Spanish Technical Dictionary
Diccionario Técnico Inglés

Routledge
Spanish Technical Dictionary
Diccionario Técnico Inglés

Volume/Tomo
1

SPANISH–ENGLISH
ESPAÑOL–INGLÉS

London and New York

First published 1997
by Routledge
11 New Fetter Lane, London EC4P 4EE

Simultaneously published in the USA and Canada
by Routledge
29 West 35th Street, New York, NY 10001

Conversion tables adapted from *Dictionary of Scientific Units*, H. G. Jerrard and D. B. McNeill, London: Chapman & Hall, 1992.

Typeset in Monotype Times, Helvetica 55 and Bauer Bodoni
by Routledge

Printed in England by TJ Press (Padstow) Ltd., Cornwall

Printed on acid-free paper

British Library Cataloguing-in-Publication Data
A catalogue record for this book is available from the British Library

Library of Congress Cataloging-in-Publication Data
Applied for

Vol 1 Spanish–English 0–415–11272–9
Vol 2 English–Spanish 0–415–11273–7
2–volume set 0–415–11274–5

Registered trademarks ®

Marcas registradas ®

Contents/Índice de contenidos

Spanish Technical Dictionary
Diccionario Técnico Inglés

Project Manager/Dirección del Proyecto
Gemma Belmonte Talero

Managing Editor/Dirección Editorial
Sinda López

Programme Manager/Dirección del Programa
Elizabeth White

Editorial/Redacción
Martin Barr   Justine Bird
Lisa Carden   Janice McNeillie
Jessica Ramage   Robert Timms

Marketing
Vanessa Markey   Rachel Miller

Systems/Sistemas Informáticos
Omar Raman   Simon Thompson

Administration/Administración
Amanda Brindley

Production/Producción
Maureen James   Nigel Marsh   Joanne Tinson

Specialist Consultants/
Asesores Técnicos Especialistas

Augustín Catalán
Agencia COVER

Antonio Castillo González
Agencia Industrial del Estado (INI)

John Camm
British Ceramic Research Limited

Dr Juan Domínguez
Centro de Investigación y Desarrollo
Agrario, Córdoba

Manuel Estany Segalás
Consejo Intertextil Español

Enrique Cuñetti
Facultad de Ingeniería, Universidad de la
República, Uruguay

Philip Maylor
Brian Patterson
IDPM, Institute for the Management of
Information Systems

Juan Carlos Miguez
IEEE, Uruguay

Dr A. Lopez-Soler
Dr X. Querol
Institute of Earth Sciences "Jaume Almera"
(CSIC), Barcelona

J.E.H. Leach
Institute of Energy

Christopher Wolfe
Institute of Physics

Celia Kirby
D.B. Smith
Institute of Hydrology

Malcolm Horlick
Institute of Refrigeration

Don Goodsell
Institution of Mechanical Engineers and Society
of Automotive Engineers

J. Salvador Santiago Páez
Instituto de Acústica (CSIC), Madrid

Dra Menchu Comas
Instituto Andaluz de Ciencias de la Tierra
(CSIC and Universidad de Granada)

María Ofelia Cirone
Dr Adan Edgardo Pucci
Instituto Argentino de Oceanografía
(CONICET)

Jesús Espinosa
Instituto del Frío (CSIC), Madrid

José Manuel Olías Jiménez
Instituto de la Grasa (CSIC), Sevilla

Dr Jaime Tornos Cubillo
Alejandro Mira Monerris
Dr Pedro Osuna Rey
Dr Fernando Hevia
(*team co-ordinator/coordinador de equipo*)
Marinela García Fernández
(*team co-ordinator/coordinadora de equipo*)
Ismael Arinas
M. Paloma M. Fradejas
Victoria Machuca
Carmen Sancho
Instituto de la Ingeniería de España

Nora Susana Fígoli
José Miguel Parera
Instituto de Investigaciones en Catálisis y
Petroquímica (FIQ-UNL-CONICET),
Argentina

Dr Manuel de León
Instituto de Matemáticas y Física Fundamental
(CSIC), Madrid

Dr Jesús A. Pajares
Instituto Nacional del Carbón (CSIC), Oviedo

José Torres Riera
Instituto Nacional de Técnica Aeroespacial

Miguel A. Gaspar Paricio
Instituto Papelero Español

Guillermo González Gómez
Katholieke Universiteit Leuven

Beni Ortiz
LA REVISTA de EL MUNDO

J.E. Lunn
Locomotive and Carriage Institution, Surrey

Pablo A. Fossati Fischer
Ministerio de Defensa Nacional de Uruguay

Contributors/Colaboradores

Lucía Alvarez de Toledo
Rodolfo Antonelli
Emilio Aparicio Fernández
Beatriz A. Bonnet
Carolina Bovino de Morgan
Mª Luisa Casado López
María Fernanda Cid
Julio Alejandro Cid
Alfonso Cuevas y Fajardo
Laura Viviana Demoy
Dr Marco A. Díaz
Enric S. Dolz-Ferrer
Carmen Fernández-Marsden
Amelia Garcia-Leigh
Margarita García de Cortázar Nebreda
Lourdes García de Reece
M. Antonio Gavilanez
Jorge Hernández Osuna
John E. Kennedy
Edmund Knox

Patricia S. Luna
Eduardo Maleubre Nogales
Montserrat Maleubre Pablos
Dr Juan M. Nieves Pamplona
Carlos Novi
Carlos de Osma Calvo
María Cristina Plencovich
Eugene Polisky
Michael Rawson
Estefanía Rodríguez de Santos
Juan Rojo
Dr Sonia Roquet de Savage
Amparo Ruiz-Vera
Joaquín M. Sallarés
Norma Sánchez Meana
Oscar L. Spraggon
Mónica Liliana Suárez
Marina Tei
Daniel Ricardo Yagolkowski
Beatriz R. Zylberfisz de Yagolkowski

The English term list is based on our database of terminology first published in the *Routledge French Technical Dictionary*, 1994. We gratefully acknowledge the original contribution of the following:

La lista de términos ingleses está extraída de nuestra base de datos terminológicos que fue publicada por primera vez en el *Diccionario Técnico Francés Routledge*, en 1994. Agradecemos la contribución original de las siguientes personas:

Yves Arden, Réjane Amery, Josephine Bacon, John P. Bryon, Michael Carpenter, Anna Cordon, Maguy Couette, Elisabeth Coyne, P.J. Doyle, J.V. Drazil, Bill Duffin, James Dunster, Christopher Freeland, Crispin Geoghegan, Susan Green, Freda Klipstein, C.A. Lagall, David Larcher, Virginia Lester, Pamela Mayorcas, James Millard, Charles Polley, Michael Rawson, Louis Rioual, Tom Williams, Stephen Wilson, Stewart Wittering

Lexicographers & Proofreaders/Lexicógrafos y Correctores de pruebas

Tom Barlett
Soraya Bermejo
Michael Britton
Harry Campbell
Ximena Castillo
Yolanda Cerdá
Anna Cooke
Alison Crann
Mel Fraser
Fiona Greenall
Zoë Hambling
Kenneth Hillman
Margaret Jull Costa
Fiona Mackintosh
Isabel Mancebo-Portela
Philip Maxwell

Suzanne McCloskey
Héloïse McGuinness
Karen Miller
Julie Muleba
Jeremy Munday
Stephanie Parker
Zoë Petersen
Kathryn Phillips-Miles
Anna Reid
Mary Rigby
Jonathan Roper
Alisa Salamon
Malihe Sanatian
Martin Stark
Carmen Suárez
Gillian Wolfe

Keyboarders/Mecanógrafos

Debbie Thomas
Carmen Alpin, Emmanuelle Bels, Kristoffer Blegvad, Pedro Contreiras,
Antonio Fernández Entrena, Beatriz Fernández Martínez, Rosa Gálvez López,
Christiane Grosskopf, Michael Jopling, Ute Krebs, Ilona Lehmann, Géraldine Monnereau,
Geir Moulson, Roger Pena Muiño, Fabienne Rangeard, Beate Schmitt, Robert Timms, Jeremy Vine

Acknowledgements/Agradecimientos

We wish to acknowledge the valuable contribution of Gonzalo Álvarez Pineda, Esteban Gozalo Crespo, John Thristan and Enrique Cama Gómez, who checked the subject areas of Military Technology, Petroleum Technology, Safety Engineering and Cinematography respectively.

We are also particularly grateful to Jaime Tornos Cubillo of the Instituto de la Ingeniería de España; Ruperto Belmonte Monforte from the Spanish Armed Forces; the Universidad de Castilla-La Mancha, the Universidad de Extremadura and Carlos Márquez Linares for their kind assistance during the compilation of this dictionary.

Queremos agradecer la valiosa contribución de Gonzalo Álvarez Pineda, Esteban Gozalo Crespo, John Thristan y Enrique Cama Gómez, quienes revisaron las materias de Defensa y Armamento, Tecnología del Petróleo, Sistemas de Seguridad y Cinematografía respectivamente.

También expresamos nuestro agradecimiento a Jaime Tornos Cubillo, del Instituto de la Ingeniería de España; Ruperto Belmonte Monforte, de las Fuerzas Armadas Españolas; a la Universidad de Castilla-La Mancha, a la Universidad de Extremadura y a Carlos Márquez Linares por su grata colaboración en la compilación de este diccionario.

Introduction/Introducción

The *Routledge Spanish Technical Dictionary* meets a specific need in an important field in view of the rapid developments currently taking place across the world in the spheres of technology and science. While it is true that there have been technological advances throughout the history of humanity, those which have come about in the last decade have impacted on every facet of human knowledge and understanding and hold sway over virtually every sphere of human life. They do so especially in industry, which has become increasingly mechanized, as well as in business and science. Technological know-how has become central to the acquisition of skills and the adoption of new working methods.

It could be argued that a language only comes of age when it is capable of responding to any new realities which arise in the living environment inhabited by its users; in other words, when the terminology of a language accounts for any changes in contingent reality which may result from factors, such as human inventions, technological or scientific advances or changing attitudes. A language must be able to demonstrate its adaptability to any changes which may occur, which might require corresponding shifts in expression or definition.

Spanish is gaining widespread usage and is demonstrably well-equipped to express and define existing scientific fact, future technological advances and inventions worthy of the new millennium. The Latin-American subcontinent is also becoming increasingly influential in the areas of trade and communication. If we take all these factors as a whole, the need for a new technical dictionary, designed to meet a variety of needs, becomes apparent.

The creation of this dictionary has involved the participation of over one hundred specialists from industry, the academic world and the corporate sector. The quality and accuracy of translations between the two languages has been tightly controlled by a team of professional translators from Spain, Latin America, the United States and the United Kingdom.

El *Diccionario Técnico Inglés Routledge* responde a una necesidad concreta en un área digna de reseñarse en vista del avance y la rápida evolución de la tecnología y de la ciencia en el mundo de hoy. Bien es cierto que avances tecnológicos los hubo siempre en la historia de la Humanidad; sin embargo, ha sido en los últimos decenios cuando tales avances han invadido todos los ámbitos del saber y del conocimiento humano y se han adueñado, valga la expresión, de casi todas las esferas o sectores de la vida humana. Estos avances se hacen patentes en la industria, cada vez más tecnificada así como en los negocios y la ciencia. Por esta razón, el conocimiento tecnológico se hace totalmente necesario en la transmisión de los saberes y en el aprendizaje de nuevos métodos.

Se podría argumentar que una lengua adquiere su madurez cuando es capaz de responder a las nuevas realidades que surgen en el entorno de sus hablantes. Es decir, cuando se forma una terminología adecuada a las realidades, que sea producto de la invención humana, del avance tecnológico y científico o de los cambios de mentalidades. Una lengua ha de ser capaz de demostrar día a día su capacidad de adaptación y adecuación a los cambios que requieran nuevas formas de expresión o definición.

El español está ganando auge y sirve, como se puede demostrar, para definir y determinar nuevas realidades científicas, experimentaciones técnicas e invenciones propias ya de un nuevo milenio. La realidad iberoamericana se torna cada vez más influyente en las esferas del comercio y de la comunicación. Cuando todas estas realidades se conjugan, surge la necesidad de elaborar un nuevo diccionario técnico que responda con su realización a los requerimientos que desde ámbitos distintos se están llevando a cabo.

Para la realización de este diccionario se ha contado con la colaboración de más de cien especialistas, pertenecientes a la industria, al mundo académico universitario y al de los negocios. La calidad y la precisión de las traducciones han sido revisadas por traductores profesionales procedentes de España, América

The *Routledge Spanish Technical Dictionary*, which is the result of careful initial planning and skilled execution on the part of translators, academics and experts, meets the demand for an efficient and easy-to-use aid to the translation of technical terms between Spanish and English.

Latina, Estados Unidos y Reino Unido.

El *Diccionario Técnico Inglés Routledge*, que es el resultado de una cuidadosa planificación inicial y especializada realización por parte de los traductores, académicos, expertos e investigadores que en él aparecen, viene a llenar un vacío existente: el de no poseer un instrumento eficaz y útil en la traducción de términos técnicos entre el español y el inglés.

César Chaparro Gómez
Vice Chancellor of the University of Extremadura/
Rector de la Universidad de Extremadura

Preface/Prólogo

This is the third dictionary to be published from Routledge's programme of bilingual specialist and technical dictionaries, following on from the Routledge French Technical Dictionary in 1994 and the Routledge German Technical Dictionary in 1995.

The two main factors that have enabled us to create this generation of new bilingual technical dictionaries are the database system and the method of compilation.

It would not have been possible to compile this dictionary within a realistic timescale, and to the standard achieved, without the use of a custom-designed database.

The database's most significant feature is that it is designed as a relational database: term records for each language are held in separate files, with further files consisting only of link records. Links between terms in different language files represent translations which enable us to handle, in a complex way, various types of one-to-many and many-to-one translation equivalences. Links between terms within a single language file represent cross-references of various types: spelling variants, geographical variants and abbreviations.

The content of the database for this dictionary was created in three principal phases. A considerable proportion of the English term list was already available following the publication of our French and German Technical Dictionaries. The Spanish terminology was then solicited from specialist translators with current practical experience of a narrowly-defined specialist subject area and an interest in the collection and dissemination of technology. The specialist translators targeted the coverage to the Spanish market. The terms in each language were then vetted by native-speaker subject specialists working at the leading edge of the respective technology in order to ensure their currency, the accuracy of contextual information, and the adequacy of coverage. Finally, each language file was reviewed by editors to ensure coverage of American English and Latin American terms and

El diccionario técnico bilingüe español–inglés que presenta la editorial Routledge es el tercero de una serie especializada que se inició en 1994 con el Diccionario Técnico Francés y se continuó al año siguiente con el Diccionario Técnico Alemán.

Dos son los factores que han intervenido para hacer posible la creación de esta serie de diccionarios: la red de la base de datos y el método de compilación del repertorio léxico.

Podemos afirmar que no habría sido posible realizar una obra de estas características en un tiempo razonable si no hubiera sido por el uso de una base de datos que se creó específicamente para esta obra.

El rasgo que distingue a esta base de datos es su estructura relacional, mediante la cual el léxico de cada lengua se almacena en un fichero y las relaciones entre los términos en otros diferentes. Las conexiones entre los términos que se encuentran en los ficheros de cada lengua funcionan como traducciones gracias a las cuáles podemos tratar de manera compleja diversos tipos de equivalencias traducibles de una a muchas acepciones y de muchas a una. Las relaciones entre los vocablos dentro de una misma lengua actúan como remisiones, que cubren a su vez una gran variedad de categorías: variantes ortográficas o geográficas y abreviaturas.

El contenido de la base de datos se gestó en tres fases. Una proporción muy considerable del vocabulario inglés ya se había elaborado para editar el Diccionario Técnico Francés y el Diccionario Técnico Alemán. Un equipo de traductores con experiencia demostrada en las distintas especialidades y con un interés particular en la lexicografía cientificotécnica se ocupó de la terminología española. Los traductores se encargaron de delimitar los centros de interés para el público español. El léxico de cada lengua se sometió posteriormente a la consideración de diversos especialistas nativos en contacto directo con los últimos avances en cada ámbito tecnológico para asegurarse de su pertinencia, la exactitud de la información contextualizada y la extensión del vocabulario. Por último, los editores spelling variants; these are clearly labelled and distinguished.

The creation and editing of the database of terms was, however, only the first stage in the making of the dictionary. Within the database the distinction between source and target languages is not meaningful, but for this printed dictionary it has been necessary to format the data to produce separate Spanish–English and English–Spanish volumes. The data was processed by a further software module to produce two alphabetic sequences, of Spanish headwords with English translations and vice versa, each displaying the nesting of compounds, ordering of translations, style for cross-references of different types, and other features according to a complex algorithm.

At this stage the formatted text was edited by a team of experienced Spanish and English lexicographers whose task it was to eliminate duplication or inconsistency; edit the contextual information and explanations; and remove terms that were on the one hand too general, or on the other, too specialized for inclusion in a general technical dictionary. This phased method of working has enabled us to set extremely high standards of quality control.

The editorial team

revisaron los ficheros de cada lengua para cerciorarse de la inclusión del vocabulario específico y las variantes léxicas y fonéticas, tanto del español latino-americano como del inglés norte-americano, que aparecen claramente especificados.

Con todo, la creación y edición de una base de datos léxicos fue sólo el primer peldaño en la elaboración del diccionario. En una base de datos no es pertinente la distinción entre lengua de origen y de destino, así que a la hora de imprimir el diccionario fue necesario organizar los datos para elaborar de manera separada los volúmenes español–inglés e inglés–español. Los datos se elaboraron en otro módulo de software para obtener las listas alfabéticas del español al inglés y viceversa, sin olvidar los cuadros de variantes de palabras compuestas, los diferentes tipos de remisiones y demás características, siguiendo un algoritmo complejo.

Llegados a este punto un experimentado equipo de lexicógrafos españoles y británicos, cuya tarea consistía en eliminar redundancias e incoherencias, revisó el texto. Dicho equipo se encargó asimismo de redactar las explicaciones e informaciones contextuales y de suprimir aquellos términos cuyo carácter pecaba de demasiado general o demasiado especializado para ser incluído en un diccionario general técnico. Este método de trabajo por etapas nos ha permitido establecer altísimas cotas de control de calidad.

El equipo editorial

Features of the dictionary/Estructura del diccionario

The main features of the dictionary are high-lighted in the text extract on the opposite page. For a more detailed explanation of each of these features and information on how to get the most out of the dictionary, see pages xxi-xxiii.

Los principales rasgos del diccionario están señalados en el extracto que aparece en la página siguiente. Para obtener una explicación más detallada de cada uno de estos rasgos así como información sobre la óptima utilización del diccionario, consúltense las páginas xxv-xxvii.

*Genders indicated at
Spanish nouns*
— **degaussado** *m* FÍS, INFORM&PD degaussing
degaussar *vt* FÍS, INFORM&PD degauss
degeneración *f* CRISTAL, ELECTRÓN, FÍS, FÍS RAD degeneracy, ING MECÁ degradation, METAL, QUÍMICA degeneracy
degenerado *adj* ELECTRÓN, FÍS, METAL, QUÍMICA degenerate
degradable: no ~ *adj* EMB, RECICL nondegradable
— *Se indican los géneros para
los nombres españoles*

*Subject area labels given in
alphabetical order to show
appropriate translation*
— **degradación** *f* CALIDAD degradation, CINEMAT grad, gradding, FÍS, INFORM&PD, ING MECÁ degradation, NUCL degradation, thindown, P&C *defecto* degradation, shading, TEC ESP *de equipos*, TV degradation; **~ biológica** *f* CONTAM, CONTAM MAR, DETERG, EMB, HIDROL, RECICL biodegradation; **~ de calidad** *f* TV quality degradation; **~ energética** *f* NUCL energy degradation; **~ de la energía** *f* ELEC, ING ELÉC dissipation; **~ con garbo** *f* INFORM&PD graceful degradation; **~ por irradiación** *f* FÍS RAD radiation degradation; **~ parcial** *f* INFORM&PD graceful degradation; **~ del suelo** *f* CONTAM land degradation; **~ del terreno** *f* CONTAM land degradation
— *Los indicadores de materia
ordenados alfabéticamente
muestran la traducción
apropiada*

*Compound terms nested
alphabetically at the first
element*
— *Los compuestos siguen el
orden alfabético a partir de
su primer elemento*

dejar[1]: **~ inactivo** *vt* PROD lay off; **~ reproducir** *vt* PROD *fuego de un horno alto* allow to breed
dejar[2]: **~ escapar el agua** *vi* CONST leak

*Cross-references from
abbreviations to full forms
are shown for both Spanish
and English*
— **DEL** *abr* GEN (*diodo electroluminiscente, diodo emisor de luz*) LED (*light-emitting diode*)
delaminación *f* C&V, P&C *de plásticos, adhesivos, defecto* delamination
— *Las remisiones entre las
abreviaturas y sus formas
plenas aparecen tanto para
los términos españoles
como para los ingleses*

delantal *m* PAPEL, SEG apron; **~ de cuero** *m* SEG leather apron; **~ protector** *m* SEG protective apron
delantero *adj* AUTO, ELECTRÓN, FOTO, INFORM&PD, INSTR front-end, TEC ESP ahead, TRANSP AÉR, VEH front-end
delegado: ~ de seguridad *m* SEG safety representative
delessita *f* MINERAL delessite
delfinidina *f* MINERAL, QUÍMICA *antocianina* delphinin

*Contexts give supplementary
information*
— **delfinina** *f* MINERAL, QUÍMICA *alcaloide* delphinine
delfinita *f* MINERAL, QUÍMICA delphinite, epidote
delga *f* ELEC, ING ELÉC *colector dinamico* commutator bar; **~ de colector** *f* ELEC, ING ELÉC commutator bar, commutator segment
delicuescencia *f* ALIMENT, QUÍMICA *conversión en fluidos* deliquescence
delicuescente *adj* ALIMENT, QUÍMICA deliquescent
delimitación: ~ de un terreno *f* CONST *topografía* surface demarcation
delimitador *m* IMPR, INFORM&PD delimiter; **~ de campo** *m* INFORM&PD field delimiter; **~ de palabra** *m* INFORM&PD word delimiter
delineante *m* ING MECÁ, PROD, TRANSP MAR *diseño naval* draftsman (*AmE*), draughtsman (*BrE*);
— *Los contextos ofrecen
información suplementaria*

*Both Spanish and Latin
American and British English
and American English
variants labelled accordingly*
— **~ mecánico** *m* Esp (*cf dibujante mecánico AmL*) PROD mechanical draftsman (*AmE*), mechanical draughtsman (*BrE*); **~ proyectista** *m* ING MECÁ design draftsman (*AmE*), design draughtsman (*BrE*)
— *Las variantes lingüísticas
tanto de España y América
Latina como las del inglés
británico y americano
aparecen indicadas
pertinentemente*

delinear *vt* CONST line out, GEOM plot, IMPR outline, rule

Using the dictionary

Range of coverage

This is one volume (the Spanish–English volume) of a general technical dictionary that covers the whole range of modern technology and the scientific knowledge that underlies it. It contains a broad base of terminology drawn from traditional areas of technology such as mechanical engineering, the construction industry, electrical engineering and electronics, but also includes the vocabulary of newly prominent subject areas such as fuelless energy sources, safety engineering and quality assurance.

Selection of terms

We have aimed to include the essential vocabulary of each subject area, and the material has been checked by leading subject experts to ensure that both the Spanish and the English terms are accurate and current, that the translations are valid equivalents, and that there are no gaps in coverage.

We have been careful about including only genuine technical terms and not allowing general vocabulary with no technical value. At the same time, we have entered the core vocabulary of technical discourse in its totality, although some of these items may also be found in general dictionaries. Although other variant translations would often be permissible in a particular subject area, we have given the term most widely preferred by specialists in the area.

Coverage of the subject areas is given proportionally so that an established and wide-ranging area such as mechanical engineering has a count of around 7,000 terms whereas a new area in which terminology is still developing, such as fuelless energy sources, will have considerably fewer terms.

Stoplists

Terms in Spanish are not entered under the following elements:

a, al, algo, algún, alguna, algunas, algunos, allá, allí, ante, antes, aquel, aquella, aquellas, aquellos, aquí, bajo, bastante, bastantes, cabe, como, cómo, con, contra, cual, cuál, cuales, cuáles, cualesquiera, cualquier, cualquiera, cuándo, cuya, cuyas, cuyo, cuyos, de, del, demasiado, desde, después, dónde, e, el, en, entre, es, esta, está, están, estar, estas, este, esto, estos, hacia, hasta, hay, la, las, los, mediante, mucha, muchas, mucho, muchos, muy, ningún, ninguna, ningunos, no, o, otra, otras, otro, otros, para, poca, pocas, poco, pocos, por, que, qué, quién, quienes, se, según, ser, si, sí, sin, so, sobre, solo, sólo, son, tan, tanto, toda, todas, todo, todos, tras, u, un, una, unas, uno, unos, y

Compounds are listed at their first element. In these nested listings, the simple form is replaced by a swung dash (~). For example:

cercado[1]: ~ **por el hielo** *adj* TRANSP MAR icebound

When the first element is itself a headword with one or more technical senses, compounds follow the simple form in alphabetical order. For example:

potencial *m* ELEC, FÍS, ING ELÉC, TEC ESP potential; ~ **absoluto** *m* ELEC absolute potential; ~ **activo** *m* ELEC active potential

If the first element is not itself translated, a colon (:) precedes the compounds. For example:

diario: ~ **de a bordo** *m* TRANSP AÉR flight log

Ordering of terms

All terms are ordered alphabetically at their first element. This is also the policy for hyphenated compounds. Compound terms are never entered under their second or third element, regardless of the semantic structure of the unit.

Articles and prepositions are ignored in determining the sequence of nested open compounds. For example:

onda *f* ACÚST wave, CONST *de agua* ripple, ELEC,

FÍS, FÍS ONDAS wave, HIDROL ripple, ING ELÉC, TELECOM, TRANSP MAR wave; **~ acústica** *f* FÍS ONDAS, TELECOM soundwave; **~ acústica de superficie** *f (OAS)* ACÚST, ELECTRÓN, FÍS, FÍS ONDAS, TELECOM surface acoustic wave *(SAW)*; **~ acústica volumétrica** *f* ING ELÉC bulk acoustic-wave *(BAW)*; **~ aérea** *f* FÍS ONDAS airwave; **~ aérea estacionaria** *f* FÍS ONDAS stationary aereal wave; **~ de aire** *f* PETROL airwave; **~ anormal** *f* OCEAN freak wave; **~ de baja frecuencia** *f* ING ELÉC low frequency wave

In the case of very long compound nests, marginal markers have been used to make it easy to find a term more quickly. For example:

interruptor:

~ g **~ giratorio de sectores sellado** *m* ING ELÉC sealed-wafer rotary switch; **~ de graduación de la luz** *m* ELEC, ING ELÉC dimmer switch;

~ h **~ horario** *m* ELEC, ING ELÉC switch clock, time switch, TELECOM time switch;

~ i **~ ignífugo** *m* SEG, TERMO fireproof switch, flameproof switch; **~ de imagen en negativo** *m* TV reverse-image switch; **~ de impacto** *m* TRANSP AÉR crash switch; **~ inmediato** *m* ING ELÉC snap-in switch

Terms containing figures and symbols are alphabetized according to the usual expansion when written out in full:

acuático *adj* HIDROL, OCEAN aquatic
A4 *m* IMPR, PAPEL A4

Homographs

Every term is accompanied by a label indicating its part of speech. For a complete list of these labels and their expansions, please see page xxix.

When terms beginning with the same element fall into two or more part-of-speech categories, the different nests will be distinguished by a raised number immediately following the head of the nest, whether the head has technical senses of its own or is a dummy. The sequence is abbreviation, adjective, adverb, noun and verb followed by less frequent parts of speech. For example:

peligro[1]**: en ~** *adj* TRANSP MAR *buque* in distress
peligro[2] *m* AUTO, CALIDAD, SEG danger, TRANSP AÉR hazard; **~ del ambiente** *m* CALIDAD environmental hazard

Ordering of translations

Every term is accompanied by one or more labels indicating the technological area in which it is

used. For a complete list of these labels and their expansions, please see pages xxix-xxx.

Where the same term is used in more than one technological area, multiple labels are given as appropriate. These labels appear in alphabetical order.

Where a term has the same translation in more than one technological area, this translation is given after the sequence of the labels. For example:

lente *f* CINEMAT, FÍS, INSTR, LAB, TEC ESP lens

Where a term has different translations according to the technological area in which it is used, the appropriate translation is given after each label or set of labels. For example:

cesta *f* OCEAN *trampa de pescado* basket trap, TRANSP *grúas* skip, TRANSP AÉR *reaprovisionamiento en vuelo* drogue

Supplementary information

The gender is given for every Spanish noun term. In the case of compound terms this is the gender of the term as a whole (that is, its noun head) rather than the final element. For example:

precintadora: ~ de bandejas *f* EMB tray sealer

In many cases additional data is given about a term in order to show how it is used. Such contextual information can be:

(a) the typical subject or object of a verb, for example:

tratar[2] *vt* CARBÓN *aceros, aire* condition, treat, PROD handle, QUÍMICA process

(b) typical nouns used with an adjective, for example:

unidireccional *adj* ACÚST *micrófono*, ELEC *corriente*, ING ELÉC unidirectional, TELECOM *circuito de transmisión* simplex, unidirectional

(c) words indicating the reference of a noun, for example:

triturador *m* ALIMENT *molienda, panadería* kibbler, CARBÓN stamp mill, HIDROL *sedimentos* comminutor, ING MECÁ crusher, triturator, NUCL, PROC QUÍ comminutor, PROD crusher

(d) information which supplements the subject area label, for example:

tránsito *m* CONST, NUCL transit, TRANSP traffic, TRANSP MAR *navegación* passage

(e) a paraphrase or broad equivalent, for example:

edafología *f* AGRIC, CARBÓN *ciencia del suelo* edaphology, soil science

When various different translations apply in the same subject area, contextual information is also used to show which translation is appropriate in different circumstances. For example:

puerta *f* AGUA, CONST gate, CONTAM MAR trawl board, ELECTRÓN gate, INFORM&PD gate, gating, ING ELÉC gate, OCEAN trawl board, PROD *de esclusas* gate, *para observar comportamiento de mecanismos* door, TRANSP MAR *de esclusa* sluice-gate, trawl board, VEH *carrocería* door

Cross-references

Both British and North American terms are covered, and these are differentiated by regional labels. Coverage also includes terms from Latin America and Spain. For a complete list of these labels and their expansions, please see page xxix.

In the case of lexical variants, full information – including translations and cross-references to the other form – is given at each entry. For example:

adoquinado *(Esp)* *m* (*cf pavimento LAm*) CARBÓN, CONST pavement *(BrE)*, sidewalk *(AmE)*

Both abbreviations and their full forms are entered in the main body of the dictionary in alphabetical sequence. Full information – including translations and cross-references to the full form or abbreviations as appropriate – is given at each entry. For example:

PTV *abr (peso total del vehículo)* VEH GVW *(gross vehicle weight)*

Consejos para la utilización de este diccionario

Delimitación de los contenidos

Éste es el volumen español–inglés del diccionario técnico general que presentamos. En él se da cuenta de todos los ámbitos de la tecnología moderna y del conocimiento científico en que se basa. Su acervo terminológico abarca desde los dominios tradicionales de la tecnología como la ingeniería mecánica, la industria de la construcción, la ingeniería eléctrica y electrónica a las especialidades de vanguardia como las fuentes de energía renovables, la ingeniería de seguridad y el control de calidad.

Selección del vocabulario

Hemos tratado de incluir el vocabulario esencial de cada materia. Un equipo de prominentes especialistas se ha ocupado de revisar los resultados, para asegurarnos de que tanto los términos españoles como los ingleses son precisos y actuales, de que las traducciones tienen equivalentes válidos y de que no quedaban lagunas en los campos léxicos seleccionados.

Hemos tomado la precaución de incluir únicamente aquellos términos cuyo carácter técnico resultaba incuestionable y de excluir los que carecían de significado específico en este campo y, si bien el vocabulario técnico esencial se encuentra a menudo adecuadamente representado en los diccionarios generales, no por eso hemos dejado de incluirlo en toda su extensión. En aquellos casos en los que una palabra permitía varias traducciones, hemos optado por la que los especialistas han considerado más plausible.

La extensión concedida a cada materia es proporcional a su importancia, de modo que a una especialidad del peso específico de la ingeniería mecánica se le dedican 7.000 términos, en tanto que a un campo nuevo como el de las fuentes de energía renovables que sólo ahora empieza a desarrollar un léxico propio le corresponde un ámbito terminológico mucho más reducido.

Exclusiones y supresiones

A la hora de buscar una palabra en español no se tomarán en cuenta los siguientes morfemas:

a, al, algo, algún, alguna, algunas, algunos, allá, allí, ante, antes, aquel, aquella, aquellas, aquellos, aquí, bajo, bastante, bastantes, cabe, como, cómo, con, contra, cual, cuál, cuales, cuáles, cualesquiera, cualquier cualquiera, cuándo, cuya, cuyas, cuyo, cuyos, de, del, demasiado, desde, después, dónde, e, el, en, entre, es, esta, está, están, estar, estas, este, esto, estos, hacia, hasta, hay, la, las, los, mediante, mucha, muchas, mucho, muchos, muy, ningún, ninguna, ningunos, no, o, otra, otras, otro, otros, para, poca, pocas, poco, pocos, por, que, qué, quién, quienes, se, según, ser, si, sí, sin, so, sobre, solo, sólo, son, tan, tanto, toda, todas, todo, todos, tras, u, un, una, unas, uno, unos, y

Las palabras compuestas siguen el orden alfabético a partir de su primer elemento. Cuando varios compuestos participan de un lexema inicial común, éste se reemplaza por una virgulilla (~), por ejemplo:

cercado[1]: **~ por el hielo** *adj* TRANSP MAR icebound

Cuando el primer elemento es una entrada dotada de uno o varios significados técnicos, las acepciones correspondientes aparecen al principio, seguidas de los compuestos en orden alfabético. Por ejemplo:

potencial *m* ELEC, FÍS, ING ELÉC, TEC ESP potential; **~ absoluto** *m* ELEC absolute potential; **~ activo** *m* ELEC active potential

Cuando el primer elemento no se traduce, los compuestos aparecen precedidos de dos puntos(:). Por ejemplo:

diario: **~ de a bordo** *m* TRANSP AÉR flight log

Orden de las entradas

Las entradas se ordenan alfabéticamente e

idéntico criterio se aplica a las locuciones y a las palabras compuestas. A este respecto debe considerarse irrelevante la carga semántica de los diferentes lexemas.

Los artículos y las preposiciones no se toman en cuenta a la hora de determinar el orden de las palabras compuestas en un artículo. Por ejemplo:

onda *f* ACÚST wave, CONST *de agua* ripple, ELEC, FÍS, FÍS ONDAS wave, HIDROL ripple, ING ELÉC, TELECOM, TRANSP MAR wave; ~ **acústica** *f* FÍS ONDAS, TELECOM soundwave; ~ **acústica de superficie** *f (OAS)* ACÚST, ELECTRÓN, FÍS, FÍS ONDAS, TELECOM surface acoustic wave *(SAW)*; ~ **acústica volumétrica** *f* ING ELÉC bulk acoustic-wave *(BAW)*; ~ **aérea** *f* FÍS ONDAS airwave; ~ **aérea estacionaria** *f* FÍS ONDAS stationary aereal wave; ~ **de aire** *f* PETROL airwave; ~ **anormal** *f* OCEAN freak wave; ~ **de baja frecuencia** *f* ING ELÉC low frequency wave

En las entradas que registran gran cantidad de acepciones se utilizan marcadores para facilitar la localización de los términos. Por ejemplo:

interruptor:
~ g ~ **giratorio de sectores sellado** *m* ING ELÉC sealed-wafer rotary switch; ~ **de graduación de la luz** *m* ELEC, ING ELÉC dimmer switch;
~ h ~ **horario** *m* ELEC, ING ELÉC switch clock, time switch, TELECOM time switch;
~ i ~ **ignífugo** *m* SEG, TERMO fireproof switch, flameproof switch; ~ **de imagen en negativo** *m* TV reverse-image switch; ~ **de impacto** *m* TRANSP AÉR crash switch; ~ **inmediato** *m* ING ELÉC snap-in switch

Los términos que incluyen cifras y símbolos se ordenan alfabéticamente, como si estuvieran escritos en letra:

acuático *adj* HIDROL, OCEAN aquatic
A4 *m* IMPR, PAPEL A4

Homógrafos

Tras cada término se indica su categoría gramatical mediante abreviatura. Puede consultarse una lista completa de las abreviaturas en la página xxix.

En el caso de términos cuyo elemento inicial es idéntico pero pertenecen a diferentes categorías gramaticales, las acepciones se distinguen por la inclusión de un supraíndice tras la entrada, ya sea dicha entrada semánticamente plena o tenga mero valor significante. El orden seguido es el de abreviatura, adjetivo, adverbio, sustantivo y verbo, seguidos por otras categorías gramaticales de uso más restringido, por ejemplo:

peligro[1]: **en ~** *adj* TRANSP MAR *buque* in distress
peligro[2] *m* AUTO, CALIDAD, SEG danger, TRANSP AÉR hazard; ~ **del ambiente** *m* CALIDAD environmental hazard

Orden de las traducciones

Cada término se acompaña de uno o más indicadores que remiten al ámbito tecnológico en que se utiliza. En las páginas xxiii-xxiv se encuentra una lista completa de tales indicadores así como de sus formas léxicas plenas.

Cuando una acepción se utiliza en varios ámbitos tecnológicos, se consignan alfabéticamente los indicadores pertinentes.

Cuando la traducción de un término es la misma en varios ámbitos tecnológicos, dicha traducción va precedida de los indicadores pertinentes. Por ejemplo:

lente *f* CINEMAT, FÍS, INSTR, LAB, TEC ESP lens

Cuando un vocablo acepta diferentes traducciones, dependiendo del campo en que se use, la traducción pertinente se consigna tras cada indicador o conjunto de indicadores. Por ejemplo:

cesta *f* OCEAN *trampa de pescado* basket trap, TRANSP *grúas* skip, TRANSP AÉR *reaprovisionamiento en vuelo* drogue

Información complementaria

El género aparece para cada término español que posea la categoría de nombre. Con respecto a los compuestos, éstos no llevan el género del último elemento, sino el de la unidad completa (es decir, el de su radical). Por ejemplo:

precintadora: ~ **de bandejas** *f* EMB tray sealer

En muchos casos se ofrecen datos adicionales para precisar el uso de un término. Dicha información puede consistir en:

(a) El sujeto u objeto típicos de un verbo, por ejemplo:

tratar[2] *vt* CARBÓN *aceros, aire* condition, treat, PROD handle, QUÍMICA process

(b) Nombres usados habitualmente con un adjetivo, por ejemplo:

unidireccional *adj* ACÚST *micrófono*, ELEC *corriente*, ING ELÉC unidirectional, TELECOM *circuito de transmisión* simplex, unidirectional

(c) Palabras que indican la referencia de un nombre, por ejemplo:

triturador *m* ALIMENT *molienda, panadería* kibbler, CARBÓN stamp mill, HIDROL *sedimentos* comminutor, ING MECÁ crusher, triturator, NUCL, PROC QUÍ comminutor, PROD crusher

(d) Información complementaria de la abreviatura, por ejemplo:

tránsito *m* CONST, NUCL transit, TRANSP traffic, TRANSP MAR *navegación* passage

(e) una paráfrasis o equivalente funcional, por ejemplo:

edafología *f* AGRIC, CARBÓN *ciencia del suelo* edaphology, soil science

En los casos en que una palabra puede traducirse de varias maneras en un mismo ámbito, se utiliza la información contextual para mostrar cuál es la traducción más apropiada en cada caso. Por ejemplo:

puerta *f* AGUA, CONST gate, CONTAM MAR trawl board, ELECTRÓN gate, INFORM&PD gate, gating, ING ELÉC gate, OCEAN trawl board, PROD *de esclusas* gate, *para observar comportamiento de mecanismos* door, TRANSP MAR *de esclusa* sluicegate, trawl board, VEH *carrocería* door

Remisiones

Dado que este diccionario incluye tanto los términos ingleses como los americanismos, se indica siempre su procedencia geográfica mediante indicador. La cobertura de este diccionario también incluye términos de América Latina y España. Para una lista completa de estos indicadores y de sus correspondencias, por favor consúltese la página xxix.

En el caso de variantes léxicas se ofrece una información completa y detallada en cada entrada, incluyendo las traducciones y referencias a la forma alternativa. Por ejemplo:

adoquinado *(Esp)* *m* (*cf pavimento LAm*) CARBÓN, CONST pavement (*BrE*), sidewalk (*AmE*)

Las abreviaturas y sus formas plenas correspondientes se consignan alfabéticamente. La información completa, incluyendo las traducciones y las remisiones a las palabras completas o a las abreviaturas, según corresponda, aparece en cada entrada. Por ejemplo:

PTV *abbr* (*peso total del vehículo*) VEH GVW (*gross vehicle weight*)

Abbreviations used in this dictionary/
Abreviaturas utilizadas en este diccionario

Parts of speech/Categorías gramaticales

abr	abreviatura	abbreviation
adj	adjetivo	adjective
adv	adverbio	adverb
f	femenino	feminine
f pl	femenino plural	feminine plural
fra	frase	phrase
m	masculino	masculine
m pl	masculino plural	masculine plural
pref	prefijo	prefix
prep	preposición	preposition
vi	verbo intransitivo	intransitive verb
v refl	verbo reflexivo	reflexive verb
vt	verbo transitivo	transitive verb
vti	verbo transitivo e intransitivo	transitive and intransitive verb

Geographic codes/Indicadores geográficos

Esp	España	Spain
AmL	América Latina	Latin America
BrE	Inglés británico	British English
AmE	Inglés americano	American English

Subject area labels/Indicadores de materia

ACÚST	Acústica	Acoustics
AGRIC	Agricultura	Agriculture
AGUA	Abastecimiento de Agua	Water Supply Engineering
ALIMENT	Tecnología de la Alimentación	Food Technology
AUTO	Ingeniería de Automoción	Automotive Engineering
CALIDAD	Control de Calidad	Quality Assurance
CARBÓN	Tecnología del Carbón	Coal Technology
CINEMAT	Cinematografía	Cinematography
COLOR	Tecnología del Color	Colours Technology
CONST	Construcción	Construction
CONTAM	Contaminación	Pollution
CONTAM MAR	Contaminación Marina	Marine Pollution
CRISTAL	Cristalografía	Crystallography
C&V	Cerámica y Vidrio	Ceramics and Glass
DETERG	Detergentes	Detergents
D&A	Defensa y Armamento	Military Technology
ELEC	Electricidad	Electricity
ELECTRÓN	Ingeniería Electrónica	Electronic Engineering
EMB	Envasado/Embalaje	Packaging
ENERG RENOV	Fuentes de Energías Renovables	Fuelless Energy Sources
FERRO	Sistemas Ferroviarios	Railway Engineering
FÍS	Física	Physics
FÍS OND	Física de las Ondas	Wave Physics

FÍS PART	Física de las Partículas	Particle Physics
FÍS RAD	Física de las Radiaciones	Radiation Physics
FÍS FLUID	Física de los Fluidos	Fluid Physics
FOTO	Fotografía	Photography
GAS	Tecnología del Gas	Gas Technology
GEN	Tecnología General	General Technology
GEOFÍS	Geofísica	Geophysics
GEOL	Geología	Geology
GEOM	Geometría	Geometry
HIDROL	Hidrología	Hydrology
INFORM&PD	Informática y Procesamiento de Datos	Computing and Data Processing
IMPR	Imprenta	Printing
ING ELÉC	Ingeniería Eléctrica	Electrical Engineering
ING MECÁ	Ingeniería Mecánica	Mechanical Engineering
ING TERM	Componentes de la Ingeniería Térmica	Heating Engineering Components
INSTAL HIDRÁUL	Instalaciones Hidráulicas	Hydraulic Equipment
INSTAL TERM	Instalaciones Térmicas	Heat Equipment
INSTR	Instrumentación	Instrumentation
LAB	Equipamiento de Laboratorio	Laboratory Equipment
MATEMAT	Matemáticas	Mathematics
MECÁ	Mecánica	Mechanics
METAL	Metalurgia	Metallurgy
METEO	Meteorología	Meteorology
METR	Metrología	Metrology
MINAS	Minería	Mining
MINERAL	Minerología	Mineralogy
NUCL	Tecnología Nuclear	Nuclear Technology
OCEAN	Oceanografía	Oceanography
ÓPT	Óptica	Optics
PAPEL	Tecnología del Papel	Paper Technology
PETROL	Petrología	Petrology
PROC QUÍ	Procesos de Tecnología Química	Chemical Technology Processes
PROD	Gestión de la Producción	Production Engineering
P&C	Plásticos y Cauchos	Plastics and Rubber
QUÍMICA	Química	Chemistry
RECICL	Reciclaje	Recycling
REFRIG	Refrigeración	Refrigeration
REVEST	Revestimientos	Coatings Technology
SEG	Sistemas de Seguridad	Safety Engineering
TEC PETR	Tecnología del Petróleo	Petroleum Technology
TEC ESP	Tecnología Espacial	Space Technology
TELECOM	Telecomunicaciones	Telecommunications
TERMO	Termodinámica	Thermodynamics
TERMOTEC	Componentes de la Termotecnia	Heating Engineering Components
TEXTIL	Textiles	Textiles
TRANSP	Transporte	Transport
TRANSP AÉR	Transporte Aéreo	Air Transport
TRANSP MAR	Transporte Marítimo	Water Transport Engineering
TV	Televisión	Television
VEH	Componentes de Vehículos	Vehicle Components

SPANISH–ENGLISH DICTIONARY
DICCIONARIO ESPAÑOL–INGLÉS

A

a- *abr* (*ato*) METR a- (*atto-*)

A *abr* ELEC, ELECTRÓN, FÍS (*ampere, amperio*) A (*ampere*), FÍS PART (*masa atómica, peso atómico*) A (*atomic mass, atomic weight, mass number*), ING ELÉC, METR (*ampere, amperio*), A (*ampere*)

AAL *abr* (*capa adaptable de ATM*) TELECOM AAL (*ATM adaptative layer*)

Aaleniense *adj* GEOL Aalenian

abacá *m* AGRIC, PAPEL Manila hemp

ábaco *m* ACÚST chart, MINAS abacus major; **~ de inclinación** *m* OCEAN dip net

abaleo *m* ALIMENT dockage

abanderado *adj* TRANSP AÉR feathered

abanderamiento *m* TRANSP AÉR feathering

abandonar[1] *vt* ING MECÁ discard

abandonar[2]: **~ el ancla** *vi* TRANSP MAR slip one's cable; **~ el buque** *vi* TRANSP MAR abandon ship

abandono: ~ del buque *m* TRANSP MAR abandonment of ship

abanico *m* GEN fan, TRANSP MAR *construcción naval* fantail, *de buque de carga* derrick; **~ abisal** *m* GEOL, OCEAN deep-sea fan; **~ aluvial** *m* GEOL alluvial fan; **~ axial** *m* SEG axial blower; **~ extractor de polvos** *m* SEG dust exhaust fan; **~ movido por motor** *m* SEG motor-driven fan; **~ oceánico** *m* GEOL, OCEAN deep-sea fan; **~ submarino** *m* GEOL, OCEAN deep-sea fan

abarbetar *vt* TRANSP MAR lash, *cabo* seize

abarloado *adj* TRANSP MAR *a otro buque* alongside

abarloar: ~ a *vt* TRANSP MAR go alongside

abarloarse *v refl* TRANSP MAR lie alongside

abarquillado[1] *adj* EMB, IMPR, PAPEL curling

abarquillado[2] *m* EMB, IMPR, PAPEL curl, curling

abastecedor *m* EMB deliverer; **~ de agua** *m* AGUA, CONTAM, HIDROL water supplier; **~ de buques** *m* TRANSP MAR *provisiones* ship chandler; **~ de combustible** *m* TEC PETR fueler (*AmE*), fueller (*BrE*)

abastecedora *f* EMB deliverer

abastecer *vt* AGUA, ELEC, ENERG RENOV, GEOL, TEC ESP, TRANSP MAR supply

abastecimiento *m* AGUA, ELEC, ENERG RENOV, GEOL supply, ING MECÁ, PROD delivery, TEC ESP supply, TEC PETR, TRANSP AÉR fueling (*AmE*), fuelling (*BrE*), TRANSP MAR supply; **~ de agua** *m* AGUA, CONTAM, HIDROL water supply; **~ de agua de altura** *m* AGUA, HIDROL distant water supply; **~ de agua asegurado** *m* AGUA, HIDROL assured water supply; **~ de agua doméstica** *m* AGUA, HIDROL domestic-water supply; **~ de agua individual** *m* AGUA, HIDROL individual water supply; **~ de agua potable** *m* AGUA, HIDROL drinking-water supply; **~ de agua rural** *m* AGRIC, AGUA, HIDROL rural water supply; **~ de agua subterránea** *m* AGRIC, AGUA, CARBÓN, HIDROL ground-water supply; **~ de combustible para la máquina** *m* ING MECÁ, TRANSP MAR engine fuel supply; **~ de energía** *m* ELEC, ENERG RENOV, ING ELÉC, ING MECÁ, NUCL, TEC ESP, TERMO energy supply; **~ de energía nuclear** *m* NUCL, TEC ESP nuclear power supply; **~ público de agua** *m* AGUA public water supply; **~ de reserva** *m* PROD backup power

abatible *adj* ING MECÁ, PROD collapsible, hinged

abatimiento *m* INFORM&PD, ING MECÁ disassembly, TEC ESP *navegación* drift, *propulsión* casting, TRANSP MAR *navegación* drift, *navegación a vela* leeway

abatir[1] *vt* INFORM&PD, ING MECÁ disassemble, TRANSP MAR *velas* haul down

abatir[2] *vi* TRANSP MAR drift, *a babor, a estribor* cast

abeliano *adj* MATEMÁT abelian

aberración *f* GEN aberration; **~ cromática** *f* FÍS, FÍS RAD, ÓPT, TELECOM, TV chromatic aberration; **~ cromática longitudinal** *f* FÍS, FÍS RAD, ÓPT, TELECOM, TV longitudinal chromatic aberration; **~ cromática transversal** *f* FÍS, FÍS RAD, ÓPT, TELECOM, TV transverse chromatic aberration; **~ de cromaticidad** *f* FÍS, FÍS RAD, ÓPT, TELECOM, TV chromaticity aberration; **~ esférica** *f* CINEMAT, FÍS, FOTO, TELECOM spherical aberration; **~ de esfericidad** *f* CINEMAT, FÍS, FOTO, TELECOM spherical aberration; **~ newtoniana** *f* FÍS Newtonian aberration; **~ óptica** *f* ÓPT, TELECOM optical aberration

abertura *f* ACÚST mouth, C&V *de un cilindro* opening, CINEMAT aperture, lens stop, CONST *tejado* break, opening, CONTAM MAR aperture, ELEC gap, EMB eyelet, FÍS aperture, FOTO aperture, lens stop, INFORM&PD aperture, ING ELÉC gap, ING MECÁ cutout, aperture, ear, daylight, opening, port, orifice, INSTAL HIDRÁUL *en el cilindro de vapor* port, *bajo el agua* orifice, MECÁ aperture, port, METAL aperture, MINAS *de un pozo* sinking, NUCL port, P&C *término general* vent, *prensa* daylight, PROD hole, REFRIG opening, TEC ESP slit, aperture, VEH *carrocería* louvre (*BrE*), louver (*AmE*), *motor* port; **~ académica** *f* CINEMAT academy aperture; **~ aislada** *f* REFRIG insulated opening; **~ de aspersión** *f* CONTAM MAR spray aperture; **~ de carga** *f* MECÁ filling hole, PROD *bóveda* charging hole; **~ de chispas en paralelo** *f* ELEC, ING ELÉC parallel spark gap; **~ del cilindro** *f* C&V splitting of the cylinder; **~ circular** *f* FÍS circular aperture; **~ de colada** *f* P&C *moldeo por inyección* sprue opening; **~ de conexión** *f* ING MECÁ connection port; **~ delantera** *f* ACÚST front gap; **~ de diafragma** *f* CINEMAT, FÍS, FOTO, TV aperture stop, F-stop; **~ eficaz** *f* CINEMAT, FOTO, TV effective aperture; **~ de entrada** *f* AUTO, VEH inlet port; **~ del haz** *f* ÓPT *angular* beamwidth; **~ del iris** *f* CINEMAT iris-out; **~ de limpieza** *f* PROD cleaning hole; **~ de la lumbrera** *f* INSTAL HIDRÁUL port opening; **~ del objetivo** *f* CINEMAT, FÍS, FOTO, TV lens aperture; **~ del piso** *f* MINAS bench hole; **~ de la positivadora** *f* CINEMAT, FOTO, TV printer aperture; **~ real** *f* CINEMAT, FOTO, TV actual aperture; **~ relativa** *f* CINEMAT aperture ratio, FÍS, FOTO F-number; **~ T** *f* CINEMAT, FOTO, TV T-stop; **~ de transporte** *f* NUCL transfer port; **~ útil** *f* CINEMAT, FOTO, TV *de un objetivo* effective aperture; **~ de ventilación** *f* MINAS blast-hole

abierto[1] *adj* ING MECÁ open-type, TRANSP MAR *barco* open

abierto[2] *m* TRANSP *semáforo* green

abietato *m* QUÍMICA abietate

abigarrado *adj* GEOL, MINERAL variegated

abisagrado *adj* ING MECÁ, MECÁ hinged

abisal *adj* GEOL, OCEAN abyssal

ablación *f* FÍS, GEOL, HIDROL, METEO, TEC ESP ablation; **~ lateral** *f* GEOL lateral planation

ablandador *m* ALIMENT tenderizer, CONST, GAS, IMPR, P&C, PROC QUÍ softener; **~ de agua** *m* AGUA, CONST, HIDROL, IMPR, P&C, PROC QUÍ water softener

ablandamiento *m* CONST, GAS, IMPR, P&C, PROC QUÍ, TERMOTEC softening; **~ del agua** *m* AGUA, CONST, HIDROL, IMPR, P&C, PROC QUÍ water softening; **~ del agua de alimentación** *m* AGUA, HIDROL, TERMOTEC feedwater softening

ablandar *vt* GEN soften, TERMO melt

ablativo *adj* TEC ESP ablative

ablución *f* PROC QUÍ washing out

abocardar *vt* ING MECÁ, MECÁ, PROD countersink

abocinado *adj* C&V flaring, MECÁ flared, PROD flaring, TEXTIL flared

abocinamiento *m* ING MECÁ bell mouth, PROD flaring, TEC ESP bell mouth

abocinar *vt* C&V, MECÁ, PROD, TERMO, TERMOTEC, TEXTIL flare

abolladura *f* ING MECÁ, MECÁ, METAL dent, NUCL *de tubos* denting

abollar *vt* ING MECÁ, MECÁ, METAL, NUCL dent

abombado *adj* MECÁ dished

abombamiento *m* ALIMENT *tarros, latas* blowing, CONST *deformación* camber, PROD *de polea bombada* crowning

abombar *vt* TEXTIL swell

abombarse *v refl* CONST bulge

abonado *m Esp* (*cf afiliado AmL*) ELEC, INFORM&PD, TELECOM, TV affiliate, subscriber; **~ ausente** *m* TELECOM absent subscriber; **~ a bordo** *m* TELECOM on-board subscriber; **~ dado de baja** *m* TELECOM ceased subscriber; **~ llamado** *m* TELECOM called party; **~ de prueba** *m* TELECOM test customer; **~ que llama** *m* TELECOM caller; **~ de vagabundeo** *m* TELECOM roaming subscriber

abonadura: **~ sobre la superficie del suelo** *f* AGRIC top dressing

abonamiento: **~ lateral** *m* AGRIC side dressing

abonar *vt* AGRIC, CONTAM, RECICL fertilize, manure

abono *m* AGRIC, CONTAM, RECICL fertilizer; **~ artificial** *m* AGRIC, CONTAM, RECICL artificial manure; **~ de base** *m* AGRIC bottom dressing; **~ completo** *m* AGRIC, CONTAM, RECICL complete fertilizer; **~ compuesto** *m* AGRIC, CONTAM, RECICL composting; **~ de corral** *m* AGRIC barnyard manure; **~ de excrementos humanos** *m* AGRIC, RECICL night soil; **~ de excrementos pútridos** *m* AGRIC, RECICL night soil; **~ fosfatado** *m* AGRIC phosphatic fertilizer; **~ verde** *m* AGRIC, CONTAM, RECICL green manure

aboquillar *vt* PROD crimp

aborcadar *vt* MECÁ, PROD, TERMO, TERMOTEC flare

abordaje *m* TRANSP MAR *accidente* collision, *embarque* boarding, boarding party; **~ deliberado** *m* TRANSP MAR ram; **~ de proa** *m* TRANSP MAR *colisión* ram

abordar[1] *vt* TRANSP MAR *embarque* board; **~ accidentalmente** *vt* TRANSP MAR *embarcación* run down; **~ de proa** *vt* TRANSP MAR *colisión* ram

abordar[2] *vi* TRANSP MAR dock

abortar *vt* INFORM&PD, TEC ESP abort

aborto *m* INFORM&PD, TEC ESP abort

abovedarse *v refl* MINAS arch

aboyado: **~ con punzón especial** *m* ING MECÁ dimpling

abozar: **~ delante** *vt* TRANSP MAR *estacha* stop off

abradante *m* CARBÓN abradant

abrasado *adj* TERMO scorched

abrasamiento *m* NUCL, TEC ESP *vehículos espaciales, cuerpos celestes* burnout; **~ neutral** *m* NUCL neutral burn-out

abrasar *vt* TERMO burn, scorch

abrasarse *v refl* TERMO glow

abrasímetro *m* CARBÓN abrasion tester, GEOL abrasive tester, HIDROL, INSTR, MECÁ, P&C abrasion tester, PAPEL *telas metálicas* abraser

abrasión *f* CARBÓN, GEOL, HIDROL, MECÁ, P&C, PAPEL abrasion; **~ Taber** *f* P&C Taber abrasion

abrasividad *f* GEN abrasiveness

abrasivo[1] *adj* GEN abrasive

abrasivo[2] *m* CARBÓN abradant, ING MECÁ, MECÁ abrasive, METAL grinding, REVEST, SEG abrasive; **~ metálico** *m* MECÁ abrasive shot; **~ de recubrimiento** *m* ING MECÁ, REVEST coated abrasive

abrazadera *f* ALIMENT thimble, AUTO bolt, strap, CONST brace, keeper, pipe collar, strap, band, ELEC *cable* clamp, ING ELÉC flange, clip, ING MECÁ strap, clamp, clevis, clip, fastener, shackle, cleat, threaded fastener, INSTAL HIDRÁUL brace, *de sujeción de la caldera* bracket, LAB *soporte* clamp, MECÁ clamp, clip, clevis, fastener, MINAS *barrenado* clamp, *jaula minas* cutting ring, shoe, OCEAN clamp, PETROL buckle, PROD strap, fastener, clamp, TEC ESP *naves espaciales, cohetes portadores* Faraday rotation, TEXTIL tieback, TRANSP *ruedas* clamp, TRANSP AÉR brace, VEH *término genérico, piezas internas del coche* clamp; **~ de acero inoxidable anti-corrosivo** *f* ING MECÁ corrosion-resistant stainless steel fastener; **~ aislante** *f* ING ELÉC insulator clamp; **~ de alineación** *f* PETROL line-up clamp; **~ de batería** *f* ING ELÉC battery clip; **~ de brida** *f* ING MECÁ stirrup strap; **~ de cabeza ovalada** *f* ING MECÁ oval-head fastener; **~ de cables** *f* ELEC, ING ELÉC cable clamp; **~ para cables** *f* ING MECÁ rope cleat; **~ de conducción de plato de presión** *f* AUTO pressure-plate drive strap; **~ de conexión a tierra** *f* ELEC, ING ELÉC earth clamp (*BrE*), ground clamp (*AmE*); **~ del distribuidor** *f* AUTO distributor-clamp bolt; **~ doble** *f* ING MECÁ dual clevis; **~ de extracción** *f* ALIMENT extraction thimble; **~ de extremo** *f* MINAS end bracket; **~ de guía** *f* ING MECÁ guide clamp; **~ de junta de vidrio esmerilado** *f* LAB *soporte* ground-glass joint clamp; **~ de manguera** *f* ING MECÁ, MECÁ hose clamp; **~ de palanca de funcionamiento antirresbaladizo** *f* PROD antislip operating lever clamp; **~ de plato** *f* AUTO plate strap; **~ de resorte** *f* ING MECÁ spring clip; **~ de retención de la pala** *f* TRANSP AÉR blade retention-strap; **~ revestida** *f* ING MECÁ lined clamp; **~ del rodamiento de bolas** *f* AUTO, VEH piston-ring clamp; **~ roscada** *f* ING MECÁ, MECÁ clevis bolt; **~ para sujetar tubos** *f* MINAS pipe clamp; **~ de tierra** *f* ELEC, ING ELÉC earth clamp (*BrE*), ground clamp (*AmE*); **~ de tornillo sin fin para manguera** *f* ING MECÁ worm-drive clamp; **~ del tubo** *f* CONST tube clip; ~ **de tubos** *f* CONST pipe joint; ~ **para tubos** *f* ING MECÁ pipe clip

abrazar *vt* ING MECÁ, MECÁ brace, clasp

abresurcos *m* AGRIC deep-furrow drill, runner; ~ **tipo zapata** *m* AGRIC shoe furrow opener

abretubos *m* PROD swage

abrevadero *m* AGRIC drinking trough

abrevar *vt* AGRIC, AGUA, HIDROL irrigate, water

abridor: ~ **de la bolsa** *m* EMB bag opener; ~ **de puerta de ventilación** *m* SEG ventilation door opener

abridora *f* TEXTIL opener

abrigo *m* D&A dugout, shield, GEOL, HIDROL screen, TELECOM dugout, TRANSP MAR shelter; ~ **del cañón** *m* D&A gun shield; ~ **subterráneo** *m* D&A, TELECOM dugout; ~ **subterráneo telefónico** *m* D&A, TELECOM telephone dugout

abrillantado *m* ING MECÁ *operación*, METAL, P&C, PAPEL polishing; ~ **a cepillo** *m* PAPEL brush polishing; ~ **mecánico** *m* ING MECÁ, METAL mechanical polishing

abrillantador: ~ **óptico** *m* P&C *materia prima* optical brightener

abrillantamiento *m* PROD brightening

abrir[1] *vt* AGUA turn on, CARBÓN blast, CINEMAT open up, CONST *tuberías* unstop, D&A unlock, FOTO *diafragma* open, ING ELÉC break, ING MECÁ bore, MINAS *un barreno con el pistolete* jump, *canteras* open up, crossdrive, *pozos* bore, cut down, OCEAN *frutos del mar* shuck, PROD *ordenadores* log in, *grifos, llaves, interruptores* turn on, SEG break open; ~ **completamente** *vt* ING MECÁ turn full on; ~ **de costado** *vt* TRANSP MAR *bote* fend off

abrir[2]: ~ **por aquí** *fra* EMB open here; ~ **por este extremo** *fra* EMB open this end

abrirse *v refl* CONST break open; ~ **de** *v refl* TRANSP MAR *muelle* clear

abrochado: ~ **con cierre** *m* TEXTIL plugging

abrochar *vt* TEXTIL *con botones* button, *con broches* fasten, *con hebilla* clasp

abrumadero: ~ **de peces** *m* OCEAN fish smoking

ABS *abr* AUTO *(sistema antibloqueo)* ABS *(antiblocking system)*, P&C *(acrilonitrilo-butadieno-estireno)* materia prima ABS *(acrylonitrile-butadiene-styrene)*

abscisa *f* INFORM&PD, MATEMÁT abscissa

absorbedor *m* GEN absorber; ~ **de cavidad** *m* ACÚST cavity absorber; ~ **de ondas** *m* ELEC *transmisión* surge absorber; ~ **solar** *m* TEC ESP solar absorber; ~ **de sonido** *m* ACÚST sound absorber

absorbencia *f* GEN absorbance, absorbency, IMPR holdout; ~ **solar** *f* ENERG RENOV solar absorbency; **poca** ~ **de la tinta** *f* IMPR ink holdout

absorbente[1] *adj* GEN absorbent, absorber; ~ **de rayos Ultra Violeta** *adj* C&V UV-absorbing

absorbente[2] *m* ACÚST absorber, AGUA, ALIMENT absorbent, absorber, CARBÓN, CONTAM, CONTAM MAR absorbent, ELEC, EMB absorbent, ENERG RENOV, FÍS ONDAS absorber, absorbent, FÍS PART absorbent, absorber, MECÁ absorbent, NUCL, PROC QUÍ absorber, QUÍMICA absorbent, absorber, REFRIG, TEC ESP, TEC PETR, TEXTIL absorber; ~ **del calor** *m* TERMO, TEXTIL heat absorber; ~ **compuesto** *m* NUCL composite absorber; ~ **de humedad** *m* EMB humidity absorber; ~ **de neutrones** *m* NUCL neutron absorber

absorber *vt* GEN absorb

absorbible *adj* QUÍMICA absorbable; **no** ~ *adj* QUÍMICA nonabsorbable

absorciometría *f* NUCL absorptiometry

absorciómetro *m* PAPEL absorptiometer

absorción *f* GEN absorption, ING MECÁ indraft *(AmE)*, indraft of air *(AmE)*, indraught *(BrE)*, indraught of air *(BrE)*, TEXTIL *del tinte* uptake; ~ **de aceite** *f* P&C oil absorption; ~ **atmosférica** *f* FÍS RAD, METEO atmospheric absorption; ~ **del baño** *f* TEXTIL *tintura* dip pickup; ~ **de calor** *f* TERMO heat absorption; ~ **por carbón activado** *f* CONTAM *tratamiento físico-químico* active carbon absorption; ~ **de carga** *f* TELECOM energy absorption; ~ **de criptón en anhídrido carbónico líquido** *f* NUCL krypton absorption in liquid carbon dioxide *(KALC process)*; ~ **dieléctrica** *f* ELEC, ING ELÉC dielectric absorption; ~ **específica** *f* METEO specific absorption; ~ **de fondo** *f* FÍS RAD background absorption; ~ **fotón-fotón** *f* FÍS PART photon-photon absorption; ~ **fotosférica** *f* FÍS RAD photospheric absorption; ~ **de gas** *f* GAS gas absorption, QUÍMICA *por metal* gassing, TEC PETR *refino* gas absorption; ~ **de luz** *f* FÍS RAD absorption of light; ~ **máxima** *f* ING ELÉC absorption peak; ~ **de microondas** *f* TELECOM microwave absorption; ~ **óptica** *f* FÍS RAD, ÓPT optical absorption; ~ **ozónica** *f* FÍS RAD ozone absorption; ~ **de radiación** *f* FÍS ONDAS, FÍS RAD absorption of radiation; ~ **de radiación ionizante** *f* FÍS ONDAS, FÍS RAD absorption of ionizing radiation; ~ **de rayos X** *f* FÍS RAD X-ray absorption; ~ **de resonancia** *f* TELECOM resonant absorption; ~ **Sabine** *f* ACÚST Sabine absorption; ~ **sonora** *f* ACÚST sound absorption; ~ **del suelo** *f* AGUA absorption in the soil; ~ **del terreno** *f* ACÚST ground absorption

abstracción *f* INFORM&PD abstraction; ~ **de datos** *f* INFORM&PD data abstraction

abstracto *adj* INFORM&PD abstract

abstraer *vt* INFORM&PD abstract

abullonado *m* TEXTIL swelling

abultamiento *m* AGUA, CONST bulking; ~ **de cienos** *m* AGUA sludge bulking

abundancia *f* FÍS, NUCL abundance; ~ **isotópica** *f* FÍS, NUCL isotopic abundance; ~ **nuclear** *f* FÍS, NUCL nuclear abundance; ~ **relativa** *f* FÍS, NUCL relative abundance

Ac *abr* *(actinio)* FÍS RAD, METAL, QUÍMICA Ac *(actinium)*

AC *abr* *(altocúmulo)* METEO, TRANSP AÉR AC *(altocumulus)*

acabado[1]: **sin** ~ *adj* PAPEL without finish; ~ **de cromado de bronce** *adj* PROD bronze-chromate finished

acabado[2] *m* C&V, COLOR finish, finishing, CONST *piedra* dressing, IMPR finish, finishing, ING MECÁ finishing, surfacing, MECÁ, P&C, PAPEL, PROD, REFRIG, REVEST, TEXTIL finish, finishing; ~ **abollado** *m* P&C *pintura* hammer finish; ~ **en agua** *m* PROD water finish, water finishing; ~ **por aire caliente** *m* REVEST hot-air finish; ~ **aislante** *m* REFRIG insulation finish; ~ **antideslizante** *m* REVEST slip-proof finish; ~ **arpillera** *m* PAPEL burlap finish; ~ **con aspecto de papel rizado** *m* TEXTIL crisp paper-like finish; ~ **avitelado** *m* PROD vellum finish; ~ **brillante** *m* COLOR, REVEST gloss finish; ~ **a casquillo de bayoneta** *m* C&V bayonet cap finish; ~ **champaña** *m* C&V champagne finish; ~ **de color** *m* COLOR color finish *(AmE)*, colour finish *(BrE)*; ~ **al corcho** *m* C&V

cork finish; ~ **craquelado** *m* PROD crackled finish; ~ **en cromo duro** *m* ING MECÁ hard chrome finish; ~ **cuadriculado** *m* C&V checked finish; ~ **del cuarzo** *m* ELECTRÓN finished quartz; ~ **con efecto de crujido** *m* REVEST scroop finish; ~ **encorvado** *m* C&V bent finish; ~ **especular** *m* ING MECÁ, MECÁ, PROD mirror finish; ~ **a espejo** *m* ING MECÁ, MECÁ, PROD mirror finish; ~ **de estañocincado** *m* REVEST tin-zinc finish; ~ **estucado** *m* REVEST non-iron finish; ~ **exterior** *m* COLOR exterior finish; ~ **final por galvanizado** *m* REVEST finally galvanized coating; ~ **a fondo** *m* REVEST full finish; ~ **a fuego** *m* C&V fire finish; ~ **granulado** *m* REVEST soil-release finish; ~ **hidrófugo** *m* REVEST, TEXTIL water-repellent finish; ~ **interior** *m* REVEST back finish; ~ **ligado** *m* REVEST bonded finish; ~ **de lustre** *m* REVEST luster finish (*AmE*), lustre finish (*BrE*); ~ **con lustre permanente** *m* REVEST permanent-sheen finish; ~ **martillado** *m* COLOR hammer tone finish; ~ **mate** *m* COLOR dull finish, matt finish, P&C *pintura* matting, TEXTIL dull finish; ~ **mate antiguo** *m* PAPEL velvet finish; ~ **mecánico** *m* ING MECÁ machine finishing; ~ **en negro** *m* PROD black finishing; ~ **no redondo** *m* C&V out-of-round finish; ~ **protector** *m* REVEST protective finish; ~ **a prueba de arrugas** *m* REVEST, TEXTIL crease-resistant finish, wrinkle-resistant finish; ~ **a prueba de contracciones** *m* REVEST shrinkproof finish; ~ **a prueba de deslizamiento** *m* REVEST slip-proof finish; ~ **a prueba de hinchazón** *m* REVEST swell-resistant finish; ~ **a prueba de niños** *m* C&V, REVEST, SEG childproof finish; ~ **a prueba de rugosidades** *m* REVEST wrinkle-resistant finish; ~ **resistente a las arrugas** *m* REVEST, TEXTIL crease-resistant finish; ~ **del reverso** *m* REVEST back finish; ~ **de revestimiento** *m* REVEST coating finish; ~ **satinado** *m* COLOR, REVEST satin finish; ~ **semimate** *m* COLOR ecru finish; ~ **de la superficie** *m* PAPEL surface finish; ~ **de superficie** *m* CONST surface dressing; ~ **en tambor giratorio** *m* PROD barrel finishing; ~ **total** *m* REVEST full finish

acabador *m* C&V, REVEST fire finisher

acaballonado *adj* AGRIC ridging

acabar[1]: **sin ~** *adj* PAPEL unfinished

acabar[2] *vt* GEN finish

acalabrotado *adj* TRANSP MAR *guindaleza* cable-laid, warped

acampanado[1] *adj* C&V flaring, MECÁ flared, PROD flaring, TEXTIL flared

acampanado[2] *m* C&V, PROD flaring

acampanar *vt* C&V, MECÁ, PROD, TEXTIL flare

acanalado[1] *adj* CONST ribbed, ING MECÁ corrugated, ribbed, MECÁ, TEXTIL *género de punto* ribbed

acanalado[2] *m* C&V fluting, ING MECÁ ribbing, MECÁ rib, ribbing, PAPEL flute, TEXTIL rib, *ligamento* ribbing; ~ **en espiral** *m* ING MECÁ involute serration; ~ **de involuta** *m* ING MECÁ involute serration; ~ **en V** *m* ELECTRÓN *transistores* V-groove

acanaladora *f* CARBÓN cutter arm, ING MECÁ routing machine, MINAS *canteras* channeler (*AmE*), channeling machine (*AmE*), channeller (*BrE*), channelling machine (*BrE*); ~ **para roca** *f* CONST, MINAS rock channeler (*AmE*), rock channeller (*BrE*); ~ **de rocas** *f* CONST, MINAS rock channeler (*AmE*), rock channeller (*BrE*)

acanaladura *f* CONST groove, ING MECÁ gash, spire, MECÁ cable, flute, groove, PROD groove, TEC ESP spline; ~ **abierta** *f* PROD *laminador* open pass

acanalar *vt* GEOL channel, PROD *ranuras* hollow out

acanticona *f* MINERAL acanticone

acantilado *m* CONST, GEOL, OCEAN, TRANSP MAR bluff, cliff

acantita *f* MINERAL acanthite

ácaro *m* AGRIC mite

acarreador *m* AGRIC feed conveyor; ~ **de pie de copa** *m* C&V stem carrier

acarrear *vt* INFORM&PD, ING MECÁ carry; ~ **hacia el mar** *vt* HIDROL carry out to sea

acarreo *m* EMB carriage, INFORM&PD carry, MINAS *geología* drifting, TRANSP carriage; ~ **anticipado** *m* INFORM&PD carry lookahead; ~ **circular** *m* INFORM&PD end-around carry; ~ **glaciar** *m* CARBÓN, GEOL boulder clay; ~ **hidráulico** *m* AGUA sluicing; ~ **parcial** *m* INFORM&PD partial carry

acarreos *m pl* CARBÓN *hidráulica* debris

acceder *vti* INFORM&PD access

accesibilidad *f* CALIDAD, MECÁ accessibility

acceso *m* CARBÓN access, adit, CONST ingress, ELECTRÓN, FERRO, INFORM&PD access, ING ELÉC *sistema*, ING MECÁ port, OCEAN approach, PROD port, TEC ESP access, TELECOM access, gateway; ~ **aleatorio** *m* ELECTRÓN, INFORM&PD, ING ELÉC, TEC ESP random access; ~ **aleatorio a la memoria** *m* INFORM&PD memory random access; ~ **de almacén frigorífico** *m* REFRIG port coldstore; ~ **a los andenes** *m* FERRO access to platforms; ~ **a archivos** *m* INFORM&PD file access; ~ **básico** *m* TELECOM basic access; ~ **de cabecera** *m* TELECOM overhead access (*OHA*); ~ **casual** *m* INFORM&PD, TEC ESP random access; ~ **directo** *m* ELECTRÓN direct access, INFORM&PD direct access, random access, ING ELÉC direct access, TEC ESP *información* random access; ~ **directo a la memoria** *m* (*DMA*) ELECTRÓN, INFORM&PD direct memory access (*DMA*); ~ **a distancia** *m* INFORM&PD remote access; ~ **al fichero** *m* INFORM&PD file access; ~ **a ficheros** *m* INFORM&PD file access; ~ **por fila en espera** *m* INFORM&PD queued access; ~ **fortuito** *m* INFORM&PD, TEC ESP random access; ~ **inicial** *m* TELECOM overhead access (*OHA*); ~ **inmediato** *m* INFORM&PD immediate access; ~ **integrado** *m* INFORM&PD, TELECOM integrated access; ~ **léxico** *m* INFORM&PD, TELECOM lexical access; ~ **mediante cola** *m* INFORM&PD queued access; ~ **a la memoria** *m* INFORM&PD, ING ELÉC memory access; ~ **múltiple** *m* INFORM&PD, TEC ESP, TELECOM multiple access; ~ **múltiple aleatorio** *m* INFORM&PD, TELECOM random multiple access; ~ **múltiple por detención de portadora** *m* TELECOM carrier sense multiple access (*CSMA*); ~ **múltiple en dirección a la onda portadora con detección de colisiones** *m* (*AMDOP/DC*) ELECTRÓN, INFORM&PD carrier sense multiple access with collision detection (*CSMA/CD*); ~ **múltiple por división de código** *m* TEC ESP code-division multiple access (*CDMA*); ~ **múltiple por división de frecuencias** *m* (*AMDF*) TEC ESP, TELECOM frequency-division multiple access (*FDMA*); ~ **múltiple por división de tiempo** *m* ELECTRÓN, INFORM&PD, TEC ESP, TELECOM time-division multiple access (*TDMA*); ~ **múltiple por espectro ensanchado** *m* TEC ESP spread spectrum multiple access (*SSMA*); ~ **a nivel de aplicaciones** *m* TELECOM application level gateway; ~ **al paquete**

m TELECOM packet port; ~ **en paralelo** *m* INFORM&PD parallel access; ~ **de periféricos** *m* PROD peripheral port; ~ **primario** *m* TELECOM primary access; ~ **a la RDSI** *m* TELECOM ISDN access; ~ **remoto** *m* INFORM&PD, TELECOM remote access; ~ **secuencial** *m* INFORM&PD sequential access; ~ **secuencial indexado** *m* INFORM&PD indexed sequential access; ~ **secuencial indicado** *m* INFORM&PD indexed sequential access; ~ **secuencial selectivo** *m* INFORM&PD selective sequential access; ~ **en serie** *m* INFORM&PD serial access; ~ **superior** *m* TELECOM path overhead (*POH*); ~ **a la tarifa primaria** *m* TELECOM primary-rate access; ~ **de transferencia de ficheros y elemento del servicio de manipulación** *m* INFORM&PD, TELECOM file transfer access and manipulation service element (*FTAMSE*); ~ **universal S** *m* TELECOM S-universal access; ~ **del usuario** *m* INFORM&PD, TELECOM user access; ~ **para un solo usuario** *m* INFORM&PD, TELECOM single-user access; ~ **en velocidad primaria a la RDSI** *m* TELECOM ISDN primary rate access

accesorio *m* CONST fitting, furnishing, ING MECÁ accessory, fitment, fitting, fixture, implement, INSTR *máquinas* attachment, MECÁ, PROD fitting, TEXTIL *para una máquina* attachment, TRANSP MAR *buque* fitting; ~ **de actuador** *m* ING MECÁ actuator attachment; ~ **actuador de giro parcial** *m* ING MECÁ part-turn actuator attachment; ~ **del automóvil** *m* AUTO, VEH car accessory; ~ **de bayoneta** *m* CONST bayonet fitting; ~ **de caldera** *m* INSTAL HIDRÁUL, PROD, TERMO boiler fitting; ~ **de cañería** *m* ING MECÁ pipe fitting; ~ **de captación** *m* ING MECÁ pick-up fitting; ~ **para cielo raso** *m* ELEC *alumbrado* ceiling fitting; ~ **de conducción de energía aérea** *m* ING ELÉC overhead power line fitting; ~ **de contacto auxiliar** *m* ELEC, PROD auxiliary contact accessory; ~ **para copiado de diapositivas** *m* FOTO slide-copying attachment; ~ **de cubierta** *m* TRANSP MAR deck gear; ~ **desmontable** *m* PROD removable insert; ~ **forjado de unión del ala** *m* TRANSP AÉR forged wing attachment; ~ **de fresado** *m* ING MECÁ cherrying attachment; ~ **para fresar** *m* ING MECÁ milling attachment; ~ **para fresar circularmente** *m* ING MECÁ circular milling attachment; ~ **de fuelle** *m* FOTO bellows attachment; ~ **para gas** *m* GAS, PROD gas fitting; ~ **hidráulico** *m* MINAS hydraulic prop; ~ **de inserción** *m* PROD removable insert; ~ **para la instalación** *m* PROD installation accessory; ~ **para lapidado** *m* ING MECÁ lapping fixture; ~ **lubricante** *m* ING MECÁ lubrication fitting; ~ **de luz incidente** *m* FOTO incident-light attachment; ~ **de minas** *m* MINAS mining appliance; ~ **para montaje en el techo** *m* ELEC *alumbrado* ceiling fitting; ~ **del motor** *m* ING MECÁ engine fitting; ~ **para muela de esmeril** *m* MECÁ, PROD emery-wheel attachment; ~ **plano** *m* ING MECÁ plain fitting; ~ **a presión** *m* ING MECÁ compressed fitting; ~ **de radar** *m* D&A radar equipment; ~ **reductor de tubería** *m* CONST reducing pipe fitting; ~ **sencillo** *m* ING MECÁ plain fitting; ~ **soldado** *m* ING MECÁ welded fitting; ~ **de suelda capilar** *m* ING MECÁ capillary solder fitting; ~ **de taladradora** *m* ING MECÁ drilling attachment; ~ **telescópico** *m* INSTR telescope mounting; ~ **termoencogible** *m* TERMO heat shrink fitting; ~ **para tubos** *m* ING MECÁ pipe fitting

accidente *m* GEOL, MINAS leap, SEG accident; ~ **al** **caminar** *m* SEG walking accident; ~ **de despresurización** *m* NUCL, SEG depressurization accident; ~ **eléctrico** *m* ELEC, ING ELÉC, SEG electrical accident; ~ **de eyección de barras** *m* NUCL, SEG control-rod ejection accident; ~ **fatal** *m* SEG fatal accident; ~ **industrial** *m* SEG industrial accident; ~ **laboral** *m* CONST, SEG accident at work; ~ **al levantar algo** *m* SEG lifting accident; ~ **de pérdida de refrigerante** *m* (*LOCA*) NUCL, REFRIG, SEG loss of coolant accident (*LOCA*); ~ **potencial** *m* SEG potential accident; ~ **de purga** *m* NUCL, SEG blowdown accident; ~ **reportable** *m* SEG notifiable accident, reportable accident

acción[1]: **de ~ doble** *adj* ING MECÁ double-acting; **de ~ rápida** *adj* ING ELÉC, ING MECÁ fast-acting, MECÁ high-speed, PROD fast-acting, snap-action

acción[2] *f* FÍS action, ING MECÁ action, movement, MECÁ action; ~ **de agrupar por pares** *f* ELEC *cable* cable; ~ **amortiguadora** *f* DETERG, QUÍMICA buffer action; ~ **de un buffer** *f* DETERG, QUÍMICA buffer action; ~ **capilar** *f* FÍS FLUID, TEC PETR *mecánica de fluidos* capillary action; ~ **de la capilaridad** *f* FÍS FLUID, TEC PETR capillary action; ~ **de casar por pares** *f* ELEC cable; ~ **de cebar** *f* MINAS *voladuras* priming; ~ **correctiva** *f* CALIDAD corrective action; ~ **de emergencia final** *f* TRANSP AÉR last-emergency action; ~ **de entidad funcional** *f* TELECOM functional entity action (*FEA*); ~ **de formar pares** *f* ELEC *cable* cable; ~ **de frenado** *f* AUTO, FERRO, VEH brake application; ~ **frenante de la corriente de Foucault** *f* ELEC *motor* eddy-current braking; ~ **impeditiva** *f* CALIDAD preventive action; ~ **insegura** *f* SEG unsafe act; ~ **interior** *f* ING MECÁ interaction; ~ **inversa** *f* CINEMAT, TV reverse action; ~ **con láser** *f* ELECTRÓN laser action; ~ **lenta** *f* CINEMAT, TV slow motion; ~ **de limpiar** *f* PROD *gasógenos* poking; ~ **momentánea** *f* ING ELÉC momentary action; ~ **de los muros** *f* CARBÓN wall effect; ~ **de poner aros** *f* PROD *barriles* hooping; ~ **de ponerse los cinturones de seguridad** *f* SEG, TRANSP AÉR belting-in; ~ **de probar el techo golpeándolo** *f* CARBÓN dynamic sounding; ~ **de rebasar los topes de fin de carrera** *f* MINAS *jaula de extracción* pulleying; ~ **recíproca** *f* ING MECÁ interaction; ~ **de reembarcar** *f* TRANSP MAR reshipping; ~ **de sobrepasar el fin de carrera** *f* MINAS *jaula de extracción* overwind, overwinding; ~ **de soltar los frenos** *f* AUTO, FERRO, VEH brake release, brakes off; ~ **tampón** *f* DETERG, QUÍMICA buffer action; ~ **tamponadora** *f* DETERG, QUÍMICA buffer action; ~ **de telescopiaje** *f* ING MECÁ telescoping; ~ **de timbrar una caldera** *f* C&V badging; ~ **viscosa** *f* FÍS FLUID viscous action; ~ **de la viscosidad entre la pared y el fluido** *f* FÍS FLUID action of viscosity between wall and fluid; ~ **en vivo** *f* CINEMAT, TV live action

accionado: ~ **por batería** *adj* ING ELÉC battery-powered; ~ **por CA** *adj* PROD AC operated; ~ **por cadena** *adj* ING MECÁ chain-driven; ~ **por correa** *adj* PROD belt-driven; ~ **por la corriente** *adj* ELEC, ING ELÉC current-operated (*AmE*), mains-operated (*BrE*); ~ **por debajo** *adj* ING MECÁ underdriven; ~ **eléctricamente** *adj* ELEC, FOTO, ING ELÉC, MECÁ electrically-driven; ~ **electromagnéticamente** *adj* ELEC, ING ELÉC electromagnetically-operated; ~ **por electromotor** *adj* ELEC, ING ELÉC, MECÁ motor-driven; ~ **por engranaje** *adj* ING MECÁ, MECÁ geared; ~ **por leva** *adj* ING MECÁ cam action; ~ **a mano** *adj* ING MECÁ, MECÁ hand-operated, PROD by hand power; ~ **por máquina** *adj* PROD engine-driven; ~ **por motor** *adj* ING MECÁ, MECÁ, PROD motor-driven; ~ **por muelle** *adj* PROD spring-loaded; ~ **por resorte** *adj* PROD spring-loaded; ~ **a tecla** *adj* INFORM&PD key-driven; ~ **y temporizado** *adj* PROD timed-driven

accionador *m* ELEC, INFORM&PD, ING ELÉC, ING MECÁ, MECÁ, PROD actuator, TEC ESP, TRANSP AÉR actuator; ~ **analógico** *m* ING ELÉC analog actuator; ~ **axial** *m* TEC ESP axial actuator; ~ **de la cerradura del cilindro** *m* PROD cylinder-lock operator; ~ **conductivo** *m* ING ELÉC conduction pump; ~ **digital** *m* ING ELÉC digital actuator; ~ **de extintor** *m* TRANSP AÉR extinguisher striker; ~ **numérico** *m* ING ELÉC digital actuator; ~ **oleoneumático** *m* PROD air-oil actuator; ~ **de ranura de monedas** *m* PROD coin-slot operator

accionamiento *m* MECÁ action, drive, TRANSP AÉR actuation; ~ **asincrónico** *m* ELEC *motor* asynchronous operation; ~ **asíncrono** *m* ELEC *motor* asynchronous operation; ~ **por baterías** *m* CINEMAT battery drive; ~ **por cadena** *m* ING MECÁ, MECÁ, VEH *motocicletas* chain drive; ~ **de centrífuga** *m* ING MECÁ centrifuge drive; ~ **centrífugo** *m* ING MECÁ centrifuge drive; ~ **directo** *m* ING MECÁ direct drive; ~ **con disparador** *m* FOTO push-button operation; ~ **a distancia** *m* GEN remote control; ~ **eléctrico** *m* ELEC, FOTO, ING ELÉC, NUCL electrical drive; ~ **de engranajes cónicos** *m* ING MECÁ bevel-gear drive; ~ **flexible** *m* MECÁ flexible drive; ~ **hidráulico** *m* ING MECÁ hydraulic drive, MECÁ fluid drive; ~ **individual** *m* ING MECÁ individual drive; ~ **con láser continuo** *m* ELECTRÓN continuous-laser action; ~ **sobre la leva** *m* VEH *motor* cam following; ~ **manual** *m* CINEMAT hand drive, ELEC manual control; ~ **de la máquina de papel** *m* PAPEL paper-machine drive; ~ **mecánico** *m* ING MECÁ, MECÁ mechanical drive; ~ **por monopolea y caja de velocidades** *m* ING MECÁ all-gear single-pulley drive; ~ **por motor** *m* FOTO, ING ELÉC motor drive; ~ **múltiple** *m* ING MECÁ multiple of gearing; ~ **por polea escalonada** *m* ING MECÁ step cone drive; ~ **principal** *m* ING MECÁ main drive; ~ **seccional** *m* PAPEL sectional drive; ~ **síncrono** *m* TV synchronous drive; ~ **por vapor** *m* INSTAL HIDRÁUL steam power

accionar *vt* CINEMAT run, ING MECÁ actuate, run, trigger, *una carga* run, MECÁ drive, actuate, PROD *mecanismos* run, *máquinas* operate, TEC ESP *dispositivos, mecanismos* drive, TELECOM *alarma* trigger; ~ **por aire comprimido** *vt* ING MECÁ run by compressed air; ~ **con palanca** *vt* ING MECÁ lever

aceitador *m* ING MECÁ oil cup

aceitadora *f* ING MECÁ oil cup

aceite *m* GEN oil; ~ **de absenta** *m* QUÍMICA absinthole; ~ **aislador** *m* ELEC, ING ELÉC, TEC PETR insulating oil; ~ **aislante** *m* ELEC, ING ELÉC, TEC PETR *lubricante* insulating oil; ~ **de ajonjolí** *m* AGRIC, ALIMENT sesame oil; ~ **de alcanfor** *m* QUÍMICA camphor oil; ~ **alcanforado** *m* QUÍMICA camphor oil; ~ **de anacardo** *m* ALIMENT, P&C *aglutinante de pintura* cashew-nut oil; ~ **anticorrosivo** *m* PROD anticorrosion oil; ~ **de antraceno** *m* QUÍMICA anthracene oil; ~ **de ballena** *m* DETERG sperm oil; ~ **de blanco de ballena** *m* DETERG spermaceti oil; ~ **de cacahuete** *m*

Esp (*cf aceite de maní AmL*) AGRIC, ALIMENT, QUÍMICA arachis oil, peanut oil; ~ **de calefacción** *m* TERMO furnace oil, heating oil; ~ **de cardamomo** *m* ALIMENT cardamom oil; ~ **de cártamo** *m* ALIMENT safflower oil; ~ **de cassia** *m* ALIMENT cassia oil; ~ **de colza** *m* AGRIC, ALIMENT canola oil, rapeseed oil; ~ **de conversión** *m* TEC PETR *refino* conversion oil; ~ **detergente** *m* DETERG, VEH *lubricación* detergent oil; ~ **de dos tiempos** *m* AUTO two-stroke oil; ~ **de encofrado** *m* CONST mold oil (*AmE*), mould oil (*BrE*); ~ **epoxi** *m* P&C *aglutinante de pintura* epoxidized oil; ~ **epoxídico** *m* P&C *aglutinante de pintura* epoxidized oil; ~ **epóxido** *m* P&C *aglutinante de pintura* epoxidized oil; ~ **esencial** *m* QUÍMICA essential oil; ~ **para extender el caucho** *m* TEC PETR extender oil; ~ **fijo** *m* QUÍMICA fixed oil; ~ **fuel** *m Esp* (*cf fueloil AmL*) QUÍMICA fuel oil; ~ **de fusel** *m* ALIMENT *fermentación* fusel oil; ~ **hidrogenado** *m* PROD hardened oil; ~ **de hígado de bacalao** *m* ALIMENT cod-liver oil; ~ **de inmersión** *m* METAL immersion oil; ~ **de linaza** *m* CONST, QUÍMICA, REVEST, TRANSP MAR *mantenimiento* linseed oil; ~ **de linaza cocido** *m* CONST, QUÍMICA, REVEST, TRANSP MAR boiled linseed oil; ~ **litográfico** *m* REVEST lithographic oil; ~ **de lubricación** *m* AUTO, PROD, VEH lubricating oil; ~ **lubricante** *m* AUTO, PROD, VEH lubricating oil; ~ **de macis** *m* ALIMENT mace oil; ~ **de maní** *m AmL* (*cf aceite de cacahuete Esp*) AGRIC, ALIMENT, QUÍMICA arachis oil, peanut oil; ~ **de mantequilla** *m* ALIMENT butter oil; ~ **de menta perilla** *m* QUÍMICA perilla-seed oil; ~ **mineral** *m* AUTO, CARBÓN, MINERAL, P&C mineral oil, TEC PETR stone oil; ~ **de mostaza** *m* ALIMENT mustard oil; ~ **del motor** *m* AUTO, ING MECÁ, MECÁ, VEH engine oil; ~ **multigrado** *m* AUTO, CARBÓN, VEH multigrade oil; ~ **neutralizado químicamente** *m* TEC PETR *refino* chemically-neutral oil; ~ **de nuez** *m* ALIMENT walnut oil; ~ **de orujo** *m* ALIMENT grape-seed oil; ~ **de palma** *m* ALIMENT, QUÍMICA palm nut oil, palm oil; ~ **de palmiste** *m* ALIMENT palm-kernel oil; ~ **de parafina** *m* PETROL, QUÍMICA, TEC PETR, TERMO liquid paraffin, paraffin oil; ~ **pesado** *m* TEC PETR *refino* fuel oil; ~ **de pescado** *m* QUÍMICA fish oil; ~ **de pino** *m* DETERG pine oil; ~ **de pulpa de madera** *m* P&C, QUÍMICA tall-oil; ~ **de refrigeración** *m* ING MECÁ, REFRIG, TERMO cooling oil; ~ **refrigerante** *m* ING MECÁ, REFRIG, TERMO cooling oil; ~ **de resina** *m* P&C, QUÍMICA tall-oil; ~ **de ricino** *m* P&C *aglutinante de pintura* castor oil; ~ **de sebo** *m* ALIMENT, QUÍMICA, TEXTIL tallow oil; ~ **de secado** *m* ALIMENT drying oil; ~ **secante** *m* P&C *aglutinante de pintura*, PROD *pinturas* drying oil; ~ **de semilla** *m* ALIMENT oil seed; ~ **de semilla de algodón** *m* ALIMENT cotton-seed oil; ~ **de semilla de mostaza** *m* ALIMENT mustard-seed oil; ~ **sulfatado** *m* DETERG sulfated oil (*AmE*), sulphated oil (*BrE*); ~ **del transformador** *m* ELEC, ING ELÉC transformer oil; ~ **para transformadores** *m* ELEC, ING ELÉC transformer oil; ~ **vegetal** *m* ALIMENT vegetable oil; ~ **de vitriolo** *m* QUÍMICA oil of vitriol

aceitera *f* PROD *lata* oiler, *para lubricación* oilcan

acelaico *adj* QUÍMICA azelaic

aceleración *f* GEN acceleration; ~ **acústica de referencia** *f* ACÚST reference sound acceleration; ~ **brusca** *f* ING MECÁ shock; ~ **centrífuga** *f* TRANSP AÉR centrifugal acceleration; ~ **centrípeta** *f* TRANSP

AÉR centripetal acceleration; **~ de Coriolis** *f* MECÁ, TEC ESP Coriolis acceleration; **~ debida a la gravedad** *f* FÍS, GEOFÍS, GEOL, TEC ESP acceleration due to gravity; **~ del giro** *f* TEC ESP spin-up; **~ de la gravedad** *f* FÍS gravity acceleration, GEOFÍS, GEOL acceleration due to gravity, TEC ESP gravity acceleration; **~ por la gravedad** *f* FÍS, GEOFÍS, GEOL, TEC ESP acceleration due to gravity; **~ gravitacional** *f* FÍS, GEOFÍS, GEOL gravitational acceleration; **~ de haz de electrones** *f* ELEC, ELECTRÓN, NUCL electron-beam acceleration; **~ larga** *f* PROD long acceleration; **~ lineal temporizada** *f* PROD linear timed acceleration; **~ negativa** *f* ING MECÁ, MECÁ deceleration, minus acceleration, retarded acceleration; **~ de partículas** *f* ACÚST, ELECTRÓN, FÍS, FÍS PART, NUCL particle acceleration; **~ retardada** *f* ING MECÁ retarded acceleration; **~ retardatriz** *f* ING MECÁ retarded acceleration; **~ sonora** *f* ACÚST sound acceleration; **~ tangencial** *f* FÍS, MECÁ tangential acceleration

acelerado *adj* AUTO, FÍS accelerated

acelerador *m* GEN accelerator, AUTO accelerator pedal (*BrE*), gas pedal (*AmE*), PROD *autos* throttle valve, VEH accelerator pedal (*BrE*), gas pedal (*AmE*); **~ de choque frontal** *m* ELECTRÓN, NUCL colliding-beam accelerator; **~ cíclico** *m* ELECTRÓN, FÍS, FÍS PART, NUCL cyclotron; **~ continuo** *m* ELECTRÓN, FÍS, FÍS PART, NUCL cyclotron; **~ de electrones** *m* ELECTRÓN, FÍS, FÍS PART, NUCL cyclotron, electron accelerator, synchrotron; **~ de fragrado** *m* CONST *hormigones* accelerator; **~ por inducción** *m* FÍS RAD induction accelerator; **~ intermedio** *m* TEC ESP booster; **~ de iones** *m* ELECTRÓN, FÍS, FÍS PART, NUCL cyclotron, ion accelerator, synchrotron; **~ iónico** *m* ELECTRÓN, FÍS PART, NUCL ion accelerator; **~ lineal** *m* ELECTRÓN, FÍS, FÍS PART, NUCL linear accelerator (*LINAC*); **~ de partículas** *m* ELECTRÓN, FÍS, FÍS PART, NUCL particle accelerator; **~ por propagación de ondas** *m* FÍS RAD wave-propagating accelerator; **~ por pulsos** *m* ELECTRÓN, FÍS RAD pulse-accelerator; **~ reproductor** *m* NUCL accelerator breeder

acelerar *vt* GEN accelerate, VEH *el motor* rev up

acelerómetro *m* GEN accelerometer; **~ de inercia** *m* TEC ESP inertial accelerometer; **~ inercial** *m* TEC ESP inertial accelerometer; **~ lateral** *m* TRANSPAÉR lateral accelerometer

acemite *m* ALIMENT *molienda, panadería, repostería* middlings

acento *m* IMPR accent; **~ flotante** *m* IMPR floating accent; **~ postizo** *m* IMPR loose accent (*BrE*), piece accent (*AmE*)

acentuación: **~ de los contrastes** *f* ACÚST, TEC ESP pre-emphasis

acentuador: **~ de bebedero** *m* ING MECÁ gate accentuator

acentuar *vt* IMPR accentuate

aceotrópico *adj* QUÍMICA, REFRIG, TERMO azeotropic

aceótropo *m* QUÍMICA, REFRIG, TERMO azeotrope

aceptación *f* ACÚST recognition, CALIDAD acceptance, INFORM&PD support, MECÁ acceptance, TRANSP AÉR approval; **~ de direcciones extensas** *f* TELECOM long-address acceptance; **~ de un haz** *f* ELECTRÓN, NUCL acceptance of a beam; **~ de productos** *f* CALIDAD receipt of goods; **~ de sobrecarga** *f* TELECOM overflow accept

aceptador *m* INFORM&PD acceptor

aceptar *vt* INFORM&PD support, ING MECÁ take in

aceptor *m* ELEC, ELECTRÓN, FÍS, FÍS PART, INFORM&PD, QUÍMICA *átomo, molécula* acceptor; **~ de ácidos** *m* P&C acid acceptor; **~ de impurezas** *m* ELECTRÓN acceptor impurity

acequia *f* AGUA, CARBÓN ditch, drain

acera *f* *Esp* (*cf vereda AmL*) CONST sidewalk (*AmE*), pavement (*BrE*), *puente* catwalk; **~ móvil** *f* *Esp* (*cf vereda móvil AmL*) TRANSP moving pavement (*BrE*), moving sidewalk (*AmE*), traveling sidewalk (*AmE*); **~ móvil capotada** *f* *Esp* (*cf vereda móvil capotada AmL*) TRANSP cabin-type moving pavement (*BrE*), cabin-type moving sidewalk (*AmE*); **~ para transporte** *f* *Esp* (*cf vereda para transporte AmL*) CONST moving platform (*BrE*), moving sidewalk (*AmE*)

acercamiento *m* INFORM&PD zoom-in, TEC ESP approach; **~-alejamiento** *m* INFORM&PD zooming

acercar *vt* INFORM&PD zoom in, MINAS approach

acería *f* METAL, PROD steelworks

acero *m* METAL steel; **~ aleado** *m* CARBÓN, METAL, MINAS alloyed steel; **~ austenítico** *m* MECÁ, METAL austenitic steel; **~ azulado con nitrato potásico** *m* ING MECÁ, MECÁ, METAL niter-blued steel (*AmE*), nitre-blued steel (*BrE*); **~ de baja aleación** *m* METAL low-alloy steel; **~ bajo en carbono** *m* ING MECÁ, METAL low-carbon steel, mild steel; **~ para barrenas** *m* METAL, MINAS drill steel; **~ Bessemer** *m* METAL, PROD Bessemer steel; **~ para brocas** *m* ING MECÁ, METAL, MINAS drill steel; **~ calmado con aluminio** *m* MECÁ, METAL aluminium-killed steel (*BrE*), aluminum-killed steel (*AmE*); **~ al carbono** *m* CARBÓN, ING MECÁ, MECÁ, METAL, NUCL carbon steel; **~ de cementación** *m* ING MECÁ, MECÁ, METAL case-hardened steel; **~ colado** *m* MECÁ, METAL cast steel; **~ de corte rápido** *m* MECÁ, METAL high-speed steel; **~ diamagnético** *m* ELEC, ING ELÉC, METAL nonmagnetic steel; **~ dulce** *m* ING MECÁ, METAL mild steel; **~ duro** *m* CONST, METAL breaker steel; **~ estructural** *m* CONST, METAL structural steel; **~ fundido** *m* MECÁ, METAL cast steel; **~ de gran resistencia a la tracción** *m* MECÁ, METAL high-tensile steel; **~ hipereutectoide** *m* METAL hypereutectoid steel; **~ inoxidable** *m* AUTO, ING MECÁ, METAL, PROD, VEH stainless steel; **~ inoxidable austenítico** *m* ING MECÁ, METAL, PROD austenitic stainless steel; **~ inoxidable cementado** *m* ING MECÁ, METAL, PROD hardened stainless steel; **~ inoxidable ferrítico** *m* ING MECÁ, MECÁ, METAL, PROD, TEC ESP ferritic stainless steel; **~ al manganeso** *m* MECÁ, METAL, PROD manganese steel; **~ moldeado** *m* MECÁ, METAL cast steel; **~ nitrurado** *m* MECÁ, METAL nitrided steel; **~ para picos** *m* METAL, MINAS drill steel; **~ para la producción de martensita exenta de carbono** *m* METAL maraging steel; **~ al silicio** *m* ING ELÉC, METAL silicon steel; **~ suave** *m* ING MECÁ, METAL low-carbon steel, mild steel

acertar *vt* INFORM&PD hit

acetal[1] *adj* ALIMENT acetal

acetal[2] *m* P&C, QUÍMICA *polímero* acetal; **~ de polivinilo** *m* P&C, QUÍMICA *polímero* polyvinyl acetal

acetaldehído *m* ALIMENT, QUÍMICA acetaldehyde, ethanal

acetaldol *m* QUÍMICA aldol

acetanilida *f* QUÍMICA acetanilide

acetato *m* ALIMENT, P&C, QUÍMICA, TEXTIL acetate; **~ de amilo** *m* P&C, QUÍMICA amyl acetate; **~ básico**

de cobre *m* QUÍMICA verdigris; **~ de bornilo** *m* QUÍMICA bornyl acetate; **~ de butilo** *m* P&C, QUÍMICA butyl acetate; **~~butirato de celulosa** *m* P&C, QUÍMICA cellulose acetobutyrate; **~ de celulosa** *m* CINEMAT, EMB, P&C, QUÍMICA cellulose acetate; **~ de etilo** *m* ALIMENT, P&C, QUÍMICA ethyl acetate; **~ de metilo** *m* DETERG, P&C, QUÍMICA methyl acetate; **~ de polivinilo** *m* P&C *polímero* polyvinyl acetate; **~ secundario** *m* TEXTIL secondary acetate

acético *adj* QUÍMICA *agente acidificante* acetic

acetificación *f* QUÍMICA acetification

acetificar *vt* QUÍMICA acetify

acetilénico *adj* CONST, GAS, MECÁ, QUÍMICA, SEG, TERMO acetylenic

acetileno *m* CONST, GAS, MECÁ, QUÍMICA TERMO, SEG acetylene, ethyne

acetilo *m* QUÍMICA acetyl

acetilsalicílico *adj* QUÍMICA acetylsalicylic

acetina *f* QUÍMICA acetin

acetobacteria *f* ALIMENT acetobacterium

acetoína *f* ALIMENT *producto para fermentación*, QUÍMICA acetoin

acetólisis *f* ALIMENT, QUÍMICA acetolysis

acetona *f* ALIMENT, P&C, QUÍMICA, REVEST acetone, propanone

acetonémico *adj* AGRIC ketotic

acetonitrilo *m* QUÍMICA acetonitrile

acetonuria *f* QUÍMICA acetonuria

acetosil *m* ALIMENT, QUÍMICA acetoxyl, benzoyl peroxide

acetoxi- *pref* QUÍMICA acetoxy-

ACF *abr (función avanzada de comunicaciones)* INFORM&PD ACF *(advanced communications function)*

achacamiento *m* PROD flattening

achaflanar *vt* C&V bevel, CONST bevel, chamfer, IMPR bevel, ING MECÁ chamfer, bevel, INSTR, MECÁ, PETROL, PROD, TEC PETR bevel

achaparrado *adj* AGRIC stunted

achatado *adj* GEOM oblate

achicador *m* FÍS FLUID, PETROL bailer, TRANSP MAR bailer, dipper

achicar *vt* AGUA drain, FÍS FLUID bail, bail out, INSTAL HIDRÁUL dewater, drain, dry up, MINAS dewater, TRANSP MAR bail out, *con bomba* pump out

achiote *m* ALIMENT anatto

achique *m* MINAS drainage, draining, PETROL bailing, TEC PETR swabbing; **~ de sedimentos** *m* AGUA, HIDROL sludge dewatering

acicular *adj* AGRIC narrow-leaved, GEOL acicular, METAL acicular, needle-shaped, QUÍMICA aciform

aciculita *f* MINERAL aciculite

acidez *f* GEN acidity; **~ atmosférica** *f* CONTAM, METEO atmospheric acidity; **~ libre** *f* QUÍMICA free acidity

acídico *adj* CONTAM, HIDROL, QUÍMICA acidic

acidífero *adj* QUÍMICA acidiferous

acidificable *adj* QUÍMICA acidifiable

acidificación *f* GEN acidification; **~ acuática** *f* AGUA, CONTAM, CONTAM MAR, HIDROL aquatic acidification; **~ del agua** *f* AGUA, CONTAM, CONTAM MAR, HIDROL aquatic acidification; **~ antropogénica** *f* CONTAM anthropogenic acidification; **~ artificial** *f* CONTAM artificial acidification; **~ natural** *f* CONTAM natural acidification; **~ del pozo** *f* PETROL, TEC PETR *producción* acid well treatment *(AWT)*

acidificado *adj* QUÍMICA acidulated

acidificador[1] *adj* CONTAM, TEC PETR *características químicas* acidifying

acidificador[2] *m* CONTAM, CONTAM MAR, QUÍMICA, TEC PETR, TEXTIL acidifier

acidificante *adj* CONTAM *características químicas* acidifying

acidificar *vt* GEN acidify

acidimetría *f* PAPEL, PROC QUÍ, QUÍMICA acidimetry

acidimétrico *adj* PAPEL, PROC QUÍ, QUÍMICA acidimetric

acidímetro *m* PAPEL, PROC QUÍ, QUÍMICA acidimeter

ácido *m* AGUA, ALIMENT, CONTAM, QUÍMICA, TEXTIL acid; **~ abiético** *m* PAPEL, QUÍMICA abietic acid; **~ acético** *m* ALIMENT acetic acid, acetic anhydride, P&C *materia prima*, QUÍMICA acetic acid; **~ acético glacial** *m* ALIMENT, QUÍMICA glacial acetic acid; **~ aconítico** *m* QUÍMICA aconitic acid; **~ agotado** *m* DETERG spent acid; **~ de aldehído** *m* ALIMENT, QUÍMICA aldehyde acid; **~ algínico** *m* ALIMENT, QUÍMICA alginic acid; **~ alquilsulfónico** *m* DETERG, QUÍMICA alkyl sulphonic acid; **~ amíctico** *m* QUÍMICA amic acid; **~ aminoacético** *m* ALIMENT, QUÍMICA glycine; **~ araquidónico** *m* ALIMENT, QUÍMICA arachidonic acid; **~ aspártico** *m* QUÍMICA aspartic acid; **~ barbitúrico** *m* QUÍMICA barbituric acid; **~ behénico** *m* QUÍMICA behenic acid; **~ benzoico** *m* ALIMENT, QUÍMICA benzoic acid; **~ biliar** *m* QUÍMICA bile acid; **~ bisulfuroso** *m* QUÍMICA hydrosulfurous acid *(AmE)*, hydrosulphurous acid *(BrE)*; **~ bórico** *m* C&V, QUÍMICA, REVEST boric acid; **~ butírico** *m* AGRIC, ALIMENT, QUÍMICA butyric acid; **~ cafeico** *m* ALIMENT, QUÍMICA caffeic acid; **~ de cámaras** *m* PROC QUÍ, QUÍMICA chamber acid; **~ cáprico** *m* QUÍMICA capric acid; **~ carbámico** *m* QUÍMICA carbamic acid; **~ de caseína** *m* ALIMENT, QUÍMICA casein acid; **~ cítrico** *m* ALIMENT, DETERG, QUÍMICA citric acid; **~ clorhídrico** *m* DETERG, QUÍMICA hydrochloric acid; **~ clórico** *m* ALIMENT, QUÍMICA chloric acid; **~ comestible** *m* ALIMENT edible acid; **~ decanoico** *m* QUÍMICA decanoic acid; **~ desoxirribonucleico** *m* *(ADN)* QUÍMICA deoxyribonucleic acid *(DNA)*; **~ diacético** *m* QUÍMICA diacetic acid; **~ dialúrico** *m* QUÍMICA dialuric acid; **~ dibromosuccínico** *m* QUÍMICA dibromosuccinic acid; **~ dicloroacético** *m* QUÍMICA dichloracetic acid; **~ dietilénico** *m* QUÍMICA diethylenic acid; **~ ditiobenzoico** *m* QUÍMICA dithiobenzoic acid; **~ ditiónico** *m* QUÍMICA dithionic acid; **~ embélico** *m* QUÍMICA *producto natural* embelin; **~ empleado** *m* TEXTIL spent acid; **~ estífnico** *m* QUÍMICA styphnic acid; **~ etanoico** *m* ALIMENT, QUÍMICA ethanoic acid; **~ etilendiamino tetra-acético** *m* *(EDTA)* DETERG, QUÍMICA ethylenediamin etra-acetic acid *(EDTA)*; **~ fítico** *m* ALIMENT, QUÍMICA phytic acid; **~ fluobórico** *m* QUÍMICA fluoroboric acid; **~ fólico** *m* ALIMENT, QUÍMICA folic acid; **~ fosfórico** *m* QUÍMICA phosphoric acid; **~ fulmínico** *m* QUÍMICA fulminic acid; **~ glucónico** *m* DETERG, QUÍMICA gluconic acid; **~ graso** *m* AGRIC, ALIMENT, DETERG, QUÍMICA fatty acid; **~ graso esencial** *m* *(EFA)* ALIMENT, QUÍMICA essential fatty acid *(EFA)*; **~ graso volátil** *m* AGRIC, QUÍMICA volatile fatty acid *(VFA)*; **~ hidantoico** *m* QUÍMICA hydantoic acid; **~ hidracrílico** *m* QUÍMICA hydracrylic acid; **~ hidrazoico** *m* QUÍMICA hydrazoic acid; **~ hidrosulfuroso** *m* QUÍMICA hydrosulfurous acid *(AmE)*, hydrosulphurous acid *(BrE)*; **~ hipoclórico**

m QUÍMICA hypochloric acid; ~ **hipocloroso** *m* QUÍMICA hypochlorous acid; ~ **hipofosfórico** *m* QUÍMICA hypophosphoric acid; ~ **hipofosforoso** *m* QUÍMICA hypophosphorous acid; ~ **hiposulfuroso** *m* QUÍMICA hyposulfurous acid (*AmE*), hyposulphurous acid (*BrE*); ~ **húmico** *m* QUÍMICA humic acid; ~ **isopropílico** *m* (*IPA*) DETERG, QUÍMICA isopropyl acid (*IPA*); ~ **láctico** *m* ALIMENT, QUÍMICA lactic acid; ~ **linoleico** *m* ALIMENT, P&C, QUÍMICA linoleic acid; ~ **maleico** *m* QUÍMICA maleic acid; ~ **málico** *m* ALIMENT, QUÍMICA malic acid; ~ **nítrico** *m* CONTAM *producto químico*, QUÍMICA nitric acid; ~ **oléico** *m* ALIMENT, DETERG, PROD, QUÍMICA oleic acid; ~ **orgánico** *m* QUÍMICA organic acid; ~ **ortofosfórico** *m* DETERG, QUÍMICA orthophosphoric acid; ~ **oxálico** *m* FOTO, QUÍMICA oxalic acid; ~ **parabánico** *m* QUÍMICA parabanic acid; ~ **periódico** *m* QUÍMICA periodic acid; ~ **pirofosfórico** *m* DETERG, QUÍMICA pyrophosphoric acid; ~ **polifosfórico** *m* DETERG polyphosphoric acid; ~ **precursor** *m* QUÍMICA parent acid; ~ **prúsico** *m* QUÍMICA prussic acid; ~ **silícico** *m* QUÍMICA silicic acid; ~ **sórbico** *m* QUÍMICA sorbic acid; ~ **sulfámico** *m* DETERG, QUÍMICA sulfamic acid (*AmE*), sulphamic acid (*BrE*); ~ **sulfhídrico** *m* QUÍMICA hydrogen sulfide (*AmE*), hydrogen sulphide (*BrE*); ~ **sulfúrico** *m* CONTAM, DETERG, QUÍMICA sulfuric acid (*AmE*), sulphuric acid (*BrE*), vitriolic acid; ~ **sulfúrico fumante** *m* DETERG, QUÍMICA fuming sulfuric acid (*AmE*), fuming sulphuric acid (*BrE*), oleum; ~ **sulfuroso** *m* CONTAM *producto químico*, QUÍMICA sulfurous acid (*AmE*), sulphurous acid (*BrE*); ~ **tartárico** *m* DETERG, QUÍMICA tartaric acid; ~ **titánico** *m* QUÍMICA titanic acid; ~ **túngstico** *m* QUÍMICA tungstic acid; ~ **vitriólico** *m* CONTAM, DETERG, QUíMICA sulfuric acid (*AmE*), sulphuric acid (*BrE*), vitriolic acid

acidólisis *m* ALIMENT, PROC QUÍ, QUÍMICA acidolysis

acidómetro *m* ALIMENT, INSTR, PROC QUÍ, QUÍMICA acidimeter

acidular *vt* ALIMENT, CONTAM acidify, PROC QUÍ, QUÍMICA acidulate

acierto *m* INFORM&PD hit

acilación *f* PROC QUÍ, QUÍMICA acylation

acilar *vt* QUÍMICA acylate

acílico *adj* QUÍMICA acyclic

acilo *m* QUÍMICA acyl

aclimatación *f* METEO acclimatization

acmita *f* MINERAL acmite

acodado: ~ **doble** *m* ING MECÁ duplex crank

acodamiento *m* CONST, MECÁ bend

acodar *vt* CONST, MECÁ bend

acodo *m* AGRIC layering, GEOFÍS offset, MECÁ knee bend

acolchado[1] *adj* EMB, ING MECÁ, MECÁ padded

acolchado[2] *m* EMB, ING MECÁ, MECÁ padding

acolchador *m* TEXTIL quilter

acolchar *vt* TEXTIL quilt, TRANSP *contenedores* stuff

acolchonado *m* TEC ESP padding

acolchonamiento: ~ **de suelos** *m* AGRIC mulching

acollador *m* TRANSP MAR *construcción naval* lanyard, rigging screw

acolmatado *adj* PAPEL *fieltros, telas* clogging

acometida *f* ELEC, ING ELÉC connection, tapping, TEC ESP *electricidad* service; ~ **de agua** *f* AGUA water fitting; ~ **de entrada** *f* ELEC, ING ELÉC input tapping

acomodamiento *m* D&A *artillería* bedding

acomodo: ~ **en estibas** *m* C&V stacking

acompaña: ~ **al embalaje** *fra* EMB *documentación* add-on to the packaging

acondicionado[1] *adj* INSTAL TERM, REFRIG, TERMO air-conditioned, TEXTIL conditioned

acondicionado[2]: ~ **del papel para ensayos** *m* PAPEL paper conditioning

acondicionador: ~ **de aire** *m* AUTO, CONST, ING MECÁ, INSTAL TERM, REFRIG, TERMO air-conditioning unit; ~ **de aire con condensador de agua** *m* INSTAL TERM, REFRIG, TERMO water-cooled air-conditioning unit; ~ **por aire enfriado** *m* INSTAL TERM, REFRIG, TERMO air-cooled air conditioning unit; ~ **autónomo** *m* INSTAL TERM, REFRIG, TERMO self-contained air-conditioning unit; ~ **del fieltro** *m* PAPEL felt conditioner; ~ **de habitaciones** *m* CONST, INSTAL TERM, REFRIG, TERMO room air-conditioning unit; ~ **de insuflación directa** *m* INSTAL TERM, REFRIG, TERMO free-flow air-conditioning unit; ~ **mural** *m* REFRIG console air conditioning unit; ~ **de señal** *m* ELECTRÓN signal conditioner; ~ **de techo** *m* CONST, INSTAL TERM, REFRIG, TERMO rooftop air-conditioning unit; ~ **a través de la pared** *m* CONST, INSTAL TERM, REFRIG, TERMO through-the-wall air-conditioning unit; ~ **de ventana** *m* CONST, INSTAL TERM, REFRIG, TERMO window air conditioning unit; ~ **de zona** *m* REFRIG zone air conditioning unit

acondicionadora: ~ **de heno** *f* AGRIC hay conditioner

acondicionamiento *m* C&V conditioning, CARBÓN *del aire* air conditioning, IMPR, P&C conditioning; ~ **de aire** *m* GEN air conditioning; ~ **de aire para invierno y verano** *m* CONST, INSTAL TERM, REFRIG all year air conditioning; ~ **de ambientes** *m* ING MECÁ air conditioning; ~ **de cienos** *m* AGUA sludge conditioning; ~ **de señal** *m* ELECTRÓN signal conditioning; ~ **de señales de entrada** *m* ELECTRÓN input signal conditioning

acondicionar *vt* CONST, INSTAL TERM air-condition, MINAS lay out, REFRIG, TERMO air-condition

aconitasa *f* QUÍMICA aconitase

aconitato *m* QUÍMICA aconitate

aconítico *adj* QUÍMICA aconitic

aconitina *f* ALIMENT aconitine

acontecimiento *m* ELECTRÓN conditioning, GEN event; ~ **fortuito** *m* ELECTRÓN random event; ~ **de telefonía entrelazada hacia atrás** *m* TELECOM backward-interworking telephony event

acopio *m* CONST stockyard, ENERG RENOV pondage

acoplado[1] *adj* ELECTRÓN *base, diodo, transistor* implanted, ING MECÁ connected; ~ **por carga** *adj* INFORM&PD load-coupled; ~ **directamente** *adj* INFORM&PD direct-coupled; ~ **lateralmente** *adj* ING ELÉC sidewall-coupled; ~ **en paralelo** *adj* ING ELÉC parallel-connected

acoplado[2]: **no** ~ **al sistema** *adv* INFORM&PD, TELECOM off-line

acoplado[3]: ~ **alimentador de comedero** *m* AGRIC self-feeding rack wagon

acoplador *m* GEN coupler, coupling; ~ **acústico** *m* ACÚST, ELECTRÓN, INFORM&PD acoustic coupler; ~ **de barra colectora** *m* ELEC *conmutador* busbar coupler; ~ **de corriente alterna** *m* TELECOM alternating-current coupler; ~ **de corriente continua** *m* TELECOM direct-current coupler; ~ **direccional** *m* FÍS, ING ELÉC, TELECOM directional coupler; ~ **de enchufe**

hembra *m* ELEC *conexión* socket coupler; ~ **en estrella** *m* ÓPT, TELECOM star coupler; ~ **de fibra óptica** *m* ÓPT, TELECOM optical fiber coupler (*AmE*), optical fibre coupler (*BrE*); ~ **de haz longitudinal** *m* TRANSP AÉR longitudinal-beam coupler; ~ **lateral de vigas** *m* TRANSP AÉR lateral-beam coupler; ~ **óptico** *m* ING ELÉC, ÓPT, TELECOM optical coupler; ~ **óptico direccional** *m* ING ELÉC, ÓPT, TELECOM directional optical coupler; ~ **optoelectrónico** *m* ING ELÉC, ÓPT, TELECOM optoelectronic coupler; ~ **rápido** *m* TEC ESP quick coupler; ~ **en T** *m* ÓPT, TELECOM tee coupler; ~ **variable** *m* ING ELÉC variocoupler; ~ **de vuelo estacionario** *m* TRANSP AÉR hover flight coupler: ~ **en Y** *m* LAB Y-piece, ÓPT, TELECOM y-coupler

acoplamiento *m* CONST *tuberías* connection, joint, strutting, ELEC *transformador* vector group, *conexión, inducción* coupling, ELECTRÓN implant, implantation, FERRO, FÍS, INFORM&PD coupling, ING ELÉC coupling, grouping, ING MECÁ link, linkage, coupling, attachment, connection, MECÁ mating, coupling, interlocking, NUCL linkage, QUÍMICA coupling, TEC ESP docking, attachment, TELECOM grouping, coupling, TRANSP, TRANSP AÉR, TRANSP MAR, VEH coupling; ~ **del acelerador** *m* FERRO *carburador* accelerator linkage; ~ **acodado** *m* ING MECÁ bend coupling; ~ **acústico** *m* ACÚST, ELECTRÓN, TELECOM acoustic coupling; ~ **ajustado** *m* FÍS tight coupling; ~ **antiparalelo** *m* ELEC antiparallel connection; ~ **automático rígido** *m* AUTO, TRANSP, VEH rigid automatic coupling; ~ **automático semi-rígido** *m* TRANSP semi-rigid automatic coupling; ~ **de bayoneta** *m* ING ELÉC bayonet coupling; ~ **de bridas** *m* ING MECÁ flange coupling, flanged union, plate coupling; ~ **de cables** *m* ELEC, ING ELÉC cable coupling; ~ **de cadena** *m* ING MECÁ chain coupling; ~ **en cantidad** *m* MINAS parallel connection; ~ **de capacidad** *m* ELEC, ING ELÉC capacitance coupling; ~ **por capacitancia y resistencia** *m* ELEC, ELECTRÓN, ING ELÉC resistance-capacitance coupling; ~ **capacitivo** *m* ELEC, ING ELÉC *capacitancia*, TELECOM capacitive coupling; ~ **cardan** *m* ING MECÁ cardan coupling; ~ **por carga** *m* INFORM&PD load coupling; ~ **cerrado** *m* ELEC *inductancia*, ING ELÉC close coupling, TV tight coupling; ~ **de charnela** *m* ING MECÁ swing-bolt coupling; ~ **de cinta** *m* ING MECÁ band coupling; ~ **de codo en ángulo recto** *m* ING MECÁ right-angled bend coupling; ~ **de collar** *m* ING MECÁ collared coupling; ~ **compacto** *m* INFORM&PD close coupling; ~ **de conexión en U** *m* ING MECÁ, MECÁ clevis link; ~ **coordinado** *m* METAL coordinate linkage; ~ **copiador** *m* ING MECÁ copying attachment; ~ **en cruce** *m* ING ELÉC cross coupling; ~ **cruzado** *m* ING ELÉC cross coupling; ~ **débil** *m* FÍS loose coupling, NUCL weak coupling; ~ **en derivación** *m* ING ELÉC paralleling, MINAS parallel connection; ~ **de desembrague rápido** *m* ING MECÁ, PROD quick-release coupling; ~ **desviador** *m* ELEC, ING ELÉC pull-off coupling; ~ **direccional** *m* ELECTRÓN, ING ELÉC, TELECOM directional coupling; ~ **directo** *m* ELEC, ELECTRÓN, ENERG RENOV, ING ELÉC direct coupling, ING MECÁ direct drive; ~ **de discos** *m* ING MECÁ flange coupling, plate coupling; ~ **de eje** *m* ING ELÉC, ING MECÁ shaft coupling; ~ **de eje flexible** *m* ING MECÁ flexible-shaft coupling, resilient shaft-coupling; ~ **elástico** *m* ING MECÁ compensating coupling; ~ **electromagnético** *m* ELEC, ING ELÉC,

TRANSP electromagnetic coupling; ~ **de electrones** *m* ELECTRÓN, FÍS PART, ING ELÉC electron coupling; ~ **electrónico** *m* ELECTRÓN, FÍS PART, ING ELÉC electron coupling; ~ **del embrague** *m* ING MECÁ clutch coupling; ~ **escalonado** *m* ING ELÉC stage coupling; ~ **estrecho** *m* ELEC *inductancia*, INFORM&PD, ING ELÉC close coupling; ~ **y fijación** *m* ELECTRÓN implant and anneal; ~ **flexible** *m* ING MECÁ flexible coupling, flexible joint; ~ **flojo** *m* FERRO loosening, ING ELÉC loose coupling; ~ **de la frecuencia del campo con la de la red** *m* TV locking; ~ **de fresado vertical** *m* ING MECÁ vertical milling attachment; ~ **por fricción** *m* ING MECÁ friction coupling; ~ **fuerte** *m* ELEC *inductancia*, ING ELÉC close coupling, tight coupling; ~ **de garras** *m* ING MECÁ dog clutch; ~ **de guía de ondas** *m* ING ELÉC waveguide coupling; ~ **de guíaondas** *m* ING ELÉC waveguide coupling; ~ **hidráulico** *m* PROD, VEH *transmisión* fluid coupling; ~ **por impedancia** *m* ING ELÉC impedance coupling; ~ **por impulso** *m* ING MECÁ impulse coupling; ~ **impulsor** *m* ING MECÁ impulse coupling; ~ **por inducción** *m* ELEC, FÍS, ING ELÉC inductive coupling; ~ **por inducción mutua** *m* ELEC, FÍS, ING ELÉC mutual inductive coupling; ~ **por inductancia** *m* ELEC, FÍS, ING ELÉC inductance coupling; ~ **de inductancia mutua** *m* ELEC, FÍS, ING ELÉC mutual inductance coupling; ~ **inductivo** *m* ELEC, FÍS, ING ELÉC inductive coupling, loose coupling; ~ **inicial a la red** *m* NUCL first connection to grid; ~ **j-j** *m* FÍS j-j coupling; ~ **de línea** *m* ING ELÉC line coupling; ~ **por líquido** *m* VEH *transmisión* fluid coupling; ~ **L-S** *m* FÍS L-S coupling; ~ **magnético** *m* ELEC *transformador* magnetic coupling; ~ **en malla** *m* ING ELÉC *redes* loop coupling; ~ **de mamparo** *m* ING MECÁ bulkhead coupling; ~ **de manguera** *m* CONST hose coupler; ~ **de manguito** *m* ING MECÁ box coupling, butt coupling; ~ **mecánico** *m* ELEC *resistencias*, ING ELÉC ganging; ~ **metálico flexible** *m* ING ELÉC pigtail; ~ **de modos** *m* ÓPT, TELECOM mode coupling; ~ **mutuo** *m* ING MECÁ interface, TELECOM mutual coupling; ~ **neumático** *m* ING MECÁ pneumatic coupling; ~ **normal** *m* NUCL normal coupling; ~ **Oldham** *m* MECÁ Oldham coupling; ~ **de onda** *m* FÍS ONDAS, TELECOM wave coupling; ~ **óptico** *m* ING ELÉC, ÓPT optical coupling; ~ **óptico direccional** *m* ING ELÉC, ÓPT directional optical coupler; ~ **en paralelo** *m* ELEC *circuito*, ING ELÉC parallel connection; ~ **parásito** *m* ING ELÉC, TELECOM parasitic coupling, stray coupling; ~ **por pasador** *m* ING MECÁ pin coupling; ~ **de perno articulado** *m* ING MECÁ swing-bolt coupling; ~ **de platillos** *m* ING MECÁ plate coupling; ~ **de platos** *m* ING MECÁ faceplate coupling, flange coupling, plate coupling; ~ **de platos con pernos de cabeza embutida** *m* ING MECÁ recessed-flange coupling; ~ **de platos con pernos a paño** *m* ING MECÁ pulley coupling; ~ **proyector** *m* INSTR projector attachment; ~ **Racah** *m* NUCL Racah coupling; ~ **RC** *m* ELECTRÓN, ING ELÉC RC coupling; ~ **de referencia** *m* ELEC, ING ELÉC reference coupling; ~ **por resistencia eléctrica** *m* ELEC, ING ELÉC, TELECOM resistive coupling; ~ **por resistencia-capacitancia** *m* ELEC, ELECTRÓN, ING ELÉC resistance-capacitance coupling; ~ **rígido** *m* ELEC *inductancia*, ING ELÉC close coupling, ING MECÁ rigid coupling; ~ **a rosca de amortiguador lateral** *m* TRANSP side-buffer screw coupling;

~ roscado *m* ING MECÁ screw coupling, screw socket, tapped fitting; **~ de Russell-Saunders** *m* FÍS Russell-Saunders coupling; **~ en serie** *m* ELEC *circuito*, ING ELÉC cascade arrangement, series connection; **~ simétrico de tubo** *m* ING MECÁ symmetric pipe coupling; **~ sincrónico** *m* ELEC synchronous coupling; **~ spin-órbita** *m* FÍS spin-orbit coupling; **~ de transformador** *m* ELEC transformer coupling; **~ en triángulo** *m* ELEC mesh connection; **~ de trinquete** *m* ING MECÁ pawl coupling; **~ de tuberías** *m* CONST, ING MECÁ, PROD pipe coupling; **~ de tubo roscado** *m* ING MECÁ screwed-pipe coupling; **~ de tubos** *m* CONST, ING MECÁ, PROD pipe coupling; **~ de tubos de desembrague rápido** *m* ING MECÁ, PROD quick-release pipe coupling; **~ universal** *m* AUTO universal joint; **~ de vástago** *m* MINAS rod coupling; **~ vectorial** *m* FÍS RAD vector coupling; **~ Willison** *m* TRANSP Willison coupling

acoplar *vt* CINEMAT patch, ELEC, ING ELÉC connect, couple, group, ING MECÁ, MECÁ attach, couple, engage, MINAS *sondas* couple, PROD *cinta del transportador* lay on, TEC ESP attach; **~ en derivación** *vt* ELEC, ING ELÉC couple in parallel; **~ mutuamente** *vt* INFORM&PD interface; **~ en paralelo** *vt* ELEC, ING ELÉC connect in parallel; **~ en serie** *vt* ELEC, ING ELÉC cascade, connect in series

acoplarse *v refl* TEC ESP *vehículos espaciales* dock

acople *m* ING MECÁ connection; **~ por transformador** *m* ING ELÉC transformer coupling

acoplo[1]: **de ~ directo** *adj* ING MECÁ, MECÁ gearless

acoplo[2] *m* GEOM join; **~ inductivo** *m* ING ELÉC flux linkage

acorazado *adj* D&A, TRANSP MAR *buque de guerra* armored (*AmE*), armoured (*BrE*)

acorazar *vt* D&A, TRANSP MAR *buque de guerra* armor (*AmE*), armour (*BrE*), plate

acordar *vt* ACÚST, TELECOM tune

acorde *m* ACÚST chord, tone, tono; **~ combinado** *m* ACÚST combination tone; **~ fundamental** *m* ACÚST *música* fundamental chord; **~ perfecto mayor** *m* ACÚST *música* major common chord

acordonado *m* TRANSP cordon line

acortado *m* TEXTIL *de talla* take-up

acortar *vt* D&A shorten, TEC ESP dock, TEXTIL shorten

acotación *f* ING MECÁ, PROD dimensioning

acotar *vt* ING MECÁ, PROD dimension

ACP *abr* (*adsorción por cambios de presión*) GAS PSA (*pressure swing adsorption*)

acre *m* METR acre

acreación *f* AGUA aggradation

acreción *f* AGUA *de sedimento*, GEOL, METEO accretion

acreditación *f* INFORM&PD *seguridad* clearance

acrematita *f* MINERAL achrematite

acrilato *m* P&C, QUÍMICA acrylate

acrílico[1] *adj* CONST, MECÁ, QUÍMICA, TEXTIL acrylic

acrílico[2] *m* CONST, MECÁ, TEXTIL acrylic

acrilonitrilo *m* QUÍMICA acrylonitrile; **~-butadieno-estireno** *m* (*ABS*) P&C *materia prima* acrylonitrile-butadiene-styrene (*ABS*)

acritud *f* CRISTAL, MECÁ work hardening, METAL tempering, work hardening

acroíta *f* MINERAL achroite

acroleína *f* TRANSP AÉR acrolein

acromático *adj* GEN achromatic

acrómetro *m* METR acrometer

acronecrosis *f* AGRIC dieback

Acta: **~ del Aire Limpio** *f* CONTAM Clean Air Act

actina *f* ALIMENT, FÍS, FOTO actin

actínico *adj* ALIMENT, FÍS, FOTO actinic

actínido *m* FÍS, FÍS RAD, NUCL, QUÍMICA actinide, actinide element

actinio *m* (*Ac*) FÍS RAD, METAL, QUÍMICA actinium (*Ac*); **~ radioactivo** *m* FÍS RAD, QUÍMICA, TRANSP MAR radioactinium

actinismo *m* FÍS, FÍS RAD, FOTO actinism

actinolita *f* MINERAL *anfíbol* actinolite

actinometría *f* FÍS, FÍS RAD, METEO actinometry

actinómetro *m* FÍS, FÍS RAD, INSTR, METEO actinometer

actinón *m* (*An*) FÍS RAD, QUÍMICA actinon (*An*)

actinota *f* MINERAL *anfíbol* actinote

actitud[1]: **~ de cabeceo** *f* TRANSP AÉR pitch attitude; **~ de vuelo** *f* TRANSP AÉR flight attitude

actitud[2]: **en ~ encabritada** *fra* TRANSP AÉR in a nose-up attitude

activación *f* CARBÓN activation, ELECTRÓN activation, firing, FÍS RAD, HIDROL activation, ING ELÉC firing, activation, METAL, TELECOM activation; **~ electrónica** *f* ELECTRÓN, FÍS RAD electron-induced activation; **~ por fotones gamma** *f* FÍS RAD gamma-photon activation; **~ inducida por electrones** *f* ELECTRÓN, FÍS RAD electron-induced activation; **~ inducida por fotones gamma** *f* FÍS RAD gamma-photon activation; **~ inducida por irradiación** *f* FÍS RAD radiation-induced activation; **~ inducida por radiación** *f* FÍS RAD radiation-induced activation; **~ térmica** *f* METAL, TERMO thermal activation; **~ con tres niveles sucesivos de decisión** *f* FÍS RAD activation comprising three successive levels of decision; **~ vocal** *f* TELECOM voice activation; **~ por la voz** *f* TELECOM voice activation

activado *adj* ELEC, ING ELÉC, PROD energized; **~ por la voz** *adj* TELECOM voice-activated

activador *m* AUTO governor, CARBÓN activator, CINEMAT, ENERG RENOV governor, ING ELÉC activator, governor, ING MECÁ, MECÁ governor, P&C, QUÍMICA activator, TRANSP MAR, VEH governor; **~ de la sinterización** *m* PROC QUÍ sintering activator

activar *vt* CARBÓN activate, ELEC energize, IMPR actuate, INFORM&PD enable, ING ELÉC energize, turn on, ING MECÁ start, trigger, MECÁ drive, PROD energize, PROD *entrega* expedite, TERMO *los fuegos de una caldera* fire up

actividad *f* FÍS, INFORM&PD activity; **~ absoluta** *f* FÍS absolute activity; **~ del agua** *f* ALIMENT Atwater factor; **~ colectiva** *f* TELECOM ensemble activity; **~ desclasificada** *f* PROD out-of-sequence activity; **~ específica** *f* FÍS, NUCL specific activity; **~ específica del elemento** *f* FÍS, NUCL element specific activity; **~ lineal** *f* NUCL linear activity; **~ naviera** *f* TRANSP MAR shipping trade; **~ nuclear** *f* FÍS RAD, NUCL nuclear activity; **~ óptica** *f* FÍS, ÓPT, QUÍMICA optical activity; **~ radicular** *f* AGRIC root activity; **~ sísmica** *f* GEOL seismic activity; **~ solar** *f* FÍS, METEO solar activity

activo *adj* ELEC *circuito* alive, *componente* live, INFORM&PD active

acto: **~ de forzar** *m* CONST forcing; **~ inseguro** *m* SEG unsafe act

actomiosina *f* ALIMENT actomyosin

actuación: **~ involuntaria** *f* PROD unintended actuation; **~ según carga** *f* TRANSP range of action per charge

actuador *m* ELEC transductor, INFORM&PD, MECÁ, PROD, TEC ESP actuator, TELECOM power feed, TRANSP AÉR actuator; **~ axial** *m* TEC ESP axial actuator; **~ eléctrico** *m* ELEC, TEC ESP electric actuator; **~ hidráulico de flaps** *m* TRANSP AÉR flap jack; **~ oscilatorio** *m* PROD semirotary actuator; **~ de varillas** *m* PROD rod actuator

actualización *f* INFORM&PD, PROD, TELECOM update, updating; **~ de archivos** *f* INFORM&PD file updating; **~ de ficheros** *f* INFORM&PD file updating; **~ inmediata** *f* PROD immediate update; **~ rápida** *f* INFORM&PD, TELECOM fast update; **~ de registros** *f* INFORM&PD record updating; **~ retroactiva** *f* INFORM&PD retrofit

actualizar *vt* INFORM&PD, ING MECÁ, PROD, TELECOM update

actuante *m* ELEC transductor actuator

actuar[1] *vt* ELEC, ING ELÉC, ING MECÁ actuate

actuar[2] *vi* FÍS act

acuacultura *f* AGRIC, AGUA, OCEAN aquaculture, aquiculture; **~ marina** *f* OCEAN marine aquaculture

acuanauta *m* OCEAN aquanaut

acuaplaning *m* AUTO, TRANSP AÉR aquaplaning

acuaplano *m* OCEAN, TRANSP MAR aquaplane

acuático *adj* HIDROL, OCEAN aquatic

acueducto *m* AGUA aqueduct, conduit, water supply line, CONST, ENERG RENOV, HIDROL aqueduct, conduit, ING MECÁ pipeline; **~ de fluidos** *m* ING MECÁ fluid pipeline

acuerdo: ~ de comunicaciones *m* TELECOM communications agreement

Acuerdo: ~ General sobre Aranceles y Comercio Aduaneros *m* (*GATT*) AGRIC General Agreement on Tariffs and Trade (*GATT*)

acúfeno *m* ACÚST tinnitus

acuicludo *m* AGUA aquiclude

acuiclusa *f* AGUA aquiclude

acuicultivo *m* AGRIC, AGUA, OCEAN aquaculture, aquiculture

acuicultor *m* AGRIC, AGUA aquaculturist, OCEAN aquaculturist, fish farmer

acuicultura *f* AGRIC, AGUA aquaculture, aquiculture, OCEAN aquaculture, aquiculture, fish farming; **~ marina** *f* OCEAN marine aquaculture

acuífero[1] *adj* AGUA, GAS, GEOL, HIDROL, MINAS, TEC PETR aquiferous, water-bearing

acuífero[2] *m* AGUA, GAS, GEOL, HIDROL aquifer, MINAS water reservoir, TEC PETR aquifer; **~ cárstico** *m* AGUA, GEOL, HIDROL karstic aquifer; **~ cerrado** *m* AGUA, GEOL, HIDROL confined aquifer; **~ ilimitado** *m* AGUA, HIDROL unconfined water; **~ lateral** *m* HIDROL, TEC PETR edge water; **~ de varias capas** *m* AGUA, GEOL, HIDROL multilayer aquifer

acuifugo *m* AGUA, GEOL aquifuge

acuitardo *m* AGUA aquitard

acumulación *f* C&V gathering, GAS pool, GEOL, METEO accumulation, PETROL pool, PROD stacking, RECICL *de fango* bulking; **~ calcárea** *f* ALIMENT boiler scale; **~ de calor** *f* TERMO heat accumulation; **~ de carga** *f* ING ELÉC charge storage; **~ de datos** *f* INFORM&PD data gathering; **~ de emulsión** *f* CINEMAT emulsion pile-up; **~ de energía** *f* ING ELÉC, TEC ESP, TERMO energy storage; **~ de frío para su distribución** *f* REFRIG dispatching cold store; **~ de hielo** *f* AGUA ice accretion; **~ de información parasítica** *f* INFORM&PD garbage; **~ de materiales** *f* MECÁ back-

log; **~ de óxido** *f* CINEMAT, TV oxide buildup; **~ de partículas de tinta** *f* IMPR caking; **~ de presión** *f* TRANSP AÉR pressure buildup; **~ térmica** *f* TERMO heat buildup; **~ de tierra** *f* AGUA land accretion; **~ de trabajo** *f* PROD backlog; **~ de xenón** *f* NUCL xenon buildup

acumulador *m* AUTO, ELEC, ELECTRÓN, FÍS accumulator, storage battery, INFORM&PD accumulator, ING ELÉC accumulator, storage battery, INSTAL HIDRÁUL, NUCL, QUÍMICA, REFRIG accumulator, TEC ESP battery cell, storage battery, TERMO, TERMOTEC, TRANSP MAR *aparato eléctricos*, VEH accumulator, storage battery; **~ de agua** *m* NUCL water accumulator; **~ alcalino** *m* ELEC, ING ELÉC alkaline storage battery; **~ de aspiración** *m* REFRIG suction accumulator; **~ de calor** *m* TERMO, TERMOTEC heat accumulator; **~ de calor nocturno** *m* INSTAL TERM, TERMO, TERMOTEC night storage heater; **~ completo** *m* INSTAL HIDRÁUL self-contained accumulator; **~ eléctrico** *m* ELEC, ING ELÉC battery; **~ de estructura invertida** *m* INSTAL HIDRÁUL inverted-pattern accumulator; **~ de frío** *m* REFRIG hold-over coil; **~ a gas** *m* GAS, ING MECÁ gas-loaded accumulator; **~ hidráulico** *m* ING MECÁ, P&C, TRANSP AÉR hydraulic accumulator; **~ hidroneumático** *m* ING MECÁ hydropneumatic accumulator; **~ intermedio** *m* ING ELÉC buffer; **~ para mineral** *m* MINAS ore bin; **~ de níquel-cadmio** *m* CINEMAT, ING ELÉC, TEC ESP, TRANSP nickel-cadmium battery; **~ de níquel-hierro** *m* ING ELÉC, TEC ESP nickel-iron battery; **~ de plomo** *m* ELEC lead-acid accumulator, lead-acid battery, FÍS lead accumulator, lead-acid battery, ING ELÉC, TRANSP, VEH lead-acid accumulator, lead-acid battery; **~ recargable** *m* CINEMAT, ELEC, FÍS, FOTO, ING ELÉC, TV rechargeable battery; **~ regenerable** *m* TRANSP regenerative cell; **~ de salmuera** *m* REFRIG brine drum; **~ seco** *m* ELEC, ING ELÉC dry storage battery; **~ simple** *m* ELEC, ING ELÉC battery; **~ de succión** *m* REFRIG suction accumulator; **~ térmico** *m* ING MECÁ, INSTAL TERM, TERMO, TERMOTEC storage heater; **~ para el tubo de aspiración** *m* REFRIG suction-line accumulator; **~ de vapor de agua** *m* INSTAL HIDRÁUL steam accumulator

acuñado[1] *adj* ING MECÁ, MECÁ keyed, wedged

acuñado[2] *m* ING MECÁ, MECÁ scotching, wedging

acuñadura *f* ING MECÁ, MECÁ wedging

acuñamiento *m* ING MECÁ jam, shimming, wedging, PROD keying

acuñar *vt* AUTO cog, ING MECÁ, MECÁ cog, jam, wedge, PROD stamp

acuñarse *v refl* INSTAL HIDRÁUL, MECÁ jam

acuófeno *m* ACÚST acouphene

acuoso *adj* AGUA, FÍS FLUID, GEOL, HIDROL, QUÍMICA, TEC PETR aqueous, hydrous; **no ~** *adj* QUÍMICA nonaqueous

acusar: ~ recibo de *vt* INFORM&PD acknowledge

acuse: ~ de recibo *m* INFORM&PD, TELECOM acknowledgement; **~ de recibo negativo** *m* INFORM&PD negative acknowledgement (*NAK*)

acustiaislamiento *m* ACÚST acoustic isolation

acústica *f* ACÚST, FÍS, SEG acoustics; **~ arquitectónica** *f* ACÚST architectural acoustics; **~ de edificios** *f* ACÚST building acoustics; **~ fisiológica** *f* ACÚST physiological noise; **~ de salas** *f* ACÚST room acoustics; **~ subacuática** *f* ACÚST, OCEAN underwater acoustics; ~ **submarina** *f* ACÚST, OCEAN marine acoustics

acústico *adj* TEC PETR sonic

acutancia *f* FOTO acutance

acutangular *adj* GEOM acute-angular

acutángulo *m* GEOM acute angle

adamina *f* MINERAL adamine

adamita *f* MINERAL adamite

adamsita *f* MINERAL adamsite

adaptable *adj* INFORM&PD, MECÁ adaptive

adaptación *f* ELECTRÓN customization, IMPR adaptation, marring, ING MECÁ accommodation, TEC ESP *circuito* interfacing; ~ **de antenas** *f* TELECOM aerial matching; ~ **de cables** *f* ELEC, ING ELÉC cable fitting; ~ **de cámaras** *f* CINEMAT, TV camera matching; ~ **de camino de menor orden** *f* TELECOM lower-order path adaptation (*LPA*); ~ **de corte** *f* TV match cut; ~ **del desvanecido de imagen** *f* TV match dissolve; ~ **de imagen** *f* TV picture match; ~ **de impedancias** *f* ING ELÉC impedance matching; ~ **del pedestal** *f* TV pedestal adjustment; ~ **de sección** *f* TELECOM section adaptation (*SA*); ~ **del software** *f* INFORM&PD software adaptation; ~ **del soporte lógico** *f* INFORM&PD software adaptation; ~ **de trayecto de orden superior** *f* TELECOM higher-order path adaptation (*HPA*)

adaptado *adj* ELECTRÓN customized

adaptador *m* ELEC adaptor, ELECTRÓN adaptor, pad, INFORM&PD adaptor, ING ELÉC adaptor, pad, ING MECÁ adaptor, fitting, LAB adaptor, MECÁ adaptor, fitting, TELECOM adaptor, pad, TEXTIL adaptor; ~ **para acoplamiento** *m* TEC ESP docking adaptor; ~ **ajustable** *m* ING MECÁ adjustable adaptor; ~ **de bayoneta** *m* ELEC, ING ELÉC socket adaptor; ~ **de bisagra** *m* ING MECÁ hinge fitting; ~ **de la bobina plana** *m* TEXTIL cheese adaptor; ~ **de boquilla** *m* ING MECÁ nozzle adaptor; ~ **del borne** *m* TELECOM terminal adaptor; ~ **de brida** *m* ING MECÁ flanged fitting; ~ **de bus del ratón** *m* INFORM&PD bus-mouse adaptor; ~ **bus-ratón** *m* INFORM&PD bus-mouse adaptor; ~ **de canal asíncrono** *m* INFORM&PD asynchronous channel adaptor; ~ **de color** *m* INFORM&PD color adaptor (*AmE*), colour adaptor (*BrE*); ~ **de conductor común** *m* INFORM&PD bus-mouse adaptor; ~ **en cruz** *m* LAB *soporte, pinza* boss head; ~ **de cuchilla** *m* ING MECÁ cutter adaptor; ~ **de eje flexible** *m* ING MECÁ flexible-shaft adaptor; ~ **de enchufe** *m* ELEC, ING ELÉC plug adaptor, socket adaptor; ~ **de E/S** *m* PROD I/O adapter; ~ **de fase** *m* ELEC *corriente alterna* phase adaptor; ~ **de fijación de la pala** *m* TRANSP AÉR blade-attachment fitting; ~ **de fresa** *m* ING MECÁ cutter adaptor; ~ **gráfico de colores** *m* (*CGA*) INFORM&PD color graphics adaptor (*AmE*) (*CGA*), colour graphics adaptor (*BrE*) (*CGA*); ~ **de gráficos de video** *m AmL*, ~ **de gráficos de vídeo** *m Esp* (*VGA*) INFORM&PD video graphics adaptor (*VGA*); ~ **de interfaz de periféricos** *m* (*PIA*) INFORM&PD peripheral interface adaptor (*PIA*); ~ **intermedio** *m* TELECOM medium adaptor (*MA*); ~ **de línea** *m* INFORM&PD line adaptor; ~ **medio** *m* TELECOM medium adaptor (*MA*); ~ **de microscopio** *m* FOTO microscope adaptor; ~ **para montaje C** *m* CINEMAT C-mount adaptor; ~ **multiclavija** *m* ING ELÉC multiplug adaptor; ~ **de pantalla** *m* INFORM&PD display adaptor; ~ **de perno articulado** *m* ING MECÁ hinged bolt fitting; ~ **de**

plato *m* ING MECÁ flanged fitting; ~ **portaherramienta** *m* ING MECÁ chuck adaptor; ~ **de potencia** *m* CINEMAT power adaptor; ~ **de ratón paralelo** *m* INFORM&PD parallel mouse-adaptor; ~ **reductor** *m* GEOFÍS, LAB *material de vidrio* reduction tube; ~ **de representación gráfica** *m* INFORM&PD graphic display adaptor; ~ **roscado** *m* ING MECÁ threaded fitting; ~ **terminal** *m* TELECOM terminal adaptor (*TA*); ~ **terminal B de la red de transmisión digital de RDSI-BA** *m* TELECOM B-ISDN terminal adaptor

adaptar *vt* ELECTRÓN customize, ING MECÁ fit, set, MECÁ fit, PROD, TV match; ~ **a las condiciones del espacio** *vt* TEC ESP adapt to space conditions; ~ **al tiempo** *vt* PROD time

adaptativo *adj* INFORM&PD adaptive

ADC: ~ **automático** *m* TELECOM automatic ADC

addendum *m* ING MECÁ addendum circle

adecuación: ~ **de iluminación** *f* ING ELÉC light fitting

adecuarse *v refl* IMPR fit

adelantador: ~ **de fase** *m* ELEC *motor* phase advancer

adelantamiento *m* INFORM&PD look ahead

adelantar[1]: ~ **con cuidado** *vt* TRANSP AÉR ease forward

adelantar[2]: ~ **la palanca de gases** *vi* TRANSP AÉR advance the throttle; ~ **el regulador de gases** *vi* TRANSP AÉR advance the throttle

adelante[1] *adv* TEC ESP ahead, TELECOM go ahead (*GA*)

adelante[2] *m* TELECOM go ahead (*GA*)

adelanto *m* CINEMAT, FÍS advance, OCEAN *de la marea* priming; ~ **de la chispa de encendido** *m* TRANSP AÉR spark advance; ~ **de fase** *m* ELEC *corriente alterna* phase lead; ~ **de información** *m* TRANSP advance information; ~ **de la lámpara de carbón** *m* CINEMAT carbon advancing

adelfa *f* AGRIC oleander

adelfolita *f* MINERAL adelpholite

adelgazador *m* QUÍMICA thinner

adelgazamiento *m* MINAS *de un filón* pinch

adelgazar *vt* ING MECÁ, MATEMÁT reduce, PROD reduce, thin

adema *f* CARBÓN, MINAS post, prop, shore

adenina *f* ALIMENT, QUÍMICA adenine

adenosina *f* QUÍMICA adenosine;

adentro: **hacia ~** *adj* TRANSP AÉR inboard

aderezado *adj* ALIMENT dressed

aderezar *vt* ALIMENT flavor (*AmE*), flavour (*BrE*)

aderezo *m* ALIMENT flavoring (*AmE*), flavouring (*BrE*), topping

ADF *abr* (*goniometría automática*) TELECOM, TRANSP AÉR ADF (*automatic direction finding*)

adherencia *f* CONST bond, EMB adherence, ING MECÁ traction, P&C adhesion, *adhesivos* bonding, PROD sticking; ~ **covalente** *f* METAL covalent bond; ~ **entre capas** *f* P&C blocking; ~ **específica** *f* P&C specific adhesion; ~ **metálica** *f* METAL metallic bond; ~ **a la pared** *f* FÍS FLUID *viscosidad* wall attachment

adherir *vt* CONST bond

adherirse *v refl* EMB adhere, GEOL cohere, P&C adhere, PROD stick

adherómetro *m* INSTR, P&C adherometer

adhesión *f* CARBÓN, CONST, EMB, P&C adhesion, PROD bond, sticking; ~ **del electrón** *f* ELECTRÓN, FÍS PART, NUCL electron attachment; ~ **específica** *f* P&C specific adhesion; ~ **a la pared** *f* FÍS FLUID *viscosidad* wall attachment

adhesividad *f* PAPEL tackiness

adhesivo[1] *adj* ALIMENT sticky, EMB, P&C, PAPEL, PROD adhesive

adhesivo[2] *m* CONST binder, cementing material, EMB, MECÁ, P&C, PROD, SEG adhesive, bonding agent; **~ de acetato** *m* EMB acetate adhesive; **~ anaeróbico** *m* EMB, P&C anaerobic adhesive; **~ de base acuosa** *m* EMB water-based backing adhesive; **~ a base de fusiones en caliente** *m* EMB hot-melt adhesive; **~ de cola** *m* COLOR glue size; **~ de contacto** *m* EMB, P&C contact adhesive; **~ en emulsión** *m* EMB, P&C emulsion adhesive; **~ de endurecimiento en frío** *m* EMB, P&C cold-setting adhesive; **~ para estucado** *m* PAPEL, REVEST coating binder; **~ que se fija a temperatura ambiente** *m* EMB room-temperature setting adhesive; **~ formulado para un fin determinado** *m* EMB, P&C purpose-formulated adhesive; **~ de fraguado en caliente** *m* EMB, P&C hot-setting adhesive; **~ para madera contrachapada** *m* EMB plywood adhesive; **~ multiuso** *m* EMB all-purpose adhesive; **~ natural** *m* EMB, SEG natural adhesive; **~ de paletizado** *m* EMB palletizing adhesive; **~ resistente al agua** *m* EMB, P&C waterproof adhesive; **~ que sigue actuando a baja temperatura** *m* EMB chill-permanent adhesive; **~ termoendurecible** *m* EMB, P&C hot-setting adhesive; **~ termosellable** *m* EMB heat-sealing adhesive

adhiere: **que se ~ fuertemente al carbón** *adj* PROD burned

adiabáticamente *adv* FÍS, FÍS FLUID, MECÁ, TERMO adiabatically

adiabático *adj* FÍS, FÍS FLUID, MECÁ, TERMO adiabatic

adiabatismo *m* FÍS, FÍS FLUID, MECÁ, TERMO adiabatism

adición *f* IMPR addendum, INFORM&PD, ING MECÁ, MATEMÁT addition; **~ de alcohol** *f* QUÍMICA *al vino* alcohol addition; **~ de dosis elevada** *f* ELECTRÓN high-dose implant

adinola *f* GEOL adinole

adípico *adj* QUÍMICA *ácido* adipic

adipocerito *m* FÍS adipocerite

aditamento *m* INFORM&PD attachment; **~ de aire limpio** *m* SEG clean-air device; **~ de autorrescate con oxígeno** *m* SEG oxygen self-rescuer; **~ cónico** *m* CINEMAT conical snoot, snoot; **~ de corte** *m* SEG cutout device; **~ de escape** *m* SEG escape device; **~ para manejo remoto** *m* SEG remote-handling device; **~ de la máscara** *m* CINEMAT mask attachment; **~ mecánico giratorio** *m* CINEMAT mechanical spin attachment; **~ de paro** *m* SEG cutout device; **~ para el paro de emergencia** *m* SEG emergency-stopping device; **~ para prevenir crímenes** *m* SEG crime prevention device; **~ para retorno seguro de gases** *m* SEG gas return safety device; **~ de seguridad** *m* SEG fail-safe device, safety device; **~ de seguridad en el camino** *m* AUTO, SEG, TRANSP, VEH road-safety device

aditivo[1]: **sin ~** *adj* ALIMENT additive-free

aditivo[2] *m* ALIMENT additive, COLOR dope, easer, CONST admixture, CONTAM, DETERG, MECÁ additive, P&C additive, admixture, PETROL dope, QUÍMICA, TEC PETR additive; **~ alimenticio** *m* ALIMENT food additive; **~ anticontracción** *m* CONST antishrinkage admixture; **~ anticorrosión** *m* TEC PETR *refino* anticorrosion additive; **~ antidetonante** *m* AUTO, CONTAM, VEH antiknock additive, antiknock agent; **~ para curado** *m* CONST curing compound;

~ detergente *m* DETERG, QUÍMICA detergent additive; **~ endurecedor** *m* COLOR stiffening dope; **~ graso** *m* ALIMENT shortening; **~ de plomo** *m* CONTAM lead additive; **~ de presión extrema** *m* TEC PETR *refino* extreme-pressure additive

adjudicatario: **~ principal** *m* TEC ESP prime contractor

adjunta: **se ~** *fra* EMB enclosed

adjuvante *m* CARBÓN adjuvant

administración: **~ agrícola** *f* AGRIC farm management; **~ de las aguas superficiales** *f* AGUA surface-water management; **~ de aguas urbanas** *f* AGUA urban-water management; **~ de almacén** *f* PROD warehouse management; **~ de archivos** *f* INFORM&PD file management; **~ de la base de datos** *f* INFORM&PD, TELECOM database management (*DBM*); **~ de calidad** *f* CALIDAD, CONST, ING MECÁ, PROD quality management; **~ de colas** *f* INFORM&PD, TELECOM queue management; **~ de configuración** *f* INFORM&PD configuration management (*CM*); **~ de datos** *f* (*DM*) INFORM&PD, TELECOM data management (*DM*); **~ de errores** *f* INFORM&PD, TELECOM error management; **~ de fichero** *f* INFORM&PD file management; **~ integral de aguas** *f* AGUA integral water management; **~ de memoria** *f* INFORM&PD memory management; **~ de red** *f* INFORM&PD, TELECOM network management

Administración: **~ Nacional de Aeronáutica y del Espacio** *f* (*NASA*) TEC ESP National Aeronautics and Space Administration (*NASA*)

administrador: **~ de la base de datos** *m* INFORM&PD, TELECOM database administrator (*DBA*); **~ de red** *m* INFORM&PD, TELECOM network manager

admisión[1] *f* AGUA inlet, intake, AUTO inlet, ELEC induction, INFORM&PD support, ING MECÁ intake, INSTAL HIDRÁUL drawing off, entrance, induction, input, MECÁ intake, PROD inlet, TRANSP AÉR intake; **~ de aire** *f* ING MECÁ air inlet; **~ de aire con forma variable envuelo** *f* TRANSP variable geometry inlet; **~ anticipada** *f* PROD *válvula de deslizamiento* early admission; **~ retardada** *f* HIDROL, INSTAL HIDRÁUL, TRANSP retarded admission; **~ de vapor** *f* INSTAL HIDRÁUL steam admission

admisión[2]: **con ~ delantera** *fra* ING MECÁ head end; **con ~ posterior** *fra* ING MECÁ head end

admitancia *f* ACÚST, ELEC, FÍS, ING ELÉC, TELECOM admittance; **~ acústica** *f* ACÚST, ELEC, FÍS, ING ELÉC acoustic admittance; **~ característica** *f* ACÚST, ELEC, FÍS, ING ELÉC characteristic admittance; **~ compleja** *f* ELEC, FÍS, ING ELÉC complex admittance; **~ eléctrica** *f* ACÚST, ELEC, FÍS, ING ELÉC, TELECOM electrical admittance; **~ de electrodo** *f* ELEC, FÍS, ING ELÉC electrode admittance; **~ de entrada** *f* ELEC, FÍS, ING ELÉC input admittance; **~ mecánica** *f* ACÚST, ELEC, FÍS, ING ELÉC, MECÁ mechanical admittance; **~ de salida** *f* ELEC, FÍS, ING ELÉC output admittance

admítese *m* TRANSP MAR *marina mercante* mate's receipt

admitir *vt* INFORM&PD support

ADN *abr* (*ácido desoxirribonucleico*) QUÍMICA DNA (*deoxyribonucleic acid*)

adobe *m* C&V, CONST adobe

adobo *m* OCEAN pickle

adoquín *m* CONST paving stone, *carreteras* sett

adoquinado *m* *Esp* (*cf pavimento AmL*) CARBÓN, CONST, TRANSP pavement (*BrE*), sidewalk (*AmE*); **~ móvil cubierto** *m* *Esp* (*cf pavimento móvil cubierto*

AmL) TRANSP cabin-type moving pavement (*BrE*), cabin-type moving sidewalk (*AmE*); ~ **rodante cubierto** *m Esp* (*cf pavimento rodante cubierto AmL*) TRANSP cabin-type moving pavement (*BrE*), cabin-type moving sidewalk (*AmE*); ~ **rodante tipo cinta** *m Esp* (*cf pavimento rodante tipo cinta AmL*) TRANSP belt-type moving pavement (*BrE*), belt-type moving sidewalk (*AmE*)

adornar: ~ **con dibujos** *vt* TEXTIL pattern

adorno *m* IMPR adornment, TEXTIL trimming; ~ **floral** *m* IMPR fleuron, rosette; ~ **al principio de página** *m* IMPR headband

adosado *adj* PROD back-to-back

adosar *vt* CONST abut

ADPCM *abr* (*modulación adaptable diferencial de impulsos en código*) TELECOM ADPCM (*adaptive differential pulse coded modulation*)

adquisición: ~ **de datos** *f* ELECTRÓN, INFORM&PD data acquisition; ~ **de un espectro** *f* FÍS ONDAS scan; ~ **de la portadora** *f* ELECTRÓN carrier acquisition; ~ **de posición de vuelo** *f* TEC ESP attitude acquisition

adragante *f* ALIMENT tragacanth

adsorbedor *m* AGUA, ALIMENT, CARBÓN, GAS, P&C, QUÍMICA adsorber

adsorbente[1] *adj* AGUA, ALIMENT, CARBÓN, GAS, P&C, QUÍMICA adsorbent

adsorbente[2] *m* AGUA, ALIMENT, CARBÓN, GAS, P&C, QUÍMICA adsorbent

adsorber *vt* AGUA, ALIMENT, CARBÓN, GAS, P&C, QUÍMICA adsorb

adsorbible *adj* AGUA, ALIMENT, CARBÓN, GAS, P&C, QUÍMICA adsorbable

adsorción *f* AGUA, ALIMENT, CARBÓN, GAS, P&C, QUÍMICA adsorption; ~ **por cambios de presión** *f* (*ACP*) GAS pressure swing adsorption (*PSA*)

aducción *f* HIDROL, QUÍMICA adduction

aducto *m* HIDROL, QUÍMICA adduct

aduja *f* ING MECÁ, TRANSP MAR *cabo* coil

adujada: ~ **de refuerzo** *f AmL* (*cf bobina de inducción cebada con acumulador para el arranque Esp*) AUTO, TRANSP, TRANSP AÉR, VEH booster coil

adularia *f* MINERAL adularia

adulteración *f* ALIMENT adulteration, doping, ELECTRÓN, FÍS doping; ~ **con silicio** *f* ELECTRÓN, FÍS silicon doping

adulterado *adj* ALIMENT adulterated, ELECTRÓN, FÍS doped

adulterante *m* ALIMENT adulterant, ELECTRÓN, FÍS dopant

adulterar *vt* ALIMENT adulterate, ELECTRÓN, FÍS dope

advección *f* FÍS, METEO advection

advertencia *f* PROD warning; ~ **meteorológica** *f* METEO, TRANSP AÉR, TRANSP MAR weather warning; ~ **prearranque** *f* ING MECÁ pre-start warning

adyacencia *f* GEOM adjacence, adjacency

adyacente *adj* GEOM adjacent

adyuvante: ~ **de lavado** *m* DETERG washing adjuvant

AENOR *abr* (*Asociación Española de Normalización*) PROD AENOR (*Spanish Association for Normalization and Certification*)

aeráulica *f* GAS aeraulics

aéreo *adj* CARBÓN overground

aerificación *f* QUÍMICA aerification

aeroacústica *f* ACÚST aeroacustics

aeróbico *adj* ALIMENT, HIDROL, RECICL aerobic

aerobio: ~ **facultativo** *m* ALIMENT facultative aerobe; ~ **obligado** *m* ALIMENT obligate aerobe

aerobiología *f* METEO aerobiology

aerobiosis *f* RECICL aerobiosis

aerodeslizador *m* TRANSP, TRANSP AÉR, TRANSP MAR air-cushion vehicle (*TACV*), hovercraft; ~ **de costados rígidos** *m* TRANSP, TRANSP MAR rigid sidewall hovercraft; ~ **guiado** *m* TRANSP, TRANSP MAR guided air-cushion vehicle; ~ **marítimo** *m* TRANSP marine air-cushion vehicle, TRANSP MAR maritime air-cushion vehicle; ~ **de plinto rígido** *m* TRANSP, TRANSP MAR rigid-skirt hovercraft; ~ **semi-anfibio** *m* TRANSP, TRANSP MAR semiamphibious air-cushion vehicle; ~ **terrestre** *m* TRANSP land air-cushion vehicle; ~ **de tipo aerodinámico** *m* TRANSP, TRANSP AÉR aerodynamic-type air-cushion vehicle; ~ **sobre vía** *m* TRANSP tracked air cushion vehicle (*TACV*)

aerodeslizamiento *m* TRANSP aeroglide

aerodinámica *f* GEN aerodynamics

aerodinámico *adj* GEN aerodynamic, streamlined

aeródromo *m* TRANSP AÉR aerodrome (*BrE*), airdrome (*AmE*)

aeroducto *m* CARBÓN air duct

aeroelasticidad *f* ENERG RENOV *energía de viento*, NUCL aeroelasticity

aeroembolismo *m* TRANSP MAR *buceo* bends

aeroespacial *adj* TEC ESP aerospatial

aeroespacio *m* TEC ESP aerospace

aerofotografía *f AmL* (*cf fotografía aérea Esp*) FOTO aerial photography

aerofrenar *vi* ING MECÁ, TEC ESP, TRANSP AÉR aerobrake (*BrE*), air brake (*AmE*)

aerofreno *m* ING MECÁ, TEC ESP, TRANSP AÉR *aviones* aerobrake (*BrE*), air brake (*AmE*)

aerofumigación: ~ **de cosechas** *f* AGRIC, TRANSP AÉR crop dusting

aerogel: ~ **de sílice** *m* REFRIG, TERMOTEC silica aerogel; ~ **de sílice opalizado** *m* REFRIG, TERMOTEC opacified silica aerogel

aerogenerador *m* ELEC, ENERG RENOV wind-powered generator

aerogiroscopio *m* TRANSP AÉR aerogyro

aerografía *f* C&V, EMB, FOTO, IMPR aerography, air-brushing

aerógrafo *m* C&V, EMB, FOTO, IMPR aerograph, airbrush

aerograma *m* PAPEL lettercard; ~ **ilustrado** *m* PAPEL illustrated lettercard

aerolínea *f* TRANSP AÉR airline

aerolito *m* TEC ESP aerolite

aerología *f* METEO aerology

aeromagnético *adj* GEOL, PETROL aeromagnetic

aerometría *f* ELEC, FÍS, HIDROL, TEC PETR aerometry

aerómetro *m* ELEC, FÍS, HIDROL, INSTR, TEC PETR aerometer

aeronáutica *f* TRANSP AÉR aeronautics

aeronaval *adj* TRANSP MAR *fuerzas* air-and-sea

aeronave *f* TEC ESP spacecraft, TRANSP AÉR aircraft; ~ **de búsqueda para todo tiempo** *f* D&A, TRANSP AÉR all-weather search aircraft

aeronavegabilidad *f* TRANSP AÉR airworthiness

aeronavegable *adj* TRANSP AÉR airworthy

aeronomía *f* METEO, TEC ESP, TRANSP AÉR aeronomy

aeropuente *m* TRANSP AÉR airbridge; ~ **de pasajeros** *m* TRANSP AÉR passenger bridge

aeropuerto *m* TEC ESP *aviación* port, TRANSP AÉR airport; ~ **de despegue y aterrizaje corto** *m* TRANSP

AÉR stolport; **~ internacional** *m* TRANSP AÉR international airport; **~ regional** *m* TRANSP AÉR regional airport

aerorrefrigerado *adj* REFRIG, TERMO air-cooled

aerorrefrigerar *vt* REFRIG, TERMO air-cool

aerosita *f* MINERAL aerosite

aerosol *m* CONTAM, EMB, FÍS, FÍS FLUID, METEO, PROD aerosol; **~ ácido** *m* CONTAM acid aerosol; **~ antiestático** *m* IMPR antistatic spray; **~ opacificante** *m* CINEMAT dulling spray

aerostática *f* METEO aerostatics

aerotermo *m* INSTAL TERM, TERMO unit heater

aerotransportado *adj* TRANSP AÉR airborne

aerotren *m* FERRO, TRANSP aerotrain, hovertrain

aeroturbina *f* ING MECÁ wind turbine

AERR *abr* (*Asociación Europea de Reciclado y Recuperación*) EMB, RECICL ERRA (*European Recycling and Recovery Association*)

AF *abr* (*alta frecuencia*) ELEC, ELECTRÓN, ING ELÉC, TELECOM HF (*high frequency*)

afanesita *f* MINERAL aphanesite

afanita *f* GEOL, PETROL aphanite

afanítico *adj* GEOL, PETROL aphanitic

afelio *m* FÍS *astronomía*, TEC ESP aphelion

aferramiento *m* ING MECÁ seizing

aferrar *vt* CONST bite, TRANSP MAR *vela* furl

afianzar *vt* TEC ESP anchor

afiche *m* ING MECÁ placard

áfido *m* AGRIC aphid, plant louse

afilado[1] *adj* FOTO, IMPR sharp, ING MECÁ, MECÁ sharp, sharpened, PROC QUÍ sharpened, PROD sharp, sharpened

afilado[2] *m* ING MECÁ, MECÁ, PROD grinding; **~ de herramientas** *m* ING MECÁ, MECÁ, PROD tool grinding, tool sharpening

afilador *m* ING MECÁ, MECÁ sharpener, PROD *persona* grinder

afiladora *f* ING MECÁ, MECÁ, PROD grinder, grinding machine, sharpener, sharpening machine; **~ de brocas** *f* ING MECÁ drill grinder, drill sharpener; **~ de brocas helicoidales** *f* ING MECÁ twist-drill grinder; **~ de cuchillas** *f* ING MECÁ tool grinder; **~ de cuchillos** *f* PROD knife-grinding machine; **~ de dos muelas** *f* ING MECÁ two-wheel grinding machine; **~ de fresas** *f* PROD milling-cutter sharpening machine; **~ de fresas de refrentar** *f* ING MECÁ face-milling grinder; **~ de herramientas** *f* ING MECÁ, MECÁ, PROD tool grinder, tool sharpener; **~ de sierra** *f* ING MECÁ, MECÁ, PROD saw-sharpening machine; **~ universal de herramientas de corte** *f* ING MECÁ, MECÁ, PROD universal tool-and-cutter sharpener

afiladura *f* ING MECÁ, MECÁ, PROD sharpening

afilar *vt* CONST point, ING MECÁ, MECÁ, PROC QUÍ, PROD grind, sharpen

afiliado *m* AmL (*cf abonado Esp*) ELEC, INFORM&PD, TELECOM, TV affiliate, subscriber

afinación *f* ACÚST tuning

afinado *adj* ACÚST *música* sharp

afinadora *f* CONST *máquina* finisher

afinamiento: ~ de órbita *m* TEC ESP orbit trimming; **~ progresivo** *m* ING MECÁ tapering

afinar *vt* ACÚST, FÍS ONDAS tune, ING MECÁ true up

afinidad *f* ELECTRÓN, FÍS PART, NUCL affinity, P&C *enlace químico* linkage, QUÍMICA, TEXTIL affinity;

~ electrónica *f* ELECTRÓN, FÍS PART, NUCL electron affinity; **~ a la tintura** *f* TEXTIL dyeing affinity

afino: ~ por chorro de aire *m* METAL air-blast refining; **~ por corriente de aire** *m* METAL air-blast refining; **~ por corriente de oxígeno** *m* METAL, PROD *aceros* oxygen lancing; **~ del grano** *m* METAL grain refinement; **~ del plomo** *m* PROD lead refining

afírico *adj* GEOL aphyric

aflatoxina *f* ALIMENT aflatoxin

aflojado *m* TEC PETR *tuberías* screwing-out

aflojamiento: ~ rápido *m* ING MECÁ, PROD, TEC ESP quick release

aflojar *vt* GAS release, ING MECÁ *tornillo* back off, relieve, unclamp, INSTAL HIDRÁUL, MECÁ relieve, PROD relieve, *tuerca* loosen, TRANSP MAR *cabo* loosen

aflojarse *v refl* ING MECÁ work loose

afloramiento *m* GAS seepage, GEOL, MINAS breakthrough, outcrop, OCEAN upwelling; **~ de rocas antiguas** *m* GEOL inlier

aflorar *vt* MINAS *un filón* strike, PROD flush up

afluencia *f* AGUA, HIDROL concourse, confluence, inflow; **~ ininterrumpida** *f* TRANSP uninterrupted flow; **~ normal** *f* TRANSP normal flow; **~ de tráfico** *f* TRANSP traffic flow; **~ de tráfico encontrado** *f* TRANSP conflicting traffic flow; **~ de tráfico no conflictivo** *f* TRANSP nonconflicting traffic flow; **~ de tráficos opuestos** *f* TRANSP conflicting traffic flow; **~ transversal** *f* TRANSP crossflow

afluente *m* AGUA, HIDROL confluent, PETROL *río* arm, TRANSP MAR *río* tributary

aflujo *m* AGUA inflow, runoff; **~ de aire** *m* MINAS rush of air

AFNOR *abr* (*Association Française de Normalisation*) AFNOR (*French Association for Normalization*)

aforador *m* GEN flowmeter

aforar *vt* GEN gage (*AmE*), gauge (*BrE*)

aforo *m* AGUA, INSTAL HIDRÁUL, PROD, TELECOM gaging (*AmE*), gauging (*BrE*), metering; **~ en vertedero** *m* INSTAL HIDRÁUL notch gaging (*AmE*), notch gauging (*BrE*)

afrecho *m* AGRIC bran; **~ de trigo** *m* AGRIC wheat bran

afrita *f* MINERAL aphrite

afrizita *f* MINERAL aphrizite

afrodita *f* MINERAL aphrodite

afrosiderita *f* MINERAL aphrosiderite

aftalosa *f* MINERAL aphthalose

aftitalita *f* MINERAL aphthitalite

aftonita *f* MINERAL aphthonite

afuera: hacia ~ *adj* TRANSP MAR *navegación* seaward

afusión *f* QUÍMICA affusion

afuste *m* MINAS *perforaciones* bar

afwilita *f* MINERAL afwillite

AFWS *abr* (*sistema de agua de alimentación auxiliar*) NUCL AFWS (*auxiliary feedwater system*)

Ag *abr* (*plata*) C&V, FOTO, ING ELÉC, METAL, QUÍMICA Ag (*silver*)

agafita *f* MINERAL agaphite

agalita *f* MINERAL agalite

agalla *f* AGRIC cyst, gall, C&V *defecto del vidrio* gall; **~ de la corona** *f* AGRIC crown gall

agalmatolita *f* MINERAL agalmatolite

agar: ~~agar *m* ALIMENT agar-agar, QUÍMICA gelose; **~ inclinado** *m* ALIMENT agar slant

agarradera *f* CONST grab, ING MECÁ handling lug, lifting lug, lug, MECÁ, PROD lug, TEC ESP handle,

TRANSP AÉR hold; ~ **revestida** ƒ ING MECÁ lined clamp; ~ **de tubería** ƒ CONST pipe grab

agarradero *m* ING MECÁ handgrip, handle, PROD *para martinete* die, TRANSP AÉR hold; ~ **de marco de ventana** *m* CONST casing grab; ~ **de tubería de pozo** *m* CONST *petróleo, sondeos* casing grab

agarrar[1] *vt* CONST bite, PROD grip, hold, TRANSP MAR *ancla* grip

agarrar[2] *vi* TRANSP MAR *ancla* bite

agarre *m* C&V, CONST, PROD, TRANSP MAR grip

agarrotado *adj* GEN jammed

agarrotamiento *m* AUTO seizing, ING MECÁ jam, jamming, seizing, PROD *cojinete, pistón* sticking, TRANSP AÉR binding, VEH seizing; ~ **hidráulico** *m* INSTAL HIDRÁUL *bomba, válvula* hydraulic lock, hydraulic locking; ~ **del pistón** *m* AUTO, VEH piston freezing

agarrotar[1] *vt* GEN jam

agarrotar[2] *vi* AUTO, ING MECÁ, INSTAL HIDRÁUL, PROD, TRANSP AÉR seize

agarrotarse *v refl* NUCL seize, TEC ESP *válvulas* jam; ~ **en su asiento** *v refl* INSTAL HIDRÁUL *válvulas* jam

ágata ƒ C&V, MINERAL agate; ~ **musgosa** ƒ C&V, MINERAL moss agate

agatino *m* MINERAL agaty

agavilladora ƒ AGRIC binder; ~ **de arroz** ƒ AGRIC rice binder

agencia ƒ TEC ESP, TELECOM, TEXTIL, TRANSP AÉR agency; ~ **espacial** ƒ TEC ESP space agency; ~ **explotadora privada reconocida** ƒ TELECOM recognised private operating agency; ~ **de transportes marítimos** ƒ TRANSP MAR shipping agency

Agencia: ~ **Espacial Europea** ƒ (*ESA*) TEC ESP European Space Agency (*ESA*); ~ **para la Protección del Medio Ambiente** ƒ CONTAM environmental protection agency (*EPA*)

agente *m* GEN agent, medium; ~ **abrasivo** *m* QUÍMICA abrasive agent; ~ **de acabado** *m* TEXTIL finishing agent; ~ **acidificante** *m* QUÍMICA, TEXTIL acidifying agent; ~ **acomplejante** *m* QUÍMICA sequestering agent; ~ **de acoplamiento** *m* P&C coupling agent; ~ **activador** *m* CARBÓN activating agent; ~ **de aduanas** *m* FERRO, TRANSP, TRANSP MAR forwarding agent (*BrE*), freight agent (*AmE*), goods agent (*BrE*); ~ **aduanero** *m* FERRO, TRANSP, TRANSP MAR forwarding agent (*BrE*), freight agent (*AmE*), goods agent (*BrE*); ~ **adherente** *m* P&C tackifying agent, QUÍMICA tackfying agent; ~ **adhesivo** *m* P&C tackness agent, QUÍMICA tackiness agent; ~ **adulterador** *m* ELECTRÓN doping agent; **agente** *m* GEN agent, medium; ~ **aglomerante** *m* CONTAM MAR, QUÍMICA herding agent; ~ **aglutinante** *m* ALIMENT, QUÍMICA binding agent; ~ **antiapelmazante** *m* ALIMENT anticaking agent; ~ **antibloqueante** *m* P&C *adhesivo, aditivo*, QUÍMICA antiblocking agent; ~ **anticongelante** *m* EMB antifreeze agent, frost-preventive agent, QUÍMICA frost-preventive agent, REFRIG antifreeze agent, frost-preventive agent; ~ **anticorrosivo** *m* QUÍMICA, TEC PETR anticorrosion agent; ~ **antidespellejante** *m* P&C *pinturas*, QUÍMICA antiskinning agent; ~ **antienvejecimiento** *m* ALIMENT, QUÍMICA antistaling agent; ~ **antiespumante** *m* CARBÓN, P&C, QUÍMICA, TEC PETR antifoaming agent; ~ **antiestático** *m* DETERG, P&C, QUÍMICA, TEXTIL antistatic agent; ~ **antimancha** *m* CINEMAT, QUÍMICA antiscratch solution; ~ **antipelmazante** *m* QUÍMICA

anticaking agent; ~ **de antirredeposición** *m* DETERG, QUÍMICA antiredeposition agent; ~ **antisalpicaduras** *m* ALIMENT, QUÍMICA antispattering agent; ~ **antivelo** *m* CINEMAT, FOTO, QUÍMICA antifogging agent; ~ **de apresto** *m* COLOR, QUÍMICA *almidón, ceras, gelatina* sizing agent; ~ **atmosférico** *m* QUÍMICA, TEC PETR atmospheric agent; ~ **biológico** *m* CONTAM biological agent; ~ **blanqueador** *m* ALIMENT, DETERG, QUÍMICA, TEXTIL bleaching agent; ~ **de blanqueo** *m* DETERG, QUÍMICA, TEXTIL bleaching agent; ~ **cargador** *m* ALIMENT, FERRO, TRANSP, TRANSP MAR forwarding agent (*BrE*), freight agent (*AmE*), goods agent (*BrE*); ~ **colorante** *m* C&V, COLOR, QUÍMICA coloring agent (*AmE*), colouring agent (*BrE*); ~ **concentrador de aceite** *m* CONTAM oil-concentrating agent; ~ **congelante** *m* REFRIG freezing medium; ~ **contaminante** *m* CONTAM, CONTAM MAR, QUÍMICA, SEG polluting agent; ~ **de control de llamadas** *m* TELECOM call control agent (*CCA*); ~ **cremante** *m* P&C, QUÍMICA creaming agent; ~ **de descomposición** *m* QUÍMICA decomposing agent; ~ **desecante** *m* PROC QUÍ, QUÍMICA desiccative agent; ~ **desengrasante** *m* MECÁ degreasing agent, QUÍMICA degrasing agent, degreasing agent; ~ **desespumante** *m* ALIMENT, QUÍMICA, TEC PETR *ingeniería de lodos* defoaming agent; ~ **deshidratante** *m* PROC QUÍ, QUÍMICA dehydrating agent; ~ **de desmoldeado** *m* P&C, QUÍMICA release agent; ~ **desmulsificante** *m* ALIMENT, QUÍMICA de-emulsifying agent; ~ **de desoxidación** *m* METAL, QUÍMICA killing agent; ~ **despachador** *m* TRANSP, TRANSP MAR forwarding agent (*BrE*), freight agent (*AmE*), goods agent (*BrE*); ~ **despolarizador** *m* ELEC, FÍS *de pila* depolarizing agent; ~ **despolarizante** *m* ELEC, FÍS *de pila* depolarizing agent; ~ **de dilución** *m* PROC QUÍ diluting agent; ~ **dispersante** *m* ALIMENT, CONTAM, HIDROL, PROC QUÍ, QUÍMICA, TEC PETR dispersant, dispersing agent; ~ **dopante** *m* ELECTRÓN doping agent; ~ **eliminador de espuma** *m* QUÍMICA defoaming agent; ~ **eluyente** *m* NUCL, QUÍMICA eluting agent; ~ **emulgente** *m* ALIMENT, EMB, P&C, PAPEL, PROC QUÍ, PROD, QUÍMICA emulsifying agent; ~ **emulsionante** *m* ALIMENT, EMB, P&C, PAPEL, PROC QUÍ, PROD, QUÍMICA emulsifying agent; ~ **encolador** *m* TEXTIL sizing agent; ~ **endurecedor** *m* P&C, REVEST curing agent; ~ **endurecedor de amina** *m* P&C, REVEST amine curing agent; ~ **esmerilador** *m* C&V grinding agent; ~ **de espesamiento** *m* ALIMENT, DETERG, QUÍMICA thickening agent; ~ **espesante** *m* ALIMENT, DETERG, QUÍMICA thickening agent; ~ **espumante** *m* C&V, DETERG, PROC QUÍ, PROD, QUÍMICA, TEC PETR foamer, foaming agent; ~ **espumoso** *m* ALIMENT, C&V, DETERG, PROC QUÍ, PROD, QUÍMICA, TEC PETR foaming agent; ~ **estabilizante** *m* QUÍMICA stabilizing agent; ~ **para evitar la corrosión** *m* EMB corrosion preventive; ~ **expedidor** *m* FERRO, TRANSP, TRANSP MAR forwarding agent (*BrE*), freight agent (*AmE*), goods agent (*BrE*); ~ **de extensión** *m* AGRIC county agent; ~ **extintor de incendios** *m* SEG fire-extinguishing agent; ~ **favorecedor del drenado** *m* PAPEL drainage aid; ~ **favorecedor de la retención** *m* PAPEL retention aid; ~ **físico** *m* CONTAM physical agent; ~ **de gelatinización** *m* ALIMENT, CONTAM MAR, P&C, QUÍMICA gelling agent; ~ **gelificante** *m* ALIMENT, CONTAM MAR, P&C, QUÍMICA gelling agent;

~ **grasiento** *m* ALIMENT, QUÍMICA greasing agent; ~ **humectante** *m* GEN, wetting agent; ~ **hundidor** *m* CONTAM MAR sinking agent; ~ **igualador** *m* QUÍMICA, TEXTIL leveling agent (*AmE*), levelling agent (*BrE*); ~ **impermeabilizador** *m* REVEST waterproofing agent; ~ **de impregnación** *m* EMB impregnating agent; ~ **de lavado** *m* DETERG, QUÍMICA washing agent; ~ **leudante** *m* ALIMENT *panadería, repostería* raising agent; ~ **de limpieza en seco** *m* DETERG, QUÍMICA dry-cleaning agent; ~ **lixiviador** *m* NUCL leaching agent; ~ **neutralizador** *m* QUÍMICA, SEG neutralizing agent; ~ **neutralizante** *m* QUÍMICA, SEG neutralizing agent; ~ **odorante** *m* PROC QUÍ odorizer; ~ **oxidante** *m* CONTAM, QUÍMICA oxidizing agent, TEC ESP fuel; ~ **de patentes** *m* ING MECÁ patent agent; ~ **de pegajosidad** *m* QUÍMICA tackiness agent; ~ **de permanencia** *m* QUÍMICA stayput agent; ~ **de precipitación** *m* PROC QUÍ, QUÍMICA precipitating agent; ~ **precipitante** *m* PROC QUÍ, QUÍMICA precipitating agent; ~ **protector** *m* EMB, SEG protective agent; ~ **protector que forma capas de espuma contra incendios** *m* SEG foam layer-forming flameproofing agent; ~ **pulidor** *m* C&V polishing agent; ~ **purificador** *m* AGUA, GAS, HIDROL, PROC QUÍ, QUÍMICA, TEC PETR purifying agent; ~ **quelatante** *m* DETERG, HIDROL, QUÍMICA chelating agent; ~ **químico** *m* CONTAM, D&A, QUÍMICA, TEXTIL chemical agent, reducing agent; ~ **recogedor de aceite** *m* CONTAM oil-concentrating agent; ~ **reductor** *m* CARBÓN *química*, CONTAM, QUÍMICA reducing agent; ~ **de refuerzo** *m* P&C, QUÍMICA reinforcing agent; ~ **retardante** *m* CONST *hormigón*, TEXTIL retarding agent; ~ **de reticulación** *m* P&C *polimerización, vulcanización* cross-linking agent; ~ **de revestido** *m* EMB, REVEST coating compound; ~ **salino** *m* NUCL salting agent; ~ **salpicaduras** *m* TEC ESP antispattering agent; ~ **secuestrador** *m* QUÍMICA sequestering agent; ~ **de separación** *m* PROC QUÍ, QUÍMICA separating agent; ~ **de solidificación** *m* P&C *aditivo* gelling agent; ~ **suavizante** *m* TEXTIL softening agent; ~ **superficiactivo** *m* CONTAM MAR, P&C surface-active agent, surfactant; ~ **supresor** *m* COLOR inhibitor dye; ~ **surfactante** *m* CONTAM MAR surface-active agent, P&C *aditivo* surface-active agent, surfactant; ~ **de tamponación** *m* CINEMAT buffering agent; ~ **telemático** *m* TELECOM telematic agent (*TLMA*); ~ **tensoactivo** *m* ALIMENT, CARBÓN, CONTAM, CONTAM MAR, DETERG, P&C, QUÍMICA surface-active agent, surfactant; ~ **tensoactivo aniónico** *m* DETERG anionic surface-active agent; ~ **tranquilizador del baño líquido** *m* METAL killing agent; ~ **de transferencia de mensajes** *m* TELECOM message transfer agent (*MTA*); ~ **de transporte** *m* FERRO, TRANSP, TRANSP MAR forwarding agent (*BrE*), freight agent (*AmE*), goods agent (*BrE*); ~ **de transporte marítimo** *m* TRANSP MAR shipping agent; ~ **usuario** *m* INFORM&PD, TELECOM user agent (*UA*); ~ **usuario de intercambio de datos electrónicos** *m* INFORM&PD, TELECOM electronic data interchange user agent (*EDI-UA*)

agilidad: ~ **de frecuencia** *f* ELECTRÓN frequency agility; ~ **de la señal** *f* ELECTRÓN signal agility

agitación *f* INFORM&PD gating, ING MECÁ shaking, METEO, OCEAN choppiness, QUÍMICA slugging, *proceso* stirring, TRANSP AÉR, TRANSP MAR choppiness; ~ **intermitente** *f* FOTO intermittent agitation;

~ **térmica** *f* FÍS, ING ELÉC, METAL, TERMO thermal agitation

agitador *m* AGRIC, ALIMENT, CARBÓN, ING MECÁ agitator, LAB, P&C, PROC QUÍ shaker, stirrer, TEC PETR *ingeniería de lodos, perforación* agitator; ~ **para limpiar piezas** *m* PROD *fundición* shaking machine; ~ **magnético** *m* LAB magnetic stirrer; ~ **refractario** *m* LAB fireproof stirrer, flameproof stirrer; ~ **de salmuera por inyección de aire comprimido** *m* REFRIG brine sparge; ~ **de semilla** *m* AGRIC seed agitator; ~ **vibratorio** *m* LAB vibrating stirrer; ~ **de vidrio** *m* C&V, LAB glass stirring-rod

agitar *vt* FOTO, PROC QUÍ agitate, PROD *funderías* churn

aglomeración *f* CARBÓN balling, congestion, sintering, GEOL agglomerate, ING MECÁ packing, cluster, P&C *cargas, pigmentos*, PROC QUÍ agglomeration, TRANSP cluster; ~ **en caliente** *f* PROD hot set; ~ **continental** *f* GEOL continental accretion; ~ **de goma en caliente** *f* EMB hot-setting glue; ~ **en lecho fluidizado** *f* PROC QUÍ fluidized-bed sintering

aglomerado¹ *adj* MECÁ sintered, TRANSP clustered

aglomerado² *m* C&V, CARBÓN aggregate, CONST, ENERG RENOV conglomerate, GEOL conglomerate, *brecha volcánica áspera* agglomerate, METAL aggregate, PETROL, PROC QUÍ agglomerate; ~ **de carbón** *m* CARBÓN briquette; ~ **descompuesto** *m* GEOL blueground

aglomerante *m* CONST *albañilería, obra civil* binder, CONTAM MAR herding agent, TEC ESP *revestimientos ignífugos* binder; ~ **carbón-carbón insaturado** *m* TEC PETR unsaturated carbon-to-carbon bond

aglomerar *vt* CONST *albañilería*, MECÁ bond, PROC QUÍ agglomerate

aglomerarse *v refl* CARBÓN agglomerate

aglutinación *f* ALIMENT agglutination, CARBÓN sintering, P&C, PROC QUÍ, QUÍMICA agglutination, caking, REFRIG *tratamiento de fibras aislantes* bonding; ~ **del grano** *f* FOTO grain clumping

aglutinamiento: ~ **electrónico** *m* TV electronic matting

aglutinante¹ *adj* GEN agglutinant

aglutinante² *m* C&V binder, CONST cementing material, CONTAM MAR binding agent, P&C *pintura*, TEC ESP *revestimientos ignífugos* binder

aglutinar¹ *vt* ALIMENT cake, PROC QUÍ agglutinate, cake, QUÍMICA agglutinate

aglutinar² *vi* ALIMENT agglutinate, cake, PROC QUÍ, QUÍMICA agglutinate

aglutinarse *v refl* ALIMENT, PROC QUÍ, QUÍMICA cake

aglutinina *f* ALIMENT, PROC QUÍ, QUÍMICA agglutinin

agotadas: ~ **las existencias** *f pl* PROD stock-out

agotado *adj* QUÍMICA spent

agotamiento *m* AGUA depletion, drainage, pumping out, pumping, CARBÓN drying, ELECTRÓN depletion, HIDROL *de una corriente de agua* drying-up, ING ELÉC *acumuladores* overdischarging, METAL exhaustion, MINAS dewatering, drainage, draining, pumping, NUCL exhaustion, QUÍMICA depletion, TEC PETR pumping, TRANSP MAR pumping out; ~ **del aire en oxígeno** *m* MINAS oxygen depletion of the air; ~ **en una sola fase** *m* MINAS single-stage pumping ~ **de los recursos del subsuelo** *m* AGRIC depletion of subsoil resources; ~ **por repeticiones** *m* MINAS stage pumping; ~ **repetido** *m* MINAS multistage pumping

agotar *vt* AGUA deplete, drain, drain off, CARBÓN dry, ELECTRÓN deplete, METAL exhaust, MINAS dewater, drain, *frente de arranque* spend ground NUCL deplete, exhaust, QUÍMICA deplete, RECICL waste, TEC PETR pump, TRANSP MAR pump out

agotarse *v refl* AGUA *pozo* run dry, CARBÓN dry, HIDROL, OCEAN run dry

agradación *f* AGUA aggradation

agramadera *f* PROD *funderías* stripping machine

agramado *m* MECÁ braking

agramiladora *f* ING MECÁ trimming machine

agrandamiento *m* CINEMAT, ING MECÁ, TV wipe; ~ **gradual circular de la imagen** *m* CINEMAT, TV circle wipe; ~ **gradual de la imagen** *m* CINEMAT, TV wipe

agregación: ~ **de datos** *f* INFORM&PD data aggregate

agregado *m* CARBÓN aggregate, P&C admixture; ~ **de mar** *m* OCEAN marine aggregate; ~ **triturado** *m* CARBÓN crushed aggregate

agregar[1] *vt* CONST tail; ~ **y eliminar** *vt* IMPR add and delete; ~ **espacio fino** *vt* IMPR *justificación de textos* add thin space

agregar[2] *vi* GEOL *partículas* aggregate

agricolita *f* MINERAL agricolite

agricultor *m* AGRIC farmer; ~ **de maíz** *m* AGRIC corn grower (*AmE*), maize grower (*BrE*)

agricultura *f* AGRIC farming; ~ **de secano** *f* AGRIC rain-fed agriculture; ~ **en tierras de aluvión** *f* AGRIC floodplain agriculture

agrietado *adj* ING MECÁ, MECÁ cracked

agrietamiento *m* C&V, CONST crazing, CRISTAL, ING MECÁ, NUCL, P&C cracking, crazing TERMO cracking; ~ **por contracción** *m* CONST shrinkage cracking; ~ **intergranular por tensocorrosión** *m* NUCL intergranular stress corrosion cracking (*IGSCC*); ~ **por tensocorrosión en medio cáustico** *m* NUCL caustic stress corrosion cracking; ~ **por tratamiento térmico** *m* TERMO heat-treatment crack

agrietar *vt* C&V, CONST craze

agrietarse *v refl* CRISTAL, ING MECÁ, METAL, P&C crack

agrimensor *m* AGRIC, CONST land surveyor, surveyor

agrimensura *f* AGRIC, CONST land survey, surveying

agrio *adj* MECÁ, METAL brittle

agro *m* AGRIC farming sector, rural sector

agroindustria *f* AGRIC, ALIMENT agribusiness

agropiro *m* AGRIC crested wheatgrass

agrumación *f* TEC PETR *ingeniería de lodos* flocculation

agrupación *f* ELECTRÓN *klystron* bunching, GEOL *de arribos consecutivos* line-up, ING MECÁ cluster, METAL clustering; ~ **celular** *f* ELECTRÓN cellular array; ~ **excesiva** *f* ELECTRÓN overbunching; ~ **de imágenes** *f* ELECTRÓN imaging array; ~ **local de iones** *f* METAL crowdion; ~ **naval** *f* D&A, TRANSP MAR navy task force; ~ **óptima** *f* ELECTRÓN optimum bunching; ~ **de reflejos** *f* ELECTRÓN reflex bunching; ~ **de la tobera del motor** *f* TRANSP AÉR engine-nozzle cluster

agrupado *adj* TRANSP clustered

agrupamiento *m* INFORM&PD cluster, TRANSP AÉR bunching; ~ **en un solo control** *m* ELEC, ING ELÉC ganging

agrupar *vt* AGUA bulk, GEOL *partículas* cluster together; ~ **para ser impresos a la vez** *vt* IMPR *moldes* gang

agrura *f* MECÁ, METAL brittleness

agua[1]: **de ~ dulce** *adj* GEOL *fauna* freshwater

agua[2]: **al ~** *adv* TRANSP MAR overboard

agua[3]: ~ **absorbida por capilaridad** *f* AGUA, HIDROL capillary soil water; ~ **ácida** *f* AGUA, CONTAM, HIDROL acid water; ~ **activa** *f* AGUA, HIDROL active water; ~ **agresiva** *f* AGUA, HIDROL aggressive water; ~ **de alimentación** *f* AGUA, HIDROL, ING MECÁ, NUCL, PAPEL, PROD, TEC PETR, TERMOTEC, TRANSP MAR *calderas* feedwater; ~ **artesiana** *f* AGUA confined ground water, CARBÓN, HIDROL artesian water, confined ground water; ~ **blanda** *f* AGUA, HIDROL soft water; ~ **bombeada** *f* AGUA, HIDROL, INSTAL HIDRÁUL pump water; ~ **bruta** *f* AGUA, HIDROL raw water; ~ **de cal** *f* QUÍMICA *medicina*, REVEST limewater; ~ **de calado** *f* OCEAN draw water; ~ **calcárea** *f* AGUA, HIDROL hard water; ~ **caliente** *f* TERMO, TERMOTEC hot water; ~ **capilar** *f* AGRIC, CARBÓN, HIDROL capillary water; ~ **de cierres** *f* NUCL seal water; ~ **de ciudad** *f* AGUA, HIDROL town water; ~ **clarificada** *f* AGUA, HIDROL clear water; ~ **clorada** *f* AGUA, HIDROL, PROD chlorinated water; ~ **combinada** *f* AGUA, HIDROL combined water; ~ **combinada químicamente** *f* AGUA, GEOL, HIDROL, PETROL, QUÍMICA, TEC PETR bound water; ~ **de condensación** *f* AGUA, HIDROL, RECICL wastewater; ~ **de constitución** *f* AGUA, HIDROL combined water; ~ **de consumo** *f* AGUA, HIDROL consumption water; ~ **contaminada** *f* AGUA, CONTAM, CONTAM MAR, HIDROL polluted water; ~ **corrosiva** *f* AGUA, HIDROL corrosive water; ~ **cruda** *f* AGUA, HIDROL raw water; ~ **de descarga** *f* AGUA, ENERG RENOV, HIDROL, INSTAL HIDRÁUL tailwater; ~ **desionizada** *f* AGUA, ELEC de-ionized water; ~ **desmineralizada** *f* AGUA, CONST, HIDROL demineralized water; ~ **de desperdicios industriales** *f* AGUA, CONTAM, HIDROL industrial waste water; ~ **dormida** *f* AGUA, HIDROL quiet water; ~ **dulce** *f* AGRIC, AGUA, GAS, HIDROL, PETROL, TRANSP MAR freshwater; ~ **dura** *f* AGUA, HIDROL hard water; ~ **de elaboración** *f* AGUA, HIDROL process water; ~ **de entrada** *f* AGUA, HIDROL influent water; ~ **estancada** *f* AGUA, HIDROL backwash water, stagnant water, standing water, still water; ~ **de extracción** *f* MINAS tapping water; ~ **filtrada** *f* HIDROL percolating water; ~ **fluvial** *f* AGRIC, AGUA, HIDROL river water; ~ **de fondo** *f* HIDROL, OCEAN bottom water; ~ **de formación** *f* AGUA bound water, GEOL, HIDROL bound water, connate water, formation water, OCEAN connate water, PETROL, TEC PETR bound water, connate water, formation water; ~ **fósil** *f* GEOL, HIDROL connate water, fossil water, OCEAN connate water, PETROL, TEC PETR connate water, fossil water; ~ **freática** *f* AGRIC, AGUA, CARBÓN, HIDROL ground water, well water; ~ **freática libre** *f* AGUA, HIDROL free groundwater; ~ **freática profunda** *f* AGUA deep groundwater; ~ **sin gas** *f* ALIMENT still water; ~ **gorda** *f* AGUA hard water; ~ **de gravedad** *f* AGUA, CARBÓN, HIDROL gravitational water, gravity water; ~ **de grieta** *f* CARBÓN cleft water; ~ **gruesa** *f* AGUA, HIDROL hard water; ~ **helada** *f* REFRIG ice water; ~ **higroscópica** *f* AGUA, HIDROL hygroscopic water; ~ **hypersalina** *f* OCEAN hypersaline water; ~ **para la industria** *f* AGUA, HIDROL industrial water; ~ **industrial** *f* AGUA, HIDROL city water (*AmE*), mains water (*BrE*), commercial water; ~ **de infiltración** *f* AGUA, HIDROL infiltration water, leak water, percolating water; ~ **infiltrada** *f* HIDROL percolating water, QUÍMICA seepage water; ~ **intersticial** *f* AGUA, CARBÓN connate water, interstitial water, pore water, GEOL connate water, HIDROL, OCEAN connate water, interstitial water, pore water,

PETROL connate water, pore water, TEC PETR connate water, interstitial water, pore water; ~ **juvenil** *f* AGUA, GEOL *magmática* juvenile water, mineral water; ~ **lacustre** *f* CONTAM lake water; ~ **de lago** *f* CONTAM lake water; ~ **lanzada hacia atrás** *f* AGUA backwash water; ~ **ligada** *f* AGUA, GEOL, HIDROL, PETROL, TEC PETR bound water; ~ **limpia** *f* AGUA, CONTAM MAR, HIDROL clean water; ~ **de lluvia** *f* AGRIC, AGUA, CARBÓN, HIDROL, METEO rainwater, storm water; ~ **de lluvia contaminada** *f* AGUA, HIDROL polluted rainwater; ~ **de lluvia impura** *f* AGUA polluted water, CONTAM, HIDROL polluted rainwater; ~ **magmática** *f* AGUA, GEOL juvenile water; ~ **de manantial** *f* AGUA, GEOL, HIDROL spring water; ~ **mancomunada** *f* CARBÓN joint water; ~ **de mar** *f* OCEAN, TRANSP MAR sea water; ~ **de mar normal** *f* OCEAN, TRANSP MAR standard sea water; ~ **de mar de referencia** *f* OCEAN, TRANSP MAR standard sea water; ~ **de marea** *f* AGUA, HIDROL tidewater; ~ **con materia corrosiva** *f* AGUA, HIDROL aggressive water; ~ **mineral** *f* AGUA, ALIMENT, HIDROL mineral water; ~ **nativa** *f* AGUA, GEOL, HIDROL, OCEAN, PETROL, TEC PETR connate water; ~ **necesaria** *f* ALIMENT make-up water; ~ **nerítica** *f* OCEAN neritic water; ~ **normal** *f* NUCL normal water, ordinary water; ~ **normal de mar** *f* OCEAN normal sea water; ~ **de origen antiguo** *f* AGUA, GEOL, HIDROL fossil water; ~ **oxidada** *f* OCEAN red tide; ~ **oxigenada** *f* AGUA, HIDROL oxygenated water, QUÍMICA, TEC ESP hydrogen peroxide; ~ **ozonizada** *f* AGUA, HIDROL ozonized water; ~ **pelicular** *f* CARBÓN pellicular water; ~ **pesada** *f* FÍS, NUCL, QUÍMICA heavy water; ~ **potable** *f* AGUA, ALIMENT, CONST, HIDROL drinking water; ~ **de pozo** *f* AGRIC, AGUA, CARBÓN, HIDROL ground water, well water; ~ **precipitable** *f* METEO precipitable water; ~ **a presión** *f* AGUA water under pressure; ~ **de proceso industrial** *f* HIDROL industrial process water; ~ **de profundidad limitada** *f* AGUA, CARBÓN, HIDROL confined ground water; ~ **pulverizada** *f* AGUA pulverizing water, HIDROL, PROD spray; ~ **pura** *f* AGUA, HIDROL pure water; ~ **purificada** *f* AGUA, HIDROL purified water; ~ **de recepción** *f* AGUA, HIDROL receiving water; ~ **reciente** *f* AGUA, HIDROL recent water; ~ **de referencia de mar** *f* OCEAN normal sea water; ~ **refrigerada** *f* HIDROL, REFRIG chilled water; ~ **regia** *f* QUÍMICA aqua regia; ~ **de relleno** *f* HIDROL, REFRIG *perdida en la evaporación* make-up water; ~ **remansada** *f* OCEAN dead water; ~ **represada** *f* OCEAN slack water; ~ **residual industrial** *f* CONTAM, RECICL industrial waste water; ~ **retenida** *f* ALIMENT bound water; ~ **de retorno** *f* PAPEL back water; ~ **salada** *f* AGRIC, AGUA, GEOL, HIDROL brackish water, saline water, salt water; ~ **salobre** *f* AGRIC, AGUA, GEOL, HIDROL brackish water, saline water; ~ **de sentina** *f* TRANSP MAR bilge water; ~ **para servicio propio** *f* HIDROL domestic water; ~ **de servicios esenciales** *f* NUCL essential-services water; ~ **singenética** *f* AGUA, GEOL, HIDROL, OCEAN, PETROL, TEC PETR connate water; ~ **sobrante** *f* AGUA, HIDROL waste water; ~ **subfundida** *f* HIDROL supercooled water; ~ **del subsuelo** *f* AGRIC, AGUA, CARBÓN, HIDROL ground water, well water; ~ **subterránea** *f* AGRIC ground water, AGUA ground water, subsoil water, subterranean water, well water, CARBÓN, CONTAM, HIDROL ground water, subsoil water, underground water; ~ **subterránea libre** *f*

CARBÓN unconfined ground water; ~ **subterránea más arriba de la capa freática** *f* AGUA vadose water; ~ **subyacente** *f* TEC PETR *geología* edge water; ~ **superficial** *f* AGUA, CARBÓN, CONST, HIDROL surface water; ~ **termal** *f* OCEAN hot spring water; ~ **tranquila** *f* AGUA, HIDROL, OCEAN slack water, still water; ~ **turbia** *f* AGUA, CONTAM turbid water; ~ **de uso local** *f* AGUA, HIDROL municipal water
agua[4]: **sin** ~ *fra* TEXTIL water-free
aguacero *m* CONST shower
aguaje *m* ENERG RENOV *energía del agua* bore; ~ **producido por la resistencia de apéndices** *m* OCEAN *hidrodinámica* dead water
aguamarina *f* MINERAL aquamarine
aguantar *vt* TRANSP MAR *cadena* check, *chubasco* weather
aguante *m* MINAS *de filón o galería* bearing
aguar *vt* AGUA irrigate, water
aguarrás *m* COLOR, P&C, QUÍMICA turpentine, REVEST turpentine varnish
aguas[1]: ~ **abajo** *adv* GEN downstream; ~ **arriba** *adv* AGUA, CONST *hidráulica*, FÍS, FÍS FLUID, GAS, HIDROL, MECÁ, TRANSP MAR upstream
aguas[2] *f pl* PROD *tejidos*, TEXTIL glossing; ~ **abiertas** *f pl* OCEAN open water; ~ **adyacentes** *f pl* Esp (*cf aguas colindantes AmL*) OCEAN adjacent waters; ~ **de albañal** *f pl* AGUA sewage, CONTAM black water, sewage, sewerage, waste water, RECICL sewage; ~ **de alcantarilla** *f pl* AGUA, CONTAM, RECICL sewage; ~ **de alcantarillado** *f pl* AGUA, HIDROL, RECICL sanitary waste water; ~ **blancas** *f pl* PAPEL white waters; ~ **de cabecera** *f pl* HIDROL *de un curso de agua* headwater; ~ **de cabeza** *f pl* ALIMENT *fermentación* first runnings, fore-runnings; ~ **cloacales** *f pl* AGUA, HIDROL, RECICL sewage; ~ **cloacales de lluvia** *f pl* AGUA, HIDROL, RECICL storm sewage; ~ **cloacales sanitarias** *f pl* AGUA, HIDROL, RECICL domestic water; ~ **colgadas** *f pl* AGUA perched water; ~ **colindantes** *f pl AmL* (*cf aguas adyacentes Esp*) OCEAN adjacent waters; ~ **coloradas** *f pl* OCEAN red tide; ~ **contaminadas** *f pl* AGUA, CONTAM, HIDROL, RECICL foul water; ~ **contiguas** *f pl* OCEAN contiguous waters; ~ **costeras** *f pl* AGUA, HIDROL, OCEAN, TRANSP MAR coastal waters, inshore waters; ~ **de descarga industrial** *f pl* RECICL industrial waste water; ~ **de desecho** *f pl* AGUA black water, waste water, CARBÓN, CONST black water, CONTAM, HIDROL black water, waste water, RECICL black water; ~ **epicontinentales** *f pl* OCEAN continental-shelf waters; ~ **estancadas** *f pl* AGUA, CONTAM, HIDROL, RECICL foul water; ~ **fecales** *f pl* AGUA, CONTAM, RECICL sewage; ~ **del interior** *f pl* AGUA inland waters; ~ **interiores** *f pl* OCEAN internal waters; ~ **intermedias** *f pl* OCEAN intermediate waters; ~ **internacionales** *f pl* OCEAN, TRANSP MAR *zonas marítimas* international waters; ~ **libres** *f pl* PETROL open water; ~ **litorales** *f pl* AGUA, HIDROL, OCEAN, TRANSP MAR coastal waters; ~ **de mareas** *f pl* TRANSP MAR equinoctial tide, water tide; ~ **meteóricas** *f pl* HIDROL meteoric water; ~ **muertas** *f pl* OCEAN dead water; ~ **navegables** *f pl* OCEAN, TRANSP MAR safe water; ~ **negras** *f pl* AGUA black water, domestic waste water, sewage, CARBÓN, CONST black water, CONTAM black water, domestic waste water, sewage, waste water, HIDROL black water, domestic waste water sewage, RECICL black water, domestic waste

water, sewage; ~ **negras crudas** *f pl* AGUA, CONTAM, HIDROL, RECICL raw sewage; ~ **negras decantadas** *f pl* AGUA, CONTAM, HIDROL, RECICL settled sewage; ~ **negras sin depurar** *f pl* AGUA, CONTAM, HIDROL, RECICL raw sewage; ~ **pelágicas** *f pl* OCEAN pelagic waters; ~ **de la plataforma continental** *f pl* OCEAN continental-shelf waters; ~ **pluviales** *f pl* CARBÓN storm water, CONTAM, HIDROL surface waters; ~ **residuales** *f pl* AGUA black water, CARBÓN, CONST black water, sewerage, waste water, CONTAM black water, HIDROL, RECICL black water, sewerage, waste water; ~ **residuales de alcantarilla** *f pl* AGUA, CONTAM, HIDROL, RECICL sewage waste water; ~ **residuales no tratadas** *f pl* AGUA, CONTAM, HIDROL, RECICL raw sewage, untreated sewage; ~ **residuales de servicios propios** *f pl* AGUA, CONTAM, HIDROL, RECICL domestic waste water; ~ **retenidas en el cuenco** *f pl* TRANSP MAR *canal* sluice; ~ **rojas** *f pl* OCEAN red waters; ~ **seguras** *f pl* TRANSP MAR safe water; ~ **someras** *f* OCEAN shallow water; ~ **sucias** *f pl* CONTAM, HIDROL, RECICL foul water; ~ **territoriales** *f pl* GEOL marginal sea, OCEAN territorial waters

agudeza *f* CONST keenness, FÍS *de una resonancia* sharpness, GEOM acuteness, ING MECÁ sharpness; ~ **auditiva** *f* ACÚST *audición* auditory acuity

agudo *adj* ING MECÁ sharp

aguilón *m* CONST *grúa*, MECÁ boom, jib; ~ **de la pulverizadora** *m* AGRIC spray boom

aguja *f* CONST, FERRO points (*BrE*), set of points (*BrE*), switch (*AmE*), ING MECÁ needle, pointer, tongue, LAB needle, *indicadora de un instrumento* hand, MINAS lagging piece, TEC ESP *cuadrante indicador* index, TELECOM pointer, TEXTIL needle, TRANSP MAR compass; ~ **de la bitácora** *f* TRANSP MAR binnacle compass; ~ **de cadeneo** *f* CONST *topografía* arrow, MINAS *topografía* chain pin; ~ **de cambio** *f* FERRO *equipo inamovible* switch blade, switch tongue; ~ **de cambio de vía** *f* FERRO point switch; ~ **compuesta** *f* CARBÓN composite pile; ~ **cónica** *f* ING MECÁ taper pin, tapered needle; ~ **de contacto** *f* ING MECÁ touch needle; ~ **descarriladora** *f* FERRO *equipo inamovible* derailing points (*BrE*), derailing switch (*AmE*); ~ **de descarrilamiento** *f* FERRO *equipo inamovible* derailing points (*BrE*), derailing switch (*AmE*); ~ **de desmoldear** *f* PROD *herramienta* draw spike, *herramienta de moldería* draw stick; ~ **de disección** *f* LAB *biología* dissection needle; ~ **de encuadernar** *f* IMPR bookbinder's needle; ~ **de ensayar** *f* ING MECÁ touch needle; ~ **del exposímetro** *f* CINEMAT, FÍS, FOTO exposure-meter needle; ~ **del flotador** *f* AUTO float needle; ~ **giroscópica** *f* TRANSP MAR gyroscopic compass; ~ **de gobierno** *f* TRANSP MAR steering compass; ~ **para hacer agujeros** *f* PROD *moldes* pricker, vent wire; ~ **imanada** *f* FÍS, GEOFÍS, ING MECÁ, INSTR magnetic needle; ~ **imantada** *f* FÍS, GEOFÍS, ING MECÁ, INSTR magnetic needle; ~ **de inclinómetro** *f* GEOFÍS dip circle needle; ~ **indicadora** *f* ELEC indicator needle, needle, pointer, FOTO, ING ELÉC indicator needle, pointer, ING MECÁ, LAB indicator needle, needle, pointer; ~ **de lengüeta** *f* TEXTIL latch needle; ~ **de líquido** *f* TRANSP MAR liquid compass; ~ **loca** *f* TRANSP MAR disturbed compass; ~ **magistral** *f* TRANSP MAR standard compass; ~ **magnética** *f* FÍS, GEOFÍS, ING MECÁ, INSTR, TRANSP AÉR, TRANSP MAR magnetic compass, mag-

netic needle; ~ **de marcar** *f* ING MECÁ scriber, TRANSP MAR *navegación* bearing compass; ~ **de marear** *f* TRANSP MAR compass, mariner's compass; ~ **de marinero** *f* GEOFÍS mariner's needle; ~ **de mina** *f* MINAS nail, needle, picker; ~ **náutica** *f* TRANSP MAR *navegación* compass; ~ **de polvorero** *f* MINAS picker, pricker, shooting needle; ~ **de proyección** *f* TRANSP MAR projector compass; ~ **de prueba** *f* ING MECÁ touch needle, test needle; ~ **repetidora** *f* TRANSP MAR compass repeater; ~ **de terraplén** *f* CARBÓN embankment pile; ~ **terraplén** *f* CARBÓN, MINAS embankment pile; ~ **de ventear** *f* PROD *moldes* piercer, vent wire

agujas: ~ **chapitel** *f pl* CONST *arquitectura* broach

agujereado *adj* MECÁ, PROD holed

agujereadora *f* MECÁ *máquina herramienta*, MINAS, PROD drill

agujerear *vt* MECÁ bore, drill, MINAS hole, PROD *moldes* pierce

agujero *m* C&V, ING ELÉC, ING MECÁ, MECÁ, PROD aperture, hole, opening, orifice, TEC PETR *perforación* hole, well bore; ~ **de agua** *m* PROD passage of water, water port; ~ **de aire** *m* PROD vent, vent hole; ~ **alargado** *m* ING MECÁ, MECÁ elongated hole; ~ **de alfiler** *m* PAPEL *defecto* pinhole; ~ **de aligeramiento** *m* MECÁ, TRANSP AÉR lightening hole; ~ **de alivio** *m Esp* (*cf agujero de desahogo AmL*) MINAS easer hole; ~ **avellanado** *m* ING MECÁ countersunk hole; ~ **del bitoque** *m* ALIMENT *fermentación* bunghole; ~ **de caldeo** *m* PROD stoke hole; ~ **de carga** *m* MECÁ filling hole; ~ **de centrado** *m* ING MECÁ center hole (*AmE*), centre hole (*BrE*); ~ **de centrado con chaflán protector** *m* ING MECÁ center-hole with protecting chamfer (*AmE*), centre hole with protecting chamfer (*BrE*); ~ **de colada** *m* MINAS *hornos* mouth, P&C *moldeo por inyección* sprue opening, PROD sprue hole, *fundición* running gate, *horno* draught hole (*BrE*), taphole, cast gate, tapped hole, *moldes* down runner, pouring-hole, pouring-gate, *piquera de horno* drawhole, mouth; ~ **de derrame** *m* AUTO overflow hole; ~ **de desagüe de seguridad** *m* AUTO overflow hole; ~ **de desahogo** *m AmL* (*cf agujero de alivio Esp*) MINAS easer hole; ~ **de desarenar** *m* ING MECÁ cored passage; ~ **de desbordamiento** *m* AUTO overflow hole; ~ **de desviación** *m* AUTO bypass bore; ~ **de diámetro demasiado pequeño** *m* PETROL, TEC PETR undergage hole (*AmE*), undergauge hole (*BrE*); ~ **de diámetro excesivo** *m* PETROL, TEC PETR overgage hole (*AmE*), overgauge hole (*BrE*); ~ **de diámetro pequeño** *m* PETROL, TEC PETR slim hole; ~ **de drenaje** *m* TEC PETR *conductos, tuberías* drain tap; ~ **de drenaje del aceite** *m* AUTO oil drain hole; ~ **de escariado** *m* ING MECÁ counterbore; ~ **de evacuación de escorias** *m* MINAS *al deshornar*, PROD breast hole; ~ **extendido** *m* ING MECÁ, MECÁ elongated hole; ~ **de fondo** *m* MINAS *sondeos*, TEC PETR *perforación* bottom hole; ~ **guía** *m* ING MECÁ, PETROL pilot hole; ~ **de inspección** *m* ING MECÁ inspection hole, PROD hand hole, inspection cover; ~ **de lavado** *m* PROD *calderas* hand hole; ~ **de limpieza** *m* PROD *calderas* mud door; ~ **de llamada** *m* CONST draw bore; ~ **de llenado** *m* TEC ESP filling hole; ~ **de lubricación** *m* ING MECÁ oilhole; ~ **de mano** *m* PROD *calderas* washout hole; ~ **moldeado** *m* ING MECÁ cored passage; ~ **negro** *m* FÍS, TEC ESP black hole; ~ **de**

observación *m* ING MECÁ spy hole; ~ **pasante** *m* CONST thoroughfare hole; ~ **de paso** *m* ING MECÁ clearance hole; ~ **de perforación** *m* GAS, GEOFÍS, GEOL, HIDROL borehole; ~ **de perno** *m* ING MECÁ bolt hole; ~ **piloto** *m* ING MECÁ pilot hole; ~ **de prospección** *m* MINAS prospect hole; ~ **puntual** *m* FÍS, ÓPT pinhole; ~ **rayado** *m* ING MECÁ notched hole; ~ **de rebose** *m* AUTO overflow hole; ~ **rectificado** *m* AUTO, ING MECÁ, VEH reboring; ~ **de referencia** *m* ING MECÁ locating hole; ~ **de salida** *m* AGUA spout hole; ~ **sondeador de pozos** *m* CARBÓN, MINAS well drill hole; ~ **de sondeo** *m* MINAS drill hole, prospect hole; ~ **de sondeo hecho en seco** *m* MINAS dry hole; ~ **de sondeo profundo** *m* MINAS deep drillhole; ~ **del tapón** *m* ALIMENT bunghole; ~ **de ventilación** *m* ING MECÁ air vent
aguzado *adj* ING MECÁ sharpened, METAL needle-shaped
aguzar *vt* ING MECÁ grind, sharpen, MECÁ, PROD grind
aherrojado *adj* PROD ironbound
ahogado[1] *adj* AUTO *carburador*, MECÁ, VEH flooded
ahogado[2] *m* AUTO *carburador*, MECÁ, VEH flood
ahogador *m* *Esp* (*cf cebador AmL*) AUTO, MECÁ, VEH choke
ahogar *vt* AUTO flood, ENERG RENOV stall, MECÁ quench, VEH flood
ahondar *vt* PROD hollow, hollow out
ahorquillar *vt* D&A range
ahorro: ~ **de energía** *m* PROD, TERMO energy saving; ~ **de espacio** *m* EMB economy of space, space saving
ahuecado *adj* MECÁ dished
ahuecamiento *m* CONST hollowing, PROD coring-out, *moldería* hollowing
ahuecar *vt* CONST hollow, hollow out, PROD hollow, hollow out, *moldería* core, core out; ~ **con macho** *vt* PROD core
ahumado[1]: ~ **en caliente** *adj* ALIMENT hot-smoked; ~ **en frío** *adj* ALIMENT cold-smoked
ahumado[2] *m* ALIMENT smoking
ahumar *vt* ALIMENT smoke
ahusamiento *m* ING MECÁ, MECÁ, PROD taper, tapering
AI *abr* (*índice de articulación*) ACÚST *palabra* AI (*articulation index*)
aikinita *f* MINERAL aikinite
ainalita *f* MINERAL ainalite
airbus *m* TRANSP AÉR airbus
aire[1]: **con ~ acondicionado** *adj* AUTO, CONST, INSTAL TERM, REFRIG, TERMO, VEH air-conditioned
aire[2]: **en el ~** *adv* TRANSP AÉR airborne, TV on air; **al ~ libre** *adv* MINAS open-cast
aire[3] *m* GAS air gas, TV air; ~ **acondicionado** *m* ING MECÁ, INSTAL TERM, REFRIG air conditioning, VEH *habitáculo* air conditioner; ~ **admitido en sentido de la marcha** *m* ING MECÁ ram air; ~ **ambiente** *m* CONTAM, REFRIG ambient air; ~ **aspirante** *m* ING MECÁ induced air; ~ **calentador** *m* TERMO, TRANSP AÉR heating air; ~ **de calentamiento** *m* TERMO, TRANSP AÉR heating air; ~ **caliente forzado** *m* INSTAL TERM, TERMO forced hot air, forced warm air; ~ **de combustión** *m* TERMO combustion air; ~ **comprimido** *m* CONST, ING MECÁ, MINAS compressed air; ~ **continental** *m* METEO continental air; ~ **de derivación del motor** *m* TRANSP AÉR engine bypass air; ~ **desecante** *m* CARBÓN drying air; ~ **deshelador** *m* TRANSP AÉR de-icing air; ~ **de desvío del motor** *m* TRANSP AÉR engine bypass air;

~ **ecuatorial marítimo** *m* METEO maritime equatorial air; ~ **de enfriamiento** *m* ING MECÁ, REFRIG, TERMO cooling air; ~ **estable** *m* METEO stable air; ~ **húmedo** *m* METEO humid air; ~ **inducido** *m* ING MECÁ induced air; ~ **inestable** *m* METEO unstable air; ~ **limpio respirable** *m* SEG air fit to breathe; ~ **líquido** *m* TERMO liquid air; ~ **marítimo** *m* METEO maritime air; ~ **no respirable** *m* SEG air unfit for respiration; ~ **ocluido** *m* EMB entrapped air; ~ **polar** *m* METEO polar air; ~ **a presión** *m* CONST, ING MECÁ, MINAS compressed air, REFRIG ram air; ~ **quieto** *m* ENERG RENOV still air; ~ **refrigerante** *m* ING MECÁ, REFRIG, TERMO cooling air; ~ **respirable** *m* SEG breathable air; ~ **saturado** *m* METEO, REFRIG saturated air; ~ **secante** *m* CARBÓN drying air; ~ **seco** *m* METEO dry air; ~ **secundario** *m* AUTO, TERMOTEC secondary air; ~ **sobresaturado** *m* METEO supersaturated air; ~ **tropical** *m* METEO tropical air; ~ **viciado** *m* MINAS bad air, TV dead air
aireación *f* CARBÓN, CONST, EMB, HIDROL, ING MECÁ, RECICL aeration
aireado *m* IMPR winding and fanning; ~ **forzado** *m* ING MECÁ forced draft (*AmE*), forced draught (*BrE*)
aireador *m* AUTO, HIDROL, TRANSP AÉR, VEH aerator, demister; ~ **de cascada** *m* HIDROL cascade aerator; ~ **por contacto** *m* HIDROL *aguas residuales* contact aerator
airear *vt* GEN aerate, air, ventilate
aislable *adj* QUÍMICA isolable
aislación *f* ACÚST, AGRIC isolation, ELEC, ING ELÉC, MECÁ, P&C insulation, PROD isolation; ~ **de aire** *f* ELEC, ING ELÉC air insulation; ~ **de algodón** *f* ELEC, ING ELÉC cotton insulation; ~ **a alta temperatura** *f* ELEC, ING ELÉC high-temperature insulation; ~ **a alta tensión** *f* ELEC, ING ELÉC high-voltage insulation; ~ **del cable** *f* ELEC, ING ELÉC cable insulation; ~ **por capa de aire** *f* TERMOTEC airspace insulation; ~ **de capas** *f* ELEC, ING ELÉC layer insulation; ~ **defectuosa** *f* ELEC, ING ELÉC faulty insulation; ~ **del devanado** *f* ELEC *bobina*, ING ELÉC winding insulation; ~ **dieléctrica** *f* ELEC, ING ELÉC dielectric insulation; ~ **de entreturno** *f* ELEC *bobina* interturn insulation; ~ **exterior** *f* ELEC outer insulation, ING ELÉC outer insolation; ~ **externa** *f* ELEC, ING ELÉC external insulation; ~ **de fase** *f* ELEC, ING ELÉC *corriente alterna* phase insulation; ~ **interior** *f* ELEC, ING ELÉC indoor insulation; ~ **por material fibroso** *f* TERMOTEC fibrous insulation; ~ **de papel** *f* ELEC, ING ELÉC paper insulation; ~ **de papel impregnado** *f* ELEC, ING ELÉC *conductor de cable* impregnated paper insulation; ~ **de PVC** *f* ELEC, ING ELÉC PVC insulation; ~ **reflectora** *f* REFRIG, TERMO, TERMOTEC reflective insulation; ~ **térmica** *f* TERMO, TERMOTEC heat insulation, thermal insulation
aislado *adj* GEN insulated; ~ **al vacío** *adj* ING ELÉC vacuum-insulated
aislador *m* GEN insulator, isolator; ~ **acampanado** *m* ELEC, ING ELÉC bell-shaped insulator; ~ **de aceite** *m* ELEC, ING ELÉC oil insulator; ~ **de alimentación** *m* ELEC, ING ELÉC feedthrough insulator; ~ **del cable** *m* ELEC, ING ELÉC cable insulator; ~ **de campana** *m* ELEC, ING ELÉC petticoat insulator; ~ **cerámico** *m* ELEC, ING ELÉC ceramic insulator; ~ **de crisol** *m* ELEC, ING ELÉC pot insulator; ~ **óptico** *m* ING ELÉC, ÓPT, TELECOM optical isolator; ~ **de parada** *m* ELEC, ING ELÉC shackle insulator; ~ **de porcelana** *m* C&V,

ELEC, ING ELÉC porcelain insulator; ~ **resilente** *m* SEG *contra vibraciones*, TEC ESP, TRANSP AÉR resilient isolator; ~ **rígido** *m* ELEC, ING ELÉC pin insulator; ~ **de vibración** *m* GEOFÍS vibration isolator; ~ **de vidrio** *m* C&V, ELEC, ING ELÉC glass insulator

aislamiento *m* ACÚST isolation, insulation, C&V, CONST, ELEC insulation, isolation, ENERG RENOV *de colector*, FÍS isolation, insulation, INFORM&PD *de información*, ING ELÉC, MECÁ, P&C *calor, electricidad* isolation, insulation, PROD, REFRIG, SEG, TEC ESP, TELECOM, TERMO, TRANSP MAR insulation, isolation; ~ **acústico** *m* ACÚST, TRANSP MAR acoustic isolation, sound insulation; ~ **acústico bruto** *m* ACÚST average sound pressure level difference; ~ **acústico normalizado** *m* ACÚST standardized sound insulation; ~ **de aire** *m* ELEC, ING ELÉC air insulation; ~ **de algodón** *m* ELEC, ING ELÉC cotton insulation; ~ **para baja temperatura** *m* TERMO low-temperature insulation; ~ **del cable** *m* ELEC, ING ELÉC cable insulation; ~ **al calor** *m* EMB heat insulation; ~ **calorífugo** *m* TERMO heat insulation; ~ **de capas** *m* ELEC, ING ELÉC layer insulation; ~ **de caucho** *m* ELEC, ING ELÉC, P&C, REVEST rubber insulation; ~ **de CC** *m* ELEC, ING ELÉC DC isolation; ~ **defectuoso** *m* ELEC, ING ELÉC insulation defect; ~ **del devanado** *m* ELEC *bobina* winding insulation; ~ **eléctrico** *m* ELEC, ING ELÉC electrical insulation; ~ **eléctrico y óptico** *m* ELEC, ING ELÉC, ÓPT, PROD electrical-optical isolation; ~ **electromagnético** *m* ELEC, ING ELÉC electromagnetic isolation; ~ **de entreturno** *m* ELEC *bobina* interturn insulation; ~ **para evitar la ganancia térmica** *m* TERMO insulation against heat gain; ~ **para evitar la pérdida de calor** *m* TERMO insulation against heat loss; ~ **exterior** *m* ELEC, ING ELÉC outer insulation; ~ **externo** *m* ELEC, ING ELÉC external insulation; ~ **extruido** *m* ELEC, ING ELÉC *cable* extruded insulation; ~ **de fase** *m* ELEC, ING ELÉC *corriente alterna* phase insulation; ~ **fibroso** *m* TERMOTEC fibrous insulation; ~ **fónico** *m* ACÚST sound insulation, sound isolation; ~ **frigorífico** *m* TERMO insulation against heat gain; ~ **galvánico** *m* ELEC, ING ELÉC galvanic isolation; ~ **interior** *m* ELEC, ING ELÉC indoor insulation; ~ **isotermo** *m* TERMO, TRANSP MAR thermal insulation; ~ **mecánico** *m* SEG *contra vibraciones* mechanical isolation; ~ **mediante paneles en sandwich** *m* REFRIG sandwich-panel insulation; ~ **de minerales** *m* ELEC, ING ELÉC *conductor de cable* mineral insulation; ~ **óptico** *m* ELEC, ING ELÉC, ÓPT optical isolation; ~ **de papel** *m* ELEC, ING ELÉC paper insulation; ~ **de papel impregnado** *m* ELEC, ING ELÉC *conductor del cable* mass-impregnated paper insulation; ~ **de polarización** *m* TEC ESP polarization isolation; ~ **de porcelana** *m* ELEC, ING ELÉC porcelain insulation; ~ **de PVC** *m* ELEC, ING ELÉC PVC insulation; ~ **con recubrimiento** *m* ELEC, ING ELÉC *conductor de cable* lapped insulation; ~ **reflector** *m* REFRIG, TERMO, TERMOTEC reflective insulation; ~ **reforzado** *m* REFRIG superinsulation; ~ **sónico** *m* ACÚST sound insulation, sound isolation; ~ **térmico** *m* CONST, EMB, FERRO, MECÁ, TEC PETR heat insulation, thermal insulation, TERMO, TERMOTEC thermal insulation, heat insulation; ~ **de transformador** *m* ELEC, ING ELÉC transformer isolation; TRANSP MAR; ~ **al vacío** *m* ELEC, ING ELÉC, ING MECÁ vacuum insulation

aislante[1] *adj* GEN insulating

aislante[2] *m* CONST insulator, ELEC insulant, insulator, isolator, ING ELÉC insulating, insulator, PROD, QUÍMICA, REFRIG insulant, insulator, isolator, TELECOM insulator; ~ **acampanado** *m* ELEC, ING ELÉC bell-shaped insulator; ~ **bajo carpa** *m* ELEC, ING ELÉC bell-shaped insulator; ~ **del borne del acumulador** *m* ELEC, ING ELÉC post insulator; ~ **de cable** *m* ELEC, ING ELÉC cable insulation, cable isolator; ~ **para cable de retenida** *m* ELEC guy insolator, ING ELÉC guy insulator; ~ **de campana** *m* ELEC, ING ELÉC mushroom insulator; ~ **de carrete** *m* ELEC, ING ELÉC reel insulator; ~ **de caucho** *m* ELEC, ING ELÉC, P&C, REVEST rubber insulation; ~ **cerámico** *m* ELEC, ING ELÉC ceramic insulator; ~ **de clavija** *m* ELEC, ING ELÉC pin insulator; ~ **comunicado** *m* ELEC, ING ELÉC insulation breakdown; ~ **contra calor y frío** *m* SEG insulation against heat and cold; ~ **de crisol** *m* ELEC, ING ELÉC pot insulator; ~ **doble** *m* ELEC, ING ELÉC double insulator; ~ **de fijación mural** *m* ELEC, ING ELÉC wall-entrance insulator; ~ **de guía de ondas** *m* ELEC, ING ELÉC waveguide isolator; ~ **de guíaondas** *m* ELEC, ING ELÉC waveguide isolator; ~ **de línea** *m* ELEC, ING ELÉC line insulator; ~ **en mantas flexible** *m* REFRIG blanket-type insulant; ~ **por núcleo sólido** *m* ELEC, ING ELÉC solid-core-type insulator; ~ **en paneles** *m* REFRIG block-type insulant, board-type insulant; ~ **en planchas** *m* REFRIG block-type insulant, board-type insulant; ~ **de poca pérdida** *m* ELEC, ING ELÉC low-loss insulator; ~ **en polvo** *m* REFRIG powdered insulant; ~ **de porcelana** *m* C&V, ELEC, ING ELÉC china insulator, porcelain insulator; ~ **del poste** *m* ING ELÉC post insulator; ~ **reflector** *m* REFRIG, TERMO, TERMOTEC reflective insulant; ~ **rígido** *m* ELEC, ING ELÉC pin insulator; ~ **Schottky** *m* ELECTRÓN Schottky barrier; ~ **de suspensión** *m* ELEC, ING ELÉC suspension insulator; ~ **del terminal** *m* ELEC, ING ELÉC terminal insulator

aislar *vt* CONST, ELEC, FÍS, ING ELÉC insulate, isolate, MINAS insulate, isolate, seal, PROD, QUÍMICA, REFRIG, SEG insulate, isolate

AISM *abr* (*Asociación Internacional de Señalización Marítima*) TRANSP MAR IALA (*International Association of Lighthouses Authorities*)

ajedrezados *m pl* C&V checkers (*AmE*), chequers (*BrE*)

ajeno *m* ELECTRÓN aliasing

ajustable *adj* ELECTRÓN, ING MECÁ, MECÁ adjustable, adjusting; ~ **continuamente** *adj* ELECTRÓN continuously adjustable; ~ **a voluntad** *adj* ING MECÁ adjustable at will

ajustable[1] *adj* ING MECÁ set; ~ **de fábrica** *adj* MECÁ factory-adjusted

ajustado[2]: ~ **en caliente** *m* ING MECÁ hot shrink fit

ajustador: ~ **del cero eléctrico** *m* ELEC, INSTR electrical zero adjuster; ~ **dióptrico** *m* CINEMAT, TV dioptric adjuster; ~ **de imagen digital** *m* CINEMAT digital frame, TV digital framer; ~ **del impulsor** *m* ING MECÁ tappet adjuster; ~ **de tetilla para bicicleta** *m* TRANSP bicycle nipple adjuster

ajustadora *f* MECÁ, PROD lapping machine

ajustar[1] *vt* AUTO tune, C&V fit, CARBÓN size, CINEMAT *obturador* set, CONST adjust, bite, ELECTRÓN tune, FÍS ONDAS focus, IMPR, INFORM&PD set, ING MECÁ fit to, fit, MECÁ adjust, fit, trim, METR adjust, PROD *madera* trim, SEG adjust, TRANSP MAR *cabo* bend; ~ **a cero** *vt* QUÍMICA zero; ~ **dentro de** *vt* ING MECÁ fit into; ~ **la**

forma de *vt* IMPR lock up; **~ a resonancia** *vt* TELE-COM tune

ajustar[2]: **~ la abertura** *vi* CINEMAT, FOTO, TV follow F-stop; **~ el cepillado** *vi* CONST *carpintería* shoot

ajustarse: **~ a la forma de** *v refl* PROD conform in shape to

ajuste *m* AUTO *de ruedas delanteras* tracking, tuning, CONST *instrumentos* setting, C&V, CRISTAL, ELECTRÓN adjustment, IMPR adjustment, make-up, matching, ING MECÁ, MECÁ fitting, regulating, adjustment, METR adjustment, PROD *tipografía* setup, *moldes* setting up, TEC PETR *perforación* making-up, TRANSP AÉR setting, TRANSP MAR *cabo* splice; **~ de abertura** *m* CINEMAT, FÍS, FOTO, TV F-stop; **~ accionado con el dedo pulgar** *m* PROD thumb wheel setting; **~ acoplado de abertura y velocidad del diafragma** *m* CINEMAT, FOTO, TV coupled speed and F-stop setting; **~ adosado añadido** *m* ING ELÉC back-to-back arrangement; **~ de alarma** *m* TELECOM alarm setting; **~ del altímetro** *m* FÍS, GEOFÍS, TEC ESP, TRANSP AÉR altimeter setting; **~ en altura** *m* INSTR height adjustment; **~ de amplitud** *m* ELECTRÓN amplitude adjustment; **~ del ángulo del ala** *m* TRANSP AÉR angle of wing setting; **~ de anillo** *m* ING MECÁ collar joint; **~ con apriete** *m* ING MECÁ interference fit; **~ aproximado** *m* INSTR coarse adjustment; **~ aproximativo** *m* ELEC coarse adjustment; **~ de aro** *m* ING MECÁ collar joint; **~ automático a la intensidad de la luz** *m* SEG automatic adjustment to light intensity; **~ de azimut** *m* TV azimuth adjustment; **~ azimutal** *m* TV azimuth adjustment; **~ de la cabeza** *m* TV head adjustment; **~ de cable** *m* ELEC, ING ELÉC cable fitting, cable splicing; **~ en caliente** *m* MECÁ shrink fit; **~ de la carrera sin parar la máquina** *m* ING MECÁ adjustment of the stroke without stopping the machine; **~ cero** *m* NUCL zero setting; **~ de cerradura** *m* CONST lock fitting; **~ de collar** *m* ING MECÁ collar joint; **~ de color** *m* COLOR, IMPR color matching (*AmE*), colour matching (*BrE*); **~ del color de la imagen** *m* TV color framing (*AmE*), colour framing (*BrE*); **~ combinatorio** *m* INFORM&PD combinational setting; **~ continuo** *m* ELECTRÓN continuous adjustment; **~ por contracción** *m* MECÁ shrink fit; **~ corredizo** *m* ING MECÁ, MECÁ running fit, sliding fit; **~ corredizo cerrado** *m* ING MECÁ, MECÁ close-sliding fit; **~ cromático** *m* COLOR color match (*AmE*), color matching (*AmE*), colour match (*BrE*), colour matching (*BrE*), IMPR color print (*AmE*), colour print (*BrE*); **~ en delta** *m* ING MECÁ delta fitting; **~ de deslizamiento** *m* ING MECÁ running fit; **~ deslizante** *m* ING MECÁ running fit; **~ deslizante cerrado** *m* ING MECÁ, MECÁ close-sliding fit; **~ de dimensión** *m* INFORM&PD dimensioning, sizing; **~ de dos máquinas eléctricas acopladas** *m* ING ELÉC back-to-back arrangement; **~ de eje** *m* ING MECÁ axle fit; **~ de encendido** *m* AUTO ignition setting, ignition timing; **~ del enfoque** *m* CINEMAT, FOTO, TV focus setting; **~ de espacio a espacio** *m* IMPR gap-to-gap adjustment; **~ estándar frontal** *m* FOTO front standard adjustment; **~ de fase** *m* ELEC *corriente alterna* phasing, ELECTRÓN phase adjustment, TRANSP phasing, TV phase adjustment, phasing; **~ final de posición** *m* NUCL final position setting; **~ fino** *m* ELECTRÓN, INSTR, NUCL, PROD, TV fine adjustment; **~ fino de sintonización** *m* ELEC-

TRÓN fine tuning; **~ flojo** *m* ING MECÁ loose fit; **~ del foco** *m* CINEMAT, FOTO, TV focus pulling, focus setting; **~ forzado** *m* ING MECÁ, MECÁ force fit, PROD press fit; **~ fraccionado del ciclo** *m* TRANSP *control de tráfico* cycle split adjustment; **~ de frecuencia** *m* ELECTRÓN frequency adjustment; **~ al gálibo** *m* FERRO tight-to-gage (*AmE*), tight-to-gauge (*BrE*); **~ de ganancia** *m* ELECTRÓN gain adjustment; **~ de gaza** *m* TRANSP MAR eyesplice; **~ GFI** *m* ELECTRÓN YIG tuning; **~ grueso** *m* ELEC, INSTR coarse adjustment; **~ de guiñadas** *m* ENERG RENOV yaw adjustment; **~ holgado** *m* ING MECÁ loose fit; **~ con holgura** *m* ING MECÁ loose fit; **~ sin holgura** *m* ING MECÁ push fit; **~ con huelgo** *m* ING MECÁ loose fit; **~ sin huelgo** *m* ING MECÁ push fit; **~ de imagen** *m* CINEMAT, TV framing; **~ isostático** *m* GEOL isostatic adjustment; **~ de lastre** *m* OCEAN buoyancy adjustment; **~ lateral** *m* ING MECÁ side fit; **~ mecánico de cero** *m* ELEC *de instrumento*, INSTR mechanical zero adjustment; **~ de nivel** *m* ELECTRÓN level adjustment; **~ oscilante** *m* INSTAL HIDRÁUL *de válvula* dancing seat; **~ del paso** *m* TRANSP AÉR pitch setting; **~ entre piezas** *m* ING MECÁ interference fit; **~ de pista** *m* TV track adjustment; **~ posterior** *m* FOTO back adjustment; **~ de precisión** *m* ING MECÁ precision setting; **~ preciso** *m* ELECTRÓN, INSTR, NUCL, PROD, TV fine adjustment; **~ a presión** *m* ING MECÁ compressed fitting, driving fit; **~ de la presión** *m* PROD pressure setting; **~ previo** *m* ELECTRÓN preset; **~ de resistencia** *m* ING ELÉC resistor trimming; **~ roscado** *m* ING MECÁ screwed fitting; **~ simple** *m* ING MECÁ plain fitting; **~ de sincronización** *m* AUTO timing adjustment; **~ suave** *m* ING MECÁ push fit, MECÁ sliding fit; **~ suelto** *m* ING MECÁ running fit; **~ para techo** *m* ELEC *alumbrado* ceiling fitting; **~ de tono** *m* TRANSP AÉR beeper trim; **~ de válvula** *m* AUTO valve setting; **~ del varactor** *m* ELECTRÓN varactor tuning; **~ de velocidad del obturador** *m* CINEMAT, FOTO, TV shutter speed setting; **~ de visualización** *m* INFORM&PD display setting

ajustes: **~ sucesivos** *m pl* INFORM&PD stepwise refinement; **~ y tolerancias** *m pl* ING MECÁ fits and clearances

ancla[2]: **estar al ~** *fra* TRANSP MAR lie at anchor; **estar al ~ en la rada** *fra* TRANSP MAR anchor in the roads

Al *abr* (*aluminio*) METAL, QUÍMICA Al (*aluminium BrE, aluminum AmE*)

ala *f* CONST flange, wing, FÍS, ING MECÁ, OCEAN, TRANSP, TRANSP AÉR wing, TRANSP MAR faceplate, wing; **~ de cantilever** *f* TRANSP, TRANSP AÉR cantilever wing; **~ crítica** *f* TRANSP, TRANSP AÉR critical wing; **~ decreciente** *f* TRANSP, TRANSP AÉR tapered wing; **~ en delta** *f* TRANSP, TRANSP AÉR delta wing; **~ de doble delta** *f* TRANSP, TRANSP AÉR double delta wing; **~ fija** *f* TRANSP, TRANSP AÉR fixed wing; **~ en flecha** *f* TRANSP, TRANSP AÉR swept wing; **~ de flecha negativa** *f* TRANSP, TRANSP AÉR forward-swept wing; **~ de flecha positiva** *f* TRANSP, TRANSP AÉR swept-back wing; **~ media** *f* TRANSP, TRANSP AÉR mid wing; **~ de punta recortada** *f* TRANSP, TRANSP AÉR clipped wing; **~ punta-flecha** *f* TRANSP, TRANSP AÉR arrowhead wing; **~ con ranura** *f* TRANSP, TRANSP AÉR slotted wing; **~ ranurada** *f* TRANSP, TRANSP AÉR slotted wing; **~ de la red de arrastre** *f* CONTAM MAR, OCEAN, TRANSP MAR trawl wing; **~ de red barredera** *f* CONTAM MAR, OCEAN, TRANSP MAR trawl wing; ~ **de rotor de ciclo caliente** *f* TRANSP, TRANSP AÉR hot-cycle rotor wing; ~ **semialta** *f* TRANSP, TRANSP AÉR shoulder wing; ~ **superior** *f* CONST top flange; ~ **trapecial** *f* TRANSP, TRANSP AÉR tapered wing

alabante: ~ **de rodillos** *m* TRANSP MAR *instalación de cubierta* roller fairlead

alabastro *m* MINERAL alabaster; ~ **oriental** *m* MINERAL oriental alabaster

álabe *m* CONST bucket, vane, ELEC, ING MECÁ, INSTAL HIDRÁUL, MECÁ, NUCL, PROD, REFRIG, TEC ESP, TRANSP AÉR blade, vane; ~ **de admisión de paso variable** *m* NUCL variable-pitch inlet vane; ~ **de cascada** *m* TRANSP AÉR cascade vane; ~ **de contacto** *m* ELEC *relé* contact blade; ~ **director** *m* INSTAL HIDRÁUL, MECÁ, REFRIG guide vane; ~ **del distribuidor** *m* INSTAL HIDRÁUL, MECÁ, REFRIG guide vane; ~ **de entrada múltiple** *m* TRANSP AÉR multi-throat vane; ~ **fijo** *m* INSTAL HIDRÁUL stationary vane; ~ **de garganta múltiple** *m* TRANSP AÉR multi-throat vane; ~ **giratorio** *m* INSTAL HIDRÁUL, MECÁ *turbina* gate; ~ **guía** *m* INSTAL HIDRÁUL, MECÁ, REFRIG *del ventilador* guide vane; ~ **del rodete** *m* INSTAL HIDRÁUL *de turbina hidráulica* runner vane; ~ **de turbina** *m* TRANSP MAR turbine blade

alabeado *adj* ING MECÁ, MECÁ warped

alabeamiento: ~ **por dislocación** *m* METAL dislocation kink

alabear *vt* CONST warp, ING MECÁ, MECÁ *distorsión* buckle

alabearse *v refl* ING MECÁ, MECÁ warp, PROD *madera* kink out of line, TRANSP MAR warp

alabeo *m* *AmL* (*cf combado Esp*) ACÚST warp, CONST *deformaciones* winding, *madera* warping, FOTO buckling, IMPR *encuadernación* warping, ING MECÁ *distorsión* buckling, PAPEL twist, warp, TEXTIL warp; ~ **en tres puntos** *m* METAL three-point bending

alacrán *m* ING MECÁ swivel hook

alactita *f* MINERAL allactite

alagita *f* MINERAL allagite

alalita *f* MINERAL alalite

alambique *m* ALIMENT, INSTAL TERM, LAB, PROC QUÍ, QUÍMICA, TEC PETR retort, still; ~ **pirolizador** *m* CARBÓN cracker; ~ **secundario** *m* ALIMENT secondary still

alambrada *f* AGRIC, CONST wire fence, D&A barbed-wire entanglement

alambrado *m* ELEC, TEC ESP wiring; ~ **de punto a punto** *m* ELEC *fuente de alimentación* point-to-point wiring

alambrar *vt* AGRIC, CONST fence in

alambre *m* GEN wire; ~ **acanalado** *m* ÓPT, TELECOM grooved cable; ~ **de acero para guardines** *m* TRANSP MAR *construcción naval* steering wire; ~ **aislado** *m* ELEC, ING ELÉC insulated wire; ~ **bimetálico** *m* ELEC, ING ELÉC bimetallic wire; ~ **de blindaje** *m* PROD shield drain wire; ~ **brillante estirado en frío** *m* ING MECÁ bright hard-drawn wire; ~ **de cobre** *m* CONST, ELEC, ING ELÉC, REVEST *conductor* copper wire; ~ **de cobre esmaltado** *m* CONST, ELEC, ING ELÉC, REVEST enameled copper wire (*AmE*), enamelled copper wire (*BrE*); ~ **común** *m* PROD common wire; ~ **descubierto** *m* ING ELÉC, PROD open wire; ~ **deslizante** *m* ELEC skid wire; ~ **desnudo** *m* ING ELÉC, PROD bare wire, open wire; ~ **desnudo de drenaje** *m* PROD bare drain-wire; ~ **del detonador** *m*

MINAS leg wire; ~ **de drenaje** *m* PROD drain wire; ~ **de envoltura** *m* ING MECÁ wire wrap; ~ **esmaltado** *m* ELEC, ING ELÉC, REVEST *conductor* enameled wire (*AmE*), enamelled wire (*BrE*); ~ **de espino** *m* AGRIC, CONST, D&A barbed wire; ~ **estañado** *m* ING ELÉC tinned wire; ~ **fijador** *m* ING MECÁ lock wire; ~ **forrado** *m* REVEST coated wire; ~ **freno** *m* ING MECÁ lock wire; ~ **fusible** *m* ELEC, ING ELÉC fuse wire; ~ **para grapar** *m* EMB stapling wire; ~ **para hacer tornillos** *m* MECÁ blank; ~ **de hierro** *m* PROD iron wire; ~ **macizo** *m* ELEC, ING ELÉC *conductor* solid wire; ~ **de malla** *m* TEXTIL heald wire; ~ **para máquina de coser con grapas** *m* EMB, TEXTIL stitching wire; ~ **neutro** *m* ELEC, ING ELÉC, PROD *circuito* neutral wire; ~ **plano** *m* ELEC, ING ELÉC flat wire; ~ **de platino** *m* LAB platinum wire; ~ **para precinto de plomo** *m* PROD lead-seal wire; ~ **de protección** *m* ELEC *línea aérea* guard wire; ~ **de púas** *m* AGRIC, CONST, D&A barbed wire; ~ **recocido brillante** *m* ING MECÁ bright-annealed wire; ~ **de relleno** *m* MECÁ filler wire; ~ **de resistencia** *m* ELEC, ING ELÉC, METAL resistance wire; ~ **de seguridad** *m* ING MECÁ, SEG lock wire; ~ **para soldadura** *m* CONST welding wire; ~ **de sujeción del cable** *m* ELEC, ING ELÉC cable suspension wire; ~ **de suspensión del cable** *m* ELEC, ING ELÉC cable suspension wire; ~ **de sustentación del cable** *m* ELEC, ING ELÉC cable suspension wire; ~ **de talón** *m* AUTO bead core; ~ **trenzado** *m* ING ELÉC braided wire, PROD stranded wire

alámbrico[1] *adj* GEN wire

alámbrico[2] *m* ING ELÉC wire

álamo *m* PAPEL poplar

alanina *f* QUÍMICA alanine

alanita *f* MINERAL allanite

alantoína *f* QUÍMICA *oxidación de urea* allantoin

alargabilidad *f* INFORM&PD, ING MECÁ, P&C extensibility

alargadera *f* ELEC *conexiones* adaptor, ING MECÁ extension piece, lengthening bar, lengthening tube; ~ **de sonda** *f* ING MECÁ lengthening rod

alargado *adj* GEOM oblong, MECÁ elongated; ~ **hacia los polos** *adj* GEOM prolate

alargamiento *m* C&V elongation, CONST stretching, FÍS elongation, extension, ING MECÁ elongation, fineness ratio, stretch, MECÁ strain, METAL, P&C elongation, PAPEL stretch, QUÍMICA *enlace* stretching, TRANSP AÉR aspect ratio; ~ **debido a la humedad** *m* PAPEL damping stretch; ~ **elástico** *m* EMB elastic elongation, elastic stretch; ~ **al fallar** *m* P&C *propiedad física* ultimate elongation; ~ **geométrico** *m* FÍS *de un perfil aerodinámico* aspect ratio; ~ **del hilo en el encolado** *m* TEXTIL *en la operación de apresto* stretch of yarn in sizing; ~ **de la imagen** *m* TV pulling; ~ **en límite elástico** *m* P&C yield strength; ~ **en el punto de rotura** *m* PAPEL stretch at breaking point; ~ **al romperse** *m* P&C *propiedad física* elongation at break; ~ **de rotura** *m* P&C *propiedad física* ultimate elongation; ~ **hasta la rotura** *m* PAPEL stretch at break; ~ **de la rotura por fluencia** *m* METAL creep rupture elongation; ~ **de ruptura** *m* FÍS elongation at rupture

alargarse *v refl* TRANSP MAR veer aft

alarma *f* GEN alarm, warning; ~ **antirrobo** *f* SEG burglar alarm; ~ **audible** *f* SEG, TELECOM audible alarm; ~ **automática contraincendios** *f* CONST, SEG,

TERMO automatic fire alarm; ~ **por baja presión** *f* PROD low-pressure alarm; ~ **contra humos** *f* SEG smoke alarm; ~ **contra ladrones** *f* SEG burglar alarm; ~ **contraincendios** *f* ELEC, SEG, TERMO, TRANSP AÉR, TRANSP MAR fire alarm; ~ **de cortocircuito eléctrico** *f* ELEC, SEG electric wire-break alarm; ~ **de escape de agua** *f* MINAS water-leakage alarm; ~ **de fuego** *f* ELEC, SEG fire alarm; ~ **de fugas de agua** *f* MINAS water-leakage alarm; ~ **importante** *f* TELECOM major alarm; ~ **de incendios** *f* ELEC, SEG, TERMO, TRANSP MAR fire alarm; ~ **de incendios automática** *f* CONST, SEG automatic fire alarm; ~ **infrarroja detectora de movimiento** *f* SEG infrared motion alarm, infrared movement-sensing alarm; ~ **de interrupción de cable eléctrico** *f* ELEC, SEG electric wire-break alarm; ~ **de intervalos de tiempo** *f* LAB sub timer alarm; ~ **luminosa intermitente** *f* SEG alarm flashing light; ~ **menor** *f* TELECOM minor alarm; ~ **pasiva** *f* TELECOM passive alerting; ~ **por presión** *f* PROD pressure alarm; ~ **por temperatura** *f* PROD temperature alarm; ~ **visual** *f* TELECOM visual alarm

alascaíta *f* MINERAL, PETROL alaskaite

alavante *m* TRANSP MAR *equipamiento de cubierta* chock, fairlead

albahaca: ~ **silvestre** *f* AGRIC quickweed

albán *m* QUÍMICA *resina* alban

albana *f* QUÍMICA *resina* alban

albañal *m* AGUA open drain, CONST, CONTAM open drain, sewer, RECICL open drain

albañil *m* C&V bricklayer, CONST bricklayer, mason

albañilería *f* C&V, CONST brickwork, masonry

albarán *m* PROD delivery ticket; ~ **de entrega** *m* PROD internal delivery slip; ~ **de expedición de materiales** *m* PROD material issue note

albardilla *f* CONST coping, square-to roof

albedo *m* GEOFÍS, METEO, TEC ESP albedo; ~ **geométrico** *m* GEOM, TEC ESP geometric albedo; ~ **neutrónico** *m* TEC ESP albedo

alberca *f* CARBÓN, CRISTAL, PROC QUÍ pond; ~ **de cristalización** *f* CARBÓN, CRISTAL, PROC QUÍ crystallizing pond

albertita *f* MINERAL albertite

albiense *m* GEOL albian

albita *f* C&V, MINERAL albite

albufera *f* GEOL, OCEAN lagoon

albúmina *f* ALIMENT, QUÍMICA albumen, albumin; ~ **sanguínea** *f* ALIMENT blood albumen, blood albumin

albuminato *m* ALIMENT, QUÍMICA albuminate

albuminoide[1] *adj* QUÍMICA albuminoid

albuminoide[2] *m* QUÍMICA albuminoid

albumosa *f* ALIMENT, QUÍMICA albumose

ALC *abr* (*árbol de levas en culata*) AUTO, MECÁ, VEH OHC (*overhead camshaft*)

alcachofa *f* AGUA rose; ~ **de aspiración** *f* AGUA strainer, tailpiece, wind bore, PROD *bombas* strainer; ~ **roscada** *f* ING ELÉC screw cap; ~ **de toma** *f* ING ELÉC rose

alcadieno *m* QUÍMICA *compuesto* alkadiene

álcali *m* GEN alkali; ~ **cáustico** *m* DETERG, QUÍMICA caustic alkali; ~ **-celulosa** *m* PAPEL alkali cellulose

alcalimetría *f* ALIMENT, DETERG, QUÍMICA alkalimetry

alcalímetro *m* ALIMENT, DETERG, QUÍMICA alkalimeter

alcalinidad *f* GEN alkalinity; ~ **carbonatada** *f* HIDROL carbonate alkalinity; ~ **preacidificación** *f* CONTAM *tratamiento químico* pre-acidification alkalinity; ~ **previa acidificación** *f* CONTAM pre-acidification alkalinity

alcalinizar *vt* HIDROL, QUÍMICA alkalize

alcalino *adj* GEN alkalic, alkaline

alcalinocalcífera *adj* GEOL *rocas ígneas o similares* calc-alkaline

alcalirresistente *adj* DETERG, QUÍMICA alkali-fast, alkali-proof

alcalisoluble *adj* DETERG, QUÍMICA alkali-soluble

alcalización *f* HIDROL, QUÍMICA alkalization

alcalizar *vt* HIDROL, QUÍMICA alkalize

alcaloide *m* QUÍMICA alkaloid; ~ **de lobelia** *m* QUÍMICA lobelia alkaloid; ~ **de pimienta** *m* QUÍMICA pepper alkaloid

alcaloides: ~ **del ergot** *m pl* QUÍMICA ergot alkaloids

alcance *m* D&A *balística*, ELECTRÓN range, reach, scope, GEOM range, INFORM&PD, ING MECÁ, PROD TEC ESP, TEXTIL TRANSP AÉR, TRANSP MAR reach, reach, scope; ~ **de alimentación** *m* ING MECÁ feed range; ~ **del balanceo hacia la derecha** *m* ING MECÁ swivel range clockwise; ~ **del balanceo hacia la izquierda** *m* ING MECÁ swivel range counterclockwise; ~ **de crucero** *m* TEC ESP cruising range; ~ **efectivo** *m* D&A effective range, TEC ESP *altavoz* beam; ~ **eficaz** *m* D&A operating range, PROD operational range; ~ **del fusil** *m* D&A rifle range; ~ **geográfico** *m* TRANSP MAR geographical range; ~ **inclinado** *m* D&A slant angle; ~ **inferior al normal** *m* PROD underrange; ~ **de marea** *m* ENERG RENOV, GEOFÍS, HIDROL, TRANSP MAR tidal range; ~ **máximo eficaz** *m* D&A maximum effective range; ~ **de medida** *m* PROD metering land; ~ **nocturno** *m* TRANSP AÉR night range; ~ **orbital** *m* TEC ESP orbital catch-up; ~ **próximo** *m* TEC ESP near range; ~ **del radar** *m* D&A, TRANSP, TRANSP AÉR, TRANSP MAR radar range; ~ **de radio** *m* TELECOM, TRANSP AÉR, TRANSP MAR radio range; ~ **de recepción** *m* FÍS RAD, TELECOM, TV receiving range; ~ **útil de funcionamiento** *m* TRANSP useful working range; ~ **visual** *m* D&A range of vision; ~ **visual de pista** *m* TRANSP AÉR runway visual range; ~ **de la voz** *m* TRANSP MAR hailing distance

alcancía *f* AmL (*cf tolua Esp*) CARBÓN, MINAS chute, ore pass; ~ **principal** *f* AmL (*cf puerta principal Esp*) CARBÓN, MINAS, TERMO main chute,

alcanfor *m* QUÍMICA camphor

alcanforado *adj* QUÍMICA camphorated

alcano *m* DETERG, PETROL, QUÍMICA, TEC PETR alkane

alcantarilla *f* AGUA, CARBÓN, CONST, HIDROL, ING MECÁ, RECICL culvert, drain, sewer; ~ **para aguas negras y de lluvia** *f* AGUA, HIDROL combined sewer, combined sewer system; ~ **aplanada** *f* CONST flat-top culvert; ~ **de fondo** *f* AGUA, HIDROL bottom culvert; ~ **principal** *f* AGUA, HIDROL main sewer; ~ **rectangular** *f* AGUA, CONST, HIDROL box culvert

alcantarillado *m* AGUA, CARBÓN, CONTAM, HIDROL, RECICL domestic sewage system, drainage system, grid

alcantarillaje *m* AGUA, CONTAM, HIDROL, RECICL sewerage

alcanzar *vt* TRANSP MAR fetch, *navegación* overhaul

alcaptona *f* QUÍMICA alcaptone

alcohol *m* GEN alcohol, spirit; ~ **alílico** *m* QUÍMICA allyl alcohol; ~ **de azufre** *m* ALIMENT, QUÍMICA thiol; ~ **bencílico** *m* QUÍMICA benzyl alcohol; ~ **bornílico**

m QUÍMICA bornyl alcohol; ~ **butílico** *m* DETERG, QUÍMICA butanol, butyl alcohol; ~ **de cereales** *m* ALIMENT grain alcohol; ~ **decílico** *m* QUÍMICA *perfumería* decyl alcohol; ~ **desnaturalizado** *m* ALIMENT, COLOR, QUÍMICA denatured alcohol, methylated spirit; ~ **etílico** *m* ALIMENT, FOTO, QUÍMICA, TEC PETR ethanol, ethyl alcohol; ~ **de grano** *m* ALIMENT grain alcohol; ~ **graso** *m* DETERG, QUÍMICA fatty alcohol; ~ **hexílico** *m* QUÍMICA hexanol, hexyl alcohol; ~ **icosílico** *m* QUÍMICA eicosyl alcohol; ~ **industrial** *m* ALIMENT industrial alcohol; ~ **isobutílico** *m* QUÍMICA isobutyl alcohol, isopropanol; ~ **isopropílico** *m* ALIMENT, QUÍMICA isopropanol, isopropyl alcohol; ~ **laurílico** *m* DETERG, QUÍMICA lauryl alcohol; ~ **de madera** *m* TERMO wood alcohol; ~ **metilado** *m* ALIMENT, COLOR, QUÍMICA methylated spirit; ~ **metílico** *m* GAS, P&C, QUÍMICA, TERMO methanol, methyl alcohol; ~ **mirístico** *m* DETERG, QUÍMICA myristic alcohol; ~ **nonílico** *m* DETERG, QUÍMICA nonyl alcohol; ~ **polivinílico** *m* P&C, QUÍMICA *polímero* polyvinyl alcohol; ~ **propenílico** *m* QUÍMICA allyl alcohol; ~ **tuyílico** *m* QUÍMICA thujyl alcohol; ~ **de Ziegler** *m* DETERG Ziegler alcohol
alcoholato *m* QUÍMICA *farmacéutico* alcoholate
alcohólico *adj* ALIMENT alcoholic
alcohólisis *f* QUÍMICA alcoholysis
alcoxi- *pref* QUÍMICA alkoxy-
alcóxido *m* QUÍMICA alkoxide
aldaba *f* CONST, ING MECÁ, MECÁ latch
aldehídico *adj* ALIMENT, P&C, QUÍMICA aldehydic
aldehído *m* ALIMENT, P&C, QUÍMICA aldehyde; ~ **ácido** *m* QUÍMICA aldehyde acid; ~ **crotónico** *m* QUÍMICA crotonaldehyde; ~ **cumínico** *m* QUÍMICA cumaldehyde; ~ **fórmico** *m* ALIMENT, P&C, QUÍMICA, TEXTIL formaldehyde
aldohexosa *f* ALIMENT, QUÍMICA aldohexose
aldol *m* QUÍMICA aldol
aldosa *f* ALIMENT, QUÍMICA aldose
aldosterona *f* ALIMENT, QUÍMICA aldosterone
aleación *f* GEN *metales* alloy; ~ **de acero** *f* CARBÓN, METAL steel alloy; ~ **de alta temperatura** *f* MECÁ, METAL high-temperature alloy; ~ **de aluminio** *f* ING MECÁ, METAL aluminium alloy (*BrE*), aluminum alloy (*AmE*); ~ **de banda lateral** *f* METAL side-band alloy; ~ **de circonio** *f* METAL, NUCL zirconium base alloy; ~ **de cobre para cobresoldar** *f* PROD brazing metal; ~ **de cobre y estaño** *f* METAL, QUÍMICA bronze; ~ **de cobre forjado para cojinetes** *f* ING MECÁ wrought-copper alloy for plain bearings; ~ **liviana** *f* MECÁ, METAL light alloy; ~ **ordenada** *f* METAL ordered alloy; ~ **de plomo y estaño** *f* ING MECÁ, METAL lead-tin alloy; ~ **de relleno** *f* CONST, METAL filler alloy; ~ **para soldar** *f* PROD brazing solder; ~ **ternaria** *f* METAL ternary alloy, three-component alloy; ~ **de titanio** *f* METAL, TEC ESP titanium alloy, titanium forging; ~ **de tres componentes** *f* METAL ternary alloy, three-component alloy; ~ **ultraligera** *f* TEC ESP ultralight alloy
alear *vt* GEN alloy
aleatorio *adj* ACÚST, ELEC, ELECTRÓN, INFORM&PD, ING ELÉC random, MATEMÁT random, stochastic, QUÍMICA, TEC ESP random
alejado *adj* TELECOM spaced-out
alejamiento *m* INFORM&PD zoom-out
alejarse *v refl* CINEMAT, TV *cámara* pull back

alelótropo *adj* QUÍMICA allelotropic
alemontita *f* MINERAL allemontite
alero *m* CONST eave, *tejado* side
alerón *m* TRANSP *vehículos de carretera* spoiler, TRANSP AÉR aileron, VEH *carrocería* overhang; ~ **equilibrado** *m* TRANSP AÉR balanced aileron; ~ **multi-velocidad** *m* TRANSP all-speed aileron; ~ **trasero** *m* VEH *carrocería* spoiler; ~ **de varias velocidades** *m* TRANSP all-speed aileron
alerta *f* TELECOM alerting; ~ **barco-tierra** *f* TELECOM, TRANSP MAR ship-to-shore alerting; ~ **entre barcos** *f* TELECOM, TRANSP MAR ship-to-ship alerting; ~ **de gases** *f* D&A gas alert; ~ **pasiva** *f* TELECOM passive alerting; ~ **de una situación de peligro** *f* TRANSP MAR distress alert; ~ **tierra-barco** *f* TELECOM, TRANSP MAR shore-to-ship alerting
aleta *f* AUTO fender (*AmE*), mudguard (*BrE*), C&V fin, CONST, EMB flap, ING MECÁ blade, flange, flap, fly, wing, MECÁ blade, fin, flap, NUCL fin, PROD *de tubo* flange, REFRIG fin, TRANSP mudguard (*BrE*), fender (*AmE*), TRANSP AÉR flap, TRANSP MAR *construcción naval* fin, quarter, VEH *carrocería* fender (*AmE*), mudguard (*BrE*), wing (*BrE*); ~ **adelantada** *f* PROD flying lead; ~ **compensadora** *f* TRANSP AÉR balance tab, trim tab; ~ **delantera** *f* AUTO front fender (*AmE*), front wing (*BrE*); ~ **disipadora de calor** *f* NUCL cooling fin; ~ **dorsal del fuselaje** *f* TRANSP AÉR fuselage dorsal fin; ~ **de enfriamiento** *f* CINEMAT cooling rib, VEH *refrigeración* cowling flaps; ~ **estabilizadora** *f* TRANSP MAR stabilizing fin; ~ **de estátor** *f* TRANSP AÉR stator vane; ~ **de estátor fija** *f* TRANSP AÉR fixed-stator vane; ~ **fondo** *f* EMB bottom flap; ~ **guardabarros** *f* VEH *accesorio* mudflap; ~ **de guía de entrada** *f* TRANSP AÉR intake-guide vane; ~ **guiadora** *f* MECÁ guide vane; ~ **hipersustentadora de aterrizaje** *f* TRANSP landing flap; ~ **hipersustentadora del borde de ataque** *f* TRANSP leading-edge flap; ~ **hipersustentadora de inclinación** *f* TRANSP droop flap; ~ **de montaje** *f* TEC ESP flange mounting; ~ **de popa** *f* TRANSP MAR *yates* skeg; ~ **del radiador** *f* AUTO, VEH *sistema de refrigeración* radiator fin; ~ **de ranura** *f* TRANSP AÉR spoiler; ~ **de ranura del borde de ataque** *f* TRANSP AÉR leading-edge slat; ~ **recubierta de vinilo** *f* REFRIG vinyl-coated fin; ~ **de refrigeración** *f* ING MECÁ, REFRIG, TERMO, TRANSP AÉR cooling fin; ~ **refrigeradora** *f* ING MECÁ, REFRIG, TERMO, TRANSP AÉR cooling fin; ~ **reguladora** *f* TRANSP AÉR trimming tab; ~ **para remeter** *f* EMB tuck-in flap; ~ **de sustentación** *f* TEC ESP flange mounting; ~ **trasera** *f* AUTO, VEH rear fender (*AmE*), rear wing (*BrE*); ~ **al tresbolillo** *f* REFRIG staggered fin; ~ **de turbina** *f* TRANSP MAR turbine blade
aletas: con ~ *adj* GEN flanged
aleteo *m* TRANSP AÉR flapping
alexandrita *f* MINERAL alexandrite
alfa: ~ **celulosa** *f* PAPEL alpha cellulose
alfabetismo: ~ **en computadoras** *m* AmL (*cf alfabetismo en ordenadores Esp*) INFORM&PD computer literacy; ~ **en ordenadores** *m* Esp (*cf alfabetismo en computadoras AmL*) INFORM&PD computer literacy
alfabeto: ~ **telegráfico internacional** *m* TELECOM, TRANSP AÉR, TRANSP MAR international telegraph alphabet
alfalfa *f* AGRIC alfalfa (*AmE*), lucerne (*BrE*)
alfamosaico *adj* INFORM&PD alphamosaic

alfanumérico *adj* INFORM&PD, TELECOM alphameric, alphanumeric

alfaque *m* OCEAN, TRANSP MAR bar, shoal

alfarda *f* CONST angle rafter

alfarería *f* C&V, PROD crockery ware, earthenware, pottery; ~ **artesanal** *f* C&V, PROD craft pottery; ~ **artística** *f* C&V, PROD artistic pottery; ~ **cocida** *f* C&V, PROD fired earthenware; ~ **vidriada** *f* C&V, PROD earthenware glazing

alfarero *m* C&V clay worker, crockery maker, pottery maker

alfiler *m* ING MECÁ, PROD pin; ~ **de corte** *m* IMPR shearing pin; ~ **de contacto** *m* *conectores* pin

alfombra: ~ **aisladora** *f* ELEC insulating mat; ~ **de cuerda** *f* TEXTIL cord carpet; ~ **de cuerda de pelo** *f* TEXTIL haircord carpet; ~ **empenachada** *f* TEXTIL tufted carpet; ~ **de felpa de bucle** *f* TEXTIL loop-pile carpet; ~ **móvil** *f* TRANSP moving carpet; ~ **en rollo** *f* TEXTIL broadloom carpet; ~ **tejida** *f* TEXTIL woven carpet; ~ **de telar de agujas** *f* TEXTIL needloom carpet; ~ **texturada** *f* TEXTIL textured carpet

alforfón *m* AGRIC buckwheat

alga *f* CONTAM MAR, HIDROL, PETROL alga, seaweed; ~ **calcárea** *f* GEOL calcareous alga; ~ **marina** *f* ALIMENT kelp

algarroba *m* ALIMENT carob

algazos *m pl* OCEAN algal mat

álgebra *f* INFORM&PD, MATEMÁT algebra; ~ **de Boole** *f* INFORM&PD, MATEMÁT Boolean algebra; ~ **booleana** *f* INFORM&PD, MATEMÁT Boolean algebra; ~ **lineal** *f* INFORM&PD, MATEMÁT linear algebra; ~ **de proposiciones lógicas** *f* ELECTRÓN logic algebra

algebraico *adj* INFORM&PD, MATEMÁT algebraic

algibe *m* PROD tank

alginato *m* P&C *aditivo* alginate; ~ **sódico** *m* ALIMENT sodium alginate

algocultura *f* OCEAN algoculture

algodón *m* AGRIC, TEXTIL cotton; ~ **pólvora** *m* MINAS guncotton, nitrocotton explosive; ~ **en rama** *m* AGRIC, TEXTIL raw cotton

algodonal *m* AGRIC, TEXTIL cotton field

algodonita *f* MINERAL algodonite

algología *f* OCEAN algology

algoritmia *f* INFORM&PD, MATEMÁT, TEC ESP algorithmics

algorítmico *adj* INFORM&PD, MATEMÁT, TEC ESP algorithmic

algoritmo *m* INFORM&PD, MATEMÁT, TEC ESP algorithm; ~ **paralelo** *m* INFORM&PD, MATEMÁT, TEC ESP parallel algorithm; ~ **de planificación** *m* INFORM&PD, MATEMÁT, TEC ESP scheduling algorithm

alguicida[1] *adj* HIDROL algicide

alguicida[2] *m* HIDROL algicide

alguicultura *f* OCEAN algoculture

alias *m* INFORM&PD alias

alicatador *m* CONST *persona* tiler

alicates *m pl* CONST *herramienta* pincers, pliers, ELEC *herramienta* pliers, ING MECÁ, MECÁ, PROD, VEH *herramienta* pincers, pliers; ~ **ajustables** *m pl* ING MECÁ combination pliers; ~ **anulares** *m pl* ING MECÁ ring pliers; ~ **de boca plana** *m pl* ING MECÁ, MECÁ, PROD duckbill pliers, flat-nose pliers, flat-nosed pliers; ~ **de boca redonda** *m pl* ING MECÁ, MECÁ, PROD round-nosed pliers; ~ **para colocar precintos de plomo** *m pl* EMB lead-sealing pliers; ~ **cónicos** *m pl* ING MECÁ cone pliers; ~ **de corte** *m pl* CONST, ING MECÁ, PROD cutting pliers, nippers; ~ **de corte diagonal** *m pl* ING MECÁ, PROD diagonal cutting nippers; ~ **de desenfundar cables** *m pl* ING MECÁ wire-stripping pliers; ~ **de engarzar** *m pl* ING MECÁ crimping pliers; ~ **estriados** *m pl* ING MECÁ multigrip pliers; ~ **de grapar** *m pl* EMB stapling pliers; ~ **de junta de labios** *m pl* ING MECÁ lip-joint pliers; ~ **de mordazas curvadas** *m pl* ING MECÁ bent-nose pliers; ~ **pelacables** *m pl* ING MECÁ wire-stripping pliers; ~ **planos** *m pl* ING MECÁ, MECÁ, PROD flat-nose pliers, flat-nosed pliers; ~ **de precintar** *m pl* ING MECÁ sealing pliers; ~ **de punta larga** *m pl* ING MECÁ, PROD long-nose pliers, long-nosed pliers; ~ **de punta redonda** *m pl* ING MECÁ, MECÁ, PROD round-nosed pliers; ~ **punzonadores revólver** *m pl* ING MECÁ revolving punch pliers; ~ **para resortes circulares** *m pl* ING MECÁ circlip pliers; ~ **de sujeción** *m pl* ING MECÁ lock-grip pliers (*BrE*), locking pliers (*AmE*), vice grips (*BrE*), vise grips (*AmE*); ~ **terminales** *m pl* ING MECÁ battery terminal pliers; ~ **torcedores para empalmes** *m pl* PROD *electricidad* joint-twisting pliers

alidada *f* CONST, INSTR *topografía* alidade, TRANSP MAR *sextante* sight bar; ~ **de eclímetro** *f* CONST, INSTR clinometer alidade; ~ **con mira de ranura** *f* CONST, INSTR open-sight alidade; ~ **de nivelación** *f* CONST, INSTR leveling alidade (*AmE*), levelling alidade (*BrE*); ~ **de pínulas** *f* CONST, INSTR sighted alidade; ~ **telescópica** *f* CONST, INSTR telescopic alidade

alifático *adj* TEC PETR *petroquímica* aliphatic

aligeramiento *m* CONTAM MAR lightening

aligerar *vt* PROD *chapa* thin down, TRANSP MAR *amarras* single up

aligerarse *v refl* PROD thin out

alijar *vt* TRANSP MAR *para desencallar* jettison

alijo *m* CONTAM MAR lightering

alilglicidiléter *m* P&C *compuesto químico*, QUÍMICA allyl glycidyl ether (*AGE*)

alilo *m* QUÍMICA allyl

alimentación *f* AGRIC feeding, AUTO feed, C&V method of feeding, ELEC *red de distribución* feed, power lead, ELECTRÓN input, *de circuitos híbridos de filamento grueso* firing, INFORM&PD, ING ELÉC feed, ING MECÁ infeed, INSTAL HIDRÁUL feed, PROD feed, infeed, feeding, TEC PETR feed, TELECOM power-feeding, power supply, power feed, TEXTIL *mecánica* feeding, TV feed; ~ **de acción automática** *f* ING MECÁ self-acting feed; ~ **de aceite** *f* AUTO, VEH *lubricación* oil feed; ~ **ácida** *f* CONTAM *proceso industrial* acid loading; ~ **aislada de la entrada** *f* ING ELÉC isolated feed through input; ~ **de alta tensión** *f* ING ELÉC high-tension power supply; ~ **automática** *f* C&V automatic feeding, ING MECÁ self-acting feed, TEC PETR *material* automatic feed; ~ **por bobina** *f* IMPR reel-feed; ~ **de CA** *f* ING ELÉC AC input; ~ **de la caldera** *f* INSTAL HIDRÁUL boiler feeding; ~ **con carbón pulverizado** *f* CARBÓN, INSTAL TERM *hornos* coal dust feeder; ~ **por choque** *f* ELEC choke feed; ~ **de correa** *f* PROC QUÍ belt feed; ~ **de corriente** *f* ELEC, ELECTRÓN, ING ELÉC *red*, TELECOM power supply; ~ **en derivación** *f* ELEC, ING ELÉC shunt feed; ~ **directa** *f* ING ELÉC feedthrough; ~ **eléctrica** *f* ELEC *red de distribución*, ING ELÉC current supply (*AmE*), mains (*BrE*), mains supply (*BrE*); ~ **de emergencia** *f* ELEC emergency power supply; ~ **de energía regulada por voltaje** *f* ELEC,

ING ELÉC voltage-regulated power supply; ~ **de energía de salida única** *f* ELEC, ING ELÉC single-output power supply; ~ **de entrada** *f* TV incoming feed; ~ **de fuerza** *f* TELECOM power supply; ~ **por gas** *f* GAS, INSTAL TERM, TERMO gas fire; ~ **hacia delante AGC** *f* TELECOM feed-forward AGC; ~ **de hojas** *f* IMPR, INFORM&PD sheet feeding; ~ **por husillo principal** *f* ING MECÁ screw feed; ~ **de inducción** *f* ING MECÁ induction pickup; ~ **inhibidora** *f* ELECTRÓN inhibiting input; ~ **interlineal** *f* INFORM&PD line feed (*LF*); ~ **inversa de cola de cuatro rodillos** *f* IMPR four-roll reverse pan feed; ~ **por inyección** *f* AUTO pressure-feed; ~ **de la línea** *f* TELECOM line feed; ~ **a mano** *f* EMB, IMPR, ING MECÁ hand feed; ~ **manual** *f* EMB, IMPR, ING MECÁ hand feed; ~ **mecánica** *f* ING MECÁ power feed; ~ **mixta** *f* TRANSP mixed power supply; ~ **por palanca de mano** *f* ING MECÁ hand-lever feed; ~ **de papel** *f* IMPR *informática*, INFORM&PD form feed (*FF*); ~ **en paralelo** *f* ELEC, ING ELÉC shunt feed; ~ **por peso** *f* PROD weight feeding; ~ **con petróleo** *f* CONST, PETROL oil firing; ~ **por ranura** *f* IMPR slot feeding; ~ **por la red** *f* ELEC current supply (*AmE*), mains (*BrE*), mains supply (*BrE*); ~ **de salida** *f* TV outgoing feed; ~ **en serie** *f* ING ELÉC series feed; ~ **simple** *f* ING ELÉC single supply; ~ **por succión** *f* C&V suction feeding; ~ **suplementaria** *f* AGRIC supplementary feeding; ~ **suplementaria de terneros** *f* AGRIC creep feeding; ~ **trifásica** *f* ELEC, ING ELÉC three-phase supply; ~ **única** *f* ING ELÉC single supply; ~ **de vela** *f* C&V gob feeding; ~ **de vela sencilla** *f* C&V single-gob feeding; ~ **en el vértice** *f* TELECOM *de antena de reflector* vertex feed

alimentado: ~ **por baterías** *adj* CINEMAT, FOTO, PROD, TV battery-powered; ~ **por correa** *adj* PROC QUÍ belt-fed; ~ **por energía solar** *adj* ELEC, ENERG RENOV, ING ELÉC solar-powered; ~ **por gas** *adj* GAS, PROD, TERMO, TERMOTEC gas-fired; ~ **en paralelo** *adj* TRANSP AÉR parallel-fed; ~ **por la red** *adj* ELEC *dispositivo*, ING ELÉC current-operated (*AmE*), mains-operated (*BrE*)

alimentador *m* C&V feeder, *pequeños candales* fore-hearth, ELEC *red de distribución*, EMB, GEOL, IMPR, ING MECÁ feeder, PROD *moldes* riser, TEC ESP *antenas*, TEXTIL *mecánica* feeder; ~ **en anillo** *m* ELEC *red de distribución* ring feeder; ~ **automático** *m* AGRIC self feeder, PROD automatic feeder; ~ **de caldera auxiliar** *m* INSTAL HIDRÁUL, TERMO, TERMOTEC auxiliary boiler feeder; ~ **de cavidad simple** *m* C&V single-gob feeding (*BrE*), single-gob process (*AmE*); ~ **de la cinta** *m* TV ribbon guide; ~ **de compensación** *m* ELEC *red de distribución* equalizing feeder; ~ **doble** *m* ELEC duplicate feeder; ~ **de doble vía** *m* ELEC two-way feed; ~ **de dos vías** *m* ELEC two-way feed; ~**-dosificador** *m* EMB dosing feeder; ~ **de esclusa neumática** *m* PAPEL airlock feeder; ~ **extremo** *m* ELEC dead end feeder, dead-ended feeder; ~ **de filtrado** *m* RECICL filter feeder; ~ **de hojas** *m* IMPR, INFORM&PD sheet feeder; ~ **independiente** *m* ELEC independent feeder; ~ **de interconexión** *m* ELEC interconnecting feeder; ~ **múltiple** *m* ELEC multiple feeder; ~ **negativo** *m* ELEC negative feeder; ~ **de la noria** *m* TRANSP scoop wheel feeder; ~ **en paralelo** *m* ELEC parallel feeder; ~ **principal** *m* ELEC trunk feeder; ~ **radial** *m* ELEC radial feeder; ~ **rotativo y mesa de recogida** *m* EMB rotary feeder and collect-

ing table; ~ **rotativo de paletas** *m* ING MECÁ rotary vane feeder; ~ **rotatorio** *m* ING MECÁ rotary feeder; ~ **en shunt** *m* ELEC parallel feeder; ~ **de tambor** *m* ING MECÁ, TEXTIL drum feeder; ~ **de tambor rotativo** *m* ING MECÁ, TEXTIL rotary-drum feeder; ~ **a través** *m* FÍS, TEC ESP feedthrough; ~ **único** *m* ELEC single feeder; ~ **vibrante** *m* PROD vibrating feeder; ~ **vibratorio** *m* EMB, MECÁ vibratory feeder

alimentadores: ~ **y transportadores vibratorios** *m pl* ING MECÁ vibrating feeders and conveyors

alimentar *vt* CINEMAT feed, ELECTRÓN input, INFORM&PD, ING MECÁ, MECÁ, PROD feed, TERMO charge, TV feed; ~ **con carbón** *vt* CARBÓN, MINAS, TERMO coal

alimenticio *adj* ALIMENT nutritious, nutritive

alimento *m* AGRIC feed, ALIMENT food, P&C *procesamiento* feed; ~ **bajo en sal** *m* ALIMENT reduced-salt food; ~ **básico** *m* ALIMENT staple food; ~ **congelado** *m* ALIMENT, REFRIG, TERMO frozen food; ~ **en conserva** *m* ALIMENT, EMB canned food (*AmE*), tinned food (*BrE*); ~ **enlatado** *m* ALIMENT, EMB canned food (*AmE*), tinned food (*BrE*); ~ **para ganado vacuno** *m* AGRIC cattle feed; ~ **de gluten de maíz** *m* AGRIC corn gluten feed (*AmE*), maize gluten feed (*BrE*); ~ **natural** *m* ALIMENT health food; ~ **para peces** *m* OCEAN fish meal; ~ **principal** *m* ALIMENT staple food

alineación *f* C&V truing, CONST alignment, boning, IMPR alignment, line-up, MECÁ alignment, lining, PETROL line-up, TEC ESP tracking, TELECOM, TV alignment; ~ **básica de imagen** *f* TELECOM, TV basic frame alignment (*BFA*); ~ **de la cabeza** *f* TV head alignment; ~ **del cabezal de video** *f* AmL, ~ **del cabezal de vídeo** *f* Esp TV video-head alignment; ~ **de frecuencias** *f* TELECOM frequency alignment; ~ **del haz** *f* ELECTRÓN, TV beam alignment; ~ **horizontal-vertical** *f* TV X-Y alignment; ~ **láser** *f* ELECTRÓN, ING MECÁ laser alignment; ~ **local** *f* ELECTRÓN local alignment; ~ **de ruedas** *f* AUTO, TRANSP, VEH wheel alignment; ~ **de las ruedas delanteras** *f* AUTO, TRANSP, VEH front-wheel alignment; ~ **de rumbo** *f* TRANSP AÉR course alignment; ~ **de sincronización** *f* TV sync line-up; ~ **de la sincronización de imagen** *f* TV field-sync alignment; ~ **de las tijeras** *f* C&V shear alignment; ~ **de trama** *f* TELECOM frame alignment; ~ **vertical** *f* CONST vertical alignment; ~ **X-Y** *f* TV X-Y alignment

alineado[1] *adj* GEN aligned, in-line

alineado[2]: ~ **de palabras** *m* IMPR word wrap

alineador *m* IMPR aligning tool, liner

alineal *adj* ELEC nonlinear

alineamiento *m* GEOL lineament, IMPR, TELECOM, TV alignment; ~ **de fase** *m* TELECOM phase alignment; ~ **de múltiples bastidores** *m* TELECOM multiframe alignment (*MFA*)

alinear *vt* C&V true up, CINEMAT align, line up, CONST align, bone, line, ELECTRÓN collimate, FOTO, IMPR align, ING MECÁ align, true, INSTR collimate, MECÁ, TELECOM, TV align

alinearse *v refl* CONST align

alisado[1] *adj* C&V, PAPEL, PROD smooth

alisado[2] *m* C&V smoothing, PAPEL smooth finish, *acabado del papel en máquina* glazing, PROD smoothing; ~ **de escuadra** *m* PROD *funderías* corner slick; ~ **de picadura basta** *m* ING MECÁ, PROD rough-cut capacity planning

alisador *m* ING MECÁ planisher, PROD *moldes* slicker; ~ **de escuadra** *m* PROD corner smoother; ~ **de escuadra de bordes redondos** *m* PROD round-edge corner smoother; ~ **de seta** *m* ING MECÁ *moldeo* bacca-box smoother; ~ **de tubo** *m* PROD *fundería* pipe slick, pipe smoother

alisadora *f* EMB calender, calender unit

alisadura *f* PROD smoothing

alisamiento *m* INFORM&PD smoothing, ING MECÁ surfacing

alisar *vt* CONST *carpintería* trim, try up, ING MECÁ true up, MECÁ plane, lap, PROD lap, TEXTIL surface

alíscafo *m* TRANSP, TRANSP MAR hydrofoil

aliso *m* PAPEL alder

aliviadero *m* AGUA, CONST, HIDROL, INSTAL HIDRÁUL channel, overflow, spillway, weir, TRANSP MAR channel; ~ **de avenidas** *m* AGUA, HIDROL flood spillway; ~ **sin compuertas** *m* AGUA, HIDROL ungated spillway; ~ **controlado** *m* AGUA, INSTAL HIDRÁUL controlled spillway; ~ **de crecidas** *m* AGUA, HIDROL spillway; ~ **entallado** *m* AGUA, HIDROL notched weir; ~ **evacuador de crecidas** *m* AGUA *presas* spillway; ~ **flotante autorregulado** *m* CONTAM MAR self-adjusting floating weir; ~ **inclinado** *m* AGUA, HIDROL diagonal weir; ~ **de inmersión regulable** *m* CONTAM *instalaciones para fluidos* adjustable submersion weir; ~ **en pared delgada** *m* AGUA, INSTAL HIDRÁUL *presas* narrow weir, thin-edged weir; ~ **principal** *m* AGUA, HIDROL main spillway; ~ **con rasera vortical** *m* CONTAM MAR weir with vortex skimmer; ~ **de seguridad** *m* AGUA, HIDROL, INSTAL HIDRÁUL, SEG emergency spillway

aliviar *vt* ING MECÁ, INSTAL HIDRÁUL, MECÁ, PROD relieve

alivio *m* ING MECÁ, MECÁ, PROD relief; ~ **térmico** *m* PROD thermal relief

aljibe *m* AGUA cistern; ~ **sin propulsión** *m* TRANSP MAR *transporte de agua* dumb barge

allanamiento *m* GEOL flattening

allanar *vt* CONST grade, TEXTIL iron out

alleno *m* QUÍMICA *compuesto* allene

alma *f* CONST core, *de viga* stem, web, ELEC, FÍS, ING ELÉC core, MECÁ web, ÓPT, TELECOM, TRANSP MAR, TV *de cabo* core; ~ **aislada** *f* ELEC *cable* insulated conductor (*AmE*), insulated core (*BrE*); ~ **del andamio** *f* CONST scaffold pole; ~ **del cable** *f* ELEC, ING ELÉC, P&C cable core; ~ **de fibra** *f* TV fiber core (*AmE*), fibre core (*BrE*); ~ **torcida** *f* TEXTIL *núcleo del hilado* twisted core

almacén[1] *m* CONTAM storage facility, storage tank, tank, FERRO depot, ING ELÉC store, PROD store, warehouse, SEG *de materiales peligrosos* storage, TRANSP MAR warehouse; ~ **de alimentación** *m* CONST, EMB, MECÁ, PROD feed hopper; ~ **de archivo** *m* INFORM&PD file storage (*AmE*), file store (*BrE*); ~ **auxiliar** *m* INFORM&PD auxiliary storage (*AmE*), auxiliary store (*BrE*); ~ **de bombas** *m* D&A bomb bay; ~ **con cámaras de gran altura** *m* REFRIG high-rise cold store; ~ **de carga** *m* TRANSP cargo warehouse; ~ **de conmutación progresiva** *m* TELECOM progressive switching magazine; ~ **de desestibado** *m* EMB denesting magazine; ~ **de estación** *m* FERRO *equipo inamovible* depot; ~ **de fichero** *m* INFORM&PD file storage (*AmE*), file store (*BrE*); ~ **frigorífico** *m* ALIMENT, ING MECÁ, REFRIG, TERMO cold store, refrigerated warehouse; ~ **frigorífico especializado**

m REFRIG, TERMO specialized cold store; ~ **frigorífico polivalente** *m* REFRIG, TERMO multipurpose cold store; ~ **frío** *m* ALIMENT, REFRIG, SEG, TERMO cold storage; ~ **de hielo** *m* REFRIG ice cellar; ~ **individual** *m* TELECOM individual store; ~ **de maderas** *m* CONST, MINAS timberyard; ~ **de mensajes** *m* TELECOM message store (*MS*); ~ **de mercancías** *m* TRANSP freight depot (*AmE*), freight station (*AmE*), goods depot (*BrE*), goods station (*BrE*); ~ **de mezclas** *m* C&V mixed-batch store; ~ **en los muelles** *m* TRANSP MAR dock warehouse; ~ **por núcleos** *m* INFORM&PD core storage (*AmE*), core store (*BrE*); ~ **de productos acabados** *m* EMB finished-goods store; ~ **seguro de líquidos inflamables** *m* SEG safe storage of flammable liquids; ~ **telescópico** *m* EMB nesting magazine

almacén[2]: **en ~** *fra* PROD ex-store

almacenado: ~ **de explosivos** *m* MINAS explosives magazine; ~ **de imágenes** *m* ING ELÉC image storage

almacenaje *m* CONST stockyard, FOTO storage, PROD storage, warehousing, TEC ESP stowage; ~ **de datos** *m* ING ELÉC data storage; ~ **a granel** *m* CONTAM bulk deposition; ~ **en tanques** *m* PROD tank storage

almacenamiento *m* CONTAM MAR storage, ENERG RENOV pondage, INFORM&PD storage, store (*BrE*), PROD storage, warehousing; ~ **de acceso directo** *m* INFORM&PD direct access storage (*AmE*), direct access store (*BrE*); ~ **de acceso inmediato** *m* INFORM&PD immediate access storage (*IAS*) (*AmE*), immediate access store (*IAS*) (*BrE*); ~ **de archivos** *m* INFORM&PD file storage (*AmE*), file store (*BrE*); ~ **asociativo** *m* INFORM&PD associative storage (*AmE*), associative store (*BrE*); ~ **auxiliar** *m* INFORM&PD auxiliary storage (*AmE*), auxiliary store (*BrE*); ~ **borrable** *m* INFORM&PD erasable storage (*AmE*), erasable store (*BrE*); ~ **por capacitor** *m* INFORM&PD capacitor storage (*AmE*), capacitor store (*BrE*); ~ **de carbón** *m* CARBÓN coal stockyard, coal yard, MINAS coal yard; ~ **central** *m* INFORM&PD main storage (*AmE*), main store (*BrE*), primary storage (*AmE*), primary store (*BrE*); ~ **de compensación** *m* TELECOM translation store; ~ **y conmutación hacia adelante** *m* IMPR storage and forwarding; ~ **de contenido direccionable** *m* INFORM&PD content-addressable storage (*AmE*), content-addressable store (*BrE*); ~ **de corrección** *m* TELECOM translation store; ~ **de datos** *m* INFORM&PD data storage (*DS*) (*AmE*), data store (*DS*) (*BrE*), NUCL data storage (*DS*), TELECOM data store (*DS*), TV memory store; ~ **definitivo de residuos** *m* NUCL ultimate waste disposal, waste disposal; ~ **en directorio** *m* TELECOM directory store; ~ **en disco** *m* TELECOM disk store; ~ **en doble buffer** *m* INFORM&PD double buffering; ~ **en el emplazamiento del reactor** *m* NUCL on-site storage; ~ **de energía** *m* ELEC, ING ELÉC, NUCL, TEC ESP, TERMO energy storage; ~ **de energía en horas de baja demanda** *m* ELEC, ING ELÉC, NUCL, TERMO off-peak energy storage; ~ **externo** *m* INFORM&PD external storage (*AmE*), external store (*BrE*); ~ **de ficheros** *m* INFORM&PD file storage (*AmE*), file store (*BrE*); ~ **en frío** *m* ALIMENT, REFRIG, SEG, TERMO cold storage; ~ **de gas** *m* ALIMENT, GAS, PROD gas storage; ~ **geológico de residuos a gran profundidad** *m* NUCL deep geological waste disposal; ~ **en húmedo** *m* NUCL wet storage; ~ **de imagen** *m* ELECTRÓN image storage; ~ **de imágenes** *m*

CINEMAT, TV frame store; ~ **de información** *m* TV information storage (*AmE*), information store (*BrE*); ~ **de la información** *m* INFORM&PD information storage (*AmE*), information store (*BrE*); ~ **intermedio** *m* INFORM&PD intermediate storage (*AmE*), intermediate store (*BrE*); ~ **interno** *m* INFORM&PD, PROD internal storage (*AmE*), internal store (*BrE*); ~ **en línea de retardo** *m* NUCL *de memoria* delay-line storage; ~ **masivo** *m* INFORM&PD mass storage (*AmE*), mass store (*BrE*); ~ **en memoria intermedia** *m* INFORM&PD buffering; ~ **en memoria intermedia doble** *m* INFORM&PD double buffering; ~ **de mensajes** *m* TELECOM message storing (*MS*); ~ **de mensajes por intercambio de datos electrónicos** *m* ELECTRÓN data interchange message store, TELECOM electronic data interchange message store (*EDI-MS*); ~ **no borrable** *m* INFORM&PD nonerasable storage (*AmE*), nonerasable store (*BrE*); ~ **por núcleos** *m* INFORM&PD core storage (*AmE*), core store (*BrE*); ~ **óptico** *m* INFORM&PD optical storage (*AmE*), optical store (*BrE*), ÓPT optical storage; ~ **óptico borrable** *m* ÓPT erasable optical storage; ~ **óptico para escribir una sola vez** *m* ÓPT write-once optical storage; ~ **de partículas** *m* FÍS PART particle storage; ~ **de persistencia variable** *m* ELECTRÓN variable-persistance storage; ~ **a poca profundidad** *m* NUCL shallow burial; ~ **principal** *m* INFORM&PD main storage (*AmE*), main store (*BrE*), primary storage (*AmE*), primary store (*BrE*); ~ **de programas** *m* INFORM&PD program storage; ~ **provisional** *m* NUCL temporary storage; ~ **en registro intermedio** *m* INFORM&PD buffering; ~ **de reserva** *m* NUCL buffer storage; ~ **de residuos en el emplazamiento** *m* NUCL on-site waste disposal; ~ **de respaldo** *m* INFORM&PD backing storage (*AmE*), backing store (*BrE*); ~ **y retransmisión** *m* INFORM&PD storage and forwarding; ~ **en seco** *m* NUCL dry storage; ~ **secundario** *m* INFORM&PD secondary storage (*AmE*), secondary store (*BrE*); ~ **en serie** *m* INFORM&PD serial storage (*AmE*), serial store (*BrE*); ~ **subterráneo** *m* AGUA, GAS, TERMO underground storage; ~ **de tambor** *m* TELECOM drum store; ~ **temporal** *m* INFORM&PD temporary storage (*AmE*), temporary store (*BrE*); ~ **de trabajo** *m* INFORM&PD working storage (*AmE*), working store (*BrE*); ~ **de traducción** *m* TELECOM translation store; ~ **de traslación** *m* TELECOM translation store; ~ **vigilado con posibilidad de recuperación** *m* (*AVR*) NUCL monitored retrievable storage (*MRS*)

almacenar *vt* CARBÓN store, CONST stockpile, ELEC, GAS, INFORM&PD, ING ELÉC store, TEC ESP, TRANSP AÉR stow; ~ **en la memoria intermedia** *vt* INFORM&PD buffer

almacenero *m* PROD storeman

almacenista *m* PROD storeperson

almadraba *f* OCEAN *artes de pesca* tunny fish net, tunny fishery, tunny net

almanaque: ~ **náutico** *m* TRANSP MAR nautical almanac

almandino *m* MINERAL almandine

almandita *f* MINERAL almandite

almártaga *f* P&C *materia prima* litharge

almecenamiento: ~ **de carbón** *m* MINAS coal stockyard

almidón *m* ALIMENT, IMPR, PAPEL, QUÍMICA, TEXTIL starch; ~ **animal** *m* ALIMENT animal starch; ~ **modificado** *m* ALIMENT modified starch; ~ **reducido** *m* ALIMENT boiled starch

almidonador *m* TEXTIL stiffener

almidonar *vt* TEXTIL stiffen

almirante *m* TRANSP MAR *armada* admiral, flag officer

almohadilla *f* ELEC *cable* bedding, EMB, IMPR, ING MECÁ, MECÁ, PROD, TEC PETR pad; ~ **antirresbaladiza** *f* PROD antislide pad; ~ **de impresión** *f* EMB impression pad; ~ **de pared lateral** *f* TEC PETR sidewall pad; ~ **sujetadora** *f* PROD gripping pad; ~ **de toque** *f* INFORM&PD touchpad

almohadillado[1] *adj* EMB, IMPR, ING MECÁ, MECÁ, PROD, TEC PETR padded

almohadillado[2] *m* EMB, IMPR, ING MECÁ, MECÁ, PROD, TEC PETR padding; ~ **del asiento** *m* VEH seat cushion; ~ **del borde** *m* EMB edge cushion

almojaya *f* CONST *andamio* putlock, putlog; ~ **del andamio** *f* CONST putlock, putlog

alnoeíta *f* MINERAL alnoeite, alnoite

alnoíta *f* MINERAL alnoeite, alnoite

alocinesis *f* TEC PETR *geología* halokinesis

alóctono *adj* CONTAM, PETROL allochthonous

alófana *f* MINERAL allophane

aloína *f* QUÍMICA aloin

alojamiento *m* AUTO housing, ING MECÁ accommodation, housing, INSTAL HIDRÁUL *de válvula* seat, MECÁ accommodation, housing, TRANSP MAR *contrucción naval* accommodation, *de pasajeros* berthing, VEH *motor, transmisión* casing, housing; ~ **del árbol de mando** *m* AUTO, VEH *transmisión* drive-shaft tunnel (*AmE*), propeller-shaft tunnel (*BrE*); ~ **para barra** *m* ING MECÁ bar hole; ~ **de la bomba** *m* AUTO, VEH pump housing; ~ **de la bujía** *m* AUTO, ELEC, VEH spark plug hole; ~ **del cojinete** *m* ING MECÁ bearing cage; ~ **del convertidor del par** *m* AUTO, VEH *motor* torque-converter housing; ~ **del convertidor de torsión** *m* AUTO, VEH torque-converter housing; ~ **del distribuidor** *m* AUTO, VEH distributor housing; ~ **dividido** *m* AUTO, VEH split housing; ~ **del eje** *m* AUTO, MECÁ, VEH axle casing, axle housing; ~ **del eje trasero** *m* AUTO, MECÁ, VEH rear-axle housing; ~ **del engranaje de sincronización** *m* AUTO, VEH timing-gear housing; ~ **en espiral** *m* AUTO, VEH volute casing; ~ **del freno** *m* AUTO, VEH brake housing; ~ **del macho** *m* PROD *moldería* seating; ~ **para machos** *m* PROD *moldería* core print; ~ **de la marinería** *m* TRANSP MAR *embarcación* mess deck; ~ **del mecanismo de accionamiento de las barras de control** *m* (*CRDH*) NUCL control rod drive housing (*CRDH*); ~ **de mercancías** *m* TRANSP freight house (*AmE*), goods house (*BrE*); ~ **óptico** *m* CINEMAT, ÓPT optical house; ~ **para pulsaciones hidráulicas del líquido circulante** *m* INSTAL HIDRÁUL fluttering seat; ~ **para radio-casette** *m* AUTO, PROD, VEH radio-cassette deck; ~ **de las ruedas** *m* TRANSP AÉR wheel well; ~ **del segmento** *m* AUTO, VEH *motor* piston-ring groove; ~ **del tren de aterrizaje** *m* TRANSP AÉR landing-gear well; ~ **de la unidad** *m* REFRIG unit housing; ~ **de válvula** *m* INSTAL HIDRÁUL valve guide, valve seat

alomado *adj* AGRIC ridging

alomador *m* AGRIC ridger

alomorfita *f* MINERAL allomorphite

alopaladio *m* MINERAL allopalladium

aloquema *m* PETROL allochem
aloquímico *m* GEOL allochem
alotriomórfico *adj* GEOL, PETROL allotriomorphic
alotropía *f* CRISTAL, METAL, QUÍMICA allotropy, polymorphism
alotrópico *adj* CRISTAL, METAL, QUÍMICA allotropic
alótropo *m* CRISTAL, METAL, QUÍMICA allotrope
alquenilo *adj* QUÍMICA alkenyl
alqueno *m* DETERG, QUÍMICA, TEC PETR *petroquímica* alkene
alquid *m* QUÍMICA alkyd
alquifol *m* MINERAL alquifou, alquifoux
alquil: **~ benceno** *m* *Esp* (*cf alquilobenceno AmL*) DETERG, QUÍMICA alkyl benzene
alquilación *f* DETERG, QUÍMICA, TEC PETR alkylation
alquilamina *f* DETERG, QUÍMICA alkylamine
alquileno *m* DETERG, QUÍMICA alkylene
alquiler: **~ de líneas** *m* TELECOM line rental
alquilo *m* DETERG, QUÍMICA, TEC PETR alkyl
alquiloaromáticos *m* DETERG alkylene, TEC PETR alkylaromatics
alquilobenceno *m* *AmL* (*cf alquil benceno Esp*) DETERG, QUÍMICA alkyl benzene
alquino *m* QUÍMICA, TEC PETR *petroquímica* alkyne
alquitrán *m* GEN bitumen, tar; **~ de gas** *m* CARBÓN gas tar; **~ de hulla** *m* CARBÓN coal tar, gas tar; **~ de petróleo** *m* PETROL, TEC PETR rock tar
alquitranado *m* GEN tarring
alquitranar *vt* GEN tar
ALS *abr* (*fuente luminosa avanzada*) FÍS PART ALS (*advanced light source*)
alstonita *f* MINERAL alstonite
alta¹: **de ~ carga** *adj* PROD highly-loaded; **de ~ densidad** *adj* INFORM&PD high-density; **de ~ flexibilidad** *adj* PROD highly-flexible; **en ~ mar** *adj* GEOL, OCEAN, TEC PETR, TRANSP MAR offshore; **de ~ presión** *adj* AGUA, FÍS, GAS high-pressure; **de ~ producción** *adj* ING MECÁ heavy-duty; **a ~ temperatura** *adj* FÍS, TERMO, TERMOTEC high-temperature; **de ~ temperatura** *adj* FÍS high-temperature; **de ~ tensión** *adj* FÍS high-voltage; **de ~ velocidad** *adj* ING MECÁ, MECÁ high-speed
alta²: **en ~ mar** *adv* GEOL, OCEAN, TEC PETR, TRANSP MAR offshore
alta³ *f* METEO *presión atmosférica* anticyclone, high; **~ frecuencia** *f* (*AF*) ELEC, ELECTRÓN, ING ELÉC, TELECOM high frequency (*HF*); **~ luz** *f* FOTO highlight; **~ mar** *f* GEOL, OCEAN, TRANSP MAR deep sea, high seas, offshore, open sea; **~ de Ogasawara** *f* METEO permanent anticyclone; **~ presión** *f* AGUA, FÍS, GAS high pressure; **~ resistencia** *f* TELECOM high resistance (*HR*); **~ resolución** *f* INFORM&PD high resolution (*HR*); **~ temperatura** *f* FÍS, TERMO, TERMOTEC high temperature; **~ tensión** *f* ELEC, FÍS, ING ELÉC *fuente de alimentación* high tension, high voltage; **~ velocidad** *f* ING MECÁ, MECÁ high gear, high speed
altaíta *f* MINERAL altaite
altar *m* SEG, TERMO fire stop; **~ del horno** *m* PROD furnace bridge
altavoz *m* ACÚST, CINEMAT, ELEC, FÍS, TELECOM, TV loudspeaker; **~ de bocina** *m* ACÚST horn loudspeaker; **~ de cinta** *m* ACÚST ribbon loudspeaker; **~ coaxial** *m* ACÚST coaxial loudspeaker; **~ de control** *m* ACÚST, CINEMAT monitoring loudspeaker; **~ dinámico** *m* ACÚST dynamic loudspeaker;

~ directo *m* ACÚST direct loudspeaker; **~ electrodinámico** *m* ACÚST, ELEC electrodynamic loudspeaker; **~ electromagnético** *m* ACÚST, ELEC electromagnetic loudspeaker; **~ electrostático** *m* ACÚST, ELEC electrostatic loudspeaker; **~ elemental** *m* ACÚST elementary loudspeaker; **~ elemental multicanal** *m* ACÚST, TELECOM multichannel elementary loudspeaker; **~ de imán permanente** *m* ACÚST permanent-magnet loudspeaker; **~ iónico** *m* ACÚST ionic loudspeaker; **~ de magnetoestricción** *m* ACÚST magnetostriction loudspeaker; **~ de membrana** *m* ACÚST membrane loudspeaker; **~ neumático** *m* ACÚST pneumatic loudspeaker
alterable *adj* ELECTRÓN, GEOL, QUÍMICA labile
alteración *f* ACÚST alteration, CARBÓN deformation; **~ accidental** *f* ACÚST accidental alteration; **~ por acción de los agentes atmosféricos** *f* P&C, REVEST weathering; **~ alveolar** *f* GEOL *roca* honeycomb structure; **~ atmosférica** *f* CARBÓN weathering; **~ de base** *f* ELECTRÓN *de transistores* base doping; **~ de canal** *f* ELECTRÓN *de transistores* channel doping; **~ del captador** *f* ELECTRÓN *de transistores* collector doping; **~ del color** *f* P&C *defecto* color removal (*AmE*), colour removal (*BrE*); **~ cromática de los extremos** *f* TV fringing; **~ deutérica** *f* GEOL deuteric alteration; **~ estructural** *f* ACÚST constitutive alteration; **~ por exposición a la intemperie artificial** *f* P&C *prueba* accelerated weathering, artificial weathering; **~ a la intemperie** *f* P&C weathering; **~ mecánica** *f* GEOL *de rocas* mechanical weathering; **~ mediante fósforo** *f* ELECTRÓN *de transistores* phosphorus doping; **~ meteórica** *f* CARBÓN weathering; **~ del negro por parásitos** *f* TV noisy blacks; **~ en profundidad** *f* REFRIG bone taint; **~ química** *f* GEOL *de rocas* chemical weathering, PETROL, QUÍMICA chemical altering; **~ de rocas** *f* GEOL weathering; **~ del suelo** *f* CONTAM land disturbance; **~ del terreno** *f* CONTAM land disturbance; **~ de la tinta durante su secado** *f* IMPR dry back; **~ de la trayectoria** *f* TEC ESP alteration of course; **~ en el valor de una variable** *f* TEC ESP pulse
alterar *vt* GEOL weather
alternación *f* ING MECÁ alternation; **~ del movimiento** *f* ING MECÁ alternation of movement
alternado *adj* ING MECÁ alternating
alternador *m* ELEC, ENERG RENOV, FÍS, ING ELÉC, VEH alternating-current generator, alternator, synchronous generator; **~ de ampollas** *m* ELEC bulb alternator; **~ asincrónico** *m* ELEC, ING ELÉC induction generator; **~ asíncrono** *m* ELEC, ING ELÉC asynchronous alternator; **~ bifásico** *m* ELEC, ING ELÉC two-phase alternator; **~ de bombillas** *m* ELEC bulb alternator; **~ de eje vertical con rangua inferior** *m* ELEC, ING ELÉC umbrella-type alternator; **~ fijo** *m* TEC ESP *aviones* fixed generator; **~ hidráulico** *m* ELEC, ING ELÉC hydroelectric generator; **~ de hierro giratorio** *m* ELEC, ING ELÉC inductor alternator; **~ de inducción** *m* ELEC, ING ELÉC induction generator; **~ de RF** *m* ELECTRÓN, ING ELÉC, TELECOM RF alternator; **~ sincrónico** *m* ELEC, ING ELÉC synchronous alternator; **~ trifásico** *m* ELEC, ING ELÉC three-phase alternator
alternante¹ *adj* ING MECÁ, MECÁ alternating
alternante² *m* ING MECÁ, MECÁ alternation
alternaria *f* AGRIC *enfermedad* alternaria

alternativo *adj* ING MECÁ, MECÁ alternating, reciprocating

alterno *adj* ING MECÁ, MECÁ alternating

alternomotor *m* ELEC AC motor, alternating-current motor, TRANSP alternating-current motor; ~ **trifásico** *m* ELEC, TRANSP three-phase alternomotor

altimetría *f* FÍS, GEOFÍS, METR, TEC ESP, TRANSP, TRANSP AÉR altimetry

altímetro *m* FÍS, GEOFÍS altimeter, height gage (*AmE*), height gauge (*BrE*), METR altimeter, height gauge (*BrE*), height gage (*AmE*), TEC ESP height gage (*AmE*), altimeter, height gauge (*BrE*), TRANSP AÉR altimeter, height gage (*AmE*), height gauge (*BrE*); ~ **aneroide** *m* TRANSP AÉR aneroid altimeter; ~ **de cabina** *m* TRANSP AÉR cabin altimeter; ~ **cifrado** *m* TRANSP AÉR encoding altimeter; ~ **codificado** *m* TRANSP AÉR encoding altimeter; ~ **de precisión** *m* INSTR, TRANSP AÉR sensitive altimeter; ~ **de presión** *m* TRANSP AÉR pressure altimeter; ~ **radárico** *m* TRANSP AÉR radar altimeter; ~ **radioeléctrico** *m* TRANSP AÉR radio altimeter; ~ **de tambor** *m* TRANSP AÉR drum altimeter; ~ **de vernier** *m* METR vernier height gage (*AmE*), vernier height gauge (*BrE*)

altiplano: ~ **continental** *m* GEOFÍS, GEOL continental plate

altitud *f* CONST elevation, FÍS, FÍS ONDAS, ING MECÁ, TEC ESP, TRANSP AÉR altitude, height; ~ **barométrica** *f* TRANSP AÉR pressure altitude; ~ **de cabina** *f* TRANSP AÉR cabin altitude; ~ **crítica** *f* TRANSP AÉR critical altitude; ~ **de crucero** *f* TRANSP AÉR cruising altitude; ~ **de descenso mínima** *f* TRANSP AÉR minimum descent altitude; ~ **nominal** *f* TRANSP AÉR rated altitude; ~ **de operación máxima** *f* TRANSP AÉR maximum-operating altitude; ~ **operacional** *f* TEC ESP operating altitude; ~ **para reingreso en la tierra** *f* TEC ESP earth-reentry altitude; ~ **relativa** *f* TRANSP AÉR relative altitude; ~ **de seguridad mínima** *f* TRANSP AÉR minimum safe altitude; ~ **de techo** *f* TEC ESP, TRANSP AÉR ceiling altitude; ~ **de visibilidad** *f* TEC ESP, TRANSP AÉR ceiling altitude

alto[1]: ~ **en fibra** *adj* ALIMENT high-fibre; **de ~ grado de vacío** *adj* MECÁ high-vacuum; **de ~ nivel** *adj* CARBÓN high-grade; ~ **en proteína** *adj* ALIMENT high-protein; **de ~ rendimiento** *adj* FÍS high-performance; **de ~ vacío** *adj* MECÁ high-vacuum

alto[2] *m* ING MECÁ stop, standstill; ~ **brillo** *m* P&C, REVEST *pintura* high gloss; ~ **contenido de azufre** *m* PETROL, QUÍMICA, TEC PETR high sulfur content (*AmE*), high sulphur content (*BrE*); ~ **explosivo** *m* D&A, MINAS high explosive; ~ **explosivo anticarro** *m* D&A high explosive anti-tank (*HEAT*); ~ **grado** *m* QUÍMICA high grade; ~ **grado del flujo** *m* PROD high flow rate; ~ **grado de protección** *m* PROD high degree of protection; ~ **horno** *m* C&V, CARBÓN, ING MECÁ, INSTAL TERM, TERMO blast furnace; ~ **nivel de imagen** *m* TV high picture level; ~ **número de Reynolds** *m* FÍS FLUID high Reynolds number; ~ **vacío** *m* ELECTRÓN hard vacuum, FÍS high vacuum; ~ **voltaje** *m* ELEC *fuente de alimentación*, FÍS high voltage

altocúmulo *m* (*AC*) METEO, TRANSP AÉR altocumulus (*AC*)

altoestrato *m* METEO, TRANSP AÉR altostratus (*AS*)

altoparlante *m* ACÚST, ELEC, FÍS, TELECOM loudspeaker

altostratus *m* METEO, TRANSP AÉR altostratus (*AS*)

altura[1] *f* ACÚST *escala musical* pitch, CONST rise, height, level, ENERG RENOV *de una onda* height, altitude, FÍS altitude, height, FÍS ONDAS height, altitude, GEOFÍS, GEOM altitude, height, IMPR *de mayúsculas* height, INFORM&PD height, altitude, ING MECÁ altitude, height, *de diente de engranaje* addendum, head, INSTR height, MECÁ addendum, METR, TEC ESP height, altitude, TEC PETR *de fluidos* head, TRANSP altitude, height, TRANSP AÉR altitude, TRANSP MAR *de la marea* height, *de luces exhibidas como ayudas a la navegación* elevation, *astronavegación* altitude; ~ **aparente** *f* TRANSP MAR *navegación* apparent altitude; ~ **de la banda lateral de apoyo** *f* IMPR bearer height; ~ **barométrica de la presión atmosférica** *f* METEO barometric height of pressure; ~ **de la bomba** *f* INSTAL TERM pump head; ~ **de la cabeza del diente** *f* ING MECÁ height above pinch line; ~ **de caída** *f* CARBÓN drop height, EMB drop height, height of fall; ~ **de la caída de agua** *f* INSTAL HIDRÁUL water fall height; ~ **de caída efectiva** *f* CARBÓN effective drop height; ~ **de caída libre** *f* TRANSP AÉR free-drop height; ~ **calculada** *f* TRANSP AÉR rated altitude; ~ **del centro del eje** *f* ING MECÁ shaft center height (*AmE*), shaft centre height (*BrE*); ~ **de centros** *f* ING MECÁ height of centers (*AmE*), height of centres (*BrE*); ~ **cinética** *f* INSTAL HIDRÁUL velocity head; ~ **de coronamiento** *f* ENERG RENOV crest height; ~ **debida a la velocidad** *f* INSTAL HIDRÁUL velocity head; ~ **de decisión** *f* TRANSP AÉR decision height; ~ **de descarga** *f* AGUA delivery lift, HIDROL, NUCL *bombas* discharge head; ~ **de descenso mínima** *f* TRANSP AÉR minimum descent height; ~ **dinámica** *f* INSTAL HIDRÁUL velocity head; ~ **de elevación** *f* AGUA lift, static head, static lift; ~ **estática de aspiración** *f* AGUA static suction lift; ~ **de extracción** *f* AmL (*cf altura fija Esp*) MINAS drawing height, fixed lift; ~ **fija** *f* Esp (*cf altura de extracción AmL*) MINAS drawing height, fixed lift; ~ **geopotencial** *f* GEOFÍS geopotential height; ~ **hidrostática** *f* TEC PETR *mecánica de fluidos* hydrostatic head, mud column; ~ **de impulsión** *f* AGUA delivery head, delivery lift, discharge lift, lift, static discharge head, HIDROL *bombas* discharge head, MINAS *bombas de mina* stage, NUCL delivery head, *de una bomba* discharge head; ~ **del impulso** *f* ELECTRÓN, INFORM&PD pulse height; ~ **de inclinación** *f* TV tip height; ~ **inclinada** *f* GEOM slant height; ~ **instrumental** *f* TRANSP MAR *navegación* sextant altitude; ~ **interna total** *f* ING MECÁ overall internal height; ~ **de levitación** *f* TRANSP hover height; ~ **libre** *f* CONST, TRANSP MAR *construcción* headroom; ~ **de la línea de deslizamiento** *f* METAL *deformación* slip step height; ~ **manométrica** *f* AGUA, INSTAL HIDRÁUL *bombas* static head, static lift; ~ **manométrica de aspiración** *f* AGUA, INSTAL HIDRÁUL *bombas* suction head; ~ **de la marea** *f* OCEAN, TRANSP MAR tidal height; ~ **máxima de caída del agua** *f* AGUA, INSTAL HIDRÁUL full head of water; ~ **media** *f* IMPR z-height; ~ **metacéntrica** *f* TRANSP MAR *arquitectura naval* metacentric height; ~ **de la mira del instrumento** *f* CONST *topografía* height of instrument; ~ **sobre el nivel de imposta** *f* CONST *arquitectura* height above impost level; ~ **normal** *f* IMPR standard height; ~ **del núcleo de la letra** *f* IMPR X-height; ~ **observada** *f* TRANSP MAR *navegación* observed altitude; ~ **de la ola** *f* OCEAN, TRANSP MAR wave height; ~ **del oleaje** *f* OCEAN, TRANSP MAR height of swell; ~ **de onda** *f* ENERG RENOV, FÍS, FÍS ONDAS wave height; ~ **de página** *f* IMPR page depth; ~ **de paso** *f* TRANSP MAR *construcción* headroom, *yate* cabin headroom; ~ **de pelo** *f* TEXTIL *tejido* pile height; ~ **del penacho de humos** *f* CONTAM *evacuación de gases* plume rise; ~ **en perpendicular** *f* GEOL normal throw; ~ **del pie del diente** *f* ING MECÁ, MECÁ dedendum; ~ **piezométrica** *f* CARBÓN *hidráulica* pressure head, INSTAL HIDRÁUL head, TEC PETR *mecánica de fluidos* piezometric head; ~ **de la pila** *f* EMB stacking height; ~ **del piso** *f* MINAS bench height; ~ **de pleamar** *f* TRANSP MAR rise; ~ **potenciométrica** *f* TEC PETR *mecánica de fluidos* potentiometric head; ~ **de la presión** *f* TRANSP AÉR pressure height; ~ **pseudopotenciométrica** *f* TEC PETR *mecánica de fluidos* pseudo-potentiometric head; ~ **de puntos** *f* ING MECÁ height of centers (*AmE*), height of centres (*BrE*), swing; ~ **de puntos sobre la bancada** *f* ING MECÁ swing of the bed; ~ **de rebote** *f* TEC ESP, TRANSP AÉR float altitude; ~ **de la red** *f* OCEAN net depth; ~ **de la regla** *f* CONST screed height; ~ **de salto** *f* GEOL normal throw; ~ **de seguridad** *f* MINAS, SEG *castillete de extracción en pozos mineros* overwinding allowance; ~ **sobre el nivel del mar** *f* FÍS altitude; ~ **del sol** *f* ENERG RENOV, TEC ESP solar altitude; ~ **del sol sobre el horizonte** *f* ENERG RENOV, TEC ESP solar altitude angle; ~ **de succión** *f* AGUA suction lift; ~ **en suspenso** *f* TRANSP *aerodeslizadores* hover height; ~ **del tipo** *f* IMPR height of type, type height; ~ **tipográfica** *f* IMPR height-to-paper; ~ **tonal** *f* ACÚST *de instrumento musical* pitch; ~ **total de dragado** *f* AGUA full dredging depth; ~ **total de la vasija del reactor** *f* NUCL total height; ~ **de trabajo del diente de engranaje** *f* ING MECÁ working depth of tooth; ~ **útil** *f* PROD operating head

altura²: **a la ~ de los ojos** *fra* CONST at eye level; **a la ~ del terreno** *fra* CONST even with the ground

alturas: ~ **tonales conjuntas** *f pl* ACÚST conjoined pitches

aluaudita *f* MINERAL alluaudite

alud *m* METEO avalanche

alumbrado *m* ELEC, ING ELÉC, TEC ESP illumination; ~ **eficiente** *m* TEC ESP illumination efficiency; ~ **de emergencia** *m* ELEC emergency lighting; ~ **instrumental** *m* INSTR instrument lighting; ~ **de posición** *m* TEC ESP position light; ~ **por proyección** *m* ELEC floodlighting; ~ **de socorro** *m* ELEC emergency lighting; ~ **vertical incidente** *m* INSTR incident top lighting

alumbrador: ~ **eléctrico** *m* ELEC electric lighter

alumbre *m* MINERAL, PAPEL, QUÍMICA alum; ~ **crómico** *m* QUÍMICA chrome alum; ~ **de hierro** *m* QUÍMICA iron alum; ~ **potásico** *m* QUÍMICA potash alum, soda alum; ~ **sódico** *m* QUÍMICA soda alum

alúmina *f* ALIMENT, CARBÓN, MINERAL alumina, PAPEL alum, QUÍMICA alumina; ~ **activada** *f* ALIMENT, QUÍMICA activated alumina

aluminación *f* QUÍMICA alumination

aluminar *vt* QUÍMICA aluminate

aluminato *m* QUÍMICA aluminate; ~ **de magnesio** *m* MINERAL, QUÍMICA spinel

alumínico *adj* QUÍMICA aluminic

aluminífero *adj* QUÍMICA aluminiferous

aluminio *m* (*Al*) METAL, QUÍMICA aluminium (*BrE*) (*Al*), aluminum (*AmE*) (*Al*)

aluminita *f* MINERAL aluminite, websterite

aluminizado *adj* METAL, REVEST aluminium-coated (*BrE*), aluminum-coated (*AmE*)

alumino- *pref* QUÍMICA alumino-

aluminosilicato *m* MINERAL, QUÍMICA aluminosilicate

aluminoso *adj* QUÍMICA aluminous

alunita *f* MINERAL alunite

alunógeno *m* MINERAL alunogen

alurgita *f* MINERAL alurgite

aluvión *m* AGUA, HIDROL, MINAS, OCEAN, PETROL alluvial deposit, alluvium; ~ **aurífero** *m* CARBÓN, MINAS gold diggings, gravel mine; ~ **explotable** *m* MINAS pay dirt; ~ **fluvial** *m* AGUA fluvial alluvium; ~ **productivo** *m* *Esp* (*cf aluvión remunerador AmL*) MINAS pay dirt; ~ **remunerador** *m* *AmL* (*cf aluvión productivo Esp*) MINAS pay dirt

álveo *m* CARBÓN *río* bed

alveolar: ~ **de nido de abejas** *m* TRANSP AÉR honeycomb

alvéolo: ~ **conector para remolques** *m* AUTO, ELEC *automotor* connector socket for trailer

alvita *f* MINERAL albite

alza *f* AGUA flashboard, D&A *de fusiles o cañones* range, sight hole, ING MECÁ lifting, INSTAL HIDRÁUL overflow, shutter, MINAS lifting, PROD *de presas* gate, TRANSP lifting; ~ **abatida** *f* D&A battle sight; ~ **circular** *f* D&A dial sight; ~ **conjugada** *f* D&A combined sight; ~ **giroscópica** *f* D&A gyroscopic sight; ~ **nocturna** *f* D&A night sight; ~ **óptica** *f* TEC ESP optical sight; ~ **panorámica** *f* D&A dial sight; ~ **reguladora** *f* D&A setter sight; ~ **removible** *f* AGUA flashboard

alzada *f* ING MECÁ throw; ~ **del caballo** *f* AGRIC height of withers

alzado *m* (*cf espaciado Esp*) CONST elevation, *planos* vertical section, IMPR make-ready, gathering, ING MECÁ front elevation, MINAS *de los cables de extracción* clearing; ~ **de la jaula** *m* MINAS clearing of the cage

alzadora *f* IMPR collating machine, gatherer; ~ **de doble flujo** *f* IMPR twin-stream collator; ~ **de papel** *f* IMPR collating machine

alzamiento *m* MINAS raise, raising

alzar *vt* IMPR collate, ING MECÁ hold up, MECÁ boost, lift, MINAS hoist; ~ **a la superficie** *vt* MINAS hoist to the surface

Am *abr* (*americio*) FÍS RAD, NUCL, QUÍMICA Am (*americium*)

AM *abr* (*amplitud modulada, modulación de amplitud*) INFORM&PD, TELECOM, TV AM (*amplitude modulation*)

amagnético *adj* ELEC nonmagnetic

amainar *vi* TRANSP MAR *del tiempo* drop, lower, shorten

amalgama *f* GEN amalgam; ~ **de criptonato de cadmio** *f* NUCL kryptonate of cadmium amalgam; ~ **para espejos** *f* PROD, REVEST quicksilvering

amalgamación *f* GEN amalgamation; ~ **en barriles** *f* PROD barrel amalgamation; ~ **de placa** *f* CARBÓN plate amalgamation

amalgamador *m* GEN amalgamator; ~ **en cubetas** *m* PROD pan amalgamator

amalgamar *vt* GEN amalgamate

amante: ~ **de rizos** *m* TRANSP MAR *velero clásico* reefing pennant

amantillo *m* TRANSP MAR *jarcia de labor* topping lift, *yates* reefing pennant; ~ **de verga para botes** *m* TRANSP boat lift

amarilleamiento *m* COLOR, PAPEL *de las fibras o tejidos* yellowness, yellowing, TEXTIL yellowing, yellowness

amarillear *vi* COLOR, PAPEL, TEXTIL yellow

amarillento *adj* COLOR, PAPEL, TEXTIL yellowish

amarillo *adj* COLOR, PAPEL, TEXTIL yellow

amarina *f* QUÍMICA amarine

amarra *f* CARBÓN cable, martingale, mooring lane, TRANSP MAR lashing; ~ **de proa** *f* TRANSP MAR headline; ~ **de tierra** *f* TRANSP MAR shoreline; ~ **de través** *f* TRANSP MAR breast line

amarrado *adj* TRANSP MAR *buque* laid-up

amarrador *m* TRANSP MAR *de pasajeros* boatman

amarradura *f* TRANSP MAR *cabo, cuerda, equipamiento* mooring

amarraje *m* ELEC *cable* serving

amarrar *vt* CONST anchor, *carpintería* brace, CONTAM MAR moor, MINAS *jaulas* hitch, TRANSP secure, TRANSP MAR moor, *buque* lay up, *cabo* lash, *hidroavión* land, *bote, cabo* make fast

amarre *m* CONTAM MAR mooring, ING MECÁ trunnion, tie, PROD *de cables* lashing, TEC ESP binder, anchoring, TEC PETR, TRANSP MAR mooring; ~ **simple** *m* TEC PETR *producción costa-fuera* single-point mooring

amasado *m* ALIMENT, C&V, PROC QUÍ kneading, mixing

amasador: ~ **de barro** *m* C&V clay kneader; ~ **dispersador** *m* PROC QUÍ dispersion kneader

amasadora *f* CONST mortar mixer, INSTAL HIDRÁUL paddle mixer, pug mill, PROC QUÍ mixer; ~ **de cilindros** *f* PROC QUÍ mixing mill; ~ **de platos** *f* PROC QUÍ mixing pan mill

amasadura *f* PROD kneading

amasamiento *m* PROD kneading

amasar *vt* CONST *mortero* mix

amatista *f* MINERAL amethyst

amausita *f* MINERAL amausite

amazonita *f* MINERAL Amazon stone, amazonite

ámbar *m* MINERAL amber, succinite; ~ **prensado** *m* MINERAL amberoid

amberoide *m* MINERAL amberoid, ambroid

ambiental *adj* GEN environmental

ambiente[1] *adj* GEN ambient

ambiente[2] *m* GEN environment; ~ **electromagnético** *m* TEC ESP electromagnetic environment; ~ **espacial** *m* TEC ESP space environment; ~ **geológico** *m* GAS, GEOL geological environment; ~ **hiperbárico** *m* OCEAN hyperbaric environment; ~ **natural** *m* AGUA, CONTAM natural environment; ~ **operacional del usuario** *m* INFORM&PD user operating environment; ~ **palustre** *m* GEOL paludal environment; ~ **pantanoso** *m* GEOL paludal environment; ~ **peligroso** *m* FÍS RAD, SEG dangerous environment; ~ **de plasma** *m* TEC ESP plasma environment; ~ **sedimentario** *m* GEOL sedimentary environment

ámbito: ~ **faunal** *m* GEOL faunal province; ~ **faunístico** *m* GEOL faunistic province; ~ **de frecuencia** *m* ELECTRÓN frequency domain; ~ **de temperatura intrínseca** *m* ELECTRÓN intrinsic temperature range

ambligonita *f* MINERAL amblygonite, hebronite

ambos: ~ **justificados** *adj* IMPR both justified

ambreína *f* QUÍMICA ambrain

ambrita *f* MINERAL ambrite

ambroide *m* MINERAL amberoid, ambroid

ambulancia *f* AUTO, VEH ambulance

AMDF *abr* (*acceso múltiple por división de frecuencias*) TEC ESP, TELECOM FDMA (*frequency-division multiple access*)

AMDOP/DC *abr* (*acceso múltiple en dirección a la onda portadora con detección de colisiones*) ELECTRÓN, INFORM&PD CSMA/CD (*carrier sense multiple access with collision detection*)

amenaza: ~ **activa** *f* TELECOM active threat; ~ **pasiva** *f* TELECOM passive threat

americio *m* (*Am*) FÍS RAD, NUCL, QUÍMICA americium (*Am*)

amerizar *m* TRANSP AÉR landing on water

amesita *f* MINERAL amesite

ametralladora *f* D&A gun, machine gun; ~ **antiaérea** *f* D&A air defence gun (*BrE*), air defense gun (*AmE*)

amiantinita *f* MINERAL amianthinite

amianto *m* C&V, MECÁ asbestos, MINERAL amiant, amianthus, P&C, TEXTIL asbestos; ~ **-cemento** *m* CONST asbestos cement

amida *f* DETERG, P&C, QUÍMICA amide

amidina *f* QUÍMICA amidine

amidógeno *m* QUÍMICA amidogen

amiláceo *adj* ALIMENT amylaceous, starchy

amilasa *f* AGRIC *enzima* amylase

amilo *m* QUÍMICA amyl

amilopectina *f* TEXTIL amylopectin

amina *f* DETERG, P&C, QUÍMICA amine; ~ **cicloalifática** *f* P&C, QUÍMICA cycloaliphatic amine; ~ **grasa** *f* DETERG, QUÍMICA fatty amine

aminación *f* QUÍMICA amination

amino- *pref* QUÍMICA amino-

aminoácido *m* ALIMENT, QUÍMICA amino acid

aminoazo *adj* QUÍMICA aminoazo

aminoetano *m* QUÍMICA ethylamine

aminorador *m* TRANSP spoiler

amiolita *f* MINERAL ammiolite

amojonamiento *m* CONST *topografía* monumenting

amojonar *vt* CONST mark out

amolado *m* ING MECÁ, PROD *herramientas* grinding, whetting; ~ **de herramientas** *m* ING MECÁ, PROD tool grinding, tool sharpening

amolador *m* ING MECÁ, PROD sharpener; ~ **de cuchillos** *m* ING MECÁ, PROD knife-grinding machine

amoladora *f* ING MECÁ, MECÁ, PROC QUÍ, PROD grinding device, grinding machine; ~ **de superficie** *f* ING MECÁ, MECÁ surface grinder

amolar *vt* ING MECÁ, MECÁ, PROC QUÍ, PROD grind

amoldadora: ~ **-sacadora** *f* IMPR former

amoniacal *adj* QUÍMICA ammoniacal

amoniaco *m* (*NH₃*) GEN ammonia (*NH₃*); ~ **anhidro** *m* QUÍMICA, TEC PETR anhydrous ammonia; ~ **líquido** *m* QUÍMICA, TERMO liquid ammonia

amonio *m* QUÍMICA ammonium

amonite *m* GEOL *paleontología* ammonite

amontonado *adj* CONST heaped, TERMO *fuegos de un horno* banked up

amontonamiento *m* CONST piling, PROD piling, stacking, RECICL *de fango* bulking; ~ **en terraplén** *m* CARBÓN embankment piling

amontonar *vt* (*cf taluzar Esp*) CARBÓN pile, CONST, MINAS bank up, PROD stack

amontonarse *v refl* METEO *nubes* bank up

amor: ~ **de hortelano** *m* ALIMENT lady's bedstraw

amorfismo *m* QUÍMICA amorphism

amorfo *adj* CARBÓN, CRISTAL, ELECTRÓN amorphous,

noncrystalline, GEOL amorphous, non-crystalline, P&C, QUÍMICA amorphous, noncrystalline

amortiguación f ACÚST, ELEC damping, ELECTRÓN falloff, FÍS, INFORM&PD, ING ELÉC, METAL, TV damping; ~ **de aire por chorros periféricos** f TRANSP peripheral-jet air cushion; ~ **por aire estático** f TRANSP static air cushion; ~ **en altura** f TRANSP height-on cushion; ~ **crítica** f FÍS, ING ELÉC critical damping; ~ **electromagnética** f ELEC *oscilación*, GEOFÍS electromagnetic damping; ~ **sin elevación** f TRANSP height-off cushion; ~ **exponencial** f ELECTRÓN exponential decay; ~ **interna** f TRANSP AÉR internal damping; ~ **magnética** f ING ELÉC magnetic damping, TRANSP magnetic cushion; ~ **de la nutación** f TEC ESP nutation damper; ~ **de olas** f OCEAN wave damping; ~ **óptima** f ING ELÉC optimum damping; ~ **rígida por aire lateral** f TRANSP rigid sidewall air cushion

amortiguador m ACÚST, AUTO damper, CONST shock absorber, D&A pad, ELEC shock absorber, *antiinterferencia* suppressor choke, FERRO *vehículo*, ING ELÉC damper, ING MECÁ cushion, dashpot, damper, *de choques* buffer, MECÁ dashpot, snubber, shock absorber, buffer, PROD suppressor, QUÍMICA buffer, TEC ESP *dispositivos guiadores de ondas* attenuator, nutation damper, snubber, shock absorber, damper, TERMOTEC damper, TRANSP dashpot, VEH *componente* dashpot, *suspensión* damper, shock absorber; ~ **abovedado** m ING MECÁ dome pad; ~ **de acoplamiento** m TRANSP AÉR coupling buffer; ~ **de aire** m ING MECÁ, TRANSP, TRANSP AÉR air cushion, air dashpot; ~ **de aire con costados rígidos** m TRANSP rigid sidewall air cushion; ~ **de aire inverso** m TRANSP negative air cushion; ~ **de arrastre** m TRANSP AÉR drag damper; ~ **del cabeceo** m TRANSP AÉR pitch damper; ~ **de caucho** m ING MECÁ, P&C rubber buffer; ~ **de chispas** m ELEC *relé* spark absorber, spark arrester, spark-quencher; ~ **de choques** m FERRO *vehículos* shock-absorbing body; ~ **de control** m TRANSP AÉR control damper; ~ **delantero** m AUTO front shock-absorber; ~ **de deslizamiento** m TERMOTEC slide damper; ~ **electrodinámico pasivo** m TEC ESP passive electrodynamic damper; ~ **electromagnético** m ELEC, GEOFÍS electromagnetic damper; ~ **esférico** m ING MECÁ ball pad; ~ **de frenado** m TRANSP AÉR drag damper; ~ **de fricción** m MECÁ, TRANSP AÉR friction damper; ~ **-guía** m TRANSP guidance cushion; ~ **hidráulico** m ING MECÁ oil dashpot; ~ **de impactos accidentales** m TERMOTEC hit-and-miss damper; ~ **de impactos aleatorios** m TERMOTEC hit-and-miss damper; ~ **de líquido viscoso** m MECÁ viscous damping; ~ **de luz** m ELEC *control* dimmer; ~ **manual** m TERMOTEC manual damper; ~ **de mariposa** m TERMOTEC butterfly damper; ~ **mecánico** m AUTO shock absorber; ~ **de múltiples ballestas** m ING MECÁ multiple-leaf damper; ~ **de neopreno** m CONST pad inside rubber boot; ~ **de olas** m OCEAN wave damper; ~ **de oscilaciones** m ELEC *corriente alterna, galvanómetro* oscillating damper, TRANSP AÉR shimmy damper; ~ **de resonancia** m AUTO resonance damper; ~ **de retroceso** m D&A recoil brake; ~ **de shimmy** m TRANSP AÉR shimmy damper; ~ **de sonido catalítico** m Esp (*cf mofle catalítico AmL, silenciador catalítico AmL*) CONTAM catalytic muffler (*AmE*), catalytic silencer (*BrE*); ~ **telescópico** m AUTO, VEH telescopic shock absor-

ber; ~ **de la válvula de mariposa** m AUTO, VEH *carburador* throttle dashpot; ~ **de vibraciones** m AUTO, ING MECÁ, MECÁ, REFRIG vibration damper, TRANSP AÉR shimmy damper, vibration damper, VEH *motor* vibration damper; ~ **de vibraciones y absorbedor de choque** m SEG vibration damper and shock absorber

amortiguamiento m ACÚST, ELEC, FÍS, INFORM&PD, PETROL damping, PROD cushioning, TEC ESP fading, TRANSP AÉR lagging, TV damping, decay; ~ **crítico** m ACÚST, PETROL, TEC PETR critical damping; ~ **electromagnético** m ELEC *oscilación*, GEOFÍS electromagnetic damping; ~ **magnético** m GEOFÍS magnetic damping

amortiguar vt ELEC *oscilación*, FÍS, ING ELÉC damp, ING MECÁ absorb, deaden, INSTAL HIDRÁUL cushion

amp: ~ **op funcionando como comparador** m ELECTRÓN operational amplifier comparator

amparo m CARBÓN prop

ampelita f PETROL ampelite

amperaje m ELEC, ING ELÉC amperage, PROD *de fusibles* size

ampere m (*A*) ELEC, ELECTRÓN, FÍS, ING ELÉC, METR, QUÍMICA ampere (*A*); ~ **-vuelta** m ELEC *devanado* ampere turn

amperímetro m GEN ammeter; ~ **de alambre caliente** m ELEC, ING ELÉC hot-wire ammeter; ~ **analógico** m ELEC, ING ELÉC, INSTR analog ammeter; ~ **astático** m ELEC, ING ELÉC, INSTR astatic ammeter; ~ **CA** m ELEC, ING ELÉC AC ammeter; ~ **de CC** m ELEC, ING ELÉC DC ammeter; ~ **de cuadro móvil** m ELEC, ING ELÉC moving-coil ammeter; ~ **diferencial** m ELEC, ING ELÉC, INSTR differential ammeter; ~ **digital** m ELEC, ING ELÉC, INSTR digital ammeter; ~ **de hilo caliente** m ELEC, ING ELÉC, INSTR thermal ammeter; ~ **de núcleo de hierro** m ELEC, ING ELÉC, INSTR iron core ammeter; ~ **térmico** m ELEC, ING ELÉC, INSTR hot-wire ammeter, thermal ammeter, thermoammeter; ~ **termoeléctrico** m ELEC, ING ELÉC, INSTR thermoammeter; ~ **de tipo empotrable** m PROD clamp-on type amp probe

amperio m (*A*) ELEC, ELECTRÓN, FÍS, ING ELÉC, METR, QUÍMICA ampere (*A*); ~ **-conductor** m ELEC ampere conductor; ~ **-hora** m ELEC *cantidad*, FÍS ampere-hour; ~ **-segundo** m ELEC, FÍS *cantidad* ampere-second; ~ **-vuelta** m ELEC *devanado*, FÍS ampere turn

ampervuelta m ELEC, FÍS ampere turn

ampliación f CINEMAT, FOTO, IMPR blowup, enlargement, enlarging, INFORM&PD zoom, zoom-in, ING MECÁ development, INSTR enlarging, *fotografía* enlargement; ~ **por acercamiento de la lente** f INFORM&PD zoom-in; ~ **con bajo nivel de ruidos** f ELECTRÓN low-noise amplification; ~ **fotográfica** f FOTO, INSTR magnifier enlargement; ~ **del hardware** f INFORM&PD hardware upgrade; ~ **óptica** f CINEMAT optical enlargement; ~ **de un solo cristal** f ELECTRÓN single-crystal growth; ~ **-reducción** f INFORM&PD zooming

ampliador: ~ **óptico** m INSTR, METR comparator

ampliadora f CINEMAT, FOTO, IMPR, INSTR enlarger, enlarger camera, enlarging camera; ~ **automática** f FOTO, IMPR automatic enlarger; ~ **fotográfica** f FOTO, IMPR enlarging camera

ampliar vt CINEMAT blow up, enlarge, FÍS, FÍS ONDAS amplify, FOTO, IMPR blow up, enlarge, TELECOM, TRANSP MAR amplify

amplidina: ~ **amplificador magnético rotativo** *f* ING ELÉC amplidyne

amplificación[1]: **de** ~ *adj* ELEC *intensidad*, ELECTRÓN, FÍS, FÍS ONDAS, INFORM&PD, TELECOM, TRANSP MAR amplifying

amplificación[2] *f* ELEC *señal* amplification, gain, ELECTRÓN gain, amplification, ING ELÉC gain, INSTR *óptica* magnifying, magnification, TEC ESP gain; ~ **alineal** *f* ELECTRÓN, TELECOM nonlinear amplification; ~ **de alta frecuencia** *f* ELECTRÓN high-frequency amplification; ~ **de baja frecuencia** *f* ELECTRÓN low-frequency amplification; ~ **de bajo nivel** *f* ELECTRÓN low-level amplification; ~ **de banda ancha** *f* ELECTRÓN wideband amplification; ~ **de corriente** *f* ELEC, ELECTRÓN current amplification; ~ **elemental** *f* METAL ultimate magnification; ~ **factorial** *f* INSTR, ÓPT factorial magnification; ~ **de FI** *f* ELECTRÓN, TELECOM IF amplification; ~ **final** *f* ELECTRÓN final amplification; ~ **de fotones** *f* FÍS PART, FÍS RAD photon amplification; ~ **fotónica** *f* FÍS PART, FÍS RAD photon amplification; ~ **de frecuencia intermedia** *f* ELECTRÓN, TELECOM intermediate-frequency amplification; ~ **de impulso** *f* ELECTRÓN pulse amplification; ~ **lateral** *f* FÍS, INSTR lateral magnification; ~ **lineal** *f* ELECTRÓN, TELECOM linear amplification; ~ **longitudinal** *f* INSTR longitudinal magnification; ~ **de la luz por estímulo en la emisión de radiaciones** *f* (*LASER*) ELECTRÓN, FÍS ONDAS, FÍS RAD light amplification by stimulated emission of radiation (*LASER*); ~ **máxima** *f* ELECTRÓN singing point; ~ **de microondas** *f* ELECTRÓN microwave amplification; ~ **no lineal** *f* ELECTRÓN, TELECOM nonlinear amplification; ~ **de onda** *f* ELECTRÓN, FÍS ONDAS TELECOM wave amplification; ~ **de onda milimétrica** *f* ELECTRÓN, FÍS ONDAS, TELECOM millimeter-wave amplification (*AmE*), millimetre-wave amplification (*BrE*); ~ **paramétrica** *f* ELECTRÓN, FÍSICA, TEC ESP, TELECOM parametric amplification; ~ **de pequeña señal** *f* ELECTRÓN small-signal amplification; ~ **de potencia** *f* ELECTRÓN power amplification; ~ **de potencia de microondas** *f* ELECTRÓN microwave-power amplification; ~ **de pulsos** *f* ELECTRÓN pulse amplification; ~ **reactiva** *f* ELECTRÓN regenerative amplification; ~ **regenerativa** *f* ELECTRÓN regenerative amplification; ~ **de RF** *f* ELECTRÓN, ING ELÉC, TELECOM RF amplification; ~ **con transistor** *f* ELECTRÓN transistor amplification; ~ **de tubo de vacío** *f* ELECTRÓN vacuum-tube amplification; ~ **del voltaje** *f* ELECTRÓN voltage amplification, ING ELÉC voltage gain

amplificado *adj* FÍS ONDAS amplifying; **no** ~ *adj* TELECOM *circuito* unamplified

amplificador[1] *adj* ELEC *intensidad* amplifying, INSTR *óptica* magnifying

amplificador[2] *m* ELEC, ELECTRÓN amplifier, voltage booster, FÍS, FÍS ONDAS, INFORM&PD amplifier, INSTR *óptica* magnifier, TEC ESP repeater, TELECOM, TRANSP MAR amplifier;

a ~ **de acción directa** *m* ELECTRÓN forward amplifier; ~ **de acoplo directo** *m* ELECTRÓN direct-coupled amplifier; ~ **de adaptación** *m* ELECTRÓN matching amplifier; ~ **de aislamiento** *m* ELECTRÓN isolation amplifier; ~ **de alimentación** *m* ELECTRÓN input amplifier; ~ **alineal** *m* ELECTRÓN, TELECOM nonlinear amplifier; ~ **de alta frecuencia** *m* ELECTRÓN high-frequency amplifier; ~ **de alta ganancia** *m* ELECTRÓN high-gain amplifier; ~ **de alta potencia** *m* ELECTRÓN, TEC ESP high-power amplifier; ~ **de audio** *m* ACÚST, ELECTRÓN audio amplifier, audio power amplifier; ~ **de audio frecuencia** *m* ACÚST, ELECTRÓN audio-frequency amplifier;

b ~ **de baja frecuencia** *m* ELECTRÓN low-frequency amplifier; ~ **de bajas frecuencias** *m* ELECTRÓN bass boost; ~ **de bajo nivel** *m* ELECTRÓN low-level amplifier; ~ **con bajo nivel de ruidos** *m* ELECTRÓN, TEC ESP, TELECOM low-noise amplifier (*LNA*); ~ **de banda ancha** *m* ELECTRÓN broadband amplifier, wideband amplifier; ~ **de banda estrecha** *m* ELECTRÓN narrow-band amplifier; ~ **en base común** *m* ELECTRÓN *transistores* common-base amplifier; ~ **básico** *m* ELECTRÓN basic amplifier; ~ **bilateral** *m* ELECTRÓN bilateral amplifier; ~ **bipolar** *m* ELECTRÓN bipolar amplifier; ~ **de bolsillo por desdoblamiento** *m* INSTR folding pocket-magnifier;

c ~ **de CA** *m* ELECTRÓN AC amplifier; ~ **de campo cruzado** *m* ELECTRÓN, FÍS, TELECOM crossed-field amplifier; ~ **de canales** *m* TV channel amplifier; ~ **de carga** *m* ELECTRÓN charge amplifier; ~ **de cascada** *m* ELECTRÓN cascade amplifier; ~ **del casco con auriculares** *m* INSTR headset magnifier; ~ **catódico** *m* ELECTRÓN, FÍS, ING ELÉC cathode amplifier; ~ **en chip** *m* ELECTRÓN on-chip amplification; ~ **de clase A** *m* ELECTRÓN, FÍS class A amplifier; ~ **de colector común** *m* ELECTRÓN common-collector amplifier; ~ **comercial** *m* ELECTRÓN commercial amplifier; ~ **compensado** *m* ELECTRÓN balanced amplifier, compensated amplifier; ~ **compensador** *m* ELECTRÓN buffer amplifier; ~ **compensador de potencia de entrada** *m* ELECTRÓN input buffer amplifier; ~ **en contrafase** *m* ELEC, ELECTRÓN, TELECOM push-pull amplifier; ~ **de corrección** *m* ELECTRÓN shaping amplifier; ~ **de corriente** *m* ELEC, ELECTRÓN current amplifier; ~ **de corriente continua** *m* ELECTRÓN DC amplifier; ~ **de corriente continua acoplada** *m* ELECTRÓN DC-coupled amplifier; ~ **cromeado** *m* ELECTRÓN chrominance amplifier; ~ **cuádruple** *m* ELECTRÓN quad operational amplifier; ~ **de cuatro canales** *m* ELECTRÓN four-channel amplifier;

d ~ **de datos** *m* ELECTRÓN data amplifier; ~ **DCME** *m* TELECOM DCME gain; ~ **de desconexión periódica** *m* PROD gate amplifier; ~ **desfasador** *m* ELECTRÓN, TV paraphase amplifier; ~ **de desviación** *m* ELECTRÓN deflection amplifier; ~ **de detección** *m* ELECTRÓN sense amplifier; ~ **diferencial** *m* ELECTRÓN, FÍS, TELECOM differential amplifier; ~ **de diodo** *m* ELECTRÓN diode amplifier; ~ **por diodo de efecto túnel** *m* TEC ESP tunnel-diode amplifier; ~ **directamente acoplado** *m* ELECTRÓN direct-coupled amplifier; ~ **discontinuo** *m* ELECTRÓN discontinuous amplifier; ~ **discreto** *m* ELECTRÓN discrete amplifier; ~ **con dispositivos semiconductores** *m* ELECTRÓN, TELECOM semiconductor amplifier; ~ **de distancias** *m* INSTR ranging magnifier; ~ **de distribución** *m* ELECTRÓN, TELECOM, TV distribution amplifier; ~ **de distribución de impulsos** *m* ELECTRÓN, TELECOM pulse-distribution amplifier (*PDA*); ~ **de distribución de video** *m* *AmL*, ~ **de distribución de vídeo** *m* *Esp* TV video distribution amplifier (*VDA*); ~ **de doble sintonización** *m* ELECTRÓN double-tuned amplifier;

~ en drenaje común *m* ELECTRÓN *transistores* common-drain amplifier;
■ e **~ de ecualización** *m* ELECTRÓN, TV equalizing amplifier; **~ de efecto de campo** *m* ELECTRÓN, TELECOM field-effect amplifier; **~ electrométrico** *m* ELEC, ELECTRÓN, ING ELÉC electrometer amplifier; **~ de emisor común** *m* ELECTRÓN common-emitter amplifier; **~ de enfoque** *m* INSTR focusing magnifier; **~ estabilizado de interrupción** *m* ELECTRÓN chopper-stabilized amplifier; **~ estabilizado de muestreo** *m* ELECTRÓN chopper-stabilized amplifier; **~ de estado sólido** *m* ELECTRÓN, TELECOM solid-state amplifier; **~ estellante** *m* ELECTRÓN, TV burst amplifier; **~ exponencial** *m* ELECTRÓN exponential amplifier;
■ f **~ ferromagnético** *m* ELECTRÓN ferromagnetic amplifier; **~ de FI** *m* ELECTRÓN, TELECOM IF amplifier; **~ de FI secundaria** *m* ELECTRÓN, TELECOM second IF amplifier; **~ de filtro** *m* ELECTRÓN filter amplifier; **~ final** *m* ELECTRÓN final amplifier; **~ final de etapa** *m* ELECTRÓN final amplifier; **~ fotoeléctrico** *m* ELECTRÓN photoelectric amplifier; **~ de frecuencia intermedia** *m* ELECTRÓN, TELECOM, TRANSP MAR *radar* intermediate-frequency amplifier; **~ de frecuencia de sonido** *m* ELECTRÓN audio-frequency amplifier; **~ en fuente común** *m* ELECTRÓN common-source amplifier;
■ g **~ de ganancia unidad** *m* ELECTRÓN unity-gain amplifier; **~ de ganancia variable** '*m* ELECTRÓN variable-gain amplifier; **~ de gema** *m* INSTR gem magnifier; **~ giratorio** *m* ELECTRÓN, ING ELÉC rotary amplifier; **~ giroscópico** *m* TRANSP AÉR gyro amplifier; **~ de las guías de ondas progresivas** *m* ELECTRÓN, TELECOM traveling waveguide amplifier (*AmE*), travelling waveguide amplifier (*BrE*); **~ Gunn** *m* ELECTRÓN, FÍS Gunn amplifier;
■ h **~ de haz lineal** *m* ELECTRÓN linear-beam amplifier; **~ horizontal** *m* ELECTRÓN horizontal amplifier;
■ i **~ de imagen** *m* ELECTRÓN, TEC ESP head amplifier; **~ de imagen de rayos X** *m* INSTR X-ray image amplifier; **~ de impulso lineal** *m* ELECTRÓN linear-pulse amplifier; **~ de impulsos** *m* TELECOM pulse amplifier; **~ industrial** *m* ELECTRÓN commercial amplifier; **~ de instrumentación** *m* ELECTRÓN instrumentation amplifier; **~ instrumental** *m* ELECTRÓN instrumentation amplifier; **~ integrador** *m* TRANSP AÉR integrator amplifier; **~ de interrupción** *m* ELECTRÓN chopper amplifier; **~ de inversión** *m* ELECTRÓN inverting amplifier; **~ inversor** *m* ELECTRÓN inverting amplifier;
■ k **~ klistron** *m* ELECTRÓN klystron amplifier;
■ l **~ de lámpara de vacío** *m* ELECTRÓN vacuum-tube amplifier; **~ de lectura** *m* ELECTRÓN read amplifier; **~ de limitación** *m* ELECTRÓN clipper amplifier, limiting amplifier; **~ limitador de volumen** *m* ELECTRÓN volume-limiting amplifier; **~ de línea** *m* ELECTRÓN line amplifier; **~ lineal** *m* ELECTRÓN, TELECOM linear amplifier; **~ lineal de potencia** *m* ELECTRÓN, TELECOM linear power amplifier; **~ logarítmico** *m* ELECTRÓN logarithmic amplifier; **~ logarítmico de imagen** *m* ELECTRÓN logarithmic video amplifier; **~ de luminancia** *m* ELECTRÓN luminance amplifier; **~ de luz** *m* ELECTRÓN light amplifier;
■ m **~ magnético** *m* ELECTRÓN, FÍS, ING ELÉC, TEC ESP magnetic amplifier (*magamp*); **~ de magnetrón**

m ELECTRÓN magnetron amplifier; **~ de matrizado** *m* ELECTRÓN, TV matting amplifier; **~ de medición** *m* ELECTRÓN measuring amplifier; **~ de mezcla** *m* ELECTRÓN mixing amplifier; **~-mezclador** *m* TV mixer-amplifier; **~ de microondas** *m* ELECTRÓN, FÍS, TELECOM microwave amplifier; **~ de microondas por emisión estimulada de radiación** *m* (*MASER*) ELECTRÓN, TEC ESP microwave amplification by stimulated emission of radiation (*MASER*); **~ de microondas transistorizado** *m* ELECTRÓN microwave transistor amplifier; **~ de modulación** *m* ELECTRÓN modulation amplifier; **~ monolítico** *m* ELECTRÓN monolithic amplifier; **~ de muestreo** *m* ELECTRÓN sampling amplifier, chopper amplifier; **~ multicanal** *m* ELECTRÓN multichannel amplifier; **~ multigradual** *m* ELECTRÓN multistage amplifier;
■ n **~ neutralizado** *m* ELECTRÓN neutralized amplifier; **~ de nivelación** *m* ELECTRÓN leveling amplifier (*AmE*), levelling amplifier (*BrE*);
■ o **~ de onda milimétrica** *m* ELECTRÓN millimeter-wave amplifier (*AmE*), millimetre-wave amplifier (*BrE*); **~ de ondas progresivas** *m* ELECTRÓN, TELECOM traveling-wave amplifier (*AmE*), travelling-wave amplifier (*BrE*); **~ operacional** *m* (*AO*) ELECTRÓN, FÍS, INFORM&PD, TELECOM operational amplifier (*op amp*); **~ óptico** *m* ELECTRÓN, ÓPT, TELECOM optical amplifier; **~ óptico-electrónico** *m* ELECTRÓN, ÓPT, TELECOM optoelectronic amplifier; **~ optoelectrónico de impulsos** *m* ELECTRÓN, ÓPT, TELECOM optoelectronic pulse amplifier (*OPA*);
■ p **~ paramagnético** *m* ELECTRÓN paramagnetic amplifier; **~ paramétrico** *m* ELECTRÓN, FÍS, TEC ESP, TELECOM parametric amplifier (*paramp*); **~ paramétrico del haz de electrones** *m* ELECTRÓN electron-beam parametric amplifier; **~ de partición de fase** *m* ELECTRÓN phase-splitter amplifier; **~ pasabanda** *m* ELECTRÓN band-pass amplifier; **~ de paso de banda** *m* ELECTRÓN band-pass amplifier; **~ de pequeña señal** *m* ELECTRÓN small-signal amplifier; **~ plano** *m* ELECTRÓN flat amplifier; **~ de poca ganancia** *m* ELECTRÓN low-gain amplifier, small-gain amplifier; **~ de la portadora** *m* ELECTRÓN carrier amplifier; **~ de potencia** *m* ELECTRÓN, FÍS, ING ELÉC, PROD, TEC ESP, TELECOM power amplifier; **~ de potencia de alta ganancia** *m* ELECTRÓN high-gain power amplifier; **~ de potencia de banda ancha** *m* ELECTRÓN wide-band power amplifier; **~ de potencia lineal** *m* ELECTRÓN, TELECOM linear power amplifier; **~ de potencia media** *m* ELECTRÓN medium-power amplifier; **~ de potencia de microondas** *m* ELECTRÓN microwave power amplifier; **~ de potencia de sonido** *m* ELECTRÓN audio power amplifier; **~ de potencia de transistor** *m* ELECTRÓN transistor power amplifier; **~ previo** *m* TEC ESP head amplifier; **~ de primera FI** *m* ELECTRÓN first IF amplifier; **~ de puente** *m* ELECTRÓN bridge amplifier; **~ en puerta común** *m* ELECTRÓN *transistores* common-gate amplifier; **~ en push-pull** *m* ELEC, ELECTRÓN, TELECOM push-pull amplifier;
■ r **~ de ráfagas** *m* ELECTRÓN burst amplifier; **~ para RCS** *m* ELECTRÓN SCR amplifier; **~ de reacción** *m* ELECTRÓN regenerative amplifier; **~ de realimentación selectiva** *m* ELECTRÓN selective feedback amplifier; **~ realimentado** *m* ELECTRÓN feedback amplifier; **~ regenerativo** *m* ELECTRÓN regenerative amplifier; **~ remoto** *m* ELECTRÓN

remote amplifier; ~ **de resistencia negativa** *m* ELEC, ELECTRÓN, ING ELÉC negative-resistance amplifier; ~ **de retroacción** *m* ELECTRÓN feedback amplifier; ~ **de RF** *m* ELECTRÓN, ING ELÉC, TELECOM RF amplifier;

~ s ~ **de salida** *m* ELECTRÓN, TV output amplifier; ~ **selectivo de frecuencia** *m* ELECTRÓN frequency-selective amplifier; ~ **de semiconductores** *m* TELECOM semiconductor amplifier; ~ **de la señal de sincronismo de color** *m* ELECTRÓN, TV burst amplifier; ~ **simétrico** *m* ELEC, ELECTRÓN, TELECOM push-pull amplifier; ~ **sincrónico** *m* ELECTRÓN, TV sync amplifier; ~ **sincronizado** *m* ELECTRÓN lock-in amplifier; ~ **del sincronizador de error de rumbo** *m* TRANSP AÉR heading error synchronizer amplifier; ~ **de sintonía escalonada** *m* ELECTRÓN stagger-tuned amplifier; ~ **sintonizado** *m* ELECTRÓN tuned amplifier; ~ **del sistema público de altavoces** *m* ACÚST, ELECTRÓN, TRANSP AÉR public address system amplifier; ~ **de una sola etapa** *m* ELEC, ELECTRÓN single-stage amplifier; ~ **de una sola fase** *m* ELEC, ELECTRÓN single-stage amplifier; ~ **de un solo canal** *m* ELECTRÓN single-channel amplifier;

~ t ~ **de TEC** *m* ELECTRÓN, TEC ESP FET amplifier; ~ **telescópico** *m* INSTR telescope magnifier; ~ **de TOP** *m* ELECTRÓN TWT amplifier; ~ **sin transformador de salida** *m* ELECTRÓN single-ended amplifier; ~ **con transistor** *m* ELECTRÓN transistor amplifier; ~ **del transistor de efecto de campo** *m* ELECTRÓN, FÍS, ÓPT, TEC ESP field-effect transistor amplifier; ~ **de tres etapas** *m* ELECTRÓN three-stage amplifier; ~ **de tubo de ondas progresivas** *m* (*ATOP*) ELECTRÓN, TEC ESP, TELECOM traveling-wave tube amplifier (*AmE*) (*TWTA*), travelling-wave tube amplifier (*BrE*) (*TWTA*); ~ **de tubo de vacío** *m* ELECTRÓN vacuum-tube amplifier;

~ v ~ **de velocidad modulada** *m* ELECTRÓN velocity-modulated amplifier; ~ **vertical** *m* ELECTRÓN vertical amplifier; ~ **vertical de muestreo** *m* ELECTRÓN sampling vertical amplifier; ~ **de video** *m* AmL, ~ **de vídeo** *m* Esp FÍS, TV video amplifier; ~ **de voltaje** *m* AUTO power booster, ELECTRÓN voltage amplifier; ~ **de voz** *m* TELECOM voice amplifier

amplificar *vt* ELECTRÓN amplify, INFORM&PD zoom in, TEC ESP *radiocomunicaciones* boost

amplitrón *m* ELECTRÓN, FÍS *tubo de microondas* amplitron

amplitud *f* ACÚST, ELEC, ELECTRÓN, ENERG RENOV, FÍS, FÍS ONDAS amplitude, INFORM&PD amplitude, width, MINAS extent, REFRIG *de la corriente de aire* spread, TEXTIL scope, TRANSP MAR *astronomía* amplitude, *marea* range; ~ **angular del haz** *f* ÓPT beamwidth; ~ **de banda de video** *f* AmL, ~ **de banda de vídeo** *f* Esp INFORM&PD, TV video bandwidth; ~ **de cresta a cresta** *f* ELECTRÓN peak-to-peak amplitude; ~ **de un cuarto de onda** *f* ELECTRÓN quarter wavelength; ~ **definitiva** *f* ELECTRÓN fixed amplitude; **Doppler** *f* FÍS RAD Doppler width; ~ **de enfriamiento** *f* REFRIG *líquidos* cooling range; ~ **de la escala** *f* METR scale range; ~ **de filtro de banda de porcentaje constante** *f* ELECTRÓN constant percentage band width filter; ~ **del gluten** *f* ALIMENT *repostería, panadería* gluten extensibility; ~ **del haz** *f* ELECTRÓN beam width; ~ **de imagen** *f* TELECOM *TV* picture amplitude; ~ **de impulso** *f* ELECTRÓN pulse ampli-

tude; ~ **de impulso comprimido** *f* ELECTRÓN compressed-pulse width; ~ **de mareas** *f* ENERG RENOV, GEOFÍS, HIDROL, TRANSP MAR tidal range; ~ **máxima** *f* ELECTRÓN peak amplitude; ~ **máxima de impulso** *f* ELECTRÓN peak pulse amplitude; ~ **máxima de señal** *f* ELECTRÓN peak signal amplitude; ~ **media de las mareas** *f* ENERG RENOV, GEOFÍS, HIDROL, TRANSP MAR mean tidal range; ~ **de modulación** *f* ELECTRÓN, FÍS modulation depth; ~ **de onda** *f* ENERG RENOV, FÍS, OCEAN wave amplitude; ~ **de la órbita atómica** *f* FÍS, FÍS PART, QUÍMICA atomic orbit width; ~ **de oscilación total** *f* ACÚST total oscillation amplitude; ~ **de presión** *f* ACÚST pressure amplitude; ~ **del pulso** *f* ELECTRÓN pulse amplitude, pulse height; ~ **reducida a la mitad** *f* ELECTRÓN half-power width; ~ **de señal** *f* ELECTRÓN signal amplitude; ~ **de señal de máxima** *f* ELECTRÓN maximum-signal amplitude; ~ **de señal relativa** *f* ELECTRÓN relative signal amplitude; ~ **de señales pico a pico** *f* TV peak-to-peak signal amplitude; ~ **vertical** *f* TV vertical amplitude

ampolla *f* C&V ampoule (*BrE*), ampule (*AmE*), *defecto* blister, EMB, LAB ampoule (*BrE*), ampule (*AmE*), P&C blub, *defecto* blister, TRANSP MAR *pintura* blister; ~ **hermética** *f* ING ELÉC plunger; ~ **de superficie** *f* C&V skin blister

ampolleta *f* C&V phial (*BrE*), vial (*AmE*), ELEC *luz* bulb, LAB, QUÍMICA phial (*BrE*), vial (*AmE*); ~ **para penicilina** *f* C&V penicillin phial (*BrE*), penicillin vial (*AmE*)

amura *f* TRANSP MAR *cabo* bow, luff, tack

amurada *f* TRANSP MAR *buque* bulwark

An *abr* (*actinón*) FÍS RAD, QUÍMICA An (*actinon*)

anabiosis *f* AGRIC anabiosis

añadido: ~ **de escoria porosa** *m* TERMOTEC foamed slag aggregate; ~ **en montaje** *m* CINEMAT, TV edit-in

añadir *vt* CONST tail, INFORM&PD add; ~ **abonos a** *vt* AGRIC fertilize; ~ **un aditivo a** *vt* FÍS dope

anaeróbico *adj* ALIMENT, HIDROL, OCEAN, P&C, RECICL anaerobic

anaerobio *m* ALIMENT, HIDROL, OCEAN, P&C, RECICL anaerobe (*BrE*), anaerobium (*AmE*)

anaerobiosis *f* RECICL anaerobiosis

anafrente *m* METEO anafront

añagaza *f* OCEAN lure

anaglifo *m* REVEST anaglypta

análisis *m* GEN analysis, TEC ESP review, scanning; ~ **de absorción atómica** *m* FÍS RAD atomic absorption analysis; ~ **de absorción gamma** *m* FÍS RAD gamma-ray absorption analysis; ~ **de absorción de partículas beta** *m* FÍS RAD beta-particle absorption analysis; ~ **de la absorción de radiaciones** *m* FÍS RAD, NUCL radiation-absorption analysis; ~ **de absorción de rayos gamma** *m* FÍS RAD gamma-ray absorption analysis; ~ **de absorción de rayos X** *m* FÍS RAD X-ray absorption analysis; ~ **de accidentes** *m* NUCL accident analysis; ~ **de activación** *m* FÍS activation analysis; ~ **de activación gamma** *m* FÍS, FÍS RAD gamma-ray activation analysis; ~ **de activación por partículas cargadas** *m* FÍS PART, FÍS RAD charged-particle activation analysis; ~ **de agua de alcantarillas** *m* AGUA sewage analysis; ~ **de aguas negras** *m* AGUA sewage analysis; ~ **armónico** *m* ELECTRÓN, FÍS, MECÁ, TELECOM harmonic analysis; ~ **en campo lejano** *m* ÓPT, TELECOM far-field analysis; ~ **de campo próximo** *m* ÓPT, TELECOM near-field analysis; ~ **por carbono 14** *m* FÍS RAD, GEOL carbon-14 analysis; ~ **de carga de la red** *m* INFORM&PD network load analysis; ~ **del circuito** *m* ING ELÉC circuit analysis; ~ **y codificación de la palabra** *m* TELECOM vocoding; ~ **de color** *m* CINEMAT, FOTO color analysis (*AmE*), colour analysis (*BrE*); ~ **de colores** *m* TV color analysis (*AmE*), colour analysis (*BrE*); ~ **combinatorio** *m* MATEMÁT combinatorial analysis; ~ **de combustión** *m* TERMO combustion analysis; ~ **por correlación de tiempo** *m* NUCL time-correlation analysis; ~ **de correspondencia** *m* MATEMÁT correspondence analysis; ~ **críptico** *m* INFORM&PD, TELECOM cryptanalysis; ~ **de criticidad** *m* CALIDAD criticality analysis; ~ **de crudo** *m* PETROL *petróleo* crude assay, TEC PETR *evaluación de la formación* crude-oil analysis, *petróleo* crude assay; ~ **cualitativo** *m* AGUA, QUÍMICA qualitative analysis; ~ **cuantitativo** *m* AGUA, QUÍMICA quantitative analysis; ~ **de datos de tiempo de vuelo** *m* FÍS RAD, TRANSP AÉR time-of-flight data analysis (*TOF data analysis*); ~ **de defectos** *m* FÍS RAD defect analysis; ~ **por difracción de rayos X** *m* FÍS RAD X-ray diffraction analysis; ~ **dispersivo de energía por rayos X** *m* QUÍMICA energy dispersive analysis by X-rays (*EDAX*); ~ **de distribución de intervalos de salida** *m* TRANSP headway distribution analysis; ~ **electrográfico** *m* ELEC, FÍS RAD electrographic analysis; ~ **elemental** *m* GEOL ultimate analysis; ~ **de errores** *m* INFORM&PD error analysis; ~ **de esfuerzo y deformación** *m* ING MECÁ stress and strain analysis; ~ **de esfuerzo y fatiga** *m* ING MECÁ stress and strain analysis; ~ **espacial** *m* CONTAM spatial resolution; ~ **espectral** *m* ELECTRÓN, FÍS, INFORM&PD, TELECOM spectral analysis, spectrum analysis; ~ **espectral óptico** *m* ELECTRÓN, FÍS, INFORM&PD, ÓPT, TELECOM optical spectral analysis; ~ **espectral en tiempo real** *m* ELECTRÓN, FÍS, INFORM&PD, TELECOM real-time spectral analysis; ~ **del espectro masa** *m* FÍS, FÍS RAD, QUÍMICA mass spectrum analysis; ~ **del espectro sísmico** *m* GEOFÍS *prospección* seismic spectral analysis; ~ **espectrográfico** *m* FÍS, FÍS RAD spectrographic analysis; ~ **espectrométrico** *m* FÍS RAD spectrometric analysis; ~ **espectroscópico** *m* FÍS RAD spectrum analysis; ~ **estadístico** *m* INFORM&PD statistical analysis; ~ **estructural** *m* CONST structural analysis; ~ **factorial** *m* MATEMÁT factor analysis; ~ **de fenómenos peligrosos** *m* CALIDAD, SEG hazard analysis; ~ **de fiabilidad** *m* SEG, TEC ESP reliability analysis; ~ **por fluorescencia** *m* FÍS, FÍS RAD fluorescence analysis; ~ **por fluorescencia atómica** *m* FÍS, FÍS RAD atomic fluorescence analysis; ~ **por fluorescencia X** *m* FÍS, FÍS RAD X-ray fluorescence analysis; ~ **foliar** *m* AGRIC four-cylinder motorcycle; ~ **por fotoactivación** *m* FÍS RAD photoactivation analysis; ~ **de Fourier** *m* ELECTRÓN, FÍS, MATEMÁT Fourier analysis; ~ **funcional** *m* MATEMÁT functional analysis, TEC ESP reliability analysis; ~ **de gas** *m* GAS, TEC PETR gas analysis; ~ **granulométrico** *m* C&V, CARBÓN granulometric analysis (*BrE*), screen analysis (*AmE*), sieve analysis, GEOL granulometric analysis (*BrE*), screen analysis, sieve analysis, NUCL, PROC QUÍ granulometric analysis (*BrE*), screen analysis (*AmE*), sieve analysis; ~ **gravimétrico** *m* CARBÓN, GEOFÍS, INSTR, PROC QUÍ, TEC PETR *exploración, geofísica* gravimetric analysis; ~ **gravimétrico**

térmico *m* CARBÓN, GEOFÍS, PROC QUÍ, TEC PETR, TERMO thermal gravimetric analysis; ~ **de hipótesis** *m* PROD assumption analysis; ~ **de imagen** *m* TELECOM image analysis; ~ **inmediato** *m* GEOL, QUÍMICA proximate analysis; ~ **isotópico** *m* FÍS RAD isotopic analysis; ~ **léxico** *m* INFORM&PD lexical analysis; ~ **de lógica** *m* ELECTRÓN, INFORM&PD logic analysis; ~ **lógico** *m* ELECTRÓN, INFORM&PD logic analysis; ~ **lógico de estado** *m* ELECTRÓN logic state analysis; ~ **lógico de fase** *m* ELECTRÓN logic state analysis; ~ **lógico de temporización** *m* ELECTRÓN logic timing analysis; ~ **mecánico** *m* GEOL mechanical analysis; ~ **metalúrgico** *m* MINERAL, QUÍMICA mineral analysis; ~ **de movimientos** *m* CINEMAT motion analysis; ~ **de núcleos** *m* AmL (*cf análisis de testigos Esp*) TEC PETR core analysis; ~ **numérico** *m* INFORM&PD numerical analysis; ~ **de operación** *m* PROD operation analysis; ~ **paramétrico** *m* ELECTRÓN parametric analysis; ~ **petrofábrico** *m* GEOL petrofabric analysis; ~ **de portadores** *m* NUCL carrier analysis; ~ **de presión, volumen y temperatura** *m* PETROL PVT analysis; ~ **probabilista del riesgo** *m* (*APR*) NUCL probabilistic risk assessment (*PRA*); ~ **probabilista de la seguridad** *m* (*APS*) NUCL probabilistic safety assessment (*PSA*); ~ **de prueba** *m* QUÍMICA check analysis; ~ **químico** *m* CARBÓN chemical analysis; ~ **por radiactivación** *m* FÍS RAD radioactivation analysis; ~ **radiométrico** *m* FÍS RAD radiometric analysis; ~ **de rayos canales** *m* FÍS, NUCL canal-ray analysis; ~ **por rayos X** *m* FÍS RAD, INSTR X-ray analysis; ~ **de la red en alta frecuencia** *m* ELEC, ING ELÉC high-frequency network analysis; ~ **de red escalar** *m* ELEC, ING ELÉC scalar-network analysis; ~ **de red vectorial** *m* ELEC, ING ELÉC vector-network analysis; ~ **de redes** *m* ELEC, ING ELÉC network analysis; ~ **de retrodispersión de partículas beta** *m* FÍS RAD beta-particle backscattering analysis; ~ **de riesgo** *m* CALIDAD, SEG hazard analysis; ~ **de la roca intacta** *m* PETROL whole-rock analysis; ~ **del ruido** *m* ACÚST, PETROL noise analysis; ~ **rutinario** *m* CALIDAD *de datos cuantitativos*, INFORM&PD routine analysis; ~ **de sedimentación** *m* CARBÓN sedimentation analysis; ~ **semántico** *m* INFORM&PD semantic analysis; ~ **de señal digital** *m* ELECTRÓN digital signal analysis; ~ **de señales** *m* ELECTRÓN, FÍS RAD, TELECOM signal analysis; ~ **de sensibilidad** *m* PROD sensitivity analysis; ~ **de sincronismo** *m* INFORM&PD timing analysis; ~ **sintáctico** *m* INFORM&PD parsing, syntax analysis, TELECOM syntactic analysis; ~ **de sistemas** *m* INFORM&PD systems analysis; ~ **del tamaño del grano** *m* ING MECÁ grain-size analysis; ~ **por tamizado** *m* CARBÓN, PROC QUÍ sieve analysis; ~ **temporal** *m* CONTAM temporal resolution, ELECTRÓN time-series analysis; ~ **de temporización** *m* INFORM&PD timing analysis; ~ **de tensión** *m* TEC ESP stress analysis; ~ **térmico** *m* QUÍMICA, TERMO thermal analysis; ~ **térmico diferencial** *m* CONTAM, P&C, QUÍMICA, TERMO differential thermal analysis (*DTA*); ~ **termogravimétrico** *m* (*ATG*) GEOFÍS, LAB, TERMO thermogravimetric analysis (*TGA*); ~ **de testigos** *m* Esp (*cf análisis de núcleos AmL*) TEC PETR core analysis; ~ **de tiempo** *m* INFORM&PD timing analysis; ~ **en tiempo real** *m* ACÚST, ELECTRÓN, INFORM&PD, TELECOM real-time analysis; ~ **por titulación** *m* QUÍMICA titrimetry; ~ **de tráfico**

m TELECOM traffic analysis; ~ **transitorio** *m* ELEC-
TRÓN transient analysis; ~ **de trayectoria crítica** *m*
TEXTIL critical-path analysis; ~ **de trazas** *m* AGUA
trace analysis; ~ **del valor** *m* PROD value analysis;
~ **de varianza** *m* (*ANOVA*) INFORM&PD analysis of
variance (*ANOVA*); ~ **por vía húmeda** *m* CARBÓN
química chemical analysis, QUÍMICA wet assay; ~ **de**
vibración *m* MECÁ vibration analysis; ~ **del vidrio** *m*
C&V glass analysis; ~ **volumétrico** *m* CARBÓN,
DETERG, QUÍMICA titration, volumetric analysis;
~ **de voz** *m* TELECOM speech analysis
analista *m* GEN analyst
analítico *adj* GEN analytic, analytical
analizador *m* GEN analyser (*BrE*), analyzer (*AmE*),
scanner, TV prompter, scanner; ~ **armónico** *m*
ELECTRÓN, FÍS harmonic analyser (*BrE*), harmonic
analyzer (*AmE*); ~ **de color** *m* CINEMAT, FOTO, TV
color analyzer (*AmE*), colour analyser (*BrE*); ~ **de**
colores de video *m AmL*, ~ **de colores de vídeo** *m*
Esp CINEMAT video color analyzer (*AmE*), video
colour analyser (*BrE*); ~ **de E/S** *m* PROD I/O scanner;
~ **espectral óptico** *m* ELECTRÓN optical spectral
analyser (*AmE*), optical spectral analyzer (*AmE*),
ÓPT, TELECOM optical spectral analyser (*BrE*), optical
spectral analyzer (*AmE*); ~ **espectral en tiempo real**
m ELECTRÓN real-time spectral analyser (*BrE*), real-
time spectral analyzer (*AmE*), INFORM&PD real-time
spectral (*AmE*), real-time spectral analyser (*BrE*),
real-time spectral analyzer (*AmE*), TELECOM real-
time spectral analyser (*BrE*), real-time spectral
analyzer (*AmE*); ~ **de espectro** *m* ELECTRÓN, TELE-
COM spectrum analyser (*BrE*), spectrum analyzer
(*AmE*); ~ **del espectro de la muestra** *m* ELECTRÓN,
TELECOM sampling spectrum analyser (*BrE*), sampling spectrum analyzer (*AmE*); ~ **de gas** *m* GAS, TEC
PETR gas analyser (*BrE*), gas analyzer (*AmE*); ~ **de**
gases por ionización alfa *m* NUCL alpha ionization
gas analyser (*BrE*), alpha ionization gas analyzer
(*AmE*); ~ **de imagen** *m* TELECOM image analyser
(*BrE*), image analyzer (*AmE*); ~ **infrarrojo de gases**
de escape *m* AUTO infrared exhaust-gas analyser
(*BrE*), infrared exhaust-gas analyzer (*AmE*); ~ **de**
lógica *m* ELECTRÓN, INFORM&PD logic analyser
(*BrE*), logic analyzer (*AmE*); ~ **lógico** *m* ELECTRÓN,
INFORM&PD logic analyser (*BrE*), logic analyzer
(*AmE*); ~ **lógico de estado y temporización** *m*
ELECTRÓN logic state and timing analyser (*BrE*),
logic state and timing analyzer (*AmE*); ~ **multicanal**
m FÍS multichannel analyser (*BrE*), multichannel
analyzer (*AmE*); ~ **de onda** *m* ACÚST, ELECTRÓN
wave analyser (*BrE*), wave analyzer (*AmE*); ~ **de**
película a color *m* CINEMAT, FOTO, TV color film
analyzer (*AmE*), colour film analyser (*BrE*); ~ **de red**
escalar *m* ELEC, ING ELÉC scalar-network analyser
(*BrE*), scalar-network analyzer (*AmE*); ~ **de red**
vectorial *m* ELEC, ING ELÉC vector-network analyser
(*BrE*), vector-network analyzer (*AmE*); ~ **de redes**
m ELEC, INFORM&PD, ING ELÉC, TELECOM network
analyser (*BrE*), network analyzer (*AmE*); ~ **de redes**
eléctricas *m* ELEC, ING ELÉC network analyser (*BrE*),
network analyzer (*AmE*); ~ **de señal digital** *m* ELEC-
TRÓN digital signal analyser (*BrE*), digital signal
analyzer (*AmE*); ~ **de señales** *m* ELECTRÓN, FÍS RAD,
TELECOM signal analyser (*BrE*), signal analyzer
(*AmE*); ~ **de señales de petróleo** *m* TEC PETR
evaluación de la formación oil show analyser (*BrE*),

oil show analyzer (*AmE*); ~ **sintáctico** *m* INFORM&PD
syntax analyser (*BrE*), syntax analyzer (*AmE*), TELE-
COM syntactic analyser (*BrE*), syntactic analyzer
(*AmE*); ~ **sonoro** *m* ACÚST sound analyser (*BrE*),
sound analyzer (*AmE*); ~ **en tiempo real** *m* ACÚST,
ELECTRÓN, INFORM&PD, TELECOM real-time analyser
(*BrE*), real-time analyzer (*AmE*); ~ **del tráfico** *m*
TRANSP traffic analyser (*BrE*), traffic analyzer
(*AmE*); ~ **transitorio** *m* ELECTRÓN transient analyser
(*BrE*), transient analyzer (*AmE*); ~ **trocoidal de**
masas *m* NUCL trochoidal mass analyser (*BrE*),
trochoidal mass analyzer (*AmE*)
analizar *vt* GEN analyse (*BrE*), analyze (*AmE*);
~ **sintácticamente** *vt* INFORM&PD parse
analógico *adj* ELECTRÓN, FÍS, INFORM&PD, ING ELÉC,
PETROL, TELECOM analog; ~ **a digital** *adj* (*A-a-D*) TV
analog-to-digital (*A-to-D*)
anamesita *f* PETROL anamesite
anamórfico *adj* CINEMAT, FOTO, INSTR, TELECOM
anamorphic
anamorfosis *m* CINEMAT, FOTO, TELECOM anamor-
phosis
anaranjado: ~ **de metilo** *m* QUÍMICA *indicador*
helianthine
anatexia *f* GEOL anatexis
anavear *vi* TRANSP MAR land
anca *f* TRANSP MAR *construcción naval* buttocks
ancho *m* GEN width; ~ **de la abertura** *m* ING MECÁ
port width; ~ **acabado de la tela** *m* TEXTIL finished
width of cloth; ~ **del arco** *m* CINEMAT arc width; ~ **de**
banda *m AmL*(*cf anchura debanda Esp*) ACÚST,
ELECTRÓN, FÍS, FÍS RAD, INFORM&PD, ÓPT, TEC ESP,
TELECOM, TV bandwidth; ~ **de banda de amplifica-**
dor vertical *m* ELECTRÓN vertical amplifier
bandwidth; ~ **de banda del canal** *m* TV channel
bandwidth; ~ **de banda de coherencia** *m* TELECOM
coherence bandwidth; ~ **de banda de crominancia**
m TV chrominance bandwidth; ~ **de banda Doppler**
m ELECTRÓN Doppler bandwidth; ~ **de banda**
espectral *m* TELECOM spectral bandwidth; ~ **de**
banda con láser *m* ELECTRÓN laser bandwidth;
~ **de banda de la portadora** *m* ELECTRÓN carrier
bandwidth; ~ **de banda según la regla de Carson** *m*
TEC ESP Carson's rule bandwidth; ~ **de banda de**
señal *m* ELECTRÓN signal bandwidth; ~ **de banda**
con señal ampliada *m* ELECTRÓN large-signal bandwidth; ~ **de banda de video** *m AmL*, ~ **de banda de**
vídeo *m Esp* INFORM&PD, TV video bandwidth; ~ **de**
la bobina de reactancia *m* TV width choke; ~ **efec-**
tivo de la línea de sonido *m* TV effective slit-width;
~ **del impulso** *m* TELECOM pulse width; ~ **de**
intervalo *m* TV gap width; ~ **de línea espectral** *m*
ÓPT spectral line width; ~ **máximo de fabricación** *m*
PAPEL machine fill; ~ **máximo de máquina tras**
cortar *m* EMB maximum trimmed-machine width;
~ **normal de vía** *m* FERRO standard gage (*AmE*),
standard gauge (*BrE*); ~ **en el peine** *m* TEXTIL width
in reed; ~ **plano y ajustado** *m* TEXTIL flat-and-fitted
width; ~ **del pulso** *m* TV pulse width; ~ **de raya**
espectral *m* TELECOM spectral-line width; ~ **del**
rodillo *m* PAPEL face roll; ~ **de tabla** *m* PAPEL *rodillo,*
cilindro face roll; ~ **de la tela** *m* TEXTIL fabric width;
~ **de tela utilizado** *m* PAPEL machine deckle; ~ **total**
m ING MECÁ overall width; ~ **de la urdimbre**
encolado *m* TEXTIL dressed width of warp; ~ **útil**
de la máquina *m* PAPEL untrimmed machine width;

~ **útil máximo de máquina** *m* PAPEL maximum trimmed-machine width; ~ **útil de la tela** *m* PAPEL maximum deckle; ~ **de vía** *m Esp* (*cf gramil AmL*) AUTO track width, CONST, FERRO *equipo inamovible* gage (*AmE*), gauge (*BrE*)

anchura *f* IMPR, INFORM&PD, PAPEL breadth, width; ~ **de banda** *f Esp* (*cf ancho de banda AmL*) ACÚST, ELECTRÓN, FÍS, FÍS RAD, INFORM&PD, ÓPT, TEC ESP, TELECOM, TV bandwidth; ~ **de banda espectral** *f* TELECOM spectral bandwidth; ~ **de banda a mitad de potencia** *f* TEC ESP half-power bandwidth; ~ **de base** *f* ELECTRÓN base width; ~ **del carácter** *f* IMPR character width; ~ **del corte** *f* ING MECÁ saw kerf, METAL width of splitting; ~ **efectiva de entrehierro** *f* ACÚST effective gap length; ~ **de entrehierro** *f* ACÚST gap length; ~ **espectral** *f* ÓPT, TELECOM spectral width; ~ **de haz a mitad de potencia** *f* TEC ESP half-power beamwidth; ~ **de impulso** *f* INFORM&PD, TELECOM pulse width; ~ **de línea espectral** *f* CRISTAL, ÓPT spectral line width, TELECOM spectral-line width; ~ **de la máquina** *f* IMPR machine width, web width; ~ **máxima** *f* PAPEL maximum width; ~ **máxima de urdimbre encolada** *f* TEXTIL maximum-dressed width of warp; ~ **mínima de urdimbre encolada** *f* TEXTIL minimum-dressed width of warp; ~ **a mitad de altura** *f* ÓPT, TELECOM full-width half maximum (*FWHM*); ~ **natural** *f* NUCL natural width; ~ **de pasado** *f* CONTAM MAR swath; ~ **de un pico** *f* FÍS RAD half width; ~ **proporcional** *f* IMPR proportional width; ~ **del pulso** *f* FÍS pulse width; ~ **de raya** *f* FÍS RAD line width; ~ **de raya espectral** *f* FÍS RAD spectral-line width; ~ **de raya natural** *f* FÍS RAD natural line width; ~ **reducida a la mitad** *f* ELECTRÓN half-power width; ~ **total** *f* PAPEL total width; ~ **trabajada** *f* CARBÓN worked thickness; ~ **de vía** *f* FERRO *equipo inamovible* rail gage (*AmE*), rail gauge (*BrE*)

anchurón *m AmL* (*cf emplazamiento Esp*) CARBÓN room, stall, MINAS *galerías* stall, *sala* chamber, room; ~ **de enganche** *m AmL* (*cf plataforma de enganche Esp*) MINAS pit landing, platt

ancla *f* CONST, TRANSP MAR anchor; ~ **de arado** *f* TRANSP MAR plough anchor (*BrE*), ploughshare (*BrE*), plow anchor (*AmE*), plowshare (*AmE*); ~ **con cepo** *f* TRANSP MAR stock anchor; ~ **sin cepo** *f* TRANSP MAR stockless anchor; ~ **CQR de seguridad** *f* SEG, TRANSP MAR CQR anchor; ~ **Danforth** *f* TRANSP MAR Danforth anchor; ~ **encepada** *f* TRANSP MAR foul anchor, fouled anchor; ~ **flotante** *f* OCEAN sea anchor, TRANSP MAR sea anchor, *con balde* bucket drogue, *de emergencia* drogue; ~ **de hongo** *f* TRANSP MAR mushroom anchor; ~ **de leva** *f* TRANSP MAR bow anchor

anclado *m* ING MECÁ fixing

anclaje *m* CONST anchor, brace, INSTAL HIDRÁUL brace, MECÁ anchor, TEC ESP anchoring, TEC PETR deadman; ~ **autoperforador** *m* CONST self-drill anchor; ~ **de baliza simple de localización descubierta** *m* TEC PETR *operación costa-fuera* exposed-location single-buoy mooring (*ELSBM*); ~ **del batán** *m* TEXTIL batt anchorage; ~ **a boya simple** *m* (*SBM*) TEC PETR *producción costa-fuera* single-buoy mooring (*SBM*); ~ **para cañería perdida** *m* PETROL liner hanger; ~ **del cinturón de seguridad** *m* AUTO, SEG, VEH safety belt anchorage (*BrE*), seat belt attachment (*AmE*); ~ **en**

forma de S *m* CONST anchor in the form of an S; ~ **en forma de X** *m* CONST anchor in the form of an X; ~ **de línea muerta** *m* TEC PETR *perforación* dead line anchor; ~ **de manguito** *m* CONST sleeve anchor; ~ **perforado con escudo** *m* CONST shield-driven anchor; ~ **del punto medio** *m* FERRO *infraestructura* midpoint anchor; ~ **a rosca** *m* ING MECÁ screw fixing; ~ **de tubería** *m* ING MECÁ pipe anchor; ~ **de la tubería de producción** *m* TEC PETR *pozos de extracción mecánica* tubing anchor; ~ **de la tubería de revestimiento** *m* TEC PETR casing set; ~ **de tubo electrónico** *m* ELECTRÓN electron-tube holder; ~ **de viento** *m* TEC PETR *perforación* guy anchor

anclar *vti* CONST, TEC ESP, TRANSP MAR anchor

anclote *m* ING MECÁ grapnel, TRANSP MAR kedge anchor

ancón *m* TRANSP MAR *geografía* bay, cove, inlet

andamiaje *m* CONST, PROD, SEG scaffolding, staging; ~ **común** *m* CONST common scaffold; ~ **tubular** *m* CONST tubular scaffolding

andamio *m* CONST, PROD scaffold, stage; ~ **de carga** *m* PROD charging scaffold; ~ **colgante** *m* CONST hanging scaffold, hanging stage, suspended scaffold; ~ **común** *m* CONST common scaffold; ~ **móvil** *m* CONST traveling cradle (*AmE*), travelling cradle (*BrE*); ~ **suspendido y oscilante** *m* CONST flying scaffold

andana: ~ **de forraje** *f* AGRIC swath

andanada *f* D&A burst, round

andar *m* TRANSP MAR *navegación* way out; ~ **de un velero** *m* TRANSP MAR rate of sailing

andén *m* CARBÓN berm, CONST causeway, footway, pavement (*BrE*), sidewalk (*AmE*), platform, FERRO platform, TEC ESP walkway; ~ **de abordaje** *m AmL* (*cf andén de embarque Esp*) TRANSP boarding platform; ~ **de carga** *m* CONST, FERRO, TRANSP loading platform; ~ **de embarque** *m Esp* (*cf andén de abordaje AmL*) TRANSP boarding platform; ~ **de entrevía** *m* FERRO island platform; ~ **móvil** *m* TRANSP moving pavement (*BrE*), moving sidewalk (*AmE*); ~ **móvil tipo cinta** *m* TRANSP belt-type moving pavement (*BrE*), belt-type moving sidewalk (*AmE*)

andesita *f* PETROL andesite

andesítico *adj* PETROL andesitic

andradita *f* MINERAL andradite, aplome, melanite, topazolite

androestéril *m* AGRIC male esterile

anegación *f* AGUA inundation

anegado *adj* AGUA, HIDROL, TRANSP MAR flooded, waterlogged; ~ **en agua** *adj* HIDROL waterlogged

anegar *vt* HIDROL, TRANSP MAR flood

añejamiento *m* ALIMENT *panadería, repostería, molienda* ageing

anejo *m* CONST *de edificio* annex (*AmE*), annexe (*BrE*), outbuilding

anemógrafo *m* ENERG RENOV, FÍS, HIDROL, METEO, TRANSP MAR anemograph

anemometría *f* ENERG RENOV, FÍS, HIDROL, LAB, METEO, TRANSP MAR anemometry

anemómetro *m* ENERG RENOV, FÍS, HIDROL, LAB *de hilo candente, de molinete, de presión, de rotación*, METEO, TRANSP MAR anemometer, wind gauge (*BrE*), ventimeter, wind gage (*AmE*); ~ **de hilo electrocalentado** *m* FÍS, METEO hot wire anemometer; ~ **de placa** *m* FÍS, INSTR, METEO plate anemometer; ~ **de presión** *m*

FÍS, INSTR, METEO pressure anemometer; ~ **registrador** *m* FÍS, INSTR, METEO recording anemometer

anemoscopio *m* TRANSP MAR wind telltale

anetol *m* QUÍMICA anethole

aneurina *f* QUÍMICA aneurin

anexo *m* CONST *de edificio* annex (*AmE*), annexe (*BrE*), outbuilding

anfibio *adj* TRANSP MAR amphibian

anfíbol *m* MINERAL, QUÍMICA amphibole

anfibolita *f* PETROL amphibolite

anfiboloide *adj* MINERAL amphiboloid

anfifílico *adj* DETERG amphiphilic

anfimixis *f* AGRIC amphimixis

anfitalita *f* MINERAL amphitalite

anfitrión *m* INFORM&PD host

anfodelita *f* MINERAL amphodelite

anfólito *m* QUÍMICA ampholyte

anfotérico *adj* DETERG, QUÍMICA amphoteric

angarillas *f pl* CONST handbarrow

angina *f* C&V bulged finish

angiografía *f* INSTR angiography

angostura *f* OCEAN narrows, TRANSP MAR *geografía* inlet; ~ **mareal** *f* OCEAN tidal inlet

angstrom *m* METR angstrom

anguilulosis *f* AGRIC nematode disease

angulación *f* IMPR angling

angulado: ~ **recto** *adj* FÍS right-angled

angular[1] *adj* CONST, GEOM, MECÁ angular

angular[2] *m* CONST knee bracket, angle, TRANSP MAR *construcción* angular, heel plate, *metal* knee; ~ **de acero** *m* MECÁ angle steel; ~ **de cuaderna** *m* TRANSP MAR *contrucción* frame angle; ~ **de hierro** *m* CONST angle iron, L-iron; ~ **de lados iguales** *m* CONST equal-sided angle; ~ **de trancanil** *m* TRANSP MAR *construcción* stringer angle

angularse *v refl* TRANSP MAR *viento* free

ángulo[1]: **de** ~ **agudo** *adj* ING MECÁ sharp-edged; **de** ~ **obtuso** *adj* GEOM obtuse-angled; **de** ~ **recto** *adj* GEOM right-angled; **en** ~ **recto a la quilla** *adj* TEC ESP abeam

ángulo[2] *m* CINEMAT angle, CONST break, inclination, FÍS *de contacto* angle, FOTO corner mount, GEOM, ING MECÁ, TRANSP MAR angle; **~ a** ~ **de abanderamiento** *m* TRANSP AÉR feathering angle; ~ **de abertura** *m* CINEMAT, FOTO, ÓPT, TV aperture angle; ~ **de abertura del haz** *m* ÓPT beamwidth; ~ **abierto** *m* GEOM open angle; ~ **de abrazamiento** *m* CARBÓN contact angle; ~ **de aceptación** *m* CINEMAT acceptance angle, NUCL *de iones* angle of acceptance, ÓPT acceptance angle; ~ **de acometida** *m* MINAS attack angle; ~ **de acoplamiento** *m* ING MECÁ pick-up angle; ~ **de adelanto** *m* ELEC *escobilla del motor* advance angle, angle of lead; ~ **de adelanto de escobillas** *m* ELEC *motor* angle of lead of brushes; ~ **de adelanto de fase** *m* ELEC *corriente alterna* phase angle, *escobilla del motor* angle of lead; ~ **de admisión** *m* ING ELÉC acceptance angle; ~ **adyacente** *m* GEOM adjacent angle; ~ **agudo** *m* GEOM acute angle; ~ **de ahusamiento** *m* MECÁ angle of taper; ~ **de apertura de los contactos del ruptor** *m* AUTO, VEH dwell angle; ~ **de arrastre** *m* TRANSP MAR angle of pull; ~ **de ascenso** *m* TEC ESP, TRANSP AÉR climb gradient; ~ **de ascenso óptimo** *m* TRANSP AÉR best-climb angle; ~ **de ataque** *m* FÍS, TRANSP AÉR angle of

attack, attack angle; ~ **de ataque frontal** *m* ING MECÁ front rake angle; ~ **de ataque inducido** *m* FÍS, TRANSP AÉR induced angle of attack; ~ **de ataque de la pala** *m* FÍS, TRANSP AÉR blade angle of attack; ~ **de aterrizaje** *m* TRANSP AÉR ground angle; ~ **de atraso** *m* ELEC *escobilla del motor* angle of lag; ~ **de atraso de escobilla** *m* ELEC *motor* angle of brush lag; ~ **de avance** *m* AUTO angle of advance, ELEC, ING ELÉC, ING MECÁ angle of advance, *distribuidor* angle of lead; ~ **de avance a la admisión** *m* ING MECÁ angular lead, *distribuidor* angle of lead; ~ **de avance al escape** *m* ING MECÁ *máquinas vapor* angle of prerelease; ~ **de avance de escobillas** *m* ELEC *motor* angle of lead of brushes; ~ **de avance de la rosca** *m* ING MECÁ thread lead angle; ~ **azimutal** *m* ENERG RENOV azimuth angle; **~ b** ~ **de la barra de control del cabeceo** *m* TRANSP AÉR pitch-control rod angle; ~ **de la barra de control del paso** *m* TRANSP AÉR pitch-control rod angle; ~ **de batimiento** *m* TRANSP AÉR flapping angle; ~ **de batimiento cíclico** *m* TRANSP AÉR cyclic-flapping angle; ~ **de batimiento de la pala** *m* TRANSP AÉR blade flapping angle; ~ **de Bragg** *m* CRISTAL Bragg angle; ~ **de Brewster** *m* ÓPT Brewster's angle; **~ c** ~ **de cabeceo** *m* TRANSP AÉR, TRANSP MAR angle of pitch, pitch angle; ~ **de caída** *m* ELECTRÓN falling edge, VEH *rueda, vía* camber; ~ **de calaje** *m* ING MECÁ angle of keying; ~ **de cámara** *m* CINEMAT, FOTO, TV camera angle; ~ **de campo** *m* CINEMAT angle of view; ~ **de cangilón** *m* ENERG RENOV *hidroelectricidad, turbinas* bucket angle; ~ **de carga** *m* ELEC *máquina eléctrica* load angle; ~ **de cenit** *m* ENERG RENOV zenith angle; ~ **ciego** *m* TRANSP AÉR blind angle; ~ **de cobertura** *m* CINEMAT *del objetivo* angle of coverage; ~ **de colocación de la pala** *m* TRANSP AÉR blade-setting angle; ~ **complementario** *m* GEOM complementary angle; ~ **cóncavo** *m* GEOM re-entrant, reflex angle; ~ **conductivo** *m* ING ELÉC conduction angle; ~ **de conicidad** *m* TRANSP AÉR coning angle; ~ **de cono del engranaje** *m* ING MECÁ gear cone angle; ~ **de contacto** *m* PROD angle of contact; ~ **contiguo** *m* GEOM contiguous angle; ~ **de convergencia** *m* TRANSP AÉR toe-in angle; ~ **de convergencia dinámica** *m* VEH *dirección* dynamic toe angle; ~ **convexo** *m* GEOM outward angle; ~ **correspondiente** *m* GEOM corresponding angle; ~ **de corte** *m* ALIMENT cutting angle, CINEMAT angle of cut-off, ING MECÁ ground-in cutting rake, cutting angle; ~ **crítico** *m* FÍS, ÓPT, PETROL, QUÍMICA, TELECOM critical angle; **~ d** ~ **de declinación** *m* FÍS, TRANSP AÉR declination angle; ~ **de declinación magnética** *m* FÍS angle of magnetic declination; ~ **de deriva** *m* TRANSP AÉR crab angle, drift angle, TRANSP MAR *navegación a vela* leeway; ~ **de derrape** *m* TEC ESP *aviones*, TRANSP yaw angle; ~ **de descenso normal** *m* TRANSP AÉR normal descent angle; ~ **de desfasaje** *m* ELEC *corriente alterna* angle of phase difference, phase angle, FÍS, FÍS ONDAS *corriente alterna* angle of phase difference; ~ **de desfasamiento** *m* ELEC *corriente alterna* phase angle; ~ **de desfase** *m* ELEC, FÍS, FÍS ONDAS *corriente alterna* angle of phase difference; ~ **de desplazamiento** *m* TRANSP AÉR drift angle; ~ **de despojo** *m* ING MECÁ, PROD *máquinas* angle of clearance; ~ **de despojo anterior** *m* ING MECÁ *salida de corte* clearance angle; ~ **de despojo superior** *m*

ING MECÁ top rake; ~ **de destalonado** *m* ING MECÁ relief angle; ~ **de desviación** *m* TV deflection angle; ~ **diedro** *m* GEOM, METAL dihedral angle; ~ **de diferencia de fases** *m* ELEC *corriente alterna* phase-difference angle; ~ **de difusión** *m* FÍS scattering angle; ~ **del dwell** *m* AUTO, VEH *encendido* dwell angle;

■ **e** ~ **de elevación** *m* CONST, GEOM, TEC ESP, TELECOM, TRANSP AÉR angle of elevation, elevation angle; ~ **de elevación nula** *m* TEC ESP, TRANSP AÉR zero lift angle; ~ **de enderezamiento** *m* METR angle plate; ~ **de enlace** *m* CRISTAL bond angle; ~ **de la entalla** *m* METAL *deformación, ensayos* notch angle; ~ **de entrada del flujo** *m* TRANSP AÉR inflow angle; ~ **entrante** *m* GEOM re-entrant; ~ **entre caras** *m* CRISTAL interfacial angle; ~ **equilátero** *m* GEOM, PROD even-sided angle; ~ **de escobilla** *m* ELEC *para máquina* brush angle; ~ **de escora** *m* TRANSP MAR *diseño de barcos* angle of heel; ~ **de escora de transición** *m* TRANSP MAR *diseño de barcos* angle of loll; ~ **de Euler** *m* FÍS Euler angle; ~ **exterior** *m* GEOM exterior angle; ~ **exterior alterno** *m* GEOM alternate exterior angle;

■ **f** ~ **de fase** *m* ELEC *corriente alterna*, FÍS, FÍS ONDAS, TRANSP AÉR phase angle; ~ **fijo** *m* CONST tight corner; ~ **de flecha** *m* TRANSP AÉR sweep angle; ~ **de fricción** *m* CARBÓN, FÍS angle of friction;

■ **g** ~ **de giro** *m* AUTO, VEH steer angle; ~ **de guiñada** *m* TEC ESP *buques*, TRANSP yaw angle;

■ **i** ~ **de iluminación** *m* IMPR illumination angle; ~ **de incidencia** *m* CINEMAT, ENERG RENOV, FÍS, FÍS ONDAS angle of incidence, incidence angle, ING MECÁ *máquinas herramientas* angle of relief, relief angle, bank clearance angle, ÓPT, PROD, TELECOM, TRANSP AÉR angle of incidence, incidence angle; ~ **de incidencia límite** *m* ÓPT critical angle; ~ **de inclinación** *m* C&V angle of pitch, CINEMAT angle of rake, *cámara* angle of tilt, ENERG RENOV *hélice,órbita* pitch angle, FÍS angle of dip, ING MECÁ *útil de cortar* angle of rake, PROD angle of inclination; ~ **de inclinación de la aguja magnética** *m* GEOFÍS *magnetometría* angle of dip; ~ **de inclinación del cilindro** *m* AUTO, VEH cylinder bank angling; ~ **de inclinación de colector solar** *m* ENERG RENOV collector tilt angle; ~ **de inclinación longitudinal** *m* TEC ESP pitch attitude; ~ **de inclinación magnética** *m* FÍS angle of magnetic inclination; ~ **interfacial** *m* CRISTAL interfacial angle; ~ **interior** *m* GEOM interior angle, internal angle; ~ **interior alterno** *m* GEOM alternate interior angle; ~ **interlabial** *m* GEOL interlimb angle; ~ **interno** *m* GEOM internal angle; ~ **de intersección** *m* CONST *topografía* intersection angle; ~ **inverso** *m* CINEMAT, TV reverse angle;

■ **l** ~ **de leva** *m* AUTO, VEH cam angle; ~ **llano** *m* GEOM flat angle; ~ **de llegada** *m* TELECOM angle of arrival;

■ **m** ~ **magnético** *m* GEOFÍS dip angle; ~ **máximo de giro de las ruedas delanteras** *m* AUTO, VEH steering lock; ~ **de mira** *m* TELECOM angle of sight; ~ **de modulación** *m* ACÚST modulation angle; ~ **de muro** *m* CONST *albañilería* angle;

■ **n** ~ **negativo** *m* GEOM negative angle;

■ **o** ~ **de obicuidad** *m* ING MECÁ angle of obliquity, TEC ESP yaw; ~ **oblicuo** *m* GEOM oblique angle; ~ **del obturador** *m* CINEMAT, FOTO, TV shutter angle;

~ **obtuso** *m* GEOM obtuse angle; ~ **óptimo de ascenso** *m* TRANSP AÉR best-climb angle;

■ **p** ~ **de pala** *m* TRANSP AÉR blade angle; ~ **de paso** *m* TRANSP AÉR pitch angle; ~ **de paso del colectivo** *m* TRANSP AÉR collective-pitch angle; ~ **de paso indicado** *m* TRANSP AÉR indicated pitch angle; ~ **de paso medio** *m* TRANSP AÉR mean pitch angle; ~ **de paso de la pala** *m* TRANSP AÉR blade-pitch angle; ~ **de la pendiente** *m* CONST slope level; ~ **de pérdida** *m* TRANSP AÉR angle of stall; ~ **de pérdida del dieléctrico** *m* ELEC dielectric loss angle; ~ **de pérdidas** *m* ELEC *capacitor*, FÍS loss angle; ~ **perforado** *m* MECÁ perforated angle; ~ **de picado** *m* CINEMAT, FOTO, TV angle of tilt; ~ **de planeo** *m* TRANSP AÉR gliding angle; ~ **plano** *m* GEOM flat angle, plane angle; ~ **de plegado** *m* PROD bend angle; ~ **de polarización** *m* ÓPT polarization angle; ~ **polarizante** *m* ÓPT polarizing angle; ~ **poliédrico** *m* GEOM polyhedral angle; ~ **positivo** *m* GEOM positive angle; ~ **de precesión** *m* ENERG RENOV angle of precession; ~ **de presión** *m* ING MECÁ, MECÁ angle of pressure, pressure angle; ~ **de puesta en bandera** *m* TRANSP AÉR feathering angle;

■ **r** ~ **de radiación** *m* ELECTRÓN, FÍS RAD, ÓPT, TELECOM beam angle, radiation angle; ~ **en radianes** *m* GEOM angle in radians; ~ **en radio** *m* GEOM angle in radians; ~ **de rebaje frontal** *m* ING MECÁ front rake angle; ~ **de rebaje negativo** *m* ING MECÁ negative rake; ~ **de rebaje superior** *m* ING MECÁ top rake; ~ **de rebajo** *m* ING MECÁ *máquinas herramientas* angle of relief; ~ **de recepción** *m* ELECTRÓN, IMPR, TELECOM acceptance angle; ~ **recto** *m* GEOM right angle; ~ **de recubrimiento** *m* ELEC *rectificador* angle of overlap, ING MECÁ *distribuidor* angle of lap; ~ **de reflexión** *m* FÍS, FÍS ONDAS angle of reflection; ~ **de refracción** *m* FÍS, FÍS ONDAS angle of refraction; ~ **de reglaje** *m* AUTO, VEH timing angle; ~ **de la resistencia** *m* TRANSP AÉR drag angle; ~ **de la resistencia aerodinámica** *m* TRANSP AÉR drag angle; ~ **de retardo de escobilla** *m* ELEC *motor* angle of brush lag; ~ **de retiro** *m* ING MECÁ, MECÁ, PROD draft (*AmE*), draught (*BrE*), draw; ~ **de retraso** *m* ELEC *escobilla del motor* angle of lag; ~ **de retraso de escobilla** *m* ELEC *motor* angle of brush lag; ~ **de retraso de fase** *m* ELEC *corriente alterna*, ENERG RENOV phase angle; ~ **de rozamiento** *m* CONST, PROD angle of friction; ~ **de rumbo** *m* TRANSP AÉR course angle;

■ **s** ~ **de salida** *m* INSTAL HIDRÁUL, ÓPT, PROD, TELECOM angle of departure, output angle; ~ **saliente** *m* GEOM outward angle, reflex angle; ~ **de situación** *m* D&A angle of sight; ~ **de solapamiento** *m* ELEC *rectificador* angle of overlap; ~ **sólido** *m* ELEC, FÍS, GEOM solid angle; ~ **de la superficie de control** *m* TRANSP AÉR control-surface angle; ~ **de superposición** *m* ELEC *rectificador* angle of overlap;

■ **t** ~ **de talud natural** *m* CONST, EMB angle of repose; ~ **del timón** *m* TRANSP MAR steering angle; ~ **tomador** *m* IMPR pick-up angle; ~ **de torsión** *m* MECÁ angle of twist; ~ **de tracción** *m* MECÁ, TRANSP MAR angle of pull, angle of traction; ~ **de la trama** *m* IMPR screen angle; ~ **de 30 grados** *m* GEOM angle of 30 degrees;

■ **u** ~ **de unión** *m* PROD bonding angle;

■ **v** ~ **vertical** *m* GEOM vertical angle; ~ **del vértice** *m* IMPR, TELECOM vertex angle; ~ **de visión** *m* CINE-

MAT viewing angle, TELECOM angle of sight; **~ de vista** *m* CINEMAT angle of view, *del objetivo* angle of coverage; **~ visual** *m* INSTR sighting line

ángulos: en ~ rectos *fra* GEOM at right angles

anguloso *adj* MECÁ angular

anhédrico *adj* GEOL anhedral

anhidración *f* ALIMENT, QUÍMICA anhydration

anhídrico *adj* ALIMENT, QUÍMICA anhydrous

anhídrido *m* DETERG, P&C, PAPEL, QUÍMICA anhydride; **~ alquenil succínico** *m* PAPEL *agentes de encolado*, QUÍMICA alkenyl succinic anhydride (*ASA*); **~ carbónico** *m* CARBÓN, QUÍMICA carbon dioxide; **~ ftálico** *m* P&C *materia prima*, QUÍMICA phthalic anhydride; **~ sulfúrico** *m* CONTAM *producto químico*, QUÍMICA sulfur trioxide (*AmE*), sulfuric anhydride (*AmE*), sulphur trioxide (*BrE*), sulphuric anhydride (*BrE*); **~ sulfuroso** *m* QUÍMICA sulfur dioxide (*AmE*), sulphur dioxide (*BrE*)

anhidrita *f* GAS, PETROL, TEC PETR anhydrite

anhidro *adj* AGUA, ALIMENT, CONTAM, HIDROL, QUÍMICA, TEC PETR anhydrous, water-free

anhidrosis *f* AGRIC *zootecnia* anhidrosis

anidado *adj* INFORM&PD nested

anidamiento *m* INFORM&PD nesting

anidar *vti* INFORM&PD nest

anilida *f* QUÍMICA anilide

anilla *f* CONST bow, hoop, ING MECÁ hoop, eye, PROD hoop; **~ de amarre** *f* TRANSP AÉR mooring ring; **~ de apertura del paracaídas** *f* TRANSP AÉR parachute release handle; **~ final** *f* ING MECÁ eye end; **~ del izador** *f* TRANSP AÉR hoisting eye, hoisting ring; **~ de soporte** *f* ING MECÁ droop-restraining ring; **~ de tirar** *f* EMB *sistemas para abrir botes, latas* pull-ring; **~ para tubos** *f* CONST pipe collar

anillado *adj* GEOM, ING MECÁ, MECÁ ring-shaped

anillar *vt* TEXTIL ring

anillo *m* AUTO, C&V ring, ELEC *bobina* torus, *de sistema* mesh, FOTO ring, GEOM annulus, INFORM&PD ring, ING MECÁ lug, hoop, collar, handling lug, INSTAL HIDRÁUL *de pistón* ring, LAB *soporte de filtración* filter support, MECÁ ring, lug, collar, PROD lug, QUÍMICA *estructura* ring, TEC ESP ring, annulus, TEC PETR annulus, TELECOM *cable*, TEXTIL *hilatura* ring, TRANSP AÉR annulus, VEH *pistón* ring; **~ de abertura** *m* CINEMAT, FOTO, TV aperture ring; **~ de acceso** *m* FÍS slip ring; **~ aceitador** *m* ING MECÁ oil ring; **~ de acoplamiento** *m* ING MECÁ coupling ring; **~ adaptador** *m* CINEMAT, FOTO adaptor ring; **~ de ajuste del valor luminoso** *m* CINEMAT, FOTO light value setting ring; **~ de almacenamiento de electrones** *m* ELECTRÓN, FÍS PART electron-storage ring; **~ antiprotón de baja energía** *m* (*LEAR*) FÍS PART low-energy antiproton ring (*LEAR*); **~ anual de crecimiento** *m* AGRIC growth ring; **~ de apriete** *m* ING MECÁ necking; **~ de aumento** *m* CINEMAT, FOTO step-up ring; **~ bajo a tierra** *m* TELECOM low ring to ground; **~ bencénico** *m* QUÍMICA benzene ring; **~ en bisel** *m* ING MECÁ bevel ring; **~ de bolas** *m* ING MECÁ, MECÁ race; **~ de centrado** *m* TV centering ring (*AmE*), centring ring (*BrE*); **~ de centraje magnético** *m* TEC ESP magnetic centering ring (*AmE*), magnetic centring ring (*BrE*); **~ central** *m* INFORM&PD *disco* hub; **~ de centrar** *m* PROD *moldes* bead; **~ cerrado** *m* QUÍMICA *estructura* closed ring; **~ de cierre hermético toroidal** *m* ING MECÁ toroidal sealing ring; **~ de cojinete falso** *m* AUTO oil control

ring, oil ring, ING MECÁ dummy bearing race; **~ colector** *m* ELEC, ING ELÉC, ING MECÁ collector ring, VEH slip ring; **~ del collector del arrancador** *m* AUTO starter collector ring; **~ de compresión** *m* AUTO, VEH compression ring; **~ de compresión superior** *m* AUTO, VEH top compression ring; **~ de concreto** *m* AmL (*cf anillo de hormigón Esp*) CONST concrete ring; **~ condensado** *m* QUÍMICA condensed ring; **~ de conexión** *m* TELECOM jumper ring; **~ cónico de compresión** *m* AUTO tapered compression ring; **~ contador** *m* TV ring counter; **~ de control de engrase** *m* AUTO, VEH oil control ring, oil ring; **~ de control de lubricación de tres piezas** *m* AUTO, VEH three-piece oil ring; **~ cortante** *m* MINAS *sondeos* cutting ring, shoe; **~ de crecimiento** *m* AGRIC annual ring; **~ deflector** *m* TEC PETR baffle ring; **~ dentado** *m* AUTO, VEH ring gear; **~ deslizante** *m* ING ELÉC slip ring; **~ de distancia** *m* TRANSP MAR *radar* calibration ring; **~ de distribución** *m* NUCL distribution ring; **~ del eje** *m* ING MECÁ shaft collar; **~ elástico para piñón de puesta en marcha** *m* AUTO, VEH *del motor* starter ring gear; **~ elástico de retención** *m* AUTO, VEH circlip; **~ de émbolo** *m* AUTO, ING MECÁ, MECÁ, VEH piston ring; **~ de empaquetadura** *m* PROD junk ring; **~ de enfoque** *m* CINEMAT, FOTO, INSTR, TV focusing ring; **~ de engrase** *m* AUTO, ING MECÁ, PROD, VEH oil ring; **~ para ensayos** *m* CONST building ring; **~ de escritura** *m* INFORM&PD write ring; **~ esmerilador** *m* PROC QUÍ grinding ring; **~ estanco** *m* ING MECÁ seal; **~ de estanqueidad** *m* ING MECÁ, VEH sealing ring; **~ excéntrico** *m* ING MECÁ eccentric hoop; **~ de expansión** *m* TEC PETR *tuberías* expansion ring; **~ de exploración** *m* TV scan ring; **~ de extensión** *m* FOTO extension ring; **~ exterior** *m* ING MECÁ outer race; **~ facetado** *m* NUCL faceted ring; **~ de fijación** *m* EMB clamping ring; **~ para fijar la abertura** *m* CINEMAT, FOTO, TV aperture-setting ring; **~ de fondo de alimentación** *m* ING MECÁ feed bush; **~ de freno** *m* ING MECÁ brake ring; **~ de fricción** *m* ING MECÁ friction ring, VEH *generador* slip ring; **~ frontal** *m* CINEMAT front scanning, INSTR front ring; **~ Gramme** *m* ELEC *bobina* Gramme ring; **~ de guarda** *m* ELEC *campo eléctrico*, FÍS, ING ELÉC, PROD guard ring; **~ de guía** *m* INSTAL HIDRÁUL guide ring; **~ guía** *m* IMPR ball race, ING MECÁ ball race, race, MECÁ race; **~ guía del cojinete** *m* ING MECÁ bearing race; **~ de hormigón** *m* Esp (*cf anillo de concreto AmL*) CONST concrete ring; **~ de inserción de bronce** *m* PROD brass insert-ring; **~ de intensificación** *m* TV intensifier ring; **~ inversor** *m* FOTO reversing ring; **~ de junta** *m* NUCL joint ring; **~ laminado sin costura** *m* ING MECÁ seamless rolled ring; **~ laminado sin soldadura** *m* ING MECÁ seamless rolled ring; **~ de llama** *m* GAS, TERMO ring of flame; **~ para llaves** *m* ING MECÁ split ring; **~ de lubricación** *m* AUTO, ING MECÁ, VEH oil ring; **~ de lubricación de tres piezas** *m* AUTO, ING MECÁ, VEH three-piece oil ring; **~ lubricante** *m* AUTO, ING MECÁ, VEH oil ring; **~ de maniobra** *m* MINAS spider and slips; **~ metálico** *m* VEH *de motor, válvula* collet; **~ de montaje** *m* CINEMAT, FOTO cocking ring; **~ de Newton** *m* FÍS, FOTO Newton's ring; **~ obturador** *m* ING MECÁ sealing ring; **~ obturador toroidal** *m* ING MECÁ toroidal sealing ring; **~ partido** *m* ING MECÁ split ring; **~ de permiso de escritura** *m* INFORM&PD

write permit ring; **~ de pistón** *m* AUTO, ING MECÁ, MECÁ, VEH circlip, piston ring; **~ de pivote** *m* AUTO pivot ring; **~ portador** *m* MINAS *pozos* curb ring; **~ de precintado** *m* ING MECÁ sealing ring; **~ de protección** *m* ELEC, FÍS, PROD guard ring; **~ protector** *m* ELEC *campo eléctrico*, FÍS, ING ELÉC, PROD guard ring; **~ que permite escribir** *m* INFORM&PD write permit ring; **~ rascador de aceite** *m* AUTO, VEH oil control ring; **~ de rectificar** *m* PROC QUÍ grinding ring; **~ reductor** *m* CINEMAT step-down ring; **~ refractorio flotante** *m* C&V *horno de vidrio* blowing ring; **~ de refuerzo** *m* TRANSP AÉR shroud ring; **~ de relés eléctricos** *m* ELEC, ELECTRÓN, FÍS, INFORM&PD flip-flop circuit; **~ de retención** *m* CINEMAT locking ring, ING MECÁ, MECÁ retaining ring; **~ retenedor** *m* ING MECÁ, MECÁ retaining ring; **~ de retenidas** *m* TEC PETR *perforación* guy ring; **~ de rodadura** *m* ING MECÁ, MECÁ, PROD race, raceway, VEH ball-bearing race; **~ de rodadura exterior** *m* ING MECÁ, MECÁ, PROD outer race; **~ de rodadura interno** *m* ING MECÁ, MECÁ, PROD ball inner race; **~ de roscar** *m* TELECOM threading ring; **~ rotatorio** *m* FÍS FLUID rotating annulus; **~ de rozamiento** *m* ING MECÁ friction ring; **~ rozante** *m* ELEC *motor* slip ring, *de conmutador* collector ring, ING ELÉC, ING MECÁ collector ring; **~ rozante del arrancador** *m* AUTO starter slip ring; **~ de seguridad** *m* AUTO circlip, ELEC, FÍS, ING ELÉC guard ring, ING MECÁ *para cojinetes de bola* lock ring, PROD guard ring, VEH *motor* circlip; **~ de seguridad partido** *m* ING MECÁ split-set collar; **~ de seis carbonos** *m* QUÍMICA six carbon ring; **~ de seis miembros** *m* QUÍMICA *compuesto* six-membered ring; **~ de sellado** *m* AUTO, VEH sealing ring; **~ de sello toroidal** *m* ING MECÁ toroidal sealing ring; **~ de la semicaja superior** *m* PROD *funderías* cope ring; **~ de sujeción** *m* FOTO retaining ring; **~ sujetador** *m* ING MECÁ, MECÁ retaining ring, shell chuck; **~ tensor** *m* PROD belt idler; **~ de tolerancias** *m* ING MECÁ limit external gage (*AmE*), limit external gauge (*BrE*); **~ de tracción** *m* ING MECÁ draw ring; **~ de triazina** *m* QUÍMICA triazine ring; **~ de trinquete** *m* ING MECÁ pawl ring; **~ variable de distancia** *m* TRANSP MAR *radar* variable-range marker; **~ de zoom** *m* CINEMAT, FOTO, TV zoom ring

ánima *f* ING ELÉC filler, ING MECÁ bore

animación: **~ por computadora** *f* AmL (*cf animación por ordenador Esp*) CINEMAT, INFORM&PD, TV computer animation; **~ de imagen aérea** *f* CINEMAT aerial image animation; **~ multiplano** *f* CINEMAT multiplane animation; **~ por ordenador** *f* (*cf animación por computadora AmL*) CINEMAT, INFORM&PD, TV computer animation

animador *m* CINEMAT animator

animagtita *f* MINERAL aenigmatite

animal: **~ de un año** *m* AGRIC yearling; **~ recién destetado** *m* AGRIC weanling; **~ de trabajo** *m* AGRIC farm animal

animar *vt* CINEMAT animate

anión *m* ALIMENT, CARBÓN anion, ELEC *partícula cargada* anion, negative ion, ELECTRÓN, FÍS, FÍS PART, FÍS RAD, ING ELÉC anion, negative ion, QUÍMICA anion; **~ predominante** *m* CONTAM dominant anion

aniónico *adj* GEN anionic

anionotropía *f* QUÍMICA anionotropy

aniquilación *f* FÍS, FÍS PART, NUCL annihilation; **~ de la**

dislocación *f* METAL dislocation annihilation; **~ electrón positrón** *f* ELECTRÓN, FÍS PART, NUCL electron-positron annihilation; **~ de pares** *f* FÍS PART, NUCL pair annihilation

anisaldehído *f* QUÍMICA *perfumería* anisaldehyde

anisentrópico *adj* FÍS, TERMO nonisentropic

anisidina *f* QUÍMICA anisidine

anisiense *m* GEOL *estratigrafía* anisian

anisoelasticidad *f* TEC ESP *mecánica* anisoelasticity

anisoelástico *adj* TEC ESP *giróscopos* anisoelastic

anisol *m* ALIMENT, QUÍMICA anisole, methoxybenzene

anisotropía *f* GEN anisotropy; **~ diamagnética** *f* FÍS, FÍS RAD diamagnetic anisotropy; **~ de turbulencia** *f* FÍS FLUID anisotropy of turbulence

anisotrópico *adj* GEN anisotropic

anisótropo *adj* GEN anisotropic

ankaramita *f* PETROL ankaramite

ankerita *f* MINERAL ankerite

annabergita *f* MINERAL nickel bloom

año: **~ hidrológico** *m* HIDROL hydrologic year; **~ luz** *m* FÍS light year; **~ planificado** *m* PROD planned year; **~ programado** *m* PROD planned year; **~ sideral** *m* FÍS, TEC ESP sidereal year

anódico *adj* GEN anodic

anodizar *vt* IMPR, QUÍMICA, REVEST, TRANSP MAR anodize

ánodo *m* GEN anode; **~ acelerador** *m* FÍS RAD, ING ELÉC accelerating anode; **~ alumínico** *m* ING ELÉC aluminium anode (*BrE*), aluminum anode (*AmE*); **~ de aluminio** *m* ING ELÉC aluminium anode (*BrE*), aluminum anode (*AmE*); **~ de batería** *m* ELEC battery plate; **~ característico** *m* ING ELÉC anode characteristic; **~ de derivación** *m* ING ELÉC bypass anode; **~ del diodo** *m* ELECTRÓN diode anode; **~ de disco perforado** *m* TV perforated disc anode (*AmE*), perforated disk anode (*BrE*); **~ de enfoque** *m* ING ELÉC focusing anode; **~ de excitación** *m* ING ELÉC excitation anode; **~ final** *m* TV final anode; **~ hueco** *m* ELEC *electrodo* hollow anode; **~ de ionización** *m* ING ELÉC holding anode; **~ nulo** *m* ING ELÉC zero anode; **~ pesado** *m* ING ELÉC heavy anode; **~ primario** *m* ING ELÉC first anode; **~ principal** *m* ING ELÉC main anode; **~ rectificador** *m* ELECTRÓN, ING ELÉC rectifier anode; **~ de rejilla** *m* FÍS RAD wire-mesh target; **~ rígido** *m* ING ELÉC rough anode; **~ de sacrificio** *m* TEC PETR *operaciones costa-fuera* sacrificial anode; **~ secundario** *m* ING ELÉC second anode; **~ sinterizado** *m* ING ELÉC sintered anode; **~ sólido de tantalio** *m* ING ELÉC tantalum slug; **~ de tantalio** *m* ING ELÉC tantalum anode; **~ tipo álabes** *m* ING ELÉC vane-type anode

anomalía *f* GEN anomaly; **~ de aire libre** *f* GEOFÍS, GEOL free-air anomaly; **~ de atmósfera libre** *f* GEOFÍS, GEOL free-air anomaly; **~ de Bouguer** *f* GEOFÍS, GEOL Bouguer anomaly, free-air anomaly; **~ del bucle** *f* PROD loop fault; **~ climática** *f* METEO climatic anomaly; **~ del equipo procesador** *f* PROD processor hardware fault; **~ excéntrica** *f* TEC ESP *orbitografía* eccentric anomaly; **~ gravitacional** *f* GEOFÍS, GEOL gravity anomaly; **~ isotópica** *f* FÍS isotopic anomaly; **~ magnética** *f* GEOFÍS, GEOL magnetic anomaly; **~ principal** *f* TEC ESP mean anomaly; **~ de propagación** *f* TV propagation anomaly; **~ real** *f* TEC ESP true anomaly; **~ termostérica** *f* OCEAN thermosteric anomaly

anómero *m* QUÍMICA anomer

anorogénico *adj* GEOL *composición tectónica* anorogenic
anortita *f* C&V anorthite
anortosita *f* PETROL anorthosite
anotación *f* INFORM&PD, PROD, TELECOM annotation, notation; **~ cronológica** *f* TELECOM logging; **~ vocal** *f* TELECOM voice annotation
ANOVA *abr* (*análisis de varianza*) INFORM&PD ANOVA (*analysis of variance*)
anoxibiótico *m* ALIMENT, HIDROL, OCEAN, P&C, RECICL anaerobe (*BrE*), anaerobium (*AmE*)
ANSI *abr* (*Instituto Americano de Normalización Nacional*) PROD ANSI (*American National Standards Institute*)
antecámara *f* AUTO antechamber; **~ de combustión** *f* TERMO, TRANSP combustion prechamber; **~ pequeña** *f* REFRIG airlock
antecapa *f* PAPEL underliner
antecrisol *m* PROD *de cubilote* receiver
antena *f* GEN aerial (*BrE*), antenna (*AmE*); **~ acromática** *f* TV achromatic aerial (*BrE*), achromatic antenna (*AmE*); **~ activa** *f* TELECOM active aerial (*BrE*), active antenna (*AmE*); **~ acústica** *f* TELECOM acoustic aerial (*BrE*), acoustic antenna (*AmE*); **~ adaptada en delta** *f* TV delta-matched aerial (*BrE*), delta-matched antenna (*AmE*); **~ adaptativa** *f* TELECOM adaptive aerial (*BrE*), adaptive antenna (*AmE*); **~ aérea** *f* FÍS aerial; **~ aérea transmisora** *f* FÍS transmitting aerial antenna; **~ de alimentador coaxial** *f* TELECOM, TV coaxial aerial (*BrE*), coaxial antenna (*AmE*); **~ con alimentador descentrado** *f* TEC ESP offset antenna; **~ amplificadora** *f* TV antennafier; **~ apantallada** *f* TV screened aerial (*BrE*), screened antenna (*AmE*); **~ aperiódica** *f* TV aperiodic aerial (*BrE*), aperiodic antenna (*AmE*); **~ de apertura** *f* TELECOM aperture aerial (*BrE*), aperture antenna (*AmE*); **~ en array** *f* TELECOM array aerial (*BrE*), array antenna (*AmE*); **~ de array con radiación máxima en la dirección del eje** *f* TELECOM end fire array aerial (*BrE*), end fire array antenna (*AmE*); **~ en array de radiación transversal** *f* TELECOM broadside array aerial (*BrE*), broadside array antenna (*AmE*); **~ artificial** *f* TV artificial aerial (*BrE*), artificial antenna (*AmE*), dummy aerial (*BrE*), dummy antenna (*AmE*); **~ autoadaptable** *f* TV self-phased array; **~ de banda ancha** *f* TEC ESP, TV broadband aerial (*BrE*), broadband antenna (*AmE*); **~ de barrido electrónico** *f* TELECOM sweep aerial (*BrE*), sweep antenna (*AmE*); **~ de barrido radar** *f* D&A, FÍS RAD, TRANSP, TRANSP AÉR, TRANSP MAR radar scanner; **~ bicónica** *f* TELECOM biconical aerial (*BrE*), biconical antenna (*AmE*); **~ bidireccional** *f* AmL TV directional aerial (*BrE*), directional antenna (*AmE*); **~ bilateral** *f* TV bilateral aerial (*BrE*), bilateral antenna (*AmE*); **~ blindada** *f* AmL TV screened aerial (*BrE*), screened antenna (*AmE*); **~ de bocina** *f* TEC ESP *radar* horn antenna; **~ de campo magnético** *f* GEOFÍS magnetic-field aerial (*BrE*), magnetic-field antenna (*AmE*); **~ circular** *f* TV circular aerial (*BrE*), circular antenna (*AmE*); **~ colectiva** *f* TELECOM, TV collective aerial (*BrE*), collective antenna (*AmE*), community aerial (*BrE*), community antenna (*AmE*); **~ colineal** *f* TELECOM collinear aerial (*BrE*), collinear antenna (*AmE*); **~ colineal de dipolos** *f* TV lazy H aerial (*BrE*), lazy H antenna (*AmE*); **~ común** *f* TELECOM,

TV common aerial (*BrE*), common antenna (*AmE*); **~ conformada** *f* TEC ESP conformal aerial (*BrE*), conformal antenna (*AmE*); **~ conforme** *f* TEC ESP conformal aerial (*BrE*), conformal antenna (*AmE*); **~ cruzada** *f* TV turnstile aerial (*BrE*), turnstile antenna (*AmE*); **~ de cuadro** *f* TELECOM loop aerial (*BrE*), loop antenna (*AmE*); **~ desplegable** *f* TEC ESP deployable aerial (*BrE*), deployable antenna (*AmE*); **~ de diedro** *f* TELECOM dihedral aerial (*BrE*), dihedral antenna (*AmE*); **~ dieléctrica** *f* TEC ESP, TELECOM dielectric aerial (*BrE*), dielectric antenna (*AmE*); **~ dipolo** *f* TRANSP MAR *comunicación por radio*, TV dipole aerial (*BrE*), dipole antenna (*AmE*), doublet aerial (*BrE*), doublet antenna (*AmE*); **~ dipolo doblado** *f* TV folded-dipole aerial (*BrE*), folded-dipole antenna (*AmE*); **~ dipolo magnético de trébol** *f* TV clover-leaf aerial (*BrE*), clover-leaf antenna (*AmE*); **~ dipolo de media onda** *f* FÍS RAD, TV half-wave dipole aerial (*BrE*), half-wave dipole antenna (*AmE*); **~ direccional** *f* TEC ESP, TELECOM, TV directional aerial (*BrE*), directional antenna (*AmE*); **~ direccional giratoria** *f* FÍS RAD *radar* scanner; **~ directiva** *f* TEC ESP, TELECOM, TV directional aerial (*BrE*), directional antenna (*AmE*); **~ dirigida** *f* TEC ESP, TELECOM, TV directional aerial (*BrE*), directional antenna (*AmE*); **~ de disco** *f* TELECOM, TV disc aerial (*BrE*), disk antenna (*AmE*); **~ de doble reflector** *f* FÍS, TEC ESP microwave aerial (*BrE*), microwave antenna (*AmE*), TELECOM double-reflector aerial (*BrE*), double-reflector antenna (*AmE*); **~ de doblete** *f* TV dipole aerial (*BrE*), dipole antenna (*AmE*), doublet aerial (*BrE*), doublet antenna (*AmE*); **~ con elementos controlados por face** *f* TEC ESP phased-array aerial (*BrE*), phased-array antenna (*AmE*); **~ de elementos múltiples desfasados** *f* TEC ESP phased-array aerial (*BrE*), phased-array antenna (*AmE*); **~ empotrada** *f* TRANSP AÉR flush aerial (*BrE*), flush antenna (*AmE*); **~ encastrada** *f* TEC ESP flush-mounted aerial (*BrE*), flush-mounted antenna (*AmE*); **~ esférica** *f* TELECOM spherical aerial (*BrE*), spherical antenna (*AmE*); **~ espiral** *f* TELECOM spiral aerial (*BrE*), spiral antenna (*AmE*); **~ exploradora** *f* TV scanner; **~ de fase progresiva** *f* TEC ESP endfire aerial (*BrE*), endfire antenna (*AmE*); **~ de fosa** *f* TRANSP AÉR flush aerial (*BrE*), flush antenna (*AmE*); **~ fusiforme** *f* TEC ESP cigar aerial (*BrE*), cigar antenna (*AmE*); **~ giratoria** *f* TELECOM, TV rotatable aerial (*BrE*), rotatable antenna (*AmE*); **~ de guía de ondas** *f* TELECOM, TV waveguide aerial (*BrE*), waveguide antenna (*AmE*); **~ de guiado** *f* TEC ESP guidance aerial (*BrE*), guidance antenna (*AmE*); **~ de guíaondas** *f* TELECOM waveguide aerial (*BrE*), waveguide antenna (*AmE*); **~ de haces múltiples** *f* FÍS, TEC ESP, TELECOM multibeam aerial (*BrE*), multibeam antenna (*AmE*); **~ de haz** *f* TV beam aerial (*BrE*), beam antenna (*AmE*); **~ de haz en abanico** *f* TV fanned-beam aerial (*BrE*), fanned-beam antenna (*AmE*); **~ de haz conformado** *f* TEC ESP contoured beam aerial (*BrE*), contoured beam antenna (*AmE*), shaped beam aerial (*BrE*), shaped beam antenna (*AmE*); **~ de haz contorneado** *f* TEC ESP contoured beam aerial (*BrE*), contoured beam antenna (*AmE*), shaped beam aerial (*BrE*), shaped beam antenna (*AmE*); **~ de haz dirigido** *f* TV beam aerial (*BrE*), beam antenna (*AmE*); **~ de haz fino** *f*

TEC ESP spot beam aerial (*BrE*), spot beam antenna (*AmE*); ~ **de haz orientable** *f* TELECOM steerable beam aerial (*BrE*), steerable beam antenna (*AmE*); ~ **de haz perfilado** *f* TEC ESP shaped beam aerial (*BrE*), shaped beam antenna (*AmE*); ~ **en hélice** *f* TEC ESP helix aerial (*BrE*), helix antenna (*AmE*); ~ **helicoidal** *f* TEC ESP helix aerial (*BrE*), helix antenna (*AmE*), TELECOM corkscrew aerial (*BrE*), corkscrew antenna (*AmE*), helical aerial (*BrE*), helical antenna (*AmE*); ~ **de hilo** *f* TELECOM, TV wire aerial (*BrE*), wire antenna (*AmE*); ~ **interior** *f* ING ELÉC, TV indoor aerial (*BrE*), indoor antenna (*AmE*); ~ **isotrópica** *f* TEC ESP isotropic aerial (*BrE*), isotropic antenna (*AmE*); ~ **de látigo** *f* TEC ESP, TRANSP MAR, TV whip aerial (*BrE*), whip antenna (*AmE*); ~ **de látigo de cuarto de onda** *f* FÍS, TELECOM, TV quarter-wave whip aerial (*BrE*), quarter-wave whip antenna (*AmE*); ~ **de lente** *f* TEC ESP, TELECOM lens aerial (*BrE*), lens antenna (*AmE*); ~ **logarítmica** *f* TV log-periodic aerial (*BrE*), log-periodic antenna (*AmE*); ~ **de mástil** *f* TELECOM mast aerial (*BrE*), mast antenna (*AmE*); ~ **matricial** *f* TELECOM array aerial (*BrE*), array antenna (*AmE*); ~ **de microondas** *f* FÍS, TEC ESP, TELECOM, TV microwave aerial (*BrE*), microwave antenna (*AmE*); ~ **monofilar** *f* TELECOM wire aerial (*BrE*), wire antenna (*AmE*); ~ **monopolo** *f* TELECOM monopole aerial (*BrE*), monopole antenna (*AmE*), unipole aerial (*BrE*), unipole antenna (*AmE*); ~ **multibanda** *f* FÍS multiband antenna, TEC ESP wideband aerial (*BrE*), wideband antenna (*AmE*); ~ **multifrecuencia** *f* TELECOM multifrequency aerial (*BrE*), multifrequency antenna (*AmE*); ~ **multihaz** *f* FÍS, TELECOM multibeam aerial (*BrE*), multibeam antenna (*AmE*); ~ **multireflector** *f* TELECOM multiple reflector aerial (*BrE*), multiple reflector antenna (*AmE*); ~ **omnidireccional** *f* TEC ESP, TELECOM omnidirectional aerial (*BrE*), omnidirectional antenna (*AmE*); ~ **de onda progresiva** *f* TELECOM, TV traveling-wave antenna (*AmE*), travelling-wave aerial (*BrE*); ~ **de ondas gravitacionales** *f* FÍS, FÍS RAD gravitational-wave aerial (*BrE*), gravitational-wave antenna (*AmE*); ~ **orientable** *f* TELECOM, TV rotatable aerial (*BrE*), rotatable antenna (*AmE*); ~ **parabólica** *f* TEC ESP parabolic aerial (*BrE*), parabolic antenna (*AmE*), dish aerial (*BrE*), dish antenna (*AmE*), TELECOM dish aerial (*BrE*), dish antenna (*AmE*), parabolic aerial (*BrE*), parabolic antenna (*AmE*), TV parabolic aerial (*BrE*), parabolic antenna (*AmE*); ~ **parabólica receptora** *f* TELECOM, TV receiving-dish aerial (*BrE*), receiving-dish antenna (*AmE*); ~ **paraboloidal** *f* TEC ESP paraboloidal aerial (*BrE*), paraboloidal antenna (*AmE*); ~ **parásita** *f* FÍS parasitic aerial (*BrE*), parasitic antenna (*AmE*); ~ **pasiva** *f* FÍS passive aerial (*BrE*), passive antenna (*AmE*); ~ **de período logarítmico** *f* TV log-periodic aerial (*BrE*), log-periodic antenna (*AmE*); ~ **periscópica** *f* FÍS periscope aerial (*BrE*), periscope antenna (*AmE*); ~ **plana** *f* TELECOM flat aerial (*BrE*), flat antenna (*AmE*); ~ **de planeo** *f* TRANSP AÉR glide aerial (*BrE*), glide antenna (*AmE*); ~ **de plato** *f* TEC ESP, TELECOM dish aerial (*BrE*), dish antenna (*AmE*); ~ **plegable** *f* TEC ESP collapsible aerial (*BrE*), collapsible antenna (*AmE*); ~ **de radar** *f* FÍS, FÍS RAD, TELECOM, TRANSP, TRANSP AÉR, TRANSP MAR, TV radar aerial (*BrE*), radar antenna (*AmE*); ~ **radárica** *f* FÍS, FÍS RAD, TELECOM, TRANSP, TRANSP AÉR, TRANSP MAR, TV radar aerial (*BrE*), radar antenna (*AmE*); ~ **de radiación lateral** *f* TEC ESP broad-side aerial (*BrE*), broad-side antenna (*AmE*); ~ **de radiación longitudinal** *f* TELECOM end-fire aerial (*BrE*), end-fire antenna (*AmE*); ~ **de radiación regresiva** *f* TEC ESP back-fire aerial (*BrE*), back-fire antenna (*AmE*); ~ **de radiación transversal** *f* TEC ESP broad-side aerial (*BrE*), broad-side antenna (*AmE*); ~ **de radio** *f* AUTO, TELECOM, TRANSP, VEH radio aerial (*BrE*), radio antenna (*AmE*); ~ **radioastronómica** *f* TELECOM radioastronomical aerial (*BrE*), radioastronomical antenna (*AmE*); ~ **de radiogoniómetro** *f* TRANSP direction finder aerial (*BrE*), direction finder antenna (*AmE*); ~ **de ranura** *f* TEC ESP, TELECOM slot aerial (*BrE*), slot antenna (*AmE*); ~ **ranurada** *f* TEC ESP, TELECOM slot aerial (*BrE*), slot antenna (*AmE*); ~ **rasa** *f* TRANSP AÉR flush aerial (*BrE*), flush antenna (*AmE*); ~ **de rastreo** *f* TEC ESP tracking aerial (*BrE*), tracking antenna (*AmE*); ~ **de recepción** *f* FÍS RAD, TELECOM, TV receiving aerial (*BrE*), receiving antenna (*AmE*), reception aerial (*BrE*), reception antenna (*AmE*); ~ **receptora** *f* FÍS RAD, TELECOM, TV receiving aerial (*BrE*), receiving antenna (*AmE*); ~ **rectilínea** *f* TELECOM rectilinear aerial (*BrE*), rectilinear antenna (*AmE*); ~ **de reflector** *f* TEC ESP, TELECOM, TV reflector aerial (*BrE*), reflector antenna (*AmE*); ~ **con reflector angular** *f* TV corner reflector aerial (*BrE*), corner reflector antenna (*AmE*); ~ **de reflector Gregoriana** *f* TELECOM Gregorian reflector aerial (*BrE*), Gregorian reflector antenna (*AmE*); ~ **con reflector múltiple** *f* TELECOM multiple reflector aerial (*BrE*), multiple reflector antenna (*AmE*); ~ **de reflector parabólico** *f* TELECOM parabolic reflector aerial (*BrE*), parabolic reflector antenna (*AmE*); ~ **reflectora** *f* TEC ESP, TELECOM, TV reflector aerial (*BrE*), reflector antenna (*AmE*); ~ **reflectora cilíndrica** *f* TELECOM cylindrical reflecting aerial (*BrE*), cylindrical reflecting antenna (*AmE*); ~ **reflectora cónica** *f* TEC ESP umbrella reflector aerial (*BrE*), umbrella reflector antenna (*AmE*); ~ **retractible** *f* TV retractable aerial (*BrE*), retractable antenna (*AmE*); ~ **retráctil** *f* TV retractable aerial (*BrE*), retractable antenna (*AmE*); ~ **de seguimiento** *f* TEC ESP tracking aerial (*BrE*), tracking antenna (*AmE*); ~ **en serie** *f* TEC ESP array aerial (*BrE*), array antenna (*AmE*); ~ **toroidal** *f* TELECOM toroidal aerial (*BrE*), toroidal antenna (*AmE*); ~ **de transmisión** *f* TELECOM, TV transmit aerial (*BrE*), transmit antenna (*AmE*); ~ **transmisora** *f* TELECOM, TV transmitting aerial (*BrE*), transmitting antenna (*AmE*); ~ **de la trayectoria de planeo** *f* TRANSP, TRANSP AÉR glide-slope aerial (*BrE*), glide-slope antenna (*AmE*); ~ **en V** *f* TELECOM V-shaped aerial (*BrE*), V-shaped antenna (*AmE*); ~ **vertical flexible** *f* TV whip aerial (*BrE*), whip antenna (*AmE*); ~ **Yagi** *f* TEC ESP, TELECOM Yagi aerial (*BrE*), Yagi antenna (*AmE*)

anteojo: ~ **buscador** *m* INSTR *telescopio* finder; ~ **estadiómetro** *m* CONST *topografía* stadimeter; ~ **de puntería** *m* TEC ESP optical sight

anteojos *m pl* AmL (*cf gafas Esp*) INSTR glasses, spy glass, ÓPT glasses; ~ **de alto rendimiento para visión nocturna** *m pl* AmL D&A high-performance night vision goggles; ~ **industriales** *m pl* AmL LAB, PROD, SEG industrial eye protectors; ~ **de lectura** *m pl* INSTR reading glasses, reading spectacles, ÓPT reading glasses; ~ **multifocales** *m pl AmL* INSTR, ÓPT multifocal glasses; ~ **polarizantes** *m pl AmL* FOTO polarizing spectacles; ~ **de protección** *m pl AmL* ING MECÁ, LAB, ÓPT, PROD, SEG glasses, goggles, protective glasses, protective spectacles; ~ **protectores de radiaciones X** *m pl AmL* INSTR X-ray protective glasses; ~ **protectores de rayos X** *m pl AmL* ÓPT, SEG X-ray protective glasses; ~ **de seguridad** *m pl AmL* (*cf lentes de seguridad Esp*) C&V, INSTR, LAB, ÓPT, PROD, SEG eye-protection glasses, safety glasses, safety goggles, safety spectacles; ~ **de sol** *m pl AmL* INSTR, ÓPT sun spectacles; ~ **de soldador** *m pl AmL* INSTR, SEG welder's goggles; ~ **para soldadores** *m pl AmL* INSTR, SEG welding goggles; ~ **para soldar** *m pl AmL* SEG welding visor (*BrE*), welding vizor (*AmE*); ~ **de visión nocturna** *m pl AmL* D&A, ÓPT night vision goggles (*NVG*)

antepaís *m* GEOL foreland, TRANSP MAR headland

antepasado *m* INFORM&PD ancestor

antepecho: ~ **de ventana** *m* CONST windowsill

anteposo *m* GEOL basement

antepozo *m* PETROL basement, cellar, derrick cellar, TEC PETR basement, cellar, *perforación* derrick cellar

anteproyecto *m* CONST preliminary design, survey

antepuerto *m* OCEAN, TRANSP MAR outer harbor (*AmE*), outer harbour (*BrE*), outer port, outport

anterior *adj* AUTO, ELECTRÓN, INFORM&PD, INSTR front-end, TEC ESP ahead, VEH front-end

antiabrasivo *adj* MECÁ abrasion-proof

antiaceleración: ~ **de la gravedad** *f* TEC ESP antigravity

antiácido *m* QUÍMICA antacid

antiadherente *m* REVEST bond-breaker (*BrE*)

antiadhesivo *m* PAPEL release

antibarión *m* FÍS PART antibaryon

antibloqueante *m* P&C *adhesivos, aditivo* antiblocking agent

anticatalizador *m* ALIMENT anticatalyst

anticatión *m* FÍS RAD antication

anticátodo *m* ELEC, FÍS, FÍS PART anticathode, target

antichoque *adj* EMB shockproof

anticiclogénesis *f* METEO anticyclogenesis

anticiclón *m* METEO anticyclone; ~ **de bloqueo** *m* METEO blocking anticyclone; ~ **continental** *m* METEO continental anticyclone; ~ **permanente** *m* METEO permanent anticyclone; ~ **semipermanente** *m* METEO semipermanent anticyclone; ~ **subtropical** *m* METEO subtropical anticyclone

anticipación[1]: **con** ~ *adv* TEC ESP ahead

anticipación[2] *f* INFORM&PD lookahead

anticipado *adj* ELEC antiparallel

anticipador: ~ **del paso del colectivo** *m* TRANSP AÉR collective-pitch anticipator

anticlinal[1] *adj* GAS, GEOL anticlinal

anticlinal[2] *m* GEOL, MINAS, TEC PETR anticline; ~ **de crecimiento** *m* GEOL growth anticline; ~ **exhumado** *m* GEOL exhumed anticlinal fold; ~ **de inversión** *m* GEOL roll-over anticline

anticlinical *adj* MINAS anticlinical

anticlinorio *m* GEOL *tectónica* anticlinorium

anticoincidencia *f* ELECTRÓN, FÍS, NUCL anticoincidence

anticongelamiento *m* TEC ESP anti-icing

anticongelante[1] *adj* GEN antifreezing

anticongelante[2] *m* GEN antifreeze, frost-preventive agent

anticorrosivo *adj* COLOR, EMB, MECÁ, P&C anticorrosive

antideflagrante *adj* GEN fireproof, flameproof

antideslizante *adj* QUÍMICA skidproof, TRANSP MAR *superficie de cubierta* nonslip

antideslizantes *m pl* ING MECÁ, MECÁ anti-drift units

antideslumbrante[1] *adj* C&V, CINEMAT, INSTR, SEG, VEH antidazzle

antideslumbrante[2] *m* C&V, CINEMAT, SEG, VEH antiglare

antidesplazantes *m pl* ING MECÁ, MECÁ anti-drift units

antidetonante *adj* MECÁ explosion-proof

antídoto *m* QUÍMICA, SEG antidote

antierosionante: ~ **marino** *m* METEO sea defence (*BrE*), sea defense (*AmE*)

antiespumante *m* CARBÓN, EMB, P&C, PAPEL, QUÍMICA, TEC PETR antifoamer, antifoaming agent, antifroth, defoaming agent

antiestático *adj* GEN antistatic

antiexplosivo *adj* MECÁ explosion-proof

antifase *adv* TV antiphase

antiferromagnético *adj* ELEC, FÍS antiferromagnetic

antiferromagnetismo *m* ELEC, FÍS antiferromagnetism

antifricción *adj* ING MECÁ, MECÁ, PROD antifriction

antifúngico *adj* PROD fungus-proof

antígeno *m* QUÍMICA *anticuerpos* antigen

antigravitatorio *adj* TEC ESP antigravity (*anti-G*)

antigüedad *f* GEOL age dating; ~ **del agua freática** *f* AGUA, GEOL, HIDROL age of groundwater

antihalo *m* FOTO, IMPR nonhalation

antiherrumbroso *adj* EMB, MECÁ, P&C, REVEST antirust

antihielo *m* TEC ESP anti-icing, TRANSP AÉR antifreeze

antiincrustante *m* REFRIG *condensadores, torres de refrigeración* scale inhibitor

antiinflamable *adj* GEN flame-resistant

antiinterferencia *f* ELEC *protección* anti-interference

antillama *adj* GEN fireproof, flameproof

antilogaritmo *m* MATEMÁT antilogarithm

antimagnético *adj* ELEC, GEOFÍS nonmagnetic

antimateria *f* FÍS, FÍS PART, NUCL antimatter

antimoniado *adj* QUÍMICA antimonial

antimoniato *m* QUÍMICA antimoniate

antimónico *adj* QUÍMICA antimonic

antimonio *m* (*Sb*) C&V, METAL, QUÍMICA antimony (*Sb*)

antimonita *f* MINERAL, QUÍMICA antimonite, stibnite

antimoniuro *m* QUÍMICA antimonide

antineutrino *m* FÍS PART, NUCL antineutrino

antineutrón *m* FÍS antineutron

antinodo *m* ACÚST, ELEC antinode, loop, FÍS, FÍS ONDAS, ING ELÉC loop, antinode, PROD *oscilaciones* loop; ~ **de la corriente** *m* FÍS current loop

antioxidante[1] *adj* ALIMENT, AUTO, EMB, MECÁ, P&C, REVEST antioxidant, antirust

antioxidante[2] *m* ALIMENT, AUTO, EMB, MECÁ, P&C, REVEST antioxidant, corrosion inhibitor, TEC ESP, VEH corrosion inhibitor

antióxido *m* CINEMAT antistain agent

antiparalelo[1] *adj* ELECTRÓN, GEOM antiparallel

antiparalelo[2] *m* GEOM antiparallel

antiparasitario *adj* ACÚST *radio*, SEG antinoise

antiparásito *m* ACÚST, SEG *radio* antinoise

antipartícula *f* FÍS, FÍS PART antiparticle

antipersonal *adj* D&A antipersonnel (*AP*)

antipertita *f* MINERAL antiperthite
antiplástico *m* GEOL shortening
antípoda: ~ **óptico** *m* CRISTAL optical antipode
antípodas *m pl* GEOM antipodes
antiprotón *m* FÍS, FÍS PART antiproton
antiquark *m* FÍS, FÍS PART antiquark
antirreactividad *f* FÍS, NUCL negative reactivity
antirredeposición *f* DETERG antiredepositing
antirresbaladizo *adj* TRANSP MAR *superficie de cubierta* nonslip
antirresonancia *f* ACÚST, ELECTRÓN antiresonance
antirreventón *m* PETROL blowout preventer
antirrobo *adj* AUTO, CALIDAD, SEG antitheft
antirruido *m* ACÚST antinoise
antisedimentante *adj* P&C *aditivo de pinturas* antisettling
antisolapamiento *m* INFORM&PD anti-aliasing
antisubmarino *adj* D&A antisubmarine (*A/S*)
antivelo *m* CINEMAT, FOTO antifoggant
antiviento *m* ACÚST, CINEMAT, TV *protector para micrófonos* windshield
antorcha *f* CONST, GAS, MECÁ, PROD torch, TEC PETR *producción, refino* flare, flare stack; ~ **para gases** *f* GAS, TERMO gas flare
antozonita *f* MINERAL antozonite
antraceno *m* QUÍMICA anthracene
antracita *f* C&V culm, CARBÓN glance coal, hard coal, MINAS stone coal
antracnosis *f* AGRIC *patología vegetal* anthracnose; ~ **de la cebada** *f* AGRIC barley anthracnose; ~ **del guisante** *f* AGRIC leaf-and-pod spot of pea
antragalol *m* QUÍMICA anthragallol
antril *m* QUÍMICA anthryl
antrilo *m* QUÍMICA anthryl
anuario: ~ **hidrológico** *m* HIDROL hydrologic yearbook; ~ **de mareas** *m* OCEAN, TRANSP MAR *navegación* tide table
anubarrado *adj* TV clouding
anublarse *v refl* METEO cloud over
añublo *m* AGRIC *patología vegetal* blight, late blight; ~ **de la hoja** *m* AGRIC leaf blight; ~ **de la papa** *m* AmL (*cf añublo de la patata Esp*) ALIMENT potato blight; ~ **de la patata** *m* Esp (*cf añublo de la papa AmL*) ALIMENT potato blight; ~ **precoz** *m* AGRIC early blight
anudadura *f* TEXTIL knotting
anudamiento *m* TEXTIL knotting
anudar *vt* TEXTIL knot
anulación *f* TELECOM deletion; ~ **de filtro** *f* ELECTRÓN filter zero; ~ **manual** *f* PROD manual override
anulador *m* FÍS nullator
anular[1] *adj* GEN annular
anular[2] *m* TEC PETR *perforación, producción* annulus
ánulo *m* GEOM annulus
anuncio *m* IMPR advertisement, ING MECÁ placard; ~ **de cambio de especificación** *m* TRANSP specification change notice (*SCN*); ~ **a toda página** *m* IMPR full-page advertisement; ~ **publicitario** *m* Esp (*cf propaganda AmL*) TV commercial; ~ **supervisor** *m* TELECOM supervisory announcement
anuncios: ~ **contiguos** *m pl* TV back-to-back commercials
anverso *m* IMPR, PAPEL recto
anzuelo *m* ING MECÁ hook
apagachispas *m* ELEC *relé* spark arrester, spark blowout, spark extinguisher, spark-quencher

apagado[1] *adj* COLOR dull, IMPR mute, ING ELÉC turned-off, TEXTIL *color* dull; **no** ~ *adj* QUÍMICA *cal* unslaked
apagado[2] *m* C&V shut-off, shutdown, ELEC, TEC ESP shutdown; ~ **del arco** *m* ELEC arc extinction; ~ **de gas** *m* GAS, INSTAL TERM gas quench; ~ **por mezcla pobre** *m* TRANSP AÉR lean die-out; ~ **del motor** *m* TEC ESP, TRANSP AÉR engine flameout; ~ **del motor en vuelo** *m* TRANSP AÉR engine shutdown in flight
apagador: ~ **del arco** *m* ELEC *aparamenta eléctrica* arc breaker; ~ **de chispas** *m* ELEC *relé* spark-quencher
apagafuegos *m* CONST, SEG, TERMO firefighter
apagallamas *m* CONST, SEG, TEC ESP, TEC PETR, TERMO flame arrester, flame trap
apagamiento: ~ **del arco** *m* ELEC arc extinction
apagar *vt* AGRIC *fuego de bosques* mop up, CINEMAT kill, turn off, cut, *la luz* black out, ING ELÉC turn off, switch off, ING MECÁ switch off, deaden, PROD *horno alto* blowout, QUÍMICA slake, SEG *un incendio o fuego* put out, TERMO smother, TRANSP AÉR *motor* shut down
apagarse *v refl* ING ELÉC go out
apagón *m* ELEC, ING ELÉC, NUCL, TELECOM *servicio eléctrico* blackout; ~ **de la central** *m* NUCL station blackout
apague: ~ **del motor** *m* TEC ESP, TRANSP AÉR engine flameout
apaisado *adj* IMPR elongated
apalancar *vt* CONST prise (*BrE*), pry (*AmE*), ING MECÁ lever, lever up
apantallado *m* ELEC *campo electromagnético* screening, INFORM&PD *protección eléctrica* screen
apantallamiento *m* ELEC, FÍS, FÍS RAD, NUCL, TEC ESP *electrónica* screening, shielding; ~ **diamagnético** *m* FÍS, FÍS RAD *del núcleo* diamagnetic shielding
apantallar *vt* ELEC, FÍS, FÍS RAD, NUCL, TEC ESP screen, shield
aparadura *f* TRANSP MAR *construcción* garboard
aparamenta: ~ **eléctrica** *f* ING ELÉC switchgear
aparatista *m* AmL INSTR instrument maker
aparato *m* ELEC, ELECTRÓN, FOTO apparatus, device, mechanism, ING MECÁ apparatus, device, fixture, mechanism, LAB, MECÁ, TRANSP MAR apparatus, device, mechanism; ~ **acoplado de carga** *m* ELECTRÓN buried channel charge-coupled device (*buried channel CCD*); ~ **de acoplamiento cargado** *m* ELECTRÓN buried channel CCD (*buried channel charge-coupled device*); ~ **de aireación** *m* RECICL aerator; ~ **de ala rotatoria** *m* TRANSP AÉR rotating-wing aircraft; ~ **alimentador** *m* PROD feeder; ~ **amortiguador** *m* ING MECÁ damping device; ~ **analizador codificador de las palabras** *m* TELECOM vocoder; ~ **de arranque eléctrico** *m* AUTO, ELEC *automotor* electric starter; ~ **aspirador de polvo** *m* EMB dust-exhausting device; ~ **autorregistrador** *m* ING MECÁ self-registering apparatus; ~ **para autorrescate** *m* SEG *con filtro para monóxido de carbono* self-rescue apparatus; ~ **de autorrescate con oxígeno** *m* SEG oxygen self-rescue apparatus; ~ **de calefacción** *m* TERMO, TERMOTEC heating appliance; ~ **casero** *m* ELEC domestic appliance, home appliance; ~ **de centraje universal** *m* INSTR universal centering apparatus (*AmE*), universal centring apparatus (*BrE*); ~ **para cerrar o precintar con bandas** *m* EMB tape sealer; ~ **para cerrar puertas resistentes al fuego** *m* SEG locking device for fire-resisting doors; ~ **de choque** *m* FERRO buffing gear;

~ **climatizador** *m* ING MECÁ air conditioner; ~ **compacto** *m* ELECTRÓN solid-state device; ~ **de comprobación** *m* CALIDAD, EMB checking apparatus; ~ **computadora de distancias** *f AmL* (*cf ordenador Esp*) ELEC computer; ~ **conductor** *m* ING MECÁ driving gear; ~ **de conexión** *m* ELEC switchgear; ~ **conmutador** *m* TELECOM switching device; ~ **de contestación automática de llamadas telefónicas** *m* TELECOM telephone-answering machine; ~ **copiador** *m* ING MECÁ copying attachment; ~ **de cortar diamante circular** *m* ING MECÁ circular diamond-cutting apparatus; ~ **de cromatografía sobre papel** *m* LAB, QUÍMICA paper-chromatography apparatus; ~ **para curvar carriles** *m* FERRO rail-bending device; ~ **de Dean y Stark** *m* LAB Dean and Stark apparatus; ~ **de desprendimiento** *m* ING MECÁ detachment device; ~ **de destilación** *m* LAB, PROC QUÍ, QUÍMICA distiller, distilling apparatus; ~ **de digestión** *m* LAB digestion apparatus; ~ **de digestión de Kjeldahl** *m* LAB *análisis de nitrógeno* Kjeldahl digestion apparatus; ~ **digital** *m* ELECTRÓN digital device; ~ **disgregador de suelos** *m AmL* (*cf escarificador Esp*) MINAS ripper; ~ **de disparo a resorte** *m* TEC ESP spring-release device; ~ **de distribución de gas** *m* GAS, PROD gas fitting; ~ **distribuidor de la carga** *m* PROD *horno alto* stock distributor; ~ **doméstico** *m* ELEC domestic appliance, home appliance; ~ **doméstico a gas** *m* GAS, INSTAL TERM domestic gas appliance; ~ **eléctrico** *m* ELEC, TEXTIL electrical appliance; ~ **eléctrico para mediciones** *m* ELEC, SEG electrical measuring-apparatus; ~ **electrodinámico** *m* ELEC, INSTR electrodynamic instrument; ~ **de elevación** *m* ING MECÁ lifting apparatus; ~ **para elevar** *m* ING MECÁ lifting apparatus, SEG lifting tackle; ~ **enrollador** *m* TEXTIL wind-up apparatus; ~ **para enseñanza** *m* ELECTRÓN, SEG, TRANSP AÉR trainer; ~ **equipado de canal n** *m* ELECTRÓN n-channel device; ~ **de estado sólido** *m* FÍS solid-state device; ~ **de extracción** *m* ING MECÁ, PROD hoisting gear, lifting gear; ~ **gasífero** *m* GAS, PROD gas-making apparatus; ~ **de gobierno** *m* TRANSP MAR *buque* helm; ~ **grabador** *m* ING MECÁ recorder; ~ **de granulometría** *m* PROC QUÍ particle-size analyser; ~ **de hierro móvil** *m* ELEC, ING ELÉC moving-iron instrument; ~ **de iluminación** *m* LAB illuminating apparatus; ~ **de incidencia** *m* QUÍMICA impinger; ~ **de izar** *m* ING MECÁ, PROD, SEG hoisting gear, lifting tackle; ~ **de Kipp** *m Esp* (*cf sulfhidrador AmL*) LAB, QUÍMICA *generación de ácido sulfhídrico* Kipp's apparatus; ~ **de lavado** *m AmL* (*cf canal de lavado Esp*) MINAS sluice; ~ **de lectura directa** *m* ELEC, INSTR direct reading instrument; ~ **para levantar** *m* SEG lifting tackle; ~ **de liberización por resorte** *m* TEC ESP spring-release device; ~ **licuante refrigerante** *m* ING MECÁ, REFRIG refrigerant liquefying set; ~ **de limpieza** *m* PROD *herramienta de moldeo* cleaner; ~ **de mano** *m* TRANSP MAR *jarcia* handy billy; ~ **de máximo de corriente continua** *m* ELEC direct overcurrent release; ~ **de máximo de corriente indirecta** *m* ELEC indirect overcurrent release; ~ **de máximo de tensión** *m* ELEC *disyuntor* overvoltage release; ~ **mecánico** *m* ING MECÁ, PROD power unit; ~ **de medición** *m* CONST, ELEC, ELECTRÓN, INSTR, METR measuring apparatus; ~ **de medida** *m* CONST, ELEC, ING ELÉC, INSTR, METR measuring apparatus, measuring device, measuring instrument; ~ **de medida para calcular el grosor de una capa** *m* INSTR, METR, REVEST coating thickness measurement apparatus; ~ **de medida coordinado** *m* METR coordinate measuring machine; ~ **de medida de varias sensibilidades** *m* ELEC *instrumento*, INSTR multirange meter; ~ **para medir y analizar aerosoles y polvos** *m* INSTR, SEG aerosol and dust measuring and analysis apparatus; ~ **para medir el brillo** *m* PAPEL gloss meter; ~ **para medir la compresión** *m* PAPEL compression tester; ~ **para medir el desgarro** *m* PAPEL tearing tester; ~ **para medir el grado de encolado** *m* PAPEL sizing tester; ~ **para medir la lisura** *m* PAPEL smoothness tester; ~ **para medir la opacidad** *m* PAPEL opacity tester; ~ **para medir la resistencia al desgarro por torsión** *m* PAPEL torsional tear tester; ~ **para medir la resistencia a la flexión** *m* PAPEL bending stiffness tester; ~ **para medir la resistencia a la perforación** *m* PAPEL puncture tester; ~ **para medir la rugosidad** *m* PAPEL roughness tester; ~ **monitor** *m* TELECOM monitor unit; ~ **motor** *m* ING MECÁ driving gear; ~ **motor superior** *m* ING MECÁ top-driving apparatus; ~ **motriz** *m* ING MECÁ driving gear; ~ **Mullen para medir la resistencia al estallido** *m* PAPEL Mullen tester; ~ **portátil** *m* ELEC portable appliance; ~ **de propulsión** *m* ING MECÁ propelling gear; ~ **protector** *m* ING MECÁ, MINAS, SEG safety apparatus, safety device; ~ **para pulir con chorro de arena** *m* C&V sandblast apparatus; ~ **de punto de goteo** *m* LAB drop point apparatus; ~ **de punto de inflamación** *m* LAB *líquidos inflamables* flash point apparatus; ~ **de purificación** *m* LAB, PROC QUÍ, QUÍMICA purifying apparatus; ~ **de radiocomunicación** *m* TELECOM, TRANSP AÉR, TRANSP MAR radio facility; ~ **de rayos X** *m* INSTR X-ray apparatus; ~ **de rayos X industrial** *m* INSTR industrial X-ray apparatus; ~ **para realizar el ensayo de resistencia a la compresión** *m* EMB compression test machine; ~ **reanimador** *m* SEG resuscitator; ~ **para reavivar muelas de esmeril** *m* PROD emery-wheel dresser; ~ **de recuperación** *m* CONTAM, CONTAM MAR recovery device; ~ **registrador** *m* INSTR recording device; ~ **de registro automático** *m* ING MECÁ self-registering apparatus; ~ **de registro gráfico en coordenadas cartesianas** *m* ELEC X-Y recorder; ~ **para rescate en incendios** *m* SEG, TERMO fire rescue appliance; ~ **de respirar** *m* CARBÓN, MINAS, OCEAN, SEG, TEC PETR *seguridad personal* breathing apparatus; ~ **para respirar con inyección de aire** *m* SEG supplied-air breathing apparatus; ~ **para respirar oxígeno** *m* SEG oxygen breathing apparatus; ~ **respiratorio** *m* CARBÓN, MINAS, OCEAN, SEG, TEC PETR breathing apparatus, respiratory system; ~ **respiratorio autónomo** *m* OCEAN, TRANSP MAR *submarinismo* scuba; ~ **de rodadura** *m* ING MECÁ running gear; ~ **salvavidas** *m* SEG, TRANSP MAR life-saving apparatus; ~ **de seguridad** *m* ING MECÁ, MINAS, SEG safety apparatus, safety device; ~ **de separación** *m* ING MECÁ detachment device; ~ **separador** *m* PROD *para tuercas hexagonales y octogonales* dividing apparatus; ~ **de sondeo a mano** *m* MECÁ hand drill; ~ **superconductor** *m* ELECTRÓN superconducting device; ~ **de suspensión de barra** *m* ING MECÁ bar hanger; ~ **telefónico** *m* TELECOM handset; ~ **de telemedida** *m* ELEC *medición* telemeter; ~ **terminal** *m* TELECOM terminating equipment; ~ **termosellante** *m* LAB heat seal apparatus; ~ **para el tope de líquidos** *m* CARBÓN liquid limit device; ~ **de tracción por fricción** *m* ING MECÁ friction draft gear (*AmE*), friction draught gear (*BrE*); ~ **de uso doméstico** *m* ELEC domestic appliance, home appliance; ~ **de varias sensibilidades** *m* ELEC *instrumento*, INSTR multirange meter; ~ **ventilador de cochera** *m* SEG garage ventilating apparatus; ~ **verificador de transmisiones** *m* METR gear-testing machine; ~ **visor** *m* INSTR sighting device

aparatos *m pl* CONST fixtures and fittings; ~ **de conexión** *m pl* ING ELÉC switchgear; ~ **individuales para remoción de polvos** *m pl* SEG individual dust removal apparatus; ~ **para izar** *m pl* SEG lifting gear; ~ **para levantar** *m pl* SEG lifting gear; ~ **sanitarios** *m pl* C&V sanitary ware

aparcamiento *m* CONST lay-by, TRANSP car pool; ~ **de coches** *m Esp* (*cf utilización en común de automóviles AmL*) TRANSP car pooling; ~ **de helicópteros civiles** *m* TRANSP AÉR helistop

aparcar *vt* TV park

apareado *adj* ELECTRÓN, FÍS, ING ELÉC, QUÍMICA paired; **no** ~ *adj* ELECTRÓN, FÍS, ING ELÉC, QUÍMICA unpaired

apareamiento *m* ELEC, ING ELÉC *cable* pairing

aparejador *m* (*cf maestro de obras, constructor*) CONST house builder, master builder, TRANSP MAR *jarcia* rigger

aparejamiento *m* QUÍMICA coupling

aparejar *vt* PROD *cables* rig, TRANSP MAR *buque* equip, *jarcia* rig

aparejería *f* PROD *mecanismo de desplazamiento* tackle

aparejo *m* COLOR filler, CONST *albañería* bond, bonding, tackle, ING MECÁ purchase, MECÁ tackle, hoist, NUCL kit, OCEAN *artes de pesca* fishing gear, rigging, rig, PROD *mecanismo de desplazamiento* tackle, kit, TRANSP MAR *arte de pesca, motonería* tackle, rig, boat tackle, VEH *carrocería* undercoat (*AmE*), underseal (*BrE*); ~ **de amantillo** *m* TRANSP MAR lifting tackle; ~ **de bloques** *m* TEC PETR *perforación* blocks; ~ **de cabria** *m* PROD gin tackle; ~ **de cadena** *m* ING MECÁ chain pulley block; ~ **de combés** *m* CONST *navegación* luff tackle; ~ **compuesto** *m* TRANSP MAR *motonería, arte de pesca* purchase rig; ~ **diferencial de cadena** *m* ING MECÁ chain block, NUCL differential chain block; ~ **doble** *m* TRANSP MAR double tackle; ~ **para elevar** *m* ING MECÁ lifting tackle; ~ **de estrellera** *m* TRANSP MAR *veleros* winding tackle; ~ **de estrinque** *m* TRANSP MAR Spanish burton; ~ **de fuerza de un solo motón** *m* TRANSP MAR *jarcia de labor* runner; ~ **de izada** *m* PROD hoisting appliance; ~ **izador** *m* PROD hoisting tackle; ~ **para levantar** *m* ING MECÁ lifting tackle; ~ **de motón movible** *m* ING MECÁ runner; ~ **móvil** *m* PETROL traveling block (*AmE*), travelling block (*BrE*); ~ **con polea de gancho** *m* PROD tackle with hook block; ~ **de poleas** *m* ING MECÁ block-and-tackle, SEG pulley block; ~ **de poleas de seguridad** *m* SEG safety pulley block; ~ **a tizón** *m* CONST *albañilería* heading bond

aparición: ~ **gradual de la imagen** *f* CINEMAT, TV fade-in; ~ **en silueta** *f* CINEMAT, TV silhouetting

apariencia: ~ **fría** *f* C&V cold appearance

apartadero *m* FERRO *equipo inamovible*, TRANSP *ferrocarriles* siding, sidings; ~ **de almacenaje** *m* FERRO, TRANSP storage siding; ~ **de carga** *m* FERRO, TRANSP

loading siding; ~ **de clasificación** *m* FERRO, TRANSP *vehículos* sorting siding; ~ **en curva** *m* FERRO *equipo inamovible* turnout on the curve; ~ **industrial** *m* FERRO, TRANSP factory siding; ~ **para locomotoras** *m* FERRO, TRANSP *equipo inamovible* locomotive-holding siding; ~ **de retención** *m* FERRO, TRANSP *equipo inamovible* holding siding; ~ **para vagones vacíos** *m* FERRO, TRANSP empties siding

apartamiento: ~ **de la ebullición nucleada** *m* NUCL departure from nuclear boiling (*DNB*)

apartar *vt* ING MECÁ deflect, set over

apastro *m* TEC ESP apastron

apatelita *f* MINERAL apatelite

apatito *m* MINERAL apatite

apea *f* CARBÓN, MINAS prop, tree

apelmazamiento *m* ALIMENT caking

apelmazar[1] *vt* PROC QUÍ, QUÍMICA cake

apelmazar[2] *vi* ALIMENT cake

apéndice *m* IMPR addendum, appendix, back matter, MECÁ, PROD tab; ~ **geomagnético** *m* GEOFÍS, TEC ESP geomagnetic tail

apeo *m* CONST flying shore, prop; ~ **de mina** *m AmL* (*cf asiento de mina Esp*) MINAS headgear

aperiódico *adj* ELEC *galvanómetro*, ELECTRÓN, FÍS, TV aperiodic

apero *m* AGRIC farm implement

apertura *f* ELECTRÓN, ING ELÉC, INSTR aperture, opening, MINAS cross driving, *de galería de mina* driving, forcing, *del manto, veta, filón* opening up, NUCL *de una válvula* lift, TEC ESP, TELECOM, TRANSP AÉR, TV aperture; ~ **del cierre de la puerta del tren de aterrizaje** *f* TRANSP AÉR landing gear door unlatching; ~ **de galerías con perforadoras neumáticas** *f* CARBÓN *minas*, MINAS churn drilling; ~ **numérica** *f* CONST, ÓPT, TELECOM numerical aperture (*NA*); ~ **numérica de lanzamiento** *f* ÓPT, TELECOM launch numerical aperture (*LNA*); ~ **numérica máxima** *f* ÓPT, TELECOM maximum numerical aperture; ~ **numérica de salida** *f* ÓPT, TELECOM launch numerical aperture (*LNA*); ~ **relativa** *f* FÍS relative aperture; ~ **superior del cursor** *f* INSTR aperture top slide; ~ **de zanja** *f* CONST trench work, trenching

ápice *m* AGRIC apex, CARBÓN apex, *botánica* tip, GEOM, TEC ESP apex

apicultura *f* AGRIC apiculture, beekeeping

apiína *f* QUÍMICA apiin

apilable *adj* EMB stackable

apilado[1] *adj* CONST drifting, heaped, TERMO *fuegos de un horno* banked up

apilado[2] *m* EMB, IMPR, INFORM&PD, MINAS, PAPEL stack, stacking; ~ **múltiple** *m* TRANSP multiple pile-up

apilador: ~ **de carbón** *m AmL* (*cf extendedura de carbón Esp*) CARBÓN, MINAS trimmer

apiladora: ~ **y atadora con alambre** *f* EMB wire stacking machine

apilamiento *m* CONST, CRISTAL, GEOM, INFORM&PD, ING ELÉC, PROD piling, piling up, stacking, TRANSP MAR *de carga* overstowage; ~ **por compactación** *m* CARBÓN compaction piling; ~ **estrecho** *m* CONST close piling

apilar *vt* CONST bank, bank up, pile, stack, INFORM&PD stack, MINAS bank, bank up, PAPEL stack up, PROD stack

apiñamiento *m* MINAS squeezing

apionol *m* QUÍMICA apionol

apiro *adj* QUÍMICA apyrous
apisonado *m* ING MECÁ packing, PROD *de arena, fundería* packing, ramming, tamping
apisonador *m* CONST rammer, PROD *persona* tamper
apisonadora *f* AGRIC tamper, AUTO road locomotive, road roller, CONST earth-rammer, ram, *carreteras* roller, road roller, MINAS tamping material, tamper, PROD *máquina* tamper, TRANSP, VEH road roller; **~ con neumáticos** *f* CONST pneumatic-tired roller (*AmE*), pneumatic-tyred roller (*BrE*); **~ de neumáticos** *f* CONST rubber-tyred roller (*BrE*), *maquinaria* rubber-tired roller (*AmE*); **~ de rodillo vibratorio** *f* CONST vibrating sheepsfoot roller; **~ de sílex** *f* CARBÓN chat roller; **~ vibradora** *f* CONST tandem vibrating roller
apisonamiento *m* CONST, MINAS, PROD ramming, tamping
apisonar *vt* CONST, MINAS, PROD ram, tamp
apjohnita *f* MINERAL apjohnite
aplanado *m* C&V, ING MECÁ, MECÁ planing, surfacing
aplanador *m* C&V, ING MECÁ planer, planing machine, MECÁ planer (*AmE*), planing machine (*BrE*)
aplanadora *f* CARBÓN dresser, CONST leveling machine (*AmE*), levelling machine (*BrE*), ING MECÁ straightening machine; **~ y curvadora de chapas** *f* ING MECÁ plate-flattening-and-bending machine
aplanamiento *m* INFORM&PD smoothing, PROD flattening, smoothing
aplanar *vt* AGRIC *el suelo* level, CONST grade, MECÁ level, plane
aplastamiento *m* CONST crushing, ELECTRÓN implosion, GEOL flattening, P&C crushing, PAPEL battering, crushing, PROD flattening, implosion, SEG crushing; **~ de la calandra** *m* PAPEL calender smash
aplastar *vt* CARBÓN, CONST, PAPEL crush
aplazamiento *m* TEC ESP *lanzamiento de misiones* delay
aplicación *f* INFORM&PD, MECÁ application; **~ de un acabado nuevo** *f* P&C *pintura*, REVEST refinishing; **~ del adhesivo en líneas paralelas** *f* EMB parallel glueing; **~ en banda** *f* AGRIC *de fertilizantes* band application; **~ de bridas de angular** *f* FERRO angle fishplating; **~ de carga** *f* TRANSP AÉR load application; **~ después de siembra** *f* AGRIC post-sowing application; **~ directa del hielo** *f* REFRIG contact icing; **~ directa del hielo en seno de carga** *f* REFRIG body-icing; **~ esclava** *f* INFORM&PD slave application; **~ espontánea de frenos** *f* FERRO, VEH spontaneous brake application; **~ gestionada por menús** *f* INFORM&PD menu-driven application; **~ con láser** *f* ELECTRÓN lasing medium; **~ de microprocesores a sistemas de calibración** *f* ING MECÁ application of microprocessors to gaging systems (*AmE*), application of microprocessors to gauging systems (*BrE*); **~ normal del freno de vacío** *f* FERRO normal vacuum-brake application; **~ normal de frenos** *f* FERRO normal brake application; **~ normal del vacuofreno** *f* FERRO normal vacuum-brake application; **~ de plasma reactivo** *f* ELECTRÓN reactive-plasma etching; **~ en postemergencia** *f* AGRIC post-emergence application; **~ en preemergencia** *f* AGRIC pre-emergence application; **~ en presiembra** *f* AGRIC pre-sowing application; **~ regida por menús** *f* INFORM&PD menu-driven application; **~ de revestimiento con cuchilla** *f* P&C, REVEST knife spreading; **~ de revestimiento por inmersión** *f* P&C *proceso*, REVEST dip coating;

~ subordinada *f* INFORM&PD slave application; **~ tipo perro guardián** *f* *AmL* (*cf aplicación de vigilancia Esp*) INFORM&PD watchdog application; **~ de vidrio** *f* ELECTRÓN glassivation; **~ de vigilancia** *f* *Esp* (*cf aplicación tipo perro guardián AmL*) INFORM&PD watchdog application; **~ a voleo** *f* AGRIC broadcast
aplicaciones: para ~ diversas *adj* DETERG, ING ELÉC, ING MECÁ, MECÁ, PROD general-purpose
aplicador: ~ de adhesivos *m* EMB adhesive applicator; **~ de bandas** *m* EMB band sealer; **~ de cubierta de protección** *m* C&V apron applicator; **~ de filamento** *m* C&V yarn applicator; **~-impresor** *m* EMB printer-applicator; **~ de precintos** *m* EMB band sealer
aplicar *vt* CINEMAT *emulsion*, COLOR *tintura* stain, ING ELÉC, MECÁ apply, PROD apportion, REVEST apply; **~ una capa provisional** *vt* PROD *moldes* thicken; **~ energía a** *vt* ING ELÉC apply power to
aplique *m* CONST, ELEC, ING ELÉC wall fitting
aplita *f* GEOL, PETROL aplite, haplite
aplítico *adj* GEOL, PETROL aplitic
aplomar *vt* CONST plumb
aplomo *m* MINERAL aplome
apnea *f* OCEAN apnea (*AmE*), apnoea (*BrE*); **~ voluntaria** *f* OCEAN *buceo* breath-holding
apocromático *adj* IMPR apochromatic
apofilita *f* MINERAL apophyllite
apogeo *m* FÍS *astronomía*, TEC ESP apogee
apomazado *m* PROD pumicing
aporcado *adj* AGRIC ridging
aporcador *m* AGRIC hiller, ridger
aporcamiento *m* AGRIC *plantas* earthing-up
aportación *f* AGUA make-up, *cuencas* annual runoff; **~ de agua** *f* AGUA inflow of water
aporte *m* GEOL supply; **~ de agua al primario** *m* AGUA, HIDROL, NUCL primary makeup
apóstrofo *m* IMPR apostrophe
apotema *f* GEOM apothem
apótome *m* ACÚST *medio tono cromático* apotome
apoyabrazo *m* AUTO, VEH *asiento* armrest
apoyacabeza *m* AUTO, VEH *asiento* headrest
apoyamanos: ~ del torno *m* ING MECÁ turning rest
apoyar *vt* CONST bed, INFORM&PD support, ING MECÁ bear, hold up, support
apoyo[1] *m* *AmL* (*cf poste Esp*) CARBÓN prop, CONST bearing, prop, support, supporting, INFORM&PD support, ING MECÁ holder, prop, standard, upright, INSTR, MECÁ cradle, MINAS *construcción* upright, prop, PROD backup, TRANSP AÉR backing; **~ aéreo cercano** *m* D&A close air support (*CAS*); **~ para caños** *m* ING MECÁ pipe support; **~ de la cimentación** *m* CONST footing; **~ de control cíclico longitudinal** *m* TRANSP AÉR fore-and-aft cyclic control support; **~ empotrado** *m* MINAS *construcción*, TELECOM wall bracket; **~ empotrado de la galería** *m* MINAS end-wall bracket; **~ logístico** *m* TRANSP AÉR logistic support; **~ de mano** *m* ING MECÁ handrest; **~ de oscilación** *m* ING MECÁ rocker bearing; **~ postventa** *m* CALIDAD after sales servicing; **~ protector** *m* MINAS shield support; **~ del transporte de carga** *m* TRANSP AÉR cargo-carrier support; **~ en V para trazar** *m* ING MECÁ V-block
apoyo[2]: **de ~** *fra* INFORM&PD backup
APR *abr* (*análisis probabilista del riesgo*) NUCL PRA (*probabilistic risk assessment*)

APRA *abr* (*ayuda de punteo radar automático*) TRANSP MAR ARPA (*automatic radar plotting aid*)

apreciación *f* GEN assessment; **~ de los riesgos** *f* CALIDAD, CONTAM, CONTAM MAR, SEG risk assessment, risk evaluation

apreciar *vt* CALIDAD assess

apremio *m* INFORM&PD, PROD *ordenador* prompt

aprendizaje: **~ asistido por ordenador** *m Esp* (*cf aprendizaje con ayuda de computadora AmL*) INFORM&PD machine learning; **~ con ayuda de computadora** *m AmL* (*cf aprendizaje asistido por ordenador Esp*) INFORM&PD machine learning

aprestador *m* P&C primer

apresto *m* P&C, PAPEL size, TEXTIL sizing, *acabado, encolado* finish, size; **~ acrílico** *m* TEXTIL acrylic size; **~ de agua sobre calandria** *m* PAPEL water finish; **~ de alcohol de polivinilo** *m* TEXTIL polyvinyl alcohol size; **~ de carbohidrato** *m* TEXTIL carbohydrate size; **~ por encoladora** *m* TEXTIL slasher sizing; **~ del hilado** *m* TEXTIL spun yarn sizing; **~ proteínico** *m* TEXTIL protein size; **~ sintético** *m* TEXTIL synthetic size

apretador *m* ING MECÁ tightener

apretadora *f* CONST pinch cock

apretamiento *m* MINAS *del techo* squeeze

apretar *vt* ING MECÁ tighten, *laminadoras* pinch, PROD grip, *fundición* ram, TEC ESP *tanques* top up; **~ al máximo** *vt* ING MECÁ tighten up hard

apretura *f* FERRO, ING MECÁ tightness

aprietacable: **~ para armaduras** *m* ELEC *cable* armor clamp (*AmE*), armour clamp (*BrE*)

aprietatuercas: **~ neumático de percusión** *m* ING MECÁ pneumatic impact-wrench

aprietavacío *m* AGRIC *manga* squeeze gate

apriete *m* CONST *tornillo* driving, driving-in, ELECTRÓN clamping, ING MECÁ screwing, jamming, clamping, tightening, MECÁ, PROD clamping; **~ hidráulico** *m* ING MECÁ hydraulic clamping; **~ limitado** *m* ING MECÁ limited tightness

aprisionado: **~ por la nieve** *adj* METEO snowed-up

aproar: **~ a** *vt* TRANSP MAR head for, stand for

aprobación *f* CALIDAD, TRANSP AÉR, VEH *normas establecidas* approval; **~ de aptitud** *f* CALIDAD capability approval; **~ de calificación** *f* CALIDAD qualification approval; **~ de capacidad** *f* CALIDAD capability approval; **~ de tipo** *f* CALIDAD type approval

aprobado *adj* CALIDAD, SEG, TELECOM, TRANSP AÉR approved; **no ~** *adj* CALIDAD, SEG, TELECOM *aparato* nonapproved

aprobar *vt* CALIDAD approve, IMPR *visto bueno para imprimir* pass, SEG, TRANSP AÉR approve

aprovechamiento: **~ industrial** *m* AGUA harnessing; **~ de un río** *m* AGUA, HIDROL river capture; **~ de tierras** *m* AGRIC land development

aprovisionamiento *m* AGUA supply, PROD store, supplying; **~ de buques** *m* TRANSP MAR *provisiones* shipchandling; **~ en vuelo** *m* D&A air-to-air refueling (*AmE*), air-to-air refuelling (*BrE*)

aprovisionar *vt* CARBÓN store, TERMO *de combustible* fuel; **~ de combustible** *vt* PROD stoke

aproximación *f* MATEMÁT approximation, TEC ESP, TRANSP AÉR, TRANSP MAR *navegación* approach; **~ asintomática** *f* TELECOM asymptomatic approximation, asymptomatical approximation; **~ a ciegas** *f* TRANSP AÉR instrument approach; **~ circular** *f*

TRANSP AÉR circling approach; **~ continuada** *f* TRANSP AÉR steady approach; **~ controlada desde tierra** *f* TRANSP AÉR ground-controlled approach; **~ cuasiquímica** *f* METAL quasi-chemical approximation; **~ curvada** *f* TRANSP AÉR curved approach; **~ directa** *f* TRANSP AÉR straight-in approach; **~ final** *f* TRANSP AÉR final approach; **~ inicial** *f* TRANSP AÉR initial approach; **~ por instrumentos** *f* TRANSP AÉR instrument approach; **~ intermedia** *f* TRANSP AÉR intermediate approach; **~ interrumpida** *f* TRANSP AÉR discontinued approach; **~ al límite de la pista** *f* TRANSP AÉR end of runway approach; **~ lineal** *f* TELECOM linear approximation; **~ al milímetro más cercano** *f* MATEMÁT approximation to the nearest millimeter (*AmE*), approximation to the nearest millimetre (*BrE*); **~ de precisión** *f* TRANSP AÉR precision approach; **~ por radar** *f* TRANSP, TRANSP AÉR, TRANSP MAR radar approach, radar homing; **~ semiclásica** *f* NUCL semiclassical approximation; **~ visual** *f* TEC ESP visual approach

aproximaciones: **~ sucesivas** *f pl* INFORM&PD stepwise refinement

aproximado *adj* ING MECÁ coarse, MATEMÁT approximate

aproximar *vt* MATEMÁT approximate, MINAS, TRANSP AÉR, TRANSP MAR approach

APS *abr* (*análisis probabilista de la seguridad*) NUCL PSA (*probabilistic safety assessment*)

APT *abr* (*herramienta de programación automática*) INFORM&PD APT (*automatic programming tool*)

áptero *adj* AGRIC apterous

Aptiense *m* GEOL *estratigrafía* Aptian

aptitud *f* CALIDAD capability; **~ al desgote** *f* PAPEL drainability; **~ de despegue** *f* TRANSP AÉR takeoff ability; **~ para la impresión** *f* IMPR, PAPEL printability; **~ al plegado** *f* PAPEL creaseability; **~ al refino** *f* PAPEL beatability; **~ a volver a ser apilado** *f* EMB restackability

apto: **no ~ para consumo humano** *adj* ALIMENT unfit for human consumption

apuntador *m* ELEC *instrumento*, TELECOM pointer; **~ de elevación** *m* TEC ESP *cañones* layer

apuntalamiento *m* AGRIC inarching, CARBÓN shoring-up, CONST bracing, buttressing, flying shore, shoring, strutting, underpinning, PROD staying

apuntalar *vt* CARBÓN shore up, CONST, ING MECÁ, MECÁ brace, TRANSP MAR *construcción* shore up

apuntar *vt* CONST point, FOTO sight

apunte: **~ de datos** *m* ELEC data recording

aquerita *f* PETROL akerite

Ar *abr* (*argón*) ELECTRÓN, METAL, QUÍMICA Ar (*argon*)

arabinosa *f* QUÍMICA *azúcar* arabinose

arabitol *m* QUÍMICA *azúcar* arabitol

arada: **~ en curva de nivel** *f* AGRIC contour ploughing (*BrE*), contour plowing (*AmE*); **~ superficial** *f* AGRIC shallow ploughing (*BrE*), shallow plowing (*AmE*)

arado *m* AGRIC plough (*BrE*), plow (*AmE*); **~ alternativo** *m* AGRIC two-way plough (*BrE*), two-way plow (*AmE*); **~ con antetrén** *m* AGRIC gallow plough (*BrE*), gallow plow (*AmE*); **~ aporcador** *m* AGRIC middle breaker; **~ para arrozales** *m* AGRIC riceland plough (*BrE*), riceland plow (*AmE*); **~ báscula** *m* AGRIC balance plough (*BrE*), balance plow (*AmE*); **~ cincel** *m* AGRIC chisel plough (*BrE*), chisel plow (*AmE*); **~ de desfonde** *m* AGRIC breaker plough (*BrE*), breaker plow (*AmE*); **~ de discos** *m*

AGRIC disc plough (*BrE*), disk plow (*AmE*); ~ **escardador** *m* AGRIC sweep plough (*BrE*), sweep plow (*AmE*); ~ **de ida y vuelta** *m* AGRIC roll-over plough (*BrE*), roll-over plow (*AmE*); ~ **rastra** *m* AGRIC tiller plough (*BrE*), tiller plow (*AmE*); ~ **reversible** *m* AGRIC reversible plough (*BrE*), reversible plow (*AmE*); ~ **reversible de ida y vuelta** *m* AGRIC two-way plough (*BrE*), two-way plow (*AmE*); ~ **rotativo** *m* AGRIC rotary plough (*BrE*), rotary plow (*AmE*); ~ **de subsuelo** *m* AGRIC chisel subsoiler; ~ **topo** *m* AGRIC subsoiler plough with mole drain (*BrE*), subsoiler plow with mole drain (*AmE*); ~ **topo de drenaje** *m* AGRIC mole plough (*BrE*), mole plow (*AmE*); ~ **viñero** *m* AGRIC vineyard plough (*BrE*), vineyard plow (*AmE*)

aragonito *m* MINERAL, PETROL aragonite

arámida *f* P&C *tipo de polímero* aramid

araña *f* C&V spider, OCEAN *ictiología* weaver, TEC PETR spider, TRANSP AÉR spider unit; ~ **de cambio de paso** *f* TRANSP AÉR pitch-change spider; ~ **de luces** *f* C&V, ELEC *alumbrado* chandelier

arandela *f* CONST drum washer, washer, ING MECÁ washer, INSTAL HIDRÁUL, MECÁ ring, PROD lute, VEH *de tornillo, perno* washer; ~ **de abrazadera** *f* PROD clamp washer; ~ **achaflanada** *f* ING MECÁ beveled washer (*AmE*), bevelled washer (*BrE*); ~ **acopada** *f* ING MECÁ cup washer, dished washer; ~ **aisladora** *f* ELEC, ING ELÉC insulating washer; ~ **aislante** *f* ELEC, ING ELÉC insulating washer, ING MECÁ grommet; ~ **de amianto** *f* ING MECÁ asbestos washer; ~ **de asiento esférico** *f* ING MECÁ tap with metric thread; ~ **de la base** *f* ING MECÁ base washer; ~ **bimetálica de semi-empuje prensada** *f* ING MECÁ pressed bimetallic half-thrust washer; ~ **de blocaje reforzada** *f* ING MECÁ heavy-duty lock washer; ~ **de canilla** *f* ING MECÁ tap washer; ~ **de caras no paralelas** *f* ING MECÁ beveled washer (*AmE*), bevelled washer (*BrE*); ~ **ciega** *f* ING MECÁ blind washer; ~ **para columnas de guías cónicas** *f* ING MECÁ washer for tapered guide pillar; ~ **de compensación** *f* PROD trim washer; ~ **de compensador** *f* TRANSP AÉR balance washer; ~ **común** *f* ING MECÁ plain washer; ~ **cóncava** *f* ING MECÁ cup washer; ~ **cónica** *f* ING MECÁ taper washer; ~ **de corcho** *f* ING MECÁ cork washer; ~ **cuadrada** *f* ING MECÁ square washer; ~ **dentada de fijación** *f* ING MECÁ serrated lock-washer, tooth lock washer; ~ **elástica** *f* ING MECÁ disk spring, spring washer; ~ **de empuje** *f* AUTO, ING MECÁ, PROD, VEH thrust washer; ~ **engarzadora** *f* ING MECÁ crimping washer; ~ **de estrella** *f* PROD star washer; ~ **estriada de fijación** *f* ING MECÁ serrated lockwasher, tooth lock washer; ~ **fiador** *f* ING MECÁ, MECÁ locking washer; ~ **fiadora** *f* ING MECÁ, MECÁ locking washer; ~ **de fieltro** *f* ING MECÁ felt washer; ~ **de fijación acanalada** *f* ING MECÁ serrated lock-washer; ~ **de grifo** *f* ING MECÁ tap washer; ~ **Groover** *f* ING MECÁ, MECÁ lock washer; ~ **húmeda ionizante** *f* SEG ionizing wet washer; ~ **para impedir el paso del lubricante al eje** *f* ING MECÁ oil slinger; ~ **con lengüeta** *f* ING MECÁ, MECÁ tab washer; ~ **de medio empuje** *f* ING MECÁ half-thrust washer; ~ **muelle** *f* ING MECÁ disk spring; ~ **del muelle de válvula** *f* ING MECÁ, INSTAL HIDRÁUL valve-spring washer; ~ **de obturación** *f* ING MECÁ blind washer; ~ **obturadora** *f* ING MECÁ blank washer; ~ **plana** *f* ING MECÁ flat washer, plain washer, VEH plain washer; ~ **plana para**

tornillos y tuercas métricas *f* ING MECÁ plain washer for metric bolts screws and nuts; ~ **de presión** *f* AUTO, ING MECÁ, MECÁ, PROD, VEH lock washer, thrust washer; ~ **de presión de anillo** *f* ING MECÁ ring-type thrust washer; ~ **de presión del tipo anular** *f* ING MECÁ ring-type thrust washer; ~ **rectangular** *f* ING MECÁ strip washer; ~ **de refuerzo** *f* TRANSP AÉR shroud ring; ~ **de resorte** *f* ING MECÁ spring washer; ~ **de resorte de válvula** *f* ING MECÁ, INSTAL HIDRÁUL valve-spring washer; ~ **rizada** *f* ING MECÁ crinkle washer; ~ **de seguridad** *f* ING MECÁ, MECÁ lock washer, locking washer; ~ **de sujeción** *f* EMB clamping ring; ~ **de taza** *f* ING MECÁ bend-up lock washer; ~ **torneada** *f* ING MECÁ turned washer; ~ **trabante** *f* ING MECÁ, MECÁ locking washer

aráquico *adj* QUÍMICA arachic

árbol *m* AUTO sleeve, INFORM&PD tree, ING MECÁ, MECÁ shaft, PETROL mast, PROD *fundería* spindle, TRANSP MAR *hélice, motor* shaft, VEH *motor, transmisión* axle, drive, shaft; ~ **acamado** *m* ING MECÁ lying shaft; ~ **auxiliar con engranajes** *m* ING MECÁ countershaft, lay shaft; ~ **binario** *m* INFORM&PD binary tree; ~ **de bombeo** *m* ENERG RENOV *bomba de molino de viento* sucker rod; ~ **de búsqueda** *m* INFORM&PD search tree; ~ **del cigüeñal** *m* AUTO, ING MECÁ, MECÁ, TRANSP MAR, VEH crankshaft; ~ **de conexiones** *m* PETROL Christmas tree; ~ **contraeje** *m* AUTO countershaft; ~ **de contramarcha** *m* ING MECÁ jackshaft; ~ **de decisión** *m* INFORM&PD decision tree; ~ **desordenado** *m* INFORM&PD unordered tree; ~ **de la dirección** *m* AUTO, VEH steering shaft; ~ **de distribución** *m* AUTO, ELEC, ING MECÁ, MECÁ, TRANSP MAR, VEH *motor* camshaft; ~ **húmedo** *m* TEC PETR *producción costa-fuera* wet tree; ~ **impulsor** *m* AUTO, VEH drive shaft (*AmE*), propeller shaft (*BrE*); ~ **de información del directorio** *m* TELECOM directory information tree (*DIT*); ~ **intermedio de marcha atrás** *m* AUTO, VEH reverse idler shaft; ~ **de latón** *m* ENERG RENOV *bomba de molino de viento* brass rod; ~ **de levas** *m* AUTO, ELEC, ING MECÁ, MECÁ, TRANSP MAR camshaft; ~ **de levas colocado en la culata** *m* AUTO, MECÁ, VEH overhead camshaft (*OHC*); ~ **de levas en culata** *m* (*ALC*) AUTO, MECÁ, VEH overhead camshaft (*OHC*); ~ **de levas en culata de acción directa** *m* AUTO, MECÁ, VEH direct-acting overhead camshaft; ~ **de mando** *m* AUTO, VEH drive shaft (*AmE*), propeller shaft (*BrE*); ~ **de manivelas** *m* ING MECÁ main shaft; ~ **de marcha atrás** *m* AUTO, VEH reverse idler shaft; ~ **de Navidad** *m* TEC PETR *producción* Christmas tree; ~ **ordenado** *m* INFORM&PD ordered tree; ~ **para película** *m* CINEMAT film tree; ~ **de plomo** *m* QUÍMICA lead tree; ~ **posicionador** *m* ING MECÁ locating arbor; ~ **de productos** *m* PROD product tree; ~ **propulsor** *m* AUTO, VEH *transmisión* shaft drive; ~ **seco** *m* TEC PETR *producción costa-fuera* dry tree; ~ **de tercer movimiento** *m* AUTO, VEH third-motion shaft; ~ **de transmisión** *m* AUTO, ING MECÁ, VEH drive shaft (*AmE*), propeller shaft (*BrE*)

arboladura *f* TRANSP MAR rig, rigging

arbolar *vt* TRANSP MAR *bandera* fly, raise, set up

arbolito: ~ **de producción** *m* TEC PETR Christmas tree

arboricultura *f* AGRIC arboriculture

arborización *f* QUÍMICA arbor

arbotante *m* CONST flying buttress, TRANSP MAR propeller bracket, *del eje* strut

arca f C&V chest, PROD safe; ~ **de viento** f ING MECÁ air box

arcada f CONST arcade

Arcaica f GEOL *estratigrafía* Archaean

arcanita f MINERAL arcanite

arcén m AUTO, CONST *carretera*, TRANSP, VEH shoulder

archipiélago m GEOL, OCEAN, TRANSP MAR archipelago

archivado adj TELECOM *télex* put on file

archivador: ~ **de documentos** m INFORM&PD file folder

archivar vt INFORM&PD archive, PROD, TEXTIL *solicitud de patente* file a patent application

archivo m CINEMAT archive, INFORM&PD, PROD archive, file, filing; ~ **de acceso aleatorio** m INFORM&PD random-access file; ~ **aleatorio** m INFORM&PD random file; ~ **borrador** m INFORM&PD scratch file; ~ **de cambios** m INFORM&PD change file; ~ **de cinta** m INFORM&PD tape file; ~ **compartido** m INFORM&PD shared file; ~ **de contabilidad** m INFORM&PD accounting file; ~ **contable** m INFORM&PD accounting file; ~ **corrupto** m INFORM&PD corrupt file; ~ **de datos** m INFORM&PD data file; ~ **directo** m INFORM&PD direct file; ~ **de disco** m INFORM&PD disk file; ~ **electrónico** m INFORM&PD electronic filing; ~ **encadenado** m INFORM&PD chained file; ~ **para enmiendas** m INFORM&PD amendment file; ~ **de entrada** m INFORM&PD input file; ~ **de entrada/salida** m INFORM&PD input/output file; ~ **físico** m INFORM&PD physical file; ~ **guardado en memoria** m INFORM&PD buffered file; ~ **de imágenes** m CINEMAT stock shot library; ~ **indexado** m INFORM&PD indexed file; ~ **indicado** m INFORM&PD indexed file; ~ **inverso** m INFORM&PD inverted file; ~ **invertido** m INFORM&PD inverted file; ~ **de lectura** m INFORM&PD input file; ~ **de lectura/escritura** m INFORM&PD input/output file; ~ **lógico** m INFORM&PD logical file; ~ **maestro** m INFORM&PD master file; ~ **de movimientos** m INFORM&PD movement file, transaction file; ~ **original** m INFORM&PD father file; ~ **padre** m INFORM&PD father file; ~ **de partidas de material** m PROD material item file; ~ **de programa** m INFORM&PD program file; ~ **de salida** m INFORM&PD output file; ~ **secuencial** m INFORM&PD sequential file; ~ **secuencial indexado** m INFORM&PD indexed sequential file; ~ **de texto** m INFORM&PD text file; ~ **de tomas** m CINEMAT stock shot library; ~ **de trabajo** m INFORM&PD scratch file, work file; ~ **de transacciones** m INFORM&PD movement file, transaction file

arcilla f GEN clay; ~ **de alfarería** f C&V, CONST pottery clay; ~ **amasada** f MINAS pug; ~ **arenosa** f C&V, CONST, GEOL sandy clay, sandy loam; ~ **azul** f C&V blue clay; ~ **de bajo rendimiento** f GEOL low-yield clay; ~ **batida** f INSTAL HIDRÁUL puddling, pug, MINAS pug; ~ **blanca muy pura** f C&V, GEOL, MINERAL, P&C, REVEST china clay; ~ **cerámica** f C&V, CONST argil; ~ **decoladora** f C&V bleaching clay; ~ **dilatable** f C&V, TEC PETR swelling clay; ~ **dura** f C&V, CONST, GEOM Gault clay; ~ **esquistosa** f TEC PETR shale; ~ **expansible** f TEC PETR swelling clay; ~ **glaciar** f CARBÓN, GEOL boulder clay; ~ **glaciárica** f CARBÓN, GEOL glacial clay, till; ~ **gredosa** f AGUA chalky clay; ~ **para junturas** f PROD lute; ~ **para ladrillo** f C&V, CONST brick clay; ~ **laterítica** f CONST, GEOL, PETROL laterite; ~ **margosa** f AGUA marly clay; ~ **de modelar** f C&V,

PROD modeling clay (*AmE*), modelling clay (*BrE*); ~ **para moldear** f C&V, CONST clay loam; ~ **moviente** f CARBÓN quick clay; ~ **muy plástica** f TEC PETR fat clay; ~ **pegajosa** f PETROL, TEC PETR gumbo, sticky clay; ~ **plástica** f C&V, GEOL plastic clay, PETROL gumbo, PROD modeling clay (*AmE*), modelling clay (*BrE*), TEC PETR ball clay, *perforación* gumbo; ~ **posglaciar** f CARBÓN, GEOL postglacial clay; ~ **prodeltaica** f GEOL prodelta clay; ~ **refractaria** f GEOL, TERMO fireclay; ~ **roja** f PETROL red clay; ~ **con sílex** f GEOL clay with flints; ~ **de tubería** f CONST pipeclay; **placa de** ~ f GEOL tile clay

arcilloso adj ENERG RENOV *rocas*, GEOL argillaceous

arco m CONST bow, arch, ELEC *descarga* arc, arc-over, FERRO *vehículos* bow, GEOM arc, ING MECÁ *berbiquí* bow, TEC ESP rib; ~ **apainelado** m CONST three-centered arch (*AmE*), three-centred arch (*BrE*); ~ **ascendente** m CONST rising arch; ~ **botarel** m CONST *arquitectura* flying buttress; ~ **de la cámara** m C&V rider arch; ~ **choque** m TEC ESP bow shock; ~ **circular** m CONST elliptical arch, GEOM circular arc; ~ **cóncavo** m C&V concave bow; ~ **de contacto** m ING MECÁ *de correa* arc of contact; ~ **convexo** m C&V convex bow; ~ **de corriente alterna** m ELEC alternating-current arc; ~ **de cuatro centros** m CONST four-centered arch (*AmE*), four-centred arch (*BrE*); ~ **de descarga** m CONST arch of discharge, discharging arch, flat arch, safety arch; ~ **de descarga para las crecidas** m AGUA flood arch; ~ **descubierto** m ING ELÉC open arc; ~ **de diseño** m CONST scheme arch; ~ **eléctrico** m CARBÓN, ELEC, FÍS, GAS, ING ELÉC electric arc; ~ **eléctrico de alta intensidad** m ELEC, ING ELÉC high-intensity electric arc; ~ **eléctrico con electrodos de carbón** m ELEC, ING ELÉC carbon arc; ~ **con electrodos de carbón** m ELEC, ING ELÉC carbon arc; ~ **elevadizo** m CONST rising arch; ~ **elíptico** m CONST elliptical arch; ~ **de encendido** m ELEC, ING ELÉC flame arc; ~ **de encuentro** m CONST *arquitectura* groined vault; ~ **de engrane** m ING MECÁ pitch arc; ~ **escarzano** m CONST skene arch; ~ **de ferrocarril** m CONST, FERRO railroad arch (*AmE*), railway arch (*BrE*); ~ **en forma de herradura** m CONST horseshoe arch; ~ **frontal** m C&V front arch; ~ **de humo** m TERMOTEC smoke arch; ~ **de ignición** m TERMOTEC ignition arc; ~ **insular** m OCEAN island arc; ~ **inverso** m ING ELÉC arc back; ~ **invertido** m CONST inflected arch, inverted arch, reversed arch, INSTAL HIDRÁUL reversed arch; ~ **de involuta** m GEOM involute arc; ~ **involuto** m GEOM involute arc; ~ **iris** m FÍS, METEO rainbow; ~ **-isla** m GEOL, OCEAN island arc; ~ **de ladrillo** m CONST brick arch; ~ **mayor** m GEOM major arc; ~ **de medio punto** m CONST Roman arch; ~ **menor** m GEOM minor arc; ~ **de mercurio** m ELEC, ING ELÉC mercury arc; ~ **oblicuo** m CONST *puentes* oblique arch; ~ **peraltado** m CONST stilted arch; ~ **primario** m GEOL internides; ~ **proyectado** m CONST *arquitectura* scheme arch; ~ **realzado** m CONST basket handle arch, stilted arch; ~ **rebajado** m CONST diminished arch, skene arch, *estructura* segmental arch; ~ **de ruptura** m ELEC, ING ELÉC breaking arc; ~ **secundario** m GEOL externides; ~ **de segueta** m ING MECÁ hacksaw frame; ~ **semicircular** m CONST semicircular arch; ~ **de la sierra** m ING MECÁ saw frame; ~ **de sierra para metales** m ING MECÁ hacksaw frame; ~ **simulado** m CONST *arquitectura* blind arch; ~ ·**de soporte del**

motor *m* AUTO, TRANSP AÉR, VEH engine support arch; **~ de suspensión** *m* PROD bow hanger; **~ de tres centros** *m* CONST *estructura* three-centered arch (*AmE*), three-centred arch (*BrE*); **~ triangular** *m* CONST *estructura* triangular arch; **~ de vacío** *m* ELEC, ING ELÉC vacuum arc; **~ de viga de celosía** *m* CONST lattice-girder arch; **~ voltaico** *m* CARBÓN, ELEC, FÍS, GAS, ING ELÉC electric arc

arcón: **~ congelador** *m* ALIMENT, REFRIG, TERMO chest freezer

arcos: **~ intersectantes** *m pl* GEOM intersecting arcs; **~ que se cortan** *m pl* GEOM intersecting arcs

arcosa *f* C&V, GEOL, PETROL arkose

arcosoldadura *f* MECÁ, PROD, TERMO arc welding

ardenita *f* MINERAL ardennite

arder[1] *vt* TERMO burn

arder[2] *vi* TERMO glow; **~ sin llama** *vi* TERMO smolder (*AmE*), smoulder (*BrE*)

ardiente *adj* QUÍMICA, TERMO glowing; **no ~** *adj* MINAS nonfiery

ardor *m* TERMO burning

área *f* GEN area; **~ de absorción equivalente** *f* ACÚST equivalent absorption area; **~ de acción segura** *f* CINEMAT *del encuadre* safe action area; **~ ácida** *f* CONTAM *caracterización del medio* acidic area; **~ de activación** *f* METAL activation area; **~ de actuación** *f* CINEMAT acting area; **~ del ala de diseño** *f* TRANSP AÉR design wing area; **~ del ala neta** *f* TRANSP AÉR net wing area; **~ de anaveaje** *f* TRANSP MAR *para helicópteros* helicopter pad; **~ anticiclónica** *f* GEOL, METEO high; **~ de aparcamiento** *f* TEC ESP, TRANSP AÉR parking area; **~ de aterrizaje** *f* TEC ESP, TRANSP AÉR landing area; **~ de baja presión** *f* GEOL, METEO low-pressure area; **~ de bombardeo** *f* D&A bombing area; **~ bruta** *f* ENERG RENOV gross area; **~ de búsqueda** *f* INFORM&PD seek area, TV scanning area; **~ de carga** *f* TRANSP AÉR loading area; **~ del chip** *f* ELECTRÓN chip area; **~ de clasificación** *f* RAIL, TRANSP *para contenedores* marshaling area (*AmE*), marshalling area (*BrE*); **~ de coherencia** *f* FÍS, ÓPT, TELECOM coherence area; **~ coherente** *f* ÓPT, TELECOM coherent area; **~ colectora** *f* AGUA drainage area; **~ comprobada** *f* TEC PETR *yacimientos* proven area; **~ común** *f* INFORM&PD common area; **~ de cruce** *f* ELECTRÓN crossover area; **~ del depósito de difusión** *f* TEC ESP nozzle area; **~ de descarga** *f* PROD dump area; **~ de difusión** *f* REFRIG *de la corriente de aire acondicionado* diffusion area; **~ del disco** *f* TRANSP AÉR *de hélice* disc area (*BrE*), disk area (*AmE*); **~ para disposición de aguas negras** *f* AGUA, CONTAM, HIDROL, RECICL sewage farm; **~ de drenaje** *f* AGUA, PETROL drainage area; **~ de drenaje ciega** *f* AGUA blind drainage area; **~ efectiva** *f* TEC ESP *antenas* effective area; **~ de ejercitación** *f* OCEAN exercise area; **~ del emisor** *f* ELEC, ING ELÉC emitter region; **~ de enfoque** *f* CINEMAT, FOTO focusing stage; **~ de entrada** *f* INFORM&PD input area; **~ de escurrimiento** *f* AGUA drainage area; **~ de espera** *f* TRANSP AÉR holding bay; **~ de estériles** *f* MINAS tailings area; **~ de exfoliación relativa** *f* CARBÓN relative grain area; **~ de frenado** *f* FÍS slowing-down area; **~ frontal** *f* ING MECÁ frontal area; **~ de la fuente** *f* CONTAM, GEOL source area; **~ hundida en la mantilla** *f* IMPR blanket low spot; **~ de la imagen** *f* CINEMAT, IMPR image area; **~ de influencia** *f* AGUA area of influence; **~ de intercambio del centro de**

conmutación de grupo *f* TELECOM group-switching centre exchange area; **~ del inyector** *f* TEC ESP nozzle area; **~ de lanzamiento** *f* TEC ESP *vehículos* launch area; **~ lateral** *f* GEOM lateral area; **~ de mantenimiento** *f* TRANSP AÉR holding bay; **~ minera** *f* MINAS mining area; **~ de músculo dorsal** *f* AGRIC rib eye area; **~ del núcleo** *f* ÓPT core area; **~ de ordenamiento** *f* RAIL, TRANSP *para contenedores* marshaling area (*AmE*), marshalling area (*BrE*); **~ peligrosa** *f* SEG danger area; **~ del pistón** *f* ING MECÁ piston area; **~ de precipitación** *f* AGUA, METEO, PROC QUÍ precipitation area; **~ del radiómetro** *f* ACÚST label area; **~ de recogida del centro de conmutación de grupos** *f* TELECOM group-switching centre catchment area; **~ recuperada** *f* AGRIC, CONTAM, RECICL reclaimed area; **~ restaurada** *f* AGRIC, CONTAM reclaimed area; **~ de sección eficaz** *f* MECÁ effective cross-sectional area; **~ de seguridad de la imagen** *f* CINEMAT picture safety area; **~ de sensación auditiva** *f* ACÚST auditory sensation area; **~ de servicio** *f* TEC ESP, TRANSP AÉR service area; **~ superficial** *f* CONTAM surface area; **~ de la terminal** *f* TRANSP AÉR terminal area; **~ del texto** *f* IMPR text area; **~ de la tobera** *f* TEC ESP nozzle area; **~ de la tolva** *f* PROD boot area; **~ de trabajo** *f* INFORM&PD work area, working area; **~ de usuario** *f* INFORM&PD, TELECOM user area (*UA*); **~ vélica** *f* TRANSP MAR sail area; **~ de viraje** *f* TRANSP AÉR turnaround

arecaína *f* QUÍMICA arecaine

arena *f* CARBÓN sand, CONST grit, PROD, TEC PETR sand; **~ aurífera** *f* CARBÓN gravel; **~ basta** *f* *Esp* (*cf arena granugienta AmL*) MINAS coarse sand; **~ bituminosa** *f* PETROL bituminous sand; **~ de cantera** *f* MINAS pit sand; **~ de desecho** *f* C&V spent grinding sand; **~ de esmerilado** *f* C&V grinding sand; **~ de espolvorear** *f* PROD *fundería* parting dust; **~ para espolvorear moldes** *f* PROD parting sand; **~ estufada** *f* PROD *fundición* baked sand; **~ fina** *f* CARBÓN, CONST, PETROL, TEC PETR fine sand; **~ fluyente** *f* CARBÓN, HIDROL, OCEAN quicksand, running sand; **~ de fundición** *f* CONST foundry sand, PROD molding sand (*AmE*), moulding sand (*BrE*); **~ glauconítica** *f* GEOL *roca* greensand; **~ granugienta** *f* *AmL* (*cf arena basta Esp*) MINAS coarse sand; **~ gris** *f* PROD *fundición* burned sand, dead sand; **~ gruesa** *f* CONST, GEOL, MINAS coarse sand; **~ húmeda** *f* PROD greensand; **~ impregnada de brea** *f* TEC PETR *geología* tar sand; **~ para machos** *f* PROD *fundición* core sand; **~ de mina** *f* MINAS pit sand; **~ de moldear** *f* PROD molding sand (*AmE*), moulding sand (*BrE*); **~ de moldurar** *f* PROD molding sand (*AmE*), moulding sand (*BrE*); **~ en montón** *f* PROD *funderías* heap sand; **~ movediza** *f* CARBÓN, HIDROL, OCEAN quicksand, running sand; **~ negra ferruginosa** *f* PROD black iron sand; **~ petrolífera** *f* PETROL oil sand; **~ productiva** *f* PETROL pey sand; **~ reforzada con fibra** *f* CONST fiber-reinforced sand (*AmE*), fibre-reinforced sand (*BrE*); **~ silícea** *f* C&V glassmaking sand, silica; **~ usada** *f* PROD *fundición* black sand; **~ verde** *f* GEOL *roca* greensand; **~ vieja** *f* PROD *funderías* floor sand, heap sand

arenadora *f* ING MECÁ surface sander

arendalita *f* MINERAL arendalite

arenero *m* PAPEL riffler, sandtable

arenífero *adj* GEOL sandy

arenilla *f* CARBÓN, CONST, PETROL, TEC PETR fine sand
arenisca *f* CONST, GEOL, OCEAN, TEC PETR sandstone;
~ **calcárea** *f* GEOL calcareous sandstone; ~ **caliza** *f*
GEOL calcareous sandstone, lime sandstone;
~ **conchífera** *f* GEOL shelly sandstone; ~ **cuarcítica**
f GEOL quartzitic sandstone; ~ **de cuarzo** *f* MINERAL
quartz sandstone; ~ **pizarrosa** *f* CARBÓN fakes;
~ **silícea** *f* CONST, GEOL grit, millstone grit
arenita *f* GEOL beach rock, OCEAN arenite, beach rock,
sandstone
arenoso *adj* ENERG RENOV *rocas* arenaceous, GEOL
sandy
areometría *f* CARBÓN, FÍS, HIDROL araeometry
areómetro *m* CARBÓN, FÍS araeometer, hydrometer,
HIDROL araeometer
arete *m* AGRIC *ganado* ear tag
arfvedsonita *f* MINERAL arfvedsonite
argamasa *f* CONST mortar, HIDROL puddle, PROC QUÍ,
QUÍMICA mortar
arganeo *m* TRANSP MAR anchor ring
argazos *m pl* OCEAN algal mat
argéntico *adj* QUÍMICA argentic
argentífero *adj* MINERAL argentiferous
argentina *f* COLOR, MINERAL argentine
argentita *f* MINERAL, QUÍMICA argentite, silver glance
argentopirita *f* MINERAL argentopyrite
argilita *f* GEOL claystone, PETROL argillite
arginasa *f* QUÍMICA arginase
arginina *f* QUÍMICA arginine
argírico *adj* QUÍMICA argyric
argirita *f* MINERAL argyrose
argiritrosa *f* MINERAL argyrythrose
argirodita *f* MINERAL argyrodite
argiropirita *f* MINERAL argyropyrite
argirosa *f* MINERAL argyrose
argolla *f* AUTO shackle, CONST staple, ING MECÁ, MECÁ
eye, hoop, lug, ring, PETROL buckle, PROD hoop, VEH
shackle; ~ **de izada** *f* TRANSP MAR lifting eye; ~ **de
izada a paño** *f* TRANSP MAR *en la cubierta* flush lifting
ring; ~ **del izador** *f* TRANSP AÉR hoisting ring
argón *m* (*Ar*) ELECTRÓN, METAL, QUÍMICA argon (*Ar*)
argumento *m* CINEMAT outline, INFORM&PD, MATEMÁT
argument
aridez *f* AGUA, CARBÓN, GEOL dryness
árido[1] *adj* CARBÓN barren
árido[2] *m* CONST *hormigón*, PETROL aggregate; ~ **fino** *m*
CONST fine aggregate; ~ **de tamaño uniforme** *m*
CONST single-size gravel aggregate
ariete *m* AGUA, CONST, MECÁ, PROD ram; ~ **aspirador**
m PROD suction ram; ~ **hidráulico** *m* AGUA, CARBÓN,
CONST, INSTAL HIDRÁUL hydraulic conveyor ram,
hydraulic impulse ram, hydraulic ram, ram pump,
water ram
arilamina *f* QUÍMICA arylamine
arilo *m* AGRIC *semilla* aril, QUÍMICA *grupo* aryl
arista[1]: **de ~ cortante** *adj* ING MECÁ sharp-edged
arista[2] *f* AGRIC awn, CARBÓN *de una pirámide* rib,
CRISTAL, GEOM, ING MECÁ, INSTR edge, METAL *de
dislocación* cusp, TRANSP AÉR chine; ~ **del borde de
ataque** *f* TRANSP AÉR leading-edge rib; ~ **cortante** *f*
ING MECÁ, MECÁ cutting edge; ~ **de corte** *f* ING
MECÁ, MECÁ lip; ~ **exterior** *f* INSTAL HIDRÁUL *válvula
de corredera* steam edge; ~ **interior** *f* INSTAL HIDRÁUL
válvula de corredera exhaust edge
arista[3]: **de ~ a arista** *fra* ING MECÁ edge-to-edge

aristón *m* CONST *arquitectura* groin (*AmE*), groyne
(*BrE*)
aritmética: ~ **binaria** *f* ELECTRÓN, INFORM&PD, MATE-
MÁT binary arithmetic; ~ **de coma fija** *f* INFORM&PD,
MATEMÁT fixed-point arithmetic; ~ **de coma
flotante** *f* INFORM&PD, MATEMÁT floating-point
arithmetic; ~ **congruente** *f* MATEMÁT congruence
arithmetic; ~ **de doble precisión** *f* INFORM&PD,
MATEMÁT double precision arithmetic; ~ **modular** *f*
INFORM&PD, MATEMÁT modular arithmetic; ~ **en
paralelo** *f* INFORM&PD, MATEMÁT parallel arithmetic;
~ **de punto flotante** *f* INFORM&PD, MATEMÁT floating-point arithmetic; ~ **del residuo** *f* INFORM&PD
residue arithmetic
arizonita *f* MINERAL arizonite
arkansita *f* MINERAL arkansite
arksutita *f* MINERAL arksutite
arma: ~ **atómica** *f* D&A, NUCL atomic weapon; ~ **de
fuego** *f* D&A gun; ~ **de implosión** *f* D&A, NUCL
implosion weapon; ~ **inteligente** *f* D&A smart
weapon; ~ **nuclear** *f* D&A, NUCL atomic weapon;
~ **nuclear de implosión** *f* D&A, NUCL implosion
weapon; ~ **de pequeño calibre contra blindaje** *f*
D&A, NUCL light anti-armor weapon (*AmE*) (*LAW*),
light anti-armour weapon (*BrE*) (*LAW*)
ARMA *abr* (*media móvil autorregresiva*) INFORM&PD
ARMA (*autoregressive moving average*)
Arma: ~ **Aérea Naval** *f* D&A, TRANSP MAR Fleet Air
Arm (*BrE*), Naval Air Service (*AmE*)
armada *f* D&A, TRANSP MAR fleet, navy; ~ **nacional** *f*
D&A, TRANSP MAR national navy
armadía *f* TRANSP MAR raft
armado[1] *adj* ELEC *cable*, TELECOM armored (*AmE*),
armoured (*BrE*)
armado[2] *m* CONST *hormigón* reinforcing, ELEC assembly, ING MECÁ assembling, OCEAN *arte de pesca*
setting, TEC ESP shielding
armador *m* TRANSP MAR shipowner; ~ **de bandejas** *m*
EMB tray erector; ~ **y cargador de bandejas** *m* EMB
tray erector and loader; ~**-desarmador de paquetes
de videotex** *m* *Esp* (*cf armador-desarmador de
paquetes de videtex AmL*) TELECOM videotex packet
assembler-disassembler; ~**-desarmador de paquetes de videtex** *m* *AmL* (*cf armador-desarmador de
paquetes de videotex Esp*) TELECOM videotex packet
assembler-disassembler
armadora *f* TRANSP MAR shipowner
armadura *f* C&V bracing, CONST framework, reinforcement, trussing, ELEC *conductor del cable* armor
(*AmE*), armour (*BrE*), *imán* keeper, FÍS *imán permanente* armature, keeper, MECÁ anchor, OCEAN *arte de
pesca* net frame, netting frame, PROD *fundería* grate,
spider, grating, tie rod, strap, grid; ~ **de anillo** *f* ELEC
generador ring armature; ~ **de anillo dentado** *f* ELEC
generador toothed ring armature; ~ **de la bomba** *f*
AGUA, PROD pump gear; ~ **de cercha de dos pares** *f*
CONST *edificación* close-couple truss; ~ **de
condensador** *f* ELEC, FÍS, ING ELÉC capacitor plate;
~ **conectada en estrella** *f* ELEC *generador* star-
connected armature; ~ **con conexión en estrella** *f*
ELEC *generador* star-connected armature; ~ **de
corriente trifásica** *f* ELEC *generador* rotary-current
armature, three-phase current armature, ING ELÉC
rotary-current armature; ~ **en cortocircuito** *f* ELEC
generador short-circuit armature; ~ **de cubierta** *f*
CONST roof truss, truss; ~ **dentada** *f* ELEC *generador*

slotted armature; ~ **devanada** f ELEC *generador* wire-wound armature; ~ **de distribución** f CONST *hormigón armado* distribution steel; ~ **estacionaria** f ELEC *generador* stationary armature; ~ **del estator** f ING ELÉC stator frame; ~ **exterior** f ELEC, ING ELÉC *cables* armor (*AmE*), armour (*BrE*); ~ **de falso tirante** f CONST *edificación* collar-beam truss; ~ **giratoria** f ELEC *generador* rotating armature; ~ **laminar** f ELEC *motor* laminated armature; ~ **longitudinal** f CONST longitudinal reinforcement; ~ **del macho** f PROD *fundición* core grid, core iron; ~ **magnética** f ELEC, ING ELÉC relay armature; ~ **de malla** f CONST mat reinforcement; ~ **metálica del neumático** f AUTO bead core; ~ **para moldeo al barro** f PROD *fundición* loam plate; ~ **del motor** f ELEC motor armature; ~ **móvil** f ACÚST, ELEC moving armature; ~ **multipolar** f ELEC *generador* multipolar armature; ~ **neutra** f ELEC *relé* neutral armature; ~ **de núcleo liso** f ELEC *generador* smooth core armature; ~ **del obturador** f ELEC *generador* shuttle armature; ~ **de parilla** f C&V wire mesh reinforcement; ~ **primaria** f ELEC *generador* primary armature; ~ **principal** f CONST main reinforcement; ~ **de ranura abierta** f ELEC *generador* open-slot armature; ~ **de la sierra** f ING MECÁ saw frame; ~ **suave del núcleo** f ELEC *generador* smooth core armature; ~ **de surgencia** f PETROL Christmas tree

armamento m TRANSP MAR *construcción* armament, fitting, *de barco* fitting out; ~ **de grueso calibre** m D&A heavy armament; ~ **de pequeño calibre** m D&A light armament

armar vt CONST *vigas* brace, ING MECÁ fit out, *vigas* brace, MECÁ brace, OCEAN *tecnología pesquera* set, PROD mount, TRANSP MAR *buque* put into commission, ship, equip, fit out; ~ **con** vt ING MECÁ fit with; ~ **y escorar** vt TRANSP MAR *cuadernas* frame

armario m (cf *gabinete AmL*) LAB *mobiliario* cupboard, MECÁ, P&C cabinet, PROD enclosure, TELECOM cabinet; ~ **de almacenamiento** m LAB *mobiliario* storage cupboard; ~ **de almacenamiento de comestibles congelados** m ALIMENT, ING MECÁ, REFRIG, TERMO frozen-food storage cabinet; ~ **casero de almacenaje de comestibles congelados** m ING MECÁ, REFRIG household frozen-food storage cabinet; ~ **del conversor** m TELECOM converter cabinet; ~ **del convertidor** m TELECOM converter cabinet; ~ **de distribución** m ING ELÉC distribution cabinet; ~ **de fermentación** m ALIMENT *panadería, repostería* proving cabinet; ~ **frigorífico** m ALIMENT, ING MECÁ, REFRIG, TERMO cold store, refrigerated cabinet; ~ **interconectado** m TELECOM cross-connect cabinet; ~ **metálico** m TELECOM rack; ~ **secadero** m ING MECÁ drying cabinet; ~ **de secado** m ALIMENT, CINEMAT drying cabinet, drying cupboard; ~ **de secado al vacío** m ING MECÁ vacuum-drying cabinet; ~ **secador** m ING MECÁ drying cabinet; ~ **de sustancias tóxicas** m LAB, SEG poisons cupboard; ~ **vacuosecador** m ING MECÁ vacuum-drying cabinet

armas: ~ **ligeras** f pl PROD small arms

armazón m CONST skeleton, frame, framework, *carpintería* carcass, ING ELÉC frame, ING MECÁ jacket, frame, MECÁ frame, OCEAN *de red* netting frame, net frame, TEC ESP casing, TEXTIL carcass, TRANSP AÉR coaming, TRANSP MAR *luces* lantern casing, VEH *carrocería* frame; ~ **de la cabina** m VEH chassis cab;

~ **cortocircuitado** m ING ELÉC short-circuited armature; ~ **en cortocircuito** m ING ELÉC short-circuited armature; ~ **4-D** m TEC ESP *astronave* 4-D reinforcement; ~ **diagonal del tren de aterrizaje** m TRANSP AÉR landing-gear diagonal truss; ~ **del farol** m TRANSP MAR *luces* lantern casing; ~ **fijo** m ING ELÉC stationary armature; ~ **para machos** m PROD *fundición* core trestle; ~ **móvil** m CONST traveling cradle (*AmE*), travelling cradle (*BrE*); ~ **polar** m ELEC *de máquina* field spider; ~ **del radiador** m AUTO, VEH radiator frame; ~ **con ranuras** m ING ELÉC slotted armature; ~ **de señales** m FERRO *equipo inamovible* signal structure; ~ **de sustentación** m ING MECÁ underframe; ~ **de tamices** m PROC QUÍ sieve frame; ~ **del tejado** m CONST roof frame

armella: ~ **roscada** f CONST screw eye

armería f D&A armory (*AmE*), armoury (*BrE*)

armonía f ACÚST harmony, ING MECÁ accordance; ~ **esférica** f TEC ESP spherical harmonic; ~ **de tonalidades de un color** f COLOR self-color (*AmE*), self-colour (*BrE*)

armónico[1] adj GEN harmonic

armónico[2] m GEN harmonic; ~ **aural** m ACÚST, TEC ESP, TELECOM aural harmonic; ~ **de bajo nivel** m ELECTRÓN low-order harmonic; ~ **fundamental** m ELECTRÓN, FÍS first harmonic; ~ **impar** m ELECTRÓN, FÍS odd harmonic; ~ **inferior** m ELECTRÓN lower harmonic; ~ **natural** m ACÚST, FÍS ONDAS natural harmonic; ~ **de orden superior** m ACÚST, ELECTRÓN high-order harmonic, ripple, FÍS high-order harmonic, ripple, overtone, TEC ESP *corriente ondulatoria* high-order harmonic, ripple; ~ **superior** m ELECTRÓN overtone; ~ **teseral** m TEC ESP tesseral harmonic; ~ **zonal** m TEC ESP zonal harmonic

arnés m AGRIC stud, D&A, SEG, TEC ESP, TRANSP AÉR harness; ~ **de amarre** m TRANSP AÉR mooring harness; ~ **de detección de incendios** m SEG, TRANSP AÉR fire-detection harness; ~ **de encendido** m TRANSP AÉR ignition harness; ~ **de seguridad** m CINEMAT, SEG, TRANSP AÉR safety harness

aro m AUTO ring, C&V loop, CONST hoop, INFORM&PD ring, ING MECÁ hoop, collar, MECÁ collar, hoop, PETROL collar, PROD hoop, TRANSP, TRANSP AÉR ring; ~ **de abertura** m CINEMAT aperture ring; ~ **adaptador** m CINEMAT adaptor ring; ~ **de admisión de la góndola** m TRANSP AÉR nacelle intake ring; ~ **de aumento** m CINEMAT step-up ring; ~ **de balancín** m EMB lever ring; ~ **cónico** m ING MECÁ conical spacer; ~ **de diafragma** m CINEMAT aperture ring; ~ **del eje** m ING MECÁ shaft collar; ~ **de émbolo** m MECÁ piston ring; ~ **empaquetador** m ING MECÁ packing ring; ~ **de enfoque** m CINEMAT FOTO, TV focusing ring; ~ **excéntrico** m ING MECÁ eccentric hoop; ~ **de freno** m ING MECÁ brake ring, *para cojinetes de bola* lock ring; ~ **de fundición para hinca de pozos** m MINAS curb; ~ **de guarnición** m ING MECÁ packing ring; ~ **del pistón** m AUTO piston ring, C&V plunger ring, ING MECÁ, MECÁ piston ring; ~ **reductor** m CINEMAT step-down ring; ~ **de retención** m CINEMAT locking ring; ~ **de rodaduras** m ING MECÁ, MECÁ race; ~ **de rodamiento** m ING MECÁ, MECÁ race; ~ **de rodamiento exterior** m ING MECÁ, MECÁ outer race; ~ **salvavidas** m SEG, TRANSP MAR *emergencia* lifebuoy; ~ **de seguridad** m ING MECÁ *para cojinetes de bola* lock ring; ~ **de soporte** m ING MECÁ droop-restraining ring; ~ **de tope** m

MINAS retainer; ~ **tórico** *m* ING MECÁ O-ring; ~ **de zoom** *m* CINEMAT, FOTO zoom ring

aroma: ~ **desagradable** *m* ALIMENT off-flavor (*AmE*), off-flavour (*BrE*)

aromaticidad *f* ALIMENT, QUÍMICA aromaticity

aromático *adj* ALIMENT, TEC PETR *compuestos del petróleo* aromatic

aromatización *f* ALIMENT, QUÍMICA, TEC PETR aromatization

aromatizar *vt* PROC QUÍ, QUÍMICA aromatize

arpeo *m* TRANSP MAR grapple, grappling hook

arpillera *f* ING MECÁ crocus cloth, TEXTIL hessian

arpón *m* OCEAN harpoon; ~ **pescador** *m* MINAS *sondeos* spear rod; ~ **pescaherramientas** *m* MINAS *sondeos* spear rod; ~ **pescatubos** *m* CONST casing spear, MINAS *sondeos* bulldog-casing spear

arqueado[1] *adj* CONST, MINAS arched

arqueado[2] *m* C&V roving, CONST, MINAS arching; ~ **alrededor** *m* C&V spun roving; ~ **sin torcedura** *m* C&V no-twist roving

arquear *vt* CONST arch, MINAS arch, bend

arquearse *v refl* CONST, MINAS bend

arqueo *m* TRANSP MAR tonnage; ~ **bruto** *m* TRANSP MAR gross register, gross tonnage; ~ **neto** *m* TRANSP MAR net tonnage; ~ **de registro** *m* TRANSP MAR register tonnage

arqueozoica *f* GEOL *estratigrafía* Archaean (*BrE*), Archean (*AmE*)

arqueta *f* MINAS *galería* manhole

arquitecto: ~ **naval** *m* TRANSP MAR naval architect; ~ **naval e ingeniero de máquinas** *m* TRANSP MAR marine architect and engineer

arquitectura: ~ **distribuida** *f* INFORM&PD distributed architecture; ~ **naval** *f* TRANSP MAR naval architecture; ~ **orientada al objeto** *f* INFORM&PD object-oriented architecture; ~ **de pila** *f* INFORM&PD stack architecture; ~ **de porciones** *f* INFORM&PD slice architecture; ~ **de red** *f* ING ELÉC network architecture; ~ **de red de computadoras** *f* AmL (*cf arquitectura de red de ordenadores Esp*) INFORM&PD computer-network architecture; ~ **de red de ordenadores** *f* Esp (*cf arquitectura de red de computadoras AmL*) INFORM&PD computer-network architecture; ~ **por registros encadenados** *f* INFORM&PD slice architecture; ~ **del sistema de división de funciones** *f* TELECOM function-division system architecture; ~ **de sistemas abiertos** *f* INFORM&PD, TELECOM open systems architecture (*OSA*); ~ **de sistemas de redes** *f* INFORM&PD, TELECOM systems network architecure (*SNA*); ~ **sistólica** *f* TELECOM systolic architecture

arrabio *m* METAL pig iron, PROD pig iron, *hierro fundido* pig iron, metal; ~ **para moldería** *m* PROD foundry iron

arracimado *adj* MINERAL botryoidal

arraigada: ~ **de la cadena del ancla** *f* TRANSP MAR anchor cable attachment; ~ **de metal** *f* TRANSP MAR eyeplate

arraigado: ~ **del estay del trinquete** *m* TRANSP MAR forestay pin

arranca: ~~**apeas** *m* MINAS prop drawing; ~~**estemples** *m* MINAS prop drawing

arrancada *f* OCEAN, TRANSP MAR *navegación* way out; ~ **avante** *f* TRANSP MAR headway

arrancado *m* IMPR tear-off, PAPEL picking; ~ **superficial** *m* PAPEL picking

arrancador *m* AUTO, ELEC, ING ELÉC, VEH starter; ~ **contactor** *m* AUTO, ELEC, ING ELÉC, VEH contactor starter; ~ **diferido** *m* ELEC time-delay starter; ~ **electromagnético** *m* AUTO, ELEC, ING ELÉC, VEH magnetic starter; ~ **entre los lados de la línea** *m* ELEC across-the-line starter; ~ **de estrella-triángulo** *m* ELEC *conmutador*, ING ELÉC star-delta starter; ~ **hidráulico de aceite** *m* ELEC oil hydraulic starter; ~ **de inercia** *m* MECÁ inertial starter; ~ **de línea** *m* ELEC *conmutador* line starter (*AmE*), starter (*BrE*); ~ **en la línea** *m* ELEC across-the-line starter; ~ **magnético** *m* PROD magnetic starter; ~ **de n pasos** *m* ELEC n-step starter; ~ **de placa frontal** *m* ELEC *motor* faceplate starter; ~ **de pulsador de rosca** *m* AUTO, ELEC, ING ELÉC screw push starter, VEH screw push; ~ **regulador** *m* ELEC regulating starter; ~ **reostático** *m* ELEC, ING ELÉC rheostat starter; ~ **del rotor** *m* ELEC, ING ELÉC rotor starter; ~ **de tambor** *m* ELEC *interruptor*, ING ELÉC drum starter; ~ **tipo Béndix**® *m* AUTO bendix-type starter; ~ **de volante** *m* MECÁ inertial starter; ~ **Y-delta** *m* ELEC Y-delta starter

arrancadora *f* AGRIC, CARBÓN, MINAS digger; ~~**cargadora para diversos usos** *f* MINAS universal ripper-loader; ~~**cargadora de papas** *f* AmL (*cf arrancadora-cargadora de patatas Esp*) AGRIC load potato digger; ~~**cargadora de patatas** *f* Esp (*cf arrancadora-cargadora de papas AmL*) AGRIC load potato digger; ~~**hileradora** *f* AGRIC lifter windrower

arrancapilotes *m* CONST pile extractor

arrancar *vt* CONST pull out, INFORM&PD start, boot, ING MECÁ start up, start, MINAS *el mineral* stope, PAPEL pick, PROD *filetes de un tornillo* strip, *máquinas* start; ~ **con arado** *vt* AGRIC plough out (*BrE*), plough up (*BrE*), plow out (*AmE*), plow up (*AmE*)

arrancasonda *m* MINAS *sondeos* drill rod grab

arrancatubos *m* AmL (*cf retén Esp*) MINAS *entubación* dog

arranchar *vt* TRANSP MAR *velero* trim

arranque *m* CARBÓN *minas* digging, broken working, *del carbón* breaking down, CONST spring, *arquitectura* springing, HIDROL *canal* intake, INFORM&PD start, ING MECÁ starting, MINAS (*cf corte mineral, recorte Esp*) *chimenea atascada, obstruida* starting, *de mineral* drawing, holing, breakage, stoping, breaking, *mineral* cut, TRANSP AÉR ignition, VEH *motor* starter; ~ **de los áridos** *m* CONST aggregate stripping; ~ **Béndix**® *m* AUTO, VEH Bendix starter®; ~ **en caliente** *m* INFORM&PD warm start, TRANSP AÉR hot start; ~ **de carbón** *m* CARBÓN, MINAS coal getting, coal winning; ~ **de carbón mecánico** *m* CARBÓN, MINAS mechanized coal-winning; ~ **de la central** *m* NUCL plant start-up; ~ **con condensador** *m* AUTO, VEH capacitor ignition; ~ **del devanado de piezas** *m* AUTO, ELEC, ING ELÉC, ING MECÁ part-winding starting; ~ **directo** *m* ELEC, ING ELÉC, ING MECÁ *de motor* direct starting; ~ **electromagnético** *m* AUTO, ELEC, ING ELÉC, VEH electromagnetic ignition; ~ **electrónico** *m* AUTO, ELEC, ELECTRÓN, VEH electronic ignition; ~ **de filones** *m* MINAS stoping of the seam; ~ **en frío** *m* INFORM&PD, TERMO, VEH cold start; ~ **inicial** *m* INFORM&PD cold start; ~ **lateral** *m* MINAS *explotación* side stoping; ~ **de la máquina** *m* C&V machine start-up; ~ **del motor por medio de manivela** *m* TRANSP AÉR cranking; ~ **por pedal** *m* VEH *motocicletas* kick-starter; ~ **por realce** *m* Esp (*cf* *labor de realce AmL*) MINAS *explotación* overhand stoping; ~ **secundario** *m Esp* (*cf arranque tibio AmL*) INFORM&PD warm start; ~ **suave** *m* PROD soft start; ~ **por tensión parcial** *m* ELEC *motor* partial-voltage starting; ~ **tibio** *m AmL* (*cf arranque secundario Esp*) INFORM&PD warm start; ~ **transistorizado sin platinos** *m* AUTO contactless transistorized ignition (*BrE*), pointless transistorized ignition (*AmE*); ~ **en vacío** *m* ELEC *de motor* no-load start

arrastrado: ~ **por las aguas** *adj* HIDROL waterborne

arrastrador: ~ **de lodos** *m* HIDROL *aguas residuales* sludge scraper; ~ **de partículas** *m* OCEAN air exhauster

arrastrar *vt* HIDROL *materiales flotantes* drift, INFORM&PD *aritmética* carry

arrastre *m* AGUA scour, C&V elutriation, CARBÓN *erosión de suelo* creep, FÍS drag, FÍS FLUID entrainment, HIDROL scour, INFORM&PD *aritmética* carry, ING MECÁ traction, pulling, METAL *soldadura oxiacetilénica* drag, NUCL scavenging, *de productos de corrosión* carry-off, OCEAN haul, TEC ESP *vuelo en atmósfera* drag, tracking, TEC PETR *perforación* drill string drag, *geotécnia* creep; ~ **de accesorios** *m* TRANSP AÉR accessory drive; ~**por la acción del agua** *f* CONTAM washout; ~ **aerodinámico** *m* TRANSP aerodynamic drag; ~ **por cable** *m* CARBÓN rope hauling; ~ **con chorros de agua** *m* AGUA sluicing; ~ **circular** *m* INFORM&PD end-around carry; ~ **de dientes** *m* IMPR, INFORM&PD tractor feed; ~ **electrónico** *m* ELECTRÓN, NUCL electron drift; ~ **eólico alto de nieve** *m* METEO blowing snow; ~ **de falla** *m* GEOL fault drag; ~ **hidráulico** *m* CARBÓN *minas* piping; ~ **litoral** *m* HIDROL littoral drift; ~ **del motor** *m* AUTO, VEH overrun; ~ **de polvos en la mezcla** *m* C&V carry-over; ~ **y relleno** *m* GEOL scouring-and-filling; ~ **en serie** *m* ELECTRÓN cascaded carry; ~ **en suspensión** *m* HIDROL suspended load; ~ **tierra adentro** *m* TRANSP inland haulage; ~ **de unidades** *m* INFORM&PD *aritmética* carry; ~ **de vagonetas** *m* CARBÓN shuttle haulage; ~ **de vagonetas en cadena** *m* CARBÓN chain haulage; ~ **de vapor** *m* PROC QUÍ steam entraining

arrastrero *m* OCEAN, TRANSP MAR trawler; ~ **congelador** *m* OCEAN, REFRIG, TRANSP MAR *para pescado* freezing trawler

array *m* TELECOM array; ~ **de diodos** *m* TELECOM diode array; ~ **lógico preformable** *m* (*PAL*) TELECOM, TV programmable array logic (*PAL*); ~ **puertas** *m* TELECOM gate array

arrecife *m* GEOL, OCEAN fringing reef, reef; ~ **de algas** *m* GEOL, OCEAN algal reef; ~ **artificial** *m* GEOL, OCEAN artificial reef; ~ **barrera** *m* GEOL, OCEAN barrier reef; ~ **de coral** *m* GEOL, OCEAN coral reef; ~ **coralígeno** *m* GEOL, OCEAN coral reef

arreglar *vt* CONST fix

arreglo *m* GAS *de equipos averiados* fixing, IMPR makeready, INFORM&PD array, ING MECÁ arrangement, MATEMÁT *de cifras, figuras*, TEC ESP, TELECOM array; ~ **y compaginación de libros** *m* IMPR book makeup; ~ **de datos** *m* INFORM&PD data array; ~ **de frente de onda** *m* INFORM&PD wavefront array; ~ **gráfico extendido** *m* (*EGA*) INFORM&PD extended graphic arrangement (*EGA*); ~ **lógico programable** *m* (*PAL*) TELECOM, TV programmable array logic (*PAL*); ~ **lógico programable de campo** *m* (*FPLA*) INFORM&PD field-programmable logic array

(*FPLA*); ~ **de puerta** *m* INFORM&PD, TELECOM gate array; ~ **en serie** *m* INFORM&PD string array; ~ **sistólico** *m* INFORM&PD systolic array

arreico *adj* AGUA, HIDROL arheic

arremetida *f* TEC PETR *perforación* kick

arremolinarse *v refl* TRANSP MAR eddy

arrendamiento: ~ **a casco desnudo** *m* TRANSP MAR bare-boat charter

arrendar *vt* AGRIC farm out

arrhenita *f* MINERAL arrhenite

arriado *m* TRANSP MAR *de buque* launching

arriar[1] *vt* TRANSP MAR draw, *bandera* haul down, *bote* lower, *cadena* slip

arriar[2]: ~ **bandera del país** *vi* TRANSP MAR strike colors (*AmE*), strike colours (*BrE*)

arriate: ~ **marginal** *m* AGRIC, AUTO, VEH rear-wheel drive

arriba *adv* TEC ESP upstream

arribada: **de** ~ *adj* TRANSP MAR *tráfico portuario* inwardbound

arribar *vi* TRANSP MAR bear away, cast off

arribazón *f* OCEAN run

arriostramiento *m* CONST bracing, bridging, PROD *virotillo roscado* staying, REFRIG stay block; ~ **de celosía** *m* CONST lattice bracing; ~ **contra vientos** *m* CONST wind bracing; ~ **en cruz** *m* TRANSP MAR *construcción naval* cross brace; ~ **longitudinal entre pies derechos de marcos** *m* MINAS sill piece

arrizar *vt* TEC ESP *el ancla* stow

arrojable *adj* TEC ESP jettisonable

arrojar: ~ **por la borda** *vt* TEC ESP jettison

arrollado *m* TEXTIL beaming

arrollador *m* TEXTIL beamer

arrollamiento *m* ELEC *generador* winding, ELECTRÓN convolver, FÍS coil, VEH winding; ~ **de alta tensión** *m* ELEC *transformador* high-voltage winding; ~ **en anillo** *m* ELEC *bobina* ring winding; ~ **anular** *m* ELEC *bobina* banked winding, ring winding; ~ **bifilar** *m* ELEC *bobina* double-layer winding, *inductor* bifilar winding; ~ **bipolar** *m* ELEC *máquina* bipolar winding; ~ **en capas superpuestas** *m* ELEC *bobina* bank winding, banked winding; ~ **en circuito abierto** *m* ELEC *máquina eléctrica, transformador* open-circuit winding; ~ **en doble capa** *m* ELEC *de bobina* double-layer winding; ~ **de drenaje** *m* ELEC *regulación* bleeder winding; ~ **dúplex** *m* ELEC *máquina eléctrica* duplex lap winding; ~ **del inducido** *m* ELEC *máquina* armature winding; ~ **múltiple** *m* ELEC *de transformador*, ING ELÉC multiple winding; ~ **primario** *m* AUTO, ELEC, ING ELÉC, VEH primary winding; ~ **recorrido por la corriente** *m* ELECTRÓN, FÍS current-carrying coil; ~ **secundario** *m* AUTO, ELEC, ING ELÉC, VEH secondary winding; ~ **en serie** *m* ELEC series winding; ~ **superpuesto** *m* ELEC *bobina* bank winding, banked winding; ~ **térmico** *m* ÓPT, TERMO thermal wrap; ~ **uniforme en capas superpuestas** *m* ELEC uniform layer winding; ~ **de unión** *m* PROD tie wrap

arroyo *m* GEOL brook, rill, stream, HIDROL brook, stream, rill; ~ **en movimiento** *m* GEOL, HIDROL running brook; ~ **pequeño** *m* HIDROL small stream

arroyuelo *m* GEOL, HIDROL brook, rivulet, streamlet; ~ **de marea** *m* OCEAN tidal creek

arroz: ~ **descascarado** *m* AGRIC, ALIMENT husked rice; ~ **descascarillado** *m* AGRIC, ALIMENT hulled rice; ~ **machacado** *m* AGRIC, ALIMENT broken rice; ~ **partido** *m* AGRIC, ALIMENT broken rice

arrozal *m* AGRIC rice field

arrufo: ~ **a proa** *m* TRANSP MAR *diseño de barcos* sheer forward; ~ **de proa** *m* TRANSP MAR *diseño de barcos* sheer aft; ~ **reglamentario** *m* TRANSP MAR *diseño de barcos* sheer

arruga *f* C&V crizzle, IMPR crease, OCEAN sand ripple, TEXTIL crease; ~ **del pliegue** *f* IMPR gusset wrinkle

arrugado *adj* ING MECÁ corrugated

arrugamiento *m* P&C *pintura* curtaining; ~ **de la película** *m* CINEMAT film wrinkling

arrugar *vt* IMPR, TEXTIL crease

arrugarse *v refl* TEXTIL crease

arrumaje *m* TRANSP MAR stowage, trim

arrumar *vt* TRANSP MAR stow

arrumazón *m* TRANSP MAR stowage, *de carga* trim

arrumbamiento *m* *AmL* (*cf dirección de buzamiento Esp*) MINAS level, *geología* strike, TRANSP MAR *dirección del buque* heading

arrumbarse *v refl* TRANSP MAR *navegación* steer for

arruzuz *m* AGRIC, ALIMENT arrowroot

arsenal *m* TRANSP MAR dockyard; ~ **de marina** *m* TRANSP MAR dockyard; ~ **naval** *m* D&A, TRANSP MAR naval dockyard

arseniato *m* QUÍMICA arsenate

arsénico *m* (*As*) C&V, METAL, QUÍMICA arsenic (*As*); ~ **rojo** *m* MINERAL, QUÍMICA red arsenic

arsenido *m* QUÍMICA arsenide

arsenífero *adj* MINERAL arseniferous

arseniopleíta *f* MINERAL arseniopleite

arseniosiderita *f* MINERAL arseniosiderite

arsenita *f* QUÍMICA arsenite

arseniuro: ~ **de galio** *m* ELECTRÓN, FÍS, ÓPT gallium arsenide; ~ **de níquel** *m* QUÍMICA nickel arsenide

arsenolita *f* MINERAL arsenolite

arsenopirita *f* MINERAL arsenopyrite; ~ **cobaltífera** *f* MINERAL glaucodote

arsina *f* QUÍMICA arsine

arte: ~ **de arrastre** *m* CONTAM MAR, OCEAN, TRANSP MAR *pesca* trawl; ~ **de arrastre y rastreo** *m* OCEAN, TRANSP MAR *pesca* trawling and dredging gear; ~ **de arrastre de Vigneron-Dahl** *m* OCEAN, TRANSP MAR *pesca* Vigneron-Dahl trawl; ~ **cerámico** *m* C&V ceramic art; ~ **por computadora** *f* *AmL* (*cf arte por ordenador Esp*) INFORM&PD computer art; ~ **de deriva** *m* OCEAN, TRANSP MAR *red* fishing drift; ~ **de enmalle** *m* OCEAN, TRANSP MAR *red* fishing drift; ~ **de estacas** *m* OCEAN, TRANSP MAR *tecnología pesquera* stake net, tide net; ~ **floral** *m* AGRIC florestry; ~ **de imprimir** *m* IMPR art of printing; ~ **de la maniobra** *m* TRANSP MAR *del buque* shiphandling; ~ **marinera** *m* TRANSP MAR seamanship; ~ **por ordenador** *m* *Esp* (*cf arte por computadora AmL*) INFORM&PD computer art; ~ **de playa** *m* OCEAN beach seine

artefacto *m* ING MECÁ gear, contrivance, device, PROD appliance, TEC ESP, TRANSP MAR *buque* craft; ~ **de arranque** *m* ELEC *motor* starter; ~ **de calefacción** *m* ING MECÁ, INSTAL TERM, TERMO heating device; ~ **espacial** *m* TEC ESP spacecraft; ~ **mecánico** *m* ING MECÁ, MECÁ engine; ~ **de salvamento** *m* SEG, TRANSP MAR life-saving apparatus

arteria *f* TELECOM route, routing, trunk; ~ **principal** *f* TRANSP arterial highway (*AmE*), arterial motorway (*BrE*)

artes: ~ **gráficas** *f pl* IMPR graphic arts; ~ **de pesca** *f pl* OCEAN fish traps

artesa *f* CARBÓN pan, *lavado de minerales* buddle, CONST trough, MINAS *lavado de minerales* trough, cradle, buddle, PROD trough, vat; ~ **de barro trabajado** *f* MINAS puddling trough; ~ **colectora** *f* *Esp* (*cf cuba colectora Esp*) MINAS collecting vat; ~ **de encofrado** *f* MINAS puddling trough; ~ **instrumental** *f* INSTR instrument basin; ~ **para lavado** *f* MINAS hutch; ~ **de pudelado** *f* MINAS puddling trough; ~ **de sacudidas** *f* PROD shaking tray; ~ **de sedimentación** *f* PROC QUÍ settling basin

artesano: ~ **en bronce** *m* CONST brasssmith

artesiano[1] *adj* AGUA, TEC PETR artesian

artesiano[2] *m* HIDROL artesian

articulación *f* ING MECÁ hinge, kunckle, link, linkage, MECÁ hinge, knuckle, link, linkage, NUCL jointing, linkage; ~ **del acelerador** *f* TRANSP accelerator linkage; ~ **con bisagra** *f* CONST hinge joint; ~ **de codo** *f* ING MECÁ, MECÁ elbow joint; ~ **esférica** *f* ING MECÁ, MECÁ ball-and-socket joint; ~ **de extremidades esféricas** *f* ING MECÁ, MECÁ ball-ended linkage; ~ **giratoria** *f* ING MECÁ, MECÁ swivel joint; ~ **de horquilla** *f* ING MECÁ, MECÁ fork head; ~ **lateral** *f* INSTR side joint; ~ **de logátomos** *f* TELECOM logatom articulation; ~ **de la placa de apoyo** *f* PROD crank link; ~ **de rótula** *f* ING MECÁ, MECÁ ball-and-socket joint, knuckle joint; ~ **a rótula y pernio** *f* ING MECÁ hinge-and-ball joint; ~ **tipo rótula** *f* AUTO kuuckle

articulado *adj* ING MECÁ, MECÁ, TRANSP articulated, hinged, pivoted

artículo *m* ING MECÁ, TEXTIL item; ~ **final** *m* TEXTIL end use; ~ **lacado** *m* COLOR lacquered work

artículos: ~ **de consumo** *m pl* EMB consumer goods; ~ **para el hogar** *m pl* TEXTIL home furnishings; ~ **de hojalata** *m pl* PROD tinware; ~ **navieros** *m pl* TEC PETR *transporte*, TRANSP MAR ship's articles; ~ **de papelería** *m pl* INFORM&PD stationery; ~ **de segunda** *m pl* TEXTIL seconds

artificiero *m* D&A artificer

artigar *vt* AGRIC burn

artillería *f* D&A artillery; ~ **a caballo** *f* D&A mounted artillery; ~ **de campaña** *f* D&A field artillery; ~ **de costa** *f* D&A coastal artillery; ~ **de grueso calibre** *f* D&A heavy artillery; ~ **de pequeño calibre** *f* D&A light artillery; ~ **pesada** *f* D&A heavy artillery; ~ **de sitio** *f* D&A siege artillery

artillero *m* *Esp* MINAS shot firer

Artinskiense *adj* GEOL *estratigrafía* Artinskian

artnalgia: ~ **hiperbárica** *f* OCEAN hyperbaric arthralgia

artrópodo *m* AGRIC arthropod

arvejilla *f* AGRIC hairy vetch

as: ~ **de guía** *m* TRANSP MAR *nudo* bowline

As *abr* (*arsénico*) C&V, METAL, QUÍMICA As (*arsenic*)

asa *f* CONST bow, grab, *cajón* handle, EMB handgrip, handle, ING MECÁ *fundición* bow, MECÁ, MINAS handle, PROD *de canasta, cesto* grip, bow handle, handle, bow, TRANSP AÉR hold; ~ **de cazo mayor** *f* PROD *fundición* bull handle; ~ **para elevación** *f* CONST lifting bow; ~ **para facilitar el transporte** *f* EMB, INSTR, MECÁ carrying handle; ~ **del portaequipajes** *f* VEH boot handle (*BrE*), trunk handle (*AmE*); ~ **de sujeción** *f* ING MECÁ handling lug; ~ **de transporte** *f* EMB, INSTR, MECÁ carrying handle

asbesto *m* GEN asbestos; **~-cemento** *m* CONST asbestos cement

asbestosis *f* CONST asbestosis

asbolana *f* MINERAL asbolane

asbolita *f* MINERAL asbolite

ascaricida *m* QUÍMICA *intestinal* ascaricide

ascendente[1] *adj* FÍS FLUID upstream, GEOL upward, HIDROL upstream, TEC PETR updip

ascendente[2] *m* CONST upgrade

ascender[1] *vt* CONST upgrade

ascender[2] *vi* TEC ESP, TRANSP AÉR climb

ascensión *f* CONST *del magma* upright, upgrade, CRISTAL climb, OCEAN ascension; **~ capilar** *f* CARBÓN capillary rise; **~ recta** *f* TEC ESP right ascension

ascenso *m* HIDROL, OCEAN rise, TEC ESP climb, TRANSP AÉR climb, climb-out; **~ escalonado** *m* TEC ESP, TRANSP AÉR stepped climb; **~ inicial** *m* TEC ESP stepped climb, TRANSP AÉR initial climb-out; **~ vertical** *m* TEC ESP vertical ascent

ascensor *m* CONST, MECÁ elevator (*AmE*), lift (*BrE*); **~ hidráulico** *m* CONST plunger elevator (*AmE*), plunging lift (*BrE*); **~ de pasajeros** *m* TRANSP passenger elevator (*AmE*), passenger lift (*BrE*)

aschynita *f* MINERAL aeschynite

ASCII *abr* (*Código General Americano de Intercambio de Información*) IMPR ASCII (*American Standard Code for Information Interchange*)

ascórbico *adj* QUÍMICA ascorbic

ascua *f* CARBÓN live coal

asegurador *m* TRANSP MAR underwriter; **~ marítimo** *m* TRANSP MAR underwriter

asegurar[1] *vt* MECÁ fasten, TRANSP MAR make fast

asegurar[2]: **~ la salud, la seguridad y el bienestar de** *fra* SEG ensure the health, safety and welfare of

asentador *m* ING MECÁ strop

asentamiento *m* *AmL* (*cf apretamiento Esp*) CARBÓN *de un terreno* settlement, CONST settlement, seating, GEOL settling, MINAS *del techo* squeeze, PROD *fundición* sinking, QUÍMICA settling; **~ del cañón** *m* D&A gun pit; **~ diferencial** *m* CONST, GEOL differential settlement; **~ de popa debido a la velocidad** *m* TRANSP squat; **~ preliminar** *m* ING MECÁ preliminary seating; **~ de refractarios en un horno** *m* PROD setting

asentar *vt* CONST bed out, set, subside, D&A *un fusil* bed in, MINAS *el polvo* lay, PROC QUÍ settle on, TEC ESP trim, TRANSP MAR *libro diario* enter

asentarse *v refl* CARBÓN, CONST settle, MINAS sink; **~ el terreno** *v refl* MINAS squeeze

aséptico *adj* ALIMENT, EMB, QUÍMICA aseptic

aserradero *m* *Esp* (*cf serrería mecánica AmL*) AGRIC, CONST, MECÁ, MINAS sawmill

aserradura *f* AGRIC, CONST, MECÁ, MINAS sawing

aserrar *vt* AGRIC, CONST, MECÁ, MINAS saw

aserrín *m* CONST *carpintería* sawdust

asesor: **~ en producción animal** *m* AGRIC livestock adviser; **~ sobre seguridad** *m* CONST, SEG safety advisor; **~ técnico de turno** *m* NUCL shift technical adviser

asesora: **~ en producción animal** *f* AGRIC livestock adviser; **~ sobre seguridad** *f* CONST, SEG safety advisor; **~ técnica de turno** *f* NUCL shift technical adviser

asesoramiento: **~ por láser** *m* ELECTRÓN laser guidance

asfaltado[1] *adj* CONST, TRANSP bituminized

asfaltado[2] *m* CONST, TRANSP asphalting, bituminization

asfaltadora *f* CONST, TRANSP asphalt-spreading machine (*AmE*), pavement-spreading machine, road-metal-spreading machine (*BrE*)

asfaltar *vt* CONST, TRANSP asphalt, bituminize

asfalteno *m* MINERAL asphaltene

asfaltita *f* MINERAL asphaltite

asfalto[1] *adj* TRANSP bituminized

asfalto[2] *m* CONST asphalt, bitumen, ELEC *aislación*, GEOL, ING MECÁ asphalt, bitumen, MINERAL bitumen, asphalt, P&C asphalt, bitumen, PETROL bitumen, asphalt, PROC QUÍ, TEC PETR asphalt, bitumen, TRANSP asphalting, bituminization, asphalt, bitumen; **~ mineral** *m* GEOL, PETROL, TEC PETR rock asphalt

asidero *m* EMB, ING MECÁ, MINAS, PROD, TEC ESP, VEH handgrip, handle; **~ de bayoneta** *m* EMB bayonet catch; **~ reforzado que forma parte del envase** *m* EMB integral reinforced handle

asiento[1]: **con ~ a popa** *adv* TRANSP MAR trimmed by the stern; **con ~ a proa** *adv* TRANSP MAR trimmed by the head

asiento[2] *m* AUTO, C&V seat, CARBÓN *del terreno* settling, CONST settlement, bed, subsidence, seat, FERRO *equipo inamovible*, GEOL subsidence, ING MECÁ seat, seating, INSTAL HIDRÁUL *válvula* seating, seat, MECÁ, NUCL, PROD seat, seating, TEC ESP *aeróstatos, buques, hidroaviones* trim, TEC PETR *geología* subsidence, TRANSP seat, TRANSP MAR *de la embarcación* trim, VEH seat; **~ abatible** *m* AUTO, VEH folding seat (*AmE*), reclining seat, tip-up seat (*BrE*); **~ bajo** *m* AUTO, VEH bucket seat; **~ de charnela** *m* ING MECÁ clapper seat; **~ de clavija** *m* ING MECÁ plug seat; **~ corrido** *m* VEH bench seat; **~ de cubeta** *m* AUTO, VEH bucket seat; **~ delantero** *m* AUTO, VEH front seat; **~ de eje** *m* ING MECÁ axle seat; **~ eyectable** *m* D&A, TEC ESP, TRANSP AÉR ejection seat, ejector seat; **~ del ingeniero de vuelo** *m* TRANSP AÉR flight-engineer's seat; **~ de máquinas** *m* TRANSP MAR engine seating; **~ del mecánico de vuelo** *m* TRANSP AÉR flight-engineer's seat; **~ de mina** *m* *Esp* (*cf apeo de mina AmL*) MINAS headgear; **~ de montaje** *m* ING MECÁ mounting pad; **~ de pasajero** *m* TRANSP AÉR passenger seat; **~ de la platina** *m* PAPEL bedplate box; **~ plegable** *m* AUTO, VEH folding seat (*AmE*), reclining seat, tip-up seat (*BrE*); **~ posterior** *m* VEH bench seat; **~ reclinable** *m* AUTO, VEH reclining seat; **~ de resorte** *m* AUTO, VEH spring seat; **~ de rótula** *m* ING MECÁ ball-socket seat; **~ de tapón** *m* ING MECÁ plug seat; **~ del terreno** *m* MINAS *construcción* squeezing; **~ trasero** *m* VEH *motocicleta* pillion; **~ trasero descubierto** *m* AUTO, VEH rumble seat; **~ de válvula** *m* AUTO, INSTAL HIDRÁUL, VEH *motor* valve lap, valve seat; **~ de válvula de corredera** *m* INSTAL HIDRÁUL seat of slide-valve; **~ vibrante** *m* INSTAL HIDRÁUL *válvula* loose seat

asignación *f* INFORM&PD, PROD allocation, allotment; **~ de almacenamiento** *f* INFORM&PD storage allocation; **~ de canal adaptivo** *f* INFORM&PD adaptive-channel allocation; **~ de clavijas** *f* PROD pin assignment; **~ dinámica** *f* INFORM&PD dynamic allocation; **~ de espectro** *f* TRANSP MAR *comunicación por satélite* spectrum allocation; **~ estática** *f* INFORM&PD static allocation; **~ en firme** *f* PROD firm allocation; **~ de función** *f* TEC ESP demand assignment; **~ de**

masas *f* NUCL mass assignment; ~ **de memoria** *f* INFORM&PD storage allocation; ~ **del número de identidad personal** *f* PROD PIN assignment; ~ **de recursos** *f* INFORM&PD resource allocation; ~ **de ruta adaptable** *f* INFORM&PD adaptive routing; ~ **de ruta alternativa** *f* PROD alternate routing; ~ **de tráfico** *f* TRANSP traffic assignment; ~ **de vía** *f* FERRO track allocation

asignar *vt* INFORM&PD, PROD allocate, assign

asimetría *f* ELEC unbalance, GEOM, INFORM&PD asymmetry, ING MECÁ unbalance, MATEMÁT, QUÍMICA *sistema optico* asymmetry

asimétrico *adj* ELEC unbalanced, GEOM, INFORM&PD, MATEMÁT asymmetric, asymmetrical, TEC ESP unbalanced

asimilable *adj* AGRIC *nutrientes* available

asincrónico *adj* CINEMAT asynchronous, nonsync, ELEC, FÍS, INFORM&PD asynchronous, TV asynchronous, nonsync

asincronismo *m* CINEMAT, ELEC, FÍS, INFORM&PD, TV asynchronism

asíncrono *adj* CINEMAT asynchronous, nonsync, ELEC, FÍS, INFORM&PD asynchronous, TV asynchronous, nonsync

asíntota *f* GEOM, MATEMÁT asymptote

asintótico *adj* GEOM, MATEMÁT asymptotic

asir *vt* PROD grip

asistencia *f* INFORM&PD support; ~ **al diseño** *f* INFORM&PD design aid; ~ **de vídeo** *f* AmL, ~ **de vídeo** *f Esp* CINEMAT video assist

asistente: ~ **de cabina** *m* TRANSP, TRANSP AÉR cabin attendant

asistido: ~ **por computadora** *adj AmL* (*cf asistido por ordenador Esp*) INFORM&PD computer-assisted; ~ **por ordenador** *adj Esp* (*cf asistido por computadora AmL*) INFORM&PD computer-assisted

asistir *vt* INFORM&PD support

askarel *m* ING ELÉC askarel

asmanita *f* MINERAL asmanite

ASME *abr* (*Sociedad Norteamericana de Ingenieros Mecánicos*) MECÁ ASME (*American Society of Mechanical Engineers*)

asociación *f* GEOL, QUÍMICA, TELECOM association; ~ **de aplicaciones** *f* TELECOM application association; ~ **mineral** *f* GEOL mineral assemblage; ~ **mineralógica** *f* GEOL mineral assemblage; ~ **ofiolítica** *f* GEOL ophiolite suite; ~ **de rocas** *f* GEOL rock association; ~ **de rocas ígneas** *f* GEOL igneous suite

Asociación: ~ **Europea de Reciclado y Recuperación** *f* (*AERR*) EMB, RECICL European Recycling and Recovery Association (*ERRA*); ~ **Internacional de Señalización Marítima** *f* (*AISM*) TRANSP MAR International Association of Lighthouses Authorities (*IALA*); ~ **Mundial de Operadores Nucleares** *f* (*WANO*) NUCL World Association of Nuclear Operators (*WANO*); ~ **de Protección e Indemnización** *f* CONTAM MAR Protection and Indemnity Association

aspa *f* CONST flight, INSTR, PROD *molino* arm, vane; ~ **del ventilador** *f* AUTO, MECÁ, REFRIG, TERMO, TRANSP, VEH fan blade

aspartamo *m* ALIMENT aspartame

aspártico *adj* ALIMENT, QUÍMICA aspartic

aspatrón *m* NUCL aspatron

aspecto[1]: **de ~ inútil** *adj* MINAS hungry

aspecto[2] *m* PAPEL appearance, TEXTIL *acabado* look; ~ **acabado** *m* TEXTIL finished appearance; ~ **blancuzco y turbio** *m* P&C *defecto de la pintura* blushing; ~ **brillante** *m* COLOR, PROD glossiness; ~ **monocromático** *m* METAL monochromatic light; ~ **sedoso** *m* P&C *pintura* silking; ~ **tras el almacenado** *m* EMB shelf appeal

aspereza *f* ING MECÁ, MECÁ roughness; ~ **superficial** *f* ING MECÁ, MECÁ surface roughness

asperjar *vt* AGUA, CONST, CONTAM MAR, QUÍMICA spray, sprinkle

áspero *adj* ING MECÁ, MECÁ coarse

asperolita *f* MINERAL asperolite

asperón *m* GEOL flag, PETROL grindstone

aspersar *vt* AGRIC, AGUA, CONST, HIDROL, SEG sprinkle

aspersión *f* AGUA, HIDROL spraying, sprinkling; ~ **del dispersante** *f* CONTAM MAR dispersant spraying; ~ **del núcleo a baja presión** *f* (*LPCS*) NUCL low pressure core spray (*LPCS*); ~ **seca** *f* C&V dry spray

aspersor *m* AGUA, CONST, HIDROL, SEG *extintor automático de incendios* sprinkler; ~ **para césped** *m* AGRIC, AGUA, HIDROL lawn sprinkler; ~ **contra incendios** *m* SEG, TERMO fire sprinkler; ~ **giratorio** *m* AGUA rotating sprayer

aspidolita *f* MINERAL aspidolite

aspiración *f* AGUA drawing, CONST draft (*AmE*), draught (*BrE*), ING MECÁ indraft (*AmE*), indraft of air (*AmE*), indraught (*BrE*), indraught of air (*BrE*), suction, INSTAL HIDRÁUL drawing off, *máquinas* induction, intake, MECÁ suction, MINAS *del aire* exhausting, REFRIG, TEXTIL suction; ~ **adicional** *f* AGUA *pozo* draw-down; ~ **de aire** *f* ING MECÁ indraft (*AmE*), indraught (*BrE*); ~ **inducida** *f* ING MECÁ induced draft (*AmE*), induced draught (*BrE*)

aspirador *m* ALIMENT, ING MECÁ exhauster, LAB *material de vidrio* aspirator, MINAS (*cf canaleta AmL, chimenea AmL*) chute, PROD aspirator; ~ **de partículas** *m* OCEAN air exhauster; ~ **de polvo** *m* CARBÓN deduster, EMB dust aspirator; ~ **del polvo por rodillos** *m* ING ELÉC roller dust collector

aspiradora: ~ **para propósitos industriales** *f* SEG vacuum cleaner for industrial purposes; ~ **al vacío de polvos peligrosos para la salud** *f* SEG vacuum cleaner for dusts hazardous to health

aspirar *vt* CARBÓN dedust, TEXTIL suck

ASR *abr* (*terminal emisor-receptor automático*) INFORM&PD ASR (*automatic send-receive*)

asta *f* PETROL mast; ~ **de bandera** *f* TRANSP MAR flagstaff; ~ **de pararrayos** *f* GEOFÍS lightning rod; ~ **del viento** *f* METEO wind arrow, wind shaft

ástaco *m* AGUA crawfish, crayfish

astático *adj* FÍS astatic

astato *m* (*At*) FÍS RAD, QUÍMICA astatine (*At*)

astenosfera *f* GEOL, PETROL asthenosphere

asteroide *m* TEC ESP minor planet, asteroid

astigmatismo *m* FÍS, ÓPT, TV astigmatism

astil *m* PROD helve; ~ **de balanza** *m* ING MECÁ, METR balance beam

astilla *f* C&V sliver, splinter, MECÁ *madera* chip; ~ **de madera** *f* CONST splinter; ~ **muerta** *f* TRANSP MAR *buque* deadrise, *diseño de embarcaciones* rise of floor

astillado: ~ **en estrella** *m* C&V star crack

astillamiento *m* C&V spalling

astillero *m* TRANSP MAR *construcción naval* construction yard, dockyard, shipyard

astrágalo *m* CONST *arquitectura* astragal

astrakanita *f* MINERAL astrakanite

astrobrújula *f* TEC ESP astrocompass

astrodinámica *f* TEC ESP astrodynamics

astrodomo *m* TEC ESP astrodrome

astrofilita *f* MINERAL astrophyllite

astrofísica *f* FÍS, TEC ESP astrophysics

astrometría *f* TEC ESP astrometry

astronauta *m* TEC ESP astronaut

astronave *f* TEC ESP spacecraft; ~ **de lanzadera** *f* TEC ESP shuttle

astronavegación *f* TRANSP MAR astronavigation

astronomía: ~ **espacial** *f* TEC ESP space astronomy; ~ **de rayos gamma** *f* FÍS RAD, TEC ESP gamma-ray astronomy

astrónomo *m* TEC ESP astronomer

astropuerto *m* TEC ESP astrodome

At *abr* (*astato*) FÍS RAD, QUÍMICA At (*astatine*)

atacadera *f* MINAS *barrenos* stemmer; ~ **de barrenos** *f* MINAS *explosivos* ramming bar

atacado[1]: ~ **por ácidos** *adj* CONTAM, PROC QUÍ, QUÍMICA attacked by acids

atacado[2] *m Esp* (*cf atraque AmL*) MINAS *barrenos* tamping, stemming

atacador *m* D&A *de un cañón* rammer, MINAS *barrenos* tamping rod, tamping stick; ~ **automático** *m* D&A *de un cañón* automatic rammer; ~ **para banco** *m* PROD bench rammer; ~ **cónico** *m* PROD *moldería* pegging rammer

atacamita *f* MINERAL atacamite

atacar *vt Esp* (*cf atracar AmL*) MINAS *barrenos, galería de voladuras* tamp, *cargas explosivas* ram, *una carga* stem; ~ **con un ácido** *vt* CRISTAL, ELECTRÓN, IMPR, METAL, QUÍMICA etch

atacolita *f* MINERAL attacolite

atáctico *adj* QUÍMICA *polímeros* atactic

atado *m* C&V bundle; ~ **de la bolsa con alambre** *m* EMB wire bag tie

atadora: ~ **de paquetes** *f* EMB bundle-tying machine

atadura *f* ING MECÁ tie, TRANSP AÉR brace

ataduras: ~ **por pulgada** *f pl* IMPR ties per inch

ataguía *f* AGUA, CONST, HIDROL, INSTAL HIDRÁUL, TRANSP MAR batardeau, caisson, cofferdam; ~ **celular** *f* AGUA, CONST, HIDROL, TRANSP MAR cellular cofferdam

atajo *m* CONST, TRANSP byroad

ataludar *vt* ING MECÁ batter

atapulgita *f* PETROL attapulgite

ataque[1]: **de ~ directo** *adj* ING MECÁ, MECÁ gearless

ataque[2] *m* ACÚST, CARBÓN attack; ~ **catódico** *m* METAL *corrosión* cathodic etching; ~ **por corrosión en la línea de nivel de vidrio** *m* C&V flux-line attack; ~ **electrolítico** *m* ELEC, NUCL electrolytic etching; ~ **con granadas de mano** *m* D&A bombing; ~ **a las piedras del horno por movimiento del vidrio** *m* C&V upward drilling; ~ **de presión** *m* OCEAN bends; ~ **químico** *m* QUÍMICA etching; ~ **de sulfato** *m* CONST sulfate attack (*AmE*), sulphate attack (*BrE*); ~ **térmico** *m* METAL thermal etching

atar *vt* CONST, INFORM&PD bind, MECÁ fasten, lock, MINAS *jaulas* hitch

atarazana *f* TRANSP MAR *construcción naval* dockyard

atascado *adj* ING MECÁ, MECÁ clogged

atascamiento *m* AGUA, CARBÓN clogging, IMPR jam, ING MECÁ clogging, jam, seizing, MINAS *de la chimenea* clogging, PROD *horno alto* bunging-up

atascar *vt* AGUA clog, ING MECÁ clog, foul, jam

atascarse *v refl* ING MECÁ seize, PROD choke

atasco *m* AUTO jamming, road jam, CINEMAT jam, MECÁ, PROD bottleneck, stoppage, TRANSP *tráfico* bottleneck, road jam, VEH road jam; ~ **de tráfico** *m* TRANSP traffic jam

atelesita *f* MINERAL atelestite

atelita *f* MINERAL atelite

atenuación *f* ACÚST, ELEC attenuation, damping, fading, ELECTRÓN attenuation, decrement, fading, FÍS attenuation, damping, fading, FÍS ONDAS *de onda* attenuation, INFORM&PD attenuation, damping, fading, ING ELÉC damping, loss, ÓPT loss, attenuation, PETROL attenuation, fading, damping, TEC ESP, TELECOM attenuation, TV attenuation, damping, fading; ~ **acústica** *f* ELECTRÓN acoustic attenuation; ~ **de armonía** *f* ELECTRÓN matching attenuation; ~ **armónica** *f* ELECTRÓN harmonic attenuation; ~ **de banda lateral** *f* ELECTRÓN sideband attenuation; ~ **copolar** *f* TEC ESP copolar attenuation; ~ **cruzada** *f* ACÚST, ELECTRÓN cross fade; ~ **diferencial de modos** *f* ELECTRÓN, TELECOM differential mode attenuation; ~ **específica** *f* TEC ESP specific attenuation; ~ **de filtro** *f* ELECTRÓN filter attenuation; ~ **infinita** *f* ELECTRÓN infinite attenuation; ~ **por lluvia** *f* TEC ESP rain attenuation; ~ **de microondas** *f* ELECTRÓN, TELECOM microwave attenuation; ~ **en modo diferencial** *f* ÓPT, TELECOM differential mode attenuation; ~ **de onda** *f* CONST ripple attenuation; ~ **paso banda** *f* ELECTRÓN *filtros* pass-band attenuation; ~ **de radiación** *f* ELECTRÓN beam attenuation; ~ **de salida** *f* ELECTRÓN output attenuation; ~ **de tensión** *f* INSTAL TERM *mecánica* stress relief; ~ **de la transmisión** *f* TEC ESP transmission attenuation; ~ **de la trayectoria** *f* TV path attenuation; ~ **por unión intrínseca** *f* ÓPT intrinsic joint loss; ~ **variable** *f* ELECTRÓN variable attenuation

atenuado *adj* METAL *aleación, solución* dilute

atenuador *m* CINEMAT dimmer, ELEC, ELECTRÓN, FÍS, FÍS ONDAS, INFORM&PD, ING ELÉC, ÓPT attenuator, fader, TEC ESP attenuator, TELECOM attenuator, pad; ~ **absorbente** *m* ELECTRÓN, TELECOM absorptive attenuator; ~ **de álabe en paralelo** *m* ELECTRÓN parallel vane-attenuator; ~ **de aleta** *m* ELECTRÓN flap attenuator; ~ **de audio** *m* ELECTRÓN audio attenuator; ~ **de audiofrecuencia** *m* ELECTRÓN audio attenuator; ~ **coaxial** *m* ELECTRÓN coaxial attenuator; ~ **coaxial variable** *m* ELECTRÓN variable coaxial attenuator; ~ **de décadas** *m* ELECTRÓN decade attenuator; ~ **digital** *m* ELECTRÓN digital attenuator; ~ **de diodo NIP** *m* ELECTRÓN PIN-diode attenuator; ~ **en escalera** *m* ELECTRÓN ladder attenuator; ~ **fijo** *m* TEC ESP *guía de ondas* pad; ~ **manual** *m* CINEMAT manual dimmer; ~ **de microonda variable** *m* ELECTRÓN, TELECOM variable-microwave attenuator; ~ **de microondas** *m* ELECTRÓN, TELECOM microwave attenuator; ~ **óptico** *m* ELECTRÓN, ÓPT, TELECOM optical attenuator; ~ **permanente** *m* ELECTRÓN fixed attenuator; ~ **de pistón** *m* ELECTRÓN, FÍS piston-attenuator; ~ **de potencia de entrada** *m* ELECTRÓN input attenuator; ~ **de reactancia** *m* ELEC, ELECTRÓN, FÍS, ING ELÉC reactance attenuator; ~ **resistivo** *m* ELEC, ELECTRÓN, ING ELÉC, TELECOM resistive attenuator; ~ **de salida** *m* ELECTRÓN output attenuator; ~ **solar** *m* TEC ESP solar absorber; ~ **variable** *m* ELECTRÓN variable attenuator; ~ **variable continuamente** *m* ELECTRÓN continuously-variable attenuator

atenuar *vt* CINEMAT, FOTO *luz* dim

aterciopelado: ~ **artificial** *m* PAPEL *acabado* velvet finish

atermancia *f* FÍS athermancy

atérmico *adj* QUÍMICA athermal

aterrada *f* OCEAN approach

aterrajado *m* ING MECÁ tapping

aterrajadora *f* ING MECÁ tapping machine

aterrajamiento *m* PROD strickling, sweeping up, *fundería* sweeping

aterrajar *vt* PROD thread, tap, *fundería* sweep up, *moldería* sweep

aterraje *m* TRANSP AÉR, TRANSP MAR landing

aterrizador *m* TEC ESP, TRANSP AÉR landing gear

aterrizaje[1] *m* TEC ESP landing, touchdown, TRANSP AÉR landing; ~ **abocardado** *m* TRANSP AÉR flared landing; ~ **abortado** *m* TRANSP AÉR aborted landing; ~ **automático** *m* TRANSP AÉR autoland; ~ **con baja visibilidad** *m* TRANSP AÉR low-visibility landing; ~ **brusco** *m* TRANSP AÉR hard landing; ~ **a ciegas** *m* TRANSP AÉR blind landing, instrument landing; ~ **corto** *m* TRANSP AÉR undershoot; ~ **y despegue convencional** *m* (*CTOL*) D&A, TRANSP AÉR conventional take-off and landing (*CTOL*); ~ **con desviación lateral** *m* TRANSP AÉR lateral-drift landing; ~ **difícil** *m* TEC ESP hard landing; ~ **duro** *m* TRANSP AÉR hard landing; ~ **de emergencia** *m* TEC ESP, TRANSP AÉR crash landing, emergency landing; ~ **forzoso** *m* TRANSP AÉR crash landing, ditch, ditching, emergency landing, splashdown; ~ **guiado** *m* TEC ESP, TRANSP AÉR homing; ~ **por instrumentos** *m* TRANSP AÉR instrument landing; ~ **sin motor** *m* TRANSP AÉR dead-stick landing; ~ **en paracaídas** *m* D&A, TRANSP AÉR parachute landing; ~ **por radiogoniómetro automático** *m* TRANSP AÉR ADF letdown; ~ **rebotado** *m* TRANSP AÉR bounced landing; ~ **de rebote** *m* TRANSP AÉR bounced landing; ~ **suave** *m* TEC ESP soft landing; ~ **de toma de contacto y despegue** *m* TRANSP AÉR touch-and-go landing; ~ **el tren replegado** *m* TRANSP AÉR wheels-up landing; ~ **de tres puntos** *m* TRANSP AÉR three-point landing; ~ **sobre el vientre del avión** *m* TRANSP AÉR belly landing; ~ **con vientos cruzados** *m* TRANSP AÉR crosswind landing; ~ **violento** *m* TRANSP AÉR rough landing; ~ **con visibilidad pobre** *m* TRANSP AÉR low-visibility landing

aterrizaje[2]: ~ **suave** *f* TRANSP AÉR soft landing

aterrizar *vi* TEC ESP land, touch down, TRANSP AÉR land

atesador *m* ING MECÁ stretcher

atesamiento *m* ING MECÁ tightening

ATDM *abr* (*multiplexión por división en tiempo asíncrono*) TELECOM ATDM (*asynchronous time division multiplexing*)

ATG *abr* (*análisis termogravimétrico*) GEOFÍS, LAB, TERMO TGA (*thermogravimetric analysis*)

atierre *m* CARBÓN *minas* slide, MINAS *relleno de desechos* gob

atíncar *m* MINERAL tinkal

atirantador *m* ING MECÁ straightener

atirantamiento *m* ING MECÁ stiffening, stiffness

atirantar *vt* CONST brace

atizador: ~ **de ignición** *m* ING MECÁ ignition poker

atizar *vt* PROD stoke

atlas: ~ **de corrientes** *m* TRANSP MAR tidal stream

atlas; ~ **de corrientes de marea** *m* TRANSP MAR *navegación* current chart

atlasita *f* MINERAL atlasite

ATM *abr* INFORM&PD (*cajero automático Esp, máquina medidora automática AmL*) ATM (*automatic teller machine*), TELECOM (*modo de transferencia asíncrono m*) ATM (*asynchronous transfer mode*), (*método de ensayo alternativo*) ATM (*alternative test method*)

atmólisis *f* QUÍMICA atmolysis

atmósfera[1]: **de** ~ **gaseosa** *adj* GAS, TERMO *lámpara eléctrica* gas-filled

atmósfera[2] *f* FÍS, FÍS RAD, ING MECÁ, METEO, METR, PROD atmosphere; ~ **controlada** *f* INSTAL TERM controlled atmosphere; ~ **corrosiva** *f* PROD corrosive atmosphere; ~ **explosiva** *f* SEG explosive atmosphere; ~ **de gas inerte** *f* NUCL inert gas blanketing; ~ **hiperbárica** *f* OCEAN hyperbaric atmosphere; ~ **inflamable** *f* SEG, TERMO flammable atmosphere; ~ **iónica** *f* FÍS RAD ionic atmosphere; ~ **libre** *f* METEO free atmosphere; ~ **potencialmente explosivas** *f* CARBÓN potentially explosive atmosphere; ~ **de referencia** *f* METEO standard atmosphere, TEC ESP reference atmosphere; ~ **de referencia estándar** *f* ING MECÁ standard reference atmosphere; ~ **salada** *f* REFRIG salted atmosphere; ~ **terrestre** *f* METEO, TEC ESP earth's atmosphere; ~ **tipo** *f* METEO standard atmosphere

atmosférico[1] *adj* FÍS, METEO atmospheric

atmosférico[2]: ~ **silbante** *m* TEC ESP whistler

atmosféricos *m pl* ELECTRÓN, ING ELÉC atmospherics, METEO sferics (*AmE*), spherics (*BrE*)

ato *pref* (*a-*) METR atto- (*a-*)

atoaje *m* TRANSP MAR *embarcaciones* towing

atolón *m* OCEAN, TRANSP MAR atoll

atomicidad *f* FÍS, FÍS PART, INFORM&PD, QUÍMICA atomicity

atómico *adj* FÍS, FÍS PART, NUCL, QUÍMICA atomic

atomización *f* FÍS PART, NUCL, PROC QUÍ, QUÍMICA, TEC PETR *refino* atomization

atomizador *m* AGRIC mist blower, EMB, LAB, PROC QUÍ, QUÍMICA, TEC PETR, TRANSP atomizer, nebulizer, sprayer; ~ **de aerosoles** *m* ING MECÁ aerosol spray container; ~ **ultrasónico de combustible** *m* TRANSP ultrasonic fuel atomizer

atomizar *vt* FÍS, FÍS PART, NUCL, QUÍMICA, TEC PETR atomize

átomo *m* GEN atom; ~ **aceptor** *m* ELECTRÓN *semiconductores*, FÍS, FÍS PART acceptor atom; ~ **donador** *m* ELECTRÓN, FÍS, FÍS PART donor atom; ~ **dopante** *m* ELECTRÓN dopant; ~ **excitado** *m* FÍS, FÍS PART excited atom, ion, ionized atom; ~ **excitado positivamente** *m* FÍS PART positive ion; ~ **de impureza** *m* CRISTAL impurity atom; ~ **intersticial** *m* CRISTAL interstitial atom; ~ **ionizado** *m* FÍS PART excited atom, ion, ionized atom; ~ **ionizado negativamente** *m* FÍS PART anion; ~ **marcado** *m* NUCL labeled atom (*AmE*), labelled atom (*BrE*), tagged atom; ~ **metaestable** *m* FÍS PART, FÍS RAD metastable atom; ~ **neutro** *m* FÍS PART neutral atom; ~ **percutado** *m* FÍS PART, NUCL knocked-on atom; ~ **receptor** *m* FÍS, FÍS PART acceptor atom; ~ **terminal** *m* QUÍMICA *de molécula* terminal atom; ~ **trazador** *m* FÍS PART, NUCL tracer atom

atopita *f* MINERAL atopite

atorado *adj* ING MECÁ, MECÁ clogged, fouled, PROD *bombas, tuberías* fouled

atoramiento *m* ING MECÁ, MECÁ jam

atorar *vt* ING MECÁ, MECÁ jam

atorarse *v refl* ING MECÁ, MECÁ seize

atornillado *m* TEC PETR *tuberías* screwing-in; **~ de tubería** *m* CONST pipe screwing

atornillamiento *m* ING MECÁ, MECÁ screwing, SEG bolting

atornillar *vt* CONST bolt, drive, ING MECÁ, MECÁ screw

ATP *abr* (*trifosfato de adenosita*) ALIMENT, PETROL, QUÍMICA ATP (*adenosine triphosphate*)

atracada *f* TRANSP MAR drawing alongside, *navegación* docking

atracado *adj* TRANSP MAR *al muelle* alongside

atracamiento *m* TEC ESP *naves espaciales* berthing

atracar[1] *vt* (*cf atacar Esp*) MINAS *barrenos, galería de voladuras, terrenos* tamp, *cargas explosivas* ram, *barrenos* ram home, *una carga* stem, TEC ESP *instalaciones espaciales* dock, TRANSP MAR dock, haul alongside, berth

atracar[2] *vi* TRANSP MAR berth; **~ terrenos** *vt* PROD *minas* tamp

atracción *f* ELEC, ING ELÉC, TEC ESP attraction; **~ electrostática** *f* ELEC, ING ELÉC, TEC ESP electrostatic attraction; **~ magnética** *f* ELEC, FÍS, ING ELÉC, TEC ESP magnetic attraction

atractivo *adj* ELEC, FÍS, ING ELÉC, TEC ESP attractive

atraerse *v refl* ELEC, FÍS, ING ELÉC, TEC ESP attract

atrapado[1]: **~ por el hielo** *adj* TRANSP MAR icebound

atrapado[2] *m* SEG trapping; **~ rápido** *m* TRANSP AÉR fast-slaving

atrapador: ~ de llamas *m* TRANSP AÉR flame trap

atrapallamas *m* TRANSP AÉR flame trap

atraque *m* *AmL* (*cf atacado Esp*) MINAS *barrenos* tamping, stemming TEC ESP *del proyectil en la recámara* berthing, *misiones espaciales* docking, TRANSP MAR *de pasajeros* berthing, *situación* berth

atrás *adv* TEC ESP aft; **~ media** *adv* TRANSP MAR half astern; **~ a poca** *adv* TRANSP MAR *motor* slow astern; **~ poco a poco** *adv* TRANSP MAR *motor* dead slow astern; **~ toda** *adv* TRANSP MAR *motor* full astern

atrasado *adj* PROD outstanding

atraso *m* ELEC, ING MECÁ, MECÁ, QUÍMICA, TEC ESP, TRANSP lagging, retardation; **~ de la marea** *m* OCEAN *mareas máximas* tidal epoch

atravesar *vt* CONST cross, HIDROL run through; **~ mediante un puente** *vt* CONST bridge over

atributo *m* INFORM&PD, TELECOM attribute; **~ de estilo** *m* IMPR style attribute; **~ de valor de consigna** *m* TELECOM set-valued attribute; **~ de valor predeterminado** *m* TELECOM set-valued attribute

atril *m* ACÚST *música* rest, stand, ING MECÁ rack

atrópico *adj* QUÍMICA atropic

atuendo *m* TEXTIL turnout; **~ de tela** *m* TEXTIL cloth turnout

atunero *m* OCEAN *barcos de pesca* tuna boat, tunny boat

ATWS *abr* (*transitorios previstos sin parada de emergencia*) NUCL ATWS (*anticipated transient without scream*)

Au *abr* (*oro*) METAL, QUÍMICA Au (*gold*)

audición *f* ACÚST hearing; **~ digital** *f* ELECTRÓN digital speech

audífono *m* ACÚST earphone, hearing aid; **~ inserto** *m* ACÚST insert earphone

audio[1] *adj* ACÚST, ELECTRÓN, TELECOM audio

audio[2] *m* INFORM&PD audio; **~ amplificador** *m* ELEC-TRÓN audio amplifier; **~ atenuador** *m* ELECTRÓN audio attenuator; **~ videotexto** *m* TELECOM audio-videotext

audiofrecuencia *f* ACÚST, ELECTRÓN, FÍS ONDAS, TELE-COM audio frequency

audiograma *m* ACÚST audiogram; **~ de nivel de enmascaramiento** *m* ACÚST masking-level audiogram; **~ de tonos puros** *m* ACÚST pure-tone audiogram; **~ vocal** *m* ACÚST speech audiogram

audiometría *f* ACÚST, FÍS ONDAS, METR audiometry; **~ tonal** *f* ACÚST pure-tone audiometry; **~ de umbral** *f* ACÚST, SEG threshold audiometry; **~ por vía ósea** *f* ACÚST bone audiometry; **~ vocal** *f* ACÚST speech audiometry

audiómetro *m* ACÚST, FÍS ONDAS, INSTR, SEG audio-meter; **~ vocal** *m* ACÚST speech audiometer

audiooscilador *m* FÍS RAD audio oscillator

auditoría: ~ en proceso de inspección *f* METR audit of inspection procedure

auge *m* TEC ESP apogee

augelita *f* MINERAL augelite

augita *f* MINERAL augite

aulacógeno *m* GEOL aulacogen

aullido *m* ACÚST howling, TELECOM *acústica* warble

aumentador: ~ de espuma *m* ALIMENT foam booster

aumentar *vt* ELECTRÓN boost, FOTO, TEC ESP *fotografía* blow up

aumento[1]: **de ~** *adj* INSTR *óptica* magnifying

aumento[2] *m* FÍS *de una lente* power, HIDROL rise, INFORM&PD zoom, INSTR *óptica* magnification, OCEAN *presión barométrica* rise, TEC ESP gain, power, access, spurt; **~ absoluto** *m* TEC ESP isotropic gain, *comunicaciones* absolute gain; **~ axial** *m* FÍS axial magnification; **~ brusco del nivel de voz** *m* TEC ESP *radio* burst word; **~ brusco de la presión** *m* INSTAL HIDRÁUL pressure surge; **~ brusco del tráfico** *m* TEC ESP traffic burst; **~ de la competencia** *m* METAL competition growth; **~ de contraste** *m* FOTO increase in contrast; **~ factorial** *m* INSTR *óptica*, ÓPT factorial magnification; **~ instantáneo** *m* TEC ESP burst; **~ isotrópico** *m* TEC ESP isotropic gain; **~ lateral** *m* FÍS, INSTR lateral magnification; **~ lineal** *m* FÍS, ÓPT linear magnification; **~ de la masa de lodo** *m* HIDROL *aguas residuales* sludge bulking; **~ del oleaje** *m* HIDROL upsurge; **~ de pasos** *m* TEXTIL step increment; **~ de peso en vivo** *m* AGRIC gain in live weight; **~ relativo** *m* TEC ESP relative gain; **~ repentino del tráfico** *m* TEC ESP traffic burst; **~ de tamaño de grano** *m* METAL coarsening; **~ de temperatura** *m* CONTAM, FÍS, TERMO temperature rise; **~ de la velocidad** *m* TRANSP progression speed; **~ de volumen debido a la humedad** *m* PAPEL moisture expansion

áureo *adj* QUÍMICA aurous

aureola *f* MINAS *lámpara* gas cap, show, TEC ESP halo, TEXTIL *vestimenta* ring; **~ por contacto** *f* GEOL *metamórfico* contact aureole

auricalcita *f* MINERAL aurichalcite

auricular *m* ACÚST, TELECOM earpiece; **~ circumaural** *m* ACÚST circumaural earphone; **~ de pistón** *m* pistonphone; **~ telefónico** *m* ACÚST, TELECOM tele-phone earphone

auriculares *m pl* ACÚST, TELECOM headphones

aurífero *adj* METAL, MINAS *yacimiento* gold-bearing, auriferous, MINERAL auriferous, gold-bearing

aurocianuro *m* QUÍMICA aurocyanide

aurolita *f* MINERAL auralite

aurora *f* GEOFÍS, METEO, TEC ESP aurora; **~ austral** *f* GEOFÍS, METEO, TEC ESP, TRANSP MAR aurora australis; **~ boreal** *f* GEOFÍS, METEO, TEC ESP, TRANSP MAR aurora borealis, northern lights; **~ polar** *f* GEOFÍS, METEO, TEC ESP, TRANSP MAR polar aurora; **~ polaris** *f* GEOFÍS, METEO, TEC ESP, TRANSP MAR aurora polaris

ausencia *f* ACÚST *de armónicas*, TEC ESP *de convección* absence; **~ sistemática** *f* CRISTAL systematic absence

austenita: **~ residual** *f* METAL residual austenite

austenítico *adj* MECÁ austenitic

autenticación *f* CALIDAD, INFORM&PD, TELECOM authentication; **~ por entidades nobles** *f* TELECOM peer-entity authentication; **~ del origen de datos** *f* INFORM&PD, TELECOM data origin authentication

autenticar *vt* CALIDAD, INFORM&PD, TELECOM authenticate

autentificación *f* CALIDAD, INFORM&PD, TELECOM authentication; **~ inversa** *f* INFORM&PD reverse authentication

autigénesis *f* GEOL *de rocas sedimentarias*, PETROL authigenesis

autígeno *adj* GEOL, PETROL authigenic

auto *m* ING MECÁ motor

autoabsorción *f* FÍS RAD, QUÍMICA self-absorption

autoadaptable *adj* INFORM&PD self-adapting

autoadaptado *adj* INFORM&PD self-adapting

autoadhesivo *m* EMB, P&C contact adhesive; **~ con elevado tiro** *m* EMB high-tackpressure sensitive adhesive

autoadrizable *adj* TRANSP MAR self-righting

autoajuste *m* TV self-adjustment

autoalimentado *adj* ING ELÉC self-powered

autoalineación *f* ELECTRÓN self-tracking

autoapantallado *adj* PROD self-shielding

autoascensible *adj* PROD self-lifting

autoblindado *adj* CINEMAT self-blimped

autobloqueo: **de ~** *adj* EMB, VEH self-locking

autobús: **~ articulado** *m* Esp (*cf bus articulado AmL, cf omnibús articulado AmL*) AUTO bimodal bus, TRANSP articulated bus, bimodal bus; **~ bimodal** *m* Esp (*cf omnibús bimodal AmL, bus bimodal AmL*) AUTO bimodal bus, TRANSP bimodal bus, dual-mode bus; **~ eléctrico** *m* ELEC, TRANSP electrobus; **~ eléctrico de batería** *m* Esp (*cf bus eléctrico AmL*) TRANSP battery bus; **~-ferrobús** *m* Esp (*cf bus-ferrobús, cf omnibús-ferrobús AmL*), TRANSP VEH road-rail bus; **~ de gas licuado de petróleo** *m* Esp (*cf bus de gas-petróleo líquido AmL*) TRANSP liquid-petroleum-gas bus; **~ de gas natural licuado** *m* TRANSP liquid-natural-gas bus (*LNG bus*); **~ de gas natural a presión** *m* Esp (*cf bus por gas natural a presión AmL*) TRANSP pressurized natural-gas bus; **~ de gasolina** *m* AUTO, VEH gas-fueled bus (*AmE*), petrol-fuelled bus (*BrE*); **~ giroscópico** *m* Esp TRANSP, VEH (*cf ómnibus giroscópico AmL, cf bus giroscópico AmL*) gyrobus; **~ de línea** *m* TRANSP coach; **~ mixto** *m* Esp (*cf bus híbrido AmL, ómnibus híbrido AmL*) TRANSP hybrid bus; **~ montado sobre vagón de ferrocarril** *m* Esp (*cf ómnibus AmL, cf bus montado en vagón de ferrocarril AmL*) TRANSP bus on railroad car (*AmE*), bus on railway wagon (*BrE*); **~ oruga** *m* Esp (*cf bus oruga AmL, ómnibus oruga AmL*) AUTO, TRANSP bimodal bus; **~ subterráneo** *m* TRANSP underground bus; **~ sobre la vía de ferrocarril** *m* Esp (*cf bus sobre la vía de ferrocarril AmL*) TRANSP bus on railroad tracks (*AmE*), bus on railway tracks (*BrE*); **~ con tacógrafo** *m* TRANSP dial-a-bus; **~ con turbina de combustión interna** *m* TRANSP gasturbine bus; **~ de vapor** *m* Esp (*cf omnibús de vapor AmL*) VEH steam bus

autocalentable *adj* ING ELÉC self-heating

autocancelación *f* ELEC self-canceling (*AmE*), self-cancelling (*BrE*)

autocapacitancia *f* ELEC *bobina*, FÍS self-capacitance

autocar *m* TRANSP coach

autocarga *f* IMPR, INFORM&PD bootstrap, bootstrapping

autocargado *m* IMPR, INFORM&PD bootstrapping

autocargador *m* IMPR, INFORM&PD bootstrap

autocatálisis *f* PROC QUÍ, QUÍMICA autocatalysis

autocerrable *adj* EMB self-sealing, self-locking

autocerrado *adj* EMB self-closing

autoclástico *adj* GEOL autoclastic

autoclave *f* ALIMENT, C&V, CARBÓN autoclave, digester, LAB *calor, presión*, PROD digester, autoclave, QUÍMICA *cocina*, esterilización, RECICL, TERMO autoclave, digester, TEXTIL *máquina* kier; **~ a gas** *f* RECICL digester gas

autocolimador *m* INSTR, METR *para medir ángulos* autocollimator

autocontenido *adj* FÍS self-contained

autocontrol *m* CALIDAD self-inspection

autocopiativo: **~ no carbonado** *m* (*NCR*) PAPEL no carbon required (*NCR*)

autocorrelación *f* ELECTRÓN, TELECOM autocorrelation

autóctono[1] *adj* CONTAM, GEOL, PETROL autochthonal, autochthonous

autóctono[2] *m* CONTAM, GEOL, PETROL autochthon

autodepuración *f* AGUA self-purification

autodescarga *f* TEC ESP self-discharge

autodestrucción *f* D&A self-destruction

autodifusión *f* FÍS self-scattering

autodino *m* ELECTRÓN autodyne

autodireccionamiento: **~ relativo** *m* INFORM&PD self-relative addressing

autodisparador *m* FOTO self-timer

autodocumentado *adj* INFORM&PD self-documenting

autodrenante *adj* TRANSP MAR self-draining

autoedición *f* IMPR, INFORM&PD desktop publishing (*DTP*); **~ de textos** *f* IMPR, INFORM&PD desktop publishing (*DTP*)

autoelevación *f* TV bootstrap

autoencender *vi* ING MECÁ knock

autoencendido *m* AUTO dieseling, pinging (*AmE*), pinking (*BrE*), TEC PETR *motores de combustión interna* knock, TRANSP AÉR autoignition, VEH *motor* autoignition, pinking (*BrE*), pinging (*AmE*)

autoendurecimiento *m* METAL *temple de precipitación* self-hardening

autoenhebrante *adj* CINEMAT self-threading

autoestable *adj* EMB freestanding

autoexcitación *f* ING ELÉC, NUCL self-excitation

auto-flujo *m* FÍS self-flux

autofunción *f* ACÚST, FÍS eigenfunction

autógeno *adj* CARBÓN, ING MECÁ *soldadura* autogenous

autogiro *m* TRANSP AÉR autogiro, autogyro

autoguía: **~ activa** *f* TRANSP AÉR homing active guidance; **~ pasiva** *f* TRANSP AÉR homing passive guidance; ~ **semiactiva** *f* TRANSP AÉR homing semi-active guidance

autoguiado *m* D&A, TEC ESP *misiles* homing

autohinchable *adj* CONTAM MAR self-inflating

autoidentificación *f* TELECOM self-identification

autoinducción *f* AUTO, ELEC, FÍS, ING ELÉC self-induction; ~ **distribuida** *f* ING ELÉC distributed inductance

autoinducido *adj* AUTO, ELEC, FÍS, ING ELÉC self-induced

autoinductancia *f* ELEC, FÍS, ING ELÉC self-inductance

autoinflable *adj* CONTAM MAR self-inflating

autoinspección *f* CALIDAD self-inspection

autoionización *f* ELECTRÓN, FÍS autoionization

autólisis *m* ALIMENT *de célula* autolysis

autoluminosidad *f* TEC ESP self-luminosity

autómata *m* INFORM&PD automaton

automático[1] *adj* ING ELÉC self-powered, TEC ESP unattended, unmanned

automático[2]: ~ **de sobreintensidad máxima** *m* ING ELÉC maximum cutout

automatimonel *m* TRANSP MAR automatic helm, automation

automatismo *m* TEC ESP automation

automatización *f* CONST, INFORM&PD, ING MECÁ, PROD, TEC ESP, TRANSP MAR automation; ~ **de biblioteca** *f* INFORM&PD library automation; ~ **del buque** *f* TRANSP MAR ship automation; ~ **del diseño** *f* INFORM&PD design automation; ~ **en el hogar** *f* GAS home automation; ~ **industrial** *f* ING MECÁ industrial automation; ~ **de oficina** *f* (*OA*) INFORM&PD, PROD office automation (*OA*); ~ **de procesos** *f* INFORM&PD process automation; ~ **del proyecto** *f* INFORM&PD design automation

automatizado *adj* INFORM&PD, ING MECÁ, PROD automated

automatizar *vt* CONST, INFORM&PD, ING MECÁ, PROD, TEC ESP automate, automatize

autometamorfismo *m* ENERG RENOV autometamorphism

automodulación *f* TELECOM automodulation

automórfico *adj* GEOL *textura*, PETROL automorphic

automorfo *adj* GEOL, PETROL automorphous

automotor[1] *adj* TRANSP automotive

automotor[2] *m* FERRO *vehículos* railcar, TRANSP rail motor unit; ~ **por acumuladores** *m* FERRO accumulator railcar; ~ **diesel-eléctrico** *m* FERRO diesel-electric railcar

automotriz *adj* TRANSP automotive

automóvil[1] *adj* TRANSP automotive

automóvil[2] *m* AUTO motorcar, ING MECÁ, MECÁ motor; ~ **eléctrico** *m* AUTO, VEH electric road vehicle; ~ **policial con radioteléfono** *m* AUTO, TRANSP cruiser; ~ **con portón trasero** *m* AUTO, VEH hatchback automobile (*AmE*), hatchback car (*BrE*)

autonomía *f* D&A range, PROD operating radius, TEC ESP *vuelo espacial* autonomy, endurance, TRANSP AÉR endurance

autónomo *adj* TEC ESP autonomous, self-contained

autooruga *m* D&A half-track vehicle

autooxidación *f* ALIMENT, QUÍMICA autoxidation

autopiloto *m* TEC ESP, TRANSP AÉR autopilot

autopista *f* AUTO, CONST freeway (*AmE*), highway, motorway (*BrE*), roadway (*AmE*), INFORM&PD, TELECOM highway, TRANSP, VEH freeway (*AmE*), highway, motorway (*BrE*), roadway (*AmE*); ~ **de doble**

calzada *f* CONST, TRANSP divided highway (*AmE*), dual carriageway (*BrE*); ~ **de dos carriles por sentido** *f* CONST, TRANSP divided highway (*AmE*), dual carriageway (*BrE*); ~ **principal** *f* TRANSP arterial highway; ~ **de transmisión** *f* TELECOM transmission highway

autopolarización *f* ING ELÉC self-bias

autopolinización *f* AGRIC self-pollination

autopropulsado *adj* CONTAM MAR, TEC ESP, TRANSP MAR self-propelled

autopropulsión *f* TEC ESP, TRANSP MAR self-propulsion

autoprotección *f* ELEC fail-safe

autopurificación *f* AGUA self-purification

autoridad *f* PROD authority; ~ **de aguas** *f* AGUA water authority; ~ **fluvial** *f* AGUA river authority; ~ **rectora del puerto** *f* TRANSP MAR port authority

autorización *f* ELECTRÓN *circuito lógico* enabling, INFORM&PD authorization, *seguridad* clearance, TEC PETR *permiso de exploración* license block (*AmE*), licence block (*BrE*), TELECOM authorization; ~ **del control de tráfico aéreo** *f* TRANSP AÉR air traffic control clearance

autorizado: ~ **para despegar** *adj* TRANSP AÉR cleared for take-off

autorizar *vt* ELECTRÓN *circuito lógico* enable

autorradiografía *f* FÍS, FÍS RAD autoradiography; ~ **cualitativa** *f* NUCL qualitative autoradiography

autorradiólisis *f* FÍS, FÍS RAD autoradiolysis

autorrealimentación *f* ELECTRÓN, ING ELÉC, NUCL inherent feedback

autorregenerable *adj* ING ELÉC self-healing

autorreglaje *m* ELECTRÓN self-tracking

autorregresión *f* INFORM&PD autoregression

autorregulación *f* ELEC automatic control, ELECTRÓN inherent regulation, ING MECÁ automatic regulation

autorregulador *m* ING MECÁ automatic regulator; ~ **de gases** *m* TRANSP AÉR autothrottle; ~ **de tensión** *m* ELEC automatic voltage control; ~ **de voltaje** *m* ELEC automatic voltage control

autorrotación: ~ **de la hélice del motor** *f* TRANSP AÉR engine windmilling

autosincronización *f* TELECOM autosynchronization

autosoldable *adj* TEC ESP self-sealing

autosuficiencia: ~ **de petróleo** *f* TEC PETR oil self-sufficiency

autotensión *f* TV autotension

autotimonel *m* TRANSP MAR autopilot

autotipia *f* IMPR halftone

autotransformador *m* ELEC *circuitos de corriente alterna* compensator, FÍS, ING ELÉC autotransformer, compensator, TEC PETR compensator; ~ **de arranque** *m* ELEC *motor*, ING ELÉC autotransformer starter; ~ **elevador** *m* ELEC, ING ELÉC step-up autotransformer; ~ **reductor** *m* ELEC, ING ELÉC step-down autotransformer

autótrofo *m* ALIMENT autotroph

autovalor *m* ELECTRÓN, FÍS eigenvalue; ~ **geométrico** *m* NUCL geometric buckling

autovector *m* ELECTRÓN, FÍS eigenvector

autovía *f* TRANSP rail motor unit; ~ **eléctrica** *f* ELEC, VEH electric railcar

autovulcanización *f* P&C self-vulcanization

autunita *f* MINERAL autunite

auxiliado: ~ **por computadora** *adj* AmL (*cf auxiliado por ordenador Esp*) INFORM&PD computer-aided, computer-assisted; ~ **por ordenador** *adj Esp* (*cf* *auxiliado por computadora AmL*) INFORM&PD computer-aided, computer-assisted

auxiliar[1] *adj* INFORM&PD standby, ING MECÁ, MECÁ, TEC ESP ancillary, auxiliary

auxiliar[2]: **~ de lavado** *m* DETERG washing adjuvant, washing auxiliary

auxilio: **~ de diseño** *m* INFORM&PD design aid; **~ de proyecto** *m* INFORM&PD design aid

auxina *f* AGRIC *hormonas*, QUÍMICA auxin

auxocromo *m* QUÍMICA auxochrome

avalancha *f* GEOL freshet, ING ELÉC, METEO avalanche

avalita *f* MINERAL avalite, fuchsite

avance[1] *m* CARBÓN *minas* break, ELEC *fase* advance, IMPR infeed, ING MECÁ sliding, advance, lead, INSTAL HIDRÁUL *válvula de corredera* lead, MECÁ *herramientas* feed, MINAS *galería* cross driving, driving, break, heading, PROD progress, feed, feeding, feed-forward, *máquinas herramientas* infeed, TV preview, VEH *encendido* advance; **~ a la admisión** *m* INSTAL HIDRÁUL preadmission, PROD *válvula de deslizamiento* early admission; **~ con agujas** *m* CARBÓN, MINAS piling; **~ con agujas de sondeo** *m* CARBÓN, MINAS test piling; **~ con agujas de terraplén** *m* CARBÓN, MINAS embankment piling; **~ angular** *m* ING MECÁ *distribuidor* angle of advance; **~ angular a la admisión** *m* ING MECÁ angular lead, *distribuidor* angle of lead; **~ angular al escape** *m* ING MECÁ angular pre-release, *máquinas vapor* angle of pre-release; **~ automático** *m* ING MECÁ power feed, *máquinas herramientas* automatic feed; **~ automático de encendido** *m* PROD *motores* timer; **~ automático del encendido** *m* TRANSP AÉR ignition advance; **~ automático por tornillo sin fin** *m* ING MECÁ screw feed; **~ automático vertical hacia abajo** *m* ING MECÁ *máquinas herramientas* automatic down feed; **~ del cabezal** *m* ING MECÁ head feed; **~ de cinta** *m* TV tape advance; **~ por cremallera** *m* ING MECÁ rack feed; **~ del encendido** *m* PROD *motores* timing range, TRANSP AÉR spark advance; **~ al encendido** *m* AUTO, VEH advanced ignition; **~ al escape** *m* PROD *válvula de corredera* early release; **~ de escape** *m* INSTAL HIDRÁUL *máquina de vapor* exhaust lead; **~ de fase** *m* ELEC *corriente alterna*, ELECTRÓN phase lead; **~ filoniano** *m* MINAS lode drive; **~ sin fin** *m* ING MECÁ screw feed; **~ de hoja** *m* IMPR, INFORM&PD form feed (*FF*); **~ informativo** *m* TRANSP advance information; **~ inicial** *m* AUTO initial advance, initial feed; **~ lento** *m* CINEMAT, PROD inching; **~ de línea** *m* IMPR, INFORM&PD line feed (*LF*); **~ lineal a la admisión** *m* INSTAL HIDRÁUL *válvula de corredera* linear lead, outside lead; **~ longitudinal** *m* ING MECÁ traverse feed, *tornos* longitudinal traverse; **~ a mano** *m* ING MECÁ hand feed; **~ manual** *m* ING MECÁ hand feed; **~ normal** *m* ING MECÁ infeed; **~ de página** *m* IMPR, INFORM&PD form feed (*FF*); **~ por palanca** *m* ING MECÁ lever feed; **~ por palanca de mano** *m* ING MECÁ hand-lever feed; **~ de papel** *m* IMPR, INFORM&PD form feed (*FF*); **~ de penetración** *m* PETROL drilling break; **~ radial** *m* ING MECÁ infeed; **~ rápido** *m* CINEMAT fast forward, ING MECÁ fast feed, coarse feed, TV fast forward; **~ de seguridad** *m* SEG, TRANSP *tráfico* safety headway; **~ con tablestacas** *m* MINAS *galerías, túneles* forepoling, lagging; **~ técnico** *m* PROD technical breakthrough; **~ transversal** *m* ING MECÁ crossfeed; **~ transversal rápido automático** *m* ING MECÁ power rapid traverse; **~ vertical** *m* ING MECÁ vertical traverse; **~ vertical descendente** *m* ING MECÁ downfeed; **~ vertical hacia abajo** *m* ING MECÁ downfeed

avance[2]: **estar en ~ de fase de medio pi** *vi* FÍS lead in phase by half pi

avante *adv* TEC ESP, TRANSP MAR ahead; **tanto ~** *adv* TRANSP MAR abreast; **~ media** *adv* TRANSP MAR half ahead; **~ poca** *adv* TRANSP MAR *motor* slow ahead; **~ poco a poco** *adv* TRANSP MAR *motor* dead slow ahead; **~ toda** *adv* TRANSP MAR *motor* full ahead

avanzado *adj* FOTO front-end

avanzar *vt* ING MECÁ, MECÁ feed, MINAS *túnel* crossdrive, PROD feed; **~ con rapidez** *vt* CINEMAT, TV fast forward

avellanado *m* GEN countersink, countersinking, reaming

avellanador *m* ING MECÁ, MECÁ, PROD countersink bit, octagonal reamer, reamer; **~ cónico** *m* ING MECÁ, MECÁ, PROD cone countersink

avellanar *vt* ING MECÁ, MECÁ, PROD countersink, ream

avena: **~ fatua** *f* AGRIC wild oat; **~ a medio moler** *f* AGRIC groats

avenado *adj* ELEC, ING MECÁ out-of-action

avenamiento *m* AGUA, CARBÓN, HIDROL drain, draining

avenar *vt* AGUA, HIDROL drain

avenida: **~ de agua** *f* AGUA, HIDROL inflow of water, influx of water, irruption

avenina *f* QUÍMICA *proteína* avenine

aventurina: **~ oriental** *f* MINERAL aventurine orientale

avería *f* GEN breakdown, damage, failure, fault; **~ en el aparato de gobierno** *f* TRANSP MAR helm damage; **~ de la base del emisor** *f* ELECTRÓN emitter-base breakdown; **~ del bucle** *f* PROD loop fault; **~ por caída del cabezal** *f* INFORM&PD *unidades de disco* head crash; **~ por calor excesivo** *f* TEC ESP burnout; **~ por cavitación** *f* METAL *corrosión* cavitation failure; **~ por desgaste** *f* ING ELÉC wear-out failure; **~ de la energía** *f* ELEC, ING ELÉC power failure; **~ del equipo procesador** *f* PROD processor hardware fault; **~ por el incendio** *f* CALIDAD, SEG fire damage; **~ inducida** *f* INFORM&PD induced failure; **~ interna** *f* REFRIG internal breakdown; **~ de la máquina** *f* TRANSP MAR engine failure; **~ del motor** *f* TRANSP MAR, VEH engine breakdown, engine failure; **~ relativa a tiempo** *f* ING ELÉC time-related failure; **~ del sistema** *f* INFORM&PD system crash; **~ térmica** *f* ING ELÉC, TERMO temperature-related failure; **~ transitoria** *f* ELECTRÓN transient fault; **~ Zener** *f* ELECTRÓN Zener breakdown

averiado *adj* GEN broken, faulty, out of order; **~ por el agua de mar** *adj* TRANSP MAR sea-damaged

averiarse *v refl* CONST break down, MECÁ, PROD fail, TEXTIL break down

averías: **sin ~** *fra* CALIDAD, SEG, TELECOM fault-free

aves: **~ limpias** *f pl* AGRIC dressed poultry

aviación: **~ embarcada** *f* D&A, TRANSP AÉR carrier-borne aircraft

aviador *m* TRANSP AÉR pilot

avicultor *m* AGRIC poultry keeper

avión *m* TEC ESP craft, TRANSP AÉR aeroplane (*BrE*), aircraft, airplane (*AmE*), plane; **~ de ala baja** *m* TRANSP AÉR low-wing plane; **~ de ala fija** *m* TRANSP AÉR fixed-wing aircraft; **~ de ala plegable** *m* TRANSP AÉR folding-wing aircraft; **~ de ala rotatoria** *m*

TRANSP AÉR rotary-wing aircraft; **~ alargado** *m* TRANSP AÉR stretched aircraft; **~ con alas de canard** *m* TRANSP AÉR canard wing aircraft; **~ de alcance medio** *m* TRANSP AÉR medium-range aircraft; **~ anfibio** *m* TRANSP MAR amphibian; **~ de aprovisionamiento** *m* D&A, TRANSP AÉR supply aircraft; **~ bimotor a reacción** *m* TRANSP AÉR twin-engine jet aircraft; **~ blanco sin piloto** *m* D&A, TRANSP AÉR pilotless target aircraft; **~ de búsqueda para todo tiempo** *m* D&A all-weather search aircraft; **~ de cambio rápido** *m* TRANSP AÉR quick-change aircraft; **~ de carga** *m* TEC ESP cargo; **~ de caza** *m* D&A, TRANSP AÉR pursuit plane, strike aircraft; **~ cisterna de reaprovisionamiento en vuelo** *m* TRANSP AÉR tanker; **~ de combate** *m* D&A, TRANSP AÉR combat aircraft; **~ compuesto** *m* TRANSP AÉR composite aircraft; **~ de cuatro plazas** *m* TRANSP AÉR four-seat aircraft; **~ de despegue y aterrizaje convencional** *m* D&A, TRANSP AÉR conventional takeoff and landing aircraft (*CTOL aircraft*); **~ de despegue y aterrizaje corto** *m* TRANSP, TRANSP AÉR short take-off and landing aircraft (*STOL aircraft*); **~ de despegue y aterrizaje reducido** *m* TRANSP AÉR reduced take-off and landing aircraft (*RTOL aircraft*); **~ de despegue y aterrizaje silencioso** *m* TRANSP AÉR quiet take-off and landing aircraft (*QTOL aircraft*); **~ de despegue y aterrizaje vertical** *m* D&A, TRANSP AÉR vertical take-off and landing aircraft (*VTOL aircraft*); **~ de dos plazas** *m* TRANSP AÉR two-seat aircraft; **~ de ejecutivo** *m* TRANSP AÉR executive aircraft; **~ explorador** *m* CONTAM MAR, TRANSP AÉR spotter plane; **~ de fuselaje ancho** *m* TRANSP AÉR wide-body aircraft; **~ de fuselaje estrecho** *m* TRANSP AÉR single-aisle aircraft; **~ jumbo** *m* TRANSP AÉR jumbo jet; **~ de morro caído** *m* TRANSP AÉR droop-nose aircraft; **~ de pasajeros** *m* TRANSP AÉR airliner; **~ de pasajeros de alcance medio** *m* TRANSP AÉR medium-range airliner; **~ de pasajeros de corto alcance** *m* TRANSP AÉR short-haul airliner; **~ de pasillo único** *m* TRANSP AÉR single-aisle aircraft; **~ de preproducción** *m* TRANSP AÉR preproduction aircraft; **~ de producción en serie** *m* TRANSP AÉR production aircraft; **~ a reacción** *m* TRANSP AÉR jet aeroplane (*BrE*), jet airplane (*AmE*), jet plane; **~ a reacción cuatrimotor** *m* TRANSP AÉR four-engine jet aircraft; **~ a reacción tetramotor** *m* TRANSP AÉR four-engine jet aircraft; **~ reactor de fuselaje ancho** *m* TRANSP wide-bodied aircraft; **~ de reaprovisionamiento en vuelo** *m* TRANSP AÉR refueling tanker (*AmE*), refuelling tanker (*BrE*); **~ remolcador** *m* TRANSP AÉR tractor aircraft, tug aircraft; **~ de repostaje** *m* D&A, TRANSP AÉR refueling aircraft (*AmE*), refuelling aircraft (*BrE*); **~ para repostar en vuelo** *m* TRANSP AÉR refueler (*AmE*), refueller (*BrE*); **~ subsónico** *m* TRANSP AÉR subsonic aircraft; **~ teledirigido** *m* TEC ESP, TRANSP AÉR drone; **~ transportado en portaviones** *m* D&A, TRANSP AÉR carrier-borne aircraft; **~ de transporte de carga** *m* TRANSP AÉR cargo aircraft; **~ de transporte supersónico** *m* TRANSP AÉR supersonic transport aircraft (*SST*); **~ turbo-hélice** *m* TRANSP, TRANSP AÉR propeller-turbine plane

aviónica *f* TEC ESP, TRANSP AÉR avionics

avíos: **~ del buque** *m pl* OCEAN vessel equipment

avisador *m* CINEMAT cuer, ING MECÁ telltale, TEC ESP annunciator, TV cuer; **~ acústico** *m* AUTO horn, PROD hooter, VEH horn; **~ alfanumérico** *m* TELECOM alphanumeric pager; **~ de flotador** *m* PROD flow alarm; **~ de incendios** *m* ELEC, SEG, TERMO, TRANSP MAR fire alarm; **~ de intercepción** *m* TELECOM intercept announcer

aviso *m* PROD warning, TEC ESP signal, TELECOM message; **~ de asignaciones** *m* TELECOM assignment message; **~ de averías** *m* PROD fault reporting; **~ de cambio** *m* CINEMAT, TV changeover cue; **~ de cambio de especificación** *m* TRANSP specification change notice (*SCN*); **~ falso** *m* TRANSP AÉR false warning; **~ de fugas** *m* SEG leakage warning; **~ para interceptar un vehículo** *m* AUTO vehicle intercept survey; **~ de mal tiempo** *m* METEO, TRANSP MAR weather warning; **~ de mensaje entrante** *m* TELECOM input message acknowledgment (*IMA*); **~ de niebla** *m* METEO, TRANSP MAR fog warning; **~ previo de entrega** *m* PROD predelivery reminder; **~ de proximidad a tierra** *m* TRANSP AÉR ground-proximity warning; **~ de recepción** *m* TELECOM acknowledgement; **~ de seguridad** *m* LAB, PROD, SEG safety placard, safety warning; **~ sonoro de la cadena** *m* TV network cue; **~ supervisor** *m* TELECOM supervisory announcement; **~ de tempestad** *m* METEO, TRANSP MAR storm warning; **~ de temporal** *m* METEO, TRANSP MAR gale warning; **~ de viento duro** *m* METEO, TRANSP MAR gale warning

avituallar *vt* TRANSP MAR *barco* supply with provisions

avivador: **~ de colores** *m* DETERG optical brightener

AVR *abr* (*almacenamiento vigilado con posibilidad de recuperación*) NUCL MRS (*monitored retrievable storage*)

awaruíta *f* MINERAL awaruite

axial *adj* GEOM, MATEMÁT axial

axialmente: **~ simétrico** *adj* CRISTAL, GEOL axially symmetric

axila: **~ de la hoja** *f* AGRIC leaf axil

axinita *f* MINERAL axinite

axioma *m* GEOM, MATEMÁT axiom, postulate; **~ paralelo de Euclides** *m* GEOM, MATEMÁT Euclid's parallel postulate

axiomático *adj* GEOM, MATEMÁT axiomatic

axiómetro *m* FÍS, TRANSP AÉR rudder-angle indicator, TRANSP MAR helm indicator, rudder-angle indicator

ayuda: **~ de aproximación por radio** *f* TELECOM, TRANSP AÉR radio-approach aid; **~ en carretera** *f* TRANSP *tráfico* passing aid; **~ al diagnóstico** *f* TELECOM diagnostic aid; **~ electrónica para el tráfico** *f* TRANSP electronic traffic aid; **~ de emergencia** *f* TRANSP emergency aid; **~ de enfoque** *f* CINEMAT, FOTO, TV focusing aid; **~ a la navegación** *f* TRANSP MAR navigation aid; **~ de navegación autónoma** *f* TRANSP AÉR self-contained navigational aid; **~ navegacional autónoma** *f* TRANSP AÉR self-contained navigational aid; **~ para prevenir los abordajes** *f* SEG, TRANSP MAR collision avoidance aid; **~ de punteo radar automático** *f* (*APRA*) TRANSP MAR automatic radar plotting aid (*ARPA*); **~ supervisora** *f* TELECOM supervisory aid

ayudante: **~ de cabina** *m* TRANSP, TRANSP AÉR cabin attendant; **~ de perforación** *m* PETROL, TEC PETR roughneck; **~ de templador** *m* C&V lehr assistant

azabache *m* CARBÓN pile point, pitch coal, GEOL, MINERAL, PETROL jet

azada *f* CONST poll adze, *agricultura* hoe, *herramienta* spade; ~ **compactadora** f AGRIC trash treader; ~ **mecánica** f AGRIC hoeing machine

azadón m AGRIC scoop; ~ **rodante** m AGRIC roller mill; ~ **rotativo** m AGRIC rotary hoe

azafrán m TRANSP MAR *construcción naval* rudder plane

azafrol m QUÍMICA safrol

azelaico adj QUÍMICA azelaic

azeotropía f QUÍMICA, REFRIG azeotropy

azeotrópico adj QUÍMICA, REFRIG azeotropic

azeótropo m QUÍMICA, REFRIG, TERMO azeotrope

azida f QUÍMICA azide

azimino adj QUÍMICA azimino

azimut m GEN azimuth; ~ **de deriva** m PETROL *topografía* drift azimuth; ~ **inverso** m TRANSP AÉR *radiogoniometría* reciprocal bearing; ~ **de lanzamiento** m TEC ESP launch azimuth; ~ **magnético** m GEOFÍS, MECÁ, TEC ESP, TRANSP AÉR magnetic azimuth, magnetic bearing; ~ **solar** m ENERG RENOV, TEC ESP solar azimuth

azobenceno m QUÍMICA *insecticidas* azobenzene

azobenzoico adj QUÍMICA azobenzoic

azogamiento m PROD *espejos* silvering

azoico adj GEOL *paleontología* barren

azorita f MINERAL azorite

azotar vt TRANSP MAR *viento* lash

azotímetro m QUÍMICA nitrometer

azúcar¹ m AGRIC, ALIMENT, C&V, QUÍMICA sugar; ~ **de caña** m AGRIC, ALIMENT cane sugar; ~ **caramelizado** m ALIMENT caramelized sugar; ~ **en cubitos** m ALIMENT lump sugar; ~ **dietético** m ALIMENT dietary sugar; ~ **de fruta** m ALIMENT, QUÍMICA fruit sugar; ~ **invertido** m ALIMENT, QUÍMICA invert sugar; ~ **de**

plomo m QUÍMICA *acetato básico de plomo* white lead; ~ **sin refinar** m AGRIC raw sugar

azúcar²: **sin** ~ fra ALIMENT no added sugar

azuche m CARBÓN *pilotes* jet, pile shoe, pile point, CONST pile shoe

azud m AGUA barrage, irrigation water wheel; ~ **sumergido** m AGUA, HIDROL, INSTAL HIDRÁUL drowned weir

azuela f CONST adze

azufrado m C&V sulfuring (*AmE*), sulphuring (*BrE*)

azufre m (*S*) METAL, P&C, QUÍMICA, TEC PETR sulfur (*AmE*) (*S*), sulphur (*BrE*) (*S*); ~ **atmosférico** m CONTAM atmospheric sulfur (*AmE*), atmospheric sulphur (*BrE*); ~ **cilindrado** m QUÍMICA roll sulfur (*AmE*), roll sulphur (*BrE*); ~ **combinado** m P&C combined sulfur (*AmE*), combined sulphur (*BrE*); ~ **extraíble** m P&C extractable sulfur (*AmE*), extractable sulphur (*BrE*); ~ **libre** m P&C *vulcanización* free sulfur (*AmE*), free sulphur (*BrE*); ~ **total** m P&C *vulcanización* total sulfur (*AmE*), total sulphur (*BrE*)

azul: ~ **chino** m C&V, COLOR Chinese blue; ~ **primario** m TV blue primary; ~ **de Prusia** m QUÍMICA Prussian blue

azulado m COLOR, PAPEL blueing

azular vt IMPR blue

azulejero m C&V, CONST tile maker, tiler

azulejo m C&V, CONST glazed tile, tile; ~ **cerámico** m C&V ceramic wall tile; ~ **a prueba de sonidos** m C&V, SEG soundproof tile

azuleno m QUÍMICA azulene

azulina f QUÍMICA azulin

azulmín m QUÍMICA azulmin

azurita f MINERAL, QUÍMICA azurite, chessylite

B

b *abr* ACÚST (*belio*) b (*bel*), FÍS, FÍS PART (*barn, barnio, belio*) b (*barn,bel*), INFORM&PD, ING ELÉC (*belio*) b (*bel*), NUCL (*bel, barn, barnio*) b (*bel, barn*)

B *abr* (*boro*) QUÍMICA B (*boron*); **~ y N** *abr* (*blanco y negro*) CINEMAT, FOTO, IMPR, TV B&W (*black and white*)

Ba *abr* (*bario*) QUÍMICA Ba (*barium*)

babaza *f* AGUA *ductos agua*, CARBÓN *del agua*, HIDROL slime

babingtonita *f* MINERAL babingtonite

babor[1] *m* TEC ESP *naves*, TRANSP MAR *parte del buque* port

babor[2]: **de ~ a estribor** *fra* TRANSP MAR athwartships

baca *f* AUTO, VEH roof rack

bache *m* CONST, VEH *carreteras* pothole, rut; **~ de coincidencia** *m* ACÚST coincidence dip

bacheo *m* PROD *carreteras* patching, patching-up

bacilo *m* ALIMENT bacillus; **~ coliforme** *m* ALIMENT coliform bacterium; **~ del colon** *m* HIDROL colon bacillus

bacteria *f* AGUA, ALIMENT, HIDROL, QUÍMICA, RECICL bacterium; **~ del ácido acético** *f* AGUA, ALIMENT, HIDROL, RECICL acetic acid bacteria; **~ olefaga** *f* RECICL oil-eating bacterium; **~ viable** *f* AGUA, HIDROL, RECICL viable bacterium

bactericida[1] *adj* AGUA, ALIMENT, HIDROL, QUÍMICA, RECICL bactericidal

bactericida[2] *m* AGUA, ALIMENT, HIDROL, QUÍMICA, RECICL bactericide

bacteriófago *m* ALIMENT, QUÍMICA bacteriophage

bacteriolisis *f* ALIMENT, QUÍMICA bacteriolysis

bacteriostato *m* ALIMENT, QUÍMICA bacteriostat

bacteriotoxina *f* ALIMENT, QUÍMICA bacteriotoxin

bagazo: **~ de la caña de azúcar** *m* PAPEL bagasse

bahía *f* ENERG RENOV, HIDROL, TRANSP MAR bay

baikalita *f* MINERAL baikalite

bailoteo: **~ del líquido** *m* TEC ESP *depósitos parcialmente llenos* liquid sloshing, sloshing

bainita *f* METAL bainite; **~ inferior** *f* METAL lower bainite; **~ superior** *f* METAL upper bainite

baivel *m* INSTR bevel, METR bevel square

baja *f* METEO *presión atmosférica* low; **~ capacidad** *f* ING ELÉC low capacitance; **~ capacitancia** *f* ING ELÉC low capacitance; **~ concentración** *f* ELECTRÓN low concentration; **~ frecuencia** *f* (*BF*) ELEC *onda*, ELECTRÓN, FÍS, FÍS ONDAS, FÍS RAD, ING ELÉC, TELECOM low frequency (*LF*); **~ de potencia** *f* ING ELÉC power-down; **~ resistencia** *f* FÍS, TELECOM low resistance; **~ temperatura** *f* FÍS, REFRIG, TERMO low temperature; **~ tensión** *f* ELEC, ING ELÉC, TELECOM low tension, low voltage; **~ velocidad** *f* ING MECÁ, TELECOM low speed

bajada *f* TRANSP MAR companionway; **~ de antena** *f* TV aerial lead (*BrE*), antenna lead (*AmE*); **~ a la cámara** *f* TRANSP MAR companionway; **~ manual del tren de aterrizaje** *f* TRANSP AÉR landing gear manual release; **~ de la marea** *f* OCEAN, TRANSP MAR fall of the tide, tidal fall; **~ del pH** *f* CONTAM *tratamiento químico* pH depression, pH drop; **~ del potencial**

hidrógeno *f* CONTAM *tratamiento químico* pH depression, pH drop; **~ de presión adiabática** *f* TRANSP AÉR adiabatic pressure drop

bajamar *m* AGUA, ENERG RENOV, HIDROL, OCEAN, TRANSP MAR ebb tide, low tide, low water

bajante *m* CONST downpipe (*BrE*), downspout (*AmE*), *tuberías* stack pipe; **~ de aguas** *m* AGUA *tubería* leader, CONST rainwater downpipe (*BrE*), leader, rainwater downspout (*AmE*); **~ de aguas sucias** *m* AGUA soil pipe

bajar[1] *vt* ING ELÉC *tensión* drop, lower, *potencia* power down SEG *una carga* lower

bajar[2] *vi* AGUA, ENERG RENOV, HIDROL, TRANSP MAR *la marea* ebb, *las aguas* subside; **~ la potencia** *vi* ING ELÉC power down; **~ la tensión** *vi* ING ELÉC drop the voltage

bajas: **~ pérdidas** *f pl* TELECOM low loss; **~ pérdidas de inserción** *f pl* TELECOM low insertion loss

bajío *m* OCEAN, TRANSP MAR flat, shallow, *accidentes geográficos para navegación* shoal; **~ del arrecife** *m* GEOL, OCEAN reef flat

bajo *m* ACÚST bass, OCEAN *geología submarina* shelf, TRANSP MAR middle ground

Bajociense *adj* GEOL *estratigrafía* Bajocian

bajocorrimiento *m* GEOL underthrust

bajos: **~ de carrocería** *m pl* VEH underbody

bajura: **de ~** *adj* TRANSP MAR inshore

bakelita® *f* ELEC *aislación*, P&C *polímero*, REVEST Bakelite®

bala *f* AGRIC, EMB, PAPEL, TEXTIL *de algodón* bale; **~ del fusil** *f* D&A rifle bullet; **~ perdida** *f* D&A lost bullet; **~ perforante** *f* D&A armor-piercing bullet (*AmE*), armour-piercing bullet (*BrE*); **~ trazadora** *f* D&A tracer bullet

balance *m* ING MECÁ offset, TRANSP MAR rolling; **~ de blanco** *m* TV white balance; **~ calorífico** *m* REFRIG calorific balance, TERMO calorific balance, *metalurgia* heat balance; **~ de cargas** *m* CONTAM *procesos industriales* ion budget; **~ cromático** *m* TV chromatic balance; **~ energético** *m* AGUA energy balance, CONTAM *procesos industriales* energy budget, TERMO energy balance; **~ de energía** *m* CONTAM *procesos industriales* energy budget; **~ de enlace** *m* TEC ESP link budget; **~ forrajero** *m* AGRIC feed balance; **~ de gas** *m* GAS gas balance; **~ hídrico** *m* AGUA water balance; **~ hidrológico** *m* AGUA hydrologic balance; **~ de intensidad** *m* ELEC ampere balance; **~ iónico** *m* CONTAM *procesos industriales* ion budget; **~ de masas** *m* NUCL mass balance, TEC ESP mass budget; **~ de la onda portadora** *m* TV carrier balance; **~ salino** *m* OCEAN salt balance; **~ térmico** *m* REFRIG, TERMO calorific balance, heat balance, thermal balance

balanceado *adj* ELEC, TELECOM *electricidad* push-pull

balancearse *v refl* MECÁ *barcos* slew, TEC ESP wobble

balanceo *m* ING MECÁ seesawing, swaying, swinging, TEC ESP *aviones* roll

balancera *f* TRANSP MAR *buque* fiddle

balancín *m* AUTO rocker, rocker arm, tappet, C&V

rocker, FÍS balance, GAS *mecánica* beam, ING MECÁ *motor* arm rocker, lever, logging head, yoke, working beam, balance lever, walking beam, rocker, link, MECÁ rocker, pendulum, METR *de tipo resorte o palanca* balance, PROD *fundería* lifting beam, sling beam, TEC ESP *mando de aviones* horn, VEH *motor, válvulas* rocker arm, *válvula de motor* tappet; ~ **a bolas** *m* PROD fly-bar with two balls; ~ **compensador** *m* ING MECÁ balance beam, compensating beam, METR balance beam; ~ **de contrapeso** *m* MINAS *del vástago de la bomba de sondeo* balance bob; ~ **de cuello de cisne** *m* ING MECÁ swanneck fly press; ~ **de escuadra** *m* ING MECÁ quadrant, V-bob, MINAS *bombas de mina* quadrant; ~ **graduado** *m* METR divided beam; ~ **de perforación** *m* ING MECÁ, MINAS *sondeos* walking beam; ~ **transversal del armazón** *m* ING MECÁ frame crossbeam; ~ **de válvula** *m* INSTAL HIDRÁUL valve rocker, valve tappet

balandra *f* TRANSP MAR sloop
balandro *m* TRANSP MAR sloop
balanza *f* EMB check weighing machine, FÍS balance, ING MECÁ scale, scales, LAB *análisis*, METR balance, PROD weighing machine; ~ **analítica** *f* LAB analytical balance; ~ **comprobadora de pesos** *f* EMB check-weigher; ~ **de Cotton** *f* FÍS Cotton balance; ~ **electrodinámica** *f* ELEC, FÍS, INSTR current balance; ~ **electrónica** *f* ELEC, EMB electronic weighing scales, LAB electronic balance; ~ **de ensayos** *f* QUÍMICA assay balance; ~ **de Eötvös** *f* FÍS Eötvös balance; ~ **física** *f* LAB *determinación de peso* physical balance; ~ **hidrostática** *f* LAB *medida de densidad* hydrostatic balance; ~ **de Kelvin** *f* ELEC *medición* Kelvin balance; ~ **de laboratorio** *f* FÍS, LAB analytical balance, chemical balance, QUÍMICA chemical balance; ~ **magnética** *f* GEOFÍS magnetic balance; ~ **de máquina herramienta dividida en líneas** *f* ING MECÁ linear-divided machine tool scale; ~ **química** *f* FÍS, LAB *análisis*, QUÍMICA chemical balance; ~ **de resorte** *f* ING MECÁ spring balance, METR spiral balance; ~ **de torsión** *f* FÍS, GEOFÍS *magnetometría*, ING MECÁ torsion balance; ~ **de torsión de Coulomb** *f* FÍS Coulomb's torsion balance; ~ **del túnel de viento** *f* TRANSP AÉR wind tunnel balance

balanzas: ~ **y básculas** *f pl* METR beam and scales
balastaje: ~ **suelto** *m* FERRO *equipo inamovible* loose ballasting
balasto *m* CONST paving material (*AmE*), ballasting, roadstones, road metal (*BrE*), *ferrocarriles* ballast, FERRO *equipo inamovible* ballast; ~ **suelto** *m* FERRO *equipo inamovible* loose ballasting
balata *f* P&C *resina natural* balata
balaustrada *f* CONST balustrade, railing
balaústre *m* CONST baluster
balcón *m* C&V balcony, TRANSP MAR *equipamiento de cubierta* pulpit; ~ **de popa** *m* TRANSP MAR pushpit, stern pulpit
baldada *f* CONST pailful
balde *m* CONST *recipiente*, CONTAM MAR, INFORM&PD bucket; ~ **de acero** *m* CONST bucket, CONTAM MAR, INFORM&PD bucket, pail; ~ **basculante** *m* CONST dump bucket (*AmE*), tipping bucket (*BrE*); ~ **embragado** *m* CONTAM MAR *bajo la panza del helicóptero* slung bucket; ~ **estañado** *m* PROD tin-plate pail; ~ **de mandíbulas** *m* CONST grab bucket
baldeo *m* AGUA flushing

baldosa *f* C&V floor tile, flooring tile, CONST flag, flagstone, flooring tile, *pavimentación* floor tile, tile; ~ **de cerámica** *f* C&V, GAS ceramic tile
balín *m* CARBÓN pellet
balística *f* TEC ESP *mecánica* ballistics
balístico *adj* TEC ESP ballistic
baliza *f* CONST, D&A, FERRO *equipo imamovible* TEC ESP *señalización de vehículos espaciales* beacon, flare, TRANSP *aviación*, TRANSP AÉR marker, TRANSP MAR *señales de navegación* buoy, beacon, seamark; ~ **de canal** *f* TRANSP channel marking; ~ **para espacio profundo** *f* TEC ESP deep-space transponder; ~ **de espeque** *f* TRANSP MAR *señales de navegación* spar buoy; ~ **fija** *f* TRANSP MAR fixed beacon; ~ **flotante** *f* TRANSP MAR *señales de navegación* floating beacon; ~ **de fondos limpios** *f* TRANSP MAR *señales de navegación* safe-water mark; ~ **giratoria** *f* TRANSP AÉR rotating beacon; ~ **interna** *f* TRANSP AÉR inner marker; ~ **de llamada de atención** *f* TEC ESP *salvamento* distress beacon; ~ **luminosa de aproximación** *f* TRANSP AÉR approach light beacon; ~ **luminosa de guía para vuelos de circunvalación** *f* TRANSP AÉR circling guidance light; ~ **de pista de aterrizaje** *f* TRANSP AÉR landing-strip marker; ~ **de radar** *f* D&A, FÍS, FÍS RAD radar beacon, TEC ESP transponder, TELECOM, TRANSP, TRANSP AÉR radar beacon, TRANSP MAR *señales de navegación* racon, radar beacon; ~ **radárica** *f* D&A, FÍS, FÍS RAD, TELECOM, TRANSP, TRANSP AÉR, TRANSP MAR radar beacon; ~ **respondedora radar** *f* D&A, TRANSP, TRANSP AÉR, TRANSP MAR radar transponder beacon; ~ **retrorreflectiva** *f* TRANSP AÉR retroreflective marker; ~ **de señalización** *f* TEC ESP distress beacon; ~ **transpondedora** *f* TEC ESP transponder beacon; ~ **de trayectoria de planeo** *f* TRANSP, TRANSP AÉR glide-path beacon
balizado: ~ **de los bordes de la pista de rodaje** *m* TRANSP AÉR taxiway-edge marker; ~ **de cruce de la pista de rodaje** *m* TRANSP AÉR taxiway intersection marking; ~ **del eje de la pista** *m* TRANSP AÉR runway-centerline marking (*AmE*), runway-centre-line marking (*BrE*); ~ **del eje de la pista de rodaje** *m* TRANSP AÉR taxiway centerline marking (*AmE*), taxiway centreline marking (*BrE*)
balizador *m* TRANSP MAR *señales de navegación* buoy tender
balizaje: ~ **del umbral de pista** *m* TRANSP AÉR runway-threshold marking
balizamiento *m* CONST beaconing, TRANSP MAR *navegación* seamarking
balizar *vt* CONST beacon, CONTAM MAR *barreras flotantes* mark, TRANSP MAR buoy
ballenera *f* OCEAN *barcos de pesca* whale boat
ballenero *m* OCEAN *barcos de pesca* whale boat; ~ **factoría** *m* OCEAN *barcos de pesca* whale factory ship
ballesta *f* ING MECÁ plate spring; ~ **de coche** *f* ING MECÁ carriage spring; ~ **de hojas** *f* ING MECÁ carriage spring, VEH *suspensión* leaf spring; ~ **posterior** *f* D&A rear spring
ballestrinque *m* TRANSP MAR clove hitch
ballico *m* AGRIC ryegrass
balón *m* TRANSP MAR *navegación a vela* spinnaker; ~ **de tres bocas** *m* LAB *material de vidrio* three-necked flask
balona *f* TEXTIL flange

balonet: ~ **neumático** _m_ CARBÓN _dirigibles_ pneumatic cell

balsa _f_ AGUA reservoir, CARBÓN pond, raft, TEC PETR _perforación_ pit, well bore, TRANSP MAR _embarcación_ raft; ~ **auto-emergente** _f_ TRANSP, TRANSP MAR emerging-foil craft; ~ **de caucho** _f_ D&A, P&C, TRANSP MAR rubber dinghy; ~ **de clarificación** _f_ TEC PETR clarifying basin; ~ **de decantación** _f_ CARBÓN, MINAS clear pond, settling basin; ~ **de lodos** _f_ CARBÓN _glaciar_ slurry basin, PETROL _perforación_ mud box, TEC PETR _perforación_ mud box, mud pit; ~ **de madera** _f_ CONST timber raft; ~ **salvavidas** _f_ SEG, TRANSP MAR _emergencia_ life raft; ~ **de sedimentación** _f_ CARBÓN sedimentation pond, settling pond

baltimorita _f_ MINERAL baltimorite

baluma: ~ **de popa** _f_ TRANSP MAR _vela trapezoidal_ drop

bamboleo _m_ ING MECÁ shaking

bañado: **no** ~ _adj_ TEXTIL _tejido_ undipped; ~ **de cinc** _adj_ REVEST zinc-coated

bañar _vt_ REVEST _con metales o líquidos_ coat

banca _f_ FOTO, OCEAN _de pescadería_ bank

bancada _f_ ING MECÁ mount, base, shears, TRANSP MAR _máquina_ bed plate; ~ **escotada** _f_ ING MECÁ gap bed; ~ **de escote** _f_ ING MECÁ gap bed; ~ **de medio escote** _f_ ING MECÁ half-gap bed; ~ **del motor** _f_ AUTO, TRANSP AÉR, VEH engine cradle, engine support; ~ **de motores** _f_ TEC ESP _vehículos_ engine mountings; ~ **partida** _f_ ING MECÁ gap bed; ~ **de popa** _f_ OCEAN, TRANSP MAR bench, terrace; ~ **del torno** _f_ ING MECÁ, MECÁ lathe bed; ~ **transversal** _f_ ING MECÁ cross girth

bancal _m_ ING MECÁ bed, OCEAN _playas_ berm

banco _m_ CONST bank, bench, ELECTRÓN bank, ING MECÁ seat, shears, LAB laboratory stool, MINAS _AmL_ (_cf destroza Esp_) _de minerales_ bunker, _galería_ stope, OCEAN _de peces_ shoal, _meseta submarina_ bank, _taller mecánico a bordo_ terrace, bench, TRANSP MAR _meseta submarina_ bank, _taller mecánico a bordo_ bench, terrace; ~ **de ajustador** _m_ INSTR workbench; ~ **de arena** _m_ GEOL, HIDROL, OCEAN, TRANSP MAR sandbank, spit, _en la desembocadura de un río_ bar; ~ **de arena de alta mar** _m_ HIDROL, OCEAN, TRANSP MAR offshore bar; ~ **de arena inundado** _m_ AGUA, HIDROL submerged bar, OCEAN shoal, submerged bar; ~ **de arena sumergido** _m_ AGUA, HIDROL, OCEAN shoal, submerged bar; ~ **ascendente** _m_ _Esp_ (_cf gradería AmL_) MINAS stoping; ~ **de barras de compensación** _m_ NUCL shim rod bank; ~ **de calibración de medidores** _m_ GAS meter-calibration bench; ~ **para calibrar medidores** _m_ GAS meter-calibration bench; ~ **de carpintero** _m_ CONST carpenters' bench; ~ **central** _m_ TRANSP MAR _geografía_ middle ground; ~ **de datos** _m_ ELECTRÓN, INFORM&PD, TELECOM databank; ~ **de ebanista** _m_ CONST joiner's bench; ~ **de efectos** _m_ CINEMAT, TV effects bank; ~ **de ensayos** _m_ ING MECÁ test bench, TEC ESP test bed; ~ **de estiraje** _m_ PROD drawing bench; ~ **de germoplasma** _m_ AGRIC gene bank; ~ **de grava** _m_ AGUA, OCEAN, TRANSP MAR shingle bank; ~ **de guijarros** _m_ OCEAN, TRANSP MAR shingle bank; ~ **de huesos** _m_ REFRIG bone bank; ~ **de luces de techo** _m_ CINEMAT, TV overhead bank; ~ **de madera** _m_ CONST _carpintería_ buck; ~ **de mediciones** _m_ ELEC _instrumento_, INSTR measuring system; ~ **de memoria** _m_ INFORM&PD computer bank, memory bank, TELE-

COM computer bank; ~ **de moldear** _m_ PROD molding bench (_AmE_), moulding bench (_BrE_); ~ **de moldeo** _m_ PROD molding bench (_AmE_), moulding bench (_BrE_); ~ **del motor** _m_ TRANSP AÉR engine stand; ~ **de niebla** _m_ METEO fog bank; ~ **de nieve** _m_ METEO snowdrift; ~ **óptico** _m_ CINEMAT, FÍS, FOTO, IMPR, METR, ÓPT optical bench; ~ **portátil de tornillo** _m_ CONST portable-vice bench (_BrE_), portable-vise bench (_AmE_); ~ **de pruebas** _m_ ELEC, INFORM&PD, ING MECÁ, MECÁ, NUCL, TELECOM test bed, test bench; ~ **de pruebas del motor** _m_ TEC ESP _aviones_, TRANSP AÉR engine test stand; ~ **de pruebas de vuelo** _m_ TRANSP AÉR flying test bench; ~ **de roca natural** _m_ MINAS _geología_ wall; ~ **de sangre** _m_ REFRIG blood bank; ~ **de semen** _m_ AGRIC semen bank; ~ **de sierra** _m_ CONST saw bench; ~ **de taller** _m_ INSTR, PROD workbench, workbench unit; ~ **de tejidos** _m_ REFRIG tissue bank; ~ **de tornero** _m_ ING MECÁ, MECÁ frame; ~ **de torno** _m_ ING MECÁ, MECÁ lathe bed; ~ **de torno asegurado con cinchas cruzadas** _m_ ING MECÁ, MECÁ lathe bed braced by cross-girths; ~ **de trefilación** _m_ PROD wiredrawing bench; ~ **de trefilar** _m_ PROD draw bench

bancos: ~ **de enfriamiento y transportadores de rodillos** _m pl_ ING MECÁ, REFRIG, TERMO cooling banks and roller conveyors; ~ **de refrigeración y transportadores de rodillos** _m pl_ ING MECÁ, REFRIG, TERMO cooling banks and roller conveyors

banda[1]: **de** ~ **ancha** _adj_ ELECTRÓN, INFORM&PD, TELECOM wideband; **de** ~ **estrecha** _adj_ ELECTRÓN, INFORM&PD, TELECOM, TV narrow-band

banda[2] _f_ CINEMAT stripe, _de sonido_ track, CONST hoop, FERRO _de rueda_ tread, FÍS, FÍS ONDAS, FÍS RAD _debido a interferencias_ ban, band, fringe, IMPR _de papel_ band, web, INFORM&PD band, strip, ING MECÁ band, PAPEL web, PROD strap, strip, TRANSP AÉR fillet, strip; ~ **de absorción** _f_ FÍS, FÍS RAD absorption band; ~ **de acabado** _f_ C&V finishing belt; ~ **de acuñamiento** _f_ PROD keying band; ~ **alta** _f_ TV high band; ~ **de aluminio** _f_ EMB aluminium tape (_BrE_), aluminum tape (_AmE_); ~ **ancha** _f_ INFORM&PD, TV broadband, wideband; ~ **de atenuación** _f_ ELECTRÓN _filtros_ attenuation band; ~ **base** _f_ ELECTRÓN, INFORM&PD, TELECOM, TV baseband; ~ **cercana al infrarrojo** _f_ FÍS near infrared; ~ **cercana al ultravioleta** _f_ FÍS near ultraviolet; ~ **cero** _f_ NUCL zero band; ~ **ciudadana** _f_ TELECOM citizen band (_CB_); ~ **clasificadora** _f_ PROD sorting belt; ~ **de comentarios** _f_ CINEMAT commentary track; ~ **completa** _f_ FÍS full band; ~ **de conducción** _f_ FÍS, FÍS RAD (_BC_) conduction band (_CB_), METEO (_Cb_) cumulonimbus (_Cb_); ~ **de conexión a tierra** _f_ ELEC, ING ELÉC, PROD earth-fault tray (_BrE_), ground-fault tray (_AmE_); ~ **de control** _f_ CINEMAT, TV control band; ~ **crítica** _f_ ACÚST critical band; ~ **de deformación** _f_ GEOL, METAL kink band; ~ **de deslizamiento** _f_ CRISTAL slip band, METAL glide band; ~ **de desperdicio** _f_ C&V scrap return; ~ **de desplazamiento** _f_ NUCL _de una barra de control_ range of movement; ~ **de diálogos** _f_ CINEMAT dialog track (_AmE_), dialogue track (_BrE_); ~ **de emisión** _f_ FÍS emission band; ~ **energética permitida** _f_ FÍS, FÍS RAD, NUCL allowed energy band; ~ **energética prohibida** _f_ FÍS, FÍS RAD, NUCL forbidden energy band; ~ **entrante** _f_ IMPR incoming web; ~ **de equilibrio** _f_ CINEMAT balance stripe; ~ **de espectro** _f_ FÍS, FÍS ONDAS band, fringe; ~ **fosforescente** _f_ TV phosphor strip; ~ **de frecuencias** *f* ELEC *de corriente alterna*, ELECTRÓN, FÍS ONDAS, FÍS RAD, INFORM&PD, TELECOM, TRANSP AÉR, TRANSP MAR, TV frequency band, frequency range, waveband; ~ **de frecuencias sonoras** *f* ACÚST, OCEAN *transmisión del sonido* acoustic waveband; ~ **de frecuencias telefónicas** *f* TELECOM voiceband; ~ **de frecuencias vocales** *f* TELECOM voiceband; ~ **de gas** *f* GAS gas band; ~ **de gradación** *f* CINEMAT grading band; ~ **de guardar** *f* INFORM&PD guard band; ~ **húmeda** *f* PAPEL *defectos* damp streak; ~ **inactiva** *f* ING ELÉC dead band; ~ **de incertidumbre de frecuencias** *f* TELECOM frequency-uncertainty band; ~ **de interferencia** *f* FÍS ONDAS interference band; ~ **L** *f* ELECTRÓN, TEC ESP, TRANSP MAR L-band; ~ **L de las frecuencias** *f* ELECTRÓN, TEC ESP, TRANSP MAR *comunicaciones por satélite* L-band frequency; ~ **lateral** *f* ELECTRÓN, FÍS, INFORM&PD, TELECOM, TV sideband; ~ **lateral de apoyo** *f* IMPR bearer; ~ **lateral inferior** *f* ELECTRÓN, FÍS, INFORM&PD, TELECOM, TV lower sideband; ~ **lateral residual** *f* ELECTRÓN, FÍS, INFORM&PD, TELECOM, TV residual sideband, vestigial sideband; ~ **lateral superior** *f* ELECTRÓN, FÍS, INFORM&PD, TELECOM, TV upper sideband; ~ **lateral única** *f* ELECTRÓN, FÍS, TELECOM, TV single sideband (*SSB*); ~ **lateral vestigial** *f* ELECTRÓN, FÍS, TELECOM, TV vestigial sideband; ~ **libre** *f* CINEMAT wild track; ~ **límite** *f* ING MECÁ limit strip; ~ **magnética** *f* GEOL magnetic stripe; ~ **magnética de balance** *f* CINEMAT magnetic balance track; ~ **magnética central** *f* CINEMAT magnetic center track (*AmE*), magnetic centre track (*BrE*); ~ **magnética individual** *f* CINEMAT, TV sepmag; ~ **magnética de sonido** *f* CINEMAT magnetic sound stripe, magnetic soundtrack; ~ **marginal** *f* CINEMAT edge stripe, edge track; ~ **marina** *f* GEOL marine band; ~ **de media octava** *f* ACÚST half-octave band; ~ **de Möbius** *f* GEOM Möbius strip; ~ **de modulación** *f* ELECTRÓN modulation band; ~ **de Moebius** *f* GEOM Möbius strip; ~ **de música y efectos sonoros** *f* CINEMAT music and effects track (*M and E track*); ~ **de muy alta frecuencia** *f* TV VHF band; ~ **de narración** *f* CINEMAT narration track; ~ **del negativo** *f* CINEMAT negative track; ~ **de octava** *f* ACÚST, ELECTRÓN octave band; ~ **de ondas sonoras** *f* ACÚST, FÍS ONDAS, OCEAN *transmisión del sonido* acoustic waveband; ~ **óptica** *f* CINEMAT optical track; ~ **óptica individual** *f* CINEMAT sepopt; ~ **de papel** *f* IMPR paper web; ~ **de papel para registro gráfico** *f* ELEC chart strip; ~ **paralela** *f* FÍS RAD parallel band; ~ **de paro** *f* C&V stop belt; ~ **pasante** *f* INFORM&PD, ING MECÁ, TELECOM pass band; ~ **de paso** *f* INFORM&PD, ING MECÁ, TELECOM pass band; ~ **de paso bajo** *f* TELECOM low-pass band; ~ **de paso compuesta** *f* ELECTRÓN composite pass band; ~ **de paso de video** *f* *AmL*, ~ **de paso de vídeo** *f* *Esp* TELECOM picture pass band; ~ **de película de control** *f* CINEMAT, TV control film band; ~ **prohibida** *f* FÍS, FÍS RAD, NUCL forbidden band; ~ **de protección** *f* INFORM&PD guard band; ~ **del radar marino** *f* TRANSP MAR marine radar band; ~ **del radar náutico** *f* TRANSP MAR marine radar band; ~ **de rodadura** *f* AUTO, TRANSP AÉR, VEH tread; ~ **de rodamiento** *f* ING MECÁ crawler track, MECÁ track; ~ **de rozamiento** *f* TRANSP AÉR chafing strip; ~ **S** *f* ELECTRÓN S band; ~ **saliente** *f* IMPR outgoing web; ~ **satélite intermedia** *f* TELECOM

intermediate satellite band; ~ **de sonido ambiental** *f* CINEMAT ambience track; ~ **sonora** *f* ACÚST, CINEMAT, TV soundtrack; ~ **sonora internacional** *f* ACÚST, CINEMAT international soundtrack; ~ **sonora simétrica** *f* ACÚST, CINEMAT symmetrical soundtrack; ~ **de sujeción** *f* ING MECÁ clamping band; ~ **de supresión** *f* ELECTRÓN rejection band; ~ **de templador** *f* C&V lehr belt; ~ **de tercio de octava** *f* ACÚST one-third octave band; ~ **de tierra** *f* AGRIC furrow slice; ~ **de transferencia de carga** *f* FÍS RAD, NUCL charge-transfer band; ~ **de tránsito viario** *f* TRANSP lane; ~ **de transmisión libre** *f* ING MECÁ pass band; ~ **transversal** *f* TRANSP *control de tráfico* through band; ~ **unilateral** *f* ACÚST unilateral track; ~ **vacía** *f* FÍS empty band; ~ **de valencia** *f* FÍS, FÍS RAD valence band (*VB*); ~ **VHF** *f* TV VHF band; ~ **vocal** *f* INFORM&PD voiceband; ~ **de voz** *f* INFORM&PD voiceband; ~ **X** *f* ELECTRÓN X band

banda[3]: **de ~ a banda** *fra* TRANSP MAR athwartships

bandaje *m* FERRO *vehículos* tire (*AmE*), tyre (*BrE*); ~ **acanalado** *m* FERRO *vehículos* hollow tread; ~ **de rueda** *m* FERRO tire groove (*AmE*), tyre groove (*BrE*)

bandas: ~ **de estrías paralelas** *f pl* C&V *vidrio plano* ream

bandazo *m* TRANSP MAR *movimiento de barco* lurch, seeling; ~ **a sotavento** *m* TRANSP MAR lee lurch

bandeado[1] *adj* GEOL banded

bandeado[2]: ~ **fluidal** *m* GEOL flow banding

bandeador: ~ **para roscar** *m* ING MECÁ tap wrench

bandeja *f* *Esp* (*cf pálet AmL, paleta Esp*) C&V pan, tray, EMB pallet, tray, FOTO tray, ING MECÁ pan, tray, LAB, PAPEL, QUÍMICA tray, TRANSP, TRANSP AÉR pallet; ~ **de aguas blancas** *f* PAPEL save-all tray; ~ **de alimentos** *f* EMB food tray; ~ **apilable** *f* (*cf pálet apilable AmL*) TRANSP stacking pallet; ~ **para botellas con tapa movible incorporada** *f* EMB coupled full-base bottle tray; ~ **para cables** *f* ELEC, ING ELÉC cable rack; ~ **de carga** *f* (*cf pálet de carga AmL*) TRANSP AÉR pallet; ~ **de cartulina** *f* EMB, PAPEL cardboard tray; ~ **cilíndrica** *f* (*cf pálet cilíndrico AmL*) TRANSP roller pallet; ~ **de compartimientos** *f* EMB compartmented tray; ~ **desechable** *f* (*cf pálet desechable AmL*) TRANSP expendable pallet; ~ **del disco** *f* ÓPT disc platter (*BrE*), disk platter (*AmE*); ~ **de disección** *f* LAB *biología* dissecting tray; ~ **de dos accesos** *f* (*cf pálet de dos accesos AmL*) TRANSP two-way pallet; ~ **con embalaje envolvente** *f* EMB wraparound tray; ~ **de entrada** *f* PROD in-tray; ~**envase** *f* EMB punnet tray; ~ **estiba** *f* (*cf pálet estiba AmL*) TRANSP stevedore-type pallet; ~ **para hielo** *f* REFRIG ice-cube tray; ~**jaula** *f* (*cf pálet-jaula AmL*) TRANSP crate pallet; ~ **de la máquina** *f* C&V machine tray; ~ **de pérdida a tierra** *f* ELEC, ING ELÉC, PROD earth-fault tray (*BrE*), ground-fault tray (*AmE*); ~ **para piezas** *f* EMB tote box; ~ **recogedora** *f* PAPEL collection tray; ~ **para recoger el goteo** *f* PROD drip tray, drop pan; ~ **no retornable** *f* (*cf pálet no retornable AmL*) TRANSP one-way pallet; ~ **reversible** *f* (*cf pálet invertible AmL*) TRANSP reversible pallet; ~ **del rodillo** *f* C&V roller tray; ~ **de salida** *f* PROD out-tray; ~ **sencilla** *f* (*cf pálet sencillo AmL*) TRANSP single-decked pallet; ~ **con separación desmontable** *f* (*cf pálet con separación desmontable AmL*) TRANSP pallet with loose partition; ~ **de una sola cara** *f* (*cf pálet de una sola cara AmL*) TRANSP single-faced pallet; ~ **de un solo acceso** *f* TRANSP one-way pallet; ~ **de un solo tablero** *f* (*cf pálet de un solo tablero AmL*) TRANSP single-platform pallet; ~ **de sujeción** *f* (*cf pálet de sujeción AmL*) TRANSP post pallet; ~ **de tela** *f* PAPEL wire tray; ~ **uniflex** *f* QUÍMICA uniflex tray; ~ **de vacío** *f* LAB vacuum pan

bandejón *m* METEO, TRANSP MAR ice floe

bandeo *m* C&V banding

bandera *f* CINEMAT *para cortar un haz de luz*, TRANSP MAR flag; ~ **alfabética Q** *f* TRANSP MAR Q flag; ~ **amarilla** *f* TRANSP MAR yellow flag; ~ **de confirmación** *f* TELECOM acknowledgement flag; ~ **de conveniencia** *f* TRANSP MAR flag of convenience; ~ **de cortesía** *f* TRANSP MAR *del país visitado* courtesy ensign, courtesy flag; ~ **de datos nuevos** *f* INFORM&PD, TELECOM new data flag (*NDF*); ~ **de dibujo** *f* TEXTIL *tejidos* pattern length; ~ **de luto** *f* TRANSP MAR mourning flag; ~ **nacional** *f* TRANSP MAR colors (*AmE*), colours (*BrE*); ~ **de la naviera** *f* TRANSP MAR *marina mercante* house flag; ~ **de popa** *f* TRANSP MAR stern flag; ~ **de práctico** *f* TRANSP MAR pilot flag; ~ **reglamentaria de segmentación** *f* TELECOM segmentation permitted flag (*SPF*); ~ **roja** *f* TRANSP MAR red ensign; ~ **de sanidad** *f* TRANSP MAR quarantine flag; ~ **de señales** *f* TRANSP MAR code flag, signal flag

banderín: ~ **de error del programa interno** *m* PROD internal program error flag (*AmE*), internal programme error flag (*BrE*); ~ **de errores del programa** *m* PROD program error flag (*AmE*), programme error flag (*BrE*); ~ **de señales** *m* TEC ESP weft

bañera *f* TRANSP MAR *embarcación* cockpit

baño *m* AGRIC cattle dip, dip, C&V, CINEMAT, IMPR bath, PETROL coating, dip, REVEST, TEC PETR coating, TEXTIL *de las telas* dip, bath; ~ **de aceite** *m* LAB *calentamiento* oil bath; ~ **ácido** *m* PROC QUÍ acid bath; ~ **de ácido** *m* C&V frosting bath; ~ **anticorrosivo** *m* P&C *pintura*, REVEST anticorrosive coating; ~ **de apresto** *m* TEXTIL size bath; ~ **de arena** *m* LAB *calentamiento* sand bath; ~ **de blanqueo** *m* CINEMAT, FOTO bleach bath, bleaching bath; ~ **clarificador** *m* CINEMAT, FOTO clearing bath; ~ **desensibilizador** *m* CINEMAT, FOTO desensitizing bath; ~ **detenedor** *m* AmL (*cf baño de paro Esp*) CINEMAT, FOTO stop bath; ~ **detenedor ácido** *m* AmL (*cf baño de paro ácido Esp*) CINEMAT, FOTO acid stop bath; ~ **electrolítico** *m* ELEC, FÍS, ING ELÉC electrolytic bath; ~ **endurecedor** *m* IMPR hardening bath; ~ **endurecedor ácido** *m* CINEMAT, FOTO acid hardening bath; ~ **de enfriamiento** *m* METAL quenching bath, REFRIG cooling bath; ~ **estabilizador** *m* CINEMAT, FOTO stabilizing bath; ~ **fijador** *m* CINEMAT, FOTO fixer, fixing bath, PROC QUÍ fixing bath, fixer; ~ **fijador ácido** *m* CINEMAT, FOTO, PROC QUÍ acid fixing bath; ~ **de fusión** *m* PROD *soldadura* molten pool, TERMO melting bath; ~ **de galvanoplastia** *m* ELEC, ING ELÉC, REVEST electroplating bath; ~ **para grabado con ácido** *m* C&V clear-etching bath; ~ **para grabado profundo** *m* C&V deep-etching bath; ~ **histológico** *m* LAB histology bath; ~ **de inversión** *m* CINEMAT, FOTO reversing bath; ~ **de lavado** *m* CINEMAT wash, DETERG washing bath; ~ **María** *m* LAB *calentamiento*, QUÍMICA water bath; ~ **de montaje** *m* LAB *microscopio* mounting bath; ~ **de paro** *m* Esp (*cf baño detenedor AmL*) CINEMAT, FOTO stop bath; ~ **de paro ácido** *m* Esp (*cf baño detenedor ácido AmL*) CINEMAT, FOTO acid stop

bath; ~ **preliminar** *m* CINEMAT, FOTO preliminary bath; ~ **de recocido** *m* TERMO annealing bath; ~ **reforzador** *m* CINEMAT, FOTO, TV replenisher; ~ **refractario** *m* REVEST refractory wash; ~ **regenerador** *m* CINEMAT, FOTO, TV replenisher; ~ **con regulación termostática** *m* LAB *equipo general* thermostatically-controlled bath; ~ **revelador** *m* CINEMAT, FOTO developing bath; ~ **de sensibilización** *m* CINEMAT, FOTO sensitizing bath; ~ **de temple** *m* METAL quenching bath; ~ **ultrasónico** *m* LAB ultrasonic bath; ~ **de vapor** *m* LAB, PROC QUÍ vapor bath (*AmE*), vapour bath (*BrE*); ~ **de viraje** *m* CINEMAT, FOTO toning bath; ~ **de zinc** *m* P&C *pintura*, REVEST zinc coating

banqueta *f* CARBÓN *tierras* bank, CONST banquette, haunch

bantea: ~ **de carga** *f* ING MECÁ pan

bao *m* TRANSP MAR *construcción naval* beam; ~ **de cubierta** *m* TRANSP MAR *construcción naval* deck beam; ~ **maestro** *m* TRANSP MAR main beam, midship beam; ~ **reforzado** *m* TRANSP MAR *construcción naval* deck transverse; ~ **transversal** *m* TRANSP MAR *construcción naval* transverse beam

baqueta *f* D&A *de un fusil* rammer

bar *m* TEC PETR *unidad de presión* bar

baranda *f* AmL (*cf barandilla Esp*) CONST rail, railing, *escalera* side bar, MECÁ handrail, SEG guardrail, TRANSP MAR *construcción naval* topgallant bulwark, *equipamiento de cubierta* guardrail, *a popa* taffrail; ~ **del puente** *f* AmL (*cf barandilla del puente Esp*) CONST bridge rail; ~ **con punta de flecha** *f* AmL (*cf barandilla con punta de flecha AmL*) CONST spearheaded railing; ~ **de seguridad al pie del palo** *f* AmL (*cf barandilla de seguridad al pie de palo Esp*) TRANSP MAR *construcción, equipamiento de cubierta* mast foot safety rail

barandal *m* SEG guardrail

barandilla *f* Esp (*cf baranda AmL*) CONST rail, railing, *escalera* side bar, MECÁ handrail, SEG guardrail, TRANSP MAR *construcción naval* topgallant bulwark, *equipamiento de cubierta* guardrail, *a popa* taffrail; ~ **del puente** *f* Esp (*cf baranda del puente AmL*) CONST bridge rail; ~ **con punta de flecha** *f* Esp (*cf baranda con punta de flecha AmL*) CONST spearheaded railing; ~ **de seguridad al pie del palo** *f* Esp (*cf baranda de seguridad al pie de palo AmL*) TRANSP MAR *construcción, equipamiento de cubierta* mast foot safety rail

barba *f* PAPEL deckle; ~ **de espiga de los cereales** *f* AGRIC awn; ~ **de succión** *f* TEC PETR suction pit

barbecho *m* AGRIC fallow; ~ **forestal** *m* AGRIC forest fallow; ~ **en maleza** *m* AGRIC bush fallow; ~ **negro** *m* AGRIC bare fallow; ~ **de verano** *m* AGRIC summer fallow; ~ **verde** *m* AGRIC green fallow

barbeta *f* ELEC *cable* serving

barbilla *f* CONST *carpintería*, ING MECÁ, PROD feather

barbiquejo *m* TRANSP MAR bobstay

barbiquí: ~ **de mano** *m* ING MECÁ hand brace; ~ **de pecho** *m* ING MECÁ fiddle drill

barbiturato *m* QUÍMICA barbiturate

barbotar *vi* QUÍMICA *líquido* sputter

barbotén *m* TRANSP MAR chain lifter

barboteo *m* QUÍMICA *de líquido* sputtering

barbotún *m* TRANSP MAR chain lifter

barca: ~ **acidificante** *f* TEXTIL *para tintes* acidifying

beck; ~ **de torniquete** *f* TEXTIL winch; ~ **de torniquete para hilados** *f* TEXTIL spun yarn winch
barcada *f* TRANSP MAR *de pasajeros* boatload
barcaje *m* TRANSP MAR waterage
barcaza *f* PETROL, TRANSP, TRANSP MAR barge; ~ **algibe** *f* TRANSP tank barge; ~ **a bordo** *f* TRANSP ship-borne lighter; ~**-contenedor** *f* TRANSP container lighter; ~ **de desembarco** *f* D&A, TRANSP landing barge; ~ **de perforación** *f* PETROL drilling barge; ~ **portacontenedores** *f* TRANSP container lighter; ~ **de reparaciones** *f* TEC PETR *perforación costa-fuera* workover barge; ~**-taller** *f* PETROL work barge; ~ **para tender oleoductos submarinos** *f* PETROL, TEC PETR *producción costa-fuera* lay barge; ~ **de tendido submarino** *f* PETROL, TEC PETR *producción costa-fuera* bury barge; ~ **para tendido de tuberías** *f* PETROL, TEC PETR *producción costa-fuera* pipelaying barge
barco *m* CONST boat, CONTAM MAR, OCEAN, TRANSP MAR ship, vessel; ~ **de abastecimiento** *m* TEC PETR, TRANSP MAR supply boat; ~ **de aleta hidrodinámica sustentadora mixta** *m* TRANSP hybrid foil craft; ~ **de apoyo** *m* TEC PETR *perforación costa-afuera* support vessel; ~ **de apoyo al buceo** *m* OCEAN diving support barge; ~ **ballenero** *m* OCEAN *barcos de pesca* whaling ship; ~ **de bandejas de carga** *m* Esp (cf *barco de pálets de carga AmL*) TRANSP pallet ship; ~ **butanero** *m* TEC PETR *transporte*, TRANSP butane carrier; ~ **de cabotaje mixto** *m* TRANSP combined vessel; ~ **con casco tipo hidroavión** *m* TRANSP planing hull-type ship; ~ **catamarán** *m* TRANSP twin-hull ship; ~ **descontaminador** *m* CONTAM, CONTAM MAR depolluting ship; ~ **factoría** *m* REFRIG factory ship; ~ **frigorífico** *m* REFRIG, TRANSP, TRANSP MAR freezer vessel, refrigerated-cargo vessel; ~ **de investigaciones oceanográficas** *m* OCEAN, TRANSP MAR oceanographic research ship, oceanographic research vessel; ~ **mejillonero** *m* OCEAN *tecnología de pesca* mussel boat; ~ **multiusos** *m* TEC PETR *perforación costa-fuera* multipurpose vessel; ~ **oceanográfico** *m* OCEAN, TRANSP MAR oceanographic research ship, oceanographic research vessel; ~ **de paletas de carga** *m* Esp (cf *barco de pálets de carga AmL*) TRANSP pallet ship; ~ **de pálets de carga** *m* AmL (cf *barco de paletas de carga Esp*) TRANSP pallet ship; ~ **de perforación** *m* TEC PETR *perforación costa-fuera* drill ship; ~**-plataforma** *m* TEC PETR ship-type rig; ~ **rastreador** *m* OCEAN trawler; ~ **rastreador frigorífico** *m* OCEAN, REFRIG freezer trawler; ~ **recuperador de aceites** *m* TEC PETR *perforación costa-afuera* oil-recovery vessel; ~ **reflotador** *m* TRANSP lift on-off ship; ~ **de suministro** *m* TEC PETR supply boat; ~ **de transporte de materiales mixtos** *m* TRANSP oil-coal-ore carrier (*OCO carrier*); ~ **TRISEC** *m* TRANSP TRISEC ship
bardana: ~ **menor** *f* AGRIC *maleza* cocklebur
baremo *m* PROD schedule
baria *f* ACÚST, CARBÓN, FÍS, OCEAN *unidad de presión* bar
baricentro *m* MATEMÁT, TEC ESP barycenter (*AmE*), barycentre (*BrE*)
bárico *adj* QUÍMICA baric
barilita *f* MINERAL barylite
bario *m* (*Ba*) QUÍMICA barium (*Ba*)
barión *m* FÍS, FÍS PART baryon
barisfera *f* GEOFÍS barysphere, centrosphere

barita *f* C&V, MINERAL, P&C, PETROL, QUÍMICA, TEC PETR barite, barytes
baritina *f* C&V, MINERAL, P&C, PETROL, QUÍMICA, TEC PETR *ingeniería de lodos* barite, barytes
baritocalcita *f* MINERAL barytocalcite
baritocelestina *f* MINERAL barytocelestine, barytocelestite
barkevikita *f* MINERAL barkevikite
barlovento: **a ~** *adv* METEO, TRANSP MAR *navegación* windward; **de ~** *adv* METEO, TRANSP MAR *navegación* windward
barn *m* (*b*) FÍS, FÍS PART, NUCL barn (*b*)
barnhardtita *f* MINERAL barnhardite
barnio *m* (*b*) FÍS, FÍS PART, NUCL barn (*b*)
barniz *m* COLOR, CONST lacker, lacquer, varnish, ELEC *aislación*, IMPR varnish, MECÁ lacker, lacquer, P&C, QUÍMICA varnish, REVEST lacker, lacquer, polish; ~ **de acabado** *m* C&V, COLOR, MECÁ, REVEST finishing varnish; ~ **de aceite de resina** *m* REVEST resin-oil varnish; ~ **adhesivo** *m* COLOR adhesive varnish; ~ **al aguarrás** *m* COLOR turpentine varnish; ~ **aislador** *m* ELEC, REVEST insulating varnish; ~ **aislante** *m* ELEC, REVEST insulating varnish; ~ **de alcohol** *m* COLOR, REVEST French polish, spirit varnish; ~ **anímico** *m* COLOR, REVEST anime varnish; ~ **antiveteado** *m* COLOR, REVEST antiflash varnish; ~ **de asfalto** *m* COLOR, REVEST asphalt varnish; ~ **bajo en aceite** *m* COLOR, REVEST short oil varnish; ~ **de bajo porcentaje de aceite** *m* COLOR, REVEST short oil varnish; ~ **de base** *m* COLOR flatting varnish; ~ **brillante** *m* COLOR brilliant varnish; ~ **bronceado** *m* COLOR bronze varnish; ~ **de caucho** *m* P&C, REVEST rubber varnish; ~ **de China** *m* PROD lacquer; ~ **de cobertura** *m* AmL (cf *barniz de recubrimiento Esp*) COLOR, REVEST coating varnish; ~ **color ámbar** *m* COLOR, REVEST amber varnish; ~ **de copal** *m* COLOR copal varnish; ~ **corto en aceite** *m* COLOR, REVEST short oil varnish; ~ **decorativo** *m* P&C *pintura* decorative varnish; ~ **endurecedor** *m* COLOR, REVEST stiffening varnish; ~ **esmaltado** *m* COLOR, REVEST enamel varnish; ~ **esmaltado para secar en estufa** *m* P&C *pintura*, REVEST stoving enamel varnish; ~ **al esmalte** *m* COLOR, REVEST enamel varnish; ~ **de estampación** *m* COLOR, REVEST stamping varnish; ~ **estañado** *m* COLOR, REVEST tin-plate varnish; ~ **estructural** *m* REVEST structural varnish; ~ **en frío** *m* COLOR, REVEST cold-cut varnish; ~ **fungicida** *m* COLOR, REVEST fungicidal varnish; ~ **de impregnación** *m* ELEC *aislación*, REVEST impregnating varnish; ~ **impregnado** *m* ELEC *aislación*, REVEST impregnating varnish; ~ **de impresión** *m* COLOR, IMPR, REVEST printing varnish; ~ **de inmersión** *m* COLOR, REVEST dipping varnish; ~ **japonés** *m* COLOR, REVEST black japan, japan; ~ **de laca** *m* COLOR lac varnish, REVEST shellac varnish; ~ **largo en aceite** *m* COLOR, REVEST long-oil varnish; ~ **litográfico** *m* COLOR, REVEST lithographic varnish; ~ **de lustre** *m* C&V, COLOR, REVEST finishing varnish; ~ **mate** *m* COLOR, REVEST flatting varnish; ~ **metálico** *m* COLOR, REVEST bronze varnish; ~ **de muñeca** *m* COLOR, REVEST French polish; ~ **negro** *m* COLOR, REVEST black varnish; ~ **oleoresinoso** *m* COLOR, REVEST oleoresinous varnish; ~ **de porcelana** *m* C&V, REVEST porcelain varnish; ~ **con propiedades conductoras** *m* REVEST conductive varnish; ~ **protector** *m* COLOR, REVEST covering varnish; ~ **a prueba de ácidos** *m* EMB,

REVEST acid-proof varnish; **~ de recubrimiento** *m Esp* (*cf barniz de cobertura AmL*) COLOR, REVEST coating varnish; **~ de relleno** *m* COLOR, REVEST knot varnish; **~ de resina alkídica** *m* COLOR, REVEST alkyde-resin varnish; **~ de resina alquídica** *m* COLOR, REVEST alkyde-resin varnish; **~ resistente a los ácidos** *m* COLOR, EMB, REVEST acid-proof varnish; **~ de revestimiento** *m* COLOR, REVEST coating varnish; **~ de secado** *m* REVEST drying varnish; **~ secado al horno** *m* COLOR, REVEST baking varnish; **~ sintético** *m* COLOR, REVEST synthetic varnish; **~ sintético de alto porcentaje en aceite** *m* COLOR, REVEST synthetic long oil varnish; **~ sintético largo en aceite** *m* REVEST synthetic long oil varnish; **~ de sobreimpresión** *m* IMPR, REVEST overprint varnish; **~ para suelos** *m* REVEST floor varnish; **~ transparente** *m* COLOR, REVEST clear varnish, glazing varnish; **~ con tratamiento químico** *m* REVEST screen varnish; **~ de trementina** *m* COLOR, REVEST turpentine varnish

barnizado[1] *adj* COLOR, CONST, IMPR, REVEST varnished, **~ a la sal** *adj* COLOR salt-glazed

barnizado[2] *m* CINEMAT lacquering, COLOR, CONST, IMPR, PROD varnishing, REVEST varnish run, varnishing; **~ de base plomo** *m* REVEST lead glazing; **~ con laca** *m* PROD japanning

barnizador *m* COLOR, CONST, IMPR, REVEST varnisher

barnizar *vt* CINEMAT, COLOR, CONST, IMPR, PROD, REVEST varnish; **~ por inmersión** *vt* COLOR, REVEST dip varnish; **~ con muñequilla** *vt* COLOR, REVEST French polish

baroaltímetro *m* FÍS, LAB, METEO, TRANSP AÉR pressure altimeter

barógrafo *m* FÍS, LAB, METEO, TRANSP AÉR, TRANSP MAR barograph

barométrico *adj* GEN barometric

barómetro *m* CONST, FÍS, LAB barometer, pressure gage (*AmE*), pressure gauge (*BrE*), METEO barometer, METR, TRANSP AÉR, TRANSP MAR barometer, pressure gage (*AmE*), pressure gauge (*BrE*); **~ aneroide** *m* CONST, FÍS, LAB, METEO, TRANSP AÉR aneroid barometer; **~ de Fortin** *m* FÍS Fortin barometer; **~ de mercurio** *m* FÍS, METEO, METR mercury barometer; **~ registrador** *m* FÍS, LAB, METEO, TRANSP MAR barograph, recording barometer

baroscopio *m* FÍS baroscope

barostato *m* REFRIG snap-action valve, TRANSP AÉR barostat

barotermógrafo *m* FÍS barothermograph

barotrauma *m* ACÚST barotrauma

barquero *m* TRANSP MAR waterman

barquilla *f* TEC ESP gondola, *de un globo dirigible* basket; **~ aerodinámica portamotor colgada del ala** *f* TEC ESP *avión* pod; **~ auxiliar suspendida** *f* TRANSP sidecar

barquillero *m* TRANSP MAR waterman

barra *f* CARBÓN *de oro o de plata* ingot, *banco de arena* bar, CONST, GEOL bar, HIDROL *en la entrada de un puerto* bar, ING MECÁ tie, lever, INSTR *cabrestante* arm, MECÁ *herramientas* shaft, ingot, MINAS *de plata* bar, *gravas auríferas* bank, *de máquina* rod, OCEAN *banco de arena*, PAPEL bar, PROD *oro o plata* ingot, TRANSP MAR *banco de arena* bar;

▪ **a** ~ **absorbente** *f* NUCL absorber rod; **~ absorbente de caída por la gravedad** *f* NUCL gravity-drop absorber rod; **~ absorbente de caída libre** *f* NUCL

gravity-drop absorber rod; **~ accionadora** *f* TRANSP AÉR actuating rod; **~ de acoplamiento** *f* AUTO, VEH *dirección* drag link, tie rod; **~ de acoplamiento de las ruedas motrices** *f* AUTO, VEH tie rod; **~ de afilar** *f* ALIMENT *cuchillería* sharpening steel; **~ de agujas** *f* TEXTIL needle bar; **~ de aire escalonada** *f* IMPR staggered air bar; **~ de alargamiento** *f* ING MECÁ lengthening rod; **~ de alimentación** *f* ELEC feeder bar; **~ de anclaje** *f* ING MECÁ tie bar; **~ de anclaje a tierra** *f* TRANSP AÉR grounding bar; **~ antialabeo** *f* TRANSP AÉR antiroll bar; **~ antibalanceo** *f* MECÁ, TEC ESP antiroll bar; **~ antivuelco** *f* VEH *carrocería* rollover bar; **~ de apoyo** *f* MINAS *construcción* rest, TRANSP AÉR backing bar; **~ de arena** *f* GEOL, HIDROL, OCEAN, TRANSP MAR sand bar, sandbank; **~ de arena de alta mar** *f* GEOL, OCEAN offshore bar; **~ de arranque** *f* CONST starter bar; **~ de ataque** *f Esp* (*cf barra de atraque AmL*) MINAS stemming rod, stemming stick; **~ de atraque** *f AmL* (*cf barra de ataque Esp*) MINAS stemming rod, stemming stick; **~ de azufre** *f* QUÍMICA stick of sulfur (*AmE*), stick of sulphur (*BrE*);

▪ **b** ~ **de bahía** *f* GEOL, OCEAN, TRANSP MAR baymouth bar; **~ de bloqueo** *f* FERRO locking bar;

▪ **c** ~ **de calibrado** *f* CONST gage bar (*AmE*), gauge bar (*BrE*); **~ para calzar** *f* CONST shoeing bar; **~ de cambio de marcha** *f* FERRO, ING MECÁ reversing lever rod, reversing rod; **~ de cambio de paso** *f* TRANSP AÉR pitch-change rod; **~ de cambio de paso de la pala** *f* TRANSP AÉR blade pitch-change rod; **~ de carga** *f AmL* (*cf lastrabarrena Esp*) TEC PETR *perforación* drill collar; **~ de cierre para la piquera** *f* PROD bott stick; **~ del cilindro** *f* ING MECÁ cylinder rod; **~ colectora** *f* ELEC *conexión*, ING ELÉC, ING MECÁ, TEC ESP TELECOM TV bus, busbar; **~ colectora de alta tensión** *f* ELEC *fuente de alimentación* high-voltage bus; **~ colectora principal** *f* ELEC, TELECOM main busbar; **~ colectora de puesta a tierra** *f* ELEC, ING ELÉC, PROD **~ colectora subdividida** *f* ELEC *fuente de alimentación* sectionalized busbar; **~ colectora de tierra** *f* ELEC *conexión* earth bus (*BrE*), ground bus (*AmE*), ING ELÉC, PROD earth bus (*BrE*), ground bus (*AmE*); **~ de color** *f* TV color bar (*AmE*), colour bar (*BrE*); **~ de compensación del tren de aterrizaje** *f* TRANSP AÉR landing-gear compensation rod; **~ compensadora** *f* AGUA balance bar; **~ compensadora de seguridad** *f* NUCL shim safety rod; **~ de conductor común** *f* INFORM&PD bus-mouse adaptor; **~ conductora** *f* ELEC *conexión*, ING ELÉC, ING MECÁ, TELECOM busbar; **~ de conexión** *f* ING MECÁ connecting rod, pitman, tie bar, MECÁ, TRANSP AÉR connecting rod; **~ de conexión de longitud infinita** *f* ING MECÁ connecting-rod of infinite length; **~ de conmutación** *f* TV switching bar; **~ de contacto del pantógrafo** *f* FERRO, ING MECÁ pantograph contact-strip; **~ de control** *f* AUTO, FÍS, TRANSP AÉR control rod; **~ de control de la rueda del morro** *f* TRANSP AÉR nose-wheel steering bar; **~ corrediza** *f* ING MECÁ slide bar; **~ corta** *f* ING MECÁ jimmy bar; **~ de corte** *f* ING MECÁ cutter, cutter bar; **~ costera** *f* GEOL, OCEAN, TRANSP MAR longshore bar; **~ cruzada** *f* PROD, TRANSP AÉR crossbar; **~ cruzada de contacto móvil** *f* PROD, TRANSP AÉR moveable-contact crossbar; **~ curva** *f* MECÁ gooseneck;

▪ **d** ~ **depresora** *f* C&V depression bar; **~ desarrugadora** *f* PAPEL camber bar; **~ de**

desconexión *f* NUCL disconnect rod; ~ **deslizante** *f* CINEMAT slide bar; ~ **desnatadora** *f* C&V skim bar, skimming rod; ~ **para despegar del molde** *f* PROD *fundición* loosening bar; ~ **de desplazamiento** *f* INFORM&PD scroll bar, ING MECÁ shifting rod; ~ **desviadora** *f* IMPR angle bar, turn bar; ~ **de la dirección** *f* AUTO, VEH steering rod; ~ **directriz** *f* ING MECÁ guide bar; ~ **de distancia** *f* ING MECÁ distance bar; ~ **de distribución** *f* ELEC distribution bus, *conexión* busbar, ING ELÉC, ING MECÁ, TELECOM busbar; ~ **del distribuidor** *f* INSTAL HIDRÁUL slide rod; ~ **distribuidora** *f* INFORM&PD bus; ~ **doble** *f* TRANSP AÉR dual rod;

▪ **e** ~ **eliminadora de arrugas** *f* PAPEL spreader bar; ~ **de emparrillado** *f* PROD grate bar; ~ **de empuje de horquilla** *f* AUTO, VEH *embrague* fork push rod; ~ **engalonada** *f* C&V chevron runner bar; ~ **de enganche** *f* FERRO *vehículos*, VEH *remolque* drawbar; ~ **de enrollamiento** *f* INFORM&PD scroll bar; ~ **en la entrada de una bahía** *f* GEOL, OCEAN, TRANSP MAR baymouth bar; ~ **de eorte** *f* AGRIC sickle bar; ~ **de escaparate** *f* CONST glazing bar; ~ **escardadora** *f* AGRIC rod weeder; ~ **de espina** *f* PROD herringbone bar; ~ **esplegadora** *f* PAPEL spreader bar; ~ **estabilizadora** *f* AUTO, VEH antiroll bar, equalizer bar, stabilizer bar;

▪ **f** ~ **para fabricar barrenas** *f* MINAS *perforación* drill rod; ~ **de ferrita** *f* FÍS ferrite rod; ~ **flexible** *f* AUTO, VEH *carrocería* underrun bar;

▪ **g** ~ **guía** *f* TEXTIL *hilado* guide bar; ~ **de guía** *f* CINEMAT slide bar, ING MECÁ guide bar; ~ **de guía al aire libre** *f* ING MECÁ overhead pilot bar; ~ **de guía alta** *f* ING MECÁ overhead pilot bar;

▪ **h** ~ **en H** *f* CONST H-bar; ~ **para hacer la cruz** *f* TEXTIL lease rod; ~ **de hierro pudelado** *f* PROD puddle bar; ~ **de hinca** *f* MINAS *sondeos* sinker bar; ~ **de la holandesa** *f* PAPEL beater bar; ~ **horizontal** *f* INSTR horizontal arm, TV horizontal bar; ~ **horizontal y vertical de la dirección de vuelo** *f* TRANSP AÉR horizontal and vertical bar of flight direction; ~ **de horquilla** *f* ING MECÁ fork bar; ~ **de horquilla del tren de aterrizaje** *f* TRANSP AÉR landing-gear fork rod;

▪ **i** ~ **indicatoria de la intensidad del viento** *f* METEO wind-speed barb; ~ **del inducido** *f* ELEC armature bar; ~ **introducida en el núcleo** *f* NUCL scrammed rod;

▪ **l** ~ **lateral** *f* AUTO, VEH *chasis* side member; ~ **litoral** *f* GEOL, OCEAN, TRANSP MAR longshore bar; ~ **longitudinal** *f* METR length bar;

▪ **m** ~ **maestra** *f* MINAS sinker bar, *de máquina* rod; ~ **de mandrinar** *f* ING MECÁ boring bar; ~ **de Meyer** *f* IMPR Meyer bar; ~ **de mina** *f* *AmL* (*cf barra de percusión Esp*) MINAS *sondeos* jumping drill, miners' bar, percussion drill;

▪ **o** ~ **de obturación** *f* ING MECÁ plugging bar; ~ **ojalada** *f* PROD looped rod; ~ **ómnibus** *f* ELEC *conexión*, ING ELÉC, ING MECÁ, TELECOM busbar; ~ **ómnibus de alta tensión** *f* ELEC *fuente de alimentación* high-voltage bus; ~ **ómnibus de alto voltaje** *f* ELEC *fuente de alimentación* high-voltage bus; ~ **ómnibus subdividida** *f* ELEC *fuente de alimentación* sectionalized busbar;

▪ **p** ~ **de panorámica** *f* CINEMAT *de cabeza de cámara* pan handle; ~ **de panorámica horizontal y vertical** *f* CINEMAT *de cabeza de cámara* pan-and-tilt

handle; ~ **del par** *f* AUTO torsion bar; ~ **de parada** *f* NUCL scram rod; ~ **de parada de emergencia** *f* NUCL emergency-shutdown rod; ~ **de parada insertada** *f* NUCL inserted scram rod; ~ **pararrayos** *f* ELEC, SEG lightning rod; ~ **de pasadores** *f* TEXTIL guide bar; ~ **patrón** *f* ING MECÁ test bar; ~ **de percusión** *f* *Esp* (*cf barra de mina AmL*) MINAS *sondeos* jumping drill, miners' bar, percussion drill; ~ **de perforación** *f* MINAS *sondeos*, PROD jar; ~ **perforadora** *f* MINAS *sondeos* sinker bar; ~ **del pistón** *f* FERRO, ING MECÁ piston rod; ~ **del pistón del cilindro** *f* FERRO, ING MECÁ cylinder piston rod; ~ **de platinas** *f* TEXTIL sinker bar; ~ **portacuchillas** *f* CARBÓN cutter arm; ~ **prensatelas** *f* TEXTIL presser bar; ~ **de prueba** *f* ING MECÁ test bar; ~ **de la puerta** *f* CONST door bar; ~ **de puesta a tierra** *f* ELEC, ING ELÉC, PROD earth bus (*BrE*), ground bus (*AmE*); ~ **de punta** *f* MECÁ crowbar; ~ **puntiaguda** *f* *AmL* (*cf punterola Esp*) MINAS *entibación* gad;

▪ **r** ~ **de reacción** *f* AUTO, VEH *suspensión de las ruedas* torque arm; ~ **de reacción del puente trasero** *f* AUTO, VEH torque arm; ~ **de refuerzo interior** *f* EMB interior strengthening bar; ~ **registradora** *f* CINEMAT register bar; ~ **de remolque** *f* MECÁ towbar; ~ **para romper el tapón de la piquera** *f* PROD runner pin, *fundería* sprue; ~ **para roscar** *f* ING MECÁ screw stock;

▪ **s** ~ **de semiarrastre** *f* AUTO, VEH *eje trasero* semitrailing arm; ~ **de senos** *f* METR sine bar; ~ **separadora** *f* ING MECÁ distance bar; ~ **de sobrerreactividad** *f* NUCL booster rod; ~ **de soporte** *f* TRANSP AÉR backing bar; ~ **de suspensión** *f* CONST hanger;

▪ **t** ~ **taladradora** *f* ING MECÁ boring bar; ~ **de taponamiento** *f* ING MECÁ plugging bar; ~ **tensada** *f* TRANSP MAR *construcción de motor* tie rod; ~ **de tensión** *f* IMPR tensioning bar, TEXTIL tension bar; ~ **tensora** *f* TRANSP MAR *construcción de motor* tie rod; ~ **testigo** *f* ING MECÁ test bar; ~ **de tierra** *f* ELEC, ING ELÉC, PROD earth bar (*BrE*), earth rod (*BrE*), ground bar (*AmE*), ground rod (*AmE*); ~ **del timón de dirección** *f* TRANSP AÉR, TRANSP MAR rudder bar; ~ **tirante** *f* ING MECÁ tie rod; ~ **de tiro** *f* ING MECÁ drawbar; ~ **de torsión** *f* AUTO, VEH torque arm, torsion bar; ~ **de torsión laminada** *f* TRANSP AÉR laminated torsion bar; ~ **de tracción** *f* FERRO *vehículos* draft bar (*AmE*), draught bar (*BrE*), drawbar, ING MECÁ drawbar, VEH *remolque* towbar; ~ **tractora** *f* ING MECÁ drawbar; ~ **de transmisión** *f* ING MECÁ string rod; ~ **transversal** *f* CONST crossbar, ING MECÁ bar, VEH *carrocería* roll-over bar; ~ **de trefilar** *f* ING MECÁ die bar;

▪ **u** ~ **unida al cable** *f* MINAS sinker bar; ~ **de unión** *f* ING MECÁ, MECÁ connecting rod, TRANSP AÉR tie bar;

▪ **v** ~ **de la ventana** *f* CONST window bar; ~ **de vidriera** *f* C&V, CONST glazing bar; ~ **de vidrio** *f* C&V, CONST glass bar

barracas *f pl* CONST barracks
barracones *m pl* CONST barracks
barrado *m* TEXTIL barriness; ~ **por trama** *m* TEXTIL barriness in the weft; ~ **por urdimbre** *m* TEXTIL stripiness in the warp
barraescota: ~ **de la mayor** *f* TRANSP MAR *yates* mainsheet track
barrandita *f* MINERAL barrandite

barras *f pl* PAPEL *variaciones cíclicas del calibre* barring
barreduras *f pl* PROD sweepings
Barremiense *adj* GEOL *estratigrafía* Barremian
barreminas *m* D&A, TRANSP MAR minesweeper
barrena *f* CARBÓN, CONST, ING MECÁ, MECÁ, MINAS, PETROL boring tool, drilling bit, auger, PROD perforator, TEC PETR *Esp* (*cf trépano AmL*) rock bit, rotary bit, TRANSP AÉR spin; ~ **para abrir el agujero de colada** *f* PROD git cutter; ~ **para abrir agujeros para postes** *f* MINAS *sondeos* earth-auger; ~ **acústica** *f* ACÚST, CONST noise barrier; ~ **de arrastre** *f* PETROL bit, TEC PETR *perforación* crown bit, drag bit, drill bit, hard of formation bit, bit, reaming bit, roller bit, spiral bit, tricone bit, pilot bit; ~ **para berbiquí** *f* ING MECÁ bit-stock drill; ~ **de cable** *f* MINAS *sondeos* churn drill; ~ **de cáliz** *f Esp* (*cf barrena tubular AmL*) MINAS *sondeos* calyx drill; ~ **de cateo** *f* MINAS *prospección* earth-borer; ~ **cilíndrica** *f* CONST shell gimlet; ~ **de cincel** *f* MINAS *perforación* chisel bit; ~ **de colisión** *f* SEG crush barrier; ~ **cónica** *f Esp* (*cf trépano cónico AmL*) TEC PETR *perforación* cone bit; ~ **controlada** *f* TRANSP AÉR controlled spin; ~ **cortanúcleos** *f AmL* (*cf barrena sacatestigos Esp*) TEC PETR *perforación* core bit; ~ **cortatestigos** *f* TEC PETR *perforación* core bit; ~ **de cuchara** *f* CONST spoon auger, MINAS *sondeos* shell auger; ~ **de diamantes** *f* TEC PETR *perforación* diamond bit; ~ **de ensanchar** *f* ING MECÁ opening bit; ~ **espiral** *f* ING MECÁ twist drill; ~ **excéntrica** *f* MINAS *sondeos*, TEC PETR eccentric bit; ~ **gastada** *f* TEC PETR *perforación* worn bit; ~ **de grueso calibre** *f* MINAS *sondeos* large hole-cut; ~ **helicoidal** *f Esp* (*cf barrena para tirafondos AmL*) CONST, ING MECÁ screw auger, MINAS *barrenado* auger; ~ **de insertos** *f Esp* (*cf trépano de insertos AmL*) TEC PETR *perforación* insert bit; ~ **de mano** *f* MECÁ hand drill; ~ **para martillo perforador** *f AmL* (*cf boca de barrena Esp*) MINAS *sondeos* drill bit; ~ **de percusión** *f* CARBÓN jumper boring, MINAS *sondeos* jumper, TEC PETR *perforación* chisel bit, crossbit; ~ **de perforación con cadena** *f* PETROL, TEC PETR cable-drilling bit; ~ **de perforación de mortaja** *f* CONST mortise-boring bit; ~ **plana** *f* TRANSP AÉR flat spin; ~ **con punta de diamante** *f* GAS diamond drill; ~ **de punzonar** *f* ING MECÁ, MECÁ punch shank; ~ **para roca** *f* TEC PETR rock bit, rotary bit; ~ **sacamuestras** *f* ING MECÁ core drill, MINAS *sondeos* core drilling; ~ **sacatestigos** *f Esp* (*cf barrena cortanúcleos AmL*) TEC PETR *perforación* core bit; ~ **sacatestigos de diamantes** *f* TEC PETR *perforación* diamond core drill; ~ **salomónica** *f* ING MECÁ screw auger; ~ **tetracónica** *f Esp* (*cf trépano tetracónico AmL*) TEC PETR *perforación* quadricone bit; ~ **para tierra** *f* CARBÓN earthdrill; ~ **para tirafondos** *f AmL* (*cf barrena helicoidal Esp*) MINAS auger; ~ **tricónica** *f Esp* (*cf tricono AmL*) TEC PETR three-cone bit, tricone bit; ~ **tubular** *f AmL* (*cf barrena de cáliz Esp*) MINAS *sondeos* calyx drill
barrenación *f* CARBÓN, CONST, ING MECÁ, MINAS, PROD boring
barrenado *m* CARBÓN *minas* drilling, CONST *carpintería* boring bit, MINAS *de rocas o tierra* drilling; ~ **por agua** *m* AGUA boring against water; ~ **de alcance** *m* MINAS *galerías* breaking-in shot; ~ **por chorro de fueloil con oxígeno** *m* CARBÓN *minas* jet piercing;

~ **de franqueo** *m* MINAS *pega* breaking-in shot; ~ **profundo** *m* MINAS *pega* shaft sinking
barrenador *m* AGRIC *insecto* borer; ~ **europeo de maíz** *m* AGRIC European corn-borer (*AmE*), European maize-borer (*BrE*); ~ **del maíz** *m* AGRIC corn-borer (*AmE*), maize-borer (*BrE*)
barrenadora *f* TEC PETR *perforación* driller
barrenar[1] *vt* CARBÓN, CONST, ING MECÁ, MECÁ, MINAS, TEC PETR drill, bore, TRANSP MAR *buque* scuttle
barrenar[2] *vi* MINAS bore
barrenero *m* MINAS *persona* borer
barrenista *m* CARBÓN shotfirer
barreno *m Esp* (*cf trépano AmL*) CARBÓN borehole, *perforación, sondeo* bore bit, drill pin, jumper drill, drill, percussion drill, jumper, borer, jumping drill, *pega* bore, *taladro pepa* blast-hole, *voladuras* drill hole; ~ **de alcance** *m* MINAS *voladuras* breaking shot, buster shot; ~ **de alivio** *m AmL* (*cf barreno de descarga Esp*) MINAS *voladuras* relief hole; ~ **auxiliar** *m* MINAS *voladuras* relief hole, reliever; ~ **de ayuda** *m* MINAS *voladuras* easer; ~ **de calado** *m Esp* (*cf barreno de franqueo AmL*) MINAS *pega* breaking shot, bursting shot, busting shot; ~ **de cara** *m* MINAS *voladuras* breast face, breast hole, PROD *voladuras* breast hole; ~ **de costado** *m* MINAS, PROD breast hole; ~ **de cuele** *m* MINAS *voladuras* center cut hole (*AmE*), centre cut hole (*BrE*); ~ **de cuña** *m* MINAS *túneles* center cut hole (*AmE*), centre cut hole (*BrE*); ~ **de descarga** *m Esp* (*cf barreno de alivio AmL*) MINAS *voladuras* relief hole, reliever; ~ **para desprender trozos secundarios de la cara del frente** *m AmL* (*cf saneo de frentes Esp*) MINAS *túneles* lifter hole; ~ **de excavación** *m* MINAS *túneles* opening shot; ~ **con el fondo ensanchado** *m* MINAS *voladuras* chambered hole; ~ **de franqueo** *m AmL* (*cf barreno de calado Esp*) MINAS *pega* bursting shot, busting shot, breaking shot, *voladuras* center cut hole (*AmE*), centre cut hole (*BrE*); ~ **húmedo** *m* MINAS *voladuras* wet hole; ~ **limpiador** *m* MINAS *voladuras* scaling bar; ~ **liso** *m* MINAS *voladuras* plain bit; ~ **de pie** *m* MINAS *voladuras* lifter; ~ **de pola** *m* MINAS *voladuras* shell auger; ~ **seco** *m Esp* (*cf barreno sin utilizar agua para mitigar el polvo AmL*) MINAS *voladuras* dry hole; ~ **sísmico** *m* GEOFÍS seismic borehole; ~ **de suelo** *m Esp* (*cf barreno tendido AmL*) CARBÓN floor shots, MINAS *pega* bottom hole, bottom shot; ~ **de techo** *m* MINAS *pega* backhole, *voladuras* roadheader; ~ **tendido** *m AmL* (*cf barreno de suelo Esp*) CARBÓN floor shots, MINAS *pega* bottom hole, bottom shot; ~ **sin utilizar agua para mitigar el polvo** *m AmL* (*cf barreno seco Esp*) MINAS *voladuras* dry hole
barrer *vt* CONST sweep, PROD, TRANSP scan
barrera *f* CONST barrage, barrier, barricade, ING MECÁ, MECÁ guard, MINAS clay pit, *de sostenimiento* barrier, OCEAN *protección de un río contra las grandes mareas* barrage, PROD *de paso a nivel* gate, SEG *para evitar la entrada* guardrail; ~ **acústica** *f* ACÚST acoustic barrier, acoustic screen; ~ **de agua** *f* MINAS *previene la propagación de la costra de coque* water barrier, water curtain; ~ **de burbujas** *f* CONTAM MAR bubble barrier; ~ **de choque** *f* SEG crush barrier; ~ **de contención** *f* TEC PETR *derrames de crudo* containment boom; ~ **contra el chorro** *f* TRANSP AÉR blast fence; ~ **contra escombros** *f* ING MECÁ guard against debris; ~ **contra fuegos** *f* SEG, TRANSP AÉR firewall; ~ **de control** *f* CONST control barrier; ~ **de**

emergencia anti-choque *f* SEG, TRANSP emergency crash barrier; ~ **enfardada** *f* CONTAM MAR boom pack; ~ **flotante** *f* CONTAM MAR boom; ~ **flotante con aliviadero** *f* CONTAM MAR weir boom; ~ **flotante con faldilla** *f* CONTAM MAR curtain boom; ~ **frente al vapor de agua** *f* EMB water vapor barrier (*AmE*), water vapour barrier (*BrE*); ~ **de fuego** *f* D&A curtain of fire; ~ **de hielo** *f* HIDROL, OCEAN ice barrier; ~ **de hielo suspendida** *f* HIDROL, OCEAN ice jam; ~ **de impacto** *f* SEG crash barrier; ~ **inicial** *f* PROD start fence; ~ **micrométrica** *f* ELECTRÓN micron barrier; ~ **de potencial** *f* CONTAM, FÍS, ING ELÉC, TELECOM barrier layer, potential barrier; ~ **de protección acústica** *f* ACÚST, CONST acoustic fencing; ~ **de Schottky** *f* ELECTRÓN, FÍS Schottky barrier; ~ **de seguridad** *f Esp* (*cf paradera AmL*) MINAS safety drag bar, SEG safety barrier; ~ **superficial** *f* CONTAM MAR skimming barrier; ~ **tecnológica** *f* NUCL engineered barrier; ~ **térmica** *f* C&V, NUCL, TEC ESP, TERMO, TERMOTEC, TRANSP heat barrier, thermal barrier; ~ **de tiempo** *f* PROD time fence
barreras: ~ **y escáners fotoeléctricos** *m pl* EMB photoelectric-light barriers and scanners
barrero *m* MINAS clay pit
barreta *f* MECÁ crowbar, PROD *válvula distribuidora* face flange; ~ **abrasiva** *f* ING MECÁ honing stone, PROD honestone
barrica *m* TRANSP barrel
barricada *f* CONST, SEG, TEC ESP barricade
barrido[1]: ~ **por la mar** *adj* TRANSP MAR washed overboard
barrido[2] *m* AGUA scavenging, ELECTRÓN sweep, FÍS ONDAS *médica* scan, FÍS RAD scanning, INFORM&PD interrogation, raster, MECÁ, NUCL scavenging, TEC ESP *alas de los aviones* sweep, TEC PETR *yacimientos* sweeping, scan, scanning, scavenging, TELECOM scanning, TRANSP AÉR *máquina* scan, TRANSP MAR *máquina* scavenging, sweep; ~ **de alta definición** *m* FÍS RAD high-resolution scan; ~ **de alta resolución** *m* FÍS RAD high-resolution scan; ~ **axial** *m* ÓPT *en espectrometría* axial scanning; ~ **de cable lateral** *m* ING ELÉC side wiping; ~ **electrónico** *m* ELECTRÓN, NUCL electron scanning; ~ **de frecuencia** *m* ELECTRÓN, TELECOM frequency sweep; ~ **gamma del combustible** *m* NUCL gamma fuel scanning; ~ **horizontal** *m* TV horizontal sweep; ~ **de izquierda a derecha** *m* TV horizontal sweep; ~ **de línea** *m* IMPR line scanning; ~ **lineal** *m* TV line sweep; ~ **logarítmico** *m* ELECTRÓN logarithmic sweep; ~ **de remanencia de la imagen** *m* TV scan burn; ~ **repetitivo** *m* ELECTRÓN repetitive sweep; ~ **ultrasónico** *m* FÍS RAD ultrasound scan; ~ **vertical** *m* ELECTRÓN vertical sweep
barril *m* ALIMENT barrel, EMB tub, GEOL, LAB *de jeringa* MECÁ, PETROL barrel, QUÍMICA keg, TRANSP barrel; ~ **de amalgamación** *m* CARBÓN amalgam barrel; ~ **del carburador** *m* AUTO, VEH carburetor barrel (*AmE*), carburettor barrel (*BrE*); ~ **del cilindro** *m* ING MECÁ, INSTAL HIDRÁUL, PROD, VEH cylinder barrel; ~ **cortanúcleos** *m* TEC PETR *perforación* core barrel; ~ **mezclador** *m* ALIMENT barrel mixer; ~ **de testigos** *m Esp* (*cf testiguero AmL*) TEC PETR *perforación* core barrel
barrilete *m* INSTR *objetivo de anteojo* body tube, NUCL *de la vasija del reactor* baffle, TRANSP MAR *nudo* stopper knot; ~ **vertical** *m* INSTR vertical clamp

barrilito *m* ALIMENT firkin
barrio: ~ **de los muelles** *m* TRANSP MAR waterfront
barrita: ~ **para rectificar** *f* ING MECÁ honing stone
barrizal *m* MINAS clay pit
barro *m* AGUA mud, ALIMENT sludge, C&V common clay, clay, CARBÓN clay, dirt, CONST, GAS, GEOL clay, MINAS slush, slurry, MINERAL clay, OCEAN mud, PETROL clay, QUÍMICA clay, sludge, RECICL sludge, TEC PETR *perforación, ingeniería de lodos* clay, mud; ~ **altamente plástico** *m* C&V, P&C highly plastic clay; ~ **de amolado** *m* PROD *muelas abrasivas* swarf; ~ **calcáreo** *m* GEOL *sedimento pelágico* calcareous ooze; ~ **cocido** *m* C&V burnt clay; ~ **de colores** *m* C&V colored clay (*AmE*), coloured clay (*BrE*); ~ **para crisoles** *m* C&V pot clay; ~ **fundible** *m* C&V fusible clay; ~ **de fundición** *m* C&V, PROD loam; ~ **de globigerinas** *m* GEOL, OCEAN Globigerina ooze; ~ **grasoso** *m* C&V fatty clay; ~ **líquido** *m* DETERG slurry; ~ **magro** *m* C&V meager clay (*AmE*), meagre clay (*BrE*); ~ **molido** *m* C&V clay powder; ~ **de plasticidad baja** *m* C&V low-plasticity clay; ~ **de plasticidad pobre** *m* C&V lean-plasticity clay; ~ **quemado** *m* C&V fired clay; ~ **refractario** *m* C&V fireclay; ~ **rico** *m* C&V rich clay; ~ **para tubos** *m* C&V pipeclay
barrón: ~ **portaherramientas** *m* ING MECÁ cutter mandrel
barros *m pl* PROD sludge; ~ **arenosos** *m pl* OCEAN ooze
barrote *m* CONST *ventana* window bar, *reja* screen bar, PROD *de escalera de mano* step; ~ **de enrejado** *m* CONST trellis post; ~ **de parrilla** *m* PROD grate bar, grid bar; ~ **redondo** *m* MINAS *escalas* round
basado: ~ **en máquina** *adj AmL* (*cf basado en ordenador Esp*) INFORM&PD machine-oriented; ~ **en ordenador** *adj Esp* (*cf basado en máquina AmL*) INFORM&PD machine-oriented; ~ **en pantalla** *adj* INFORM&PD screen-based
basaltina *f* MINERAL basaltine
basalto *m* CONST, ENERG RENOV, GEOL, PETROL basalt; ~ **de colada** *m* GEOL flood basalt; ~ **columnar** *m* GEOL columnar basalt
basamento *m* GEOL basement, ING MECÁ pedestal, PETROL, TEC PETR basement, TEXTIL *capa inferior de tejido* underlayer; ~ **de alfombra** *m* TEXTIL carpet underlay; ~ **cristalino** *m* CRISTAL, GEOL crystalline basement
basamiento *m* GEOL basement complex
basanita *f* PETROL basanite
báscula *f* ELECTRÓN flip-flop, EMB check weighing machine, ING MECÁ bascule, scale, scales, LAB, METR *balanza* balance, pan scales, scale, PROD weighing machine; ~ **de columna** *f* LAB *balanza* pillar scales; ~ **de pesar** *f* METR weighing scale
basculación *f* ING MECÁ seesaw motion
basculador *m* CARBÓN dump truck (*AmE*), tipper (*BrE*), toggle, ING MECÁ rocker, tripper, tripping device, MECÁ rocker, PROD tilter; ~ **cronometrado** *m* ELECTRÓN clocked flip-flop; ~ **en T** *m* ELECTRÓN T-flip-flop; ~ **tipo D** *m* ELECTRÓN D-type flip-flop; ~ **de vagones** *m* CARBÓN dump truck (*AmE*), tipper (*BrE*)
basculadora *f* CARBÓN tip
basculamiento *m AmL* (*cf descarga Esp*) MINAS dumping, PROD dumping, tilting
basculante[1] *adj* ING MECÁ hinged, MECÁ pivoted
basculante[2] *m* CONST dump truck (*AmE*), dumper (*AmE*), dumper truck (*AmE*), tipper (*BrE*)

bascular[1] *vt* INFORM&PD toggle, MINAS *tolva de mineral* dump, PROD *interruptor* toggle

bascular[2] *vi* CONST tip, tip up

base[1]: **hacia la ~** *adv* TRANSP AÉR inbound-heading

base[2] *f* CONST base, bed, foot, toe, sole, ELECTRÓN *de tubo electrónico*, FÍS, GEOM base, IMPR base, radix, substrate, INFORM&PD root, radix, base, ING ELÉC base, ING MECÁ seating, MATEMÁT *espacio vectorial* base, MINAS *de una mina* seat, QUÍMICA *álcali*, TELECOM base; **~ aérea** *f* TRANSP AÉR air base; **~ de apoyo** *f* CARBÓN spot footing, CONST foothold; **~ de apoyo de las pinzas** *f* IMPR gripper pad; **~ de aprovisionamiento** *f* TEC PETR *operaciones de perforación* supply base; **~ de calzo** *f* EMB skid base; **~ de capa** *f* GEOL sole; **~ del carril** *f* FERRO *equipo inamovible* rail base, railbed; **~ de cinta** *f* TV tape base; **~ de compensación de la brújula** *f* TRANSP AÉR compass compensation base; **~ de comprobación** *f* CONST base of verification; **~ concreta** *f* GEOFÍS concrete base; **~ de conocimiento** *f* INFORM&PD knowledge base; **~ costera** *f* TEC PETR *operaciones* onshore base; **~ de la curva característica** *f* FOTO toe region of characteristic curve; **~ de datos** *f* ELECTRÓN, INFORM&PD, TELECOM database; **~ de datos distribuida** *f* ELECTRÓN, INFORM&PD, TELECOM distributed database; **~ de datos en línea** *f* ELECTRÓN, INFORM&PD, TELECOM on-line database; **~ de datos óptica** *f* ÓPT optical database; **~ de datos de red** *f* ELECTRÓN, INFORM&PD, TELECOM network database; **~ de datos relacional** *f* ELECTRÓN, INFORM&PD, TELECOM relational database (*RDB*); **~ deslizante** *f* CINEMAT *de cámara* sliding base plate; **~ de enchufe** *f* TELECOM socket, VEH *electricidad* plug socket; **~ enteriza** *f* PROD one-piece base; **~ esmerilada** *f* C&V ground base; **~ espacial** *f* TEC ESP *exploración del espacio* exobase; **~ estructural permanente** *f* PETROL permanent guide base, permanent guide structure; **~ de un glaciar** *f* GEOL sole; **~ de guía transitoria** *f* PETROL temporary guide base; **~ de helicópteros** *f* TRANSP AÉR helicopter station; **~ de hidroaviones** *f* TRANSP AÉR seaplane base; **~ horizontal** *f* INSTR horizontal base; **~ de información para dirección de empresas** *f* TELECOM management information base (*MIB*); **~ de información para la gestión de la seguridad** *f* TELECOM security management information base (*SMIB*); **~ de instalación** *f* ING MECÁ mounting base; **~ de la lámpara** *f* CINEMAT lamp base; **~ de lanzamiento** *f* TEC ESP launching base; **~ loctal** *f* ELECTRÓN loctal base; **~ de montaje** *f* ING MECÁ mounting base; **~ de Mylar** *f* TV mylar base; **~ naval** *f* D&A, TRANSP MAR naval base; **~ niveladora de cámara** *f* CINEMAT four-plate, four-way; **~ de nubes** *f* METEO, TRANSP AÉR cloudbase; **~ octal** *f* ING MECÁ octal base; **~ de operaciones** *f* PROD base of operations; **~ orgánica** *f* QUÍMICA organic base; **~ de pago por uso** *f* TELECOM pay-by-use basis; **~ del pedestal** *f* INSTR pedestal base; **~ de la polea de la cabina** *f* TRANSP cabin pulley cradle; **~ portátil** *f* ELEC *conexión* portable-socket outlet; **~ de recursos accesibles** *f* ENERG RENOV accessible resource base; **~ de ruedas** *f* PROD stand on wheels, TRANSP AÉR wheel base; **~ de la soldadura** *f* CONST root of weld; **~ en T** *f* ING MECÁ wing base; **~ de tiempo** *f* ELECTRÓN, FÍS, NUCL, PROD, TELECOM, TV time base; **~ de tiempo de cristal** *f* CRISTAL, ELECTRÓN crystal time base; **~ de tiempo de línea** *f* TV line output; **~ de tiempo lineal** *f* ELECTRÓN linear time base; **~ de tiempo retardada** *f* ELECTRÓN delaying time base; **~ tipo p** *f* ELECTRÓN p-type base; **~ de tornillo** *f* CINEMAT screw base; **~ de transistor** *f* ELECTRÓN, TELECOM transistor base; **~ de tubo electrónico** *f* ELECTRÓN electron-tube base

BASIC *abr* (*código de instrucciones simbólicas de carácter general para principiantes*) IMPR, INFORM&PD *lenguaje de programación* BASIC (*Beginner's All-purpose Symbolic Instruction Code*)

basicidad *f* QUÍMICA basicity

basificación *f* GEOL, QUÍMICA *proceso* basification

basificar *vt* GEOL, QUÍMICA basify

bastidor *m* AUTO perimeter frame, C&V casement, CINEMAT flat, D&A rack, EMB frame, rack mount, FERRO *vehículos* frame, underframe, FOTO, IMPR frame, ING MECÁ rack, frame, plate, case, housing, INSTR, MECÁ cradle, PAPEL frame, PROD rack, TEC ESP undercarriage, TEXTIL frame, TRANSP *camiones* undercarriage, VEH *carrocería* frame, chassis; **~ de agujas de ganchillo** *m* TEXTIL bearded needle frame; **~ de almacenamiento de combustible gastado** *m* NUCL spent-fuel rack; **~ para armas** *m* D&A arms rack; **~ auxiliar** *m* VEH *carrocería* subframe; **~ auxiliar múltiple** *m* TELECOM submultiframe (*SMF*); **~ del bogie** *m* FERRO bogie frame (*BrE*), truck frame (*AmE*); **~ para bombas** *m* D&A bomb rack; **~ de cabina** *m* VEH chassis cab; **~ de cajón** *m* ING MECÁ box frame; **~ del carretón** *m* FERRO bogie frame (*BrE*), truck frame (*AmE*); **~ del castillete de extracción** *m* MINAS head frame; **~ de clavijas** *m* TEXTIL pin frame; **~ de conductores** *m* ELEC, ING ELÉC lead frame; **~ de E/S** *m* PROD I/O rack; **~ de filtro** *m* PROC QUÍ filter frame; **~ de flexibilidad óptica** *f* ÓPT, TELECOM optical flexibility frame (*OFF*); **~ de hierro** *m* CONST sash iron; **~ para la impresión por contacto** *m* CINEMAT, FOTO, IMPR contact-printing frame; **~ intermedio** *m* TEXTIL intermediate frame; **~ magnético** *m* ING MECÁ magnetic rack; **~ de la manguera** *m* TEC PETR *perforación* standpipe; **~ de mecha** *m* TEXTIL slubbing frame; **~ de moldeo** *m* TRANSP MAR *construcción* molding frame (*AmE*), moulding frame (*BrE*); **~ de montaje** *m* CINEMAT editing rack, pegboard, ING MECÁ assembly jig, TV editing rack; **~ permanente** *m* PETROL permanent guide structure; **~ de pinzas** *m* TEXTIL clip frame; **~ de rame** *m* TEXTIL stenter frame; **~ reversible** *m* OCEAN reversing frame; **~ de secado** *m* FOTO *para negativos, pruebas o ampliaciones* drying frame; **~ de sierra** *m* ING MECÁ saw frame; **~ de soporte** *m* NUCL underframe; **~ de la tela** *m* PAPEL wire frame; **~ de terraja** *m* ING MECÁ screw plate stock; **~ de transposición** *m* FOTO transposing frame; **~ tubular** *m* VEH *de motocicleta* tubular frame; **~ de varillas** *m* TEXTIL bar frame

bastita *f* MINERAL bastite, schiller spar

bastnasita *f* MINERAL bastnaesite

basto *adj* ING MECÁ coarse

basura[1] *f* AGUA refuse, CARBÓN dust, refuse, EMB garbage (*AmE*), refuse, rubbish (*BrE*), NUCL, PROD, RECICL, SEG garbage (*AmE*), refuse, rubbish (*BrE*), waste; **~ de alto nivel** *f* CONTAM, NUCL, RECICL high-level waste (*HLW*); **~ casera** *f* RECICL household waste; ~ **de cocina** f ALIMENT, CONTAM, RECICL kitchen waste; ~ **doméstica** f AGUA, CONTAM, RECICL domestic sewage; ~ **indeterminada** f RECICL indeterminate waste; ~ **nuclear** f CONTAM, NUCL, RECICL, SEG nuclear waste; ~ **residual** f RECICL sewage waste; ~ **voluminosa** f RECICL bulky waste

basura[2]: ~ **entra/basura sale** fra INFORM&PD garbage-in/garbage-out (*GIGO*)

basurero m AGUA, CONTAM, RECICL refuse dump, tip (*BrE*); ~ **a cielo abierto** m Esp (*cf pepena a cielo abierto AmL*) CONTAM below-cloud scavenging; ~ **municipal** m CONTAM, RECICL municipal dump

bata: ~ **de laboratorio** f LAB, SEG laboratory coat; ~ **protectora** f LAB, SEG protective gown; ~ **de trabajo** f Esp (*cf overoles AmL*) PROD overalls, SEG coveralls (*AmE*), industrial overalls (*BrE*), overalls (*BrE*)

batalla f TRANSP AÉR wheel base, VEH wheelbase

batán m PAPEL beating mill, fulling mill; ~ **de fieltro** m PAPEL felt whipper; ~ **fino** m TEXTIL fine batt; ~ **grueso** m TEXTIL coarse batt; ~ **posterior** m TEXTIL backside batt

batanado m TEXTIL fulling, *operación de acabado* milling

bataneo m TEC ESP *aviones* buffeting

batayola f TRANSP MAR raft, *construcción* topgallant bulwark, *equipamiento de cubierta* pulpit

batch m INFORM&PD batch

batea f CARBÓN *minería* strake, LAB trough; ~ **de carga** f CARBÓN *horno alto* pan; ~ **de lodo** f TEC PETR slurry trough

bateadora: ~ **de aluviones** f MINAS *prospección minera* dirt-packing machine; ~ **de traviesas** f FERRO *equipo inamovible* tamping machine

batear vt FERRO *traviesas* tamp

batería f AUTO battery, CARBÓN bank of cells, ELEC *pila* FÍS, FOTO, INFORM&PD, ING ELÉC battery, storage battery, ING MECÁ battery, cluster, INSTAL HIDRÁUL *de calderas* battery, set, MINAS *de hornos*, PAPEL battery, PROD cluster, QUÍMICA, REFRIG, TEC ESP, TELECOM, TRANSP MAR, VEH battery; ~ **de accionamiento** f TRANSP drive battery; ~ **de acumuladores** f AUTO, ELEC, FÍS, ING ELÉC, TEC ESP, TELECOM, TRANSP, TRANSP MAR, VEH *instalación eléctrica* accumulator, lead-acid battery, secondary battery, storage battery; ~ **de acumuladores de aire y zinc** f TRANSP zinc-air storage battery; ~ **de acumuladores alcalinos** f ING ELÉC, TRANSP alkaline storage battery; ~ **de acumuladores de azufre-sodio** f ING ELÉC, TRANSP sodium sulfur storage battery (*AmE*), sodium sulphur storage battery (*BrE*); ~ **de acumuladores de hierro-níquel** f ING ELÉC, TRANSP nickel-iron storage battery; ~ **de acumuladores de litio-cloro** f ING ELÉC, TRANSP lithium-chlorine storage battery; ~ **de acumuladores de óxido-plata** f ING ELÉC, TRANSP silver-oxide storage battery; ~ **de acumuladores de plata-zinc** f ING ELÉC, TRANSP silver-zinc storage battery; ~ **de acumuladores plata-zinc** f TRANSP silver-zinc storage battery; ~ **de acumuladores secos** f ELEC, ING ELÉC dry storage battery; ~ **de acumuladores de zinc-níquel** f ING ELÉC, TRANSP nickel-zinc storage battery; ~ **alcalina** f ELEC, ING ELÉC alkaline battery, alkaline storage battery; ~ **de altas prestaciones** f TRANSP high-performance battery; ~ **de arranque** f ELEC *automotor*, ING ELÉC, TRANSP starter battery; ~ **de artillería de campaña** f D&A field battery;

~ **blindada** f TRANSP armored battery (*AmE*), armoured battery (*BrE*); ~ **de bocartes** f PROD stamp battery; ~ **de campaña** f D&A field battery; ~ **de cinturón** f CINEMAT battery belt; ~ **de compensación** f ING ELÉC, TRANSP buffer battery; ~ **compensadora** f ING ELÉC, TRANSP buffer battery; ~ **de condensadores** f ING ELÉC capacitor bank; ~ **de costa** f D&A coast battery; ~ **despolarizada por aire** f ING ELÉC air-depolarized battery; ~ **con electrólito** f TRANSP electrolyte battery; ~ **con electrólito anhidro** f TRANSP nonaqueous electrolyte battery; ~ **elevadora de voltaje** f TRANSP booster battery; ~ **de emergencia** f ELEC *acumulador* ING ELÉC, PROD backup battery, emergency battery; ~ **ferroníquel** f ING ELÉC, TEC ESP nickel-iron battery; ~ **de filtros** f ELECTRÓN filter bank; ~ **de fuel de sales fundidas a alta temperatura** f TRANSP high-temperature molten salts fuel battery; ~ **hermética** f VEH *instalación eléctrica* nonspill battery (*BrE*), sealed battery (*AmE*); ~ **hidráulica** f TRANSP AÉR hydraulic battery; ~ **hidroactivada** f ING ELÉC water-activated battery; ~ **interior** f ING ELÉC internal battery; ~ **interna** f ING ELÉC internal battery; ~ **de lámparas** f CINEMAT bank of lights; ~ **de lámparas múltiples** f CINEMAT multibroad; ~ **de lámparas simples** f CINEMAT single broad; ~ **de larga duración** f ING ELÉC long-life battery; ~ **metálica** f TRANSP metal-air battery; ~ **negativa** f TELECOM negative battery; ~ **de níquel-cadmio** f CINEMAT, ING ELÉC, TEC ESP, TRANSP nickel-cadmium battery; ~ **de níquel-hierro** f ING ELÉC, TEC ESP, TRANSP nickel-iron battery; ~ **nuclear** f CONST, NUCL nuclear battery; ~ **de óxido de plata** f ING ELÉC silver oxide battery; ~ **de pilas** f ING ELÉC primary battery; ~ **plana** f TELECOM flat battery; ~ **de plata** f ING ELÉC silver battery; ~ **en plata y cadmio** f ING ELÉC silver-cadmium battery; ~ **de plata y zinc** f ING ELÉC silver-zinc battery; ~ **PLD** f ING ELÉC DIP battery; ~ **de plomo** f FÍS, ING ELÉC, TRANSP, VEH lead-acid battery; ~ **de plomo-ácido** f FÍS, ING ELÉC, TRANSP, VEH lead-acid battery; ~ **de polarización** f ING ELÉC bias battery; ~ **positiva** f TELECOM positive battery; ~ **primaria de plata y zinc** f ING ELÉC silver-zinc primary battery; ~ **principal** f TRANSP AÉR main battery; ~ **recargable** f CINEMAT, FÍS, FOTO, ING ELÉC, TV rechargeable battery, recharging battery; ~ **refrigerante** f REFRIG cooling battery; ~ **regenerable** f TRANSP regenerative cell; ~ **de relés** f PROD bank of relays; ~ **de repuesto** f ING ELÉC battery backup; ~ **de reserva** f ING ELÉC battery backup, reserve battery; ~ **de reserva sulfuro-sódica** f ING ELÉC, TRANSP sodium sulfur storage battery (*AmE*), sodium sulphur storage battery (*BrE*); ~ **seca** f ING ELÉC dry battery, TRANSP nonaqueous electrolyte battery; ~ **seca de mercurio** f ING ELÉC mercury battery; ~ **secundaria** f TEC ESP secondary battery; ~ **de socorro** f ELEC *acumulador*, ING ELÉC, SEG emergency battery; ~ **solar** f ENERG RENOV, TV solar battery; ~ **térmica** f ING ELÉC, TERMO thermal battery; ~ **de urgencia** f ELEC, ING ELÉC emergency battery, TRANSP booster battery; ~ **de zinc** f ING ELÉC zinc battery

baterías: ~ **de enfriamiento y transportadores de rodillos** f pl ING MECÁ, REFRIG, TERMO cooling banks and roller conveyors; ~ **de refrigeración y transpor-**

tadores de rodillos *f pl* ING MECÁ, REFRIG, TERMO cooling banks and roller conveyors

batido *m* ACÚST beat, ALIMENT churn, churning, FÍS beat, PROD beating, TRANSP AÉR beat

batidor *m* CONST beater; ~ **del fieltro** *m* PAPEL thrasher

batidora: ~ **eléctrica** *f* ALIMENT, ELEC electric mixer; ~ **de paletas** *f* MINAS *preparación de minas* paddle mixer

batiduras: ~ **de forja** *f pl* PROD forge scale, hammer scale, nill; ~ **de hierro** *f pl* PROD iron scales; ~ **de laminado** *f pl* PROD mill scale; ~ **de martillado** *f pl* PROD hammer scale

batiente *m* AGUA miter post (*AmE*), miter sill (*AmE*), mitre post (*BrE*), mitre sill (*BrE*), ING MECÁ frame, rack; ~ **de concrete** *m* *AmL* (*cf plataforma de hormigón Esp*) HIDROL concrete apron; ~ **de hormigón** *m* *Esp* (*cf batiente de concrete AmL*) HIDROL *diques* concrete apron; ~ **de la máquina** *m* ING MECÁ engine frame; ~ **del motor** *m* ING MECÁ engine frame; ~ **de puerta** *m* AGUA miter post (*AmE*), miter sill (*AmE*), mitre post (*BrE*), mitre sill (*BrE*)

batimetría *f* ENERG RENOV, FÍS, OCEAN, TRANSP MAR bathymetry

batímetro *m* ENERG RENOV, FÍS, OCEAN, TRANSP MAR bathometer, bathymeter

batimiento *m* TRANSP AÉR flapping

batir *vt* ALIMENT churn, IMPR agitate, OCEAN *barcos de guerra* strike, PAPEL beat, TRANSP MAR *olas* lash

batiscafo *m* OCEAN bathyscaph

batisfera *f* OCEAN bathysphere

batitaquimetría *f* OCEAN bathytachymetry

batitermia *f* OCEAN bathythermy

batitermografía *f* OCEAN bathythermography

batitermógrafo *m* OCEAN bathythermograph

batolito *m* PETROL batholith; ~ **pequeño** *m* CARBÓN *geología* stock

batómetro *m* TRANSP MAR *navegación* sounder

Batoniense *adj* GEOL *estratigrafía* Bathonian

baudio *m* INFORM&PD baud

baulita *f* PETROL baulite

bauprés *m* TRANSP MAR bowsprit

bauxita *f* C&V, GEOL, MINERAL, PETROL, QUÍMICA bauxite

bayoneta *f* ING ELÉC bayonet; ~ **del fusible** *f* ING ELÉC fuse base

BC *abr* (*banda de conducción*) FÍS, FÍS RAD CB (*conduction band*)

Be *abr* (*berilio*) QUÍMICA Be (*beryllium*)

beaumontita *f* PETROL beaumontite

beauty: ~ **quark** *m* FÍS PART beauty quark (*b-quark*)

bebedero *m* AGRIC trough, P&C *parte de equipo* sprue bush, PROD running gate, trough, main gate, runner, sow, riser, jet; ~ **alimentador** *m* PROD *molde* pouring-gate, pouring-hole; ~ **de boquilla** *m* AGRIC nipple-drinker; ~ **de colada** *m* PROD feed head; ~ **de despumar** *m* PROD *fundería* skim gate

bedano *m* ING MECÁ parting tool

behénico *adj* QUÍMICA behenic

beidelita *f* MINERAL beidellite

belio *m* (*b*) ACÚST, FÍS, FÍS PART, FÍS RAD, INFORM&PD, ING ELÉC bel (*b*)

belonita *f* MINERAL belonite

bemol *m* ACÚST flat

benceno *m* QUÍMICA benzene, benzole

bencenoide *adj* QUÍMICA benzenoid

bencidina *f* QUÍMICA benzidine

bencilo *m* QUÍMICA benzil

bencina *f* QUÍMICA ligroin, benzine

Béndix® *m* AUTO, VEH Bendix starter®

beneficiación *f* C&V beneficiation

beneficio *m* PROD yield; ~ **bruto** *m* PROD gross margin, gross profit; ~ **de minerales** *m* MINAS *preparación de minas* ore dressing; ~ **sobre recursos** *m* TEC PETR return on asset (*ROA*)

bengala *f* D&A, TRANSP AÉR flare, TRANSP MAR flare, rocket; ~ **para casos de emergencia** *f* TRANSP MAR emergency rocket; ~ **de emergencia** *f* TRANSP MAR emergency rocket; ~ **de mano** *f* TRANSP MAR hand flare; ~ **con paracaídas** *f* D&A, TRANSP MAR *pirotecnia, señal de peligro* parachute flare; ~ **de señales** *f* TRANSP MAR *sistema Very* Very light; ~ **para señales de socorro** *f* TRANSP MAR distress flare

bentónico *adj* GEOL benthonic

bentonita *f* CARBÓN, DETERG, MINERAL, PETROL, TEC PETR *perforación, ingeniería de lodos* bentonite

benzaldehído *m* QUÍMICA benzaldehyde

benzaldoxima *f* QUÍMICA benzaldoxime

benzamida *f* QUÍMICA benzamide

benzanilida *f* QUÍMICA benzanilide

benzina *f* TEC PETR *refino* benzene

benzoato *m* QUÍMICA benzoate

benzofenona *f* QUÍMICA *perfumería* benzophenone

benzohidrol *m* QUÍMICA benzohydrol

benzoico *adj* QUÍMICA benzoic

benzoína *f* QUÍMICA benzoin

benzol *m* CARBÓN coal naphtha, coal-tar naphtha, QUÍMICA benzene, benzol, benzole

benzonaftol *m* QUÍMICA benzonaphthol

benzonitrilo *m* QUÍMICA benzonitrile

benzopireno *m* QUÍMICA benzopyrene

benzoquinona *f* QUÍMICA benzoquinone

bequerelio *m* (*Bq*) FÍS, FÍS RAD, METR, NUCL becquerel (*Bq*)

BER *abr* (*porcentaje de error de bit*) ELECTRÓN, INFORM&PD BER (*bit error rate*)

berbiquí *m* CARBÓN auger, CONST brace, ING MECÁ bit brace, bit stock, screw auger, MECÁ *carpintería* brace; ~ **de barrena** *m* MINAS *sondeos* stock; ~ **de clavija** *m* ING MECÁ plug center bit (*AmE*), plug centre bit (*BrE*); ~ **de pecho** *m* ING MECÁ breast drill; ~ **de trinquete** *m* CONST, ING MECÁ ratchet brace; ~ **de violín** *m* ING MECÁ bow drill

bergamaskita *f* MINERAL bergmannite

berilea *f* C&V beryllia

berilio *m* (*Be*) QUÍMICA beryllium (*Be*)

berilo *m* C&V, MINERAL beryl; ~ **dorado** *m* MINERAL golden beryl, heliodor

berilómetro *m* NUCL beryllium-prospecting meter

berilonita *f* MINERAL beryllonite

berlinga *f* TRANSP MAR *palos* spar

berlingado *m* MINAS *metalurgia* piling

berma *f* CARBÓN *construcción* berm, GEOL, OCEAN berm, beach berm

berquelio *m* (*Bk*) FÍS RAD, QUÍMICA berkelium (*Bk*)

Berriasiense *adj* GEOL Berriasian

berthierita *f* MINERAL berthierite

bertrandita *f* MINERAL bertrandite

berzelianita *f* MINERAL berzelianite

beta: ~ **amilasa** *f* ALIMENT beta-amylase; ~ **de aparejo** *f* ING MECÁ purchase fall; ~ **-celulosa** *f* PAPEL beta-cellulose

betatrón _m_ FÍS, FÍS RAD, NUCL betatron

betumen: ~ **elástico** _m_ AmL (_cf betún elástico Esp_) MINERAL elastic bitumen

betún _m_ CONST pitch, bitumen, ELEC _aislación_, GEOL, ING MECÁ, P&C, PETROL, PROC QUÍ, TEC PETR _compuestos del petróleo_ bitumen; ~ **elástico** _m_ Esp (_cf betumen elástico AmL_) MINERAL elastic bitumen

beudantina _f_ MINERAL beudantine

beudantita _f_ MINERAL beudantite

BF _abr_ (_baja frecuencia_) ELEC _onda_, ELECTRÓN, FÍS, FÍS ONDAS, FÍS RAD, ING ELÉC, TELECOM LF (_low frequency_)

BHA _abr_ (_butilhidroxianisol_) ALIMENT BHA (_butylated hydroxyanisole_)

BHP _abr_ (_potencia al freno en caballos de fuerza_) ING MECÁ, PROD, TRANSP BHP (_brake horsepower_)

BHT _abr_ (_butilhidroxitolueno_) ALIMENT BHT (_butylated hydroxytoluene_)

Bi _abr_ (_bismuto_) METAL, QUÍMICA Bi (_bismuth_)

biblioteca: ~ **de cintas** _f_ INFORM&PD tape library; ~ **de discos ópticos** _f_ INFORM&PD, ÓPT optical disk library; ~ **de fotografías** _f_ FOTO picture library; ~ **de procedimientos** _f_ INFORM&PD procedure library; ~ **de programas** _f_ INFORM&PD program library; ~ **de sistemas** _f_ IMPR, INFORM&PD systems library; ~ **de subrutinas** _f_ INFORM&PD subroutine library

bicapa _f_ REVEST _pintura para automóviles_ double layer

bicarbonato: ~ **sódico** _m_ ALIMENT, DETERG, QUÍMICA bicarbonate of soda, sodium bicarbonate; ~ **de sodio** _m_ ALIMENT, DETERG, QUÍMICA bicarbonate of soda, sodium bicarbonate

bichero _m_ TRANSP MAR _ayuda atraque y navegación_ boat-hook

bicicleta _f_ TRANSP bicycle; ~ **con motor** _f_ AUTO, TRANSP motor-assisted bicycle; ~ **plegable** _f_ TRANSP folding bicycle

bicíclico _adj_ QUÍMICA bicyclic

bicolor _adj_ COLOR bicolored (_AmE_), bicoloured (_BrE_)

bicristal _m_ METAL bicrystal

bidimensional _adj_ FÍS two-dimensional

bidireccional _adj_ ACÚST, INFORM&PD, PROD, TELECOM bidirectional; ~ **simultáneo** _adj_ ING ELÉC duplex

bidón _m_ CONTAM MAR drum, INSTAL HIDRÁUL _de indicador de máquina de vapor de agua_ can, drum, tin; ~ **de almacenamiento de elementos nuevos** _m_ NUCL new-element storage drum; ~ **de fibra** _m_ NUCL _para residuos radiactivos_ fiber drum (_AmE_), fibre drum (_BrE_); ~ **de gasolina de 20 litros** _m_ TRANSP jerry can; ~ **de metal** _m_ EMB metal drum; ~ **de reserva de gasolina** _m_ TRANSP jerry can; ~ **de residuos** _m_ NUCL waste drum

bieberita _f_ MINERAL bieberite

biela _f_ AUTO connecting rod, ING MECÁ connecting rod, pitman, INSTAL HIDRÁUL _de válvula_ rod, MECÁ, TRANSP MAR, VEH connecting rod; ~ **de acoplamiento** _f_ FERRO _vehículos_ coupling rod, MECÁ link, PROD side push rod; ~ **articulada** _f_ ING MECÁ link rod, knuckle-jointed connecting rod; ~ **de cabeza ahorquillada** _f_ ING MECÁ fork-end connecting rod; ~ **con cabeza de caja** _f_ AUTO, ING MECÁ, VEH connecting-rod with strap end; ~ **con cabeza de caja cerrada** _f_ AUTO, ING MECÁ, VEH connecting-rod with box end; ~ **con cabeza en horquilla** _f_ AUTO, ING MECÁ, VEH connecting-rod with fork end; ~ **de la cuchilla** _f_ AGRIC sickle pitman; ~ **del distribuidor** _f_ INSTAL HIDRÁUL slide rod; ~ **de empuje** _f_ AUTO, VEH

suspensión radius arm; ~ **de excéntrica** _f_ ING MECÁ, MECÁ, TRANSP AÉR cam follower, eccentric rod; ~ **excéntrica** _f_ AUTO, VEH offset connecting rod; ~ **de mando de la dirección** _f_ AUTO, VEH pitman arm; ~ **del pistón** _f_ AUTO, FERRO, ING MECÁ, PROD, VEH piston rod; ~ **de suspensión** _f_ FERRO _vehículos_ suspension rod; ~ **de suspensión de la corredera** _f_ ING MECÁ link hanger, link support

bieleta _f_ ING MECÁ link rod

bienes: ~ **de consumo** _m pl_ EMB consumer goods; ~ **salvados** _m pl_ TRANSP MAR salvage

bies _m_ CARBÓN bias

biestabilidad _f_ ELECTRÓN, FÍS, INFORM&PD, ING ELÉC, TELECOM bistability; ~ **óptica** _f_ ÓPT, TELECOM optical bistability

biestable[1] _adj_ ELECTRÓN, FÍS, INFORM&PD, ING ELÉC, TELECOM bistable

biestable[2] _m_ ELECTRÓN flip-flop; ~ **asíncrono** _m_ ELECTRÓN unclocked flip-flop; ~ **D** _m_ ELECTRÓN D-type flip-flop; ~ **ordenador-seguidor** _m_ ELECTRÓN master-slave flip-flop; ~ **sincrono activado por cambios de nivel** _m_ ELECTRÓN edge-triggered flip-flop; ~ **tipo T** _m_ ELECTRÓN T-flip-flop

bifásico _adj_ ELEC, ING ELÉC biphase, diphase, two-phase

bifenilo _m_ QUÍMICA biphenyl

bifilar _adj_ ELEC _devanado_, FÍS bifilar

bifocales: ~ **fundidos** _m pl_ C&V fused bifocals; ~ **sólidos** _m pl_ C&V solid bifocals

bifurcación _f_ AGUA branch pipe, CONST branch, fork, y-branch, ELEC _conexión_ tapping, ELECTRÓN branch, FÍS branching, INFORM&PD jump, _circuito_ branch, ING MECÁ parting, PROD _carreteras, ferrocarril_ junction, _ventilación minas_ split, TRANSP bifurcation; ~ **abierta** _f_ PROD branch open; ~ **incondicional** _f_ INFORM&PD unconditional jump; ~ **negativa** _f_ ELEC minus tapping

bifurcador _m_ ING MECÁ divider

bifurcar _vt_ CONST, INFORM&PD _programa_ branch

bifurcarse _vt_ CONST branch, branch off

big: **el ~ bang** _m_ FÍS TEC ESP _cosmogénesis_ the Big Bang

Big: **el ~ Bang** _m_ FÍS Big Bang

bigorneta _f_ CONST _metales_ stake; ~ **de banco** _f_ CONST bench stake

bigotera _f_ CONST bow dividers, bow compass

biguanida _f_ QUÍMICA biguanide

biimpelente _m_ TEC ESP bipropellant

bija _f_ ALIMENT _planta de origen americano intertropical_ annatto

bilirrubina _f_ QUÍMICA _sangre, urea_ bilirubin

billete: ~ **de avión** _m_ TRANSP AÉR airway bill; ~ **de pasaje aéreo** _m_ TRANSP AÉR airway bill; ~ **de recorrido** _m_ EMB swing ticket (_BrE_)

bimetal _m_ ELEC _en termostatos, termómetros_ bimetallic strip

bimolecular _adj_ QUÍMICA bimolecular

bimotor _adj_ ING MECÁ, TRANSP AÉR twin-engined

binario _adj_ INFORM&PD, MATEMÁT, METAL, QUÍMICA binary

binaural _adj_ ACÚST binaural

binocular _adj_ FÍS, INSTR, LAB binocular

binoculares: ~ **miniatura** _m pl_ INSTR miniature binoculars; ~ **prismáticos** _m pl_ INSTR prism binoculars, prismatic binoculars

binomio _m_ MATEMÁT binomial

bioblanqueo _m_ PAPEL biobleaching

bioclástico *adj* GEOL, PETROL bioclastic

bioclasto *m* GEOL, PETROL bioclast

biodegradabilidad *f* CONTAM, CONTAM MAR, DETERG, EMB, HIDROL, RECICL biodegradability

biodegradable *adj* CONTAM, CONTAM MAR, DETERG, EMB, HIDROL, RECICL biodegradable

biodegradación *f* CONTAM, CONTAM MAR, DETERG, EMB, HIDROL, RECICL biodegradation

bioesparita *f* PETROL biosparite

bioestratigrafía *f* GEOL biostratigraphy

biofiltro *m* CONTAM biofilter

biofísica *f* FÍS biophysics

biogás *m* GAS biogas

biogénico *adj* TEC PETR *generación de hidrocarburos* biogenic

bioihermo *m* GEOL, OCEAN reef knoll

biomasa *f* AGUA, ENERG RENOV, GAS, HIDROL *aguas residuales* biomass; **~ arbórea** *f* AGRIC biomass from trees

biomicrita *f* PETROL biomicrite

biorrotor *m* HIDROL *aguas residuales* biorotor

biosa *f* QUÍMICA biose

biosfera *f* CONTAM, METEO biosphere

biosíntesis *f* QUÍMICA biosynthesis

biostroma *m* GEOL biostrome

biotecnología *f* AGRIC, ALIMENT biotech, biotechnology

biotina *f* AGRIC, ALIMENT biotin

biotita *f* C&V, MINERAL, PETROL biotite

bioturbación *f* GEOL bioturbation

biovidrio *m* C&V bioglass

bióxido *m* QUÍMICA dioxide

biozona *f* GEOL *unidad bioestratigráfica* biozone

bipolar *adj* ELEC *motor*, ELECTRÓN, FÍS, INFORM&PD bipolar

biprisma: **~ de Fresnel** *m* FÍS Fresnel biprism

bipropelante *m* TEC ESP bipropellant

biquinario *adj* INFORM&PD, MATEMÁT biquinary

birrefringencia *f* CRISTAL, FÍS, FÍS RAD, ÓPT birefringence, double refraction

birrefringente *adj* CRISTAL, FÍS, FÍS RAD, ÓPT birefringent

bisagra *f* CONST strap hinge, hinge, *carpintería* butt, joint, ING MECÁ, INSTR, MECÁ hinge; **~ continua** *f* CONST garnet hinge; **~ cubrejuntas** *f* CONST butt-and-strap hinge; **~ embutida** *f* CONST flap hinge; **~ en forma de H** *f* CONST H hinge; **~ con granata transversal** *f* CONST cross-garnet hinge; **~ de guía cruzada** *f* CONST crosstailed hinge; **~ de guía transversal** *f* CONST crosstail hinge; **~ con pasador con pomo** *f* CONST hinge with knobbed pin; **~ de pasador suelto** *f* CONST loose-pin hinge; **~ de la puerta** *f* CONST, VEH door hinge; **~ en T** *f* CONST T-hinge; **~ de tope** *f* CONST butt hinge

bisecado *adj* GEOM bisecting

bisecar *vt* GEOM bisect

bisección *f* GEOM bisection

bisectado *adj* GEOM bisecting

bisectar *vt* GEOM bisect

bisector *m* GEOM bisector

bisel *m* CONST chamfer, wane, IMPR miter (*AmE*), mitre (*BrE*), ING MECÁ bevel edge, edge, INSTR, MECÁ, PROD bevel, bezel; **~ agudo** *m* C&V steep bevel; **~ anterior** *m* ING MECÁ leading chamfer; **~ de bajo perfil** *m* PROD low profile bezel; **~ desviador** *m* TEC PETR *perforación* whipstock; **~ escalonado** *m* CONST

stopped chamfer; **~ con forma** *m* C&V shaped bevel; **~ graduado** *m* C&V taper bevel; **~ de protección** *m* ING MECÁ protecting chamfer; **~ Vauxhall** *m* C&V Vauxhall bevel

biselación *f* GEN beveling (*AmE*), bevelling (*BrE*)

biselado *m* CONST beveling (*AmE*), bevelling (*BrE*), chamfering, INSTR beveling (*AmE*), bevelling (*BrE*), PROD *soldadura* scarfing; **~ por ambos lados** *m* C&V miter bevel both sides (*AmE*), mitre bevel both sides (*BrE*); **~ a inglete** *m* C&V miter bevel (*AmE*), mitre bevel (*BrE*)

biselar *vt* GEN bevel, chamfer

bisfenol: **~ A** *m* P&C *compuesto químico, materia prima* bisphenol A

bisilicato *m* QUÍMICA bisilicate

bismita *f* MINERAL bismite

bismutato *m* QUÍMICA bismuthate

bismutina *f* MINERAL bismuthine

bismutinita *f* MINERAL bismuthinite

bismutita *f* MINERAL bismutite

bismuto *m* (*Bi*) METAL, QUÍMICA bismuth (*Bi*)

bisolita *f* MINERAL byssolite

bisturí *m* LAB *disección* scalpel; **~ térmico** *m* TERMO thermic lance

bisulfato *m* QUÍMICA bisulfate (*AmE*), bisulphate (*BrE*)

bisulfito *m* QUÍMICA bisulfite (*AmE*), bisulphite (*BrE*)

bisulfuro *m* QUÍMICA bisulfide (*AmE*), bisulphide (*BrE*), disulfide (*AmE*), disulphide (*BrE*), hydrosulfide (*AmE*), hydrosulphide (*BrE*); **~ de carbono** *m* QUÍMICA carbon bisulfide (*AmE*), carbon bisulphide (*BrE*)

bit[1] *abr* ELECTRÓN, IMPR, INFORM&PD, TELECOM (*cifra binaria, dígito binario*) bit (*binary digit*)

bit[2] *m* ELECTRÓN, IMPR, INFORM&PD, PETROL, TEC PETR, TELECOM bit; **~ adicional** *m* TELECOM more bit; **~ de alarma del rango individual de herramientas** *m* PROD individual tool range alarm bit; **~ de arranque** *m* INFORM&PD start bit; **~ breaker** *m* TEC PETR *perforación* bit breaker; **~ de cabeza** *m* TELECOM overhead bit; **~ de cierre del temporizador** *m* PROD timer-off done bit; **~ de comienzo** *m* INFORM&PD start bit; **~ de control** *m* INFORM&PD control bit; **~ de cronometraje del temporizador** *m* PROD timer-timing bit; **~ de detenida** *m* INFORM&PD stop bit; **~ de entrada ejecutada** *m* INFORM&PD, PROD input satisfied status bit; **~ de entrada forzada** *m* INFORM&PD, PROD forced-input bit; **~ de estado del temporizador** *m* INFORM&PD, PROD timer-status bit; **~ de fin de ejecución** *m* INFORM&PD, PROD step completion bit; **~ de fin de instrucciones** *m* INFORM&PD, PROD step completion bit; **~ habilitador de decrecimiento** *m* INFORM&PD, PROD down-enable bit; **~ indicador** *m* INFORM&PD, PROD flag bit; **~ de información** *m* INFORM&PD information bit; **~ inicial** *m* TELECOM overhead bit; **~ a la izquierda** *m* INFORM&PD high-order bit; **~ M** *m* INFORM&PD, TELECOM M bit; **~ más significativo** *m* (*BMS*) INFORM&PD, PROD, TELECOM most significant bit (*MSB*); **~ de mayor significación** *m* (*BMS*) INFORM&PD, PROD, TELECOM most significant bit (*MSB*); **~ de memoria** *m* INFORM&PD, PROD store bit; **~ de menor significación** *m* (*BMS*) INFORM&PD, PROD, TELECOM least significant bit (*LSB*); **~ menos significativo** *m* (*BMS*) INFORM&PD, PROD, TELECOM least significant bit (*LSB*); **~ de orden superior** *m* INFORM&PD high-order bit; **~ de parada** *m*

INFORM&PD stop bit; ~ **de paridad** *m* INFORM&PD parity bit; ~ **de ruptura a nivel superior** *m* INFORM&PD, PROD major fault bit; ~ **señalador** *m* INFORM&PD flag bit; ~ **de servicio** *m* INFORM&PD service bit; ~ **de signo** *m* INFORM&PD sign bit; ~ **de temporizador** *m* PROD timer bit, timer clock bit; ~ **de terminación del recuento** *m* PROD count-up done bit; ~ **de verificación** *m* ELECTRÓN check bit

bita *f* ING MECÁ cleat, TRANSP MAR bitt; ~ **de amarre** *f* TRANSP MAR mooring bitt; ~ **de los linguetes** *f* ING MECÁ pawl bitt

bitácora *f* TRANSP MAR binnacle

bitadora *f* TRANSP MAR *del cable de fondeo* range

bitartrato: ~ **potásico** *m* QUÍMICA cream of tartar; ~ **de potasio** *m* QUÍMICA tartar

bitoque *m* ALIMENT *fermentación* bung

bitownita *f* MINERAL bytownite

bits: ~ **por pulgada** *m pl* (*BBP*) INFORM&PD bits per inch (*BPI*); ~ **por segundo** *m pl* (*BPS*) INFORM&PD bits per second (*BPS*)

bitumen *m* GEN *compuestos del petróleo* bitumen; ~ **de granulación** *m* PROC QUÍ granulation pitch

bituminoso *adj* GEN bituminous

bivalencia *f* INFORM&PD, QUÍMICA bivalence

bivalente *adj* INFORM&PD, QUÍMICA bivalent

biveleta *f* METEO bidirectional wind vane

bixina *f* ALIMENT bixin

bizcocho: ~ **de porcelana de color de caña** *m* C&V bamboo

Bk *abr* (*berquelio*) FÍS RAD, QUÍMICA Bk (*berkelium*)

blancarte *m* AmL (*cf ripios Esp*) MINAS attle

blanco[1]: ~ **y negro** *adj* (*B y N*) CINEMAT, FOTO, IMPR, TV black and white (*B&W*)

blanco[2]: *m* INFORM&PD blank, TEC ESP rover, target, TV target; ~ **colocado a una distancia alejada** *m* TEC ESP *arquería* rover; ~ **de emisión secundaria** *m* ELECTRÓN secondary emission target; ~ **de España** *m* QUÍMICA whiting; ~ **hueco** *m* NUCL hollow target; ~ **de plomo** *m* QUÍMICA ceruse; ~ **1ª bis** *m* PAPEL *tipo de papel de recuperación* mixed white shavings; ~ **de prueba** *m* D&A dummy target; ~ **de referencia** *m* TV reference white; ~ **reticular** *m* FÍS RAD wire-mesh target; ~ **2ª con pasta mecánica** *m* PAPEL *papelote* mixed light-coloured printer shavings; ~ **3ª con alcance** *m* PAPEL *papel de recuperación* bookbinder's shavings; ~ **terrestre** *m* D&A ground target; ~ **de yeso** *m* QUÍMICA whiting

blancos *m pl* CONTAM MAR *productos de hidrocarburos* white product; ~ **sangrientes** *m pl* TV bleeding whites

blancura *f* IMPR, PAPEL brightness

blandura *f* IMPR, PAPEL softness

blanqueado *m* ALIMENT *panadería, repostería y molienda* bleaching, CONST whitewash, whitewashing, QUÍMICA whitewash, bleaching, REVEST *con cal* lime washing, whitewashing

blanqueador *m* DETERG, PAPEL bleacher, bleaching agent; ~ **clorado** *m* DETERG chlorine bleaching agent

blanqueante: ~ **fluorescente** *m* PAPEL fluorescent whitening; ~ **óptico** *m* PAPEL optical brightener

blanquear *vt* ALIMENT bleach, blanch, CINEMAT bleach, COLOR whitewash, FOTO bleach out, IMPR, PAPEL, QUÍMICA, TEXTIL bleach

blanqueo *m* DETERG, PAPEL bleaching; ~ **sin compuesto clorado** *m* PAPEL *pasta* TCF bleaching;

~ **óptico** *m* DETERG optical bleaching; ~ **con polvos de blanqueo** *m* DETERG chloride of lime bleaching

blastomilonita *f* GEOL blastomylonite

blenda *f* MINAS *mineral* jack, MINERAL, QUÍMICA *sulfuro de cinc* blende, sphalerite, zinc blende; ~ **cadmífera** *f* MINERAL cadmium blende

blendoso *adj* PROD blendous, blendy

blindado *adj* ELEC *cable*, ING ELÉC armored (*AmE*), armour-clad (*BrE*), armoured (*BrE*), armor-clad (*AmE*), ING MECÁ, MECÁ metal-clad; ~ **con hoja de cobre** *adj* ELEC *cable*, REVEST copper-clad

blindaje *m* D&A armor (*AmE*), armour (*BrE*), ELEC *cable* shielding, armor (*AmE*), screening, armour (*BrE*), shield, INFORM&PD shield, ING ELÉC armor (*AmE*), armour (*BrE*), LAB *seguridad* shield, MINAS *pozos* piling, NUCL shield, PROD *electricidad* shield, metal coating, SEG *espectrómetro de neutrones* shielding, TV screening, shield, screen; ~ **antimagnético** *m* ELEC *campo* magnetic screening; ~ **antimisiles** *m* NUCL guard vessel; ~ **axial** *m* NUCL axial shield; ~ **biológico** *m* NUCL biological shield; ~ **del cable** *m* ELEC, ING ELÉC *conductor* cable screen; ~ **electromagnético** *m* ELEC electromagnetic shielding; ~ **de grafito** *m* NUCL graphite shielding; ~ **insonoro para la cámara** *m* CINEMAT blimp; ~ **interno** *m* ELECTRÓN internal shield; ~ **magnético** *m* ING ELÉC magnetic shielding; ~ **del núcleo** *m* ELEC *conductor del cable* core screen; ~ **protector** *m* D&A protective armor (*AmE*), protective armour (*BrE*); ~ **térmico** *m* SEG, TEC ESP, TERMO heat shield, thermal shield; ~ **de trenza de cobre** *m* ELEC *cable*, ELECTRÓN, PROD copper-braid shielding

blindar *vt* CINEMAT blimp, MECÁ plate, MINAS line, line with metal, tub, QUÍMICA screen

blip[1] *m* TRANSP MAR *señal radioeléctrica* bleep

blip[2] *vi* TRANSP MAR bleep

blizzard *f* METEO blizzard

blocaje: ~ **de aire** *m* TRANSP AÉR airlock; ~ **antirráfaga** *m* TEC ESP, TRANSP AÉR, TRANSP MAR gust lock; ~ **del control del tren de aterrizaje del morro** *m* TRANSP AÉR nose gear steer lock; ~ **de la superficie de control** *m* TRANSP AÉR control-surface locking; ~ **del tren de aterrizaje** *m* TRANSP AÉR landing gear uplock; ~ **de zona de Fresnel** *m* TELECOM Fresnel zone blockage

blocao: ~ **de cemento** *m* D&A *para ametralladoras* pillbox

block: ~ **para hacer caras** *m* C&V facing block

blödita *f* MINERAL blödite

bloedita *f* MINERAL bloedite

blooming *m* FOTO blooming

bloque *m* AUTO cylinder block, CONST block, ELECTRÓN *de filtros* bank, INFORM&PD, PROD, TEC PETR *licencias de explotación* block; ~ **de alimentación** *m* ING ELÉC power pack; ~ **de alto nivel** *m* ELECTRÓN high-logic level; ~ **de anclaje** *m* CONST anchorage block; ~ **de bornas** *m* PROD terminal block; ~ **calibrador** *m* MECÁ, METR *carburo* gage block (*AmE*), gauge block (*BrE*); ~ **del cierre** *m* D&A breech block; ~ **de cilindros** *m* AUTO, VEH cylinder block; ~ **de cimentación** *m* CONST foundation block; ~ **de circuito integrado** *m* ELECTRÓN integrated circuit package; ~ **de concreto** *m* AmL (*cf bloque de hormigón Esp*) C&V, CONST concrete block; ~ **de conectores** *m* ELEC terminal block; ~ **de conexiones** *m* ELEC terminal block; ~ **de contacto**

m ING ELÉC contact block; ~ **de contacto moldeado** *m* PROD molded contact block (*AmE*), moulded contact block (*BrE*); ~ **de corona** *m* TEC PETR crown block; ~ **corrido** *m* GEOL overthrust block; ~ **de datos** *m* INFORM&PD data tablet; ~ **deslizado** *m* GEOL landslide block; ~ **desnatador** *m* C&V skimmer block; ~ **de dislocación ascendente** *m* GEOL upfaulted block; ~ **de dislocación descendente** *m* GEOL downfaulted block; ~ **de enlace ascendente** *m* TELECOM uplink block; ~ **errático** *m* CARBÓN boulder, GEOL *sedimento glacial* erratic block; ~ **erróneo** *m* TELECOM erroneous block; ~ **de erro-res consecutivos** *m* TELECOM consecutive error block (*CEB*); ~ **de fresado** *m* IMPR routing block; ~ **de fundación del mortero** *m* PROD mortar block; ~ **de grafito** *m* NUCL graphite block; ~ **guía** *m* ING MECÁ guide block; ~ **de hielo** *m* REFRIG ice block, ice cake; ~ **de hormigón** *m* *Esp* (*cf bloque de concreto AmL*) C&V, CONST concrete block; ~ **hueco de vidrio** *m* C&V hollow glass block; ~ **hundido** *m* GEOL downfaulted side; ~ **inclinado** *m* GEOL tilted block; ~ **del izador** *m* TRANSP AÉR hoisting block; ~ **en L** *m* C&V L-block; ~ **de lengüetas de conexión** *m* ELEC terminal block; ~ **de lógica** *m* INFORM&PD logical block; ~ **lógico** *m* INFORM&PD logical block; ~ **lógico de láminas** *m* PROD logic reed block; ~ **de longitud fija** *m* INFORM&PD fixed-length block; ~ **macizo** *m* PROD solid block; ~ **de media tinta** *m* FOTO halftone block; ~ **de medición** *m* ING MECÁ measuring block; ~ **de mensajes CRC** *m* TELECOM CRC message bloc (*CMB*); ~ **metálico con ranura en V** *m* ING MECÁ, METR V-block; ~ **de mineral** *m* CARBÓN boulder; ~ **del motor** *m* AUTO, ING MECÁ, MECÁ, VEH cylinder block, engine block; ~ **motor** *m* ING MECÁ, PROD power unit; ~ **móvil** *m* TEC PETR *perforación* traveling block (*AmE*), travelling block (*BrE*); ~ **de organigrama** *m* INFORM&PD flowchart block; ~ **de osciladores** *m* ELECTRÓN bank of oscillators; ~ **de palancas acodadas** *m* ING MECÁ bellcrank block; ~ **pequeño** *m* CONST *de piedra* chip; ~ **de piso** *m* C&V paving block; ~ **de pistones** *m* IMPR air pin block; ~ **de polea** *m* ING MECÁ sheave block, sheave pulley block; ~ **de posicionamiento** *m* ING MECÁ positioning block; ~ **para pulir mármol** *m* CARBÓN float; ~ **de recalcar** *m* PROD swage block; ~ **refractario del quemador** *m* C&V burner block; ~ **de sal** *m* ALIMENT salt cake; ~ **separador** *m* CONST spacer block; ~ **de sintonía** *m* TELECOM tuner; ~ **de taponamiento de pozo** *m* GAS well-plugging block; ~ **de tensión** *m* ING MECÁ tension block; ~ **terminal** *m* ELEC *conexión*, ING ELÉC, PROD, VEH *instalación eléctrica* terminal block; ~ **de terminales** *m* ELEC *conexión*, ING ELÉC, PROD, VEH terminal block, terminal barrier strip; ~ **de transmisión** *m* INFORM&PD transmission block; ~ **de tres poleas** *m* ING MECÁ three-sheave block; ~ **de vaciado** *m* C&V tapout block; ~ **viajero** *m* TEC PETR *perforación* traveling block (*AmE*), travelling block (*BrE*); ~ **de vidrio** *m* C&V glass block

bloqueado[1]: ~ **por el hielo** *adj* TRANSP MAR icebound
bloqueado[2] *m* INFORM&PD lockout, blocking
bloqueador *m* ING MECÁ detent; ~ **de ruedas** *m* AUTO *mantenimiento* wheel clamp
bloquear *vt* CINEMAT lock, CONST *carretera* close, INFORM&PD lock, ING MECÁ block, INSTAL HIDRÁUL

máquina de vapor, motor de vapor bar, TRANSP MAR *puerto* block, blockade, *timón* lash
bloqueo *m* AGUA blocking, AUTO choking, CARBÓN blocking, CONST closing, ELECTRÓN blocking state, *oscilador* lock, FERRO *de ruedas* locking, IMPR lock, INFORM&PD *de canal* interlock, deadlock, lock, lock-out, ING ELÉC sticking, interlock, blocking, TELECOM blocking, TRANSP AÉR choking, TRANSP MAR *puerto* blockade, TV blocking; ~ **automático** *m* CINEMAT, TV automatic lock; ~ **de banda magnética individual** *m* CINEMAT, TV sepmag lock; ~ **de la caja** *m* CINEMAT housing lock; ~ **de chimenea** *m* MINAS chute hang-up; ~ **del circuito** *m* ELECTRÓN loop lock; ~ **de la columna de dirección** *m* AUTO, VEH steering-column lock; ~ **del diferencial** *m* AUTO, VEH *transmisión* differential lock; ~ **exterior** *m* TELECOM external blocking; ~ **de fase** *m* ELECTRÓN phase locking; ~ **fatal** *m* INFORM&PD deadly embrace; ~ **inferior** *m* TRANSP AÉR down lock; ~ **interno** *m* TELECOM internal blocking; ~ **mecánico** *m* ING MECÁ mechanical locking; ~ **momentáneo** *m* TELECOM freeze-out; ~ **de puertas** *m* FERRO *vehículos* door blocking; ~ **y rebobinado automático** *m* FOTO auto-stop and rewind; ~ **de sincronización** *m* CINEMAT, TV synchrolock; ~ **del tacómetro** *m* AUTO, CINEMAT, FÍS, TRANSP, TV, VEH tachometer lock; ~ **del teclado** *m* INFORM&PD keyboard lockout; ~ **de transmisión** *m* TELECOM blackout
bloques: ~ **lógicos** *m pl* ELECTRÓN logic array; ~ **no separables** *m pl* TELECOM nondroppable blocks; ~ **separables optativamente** *m pl* TELECOM optionally droppable blocks
blusa: ~ **para polvo** *f* SEG dust blouse
BMS *abr* INFORM&PD, PROD, TELECOM (*bit de mayor significación, bit más significativo*) MSB (*most significant bit*); (*bit de menor significación, bit menos significativo*) LSB (*least significant bit*)
bobierrita *f* MINERAL bobierrite
bobina *f* ACÚST spool, C&V, CINEMAT bobbin, spool, CONST, ELEC coil, EMB reel, FÍS coil, IMPR *cartón, papel* reel, ING ELÉC, ING MECÁ, MECÁ coil, PAPEL reel, PROD spool, *de alambre* coil, TEXTIL bobbin, TV reel, bobbin, VEH coil; ~ **accionada en la superficie** *f* PAPEL surface-drive reel; ~ **de acoplamiento** *f* ELEC coupling coil, mutual inductor, FÍS, ING ELÉC mutual inductor; ~ **adyacente** *f* ELEC adjacent coil; ~ **de alimentación** *f* CINEMAT feed reel, feed spool; ~ **de amortiguamiento** *f* ELEC, ING ELÉC damping coil; ~ **de apagado de arco** *f* ELEC *relé*, ING ELÉC arc suppression coil; ~ **de apagado por rejillas** *f* ELEC *relé*, ING ELÉC arc suppression coil; ~ **apagadora del arco** *f* ELEC *relé*, ING ELÉC arc suppression coil; ~ **de apertura-cierre** *f* ELEC make-and-break coil; ~ **arro-llada sobre mandril** *f* ELEC form-wound coil; ~ **de arrollamiento al azar** *f* ELEC random winding; ~ **de arrollamiento desordenado** *f* ELEC random winding; ~ **para arrollar cable eléctrico** *f* ELEC cable drum; ~ **de autoinducción** *f* ING ELÉC inductance, choke; ~ **de bolsas de polietileno** *f* EMB polyethylene bags on the roll; ~ **de bucle** *f* ELEC loop coil; ~ **de cable** *f* EMB cable reel, ING ELÉC cable drum; ~ **cabo** *f* PAPEL side run; ~ **de campo** *f* ELEC *de máquina*, ING ELÉC field coil; ~ **de carga** *f* CINEMAT loading spool, ELEC, FÍS loading coil; ~ **chata** *f* ELEC slab coil; ~ **de choque** *f* ELEC *inductor* choke, choke coil, ING ELÉC choke coil, choke, PETROL thumper,

TELECOM choke; ~ **de cierre** *f* PROD lockout coil; ~ **de clavija** *f* ING ELÉC plug-in coil; ~ **colectora** *f* PROD bus choke; ~ **compensadora** *f* ING ELÉC bucking coil; ~ **de concentración** *f* ING ELÉC focusing coil; ~ **conformada** *f* ELEC form-wound coil; ~ **correctora de zumbido** *f* ING ELÉC, TV humbucking coil; ~ **deflectora** *f* FÍS, NUCL deflecting coil; ~ **de deflexión vertical** *f* ING ELÉC vertical deflection coil; ~ **en derivación** *f* ING ELÉC tapped coil; ~ **desmagnetizante** *f* TV degaussing coil; ~ **de desviación** *f* ELECTRÓN deflection yoke, ING ELÉC yoke coil; ~ **de desviación electromagnética** *f* ING ELÉC deflection coil; ~ **de desviación horizontal** *f* ING ELÉC horizontal deflection coil; ~ **devanada** *f* ELEC wire-wound coil; ~ **diferencial** *f* ELEC differential coil; ~ **de disparo** *f* ELEC trip coil; ~ **de disyuntor** *f* ING ELÉC relay coil; ~ **eléctrica** *f* ELEC, ING ELÉC electric coil; ~ **de electroimán** *f* ELEC, ING ELÉC magnet coil; ~ **de encendido** *f* AUTO, ING ELÉC, TRANSP AÉR, VEH ignition coil; ~ **de encuadre de la imagen** *f* TELECOM picture-control coil; ~ **de enfoque** *f* CINEMAT, TV focusing coil; ~ **para enrollar cables** *f* EMB cable drum; ~ **con entrehierro** *f* ELEC air-gap coil; ~ **de equilibrio** *f* ELEC balance coil; ~ **con espacio de aire** *f* ELEC air-gap coil; ~ **del estator** *f* ING ELÉC stator coil; ~ **estropeada en los bordes** *f* PAPEL starred roll; ~ **de excitación** *f* ELEC, ING ELÉC magnetizing coil, ING ELÉC *relé* drive coil; ~ **de exploración** *f* CINEMAT, TV scanning coil; ~ **exploradora** *f* ELEC exploring coil, search coil, ING ELÉC search coil, *electrotecnia* flip coil; ~ **extensible** *f* ING ELÉC extension reel; ~ **de extinción** *f* ING ELÉC blowout coil; ~ **extraplana** *f* ELEC slab coil; ~ **fantasma** *f* ING ELÉC phantom coil; ~ **fija** *f* ELEC fixed coil; ~ **de final de máquina** *f* PAPEL batch roll; ~ **de fondo de cesta** *f* ING ELÉC basket coil; ~ **de fricción** *f* PAPEL friction reel; ~ **de Helmholtz** *f* FÍS Helmholtz coil; ~ **de imagen** *f* CINEMAT, TV picture reel; ~ **imanante** *f* ELEC, ING ELÉC magnetizing coil; ~ **de impedancia** *f* ELEC, ING ELÉC *inductor* choke coil; ~ **de impedancia de filtrado** *f* ELEC, ING ELÉC smoothing choke; ~ **impregnada** *f* ING ELÉC impregnated coil; ~ **de inducción** *f* AUTO, ELEC, ING ELÉC, VEH *sistema de encendido* induction coil, spark coil; ~ **de inducción cebada con acumulador para el arranque** *f Esp* (*cf adujada de refuerzo AmL*) AUTO, TRANSP, TRANSP AÉR, VEH booster coil; ~ **de inducción telefónica** *f* ING ELÉC, TELECOM telephone induction coil; ~ **del inducido** *f* ELEC, ING ELÉC armature coil; ~ **de inductancia** *f* ELEC, ING ELÉC inductance coil, inductor; ~ **de inductancia ajustable** *f* ELEC, ING ELÉC adjustable inductance coil; ~ **de inductancia con devanado** *f* ELEC, ING ELÉC pancake coil; ~ **inductora del arrancador** *f* AUTO starter field coil; ~ **inductora de sonido** *f* CINEMAT, TV humbucking coil; ~ **inductora de zumbido** *f* CINEMAT, TV hum-bucking coil; ~ **de interrupción** *f* ELEC trip coil; ~ **jumbo** *f* PAPEL jumbo roll; ~ **madre** *f* PAPEL jumbo roll; ~ **magnetizante** *f* ELEC, ING ELÉC magnetizing coil; ~ **con mandril** *f* PAPEL center-wind reel (*AmE*), centre-wind reel (*BrE*); ~ **móvil** *f* ELEC, ING ELÉC moving coil; ~ **muy estrecha** *f* PAPEL coil; ~ **no inductiva** *f* CONST, ELEC noninductive winding; ~ **en oposición** *f* ELEC, ING ELÉC bucking coil; ~ **osciladora** *f* CINEMAT, ING ELÉC oscillator coil;

~ **de panal** *f* ELEC, ING ELÉC lattice-wound coil; ~ **de papel** *f* PAPEL paper roll; ~ **de papel de mitad de anchura** *f* IMPR, PAPEL dinky reel; ~ **de paralelo** *f* ING ELÉC shunt coil; ~ **de paso entero** *f* ELEC full-pitch coil; ~ **perezosa** *f* ELEC *motor* lazy coil; ~ **perforada** *f* EMB perforated on the reel; ~ **en pilas** *f* ELEC pile-wound coil; ~ **plana** *f* ELEC, ING ELÉC pancake coil, slab coil; ~ **de prueba** *f* CINEMAT test reel, ING ELÉC search coil; ~ **de reacción** *f* AUTO, VEH *instalación eléctrica* choke coil; ~ **de reactancia** *f* ELEC, FÍS, ING ELÉC choke coil, impedance coil, reactance coil; ~ **de reactancia de absorción** *f* ELEC, ING ELÉC smoothing choke; ~ **de reactancia de hierro saturado** *f* ELEC, ING ELÉC swinging choke; ~ **de reactancia limitadora de corriente** *f* ELEC, ING ELÉC current-limiting reactor; ~ **de reactancia de línea** *f* ELEC, ING ELÉC line-choking coil; ~ **de realimentación** *f* ELEC, ING ELÉC feedback coil; ~ **de recogida** *f* CINEMAT take-up; ~ **reemplazable** *f* ING ELÉC plug-in coil; ~ **de regulación de la imagen** *f* CINEMAT, TV image-control coil; ~ **reguladora de la exploración** *f* TV scanning yoke; ~ **de relé** *f* PROD relay coil; ~ **de relevador** *f* ELEC, TELECOM relay winding; ~ **de repetición** *f* ING ELÉC repeating coil; ~ **de resistencia** *f* ELEC, ING ELÉC resistance coil; ~ **de retardo** *f* ELEC retardation coil; ~ **de retención** *f* ING ELÉC holding coil; ~ **de revelado** *f AmL* (*cf espiral de relevado Esp*) FOTO developing reel (*AmE*), developing spiral (*BrE*), developing spool (*AmE*); ~ **de RF** *f* ELEC, ELECTRÓN, ING ELÉC, TELECOM RF coil; ~ **de Ruhmkorff** *f* ELEC induction coil; ~ **secundaria** *f* ING ELÉC secondary coil; ~ **en serie** *f* ELEC series coil; ~ **de simetría** *f* ELEC balance coil; ~ **de sintonía** *f* D&A tuning coil; ~ **de sintonización** *f* ELECTRÓN tuning coil; ~ **de solenoide** *f* REFRIG solenoid coil; ~ **superconductora** *f* TRANSP superconducting coil; ~ **del tanque de unidades múltiples** *f AmL* (*cf espiral del tanque de unidades múltiples Esp*) FOTO multiunit tank spiral; ~ **de telar** *f* C&V pirn; ~ **Tesla** *f* ING ELÉC Tesla coil; ~ **toroidal** *f* ELEC, NUCL toroidal coil; ~ **de varias capas** *f* ELEC multilayer coil

bobinado *m AmL* (*cf rebobinado Esp*) CONST, ELEC, FÍS winding, FOTO spooling, ING ELÉC, ING MECÁ winding, TEXTIL wind-up, VEH *generador* winding; ~ **apretado** *m AmL* (*cf rebobinado apretado Esp*) CINEMAT tight winder, FOTO tight spooling; ~ **automático** *m AmL* (*cf rebobinado automático Esp*) CINEMAT, FOTO auto winding; ~ **B** *m AmL* (*cf rebobinado B Esp*) CINEMAT, TV B-wind; ~ **cilíndrico** *m* ELEC cylindrical winding; ~ **de derivación** *m* ELEC, ING ELÉC bias winding; ~ **diametral** *m* ELEC, ING ELÉC diametrical winding; ~ **en disco aplanado** *m* ING ELÉC pie winding; ~ **sobre mandril** *m* PAPEL center winding (*AmE*), centre winding (*BrE*); ~ **de motores** *m* TEC ESP wiring; ~ **en pilas** *m* ELEC pile-wound coil; ~ **primario** *m* AUTO, FÍS primary winding; ~ **de relé** *m* ELEC, TELECOM relay winding; ~ **secundario** *m* AUTO, FÍS secondary winding

bobinador *m* INFORM&PD spooler, PROD coiler

bobinadora *f* CINEMAT, EMB winding machine

bobinados: ~ **A y B** *m pl* CINEMAT A and B windings

bobinando *m* INFORM&PD spooling

bobinar *vt* CINEMAT spool, wind, FÍS wind, IMPR, ING ELÉC reel, ING MECÁ wind, PAPEL reel, wind

boca *f* ACÚST mouth, CARBÓN *ríos, desfiladeros* entry,

CONST mouth, D&A *de un arma de fuego* muzzle, ING MECÁ orifice, nozzle, nose, jaw, *micrómetro* anvil, LAB *de matraz* neck, MECÁ *frascos* neck, jaw, MINAS *hornos* mouth, *pozos* top hole, *instrumento de medir* anvil, PROD *martillo* face, *hornos altos* mouth, pane, nozzle, *de tenazas* nose, TELECOM *de conducto* mouthpiece, TRANSP MAR *desembocadura de río, estrecho* mouth; ~ **acampanada** *f* ING MECÁ, TEC ESP bell mouth; ~ **de acceso** *f* ING MECÁ manhole; ~ **de agua** *f* AGUA plug; ~ **de agujero** *f* PROD stop gap; ~ **de alcantarilla** *f* AGUA, CONTAM, RECICL drain; ~ **artificial** *f* ACÚST artificial mouth; ~ **de barrena** *f Esp* (*cf barrena para martillo perforador AmL*) MINAS *sondeos* drill bit; ~ **bombeada** *f* ING MECÁ ball peen; ~ **de carbonera** *f* CARBÓN, MINAS coal chute; ~ **contraincendios** *f* AGUA, CONST, SEG, TERMO, TRANSP MAR fire hydrant (*BrE*), fireplug (*AmE*); ~ **del crisol** *f* C&V pot mouth; ~ **de entrada** *f* HIDROL, MECÁ inlet; ~ **del hogar** *f* PROD fire hole; ~ **de hombre** *f* NUCL manhole; ~ **del horno** *f* C&V gathering hole; ~ **de incendios** *f* AGUA, CONST, SEG, TERMO, TRANSP MAR fire hydrant (*BrE*), fireplug (*AmE*); ~ **de inspección** *f* ING MECÁ, MINAS, TEC ESP manhole; ~ **de lobo** *f* METEO, TRANSP MAR catspaw; ~ **de un pico de cangrejo** *f* TRANSP MAR *veleros* horn; ~ **de pozo** *f* CARBÓN bank, ENERG RENOV, GAS wellhead; ~ **del pozo** *f* MINAS pit mouth, pithead, *pozo* pit eye; ~ **del puerto** *f* C&V port mouth; ~ **de la red** *f* OCEAN net mouth; ~ **de riego** *f* AGUA hydrant, water hydrant, water plug; ~ **del río** *f* AGUA, HIDROL, MINAS, OCEAN, TRANSP MAR river mouth; ~ **de salida** *f* AGUA outlet, HIDROL *presas* bottom outlet, ING MECÁ outlet; ~ **del socavón** *f* MINAS bankhead; ~ **de vaciado** *f* C&V casting lip

bocabarras *f* ING MECÁ poppet holes

bocal *m* OCEAN narrows

bocallave *f* CONST key plate, ING MECÁ angled socket wrench, keyway

bocamina *f* MINAS bank, pithead, mine entrance

bocarte *m AmL* (*cf troquel Esp*) MINAS ore stamp, die, PROD stamp; ~ **de cabezal triple** *m* MINAS *molde metálico* three-head battery; ~ **de caída libre** *m* PROD gravity stamp; ~ **de mineral** *m AmL* (*cf triturado de mineral Esp*) MINAS *preparación de minas* ore crusher; ~ **de pilón simple** *m* PROD single-stamp mill

bocartear *vt* MINAS *preparación* squeeze, PROD mill, stamp

bocarteo *m AmL* (*cf fresado Esp*) MINAS crushing, milling, PROD stamp milling

bocatoma *f* AGUA, HIDROL inlet

bocazo *m Esp* (*cf fallo en el tiro AmL*) CARBÓN *barrenos* fizzle, weathering, MINAS misfire

boceto *m* CONST sketch, IMPR design, dummy, layout, outline; ~ **del anuncio** *m* IMPR advertisement layout; ~ **definitivo** *m* IMPR comprehensive

bocha *f AmL* (*cf compresor hermético Esp*) ING MECÁ, REFRIG hermetic compressor, hermetically-sealed compressor unit

bocín ~ **de tobera** *m* PROD *horno alto* tuyere nozzle, *tobera* blast nozzle

bocina *f* ACÚST *automóviles*, AUTO horn, TRANSP MAR *construcción* stern tube, VEH horn; ~ **de boquilla** *f AmL* (*cf nautófono Esp*) TRANSP MAR *ayuda a la navegación* reed; ~ **cónica** *f* ACÚST conical horn; ~ **del escobén** *f* TRANSP MAR *construcción* hawse pipe; ~ **exponencial** *f* ACÚST exponential horn; ~ **de**

la limera *f* TRANSP MAR *construcción* rudder trunk; ~ **de niebla** *f* TRANSP MAR *navegación* horn

bodega *f* OCEAN fish hold, PROD warehouse, TEC ESP bay, TRANSP AÉR, TRANSP MAR *buque* hold; ~ **almacenamiento cables** *f* TELECOM cable storage hold; ~ **de carga** *f* TEC ESP, TRANSP, TRANSP MAR cargo hold; ~ **de empaque** *f* C&V packer's bay; ~ **de estibado de la carga** *f* TRANSP, TRANSP MAR cargo hold; ~ **para estibar la carga** *f* TRANSP, TRANSP MAR cargo hold; ~ **refrigerada** *f* REFRIG, TRANSP refrigerating hold

boehmita *f* MINERAL boehmite

Boeing ~ **747** *m* TRANSP AÉR jumbo jet

bofetada ~ **de la pala** *f* TRANSP AÉR blade slap

bogar *vi* TRANSP MAR row

bogie *m Esp* (*cf boje AmL*) CONST *ferrocarril*, FERRO bogie (*BrE*), truck (*AmE*), ING MECÁ truck (*AmE*), trailer (*AmE*), bogie (*BrE*), MECÁ, VEH bogie (*BrE*), truck (*AmE*); ~ **motor** *m Esp* (*boje motor AmL*) FERRO motor bogie (*BrE*), motor truck (*AmE*)

boje *m AmL* (*cf bogie Esp*) CONST *ferrocarril*, FERRO *vehículos* truck (*AmE*), bogie (*BrE*), ING MECÁ truck (*AmE*), trailer (*AmE*), bogie (*BrE*), MECÁ, VEH bogie (*BrE*), truck (*AmE*); ~ **motor** *m AmL* (*cf bogie motor Esp*) FERRO motor bogie (*BrE*), motor truck (*AmE*)

bola *f* CONST ball, ING MECÁ *termómetro* bulb, flyweight, *regulador de máquina* ball, OCEAN net roller; ~ **de alquitrán** *f* CONTAM MAR tar ball; ~ **de control del cursor** *f* INFORM&PD trackball, trackerball; ~ **de fricción** *f* ING MECÁ friction ball; ~ **de torsión** *f* AUTO torque ball

bolardo *m* TRANSP MAR bollard

boleador *m* INFORM&PD trackball, trackerball

boleíta *f* MINERAL boleite

boletín ~ **electrónico** *m* INFORM&PD *módem* bulletin board; ~ **de información** *m* TEC ESP report; ~ **informativo** *m Esp* (*cf noticioso de televisión AmL*) TV newscast; ~ **meteorológico** *m* METEO weather report

bolina **de** ~ *adj* TRANSP MAR *navegación a vela* close-hauled

bolita *f* P&C *operación, plásticos* pellet

bolo *m AmL* (*cf pastilla Esp*) CARBÓN pellet; ~ **alimenticio** *m* AGRIC cud

bolómetro *m* FÍS, ING ELÉC, REFRIG, TEC ESP bolometer

bolsa *f* GEN bag, CARBÓN *de mineral* crevice; ~ **para accesorios** *f* FOTO gadget bag; ~ **con acolchado burbujas de aire** *f* EMB air-bubble bag; ~ **de agua caliente** *f* INSTAL TERM hot-water bottle; ~ **de aire** *f* AUTO air bag, MINAS *sondeos* airlock, TERMOTEC *sistemas de calor por agua caliente* air lock, air pocket, TRANSP AÉR air pocket; ~ **de arena** *f* SEG sandbag; ~ **de basura** *f* EMB garbage bag (*AmE*), rubbish bag (*BrE*); ~ **con cierre de cremallera** *f* EMB zip lock bag; ~ **de colada** *f* ING MECÁ teeming pouch; ~ **elevadora** *f* TRANSP AÉR lifting bag; ~ **de fondo cuadrado de fuelle** *f* EMB square bag with gussets; ~ **de fondo redondo** *f* EMB round-ended pouch; ~ **de herramientas** *f* ING MECÁ kit; ~ **lateral** *f* C&V side pocket; ~ **de líquido** *f* REFRIG liquid slug; ~ **oscura de carga** *f* FOTO changing bag; ~ **de papel** *f* EMB, PAPEL paper bag; ~ **del paracaídas** *f* D&A parachute pack; ~ **perforada formando una bobina** *f* EMB perforated bag on a roll; ~ **de petróleo** *f* GEOL, PETROL, TEC PETR oil basin; ~ **plana de plástico** *f* EMB, P&C lay-flat film bag; ~ **para polvos** *f* SEG dust bag; ~ **postal** *f* EMB, PAPEL correspondence envelope; ~ **postal autocerrable** *f* EMB, PAPEL self-seal pocket envelope; ~ **que contiene un desecante** *f* EMB desiccant bag; ~ **que contiene un producto absorbente de la humedad** *f* EMB moisture-absorbent bag; ~ **que se puede hervir** *f* EMB boilable pouch; ~ **que se suministra en bobinas** *f* EMB reel feed bag; ~ **recubierta** *f* EMB lined bag; ~ **de reparaciones** *f* PROD repair outfit; ~ **secadora** *f* PAPEL dryer pocket; ~ **separable unida formando una bobina** *f* EMB bag reel; ~ **de vapor** *f* AUTO vapor lock (*AmE*), vapour lock (*BrE*); ~ **de vidrio** *f* C&V glass pocket

bolsada *f* CARBÓN, MINAS *de minerales*, MINERAL pocket; ~ **de mineral** *f* CARBÓN, MINAS patch

bolsillo: ~ **de la pala** *m* TRANSP AÉR blade pocket

bolsita *f* EMB sachet; ~ **dosis unitarias** *f* EMB unit dose sachet; ~ **soluble** *f* EMB soluble sachet

bolso *m* CARBÓN bag

bolsón *m* CARBÓN, MINAS *de minerales* pocket; ~ **de gas** *m* GAS gas pocket; ~ **de mineral** *m* *AmL* (*cf parcela de mineral Esp*) MINAS patch

bomba *f* GEN pump, NUCL H bomb;

~ a ~ **de abastecimiento** *f* AGUA, CONTAM MAR supply pump; ~ **de absorción** *f* FÍS sorption pump; ~ **de acción directa** *f* AGUA, INSTAL HIDRÁUL direct action pump, direct-acting pump; ~ **accionada por correa** *f* PROD belt-driven pump; ~ **accionada por motor** *f* TRANSP AÉR engine-driven pump; ~ **de aceite** *f* AUTO, CONST, REFRIG, VEH oil pump; ~ **de aceite de engranajes** *f* AUTO, VEH gear-type oil pump; ~ **de aceleración** *f* AUTO, VEH *carburador* accelerator pump; ~ **de achique** *f* AGUA, MINAS *sondeos* dewatering pump; ~ **de achique portátil** *f* TRANSP MAR handy billy; ~ **de agotamiento** *f* AGUA drainage pump, draining engine, draining pump, *minas* pumping engine, CONTAM MAR *de la carga* stripping pump, MINAS, TEC PETR pumping engine; ~ **de agotamiento ajustable** *f* ING MECÁ adjustable-discharge pump; ~ **de agua** *f* AUTO, VEH *sistema de refrigeración* water pump; ~ **de agua de alimentación** *f* HIDROL, ING MECÁ feedwater pump; ~ **de aguas residuales** *f* NUCL dirty-water pump; ~ **de aire** *f* FÍS, ING MECÁ *para compresión, succión de aire* air pump; ~ **de alimentación** *f* AUTO, ING MECÁ fuel pump, gas pump (*AmE*), gasoline pump (*AmE*), petrol pump (*BrE*), INSTAL TERM feed pump, MECÁ, TEC ESP, TRANSP AÉR, TRANSP MAR, VEH fuel pump, gas pump (*AmE*), gasoline pump (*AmE*), petrol pump (*BrE*); ~ **de alimentación de caldera** *f* INSTAL TERM boiler feed pump; ~ **de alimentación de combustible** *f* AUTO, ING MECÁ, MECÁ, TEC ESP, TRANSP AÉR, TRANSP MAR, VEH fuel pump; ~ **alternativa de combustible** *f* TEC ESP fuel backup pump; ~ **de arena** *f* MINAS sludger, *sondeos* sludge pump; ~ **de aspiración** *f* AGUA, INSTAL HIDRÁUL aspiration pump, aspiring pump, exhaust pump, suction pump, TEXTIL suction pump; ~ **aspirante** *f* AGUA, ALIMENT, ING MECÁ, INSTAL HIDRÁUL aspiration pump, aspiring pump, suction pump; ~ **aspirante e impelente** *f* AGUA reciprocating pump; ~ **aspirante-impelente** *f* AGUA bucket pump, lift-and-force pump, ING MECÁ, INSTAL HIDRÁUL bucket pump; ~ **aspirante-impelente con válvulas en el pistón** *f* AGUA, ING MECÁ, INSTAL HIDRÁUL bucket pump; ~ **atómica** *f* NUCL atomic bomb; ~ **con aumento y disminución de volumen** *f* INSTAL HIDRÁUL positive-displacement

pump; ~ **autocebable** *f* INSTAL HIDRÁUL self-priming pump; ~ **autocebadora** *f* INSTAL HIDRÁUL self-priming pump; ~ **autocebadora de aguas residuales** *f* INSTAL HIDRÁUL, NUCL self-priming dirty-water pump; ~ **axial** *f* ING MECÁ axial pump;

~ b ~ **de barrido** *f* PROD scavenge pump; ~ **de bicicleta** *f* TRANSP bicycle pump; ~ **bicilíndrica** *f* INSTAL HIDRÁUL duplex pump;

~ c ~ **de cabeza de máquina** *f* PAPEL fan pump; ~ **de cadena** *f* AGUA, ING MECÁ chain pump; ~ **de cadena sin fin** *f* AGUA chapelet, paternoster pump; ~ **de calor** *f* FÍS, ING MECÁ, MECÁ, TERMO, TERMOTEC heat pump; ~ **calorífica** *f* FÍS, ING MECÁ, MECÁ, TERMO, TERMOTEC heat pump; ~ **calorimétrica** *f* FÍS, LAB calorimetric bomb; ~ **de carga** *f* CONTAM MAR cargo pump, ING MECÁ, MECÁ booster pump, NUCL charging pump, TEC PETR *perforación, oleoductos*, TRANSP, TRANSP AÉR booster pump, TRANSP MAR *petrolero* cargo pump; ~ **catalítica** *f* PROC QUÍ catalytic pump; ~ **de caudal constante** *f* ING MECÁ constant flow pump; ~ **de caudal regulable** *f* ING MECÁ variable-discharge pump, PROD variable-delivery pump; ~ **de cebado** *f* TRANSP AÉR primer pump; ~ **de cebado automático** *f* AUTO, VEH *carburante* self-priming pump; ~ **cebadora** *f* ING MECÁ, MECÁ, TEC PETR, TRANSP, TRANSP AÉR booster pump; ~ **centrífuga** *f* CONTAM MAR, ING MECÁ, PROC QUÍ, REFRIG centrifugal pump; ~ **centrífuga para arenas y lodos** *f* MINAS *sondeos* sludger; ~ **centrífuga de aspiración final** *f* ING MECÁ end suction centrifugal pump; ~ **centrífuga con difusor** *f* INSTAL HIDRÁUL turbine pump; ~ **centrífuga de succión final** *f* ING MECÁ end suction centrifugal pump; ~ **de chorro** *f* AGUA, NUCL jet pump; ~ **cinética** *f* ING MECÁ kinetic pump; ~ **de circulación** *f* ALIMENT, INSTAL HIDRÁUL, REFRIG, TERMOTEC circulating pump, circulation pump; ~ **de combustible** *f* AUTO, ING MECÁ, MECÁ, TEC ESP, TRANSP AÉR fuel pump, gas pump (*AmE*), gasoline pump (*AmE*), petrol pump (*BrE*), TRANSP MAR *motor* fuel pump, gas pump (*AmE*), gasoline pump (*AmE*), VEH fuel pump, gas pump (*AmE*), gasoline pump (*AmE*), petrol pump (*BrE*); ~ **de combustible de alta presión** *f* AUTO, VEH fuel high-pressure pump; ~ **de combustible de diafragma** *f* AUTO, VEH diaphragm fuel pump; ~ **de combustible eléctrica** *f* AUTO, ELEC, VEH electric fuel-pump; ~ **de combustible de émbolo** *f* AUTO, VEH plunger fuel-pump; ~ **de combustible de pistón** *f* AUTO, VEH plunger fuel-pump; ~ **de combustible de vacío** *f* AUTO, VEH vacuum fuel pump; ~ **de compresión** *f* INSTAL HIDRÁUL compression pump; ~ **de condensación** *f* PROC QUÍ condensating pump; ~ **del condensado de residuos líquidos** *f* NUCL waste condensate pump; ~ **y conexiones** *f pl* ING MECÁ pump and connections; ~ **de contraincendios** *f* TERMO, VEH fire-engine; ~ **criogénica** *f* TEC ESP *motores* cryopump;

~ d ~ **de desagüe** *f* AGUA drainage pump, draining engine, draining pump; ~ **de descarga constante** *f* PROD fixed-delivery pump; ~ **de descarga variable** *f* ING MECÁ adjustable-discharge pump, variable-discharge pump; ~ **deshieladora** *f* TRANSP AÉR de-icing pump; ~ **de desplazamiento positivo** *f* ING MECÁ positive-displacement pump, constant flow pump; ~ **sin detonar** *f* D&A unexploded bomb (*UXB*); ~ **de diafragma** *f* AGUA, CONTAM MAR diaphragm pump;

~ **de difusión** *f* FÍS, ING MECÁ, INSTR diffusion pump; ~ **dinámica** *f* ING MECÁ kinetic pump; ~ **doble** *f* PROD double pump; ~ **de doble efecto** *f* PROD *cilindro* double-acting pump; ~ **dosificadora** *f* EMB dosing pump; ~ **de dragado** *f* TRANSP MAR dredge pump;

■ **e** ~ **de efecto retardado** *f* D&A time bomb; ~ **electromagnética** *f* ELEC, ELECTRÓN, NUCL *para metales líquidos* electromagnetic pump; ~ **de elevación por aire comprimido** *f* PROC QUÍ airlift pump; ~ **de émbolo aspirante** *f* PROD free-fall pump; ~ **de émbolo buzo** *f* AGUA plunger pump; ~ **de engranaje tetracilíndrica** *f* PROD four-throw geared pump; ~ **de engranajes** *f* AUTO, P&C, REFRIG, VEH gear pump; ~ **eólica** *f* ENERG RENOV windmill pump; ~ **de eslinga** *f* AmL (*cf bomba de tiro Esp*) MINAS slung pump; ~ **de estribo para agua** *f* AGUA, HIDROL, SEG stirrup pump for water; ~ **de evaporación** *f* PROC QUÍ evaporating pump; ~ **sin explotar** *f* D&A unexploded bomb (*UXB*); ~ **expulsora** *f* ING MECÁ ejector pump; ~ **de extracción de agua** *f* GEOL degasser; ~ **de extracción neumática** *f* PROC QUÍ pneumatic extraction-pump; ~ **eyectora** *f* ING MECÁ ejector pump;

■ **f** ~ **para fangos** *f* MINAS mud pump; ~ **filtrante** *f* AmL (*cf trompa de vacío Esp*) CONTAM MAR strainer pump, LAB filter pump; ~ **de flujo axial** *f* ING MECÁ, MINAS *aguas* axial flow pump; ~ **de flujo constante** *f* ING MECÁ constant flow pump; ~ **de flujo mixto** *f* ING MECÁ, MINAS *aguas* mixed-flow pump; ~ **de fuelle** *f* LAB bellows pump; ~ **de fugas** *f* NUCL leakage water pump; ~ **fumígena** *f* D&A smoke bomb;

■ **g** ~ **de gasolina** *f* Esp (*cf bomba de nafta AmL*) AUTO, ING MECÁ, MECÁ, TEC ESP, TRANSP AÉR, TRANSP MAR, VEH fuel pump, gas pump (*AmE*), gasoline pump (*AmE*), petrol pump (*BrE*); ~ **de gasolina mecánica** *f* AUTO *alimentación*, VEH mechanical fuel pump; ~ **guiada por láser** *f* D&A laser-guided bomb (*LGB*);

■ **h** ~ **helicoidal** *f* RECICL screw pump;

■ **i** ~ **impelente** *f* AGUA lift pump, force pump, plunger pump, ram pump, INSTAL HIDRÁUL blast pump, force pump, lift pump, pressing pump, ram pump, PROD force pump; ~ **impelente de doble asiento** *f* INSTAL HIDRÁUL duplex pump; ~ **de impulsión** *f* TEC ESP boost pump; ~ **sin impulsión** *f* PROD free delivery pump; ~ **de incendios** *f* CONST, SEG fire pump, TERMO fire-engine, fire pump; ~ **de inducción** *f* ING ELÉC induction pump; ~ **inteligente** *f* D&A smart bomb; ~ **de inyección** *f* CONST, ING MECÁ, TRANSP MAR *motor* injection pump; ~ **de inyección de combustible** *f* AUTO, TERMO, TERMOTEC, VEH fuel-injection pump; ~ **de inyección con distribuidor** *f* AUTO distributor injection pump; ~ **de inyección de lodos** *f* PETROL mud pump; ~ **de inyección multicilíndrica** *f* AUTO multicylinder injection pump; ~ **iónica** *f* FÍS, ING MECÁ ion pump; ~ **del izador** *f* TRANSP AÉR hoist pump;

■ **l** ~ **de lodo** *f* TEC PETR *perforación* mud pump; ~ **del lodo** *f* MINAS *sondeos* mud pump, TEC PETR *perforación* slush pump; ~ **para lodos** *f* MINAS *desagües* slurry pump; ~ **de lubricación** *f* MECÁ, PROD lubricating pump; ~ **de lubricación con sus tuberías** *f* PROD *tornos* lubricating pump and pipe connections;

■ **m** ~ **magnetohidrodinámica** *f* ING ELÉC magneto-hydrodynamic pump; ~ **de mano** *f* PROD, SEG hand pump; ~ **manual** *f* PROD, SEG hand pump; ~ **mecánica** *f* AGUA, MINAS, TEC PETR pumping engine; ~ **mecánica de combustible** *f* AUTO, VEH mechanical fuel pump; ~ **de membrana** *f* AGUA, CONTAM MAR diaphragm pump; ~ **mezcladora** *f* PAPEL mixing pump; ~ **MHD** *f* ING ELÉC MHD pump; ~ **molecular** *f* FÍS molecular pump; ~ **de molino de viento** *f* ENERG RENOV windmill pump; ~ **motorizada** *f* CONTAM MAR motor pump; ~ **de movimiento alternativo** *f* AGUA reciprocating pump; ~ **de muestreo** *f* LAB sampling pump;

■ **n** ~ **de nafta** *f* AmL (*cf bomba de gasolina Esp*) AUTO, ING MECÁ, MECÁ, TEC ESP, TRANSP AÉR, TRANSP MAR, VEH fuel pump, gas pump (*AmE*), gasoline pump (*AmE*), petrol pump (*BrE*); ~ **neumática** *f* FÍS, ING MECÁ pneumatic pump, vacuum pump, *para compresión, succión* air pump;

■ **o** ~ **oscilatoria** *f* AGUA semirotary pump;

■ **p** ~ **de paletas** *f* AGUA, PROD vane pump; ~ **peristáltica** *f* LAB *manejo de líquidos* peristaltic pump; ~ **de pie** *f* AUTO, ING MECÁ foot pump; ~ **de pistón** *f* AGUA, PROD piston pump; ~ **de pistón axial** *f* MECÁ axial piston pump; ~ **de pistón tubular** *f* AGUA plunger pump; ~ **portátil** *f* LAB *muestreo* portable pump; ~ **con prensaestopas** *f* NUCL glanded pump; ~ **de presión** *f* AGUA, CARBÓN pressure pump, INSTAL HIDRÁUL compression pump, pressure pump, PETROL pressure bomb; ~ **principal** *f* NUCL main coolant pump; ~ **de prueba** *f* AGUA test pump; ~ **de prueba del inyector** *f* AUTO injector test pump; ~ **para purgar agua** *f* AGUA drip pump;

■ **q** ~ **que explota cerca del blanco** *f* TEC ESP near miss; ~ **que puede funcionar sumergida** *f* AGUA submerged pump;

■ **r** ~ **rafadora de alta presión** *f* MINAS high-pressure shearer pump; ~ **de reacción** *f* LAB reaction bomb; ~ **reforzadora** *f* ING MECÁ, MECÁ, TEC PETR, TRANSP, TRANSP AÉR booster pump; ~ **del refrigerante del reactor** *f* NUCL reactor coolant pump, main coolant pump, REFRIG, TERMO reactor-coolant pump; ~ **de régimen** *f* PROD duty pump; ~ **reguladora** *f* LAB *manejo de fluidos* metering pump; ~ **de repuesto de combustible** *f* TEC ESP *motores* fuel backup pump; ~ **del residuo de la condensación** *f* NUCL waste condensate pump; ~ **de rosario** *f* AGUA chain pump, chapelet, paternoster pump, ING MECÁ chain pump; ~ **a rosca** *f* TRANSP screw pump; ~ **de rotación** *f* AGUA, FÍS rotary pump; ~ **rotativa** *f* AGUA, FÍS rotary pump; ~ **rotativa de paletas** *f* PROD vane pump;

■ **s** ~ **de salmuera** *f* REFRIG brine pump; ~ **de sentina** *f* AGUA sump pump, TRANSP MAR bilge pump; ~ **de servodirección** *f* AUTO, VEH power-steering pump; ~ **de un solo émbolo** *f* INSTAL HIDRÁUL solo-piston pump; ~ **de la sonda** *f* CONST borehole pump; ~ **de sondeos** *f* AGUA borehole pump; ~ **submarina** *f* AGUA subaqueous pump; ~ **de succión** *f* ALIMENT *maquinaria para proceso de alimentos* suction pump; ~ **sumergible** *f* MINAS *aguas* submersible pump; ~ **de sumidero** *f* AGUA sump pump; ~ **suspendida** *f* Esp (*cf bomba volante AmL*) MINAS *aguas sondeos* suspended pump;

■ **t** ~ **de tiro** *f* Esp (*bomba de eslinga AmL*) MINAS slung pump; ~ **de tornillo** *f* PROD screw pump; ~ **de trasiego** *f* CONTAM MAR *de la carga* transfer pump;

~ **tricilíndrica** *f* AGUA three-throw pump; ~ **de turbina** *f* INSTAL HIDRÁUL turbine pump; ~ **v** ~ **de vacío** *f* ALIMENT vacuum air pump, vacuum pump, CONTAM MAR vacuum pump, FÍS air pump, vacuum pump, ING MECÁ vacuum pump, *para compresión, succión* air pump, LAB vacuum pump; ~ **de vacío de alta presión** *f* ING MECÁ high-pressure vacuum pump; ~ **de vacío de baja presión** *f* ING MECÁ low-pressure vacuum pump; ~ **de vacío cinética** *f* MECÁ kinetic vacuum pump; ~ **de vacío de desplazamiento positivo** *f* ING MECÁ positive-displacement vacuum pump; ~ **de válvula de charnela** *f* PROD flap valve pump; ~ **de vapor de acción simple** *f* MINAS bull pump; ~ **de ventilación** *f* FÍS, ING MECÁ *para compresión, succión* air pump; ~ **de volante** *f* PROD hand flywheel pump; ~ **volante** *f* AmL (*cf bomba suspendida Esp*) MINAS *aguas sondeos* suspended pump; ~ **volumétrica** *f* ING MECÁ, INSTAL HIDRÁUL, PROD positive-displacement pump

bombardeo *m* D&A bombardment, bombing, ELECTRÓN, FÍS RAD, GAS, METAL bombardment; ~ **cruzado** *m* FÍS RAD cross bombardment; ~ **de electrones** *m* ELECTRÓN, FÍS PART, NUCL electron bombardment; ~ **de iones** *m* ELECTRÓN, METAL ion bombardment; ~ **iónico** *m* ELECTRÓN ionic bombardment, QUÍMICA *de espécimen* ion sputtering; ~ **en lanzadera** *m* TEC ESP *aviación* shuttle; ~ **múltiple** *m* FÍS RAD cross bombardment

bombear *vt* AGUA, CONST pump, pump out, ING MECÁ feed, pump, pump out, MECÁ feed, PROD feed, pump, pump out, TRANSP MAR pump

bombeo *m* AGUA pumping, pumping out, CONST *carretera* cambering, crossfall, camber, ING MECÁ bulge, MINAS *aguas* pumping, PROD *polea de cara combada* crown, TEC PETR pumping, TRANSP MAR pumping out; ~ **de agotamiento** *m* MINAS *aguas* mine pumping; ~ **de desagüe** *m* AGUA, MINAS *vaciado de pozos*, TEC PETR pumping; ~ **de extracción** *m* AmL (*cf bombeo de limpieza Esp*) MINAS *pozos* fall cleanup; ~ **de haz de electrones** *m* ELECTRÓN electron-beam pumping; ~ **de limpieza** *m* Esp (*cf bombeo de extracción AmL*) MINAS *pozos* fall cleanup; ~ **neumático** *m* MINAS *aguas* plunger lift; ~ **óptico** *m* ELECTRÓN, FÍS, FÍS RAD, NUCL optical pumping

bombero *m* CONST, SEG, TERMO firefighter, fireman

bombilla *f* AUTO, CINEMAT, ELEC, ING ELÉC, ING MECÁ bulb; ~ **eléctrica** *f* ING ELÉC lamp; ~ **del faro delantero** *f* AUTO headlamp bulb; ~ **incandescente pequeña** *f* CINEMAT baby

bombillo *m* C&V parison mold (*AmE*), parison mould (*BrE*)

bombín: ~ **de freno** *m* AUTO, ING MECÁ, VEH brake cylinder

bombona *f* ALIMENT, LAB carboy

bono: ~ **de trabajo** *m* PROD job ticket

boom *m* CINEMAT *soporte extensible para micrófonos* boom; ~ **de la cámara** *m* CINEMAT, TV camera boom

booster *m* REFRIG booster compressor

boquerel *m* ING MECÁ nozzle, PROD *de manguera o similar* nosepiece

boquilla *f* AGUA jet, AUTO nozzle, C&V nozzle, mouth tool, CARBÓN *orificio* nozzle, CONST nozzle, nipple, ENERG RENOV nozzle, FÍS jet, GAS nozzle, ING MECÁ die, nipple, collet, nozzle, bushing, LAB *material de vidrio* nozzle, MECÁ *de tubo* faucet, nozzle, bushing, PROD *fuelle, inyector, soplete oxiacetilénico*, TEC ESP nozzle; ~ **acopladora** *f* CONST, ING MECÁ nipple; ~ **ajustable** *f* ING MECÁ adjustable nozzle; ~ **bloqueada** *f* TRANSP AÉR choked nozzle; ~ **de borda** *f* INSTAL HIDRÁUL *depósito hidráulico* Borda mouthpiece; ~ **cilíndrica** *f* INSTAL HIDRÁUL cylindrical mouthpiece; ~ **de combustible** *f* AUTO, ING MECÁ, VEH fuel nozzle; ~ **convergente** *f* ING MECÁ combining nozzle; ~ **de corrección de aire** *f* AUTO air correction jet; ~ **de descarga** *f* PROD *de inyector* delivery nozzle; ~ **de engrase** *f* ING MECÁ oilhole; ~ **de flujo** *f* ING MECÁ flow nozzle; ~ **giratoria** *f* TEC ESP rotatable nozzle; ~ **de inyección** *f* AUTO injection nozzle; ~ **de inyección de aguja** *f* AUTO pintle injection nozzle; ~ **de inyección de chorro doble** *f* AUTO twin-jet injection nozzle; ~ **de inyección de surtidor único** *f* AUTO single-jet injection nozzle; ~ **inyectora** *f* ING MECÁ combining nozzle; ~ **de llenado** *f* EMB filling nozzle; ~ **de lubricación** *f* MECÁ lubricating nipple; ~ **para macho roscador** *f* ING MECÁ tapping chuck; ~ **de manguera** *f* ING MECÁ hose nozzle; ~ **de manguera contra incendios** *f* CONST, SEG, TERMO fire-hose nozzle; ~ **medidora de flujo** *f* ING MECÁ flow nozzle; ~ **de mordazas** *f* ING MECÁ collet chuck; ~ **de pivote del estrangulador** *f* AUTO throttle pintle nozzle; ~ **para pulir con chorro de arena** *f* C&V sandblasting nozzle; ~ **pulverizadora** *f* ALIMENT atomizing nozzle, ING MECÁ sprayer nozzle; ~ **para pulverizar** *f* ALIMENT atomizing nozzle; ~ **de regadera** *f* ING MECÁ sprayer nozzle; ~ **de rociado con agua pesada** *f* NUCL heavy water spray nozzle; ~ **rociadora** *f* ING MECÁ sprayer nozzle; ~ **para taladro** *f* ING MECÁ drill socket; ~ **de la tobera** *f* PROD nozzle tip; ~ **de ventilación** *f* CARBÓN, ING MECÁ air nozzle

boracita *f* MINERAL boracite

borano *m* QUÍMICA borane

boratera *f* DETERG borax lake

borato *m* QUÍMICA borate; ~ **sódico** *m* QUÍMICA sodium borate

bórax *m* AGRIC, C&V, CONST, DETERG, MINERAL, QUÍMICA borax

borazón *m* QUÍMICA borazon

borbolleo: ~ **resonante** *m* NUCL chugging

borboteador *m* FÍS bubbler

borda: **por la** ~ *adv* TRANSP MAR overboard

bordada *f* TRANSP MAR leg, tack

bordadora *f* TEXTIL wrapper

borde[1]: **de** ~ **afilado** *adj* ING MECÁ sharp-edged; **de** ~ **dorado** *adj* IMPR gilt edge

borde[2] *m* C&V edging, CONST lip, selvage, CRISTAL, EMB, FÍS edge, GEOL margin, INFORM&PD edge, ING MECÁ edging, INSTR edge, LAB *de vaso de precipitado* lip, *abertura, acceso* flange, MATEMÁT limb, MECÁ edge, MINAS *geología* brow, TEXTIL edge, TRANSP AÉR chine; ~ **de absorción** *m* CRISTAL, NUCL absorption edge; ~ **de absorción K** *m* NUCL K-absorption edge; ~ **afilado** *m* C&V arrissed edge, CONST *herramientas* cutting edge, keen edge; ~ **alisado** *m* C&V smoothed edge; ~ **anterior** *m* IMPR fore edge; ~ **arrugado** *m* C&V wrinkled rim; ~ **de ataque** *m* CINEMAT, ELECTRÓN, FÍS, PROD, TELECOM, TRANSP AÉR leading edge; ~ **de ataque de la pala** *m* TRANSP AÉR blade leading edge; ~ **de ataque del plano de deriva** *m* TRANSP AÉR fin leading edge; ~ **de ataque único** *m*

PROD leading-edge one shot; ~ **burdo** *m* C&V rough edge; ~ **de caída** *m* ING ELÉC trailing edge; ~ **continental** *m* OCEAN continental borderland; ~ **de convección forzada** *m* IMPR forced-convection edge; ~ **de la cortadura** *m* METAL shear lip; ~ **cortante** *m* CONST cutting edge, keen edge; ~ **cortante de la vertedera** *m* AGRIC shin; ~ **del corte** *m* TEXTIL cutting edge; ~ **craso** *m* P&C *defecto en la pintura* fat edge; ~ **delantero** *m* CINEMAT, ELECTRÓN, FÍS, PROD, TELECOM, TRANSP AÉR leading edge; ~ **dorado** *m* IMPR gilt edge; ~ **encolado** *m* EMB gummed edge; ~ **de entrada** *m* CINEMAT, ELECTRÓN, FÍS, PROD, TELECOM, TRANSP AÉR leading edge; ~ **entrante** *m* CINEMAT, ELECTRÓN, FÍS, PROD, TELECOM, TRANSP AÉR leading edge; ~ **de la esquina interior** *m* EMB inside corner edge; ~ **festoneado** *m* ING MECÁ pinked edge; ~ **filoso** *m* C&V sharp edge; ~ **frontal** *m* IMPR fore edge; ~ **de fuga de una pala** *m* TRANSP AÉR blade trailing edge; ~ **de grano** *m* NUCL grain boundary; ~ **de la hoja** *m* AGRIC leaf margin, C&V edge of the sheet; ~ **inferior** *m* IMPR tail edge, TRANSP MAR *portillos de luz* sill; ~ **irregular** *m* C&V irregular edge, PAPEL deckle; ~ **de media caña** *m* C&V half round edge; ~ **penetrante** *m* CONST *herramientas* cutting edge, keen edge; ~ **perchado** *m* TEXTIL raised edge; ~ **plano** *m* C&V flat edge; ~ **plano y bisel** *m* C&V flat edge and bevel; ~ **posterior** *m* ELECTRÓN, PROD, TRANSP AÉR *de la pala* trailing edge; ~ **pulido** *m* C&V polished edge; ~ **pulido a fuego** *m* C&V fire-polished edge; ~ **recortado** *m* ING MECÁ pinked edge; ~ **redondeado** *m* C&V rounded edge; ~ **de referencia** *m* TV reference edge; ~ **sin rematar** *m* CONST cut end; ~ **de salida** *m* ELECTRÓN, PROD, TRANSP AÉR *de la pala* trailing edge; ~ **de sellado** *m* C&V sealing edge; ~ **serrado** *m* SEG jagged edge; ~ **soldado** *m* ALIMENT *latería* soldered seam; ~ **superpuesto** *m* IMPR foldover edge; ~ **trasero** *m* FÍS trailing edge

bordeado *m* ING MECÁ edging; ~ **del armazón** *m* ING MECÁ frame edging

bordeamiento: ~ **de color** *m* TV color fringing (*AmE*), colour fringing (*BrE*)

bordear[1]: ~ **con ranura** *vt* C&V edge with a groove

bordear[2] *vi* CONST set boards edgeways, TRANSP MAR tack

bordes: ~ **cerrados** *m pl* IMPR yapp edges; ~ **coloreados** *m pl* IMPR *en cuadernación* colored edges (*AmE*), coloured edges (*BrE*); ~ **de filones** *m pl* MINAS ledger; ~ **jaspeados** *m pl* IMPR *en cuadernación* sprinkled edges

bordillo *m* CONST curb (*AmE*), kerb (*BrE*); ~ **de la acera** *m* CONST curb (*AmE*), kerb (*BrE*); ~ **osculante** *m* GEOM osculating curve

bordo[1]: **a** ~ *adv* D&A, TEC ESP on board; **a** ~ **de** *adv* TRANSP MAR on board

bordo[2] *m* TRANSP MAR leg, board; ~ **del buque** *m* TRANSP MAR shipboard

bordón *m* TEXTIL *tejidos* rib

borna *f* AUTO post, PROD *electricidad* stud, TELECOM terminal; ~ **positiva** *f* ELEC *conexión* positive terminal; ~ **de salida** *f* ELEC, PROD output terminal; ~ **de tornillo** *f* PROD screw terminal

borne *m* GEN terminal; ~ **de antena** *m* TV aerial terminal (*BrE*), antenna terminal (*AmE*); ~ **de batería** *m* AUTO, ING ELÉC, VEH battery terminal; ~ **de conexión** *m* ING ELÉC binding post; ~ **de fase** *m* ELEC *conexión*, ING ELÉC phase terminal; ~ **del inducido** *m* ELEC *máquina* armature end connection, armature end plate; ~ **de línea** *m* ELEC *conexión*, PROD *telegrafía*, TELECOM line terminal; ~ **de masa** *m* ELEC, ING ELÉC earth terminal (*BrE*) ground terminal (*AmE*); ~ **de montaje lateral** *m* AUTO, VEH side-mounted terminal; ~ **negativo** *m* AUTO, ELEC, ING ELÉC, VEH negative terminal; ~ **neutro** *m* ELEC *conexión*, ING ELÉC neutral terminal; ~ **de perforaciones** *m* PROD stab terminal; ~ **positivo** *m* AUTO, ELEC, ING ELÉC, VEH positive terminal; ~ **positivo de tierra** *m* AUTO, ELEC, ING ELÉC, VEH positive earthed terminal (*BrE*), positive grounded terminal (*AmE*); ~ **principal** *m* AUTO main terminal; ~ **de puesta a masa** *m* ELEC, ING ELÉC, TRANSP MAR electrical earth connector (*BrE*), electrical ground connector (*AmE*); ~ **de puesta a tierra** *m* ELEC, ING ELÉC, TRANSP MAR earth terminal (*BrE*), electrical earth connector (*BrE*), electrical ground connector (*AmE*), ground terminal (*AmE*); ~ **de salida** *m* ELEC output terminal; ~ **de tierra** *m* ELEC *conexión*, ING ELÉC earth terminal (*BrE*), ground terminal (*AmE*); ~ **de tierra aterrado** *m* ELEC, ING ELÉC, PROD earthed earth terminal (*BrE*), grounded earth terminal (*AmE*); ~ **umbilical** *m* TEC ESP umbilical connector

borneando: ~ **proa al viento** *adv* TRANSP MAR *buque fondeado* swinging

borneol *m* QUÍMICA borneol, camphol

bornilo *m* QUÍMICA bornyl

bornita *f* MINERAL bornite, erubescite, variegated copper

boro *m* (*B*) QUÍMICA boron (*B*)

borofluoruro *m* QUÍMICA borofluoride

borosilicato *m* QUÍMICA borosilicate

borra: ~ **de algodón** *f* P&C *carga* cotton linter, PROD, TEXTIL cotton waste; ~ **fibrosa** *f* TEXTIL fibrous waste

borrado *m* ACÚST erasure, ELECTRÓN *C.I. digitales* blanking, IMPR deleatur, deletion, INFORM&PD erasure, deletion, clear, PETROL muting, TV blanking, erasure; ~ **final** *m* TV final blanking; ~ **por rayos ultravioleta** *m* INFORM&PD ultraviolet erasing; ~ **selectivo** *m* INFORM&PD selective erasure; ~ **síncrono** *m* TV sync blanking; ~ **ultravioleta** *m* INFORM&PD ultraviolet erasing; ~ **vertical** *m* INFORM&PD, TV vertical blanking

borrador *m* CONST draft, INFORM&PD eraser, ING MECÁ draft, TV eraser; ~ **de norma internacional** *m* INFORM&PD, TELECOM draft international standard (*DIS*); ~ **volumétrico** *m* TV bulk eraser

borrar *vt* ELECTRÓN reset, IMPR delete, INFORM&PD clear, delete, erase, ÓPT erase, PROD clear, TV erase

borrasca *f* METEO, TRANSP MAR rough weather

borrosidad *f* TV smearing

borroso *adj* C&V, FÍS blurred

bort *m* MINAS bort

boruro *m* QUÍMICA boride

boruros: ~, **carburos, nitruros y siliciliros cementados** *m pl* ING MECÁ hard metal

bosón *m* FÍS, FÍS PART boson; ~ **de Higgs** *m* FÍS, FÍS PART Higgs boson, Higgs particle; ~ **vectorial intermedio** *m* FÍS, FÍS PART intermediate vector boson; ~ **Z** *m* FÍS Z boson

bosque: ~ **mixto** *m* AGRIC mixed forest

bosquejo *m* ING MECÁ outline, scheme

BOT *abr* (*inicio de la cinta*) INFORM&PD BOT (*beginning of tape*)

bota f TRANSP MAR line; ~ **de agua** f CONST water butt; ~ **de chavetas** f ING MECÁ key drift; ~ **de seguridad** f SEG safety boot

botador m ING MECÁ, MECÁ pin driver, punch, tappet; ~ **de chavetas** m ING MECÁ key drift, drift; ~ **de pernos** m ING MECÁ, MECÁ drift

botalón m MECÁ boom; ~ **de foque** m CONTAM MAR *veleros* jib boom

botapasador m ING MECÁ pin drift, pin extractor

botar[1] vt ING MECÁ, MECÁ *pasador* punch out, TRANSP MAR *barco* launch

botar[2] vi PROD *remaches* knock out; ~ **un pasador** vi ING MECÁ, MECÁ punch out

botas: ~ **de goma** f pl P&C rubber boots; ~ **de trabajo** f pl SEG work boots

botavara: ~ **con dispositivo de arrollamiento** f TRANSP MAR roller boom

bote m CONST, CONTAM MAR, OCEAN, TRANSP MAR boat; ~ **con abertura automática** m EMB key-opening can; ~ **aerosol** m EMB aerosol can; ~ **de aluminio** m EMB aluminium can (*BrE*), aluminum can (*AmE*); ~ **auxiliar inflable** m TRANSP MAR inflatable dinghy; ~ **de a bordo** m TRANSP MAR ship's boat; ~ **de caucho** m P&C, TRANSP MAR rubber boat; ~ **de goma** m P&C, TRANSP MAR rubber boat; ~ **de humo** m CINEMAT smoke candle; ~ **inflable** m TRANSP AÉR, TRANSP MAR inflatable boat, inflatable dinghy; ~ **langostero** m OCEAN lobster boat; ~ **metálico** m EMB metal can; ~ **de pesca** m OCEAN fish gig; ~ **plegable** m TRANSP MAR collapsible boat; ~ **de rescate** m SEG, TRANSP AÉR, TRANSP MAR lifeboat, rescue boat; ~ **salvavidas** m SEG, TRANSP AÉR, TRANSP MAR lifeboat, rescue boat; ~ **salvavidas inflable** m SEG, TRANSP AÉR, TRANSP MAR inflatable dinghy; ~ **velero de recreo sin cámara** m TRANSP MAR day boat

botella: ~ **de aire** f OCEAN *buzos autónomos* scuba tank; ~ **de aluminio** f EMB aluminium bottle (*BrE*), aluminum bottle (*AmE*); ~ **de boca ancha** f LAB *material de vidrio* wide-mouth bottle; ~ **de Borgoña** f C&V burgundy bottle; ~ **para cerveza europea** f C&V Eurobeer bottle; ~ **para champaña** f C&V champagne bottle; ~ **color cobalto** f C&V cobalt bottle; ~ **para conserva** f ALIMENT, C&V, EMB canning jar (*AmE*), preserving jar (*BrE*); ~ **de cuello angosto** f C&V, LAB *material de vidrio* narrow-necked bottle; ~ **con cuello moldeado** f C&V, LAB *material de vidrio* bottle with molded neck (*AmE*), bottle with moulded neck (*BrE*); ~ **cuentagotas** f C&V, LAB *material de vidrio* dropping bottle; ~ **de depósito** f EMB deposit bottle; ~ **de Dreschel** f LAB *material de vidrio* Dreschel bottle; ~ **fabricada por soplado** f C&V, EMB blown bottle; ~ **de gas comprimido** f GAS, ING MECÁ, MECÁ, NUCL, PROD, TERMO gas cylinder; ~ **de gas a presión** f GAS, NUCL, PROD, TERMO gas bottle; ~ **con gotero** f C&V eye-drop bottle; ~ **inclinada** f C&V leaner; ~ **inderramable** f EMB antispill bottle; ~ **para infusión** f C&V infusion bottle; ~ **de Klein** f GEOM Klein bottle; ~ **lavaojos** f SEG eye-rinse bottle; ~ **de Leyden** f ING ELÉC Leyden jar; ~ **magnética** f FÍS magnetic bottle; ~ **no retornable** f C&V single-trip bottle, EMB no-return bottle; ~ **de oxígeno** f PROD oxygen cylinder; ~ **para pastillas** f C&V tablet bottle; ~ **perfumera** f C&V miniature bottle; ~ **plástica** f FOTO plastic bottle; ~ **de polietileno** f EMB PET bottle; ~ **de reactivo** f LAB reagent bottle; ~ **de PVC** f EMB PVC bottle; ~ **retornable** f EMB, RECICL returnable bottle; ~ **sin retorno** f EMB, RECICL nonreturnable bottle, one-way bottle; ~ **semirrígida** f FOTO collapsible bottle; ~ **con tapón de rosca** f EMB screw cap bottle; ~ **para transfusión** f C&V transfusion bottle

botellita f C&V phial (*BrE*), vial (*AmE*)

botellón: ~ **de amoníaco** m DETERG ammonia carboy; ~ **de oxígeno** m TEC ESP oxygen cylinder

botero m TRANSP MAR waterman

botiquín: ~ **de primeros auxilios** m SEG first-aid box, first-aid kit

botón m GEN button, TEXTIL *hilatura* nep; ~ **de ajuste** m CINEMAT, FOTO, GEOFÍS, MECÁ adjusting knob, adjustment knob; ~ **de ajuste ocular** m FOTO, INSTR eyepiece focusing knob; ~ **de arranque** m CINEMAT start button, VEH *del motor* starter button; ~ **de avance lento** m CINEMAT inching knob; ~ **de contacto** m CINEMAT, ELEC, FOTO, ING ELÉC, ING MECÁ, TELECOM push-button; ~ **de control del iris** m TV iris control button; ~ **de control de movimiento lento** m INSTR slow-motion control knob; ~ **corrector** m MECÁ adjusting knob; ~ **de cruce** m TEXTIL jigger; ~ **de descarga** m ING MECÁ discharge button; ~ **de desconexión del piloto automático** m TRANSP AÉR autopilot disengage push button; ~ **del disparador** m CINEMAT, FOTO, TV release button, shutter release button; ~ **de emergencia** m ELEC, SEG emergency button; ~ **de encendido y apagado** m CINEMAT, FOTO on-off switch; ~ **de enfoque** m CINEMAT, FOTO, INSTR, TV focus knob, focusing knob; ~ **de escala luminosa** m FOTO light scale switch; ~ **esférico** m ING MECÁ ball knob; ~ **estriado** m ING MECÁ milled knob; ~ **del flash** m FOTO flash switch; ~ **giratorio** m ING ELÉC rotary knob; ~ **de grabación** m CINEMAT, TV record button; ~ **de iluminación a contraluz** m PROD backlit push button; ~ **de iluminación reticular** m INSTR reticle illumination knob; ~ **impulsor** m ING MECÁ push-button; ~ **con índice** m ING MECÁ pointer knob; ~ **inversor** m INSTR inverter knob; ~ **de llamada** m ING ELÉC call button; ~ **de mando** m ELEC control knob, INSTR *limbo horizontal* drive knob; ~ **de parada** m TV stop key; ~ **de porcelana** m C&V porcelain button; ~ **a presión** m TEXTIL press stud; ~ **de presión** m QUÍMICA push-button; ~ **pulsador** m ELEC *control*, ING ELÉC, TELECOM, TV push-button; ~ **de reajuste** m CINEMAT, TV reset knob; ~ **de reencendido del motor** m TRANSP AÉR engine-relight push-button; ~ **de reglaje** m CINEMAT, FOTO, INSTR adjustment knob; ~ **de reglaje de la lente condensadora** m FOTO, INSTR condenser adjustment knob; ~ **de regulación** m GEOFÍS *instrumentos* adjustment knob; ~ **de reposición** m INFORM&PD reset button; ~ **de reset** m INFORM&PD reset button; ~ **de reseteado** m CINEMAT, TV reset knob; ~ **de restauración** m INFORM&PD reset button

botonera: ~ **de discado** f *AmL* INFORM&PD *teléfono* keypad

botriógeno m MINERAL botryogen

botroidal adj MINERAL botryoidal

botulismo m ALIMENT botulism

Bouguer: **anomalía de** ~ f GEOL Bouguer anomaly, free-air anomaly

boulangerita f MINERAL boulangerite

bournonita *f* MINERAL bournonite, cogwheel ore, wheel ore

bóveda *f* C&V *del horno* crown, CARBÓN *geología* cupola, *de túnel* roof, CONST arch, *arquitectura* vault, GAS dome, GEOL *estructura ígnea intrusiva* cupola, INSTAL HIDRÁUL *de horno* dome, glut, MINAS *geología* anticline, roof, arch, PROD chimney, dome, crown, cupola; ~ **de arista** *f* CONST *arquitectura* groined vault; ~ **de cañón** *f* CONST barrel vault, wagon vault; ~ **colectora de humos** *f* TERMOTEC smoke arch; ~ **laminar** *f* TEC ESP membrane; ~ **rebajada** *f* CONST diminished arch; ~ **segmentada** *f* CONST *estructura* segmental arch; ~ **del túnel** *f* CONST tunnel vault; ~ **de vapor** *f* INSTAL HIDRÁUL steam dome

bovedilla *f* CONST overhang, TRANSP MAR *buques de popa cuadra* transom, *construcción naval* fantail, *estructura de popa* counter

bowenita *f* MINERAL bowenite

bowlingita *f* MINERAL bowlingite

boya *f* AUTO, CONST float, PETROL buoy, TRANSP MAR float, *marcas de navegación* buoy; ~ **de alarma en la caldera** *f* INSTAL HIDRÁUL boiler float; ~ **de amarre** *f* CONTAM MAR, PETROL, TRANSP MAR mooring buoy; ~ **de anclaje** *f* CONTAM MAR, PETROL, TRANSP MAR mooring buoy; ~ **de campana** *f* TRANSP MAR *marcas de navegación* bell buoy; ~ **del carburador** *f* AUTO, VEH carburetor float (*AmE*), carburettor float (*BrE*); ~ **cilíndrica** *f* TRANSP MAR *marcas de navegación* barrel buoy, can buoy; ~ **cónica** *f* TRANSP MAR conical buoy; ~ **de espeque** *f* TRANSP MAR *marcas de navegación* spar buoy; ~ **largada a la deriva** *f* METR drifting buoy; ~ **luminosa** *f* TRANSP MAR *marcas de navegación* light buoy; ~ **luminosa de castillete** *f* TRANSP MAR *marcas de navigación* pillar buoy; ~ **marcadora** *f* TRANSP MAR marker buoy; ~ **para marcar** *f* TRANSP MAR marker buoy; ~ **oceanográfica** *f* OCEAN oceanographic buoy; ~ **de recalada** *f* TRANSP MAR *marcas de navegación* sea buoy; ~ **reflector de radar** *f* TRANSP, TRANSP MAR radar reflector buoy; ~ **de silbato** *f* TRANSP MAR *marcas de navegación, señales de sonido* whistle buoy; ~ **simple de amarre de localización descubierta** *f* TEC PETR *producción costa-fuera* exposed-location single-buoy mooring (*ELSBM*); ~ **de superficie** *f* TRANSP MAR *investigación oceanográfica* surface float

boyante *adj* TRANSP MAR buoyant

boyar *vi* TRANSP MAR *barco perdido* float off

boyarín *m* TRANSP MAR beacon; ~ **del ancla** *m* TRANSP MAR beacon

boza *f* TRANSP MAR painter; ~ **de mal tiempo** *f* TRANSP MAR *jarcias* jumper stay

BPP *abr* (*bits por pulgada*) INFORM&PD BPI (*bits per inch*)

BPS *abr* (*bits por segundo*) INFORM&PD BPS (*bits per second*)

Bq *abr* (*bequerelio*) FÍS, FÍS RAD, METR, NUCL Bq (*becquerel*)

b-quark *m* (*quark belleza*) FÍS, FÍS PART b-quark (*beauty quark*)

Br *abr* (*bromo*) QUÍMICA Br (*bromine*)

bracear *vt* TRANSP MAR *velas* handle, *vergas* brace

bracero *m* AGRIC farm-hand

bráctea *f* AGRIC bract

braga *f* CONTAM MAR, TRANSP MAR sling

bragas: ~ **de la tobera de escape** *f pl* TRANSP AÉR exhaust-nozzle breeches

brandisita *f* MINERAL brandisite

braquisinclinal *f* CARBÓN, GEOL, MINAS basin

braquistócrono *m* GEOM brachistochrone

brasa *f* CARBÓN burning-coal, live coal; ~ **del horno de arco eléctrico** *f* CARBÓN, TERMO EAF dust

brasilina *f* QUÍMICA brasilin

braunita *f* MINERAL braunite

braza *f* METR, OCEAN, TRANSP MAR fathom; ~ **mayor** *f* TRANSP MAR *jarcias* mainbrace

brazar *vt* TRANSP MAR *vergas* brace

brazo *m* (*cf tentáculo Esp*) CINEMAT *de grúa o pluma* jib, CONST boom, ELEC *de red de distribución* branch, FÍS lever, ING MECÁ web, lever, *de un tren epicicloidal* arm, MECÁ boom, MINAS horn, PETROL arm, TEC ESP bonding strap, TRANSP AÉR lever, TRANSP MAR *ancla* arm, VEH *cigüeñal* web; ~ **de acceso** *m* INFORM&PD access arm; ~ **adrizante** *m* TRANSP MAR *arquitectura naval* righting lever; ~ **de ancla** *m* TRANSP MAR anchor arm; ~ **de balanza** *m* ING MECÁ, METR balance beam, TRANSP AÉR balance arm; ~ **basculante** *m* AUTO rocker arm, INSTR swivel arm, VEH rocker arm; ~ **del boom** *m* CINEMAT boom arm; ~ **de búsqueda** *m* INFORM&PD seek arm; ~ **de cableado** *m* PROD wiring arm; ~ **de cableado del inductor** *m* PROD field wiring arm; ~ **de caída** *m* TRANSP AÉR drop arm; ~ **de carga para combustibles** *m* TRANSP AÉR refueling boom (*AmE*), refuelling boom (*BrE*); ~ **central** *m* ELEC *fuente de alimentación*, ING ELÉC common branch; ~ **de cigüeñal** *m* AUTO, ING MECÁ, VEH crankweb; ~ **de compás** *m* METR beam compasses; ~ **de control** *m* CINEMAT, TV control arm; ~ **de control de cabeceo** *m* TRANSP AÉR pitch-control arm; ~ **de control de paso** *m* TRANSP AÉR pitch-control arm; ~ **criotécnico** *m* TEC ESP cryotechnic arm; ~ **del cultivador** *m* AGRIC shank; ~ **empujaválvulas** *m* AUTO, VEH rocker arm; ~ **del estibador** *m* C&V stacker arm; ~ **flexible** *m* CINEMAT, TV flexi-arm; ~ **del fonocaptor** *m* ACÚST pick-up arm; ~ **de fuerza** *m* ING MECÁ leverage of a force; ~ **con garras para sujetar el tronco del árbol** *m* ING MECÁ knee; ~ **girable** *m* CONTAM MAR rotatable arm; ~ **de gradiente de gravedad** *m* TEC ESP gravity-gradient boom; ~ **de grúa** *m* ING MECÁ, MECÁ crane jib, TEC PETR *mecanismos* boom, TRANSP AÉR hoist arm; ~ **de horquilla** *m* AUTO, ING MECÁ, VEH fork arm; ~ **del huso** *m* AUTO, VEH spindle arm; ~ **idéntico del puente** *m* ELEC *circuito* equal-arm bridge; ~ **del indicador del nivel en las balsas de lodo** *m* TEC PETR *perforación* pit lever; ~ **intermedio de la dirección** *m* AUTO, VEH steering idler arm; ~ **largo** *m* PROD *de cazo de funderías* bull; ~ **del limpiaparabrisas** *m* AUTO, VEH *accesorio* wiper arm; ~ **del magnetómetro** *m* TEC ESP magnetometer boom; ~ **de mando** *m* AUTO, VEH pitman arm, *dirección* drop arm, steering arm; ~ **de manipulación a distancia** *m* TEC ESP remote manipulator arm; ~ **de manivela** *m* AUTO, ING MECÁ, VEH crank, crank arm, crank cheek, crankweb; ~ **de mar** *m* OCEAN, TRANSP MAR *navegación* firth, sea arm, sound; ~ **del momento** *m* ING MECÁ moment arm; ~ **montador telescópico** *m* CONST telescopic erector arm; ~ **muerto** *m* AGUA cutoff; ~ **mural** *m AmL* (*cf soporte de pared Esp*) MINAS wall bracket; ~ **oscilante** *m* AUTO rocker arm, PROD swing arm, VEH rocker arm, *suspensión* control arm; ~ **oscilante terminal** *m* PROD terminal swing arm; ~ **de palanca** *m* CONST, ING MECÁ lever arm, leverage; ~ **del par** *m* AUTO, VEH torque arm; ~ **de pared** *m* TELECOM wall bracket; ~ **de pared de la galería** *m* MINAS end-wall bracket; ~ **de piñón** *m* ING MECÁ pinion web; ~ **de pitman** *m* AUTO, VEH pitman arm; ~ **portaplaca** *m* PROD *moldeo al barro* spindle arm; ~ **proporcionado** *m* ELEC *puente* proportionate arm; ~ **del puente** *m* INSTR bridge arm; ~ **regulador** *m* ING MECÁ adjusting arm; ~ **de relación** *m* ELEC *puente* ratio arm; ~ **robótico** *m* PROD robotic arm; ~ **del rotor** *m* AUTO, VEH *encendido* rotor arm; ~ **sacador** *m* C&V takeout arm; ~ **secundario libre** *m* AUTO idler arm; ~ **de la silla** *m* C&V chair arm; ~ **de suspensión** *m* AUTO, VEH suspension arm; ~ **telescópico** *m* CINEMAT telescopic arm; ~ **de tensión** *m* INFORM&PD tension arm; ~ **tensor** *m* INFORM&PD tension arm; ~ **de torsión** *m* AUTO, VEH torque arm

brazola *f* TRANSP AÉR, TRANSP MAR coaming; ~ **de escotilla** *f* TRANSP MAR hatch coaming

brea *f* CARBÓN coal-tar pitch, CONST mineral pitch, tar, P&C *materia prima, pintura*, QUÍMICA tar; ~ **de alquitrán** *f* CARBÓN coal-tar pitch; ~ **depositable** *f* PAPEL pitch; ~ **mineral** *f* MINERAL maltha

brecha *f* AGUA breach, GEOL crush rock, ING MECÁ opening, PETROL *sedimentología* breccia; ~ **abisal** *f* OCEAN abyssal pass; ~ **en capa de roca dura** *f* GEOL shell breccia; ~ **de explosión** *f* GEOL *volcánica* explosion breccia; ~ **de fricción** *f* INSTAL HIDRÁUL *en accionamiento hidráulico* crush breccia, fault breccia, rock rubble

brechificado: anticlinal ~ *m* GEOL breached anticline

breithauptita *f* MINERAL breithauptite

breunnerita *f* MINERAL breunnerite

brewsterita *f* MINERAL brewsterite

brezel *m* GEOM pretzel

brida *f* (*cf reborde AmL*) ALIMENT *resortes hojas* buckle, martingale, stirrup, *tuberías* flange, strap, clip, trunnion, grip, yoke, *máquina de taladrar* bridge, *mecánica* bridle, fastener; ~ **de acoplamiento** *f* MECÁ coupling flange; ~ **de angular** *f* FERRO *equipo inamovible de carriles* angle fishplate; ~ **de ballesta** *f* VEH *suspensión* U-bolt; ~ **de ballestas** *f* ING MECÁ spring washer; ~ **basculante** *f* ING MECÁ swivel stirrup; ~ **de base** *f* CONST *tuberías* bottom flange; ~ **de campana** *f* ING MECÁ hub-type flange; ~ **con cara revestida** *f* PROD faced flange; ~ **de carril** *f* FERRO *equipo inamovible* rail bond (*BrE*), rail splice (*AmE*); ~ **de carriles** *f* FERRO strap rail; ~ **de choque** *f* ELEC, ING ELÉC choke flange; ~ **ciega** *f* ING MECÁ blank flange, blind flange; ~ **de cocodrilo** *f* ELEC, ING ELÉC, ING MECÁ crocodile clip; ~ **de collar** *f* ING MECÁ neck flange; ~ **conectora** *f* TEC ESP connecting flange; ~ **doble de conexión** *f* TRANSP AÉR connecting twin-yoke; ~ **Dzus** *f* TEC ESP *aviones* Dzus fastener; ~ **del eje** *f* AUTO, VEH *ruedas* axle flange; ~ **de empalme** *f* MECÁ coupling flange; ~ **de enganche desmontable** *f* FERRO *vehículos* removable coupling link; ~ **del objetivo** *f* FOTO lens flange; ~ **de obturación** *f* ING MECÁ blank flange, blind flange; ~ **del radiador** *f* AUTO, VEH radiator flange; ~ **rebatida** *f* PROD *chapistería* turned-up flange; ~ **de rebote** *f* AUTO, VEH rebound clip; ~ **de sierra** *f* ING MECÁ saw clamp; ~ **de tornillo** *f* ING MECÁ screw clamp; ~ **de unión** *f* MECÁ coupling flange

bridas[1]: **con** ~ *adj* GEN flanged

bridas[2]: ~ **de contacto** *f pl* NUCL mating flanges; ~ **que casan** *f pl* MECÁ mating flanges

brigada: ~ **de incendios** *f* SEG, TERMO firefighting personnel; ~ **en voladizo** *f* CONST flying squad

brillante *adj* CINEMAT, FOTO glossy, ÓPT bright, P&C *pintura* glossy, TEXTIL glossy, lustrous, TV glossy

brillo[1]: **sin** ~ *adj* SEG glare-free

brillo[2] *m* (*cf luminosidad Esp*) COLOR glossiness, CONTAM MAR sheen, ELECTRÓN, FÍS brightness, GEOL luster (*AmE*), lustre (*BrE*), IMPR gloss, METAL illumination, MINERAL luster (*AmE*), lustre (*BrE*), ÓPT brightness, P&C *propiedad física*, PAPEL gloss, PROD gloss, glossiness, REVEST luster finish (*AmE*), lustre finish (*BrE*), polish, SEG glare, TEC ESP radiance, TRANSPAÉR, TRANSP MAR display brilliance, TV brightness; ~ **adamantino** *m* MINERAL adamantine luster (*AmE*), adamantine lustre (*BrE*); ~ **catódico** *m* FÍS cathode glow; ~ **fotométrico** *m* FÍS RAD photometric brightness; ~ **de la imagen** *m* TELECOM *TV* picture brightness, TRANSPAÉR, TRANSP MAR display brilliance; ~ **metálico** *m* QUÍMICA metallic luster (*AmE*), metallic lustre (*BrE*); ~ **negativo** *m* FÍS negative glow; ~ **de pantalla** *m* CINEMAT screen brightness; ~ **de la tinta** *m* IMPR ink gloss

brinelación *f* ING MECÁ brinelling

briqueta *f* C&V briquette, CARBÓN boulet, briquette

brisa *f* METEO breeze; ~ **débil** *f* METEO gentle breeze; ~ **fuerte** *f* METEO strong breeze; ~ **de lago** *f* METEO onshore breeze; ~ **ligera** *f* METEO light breeze; ~ **de mar** *f* METEO onshore breeze, sea breeze; ~ **de montaña** *f* METEO mountain breeze; ~ **muy débil** *f* METEO light breeze; ~ **suave** *f* METEO gentle breeze; ~ **de tierra** *f* METEO, TRANSP MAR land breeze; ~ **de valle** *f* METEO valley breeze

brizna *f* CONST splinter, MECÁ chip; ~ **de silicio** *f* ELEC silicon chip

broca *f* CONST *sondeo* bit, bore bit, GEOL auger, ING MECÁ, MECÁ bit, cutter, boring tool, drill, tool bit, MINAS *barreno* boring tool, *sondeo* drill, PETROL drilling bit, TEC PETR *perforación* pilot bit, hard of formation bit, crown bit, bit, core bit, spudding bit, roller bit, reaming bit, rock bit, rotary bit, drag bit, spiral bit, drill bit; ~ **abierta** *f* Esp (*cf brochadora AmL*) MINAS *mandrinado* broaching bit; ~ **de agujeros a roscar** *f* ING MECÁ tapping drill; ~ **ajustable** *f* ING MECÁ extension bit; ~ **de arrastre** *f* PETROL bit; ~ **y avellanadora combinadas** *f* ING MECÁ, PROD combined drill and countersink; ~ **de avellanar** *f* ING MECÁ, MECÁ, PROD countersink bit; ~ **de berbiquí** *f* ING MECÁ screw auger, *dispositivo de taladrar* auger bit; ~ **de botar** *f* ING MECÁ center key (*AmE*), centre key (*BrE*); ~ **de cabeza de flecha** *f* ING MECÁ flat drill; ~ **de carburo al tungsteno descartable** *f* ING MECÁ throwaway carbide drill; ~ **de carburo al tungsteno desechable** *f* ING MECÁ throwaway carbide drill; ~ **de centrar** *f* ING MECÁ center bit (*AmE*), centre bit (*BrE*), center bit for bit stock (*AmE*), centre bit for bit stock (*BrE*); ~ **para centrar** *f* ING MECÁ center drill (*AmE*), centre drill (*BrE*); ~ **de centro** *f* ING MECÁ center brace bit (*AmE*), centre brace bit (*BrE*); ~ **corriente** *f* ING MECÁ jobber drill; ~ **corta** *f* ING MECÁ stub drill; ~ **de corte** *f* ING MECÁ cutting driftpin; ~ **de diamante** *f* TEC PETR diamond bit; ~ **enganchada** *f* PROD latched bit; ~ **escalonada** *f* ING MECÁ step drill; ~ **con espiga**

de cono Morse *f* ING MECÁ Morse taper shank drill; **~ de estrías rectas** *f* ING MECÁ straight-fluted drill; **~ extensible** *f* ING MECÁ extension bit; **~ con guía** *f* ING MECÁ pin drill; **~ de guía** *f* ING MECÁ center bit (*AmE*), centre bit (*BrE*), center brace bit (*AmE*), centre brace bit (*BrE*); **~ helicoidal** *f* ING MECÁ, MECÁ twist drill; **~ helicoidal con canal lubricador** *f* ING MECÁ oil tube twist drill; **~ helicoidal de cola cilíndrica** *f* ING MECÁ parallel-shank twist drill; **~ helicoidal con mango cilíndrico** *f* ING MECÁ twist drill with parallel shank, twist drill with straight shank; **~ helicoidal con mango cuadrado cónico** *f* ING MECÁ twist drill with taper square shank; **~ helicoidal multidiámetro** *f* ING MECÁ subland twist drill; **~ hueca** *f* ING MECÁ core drill, shell drill, MINAS *sondeo testiquero* core drilling; **~ hueca cilíndrica de bordes cortantes de sierra** *f* ING MECÁ hole saw; **~ hueca cilíndrica con bordes de sierra de carburo al tungsteno** *f* ING MECÁ tungsten-carbide grit hole saw; **~ hueca de plaquitas de carburo para hormigón** *f* ING MECÁ core drill carbide-tipped for concrete; **~ de labios rectos** *f* ING MECÁ straight-fluted drill; **~ de lanzadera** *f* TEXTIL shuttle spindle; **~ de madera** *f* ING MECÁ wood bit; **~ de mango cilíndrico** *f* ING MECÁ straight-shank twist drill; **~ de mango cuadrado** *f* ING MECÁ square-shank drill; **~ de mango recto** *f* ING MECÁ straight-shank twist drill; **~ de manguito** *f* ING MECÁ shell drill; **~ para metal** *f* ING MECÁ twist drill; **~ de múltiples diámetros** *f* ING MECÁ step drill; **~ plana** *f* ING MECÁ flat drill; **~ de punzón** *f* ING MECÁ pin drill; **~ de retenida** *f* PROD latch bit; **~ de rodillos** *f* TEC PETR roller bit; **~ salomónica** *f* ING MECÁ, MECÁ twist drill; **~ salomónica con espiga de cono Morse** *f* ING MECÁ, MECÁ Morse taper shank twist drill; **~ de sondeo** *f* MINAS jumper; **~ de tetón cilíndrico** *f* ING MECÁ pin drill; **~ de tres puntas** *f* ING MECÁ center bit for bit stock (*AmE*), centre bit for bit stock (*BrE*); **~ tricona** *f* TEC PETR tricone bit; **~ de tubos y acero** *f* ING MECÁ tube-and-steel drill; **~ para vidrio** *f* C&V glass drill, ING MECÁ glass bit, glass drill
brocado *m* TEXTIL brocade
brocal *m* C&V wide mouth container
brocha *f* COLOR, CONST brush, ING MECÁ, MECÁ, MINAS broach, P&C *pintura* paintbrush; **~ de dientes insertados** *f* ING MECÁ inserted-tooth broach; **~ para encalar** *f* COLOR whitewash brush; **~ para pintura a la cal** *f* COLOR whitewash brush; **~ plana** *f* COLOR flat brush
brochado *m* ING MECÁ, MECÁ, MINAS broaching
brochadora *f* *AmL* (*cf broca abierta Esp*) ING MECÁ, MECÁ broaching machine, MINAS *mandrinado* broaching bit; **~ de correderas múltiples** *f* ING MECÁ multiple-ram broaching machine; **~ vertical por presión** *f* ING MECÁ press-type vertical broaching machine
brochal *m* CONST trimmed joist, trimmer
brochantita *f* MINERAL brochantite
brochar *vt* ING MECÁ, MECÁ, MINAS broach
broche: **~ y corchete** *m* ING MECÁ hook and eye; **~ pentagonal** *m* ING MECÁ five-sided broach
bromal *m* QUÍMICA bromal
bromargirita *f* MINERAL bromargyrite, bromyrite
bromato *m* QUÍMICA bromate
bromelina *f* ALIMENT bromelain

bromhidrato *m* QUÍMICA hydrobromide
bromhídrico *adj* QUÍMICA bromhidric
brómico *adj* QUÍMICA bromic
bromita *f* MINERAL bromite
bromo *m* (*Br*) AGRIC brome grass, QUÍMICA bromine (*Br*)
bromoacético *adj* QUÍMICA bromoacetic
bromoacetona *f* QUÍMICA bromoacetone
bromobenceno *m* QUÍMICA bromobenzene
bromofenol *m* QUÍMICA bromophenol
bromoformo *m* QUÍMICA bromoform
bromometano *m* QUÍMICA bromomethane
bromuro *m* DETERG, FOTO, QUÍMICA bromide; **~ decílico** *m* QUÍMICA decyl bromide; **~ de metilo** *m* QUÍMICA methyl bromide; **~ de plata** *m* MINERAL bromyrite, QUÍMICA silver bromide
bronce *m* COLOR, MECÁ, PROD, QUÍMICA brass, bronze; **~ de aluminio** *m* MECÁ aluminium bronze (*BrE*), aluminum bronze (*AmE*); **~ de cañón** *m* D&A, MECÁ gunmetal; **~ fosforoso** *m* ELEC, ING ELÉC, QUÍMICA phosphor bronze; **~ fundido** *m* MECÁ, METAL cast bronze; **~ industrial** *m* MECÁ gunmetal; **~ manganoso** *m* MECÁ manganese bronze; **~ mecánico** *m* MECÁ gunmetal
broncesoldado *adj* CONST, ING MECÁ, PROD, TERMO brazed
broncesoldadura *f* CONST, ING MECÁ, PROD, TERMO brazing; **~ de inducción de alta frecuencia** *f* CONST high-frequency induction brazing
broncesoldar *vt* CONST, ING MECÁ, PROD, TERMO braze
broncita *f* MINERAL bronzite
brookita *f* MINERAL brookite
brotar *vi* AGRIC bud
brote *m* AGRIC *de enfermedad* outbreak, HIDROL spurting out, TERMO *de fuego* outbreak; **~ de cebada germinada** *m* AGRIC malt sprout
broza *f* AGRIC slash
brucelosis *f* AGRIC Bang's disease, brucellosis, undulant fever
brucina *f* ALIMENT, QUÍMICA brucine
brucita *f* MINERAL brucite
brujido *m* MINAS diamond dust, diamond powder
brujir *vt* ING MECÁ, MECÁ, PROD grind
brújula *f* CONST, D&A, FÍS compass, GEOFÍS compass, magnetic compass, mariner's needle, INSTR magnetic compass, TEC ESP compass, TRANSP AÉR, TRANSP MAR magnetic compass; **~ de agrimensor** *f* CONST surveyor's compass; **~ de bolsillo** *f* D&A pocket compass; **~ declinatoria** *f* PROD trough compass; **~ giroscópica** *f* TRANSP directional gyro, TRANSP AÉR gyrocompass, gyroscopic compass; **~ de giróscopo direccional sincronizado** *f* TRANSP AÉR gyrosyn compass; **~ de inclinación** *f* CARBÓN inclinometer, GEOFÍS dip compass; **~ magnética** *f* GEOFÍS, INSTR, TRANSP AÉR magnetic compass; **~ de mina** *f* MINAS mine surveying compass; **~ de minero** *f* GEOFÍS dip compass, MINAS mine surveying compass; **~ radiogoniométrica** *f* TRANSP AÉR, TRANSP MAR radio compass; **~ de senos** *f* ELEC *instrumento* sine galvanometer; **~ de trípode** *f* MINAS *herramienta, geológica* dial; **~ de variación** *f* GEOFÍS variation compass
brulote *m* TERMO fireship
bruma *f* CONTAM, METEO mist; **~ atmosférica** *f* FOTO atmospheric haze; **~ cargada de humo** *f* CONTAM, METEO smog

brumoso *adj* TRANSP MAR *tiempo atmosférico* hazy

bruñido *m* C&V, IMPR burnishing, ING MECÁ honing, MECÁ burnishing, P&C *operación*, PROD polishing; **~ profundo** *m* C&V deep sleek

bruñidor *m* ING MECÁ, MECÁ planisher, PROD polisher

bruñidora *f* C&V burnisher

bruñir *vt* C&V, IMPR *papel de estaño* burnish, ING MECÁ, MECÁ planish, polish, P&C, PROD polish, TV burn in

brusca *f* TRANSP MAR *cubierta* round; **~ reglamentaria** *f* TRANSP MAR *construcción* camber

brushita *f* MINERAL brushite

bruto[1]: **en ~** *adj* CARBÓN raw

bruto[2] *m* CINEMAT *lámpara de iluminación* brute

BSE *abr* (*encefalopatía espongiforme bovina*) AGRIC BSE (*bovine spongiform encephalopathy*)

BSI *abr* (*Instituto Británico de Normalización*) PROD BSI (*British Standards Institution*)

B3ZS *abr* (*código bipolar con sustitución de tres ceros*) TELECOM B3ZS (*bipolar code with three-zero substitution*)

bucarán *m* IMPR art canvas, *lona* buckram

buceador: **~ en aguas profundas** *m* PETROL, TRANSP MAR deep-sea diver

buceadora: **~ en aguas profundas** *f* PETROL, TRANSP MAR deep-sea diver

buceo *m* OCEAN diving, skin dive, PETROL diving, TEC ESP dive, TEC PETR scuba diving; **~ en aguas profundas** *m* TEC PETR, TRANSP MAR deep-water diving; **~ con aparato respiratorio autónomo** *m* TRANSP MAR scuba diving; **~ hacia abajo** *m* TEC PETR *geología* down dip; **~ a saturación** *m* OCEAN saturation dive; **~ en saturación** *m* OCEAN saturation diving; **~ por saturación** *m* TEC PETR saturated diving; **~ sin traje** *m* OCEAN skin diving

bucle[1]: **en ~** *adj* ELEC *circuito* looped

bucle[2] *m* ACÚST, CINEMAT, ELEC, ELECTRÓN, INFORM&PD, ING ELÉC loop, ING MECÁ eye, PETROL, PROD, TELECOM, TEXTIL *hilatura* loop; **~ abierto** *m* ELEC *circuito* open loop; **~ anidado** *m* INFORM&PD nested loop; **~ de audio** *m* TELECOM audio loop; **~ de autorrestauración** *m* INFORM&PD self-resetting loop; **~ cerrado** *m* ELEC *circuito*, ELECTRÓN, ING ELÉC closed loop; **~ de cinta de papel** *m* INFORM&PD paper-tape loop; **~ continuo** *m* CINEMAT continuous loop; **~ de corriente de inducción magnética** *m* TELECOM magnetic induction current loop; **~ de Costas** *m* TELECOM Costas loop; **~ digital** *m* TELECOM digital loop; **~ de dislocación** *m* CRISTAL dislocation loop; **~ enganchado** *m* TELECOM *oscilador* locked loop; **~ de enganche de fase** *m* TEC ESP *telecomunicación*, TELECOM phase-locked loop (*PLL*); **~ de espera** *m* INFORM&PD wait loop; **~ de histéresis** *m* METAL hysteresis loop; **~ de histéresis magnética** *m* ELEC magnetic hysteresis loop; **~ inferior** *m* CINEMAT lower loop; **~ infinito** *m* INFORM&PD infinite loop; **~ integrado** *m* TRANSP embedded loop; **~ del izador** *m* TRANSP AÉR hoisting eye; **~ de realimentación** *m* PROD feedback loop; **~ de tierra** *m* PROD ground loop; **~ de video** *AmL*, **~ de vídeo** *m Esp* TELECOM video loop

bucles: **~ engarzados** *m pl* TEXTIL meshed loops; **~ enredados** *m pl* TEXTIL intermeshed loops

buen[1]: **de ~ rendimiento** *adj* TERMO fuel-efficient

buen[1]: **en ~ estado de funcionamiento** *fra* PROD in working order; **en ~ estado para navegar** *fra* TRANSP MAR in navigable condition

buena: **en ~ vela** *adj* TRANSP MAR *navegación* full and by

bufadero *m* OCEAN *erosión de costas* blowhole

bufanda *f* TEXTIL comforter

búfer: **~ de entrada** *m* INFORM&PD input buffer; **~ de salida** *m* INFORM&PD output buffer

buffer *m* QUÍMICA buffer; **~ de epoxi** *m* TELECOM epoxy buffer

bufotoxina *f* QUÍMICA bufotoxin

bujarda *f* PROC QUÍ granulating hammer

buje *m* ING MECÁ pillow, TRANSP AÉR spinner, VEH *rueda* hub; **~ de aislamiento** *m* ING MECÁ insulation bush; **~ de aleación de cobre** *m* ING MECÁ copper alloy bush; **~ de alimentación** *m* ING MECÁ feed bush; **~ de cojinete principal** *m* AUTO main-bearing bushing; **~ cónico** *m* VEH *rueda* tapered hub; **~ del cuadrante** *m AmL* (*cf buje del vástago Esp*) TEC PETR *perforación* kelly bushing; **~ de desenroscar** *m* ING MECÁ unscrewing bush; **~ flotante** *m* ING MECÁ floating bush; **~ de kelly** *m AmL* (*cf buje del vástago Esp*) TEC PETR *perforación* kelly bushing; **~ de la palanca de control** *m* TRANSP AÉR control-column boss; **~ del pie de biela** *m* AUTO small end bushing; **~ soporte del pie de biela** *m* AUTO piston-boss bushing; **~ del vástago** *m Esp* (*cf buje de kelly AmL, cf buje del cuadrante AmL*) TEC PETR *perforación* kelly bushing

bujía *f* AUTO, MECÁ, VEH *encendido* plug; **~ caliente** *f* AUTO heat plug, hot spark plug; **~ de encendido** *f* AUTO, ELEC spark plug, MECÁ ignition plug, VEH ignition plug, spark plug; **~ de ignición** *f* AUTO ignition plug, spark plug; **~ de incandescencia** *f* AUTO resistor-type spark plug; **~ incandescente** *f* AUTO, TERMO glow plug; **~-pie** *f* CINEMAT *iluminación* foot candle

bulárcama *f* OCEAN *construcción naval* web frame, TRANSP MAR web frame, *construcción metálica* plate web

bulbo *m* C&V bulb; **~ de caucho** *m AmL* (*cf pera de caucho Esp*) LAB *pipeta*, P&C rubber bulb; **~ desecador** *m* FOTO, LAB *material de vidrio* potash bulb; **~ lateral** *m* AGRIC offset; **~ térmico** *m* AUTO thermal bulb

bulldozer *m* AUTO, CONST bulldozer

bullidor *m* ING MECÁ back boiler

bullir *vi* TERMO boil

bulón *m* MECÁ bolt, MINAS *del bocarte* boss; **~ de anclaje** *m* MECÁ anchor bolt; **~ de cabeza hexagonal** *m* ING MECÁ hexagonal-head bolt; **~ explosivo** *m* MECÁ, TEC ESP *vehículos* explosive bolt; **~ ordinario** *m* ING MECÁ machine bolt; **~ del pie de biela unido al pistón** *m* AUTO, VEH piston-pin locked to piston; **~ del pistón** *m* AUTO, VEH *motor* piston pin; **~ de saneo** *m Esp* (*cf redondo de saneado del techo AmL*) MINAS *galerías* scaling bar

bulonado *m* MINAS *galerías* resin roof bolting

bulto *m* EMB bundle, TRANSP parcel; **~ de residuos de baja actividad** *m* NUCL low-level waste package

bumerán *m* FÍS FLUID boomerang

búnker: **~ C** *m* CONTAM MAR, TRANSP AÉR bunker C

bunsenina *f* MINERAL bunsenine

bunsenita *f* MINERAL bunsenite

buque *m* CONTAM MAR *pesqueros, veleros* vessel, *lenguaje profesional y jurídico* ship, OCEAN ship, TRANSP, TRANSP MAR ship, vessel; **~ almirante** *m* TRANSP MAR flagship; **~ de altura** *m* TRANSP MAR ocean-going ship, seagoing vessel; ~ **de apoyo** *m* TRANSP MAR support vessel; ~ **auxiliar** *m* CONTAM MAR tender; ~ **balizador** *m* TRANSP MAR buoy tender; ~ **butanero** *m* TEC PETR *transporte*, TRANSP butane carrier, butane tanker; ~ **cablero** *m* TELECOM, TRANSP MAR cable ship; ~ **de cabotaje** *m* TRANSP MAR coaster; ~ **carbonero** *m* CARBÓN, MINAS, TRANSP MAR coal ship, collier; ~ **de carga** *m* *Esp* (*cf fletero AmL*) TRANSP MAR cargo ship, freighter; ~ **de carga mixta** *m* TRANSP MAR mixed cargo ship; ~ **de carga y pasaje** *m* TRANSP MAR cargo and passenger ship; ~ **de carga politérmica** *m* TRANSP polythermal cargo ship; ~ **de carga refrigerada** *m* REFRIG, TRANSP, TRANSP MAR refrigerated-cargo ship; ~ **de carga rompe-hielos** *m* TRANSP MAR icebreaking cargo ship; ~ **de carga seca** *m* TRANSP MAR dry-cargo ship; ~ **de carga seca a granel** *m* TRANSP MAR dry-bulk carrier; ~ **de carga para usos varios** *m* TRANSP MAR multipurpose carrier; ~ **carguero tipo Seabee** *m* TRANSP MAR Seabee carrier; ~ **cisterna** *m* TEC PETR tanker, TRANSP MAR butane tanker; ~ **cisterna de butano** *m* TEC PETR, TRANSP, TRANSP MAR butane tanker; ~ **cisterna de crudo** *m* PETROL, TEC PETR, TRANSP MAR crude-oil tanker; ~ **cisterna de entregas** *m* PROD, TRANSP MAR delivery tanker; ~ **cisterna de LGN** *m* TEC PETR, TRANSP MAR NGL tanker; ~ **cisterna para líquidos espesos** *m* TRANSP MAR ore-slurry-oil tanker (*OSO tanker*); ~ **-cisterna de propano** *m* TEC PETR, TRANSP MAR propane tanker; ~ **para cruceros** *m* TRANSP MAR cruise ship; ~ **de cuatro palos** *m* TRANSP MAR four-master; ~ **de una cubierta** *m* TRANSP MAR single-decked ship; ~ **dedicado al tendido de cables** *m* TRANSP MAR cable-laying ship; ~ **descargador** *m* CONTAM MAR unloader; ~ **descontaminador** *m* CONTAM, CONTAM MAR depolluting ship; ~ **de dos hélices** *m* TRANSP MAR twin-screw steamer; ~ **dragaminas** *m* D&A, TRANSP MAR minesweeper; ~ **escolta** *m* TRANSP MAR *marina* escort ship; ~ **escuela** *m* TRANSP MAR training ship; ~ **para la extracción de petróleo** *m* CONTAM *combustibles* oil-recovery vessel; ~ **factoría** *m* OCEAN, TRANSP MAR factory ship; ~ **faro** *m* TRANSP MAR *carguero* light vessel, *marcas de navegación* lightship; ~ **frigorífico** *m* OCEAN deep-freeze ship, TRANSP MAR cold storage ship, reefer ship; ~ **frutero** *m* TRANSP MAR fruit carrier; ~ **gasero** *m* TRANSP MAR liquefied-natural-gas carrier (*LNG carrier*); ~ **gemelo** *m* TRANSP MAR sister ship; ~ **de guerra** *m* D&A, TRANSP MAR man-of-war, warship; ~ **hidrográfico** *m* TRANSP MAR hydrographic survey vessel; ~ **hospital** *m* TRANSP MAR hospital ship; ~ **de iguales características** *m* TRANSP MAR sister ship; ~ **insignia** *m* TRANSP MAR flagship; ~ **de investigaciones oceanográficas** *m* OCEAN, TRANSP MAR oceanographic research ship; ~ **en lastre** *m* TRANSP MAR ship in ballast; ~ **de línea** *m* TRANSP MAR liner; ~ **mercante** *m* TRANSP MAR merchant ship, merchantman; ~ **meteorológico** *m* METEO, TRANSP MAR weather ship; ~ **minador** *m* D&A, TRANSP MAR minelayer, minelaying ship; ~ **con las mismas características** *m* TRANSP MAR sister ship; ~ **multicasco** *m* TRANSP MAR multihulled ship; ~ **multiservicios** *m* TEC PETR *operaciones marinas*, TRANSP MAR multiservice vessel (*MSV*); ~ **naufragado** *m* TRANSP MAR *navegación* wreck;

~ **náufrago** *m* TRANSP MAR *navegación* wreck; ~ **de navegación marítima** *m* CONTAM MAR seagoing vessel; ~ **nodriza** *m* TEC ESP mother ship, TRANSP MAR mother ship, tender; ~ **oceanográfico** *m* TRANSP MAR ocean survey vessel; ~ **de pasaje** *m* TRANSP MAR liner, passenger ship; ~ **patrullero** *m* CONTAM MAR patrol boat, D&A patrol boat, patrol craft, patrol vessel; ~ **patrullero impulsado por motor** *m* D&A motor-propelled patrol boat; ~ **en peligro** *m* TRANSP MAR ship in distress; ~ **de perforación** *m* PETROL, TEC PETR drill ship; ~ **para perforaciones** *m* TRANSP MAR drill ship; ~ **perforador desde superficie** *m* PETROL, TEC PETR surface-piercing craft; ~ **pesquero** *m* OCEAN, TRANSP MAR fishing vessel; ~ **petrolero** *m* MINAS oiler, PETROL, TEC PETR tanker, TRANSP MAR oiler; ~ **petrolero de butano** *m* TEC PETR, TRANSP, TRANSP MAR butane tanker; ~ **petrolero de crudo** *m* PETROL, TEC PETR, TRANSP MAR crude-oil tanker; ~ **petrolero rompe-hielos** *m* PETROL, TEC PETR, TRANSP MAR icebreaking oil tanker; ~ **planero** *m* TRANSP MAR hydrographic survey vessel; ~ **polivalente** *m* TRANSP MAR *marina mercante* multipurpose ship; ~ **portaviones** *m* D&A, TRANSP MAR aircraft carrier; ~ **portabarcazas** *m* TRANSP MAR barge carrier; ~ **portacontenedores** *m* TRANSP MAR container ship; ~ **portagabarras** *m* TRANSP lighter aboard ship carrier (*LASH carrier*), TRANSP MAR barge carrier; ~ **propulsado por turbina de gas** *m* TRANSP MAR gas-turbine ship; ~ **de propulsión mecánica** *m* TRANSP MAR mechanically-propelled ship; ~ **de prospección de hidrocarburos** *m* PETROL, TEC PETR, TRANSP MAR drill ship; ~ **puerta** *m* AGUA, HIDROL, INSTAL HIDRÁUL, TRANSP MAR *diques, esclusas* batardeau; ~ **que cede el paso** *m* TRANSP MAR give-way vessel; ~ **en que se halla el mando naval** *m* D&A, TRANSP MAR command ship; ~ **remolcador de gabarras** *m* TRANSP lighter carrier; ~ **repostador** *m* TRANSP MAR refueler (*AmE*), refueller (*BrE*); ~ **rolón** *m* TRANSP, TRANSP MAR roll-on vessel, roll-on/roll-off vessel (*ro-ro vessel*); ~ **rompehielos** *m* D&A, TRANSP MAR icebreaker; ~ **en rosca** *m* TRANSP MAR *carguero* light ship; ~ **de rueda** *m* TRANSP MAR paddle boat; ~ **de salvamento** *m* D&A, TRANSP wrecker, TRANSP MAR salvage vessel, *diseño de barcos* reserve buoyancy; ~ **siniestrado** *m* TRANSP MAR ship in distress; ~ **con sistema de transflotación de gabarras** *m* TRANSP MAR float-on/float-off vessel; ~ **de una sola hélice** *m* TRANSP MAR single-screw ship; ~ **de suministro** *m* TRANSP MAR supply vessel; ~ **de sustentación hidrodinámica** *m* TRANSP MAR surface effect ship (*SES*); ~ **tanque** *m* CONTAM MAR, TRANSP MAR tanker; ~ **tanque de gas natural líquido** *m* TEC PETR natural-gas liquids tanker (*NGL tanker*); ~ **tanque de metano** *m* TEC PETR methane tanker; ~ **tanque petrolero** *m* TEC PETR oil tanker; ~ **tanque quimiquero** *m* TRANSP MAR chemical tanker; ~ **tanque con sistema de vacío** *m* CONTAM MAR *limitación del derrame de hidrocarburos en caso de brecha* vacuum tanker; ~ **tanque para el transporte de gas** *m* TRANSP MAR gas carrier; ~ **tanque para el transporte de GPL** *m* TRANSP MAR LPG carrier; ~ **tanque para el transporte de vino** *m* TRANSP MAR wine carrier; ~ **torpedero** *m* D&A torpedo boat; ~ **transbordador** *m* TRANSP MAR ferry; ~ **transbordador de trenes y pasaje** *m* TRANSP MAR ferryboat;

~ de transbordo rodado *m* TRANSP, TRANSP MAR roll-on vessel, roll-on/roll-off vessel (*ro-ro vessel*); **~ de transbordo vertical** *m* TRANSP MAR lift-on/lift-off vessel; **~ de transbordo vertical por izada** *m* TRANSP MAR lift-on/lift-off vessel; **~ transportador** *m* TRANSP MAR ferry; **~ transportador mixto** *m* TRANSP MAR oil-bulk-ore carrier (*OBO carrier*); **~ transportador de vehículos carreteros y vagones cargados** *m* TRANSP, TRANSP MAR roll-on/roll-off ship (*ro-ro ship*); **~ para transporte de barcazas** *m* TRANSP MAR lighter aboard ship (*LASH*); **~ para transporte de butano líquido** *m* TEC PETR, TRANSP MAR butane carrier, butane tanker; **~ de transporte combinado** *m* TRANSP MAR oil-coal-ore carrier (*OCO carrier*); **~ para el transporte de crudos** *m* CONTAM, CONTAM MAR, PETROL, TEC PETR, TRANSP MAR crude carrier; **~ para transporte de gas natural licuado y gas de petróleo licuado** *m* TERMO, TRANSP MAR liquefied-natural-gas and liquefied-petroleum-gas carrier (*LNG-LPG carrier*); **~ de transporte mixto** *m* TRANSP MAR oil-bulk-ore carrier (*OBO carrier*); **~ de transporte de tropas** *m* D&A troopship; **~ de turbinas** *m* TRANSP MAR turbine vessel; **~ vacío** *m* TRANSP MAR light vessel, lightship; **~ de vapor** *m* TRANSP MAR mechanically-propelled ship; **~ de vela con motor auxiliar** *m* TRANSP MAR auxiliary engine sailing ship; **~ viga** *m* TRANSP MAR *construcción naval* hull girder

buques *m pl* TRANSP MAR shipping

buratita *f* MINERAL buratite

burbuja *f* C&V air bell, blister, bubble, CINEMAT air bell, FÍS bubble, P&C *defecto de envasado, pintura,* blister, bubble, PROC QUÍ bubble, TRANSP MAR *metacrilato, pintura* blister; **~ de aire** *f* FÍS air bubble, P&C entrapped air; **~ alargada en tubos de vidrio** *f* C&V airline; **~ ampollada** *f* C&V broken seed; **~ de debituse** *f* C&V debituse bubble; **~ de sal** *f* C&V salt bubble; **~ de sobrecalentamiento** *f* C&V reboil bubble; **~ al vacío** *f* C&V vacuum bubble; **~ de vapor** *f* PROC QUÍ vapor bubble (*AmE*), vapour bubble (*BrE*)

burbujeado *m* C&V bubbling; **~ del esmalte** *m* C&V bubbling

burbujeador *f* C&V bubbler

burbujeante *adj* FÍS bubbling

burbujear: ~ a través *vi* FÍS bubble through

burbujeo *m* FÍS, OCEAN, QUÍMICA bubbling

burda *f* TRANSP MAR *jarcia* backstay

bureta *f* LAB, QUÍMICA *material de vidrio* burette; **~ automática** *f* LAB automatic burette; **~ de gas** *f* LAB gas burette; **~ de separación** *f* LAB, PROC QUÍ separating burette

buril *m* CONST chisel, ING MECÁ burr, cold chisel, scriber, PROD crosscut chisel, *molde de arcilla* cold chisel, pricker; **~ de acanalar** *m* MECÁ groove-cutting chisel

burilado *m* PROD chiseling (*AmE*), chiselling (*BrE*)

burilador *m* PROD *persona* chiseler (*AmE*), chiseller (*BrE*)

buriladora *f* ING MECÁ router

burlete *m* MECÁ gasket

burro: ~ de arranque *m* AmL (*cf motor de arranque Esp*) ING MECÁ *automóviles* starter motor

burst: ~ alternante *m* AmL (*cf estallido alternante Esp*) TV alternating burst

bus *m* INFORM&PD bus, highway, trunk; **~ articulado** *m* AmL (*cf autobús articulado Esp*) AUTO, TRANSP bimodal bus; **~ bimodal** *m* AmL (*cf autobús bimodal Esp*) AUTO, TRANSP bimodal bus, dual-mode bus; **~ de control** *m* INFORM&PD control bus; **~ de datos** *m* INFORM&PD data bus; **~ de datos ópticos** *m* TELECOM optical data bus; **~ de direcciones** *m* INFORM&PD address bus; **~ de efectos** *m* CINEMAT, TV effects bus; **~ de efectos especiales** *m* TV special effects bus; **~ eléctrico** *m* AmL (*cf autobús eléctrico de batería Esp*) TRANSP battery bus; **~ de encadenamiento** *m* INFORM&PD daisychain bus; **~ de encadenamiento mariposa** *m* INFORM&PD daisychain bus; **~ de entrada/salida** *m* INFORM&PD input/output bus; **~-ferrobús** *m* AmL (*cf autobús-ferrobús Esp*) TRANSP road-rail bus; **~ de gas natural licuado** *m* AmL (*cf autobús de gas natural licuado Esp*) TRANSP liquid-natural-gas bus (*LNG bus*); **~ por gas natural a presión** *m* AmL (*cf autobús de gas natural a presión Esp*) TRANSP pressurized natural-gas bus; **~ de gas-petróleo líquido** *m* AmL (*cf autobús de gas licuado de petróleo Esp*) TRANSP liquid-petroleum-gas bus; **~ giroscópico** *m* AmL (*cf autobús giroscópico Esp, cf girobús Esp*) TRANSP, VEH gyrobus; **~ híbrido** *m* AmL (*cf autobús híbrido Esp*) TRANSP hybrid bus; **~ montado en vagón de ferrocarril** *m* AmL (*cf autobús montado sobre vagón de ferrocarril Esp*) TRANSP bus on railroad car (*AmE*), bus on railway wagon (*BrE*); **~ oruga** *m* AmL (*cf autobús oruga Esp*) AUTO, TRANSP bimodal bus; **~ principal** *m* INFORM&PD backbone bus; **~ con tacógrafo** *m* AmL (*cf autobús con tacógrafo Esp*) TRANSP dial-a-bus; **~ del testigo** *m* INFORM&PD red en anillo token bus; **~ sobre la vía de ferrocarril** *m* AmL (*cf autobús sobre la vía de ferrocarril Esp*) TRANSP bus on railroad tracks (*AmE*), bus on railway tracks (*BrE*)

busa *f* PROD *de fuelles* nosepiece, *de tobera* nose, *horno alto* bustle pipe

busca: ~ de minerales *f* AmL (*cf prospección Esp*) MINAS cutting, prospecting

buscador[1]**: ~ de calor** *adj* TEC ESP heat-seeking; **~ por infrarrojo** *adj* TERMO heat-seeking

buscador[2] *m* PROD cat whisker, TEC ESP viewfinder; **~ del blanco** *m* D&A homing device; **~ de la dirección del emisor de ondas de radio** *m* FÍS radio direction finding; **~ de núcleo** *m* METEO center finder (*AmE*), centre finder (*BrE*); **~ y reparador de averías** *m* ING MECÁ troubleshooter

buscafugas: ~ espectrómetro de masa *m* ING MECÁ mass spectrometer-type leak detector

buscapiés *m* TEC ESP *cohetes* squib

buscapolos *m* ELEC *conexiones* polarity tester

buscar *vt* INFORM&PD seek

busco *m* AGUA base of a sluice gate, miter post (*AmE*), miter sill (*AmE*), mitre post (*BrE*), mitre sill (*BrE*), sill; **~ de aguas abajo** *m* AGUA tail miter sill (*AmE*), tail mitre sill (*BrE*); **~ de aguas arriba** *m* AGUA head miter sill (*AmE*), head mitre sill (*BrE*)

bushel *m* METR *medida de volumen* bushel

búsqueda *f* INFORM&PD look-up, search, searching, TEC ESP tracking, TV scanning; **~ de agua subterránea** *f* AGRIC, AGUA, CARBÓN, HIDROL ground-water investigation; **~ en árbol** *f* INFORM&PD tree search; **~ binaria** *f* INFORM&PD binary chop, binary search; **~ de Boole** *f* INFORM&PD Boolean search; **~ dicotómica** *f* INFORM&PD binary chop, binary search, dichotomizing search; **~ encadenada** *f* INFORM&PD chaining search; **~ de**

Fibonacci *f* INFORM&PD Fibonacci search; ~ **de radio de acción** *f* ELECTRÓN range finding; ~ **y rescate** *f* (*SAR*) D&A, TELECOM, TRANSP AÉR *tipo de misión* search and rescue (*SAR*); ~ **y salvamento** *f* TRANSP MAR search and rescue; ~ **secuencial** *f* INFORM&PD sequential search; ~ **en tabla** *f* INFORM&PD table search

bustamita *f* MINERAL bustamite

butadieno *m* QUÍMICA, TEC PETR butadiene

butanero *m* TEC PETR, TRANSP butane carrier

butano *m* QUÍMICA, TEC PETR *petroquímica* butane

butanol *m* QUÍMICA butanol

butanona *f* P&C butanone

butenal *m* QUÍMICA butenal

buteno *m* DETERG, QUÍMICA butene

buterola *f* CONST *remaches* rivet set, set

butileno *m* DETERG butylene

butilhidroxianisol *m* (*BHA*) ALIMENT *BHA* butylated hydroxyanisole (*BHA*)

butilhidroxitolueno *m* (*BHT*) ALIMENT butylated hydroxytoluene (*BHT*)

butiraldehído *m* QUÍMICA butyraldehyde

butirato *m* QUÍMICA butyrate

butirina *f* QUÍMICA butyrin

butirita *f* CARBÓN butyrite

buza *f* CARBÓN *horno metalúrgico* nozzle, METAL jet

buzamiento *m* CARBÓN *filones, geología* dip, CONST *geología* inclination, ENERG RENOV, GEOL, MINAS, PETROL dip, TEC PETR slope shaft; ~ **ascendente** *m* TEC PETR *geología* up dip; ~ **descendente** *m* TEC PETR *geología* down dip; ~ **de gran ángulo** *m* TEC PETR *geología* high-angle dip; ~ **hacia abajo** *m* TEC PETR *geología* down dip; ~ **hacia arriba** *m* TEC PETR *geología* up dip; ~ **inicial** *m* GEOL initial dip; ~ **en sentido contrario a la inclinación** *m* GEOL up dip

buzar *vi* GEOL dip

buzo *m* OCEAN, TEC PETR diver; ~ **de saturación** *m* TEC PETR *operaciones marinas* saturation diver

buzón *m* AGUA *bajantes de agua* cesspool; ~ **de alcancía** *m* MINAS chute door, chute gate; ~ **de chimenea** *m* MINAS chute door, chute gate; ~ **de correo electrónico** *m* ELECTRÓN, INFORM&PD, TELECOM electronic mailbox (*e-mail box*); ~ **electrónico** *m* ELECTRÓN, INFORM&PD, TELECOM electronic mailbox (*e-mail box*); ~ **radiotelefónico** *m* TELECOM voice mailbox; ~ **telefónico** *m* TELECOM voice mailbox; ~ **para textos** *m* TELECOM text mailbox

BV *abr* (*banda de valencia*) FÍS, FÍS RAD VB (*valence band*)

byte *m* INFORM&PD, PROD, TELECOM byte; ~ **de salida** *m* ELECTRÓN, INFORM&PD eight-bit output

C

c *abr* (*centi-*) METR c (*centi-*)

C *abr* ELEC, ELECTRÓN (*culombio*) C (*coulomb*), FÍS (*Celsio, culombio*) C (*Celsius, coulomb*), ING ELÉC (*culombio*) C (*coulomb*), MECÁ, METEO (*Celsio*) C (*Celsius*), METR (*Celsio, culombio*) C (*Celsius, coulomb*), QUÍMICA (*carbono, Celsio*) C (*carbon, Celsius*), TEC PETR (*carbono*) C (*carbon*)

Ca *abr* (*calcio*) ALIMENT, METAL, QUÍMICA Ca (*calcium*)

CA *abr* GEN (*corriente alterna, corriente alternativa*) AC (*alternating current*)

CAB *abr* (*control automático de brillo*) TV ABC (*automatic brightness control*)

cabalgamiento *m* GEOL overthrust, thrusting, TEC PETR *geología* overthrust; **~ ciego** *m* GEOL blind thrust; **~ por cizalla en un pliegue** *m* GEOL stretch thrust; **~ de desplazamiento direccional** *m* GEOL strike-slip thrust; **~ transcurrente** *m* GEOL transcurrent thrust; **~ en zonas con comprensión oblícua** *m* GEOL transpressive thrust

caballaje *m* AmL (*cf caballo de vapor Esp*) AUTO, ING MECÁ horsepower (*hp*); **~ hora** *m* ING MECÁ horsepower hour

caballero *m* TEXTIL drop wire

caballete *m* CONST trestle, *tejados* ridge capping, ING MECÁ, MECÁ pedestal, rack, standard, PROD *trabajo con arcilla* horse, rig; **~ de aserrar** *m* CONST sawbuck, sawhorse; **~ de extracción** *m* MINAS winding tower; **~ portapoleas** *m* PROD *de torre de perforación* crown block

caballo *m* C&V horse; **~ americano** *m* AGRIC quarter horse; **~ castrado** *m* AGRIC gelding; **~-hora efectivo** *m* ING MECÁ brake horsepower hour, MECÁ actual horsepower hour; **~ métrico** *m* METR *potencia* metric horsepower; **~ de Troya** *m* INFORM&PD Trojan horse; **~ troyano** *m* INFORM&PD Trojan horse; **~ de vapor** *m* Esp (*cf caballaje AmL*) AUTO, ING MECÁ horsepower (*hp*)

caballos *m pl* ING MECÁ horsepower (*hp*); **~ efectivos** *m pl* ING MECÁ effective horsepower; **~ hora** *m pl* ING MECÁ horsepower hour; **~ efectivos hora** *m pl* ING MECÁ effective horsepower hour

cabaña *f* AmL (*cf explotación ganadera Esp*) AGRIC livestock farm

cabeceado *m* TV *cabeza de grabación fuera de azimut* head banding

cabecear[1] *vt* CINEMAT pan down

cabecear[2] *vi* TRANSP MAR pitch

cabeceo *m* FÍS *movimiento de una embarcación o de un aeroplano*, TEC ESP *satélite* pitch, TRANSP AÉR pitch, *satélite* pitching, TRANSP MAR *movimiento buque* pitching; **~ de frenado** *m* TRANSP AÉR braking pitch; **~ del morro hacia arriba** *m* TRANSP AÉR pitch-up

cabecera *f* AGUA, HIDROL *canal* intake, IMPR, INFORM&PD header, MINAS headboard, TELECOM header; **~ del aparato de transacción** *f* TELECOM transactional set-header; **~ de cinta** *f* INFORM&PD tape header; **~ de grupo funcional** *f* TELECOM functional-group header; **~ de guía** *f* INFORM&PD *cinta* leader; **~ de intercambio** *f* TELECOM inter-change header; **~ de un manantial** *f* HIDROL spring head; **~ de mensaje** *f* INFORM&PD, TELECOM message header

cabecero *m* CONST head, head beam, *obra civil* waling

cabeza[1]: **a la ~** *adv* TEC ESP ahead

cabeza[2] *f* CINEMAT, CONST, IMPR head, ING MECÁ addendum, butt, head, MINAS *sondeos* cap piece, PROD *de clavo, perno, tornillo* head, TRANSP AÉR hub, TRANSP MAR *de mecha del timón* head; **~ de anillo** *f* TV ring head; **~ antisalpicadura** *f* LAB *destilación* antisplash head; **~ aplicadora** *f* EMB applicator head; **~ de arrastre** *f* TV head tracking; **~ de avance continuo** *f* ING MECÁ continuous feed head; **~ de barra de conexión** *f* ING MECÁ stub end; **~ basculante** *f* CINEMAT cradle head; **~ de biela** *f* AUTO big end, connecting-rod end, connecting-rod big end, ING MECÁ connecting-rod big end, connecting-rod end, large end of connecting rod, pitman head, stub end, VEH big end, connecting-rod big end, connecting-rod end; **~ de biela de caja cerradas** *f* ING MECÁ box-connecting rod end; **~ de borrado volante** *f* TV flying erase head; **~ buscadora** *f* D&A, TEC ESP *torpedos* homing head; **~ de cable** *f* ELEC, ING ELÉC *conexión* cable termination; **~ del cable** *f* ELEC, ING ELÉC cable head; **~ caliente** *f* CINEMAT *de control remoto para grúa* hot head; **~ de la caña** *f* C&V gathering end; **~ de carril** *f* FERRO *vehículos* rail head; **~ chanfleada** *f* CONST, ING MECÁ, PROD pan head; **~ cilíndrica taladrada** *f* ING MECÁ drilled fillister head screw; **~ de cilindro** *f* REFRIG cylinder head; **~ del cilindro** *f* TRANSP MAR *motores* cylinder head; **~ de combate** *f* D&A, TEC ESP *de un torpedo o proyectil nuclear* warhead; **~ de combate contra blindaje** *f* D&A anti-armor warhead (*AmE*), anti-armour warhead (*BrE*); **~ cortadora** *f* CONST *perforación* bit; **~ cortante** *f* CONST cutter head; **~ cuadrada** *f* CONST square head; **~ cubretornillos** *f* ING MECÁ screw casing head; **~ dentada** *f* ING MECÁ dented knob; **~ del disco** *f* ÓPT disc head (*BrE*), disk head (*AmE*); **~ doble de descarga** *f* ING MECÁ dual discharge head; **~ elevada** *f* ING MECÁ raised head; **~ del émbolo** *f* AUTO piston head; **~ embutida** *f* ING MECÁ flush head, MECÁ dished head, PROD flush head; **~ de encendido** *f* TEC ESP *nave espacial* fire bulkhead; **~ de engranaje basculante** *f* CINEMAT cradle gear head; **~ en espiral** *f* TV head winding; **~ espiralada** *f* TV head winding; **~ de extrusión** *f* ING MECÁ extrusion head; **~ fija** *f* INFORM&PD fixed head, ING MECÁ standing end; **~ fluida** *f* CINEMAT fluid head; **~ del fonocaptor** *f* ACÚST pick-up head; **~ fresada** *f* ING MECÁ milled head; **~ de fricción** *f* CINEMAT friction head; **~ de ganado** *f* AGRIC head; **~ giratoria** *f* MINAS *sondeos* water swivel; **~ giratoria del torno** *f* ING MECÁ live head; **~ giroscópica** *f* CINEMAT gyroscopic head; **~ golpeadora de hinca** *f* MINAS *sondeos* drive head; **~ grabadora** *f* IMPR engraving head; **~ hacia afuera** *f* CINEMAT head-out; **~ hexagonal** *f* ING MECÁ hexagon head, MECÁ hex nut; **~ de horquilla** *f* ING MECÁ fork head; **~ del**

husillo *f* ING MECÁ spindle, spindle nose; ~ **de impresión** *f* IMPR *informática* printing head; ~ **de inyección** *f* MINAS *sondeos* water swivel, TEC PETR swivel; ~ **lectora** *f* CINEMAT playback head; ~ **de lectura** *f* INFORM&PD, ÓPT read head; ~ **de línea** *f* CARBÓN *ferrocarril* road head; ~ **magnética** *f* ACÚST, INFORM&PD, TV magnetic head; ~ **magnética borradora** *f* ACÚST erasing magnetic head, INFORM&PD, TV magnetic erasing head; ~ **magnética grabadora** *f* ACÚST recording magnetic head; ~ **magnética reproductora** *f* ACÚST reproducing magnetic-head; ~ **magnética reproductora o grabadora** *f* ACÚST recording-reproducing magnetic head; ~ **de maniobra** *f* MINAS brace head; ~ **de manivelas** *f* CINEMAT geared head; ~ **de mazo** *f* CONST beetle head; ~ **moleteada** *f* ING MECÁ milled head, ratchet stop; ~ **móvil de torno** *f* ING MECÁ poppet head; ~ **Moy** *f* CINEMAT Moy head; ~ **de muñón** *f* ING MECÁ stub end; ~ **nodal** *f* CINEMAT nodal head; ~ **obstruida** *f* TV clogged head; ~ **de página** *f* IMPR headline; ~ **de percursión** *f* CONST driving block; ~ **de perno** *f* ING MECÁ, LAB, MECÁ bolt head; ~ **de pilón** *f* PROD stamp head, *bocarte* stamp boss; ~ **de pilote** *f* CARBÓN pile head; ~ **de pistón** *f* MECÁ crown; ~ **del pistón** *f* AUTO piston top, ING MECÁ piston head; ~ **potenciométrica** *f* TEC PETR *mecánica de fluidos* potentiometric head; ~ **de pozo** *f* GAS, PETROL, TEC PETR *perforación, producción* wellhead; ~ **de pozo submarina** *f* GAS, PETROL, TEC PETR *perforación, producción costa-fuera* subsea wellhead; ~ **de la rasera** *f* CONTAM MAR skimming head; ~ **redonda** *f* PROD *operador* mushroom head; ~ **redonda taladrada** *f* ING MECÁ drilled fillister head screw; ~ **de registro** *f* INFORM&PD, TV record head, recording head; ~ **reproductora** *f* CINEMAT, TV playback head, reproducing head; ~ **rotativa de video** *AmL*, ~ **rotativa de vídeo** *Esp* CINEMAT, TV rotary video head; ~ **de rotor** *f* TRANSP AÉR rotor head; ~ **de rotor principal** *f* TRANSP AÉR main-rotor head; ~ **de sonda** *f* MINAS *sondeos* temper screw; ~ **de sonido óptico** *f* CINEMAT optical sound head; ~ **de soplado** *f* C&V *vidrio soplado* blow-head; ~ **de soplar** *f* C&V *vidrio soplado* blowing-crown; ~ **de tercereje** *f* CINEMAT gimbal head; ~ **de tolva** *f* CONST hopper head; ~ **de tornillo** *f* ING MECÁ bolt head, screw head; ~ **tractora del remolque** *f* TRANSP driving trailer car; ~ **tractora del trailer** *f* TRANSP driving trailer car; ~ **de transporte** *f* TV head wheel; ~ **trasera** *f* ING MECÁ *barrenos* backhead; ~ **del trípode** *f* CINEMAT tripod head; ~ **de trípode giroscópico** *f* CINEMAT gyroscopic tripod head; ~ **troncocónica** *f* CONST, ING MECÁ, PROD pan head; ~ **de válvula** *f* AUTO valve head

cabezada *f* TRANSP MAR *de la hélice* pitch

cabezal[1]: **de ~ doble** *adj* CINEMAT double-headed

cabezal[2] *m* AUTO headstock, CINEMAT head, CONST bolster, casing head, pile cap, ram, IMPR, INFORM&PD, ING MECÁ head, MECÁ headstock, MINAS crown tree, *entibación de galerías* crown piece, crown, *sondeos* cap piece, TEXTIL *géneros de punto* section; ~ **de la bajante de aguas** *m* CONST rainwater head; ~ **barrenador** *m* ING MECÁ, MINAS *sondeos* boring head; ~ **binocular** *m* INSTR binocular head; ~ **blindado** *m* ING MECÁ ironclad headstock, totally-enclosed headstock; ~ **de borrado** *m* INFORM&PD, TV erase head, erasing head; ~ **de**

buscador *m* PROD cat-whisker head; ~ **de captación láser** *m* ÓPT laser pick-up head; ~ **de color** *m* CINEMAT *fotografía*, FOTO, TV color head (*AmE*), colour head (*BrE*); ~ **conducido por rueda libre** *m* ING MECÁ freewheel-driven head; ~ **conductor** *m* ING MECÁ driving head; ~ **cortador** *m* MINAS *barrenos* cutter head; ~ **cortante** *m* MINAS *perforación* cutting head; ~ **divisor** *m* ING MECÁ index; ~ **divisor mecánico** *m* ING MECÁ mechanical dividing head; ~ **de emulsionado** *m* CINEMAT, FOTO coating head; ~ **de engranajes** *m* ING MECÁ gear head; ~ **de escritura** *m* INFORM&PD write head; ~ **de estirado** *m* ING MECÁ engineering drawing block; ~ **estirador** *m* ING MECÁ bull block; ~ **de estirar** *m* PROD *trefilería* drawing block; ~ **de exploración** *m* TV scanning head; ~ **explorador** *m* IMPR scanning head; ~ **del extrusor** *m* ING MECÁ extrusion head; ~ **fijo** *m* INFORM&PD fixed head, PROD fast head; ~ **fijo de torno** *m* ING MECÁ lathe headstock; ~ **para fresar** *m* ING MECÁ milling head; ~ **de fricción** *m* ING MECÁ friction headstock; ~ **de funcionamiento oscilante** *m* PROD wobble stick operating head; ~ **giratorio** *m* TEC PETR swivel; ~ **de grabación** *m* ACÚST, INFORM&PD, TV record head, recording head; ~ **de impresión** *m* INFORM&PD print head; ~ **inclinable** *m* FOTO tilting head; ~ **de inyección** *m* MINAS *sondeos* water swivel; ~ **láser** *m* ÓPT laser head; ~ **de lectura** *m* INFORM&PD, ÓPT read head; ~ **de lectura-escritura** *m* INFORM&PD, ÓPT read-write head; ~ **magnético** *m* INFORM&PD magnetic head; ~ **de marcha oscilante** *m* PROD wobble stick operating head; ~ **de margarita** *m* IMPR *impresora* daisywheel; ~ **medidor** *m* CARBÓN, ELEC *medición* probe; ~ **micrométrico del indicador electrónico** *m* METR electronic-display micrometric head; ~ **de monopolea** *m* ING MECÁ *tornos* all-gear head, all-geared headstock, PROD constant-speed belt head; ~ **móvil** *m* ACÚST, CINEMAT, ING ELÉC capstan, ING MECÁ capstan, tailstock, MECÁ, TRANSP MAR, TV capstan; ~ **con movimiento panorámico vertical** *m* FOTO tilt head; ~ **óptico** *m* ÓPT optical head; ~ **de panorámica horizontal y vertical** *m* CINEMAT, FOTO pan-and-tilt head; ~ **portabrocas** *m* ING MECÁ drill head; ~ **portacuchillas** *m* ING MECÁ cutter head; ~ **portafresas** *m* ING MECÁ cutter head; ~ **portaherramientas** *m* ING MECÁ tool head; ~ **portamuelas** *m* PROD wheel head; ~ **de portaobjetivos** *m* INSTR objective turret; ~ **de pozo** *m* GAS, PETROL wellhead; ~ **de proyección** *m* CINEMAT picture head, projection head; ~ **pseudopotenciométrico** *m* TEC PETR *hidrogeología, mecánica de fluidos* pseudo-potentiometric head; ~ **pulidor** *m* ING MECÁ polishing head; ~ **de rayos X** *m* INSTR X-ray head; ~ **de retención oscilante** *m* PROD wobble stick head; ~ **de retención oscilante del alambre** *m* PROD wire wobble stick head; ~ **del revestimiento** *m* TEC PETR casing head; ~ **roscador** *m* ING MECÁ screwing head; ~ **roscador de carraca** *m* ING MECÁ ratchet screwing stock; ~ **para roscar** *m* ING MECÁ screwing chuck; ~ **para roscar radial** *m* ING MECÁ radial diehead; ~ **del serpentín** *m* REFRIG *del evaporador* coil header; ~ **de sonda** *m* *AmL* (*cf giratoria Esp*) MINAS *sondeos* swivel, swivel rod; ~ **de sonido para banda magnética** *m* CINEMAT magnetic strip sound head; ~ **sonoro** *m* CINEMAT sound film head; ~ **de taladrar de husillos múltiples** *m* ING

MECÁ multiple-spindle drilling head; ~ **de taladrar de polihusillos** *m* ING MECÁ multiple-spindle drilling head; ~ **de torno** *m* ING MECÁ lathe head; ~ **del torno** *m* ING MECÁ lathe headstock; ~ **totalmente cerrado** *m* ING MECÁ totally-enclosed headstock; ~ **totalmente encastrado** *m* ING MECÁ totally-enclosed headstock; ~ **del trípode** *m* FOTO tripod head; ~ **de tubería de revestimiento** *m* TEC PETR *perforación* casing head; ~ **de tubos** *m* PROD *caldera acuotubular* header; ~ **de video** *AmL*, ~ **de vídeo** *m Esp* TV video head

cabezales: ~ **divisores** *m pl* ING MECÁ dividing heads
cabezas: ~ **apiladas** *f pl* TV stacked heads; ~ **escalonadas** *f pl* TV staggered heads
cabida *f* ING MECÁ carrying capacity; ~ **de caja** *f* TRANSP *camiones*, TRANSP AÉR, TRANSP MAR, VEH cargo space
cabilla *f* PROD drift bolt, pin; ~ **de maniobra** *f* TRANSP MAR *equipamiento de cubierta* belaying pin
cabina *f* CINEMAT lamp house, *de un estudio* booth, CONST *grúa* cabin, cab, FERRO *vehículos* cabin, IMPR booth, INSTR *cine* lamp house, P&C cabinet, TEC ESP cabin, TRANSP cab, TRANSP AÉR cabin, VEH *carrocería* cab; ~ **de ascensor** *f* TRANSP passenger car (*BrE*), passenger coach (*AmE*); ~ **de cambio de agujas controlada por computadora** *f AmL* (*cf cabina de cambio de agujas controlada por ordenador Esp*) FERRO computerized signal box; ~ **de cambio de agujas controlada por ordenador** *f Esp* (*cf cabina de cambio de agujas controlada por computadora AmL*) FERRO computerized signal box; ~ **de control sonoro** *f* CINEMAT sound control room; ~ **cósmica** *f* TEC ESP cabin; ~ **con indicadores a base de pantallas electrónicas** *f* TRANSP AÉR all-glass cockpit; ~ **insonorizada** *f* CINEMAT sound booth; ~ **de mando** *f* TRANSP AÉR cockpit; ~ **del maquinista** *f* FERRO, VEH driver's cab, driver's cabin, engineer's cab; ~ **sobre el motor** *f* VEH *carrocería* cab-over-engine; ~ **de pasajeros** *f* TRANSP passenger cabin; ~ **del piloto** *f* TEC ESP cockpit; ~ **de proyección** *f* CINEMAT screening booth; ~ **de secado** *f* P&C drying cabinet; ~ **telefónica** *f* TELECOM telephone booth (*AmE*), telephone kiosk (*BrE*); ~ **de vuelo** *f* TRANSP AÉR flight deck
cabio *m* CONST rafter; ~ **bajo** *m* CONST *carpintería* bottom rail; ~ **común** *m* CONST *cubiertas* common rafter; ~ **corto** *m* CONST jack rafter
cabirón *m* TRANSP MAR gypsy, warping head
cable *m* ELEC *conductor* cable, cord, ELECTRÓN *fase* lead, INFORM&PD cable, ING ELÉC lead, cable, ING MECÁ cable, cord, MECÁ rope, cable, P&C, PETROL cable, TELECOM wire, cable, TRANSP MAR *comunicaciones*, VEH *controles, sistema eléctrico* cable;

◼ a ~ **en abanico** *m* ING ELÉC fanned cable; ~ **acabado en masa** *m* ING ELÉC mass-terminated cable; ~ **acanalado** *m* ÓPT, TELECOM grooved cable; ~ **de aceite** *m* ELEC oil-filled cable; ~ **con aceite circulante** *m* ELEC oil-filled cable; ~ **de aceite fluido** *m* ELEC oil-filled cable; ~ **de acometida** *m* ELEC lead-in cable; ~ **de acoplamiento** *m* TELECOM jumper wire; ~ **aéreo** *m* ELEC *red de distribución* overhead cable, ING ELÉC overhead cable, aerial cable, TELECOM aerial cable, TRANSP cableway; ~ **aéreo aislado** *m* ELEC aerial insulated cable; ~ **aislado** *m* ELEC, ING ELÉC insulated cable; ~ **aislado por**

caucho *m* ELEC, ING ELÉC, P&C, REVEST rubber-insulated cable; ~ **aislado de minerales** *m* ELEC, ING ELÉC mineral-insulated cable; ~ **aislado con papel** *m* ELEC, -ING ELÉC paper-insulated cable; ~ **aislado con plástico** *m* ING ELÉC, TELECOM plastic-insulated cable; ~ **con aislamiento de caucho** *m* ING ELÉC, P&C rubber cable; ~ **con aislamiento de material mineral** *m* ELEC, ING ELÉC mineral-insulated cable; ~ **de alambres** *m* MINAS cable chain; ~ **alargador** *m* CINEMAT extension cable; ~ **de alimentación** *m* ELEC *red de distribución* feeder, feeder cable, feed cable, power cable, INFORM&PD feeder cable, ING ELÉC feeder cable, feed cable, service line, ING MECÁ feed cable, PROD supply cable, TV feeder cable; ~ **de alimentación aéreo** *m* ING ELÉC open-wire feeder; ~ **alimentador** *m* ELEC *red de distribución* feeder cable, feed cable, power cable, INFORM&PD feeder cable, ING ELÉC feeder cable, feed cable, TEC ESP *comunicaciones* feeder, TV feeder cable; ~ **de alta frecuencia** *m* ELEC *fuente de alimentación de c.a.* high-frequency cable; ~ **de alta tensión** *m* ELEC *fuente de alimentación*, ING ELÉC extra-high-voltage cable, high-voltage cable; ~ **de alto voltaje** *m* INSTR high-voltage cable, VEH *instalación eléctrica* jump lead; ~ **anular** *m* ELEC *circuito* ING ELÉC ring current (*AmE*), ring main (*BrE*); ~ **apantallado** *m* ELEC, FÍS, INFORM&PD screened cable; ~ **de apertura manual** *m* D&A, TRANSP AÉR ripcord; ~ **armado** *m* ING ELÉC, ING MECÁ armored cable (*AmE*), armoured cable (*BrE*), PROD shielded cable, TELECOM armored cable (*AmE*), armoured cable (*BrE*); ~ **de arrastre** *m* CONTAM MAR, OCEAN trawl warp, TRANSP haulage cable, TRANSP MAR trawl warp; ~ **arterial** *m* ING ELÉC trunk cable;

◼ b ~ **de baja tensión** *m* ELEC *fuente de alimentación* low voltage cable; ~ **bajo plomo** *m* ELEC, ING ELÉC lead-covered cable, lead-sheathed cable; ~ **de las baterías** *m* CINEMAT battery cable; ~ **biaxial** *m* ELEC biaxial cable; ~ **blindado** *m* ELEC screened cable, shielded cable, FÍS, ING ELÉC shielded cable, ING MECÁ armored cable (*AmE*), armoured cable (*BrE*), TV shielded cable; ~ **Bowden** *m* MECÁ, VEH *de embrague, freno* Bowden cable; ~ **de bronce** *m* CONST brass wire; ~ **de la bujía** *m* AUTO, ELEC, VEH spark plug wire, spark-plug cable;

◼ c ~ **de cadena** *m* ING MECÁ chain cable; ~ **de cadena de eslabones con pasador** *m* CONST stud-link chain cable; ~ **de caída de voltaje** *m* TV drop cable; ~ **de calefacción** *m* ELEC heating cable; ~ **de calefacción del pavimento** *m* ING MECÁ floor-warming cable; ~ **de calefacción del suelo** *m* ING MECÁ floor-warming cable; ~ **de carga por bobinas** *m* ING ELÉC coil-loaded cable; ~ **cargado** *m* ING ELÉC, TELECOM loaded cable; ~ **cargado continuamente** *m* ING ELÉC, TELECOM continuously-loaded cable; ~ **cargado de continuo** *m* ING ELÉC, TELECOM continuously-loaded cable; ~ **de cierre** *m* ELEC *conexión* jumper; ~ **cinta** *m* ELEC, ÓPT, PROD, TELECOM ribbon cable; ~ **coaxial** *m* ELEC, FÍS, INFORM&PD, ING ELÉC, TELECOM, TV coax, coaxial cable; ~ **de cobre** *m* CONST copper wire, ELEC copper cable, copper wire, ELECTRÓN copper cable, ING ELÉC, REVEST copper wire, TELECOM copper cable; ~ **de comunicaciones** *m* ING ELÉC, TELECOM communications cable; ~ **concéntrico** *m* ELEC, FÍS, INFORM&PD coaxial cable, ING ELÉC coax, coaxial cable, TELECOM,

TV coaxial cable; ~ **conductor de conexión con la célula** m TRANSP AÉR airframe bonding lead; ~ **conductor de corriente** m ELEC, ING ELÉC current lead (*AmE*), mains lead (*BrE*); ~ **conductor electroaislado** m ING ELÉC insulated conductor (*AmE*), insulated core (*BrE*); ~ **conductor de entrada** m ING ELÉC input lead; ~ **conductor de una fibra** m ING ELÉC single-fiber cable (*AmE*), single-fibre cable (*BrE*); ~ **conductor de llegada** m ING ELÉC pigtail; ~ **conductor del rayo** m ING ELÉC lightning conductor; ~ **conductor de la red** m ELEC, ING ELÉC current lead (*AmE*), mains lead (*BrE*); ~ **de conductores axiales retorcidos** m PROD twin-axial cable; ~ **de conductores emplomados** m ELEC separate lead cable (*SL cable*), separately lead-sheathed cable; ~ **de conductores metálicos** m TELECOM metal-conductor cable, metallic conductor cable; ~ **de conductores múltiples** m ELEC, ELECTRÓN, ING ELÉC bunched cable (*BrE*), bundled cable (*AmE*); ~ **de conductores pareados** m ING ELÉC paired cable; ~ **de conductores con vaina de plomo** m ELEC separately lead-sheathed cable; ~ **de conexión** m CINEMAT connecting cable, connecting cord, ELEC connecting wire, FOTO connecting cable, connecting cord, INFORM&PD patch cord, ING ELÉC connecting cable, connecting lead, jumper, patch cord, connecting wire, TV connecting cable; ~ **de conexionado** m MECÁ harness cable; ~ **de construcción apretado** m ÓPT tight construction cable, tight-jacketed cable; ~ **construido flotante** m ÓPT loose construction cable; ~ **de contrapeso** m AmL (*cf cable de tracción Esp*) MINAS load rope; ~ **de control** m ING MECÁ control cable; ~ **de control del carburador** m AUTO, VEH carburetor control cable (*AmE*), carburettor control cable (*BrE*); ~ **convencional** m ÓPT conventional cable; ~ **de corriente** m ELEC, ING ELÉC current lead (*AmE*), mains lead (*BrE*); ~ **de la corriente** m ELEC, ING ELÉC current cable (*AmE*), mains cable (*BrE*); ~ **de corriente de corto alcance** m ING ELÉC short-haul cable; ~ **en cuadretes** m ING ELÉC quadded cable; ~ **cuádruple de pares** m ING ELÉC quadded cable; ~ **con cuatro conductores aislados** m ING ELÉC quad; ~ **de cuatro pares** m ING ELÉC quadruple-pair cable; ~ **con cubierta de plomo** m ELEC, ING ELÉC lead-covered cable; ~ **de cuchareo** m TEC PETR sand line;

~ **d** ~ **de detección de incendio** m SEG, TRANSP AÉR fire-detecting wire; ~ **del disparador** m CINEMAT shutter release cable; ~ **de distribución** m ELEC distribution cable, ING ELÉC distribution cable, supply main; ~ **de distribución local** m TELECOM local distribution cable; ~ **de doble par** m ELEC double core cable; ~ **de doble-conductor plano** m TRANSP twin-ribbon cable; ~ **de dos almas** m ELEC double core cable; ~ **de dos conductores aislados** m ELEC, ING ELÉC twin cable; ~ **de dos conductores trenzados** m ELEC, ING ELÉC duplex cable; ~ **dúplex** m ELEC, ING ELÉC duplex cable;

~ **e** ~ **elástico** m TRANSP sandow; ~ **eléctrico impregnado de aceite** m ING ELÉC oil-filled cable; ~ **eléctrico para servicio de un equipo** m TEC ESP umbilical cable; ~ **de embrague** m AUTO, VEH clutch cable; ~ **de emergencia** m TRANSP emergency cable; ~ **de empalme** m ING ELÉC junction cable, VEH *instalación eléctrica* jumper; ~ **emplomado** m ELEC,

ING ELÉC lead-covered cable, lead-sheathed cable; ~ **enfundado** m ELEC, ING ELÉC sheathed cable; ~ **de enlace** m ING ELÉC trunk cable; ~ **enroscatubos** m TEC PETR *perforación* spinning line; ~ **entubado** m ELEC pipe-type cable; ~ **envainado** m ING ELÉC sheathed cable; ~ **envainado en plomo** m ELEC, ING ELÉC lead-covered cable, lead-sheathed cable; ~ **del estárter** m AUTO, VEH starter cable; ~ **de estrella** m FÍS star quad; ~ **estrella-cuadrete** m ING ELÉC star-quad cable; ~ **de expansión** m PROD expansion cable; ~ **de extensión** m LAB *alimentación eléctrica* extension lead; ~ **de extensión telefónica** m TELECOM telephone extension cable; ~ **de exterior** m ELEC outdoor cable; ~ **exterior con gas a presión** m ELEC external gas-pressure cable; ~ **exterior de planta** m AmL TV outside plant cable; ~ **de extracción** m MINAS *sondeos y pozos* hoisting rope;

~ **f** ~ **de fibra óptica** m INFORM&PD, ING ELÉC, ÓPT, TELECOM fiber-optic cable (*AmE*), fibre-optic cable (*BrE*), optical fiber cable (*AmE*), optical fibre cable (*BrE*); ~ **de fibra plástica** m ING ELÉC plastic-fiber cable (*AmE*), plastic-fibre cable (*BrE*); ~ **de fibroóptica** m ÓPT fiber-optic cable (*AmE*), fibre-optic cable (*BrE*), optical fiber cable (*AmE*), optical fibre cable (*BrE*); ~ **fibroplástico** m ING ELÉC plastic-fiber cable (*AmE*), plastic-fibre cable (*BrE*); ~ **fibroóptico** m INFORM&PD, ING ELÉC, ÓPT, TELECOM fiber-optic cable (*AmE*), fibre-optic cable (*BrE*); ~ **flexible** m ELEC, ING ELÉC flexible cable; ~ **flexible aislado** m ELEC, ING ELÉC field wire; ~ **flexible de conexión** m ELEC, ING ELÉC pigtail; ~ **flexible de conexión de fibra óptica** m INFORM&PD, ING ELÉC, ÓPT, TELECOM optical fiber pigtail (*AmE*), optical fibre pigtail (*BrE*); ~ **flexible de remolque** m ING ELÉC trailing cable; ~ **forrado** m REVEST coated wire; ~ **forrado de metal** m ING ELÉC metal-sheathed cable; ~ **forrado de plomo** m ELEC, ING ELÉC lead-covered cable; ~ **de freno** m AUTO, VEH brake cable; ~ **del freno de mano** m AUTO, VEH handbrake cable;

~ **g** ~ **con gas a presión interna** m ELEC internal gas pressure cable; ~ **gemelo** m TV paired cable; ~ **guía** m MINAS *pozos* gland, PROD guide rope;

~ **h** ~ **de henequén** m ING MECÁ sisal rope; ~ **hertziano** m TELECOM radio link; ~ **hilado** m ÓPT spun cable; ~ **con hilos desnudos** m ING MECÁ piano wire;

~ **i** ~ **de iluminación** m ELECTRÓN lighting cable; ~ **impregnado** m ELEC impregnated cable; ~ **de impulsos de sincronización** m CINEMAT sync pulse cable; ~ **de incendios** m TRANSP AÉR fire wire; ~ **interconector del programador de reposición** m PROD replacement programer interconnect cable (*AmE*), replacement programmer interconnect cable (*BrE*); ~ **de interconexión** m CONST interconnecting cable, ING ELÉC interconnection cable; ~ **interior** m ELEC indoor cable; ~ **izador** m MINAS, PROD hoisting rope;

~ **l** ~ **a larga distancia** m ING ELÉC long-distance cable; ~ **para líneas auxiliares** m TELECOM trunk cable; ~ **de llegada de fibra óptica** m INFORM&PD, ING ELÉC, ÓPT, TELECOM optical fiber pigtail (*AmE*), optical fibre pigtail (*BrE*); ~ **lleno de gas** m ELEC, GAS, ING ELÉC gas-filled cable; ~ **luminoso** m ING ELÉC light cable;

~ **m** ~ **de masa** m ELEC, ING ELÉC, VEH earth cable (*BrE*), ground cable (*AmE*); ~ **de masa con la célula** *m* TRANSP AÉR airframe bonding lead; ~ **metálico** *m* PROD wire rope; ~ **en minihaz** *m* ÓPT mini-bundle cable; ~ **mixto** *m* ING ELÉC composite cable; ~ **monoconductor** *m* ELEC, ING ELÉC single-conductor cable, single-core cable; ~ **monofibra** *m* ING ELÉC single-fiber cable (*AmE*), single-fibre cable (*BrE*); ~ **monofilar** *m* ELEC, ING ELÉC single-conductor cable, single-core cable, PROD single-conductor wire; ~ **móvil** *m* ING ELÉC trailing cable; ~ **de muchas fibras** *m* ING ELÉC, ÓPT, TELECOM multifiber cable (*AmE*), multifibre cable (*BrE*); ~ **multiconductor** *m* ELEC, ING ELÉC, TV multicore cable; ~ **multifibra** *m* ING ELÉC, ÓPT, TELECOM multifiber cable (*AmE*), multifibre cable (*BrE*); ~ **multifilar** *m* ELEC, ING ELÉC, TV multicore cable; ~ **de múltiples conductores** *m* ELEC, ING ELÉC, TV multiconductor cable; ~ **de múltiples fibras** *m* ING ELÉC, ÓPT, TELECOM multifiber cable (*AmE*), multifibre cable (*BrE*); ~ **multitorónico** *m* TRANSP MAR *cuerda* multistrand;

■ **n** ~ **no cargado** *m* TELECOM unloaded cable; ~ **con núcleo de fundente** *m* CONST *soldadura* flux-cored wire; ~ **de núcleo ranurado** *m* ÓPT slotted-core cable;

■ **o** ~ **óptico** *m* ÓPT, TELECOM, TV optical cable; ~ **óptico concéntrico** *m* ÓPT, TELECOM, TV concentric optical cable;

■ **p** ~ **de un par** *m* INFORM&PD cable pair; ~ **de par coaxial** *m* ING ELÉC coaxial pair cable; ~ **de par trenzado** *m* INFORM&PD, TELECOM twisted-pair cable; ~ **de par único** *m* ING ELÉC single-pair cable; ~ **de parada de seguridad** *m* TRANSP safety stop cable; ~ **de pares** *m* ELEC, ING ELÉC, TV paired cable; ~ **de pares simétricos** *m* TELECOM symmetrical pair cable; ~ **de pequeñas pérdidas** *m* ELEC *fuente de alimentación*, TELECOM low-loss cable; ~ **de pérdidas de acoplamiento de mejora** *m* ELEC, TELECOM grading-coupling loss cable; ~ **de perforación** *m* Esp (*cf* guaya *AmL*) MINAS *sondeos* drilling cable, wire rope, PETROL, TEC PETR *perforación* drilling line; ~ **plano** *m* ELEC flat cable, ribbon cable, ING ELÉC flat cable, ÓPT, PROD, TELECOM ribbon cable; ~ **plano doble** *m* TRANSP twin-ribbon cable; ~ **plano flotante** *m* ÓPT loose-ribbon cable; ~ **plano de par trenzado** *m* ING ELÉC twisted-pair flat cable; ~ **poco disipativo** *m* ELEC *fuente de alimentación* low-loss cable; ~ **policonductor** *m* ELEC, ING ELÉC multiconductor cable; ~**portante** *m* TRANSP track cable; ~ **preformado** *m* ELEC harness cable; ~ **a presión** *m* ELEC pressure cable; ~ **a presión autónomo** *m* ELEC self-contained pressure cable; ~ **de prolongación** *m* ELEC, ING ELÉC extension cable; ~ **de puente** *m* ELEC *conexión* jumper; ~ **de PVC** *m* ING ELÉC PVC sheath;

■ **r** ~ **R** *m* TELECOM R-wire; ~ **de red** *m* TELECOM network cable; ~ **de la red eléctrica** *m* ING ELÉC current cable (*AmE*), mains cable (*BrE*); ~ **de remolque** *m* CONTAM MAR towrope, ING MECÁ hawser, TRANSP MAR towrope, warp; ~ **de retardo** *m* TV delay cable; ~ **de retenida** *m* ING ELÉC guy wire; ~ **revestido compacto** *m* TELECOM tight-jacketed cable;

■ **s** ~ **de la señal de identificación** *m* CINEMAT pilot-tone cable; ~ **sensible** *m* ING ELÉC sensing lead; ~ **de sincronización** *m* CINEMAT sync cable; ~ **sinfín** *m* TRANSP, VEH endless cable; ~ **del soldador** *m* TELECOM splicer cable; ~ **de sondeo** *m* MINAS drill rope; ~ **soportado por catenaria** *m* FERRO *equipo*

inamovible catenary support; ~ **de soporte** *m* Esp (*cf* línea de soporte *AmL*, línea de sujeción *AmL*) TEC PETR *perforación* backup line; ~ **soterrado** *m* ELEC *red de distribución* underground cable; ~ **submarino** *m* FÍS, ING ELÉC, TELECOM submarine cable, TRANSP MAR submarine cable, *en aguas profundas* deep-sea cable; ~ **subterráneo** *m* ELEC *red de distribución*, ING ELÉC underground cable; ~ **suelto** *m* PROD loose cable; ~ **de sujeción** *m* Esp (*cf* línea de soporte *AmL*, línea de sujeción *LAm*) TEC PETR *perforación* backup line; ~ **sumergible** *m* FÍS, ING ELÉC, TELECOM submarine cable; ~ **superconductor** *m* TELECOM superconductor cable; ~ **de suspensión** *m* TRANSP MAR boat sling;

■ **t** ~ **de telecomunicaciones** *m* TELECOM telecommunications cable; ~ **telefónico** *m* ING ELÉC, TELECOM telephone cable; ~ **telegráfico** *m* ING ELÉC telegraph cable; ~ **de televisión** *m* ELECTRÓN, FÍS, ING ELÉC, TELECOM, TV television cable; ~ **tendido bajo tierra** *m* ELEC *red de distribución* underground cable; ~ **terrestre** *m* TELECOM land cable; ~ **a tierra** *m* ELEC, FÍS, ING ELÉC, PROD, TELECOM earth lead (*BrE*), ground lead (*AmE*); ~ **de tierra** *m* TV earth lead (*BrE*), ground lead (*AmE*); ~ **de timbre** *m* ING ELÉC bell wire; ~ **tipo tubo de lubricación de aceite** *m* ELEC oil-filled pipe-type cable; ~ **de tipo unitario** *m* ÓPT unit-type cable; ~ **de tiras** *m* ELEC, ÓPT, PROD, TELECOM ribbon cable; ~ **de toma a tierra** *m* ELEC, ING ELÉC, VEH earth cable (*BrE*), ground cable (*AmE*); ~ **tomacorriente** *m* ELEC *red de distribución*, ING ELÉC power cable; ~ **de tracción** *m* Esp (*cf* cable de contrapeso *AmL*) ING MECÁ pulling rope, MINAS load rope, TRANSP haulage cable, traction cable, traction rope; ~ **tractor** *m* ING MECÁ pulling rope; ~ **de transmisión** *m* TRANSP MAR *del timón* transmission wire; ~ **para transporte de energía** *m* ING ELÉC power cable; ~ **trenzado** *m* ELEC *conductor*, ING ELÉC stranded cable, PROD stranded wire; ~ **trenzado directo** *m* ÓPT direct strand cable; ~ **de tres almas** *m* ELEC triple-core cable; ~ **de tres conductores** *m* ING ELÉC three-conductor cable; ~ **troncal** *m* TELECOM trunk cable; ~ **tubo flotante** *m* ÓPT loose-tube cable;

■ **u** ~ **umbilical** *m* TEC ESP umbilical cable; ~ **unimodal** *m* ING ELÉC single-mode cable; ~ **de unión** *m* TELECOM trunk cable; ~ **de unión de dos centrales generadoras** *m* ELEC *fuente de alimentación* trunk feeder;

■ **v** ~ **de varias fibras** *m* ING ELÉC multifiber cable (*AmE*), multifibre cable (*BrE*); ~ **de varios conductores no concéntricos** *m* ELEC, ING ELÉC, TV multicore cable; ~ **de varios núcleos** *m* ING ELÉC n-core cable; ~ **del velocímetro** *m* AUTO, VEH speedometer cable; ~**vía** *m* TRANSP track cable; ~ **video** *AmL*, ~ **vídeo** *Esp* *m* ING ELÉC video cable; ~ **de vitrofibra** *m* C&V, CONST, ELEC, ING ELÉC, TELECOM glass-fiber cable (*AmE*), glass-fibre cable (*BrE*); ~ **vitrofibra** *m* C&V, CONST, ELEC, ING ELÉC, TELECOM glass-fiber cable (*AmE*), glass-fibre cable (*BrE*);

■ **y** ~ **Y** *m* TV y-cable

cableado[1] *adj* INFORM&PD hard-wired

cableado[2] *m* ELEC *fuente de alimentación* electric wiring, cable network, *conexión* wiring, electrical wiring, *componente del cable* lay, ING ELÉC electrical wiring, cable network, electric wiring, wiring, cabling, ING MECÁ lay, TEC ESP wiring, *aviones* harness, TELECOM, TV cabling; **~ del circuito principal** *m* PROD *telefonía*, TELECOM trunkline wiring; **~ embutido** *m* ELEC *red de distribución*, ING ELÉC flush wiring; **~ de fuerza** *m* PROD power wiring; **~ del inductor** *m* PROD field wiring; **~ de la instalación eléctrica** *m* VEH electric wiring; **~ interior** *m* ING ELÉC indoor wiring; **~ interno** *m* TELECOM internal wiring; **~ de línea** *m* PROD line wiring; **~ preformado** *m* TEC ESP wiring harness; **~ de punto a punto** *m* ING ELÉC point-to-point wiring; **~ rectangular** *m* ING ELÉC rectangular wiring; **~ en serie** *m* PROD serial wiring; **~ subterráneo** *m* CONST underground cabling

cableaje *m* ELEC cabling, *conexión* wiring, ING ELÉC, TELECOM, TV cabling

cablear *vt* ELEC, TV wire

cablegrama *m* ELEC, ING ELÉC cablegram

cablero *m* TELECOM cable ship, TRANSP MAR cable ship, cable-laying ship

cabo[1]: **a ~ abierto** *adj* TEXTIL *hilo* open-ended

cabo[2] *m* FÍS *guía de ondas* stub, INFORM&PD tail, MECÁ rope, OCEAN *accidentes geográficos* foreland, cape, PAPEL offcut, TRANSP MAR cape, line, rope; **~ acollador** *m* TRANSP MAR *construcción naval* lanyard; **~ ayustado** *m* TRANSP MAR spliced rope; **~ colchado con tres cordones** *m* TRANSP MAR three-stranded line; **~ de disparo** *m* OCEAN *zoología marina* tripping line; **~ guía** *m* TRANSP MAR *del ancla flotante* tripping line; **~ del hilo** *m* TEXTIL yarn end; **~ de jarcia** *m* TRANSP MAR *para enjarciar* rope; **~ de maniobra** *m* *Esp* (*cf timonel de combate AmL*) TRANSP MAR quartermaster; **~ de mar** *m* *Esp* (*cf timonel de combate AmL*) TRANSP MAR quartermaster; **~ multitorónico** *m* TRANSP MAR *cuerda* multistrand; **~ de nylon** *m* TRANSP MAR nylon line; **~ que no hace cocas** *m* TRANSP MAR nonkinking rope; **~ de remolque** *m* CONTAM MAR towrope, TRANSP MAR towrope, warp; **~ de retenida** *m* TRANSP MAR *cuerda* guy; **~ roto** *m* TEXTIL broken end; **~ salvavidas** *m* SEG, TRANSP MAR *equipamiento de cubierta, de seguridad* lifeline; **~ salvavidas flotante** *m* TRANSP MAR *para el rescate* floating line; **~ de señales** *m* TRANSP MAR *tripulación* yeoman of signals; **~ tenso** *m* TEXTIL tight end

cabos: **~ cruzados** *m pl* TEXTIL crossed ends; **~ dobles** *m pl* TEXTIL twin ends; **~ perdidos** *m pl* TEXTIL lost ends

cabotaje *m* TRANSP cabotage, TRANSP MAR home trade, *marina mercante* coasting trade

cabotero *m* TRANSP MAR coaster

cabrerita *f* MINERAL cabrerite

cabrestante *m* CINEMAT *herramienta* capstan, CONST *herramienta* axle pulley, winch, windlass, ING ELÉC, ING MECÁ, MECÁ capstan, MINAS *sondeos* hoist, winding engine, TRANSP MAR, TV capstan; **~ del cable** *m* ELEC, ING ELÉC cable winch; **~ de maniobras** *m* CONST shunting winch; **~ portátil** *m* MINAS *sondeos* slusher; **~ de sondeo** *m* MINAS *sondeos* drilling winch, *sondeos y castilletes de mina* hoist; **~ al vacío** *m* TV vacuum capstan

cabria *f* CONST *mecánica* shear legs, gin, TEC PETR *perforación* derrick; **~ de caballo** *f* ING MECÁ horsepower (*hp*)

cabrillas *f pl* OCEAN white horses, *clasificación de olas* white caps

cabrilleo *m* OCEAN ripples

cabriolé *m* TRANSP, VEH cabriolet

cabuyería *f* TRANSP MAR *aparejos* cordage, rigging; **~ alquitranada** *f* TRANSP MAR tarred cordage

CA/CC: **de ~ a pleno voltaje** *adj* PROD full-voltage AC/DC

cacha *f* ING MECÁ handle plate; **~ de la corredera** *f* ING MECÁ link plate

cacharro *m* C&V coarse pottery

cachaza *f* CARBÓN *fabricación de azúcar* foam

caché *m* INFORM&PD *memoria* cache

cacoxenita *f* MINERAL cacoxenite

cacoxeno *m* MINERAL cacoxenite

CAD *abr* ELEC, IMPR, INFORM&PD, MECÁ, PROD, TELECOM (*diseño asistido por computador, diseño asistido por computadora, diseño asistido por ordenador*) CAD (*computer-aided design*)

cadabina *f* QUÍMICA stachydrine

cadaverina *f* QUÍMICA cadaverine

cadena *f* ELECTRÓN daisychain, EMB chain, line, GEOL chain, INFORM&PD chain, string, ING ELÉC network, ING MECÁ chain, series, MECÁ, METR chain, PROD daisychain, QUÍMICA, TEXTIL chain, TRANSP network, TRANSP MAR cable chain, chain, *amarradero* chain cable, TV network, VEH *de la distribución, amaradero* chain; **~ abierta** *f* QUÍMICA open chain; **~ de agrimensor** *f* CONST Gunter's chain, measuring chain, *topografía* surveyor's chain; **~ de amarre** *f* CONTAM MAR, TRANSP MAR mooring chain; **~ de ancla** *f* MINAS *buques* cable chain; **~ del ancla** *f* TRANSP MAR anchor chain; **~ apiladora** *f* PROD *troncos* load chain; **~ de arrastre** *f* *Esp* (*cf escala AmL*) MINAS *dragas de rosario* digging ladder; **~ de articulaciones** *f* ING MECÁ flat-top chain; **~ articulada** *f* ING MECÁ block chain, pitch chain, pitched chain, stud chain, MECÁ sprocket chain; **~ de aviso de servicio** *f* TELECOM *transmisión de caracteres* service string advice; **~ de barbada** *f* PROD curbed chain; **~ binaria** *f* ELECTRÓN bit string; **~ de bits** *f* INFORM&PD bit string; **~ de buje de transmisión** *f* ING MECÁ transmission bush chain; **~ de cable de eslabones con pasador** *f* CONST studded link cable chain; **~ calibrada de rodillos y polea de cadena** *f* ING MECÁ precision roller-chain and chain-wheel; **~ de cangilón** *f* CONST bucket chain; **~ de caracteres** *f* INFORM&PD character string; **~ de carga** *f* ING MECÁ load chain, PROD *aparejos* lifting chain, *grúas, polipastos* load chain, SEG *de una grúa* load chain; **~ cerrada** *f* QUÍMICA closed chain; **~ de los chapones** *f* ING MECÁ flat-top chain; **~ cinemática** *f* MECÁ kinematic chain; **~ con cinta de acero** *f* CONST *topografía* steel band chain; **~ de contrapeso** *f* ING MECÁ balance chain; **~ dentada** *f* ING MECÁ indented chain; **~ de desintegración** *f* FÍS, FÍS PART decay chain; **~ de desintegración radiactiva** *f* FÍS PART, FÍS RAD radioactive-decay series; **~ desordenada** *f* QUÍMICA *de polímero* disordered chain; **~ diferencial** *f* MECÁ differential chain block; **~ de diodos** *f* ELECTRÓN diode string; **~ de distribución** *f* TRANSP distribution chain; **~ de la distribución** *f* AUTO, VEH timing chain; **~ electroacústica** *f* ACÚST, ING ELÉC electroacoustic chain; **~ de elevación de pesos** *f* PROD *aparejos* lifting chain; **~ de embalaje** *f* EMB packaging line; **~ de embotellado** *f* EMB bottling line; **~ de envasado en láminas al vacío** *f* EMB blister-packing line; **~ para envasar líquidos** *f* EMB liquid-packaging line;

~ **de eslabones** *f* ING MECÁ link chain, MECÁ chain block; ~ **de eslabones abiertos** *f* PROD open-link chain; ~ **de eslabones macizos** *f* ING MECÁ block chain; ~ **de estaciones de radio** *f* TV radio link; ~ **sin fin** *f* VEH endless chain; ~ **del frío** *f* REFRIG cold chain; ~ **Galle** *f* ING MECÁ pitch chain, pitched chain, stud chain; ~ **de gorrones** *f* ING MECÁ stud chain; ~ **de grabación** *f* TV recording chain; ~ **de grapas sin fin** *f* TEXTIL *mecánica* endless pin-chain; ~ **para guardines** *f* TRANSP MAR steering chain; ~ **de huesecillos** *f* ACÚST ossicular chain; ~ **de identificación de ganado** *f* AGRIC neck chain; ~ **indentada** *f* ING MECÁ indented chain; ~ **irregular** *f* QUÍMICA *de polímero* disordered chain; ~ **para izar** *f* SEG lifting chain; ~ **de láminas** *f* ING MECÁ leaf chain; ~ **lateral** *f* QUÍMICA side chain; ~ **para levantar** *f* CONST, SEG lifting chain; ~ **lineal** *f* QUÍMICA *estructura* straight chain; ~ **de llenado** *f* EMB filling line; ~ **de llenado de latas** *f* EMB can-filling line (*AmE*), tin-filling line (*BrE*); ~ **de llenado de sacos** *f* EMB sack-filling line; ~ **Loran** *f* TRANSP AÉR Loran chain; ~ **de mando de la distribución** *f* AUTO, VEH *motor* camshaft drive chain; ~ **de maniobra** *f* PROD *aparejos de izar* hand chain; ~ **de mano** *f* CONST hand chain; ~ **para medición del terreno** *f* CONST, METR land measuring chain; ~ **de montaje** *f* EMB, ING MECÁ, MECÁ, PROD, TRANSP MAR *de taller* assembly line, assembly list; ~ **de montañas** *f* GEOL mountain chain; ~ **de noticias** *f* TV news network; ~ **nula** *f* INFORM&PD null string; ~ **ordenada** *f* QUÍMICA ordered chain; ~ **ordinaria de eslabones** *f* ING MECÁ link chain; ~ **osicular** *f* ACÚST ossicular chain; ~ **de perforación** *f* MINAS drill chain; ~ **portacuchillas** *f* CARBÓN *rafadoras* cutter chain; ~ **probada** *f* ING MECÁ tested chain; ~ **de producción** *f* EMB assembly line, production line, ING MECÁ, MECÁ, PROD, TRANSP MAR assembly line; ~ **de producción de cierres** *f* EMB closure production line; ~ **de producción continua** *f* EMB, PROD continuous production line; ~ **de producción en serie** *f* EMB, ING MECÁ, PROD production line; ~ **puntero** *f* INFORM&PD pointer chain; ~ **con raederas** *f* MINAS scraper chain assembly; ~ **ramificada** *f* QUÍMICA branched chain; ~ **de Ramsden** *f* CONST Ramsden's chain; ~ **con rastras** *f* MINAS scraper chain assembly; ~ **recta** *f* QUÍMICA *estructura* straight chain; ~ **regional en A** *f* TELECOM A-leg; ~ **sin remaches estampada en caliente** *f* ING MECÁ drop-forged rivetless chain; ~ **reproductora** *f* TV reproducing chain, reproducing network; ~ **de retención** *f* MINAS *jaula de extracción en minas* bridle chain; ~ **de rodillos** *f* ING MECÁ block chain, roller chain; ~ **de rodillos doble** *f* AUTO double roller chain; ~ **de rosario** *f* ING MECÁ beaded chain; ~ **y ruedas dentadas** *f* ING MECÁ chain and sprocket wheels; ~ **de seguridad** *f* ING MECÁ, SEG safety chain; ~ **de símbolos** *f* INFORM&PD symbol string; ~ **de sincronización** *f* AUTO timing chain; ~ **de sincronización silenciosa** *f* AUTO noiseless timing chain; ~ **sincronizadora de rodillos** *f* AUTO roller timing chain; ~ **de suspensión** *f* TRANSP MAR boat sling; ~ **de telecine** *f* CINEMAT telecine chain; ~ **de texto** *f* IMPR text stream; ~ **de transmisión** *f* AUTO, VEH *motor* drive chain; ~ **de transmisión de enlace acodada** *f* ING MECÁ cranked-link transmission chain; ~ **de transmisión de unión acodada** *f* ING MECÁ cranked-link transmission chain;

~ **transportadora** *f* ING MECÁ chain conveyor, conveyor chain, MINAS *tracción* creeper chain; ~ **de unión** *f* CONST band chain; ~ **unitaria** *f* INFORM&PD unit string; ~ **vacía** *f* *Esp* (*cf serie vacía AmL*) INFORM&PD empty string, null string; ~ **Vaucanson** *f* PROD ladder chain

cadencia *f* INFORM&PD latency; ~ **de digitalización** *f* ELECTRÓN, INFORM&PD, TELECOM digitizing rate; ~ **elástica** *f* CARBÓN creep; ~ **de emisión secundaria** *f* ELECTRÓN secondary emission ratio; ~ **de la emisora** *f* TV station timing; ~ **de fuego** *f* D&A rate of fire; ~ **de miscibilidad** *f* METAL miscibility gap; ~ **de tiro** *f* D&A rate of fire

cadeneo *m* CONST chainage, chaining

cadenero *m* PROD *persona* chain maker

cadenote *m* TRANSP MAR chain plate

cadmiado *adj* MECÁ, METAL, QUÍMICA cadmium-plated

cadmio *m* (*Cd*) ELEC, ING ELÉC, METAL, MINERAL, QUÍMICA cadmium (*Cd*)

caducidad *f* EMB shelf life, storage durability, PROD obsolescence; ~ **del temporizador** *f* TELECOM expiration of timer (*AmE*), expiry of timer (*BrE*)

caer: ~ **al agua** *vi* TRANSP MAR fall overboard; ~ **a sotavento** *vi* TRANSP MAR *navegación a vela* cast off

CAF *abr* (*control automático de frecuencia*) ELEC, ELECTRÓN, PROD, TELECOM, TV AFC (*automatic frequency control*)

cafeico *adj* ALIMENT, QUÍMICA caffeic

cafeína *f* ALIMENT, QUÍMICA caffeine

cafetánico *adj* ALIMENT, QUÍMICA caffetannic

CAG *abr* ELECTRÓN (*control automático de ganancia*) AGC (*automatic gain control*)

CAI *abr* (*enseñanza asistida por computadora, enseñanza asistida por ordenador*) INFORM&PD CAI (*computer-assisted instruction*)

caída *f* CONST drop, FÍS fall, GEOL drop, INSTAL HIDRÁUL fall, REFRIG drop, TEC ESP *hidrología, vuelo orbital* drop, chute, TRANSP MAR *palo* rake, *vela cuadrilátera* drop, TV drop-out, VEH *geometría de las ruedas* camber; ~ **adiabática de presión** *f* FÍS FLUID *gases* adiabatic pressure drop; ~ **anódica de voltaje** *f* GAS anode drop; ~ **bruta** *f* ING MECÁ *hidráulica* bulkhead; ~ **en cascada** *f* CARBÓN cascading; ~ **en catarata** *f* CARBÓN *hidráulica* cataracting; ~ **catódica de voltaje** *f* ELECTRÓN, GAS cathode drop; ~ **efectiva** *f* INSTAL HIDRÁUL effective head, operating head, working head; ~ **de frecuencia** *f* ELEC frequency fall-off; ~ **inductiva** *f* ELEC *de tensión* inductive drop; ~ **libre** *f* EMB, TRANSP AÉR freefall; ~ **de la línea** *f* ELEC *red de distribución* line break; ~ **del listón** *f* C&V loss of sheet; ~ **de nieve** *f* METEO *precipitación de nieve* snowfall; ~ **de nivel** *f* ING MECÁ level drop; ~ **óhmica** *f* ELEC *resistencia* ohmic drop; ~ **del palo** *f* TRANSP MAR mast rake; ~ **del pH** *f* CONTAM, QUÍMICA *tratamiento químico* pH drop; ~ **de piedras** *f* CONST fall of rock; ~ **de popa** *f* TRANSP MAR *vela trapezoidal* drop; ~ **de potencial** *f* ELEC *de voltaje* fall of potential, potential drop, FÍS, ING ELÉC *de voltaje* potential drop; ~ **de presión** *f* ENERG RENOV pressure drop, INSTAL HIDRÁUL drop of pressure, pressure declination, pressure drop, INSTAL TERM, PROD, TEC PETR *mecánica de fluidos* pressure drop; ~ **de proa** *f* TRANSP MAR *vela trapezoidal* luff; ~ **en la reactancia** *f* ELEC, ING ELÉC reactance drop; ~ **del sistema** *f* INFORM&PD

crash; **~ de temperatura** *f* GAS, TERMO temperature drop; **~ de tensión** *f* CINEMAT power failure, ELEC potential drop, voltage drop, ING ELÉC potential drop, PROD voltage drop, TELECOM power fail; **~ de tensión anódica** *f* GAS anode drop; **~ de tensión de cátodo** *f* ELECTRÓN, GAS cathode drop; **~ de tensión debida a la impedancia** *f* ELEC impedance drop; **~ de tensión de línea** *f* ELEC *voltaje*, ING ELÉC line drop; **~ de tensión en una línea** *f* PROD line drop; **~ de tensión en la reactancia** *f* ELEC, ING ELÉC reactance drop; **~ de tensión por resistencia** *f* ELEC resistance drop, ING ELÉC resistance drop, voltage drop; **~ térmica** *f* TERMO heat drop; **~ útil** *f* INSTAL HIDRÁUL effective head, operating head, working head, PROD operating head; **~ de voltaje** *f* ELEC *resistencia* IR-drop, potential drop, voltage drop, FÍS voltage drop; **~ de voltaje en línea** *f* ELEC, ING ELÉC line drop; **~ de voltaje por resistencia** *f* ELEC resistance drop, ING ELÉC resistance drop, voltage drop

cainita *f* MINERAL kainite

caja *f* ACÚST *de un aparato* enclosure, AGUA body, CONST case, ELEC *del motor* casing, FERRO *vehículos* body, ING MECÁ housing, bearer, frame, case, INSTR *carruajes, vagones* body, MECÁ casing, cabinet, case, chamber, MINAS *bocartes* cutting ring, shoe, QUÍMICA tray, TRANSP AÉR case;

~ a ~ **de acceso** *f* ELEC *instalación* pull box; **~ accionadora** *f* ING MECÁ kicker box; **~ acústica** *f* ACÚST *radio* baffle; **~ acústica reflectora** *f* ACÚST reflex baffle; **~ de admisión** *f* ING MECÁ inlet case; **~ de adorno** *f* EMB fancy box; **~ de agua** *f* C&V, NUCL water box; **~ de agua de la calandra** *f* PAPEL calender water box; **~ de aire** *f* MINAS air blast, air box; **~ aislada** *f* REFRIG insulated body; **~ alar** *f* TRANSP AÉR wing box; **~ de alimentación** *f* ING MECÁ change feed box, feed box, PAPEL, PROD feed box; **~ alineadora-ensanchadora** *f* TRANSP AÉR align-reaming box; **~ alta** *f* IMPR upper case; **~ de ampollas** *f* EMB ampoule box (*BrE*), ampule box (*AmE*); **~ de anclas** *f* TRANSP MAR chain locker; **~ anular de la cámara de combustión** *f* TRANSP AÉR combustion-chamber annular case; **~ apilable** *f* EMB stacking box; **~ de arenque** *f* OCEAN herring boat; **~ de armón** *f* D&A ammunition box; **~ aspirante** *f* PAPEL suction box, vacuum box; **~ aspirante desplegadora** *f* PAPEL cambered suction box; **~ aspirante plana** *f* PAPEL flat box; **~ de avances** *f* ING MECÁ, PAPEL, PROD feed box;

~ b ~ **de banderas** *f* TRANSP MAR signal locker; **~ de batería** *f* AUTO, PAPEL battery box; **~ Baum** *f* CARBÓN Baum jig; **~ de blocaje del tren de aterrizaje** *f* TRANSP AÉR landing gear uplock box; **~ de bornes** *f* ING ELÉC terminal box;

~ c ~ **de cabeza** *f* AGUA head box; **~ de cabeza de máquina** *f* PAPEL head box; **~ de cables** *f* ELEC *accesorio de cable*, ING ELÉC terminal box, TRANSP MAR cable locker; **~ de cadenas** *f* TRANSP MAR cable locker; **~ de la cámara** *f* CINEMAT, FOTO, TV camera housing; **~ de cambio de velocidades** *f* AUTO, ING MECÁ, MECÁ, PROD, VEH gearbox; **~ de cambios** *f* AUTO gearbox, ING MECÁ *automóviles* gear case, MECÁ, PROD, TRANSP MAR, VEH *transmisión* gearbox; **~ del carrete inferior** *f* CINEMAT bottom spool box; **~ de cartón ondulado** *f* EMB, PAPEL corrugated-board box; **~ de cartón usada, aplastada y limpia** *f*

EMB washed and squashed consumer waste carton; **~ de caudales** *f* PROD, SEG safe; **~ de cementación** *f* PROD *horno* converting pot; **~ central del fuselaje** *f* TRANSP AÉR fuselage center box (*AmE*), fuselage centre box (*BrE*); **~ de cepillo** *f* CONST plane stock; **~ del cepillo** *f* ING MECÁ stock; **~ de la cerradura** *f* CONST lock casing; **~ del cerrojo** *f* CONST box staple; **~ de cervezas** *f* EMB beer crate; **~ de chapa fina de acero** *f* PROD sheet steel case; **~ del cigüeñal** *f* AUTO, MECÁ, VEH crankcase; **~ de cola** *f* AGUA end box, tail box; **~ colectora del polvo** *f* PROC QUÍ dust catcher; **~-compartimiento** *f* EMB compartment case; **~ de concentrados** *f* METAL, MINAS *metalurgia* concentrates box; **~ de conexión** *f* ING MECÁ jack box; **~ de conexiones** *f* ELEC junction box, ING ELÉC connection box, junction box, TEC ESP junction box; **~ de contacto** *f* ELEC *conexión* plug receptacle; **~ control** *f* AmL (*cf jaula Esp*) TRANSP box; **~ de control** *f* ELEC, ING ELÉC control unit, ING MECÁ control box; **~ de control de arranque del motor** *f* TRANSP AÉR engine-starting control box; **~ de control digital** *f* AUTO digital control box; **~ de control de encendido del motor** *f* TRANSP AÉR engine-starting control box; **~ de corte** *f* ELEC, ING ELÉC *conexión* cable termination; **~ cosida con grapas** *f* EMB stitched box; **~ de cribado** *f* PROD screening box;

~ d ~ **de décadas** *f* ING ELÉC, TRANSP AÉR decade box; **~ de décadas de condensadores** *f* ING ELÉC, TRANSP AÉR decade capacitance box; **~ de décadas de inductancias** *f* ING ELÉC, TRANSP AÉR decade inductance box; **~ de décadas de resistencias** *f* ING ELÉC, TRANSP AÉR decade resistance box; **~ de depuración neumática** *f* CARBÓN air jig; **~ de derivación** *f* ELEC distribution box, *accesorio del cable* dividing box, *conexión* branch box, junction box, *instalación* pull box; **~ de derivación trifurcada** *f* ELEC *accesorio de cable* trifurcating box, *conexión de cable* trifurcating joint; **~ de derivaciones** *f* ING ELÉC conduit box, TEC ESP junction box; **~ de descarga** *f* PROD *de bomba* delivery box; **~ de desgotado** *f* PAPEL straining chest; **~ de desgranamiento** *f* AGRIC shelling cage; **~ deslizante** *f* EMB slip case; **~ de deslustrado** *f* FOTO matte box; **~ desmontable** *f* EMB collapsible case; **~ diafragma** *f* MINAS diaphragm jig; **~ del diferencial** *f* AUTO differential case; **~ de dilatación** *f* AUTO expansion tank, INSTAL HIDRÁUL expansion box, expansion tank, MECÁ, TERMOTEC expansion tank; **~ de la dirección** *f* AUTO, VEH steering gearbox; **~ disparadora** *f* AUTO, VEH trigger box; **~ de distribución** *f* CINEMAT switch box, ELEC *conexión* distribution box, outlet box, ING ELÉC conduit box, distribution box, TEC PETR valve chest; **~ del distribuidor** *f* INSTAL HIDRÁUL *válvula de corredera* slide box, valve liner; **~ divisoria de fases** *f* ELEC *accesorio del cable* splitter box; **~ de doble hilera de conexiones** *f* INFORM&PD dual-in-line package (*DIP*); **~ doblemente reforzada** *f* EMB double-battened case;

~ e ~ **de efectos sonoros** *f* CINEMAT, TV effects box; **~ del eje portador** *f* FERRO bearing axle box; **~ de embalaje** *f* EMB case packing, TEXTIL packing case; **~ de émbolo** *f* CARBÓN plunger-type jig; **~ de embutir** *f* ELEC *conexión* outlet box; **~ de empalme** *f* CONST joint box, ELEC *accesorio de cable* terminal box, *conexión* joint box; **~ de empalme para cable tripolar** *f* ELEC *accesorio de cable* trifurcating box;

~ **de empalme de cables** *f* ELEC *conexión*, ING ELÉC cable box; ~ **de empalmes** *f* ELEC *conexión* junction box, splice box, ING ELÉC conduit box, TEC ESP junction box; ~ **de empalmes para cables** *f* ELEC, ING ELÉC *conexión* cable junction box; ~ **de empaquetadura** *f* ING MECÁ stuffing box; ~ **de enchufe** *f* PROD, TELECOM socket; ~ **de enchufes** *f pl* ING ELÉC plug box; ~ **de engranajes** *f* AUTO gearbox, ING MECÁ gear case, gear casing, gear cover, gearbox, axle box (*BrE*), gear box (*AmE*), MECÁ, PROD, TRANSP MAR, VEH gearbox; ~ **de engranajes cónicos** *f* TRANSP AÉR bevel-gear housing; ~ **de engranajes de la dirección** *f* AUTO, VEH steering gearbox; ~ **de engranajes intermedia** *f* TRANSP AÉR intermediate gearbox; ~ **de engranajes principal** *f* TRANSP AÉR main gearbox; ~ **de engrase** *f* ING MECÁ pedestal box, PROD oil box; ~ **de entrada** *f* ING MECÁ inlet case, PAPEL head box; ~ **de entrada abierta** *f* PAPEL open headbox; ~ **de entrada multicapa** *f* PAPEL *máquina de papel o de cartón* multilayer headbox; ~ **de entrada de pasta** *f* PAPEL breaking-strength tester; ~ **de entrada principal** *f* PAPEL primary headbox; ~ **de entrada secundaria** *f* PAPEL secondary headbox; ~ **de entradas con regletas de nivel** *f* PAPEL lath breast; ~ **del escape** *f* TRANSP AÉR exhaust case; ~ **de esclusa** *f* AGUA sluice box; ~ **espiral** *f* AGUA volute chamber; ~ **estancadora** *f* ING MECÁ, PROD gland; ~ **de estopas de relleno** *f* ING MECÁ stuffing box; ~ **para examinar diapositivas** *f* FOTO slide box; ~ **de expansión del combustible** *f* TRANSP AÉR fuel expansion box; ~ **de expansión del combustible de punta del ala** *f* TRANSP AÉR fuel ullage box; ~ **expositora** *f* EMB display box; ~ **de expulsión** *f* ALIMENT exhaustion box; ~ **externa** *f* EMB outer case;

■ **f** ~ **de fangos** *f* PROD *compuertas* mud box; ~ **de fantasía** *f* EMB fancy box; ~ **de fondo plegable** *f* EMB folded-bottom box; ~ **de fuegos** *f* FERRO *vehículos*, TERMO *de un horno o caldera* firebox; ~ **fuerte** *f* PROD, SEG safe; ~ **del fuselaje** *f* TRANSP AÉR fuselage box; ~ **de fusibles** *f* AUTO, ELEC, ING ELÉC, VEH *instalación eléctrica* fuse box; ~ **del fusil** *f* D&A rifle stock; ~ **de fusores** *f* AUTO, ELEC, ING ELÉC, VEH fuse box;

■ **g** ~ **giroscópica** *f* TRANSP AÉR gyro caging; ~ **de grasa** *f* ING MECÁ pedestal box, PROD *ejes* grease box, oil box; ~ **de grasa recalentada** *f* FERRO *vehículos* hot box; ~ **de grasas** *f* FERRO *vehículos*, ING MECÁ axle box (*BrE*), journal box (*AmE*); ~ **de guantes** *f* NUCL glove box;

■ **h** ~ **de herramientas** *f* ING MECÁ kit, tool box; ~ **hidráulica** *f* MINAS jig; ~ **de hojalata** *f* EMB canister; ~ **de humos** *f* FERRO *vehículos*, TERMOTEC smoke box;

■ **i** ~ **de inductancias** *f* ING ELÉC inductance box; ~ **inferior de descarga** *f* AGUA drop box; ~ **a inglete** *f* CONST miter box (*AmE*), mitre box (*BrE*); ~ **de inspección** *f* PROD inspection fitting; ~ **intermedia** *f* TRANSP AÉR intermediate case; ~ **isoterma** *f* REFRIG insulated body;

■ **l** ~ **del larguero del estabilizador** *f* TRANSP AÉR fin spar box; ~ **de lavado** *f* CARBÓN *de minerales* cleaner jig; ~ **de lujo** *f* EMB fancy box; ~ **de luz** *f* FOTO light box;

■ **m** ~ **de machos** *f* PROD core stock, *fundición* core box; ~ **de machos de dos partes** *f* PROD *moldería* split core box; ~ **de madera contrachapada** *f* EMB plywood case; ~ **metálica** *f* EMB metal box; ~ **metálica de bordes canteados** *f* EMB metal edging case; ~ **de mezclas** *f* C&V mixing box; ~ **de moldear** *f* PROD molding box (*AmE*), molding flask (*AmE*), moulding box (*BrE*), moulding flask (*BrE*); ~ **de moldeo** *f* PROD casting box, foundry flask; ~ **del motor** *f* TEC ESP motor case; ~ **de municiones** *f* D&A ammunition box; ~ **en un muro** *f* CONST wall box;

■ **n** ~ **negra** *f* INFORM&PD, INSTR, TRANSP AÉR black box;

■ **p** ~ **de paredes de cristal** *f* AGRIC *para cubrir plantas jóvenes* cold frame; ~ **de paso** *f* ELEC *instalación* pull box; ~ **pegada** *f* EMB glued box; ~ **Petri** *f* AGRIC, LAB *bacteriología* culture plate, *material de vidrio, microbiología* Petri dish; ~ **del piñón planetario** *f* TRANSP AÉR planet-pinion cage; ~ **de pinturas** *f* TV paintbox, palette; ~ **de pistón** *f* CARBÓN piston jig, plunger-type jig; ~ **plana** *f* ELECTRÓN flat pack; ~ **plana de banda móvil** *f* PAPEL moving-belt flat box; ~ **plantilla de alineación** *f* TRANSP AÉR align-reaming box; ~ **plegable** *f* EMB fold carton, folding box, *de cartón* folding carton, PAPEL folding box; ~ **plegable de cartón** *f* EMB fold carton; ~ **plegable impresa** *f* EMB printed folding-carton; ~ **de polea** *f* ING MECÁ shell; ~ **de presión** *f* PROD pressure box; ~ **de pulsación** *f* CARBÓN pulsator jig; ~ **de pulsación neumática** *f* CARBÓN pneumatic jig; ~ **de puros** *f* EMB cigar box;

■ **r** ~ **reflex** *f* CINEMAT, FOTO reflex housing; ~ **refrigerante** *f* PROD jumbo; ~ **de registro de porcelana** *f* C&V porcelain junction-box; ~ **registro de porcelana** *f* C&V *para instalaciones eléctricas* porcelain conduit-box; ~ **de relé de fuerza e interconexión** *f* TRANSP AÉR power-and-interlock relay box; ~ **de reloj** *f* ING MECÁ watch casing; ~ **de resistencia** *f* ELEC, ING ELÉC resistance box; ~ **de resonancia** *f* ELECTRÓN, FÍS PART, TELECOM *en telegrafía* resonator; ~ **reutilizable** *f* EMB, RECICL reusable box; ~ **rígida** *f* EMB rigid box; ~ **de ruedas** *f* ING MECÁ wheel guard; ~ **de ruedas interna** *f* TRANSP AÉR internal wheel case; ~ **con ruedas para transportar libros** *f* IMPR trolley;

■ **s** ~ **de salida** *f* ELEC *conexión* outlet box; ~ **de seguridad** *f* PROD, SEG safe; ~ **de señales** *f* FERRO *equipo inamovible* signal box (*BrE*), signal tower (*AmE*); ~ **de sonda** *f* MINAS *sondeos* driller box, pole-box; ~ **subterránea** *f* ELEC *fuente de alimentación* manhole; ~ **de succión** *f* ING MECÁ suction box;

■ **t** ~ **telescópica** *f* EMB nesting box; ~ **del televisor** *f* TELECOM, TV television cabinet; ~ **terminal** *f* ELEC, ING ELÉC *conexión* cable termination; ~ **terminal para cables** *f* CINEMAT cable terminal box; ~ **de terminales** *f* ELEC *accesorio del cable* terminal box; ~ **de terminales del compresor** *f* REFRIG *de terminales eléctricos* compressor terminal box; ~ **de terraja** *f* MECÁ die; ~ **de transporte** *f* EMB carrier box; ~ **de tres partes** *f* PROD three-part flask, *moldería* three-parted box; ~ **para tubos** *f* PROD *fundería* pipe box;

■ **u** ~ **de unión** *f* ELEC *conexión* joint box, TEC ESP junction box;

■ **v** ~ **de válvulas** *f* ING MECÁ clack box, INSTAL HIDRÁUL valve box, valve chest, TEC PETR valve chest; ~ **de la varilla de obturación** *f* ING MECÁ rod-seal housing; ~ **de velocidades** *f* AUTO gearbox, ING MECÁ gear case, gearbox, MECÁ, PROD, TRANSP MAR,

VEH *transmisión* gearbox; ~ **de vidrio** *f* INFORM&PD glass box; ~ **de viento** *f* ING MECÁ *hornos* air box, MINAS air blast, air box

cajas: ~ **de cartón ondulado usadas** *f pl* PAPEL old corrugated cartons (*OCC*); ~ **de cartón rígidas y plegables** *f pl* EMB rigid and folding cartons; ~ **de humo** *f pl* PROD *caldera* breeching

cajeado *m* CONST *carpintería* mortising

cajeadora *f* CONST, PROD mortising machine

cajear *vt* MINAS hew

cajera *f* PROD groove

cajero: ~ **automático** *m Esp* (*ATM, cf máquina medidora automática AmL*) INFORM&PD automatic teller machine (*ATM*)

cajeta *f* TRANSP MAR sennet; ~ **de iluminación** *f AmL* (*cf negatoscopio Esp*) FOTO light box

cajetilla *f* EMB cigarette pack (*AmE*), cigarette packet (*BrE*); ~ **de cigarrillos** *f* EMB cigarette pack (*AmE*), cigarette packet (*BrE*)

cajetín *m* IMPR *tipografía* box, ING ELÉC molding (*AmE*), moulding (*BrE*); ~ **de alimentación** *m* PAPEL breast box; ~ **de capacitancia** *m* ELEC, ING ELÉC, INSTR capacitance box; ~ **de iluminación** *m* CINEMAT light box; ~ **de madera** *m* ING ELÉC *instalación eléctrica* molding (*AmE*), moulding (*BrE*)

cajillo: ~ **para tuberías** *m* CARBÓN *buques* pipe casing

cajista *m* IMPR compositor, typesetter

cajón *m* AGUA, CONST caisson, CONTAM MAR skip, EMB crate, HIDROL, INSTAL HIDRÁUL caisson, TRANSP skip, TRANSP MAR caisson; ~ **de apilar** *m* CONST pile drawer; ~ **de avena** *m* AGRIC oat bin; ~ **de fuegos** *m* FERRO *vehículos* firebox; ~ **de ventilación** *m* ING MECÁ *minas* air box

cajonera *f* AGRIC *horticultura* flat, *para cubrir plantas jóvenes* cold frame; ~ **para filtros** *f* FOTO sliding filter drawer

cal *f* MINERAL, CONST lime; ~ **apagada** *f* CONST, QUÍMICA slaked lime; ~ **clorada** *f* ALIMENT chlorinated lime; ~ **soda** *f* QUÍMICA *cal-hidróxido sódico* soda lime; ~ **viva** *f* C&V quicklime, CARBÓN *industrias, metalurgia, residuos* lime, CONST, CONTAM MAR quicklime, QUÍMICA quicklime, *industrias, metalurgia, residuos* lime; ~ **viva de blanqueo** *f* DETERG bleaching lime

CAL *abr* (*enseñanza auxiliada por ordenador*) INFORM&PD CAL (*computer-assisted learning*)

cala *f* MINAS *túneles* holing, OCEAN, TRANSP MAR *geografía* cove

calabrote *m* ING MECÁ hawser

calada *f* GEOL, MINAS breakthrough, OCEAN *pesca* catch, TEXTIL *tejedura* shed

caladero *m* TRANSP MAR fishing ground

calado *m* MINAS *túneles* holing, PROD fretting, fretwork, fret sawing, TEXTIL *género de punto* lacy fabric, TRANSP MAR *buque* draught (*BrE*), draft (*AmE*); **poco** ~ *m* TRANSP MAR *buque* shallow draft (*AmE*), shallow draught (*BrE*); ~ **máximo** *m* TRANSP MAR *diseño naval* extreme draft (*AmE*), extreme draught (*BrE*); ~ **máximo en agua salada** *m* TRANSP MAR deepest seagoing draft (*AmE*), deepest seagoing draught (*BrE*); ~ **medio** *m* TRANSP MAR *diseño naval* mean draft (*AmE*), mean draught (*BrE*); ~ **de trazado** *m* TRANSP MAR rabbet draft (*AmE*), rabbet draught (*BrE*), *diseño naval* molded draft (*AmE*), moulded draught (*BrE*)

calafate *m* ING MECÁ, MECÁ, PROD, TRANSP MAR caulker

calafateado *m* ING MECÁ, MECÁ, PROD, TRANSP MAR caulking

calafatear *vt* ING MECÁ, MECÁ caulk, PROD *hierro* caulk, jag, TRANSP MAR *mantenimiento, construcción de barco* caulk

calafateo *m* ING MECÁ, MECÁ caulking, PROD *de hierro* caulking, jagging, TRANSP MAR caulking

calaje *m* ING ELÉC set; ~ **en cascada** *m* ING ELÉC cascade set; ~ **de las escobillas plegable** *m* PROD fold brush setting

calambres *m pl* OCEAN diver's cramp

calamina *f* CARBÓN, MINAS calamine, MINERAL calamine, electric calamine

calamita *f* MINERAL calamite, tremolite

calandra *f* PAPEL calender; ~ **abrillantadora** *f* PAPEL gloss calender; ~ **de bastidor abierto** *f* PAPEL openframe calender; ~ **calibradora** *f* PAPEL caliper calender (*AmE*), calliper calender (*BrE*), thickness calender; ~ **de construcción abierta** *f* PAPEL openface calender; ~ **de construcción cerrada** *f* PAPEL closed-face calender; ~ **friccionadora** *f* PAPEL friction calender; ~ **gofradora** *f* EMB, PAPEL embossing calender; ~ **de hojas** *f* PAPEL sheet calender; ~ **satinadora de placas** *f* PAPEL plate glazing-calender; ~ **de vaporización** *f* TEXTIL steam calender

calandrado *m* P&C, PAPEL *procesamiento* calendering; ~ **en caliente** *m* PAPEL hot calendering

calandrar *vt* P&C, PAPEL *operación* calender

calandria *f* P&C *equipo* calender, PROC QUÍ, PROD rolling press, TEXTIL *acabado de las telas* calender

calar[1] *vt* AUTO stall, OCEAN *red* shoot, VEH stall

calar[2] *vi* TRANSP MAR draw

calarse *v refl* AUTO, VEH *motor* stall

calaverita *f* MINERAL calaverite

calcado *m* IMPR, ING MECÁ tracing

calcantita *f* MINERAL chalcanthite

calcar *vt* ING MECÁ trace

calcarenita *f* GEOL calcarenite

calcáreo *adj* GEOL calcareous, *roca* limey, QUÍMICA calcareous

calce *m* ING MECÁ wedging; ~ **de centrado** *m* IMPR centering wedge (*AmE*), centring wedge (*BrE*)

calcedonia *f* MINERAL chalcedony; ~ **parda** *f* MINERAL sardonyx

calcedonix *m* MINERAL chalcedonyx

calcés *m* TRANSP MAR *del palo* masthead

calcetería: ~ **y medias** *f pl* TEXTIL *actividad comercial* hosiery

cálcico *adj* GEOL, QUÍMICA calcic

calcífero *adj* GEOL, QUÍMICA *vitamina* calciferous

calciferol *m* GEOL, QUÍMICA calciferol

calcificar[1] *vt* GEOL calcify

calcificar[2] *vi* QUÍMICA calcify

calcimina *f* COLOR calcimine

calcín *m* C&V roasting kiln, INSTAL TERM, PROD, TERMO roaster, roasting furnace, roasting kiln, roasting oven

calcinación *f* CARBÓN calcination, calcining, roasting, INSTAL TERM calcination, calcining, PROD *cocción, tostación* burning, roasting, QUÍMICA calcination, calcining

calcinado[1] *adj* TERMO burnt, dead-burned

calcinado[2] *m* CARBÓN, INSTAL TERM, QUÍMICA calcining

calcinar *vt* CARBÓN, INSTAL TERM, QUÍMICA calcine, TERMO burn

calcio *m* (*Ca*) ALIMENT, METAL, QUÍMICA calcium (*Ca*)

calciocelestina *f* MINERAL calciocelestine

calcioferrita *f* MINERAL calcioferrite

calciotorita *f* MINERAL calciothorite

calcita *f* MINERAL calcite

calco: ~ **azul de dibujo** *m* ING MECÁ blueprint; ~**-esquisto** *m* GEOL calc-schist

calcoestibina *f* MINERAL chalcostibite

calcofanita *f* MINERAL chalcophanite

calcofilita *f* MINERAL chalcophyllite

calcolita *f* MINERAL chalcolite

calcomanía *f* C&V decal, transfer (*BrE*); ~ **térmica** *f* C&V heat-release decal (*AmE*), heat-release transfer (*BrE*)

calcomenita *f* MINERAL chalcomenite

calcomorfita *f* MINERAL chalcomorphite

calcona *f* QUÍMICA chalcone

calcopirita *f* MINERAL chalcopyrite

calcopirrotina *f* MINERAL chalcopyrrhotine

calcosiderita *f* MINERAL chalcosiderite

calcosina *f* MINERAL chalcosine

calcosita *f* MINERAL chalcocite

calcotriquita *f* MINERAL chalcotrichite

calcreta *f* GEOL calcrete

calculador *m* ELECTRÓN, FÍS, MATEMÁT calculator

calculadora *f* ELECTRÓN, FÍS, INFORM&PD, MATEMÁT calculator; ~ **de bolsillo** *f* TELECOM pocket calculator; ~ **electrónica** *f* ELEC computer; ~ **de exposición** *f* CINEMAT, FOTO, TV exposure calculator; ~ **de tiempo y metraje** *f* CINEMAT time-footage calculator

calcular[1] *vt* FÍS, INFORM&PD, MATEMÁT calculate, MECÁ design, TEC ESP time

calcular[2]: ~ **el espacio tipográfico** *vi* IMPR *fotocomposición, tipografía* cast off

cálculo *m* FÍS calculation, INFORM&PD computation, computing, calculation, MATEMÁT calculation, TELECOM computation; ~ **analógico** *m* INFORM&PD analog calculation; ~ **por aproximaciones sucesivas** *m* GEOM trial and error calculation; ~ **del centro de gravedad** *m* TRANSP AÉR computation of center of gravity (*AmE*), computation of centre of gravity (*BrE*); ~ **diferencial** *m* MATEMÁT calculus, differential calculus; ~ **de escalas lineales** *m* TELECOM linear scaling calculation; ~ **de estructuras** *m* CONST structural analysis; ~ **justificativo del puntero** *m* TELECOM pointer justification-count (*PJC*); ~ **de magnitudes** *m* GEOM medida mensuration; ~ **operacional** *m* ELECTRÓN operational calculus; ~ **de originales** *m* IMPR copy fitting; ~ **de la potencia de enlace** *m* TELECOM link power budget; ~ **tensorial** *m* MATEMÁT tensor calculus; ~ **de variaciones** *m* MATEMÁT calculus of variations, variational calculus

calda *f* MECÁ heating; ~ **sudante** *f* PROD sparkling heat

caldeado: ~ **con coque** *adj* PROD coke-fired; ~ **con fueloil** *adj* TERMO oil-burning, oil-fired

caldear *vt* PROD stoke, TERMO heat up

caldecín *m* MINAS air reservoir

caldedera *f* INSTAL HIDRÁUL boilermaker

caldeo *m* MECÁ heating, PROD stoking, heating; ~ **dieléctrico** *m* ELEC, ING ELÉC dielectric heating; ~ **con gas** *m* GAS, INSTAL TERM, TERMO gas fire; ~ **por histéresis dieléctrica** *m* ELECTRÓN, ING ELÉC electronic heating; ~ **por radiación** *m* FÍS RAD, TERMO, TERMOTEC radiation heating; ~ **por resistencia** *m* ELEC, ING ELÉC, TERMO resistance heating

caldera *f* ELEC *fuente de alimentación* generator, FERRO *vehículos*, FÍS boiler, GEOL *volcánica* caldera, ING MECÁ pan, boiler, INSTAL HIDRÁUL *sistema generador de vapor*, MECÁ boiler, MINAS cauldron, PAPEL boiler, PROD cauldron, TERMO, TRANSP MAR *motor* boiler; ~ **acuotubular** *f* INSTAL TERM, TERMOTEC water tube boiler; ~ **de agua caliente** *f* ING MECÁ, TERMO hot water boiler; ~ **de agua caliente a baja presión** *f* INSTAL TERM low-pressure hot-water boiler; ~ **de asfalto** *f* PROD devil; ~ **para asfalto** *f* CONST asphalt boiler; ~ **auxiliar** *f* INSTAL TERM auxiliary boiler, TERMOTEC standby boiler; ~ **para calefacción central** *f* INSTAL TERM, TERMO central-heating boiler; ~ **para calentar alquitrán** *f* CONST tar boiler; ~ **de calor de desecho** *f* INSTAL TERM waste heat boiler; ~ **a carbón** *f* INSTAL TERM coal-fired boiler; ~ **de casco** *f* INSTAL TERM shell-type boiler; ~ **cilíndrica** *f* INSTAL HIDRÁUL barrel boiler, cylinder boiler, cylindrical boiler, shell boiler; ~ **de circulación** *f* INSTAL HIDRÁUL circulating boiler; ~ **de circulación forzada** *f* INSTAL TERM forced-circulation boiler; ~ **de un conducto de humos** *f* PROD single-flue boiler; ~ **de cuerpo cilíndrico** *f* INSTAL TERM shell-type boiler; ~ **desmontable** *f* INSTAL TERM sectional boiler; ~ **de doble paso** *f* INSTAL TERM double pass boiler; ~ **doméstica** *f* ING MECÁ domestic boiler; ~ **eléctrica** *f* ELEC, ING MECÁ electric steam-boiler; ~ **de electrodos** *f* ELEC, TERMOTEC electrode boiler; ~ **de entrada única** *f* INSTAL TERM once-through boiler; ~ **fija** *f* PROD stationary boiler; ~ **de fondo abombado** *f* INSTAL HIDRÁUL dish-ended boiler; ~ **de fusión** *f* PROD melter, melting pot, TERMO melting pot; ~ **de gas** *f* GAS, INSTAL TERM, TERMO gas boiler; ~ **de gasóleo** *f* INSTAL TERM, TERMOTEC oil-fired boiler; ~ **de hervidores** *f* INSTAL HIDRÁUL elephant boiler, French boiler; ~ **de hogar exterior** *f* PROD outside-fired boiler; ~ **de hogar interior** *f* PROD inside-fired boiler; ~ **de hogar interior cilíndrico** *f* INSTAL HIDRÁUL cylindrical flue boiler; ~ **de hogar tubular** *f* INSTAL HIDRÁUL tubular furnace boiler; ~ **de horno** *f* PROD flue boiler; ~ **con humero de retorno** *f* PROD return-flue boiler; ~ **humotubular** *f* INSTAL TERM *multitubular* fire-tube boiler; ~ **de llama directa** *f* PROD direct draught boiler; ~ **de locomotora** *f* INSTAL TERM locomotive boiler; ~ **marina** *f* INSTAL TERM marine boiler; ~ **multitubular** *f* PROD multiflue boiler; ~ **de pequeños elementos** *f* PROD sectional boiler; ~ **a presión** *f* ING MECÁ fired pressure vessel; ~ **que quema cortezas** *f* PAPEL bark boiler; ~ **radiante** *f* INSTAL TERM, TERMO, TERMOTEC radiant boiler; ~ **radiante marina** *f* INSTAL TERM, TERMO marine radiant boiler; ~ **radiante de recalentamiento marina** *f* INSTAL TERM, TERMO, TERMOTEC marine radiant reheat boiler; ~ **de recuperación** *f* TERMO waste heat boiler; ~ **de reserva** *f* TERMOTEC standby boiler; ~ **de seguridad** *f* PROD safety boiler; ~ **de tambor doble** *f* INSTAL TERM bi-drum boiler; ~ **de triple paso** *f* INSTAL TERM treble-pass boiler, triple-pass boiler; ~ **de tubos de agua** *f* INSTAL TERM, TERMOTEC water tube boiler; ~ **de tubos cruzados** *f* INSTAL TERM, TERMO cross-tube boiler; ~ **con tubos de humos** *f* INSTAL TERM *multitubular* fire-tube boiler; ~ **tubular** *f*

ALIMENT *maquinaria para proceso de alimentos* tubular boiler; **~ tubular con retorno de llama** *f* PROD return tube boiler, return tubular boiler; **~ uniformemente cilíndrica** *f* INSTAL HIDRÁUL plain cylindrical boiler; **~ unitaria** *f* TERMOTEC packaged boiler; **~ de vapor** *f* INSTAL HIDRÁUL power boiler, steam boiler, steamer, TERMO steam boiler; **~ de vapor que quema cortezas** *f* PAPEL bark power boiler; **~ de vaporización instantánea** *f* INSTAL HIDRÁUL, TERMO, TERMOTEC flash boiler; **~ de vaporización muy rápida** *f* INSTAL HIDRÁUL, TERMO, TERMOTEC flash boiler; **~ de vaporización rápida** *f* INSTAL HIDRÁUL flasher

calderería *f* INSTAL TERM boilermaking

calderero *m* INSTAL HIDRÁUL boilermaker

calderilla *m* MINAS winze

calderín *m* TEC PETR *refino* reboiler

caldero *m AmL* (*cf espuerta Esp*) MINAS *minerales o carbón* tub, cauldron, kibble, *laminación* chock; **~ estampado** *m* CONST stamped bucket; **~ grande** *m* MINAS *minerales o carbón* cauldron

caldo *m* AGUA *de cultivos* medium; **~ defectuoso** *m* TERMO off heat

caledonita *f* MINERAL caledonite

calefacción[1]: **de ~** *adj* PROD heating

calefacción[2] *f* GAS, MECÁ, PROD, TERMO, TERMOTEC, TRANSP AÉR heating; **~ acoplada a la red eléctrica** *f* INSTAL TERM power frequency heating; **~ del ambiente** *f* GAS space heating; **~ de automóvil** *f* AUTO, INSTAL TERM automotive heating; **~ a baja presión** *f* TERMOTEC low-pressure heating; **~ central** *f* CONST, INSTAL TERM, TERMO central heating; **~ centralizada** *f* TERMO district heating; **~ debajo del piso** *f* ELEC, TERMO underfloor heating; **~ dieléctrica** *f* INSTAL TERM dielectric heating; **~ por distritos** *f* INSTAL TERM district heating; **~ eléctrica** *f* ELEC, TERMOTEC electric heating; **~ empotrada en el suelo** *f Esp* (*cf losa radiante AmL*) TERMOTEC underfloor heating; **~ a gas** *f* GAS, PROD, TERMO, TERMOTEC gas heating; **~ por gasóleo** *f* INSTAL TERM, TERMOTEC oil-fired heating; **~ por infrarrojos** *f* TERMOTEC infrared heating; **~ por panel de infrarrojos** *f* INSTAL TERM infrared panel heating; **~ radiante** *f* FÍS RAD, INSTAL TERM, TERMO, TERMOTEC radiant heating; **~ por recuperación** *f* INSTAL TERM, REFRIG, TERMO regenerative heating; **~ por vapor** *f* CONST steam heating

calefactor *m* MECÁ, TERMO, VEH heater; **~ de aletas** *m* TERMO finned heater; **~ de calor radiante** *m* FÍS RAD, ING MECÁ, TERMO, TERMOTEC radiant heater; **~ por radiación** *m* FÍS RAD, ING MECÁ, TERMO, TERMOTEC radiant heater; **~ unidad** *m* TERMO unit heater; **~ de ventilador** *m* ING MECÁ, INSTAL TERM, TERMO fan heater

calentado: **~ al blanco** *adj* QUÍMICA *condición*, TERMO white-hot; **~ por carbón** *adj* CARBÓN coal-fired; **~ con gas** *adj* GAS, PROD, TERMO, TERMOTEC gas-fired; **~ al rojo** *adj* TERMO, TERMOTEC red-hot;

calentador *m* GAS, ING MECÁ, INSTAL TERM, LAB, MECÁ heater, NUCL *del presionador* heater rod, PROD, TERMO, TERMOTEC heater; **~ del aglomerante** *m* CONST binder heater; **~ de agua** *m* ING MECÁ hot water heater, water heater, LAB water heater; **~ de agua de alimentación** *m* PROD, TERMOTEC feedwater heater; **~ de agua caliente** *m* ING MECÁ hot water heater; **~ de agua a gas** *m* GAS, HIDROL, ING MECÁ,

INSTAL TERM, TERMO gas-fired water heater; **~ de agua por medio de gas** *m* GAS, INSTAL TERM, TERMO gas water heater; **~ de aire** *m* ING MECÁ, INSTAL TERM air heater, TRANSP AÉR heater blower; **~ por aire caliente** *m* TERMOTEC hot-air heater; **~ de aire giratorio** *m* INSTAL TERM, TERMO rotary air heater; **~ de aire regenerativo** *m* INSTAL TERM, SEG, TERMO regenerative airheater; **~ de aire tubular** *m* INSTAL TERM tubular air heater; **~ para almacenamiento térmico** *m* TERMO storage heater; **~ para almacenamiento térmico nocturno** *m* TERMO night storage heater; **~ de alta frecuencia** *m* ING ELÉC high-frequency heater; **~ de arco eléctrico** *m* ELEC, ING MECÁ electric-arc heater; **~ al arco eléctrico** *m* ING MECÁ arc heater; **~ por arco eléctrico** *m* ING MECÁ arc heater; **~ de calor radiante** *m* FÍS RAD, ING MECÁ, TERMO, TERMOTEC radiant heater; **~ del cárter** *m* REFRIG crankcase heater; **~ de cartuchos** *m* ING MECÁ cartridge heater; **~ de cinta** *m* ING MECÁ band heater; **~ de combustión limpia** *m* TERMOTEC flueless heater; **~ por conducción** *m* ING MECÁ, TERMO conduction heater; **~ de contacto** *m* ING MECÁ, TERMO contact heater; **~ convector** *m* INSTAL TERM, TERMO, TERMOTEC convector heater; **~ por corrientes de inducción** *m* ING ELÉC induction heater; **~ de cubeta** *m* FOTO dish heater; **~ dieléctrico** *m* ING MECÁ dielectric heater; **~ eléctrico** *m* ELEC electric hot-plate, ING MECÁ electric heater, TERMO electric hot-plate, electric heater; **~ eléctrico de superficies** *m* ELEC, ING MECÁ electric surface-heater; **~ de embudo** *m* LAB funnel heater; **~ entubado** *m* ING MECÁ, TERMOTEC flued heater; **~ del evaporador** *m* PROC QUÍ evaporating boiler; **~ a gas** *m* GAS, ING MECÁ, INSTAL TERM, TERMO gas-fired heater; **~ de inducción de baja frecuencia** *m* ING ELÉC low-frequency induction heater; **~ de inmersión** *m* ELEC *calefacción*, ING ELÉC, ING MECÁ, INSTAL TERM, LAB, MECÁ immersion heater; **~ instantáneo de agua** *m* INSTAL TERM instantaneous water heater; **~ de panel** *m* TERMO panel heater; **~ de parabrisas** *m* AUTO, VEH *accesorio* defroster; **~ radiante** *m* FÍS RAD, ING MECÁ, TERMO, TERMOTEC radiant heater; **~ refrigerado por vapor saturado** *m* NUCL saturated steam-cooled heater; **~ sumergible** *m* ING ELÉC immersion heater; **~ superficial eléctrico** *m* ELEC, ING MECÁ electric surface-heater; **~ de tanque** *m* FOTO tank heater; **~ sin tubos** *m* ING MECÁ flueless heater; **~ de tubos de aletas** *m* TERMO finned heater; **~ de varias clases de combustibles** *m* ING MECÁ multifuel heater; **~ de ventilador** *m* ING MECÁ, INSTAL TERM, TERMO fan heater

calentamiento[1]: **de ~** *adj* PROD heating

calentamiento[2] *m* C&V heating-up, warming-in, CONTAM temperature rise, GAS, MECÁ, PROD, TERMO, TEXTIL heating, TRANSP AÉR run-up area; **~ por aerodinámica** *m* TEC ESP aerodynamic heating; **~ aerodinámico** *m* TEC ESP kinetic heating, *vehículos espaciales* aerodynamic heating; **~ de agujas** *m* FERRO *equipo inamovible* points heating; **~ por alta frecuencia** *m* ELEC, ING ELÉC high-frequency heating; **~ anormal** *m* ING ELÉC overheating; **~ por arco** *m* ING ELÉC arc heating; **~ cinético** *m* FÍS, TEC ESP kinetic heating; **~ dieléctrico** *m* ELEC, ING ELÉC, P&C *operación* dielectric heating; **~ dinámico** *m* TERMO, TERMOTEC dynamic heating; **~ por efecto Joule** *m*

ELEC, ING ELÉC, TERMO resistance heating; ~ **eléctrico del suelo** *m* AGRIC, ELEC, INSTAL TERM electrical soil-heating; ~ **electrónico** *m* ELECTRÓN, ING ELÉC electronic heating; ~ **por fricción con el aire** *m* TEC ESP air friction heating; ~ **gamma** *m* FÍS RAD gamma-ray heating; ~ **por inducción** *m* MECÁ, P&C *operación* induction heating; ~ **por inducción de baja frecuencia** *m* ING ELÉC low-frequency induction heating; ~ **inductivo** *m* TERMO inductive heating; ~ **instantáneo** *m* ALIMENT, TERMOTEC flash heating; ~ **por manta eléctrica** *m* ELEC, TERMOTEC electric-blanket heating; ~ **de media frecuencia** *m* ING ELÉC, INSTAL TERM medium-frequency heating; ~ **con microondas** *m* P&C *operación* dielectric heating; ~ **moderado** *m* INSTAL TERM, TERMO gentle heat; ~ **del motor** *m* TRANSP AÉR engine run-up; ~ **de panel** *m* TERMO panel heating; ~ **radiante** *m* FÍS RAD, INSTAL TERM, TERMO, TERMOTEC radiant heating; ~ **por rayos infrarrojos** *m* FÍS RAD infrared heating; ~ **por resistencia** *m* ELEC, ING ELÉC, TERMO resistance heating; ~ **por RF** *m* ELECTRÓN, INSTAL TERM, P&C, TERMO *operación* RF heating; ~ **suave** *m* INSTAL TERM, TERMO gentle heat

calentar *vt* C&V heat up, FÍS heat, PROD *caldera* serve, QUÍMICA heat, heat up, TERMO heat, warm up, *vidrio* heat up, TEXTIL heat; ~ **en seco** *vt* TEXTIL bake

calentarse *v refl* ING MECÁ run hot, QUÍMICA, TERMO heat up

calera *f* MINAS lime pit, *explotación* limestone quarry, PROD lime pit

caleta *f* MINAS *geografía* inlet, OCEAN cove, inlet, TRANSP MAR *geografía* creek

cali *m* QUÍMICA *álcali* kali

calibita *f* MINERAL chalybite

calibración *f* AGUA gaging (*AmE*), gauging (*BrE*), CARBÓN calibration, sizing, *instrumentos* setting, ELEC *instrumento*, ELECTRÓN, FÍS RAD, INFORM&PD calibration, ING MECÁ verification, calibration, gauging (*BrE*), gaging (*AmE*), INSTR, LAB calibration, METR calibration, gaging (*AmE*), gauging (*BrE*), NUCL, PAPEL calibration, PROD gaging (*AmE*), gauging (*BrE*), calibration, TEC ESP, TEC PETR calibration, TELECOM metering, TRANSP AÉR calibration; ~ **de ajuste previo** *f* METR presetting gage (*AmE*), presetting gauge (*BrE*); ~ **de alarma** *f* TELECOM alarm setting; ~ **de la amplitud** *f* ELECTRÓN amplitude calibration; ~ **por fuerza dinámica** *f* ING MECÁ dynamic force calibration; ~ **por fuerza viva** *f* ING MECÁ dynamic force calibration; ~ **del generador de señal** *f* ELECTRÓN, FÍS signal generator calibration; ~ **geométrica** *f* TV geometric calibration; ~ **del giróscopo** *f* TRANSP AÉR gyro resetting; ~ **de radar** *f* FÍS ONDAS, FÍS RAD radar calibration

calibrado *m* CARBÓN, ELEC, ELECTRÓN, FÍS RAD, INFORM&PD calibration, ING MECÁ gaging (*AmE*), gauging (*BrE*), calibration, INSTR, LAB calibration, METR *de un instrumento de medida* gaging (*AmE*), calibration, gage (*AmE*), sizing, gauge (*BrE*), gauging (*BrE*), NUCL, PAPEL, PROD, TEC ESP, TEC PETR, TRANSP AÉR calibration; ~ **del freno de disco** *m* FERRO *vehículos* disc brake calliper (*BrE*), disk brake caliper (*AmE*); ~ **en sondeos** *m* GEOFÍS *propiedades de materiales* borehole logging

calibrador *m* CALIDAD *herramientas* CARBÓN gage (*AmE*), gauge (*BrE*), ING MECÁ gage (*AmE*), gauge (*BrE*), template, templet INSTR gage (*AmE*), gauge

(*BrE*), LAB, MECÁ caliper (*AmE*), calliper (*BrE*), gage (*AmE*), gauge (*BrE*), METR *instrumento*, PROD *hornos altos* gauge (*BrE*), gage (*AmE*), TRANSP MAR *medidas* template, VEH *freno*, caliper (*AmE*), calliper (*BrE*); ~ **accionado por aire** *m* METR air-operated gage (*AmE*), air-operated gauge (*BrE*); ~ **acústico** *m* ACÚST acoustic calibrator; ~ **de ajuste por resorte** *m* METR spring adjusting caliper (*AmE*), spring adjusting calliper (*BrE*); ~ **de alambres** *m* METR wire gage (*AmE*), wire gauge (*BrE*); ~ **de amplitud** *m* ELECTRÓN amplitude calibrator; ~ **del ánima** *m* METR *cañones* bore gage (*AmE*), bore gauge (*BrE*); ~ **del balancín** *m* METR beam caliper (*AmE*), beam calliper (*BrE*); ~ **de brocas** *m* ING MECÁ drill gage (*AmE*), drill gauge (*BrE*); ~ **de burbuja** *m* PROC QUÍ bubble gage (*AmE*), bubble gauge (*BrE*); ~ **de carbón** *m* CARBÓN, MINAS coal sizer; ~ **de cerámico de espesores** *m* ING MECÁ ceramic slip gage (*AmE*), ceramic slip gauge (*BrE*); ~ **cilíndrico** *m* PROD cylindrical gage (*AmE*), cylindrical gauge (*BrE*); ~ **cilíndrico de exteriores** *m* ING MECÁ external cylindrical gage (*AmE*), external cylindrical gauge (*BrE*); ~ **cilíndrico de interiores** *m* ING MECÁ internal cylindrical gage (*AmE*), internal cylindrical gauge (*BrE*); ~ **de clisés** *m* METR *topografía* plate gage (*AmE*), plate gauge (*BrE*); ~ **para clisés** *m* METR *topografía* plate gage (*AmE*), plate gauge (*BrE*); ~ **de compás interior** *m* ING MECÁ internal caliper gage (*AmE*), internal calliper gauge (*BrE*); ~ **de cursor de corredera** *m* ING MECÁ sliding caliper gage (*AmE*), sliding calliper gauge (*BrE*); ~ **doble** *m* CONST double calipers (*AmE*), double callipers (*BrE*); ~ **de espesores** *m* ING MECÁ, MECÁ thickness gage (*AmE*), thickness gauge (*BrE*), METR slip gage (*AmE*), slip gauge (*BrE*); ~ **de espesores de cerámica** *m* ING MECÁ ceramic slip gage (*AmE*), ceramic slip gauge (*BrE*); ~ **para espesores de chapas** *m* ING MECÁ plate gage (*AmE*), plate gauge (*BrE*); ~ **de exteriores de tres puntos** *m* ING MECÁ three-point snap gage (*AmE*), three-point snap gauge (*BrE*); ~ **fijo de medida exacta** *m* ING MECÁ go gage (*AmE*), go gauge (*BrE*); ~ **de filetes de tornillos** *m* ING MECÁ thread gage (*AmE*), thread gauge (*BrE*); ~ **de frecuencia** *m* ELECTRÓN frequency calibrator; ~ **de fuerza centrífuga** *m* FÍS, PROC QUÍ centrifugal-force calibrator; ~ **de herradura de tres puntos** *m* ING MECÁ three-point snap gage (*AmE*), three-point snap gauge (*BrE*); ~ **del hilo de rosca** *m* ING MECÁ screw gage (*AmE*), screw gauge (*BrE*), screw pitch gage (*AmE*), screw pitch gauge (*BrE*), screw thread gage (*AmE*), screw thread gauge (*BrE*), METR *tornillos* screw thread gage (*AmE*), screw thread gauge (*BrE*); ~ **de horquilla de tres puntos** *m* ING MECÁ three-point snap gage (*AmE*), three-point snap gauge (*BrE*); ~ **de huelgo mínimo** *m* ING MECÁ no-go gage (*AmE*), no-go gauge (*BrE*); ~ **de inspección** *m* ING MECÁ inspection gage (*AmE*), inspection gauge (*BrE*); ~ **Johansson** *m* MECÁ Johansson gage (*AmE*), Johansson gauge (*BrE*); ~ **de longitud** *m* METR length gage (*AmE*), length gauge (*BrE*); ~ **macho** *m* CONST plug gage (*AmE*), plug gauge (*BrE*), ING MECÁ internal caliper gage (*AmE*), internal calliper gauge (*BrE*); ~ **de máxima y mínima** *m* ING MECÁ in-and-out calipers (*AmE*), in-and-out callipers (*BrE*); ~ **mínimo** *m* ING MECÁ no-go gage (*AmE*), no-go gauge (*BrE*); ~ **neumático** *m* METR air gage (*AmE*), air gauge (*BrE*); ~ **no pasa** *m* ING MECÁ no-go gage (*AmE*), no-go gauge (*BrE*); ~ **en ocho** *m* ING MECÁ figure-of-eight calipers (*AmE*), figure-of-eight callipers (*BrE*); ~ **de paso** *m* METR *hilo de rosca* gap gage (*AmE*), gap gauge (*BrE*); ~ **patrón** *m* ING MECÁ standard gage (*AmE*), standard gauge (*BrE*); ~ **de pieza** *m* ING MECÁ detail gage (*AmE*), detail gauge (*BrE*); ~ **plano de exteriores** *m* ING MECÁ external caliper gage (*AmE*), external calliper gauge (*BrE*); ~ **que no pasa** *m* ING MECÁ no-go gage (*AmE*), no-go gauge (*BrE*); ~ **que pasa** *m* ING MECÁ go gage (*AmE*), go gauge (*BrE*); ~ **de ranura** *m* METR spline gage (*AmE*), spline gauge (*BrE*); ~ **de regulación** *m* ING MECÁ setting master; ~ **remoto de temperatura** *m* INSTR, REFRIG, TERMOTEC remote temperature gage (*AmE*), remote temperature gauge (*BrE*); ~ **de rodillos** *m* C&V roller gage (*AmE*), roller gauge (*BrE*); ~ **de roscas** *m* ING MECÁ thread gage (*AmE*), thread gauge (*BrE*), METR screw gage (*AmE*), screw gauge (*BrE*); ~ **de sedimentación** *m* CARBÓN settlement gage (*AmE*), settlement gauge (*BrE*); ~ **de separaciones** *m* METR feeler gage (*AmE*), feeler gauge (*BrE*); ~ **de tapón** *m* ING MECÁ internal cylindrical gauge (*AmE*), internal cylindrical gauge (*BrE*); ~ **de tapón para agujeros** *m* ING MECÁ plug gage (*AmE*), plug gauge (*BrE*); ~ **de tolerancia mínima** *m* ING MECÁ go gage (*AmE*), go gauge (*BrE*); ~ **de tubos** *m* ING MECÁ tube gage (*AmE*), tube gauge (*BrE*); ~ **de verificación de piezas** *m* METR feeler gage (*AmE*), feeler gauge (*BrE*); ~ **de vibraciones** *m* ACÚST vibration calibrator

calibradores *m pl* C&V calipers (*AmE*), callipers (*BrE*); ~ **de entrada y salida** *m pl* ING MECÁ in-and-out calipers (*AmE*), in-and-out callipers (*BrE*); ~ **de interiores** *m pl* ING MECÁ inside calipers (*AmE*), inside callipers (*BrE*); ~ **de máxima y mínima** *m pl* ING MECÁ in-and-out calipers (*AmE*), in-and-out callipers (*BrE*); ~ **micrométricos** *m pl* ING MECÁ micrometer calipers (*AmE*), micrometer callipers (*BrE*)

calibrar *vt* CARBÓN size, calibrate, ELEC, ELECTRÓN, FÍS, FÍS RAD, INFORM&PD calibrate, ING MECÁ calibrate, caliper (*AmE*), calliper (*BrE*), measure, INSTR, LAB, METR, NUCL, PAPEL calibrate, PROD calibrate, *alambres metálicos, chapas metálicas, toneles* gage (*AmE*), gauge (*BrE*), TEC ESP, TEC PETR calibrate, TEXTIL gage (*AmE*), gauge (*BrE*), TRANSP AÉR calibrate

calibre *m* AUTO bore, caliper (*AmE*), calliper (*BrE*), CALIDAD *alambre*, ELEC *medición, fabricación* gage (*AmE*), gauge (*BrE*), ING MECÁ caliper (*AmE*), calliper (*BrE*), gage (*AmE*), gauge (*BrE*), standard gage (*AmE*), standard gauge (*BrE*), INSTR, LAB gage (*AmE*), gauge (*BrE*), MECÁ caliper (*AmE*), calliper (*BrE*), gage (*AmE*), gauge (*BrE*), template, METEO *medición* vernier caliper (*AmE*), vernier calliper (*BrE*), PAPEL, PETROL caliper (*AmE*), calliper (*BrE*), PROD gauge (*AmE*), gauge (*BrE*), REVEST size, TEC PETR *perforación, petrofísica* caliper log (*AmE*), calliper log (*BrE*), TEXTIL *mecánica* gage (*AmE*), gauge (*BrE*), VEH *utillaje* bore, gage (*AmE*), gauge (*BrE*); ~ **por aguja** *m* VEH *carburador* needle jet; ~ **para ángulos de tuercas octogonales** *m* ING MECÁ octagonal nut angle gage (*AmE*), octagonal nut angle gauge (*BrE*); ~ **de anillo** *m* METR ring gage (*AmE*), ring gauge (*BrE*); ~ **de anillos** *m* PROD cylindrical gage (*AmE*), cylindrical gauge (*BrE*); ~ **anular** *m* METR ring gage

(*AmE*), ring gauge (*BrE*); ~ **de carpintero** *m* CONST carpenters' gage (*AmE*), carpenters' gauge (*BrE*); ~ **cilíndrico** *m* ING MECÁ, METR plug gage (*AmE*), plug gauge (*BrE*); ~ **del cilindro** *m* ING MECÁ cylinder bore; ~ **de comparación** *m* ING MECÁ master gage (*AmE*), master gauge (*BrE*); ~ **de comprobación** *m* ING MECÁ reference gage (*AmE*), reference gauge (*BrE*); ~ **crítico** *m* MINAS *barreno* critical diameter; ~ **de diferencia de presión** *m* MECÁ differential pressure gage (*AmE*), differential pressure gauge (*BrE*); ~ **doble** *m* CONST double calipers (*AmE*), double callipers (*BrE*); ~ **de estirar** *m* PROD drawplate; ~ **exterior** *m* ING MECÁ three-point snap gage (*AmE*), three-point snap gauge (*BrE*); ~ **de herradura** *m* METR snap gage (*AmE*), snap gauge (*BrE*); ~ **de horquilla** *m* METR snap gage (*AmE*), snap gauge (*BrE*); ~ **interior plano** *m* ING MECÁ male caliper gage (*AmE*), male calliper gauge (*BrE*); ~ **de límites** *m* ING MECÁ, METR limit gage (*AmE*), limit gauge (*BrE*); ~ **de macho** *m* ING MECÁ plug gage (*AmE*), plug gauge (*BrE*); ~ **macho cilíndrico** *m* ING MECÁ plug gage (*AmE*), plug gauge (*BrE*); ~ **macho plano** *m* ING MECÁ male caliper gage (*AmE*), male calliper gauge (*BrE*); ~ **maestro** *m* ING MECÁ master gage (*AmE*), master gauge (*BrE*); ~ **de mordaza** *m* METAL *metrología* clip gage (*AmE*), clip gauge (*BrE*), METR snap gage (*AmE*), snap gauge (*BrE*); ~ **de nonio** *m* MECÁ, METR vernier caliper (*AmE*), vernier calliper (*BrE*); ~ **normal** *m* CONST standard gage (*AmE*), standard gauge (*BrE*); ~ **patrón** *m* ING MECÁ master gage (*AmE*), master gauge (*BrE*), reference gage (*AmE*), reference gauge (*BrE*); ~ **plano** *m* ING MECÁ caliper gage (*AmE*), calliper gauge (*BrE*); ~ **de profundidad** *m* MECÁ, METR depth gage (*AmE*), depth gauge (*BrE*); ~ **de referencia** *m* ING MECÁ master gage (*AmE*), master gauge (*BrE*); ~ **de reloj de arena** *m* ING MECÁ hourglass calipers (*AmE*), hourglass callipers (*BrE*); ~ **reproductor** *m* PROD former; ~ **para la separación entre ejes de carriles** *m* FERRO track gage (*AmE*), track gauge (*BrE*); ~ **simple** *m* ING MECÁ plain gage (*AmE*), plain gauge (*BrE*); ~ **de superficies** *m* MECÁ surface gage (*AmE*), surface gauge (*BrE*); ~ **de tapón** *m* METR plug gage (*AmE*), plug gauge (*BrE*); ~ **de tolerancia** *m* ING MECÁ, METR limit gage (*AmE*), limit gauge (*BrE*); ~ **de topes** *m* PROD cylindrical gage (*AmE*), cylindrical gauge (*BrE*)

calicata *f* AGRIC soil profile pit, GEOL test pit, MINAS *prospección* cutting, prospecting, PETROL *minería* ditch

calicó *m* IMPR *tela para encuadernar*, TEXTIL calico

calidad¹: **de ~ media** *adj* MINAS *mineral* medium grade

calidad² *f* GEN grade, quality; ~ **del agua** *f* AGUA, CALIDAD, CONTAM, ENERG RENOV, HIDROL water quality; ~ **de agua potable** *f* AGUA drinking water quality; ~ **del aire** *f* TV air quality; ~ **apreciada** *f* CALIDAD assessed quality; ~ **de los áridos** *f* CONST quality of aggregate; ~ **de boceto** *f* IMPR draft quality (*AmE*), draught quality (*BrE*); ~ **de carta** *f* (*LQ*) INFORM&PD letter quality (*LQ*); ~ **cercana a la de carta** *f* (*NLQ*) INFORM&PD near-letter-quality (*NLQ*); ~ **de conservación** *f* ALIMENT keeping quality; ~ **corriente** *f* PROD common quality; ~ **económica** *f* CALIDAD economic quality; ~ **de la emisión** *f* TV broadcast quality; ~ **de horneado** *f* ALIMENT *molienda* baking quality; ~ **inaceptable** *f*

TELECOM unacceptable quality; ~ **inferior de servicio** *f* CALIDAD, TELECOM lower quality of service; ~ **de letra** *f* (*LQ*) IMPR, INFORM&PD letter quality (*LQ*); ~ **media después de control** *f* CALIDAD average outgoing quality (*AOQ*); ~ **de miembro** *f* INFORM&PD membership; ~ **de planeidad** *f* ING MECÁ flatness quality; ~ **de prueba** *f* PROD draft quality (*AmE*), draught quality (*BrE*); ~ **de reproducción** *f* CALIDAD, CINEMAT, FOTO, TV quality of reproduction; ~ **del servicio** *f* CALIDAD, TELECOM quality of service (*QOS*); ~ **de servicio de la red** *f* TELECOM network performance (*NP*); ~ **del sonido** *f* TELECOM sound quality; ~ **telefónica** *f* TELECOM *circuito* voice grade; ~ **de transmisión** *f* TELECOM transmission quality; ~ **a través del diseño** *f* CALIDAD, ING MECÁ, PROD quality through design; ~ **del vapor** *f* INSTAL HIDRÁUL steam quality; ~ **de vidriado** *f* C&V glazing quality

caliente *adj* TERMO warm

calificador: ~ **del código de identificación** *m* TELECOM identification code qualifier; ~ **de información de autorización** *m* TELECOM authorization information qualifier; ~ **de referencia del recibidor** *m* TELECOM recipient reference qualifier

californio *m* (*Cf*) FÍS RAD, QUÍMICA californium (*Cf*)

calima *f* METEO haze, TERMO heat haze, TRANSP MAR haze

calinoso *adj* METEO, TERMO, TRANSP MAR *tiempo atmosférico* hazy

caliza *f* C&V, CONST, GEOL, PETROL, QUÍMICA limestone; ~ **arenosa** *f* GEOL sandy limestone; ~ **bioconstruida** *f* GEOL constructed limestone; ~ **biomicrítica** *f* PETROL algal limestone; ~ **conchífera** *f* GEOL shelly limestone; ~ **cristalina** *f* CRISTAL, GEOL crystalline limestone; ~ **dolomítica** *f* GEOL dolomitic limestone; ~ **encrinítica** *f* GEOL crinoidal limestone, encrinitic limestone; ~ **espumosa** *f* MINERAL aphrite; ~ **granular** *f* GEOL lump limestone; ~ **lumaquélica** *f* GEOL coquina limestone; ~ **magnésica** *f* GEOL magnesian limestone; ~ **micrítica** *f* GEOL mudstone; ~ **microcristalina** *f* GEOL microcrystalline limestone; ~ **nodular** *f* GEOL nodular limestone; ~ **numulítica** *f* GEOL nummulitic limestone; ~ **oolítica** *f* GEOL oolitic limestone; ~ **ortoquímica** *f* GEOL orthochemical limestone; ~ **peloide** *f* GEOL pellet limestone; ~ **de pisolita** *f* GEOL pisolite limestone; ~ **pisolítica** *f* GEOL pisolitic limestone; ~ **silícea** *f* GEOL cherty limestone; ~ **terrosa** *f* MINERAL rock milk

calizo *adj* GEOL, QUÍMICA calcareous

callainita *f* MINERAL callainite

calle: ~ **de acceso** *f* CONST, TRANSP access road (*AmE*), slip road (*BrE*); ~ **peatonal sin salida** *f* TRANSP blind sector without traffic rights; ~ **de torbellinos** *f* FÍS FLUID *patrón de flujo* vortex street

callejón: ~ **sin salida** *m* CONST blind alley

Calloviense *adj* GEOL Callovian

calma[1]: **en** ~ *adj* HIDROL, METEO, OCEAN calm

calma[2] *f* HIDROL *del mar*, METEO, OCEAN calm; ~ **chicha** *f* HIDROL *marina*, TRANSP MAR *meteorología* dead calm; ~ **profunda** *f* HIDROL *marina* dead calm; ~ **tropical** *f* OCEAN tropical calm

calmado *adj* REVEST killed

calmar *vt* TRANSP MAR *barco* calm

calmas: ~ **ecuatoriales** *f pl* METEO doldrums, equatorial calms, OCEAN equatorial calms, TRANSP MAR doldrums; ~ **subtropicales** *f pl* METEO, OCEAN subtropical calms; ~ **tropicales** *f pl* METEO tropical calms

calmazo *m* TRANSP MAR *meteorología* dead calm

calomel *m* MINERAL, QUÍMICA calomel

calomelano *m* MINERAL, QUÍMICA calomel

calón *m* OCEAN seine staff

calor *m* GEN heat, NUCL warmth, TERMO off heat, warmth; ~ **de absorción** *m* TERMO heat of absorption; ~ **de activación** *m* TERMO heat of activation; ~ **de adsorción** *m* NUCL adsorption heat; ~ **de alta calidad** *m* NUCL high grade heat; ~ **blanco** *m* TERMO white heat; ~ **causado por fricción** *m* TERMO heat caused by friction; ~ **cedido** *m* REFRIG heat removed; ~ **cedido en el condensador** *m* REFRIG condenser heat; ~ **de combinación** *m* TERMO heat of combination; ~ **de combustión** *m* FÍS heat, QUÍMICA *de sustancia* heat of combustion, TERMO burning-heat, heat of combustion; ~ **de compresión** *m* TERMO heat of compression; ~ **de condensación** *m* TERMO heat of condensation; ~ **de convección** *m* INSTAL TERM convected heat, TERMO convected heat, convection heat; ~ **de descomposición** *m* QUÍMICA heat of decomposition; ~ **de desecho** *m* TERMOTEC waste heat; ~ **de disociación** *m* TERMO heat of dissociation; ~ **de disolución** *m* TERMO heat of solution; ~ **disponible** *m* TERMO available heat; ~ **específico** *m* P&C *propiedad física, prueba*, REFRIG, TEC ESP, TERMO, TERMOTEC *física* specific heat; ~ **específico a presión constante** *m* TERMO specific heat at constant pressure; ~ **excesivo** *m* PROD excessive heat; ~ **de expansión** *m* TERMO heat of expansion; ~ **del fondo** *m* C&V *hornos* bottom heat; ~ **de formación** *m* QUÍMICA *de compuesto*, TERMO heat of formation; ~ **de fusión** *m* TERMO heat of fusion, melting heat; ~ **de los gases de desecho** *m* TERMO waste gas heat; ~ **de hidratación** *m* CONST *cemento*, TERMO heat of hydration; ~ **latente** *m* CONST, FÍS, MECÁ, METEO, REFRIG, TEC PETR, TERMO, TERMOTEC latent heat; ~ **latente de compresión** *m* TERMO latent heat of compression; ~ **latente específico** *m* FÍS specific latent heat; ~ **latente de evaporación** *m* FÍS latent heat of vaporization, TERMO latent heat of evaporation, latent heat of vaporization; ~ **latente de expansión** *m* TERMO latent heat of expansion; ~ **latente de formación** *m* TERMO enthalpy of formation; ~ **latente de fusión** *m* FÍS, TERMO latent heat of fusion; ~ **latente de fusión efectivo** *m* TERMO effective latent heat of fusion; ~ **latente de hielo** *m* REFRIG ice-melting equivalent; ~ **latente de solidificación** *m* TERMO latent heat of solidification; ~ **latente de transformación** *m* TERMO latent heat of transformation; ~ **latente de vaporización** *m* PROC QUÍ, TERMO enthalpy of vaporization; ~ **libre** *m* TERMO, TERMOTEC free heat; ~ **luminiscente** *m* TERMO, TERMOTEC glowing heat; ~ **de mezclado** *m* TERMO heat of mixing; ~ **moderado** *m* TERMO warmth; ~ **molecular** *m* TERMO molecular heat; ~ **de neutralización** *m* TERMO heat of neutralization; ~ **perceptible** *m* FÍS sensible heat; ~ **de radiación** *m* FÍS RAD, TERMO, TERMOTEC radiation heating; ~ **de radiactividad** *m* FÍS RAD heat of radioactivity; ~ **radiante** *m* FÍS RAD, INSTAL TERM radiant heat, TERMO, TERMOTEC glowing heat, radiant heat; ~ **de reacción** *m* FÍS heat, *en procesos nucleares* Q-value, QUÍMICA, TERMO heat of reaction; ~ **reflejado** *m* ÓPT, TERMO reflected heat; ~ **residual** *m* CONTAM *procesos*

industriales, TERMO, TERMOTEC waste heat;
~ **rutilante** *m* TERMO, TERMOTEC glowing heat;
~ **seco** *m* TEXTIL dry heat; ~ **sensible** *m* GEOFÍS,
TERMOTEC sensible heat; ~ **sobrante** *m* CONTAM
procesos industriales, NUCL waste heat; ~ **solar** *m*
ENERG RENOV solar heat; ~ **del soldeo** *m* PROD
sparkling heat; ~ **útil** *m* TERMO heat output, TERMO-
TEC useful heat; ~ **de vaporización** *m* TERMO heat of
vaporization
caloría *f* ALIMENT, FÍS, TERMO, TERMOTEC calorie;
~~**-gramo** *f* FÍS, METR *unidad de calor* gram calorie,
gramme calorie
calorías: bajo en ~ *adj* ALIMENT low calorie
calórico *adj* TERMO caloric
calorífero[1] *adj* TERMO heat-carrying
calorífero[2] *m* ING MECÁ calorifier, MECÁ heater, TERMO
radiator, TERMOTEC *transferencia de calor* heater,
radiator
calorífico *adj* ALIMENT, FÍS, REFRIG calorific, TERMO
calorific, caloric, TERMOTEC calorific; **de** ~ *adj* PROD
heating
calorifugacia *f* TERMO heat resistance
calorifugado *adj* MECÁ, TERMO heat-insulated
calorífugo *adj* FÍS heat-resistant, MECÁ heat insulation,
TEC ESP heat-resisting, TERMO heat insulation
calorimetría *f* GEN calorimetry, TERMO science of heat;
~ **de exploración diferencial** *f* FÍS RAD, P&C, TERMO
differential scanning calorimetry
calorimétrico *adj* GEN calorimetric
calorímetro *m* GEN calorimeter; ~ **adiabático** *m*
TERMO adiabatic calorimeter; ~ **de bomba** *m* FÍS
bomb calorimeter; ~ **por condensación** *m* PROC
QUÍ condensation calorimeter; ~ **cuasiadiabático** *m*
NUCL quasi-adiabatic calorimeter; ~ **electromagnético** *m* ELEC, FÍS PART, FÍS RAD *detector de
partículas* electromagnetic calorimeter; ~ **hadrónico**
m FÍS PART, FÍS RAD *detector de hadrones* hadronic
calorimeter; ~ **a volumen constante** *m* FÍS bomb
calorimeter
calza *f* CARBÓN *minas* block, MINAS *sondeos* drive
block; ~ **de precisión** *f* ING MECÁ precision shim
calzada *f* AUTO freeway (*AmE*), motorway (*BrE*), road,
roadway (*AmE*), CARBÓN pavement (*BrE*), sidewalk
(*AmE*), CONST causeway, footway, pavement (*BrE*),
road, sidewalk (*AmE*), freeway (*AmE*), motorway
(*BrE*), roadway (*AmE*), TRANSP freeway (*AmE*),
motorway (*BrE*), road, roadway (*AmE*), TRANSP
MAR *para cruzar zonas pantanosas* causeway, VEH
road, motorway (*BrE*), roadway (*AmE*), freeway
(*AmE*); ~ **para ciclistas** *f* AmL (*cf carril para
bicicletas Esp*) TRANSP cycle path; ~ **elevada** *f*
TRANSP MAR *para cruzar zonas pantanosas* causeway
calzado *m* SEG *de trabajo, seguridad* footwear;
~ **antiestático** *m* SEG antistatic footwear;
~ **aprobado** *m* SEG approved footwear; ~ **de plástico moldeado** *m* SEG plastic-molded footwear
(*AmE*), plastic-moulded footwear (*BrE*);
~ **protector** *m* SEG protective footwear; ~ **protector
para evitar quemaduras** *m* SEG footwear for protection against burns; ~ **de trabajo** *m* SEG industrial
footwear
calzar[1] *vt* MECÁ wedge, PROD tip, *galga* chock
calzar[2]: ~ **la matriz** *vi* IMPR back up
calzo *m* AUTO brake shoe, ING MECÁ brake shoe,
packing piece, scotch, wedge, MECÁ brake shoe,
chock, wedge, MINAS chuck, PROD *martillo pilón*

tup, TRANSP skid, TRANSP MAR cradle, *construcción*
chock, VEH brake shoe, skid pad; ~ **de bote** *m*
TRANSP MAR *construcción* boat chock; ~ **de maza** *m*
PROD *martillo pilón* pallet-for-tup; ~ **de maza de
martinete** *m* PROD *martillo pilón* tup die; ~ **superior**
m PROD top die
CAM *abr* ELEC, INFORM&PD, PROD, TELECOM (*fabricación asistida por computadora AmL, fabricación
asistida por ordenador Esp*) CAM (*computer-aided
manufacturing*)
cama *f* CONST cradle, IMPR overlay, OCEAN bed;
~ **bacteriana** *f* AGUA bacteria bed; ~ **de
construcción** *f* TRANSP MAR *construcción* building
berth; ~ **fría** *f* AGRIC cold frame; ~ **de varada** *f*
TRANSP MAR *construcción* building slip
camada *f* GAS, GEOL bed
camafeo *m* C&V cameo, intaglio
cámara[1] *m* Esp (*cf camarógrafo AmL*) TV cameraman
cámara[2] *f* AGUA coffer, CARBÓN *minas* breast, room,
stall, CINEMAT camera, CONST cabin, ELEC *fuente de
alimentación* manhole, FÍS, FOTO, IMPR camera, ING
MECÁ chamber, INSTAL HIDRÁUL *bombas centrífugas*
clearance space, INSTR camera, MECÁ chamber,
MINAS stall, *de mineral* chamber, *habitación* room,
QUÍMICA *proceso* chamber, camera, TRANSP MAR
alojamiento de pasajeros berth, cabin, *buques* saloon,
TV camera, VEH *de neumático* inner tube;
■ **a** ~ **accionada manualmente** *f* CINEMAT hand-
cranked camera; ~ **accionada por resorte** *f* CINE-
MAT spring-drive camera; ~ **de aceleración** *f* PROC
QUÍ *acelerador de partículas* accelerating chamber;
~ **de aclarado** *f* EMB chambre blanche; ~ **para
aerofotografía** *f* CINEMAT aerial camera; ~ **de
aerofotogrametría** *f* FOTO aerial mapping camera;
~ **de agua** *f* INSTAL HIDRÁUL *turbinas* forebay; ~ **de
agua abierta** *f* AGUA *turbinas* open-turbine chamber,
PROD *hidroturbinas* open flume; ~ **de aire** *f* ING MECÁ
air receiver; MINAS airspace; TEC PETR air box; ~ **de
aire del bote** *f* TRANSP boat tank; ~ **de aire
comprimido** *f* ING MECÁ air receiver; ~ **aislada** *f*
MINAS panel; ~ **de almacenamiento** *f* OCEAN fish
room; ~ **de almacenamiento frigorífico** *f* ALIMENT,
REFRIG, TERMO cold storage room; ~ **de almacenamiento de hielo** *f* REFRIG ice storage room; ~ **de
almacenamiento para productos congelados** *f*
ALIMENT, REFRIG, TERMO frozen-food storage room;
~ **ampliadora** *f* FOTO, IMPR enlarger camera, enlarging camera, INSTR enlarger camera; ~ **para análisis
de movimientos** *f* CINEMAT motion analysis camera;
~ **anecoica** *f* ACÚST anechoic room, dead room,
echoless chamber, FÍS dead room, ING MECÁ echoless
chamber, TELECOM anechoic chamber, echoless
chamber; ~ **para animación** *f* CINEMAT animation
camera; ~ **de apagado de arco** *f* ING ELÉC arc
quench chamber; ~ **aséptica** *f* REFRIG clean room;
~ **de astrofotografía** *f* Esp (*cf cámara astronómica
AmL*) FOTO astronomical camera; ~ **astronómica** *f*
AmL (*cf cámara astrofotográfica Esp*) FOTO astronomical camera; ~ **audiométrica** *f* ACÚST, FÍS ONDAS
audiometric room;
■ **b** ~ **de baja velocidad** *f* CINEMAT low-speed
camera; ~ **de balas** *f* AGRIC bale chamber; ~ **de
bocarte** *f* MINAS stamp house; ~ **de bombas** *f* AGUA
pump room; ~ **de burbujas** *f* FÍS, FÍS PART *detección
de partículas* bubble chamber;
■ **c** ~ **de cajón** *f* FOTO box camera; ~ **de calderas** *f*

INSTAL HIDRÁUL boiler-room, stokehold, INSTAL TERM boiler-room; **~ de calefacción** *f* TERMO heating chamber; **~ de calentamiento** *f* TERMO heating chamber; **~ caliente del condensador** *f* NUCL condenser hotwell; **~ de campaña** *f* FOTO *máquina fotográfica con pie* field camera, folding camera; **~ de campo** *f* FOTO field camera; **~-car** *f* CINEMAT, TV *automóvil para tomas en movimiento* camera car; **~ de caracol** *f* PROD spiral casing; **~ con chasis** *f* CINEMAT magazine camera; **~ de chasis paralelo** *f* CINEMAT parallel magazine camera; **~ de chispas** *f* FÍS spark chamber; **~ de chispas seguidas** *f* FÍS PART *detección de partículas* streamer chamber; **~ de choque térmico** *f* LAB thermal shock chamber; **~ de cine** *f* CINEMAT film camera; **~ cinematográfica** *f* CINEMAT motion picture camera; **~ climática** *f* EMB, REFRIG, TERMO climatic chamber; **~ climatizada** *f* EMB, REFRIG climatic chamber; **~ de combustión** *f* AUTO, ING MECÁ combustion chamber, INSTAL HIDRÁUL *horno de caldera de vapor* fire chamber, combustion chamber, boiler, firebox, INSTAL TERM combustion chamber, MECÁ boiler, PROD fire chamber, RECICL can bank, TEC ESP combustor, combustion chamber, burner can, TERMO *motores de pistón* combustion chamber, firebox, fire chamber, combustor, TRANSP AÉR, TRANSP MAR *motor* combustion chamber; **~ de combustión de aluminio** *f* RECICL aluminium can bank (*BrE*), aluminum can bank (*AmE*); **~ de combustión del motor** *f* ING MECÁ engine combustion-chamber; **~ de combustión semiesférica** *f* AUTO hemispherical combustion chamber; **~ de combustión tipo cuña** *f* AUTO wedge-type combustion chamber; **~ de combustión de turbulencia** *f* AUTO turbulence combustion chamber; **~ compacta** *f* ELECTRÓN solid-state camera; **~ de compensación** *f* HIDROL *central hidroeléctrica* surge tank; **~ de compresión** *f* ACÚST pistonphone, AUTO, CARBÓN, INSTAL HIDRÁUL compression chamber, TEC ESP thrust chamber, TEC PETR *buceo*, VEH compression chamber; **~ de compuertas** *f* HIDROL valve chamber; **~ de congelación** *f* ALIMENT, REFRIG freezing room; **~ de contacto de cloro** *f* HIDROL chlorine contact chamber; **~ contenida en blindaje insonoro** *f* CINEMAT blimped camera; **~ de control remoto** *f* CINEMAT, FOTO, TV remote-controlled camera; **~ de conversión de imágenes unidimensionales** *f* CINEMAT streak image converter camera; **~ para copiado** *f* FOTO copy camera; **~ con corriente** *f* TV live camera; **~ de cuerda** *f* CINEMAT clockwork camera, spring-drive camera; **~ de curado** *f* REFRIG *de carne* curing cellar;

~ d **~ de decantación** *f* NUCL settler chamber; **~ de depuración** *f* PROD skimming chamber, *fundería* skim gate; **~ desacoplada** *f* TV mismatched camera; **~ de descompresión** *f* PETROL, TEC PETR, TRANSP MAR *actividades subacuáticas* decompression chamber; **~ de desecación** *f* PROD drying chamber; **~ de desecar** *f* PROD drying chamber; **~ de destilación** *f* PROC QUÍ distillation chamber; **~ de difracción de rayos X** *f* NUCL X-ray diffraction camera; **~ digital** *f* TV digital camera; **~ directa** *f* TV live camera; **~ con dispositivo de acoplamiento de carga** *f* EMB charge couple device camera; **~ de distribución** *f* AUTO plenum chamber, ING MECÁ distribution chamber, INSTAL HIDRÁUL steam chamber, steam chest, TRANSP

AÉR plenum chamber; **~ de distribución de vapor** *f* INSTAL HIDRÁUL valve chest, *distribuidor máquina de vapor* steam box, steam case; **~ de distribución del vapor** *f* TEC PETR valve chest; **~ distribuidora de aire** *f* AUTO, TRANSP AÉR plenum chamber; **~ para documentales de actualidades** *f* CINEMAT newsreel camera; **~ de dos formatos** *f* FOTO dual-format camera;

~ e **~ electronográfica** *f* FÍS PART electronographic camera; **~ de elutriación** *f* PROC QUÍ, QUÍMICA elutriation chamber; **~ de emisión** *f* INSTR emission chamber; **~ de enganche** *f* MINAS plat, *pozo de minas* lodge room; **~ de enganche de vagonetas** *f* MINAS *pozo* plat; **~ de ensayo en condiciones ambientales** *f* TEC ESP *equipos* environmental test chamber; **~ de esclusa** *f* AGUA, TRANSP MAR lock chamber; **~ de espejo rotativo para imágenes unidimensionales** *f* CINEMAT, FOTO, TV rotating-mirror streak camera; **~ espiral** *f* PROD spiral casing; **~ de estado sólido** *f* ELECTRÓN solid-state camera; **~ estereométrica** *f* INSTR stereometric camera; **~ estereoscópica** *f* FOTO stereoscopic camera; **~ de estudio** *f* FOTO studio camera; **~ de expansión** *f* ING MECÁ expansion chamber; **~ de explosión** *f* AUTO, ING MECÁ, INSTAL HIDRÁUL, INSTAL TERM, TEC ESP, TERMO, TRANSP AÉR, TRANSP MAR combustion chamber; **~ de extracción** *f* MINAS *cuartel de explotación* winder house;

~ f **~ de fardos** *f* AGRIC bale chamber; **~ de fermentación** *f* ALIMENT *panadería, repostería* proving cabinet; **~ de filtros de bolsas o mangas** *f* CONTAM *tratamiento de gases* baghouse; **~ de filtros de sacos** *f* CARBÓN baghouse; **~ de flotabilidad** *f* CONTAM MAR *de la barrera flotante* flotation chamber; **~ de formato ancho** *f* INSTR large-format camera; **~ de fotografía fija** *f* FOTO still camera; **~ fotográfica** *f* CINEMAT, FÍS, FOTO, IMPR, INSTR, QUÍMICA, TV camera; **~ fotográfica con enfoque reflex por espejos** *f* FOTO camera with mirror-reflex focusing; **~ fotográfica con exposímetro incorporado** *f* FOTO camera with built-in exposure meter, camera with coupled exposure meter; **~ fotográfica con extensión de fuelle corto** *f* FOTO camera with short bellows extension; **~ fotográfica con extensión de fuelle grande** *f* FOTO camera with large bellows extension; **~ fotográfica de foco fijo** *f* CINEMAT, FOTO, TV fixed-focus camera; **~ fotográfica de fuelle cuadrado** *f* FOTO square-bellows camera; **~ fotográfica de medio clisé** *f* FOTO half-plate camera; **~ fotográfica con objetivo descentrable** *f* FOTO camera with rising and swinging front; **~ fotográfica con objetivo intercambiable** *f* FOTO camera with interchangeable lens; **~ fotográfica con obturador entre lentes** *f* FOTO camera with diaphragm shutter; **~ fotográfica plegable** *f* FOTO folding camera; **~ fotográfica submarina** *f* FOTO underwater camera; **~ fotográfica subminiatura** *f* FOTO subminiature camera; **~ fotográfica con visor reflex desmontable** *f* FOTO camera with detachable reflex viewfinder; **~ fotogramétrica** *f* INSTR photogrammetric camera; **~ frigorífica** *f* ING MECÁ coldroom, REFRIG cold chamber, coldroom, TERMO coldroom; **~ frigorífica desmontable** *f* REFRIG portable coldroom; **~ frigorífica de doble pared** *f* REFRIG jacket coldroom; **~ frigorífica inflable** *f* REFRIG *por presión del aire* inflatable coldroom; **~ frigorífica pequeña** *f*

REFRIG walk-in refrigerator; ~ **frigorífica para plantas de vivero** *f* REFRIG *para plantas* nursery cold store; ~ **frigorífica prefabricada** *f* REFRIG sectional cold room; ~ **de frío negativo** *f* REFRIG negative-cold chamber; ~ **de frío positivo** *f* REFRIG positive-cold chamber;

~ h   ~ **de haz dividido** *f* CINEMAT split-beam camera; ~ **hiperbárica** *f* OCEAN, TEC PETR *buceo* hyperbaric chamber; ~ **húmeda** *f* OCEAN *experiencias hiperbáricas* wet chamber; ~ **de humos** *f* CONST, FERRO smoke box;

~ i   ~ **para imágenes unidimensionales** *f* CINEMAT streak camera; ~ **impelente** *f* AUTO, TRANSP AÉR plenum chamber; ~ **impelente multiplinto** *f* TRANSP multiple-skirted plenum chamber; ~ **insonora** *f* CINEMAT blimped camera, sound camera; ~ **intensificadora de silicio** *f* ELECTRÓN silicon-intensifier-target camera tube; ~ **de ionización** *f* FÍS ionization chamber, FÍS ONDAS *para detectar la radiación* cloud chamber, ionization chamber, FÍS PART *detección de partículas* ionization chamber, ING MECÁ expansion chamber; ~ **de ionización de fisión** *f* FÍS, FÍS PART, NUCL fission ionization chamber; ~ **de ionización intranuclear** *f* NUCL in-core ionization chamber; ~ **de irradiación** *f* FÍS RAD irradiation chamber;

~ k   ~ **de Kanne** *f* NUCL *para la monitorización de gases radiactivos* Kanne chamber;

~ l   ~ **limpiadora** *f* CARBÓN scavenger cell; ~ **de limpieza con chorro de arena** *f* PROD sandblast cleaning room; ~ **lúcida** *f* CINEMAT, FOTO, TV camera lucida;

~ m   ~ **de maduración de carnes** *f* REFRIG ageing room; ~ **de mando** *f* TRANSP MAR *de barco* control room; ~ **con mando a distancia** *f* CINEMAT, FOTO, TV remote-controlled camera; ~ **de mano** *f* CINEMAT hand-held camera; ~ **de máquinas** *f* TRANSP MAR engine room; ~ **de mezcla** *f* AUTO, GAS, VEH *del carburador* mixing chamber; ~ **de mezclado** *f* AUTO, GAS, VEH mixing chamber; ~ **microscópica automática** *f* INSTR automatic microscope camera; ~ **del microscopio** *f* INSTR microscope camera; ~ **miniatura** *f* FOTO *para película de 35 mm*, INSTR miniature camera, TV minicam; ~ **de montaje** *f* TV insert camera; ~ **con montura plegable** *f* FOTO camera with collapsible mount; ~ **móvil** *f* TV mobile camera;

~ n   ~ **de niebla** *f* FÍS ONDAS *para detectar partículas cargadas* cloud chamber, FÍS PART *para detectar partículas ionizadas* cloud chamber, ionization chamber; ~ **de niebla de Wilson** *f* FÍS Wilson cloud chamber; ~ **no apta para sonido directo** *f* CINEMAT wild camera;

~ o   ~ **de observación** *f* OCEAN observation chamber, TEC ESP tracking camera; ~ **de oficiales** *f* TRANSP MAR *Marina Real* wardroom; ~ **de oscilaciones** *f* CRISTAL oscillation camera;

~ p   ~ **panorámica** *f* FOTO panoramic camera; ~ **para película de tres capas de emulsión** *f* CINEMAT three-strip camera; ~ **de placas** *f* FOTO plate camera; ~ **plegable de formato grande** *f* FOTO large-format folding camera; ~ **plegable de gran formato** *f* FOTO large-format folding camera; ~ **plegable tipo fuelle** *f* FOTO bellows-type folding camera; ~ **de pleno** *f* TRANSP AÉR plenum chamber; ~ **plenum** *f* ACÚST plenum chamber; ~ **de polvo** *f* CRISTAL powder camera; ~ **para polvos** *f* CARBÓN dust

chamber; ~ **con portaobjetivos** *f* CINEMAT turret camera; ~ **portátil** *f* TV camcorder; ~ **de poscombustión** *f* CARBÓN postcombustion chamber; ~ **de precalentamiento de gases** *f* C&V checker chamber; ~ **de precesión** *f* CRISTAL precession camera; ~ **de precombustión** *f* TERMO, VEH *motor diesel* precombustion chamber; ~ **de presión piloto** *f* TRANSP AÉR pilot pressure chamber; ~ **para productos congelados** *f* ALIMENT, REFRIG, TERMO frozen-food storage room; ~ **de puertas** *f* AGUA, HIDROL *esclusas* gate chamber; ~ **de pulverización** *f* PROC QUÍ pulverizing chamber, REFRIG spray chamber;

~ r   ~ **de rayos X** *f* FÍS RAD, NUCL X-ray camera; ~ **de reacción electrón-positrón** *f* FÍS PART electron-positron collider; ~ **de reacción protón-antiprotón** *f* FÍS PART proton-antiproton collider; ~ **de reacondicionado** *f* TERMO humidifier; ~ **de recalentamiento** *f* REFRIG warming room; ~ **rectangular** *f* FOTO box camera; ~ **de recuento de fotones** *f* FÍS PART photon-counting camera; ~ **reflex** *f* CINEMAT, FOTO, TV reflex camera; ~ **reflex de dos objetivos** *f* FOTO twin-lens reflex, twin-lens reflex camera; ~ **reflex de un objetivo** *f* FOTO single-lens reflex camera (*SLR*); ~ **reflex con objetivo acromático** *f* FOTO single-lens reflex camera (*SLR*); ~ **de refrigeración** *f* REFRIG, TERMO cooler; ~ **de reportero de prensa** *f* FOTO press camera; ~ **de reproducción** *f* FOTO reproduction camera, IMPR copy camera, reproduction camera; ~ **para reproducción fotomecánica** *f* CINEMAT process camera; ~ **de resonancia** *f* ACÚST, ING MECÁ, TELECOM echo chamber; ~ **de reverberación** *f* ACÚST, ING MECÁ echo chamber, TEC ESP reverberation chamber, TELECOM echo chamber; ~ **sin reverberaciones** *f* ACÚST, ING MECÁ, TELECOM echoless chamber; ~ **reverberante** *f* ACÚST reverberation room; ~ **rígida** *f* FOTO box camera;

~ s   ~ **de sacos para filtrar gases** *f* CARBÓN *hornos* baghouse; ~ **de salazón** *f* REFRIG *de carne* curing cellar; ~ **de salida** *f* AGUA afterbay; ~ **de salvamento** *f* TRANSP MAR *para personal de submarinos* diving bell; ~ **seca** *f* AGUA dry well; ~ **de secado** *f* PROD drying house, TERMO drying chamber; ~ **de sedimentación** *f* AUTO sediment chamber, PROC QUÍ sedimentation basin; ~ **seguidora** *f* TEC ESP tracking camera; ~ **sellada impermeable** *f* FOTO waterproof sealed camera; ~ **semirreverberante** *f* ACÚST semi-reverberant room; ~ **de simulación espacial** *f* TEC ESP space simulation chamber, *ensayos* environmental test chamber; ~ **de sobrepresión** *f* AUTO plenum chamber; ~ **de sonido óptico** *f* CINEMAT optical sound camera; ~ **sorda** *f* ACÚST dead room; ~ **subjetiva** *f* CINEMAT subjective camera; ~ **subterránea** *f* MINAS *sala* chamber, TELECOM underground chamber; ~ **de succión** *f* TRANSP suction chamber;

~ t   ~ **de tambor rotativo para imágenes unidimensionales** *f* CINEMAT, FOTO, TV rotating-drum streak camera; ~ **con telémetro acoplado** *f* FOTO camera with coupled rangefinder; ~ **de televisión** *f* ELECTRÓN, FÍS, ING ELÉC, TELECOM, TV television camera; ~ **de toma** *f* HIDROL intake chamber; ~ **tomavistas con corriente** *f* TV hot camera; ~ **de tostación** *f* AmL (*cf caseta de tostación Esp*) CARBÓN *metalurgia*, MINAS *hornos* stall; ~ **de tres tubos** *f* TV three-tube camera; ~ **sin**

trípode *f* CINEMAT hand-held camera; ~ **de turbina** *f* INSTAL HIDRÁUL turbine chamber, wheel pit; ~ **de turbina cerrada** *f* PROD closed-turbine chamber; ~ **de turbinas** *f* INSTAL HIDRÁUL turbine pit; ~ **de turbulencia** *f* AUTO turbulence chamber;

■ **~ u** ~ **ultrarrápida** *f* CINEMAT high-speed camera;

■ **~ v** ~ **de vacío** *f* TV vacuum chamber; ~ **de vacío con conexiones múltiples** *f* REFRIG manifold drying apparatus; ~ **de vacío con estanterías** *f* REFRIG tray drying chamber; ~ **de vapor** *f* INSTAL HIDRÁUL *distribuidor* steam chamber; ~ **de velocidad variable** *f* CINEMAT variable-speed camera; ~ **de velocidades múltiples** *f* CINEMAT multiple-speed camera; ~ **de ventilación** *f* TEC PETR air box; ~ **de video de mano** *f* AmL, ~ **de vídeo de mano** *f* Esp TV camcorder; ~ **vidicón** *f* TV vidicon camera; ~ **de visita** *f* CONST inspection chamber, ING MECÁ manhole;

■ **~ w** ~ **de Weissenberg** *f* CRISTAL Weissenberg camera

cámaras: ~ **de la tripulación** *f pl* TRANSP MAR crew's quarters

camareta *f* TRANSP MAR *de suboficiales y empleos equivalentes* warrant officers' wardroom

camarógrafo *m* AmL (*cf cámara Esp*) TV cameraman

camaronero *m* OCEAN shrimp boat, *tecnología pesquera* shrimper, TRANSP MAR shrimp boat, shrimp trawler

camarote *m* TRANSP MAR *alojamiento* cabin; ~ **de pasaje** *m* TRANSP passenger cabin

cambiable *adj* GEN removable

cambiacorreas *m* PROD belt shifter

cambiadiscos *m* ACÚST *gramófono* record changer

cambiador: ~ **de calor** *m* REFRIG *entre fases de fluido frigorífico* liquid suction heat exchanger; ~ **de calor de caudales en paralelo** *m* NUCL parallel-flow heat exchanger; ~ **de calor de corrientes cruzadas** *m* REFRIG counterflow heat exchanger; ~ **de calor de haz tubular** *m* NUCL shell-and-tube heat exchanger; ~ **de diapositivas** *m* FOTO slide changer; ~ **de fase de ferrita** *m* ING ELÉC ferrite phase-shifter; ~ **de frecuencia** *m* ELEC *convertidor*, ELECTRÓN frequency changer; ~ **de muestras** *m* NUCL sample changer; ~ **de relación de transformación** *m* ELEC tap change operation; ~ **de toma de carga** *m* ELEC load tap changer

cambiaplacas: ~ **a plena luz** *m* AmL (*cf bolsa oscura de carga Esp*) FOTO changing bag

cambiar[1] *vt* CINEMAT change over; ~ **la derrota habitual de** *vt* TRANSP MAR *navegación* reroute; ~ **de proporción** *vt* INFORM&PD scale; ~ **la ruta habitual de** *vt* TRANSP MAR *transporte marítimo* reroute

cambiar[2] *vti* INFORM&PD shift

cambiar[3] *vi* ELEC fluctuate; ~ **la proa y tomar la otra vuelta** *vi* TRANSP MAR *buques* wear; ~ **de rumbo virando por redondo** *vi* TRANSP MAR *navegación* veer off course; ~ **velocidades** *vi* PROD *autos* shift

cambiavía *m* CONST *ferrocarril*, FERRO points (*BrE*), set of points (*BrE*), switch (*AmE*)

cambio *m* CINEMAT changeover, ELEC fluctuation, FÍS PART *en la forma del núcleo* change, INFORM&PD shift, ING MECÁ exchange, PROD exchange, trade-in, TEC ESP rotation, TV changeover; ~ **de aceite** *m* VEH *lubricación* oil change; ~ **adiabático** *m* TERMO adiabatic change; ~ **de agujas** *m* FERRO *equipo inamovible*

points switching; ~ **de agujas semi-automático** *m* FERRO *equipo inamovible* semiautomatic points switching; ~ **de ángulo por unidad de tiempo** *m* ING MECÁ angular velocity; ~ **automático** *m* AUTO automatic clutch, TELECOM automatic changeover; ~ **automático de canilla** *m* TEXTIL automatic pirn change; ~ **automático de línea** *m* INFORM&PD word wrap, wrap around; ~ **de bandas** *m* TV waveband switching; ~ **de barrena** *m* TEC PETR *perforación* bit change (*BC*); ~ **brusco de la acidez** *m* CONTAM *tratamiento químico* acid shock; ~ **brusco del pH** *m* CONTAM *tratamiento químico* acid shock; ~ **brusco de temperatura** *m* TERMO thermal shock; ~ **de carga en rampa** *m* NUCL ramp change of load; ~ **de carril** *m* TRANSP lane switching; ~ **a cifras** *m* TELECOM figure shift; ~ **de color** *m* C&V color change (*AmE*), colour change (*BrE*), IMPR color shift (*AmE*), colour shift (*BrE*); ~ **de color por exposición al fuego** *m* MINAS *gemas* firing; ~ **de contexto** *m* INFORM&PD context switching; ~ **de la derrota habitual** *m* TRANSP MAR rerouting; ~ **de diálogo** *m* CINEMAT dialog replacement (*AmE*), dialogue replacement (*BrE*); ~ **de elevación** *m* CONST lifting gear; ~ **de escala** *m* INFORM&PD scaling; ~ **de espín** *m* NUCL spin exchange; ~ **de estado** *m* FÍS, TERMO change of state; ~ **de fase** *m* ELEC *corriente alterna*, TRANSP phase shift; ~ **de forma** *m* METAL shape change; ~ **de frecuencia** *m* ELECTRÓN frequency change; ~ **de frecuencia del oscilador** *m* ELECTRÓN oscillator drift; ~ **de inventario** *m* PROD inventory change; ~ **a letras** *m* TELECOM letter shift; ~ **de línea automático** *m* INFORM&PD word wrap, wrap around; ~ **lógico** *m* INFORM&PD logical shift; ~ **de marcha** *m* ING MECÁ reversing, reversing shaft; ~ **de marcha de husillo** *m* ING MECÁ screw reversing gears; ~ **de marcha por tornillo sin fin** *m* ING MECÁ screw reversing gears; ~ **de modo** *m* INFORM&PD mode change; ~ **por pedal** *m* VEH *en motocicletas* foot change; ~ **de pendiente** *m* GEOL break-of-slope; ~ **de plancha** *m* IMPR plate changeover; ~ **con preselección de marchas** *m* VEH *caja de cambio* preselection gear change; ~ **de presión** *m* GAS pressure change; ~ **principal** *m* TRANSP master change (*MC*); ~ **de rasante** *m* GEOL break-of-slope, VEH *superficie* road camber; ~ **de rumbo** *m* TRANSP MAR *navegación* alteration of course; ~ **de la ruta habitual** *m* TRANSP MAR rerouting; ~ **secular** *m* GEOFÍS secular change; ~ **de trépano** *m* TEC PETR *perforación* bit change (*BC*); ~ **de turno** *m* NUCL *del equipo de operación de una central nuclear* handover; ~ **de velocidad** *m* ING MECÁ change speed rate; ~ **en la velocidad del aire** *m* INSTAL TERM air rate change, rate of air change; ~ **de velocidad de avance de la perforación** *m* PETROL drilling break; ~ **de velocidad por pedal** *m* AUTO, VEH *en motocicletas* gear change (*BrE*), gear shift (*AmE*); ~ **de velocidades** *m* AUTO, VEH *caja de cambio* gear change (*BrE*), gear shift (*AmE*); ~ **de velocidades automático** *m* VEH automatic gear shift, *caja de cambio* automatic gear change; ~ **de velocidades manual** *m* AUTO, VEH manual gearbox; ~ **de velocidades de pedal** *m* VEH *en motocicletas* foot change; ~ **de vía** *m* CONST turnout; ~ **de vía completo** *m* FERRO turnout; ~ **de vía a la derecha** *m* FERRO *equipo inamovible* right-hand turnoff; ~ **de volumen** *m* METAL volume change

cambios: ~ **de vía** *m pl* CONST *ferrocarriles* points (*BrE*), switch (*AmE*); ~ **de vía ordinarios** *m pl* FERRO ordinary points (*BrE*), ordinary switch (*AmE*)

cambray *m* TEXTIL cambric

Cámbrico *adj* GEOL Cambrian

camello *m* AGUA, CONST, HIDROL, INSTAL HIDRÁUL caisson, TRANSP MAR caisson, camel

camellón *m* AGRIC ridge

camilla *f* SEG folding bed, *para lesionados* stretcher; ~ **con ruedas** *f* SEG stretcher cart

caminata: ~ **aleatoria** *f* INFORM&PD random walk

camino *m* AUTO road, CONST *carreteras* road, way, FÍS, GEOM path, QUÍMICA way, TRANSP, VEH road; ~ **de acceso** *m* CONST access road (*AmE*), slip road (*BrE*), INFORM&PD access path, TRANSP access road (*AmE*), slip road (*BrE*); ~ **asfaltado** *m* CONST paved road (*AmE*); ~ **crítico** *m* CONST *planificaciones*, INFORM&PD critical path, NUCL critical pathway; ~ **crítico de la recarga de combustible** *m* NUCL refueling outage critical path (*AmE*), refuelling outage critical path (*BrE*); ~ **de fuga** *m* ELEC, ING ELÉC leakage path; ~ **medio libre** *m* METAL mean free path; ~ **óptico** *m* FÍS optical path, ÓPT optical path length, TELECOM optical path; ~ **óptimo** *m* TELECOM optimal path; ~ **para pasar** *m* CONST passing lane; ~ **en pendiente** *m* CONST climbing lane; ~ **principal** *m* CONST, TRANSP main road; ~ **de rodadura** *m* ING MECÁ ball race; ~ **secundario** *m* CONST, TRANSP byroad, byway; ~ **de transmisión** *m* TELECOM transmission highway

camión *m* AUTO, CONST, MECÁ, REFRIG, TRANSP, VEH lorry (*BrE*), truck (*AmE*); ~ **aislado** *m* TERMO insulated lorry (*BrE*), insulated truck (*AmE*); ~ **articulado** *m* VEH articulated lorry (*BrE*), articulated truck (*AmE*); ~ **basculante** *m* CONST dump truck (*AmE*), dumper (*AmE*), dumper truck (*AmE*), tipper (*BrE*), TRANSP dump truck (*AmE*), tipper (*BrE*); ~ **de la basura** *m* RECICL dustbin lorry (*BrE*), garbage truck (*AmE*), refuse collection vehicle; ~ **butanero** *m* TEC PETR *transporte*, TRANSP butane carrier; ~ **de caja basculante** *m* TRANSP skip lorry (*BrE*), skip truck (*AmE*); ~~**cisterna** *m* CONST tanker lorry (*BrE*), tanker truck (*AmE*), CONTAM MAR road tanker, TRANSP fuel tanker, road tank car (*RTC*), tanker, TRANSP AÉR bowser, VEH tanker; ~~**cisterna de butano** *m* TEC PETR, TRANSP butane tanker; ~~**cisterna de combustible** *m* TRANSP refueling tanker (*AmE*), refuelling tanker (*BrE*); ~~**cisterna de leche** *m* AUTO milk tanker; ~~**cisterna para transporte de agua** *m* CONST water lorry (*BrE*), water truck (*AmE*); ~~**cuba** *m* TRANSP tank lorry (*BrE*), tank truck (*AmE*); ~ **de distribución eléctrica** *m* AUTO electric delivery lorry (*BrE*), electric delivery truck (*AmE*); ~ **eléctrico** *m* TRANSP battery lorry (*BrE*), battery truck (*AmE*); ~ **eléctrico de acumuladores** *m* TRANSP battery lorry (*BrE*), battery truck (*AmE*); ~~**espuerta** *m* TRANSP skip lorry (*BrE*), skip truck (*AmE*); ~ **con estanterías** *m* TRANSP rack body lorry (*BrE*), rack body truck (*AmE*); ~ **frigorífico** *m* AUTO, REFRIG, TRANSP refrigerated lorry (*BrE*), refrigerated truck (*AmE*); ~ **frigorífico articulado** *m* REFRIG, TRANSP refrigerated trailer; ~ **de gran capacidad** *m* TRANSP large-capacity truck; ~ **de gran tonelaje** *m* AUTO heavy goods vehicle (*HGV*), heavy lorry (*BrE*), heavy truck (*AmE*); ~ **hormigonera** *m* CONST mixer truck;

~ **industrial** *m* SEG industrial truck; ~ **isotermo** *m* REFRIG insulated lorry (*BrE*), insulated truck (*AmE*); ~ **ligero** *m* AUTO light lorry (*BrE*), light truck (*AmE*); ~ **de limpieza urbana** *m* AUTO street cleaning lorry (*BrE*), street cleaning truck (*AmE*); ~ **de mercancías** *m* AUTO freight truck (*AmE*), goods lorry (*BrE*); ~ **con motor muy potente** *m* TRANSP heavy motor lorry (*BrE*), heavy motor truck (*AmE*); ~ **con motor de turbina** *m* AUTO turbine-engined lorry (*BrE*), turbine-engined truck (*AmE*); ~ **de mudanzas** *m* AUTO, VEH removal truck (*AmE*), removal van (*BrE*); ~ **plataforma** *m* AUTO platform lorry (*BrE*), platform truck (*AmE*); ~ **de plataforma estática** *m* TRANSP passive flat car; ~ **de plataforma pasiva** *m* TRANSP passive flat car; ~ **portacontenedores** *m* TRANSP container-carrier lorry (*BrE*), container-carrier truck (*AmE*); ~ **portavolquetes** *m* CONTAM MAR skip lorry (*BrE*), skip truck (*AmE*); ~ **de propulsión eléctrica** *m* AUTO, ELEC electric propulsion lorry (*BrE*), electric propulsion truck (*AmE*); ~ **de recogida de basuras** *m* VEH refuse-collection lorry (*BrE*), refuse-collection truck (*AmE*); ~ **de recolección de basura** *m* RECICL garbage truck (*AmE*), refuse-collection lorry (*BrE*), VEH refuse-collection vehicle; ~ **refrigerante** *m* AUTO, REFRIG, TRANSP refrigerated lorry (*BrE*), refrigerated truck (*AmE*); ~ **repostador** *m* VEH *transporte de combustible* refueler (*AmE*), refueller (*BrE*); ~ **para repostar** *m* TEC ESP bowser; ~ **de salvamento** *m* AUTO, TRANSP salvage lorry (*BrE*), salvage truck (*AmE*); ~ **semioruga** *m* VEH *tipo de vehículo* half-track lorry (*BrE*), half-track truck (*AmE*); ~ **semirremolque** *m* TRANSP semitrailer lorry (*BrE*), semitrailer truck (*AmE*); ~ **semi-trailer** *m* TRANSP semitrailer lorry (*BrE*), semitrailer truck (*AmE*); ~ **sustentado por la cola** *m* TRANSP tail-lift lorry (*BrE*), tail-lift truck (*AmE*); ~ **tanque** *m* CONST tanker lorry (*BrE*), tanker truck (*AmE*); ~ **tanque con sistema de vacío** *m* CONTAM MAR vacuum lorry (*BrE*), vacuum truck (*AmE*); ~ **termoaislado** *m* TERMO heat-insulated lorry (*BrE*), heat-insulated truck (*AmE*); ~ **todo-terreno** *m* TRANSP cross-country lorry (*BrE*), cross-country truck (*AmE*); ~ **tractor de remolque** *m* VEH tractor-trailer lorry (*BrE*), tractor-trailer truck (*AmE*); ~ **para transportar ganado** *m* AGRIC, TRANSP cattle lorry (*BrE*), cattle truck (*AmE*); ~ **de transporte de ganado** *m* AGRIC, TRANSP cattle lorry (*BrE*), cattle truck (*AmE*); ~ **volquete** *m* AGRIC, CARBÓN dump truck (*AmE*), tipper (*BrE*)

camioneta *f* AGRIC pick-up lorry (*BrE*), pick-up truck (*AmE*), AUTO light lorry (*BrE*), light truck (*AmE*), pick-up lorry (*BrE*), pick-up truck (*AmE*), VEH pick-up lorry (*BrE*), pick-up truck (*AmE*); ~ **eléctrica de reparto** *f Esp* (*cf chatita AmL*) TRANSP electric pick-up; ~ **con equipo de TV de exteriores** *f* TRANSP shooting brake; ~ **montacargas** *f* MECÁ lifting truck

camisa[1]: **con** ~ **exterior** *adj* MECÁ jacketed

camisa[2] *f* AGUA lining, ING ELÉC *cable, cilindro*, ING MECÁ, LAB jacket, MECÁ case, sleeve, jacket, NUCL jacket, PROD *de ladrillos refractarios* jacket, sleeve, lining, TEC PETR *perforación* liner; ~ **de agua** *f* AUTO, TERMO water jacket; ~ **aislante de la caldera** *f* INSTAL HIDRÁUL boiler lagging; ~ **del cable** *f* ELEC, ING ELÉC, ING MECÁ cable sheath; ~ **de calefacción** *f* TERMO heating jacket; ~ **calefactora** *f* TERMO heating jacket; ~ **del carburador** *f Esp* (*cf chaqueta del* *carburador*) *AmL* AUTO, VEH carburetor jacket (*AmE*), carburettor jacket (*BrE*); ~ **de cilindro** *f* AUTO, MECÁ, VEH *motor* cylinder liner; ~ **del cilindro** *f* AUTO, VEH carburetor jacket (*AmE*), carburettor jacket (*BrE*); ~ **del cilindro húmeda** *f* AUTO wet cylinder liner; ~ **de cilindro húmedo** *f* AUTO wet cylinder liner; ~ **de colada** *f* PROD slip jacket; ~ **exterior** *f* PROD *horno alto* mantle; ~ **exterior de agua** *f* PROD, VEH *sistema de refrigeración* water jacket; ~ **exterior del cilindro** *f* PROD cylinder jacket; ~ **exterior de vapor** *f* INSTAL HIDRÁUL steam jacket; ~ **de extracción** *f* AGUA pulling casing; ~ **de fibra** *f* TELECOM fiber jacket (*AmE*), fibre jacket (*BrE*); ~ **interior** *f* CARBÓN *de cilindros* liner, MECÁ liner, lining; ~ **interior del cilindro** *f* AUTO, MECÁ, VEH cylinder liner; ~ **lisa** *f* TEC PETR *perforación* blank liner; ~ **de refrigeración** *f* TERMO water jacket; ~ **del rodillo manchón** *f* PAPEL couch roll jacket; ~ **termoaislante** *f* TERMO heat-insulating jacket; ~ **de vapor** *f* TERMOTEC steam jacket

camisería *f* TEXTIL *confección y artículos* shirting

campana *f* ING MECÁ, PAPEL, PROD hood, QUÍMICA bell jar; ~ **abierta** *f* PAPEL open hood; ~ **de advertencia** *f* SEG warning bell; ~ **de alarma** *f* SEG alarm bell; ~ **de buceo** *f* OCEAN, TRANSP MAR diving bell; ~ **burbujeadora** *f* TEC PETR *refino* bubble cap; ~ **de buzo** *f* PETROL, TEC PETR *operaciones marinas* diving bell; ~ **eléctrica** *f* ELEC electric bell; ~ **del embrague** *f* AUTO, VEH bell housing, clutch housing; ~ **de extracción** *f* LAB *mobiliario, seguridad* extraction hood, fume cupboard, fume hood, PROC QUÍ fume cupboard, fume hood; ~ **de laboratorio** *f* LAB, PROC QUÍ fume cupboard, fume hood; ~ **de pesca** *f* MINAS *sondeos* horn socket, socket, TEC PETR *perforación* overshot; ~ **protectora** *f* INSTR, SEG protective hood; ~ **protectora contra ruidos** *f* SEG noise-protective hood; ~ **recta** *f* PROD straight hood; ~ **de vidrio** *f* INSTR *para proteger plantas* hand glass, LAB *material de vidrio* bell jar

campaña *f* AGRIC crop year, season, C&V *duración de un horno de vidrio*, CARBÓN *horno alto*, D&A campaign, PROD *de horno* campaign, *horno alto* run; ~ **cerealista** *f* AGRIC grain season, grain year; ~ **de horno alto** *f* PROD blast-furnace campaign; ~ **"La Seguridad Primero"** *f* SEG safety first campaign

Campaniense *adj* GEOL Campanian

campanilla *f* AGRIC convulvulus sepium, hedge bindweed, ING ELÉC call bell; ~ **de alarma** *f* ELEC alarm bell

campilita *f* MINERAL campylite

campo *m* ACÚST, ELEC field, ELECTRÓN domain, FOTO background, ING ELÉC, PETROL, TEC PETR field, TELECOM *de aplicación, de atracción* range; ~ **de acción** *m* CINEMAT field of action; ~ **de acción de la fuente de emisión** *m* CONTAM area emission source; ~ **de acoplamiento** *m* CINEMAT patch field; ~ **acústico** *m* ACÚST acoustical field, acoustic field; ~ **administrativo** *m* CONST administrative area; ~ **alternativo** *m* ING ELÉC alternating field; ~ **de aplicación** *m* TEC ESP range; ~ **auxiliar superior** *m* TELECOM upper subfield; ~ **de batalla** *m* D&A battlefield; ~ **cercano** *m* TEC ESP near range; ~ **de clasificación** *m* INFORM&PD sort field; ~ **clave** *m* INFORM&PD key field; ~ **comercial** *m* TEC PETR *producción* commercial field; ~ **conductivo** *m* ING ELÉC field; ~ **constante** *m* ELEC constant field, ING

ELÉC stationary field; ~ **de control** *m* INFORM&PD control field; ~ **de control UIH** *m* TELECOM UIH control field; ~ **de corriente alterna** *m* ELEC *electromagnetismo* alternating-current field; ~ **cristalino** *m* CRISTAL, FÍS RAD crystal field; ~ **cruzado** *m* ELECTRÓN, FÍS, TELECOM crossed field; ~ **cuadripolar** *m* NUCL quadrupole field; ~ **de datos** *m* ELECTRÓN data domain, INFORM&PD data field; ~ **desimanante** *m* ELEC, FÍS demagnetizing field; ~ **desmagnetizante** *m* ELEC, FÍS demagnetizing field; ~ **despolarizante** *m* FÍS depolarizing field; ~ **diamantífero** *m* MINAS diamond field; ~ **difuso** *m* ACÚST diffuse field; ~ **digital** *m* ELECTRÓN digital domain; ~ **de dirección de arranque** *m* PROD start address field; ~ **de direcciones de palabra** *m* PROD word address field; ~ **de dispersión** *m* ELEC *de máquina eléctrica, transformador* stray field; ~ **de dispersión magnética** *m* ELEC *de máquina eléctrica, transformador* stray field; ~ **distante** *m* TELECOM distant field; ~ **eléctrico** *m* ELEC *electromagnetismo*, FÍS, FÍS RAD, ING ELÉC, PETROL, TELECOM, TV electric field; ~ **eléctrico alternativo** *m* ING ELÉC alternating electric field; ~ **eléctrico atmosférico** *m* GEOFÍS atmospheric electric field; ~ **eléctrico estático** *m* ING ELÉC static electric field; ~ **electromagnético** *m* ELEC, FÍS, FÍS ONDAS, ING ELÉC, TELECOM electromagnetic field; ~ **electrostático** *m* ELEC, FÍS electrostatic field; ~ **de esparcimiento** *m* AGUA spreading field; ~ **estable** *m* METEO stable field; ~ **estacionario** *m* ING ELÉC stationary field; ~ **estático** *m* ING ELÉC, TELECOM static field; ~ **evanescente** *m* ÓPT, TELECOM evanescent field; ~ **de excitación** *m* ING ELÉC excitation field; ~ **experimental** *m* AGRIC experimental field; ~ **de exploración** *m* TV scanning field; ~ **fijo** *m* INFORM&PD fixed field; ~ **de fuerza** *m* ING ELÉC field of force; ~ **de fuerzas nucleares** *m* NUCL field of nuclear forces; ~ **de fugas** *m* TELECOM leakage field; ~ **de gas** *m* GAS, TEC PETR *producción* gas field; ~ **geomagnético externo** *m* FÍS, GEOFÍS external geomagnetic field; ~ **geotérmico** *m* ENERG RENOV geothermal field; ~ **giratorio** *m* ELEC *electromagnetismo*, FÍS, ING ELÉC, TELECOM rotating field; ~ **gravitacional** *m* FÍS, GEOFÍS, GEOL gravitational field; ~ **gravitacional terrestre** *m* FÍS, GEOFÍS, GEOL gravitational field; ~ **de Hall** *m* FÍS Hall field; ~ **hipertérmico** *m* ENERG RENOV hyperthermal field; ~ **de imagen** *m* CINEMAT image field; ~ **de la imagen** *m* CINEMAT field of image; ~ **de imagen eficaz** *m* *AmL* (*cf profundidad de cambio Esp*) FOTO effective image-field; ~ **de imagen nítida** *m* CINEMAT *de la lente* lens coverage; ~ **de inducción** *m* ELEC *electromagnetismo* induced field; ~ **del inducido** *m* ELEC *máquina* armature field; ~ **inductor** *m* ING ELÉC, TV induction field; ~ **inductor constante** *m* ELEC constant field; ~ **inductor crítico** *m* ELEC, ING ELÉC *microondas* critical field; ~ **inestable** *m* METEO disturbed field; ~ **de influencia de la fuente de emisión** *m* CONTAM area emission source; ~ **de interferencia magnética** *m* GEOFÍS magnetic interference field; ~ **inverso** *m* GEOL field reversal; ~ **de investigación** *m* CARBÓN investigation field, research area; ~ **iónico** *m* FÍS RAD ionic yield; ~ **irrotacional** *m* FÍS irrotational field; ~ **de lava** *m* GEOL lava plateau; ~ **de lectura** *m* TV scanning area; ~ **lejano de radiación** *m* TEC ESP *radiofrecuencia* Fraunhofer region; ~ **libre** *m*

ACÚST, INFORM&PD free field; ~ **magnético** *m* ELEC, FÍS, GEOL, ING ELÉC, TELECOM, TV magnetic field; ~ **magnético alternativo** *m* ING ELÉC alternating magnetic field; ~ **magnético crítico** *m* ELEC, ING ELÉC *microondas* critical field; ~ **magnético externo** *m* ELEC external magnetic field; ~ **magnético giratorio** *m* ELEC, ING ELÉC rotary field; ~ **magnético terrestre** *m* ELEC, FÍS, GEOFÍS, TEC ESP earth's magnetic field; ~ **magnetizante** *m* ING ELÉC magnetizing field; ~ **marginal** *m* TEC PETR *producción* marginal field; ~ **de minas** *m* D&A minefield; ~ **molecular** *m* FÍS molecular field; ~ **de nitidez** *m* CINEMAT field of sharpness; ~ **nominal de medida** *m* NUCL rated range; ~ **opuesto** *m* ELEC *electromagnetismo* opposing field; ~ **parásito** *m* ELEC *de máquina eléctrica, transformador* stray field; ~ **de periodicidad temporal** *m* ING ELÉC time-periodic field; ~ **perturbado** *m* METEO disturbed field; ~ **perturbador** *m* TELECOM, TV noise field; ~ **de petróleo** *m* TEC PETR *geología del petróleo* petroleum province; ~ **de petróleo con presión de gas** *m* TEC PETR *producción*, TERMO gas field; ~ **petrolero costa-fuera** *m* TEC PETR *producción costa-fuera* offshore field; ~ **petrolífero** *m* PETROL oilfield, TEC PETR oilfield; ~ **petrolífero marino** *m* TEC PETR *producción costa-fuera* offshore field; ~ **de propiedades estadísticas** *m* TELECOM random field; ~ **protegido** *m* INFORM&PD protected field; ~ **próximo** *m* TELECOM near field; ~ **de radiación** *m* FÍS RAD radiation field; ~ **reservado** *m* TELECOM reserved field (*RES*); ~ **reverberante** *m* ACÚST reverberant field; ~ **rotacional** *m* ING ELÉC curl field; ~ **rotórico** *m* ELEC *generador* rotor field; ~ **de ruido** *m* TELECOM, TV noise field; ~ **sinusoidal** *m* ING ELÉC sinusoidal field; ~ **solenoidal** *m* FÍS solenoidal field; ~ **sonoro** *m* ACÚST sound field, CONTAM *acústica* source area; ~ **de tiro** *m* D&A shooting range, firing range; ~ **de tolerancia del alma** *m* TELECOM core tolerance field; ~ **de tolerancia del chapado** *m* ÓPT, TELECOM cladding tolerance field; ~ **de tolerancia del núcleo** *m* ÓPT core tolerance field; ~ **de tolerancia del revestimiento** *m* ÓPT *fibras ópticas*, TELECOM cladding tolerance field; ~ **de tolerancia de la superficie de referencia** *m* ÓPT, TELECOM reference-surface tolerance field; ~ **uniforme** *m* FÍS uniform field; ~ **de variabilidad** *m* TEC ESP range; ~ **variable** *m* INFORM&PD, ING ELÉC variable field; ~ **vectorial** *m* ELEC *electromagnetismo*, ING ELÉC vector field; ~ **visible** *m* ENERG RENOV visible region; ~ **visual** *m* CINEMAT field of view

campos: ~ eléctricos radiales *m pl* ELEC, FÍS RAD radial electrical fields

can *m* CONST *arquitectura* corbel; ~ **lanzaespuma** *m* CONTAM MAR *equipo contra incendios* fire monitor

caña *f* C&V *para sacar vidrio del horno manualmente* draw rod, gathering iron, CINEMAT *soporte extensible para micrófonos* fishpole, CONST *carpintería* reed, *llave* stem, shank, ING MECÁ shank, MINAS *pozo de minas* shaft; ~ **de soplar** *f* ING MECÁ *vidriero* blowpipe; ~ **del timón** *f* TRANSP MAR tiller, *construcción* helm; ~ **de vidriero** *f* C&V *vidrio soplado* blowpipe

canal *m* ACÚST channel, AGUA waterway, channel, duct, C&V chute, flute, trough, CARBÓN ditch, CONST *hidráulica* chute, eaves trough, ELEC, ELECTRÓN channel, ENERG RENOV fairway, channel, GEOL *fluvial*

channel, HIDROL *de un río* discharge, channel, navigation channel, INFORM&PD, ING ELÉC channel, ING MECÁ duct, groove, INSTAL HIDRÁUL channel, MECÁ flute, channel, duct, chute, MINAS *ventilación* air pipe, OCEAN navigation channel, ÓPT *electroacústica* track, PROD pass, REFRIG duct, TEC ESP *telecomunicaciones* channel, TRANSP navigation channel, TRANSP MAR channel, *canalón de trancanil* canal, fairway, TV channel;

~ a ~ **abierto para agua** *m* AGUA flume; ~ **de acceso** *m* TELECOM *telefonía* access channel, TRANSP MAR approach channel; ~ **activo** *m* TELECOM working channel; ~ **acústico** *m* OCEAN *zoología marina: oído de peces y mamíferos* sound channel; ~ **adyacente** *m* TV adjacent channel; ~ **de agotamiento** *m* OCEAN drainage channel; ~ **agrupador de señalización** *m* TELECOM signaling grouping channel (*AmE*) (*SGC*), signalling grouping channel (*BrE*) (*SGC*); ~ **de agrupamiento de señalización** *m* TELECOM signaling grouping channel (*AmE*) (*SGC*), signalling grouping channel (*BrE*) (*SGC*); ~ **de agua jabonosa** *m* ING MECÁ sud-channel; ~ **de aguas arriba** *m* AGUA penstock; ~ **de aire** *m* PROD *de horno* air flue; ~ **aislado** *m* TELECOM blanked-off channel; ~ **de ajuste inicial** *m* TELECOM setup channel; ~ **de ajuste preliminar** *m* TELECOM setup channel; ~ **de alabeo** *m* TRANSP AÉR roll channel; ~ **de albufera** *m* OCEAN lagoon channel; ~ **aleatorio** *m* TELECOM random channel; ~ **de alimentación** *m* AGUA headrace, pentrough, ENERG RENOV headrace canal; ~ **alimentador** *m* TV head channel; ~ **aliviadero** *m* AGUA spillway; ~ **analógico** *m* INFORM&PD analog channel; ~ **anterior** *m* AGUA head bay; ~ **anular** *m* NUCL annular channel; ~ **apto para buques de navegación marítima** *m* TRANSP MAR ship canal; ~ **de arrastre** *m* CARBÓN flume, MINAS drive tube;

~ b ~ **B del servicio del circuito virtual** *m* TELECOM B-channel virtual circuit service; ~ **de banda limitada** *m* INFORM&PD band-limited channel; ~ **de bolas** *m* ING MECÁ ball race;

~ c ~ **de cabeceo** *m* TRANSP AÉR pitch channel; ~ **de cables** *m* ELEC, ING ELÉC *fuente de alimentación* cable trench; ~ **de calefacción** *m* EMB heating channel; ~ **de cámara** *m* CINEMAT, TV camera channel; ~ **de carga** *m* AGUA headrace, pentrough, INSTAL HIDRÁUL feeder, headrace; ~ **de colada** *m* C&V *hornos* baffle, CARBÓN *funderías* main gate, P&C *moldeo, parte de equipo* sprue, PROD *fundería* runner, main gate, deadhead, sow; ~ **de colada en caída directa** *m* PROD drop runner, *en encofrado* direct-pouring gate, *fundería* plump gate, pop gate; ~ **de colada de la carga** *m* P&C *parte de equipo* feed runner; ~ **de colada en fuente** *m* PROD horn gate, *molde* fountain runner; ~ **de colada en talón** *m* PROD *fundería* side gate; ~ **colectivo** *m* ELECTRÓN *transistores* bulk channel; ~ **colector** *m* AGUA *bomba centrífuga* volute chamber; ~ **de comunicación** *m* INFORM&PD link (*BrE*), trunk (*AmE*), TRANSP AÉR message chute; ~ **de comunicación de datos** *m* TELECOM data communication channel (*DCC*); ~ **de comunicación entre galerías adyacentes** *m* GEOL, MINAS breakthrough; ~ **de comunicaciones** *m* INFORM&PD communication channel; ~ **de conducción** *m* HIDROL, ING MECÁ, INSTAL HIDRÁUL, PROD *hidráulica* raceway; ~ **de control** *m* TELECOM control channel (*CC*); ~ **de control embebido** *m* TELECOM embedded control channel (*ECC*); ~ **de control por inversión** *m* TELECOM reversal control channel; ~ **de control de la red** *m* INFORM&PD network control channel; ~ **de conversación** *m* TELECOM speech path; ~ **de corriente portadora** *m* TELECOM bearer channel (*BC*); ~ **corto** *m* ELECTRÓN short channel;

~ d ~ **D** *m* TELECOM D-channel; ~ **de datos** *m* INFORM&PD data channel, TEC ESP *comunicaciones* data link, *comunicaciones informatizadas* data bus; ~ **dedicado** *m* TELECOM, TV dedicated channel; ~ **deflector** *m* C&V deflector chute; ~ **de derivación** *m* AGUA diversion canal, leat, HIDROL diversion canal, diversion channel, PROD leat; ~ **de desagüe** *m* AGUA tailrace, HIDROL drainage ditch, tailrace, OCEAN drainage channel, RECICL sluice; ~ **de desbaste** *m* CARBÓN, PROD roughing pass; ~ **de descarga** *m* AGRIC chute, AGUA delivery channel, discharge canal, *turbina hidráulica* tailrace, CARBÓN flume, HIDROL tailrace, discharge channel, INSTAL HIDRÁUL spillway channel, wasteway, spillway canal, PROD flume; ~ **descubierto** *m* INSTAL HIDRÁUL open channel; ~ **de deslizamiento** *m* ING MECÁ track roller; ~ **despejado** *m* TELECOM clear channel; ~ **de desviación** *m* AGUA diversion canal; ~ **desviador** *m* AGUA diversion canal; ~ **discreto** *m* TELECOM discrete channel; ~ **de distribución** *m* AGUA distributing canal, C&V *hidrodinámica* forehearth; ~ **distribuidor** *m* AGUA delivery race; ~ **distributario** *m* GEOL distributary channel; ~ **de drenaje** *m* AGRIC, AGUA drainage channel, RECICL drain;

~ e ~ **de efluente** *m* AGUA effluent channel; ~ **embutido** *m* PROD *laminador* closed pass; ~ **de engrase** *m* ING MECÁ oil groove; ~ **de entrada** *m* AGUA inflow canal; ~ **de entrada/salida** *m* INFORM&PD input/output channel; ~ **de esclusa Venturi-Parshall** *m* HIDROL Venturi-Parshall flume; ~ **con esclusas** *m* AGUA locked canal; ~ **de esclusas y de lago interior** *m* AGUA lock-and-inland-lake canal; ~ **de evacuación** *m* HIDROL discharge channel; ~ **de evacuación de aguas de lluvia** *m Esp* (*cf canal de evacuación de aguas pluviales AmL*) RECICL storm sewer; ~ **de evacuación de aguas pluviales** *m AmL* (*cf canal de evacuación de aguas de lluvia Esp*) RECICL storm sewer; ~ **evacuador** *m* INSTAL HIDRÁUL spillway canal, spillway channel, wasteway; ~ **evacuador de crecidas** *m* INSTAL HIDRÁUL wasteway; ~ **de experiencias hidrodinámicas** *m* TRANSP MAR *casco, hélice* testing tank, towing tank, *construcción* ship model towing tank;

~ f ~ **filoniano** *m* MINAS *placeres* lode channel; ~ **fluvial** *m* HIDROL stream channel; ~ **de frecuencias** *m* TELECOM frequency channel; ~ **de fuga** *m* AGUA *turbina hidráulica*, HIDROL tailrace;

~ h ~ **hembra** *m* PROD *laminador* box pass; ~ **hidrodinámico con generador de olas** *m* TRANSP MAR wave generating towing tank; ~ **hidrodinámico de oscilación** *m* FÍS ONDAS *para experimentos ondulatorios* ripple tank; ~ **de humos** *m* PROD *de hornos* flue;

~ i ~ **de instalación** *m* TELECOM setup channel; ~ **interoceánico** *m* OCEAN interocean channel, sea canal; ~ **inverso** *m* INFORM&PD reverse channel; ~ **de irrigación** *m* AGUA catch feeder;

~ l ~ **de lavado** *m Esp* (*cf aparato de lavado AmL*)

MINAS sluice; ~ **de lectura-escritura** *m* INFORM&PD, ÓPT read-write channel; ~ **libre marcado** *m* TELECOM marked idle channel; ~ **limpia** *m* OCEAN *navegación en aguas restringidas* sound channel; ~ **lineal** *m* TELECOM linear channel; ~ **litoral** *m* OCEAN side channel; ~ **de llamadas** *m* TELECOM calling channel; ~ **de llegada** *m* AGUA headrace, leat, penstock, INSTAL HIDRÁUL feeder, headrace, PROD leat;

~ m ~ **de mando** *m* D&A, TEC ESP *organizaciones militares* command channel; ~ **de mantenimiento de operadores de red** *m* TELECOM network operators maintenance channel (*NOMC*); ~ **de mareas** *m* TRANSP MAR tide gate, tideway; ~ **marítimo** *m* OCEAN *navegación comercial* sea canal, seaway, TRANSP MAR ship canal; ~ **de medición del período** *m* NUCL period-measuring channel; ~ **de mensajes** *m* TRANSP AÉR message chute; ~ **de molino** *m* AGUA, PROD leat; ~ **múltiple** *m* ING MECÁ gang channel; ~ **multiplexor** *m* INFORM&PD multiplexer channel;

~ n ~ **n** *m* ELECTRÓN n-channel; ~ **navegable** *m* HIDROL navigable channel, OCEAN *barcos de alta mar* ship canal, TRANSP MAR navigable channel; ~ **de navegación** *m* HIDROL navigation channel, OCEAN navigation channel, *para paso* shipping channel, TRANSP navigation channel; ~ **a nivel** *m* AGUA ditch canal;

~ o ~ **operativo embebido** *m* TELECOM embedded operations channel (*EOC*); ~ **operativo libre** *m* TELECOM idle working channel;

~ p ~ **p** *m* ELECTRÓN p-channel; ~ **de paginación** *m* TELECOM paging channel; ~ **de paso** *m* TRANSP AÉR pitch channel; ~ **en pendiente** *m* NUCL inclined channel; ~ **periodimétrico** *m* NUCL period-measuring channel; ~ **de preparación inicial** *m* TELECOM setup channel; ~ **principal** *m* TELECOM trunk channel; ~ **principal de una red** *m* TELECOM trunk channel; ~ **probabilístico** *m* TELECOM random channel; ~ **de pruebas con generador de olas** *m* TRANSP MAR *diseño de buques* wave-generating test tank; ~ **de pruebas hidrodinámicas** *m* TRANSP MAR *casco, hélice* testing tank, towing tank;

~ q ~ **Q** *m* ELECTRÓN Q-channel;

~ r ~ **de radiación** *m* FÍS RAD, NUCL, QUÍMICA radiation channel; ~ **de radio** *m* TELECOM, TRANSP AÉR radio channel; ~ **de recepción** *m* TELECOM receiving channel, reception channel, TV reception channel; ~ **de registro** *m* CINEMAT recording channel; ~ **remoto** *m* PROD remote channel; ~ **de resaca** *m* OCEAN tidal channel; ~ **de retorno** *m* INFORM&PD return channel, reverse channel; ~ **de retroceso** *m* INFORM&PD backward channel; ~ **de riego** *m* AGUA catch feeder, irrigation canal;

~ s ~ **de salida** *m* AGUA outlet channel, TV outgoing channel; ~ **por satélite** *m* TV satellite channel; ~ **secundario** *m* TRANSP AÉR secondary duct; ~ **selector** *m* INFORM&PD selector channel; ~ **semicircular** *m* ACÚST semicircular channel; ~ **para señales de control** *m* TEC ESP command channel; ~ **de señalización** *m* TELECOM signaling channel (*AmE*), signalling channel (*BrE*); ~ **para señalización** *m* TELECOM signaling channel (*AmE*), signalling channel (*BrE*); ~ **de señalización dedicado** *m* TELECOM dedicated signaling channel (*AmE*), dedicated signalling channel (*BrE*); ~ **de señalización no dedicado** *m* TELECOM nondedicated signaling channel (*AmE*), nondedicated signalling channel (*BrE*); **~ de sobrecarga** *m* TELECOM overload channel; **~ de sonido** *m* TV sound channel; **~ sonoro** *m* OCEAN sound channel, *acústica submarina* acoustic channel; **~ subterráneo AAC** *m* ELECTRÓN buried channel charge-coupled device (*buried channel CCD*); **~ de superficie** *m* ELECTRÓN surface channel;

~ t **~ de toma** *m* AGUA headrace, intake canal, leat, penstock, INSTAL HIDRÁUL headrace, PROD leat; **~ de trabajo** *m* HIDROL race, raceway, INSTAL HIDRÁUL feeder, headrace, race, raceway, TELECOM working channel; **~ de trancanil** *m* TRANSP MAR waterway; **~ de transmisión** *m* INFORM&PD transmission channel, MINAS drive tube, TELECOM transmission channel, transmit channel; **~ de transmisión acústica** *m* OCEAN acoustic channel; **~ de transmisión de datos** *m* INFORM&PD data transmission channel, TELECOM data communication channel (*DCC*); **~ transversal** *m* TEXTIL crosswise ribs; **~ tributario** *m* HIDROL tributary channel;

~ u **~ único por portadora** *m* (*SCPC*) TEC ESP, TELECOM single channel per carrier (*SCPC*); **~ utilizando banda lateral inferior** *m* TV channel using lower sideband; **~ utilizando banda lateral superior** *m* TV channel using upper sideband;

~ v **~ en V** *m* ING MECÁ V-groove; **~ de vertedero** *m* INSTAL HIDRÁUL spillway canal, spillway channel, wasteway; **~ vertedor** *m* ENERG RENOV *presas* spillway channel; **~ de video** *m* AmL, **~ de vídeo** *m* Esp TV video channel; **~ para vidrio de deshecho** *m* C&V cullet chute; **~ virtual** *m* TELECOM virtual channel (*VC*); **~ virtual de señalización** *m* TELECOM signalling virtual channel (*AmE*) (*SVC*), signalling virtual channel (*BrE*) (*SVC*); **~ virtual de señalización de radiodifusión** *m* TELECOM broadcast-signaling virtual channel (*AmE*) (*BSVC*), broadcast-signalling virtual channel (*BrE*) (*BSVC*); **~ vocal** *m* INFORM&PD, TELECOM voice channel; **~ de voz** *m* INFORM&PD speech channel, voice channel, TELECOM voice channel; **un solo ~ por portadora** *m* (*SCPC*) TEC ESP, TELECOM single channel per carrier (*SCPC*)

canales[1]: **sin ~** *adj* ING MECÁ, MECÁ fluteless
canales[2]: **en ~** *fra* FÍS FLUID in channels
canaleta *f* AmL (*cf aspirador Esp*) CARBÓN *minas* chute, CONST channel, gutter, MECÁ, MINAS chute, PETROL ditch; **~ alimentadora del filtro** *f* CARBÓN filter feed trough; **~ de carga** *f* MINAS loading chute; **~ de distribución del hormigón** *f* CARBÓN flume; **~ por la que pasa el cableado** *f* ELEC *instalación* conduit

canalización *f* AGUA canalization, ELECTRÓN pipelining, ENERG RENOV channeling (*AmE*), channelling (*BrE*), HIDROL piping, raceway, ING MECÁ ducting, pipeline, INSTAL HIDRÁUL line, pipeline, pipework, raceway, boosting main, inlet, PROD duct work, TELECOM channeling (*AmE*), channelling (*BrE*), TRANSP channel track, channelization; **~ del aire** *f* ING MECÁ air ducting; **~ de aire comprimido** *f* CONST compressed-air line, ING MECÁ air main, compressed-air line; **~ ascendente** *f* ELEC *fuente de alimentación* rising main; **~ automática** *f* HIDROL mechanical piping; **~ circular** *f* ELEC, ING ELÉC ring circuit (*AmE*), ring current (*AmE*), ring main (*BrE*); **~ de datos** *f* ELECTRÓN data path; **~ eléctrica** *f* ELEC *conexión* wiring; **~ flexible** *f* ELEC, ING ELÉC flexible conductor, ING MECÁ flexible tubing; **~ de fluidos** *f*

ING MECÁ fluid pipeline; **~ de gas** *f* CONST gas pipeline, GAS gas main, gas pipeline, PETROL gas pipe, gas pipeline, PROD gas main, TEC PETR gas pipe, gas pipeline, TERMO gas main, gas pipeline; **~ metálica flexible** *f* ING MECÁ flexible metal piping; **~ neumática** *f* ING MECÁ air main; **~ principal** *f* AGUA main (*AmE*), mains (*BrE*); **~ principal de aguas** *f* AGUA, HIDROL city water (*AmE*), mains water (*BrE*)
canalizado *m* OCEAN fairway
canalizar *vt* AGUA canalize, CONST, ENERG RENOV channel
canalizo *m* CARBÓN flume
canalón *m* AGUA flume, CONST eaves trough, flume, gutter, trough gutter, ING ELÉC conductor, TRANSP AÉR gutter; **~ del alero** *m* CONST eaves gutter; **~ de arista** *m* CONST *edificación* fillet gutter; **~ del cable** *m* ELEC, ING ELÉC cable trough; **~ de calibración** *m* AGUA calibration flume; **~ eléctrico** *m* ELEC, ING ELÉC electrical conductor; **~ oculto** *m* CONST secret gutter; **~ de pretil** *m* CONST parapet gutter; **~ de prueba** *m* AGUA test flume; **~ rectangular** *m* CONST box gutter
cañamazo *m* AmL (*cf tabique de lona Esp*) CONST canvas brattice, TEXTIL *material textil* canvas
cáñamo *m* TRANSP MAR hemp
canasta *f* OCEAN *tecnología pesquera* shellfish basket; **~ de calentamiento** *f* AmL (*cf manto de calentamiento Esp*) LAB heating mantle; **~ del extractor** *f* PROC QUÍ extractor basket; **~ salvavidas** *f* TRANSP MAR *para rescate* breeches buoy
canastilla *f* OCEAN pad
cañaveral *m* AGRIC cane plantation
cáncamo *m* ING MECÁ *equipamiento de cubierta*, MECÁ, TRANSP MAR eyebolt; **~ de argolla** *m* TRANSP MAR *accesorios de cubierta* ring bolt; **~ para izar** *m* ING MECÁ lifting eyebolt; **~ de maniobra** *m* TRANSP MAR *equipamiento de cubierta* lifting eye; **~ de ojo** *m* PROD eyebolt
cancelación *f* INFORM&PD abort; **~ de existencias** *f* PROD inventory wipe-off; **~ de lóbulo lateral** *f* ELECTRÓN side lobe cancellation; **~ de salida** *f* PROD output override
cancelar: **~ el efecto de** *vt* PROD override
cancerígeno[1] *adj* ALIMENT, QUÍMICA, SEG carcinogenic
cancerígeno[2] *m* ALIMENT, QUÍMICA, SEG carcinogen
cancha: **~ de mineral** *f* AmL (*cf vertedero de mineral Esp*) MINAS ore dump
cancrinita *f* MINERAL cancrinite
candado *m* CONST padlock; **~ de la puerta** *m* CONST gate latch
candal: **~ intrínseco** *m* PAPEL baseflow; **~ de pasta** *m* PAPEL *que llega a la máquina de papel* approach flow
candela *f* (*cd*) ELEC *unidad luminosa*, FÍS, ING ELÉC, METR, ÓPT candela (*cd*); **~ decimal** *f* METR decimal candle
candelabro *m* C&V, ELEC *alumbrado* chandelier
candelero *m* C&V, ELEC chandelier, TRANSP MAR *cubierta* stanchion
candeletón *m* TRANSP MAR *jarcia* Spanish burton
candente *adj* TERMO, TERMOTEC red-hot
candileja *f* CINEMAT, TV footlight
candita *f* MINERAL candite
cañería *f* AGUA conduit, water supply line, CONTAM pipeline, GAS piping, ING MECÁ pipeline, MECÁ pipe, pipework, TEC PETR *perforación* casing; **~ de abaste-**

cimiento de agua *f* AGUA water supply pipe; **~ de agua** *f* AGUA water pipe; **~ de aire** *f* ING MECÁ air pipe; **~ cloacal** *f* RECICL main sewer; **~ de desagüe** *f* CONST waste cock, RECICL outfall pipe; **~ de distribución** *f* AGUA distributing pipe; **~ de entubación** *f* GAS casing; **~ flexible** *f* ING MECÁ flexible pipe; **~ de fluidos** *f* ING MECÁ fluid pipeline; **~ de hierro gris** *f* ING MECÁ gray-iron pipe (*AmE*), grey-iron pipe (*BrE*); **~ metálica flexible** *f* ING MECÁ flexible metal conduit, flexible metal piping; **~ perdida** *f* PETROL liner; **~ perforada** *f* PETROL perforated pipe; **~ de plomo** *f* CONST lead piping; **~ principal del agua** *f* AGUA water main; **~ principal de gas** *f* GAS, PROD, TERMO gas main; **~ de revestimiento** *f* GAS casing

canfano *m* QUÍMICA camphane

canfeno *m* QUÍMICA *terpeno insecticida* camphene

canfol *m* QUÍMICA *borneal* camphol

canforato *m* QUÍMICA camphorate

canfórico *adj* QUÍMICA camphoric

cangilón *m* AGRIC *elevador*, AGUA, CONST bucket, CONTAM *maquinaria industrial* scoop, ENERG RENOV *turbina* bucket, MINAS *recipiente* tub, TRANSP MAR *pesca* bucket; **~ de acero embutido** *m* PROD pressed-steel bucket; **~ de agua de descarga automática** *m* AGUA self-discharging water bucket; **~ de draga** *m* TRANSP dredger bucket; **~ del elevador** *m* PROD elevator bucket, elevator cup; **~ suspendido** *m* PROD hanging bucket

cangreja *f* TRANSP MAR *mástil, vela* mizzen, mizzen sail

cangrejo *m* CINEMAT spreader, *base para patas de trípode* spider, TRANSP MAR *mástil, vela* mizzen, mizzen sail; **~ de río** *m* AGUA crawfish, crayfish

canguro *m* TRANSP, TRANSP MAR car ferry, ferry

canica: **~ de vidrio** *f* C&V glass marble

canilla *f* CONST cock, spigot, bib, MECÁ *tuberías* tap, faucet, PROD spout, TEXTIL pirn; **~ del agua** *f* AmL (*cf llave del agua Esp*) CONST water cock; **~ atomizadora** *f* ING MECÁ spray tap; **~ de base** *f* CONST butt cock; **~ de bebedero** *f* PROD gate stick; **~ de botón** *f* CONST push-button faucet (*AmE*), push-button tap (*BrE*); **~ de bronce** *f* CONST cock brass; **~ de cierre automático** *f* CONST self-closing cock (*BrE*), self-closing faucet (*AmE*); **~ cilíndrica** *f* CONST cylinder cock; **~ curva** *f* AmL CONST bib nozzle; **~ de metal** *f* CONST cock metal; **~ mezcladora de seguridad** *f* ING MECÁ safety mixing tap; **~ principal** *f* ING MECÁ main tap; **~ pulverizadora** *f* ING MECÁ spray tap; **~ recta** *f* AmL (*cf grifo de boca recta Esp*) CONST straight-nose cock

canillera *f* TEXTIL *tejedura* pirn-winding machine

canje *m* PROD trade-in; **~ catiónico** *m* HIDROL, TEC PETR cation exchange

canoa *f* TRANSP MAR *embarcación* canoe

canon: **~ de emisiones** *m* CONTAM polluter pays principle

caño *m* CONST pipe, GAS piping, MECÁ pipe, TRANSP MAR creek; **~ de aire** *m* MINAS brattice; **~ cónico** *m* ING MECÁ taper pipe; **~ con costura** *m* ING MECÁ seamed pipe; **~ sin costura** *m* ING MECÁ seamless pipe; **~ del desagüe** *m* CONST drainpipe; **~ de hierro dúctil** *m* ING MECÁ ductile iron pipe; **~ de subida** *m* PETROL, TEC PETR riser pipe

cañón *m* CONST barrel, *llave* tube, D&A cannon, gun, ING MECÁ barrel, TEC PETR *perforación* gun perforator, TERMOTEC *para chimeneas* flue; **~ de ánima lisa**

m D&A smooth-bore gun; **~ antiaéreo de grueso calibre** *m* D&A heavy anti-aircraft gun; **~ antiaéreo de pequeño calibre** *m* D&A light air defence gun (*BrE*), light air defense gun (*AmE*); **~ anticarro** *m* D&A anti-tank gun; **~ anticarro de grueso calibre** *m* D&A heavy anti-tank gun; **~ anticarro de pequeño calibre** *m* D&A light anti-tank gun; **~ de artillería** *m* D&A ordnance gun; **~ automático** *m* D&A automatic gun; **~ azul** *m* TV blue gun; **~ de chimenea** *m* CONST chimney flue; **~ para disparar el arpón** *m* OCEAN harpoon gun; **~ de electrones** *m* ELECTRÓN, FÍS, FÍS RAD, INSTR, TV electron gun; **~ de electrones pulsados** *m* ELECTRÓN, FÍS RAD pulsed-electron gun; **~ electrónico** *m* ELECTRÓN, FÍS, FÍS RAD, INSTR electron gun, PROD ion gun, TV electron gun; **~ de emisiones** *m* CONTAM flue; **~ de fusil** *m* D&A gun barrel; **~ del fusil** *m* D&A rifle barrel; **~ de información** *m* ELECTRÓN *tubos de almacenamiento de memoria* writing gun; **~ de lectura** *m* ELECTRÓN reading gun; **~ montado en torreta** *m* D&A turret gun; **~ perforador** *m* TEC PETR *perforación* perforating gun; **~ rayado** *m* D&A rifled gun; **~ sin retroceso** *m* D&A recoilless gun; **~ de tiro rápido** *m* D&A rapid-fire gun; **~ toroidal de electrones** *m* TV toroidal electron gun; **~ verde** *m* TV green gun

cañoneo *m Esp* (*cf punzado de la cañería AmL*) D&A gunfire, TEC PETR *perforación* casing perforation; **~ del revestimiento** *m Esp* (*cf punzado de la cañería AmL*) TEC PETR *perforación* casing perforation

cántara *f* TEXTIL *tejido* creel

cantaridina *f* QUÍMICA cantharidine

canteador: **~ en bisel** *m* C&V beaded bevel

cantera *f* CARBÓN stone pit, CONST pit, quarry, GEOL quarry, MINAS pit, quarry; **~ abandonada** *f* MINAS disused quarry; **~ de arcilla** *f* MINAS clay pit; **~ de caliza** *f* MINAS lime pit, limestone quarry, PROD lime pit; **~ a cielo abierto** *f* MINAS open quarry; **~ de grava** *f* CONST gravel pit; **~ de mármol** *f* MINAS marble quarry; **~ de piedra** *f* CONST cut-stone quarry, MINAS stone pit; **~ de piedra de cal** *f* MINAS lime pit, limestone quarry, PROD lime pit; **~ de piedra natural** *f* MINAS stone quarry; **~ de pizarra** *f* MINAS slate quarry; **~ de préstamo** *f* CARBÓN *excavaciones* borrow pit

cantería *f* MINAS quarrying

cantero *m* CONST hewer, stone dresser, MINAS *persona* cutter, quarryman

cantidad: **~ analógica** *f* INFORM&PD analog quantity; **~ de carne** *f* ALIMENT *carnicería* carcass yield; **~ de chatarra** *f* PROD scrap quantity; **~ crítica** *f* NUCL critical amount; **~ cuantificada** *f* ELECTRÓN quantized quantity; **~ declarada defectuosa** *f* PROD quantity declared unfit; **~ embarcada** *f* PROD shipped quantity; **~ de entrada de energía desequilibrada** *f* ING ELÉC unbalanced input; **~ entregada por completo** *f* PROD quantity completed; **~ extensiva** *f* FÍS extensive quantity; **~ fija** *f* TEC PETR *cementación, perforación* batch; **~ fija de pedido** *f* PROD fixed-order quantity; **~ de flujo** *f* FÍS flux quantum; **~ de flujo calorífico** *f* FÍS heat flow rate; **~ intensiva** *f* FÍS intensive quantity; **~ irracional** *f* MATEMÁT surd; **~ de lluvia que se evapora** *f* HIDROL fly-off; **~ de materiales por unidad** *f* PROD material quantity per unit; **~ medida** *f* ELECTRÓN measured quantity; **~ mensurable** *f* METR measurable quantity; ~ **de minerales** *f AmL* (*cf porción de minerales Esp*) MINAS parcel of ore; ~ **de movimiento** *f* ING MECÁ, MECÁ momentum; ~ **de movimiento angular-orbital** *f* FÍS RAD orbital angular momentum; ~ **no crítica** *f* NUCL off-critical amount; ~ **de orden económico** *f* PROD economic order quantity; ~ **pedida de nuevo** *f* PROD reorder quantity; ~ **pendiente** *f* PROD quantity back-order; ~ **periódica** *f* ELECTRÓN periodic quantity; ~ **de primer orden** *f* GEOM first-order quantity; ~ **radiante** *f* FÍS RAD, NUCL, TERMO, TERMOTEC radiant quantity; ~ **rechazada** *f* CALIDAD, PROD rejected quantity; ~ **de salida** *f* ING ELÉC output quantity; ~ **sinusoidal** *f* ELECTRÓN, ING ELÉC sinusoidal quantity; ~ **sinusoidal amortiguada** *f* ING ELÉC damped sinusoidal quantity; ~ **de sustancia** *f* FÍS amount of substance; ~ **de toma** *f* ELEC tapping quantity; ~ **variable** *f* ING ELÉC variable quantity; ~ **vuelta a pedir** *f* PROD reorder quantity

cantil *m* CONST, GEOL, OCEAN, TRANSP MAR cliff

canto[1]: **de ~ vivo** *adj* ING MECÁ sharp-edged

canto[2] *m* CONST *carpintería* feather, EMB, GEOM edge, ING MECÁ feather, MECÁ, PETROL edge, PROD feather; ~ **arcilloso** *m* GEOL, TEC PETR ball clay; ~ **en bisel** *m* CONST *carpintería* feather edge; ~ **doblado** *m* IMPR *de un libro* bolt; ~ **errático** *m* GEOL *sedimento glacial* erratic block; ~ **en línea recta** *m* ING MECÁ straight edge; ~ **rodado** *m* CONST boulder, rolling, GEOL cobble, PETROL boulder; ~ **rodado pequeño** *m* PETROL pebble

cantonita *f* MINERAL cantonite

cantos: ~ **doblados** *m pl* IMPR closed folds

cañuela *f* AGRIC fescue

caoba: ~ **de Guinea** *f* TRANSP MAR *construcción naval* sapele

caolín *m* C&V china clay, porcelain clay, GEOL china clay, MINERAL china clay, kaolin, P&C *material de carga, mineral, pigmento, refuerzo* kaolin, china clay, QUÍMICA kaolin, REVEST china clay, TEC PETR *geología* kaolinite

caolinita *f* CARBÓN, MINERAL kaolinite

caolinización *f* ENERG RENOV, GEOL kaolinization

capa[1]: **de ~ única** *adj* REVEST single-layer

capa[2] *f* AGUA layer, AUTO ply, CARBÓN layer, CONST bed, coat, layer, GAS bed, GEOFÍS mantle, GEOL bed, crust, marker bed, stratum, GEOM nappe, INFORM&PD layer, P&C *adhesivo, de pintura, plástico* coating, layer, PETROL crust, PROD coating, *agua, petróleo* body, REVEST *de pintura* coating, film, TEC ESP *revestimiento exterior de vehículos espaciales* coat, layer, TEXTIL ply, layer, *de fibras* cap, TRANSP MAR layer;

~ **a** ~ **acuífera** *f* AGUA aquifer; ~ **acuífera subterránea** *f* AGRIC ground water, AGUA ground water, well water, CARBÓN, HIDROL ground water; ~ **adaptable de ATM** *f* (*AAL*) TELECOM ATM adaptive layer (*AAL*); ~ **adherente** *f* HIDROL adhering nappe, REVEST adhesive coat, bonding layer; ~ **adhesiva** *f* CONST tack coat, REVEST adhesive coat; ~ **de adhesivo entre dos superficies** *f* P&C glue line; ~ **de aglomerante** *f* CONST binder course; ~ **agotada** *f* ELECTRÓN *transistores* depletion layer; ~ **de agua** *f* HIDROL nappe, sheet of water, INSTAL HIDRÁUL nappe; ~ **de aire** *f* TEC PETR *perforación costa-fuera* air gap; ~ **aislante** *f* EMB, ING ELÉC, REVEST insulating layer; ~ **aluvial** *f* HIDROL alluvial nappe; ~ **anegada** *f* HIDROL drowned nappe;

~ **anódica** *f* REVEST anodic coat; ~ **antiabrasiva** *f* REVEST anti-abrasion layer; ~ **anticapilar** *f* CONST anticapillary course; ~ **antihalo** *f AmL* (*cf revestimiento antihalo Esp*) CINEMAT antihalation layer, FOTO antihalo layer; ~ **de aparejo** *f AmL* (*cf tapagrietas Esp*) COLOR filler coat; ~ **de aplicación** *f* INFORM&PD, TELECOM application layer; ~ **de Appleton** *f* FÍS Appleton layer, GEOFÍS F-layer; ~ **arable** *f* AGRIC topsoil; ~ **de arcilla** *f* AGRIC claypan; ~ **de asiento** *f* CONST base course, FERRO *vía férrea* subgrade;

~ **b** ~ **de barniz** *f* REVEST coat of varnish; ~ **barrera** *f* CONTAM *procesos industriales*, FÍS, TELECOM barrier layer; ~ **de base** *f* P&C *pintura*, PROD base coat, REVEST ground coat; ~ **básica sobre la película** *f* IMPR key coating on film; ~ **al bies** *f* TEXTIL bias ply;

~ **c** ~ **de calcita fibrosa** *f* GEOL beef; ~ **calefactora** *f* REFRIG *para evitar que el suelo se hiele* heater mat; ~ **caliente** *f* METEO warm layer; ~ **de carbón** *f* CARBÓN *minas* coal seam, coal measure, seam, GEOL coal measure, MINAS *geología* seam, coal measure, coal seam, seam coal; ~ **catalizadora** *f* NUCL catalyst bed; ~ **cerca del suelo** *f* METEO ground layer; ~ **cerrada** *f* FÍS closed shell; ~ **cilíndrica** *f* TEC ESP cylindrical shell; ~ **de cola** *f* REVEST glue layer; ~ **condicionalmente inestable sin saturar** *f* METEO conditionally unstable unsaturated layer; ~ **de contacto** *f* ELECTRÓN contact mask; ~ **de conversión** *f* COLOR conversion layer; ~ **de coque** *f* CARBÓN coke bed; ~ **de cortadura** *f* FÍS FLUID *fluidos* shear layer; ~ **cristalina** *f* CRISTAL crystal face;

~ **d** ~ **D** *f* FÍS, METEO D-layer; ~ **delgada** *f* ELECTRÓN *fabricación de circuitos integrados* thin film, MECÁ film, METAL *cristalografía* stratification, layer, P&C *pintura, plásticos* film; ~ **de desgaste** *f* CONST wearing course, ING MECÁ wearing surface; ~ **de detención** *f* CONTAM, ELECTRÓN, FÍS, ÓPT, TELECOM barrier layer; ~ **de discontinuidad** *f* OCEAN discontinuity layer;

~ **e** ~ **E** *f* FÍS, GEOFÍS *atmósfera* E-layer; ~ **eléctrica doble** *f* QUÍMICA electrical double-layer; ~ **de electrón** *f* FÍS M shell; ~ **del electrón** *f* ELECTRÓN, FÍS, FÍS PART, NUCL, QUÍMICA electron shell; ~ **electrónica** *f* ELECTRÓN, FÍS, FÍS PART, NUCL, QUÍMICA electron shell; ~ **de emulsión** *f* REVEST emulsion layer; ~ **endurecida** *f* GEOL hard ground, *en suelos carbonatados* hardpan; ~ **de enlace** *f* TELECOM link layer; ~ **de enlace de datos** *f* INFORM&PD data link layer; ~ **epitaxial** *f* CRISTAL, ELECTRÓN, TELECOM epitaxial layer; ~ **de equipos** *f* TELECOM equipment layer; ~ **de escorias** *f* PROD cinder bed; ~ **de espuma** *f* TRANSP AÉR foam blanket; ~ **estable totalmente sin saturar** *f* METEO unsaturated completely stable layer; ~ **de estériles** *f* CARBÓN dirt band; ~ **de estucado de bajo gramaje** *f* PAPEL lowweight coating (*LWC*); ~ **eufótica** *f* OCEAN photosynthetic layer; ~ **exterior** *f* TEC ESP skin; ~ **exterior delgada** *f* P&C *pintura, plásticos* skin; ~ **de extracción** *f* PROC QUÍ separation layer;

~ **f** ~ **F** *f* FÍS, GEOFÍS F-layer; ~ **fangosa** *f* CARBÓN slurry basin; ~ **fértil** *f* TEC ESP *reactor nuclear* blanket; ~ **fibrosa** *f* PAPEL fibrous layer, furnish layer; ~ **de filtración** *f* PROC QUÍ filtering layer; ~ **del filtro** *f* PROC QUÍ filtering layer; ~ **de filtros** *f* CINEMAT, TV filter layer; ~ **fina** *f* ELECTRÓN thin film; ~ **fina de arena** *f* CONST *carreteras* sand seal; ~ **final** *f* COLOR final coat,

REVEST appearance cover; ~ **física** _f_ INFORM&PD _interconexión de sistemas abiertos_, TELECOM physical layer; ~ **fotoconductora** _f_ ÓPT photoconducting layer; ~ **fotoeléctrica** _f_ REVEST photoelectric layer; ~ **de fotosíntesis** _f_ OCEAN photosynthetic layer; ~ **freática** _f_ AGRIC ground-water level, ground-water table, AGUA ground-water table, phreatic bed, ground-water level, CARBÓN ground-water level, ground-water table, CONST ground-water level, GEOL phreatic ground water, HIDROL ground-water level, ground-water table, underground nappe, TEC PETR _hidrogeología_ water table; ~ **de fricción** _f_ METEO planetary boundary layer; ~ **frontal** _f_ GEOL foreset bed;

~ g ~ **gasífera** _f_ MINAS _pozo petróleo_, TEC PETR _geología_ gas cap; ~ **gasífera secundaria** _f_ TEC PETR secondary gas cap; ~ **de gel** _f_ REVEST, TRANSP MAR gel coat; ~ **geológica** _f_ GAS, GEOL geological layer; ~ **de grava** _f_ CONST ballast; ~ **gruesa** _f_ ELECTRÓN _fabricación de circuitos integrados_ thick film; ~ **de guía** _f_ GEOL key bed;

~ h ~ **hidratada** _f_ C&V hydrated layer; ~ **con hidrocarburos** _f_ TEC PETR _geología_ production layer; ~ **de hielo** _f_ REFRIG glaze; ~ **horizontal sobre la capa frontal del lecho** _f_ GEOL topset bed; ~ **de humus** _f_ CARBÓN duff;

~ i ~ **ignífuga** _f_ REVEST fire-resistant layer, fireproof coat, flameproof coat, TERMO fire-resistant layer; ~ **impermeable** _f_ ENERG RENOV _geología_ cap rock, GAS impermeable layer, GEOL _geología_, MINAS _encima de una roca petrolífera_, TEC PETR _geología_ cap rock; ~ **de imprimación** _f_ CONST _carreteras, pintura_ prime coat, _pintura_ undercoat, PROD _pinturas_ priming coat; ~ **inferior** _f_ REVEST lower ply, TEXTIL sublayer; ~ **infinitamente gruesa** _f_ FÍS RAD infinitely thick layer; ~ **de insonorización** _f_ CONST noise barrier; ~ **de interconexión de sistemas abiertos** _f_ TELECOM open systems interconnection layer (_OSI layer_); ~ **interior** _f_ QUÍMICA _pintura_, REVEST undercoat; ~ **intermedia** _f_ REVEST intermediate layer; ~ **interna de dos electrones** _f_ NUCL two-electron innermost shell; ~ **de inversión** _f_ CONTAM _atmósfera_, METEO, TELECOM inversion layer; ~ **ionizante** _f_ ELEC _carga_ ionizing layer; ~ **de la ionosfera** _f_ GEOFÍS ionosphere layer; ~ **isotérmica** _f_ METEO isothermal layer;

~ k ~ **K** _f_ FÍS, NUCL K-shell; ~ **de Kennelly-Heaviside** _f_ FÍS Kennelly-Heaviside layer, GEOFÍS _atmósfera_ E-layer;

~ l ~ **L** _f_ FÍS L shell; ~ **libre** _f_ HIDROL free nappe; ~ **ligante** _f_ CONST tack coat; ~ **límite** _f_ ACÚST, ENERG RENOV, FÍS, FÍS FLUID, MECÁ, OCEAN, REFRIG boundary layer, REVEST boundary layer, boundary film, TRANSP AÉR boundary layer; ~ **límite atmosférica** _f_ METEO planetary boundary layer; ~ **límite planetaria** _f_ METEO planetary boundary layer; ~ **límite superficial** _f_ METEO ground layer, surface boundary layer; ~ **límite turbulenta** _f_ FÍS FLUID turbulent boundary layer;

~ m ~ **M** _f_ FÍS M shell; ~ **magnética** _f_ FÍS magnetic shell; ~ **metálica** _f_ REVEST metal coat, metal coating; ~ **monomolecular** _f_ QUÍMICA monomolecular layer; ~ **muy inclinada** _f_ MINAS highly-inclined seam;

~ n ~ **N** _f_ TELECOM N-layer; ~ **de nieve** _f_ METEO snow cover; ~ **de nitruro de titanio** _f_ ING MECÁ titanium nitride coating; ~ **de nubes** _f_ METEO, TEC ESP cloud layer;

~ o ~ **obtenida por deposición de vapor** _f_ PROC QUÍ vapor-deposited layer (_AmE_), vapour-deposited layer (_BrE_); ~ **oculta** _f_ TEC PETR hidden layer; ~ **de óxido** _f_ INFORM&PD oxide layer, MECÁ scale; ~ **de óxido catódico** _f_ REVEST cathodic-oxide coating; ~ **de óxido de la cinta** _f_ TV tape oxide layer; ~ **de ozono** _f_ CONTAM _atmósfera_ ozone layer, ozonosphere, METEO, TEC ESP ozone layer;

~ p ~ **pegajosa** _f_ REFRIG _en carne y pescado atrasados_ slime; ~ **permeable** _f_ GAS permeable layer; ~ **de pintura al barniz** _f_ COLOR varnish paint coat; ~ **porosa** _f_ CARBÓN _galvanotecnia_, GAS porous layer; ~ **porosa de extracción** _f_ PROC QUÍ porous extracting cup; ~ **de presentación** _f_ INFORM&PD presentation layer; ~ **de producción** _f_ TEC PETR _geología_ production layer; ~ **productiva** _f_ TEC PETR _geología_ production layer; ~ **productora** _f_ PETROL carrier bed; ~ **profunda de dispersión** _f_ OCEAN deep scattering layer (_DSL_); ~ **de protección** _f_ EMB protection layer; ~ **de protección al ácido** _f_ IMPR top; ~ **protectora** _f_ COLOR top coat, EMB coating, IMPR _encuadernación_ resist, P&C _pintura, plásticos_ resistant coating, protective coating, PROD protection coating, REVEST protective layer, resist, resistant coating; ~ **protectora antihalo** _f_ IMPR antihalation backing; ~ **protectora de cintas** _f_ TV tape-coating material; ~ **protectora de pintura** _f_ REVEST protective coat of paint, protective coating of paint; ~ **pulverizada** _f_ REVEST powder coating;

~ q ~ **Q** _f_ NUCL Q-shell; ~ **que actúa de filtro amarillo** _f_ IMPR yellow filter layer;

~ r ~ **de red conmutada** _f_ TELECOM switched network layer; ~ **de referencia** _f_ GEOL _estrato_ marker, NUCL key bed; ~ **reflectante** _f_ REVEST reflective coat; ~ **reflectora** _f_ REVEST reflective coat; ~ **de refuerzo** _f_ REVEST tie coat; ~ **resistente** _f_ REVEST resist; ~ **resistente a la abrasión** _f_ REVEST anti-abrasion layer; ~ **de revestimiento** _f_ REVEST coating film; ~ **rizada** _f_ REVEST kiss roll coating; ~ **de roca debajo del terreno blando** _f_ GEOL _edafología_ hardpan; ~ **de roca dura** _f_ CARBÓN _sondeos_ shell; ~ **de roca dura y barrena** _f_ MINAS shell and auger; ~ **rocosa** _f_ CARBÓN stone bed; ~ **de rodadura** _f_ CONST _carreteras_ carpet; ~ **roja** _f_ GEOL red bed; ~ **de rotura de capilaridad** _f_ CARBÓN capillarity-breaking layer;

~ s ~ **saturada** _f_ METEO saturated layer; ~ **sedimentaria** _f_ GAS sediment layer; ~ **de sellado** _f_ C&V covercoat, CONST _carreteras_ seal coat; ~ **semiimpermeable** _f_ AGUA semi-impermeable layer; ~ **de separación** _f_ PROC QUÍ separation layer; ~ **separadora** _f_ ELEC _cable_ separator; ~ **de servicio** _f_ TELECOM service layer; ~ **de sesión** _f_ INFORM&PD session layer; ~ **de silicato** _f_ GEOL silicate-bearing sheet; ~ **silícea** _f_ GEOL cherty bed; ~ **de silicio** _f_ ELECTRÓN silicon layer; ~ **subterránea de hielo** _f_ CARBÓN, TEC PETR permafrost; ~ **superficial** _f_ CONTAM MAR, OCEAN surface layer; ~ **superficial protectora** _f_ MECÁ finish; ~ **superficial del suelo** _f_ CARBÓN topsoil; ~ **de superficie** _f_ CONTAM MAR surface layer; ~ **superior** _f_ TEXTIL top ply; ~ **superior de pintura** _f_ P&C top coat of paint; ~ **superior del suelo** _f_ CONST topsoil; ~ **superpuesta** _f_ CONST, TV overlay;

~ t ~ **de tejido** _f_ TEXTIL _bucarán_ layer; ~ **de tejido intermedia** _f_ TEXTIL middle layer; ~ **de tejido superior** _f_ TEXTIL top layer; ~ **totalmente estable** _f_

METEO completely stable layer; ~ **de transición** *f*
ELECTRÓN, FÍS depletion layer, GEOL passage bed;
~ **de transmisión** *f* INFORM&PD transport layer,
TELECOM transmission layer; ~ **de turbidez** *f* OCEAN
turbidity layer; ~ **turbulenta** *f* METEO turbulent layer;
~ u ~ **de unión** *f* REVEST tie coat;
~ v ~ **de velocidad baja** *f* PETROL low-velocity layer;
~ **de vidrio metálico** *f* REVEST metal glaze film;
~ **viscosa** *f* REFRIG *en carne y pescado atrasados*
slime;
~ y ~ **yesífera** *f* CONST chalk stratum; ~ de yeso *f*
CONST chalk stratum
capacete *m* ING ELÉC end cap
capacidad *f* AGUA capacity, CALIDAD capability,
CARBÓN content, CONST capacity, ELEC capacitance,
ELECTRÓN contention, ENERG RENOV *tubería* carrying
capacity, FÍS capacitance, ING ELÉC capacitance,
capacity, power output, ING MECÁ capacity, output,
performance, *bombas* delivery, METAL bulk property,
METR *de un micrómetro*, NUCL *de una central* capacity,
ÓPT capacitance, PROD *ventilador* capacity of output,
output, *energía productiva* capacity, REFRIG capacity,
TEC ESP *instrumentos* range, TELECOM *posibilidades*
facility, capacitance, capability, TRANSP *carretera*
capacity, TRANSP MAR *de algibe* volume, VEH *motor*
capacity; ~ **absoluta** *f* TRANSP AADT (*average
annual daily traffic*); ~ **de absorción** *f* EMB, HIDROL
absorptive capacity, MECÁ absorbing capacity; ~ **de
absorción de protones** *f* CONTAM, FÍS RAD *propiedad
química* proton-absorptive capacity; ~ **de absorción
protónica** *f* CONTAM, FÍS RAD *propiedad química*
proton-absorptive capacity; ~ **de aceleración** *f* VEH
engine pick-up; ~ **de aceptación de mensajes** *f*
TELECOM facility accepted message (*FAA*); ~ **de
adherencia** *f* METAL *enlaces* bond energy;
~ **agotada** *f* PROD spent capacity; ~ **agudizadora** *f*
ELEC *de red de distribución* peaking capacity; ~ **de
almacenamiento** *f* AGUA, INFORM&PD storage capacity; ~ **en amperios-hora** *f* FOTO ampere-hour
capacity; ~ **de apoyo** *f* TEC PETR support capacity;
~ **asignada** *f* AGUA, CARBÓN rated capacity;
~ **básica** *f* TRANSP basic capacity; ~ **del bobinado** *f*
ING ELÉC winding capacity; ~ **para cálculos
aritméticos** *f* PROD arithmetic capability; ~ **de la
caldera** *f* INSTAL TERM boiler capacity; ~ **calefactora**
f TERMO *de una bomba* heating capacity; ~ **de
calentamiento** *f* TERMOTEC heating capacity; ~ **calórica específica** *f* FÍS specific heat capacity;
~ **calórica específica a presión constante** *f* FÍS
specific heat capacity at constant pressure; ~ **calórica específica a volumen constante** *f* FÍS specific
heat capacity at constant volume; ~ **calórica molar** *f*
FÍS molar heat capacity; ~ **calorífica** *f* FÍS heat
capacity, TERMO heat capacity, heat content; ~ **calorífica atómica** *f* FÍS RAD atomic heat capacity; ~ **de
campo** *f* AGRIC field capacity; ~ **del canal** *f*
INFORM&PD, TELECOM channel capacity; ~ **de canales despejados** *f* TELECOM clear-channel capability;
~ **de canje catiónico** *f* TEC PETR *química* cation
exchange capacity (*CEC*); ~ **de carga** *f* CARBÓN
load-bearing capacity, bearing capacity, EMB, ING
MECÁ carrying capacity, TRANSP carriage, loading
capacity, TRANSP AÉR load capacity, TRANSP MAR
diseño naval cargo capacity; ~ **de carga breve** *f*
ELEC *equipo* short-time rating; ~ **cargada** *f* CARBÓN
loaded capacity; ~ **del carrete** *f* CINEMAT, TV reel

capacity; ~ **de una carretilla** *f* CONST barrowful; ~ **en
caso de afluencia masiva** *f* TRANSP *control de tráfico*
tidal capacity; ~ **de cierre** *f* ELEC *de relé* making
capacity; ~ **del cilindro** *f* AUTO, VEH cylinder capacity; ~ **comprobada** *f* PROD demonstrated capacity;
~ **en condiciones predominantes** *f* TRANSP capacity
under prevailing conditions; ~ **de conducción de
corriente** *f* ELEC, ELECTRÓN, FÍS, ING ELÉC current-
carrying capacity; ~ **de conducto** *f* ENERG RENOV
conduit capacity; ~ **de conexión** *f* ING ELÉC making
capacity; ~ **de congelación** *f* ALIMENT, REFRIG
freezing capacity; ~ **del congelador** *f* ALIMENT,
REFRIG freezer capacity; ~ **de contacto** *f* ELEC
contact bounce; ~ **de corrección** *f* ELEC *de red de
distribución* peaking capacity; ~ **para corregir errores impulsivos** *f* TELECOM burst-error correcting-
capability; ~ **de corriente** *f* ELEC *línea*, ELECTRÓN,
FÍS, ING ELÉC current-carrying capacity; ~ **cúbica** *f*
MECÁ, METR, TRANSP MAR *de área de estibaje* cubic
capacity; ~ **de desconexión** *f* ELEC, ELECTRÓN, ING
ELÉC breaking capacity; ~ **encubridora** *f* P&C *propiedad, prueba* hiding-power; ~ **específica** *f* ENERG
RENOV *de pozo* specific capacity; ~ **estipulada** *f*
AGUA, CARBÓN rated capacity; ~ **de extensión de
la grieta** *f* METAL crack extension force; ~ **de las
funciones locales** *f* TELECOM local function capabilities (*LFC*); ~ **generadora** *f* ELEC, ENERG RENOV,
ING ELÉC, NUCL *fuente de alimentación* generating
capacity; ~ **hidroscópica** *f* AGUA water absorption
capacity; ~ **incremental** *f* ING ELÉC differential
capacitance; ~ **indicada** *f* AGUA, CARBÓN *motores*
rated capacity; ~ **inductiva específica** *f* ING ELÉC
specific inductive capacity, P&C *propiedad física,
prueba* dielectric constant, TV specific inductive
capacity; ~ **de intercambio de cationes** *f* CONTAM
características físico-químicas cation exchange capacity; ~ **de intercambio catiónico** *f* CONTAM cation
exchange capacity; ~ **de intercambio iónico** *f* GEOL
ion exchange capacity; ~ **interruptora** *f* ELEC, ING
ELÉC breaking capacity; ~ **de introducción de la
información** *f* ELECTRÓN write-through capability;
~ **de izado del gancho** *f* NUCL lifting capacity with
hook; ~ **de levantamiento** *f* ING MECÁ lifting power;
~ **local** *f* TRANSP local capacity; ~ **máxima** *f* C&V
hasta el tope brim capacity, CONST peak capacity;
~ **máxima de retención de agua** *f* AGRIC maximum
water-holding capacity; ~ **de la memoria** *f*
INFORM&PD, ING ELÉC memory capacity; ~ **de
memoria** *f* INFORM&PD storage capacity; ~ **de un
molino** *f* PROD mill run; ~ **momentánea** *f* TRANSP
momentaneous capacity; ~ **de mordaza paralela** *f*
ING MECÁ parallel jaw capacity; ~ **neutralizadora de
ácidos** *f* CONTAM *propiedades o características
químicas* acid-neutralizing capacity; ~ **nominal** *f*
AGUA rated capacity, C&V nominal capacity,
CARBÓN *motores* rated capacity; ~ **de ordeño** *f*
AGRIC milking ability; ~ **de oxigenación** *f* HIDROL
oxygenation capacity; ~ **de petición de mensajes** *f*
TELECOM facility request message (*FAR*); ~ **del
portador** *f* TELECOM bearer capacity; ~ **portante** *f*
CONST bearing capacity; ~ **posible** *f* TRANSP possible
capacity; ~ **de potencia** *f* MECÁ carrying capacity;
~ **práctica** *f* TRANSP practical capacity; ~ **de
predicción** *f* CONTAM predictive capability;
~ **procesamiento** *f* TELECOM processing power;
~ **de producción** *f* PROD throughput; ~ **para produ-**

cir hielo f REFRIG ice-making capacity; ~ **de producir semillas** f C&V seeding potential; ~ **productiva de la tierra** f AGRIC land capability; ~ **real en el medio rural** f TRANSP practical capacity under rural conditions; ~ **real en el medio urbano** f TRANSP practical capacity under urban conditions; ~ **de rebrote** f AGRIC recovery ability; ~ **de rechazo de mensajes** f TELECOM facility rejected message (*FRJ*); ~ **de recubrimiento de la tinta** f PAPEL ink coverage; ~ **de refrigerante** f REFRIG refrigerant capacity; ~ **de régimen** f AGUA, CARBÓN *motores* rated capacity; ~ **reproductora** f AGRIC reproductive ability; ~ **de reserva** f TRANSP reserve capacity; ~ **residual** f ELEC, ING ELÉC residual capacitance; ~ **de retención** f HIDROL *terreno* retentive capacity; ~ **de retención de agua aprovechable** f AGRIC available water-holding capacity; ~ **de retener** f FÍS, TV retentivity; ~ **de ruptura** f ELEC *de fusible, interruptor*, ELECTRÓN, ING ELÉC breaking capacity; ~ **de soporte** f TEC PETR support capacity; ~ **de la superficie de contacto** f METAL interface energy; ~ **térmica** f TERMO heat capacity, thermal capacity, TERMOTEC thermal capacity; ~ **térmica electrónica** f ELECTRÓN, FÍS electronic heat capacity; ~ **en los terminales** f PROD terminal capacity; ~ **de torneado** f ING MECÁ swing; ~ **de transporte de agua** f TEXTIL water-carrying capacity; ~ **de transporte de sólidos** f HIDROL *de un curso de agua* transport capacity for solids; ~ **de tratamiento** f PROD throughput; ~ **de tratamiento útil** f INFORM&PD throughput; ~ **de vuelo estacionario** f TRANSP AÉR hovering capability
capacidades: ~ **para realizar muestras** f pl TEXTIL patterning capacities
capacitación f ING MECÁ training; ~ **térmica** f ENERG RENOV thermal capacitance
capacitador: ~ **electrolítico no polarizado** m CONST nonpolarized electrolytic capacitor; ~ **variable** m ELEC trim capacitor, trimmer, variable capacitor
capacitancia f ELEC *propiedad básica, capacitor*, FÍS, ING ELÉC, ÓPT, TELECOM capacitance; ~ **acústica** f ACÚST acoustic capacitance; ~ **del cátodo de rejilla** f ING ELÉC grid cathode capacitance; ~ **centralizada** f ELEC *capacitor* lumped capacitance; ~ **concentrada** f ING ELÉC lumped capacitance; ~ **en derivación** f ING ELÉC shunt capacitance; ~ **de devanado** f ELEC *de bobina*, ING ELÉC winding capacitance; ~ **distribuida** f ING ELÉC distributed capacitance; ~ **eléctrica** f ELEC, TELECOM electrical capacitance; ~ **de entrada** f ING ELÉC input capacitance; ~ **entre electrodos** f ING ELÉC interelectrode capacitance; ~ **entre patillas** f ING ELÉC pin-to-pin capacitance; ~ **de entreturno** f ELEC *de bobina* interturn capacitance; ~ **específica** f ELEC specific capacitance; ~ **incremental** f ELEC incremental capacitance; ~ **inductiva específica** f, FÍS, ING ELÉC, TEC ESP, TELECOM permittivity; ~ **interelectródica** f ING ELÉC interelectrode capacitance; ~ **de la juntura** f ING ELÉC junction capacitance; ~ **localizada** f ELEC *capacitor* lumped capacitance; ~ **nominal** f PROD nominal capacitance; ~ **parásita** f ING ELÉC parasitic capacitance; ~ **puerta a drenaje** f ING ELÉC gate-to-drain capacitance; ~ **puerta a generador** f ING ELÉC gate-to-source capacitance; ~ **puerta a substrato** f ING ELÉC gate-to-substrate capacitance; ~ **residual** f ELEC residual capacitance, ING ELÉC residual capacitance, zero capacitance; ~ **de salida** f ING ELÉC output capaci-

tance; ~ **en serie** f ING ELÉC series capacitance; ~ **de variación lineal** f ING ELÉC straight-line capacitance
capacitator: ~ **de dieléctrico cerámico** m TELECOM ceramic capacitor
capacitivo: **no** ~ adj ELEC *de carga*, ING ELÉC noncapacitive
capacitor m GEN capacitor; ~ **de acoplamiento** m ELEC, FÍS coupling capacitor; ~ **de aire** m ELEC, FÍS air capacitor; ~ **ajustable** m ELEC adjustable capacitor; ~ **de aplanamiento** m ELEC smoothing capacitor; ~ **de arranque** m ELEC starting capacitor; ~ **en baño de aceite** m ELEC oil-immersed capacitor; ~ **de bloqueo** m ELEC, FÍS, ING ELÉC blocking capacitor; ~ **de CA** m *AmL* ING ELÉC AC capacitor; ~ **calculable** m FÍS calculable capacitor; ~ **de capa delgada** m ELECTRÓN thin-film capacitor; ~ **de capa gruesa** m ELECTRÓN thick-film capacitor; ~ **cerámico** m C&V, ELEC, FÍS, ING ELÉC, TELECOM ceramic capacitor; ~ **cilíndrico** m FÍS cylindrical capacitor; ~ **desfasador** m ELEC phase-shifting capacitor; ~ **de dieléctrico cerámico** m C&V, ELEC, FÍS, ING ELÉC ceramic capacitor; ~ **dieléctrico de mica** m ELEC, ING ELÉC mica dielectric capacitor; ~ **de dos baterías** m ELEC two-cell capacitor; ~ **electrolítico** m ELEC, FÍS, ING ELÉC electrolytic capacitor; ~ **de energía** m ELEC, ING ELÉC power capacitor; ~ **de impulsos** m ELEC pulse capacitor; ~ **interdigital** m FÍS interdigital capacitor; ~ **de Lampard y Thomson** m FÍS Lampard and Thomson capacitor; ~ **de múltiples placas** m ELEC multiple plate capacitor; ~ **no regulable** m ELEC, FÍS, ING ELÉC fixed capacitor; ~ **oscilante** m ELEC oscillating capacitor; ~ **de papel** m ELEC, ING ELÉC paper capacitor; ~ **de paso** m FÍS bypass capacitor; ~ **de película delgada** m ELEC thin-film capacitor; ~ **para película metalizada** m ING ELÉC metalized film capacitor (*AmE*), metallized film capacitor (*BrE*); ~ **de placa** m ELEC plate capacitor; ~ **de referencia** m ELEC, ING ELÉC reference capacitor; ~ **rotativo** m ELEC, ING ELÉC rotary capacitor; ~ **supresor** m ELEC *antiinterferencia* suppressor capacitor
capado: ~ **del tabaco** m AGRIC topping
capar vt C&V cap
caparazón[1]: **con** ~ adj ALIMENT *mariscos* shell-on
caparazón[2] m TEC ESP shell
caparrosa f QUÍMICA copperas
capas[1]: **en** ~ **horizontales** adv FÍS FLUID in horizontal layers
capas[2]: ~ **empinadas de carbón** f pl CARBÓN edge coals; ~ **estratificadas Ekman** f pl HIDROL Ekman layer; ~ **múltiples** f pl TEC ESP multilayer
capataz m C&V gaffer, CONST foreman, PROD charge hand, foreman; ~ **encargado de las voladuras** m CONST blasting foreman; ~ **de maniobras** m FERRO foreman shunter; ~ **del patio de carga** m FERRO freight-yard foreman (*AmE*), goods yard foreman (*BrE*); ~ **del patio de mercancías** m FERRO freight-yard foreman (*AmE*), goods yard foreman (*BrE*); ~ **de perforación** m PETROL tool pusher; ~ **de sondeo** m TEC PETR *perforación* tool pusher; ~ **del tendido de la vía** m FERRO tracklaying foreman
caperuza f FOTO, ING ELÉC hood, TRANSP MAR *del ventilador* ventilator cowl
capilar[1] adj ING MECÁ capillary
capilar[2] m FÍS *del termómetro* stem

capilaridad *f* CARBÓN, CONST, FÍS, QUÍMICA *de diámetro pequeño* capillarity

capilarímetro *m* CARBÓN capillarimeter

capilla *f* IMPR advance sheet, press proof, show sheet

capital: ~ **en acciones ordinarias** *m* TEC PETR *finanzas* equity capital; ~ **ajeno** *m* TEC PETR *comercio, finanzas* loan capital

capitán *m* D&A, TRANSP AÉR captain, TRANSP MAR *marina* captain, *marina mercante* shipmaster; ~ **de fragata** *m* D&A, TRANSP MAR *marina* commander; ~ **general de la armada** *m* D&A, TRANSP MAR fleet admiral; ~ **del puerto** *m* TRANSP MAR harbor master (*AmE*), harbour master (*BrE*)

capitanía *f* D&A, TRANSP AÉR, TRANSP MAR captaincy; ~ **de puerto** *f* TRANSP MAR harbor master's office (*AmE*), harbour master's office (*BrE*)

capitel *m* CONST cap, MINAS *columnas* head

capítulo *m* AGRIC head

capó *m* AUTO, ING MECÁ hood, MECÁ *automóvil* cowl, TRANSP AÉR cowl, cowling, VEH *carrocería* bonnet (*BrE*), hood (*AmE*); ~ **anti-choque** *m* TRANSP *camiones* crushproof safety bonnet (*BrE*), crushproof safety hood (*AmE*); ~ **anti-golpes** *m* TRANSP *camiones* crushproof safety bonnet (*BrE*), crushproof safety hood (*AmE*); ~ **de seguridad** *m* TRANSP safety bonnet (*BrE*), safety hood (*AmE*); ~ **de tobera** *m* TRANSP AÉR nozzle cowl

capota *f* ING MECÁ, VEH *coches* hood; ~ **abatible** *f* TRANSP MAR *botes salvavidas* canopy; ~ **contra rociones** *f* TRANSP MAR spray hood; ~ **de descapotable** *f* AUTO top of convertible; ~ **plegable** *f* AUTO top of convertible

capotaje *m* SEG overturning, TRANSP *aviones* overturning, roll-over

cáprico *adj* QUÍMICA capric

caprílico *adj* QUÍMICA caprylic

caprilo *m* QUÍMICA capryl

caproico *adj* QUÍMICA caproic

caproilo *m* QUÍMICA caproyl

caproína *f* QUÍMICA caproin

caprolactama *f* QUÍMICA caprolactam

capsaicina *f* QUÍMICA capsaicin

capsicina *f* QUÍMICA *agente terapéutico* capsicin

cápsula *f* LAB *análisis* boat, *incineración* dish, MINAS *detonación*, TEC ESP (*cf navecilla AmL*) *vehículos* capsule; ~ **que se abre por fricción** *f* EMB friction snap-on cap; ~ **aerosol** *f* EMB aerosol cap; ~ **de algodón** *f* AGRIC boll; ~ **de aluminio** *f* EMB aluminium capsule (*BrE*), aluminum capsule (*AmE*); ~ **de aterrizaje** *f* TEC ESP landing capsule; ~ **del cartucho** *f* D&A percussion cap; ~ **detonadora** *f* MINAS *explosivos* detonator cap; ~ **detonante** *f* D&A fuse cap; ~ **de emergencia** *f* TEC ESP *vehículos* emergency capsule; ~ **de escape** *f* TEC PETR *operaciones costa-fuera* escape capsule; ~ **espacial** *f* TEC ESP 'space capsule; ~ **de evaporación** *f* LAB *material de vidrio* evaporating basin, evaporating dish; ~ **expulsable** *f* TEC ESP pod; ~ **extrema** *f* ING ELÉC end cap; ~ **eyectable** *f* TEC ESP ejectable capsule; ~ **fonocaptora coplanar** *f* ACÚST coplanar cartridge; ~ **fulminante** *f* (*cf cebo Esp*) D&A detonator, MINAS *explosivos* primer; ~ **de gelatina** *f* EMB gelatine capsule; ~ **marina de observación** *f* TEC PETR *buceo* observation chamber; ~ **del motor** *f* TRANSP AÉR engine nacelle, engine pod; ~ **de porcelana** *f* LAB basin, PROC QUÍ porcelain capsule; ~ **protectora contra ruidos** *f* SEG noise-protective capsule; ~ **a prueba de sonidos** *f* SEG soundproof capsule; ~ **recipiente** *f* TRANSP container capsule; ~ **retráctil** *f* EMB shrink capsule; ~ **roscada que se rompe al abrirse por primera vez** *f* EMB snap cap; ~ **de seguridad** *f* MECÁ bursting disc (*BrE*), bursting disk (*AmE*); ~ **termoformada en forma concha** *f* EMB clamshell blister; ~ **termoformada a medida del objeto a envasar** *f* EMB contract blister packaging service

captación *f* AGUA capture, catching, catchment, collection, uptake, ING MECÁ pick-up, NUCL pick-up, sensing, TELECOM pick-up, TRANSP AÉR lock-on; ~ **en abanico** *f* GAS arc discharge; ~ **en abanico con transferencia** *f* GAS arc discharge with transfer; ~ **de agua** *f* AGUA water catchment; ~ **de agua de lluvia** *f* AGRIC, AGUA rainwater catchment; ~ **del enlace entre satélites** *f* TEC ESP intersatellite link acquisition; ~ **de hidrógeno** *f* NUCL hydrogen uptake; ~ **metálica** *f* PROD metal pick-up; ~ **de modo normal** *f* TEC ESP acquisition of normal mode; ~ **del objetivo** *f* TEC ESP acquisition; ~ **optoelectrónica** *f* ING MECÁ optoelectronic pick-up; ~ **de órbita** *f* TEC ESP acquisition of orbit; ~ **de posición de vuelo** *f* TEC ESP acquisition of attitude; ~ **urbana** *f* AGUA urban catchment

captador *m* ACÚST pick-up, ELECTRÓN, METR, TEC ESP sensor, TELECOM pick-up, sensor; ~ **activo** *m* TEC ESP active sensor; ~ **estereofónico** *m* ACÚST stereophonic pick-up; ~ **fuera del aire** *m* TV off-air pick-up; ~ **de iones** *m* ELECTRÓN, TV ion trap; ~ **para medidor de contacto** *m* ING MECÁ pick-up for contact meter; ~ **monofónico** *m* ACÚST monophonic pick-up; ~ **piezoeléctrico** *m* ING MECÁ piezoelectric pick-up; ~ **de polvo** *m* CARBÓN, P&C *equipo* dust collector; ~ **que no transmite** *m* TV off-air pick-up

captar *vt* AGUA collect; ~ **aire** *vt* PROD trap air

captura *f* FÍS capture, OCEAN *pesca* catch, fish stock, TEC ESP *satélites, satélites militares* capture, TRANSP AÉR capture, lock-on, TRANSP MAR *pesca* landing; ~ **en la capa K** *f* FÍS K-capture; ~ **de datos** *f* INFORM&PD data capture; ~ **electrónica** *f* ELECTRÓN, FÍS, FÍS RAD, NUCL electron capture; ~ **de errores** *f* INFORM&PD error report; ~ **de haz** *f* TRANSP AÉR beam capture; ~ **máxima autorizada** *f* OCEAN total authorized catch; ~ **máxima permitida** *f* (*CMP*) OCEAN total allowable catch (*TAC*); ~ **de neutrones** *f* FÍS neutron capture; ~ **parásita** *f* NUCL parasitic capture; ~ **radiante de neutrones** *f* FÍS RAD neutron radiative capture; ~ **radiativa** *f* FÍS RAD radiative capture; ~ **radiativa de neutrones** *f* FÍS RAD neutron radiative capture; ~ **de un río** *f* AGUA, HIDROL river capture; ~ **río arriba** *f* HIDROL *de un río* headwater capture; ~ **por unidad de esfuerzo** *f* OCEAN catch per unit effort (*CUE*)

capucha *f* ING MECÁ hood; ~ **metálica** *f* PROD *muelas abrasivas* hood for emery wheels; ~ **de neoprene** *f* OCEAN diving hood; ~ **para polvos** *f* SEG dust hood; ~ **protectora** *f* D&A, SEG *del trabajador* protective hood; ~ **protectora contra polvo y aspersión** *f* SEG dust and spray protective hood; ~ **protectora contra ruidos** *f* SEG noise-protective hood

capuchón *m* CONST pile hood, GEOL cap, TRANSP MAR *del ventilador* ventilator cowl; ~ **roscado** *m* ING MECÁ, MECÁ cap nut; ~ **del teclado** *m* IMPR *informática* keycap

capullo *m* AGRIC cocoon

cara *f* CARBÓN *cuñas, cristales* face, CONST cheek, face, facing, GEOM face, ING MECÁ pane, head, INSTR *de un sólido*, MINAS *de filón* side; ~ **activa** *f* CONTAM MAR front blade; ~ **de un cristal** *f* CRISTAL crystal face; ~ **de empuje de la válvula** *f* VEH *del motor* valve face; ~ **fieltro** *f* PAPEL felt side, top side; ~ **firme** *f* ING MECÁ *mortaja de carpintería* cheek; ~ **de fricción** *f* AUTO friction facing; ~ **frontal** *f* ING MECÁ front face; ~ **del grano** *f* PROD grain side; ~ **magnética** *f* ING MECÁ magnetic face; ~ **del pelo** *f* PROD *correas* grain side; ~ **plana** *f* MECÁ flat; ~ **del reloj** *f* ING MECÁ dial; ~ **tela** *f* PAPEL wire side; ~ **de trabajo** *f AmL* (*cf frente de trabajo Esp*) MINAS *laboreo* face

caracol *m* ING MECÁ spiral

caracola *f AmL* (*cf gancho Esp*) MINAS *sondeos* pole hook

carácter *m* ELECTRÓN, IMPR, INFORM&PD, TELECOM, TV character; ~ **de acuse positivo de recepción** *m* INFORM&PD positive acknowledgement; ~ **alfanumérico** *m* INFORM&PD, TELECOM alphanumeric character; ~ **de alimentación de papel** *m* INFORM&PD form-feed character; ~ **de atención** *m* INFORM&PD bell character; ~ **de avance de página** *m* INFORM&PD form-feed character; ~ **de aviso** *m* INFORM&PD bell character; ~ **de borrado** *m* INFORM&PD delete character (*DEL*); ~ **cambiador de código** *m* (*carácter SI*) INFORM&PD shift-in character (*SI character*); ~ **de la categoría** *m* INFORM&PD status character; ~ **comodín** *m* INFORM&PD wild card character; ~ **de control** *m* INFORM&PD check character, control character; ~ **de control de errores** *m* TELECOM error-check character; ~ **en cursiva** *m* INFORM&PD italic character; ~ **descendiente** *m* IMPR long descender; ~ **de desplazamiento** *m* INFORM&PD shift character; ~ **de desplazamiento hacia dentro** *m* INFORM&PD shift-in character (*SI character*); ~ **de desplazamiento hacia fuera** *m* INFORM&PD shift-out character; ~ **doble** *m* IMPR *letras enlazadas* quaint character; ~ **de empaquetadura** *m* TELECOM stuffing character; ~ **de enlace** *m* PROD break character; ~ **de espacio** *m* INFORM&PD space character; ~ **especial** *m* INFORM&PD special character; ~ **de estado** *m* INFORM&PD status character; ~ **de extensión de código** *m* INFORM&PD code-extension character; ~ **de formato** *m* INFORM&PD layout character; ~ **funcional** *m* IMPR functional character; ~ **gráfico** *m* INFORM&PD graphic character; ~ **de identificación** *m* INFORM&PD identification character; ~ **ilegal** *m* INFORM&PD illegal character; ~ **de la imagen** *m* CINEMAT key; ~ **de instrucción que indica dejar sin efecto** *m* INFORM&PD cancel (*CAN*); ~ **inválido** *m* INFORM&PD illegal character; ~ **más significativo** *m* INFORM&PD most significant character; ~ **de omisión** *m* INFORM&PD ignore character; ~ **privado** *m* INFORM&PD privacy; ~ **de reconocimiento** *m* INFORM&PD acknowledge character; ~ **de relleno** *m* INFORM&PD idle character, filler character; ~ **de retorno del carro** *m* INFORM&PD carriage return (*CR*); ~ **de retroceso** *m* INFORM&PD backspace character; ~ **SI** *m* (*carácter cambiador de código*) INFORM&PD SI character (*shift-in character*); ~ **sincrónico** *m* INFORM&PD synchronous idle (*SYN*), synchronous character; ~ **de sincronismo** *m* INFORM&PD synchronous idle (*SYN*); ~ **SO** *m* INFORM&PD SO character; ~ **de supresión** *m*

INFORM&PD ignore character, delete character (*DEL*); ~ **de sustitución** *m* INFORM&PD substitute character (*SUB character*); ~ **sin uso** *m* INFORM&PD idle character

caracteres: ~ **inclinados** *m pl* IMPR slanted letters; ~ **por pulgada** *m pl* (*CPP*) IMPR characters per inch (*CPI*); ~ **raros** *m pl* IMPR pi characters; ~ **por segundo** *m pl* (*CPS*) IMPR, INFORM&PD characters per second (*CPS*)

característica *f* GEN characteristic, feature; ~ **del ánodo** *f* ING ELÉC anode characteristic; ~ **de control** *f* ING ELÉC control characteristic; ~ **corriente de voltaje** *f* TEC ESP current-voltage characteristic; ~ **de corriente y voltaje** *f* ING ELÉC voltage-current characteristic; ~ **cualitativa** *f* CALIDAD qualitative characteristic; ~ **cuantitativa** *f* CALIDAD quantitative characteristic; ~ **de descenso de potencia** *f* ING ELÉC power-down feature; ~ **del diodo** *f* ELECTRÓN diode characteristic; ~ **de disminución de intensidad lumínica** *f* ING ELÉC decay characteristic; ~ **de drenaje dendrítico** *f* HIDROL dendritic drainage pattern; ~ **eléctrica** *f* ELEC, ELECTRÓN electrical characteristic, ING ELÉC electric image, electrical characteristic; ~ **electródica** *f* ELEC, FÍS, ING ELÉC electrode characteristic; ~ **del embarque** *f* TEC ESP handling characteristic; ~ **espectral** *f* ELECTRÓN spectral characteristic; ~ **estática** *f* ING ELÉC static characteristic; ~ **de fase y amplitud** *f* TELECOM phase-amplitude characteristic; ~ **fase-amplitud** *f* TELECOM phase-amplitude characteristic; ~ **de frecuencia** *f* TELECOM frequency characteristic; ~ **de frecuencia de ganancia** *f* ELECTRÓN gain frequency characteristic; ~ **de grabación** *f* ACÚST, TV recording characteristic; ~ **de impedancia** *f* ING ELÉC impedance characteristic; ~ **de impedancia con rotor enclavado** *f* ELEC *máquina asincrónica* locked-rotor impedance characteristic; ~ **lineal** *f* ELECTRÓN linear characteristic; ~ **logarítmica** *f* ELECTRÓN logarithmic characteristic; ~ **de la maniobra** *f* TEC ESP handling characteristic; ~ **óptica** *f* TELECOM optical characteristic; ~ **de orden** *f* ING ELÉC control characteristic; ~ **de persistencia** *f* ELECTRÓN persistence characteristic; ~ **de la presión** *f* METEO pressure characteristic; ~ **de radiación** *f* FÍS, FÍS RAD, ÓPT, TEC ESP, TELECOM, TV radiation pattern; ~ **de rayos gamma** *f* ELECTRÓN *tubos de cámara* gamma characteristic; ~ **de la rejilla** *f* ELECTRÓN grid characteristic; ~ **de la repetición** *f* TV replay characteristic; ~ **de reproducción** *f* ACÚST reproducing characteristic; ~ **de reproducción o de grabación** *f* ACÚST reproducing-recording characteristic; ~ **de resistencia negativa** *f* ING ELÉC negative resistance characteristic; ~ **de respuesta** *f* TV response characteristic; ~ **de respuesta armónica** *f* FÍS ONDAS harmonic response characteristic; ~ **de sintonización** *f* FÍS RAD tuning characteristic; ~ **temporal** *f* TELECOM time characteristic; ~ **térmica** *f* TELECOM thermal characteristic; ~ **de transferencia** *f* ELECTRÓN, FÍS transfer characteristic; ~ **del transistor** *f* ELECTRÓN transistor characteristic; ~ **de transmisión** *f* TELECOM transmission characteristic

características: ~ **de la caída** *f pl* TEXTIL *del tejido* draping properties; ~ **de carga** *f pl* TRANSP AÉR load characteristic; ~ **sin carga** *f pl* TRANSP AÉR no-load characteristic; ~ **en circuito abierto** *f pl* ELEC *equipo* open-circuit characteristics; ~ **constructivas** *f pl* TRANSP MAR *construcción naval* constructional features; ~ **directas** *f pl* ELECTRÓN forward characteristic; ~ **de fatiga** *f pl* MECÁ fatigue properties; ~ **funcionales** *f pl* ING MECÁ performance properties, performance specification; ~ **de funcionamiento** *f pl* ING MECÁ performance properties; ~ **hidráulicas** *f pl* HIDROL *cuenca de captación* hydraulic characteristics; ~ **de impulso** *f pl* ELECTRÓN pulse characteristics; ~ **mecánicas** *f pl* CONST, FÍS FLUID, ING MECÁ, P&C mechanical properties; ~ **de operación** *f pl* TELECOM performance data; ~ **del pulso** *f pl* ELECTRÓN pulse characteristics; ~ **de la red** *f pl* TELECOM network performance (*NP*); ~ **de reproducción** *f pl* CINEMAT, TV reproduction characteristics; ~ **de saturación** *f pl* FÍS RAD saturation characteristics; ~ **de viscosidad y temperatura** *f pl* TERMO viscosity-temperature characteristics; ~ **del volumen de información** *f pl* ELECTRÓN bulk properties

característico *adj* INFORM&PD characteristic

caramelo *m* ALIMENT caramel, caramel sugar, COLOR caramel

caras: **entre ~** *adj* ING MECÁ across flats

carate *m AmL* (*cf quilate Esp*) METR *de diamante* carat, carat fine

carátula: ~ **de aguja** *f* LAB *de instrumento* needle dial

caravana *f* AUTO TRANSP, VEH line of cars (*AmE*), line of traffic (*AmE*), queue of cars (*BrE*), queue of traffic (*BrE*), TRANSP caravan (*BrE*), trailer (*AmE*), camper

carbamato *m* QUÍMICA carbamate

carbamila *f* QUÍMICA carbamyle

carbamilo *m* QUÍMICA carbamyl

carbanilida *f* QUÍMICA carbanilide

carbanilo *m* QUÍMICA carbanil

carbanión *m* QUÍMICA carbanion

carbazida *f* QUÍMICA carbazide

carbazol *m* QUÍMICA carbazole

carbeno *m* QUÍMICA carbene

carbilamina *f* QUÍMICA carbylamine, isonitrile

carbinol *m* QUÍMICA *metanol* carbinol

carbocementación *f* ING MECÁ carburizing, *aceros* case hardening, MECÁ case hardening; ~ **de gas** *f* GAS, TERMO gas carburizing

carbocíclico *adj* QUÍMICA carbocyclic

carboestirilo *m* QUÍMICA carbostyril

carbohidrasa *f* ALIMENT, QUÍMICA carbohydrase

carbohidrato *m* ALIMENT carbohydrate, QUÍMICA carbohydrate, glucide

carbólico *adj* QUÍMICA carbolic

carbón *m* CARBÓN coal, ELEC carbon, GEOL coal, ING ELÉC carbon, MINAS, PETROL coal, QUÍMICA charcoal, TERMO coal; ~ **activado** *m* AGUA activated carbon, ALIMENT activated charcoal, GAS, HIDROL activated carbon, P&C *carga* activated charcoal, activated carbon, PROC QUÍ activated carbon, QUÍMICA activated carbon, activated charcoal; ~ **activo** *m* PROC QUÍ activated carbon; ~ **aglomerado** *m* QUÍMICA tamped carbon; ~ **aglomerante** *m* CARBÓN close-burning coal; ~ **aglutinante** *m* CARBÓN caking coal, close-burning coal; ~ **de algas** *m* CARBÓN boghead; ~ **almacenado** *m* CARBÓN stock coal; ~ **de antracita** *m* CARBÓN anthracite coal; ~ **ardiente** *m* CARBÓN live coal; ~ **autóctono** *m* CARBÓN in situ pile; ~ **bituminoso** *m* CARBÓN bituminous coal, pit coal, soft coal, MINAS pit coal, soft coal, TERMO brown

coal; ~ **no bituminoso** *m* CARBÓN nonbituminous coal; ~ **brillante** *m* CARBÓN *petrografía* bright coal; ~ **bruto** *m* MINAS run-of-mine coal; ~ **de bujía** *m* CARBÓN, MINAS cannel coal; ~ **de caldera** *m* CARBÓN boiler coal; ~ **calibrado** *m* CARBÓN graded coal; ~ **de calidad inferior** *m* CARBÓN dross; ~ **de carboneras** *m* CARBÓN *para el consumo de buques* bunker coal; ~ **clasificado** *m* CARBÓN graded coal, sized coal; ~ **comercial** *m* CARBÓN commercial coal; ~ **de coque** *m* CARBÓN coking coal; ~ **de coque hinchado** *m* CARBÓN swelled coking coal; ~ **para craqueo** *m* C&V *en destilación de hidrocarburos* cracking coal; ~ **dióxido** *m* (CO_2) ELECTRÓN carbon dioxide (CO_2); ~ **doméstico** *m* CARBÓN domestic coal, household coal; ~ **de electrodos** *m* ING ELÉC electrode carbon; ~ **esquistoso** *m* CARBÓN *minería* foliated coal, schistous coal, MINAS foliated coal; ~ **explotado en profundidad** *m* CARBÓN deep-mined coal; ~ **en galleta** *m* CARBÓN cobbles; ~ **de gas** *m* CARBÓN cannel coal, gas coal, lean coal, MINAS cannel coal, TERMO lean coal; ~ **de gasógeno** *m* CARBÓN generator coal; ~ **granza** *m* CARBÓN pea coal; ~ **granza a granel** *m* CARBÓN, rough-pea coal; ~ **graso** *m* CARBÓN fat coal, rich coal, MINAS fat coal; ~ **grueso** *m* CARBÓN large coal, lump coal, MINAS lump coal; ~ **inaglutinable** *m* CARBÓN noncaking coal; ~ **no inflamable** *m* CARBÓN uninflammable coal; ~ **de la inflorescencia** *m* AGRIC head smut; ~ **lavado** *m* CARBÓN clean coal, cleaned coal, washed coal; ~ **no lavado** *m* CARBÓN, MINAS raw coal; ~ **de leña** *m* CARBÓN charcoal; ~ **limpiado a mano** *m* CARBÓN wale; ~ **limpio** *m* CARBÓN clean coal; ~ **de llama** *m* CARBÓN flaming coal; ~ **de llama larga** *m* CARBÓN candel coal; ~ **de madera** *m* CARBÓN wood charcoal, wood coal; ~ **magro** *m* CARBÓN, TERMO lean coal; ~ **mate** *m* CARBÓN, MINAS dull coal; ~ **materias volátiles** *m* CARBÓN low-volatile coal; ~ **medio en volátiles** *m* CARBÓN medium-volatile coal; ~ **menudo** *m* CARBÓN fine coal, slack, pea coal, small coal; ~ **menudo sin finos de minerales** *m* CARBÓN small coal without fines; ~ **mineral** *m* CARBÓN bituminous coal, mineral, mineral coal, pit coal, MINAS pit coal; ~ **ordinario** *m* CARBÓN inferior coal; ~ **de pantanos** *m* CARBÓN mud coal; ~ **en pedazos pequeños** *m* CARBÓN cobbles; ~ **de piedra** *m* CARBÓN bituminous coal, coal, pit coal, GEOL coal, MINAS coal, pit coal, TERMO coal; ~ **pobre** *m* CARBÓN producer coal; ~ **en polvo** *m* CARBÓN dust coal; ~ **pulverizado** *m* CARBÓN pulverized coal; ~ **puro** *m* CARBÓN, MINAS pure coal; ~ **de raíz** *m* AGRIC charcoal root; ~ **de retorta** *m* CARBÓN gas carbon, retort coal; ~ **de roca** *m* CARBÓN best coal; ~ **seleccionado a mano** *m* CARBÓN cob coal; ~ **semibituminoso** *m* CARBÓN semibituminous coal; ~ **semigraso** *m* CARBÓN semibituminous coal; ~ **sinterizado** *m* CARBÓN, MINAS sintering coal; ~ **tamizado** *m* CARBÓN screened coal, sifted coal; ~ **sin tamizar** *m* CARBÓN unscreened coal; ~ **térmico** *m* CARBÓN steam coal; ~ **terroso** *m* CARBÓN, MINAS bone coal; ~ **todo en uno** *m AmL*, ~ **todouno** *m Esp* MINAS run-of-mine coal; ~ **ultra limpio** *m* CARBÓN super clean coal; ~ **vegetal** *m* CARBÓN charcoal, filtering charcoal; ~ **vegetal pulverizado** *m* CARBÓN pulverized charcoal; ~ **vendible** *m* CARBÓN saleable coal

carbonáceo *adj* GEOL, QUÍMICA carbonaceous

carbonación *f* DETERG, HIDROL carbonation
carbonado *m* MINERAL carbonado
carbonatación *f* PAPEL, QUÍMICA carbonatation
carbonatado *adj* ALIMENT carbonated, GEOL carbonated, chalky, QUÍMICA carbonated
carbonatar *vt* ALIMENT, QUÍMICA carbonate
carbonatita *f* PETROL carbonatite
carbonatización *f* GEOL carbonatization
carbonato *m* DETERG, PAPEL, QUÍMICA carbonate; ~ **cálcico** *m* ALIMENT calcium carbonate, GEOL chalk; ~ **de hierro** *m* QUÍMICA iron carbonate; ~ **de magnesio** *m* TERMOTEC magnesium carbonate; ~ **potásico** *m* QUÍMICA potash; ~ **de potasio** *m* DETERG potassium carbonate; ~ **sódico** *m* PAPEL soda ash, QUÍMICA sodium carbonate; ~ **de sodio** *m* DETERG, PAPEL, QUÍMICA sodium carbonate
carbonera *f* CARBÓN *buques* bunker, coal bunker, FERRO *vehículos* coal bunker, MINAS *buques* bunker, coal bunker; ~ **de carbón no lavado** *f* CARBÓN, MINAS raw-coal bunker; ~ **de carbón todouno** *f* CARBÓN, MINAS raw-coal bunker
carbonero *m* CARBÓN, MINAS coal ship, TRANSP MAR coal ship, collier
carbónico *adj* QUÍMICA carbonic
carbonífero *adj* CARBÓN carboniferous, coal-bearing, MINERAL, TEC PETR carboniferous
carbonificación *f* CARBÓN, GEOL, MINAS coalification
carbonilo *m* QUÍMICA carbonyl
carbonización *f* CARBÓN carbonization, coalification, coking, GEOL, MINAS coalification, QUÍMICA charring, TERMO carbonization
carbonizado[1] *adj* CARBÓN carbonized, QUÍMICA charred, TERMO carbonized
carbonizado[2] *m* TELECOM char
carbonizar *vt* CARBÓN carbonize, char, TERMO carbonize
carbono *m* (*C*) QUÍMICA, TEC PETR *geología* carbon (*C*); ~ **del arco** *m* CINEMAT arc carbon; ~ **para arco voltaico** *m* ING MECÁ arc lamp carbon; ~ **combinado** *m* QUÍMICA fixed carbon; ~ **fijo** *m* GEOL fixed carbon; ~ **inorgánico disuelto** *m* CONTAM *caracterización de aguas* dissolved inorganic carbon; ~ **orgánico disuelto** *m* CONTAM *caracterización de aguas* dissolved organic carbon
carbonoso *adj* GEOL, QUÍMICA carbonaceous
carborundo *m* ING MECÁ carborundum®
carborundum® *m* QUÍMICA *regulador de ebullición* carborundum®
carboxilato *m* QUÍMICA carboxylate
carboxílico *adj* P&C *término químico* carboxylated, QUÍMICA carboxylic
carboxilo *m* QUÍMICA carboxyl
carboximetilcelulosa *f* ALIMENT, DETERG, P&C carboxymethyl cellulose (*CMC*)
carburación *f* ING MECÁ carburizing, INSTAL TERM *combustión* carburation
carburador *m* AUTO, C&V, ING MECÁ, MECÁ carburetor (*AmE*), carburettor (*BrE*), PROD gas machine, TRANSP MAR *motor*, VEH *alimentación de combustible* carburetor (*AmE*), carburettor (*BrE*); ~ **de aspiración lateral** *m* AUTO, VEH sidedraft carburetor (*AmE*), sidedraught carburettor (*BrE*); ~ **bifásico** *m* AUTO, VEH two-phase carburetor (*AmE*), two-phase carburettor (*BrE*); ~ **cuádruple** *m* AUTO, VEH quad carburetor (*AmE*), quad carburettor (*BrE*); ~ **de cuatro cubas** *m* AUTO, VEH four-barrel car-

buretor (*AmE*), four-barrel carburettor (*BrE*); ~ **de cuba doble** *m* AUTO, VEH twin-barreled carburetor (*AmE*), twin-barrelled carburettor (*BrE*); ~ **doble** *m* AUTO, MECÁ, VEH dual carburetor (*AmE*), dual carburettor (*BrE*), twin carburetor⁺ (*AmE*), twin carburettor (*BrE*); ~ **de doble cuerpo** *m* AUTO, VEH twin carburetor (*AmE*), twin carburettor (*BrE*); ~ **de dos fases** *m* AUTO, VEH two-phase carburetor (*AmE*), two-phase carburettor (*BrE*); ~ **electrónico** *m* ELECTRÓN, TRANSP electronic carburetor (*AmE*), electronic carburettor (*BrE*); ~ **de estrangulador doble** *m* AUTO, VEH twin-choke carburetor (*AmE*), twin-choke carburettor (*BrE*); ~ **horizontal** *m* AUTO, VEH horizontal carburetor (*AmE*), horizontal carburettor (*BrE*); ~ **invertido** *m* AUTO, VEH downdraft carburetor (*AmE*), downdraught carburettor (*BrE*); ~ **de mariposa doble** *m* AUTO, VEH twin-choke carburetor (*AmE*), twin-choke carburettor (*BrE*); ~ **de succión** *m* AUTO, VEH suction carburetor (*AmE*), suction carburettor (*BrE*); ~ **de succión lateral** *m* AUTO, VEH sidedraft carburetor (*AmE*), sidedraught carburettor (*BrE*); ~ **de tiro descendente** *m* AUTO, VEH downdraft carburetor (*AmE*), downdraught carburettor (*BrE*); ~ **de tiro invertido** *m* AUTO, VEH downdraft carburetor (*AmE*), downdraught carburettor (*BrE*); ~ **de tiro lateral** *m* AUTO, VEH sidedraft carburetor (*AmE*), sidedraught carburettor (*BrE*); ~ **de venturi variable** *m* AUTO, VEH variable venturi carburetor (*AmE*), variable venturi carburettor (*BrE*); ~ **zenit** *m* AUTO, VEH zenith carburetor (*AmE*), zenith carburettor (*BrE*)
carburar *vt* QUÍMICA carburet
carburo *m* ING MECÁ, MECÁ, METAL, QUÍMICA carbide; ~ **cálcico** *m* QUÍMICA calcium carbide; ~ **de calcio** *m* QUÍMICA calcium carbide; ~ **metálico de aleación** *m* METAL alloy carbide; ~ **metálico primario** *m* METAL primary carbide; ~ **de silicio** *m* FÍS, ING ELÉC silicon carbide, QUÍMICA carborundum®; ~ **al tungsteno** *m* ING MECÁ hard metal, tungsten carbide; ~ **de volframio** *m* TEC PETR *perforación* tungsten carbide
carcasa *f* C&V casing, ING ELÉC carcass, MECÁ casing, NUCL, PAPEL shell, TEC ESP *protección de vehículos espaciales* casing; ~ **de la bujía** *f* AUTO, ELEC, VEH spark plug shell; ~ **del motor** *f* TRANSP motor home; ~ **submarina** *f* *Esp* (*cf chasis submarino AmL*) CINEMAT *para cámara*, FOTO underwater housing; ~ **de turbina** *f* TRANSP MAR turbine casing
carcaza *f* ELEC *del motor* casing, ING MECÁ *máquinas* case
carcinotrón *m* ELECTRÓN, FÍS, TELECOM carcinotron; ~ **tipo O** *m* ELECTRÓN O-type carcinotron
carda *f* TEXTIL *hilatura* card
cardado *m* TEXTIL *carda* carding
cardadura *f* TEXTIL carding
cardan *m* AUTO, ING MECÁ, TRANSP AÉR cardan; ~ **del eje exterior** *m* TRANSP AÉR outer axis gimbal
cardar *vt* TEXTIL card
cardas: ~ **dobles** *f pl* TEXTIL twin cards
cardenillo *m* QUÍMICA verdigris
cardiode *n* ING MECÁ cardioid
cardo: ~ **ruso** *m* AGRIC Russian thistle
cardumen *m* OCEAN *ecología-etología marinas* shoal
carena *f* TRANSP MAR quickwork, ship's bottom, *diseño naval* underwater hull; ~ **líquida** *f* TRANSP MAR *diseño naval* free surface
carenado *m* REFRIG *del ventilador de aire*

acondicionado, TRANSP AÉR fairing; **~ corriente abajo** *m* REFRIG downstream fairing; **~ corriente arriba** *m* REFRIG upstream fairing; **~ del extremo de la pala** *m* TRANSP AÉR blade-tip fairing; **~ de plexiglas** *m* TRANSP AÉR plexiglass fairing; **~ de unión** *m* TRANSP AÉR fillet

carenadura *f* TRANSP MAR careening

carenaje *m* TEC ESP fairing

carenar *vt* TRANSP MAR repair, *mantenimiento de buques* grave, overhaul, *para limpiar fondos de buques* careen

carencia *f* ACÚST *de armónicas*, TEC ESP *de convección* absence; **~ de humedad** *f* AGRIC moisture deficiency; **~ de minerales** *f* AGRIC mineral deficiency

carenero *m* TRANSP MAR careening grid

careno *m* QUÍMICA carene

careta *f* D&A, PROD, SEG face shield; **~ antigás** *f* D&A gas mask; **~ contra polvos** *f* SEG dust mask; **~ de protección** *f* PROD, SEG face shield; **~ de soldador** *f* SEG welder's hood, welding helmet; **~ de soldador de mano** *f* SEG welder's hand shield

carfolita *f* MINERAL carpholite

carga[1]: **bajo ~** *adj* ING MECÁ under load; **sin ~** *adj* ELEC uncharged, PROD empty, TRANSP AÉR no-load; **sin ~ eléctrica** *adj* ELEC uncharged

carga[2] *f* C&V charge, charging, *del horno antes de empezar fundición* filling, CARBÓN *mina* fill, load, stress, CONST load, ELEC *capacitor* charge, *fuente de alimentación* load, ELECTRÓN charge, FÍS loading, charge, FÍS PART charge, HIDROL load, *aguas residuales* loading, INFORM&PD loading, load, ING ELÉC charging, charge, ING MECÁ head, load, INSTAL HIDRÁUL *de la válvula* load, MECÁ stress, pressure, METAL load, MINAS *pega* charge, charging, P&C *procesamiento* stress, filler, feed, PAPEL filler, load, PROD *horno* charging, load, stock, REFRIG load, TEC ESP cargo, TELECOM charge, load, TRANSP AÉR payload, TRANSP MAR load, cargo, freight, loading, goods (*BrE*);
~ ácida *f* CONTAM *proceso industrial* acid loading; **~ activa** *f* ELEC *resistencia óhmica*, ING ELÉC active load, P&C *ingrediente* active filler; **~ de acumulador** *f* ELEC accumulator charge, battery charge, ING ELÉC battery charge; **~ admisible** *f* CARBÓN bearing capacity, permissible load, safe load, ELEC *equipo* allowable load, ING MECÁ safe load; **~ aérea** *f* EMB air cargo, TRANSP AÉR air freight; **~ aerodinámica** *f* TRANSP AÉR aerodynamic load; **~ de agotamiento** *f* NUCL once-through charge; **~ de agua** *f* CARBÓN *hidráulica* pressure head, INSTAL HIDRÁUL pressure head, head of water, static head, head; **~ ajustada** *f* FÍS matched load; **~ alar** *f* TRANSP AÉR *aerodinámica* wing loading; **~ para alumbrado** *f* ING ELÉC light loading; **~ antisubmarinos** *f* D&A depth charge; **~ anual** *f* AGUA annual load; **~ de aspiración neta positiva** *f* NUCL net-positive-suction head; **~ atmosférica** *f* CONTAM *procesos industriales* atmospheric loading; **~ automática** *f* INFORM&PD autoload; **~ axial** *f* CONST axle load, METAL axial loading;
~ de la bandeja *f* *Esp* (*cf carga de la pálet AmL*) EMB pallet load; **~ de bandejas** *f* *Esp* (*cf carga de pálets AmL*) EMB pallet loading; **~ de base** *f* CONST *movimiento de tierras* toe weighting; **~ de batería** *f* ELEC, ING ELÉC battery charge; **~ biaxial** *f* METAL biaxial loading; **~ de bobinas** *f* ING ELÉC coil

loading; **~ del buque** *f* TRANSP MAR *acción* shiploading;
~ de calada *f* PROD stall load; **~ del canal** *f* TELECOM channel loading; **~ sin capacidad** *f* ING ELÉC noncapacitive load; **~ capacitiva** *f* ELEC *circuito de corriente alterna*, ING ELÉC, TELECOM capacitive load; **~ de cartucho** *f* INFORM&PD cartridge loading; **~ de casettes** *f* CINEMAT cassette loading; **~ de cebado** *f* MINAS *voladuras*, TEC ESP priming charge; **~-cebo** *f* MINAS *voladuras*, TEC ESP priming charge; **~ cíclica** *f* METAL repeated loading; **~ de cinta máxima permisible** *f* ING MECÁ maximum allowable belt stress; **~ circulante** *f* CARBÓN circulating load; **~ de columna** *f* MINAS *voladuras* column charge; **~ columnar** *f* MINAS *voladuras* columnar charge; **~ de combustible** *f* NUCL fuel charge, TRANSP AÉR fuel load; **~ de combustible en condiciones de operación** *f* NUCL hot refueling (*AmE*), hot refuelling (*BrE*), on-load charging, on-load fueling (*AmE*), on-load fuelling (*BrE*), on-load refueling (*AmE*), on-load refuelling (*BrE*); **~ de combustible por lotes** *f* NUCL batch fuel-loading; **~ de combustible por zonas** *f* NUCL zoned fuel-loading; **~ compensada** *f* ING ELÉC matched load; **~ completa de un buque** *f* TRANSP MAR shipload; **~ completa de mineral** *f* MINAS *alto horno* round; **~ de compresión** *f* EMB compression load; **~ concéntrica** *f* CONST concentric loading, ING ELÉC coaxial load; **~ de conducción** *f* ELEC, FÍS conduction charge; **~ constante** *f* ELEC *máquina eléctrica* constant load, ING ELÉC fixed load; **~ continua** *f* ELEC continuous load, continuous loading, ING ELÉC continuous load, continuous loading, floating charge; **~ continua submarina** *f* ING ELÉC krarup loading; **~ de control de cabeceo** *f* TRANSP AÉR pitch-control load; **~ de control del paso** *f* TRANSP AÉR pitch-control load; **~ de correa máxima admisible** *f* ING MECÁ maximum allowable belt stress; **~ crítica** *f* METAL critical stress; **~ de crudo sobre residuos de limpieza de tanques** *f* MINAS *petroleros*, TEC PETR load on top; **~ cuasiestática** *f* TEC ESP quasi-statical loading; **~ de cursor** *f* ING ELÉC sliding load;
~ de deformación remanente *f* METAL yield stress; **~-descarga de la memoria principal** *f* INFORM&PD roll in-roll out; **~-descarga de la RAM** *f* INFORM&PD roll in-roll out; **~ desplazable guía de ondas** *f* ING ELÉC waveguide sliding load; **~ desplazable guíaondas** *f* ING ELÉC waveguide sliding load; **~ dieléctrica** *f* TELECOM dielectric charge; **~ dinámica** *f* CONST, METAL dynamic loading; **~ de disco** *f* TRANSP AÉR disc loading (*BrE*), disk loading (*AmE*); **~ de diseño** *f* TRANSP AÉR design load; **~ de diseño por rueda** *f* TRANSP AÉR design wheel load; **~ dispersa** *f* INFORM&PD scatter load; **~ de dispersión** *f* INFORM&PD scatter load; **~ disponible** *f* TRANSP quick charge; **~ a distancia** *f* INFORM&PD, TELECOM remote loading; **~ distribuida de forma arbitraria** *f* METAL random loading;
~ efectiva *f* CARBÓN effective stress, EMB effective load, TEC PETR *geología* effective stress; **~ del eje** *f* CONST axle load; **~ por eje** *f* CONST axle load; **~ ejercida por el viento** *f* OCEAN wind stress; **~ elástica general** *f* METAL general yield load; **~ eléctrica** *f* ELEC *electroquímica, electrostática* electric charge, electrical charge, FÍS charge, electrical charge, electric charge, FÍS PART electric charge, ING

ELÉC, TELECOM electric charge, electrical charge; ~ **electrostática** *f* ELEC electrostatic charge; ~ **elemental** *f* FÍS elementary charge; ~ **de ensayo** *f* CARBÓN test loading; ~ **de entretenimiento** *f* ING ELÉC trickle charge; ~ **equilibrada** *f* ELEC, ING ELÉC balanced load; ~ **espacial** *f* FÍS, TEC ESP space-time, TRANSP AÉR space charge; ~ **específica** *f* FÍS *cociente carga-masa* specific charge, *de un portador electrizado* charge mass ratio, FÍS PART *de un electrón* specific charge; ~ **específica del electrón** *f* NUCL electron specific charge; ~ **estática** *f* ING ELÉC static charge; ~ **estocástica** *f* METAL stochastic loading; ~ **excéntrica** *f* CONST eccentric loading; ~ **explosiva** *f* AmL MINAS *voladura* blasting-charge; ~ **explosiva dentro de un recipiente de paredes** *f* TEC ESP *aviones* squib; ~ **de explotación** *f* TRANSP traffic load; ~ **de los extinguidores de incendios** *f* AmL (*cf carga de los extintores de incendios Esp*) CONST, SEG, TERMO, TRANSP AÉR, TRANSP MAR fire-extinguisher filling; ~ **de los extintores de incendios** *f* Esp (*cf carga de los extinguidores de incendios AmL*) CONST, SEG, TERMO, TRANSP AÉR, TRANSP MAR fire-extinguisher filling;

~f ~ **de fatiga** *f* TEC ESP fatigue strength; ~ **fija coaxial** *f* ING ELÉC coaxial fixed load; ~ **fija guía de ondas** *f* ING ELÉC waveguide fixed load; ~ **fija guíaondas** *f* ING ELÉC waveguide fixed load; ~ **de fondo** *f* Esp (*cf carga inferior AmL*) GEOL bed load, MINAS bottom charge; ~ **de formación inicial** *f* AUTO initial forming charge; ~ **de fractura** *f* CARBÓN, EMB, ING MECÁ, MECÁ, PAPEL, TRANSP MAR *de cuerda, cadena* breaking load; ~ **de fuego** *f* TERMO fire load; ~ **fuera de la hora punta** *f* ING ELÉC off-peak load; ~ **fundida** *f* PROD *encofrado* blow;

~g ~ **ganadera** *f* AGRIC stocking rate; ~ **en el gancho** *f* TEC PETR *perforación* hook load; ~ **general** *f* C&V blanket feed, EMB general cargo; ~ **geodésica potencial** *f* NUCL potential geodesic head; ~ **de gran potencia** *f* ING ELÉC high-power load; ~ **de guía de ondas** *f* ING ELÉC waveguide load; ~ **de guíaondas** *f* ING ELÉC waveguide load;

~h ~ **hasta la línea de carga máxima** *f* TRANSP MAR *cargamento* maximum load; ~ **del haz** *f* TV beam loading; ~ **hidráulica** *f* HIDROL hydraulic head, hydraulic load, INSTAL HIDRÁUL headwater level, hydraulic charging, hydraulic head, NUCL elevation head, TEC PETR *mecánica de fluidos* hydraulic head; ~ **hidrodinámica** *f* HIDROL hydrodynamic load; ~ **hidrostática** *f* HIDROL hydrostatic load, INSTAL HIDRÁUL head, velocity head, METAL hydrostatic stress; ~ **hidrostática debida a la presión** *f* CARBÓN pressure head; ~ **del horno** *f* C&V furnace charge;

~i ~ **de ignición** *f* D&A rocket igniter; ~ **de imagen** *f* ING ELÉC image charge; ~ **de impacto** *f* TRANSP AÉR impact load; ~ **impulsada** *f* D&A propelled charge; ~ **de impulso** *f* ING ELÉC moving charge; ~ **impulsora** *f* ING ELÉC active load; ~ **inducida** *f* ELEC *electrostática*, ING ELÉC, QUÍMICA induced charge; ~ **inductiva** *f* ELEC *corriente alterna*, ING ELÉC, TELECOM inductive load; ~ **inestable** *f* ING ELÉC charge pump; ~ **inferior** *f* AmL (*cf carga de fondo Esp*) MINAS bottom charge; ~ **de inflamación** *f* TEC ESP *en el interior del proyectil* igniter; ~ **iniciadora** *f* MINAS *voladuras* priming charge, TEC ESP priming charge, squib, primer charge; ~ **inicial de material**

fisionable *f* NUCL initial fissile charge; ~ **integrada** *f* ING ELÉC integrated charge; ~ **intermitente** *f* ELEC *de generador* intermittent load;

~l ~ **lenta** *f* CINEMAT trickle charge, ING ELÉC trickle charge, *acumuladores* float life; ~ **levantada** *f* TRANSP AÉR lifted load; ~ **libre** *f* FÍS free charge; ~ **límite** *f* CARBÓN ultimate load, ELEC limit load, ING MECÁ safe load, TRANSP AÉR limit load; ~ **líquida** *f* REFRIG *en termostato* liquid charge; ~ **longitudinal de la barra cíclica** *f* TRANSP AÉR longitudinal cyclic stick load; ~ **por lotes** *f* TRANSP package freight (*AmE*), package goods (*BrE*); ~ **a la luz del día** *f* CINEMAT daylight loading;

~m ~ **máxima** *f* ELEC *red de distribución* load peak, ING ELÉC peak load, TRANSP AÉR maximum load, full load, maximum payload; ~ **máxima admisible** *f* ELEC, ING ELÉC rated load; ~ **máxima permitida** *f* VEH *normas* total permissible laden weight (*BrE*), total permissible loaded weight (*AmE*); ~ **máxima total** *f* CONST maximum total load; ~ **máxima de tracción sin deformación** *f* METAL creep strength; ~ **máxima de tracción sin deformación apreciable** *f* CARBÓN creep strength; ~ **máxima unitaria a la tracción** *f* ING MECÁ, MECÁ, METAL *fluencia* ultimate tensile strength; ~ **máxima vertical de la rueda** *f* TRANSP AÉR maximum wheel vertical load; ~ **media** *f* ELEC average load; ~ **en la memoria principal** *f* INFORM&PD roll-in; ~ **motriz** *f* PROD *hidráulica* operating head; ~ **móvil** *f* ING ELÉC moving charge; ~ **muerta** *f* CONST dead load;

~n ~ **negativa** *f* ELEC *electrostática, etc*, FÍS, ING ELÉC negative charge; ~ **neta** *f* ING ELÉC net charge; ~ **no inductiva** *f* ELEC *corriente alterna* noninductive load; ~ **no reactiva** *f* ELEC *corriente alterna* non-reactive load; ~ **nominal** *f* ELEC *equipo* allowable load; ~ **normal prevista** *f* TRANSP AÉR estimated normal payload; ~ **nuclear** *f* FÍS PART, NUCL nuclear charge;

~o ~ **opuesta** *f* ELEC, QUÍMICA opposite charge;

~p ~ **de pago** *f* TEC ESP payload; ~ **de la pala** *f* TRANSP AÉR blade loading; ~ **de la pálet** *f* AmL (*cf carga de la bandeja Esp*, cargade paletas *f* Esp) EMB pallet load; ~ **de la paleta** *f* Esp (*cf carga de la pálet AmL*) EMB pallet load; ~ **de paletas** *f* Esp (*cf carga de pálets AmL*) EMB pallet loading; ~ **de pálets** *f* AmL (*cf carga de bandejas Esp*) EMB pallet loading; ~ **de pandeo** *f* MECÁ buckling load; ~ **parcial** *f* AUTO part load; ~ **parcial de mercancías** *f* TRANSP part-load freight (*AmE*), part-load goods (*BrE*); ~ **pasiva** *f* ING ELÉC passive load; ~ **de pasta en la pila holandesa** *f* PAPEL pulp batch; ~ **permanente** *f* CARBÓN permanent load, CONST dead load; ~ **permisible** *f* ELEC *equipo*, TRANSP AÉR allowable load; ~ **plena** *f* ELEC *de generador*, PROD full load; ~ **de polarización** *f* FÍS polarization charge; ~ **positiva** *f* ELEC *electroestática*, FÍS, ING ELÉC positive charge; ~ **a presión atmosférica** *f* CONTAM *procesos industriales* atmospheric load; ~ **previa** *f* ING MECÁ preload; ~ **principal** *f* CONST main load; ~ **de procesamiento** *f* INFORM&PD processing load; ~ **de profundidad** *f* D&A depth charge; ~ **de programa inicializado** *f* INFORM&PD initial program load (*IPL*); ~ **promedio** *f* ELEC average load; ~ **propulsora** *f* D&A propelling charge; ~ **de proyección** *f* TEC ESP *cañones* igniter; ~ **de prueba** *f* CARBÓN test loading, ING MECÁ proof load, test load, TRANSP AÉR proof load;

~ **de punta** *f* ELEC *electroestática* point charge, *red de distribución* load peak, peak load; ~ **puntual** *f* ELEC *electroestática*, FÍS point charge;

~ r ~ **reactiva** *f* ELEC, ELECTRÓN, FÍS, ING ELÉC, TELECOM reactive load; ~ **real** *f* METAL true stress; ~ **recuperada** *f* ING ELÉC recovered charge; ~ **de refrigerante** *f* REFRIG refrigerant charge; ~ **de refuerzo** *f* P&C reinforcing filler, TRANSP boost charge; ~ **de régimen** *f* METAL nominal stress; ~ **relativa a rueda portante** *f* TRANSP trailing load; ~ **remanente** *f* ELEC *electroestática* residual charge, remanent charge, FÍS remanent charge, ING ELÉC remanent charge, residual charge, PETROL remanent charge, TV residual charge; ~ **remolcada** *f* TRANSP trailing load; ~ **remota** *f* INFORM&PD, TELECOM remote loading; ~ **repartida** *f* ELEC, ING ELÉC partitioned charge; ~ **residual** *f* ELEC *electroestática*, ING ELÉC, TV residual charge; ~ **resistiva** *f* ELEC *máquina*, ING ELÉC, TELECOM resistive load; ~ **retrógrada** *f* TEC ESP retropack; ~ **al romperse** *f* P&C *propiedad física, prueba* load at break; ~ **de rotura** *f* CARBÓN failure load, breaking load, EMB breaking load, ING MECÁ, MECÁ breaking load, breaking stress, PAPEL, TRANSP MAR *de cadena, cuerda* breaking load; ~ **de rotura por tracción** *f* ING MECÁ, METAL ultimate tensile strength; ~ **por rueda** *f* CONST *estructura* wheel load; ~ **de ruptura** *f* CARBÓN, EMB, ING MECÁ, MECÁ, PAPEL, TRANSP MAR *de cuerda, cadena* breaking load;

~ s ~ **de salida** *f* ING ELÉC output charge; ~ **del secuenciador** *f* PROD sequencer load; ~ **segura de trabajo** *f* SEG safe working load; ~ **de seguridad admisible** *f* ING MECÁ safe load; ~ **de servicio** *f* AGUA service load; ~ **del sistema** *f* TELECOM system load; ~ **de un solo paso** *f* NUCL once-through charge; ~ **superficial** *f* ELEC *electroestática*, FÍS surface charge; ~ **superior** *f* ING ELÉC topping charge; ~ **en suspensión** *f* HIDROL load in suspension;

~ t ~ **temporal** *f* CARBÓN temporary load; ~ **térmica** *f* REFRIG cooling load, TERMO heat load, cooling load, thermal head; ~ **térmica debida al producto** *f* REFRIG product load; ~ **térmica debida al servicio** *f* REFRIG service load; ~ **tolerable** *f* ELEC *equipo* allowable load; ~ **total** *f* NUCL total charge; ~ **de trabajo** *f* ING MECÁ working load, TEXTIL work load; ~ **de tracción** *f* GEOL bed load; ~ **del trépano** *f* PETROL bit load; ~ **de tungsteno** *f* TEC PETR *perforación* tungsten carbide;

~ u ~ **por unidad de potencia** *f* METR power loading; ~ **unitaria de rotura** *f* ING MECÁ ultimate stress; ~ **unitaria de rotura a la tracción** *f* MECÁ tensile strength; ~ **unitaria a la tracción** *f* ING MECÁ, MECÁ tensile stress; ~ **unitaria de tracción** *f* METAL tensile stress; ~ **útil** *f* CONTAM MAR payload, EMB, ING MECÁ, MECÁ carrying capacity, TELECOM *comunicación por satélite*, VEH payload; ~ **útil del satélite** *f* TEC ESP useful satellite load;

~ v ~ **de vagón** *f* TRANSP carload (*AmE*), wagonload (*BrE*); ~ **de vagones incompletos** *f* FERRO *vehículos* less-than-carload freight (*AmE*), less-than-carload goods (*BrE*); ~ **de vagones en la jaula** *f* MINAS onsetting; ~ **variable** *f* CONST live load, METAL varying loading; ~ **de vidrio de desecho solo** *f* C&V charging cullet only; ~ **del viento** *f* METEO wind pressure; ~ **de voladura** *f* MINAS blasting-charge

cargadera *f* TRANSP MAR *cabo* downhaul

cargadero: ~ **de dragados** *m* CARBÓN spoil tip; ~ **de**

mineral *m* CARBÓN spoil tip; ~ **de rocas** *m* CARBÓN rock tip

cargado[1] *adj* D&A *granada, munición, proyectil*, ELEC live, INFORM&PD loaded, PROD *proyectil* live, TRANSP MAR *barco de cargamento* loaded, VEH under load; **no** ~ *adj* FÍS RAD uncharged; ~ **de baterías** *adj* TEC ESP battery-charged; ~ **por muelle** *adj* PROD spring-loaded; ~ **de pilas** *adj* TEC ESP battery-charged; ~ **de polvo** *adj* CARBÓN, SEG dust-laden; ~ **por resorte** *adj* PROD spring-loaded

cargado[2] *m* INFORM&PD loading

cargador *m* INFORM&PD loader, ING ELÉC charging, PROD *persona, máquina* loader, *hornos* charger, TRANSP loader, TRANSP MAR shipper; ~ **absoluto** *m* INFORM&PD absolute loader; ~ **de balas** *m* AGRIC bale loader; ~ **de banda** *m* Esp (*cf cargador de cinta AmL*) MINAS *ametralladoras* loading belt; ~ **de baterías** *m* ING ELÉC, TRANSP battery charger; ~ **de cajas** *m* EMB case loader; ~ **de cajas en horizontal** *m* EMB horizontal case loader; ~ **de cinta** *m* AmL (*cf cargador de banda Esp*) MINAS *ametralladoras* loading belt; ~ **por correa** *m* PROD *con mango de madera* belt shipper; ~ **de cuchillas** *m* CARBÓN cutter loader; ~ **de energía** *m* CONST *maquinaria, obras públicas* power loader; ~ **del enlace** *m* INFORM&PD link loader; ~ **de equipaje** *m* TRANSP baggage loader; ~ **de estiércol** *m* AGRIC manure crane; ~ **por etapas** *m* MINAS stage loader; ~ **de fardos** *m* AGRIC bale loader; ~ **general** *m* C&V blanket charger; ~ **de goteo** *m* ELEC *acumulador* trickle charger; ~ **por goteo** *m* ELEC *acumulador* trickle charger; ~ **gradual** *m* MINAS stage loader; ~ **del horno** *m* C&V doghouse (*BrE*), filling end (*AmE*); ~ **incorporado** *m* TRANSP built-in charger; ~ **lento** *m* ING ELÉC trickle charger; ~ **mecánico** *m* INSTAL TERM *horno* mechanical stoker, PROD *máquina* stoker; ~ **de muelle** *m* TRANSP MAR docker (*BrE*), longshoreman (*AmE*); ~ **de pasta** *m* C&V batch charger; ~ **de película a granel** *m* AmL (*cf cargador de película en rollo Esp*) FOTO bulk film loader; ~ **de película en rollo** *m* Esp (*cf cargador de película a granel AmL*) FOTO bulk film loader; ~ **permanente** *m* TRANSP stationary charger; ~ **de las pilas** *m* FOTO battery charger; ~ **recíproco** *m* C&V reciprocating charger; ~ **trasero** *m* CONST back loader; ~ **de vagonetas** *m* MINAS *trabajador* trimmer

cargadora *f* PROD *máquina* loader, TRANSP MAR shipper; ~ **de ataque frontal** *f* CONTAM MAR front-end loader; ~ **de bandejas** *f* Esp (*cf cargadora de pálets AmL*) EMB pallet loader; ~ **de fardos** *f* TRANSP bale loader; ~ **de pacas** *f* TRANSP bale loader; ~ **de paletas** *f* Esp (*cf cargadora de pálets AmL*) EMB pallet loader; ~ **de pálets** *f* AmL (*cf cargadora de bandejas Esp, cf cargadora de paletas Esp*) EMB pallet loader

cargadura *f* TRANSP MAR *para arriar vela* tripping line

cargamento *m* ING MECÁ load, MINAS loading, TEC ESP loading, cargo, TRANSP MAR cargo, shipment; ~ **de la carbonera** *m* MINAS shaft bunker loading; ~ **cuasiestático** *m* TEC ESP quasi-statical loading; ~ **en cubierta** *m* TRANSP MAR deck load; ~ **entero** *m* TRANSP MAR shipload; ~ **a granel** *m* TRANSP MAR bulk; ~ **refrigerado** *m* REFRIG, TRANSP refrigerated cargo; ~ **de vagonetas** *m* AmL (*cf llenado de vagonetas Esp*) MINAS filling

cargar[1] *vt* CINEMAT, CONST, IMPR, INFORM&PD load, MINAS *minerales* fill, *en el piso de carga* cut a plat, PROD, TERMO charge, TRANSP MAR *mercancías* ship, load, *velas* haul down; ~ **por fricción** *vt* Fís charge by friction; ~ **la llamada a** *vt* TELECOM charge the call to; ~ **en la memoria principal** *vt* INFORM&PD roll in **cargar**[2] *vi* TELECOM charge

cargas: ~ **de igual polaridad** *f pl* ELEC like charges; ~ **iguales** *f pl* ELEC like charges; ~ **del mismo signo** *f pl* ELEC like charges; ~ **parciales** *f pl* TRANSP part loads

cargo *m* TRANSP AÉR cargo; ~ **aéreo** *m* TRANSP AÉR air freight; ~ **de conexión inicial** *m* TELECOM initial connection charge

carguero *m* TRANSP MAR cargo ship; ~ **polivalente** *m* TRANSP multipurpose carrier; ~ **tipo Seabee** *m* TRANSP MAR Seabee carrier

caries: ~ **húmeda** *f* CONST wet rot; ~ **seca** *f* TRANSP MAR *en la madera* dry rot

cariocerita *f* NUCL karyocerite

carleta: ~ **plana** *f* ING MECÁ cotter file

carlinga *f* TEC ESP *aeronavegación militar* cockpit, TRANSP MAR *mástil* mast step, step

carmada: ~ **viva** *f* OCEAN live bait

carmesí *m* ALIMENT, QUÍMICA carmine

carmín *m* ALIMENT carmine, COLOR cochineal dye, QUÍMICA carmine

carmínico *adj* QUÍMICA carminic

carnalita *f* MINERAL carnallite

carne: ~ **bovina** *f* AGRIC beef and veal; ~ **en canal** *f* AGRIC, ALIMENT carcass meat; ~ **fósil** *f* MINERAL mountain flesh; ~ **limpia de res** *f* AGRIC dressed beef; ~ **de ternera blanca** *f* AGRIC baby beef; ~ **de vacuno** *f* AGRIC beef

carnero *m* ING MECÁ tool box; ~ **castrado** *m* AGRIC wether; ~ **portaherramientas** *m* ING MECÁ tool ram, tool slide

carnitina *f* QUÍMICA carnitine

carnotita *f* MINERAL carnotite

carona *f* QUÍMICA carone

caroteno *m* ALIMENT, QUÍMICA carotene

carpe *m* PAPEL hornbeam

carpelo *m* AGRIC *pistilo* carpel

carpidor *m* AGRIC shovel cultivator

carpintería *f* CONST carpentry; ~ **de hierro** *f* PROD ironwork; ~ **de hierro pesada** *f* CONST heavy iron-work

carpintero *m* CONST carpenter, TRANSP MAR *a bordo* shipwright; ~ **de ribera** *m* TRANSP MAR shipwright

carraca *f* CONST *carpintería* ratchet brace, ING MECÁ ratchet lever

carragenato *m* ALIMENT carrageen, carrageenan

carragenina *f* ALIMENT carrageen, carrageenan

carraguín *m* CONTAM MAR *alga de utilidad medicinal y nutritiva* carageenan, carrageen

carrera *f* AUTO stroke, C&V *de ladrillos o bloques* course, CONST *de émbolo* stroke, ING ELÉC pretravel, ING MECÁ travel, run, stroke, play, INSTAL HIDRÁUL *de la válvula de corredera* stroke, *de válvula de charnela* lift, PROD *máquina* runout; ~ **de admisión** *f* VEH *motor* intake stroke; ~ **de apertura** *f* PROD *válvulas* opening travel; ~ **del ariete** *f* ING MECÁ stroke of ram; ~ **ascendente** *f* AUTO, ING MECÁ, VEH *motor, pistón* upstroke; ~ **de aspiración** *f* VEH *motor* induction stroke; ~ **de aterrizaje** *f* TRANSP AÉR landing run; ~ **del balancín de perforación** *f* MINAS *sondeos* stroke of the walking beam; ~ **del carnero** *f* ING MECÁ stroke of ram; ~ **completa** *f* PETROL, TEC PETR round trip; ~ **de compresión** *f* AUTO, TERMO, VEH *motores* compression stroke; ~ **de corte** *f* ING MECÁ cutting stroke; ~ **descendente** *f* AUTO downstroke, ING MECÁ downstroke, downward stroke, MECÁ, P&C, VEH *motor* downstroke; ~ **de despegue** *f* TRANSP AÉR takeoff run; ~ **del distribuidor** *f* INSTAL HIDRÁUL valve lift, valve travel; ~ **del émbolo** *f* ING MECÁ piston stroke; ~ **del embrague** *f* AUTO, VEH clutch throwout; ~ **de escape** *f* AUTO, VEH *motor* exhaust stroke; ~ **de expansión** *f* ING MECÁ, TERMO, VEH *motor* expansion stroke; ~ **de extracción** *f* *Esp* (*cf carrera de subida AmL*) MINAS *motor de extracción* hoist; ~ **horizontal** *f* PROD horizontal travel; ~ **del husillo de la taladradora** *f* ING MECÁ stroke of drilling spindle; ~ **de ida** *f* INSTAL HIDRÁUL *del pistón* outstroke; ~ **inactiva** *f* ING MECÁ idle stroke; ~ **inferior** *f* CONST *arquitectura* pole plate; ~ **de la marea** *f* ENERG RENOV, GEOFÍS, HIDROL, TRANSP MAR tidal range; ~ **motriz** *f* AUTO power stroke; ~ **muerta** *f* ING MECÁ idle stroke; ~ **del pistón** *f* AUTO stroke, ING MECÁ length of stroke, piston stroke; ~ **de retorno** *f* ING MECÁ return stroke, INSTAL HIDRÁUL *de cilindros, del pistón* back stroke; ~ **retrógrada** *f* ING MECÁ *motor* backward stroke; ~ **de sacada y bajada** *f* PETROL, TEC PETR round trip; ~ **de subida** *f* *AmL* (*cf carrera de extracción Esp*) MINAS *motor de extracción* hoist; ~ **superior** *f* C&V top course of tank blocks; ~ **de trabajo** *f* AUTO power stroke, VEH *motor* expansion stroke; ~ **útil** *f* ING MECÁ cutting stroke; ~ **en vacío** *f* ING MECÁ idle stroke; ~ **de vacío inactiva** *f* ING MECÁ noncutting stroke; ~ **de la válvula** *f* INSTAL HIDRÁUL valve lift, valve travel; ~ **de vuelta** *f* INSTAL HIDRÁUL *de pistón* instroke, back stroke

carrete[1]: **de ~ a carrete** *adj* CINEMAT reel-to-reel

carrete[2] *m* ACÚST spool, C&V spool, bobbin, CINEMAT bobbin, reel, ELEC coil, FOTO cartridge, INFORM&PD reel, ING ELÉC bobbin, coil, wire bundle, ING MECÁ *tornos* cat head, ratchet, MECÁ ratchet, PETROL reel, TEXTIL bobbin, TV bobbin, spool, reel; ~ **de alimentación** *m* TV supply roll; ~ **alimentador** *m* TV supply reel; ~ **de la bobina** *m* PAPEL reel spool; ~ **para cable** *m* PETROL cable reel; ~ **de la cinta** *m* TV tape roller; ~ **compensador** *m* PROD compensating spool; ~ **de demostración** *m* CINEMAT demo reel, take-off reel; ~ **desenrollador** *m* CINEMAT takeoff reel; ~ **dividido** *m* CINEMAT split reel, split spool; ~ **de enrollado** *m* INFORM&PD take-up reel; ~ **para enrollar cables** *m* CONST cable drum; ~ **de entrega** *m* INFORM&PD supply reel; ~ **extensible** *m* ING ELÉC extension reel; ~ **de extensión telefónica** *m* TELECOM telephone extension reel; ~ **de inducción** *m* ELEC induction coil; ~ **de inercia** *m* ING MECÁ inertia reel; ~ **inferior** *m* CINEMAT lower spool; ~ **para manguera contra incendios** *m* CONST, SEG, TERMO fire-hose reel; ~ **de manguera móvil** *m* SEG *equipo para combate de incendios* mobile hose reel; ~ **de película** *m* CINEMAT, FOTO film spool; ~ **proveedor** *m* CINEMAT supply reel, supply spool; ~ **receptor** *m* CINEMAT take-up reel, TV take-up spool; ~ **de relevador** *m* ELEC, TELECOM relay winding; ~ **de revelado** *m* FOTO developing spool (*AmE*), developing spiral (*BrE*); ~ **de Ruhmkorff** *m* ING ELÉC spark coil; ~ **suelto** *m* ING MECÁ loose reel

carretel *m* OCEAN, PETROL reel, PROD spool; ~ **para cable** *m* PETROL cable reel; ~ **de mangueras** *m* PROD hose reel; ~ **de maniobras** *m* PETROL cathead

carreteo *m* TRANSP AÉR taxiing
carretera *f* AUTO, CONST freeway (*AmE*), motorway (*BrE*), road, roadway (*AmE*), INFORM&PD bus, TRANSP, VEH freeway (*AmE*), motorway (*BrE*), road, roadway (*AmE*); ~ **de acceso** *f* CONST access road (*AmE*), slip road (*BrE*), PROD *minas* gateway, TRANSP access road (*AmE*), slip road (*BrE*); ~ **de acceso limitado** *f* TRANSP arterial highway (*AmE*), dual carriageway (*BrE*); ~ **de acceso válida** *f* TRANSP slip road count (*BrE*); ~ **controlada** *f* TRANSP guided road; ~ **costera de circulación** *f* CONST coastal ring road; ~ **de cruce de doble sentido** *f* CONST double crossover road; ~ **en desmonte** *f* CONST sunken road; ~ **de desvío** *f* TRANSP bypass road; ~ **de doble sentido** *f* CONST divided highway (*AmE*), dual carriageway (*BrE*), two-way road; ~ **de entrada censada** *f* TRANSP access road census (*AmE*), slip road census (*BrE*); ~ **inferior** *f* CONST *arquitectura* checkrail, groundsill; ~ **de montaña** *f* CONST mountain road; ~ **pavimentada** *f* CONST metaled road (*AmE*), metalled road (*BrE*); ~ **de peaje** *f* CONST toll road; ~ **de poco tráfico** *f* CONST low-traffic road; ~ **principal** *f* AUTO arterial road, CONST, TRANSP main road, major road; ~ **provisional** *f* CONST temporary road; ~ **rústica** *f* AUTO, CONST, VEH rough road; ~ **secundaria** *f* CONST minor road, TRANSP secondary road; ~ **de segundo orden** *f* TRANSP secondary road; ~ **troncal** *f* TRANSP arterial highway (*AmE*), divided highway (*AmE*), dual carriageway (*BrE*)
carretilla *f* CARBÓN buggy, CONST barrow, bogie (*BrE*), truck (*AmE*), wheelbarrow, FERRO *vehículos* bogie (*BrE*), truck (*AmE*), ING MECÁ, MECÁ bogie (*BrE*), truck (*AmE*), barrow, dolly, MINAS buggy, PROD *transporte de material* hand truck, dolly, TRANSP skidder, barrow, wheelbarrow, VEH bogie (*BrE*), truck (*AmE*); ~ **de alzamiento** *f* MECÁ lifting truck; ~ **eléctrica** *f* CONST squib; ~ **eléctrica de equipaje** *f* TRANSP electrovan; ~ **elevadora** *f* AGRIC forklift truck, AUTO fork truck, forklift truck, CONST forklift truck, EMB forklift truck, lift truck, ING MECÁ, PROD, VEH forklift truck; ~ **elevadora manual** *f* EMB manual lift truck; ~ **de horquilla elevadora** *f* AGRIC, AUTO, CONST, EMB, ING MECÁ, PROD, VEH forklift truck; ~ **de mano** *f* PROD hand truck; ~ **mecánica** *f* C&V mechanical boy; ~ **con motor** *f* AUTO monomotor bogie (*BrE*), monomotor truck (*AmE*); ~ **del motor** *f* TRANSP AÉR engine trolley; ~ **motorizada para apilar** *f* EMB stacking truck; ~ **de portes** *f* TRANSP short-haul skidder; ~ **de rodillos** *f* C&V dolly; ~ **para sacos** *f* EMB sack barrow; ~ **de servicio** *f* TRANSP AÉR service trolley; ~ **de torno para metales** *f* ING MECÁ turning carrier
carretón *m* CONST cart, bogie (*BrE*), truck (*AmE*), FERRO *vehículos* bogie (*BrE*), truck (*AmE*), ING MECÁ, MECÁ, VEH bogie (*BrE*), truck (*AmE*); ~ **de descarga auto-basculante** *m* TRANSP bogie open self-discharge wagon (*BrE*), truck open self-discharge car (*AmE*); ~ **de motor** *m* FERRO motor bogie (*BrE*), motor truck (*AmE*); ~ **portador** *m* FERRO carrying bogie (*BrE*), carrying truck (*AmE*); ~ **de remolque** *m* FERRO *vehículos* trailer bogie (*BrE*), trailer car (*AmE*); ~ **con techo giratorio** *m* TRANSP bogie wagon with swivelling roof (*BrE*), truck car with swiveling roof (*AmE*)

carril *m* CONST, FERRO, ING ELÉC rail, ING MECÁ plate, MECÁ track, TRANSP lane, TRANSP MAR *de la burda, de yates* track; **contra** ~ *m* ING MECÁ guiderail; ~ **acanalado** *m* FERRO grooved rail; ~ **de aceleración** *m* CONST, TRANSP acceleration lane; ~ **de adelantamiento** *m* CONST overtaking lane; ~ **aislado** *m* FERRO insulated rail; ~ **de apoyo** *m* TRANSP bearing rail; ~ **de autobús equipado con dispositivo de guiado** *m* TRANSP bus-lane equipped with guiding device; ~ **para bicicletas** *m* *Esp* (*cf calzada para ciclistas AmL*) TRANSP cycle path; ~ **bus equipado con dispositivo-guía** *m* TRANSP bus-lane equipped with guiding device; ~ **bus de tránsito rápido** *m* TRANSP busway for rapid transit; ~ **conductor** *m* ELEC *ferrocarril*, FERRO, ING ELÉC conductor rail; ~ **continuo soldado** *m* FERRO continuous welded rail (*CWR*); ~ **de desaceleración** *m* CONST, TRANSP deceleration lane; ~ **descamado** *m* FERRO flaked rail; ~ **de dirección** *m* TRANSP guiderail; ~ **doble** *m* ELECTRÓN *en circuitos lógicos* double rail; ~ **de doble cabeza** *m* FERRO double-headed rail; ~ **de doble seta** *m* FERRO bullheaded rail; ~ **de escape** *m* TRANSP escape lane; ~ **festoneado** *m* FERRO *equipo inamovible* scalloped rail; ~ **de grúa** *m* CONST crane rail; ~ **de guía** *m* ING MECÁ guiderail; ~ **guía** *m* TRANSP guiderail; ~ **guiador** *m* ING MECÁ guiderail; ~ **de montaje** *m* PROD mounting rail; ~ **paramagnético** *m* TRANSP paramagnetic rail; ~ **al ras del pavimento** *m* FERRO *equipo inamovible* sunken rail; ~ **de reacción** *m* FERRO *equipo inamovible* reaction rail; ~ **de salida** *m* TRANSP escape lane; ~ **soldado** *m* FERRO *equipo inamovible* ribbon rail; ~ **de tránsito rápido para autobús** *m* TRANSP busway for rapid transit; ~ **de translación** *m* FERRO, ING MECÁ running rail; ~ **para tranvía** *m* CONST tramtrack; ~ **de unión** *m* FERRO *vía* closure rail; ~ **de la vía** *m* FERRO *vía*, ING MECÁ running rail
carrilera *f* ING MECÁ runway; ~ **de rodamiento** *f* ING MECÁ crawler track
carrillo *m* LAB *mobiliario* trolley
carrito: ~ **eléctrico** *m* ELEC, TRANSP electric trolley; ~ **porta-equipajes** *m* TRANSP luggage trolley
carro *m* CONST cart, INFORM&PD carriage, ING MECÁ truck (*AmE*), traveler (*AmE*), traveller (*BrE*), *máquina herramienta* carriage, MECÁ *máquina herramienta* slide, carriage, NUCL trolley, PROD carrier, *elevación* crab, jenny, *puente grúa* traveler (*AmE*), traveller (*BrE*), TRANSP car, wagon; ~ **bomba contra incendios** *m* *AmL* SEG fire-engine; ~ **para camillas** *m* SEG stretcher cart; ~ **de combate** *m* D&A tank; ~ **del contrapeso** *m* MINAS *planos inclinados* counterbalance carriage; ~ **corredizo** *m* MECÁ carriage; ~ **de corte** *m* ING MECÁ cutting-off slide; ~ **para crisoles** *m* C&V pot carriage; ~ **cuello de ganso** *m* C&V goose-necked pot carriage; ~ **de estufa de machos** *m* PROD *fundición* core carriage; ~ **final** *m* PAPEL end deckle; ~ **de guiado de flaps** *m* TRANSP AÉR flap roller carriage; ~ **izador** *m* TRANSP AÉR hoisting carriage; ~ **perforador** *m* *Esp* (*cf perforadora montada sobre carillo AmL*) MINAS drill rig; ~ **de perforadoras múltiples** *m* *AmL* (*cf vagón perforador Esp*) MINAS *para hacer barrenos* jumbo including boom; ~ **pesado** *m* D&A heavy tank; ~ **portaherramienta** *m* ING MECÁ ram, saddle, MECÁ *máquina herramienta* ram; ~ **portaherramientas** *m* ING MECÁ tool box, tool

carriage, tool slide, tool-holding slide; ~ **portamuela** *m* PROD wheel slide; ~ **portamuelas** *m* ING MECÁ wheel carriage; ~ **de precisión** *m* ING MECÁ precision slide; ~ **de rodadura** *m* ING MECÁ block carriage, PROD *de puente-grúa, teleférico* monkey carriage, traveling runner (*AmE*), traveling trolley (*AmE*), travelling runner (*BrE*), travelling trolley (*BrE*), trolley; ~ **sobre ruedas** *m* D&A wheeled carriage; ~ **superior** *m* ING MECÁ cross slide; ~ **tiendepuentes** *m* D&A bridge-layer tank, bridging tank; ~ **del torno** *m* MECÁ lathe slide; ~ **de la torreta** *m* ING MECÁ turret slide; ~ **transversal** *m* ING MECÁ cross slide, MECÁ runner, PROD *máquina herramienta* cross traverse

carrocería *f* INSTR *automóvil* body, VEH body, body shell; ~ **desmontable** *f* VEH unit construction body; ~ **normalizada** *f* VEH unitized body; ~ **tratada con parafina** *f* VEH body in white

carroza *f* TRANSP coach, TRANSP MAR *falúas* canopy

carruaje *m* TRANSP *ferrocarril* coach

carrucha *f* *Esp* (*cf garrucha AmL*) MINAS pulley, *martillo pilón* drive block

carstenita *f* MINERAL karstenite

carta *f* INFORM&PD letter, TEXTIL *de un tejido* body; ~ **de ajuste** *f* TV test pattern; ~ **de ajuste de barras** *f* *Esp* (*cf patrón de ajuste de barras AmL*) TV bar pattern; ~ **del Almirantazgo** *f* TRANSP MAR Admiralty chart (*BrE*); ~ **de aproximación** *f* TRANSP AÉR approach chart; ~ **de aproximación por instrumentos** *f* TRANSP AÉR instrument approach chart; ~ **de arrumbamiento** *f* TRANSP MAR *navegación* plotting sheet, routing chart; ~ **de aterrizaje** *f* TRANSP AÉR landing chart; ~ **barimétrica** *f* TRANSP MAR *navegación* bathymetric chart; ~ **celeste** *f* TEC ESP, TRANSP MAR star chart; ~ **del contorno de aguas subterráneas** *f* AGRIC, AGUA, CARBÓN, HIDROL ground-water contour map; ~ **de control** *f* CALIDAD inspection card; ~ **de corrientes de marea** *f* TRANSP MAR *navegación* tide chart; ~ **costera** *f* TRANSP MAR coastal chart; ~ **cromática** *f* CINEMAT, IMPR, TV color chart (*AmE*), colour chart (*BrE*); ~ **cromática y escala de grises** *f* CINEMAT, IMPR, TV color chart and gray scale (*AmE*), colour chart and grey scale (*BrE*); ~ **de derrotas** *f* TRANSP MAR *navegación* track chart; ~ **de derrotas óptimas** *f* TRANSP MAR *navegación* oceanic routing chart; ~ **de desembarco** *f* TRANSP MAR *navegación* landing chart; ~ **de doblaje** *f* CINEMAT dubbing chart; ~ **electrónica** *f* TRANSP MAR *navegación* electronic chart; ~ **gnomónica** *f* TRANSP MAR *navegación* great circle chart; ~ **de grises estándar** *f* CINEMAT standard gray card (*AmE*), standard grey card (*BrE*); ~ **hidrogeológica** *f* HIDROL hydrogeological map; ~ **hidrográfica** *f* HIDROL *marina*, TRANSP MAR *navegación* hydrographic chart; ~ **de inspección** *f* CALIDAD inspection card; ~ **local** *f* TRANSP AÉR sectional chart; ~ **de marear** *f* TRANSP MAR *mapa* chart; ~ **Mercator** *f* TRANSP MAR Mercator chart; ~ **mercatoriana** *f* TRANSP MAR *navegación* Mercator chart; ~ **meteorológica** *f* TRANSP MAR weather map; ~ **náutica** *f* TRANSP MAR *mapa* pilot chart; ~ **náutica del Almirantazgo** *f* TRANSP MAR Admiralty chart (*BrE*); ~ **de navegación aeronáutica** *f* TRANSP AÉR aeronautical chart; ~ **de nitidez** *f* CINEMAT definition chart; ~ **nomónica** *f* TRANSP MAR *navegación* great circle chart; ~ **de porte** *f* TRANSP MAR waybill; ~ **de porte aéreo** *f* IMPR airway bill; ~ **de prueba** *f* CINEMAT test chart; ~ **de prueba de barras de color** *f* TV color-bar test pattern (*AmE*), colour bar test pattern (*BrE*); ~ **de punteo de la derrota** *f* TRANSP MAR *navegación* plotting chart; ~ **en punto mayor** *f* TRANSP MAR large-scale chart; ~ **de resolución** *f* CINEMAT resolution chart; ~ **de rollo** *f* ELEC *registro* chart strip; ~ **de ruta aeronáutica** *f* TRANSP AÉR aeronautical route chart; ~ **sinóptica** *f* METEO weather chart; ~ **de Smith** *f* FÍS Smith chart; ~ **del tiempo** *f* METEO weather chart; ~ **de los vientos** *f* TRANSP MAR *navegación* wind chart

cartabón *m* GEOM set square, ING MECÁ rule, set square, square, INSTR bevel, rule, METR bevel, MINAS quadrant, TRANSP MAR *construcción* gusset, heel plate; ~ **de arriostramiento** *m* TRANSP MAR *contra el pandeo* tripping bracket; ~ **de bao** *m* TRANSP MAR *construcción* beam bracket; ~ **de unión** *m* TRANSP MAR *construcción* gusset

cártamo *m* AGRIC safflower

cartear *vt* TRANSP MAR *navegación* plot

cartel *m* ING MECÁ placard

cartela *f* TRANSP MAR *construcción* heel plate

cartelas *f pl* PROD *cadena de rodillos, ruedas de esmeril* side plates

carteo *m* PAPEL *del papel* rattle

cárter *m* CARBÓN *autos* pan, ING MECÁ housing, pan, VEH *motor, transmisión* casing; ~ **de aceite** *m* AUTO oil pan, TRANSP AÉR oil sump; ~ **de la cadena** *m* SEG chain case, VEH *motocicletas, transmisión* chaincase; ~ **de la caja de cambio de velocidades** *m* ING MECÁ gearbox guard; ~ **de la caja de cambios** *m* AUTO, VEH gearbox housing; ~ **del diferencial** *m* AUTO differential case; ~ **de la dirección** *m* VEH steering gearbox; ~ **dividido** *m* VEH *puente trasero* split housing; ~ **del eje** *m* VEH axle casing, axle housing; ~ **del eje trasero** *m* AUTO, MECÁ, VEH rear-axle housing; ~ **de embrague** *m* AUTO, VEH clutch casing; ~ **del embrague** *m* AUTO, VEH bell housing; ~ **de engranajes** *m* ING MECÁ gear case, gear casing, gear cover, wheel guard; ~ **de entrada** *m* AUTO input shell; ~ **inferior** *m* VEH sump, *carrocería* floor pan; ~ **inferior del cigüeñal** *m* AUTO, MECÁ, VEH crankcase; ~ **de lubricante** *m* TRANSP AÉR oil sump; ~ **del motor** *m* AUTO, MECÁ, VEH *motor* crankcase; ~ **motor con ventilación directa** *m* VEH *motor* positive crankcase ventilation; ~ **principal de la cadena** *m* VEH *transmisión* primary chaincase; ~ **en seco** *m* VEH *motor* dry sump; ~ **superior del cigüeñal** *m* AUTO, MECÁ, VEH crankcase top half; ~ **tipo partido** *m* VEH *puente trasero* banjo-type housing; ~ **del volante** *m* VEH *motor* flywheel housing

cartera *f* TEXTIL *del bolsillo* flap

cartilla: ~ **de trazado** *f* TRANSP MAR *arquitectura naval* offsets

cartografiar *vt* HIDROL, TRANSP MAR *hidrografía, navegación* chart

cartógrafo *m* CONST mapper

cartón *m* EMB *envase conteniendo varias cajetillas de cigarrillos* carton, PAPEL board; ~ **aislante** *m* EMB insulating board; ~ **alisado** *m* PAPEL machine-finished paperboard (*MF paperboard*); ~ **alisado en húmedo** *m* PAPEL water-finished board; ~ **de amianto** *m* CONST asbestos millboard, PAPEL asbestos board; ~ **de asbesto** *m* CONST asbestos millboard; ~ **para cajas plegables** *m* PAPEL folding boxboard;

~ **calandrado** *m* PAPEL calendered paper board; ~ **para calzados** *m* PAPEL shoe board; ~ **de una capa** *m* PAPEL single-ply board; ~ **con la capa superior blanca** *m* EMB white lined board; ~ **coloreado por ambas caras** *m* PAPEL two-side colored board (*AmE*), two-side coloured board (*BrE*); ~ **compacto** *m* PAPEL carton compact; ~ **compacto para cajas** *m* PAPEL pressboard; ~ **conteniendo pasta mecánica** *m* PAPEL mechanical woodpulp board; ~ **conteniendo pulpa mecánica** *m* PAPEL mechanical-pulp board; ~ **contracolado** *m* PAPEL carton compact, pasted board; ~ **corrugado** *m* ALIMENT *embalaje* corrugated cardboard, EMB, PAPEL corrugated board; ~ **dieléctrico** *m* PAPEL electrical insulating board; ~ **de dos capas** *m* PAPEL two-layer board; ~ **dúplex** *m* EMB duplex board; ~ **dúplex resistente al agua** *m* EMB duplex waterproof board; ~ **duro** *m* PAPEL glazed millboard; ~ **para embutir** *m* PAPEL board for pressing; ~ **para encuadernación** *m* PAPEL bookbinding board; ~ **entretelado de dos caras** *m* PAPEL cloth-centered board (*AmE*), cloth-centred board (*BrE*); ~ **estucado para cajas plegables** *m* PAPEL coated folding board; ~ **fieltro** *m* PAPEL felt board; ~ **para flanes de esterotipia** *m* PAPEL flong board; ~ **forrado** *m* PAPEL pasted lined board; ~ **gris** *m* C&V gray board (*AmE*), grey board (*BrE*), EMB chipboard, millboard, PAPEL *papeles de recuperación* grey board (*BrE*), brown mixed-pulp board, millboard, gray board (*AmE*), *en máquina continua* chipboard, PROD grey board (*BrE*), gray board (*AmE*); ~ **gris forrado** *m* PAPEL lined chipboard; ~ **en hojas** *m* PAPEL paper board in the flat; ~ **homogéneo** *m* PAPEL solid board; ~ **jacquard** *m* PAPEL jacquard board; ~ **jaspeado** *m* PAPEL veined board; ~ **para maletas** *m* PAPEL suitcase board; ~ **microondulado** *m* EMB, PAPEL corrugated-board with narrowly spaced flutes; ~ **moldeado** *m* EMB, PAPEL molded board (*AmE*), moulded board (*BrE*); ~ **multicapa** *m* PAPEL combination board; ~ **multicapas** *m* PAPEL multilayer board; ~ **ondulado** *m* EMB corrugated board, corrugated fiberboard (*AmE*), corrugated fibreboard (*BrE*), PAPEL corrugated board; ~ **ondulado de doble cara** *m* EMB double face corrugated board; ~ **ondulado para envasado** *m* EMB pads; ~ **ondulado de onda ancha** *m* EMB, PAPEL corrugated-board with broadly spaced flutes; ~ **ondulado para trabajos pesados** *m* EMB *para soportar grandes pesos* heavy-duty corrugated fiberboard (*AmE*), heavy-duty corrugated fibreboard (*BrE*); ~ **de paja mixto** *m* PAPEL mixed strawboard; ~ **parafinado** *m* PAPEL waxed board; ~ **pardo** *m* PAPEL brown mixed-pulp board, *en máquina intermitente* brown millboard; ~ **piedra** *m* PAPEL tarred felt; ~ **prensado** *m* PAPEL pressboard, presspahn-transformer board; ~ **de pulpa mecánica sin blanquear** *m* PAPEL unbleached mechanical pulp board; ~ **de pulpa mecánica parda** *m* PAPEL brown mechanical pulp board; ~ **reforzado** *m* PAPEL reinforced board; ~ **resistente a las grasas** *m* EMB, PAPEL grease-resistant board; ~ **satinado en una cara** *m* PAPEL machine-glazed board (*MG board*); ~ **semikraft** *m* PAPEL halfkraft board; ~ **tipo test liner** *m* EMB test liner board

cartonaje: ~ **para películas** *m* EMB film cartoning; ~ **vertical** *m* EMB vertical cartoner

cartoncillo: ~ **para cajas plegables** *m* PAPEL folding boxboard; ~ **con la cara superior blanca** *m* PAPEL bleached lined

cartucho *m* D&A *explosivo*, INFORM&PD cartridge, MINAS *explosivo* cartridge, *pega* charge; ~ **sin bala** *m* D&A blank cartridge; ~ **de bucle magnético sin fin** *m* ACÚST endless magnetic-loop cartridge; ~ **sin carga** *m* AmL (*cf cartucho falso Esp*) MINAS dummy cartridge; ~**-cebo** *m* MINAS *explosivos* primer cartridge; ~ **de cinta** *m* INFORM&PD tape cartridge; ~ **de cinta magnética** *m* INFORM&PD magnetic tape cartridge; ~ **coplanar** *m* ACÚST coplanar cartridge; ~ **de datos** *m* INFORM&PD data cartridge; ~ **de dinamita** *m* MINAS *explosivo* dynamite cartridge; ~ **de disco** *m* INFORM&PD disk cartridge; ~ **de extracción** *m* LAB *aparato Soxhlet* extraction thimble; ~ **falso** *m* Esp (*cf cartucho sin carga AmL*) MINAS dummy cartridge; ~ **filtrante** *m* ING MECÁ cartridge filter, MECÁ, PROC QUÍ filter cartridge, VEH cartridge filter, filter cartridge; ~ **del filtro** *m* MECÁ, PROC QUÍ, VEH *aceite* filter cartridge; ~ **del filtro de aire impregnado en aceite** *m* AUTO oil-moistened air filter cartridge; ~ **de fogueo** *m* D&A blank cartridge, MINAS *explosivos* blank cartridge, dummy cartridge; ~ **para inserto rotatorio** *m* ING MECÁ cartridge for indexable inserts; ~ **de instrucción** *m* AmL (*cf cartucho de prueba Esp*) MINAS dummy cartridge; ~ **de memoria** *m* PROD memory cartridge; ~ **de prueba** *m* Esp (*cf cartucho de instrucción AmL*) MINAS dummy cartridge; ~ **de uranio natural** *m* NUCL natural-uranium slug

cartulina *f* EMB, PAPEL cardboard; ~ **bristol** *f* PAPEL ivory board; ~ **con la cara superior blanca** *f* PAPEL bleached lined; ~ **marfil** *f* PAPEL pasted-ivory board; ~ **neutra de prueba** *f* CINEMAT neutral test card; ~ **ondulada** *f* EMB, PAPEL corrugated cardboard

casa *f* CONST cabin; ~ **de bombas** *f* AGUA pump house, CONST pumping-station; ~ **desmontable** *f* CONST frame house; ~ **de esclusas** *f* AGUA lock house; ~ **de máquinas** *f* PROD engine house; ~ **de mezclas** *f* C&V mixing room; ~ **del perforador** *f* TEC PETR *perforación* doghouse; ~ **de perro** *f* TEC PETR *perforación* doghouse

cascada *f* CARBÓN *de la voz* cascade, crack, CINEMAT, ELEC, ELECTRÓN, FÍS, FÍS RAD cascade, HIDROL cascade, waterfall, INFORM&PD, ING ELÉC, NUCL cascade; ~ **electrónica** *f* ELECTRÓN, FÍS PART, NUCL electron cascade; ~ **de energía** *f* FÍS FLUID energy cascade; ~ **de enfriamiento por agua** *f* ING MECÁ water cooling cascade; ~ **genérica** *f* NUCL generic cascade; ~ **integrada** *f* REFRIG mixed refrigerant cascade; ~ **sin mezcla** *f* NUCL no-mixing cascade; ~ **radiativa** *f* FÍS RAD radiative cascade; ~ **de refrigeración por agua** *f* ING MECÁ water cooling cascade

cascajal *m* AGUA gravel pit

cáscara[1]: **de ~ dura** *adj* AGRIC hard-coated

cáscara[2] *f* AGRIC chaff, CARBÓN skin; ~ **del grano de arroz** *f* AGRIC rice hull; ~ **de naranja** *f* C&V, P&C orange peel

cascarilla: ~ **de laminación** *f* PROD mill scale, roll scale; ~ **de óxido** *f* PROD scale

cascarón *m* C&V shell

casco[1]: **de un solo ~** *adj* TRANSP MAR *embarcación* single-hull

casco[2] *m* AGUA hull, ING MECÁ hood, SEG *protector*, TEC PETR helmet, TRANSP AÉR hull, headgear, TRANSP MAR hull; ~ **aerodinámico** *m* TRANSP MAR aerofoil hull (*BrE*), airfoil hull (*AmE*); ~ **con auriculares** *m* INSTR, TELECOM headset; ~ **de bombero** *m* SEG fireman's helmet; ~ **contra humos** *m* SEG smoke helmet; ~ **desnudo** *m* TRANSP AÉR bare hull; ~ **esférico de pared delgada** *m* ING MECÁ thin-walled spherical shell; ~ **de frenado** *m* TRANSP AÉR drag cup; ~ **de grasa consistente** *m* ING MECÁ set grease cup; ~ **industrial de seguridad** *m* SEG industrial safety helmet; ~ **protector** *m* PETROL protection helmet, SEG protective helmet; ~ **protector para soldadura** *m* CONST welding helmet; ~ **del proyectil** *m* D&A shell splinter; ~ **rígido de buceo** *m* OCEAN diving helmet; ~ **de seguridad** *m* SEG, TRANSP safety helmet; ~ **de seguridad para soldador** *m* SEG welder's safety helmet; ~ **de soldador** *m* SEG welder's hood; ~ **ventilado para tripulación de carros** *m* D&A ventilated tank crew helmet

cascos *m pl* TV head wear

cascotes: de ~ *adj* CONST *piedra* rubbly

caseína *f* ALIMENT, P&C, QUÍMICA casein; ~ **de cuajo** *f* AGRIC, ALIMENT rennet casein

caseinato: ~ **sódico** *m* ALIMENT sodium caseinate

caseta *f* CONST house, cabin, TEC PETR *perforación* doghouse; ~ **de acumuladores** *f* ELEC battery housing; ~ **de bocamina** *f* MINAS head house; ~ **de derrota** *f* TRANSP MAR *construcción* chartroom; ~ **del filtro** *f* MINAS *conducción de aguas* head house; ~ **de filtros** *f* NUCL filter house; ~ **de gobierno** *f* TRANSP MAR *buques* wheel house; ~ **de herramientas** *f* CONST tool shed; ~ **protectora contra ruidos** *f* SEG noise-protection booth; ~ **a prueba de sonidos** *f* SEG soundproof booth; ~ **de tostación** *f* MINAS *hornos* stall

casette *f* CINEMAT, ÓPT, TV cassette; ~ **de cinta continua** *f* TV tape loop cassette; ~ **digital** *f* INFORM&PD digital cassette; ~ **del disco óptico** *f* INFORM&PD, ÓPT optical disk cassette; ~ **grabadora** *f* INSTR cassette recorder; ~ **miniatura de película** *f* INSTR miniature film cassette; ~ **de película** *f* INSTR film cassette; ~ **para pielograma** *f* INSTR pyelogram cassette; ~ **de rayos X** *f* INSTR *película* X-ray cassette

casi: ~ **colisión** *f* TRANSP AÉR near collision, near miss

casilla: ~ **angular** *f* IMPR angle box; ~ **electrónica** *f* INFORM&PD e-mail address; ~ **de organigrama** *f* INFORM&PD flowchart block

casita *f* CONST *arquitectura* maisonette

casiterita *f* MINERAL cassiterite, tinstone, wood tin

caso[1]: ~ **de emergencia** *m* SEG emergency case

caso[2]: **en** ~ **de descompostura** *fra* SEG in the event of breakdown; **en** ~ **de incendio** *fra* SEG in case of fire

casquete *m* ING MECÁ, INSTR, MECÁ, PETROL cap; ~ **aislador** *m* ELEC insulator cap; ~ **aislante** *m* ELEC insulator cap; ~ **de burbujeo** *m* ALIMENT, TEC PETR bubble cap; ~ **del cojinete** *m* AUTO, ING MECÁ bearing cap; ~ **cubretornillos** *m* ING MECÁ screw casing head; ~ **de gas** *m* TEC PETR gas cap; ~ **glaciar** *m* OCEAN ice cap; ~ **inferior** *m* NUCL *de una vasija a presión* dished bottom, hemiellipsoidal bottom; ~ **de soldador** *m* CONST welding helmet; ~ **superior** *m* NUCL *de una vasija a presión* hemiellipsoidal head

casquillo *m* C&V bushing, CONST socket, ING ELÉC base, cap, holder, ING MECÁ bush, ferrule, gland, collar, lug, *tuberías* bushing, MECÁ bushing, collar, ferrule, gland, lug, ÓPT ferrule, PROD *de prensaestopas* follower, lug, gland, follower bush, thimble, TEC ESP, TEC PETR gland, TELECOM ferrule, VEH bush, *transmi-*

sión axle bush; ~ **adaptador** *m* PROD socket; ~ **de aislamiento** *m* ING MECÁ insulation cap; ~ **aislante** *m* CONST cap; ~ **de ajuste forzado** *m* ING MECÁ forced-fit bush; ~ **de aleación de cobre** *m* ING MECÁ copper alloy bush; ~ **de alimentación** *m* ING MECÁ feed bush; ~ **antifricción** *m* VEH *transmisión* axle bushing; ~ **anular** *m* C&V annular bushing; ~ **del árbol de levas** *m* AUTO, VEH camshaft bushing; ~ **y barra aislantes** *m* ING ELÉC cap and rod insulator; ~ **de bayoneta** *m* AmL (*cf montura de bayoneta Esp*) ELEC *ampolla, bombilla* bayonet cap, FOTO bayonet base, bayonet socket, ING ELÉC bayonet base; ~ **de bronce de cañón** *m* ING MECÁ gunmetal bush; ~ **del bulón del pistón** *m* AUTO, VEH piston-pin bushing; ~ **del canal de colada** *m* ING MECÁ sprue bush; ~ **de cerrojo** *m* CONST lock bush; ~ **y clavija aislantes** *m* ING ELÉC cap and pin insulator; ~ **de cojinete** *m* ELEC *máquina* brass; ~ **del cojinete** *m* AUTO, ING MECÁ bearing cap; ~ **del cojinete del cigüeñal** *m* AUTO, VEH crankshaft-bearing cap; ~ **del cojinete principal** *m* AUTO main-bearing bushing; ~ **de la columna de apoyo** *m* ING MECÁ support pillar bush; ~ **conductor para taladrar** *m* ING MECÁ jig bush; ~ **del eje** *m* AUTO axle cap; ~ **de enchufe** *m* ING ELÉC jack bush; ~ **-enchufe de lámpara** *m* ING ELÉC lamp holder; ~ **excéntrico** *m* ING MECÁ eccentric bush; ~ **de extracción** *m* ING MECÁ stripper bush; ~ **del extremo de la pala** *m* TRANSP AÉR blade-tip cap; ~ **del fusible** *m* ING ELÉC fuse base; ~ **guardapolvos** *m* ING MECÁ dust cap; ~ **guía** *m* ING MECÁ guide bush, jig bush; ~ **de guía de bronce** *m* ING MECÁ bronze guide bush; ~ **guía con cabeza** *m* ING MECÁ headed guide bush; ~ **guía sin cabeza** *m* ING MECÁ headless guide bush; ~ **guía encabezado** *m* ING MECÁ headed guide bush; ~ **guía para taladrar con plantilla** *m* ING MECÁ, MECÁ drill bushing; ~ **de inyección de colada** *m* ING MECÁ sprue bush; ~ **de lámpara** *m* ING ELÉC lamp cap; ~ **para lámparas** *m* FOTO lamp socket; ~ **con ocho pistones** *m* ING MECÁ octal base; ~ **octal** *m* ING MECÁ octal base; ~ **de ondulación** *m* ING MECÁ crimping bush; ~ **de pie de biela** *m* VEH small end bush; ~ **del pie de biela** *m* AUTO, VEH piston-pin bushing; ~ **piloto** *m* AUTO pilot bushing; ~ **de plato** *m* ING MECÁ chuck bushing; ~ **de pliegue** *m* ING MECÁ crimping bush; ~ **de presión** *m* AUTO pressure cap; ~ **primario** *m* AUTO primary cup; ~ **a rosca** *m* ING ELÉC screw base; ~ **roscado** *m* ING MECÁ screw box, screw ferrule, threaded bush; ~ **secundario** *m* AUTO secondary cup; ~ **-soporte de la lámpara** *m* ING ELÉC lamp holder; ~ **del trípode** *m* FOTO tripod bush; ~ **de ventilación impermeable** *m* AUTO splash-proof vent cap

castigado: ~ **por el tiempo** *adj* TRANSP MAR *buque* weather-beaten

castillete *m* ELEC *red de distribución*, ING ELÉC pylon, ING MECÁ housing, MINAS winding tower, PETROL *perforación de sondeos* derrick, PROD *tren de laminación* holster; ~ **de celosía** *m* CONST, ELEC *línea de alimentación aérea* lattice tower; ~ **de extracción** *m* MINAS gallows frame, head frame, headgear, headstock, headwork, pithead frame, PROD *minas* poppet; ~ **de transmisión** *m* ELEC *red de distribución* transmission tower

castillo: ~ **de proa** *m* TRANSP MAR *construcción* forecastle

castorina *f* QUÍMICA castorin

castración *f* AGRIC emasculation

CAT *abr* INFORM&PD (*prueba asistida por ordenador Esp, prueba auxiliada por computadora AmL*) CAT (*computer-aided testing*), IMPR (fotocomposición asistida por computadora *AmL*, fotocomposición asistida por ordenador *Esp*) CAT (*computer assisted typesetting*)

cata *f Esp* (*cf cateo AmL*) ALIMENT *vinícola, cervecería* tasting, MINAS *investigación* cutting, prospecting

cataclasis *f* GEOL cataclasis

cataclasita *f* GEOL cataclasite

cataclismo *m* GEOL cataclysm

catadióptrico[1] *adj* FÍS catadioptric

catadióptrico[2] *m* ÓPT, VEH reflex reflector

catadrióptico *m* FOTO mirror lens

catadura *f* AGUA sampling

catafaro *m* ÓPT, VEH reflex reflector

cataforesis *f* QUÍMICA *industria automotriz* cataphoresis

catafrente *m* METEO katabatic front

catalasa *f* ALIMENT, QUÍMICA *enzima* catalase

catalejo *m* TRANSP MAR telescope

catálisis *f* ALIMENT catalysis; **~ por radiación** *f* FÍS RAD radiation catalysis

catalítico *adj* QUÍMICA catalytic

catalizador *m* GEN *sistema de escape* catalyst; **~ de Ziegler** *m* QUÍMICA Ziegler catalyst

catalizar *vt* P&C *reacción química*, QUÍMICA catalyze

catalobara *f* METEO isallobar

catalogar *vt* CARBÓN table, MECÁ index

catálogo *m* D&A catalogue, INFORM&PD catalog (*AmE*), catalogue (*BrE*)

catamarán *m* TRANSP MAR catamaran

catapleíta *f* MINERAL catapleite

catapulta *f* D&A, TRANSP AÉR *portaaviones* catapult; **~ de lanzamiento** *f* TEC ESP *dispositivo de disparo para cohetes y misiles* booster

catarata *f* C&V *en operador* cataract

catártico *adj* QUÍMICA cathartic

catastro *m* AGRIC land register

catástrofe: **~ ultravioleta** *f* FÍS ultraviolet catastrophe

catatermómetro *m* TERMOTEC *corrientes de aire* katathermometer

catavientos *m* TRANSP MAR wind telltale

catear *vt* MINAS costean

catecol *m* QUÍMICA catechol

catecolamina *f* QUÍMICA catecholamine

cátedra *f* ING MECÁ chair

categoría *f* INFORM&PD status, ING MECÁ class; **~ del aeroplano** *f* TRANSP AÉR aircraft category; **~ del avión** *f* TRANSP AÉR aircraft category; **~ del motor** *f* TRANSP AÉR engine rating; **~ de vuelo** *f* TRANSP AÉR flight status

catenación *f* QUÍMICA *formación de cadena* catenation

catenaria *f* FERRO *equipo inamovible*, FÍS, GEOM, MECÁ, TRANSP MAR catenary

catenoide *f* GEOM catenoid

cateo *AmL m* (*cf cata Esp*) ALIMENT tasting, MINAS *investigación* prospection, cutting, prospecting

catepsina *f* QUÍMICA cathepsin

catetómetro *m* FÍS, INSTR, LAB cathetometer

catión *m* ELEC, ELECTRÓN *partícula cargada*, FÍS, FÍS PART, FÍS RAD, QUÍMICA cation, positive ion; **~ base** *m* CONTAM base cation; **~ predominante** *m* CONTAM dominant cation

catiónico *adj* CARBÓN, FÍS, FÍS PART, HIDROL, QUÍMICA cationic

cátodo *m* GEN cathode; **~ caldeado directamente** *m* ING ELÉC directly-heated cathode; **~ caldeado indirectamente** *m* ING ELÉC indirectly-heated cathode; **~ de calentamiento indirecto** *m* ING ELÉC indirect heater-type cathode; **~ de calentamiento rápido** *m* TERMO, TV rapid heat-up cathode; **~ de célula fotoeléctrica** *m* ING ELÉC photocathode; **~ de cesio** *m* ING ELÉC caesium cathode (*BrE*), cesium cathode (*AmE*); **~ de charco de mercurio** *m* ING ELÉC mercury pool cathode; **~ común** *m* ING ELÉC common cathode; **~ de descarga luminiscente** *m* ELEC, ELECTRÓN, FÍS, ING ELÉC glow-discharge cathode; **~ emisor** *m* ELEC, ELECTRÓN emitter, ING ELÉC dispenser cathode, emitter, hot cathode; **~ frío** *m* ELECTRÓN, ING ELÉC cold cathode; **~ impregnado** *m* ING ELÉC impregnated cathode; **~ líquido** *m* ING ELÉC *electrotecnia* pool cathode; **~ recubierto con óxido** *m* ING ELÉC oxide-coated cathode; **~ revestido de óxido** *m* ING ELÉC oxide-coated cathode; **~ secundario** *m* ELECTRÓN, INSTR electron mirror; **~ suministrador** *m* ING ELÉC dispenser cathode; **~ termiónico** *m* ING ELÉC thermionic cathode

catodoluminiscencia *f* FÍS, ING ELÉC cathodoluminescence

catraca *f* ING MECÁ ratchet, ratchet brace, MECÁ ratchet

cauce *m* AGUA channel bed; **~ de corriente** *m* AGUA streamway; **~ de una corriente de agua** *m* AGUA channelway; **~ del río** *m* AGUA, HIDROL river bed; **~ salino** *m* HIDROL saltbed

caucho *m* IMPR india rubber, rubber, P&C *materia prima, elastómero* india rubber, natural rubber, oilcloth, rubber, QUÍMICA oilcloth, rubber, TEXTIL oilcloth; **~ acrílico** *m* EMB acrylic rubber; **~ de acrilonitrilo** *m* MECÁ, P&C *elastómero* acrylonitrile rubber; **~ antiestático** *m* P&C antistatic rubber; **~ bruto** *m* P&C *materia prima* crude rubber; **~ de butadieno** *m* P&C *elastómero* butadiene rubber; **~ de butadieno-acrilonitrilo** *m* P&C *elastómero* butadiene acrylonitrile rubber; **~ butadieno-estireno** *m* (*SBR*) P&C styrene butadiene rubber (*SBR*); **~ butílico** *m* P&C *elastómero* butyl rubber; **~ celular** *m* P&C *elastómetro* foam rubber, cellular rubber, *elastómero* foam, REFRIG, TERMOTEC cellular rubber; **~ clorado** *m* P&C chlorinated rubber; **~ de cloropreno** *m* P&C *elastómero* chloroprene rubber; **~ CR** *m* QUÍMICA CR rubber; **~ crudo** *m* P&C *materia prima* crude rubber, raw rubber; **~ endurecido** *m* P&C vulcanite; **~ esponjoso** *m* P&C *elastómero* foam, foam rubber, latex foam; **~ espumado** *m* P&C *elastómero* foam, foam rubber; **~ de etilenopropileno** *m* P&C *elastómero* ethylene-propylene rubber; **~ expandido** *m* P&C expanded rubber; **~ para grabar** *m* IMPR engraving rubber; **~ microcelular** *m* P&C microcellular rubber; **~ mineral** *m* MINERAL mineral caoutchouc; **~ nitrílico** *m* P&C *elastómero* nitrile rubber; **~ de nitrilo** *m* P&C *elastómero* nitrile rubber; **~ reblandecido con aceite** *m* P&C oil-softened rubber; **~ recuperado** *m* P&C reclaimed rubber; **~ regenerado** *m* P&C regenerated rubber; **~ silicónico** *m* ELEC *aislador* silicone rubber; **~ sintético** *m* P&C *elastómero*, TEC PETR *petroquímica* synthetic rubber; **~ suavizado con aceite** *m* P&C oil-softened rubber; ~ **termoplástico** *m* P&C *elastómero* thermoplastic rubber

caudal *m* AGUA delivery, flow, discharge, efflux, outflow, CARBÓN *fluido* discharge, *de agua, de aire* feed, *hidráulica* flow, CONTAM *fluidos* discharge, ENERG RENOV, FÍS flow, FÍS FLUID *maquinaria fluidodinámica* flow rate, flow, GEOL flow, discharge, HIDROL *de agua* flow, ING MECÁ discharge, *bombas* output, INSTAL HIDRÁUL flow, efflux, MECÁ flow, PROD *ventilador* capacity of output, *agua, gas, tuberías* outflow, *bombas* output, REFRIG rate of flow, TEC PETR throughput, TERMO, TEXTIL flow; ~ **afluente** *m* AGUA inflow; ~ **de agua** *m* AGUA, CONST, HIDROL flow of water, INSTAL HIDRÁUL feed, INSTAL TERM water flow rate; ~ **de agua diario** *m* AGUA daily water flow; ~ **de aire** *m* FÍS FLUID airflow, INSTAL HIDRÁUL feed, INSTAL TERM airflow rate; ~ **de alimentación** *m* CONST *pozo* inducing flow; ~ **anual** *m* AGUA annual flow; ~ **en aumento** *m* HIDROL *de un río* increasing flow; ~ **bruto** *m* NUCL gross flow; ~ **descendente** *m* AGUA, HIDROL receding flood; ~ **de Ekman** *m* OCEAN Ekman flow; ~ **existente** *m* TEC PETR *finanzas* asset; ~ **de infiltración** *m* INSTAL TERM infiltration rate; ~ **intrínseco** *m* HIDROL *régimen* base-flow; ~ **líquido** *m* NUCL liquid flow; ~ **másico** *m* REFRIG mass flow rate; ~ **máximo** *m* HIDROL peak flow; ~ **máximo de agua** *m* AGUA peak water flow; ~ **máximo de crecida** *m* AGUA, HIDROL flood peak; ~ **máximo específico** *m* HIDROL specific peak flow; ~ **medio** *m* AGRIC, CARBÓN *hidráulica* rate of flow; ~ **medio diario** *m* AGUA mean daily flow; ~ **retardado** *m* AGUA, HIDROL retarded flow; ~ **total** *m* HIDROL *de la capa de agua* total flow; ~ **uniforme** *m* HIDROL uniform flow; ~ **variable** *m* HIDROL variable flow; ~ **volumétrico** *m* REFRIG volume flow rate

caudalímetro *m* GEN flowmeter; ~ **de burbuja** *m* LAB *flujo de gases* bubble flowmeter, PROC QUÍ bubble gage (*AmE*), bubble gauge (*BrE*)

causa: ~ **asignable** *f* CALIDAD assignable cause; ~ **común de accidentes** *f* SEG common cause of accidents

causar *vt* CONST *asiento, hundimiento* cause to subside, ELEC induce

causticidad *f* DETERG, FÍS, QUÍMICA *corrosividad* causticity

cáustico[1] *adj* DETERG, FÍS, QUÍMICA caustic

cáustico[2] *m* QUÍMICA caustic

caustificar *vt* DETERG, FÍS, QUÍMICA causticize

cautín *m* ELEC *conexión* soldering iron

cautividad: **en** ~ *adj* TEC ESP tethered

cautivo[1] *adj* D&A cautivo, MECÁ captive

cautivo[2] *m* D&A captive

cavadora: ~ **de hoyos** *f* AGRIC post-hole digger

caverna *f* MINAS *geología* cavity

cavernoso *adj* GEOL caved, cavernous, vuggy, vughy, vugular, MINAS caved

caveto *m* CONST *arquitectura, carpintería* quirk

cavidad *f* (*cf bolsa Esp*) CARBÓN cavity, CONST hollowness, hollow, ELECTRÓN, FÍS, GAS cavity, GEOFÍS *superficie terrestre* sima, GEOL *tipo de porosidad* vug, ING MECÁ hole, METAL cavity, sink, MINAS *conteniendo gas o agua* cavity, *llena de agua o gas* bag, NUCL, PETROL, TEC ESP, TEC PETR cavity; ~ **del acelerador** *f* FÍS PART *parte de la instalación del LEP* accelerator cavity; ~ **de captación** *f* ELECTRÓN *klystron* catcher cavity; ~ **coaxial** *f* ELECTRÓN coaxial cavity; ~ **de contracción** *f* CARBÓN *metalurgia* piping; ~ **que se deja en un molde para sacar probeta** *f* PROD *fundición* runner; ~ **de desactivación** *f* NUCL decay cavity; ~ **de doble sintonización** *f* ELECTRÓN double-tuned cavity; ~ **de encastre** *f* ING MECÁ insert cavity; ~ **de enfriamiento** *f* NUCL, REFRIG cooling cavity; ~ **de entrada** *f* ELECTRÓN input cavity; ~ **de láser** *f* ELECTRÓN laser cavity; ~ **de microondas** *f* ELECTRÓN microwave cavity; ~ **óptica** *f* FÍS RAD, ÓPT, TELECOM optical cavity; ~ **óptica resonante** *f* ÓPT resonant optical cavity; ~ **a presión** *f* CARBÓN pressure cell; ~ **rellena de agua** *f* CARBÓN *minas*, MINAS pocket; ~ **rellena de gas** *f* CARBÓN *minas*, MINAS pocket; ~ **resonante** *f* ELECTRÓN, FÍS, ÓPT, TELECOM resonant cavity; ~ **de sal** *f* GAS salt cavity; ~ **de salida** *f* ELECTRÓN output cavity; ~ **toroidal** *f* FÍS RAD torus

cavitación *f* GEN cavitation

cavitante *adj* GEN cavitating

cavitar *vi* GEN cavitar

cayo *m* OCEAN cay

caz *m* AGUA flume, leat, level, canal, raceway, *molinos* mill course, race, mill race, mill tail, CARBÓN *turbina hidráulica* flume, PROD leat; ~ **de descarga** *m* AGUA, HIDROL tailrace; ~ **de traída** *m* AGUA headrace; ~ **vertedor** *m* AGUA spillway channel

caza: ~ **subacuática** *f* OCEAN underwater fishing

cazarete *m* OCEAN *paño de red de arrastre* baiting

cazo *m* ALIMENT ladle; ~ **de colada** *m* PROD foundry ladle; ~ **con dos brazos largos** *m* ING MECÁ *fundería* bull ladle; ~ **mayor** *m* ING MECÁ *fundería* bull ladle; ~ **mayor con brazos** *m* PROD *fundería* bull ladle; ~ **de plomero** *m* PROD *crisol* melter, melting pot

cazoleta: ~ **del pivote** *f* PROD *fragua* main casting

cazonete *m* TRANSP MAR toggle

Cb *abr* (*cumulonimbo, cumulonimbus*) METEO Cb (*cumulonimbus*)

CBB *abr* (*conmutador bipolar bidireccional*) ELEC, ING ELÉC DPDTS (*double-pole double-throw switch*)

CBL *abr* (*registro de adherencia del cemento*) TEC PETR CBL (*cement bond log*)

CBR *abr* (*índice de resistencia del terreno*) CONST CBR (*California Bearing Ratio*)

CC *abr* (*corriente continua*) ELEC DC (*direct current*)

Cc *abr* (*cirrocúmulo*) METEO Cc (*cirrocumulus*)

CCC *abr* (*cúbico de caras centradas*) CRISTAL, METAL, QUÍMICA FCC (*face-centered cubic AmE, face-centred cubic BrE*)

CCD *abr* INFORM&PD (*central de comunicación de datos, central de conmutación de datos*) DSE (*data-switching exchange*), LAB, QUÍMICA (*cromatografía de capa delgada*) TLC (*thin-layer chromatography*), TELECOM (*central de comunicación de datos, central de conmutación de datos*) DSE (*data-switching exchange*)

CCITT *abbr* (*Comité Consultivo Internacional de Telefonía y Telegrafía*) TELECOM CCITT (*International Telegraph and Telephone Consultative Committee*)

CCP *abr* (*central de conmutación de paquetes*) INFORM&PD, TELECOM PSE (*packet-switching exchange*)

CCWS *abr* (*sistema de agua de refrigeración de componentes*) NUCL CCWS (*component-cooling water system*)

cd *abr* (*candela*) ELEC, FÍS, ING ELÉC, METR, ÓPT cd (*candela*)

Cd *abr* (*cadmio*) ELEC, ING ELÉC, METAL, MINERAL, QUÍMICA Cd (*cadmium*)

CD *abr* (*disco compacto*) INFORM&PD, ÓPT CD (*compact disc*)

CDA *abr* (*concentración derivada en aire*) NUCL DAC (*derived air concentration*)

CD-ROM *abr* (*disco compacto de memoria de sólo lectura*) ÓPT CD-ROM (*compact disc read-only memory*)

Ce *abr* (*cerio*) QUÍMICA Ce (*cerium*)

ceaxantina *f* QUÍMICA zeaxanthin

cebada: ~ **de grano desnudo** *f* AGRIC hull-less barley; ~ **para maltear** *f* ALIMENT *fermentación, molienda* malting barley

cebadilla *f* AGRIC brome grass

cebado *m* AGUA priming, INFORM&PD bootstrap, ING ELÉC striking, INSTAL HIDRÁUL *bombas*, MINAS *voladuras*, TRANSP AÉR priming; ~ **de arco** *m* ELEC arc striking; ~ **indirecto** *m* MINAS indirect priming; ~ **magnético** *m* ING ELÉC energization

cebador *m* *AmL* (*cf iniciador Esp*) CARBÓN water channel, *electricity* priming power, starter, ING MECÁ striker, MINAS *voladuras* primer, OCEAN choke; ~ **del carburador** *m* *AmL* (*ahogador Esp*) MECÁ choke

cebadura *f* MINAS *voladuras* priming, OCEAN bait; ~ **indirecta** *f* MINAS *voladuras* indirect priming

cebar[1]: ~ **un sifón** *fra* AGUA start the flow of water in a siphon

cebar[2] *vt* AGUA prime, start in operation, ING MECÁ start, INSTAL HIDRÁUL *bomba* fang, prime, start, MINAS *un cartucho* prime, PROD *bombas, inyectores, sifones* start

cebar[3]: ~ **el arco** *vi* ING ELÉC arc

cebo *m* D&A fuse, *de explosivos* primer, MINAS (*cf fulminante Esp, detonador Esp*), blasting-cap, *voladuras* fuse, *explosivos* primer, exploder, OCEAN bait, QUÍMICA primer; ~ **de acción rápida** *m* D&A quick-acting fuse; ~ **de chispa** *m* *Esp* (*cf cebo explosivo AmL*) MINAS *pega* spark fuse; ~ **de combustión lenta** *m* MINAS *voladuras* slow-burning fuse; ~ **desfasado en milisegundos** *m* MINAS millisecond delay detonator; ~ **eléctrico** *m* MINAS *pega* electric blasting cap; ~ **eléctrico de cantidad** *m* MINAS *pega* battery fuse; ~ **eléctrico de incandescencia** *m* MINAS quantity fuse; ~ **eléctrico instantáneo** *m* MINAS *pega* instantaneous electric blasting-cap; ~ **eléctrico de retardo** *m* ELEC, MINAS *pega* electric delay detonator cap; ~ **explosivo** *m* *AmL* (*cf cebo de chispa Esp*) MINAS *pega* spark fuse; ~ **fulminante** *m* D&A percussion fuse, percussion priming; ~ **de mando** *m* MINAS master fuse; ~ **progresivo** *m* MINAS *pega* ordinary fuse; ~ **con retardo de milisegundo** *m* MINAS *voladuras* millisecond delay cap

cebolla *f* C&V onion

cebollita: ~ **para plantar** *f* AGRIC onion set

cebú *m* AGRIC zebu

cedazo *m* CARBÓN cribble, riddle, sieve, CONST screen, sieve, LAB *separación* sieve, MINAS *preparación minerales* screen, PROD riddle, screen, sieve, sifter, strainer, RECICL screen, sieve; ~ **sacudidor** *m* PROD shaking screen; ~ **para trapos** *m* PAPEL rag duster, rag thrasher

cedrene® *m* QUÍMICA cedrene®

cedrol *m* QUÍMICA *perfumería* cedrol

cefalópodo *m* GEOL cephalopod

cefalosporina *f* QUÍMICA *sustancias antibióticas* cephalosporin

céfiro *m* METEO zephyr

cegado[1] *adj* IMPR plug

cegado[2]: ~ **de los puntos** *m* IMPR plugging

cegar *vt* TERMO smother

cegesimal *m* (*CGS*) METR centimeter-gram-second (*AmE*) (*CGS*), centimetre-gramme-second (*BrE*) (*CGS*)

ceguera: ~ **de los hielos** *f* TEC ESP *oceanografía* blind; ~ **de la nieve** *f* METEO snow blindness

ceilanita *f* MINERAL ceylanite

ceilonita *f* MINERAL ceylonite

ceína *f* P&C, QUÍMICA *proteína* zein

celadonita *f* MINERAL celadonite

celda *f* CINEMAT, IMPR, INFORM&PD, INSTR, LAB *análisis, electrólisis* cell; ~ **de absorción** *f* PROC QUÍ absorption cell; ~ **de acabado** *f* CARBÓN *flotación minerales* cleaner cell; ~ **de acumulador** *f* ELEC accumulator cell; ~ **de animación** *f* CINEMAT animation cell; ~ **de bit** *f* PROD bit location; ~ **de Bragg** *f* CRISTAL Bragg cell; ~ **circular** *f* IMPR round cell; ~ **conductimétrica** *f* ELECTRÓN, LAB *análisis, electroquímica* conductance cell; ~ **cromatográfica** *f* LAB *análisis*, QUÍMICA chromatography tank; ~ **de difusión** *f* PROC QUÍ diffusion cell; ~ **de electroforesis** *f* ELEC, LAB, QUÍMICA electrophoresis cell; ~ **de ensayo de cobre** *f* FÍS RAD copper test cell; ~ **galvánica** *f* ELEC, ING ELÉC, QUÍMICA, REVEST galvanic cell; ~ **nuclear** *f* CONST nuclear cell; ~ **de oxidación-reducción** *f* LAB, QUÍMICA oxidation-reduction cell; ~ **de óxido-reducción** *f* LAB, QUÍMICA oxidation-reduction cell; ~ **de oxidorreducción** *f* LAB, QUÍMICA oxidation-reduction cell; ~ **de redox** *f* LAB, QUÍMICA redox cell; ~ **secundaria** *f* ELEC *acumulador* secondary cell; ~ **standard** *f* FÍS standard cell; ~ **unidad** *f* CRISTAL unit cell; ~ **unitaria** *f* NUCL unit cell

celdilla: ~ **solar** *f* TEC ESP solar cell

celemín *m* METR peck

celeste *adj* TEC ESP, TRANSP AÉR, TRANSP MAR celestial

celestina *f* MINERAL celestine

celestita *f* MINERAL celestite

celo[1] *m* AGRIC heat

celo[2]: **en** ~ *fra* AGRIC in heat

celosía *f* CONST lattice, ING MECÁ, MECÁ louver (*AmE*), louvre (*BrE*)

Celsio *m* (*C*) FÍS, MECÁ, METEO, METR, PROD, QUÍMICA Celsius (*C*)

célula *f* GEN cell, CARBÓN *metalografía* pressure cell; ~ **abierta** *f* P&C open cell; ~ **aerífera** *f* *AmL* (*cf conducto aerífero Esp*) ING MECÁ, MINAS air conduit, air passage; ~ **de barrido** *f* CARBÓN *aeronáutica* scavenger cell; ~ **cerrada** *f* P&C closed cell; ~ **ciliada** *f* ACÚST *audición* acoustic cell, hair cell; ~ **de combustible para altas temperaturas** *f* TRANSP high-temperature fuel cell; ~ **de combustible principal** *f* TRANSP primary fuel cell; ~ **combustible secundaria** *f* TRANSP secondary fuel cell; ~ **cónica** *f* TEC ESP conical shell; ~ **de dióxido de manganeso-zinc** *f* ING ELÉC zinc-manganese dioxide cell; ~ **directa** *f* TRANSP direct cell; ~ **directa metanol-aire** *f* TRANSP direct methanol-air cell; ~ **directa de oxígeno-hidrógeno** *f* TRANSP direct hydrogen-oxygen cell; ~ **electrolítica** *f* ELEC, FÍS, ING ELÉC electrolytic cell; ~ **energética** *f* TEC ESP battery cell, TERMO fuel cell;

~ **energética secundaria** *f* TRANSP secondary fuel cell; ~ **del exposímetro** *f* FOTO meter cell; ~ **fotoconductiva** *f* ELECTRÓN, FÍS photoconductive cell; ~ **fotoeléctrica** *f* CINEMAT photocell, photoelectric cell (*PEC*), ELEC photocell, photoelectric cell (*PEC*), electric eye, ELECTRÓN, FÍS photocell, photoelectric cell (*PEC*), FÍS RAD photoelectric cell (*PEC*), FOTO photoelectric cell (*PEC*), photocell, IMPR photocell, ING ELÉC electric eye, ING MECÁ, QUÍMICA photocell, TEC ESP photoelectric cell (*PEC*), photocell, TRANSP *contról de tráfico* photoelectric detector, TV photocell, photoelectric cell (*PEC*); ~ **fotoeléctrica alcalina** *f* FOTO alkaline photocell; ~ **fotoeléctrica de cadmio-níquel** *f* ING ELÉC cadmium-nickel cell; ~ **fotoeléctrica multiplicadora** *f* INSTR multiplier phototube; ~ **fotogalvánica** *f* ENERG RENOV photogalvanic cell; ~ **del fotómetro** *f* FOTO light meter cell; ~ **fotovoltaica** *f* ENERG RENOV, FÍS, ING ELÉC, TEC ESP photovoltaic cell; ~ **fotovoltaica con barrera anterior** *f* ING ELÉC front-wall photovoltaic cell; ~ **fotovoltaica de carrera posterior** *f* ING ELÉC back-wall photovoltaic cell; ~ **fotovoltaica solar** *f* ING ELÉC, TEC ESP solar cell; ~ **galvánica** *f* ELEC, ING ELÉC, QUÍMICA, REVEST galvanic cell; ~ **de gel** *f* ING ELÉC gel cell; ~ **de Golay** *f* FÍS Golay cell; ~ **H** *f* ING ELÉC H cell; ~ **de Kerr** *f* FÍS Kerr cell; ~ **magnética** *f* INFORM&PD magnetic cell; ~ **de memoria** *f* IMPR, ING ELÉC storage cell; ~ **de memoria de plata y zinc** *f* ING ELÉC silver-zinc storage cell; ~ **de metanol** *f* TRANSP methanol cell; ~ **de óxido de plata y zinc** *f* ING ELÉC zinc-silver oxide cell; ~ **de plata y zinc** *f* ING ELÉC zinc-silver cell; ~ **recargable** *f* FÍS rechargeable cell; ~ **rectificadora** *f* ELEC metal rectifier, ING ELÉC rectifier cell; ~ **de redox** *f* LAB, QUÍMICA redox cell; ~ **de selenio** *f* ING ELÉC selenium cell; ~ **de silicio** *f* ENERG RENOV silicon cell; ~ **solar** *f* ELEC, ENERG RENOV, FÍS, FÍS RAD, TEC ESP, TELECOM solar cell; ~ **de sulfuro de cadmio** *f* ELEC, ING ELÉC cadmium sulfide cell (*AmE*), cadmium sulphide cell (*BrE*); ~ **unielemental** *f* ING ELÉC one-element cell

celular *adj* AGUA, CONST, INSTAL HIDRÁUL, TRANSP AÉR, TRANSP MAR cellular

celuloide *m* EMB, QUÍMICA celluloid

celulosa *f* CINEMAT, EMB, P&C, QUÍMICA *polímero natural* cellulose; ~ **de carboximetilo** *f* ALIMENT, DETERG, P&C carboxymethyl cellulose (*CMC*); ~ **en copos** *f* PAPEL dusting, fluff; ~ **de etilo** *f* P&C *polímero* ethyl cellulose; ~ **hidratada** *f* QUÍMICA hydrocellulose; ~ **hidroxietílica** *f* TEC PETR *petroquímica* hydroxyethylcellulose

celulósico *adj* QUÍMICA, TEXTIL cellulosic

cementación *f* C&V cementation, CONST cementation, cementing, GAS cementation, ING MECÁ carburizing, MECÁ hard facing, NUCL cementation, TEC PETR cementing, cementation, TERMO cementation; ~ **en caja** *f* ING MECÁ *aceros*, MECÁ case hardening; ~ **por gas** *f* GAS, TERMO gas carburizing; ~ **gaseosa** *f* GAS, TERMO gas carburizing; ~ **de grietas acuíferas** *f* CARBÓN grouting

cementado *adj* ING MECÁ hard-faced

cementador *m* CONST *albañilería* binder

cementar[1] *vt* C&V cement, CARBÓN *acero* carbonize, CONST, GAS cement, ING MECÁ hard-face, NUCL cement, PROD *aceros* harden, TEC PETR, TERMO cement

cementar[2] *vi* MINAS carbonize

cementera *f* CONST cement maker

cementista *m* CONST *persona* finisher

cemento *m* C&V, CINEMAT, CONST, GEOL cement, PETROL mastic, QUÍMICA, TEC ESP cement, TRANSP MAR mastic; ~ **aislante** *m* REFRIG insulating cement; ~ **de amianto** *m* PAPEL asbestos cement; ~ **de empalmar** *m* CINEMAT splicing cement; ~ **sin finos** *m* CONST no-fines concrete; ~ **de fraguado lento** *m* CONST slow-setting cement; ~ **a granel** *m* CONST bulk cement; ~ **Portland normal** *m* CONST ordinary Portland cement; ~ **Portland resistente a los sulfatos** *m* CONST Portland sulfate-resisting cement (*AmE*), Portland sulphate-resisting cement (*BrE*); ~ **puzolánico** *m* CONST pozzolanic cement; ~ **rápido** *m* CONST quick-setting cement, rapid-hardening cement; ~ **reforzado con vidrio** *m* C&V, CONST, PROD glass-reinforced concrete; ~ **siderúrgico** *m* CONST blast-furnace cement; ~ **sintaxial** *m* GEOL rim cementation

CEN *abr* (*Comité Européen de Normalisation*) PROD CEN (*European Committee for National Standards*)

cenagal *m* AGUA marsh, MINAS *piso de galerías* mud bottom

cenagoso *adj* GEOL, HIDROL marshy

cenefa *f* TEXTIL edge trim

CENELEC *abr* (*Comité Européen de Normalisation Électronique*) ELEC, ING ELÉC CENELEC (*European Committee for Electronic Normalization*)

cenicero *m* INSTAL TERM ash box, PROD ash bin, ash pit

ceñida: ~ **a la cuadra** *f* TRANSP MAR beam reach

cenit *m* ENERG RENOV, TEC ESP, TRANSP MAR *astronavegación* zenith

ceniza[1]: **sin** ~ *adj* CONTAM *caracterización de combustibles* ash-free

ceniza[2] *f* CONTAM *combustión, caracterización de combustibles*, P&C, PETROL, TERMO ash; ~ **de carbón** *f* CARBÓN coal ash; ~ **de huesos** *f* C&V, QUÍMICA bone ash; ~ **y residuo de combustión** *f* CONTAM ash and combustion residue; ~ **de sosa** *f* DETERG soda ash

cenizas *f pl* CARBÓN ash, dust, fly ash, P&C ash, dust, PAPEL dust, PROD *batiduras de forja* cinders; ~ **de horno** *f pl* C&V breezing, CARBÓN breeze; ~ **radiactivas** *f pl* AGRIC radioactive fallout, CONTAM *en la atmósfera* radioactive fallout, rainout, FÍS, FÍS RAD, NUCL radioactive fallout; ~ **silíceas** *f pl* GEOL *escorias petrificadas de grano fino* flinty ash; ~ **volantes** *f pl* CARBÓN fly ash; ~ **voltantes** *f pl* PROD clinker

Cenomaniense *adj* GEOL Cenomanian

Cenozoico *adj* GEOL Cenozoic

censo: ~ **de carreteras de acceso** *m* TRANSP access road census (*AmE*), slip road census (*BrE*); ~ **direccional** *m* TRANSP directional census; ~ **del tráfico** *m* TRANSP traffic census; ~ **de vías de acceso** *m* TRANSP access road census (*AmE*), slip road census (*BrE*)

centellear *vi* CINEMAT flicker

centelleo[1]: **sin** ~ *adj* CINEMAT flicker-free

centelleo[2] *m* CINEMAT *AmL* (*cf disparo de flash Esp*) flash, flashing, flicker, FÍS scintillation, FOTO flashing, INFORM&PD *pantalla* flicker, ING ELÉC flashing, TEC ESP, TEXTIL scintillation, TV flicker; ~ **aurífero** *m* TEC ESP gold flashing; ~ **dorado** *m* TEC ESP gold flashing; ~ **solar** *m* TEC ESP solar flare

centi- *pref* (*c*) METR centi- (*c*)

centiárea *f* METR centiare
centígrado *adj* FÍS, MECÁ, METEO, METR, QUÍMICA, TERMO centigrade
centigramo *m* METR centigram (*AmE*), centigramme (*BrE*)
centilitro *m* METR centiliter (*AmE*), centilitre (*BrE*)
centímetro *m* METR centimeter (*AmE*), centimetre (*BrE*); **~ cuadrado** *m* METR square centimeter (*AmE*), square centimetre (*BrE*); **~ cúbico** *m* METR cubic centimeter (*AmE*), cubic centimetre (*BrE*); **~-gramo** *m* METR *unidad de calor* gram centimeter (*AmE*), gramme centimetre (*BrE*)
centinela *m* INFORM&PD sentinel
centinormal *adj* QUÍMICA centinormal
centipoise *m* P&C *unidad física* centipoise
centrado[1] *adj* GEOM in-center (*AmE*), in-centre (*BrE*)
centrado[2] *m* CONST centering (*AmE*), centring (*BrE*), striking the centering (*AmE*), striking the centring (*BrE*), ING MECÁ, MECÁ centering (*AmE*), centring (*BrE*), TRANSP AÉR trim; **~ por corriente continua** *m* TV DC centering (*AmE*), DC centring (*BrE*); **~ horizontal de la imagen** *m* TV horizontal centering control (*AmE*), horizontal centring control (*BrE*); **~ de imán permanente** *m* TV permanent-magnet centering (*AmE*), permanent-magnet centring (*BrE*); **~ vertical** *m* ELECTRÓN vertical centering (*AmE*), vertical centring (*BrE*)
centrador: **~ fijo** *m* ING MECÁ steady rest
centraje *m* ING MECÁ, MECÁ centering (*AmE*), centring (*BrE*)
central[1] *adj* AUTO, ELECTRÓN, FOTO front-end, GEOL middle, INFORM&PD front-end, host, INSTR, VEH front-end
central[2] *f* ELEC, ENERG RENOV, GAS, ING ELÉC station, TELECOM exchange, TERMO, TERMOTEC station; **~ de acceso internacional** *f* TELECOM international gateway exchange; **~ automática** *f* TELECOM unmanned exchange; **~ automática privada** *f* TELECOM private automatic exchange; **~ automática rural** *f* TELECOM rural automatic exchange; **~ automática unitaria** *f* TELECOM unit automatic exchange (*UAX*); **~ autónoma** *f* TELECOM stand-alone exchange; **~ auxiliar** *f* TELECOM satellite exchange; **~ de barras cruzadas** *f* TELECOM crossbar exchange; **~ de circuitos conmutados** *f* TELECOM circuit-switched exchange; **~ de comunicación de datos** *f* (*CCD*) INFORM&PD, TELECOM data-switching exchange (*DSE*); **~ de conmutación de datos** *f* (*CCD*) INFORM&PD, TELECOM data-switching exchange (*DSE*); **~ de conmutación de datos internacionales** *f* TELECOM international data switching exchange (*IDSE*); **~ de conmutación para mensajes internacionales** *f* TELECOM international data switching exchange (*IDSE*); **~ de conmutación de paquetes** *f* (*CCP*) INFORM&PD, TELECOM packet-switching exchange (*PSE*); **~ de control de programa almacenado** *f* TELECOM stored program control exchange (*AmE*), stored programme control exchange (*BrE*); **~ de conversión** *f* ELEC *transformador*, ING ELÉC converting station; **~ de destino** *f* TELECOM terminating exchange; **~ digital integrada** *f* TELECOM integrated digital exchange; **~ de distribución** *f* ING ELÉC distribution station; **~ eléctrica** *f* ENERG RENOV powerhouse, ING ELÉC electric-power station, FÍS, ING MECÁ, MINAS power station, TELECOM power plant, power station;

~ eléctrica activada con aceite *f* ELEC *fuente de alimentación* oil-fired power station; **~ eléctrica alimentada por gasóleo** *f* TERMOTEC oil-fired power station; **~ eléctrica de carga** *f* ELEC, ING ELÉC charging station; **~ eléctrica con carga máxima** *f* ELEC *red de distribución* peak-load power plant; **~ eléctrica de conversión** *f* ELEC *transformador*, ING ELÉC converting station; **~ eléctrica de fueloil** *f* NUCL oil-fired power plant; **~ eléctrica geotérmica** *f* ELEC *fuente de alimentación*, ENERG RENOV, ING ELÉC geothermal power station; **~ eléctrica maremotriz** *f* ENERG RENOV tidal power station; **~ eléctrica de turbina de gas** *f* ELEC *fuente de alimentación*, GAS gas-turbine power station; **~ de electricidad** *f* ING ELÉC electric-power station; **~ electromecánica** *f* ELEC, ING ELÉC, TELECOM electromechanical exchange; **~ electrónica** *f* TELECOM electronic exchange; **~ electrónica de relé de láminas** *f* ELEC, ING ELÉC, TELECOM reed-relay electronic exchange; **~ energética** *f*, FÍS, ING MECÁ, MINAS, TELECOM power station, TEC ESP power plant; **~ energética caldeada con petróleo** *f* TERMO oil-fired power station; **~ energética impulsada por combustibles fósiles** *f* ELEC, TERMO fossil-fuel power station; **~ de energía** *f* ELEC *fuente de alimentación* power plant, FÍS power station, TELECOM power plant, power station; **~ de energía diesel-eléctrica** *f* ING ELÉC diesel-electric power station; **~ de energía eléctrica a vapor** *f* ING ELÉC steam-electric power station; **~ de energía eólica** *f* ING ELÉC wind-electric power station; **~ de energía hidráulica** *f* MINAS hydraulic power pack; **~ de energía hidroeléctrica** *f* ING ELÉC hydroelectric power station, water power station; **~ de energía solar** *f* ING ELÉC solar power farm; **~ de energía termo-eléctrica** *f* ING ELÉC thermal-electric power station; **~ de energía de turbina de combustión** *f* ELEC *fuente de alimentación*, GAS gas-turbine power station; **~ de fuerza eléctrica solar** *f* ING ELÉC solar electric power station; **~ de fuerza motriz** *f* FÍS, ING MECÁ, MINAS, TELECOM power station; **~ de fuerza motriz caldeada por gasoil** *f* TERMO oil-fired power station; **~ de fuerza motriz de combustible fósil** *f* ELEC, TERMO fossil-fuel power station; **~ de generación eléctrica** *f* CONST, ELEC, ING ELÉC, NUCL electricity generating station; **~ geotérmica** *f* ELEC *fuente de alimentación*, ENERG RENOV, ING ELÉC geothermal power station; **~ hidroeléctrica** *f* ELEC *fuente de alimentación*, HIDROL hydroelectric power station; **~ holográfica** *f* TELECOM holographic exchange; **~ independiente** *f* TELECOM stand-alone exchange; **~ internacional de acceso a la conmutación de paquetes** *f* TELECOM international packet-switching gateway exchange; **~ interurbana** *f* TELECOM trunk exchange; **~ interurbana de tránsito** *f* TELECOM trunk transit exchange; **~ lechera** *f* REFRIG dairy plant; **~ local** *f* TELECOM local exchange (*LE*); **~ de luz y fuerza** *f* ELEC *fuente de alimentación* power plant; **~ manual privada** *f* TELECOM private manual exchange; **~ maremotriz** *f* ING ELÉC tidal power plant, OCEAN tidal power station; **~ no atendida** *f* TELECOM unattended exchange, unmanned exchange; **~ nodal de tránsito** *f* TELECOM tandem exchange; **~ nuclear** *f* CONST nuclear power plant, nuclear power station, ELEC *generador*, FÍS nuclear power station, NUCL nuclear power plant, nuclear power station; **~ nuclear de**

carga de pico *f* NUCL peak-load nuclear-power plant; **~ nuclear de dos grupos** *f* NUCL twin-reactor station; **~ nuclear de dos lazos** *f* NUCL two-circuit nuclear power plant; **~ nuclear con reactor refrigerado por gas** *f* NUCL gas-cooled nuclear power plant; **~ nuclear de reactores gemelos** *f* NUCL twin-reactor station; **~ nuclear de tres lazos** *f* NUCL three-circuit nuclear power plant; **~ nucleoeléctrica** *f* CONST, ELEC, FÍS, NUCL nuclear power station; **~ óptica** *f* TELECOM optical exchange; **~ sin personal** *f* TELECOM unmanned exchange; **~ sin personal de guardia** *f* TELECOM unattended exchange; **~ sin personal permanente** *f* TELECOM unattended exchange; **~ primaria** *f* TELECOM home exchange (*BrE*), host exchange (*AmE*); **~ principal** *f* TELECOM main exchange; **~ privada automática** *f* TELECOM private automatic exchange; **~ privada manual** *f* TELECOM private manual branch exchange (*PMBX*); **~ de procedencia** *f* TELECOM originating exchange; **~ de producción hidroeléctrica** *f* ING ELÉC hydroelectric generating station; **~ de la RDSI** *f* TELECOM ISDN exchange; **~ de red privada** *f* TELECOM private network exchange (*PNX*); **~ de referencia** *f* NUCL reference plant; **~ rotativa** *f* TELECOM rotary exchange; **~ rural** *f* TELECOM rural exchange; **~ satélite** *f* TELECOM satellite exchange; **~ de servicio** *f* TELECOM serving exchange; **~ de servicios digitales integrados** *f* TELECOM integrated digital services exchange; **~ de servicios integrados** *f* TELECOM integrated services exchange; **~ tándem** *f* TELECOM tandem exchange; **~ telefónica** *f* TELECOM telephone exchange, central exchange, exchange switchboard; **~ telefónica automática** *f* TELECOM automatic switchboard, automatic telephone exchange; **~ telefónica comunitaria** *f* TELECOM community dial office (*CDO*); **~ telefónica dependiente** *f* TELECOM dependent exchange; **~ telefónica digital** *f* TELECOM digital exchange; **~ telefónica discriminante de satélite** *f* TELECOM discriminating satellite exchange; **~ telefónica intermedia** *f* TELECOM tandem exchange; **~ telefónica intermedia de enlace** *f* TELECOM junction tandem exchange; **~ telefónica manual** *f* TELECOM manual exchange; **~ telefónica privada** *f* (*PBX*) TELECOM private branch exchange (*PBX*); **~ telefónica privada distribuida** *f* TELECOM distributed PBX; **~ telefónica pública** *f* TELECOM public telephone exchange; **~ telefónica subordinada** *f* TELECOM dependent exchange; **~ telegráfica** *f* ING ELÉC telegraph exchange; **~ télex** *f* TELECOM telex exchange; **~ temporizadora** *f* TELECOM time-division exchange; **~ térmica** *f* NUCL thermal power plant, TERMO thermal power station; **~ de tránsito** *f* TELECOM tandem exchange, transit exchange; **~ de tránsito internacional** *f* TELECOM international transit exchange (*INT TR*); **~ urbana** *f* TELECOM local exchange

centralita *f* INFORM&PD, TELECOM exchange; **~ automática** *f* TELECOM private automatic exchange; **~ privada** *f* TELECOM private branch exchange (*PBX*); **~ privada automática conectada a la red pública** *f* TELECOM private automatic branch exchange (*PABX*); **~ telefónica automática privada de servicios integrados** *f* TELECOM integrated services private automatic branch exchange; **~ telefónica automática privada de voz-datos**

integrados *f* TELECOM integrated voice-data private automatic branch exchange

centralizado *adj* INFORM&PD, ING MECÁ, TELECOM centralized

centralizador *m* TEC PETR *perforación* centralizer

centramiento *m* ING MECÁ, MECÁ centering (*AmE*), centring (*BrE*)

centrar[1] *vt* CONST *la burbuja* come back to the center of its run (*AmE*), come back to the centre of its run (*BrE*), ING MECÁ true, true up

centrar[2]: **~ la burbuja** *vi* CONST *topografía* keep its center (*AmE*), keep its centre (*BrE*)

céntrico *adj* GEOM in-center (*AmE*), in-centre (*BrE*)

centrífuga *f* CARBÓN centrifuge, FÍS centrifugal, centrifuge, HIDROL *aguas residuales*, LAB *separación* centrifuge, MECÁ centrifuge, centrifugal, P&C *equipo* centrifuge, PROC QUÍ centrifugal, centrifuge, centrifugal machine, QUÍMICA centrifuge, TEXTIL *mangle* hydroextractor

centrifugación *f* AGUA centrifugation, CARBÓN centrifuging, PROC QUÍ, QUÍMICA *proceso* centrifugation

centrifugado *m* TEXTIL *secado* hydroextraction

centrifugadora *f* FÍS centrifugal, centrifuge, HIDROL *aguas residuales*, LAB *separación* centrifuge, MECÁ centrifugal, centrifuge, P&C *equipo, operación* centrifuge, PROC QUÍ centrifuge, centrifugal, QUÍMICA centrifuge, TEXTIL *secadora por centrifugación* hydroextractor; **~ de cesta** *f* CARBÓN basket centrifuge; **~ manual** *f* LAB hand centrifuge; **~ de taza** *f* CARBÓN bowl centrifuge

centrifugar *vt* GEN centrifuge, TEXTIL *secado* hydroextract

centrífugo *adj* AUTO, FÍS, MECÁ, PROC QUÍ, TRANSP AÉR, VEH centrifugal

centrípeto *adj* FÍS, ING MECÁ, TRANSP AÉR centripetal

centro[1]: **del ~** *adj* GEOM in-center (*AmE*), in-centre (*BrE*)

centro[2]: **en el ~** *adv* TRANSP MAR midship; **en el ~ del buque** *adv* TRANSP MAR amidships

centro[3] *m* CONST, EMB, GEOM center (*AmE*), centre (*BrE*), INFORM&PD hub, PROD, SEG center (*AmE*), centre (*BrE*), TEC ESP core, TELECOM center (*AmE*), centre (*BrE*); **~ de accionamiento** *m* TEC ESP operation center (*AmE*), operation centre (*BrE*); **~ aerodinámico** *m* TRANSP AÉR, TRANSP MAR aerodynamic center (*AmE*), aerodynamic centre (*BrE*); **~ aerodinámico de la pala** *m* TRANSP AÉR blade aerodynamic center (*AmE*), blade aerodynamic centre (*BrE*); **~ del alma** *m* TELECOM core center (*AmE*), core centre (*BrE*); **~ del área de flotación** *m* TRANSP MAR center of waterplane area (*AmE*), centre of waterplane area (*BrE*); **~ automático conmutador de mensajes** *m* TELECOM automatic message switching center (*AmE*), automatic message switching centre (*BrE*); **~ de baja presión** *m* METEO center of low pressure (*AmE*), centre of low pressure (*BrE*); **~ de cálculo de costes** *m* PROD cost center (*AmE*), cost centre (*BrE*); **~ de carena** *m* TRANSP MAR *construcción* center of buoyancy (*AmE*), centre of buoyancy (*BrE*); **~ de chapado** *m* TELECOM cladding center (*AmE*), cladding centre (*BrE*); **~ de conexión** *m* TELECOM trunk switching center (*AmE*), trunk switching centre (*BrE*); **~ de conexiones telefónicas** *m* TELECOM trunk switching center (*AmE*), trunk switching centre (*BrE*); **~ de conmutación** *m* TELECOM, TV switching center

(*AmE*), switching centre (*BrE*); ~ **de conmutación de almacenamiento y retransmisión** *m* INFORM&PD storage and forwarding switching center (*AmE*), store and forward switching centre (*BrE*); ~ **de conmutación de circuitos** *m* INFORM&PD, TELECOM circuit-switching center (*AmE*), circuit-switching centre (*BrE*); ~ **de conmutación digital** *m* TELECOM digital switching center (*AmE*), digital switching centre (*BrE*); ~ **de conmutación digital de la red principal** *m* TELECOM digital main network switching center (*AmE*), digital main network switching centre (*BrE*) (*DMNSC*); ~ **de conmutación de enlace principal** *m* TELECOM main trunk-switching center (*AmE*), main trunk-switching centre (*BrE*); ~ **de conmutación de grupos** *m* TELECOM group-switching center (*AmE*), group-switching centre (*GSC*); ~ **de conmutación internacional** *m* TELECOM international switching center (*AmE*) (*ISC*), international switching centre (*BrE*) (*ISC*); ~ **de conmutación marítima** *m* TELECOM maritime switching center (*AmE*) (*MSC*), maritime switching centre (*BrE*) (*MSC*); ~ **de conmutación de mensajes** *m* TELECOM message-switching center (*AmE*), message-switching centre (*BrE*); ~ **de conmutación móvil** *m* TELECOM mobile switching center (*AmE*) (*MSC*), mobile switching centre (*BrE*) (*MSC*); ~ **de conmutación del satélite** *m* TEC ESP satellite switching; ~ **de conmutación de tráfico** *m* TELECOM message-switching center (*AmE*), message-switching centre (*BrE*); ~ **de conmutación de tránsito** *m* TELECOM transit switching center (*AmE*), transit switching centre (*BrE*); ~ **de control** *m* FERRO *equipo inamovible*, TEC ESP, TV control centre (*BrE*), control center (*AmE*); ~ **de control de aproximación por radar** *m* TRANSP radar approach control center (*AmE*), radar approach control centre (*BrE*); ~ **de control de red** *m* TELECOM network control center (*AmE*) (*NCC*), network control centre (*BrE*) (*NCC*); ~ **de control de satélites** *m* TELECOM satellite control center (*AmE*) (*SCC*), satellite control centre (*BrE*) (*SCC*); ~ **de control de tráfico aéreo** *m* TRANSP AÉR air traffic control center (*AmE*), air traffic control centre (*BrE*); ~ **de coordinación de rescate** *m* SEG, TRANSP AÉR, TRANSP MAR rescue coordination center (*AmE*), rescue coordination centre (*BrE*); ~ **coordinador de salvamento** *m* SEG, TRANSP AÉR, TRANSP MAR rescue coordination center (*AmE*), rescue coordination centre (*BrE*); ~ **de coste** *m* PROD cost center (*AmE*), cost centre (*BrE*); ~ **de curvatura** *m* FÍS, FÍS RAD, GEOM center of curvature (*AmE*), centre of curvature (*BrE*); ~ **de deriva** *m* TRANSP MAR *construcción* center of lateral resistance (*AmE*), centre of lateral resistance (*BrE*); ~ **de desplazamiento** *m* HIDROL center of displacement (*AmE*), centre of displacement (*BrE*); ~ **de distribución** *m* TELECOM distribution center (*AmE*), distribution centre (*BrE*); ~ **efectivo de fuente acústica** *m* ACÚST effective center of acoustic source (*AmE*), effective centre of acoustic source (*BrE*); ~ **de elección** *m* INFORM&PD hub polling; ~ **de emergencias** *m* SEG emergency center (*AmE*), emergency centre (*BrE*); ~ **de empuje** *m* HIDROL center of buoyancy (*AmE*), centre of buoyancy (*BrE*), TEC ESP center of thrust (*AmE*), centre of thrust (*BrE*); ~ **de enlace principal** *m* TELECOM main trunk-exchange area; ~ **de erupción** *m* GEOL *volcá-nica* eruption point; ~ **espacial** *m* TEC ESP space center (*AmE*), space centre (*BrE*); ~ **de explotaciones** *m* TELECOM operations center (*AmE*), operations centre (*BrE*); ~ **fabril** *m* MECÁ factory; ~ **de facturación** *m* TELECOM billing center (*AmE*), billing centre (*BrE*); ~ **de ferrita** *m* ING ELÉC ferrite rod; ~ **de ferrocarriles** *m* FERRO railroad center (*AmE*), railway centre (*BrE*); ~ **de flotación** *m* TRANSP MAR *construcción, diseño naval* center of flotation (*AmE*), centre of flotation (*BrE*); ~ **de genética ganadera** *m* AGRIC breeding unit; ~ **de gestión** *m* TELECOM management center (*AmE*), management centre (*BrE*); ~ **de gestión de red** *m* TELECOM network management center (*AmE*) (*NMC*), network management centre (*BrE*) (*NMC*); ~ **de giro** *m* ING MECÁ, INSTR, NUCL pivot; ~ **de gobierno y mando** *m* TRANSP MAR *del buque* command and control center (*AmE*), command and control centre (*BrE*); ~ **de gravedad** *m* CONST, FÍS, MECÁ, TEC ESP, TRANSP AÉR center of gravity (*AmE*), centre of gravity (*BrE*), TRANSP MAR *diseño naval* center of gravity (*AmE*), centre of gravity (*BrE*), shift of G; ~ **hiperbárico** *m* OCEAN hyperbaric center (*AmE*), hyperbaric centre (*BrE*); ~ **de iluminación** *m* CINEMAT light center (*AmE*), light centre (*BrE*); ~ **de impacto** *m* MECÁ center of impact (*AmE*), centre of impact (*BrE*); ~ **de impulsión** *m* TEC ESP center of thrust (*AmE*), centre of thrust (*BrE*); ~ **de inercia** *m* MECÁ center of inertia (*AmE*), centre of inertia (*BrE*); ~ **de información de combate** *m* D&A combat information center (*AmE*) (*CIC*), combat information centre (*BrE*) (*CIC*); ~ **de información para los viajeros** *m* FERRO passenger information center (*AmE*), passenger information centre (*BrE*); ~ **de información de vuelo** *m* TRANSP AÉR flight information center (*AmE*), flight information centre (*BrE*); ~ **informático** *m* INFORM&PD, TELECOM data processing center (*AmE*) (*DPC*), data processing centre (*BrE*) (*DPC*); ~ **del intersticio** *m* ING ELÉC gapped core; ~ **de inversión** *m* CRISTAL inversion center (*AmE*), inversion centre (*BrE*); ~ **de investigación** *m* TELECOM research center (*AmE*), research centre (*BrE*); ~ **de investigaciones** *m* AGRIC research station; ~ **de ladrillos** *m* CONST bricked-up core; ~ **de mando** *m* PROD control center (*AmE*), control centre (*BrE*); ~ **de mantenimiento acceso para abonados** *m* TELECOM customer access maintenance center (*AmE*) (*CAMC*), customer access maintenance centre (*BrE*) (*CAMC*); ~ **de mantenimiento y administración de la explotación** *m* TELECOM operation administration and maintenance center (*AmE*), operation administration and maintenance centre (*BrE*) (*OAMC*); ~ **de mantenimiento atención al cliente** *m* TELECOM customer access maintenance center (*AmE*) (*CAMC*), customer access maintenance centre (*BrE*) (*CAMC*); ~ **de maquinado** *m* ING MECÁ machine center (*AmE*), machine centre (*BrE*), MECÁ machining center (*AmE*), machining centre (*BrE*); ~ **de masa** *m* FÍS center of mass (*AmE*), centre of mass (*BrE*), MECÁ center of inertia (*AmE*), centre of inertia (*BrE*), center of mass (*AmE*), centre of mass (*BrE*), TEC ESP center of mass (*AmE*), centre of mass (*BrE*); ~ **de movimiento** *m* MECÁ center of motion (*AmE*), centre of motion (*BrE*); ~ **muerto delantero** *m* ING MECÁ crank-end dead center (*AmE*), crank-end dead centre

(*BrE*); ~ **muerto trasero** *m* ING MECÁ crank-end dead center (*AmE*), crank-end dead centre (*BrE*); ~ **del núcleo** *m* ÓPT, TELECOM core center (*AmE*), core centre (*BrE*); ~ **de operaciones** *m* TELECOM operations center (*AmE*), operations centre (*BrE*); ~ **de operadores** *m* TELECOM operator center (*AmE*), operator centre (*BrE*); ~ **óptico** *m* FÍS, TELECOM optical center (*AmE*), optical centre (*BrE*); ~ **de oscilación** *m* FÍS center of oscillation (*AmE*), centre of oscillation (*BrE*); ~ **de percusión** *m* FÍS center of percussion (*AmE*), centre of percussion (*BrE*); ~ **de personal de explotación** *m* TELECOM operator center (*AmE*), operator centre (*BrE*); ~ **de personal operador** *m* TELECOM operator center (*AmE*), operator centre (*BrE*); ~ **de presión** *m* FÍS, TRANSP AÉR center of pressure (*AmE*), centre of pressure (*BrE*); ~ **de presión de la pala** *m* TRANSP AÉR blade center of pressure (*AmE*), blade centre of pressure (*BrE*); ~ **principal de conmutación de enlaces internacionales** *m* TELECOM main international trunk-switching center (*AmE*), main international trunk-switching centre (*BrE*); ~ **principal de conmutación internacional** *m* TELECOM main international switching center (*AmE*), main international switching centre (*BrE*); ~ **principal de distribución internacional** *m* TELECOM main international switching center (*AmE*), main international switching centre (*BrE*); ~ **de procesamiento de datos** *m* (*DPC*) INFORM&PD, TELECOM data processing center (*AmE*) (*DPC*), data processing centre (*BrE*) (*DPC*); ~ **de proceso de datos** *m* (*DPC*) INFORM&PD, TELECOM data processing center (*AmE*) (*DPC*), data processing centre (*BrE*) (*DPC*); ~ **de pruebas de vuelo** *m* TEC ESP, TRANSP AÉR flight test center (*AmE*), flight test centre (*BrE*); ~ **de recepción de averías** *m* TELECOM fault-reception center (*AmE*) (*FRC*), fault-reception centre (*BrE*) (*FRC*); ~ **de recogida de datos** *m* TELECOM data collection center (*AmE*) (*DCC*), data collection centre (*BrE*) (*DCC*); ~ **de recolección de datos** *m* TELECOM data collection center (*AmE*) (*DCC*), data collection centre (*BrE*) (*DCC*); ~ **de rectificado** *m* ING MECÁ grinding center (*AmE*), grinding centre (*BrE*); ~ **de rectificado con siete ejes** *m* ING MECÁ grinding center with seven axes (*AmE*), grinding centre with seven axes (*BrE*); ~ **de relajación** *m* METAL relaxation center (*AmE*), relaxation centre (*BrE*); ~ **de reprocesado** *m* PROD rework center (*AmE*), rework centre (*BrE*); ~ **de reproducción** *m* AGRIC breeding station; ~ **del revestimiento** *m* ÓPT *fibras ópticas* cladding center (*AmE*), cladding centre (*BrE*); ~ **de rotación** *m* MECÁ pivot; ~ **de rueda** *m* FERRO *vehículos* wheel web; ~ **secundario de perturbaciones** *m* FÍS ONDAS *del frente de onda* secondary center of disturbance (*AmE*), secondary centre of disturbance (*BrE*); ~ **de simetría** *m* CRISTAL center of symmetry (*AmE*), centre of symmetry (*BrE*); ~ **de sincronización** *m* TELECOM time-division exchange; ~ **sísmico** *m* GEOFÍS *de terremotos*, GEOL earthquake focus; ~ **de sustentación** *m* TRANSP AÉR lift center (*AmE*), lift centre (*BrE*); ~ **térmico** *m* REFRIG thermal center (*AmE*), thermal centre (*BrE*); ~ **de trabajo** *m* PROD work center (*AmE*), work centre (*BrE*); ~ **de tráfico** *m* TRANSP traffic center (*AmE*), traffic centre (*BrE*); ~ **de tránsito internacional** *m* PROD *telecomunicaciones*

gateway; ~ **de vaciado** *m* CONST dumping station; ~ **vélico** *m* TRANSP MAR *diseño naval* center of wind pressure (*AmE*), centre of wind pressure (*BrE*)
Centro: ~ **de Conmutación de la Red** *m* TELECOM Network Switching Center (*AmE*) (*NSC*), Network Switching Centre (*BrE*) (*NSC*); ~ **de Control Técnico y de Operaciones** *m* (*TOCC*) TEC ESP Technical and Operational Control Center (*AmE*) (*TOCC*); ~ **de Operaciones Intelsat** *m* TEC ESP *comunicaciones* Intelsat Operations Center (*AmE*) (*IOS*), Intelsat Operations Centre (*BrE*) (*IOS*); ~ **de Tránsito Internacional** *m* TEC ESP International Transit Center (*AmE*), International Transit Centre (*BrE*)
centroide[1] *adj* FÍS, GEOM centroid
centroide[2] *m* FÍS, GEOM centroid
centrosfera *f* GEOFÍS centrosphere
centrosimétrico *adj* CRISTAL centrosymmetric
ceolita *f* MINERAL, QUÍMICA zeolite
cepellón *m* AGRIC ball
cepillado[1] *adj* MECÁ brushed
cepillado[2] *m* CONST shaving, ING MECÁ surface planing, PAPEL brush finish, TEXTIL *en las operaciones de acabado* brushing; ~ **de bebedero caliente** *m* ING MECÁ hot sprue brushing; ~ **de rebabas calientes** *m* ING MECÁ hot sprue brushing; ~ **tosco** *m* ING MECÁ, PROD rough-cut planing
cepilladora *f* CONST planing tool, ING MECÁ planomiller, shaping machine, shaping planer (*AmE*), surface planing machine (*BrE*), shaper, MECÁ *máquinas herramientas* planer (*AmE*), planing machine (*BrE*), shaper, PAPEL brush-polishing machine; ~ **de bordes de chapas** *f* ING MECÁ plate-edge planing machine; ~ **de carrera ajustable** *f* ING MECÁ adjustable stroke shaper; ~ **de carriles** *f* FERRO *equipo inamovible* rail-planing machine; ~ **de carro** *f* ING MECÁ carriage planing machine; ~ **con un costado abierto** *f* ING MECÁ, PROD open-side planing machine; ~ **de engranajes** *f* ING MECÁ gear-shaving machine; ~ **de engranajes blandos** *f* ING MECÁ gear-shaving machine; ~ **de engranajes sin templar** *f* ING MECÁ gear-shaving machine; ~ **de escote** *f* PROD side-planing machine; ~ **y esgualador** *f* CONST planing and thicknessing machine; ~ **monomontante** *f* ING MECÁ, PROD open-side planing machine (*BrE*); ~ **de puente** *f* ING MECÁ double housing planing machine
cepilladura *m* CONST *carpintería* planing
cepillar *vt* CONST clean off, trim, *madera* rough down, MECÁ plane, *carpintería* trim, PROD *madera* trim
cepillo *m* CONST brush, plane, MECÁ, PAPEL brush; ~ **de aire** *m* IMPR, PAPEL air doctor; ~ **de alambre** *m* PROD *hierro* wire brush; ~ **antiestático** *m* IMPR antistatic brush; ~ **de arañar** *m* CONST scratch brush; ~ **de balanza** *m* LAB *análisis, limpieza* balance brush; ~ **de banco** *m* CONST bench plane; ~ **de la camisa** *m* PAPEL jacket brush; ~ **con cerdas de naylon** *m* P&C nylon brush; ~ **dentado** *m* CONST tooth plane, toothing plane; ~ **para frotar arena** *m* PROD *moldería* swab brush; ~ **de hierro** *m* CONST plane iron; ~ **lateral** *m* CONST side plane; ~ **para lengüeta** *m* CONST tonguing plane; ~ **limpiador** *m* IMPR cleaning brush; ~ **machihembrado con dos extremos** *m* CONST double-ended match plane; ~ **de machihembrar** *m* CONST matching plane; ~ **para machihembrar** *m* CONST tonguing-and-grooving plane; ~ **mecánico** *m* CONST planer (*AmE*), planing machine (*BrE*); ~ **para la moldura del collarín** *m*

CONST neck molding plane (*AmE*), neck moulding plane (*BrE*); ~ **de óvolo** *m* CONST ovolo plane; ~ **con protección** *m* CONST *carpintería* nosing plane; ~ **pulidor** *m* MECÁ dolly; ~ **para quitar el polvo** *m* FOTO duster, dusting brush; ~ **de rafinar** *m* CONST *carpintería* try plane; ~ **de ranura lateral** *m* CONST side rabbet plane; ~ **ranurador** *m* CONST plough plane (*BrE*), plough (*BrE*), plow (*AmE*), *carpintería* fillister plane, plow plane (*AmE*), rebate plane, ING MECÁ rebate plane; ~ **de ranurar** *m* CONST grooving plane, rabbet plane; ~ **de ranurar cuadrado** *m* CONST square rabbet plane; ~ **raspador** *m* CONST scraping plane; ~ **redondo** *m* CONST compass plane; ~ **de talón** *m* CONST *carpintería* ogee plane

cepo *m* ING MECÁ block; ~ **del ancla** *m* TRANSP MAR anchor stock; ~ **de armadura** *m* ELEC *generador* spider-type armature; ~ **de manga** *m* AGRIC headgate; ~ **de polea** *m* ING MECÁ shell; ~ **de sierra** *m* ING MECÁ saw clamp

cepstrum *m* ELECTRÓN cepstrum

cera *f* EMB, ING ELÉC, QUÍMICA, REVEST, TEC PETR, TEXTIL wax; ~ **aislante** *f* ALIMENT wax, ING ELÉC insulating wax; ~ **de carnauba** *f* ALIMENT carnauba wax; ~ **para impregnar** *f* EMB impregnating wax; ~ **microcristalina** *f* TEC PETR *compuestos del petróleo* microcrystalline wax; ~ **mineral** *f* MINERAL hatchetine, mineral tallow; ~ **de parafina** *f* ELEC *aislador*, QUÍMICA paraffin wax; ~ **perdida** *f* C&V lost wax; ~ **de petróleo** *f* TEC PETR *compuestos del petróleo* petroleum wax; ~ **para pisos** *f AmL* (*cf cera para suelos Esp*) DETERG floor polish; ~ **para revestimiento** *f* REVEST coating wax; ~ **para suelos** *f Esp* (*cf cera para pisos AmL*) DETERG floor polish

cerametal *m* MECÁ cermet

cerámica *f* C&V delftware, ceramic, QUÍMICA *sustancia* ceramic; ~ **dental** *f* C&V dental ceramic; ~ **de Holanda** *f* C&V Delft; ~ **ordinaria** *f* C&V ordinary ceramic; ~ **vidriada** *f* C&V glazed pottery

ceramista *m* C&V ceramist

cerargirita *f* MINERAL cerargyrite

cerca[1]: ~ **del fuselaje** *adj* TRANSP AÉR inboard

cerca[2]: ~ **de la costa** *adv* TRANSP MAR inshore

cerca[3] *f* AGRIC fence, CONST enclosure, fence, screen, PROD hedge; ~ **eléctrica** *f* AGRIC, ELEC electric fence; ~ **flotante** *f* CONTAM MAR fence boom; ~ **protectora para máquina** *f* SEG machine fence

cercado[1]: ~ **por el hielo** *adj* TRANSP MAR icebound

cercado[2] *m* PROD hedging

cercanías: ~ **de costa** *f pl* OCEAN foreshore

cercar[1]: **sin** ~ *adj* CONST unfenced

cercar[2] *vt* AGRIC fence in, CONST brace, fence in

cercha *f* MECÁ A-frame, MINAS prop, *entibación* crown bearer; ~ **de dos pares** *f* CONST *edificación* close couple

cerco *m* CONST casement, stirrup, VEH *faro* rim; ~ **danés** *m* OCEAN *artificios de pesca* Danish seine; ~**-guía** *m* AGUA frame; ~ **de madera** *m* CONST timber frame

cerdo: ~ **joven castrado** *m* AGRIC barrow; ~ **libre de ciertas enfermedades específicas** *m* AGRIC specific pathogen free pig; ~ **pancetero** *m* AGRIC baconer

cereal *m* AGRIC breadstuff, small grain; ~ **forrajero** *m* AGRIC feed grain; ~ **panificador** *m* AGRIC bread grain; ~ **verde** *m* AGRIC immature grain

cerealina *f* QUÍMICA cerealin

cerealista *m* AGRIC grain dealer

cerealosa *f* QUÍMICA *glucosa* cerealose

cererita *f* MINERAL cererite

ceresina *f* QUÍMICA ceresin, ceresine

cérico *adj* QUÍMICA ceric

cerílico *adj* QUÍMICA cerylic

cerilo *m* QUÍMICA keryl

cerina *f* MINERAL cerine, orthite, NUCL orthite, QUÍMICA *mineral* cerine

cerio *m* (*Ce*) QUÍMICA cerium (*Ce*)

cerita *f* MINERAL cerite

cermet *m* MECÁ cermet

CERN *abr* (*Conseil Européen pour la Recherche Nucléaire*) FÍS PART CERN (*European Laboratory for Particle Physics*)

cernedor: ~ **giratorio** *m* AGRIC rotary cleaner; ~ **rotativo** *m* AGRIC rotary weed screen

cerner *vt* CARBÓN faredice

cernidor *m* RECICL sieve

cernidura *f* CARBÓN, QUÍMICA screening

cernir *vt* CARBÓN sieve, CONST screen, sieve, PROC QUÍ, PROD, RECICL sieve

cero *m* GEN zero; ~ **absoluto** *m* ALIMENT, FÍS, QUÍMICA, REFRIG, TERMO absolute zero; ~ **de la carta** *m* TRANSP MAR *navegación* chart datum; ~ **cruzado** *m* IMPR slashed zero; ~ **eléctrico** *m* ELEC *de instrumento*, INSTR electrical zero; ~ **de filtro** *m* ELECTRÓN filter zero; ~ **grados** *m pl* TERMO zero degrees; ~ **hidrográfico** *m* OCEAN *equipos para mediciones hidrográficas* sounding datum; ~ **a la izquierda** *m* PROD leading zero

cerolita *f* MINERAL cerolite

ceroso *adj* QUÍMICA cerous

cerótico *adj* QUÍMICA cerotic

cerovalente *adj* QUÍMICA zerovalent

cerquero *m* OCEAN *tecnología pesquera* seiner; ~ **atunero** *m* OCEAN tunny boat

cerradero *m* ING MECÁ keeper

cerrado[1] *adj* MATEMÁT *curvas, espacios topológicos, intervalos* closed, PROD clinched, locked down, TRANSP *semáforo* red

cerrado[2] *m* C&V shutdown

cerrador: ~ **automático** *m* EMB automatic capper

cerradora: ~ **de botellas** *f* EMB bottle-closing machine; ~ **de botes** *f* EMB can-closing machine (*AmE*), tin-closing machine (*BrE*); ~ **de sacos** *f* EMB sack-closing machine; ~ **de tubos** *f* EMB tube-closing machine

cerradura *f* CONST lock, locking, SEG, VEH *carrocería* lock; ~ **del cilindro** *f* PROD cylinder lock; ~ **de combinación** *f* CONST *cerrajería* combination lock, puzzle lock; ~ **de contactos para la protección de la memoria** *f* INFORM&PD, PROD memory protect keyswitch; ~ **a derechas** *f* CONST right-hand lock; ~ **embutida** *f* CONST mortise lock; ~ **entallada** *f* CONST warded lock; ~ **de funcionamiento ultrarrápido** *f* PROD snap lock; ~ **de gancho** *f* EMB hooked lock; ~ **de muelle** *f* CONST *cerrajería* spring lock; ~ **muerta** *f* CONST dead lock; ~ **a nivel** *f* CONST box lock, flush lock; ~ **con pestillo** *f* CONST tumbler lock; ~ **de puerta** *f* VEH door lock; ~ **de resorte** *f* CONST spring lock; ~ **de seguridad** *f* CONST safety lock; ~ **tubular** *f* CONST piped key lock

cerrajería *f* CONST ironworking, lock work, locksmithery, locksmithing

cerrajero *m* CONST locksmith

cerramiento *m* CONST enclosure, walled enclosure,

MINAS *embalse* dam; ~ **permanente de cemento** *m* *Esp* (*cf cerramiento permanente de concreto AmL*) CONST permanent concrete shuttering; ~ **permanente de concreto** *m AmL* (*cf cerramiento permanente de cemento Esp*) CONST permanent concrete shuttering; ~ **en superficie** *m* PROD surface-mounted enclosure

cerrar *vt* AGUA turn off, CINEMAT *diafragma* close down, CONST close, lock, CONTAM MAR *la mancha de hidrocarburos* corral, FOTO *diafragma* close down, INFORM&PD *sesión* sign off, ING ELÉC close, make, INSTAL HIDRÁUL *admisión de vapor* cut off, MECÁ lock, MINAS insulate, seal, PROD *molde, placa* close, insulate, *grifos, llaves, interruptores* turn off, TEC ESP, TRANSP AÉR shut down, TRANSP MAR *escotillas* batten down; ~ **por calor** *vt* TERMO heat-seal; ~ **con cremallera** *vt* TEXTIL zip; ~ **herméticamente** *vt* PROD insulate, seal hermetically

cerrarse *v refl* CONST, ING MECÁ lock

cerrojo *m* CONST bolt, box, keeper, latch lock, lock, lock key, latch, INFORM&PD latch, ING MECÁ bolt, fastener, latch, MECÁ latch, fastener, PROD fastener; ~ **acodado cuadrado** *m* ING MECÁ square neck bolt; ~ **de bloqueo** *m Esp* (*cf interruptor con traba AmL*) TV switch lock; ~ **de botón** *m* ING MECÁ thumb bolt; ~ **de cierre** *m* PROD locking latch; ~ **de enclavamiento de energía** *m* PROD power-interlock bar; ~ **de ignición antirrobo** *m* AUTO antitheft ignition lock; ~ **de marco de ventana** *m* CONST casement fastener; ~ **mecánico** *m* PROD mechanical latch; ~ **muerto** *m* CONST *cerrajería* dead bolt; ~ **no retentivo** *m* PROD nonretentive latch; ~ **pasador** *m* CONST barrel bolt; ~ **de la portezuela del tren de aterrizaje** *m* TRANSP AÉR landing gear door latch; ~ **de la puerta** *m* CONST door bolt; ~ **de resorte** *m* CONST *cerraduras* spring bolt, PROD spring latch; ~ **de retroceso** *m* EMB flip spout closure; ~ **de seguridad** *m* ING MECÁ safety catch; ~ **de ventana** *m* CONST sash fastener, window catch

certificación *f* VEH *reglamentaciones* certification; ~ **de ruido** *f* ACÚST noise certification

certificado: ~ **de aceptación** *m* CALIDAD acceptance certificate, TRANSP AÉR approval certificate; ~ **de aeronavegabilidad** *m* TRANSP AÉR certificate of airworthiness; ~ **de aprobación** *m* TRANSP AÉR approval certificate; ~ **aprobado de primeros auxilios** *m* SEG approved first aid certificate; ~ **de conformidad** *m* CALIDAD certificate of compliance, certificate of conformity, ING MECÁ certificate of conformity; ~ **de construcción** *m* TRANSP MAR builder's certificate; ~ **de garantía** *m* PROD warranty; ~ **de garantía de calidad** *m* CALIDAD, CONST, ING MECÁ, MECÁ, NUCL, PROD, TEC ESP quality-assurance certificate; ~ **de homologación** *m* ING MECÁ certificate of conformity; ~ **de inscripción en el Registro Mercantil** *m* TRANSP MAR *libro de buques* certificate of registration; ~ **de navegabilidad** *m* TRANSP MAR certificate of seaworthiness; ~ **de registro** *m* TRANSP AÉR certificate of registration; ~ **del Registro Marítimo** *m* TRANSP MAR certificate of registry

certificar *vt* CALIDAD, TRANSP AÉR, TRANSP MAR certify

cerúleo *m* QUÍMICA ceruleum

cerusa *f* QUÍMICA *colorante y cosmética* ceruse

cerusita *f* MINERAL cerusite

cervantita *f* MINERAL cervantite

cesación *f* ING MECÁ stop

cesar[1] *vt* ING MECÁ hold up

cesar[2]: ~ **el trabajo** *vi* MINAS cease work

cese *m* ING MECÁ knock-off

cesio *m* (*Cs*) ELECTRÓN, ING ELÉC, QUÍMICA, TEC ESP caesium (*BrE*) (*Cs*), cesium (*AmE*) (*Cs*)

cesión: ~ **de intereses** *f* TEC PETR *negociación, traspaso de licencias* farming-out; ~ **lateral** *f* NUCL lateral yielding

césped *m* AGRIC sod

cesta *f* OCEAN *trampa de pesca* basket trap, TRANSP *grúas* skip, TRANSP AÉR *reaprovisionamiento en vuelo* drogue; ~ **basculante de manejo del combustible** *f* NUCL tilting basket; ~ **de la compra** *f* EMB grocer's bag

cesto: ~ **de carbón** *m* CARBÓN *cantidad de peso de carbón* coal basket; ~ **de coque** *m* CARBÓN coke basket

cetano *m* QUÍMICA *petróleo* cetane

cetena *m* QUÍMICA ketene

cetilo *m* QUÍMICA *ceras* cetyl

cetimina *f* QUÍMICA ketimine

cetoácido *m* QUÍMICA keto-acid

cetol *m* QUÍMICA ketol

cetona *f* P&C *grupo de compuesto químico*, QUÍMICA ketone

cetónico *adj* QUÍMICA ketonic

cetosa *f* QUÍMICA ketose

cevadina *f* QUÍMICA *semillas* cevadine

Cf *abr* (*californio*) FÍS RAD, QUÍMICA Cf (*californium*)

CFC *abr* (*clorofluorocarbono*) CONTAM *productos químicos*, EMB CFC (*chlorofluorocarbon*)

CFD *abr* (*dinámica de fluidos computacional*) TRANSP AÉR CFD (*computational fluid dynamics*)

CGA[1] INFORM&PD (*adaptador gráfico de colores*) CGA (*color graphics adaptor AmE, colour graphics adaptor BrE*), TELECOM (*control automático de ganancia*) AGC (*automatic gain control*)

CGA[2]: ~ **retardada** *f* (*control de ganancia automática retardada*) ELECTRÓN delayed AGC (*delayed automatic gain control*)

CGS *abr* (*cegesimal*) METR CGS (*centimeter-gram-second AmE, centimetre-gram-second BrE*)

chabasita *f* MINERAL chabasite

chabota *f* ING MECÁ *martillo pilón* anvil block, MINAS *martillo pilón* drive block

chaflán *m* CONST chamfer, fillet, ING MECÁ bevel, bevel edge; ~ **de aflojamiento rápido** *m* ING MECÁ quick-release taper

chala *f* AGRIC, ALIMENT husk; ~ **de maíz** *f AmL* AGRIC corn husk (*AmE*), maize husk (*BrE*)

chalán *m AmL* PETROL barge

chalana *f* PETROL barge

chalcolita *f* NUCL torbernite

chaleco: ~ **antibalas** *m* D&A bulletproof jacket, bulletproof vest; ~ **ascensional** *m* OCEAN *buceo* buoyancy compensator; ~ **a prueba de balas** *m* D&A bulletproof jacket; ~ **salvavidas** *m* SEG life jacket, life preserver, lifebelt, TRANSP AÉR life jacket, life preserver, TRANSP MAR *emergencia* life jacket, lifebelt

chambrana *f* CONST jamb lining

chamico *m* AGRIC jimson weed, thorn apple

chamosita *f* MINERAL chamosite

chamuscado[1] *adj* P&C scorchy, TERMO scorched

chamuscado[2] *m* TEXTIL singeing

chamuscadora *f* TEXTIL singeing machine

chamuscar vt CARBÓN char, TERMO scorch, TEXTIL *operación de acabado* singe

chapa f CARBÓN iron, CONST veneer, EMB *botellas* crown closure, ING MECÁ plate, MECÁ plate, sheet, METAL plate, SEG lock, TEC ESP ply, TRANSP MAR *construcción naval* plate, sheet; **~ abrasiva de cono truncado** f ING MECÁ truncated-cone abrasive sheet; **~ de acero** f VEH *carrocería* sheet steel; **~ de acero para esmaltar** f REVEST enameling sheet (*AmE*), enamelling sheet (*BrE*); **~ de base en forma de U** f ING MECÁ U-shaped base plate; **~ de blindaje** f CONST armor plate (*AmE*), armour plate (*BrE*); **~ con bulbo** f MECÁ bulb plate; **~ de caldera** f INSTAL HIDRÁUL boiler test-plate; **~ para caldera** f INSTAL HIDRÁUL boiler plate; **~ de calibrar** f PROD gaging plate (*AmE*), gauging plate (*BrE*); **~ cáncamo** f TRANSP MAR *construcción naval* eyeplate, lug; **~ circular maquinada** f ING MECÁ machined circular plate; **~ cónica** f TEC ESP conical shell; **~ cortante** f CONST cutting veneer; **~ del cuadrante** f ING MECÁ quadrant plate; **~ de cubierta** f TRANSP MAR deck plate; **~ de envuelta del revestimiento calorífugo** f INSTAL HIDRÁUL clothing plate, facer, lining plate; **~ del estator** f ING ELÉC stator lamination; **~ estriada** f PROD checker plate (*AmE*), checkered plate (*AmE*), chequer plate (*BrE*), chequered plate (*BrE*); **~ extractora** f ING MECÁ stripper plate; **~ de fabricante** f ING MECÁ, MECÁ nameplate; **~ con faldilla** f PROD flange plate; **~ fina de madera revestida con cartón por ambas caras** f PROD veneer board; **~ del fondo** f TRANSP MAR inner bottom plating; **~ de forro** f PROD lining plate; **~ del forro** f PROD *de barril, calentador* shell plate; **~ de fricción** f ING MECÁ chafing plate; **~ frontal** f TV faceplate; **~ de fuego** f PROD flame plate; **~ de guarda** f CONST *cerraduras* finger plate; **~ del hogar** f PROD furnace plate; **~ horizontal de la sobrequilla** f TRANSP MAR *construcción* rider plate; **~ de laminación** f REVEST lamination sheet; **~ laminada** f CONST laminated sheet; **~ magnética** f ING ELÉC stamping; **~ de matrícula** f VEH *carrocería* license plate (*AmE*), number plate (*BrE*); **~ para moldes** f ING MECÁ mold plate (*AmE*), mould plate (*BrE*); **~ de montaje** f TEC ESP support plate; **~ de número** f ING MECÁ number plate; **~ del ojo de la llave** f CONST key plate; **~ ondulada** f CONST corrugated iron; **~ de percusión** f ING MECÁ percussion plate; **~ perforadora** f CARBÓN punched plate; **~ del piso** f MECÁ floor plate; **~ de plástico laminado** f P&C *plásticos, producto semielaborado* laminated sheet; **~ de pliegues redondos** f PROD round pinch plate; **~ posterior de la caja de fuegos** f ING MECÁ back sheet of fire box; **~ protectora** f PROD guard plate; **~ de punzonar** f ING MECÁ, MECÁ punch plate; **~ de quilla** f TRANSP MAR *construcción* keel plate; **~ rectangular maquinada** f ING MECÁ machined rectangular plate; **~ de recubrimiento** f PROD junction plate; **~ de relleno** f TRANSP MAR *en cubierta* filling plate; **~ de respaldo** f ING MECÁ backing plate; **~ de revestimiento** f PROD lining plate; **~ del revestimiento** f INSTAL HIDRÁUL *calderas* clothing plate; **~ para revestimiento** f REVEST coating sheet; **~ de rozamiento** f ING MECÁ chafing plate; **~ soporte** f ING MECÁ backing plate; **~ superior** f ING MECÁ top plate; **~ de trinca** f ING MECÁ locking plate, lockplate; **~ tubular de la caja de fuegos** f

FERRO *vehículos* firebox tube plate; **~ de varenga** f MECÁ *barcos* floor plate

chapado m CONST veneering; **~ deprimido** m TELECOM depressed cladding; **~ parcial** m REVEST partial plating; **~ plástico** m ING ELÉC plastic cladding; **~ sin recubrimiento** m ÓPT *fibras ópticas* depressed cladding

chapaleta f ING MECÁ clack

chaparrón m CONST shower, METEO cloudburst, shower; **~ electrónico** m ELECTRÓN, FÍS PART, NUCL electron shower; **~ fuerte** m METEO cloudburst, heavy shower

chapas: **~ del forro** f pl TRANSP MAR *construcción naval* shell plating; **~ perforadas y sin perforar** f pl ING MECÁ bored-and-plain plates

chapeado m MECÁ cladding; **~ de fibra óptica** m ÓPT, TELECOM fiber-optic cladding (*AmE*), fibre-optic cladding (*BrE*); **~ homogéneo** m TELECOM homogeneous cladding

chapista m PROD plater

chaqueta f ING MECÁ, LAB, MECÁ, PROD *cilindros* jacket; **~ del carburador** f *AmL* (*cf camisa del carburador Esp*) AUTO, VEH carburetor jacket (*AmE*), carburettor jacket (*BrE*); **~ del cilindro** f AUTO carburetor jacket (*AmE*), carburettor jacket (*BrE*), PROD cylinder casing, cylinder jacket, VEH carburetor jacket (*AmE*), carburettor jacket (*BrE*); **~ de la fibra** f ÓPT fiber jacket (*AmE*), fibre jacket (*BrE*); **~ del silenciador** f AUTO muffler jacket

charca f CARBÓN pond

charco m GAS pool, HIDROL puddle

charm: **~ quark** m FÍS, FÍS PART charm quark

charnela[1]: **a ~** adj ING MECÁ hinged; **de ~** adj ING MECÁ, MECÁ hinged

charnela[2] f AUTO knuckle, CONST knuckle, GEOL *de un pliegue* hinge, ING MECÁ joint, clack, hinge, knuckle, pivot, INSTR hinge, MECÁ pivot, hinge, NUCL valve flap; **~ de aleteo** f TRANSP AÉR flapping hinge; **~ de aleteo descentrada** f TRANSP AÉR offset flapping hinge; **~ de arrastre** f TRANSP AÉR drag hinge; **~ de batimiento** f TRANSP AÉR flapping hinge; **~ de cambio de paso de la pala** f TRANSP AÉR blade-pitch change-hinge; **~ de plegado de la pala** f TRANSP AÉR blade folding-hinge; **~ portacebo** f MINAS *pega* sliding shutter, TEC ESP *erizo antisubmarino* shutter; **~ de resistencia** f TRANSP AÉR drag hinge

charola f *AmL* (*cf placa Esp*) C&V pan, LAB tray; **~ de la máquina** f *AmL* (*cf placa de la máquina Esp*) C&V machine tray

charolado m COLOR japan work

charolamiento m PROD japanning

charolas f pl C&V catch pans

chárter: **~ de reserva por adelantado** m TRANSP AÉR advance booking charter; **~ de reserva anticipada** m TRANSP AÉR advance booking charter

chasis m CINEMAT magazine, FOTO dark slide, ING MECÁ frame, VEH chassis, frame; **~ de alimentación** m CINEMAT feed magazine; **~ de cabina** m VEH chassis cab; **~ coaxial** m CINEMAT coaxial magazine; **~ de columnas** m PROD column box; **~ concéntrico** m CINEMAT concentric magazine; **~ de cortinilla** m FOTO roller-blind dark slide; **~ doble** m FOTO twin magazine; **~ de E/S** m (*chasis de entrada/salida*) PROD I/O chassis (*input/output chassis*); **~ giratorio** m FOTO, TEC ESP rotary magazine; **~ inferior** m VEH underbody; ~ **para la luz del día** *m* CINEMAT daylight magazine; ~ **metálico** *m* FOTO metal dark slide; ~ **con piso descendente** *m* VEH drop bed frame; ~ **de procesador** *m* PROD processor chassis; ~ **de procesador primario** *m* PROD primary processor chassis; ~ **receptor** *m* FOTO take-up cassette; ~ **de recogida** *m* CINEMAT take-up magazine; ~ **submarino** *m* AmL (*cf carcasa submarina Esp*) CINEMAT *para cámara*, FOTO underwater housing

chasquido *m* ACÚST click, CARBÓN crack

chata *f* TRANSP MAR float, *barcaza* barge; ~ **para la colocación de oleoductos submarinos** *f* PETROL, TEC PETR lay barge; ~ **velera** *f* TRANSP MAR barge

chatarra *f* CARBÓN, MECÁ, METAL, PROD scrap, TEC PETR *pozo* junk; ~ **de bebederos** *f* PROD *fundería* sprue; ~ **de fundición** *f* PROD cast scrap; ~ **de la fundición** *f* PROD foundry scrap; ~ **de hierro** *f* PROD scrap iron; ~ **de metal** *f* PROD scrap metal

chatita *f* AmL (*cf camioneta eléctrica de reparto Esp*) TRANSP electric pick-up

chato *adj* TRANSP MAR flat-bottomed

chaveta *f* AmL (*cf cuña Esp*) ING ELÉC pin, ING MECÁ forelock, gib and cotter, key, pin, MECÁ key, peg, MINAS *entibación* gad; ~ **ajustable** *f* ING MECÁ adjustable gib; ~ **angular** *f* ING MECÁ angled key; ~ **de arrastre** *f* ING MECÁ driver; ~ **con cabeza** *f* ING MECÁ gib head key; ~ **circular** *f* ING MECÁ disc key (*BrE*), disk key (*AmE*); ~ **cóncava** *f* ING MECÁ saddle key; ~ **cónica** *f* ING MECÁ taper key; ~ **cónica delgada** *f* ING MECÁ thin taper key; ~ **cónica con tacón** *f* ING MECÁ taper key with gib head; ~ **cónica sin tacón** *f* ING MECÁ taper key without gib head; ~ **y contrachaveta** *f* ING MECÁ gib and key; ~ **corrediza** *f* CONST, ING MECÁ, PROD feather; ~ **de disco** *f* ING MECÁ disc key (*BrE*), disk key (*AmE*), VEH Woodruff key; ~ **doble** *f* ING MECÁ gib and cotter; ~ **del eje** *f* ING MECÁ shaft key; ~ **encastrada** *f* ING MECÁ sunk key; ~ **de fijación** *f* ING MECÁ set key; ~ **de fricción** *f* ING MECÁ saddle key; ~ **guía** *f* ING MECÁ, MECÁ cotter pin; ~ **hendida** *f* ING MECÁ locking pin, split cotter pin, split pin; ~ **de horquilla** *f* ING MECÁ, MECÁ clevis pin; ~ **hueca** *f* ING MECÁ saddle key; ~ **hundida** *f* ING MECÁ sunk key; ~ **de reglaje** *f* ING MECÁ adjusting gib; ~ **de resorte** *f* ING MECÁ spring cotter, spring governor; ~ **de retén** *f* ING MECÁ, MECÁ cotter pin; ~ **roscada** *f* ING MECÁ screw key; ~ **de seguridad** *f* ING MECÁ lockpin, locking pin; ~ **con talón** *f* ING MECÁ gib head key; ~ **de talón** *f* ING MECÁ gib; ~ **tangencial** *f* ING MECÁ tangent key; ~ **Woodruff** *f* VEH Woodruff key

chavetaje *m* ING MECÁ cottering

chavetas: ~ **y chaveteros cónicos** *f pl* ING MECÁ taper keys and keyways; ~ **y chaveteros de media caña** *f pl* ING MECÁ Woodruff keys and keyways; ~ **y chaveteros tangenciales** *f pl* ING MECÁ tangential keys and keyways; ~ **paralelas cuadradas** *f pl* ING MECÁ square parallel keys

chavetero *m* ING MECÁ keyway, MECÁ key slot, keyway, NUCL keyway, PROD keyslot, slot

chavibetol *m* QUÍMICA chavibetol

chavicol *m* QUÍMICA chavicol

chequeo: ~ **prepotencial** *m* PROD prepower check

chert *m* GEOL chert

chessylita *f* MINERAL, QUÍMICA chessylite

cheurón *m* ING MECÁ chevron

chibalete *m* IMPR *para cajas tipográficas* rack

chicharra *f* ING MECÁ ratchet, ratchet spanner, ratchet wrench, MECÁ ratchet

chicle *m* QUÍMICA gum; ~ **de aceleración** *m* AUTO acceleration jet; ~ **de corrección de aire** *m* AUTO air correction jet; ~ **de ralentí** *m* AUTO idle jet

chicote *m* AmL (*cf final Esp*) ING ELÉC stub, MINAS *marina* free end; ~ **de a bordo** *m* AmL (*cf final de a bordo Esp*) TRANSP MAR *de la cadena del ancla, del cabo* bitter end; ~ **del cable** *m* AmL (*cf final del cable Esp*) MINAS *marina* cable end

chiflón *m* AmL (*cf inclinación del eje en un coladero Esp*) CARBÓN *minas* dip, MINAS inclined shaft, dip, diphead, *minería a cielo abierto* slope

chigre *m* CONST, CONTAM MAR winch, TRANSP MAR *de amarre, del ancla* windlass, *de carga* winch; ~ **de amarre de tensión constante** *m* TRANSP MAR self-tensioning winch; ~ **de carga** *m* TRANSP MAR cargo winch; ~ **de tensión constante** *m* TRANSP MAR *equipamiento de cubierta* self-tensioning winch

chileíta *f* MINERAL chileite

chillido *m* TELECOM squealing

chimenea *f* AmL (*cf aspirador Esp, coladero Esp*) C&V flue, CARBÓN *minas* shaft, chute, rise face, CONST stack, chimney, hearth, GEOL *volcánica* vent, MINAS chute, raise, *minas* shaft, staple shaft, *tiro de alcancía* nail, chimney, staple pit, *comunicación de dos partes de la mina* pass, TEC PETR stack, TERMOTEC stack, chimney, TRANSP MAR *de barco* funnel; ~ **de acero** *f* CONST steel chimney; ~ **en bancos** *f* Esp (*cf contracielo escalonado AmL*) MINAS raise stope; ~ **para carbón** *f* CARBÓN *minas*, MINAS coal chute; ~ **cónica** *f* CARBÓN conical mill; ~ **de equilibrio** *f* ENERG RENOV surge shaft; ~ **metálica** *f* CONST funnel; ~ **de mineral** *f* AmL (*cf coladero Esp, cf chimenea de mineral para llenado Esp*) MINAS mill hole, fill raise, mill, *mineral* coarse concentration mill; ~ **para mineral** *f* MINAS ore pass; ~ **de mineral para llenado** *f* Esp (*cf chimenea de mineral AmL*) MINAS fill raise; ~ **principal** *f* MINAS main chute; ~ **de relleno** *f* MINAS filling raise; ~ **de ventilación** *f* CARBÓN flume, MINAS air stack; ~ **de ventilación de superficie** *f* MINAS surface ventilation chimney; ~ **volcánica** *f* GEOL volcanic vent

chimeneas *f pl* PROD *horno alto, canales* channeling (*AmE*), channelling (*BrE*)

china: ~ **de hueso** *f* C&V bone china

chinche *f* AGRIC bug, chinch bug; ~ **tintórea** *f* AGRIC, TEXTIL cotton stainer

chinchorro *m* D&A, P&C, TRANSP MAR dinghy; ~ **de caucho** *m* D&A, P&C, TRANSP MAR rubber dinghy; ~ **de goma** *m* D&A, P&C, TRANSP MAR rubber dinghy; ~ **inflable** *m* TRANSP MAR inflatable dinghy

chino *m* ALIMENT conical sieve

chip *m* ELEC, ELECTRÓN, INFORM&PD, TEC ESP, TELECOM *circuito integrado* chip; ~ **de acoplamiento** *m* ELECTRÓN companion chip; ~ **amplificador** *m* ELECTRÓN amplifier chip; ~ **de amplificador operacional** *m* ELECTRÓN operational amplifier chip; ~ **analógico** *m* ELECTRÓN analog chip; ~ **de arseniuro de galio** *m* ELECTRÓN, FÍS, ÓPT gallium arsenide chip; ~ **para la cancelación de ecos** *m* ELECTRÓN echo-canceling chip (*AmE*), echo-cancelling chip (*BrE*); ~ **codificador** *m* ELECTRÓN encryption chip; ~ **convertidor** *m* ELEC, ELECTRÓN converter chip; ~ **digital** *m* ELECTRÓN digital chip; ~ **diseñado sobre pedido** *m* ELECTRÓN custom-designed chip; ~ **especializado** *m* ELECTRÓN dedicated chip; ~ **de filtro** *m* ELECTRÓN on-chip filter; ~ **de interconexión** *m* ELECTRÓN interface chip; ~ **lógico aleatorio** *m* ELECTRÓN, INFORM&PD random-logic chip; ~ **de memoria** *m* INFORM&PD memory chip; ~ **microprocesador** *m* ELECTRÓN, INFORM&PD microprocessor chip; ~ **óptico-electrónico** *m* ELECTRÓN optoelectronic chip; ~ **personalizado** *m* INFORM&PD custom chip; ~ **procesador de señales** *m* ELECTRÓN signal-processing chip; ~ **reproductor de imágenes** *m* ELECTRÓN imaging chip; ~ **semiconductor** *m* ELECTRÓN semiconductor chip; ~ **semipersonalizado** *m* ELECTRÓN semicustom chip; ~ **de silicio** *m* ELECTRÓN, FÍS RAD, INFORM&PD silicon chip; ~ **soporte del haz** *m* ELECTRÓN *microplacas* beam lead chip; ~ **de tecnología híbrida** *m* ELECTRÓN chip-and-wire hybrid circuit; ~ **transformador** *m* ELEC, ELECTRÓN converter chip; ~ **de uso general** *m* ELECTRÓN general-purpose chip; ~ **varactor** *m* ELECTRÓN varactor chip; ~ **VLSI** *m* INFORM&PD VLSI chip; ~ **de voz** *m* INFORM&PD speech chip

chirrido *m Esp* (*cf gorjeo AmL*) INFORM&PD chirp, TELECOM chirping; ~ **de frenos** *m* ACÚST brake squeal

chispa *f* GEN spark; ~ **de colector** *f* ELEC *en máquina*, FÍS, ING ELÉC brush discharge; ~ **eléctrica** *f* ELEC *descarga electric spark*; ~ **eléctrica de escobilla** *f* ELEC *en máquina* brush sparking; ~ **de encendido** *f* ING ELÉC initiating spark; ~ **de escobilla** *f* ELEC *en máquina* brush discharge, brush sparking, FÍS, ING ELÉC brush discharge

chispas *f pl* ING MECÁ, PROD, SEG flying sparks; ~ **del colector** *f pl* ELEC, ELECTRÓN commutator sparking

chispeo *m* TV sparking

chispero *m* ING ELÉC arrester; ~ **protector** *m* ELEC protective spark discharger

chispómetro *m* ING ELÉC measuring spark gap

chisporrotear *vi* ING ELÉC spark, QUÍMICA *líquido* sputter

chisporroteo *m* ING ELÉC sparking, QUÍMICA *líquido* sputtering

chivato *m* AUTO flashing warning light, ING MECÁ telltale, SEG flashing warning light

chiviatita *f* MINERAL chiviatite

chloantita *f* MINERAL chloanthite

chocar: ~ **con** *vt* ING MECÁ, MECÁ impinge on; ~ **contra** *vt* SEG crash into

choque *m* ACÚST shock, CARBÓN collision, impact, CONST impact, ELEC *inductor* choke, ELECTRÓN *electrodeposición* striking, FÍS impact, collision, INFORM&PD crash, ING ELÉC choke, ING MECÁ impingement, MECÁ impact, PROD jar, SEG collision, TEC ESP shock, TRANSP collision; ~ **al amerizar** *m* TEC ESP *cosmonaves* splashdown; ~ **de cabeza en el disco** *m* INFORM&PD head crash; ~ **por detrás** *m* AUTO, TRANSP, VEH rear-end collision; ~ **elástico** *m* FÍS elastic collision; ~ **eléctrico** *m* ELEC, FÍS RAD, ING ELÉC electric quadrupole, SEG electric shock; ~ **de filtro** *m* ELEC *red*, ING ELÉC filter choke; ~ **importante** *m* TRANSP primary collision; ~ **inelástico** *m* FÍS inelastic collision; ~ **lateral** *m* TRANSP side collision; ~ **múltiple** *m* TRANSP multiple pile-up; ~ **oscilante** *m* ING ELÉC swinging choke; ~ **pirotécnico** *m* TEC ESP pyrotechnical shock; ~ **secundario** *m* TRANSP secondary collision; ~ **térmico** *m* TEC ESP, TERMO

thermal shock; ~ **térmico a presión** *m* NUCL pressurized thermal shock

chorlo *m* MINERAL schorl; ~ **rojo** *m* MINERAL *variedad de rutilo* red schorl

chorreado: ~~**-alimentador** *m* C&V *de un horno de vidrio* canal; ~ **con granalla** *m* MINAS shotting, NUCL shot peening

chorreador *m* C&V feeder, forehearth; ~ **de fibras** *m* C&V fiber feeder (*AmE*), fibre feeder (*BrE*)

chorreadora: ~ **de arena** *f* PROD sandblast machine; ~ **de arena de mesa rotatoria** *f* PROD *fundería* rotary-table sandblast machine

chorrear: ~ **con arena** *vt* CONST *explosivos* blast, sandblast, MECÁ sandblast

chorreo: ~ **con granalla** *m* PROD peening

chorro *m* AGUA, AUTO, FÍS, FÍS FLUID, GEOL jet, HIDROL spouting, spurting out, NUCL *fluido* jet, TEC ESP *despegue* blast, TEC PETR *perforación* jet, TRANSP AÉR blast, jet, slipstream; ~ **en abanico** *m* AGUA spreader jet; ~ **de agua** *m* AGUA water jet; ~ **de agua caliente** *m* GEOFÍS *instrumentos, géiseres, prospección* hot water jet; ~ **de agua giratorio** *m* TRANSP rotatable water jet; ~ **de agua vertical** *m* TEXTIL vertical water flow; ~ **de aire** *m* CARBÓN air blast; ~ **de arena** *m* CARBÓN sandblast, CONST, MECÁ grit blasting, PROD sand jet; ~ **de arranque** *m* AUTO starter jet, starting jet; ~ **auxiliar de alta velocidad** *m* AUTO high-speed auxiliary jet; ~ **de colada** *m* METAL *fundición* jet, PROD *funderías* header, jet; ~ **cortador de bordes** *m* PAPEL edge nozzle, edge spray; ~ **de flujo rápido** *m* CONTAM MAR flushing; ~ **de gases** *m* GAS, PROD gas jet, TERMO exhaust gases; ~ **de gran potencia** *m* TRANSP heavy jet; ~ **limpiaparabrisas** *m* AUTO windscreen washer jet (*BrE*), windshield washer jet (*AmE*); ~ **del motor a reacción** *m* TRANSP AÉR engine-jet wash; ~ **a presión** *m* TERMOTEC pressure jet; ~ **principal** *m* AUTO, VEH main jet; ~ **de ralentí** *m* AUTO idle jet; ~ **del reactor** *m* TRANSP AÉR engine-jet wash, jet stream; ~ **del rotor** *m* TRANSP AÉR rotor stream; ~ **de vapor** *m* GEOFÍS steam jet; ~ **de vapor de agua** *m* INSTAL HIDRÁUL blast of steam, steam jet; ~ **de viento** *m* PROD blast

chubasco *m* METEO shower; ~ **blanco** *m* OCEAN white squall; ~ **fuerte** *m* METEO heavy shower; ~ **seco** *m* OCEAN white squall

chumacera *f* AUTO journal, ELEC *máquina* bearing, ING MECÁ pillow block, bearing, journal bearing, axle box (*BrE*), journal, cushion, journal box (*AmE*), MECÁ, NUCL journal, PROD plummer block, TRANSP MAR *equipamiento de buques* oarlock (*AmE*), rowlock (*BrE*); ~ **del eje** *f* CONST, TRANSP MAR main-shaft bearing; ~ **de empuje** *f* TRANSP MAR *de motor* thrust bearing, thrust block; ~ **de horquilla** *f* TRANSP MAR *remo* oarlock (*AmE*), rowlock (*BrE*); ~ **reforzada** *f* PROD heavy-type plummer block

chupador *m* ING MECÁ sucker

chupón *m* AGRIC, IMPR sucker; ~ **de avance** *m* IMPR forwarding sucker

CI *abr* GEN (*circuito integrado*) IC (*integrated circuit*), (*circuito impreso*) PC (*printed circuit*)

Ci *abr* (*curie*) FÍS, FÍS RAD Ci (*curie*), METEO (*cirrus*) CI (*cirrus*)

CIAE *abr* (*circuito integrado de aplicación específica*) INFORM&PD ASIC (*application-specific integrated circuit*)

ciamélida *f* QUÍMICA cyamelid, cyamelide

cian *m* IMPR cyan
cianamida *f* QUÍMICA cyanamide
cianato *m* QUÍMICA cyanate; **~ mercúrico** *m* QUÍMICA mercury fulminate
cianhídrico *adj* QUÍMICA hydrocyanic
ciánico *adj* QUÍMICA *ácido* cyanic
cianita *f* MINERAL cyanite, disthene, kyanite
cianización *f* QUÍMICA *de materia orgánica* cyanization
cianoacético *adj* QUÍMICA *ácido* cyanacetic
cianocopia *f* IMPR, PROD blueprint
cianógeno *m* QUÍMICA cyanogen
cianosa *f* MINERAL cyanose
cianosita *f* MINERAL cyanosite
cianotipo *m* TEXTIL blueprint
cianotriquita *f* MINERAL cyanotrichite
cianuración *f* CARBÓN, PROC QUÍ, QUÍMICA *procedimiento* cyanidation, cyanide process
cianuro *m* QUÍMICA *fumigaciones* cyanide; **~ de hidrógeno** *m* QUÍMICA hydrogen cyanide; **~ de oro** *m* QUÍMICA gold cyanide; **~ de plata** *m* QUÍMICA silver cyanide; **~ potásico** *m* QUÍMICA potassium cyanide; **~ de potasio** *m* QUÍMICA potassium cyanide; **~ de vinilo** *m* QUÍMICA acrylonitrile
ciar *vi* TRANSP MAR go astern
ciberespacio *m* INFORM&PD cyberspace
cibernética *f* INFORM&PD, TEC ESP cybernetics
cicatriz: **~ de la hoja** *f* AGRIC leaf scar; **~ de vaciado** *f* C&V casting scar
cícero *m* IMPR cicero
ciclación *f* PROC QUÍ, QUÍMICA *proceso, reacción* cyclization
ciclado: **~ térmico** *m* TERMO thermal cycling
ciclamato *m* QUÍMICA cyclamate
cíclico[1] *adj* ACÚST cyclic, ELEC *corriente alterna* periodic, cyclic, GEOL, INFORM&PD, ING ELÉC, ING MECÁ cyclic, MATEMÁT recurring, QUÍMICA, TEC PETR cyclic
cíclico[2] *m* TRANSP AÉR cycling
ciclización *f* PROC QUÍ cyclization, QUÍMICA *proceso, reacción* closure, cyclization
ciclo *m* ACÚST, CONTAM, ELEC, GEOL, INFORM&PD, ING ELÉC, ING MECÁ cycle, run, NUCL, PETROL, REFRIG, TEC PETR, TV cycle; **~ abierto** *m* NUCL *de combustible* open fuel cycle; **~ de agotamiento y descarga** *m* NUCL once-through-then-out cycle (*OTTO*); **~ de agua-vapor** *m* NUCL steam-water cycle; **~ de aire con turbina de expansión** *m* REFRIG brake-turbine air-cycle; **~ alcano** *m* QUÍMICA, TEC PETR cycloalkane; **~ alqueno** *m* QUÍMICA, TEC PETR cycloalkene; **~ automatizado** *m* INFORM&PD machine cycle; **~ del azufre** *m* CONTAM sulfur cycle (*AmE*), sulphur cycle (*BrE*); **~ biogeoquímico** *m* GEOL biogeochemical cycle; **~ de Brückner** *m* METEO Brückner cycle; **~ del carbono** *m* ALIMENT carbon cycle; **~ de carga y descarga** *m* TRANSP charge discharge cycle; **~ de Carnot** *m* FÍS, INSTAL TERM, TERMO Carnot cycle; **~ de combustible** *m* TEC ESP fuel cycle; **~ de combustible avanzado** *m* NUCL advanced fuel cycle; **~ de combustible cerrado** *m* NUCL closed fuel cycle; **~ de compresión de vapor** *m* TERMO vapor compression cycle (*AmE*), vapour compression cycle (*BrE*); **~ de contrapresión** *m* NUCL topping cycle; **~ cronometrado** *m* ELECTRÓN, INFORM&PD clock cycle; **~ de cuatro tiempos** *m* AUTO, ING MECÁ, MECÁ, TRANSP MAR four-stroke cycle, VEH *motor* four-stroke cycle, Otto cycle; **~ de**

desescarche *m* REFRIG defrost cycle; **~ de exploración** *m* TV scanning cycle; **~ de fabricación** *m* PROD manufacturing cycle; **~ frigorífico** *m* REFRIG, TERMO refrigeration cycle; **~ frigorífico de absorción** *m* REFRIG absorption refrigerating cycle; **~ frigorífico de aire** *m* REFRIG air refrigeration cycle; **~ frigorífico de compresión** *m* REFRIG, TERMO compression refrigerating cycle; **~ frigorífico de eyección de vapor** *m* REFRIG vapor jet refrigerating cycle (*AmE*), vapour jet refrigerating cycle (*BrE*); **~ de funcionamiento** *m* TRANSP operating cycle; **~ geoquímico** *m* GEOL, PETROL, QUÍMICA, TEC PETR geochemical cycle; **~ hidrológico** *m* AGUA hydrologic cycle, GEOL hydrologic cycle, hydrological cycle; **~ de histéresis** *m* ELEC *magnetismo* B/H loop, hysteresis curve, hysteresis loop, FÍS, ING ELÉC, METAL, P&C *propiedad, prueba* hysteresis loop; **~ de histéresis magnética** *m* ELEC magnetic hysteresis loop; **~ de histéresis rectangular** *m* ING ELÉC rectangular hysteresis loop; **~ de inspección** *m* TRANSP AÉR inspection cycle; **~ de instrucción** *m* INFORM&PD instruction cycle; **~ de integración** *m* ING ELÉC integration period; **~ de intervalos** *m* ACÚST cycle of intervals; **~ intranuclear del combustible** *m* NUCL in-core fuel cycle; **~ de irradiación** *m* FÍS RAD irradiation loop; **~ magnético** *m* GEOFÍS magnetic cycle; **~ de mancha solar** *m* GEOFÍS sunspot cycle; **~ de máquina** *m* INFORM&PD machine cycle; **~ de memoria** *m* INFORM&PD memory cycle; **~ de moldeado** *m* P&C *procesamiento* molding cycle (*AmE*), moulding cycle (*BrE*); **~ de moldeo** *m* P&C *procesamiento* molding cycle (*AmE*), moulding cycle (*BrE*); **~ de ocupación de vagones** *m* FERRO turnout; **~ orogénico** *m* GEOL orogenic cycle; **~ de Otto** *m* AUTO, VEH *motor* Otto cycle; **~ periódico de catorce días de las mareas** *m* ENERG RENOV *mareas*, OCEAN spring neap cycle; **~ de Rankine** *m* INSTAL TERM, TERMO, TRANSP Rankine cycle; **~ de referencia** *m* REFRIG standard rating cycle; **~ de refresco** *m* ELECTRÓN, INFORM&PD refresh cycle; **~ de refrigeración por chorro de vapor** *m* REFRIG vapor jet refrigerating cycle (*AmE*), vapour jet refrigerating cycle (*BrE*); **~ de regeneración** *m* ELECTRÓN, INFORM&PD *memorias dinámicas* refresh cycle; **~ de reloj** *m* ELECTRÓN, INFORM&PD clock cycle; **~ de reproducción** *m* NUCL breeding cycle; **~ de secado por impregnación** *m* ALIMENT, C&V, CONST soaking-drying cycle; **~ de sedimentación** *m* GEOL cycle of sedimentation; **~ sedimentario** *m* GEOL sedimentary cycle; **~ de servicio** *m* PROD duty cycle, service cycle; **~ de soldadura** *m* CONST welding cycle; **~ de suministro** *m* PROD delivery cycle; **~ de temperatura** *m* TERMO temperature cycle; **~ térmico** *m* TERMO heat cycle, thermal cycle; **~ termodinámico** *m* INSTAL TERM, TERMO thermodynamic cycle; **~ de trabajo** *m* AUTO working cycle; **~ de U-Pu** *m* NUCL U-Pu cycle; **~ de uranio-plutonio** *m* NUCL uranium-plutonium cycle
ciclogiróscopo *m* TRANSP AÉR cyclogyro
ciclograma *m* MATEMÁT pie chart
ciclohexano *m* QUÍMICA, TEC PETR cyclohexane
ciclohexanol *m* QUÍMICA cyclohexanol
ciclohexeno *m* QUÍMICA cyclohexene
cicloide *f* GEOM cycloid
ciclometileno *m* QUÍMICA cyclomethylene
ciclomotor: **~ eléctrico** *m* AUTO, ELEC electric moped

ciclón *m* ALIMENT, CARBÓN, CONTAM, METEO, PROC QUÍ, TEC PETR cyclone, TRANSP twister; ~ **de alta temperatura** *m* CARBÓN hot cyclone; ~ **de sedimentación** *m* PROC QUÍ settling cyclone; ~ **tropical** *m* METEO tropical cyclone

cicloolefina *f* QUÍMICA, TEC PETR cycloolefin

cicloparafina *f* QUÍMICA, TEC PETR cycloparaffin

ciclorama *m* CINEMAT cyc

ciclotema *m* GEOL cyclothem

ciclotrón *m* ELECTRÓN, FÍS cyclotron, electron accelerator, synchrotron, FÍS PART *en posición al acelerador lineal* cyclotron, electron accelerator, ion accelerator, synchrotron, NUCL cyclotron, electron accelerator, synchrotron; ~ **de electrones** *m* ELECTRÓN, FÍS PART, NUCL electron cyclotron

cielo[1]: **a ~ abierto** *adj* MINAS open-cast

cielo[2] *m* PROD *hogar de horno* roof; ~ **falso absorbente de sonidos** *m* ACÚST, SEG sound-absorbent ceiling

ciénaga *f* AGUA marsh

ciencia: ~ **de los alimentos** *f* ALIMENT food science; ~ **de computadoras** *f* AmL (*cf informática Esp*) INFORM&PD computer science

cieno *m* AGUA mud, silt, slime, sludge, CARBÓN silt, slime, sludge, CONTAM sludge, HIDROL ooze, silt, slime, OCEAN mud, RECICL sludge, TEC PETR mud; ~ **activado** *m* CONTAM, HIDROL, RECICL activated sludge; ~ **activo** *m* CONTAM, HIDROL, RECICL activated sludge; ~ **cloacal** *m* AGUA sewage sludge, sludge, RECICL sludge; ~ **concentrado** *m* CARBÓN concentrated sludge; ~ **no tratado** *m* AGUA, CONTAM, RECICL raw sludge; ~ **reciclado activado** *m* HIDROL *aguas residuales* activated recycled sludge; ~ **de retorno activado** *m* HIDROL activated return sludge; ~ **séptico** *m* RECICL septic sludge; ~ **sin tratar** *m* AGUA, CONTAM, RECICL raw sludge

cierracircuito *m* ELEC, ING ELÉC circuit closer

cierre[1]: **de ~ automático** *adj* TEC ESP self-sealing, VEH self-locking; **de ~ propio** *adj* TEC ESP self-sealing

cierre[2] *m* AGUA shutting-off, CONST closing, sealing-up, barrage, D&A *armas de fuego* breech, ELEC *protección* seal, ELECTRÓN gating, EMB, GEOL closure, INFORM&PD *programa* close down, lock, ING ELÉC closure, make, make, fastener, INSTAL HIDRÁUL *del vapor* shutting-off, *de admisión de flujo* cutting-off, *de admisión de vapor* cutoff, MECÁ fastener, METAL locking, PROD fastener, closing, *puertas* cutting-off, QUÍMICA *de anillo* closure, TEC ESP shutdown, latch-up, TEXTIL *de prendas* fastening; ~ **de aceite** *m* ING MECÁ, MECÁ oil seal; ~ **del agua** *m* AGUA shutting-off water; ~ **de aire** *m* TRANSP AÉR airlock; ~ **antiderrame** *m* EMB pour-spout closure; ~ **apertura rápida** *m* EMB snap-off closure; ~ **por atado** *m* EMB tying closure; ~ **autoclave** *m* INSTAL HIDRÁUL pressure seal; ~ **de bayoneta** *m* MECÁ bayonet locking; ~ **blindado** *m* ING ELÉC shielded enclosure; ~ **de boca** *m* TRANSP MAR *construcción* tumblehome; ~ **de la bolsa** *m* AmL (*cf embolsado Esp*) OCEAN purse-fastener, pursing; ~ **de la botella** *m* EMB bottle closure; ~ **por calor** *m* P&C *operación, procesamiento*, TERMO heatsealing; ~ **del capó** *m* AUTO bonnet lock (*BrE*), hood lock (*AmE*); ~ **centralizado** *m* AUTO, ING MECÁ, VEH central locking; ~ **con charnela que se rompe al abrirse por primera vez** *m* EMB snap hinge closure; ~ **de la compuerta del tren de aterrizaje** *m* TRANSP AÉR landing gear door uplock; ~ **de contacto-**

membrana *m* ING ELÉC membrane keyswitch; ~ **de cremallera** *m* TEXTIL zip fastener; ~ **definitivo** *m* INFORM&PD *programa* closedown; ~ **de emergencia** *m* INFORM&PD emergency shutdown; ~ **por enfriamiento** *m* PROD *fundería* cold shot; ~ **esférico** *m* ING MECÁ ball lock; ~ **estanco a la presión** *m* INSTAL HIDRÁUL pressure seal; ~ **en falso** *m* ING ELÉC false closure; ~ **fruncido** *m* IMPR crimp lock; ~ **de garantía** *m* EMB guarantee closure; ~ **de garantía contra falsificaciones** *m* EMB tamperproof closure; ~ **hermético** *m* EMB hermetic closure, ING MECÁ vacuum seal, NUCL hermetic sealing; ~ **hermético del fuelle** *m* TEC ESP bellows seal; ~ **hermético ultrasónico** *m* EMB ultrasonic sealing, TERMO ultrasonic welding; ~ **hinchable** *m* NUCL inflatable seal; ~ **inflable** *m* NUCL inflatable seal; ~ **por jalado de la jareta** *m* OCEAN *redes de pesca* pursing; ~ **laberíntico** *m* ING MECÁ, NUCL labyrinth seal; ~ **mediante gofrado** *m* EMB embossing closure; ~ **mediante precinto** *m* EMB band sealing; ~ **mediante rizado** *m* EMB crimp-on closure; ~ **mediante tapón articulado** *m* EMB hinged plug orifice closure; ~ **del motor en vuelo** *m* TRANSP AÉR engine shutdown in flight; ~ **por pegado** *m* EMB binding closure; ~ **de polipropileno** *m* EMB polypropylene closure; ~ **precintado** *m* EMB taped closure; ~ **de prensaestopas** *m* NUCL packing seal; ~ **a presión** *m* INSTAL HIDRÁUL pressure seal; ~ **de presión** *m* EMB snap-on closure; ~ **a prueba de niños** *m* EMB, SEG child-resistant closure; ~ **de puertas** *m* FERRO, VEH door locking; ~ **que pone de manifiesto los intentos de violación** *m* EMB tamper-evident closure; ~ **que se abre estirando de un extremo** *m* EMB pull-off closure; ~ **que se abre rasgando** *m* EMB tear-off closure; ~ **remetiendo los bordes** *m* EMB tuck-in closure; ~ **de rosca sin fin** *m* EMB continuous-thread closure; ~ **roscado** *m* EMB screw closure; ~ **rotativo** *m* REFRIG rotary seal; ~ **de seguridad** *m* AUTO safety catch, EMB safety closure, VEH safety catch; ~ **de seguridad del tren de aterrizaje** *m* TRANSP AÉR landing-gear safety lock; ~ **térmico** *m* REVEST heat-sealing; ~ **de tira metálica** *m* EMB metal-strip closure; ~ **por torsión** *m* EMB twisting closure; ~ **de vapor** *m* REFRIG gas lock

cifra *f* MATEMÁT figure; ~ **binaria** *f* (*bit*) ELECTRÓN, IMPR, INFORM&PD, PETROL, TELECOM binary digit (*bit*); ~ **significativa** *f* MATEMÁT significant figure

cifrado *m* ELECTRÓN encipherment, encryption, encoding, EMB encoding, INFORM&PD, TEC ESP *comunicaciones* encoding, encryption, TELECOM encoding, encryption, encipherment, TRANSP AÉR encoding, TV encryption; ~ **de datos** *m* INFORM&PD data encryption; ~ **enlace por enlace** *m* TELECOM link-by-link encipherment; ~ **de extremo a extremo** *m* TELECOM end-to-end encipherment; ~ **de operación** *m* PROD operation code (*op code*)

cifrador *m* ELECTRÓN encoder, INFORM&PD coder, encoder, PROD encoder, TELECOM, TV coder, encoder

cifrar *vt* ELEC digitize, ELECTRÓN encipher, encrypt, INFORM&PD encrypt, ING MECÁ digitize, TELECOM encipher, encrypt, TV encrypt

cigoto *m* AGRIC zygote

cigüeña *f* AUTO crank, CONST winch, ING MECÁ crank; ~ **y sistema de varilla de conexión** *f* ING MECÁ crank and connecting rod system

cigüeñal *m* AUTO crankshaft, ING MECÁ crankshaft,

main shaft, crank, MECÁ crankshaft, main shaft, TRANSP MAR *motor* crankshaft, VEH *motor* crank, crankshaft; **~ de dos muñequillas** *m* ING MECÁ two-throw crank shaft; **~ ensamblado** *m* ING MECÁ built-up crank; **~ no enterizo** *m* ING MECÁ built-up crank; **~ de tres codos** *m* ING MECÁ three-throw crank shaft; **~ de tres muñequillas** *m* ING MECÁ three-throw crank shaft

cilindrado *m* ING MECÁ surfacing; **~ de rodillos de madera** *m* PROD *torno* rounding wooden rollers

cilindrar: **~ en el torno** *vt* PROD *electrónica* turn off

cilíndrico[1] *adj* GEOL, GEOM, ING MECÁ cylindrical

cilíndrico[2]: **~ recto** *m* ELEC *engranaje* spur

cilindro *m* AUTO, C&V cylinder, CONST barrel, GEOM cylinder, IMPR roller, INFORM&PD cylinder, ING MECÁ barrel, MECÁ barrel, cylinder, drum, roller, MINAS cylinder, P&C *de prensa, parte de equipo* cylinder, roller, PAPEL cylinder, roll, PROD roll, cylinder, TRANSP MAR *accesorios de cubierta* warping drum, *motor* cylinder, VEH *motor* cylinder; **~ acanalado** *m* ING MECÁ grooved roll, grooved roller, PROD grooved roll; **~ accionado por fluido** *m* ING MECÁ fluid-power cylinder; **~ para acetileno** *m* MECÁ acetylene cylinder; **~ de aire** *m* MECÁ air cylinder; **~ de aire comprimido** *m* ING MECÁ compressed-air cylinder; **~ alimentador** *m* ING MECÁ feed roller; **~ alzador** *m* IMPR gathering cylinder; **~ de amalgamación** *m* CARBÓN amalgam barrel; **~ de apoyo** *m* ING MECÁ top roll; **~ aspirante** *m* PAPEL suction roll; **~ auxiliar** *m* AUTO slave cylinder; **~ auxiliar del embrague** *m* AUTO, VEH clutch slave cylinder; **~ de bomba** *m* AGUA pump cylinder; **~ de buceo** *m* OCEAN diving cylinder; **~ cabrestante** *m* TRANSP MAR *accesorios de cubierta* warping drum; **~ de chapa** *m* ING MECÁ plate roll; **~ circular recto** *m* GEOM right-circular cylinder; **~ cobreado** *m* IMPR copperplated cylinder; **~ cubicado** *m* METR squareness cylinder; **~ con cuchilla plegadora** *m* PAPEL folding-knife cylinder; **~ de cuchillas de solapa** *m* IMPR tucking blades cylinder; **~ de curvar** *m* ING MECÁ bending roll; **~ desbastador** *m* CARBÓN roughing-down roll, ING MECÁ billetting roll, PROD puddle roll, roughing roll, roughing-down roll; **~ desgranador** *m* AGRIC threshing cylinder; **~ de deslizamiento** *m* METAL slip cylinder; **~ enfriador** *m* PAPEL sweat roll; **~ escalonado** *m* ING MECÁ stepped roll; **~ de Faraday** *m* ELEC *electrostática* Faraday ice pail, FÍS Faraday cylinder; **~ formador** *m* PAPEL paper-forming cylinder; **~ del formato** *m* PAPEL *máquina de cartón* liner cylinder; **~ del freno** *m* AUTO, ING MECÁ, VEH brake cylinder; **~ del freno de la rueda** *m* AUTO brake-wheel cylinder; **~ de función de templada** *m* PROD chilled-iron roll; **~ de funcionamiento hidráulico** *m* INSTAL HIDRÁUL hydraulic actuating cylinder; **~ de garganta** *m* ING MECÁ grooved roll; **~ de gas a presión** *m* GAS gas bottle, gas cylinder, ING MECÁ, MECÁ gas cylinder, NUCL, PROD, TERMO gas bottle, gas cylinder; **~ de gran potencia** *m* ING MECÁ heavy-section roll; **~ guía** *m* TEXTIL guide roller; **~ hidráulico** *m* ING MECÁ fluid-power cylinder, hydraulic cylinder; **~ para huecograbado** *m* IMPR intaglio cylinder; **~ humidificador** *m* EMB damping roll; **~ de impresión** *m* IMPR impression roll, printing cylinder; P&C *equipo* printing roll; **~ impresor** *m* P&C *equipo* impression cylinder; **~ impulsor del embrague** *m* AUTO, VEH clutch

output cylinder; **~ inferior** *m* PROD *laminador* bottom roll; **~ de laminador** *m* FERRO, ING MECÁ, PROD rolling-mill roll; **~ laminador** *m* PROD roller; **~ liso** *m* PROD plain roll; **~ de listones** *m* AGRIC rub bar cylinder; **~ macho** *m* ING MECÁ top roll, upper roll; **~ maestro** *m* AUTO, VEH *de embrague, freno* master cylinder; **~ maestro del embrague** *m* AUTO, VEH clutch master cylinder; **~ maestro del freno** *m* AUTO, VEH brake master cylinder; **~ de mando** *m* AUTO, ING MECÁ, VEH brake cylinder; **~ de la mantilla** *m* IMPR blanket cylinder; **~ mecánico** *m* AUTO power cylinder; **~ de media presión** *m* INSTAL HIDRÁUL intermediate pressure cylinder; **~ de medición** *m* FOTO measuring cylinder; **~ de medición del engranaje** *m* METR gear-measuring cylinder; **~ medidor del hilo de rosca** *m* METR screw thread-measuring cylinder; **~ mezclador** *m* PROD mixing cylinder; **~ motor del embrague** *m* AUTO, VEH clutch output cylinder; **~ neumático** *m* ING MECÁ, TRANSP AÉR pneumatic cylinder; **~ de pared delgada** *m* ING MECÁ thin-walled cylinder; **~ de la pila holandesa** *m* PAPEL beater roll; **~ plegador** *m* IMPR folding cylinder; **~ portaplanchas** *m* IMPR plate cylinder; **~ de presión** *m* ACÚST, ING MECÁ pressure roller; **~ principal** *m* AUTO brake master cylinder, master cylinder, TRANSP AÉR master cylinder, VEH brake master cylinder, master cylinder; **~ principal de freno** *m* AUTO, VEH brake master cylinder; **~ principal del freno del tren de aterrizaje** *m* TRANSP AÉR landing gear master brake cylinder; **~ para pruebas** *m* CONST test cylinder; **~ pulidor** *m* ING MECÁ planishing roll; **~ pulverizador** *m* CARBÓN cracker; **~ ralentizador** *m* IMPR slow-down cylinder; **~ recolector** *m* IMPR collect cylinder; **~ refrigerador** *m* PAPEL, REFRIG cooling roller; **~ de respaldo** *m* C&V backup roll; **~ de revolución** *m* GEOM cylindrical solid of revolution; **~ de la rueda** *m* AUTO wheel cylinder; **~ de salida de embrague** *m* AUTO, VEH clutch output cylinder; **~ satinador** *m* PAPEL machine-glazed cylinder (*MG cylinder*), yankee dryer; **~ de secado** *m* DETERG, TEXTIL drying cylinder; **~ secador** *m* DETERG, PAPEL, TEXTIL drying cylinder; **~ secador embarcador** *m* PAPEL baby dryer; **~ secundario** *m* VEH *embrague, frenos* slave cylinder; **~ superior** *m* ING MECÁ top roll, upper roll; **~ telescópico** *m* ING MECÁ telescopic cylinder; **~ tensor** *m* ING MECÁ tension roller; **~ del tren desbastador** *m* ING MECÁ blooming roll; **~ triturador** *m* PROC QUÍ crusher roll, crushing roll, PROD crusher roll; **~ en V** *m* AUTO V-shaped cylinder; **~ de vapor** *m* INSTAL HIDRÁUL steam cylinder; **~ de varilla pasante** *m* ING MECÁ through-rod cylinder; **~ de varilla única** *m* PROD simple rod cylinder; **~ de Wehnelt** *m* FÍS *electrodo*, INSTR Wehnelt cylinder; **~ yanqui** *m* PROD yankee cylinder

cilindrocónico *adj* PROD cylindroconic

cilindros: **~ en línea** *m pl* AUTO in-line cylinders; **~ opuestos** *m pl* AUTO opposed cylinders; **~ yuxtapuestos** *m pl* ING MECÁ side-by-side cylinders

CIM *abr* ELECTRÓN, FÍS (*circuito integrado de microondas*) MIC (*microwave integrated circuit*), INFORM&PD (*fabricación informatizada, fabricación integrada por computadora AmL, fabricación integrada por ordenador Esp*) CIM (*computer-integrated manufacturing*)

cima *f* CONST brow, *del tejado* crest, GEOL crest; ~ **invertida** *f* CONST *arquitectura* reversed ogee; ~ **de una loma** *f* CONST knoll

cimacio *m* CONST *de arco* ogee

cimasa *f* QUÍMICA zymase

cimbra *f* CONST *arco, edificación, obra civil* center (*AmE*), centre (*BrE*), *madera* bending, *obra civil, edificación* centering (*AmE*), centring (*BrE*); ~ **de arco** *f* CONST centering (*AmE*), centring (*BrE*); ~ **de la ola** *f* OCEAN *perfil de la ola* camber

cimbrear *vt* CONST bend

cimeno *m* QUÍMICA cymene

cimentación *f* CARBÓN, CONST foundation, PROD bedding-in; ~ **compensada** *f* CARBÓN compensated foundation; ~ **reforzada** *f* CARBÓN deep foundation; ~ **del yunque** *f* MINAS *martillo pilón* stock

cimentar *vt* CONST lay the foundations of

cimiento *m Esp* (*cf fundación AmL*) CONST footing, foundation

cimófana *f* MINERAL cymophane

cimógeno *adj* PETROL cymogene

cimohidrólisis *f* QUÍMICA zymohydrolysis

cimolita *f* MINERAL cimolite

cimómetro *m* FÍS ONDAS wavemeter; ~ **de frecuencia de pulsación** *m* FÍS ONDAS beat-frequency wavemeter

cinabrio *m* MINAS *mineral*, MINERAL cinnabar, mercury ore, QUÍMICA cinnabar

cinamato *m* ALIMENT, QUÍMICA cinnamate; ~ **de bencilo** *m* ALIMENT benzyl cinnamate; ~ **de etilo** *m* ALIMENT ethyl cinnamate

cinámico *adj* QUÍMICA *ácido* cinnamic

cinamilo *m* QUÍMICA cinnamyle

cinc[1]: **de** ~ *adj* QUÍMICA zincic

cinc[2] *m* (*Zn*) GEN zinc (*Zn*)

cincado[1] *adj* MECÁ, P&C, QUÍMICA, REVEST galvanized

cincado[2] *m* P&C *pintura* zinc coating, QUÍMICA zincking, REVEST zinc coating

cincato *m* QUÍMICA zincate

cincel *m* AGRIC cold chisel, CONST, ING MECÁ, MECÁ, MINAS, PROD *para metal* chisel; ~ **de acero templado** *m* ING MECÁ, PROD cold chisel; ~ **de boca plana** *m* MINAS *sondeos* flat chisel; ~ **de calafatear** *m* ING MECÁ, MECÁ, PROD caulking iron; ~ **para calafatear** *m* ING MECÁ, MECÁ, PROD caulking chisel; ~ **de desbarbar** *m* PROD chipping chisel; ~ **de desbastar** *m* ING MECÁ, PROD chipping chisel; ~ **ovalado** *m* ING MECÁ oval burr; ~ **de pico redondo** *m* ING MECÁ round-nose chisel; ~ **de punta de diamante** *m* ING MECÁ diamond point chisel, diamond point cold chisel; ~ **con punta rómbica** *m* ING MECÁ diamond cold chisel, diamond nose chisel; ~ **subsolador** *m* AGRIC subsoiler chisel

cincelador *m* CONST chipper

cincelar *vt* CONST chip

cincha *f* CONTAM MAR *para hacer bragas*, OCEAN *pasamanería* webbing, TV cinch, cinching

cincho *m* TEXTIL *vestimenta* webbing; ~ **de pilote** *m* CONST pile hoop

cíncico *adj* QUÍMICA zincic

cincita *f* MINERAL zincite

cincoide *adj* QUÍMICA zincoid

cincona *f* QUÍMICA zincon

cinconidina *f* QUÍMICA *alcaloide* cinchonidine

cinconina *f* QUÍMICA *alcaloide* cinchonin

cincosita *f* MINERAL zincosite, zinkosite

cincoso *adj* QUÍMICA zincky, zincous

cincum *m* QUÍMICA zincum

cine *m* CINEMAT *local de exhibición* cinema, viewing theater (*AmE*), viewing theatre (*BrE*), *películas* motion pictures

cineasta *m* CINEMAT cinematographer, film-maker; ~ **documental** *m* CINEMAT, TV documentary film-maker; ~ **independiente** *m* CINEMAT independent film-maker

Cinemascope *m* CINEMAT Cinemascope

cinemateca *f* CINEMAT stock shot library

cinemática *f* FÍS, MECÁ kinematics

cinematografía: ~ **infrarroja** *f* CINEMAT infrared cinematography; ~ **radiográfica** *f* CINEMAT X-ray cinematography

cinemicrografía *f* CINEMAT cinemicrography

ciñendo *adv* TRANSP MAR *navegación a vela* close-hauled

cineol *m* QUÍMICA *eucalipto* cineol

cineólico *adj* QUÍMICA *ácido* cineolic

cineradiografía *f* CINEMAT cineradiography

cinerama *m* CINEMAT cinerama

cinerita *f* GEOL *cenizas volcánicas* cinerite

cinescopio *m* C&V TV bulb, ELECTRÓN kinescope

cinética[1] *adj* ACÚST *impedancia* motional, TEC ESP kinetic

cinética[2] *f* FÍS, GAS, MECÁ, METAL, QUÍMICA, TERMO kinetics; ~ **de gases** *f* GAS, TERMO gas kinetics

cinglador: ~ **de quijadas** *m* PROD *hierro pudelado* crocodile squeezer

cinolina *f* QUÍMICA cinnoline

cinta *f* ACÚST ribbon, C&V tape, CONST band tape, FOTO cassette, INFORM&PD strip, ribbon, tape, ING MECÁ belt, MECÁ belt, tape, METR *de metal para tomar medidas* ribbon, P&C tape, PROD strip, TEXTIL tape; ~ **abrasiva** *f* C&V, ING MECÁ abrasive belt; ~ **adhesiva** *f* CINEMAT adhesive tape, CONST masking tape, EMB adhesive tape, masking tape, ING ELÉC, MECÁ, P&C adhesive tape; ~ **adhesiva por ambos lados** *f* EMB double-sided tape; ~ **aisladora** *f* ELEC insulating tape, ING ELÉC adhesive insulating tape, insulating tape; ~ **aislante** *f* ELEC insulating tape, *cable* serving, ING ELÉC insulating tape, adhesive insulating tape, MECÁ *electricidad* adhesive tape, PROD electrical tape; ~ **aislante adhesiva** *f* ING ELÉC adhesive insulating tape; ~ **de alineación** *f* TV alignment tape, line-up tape; ~ **de alineamiento** *f* TV alignment tape; ~ **de alta energía** *f* TV high-energy tape; ~ **de ametralladora** *f* D&A cartridge belt; ~ **de arena** *f* OCEAN sand ribbon; ~ **autoadhesiva** *f* EMB, P&C self-adhesive tape; ~ **auxiliar** *f* INFORM&PD backing tape; ~ **de borrador** *f* INFORM&PD scratch tape; ~ **de brida** *f* ING MECÁ stirrup strap; ~ **de cámara** *f* CINEMAT, TV camera tape; ~ **de cambios** *f* INFORM&PD change tape; ~ **de cartuchos** *f* D&A cartridge belt; ~ **clasificadora** *f* PROD sorting belt; ~ **conductora** *f* ING MECÁ driving belt; ~ **continua** *f* TV tape loop; ~ **de control** *f* INFORM&PD control tape; ~ **de control de la exposición** *f* CINEMAT, FOTO exposure control tape; ~ **para copia de seguridad** *f* INFORM&PD streamer; ~ **dentada** *f* ING MECÁ notched band, cog belt; ~ **de desgaste** *f* ING MECÁ wear strip; ~ **digital de audio** *f* INFORM&PD digital audio tape (*DAT*); ~ **de dióxido de cromo** *f* TV chrome-dioxide tape; ~ **para electricistas** *f* CINEMAT gaffer tape; ~ **de emergencia** *f* TV tape backing; ~ **de empalmar** *f* CINEMAT splicing tape; ~ **enrollada** *f* ÓPT tape wrap;

~ entintada f INFORM&PD inked ribbon; **~ de entrada** f IMPR leading-in tape; **~ de escogido** f AmL (cf cinta de recogida Esp) MINAS picking belt; **~ de fibras discontinuas** f TEXTIL hilatura top; **~ de filamentos continuos** f TEXTIL tow; **~ del freno** f AUTO, ING MECÁ, MECÁ, VEH embrague brake band; **~ fusible** f ELEC fuse link; **~ gruesa** f TEXTIL tejido webbing; **~ para hacer la cruz** f TEXTIL lease band; **~ laminada autoadhesiva** f EMB self-adhesive laminated tape; **~ laseróptica** f ÓPT laser-optic tape, optical tape; **~ de lija** f ING MECÁ, REVEST coated abrasive; **~ maestra** f INFORM&PD, TV master tape; **~ magnética** f ACÚST, ELEC, GEOFÍS, IMPR, INFORM&PD, TELECOM, TV magnetic tape; **~ magnética en blanco** f TV blank magnetic tape; **~ magnética uniforme** f ACÚST level magnetic tape; **~ magnética de video** AmL, **~ magnética de vídeo** f Esp TELECOM videotape; **~ magnetofónica** f TV magnetic tape, tape; **~ de medir** f CONST measuring tape, MECÁ tape measure, METR measuring tape, tape measure; **~ metálica** f CONST topografía surveyor's tape, ING MECÁ strap; **~ métrica** f MECÁ tape measure, METR instrumento measure, tape measure; **~ motriz** f ING MECÁ driving belt; **~ no elaborada** f CINEMAT, TV raw tape; **~ de notas** f INFORM&PD scratch tape; **~ olefila** f CONTAM MAR oleophilic belt; **~ óptica** f ÓPT optical tape; **~ de paja** f PROD moldería hay band, hay rope; **~ de papel** f ELEC aislación, INFORM&PD paper tape; **~ de papel engomado** f EMB gummed paper tape; **~ peinada** f TEXTIL combed top; **~ perforada** f INFORM&PD, TELECOM perforated tape, punch tape (AmE), punched tape (BrE); **~ de polietileno autoadhesiva** f EMB polyethylene self-adhesive tape; **~ pregrabada** f TV prerecorded tape; **~ protectora** f ING MECÁ wear strip, PROD protective strip; **~ protectora de la superficie** f EMB surface protection tape; **~ de prueba** f ACÚST, TV test tape; **~ de PVC autoadhesiva** f EMB PVC pressure-sensitive tape; **~ que se fija por calor** f EMB heat-fix tape; **~ de ranurar** f ING MECÁ plough belt (BrE), plow belt (AmE); **~ de recogida** f Esp (cf cinta de escogido AmL) MINAS picking belt; **~ de referencia** f ACÚST reference tape, TV reference tape, standard tape; **~ registradora de velocidad** f FERRO speed recording tape; **~ de resaltos** f ING MECÁ cog belt; **~ de respaldo** f TV tape backing; **~ de respiradero** f TV aircheck tape; **~ reutilizable** f INFORM&PD scratch tape; **~ selladora** f IMPR sealing tape; **~ sensible a la presión** f EMB precinto autoadhesivo pressure-sensitive tape; **~ de serpentina** f INFORM&PD streamer; **~ térmica** f LAB heating tape; **~ termosellable** f EMB heat seal tape; **~ transportadora** f Esp (cf máquina transportadora AmL) AGRIC belt conveyor, conveyor belt, conveyor, band conveyor, travelling apron (BrE), traveling apron (AmE), conveyor belting; **~ transportadora de carbón** f CARBÓN coal belt, GEOL cola bed, MINAS coal belt; **~ transportadora corrugada** f ING MECÁ cog belt; **~ transportadora de mandil** f TRANSP apron conveyor; **~ transportadora de pasajeros** f TRANSP en movimiento pedestrian conveyor, TRANSP AÉR moving sidewalk (AmE), passenger conveyor (BrE); **~ transportadora resistente al fuego** f ING MECÁ flame-retardant conveyor belt; **~ transportadora retardadora de llamas** f ING MECÁ flame-retardant conveyor belt;

~ transportadora de salida f PROD belt-outgoing conveyor, outgoing conveyor belt; **~ transportadora de velocidad regulable** f TRANSP variable-speed conveyor belt; **~ de video** AmL, **~ de vídeo** Esp f FÍS, TV videotape

cintas: **~ de video** AmL, **~ de vídeo** Esp f pl TV videoware

cintería f TEXTIL narrow-fabric

cintón m TRANSP MAR de casco ribband

cintoteca f INFORM&PD tape library

cintura f TEXTIL waist

cinturón m GEN de seguridad belt; **~ de asiento de tres puntos** m TRANSP three-point seat belt; **~ y bolsillo de herramientas** m ING MECÁ tool belt and pouches; **~ de calefacción** m TERMOTEC heating belt; **~ de irradiación** m FÍS RAD, TEC ESP radiation belt; **~ de lastre** m OCEAN buceo weight belt; **~ milonítico** m GEOFÍS fracture zone; **~ móvil** m GEOL mobile belt; **~ de nieve** m METEO snow belt; **~ orogénico** m GEOL orogenic belt; **~ y portacartuchera para herramientas** m ING MECÁ tool belt and pouches; **~ de radiación** m FÍS RAD, TEC ESP radiation belt; **~ de radiación de Van Allen** m FÍS RAD, GEOFÍS, TEC ESP Van Allen radiation belt; **~ de seguridad** m Esp (cf faja de seguridad AmL) SEG, TEC PETR, TRANSP, TRANSP AÉR seat belt; **~ de seguridad de tres anclajes** m TRANSP three-point seat belt; **~ de Van Allen** m FÍS RAD Van Allen belt, Van Allen radiation belt; **~ verde** m GEOL green belt

CIP abr (limpieza en el lugar) ALIMENT CIP (cleaning in place)

ciprina f MINERAL cyprine

circo: **~ glaciar** m GEOL cirque

circón m MINERAL zircon; **~ incoloro** m MINERAL jargon

circona f C&V zirconia

circonato m QUÍMICA zirconate

circonia f QUÍMICA zirconia

circónico adj QUÍMICA zirconic

circonifluoruro m QUÍMICA zirconifluoride

circonilo m QUÍMICA zirconyl

circuitería: **~ integrada** f ELECTRÓN integrated circuitry

circuito m ELEC, ELECTRÓN, FÍS, INFORM&PD circuit, ING ELÉC path, PROD de controladores loop, TELECOM circuit;

▶ **a** **~ abierto** m CARBÓN open circuit, ELEC open circuit, open loop, FÍS open circuit, ING ELÉC broken circuit, open circuit; **~ de absorción** m ELEC absorption circuit; **~ de acceso** m TELECOM access circuit; **~ de accionamiento individual** m ING ELÉC single drive circuit; **~ aceptor** m ELEC inducción acceptor circuit; **~ acoplado** m ELEC, ING ELÉC coupled circuit; **~ de activación** m ELEC trigger circuit, MINAS firing circuit; **~ activado** m ELECTRÓN triggered circuit; **~ activador** m ELEC, ELECTRÓN trigger circuit, ING ELÉC energizing circuit; **~ activo** m FÍS active circuit; **~ aditivo** m TV adder; **~ aditivo azul** m TV blue adder; **~ aditivo rojo** m TV red adder; **~ aditivo verde** m TV green adder; **~ de admisión** m ELEC inducción acceptor circuit; **~ ajustado** m FÍS tuned circuit; **~ de ajuste** m ING ELÉC adjusting circuit; **~ al aire** m ING ELÉC open circuit; **~ de alarma** m TELECOM alarm circuit; **~ de alimentación** m ELEC feed circuit, power circuit, red de distribución feeder, ING ELÉC feed circuit; **~ de**

alimentación de energía eléctrica m ING ELÉC power-supply circuit; ~ **alineal** m TELECOM nonlinear circuit; ~ **de alumbrado** m ELEC lighting circuit; ~ **amplificador** m ELECTRÓN amplifier circuit, amplifying circuit; ~ **con amplificador** m TELECOM amplified circuit; ~ **analógico** m ELECTRÓN, ING ELÉC, TELECOM analog circuit; ~ **analógico de uso privado** m TELECOM analog private wire; ~ **de anillo** m ELEC ring circuit (*AmE*), ring main (*BrE*), ING ELÉC ring current (*AmE*), ring main (*BrE*); ~ **anódico** m ING ELÉC plate circuit; ~ **del ánodo** m ELEC, ING ELÉC anode circuit; ~ **de anticoincidencia** m FÍS anticoincidence circuit; ~ **antirresonante** m ELECTRÓN, FÍS, TRANSP AÉR antiresonant circuit; ~ **anular** m ELEC ring circuit (*AmE*), ring main (*BrE*), ING ELÉC ring current (*AmE*), ring main (*BrE*); ~ **aperiódico** m ELECTRÓN aperiodic circuit; ~ **aritmético** m ELECTRÓN arithmetic circuit; ~ **de arranque** m TELECOM startup circuit; ~ **asimétrico** m ELEC asymmetric circuit, asymmetrical circuit; ~ **asíncrono** m INFORM&PD, TELECOM asynchronous circuit; ~ **astable** m ELECTRÓN astable circuit; ~ **automático de enlaces** m TELECOM automatic trunk working; **~ b** ~ **de baja impedancia** m PROD low-impedance path; ~ **con bajo nivel de señal** m *contador eléctrico* ING ELÉC dry circuit; ~ **de banda ancha** m TELECOM wideband circuit; ~ **de banda estrecha** m ELECTRÓN narrow-band circuit; ~ **basculador de bloqueo** m ELECTRÓN gated flip-flop; ~ **basculante** m ELECTRÓN flip-flop circuit, FÍS flip-flop, flip-flop circuit, INFORM&PD flip-flop circuit; ~ **bidireccional** m TELECOM both-way circuit; ~ **biestable** m ELECTRÓN bistable circuit; ~ **biestable de bloqueo** m ELECTRÓN gated flip-flop; ~ **bifilar** m ELEC two-wire circuit; ~ **bifurcado** m ELEC derived circuit; ~ **binario** m ELECTRÓN binary circuit; ~ **de bloqueo** m ING ELÉC interlock circuit; ~ **de bloqueo de fase** m TEC ESP phase-locked loop (*PLL*); ~ **de borrado** m AmL TV blanking circuit; ~ **en bucle** m FERRO loop line; ~ **Burgers** m METAL Burgers circuit; **~ c** ~ **de CA** m ING ELÉC AC circuit; ~ **de calefacción** m INSTAL TERM heating circuit; ~ **de carga** m ELEC *acumulador, batería* charging circuit, load circuit, ING ELÉC ballasting circuit, ING MECÁ load; ~ **en carga** m TELECOM load circuit; ~ **catódico** m ELECTRÓN, ING ELÉC cathode circuit; ~ **de CC** m ING ELÉC DC circuit; ~ **cerrado** m CARBÓN closed circuit, ELEC closed circuit, closed loop, ELECTRÓN closed loop, FÍS closed circuit, ING ELÉC closed circuit, loop, closed loop, ING MECÁ, TELECOM closed circuit; ~ **cerrado extenso** m TRANSP long loop; ~ **cerrado indicador de varios puntos** m TELECOM multipoint indication loop (*MIL*); ~ **cerrado de realimentación** m ELECTRÓN feedback loop; ~ **cerrado de televisión** m TELECOM, TV closed-circuit television (*CCTV*); ~ **cerrado del vídeo** AmL, ~ **cerrado del vídeo** m *Esp* TELECOM video loop; ~ **en chip** m ELECTRÓN on-chip circuit; ~ **de choque** m ELEC choke circuit; ~ **de cierre remoto** m PROD remote-shutdown circuit; ~ **clearscan** m TRANSP MAR clearscan; ~ **de coincidencia** m ELECTRÓN, FÍS coincidence circuit; ~ **de combinación** m ELECTRÓN, TV combiner circuit; ~ **combinacional** m ELECTRÓN, INFORM&PD combinational circuit (*AmE*), combinatorial circuit (*BrE*); ~ **combinador** m ELECTRÓN combiner circuit, TELE-

COM combining circuit, TV combiner circuit; ~ **combinatorio** m ELECTRÓN, INFORM&PD combinational circuit (*AmE*), combinatorial circuit (*BrE*); ~ **de combustible** m AUTO, VEH fuel line; ~ **de comparación** m TELECOM compensation circuit; ~ **comparador** m ELECTRÓN comparator circuit; ~ **de compensación** m ELEC bucking circuit, compensating circuit, ELECTRÓN buffer circuit; ~ **compensador** m ELEC bucking circuit; ~ **compensador de cable** m TV cable compensation circuit; ~ **compensador magnético** m ING ELÉC magnetic flywheel; ~ **completo** m ELEC full circuit; ~ **de comprobación** m TELECOM comparison circuit; ~ **de comunicaciones** m ING ELÉC, TELECOM communications circuit; ~ **de conductos flexibles** m TELECOM cord circuit; ~ **conectado** m ELEC linked circuit; ~ **de conexión intraoficinal** m TELECOM intraoffice junctor circuit; ~ **de conmutación** m ELECTRÓN, ING ELÉC switching circuit; ~ **conmutado** m TELECOM switched circuit; ~ **contador** m ELECTRÓN counter circuit; ~ **en contrafase** m ELEC, ELECTRÓN, TELECOM push-pull circuit; ~ **de control** m CINEMAT, ELEC, INFORM&PD, REFRIG, TELECOM control circuit; ~ **de convergencia** m TV convergence circuit; ~ **de conversación** m TELECOM speech circuit; ~ **corrector** m ING ELÉC adjusting circuit, TV corrector circuit; ~ **de corriente alterna** m ELEC alternating-current circuit; ~ **de corriente de Foucault** m ELEC, ING ELÉC eddy-current circuit; ~ **cortado** m ELEC open circuit; ~ **cortador** m TEC ESP *electrónica* chopper circuitry; ~ **para cortocircuitar carga** m ING ELÉC *auxiliar* crowbar; ~ **de cuadratura de la onda** m TV squaring circuit; ~ **de cuatro hilos** m INFORM&PD four-wire circuit; ~ **de cuatro hilos conductores** m INFORM&PD four-wire circuit; ~ **de cuatro terminales** m ING ELÉC four-terminal network; **~ d** ~ **de derivación** m ING ELÉC bias circuit; ~ **en derivación** m ELEC parallel circuit, ING ELÉC bridge circuit, MINAS *electricidad*, TELECOM parallel circuit; ~ **derivado** m FÍS branch circuit, ING ELÉC derived circuit, shunt circuit, PROD branch circuit; ~ **derivado de motor** m PROD motor branch circuit; ~ **de desbloqueo de la imagen** m TV field-gating circuit; ~ **de descarga** m ING ELÉC discharge circuit, TV dumping circuit; ~ **desconectador** m ELEC trip circuit, *cortocircuito* trigger; ~ **de desconexión de color** m TV color gate (*AmE*), colour gate (*BrE*); ~ **de desenganche** m ELEC *cortocircuito* trigger; ~ **desmultiplicador de impulsos** m ELECTRÓN scaling circuit; ~ **despuntador** m ING ELÉC despiking circuit; ~ **de detección** m TRANSP detection loop; ~ **detector** m ELECTRÓN detector circuit; ~ **diferenciado** m ELECTRÓN differentiating circuit; ~ **diferenciador** m TV peaking circuit; ~ **diferenciador de crestas** m ELECTRÓN peaking circuit; ~ **digital** m ELECTRÓN, INFORM&PD, TELECOM digital circuit; ~ **discriminador del haz** m TV beam gate; ~ **disimétrico** m ELEC asymmetric circuit, asymmetrical circuit; ~ **disparador** m ELEC, ELECTRÓN, FÍS trigger circuit; ~ **de disparo** m ELEC *cortocircuito* trigger, trigger circuit, trip circuit, FOTO triggering circuit; ~ **divisor** m ELECTRÓN dividing circuit; ~ **de doble sintonización** m ELECTRÓN double-tuned circuit; ~ **de dos B más D** m TELECOM two-B-plus-

D arrangement; ~ **de dos hilos** *m* ELEC, INFORM&PD two-wire circuit;

~ **e** ~ **eléctrico** *m* ELEC channel, circuit, electric circuit, electrical circuit, electrical path, ELECTRÓN channel, electric circuit, circuit, FÍS, INFORM&PD circuit, ING ELÉC channel, electric circuit, PROD electrical path, TELECOM circuit, electrical circuit; ~ **eléctrico principal** *m* ELECTRÓN majority gate, majority logic; ~ **electrónico** *m* ELECTRÓN, TELECOM electronic circuit; ~ **electrónico discreto** *m* ELECTRÓN discrete electronic circuit; ~ **elemental distribuido** *m* ING ELÉC distributed element circuit; ~ **de elementos localizados** *m* ING ELÉC lumped-element circuit; ~ **de encendido** *m* MINAS firing circuit; ~ **de enganche** *m* PROD latch circuit; ~ **de enlace** *m* ELECTRÓN interface circuit; ~ **de enlace entrante** *m* TELECOM incoming trunk circuit; ~ **de enlace entre centralitas telefónicas automáticas privadas** *m* TELECOM inter-PABX tie circuit; ~ **de enlace intercentral** *m* TELECOM interswitchboard tie-circuit; ~ **de enlace de salida** *m* TELECOM outgoing trunk circuit; ~ **enlazado** *m* ELEC linked circuit; ~ **enterrado** *m* TRANSP buried loop; ~ **de entrada** *m* ELEC input circuit, ELECTRÓN input gate, ING ELÉC input circuit; ~ **entrante** *m* TELECOM incoming circuit; ~ **de entrecierre** *m* ING ELÉC interlock circuit; ~ **equivalente** *m* ELEC, ELECTRÓN, FÍS, ING ELÉC equivalent circuit; ~ **equivalente del transistor** *m* ELECTRÓN transistor equivalent circuit; ~ **de espera** *m* TRANSP AÉR holding pattern; ~ **estabilizador** *m* ELEC smoothing circuit; ~ **estampado para plaquetas** *m* INFORM&PD chip card; ~ **de excitación** *m* ING ELÉC excitation circuit; ~ **excitador** *m* ELEC trigger circuit; ~ **O exclusivo** *m* ELECTRÓN exclusive OR circuit; ~ **exclusivo NO-O** *m* ELECTRÓN exclusive NOR circuit; ~ **explorador** *m* TV sweep circuit; ~ **exterior** *m* ELEC external circuit, outgoing circuit; ~ **externo** *m* ELEC external circuit;

~ **f** ~ **fallado** *m* NUCL failed circuit; ~ **fantasma** *m* ING ELÉC, TELECOM phantom circuit; ~ **de fase sincronizada** *m* ELECTRÓN, TEC ESP phase-locked loop (*PLL*); ~ **de ferrorresonancia** *m* ING ELÉC ferroresonance circuit; ~ **de fijación** *m* ELECTRÓN, TV clamping circuit; ~ **de fijación de amplitud** *m* ELECTRÓN, TV clamping circuit; ~ **fijo aeronáutico** *m* TRANSP AÉR aeronautical fixed circuit; ~ **de filtrado** *m* ELEC smoothing circuit; ~ **con filtro RC** *m* ELECTRÓN, ING ELÉC RC filter-circuit; ~ **frigorífico** *m* REFRIG refrigerating circuit;

~ **g** ~ **de gas** *m* INSTAL TERM gas circuit; ~ **de gatillado** *m* ELEC trigger circuit; ~ **gatillador** *m* ELEC trigger circuit; ~ **gatillo** *m* ELEC trigger circuit; ~ **gausiano** *m* TV Gaussian circuit; ~ **geotérmico** *m* ENERG RENOV, GEOFÍS geothermal circuit;

~ **h** ~ **híbrido** *m* ELECTRÓN, FÍS hybrid circuit; ~ **híbrido de capa delgada** *m* ELECTRÓN thin-film hybrid circuit; ~ **híbrido de capa gruesa** *m* ELECTRÓN thick-film hybrid circuit; ~ **híbrido óptico** *m* ELECTRÓN optical hybrid circuit; ~ **híbrido simple** *m* ELECTRÓN simple hybrid circuit; ~ **de hilo desnudo** *m* ING ELÉC open circuit;

~ **i** ~ **IGE** *m* FÍS, INFORM&PD, TELECOM LSI circuit; ~ **de ignición** *m* TEC ESP ignition circuit; ~ **de igualador** *m* TELECOM equalizer circuit; ~ **de imagen** *m* TELECOM *TV* picture circuit; ~ **IMGE** *m* *circuito de integración en muy gran escala* FÍS, TELE-

COM VLSI circuit; ~ **impreso** *m* (*CI*) CONST, ELECTRÓN, FÍS, INFORM&PD printed circuit (*PC*); ~ **impreso de alta frecuencia** *m* ELECTRÓN high-frequency printed circuit; ~ **impreso de capas múltiples** *m* ELECTRÓN, TELECOM multilayer printed circuit; ~ **impreso a doble cara** *m* ELECTRÓN double-sided printed circuit; ~ **impreso de extensión** *m* INFORM&PD expansion board; ~ **impreso flexible** *m* ELECTRÓN flexible printed circuit; ~ **impreso con línea de precisión** *m* ELECTRÓN fine-line printed circuit; ~ **impreso de microondas** *m* ELECTRÓN microwave printed circuit; ~ **impreso de multicapa** *m* ELECTRÓN, TELECOM multilayer printed circuit; ~ **impreso en poliamida** *m* ELECTRÓN polyamide printed circuit; ~ **impreso a una sola cara** *m* ELECTRÓN single-sided printed circuit; ~ **impreso unilateral** *m* ELECTRÓN single-sided printed circuit; ~ **de impulsos** *m* ELECTRÓN, ING ELÉC pulse circuit; ~ **del inducido** *m* ELEC *máquina* armature circuit; ~ **inductivo** *m* ELEC inductive circuit; ~ **inductor** *m* ELEC *de máquina* field circuit; ~ **de integración** *m* ING ELÉC integrating circuit; ~ **de integración a gran escala** *m* FÍS, TELECOM large-scale integration circuit; ~ **integrado** *m* (*CI*) AUTO, ELEC, FÍS integrated circuit (*IC*), INFORM&PD integrated circuit (*IC*), chip, TEC ESP chip, TELECOM integrated circuit (*IC*), TRANSP embedded loop, TV integrated circuit (*IC*); ~ **integrado analógico** *m* ELECTRÓN analog integrated circuit; ~ **integrado de aplicación específica** *m* (*CIAE*) INFORM&PD application-specific integrated circuit (*ASIC*); ~ **integrado bipolar** *m* ELECTRÓN, INFORM&PD bipolar-integrated circuit; ~ **integrado bipolar analógico** *m* ELECTRÓN analog bipolar integrated circuit; ~ **integrado bipolar Schottky** *m* ELECTRÓN Schottky bipolar integrated circuit; ~ **integrado bipolar de silicio** *m* ELECTRÓN silicon bipolar integrated circuit; ~ **integrado de capas gruesas** *m* TELECOM thick-layer integrated circuit; ~ **integrado digital** *m* ELECTRÓN digital integrated circuit; ~ **integrado electrónico** *m* ELECTRÓN electronic integrated circuit; ~ **integrado en escala ultra grande** *m* TELECOM ultra large scale integration circuit; ~ **integrado de gran densidad** *m* ELECTRÓN high-density integrated circuit; ~ **integrado en gran escala** *m* FÍS large-scale integrated circuit, INFORM&PD large-scale integration circuit, TELECOM large-scale integrated circuit; ~ **integrado de Hall** *m* AUTO Hall integrated circuit (*Hall IC*); ~ **integrado híbrido** *m* ELECTRÓN, TELECOM hybrid integrated circuit; ~ **integrado lineal** *m* ELECTRÓN linear integrated circuit; ~ **integrado de microondas** *m* (*CIM*) ELECTRÓN, FÍS microwave integrated circuit (*MIC*); ~ **integrado de microondas activo** *m* ELECTRÓN active microwave integrated circuit; ~ **integrado monolítico** *m* ELECTRÓN, TELECOM monolithic integrated circuit; ~ **integrado en muy gran escala** *m* FÍS, TELECOM very large-scale integrated circuit; ~ **integrado óptico** *m* ELECTRÓN optical integrated circuit; ~ **integrado óptico de una sola función** *m* ELECTRÓN single-mode optical integrated circuit; ~ **integrado planar** *m* ELECTRÓN planar integrated circuit; ~ **integrado de SCOM** *m* ELECTRÓN metal gate CMOS integrated circuit; ~ **integrado de semiconductor** *m* ELECTRÓN semiconductor integrated circuit; ~ **integrado de silicio** *m* ELECTRÓN silicon integrated circuit; ~ **integrado SOM de canal**

N *m* ELECTRÓN NMOS integrated circuit; ~ **integrado tridimensional** *m* ELECTRÓN three-dimensional integrated circuit; ~ **integrado unipolar** *m* ELECTRÓN unipolar integrated circuit; ~ **de intercomunicación** *m* TV talkback circuit; ~ **de interfaz** *m* TELECOM interface circuit; ~ **intermedio** *m* ING ELÉC buffer; ~ **interno de entrada** *m* ELEC input circuit; ~ **interrumpido** *m* ELEC open circuit; ~ **invertidor** *m* ELECTRÓN inverter gate; ~ **de iteración** *m* ELECTRÓN loop;

■ **l** ~ **lateral** *m* ING ELÉC side circuit; ~ **de limitación de amplitud** *m* TV amplitude limiter circuit; ~ **limitador** *m* ELECTRÓN clipper circuit, TEC ESP *electrónica* limiter, TRANSP AÉR clipper; ~ **de línea de abonado** *m* TELECOM subscriber line circuit (*SLC*); ~ **lineal** *m* ELEC, ELECTRÓN, TELECOM linear circuit; ~ **de lógica** *m* ELECTRÓN, INFORM&PD, TELECOM logic circuit; ~ **lógico** *m* ELECTRÓN logic, logic circuit, INFORM&PD, TELECOM logic circuit; ~ **O lógico** *m* ELECTRÓN inclusive OR circuit, inclusive OR gate; ~ **lógico aleatorio** *m* ELECTRÓN, INFORM&PD random logic, random-logic circuit; ~ **lógico algebraico** *m* ELECTRÓN logic algebra; ~ **lógico de alta velocidad** *m* ELECTRÓN high-speed logic, fast logic; ~ **lógico de alto nivel** *m* ELECTRÓN high-level logic; ~ **lógico de arseniuro de galio** *m* ELECTRÓN gallium arsenide logic; ~ **lógico bipolar** *m* ELECTRÓN bipolar logic; ~ **lógico combinacional** *m* ELECTRÓN combinatorial logic; ~ **lógico combinatorio** *m* ELECTRÓN combinatorial logic; ~ **lógico compatible** *m* ELECTRÓN compatible logic; ~ **lógico digital** *m* ELECTRÓN digital logic; ~ **lógico de diodo** *m* ELECTRÓN diode logic; ~ **lógico de emisor acoplado** *m* (*CLEA*) ELECTRÓN emitter-coupled logic (*ECL*); ~ **lógico de enlace** *m* ELECTRÓN interface logic; ~ **lógico de fluido** *m* ING MECÁ fluid logic circuit; ~ **lógico de gran densidad** *m* ELECTRÓN high-density logic; ~ **lógico integrado** *m* ELECTRÓN integrated logic circuit, logic integrated circuit; ~ **lógico de inyección** *m* ELECTRÓN injection logic; ~ **lógico mezclador** *m* ELECTRÓN inclusive OR gate; ~ **lógico no saturado** *m* ELECTRÓN nonsaturated logic; ~ **lógico óptico** *m* ELECTRÓN optical logic circuit, optical logic gate; ~ **lógico predifundido de EA** *m* ELECTRÓN ECL gate array; ~ **lógico predifundido de emisor acoplado** *m* ELECTRÓN emitter-coupled logic gate array; ~ **lógico rápido** *m* ELECTRÓN fast logic; ~ **lógico RCTL** *m* ELECTRÓN RCTL logic; ~ **lógico saturado** *m* ELECTRÓN saturated logic; ~ **lógico de SCOM** *m* ELECTRÓN CMOS logic; ~ **lógico de semiconductor de óxido de metal** *m* ELECTRÓN metal-oxide semiconductor logic circuit; ~ **lógico de SOM** *m* ELECTRÓN MOS logic circuit; ~ **lógico SOM de canal N** *m* ELECTRÓN NMOS logic; ~ **lógico sumador** *m* ELECTRÓN logic addition; ~ **lógico de transistor a transistor** *m* (*CLTT*) ELECTRÓN transistor-transistor logic (*TTL*); ~ **lógico transistorizado combinado** *m* (*CLTC*) ELECTRÓN merged transistor logic (*MTL*); ~ **lógico de tres estados** *m* ELECTRÓN three-state logic;

■ **m** ~ **magnético** *m* ELEC, FÍS magnetic circuit, ING ELÉC magnetic circuit, metallic circuit, TEC ESP magnetic circuit; ~ **de mando** *m* ELEC trigger circuit; ~ **de mando único** *m* ING ELÉC, TV ganged circuit; ~ **matricial** *m* TELECOM matrix circuit; ~ **de megaflujo** *m* TELECOM megastream circuit; ~ **de**

memoria *m* TELECOM memory circuit; ~ **metálico** *m* ING ELÉC physical circuit; ~ **micrométrico** *m* ELECTRÓN micron circuit; ~ **de microondas** *m* ELECTRÓN microwave circuit; ~ **de mínima potencia** *m* ING ELÉC dry circuit; ~ **monolítico integrado de microondas** *m* (*CMIM*) FÍS monolithic microwave integrated circuit (*MMIC*); ~ **MSI** *m* TELECOM MSI circuit; ~ **multietapa** *m* TELECOM multistage circuit; ~ **en multiplex** *m* PROD multiplex circuit;

■ **n** ~ **NI** *m* ELECTRÓN NAND gate; ~ **nivelador** *m* ELEC smoothing circuit; ~ **no inductivo** *m* ELEC noninductive circuit; ~ **no recíproco** *m* TELECOM nonreciprocal circuit; ~ **no simétrico** *m* ELEC asymmetric circuit, asymmetrical circuit; ~ **NO-O** *m* ELECTRÓN NOR circuit; ~ **NO-O exclusivo** *m* ELECTRÓN exclusive NOR circuit; ~ **NOT** *m* ELECTRÓN NOT circuit; ~ **NO-Y** *m* ELECTRÓN NAND circuit; ~ **NY** *m* ELECTRÓN NAND circuit;

■ **o** ~ **obturador de gran potencia** *m* ELECTRÓN deep rejection trap; ~ **óptico integrado** *m* ÓPT, TELECOM integrated optical circuit (*IOC*); ~ **opticoelectrónico integrado** *m* TELECOM integrated optoelectronic circuit (*IOC*); ~ **OR** *m* ELECTRÓN OR circuit; ~ **oscilador** *m* ELECTRÓN oscillating circuit, oscillator circuit, TV oscillating circuit; ~ **oscilante** *m* ELEC, TV oscillating circuit;

■ **p** ~ **en paralelo** *m* ELEC, MINAS, TELECOM parallel circuit; ~ **de paro de la máquina** *m* PROD machine shutdown circuit; ~ **pasivo** *m* FÍS, TELECOM passive circuit; ~ **sin pérdidas** *m* TELECOM zero loss circuit; ~ **periódico de contacto** *m* ING ELÉC contact window; ~ **de permanencia** *m* TV unblanking circuit; ~ **perturbador** *m* TV scrambler; ~ **sobre la placa** *m* ELECTRÓN on-board circuitry; ~ **de polarización de emisor** *m* ELECTRÓN emitter bias circuit; ~ **polifásico** *m* ELEC polyphase circuit; ~ **de polisilicio** *m* ELECTRÓN polysilicon gate; ~ **portamuestra** *m* ING ELÉC sample-and-hold circuit; ~ **de potencia** *m* ELEC power circuit; ~ **primario** *m* ELEC *transformador* primary circuit; ~ **principal** *m* ELEC *de red* main circuit; ~ **privado** *m* TELECOM private wire (*PW*); ~ **de producción** *m* ING MECÁ production line; ~ **de protección** *m* ELEC protective circuit, ING ELÉC, TELECOM protection circuit; ~ **en puente** *m* ELEC, ING ELÉC bridge circuit; ~ **puente** *m* ELEC bridge circuit; ~ **de puente Owen** *m* ELEC Owen bridge;

■ **r** ~ **radiante** *m* TV radiating circuit; ~ **de ralentí y baja velocidad** *m* AUTO idle and low speed circuit; ~ **RC** *m* ELECTRÓN RC circuit; ~ **de reactancia** *m* ELEC, ELECTRÓN, FÍS, ING ELÉC reactance circuit; ~ **reactivo** *m* ELEC reactive circuit, ELECTRÓN regenerative circuit, ING ELÉC reactive circuit; ~ **de realimentación** *m* MECÁ feedback loop, TV feedback circuit; ~ **de realimentación de audio** *m* TV audio feedback circuit; ~ **de realimentación de video** *AmL*, ~ **de realimentación de vídeo** *Esp m* TV video feedback circuit; ~ **recíproco** *m* FÍS *televisión* dual network, TELECOM reciprocal circuit, TV dual network; ~ **recortador** *m* ELECTRÓN clipping circuit; ~ **de rectificación** *m* ING ELÉC rectifying circuit; ~ **del refrigerante** *m* REFRIG refrigerant circuit; ~ **regenerativo** *m* ELECTRÓN regenerative circuit; ~ **de regreso del vapor** *m* AUTO vapor return line (*AmE*), vapour return line (*BrE*); ~ **de regulación** *m* ELEC control circuit; ~ **de reloj** *m* ELECTRÓN clock circuit; ~ **de reserva** *m* PROD backup circuit;

~ **resistivo** *m* ELEC, ING ELÉC, TELECOM resistive circuit; ~ **de resonancia paralelo** *m* ELECTRÓN parallel resonant circuit; ~ **resonante** *m* ELEC *inducción* acceptor circuit, resonant circuit, ELECTRÓN, FÍS, TELECOM, TRANSP resonant circuit; ~ **resonante en serie** *m* ING ELÉC series resonant circuit; ~ **retardador** *m* ELECTRÓN time-delay circuit; ~ **de retardo** *m* ELEC delay circuit, ELECTRÓN delay circuit, circuit delay, TELECOM, TV delay circuit; ~ **de retorno** *m* ING ELÉC return circuit; ~ **de retroacción** *m* TV feedback circuit;

s ~ **de salida** *m* ING ELÉC, TELECOM output circuit; ~ **de salida cortocircuitado** *m* PROD shorted output circuit; ~ **saliente** *m* TELECOM outgoing circuit; ~ **Satstream** *m* TELECOM *marca comercial BT* Satstream circuit; ~ **de seguridad** *m* ING ELÉC guard circuit, shutdown circuit, snubber circuit; ~ **selector** *m* ING ELÉC switching circuit; ~ **de semiconductor de óxido de metal** *m* ELECTRÓN metal-oxide semiconductor gate; ~ **semipersonalizado** *m* ELECTRÓN semicustom circuit; ~ **de señal** *m* TELECOM line circuit; ~ **de señalización** *m* FERRO *equipo inamovible* signaling wires (*AmE*), signalling wires (*BrE*); ~ **separador** *m* TV separation circuit; ~ **en serie** *m* ING ELÉC, MINAS, TELECOM series circuit; ~ **en serie-paralelo** *m* ING ELÉC, MINAS *electricidad* series-parallel circuit; ~ **de servomecanismo** *m* CINEMAT, TV servo loop; ~ **simétrico** *m* ELEC, ELECTRÓN, TELECOM push-pull circuit; ~ **sincrónico** *m* TELECOM synchronous circuit; ~ **de sincronización** *m* ELECTRÓN clock circuit; ~ **sincronizado** *m* ELECTRÓN clocked circuit; ~ **de sintonía** *m* ELECTRÓN, TELECOM tuning circuit; ~ **sintonizable** *m* TELECOM tuning circuit; ~ **de sintonización** *m* ING ELÉC tuning circuit; ~ **de sintonización simultánea** *m* ING ELÉC ganged tuning circuit; ~ **sintonizado** *m* ELECTRÓN tuned circuit; ~ **sintonizador** *m* TELECOM tuning circuit; ~ **de soldadura** *m* CONST welding circuit; ~ **de SOM** *m* ELECTRÓN MOS gate; ~ **de SOM con puerta de metal de óxido espeso** *m* ELECTRÓN thick-oxide metal gate MOS circuit; ~ **de succión** *m* ELEC suction circuit; ~ **de supervisión de central propia** *m* TELECOM own-exchange supervisory circuit; ~ **supresor de ondulaciones residuales** *m* ELEC smoothing circuit;

t ~ **de telefonía entrelazada hacia adelante** *m* TELECOM forward-interworking telephony event (*FITE*); ~ **telefónico privado escalonado** *m* TELECOM speech-grade private wire; ~ **de termistor** *m* PROD thermistor circuit; ~ **tipo gatillo** *m* ELEC trigger circuit; ~ **totalmente manual** *m* ELECTRÓN full-custom circuit; ~ **de tráfico** *m* TELECOM traffic circuit; ~ **de transformación de coordenadas cromáticas** *m* TELECOM matrix circuit; ~ **de tres entradas** *m* ELECTRÓN three-input gate; ~ **trifásico** *m* TELECOM three-phase circuit;

u ~ **umbral** *m* ELECTRÓN threshold gate; ~ **único** *m* ELECTRÓN one-shot circuit; ~ **unido** *m* ELEC linked circuit; ~ **de unión** *m* ING ELÉC, TELECOM trunking; ~ **de uso privado de datos** *m* TELECOM data private wire; ~ **de utilización** *m* ELEC load circuit;

v ~ **de varias etapas** *m* TELECOM multistage circuit; ~ **de la vía** *m* FERRO *equipo inamovible* track circuit, track connection; ~ **virtual** *m* INFORM&PD virtual circuit, ING ELÉC phantom circuit, TELECOM virtual circuit; ~ **virtual canal D** *m* TELECOM D-

channel virtual circuit; ~ **virtual conmutado** *m* TELECOM switched virtual circuit; ~ **virtual permanente** *m* (*PVC*) INFORM&PD, TELECOM permanent virtual circuit (*PVC*);

y ~ **Y** *m* ELECTRÓN inclusive AND circuit

circuitos: ~ **de frenado independiente** *m pl* TRANSP separated-braking circuits; ~ **de freno individuales** *m pl* TRANSP separated-braking circuits

circulación *f* AGUA circulation, efflux, EMB flow, FÍS circulation, ING MECÁ movement, INSTAL HIDRÁUL efflux, METAL flow, TEC PETR *lodo* circulation, TRANSP *vehículos* traffic; ~ **del aire** *f* TEC ESP air flow; ~ **anticiclónica** *f* METEO anticyclonic circulation; ~ **asincrónica** *f* ELEC asynchronous running; ~ **atmosférica** *f* METEO atmospheric circulation; ~ **atmosférica general** *f* METEO general atmospheric circulation; ~ **ciclónica** *f* METEO cyclonic circulation; ~ **a contravía** *f* FERRO running on wrong line; ~ **ininterrumpida** *f* TRANSP uninterrupted flow; ~ **interrumpida** *f* TRANSP interrupted flow; ~ **de inyección** *f* PETROL *perforación* mud circulation; ~ **irregular** *f* FERRO *vehículos* uneven running; ~ **del lodo** *f* TEC PETR *perforación* mud column; ~ **de los materiales** *f* PROD *operaciones, talleres* routing; ~ **del monzón** *f* METEO monsoon circulation; ~ **natural** *f* NUCL natural circulation; ~ **perdida** *f* PETROL lost circulation; ~ **con retraso** *f* FERRO out-of-course running; ~ **en sentido equivocado** *f* FERRO wrong direction running; ~ **de tráfico** *f* TELECOM, TRANSP traffic flow; ~ **ventilada** *f* INSTAL TERM fanned circulation; ~ **en vía única** *f* FERRO single-line working

circulador *m* TEC ESP *comunicaciones por microondas*, TELECOM circulator; ~ **de microondas** *m* TELECOM microwave circulator; ~ **óptico** *m* TELECOM optical circulator

circulante: ~ **Faraday** *m* ELECTRÓN, FÍS, ING ELÉC Faraday circulator

circular[1]: ~ **de información aeronáutica** *f* TRANSP AÉR aeronautical information circular

circular[2] *vti* INFORM&PD flow

circular[3] *vi* PROD *tráfico* run; ~ **en direcciones opuestas** *vi* FÍS PART *rayos* circulate in opposite directions

circularidad: **no** ~ *f* TELECOM *de revestimiento, núcleo* noncircularity

circularización *f* TEC ESP circularization

círculo *m* CINEMAT, FÍS, FOTO, GEOM circle, INFORM&PD ring, ING MECÁ circle, PROD, TEXTIL *tejidos* ring; ~ **abierto** *m* TRANSP AÉR open loop; ~ **de los agujeros de los pernos** *m* PROD bolt circle; ~ **de base** *m* ING MECÁ base circle; ~ **bordeando un texto** *m* IMPR balloon; ~ **de cabeza** *m* ING MECÁ outside circle; ~ **cerrado** *m* TRANSP AÉR closed loop; ~ **circunscrito** *m* GEOM circumscribed circle; ~ **concéntrico** *m* GEOM concentric circle; ~ **de confusión** *m* CINEMAT, FOTO circle of confusion; ~ **de contacto** *m* ING MECÁ dividing circle, pitch line; ~ **contiguo** *m* GEOM kissing circle; ~ **de la corona** *m* ING MECÁ point circle; ~ **de cristal esmerilado** *m* CINEMAT, FOTO ground-glass circle; ~ **de declinación** *m* FÍS, INSTR declination circle, METR hour circle; ~ **de deslizamiento** *m* CONST slip circle; ~ **dividido** *m* GEOM divided circle; ~ **divisorio** *m* INSTR divide circle; ~ **del eje polar** *m* INSTR polar-axis circle; ~ **de engrane** *m* ING MECÁ pitch line; ~ **exterior** *m*

ING MECÁ blank circle, outside circle, point circle; ~ **generador** *m* GEOM rolling circle; ~ **graduado** *m* INSTR graduated circle; ~ **indicador** *m* ING MECÁ dial; ~ **inscrito** *m* GEOM inscribed circle; ~ **límite** *m* ING MECÁ clearance circle; ~ **magnético** *m* GEOFÍS dip circle; ~ **máximo** *m* FÍS great circle; ~ **mayor** *m* FÍS great circle; ~ **meridiano** *m* INSTR *telescopio* meridian circle; ~ **de mínima confusión** *m* CINEMAT, FOTO circle of least confusion; ~ **del pie** *m* ING MECÁ dedendum circle, dedendum line; ~ **primitivo** *m* ING MECÁ base circle, dividing circle, pitch line; ~ **primitivo de contacto** *m* ING MECÁ pitch circle; ~ **de rotación** *m* VEH turning circle; ~ **de Rowland** *m* FÍS Rowland circle; ~ **tangente** *m* GEOM tangent circle

circumpolar *adj* TRANSP, TRANSP MAR circumpolar

circuncentro *m* GEOM circumcenter (*AmE*), circumcentre (*BrE*)

circuncírculo *m* GEOM circumcircle

circunferencia *f* GEOM circle, circumference; ~ **de base** *f* ING MECÁ *engranajes* dedendum circle; ~ **de cabeza** *f* ING MECÁ addendum circle; ~ **de la corona** *f* ING MECÁ point circle; ~ **externa** *f* ING MECÁ crown; ~ **primitiva** *f* ING MECÁ curve of contact, pitch circumference; ~ **de raíz** *f* ING MECÁ dedendum circle

circunnavegación *f* TRANSP, TRANSP MAR circumnavigation

circunscribir *vt* GEOM circumscribe

circunstancia: ~ **imprevista** *f* ELECTRÓN random event

circunvolución *f* ING MECÁ, PETROL convolution

cirrocúmulo *m* (*Cc*) METEO cirrocumulus (*Cc*)

cirrolita *f* MINERAL cirrolite

cirrostrato *m* (*Cs*) METEO cirrostratus (*Cs*)

cirrus *m* (*Ci*) METEO cirrus (*Ci*)

cis *adj* QUÍMICA *isómero* cis; ~**-trans** *adj* QUÍMICA *isómero* cis-trans

CISC *abr* (*computador de conjunto de instrucciones complejas AmL, ordenador de conjunto de instrucciones complejas Esp*) INFORM&PD CISC (*complex instruction set computer*)

cisco *m* CARBÓN coal dust, coal powder, culm, lumpless small coal, slack, MINAS coal powder; ~ **de coque** *m* CARBÓN breeze

cisterna *f* AGUA cistern, PROD, TEC ESP, TEC PETR tank, TRANSP container; ~ **de agua de condensación** *f* TEC ESP wastewater tank; ~ **de aguas residuales** *f* TEC ESP wastewater tank; ~ **de aireación** *f* RECICL aeration tank; ~ **de carga** *f* CONTAM cargo tank; ~ **de líquidos semipastosos** *f* TRANSP slurry tanker; ~ **pesada** *f* TRANSP mammoth tanker; ~ **recolectora de leche** *f* AGRIC milk collection lorry (*BrE*), milk tank; ~ **de sedimentación** *f* RECICL sedimentation tank; ~ **de seguridad para gas** *f* TRANSP safety gas tank; ~ **de seguridad para gasolina** *f* TRANSP safety gasoline tank (*AmE*), safety petrol tank (*BrE*); ~ **semirremolque** *f* TRANSP tank semitrailer; ~ **semitrailer** *f* TRANSP tank semitrailer; ~ **submarina** *f* TRANSP submarine tanker; ~ **de ventilación** *f* RECICL aeration tank

cisuras *f pl* P&C *defecto de la pintura* cissing

cita: ~ **espacial** *f* TEC ESP rendezvous, space rendezvous, spatial rendezvous

citogenética *f* AGRIC cytogenetics

citral *m* QUÍMICA citral

citrato *m* QUÍMICA citrate

cítrico *adj* ALIMENT, DETERG, QUÍMICA *ácido* citric

citrino *m* MINERAL citrine

citronela *m* ALIMENT citronella

citronelal *m* QUÍMICA *perfumería* citronellal

citronilo *m* QUÍMICA citronyl

citrulina *f* QUÍMICA *ácido* citrulline

ciudadela *f* TRANSP MAR bridge-house, *construcción naval* bridge castle

civetona *f* QUÍMICA *perfumería* civetone

cizalla *f* GEOL *estructural* shear, ING MECÁ shears, shear blade, shearing machine, MECÁ shears, shear, P&C shear, TEXTIL shears; ~ **de barras** *f* ING MECÁ bar shear; ~ **para chapa** *f* ING MECÁ plate shears; ~ **mecánica** *f* CARBÓN shearing machine; ~ **de rollos** *f* ING MECÁ scroll shears; ~ **de uña vibrante** *f* ING MECÁ nibbling machine, power nibbler (*BrE*), power nibbling machine (*AmE*); ~ **universal** *f* ING MECÁ universal shears

cizalladora *f* ING MECÁ shearing machine; ~ **de barras** *f* ING MECÁ bar-shearing machine; ~ **y perforadora** *f* ING MECÁ shearing and punching machine

cizalladura: ~ **de viento** *f* METEO wind shear

cizallamiento *m* GEOL, ING MECÁ shearing; ~ **de maclaje** *m* METAL *deformación por maclado* twinning shear

cizallar *vt* ING MECÁ shear; ~ **en frío** *vt* ING MECÁ, TERMO cold-shear

cizallas *f pl* CONST bench shears, ING MECÁ shears; ~ **de barras** *f pl* ING MECÁ bar shears; ~ **para desbastes** *f pl* PROD slab shears; ~ **de hojalatero** *f pl* PROD tinman's shears, tinman's snips; ~ **de mano** *f pl* ING MECÁ panel shears, PROD hand shears; ~ **para tochos** *f pl* ING MECÁ billet shears, bloom shears

Cl *abr* (*cloro*) CONTAM, DETERG, HIDROL, QUÍMICA, SEG Cl (*chlorine*)

CL *abr* (*cristal líquido*) GEN LC (*liquid crystal*)

CLAP *abr* (*cromatografía líquida a alta presión*) ALIMENT, LAB, QUÍMICA HPLC (*high-pressure liquid chromatography*)

claqueta *f* CINEMAT clapper board, slate; ~ **de cámara** *f* CINEMAT, TV camera slate; ~ **de cola** *f* CINEMAT tail slate; ~ **final** *f* CINEMAT end slate; ~ **invertida** *f* CINEMAT upside down slate; ~ **visual** *f* CINEMAT vision clap

claquetista *m* CINEMAT clapper person

clara: ~ **de cuadernas** *f* TRANSP MAR *construcción naval* frame spacing

claraboya *f* C&V roof light, ING MECÁ, MECÁ louver (*AmE*), louvre (*BrE*), TRANSP MAR deck light, deckhead light

claridad *f* AGUA, HIDROL clearness, IMPR bright, brightness, lightness, TELECOM clarity

clarificación *f* AGUA clarification, clarifying, ALIMENT clarifying, CALIDAD, CARBÓN clarification, clarifying, PROC QUÍ clarifying, QUÍMICA clarification, clarifying, TEC PETR clarifying; ~ **compuesta** *f* HIDROL *aguas residuales* compound clarification; ~ **primaria** *f* AGUA primary clarification

clarificador *m* ALIMENT clarifier

clarificadora *f* HIDROL clarifier

clarificar *vt* AGUA, ALIMENT, CALIDAD clarify, CARBÓN clarify, dilute, PROC QUÍ, QUÍMICA, TEC PETR clarify

clarificarse *v refl* CARBÓN *líquido* settle

clarión *m* C&V chalk

clarita *f* MINERAL clarite

clarkeíta *f* MINERAL clarkeite

claro[1] *adj* CINEMAT low-key, GEOL light-colored (*AmE*), light-coloured (*BrE*), METEO *aire* clear
claro[2] *m* AGRIC clearing; ~ **de tierra** *m* TEC ESP *astronomía* earthshine
claroscuro *m* CINEMAT chiaroscuro
clase[1]: **de ~ media** *adj* CARBÓN *tierra* medium-graded, MINAS *mineral* medium grade
clase[2] *f* CALIDAD grade, CRISTAL, INFORM&PD, ING MECÁ class, PROD grade; ~ **de aislación** *f* ELEC insulation class; ~ **de aislamiento** *f* ELEC insulation class; ~ **de amplificador** *f* ELECTRÓN amplifier class; ~ **de contacto** *f* ING ELÉC contact rating; ~ **de corriente** *f* REFRIG flow pattern; ~ **de error del protocolo** *f* TELECOM protocol error class; ~ **holoédrica** *f* CRISTAL holohedral class; ~ **holosimétrica** *f* CRISTAL holosymmetric class; ~ **no disponible de opción o servicio** *f* TELECOM service or option not available class; ~ **no disponible de recursos** *f* TELECOM resource unavailable class; ~ **no implementada de opción o servicio** *f* TELECOM service or option not implemented class; ~ **no obtenible de recursos** *f* TELECOM resource unavailable class; ~ **de objetos controlados** *f* TELECOM managed object class (*MOC*); ~ **preferente** *f* TRANSP cabin class; ~ **de primeros auxilios** *f* SEG first-aid class; ~ **de protocolo** *f* TELECOM protocol class; ~ **de registro** *f* INFORM&PD record class; ~ **de situación normal** *f* TELECOM normal situation class
clasificación *f* AGRIC grade, grading, C&V sorting, CALIDAD classification, CARBÓN sizing, classification, sorting, EMB *calidades, tamaños* grading, FERRO marshalling (*BrE*), marshaling (*AmE*), *vehículos* classification, GEOL sorting, INFORM&PD filing, sorting, sort, PAPEL screening, PROD *por calidades, tamaños* filing, grading, assortment, TELECOM sequencing, TRANSP, TRANSP MAR classification, TV rating; ~ **por aire** *f* ALIMENT air classification; ~ **alfanumérica** *f* INFORM&PD alphanumeric sort; ~ **de los animales vivos** *f* AGRIC live grades; ~ **del avión** *f* TRANSP AÉR aircraft classification; ~ **de la basura** *f* RECICL waste sorting; ~ **binaria** *f* INFORM&PD binary sort; ~ **de bloques** *f* INFORM&PD block sort; ~ **del buque** *f* TRANSP MAR *a efectos de seguro* classification of ship; ~ **por calidad** *f* AGRIC, CALIDAD quality grading; ~ **de control de aproximación** *f* TRANSP AÉR approach control rating; ~ **cromática** *f* COLOR color gradation (*AmE*), colour gradation (*BrE*); ~ **de estilo de tipo** *f* IMPR classification of type design; ~ **externa** *f* INFORM&PD external sort; ~ **granulométrica** *f* C&V, CARBÓN, PROC QUÍ granulometric classification; ~ **hidráulica** *f* CARBÓN hydraulic classification; ~ **interna** *f* INFORM&PD internal sort; ~ **inversa** *f* INFORM&PD backward sort; ~ **de llegadas** *f* GEOL picking of arrivals; ~ **a mano** *f* CARBÓN, MINAS *minerales* hand selection; ~ **del mantenimiento del avión** *f* TRANSP AÉR aircraft maintenance rating, aircraft overhaul rating; ~ **de las masas de aire** *f* METEO air mass classification; ~ **por el método de burbuja** *f* INFORM&PD bubble sort; ~ **por el método del lomo de asno** *f* FERRO *vehículos* hump shunting; ~ **del número de octano** *f* AUTO octane number rating (*ONR*); ~ **en onda** *f* INFORM&PD bubble sort; ~ **de partículas** *f* PROC QUÍ particle classification; ~ **de potencia** *f* ING ELÉC power rating; ~ **del radar** *f* FÍS RAD, TRANSP, TRANSP AÉR, TRANSP MAR radar rating;

~ **del radar de aproximación de precisión** *f* TRANSP AÉR precision-approach radar rating; ~ **del radar de aproximación de vigilancia** *f* TRANSP AÉR control early warning radar rating; ~ **de radar de búsqueda de aproximación** *f* TRANSP AÉR surveillance approach radar rating; ~ **regresiva** *f* INFORM&PD backward sort; ~ **de retroceso** *f* INFORM&PD backward channel; ~ **de ruido** *f* ING MECÁ, INSTR noise labeling (*AmE*), noise labelling (*BrE*); ~ **secuencial** *f* ING ELÉC sequencing; ~ **del servicio del avión** *f* TRANSP AÉR aircraft overhaul rating; ~ **por tamaños** *f* C&V, CARBÓN size grading; ~ **de terrenos** *f* AGRIC land classification; ~ **de vagones** *f* FERRO *vehículos* shunting; ~ **volumétrica** *f* EMB sizing
clasificador *m* C&V classifier, sorter, CINEMAT grader, PROC QUÍ classifier, TEC PETR *separador de partículas* cyclone; ~ **de aire seco** *m* CARBÓN *para materiales pulverulentos* cyclone; ~ **de botellas** *m* EMB bottle unscrambler; ~ **cónico** *m* CARBÓN *metalurgia*, PROD *mineral* cone classifier; ~ **de corriente vertical** *m* CARBÓN upward current classifier; ~ **de documentos** *m* INFORM&PD document sorter; ~ **helicoidal** *m* CARBÓN spiral classifier; ~ **mecánico** *m* CARBÓN mechanical classifier; ~ **de tambor** *m* CARBÓN *minería* drum cobber; ~ **de tamiz** *m* PAPEL screen; ~ **de taza** *m* CARBÓN bowl classifier; ~ **de trapos** *m* PAPEL rag sorter; ~ **de vasija** *m* CARBÓN bowl classifier
clasificadora *f* AGRIC spayed heifer; ~ **de cilindros giratorios** *f* AGRIC revolving-cylinder sorter; ~ **de documentos** *f* AmL (*cf archivador de documentos Esp*) INFORM&PD file folder; ~ **hidráulica** *f* MINAS *preparación* jig
clasificar *vt* ALIMENT grade, C&V, CALIDAD, CARBÓN classify, CINEMAT, ELECTRÓN, GEOL grade, INFORM&PD sort, MECÁ index, ÓPT grade, PROC QUÍ classify, PROD grade, TEC ESP index, TELECOM, TV grade; ~ **por colores** *vt* TEXTIL sort by color (*AmE*), sort by colour (*BrE*); ~ **por dimensiones** *vt* CARBÓN size; ~ **a mano** *vt* CARBÓN sort by hand, *minería* cobb (*AmE*), cob (*BrE*), MINAS *minerales* cobb (*AmE*), cob (*BrE*); ~ **por matices** *vt* TEXTIL sort by shade
clástico *adj* GEOL, PETROL, RECICL clastic
clasto *m* GEOL, PETROL, RECICL clast
claudetita *f* MINERAL claudetite
clausthalita *f* MINERAL clausthalite
cláusula: ~ **de capacidad** *f* TRANSP AÉR capacity clause; ~ **capacitoria** *f* TRANSP AÉR capacity clause
clavadura *f* CONST nailing
clavar[1] *vt* CONST nail, *tornillo* drive in, stick, drive
clavar[2] *vi* CONST batten
clave *f* ACÚST key, CONST *de arco* keystone, INFORM&PD key, TELECOM code; ~ **actual** *f* INFORM&PD actual key; ~ **de un arco** *f* CONST key; ~ **de búsqueda** *f* INFORM&PD search key; ~ **de colores** *f* IMPR color key (*AmE*), colour key (*BrE*); ~ **de control** *f* ELECTRÓN check bit, INFORM&PD control key; ~ **indefinida** *f* INFORM&PD undefined key; ~ **de interrogación** *f* TELECOM polling key; ~ **de las llaves** *f* TEC PETR *perforación* tong line; ~ **musical** *f* ACÚST key signature; ~ **nacional** *f* TELECOM national code; ~ **real** *f* INFORM&PD *COBOL* actual key
clavija *f* CINEMAT pin, CONST dowel, dowel pin, pin, pintle, *electricidad* plug, ELEC *conexión* plug, socket plug, ING ELÉC pin, ING MECÁ gudgeon, key, MECÁ *carpintería* peg, PROD dowel, pin, TELECOM plug;

~ de alineación *f* ING ELÉC aligning plug; **~ de alineamiento** *f* MECÁ alignment pin; **~ de arrastre** *f* ING MECÁ driver; **~ banana** *f* CINEMAT, FOTO, ING ELÉC banana plug; **~ bipolar** *f* ING ELÉC two-pin plug; **~ de buena coincidencia** *f* ING MECÁ locating pin; **~ cementada** *f* ING MECÁ hardened dowel pin; **~ de cierre** *f* ING MECÁ locking pin, lockpin; **~ conectora de extensión** *f* ING MECÁ extension connector-plug; **~ de conexión** *f* ELEC plug, MECÁ, PROD *electricidad* stud, TELECOM jack; **~ de conexión de flash para contactos F y X** *f* FOTO flash socket for F and X contact; **~ de conexión a la red** *f* ELEC, ING ELÉC current plug (*AmE*), mains plug (*BrE*); **~ cónica** *f* ING MECÁ taper pin; **~ de contacto** *f* ELEC *conexión* plug, ING ELÉC contact pin; **~ de corriente** *f* ELEC, ING ELÉC current socket (*AmE*), mains socket (*BrE*); **~ del disparador de cable** *f* FOTO cable release socket; **~ eléctrica** *f* ELEC, ING ELÉC, LAB *red de distribución* electric plug; **~ de embrague** *f* FERRO, ING MECÁ catch pin; **~ de enchufe** *f* ING ELÉC jack plug; **~ y enchufe** *f* TELECOM plug and socket; **~ hembra de corriente** *f* ELEC, ING ELÉC current socket (*AmE*), mains socket (*BrE*); **~ hembra de toma de corriente** *f* ING ELÉC jack plug; **~ hendida** *f* ING MECÁ cotter pin, split cotter pin, MECÁ cotter pin; **~ impolarizada** *f* ELEC *conexión* nonpolarized plug; **~ irreversible** *f* ELEC *conexión* nonreversible plug; **~ macho sobresaliente** *f* ING ELÉC pigtail; **~ maestra** *f* ING MECÁ kingbolt, kingpin; **~ no polarizada** *f* ELEC *conexión* nonpolarized plug; **~ polarizada** *f* ING ELÉC polarized plug; **~ posicionadora** *f* FOTO, ING MECÁ, MECÁ locating pin; **~ con punta cónica** *f* CINEMAT, FOTO banana plug, ING ELÉC banana jack, banana plug; **~ de reposición** *f* IMPR withdrawing pin; **~ de retención** *f* ING MECÁ catch pin; **~ con rosca para extracción** *f* ING MECÁ dowel pin with extracting thread; **~ con rosca extractora** *f* ING MECÁ dowel pin with extracting thread; **~ de situación** *f* MECÁ locating pin; **~ de terminación** *f* ELEC *accesorio del cable* plug-in termination; **~ de toma de corriente** *f* ING ELÉC power plug; **~ de ventilación** *f* CONST vent peg

clavito *m* ING MECÁ tack

clavo *m* CONST brad, nail, spike nail, MECÁ nail, MINAS chimney, nail; **~ de cabeza ancha** *m* ING MECÁ clout nail; **~ central** *m* CONST core nail; **~ de chimenea** *m* MINAS chimney, nail; **~ de fundición** *m* CONST cast nail; **~ de fundidor** *m* PROD gagger; **~ de listón** *m* CONST lath nail; **~ para mampostería** *m* CONST masonry nail; **~ de moldeador** *m* PROD gagger, sprig, *molderías* lifter; **~ para pisos** *m* CONST flooring nail; **~ de pizarrero** *m* ING MECÁ slate nail; **~ de uña** *m* FERRO *vía* dog spike

claxon *m* AUTO, TRANSP, VEH *seguridad* horn

CLEA *abr* (*circuito lógico de emisor acoplado*) ELECTRÓN ECL (*emitter-coupled logic*)

cleveíta *f* MINERAL cleveite

cliché *m* PROD stencil plate; **~ de destinatario** *m* EMB address stencil

clima *m* METEO climate; **~ continental** *m* METEO continental climate; **~ marino** *m* METEO maritime climate; **~ marítimo** *m* METEO maritime climate; **~ mediterráneo** *m* METEO Mediterranean climate; **~ de montaña** *m* METEO mountain climate; **~ del monzón** *m* METEO monsoon climate; **~ polar** *m* METEO polar climate; **~ tropical** *m* METEO tropical climate

climagrama *m* METEO climagram

climatización *f* CARBÓN, ING MECÁ, INSTAL TERM, METEO air conditioning, TERMOTEC acclimatization

climatizador *m* AUTO air conditioner, REFRIG air-conditioning unit; **~ del aire** *m* ING MECÁ air conditioner; **~ autónomo** *m* REFRIG self-contained air-conditioning unit

climatograma *m* METEO climatogram

climatología *f* METEO climatology

climograma *m* METEO climagram

clinker *m* C&V clinker cement, CONST clinker

clinoclasa *f* MINERAL clinoclase

clinoclasita *f* MINERAL clinoclasite

clinocloro *m* MINERAL clinochlore

clinoedrita *f* MINERAL clinohedrite

clinohumita *f* MINERAL clinohumite

clinometría *f* GEN clinometry

clinómetro *m* CARBÓN clinometer, inclinometer, CONST, GEOFÍS, GEOL, INSTR, METR, MINAS, TEC ESP clinometer; **~ giroscópico** *m* TEC ESP gyroclinometer; **~ para inclinación marina** *m* GEOFÍS dip compass

clinozoisita *f* MINERAL clinozoisite

clinquer *m* PROD clinker

clintonita *f* MINERAL clintonite

clip *m* EMB paperclip; **~ de junta de vidrio esmerilado** *m* LAB *soporte* ground-glass joint clamp

clipe *m* GEOL klippe

clisé *m* IMPR *fotografía* block, *serigrafía* stencil; **~ de destinatario** *m* EMB address stencil; **~ de línea** *m* IMPR line cut; **~ de trama de colores** *m* IMPR color line plate (*AmE*), colour line plate (*BrE*)

clivaje: **~ por deformación** *m* GEOL strain slip cleavage

cloaca *f* AGUA sewer, CARBÓN drain, CONST, HIDROL, RECICL sewer; **~ derivada** *f* AGUA branch sewer; **~ maestra** *f* CONST main sewer

clon *m* FÍS *biología, botánica*, INFORM&PD clone

clonación *f* FÍS *biología, botánica*, INFORM&PD cloning

clopinita *f* NUCL khlopinite

cloracetato *m* QUÍMICA chloracetate

cloracético *adj* QUÍMICA *ácido* chloracetic

cloración *f* DETERG, HIDROL, QUÍMICA *proceso* chlorination; **~ hasta el punto de aumento rápido del cloro residual** *f* HIDROL breakpoint chlorination

clorado *adj* DETERG, HIDROL, QUÍMICA chlorinated

cloral *m* QUÍMICA *aldehido grasoso* chloral; **~ hidratado** *m* QUÍMICA chloralhydrate

cloralamida *f* QUÍMICA chloralamide

cloralbenceno *m* QUÍMICA chloralbenzene

cloralbutol *m* QUÍMICA chloralbutol

cloralformatio *m* QUÍMICA chloralformate

cloralosa *f* QUÍMICA chloralose

cloranil *m* QUÍMICA *tinturas desinfectantes* chloranil

clorar *vt* DETERG, HIDROL, QUÍMICA chlorinate

clorastrolita *f* MINERAL chlorastrolite

clorato *m* QUÍMICA chlorate; **~ potásico** *m* QUÍMICA potassium chlorate; **~ de potasio** *m* QUÍMICA potassium chlorate

clorhídrico *adj* DETERG, QUÍMICA hydrochloric

clorhidrina *f* QUÍMICA chlorohydrin

clórico *adj* QUÍMICA chloric

clorita *f* CARBÓN, MINERAL, TEC PETR chlorite

cloritización *f* GEOL chloritization

clorito *m* QUÍMICA chlorite

cloritoide *m* MINERAL chloritoid

cloro *m* (*Cl*) CONTAM *producto químico*, DETERG, HIDROL, QUÍMICA, SEG chlorine (*Cl*); **~ activo** *m* DETERG active chlorine; **~ disponible** *m* PAPEL available chlorine; **~ líquido** *m* TERMO liquid chlorine

clorocalcita *f* MINERAL chlorocalcite

clorocaucho *m* P&C, QUÍMICA chlororubber

cloroespinela *f* MINERAL chlorospinel

clorofenol *m* QUÍMICA chlorophenol

clorofibra *f* TEXTIL chlorofiber (*AmE*), chlorofibre (*BrE*)

clorofila *f* QUÍMICA chlorophyl (*AmE*), chlorophyll (*BrE*)

clorofluorocarbono *m* (*CFC*) CONTAM *productos químicos*, EMB, QUÍMICA chlorofluorocarbon (*CFC*)

clorohidrato: **~ de aluminio** *m* HIDROL, QUÍMICA aluminium chlorohydrate (*BrE*), aluminum chlorohydrate (*AmE*)

cloromelanita *f* MINERAL chloromelanite

clorometano *m* QUÍMICA methyl chloride

clorópalo *m* MINERAL chloropal

cloropicrina *f* QUÍMICA *lacrimógeno vomitivo* chloropicrin

cloroplatinato *m* QUÍMICA chloroplatinate

cloropreno *m* QUÍMICA chloroprene

clororresistente *adj* DETERG, QUÍMICA chlorine-fast

clorosis *f* AGRIC chlorosis

cloroso *adj* QUÍMICA *ácido* chlorous

clorosulfonación *f* DETERG, QUÍMICA chlorosulfonation (*AmE*), chlorosulphonation (*BrE*)

clorosulfonado *adj* DETERG, QUÍMICA chlorosulfonated (*AmE*), chlorosulphonated (*BrE*)

cloruro *m* CARBÓN *química*, QUÍMICA chloride; **~ ácido** *m* QUÍMICA acidic chloride; **~ de ácido** *m* ALIMENT, QUÍMICA acid chloride; **~ amónico** *m* QUÍMICA ammonium chloride; **~ de amonio** *m* QUÍMICA ammonium chloride; **~ áurico** *m* QUÍMICA gold chloride; **~ de azufre** *m* QUÍMICA sulfur chloride (*AmE*), sulphur chloride (*BrE*); **~ de benzoílo** *m* QUÍMICA benzoyle chloride; **~ de cal** *m* PAPEL bleaching powder; **~ cálcico** *m* ALIMENT, QUÍMICA calcium chloride; **~ de calcio** *m* ALIMENT, QUÍMICA calcium chloride; **~ de carbonilo** *m* QUÍMICA carbonyl chloride; **~ de cobalto** *m* C&V, QUÍMICA cobalt chloride; **~ gelatinado** *m* QUÍMICA *utilizado en fotografía* gelatino-chloride; **~ de metilo** *m* QUÍMICA methyl chloride; **~ de plata** *m* ING ELÉC, QUÍMICA silver chloride; **~ de polivinilideno** *m* (*PVDC*) P&C polyvinylidene chloride (*PVDC*); **~ de polivinilo** *m* (*PVC*) CONST, ELEC, EMB, ING ELÉC, P&C polyvinyl chloride (*PVC*); **~ de polivinilo clorado** *m* (*CPVC*) P&C chlorinated polyvinyl chloride (*CPVC*); **~ potásico** *m* QUÍMICA potassium chloride; **~ de potasio** *m* QUÍMICA potassium chloride; **~ sódico** *m* QUÍMICA sodium chloride; **~ xenónico** *m* ELECTRÓN xenon chloride; **~ de zinc** *m* ING ELÉC zinc chloride

closet *m* C&V closet

CLTC *abr* (*circuito lógico transistorizado combinado*) ELECTRÓN MTL (*merged transistor logic*)

clupeína *f* QUÍMICA *proteína* clupeine

cluster: **~ de iones** *m* FÍS RAD ion cluster

CM *abr* (*contrato de mantenimiento*) ELECTRÓN MA (*maintenance contract*)

Cm *abr* (*curio*) FÍS RAD, QUÍMICA Cm (*curium*)

CMA *abr* (*concentración máxima admisible*) CONTAM MAC (*maximum allowable concentration*)

CMA: **~ ocupacional** *f* CONTAM *legislación* occupational MAC

CMC *abr* (*carboximetilcelulosa*) ALIMENT, DETERG, P&C CMC (*carboxymethyl cellulose*)

CME *abr* (*concentración máxima de emisión*) CONTAM MEC (*maximum emission concentration*)

CMIM *abr* (*circuito monolítico integrado de microondas*) FÍS MMIC (*monolithic microwave integrated circuit*)

CMP *abr* (*captura máxima permitida*) OCEAN TAC (*total allowable catch*)

CN *abr* (*control numérico*) ELEC, INFORM&PD, ING MECÁ, MECÁ, PROD NC (*numerical control*)

CNC *abr* (*control numérico computerizado*) INFORM&PD, ING MECÁ CNC (*computerized numerical control*)

Co *abr* (*cobalto*) C&V, COLOR, METAL, NUCL, QUÍMICA Co (*cobalt*)

coacervación *f* QUÍMICA coacervation

coacervato *m* QUÍMICA coacervate

coagulación *f* ALIMENT clotting, HIDROL, P&C, PROC QUÍ coagulation, QUÍMICA clotting, coagulation, TERMOTEC *tratamiento del agua de las calderas* coagulation

coagulado *adj* ALIMENT, HIDROL, P&C, PROC QUÍ, QUÍMICA, TERMOTEC coagulated

coagulador *m* PROC QUÍ coagulator

coagulante *m* ALIMENT coagulating, coagulant, CARBÓN flocculant, thickener, HIDROL coagulant, coagulating, P&C coagulating, coagulant, flocculant, *caucho* coagulating agent, PROC QUÍ coagulating, flocculant, coagulant, QUÍMICA coagulant, thickener, coagulating, flocculant, TERMOTEC coagulant, coagulating

coagular[1] *vt* ALIMENT, HIDROL, P&C, QUÍMICA, TERMOTEC coagulate

coagular[2] *vti* PROC QUÍ coagulate

coágulo *m* P&C coagulum

coalescencia *f* METEO coalescence

coaltitud *f* TEC ESP coaltitude

coaxial[1] *adj* GEN coaxial

coaxial[2] *m* INFORM&PD coaxial, TELECOM, TV coax

coaxil[1] *adj* GEN coaxial

coaxil[2] *m* INFORM&PD coaxial

coayudante: **~ de la filtración** *m* PROC QUÍ filter aid

cobalamina *f* QUÍMICA cobalamin

cobaltaminas *f pl* QUÍMICA cobaltammines

cobáltico *adj* QUÍMICA cobaltic

cobaltina *f* MINERAL cobalt glance, cobaltine

cobaltita *f* MINERAL cobaltite

cobalto *m* (*Co*) C&V, COLOR, METAL, NUCL, QUÍMICA cobalt (*Co*); **~ radioactivo** *m* FÍS RAD, QUÍMICA radiocobalt; **~ rojo** *m* MINERAL red cobalt

cobertizo *m* CONST shed, TEC ESP shelter

cobertor: **~ acústico** *m* TEC ESP acoustic blanket

cobertura *f* AGRIC *plaguicida* coverage, *semilla* coating, ALIMENT *repostería* couverture, topping, TEC ESP blanket, coverage, footprint, TELECOM coverage; **~ de la cadena** *f* TV network coverage; **~ de chocolate** *f* ALIMENT chocolate coating; **~ en directo** *f* TV live coverage; **~ de la emisora** *f* TV station coverage; **~ de frecuencia** *f* ELECTRÓN, TELECOM frequency coverage; **~ de gas inerte** *f* NUCL inert gas blanketing; **~ de gas nitrógeno** *f* NUCL nitrogen cover gas; **~ global** *f* TELECOM, TRANSP global coverage; **~ del haz del punto de exploración** *f* TELECOM spot beam coverage; **~ del**

haz del punto de imagen *f* TELECOM spot beam coverage; **~ hemisférica** *f* TELECOM hemispherical coverage; **~ del inducido** *f* ING ELÉC armature casing; **~ del objetivo** *f* CINEMAT lens coverage; **~ de parafina** *f* EMB paraffin coating; **~ plástica** *f* TELECOM plastic coating; **~ de rastrojo** *f* AGRIC mulch; **~ secundaria** *f* TELECOM secondary coating; **~ de sílice** *f* TELECOM silica coating

cobija *f* CONST *tejado* ridge tile; **~ contra incendios** *f* AmL (*cf manta contra incendios Esp*) SEG fire blanket; **~ de rescate** *f* AmL (*cf manta de rescate Esp*) SEG rescue blanket

cobranza *f* PROD collection

cobrar *vt* TRANSP MAR *barco, cuerda red* haul

cobre *m* (*Cu*) ELEC, METAL, QUÍMICA copper (*Cu*); **~ abigarrado** *m* MINERAL variegated copper; **~ blanco** *m* MINERAL domeykite; **~ gris** *m* MINERAL gray copper ore (*AmE*), grey copper ore (*BrE*); **~ gris antimonial** *m* MINERAL panabase; **~ rojo** *m* MINERAL, QUÍMICA red copper ore

cobreado *m* C&V copper staining

cobresoldabilidad *f* CONST, ING MECÁ, PROD, TERMO brazeability

cobresoldado *adj* CONST, ING MECÁ, PROD, TERMO brazed

cobresoldadura *f* CONST, ING MECÁ, PROD, TERMO brazing

cobresoldar *vt* CONST, ING MECÁ, PROD, TERMO braze

cobresoldeo *m* CONST, ING MECÁ, PROD, TERMO brazing, hard soldering

cobro *m* PROD collection; **~ directo** *m* PROD deferred charges, on-line data charge

cobro revertido: hacer una llamada a ~ *vi* TELECOM call collect (*AmE*), make a reverse charge call (*BrE*)

coca *f* TRANSP MAR *cabo* kink; **~ de un cable** *f* PROD kink

cocción *f* C&V baking, *en el horno de vidriado* burning, P&C *pintura* baking, TERMO cooking; **~ botulina** *f* ALIMENT botulinum cook; **~ de cal** *f* PROD lime burning; **~ química** *f* PAPEL chemical pulping

cocedor *m* TERMO cooker

cocer[1] *vt* C&V, TERMO bake; **~ parcialmente** *vt* ALIMENT parboil; **~ al vapor** *vt* ALIMENT steam

cocer[2] *vti* TERMO boil; **~ a fuego lento** *vti* TERMO boil slowly

coche *m* AUTO motorcar, FERRO coach, ING MECÁ, MECÁ motor, TRANSP car, wagon; **~ de alquiler** *m* TRANSP cab; **~ blindado** *m* D&A armored car (*AmE*), armoured car (*BrE*); **~ blindado anfibio** *m* D&A amphibious armored car (*AmE*), amphibious armoured car (*BrE*); **~ blindado de reconocimiento** *m* D&A armored reconnaissance car (*AmE*), armoured reconnaissance car (*BrE*); **~ bomba contra incendios** *m* SEG fire-engine; **~ de cabeza** *m* ING MECÁ head end; **~ cama** *m* FERRO *vehículos* sleeping car, TRANSP sleeper; **~ deportivo** *m* AUTO, VEH sports car; **~ ecológico** *m* TRANSP clean-air car; **~ electrónico** *m* AUTO, ELECTRÓN electronic car; **~ de gasolina** *m* Esp, (*cf coche de nafta AmL*) AUTO gas-fueled car (*AmE*), petrol-fuelled car (*BrE*); **~ grúa** *m* VEH towing vehicle; **~-litera** *m* TRANSP sleeper; **~ de nafta** *AmL* (*cf coche de gasolina Esp*) AUTO gas-fueled car (*AmE*), petrol-fuelled car (*BrE*); **~ particular** *m* VEH private vehicle; **~ de pasajeros** *m* TRANSP passenger automobile (*AmE*), passenger car (*BrE*); **~ policía con radioteléfono** *m* TRANSP cruiser; **~ para pruebas deportivas** *m* AUTO, VEH *tipo de vehículo* sports car; **~ con puerta trasera** *m* AUTO, VEH *vehículo* hatchback automobile (*AmE*), hatchback car (*BrE*); **~ radio-patrulla** *m* AUTO radio-patrol car; **~ de remolque** *m* MECÁ trailer; **~ restaurante** *m* FERRO *vehículos* buffet car, TRANSP saloon coach; **~ salón** *m* FERRO *vehículos* club car; **~ de turismo** *m* VEH passenger automobile (*AmE*), passenger car (*BrE*); **~ para uso particular** *m* VEH passenger automobile (*AmE*), passenger car (*BrE*); **~ para uso privado** *m* VEH private vehicle; **~ de viajeros** *m* TRANSP passenger automobile (*AmE*), passenger car (*BrE*), passenger coach

cochera *f* TRANSP bus shelter, shelter

cochinilla *f* ALIMENT, COLOR cochineal

cocido[1]: **no ~** *adj* C&V unfired, TERMO unbaked

cocido[2] *m* C&V *arcillas* bake

cociente *m* FÍS ratio, INFORM&PD quotient, MATEMÁT, METR, REFRIG ratio; **~ de armónicas** *m* GEOM harmonic ratio; **~ cruzado** *m* GEOM cross-ratio; **~ de difusión térmica** *m* FÍS, NUCL, TERMO thermal diffusion ratio; **~ dimensional** *m* INFORM&PD aspect ratio; **~ trigonométrico** *m* GEOM trigonometrical ratio; **~ de vueltas** *m* FÍS turns ratio; **~ zona de combustión superficie-garganta** *m* TEC ESP *toberas* burning surface-to-throat area ratio

cocimiento *m* C&V firing

cocina *f* TRANSP MAR *de barco* galley; **~ continua** *f* ALIMENT continuous cooker; **~ de estucado** *f* REVEST coating kitchen

cóclea *f* ACÚST cochlea

cocodrilo *m* ING MECÁ alligator clip

cocolita *f* MINERAL coccolite

cocolito *m* GEOL coccolith

codal *m* CONST *carpintería* strut

codaste *m* TRANSP MAR *construcción naval* rudder post, stern post

códec *m* ELECTRÓN, INFORM&PD, TELECOM code-decoder, codec, coder-decoder

codeína *f* QUÍMICA codeine

codera *f* SEG elbow pad

codificación *f* ELECTRÓN coding, encoding, encryption, EMB encoding, INFORM&PD coding, encryption, encoding, TEC ESP encoding, encryption, TELECOM coding, encryption, encoding, TRANSP AÉR encoding, TV encryption; **~ adaptiva** *f* TELECOM adaptive coding; **~ de la anchura** *f* ELECTRÓN *modulación por amplitud de impulso* width coding; **~ a baja velocidad** *f* TELECOM low-rate encoding (*LRE*); **~ básica** *f* INFORM&PD basic coding; **~ binaria** *f* TELECOM binary coding; **~ bit a bit** *f* TELECOM bit-by-bit encoding; **~ de bloque** *f* TELECOM block coding; **~ de contacto instalada en línea** *f* EMB in-line contact coding; **~ convolucional** *f* TELECOM convolutional coding; **~ convolucional auto-octogonal** *f* TELECOM self-orthogonal convolutional coding; **~ con corrección de errores** *f* TELECOM error-correction coding; **~ por desplazamiento de fase en banda estrecha** *f* TELECOM NBPSK (*narrow-band phase shift keying*); **~ para la detección de errores** *f* ELECTRÓN, INFORM&PD, TELECOM error-detection coding; **~ digital** *f* ELECTRÓN, TELECOM digital coding; **~ esquemática** *f* INFORM&PD skeletal coding; **~ de fase** *f* (*PE*) INFORM&PD phase encoding (*PE*); **~ de frecuencia** *f* ELECTRÓN, TELECOM frequency encoding; **~ intertrama** *f* TELECOM interframe coding; ~ **marginal** *f* CINEMAT edge coding; ~ **por mínimo desplazamiento de fase por filtrado gausiano** *f* TELECOM Gaussian-filtered minimum shift keying (*GMSK*); ~ **de predicción lineal** *f* ELECTRÓN linear predictive coding; ~ **predictiva lineal** *f* TELECOM linear predicting coding (*LPC*); ~ **predictiva vocoder lineal** *f* TELECOM linear predicting coding vocoder; ~ **propia** *f* INFORM&PD own coding; ~ **de señal compuesta** *f* ELECTRÓN, TELECOM, TV composite signal coding; ~ **por tonos secuenciales** *f* TELECOM sequential tone coding; ~ **de trama entrante** *f* TELECOM in-frame coding; ~ **vocal** *f* TEC ESP speech encoding, TELECOM vocoding; ~ **de la voz** *f* TELECOM speech coding, speech encoding

codificado: **no ~** *adj* TELECOM uncoded

codificador *m* ELECTRÓN, INFORM&PD code-decoder, coder, encoder, PROD encoder, TELECOM code-decoder, coder, encoder, TV coder, encoder; **~-decodificador** *m* ELECTRÓN, INFORM&PD, TELECOM codec, coder-decoder; ~ **del fondo** *m* EMB bottom coder; ~ **de hojas de cajas de cartón en caliente** *m* EMB hot foil carton coder; ~ **de impulsos** *m* ELECTRÓN, FÍS RAD pulse coder; ~ **de modos** *m* TELECOM mode scrambler; ~ **multiespiras** *m* PROD multiturn encoder; ~ **óptico** *m* TEC ESP optical encoder; ~ **de prioridad** *m* ELECTRÓN priority encoder; ~ **de pulsos** *m* FÍS RAD pulse coder; ~ **de señales ADPCM** *m* TELECOM ADPCM encoder; ~ **de teclado** *m* INFORM&PD keyboard encoder; ~ **para telefonía secreta** *m* TELECOM speech scrambler; ~ **vocal** *m* INFORM&PD vocoder, TELECOM vocoder, voice coder; ~ **de la voz para comunicación secreta** *m* TELECOM speech scrambler; ~ **de vuelta simple** *m* PROD single turn encoder

codificadora: ~ **de etiquetas** *f* EMB label-coding machine

codificar *vt* ELECTRÓN encode, encrypt, INFORM&PD code, encode, encrypt, TELECOM encode, encrypt, TV code, encrypt, encode; ~ **en dígito** *vt* ING MECÁ digitize

código *m* ELECTRÓN, INFORM&PD, TELECOM code; ~ **absoluto** *m* INFORM&PD absolute code; ~ **de acceso a la selección automática interurbana** *m* TELECOM subscriber trunk dialing access code (*AmE*), subscriber trunk dialling access code (*BrE*); ~ **alfabético** *m* INFORM&PD alphabetic code; ~ **alfanumérico** *m* INFORM&PD, TELECOM alphanumeric code; ~ **ampliado de caracteres decimales codificados en binario** *m* (*EBCDIC*) INFORM&PD extended binary-coded decimal-interchange code (*EBCDIC*); ~ **ASCII** *m* INFORM&PD ASCII code; ~ **ASME** *m* MECÁ ASME code; ~ **de autentificación** *m* INFORM&PD authentication code; ~ **de autocomprobación** *m* INFORM&PD self-checking code; ~ **auto-verificante** *m* INFORM&PD self-checking code; ~ **de barras** *m* EMB, IMPR, INFORM&PD, TELECOM bar code; ~ **binario** *m* ELECTRÓN, INFORM&PD binary code; ~ **bipolar** *m* TELECOM bipolar code; ~ **bipolar de alta densidad del orden 2** *m* TELECOM high-density bipolar of order 2 code (*HDB2*); ~ **bipolar de alta densidad del orden 3** *m* TELECOM high-density bipolar of order 3 code (*HDB3*); ~ **bipolar con sustitución de tres ceros** *m* (*B3ZS*) TELECOM bipolar code with three-zero substitution (*B3ZS*); ~ **biquinario** *m* INFORM&PD, MATEMÁT biquinary code; ~ **en**

cadena *m* INFORM&PD chain code; ~ **de la cadena puntera** *m* INFORM&PD chain code; ~ **de carácter** *m* INFORM&PD character code; ~ **cíclico** *m* TELECOM cyclic code; ~ **cíclico de bloque** *m* TELECOM cyclic block code; ~ **complementario** *m* TELECOM complementary code; ~ **de comprobación** *m* INFORM&PD check code, hash; ~ **de configuración estacional** *m* PROD, SEG season pattern code, seasonal pattern code; ~ **convolucional** *m* TELECOM convolution code, convolutional code; ~ **convolutivo** *m* TELECOM convolutive code; ~ **con corrección de errores** *m* ELECTRÓN, INFORM&PD error-correction coding; ~ **de corrección de errores** *m* ELECTRÓN, INFORM&PD, TEC ESP error-correcting code, TELECOM error-correcting code, error-correction code; ~ **corrector de errores** *m* ELECTRÓN, INFORM&PD error-correcting code, error-correction code, TEC ESP error-correcting code, TELECOM error-correcting code, error-correction code; ~ **de delimitación** *m* PROD fence code; ~ **de detección de errores** *m* ELECTRÓN, INFORM&PD, TELECOM error-detecting code, error-detection code; ~ **detector de errores** *m* ELECTRÓN, INFORM&PD, TELECOM error detector code; ~ **de detenida** *m* INFORM&PD stop code; ~ **digital** *m* ELECTRÓN digital code; ~ **de dirección** *m* CINEMAT address code; ~ **de dirección de la pista de aviso** *m* CINEMAT, TV cue-track address code; ~ **de EBCDIC** *m* ELECTRÓN EBCDIC code; ~ **de error** *m* INFORM&PD error code; ~ **especificado** *m* TELECOM given code; ~ **esquemático** *m* INFORM&PD skeletal coding; ~ **de fecha** *m* EMB date code; ~ **fuente** *m* INFORM&PD source code; ~ **de función** *m* INFORM&PD function code; ~ **de funcionamiento** *m* INFORM&PD function code; ~ **de Gray** *m* TEC ESP *cálculo* Gray code; ~ **de Hamming** *m* TEC ESP Hamming code; ~ **Hollerith** *m* INFORM&PD Hollerith code; ~ **de identificación** *m* INFORM&PD, TELECOM identification code; ~ **de identificación del receptor** *m* TELECOM recipient identification code; ~ **de identificación del recibidor** *m* TELECOM recipient identification code; ~ **de identificación de la red de datos** *m* TELECOM data network identification code (*DNIC*); ~ **de identificación de la red télex** *m* TELECOM telex network identification code; ~ **identificador de la red de datos** *m* TELECOM DNIC (*data network identification code*); ~ **de impulsos** *m* ELECTRÓN pulse code; ~ **de información** *m* INFORM&PD, PROD operation code (*op code*); ~ **de instrucción** *m* INFORM&PD instruction code; ~ **de instrucción de máquina** *m* INFORM&PD machine instruction code; ~ **de instrucciones simbólicas de carácter general para principiantes** *m* (*BASIC*) IMPR, INFORM&PD Beginner's All-purpose Symbolic Instruction Code (*BASIC*); ~ **interno** *m* INFORM&PD machine code; ~ **interurbano** *m* TELECOM trunk code (*TC*); ~ **láser** *m* ELECTRÓN laser code; ~ **de la línea** *m* EMB line code; ~ **en línea** *m* TELECOM line code; ~ **lineal** *m* TELECOM linear code; ~ **de longitud variable** *m* TELECOM variable-length code; ~ **de máquina** *m* INFORM&PD machine code; ~ **de marcación** *m* TELECOM dialing code (*AmE*), dialling code (*BrE*); ~ **nacional** *m* TELECOM national code; ~ **nacional de destino** *m* TELECOM national destination code (*NDC*); ~ **no binario** *m* TELECOM nonbinary code; ~ **numérico** *m* INFORM&PD numerical code; ~ **objeto** *m* FÍS, INFORM&PD object code; ~ **de ocho niveles** *m*

ELECTRÓN eight-level code; ~ **de opción** *m* INFORM&PD option code; ~ **de operación** *m* INFORM&PD, PROD operation code (*op code*); ~ **de operación prohibido** *m* PROD illegal operation code (*illegal op code*); ~ **del país** *m* TELECOM country code (*CC*); ~ **de parada** *m* INFORM&PD stop code; ~ **de la partida** *m* EMB batch code; ~ **de prioridad del circuito de transmisión** *m* TELECOM transmission priority code; ~ **de prioridad de procesamiento** *m* TELECOM processing-priority code; ~ **de prioridad de transmisión** *m* TELECOM transmission priority code; ~ **de producción** *m* CINEMAT production code; ~ **de protección de error** *m* TELECOM error protection code; ~ **de punto de origen** *m* TELECOM originating point code (*OPC*); ~ **de puntos fijos** *m* TELECOM point code; ~ **para el punzonado** *m* TEXTIL needling code; ~ **de redondeo** *m* PROD rounding code; ~ **de redundancia cíclica** *m* (*CRC*) TELECOM cyclic redundancy code (*CRC*); ~ **redundante** *m* INFORM&PD redundant code; ~ **de la remesa** *m* EMB batch code; ~ **de respuesta parcial** *m* TELECOM partial-response code; ~ **de retorno** *m* INFORM&PD return code; ~ **de ruidos pseudoaleatorios** *m* TELECOM pseudo-random noise code; ~ **de seguridad** *m* ING MECÁ, SEG safety code; ~ **de señales** *m* TRANSP MAR signal book; ~ **de situación** *m* EMB site code; ~ **de tensión de la bobina** *m* PROD coil voltage code; ~ **de tiempo** *m* CINEMAT, TV time code; ~ **de tiempos de la SMPTE** *m* CINEMAT, TV SMPTE time code; ~ **de zona** *m* TELECOM area code

codo[1]: **hasta el** ~ *adj* CONST elbow-high
codo[2] *m* AUTO crank, CONST bend, CRISTAL jog, ING MECÁ knee, tube bend, LAB elbow, MECÁ bend, elbow, NUCL *de un manipulador* elbow, PETROL dogleg, TEC ESP *tuberías* bending, TRANSP MAR *de canal, río, conducto* bend, VEH *cigüeñal* throw; ~ **en ángulo recto** *m* ING MECÁ right-angled bend; ~ **cónico** *m* ING MECÁ taper bend; ~ **de dilatación** *m* REFRIG expansion loop; ~ **doble** *m* ING MECÁ *tuberías*, MECÁ bushing; ~ **del eje** *m* ING MECÁ axle crank; ~ **de empalme** *m* CONST union elbow; ~ **de fundición** *m* CONST, NUCL cast-iron elbow; ~ **de manubrio** *m* ING MECÁ crank; ~ **de palanca** *m* ING MECÁ crank; ~ **del portavaliento** *m* PROD *altos hornos* gooseneck; ~ **purgador** *m* ING MECÁ bleed elbow; ~ **roscado y enchufado** *m* ING MECÁ screwed and socketed bend; ~ **y sistema de varilla de conexión** *m* ING MECÁ crank and connecting rod system; ~ **de tubo** *m* ING MECÁ tube bend

CO₂ *m* (*carbón dióxido*) ELECTRÓN CO_2 (*carbon dioxide*)
coeficiencia: ~ **comercial** *f* PROD commercial coefficiency
coeficiente *m* ELEC, FÍS, INSTAL HIDRÁUL, MATEMÁT, PROD coefficient;
▪ **a** ~ **de absorción** *m* ACÚST, FÍS, FÍS RAD absorption coefficient; ~ **de absorción acústica** *m* FÍS acoustic absorption coefficient; ~ **de absorción lineal** *m* FÍS linear absorption coefficient; ~ **de absorción solar** *m* ENERG RENOV solar absorption coefficient; ~ **de abundancia** *m* AGUA coefficient of abundance; ~ **de acomodación térmica** *m* TEC ESP thermal accommodation coefficient; ~ **de acoplamiento** *m* ELEC *inductor* coefficient of coupling, FÍS coupling coefficient; ~ **de acoplamiento electromagnético** *m* ELEC *transformador* magnetic coupling coefficient; ~ **de**

acoplamiento magnético *m* ELEC *transformador* magnetic coupling coefficient; ~ **de acortamiento** *m* ENERG RENOV, INSTAL HIDRÁUL contraction coefficient; ~ **de actividad** *m* FÍS activity coefficient; ~ **de adaptación** *m* ELEC, ING ELÉC *dipolos* return current coefficient; ~ **de adhesión** *m* AUTO adhesion coefficient; ~ **adiabático** *m* TERMO adiabatic coefficient; ~ **aerodinámico** *m* TEC ESP aerodynamic coefficient; ~ **de afinamiento** *m* TRANSP MAR *diseño naval* coefficient of fineness; ~ **de agitación del mar** *m* HIDROL coefficient of roughness; ~ **de aguas subterráneas utilizables** *m* HIDROL coefficient of usable groundwater; ~ **de aislamiento térmico** *m* FÍS, MECÁ, P&C, TERMO coefficient of thermal insulance, thermal insulation coefficient; ~ **de alargamiento** *m* P&C *propiedad física, prueba* modulus of elongation; ~ **de almacenamiento de agua** *m* HIDROL water storage coefficient; ~ **de amortiguación** *m* FÍS damping coefficient; ~ **del área de expansión de la tobera** *m* TEC ESP nozzle expansion area ratio; ~ **de atenuación** *m* ACÚST, ELECTRÓN, FÍS, ÓPT, TELECOM attenuation coefficient; ~ **de atenuación de imagen** *m* ELECTRÓN image attenuation coefficient; ~ **de atenuación lineal** *m* FÍS linear attenuation coefficient; ~ **de autocalentamiento** *m* ING ELÉC self-heating coefficient; ~ **de autoinducción** *m* ELEC self-inductance;
▪ **b** ~ **binómico** *m* MATEMÁT binomial coefficient; ~ **de bruma** *m* CONTAM *control de la contaminación atmosférica, legislación* coefficient of haze;
▪ **c** ~ **de calidad** *m* ELECTRÓN, TEC ESP figure of merit; ~ **de cambio de fase de imagen** *m* ELECTRÓN image phase change coefficient; ~ **de captura** *m* OCEAN coefficient of capture; ~ **de capturabilidad** *m* OCEAN catchability coefficient; ~ **de carga** *m* MECÁ load factor; ~ **de carga de avión de carga** *m* TRANSP AÉR all-cargo load factor; ~ **de caudal** *m* INSTAL HIDRÁUL coefficient of discharge; ~ **de compensación** *m* ING MECÁ make-up rate; ~ **de compresibilidad** *m* CARBÓN compressibility coefficient, compressibility modulus, FÍS compressibility coefficient; ~ **de compresión** *m* P&C *propiedad física, prueba* compression modulus, TRANSP MAR *construcción GRP* modulus of compression; ~ **de concentración** *m* CARBÓN concentration ratio; ~ **de conductividad térmica** *m* MECÁ, P&C, TERMO coefficient of thermal conductivity; ~ **de consolidación** *m* CARBÓN coefficient of consolidation, PROD shrinkage factor; ~ **de contracción** *m* ENERG RENOV contraction coefficient, INSTAL HIDRÁUL coefficient of contraction, contraction coefficient; ~ **de conversión** *m* NUCL conversion coefficient; ~ **de conversión antemeridiano AM-PM** *m* TEC ESP AM-PM conversion coefficient; ~ **de conversión de modulación de fase** *m* TEC ESP PM conversion coefficient; ~ **de correlación** *m* FÍS FLUID, INFORM&PD correlation coefficient; ~ **de corrientes reflejadas** *m* ELEC, ING ELÉC return current coefficient; ~ **de corte** *m* ING MECÁ shear modulus;
▪ **d** ~ **de desagüe** *m* CONST *hidráulica* runoff coefficient; ~ **de descarga** *m* ENERG RENOV discharge coefficient, INSTAL HIDRÁUL coefficient of discharge, coefficient of efflux; ~ **de descarga de masa** *m* TEC ESP mass discharge coefficient; ~ **de desfasador** *m* ACÚST dephasing coefficient; ~ **diferencial** *m* MATEMÁT differential coefficient; ~ **de difusión** *m*

ELECTRÓN, FÍS diffusion coefficient, FÍS FLUID diffusivity; ~ **de difusión térmica** *m* FÍS, NUCL, TERMO thermal diffusion coefficient; ~ **de dilatación** *m* MECÁ expansion coefficient, P&C *propiedad física, prueba* coefficient of expansion; ~ **de dilatación térmica** *m* TERMO thermal expansion coefficient; ~ **de disipación** *m* FÍS dissipation coefficient; ~ **de dispersión** *m* PROC QUÍ dispersion coefficient, TELECOM scattering coefficient; ~ **de dispersión magnética** *m* ELEC coefficient of magnetic dispersion; ~ **de dispersión material** *m* ÓPT, TELECOM material dispersion coefficient; ~ **de distribución** *m* QUÍMICA partition coefficient; ~ **de drenaje** *m* HIDROL coefficient of drainage;

~ e ~ **de eficiencia** *m* MECÁ coefficient of efficiency; ~ **de elasticidad** *m* CARBÓN, CONST, FÍS, ING MECÁ, MECÁ, METAL modulus of elasticity, P&C *propiedad física, prueba* modulus of elasticity, coefficient of elasticity, TEC PETR *resistencias de materiales* Young's modulus, modulus of elasticity, TRANSP AÉR, TRANSP MAR *construcción GRP* modulus of elasticity; ~ **elástico** *m* METAL elastic coefficient; ~ **de empuje** *m* TEC ESP thrust coefficient; ~ **de endurecimiento por medios mecánicos** *m* CRISTAL, MECÁ, METAL *acritud* work-hardening coefficient; ~ **de energía** *m* ENERG RENOV power coefficient; ~ **de enturbiamiento atmosférico** *m* CONTAM coefficient of haze; ~ **de esbeltez** *m* CONST slenderness ratio; ~ **de escorrentía** *m* AGUA, CONTAM, CONTAM MAR, HIDROL runoff coefficient; ~ **de escurrimiento** *m* AGUA, CONTAM, CONTAM MAR runoff coefficient, HIDROL coefficient of runoff, runoff coefficient; ~ **de expansión** *m* CONST, P&C *propiedad física, prueba* coefficient of expansion; ~ **de expansión cúbica** *m* FÍS cubic expansion coefficient; ~ **de expansión lineal** *m* FÍS linear expansion coefficient; ~ **de expansión térmica** *m* FÍS, MECÁ, TERMO coefficient of thermal expansion; ~ **de expansión volumétrica** *m* FÍS *expansividad cúbica* volume expansion coefficient; ~ **de extinción** *m* TELECOM extinction ratio (*EX*);

~ f ~ **de fase** *m* TELECOM phase coefficient; ~ **de flujo** *m* ENERG RENOV, FÍS FLUID flow coefficient; ~ **del flujo de agua subterránea** *m* HIDROL coefficient of groundwater flow; ~ **de fricción** *m* CONST, FÍS, MECÁ, METR, P&C *propiedad física, prueba* coefficient of friction; ~ **de fuerza** *m* ENERG RENOV power coefficient; ~ **de fuerza lateral** *m* TRANSP AÉR lateral-force coefficient;

~ g ~ **de gasto** *m* INSTAL HIDRÁUL coefficient of discharge;

~ h ~ **de Hall** *m* FÍS Hall coefficient; ~ **de histéresis** *m* ELEC *magnetización* hysteresis coefficient;

~ i ~ **de imanación** *m* ELEC magnetic susceptibility; ~ **de inducción múltiple** *m* ELEC coefficient of mutual induction; ~ **de inducción mutua** *m* ELEC *bobina* coefficient of mutual induction; ~ **de inducción propia** *m* ELEC self-inductance; ~ **de infiltración** *m* HIDROL coefficient of infiltration;

~ l ~ **de labranza** *m* AGRIC unit draft; ~ **de lixiviación** *m* NUCL leaching coefficient; ~ **de luminosidad** *m* FÍS RAD luminosity coefficient;

~ m ~ **de marchitamiento** *m* AGRIC wilting coefficient; ~ **másico de absorción de energía** *m* NUCL mass energy absorption coefficient; ~ **másico de atenuación** *m* NUCL mass attenuation coefficient;

~ **másico de transferencia de energía** *m* NUCL mass energy transfer coefficient; ~ **del momento** *m* ING MECÁ moment coefficient; ~ **de montaje** *m* PROD assembly coefficient;

~ o ~ **de onda estacionaria** *m* TELECOM standing-wave ratio (*SWR*);

~ p ~ **de partición** *m* METAL, QUÍMICA partition coefficient; ~ **de Peltier** *m* FÍS Peltier coefficient; ~ **de pérdidas por reflexión** *m* ING ELÉC mismatch factor; ~ **de permeabilidad** *m* CARBÓN permeability coefficient, HIDROL coefficient of permeability; ~ **pH** *m* HIDROL pH number; ~ **de planeo** *m* TRANSP AÉR lift-drag ratio (*L-D ratio*); ~ **de Poisson** *m* CARBÓN, CONST, MECÁ, PETROL Poisson's ratio; ~ **de potencia-peso** *m* TRANSP AÉR power-weight ratio; ~ **de presión** *m* ENERG RENOV, FÍS, TRANSP AÉR *aerodinámica* pressure coefficient; ~ **de presión relativa** *m* FÍS relative-pressure coefficient; ~ **de presión terrestre** *m* CARBÓN earth-pressure coefficient; ~ **de propagación** *m* ACÚST, TELECOM propagation coefficient; ~ **de propagación axial** *m* ÓPT, TELECOM axial propagation coefficient; ~ **de propulsión de la hélice** *m* TRANSP propeller-thrust coefficient;

~ r ~ **de recalcado** *m* METAL jump rate; ~ **de reciprocidad electroacústica** *m* ACÚST, ING ELÉC electroacoustical reciprocity coefficient; ~ **de recombinación** *m* ELECTRÓN, FÍS recombination coefficient; ~ **de reducción** *m* TRANSP MAR *escantillón* coefficient of reduction; ~ **de reflectividad** *m* TEC ESP reflectivity coefficient; ~ **de reflexión** *m* ACÚST, ELECTRÓN, FÍS, ÓPT reflection coefficient, TEC ESP reflectivity coefficient; ~ **de reflexión del mar** *m* TEC ESP albedo; ~ **de reflexión de potencia** *m* ÓPT power-reflection coefficient; ~ **de reflexión de presión** *m* ACÚST pressure-reflection coefficient; ~ **de relleno** *m* TELECOM filling coefficient; ~ **de rendimiento aerodinámico** *m* TRANSP AÉR lift-drag ratio; ~ **de reparto** *m* QUÍMICA partition coefficient; ~ **de resistencia aerodinámica** *m* AUTO, TRANSP AÉR drag coefficient; ~ **de resistencia al aire** *m* VEH *carrocería* drag coefficient; ~ **de resistencia al avance** *m* ENERG RENOV coefficient of drag, FÍS FLUID, TRANSP MAR drag coefficient; ~ **de resistencia al caudal** *m* HIDROL flow-resistance coefficient; ~ **de restitución** *m* FÍS restitution coefficient; ~ **de retardo** *m* FÍS drag coefficient; ~ **de rigidez** *m* CARBÓN *vigas* shear modulus; ~ **de rotura** *m* TEXTIL breakage rate; ~ **de rozamiento** *m* CONST, FÍS, MECÁ coefficient of friction, METAL friction stress, METR, P&C *propiedad física, prueba* coefficient of friction; ~ **de rozamiento debido al rodamiento** *m* FÍS rolling-friction coefficient; ~ **de rozamiento por deslizamiento** *m* FÍS sliding friction coefficient; ~ **de rozamiento estático** *m* FÍS static friction coefficient; ~ **de ruidosidad** *m* ELECTRÓN noise figure;

~ s ~ **Sabine** *m* ACÚST Sabine coefficient; ~ **de Seebeck** *m* FÍS Seebeck coefficient; ~ **de seguridad** *m* CARBÓN safety factor, CONST factor of safety, ELEC safety factor, PROD, SEG coefficient of safety; ~ **de Steinmetz** *m* FÍS Steinmetz's coefficient; ~ **de sustentación** *m* ENERG RENOV, FÍS, TRANSP AÉR *aerodinámica* lift coefficient (*Cl*); ~ **de sustentación de la pala** *m* TRANSP AÉR blade-lift coefficient;

~ t ~ **de temperatura** *m* ING ELÉC, TERMO, TV temperature coefficient; ~ **de temperatura de**

capacitancia *m* ING ELÉC temperature coefficient of capacitance; **~ de temperatura de resistencia** *m* ELEC, ING ELÉC temperature coefficient of resistance; **~ térmico** *m* TERMO temperature coefficient; **~ de termoaislamiento** *m* FÍS, TERMO thermal insulation index; **~ de termoconducción** *m* MECÁ, P&C, TERMO coefficient of thermal conduction; **~ de termotransferencia** *m* TERMO heat transfer coefficient; **~ de Thomson** *m* FÍS Thomson coefficient; **~ de torsión** *m* ENERG RENOV coefficient of torque, TEXTIL twist factor; **~ de transferencia antemeridiano AM-PM** *m* TEC ESP AM-PM transfer coefficient; **~ de transferencia de calor** *m* FÍS heat transfer coefficient; **~ de transferencia de calor por convección** *m* TERMOTEC convective heat transfer coefficient; **~ de transferencia del calor de irradiación** *m* FÍS RAD, TERMO, TERMOTEC radiation-heat transfer coefficient; **~ de transferencia de energía** *m* FÍS energy-transfer coefficient; **~ de transferencia de imagen** *m* ELECTRÓN image transfer coefficient; **~ de transferencia de masa-energía** *m* FÍS mass energy transfer coefficient; **~ de transferencia de modulación de amplitud** *m* TEC ESP AM transfer coefficient; **~ de transferencia de modulación de fase** *m* TEC ESP PM transfer coefficient; **~ de transferencia TWT** *m* TEC ESP TWT transfer coefficient; **~ de transmisión** *m* ACÚST, FÍS *del sonido* transmission coefficient; **~ de transmisión de calor** *m* TERMO heat transfer coefficient; **~ de turbiedad** *m* ENERG RENOV turbidity coefficient; **~ u** **~ de uniformidad** *m* CARBÓN uniformity coefficient; **~ de utilización** *m* MECÁ duty cycle; **~ de utilización de un almacén** *m* REFRIG *de energía* storage factor; **~ v** **~ de variación de la capacidad con la temperatura** *m* ING ELÉC temperature coefficient of capacitance; **~ de variación de la resistencia con la temperatura** *m* ING ELÉC temperature coefficient of resistance; **~ de velocidad** *m* ENERG RENOV velocity coefficient; **~ de la velocidad en la boquilla** *m* ENERG RENOV nozzle velocity coefficient; **~ de viscosidad** *m* FÍS FLUID, TERMO viscosity coefficient; **~ de viscosidad y temperatura** *m* TERMO viscosity-temperature coefficient; **~ volumétrico de emisión y absorción** *m* FÍS RAD volume emission and absorption coefficient; **~ y** **~ de Young** *m* CARBÓN Young's modulus

coeficientes: **~ de Einstein** *m pl* FÍS RAD Einstein coefficients

coenzima *f* ALIMENT coenzyme

coercitividad *f* ELEC, ELECTRÓN, FÍS coercivity, ING ELÉC coercitivity, METAL, TV coercivity

coetáneo *adj* GEOL coeval; **~ con la amplitud de marea** *adj* GEOFÍS cotidal

cofia *f* TELECOM *cohetes* fairing

coger *vt* TRANSP MAR fetch

coherencia *f* CARBÓN compatibility, ELECTRÓN, FÍS, FÍS ONDAS coherence, INFORM&PD consistency, ÓPT, TELECOM coherence; **~ espacial** *f* FÍS, TELECOM spatial coherence; **~ de espacio** *f* TELECOM space coherence; **~ de onda** *f* TELECOM wave coherence; **~ óptica** *f* TELECOM optical coherence; **~ parcial** *f* ÓPT, TELECOM partial coherence; **~ temporal** *f* FÍS temporal coherence, ÓPT temporal coherence, time coherence, TELECOM temporal coherence; **~ de tiempo** *f* TELECOM time coherence

coherente *adj* CARBÓN, CONST cohesive, ELECTRÓN coherent, FÍS coherent, cohesive, FÍS ONDAS coherent, METAL coherent, cohesive, ÓPT, TELECOM coherent

cohesión *f* CARBÓN, CONST, FÍS, METAL cohesion; **~ mecánica** *f* NUCL mechanical bond

cohesivo *adj* CARBÓN, CONST, FÍS, METAL cohesive

cohete *m* D&A, TEC ESP, TRANSP AÉR, TRANSP MAR *propulsor* rocket; **~ aprovisionado de combustible sólido** *m* TEC ESP solid-fuelled rocket; **~ atómico** *m* D&A atomic rocket; **~ atómico de aire a aire** *m* D&A atomic air-to-air rocket; **~ auxiliar** *m* TEC ESP *vehículos* booster; **~ de combustible líquido** *m* TERMO liquid-fuel rocket; **~ de control** *m* TEC ESP control rocket; **~ de despegue** *m* TEC ESP kick rocket; **~ estabilizador por giro** *m* TEC ESP spin rocket; **~ de frenado** *m* TRANSP retardation rocket; **~ giratorio** *m* TEC ESP spin rocket; **~ luminoso** *m* D&A flare; **~ multietapa** *m* TEC ESP multistage rocket; **~ en órbita** *m* TEC ESP orbital rocket; **~ orbital** *m* TEC ESP orbital rocket; **~ de propelente sólido** *m* TEC ESP solid-propellant rocket; **~ con propulsante líquido** *m* TERMO liquid-propellant rocket; **~ de propulsión iónica** *m* TEC ESP ion rocket; **~ de propulsor sólido** *m* TEC ESP solid-propellant rocket; **~ de retardo** *m* TRANSP retardation rocket; **~ de separación** *m* TEC ESP separation rocket; **~ sonda** *m* TEC ESP sounding rocket; **~ de sondeo** *m* D&A sounding rocket

cohetería *f* D&A, TEC ESP, TRANSP AÉR rocketry

cohobación *f* QUÍMICA cohobation

cohobar *vt* QUÍMICA cohobate

coincidencia: **~ retardada** *f* FÍS RAD delayed coincidence

coincidir *vi* CONST match

cojín: **~ del asiento** *m* VEH seat cushion

cojinete *m* *AmL* (*cf travesaño Esp*) AUTO journal, CONST bearing, ELEC *máquina* bearing, brass, FERRO *vehículos* axle box (*BrE*), journal box (*AmE*); ING MECÁ plummer-block bearing, bearings, bush, journal, axle box (*BrE*), journal box (*AmE*), pillow, bearing, pillow block, plummer block, cushion, INSTR cradle, MECÁ cradle, bearing, journal, pad, MINAS cap piece, TRANSP MAR engine bearing, VEH bearing, *término genérico* bush, journal; **~ ablandado** *m* AUTO run bearing; **~ aceitado** *m* ING MECÁ oiled bearing; **~ sin aceite** *m* ING MECÁ oilless bearing; **~ de agujas** *m* ING MECÁ needle bearing; **~ de aleación de cobre** *m* ING MECÁ copper alloy bush; **~ de alineación propia** *m* ING MECÁ self-aligning bearing; **~ de anillo delgado** *m* ING MECÁ thin ring bearing; **~ antifricción** *m* PROD antifriction bearing; **~ del árbol de levas** *m* VEH camshaft bushing; **~ de aro estrecho** *m* ING MECÁ thin ring bearing; **~ de aterrajar** *m* ING MECÁ screw stock; **~ autoalineable** *m* ING MECÁ, VEH self-aligning bearing; **~ autoengrasado** *m* ING MECÁ self-lubricating bearing; **~ autolubricado** *m* ING MECÁ oilless bearing, self-lubricating bearing; **~ axial de rodillos cilíndricos** *m* ING MECÁ cylindrical roller bearing; **~ de bancada** *m* ING MECÁ crank bearing; **~ de biela** *m* AUTO, ING MECÁ connecting-rod bearing; **~ de bolas** *m* ING MECÁ ball bearing, ball bushing, rolling bearing, MECÁ ball bearing; **~ de bolas acanalado** *m* ING MECÁ grooved ball bearing; **~ de bolas angular** *m* MECÁ angular ball bearing; **~ de bolas de construcción fuerte** *m* ING MECÁ heavy-duty ball bearing; **~ de bolas de doble fila** *m* ING MECÁ double row ball bearing; ~ **de bolas de empuje** *m* ING MECÁ thrust ball bearing; ~ **de bolas de guías hondas** *m* ING MECÁ deep groove ball bearing; ~ **de bolas con pista de rodadura hecha de alambre de acero duro** *m* ING MECÁ wire race ball bearing; ~ **de bolas de ranuras profundas** *m* ING MECÁ deep groove ball bearing; ~ **de bolas de tipo medio** *m* PROD medium-type ball bearing; ~ **de bolas para trabajos pesados** *m* ING MECÁ heavy-duty ball bearing; ~ **de bronce de cañón** *m* ING MECÁ gunmetal bearing; ~ **de la cabeza** *m* VEH *biela* big-end bearing; ~ **de cabeza de biela** *m* AUTO big-end bearing; ~ **central** *m* AUTO center bearing (*AmE*), centre bearing (*BrE*); ~ **de cigüeñal** *m* ING MECÁ crank bearing; ~ **del cigüeñal** *m* AUTO crankshaft bearing, TRANSP MAR main bearing, VEH *motor* crankshaft bearing, main bearing; ~ **cilíndrico** *m* ING MECÁ ball bushing, bearing bush; ~ **cilíndrico radial** *m* ING MECÁ radial cylindrical roller bearing; ~ **de contacto plano** *m* ING MECÁ, VEH *término genérico* plain bearing; ~ **en declive** *m* INSTR declination bearing; ~ **de desembrague** *m* AUTO release bearing, throw-out bearing, VEH throw-out bearing; ~ **de desgaste** *m* PETROL wear bushing; ~ **de dos mitades de pared delgada** *m* ING MECÁ thin-walled half-bearing; ~ **del eje** *m* ING MECÁ shaft-bearing; ~ **del eje del motor** *m* TRANSP AÉR engine-shaft bearing; ~ **de empuje** *m* ENERG RENOV, ING MECÁ, MECÁ thrust bearing; ~ **de empuje axial** *m* ING MECÁ axial-thrust bearing; ~ **de empuje de bolas** *m* ING MECÁ ball-thrust bearing, thrust ball bearing, MECÁ ball-thrust bearing; ~ **de empuje liso** *m* ING MECÁ plain-thrust bearing; ~ **esférico basculante** *m* ING MECÁ swivel bearing; ~ **final de empuje** *m* ING MECÁ end-thrust bearing; ~ **giratorio** *m* ING MECÁ swivel plummer block; ~ **grafitado** *m* ING MECÁ oilless bearing; ~ **guarnecido con metal antifricción** *m* ING MECÁ bearing lined with antifriction metal; ~ **de guía** *m* INSTR *turbina de eje vertical* guide bearing; ~ **hidrodinámico** *m* TEC ESP hydrodynamic bearing; ~ **hidrostático** *m* MECÁ, PROD hydrostatic bearing; ~ **de hierro fundido antifriccionado** *m* ING MECÁ babbited cast-iron bearing; ~ **inferior** *m* ING MECÁ bottom brass; ~ **Kingsbury** *m* MECÁ Kingsbury bearing; ~ **de latón rojo** *m* ING MECÁ gun-metal bearing; ~ **liso** *m* ING MECÁ journal bearing, plain bearing, NUCL journal bearing, VEH *término genérico* plain bearing; ~ **sin lubricación** *m* ING MECÁ oilless bearing; ~ **de lubricación automática por cadena** *m* ING MECÁ chain-oiled bearing; ~ **lubricado por gas** *m* ING MECÁ, NUCL gas-lubricated bearing; ~ **con lubricante sólido** *m* ING MECÁ oilless bearing; ~ **de magneto** *m* ING MECÁ magneto bearing; ~ **de metal antifricción** *m* ING MECÁ *cojinetes de bolas* antifriction bearing; ~ **de motor** *m* ING MECÁ engine bearing; ~ **de movimiento longitudinal** *m* ING MECÁ traverse motion bearing; ~ **neumático** *m* CINEMAT air bearing; ~ **no alineado** *m* ING MECÁ out-of-jig cradle; ~ **del pie de biela** *m* AUTO small end bushing; ~ **piloto** *m* AUTO pilot bearing, pilot bushing; ~ **de polea** *m* ING MECÁ, MECÁ bushing; ~ **principal** *m* AUTO, CONST, TRANSP MAR, VEH *motor* main bearing; ~ **PTC** *m* PROD RTD bearing; ~ **radial sencillo de rótula** *m* ING MECÁ tap with metric thread; ~ **revestido con metal antifricción** *m* ING MECÁ bearing lined with antifriction metal; ~ **de**

rodillos *m* ING MECÁ needle bearing, roller bearing, rolling bearing; ~ **de rodillos abarrilados** *m* ING MECÁ barrel-shaped roller bearing; ~ **de rodillos de acción radial** *m* ING MECÁ radial cylindrical roller bearing; ~ **de rodillos en aguja** *m* ING MECÁ needle roller bearing; ~ **de rodillos angular** *m* MECÁ angular roller bearing; ~ **de rodillos cilíndricos** *m* ING MECÁ barrel roller bearing; ~ **de rodillos cilíndricos axial** *m* ING MECÁ axial cylindrical roller bearing; ~ **de rodillos cónicos** *m* ING MECÁ taper roller bearing; ~ **de roscar** *m* ING MECÁ die, screw plate, screwing die, stock, MECÁ die; ~ **de roscar radial** *m* ING MECÁ radial-threading die; ~ **de roscar tangencial** *m* ING MECÁ tangential diehead, tangential threading die; ~ **de rótula** *m* ING MECÁ knuckle bearing, swivel bearing; ~ **de rueda** *m* VEH wheel bearing; ~ **de rulemán** *m* ING MECÁ roller bearing, rolling bearing; ~ **sencillo esférico** *m* ING MECÁ tap with metric thread; ~ **sencillo de rótula** *m* ING MECÁ tap with metric thread; ~ **de suspensión de la corredera** *m* ING MECÁ link bearing; ~ **de terraja** *m* MECÁ die; ~ **tipo fricción** *m* ING MECÁ friction-type bearing; ~ **para trépano** *m* PETROL bit bearing

cok: ~ **de gas** *m* CARBÓN gas coke; ~ **metalúrgico** *m* CARBÓN, METAL metallurgic coke (*BrE*), metalurgic coke (*AmE*)

col: ~ **común** *f* AGRIC kale

cola *f* AGUA stem, tail bay, C&V CINEMAT tail, COLOR size, glue, CONST tail, EMB glue, INFORM&PD queue, ING MECÁ tang, P&C *caucho, pintura* size, *adhesivo* glue, PAPEL glue, size, QUÍMICA glue, TEC ESP *torpedos* afterbody, TELECOM queue, TEXTIL *para tejidos* size, *pegamento* glue; ~ **animal** *f* PAPEL animal glue; ~ **Bewoid** *f* PAPEL *agente de encolado* Bewoid size; ~ **blanca** *f* COLOR white size; ~ **de coches** *f* TRANSP line of cars (*AmE*), line of traffic (*AmE*), traffic queue (*BrE*); ~ **de destilación** *f* PROC QUÍ, QUÍMICA, TERMO distillation tail; ~ **de dispositivo** *f* INFORM&PD device queue; ~ **de entrada** *f* INFORM&PD input queue, entry queue; ~ **de espera limitada** *f* TELECOM limited waiting queue; ~ **geomagnética** *f* GEOFÍS, TEC ESP *geofísica* geomagnetic tail; ~ **de la gota** *f* C&V gob tail; ~ **de impresión** *f* INFORM&PD spooler; ~ **de llamadas** *f* TELECOM queue; ~ **magnética** *f* TEC ESP magnetotail; ~ **de milano** *f* CONST dovetail, fantail, INSTR, MECÁ, PROD dovetail; ~ **multiservidor** *f* TELECOM multiple server queue; ~ **neutra** *f* FOTO acid-free glue; ~ **de pescado** *f* ALIMENT finings, isinglass; ~ **de presión** *f* GEOL pressure shadow; ~ **prioritaria** *f* INFORM&PD priority queue; ~ **de salida** *f* INFORM&PD output queue; ~ **de un solo servidor** *f* TELECOM single-server queue; ~ **en T** *f* ELECTRÓN tail, TRANSP AÉR T-tail; ~ **de transferencia** *f* TEXTIL *faldón* transfer tail; ~ **en V** *f* TRANSP AÉR V-tail; ~ **vegetal** *f* COLOR vegetable size; ~ **de zorro** *f* AGRIC bour foxtail, foxtail grass

colacionado *m* INFORM&PD collation

colada *f* ING MECÁ *metalurgia*, MECÁ casting, PROD pouring, casting, *hornos metalúrgicos* tap, flow, founding, *metalurgia* cast, melt; ~ **por arriba** *f* PROD *fundería* top pouring; ~ **en caída directa** *f* PROD *fundería* top pouring; ~ **a chorro** *f* PROD *lingoteras* top pouring; ~ **en fuente** *f* PROD *fundición* bottom pouring; ~ **en lingotera** *f* SEG *fundición* pouring; ~ **de llenado** *f* PROD *moldes* teeming; ~ **de metal fundido** *f* PROD tapping the metal; ~ **en molde** *f* PROD casting in molds (*AmE*), casting in moulds (*BrE*); ~ **de pie** *f* PROD *fundería* top pouring; ~ **en talón** *f* PROD *fundición* side casting

coladero *m Esp* (*cf chimenea AmL*) MINAS raise, chute, mill hole, winze, *comunicación de dos partes de la mina* pass, *mineral* coarse concentration mill; ~ **principal** *m* MINAS main chute

colador *m* GAS, PROD strainer, RECICL sieve, TEC PETR *producción, refino* strainer; ~ **cónico** *m* ALIMENT conical sieve; ~ **filtrante** *m* PROC QUÍ filtrating strainer

colágeno *m* ALIMENT, QUÍMICA *proteína* collagen

colapsar *vti* CONST *estructuras, muros, paredes* crumble, fail

colapso *m* CONST breaking, collapse, ING MECÁ collapsing, MINAS horse, *de terreno* collapse; ~ **gravitacional** *m* TEC ESP gravitation collapse

colar *vt* CARBÓN, MINAS tap, PROD *fundería* teem, *hierro* cast, *metal de un horno* run off, *metales licuados* run; ~ **en matriz** *vt* ING MECÁ die-cast

colargol *m* QUÍMICA collargol

colarse *v refl* QUÍMICA seep; ~ **por ojo** *vr* TRANSP MAR *buque* go down by the bows

colas: ~ **de componentes pesados** *f pl* TEC PETR *refino* heavy ends; ~ **de enriquecimiento** *f pl* NUCL enrichment tails

colaspis: ~ **flavida** *f* AGRIC grape colaspis

colato *m* QUÍMICA cholate

colcha *f* TEXTIL quilt

colchar *vt* TRANSP MAR *filásticas, cordones* lay

colchiceína *f* QUÍMICA colchiceine

colchicina *f* QUÍMICA colchicine

colchón *m* TEXTIL mattress; ~ **de aire** *m* TRANSP AÉR, TRANSP MAR *aerodeslizador* air cushion; ~ **amortiguador de aire estático** *m* TRANSP static air cushion; ~ **amortiguador en altura** *m* TRANSP height-on cushion; ~ **amortiguador por chorros de aire periféricos** *m* TRANSP peripheral-jet air cushion; ~ **amortiguador magnético** *m* TRANSP magnetic cushion

colchoneta: ~ **de fibra de vidrio** *f* C&V chopped strand mat, glass-fiber mat (*AmE*), glass-fibre mat (*BrE*); ~ **de filamento continuo** *f* C&V continuous strand mat; ~ **no curada** *f* C&V uncured mat

colcótar *m* PROD, QUÍMICA *limpiador de cristales, pigmentos* colcothar

colección *f* OCEAN *mediciones de la profundidad del mar* sounding pool

colecta: ~ **de basuras** *f* RECICL refuse collection, waste collection; ~ **de desperdicios** *f* RECICL waste collection; ~ **selectiva** *f AmL* RECICL selective collection; ~ **por separado** *f* RECICL separate collection

colector *m* AGRIC main drain, CARBÓN *separación de minerales* tail disposal, collecting reagent, collecting agent, CONST *tuberías* trap, CONTAM *instalaciones industriales* sink, receptor, ELEC *máquinas* commutator, FÍS collector, ING ELÉC commutator, ING MECÁ receiver, manifold, INSTAL HIDRÁUL *calderas, de turbina de vapor* collector, drum, receiver, header, LAB *destilación, material de vidrio* receiver, NUCL manifold, header, PROC QUÍ collector, PROD *calderas, de sobrecalentador* header, sump, *vapor, condensado* manifold, RECICL main sewer, TEC ESP receiver, TEC PETR *pipe* manifold, TELECOM, TRANSP collector, VEH *generador* slip ring; ~ **absorbente** *m* PROC QUÍ absorber trap; ~ **de aceite** *m* AGUA oil trap; ~ **de**

admisión *m* AUTO, ING MECÁ, MECÁ, TRANSP AÉR, VEH *motor* inlet manifold (*BrE*), intake manifold (*AmE*); ~ **de agua** *m* AGUA separator, water conduit, MINAS standage, *pozos* sinkhole; ~ **de agua de alimentación** *m* NUCL feedwater manifold; ~ **de agua condensada** *m* INSTAL HIDRÁUL *tubería de vapor* steam separator; ~ **para aguas pluviales** *m* HIDROL *aguas residuales* storm drain; ~ **de aire caliente** *m* TRANSP AÉR hot air gallery; ~ **de alimentación** *m* NUCL feeder header; ~ **de arena y grava** *m* HIDROL sand and gravel trap; ~ **de barrido** *m* ING MECÁ *motores* air box; ~ **de basura** *m* RECICL garbage chute (*AmE*), rubbish chute (*BrE*); ~ **de bebedero caliente** *m* ING MECÁ hot runner manifold; ~ **de canal de colada caliente** *m* ING MECÁ hot runner manifold; ~ **de cenizas** *m* PROD cinder pocket; ~ **de condensación** *m* PROC QUÍ condensation trap; ~ **de datos** *m* TELECOM data sink; ~ **de delgas** *m* ELEC *máquinas*, TRANSP *vehículos eléctricos* commutator; ~ **de depósito radiactivo** *m* CONTAM precipitation collector; ~ **electrostático** *m* NUCL electrostatic collector; ~ **de energía** *m* TEC ESP power bus; ~ **de entrada** *m* AUTO, ING MECÁ inlet manifold (*BrE*), intake manifold (*AmE*); ~ **de escape** *m* AUTO, MECÁ, TERMO, TRANSP AÉR, VEH exhaust manifold; ~ **de flujo** *m* PETROL, TEC PETR flow line; ~ **de goteo** *m* CONST drip cup; ~ **de imagen final** *m* INSTR final-image tube; ~ **de impurezas** *m* ING MECÁ dirt trap; ~ **de lodo** *m* TEC PETR mud ring; ~ **de lubricante** *m* PROD, VEH *lubricación del motor* sump; ~ **maestro** *m Esp* CONST main sewer; ~ **mecánico** *m* CONTAM *tratamiento de gases* mechanical collector; ~ **de muestras de suelo** *m* OCEAN *geología submarina* sediment probe; ~ **múltiple** *m* TRANSP AÉR manifold; ~ **del núcleo fundido** *m* NUCL melting core catcher; ~ **de piquera caliente** *m* ING MECÁ hot runner manifold; ~ **con placas planas** *m* ENERG RENOV *energía solar* flat-plate collector; ~ **de polvos** *m* CARBÓN, P&C dust collector, dust chamber, PROC QUÍ dust catcher, SEG dust separator; ~ **de polvos por bolsas** *m* SEG fabric dust collector; ~ **de polvos y fibras** *m* SEG filtering plant for dust and fibres; ~ **de precipitación** *m* CONTAM precipitation collector; ~ **presurizado** *m* TRANSP AÉR pressurizing manifold; ~ **del recalentador** *m* PROD superheater header, superheater manifold; ~ **de red** *m* CONTAM net receiver; ~ **de salmuera** *m* REFRIG *llenado y registro del depósito* brine header; ~ **de sedimentos** *m* ING MECÁ dirt trap; ~ **solar** *m* ENERG RENOV solar collector; ~ **de sólidos arrastrados en el agua de alcantarillas** *m* HIDROL grit trap; ~ **de sumidero** *m* HIDROL sink trap; ~ **térmico** *m* FÍS heat reservoir; ~ **tipo p** *m* ELECTRÓN p-type collector; ~ **tipo Winston** *m* ENERG RENOV Winston collector; ~ **de transistor** *m* ELECTRÓN, TELECOM transistor collector; ~ **de tubos vaciados** *m* ENERG RENOV evacuated-tube collector; ~ **de vaciadero caliente** *m* ING MECÁ hot runner manifold; ~ **de vapor** *m* ING MECÁ manifold

colemanita *f* MINERAL colemanite

coleóptilo *m* AGRIC coleoptile

cólera: ~ **aviar** *f* AGRIC fowl cholera; ~ **porcino** *f* AGRIC hog cholera

colescitografía *f* INSTR cholecystography

colestérico *adj* QUÍMICA cholesteric

colesterol *m* ALIMENT, QUÍMICA cholesterol

colgado *m* PROD hand-up, hanging-up

colgador *m* ING MECÁ suspension; **~ de películas** *m* CINEMAT, FOTO film rack

colgante¹ *adj* PROD pending

colgante² *m* CONST hanging post, PROD hanging; **~ de brida** *m* ING MECÁ stirrup hanger

colgar¹ *vt* INFORM&PD hang up, PROD hang, hang up

colgar² *vi* TELECOM *teléfono* hang up

cólico *adj* QUÍMICA *ácido* cholic

cólicos *m* OCEAN *accidentes de buzos* colics

colidina *f* QUÍMICA collidine

coligativo *adj* QUÍMICA colligative

colilla *f* ING ELÉC stub

colimación *f* GEN collimation

colimador *m* GEN collimator; **~ espectroscópico** *m* INSTR spectroscope collimator

colimar *vt* GEN collimate

colina *f* ALIMENT choline; **~ abisal** *f* OCEAN *geología submarina* abyssal hill, sea high, abyssal knoll, sea knoll

colineal *adj* GEOM collinear

colinérgico *adj* QUÍMICA cholinergic

colinesterasa *f* QUÍMICA cholinesterase

colirita *f* MINERAL collyrite

colisión *f* CARBÓN collision, impact, FÍS, INFORM&PD collision, ING MECÁ, METAL impingement, SEG collision, TELECOM clashing, TRANSP, TRANSP MAR collision; **~ aérea** *f* TRANSP AÉR aerial collision; **~ por detrás** *f* AUTO, TRANSP, VEH rear-end collision; **~ elástica** *f* FÍS elastic collision, NUCL billiard-ball collision, elastic collision; **~ electrónica** *f* ELECTRÓN, TELECOM electron collision; **~ del electrón-positrón** *f* FÍS PART electron-positron interaction; **~ entre átomos** *f* NUCL atom-atom collision; **~ importante** *f* TRANSP *tráfico* primary collision; **~ inelástica** *f* FÍS RAD inelastic collision; **~ inelástica profunda** *f* NUCL deep inelastic collision; **~ ión-ión** *f* FÍS PART ion-ion collision; **~ lateral** *f* TRANSP side collision; **~ lejana** *f* NUCL distant collision; **~ múltiple** *f* TRANSP multiple pile-up; **~ de partículas** FÍS PART particle collision; **~ de primera especie** *f* NUCL collision of the first kind; **~ protón-antiprotón** *f* FÍS PART proton-antiproton collision; **~ de protones antiprotones con energía de 500 GeV** *f* FÍS PART proton-antiproton collision at 500 GeV; **~ protón-protón** *f* FÍS PART proton-proton collision; **~ próxima** *f* NUCL close collision; **~ radiativa** *f* FÍS RAD radiative collision; **~ secundaria** *f* TRANSP secondary collision; **~ tangencial** *f* NUCL glancing collision; **~ en vuelo** *f* TRANSP AÉR midair collision

collar *m* CONST collar, ELEC brass, ING MECÁ hoop, clip, collar, MECÁ clip, collar, PETROL collar, PROD hoop, TEC PETR collar, VEH *de motor, válvula* collet; **~ de acoplamiento** *m* ING MECÁ coupling box; **~ para cable** *m* ELEC, ING ELÉC cable clip; **~ del cojinete principal** *m* AUTO main-bearing bushing; **~ de desplazamiento del engranaje planetario** *m* AUTO sun gear shift collar; **~ del eje** *m* ING MECÁ shaft collar; **~ de embrague** *m* ING MECÁ clutch collar; **~ de excéntrica** *m* ING MECÁ eccentric strap; **~ excéntrico** *m* ING MECÁ eccentric collar, eccentric hoop; **~ de fijación** *m* ING MECÁ set collar; **~ de flotación** *m* TEC PETR *perforación* flotation collar; **~ partido** *m* ING MECÁ split ring; **~ de perforación** *m* *Esp* (*cf portamechas AmL*) TEC PETR drill collar; **~ de retén** *m* ING MECÁ stop collar; **~ soldado** *m* ING

MECÁ welded collar; **~ de tope** *m* ING MECÁ stop collar

collarín *m* (*cf reborde AmL*) C&V collar, CONST, ELECTRÓN neck, FÍS necking, ING MECÁ hoop, collar, gland, LAB *abertura, acceso* flange, MECÁ *máquinas herramientas* gland, collar, PROD, TEC ESP gland, TRANSP MAR *del ventilador* ventilator socket, VEH collet; **~ de la bandeja de carga** *m* *Esp* TRANSP pallet collar; **~ de excéntrica** *m* ING MECÁ eccentric strap; **~ de flotación** *m* TEC ESP flotation collar; **~ de la pálet** *m* *AmL*, **~ de paleta** *m* *Esp* TRANSP pallet collar; **~ de seguridad del revestimiento** *m* TEC PETR *perforación* casing clamp; **~ de tubo electrónico** *m* ELECTRÓN electron-tube neck

collera *f* MECÁ shackle

colmar *vt* CONST fill up

colocación *f* CONST laying, *topografía* setting, ING MECÁ setting, fitting, positioning, location, fixing, MECÁ fitting, PROD setup; **~ de banda magnética** *f* CINEMAT magnetic striping; **~ de cables** *f* ELEC *fuente de alimentación*, ING ELÉC cable laying; **~ de la carnada** *f* OCEAN *liña, pesca con palangre, urricán* baiting; **~ del cebo** *f* OCEAN baiting; **~ de un contrafuerte** *f* CONST buttressing; **~ de las escobillas** *f* ING ELÉC brush position; **~ de instalaciones solares** *f* ENERG RENOV *por fabricante* retrofit; **~ de listones de madera** *f* CONST battening; **~ de macadam** *f* CONST macadamization; **~ de machos** *f* ING MECÁ coring, PROD coring-up, *piezas fundidas* coring; **~ de minas** *f* *AmL* (*cf explotación de minas Esp*) MINAS mining; **~ en obra** *f* PROD *funderías* floor bedding; **~ ordenada** *f* FERRO, TRANSP marshaling (*AmE*), marshalling (*BrE*); **~ en paralelo** *f* ING ELÉC parallel lay; **~ de pilotes** *f* CARBÓN sheet piling; **~ previa de la banda sonora** *f* CINEMAT prestriping; **~ de la quilla** *f* TRANSP MAR keel laying; **~ de señales** *f* CONST *topografía* setting-out; **~ de tapas** *f* IMPR *encuadernación* casing-in; **~ del tapón roscado** *f* EMB capping; **~ de tarugos** *f* ING MECÁ plugging; **~ a tope** *f* CONST *carpintería* butting; **~ de la tubería** *f* TEC PETR *perforación* casing set; **~ de ventanas** *f* C&V *en un edificio* glazing; **~ de la vía** *f* FERRO *vehículos* plate laying; **~ de vías** *f* CINEMAT tracklaying

colocado *adj* ING MECÁ, MECÁ fixed

colocar¹ *vt* CONST *caballete* ridge, *clavijas* peg, fix, lay, *topografía* set, set out, ENER RENOV *instalación solar* retrofit, ING MECÁ fit to, set, PROD, *inserciones* chill, locate; **~ en** *vt* ING MECÁ fit in; **~ a ambos lados** *vt* PROD straddle; **~ dentro de** *vt* ING MECÁ fit into; **~ en espera** *vt* TELECOM put on hold; **~ al extremo de** *vt* ING MECÁ fit on; **~ para uso** *vt* PROD *taladro* bring into position for use

colocar²: **~ el caballete** *vi* CONST ridge; **~ clavijas** *vi* CONST peg; **~ inserciones** *vi* PROD *funderías* chill; **~ una instalación solar** *vi* ENERG RENOV retrofit; **~ macadam sobre** *vi* CONST *carreteras* metal

colodión *m* QUÍMICA collodion

colofeno *m* AGRIC, QUÍMICA *fertilizantes* colophene

colofón *m* QUÍMICA colophon

colofonia *f* P&C *materia prima* colophony, resin, QUÍMICA colophony; **~ de madera** *f* P&C *materia prima* wood resin

colofonita *f* MINERAL colophonite

coloidal *adj* GEN colloidal

coloide *m* GEN colloid; **~ irreversible** *m* QUÍMICA

irreversible colloid; ~ **protector** *m* QUÍMICA protective colloid

color[1]: **de un solo** ~ *adj* COLOR single-colored (*AmE*), single-coloured (*BrE*); **a todo** ~ *adj* IMPR full-color (*AmE*), full-colour (*BrE*); ~ **de acero** *adj* COLOR steel-colored (*AmE*), steel-coloured (*BrE*); **de** ~ **cobre** *adj* COLOR copper-colored (*AmE*), copper-coloured (*BrE*); ~ **oscuro** *adj* GEOL dark-colored (*AmE*), dark-coloured (*BrE*); **de** ~ **permanente** *adj* COLOR colorfast (*AmE*), colourfast (*BrE*); **de** ~ **resistente** *adj* EMB colorfast (*AmE*), colourfast (*BrE*); **de** ~ **sólido** *adj* TEXTIL *tintura* colorfast (*AmE*), colourfast (*BrE*)

color[2] *m* GEN color (*AmE*), coloring (*AmE*), colour (*BrE*), colouring (*BrE*); ~ **acromático** *m* IMPR achromatic color (*AmE*), achromatic colour (*BrE*); ~ **aditivo** *m* COLOR additive color (*AmE*), additive colour (*BrE*); ~ **de apresto** *m* COLOR size color (*AmE*), size colour (*BrE*); ~ **cetrino** *m* COLOR citrine color (*AmE*), citrine colour (*BrE*); ~ **cromático** *m* COLOR chromatic color (*AmE*), chromatic colour (*BrE*), chrome color (*AmE*), chrome colour (*BrE*); ~ **de cromo** *m* COLOR chrome color (*AmE*), chrome colour (*BrE*); ~ **crudo** *m* COLOR ecru; ~ **desaturado** *m* TV desaturated color (*AmE*), desaturated colour (*BrE*); ~ **desvanecido** *m* COLOR desaturated color (*AmE*), desaturated colour (*BrE*); ~ **falso** *m* COLOR, INFORM&PD false color (*AmE*), false colour (*BrE*); ~ **de fondo** *m* COLOR ground color (*AmE*), ground colour (*BrE*); ~ **fugaz** *m* COLOR sighting color (*AmE*), sighting colour (*BrE*); ~ **del gluón** *m* FÍS PART *asociado con cambios en las condiciones del quark* gluon color (*AmE*), gluon colour (*BrE*); ~ **de impresión** *m* COLOR printing color (*AmE*), printing colour (*BrE*); ~ **intenso** *m* COLOR deep color (*AmE*), deep colour (*BrE*); ~ **liso** *m* COLOR solid color (*AmE*), solid colour (*BrE*); ~ **de la mancha** *m* IMPR spot color (*AmE*), spot colour (*BrE*); ~ **mate** *m* C&V flat color (*AmE*), flat colour (*BrE*); ~ **natural** *m* TEXTIL natural color (*AmE*), natural colour (*BrE*); ~ **percibido** *m* COLOR perceived color (*AmE*), perceived colour (*BrE*); ~ **permanente** *m* COLOR color fastness (*AmE*), colour fastness (*BrE*), nonfading color (*AmE*), nonfading colour (*BrE*), P&C lasting color (*AmE*), lasting colour (*BrE*), color fastness (*AmE*), colour fastness (*BrE*); ~ **en polvo** *m* C&V ground color (*AmE*), ground colour (*BrE*); ~ **de porcelana** *m* C&V porcelain color (*AmE*), porcelain colour (*BrE*); ~ **primario** *m* CINEMAT, COLOR, FÍS, FÍS RAD primary color (*AmE*), primary colour (*BrE*); ~ **principal** *m* COLOR principal color (*AmE*), principal colour (*BrE*); ~ **del quark** *m* FÍS, FÍS PART quark color (*AmE*), quark colour (*BrE*); ~ **del revestimiento** *m* REVEST coating color (*AmE*), coating colour (*BrE*); ~ **secundario** *m* COLOR, IMPR secondary color (*AmE*), secondary colour (*BrE*); ~ **semi-transparente** *m* C&V semitransparent color (*AmE*), semitransparent colour (*BrE*); ~ **sólido** *m* IMPR fast color (*AmE*), fast colour (*BrE*), P&C *materia prima de la pintura* lasting color (*AmE*), lasting colour (*BrE*); ~ **solitario** *m* IMPR spot color (*AmE*), spot colour (*BrE*); ~ **superpuesto** *m* COLOR run color (*AmE*), run colour (*BrE*); ~ **tipográfico** *m* IMPR type color (*AmE*), type colour (*BrE*); ~ **vítreo** *m* C&V, COLOR glass color (*AmE*), glass colour (*BrE*);

~ **vitrificable** *m* C&V vitrifiable color (*AmE*), vitrifiable colour (*BrE*)

coloración *f* ALIMENT coloring (*AmE*), colouring (*BrE*), CINEMAT, P&C tinting, QUÍMICA staining; ~ **anormal** *f* P&C *defecto* color removal (*AmE*), colour removal (*BrE*); ~ **rosada** *f* COLOR pink coloration (*AmE*), pink colouration (*BrE*)

coloradoíta *f* MINERAL coloradoite

colorante *m* ALIMENT color (*AmE*), coloring (*AmE*), coloring matter (*AmE*), colour (*BrE*), colouring (*BrE*), colouring matter (*BrE*), COLOR coloring (*AmE*), coloring matter (*AmE*), colouring (*BrE*), colouring matter (*BrE*), dye, dyestuff, FOTO coloring (*AmE*), colouring (*BrE*), IMPR, PAPEL color (*AmE*), coloring (*AmE*), coloring matter (*AmE*), colour (*BrE*), colouring (*BrE*), colouring matter (*BrE*), dye, dyestuff, QUÍMICA color (*AmE*), coloring matter (*AmE*), colour (*BrE*), colouring matter (*BrE*), TEXTIL color (*AmE*), coloring (*AmE*), coloring matter (*AmE*), colour (*BrE*), colouring (*BrE*), colouring matter (*BrE*), dye, dyestuff; ~ **ácido** *m* COLOR, IMPR acid color (*AmE*), acid colour (*BrE*), QUÍMICA acetylide, TEXTIL acid dye; ~ **de acridina** *m* COLOR acridine dye; ~ **adhesivo** *m* COLOR glue color (*AmE*), glue colour (*BrE*); ~ **de advertencia** *m* COLOR warning color (*AmE*), warning colour (*BrE*); ~ **de alquitrán** *m* CARBÓN coal-tar dye, COLOR coal-tar dye, tar dye; ~ **de anilina** *m* COLOR, FOTO, QUÍMICA aniline dye; ~ **de antraceno** *m* QUÍMICA anthracene dye; ~ **de apresto** *m* COLOR size color (*AmE*), size colour (*BrE*); ~ **azoico** *m* QUÍMICA *cueros, fibras, textiles* azo dye, TEXTIL azoic dye; ~ **de azúcar quemada** *m* COLOR burnt-sugar coloring (*AmE*), burnt-sugar colouring (*BrE*); ~ **básico** *m* PAPEL, QUÍMICA, TEXTIL basic dye; ~ **bis-azoico** *m* QUÍMICA bis-azo dye; ~ **cerámico** *m* COLOR ceramic color (*AmE*), ceramic colour (*BrE*); ~ **de cloruro** *m* COLOR butter coloring (*AmE*), butter colouring (*BrE*); ~ **de cobalto** *m* COLOR cobalt color (*AmE*), cobalt colour (*BrE*); ~ **al cromo** *m* COLOR chrome color (*AmE*), chrome colour (*BrE*); ~ **directo** *m* COLOR substantive dye, TEXTIL *tintura* direct dye; ~ **disperso** *m* TEXTIL *tintura* disperse dye; ~ **encapsulado** *m* COLOR encapsulated dye; ~ **ignífugo** *m* COLOR fireproof color (*AmE*), fireproof colour (*BrE*), flameproof color (*AmE*), flameproof colour (*BrE*); ~ **de imprenta** *m* COLOR printer's color (*AmE*), printer's colour (*BrE*); ~ **indicador** *m* COLOR indicator color (*AmE*), indicator colour (*BrE*); ~ **indofenólico** *m* COLOR indophenol dye; ~ **inherente** *m* COLOR inherent color (*AmE*), inherent colour (*BrE*); ~ **de interferencia** *m* COLOR interference color (*AmE*), interference colour (*BrE*); ~ **de laca** *m* COLOR lacquer dye; ~ **para lanas** *m* COLOR woollen dyer; ~ **limón** *m* COLOR lemon color (*AmE*), lemon colour (*BrE*); ~ **litográfico** *m* COLOR lithographic color (*AmE*), lithographic colour (*BrE*); ~ **mineral** *m* COLOR mineral color (*AmE*), mineral colour (*BrE*); ~ **monoazoico** *m* COLOR monoazo dye; ~ **natural** *m* COLOR natural color (*AmE*), natural colour (*BrE*); ~ **opaco** *m* COLOR opaque color (*AmE*), opaque colour (*BrE*); ~ **por oxidación** *m* COLOR oxidation dye; ~ **permanente** *m* COLOR permanent color (*AmE*), permanent colour (*BrE*); ~ **plateado** *m* COLOR silver color (*AmE*), silver colour (*BrE*); ~ **en polvo** *m* COLOR dry color (*AmE*), dry colour (*BrE*), powdered color (*AmE*), powdered colour (*BrE*); ~ **premetalizado** *m* TEXTIL premetalized dye (*AmE*), premetallized dye (*BrE*); ~ **puro** *m* COLOR plain color (*AmE*), plain colour (*BrE*); ~ **que no desvanece** *m* COLOR nonfading color (*AmE*), non-fading colour (*BrE*); ~ **reactivo** *m* COLOR, TEXTIL reactive dye; ~ **de superficie** *m* COLOR surface color (*AmE*), surface colour (*BrE*); ~ **tina** *m* TEXTIL vat dye

colorar *vt* QUÍMICA stain

coloreado[1] *adj* GEN colored (*AmE*), coloured (*BrE*)

coloreado[2]: ~ **en la superficie** *m* PAPEL surface coloring (*AmE*), surface colouring (*BrE*)

colorear *vt* COLOR tincture, tint, FOTO, PAPEL dye

colores: ~ **complementarios** *m pl* COLOR, FÍS, FOTO complementary colors (*AmE*), complementary colours (*BrE*); ~ **para cuatricromía** *m pl* IMPR process colors (*AmE*), process colours (*BrE*); ~ **de esmalte** *m pl* C&V, COLOR enamel colors (*AmE*), enamel colours (*BrE*); ~ **espectrales** *m pl* FÍS RAD spectral colors (*AmE*), spectral colours (*BrE*); ~ **del espectro** *m pl* FÍS RAD colors of the spectrum (*AmE*), colours of the spectrum (*BrE*); ~ **de fondo** *m pl* TV color balance (*AmE*), colour balance (*BrE*); ~ **de mayólica** *m pl* C&V majolica colors (*AmE*), majolica colours (*BrE*); ~ **metaméricos** *m pl* COLOR, IMPR metameric colors (*AmE*), metameric colours (*BrE*); ~ **nacionales** *m pl* TRANSP MAR colors (*AmE*), colours (*BrE*); ~ **primarios aditivos** *m pl* CINEMAT, COLOR, IMPR additive primaries, additive primary colors (*AmE*), additive primary colours (*BrE*); ~ **primarios sustractivos** *m pl* CINEMAT, IMPR, TV *amarillo, magenta y cyan* subtractive primaries, subtractive primary colors *AmE*, subtractive primary colours *BrE*; ~ **de seguridad** *m pl* SEG safety colors (*AmE*), safety colours (*BrE*); ~ **para sobreimprimir** *m pl* IMPR overprint colors (*AmE*), overprint colours (*BrE*)

colorido *adj* COLOR colorful (*AmE*), colourful (*BrE*)

colorimetría *f* GEN *análisis* colorimetry

colorímetro *m* GEN *análisis* colorimeter

coloxilina *f* QUÍMICA colloxylin

colquiceína *f* QUÍMICA colchiceine

colquicina *f* QUÍMICA *alcaloide* colchicine

columbio *m* QUÍMICA columbium

columbita *f* MINERAL, QUÍMICA *mineral de tántalo* columbite, niobite

columna *f* (*cf mamposta AmL*) CARBÓN shaft, CONST pillar, strut, FERRO *infraestructura* stanchion, GAS column, GEOFÍS pillar, IMPR column, ING MECÁ housing, standard, post, upright, MATEMÁT column, MINAS *construcción de galerías* prop, TEXTIL *géneros de punto* wale; ~ **absorbente** *f* PROC QUÍ absorber column; ~ **de absorción** *f* CARBÓN absorption column; ~ **de agotamiento** *f* REFRIG stripping column; ~ **para ampliaciones** *f* FOTO, IMPR enlarger column; ~ **angulada** *f* FOTO angled column; ~ **basáltica** *f* GEOL basaltic column; ~ **binaria** *f* INFORM&PD binary column; ~ **central engranada** *f* CINEMAT, FOTO geared center column (*AmE*), geared centre column (*BrE*); ~ **central de un trípode** *f* FOTO central column of a tripod; ~ **de concentración** *f* PROC QUÍ concentration column; ~ **de condensación** *f* PROC QUÍ condensation column; ~ **de control** *f* TRANSP AÉR control column; ~ **de cromatografía** *f* LAB *análisis*, QUÍMICA chromatography column; ~ **deformada** *f* IMPR snaking column; ~ **de**

destilación *f* PROC QUÍ, QUÍMICA distilling column; ~ **de destilación fraccionada** *f* TEC PETR fractionating column; ~ **para destilación fraccionada** *f* PROC QUÍ fraction distillation column; ~ **de difusión** *f* PROC QUÍ diffusion tower; ~ **de dirección** *f* AUTO, VEH steering column; ~ **de dirección de derrumbe axial** *f* TRANSP axially-collapsing steering column; ~ **de dirección desarmable** *f* AUTO, VEH collapsible steering column; ~ **empacada** *f* PROC QUÍ, QUÍMICA, TEC PETR packed column; ~ **de empaquetamiento** *f* PROC QUÍ packed tower; ~ **de enfoque de haces iónicos** *f* FÍS RAD ion beam focusing column; ~ **estratigráfica** *f* GEOL stratigraphic column; ~ **de fraccionamiento** *f* LAB *destilación* fractionation column; ~ **de gas uniformemente excitada** *f* FÍS RAD uniformly-excited column of gas; ~ **geológica** *f* GEOL geological column; ~ **guía** *f* ING MECÁ guide pillar; ~ **de haz de electrones** *f* ELECTRÓN electron-beam column; ~ **hidráulico** *f* *Esp* (*cf propulsor hidráulico AmL*) MINAS hydraulic prop; ~ **inclinada** *f* FOTO angled column; ~ **de lavado** *f* PROC QUÍ washing column; ~ **de pastillas** *f* NUCL pellet stack; ~ **de perforación** *f* PETROL drill column; ~ **perforadora** *f* PETROL drill stem, drill string; ~ **de placas** *f* TEC PETR *refino* plate column; ~ **de platos de burbujeo** *f* PROC QUÍ bubble-tray column; ~ **plegable** *f* TRANSP AÉR folding pylon; ~ **positiva** *f* ELECTRÓN, FÍS positive column; ~ **posterior** *f* ING MECÁ rear pillar; ~ **de producción** *f* PETROL production string; ~ **reguladora** *f* TEC PETR standpipe; ~ **de relleno** *f* PROC QUÍ, QUÍMICA, TEC PETR packed column; ~ **rica** *f* MINAS chimney, nail; ~ **de sal** *f* GEOL, TEC PETR salt column; ~ **de secado** *f* LAB *material de vidrio* drying column; ~ **de separación** *f* LAB, PROC QUÍ separating column; ~ **simple** *f* CONST plain column; ~ **sonora** *f* ACÚST sound column; ~ **de soporte** *f* ING MECÁ support pillar; ~ **de taladradora** *f* ING MECÁ drilling pillar; ~ **térmica** *f* NUCL thermal column; ~ **del tren de aterrizaje** *f* TRANSP AÉR landing gear leg; ~ **del tren de aterrizaje del morro** *f* TRANSP AÉR nose gear leg; ~ **de tuberías** *f* PETROL pipe string; ~ **tubular** *f* CONST tube column

columnas: ~ **de forma integral** *f pl* ING MECÁ integral-way columns

coluviones: ~ **de ladera** *f pl* GEOL scree

colza *f* AGRIC rapeseed

coma *f* ACÚST *música* comma, ELECTRÓN, FÍS, FOTO, IMPR comma; ~ **de Aristoxene** *f* ACÚST Aristoxene comma; ~ **decimal** *f* INFORM&PD, MATEMÁT decimal point; ~ **Didyme** *f* ACÚST Didyme comma; ~ **fija** *f* INFORM&PD fixed point; ~ **flotante** *f* INFORM&PD, TELECOM floating point; ~ **de Holder** *f* ACÚST Holder comma; ~ **de Pitágoras** *f* ACÚST, MATEMÁT Pythagorean comma; ~ **sintónica** *f* ACÚST syntonous comma

comagmático *adj* GEOL comagmatic

comandante *m* TEC ESP *misiones tripuladas*, TRANSP MAR *marina* commander; ~ **del buque insignia** *m* TRANSP MAR *marina* flag captain; ~ **de zona de defensa antiaérea** *m* TEC ESP controller

comando: ~ **del ángulo del motor** *m* TEC ESP engine angle command; ~ **de conexión** *m* ING MECÁ, PROD force-on command; ~ **de desconexión** *m* ING MECÁ, PROD force-off command; ~ **forzado de activación** *m* PROD force-enable command; ~ **forzado de desactivación** *m* PROD force-disable command;

~ multiplo de señal de liberación *m* TELECOM master control room (*MCR*); **~ multipunto de forzamiento de visualización** *m* TELECOM MCV (*multipoint command visualization forcing*); **~ multipunto para inversión MCS** *m* TELECOM MCN (*multipoint command negating MCS*); **~ multipunto de petición de autorización** *m* TELECOM MCT (*multipoint command token claim*); **~ multipunto de transmisión simétrica de datos** *m* TELECOM message control system; **~ de operador** *m* INFORM&PD operator command; **~ de tránsito digital** *m* TELECOM digital transit command

comba *f* CONST *madera, placas* bow, *deformación* bulge, ING MECÁ bulge, TRANSP AÉR camber

combado *m* *Esp* (*cf alabeo AmL*) CINEMAT, FOTO buckling; **~ del impreso al secarse la tinta** *m* IMPR ink-drying curl

combadura *f* CONST bulging, camber, *madera* warping, winding, PETROL buckle

combar *vi* ING MECÁ bulge

combarse *vr* CONST bend, bulge, *madera* wind

combate *m* TEC ESP pull; **~ de incendios** *m* CONST, SEG, TERMO firefighting

combeo *m* CINEMAT buckle, FOTO buckling

combinación *f* IMPR melding, INFORM&PD, MATEMÁT combination, P&C blend, compound, QUÍMICA, TRANSP AÉR combination; **~ de dos o más registros sonoros** *f* ACÚST dubbing; **~ de huecos** *f* METAL void coalescence; **~ de muelas abrasivas** *f* INSTR abrasive wheel combination; **~ de regulación en carga** *f* ELEC *transformador* on-load tap changing; **~ satélite-motor de apogeo** *f* TEC ESP satellite-apogee motor combination; **~ suspensión-vagón** *f* TRANSP saddle mount combination; **~ de toma** *f* ELEC *transformador* on-load tap changing

combinaciones: **~ de productos y procesos** *f pl* TEXTIL product-and-process mixes

combinado: no ~ *adj* QUÍMICA uncombined

combinador *m* ING ELÉC multiple contact switch, TELECOM combiner; **~ cilíndrico** *m* ELEC *conmutador* drum controller; **~ óptico** *m* ÓPT, TELECOM optical combiner; **~ de regulación en carga** *m* ELEC on-load tap changer; **~ de tambor** *m* ELEC *conmutador* drum controller; **~ de toma** *m* ELEC on-load tap changer

combinar *vt* INFORM&PD combine, ING MECÁ combine, blend, TEXTIL blend

combinarse *vr* *AmL* (*cf fundirse Esp*) QUÍMICA coalesce

combinational *adj* ELECTRÓN, INFORM&PD, MATEMÁT combinatorial

combinatoria *f* GEOM, MATEMÁT *teoría de probabilidades* combinatorics

combinatorio *adj* ELECTRÓN, INFORM&PD, MATEMÁT combinatorial

comburente *m* CONTAM MAR burning-agent; **~ oxidante** *m* TEC ESP oxidizer

combustibilidad *f* GEN combustibility, flammability

combustible¹ *adj* GEN combustible

combustible² *m* CARBÓN, EMB combustible, MINAS bunker, TERMO combustible, fuel, TRANSP MAR *para motor*, VEH fuel; **~ con alto contenido de azufre** *m* CONTAM sulfurous combustible (*AmE*), sulphurous combustible (*BrE*); **~ de alto octanaje** *m* AUTO premium fuel (*BrE*), premium gas (*AmE*), premium gasoline (*AmE*), premium petrol (*BrE*); **~ de aluminiuro de uranio** *m* NUCL uranium aluminide fuel;

~ de aviación *m* TRANSP AÉR aviation fuel; **~ para buques** *m* TRANSP bunker oil; **~ criogénico** *m* TEC ESP cryogenic fuel; **~ derivado** *m* TEC PETR *refino* derived fuel; **~ desechable** *m* TRANSP AÉR drainable unusable fuel; **~ diesel** *m* AUTO, TEC PETR, VEH diesel fuel; **~ de dióxido de uranio** *m* NUCL uranium dioxide fuel; **~ disperso** *m* NUCL dispersion fuel; **~ doméstico** *m* TEC PETR *derivados del petróleo* domestic fuel oil; **~ dual** *m* TEC ESP dual fuel; **~ emulsificado** *m* TRANSP AÉR emulsified fuel; **~ enriquecido** *m* NUCL enriched fuel; **~ fósil** *m* CONTAM combustible fossil fuel, PETROL, TEC PETR, TERMO fossil fuel; **~ gaseoso** *m* TEC PETR *derivados del petróleo* gaseous fuel; **~ líquido** *m* CARBÓN fuel, CONTAM MAR bunker, bunker fuel, TEC ESP liquid propellant, TERMO liquid fuel, liquid propellant, TRANSP MAR fuel; **~ matricial** *m* NUCL matrix fuel; **~ de motor a reacción** *m* TRANSP AÉR jet engine fuel, jet fuel; **~ nuclear** *m* D&A, NUCL nuclear fuel; **~ nuclear enriquecido** *m* NUCL enriched nuclear fuel; **~ nuclear sólido** *m* NUCL solid nuclear fuel; **~ nuevo** *m* NUCL fresh fuel, new fuel; **~ en pasta** *m* NUCL paste fuel; **~ propulsante** *m* TEC ESP propellant fuel, propellent fuel; **~ propulsor** *m* TEC ESP propellant fuel, propellent fuel; **~ de reactor** *m* TRANSP AÉR jet fuel; **~ de referencia** *m* TEC PETR reference fuel; **~ de reposición** *m* NUCL make-up fuel; **~ sulfuroso** *m* CONTAM sulfurous combustible (*AmE*), sulphurous combustible (*BrE*); **~ terciario** *m* AUTO tertiary fuel; **~ de turbina** *m* TEC ESP turbine fuel; **~ de UO₂** *m* NUCL UO₂ fuel; **~ de uranio natural** *m* NUCL natural-uranium fuel

combustión *f* CARBÓN, EMB, P&C, QUÍMICA, SEG, TERMO combustion; **~ completa** *f* GAS, TERMO complete combustion; **~ erosiva** *f* TEC ESP *motores propulsores, vehículos* erosive burning; **~ espontánea** *f* SEG spontaneous combustion; **~ explosiva** *f* MINAS explosive combustion; **~ de gases de escape** *f* AUTO, MECÁ, TRANSP exhaust-gas combustion; **~ in situ** *f* TEC PETR in situ combustion; **~ inversa** *f* PETROL reverse combustion; **~ irregular** *f* TEC ESP *motor del cohete* chuffing; **~ lenta** *f* AUTO slow combustion, QUÍMICA glow; **~ pobre** *f* AUTO lean-burn; **~ retardada** *f* TERMO *de un motor* afterburning; **~ sumergida** *f* GAS submerged combustion

comedero *m* AGRIC feed bunk, feeding rack, feeding trough; **~ de animales con distribución a través de sinfín** *m* AGRIC auger bank; **~ para heno** *m* AGRIC hayrack; **~ de madera** *m* AGRIC wooden trough

comedor: **~ de marinería** *m* TRANSP MAR mess

coménico *adj* QUÍMICA *ácido* comenic

comenzar *vti* INFORM&PD, ING MECÁ, TELECOM, TV start

comercial *adj* FOTO *solución* ready-made

comercialidad *f* TEXTIL marketability

comercialización: **~ de pescado fresco** *f* OCEAN fresh fish trade

comerciante: **~ de lanas** *m* TEXTIL wool merchant; **~ al por menor** *m* PROD retailer

comerciar: **~ en** *vi* PROD handle

comercio: **~ de carbón** *m* CARBÓN, MINAS colliery; **~ marítimo** *m* TRANSP MAR sea trade, seaborne trade, shipping trade; **~ del pescado** *m* TRANSP MAR fish trade; **~ del pescado fresco** *m* TRANSP MAR fresh-fish trade

cometa¹ *m* TEC ESP comet

cometa² *f* GEOM, TRANSP AÉR *juguete* kite

comida: ~ **para calentar y comer** *f* ALIMENT, EMB heat-and-eat food; ~ **cocinada y refrigerada** *f* ALIMENT, EMB *conservada por refrigeración* cook-chill meal; ~ **enlatada** *f* ALIMENT, EMB canned food (*AmE*), tinned food (*BrE*); ~ **natural completa** *f* ALIMENT whole food; ~ **preparada** *f* ALIMENT ready meal

comienzo *m* INFORM&PD start, ING MECÁ starting, MINAS initiation, TEC ESP threshold, TELECOM *de transmisión* start; ~ **anticipado** *m* TV early start; ~ **anticipado de audio** *m* TV early-start audio; ~ **anticipado de video** *m AmL*, ~ **anticipado de vídeo** *Esp m* TV early-start video; ~ **de bifurcación** *m* PROD branch start; ~ **de cabecera** *m* INFORM&PD start of header (*SOH*); ~ **indirecto** *m* MINAS indirect initiation; ~ **de llamada** *m* TELECOM beginning of call demand; ~ **de mensaje** *m* INFORM&PD start of message (*SOM*), TELECOM beginning of message (*BOM*); ~ **de texto** *m* INFORM&PD start of text (*STX*); ~ **de título** *m* INFORM&PD start of header (*SOH*)

comienzos: ~ **de vida** *m pl* NUCL *del núcleo* beginning of life; ~ **de vida del núcleo** *m pl* NUCL beginning of life of the core

comillas *f pl* IMPR French quotes, inverted commas, quotation marks, quotes

comisario *m* TRANSP MAR *marina mercante* head steward, purser

comisión: ~ **de seguridad** *f* SEG safety committee

Comisión: ~ **Internacional de Iluminación** *f* FÍS International Commission on Illumination; ~ **Internacional de Protección Radiológica** *f* FÍS RAD International Commission on Radiological Protection (*ICRP*); ~ **Reguladora Nuclear de los Estados Unidos** *f* (*US NRC*) NUCL United States Nuclear Regulatory Commission (*US NRC*)

comité: ~ **de seguridad** *m* SEG safety committee; ~ **de seguridad del aeropuerto** *m* TRANSP AÉR airport security committee

Comité: ~ **Consultivo Internacional de Radiocomunicaciones** *m* (*CCIR*) TEC ESP International Radiocommunications Consultative Committee; ~ **Consultivo Internacional de Telefonía y Telegrafía** *m* (*CCITT*) TELECOM International Telegraph and Telephone Consultative Committee; ~ **Electrotécnico Internacional** *m* PROD International Electrotechnical Committee (*IEC*); ~ **Internacional de Registro de Frecuencias** *m* (*IFRB*) TEC ESP International Frequency Registration Board (*IFRB*); ~ **Nacional de Normas de Televisión** *m* (*NTSC*) TV National Television Standards Committee (*AmE*) (*NTSC*)

cómodo *adj* INFORM&PD user-friendly

comodoro *m* TRANSP MAR *marino* commodore

compacción *f* PROD *de la arena de un molde* packing, ramming, tamping

compacidad *f* CARBÓN density, *terrenos* compatibility

compactación *f* CARBÓN, CONST compaction, GEOL compaction, compactation, INFORM&PD compactation, MINAS tamping, PROD *fundería, de arena* packing, ramming, tamping, TEC PETR *sedimentación* compaction; ~ **de bloques** *f* INFORM&PD block compaction, block compression; ~ **de datos** *f* ELECTRÓN, INFORM&PD, TELECOM data compaction, data compression; ~ **de laboratorio** *f* CARBÓN laboratory compaction; ~ **de la memoria** *f* INFORM&PD memory compaction

compactador: ~ **de suelos** *m* CONST rammer

compactadora *f* MINAS tamper

compactar *vt* GEOL, PROD compact

compacto *adj* ELECTRÓN, PROD *alambre* solid

compaginación *f* IMPR make-up

compaginar *vt* IMPR make up

compañía: ~ **aérea** *f Esp* (*cf compañía de aeronavegación AmL*) TRANSP airline; ~ **de aeronavegación** *f AmL* (*cf compañía aérea Esp*) TRANSP airline; ~ **chárter** *f* TRANSP AÉR charter company; ~ **naviera** *f* TRANSP MAR shipping company; ~ **de seguridad** *f* SEG security firm; ~ **de semillas** *f* AGRIC seed company; ~ **de televisión privada** *f* TV pay TV

compansión *f* TELECOM companding; ~ **silábica** *f* TELECOM syllabic companding

comparación: ~ **de señal** *f* ELECTRÓN signal comparison; ~ **del voltaje** *f* ING ELÉC voltage comparison

comparador *m* ELECTRÓN, FÍS comparator, INFORM&PD *electrónica* comparator, collator, INSTR, METR, TELECOM comparator; ~ **de aire comprimido** *m* METR pneumatic gage (*AmE*), pneumatic gauge (*BrE*); ~ **analógico** *m* ELECTRÓN analog comparator; ~ **de cuadrante** *m* ING MECÁ dial gage (*AmE*), dial gauge (*BrE*); ~ **diferencial** *m* ELECTRÓN differential comparator; ~ **electrónico** *m* ING MECÁ electronic comparator; ~ **de fase** *m* ELECTRÓN phase comparator, phase detector, TV phase comparator; ~ **óptico** *m* METR optical comparator; ~ **óptico de proyección para verificar perfiles** *m* METR profile projector; ~ **de señales** *m* ELECTRÓN signal comparator; ~ **de voltaje** *m* ING ELÉC voltage comparator

comparar *vt* INFORM&PD compare, collate, METR measure

compartición *f* INFORM&PD sharing

compartimentación *f* INFORM&PD compartmentalization

compartimentado *m* TRANSP MAR *arquitectura naval* compartmentation

compartimentalización *f* INFORM&PD compartmentalization

compartimiento *m* CARBÓN compartment, *minas* panel, INFORM&PD, ING MECÁ compartment, MINAS panel, REFRIG compartment, TEC ESP *vehículos aéreos y espaciales* bay, TRANSP compartment; ~ **de archivos** *m* INFORM&PD file sharing; ~ **de bombas** *m* AGUA pump compartment; ~ **para las bombas** *m* D&A *en los aviones* bomb bay; ~ **de carga** *m* TEC ESP payload bay; ~ **para la carga** *m* TEC ESP cargo bay; ~ **de cargo** *m* TRANSP AÉR cargo compartment; ~ **congelador** *m* ALIMENT freezer compartment, REFRIG freezer compartment, low-temperature compartment; ~ **desprendible de carga** *m* TEC ESP *avión* pod; ~ **de un edificio** *m* CONST bay; ~ **de elevación** *m* CONST hoisting compartment; ~ **de equipaje** *m* TRANSP baggage room (*AmE*), luggage compartment (*BrE*), baggage compartment (*AmE*); ~ **de equipajes** *m* VEH *carrocería* luggage compartment; ~ **de espera** *m* TRANSP AÉR holding bay; ~ **estanco** *m* CONST, HIDROL cofferdam; ~ **estanco celular de diafragma** *m* HIDROL diaphragm cellular cofferdam; ~ **estanco para introducción de muestras** *m* INSTR specimen insertion airlock; ~ **de fabricación hielo** *m* ING MECÁ ice-making compartment; ~ **de lodo** *m* TEC PETR *ingeniería de lodos, perforación* mud box; ~ **de**

máquinas *m* TRANSP MAR engine room; ~ **del motor** *m* TRANSP AÉR engine nacelle, engine pod, VEH engine compartment; ~ **múltiple para cohetes** *m* TEC ESP *aviones* pod; ~ **de nieve carbónica** *m* REFRIG *en vehículos frigoríficos* dry ice bunker; ~ **de pasajeros** *m* TRANSP passenger compartment; ~ **de pasajeros reforzado** *m* TRANSP strengthened passenger compartment; ~ **para pilas** *m* FOTO battery chamber; ~ **de recursos** *m* INFORM&PD resource sharing; ~ **de un tanque de combustible** *m* TERMO fuel cell; ~ **del tiempo** *m* TELECOM time sharing; ~ **del tren de aterrizaje** *m* TRANSP AER landing gear bay; ~ **de la tripulación** *m* TRANSP AÉR crew compartment; ~ **de ventilación** *m* *Esp* MINAS air brattice, brattice, stoping; ~ **de vuelo** *m* TRANSP AÉR flight compartment
compartir *vt* INFORM&PD share
compás *m* CONST bow compass, compass, D&A, FÍS, GEOFÍS compass, GEOM, ING MECÁ compasses, TEC ESP, TRANSP MAR *navegación* compass; ~ **alado** *m* ING MECÁ wing compasses; ~ **azimutal** *m* TRANSP MAR azimuth compass; ~ **de brazos curvos** *m* METR calipers (*AmE*), callipers (*BrE*); ~ **de calibración** *m* METR caliper compasses (*AmE*), calliper compasses (*BrE*); ~ **calibrador en ocho** *m* ING MECÁ figure-of-eight calipers (*AmE*), figure-of-eight callipers (*BrE*); ~ **de calibres** *m* METR calipers (*AmE*), callipers (*BrE*); ~ **de calibres indicador de cuadrante** *m* METR dial-indicating calipers (*AmE*), dial-indicating callipers (*BrE*); ~ **de carpintero** *m* CONST *carpintería* joiner's gage (*AmE*), joiner's gauge (*BrE*); ~ **de dividir** *m* ING MECÁ pair of dividers, METR dividers; ~ **de elipses** *m* ING MECÁ trammel; ~ **elíptico** *m* ING MECÁ oval compass; ~ **de espesor** *m* ING MECÁ outside calipers (*AmE*), outside callipers (*BrE*); ~ **de espesores** *m* CONST bow calipers (*AmE*), bow callipers (*BrE*), ING MECÁ bow compass, outside calipers (*AmE*), outside callipers (*BrE*); ~ **de exteriores** *m* ING MECÁ outside calipers (*AmE*), outside callipers (*BrE*); ~ **giroscópico** *m* TRANSP AÉR gyrocompass, gyroscopic compass, TRANSP MAR gyroscopic compass; ~ **de gobierno** *m* TRANSP MAR steering compass; ~ **de gruesos** *m* CONST bow calipers (*AmE*), bow callipers (*BrE*), ING MECÁ outside calipers (*AmE*), outside callipers (*BrE*), METR calipers (*AmE*), callipers (*BrE*); ~ **de interiores y exteriores** *m* ING MECÁ inside-and-outside calipers (*AmE*), inside-and-outside callipers (*BrE*), outside-and-inside calipers (*AmE*), outside-and-inside callipers (*BrE*); ~ **largo** *m* CONST long compass; ~ **de líquido** *m* TRANSP MAR liquid compass; ~ **magistral** *m* TRANSP MAR standard compass; ~ **magnético** *m* GEOFÍS, INSTR, TRANSP AÉR magnetic compass; ~ **de medición de caras externas** *m* ING MECÁ outside measuring faces; ~ **de medir** *m* METR dividers; ~ **de muelle** *m* CONST bow-spring compass; ~ **de muelle para bomba** *m* ING MECÁ pump spring-bow; ~ **de muelle de precisión** *m* ING MECÁ bow compass; ~ **de navegación** *m* TRANSP MAR mariner's compass; ~ **de nivelación** *m* CONST *topografía* leveling compass (*AmE*), levelling compass (*BrE*); ~ **de proporción** *m* TRANSP MAR *navegación* dividers; ~ **de puntas** *m* CONST bow compass, TRANSP MAR *navegación* dividers; ~ **de puntas en ocho** *m* ING MECÁ figure-of-eight calipers (*AmE*), figure-of-eight callipers (*BrE*); ~ **de puntas secas** *m* ING MECÁ pair of

dividers, METR dividers; ~ **de reducción** *m* ING MECÁ reduction compass; ~ **repetidor** *m* TEC ESP, TRANSP MAR repeater compass; ~ **rotacional** *m* ING MECÁ rotational compass; ~ **rotativo** *m* ING MECÁ rotational compass; ~ **de varas** *m* ING MECÁ trammel
compases *m pl* GEOM, ING MECÁ compasses; ~ **de calibrar interiores** *m pl* ING MECÁ inside calipers (*AmE*), inside callipers (*BrE*); ~ **de exteriores** *m pl* ING MECÁ egg calipers (*AmE*), egg callipers (*BrE*); ~ **de interiores** *m pl* ING MECÁ inside calipers (*AmE*), inside callipers (*BrE*); ~ **micrométricos** *m pl* ING MECÁ micrometer calipers (*AmE*), micrometer callipers (*BrE*); ~ **micrométricos de gruesos** *m pl* ING MECÁ micrometer calipers (*AmE*), micrometer callipers (*BrE*); ~ **con punta de bolígrafo** *m pl* ING MECÁ compasses with pen point; ~ **con punta de lápiz** *m pl* ING MECÁ compasses with pencil point; ~ **de puntas para diámetros exteriores** *m pl* ING MECÁ egg calipers (*AmE*), egg callipers (*BrE*)
compatibilidad *f* ACÚST, CALIDAD compatibility, CARBÓN compatibility, consistency, ELECTRÓN, INFORM&PD, ING MECÁ, P&C, TV *propiedad* compatibility; ~ **ambiental** *f* CONTAM environmental compatibility; ~ **ascendente** *f* INFORM&PD upward compatibility; ~ **descendente** *f* INFORM&PD downward compatibility; ~ **dimensional** *f* TELECOM dimensional compatability; ~ **electromagnética** *f* TEC ESP electromagnetic compatibility (*EMC*); ~ **de enchufes** *f* ING ELÉC plug compatibility; ~ **entre fluidos y materiales elastométricos** *f* ING MECÁ compatibility between elastometric materials and fluids; ~ **entre tintas** *f* IMPR *imprimir una tinta sobre otra* trapping; ~ **hacia abajo** *f* INFORM&PD downward compatibility; ~ **hacia arriba** *f* INFORM&PD upward compatibility; ~ **inversa** *f* TV reverse compatibility; ~ **con revisión anterior** *f* INFORM&PD downward compatibility; ~ **de sobreimpresión** *f* IMPR *imprimir una tinta sobre otra* trapping
compatible *adj* GEN compatible; ~ **hacia abajo** *adj* INFORM&PD downward compatible; ~ **hacia arriba** *adj* INFORM&PD upward compatible; ~ **con revisión anterior** *adj* INFORM&PD downward compatible
compensación *f* ACÚST, CINEMAT, COLOR, CONST, DETERG, ELEC equalization, ELECTRÓN compensation, equalization, ENERG RENOV equalization, FÍS compensation, equalization, FÍS FLUID, FÍS RAD, FOTO, GAS, GEOFÍS, GEOL equalization, HIDROL compensation, equalization, IMPR, INFORM&PD equalization, ING ELÉC compensation, ING MECÁ offset, take-up, equalization, MECÁ equalization, offset, NUCL shim, OCEAN compensation, PROD, REFRIG equalization, SEG compensation, TEC ESP, TEC PETR, TELECOM, TERMO, TERMOTEC equalization, TRANSP AÉR balancing, trimming, equalization; ~ **automática de blancos** *f* TV automatic white balance; ~ **de baja frecuencia** *f* ELECTRÓN low-frequency compensation; ~ **de cabeceo** *f* TRANSP AÉR pitch compensation; ~ **de colores** *f* CINEMAT color matching (*AmE*), colour matching (*BrE*); ~ **por desgaste** *f* MECÁ compensation for wear; ~ **dinámica** *f* ING ELÉC dynamic trimming; ~ **por dopado** *f* ELECTRÓN doping compensation; ~ **de fase** *f* ELEC *corriente alterna*, TELECOM phase compensation; ~ **de frecuencia** *f* ELECTRÓN frequency compensation; ~ **de la gravedad** *f* TEC ESP antigravity; ~ **por impurezas** *f* ELECTRÓN doping

compensation; **~ por lesiones de trabajo** *f* SEG industrial injury benefit; **~ de línea** *f* TV line tilt; **~ de órbita** *f* TEC ESP orbit trimming; **~ de paso** *f* TRANSP AÉR pitch compensation; **~ de puente de medida** *f* ING ELÉC bridge balancing; **~ de la reacción del inducido** *f* ING ELÉC armature reaction compensation; **~ de temperatura** *f* ELECTRÓN *transistores* temperature compensation; **~ térmica** *f* ING ELÉC temperature compensation, TERMO heat compensation, temperature compensation

compensado *m* TRANSP AÉR trim; **~ lateral** *m* TRANSP AÉR lateral trim

compensador *m* GEN equalizer, ELEC *circuitos de corriente alterna*, ELECTRÓN compensator, FÍS balance, compensator, balancing, ING ELÉC trim tab, balancer, compensator TEC PETR *perforación costa-fuera* compensator; **~ de Babinet** *m* FÍS Babinet compensator; **~ de brújula** *m* TRANSP AÉR compass compensating; **~ de caída** *m* TV drop-out compensator; **~ del control de paso de la pala** *m* TRANSP AÉR blade-pitch control compensator; **~ de corriente alterna** *m* ELEC alternating-current balancer; **~ de cuerno** *m* TRANSP AÉR horn balance; **~ diferencial** *m* ING MECÁ *turbinas* balance gear; **~ del error de velocidad** *m* TV velocity error compensator; **~ de fase** *m* ELEC *corriente alterna* phase equalizer; **~ de frenada** *m* VEH brake compensator; **~ de herradura** *m* TRANSP AÉR horn balance; **~ hidráulico** *m* TEC PETR *perforación costa-fuera* hydraulic compensator; **~ de impedancia** *m* ELEC impedance corrector; **~ de Mach** *m* TRANSP AÉR Mach compensator; **~ de movimiento vertical** *m* PETROL heave compensator; **~ neutral** *m* ING ELÉC neutral compensator

compensar[1] *vt* ELECTRÓN, FERRO compensate, ING MECÁ compensate, counterbalance, MECÁ counterbalance, SEG *indemnizar* compensate, TRANSP MAR *aguja* adjust, compensate

compensar[2] *vi* FÍS compensate; **~ por desgaste** *vi* MECÁ compensate for wear

competencia: **~ náutica** *f* TRANSP MAR seamanship

competente *adj* GEOL *estratos* competent

compilación *f* IMPR compilation, INFORM&PD compilation, compile, TELECOM compilation; **~ cruzada** *f* INFORM&PD cross compilation

compilado *adj* ING MECÁ embodied

compilador *m* IMPR, INFORM&PD, TELECOM compiler; **~-compilador** *m* INFORM&PD compiler-compiler; **~ cruzado** *m* INFORM&PD cross compiler

compiladora: **~ incremental** *f* INFORM&PD incremental compiler

compilancia *f* ACÚST compliance

complejidad *f* INFORM&PD complexity; **~ del chip** *f* ELECTRÓN chip complexity; **~ de microplaca** *f* ELECTRÓN chip complexity

complejo *m* GEOL complex; **~ frigorífico** *m* REFRIG cold store complex; **~ ígneo** *m* GEOL igneous complex; **~ de inclusión** *m* QUÍMICA adduct; **~ de lanzamiento** *m* TEC ESP launching complex; **~ de un orbital externo** *m* FÍS RAD outer orbital complex; **~ de un orbital interno** *m* FÍS RAD inner orbital complex; **~ petrolífero explotable** *m* PETROL pay zone; **~ de red de conmutación** *m* TELECOM switching network complex

complementariedad *f* FÍS PART, INFORM&PD complementarity

complemento *m* ELECTRÓN, INFORM&PD complement; **~ del ángulo de corte** *m* ING MECÁ top rake; **~ de base** *m* INFORM&PD radix complement; **~ de base menos uno** *m* INFORM&PD radix-minus-one complement; **~ de buzamiento** *m* GEOL hade, *inclinación del plano de falla* angle of hade; **~ a dos** *m* INFORM&PD two's complement; **~ a nueve** *m* INFORM&PD nine's complement; **~ a la refrigeración** *m* REFRIG supplement to refrigeration; **~ a uno** *m* INFORM&PD one's complement

completación *f* TEC PETR *perforación* completion; **~ en el fondo marino** *f* TEC PETR *producción costa-fuera* subsea completion; **~ submarina** *f* TEC PETR *producción costa-fuera* subsea completion

completamente: **~ abierto** *adj* CINEMAT *el diafragma del objetivo* wide open; **~ automático** *adj* ING MECÁ, INSTR, PROD fully automatic; **~ cargado hasta los calados permitidos** *adj* TRANSP MAR *buque* full and down; **~ digital** *adj* ELECTRÓN all-digital; **~ seco** *adj* TEXTIL bone-dry

complexante *m* NUCL complexing agent

complicado *adj* ING MECÁ intricate

componedor *m* IMPR setting stick; **~ para carteles** *m* IMPR poster stick

componedora *f* IMPR composing machine; **~ tipográfica** *f* IMPR typesetting machine

componente[1] *adj* ING MECÁ component

componente[2] *m* ELEC *corriente, de fuerza*, ELECTRÓN *elemento de un equipo funcional*, FÍS component, ING MECÁ part, component, MECÁ, NUCL, PROD, TELECOM component; **~ activo** *m* ELECTRÓN active component, real component, FÍS active component, ING ELÉC active component, real component, TELECOM active component; **~ de alta frecuencia** *m* ELECTRÓN high-frequency component; **~ alternado** *m* ELEC *corriente alterna* alternating component; **~ alternativo** *m* ELEC *corriente alterna* alternating component; **~ alterno** *m* ELEC *corriente alterna* alternating component; **~ de arseniuro de galio** *m* ELECTRÓN, FÍS, ÓPT gallium arsenide component; **~ en canana** *m* ING ELÉC bandoleered component; **~ capacitivo** *m* ELEC, ING ELÉC capacitive component; **~ CC** *m* TELECOM DC component; **~ consignado** *m* PROD consigned component; **~ continuo** *m* ELEC *de corriente* direct component; **~ cromático** *m* TV chromatic component, chrominance component; **~ de crominancia** *m* TV chrominance component; **~ en cuadratura** *m* ELEC *corriente alterna-componente reactivo*, ELECTRÓN quadrature component; **~ desechable** *m* ING MECÁ expendable item; **~ desmontable** *m* TRANSP AÉR removable; **~ dinámico** *m* TRANSP AÉR dynamic component; **~ discreto** *m* ING ELÉC, TELECOM discrete component; **~ eléctrico** *m* ELEC, ING ELÉC electrical component; **~ electrónico** *m* ELECTRÓN, INFORM&PD, TELECOM electronic component; **~ empacado en cinta** *m* ING ELÉC taped component; **~ enchufable** *m* ING ELÉC plug-in component; **~ encintado** *m* ING ELÉC taped component; **~ de estado sólido** *m* PROD solid-state component; **~ excitado** *m* GAS excited component; **~ del faro de charnela del tren de aterrizaje** *m* TRANSP AÉR landing gear hinge beam fitting; **~ en fase** *m* ELECTRÓN in-phase component; **~ de frecuencia** *m* ELECTRÓN frequency component; **~ de guíaondas** *m* ING ELÉC waveguide component; **~ híbrido integrado** *m* ELECTRÓN integrated hybrid component; ~ **impulsor** *m* ING ELÉC active component; ~ **inactivo** *m* ELEC *corriente alterna*, ING ELÉC idle component; ~ **instalado en superficie** *m* ELECTRÓN surface-mounted component; ~ **del izador** *m* TRANSP AÉR hoist fitting; ~ **lógico** *m* ELECTRÓN logic component, PETROL software; ~ **longitudinal** *f* FÍS longitudinal component; ~ **para máquina hidráulica** *m* INSTAL HIDRÁUL hydraulic fitting; ~ **de la marea** *m* OCEAN tidal component; ~ **mecánico** *m* ING MECÁ mechanical component; ~ **de montaje superficial** *m* TELECOM surface-mounted component (*SMC*); ~ **móvil** *m* CONTAM mobile component; ~ **negro** *m* IMPR black art; ~ **pasivo** *m* ELECTRÓN, FÍS, NUCL, TELECOM passive component; ~ **pegado con cinta** *m* ING ELÉC taped component; ~ **p-n-p-n** *m* ELECTRÓN p-n-p-n component; ~ **primario de la termofluencia** *m* METAL primary creep; ~ **químico** *m* GEOL, QUÍMICA chemical component; ~ **reactivo** *m* ELEC reactive component, ING ELÉC reactive component, wattless component; ~ **del reactor** *m* NUCL reactor component; ~ **en régimen transitorio** *m* TELECOM transient fluctuation; ~ **de retardo** *m* NUCL delay component; ~ **roscado** *m* ING MECÁ threaded component; ~ **de semiconductor** *m* ELECTRÓN semiconductor material; ~ **semiconductor** *m* ELECTRÓN semiconductor component; ~ **de señal** *m* ELECTRÓN signal component; ~ **SOM de canal N** *m* ELECTRÓN NMOS component; ~ **de la subportadora** *m* TV subcarrier component; ~ **de sustentación** *m* TRANSP AÉR lift component; ~ **térmico** *m* METAL thermal component; ~ **termoplástico** *m* ING MECÁ thermoplastic component; ~ **de tipo aditivo** *m* TEC ESP added-on component; ~ **de tipo n** *m* ELECTRÓN n-type component; ~ **para transferencia de mensajes** *m* TELECOM message transfer part; ~ **transitorio** *m* TELECOM transient fluctuation; ~ **de tubo** *m* ING MECÁ pipe component

componente[3] *f* ELEC *parte* component; ~ **de corriente alterna** *f* ELEC alternating-current component; **de corriente continua** *f* ELEC direct current component; ~ **desvatiada** *f* ELEC *corriente alterna* wattless component; ~ **fundamental** *f* FÍS fundamental component; ~ **de guía de ondas** *m* ING ELÉC waveguide component; ~ **horizontal** *f* FÍS horizontal component; ~ **horizontal de desplazo** *f* GEOFÍS offset; ~ **parásita** *f* ING ELÉC parasitic component; ~ **de potencia directa** *f* ING ELÉC direct power component; ~ **principal** *f* MATEMÁT principal component; ~ **radial** *f* FÍS radial component; ~ **reactiva** *f* ELEC *corriente alterna* wattless component, lag, ELECTRÓN, ING ELÉC lag; ~ **tangencial** *f* FÍS tangential component; ~ **transversal** *f* ELEC *corriente alterna-componente reactivo* quadrature axis component, FÍS transverse component; ~ **vertical** *f* FÍS vertical component; ~ **de viento longitudinal** *f* TRANSP AÉR longitudinal-wind component; ~ **de Zeeman** *f* FÍS Zeeman component

componer[1] *vt* IMPR set, *tipografía* assemble, compose; ~ **en cursivas** *vt* IMPR italicize; ~ **con minúsculas** *vt* IMPR keep down

componer[2]: ~ **continuamente** *vi* IMPR run on

comportamiento *m* CONST, CONTAM MAR *de la mancha de hidrocarburos* behavior (*AmE*), behaviour (*BrE*); ~ **estacional** *m* PROD seasonal behavior (*AmE*), seasonal behaviour (*BrE*); ~ **estanco** *m* AGUA, TRANSP MAR cofferdam; ~ **de la fractura** *m* METAL fracture behavior (*AmE*), fracture behaviour (*BrE*); ~ **del helicóptero** *m* TRANSP AÉR helicopter behavior (*AmE*), helicopter behaviour (*BrE*); ~ **del reactor** *m* NUCL reactor behavior (*AmE*), reactor behaviour (*BrE*); ~ **de la red** *m* TELECOM network performance (*NP*)

composición *f* IMPR composition, display, matter, set, MECÁ, NUCL, PAPEL, QUÍMICA composition; ~ **antiincrustante** *f* REVEST antifouling composition; ~ **apretada** *f* IMPR set solid; ~ **asistida por computadora** *f* AmL (*cf composición asistida por ordenador Esp*) IMPR computer-assisted setting; ~ **asistida por ordenador** *f* Esp (*cf composición asistida por computadora AmL*) IMPR computer-assisted setting; ~ **del barro** *f* C&V clay composition; ~ **de la carga** *f* C&V batch composition; ~ **en columnas** *f* IMPR column matter, columnar composition; ~ **por computadora** *f* AmL (*cf composición por ordenador Esp*) INFORM&PD computer-assisted setting; ~ **conservada para reimpresión** *f* IMPR repeat matter; ~ **continua** *f* IMPR full-out; ~ **desengrasante** *f* EMB degreasing compound; ~ **de los dientes de la sierra** *f* ING MECÁ saw set; ~ **de un edificio** *f* CONST framing; ~ **estequiométrica** *f* METAL stoichiometric composition; ~ **de fabricación** *f* PAPEL furnish; ~ **fibrosa** *f* IMPR *cartón, papel* fiber furnish (*AmE*), fibre furnish (*BrE*), PAPEL fiber composition (*AmE*), fibre composition (*BrE*); ~ **fotográfica** *f* IMPR light setting; ~ **sin interlinear** *f* IMPR solid matter; ~ **sin justificar** *f* IMPR ragged setting; ~ **con las letras apretadas** *f* IMPR kerning; ~ **llena** *f* IMPR square matter; ~ **a mano** *f* IMPR hand composition; ~ **mecánica** *f* IMPR mechanical setting; ~ **de la mezcla** *f* TEC ESP mixture composition; ~ **mineralógica** *f* GEOL *de una roca* mode; ~ **por ordenador** *f* Esp (*cf composición por computadora AmL*) INFORM&PD computer-assisted setting; ~ **pendiente de impresión** *f* IMPR standing matter, standing type; ~ **sin plomo** *f* IMPR cold type; ~ **que abarca una página doble** *f* IMPR double spread; ~ **para reutilizar** *f* IMPR alive matter; ~ **reversible** *f* IMPR reversible furniture; ~ **tabular** *f* IMPR tabular work; ~ **de tarifas** *f* TELECOM tariff structure; ~ **de texto** *f* IMPR body matter; ~ **tipográfica** *f* IMPR text matter, typesetting; ~ **todavía válida** *f* IMPR live matter; ~ **del tren** *f* FERRO *vehículos* train set

compositora: ~ **para líneas de simple distribución** *f* IMPR single-distributor line-composing machine

compostura *f* GAS fixing

compra: ~ **de componentes** *f* TEC ESP component procurement

comprendido *adj* ING MECÁ included

comprensión: ~ **de bloques** *f* INFORM&PD block compaction; ~ **de la voz** *f* INFORM&PD speech understanding

compresibilidad *f* GEN compressibility; ~ **de los gases** *f* TERMO compressibility of gases; ~ **isentrópica** *f* FÍS isentropic compressibility; ~ **isotérmica** *f* FÍS isothermal compressibility

compresible *adj* GEN compressible; **no** ~ *adj* TRANSP AÉR incompressible

compresión[1]: **a** ~ **total** *adj* ING MECÁ, PROD fully-compressed

compresión[2] *f* CONST, ELECTRÓN, FÍS, FÍS ONDAS

compression, INFORM&PD compaction, compression, ING MECÁ compression, INSTAL HIDRÁUL compression, *del vapor en el cilindro* cushion, cushioning, NUCL *de un plasma*, P&C *procesamiento*, REFRIG compression; ~ **adiabática** *f* FÍS FLUID *gases*, TERMO adiabatic compression; ~ **anamórfica** *f* CINEMAT anamorphic squeeze; ~ **del ancho de banda** *f* ELECTRÓN, TV bandwidth compression; ~ **biescalonada** *f* ING MECÁ two-space compression; ~ **bietápica** *f* ING MECÁ two-space compression; ~ **de blanco** *f* TV white compression; ~ **en caliente** *f* METAL *pulvimetalurgia* hot pressing; ~ **de circuito abierto** *f* CARBÓN open-circuit crushing; ~ **de datos** *f* ELECTRÓN, INFORM&PD, TELECOM data compaction, data compression; ~ **escalonada** *f* ING MECÁ compound compression, PROD *aire* stage compression; ~ **en escalonamiento** *f* INSTAL HIDRÁUL, REFRIG single-stage compression; ~ **en etapas** *f* PROD *aire* stage compression; ~**-expansión** *f* ELECTRÓN, TEC ESP *telecomunicaciones* companding; ~ **facsímil** *f* INFORM&PD, TELECOM facsimile compression; ~ **por fases** *f* ING MECÁ compound compression; ~ **de la gama tonal** *f* IMPR tone compression; ~ **de ganancia** *f* ELECTRÓN gain compression; ~ **húmeda** *f* REFRIG wet compression; ~ **de imagen** *f* ELECTRÓN image compression, TELECOM *TV*, TV picture compression; ~ **de impulso** *f* TELECOM pulse compression; ~ **de la memoria** *f* INFORM&PD memory compression; ~ **múltiple** *f* REFRIG multistage compression; ~ **del negro** *f* TV black compression, black crush; ~ **del puntal de choque del tren de aterrizaje** *f* TRANSP AÉR landing gear shock strut compression; ~ **seca** *f* REFRIG dry compression; ~ **de señales** *f* TELECOM signal compression; ~ **simple** *f* INSTAL HIDRÁUL, REFRIG single-stage compression; ~ **del tiempo** *f* TELECOM time compression; ~ **en el tiempo** *f* TELECOM time compression; ~**-tracción** *f* TRANSP push-pull; ~ **de vapor** *f* PROC QUÍ vapor compression (*AmE*), vapour compression (*BrE*); ~ **en varios escalonamientos** *f* REFRIG multistage compression; ~ **vocal digital** *f* TELECOM digital speech compression; ~ **de volumen** *f* ELECTRÓN *señales microfónicas* volume compression

compresómetro *m* CARBÓN compressometer

compresor *m* GEN *motores* compressor; ~ **sin aceite** *m* REFRIG oil-free compressor; ~ **de aire** *m* AUTO, ING MECÁ, INSTAL HIDRÁUL, LAB, MECÁ air compressor; ~ **de alta presión** *m* ING MECÁ high-pressure compressor; ~ **alternativo** *m* REFRIG reciprocating compressor; ~ **axial** *m* MECÁ, TRANSP AÉR axial compressor; ~ **de baja presión** *m* ING MECÁ low-pressure compressor; ~ **bicilíndrico** *m* INSTAL HIDRÁUL duplex compressor; ~ **centrífugo** *m* ING MECÁ centrifugal compressor; ~ **con cilindros en línea** *m* PROD straight-line compressor; ~ **en contracorriente** *m* REFRIG return-flow compressor; ~ **de diafragma** *m* ING MECÁ diaphragm compressor; ~ **a diesel** *m* ING MECÁ diesel-powered compressor; ~ **doble** *m* ING MECÁ dual compressor; ~ **de doble efecto** *m* ING MECÁ dual compressor; ~ **con dos cilindros de vapor** *m* INSTAL HIDRÁUL duplex compressor; ~ **de émbolo** *m* GAS piston compressor; ~ **enfriado por aire** *m* REFRIG air-cooled compressor; ~ **en equicorriente** *m* REFRIG uniflow compressor; ~**-expansor** *m* ELECTRÓN, INFORM&PD, TEC ESP compander, TELECOM compressor-expander; ~ **de**

fases escalonadas *m* ING MECÁ compound compressor, PROD stage compressor; ~ **de flujo alterno** *m* REFRIG return-flow compressor; ~ **de flujo continuo** *m* REFRIG uniflow compressor; ~ **frigorífico** *m* REFRIG refrigerating compressor; ~ **de funcionamiento sin lubricante** *m* ING MECÁ dry-running compressor; ~ **de funcionamiento en seco** *m* ING MECÁ dry-running compressor; ~ **de gas** *m* GAS, PROD gas compressor; ~ **de grupo abierto** *m* REFRIG open-type compressor unit; ~ **de haz magnético** *m* FÍS PART magnetic beam compressor; ~ **hermético** *m* *Esp* (*cf bocha AmL*) ING MECÁ, REFRIG hermetic compressor, hermetically-sealed compressor unit; ~ **de inyección de agua** *m* INSTAL HIDRÁUL spray compressor, water injection compressor, PROD spray compressor; ~ **de inyección de aire** *m* INSTAL HIDRÁUL *motor de combustión interna* air injection compressor; ~ **monocompresional** *m* INSTAL HIDRÁUL single-stage compressor; ~ **multigradual de varias etapas** *m* ING MECÁ multistage compressor; ~ **multigradual de varios pasos** *m* ING MECÁ multistage compressor; ~ **de pequeña presión** *m* ING MECÁ low-pressure compressor; ~ **de pequeña velocidad** *m* INSTAL HIDRÁUL slow-speed compressor; ~ **de pistón** *m* GAS, ING MECÁ piston compressor; ~ **de pistón de alta presión** *m* ING MECÁ high-pressure piston compressor; ~ **de pistón de baja presión** *m* ING MECÁ low-pressure piston compressor; ~ **que comprime aire desplazándolo mecánicamente** *m* ING MECÁ positive-displacement compressor; ~ **radial** *m* TRANSP AÉR radial compressor; ~ **reciprocante** *m* ING MECÁ piston compressor; ~ **reciprocante de alta velocidad** *m* ING MECÁ high-pressure compressor; ~ **refrigerado por refrigerante** *m* REFRIG refrigerant-cooled compressor; ~ **de refrigerante** *m* ING MECÁ, REFRIG refrigerant compressor; ~ **refrigerante** *m* ING MECÁ, REFRIG refrigerant compressor; ~ **rotativo** *m* ING MECÁ, REFRIG, TERMO rotary compressor; ~ **rotativo monoeje estacionario** *m* ING MECÁ rotary-compressor single shaft stationary; ~ **de segmentos del pistón** *m* AUTO, VEH piston-ring compressor; ~ **de simple efecto** *m* REFRIG single acting compressor; ~ **para la sobrealimentación** *m* MECÁ supercharger; ~ **de una sola etapa** *m* INSTAL HIDRÁUL single-stage compressor; ~ **de tornillo** *m* GAS, ING MECÁ screw compressor; ~ **de tornillos rotatorios** *m* PROD rotary-screw compressor

comprimido[1] *adj* MECÁ pressurized

comprimido[2]: ~ **crudo** *m* NUCL green compact

comprimir *vt* CINEMAT squeeze, CONST, ELECTRÓN compress, INFORM&PD pack, compress, ING MECÁ, MINAS, NUCL compress, PROD squeeze, TELECOM compress

comprobación *f* CALIDAD checking, CARBÓN assay, test, trial, EMB check, checking, INFORM&PD testing, check, ING MECÁ check, testing, verification, TEC ESP monitoring, check, TEC PETR *producción, yacimientos* gaging (*AmE*), gauging (*BrE*), TELECOM audit; ~ **y aceptación del seguro marítimo** *f* TRANSP MAR underwriting; ~ **de amplitud** *f* ELECTRÓN amplitude calibration; ~ **automática** *f* CALIDAD *de datos cuantitativos* automatic verification, INFORM&PD automatic check; ~ **de base** *f* TRANSP AÉR base check; ~ **de las baterías** *f* CINEMAT battery check; ~ **de blanco-negro** *f* TV black-white monitoring;

~ **de calidad** *f* CALIDAD quality audit, quality check, PROD quality check; ~ **estadística** *f* METR statistical check; ~ **marginal** *f* INFORM&PD marginal check; ~ **mediante hardware** *f* INFORM&PD hardware check; ~ **meticulosa** *f* CONST, SEG meticulous inspection; ~ **multicanal** *f* TRANSP MAR *radio* multichannel monitoring; ~ **de paridad par-impar** *f* INFORM&PD odd-even check; ~ **del peso** *f* EMB check weighing; ~ **de programas** *f* INFORM&PD program checkout; ~ **de la redundancia cíclica** *f* (*CRC*) ELECTRÓN cyclic redundancy check (*CRC*); ~ **de redundancia longitudinal** *f* INFORM&PD longitudinal redundancy check (*LRC*); ~ **de la remesa** *f* EMB batch tabbing; ~ **de la seguridad** *f* TELECOM *de transmisiones* security audit; ~ **del sistema** *f* INFORM&PD system testing; ~ **por tarjeta precodificada** *f* TRANSP *tráfico* precoded tag survey; ~ **de validez** *f* TELECOM validity check; ~ **vertical de la redundancia** *f* INFORM&PD vertical redundancy check (*VRC*)

comprobador *m* ELECTRÓN calibrator; ~ **de compresión de diesel** *m* AUTO diesel compression tester; ~ **de inducidos** *m* ELEC *instrumento* armature tester; ~ **de línea** *m* PAPEL line tester; ~ **de película** *m* CINEMAT film checker; ~ **de transistores** *m* ELECTRÓN transistor checker

comprobar *vt* CINEMAT, CONST check, PROD *telefonía* monitor

compromiso *m* PROD commitment; ~ **de dosis equivalente efectiva colectiva** *m* NUCL collective effective dose equivalent commitment

compuerta *f* AGRIC *riego* gate, AGUA draw gate, gate, sluice, water gate, *esclusas* shut, CARBÓN lock, ram pump, floodgate, CONST gate, *hidráulica* lock, CONTAM MAR trawl board, ENERG RENOV flap gate, FÍS, HIDROL gate, ING MECÁ overflow, INSTAL HIDRÁUL *presas* shutter, MINAS *hidráulicas* baffle, OCEAN trawl board, PROC QUÍ baffle, PROD *fundería* shut, gate, RECICL sluice, REFRIG *en conductos de aire acondicionado* damper, SEG airlock, TEC ESP hatch, TRANSP MAR trawl board, *de esclusa* floodgate, sluice, sluicegate; ~ **de abertura graduada** *f* AGUA graded sluice; ~ **de aguas abajo** *f* AGUA *esclusas* aft gate; ~ **de aguas arriba** *f* AGUA crown gate; ~ **de agujas** *f* AGUA needle dam; ~ **de aire** *f* TEC ESP *espacionaves* airlock; ~ **del alimentador** *f* C&V feeder gate; ~ **basculante** *f* ENERG RENOV *presas* tilting gate; ~ **de cabecera** *f* AGUA head gate; ~ **de carga** *f* PROD *hornos* charging door, TRANSP cargo hatchway; ~ **de cierre oscilante** *f* HIDROL *con contrapeso* flap gate; ~ **cilíndrica** *f* HIDROL *presas* roller gate, METR ring gage (*AmE*), ring gauge (*BrE*); ~ **de cola** *f* AGUA tailgate; ~ **contra incendios** *f* CONST, SEG, TERMO fire-protection gate; ~ **corrediza** *f* CONST sliding gate; ~ **de desagüe** *f* AGUA, ENERG RENOV sluicegate; ~ **de desagüe deslizable** *f* ENERG RENOV *presas* sliding sluice gate; ~ **de descarga** *f* AGUA sluicegate, tailgate, INSTAL HIDRÁUL outlet valve; ~ **deslizante** *f* CONST sliding gate; ~ **de dique** *f* ENERG RENOV, HIDROL bay; ~ **de esclusa** *f* AGUA canal lock gate, penstock, sluicegate, HIDROL *instalaciones hidroeléctricas* penstock; ~ **flotante** *f* AGUA caisson; ~ **giratoria** *f* CONST swing gate; ~ **de guillotina** *f* MINAS *comunicación en galerías* guillotine gate; ~ **inferior** *f* MECÁ tailgate; ~ **de inmersión** *f* HIDROL falling sluice; ~ **de inspección** *f* TEC ESP inspection hatch; ~ **intermedia** *f* ING ELÉC transfer gate; ~ **de**

inundación *f* AGUA, HIDROL flood gate; ~ **de limpia** *f* AGUA flush gate; ~ **de marea** *f* HIDROL tidal lock; ~ **de paso** *f* AGUA shut-off gate; ~ **de presión** *f* FOTO pressure gate; ~ **radial** *f* ENERG RENOV *presas* radial gate; ~ **de rápido** *f* CONST chute gate; ~ **reguladora** *f* AGUA regulating sluice, INSTAL HIDRÁUL regulating sluice, regulator gate; ~ **rodante** *f* ENERG RENOV fixed-roller sluice gate, *de presa* free-roller sluice gate; ~ **de rodillos** *f* HIDROL *presas* roller gate; ~ **de sector** *f* ENERG RENOV *presas* sector gate; ~ **sectorial** *f* AGUA sector weir; ~ **de sincronización cromática** *f* TV burst gate; ~ **del suelo** *f* TRANSP AÉR floor hatch; ~ **de tambor** *f* ENERG RENOV *presas* drum gate; ~ **de toma** *f* AGUA head gate, intake gate, head sluice, ENERG RENOV *de presa* headgate; ~ **de toma de agua** *f* AGUA regulating sluice, INSTAL HIDRÁUL regulating sluice, regulator gate; ~ **de trabajo** *f* AGUA head gate; ~ **trasera** *f* CINEMAT tailgate, TRANSP hatchback; ~ **de ventilación** *f* MINAS air door; ~ **Y** *f* FÍS AND gate

compuesto *m* P&C *término general* compound, *material formado por combinación de varios* composite, PROD, QUÍMICA compound; ~ **activado** *m* METAL activated complex; ~ **aislante** *m* EMB, ING ELÉC insulating compound; ~ **anfifílico** *m* DETERG amphiphilic compound; ~ **anfótero** *m* DETERG, QUÍMICA amphoteric compound; ~ **aromático** *m* ALIMENT, TEC PETR *compuestos del petróleo* aromatic compound; ~ **azimídico** *m* QUÍMICA azimino compound; ~ **de azufre** *m* GAS sulfur compound (*AmE*), sulphur compound (*BrE*); ~ **cíclico** *m* QUÍMICA closed chain; ~ **clatrático** *m* QUÍMICA clathrate compound; ~ **desengrasante** *m* EMB degreasing compound; ~ **electrónico** *m* METAL electronic compound; ~ **de espuma** *m* TRANSP AÉR foam compound; ~ **estequiométrico** *m* CRISTAL stoichiometric compound; ~ **frágil de aluminio y oro** *m* ING ELÉC purple plague; ~ **inorgánico de cromo** *m* QUÍMICA *cromato de bario* inorganic chromium compound; ~ **interhalógeno** *m* QUÍMICA interhalogen compound; ~ **intermetálico** *m* METAL intermetallic compound; ~ **intersticial** *m* QUÍMICA interstitial compound; ~ **irradiado con tritio** *m* NUCL tritiated compound; ~ **lapidado** *m* PROD lapping compound; ~ **leuco** *m* QUÍMICA *de colorante* leuco compound; ~ **lubricante** *m* ELEC *cable* filler; ~ **marcado** *m* NUCL labeled compound (*AmE*), labelled compound (*BrE*); ~ **marcado isotópicamente** *m* NUCL isotopically-tagged compound; ~ **para moldear por inyección** *m* EMB injection molding compound (*AmE*), injection moulding compound (*BrE*); ~ **para moldeo por inyección** *m* PROD injection molding compound (*AmE*), injection moulding compound (*BrE*); ~ **orgánico** *m* QUÍMICA organic compound; ~ **organomagnésico** *m* QUÍMICA organomagnesium compound; ~ **oxidante** *m* QUÍMICA oxidizing agent; ~ **premoldeado** *m* P&C *procesamiento* preform; ~ **para proteger la superficie de un metal** *m* C&V *galvanización* blanket; ~ **químico** *m* AGRIC compound fertilizer; ~ **de silicona** *m* TRANSP MAR silicone compound; ~ **silicónico** *m* TRANSP MAR silicone compound; ~ **tritiado** *m* NUCL tritiated compound; ~ **de uranio** *m* NUCL uranium compound

compuestos: ~ **azoicos** *m pl* QUÍMICA azo compounds; ~ **organoclorados** *m pl* RECICL organochlorine; ~ **organohalogenados adsorbibles** *m pl*

PAPEL adsorbable organic halogens (*AOX*); ~ **petrosulfúricos** *m pl* TEC PETR petrosulfur compounds (*AmE*), petrosulphur compounds (*BrE*)

computación *f* INFORM&PD, TELECOM *AmL* (*cf informática Esp*) *teoría* computing, computing sciences, *procesamiento de datos* computation

computador *AmL ver* computadora *AmL*, ordenador *Esp*

computadora *f AmL* (*cf ordenador Esp*) ELEC computer, INFORM&PD computer, machine; ~ **anfitrión** *f AmL* (*cf ordenador anfitrión Esp*) INFORM&PD host computer; ~ **autónoma** *f AmL* (*cf ordenador autónomo Esp*) INFORM&PD off-line computer, stand-alone, NUCL off-line computer; ~ **de a bordo** *f AmL* (*cf ordenador de a bordo Esp*) TEC ESP on-board computer; ~ **de casa** *f AmL* (*cf ordenador de casa Esp*) INFORM&PD home computer; ~ **central** *f AmL* (*cf ordenador central Esp*) INFORM&PD host computer; ~ **de conjunto de instrucciones complejas** *f AmL* (*cf ordenador de conjunto de instrucciones complejas Esp*) INFORM&PD complex instruction set computer (*CISC*); ~ **de cuarta generación** *f AmL* (*cf ordenador de cuarta generación Esp*) INFORM&PD fourth-generation computer; ~ **de datos de aire** *f AmL* (*cf ordenador de datos de aire Esp*) TRANSP AÉR air data computer; ~ **dedicada** *f AmL* (*cf ordenador dedicado Esp*) ELECTRÓN, INFORM&PD dedicated computer; ~ **de destino** *f AmL* (*cf ordenador de destino Esp*) INFORM&PD target computer; ~ **digital** *f AmL* (*cf ordenador digital Esp*) ELECTRÓN, INFORM&PD digital computer; ~ **digital de proceso** *f AmL* (*cf ordenador digital de proceso*) NUCL digital process computer system; ~ **doméstica** *f AmL* (*cf ordenador doméstico Esp*) INFORM&PD, TELECOM home computer, personal computer (*PC*); ~ **especializada** *f AmL* (*cf ordenador especializado Esp*) INFORM&PD dedicated computer, special-purpose computer; ~ **de falda** *f AmL* (*cf ordenador de falda Esp*) INFORM&PD laptop computer; ~ **fuera de línea** *f AmL* (*cf ordenador fuera de línea Esp*) INFORM&PD, NUCL off-line computer; ~ **híbrida** *f AmL* (*cf ordenador híbrido Esp*) INFORM&PD hub polling, hybrid computer; ~ **con juego reducido de instrucciones** *f AmL* (*cf ordenador con juego reducido de instrucciones Esp*) INFORM&PD reduced instruction set computer (*RISC*); ~ **laptop** *f AmL* (*cf ordenador laptop Esp*) INFORM&PD laptop computer; ~ **de longitud de palabra fija** *f AmL* (*cf ordenador de longitud de palabra fija Esp*) INFORM&PD fixed-wordlength computer; ~ **de longitud de palabra variable** *f AmL* (*cf ordenador de longitud de palabra variable Esp*) INFORM&PD variable-wordlength computer; ~ **orientada a los caracteres** *f AmL* (*cf ordenador orientado a los caracteres Esp*) INFORM&PD character-oriented machine; ~ **en paralelo** *f AmL* (*cf ordenador en paralelo Esp*) INFORM&PD parallel computer; ~ **pasarela** *f AmL* (*cf ordenador pasarela Esp*) INFORM&PD gateway computer; ~ **personal** *f AmL* (*cf ordenador personal Esp, PC*), INFORM&PD, TELECOM personal computer (*PC*); ~ **portátil** *f AmL* (*cf ordenador portátil Esp*) INFORM&PD laptop computer; ~ **de portilla** *f AmL* (*cf ordenador de portilla Esp*) INFORM&PD gateway computer; ~ **de primera generación** *f AmL* (*cf ordenador de primera generación Esp*) INFORM&PD first-generation computer; ~ **principal** *f AmL* (*cf ordenador principal Esp*) INFORM&PD host computer; ~ **con programa almacenado** *f AmL* (*cf ordenador con programa almacenado Esp*) INFORM&PD stored program computer; ~ **de programa almacenado** *f AmL* (*cf ordenador con programa almacenado Esp*) TELECOM stored program computer; ~ **con programa almacenado en memoria** *f AmL* (*cf ordenador con programa almacenado en memoria Esp*) INFORM&PD stored program computer; ~ **puente entre redes** *f AmL* (*cf ordenador puente entre redes Esp*) INFORM&PD gateway computer; ~ **de quinta generación** *f AmL* (*cf ordenador de quinta generación Esp*) INFORM&PD fifth-generation computer; ~ **satélite** *f AmL* (*cf ordenador satélite Esp*) INFORM&PD satellite computer; ~ **de secuencia** *f AmL* (*cf ordenador de secuencia Esp*) INFORM&PD, ING ELÉC, PROD, TELECOM sequencer; ~ **de segunda generación** *f AmL* (*cf ordenador de segunda generación Esp*) INFORM&PD second-generation computer; ~ **serie** *f AmL* (*cf ordenador serie Esp*) INFORM&PD serial computer; ~ **síncrona** *f AmL* (*cf ordenador síncrono Esp*) INFORM&PD synchronous computer; ~ **de sobremesa** *f AmL* (*cf ordenador de sobremesa*) INFORM&PD desktop computer; ~ **sola** *f AmL* (*cf ordenador solo Esp*) INFORM&PD stand-alone; ~ **de tercera generación** *f AmL* (*cf ordenador de tercera generación Esp*) INFORM&PD third-generation computer; ~ **para tráfico** *f AmL* (*cf ordenador para tráfico Esp*) TRANSP AÉR traffic computer; ~ **universal** *f AmL* (*cf ordenador universal Esp*) INFORM&PD general-purpose computer, universal computer; ~ **de uso general** *f AmL* (*cf ordenador de uso general Esp*) INFORM&PD general-purpose computer; ~ **virtual** *f AmL* (*cf ordenador virtual Esp*) INFORM&PD virtual machine; ~ **de vuelo** *f AmL* (*cf ordenador de vuelo Esp*) TRANSP AÉR flight computer

computadorizado *adj AmL* (*cf informatizado Esp*) PROD computer-based, computerized

cómputo *m* INFORM&PD *contador, controlador* tally, computation, PROD tally, TELECOM computation, metering; ~ **de llamadas** *m* TELECOM call metering

común *adj* INFORM&PD standard

comunicación *f* TELECOM message; ~ **amiga** *f* INFORM&PD handshaking; ~ **colectiva** *f* TELECOM conference call; ~ **para conferencia de encuentro** *f* TELECOM meet-me conference call; ~ **de datos** *f* INFORM&PD data communication; ~ **directa** *f* TELECOM *abonados* through-connection; ~ **de emergencia** *f* PROD backup communication; ~ **entre canales de E/S** *f* PROD I/O channel communication; ~ **interurbana automática en directo** *f* TELECOM direct digital interface (*DDI*); ~ **interurbana no tasada** *f AmL* (*cf llamada gratuita Esp*) TELECOM freephone call (*BrE*), toll-free call (*AmE*); ~ **por ionización meteórica** *f* TEC ESP meteor burst communication; ~ **nula de indicación para varios puntos** *f* TELECOM multipoint indication zero communication (*MIZ*); ~ **personal** *f* TELECOM personal call; ~ **punto a punto** *f* PROD point-to-point communication; ~ **por radio** *f* TELECOM radiocommunication; ~ **radioeléctrica** *f* TELECOM radiocommunication; ~ **de salida** *f* TELECOM outgoing call; ~ **telefónica colectiva de encuentro** *f* TELECOM meet-me conference call

comunicaciones[1]: ~ **por cable** *f pl* TELECOM cable communications; ~ **digitales** *f pl* INFORM&PD digital communications; **~ en escena** *f pl* TELECOM on-scene communications; **~ espaciales** *f pl* TELECOM space communications; **~ internacionales** *f pl* TELECOM worldwide communications; **~ con láser** *f pl* ELECTRÓN laser communications; **~ mediante satélites** *f pl* TELECOM satellite communications; **~ ópticas** *f pl* TELECOM optical communications; **~ por satélite** *f pl* (*SATCOM*) TELECOM, TRANSP MAR satellite communications (*SATCOM*); **~ tierra-aire** *f pl* TRANSP AÉR ground-to-air communication

comunicaciones[2]: **~ de salida no permitidas** *fra* TELECOM, outgoing calls barred (*OCB*); **~ de salida prohibidas** *fra* TELECOM outgoing calls barred (*OCB*)

comunicar *vt* ING MECÁ impart, TELECOM put through; **~ por señales** *vt* TRANSP MAR *mensajes* signal

comunicarse *v refl* TELECOM pass through

comunidad: **~ marítima** *f* TRANSP MAR maritime community; **~ de plantas** *f* AGRIC plant association; **~ vital** *f* CONTAM *mundo animal y vegetal* living community; **~ viva** *f* CONTAM *mundo animal y vegetal* living community

concatenación *f* ELECTRÓN, IMPR, INFORM&PD, ING ELÉC, TELECOM concatenation

concatenar *vt* ELECTRÓN, IMPR, INFORM&PD, ING ELÉC, TELECOM concatenate

concavidad *f* TV canoe

cóncavo[1] *adj* GEOM re-entrant, MECÁ dished

cóncavo[2] *m* CARBÓN concave

concentración *f* C&V gathering, CARBÓN *de minerales* upgrading, *solución* strength, concentration, CONTAM *soluciones* level, ELECTRÓN concentration, METR *burbuja* level, OCEAN, P&C, QUÍMICA, TEC ESP, TELECOM concentration; **~ de aceptores** *f* ELECTRÓN *semiconductores* acceptor concentration; **~ ácida** *f* CONTAM acid concentration; **~ atmosférica** *f* CONTAM atmospheric concentration; **~ de colas** *f* NUCL tail assay; **~ por congelación** *f* ALIMENT, PROC QUÍ, REFRIG freeze concentration; **~ de contaminante ambiental** *f* CONTAM *caracterización del medio* ambient pollutant concentration; **~ derivada en aire** *f* (*CDA*) NUCL derived air concentration (*DAC*); **~ de electrones de valencia** *f* NUCL valence electron concentration; **~ de fangos** *f* MINAS ragging; **~ de fondo** *f* CONTAM *legislación, medida de contaminación* background concentration, background level; **~ gaseosa** *f* ELECTRÓN gas focusing; **~ de imanes permanentes** *f* ING ELÉC permanent-magnet focusing; **~ de impurezas** *f* ELECTRÓN impurity concentration; **~ de ión hidrógeno** *f* QUÍMICA hydrogen ion concentration; **~ iónica** *f* GEOFÍS ionic concentration; **~ letal** *f* CONTAM *toxicología* lethal concentration; **~ letal media** *f* CONTAM *toxicología* median lethal concentration; **~ magnética** *f* ING ELÉC magnetic focusing; **~ masiva** *f* SEG mass concentration; **~ masiva de bióxido de azufre** *f* SEG *en el medio ambiente* mass sulfur dioxide concentration (*AmE*), mass sulphur dioxide concentration (*BrE*); **~ máxima** *f* CONTAM *análisis, legislación* peak concentration; **~ máxima de emisión** *f* (*CME*) CONTAM *legislación* maximum emission concentration (*MEC*); **~ máxima permisible ocupacional** *f* CONTAM *legislación* occupational maximum allowable concentration; **~ mecánica** *f* PROD mechanical concentration; **~ media de un elemento** *f* GEOL *en un reservorio geoquímico* abun-

dance pattern; **~ mortal** *f* CONTAM *toxicología* lethal concentration; **~ de ozono** *f* CONTAM *tratamiento de aguas* ozone concentration; **~ segura** *f* NUCL safe concentration; **~ de tráfico** *f* TRANSP traffic concentration; **~ del vapor** *f* METEO *del aire* absolute humidity; **~ de volumen crítico de pigmentos** *f* P&C *pintura* critical pigment volume concentration (*CVPC*)

concentraciones: **~ de polvo suspendido en el aire** *f pl* SEG airborne dust concentrations

concentrado[1] *adj* ALIMENT, CARBÓN, PROC QUÍ, QUÍMICA concentrated

concentrado[2] *m* ALIMENT, CARBÓN concentrate, MINAS *de minerales* hutch; **~ para panadería** *m* ALIMENT bakery concentrate; **~ de proteínas de peces** *m* OCEAN fish protein concentrate (*FPC*); **~ de uranio** *m* NUCL uranium concentrate

concentrador *m* ELECTRÓN concentrator (*CON*), INFORM&PD concentrator (*CON*), *red* hub, INSTAL TERM, TELECOM concentrator (*CON*); **~ de acceso** *m* TELECOM access concentrator; **~ antibloqueo** *m* ING ELÉC nonblocking concentrator; **~ colocado** *m* TELECOM collocated concentrator; **~ de datos** *m* INFORM&PD data concentrator; **~ de línea remota** *m* INFORM&PD, TELECOM remote-line concentrator; **~ lineal** *m* ENERG RENOV linear concentrator; **~ de líneas** *m* TELECOM line concentrator; **~ remoto** *m* TELECOM remote concentrator; **~ satélite digital** *m* TELECOM digital satellite concentrator; **~ solar** *m* INSTR solar concentrator

concentrar *vt* ALIMENT, CARBÓN concentrate, CINEMAT, ELECTRÓN, FÍS, FÍS ONDAS, FOTO, ÓPT focus, PROC QUÍ, QUÍMICA concentrate, TV focus

concentricidad *f* MECÁ concentricity

concéntrico *adj* CONST, ELEC, GEOL, GEOM, MATEMÁT, MECÁ concentric

concepción: **~ tecnológica y métodos de fabricación** *f* PROD engineering and methods

concepto: **~ de avance de la pala** *m* TRANSP AÉR advancing blade concept

conceptual *adj* INFORM&PD conceptual

concertar *vt* PROD time

concertino *m* ACÚST *música* leader

concesión *f* MINAS *Esp* (*cf pertenencia AmL*) claim, TEC PETR *licencias de exploración* license (*AmE*), licence (*BrE*), *licencias de explotación* block, concession; **~ minera** *f* MINAS *legislación* mining concession, mining claim; **~ para producción** *f* TEC PETR production licence (*BrE*), production license (*AmE*)

concha *f* C&V shell

conchar *vt* ALIMENT *cacao* conche

conciencia: **~ de consumidor** *f* TEXTIL consumer awareness

conclusión: **~ bajo el mar** *f* TEC PETR *producción costa-fuera* subsea completion

concordancia *f* OCEAN *mareas* concordance, TEC PETR *geología* conformity; **~ con un modelo** *f* PROD pattern matching

concreción *f* GEOL, PETROL, TEC PETR concretion

concretar *vt* AmL (*cf hormigonar Esp*) CONST concrete

concretera *f* AmL (*cf hormigonera Esp*) CONST mixer truck

concreto *m* AmL (*cf hormigón Esp*) C&V, CONST concrete; **~ con alta dosificación de cemento** *m* AmL (*cf hormigón con alta dosificación de cemento*

Esp) CONST fat concrete; **~ amasado en fábrica** *m AmL* (*cf hormigón amasado en fábrica Esp*) CONST ready-mixed concrete; **~ armado** *m AmL* (*cf hormigón armado Esp*) CONST, D&A reinforced concrete, TRANSP MAR *construcción naval* ferroconcrete; **~ asfáltico** *m AmL* (*cf hormigón asfáltico Esp*) CONST asphalt concrete; **~ de blindaje** *m AmL* (*cf hormigón de blindaje Esp*) CONST blinding concrete; **~ celular** *m AmL* (*cf hormigón celular Esp*) CONST honeycombing; **~ centrifugado** *m AmL* (*cf hormigón centrifugado Esp*) CONST spun concrete; **~ ciclópeo** *m AmL* (*cf hormigón ciclópeo Esp*) CONST cyclopean concrete; **~ fabricado a pie de obra** *m AmL* (*cf hormigón fabricado a pie de obra Esp*) CONST in situ concrete; **~ sin finos** *m AmL* (*cf hormigón sin finos Esp*) CONST no-fines concrete; **~ fresco** *m AmL* (*cf hormigón fresco Esp*) CONST green concrete; **~ in situ** *m AmL* (*cf hormigón in situ Esp*) CONST in situ concrete; **~ ligero** *m AmL* (*cf hormigón ligero Esp*) CONST air-entrained concrete, lightweight concrete; **~ en masa** *m AmL* (*cf hormigón en masa Esp*) CONST mass concrete; **~ mezclado a mano** *m AmL* (*cf hormigón mezclado a mano Esp*) CONST hand-mixed concrete; **~ mezclado en obra** *m AmL* (*cf hormigón mezclado en obra Esp*) CONST site concrete; **~ para pavimento** *m AmL* (*cf hormigón para pavimento Esp*) CONST pavement-quality concrete (*PQC*); **~ pobre** *m AmL* (*cf hormigón pobre Esp*) CONST lean concrete; **~ preamasado** *m AmL* (*cf hormigón preamasado Esp*) CONST ready-mixed concrete; **~ prefabricado** *m AmL* (*cf hormigón prefabricado Esp*) CONST precast concrete; **~ pretensado** *m AmL* (*cf hormigón pretensado Esp*) CONST prestressed concrete; **~ sumergido** *m AmL* (*cf hormigón sumergido Esp*) CONST submerged concrete; **~ vibrado** *m AmL* (*cf hormigón vibrado Esp*) CONST vibrated concrete; **~ zunchado** *m AmL* (*cf hormigón zunchado Esp*) CONST hooped concrete, stirruped concrete

concurso: **~ de ganado** *m* AGRIC livestock show

condensable *adj* FÍS, GAS, QUÍMICA *gas* condensable

condensación *f* ACÚST, CONST, FÍS, HIDROL condensation, IMPR sweating, PROC QUÍ, QUÍMICA condensation, REFRIG condensation, dew, *vapor sobre superficie fría* sweating, TELECOM condensing, TERMO, TERMOTEC condensation; **~ de Bose-Einstein** *f* FÍS Bose-Einstein condensation; **~ de calor** *f* METEO heat of condensation; **~ de Claisen** *f* PROC QUÍ, QUÍMICA *reacción* Claisen condensation; **~ por enfriamiento superficial** *f* PROC QUÍ condensation by surface cooling; **~ de la humedad sobre superficies metálicas** *f* MINAS *metalurgia* sweat; **~ interna** *f* C&V *en vidrio doble* condensation; **~ por inyección** *f* PROC QUÍ condensation by injection; **~ de vapor** *f* NUCL vapor condensation (*AmE*), vapour condensation (*BrE*); **~ de zinc** *f* CARBÓN zinc condensation

condensado *adj* GEN condensate, condensed

condensador *m* AUTO capacitor, CARBÓN *tubería de vapor* gage (*AmE*), gauge (*BrE*), *electricidad* permittor, CINEMAT condenser, ELEC *componente* capacitor, condenser, FÍS, INFORM&PD capacitor, ING ELÉC capacitor, condenser, ING MECÁ capacitor, LAB *elemento eléctrico* condenser, capacitor, PROC QUÍ, QUÍMICA, REFRIG, TEC PETR condenser, TELECOM capacitor, TERMOTEC, TRANSP MAR condenser, TV capacitor, VEH *encendido* condenser, capacitor; **~ de**

acoplamiento *m* ELEC coupling capacitor; **~ acumulador** *m* ING ELÉC storage capacitor; **~ de aire por convección natural** *m* REFRIG natural convection air-cooled condenser; **~ ajustable** *m* ELEC, ING ELÉC adjustable capacitor; **~ de ajuste de aire** *m* ING ELÉC air trimmer capacitor; **~ de ajuste de grupo** *m* ING ELÉC gang-tuning capacitor; **~ de ajuste de sintonía** *m* ELEC trim capacitor; **~ de alimentador** *m* ING ELÉC feedthrough capacitor; **~ de almacenaje de energía** *m* ING ELÉC, TEC ESP, TERMO energy-storage capacitor; **~ de aluminio húmedo** *m* ING ELÉC wet-aluminium capacitor (*BrE*), wet-aluminum capacitor (*AmE*); **~ de aluminio sólido** *m* ING ELÉC solid aluminium capacitor (*BrE*), solid aluminum capacitor (*AmE*); **~ del anillo de seguridad** *m* ING ELÉC guard-ring capacitor; **~ con ánodo sólido de tantalio** *m* ING ELÉC tantalum slug capacitor; **~ antiparasitario** *m* ING ELÉC spark capacitor; **~ de aplanamiento** *m* ELEC smoothing capacitor; **~ de arranque** *m* ELEC starting capacitor; **~ de atenuación** *m* ING ELÉC damping capacitor; **~ autorregenerable** *m* ING ELÉC self-healing capacitor; **~ en baño de aceite** *m* ELEC oil-immersed capacitor; **~ de bloqueo** *m* ELEC, FÍS, ING ELÉC blocking capacitor; **~ de capacidad fija de CA** *m* ING ELÉC AC capacitor; **~ cerámico** *m* C&V, ELEC, FÍS, ING ELÉC, TELECOM ceramic capacitor; **~ cerámico graduable** *m* ING ELÉC adjustable ceramic capacitor; **~ cerámico revestido de vidrio** *m* ING ELÉC glass-coated ceramic capacitor; **~ cerámico de una sola capa** *m* ING ELÉC single-layer ceramic capacitor; **~ cerámico tubular** *m* ING ELÉC tubular ceramic capacitor; **~ en chip** *m* ING ELÉC on-chip capacitor; **~ de chip cerámico** *m* C&V, ING ELÉC ceramic-chip capacitor; **~ de chorro** *m* INSTAL HIDRÁUL ejector condenser; **~ de cintas** *m* TEXTIL tape condenser; **~ de compensación térmica** *m* ING ELÉC temperature-compensating capacitor; **~ compuesto** *m* ING ELÉC lumped capacitor; **~ de conexión perpendicular** *m* ING ELÉC radial-lead capacitor; **~ de contacto directo** *m* PROD direct contact condenser; **~ controlado por la tensión** *m* ING ELÉC voltage-controlled capacitor; **~ de cubeta** *m* ING ELÉC reservoir capacitor; **~ derivado** *m* FÍS shunt capacitor; **~ de descarga** *m* ING ELÉC discharge capacitor; **~ de descargas gaseosas** *m* NUCL off-gas condenser; **~ de desconexión** *m* ING ELÉC decoupling capacitor; **~ desfasador** *m* ELEC phase-shifting capacitor; **~ devanado** *m* ING ELÉC reeled capacitor; **~ dieléctrico** *m* ELEC dielectric condenser; **~ con dieléctrico de aire** *m* ELEC air capacitor; **~ con dieléctrico de papel** *m* ELEC, ING ELÉC paper capacitor; **~ de disco** *m* ING ELÉC disc capacitor (*BrE*), disk capacitor (*AmE*); **~ discontinuo** *m* ING ELÉC discrete capacitor; **~ discreto** *m* ING ELÉC discrete capacitor; **~ de dos baterías** *m* ELEC two-cell capacitor; **~ electrolítico** *m* ELEC, FÍS, ING ELÉC electrolytic capacitor; **~ electrolítico de aluminio** *m* ING ELÉC aluminium electrolytic capacitor (*BrE*), aluminum electrolytic capacitor (*AmE*); **~ electrolítico polarizado** *m* ING ELÉC polarized electrolytic capacitor; **~ de electrólito sólido** *m* ING ELÉC solid-electrolyte capacitor; **~ electroquímico** *m* ELEC, ING ELÉC, QUÍMICA, TELECOM electrochemical capacitor; **~ de empalme** *m* ING ELÉC junction capacitor; **~ para encendido** *m* AUTO ignition capacitor;

~ **enfriado por agua** *m* TERMOTEC water-cooled condenser; ~ **enfriado por aire** *m* REFRIG air-cooled condenser; ~ **enfriado por convección forzada de aire** *m* REFRIG forced-draft air-cooled condenser (*AmE*), forced-draught air-cooled condenser (*BrE*); ~ **de entrada** *m* ING ELÉC input capacitor; ~ **de equilibrado** *m* ING ELÉC trimmer capacitor; ~ **estándar** *m* ING ELÉC standard capacitor; ~ **evaporativo** *m* REFRIG evaporative condenser; ~ **de extracción** *m* REFRIG bleeder-type condenser; ~ **por eyección** *m* INSTAL HIDRÁUL ejector condenser; ~ **de eyector** *m* INSTAL HIDRÁUL ejector condenser; ~ **fijo** *m* ELEC, FÍS, ING ELÉC fixed capacitor; ~ **de filtro** *m* ING ELÉC filter capacitor; ~ **de frío** *m* REFRIG cold trap; ~ **del generador de agua dulce** *m* TRANSP MAR freshwater condenser; ~ **graduable** *m* ING ELÉC adjustable capacitor; ~ **de gran valor** *m* ING ELÉC large-value capacitor; ~ **con hielo** *m* NUCL ice condenser; ~ **húmedo de tantalio** *m* ING ELÉC tantalum wet capacitor; ~ **para ignición** *m* AUTO ignition capacitor; ~ **de impulsos** *m* ELEC pulse capacitor; ~ **inductor** *m* ING ELÉC inductive capacitor; ~ **integrado** *m* ING ELÉC integrated capacitor; ~ **integrador** *m* ING ELÉC integrating capacitor; ~ **de inyección** *m* INSTAL HIDRÁUL *máquina de vapor* injection condenser, jet condenser; ~ **de láminas de tantalio** *m* ING ELÉC tantalum foil capacitor; ~ **de mezcla** *m* INSTAL HIDRÁUL jet condenser; ~ **de mica** *m* ELEC, ING ELÉC mica capacitor; ~ **de mica metalizada** *m* ING ELÉC metalized mica capacitor (*AmE*), metallized mica capacitor (*BrE*); ~ **de mica plateada** *m* ING ELÉC silvered mica capacitor; ~ **de microplaqueta cerámica** *m* C&V, ING ELÉC ceramic-chip capacitor; ~ **de microscopio** *m* LAB *óptica* microscope condenser; ~ **montado debajo de la turbina** *m* NUCL underfloor condenser; ~ **múltiple** *m* ELEC, ING ELÉC gang capacitor; ~ **de múltiples placas** *m* ELEC multiple plate capacitor; ~ **no regulable** *m* ELEC, FÍS, ING ELÉC fixed capacitor; ~ **óptico** *m* CINEMAT optical condenser; ~ **oscilante** *m* ELEC oscillating capacitor; ~ **de óxido de tantalio** *m* ING ELÉC tantalum oxide capacitor; ~ **de papel** *m* ELEC, ING ELÉC paper capacitor; ~ **de papel aceitado** *m* ING ELÉC oil-paper capacitor; ~ **de papel metalizado** *m* ELEC metalized-paper capacitor (*AmE*), metallized-paper capacitor (*BrE*); ~ **de película** *m* ING ELÉC film capacitor; ~ **de película delgada** *m* ELEC thin-film capacitor, TELECOM thin-layer capacitor; ~ **de película metalizada** *m* ING ELÉC metalized film capacitor (*AmE*), metallized film capacitor (*BrE*); ~ **de película plástica** *m* ING ELÉC plastic-film capacitor; ~ **con pérdidas** *m* ING ELÉC leaky capacitor; ~ **de placas articuladas** *m* ELEC book capacitor; ~ **de placas paralelas** *m* FÍS, ING ELÉC parallel-plate capacitor; ~ **de planos paralelos** *m* FÍS, ING ELÉC parallel-plate capacitor; ~ **polarizado** *m* ING ELÉC polarized capacitor; ~ **de potencia** *m* ELEC, ING ELÉC power capacitor; ~ **protector** *m* ING ELÉC protection capacitor; ~ **de referencia** *m* ELEC, ING ELÉC reference capacitor; ~ **de reflujo** *m* LAB *destilación, material de vidrio*, NUCL reflux condenser; ~ **refrigerado por aire** *m* TERMOTEC air-cooled condenser; ~ **refrigerado por corriente forzada de aire** *m* REFRIG forced-draft air-cooled condenser (*AmE*), forced-draught air-cooled condenser (*BrE*); ~ **regulable** *m* ING ELÉC adjustable

capacitor; ~ **de rejilla** *m* ING ELÉC grid capacitor; ~ **de retención** *m* ING ELÉC snubber capacitor; ~ **de rociado** *m* GAS spray condenser; ~ **rotativo** *m* ELEC, ING ELÉC rotary capacitor; ~ **de salida** *m* ING ELÉC output capacitor; ~ **semiconductor de óxido metalizado** *m* ING ELÉC metal-oxide semiconductor capacitor; ~ **en serie** *m* ELEC, ING ELÉC series capacitor; ~ **de serpentín de envolvente** *m* REFRIG shell and coil condenser; ~ **síncrono** *m* ING ELÉC synchronous capacitor; ~ **de sintonización** *m* ING ELÉC tuning capacitor; ~ **sólido de tantalio** *m* ING ELÉC tantalum solid capacitor; ~ **SOM** *m* ING ELÉC MOS capacitor; ~ **sumergido** *m* REFRIG submerged condenser; ~ **de superficie** *m* PROD *vapor* surface condenser; ~ **de superficie enfriado por aire** *m* ING MECÁ air-cooled surface condenser; ~ **supresor** *m* ELEC *antiinterferencia* suppressor capacitor; ~ **tándem** *m* ELEC, ING ELÉC gang capacitor; ~ **de tantalio** *m* ING ELÉC tantalum capacitor; ~ **de tantalio por espira húmeda** *m* ING ELÉC wet-slug tantalum capacitor; ~ **de tantalio recubierto en plata** *m* ING ELÉC silver case tantalum capacitor; ~ **de tantalio sólido** *m* ING MECÁ solid tantalum capacitor; ~ **totalmente de tantalio** *m* ING ELÉC all-tantalum capacitor; ~ **universal** *m* INSTR universal condenser; ~ **de vacío** *m* ING ELÉC vacuum capacitor; ~ **variable** *m* ELEC trim capacitor, trimmer, variable capacitor, FÍS, ING ELÉC variable capacitor; ~ **variable de aire** *m* ING ELÉC air variable capacitor; ~ **variable de armaduras articuladas** *m* ELEC book capacitor; ~ **variable de estator fraccionado** *m* ING ELÉC split-stator variable capacitor; ~ **variable tipo libro** *m* ELEC book capacitor; ~ **de vidrio** *m* C&V, ING ELÉC glass capacitor

condensadores: ~ **acoplados** *m pl* ING ELÉC ganged capacitors; ~ **de amperaje de servicio** *m pl* ING ELÉC ratioed capacitors

condensar *vt* ELECTRÓN implode, GAS, GEOL condense, INFORM&PD compress, PROC QUÍ, QUÍMICA, REFRIG, TEC PETR, TERMO, TERMOTEC condense

condición *f* INFORM&PD condition; ~ **de aislación** *f* ELEC insulation class; ~ **de colgado** *f* TELECOM *teléfono* on-hook condition; ~ **de emergencia** *f* NUCL emergency condition; ~ **estacionaria** *f* ÓPT, TELECOM steady-state condition; ~ **de estado estacionario** *f* ING ELÉC, ÓPT steady-state condition; ~ **de falla** *f* PROD fault condition; ~ **de fricción de referencia** *f* TRANSP AÉR reference-friction condition; ~ **de funcionamiento** *f* METR *de una máquina* operating condition; ~ **hidráulica** *f* GAS hydraulic status; ~ **de máximo lastre** *f* TRANSP MAR deepest ballast condition; ~ **de no deslizamiento** *f* FÍS FLUID no-slip condition; ~ **de prueba** *f* METR test condition; ~ **de ramificación** *f* PROD branching condition; ~ **de régimen permanente** *f* TELECOM steady-state condition; ~ **en reposo** *f* TELECOM *teléfono* on-hook condition; ~ **de trabajo** *f* MECÁ rating; ~ **transitoria** *f* ELEC, ING ELÉC transient condition

condicional *adj* INFORM&PD conditional

condicionalmente: ~ **obligatorio** *adj* TELECOM conditionally mandatory (*CM*)

condicionamiento: ~ **para señal** *m* PROD signal conditioning

condiciones[1] *f pl* EMB conditions of testing; ~ **de accidente** *f pl* NUCL accident conditions; ~ **ambientales** *f pl* METR, PROD environmental conditions; ~ **ambientales inseguras** *f pl* SEG unsafe environmental conditions; ~ **de amplitud de señal** *f pl* ELECTRÓN large-signal conditions; ~ **de aproximación en ralentí** *f pl* TRANSP AÉR approach idling conditions; ~ **atmosféricas** *f pl* TRANSP MAR atmospheric conditions; ~ **atmosféricas mínimas** *f pl* TRANSP AÉR minimum weather conditions; ~ **climáticas** *f pl* EMB climatic conditions; ~ **degradadas de funcionamiento** *f pl* TEC ESP *equipos* degraded operating conditions; ~ **del depósito** *f pl* PETROL, TEC PETR reservoir conditions; ~ **de deslizamiento** *f pl* TRANSP skidding conditions; ~ **dinámicas** *f pl* ING ELÉC dynamic conditions; ~ **de entrega** *f pl* PROD terms of delivery; ~ **de espera de aproximación** *f pl* TRANSP AÉR approach idling conditions; ~ **estáticas** *f pl* ING ELÉC static conditions; ~ **evaluadas** *f pl* ING ELÉC rated conditions; ~ **de una expedición** *f pl* TRANSP MAR shipping terms; ~ **en el fondo del pozo** *f pl* TEC PETR downhole conditions; ~ **frontera** *f pl* ELEC *electrostática, magnetostática, para ondas planas* boundary conditions; ~ **de funcionamiento** *f pl* ELEC *equipo*, ING ELÉC operating conditions; ~ **de funcionamiento nominal** *f pl* TEC ESP nominal operating conditions; ~ **inseguras** *f pl* SEG unsafe conditions; ~ **marineras** *f pl* TRANSP MAR *del buque* seakeeping qualities; ~ **máximas** *f pl* ELEC *fuente de alimentación* on-peak conditions; ~ **meteorológicas** *f pl* CONTAM, METEO meteorological conditions; ~ **no lineales** *f pl* CONST nonlinear conditions; ~ **nominales** *f pl* ING ELÉC rated conditions; ~ **normales** *f pl* ING MECÁ normal conditions; ~ **normales de funcionamiento** *f pl* ING MECÁ normal working conditions; ~ **de operación anormales** *f pl* NUCL off-normal operating conditions; ~ **de operación normales** *f pl* NUCL normal operating conditions; ~ **de operación previstas** *f pl* TRANSP AÉR anticipated operating conditions; ~ **óptimas** *f pl* TRANSP ideal conditions; ~ **de perforación** *f pl* TEC PETR *perforación* drilling conditions; ~ **de puesta en establecimiento del estado estacionario** *f pl* TELECOM steady-state launching conditions; ~ **de referencia** *f pl* METR reference conditions; ~ **de saturación** *f pl* ING ELÉC saturation conditions; ~ **para la semejanza dinámica de dos flujos** *f pl* FÍS FLUID conditions for dynamic similarity of two flows; ~ **sinusoidales** *f pl* ING ELÉC sinusoidal conditions; ~ **de sondeo** *f pl* TEC PETR *perforación* drilling conditions; ~ **sub-superficiales** *f pl* TEC PETR *perforación, producción* subsurface conditions; ~ **técnicas** *f pl* TEC ESP environment; ~ **de trabajo** *f pl* ELEC *equipo* operating conditions, SEG working conditions; ~ **de tráfico** *f pl* TRANSP traffic conditions; ~ **de utilización** *f pl* ELEC *equipo* operating conditions; ~ **del viento** *f pl* METEO wind conditions; ~ **del yacimiento** *f pl* PETROL, TEC PETR reservoir conditions

condiciones[2]: **bajo ~ normales de uso** *fra* METR under normal conditions of use; **en ~ de servicio** *fra* ING MECÁ in full working order, in working order

condimentar *vt* ALIMENT flavor (*AmE*), flavour (*BrE*)

condimento *m* ALIMENT flavoring (*AmE*), flavouring (*BrE*)

condrina *f* QUÍMICA chondrin

condrita *f* GEOL, TEC PETR chondrite

condrítico *adj* GEOL, TEC PETR chondritic

condroarsenita *f* MINERAL chondrarsenite

condrodita *f* MINERAL chondrodite

conducción *f* CONST, ELEC, ELECTRÓN conduction, FERRO *vehículos* driving, FÍS, HIDROL, ING ELÉC, INSTAL TERM conduction, TEC PETR *tuberías* pipe, VEH driving; ~ **aérea** *f* ACÚST air conduction; ~ **aerotimpánica** *f* ACÚST air conduction; ~ **defectuosa** *f* ING ELÉC defect conduction; ~ **eléctrica** *f* ELEC, ING ELÉC electric conduction, electrical conduction; ~ **de energía aérea** *f* ING ELÉC overhead power line; ~ **de fase** *f* FÍS phase lead; ~ **por huecos** *f* ING ELÉC hole conduction; ~ **por lagunas** *f* ING ELÉC *semiconductores* hole conduction; ~ **ósea** *f* ACÚST bone conduction; ~ **residual** *f* ING ELÉC dark conduction; ~ **térmica** *f* TERMO heat conduction, thermal conduction; ~ **de vapor** *f* INSTAL HIDRÁUL steam pipeline; ~ **para ventilar las bolsas secadoras** *f* PAPEL dryer pocket ventilation duct

conducciones *f pl* REFRIG *del aire acondicionado* duct work

conducir[1]: ~ **un canal hasta el mar** *fra* AGUA carry a canal as far as the sea

conducir[2] *vt* TEXTIL *de una persona* drive, TRANSP *coche* drive, steer; ~ **por tubería** *vt* CONST pipe

conducta *f* CONST *de estructura* behavior (*AmE*), behaviour (*BrE*)

conductancia *f* ELEC, ELECTRÓN, FÍS, ING ELÉC, TELECOM conductance; ~ **de conversión** *f* ING ELÉC conversion conductance; ~ **directa** *f* ELEC *de semiconductor*, ELECTRÓN, ING ELÉC forward conductance; ~ **eléctrica** *f* ELEC, TELECOM electrical conductance; ~ **equivalente** *f* TERMO equivalent conductance; ~ **específica** *f* ELEC specific conductance; ~ **iónica** *f* FÍS RAD ionic conductance; ~ **negativa** *f* ELEC *semiconductor* negative conductance; ~ **de superficie** *f* TERMOTEC surface conductance; ~ **térmica** *f* FÍS, TERMOTEC thermal conductance; ~ **unitaria** *f* TERMO unit conductance

conductibilidad: ~ **calórica** *f* TERMO caloric conductibility; ~ **térmica** *f* TERMO thermal conductibility

conductibilímetro *m* AmL (*cf conductímetro Esp*) LAB *análisis* conductivity meter

conductímetro *m* Esp (*cf conductibilímetro AmL, conductivímetro AmL*) LAB *análisis* conductivity meter

conductividad *f* GEN conductivity; ~ **activada** *f* ING ELÉC on-state conductivity; ~ **capilar** *f* HIDROL capillary conductivity; ~ **eléctrica** *f* ELEC, ING ELÉC electrical conductivity; ~ **electrolítica** *f* ELEC *electrólisis*, FÍS, ING ELÉC electrolytic conductivity; ~ **electrónica** *f* ELECTRÓN, FÍS RAD electron conductivity; ~ **extrínseca** *f* ING ELÉC extrinsic conductivity; ~ **hidráulica** *f* ENERG RENOV, HIDROL hydraulic conductivity; ~ **intrínseca** *f* ELEC, ING ELÉC intrinsic conductivity; ~ **molecular** *f* TERMO molecular conductivity; ~ **por portadores positivos** *f* ING ELÉC p-type conductivity; ~ **térmica** *f* CONST, FÍS, GAS, P&C, TEC PETR, TERMO, TERMOTEC thermal conductivity; ~ **térmica electrónica** *f* ELECTRÓN, FÍS, NUCL electronic heat conductivity; ~ **térmica equivalente** *f* TERMO, TERMOTEC equivalent thermal conductivity

conductivímetro *m* AmL (*cf conductímetro Esp*) LAB *análisis* conductivity meter

conducto *m* ACÚST duct, AGUA conduit, water supply line, AUTO passage, CONST channel, conduit, pipeline, *llave* pipe, CONTAM *instalación proceso industrial* pipeline, ELEC *fuente de alimentación* conduit, ELECTRÓN funnel, ENERG RENOV conduit, pipeline, HIDROL conduit, INFORM&PD pipeline, ING ELÉC conductor, ING MECÁ, MECÁ channel, chute, duct, pipe, MINAS *de descarga, de ventilación* trunk, chute, PROD, REFRIG duct, TEC ESP chute, TELECOM duct; **~ de abastecimiento** *m* AGUA *hidrología* aqueduct, supply line; **~ de aceite** *m* AUTO oil channel, oil gallery; **~ aerífero** *m Esp (cf célula aerífera AmL)* ING MECÁ air conduit, MINAS air conduit, air passage; **~ aerotermodinámico** *m* TRANSP aerothermodynamic duct; **~ de agua** *m* AGUA water pipe, AUTO water gallery, HIDROL waterway, PROD water pipe; **~ de aire** *m* ING MECÁ air conduit, air duct, air hose, air passage, INSTAL TERM air duct, MINAS air hose, PROD air flue, REFRIG air duct; **~ de aire caliente** *m* TRANSP AÉR hot air duct; **~ de aireación** *m* ING MECÁ, MINAS air passage; **~ aislado** *m* ELEC *instalación* insulated conduit; **~ para aliviar la presión** *m* FERRO pressure-relief duct; **~ para aliviar la presión del aire** *m* FERRO *equipo inamovible* air pressure relief duct; **~ ascendente** *m* AGUA uptake, CONST, ELEC *fuente de alimentación* rising main; **~ ascendente doble** *m* REFRIG double-suction riser; **~ ascendente de humos** *m* TERMO fire-rising main; **~ de cable** *m* ELEC, ING ELÉC cable duct; **~ para cables** *m* ELEC *fuente de alimentación* conduit, PROD wiring duct; **~ cárstico** *m* HIDROL karstic conduit; **~ ciego** *m* CONST blind shaft; **~ circular del viento** *m* PROD *hornos altos* circular blast aim, horseshoe main; **~ de compensación** *m* HIDROL surge shaft; **~ del condensado** *m* REFRIG condensate line; **~ corto** *m* INSTAL HIDRÁUL short pipe; **~ de desagüe** *m* AGUA sewer drain, RECICL sluice; **~ de descarga** *m* RECICL discharge pipe; **~ deshelador** *m* TRANSP AÉR de-icing duct; **~ eléctrico** *m* ELEC, ING ELÉC raceway; **~ elevador** *m* PETROL, TEC PETR riser; **~ de enfriamiento** *m* ING MECÁ, REFRIG cooling duct; **~ de escape** *m* TRANSP AÉR stack; **~ del escape** *m* AUTO exhaust passage; **~ espiral** *m* MINAS spiral chute; **~ de evacuación** *m* AGUA sluiceway, outlet pipe, ENERG RENOV *presas* sluice; **~ flexible** *m* CONST flexible ducting, CONTAM MAR flexible hose; **~ forzado** *m* AGUA, ENERG RENOV penstock, INSTAL HIDRÁUL full pipe; **~ de humos** *m* TERMOTEC *calderas* flue; **~ de humos de chimenea** *m* CONST chimney flue; **~ inferior de humos** *m* TERMOTEC bottom flue; **~ de irrigación** *m* AGUA irrigation canal; **~ de llegada** *m* AGUA head pipe, INSTAL HIDRÁUL gathering line, head pipe; **~ de lubricación** *m* ING MECÁ oil groove; **~ metálico flexible** *m* ING MECÁ flexible metal conduit; **~ oscilante** *m* CONST swinging chute; **~ de la pala** *m* TRANSP AÉR blade duct; **~ petrolero** *m* TEC PETR *transporte* oil pipeline; **~ portacables** *m* ING ELÉC, PROD conduit; **~ portacables eléctricos** *m* ELEC, ING ELÉC electric conduit; **~ principal de vapor** *m* INSTAL HIDRÁUL main steam pipe; **~ de refrigeración** *m* ING MECÁ, REFRIG cooling duct; **~ refrigerante** *m* ING MECÁ, REFRIG cooling duct; **~ de retorno del lodo** *m* TEC PETR *perforación* mud return line; **~ de salida del pistón** *m* CONST piston-relief duct; **~ de salmuera** *m* REFRIG brine line; **~ secundario** *m* TRANSP AÉR secondary duct; **~ sifónico** *m* HIDROL siphon conduit *(AmE)*, syphon conduit *(BrE)*; **~ subterráneo** *m* CONST, HIDROL

culvert; **~ para el suministro de energía** *m* FERRO power-supply duct; **~ tubular** *m* PROD passage; **~ de vapor** *m* INSTAL HIDRÁUL steamway; **~ de ventilación** *m AmL (cf galería de ventilación Esp)* CONST ventiduct, ventilation duct, ING MECÁ air duct, MINAS fang, ventilation ducting, airway, crossheading, TRANSP MAR *construcción naval* air course; **~ de ventilación de superficie** *m* MINAS surface ventilation duct; **~ del viento** *m* ING MECÁ, MINAS *alto horno* air pipe; **~ volcánico** *m* OCEAN *vulcanología submarina* volcanic vent

conductor[1]: **no ~** *adj* ING ELÉC nonconductive; **sin ~** *adj* TRANSP *coche, taxi* driverless
conductor[2] *m* CONST conveyor, ELEC conductor, ELECTRÓN *fase* lead, FERRO *persona* conductor *(AmE)*, guard *(BrE)*, FÍS conductor, IMPR carrier, ING ELÉC *escobillas* lead, ING MECÁ driver, TEC PETR *eléctrico o térmico* conductor, TELECOM conductor; **~ de accesorios** *m* TRANSP AÉR accessory drive; **~ acorazado** *m* ELEC *de cable* metal-coated conductor; **~ de aire** *m* ING MECÁ air pipe; **~ sin aislación** *m* ELEC bare conductor; **~ aislado** *m* ELEC, ING ELÉC, TELECOM insulated core *(BrE)*, insulated conductor *(AmE)*; **~ sin aislamiento** *m* ELEC bare conductor; **~ de alimentación** *m* ELEC power lead, *red de distribución* feeder; **~ alimentador** *m* ELEC feed cable, *red de distribución* feeder cable, INFORM&PD feeder cable, ING ELÉC feed cable, feeder cable, TV feeder cable; **~ de aluminio** *m* ELEC aluminium conductor *(BrE)*, aluminum conductor *(AmE)*; **~ armado** *m* ELEC *cable* metal-coated conductor; **~ de arrastre** *m* ING MECÁ driver; **~ ascendente** *m* ELEC *fuente de alimentación* riser; **~ blindado** *m* ELEC *cable* metal-clad conductor, shielding conductor; **~ de cable eléctrico monofilar** *m* ING ELÉC core; **~ de cable eléctrico multifilar** *m* ING ELÉC core; **~ de cable unifilar** *m* ELEC single-wire cable conductor, solid conductor; **~ cableado múltiple** *m* ELEC *cable* multiple-stranded conductor; **~ del calor** *m* TEXTIL heat conductor; **~ de camión de largo recorrido** *m* AUTO long-haul lorry driver *(BrE)*, long-haul truck driver *(AmE)*; **~ de capa gruesa** *m* ELECTRÓN thick-film conductor; **~ de cierre** *m* ELEC *conexión* jump lead; **~ de cilindro** *m* ING MECÁ drum drive; **~ de cinta** *m* TV tape leader; **~ circular trenzado concéntricamente** *m* ELEC *cable* concentrically-stranded circular conductor; **~ de cobre** *m* ELEC *cable*, ELECTRÓN copper conductor; **~ colectivo de transferencia para batería** *m* TEC ESP battery transfer bus; **~ compactado** *m* ELEC *cable* compacted conductor; **~ común** *m* IMPR common carrier, INFORM&PD bus, highway, trunk, TELECOM common highway; **~ común de control** *m* INFORM&PD control bus; **~ común de datos** *m* INFORM&PD data bus; **~ común de entrada/salida** *m* INFORM&PD input/output bus; **~ común de la ficha** *m* INFORM&PD token bus; **~ concéntrico** *m* ELEC *cable* concentric conductor; **~ de cono de fricción** *m* ING MECÁ friction cone drive; **~ de correa plana** *m* ING MECÁ flat belt drive; **~ descubierto** *m* ING ELÉC open conductor; **~ desnudo** *m* ELEC bare conductor; **~ doble retorcido** *m* ING ELÉC twisted pair; **~ doble torcido** *m* PROD twisted pair; **~ doble trenzado** *m* PROD twisted pair; **~ eléctrico** *m* ELEC, FÍS, ING ELÉC electric conductor, electrical conductor; **~ de electrodo de puesta a tierra** *m* ELEC, ING ELÉC, PROD

earthing electrode conductor (*BrE*), grounding electrode conductor (*AmE*); ~ **electrónico** *m* ELEC conductor; ~ **de empalme** *m* ELEC *conexión* jump lead; ~ **estañado** *m* ELEC *cable* tinned conductor; ~ **exterior** *m* TELECOM outer conductor; ~ **de filamento delgado** *m* ELECTRÓN thin-film conductor; ~ **flexible** *m* ELEC *cable*, ING ELÉC flexible conductor, TELECOM cord; ~ **forrado de metal** *m* ING ELÉC metal-sheathed conductor; ~ **de grúa** *m* TRANSP MAR crane operator; ~ **de hilos retorcidos múltiple** *m* ELEC *cable* multiple-stranded conductor; ~ **hueco** *m* ELEC hollow conductor; ~ **del inducido** *m* ELEC *máquina eléctrica* armature conductor; ~ **interior** *m* TELECOM inner conductor; ~ **interno aislado** *m* ING ELÉC core; ~ **de masa** *m* ELEC, ING ELÉC earth conductor (*BrE*), ground conductor (*AmE*); ~ **metálico** *m* ING ELÉC metallic conductor; ~ **metalizado** *m* ELEC metal-coated conductor, *cable* metal-clad conductor; ~ **Millikan** *m* ELEC *cable* Millikan conductor; ~ **negativo** *m* ING ELÉC negative conductor; ~ **neutral** *m* ING ELÉC neutral conductor; ~ **neutro** *m* ELEC *circuito* neutral wire, MECÁ neutral; ~ **nulo** *m* ING ELÉC zero conductor; ~ **óhmico** *m* FÍS ohmic conductor; ~ **ómnibus de puesta a tierra** *m* ELEC, ING ELÉC earthing bus (*BrE*), grounding bus (*AmE*); ~ **ómnibus de tierra** *m* ELEC *conexión*, ING ELÉC earth bus (*BrE*), ground bus (*AmE*); ~ **de oropel** *m* ELEC *del cable con cintas metálicas* tinsel conductor; ~ **perfilado** *m* ELEC *cable* shaped conductor; ~ **perfilado de sección sectorial** *m* ELEC *cable* sector-shaped conductor; ~ **plano** *m* ELEC *cable* plain conductor; ~ **de polea escalonada** *m* ING MECÁ step cone drive; ~ **principal** *m* ELEC *fuente de alimentación* trunk main, ING ELÉC main conductor; ~ **de protección** *m* ING ELÉC protective conductor; ~ **de puente** *m* ELEC *conexión* jump lead; ~ **de retorno** *m* D&A, ING ELÉC return conductor; ~ **con revestimiento metálico** *m* ELEC *cable* metal-clad conductor; ~ **sencillo** *m* ELEC single-wire cable conductor, solid conductor; ~ **de suministro** *m* ING ELÉC lead; ~ **de tambor** *m* ING MECÁ drum drive; ~ **de tierra** *m* ELEC earth wire (*BrE*), ground wire (*AmE*), earth conductor (*BrE*), ground conductor (*AmE*), FÍS earth wire (*BrE*), ground wire (*AmE*), ING ELÉC earth wire (*BrE*), ground wire (*AmE*), earth conductor (*BrE*), ground conductor (*AmE*), PROD, TELECOM, TV earth wire (*BrE*), ground wire (*AmE*); ~ **de tierra único** *m* ELEC *conexión* single earth (*BrE*), single ground (*AmE*); ~ **trenzado** *m* ELEC, ING ELÉC stranded conductor; ~ **trenzado múltiple** *m* ELEC *cable* multiple-stranded conductor; ~ **de tubo** *m* ING MECÁ core driver

conductores: ~ **comunes de forma encadenada** *m pl* INFORM&PD daisychain bus; ~ **de entrada** *m pl* ING ELÉC *circuitos* fan-in; ~ **de salida** *m pl* ING ELÉC *circuitos* fan-out

condurrita *f* MINERAL condurrite

conectable *adj* ING ELÉC pluggable, PROD plug-in

conectado *adj* ELEC *circuito*, FÍS connected, INFORM&PD on-line, connected, ING ELÉC, ING MECÁ connected; ~ **a la corriente** *adj* ELEC *instalación*, ING ELÉC connected to the current (*AmE*), connected to the mains (*BrE*); ~ **en derivación** *adj* ING ELÉC parallel-connected; ~ **por detrás** *adj* ELEC *conmutador* back-connected; ~ **directo a tierra** *adj* ING ELÉC directly-earthed (*BrE*), directly-grounded (*AmE*);

~ **en estrella** *adj* TELECOM star-connected; ~ **a la línea de alimentación** *adj* ELEC *instalación*, ING ELÉC connected to the current (*AmE*), connected to the mains (*BrE*); ~ **en paralelo** *adj* ELEC, FÍS, ING ELÉC connected in parallel, parallel-connected; ~ **a la red eléctrica** *adj* ELEC *instalación*, ING ELÉC connected to the current (*AmE*), connected to the mains (*BrE*); ~ **a la red principal** *adj* ELEC *instalación*, ING ELÉC connected to the current (*AmE*), connected to the mains (*BrE*); ~ **en serie** *adj* ELEC, FÍS, ING ELÉC connected in series, series-connected; ~ **a toma de tierra** *adj* ING ELÉC connected to earth (*BrE*), connected to ground (*AmE*)

conectador *m* ING MECÁ nipple, PROD *electricidad* outlet connection; ~ **acoplador** *m* ELEC *en sistema* link; ~ **acoplador asincrónico** *m* ELEC *dos sistemas de CA* asynchronous link; ~ **de acoplamiento** *m* PROD mating connector; ~ **de alimentación** *m* PROD power lead; ~ **óptico** *m* TELECOM optical connector; ~ **orientable** *m* ING MECÁ banjo union; ~ **tipo clavija** *m* TELECOM plug-type connector; ~ **de tubo flexible** *m* MINAS flexible-hose union

conectador/desconectador: ~ **de palanca** *m* PROD toggle on-off switch

conectar[1] *vt* CINEMAT connect, CONST bind, branch, bridge, ELEC *aparato* connect, connect up, *conexión* couple, plug in, switch on, FÍS connect, IMPR install, INFORM&PD attach, connect, ING ELÉC connect, connect up, couple, make, patch, plug in, switch on, turn on, PROD connect, interface, TELECOM plug in, put through, TV patch, switch on; ~ **al aire** *vt* TV switch to air; ~ **eléctricamente** *vt* REFRIG bond; ~ **a masa** *vt* AUTO, ELEC, FÍS, ING ELÉC, PROD, TELECOM, VEH *instalación eléctrica* earth (*BrE*), ground (*AmE*); ~ **mediante bridas** *vt* TEC PETR *conductos, tuberías* flange up; ~ **mediante interruptor** *vt* ELEC *aparato* switch on; ~ **en paralelo** *vt* ELEC, ING ELÉC couple in parallel; ~ **a tierra** *vt* AUTO, ELEC, FÍS, PROD, TELECOM, VEH earth (*BrE*), ground (*AmE*)

conectar[2]: ~ **correa** *vi* ING MECÁ fork a belt on; ~ **una impedancia terminal** *vi* TELECOM *cable* terminate

conectarse: ~ **al sistema** *v refl* INFORM&PD log in, log on

conectividad *f* INFORM&PD connectivity; ~ **digital de extremo a extremo** *f* TELECOM end-to-end digital connectivity

conectivo *adj* INFORM&PD connective

conector *m* CINEMAT connector, CONST nipple, INFORM&PD connector, ING MECÁ coupler, fitting, strap, LAB *material de vidrio, etc.* connector, MECÁ fitting, QUÍMICA *equipamento de laboratorio*, TELECOM, VEH connector; ~ **acoplador** *m* ELEC coupler connector; ~ **de amplificador de potencia** *m* PROD *telefonía* line drive connector; ~ **de baja fuerza de inserción** *m* ING ELÉC low-insertion force connector; ~ **basculante de tubo** *m* ING MECÁ swivel pipe connector; ~ **de bisagra** *m* ING MECÁ hinge fitting; ~ **bloqueable** *m* ING ELÉC lockable connector; ~ **de borde** *m* ING ELÉC edge connector; ~ **de brida** *m* ING MECÁ flanged fitting; ~ **en bucle** *m* PROD loop back connector; ~ **cannon** *m* CINEMAT cannon plug; ~ **de circuito impreso** *m* ING ELÉC printed-circuit connector; ~ **de clavija** *m* ING ELÉC plug connector; ~ **coaxial** *m* ING ELÉC coaxial connector; ~ **de desagüe** *m* ING MECÁ drain connector; ~ **de descarga** *m* ING MECÁ flushing connector; ~ **de**

dos contactos *m* ELEC *condensador* two-contacts connector; ~ **de dos piezas separadas** *m* ING ELÉC two-piece connector; ~ **de drenaje** *m* ING MECÁ drain connector; ~ **eléctrico** *m* ELEC, ING ELÉC electrical connector; ~ **de enchufe de borde** *m* ING ELÉC, PROD edge-socket connector; ~ **de entrada** *m* TELECOM in-connector; ~ **de expansión** *m* ELECTRÓN, INFORM&PD expansion slot; ~ **del extremo de varilla** *m* ING MECÁ strap connecting rod end; ~ **de fibra óptica** *m* ING ELÉC, ÓPT, TELECOM fiber-optic connector (*AmE*), fibre-optic connector (*BrE*), optical fiber connector (*AmE*), optical fibre connector (*BrE*); ~ **fibroóptico** *m* ING ELÉC, ÓPT, TELECOM fiber-optic connector (*AmE*), fibre-optic connector (*BrE*); ~ **de flujograma** *m* *AmL* (*cf conector de organigrama Esp*) INFORM&PD flowchart connector; ~ **frente muerto** *m* ING ELÉC dead front connector; ~ **frente protegido** *m* ING ELÉC dead front connector; ~ **fuera de circuito** *m* TELECOM out-connector; ~ **fuera de servicio** *m* TELECOM out-connector; ~ **de fuerza de inserción nula** *m* ELECTRÓN zero insertion force connector; ~ **hembra** *m* ELEC female connector, INFORM&PD jack, ING ELÉC, PROD female connector; ~ **hermafrodita** *m* ING ELÉC hermaphroditic connector; ~ **de holgura** *m* ING MECÁ self-aligning bearing; ~ **ignífugo** *m* TEC ESP fireproof bulkhead; ~ **de interfaces** *m* PROD interface connector socket; ~ **macho** *m* ELEC *conexión* socket plug, TV plug; ~ **de manguera contra incendios** *m* CONST, SEG, TERMO fire-hose coupling; ~ **monobloque** *m* ING ELÉC one-piece connector; ~ **mural** *m* ELEC *conexión* wall outlet; ~ **óptico** *m* TELECOM optical connector; ~ **de organigrama** *m* *Esp* (*cf conector de flujograma AmL*) INFORM&PD flowchart connector; ~ **de plato** *m* ING MECÁ flanged fitting; ~ **polarizado** *m* ING ELÉC polarized connector; ~ **de prueba** *m* ING ELÉC test jack; ~ **de puntas** *m* CINEMAT pin connector; ~ **roscado** *m* ING MECÁ threaded fitting; ~ **de salida** *m* INFORM&PD out-connector; ~ **seco** *m* ING ELÉC dry connector; ~ **en T para tubos** *m* ING MECÁ pipe tee; ~ **Tuchel** *m* CINEMAT Tuchel connector; ~ **umbilical** *m* TEC ESP umbilical connector; ~ **XLR** *m* CINEMAT, TV XLR connector

conejo *m* TEC PETR *perforación, producción* go devil

conexión[1]: **con ~ en estrella** *adj* TELECOM star-connected; **de ~ posterior** *adj* ELEC *conmutador* back-connected

conexión[2] *f* CONST connection, union, ELEC *circuito* connection, joint, junction, FÍS, GAS connection, INFORM&PD trunk (*AmE*), connection, link (*BrE*), attachment, ING ELÉC coupling, joint, make, turn-on, connection, ING MECÁ coupling, make, joint, connection, couple, link, tie, MECÁ bonding, coupling, link, PETROL tie-in, joint, PROD junction, TEC ESP switching, strap-on, TELECOM trunk switching, connection, TRANSP coupling, TRANSP AÉR coupling, interlock control, VEH *sistema eléctrico* coupling; ~ **ajustable magnética** *f* ING MECÁ magnetic adjustable link; ~ **alámbrica** *f* ING ELÉC tie wire; ~ **en anillo** *f* ELEC mesh connection; ~ **de ánodo común** *f* ING ELÉC common-anode connection; ~ **antiparalela** *f* ELEC antiparallel connection; ~ **axial radial** *f* INSTR axial-radial bearing; ~ **de base común** *f* ING ELÉC common-base connection; ~ **base a tierra** *f* ELEC, ING ELÉC earthed base connection (*BrE*), grounded

base connection (*AmE*); ~ **de bayoneta** *f* ELEC bayonet joint; ~ **bifurcada** *f* ELEC forked connection; ~ **blindada** *f* ELEC cross-bonding, shield-bonding; ~ **blindada subdividida** *f* ELEC sectionalized cross-bonding, sectionalized shield-bonding; ~ **de bomba** *f* AGUA pump connection; ~ **para brocas** *f* MINAS drill joint; ~ **con cable armado** *f* PROD shielded-cable connection; ~ **de cables** *f* TV cable link; ~ **de camino de menor orden** *f* TELECOM lower-order path connection (*LPC*); ~ **de canal virtual** *f* TELECOM virtual channel connection (*CRF, VCC*); ~ **en cantidad** *f* ELEC *circuito*, ING ELÉC parallel connection; ~ **del carburador** *f* AUTO, VEH carburetor linkage (*AmE*), carburettor linkage (*BrE*); ~ **de carga** *f* REFRIG charging connection; ~ **de carril** *f* FERRO rail bond (*BrE*), rail splice (*AmE*); ~ **en cascada** *f* ELEC, ING ELÉC, NUCL cascade connection; ~ **de circuito integrado** *f* ING ELÉC integrated-circuit connection; ~ **de colector común** *f* ING ELÉC common-collector connection; ~ **colector a masa** *f* ELEC, ING ELÉC earthed collector connection (*BrE*), grounded collector connection (*AmE*); ~ **cónica** *f* ING MECÁ conical clamping connection; ~ **de consumo de energía** *f* ING ELÉC common-drain connection; ~ **cruzada** *f* ELEC cross-bonding, shield-bonding, ING ELÉC cross-coupling; ~ **cruzada continua** *f* ELEC continuous cross-bonding, continuous shield-bonding; ~ **cruzada subdividida** *f* ELEC sectionalized cross-bonding, sectionalized shield-bonding; ~ **del cuerpo emisor a tierra** *f* ING ELÉC earthed emitter connection (*BrE*), grounded emitter connection (*AmE*); ~ **en delta** *f* ELEC delta connection; ~ **en delta abierta** *f* ELEC open-delta connection; ~ **en derivación** *f* ELEC *circuito*, ING ELÉC parallel connection; ~ **digital** *f* TELECOM digital connection; ~ **directa de carriles** *f* FERRO *equipo inamovible* direct rail fastening; ~ **directa de cono Morse** *f* ING MECÁ positive drive of Morse tapers; ~ **eléctrica** *f* ELEC, ING ELÉC electrical connection; ~ **de electrodo captador a masa** *f* ELEC, ING ELÉC earthed collector connection (*BrE*), grounded collector connection (*AmE*); ~ **embridada de tubería** *f* ING MECÁ clamped pipe connection; ~ **de emisor común** *f* ING ELÉC common-emitter connection; ~ **por enchufe** *f* ING ELÉC plug connection; ~ **de entrada** *f* PROD inlet connection; ~ **entre el alimentador y el carril** *f* FERRO crossbond; ~ **entre dos ejes** *f* ING MECÁ connection between two shafts; ~ **entre redes** *f* INFORM&PD internal storage (*AmE*), internal store (*BrE*); ~ **errónea** *f* TELECOM faulty connection; ~ **en estrella** *f* ELEC star connection, wye connection, Y-connection, FÍS star connection, ING ELÉC star connection, Y-connection, TELECOM star connection; ~ **estrella-estrella** *f* ELEC star-star connection; ~ **estrella-triángulo** *f* ING ELÉC star-delta connection; ~ **fácil** *f* TELECOM easy connection; ~ **de fibra óptica** *f* TELECOM fiber-optic connection (*AmE*), fibre-optic connection (*BrE*); ~ **fibroóptica** *f* ING ELÉC, ÓPT fiber-optic connection (*AmE*), fibre-optic connection (*BrE*); ~ **final del inducido** *f* ELEC *máquina* armature end connection; ~ **física** *f* TELECOM physical connection (*PhC*); ~ **flexible** *f* FERRO *vehículos* flexible connection; ~ **frontal del inducido** *f* ELEC *máquina* armature end connection; ~ **de fusible limitador de corriente** *f* ELEC current-limiting fuse link; ~ **híbrida** *f* ELECTRÓN hybrid junction;

~ **hidráulica** *f* AUTO hydraulic linkage; ~ **hiperviolenta** *f* ELECTRÓN hyperabrupt junction; ~ **inductiva** *f* FERRO *equipo inamovible* impedance bond; ~ **inductiva reglada** *f* FERRO *equipo inamovible* tuned impedance bond; ~ **inferior** *f* TRANSP AÉR *del eje del rotor* lower link; ~ **insertada** *f* ING MECÁ inset joint; ~ **intermedia** *f* ELEC tapping, *devanado* tap, ING ELÉC tap; ~ **intermedia principal** *f* ELEC principal tapping; ~ **de interrogación** *f* PROD examine-on; ~ **Josephson** *f* ELECTRÓN Josephson junction; ~ **Leblanc** *f* ELEC *conexión, transformador* Leblanc connection; ~ **de línea** *f* ELEC *abastecimiento al consumidor* line connection; ~ **mal hecha** *f* SEG faulty connection; ~ **de malla** *f* FÍS mesh connection; ~ **de manguera** *f* ING MECÁ hose connection; ~ **a masa** *f* CONST, ELEC, ELECTRÓN, FERRO, ING ELÉC, TEC ESP earthing (*BrE*), grounding (*AmE*); ~ **a masa de prueba** *f* VEH test ground; ~ **mecánica** *f* ÓPT, TELECOM, TV mechanical splice; ~ **de microonda** *f* FÍS microwave link; ~ **en paralelo** *f* ELEC *circuito*, FÍS parallel connection, ING ELÉC parallel connection, paralleling; ~ **de pérdida de energía** *f* ING ELÉC common-drain connection; ~ **con perno** *f* TEC ESP *estructura de espacionaves* bolted connection; ~ **perpendicular** *f* ING ELÉC radial lead; ~ **en polígono** *f* ELEC mesh connection, polygon connection; ~ **a presión** *f* ING ELÉC pressurized connection; ~ **a presión en línea** *f* PROD in-line pressure connection; ~ **presurizada** *f* ING ELÉC pressurized connection; ~ **provisional** *f* ING ELÉC patch; ~ **de prueba** *f* ING ELÉC test lead; ~ **en puente** *f* ELEC *en circuito* bridge connection, jumper; ~ **en puente de entrada** *f* NUCL inlet jumper; ~ **de puerta común** *f* ING ELÉC common-gate connection; ~ **de puesta a tierra superficial** *f* ELEC surface earthing connection (*BrE*), surface grounding connection (*AmE*); ~ **Q** *f* ELECTRÓN Q-switching; ~ **radial** *f* ING ELÉC radial lead; ~ **de recepción nula** *f* TRANSP aural null loop; ~ **de recorrido incompleto** *f* TELECOM lower-order path connection (*LPC*); ~ **en red** *f* INFORM&PD networking; ~ **para el refrigerante** *f* REFRIG refrigerant connection; ~ **en rizo** *f* ING ELÉC crimped connection; ~ **Scott** *f* ELEC open-delta connection, *transformador* Scott connection; ~ **en seco** *f* ING ELÉC dry connection; ~ **en serie** *f* ELEC *circuito*, FÍS series connection, ING ELÉC series connection, tandem connection, MINAS *explosivos* series connection; ~ **en shunt** *f* ELEC *circuito*, ING ELÉC parallel connection; ~ **con el sistema** *f* INFORM&PD log-in, log-on; ~ **de sobrecarga de datos** *f* TELECOM connect data overflow (*CDO*); ~ **de una sola punta** *f* ELEC single-point bonding; ~ **suelta** *f* INFORM&PD loose coupling; ~ **de superficie** *f* TELECOM surface connection; ~ **en T** *f* ING MECÁ T-connection; ~ **en tándem** *f* NUCL tandem connection; ~ **telefónica** *f* TELECOM trunk switching; ~ **a tierra** *f* GEN earth connection (*BrE*), earthing (*BrE*), ground connection (*AmE*), grounding (*AmE*); ~ **de tierra del fuselaje** *f* TRANSP AÉR fuselage ground connection; ~ **a tierra múltiple** *f* PROD multiple earthing connection (*BrE*), multiple grounding connection (*AmE*); ~ **tipo enchufe** *f* ELEC plug-type connection; ~ **tipo toma de corriente** *f* ELEC plug-type connection; ~ **de transformador monofásico sobre línea trifásica** *f* ELEC open-delta connection; ~ **de tránsito** *f* TELECOM through-connection; ~ **para la transmisión de datos** *f* TELECOM data link connection (*DLC*); ~ **de trayecto de orden superior** *f* TELECOM higher-order path connection (*HPC*); ~ **de trenes** *f* FERRO train connection; ~ **en triángulo** *f* ELEC, FÍS, ING ELÉC delta connection, mesh connection; ~ **en triángulo abierto** *f* ELEC open-delta connection; ~ **triángulo-estrella** *f* ELEC delta-star connection; ~ **de tubería flexible** *f* PROD hose coupling; ~ **de tubo** *f* ING MECÁ pipe connection; ~ **de tubo flexible** *f* MINAS *sondeos* flexible-hose connection; ~ **de tubos** *f* ING MECÁ pipe union; ~ **en V** *f* ELEC open-delta connection, V-connection; ~ **de las varillas de perforación** *f* MINAS *sondeos* drill pole joint; ~ **de la vía de transmisión virtual** *f* TELECOM virtual path connection (*VPC*); ~ **virtual** *f* INFORM&PD virtual connection; ~ **Y** *f* FÍS Y-connection; ~ **en Y** *f* ELEC *reactor, transformador*, ING ELÉC Y-connection; ~ **Y-Y** *f* ELEC star-star connection; ~ **en zigzag** *f* ELEC *reactor, transformador* zigzag connection

conexionado[1]: ~ **en estrella** *adj* TELECOM star-wired
conexionado[2] *m* ELEC, TEC ESP wiring
conexiones[1]: **sin** ~ *adj* TELECOM connectionless
conexiones[2]: ~ **eléctricas** *f pl* ELEC *conexión* wiring; ~ **externas** *f pl* ELEC external circuit
confección *f* OCEAN *de la red* netting, TEXTIL making-up; ~ **de mosquiteras** *f* TEXTIL mosquito netting
confeccionar *vt* CONST make
conferencia: ~ **colectiva de encuentro** *f* TELECOM meet-me conference call; ~ **de mando para varios puntos** *f* TELECOM multipoint command conference (*MCC*); ~ **telefónica** *f* TELECOM telephone conference; ~ **por video** *AmL*, ~ **por vídeo** *Esp* *f* TEC ESP, TELECOM videoconference
confiabilidad *f* GEN reliability, CALIDAD dependability, reliability, security; ~ **de equipo físico** *f* INFORM&PD hardware reliability
confidencialidad *f* INFORM&PD privacy, TELECOM confidentiality; ~ **en la circulación de tráfico** *f* TELECOM traffic flow confidentiality; ~ **de los datos** *f* INFORM&PD data privacy
configuración *f* ELECTRÓN configuration, patterning, INFORM&PD configuration, setup, ING MECÁ lay, outline, TEC ESP, TELECOM configuration; ~ **para almacenamiento** *f* TEC ESP storage configuration; ~ **en anillo** *f* TELECOM ring configuration; ~ **por bancos** *f* NUCL *de las barras de control* banked configuration; ~ **binaria** *f* INFORM&PD bit pattern; ~ **del cableado** *f* PROD wiring configuration; ~ **del campo magnético** *f* FÍS RAD, NUCL magnetic-field configuration; ~ **de carga máxima** *f* TEC ESP full-coverage beam; ~ **de carga reducida** *f* TEC ESP reduced-load configuration; ~ **cuadripolar** *f* NUCL quadrupolar configuration; ~ **del deslizamiento** *f* METAL geometry of glide; ~ **del electrodo** *f* ELEC, TELECOM electrode configuration; ~ **electrónica** *f* FÍS electronic configuration; ~ **del equipo físico** *f* INFORM&PD hardware configuration; ~ **especular** *f* NUCL mirror configuration; ~ **estacional** *f* PROD seasonal pattern; ~ **en estrella** *f* TELECOM star configuration; ~ **de flujo** *f* AGUA flow pattern; ~ **del fondeo de la barrera flotante** *f* CONTAM MAR boom-laying configuration; ~ **del hardware** *f* INFORM&PD hardware configuration; ~ **de lanzamiento** *f* TEC ESP launching configuration; ~ **limpia** *f* TRANSP AÉR clean configuration; ~ **de línea** *f* ING ELÉC line configuration; ~ **lógica** *f* ELECTRÓN logic pattern; ~ **matriz** *f*

TELECOM matrix configuration; ~ **de pista** *f* TV track configuration; ~ **de plena carga** *f* TEC ESP full-load configuration; ~ **de la potencia** *f* NUCL power shape; ~ **reiterada** *f* ELECTRÓN replicated pattern; ~ **de reserva** *f* PROD backup configuration; ~ **de respaldo** *f* PROD backup configuration; ~ **de la señal** *f* INFORM&PD signal shaping; ~ **del sistema** *f* ELEC *red de distribución* system configuration; ~ **del sistema bus** *f* TELECOM bus configuration; ~ **del software** *f* INFORM&PD software configuration; ~ **del soporte lógico** *f* INFORM&PD software configuration; ~ **tetrapolar** *f* NUCL quadrupolar configuration; ~ **total** *f* TELECOM total configuration; ~ **de la unidad binaria** *f* ELECTRÓN bit mapping

configurado *adj* ING MECÁ, PROD shaped

configurar *vt* INFORM&PD set up, configure

confinamiento *m* FÍS confinement, *fusión nuclear* containment, MINAS confinement; ~ **inercial** *m* FÍS inertial confinement; ~ **magnético** *m* FÍS magnetic confinement; ~ **del quark** *m* FÍS, FÍS PART quark confinement

confirmación *f* MECÁ, P&C, PROD, REVEST forming, TELECOM confirm, confirmation; ~ **de conexión** *f* TELECOM connect confirm (*CC*); ~ **de recepción** *f* INFORM&PD acknowledge, acknowledgement

confirmar *vt* TELECOM confirm; ~ **recepción de** *vt* INFORM&PD acknowledge

conflicto: ~ **laboral** *m* PROD industrial dispute

confluencia *f* AGUA confluence, CONST *carreteras, ferrocarril, ríos* junction, HIDROL confluence, PROD junction, TRANSP concourse

confluente *adj* AGUA, HIDROL confluent

confluir *vi* AGUA meet

conformación *f* PROD *estampado* shaping; ~ **aproximada a la definitiva** *f* PROD blocking; ~ **en caliente** *f* MECÁ, PROD hot forming; ~ **por explosión** *f* MECÁ, TERMO explosive forming; ~ **de impulsos** *f* INFORM&PD pulse shaping; ~ **del paisaje** *f* AGRIC landscaping; ~ **de piezas por cargas explosivas** *f* MECÁ, TERMO explosive forming; ~ **de señales** *f* INFORM&PD signal shaping

conformador *m* CINEMAT conformer

conformadora *f* ING MECÁ shaping machine; ~ **tipo banco** *f* ING MECÁ bench-type shaping machine

conformar *vt* CINEMAT conform, ING MECÁ, PROD shape

conformidad *f* CALIDAD compliance, conformity, GEOL conformity, INFORM&PD concordance, conformance; ~ **con el plazo** *f* PROD deadline conformity

confuso *adj* ING MECÁ intricate, TEXTIL *contorno, trazado* hazy

congelación *f* ALIMENT, EMB, FÍS freezing, GEOL, METEO frost, PROC QUÍ setting, REFRIG, TERMO freeze-up, freezing; ~ **en aire en reposo** *f* REFRIG still-air freezing; ~ **por chorro de aire** *f* ALIMENT blast freezing, REFRIG jet freezing; ~ **sobre cinta transportadora** *f* REFRIG belt freezing; ~ **por contacto** *f* REFRIG, TERMO contact freezing; ~ **por corriente de aire** *f* REFRIG air blast freezing; ~ **a granel** *f* REFRIG bulk freezing; ~ **por inmersión** *f* EMB dip freezing, immersion freezing, REFRIG immersion freezing; ~ **en lecho fluidificado** *f* ALIMENT, PROC QUÍ, REFRIG fluidized-bed freezing; ~ **lenta** *f* ALIMENT, REFRIG slow-freezing; ~ **de productos meltos** *f* REFRIG loose-freezing; ~ **por pulverización** *f* REFRIG spray freezing; ~ **rápida** *f*

ALIMENT quick-freezing, REFRIG deep-freezing, quick-freezing, TERMO quick-freezing; ~ **en seco** *f* EMB dry-freeze; ~ **del suelo** *f* REFRIG soil freezing; ~ **superficial** *f* REFRIG crust freezing

congelado[1] *adj* ALIMENT, REFRIG, TERMO frozen; ~ **rápidamente** *adj* ALIMENT deep-frozen, quick-frozen, REFRIG, TERMO quick-frozen

congelado[2] *m* ALIMENT, REFRIG deep-freeze

congelador *m* ELEC fridge-freezer, ING MECÁ freezer, ice-making compartment, PROC QUÍ freezer, REFRIG, TERMO freezer, fridge-freezer; ~ **de acceso total** *m* REFRIG walk-in freezer; ~ **por aire forzado** *m* REFRIG air-blast freezer; ~ **de alimentos** *m* ALIMENT, ING MECÁ, REFRIG food freezer; ~ **por aspersión** *m* REFRIG spray freezer; ~ **de comestibles** *m* ALIMENT, ING MECÁ, REFRIG food freezer; ~ **por contacto** *m* REFRIG, TERMO contact freezer; ~ **continuo** *m* REFRIG continuous freezer; ~ **discontinuo** *m* REFRIG batch-type freezer; ~ **doméstico** *m* REFRIG domestic freezer; ~ **con estanterías** *m* REFRIG shelf freezer; ~ **de lotes** *m* ALIMENT *maquinaria* batch freezer; ~ **de placas** *m* REFRIG plate-freezer; ~ **rápido** *m* ALIMENT, REFRIG, TERMO quick-freezer; ~ **con rascador** *m* REFRIG scraped-surface freezer; ~ **ventilado** *m* REFRIG ventilated froster; ~ **vertical** *m* REFRIG upright freezer

congelamiento: **de** ~ **rápido** *adj* ALIMENT, REFRIG, TERMO quick-freezing

congelar *vt* ALIMENT chill, freeze, CINEMAT, EMB, FÍS freeze, MECÁ chill, PROC QUÍ freeze, REFRIG *agua*, TERMO chill, freeze; ~ **a baja temperatura** *vt* TERMO deep-freeze; ~ **rápidamente** *vt* ALIMENT, REFRIG deep-freeze, quick-freeze, TERMO quick-freeze

congelarse *v refl* ALIMENT, FÍS, PROC QUÍ, REFRIG, TERMO freeze

congestión *f* CARBÓN, TELECOM congestion

conglomerado *m* CONST, ENERG RENOV, GEOL, PETROL conglomerate, TEC ESP *astronomía* cluster, TEC PETR conglomerate; ~ **aurífero** *m* MINAS *geología* banket, gold conglomerate; ~ **de fallas** *m* MINAS *geología* fault conglomerate; ~ **interformacional** *m* PETROL interformational conglomerate; ~ **poligenético** *m* GEOL polygenetic conglomerate; ~ **polimíctico** *m* GEOL polymictic conglomerate

congruente *adj* MATEMÁT congruent

Coniaciense *adj* GEOL Coniacian

cónica *f* GEOM conic, conics

conicidad *f* ING MECÁ tapering, MECÁ taper, PROD draft (*AmE*), draught (*BrE*), *inclinación de caras laterales del molde* strip

conicina *f* QUÍMICA coniine

cónico *adj* GEOM, TEXTIL conic

coniferina *f* QUÍMICA coniferin

conificación *f* ING MECÁ tapering

conificar: ~ **el extremo de** *vt* PROD *tubos* tag

coniína *f* QUÍMICA coniine

conina *f* QUÍMICA conine

coninquita *f* MINERAL koninckite

conjugación *f* FÍS, GEOM conjugacy, MATEMÁT conjugacy, conjugation, QUÍMICA *compuesto, enlace* conjugation

conjugado[1] *adj* MATEMÁT, QUÍMICA *compuesto, enlace* conjugated; **no** ~ *adj* QUÍMICA nonconjugated

conjugado[2] *m* FÍS, GEOM, MATEMÁT conjugate

conjugancia *f* FÍS, GEOM, MATEMÁT conjugacy

conjugar *vt* FÍS, GEOM, MATEMÁT conjugate

conjunción f INFORM&PD conjunction

conjuntivo m INFORM&PD connective

conjunto[1]: **de ~** adj ING MECÁ overall; **de ~ completo** adj ING MECÁ, PROD complete-assembly

conjunto[2] m CINEMAT assembly, CONST grupo de elementos set, ELECTRÓN bank, GEOL de capas, INFORM&PD, ING ELÉC set, ING MECÁ assembly, MATEMÁT set, METAL aggregate, PROD kit, TELECOM equipo suite, TEXTIL set, TRANSP cluster, VEH assembly; **~ de la abertura** m CINEMAT film aperture assembly; **~ de artículos** m EMB group of commodities; **~ básico** m ING ELÉC basic group; **~ de baterías** m CINEMAT battery pack, PROD battery assembly; **~ de cabezal** m TV head assembly; **~ de cables** m ÓPT, TELECOM cable assembly; **~ calorímetro** m FÍS RAD calorimeter assembly; **~ canónico** m FÍS canonical ensemble; **~ de caracteres** m INFORM&PD character set; **~ de caracteres universal** m INFORM&PD universal character set; **~ de chips** m ELECTRÓN chip set; **~ de cierre** m NUCL seal assembly; **~ de circuitos integrados** m ELECTRÓN integrated circuitry; **~ de cola** m TRANSP AÉR tail unit; **~ combustible sin vaina** m NUCL canless fuel assembly; **~ de combustibles** m FÍS fuel assembly; **~ de compensación** m NUCL shim assembly; **~ complementario** m MATEMÁT complementary set; **~ de conexiones eléctricas** m ELEC, ING ELÉC electric wiring; **~ de contextos definidos** m TELECOM defined context set (DCS); **~ de control automático** m NUCL automatic control assembly; **~ de control periférico** m NUCL edge-control assembly; **~ de convergencia** m TV convergence assembly; **~ convertidor** m ING ELÉC converter set; **~ de corona y piñón** m VEH diferencial, transmisión crown wheel and pinion; **~ de correas** m P&C belting; **~ de datos** m INFORM&PD data aggregate, data set; **~ de datos de generación** m INFORM&PD generation data set; **~ de datos listo** m TELECOM data set ready (DSR); **~ de datos preparados** m TELECOM data set ready (DSR); **~ de diaclasas** m GEOL joint set; **~ de ejes de transmisión** m ING MECÁ shafting; **~ electrónico doméstico** m ING ELÉC home electronic equipment; **~ electrónico del helicóptero** m TRANSP AÉR helicopter avionics package; **~ electrónico de rotación** m TELECOM despun control electronics (DCE); **~ de elementos de filtración** m ELECTRÓN mask set; **~ de embarcaciones de desembarco** m TRANSP tractor; **~ de embrague y marcha libre** m TRANSP AÉR freewheel and clutch unit; **~ de encerramiento montado en la pared** m PROD wall mount enclosure assembly; **~ de engranajes** m ING MECÁ train of gearing, TRANSP cluster; **~ para ensamblar** m NUCL kit; **~ de equipos** m GAS assembly; **~ de experimentos** m TEC ESP vuelo experiment package; **~ de filtros** m CINEMAT filter kit, FOTO filter set, TV filter kit; **~ formado por varios paquetes** m EMB multipack; **~ de fotodiodos** m ELECTRÓN photodiode array; **~ de fusión-fisión rápida a potencia térmica cero** m NUCL fast fusion-fission assembly at zero thermal power; **~ de giro y transferencia de potencia** m (CTCP) TEC ESP bearing and power-transfer assembly (BAPTA); **~ de herramientas** m INFORM&PD toolkit; **~ de instrucciones** m INFORM&PD instruction set, PROD set of instructions; **~ de instrumentación intranuclear** m NUCL in-core instrument assembly; **~ laminar** m ING ELÉC circuito

magnético lamination; **~ de lentes** m CINEMAT set of lenses; **~ de lentes en bruto** m C&V montadas en una herramienta para pulirlas block; **~ de máquinas** m ING MECÁ machinery; **~ monolítico** m ELECTRÓN monolithic array; **~ de osciladores** m ELECTRÓN oscillator bank; **~ de paracaídas** m D&A parachute cluster; **~ de penetración** m NUCL penetration unit; **~ de piezas** m VEH unit; **~ de piezas soldadas** m MECÁ weldment; **~ de pilas** m FOTO battery pack; **~ de piñón y corona** m VEH diferencial, transmisión crown and pinion; **~ de presión máxima** m TEC ESP full-pressure suit; **~ de puertas** m ELECTRÓN gate array; **~ de puertas en un chip** m ELECTRÓN gate array chip; **~ de RAMs** m ING ELÉC bank of RAMs; **~ de recalentado** m NUCL superheat assembly; **~ receptor** m NUCL receiving assembly; **~ rectificador** m TELECOM rectifier unit; **~ de rodillos** m PAPEL calandras, supercalandras stack; **~ de rodillos que forman la prensa** m PAPEL press stack; **~ de sellado** m NUCL seal assembly; **~ de símbolos** m INFORM&PD symbol set; **~ terminal de datos** m (DTE) INFORM&PD, TELECOM, TRANSP AÉR data terminal equipment (DTE); **~ de tipos desordenados** m IMPR pie; **~ de unidades lógicas** m ELECTRÓN logic array; **~ universal** m INFORM&PD, MATEMÁT universal set; **~ de válvulas** m GAS set of valves; **~ de velocidades** m ING MECÁ set of speeds

conjuntor m ELEC relé, ING ELÉC contactor, TELECOM originating junctor, junctor; **~-disyuntor** m ELEC conmutador line breaker, ELEC, ING ELÉC make-and-break; **~ rotatorio** m CINEMAT, ELEC, ING ELÉC, PROD, TELECOM rotary switch; **~ terminal** m TELECOM terminating junctor

conmensurable adj MATEMÁT commensurable

conmoción: **~ estructural** f TEC ESP aviones buffeting

conmutable adj TV switchable

conmutación f ELEC commutation, INFORM&PD gating, switching, ING ELÉC commutation, switching, MATEMÁT commutation, TEC ESP, TELECOM switching; **~ acelerada** f ELEC máquina accelerated commutation; **~ activada por la voz** f TELECOM voice-operated switching; **~ automática** f ELEC conmutador automatic changeover, ING ELÉC, TRANSP automatic switching; **~ automática de protección** f TELECOM automatic protection switching (APS); **~ de bancos** f INFORM&PD bank switching; **~ a bordo** f TEC ESP on-board switching; **~ de CC** f ING ELÉC DC switching; **~ de circuitos** f INFORM&PD, TELECOM circuit switching; **~ contextual** f INFORM&PD context switching; **~ cruce cero** f ING ELÉC zero-crossing switching; **~ de datos** f TELECOM data switching; **~ de desplazamiento de fase** f ELECTRÓN phase-shift keying (PSK); **~ digital** f TELECOM digital switching; **~ por división de espacio** f INFORM&PD space-division switching; **~ por división de tiempo** f INFORM&PD time-division switching; **~ por divisiones espaciadas** f TELECOM spaced division switching; **~ electromecánica** f ELEC, TELECOM electromechanical switching; **~ electrónica** f INFORM&PD, TELECOM electronic switching, TRANSP electronic commutation; **~ de entrada/salida** f INFORM&PD input/output switching; **~ de entre celdas** f TELECOM intercell switching; **~ espacial** f INFORM&PD space-division switching, TELECOM space switch; **~ de haz** f TEC ESP radar beam switching; **~ de hiperfrecuencia** f ING ELÉC high-frequency switching; ~ **de línea** *f* INFORM&PD line switching; ~ **de llamada en curso** *f* TELECOM switching call-in-progress; ~ **de longitud de onda** *f* TELECOM wavelength switching; ~ **de mando** *f* TRANSP AÉR override control; ~ **manual** *f* ING ELÉC manual switching; ~ **de mensajes** *f* INFORM&PD, TELECOM message switching; ~ **óptica** *f* INFORM&PD, ING ELÉC, TELECOM optical switching; ~ **por paquetes** *f* INFORM&PD, ING ELÉC, TELECOM packet switching; ~ **del primario del transformador** *f* PROD transformer primary switching; ~ **de relé de láminas** *f* ELEC, ING ELÉC, TELECOM reed-relay switching; ~ **de retardo de fase** *f* ELECTRÓN phase-shift keying (*PSK*); ~ **telefónica automática** *f* ING ELÉC automatic telephone switching; ~ **temporal** *f* TELECOM time switching, time-division switching; ~ **en el tiempo** *f* TELECOM time switching, time-division switching; ~ **de variación de fase cuaternaria** *f* ELECTRÓN, TELECOM quaternary phase-shift keying (*QPSK*); ~ **de voltaje nulo** *f* ING ELÉC zero voltage switching

conmutaciones: ~ **de frenado inverso** *f pl* TRANSP reversing braking switchgroup

conmutado *adj* TELECOM switched; ~ **por la voz** *adj* TELECOM voice-switched

conmutador *m* CINEMAT switch, ELEC changeover switch, commutator, selector switch, switchgear, *circuito* switch, INFORM&PD switch, ING ELÉC changeover switch, switching device, commutator, switch, TELECOM switch, telephone switchboard, TRANSP *vehículos eléctricos* commutator;

~ a ~ **de acceso flexible** *m* TELECOM flexible access switch (*FAS*); ~ **de acción rápida de dos polos** *m* ELEC double-pole snap switch (*DPSS*); ~ **accionado a mano** *m* ELEC hand-operated switch; ~ **de accionamiento con el pulgar** *m* ELEC thumbwheel switch; ~ **del acoplador direccional** *m* FÍS, ING ELÉC, TELECOM directional coupler switch; ~ **de acoplamiento** *m* ING ELÉC, TELECOM grouping switch; ~ **de acoplamiento de caracteres** *m* TELECOM character-coupling switch; ~ **activado por presión** *m* INSTAL HIDRÁUL pressure switch; ~ **activado por la voz** *m* TELECOM voice-operated switch; ~ **de aislación de línea** *m* ELEC *red de distribución* line-isolating switch; ~ **aislado** *m* ELEC isolating switch; ~ **de alta tensión** *m* ELEC high-voltage switch gear; ~ **antibloqueo** *m* ING ELÉC non-blocking switch; ~ **antideflagrante** *m* ELEC *seguridad* fireproof switch, flameproof switch; ~ **de aplicaciones** *m* ELEC function selector; ~ **del arrancador** *m* AUTO starter commutator; ~ **de arranque** *m* ELEC starting changeover switch; ~ **de arranque estrella-triángulo** *m* ELEC star-delta starting switch; ~ **de aterrizaje** *m* ELEC *de un ascensor, elevador* landing switch; ~ **de aumento** *m* INSTR magnification changer; ~ **automático** *m* ELEC automatic switch, ING ELÉC self-acting switch; ~ **automático de control** *m* ELEC *relé* automatic control switch; ~ **auxiliar** *m* ELEC, TELECOM auxiliary switch; ~ **de avance por pasos** *m* ELEC step switch;

~ b ~ **de banda ancha** *m* TELECOM broadband switch; ~ **de banda estrecha** *m* TELECOM narrow-band switch; ~ **en baño de aceite** *m* ELEC, ING ELÉC oil switch; ~ **basculante** *m* CINEMAT rocker switch, toggle switch, ELEC, ING ELÉC rocker switch; ~ **para bascular** *m* ELEC *automotor* dip selector switch, dip switch; ~ **bidireccional** *m* ING ELÉC double-throw switch; ~ **bifilar** *m* TELECOM two-wire switch; ~ **bifilar**

doble *m* TELECOM four-wire switch; ~ **bimetálico** *m* ELEC bimetallic switch; ~ **bipolar** *m* CINEMAT, ELEC, ING ELÉC double-pole switch (*DPS*); ~ **bipolar bidireccional** *m* (*CBB*) ELEC, ING ELÉC double-pole double-throw knife switch, double-pole double-throw switch (*DPDTS*); ~ **bipolar de dos vías** *m* ELEC double-pole double-throw knife switch, double-pole double-throw switch (*DPDTS*); ~ **de bit** *m* TELECOM bit switch; ~ **de botón** *m* ING ELÉC, ING MECÁ, TELECOM push-button; ~ **de botón deslizante** *m* ELEC slide switch; ~ **de bytes** *m* TELECOM byte switch;

~ c ~ **de cables** *m* ELEC cord switch; ~ **de caja** *m* ELEC box relay; ~ **de cámara** *m* CINEMAT, TV camera switching; ~ **cambiador de polos** *m* ELEC pole-changer switch; ~ **de caracteres** *m* TELECOM character switch; ~ **de carga** *m* ELEC load switch; ~ **CBB** *m* ING ELÉC DPDT switch; ~ **a chorro de aire** *m* ELEC *cortacircuito* air blast switch; ~ **de cierre retardado** *m* ELEC slow-break switch; ~ **de circuito** *m* TELECOM circuit switch; ~ **de un circuito y dos direcciones** *m* ELEC single-pole double-throw switch; ~ **del circuito virtual** *m* TELECOM virtual circuit switch (*VCS*); ~ **de clavijas** *m* ING ELÉC plug switch; ~ **de comprobación** *m* ELEC, ING ELÉC check switch; ~ **conectador-desconectador** *m* ELEC, ING ELÉC, ING MECÁ on-off switch; ~ **con conexión a tierra** *m* ELEC, ING ELÉC earthed switch (*BrE*), grounded switch (*AmE*); ~ **conjuntor** *m* ELEC, ING ELÉC circuit closer; ~ **de contactos cortocircuitantes** *m* ING ELÉC shorting contact switch; ~ **con contactos elásticos** *m* ELEC spring switch, *interruptor* spring commutator; ~ **de contactos escalonados** *m* ELEC step switch; ~ **de control** *m* ELEC check switch, control switch, ING ELÉC check switch; ~ **de control automático** *m* ELEC *relé* automatic control switch; ~ **de control de discrepancias** *m* ELEC control discrepancy switch; ~ **con cordones** *m* ELEC cord switch; ~ **corredizo** *m* ELEC slide switch; ~ **de corte** *m* CINEMAT cutoff switch, ELEC cutout switch; ~ **de cuchilla** *m* ELEC knife switch; ~ **de cuchilla de dos direcciones** *m* ELEC double-throw knife switch; ~ **de cuchilla de dos vías** *m* ELEC double-throw knife switch; ~ **del cursor** *m* ING ELÉC sliding switch;

~ d ~ **de datos** *m* TELECOM data switch; ~ **de derivación** *m* ELEC tap switch; ~ **en derivación** *m* ELEC shunt switch; ~ **deslizante** *m* ELEC slide switch; ~ **digital** *m* TELECOM digital switch; ~ **de dirección** *m* ELEC, ELECTRÓN commutator switch; ~ **de distancia** *m* ELEC distance switch; ~ **de división de longitud de onda** *m* TELECOM wavelength division switch; ~ **de doble polo** *m* CINEMAT, ELEC, ING ELÉC double-pole switch (*DPS*); ~ **de doble tiro** *m* ELEC two-way switch; ~ **de dos conductores** *m* TELECOM two-wire switch; ~ **de dos direcciones** *m* ELEC double-throw switch, two-way switch; ~ **de dos hilos** *m* TELECOM two-wire switch; ~ **de dos polos** *m* CINEMAT, ELEC, ING ELÉC double-pole switch (*DPS*); ~ **de dos polos y dos vías** *m* ELEC double-pole double-throw knife switch; ~ **de dos posiciones** *m* TRANSP MAR changeover switch; ~ **de dos posiciones de contacto** *m* ELEC double-throw switch; ~ **de dos vías** *m* ELEC double-throw switch;

~ e ~ **eléctrico de palanca bidireccional** *m* ING ELÉC toggle switch; ~ **electrónico** *m* INFORM&PD, TELECOM electronic switch; ~ **electrónico de**

mensajes *m* TELECOM electronic message switch; ~ **electroóptico** *m* ELEC, ÓPT, TELECOM electro-optic switch; ~ **de elementos amovibles** *m* ELEC draw-out switchgear; ~ **embutido** *m* ELEC, ING ELÉC flush switch; ~ **embutido en la pared** *m* ELEC, ING ELÉC flush switch; ~ **de encendido y apagado** *m* CINEMAT on-off switch; ~ **en estrella** *m* TELECOM star switch; ~ **explorador** *m* ING ELÉC scanning switch; ~ **de exterior** *m* ELEC outdoor switchgear;

~ f ~ **falso** *m* ELEC false switching; ~ **de fase coaxial** *m* ING ELÉC coaxial phase shifter; ~ **de flujo** *m* ELEC flow switch; ~ **de funcionamiento** *m* ELEC operating switch;

~ g ~ **de gamas** *m* ELEC, ING ELÉC range switch; ~ **giratorio** *m* CINEMAT, ELEC, ING ELÉC, PROD, TELECOM rotary switch; ~ **de gobierno** *m* ELEC master switch; ~ **de grupo** *m* TELECOM envelope switch;

~ h ~ **híbrido** *m* TELECOM hybrid switch;

~ i ~ **incombustible** *m* ELEC *seguridad* fireproof switch, flameproof switch; ~ **de inercia** *m* ING ELÉC inertia switch; ~ **de instalación** *m* ELEC installation switch; ~ **instantáneo de dos polos** *m* ELEC double-pole snap switch (*DPSS*); ~ **de instrumento** *m* ING ELÉC instrument switch; ~ **interferométrico de puente equilibrado** *m* TELECOM balanced-bridge interferometer switch; ~ **intermedio-n** *m* ELEC n-way switch; ~ **de interrupción** *m* PROD break contact; ~ **interurbano** *m* TELECOM toll switch; ~ **de intervalos** *m* TV gapping switch; ~ **inversor** *m* ELEC, ING ELÉC changeover switch, reversing switch; ~ **inversor de arranque** *m* ELEC starting changeover switch; ~ **inversor de marcha** *m* ELEC direction commutator, *de herramienta de máquina* direction switch; ~ **inversor de la polaridad** *m* ING ELÉC polarity-reversing switch;

~ l ~ **lento normalmente cerrado** *m* ELEC slow-break switch; ~ **de levas** *m* ELEC cam switch; ~ **de líneas** *m* ELEC line commutator; ~ **local** *m* TELECOM local switch;

~ m ~ **de mando** *m* ELEC control switch; ~ **de mando automático** *m* ELEC *relé* automatic control switch; ~ **de mando doble** *m* FERRO *equipo inamovible* dual-control switch; ~ **manual** *m* ELEC hand-operated switch; ~ **manual de monocordio** *m* TELECOM single-cord switchboard; ~ **de márgenes** *m* ELEC, ING ELÉC range switch; ~ **de memoria** *m* PROD memory store switch; ~ **de mensajes** *m* TELECOM message switch; ~ **de mercurio** *m* ING ELÉC mercury interrupter; ~ **metropolitano** *m* TELECOM metropolitan switch; ~ **mezclador de video** *AmL*, ~ **mezclador de vídeo** *Esp* *m* TELECOM video switch; ~ **monofásico** *m* ING ELÉC single-pole switch; ~ **monopolar** *m* ELEC single-pole switch; ~ **montado en el piso** *m* ELEC *ascensor, elevador* floor switch; ~ **móvil aislado con fibra** *m* TELECOM moving fiber switch (*AmE*), moving fibre switch (*BrE*); ~ **múltiple** *m* ELEC multiple switch, gang switch, ING ELÉC, PROD gang switch, TELECOM multiple switchboard; ~ **de múltiples contactos** *m* ELEC multiple contact switch;

~ o ~ **de octetos** *m* TELECOM byte switch; ~ **oficina central** *m* TELECOM central-office switch; ~ **operado con teclas** *m* ELEC key-operated switch; ~ **óptico** *m* TELECOM optical switch; ~ **óptico integrado** *m* TELECOM integrated optical switch; ~ **óptico mecánico** *m* *Esp* (*cf dispositivo de conmutación óptico mecánico*

AmL) ÓPT, TELECOM mechanical optical switch; ~ **oscilante** *m* ING ELÉC tumbler switch;

~ p ~ **de palanca** *m* ELEC lever commutator switch; ~ **de palanca acodillada** *m* ELEC toggle switch; ~ **de palanca de conexión y desconexión** *m* ELEC lever on-off switch; ~ **de paquete** *m* TELECOM packet switch (*PS*); ~ **de paquetes de voz y datos** *m* TELECOM voice data packet switch; ~ **por pasos** *m* ELEC step switch; ~ **de pedal** *m* INSTR foot switch; ~ **de pie** *m* AUTO foot dimmer (*AmE*), foot dipswitch (*BrE*); ~ **de piso** *m* ELEC *ascensor, elevador* floor switch; ~ **de un polo** *m* ELEC single-pole switch; ~ **de un polo y dos vías** *m* ELEC single-pole double-throw switch; ~ **de posición** *m* ELEC position switch; ~ **previo de ruptura** *m* TV break-before-make switch; ~ **principal** *m* TELECOM main switch; ~ **de proximidad** *m* ELEC, ING ELÉC, PROD proximity switch; ~ **de prueba** *m* ELEC, ING ELÉC check switch; ~ **de puesta a tierra** *m* ELEC, ING ELÉC earth switch (*BrE*), earthed switch (*BrE*), earthing switch (*BrE*), ground switch (*AmE*), grounded switch (*AmE*), grounding switch (*AmE*); ~ **pulsador** *m* ING ELÉC, ING MECÁ, TELECOM push-button; ~ **de pulsador de varilla** *m* PROD push-rod switch;

~ q ~ **que no produce llama** *m* ELEC *seguridad* fireproof switch, flameproof switch;

~ r ~ **rápido** *m* ELEC quick throwover switch; ~ **rápido de circuitos** *m* TELECOM fast circuit switch; ~ **rápido de dos direcciones** *m* ELEC quick throwover switch; ~ **rápido de dos vías** *m* ELEC quick throwover switch; ~ **rápido de paquetes** *m* TELECOM fast packet switch (*FPS*); ~ **de la RDSI** *m* TELECOM ISDN switch; ~ **reductor** *m* ELEC *de graduación de la luz*, ING ELÉC dimmer switch; ~ **de regulación automática** *m* ELEC *relé* automatic control switch; ~ **relé** *m* CINEMAT relay switch; ~ **de relé de láminas** *m* ELEC, ING ELÉC, TELECOM reed-relay switch; ~ **de retardo** *m* ELEC delay switch; ~ **de retardo de fase** *m* ELECTRÓN phase-delay keying; ~ **de retención en reposo** *m* ELEC lock-down switch; ~ **de retorno elástico** *m* ELEC spring return switch; ~ **de retorno por muelle** *m* ELEC spring return switch; ~ **reversible** *m* ING ELÉC reversible switch; ~ **rotativo** *m* CINEMAT, ELEC, ING ELÉC, PROD, TELECOM rotary switch; ~ **rotatorio** *m* CINEMAT, ELEC, ING ELÉC, PROD, TELECOM rotary switch; ~ **de ruptura** *m* PROD break contact; ~ **rural** *m* TELECOM rural switch;

~ s ~ **de sectores** *m* ING ELÉC wafer switch; ~ **de sectores rotatorio** *m* ELEC, ING ELÉC rotary wafer switch; ~ **secundario-n** *m* ELEC n-way switch; ~ **de seguridad** *m* ELEC safety switch; ~ **de selección de canales** *m* TELECOM, TV channel-selector switch; ~ **selectivo** *m* ELEC selective switch; ~ **selector** *m* ELEC multiple contact switch, selector switch, tap selector, tap switch, ING ELÉC, PROD selector switch; ~ **selector activado por teclado (o botón)** *m* PROD key-operated selector switch; ~ **selector para bascular** *m* ELEC *automotor* dip selector switch, dip switch; ~ **selector de modo** *m* PROD mode-setting knob; ~ **selector multiposicional** *m* ING ELÉC multi-position switch; ~ **selector rotativo** *m* ELECTRÓN rotary selector switch; ~ **de semiconductor** *m* ELEC semiconductor switching device; ~ **para semiconductor** *m* ING ELÉC semiconductor switch; ~ **en serie** *m* ELEC series switch; ~ **serie-paralelo** *m* ING ELÉC series-parallel switch; ~ **de simple polo**

doble tiro *m* ELEC single-pole double-throw switch; ~ **de sobrecorriente** *m* ELEC excess-current switch, overcurrent switch;

~ t ~ **de tambor** *m* ELEC drum switch; ~ **de tambor inversor** *m* ELEC reversing-drum switch; ~ **telefónico** *m* ING ELÉC telephone switch, TELECOM telephone switchboard, telephone switch; ~ **térmico** *m* ING ELÉC thermal switch; ~ **de tiempo** *m* ELEC, TELECOM time switch; ~ **de tomas** *m* ELEC tap changer, tap switch, ING ELÉC tap changer; ~ **de tomas de carga** *m* ELEC load tap changer; ~ **totalmente aislado** *m* ELEC all-insulated switch; ~ **TR** *m* TRANSP MAR *radar* TR cell; ~ **de tres direcciones** *m* ELEC three-point switch, three-way switch; ~ **de tres posiciones** *m* ELEC three-position switch;

~ u ~ **UB** *m* ING ELÉC DPST switch; ~ **ultrarrápido** *m* ING ELÉC snap-action switch; ~ **del umbral anterior** *m* TV front-porch switch; ~ **unipolar** *m* ELEC single-pole switch, ING ELÉC single pole double-throw switch (*SPDT switch*); ~ **unipolar bidireccional** *m* (*CUB*) ING ELÉC double-pole single-throw switch (*DPSTS*); ~ **universal** *m* TELECOM universal switch;

~ v ~ **al vacío** *m* ELEC *cortocircuito* vacuum switch; ~ **de vacío** *m* ELEC *cortocircuito* vacuum switch; ~ **velociselector** *m* PROD speed switch; ~ **de video** *AmL*, ~ **de vídeo** *Esp m* TELECOM video switch; ~ **de voz-datos integrados** *m* TELECOM integrated voice-data switch

conmutar *vt* INFORM&PD switch, ING ELÉC switch in, TELECOM switch

conmutativo *adj* ELEC, ELECTRÓN, MATEMÁT commutative

conmutatriz: ~ **síncrona** *f* ING ELÉC synchronous converter

cono *m* C&V, ELECTRÓN, GEOFÍS, GEOM, ING MECÁ, MATEMÁT cone, MECÁ taper, TEC ESP, TEXTIL, TV cone; ~ **abisal** *m* OCEAN abyssal cone; ~ **acanalado** *m* ING MECÁ grooved cone; ~ **aluvial** *m* AGUA alluvial cone, GEOL cone delta, HIDROL alluvial cone; ~ **de aproximación** *m* TRANSP AÉR approach funnel; ~ **de asentamiento** *m* CONST *ensayos, hormigón* slump cone; ~ **de avance** *m* MINAS *pozos de minas* sink; ~ **de sobre cabeza** *m* ING MECÁ cone for overhead motion; ~ **de carga** *m* D&A *artillería* warhead, TEC ESP *artillería* warhead, *proyectiles arrojadizos* ablating cone; ~ **de choque** *m* D&A nose; ~ **de cierre** *m* C&V bell cone, PROD *horno alto* bell; ~ **circular recto** *m* GEOM right-circular cone; ~ **de confusión** *m* TRANSP AÉR confusion cone; ~ **de contacto** *m* ING MECÁ pitch cone; ~ **de decantación** *m* Esp (*cf embudo de decantación AmL*) MINAS *química* separator; ~ **de depresión** *m* AGUA depression cone; ~ **de depresión de bombeo** *m* AGUA pumping depression cone; ~ **de dispersión** *m* PROC QUÍ dispersion cone; ~ **del eje conducido** *m* ING MECÁ countershaft cone; ~ **y embudo** *m* PROD *altos hornos* cup and cone; ~ **escaleno** *m* GEOM scalene cone; ~ **de escape** *m* TRANSP AÉR exhaust cone; ~ **de escorias** *m* GEOL *volcánicas* cinder cone; ~ **eyectable de la ojiva** *m* TEC ESP *vehículos espaciales, misiles* ejectable nose cone; ~ **de fractura** *m* C&V fracture cone; ~ **de hélice** *m* TRANSP AÉR spinner; ~ **hidráulico** *m* AGUA hydraucone; ~ **macho** *m* ING MECÁ male cone; ~ **de morro** *m* TRANSP AÉR nose cone; ~ **del morro** *m* TEC ESP nose cone; ~ **de morro desmontable** *m* TRANSP AÉR

detachable nose cone; ~ **Morse** *m* ING MECÁ *torno* Morse taper; ~ **oblicuo** *m* GEOM oblique cone; ~ **de la ojiva** *m* TEC ESP nose cone; ~ **primitivo** *m* ING MECÁ pitch cone; ~ **de revolución** *m* GEOM cone of revolution; ~ **de salida** *m* FÍS exit cone; ~ **de sedimentación** *m* PROC QUÍ settling cone; ~ **de sedimentación de Imhoff** *m* LAB *material de vidrio* Imhoff sedimentation cone; ~ **Seger** *m* C&V Seger cone; ~ **para señalización de tráfico** *m* CONST traffic cone; ~ **señalizador de carreteras** *m* AUTO, VEH road marker cone; ~ **de silencio** *m* ING MECÁ cone of silence; ~ **sincronizador** *m* ING MECÁ synchronizing cone; ~ **de Taylor** *m* FÍS RAD Taylor cone; ~ **para transmisión alta** *m* ING MECÁ cone for overhead motion; ~ **truncado** *m* GEOM truncated cone; ~ **de viento** *m* TRANSP AÉR wind cone, wind sock

conocimiento: ~ **de embarque** *m* TEC PETR *transporte*, TRANSP MAR *documentación* bill of lading; ~ **de embarque limpio** *m* TRANSP MAR clean bill of lading; ~ **de embarque sucio** *m* TRANSP MAR claused bill of lading

consanguinización *f* AGRIC inbreeding

consecuencia *f* CARBÓN consistency

consecutivo *adj* INFORM&PD, TELECOM consecutive

Consejo: ~ **Europeo de Investigación Nuclear** *m* (*JET*) FÍS RAD, NUCL Joint European Torus (*JET*)

conserva *f* ALIMENT preserve; ~ **de fruta** *f* AGRIC preserve

conservabilidad *f* CALIDAD, MECÁ, PROD, TEC ESP maintainability

conservación *f* MECÁ maintenance, ÓPT, PETROL, TEC PETR, TERMO conservation, TEXTIL *del tejido o de las prendas* care; ~ **de agua** *f* HIDROL water conservation; ~ **del brillo** *f* FÍS, ÓPT, TELECOM conservation of brightness; ~ **de la calidad** *f* REFRIG *de los alimentos* keeping quality; ~ **de la cantidad de movimiento lineal** *f* FÍS conservation of momentum; ~ **de la carga** *f* FÍS conservation of charge; ~ **de energía** *f* TERMO energy conservation; ~ **de la energía** *f* FÍS conservation of energy; ~ **en frío** *f* ALIMENT, REFRIG, SEG, TERMO cold storage; ~ **de la masa** *f* FÍS conservation of mass; ~ **del oído** *f* SEG hearing conservation; ~ **del paisaje** *f* AGRIC landscape conservation; ~ **de la paridad** *f* FÍS conservation of parity; ~ **de la presión** *f* PETROL pressure maintenance; ~ **de la radiancia** *f* FÍS, ÓPT, TELECOM conservation of radiance; ~ **en salmuera** *f* OCEAN *del pescado* brine pickling

conservacionista *adj* AGRIC conservationist

conservador: ~ **de aceite** *m* ING ELÉC *transformadores* oil conservator; ~ **de alimento** *m* ALIMENT food preservative

conservante *m* AGRIC, EMB, PROD preservative; ~ **para madera** *m* CONST wood preservative

conservar *vt* ALIMENT preserve

consideración: ~ **de aplicación** *f* PROD application consideration

consignación: ~ **de buques** *f* TRANSP MAR shipping agency

consignador *m* PROD consignor

consignataria: ~ **de buques** *f* TRANSP MAR shipping agent

consignatario *m* PROD consignor; ~ **de buques** *m* TRANSP MAR shipping agent

consistencia[1] *f* ALIMENT *textura de líquido y semi líquido* consistency, CARBÓN *propiedad para resistir a* *la deformación* compatibility, consistency, HIDROL *del mineral* body, INFORM&PD, PAPEL consistency, TRANSP AÉR solidity; **~ del colorante** *f* COLOR coloring body (*AmE*), colouring body (*BrE*); **~ del suelo** *f* AGRIC tilth

consistencia²: con~ metálica *fra* PROD metal -bodied

consola *f* ELEC control, INFORM&PD console, ING MECÁ rack, TRANSP AÉR console, TRANSP MAR *construcción naval* gusset, bracket; **~ de bao** *f* TRANSP MAR *construcción* beam knee; **~ de bucle conmutado** *f* TELECOM switched-loop console; **~ con una clave por línea principal** *f* TELECOM key-per-line console (*BrE*), key-per-trunk console (*AmE*); **~ de control** *f* ELEC control console, TRANSP MAR *equipo eléctrico* console, TV control console; **~ de escuadra** *f* ING MECÁ knee; **~ en escuadra** *f* ING MECÁ, MECÁ angle bracket; **~ de iluminación** *f* CINEMAT lighting console; **~ del montante** *f* PROD bearer bracket; **~ para el montaje** *f* PROD mounting bracket; **~ mural** *f* TELECOM wall bracket; **~ mural de la galería** *f* MINAS end-wall bracket; **~ de operador** *f* INFORM&PD operator console, TELECOM operating console, operator's console; **~ de operadores** *f* TELECOM operating console, operator's console; **~ principal** *f* INFORM&PD master console; **~ de producción** *f* TV production console; **~ de refuerzo** *f* TRANSP MAR *construcción naval* gusset; **~ de toma de sonido** *f* ACÚST sound take desk; **~ de trabajo** *f* TELECOM operating console, operator's console; **~ de visualización** *f* INFORM&PD display console

consolidación *f* CARBÓN compaction, consolidation, CONST, GEOL compaction, PETROL consolidation, TEC PETR compaction, consolidation, TRANSP consolidation; **~ de barras** *f* NUCL rod consolidation; **~ de laboratorio** *f* CARBÓN laboratory compaction; **~ y solidificación química** *f* MINAS chemical grouting and freezing

consolidante *m* GEOL binder

consolidar *vt* MINAS, PROD *terrenos* tamp

consonancia *f* ACÚST consonance

constantan *m* ELEC *termopar* constantan

constante¹ *adj* ELECTRÓN, FÍS, INFORM&PD, MATEMÁT, MECÁ constant

constante² *f* ELECTRÓN, FÍS constant, GEOM invariant, INFORM&PD, MATEMÁT, MECÁ constant; **~ de acoplamiento** *f* FÍS, NUCL *de interacción* coupling constant; **~ de apantallamiento** *f* NUCL screening constant, screening number; **~ arbitraria** *f* FÍS arbitrary constant; **~ de atenuación** *f* ELECTRÓN, FÍS attenuation constant; **~ de atenuación acústica** *f* ELECTRÓN acoustic attenuation constant; **~ de Boltzmann** *f* FÍS, TEC ESP Boltzmann's constant; **~ calorífica** *f* TERMO heat constant; **~ de Curie** *f* FÍS, FÍS RAD Curie constant; **~ de desintegración** *f* FÍS, FÍS RAD, GEOL, NUCL decay constant; **~ dieléctrica** *f* ELEC *capacitor* dielectric constant, permittivity, FÍS, ING ELÉC dielectric constant, permittivity, P&C *propiedad física, prueba* dielectric constant, TEC ESP, TELECOM permittivity; **~ dieléctrica absoluta** *f* ELEC *campo eléctrico* absolute permittivity; **~ dieléctrica relativa** *f* ELEC *electromagnetismo*, FÍS, ING ELÉC relative permittivity; **~ de Dirac** *f* FÍS Dirac constant; **~ de disociación** *f* QUÍMICA dissociation constant; **~ de dosis** *f* FÍS RAD dose constant; **~ elástica** *f* METAL elastic constant; **~ de elasticidad** *f* FÍS FLUID elastic

constant; **~ eléctrica** *f* FÍS, ING ELÉC electric constant; **~ de equilibrio** *f* QUÍMICA *de reacción* equilibrium constant; **~ de estructura fina** *f* FÍS, NUCL fine-structure constant; **~ de Faraday** *f* FÍS Faraday constant; **~ de fase** *f* ACÚST *de onda plana progresiva*, ELECTRÓN, FÍS, ÓPT, TELECOM, TV phase constant; **~ figurada** *f* INFORM&PD figurative constant; **~ de los gases perfectos** *f* FÍS gas constant; **~ de gravedad específica y viscosidad** *f* TERMO viscosity-gravity constant; **~ de gravitación** *f* FÍS, GEOFÍS, GEOL gravitational constant, TEC ESP gravitation constant; **~ gravitacional** *f* FÍS, GEOFÍS, GEOL gravitational constant; **~ de Hubble** *f* TEC ESP Hubble's constant; **~ de ionización** *f* QUÍMICA dissociation constant; **~ de Josephson** *f* FÍS Josephson constant; **~ de larga duración** *f* ELECTRÓN *transistores* longtime constant; **~ de Lorenz** *f* FÍS Lorenz constant; **~ magnética** *f* FÍS, ING ELÉC magnetic constant; **~ molar de los gases** *f* FÍS molar gas constant; **~ nominal** *f* METAL rate constant; **~ de Planck** *f* (*h*) FÍS Planck's constant (*h*); **~ de propagación** *f* FÍS, ÓPT, TELECOM propagation constant; **~ de propagación acústica** *f* ACÚST acoustic propagation constant; **~ de radiación** *f* TERMOTEC radiation constant; **~ de red** *f* METAL crystal-lattice parameter, lattice constant; **~ de una red** *f* ING ELÉC network constant; **~ reducida de Planck** *f* FÍS h-bar; **~ de retículo** *f* METAL crystal-lattice parameter, lattice constant; **~ de Rydberg** *f* FÍS Rydberg constant; **~ solar** *f* ENERG RENOV, GEOFÍS, TEC ESP solar constant; **~ de Stefan** *f* TERMOTEC Stefan's constant; **~ de Stefan-Boltzman** *f* FÍS Stefan-Boltzmann constant; **~ de tiempo** *f* ELEC *circuito de CA*, FÍS time constant; **~ de tiempo de la descarga automática** *f* ING ELÉC self-discharge time constant; **~ torsional** *f* FÍS torsional constant

constantes: ~ críticas *f pl* QUÍMICA critical constants

constatación *f* ING MECÁ testing, verification

constitución: ~ aislante *f* ING ELÉC insulating compound; **~ dominante** *f* METAL *cristalografía* habit

constringencia *f* FÍS constringence

construcción¹: de ~ fuerte *adj* ING MECÁ heavy-duty; **de ~ a tope** *adj* TRANSP MAR *embarcación* carvel-built

construcción² *f* CONST assembly, ING MECÁ frame, OCEAN *de la red* netting, PROD *revestimiento del horno alto* building, stuffing; **~ de arriba hacia abajo** *f* CONST top-down construction; **~ de ataguías** *f* AGUA, CONST, HIDROL coffering; **~ de calderas** *f* INSTAL HIDRÁUL boilermaking; **~ de carreteras** *f* AUTO, CONST, TRANSP, VEH road building, road making; **~ sobre cubierta en el costado** *f* TRANSP MAR *de buques especiales* side construction; **~ descendente** *f* CONST top-down construction; **~ flexible** *f* CONST *carreteras* flexible construction; **~ de máquinas** *f* ING MECÁ modular machine tool construction; **~ naval** *f* TRANSP MAR shipbuilding; **~ pesada** *f* CONST heavy engineering; **~ de puentes de vigas en voladizo** *f* CONST cantilever; **~ del tejido** *f* TEXTIL fabric construction

construcciones: ~ de cadenas *f pl* FÍS RAD string constructions; **~ en secuencia ordenada** *f pl* FÍS RAD string constructions

constructor *m* CALIDAD contractor, CONST builder, contractor, house builder, master builder MECÁ erector, PROD maker; **~ de botes** *m* TRANSP MAR boatbuilder, shipwright; **~ de buques** *m* TRANSP

MAR shipbuilder; **~ de máquinas** *m* ING MECÁ, PROD engine maker; **~ de motores** *m* ING MECÁ engine builder; **~ naval** *m* TRANSP MAR shipbuilder; **~ de tejados** *m* CONST roofer

construir[1] *vt* CONST build, make, MINAS *pozo* lay out, PROD *herramienta* make, TRANSP MAR install

construir[2]: **~ un puente sobre un río** *vi* CONST throw a bridge over a river

consulta *f* INFORM&PD look-up, enquiry, *datos* query, inquiry; **~ de la base de datos** *f* INFORM&PD database query; **~ de los datos** *f* INFORM&PD data query; **~ en tabla** *f* INFORM&PD table lookup; **~ de tablas** *f* INFORM&PD table lookup; **~ del usuario** *f* INFORM&PD user query

consultar[1] *vt* INFORM&PD query

consultar[2]: **~ duración e importe** *vi* TELECOM *telefonía* advise duration and charge

consumible *adj* TEC ESP *vehículos espaciales* expendable

consumibles: **~ de imprenta** *m pl* IMPR printer's supply

consumidor *m* CALIDAD, ELEC, EMB, GAS, TEXTIL consumer; **~ de aire** *m* PROD air breather; **~ casero** *m* ELEC domestic consumer; **~ doméstico** *m* ELEC domestic consumer

consumir[1] *vt* ING MECÁ wear

consumir[2]: **~ preferentemente antes de** *fra* EMB best before

consumirse *v refl* ALIMENT, TERMO boil away

consumo *m* AGUA, AUTO, HIDROL, MECÁ, VEH *de carburante* consumption; **~ de calor** *m* TERMO heat consumption, heat input; **~ de calor sin carga** *m* C&V no-load heat consumption; **~ calorífico** *m* TERMO heat input, heat rate; **~ del circuito** *m* ELEC circuit power requirement; **~ de combustible** *m* TERMO, TRANSP AÉR fuel consumption; **~ diario** *m* AGUA daily consumption; **~ de energía** *m* ELEC *red de alimentación* power consumption, FÍS energy consumption, ING ELÉC power consumption, drain, TERMO energy consumption; **~ de energía dinámica** *m* ING ELÉC dynamic power consumption; **~ específico diario** *m* HIDROL daily specific consumption; **~ de gasolina** *m* Esp (*cf consumo de nafta AmL*) AUTO, VEH gas consumption (*AmE*), gasoline consumption (*AmE*), petrol consumption (*BrE*); **~ por habitante** *m* AGUA per-capita consumption; **~ máximo** *m* ELEC *fuente de alimentación* maximum demand, HIDROL *de agua* maximum consumption, ING ELÉC *electricidad* maximum demand; **~ mínimo** *m* HIDROL *de agua* minimum consumption; **~ de nafta** *m* AmL (*cf consumo de gasolina Esp*) AUTO, VEH gas consumption (*AmE*), gasoline consumption (*AmE*), petrol consumption (*BrE*)

contabilidad: **~ de tareas** *f* INFORM&PD job accounting

contabilización *f* PROD posting; **~ de productos** *f* PROD product accounting

contabilizar[1]: **~ como existencias** *vt* PROD post to stock

contabilizar[2]: **~ horas** *vi* PROD post hours

contacto *m* AUTO contact, ELEC *devanado* tap, *circuito* connection, *conmutador, relé* contact, GEOL, INFORM&PD contact, ING ELÉC contact, connection, ING MECÁ make, P&C, PROD, TRANSP AÉR contact; **~ abierto** *m* ELEC *relé* break contact, open contact; **~ de acoplamiento bifurcado** *m* ELEC, PROD bifur-cated mating-contact; **~ de apertura** *m* ING ELÉC break contact; **~ auxiliar** *m* ING ELÉC, PROD auxiliary contact; **~ de banco** *m* TELECOM bank contact; **~ por barrido lateral** *m* ING ELÉC side-wipe contact; **~ de base** *m* ING ELÉC base contact; **~ bidireccional** *m* ING ELÉC double-throw contact; **~ bifilar** *m* TELECOM two-wire crosspoint; **~ bifurcado de acoplamiento** *m* ELEC, PROD bifurcated mating-contact; **~ bimetálico** *m* ING ELÉC bimetallic contact; **~ de blocaje** *m* ING ELÉC interlock contact; **~ bobinado** *m* ING ELÉC reeled contact; **~ de cabeza con cinta** *m* TV head-to-tape contact; **~ de carbón** *m* ELEC, ING ELÉC, MECÁ carbon contact; **~ del casquillo** *m* ING ELÉC base contact; **~ de cierre** *m* ELEC *relé* closing contact; **~ de cierre inmediato normalmente abierto** *m* PROD normally-open early-make contact (*NO early-make contact*); **~ de clavija** *m* ING ELÉC socket contact; **~ complementario bifurcado** *m* ELEC, PROD bifurcated mating-contact; **~ de conmutación óptica** *m* TELECOM optical-switching crosspoint; **~ corredizo de reóstato** *m* ELEC *resistencia*, ING ELÉC rheostat slider; **~ de cortocircuitar** *m* ING ELÉC shorting contact; **~ de cruce** *m* TELECOM crosspoint; **~ de cruce de banda ancha** *m* TELECOM broadband crosspoint; **~ de cruce bifilar doble** *m* TELECOM four-wire crosspoint; **~ de cruce CMOS** *m* TELECOM CMOS crosspoint; **~ de cruce de diodos** *m* TELECOM diode crosspoint; **~ de desconexión** *m* ING ELÉC break contact; **~ por deslizamiento** *m* FÍS sliding contact; **~ deslizante** *m* ING MECÁ wiper; **~ deslizante del reóstato** *m* AUTO rheostat-sliding contact; **~ del disparador** *m* ING MECÁ trigger contact; **~ disparador** *m* AUTO trigger contact; **~ de disparo** *m* PROD trip contact; **~ de dos conductores** *m* TELECOM two-wire crosspoint; **~ de dos hilos** *m* TELECOM two-wire crosspoint; **~ duro** *m* ELEC hard contact; **~ eléctrico** *m* ELEC *relé*, ING ELÉC electric contact, electrical contact; **~ eléctrico de característica óhmica** *m* ING ELÉC ohmic contact; **~ emisor** *m* ING ELÉC emitter contact; **~ fijo** *m* ELEC *relé*, ING ELÉC fixed contact, PROD stationary contact; **~ para flash** *m* FOTO flash contact; **~ gas-agua** *m* PETROL gas-water contact; **~ gas-petróleo** *m* PETROL, TEC PETR gas-oil contact (*AmE*), gasoline-oil contact (*AmE*), petrol-oil contact (*BrE*); **~ del generador** *m* ING ELÉC source contact; **~ para grandes amperajes** *m* ING ELÉC heavy-duty contact; **~ hembra** *m* ELEC, ING ELÉC female contact; **~ hermafrodita** *m* ING ELÉC hermaphroditic contact; **~ hidrófobo** *m* PROD hard contact; **~ hombre-máquina** *m* TEC ESP man-machine interface; **~ incierto** *m* ELEC *relé* contact bounce; **~ del interruptor bloqueado** *m* PROD sealed switch contact; **~ de inversión** *m* ING ELÉC, PROD reverse contact; **~ laminar** *m* ELEC *máquina*, ING ELÉC laminated brush; **~ de láminas** *m* ELEC *interruptores, relés*, TELECOM reed contact; **~ de lengüeta** *m* TELECOM remreed crosspoint; **~ de lengüetas** *m* ELEC, PROD, TELECOM reed contact; **~ lógico de láminas** *m* PROD logic reed contact; **~ macho-hembra** *m* ING ELÉC hermaphroditic contact; **~ movible bifurcado** *m* PROD bifurcated movable-contact; **~ móvil** *m* ELEC moveable contact, ING ELÉC moving contact, ING MECÁ wiper, PROD moveable contact; **~ normalmente abierto** *m* CONST, ELEC, PROD NO contact, normally-open contact; **~ nor-**

malmente cerrado *m* CONST, ELEC, PROD NC contact, normally-closed contact; ~ **óhmico** *m* ENERG RENOV, FÍS, ING ELÉC ohmic contact; ~ **optoelectrónico** *m* TELECOM optoelectronic crosspoint; ~ **de pérdida de energía** *m* ING ELÉC drain contact; ~ **petróleo-agua** *m* PETROL oil-water contact; ~ **en plata** *m* ING ELÉC silver contact; ~ **plateado** *m* ING ELÉC silver-plated contact; ~ **de presión** *m* ING MECÁ plunger; ~ **para puentear** *m* ING ELÉC shorting contact; ~ **de puerta** *m* ING ELÉC gate contact; ~ **punta** *m* ING ELÉC *electrodo de soldar* point contact; ~ **quemado** *m* VEH *sistema de encendido* burnt contact; ~ **por radar** *m* TRANSP radar contact; ~ **por radio** *m* TRANSP AÉR radio link; ~ **de recubrimiento** *m* PROD overlap contact; ~ **de relé** *m* ELEC, ING ELÉC, PROD relay contact; ~ **del relé de láminas** *m* ELEC, ING ELÉC, TELECOM reed-relay crosspoint; ~ **de reposo** *m* ELEC *relé* back contact, break contact, resting contact; ~ **de reposo-reposo** *m* ELEC break-break contact; ~ **reposo-trabajo** *m* ING ELÉC make-and-break contact; ~ **de ruptura** *m* ING ELÉC break contact; ~ **de ruptura final** *m* PROD late-break contact; ~ **de ruptura retardada normalmente abierto** *m* PROD NO late-break contact, normally-open late-break contact; ~ **de semiconductores** *m* TELECOM semiconductor crosspoint; ~ **de soplado magnético de arco** *m* ING MECÁ magnetic arc blowout contact; ~ **de la toma** *m* ING ELÉC collector contact; ~ **a tope** *m* ELEC *de relé* butt contact; ~ **de tornillo** *m* ELEC screw contact; ~ **de trabajo** *m* ELEC *relé*, ING ELÉC make contact; ~ **de trabajo-trabajo** *m* ELEC *relé* make-make contact; ~ **de transición** *m* ELEC *de transformador* transition contact; ~ **unipolar** *m* ING ELÉC single-break contact
contactor *m* ELEC *relé*, ING ELÉC contactor; ~ **de CA** *m* PROD AC contactor; ~**-disyuntor de potencia activa** *m* ELEC active power relay; ~ **de inversión** *m* ING ELÉC, PROD reverse contactor; ~ **mecánico** *m* ELEC *de relé* mechanical contactor; ~ **de posición** *m* ELEC position switch; ~ **de precarga** *m* PROD precharge contactor; ~ **con relé de máxima** *m* ELEC, ING ELÉC, PROD contactor with overload relay; ~ **con relé de sobrecarga** *m* ELEC, ING ELÉC, PROD contactor with overload relay; ~ **en vacío** *m* PROD vacuum contactor; ~ **en vacío de tensión media** *m* PROD medium-voltage vacuum contactor
contactos: ~ **acoplados** *m pl* ING ELÉC mated contacts; ~ **adhesivos** *m pl* ELEC *relé* sticking contacts; ~ **apilados** *m pl* ING ELÉC pile-up; ~ **herméticos** *m pl* ING ELÉC sealed contacts; ~ **impregnados en mercurio** *m pl* ING ELÉC mercury-wetted contacts; ~ **del interruptor general** *m pl* ELEC main switching contacts; ~ **del interruptor principal** *m pl* ELEC main switching contacts; ~ **de láminas** *m pl* ING ELÉC *de interruptor* bridging contacts; ~ **en paralelo** *m pl* ING ELÉC bridging contacts; ~ **principales** *m pl* ELEC main contacts; ~ **que se pegan** *m pl* ELEC *relé* sticking contacts; ~ **del ruptor** *m pl* VEH *del encendido* breaker contact; ~ **selectivamente revestidos** *m pl* ING ELÉC selectively-plated contacts; ~ **no simultáneos** *m pl* ING ELÉC nonbridging contacts; ~ **de soplado magnético** *m pl* ING MECÁ magnetic blowout contacts; ~ **de trabajo** *m pl* PROD operative contacts
contador *m* ELEC *instrumento* meter, ELECTRÓN counting, counter, EMB counting device, INFORM&PD

counter, ING MECÁ recorder, counter, meter, telltale, register, TELECOM register, counter; ~ **de abonado** *m* TELECOM subscriber's meter; ~ **de acumulación** *m* ELECTRÓN adding counter; ~ **acumulador** *m* ELECTRÓN accumulating counter; ~ **de agua** *m* AGUA, TEXTIL water meter; ~ **de aire** *m* ING MECÁ air meter; ~ **analógico** *m* ING ELÉC analog meter; ~ **en anillo** *m* TV ring counter; ~ **arriba-abajo** *m* ELECTRÓN increment-decrement counter; ~ **ascendente** *m* ELECTRÓN up counter; ~ **ascendente-descendente** *m* ELECTRÓN up-down counter; ~ **de aumento-disminución** *m* ELECTRÓN increment-decrement counter; ~ **de avance y retroceso** *m* TV forward-backward counter; ~ **bidireccional** *m* ELECTRÓN bidirectional counter; ~ **binario** *m* ELECTRÓN binary counter, binary scaler, INFORM&PD binary counter; ~ **de carreras** *m* ING MECÁ stroke counter; ~ **de centelleo** *m* FÍS scintillation counter, FÍS ONDAS radiation counter, FÍS PART, FÍS RAD scintillation counter; ~ **Cerenkov** *m* FÍS PART, FÍS RAD *de partículas* Cerenkov counter; ~ **de Cerenkov diferencial** *m* FÍS PART *de partículas* differential Cerenkov counter; ~ **de chispas** *m* FÍS RAD spark counter; ~ **de cinta** *m* TV tape counter; ~ **conectable en cascada** *m* INFORM&PD cascadable counter; ~ **de corrientes** *m* ENERG RENOV, INSTAL HIDRÁUL current meter; ~ **de cuenta atrás** *m* ELECTRÓN, TEC ESP countdown counter; ~ **de décadas** *m* ELECTRÓN decade scaler; ~ **decreciente** *m* PROD down counter; ~ **de demanda máxima** *m* ING ELÉC demand meter; ~ **descendente** *m* ELECTRÓN down counter; ~ **de destello** *m* FÍS PART scintillation counter, FÍS RAD ionization counter; ~ **de destellos** *m* FÍS ONDAS radiation counter; ~ **directo** *m* PROD forward contactor; ~ **de dos etapas** *m* TELECOM two-state register; ~ **de electricidad** *m* ELEC electric meter, electricity meter, ING ELÉC electricity meter; ~ **eléctrico** *m* ELEC electric meter, electricity meter, ING ELÉC electricity meter; ~ **electrónico** *m* ELECTRÓN electronic counter, electronic counting; ~ **electrónico de décadas** *m* ELECTRÓN decade scaler; ~ **de energía activa** *m* ELEC active energy meter; ~ **de energía aparente** *m* ELEC apparent energy meter, apparent-power meter; ~ **de exceso de energía** *m* ELEC excess energy meter; ~ **de exposiciones** *m* CINEMAT, FOTO exposure counter; ~ **de fotogramas** *m* *Esp* (*cf cuentafotogramas AmL*) CINEMAT, FOTO frame counter; ~ **de frecuencia** *m* ELECTRÓN, INSTR, TELECOM frequency counter; ~ **de funcionamiento** *m* ELEC *cambiador de toma* operation counter; ~ **de gas** *m* LAB *volúmenes*, PROD, TERMO gas meter; ~ **Geiger** *m* FÍS, FÍS PART *de radiaciones iónicas*, FÍS RAD, NUCL Geiger counter; ~ **de hilos** *m* IMPR linen counter; ~ **de horas trabajadas** *m* SEG working hours counter; ~ **de impulsos** *m* ELEC *instrumento* impulse counter, ELECTRÓN count-up counter, pulse counter, INSTR impulse counter; ~ **indicador de volumen** *m* TV volume indicator meter (*VI meter*); ~ **de instrucciones** *m* INFORM&PD program counter; ~ **de intervalos** *m* MECÁ elapsed-time counter; ~ **de ionización** *m* FÍS RAD ionization counter; ~ **irreversible** *m* PROD up counter; ~ **de metraje** *m* CINEMAT, TV film-footage counter; ~ **de núcleos de condensación** *m* CONTAM, METEO condensation nucleus counter; ~ **de nueve dígitos** *m* EMB nine-digit counter; ~ **de órbitas** *m* TEC ESP orbit counter; ~ **de pasos** *m* INFORM&PD step counter; ~ **de polvos** *m* INSTR dust counter; ~ **de potencia activa** *m* ELEC *instrumento* active power meter; ~ **privado de abonado** *m* TELECOM subscriber's private meter; ~ **proporcional** *m* FÍS proportional counter; ~ **que indica la cantidad de cinta utilizada** *m* TV tape counter; ~ **de radiación** *m* FÍS RAD radiation counter; ~ **de reexpedición** *m* TELECOM redirecting counter; ~ **de relámpagos** *m* GEOFÍS lightning flash counter; ~ **de retícula** *m* IMPR screen count; ~ **reversible** *m* ELECTRÓN bidirectional counter, increment-decrement counter, up-down counter, PROD up-down counter; ~ **de semiconductores** *m* FÍS PART *de partículas* semiconductor counter; ~ **de tiempo** *m* PROD timer; ~ **tipo Campbell-Stokes** *m* ENERG RENOV Campbell-Stokes recorder; ~ **totalizador** *m* ING ELÉC integrating meter; ~ **totalizador de exceso** *m* ELEC excess energy meter; ~ **de tráfico** *m* TRANSP traffic counter; ~ **de velocidad por radar** *m* FÍS RAD, TRANSP, TRANSP MAR radar speed meter; ~ **de vueltas** *m* ING MECÁ, INSTR revolution counter

contadores: ~ **en cascada** *m pl* PROD cascading counters; ~ **de silicio** *m pl* FÍS RAD silicon counters

contaje: ~ **de ciclos** *m* PROD cycle counting

contaminación[1]: **contra la** ~ *adv* MECÁ antipollution

contaminación[2] *f* CALIDAD *alimentos* contamination, pollution, CARBÓN contamination, CONTAM pollution, contamination, GEOL, HIDROL, NUCL contamination, QUÍMICA, SEG, TEC ESP contamination, pollution, TEC PETR contamination; ~ **ácida** *f* CONTAM acid pollution; ~ **acuática** *f* CONTAM water pollution; ~ **acústica** *f* ACÚST, CONTAM, SEG noise pollution; ~ **del agua** *f* AGUA, CONTAM, HIDROL water pollution; ~ **del aire** *f* SEG air pollution; ~ **ambiental por ruido** *f* ACÚST, CONTAM environmental noise pollution; ~ **atmosférica** *f* CONTAM air pollution, atmospheric pollution; ~ **excepcional concertada** *f* NUCL emergency exposure to radioactive materials; ~ **de fondo** *f* CONTAM background pollution; ~ **por lluvia radiactiva** *f* CONTAM *en la atmósfera* contamination fallout; ~ **de la mantilla** *f* IMPR *por pelusa, partículas de estucado* blanket contamination; ~ **material** *f* CONTAM material pollution; ~ **de materiales** *f* CONTAM material pollution; ~ **por polvo radiactivo** *f* CONTAM *en la atmósfera* contamination fallout; ~ **púrpura** *f* ING ELÉC purple plague; ~ **radiactiva** *f* CONTAM, FÍS, FÍS RAD radioactive contamination, radioactive pollution, NUCL, SEG radioactive pollution; ~ **por ruido** *f* ACÚST, CONTAM, SEG noise pollution; ~ **sonora** *f* ACÚST, CONTAM, SEG noise pollution; ~ **del suelo** *f* CONTAM land pollution, *tipo de contaminación* soil pollution; ~ **térmica** *f* CONTAM thermal pollution; ~ **terrestre** *f* CONTAM land pollution; ~ **de la tierra** *f* CONTAM *tipo de contaminación* soil pollution

contaminado *adj* CONTAM polluted

contaminante[1] *adj* AGUA polluted

contaminante[2] *m* CONTAM, CONTAM MAR, SEG pollutant; ~ **acuático** *m* AGUA aquatic pollutant; ~ **del agua** *m* CONTAM water pollutant; **del aire** *m* CONTAM air pollutant; ~ **atmosférico** *m* CONTAM air pollutant, air-polluting substance, airborne pollutant; ~ **precursor** *m* CONTAM precursor pollutant; ~ **del suelo** *m* CONTAM soil pollutant; ~ **del terreno** *m* CONTAM soil pollutant; ~ **terrestre** *m* CONTAM land

pollutant; ~ **de la tierra** *m* CONTAM land pollutant, soil pollutant

contaminar *vt* ALIMENT taint, CALIDAD contaminate, pollute, CARBÓN contaminate, CONTAM contaminate, pollute, GEOL, HIDROL, NUCL, QUÍMICA contaminate, SEG contaminate, pollute, TEC ESP, TEC PETR contaminate

contar: ~ **el tiempo** *vt* PROD time

contemporáneo *adj* GEOL coeval, contemporaneous, synchronous

contención *f* INFORM&PD contention, NUCL containment, TELECOM contention; ~ **secundaria** *f* NUCL secondary containment

contenedor *m* EMB container (*CT*), RECICL bank, TELECOM container (*CT*), TRANSP box, container (*CT*); ~ **con abertura superior** *m* TRANSP container with opening top; ~ **abierto** *m* TRANSP open-wall container; ~ **abovedado** *m* TRANSP igloo container; ~ **de agua** *m* TEXTIL *recipiente* water container; ~ **aislado** *m* TRANSP insulated container; ~ **de almacenamiento** *m* NUCL storage canister; ~ **alto** *m* TRANSP high-cube container (*HC*); ~ **apilable** *m* TRANSP stackable container; ~ **con armadura** *m* TRANSP skeleton container; ~ **de bandejas** *m* TRANSP pallet container; ~ **basculante** *m* TRANSP tiltainer; ~ **blindado** *m* NUCL shielded coffin; ~ **de botellas** *m* RECICL bottle bank; ~ **caldeado** *m* TRANSP heated container; ~ **de carga plegable** *m* TRANSP collapsible freight container; ~ **de cartón** *m* EMB container board box; ~ **cerrado** *m* TRANSP closed container; ~ **de cloración** *m* HIDROL chlorination tank, chlorination vessel; ~ **criogénico de presión** *m* ING MECÁ cryogenic pressure vessel; ~ **de cristal o de vidrio** *m* EMB glass container; ~ **cubierto** *m* TRANSP covered container; ~ **cubierto con lona** *m* TRANSP tarpaulin-covered container; ~ **de descarga por gravedad** *m* TRANSP bulk container with gravity discharge; ~ **con descarga por presión** *m* TRANSP bulk container with pressure discharge; ~ **descubierto** *m* TRANSP open-top container; ~ **empalmable** *m* TRANSP joinable container; ~ **enganchable** *m* TRANSP joinable container; ~ **enrejado** *m* TRANSP lattice-sided container; ~ **entre módulos** *m* TRANSP intermodal container; ~ **de equipo desmontable** *m* TRANSP AÉR detachable pod; ~ **esférico** *m* TRANSP *vehículo* spherical container; ~ **estándar** *m* TRANSP standard container; ~ **evacuador de aceites** *m* CONTAM oil-clearance vessel; ~ **frigorífico** *m* REFRIG, TRANSP refrigerated container; ~ **de gabarras** *m* TRANSP barge container; ~ **para gases transportable** *m* SEG transportable gas container; ~ **ferroviario** *m* FERRO railroad container (*AmE*), railway container (*BrE*); ~ **para gases transportable** *m* SEG transportable gas container; ~ **de gran tamaño** *m* EMB large-size container; ~ **de inclinación** *m* TRANSP tilt container; ~ **con inclinación superior** *m* TRANSP tilt-top container; ~ **intermodal** *m* TRANSP intermodal container; ~ **de lodos** *m* RECICL sludge gulper; ~ **mixto** *m* EMB composite container; ~ **múltiple** *m* EMB multiunit container; ~ **-n** *m* TELECOM container-n (*C-n*); ~ **PA** *m* TRANSP PA container; ~ **de paletas** *m Esp* TRANSP pallet container; ~ **de pálets** *m AmL* TRANSP pallet container; ~ **de polietileno** *m* EMB polyethylene container; ~ **de poliolefina** *m* EMB polyolefin container; ~ **de presión de diafragma** *m* ING MECÁ diaphragm pressure vessel; ~ **de productos a granel semielaborados** *m* EMB intermediate bulk container; ~ **de residuos** *m* NUCL waste canister; ~ **retornable** *m* EMB, RECICL returnable container; ~ **sin retorno** *m* EMB one-way container; ~ **de ruedas fijas** *m* TRANSP container with fixed wheels; ~ **singular** *m* TRANSP odd container; ~**-tanque** *m* TRANSP tank container; ~ **terrestre** *m* TRANSP land container; ~ **para transporte** *m* TRANSP transcontainer; ~ **de transporte para un único elemento de combustible** *m* NUCL single-element shipping cask; ~ **de ultramar** *m* TRANSP overseas container; ~ **de volúmenes secos** *m* TRANSP dry-bulk container

contenedorización *f* EMB, TRANSP containerization

contener *vt* PROD hold

contenido[1]: ~ **petróleo** *adj* PETROL, TEC PETR *formación con petróleo* oil-bearing

contenido[2] *m* CARBÓN, CONST content, ELECTRÓN contention, METAL content, MINAS tenor, P&C, TELECOM, TEXTIL content; ~ **en acído** *m* EMB acid content; ~ **de ácido libre** *m* ALIMENT acid value; ~ **acuoso** *m* GAS water content; ~ **de agua** *m* AGRIC moisture content, CARBÓN, CONST, FÍS, GAS, METEO, ÓPT water content, REFRIG moisture content; ~ **en agua** *m* CARBÓN water ratio; ~ **de agua residual** *m* HIDROL, ÓPT residual-water content; ~ **alcalino** *m* DETERG alkali content; ~ **alto de azufre** *m* QUÍMICA, REVEST, TEC PETR *calidad del petróleo* high sulfur content (*AmE*), high sulphur content (*BrE*); ~ **en alúmina** *m* CARBÓN aluminium content (*BrE*), aluminum content (*AmE*); ~ **en aluminio** *m* CARBÓN alumina content, aluminium content (*BrE*), aluminum content (*AmE*); ~ **en arcilla** *m* CARBÓN clay content; ~ **armónico relativo** *m* ACÚST, ELECTRÓN relative harmonic content; ~ **de azufre** *m* TEC PETR sulfur content (*AmE*), sulphur content (*BrE*); ~ **bajo de azufre** *m* TEC PETR *calidad del petróleo* low sulfur content, low sulphur content (*BrE*); ~ **bajo de sulfuro** *m* TEC PETR low sulfur content (*AmE*), low sulphur content (*BrE*); ~ **calórico** *m* TERMO caloric content; ~ **de una carretilla** *m* CONST barrowful; ~ **de ceniza** *m* ALIMENT, PAPEL, PROD ash content; ~ **de decisión** *m* INFORM&PD decision content; ~ **de diodo** *m* MINAS gold content; ~ **de energía** *m* TERMO energy content; ~ **de fango** *m* AGUA mud content; ~ **en fibras** *m* TEXTIL fiber content (*AmE*), fibre content (*BrE*); ~ **gaseoso** *m* GAS, QUÍMICA, TERMO gas content; ~ **gris** *m* IMPR gray contents (*AmE*), grey contents (*BrE*); ~ **de humedad** *m* AGUA, ALIMENT, CARBÓN, CONST, EMB, FÍS, PAPEL, QUÍMICA, REFRIG, TEXTIL moisture content; ~ **de humedad del papel** *m* IMPR paper moisture content; ~ **húmedo** *m* EMB moisture content; ~ **informático** *m* INFORM&PD information content; ~ **en materia orgánica** *m* HIDROL *aguas residuales* organic matter content; ~ **en mineral** *m* AmL (*cf ley de mineral Esp*) MINAS *evaluación yacimiento* ore contents; ~ **no volátil** *m* P&C *propiedad física, prueba* nonvolatile content; ~ **nominal** *m* EMB, IMPR nominal content; ~ **en nutriente** *m* ALIMENT nutrient content; ~ **olefínico** *m* QUÍMICA olefinic content; ~ **óptimo de humedad** *m* CONST optimum moisture content; ~ **en oro** *m* CARBÓN gold content, MINAS *evaluación* gold grade, gold tenor; ~ **de parafina** *m* REFRIG wax content; ~ **de plata** *m* FOTO silver content; ~ **de polvos** *m* CONTAM dust content; ~ **de sal** *m* GAS salt content;

~ **de sales** *m* OCEAN salt content; ~ **salino** *m* GAS salt content; ~ **en sílice** *m* MINAS silica content; ~ **en silicio** *m* MINAS *de los minerales* silica content; ~ **sólido** *m* IMPR solid content; ~ **en sólidos** *m* PAPEL dry content, dry solids content; ~ **térmico** *m* TERMO thermal content; ~ **de zumo** *m* ALIMENT juice content

conteo: ~ **bacteriano** *m* AGUA bacterial count; ~ **irreversible** *m* PROD up count

contestador *m* TEC ESP transponder, TRANSP AÉR responder

contexto: ~ **de aplicaciones** *m* TELECOM application context; ~ **tectónico** *m* GEOL tectonic setting

contextual *adj* INFORM&PD context-sensitive

contienda *f* CARBÓN contest, INFORM&PD contention, TEC ESP pull

contiene: **que** ~ **fósiles** *adj* CARBÓN, GEOL, PETROL fossil-bearing

contiguo *adj* GEOM, INFORM&PD, OCEAN contiguous

continua: ~ **de hilar de aletas** *f* TEXTIL flyer-spinning frame; ~ **de hilar de anillos** *f* TEXTIL ring-spinning frame; ~ **de hilar de campana** *f* TEXTIL cap-spinning frame; ~ **de retorcer** *f* TEXTIL twister

continuación: ~ **de mensaje** *f* TELECOM continuation of message (*COM*)

continuidad *f* ELEC *tubo de rayos catódicos* persistence; ~ **de contacto** *f* ELEC, PROD contact continuity; ~ **eléctrica** *f* ELEC, ING ELÉC electrical continuity

continuo[1] *adj* GEN continuous

continuo[2]: ~ **de Compton** *m* FÍS, FÍS RAD Compton continuum

contornear *vt* PROD round

contorno *m* CONST contour, course, lay, GEOL boundary, contour, ING MECÁ line, outline, METAL boundary, TRANSP MAR *de la costa* shoreline; ~ **de agua subterránea** *m* AGRIC, AGUA, CARBÓN, HIDROL ground-water contour; ~ **de atenuación** *m* ELECTRÓN attenuation contour; ~ **de busto** *m* TEXTIL bustline; ~ **de un carácter** *m* INFORM&PD character outline; ~ **coherente** *m* METAL coherent boundary; ~ **de excéntrica** *m* TRANSP AÉR cam contour; ~ **de fase** *m* METAL phase boundary; ~ **de grano** *m* CRISTAL, METAL grain boundary; ~ **de leva** *m* TRANSP AÉR cam contour; ~ **de líneas de puntos** *m* CONST *topografía* broken country; ~ **de la macla** *m* METAL twin boundary; ~ **de raya espectral** *m* FÍS RAD spectral line profile; ~ **de las rayas espectrales** *m pl* FÍS RAD profile of spectral lines; ~ **de raya mecánica cuántico** *m* FÍS RAD quantum mechanical line shape; ~ **de la superficie de contacto** *m* METAL interface boundary

contorsión *f* ING MECÁ twisting

contraaguja: ~ **de vía muerta** *f* CARBÓN follower

contraalza *f* AGUA tailgate

contrabalance *m* TRANSP AÉR counterbalance

contrabisagra *f* CONST leaf

contrabóveda *f* CONST inverted arch

contracabezal *m* ING MECÁ footstock

contracable: ~ **de equilibrio** *m* AmL (*cf contrapeso de equilibrio Esp*) MINAS *pozo extracción* counterbalancing rope

contracarril *m* FERRO *equipo inamovible* check rail, wing rail

contracción[1]: **tras la** ~ *adj* EMB after-shrinkage

contracción[3] *f* FÍS, HIDROL, ING MECÁ contraction, METAL constriction, shrinkage, MINAS *de un filón* pinch, *de filones* nip, contraction, P&C *defecto, procesamiento* shrinkage, PAPEL, PROD, TELECOM shrinkage, shrinking, TEXTIL contraction; ~ **por desecación** *f* CONST *hormigón* drying shrinkage; ~ **durante la descongelación** *f* REFRIG thaw rigor; ~ **de fraguado** *f* CONST *cemento, hormigón* shrinkage; ~ **por fraguado** *f* CONST *hormigón* drying shrinkage; ~ **por frío** *f* REFRIG *de los músculos de la carne* cold shortening, TERMO contraction due to cold; ~ **e hinchamiento** *f* GEOL pinch and swell; ~ **de la imagen** *f* TV image contraction; ~ **lantanoidea** *f* QUÍMICA *radio atómico/iónico* lanthanide contraction; ~ **lateral** *f* HIDROL *flujo de agua en una esclusa* end contraction, METAL lateral contraction; ~ **de longitud** *f* FÍS length contraction; ~ **de Lorentz-Fitzgerald** *f* FÍS Lorentz-Fitzgerald contraction; ~ **en el molde** *f* P&C *procesamiento* mold shrinkage (*AmE*), mould shrinkage (*BrE*); ~ **repentina de la sección transversal** *f* INSTAL HIDRÁUL sudden contraction of cross section; ~ **de secado** *f* CONST *cemento, hormigón* shrinkage; ~ **por secado defectuoso** *f* IMPR pucker; ~ **al solidificar** *f* EMB shrinkage on solidification; ~ **térmica** *f* REVEST heat shrinking, TERMO thermal contraction; ~ **volumétrica** *f* P&C shrinkage

contrachapado *m* CONST plywood

contrachaveta *f* ING MECÁ gib, gib and cotter, forelock, nose key

contracielo *m AmL* (*cf chimenea Esp*) MINAS raise; ~ **escalonado** *m AmL* (*cf chimenea en bancos Esp*) MINAS raise stope

contracolado *m* PAPEL lining, pasting

contracolar *vt* PAPEL paste

contracono *m* ING MECÁ overhead cone

contracorriente[1] *adj* TV upstream

contracorriente[2] *f* AGUA backflow, backwater, countercurrent, CARBÓN counterflow, HIDROL backwash, countercurrent, ING ELÉC countercurrent, INSTAL HIDRÁUL *agua de caldera* backset, NUCL, OCEAN, QUÍMICA countercurrent, TRANSP MAR *dirección abajo* undertow; ~ **submarina** *f* OCEAN *física hidrodinámica* backwash, TRANSP MAR *mar* undertow

contractibilidad *f* CARBÓN contractibility

contráctil *adj* TEXTIL contractile

contraeje *m* ING MECÁ jackshaft

contraembalse *m* AGUA, HIDROL compensating reservoir, equalizing reservoir

contraer *vt* AGUA, CARBÓN, ENERG RENOV, FÍS, HIDROL contract, ING MECÁ *filones* pinch, INSTAL HIDRÁUL, MINAS, PROD, TERMO contract

contraestampa *f* ING MECÁ top swage, PROD *martinete* top die, top pallet

contraestay *m* TRANSP MAR jumper stay

contrafase[1]: **en** ~ *adj* TELECOM push-pull

contrafase[2] *adv* TV antiphase

contrafase[3] *f* ELEC, ELECTRÓN, TELECOM *radio* push-pull

contrafilón *m* MINAS counter, counterlode

contraflujo *m* CARBÓN counterflow

contrafuerte *m* CONST *de pared o puente* buttress, *edificación, obra civil* close buttress, ING MECÁ prop, stiffener, TEC ESP stiffener

contragalería *f* MINAS counter gangway

contragolpe *m* CINEMAT backlash

contraíble *adj* TERMO shrinkable

contralísios *m pl* TRANSP MAR *viento* antitrades

contralomo *m* IMPR *encuadernación* back lining

contraluz[1]: **a** ~ *adv* CINEMAT against the light

contraluz[2]: ~ **principal** *f* CINEMAT counter key light

contramaestre *m* TRANSP MAR boatswain; ~ **mayor** *m* TRANSP MAR *armada* chief petty officer

contramando: ~ **de seguridad del tren de aterrizaje** *m* TRANSP AÉR landing gear safety override

contramanivela *f* PROD fly crank

contramarcha *f* ING MECÁ reversing, countershaft, countershafting, *máquinas herramientas* back gear, MECÁ reverse

contramatriz *f* IMPR *troquel hembra* counterdie

contramedidas: ~ **en las comunicaciones** *f pl* TELECOM *guerra electrónica* communication countermeasures (*CCM*); ~ **electrónicas** *f pl* D&A, TEC ESP electronic countermeasures (*ECM*)

contramovimiento: ~ **de dos velocidades** *m* ING MECÁ two-speed counter motion

contrapeldaño *m* CONST riser

contrapeso *m* CONST counterweight, ING MECÁ back balance, balance weight, flyweight, counterweight, INSTR counterweight, MECÁ balance weight, balancing weight, counterweight, VEH *carrocería* overrider; ~ **de equilibrio** *m Esp* (*cf contracable de equilibro LAm*) MINAS *pozo extracción* counterbalancing rope; ~ **de pluma de carga giratoria** *m* TRANSP MAR deadman; ~ **tensor** *m* ING MECÁ balanced tension block; ~ **volante** *m* ING MECÁ flyweight

contraplaca *f* ING MECÁ backing plate

contraplato *m* ING MECÁ chuck back, counterplate, *para montar brocas de torno* back plate

contrapolea *f* ING MECÁ tail pulley

contraportada *f* IMPR frontispiece

contrapresión *f* C&V back pressure, INSTAL HIDRÁUL backlash, counterpressure; ~ **crítica** *f* ING MECÁ critical backing pressure

contraprueba *f* ING MECÁ countertest, QUÍMICA check assay

contrapuerta: ~ **de corredera** *f* AGUA sash gate

contrapunta: ~ **del cabezal móvil** *f* ING MECÁ tailpin thrust

contrapunto *m* ING MECÁ, MECÁ loose head, loose headstock, poppet head, *torno* back centre (*BrE*), back puppet, backhead, footstock, deadhead, dead center (*AmE*), dead centre (*BrE*), headstock, back center (*AmE*), PROD *tornos* poppet; ~ **del cilindro** *m* PROD *tornos* cylinder tailstock; ~ **de fricción** *m* ING MECÁ friction headstock; ~ **de manguito** *m* PROD *torno* cylinder poppet head

contrapunzar *vt* ING MECÁ counterpunch

contrapunzón *m* ING MECÁ counterpunch

contrariamente: ~ **a la inclinación** *adv* GEOL up-dip

contrarrayado *m* IMPR crosshatching

contrarremachador *m* CONST rivet dolly

contrarroda *f* TRANSP MAR *construcción de barcos* apron

contraseña *f* INFORM&PD, TELECOM password

contrastación *f* PROD, TEC ESP monitoring

contrastar *vt* PROD, TEC ESP monitor

contraste[1]: **de poco** ~ *adj* CINEMAT low-key

contraste[2] *m* CARBÓN assay office, *estadística* test, CINEMAT, FÍS, FOTO contrast, TRANSP MAR *del viento contra la vela* eddy wind, TV contrast; ~ **de brillos** *m* TV brightness ratio; ~ **de imagen** *m* TELECOM TV picture contrast; ~ **de índice** *m* ÓPT index contrast; ~ **del índice de refracción** *m* CINEMAT, FÍS, FOTO,

ÓPT, TEC ESP, TELECOM refractive-index contrast;
~ **luminoso** *m* CINEMAT lighting contrast

contrasurco *m* AGRIC backfurrow

contrataladro *m* ING MECÁ counterbore

contratista *m* CALIDAD contractor, CONST contractor, house builder, jobbing contractor; ~ **de acarreos** *m* TRANSP haulage contractor; ~ **de perforaciones** *m* PETROL drilling contractor; ~ **principal** *m* TEC ESP prime contractor; ~ **de sondeos** *m* CONST *perforación* boring contractor

contrato: ~ **de fletamento** *m* TEC PETR *transporte*, TRANSP MAR charter party; ~ **de mantenimiento** *m* (*CM*) ELECTRÓN maintenance contract (*MA*)

contratoma *f* CINEMAT countershot

contratrama *f* CINEMAT countermatte

contratuerca *f* ING MECÁ locknut, checknut, jam nut, nut lock, pinch nut, safety nut, set nut, PROD grip nut, locknut

contravapor *m* INSTAL HIDRÁUL reversed steam

contravarilla: ~ **del pistón** *f* ING MECÁ extended piston-rod, piston with extended rod

contravástago: ~ **del pistón** *m* ING MECÁ tail piston rod

contravenir *vt* SEG *los reglamentos* contravene

contraventana *f* CONST shutter

contraventear *vt* CONST brace against wind pressure

contraviento *m* CONST bracing against wind pressure, wind brace

contrete: ~ **de apoyo** *m* CARBÓN *ingeniería minera* strut; ~ **de unión** *m* PROD bonding strut

control[1]: **de** ~ **numérico** *adj* MECÁ numerically controlled; **por** ~ **remoto** *adj* AUTO, D&A, ING MECÁ, TV remote-controlled

control[2] *m* CALIDAD inspection, ELEC, ELECTRÓN control, INFORM&PD control, check, ING MECÁ *maquinaria* check, testing, QUÍMICA regulation, TEC ESP monitoring, control, TEC PETR *producción, yacimientos* gaging (*AmE*), gauging (*BrE*), TELECOM control, TEXTIL control, monitoring, TV monitoring, VEH control;

▪ **a** ~ **de acceso** *m* TELECOM access control; ~ **de acceso al medio** *m* TELECOM media access control; ~ **de acceso al soporte** *m* TELECOM media access control; ~ **de acceso medio** *m* TELECOM medium access control (*MAC*); ~ **de acceso a la red** *m* INFORM&PD network access control; ~ **accionado por vehículos** *m* AUTO vehicle-actuated control; ~ **de aceptación** *m* CALIDAD acceptance inspection, receiving inspection; ~ **acimutal** *m* TRANSP AÉR azimuthal control; ~ **de actitud** *m* TRANSP AÉR attitude control; ~ **activo** *m* TEC ESP active control; ~ **por adaptación** *m* ING ELÉC adaptive control; ~ **de afluencia** *m* TRANSP *carreteras* flow control; ~ **del alerón** *m* TRANSP AÉR aileron control; ~ **por alimentación anticipada** *m* AmL (*cf control por alimentación hacia adelante Esp*) INFORM&PD feed-forward control; ~ **de alimentación del arco** *m* CINEMAT arc feed control; ~ **por alimentación hacia adelante** *m* Esp (*cf control por alimentación anticipada AmL*) INFORM&PD feed-forward control; ~ **de alimentación de hilo** *m* TEXTIL yarn feed control; ~ **alimentario** *m* ALIMENT, CALIDAD food control; ~ **de amplitud** *m* FÍS ONDAS amplitude control; ~ **de amplitud de la línea** *m* TV line amplitude control; ~ **de la amplitud vertical** *m* TV vertical amplitude control; ~ **de aproximación** *m* TRANSP AÉR approach

control; ~ **de aproximación automático** *m* TRANSP AÉR automatic approach control; ~ **de área** *m* TRANSP area control; ~ **de asbesto en áreas de trabajo** *m* SEG asbestos control in the workplace; ~ **por atributos** *m* CALIDAD inspection by attributes; ~ **automático** *m* CONST, ELEC, EMB automatic control; ~ **automático de avance** *m* TRANSP automatic control of headway; ~ **automático de brillo** *m* (*CAB*) TV automatic brightness control (*ABC*); ~ **automático continuo de tren** *m* FERRO, TRANSP continuous automatic train control; ~ **automático de corriente** *m* ELEC *relé* automatic current controller; ~ **automático de crominancia** *m* TV automatic chrominance control; ~ **automático de fase** *m* ELECTRÓN, TV automatic phase control (*APC*); ~ **automático de frecuencia** *m* (*CAF*) ELECTRÓN, FÍS, PROD, TELECOM, TV automatic frequency control (*AFC*); ~ **automático de ganancia** *m* (*CAG*) ELECTRÓN, GEOFÍS, TELECOM automatic gain control (*AGC*); ~ **automático de registro lateral** *m* IMPR automatic sidelay control; ~ **automático de voltaje** *m* ELEC automatic voltage control; ~ **automático de volumen** *m* FÍS, ING ELÉC, ÓPT automatic volume control (*AVC*); ~ **de avance** *m* TRANSP headway control; ~ **del avance** *m* PROD progress control; ~ **de avance mínimo** *m* TRANSP control of minimum headway; ~ **de avance de la película** *m* FOTO rapid film-advance lever;

▪ **b** ~ **en bucle cerrado** *m* ELEC, ELECTRÓN closed-loop control;

▪ **c** ~ **de cabeceo** *m* TRANSP AÉR pitch control; ~ **de la calefacción** *m* AUTO heater control; ~ **de calidad** *m* GEN quality control (*QC*); ~ **de la calidad** *m* CALIDAD quality surveillance; ~ **de calidad del aire ambiente** *m* SEG quality monitoring of ambient air; ~ **de calidad durante el proceso** *m* CALIDAD process quality control; ~ **de la calidad de la imagen** *m* TELECOM, TV picture-quality control; ~ **de calidad a tiempo real** *m* TEXTIL on-line quality monitoring; ~ **de calidad en toda la empresa** *m* CALIDAD company-wide quality control (*CWQC*); ~ **de calidad en toda la sociedad** *m* CALIDAD company-wide quality control (*CWQC*); ~ **de calidad total** *m* CALIDAD, PROD total quality control (*TQC*); ~ **del camino de acceso** *m* TRANSP access road control (*AmE*), slip road control (*BrE*); ~ **del canal** *m* TELECOM channel control; ~ **de la capa límite** *m* TRANSP boundary-layer control; ~ **de capacidad** *m* TRANSP AÉR capacity control; ~ **de carga dinámica** *m* TELECOM dynamic load control; ~ **de carretera de acceso** *m* TRANSP access road control (*AmE*), slip road control (*BrE*); ~ **de caudal** *m* ING MECÁ flow-rate controller; ~ **del caudal por canales individuales** *m* NUCL individual channel flow control; ~ **del caudal de salida** *m* AGUA outlet flow control; ~ **de centrado** *m* TV centering control (*AmE*), centring control (*BrE*); ~ **del centrado de la imagen** *m* TV field-centering control (*AmE*), field-centring control (*BrE*); ~ **de centrado vertical** *m* TV vertical-centering control (*AmE*), vertical-centring control (*BrE*); ~ **central** *m* TELECOM central control; ~ **centralizado** *m* TELECOM centralized control; ~ **de cierre o apertura** *m* ELEC, TEC ESP on-off control; ~ **de combustible** *m* TRANSP AÉR fuel control; ~ **de conexión y desconexión** *m* ELEC, TEC ESP on-off control; ~ **de configuración** *m* PROD configuration control; ~ **de consultas** *m* INFORM&PD inquiry control; ~ **de la contaminación** *m* CONTAM pollution control; ~ **de la contaminación atmosférica** *m* CONTAM air pollution control; ~ **de contención** *m* TELECOM contention control; ~ **de continuidad** *m* CINEMAT, TV continuity control; ~ **continuo** *m* INFORM&PD continuous control; ~ **de contraste** *m* CINEMAT, FÍS, FOTO, TV contrast control; ~ **del corredor** *m* TRANSP corridor control; ~ **de la corriente** *m* ING ELÉC current control; ~ **de crecidas** *m* AGUA, ENERG RENOV flood control; ~ **por cristal** *m* TRANSP AÉR crystal control; ~ **de crominancia** *m* TV chroma control; ~ **en cuadratura** *m* ELEC *corriente alterna* quadrature control;

~ d ~ **definitivo** *m* CALIDAD final inspection; ~ **de degustación** *m* AGUA taste control; ~ **de desplazamiento de imagen** *m* TV field-shift switch; ~ **de desviación horizontal de la imagen** *m* TV horizontal deflection control; ~ **de diferencia de color** *m* TV color-difference signal (*AmE*), colour-difference signal (*BrE*); ~ **digital** *m* TELECOM, TV digital control; ~ **digital directo** *m* INFORM&PD, TELECOM direct digital control (*DDC*); ~ **dimensional** *m* PROD *producción* dimensional control; ~ **directo** *m* ING MECÁ direct control; ~ **del dispositivo** *m* INFORM&PD device control (*DC*); ~ **a distancia** *m* AUTO, CINEMAT, CONST, D&A, ELEC, FOTO, ING MECÁ, MECÁ, TEC ESP, TELECOM, TRANSP, TV, VEH remote control; ~ **doble** *m* TRANSP AÉR dual control; ~ **durante la fabricación** *m* CALIDAD, PROD in-process inspection, in-process inspection in manufacturing;

~ e ~ **por eco** *m* INFORM&PD echo check; ~ **de ejecución** *m* CALIDAD performance testing; ~ **eléctrico** *m* ELEC, ING ELÉC electric control; ~ **electrónico de frenado** *m* TRANSP electronic braking-control; ~ **electrónico de velocidad** *m* TRANSP electronic speed-control; ~ **de emergencia** *m* SEG, TRANSP AÉR emergency control; ~ **de emisiones del motor** *m* ING MECÁ engine-emission control; ~ **de encaminamiento** *m* TELECOM routing control; ~ **de enclavamiento** *m* TRANSP AÉR interlock control; ~ **de encriptación** *m* TELECOM scrambling control; ~ **de enfoque** *m* CINEMAT, FOTO, TV focusing control; ~ **de enlace de datos** *m* INFORM&PD data link control; ~ **de enlace de datos de alto nivel** *m* INFORM&PD, TELECOM high-level data link control (*HDLC*); ~ **de enlace lógico** *m* TELECOM logical link control (*LLC*); ~ **de entrada** *m* ELECTRÓN input control; ~ **de entrada/salida** *m* INFORM&PD input/output control; ~ **de equipamiento** *m* PROD hardware review; ~ **de error de cabecera** *m* TELECOM header error control (*HEC*); ~ **de errores** *m* INFORM&PD, TELECOM error control; ~ **de espera** *m* TELECOM queue control; ~ **estadístico de la calidad** *m* CALIDAD statistical quality control; ~ **de etapa mecánica** *m* FOTO, INSTR mechanical stage control; ~ **de existencias** *m* FERRO, PROD stock control; ~ **de exposición a humos de soldadura** *m* SEG control of exposure to fumes from welding and brazing; ~ **de extremo a extremo** *m* INFORM&PD end-to-end control;

~ f ~ **de fase** *m* ELECTRÓN, TELECOM, TV phase control; ~ **para fijar la abertura** *m* FOTO aperture-setting lever; ~ **fino** *m* NUCL fine control; ~ **de flexibilidad** *m* ING MECÁ flexible control; ~ **de flujo** *m* INFORM&PD, INSTAL TERM, TELECOM, TRANSP flow control; ~ **fotoeléctrico del registro** *m* EMB photoelectric register control; ~ **de frecuencia de la carga** *m* ELEC *máquina eléctrica* load frequency control; ~ **de frecuencia electrónico** *m* ELECTRÓN electronic frequency-control; ~ **de frecuencias** *m* ELECTRÓN, TELECOM frequency control; ~ **de funcionamiento** *m* CALIDAD performance testing; ~ **de la fusión** *m* TRANSP merging control;

~ g ~ **de ganancia** *m* ELECTRÓN gain control; ~ **de ganancia automática retardada** *m* (*CGA retardada*) ELECTRÓN delayed automatic gain control (*delayed AGC*); ~ **de ganancia manual** *m* ELECTRÓN manual gain control; ~ **de ganancia principal** *m* ELECTRÓN master gain control; ~ **de gases** *m* TEC ESP *motores* throttle control; ~ **genérico del flujo** *m* TELECOM generic flow control (*GFC*); ~ **de guiñadas** *m* ENERG RENOV yaw control;

~ h ~ **de humos** *m* SEG smoke control;

~ i ~ **independiente** *m* NUCL independent control; ~ **indirecto** *m* ING ELÉC indirect control; ~ **individual** *m* TRANSP individual control; ~ **del inducido** *m* ELEC *máquina* armature control; ~ **infrarrojo a distancia** *m* ING MECÁ infrared remote control; ~ **infrarrojo remoto** *m* ING MECÁ infrared remote control; ~ **intercambio de datos** *m* TRANSP AÉR data link control; ~ **de interconexión** *m* TRANSP AÉR interlock control; ~ **del interruptor de datos del giróscopo** *m* TRANSP AÉR gyro data-switching control; ~ **de inundaciones** *m* AGUA, ENERG RENOV flood control; ~ **de inventario** *m* PROD inventory control; ~ **del izador** *m* TRANSP AÉR elevator control;

~ l ~ **en lazo cerrado** *m* ELECTRÓN feedback control; ~ **para limitar el flujo** *m* MINAS *pozo de petróleo* capping; ~ **lineal** *m* ELEC linear control; ~ **de linealidad** *m* TV line linearity control, linearity control; ~ **de llama** *m* INSTAL TERM flame control; ~ **de llamadas** *m* TELECOM call control; ~ **lógico** *m* ELECTRÓN logic control; ~ **lógico de enlace** *m* TELECOM logical link control (*LLC*);

~ m ~ **de macizo de anclaje** *m* SEG, TRANSP dead man's control; ~ **maestro** *m* TV master control; ~ **maestro del fundido** *m* TV master control fader; ~ **manual** *m* CONST, ELEC, FOTO, MECÁ, TRANSP AÉR manual control; ~ **del material móvil** *m* FERRO stock control; ~ **mediante pasillo** *m* TRANSP corridor control; ~ **de mezcla** *m* TRANSP AÉR *de combustible* mixture control; ~ **del microprocesador** *m* TELECOM microprocessor control; ~ **por microprocesador** *m* ELECTRÓN microprocessor control; ~ **de misión** *m* TEC ESP *misiones espaciales* control center (*AmE*), control centre (*BrE*); ~ **de montaje** *m* CINEMAT, FOTO cocking lever; ~ **de motor** *m* TV motor control; ~ **por motor** *m* TV motor control; ~ **del motor del cabezal móvil** *m* CINEMAT capstan motor control; ~ **de muestras** *m* CALIDAD sample surveillance; ~ **por muestreo** *m* CALIDAD sampling inspection;

~ n ~ **de la nitidez** *m* TV sharpness control; ~ **de nivel** *m* EMB level control; ~ **de nivel automático** *m* AUTO automatic level control; ~ **del nivel de líquido** *m* EMB liquid-level control; ~ **del nivel del pedestal** *m* TV pedestal level control; ~ **normal** *m* CALIDAD normal inspection; ~ **numérico** *m* (*CN*) ELEC *informática*, INFORM&PD, ING MECÁ, MECÁ, PROD numerical control (*NC*); ~ **numérico de cinco ejes** *m* ING MECÁ five-axis numerical control; ~ **numérico**

computerizado *m* (*CNC*) INFORM&PD, ING MECÁ computerized numerical control (*CNC*); ~ **numérico penta axial** *m* ING MECÁ five-axis numerical control; **~ o** ~ **de olores** *m* CONTAM, PROC QUÍ odor control (*AmE*), odour control (*BrE*); ~ **óptimo** *m* NUCL, TELECOM optimal control; ~ **de órbita** *m* TEC ESP orbit control; ~ **de la orientación** *m* TELECOM orientation control; **~ p** ~ **de panorámica vertical** *m* CINEMAT tilt control; ~ **por parámetros** *m* CALIDAD inspection by variables; ~ **de paridad** *m* TELECOM parity control; ~ **del pasillo** *m* TRANSP corridor control; ~ **pasivo** *m* TEC ESP passive control; ~ **de paso** *m* TRANSP AÉR pitch control; ~ **de paso cíclico** *m* TRANSP AÉR cyclic-pitch control; ~ **del paso del colectivo** *m* TRANSP AÉR collective-pitch control; ~ **paso a paso** *m* ELEC *accionamiento motorizado* step-by-step control; ~ **de pausas** *m* TV pause control; ~ **de pedal** *m* ING MECÁ foot-operated control, pedal-operated control; ~ **de petición** *m* INFORM&PD inquiry control; ~ **de pH** *m* CARBÓN *química* pH control; ~ **de picado** *m* CINEMAT tilt control; ~ **de picos** *m* TV peaking control; ~ **de plomo en el trabajo** *m* SEG lead control at work; ~ **de la polaridad** *m* TV polarity control; ~ **de polvos** *m* PROC QUÍ dust control; ~ **de posición** *m* TEC ESP station keeping; ~ **posicional** *m* ING MECÁ positioning; ~ **de presión** *m* INSTAL TERM pressure control; ~ **a presión en emergencia** *m* PROD pressure-control emergency; ~ **a presión normal** *m* PROD pressure-control normal; ~ **de la presión del primario por medio del presurizador** *m* NUCL pressurizer pressure control; ~ **de presión tensionada por resortes** *m* FOTO spring-tensioned pressure lever; ~ **en primera presentación** *m* CALIDAD original inspection; ~ **principal** *m* TV master control; ~ **principal del fundido** *m* TV master control fader; ~ **de proceso** *m* CALIDAD, ING ELÉC, TELECOM process control; ~ **de la producción** *m* MINAS *pozo de petróleo* capping, PROD production control, TEC PETR *producción, yacimiento* gaging (*AmE*), gauging (*BrE*); ~ **de programa almacenado** *m* TELECOM stored program control (*AmE*) (*SPC*), stored programme control (*BrE*) (*SPC*); ~ **programable** *m* TELECOM programmable control; ~ **programado** *m* TV programmed control; ~ **de programas** *m* INFORM&PD program control; ~ **progresivo** *m* ELEC stepless control; ~ **del proyecto** *m* CONST project monitoring; ~ **del punto de exploración** *m* TV scanning spot control; **~ r** ~ **de radar** *m* TRANSP, TRANSP AÉR, TRANSP MAR radar control; ~ **radial** *m* ÓPT radial control; ~ **de rampa de salida** *m* PROD output ramp control; ~ **del reactor** *m* NUCL, TRANSP AÉR reactor control; ~ **por realimentación** *m* ING ELÉC feedback control; ~ **de la red de tráfico** *m* TRANSP traffic network control; ~ **reducido** *m* CALIDAD reduced inspection; ~ **reforzado** *m* CALIDAD tightened inspection; ~ **de refuerzo** *m* TRANSP booster control; ~ **del registro** *m* CINEMAT, TV registration control; ~ **de reglaje** *m* TRANSP AÉR trim control; ~ **remoto** *m* GEN remote control; ~ **remoto por cámara de televisión** *m* TRANSP, TV remote-control by television camera; ~ **remoto inalámbrico** *m* CINEMAT wireless remote control; ~ **remoto infrarrojo** *m* ING MECÁ infrared remote control; ~ **remoto manual** *m* TRANSP AÉR manual remote control; ~ **remoto por radio** *m* FÍS

RAD radio remote control, TELECOM radio telecontrol, TRANSP radio remote control, radio telecontrol; ~ **remoto de temperatura** *m* REFRIG, TERMO remote temperature monitoring; ~ **de la rueda del morro** *m* TRANSP AÉR nose-wheel steering; ~ **de ruido de banda** *m* TV banding on noise; ~ **de ruidos** *m* ACÚST, SEG noise control; **~ s** ~ **de salida** *m* TV output control; ~ **de saturación de banda** *m* TV banding on saturation; ~ **de secuencia** *m* INFORM&PD sequence control; ~ **secuencial** *m* INFORM&PD sequence control, TELECOM sequencing; ~ **del seguimiento** *m* TV tracking control; ~ **de seguridad** *m* SEG, TRANSP *ferrocarriles* dead man's control; ~ **de seguridad del ciclo de trabajo** *m* CONST job cycle safety audit; ~ **selectivo** *m* CALIDAD screening inspection; ~ **de señalización de tráfico concatenada** *m* TRANSP linked traffic signal control; ~ **sensible al tacto** *m* INFORM&PD touch-sensitive control; ~ **de sincronismo horizontal** *m* TV horizontal hold control; ~ **de sincronización** *m* TV hold control; ~ **de sincronización vertical** *m* TV vertical hold control; ~ **sincronizador de imagen** *m* TV frame-synchronization control; ~ **del sistema de comunicaciones** *m* (*CSS*) TEC ESP communications system monitoring (*CSM*); ~ **de substancias peligrosas para la salud** *m* SEG control of substances hazardous to health (*COSHH*); **~ t** ~ **tangencial** *m* ÓPT tangential control; ~ **de tareas** *m* INFORM&PD job control; ~ **telemedida telemando y distancia** *m* TEC ESP telemetry command and ranging subsystem (*TCR*); ~ **de temperatura** *m* TERMO, TEXTIL temperature control; ~ **de tensión** *m* ELEC voltage control; ~ **de la tensión de la cinta** *m* TV tape tension control; ~ **térmico** *m* TEC ESP thermal control; ~ **térmico pasivo** *m* (*PTC*) TEC ESP passive thermal control (*PTC*); ~ **de termistores** *m* ING ELÉC thermistor control; ~ **termostático** *m* INSTAL TERM thermostat control; ~ **de tiempo** *m* TELECOM timing; ~ **en tiempo real** *m* ELECTRÓN, INFORM&PD, TELECOM, TRANSP real-time control; ~ **del timón de dirección** *m* TRANSP AÉR, TRANSP MAR rudder control; ~ **de todo o nada** *m* TEC ESP on-off control, all-or-nothing control; ~ **de toma intermedia** *m* ING ELÉC tapped control; ~ **de tonalidad** *m* TV hue control; ~ **de tonalidad de banda** *m* TV banding on hue; ~ **de toxicidad en el lugar de trabajo** *m* SEG control of toxicity in the workplace; ~ **de toxicidad en el trabajo** *m* SEG control of toxicity at work; ~ **del trabajo en curso de ejecución** *m* PROD work-in-progress control; ~ **de tráfico** *m* TRANSP traffic control; ~ **de tráfico aéreo** *m* TRANSP AÉR air traffic control; ~ **de transmisión** *m* INFORM&PD transmission control; ~ **de transporte de la película** *m* CINEMAT, FOTO film-transport lever; ~ **del tren de aterrizaje del morro** *m* TRANSP AÉR nose gear steering; **~ u** ~ **de la unidad** *m* INFORM&PD device control (*DC*); ~ **unidireccional** *m* FOTO single-stroke lever; **~ v** ~ **de validez** *m* INFORM&PD validity check; ~ **del vector de empuje** *m* (*TVC*) TEC ESP thrust vector control (*TVC*); ~ **de velocidad** *m* ELEC *máquina*, TEXTIL *mecánica* speed control; ~ **de velocidad automático** *m* AUTO automatic speed control; ~ **de velocidad por baliza** *m* FERRO *equipo inamovible* speed control by beacons; ~ **de la velocidad de la** cinta *m* TV tape speed control; ~ **de velocidad por imanes de seguimiento** *m* FERRO speed control with track magnets; ~ **de velocidad del obturador** *m* FOTO shutter speed control; ~ **de velocidad variable** *m* TV variable-speed control; ~ **por venenos solubles** *m* NUCL soluble poison control; ~ **vertical** *m* CINEMAT, ÓPT vertical control; ~ **del viaje** *m* TRANSP AÉR travel follow-up; ~ **vocal** *m* TELECOM voice control; ~ **de vuelo estacionario** *m* TRANSP AÉR hover control

control³: **bajo ~** *fra* ING MECÁ under control

controlabilidad *f* TRANSP AÉR controllability

controlado: ~ **por CNC** *adj* ING MECÁ CNC-controlled; ~ **por computadora** *adj* AmL (*cf controlado por ordenador Esp*) PROD, TELECOM computer-controlled; ~ **desde tierra** *adj* TEC ESP ground-controlled; ~ **a mano** *adj* ING ELÉC manually-controlled; ~ **mediante teclas** *adj* INFORM&PD key-driven; ~ **por menús** *adj* INFORM&PD menu-driven; ~ **por ordenador** *adj* Esp (*cf controlado por computadora AmL*) PROD, TELECOM computer-controlled; ~ **por radar** *adj* FÍS, TELECOM, TRANSP, TRANSP AÉR, TRANSP MAR radar-controlled

controlador *m* CALIDAD inspector, INFORM&PD driver, controller, ÓPT controller, PROD *persona* timekeeper, *inventor* controller, checker, TEC ESP controller; ~ **de agrupamientos** *m* INFORM&PD cluster controller; ~ **ajustado al tráfico** *m* TRANSP traffic-adjusted controller; ~ **de altitud** *m* TRANSP AÉR altitude controller; ~ **de altitud barométrico** *m* TRANSP AÉR barometric altitude controller; ~ **del árbol de distribución** *m* ELEC *conmutador* camshaft controller; ~ **barométrico** *m* TRANSP AÉR barometric controller; ~ **bifásico** *m* TRANSP *control de tráfico* two-phase controller; ~ **del bus** *m* TELECOM bus arbitrator; ~ **de calidad** *m* CALIDAD quality auditor; ~ **del caudal** *m* PROD flow-controller; ~ **del caudal del flujo** *m* ING MECÁ flow-rate controller; ~ **de circuito impreso para motor** *m* PROD solid-state motor controller; ~ **de comprobación** *m* PROD monitoring controller; ~ **de disco** *m* INFORM&PD disk controller; ~ **de dispositivo** *m* INFORM&PD device driver, TELECOM device controller; ~ **electrónico de velocidad** *m* TRANSP electronic speed controller; ~ **de la estación base** *m* TELECOM base station controller; ~ **gradual** *m* PROD stepper controller; ~ **independiente** *m* PROD stand-alone controller; ~ **industrial** *m* PROD industrial controller; ~ **de intersecciones locales** *m* TRANSP local intersection controller; ~ **de línea** *m* INFORM&PD line driver, ING ELÉC line controller; ~ **de línea analógica** *m* INFORM&PD analog line driver; ~ **local** *m* TRANSP local controller; ~ **lógico programable** *m* PROD programmable logic controller; ~ **de la memoria** *m* ING ELÉC memory controller; ~ **monofásico** *m* TRANSP one-phase controller; ~ **de motor** *m* PROD motor controller; ~ **de motor de estado sólido** *m* PROD solid-state motor controller; ~ **del nivel del vidrio** *m* C&V, PROD glass-level controller; ~ **óptimo** *m* NUCL, TELECOM optimal controller; ~ **de pantalla** *m* INFORM&PD display controller; ~ **de pH** *m* CARBÓN *química* pH controller; ~ **polifásico** *m* TRANSP multiphase controller; ~ **principal** *m* TRANSP master controller; ~ **de procesos** *m* ING ELÉC process controller; ~ **de profundidad** *m* TEC PETR depth controller; ~ **programable** *m* PROD programmable

controller; ~ **de proyecto** *m* TEC ESP project controller; ~ **de radar** *m* TRANSP, TRANSP AÉR, TRANSP MAR radar controller; ~ **de redes** *m* INFORM&PD network controller; ~ **de rondas** *m* PROD *vigilantes* time recorder, watchman's clock; ~ **de señales de tráfico** *m* TRANSP traffic signal controller; ~ **de tráfico aéreo** *m* TRANSP AÉR air traffic controller; ~ **transistorizado** *m* PROD solid-state motor controller; ~ **de TRC** *m* ELECTRÓN CRT controller; ~ **de visualización** *m* INFORM&PD display driver, display controller; ~ **de vuelo** *m* TRANSP AÉR flight controller

controlar *vt* AGUA overpower, INFORM&PD, TEC ESP, TELECOM control, TEXTIL monitor; ~ **el sonido de** *vt* CINEMAT monitor

controles: ~ **e indicaciones** *m pl* TRANSP AÉR controls and indicators; ~ **interconectados** *m pl* TRANSP AÉR interconnected controls; ~ **de predeterminación del nivel del combustible** *m pl* TRANSP AÉR fuel-level presetting controls; ~ **programables** *m* PROD programmable controls; ~ **tradicionales** *m pl* PROD traditional controls; ~ **de vuelo** *m pl* TRANSP AÉR flight controls; ~ **de vuelo digitales fly-by-wire** *m pl* TRANSP AÉR fly-by-wire flight controls

controversia *f* CARBÓN contest

convección *f* GEN convection; ~ **por anillo rotatorio** *f* FÍS FLUID rotating-annulus convection; ~ **forzada** *f* C&V, FÍS, FÍS FLUID, GAS, IMPR, TERMOTEC forced convection; ~ **natural** *f* FÍS, FÍS FLUID, TERMOTEC natural convection; ~ **térmica** *f* TERMO heat convection

convectivo *adj* GEN convective

convector *m* GEN convector; ~ **de aire con abanico** *m* SEG, TERMOTEC fan-assisted air heater; ~ **de aire caliente con abanico** *m* SEG fan-assisted air heater; ~ **de aire caliente con ventilador** *m* TERMOTEC fan-assisted air heater; ~ **eléctrico** *m* ELEC, TERMO electric convector

convención: ~ **de datos** *f* TRANSP AÉR data convention

convenio: ~ **sobre anotaciones** *m* PROD notation convention

convergencia *f* TRANSP concourse, VEH *de las ruedas delanteras* toe-in; ~ **dinámica horizontal** *f* TV horizontal dynamic convergence; ~ **de la imagen** *f* TV field convergence; ~ **de la onda senoidal** *f* TV sine wave convergence; ~ **de transmisión** *f* TELECOM transmission convergence; ~ **vertical** *f* ELECTRÓN vertical convergence

convergente *adj* GEN convergent

conversación: ~ **cruzada** *f* INFORM&PD crosstalk; ~ **imposible** *f* TELECOM conversation impossible (*CI*); ~ **tasable de transferencia** *f* TELECOM transfer charge call; ~ **tasada de transferencia** *f* TELECOM transfer charge call; ~ **de tres direcciones** *f* TELECOM three-way conversation; ~ **tridireccional** *f* TELECOM three-way conversation

conversacional *adj* INFORM&PD conversational

conversión *f* ELEC, ELECTRÓN, INFORM&PD conversion, ING ELÉC conversion, transformation, TEC PETR *refino* reforming; ~ **analógico-digital** *f* ELEC, ELECTRÓN, FÍS, FÍS PART, INFORM&PD, TELECOM analog-to-digital conversion (*ADC*); ~ **analógico-digital en chip** *f* ELECTRÓN on-chip analog-to-digital conversion; ~ **automática de datos** *f* INFORM&PD automatic data conversion; ~ **a bajo costo** *f* FÍS RAD low-cost conversion; ~ **de CC en CA** *f* ING ELÉC DC-to-AC conversion; ~ **de CC en CC** *f* ING ELÉC DC-to-DC conversion; ~ **de código** ƒ INFORM&PD, ING ELÉC code conversion; ~ **de datos** ƒ ELECTRÓN, INFORM&PD data conversion; ~ **digital-analógica** ƒ ELECTRÓN, INFORM&PD, TELECOM, TV digital-to-analog conversion (*DAC*); ~ **dinámica** ƒ TV dynamic convergence; ~ **dinámica de direcciones** ƒ INFORM&PD dynamic address translation; ~ **directa** ƒ ENERG RENOV direct conversion; ~ **doble** ƒ ELECTRÓN *A*/*D* double conversion; ~ **de energía** ƒ ELEC energy transformation, ING ELÉC energy conversion, energy transformation, TELECOM energy conversion, TERMO energy conversion, energy transformation; ~ **de energía directa** ƒ ING ELÉC direct energy conversion; ~ **de la energía solar** ƒ ING ELÉC solar energy conversion; ~ **de estrella a delta** ƒ ELEC *conexión* star-to-delta conversion; ~ **de estrella a triángulo** ƒ ELEC *conexión* star-to-delta conversion; ~ **de frecuencia** ƒ ELEC frequency conversion, ELECTRÓN, TELECOM frequency conversion, frequency translation, TV frequency translation; ~ **gamma** ƒ FÍS RAD gamma-ray conversion; ~ **heterodina** ƒ ING ELÉC heterodyne conversion; ~ **de imagen** ƒ ELECTRÓN, TELECOM image conversion; ~ **de impedancia** ƒ ING ELÉC impedance conversion; ~ **interna** ƒ FÍS internal conversion; ~ **magnetohidrodinámica** ƒ GEOFÍS, ING ELÉC magnetohydrodynamic conversion; ~ **MHD** ƒ GEOFÍS, ING ELÉC MHD conversion; ~ **de normas** ƒ TV standards conversion; ~ **de octeto** ƒ ELECTRÓN, INFORM&PD eight-bit conversion; ~ **paralela** ƒ ELECTRÓN parallel conversion; ~ **paralela analógica-digital** ƒ ELECTRÓN flash analog-to-digital conversion; ~ **de paralelo a serie** ƒ ING ELÉC parallel-to-serial conversion; ~ **rápida** ƒ ELECTRÓN flash conversion; ~ **de señal óptica** ƒ ING ELÉC optical signal conversion; ~ **de señales** ƒ ELECTRÓN, TELECOM signal conversion; ~ **en serie de analógico a digital** ƒ CONST, ELECTRÓN serial analog-to-digital conversion; ~ **de serie a paralelo** ƒ ING ELÉC serial-to-parallel conversion; ~ **de sistema analógico a sistema digital** ƒ TV A-to-D conversion (*analog-to-digital conversion*); ~ **de sistema analógico a sistema digital en serie** ƒ ELECTRÓN serial analog-to-digital conversion; ~ **de sistema decimal a binario** ƒ ELECTRÓN decimal-to-binary conversion; ~ **de sistema digital a sistema análogico** ƒ ELECTRÓN digital-to-analog conversion (*DAC*); ~ **temporal a digital** ƒ FÍS PART *en los detectores electrónicos* time-to-digital conversion (*TDC*); ~ **térmica océanica** ƒ ENERG RENOV ocean thermal conversion; ~ **termica solar** ƒ ENERG RENOV solar thermal conversion; ~ **termiónica** ƒ ING ELÉC, NUCL thermionic conversion; ~ **termoeléctrica** ƒ ING ELÉC thermoelectric conversion; ~ **triángulo-estrella** ƒ ELEC *conexión* delta-to-star conversion; ~ **de voltaje en frecuencia** ƒ ING ELÉC voltage-to-frequency conversion

conversor *m* TELECOM converter; ~ **analógico-digital** *m* ELEC, ELECTRÓN, FÍS, FÍS PART, INFORM&PD, TELECOM analog-to-digital converter (*ADC*); ~ **CC/CC** *m* TEC ESP DC/DC converter; ~ **digital-analógico** *m* ELECTRÓN, INFORM&PD, TELECOM, TV digital-to-analog converter (*DAC*); ~ **de energía** *m* ING ELÉC, TELECOM, TERMO energy converter; ~ **de imagen por rayos infrarrojos** *m* TV infrared image converter; ~ **de impedancia negativa** *m* ING ELÉC negative-impedance converter; ~ **de lenguaje** *m* INFORM&PD

language translator; ~ **de paralelo a serie** *m* ING ELÉC parallel-to-serial converter; ~ **pentarrejilla** *m* ING ELÉC pentagrid converter

convertidor *m* CINEMAT converter, ELEC converter, *corriente alterna* rotary converter, ELECTRÓN inverter, converter, ING ELÉC converter, inverter, selector, rotary converter, INSTAL TERM, TELECOM, VEH converter; ~ **A/D** *m* ELEC, ELECTRÓN, FÍS, FÍS PART, INFORM&PD, TELECOM analog-to-digital converter (*ADC*); ~ **A/D de alta velocidad** *m* ELECTRÓN flash analog-to-digital converter; ~ **analógico-digital** *m* ELEC, ELECTRÓN, FÍS, FÍS PART, INFORM&PD, TELECOM analog-to-digital converter (*ADC*); ~ **de arco de mercurio** *m* ELEC, ING ELÉC mercury-arc converter; ~ **ascendente** *m* TELECOM upconverter; ~ **de autoconmutación** *m* ELEC self-commutated converter; ~ **Bessemer** *m* ING MECÁ, INSTAL TERM, PROD Bessemer converter; ~ **de CA directa** *m* ELEC direct AC converter; ~ **de campo giratorio** *m* ELEC, ING ELÉC rotary-field converter; ~ **catalítico** *m* AUTO, CONTAM, VEH *sistema de escape* catalytic converter; ~ **de CC** *m* ELEC DC converter, direct DC converter; ~ **de CC en CA** *m* ING ELÉC DC-to-AC converter; ~ **de CC en CC** *m* ING ELÉC DC-to-DC converter; ~ **de cintas de filamentos continuos a discontinuos** *m* TEXTIL toe-to-top converter; ~ **de código** *m* INFORM&PD, ING ELÉC code converter; ~ **conmutado de carga** *m* ELEC load-commutated converter; ~ **del contador de carga** *m* TELECOM charge-metering converter; ~ **de corriente continua** *m* TELECOM direct current converter; ~ **de datos** *m* ELECTRÓN, INFORM&PD data converter; ~ **digital** *m* ELECTRÓN digital converter; ~ **digital-analógico** *m* ELECTRÓN, INFORM&PD, TELECOM, TV digital-to-analog converter (*DAC*); ~ **de electromotor** *m* ING ELÉC motor converter; ~ **elevador** *m* TELECOM upconverter; ~ **elevador de frecuencia** *m* TELECOM upconverter; ~ **de energía** *m* ING ELÉC, TEC ESP power converter; ~ **estático** *m* ELEC, ING ELÉC, TEC ESP static converter; ~ **de exploración** *m* ELECTRÓN scan converter; ~ **de fase** *m* ELEC *corriente alterna* phase converter, ELECTRÓN phase changer, phase converter, TV phase converter; ~ **de frecuencia** *m* ELEC, ELECTRÓN frequency changer, frequency converter, TELECOM frequency converter; ~ **de frecuencia con colector** *m* ELEC, ELECTRÓN commutator-type frequency; ~ **de la frecuencia de corriente** *m* ELEC, TELECOM frequency-current converter; ~ **de frecuencia de inducción** *m* ELEC induction frequency converter; ~ **de frecuencia rotativo** *m* ELEC, ING ELÉC rotary frequency-converter; ~ **de frecuencia de video** *AmL*, ~ **de frecuencia de vídeo** *Esp m* TV video frequency converter; ~ **giratorio** *m* ELEC *corriente alterna*, ING ELÉC rotary converter; ~ **granangular** *m* FOTO wide angle converter; ~ **de imagen** *m* CINEMAT, ELECTRÓN image converter; ~ **invertido** *m* ELEC inverted converter; ~ **magnetohidrodinámico** *m* ING ELÉC, NUCL, TEC ESP magnetohydrodynamic converter; ~ **MHD** *m* ING ELÉC, NUCL, TEC ESP MHD converter; ~ **de normas** *m* TV standards converter; ~ **de octeto** *m* ELECTRÓN, INFORM&PD eight-bit converter; ~ **de par** *m* AUTO, ING MECÁ, VEH *transmisión* torque converter; ~ **del par motor** *m* AUTO, VEH *transmisión* torque converter; ~ **paralelo** *m* ELECTRÓN parallel converter; ~ **paralelo analógico-digital** *m* ELECTRÓN flash

analog-to-digital converter; ~ **de potencia rotativo** *m* ELEC *corriente alterna*, ING ELÉC rotary converter; ~ **de protocolo** *m* INFORM&PD, TELECOM protocol converter; ~ **rápido** *m* ELECTRÓN flash converter; ~ **del RCS** *m* ING ELÉC SCR converter; ~ **reticular** *m* ING ELÉC grating converter; ~ **de RF a IF** *m* TELECOM down converter; ~ **rotativo** *m* ELEC *corriente alterna*, ING ELÉC rotary converter; ~ **de señales** *m* TELECOM signal converter; ~ **de señales lógicas** *m* NUCL logic signal converter; ~ **de señales lógicas de disparo** *m* NUCL trip logic signal converter; ~ **en serie** *m* ING ELÉC series converter; ~ **en serie de analógico a digital** *m* CONST serial analog-to-digital converter; ~ **de serie a paralelo** *m* ING ELÉC serial-to-parallel converter; ~ **sincrónico** *m* ELEC *corriente alterna* rotary converter, *suministro de corriente alterna* synchronous converter, ING ELÉC rotary converter; ~ **síncrono** *m* ING ELÉC synchronous converter, synchronous inverter; ~ **de sistema analógico a sistema digital en serie** *m* ELECTRÓN serial analog-to-digital converter; ~ **de sistema decimal a binario** *m* ELECTRÓN decimal-to-binary converter; ~ **termal** *m* ING ELÉC thermal converter; ~ **termiónico** *m* ING ELÉC thermionic converter; ~ **de termopar** *m* ING ELÉC thermocouple converter; ~ **de tiratrón** *m* ELECTRÓN thyratron inverter; ~ **de torque** *m* ING MECÁ torque converter; ~ **de torsión** *m* ING MECÁ torque converter; ~ **para UHF** *m* TV UHF converter; ~ **de voltaje en frecuencia** *m* ING ELÉC voltage-to-frequency converter
convertir *vt* ELEC, ELECTRÓN, INFORM&PD, ING ELÉC convert; ~ **en coque** *vt* CARBÓN, MINAS coke; ~ **en digitales** *vt* IMPR digitize; ~ **en valor numérico** *vt* ING MECÁ digitize
convolución *f* ING MECÁ convolution
convolvulina *f* QUÍMICA convolvulin
convoy *m* D&A, TRANSP MAR convoy; ~ **de barcazas** *m* TRANSP multiple barge convoy set; ~ **de buques graneleros** *m* TRANSP bulk-ship train; ~ **en carretera** *m* TRANSP road train; ~ **en carretera de larga distancia** *m* TRANSP long-distance road train; ~ **remolcado** *m* TRANSP towed convoy
convoyar *vt* D&A, TRANSP MAR convoy
coordenada *f* CONST, FÍS, GEOM, INFORM&PD, ING MECÁ, MATEMÁT coordinate; ~ **Cartesiana** *f* CONST, ELECTRÓN, FÍS, GEOM, MATEMÁT Cartesian coordinate; ~ **curvilínea** *f* FÍS, GEOM curvilinear coordinate; ~ **x** *f* FÍS x-coordinate; ~ **y** *f* FÍS y-coordinate; ~ **z** *f* FÍS Z coordinate
coordenadas: ~ **atómicas** *f pl* CRISTAL atomic coordinates; ~ **de centro de masa** *f pl* FÍS, MECÁ, TEC ESP center of mass coordinates (*AmE*), centre of mass coordinates (*BrE*); ~ **cilíndricas** *f pl* FÍS cylindrical coordinates; ~ **de color** *f pl* FÍS RAD color coordinates (*AmE*), colour coordinates (*BrE*); ~ **cromáticas** *f pl* FÍS chromatic coordinates; ~ **de cromaticidad** *f pl* FÍS chromaticity coordinates; ~ **esféricas** *f pl* FÍS spherical coordinates; ~ **generalizadas** *f pl* FÍS generalized coordinates; ~ **de hora universal** *f pl* (*CUT*) METEO, TELECOM universal time coordinates (*UTC*); ~ **normales** *f pl* FÍS normal coordinates; ~ **paramétricas** *f pl* GEOM parametric coordinates; ~ **polares** *f pl* FÍS polar coordinates; ~ **reducidas** *f pl* FÍS reduced coordinates
coordinación: ~ **inductiva** *f* ING ELÉC inductive coordination

coordinador: ~ **de comunicación colectiva** *m* TELECOM conference-call chairman, conference-call chairperson
coordinar *vt* CONST, TELECOM coordinate
copa *f* PROD *altos hornos* hopper; ~ **de aceite** *f* ING MECÁ oil cup; ~ **esférica** *f* ING MECÁ ball cup; ~ **de flujo** *f* LAB *viscosidad de flujo* flow cup; ~ **de lubricación** *f* AmL (*cf engrasador Esp*) ING MECÁ oil cup, MINAS *herramienta*, PROD oiler
copal: ~ **fósil** *m* MINERAL fossil copal
copalina *f* MINERAL copaline
copalita *f* MINERAL copalite
copas: ~ **con pie alto** *f pl* C&V stemware
copela *f* QUÍMICA cupel
copia *f* CINEMAT, FOTO print, IMPR copy, print, dupe, hash, manifold, INFORM&PD copy; ~ **ampliada** *f* CINEMAT *del negativo*, FOTO enlargement print; ~ **anamórfica** *f* CINEMAT anamorphic print; ~ **de aprobación** *f* CINEMAT approval print; ~ **blanda** *f* INFORM&PD soft copy; ~ **en borrada** *f* CONST blueprint; ~ **de bromoleotipia** *f* FOTO bromoil print; ~ **de bromuro** *f* FOTO bromide print; ~ **cianográfica en papel sepia** *f* IMPR brown print; ~ **de cinta** *f* INFORM&PD tape copy; ~ **a color** *f* FOTO color print (*AmE*), colour print (*BrE*); ~ **a color en papel carbro** *f* FOTO carbro color print (*AmE*), carbro colour print (*BrE*); ~ **coloreada** *f* FOTO color print (*AmE*), colour print (*BrE*); ~ **coloreada en papel carbro** *f* FOTO carbro color print (*AmE*), carbro colour print (*BrE*); ~ **combinada** *f* CINEMAT combined print, composite print, IMPR combined print; ~ **por contacto** *f* CINEMAT, FOTO, IMPR contact print; ~ **CRI** *f* CINEMAT CRI print; ~ **cronometrada** *f* CINEMAT timed print; ~ **de cuña** *f* CINEMAT wedge print; ~ **definitiva** *f* CINEMAT release positive; ~ **definitiva de prueba** *f* CINEMAT final trial print; ~ **delgada** *f* CINEMAT thin print; ~ **para el depósito legal del libro** *f* IMPR deposit copy; ~ **directa** *f* CINEMAT direct print; ~ **dupe a color** *f* CINEMAT color dupe print (*AmE*), colour dupe print (*BrE*); ~ **con efecto de sombreado** *f* CINEMAT soot-and-whitewash print; ~ **en forroprusiato** *f* CONST blueprint; ~ **fotográfica** *f* IMPR photographic print; ~ **fotográfica esmaltada** *f* FOTO glossy print; ~ **fotostática** *f* IMPR stat; ~ **de gradación** *f* CINEMAT grading copy; ~ **heliográfica** *f* CONST, IMPR, ING MECÁ, PROD blueprint; ~ **en idioma original** *f* CINEMAT original language print; ~ **impresa** *f* IMPR *informática* hard copy, INFORM&PD printout, hard copy, TEC ESP *computadores* hard copy; ~ **intermedia** *f* CINEMAT lavender, FOTO lavender print; ~ **intermedia reversible a color** *f* CINEMAT, FOTO color reversal intermediate (*AmE*), colour reversal intermediate (*BrE*); ~ **litográfica** *f* IMPR lithographic print; ~ **local** *f* TELECOM local copy; ~ **maestra de grano fino** *f* CINEMAT, FOTO fine-grain master; ~ **maestra magnética** *f* CINEMAT magnetic master; ~ **maestra de montaje** *f* CINEMAT, TV edit master; ~ **maestra de protección** *f* CINEMAT protection master; ~ **maestra reversible** *f* CINEMAT, IMPR reversal master print; ~ **maestra con separación de colores** *f* CINEMAT separation master; ~ **magnetoóptica** *f* CINEMAT mag-optical print; ~ **de memoria para restauración del sistema** *f* INFORM&PD rescue dump; ~ **monoiluminada** *f* CINEMAT one-light print; ~ **montada** *f* CINEMAT, TV edited print; ~ **negativa** *f*

FOTO negative print; ~ **de negativo completo** *f* FOTO full-frame print; ~ **numerada** *f* IMPR numbered copy; ~ **óptica** *f* CINEMAT optical print; ~ **ozalida** *f* IMPR ozalid; ~ **en pantalla** *f* INFORM&PD soft copy; ~ **en papel bromuro** *f* IMPR bromide print; ~ **de un plano** *f* CONST drawing print; ~ **positiva maestra** *f* CINEMAT master positive; ~ **positiva con separación de colores** *f* CINEMAT separation positive; ~ **positiva de sonido** *f* CINEMAT sound positive; ~ **positiva con sonido directo** *f* CINEMAT sound direct positive; ~ **positiva de sonido óptico** *f* CINEMAT optical sound positive; ~ **preliminar** *f* IMPR advance copy; ~ **de protección** *f* CINEMAT protection copy; ~ **de prueba combinada** *f* CINEMAT trial composite print; ~ **de prueba combinada definitiva** *f* CINEMAT final trial composite; ~ **rechazada** *f* IMPR kick copy; ~ **de reducción** *f* CINEMAT, FOTO, TV reduction print; ~ **de referencia** *f* CINEMAT, TV reference print; ~ **de registro** *f* CINEMAT register print; ~ **y repetición** *f* IMPR step-and-repeat; ~ **de reserva** *f* INFORM&PD backup; ~ **reversible** *f* CINEMAT, IMPR reversal print; ~ **de segunda generación** *f* CINEMAT second generation copy; ~ **sincronizada** *f* CINEMAT *con imagen y sonido* married print; ~ **de trabajo** *f* CINEMAT cutting copy, slash print, work print; ~ **de transmisión** *f* CINEMAT transmission copy; ~ **verde** *f* CINEMAT green print; ~ **virada** *f* CINEMAT toned print; ~ **de xenón** *f* CINEMAT xenon print

copiado *m* ACÚST reproducing, reproduction; ~ **uno a uno** *m* CINEMAT one-to-one printing; ~ **anamorfósico** *m* CINEMAT anamorphosing printing; ~ **A y B** *m* CINEMAT A and B printing; ~ **en bucle** *m* CINEMAT loop printing; ~ **escalonado** *m* CINEMAT step printing; ~ **por exploración de imagen** *m* CINEMAT scanning printing; ~ **de la imagen** *m* IMPR imaging; ~ **por inversión** *m* CINEMAT, FOTO, IMPR reverse printing; ~ **inverso** *m* CINEMAT, FOTO, IMPR reverse printing; ~ **multicuadro** *m* CINEMAT multiple frame printing; ~ **para obtener ampliaciones** *m* CINEMAT blow-up printing; ~ **opalino** *m* CINEMAT opal printing; ~ **por secciones** *m* CINEMAT section printing; ~ **de vaivén** *m* CINEMAT back-and-forth printing; ~ **con ventanilla fluida** *m* CINEMAT fluid-gate printing; ~ **de ventanilla líquida** *m* CINEMAT liquid-gate printing

copiapoíta *f* MINERAL copiapite

copiar *vt* CINEMAT print, strike, FOTO duplicate, ING MECÁ trace; ~ **descomprimiendo la imagen** *vt* CINEMAT stretch-frame print; ~ **dos veces** *vt* CINEMAT double-print

copiloto *m* TRANSP AÉR copilot

copión *m* CINEMAT, TV cue print, *primera copia del negativo* rushes

coplanar *adj* ACÚST, FÍS, GEOM coplanar

copo *m* AGRIC flake, TEXTIL *fibras* tuft; ~ **de maíz** *m* AGRIC corn flake (*AmE*), maize flake (*BrE*); ~ **de nieve** *m* METEO snowflake

copolimerización *f* P&C *polimerización*, QUÍMICA, TEC PETR *petroquímica*, TEXTIL copolymerization

copolímero *m* P&C *polimerización*, QUÍMICA, TEC PETR, TEXTIL copolymer; ~ **de bloque** *m* P&C *polimerización* block copolymer; ~ **butadieso-estireno** *m* TEC PETR *petroquímica* butadiene styrene copolymer; ~ **etileno** *m* (*EVA*) P&C ethylene-vinyl acetate (*EVA*)

copos: ~ **de avena** *m pl* AGRIC rolled oats

coprecipitación *f* NUCL coprecipitation

coprocesador *m* INFORM&PD coprocessor

coprolito *m* GEOL *paleontología* coprolite

copropietario *m* TRANSP MAR co-owner, part owner; ~ **de apartadero particular** *m* FERRO co-owner of private siding

coprostanol *m* QUÍMICA *esteroide* coprostanol

coque *m* CARBÓN, MINAS, PROD, QUÍMICA, TERMO coke; ~ **de alto horno** *m* CARBÓN blast-furnace coke; ~ **metalúrgico** *m* CARBÓN blast-furnace coke; ~ **pulverizado** *m* CARBÓN, MINAS coke dust

coquificación *f* CARBÓN, MINAS coking; ~ **retardada** *f* TERMO delayed coking

coquificante *adj* CARBÓN, MINAS coking

coquificar *vt* CARBÓN, MINAS coke

coquilla *f* PROD *de un molde* part; ~ **de colada** *f* PROD *fundición* chill mold (*AmE*), chill mould (*BrE*)

coquizable *adj* CARBÓN, MINAS coking; **no** ~ *adj* CARBÓN, MINAS noncoking

coquización *f* CARBÓN, MINAS coking

coquizar *vt* CARBÓN, MINAS coke

coracita *f* MINERAL, NUCL coracite

coral *m* GEOL, OCEAN coral

coralífero *adj* GEOL, OCEAN coral

coralino *adj* GEOL, OCEAN coralline

coraza *f* D&A *de un buque* armor (*AmE*), armour (*BrE*), TEC ESP shell; ~ **magnética** *f* ELEC *campo* magnetic screening; ~ **térmica** *f* ING ELÉC, TRANSP AÉR heat shield

corazón: ~ **de arrastre** *m* ING MECÁ dog

corchete *m* CONST hook, *mampostería, sillería* cramp, ING MECÁ clasp, TEXTIL *en las prendas* fastening; ~ **con extremos invertidos** *m* CONST cramp with turned-down ends; ~ **grande en forma de G** *m* CONST deep pattern G cramp; ~ **de hierro** *m* CONST cramp iron; ~ **de hierro con ganchos** *m* CONST cramp iron with stone hook; ~ **macho** *m* ING MECÁ hook

corcho *m* CONST, LAB, REFRIG *para aislar* cork; ~ **de la botella** *m* EMB bottle stopper; ~ **expandido** *m* REFRIG expanded cork; ~ **fósil** *m* MINERAL mountain cork; ~ **granulado** *m* REFRIG granulated cork; ~ **de montaña** *m* MINERAL mountain cork

cordaje *m* ING MECÁ banding, TRANSP MAR rope; ~ **de amarre** *m* TRANSP MAR lashing, mooring rope

cordel *m* PROD line, TEXTIL twine, TRANSP MAR line; ~ **de corredera** *m* TRANSP MAR log line

cordierita *f* C&V cordierite, MINERAL cordierite, dichrote, iolite

cordillera *f* GAS limits

cordón *m* CONST *vigas* boom, ELEC *cable* cord, INFORM&PD cable, ING MECÁ former, cord, OCEAN *aparejos*, TEC ESP strand, TELECOM, TEXTIL cord; ~ **de cableado** *m* TRANSP AÉR lacing cord; ~ **detonante** *m* MINAS detonating cord; ~ **de espoleta** *m* TEC ESP fuse cord; ~ **del fondo** *m* MECÁ root pass; ~ **de hielo** *m* OCEAN ice ridge; ~ **del instrumento** *m* TELECOM instrument cord; ~ **litoral** *m* GEOL barrier beach, offshore bar, OCEAN offshore bar; ~ **del microteléfono** *m* TELECOM *teléfono* handset cord; ~ **de raíz** *m* PETROL root bead, root pass; ~ **tensor** *m* ING MECÁ tightening cord

cordoncillo *m* C&V braid

corindón *m* C&V corundum, MINERAL adamantine spar, corundum

corinita *f* MINERAL corynite

cornamusa *f* TRANSP MAR *equipamiento de cubierta*

belaying cleat, cleat; ~ **de amarre** *f* TRANSP MAR *construcción naval, montaje de cubierta* mooring cleat; ~ **atochante** *f* TRANSP MAR *montaje de cubierta* jam cleat

corneana *f* GEOL, PETROL hornfels; ~ **calco-lilicatada** *f* GEOL calc-silicate hornfels

cornezuelo *m* AGRIC *cereales*, ALIMENT ergot

cornijal *m* CONST corner post

cornisa *f* CONST cornice, OCEAN seaboard, shore, TRANSP MAR *geografía* seaboard; ~ **de la cabria** *f* AmL (cf *cúpula de la torre Esp*) TEC PETR derrick crown; ~ **marina** *f* OCEAN seaboard, shore

cornubianita *f* GEOL hornfels

corola *f* AGRIC corolla

corolario *m* MATEMÁT corollary

corona *f* Esp (cf *cumbrera AmL*) AGRIC crown, ELECTRÓN, FÍS, GAS corona, GEOFÍS *instrumento de prospección* aigrette, ING ELÉC corona, ING MECÁ crown, boring head, INSTAL HIDRÁUL *de turbina*, MECÁ ring, MINAS *perforadora diamantes* boring head, *entibación de galerías* crown, P&C corona, PROD *del malacate* drumhead, TEC ESP halo, corona, TEC PETR *perforación* crown, TRANSP AÉR annulus; ~ **circular** *f* TEC ESP, TEC PETR *perforación, producción* annulus; ~ **cortante** *f* ING MECÁ, MINAS *sondeos* boring head; ~ **delgada** *f* C&V light crown; ~ **densa** *f* C&V dense crown; ~ **dentada** *f* AUTO ring gear, VEH *grupo diferencial* ring gear, *transmisión, diferencial* crown wheel; ~ **dentada cónica** *f* AUTO crown wheel; ~ **dentada de puesta en movimiento del volante** *f* ING MECÁ, VEH *motor* flywheel starter ring gear; ~ **de diamante** *f* Esp (cf *cortanúcleos de diamante AmL*) TEC PETR *perforación* diamond bit, diamond core drill; ~ **de diamantes** *f* MINAS *para sondeos* crown impregnated with diamonds; ~ **del diferencial** *f* AUTO crown wheel; ~ **directriz** *f* INSTAL HIDRÁUL *de una turbina* guide ring; ~ **de empaquetadura** *f* ING MECÁ packing gland; ~ **y piñón** *f* AUTO, VEH ring and pinion; ~ **del pistón** *f* VEH *motor* piston crown; ~ **del puerto** *f* C&V port crown (*BrE*), uptake crown (*AmE*); ~ **de puntas de diamante** *f* PROD diamond boring crown; ~ **de refuerzo interior** *f* TRANSP AÉR inner shroud; ~ **sentada** *f* C&V saddled finish; ~ **solar** *f* GEOFÍS solar corona; ~ **de sondeo** *f* MINAS core bit, core drilling; ~ **terminal** *f* GEOFÍS *instrumento de prospección* terminal aigrette; ~ **testiguera** *f* MINAS *sondeos* core cutter; ~ **para tintura** *f* TEXTIL mock cake; ~ **de la torre** *f* AmL (cf *cúpula de la torre Esp*) TEC PETR *perforación* derrick crown, crown block

coronación *f* AGUA crest, CONST *arco* crown, *arquitectura* crowning

coronado: ~ **de nieve** *adj* METEO snow-capped

coronamiento *m* AGUA coping, crest, CARBÓN *muros, presas* slope top, ENERG RENOV, HIDROL *presas* crest, TRANSP MAR *construcción naval* taffrail; ~ **de sifón** *m* ENERG RENOV siphon crest; ~ **del talud** *m* CARBÓN slope top

coronar *vt* CONST *tejidos* cap

coroneno *m* QUÍMICA coronene

corral *m* OCEAN *redes de pesca* pound net; ~ **de engorde** *m* AGRIC feedlot; ~ **de peces** *m* OCEAN fish corral

correa *f* CONST purlin, strap, band, FOTO strap, ING MECÁ belt, strap, MECÁ belt, PROD strap; ~ **abierta** *f* PROD open belt; ~ **acanalada** *f* ING MECÁ ribbed V-belt; ~ **de balata** *f* ING MECÁ balata belt;

~ **conductora** *f* ING MECÁ driving belt; ~ **conductora dentada** *f* ING MECÁ toothed drive belt; ~ **cruzada** *f* PROD *transmisiones* halved belt; ~ **de cuero** *f* PROD leather belt; ~ **en cuña** *f* ING MECÁ V-belt; ~ **dentada** *f* AUTO cogged belt, VEH *distribución* cog belt, cogged belt, notched belt; ~ **de distribución** *f* VEH *mando del árbol de levas* timing belt; ~ **sin fin** *f* PROD continuous belt, VEH *sistema de refrigeración del motor* endless belt; ~ **hexagonal sin fin** *f* ING MECÁ endless hexagonal belt; ~ **hidráulica** *f* PROD belt pump; ~ **motriz** *f* ING MECÁ driving belt; ~ **Pag** *f* CINEMAT Pag belt; ~ **de polipropileno** *f* EMB polypropylene strap; ~ **de potencia** *f* CINEMAT power belt; ~ **recta** *f* PAPEL, PROD open belt; ~ **de regulación** *f* MECÁ timing belt; ~ **de remolcar** *f* TRANSP tow strap; ~ **de rescate** *f* SEG, TRANSP AÉR, TRANSP MAR rescue strap; ~ **semicruzada** *f* PROD quarter belt, quarter-turn belt, quarter-twist belt, quartered belt; ~ **sinfín** *f* AGRIC *máquinas de clasificación* apron; ~ **de transmisión** *f* ING MECÁ, MECÁ, PROD *de máquina* belt; ~ **transmisora** *f* ING MECÁ driving belt; ~ **transmisora plana** *f* ING MECÁ flat transmission belt; ~ **transportadora** *f* AmL (cf *cinta transportadora Esp*) AGRIC, CONST, EMB, ING MECÁ, MECÁ, MINAS, PROD, TRANSP belt conveyor, conveyor belt, traveling apron (*AmE*), travelling apron (*BrE*); ~ **transportadora con armazón textil** *f* ING MECÁ conveyor-belt with a textile carcass; ~ **trapezoidal** *f* AUTO V-belt, ING MECÁ ribbed V-belt, V-belt, MECÁ, P&C V-belt, TEXTIL V-belting; ~ **trapezoidal ancha sin fin** *f* ING MECÁ endless wide V-belt; ~ **trapezoidal sin fin** *f* ING MECÁ endless V-belt; ~ **trapezoidal múltiple** *f* ING MECÁ multi-V belt; ~ **de tres capas** *f* ING MECÁ three-ply belting; ~ **triple** *f* ING MECÁ three-ply belting; ~ **para tubería** *f* CONST pipe strap; ~ **en V** *f* AUTO, ING MECÁ, MECÁ, P&C, VEH *sistema de refrigeración* V-belt; ~ **en V múltiple** *f* ING MECÁ multi-V belt; ~ **del ventilador** *f* AUTO, MECÁ, VEH *refrigeración* fan belt

correaje *m* D&A harness, PROD belting; ~ **de cuero** *m* PROD leather belting; ~ **de cuero articulado** *m* PROD leather-link belting; ~ **de goma** *m* ING MECÁ, P&C rubber belting; ~ **de seguridad** *m* TRANSP MAR safety harness

correas: ~ **articuladas** *f pl* P&C *de máquina* link belting; ~ **de cuero** *f pl* PROD leather belting; ~ **de cuero articulado** *f pl* PROD leather-link belting; ~ **de ocho capas** *f pl* PROD eight-ply belting; ~ **de seis pliegues** *f pl* PROD six-ply belting; ~ **de transmisión de caucho** *f pl* ING MECÁ, P&C rubber belting

corrección *f* ING MECÁ correction, rectification; ~ **de la aberración cromática** *f* TV color correction (*AmE*), colour correction (*BrE*); ~ **de absorción** *f* CRISTAL absorption correction; ~ **de aire libre** *f* GEOFÍS *anomalía gravimétrica*, GEOL free-air correction; ~ **de alargamiento** *f* TV aspect-ratio adjustment; ~ **de altura** *f* GEOFÍS *anomalía gravimétrica* height correction; ~ **anticipada de errores** *f* AmL (cf *corrección de errores hacia adelante Esp*) INFORM&PD, TELECOM forward error correction (*FEC*); ~ **apocromática** *f* FOTO *que elimina los defectos de lineatura* apochromatic correction; ~ **auditiva** *f* ACÚST hearing correction; ~ **automática de errores** *f* TELECOM, TRANSP AÉR automatic error correction; ~ **de Bouguer** *f* GEOFÍS *anomalía gravimétrica* Bouguer correction; ~ **de la carta** *f* TRANSP

MAR chart correction; ~ **del contorno** *f* TV edge correction; ~ **dinámica** *f* GEOL move-out correction, PETROL dynamic correction; ~ **de la distorsión de imagen** *f* TV field tilt; ~ **del error de base de tiempo** *f* TV time base error correction; ~ **de errores** *f* ELECTRÓN, INFORM&PD, TELECOM error correction; ~ **de errores hacia adelante** *f Esp* (*cf corrección anticipada de errores AmL*) INFORM&PD, TELECOM forward error correction (*FEC*); ~ **estática** *f* TEC PETR static correction; ~ **del factor de potencia** *f* ING ELÉC power-factor correction; ~ **de fase** *f* ELEC *corriente alterna*, TELECOM phase compensation; ~ **forzada de visualización de mando para varios** *f* TELECOM multipoint command visualization forcing (*MCV*); ~ **gamma** *f* CINEMAT gamma correction; ~ **por la marea** *f* OCEAN tidal correction; ~ **menor** *f* IMPR minus correction; ~ **de órbita** *f* TEC ESP orbit correction; ~ **de paralaje** *f* CINEMAT parallax correction; ~ **de pruebas** *f* IMPR proofreading; ~ **de la retrodispersión por saturación** *f* NUCL saturation back-scattering correction; ~ **de rumbo** *f* TRANSP AÉR course alignment; ~ **del tiro** *f* D&A fire correction; ~ **de trayectoria** *f* D&A, TEC ESP path correction

correcciones: ~ **diferenciales** *f pl* TRANSP MAR *navegación por satélite* differential corrections; ~ **normales de move-out** *f pl* TEC PETR *geofísica* normal move-out corrections

corrector *m* IMPR, INFORM&PD reader; ~ **de altitud** *m* AUTO altitude corrector; ~ **de amplitud** *m* TV amplitude corrector; ~ **automático de la aberración cromática** *m* TV color automatic timebase corrector (*AmE*), colour automatic time base corrector (*BrE*); ~ **de base de tiempo** *m* TV timebase corrector (*TBC*); ~ **por chorro de aire** *m* IMPR air eraser; ~ **del cuadro cromático** *m* TV color-field corrector (*AmE*), colour-field corrector (*BrE*); ~ **de distorsión de imagen** *m Esp* (*cf mezclador de la distorsión de la imagen AmL*) TV tilt mixer; ~ **gamma** *m* TV gamma corrector; ~ **de impedancia** *m* ELEC impedance corrector; ~ **de puntos y formas** *m* TV spot-shape corrector; ~ **del sombreado** *m* TV shading corrector

corredera *f* ING MECÁ runner, MECÁ *máquinas herramientas* ram, MINAS *sondeos* drilling jar, guide, jar, PROD *sondeos* jar; ~ **de ajuste** *f* ING MECÁ gib; ~ **de barras abiertas** *f* ING MECÁ link motion with open rods; ~ **de barras cruzadas** *f* ING MECÁ link motion with crossed rods; ~ **de chapa** *f* ING MECÁ plate link; ~ **de la cruceta** *f* ING MECÁ crosshead block; ~ **electrónica** *f* TRANSP MAR *equipamiento electrónico* speedometer; ~ **de esmerilado** *f* C&V grinding runner; ~ **de expansión** *f* INSTAL HIDRÁUL *distribuidor máquina de vapor* expansion plate, expansion slide; ~ **de macho angular** *f* ING MECÁ *moldes de inyección* angled core slide; ~ **mecánica** *f* TRANSP MAR passenger ship; ~ **de parrilla** *f* PROD gridiron valve; ~ **portaherramientas** *f* ING MECÁ tool slide, toolholding slide; ~ **de potenciómetro** *f* ELEC *resistor* potentiometer slider; ~ **de retroceso** *f* D&A recoil slide; ~ **del sector Stephenson** *f* ING MECÁ link block, die

corredor *m* CONST walkway; ~ **aéreo** *m* TRANSP AÉR air corridor; ~ **de ascenso** *m* TEC ESP, TRANSP AÉR climb corridor; ~ **estrecho** *m* CONST catwalk; ~ **fletador** *m* TRANSP MAR chartering broker; ~ **marítimo** *m* TRANSP MAR ship broker; ~ **de reentrada óptimo** *m*

TEC ESP optimum re-entry corridor; ~ **de salida al mar** *m* OCEAN *estado sin litoral o zona mediterránea* shipping corridor; ~ **de subida** *m* TEC ESP, TRANSP AÉR climb corridor

corregir *vt* D&A *el tiro* register, INFORM&PD *programas* patch, ING MECÁ true, amend, PROD make good, repair; ~ **provisionalmente** *vt* INFORM&PD patch

correhuela *f* AGRIC *mala hierba* hedge bindweed

correílla *f* PROD *correas* lace

correlación *f* GEN correlation; ~ **cruzada** *f* ELECTRÓN cross correlation; ~ **doble de velocidad** *f* FÍS FLUID *turbulencia* double velocity correlation; ~ **espacio-temporal** *f* TELECOM space-time correlation; ~ **múltiple** *f* TEC ESP multiplexing; ~ **temporal** *f* TELECOM time correlation

correlacionado *adj* GEN correlated

correlador *m* ELECTRÓN, TELECOM correlator; ~ **cruzado** *m* ELECTRÓN cross correlator; ~ **óptico** *m* ELECTRÓN optical correlator

correntímetro *m* TRANSP MAR *oceanografía* current meter

correntómetro *m* AGUA, HIDROL current meter; ~ **acústico** *m* OCEAN acoustic current meter; ~ **electromagnético** *m* OCEAN electromagnetic current meter

correo *m* INFORM&PD mailing; ~ **aéreo** *m* TRANSP AÉR airmail; ~ **electrónico** *m* ELECTRÓN, INFORM&PD, TELECOM electronic mail (*e-mail*); ~ **telefónico** *m* TELECOM voice mail

correr[1] *vti* ING MECÁ run

correr[2] *vi* HIDROL *las aguas* run down; ~ **un bordo** *vi* TRANSP MAR *navegación a vela* make a tack; ~ **con los costes del accionariado** *vi* TEXTIL bear stockholding costs; ~ **tabla por la red** *vi* INFORM&PD surf the net ~ **un temporal** *vi* TRANSP MAR *navegación a vela* weather a storm

corrrerse *v refl* IMPR *tinta* bleed, TRANSP MAR *carga* shift

correspondencia *f* ACÚST correspondence; ~ **radiotelefónica** *f* TELECOM voice mail; ~ **reticular** *f* METAL lattice correspondence; ~ **telefónica** *f* TELECOM voice mail

correvuela *f* AGRIC bindweed

corrida: ~ **de máquina** *f* ING MECÁ machine run

corriente[1]: **con** ~ *adj* ELEC live, *circuito* alive, FÍS, PROD *electricidad* live

corriente[2]: ~ **abajo** *adv* MECÁ downstream; **a contra** ~ *adv* TRANSP MAR upstream; **contra la** ~ *adv* TRANSP MAR upstream; **con** ~ *adv* ING ELÉC alive; ~ **abajo** *adv* FÍS, FÍS FLUID, GAS, TRANSP MAR downstream; ~ **arriba** *adv* GAS upstream

corriente[3] *f* AGUA efflux, ELEC *red de distribución* current (*AmE*), current supply (*AmE*), mains (*BrE*), mains supply (*BrE*), FÍS current, HIDROL *marítima* current, *de agua* flow, ING ELÉC current, INSTAL HIDRÁUL flow, efflux, INSTAL TERM *de aire* draught (*BrE*), draft (*AmE*), MECÁ flow, OCEAN current, REFRIG, TEC PETR flow, TRANSP MAR *mareas* current, TV current (*AmE*), mains (*BrE*);

~ **a** ~ **de absorción** *f* ELEC absorption current; ~ **activa** *f* ELEC, FÍS active current, ING ELÉC phase current; ~ **activada** *f* ING ELÉC forward current, on-state current; ~ **admisible** *f* ING ELÉC permissible current; ~ **de agua** *f* CONST, CONTAM watercourse, HIDROL stream, race, *arroyo, canal* watercourse, INSTAL HIDRÁUL *conducto de agua* race; ~ **de agua**

decreciente *f* HIDROL losing stream; ~ **de agua dulce** *f* HIDROL freshet; ~ **de agua efluente** *f* HIDROL effluent stream; ~ **de agua permanente** *f* HIDROL permanent flow; ~ **de agua subterránea** *f* HIDROL underground flow, underground stream; ~ **de aire** *f* ING MECÁ draft (*AmE*), draught (*BrE*), MINAS *ventilación* air blast, PROD blow, draft (*AmE*), draught (*BrE*); ~ **de aire caliente** *f* METEO warm air stream, TEXTIL hot air stream; ~ **de aire descendente** *f* ING MECÁ downdraft (*AmE*), down-draught (*BrE*), MINAS *ventilación* downcast, TRANSP AÉR downdraft (*AmE*), downdraught (*BrE*); ~ **de aire entrante** *f* MINAS ingoing air current; ~ **de aire exterior** *f* MINAS outgoing air current; ~ **de aire forzada** *f* INSTAL TERM forced draft (*AmE*), forced draught (*BrE*), ING MECÁ induced draft (*AmE*), induced draught (*BrE*); ~ **de aire natural** *f* INSTAL TERM natural draft (*AmE*), natural draught (*BrE*); ~ **de aire a presión** *f* ING MECÁ forced draft (*AmE*), forced draught (*BrE*); ~ **de alimentación** *f* ING ELÉC feed, supply current; ~ **de alimentación por choque** *f* ELEC *fuente de alimentación* choke feed; ~ **de alta frecuencia** *f* ELEC *CA* high-frequency current; ~ **alterna** *f* (*CA*) ELEC, ELECTRÓN, FÍS, ING ELÉC, METAL, PROD, QUÍMICA, TEC ESP, TELECOM, TV alternating current (*AC*); ~ **alterna rectificada** *f* ELEC, ING ELÉC rectified alternating-current; ~ **alternativa** *f* (*CA*) ELEC, ELECTRÓN, FÍS, ING ELÉC, METAL alternating current (*AC*), OCEAN *mareas* reversing current, PROD, QUÍMICA, TEC ESP, TELECOM, TV alternating current (*AC*); ~ **del ánodo** *f* ING ELÉC anode current; ~ **del arco** *f* ELEC arc current; ~ **de arranque** *f* ELEC breakaway starting current, OCEAN, TRANSP MAR rip current; ~ **de arrastre** *f* OCEAN entrainment current; ~ **de autoinducción** *f* ING ELÉC self-induction current;

~ b ~ **de barro con cantos gruesos** *f* GEOL *sedimentología* debris flow; ~ **de base de recombinación** *f* ELECTRÓN recombination base current; ~ **bifásica** *f* ELEC biphase current, ING ELÉC biphase current, two-phase current, REFRIG two-phase flow; ~ **bifurcada** *f* ELEC *sector* derived current; ~ **de borrado** *f* TV erasing current;

~ c ~ **de caldeo** *f* ELEC *de tubo, válvula* filament current, TERMO heating current; ~ **del campo inductor** *f* ING ELÉC field current; ~ **del cañón de electrones** *f* ELECTRÓN, FÍS, TV electron gun current; ~ **de carga** *f* ELEC *acumulador, batería* charging current; ~ **en carga** *f* ING ELÉC on-load current; ~ **sin carga** *f* ELEC no-load current; ~ **de choque** *f* ELEC impulse current; ~ **en chorro** *f* METEO jet stream; ~ **en circuito abierto** *f* ELEC open-circuit current; ~ **circular** *f* OCEAN rotary current; ~ **de compensación** *f* OCEAN compensation current; ~ **compensadora** *f* ELEC compensating current; ~ **de conducción** *f* ELEC, FÍS conduction current; ~ **conductiva** *f* ELEC, FÍS conduction current; ~ **conmutada** *f* ELEC switched current; ~ **de conservación** *f* ELEC *de relé* maintenance current; ~ **continua** *f* (*CC*) GEN continuous current (*CC*), direct current (*DC*), steady current; ~ **de convección** *f* C&V *en horno de vidrio* convection current, convection dryer, FÍS, TERMO convection current; ~ **conveccional** *f* ING ELÉC convection current; ~ **de corte** *f* ELEC *transistor* cutoff current; ~ **de cortocircuito** *f* ELEC, ELECTRÓN, ING ELÉC fault;

~ **costera** *f* OCEAN inshore current, longshore current; ~ **de crecida** *f* AGUA gaining stream; ~ **de cresta** *f* ELEC *de red de distribución* peak current;

~ d ~ **de datos** *f* INFORM&PD, TELECOM data stream; ~ **de deriva** *f* OCEAN, TRANSP MAR drift current; ~ **derivada** *f* ING ELÉC shunt current, PROD under-current; ~ **de desaccionamiento** *f* ELEC *relé* drop-out current; ~ **de descarga** *f* ELEC *de capacitor*, ING ELÉC discharge current; ~ **de descarga espontánea** *f* ELEC, ING ELÉC leakage current; ~ **de desconexión** *f* ELEC, ING ELÉC release current; ~ **de desenganche** *f* ING ELÉC drop-out current; ~ **de desplazamiento** *f* ELEC *ecuaciones de Maxwell*, FÍS, ING ELÉC displacement current; ~ **en diente de sierra** *f* TV sawtooth current; ~ **de difusión** *f* ELEC *electrólisis* diffusion current; ~ **directa** *f* C&V *del flujo del vidrio en el horno* direct current (*DC*), ELEC *de semiconductor* forward current, FERRO direct current (*DC*); ~ **de dispersión** *f* ELEC, ING ELÉC leakage current; ~ **de dispersión en empalme** *f* ING ELÉC junction leakage current; ~ **de drenaje** *f* ING ELÉC drain current;

~ e ~ **eficaz** *f* ING ELÉC effective current, RMS current; ~ **eléctrica** *f* ELEC, FÍS, FÍS RAD, ING ELÉC, TELECOM current, electric current; ~ **eléctrica monofásica** *f* CONST single-phase electric current; ~ **eléctrica monopolar** *f* CONST single-phase electric current; ~ **electrónica** *f* ELECTRÓN, ING ELÉC electron current; ~ **de emisión con campo nulo** *f* NUCL field-free emission current; ~ **de encendido** *f* ELEC *de tubo, válvula* filament current; ~ **enderezada** *f* ELEC, ING ELÉC rectified current; ~ **de entrada** *f* ELEC input current, ING ELÉC inrush current; ~ **de entrada diferencial** *f* ELECTRÓN differential input; ~ **equilibrada** *f* REFRIG *circulación, flujo* balanced flow; ~ **de escape** *f* ING MECÁ exhaust draft (*AmE*), exhaust draught (*BrE*); ~ **de estiraje** *f* C&V pull current (*BrE*), withdrawal current (*AmE*); ~ **exafásica** *f* ING ELÉC six-phase current; ~ **de excitación** *f* ELEC *electromagnetismo* excitation current, ING ELÉC energizing current; ~ **de extracción** *f* ING MECÁ exhaust draft (*AmE*), exhaust draught (*BrE*);

~ f ~ **de fango** *f* GEOL mud flow, OCEAN turbidity current; ~ **de fase** *f* ELEC *sistema*, ING ELÉC phase current; ~ **en fase** *f* ELEC in-phase current; ~ **de fase del motor** *f* PROD motor phase current; ~ **de filamento** *f* ELEC *de válvula, tubo*, ING ELÉC filament current; ~ **fluida** *f* TEC PETR *mecánica de fluidos* fluid flow; ~ **de fluidos** *f* TEC PETR *mecánica de fluidos* fluid flow; ~ **de fondo** *f* AGUA bottom flow, underflow, ENERG RENOV undercurrent, OCEAN bottom current, underflow; ~ **fotoeléctrica** *f* ELEC photoelectric current, ÓPT, TELECOM photocurrent; ~ **fotovoltaica** *f* ING ELÉC photovoltaic current; ~ **de Foucault** *f* ELEC, FERRO, FÍS eddy current, ING ELÉC eddy current, parasitic current, MECÁ, NUCL eddy current, OCEAN eddying current, TV eddy current; ~ **de fuga** *f* ELEC leakage current, GEOFÍS earth current (*BrE*), ground current (*AmE*), ING ELÉC dielectric current, leakage current, spike, stray current, TELECOM leakage current; ~ **de fuga a tierra** *f* ELEC, ING ELÉC earth-leakage current (*BrE*), ground-leakage current (*AmE*); ~ **de funcionamiento** *f* ELEC operating current, ING ELÉC operate current, operating current;

~ g ~ **galvánica** *f* ELEC, ING ELÉC galvanic current;

~ **giratoria de marea** f OCEAN rotary tidal current; ~ **de grabación** f TV *vídeo* record current; ~ **de gradiente** f OCEAN slope current;

~ h ~ **hacia adentro** f ING MECÁ indraft (*AmE*), indraft of air (*AmE*), indraught (*BrE*), indraught of air (*BrE*);

~ i ~ **de inducción** f ELEC, ING ELÉC induction current; ~ **inducida** f ELEC *electromagnetismo* induced current, *transformador* secondary current, ING ELÉC secondary current, INSTAL TERM *de aire* induced draught (*BrE*), induced draft (*AmE*), TELECOM induced current; ~ **inducida de desconexión** f ELECTRÓN, ING ELÉC break-induced current; ~ **del inducido** f ELEC, ING ELÉC armature current; ~ **del inductor** f ELEC field current; ~ **inductora** f ELEC *transformador* primary current, ING ELÉC field current, primary current; ~ **inicial** f ELEC, ING ELÉC initial current; ~ **instantánea** f ELEC, ING ELÉC instantaneous current; ~ **de instrucciones** f INFORM&PD instruction stream; ~ **intermitente** f HIDROL intermittent flow; ~ **interruptiva de cortocircuito** f PROD short-circuit interrupting current; ~ **interruptora admisible** f ELEC *cortacircuito* admissible interrupting current; ~ **de inundación** f OCEAN flood stream; ~ **inversa** f ELEC, HIDROL, ING ELÉC return current; ~ **invertida** f ELEC, ING ELÉC reverse current; ~ **iónica** f FÍS RAD ion current; ~ **de ionización** f ELECTRÓN ionization current;

~ l ~ **laminar** f GEOL sheet flow; ~ **de lava** f GEOL *estructura volcánica* lava flow, lava stream; ~ **de liberación** f ELEC, ING ELÉC release current; ~ **libre** f FÍS FLUID free-stream; ~ **limitadora** f ELEC limiting current; ~ **de línea** f ING ELÉC line current; ~ **líquida** f NUCL liquid flow; ~ **litoral** f OCEAN littoral current; ~ **de llamada** f TELECOM ringing current; ~ **de lodo y cantos rodados** f GEOL *masa de partículas volcánicas de grano fino* lahar; ~ **longitudinal** f C&V longitudinal current; ~ **luminosa** f ÓPT, TELECOM light current;

~ m ~ **magnetizante** f ING ELÉC magnetizing current; ~ **en la malla** f ING ELÉC mesh current; ~ **de mantenimiento** f ELEC *de relé* maintenance current; ~ **de la marea** f HIDROL, OCEAN, TRANSP MAR tidal stream; ~ **de marea** f ENERG RENOV, OCEAN tidal current, TRANSP MAR race, tidal current; ~ **de la marea menguante** f ENERG RENOV, HIDROL, OCEAN, TRANSP MAR *curso de navegación* ebb stream; ~ **marina** f HIDROL, OCEAN, TRANSP MAR *curso de navegación* current; ~ **máxima** f ELEC, ING ELÉC maximum current, *de red de distribución* peak current, FÍS peak current; ~ **máxima momentánea** f ELEC rated short-time current; ~ **máxima de soldadura** f CONST maximum welding current; ~ **meandriforme** f GEOL meandering stream; ~ **medida** f ELEC measured current; ~ **menguante** f ENERG RENOV, HIDROL, OCEAN, TRANSP MAR ebb current, ebb stream; ~ **mínima de soldadura** f CONST minimum welding current; ~ **momentánea** f PROD, TRANSP transient current; ~ **momentánea de chorro de arco** f PROD showering arc transient; ~ **monofásica** f CINEMAT, ING ELÉC single-phase current;

~ n ~ **natural** f FÍS streamline; ~ **de nivelación** f OCEAN gradient current; ~ **no disruptiva** f PROD withstand current; ~ **nodal** f ELEC nodal current; ~ **nominal** f ELEC, ING ELÉC rated current; ~ **nominal de soldadura** f CONST rated welding current;

~ **nominal de tránsito** f ELEC *cambiador de toma* rated through-current; ~ **normal** f ING ELÉC convection current;

~ o ~ **oceánica** f METEO, OCEAN, TRANSP MAR ocean current, oceanic current; ~ **ondulatoria** f ELEC, ING ELÉC pulsating current; ~ **operacional** f PROD operational current; ~ **de oscuridad** f ÓPT dark current; ~ **en la oscuridad** f ING ELÉC dark current;

~ p ~ **parásita** f ELEC, FERRO, FÍS eddy current, ING ELÉC eddy current, parasitic current, stray current, MECÁ, NUCL, TV eddy current; ~ **parcial** f MINAS split, *ventilación minas* split current, PROD *ventilación minas* split current; ~ **de paso al reposo** f ELEC *relé* drop-out current; ~ **de pérdida en puerta** f ING ELÉC gate leakage current; ~ **de pérdida a tierra** f PROD ground fault current; ~ **permanente** f OCEAN permanent current, stable current; ~ **de pico** f ELEC *de red de distribución* peak current; ~ **a plena carga** f ELEC, PROD full-load current; ~ **de polarización** f FÍS polarization current, TV biasing current; ~ **polarizante** f TV biasing current; ~ **polifásica** f ELEC polyphase current; ~ **portadora** f FÍS carrier; ~ **primaria** f ELEC *transformador* primary current; ~ **pulsada** f ELEC pulsed current; ~ **pulsatoria** f ELEC, ING ELÉC pulsating current, pulsatory current; ~ **de punta** f ING ELÉC peak current;

~ r ~ **reactiva** f ELEC, ING ELÉC idle current, reactive current, wattless current; ~ **de realimentación** f ELECTRÓN, ING ELÉC feedback current; ~ **rectificada** f ELEC, ING ELÉC rectified current; ~ **de la red eléctrica** f ING ELÉC mains current; ~ **de régimen** f ELEC, ING ELÉC rated current; ~ **de rejilla** f ELEC, ING ELÉC grid current; ~ **del relámpago** f ING ELÉC lightning current; ~ **remanente** f ELEC, ING ELÉC residual current; ~ **en remolino** f ELEC, FERRO, FÍS, ING ELÉC, MECÁ, NUCL, TV eddy current; ~ **de repique** f TELECOM ringing current; ~ **de resaca** f OCEAN, TRANSP MAR rip current, rip tide; ~ **residual** f ELEC *cortacircuito* cutoff current, residual current, FÍS *de reposo de una célula fotoeléctrica* dark current, ING ELÉC residual current, zero current, TELECOM dark current; ~ **de retención** f ELEC *relé* hold current, ING ELÉC holding current, latching current; ~ **de retorno** f ELEC, HIDROL, ING ELÉC return current; ~ **de retorno por tierra** f ELEC, ING ELÉC earth current (*BrE*), ground current (*AmE*); ~ **de RF** f ELEC, ELECTRÓN, ING ELÉC, TELECOM RF current, Rf current; ~ **del rotor** f TRANSP AÉR rotor stream; ~ **con rotor enclavado** f ELEC *máquina asíncrona* locked-rotor current; ~ **con rotor en reposo** f ELEC *máquina asíncrona* locked-rotor current; ~ **de ruptura** f ELEC *de relé* breaking current;

~ s ~ **de salida** f ELEC, ING ELÉC output, output current, PROD output; ~ **de salida regulada** f ING ELÉC regulated-output current; ~ **de saturación** f ELECTRÓN, FÍS saturation current; ~ **secundaria** f ELEC *transformador* secondary current; ~ **en sentido directo** f ELEC *de semiconductor* forward current; ~ **de servicio** f ELEC, ING ELÉC operating current; ~ **sinusoidal** f FÍS, ING ELÉC sinusoidal current; ~ **de sobrecarga** f ELEC *equipo, fuente de alimentación*, ING ELÉC overload current; ~ **de sobrecarga limitadora** f ELEC limiting overload current; ~ **de sobrevoltaje** f PROD transient current; ~ **subálvea** f AGUA underflow; ~ **subfluvial** f AGUA underflow;

~ **submarina** *f* AGUA, OCEAN, PROD undercurrent; ~ **submarina vaciante** *f* OCEAN *desplazamiento de masas acuáticas profundas* underwashing; ~ **subsuperficial** *f* OCEAN subsurface current, TRANSP MAR *estado de la mar* undercurrent; ~ **subterránea** *f* CARBÓN underflow, HIDROL underground flow; ~ **de suministro** *f* ING ELÉC supply current; ~ **de suministro eléctrico** *f* ING ELÉC feed; ~ **superficial** *f* TELECOM, TRANSP MAR surface current; ~ **de la superficie** *f* ING MECÁ overflow; **~ t** ~ **telúrica** *f* GEOFÍS, PROD earth current (*BrE*), ground current (*AmE*); ~ **térmica** *f* METEO thermal current; ~ **térmica nominal** *f* PROD, TERMO rated thermal current; ~ **termoeléctrica** *f* NUCL thermocurrent; ~ **terrestre** *f* GEOFÍS earth current (*BrE*), ground current (*AmE*); ~ **de tierra** *f* GEOFÍS earth current (*BrE*), ground current (*AmE*); ~ **de toma** *f* ELEC *de devanado* tapping current; ~ **de trabajo** *f* ELEC operating current, ING ELÉC make current, operating current; ~ **de trabajos** *f* INFORM&PD job stream; ~ **de tráfico** *f* TRANSP traffic stream; ~ **de tránsito máxima nominal** *f* ELEC maximum-rated through-current; ~ **transitoria** *f* ELECTRÓN, TRANSP transient current; ~ **transitoria anormal** *f* ING ELÉC surge; ~ **transversal** *f* C&V transverse current; ~ **trifásica** *f* ING ELÉC three-phase current; ~ **trifilar** *f* ING ELÉC three-wire current; ~ **de turbidez** *f* GEOL turbidity current; ~ **turbulenta** *f* CARBÓN eddy flow, GAS turbulent stream, HIDROL eddy, REFRIG eddy flow; ~ **turbulenta de retorno** *f* HIDROL back eddy; **~ u** ~ **umbral** *f* ELEC, FÍS RAD, ING ELÉC, ÓPT, TELECOM threshold current; **~ v** ~ **en vacío** *f* ELEC no-load current; ~ **variable en función de la densidad** *f* HIDROL density current; ~ **de vuelta al reposo** *f* ELEC *relé* drop-out current; ~ **de vuelta por tierra** *f* ELEC, ING ELÉC earth current (*BrE*), ground current (*AmE*)

corriente[4]: **a la ~ operacional** *fra* PROD at rated operational current; **a la ~ de régimen** *fra* PROD at rated operational current

Corriente: ~ **del Atlántico Norte** *f* OCEAN North Atlantic Current; ~ **del Pacífico Norte** *f* OCEAN North Pacific Current

corrientes: ~ **de Ampère** *f pl* FÍS Amperian currents; ~ **amperianas** *f pl* FÍS Amperian currents; ~ **balanceadas** *f pl* ING ELÉC *línea trifásica* balanced currents; ~ **litorales** *f pl* HIDROL littoral drift; ~ **neutras** *f pl* FÍS neutral currents

corrientímetro *m* INSTAL HIDRÁUL current meter

corrimiento *m* CARBÓN *de la vía férrea* creep, *de terrenos* slide, CONST *de hormigón* creep, *de tierras* thrust, GEOL overthrust, P&C *defecto de la pintura* running, PROD *geología* thrust, TEC PETR overthrust; ~ **descolgado** *m* P&C *de una capa de pintura* sagging; ~ **Doppler** *m* FÍS RAD Doppler effect, TEC ESP Doppler shift; ~ **de fase** *m* ELEC *corriente alterna*, FÍS phase shift; ~ **de frecuencia** *m* ELEC frequency drift, *de onda* frequency deviation, ELECTRÓN frequency drift, frequency deviation, TELECOM frequency drift; ~ **de tierras** *m* FERRO landslip, MINAS *geología* run of ground; ~ **de la tinta** *m* IMPR bleeding

corromper *vt* ALIMENT taint, INFORM&PD corrupt

corromperse *v refl* AGRIC, CONST, TRANSP MAR rot

corrosible *adj* GEN corrodible

corrosión *f* GEN corrosion, MECÁ fretting; ~ **circunfe-** rencial **de los tubos cerca de la placa** *f* METAL necking; ~ **por corrientes de fugas** *f* ING ELÉC stray current corrosion; ~ **fisurante** *f* TRANSP AÉR gutter, guttering; ~ **por frotamiento** *f* MECÁ fretting, NUCL galling, PROD fretting; ~ **nodular** *f* NUCL nodular corrosion; ~ **uniforme** *f* METAL uniform corrosion

corrosividad *f* GEN corrodibility

corrosivo[1] *adj* QUÍMICA, SEG corrosive

corrosivo[2] *m* QUÍMICA, SEG corrodent

corrugación *f* PAPEL corrugation

corrugado[1] *adj* ING MECÁ, PAPEL corrugated

corrugado[2] *m* C&V corrugation, crimp

corsita *f* PETROL corsite

corta *f* AGUA cutoff; ~ **duración** *f* EMB short run

cortaarandelas *m* ING MECÁ washer cutter

cortabebederos *m* PROD sprue cutter

cortabordes *m* PAPEL edge-cutter

cortacarriles *m* FERRO railroad cutting (*AmE*), railway cutting (*BrE*)

cortacircuito *m* ELEC *conmutador*, ING ELÉC circuit breaker; ~ **anódico** *m* ELEC, ING ELÉC anode circuit breaker; ~ **de ánodo** *m* ELEC, ING ELÉC anode circuit breaker; ~ **del arco** *m* ELEC arc breaker; ~ **automático** *m* ELEC, SEG automatic cut-out switch; ~ **de expulsión** *m* ELEC expulsion fuse; ~ **limitador de corriente** *m* ELEC, ING ELÉC current-limiting circuit breaker; ~ **de puesta a tierra** *m* ELEC, ING ELÉC earth arrester (*BrE*), ground arrester (*AmE*)

cortada *f* OCEAN cleft

cortadillo *m* ALIMENT lump sugar

cortado[1]: ~ **al biés** *adj* TEXTIL cut on the bias; ~ **a medida** *adj* PROD cut-to-length; ~ **transversalmente** *adj* GEOL crosscutting

cortado[2] *m* SEG cutting; ~ **y pegado** *m* IMPR, INFORM&PD cut and paste; ~ **en tiras** *m* PROD slitting

cortador *m* ING MECÁ blanking die, MECÁ, MINAS cutter, TRANSP AÉR chopper; ~ **de arandelas** *m* ING MECÁ washer cutter; ~ **de barro** *m* C&V clay cutter; ~ **de cable del izador** *m* TRANSP AÉR hoist cable cutter; ~ **de cizalla** *m* IMPR shear cutter; ~ **de esclusa** *m* ING MECÁ gate cutter; ~ **mecánico** *m* MINAS cutting machine, PETROL mechanical cutter; ~ **de negativos** *m* CINEMAT, FOTO negative cutter; ~ **de pernos** *m* MECÁ bolt cutter; ~ **de piedra** *m* MINAS *persona* cutter, quarryman; ~ **de precisión** *m* ING MECÁ fine blanking-die; ~ **rotatorio de metales** *m* ING MECÁ milling cutter; ~ **rotatorio de roscas** *m* ING MECÁ thread-milling cutter; ~ **de tacos** *m* ING MECÁ wad punch; ~ **de tubo** *m* ING MECÁ tube cutter; ~ **de tubos de tres ruedas** *m* ING MECÁ three-wheel tube cutter; ~ **de vidrio** *m* C&V, CONST *de diamante* cutting diamond, *operario* glass cutter; ~ **de vidrio estirado verticalmente** *m persona* C&V capper (*AmE*), cutoff man (*BrE*); ~ **de viento horizontal** *m* TRANSP AÉR horizontal wind shear; ~ **Woodruff** *m* ING MECÁ Woodruff cutter

cortadora *f* CONST, ING MECÁ cutter, PAPEL chopper; ~ **por arco eléctrico** *f* ING MECÁ arc cutter; ~ **automática** *f* C&V automatic cutter; ~ **de azulejos** *f* CONST tile cutter; ~ **de baldosas** *f* CONST tile cutter; ~ **de bobinas de papel** *f* IMPR paper-roll cutter; ~ **de los cantos del libro** *f* IMPR *ingenio* plough and press (*BrE*), plow and press (*AmE*); ~ **de césped de cuchillas helicoidales** *f* AGRIC reel mower; ~ **diagonal** *f* ING MECÁ angle cutter; ~ **de doble cuchilla** *f* PAPEL double blade cutter;

~-empacadora de maíz *f* AGRIC corn binder (*AmE*), maize binder (*BrE*); **~ de flejes de hierro** *f* EMB iron band cutter; **~ de forraje** *f* AGRIC chaff cutter; **~ de hojas** *f* EMB sheet-cutting machine, IMPR sheeter; **~ de hojas hidráulica** *f* PAPEL hydraulic sheet cutter; **~ de inglete** *f* CONST miter cutting machine (*AmE*), mitre cutting machine (*BrE*); **~ longitudinal** *f* IMPR, PAPEL slitter; **~ longitudinal-impresora** *f* EMB slitting and printing machine; **~ longitudinal-rebobinadora** *f* EMB slitting and rewinding machine; **~ mediante chorros de agua** *f* PAPEL spray cutter; **~ de oxígeno** *f* C&V burning-off and edge-melting machine, remelting machine; **~ de precisión** *f* ING MECÁ fine blanking-press; **~-rebobinadora** *f* PAPEL slitter rewinder; **~ redondeadora de esquinas** *f* ING MECÁ corner-rounding cutters; **~ rotativa** *f* AGRIC rotary slasher; **~ de trapos** *f* PAPEL rag cutter, rag shredder; **~ de tres cuchillas** *f* IMPR tri-blade cutting machine; **~ trilateral** *f* IMPR three-sided cutting machine; **~ de tubos** *f* CONST pipe cutter; **~ con turbina** *f* TEXTIL *de fibra* turbo-stapler

cortadores: **~ diagonales** *m pl* ING MECÁ diagonal cutters

cortadura *f* ING MECÁ shearing; **~ por soplete** *f* CONST, MECÁ, TERMO flame cutting

cortafríos *m* ING MECÁ cold chisel, MECÁ, MINAS chisel, PROD cold chisel; **~ de pico redondo** *m* ING MECÁ round-nosed cold chisel; **~ plano** *m* ING MECÁ flat chisel, flat cold chisel, MINAS flat chisel; **~ de punta de diamante** *m* ING MECÁ diamond point cold chisel; **~ con punta rómbica** *m* ING MECÁ diamond cold chisel, diamond nose chisel

cortafuegos *m* SEG, TERMOTEC *protección* fire cutoff, fire damper

cortahierros *m* ING MECÁ, PROD cold chisel

cortante[1] *adj* GEOL cutting, ING MECÁ sharp

cortante[2]: **~ del viento** *f* METEO wind shear

cortanúcleos *m* AmL (*cf utillaje del sacatestigos Esp*) TEC PETR *perforación* coring tool; **~ de diamante** *m* AmL (*cf cortatestigos de diamante Esp*) TEC PETR *perforación* diamond core drill

cortapernos *m* ING MECÁ, TRANSP MAR *reparación de barcos* bolt cropper

cortapruebas *m* FOTO trimmer; **~ con bordes dentellados** *m* FOTO jagged edge trimmer

cortar *vt* AGRIC mow, AGUA shut off, C&V *la orillas del vidrio* cut off, CINEMAT *a otra escena* cut to, cut, CONST cut, cut up, *madera* fell, GEOL intersect, GEOM *figuras geométricas* cut, intersect, IMPR cut, ING ELÉC *corriente* switch off the current, ING MECÁ punch out, cut out, shear, *bordes* pink, *corriente* switch off the current, INSTAL HIDRÁUL shut off, cut off, MECÁ cut, punch out, *bordes* break edges, PROD blank, *con el punzón sacabocados* punch out, *conexiones del servidor* cut off, TEC ESP *potencia* shut down, TRANSP *proa* cross the bows; **~ en cubitos** *vt* ALIMENT dice; **~ en dados pequeños** *vt* ALIMENT dice; **~ a medidas especificadas** *vt* PROD *conductos* cut to the required length; **~ con plantilla** *vt* SEG cut with a jig; **~ transversalmente** *vt* GEOL crosscut

cortaroscas *m* ING MECÁ threader

cortatestigos: **~ de diamante** *m* Esp (*cf cortanúcleos de diamante AmL*) TEC PETR *perforación* diamond core drill

cortatubo: **~ de revestimiento tubular de pozo** *m* CONST *petróleo, sondeos* casing cutter

cortatubos *m* CONST pipe cutter, ING MECÁ, LAB tube cutter

cortaviento *m* METEO wind break

corte[1]: **de ~ bifacial** *adj* ING MECÁ double-cutting; **de ~ doble** *adj* ING MECÁ double-cutting

corte[2] *m* C&V cut, cutout, cutting, *con sierra* sawing out, *de un cilindro* opening, CARBÓN *con sierra o con soplete* cutting, cut, kerf, CINEMAT cut, cutoff, CONST closing, cut, cutting-off, FÍS shear, GEOL shearing, section, GEOM intercept, ING ELÉC cutoff, turn-off, ING MECÁ shearing, edging, cutout, edge, section, cut, INSTAL HIDRÁUL cutting-off, MECÁ cut, notch, METAL notch, MINAS *de diamantes* cutting, P&C *reología* shear, PETROL section, PROD *rueda dentada* cutting, slicing, notch, cutoff, SEG, TEXTIL cutting, TV cutoff; **~ de acetileno** *m* MECÁ acetylene cutting; **~ por arco** *m* CONST arc cutting, *soldadura* oxygen arc cutting; **~ por arco eléctrico** *m* PROD arc cutting, TERMO electric-arc cutting; **~ por arco eléctrico y chorro de oxígeno** *m* PROD oxygen arc cutting; **~ de arco eléctrico con plasma** *m* CONST plasma arc-cutting; **~ por arco con oxígeno** *m* CONST *soldadura* oxygen arc cutting; **~ auxiliar** *m* C&V auxiliary cut; **~ biselado** *m* C&V beveling (*AmE*), bevelling (*BrE*); **~ burdo** *m* C&V rough cutting; **~ central** *m* Esp (*cf corte en Y AmL*) MINAS *pega* center cut (*AmE*), centre cut (*BrE*); **~ de césped** *m* AGRIC lawn cut; **~ de colchoneta** *m* C&V matt cutting; **~ en cono** *m* MINAS cone cut; **~ de cuchillas** *m* C&V shearcut; **~ decorativo** *m* C&V decorative cutting; **~ en diamante** *m* C&V diamond cut pattern; **~ en dirección transversal** *m* PAPEL cross-sectional cut, cross-sectional cutting; **~ con discontinuidad** *m* CINEMAT jump cut; **~ de energía** *m* PROD power outage; **~ entre campos** *m* TV interfield cut; **~ experimental** *m* CONST experimental section; **~ de film** *m* PROD film cutting; **~ geológico** *m* GEOL geological section; **~ del haz** *m* TV beam cut-off; **~ horizontal en una capa de carbón** *m* CARBÓN *minería* kerf; **~ por láser** *m* CONST, ELECTRÓN laser cutting; **~ limpio** *m* C&V clean cut; **~ por llama oxiacetilénica** *m* PROD, TERMO gas cutting; **~ longitudinal** *m* EMB slit, slitting, PAPEL slitting; **~ de madera y leña** *m* CONST cleaving; **~ en marcha** *m* EMB clip-on carrier; **~ de marquetería** *m* PROD fret cutting; **~ mineral** *m* MINAS cut; **~ del negativo** *m* CINEMAT cutting; **~ de negativos** *m* CINEMAT negative cutting; **~ núcleos** *m* AmL (*cf extracción de testigos Esp*) PETROL, TEC PETR *perforación* coring; **~ oblicuo** *m* PROD *soldeo* scarfing; **~ oxieléctrico** *m* PROD oxygen arc cutting; **~ de petróleo** *m* PETROL, TEC PETR *perforación* cut oil; **~ piramidal** *m* MINAS pyramid cut; **~ con plasma** *m* CONST plasma cutting; **~ de pliegue** *m* PAPEL blister cut; **~ profundo** *m* C&V deep cut, ING MECÁ heavy cut; **~ a ras** *m* PROD flush cut; **~ recto** *m* ING MECÁ straight edge; **~ de red** *m* TELECOM network breakdown; **~ por rodadura de las bandas de código** *m* TELECOM rolling code band splitting; **~ de rollos A y B** *m* CINEMAT A and B roll cutting; **~ de señal** *m* ELECTRÓN signal clipping; **~ sesgado** *m* ING MECÁ undercut; **~ al sesgo** *m* MINAS cut chain incline; **~ de sierra** *m* CONST saw cut; **~ por soplete** *m* CONST, MECÁ, TERMO flame cutting; **~ del suministro de gas** *m* GAS, INSTAL TERM gas quench; **~ de tráfico** *m* TRANSP traffic cut; **~ de transmisión** *m* TELECOM, TV transmission breakdown; ~ **transversal** *m* ING MECÁ crosscut, cross section, section, METAL cross section, TELECOM transverse section; ~ **transversal de captación** *m* PETROL capture cross-section; ~ **transversal de captura** *m* PETROL capture cross-section; ~ **transversal de la pala** *m* TRANSP AÉR blade cross-section; ~ **en TV** *m* TV TV cutoff; ~ **de vidrio caliente con alambre** *m* C&V hot glass wire cutting; ~ **de vidrios con diamante** *m* C&V diamond glass cutting; ~ **en Y** *m AmL* (*cf corte central Esp*) MINAS *pega* center cut (*AmE*), centre cut (*BrE*)

cortes *m pl* TEC PETR cuttings, *perforación* drill cuttings

cortesía *f* IMPR sinkage

corteza *f* CARBÓN shell, skin, GEOL crust, PAPEL bark, PETROL crust, TEC ESP shell, skin; ~ **oceánica** *f* GEOL, OCEAN ocean crust, oceanic crust; ~ **seca** *f* CARBÓN dry crust; ~ **terrestre** *f* ENERG RENOV, GEOFÍS, GEOL earth's crust, TEC ESP terrestrial crust

corticoesteroide *m* QUÍMICA corticosteroid

corticoide *m* QUÍMICA corticoid

corticosterona *f* QUÍMICA corticosterone

corticotrófico *adj* QUÍMICA corticotrophic

corticotrofina *f* QUÍMICA corticotrophin

corticotrópico *adj* QUÍMICA corticotropic

corticotropina *f* QUÍMICA corticotropin

cortijo *m* AGRIC farmstead

cortina *f* C&V, TEXTIL curtain, VEH *del radiador* blind; ~ **de aire** *f* INSTAL TERM, REFRIG air curtain, SEG air curtain installation, TRANSP air curtain; ~ **contra fuego** *f* SEG fire screen; ~ **enrollable** *f* CONST rolling shutter; ~ **de estabilidad** *f* TRANSP stability curtain; ~ **de fuego** *f* SEG, TERMO fire curtain; ~ **de humo** *f* D&A smoke screen; ~ **opaca** *f* TEXTIL blackout curtain; ~ **protectora para soldador** *f* SEG welder's protective curtain; ~ **de regadera** *f* C&V shower screen; ~ **de seguridad** *f* SEG safety curtain

cortisona *f* QUÍMICA *alergias, antinflamatorio, artritis* cortisone

corto *m* CINEMAT *publicitario* clip, short, ING ELÉC short

cortocircuitación *f* ING ELÉC, PROD shorting

cortocircuitado *adj* ELEC, FÍS, INFORM&PD, ING ELÉC, TELECOM short-circuit, short-circuited

cortocircuitar *vt* ING ELÉC, PROD short

cortocircuito *m* ELEC, FÍS, INFORM&PD, ING ELÉC, TELECOM short circuit; ~ **ajustable** *m* FÍS, ING ELÉC adjustable short circuit; ~ **de alta tensión** *m* ELEC *conmutador* high-voltage circuit breaker; ~ **con desprendimiento de chispas** *m* ELEC, ING ELÉC, TERMO flash-over; ~ **discriminante** *m* ELEC discriminating circuit breaker; ~ **con emisión de chispas** *m* ELEC *de electrodo*, ING ELÉC, TERMO flash-over; ~ **de fuga a tierra** *m* ELEC, ING ELÉC earth-leakage circuit breaker (*BrE*), ground-leakage circuit breaker (*AmE*); ~ **interfásico** *m* ELEC *falla* interphase short circuit; ~ **interfásico a la salida** *m* PROD output phase-to-phase short; ~ **momentáneo** *m* ELECTRÓN transient short-circuit; ~ **de pérdida a tierra** *m* ELEC, ING ELÉC earth-leakage circuit breaker (*BrE*), ground-leakage circuit breaker (*AmE*); ~ **perfecto** *m* ING ELÉC dead short circuit; ~ **de salida a tierra** *m* PROD output earth short (*BrE*), output ground short (*AmE*); ~ **de tensión de la falla** *m* ELEC fault voltage circuit breaker

cortometraje *m* CINEMAT short

corundelita *f* MINERAL corundellite

corundofilita *f* MINERAL corundophilite

cos *abr* (*coseno*) CONST, GEOM, INFORM&PD, MATEMÁT *trigonometría* cos (*cosine*)

cosalita *f* MINERAL cosalite

cosec *abr* (*cosecante*) CONST, GEOM, INFORM&PD, MATEMÁT *trigonometría* cosec (*cosecant*)

cosecante *f* (*cosec*) CONST, GEOM, INFORM&PD, MATEMÁT *trigonometría* cosecant (*cosec*)

cosecha *f* AGRIC crop, harvest, yield; ~ **de cereales** *f* AGRIC grain crop; ~ **destinada al corte** *f* AGRIC green feeding crop; ~ **destinada a ser enlatada** *f* AGRIC canning crop; ~ **record** *f* AGRIC record crop

cosechadora: ~ **de algodón** *f* AGRIC, TEXTIL cotton picker; ~ **autopropulsada** *f* AGRIC combine; ~ **de cebollas** *f* AGRIC onion harvester; ~**-desgranadora de maíz** *f* AGRIC corn picker-sheller (*AmE*), maize picker-sheller (*BrE*); ~**-deshojadora de maíz** *f* AGRIC maize picker-husker (*BrE*), corn picker-husker (*AmE*); ~ **de maíz** *f* AGRIC corn picker (*AmE*), maize picker (*BrE*); ~ **de papas** *f AmL* (*cf cosechadora de patatas Esp*) AGRIC potato digger; ~ **de patatas** *f Esp* (*cf cosechadora de papas AmL*) AGRIC potato digger; ~ **de peine** *f* AGRIC stripper; ~ **trilladora** *f* AGRIC combine; ~**-trilladora** *f* AGRIC harvester-thresher

cosechas: ~ **de grano grueso** *f pl* AGRIC coarse crops; ~ **principales** *f pl* AGRIC basic crops; ~ **de secano** *f pl* AGRIC dry crops

cosedora *f* IMPR sewing machine; ~ **de bolsas** *f* EMB bag-stitching machine; ~ **de sacos** *f* EMB sack-sewing machine

coseno *m* (*cos*) CONST, GEOM, INFORM&PD, MATEMÁT *trigonometría* cosine (*cos*)

coser *vt* EMB, IMPR, PROD stitch, TEXTIL seam, sew, stitch, TRANSP MAR *cordaje* lash; ~ **abajo** *vt* TEXTIL stitch down; ~ **con tiretas** *vt* PROD *correas* lace

cosido[1] *adj* EMB, IMPR, PROD, TEXTIL stitched

cosido[2] *m* TRANSP MAR stitching; ~ **remachado** *m* EMB riveted seam; ~ **de unión** *m* TEXTIL seaming

cosirita *f* MINERAL cossyrite

cosmético: ~ **de pulir** *m* C&V rouge

cosmódromo *m* TEC ESP cosmodrome

cosmogonía *f* TEC ESP cosmogony

cosmografía *f* TEC ESP cosmography

cosmología *f* TEC ESP cosmology

cosmonauta *m* TEC ESP cosmonaut

cosmonave *f* TEC ESP spacecraft, spaceship

costa[1]: ~**-fuera** *adj* GEOL, OCEAN, TEC PETR, TRANSP MAR offshore

costa[2]: **hacia la** ~ *adv* GEOL onshore, TRANSP MAR *navegación* shoreward; **sobre la** ~ *adv* TRANSP MAR onshore

costa[3] *f* GEOL coast, shore, HIDROL *marítima*, TRANSP MAR *geografía* coast; ~ **abordable** *f* TRANSP MAR accessible coast; ~ **brava** *f* OCEAN bold shore; ~ **escarpada** *f* OCEAN steep coast; ~ **marina** *f* OCEAN shore; ~ **a sotavento** *f* TRANSP MAR *navegación a vela* lee shore

costado[1]: **al** ~ *adv* TRANSP MAR *barco, muelle* alongside; **hacia el** ~ *adv* ING MECÁ, TRANSP MAR *jarcia, motor* outboard

costado[2] *m* CONST *tejado*, INSTR, TRANSP MAR *del buque* side; ~ **de barlovento** *m* TRANSP MAR *del buque* weather side; ~ **de una bóveda** *m* CONST *arquitectura* haunch; ~ **en colisión** *m* TRANSP side-on collision; ~ **izquierdo** *m* TEC ESP *aviones* port; ~ **de popa** *m* TRANSP MAR quarter

costanera f AGRIC *arado* landside, CONST coastal ring road; ~ **giratoria** f AGRIC rolling landside; ~ **de orla** f OCEAN fringing reef

costanero adj TRANSP MAR inshore

coste m PROD cost; ~ **debido al retraso** m EMB downtime cost; ~ **debido al tiempo de parada** m EMB downtime cost; ~ **de la mano de obra** m PROD labor cost (*AmE*), labour cost (*BrE*); ~ **de productos** m PROD product cost; ~ **y seguro y flete** m PROD cost insurance freight (*CIF*)

costear vi TRANSP MAR coast

costero[1] adj GEOL, HIDROL coastal, TRANSP MAR inshore, *geografía* coastal

costero[2] m *Esp* (*cf hastial AmL*) MINAS *de minas* side, *minería del carbón* cheeks; ~ **de máquina** m PAPEL machine flaw

costes: ~ **de funcionamiento** m pl ING MECÁ, PAPEL, PROD, TEXTIL running costs

costilla f CARBÓN, MECÁ, TEC ESP, TRANSP AÉR, TRANSP MAR *construcción naval* rib; ~ **enrejada de celosía** f TRANSP AÉR lattice rib; ~ **falsa** f TRANSP AÉR false rib; ~ **final** f TRANSP AÉR end rib; ~ **principal** f TRANSP AÉR main rib; ~ **de pulido** f C&V lapping rib; ~ **superior del muñón del estabilizador** f TRANSP AÉR fin stub top rib

costo: ~ **de aprobación** m CALIDAD approval cost; ~ **de la calidad** m CALIDAD quality cost; ~ **de espacio** m EMB cost of space; ~ **de funcionamiento** m CALIDAD operating cost; ~ **de inversión** m CALIDAD capital cost; ~ **de la mano de obra** m PROD labor cost (*AmE*), labour cost (*BrE*); ~ **de mantenimiento** m CALIDAD maintenance cost; ~ **de operación y de mantenimiento** m CALIDAD life-cycle cost; ~ **de prevención** m CALIDAD prevention cost; ~ **del programa** m TV program cost (*AmE*), programme cost (*BrE*); ~ **recurrente** m TEC ESP recurrent cost; ~ **relacionado con la calidad** m CALIDAD quality-related cost; ~ **de valoración** m CALIDAD appraisal cost

costos: ~ **circunstanciales** m pl TEC ESP nonrecurring cost; ~ **imprevistos** m pl TEC ESP nonrecurring cost; ~ **no recurrentes** m pl TEC ESP nonrecurring cost; ~ **ocasionales** m pl TEC ESP nonrecurring cost

costra f ALIMENT scab, CARBÓN *lingote fundido* skin, GEOL, PETROL crust; ~ **de carga** f C&V batch crust; ~ **de coque** f MINAS coked coal dust; ~ **de lodo** f CARBÓN filter cake, PETROL *perforación* mud cake

costura[1]: **sin** ~ adj IMPR unsewn

costura[2] f C&V seam, IMPR seam, sewing, ING MECÁ, MECÁ, PAPEL seam, TEXTIL seam, sewing; ~ **en bloc** f PAPEL padding; ~ **en bloque** f IMPR stab stitching; ~ **de correa** f PROD *acero* belt lacing; ~ **de folletos** f IMPR pamphlet stitching; ~ **larga** f TRANSP MAR *cordaje* long splice; ~ **por el lomo** f IMPR saddle stitching; ~ **periférica** f D&A peripheral hem; ~ **soldada** f NUCL seam weld; ~ **transversal** f PROD butt seal

costuras[1]: **sin** ~ adj TEXTIL seamless

costuras[2]: ~ **no casadas** f pl TEXTIL mismatched seams

co-sumando m INFORM&PD augend

cot abr (*cotangente*) CONST, GEOM, INFORM&PD, MATEMÁT cot (*cotangent*)

cota f CONST, MECÁ elevation

cotangente f (*cotg*) CONST, GEOM, INFORM&PD, MATEMÁT *trigonometría* cotangent (*cot*)

cotas: ~ **límite** f pl ING MECÁ boundary dimensions

cote m TRANSP MAR knot

cotejador m INFORM&PD collator

cotejar vt INFORM&PD collate, PROD match

cotg abr (*cotangente*) CONST, GEOM, INFORM&PD, MATEMÁT *trigonometría* cot (*cotangent*)

cotiledón m AGRIC cotyledon

coto: ~ **de acuicultura marina** m OCEAN sea ranch

cotunnita f MINERAL cotunnite

coulómetro m ELEC *instrumento*, FÍS, ING ELÉC, INSTR, METR coulombmeter, coulometer

Couviniense adj GEOL Couvinian

covalencia f NUCL, QUÍMICA covalence, covalency

covalente adj NUCL, QUÍMICA covalent

covariante f INFORM&PD covariant

covarianza f INFORM&PD covariance

covellina f MINERAL covelline

covellita f MINERAL covellite

coy m TRANSP MAR hammock

coz f TRANSP MAR *de mastelero, de rollizo* foot, *de palo* heel

CPP abr (*caracteres por pulgada*) IMPR CPI (*characters per inch*)

CPS abr (*caracteres por segundo*) IMPR, INFORM&PD CPS (*characters per second*)

CPVC abr (*cloruro de polivinilo clorado*) P&C CPVC (*chlorinated polyvinyl chloride*)

Cr abr (*cromo*) METAL, QUÍMICA, REVEST Cr (*chromium*)

cracking m TEC PETR *refino* cracking

craqueado: ~ **carbúrico** m METAL *pulvimetalurgia* carbide cracking

craqueo: ~ **de coque** m CARBÓN coking cracking; ~ **fluidizado** m PROC QUÍ fluid cracking

cráter m CARBÓN *soldadura* kerf, GEOL, MINAS, P&C, TEC ESP *geología* crater; ~ **de impacto** m TEC ESP impact crater; ~ **lateral** m GEOL *estructura volcánica* lateral crater; ~ **parásito** m GEOL *estructura volcánica* parasitic crater;

cratícula: ~ **de difracción** f FÍS ONDAS diffraction grating

cratón m GEOFÍS continental shield, GEOL continental shield, craton

CRC abr ELECTRÓN, INFORM&PD, TELECOM (*verificación de redundancia cíclica*) CRC (*cyclic redundancy check*), *código de redundancia cíclica* CRC (*cyclic redundancy code*)

CRDH abr (*alojamiento del mecanismo de accionamiento de las barras de control*) NUCL CRDH (*control rod drive housing*)

CRDM abr (*mecanismo de accionamiento de las barras de control*) NUCL CRDM (*control rod drive mechanism*)

creación: ~ **de archivos** f INFORM&PD file creation; ~ **de datos** f INFORM&PD data origination; ~ **de ficheros** f INFORM&PD file creation; ~ **de registro** f INFORM&PD record creation

creado adj ING MECÁ, PROD shaped

creador: ~ **de imágenes** m INFORM&PD imager

crear[1] vt INFORM&PD create

crear[2]: ~ **aspiración** vi ING MECÁ create a draft (*AmE*), create a draught (*BrE*); ~ **una corriente de aire** vi ING MECÁ create a draft (*AmE*), create a draught (*BrE*) ~ **succión** vi ING MECÁ create a draft (*AmE*), create a draught (*BrE*); ~ **un talud** vi CONST bank; ~ **un terraplén** vi CONST bank

crecer vi HIDROL *las aguas* rise

crecida *f* HIDROL flood, rise, *riada* spate, torrent, *de un río* high water, OCEAN rise; **~ de las aguas** *f* AGUA, HIDROL flood of water; **~ anual** *f* AGUA annual flood; **~ máxima** *f* AGUA high-water overflow

crecido *adj* HIDROL *río* in flood; **~ por vapor** *adj* PROC QUÍ vapor-grown (*AmE*), vapour-grown (*BrE*)

crecimiento *m* AGRIC, CRISTAL, ELECTRÓN, FÍS FLUID, METAL, NUCL growth; **~ columnar** *m* METAL oriented growth; **~ de cristales** *m* CRISTAL crystal growth; **~ cristalino** *m* CRISTAL crystal growth; **~ epitaxial** *m* CRISTAL, ELECTRÓN epitaxial growth; **~ de granos orientados** *m* METAL *cristalografía* oriented growth; **~ inicial de una grieta** *m* NUCL initial crack growth; **~ de lado a lado** *m* METAL edgewise growth; **~ de lagunas** *m* METAL void growth

credencial *m* TELECOM credential

crema: **~ de chocolate** *f* CONTAM, CONTAM MAR *hidrocarburos intemperizados* chocolate mousse; **~ para la piel** *f* SEG skin cream; **~ protectora** *f* SEG protective cream; **~ de seguridad ocupacional** *f* SEG occupational safety cream

cremallera *f* ING ELÉC rack, ING MECÁ cog rail, rack, ratch, MECÁ, PROD rack, TEXTIL zip (*BrE*), zipper (*AmE*), TRANSP *ferrocarril*, VEH *dirección* rack; **~ de alimentación** *f* ING MECÁ feed rack; **~ de avance múltiple** *f* PROD multiple feed rack; **~ base** *f* ING MECÁ basic rack; **~ circular de puntería azimutal** *f* D&A training rack; **~ sin fin** *f* ING MECÁ worm rack; **~ y piñón** *m* FERRO, ING MECÁ, INSTR, MECÁ, PROD, TRANSP rack and pinion; **~ que engancha con un piñón** *f* ING MECÁ rack which engages with a pinion; **~ de referencia** *f* ING MECÁ basic rack; **~ en segmento** *f* ING MECÁ segment gear; **~ tipo** *f* ING MECÁ basic rack

crémor: **~ tártaro** *m* ALIMENT, QUÍMICA cream of tartar

crenulación *f* GEOL crenulation

creosota *f* QUÍMICA creosote

crepé *m* P&C, QUÍMICA *caucho esponjoso* crepe; **~ de caucho** *m* P&C crepe rubber

crepitación *f* ACÚST, QUÍMICA crackle

crepitar *vi* ACÚST, QUÍMICA *petróleo* crackle

crespado[1] *adj* PAPEL creped

crespado[2] *m* PAPEL creping; **~ fuera de máquina** *m* PAPEL off-machine creping; **~ fuerte** *m* PAPEL heavy crepe; **~ en húmedo** *m* PAPEL wet creping; **~ en máquina** *m* PAPEL on-machine creping; **~ en seco** *m* PAPEL dry creping

crespón *m* TEXTIL crepe

cresta *f* CONST, GEOL, HIDROL crest, MINAS brow, OCEAN *geología submarina* ridge; **~ barométrica** *f* OCEAN pressure ridge; **~ del blanco** *f* TV peak white; **~ de blancos** *f* TV white peak; **~ de coral** *f* GEOL, OCEAN coral ridge; **~ de la corriente** *f* FÍS current loop; **~ de cúmulo volcánico** *f* GEOL *característica volcánica* cumulo dome; **~ isoclinal** *f* GEOL hogback; **~ del negro** *f* Esp (*cf pico del negro AmL*) TV black peak; **~ de una onda** *f* ENERG RENOV wave crest; **~ de playa** *f* OCEAN beach ridge, shingle spit; **~ del pliegue** *f* GEOL hinge; **~ de presión** *f* OCEAN pressure ridge; **~ de señal** *f* TELECOM pip

crestón *m* GEOL hogback, MINAS *geología* outcrop

creta *f* GEOL, TEC PETR chalk; **~ arenosa** *f* GEOL sandy chalk

cretáceo *adj* GEOL, PETROL, TEC PETR chalky, cretaceous

cría *f* AGRIC breeding, raising, rearing; **~ artificial** *f* AGRIC *ganado* rearing by hand; **~ del avestruz** *f* AGRIC ostrich breeding; **~ por cruza** *f* AGRIC outbreeding; **~ de ganado** *f* AGRIC livestock breeding; **~ de mejillones** *f* OCEAN mussel breeding; **~ piscícola** *f* ALIMENT fish-breeding; **~ y ramoneo** *f* OCEAN sea ranching; **~ de tortugas** *f* OCEAN turtle culture

criadero *m* Esp (*cf criadero menero AmL*) MINAS ore deposit; **~ de mejillones** *m* OCEAN *banco natural* mussel bed; **~ menero** *m* AmL (*cf criadero Esp*) MINAS ore deposit; **~ mineral** *m* AmL (*cf yacimiento Esp*) MINAS mineral deposit; **~ de ostras** *m* OCEAN oyster bed; **~ de productos del mar** *m* OCEAN marine farm

criar *vt* AGRIC breed

criba *f* CARBÓN cribble, griddle, riddle, screen, sieve, slug, CONST screen, sieve, CONTAM MAR screening, LAB *separación* sieve, MINAS *del bocarte* screen, PROD griddle, cribble, screen, strainer, sieve, sifter, riddle, RECICL screen; **~ de aire comprimido** *f* ING MECÁ pneumatic jig; **~ de alfarero** *f* C&V earthenware sieve; **~ para arena** *f* PROD *fundería* sand riddler, sand sifter; **~ de barrotes** *f* CARBÓN, PROD grizzly; **~ centrífuga** *f* PROC QUÍ centrifuge screen; **~ cilíndrica** *f* HIDROL drum screen; **~ de clasificación de menudos** *f* CARBÓN nut-sizing screen; **~ clasificadora** *f* PROD sizing screen; **~ corrediza** *f* PROD traveling belt screen (*AmE*), travelling belt screen (*BrE*); **~ para eliminar lodos** *f* CARBÓN desliming screen; **~ de escurrido** *f* CARBÓN running jig; **~ filtradora** *f* PROC QUÍ filtering screen; **~ filtrante** *f* PROD filter screen, plunger jig; **~ de filtro** *f* TEC ESP *equipamiento* filter jig; **~ giratoria** *f* CARBÓN, CONST, MINAS, PROD revolving screen, rotary screen; **~ de granos** *f* CARBÓN coarse screen; **~ para gruesos** *f* MINAS *preparación minerales* coarse trommel; **~ hidráulica** *f* CARBÓN, MINAS jig; **~ hidráulica de pistón** *f* PROD plunger jig; **~ de lavado** *f* CARBÓN *para el oro*, PROD riddle; **~ lavadora** *f* MINAS rocker, *para el oro, minerales* cradle, mining cradle; **~ de maíz** *f* AGRIC corn crib (*AmE*), maize drying shed (*BrE*); **~ de malla ancha** *f* CARBÓN coarse mesh screen; **~ molecular** *f* QUÍMICA molecular sieve; **~ oscilante** *f* CARBÓN swing sieve, PROD swinging screen; **~ de percusión** *f* CARBÓN impact screen, PROD impact screen, percussion sieve; **~ de pistón** *f* MINAS jig; **~ de rejilla fija** *f* PROD fixed-sieve jig; **~ para remover la ganga** *f* CARBÓN *minerales* scalping screen; **~ rotatoria** *f* CARBÓN *metalurgia*, CONST, MINAS, PROD revolving screen, rotary screen; **~ de sacudidas** *f* CARBÓN shaking screen, MINAS jigger screen, PROD bumping screen, push screen, shaking grate; **~ de tambor** *f* CARBÓN trommel screen, HIDROL drum screen; **~ de tela metálica** *f* PROD wire screen, wire sieve, wire cloth screen; **~ de tela tejida** *f* PROD wire-woven screen; **~ de trepidación** *f* PROD shaking screen; **~ vibradora** *f* CARBÓN impact screen, vibrating screen; **~ vibratoria** *f* PROD impact screen, shaking screen, QUÍMICA vibrating screen

cribado *m* CARBÓN grading, overflow of a screen, griddling, riddling, CONST screening, sieving, CONTAM *tratamiento de aguas, sólidos*, PROC QUÍ sieving, PROD screening, griddling, sifting, riddling; **~ del balasto** *m* FERRO *equipo inamovible* ballast screening; ~ **hidráulico** *m* MINAS jigging; ~ **primario** *m* CARBÓN prescreening

cribador *m* CARBÓN, PROD riddler; ~ **vibratorio** *m* EMB vibratory sifter

cribar *vt* CARBÓN faredice, sieve, CONST screen, sieve, CONTAM MAR screen, HIDROL, PROC QUÍ, PROD sieve, QUÍMICA screen, RECICL sieve

cribón *m* CARBÓN, PROD *tamiz de minerales* grizzly; ~ **oscilante** *m* CARBÓN jig sieve; ~ **vibratorio** *m* CARBÓN vibrating grizzly

cric *m* CONST *gato* jack, ING MECÁ lifting screw, ratchet, MECÁ ratchet; ~ **para elevar pesos** *m* ING MECÁ lifting jack; ~ **hidráulico** *m* CONST hydraulic jack; ~ **para posicionar** *m* ING MECÁ positioning screw jack

crictonita *f* MINERAL crichtonite

criobiología *f* REFRIG cryobiology

criocirugía *f* REFRIG cryosurgery

criodecapado *m* REFRIG cryoetching

criodeshidratación *f* ALIMENT, PROC QUÍ, REFRIG lyophilization

criofísica *f* REFRIG cryophysics

criogenia *f* FÍS, ING MECÁ, REFRIG, TEC ESP, TERMO, TERMOTEC cryogenics

criogénica *f* FÍS, ING MECÁ, REFRIG, TEC ESP, TERMO, TERMOTEC cryogenics

criogénico *adj* FÍS, REFRIG, TEC ESP, TERMO cryogenic

criolita *f* C&V, MINERAL, QUÍMICA cryolite

criología *f* REFRIG cryology

criomagnetismo *m* REFRIG cryomagnetism

criomedicina *f* REFRIG cryomedicine

criorresistencia *f* ING ELÉC cold resistance

crioscopia *f* REFRIG cryoscopy

criostato *m* FÍS, LAB cryostat, TEC ESP cryopump, cryostat

criotón *m* ING ELÉC cryotron

criotratado *adj* TERMO *congelado* deep-frozen

criotratamiento *m* TERMO deep-freezing, deep-freeze

criotratar *vt* TERMO deep-freeze

criptoanálisis *m* INFORM&PD, TELECOM cryptanalysis

criptocristalina *adj* CRISTAL, GEOL cryptocrystalline

criptografía *f* ELECTRÓN encryption, INFORM&PD cryptography, TEC ESP, TELECOM cryptography, encryption, TV encryption

criptográfico *adj* INFORM&PD, TEC ESP, TELECOM cryptographic

criptón *m* (*Kr*) QUÍMICA krypton (*Kr*)

criseno *m* QUÍMICA *cracking del petróleo* chrysene

crisis: ~ **energética** *f* TERMO energy crisis

crisoberilo *m* MINERAL cat's eye, chrysoberyl

crisócola *f* MINERAL chrysocolla

crisofánico *adj* QUÍMICA *ácido* chrysophanic

crisoidina *f* QUÍMICA chrysoidine

crisol *m* C&V day tank, CONST hearth, INSTAL TERM, LAB crucible, PROD melting pot, melter, crucible, QUÍMICA crucible, TERMO melting pot, crucible; ~ **de ajuste** *m* ELEC *resistencia* preset pot; ~ **de barro** *m* C&V clay crucible; ~ **de barro refractario** *m* C&V fireclay crucible; ~ **cerrado** *m* C&V closed pot; ~ **cubierto** *m* C&V covered pot; ~ **de filtro de vidrio sinterizado** *m* LAB *filtración* sintered glass filter crucible; ~ **de fusión** *m* QUÍMICA melting pot, TERMO melting crucible; ~ **de Gooch** *m* LAB *filtración* sintered glass filter crucible, *filtración, secado* Gooch crucible; ~ **de grafito** *m* PROD blue pot, plumbago crucible; ~ **de helio** *m* ELEC *resistencia* helipot; ~ **del**

horno *m* GAS hearth; ~ **platinado** *m* LAB *análisis, calentamiento, recipiente* platinum crucible; ~ **de plomo negro** *m* PROD black-lead crucible; ~ **de porcelana** *m* C&V, LAB *análisis, recipiente* porcelain crucible; ~ **de reducción** *m* PROD reduction crucible

crisolita *f* MINERAL chrysolite

crisoprasa *f* MINERAL chrysoprase

crisotilo *m* MINERAL chrysotile

crispado *m* PAPEL cockle finish

cristal *m* C&V crystal, crystal glass, plate glass, CONST pane, CRISTAL, ELECTRÓN, ÓPT crystal, PROD lens, QUÍMICA, TELECOM crystal; ~ **de agua** *m* DETERG water glass; ~ **de alabastro** *m* C&V alabaster glass; ~ **antideslumbrante** *m* C&V, SEG antidazzle glass; ~ **de aumento** *m* FÍS, LAB, MECÁ magnifying glass; ~ **de aumento convergente** *m* INSTR focusing magnifying glass; ~ **basal** *m* CRISTAL columnar crystal, METAL *cristalografía* basal crystal, columnar crystal; ~ **blindado** *m* C&V, CONST, SEG armored glass (*AmE*), armoured glass (*BrE*), safety glass; ~ **de Bohemia** *m* C&V Bohemian crystal; ~ **casi perfecto** *m* CRISTAL, METAL nearly perfect crystal; ~ **columnar** *m* CRISTAL columnar crystal, METAL *cristalografía* basal crystal, columnar crystal; ~ **cortado** *m* C&V cut glass; ~ **de cuarzo** *m* CINEMAT, CRISTAL, ELEC, ELECTRÓN, INFORM&PD quartz crystal; ~ **sin defectos** *m* METAL perfect crystal; ~ **doble** *m* CONST, TERMOTEC double glazing; ~ **dopado con cesio** *m* TEC ESP caesium-doped glass (*BrE*), cesium-doped glass (*AmE*); ~ **esmerilado** *m* CINEMAT, FOTO ground glass; ~ **esmerilado con lente de Fresnel** *m* CINEMAT, FOTO ground-glass with Fresnel lens; ~ **esmerilado con microprisma** *m* FOTO ground-glass screen with microprism spot; ~ **esmerilado con retícula** *m* CINEMAT, FOTO ground-glass screen with reticule; ~ **ferroeléctrico** *m* CRISTAL, ELEC, ING ELÉC ferroelectric crystal; ~ **filiforme** *m* CRISTAL whisker; ~ **de filtro** *m* ELECTRÓN filter crystal; ~ **de gafas** *m* INSTR spectacle lens; ~ **germen** *m* CRISTAL seed crystal; ~ **idiomórfico** *m* METAL, CRISTAL idiomorphic crystal; ~ **impurificado con cesio** *m* TEC ESP caesium-doped glass (*BrE*), cesium-doped glass (*AmE*); ~ **inastillable** *m* CONST, SEG safety glass; ~ **irrompible** *m* SEG, TRANSP shatterproof glass; ~ **líquido** *m* (*CL*) CRISTAL, ELEC, ELECTRÓN, FÍS, INFORM&PD, NUCL, TEC PETR, TELECOM, TV liquid crystal (*LC*); ~ **madre** *m* ELECTRÓN mother crystal; ~ **matizado mordentado al ácido** *m* C&V acid-etched frosted glass; ~ **de observación** *m* INSTR looking glass; ~ **ópticamente plano** *m* CINEMAT optical flat; ~ **óptico** *m* FOTO optical glass; ~ **de oscilador** *m* ELECTRÓN oscillator crystal; ~ **perfecto** *m* METAL perfect crystal; ~ **piezoeléctrico** *m* ING ELÉC, TELECOM piezoelectric crystal; ~ **de plomo** *m* C&V full-lead crystal glass, lead crystal glass; ~ **protector** *m* SEG protective glass; ~ **protector de gafas de soldador** *m* SEG protective glass for welder's goggles; ~ **de puerta** *m* VEH door glass; ~ **de recepción** *m* ELECTRÓN, ÓPT, TELECOM receive crystal; ~ **de refinación** *m* C&V, PROC QUÍ, SEG refining glass; ~ **reforzado** *m* SEG, TRANSP toughened glass; ~ **de roca denso** *m* C&V dense flint; ~ **de seguridad** *m* CONST, SEG, TRANSP safety glass; ~ **semiconductor** *m* ELECTRÓN semiconductor crystal; ~ **de silicio** *m* ELECTRÓN silicon crystal; ~ **soplado** *m* PROD blown glass; ~ **con temperatura controlada** *m* ELECTRÓN temperature-controlled crystal

cristalería: ~ **para hoteles** *f* C&V *copas, vasos* hotel glassware; ~ **soplada** *f* C&V antique glass

cristales: ~ **líquidos dicroicos** *m pl* ELECTRÓN dichroic liquid crystals; ~ **líquidos esmécticos** *m pl* ELECTRÓN smectic liquid crystals; ~ **líquidos nemáticos** *m pl* CRISTAL, ELECTRÓN nematic liquid crystals

cristalino *adj* CRISTAL, C&V, QUÍMICA crystalline

cristalización *f* CRISTAL crystallization, nucleation, ING MECÁ *metalurgia* coring, METAL coring, nucleation, METEO, NUCL nucleation, P&C *fenómeno químico*, PROC QUÍ crystallization; ~ **fraccionada** *f* CRISTAL, GEOL *proceso ígneo* fractional crystallization

cristalizado: ~ **después de un sobrecalentamiento** *adj* TERMO *acero* scorched

cristalizador *m* Esp (*cf vidrio de reloj AmL*) CRISTAL crystallizer, crystallizing dish, LAB crystallizing dish, PROC QUÍ, QUÍMICA crystallizer

cristalografía *f* CRISTAL, FÍS RAD, METAL crystallography; ~ **por rayos X** *f* FÍS RAD X-ray crystallography

cristianita *f* MINERAL christianite

cristobalita *f* MINERAL cristobalite

criterio: ~ **aceptación** *m* TEC ESP acceptance criterion; ~ **de búsqueda** *m* INFORM&PD search key; ~ **de defensa a ultranza** *m* NUCL defence in depth criterion; ~ **de diseño** *m* NUCL design criterion; ~ **de emplazamiento** *m* CONTAM *instalación industrial* site criteria; ~ **del fabricante** *m* ING MECÁ manufacturer's discretion; ~ **de falla** *m* CALIDAD, MECÁ criterion of failure; ~ **de fractura** *m* METAL fracture criterion; ~ **de fractura de Griffith** *m* NUCL Griffith's fracture criterion; ~ **de fuga antes de rotura** *m* NUCL leak-before-break criterion; ~ **de proyección** *m* PROD design criterion; ~ **de Rayleigh** *m* Fís Rayleigh criterion; ~ **de recepción** *m* TEC ESP acceptance criterion; ~ **de situación** *m* CONTAM site criteria; ~ **de ubicación** *m* CONTAM site criteria

criterios: ~ **de aceptación de calidad** *m pl* ING MECÁ quality acceptance criteria; ~ **de los riesgos** *m pl* CALIDAD, SEG risk criteria

criticidad *f* CALIDAD criticality; ~ **inicial** *f* NUCL first criticality, initial criticality

crítico[1] *adj* CALIDAD, ING MECÁ, NUCL, QUÍMICA, TRANSP AÉR critical

crítico[2]: ~ **con neutrones inmediatos** *m* Esp (*cf crítico con neutrones no retardados AmL*) NUCL prompt critical; ~ **con neutrones no retardados** *m* AmL (*cf crítico con neutrones inmediatos Esp*) NUCL prompt critical

croceína *f* QUÍMICA *colorante* crocein

crocidolita *f* MINERAL crocidolite

crocina *f* QUÍMICA crocin

crocoíta *f* MINERAL crocoisite, crocoite, red-lead ore

crocónico *adj* QUÍMICA *ácido* croconic

croiomarcado *m* REFRIG cryobranding

cromado[1] *adj* ING MECÁ, REVEST, VEH chromium-plated, chrome-plated

cromado[2] *m* ING MECÁ, REVEST, VEH chromium plating; ~ **duro** *m* ING MECÁ hard chrome plating, REVEST hard plating; ~ **resistente** *m* REVEST hard plating

cromaticidad *f* GEN chromaticity

cromático *adj* COLOR, FÍS, FOTO, ÓPT, QUÍMICA, TV chromatic

cromato *m* QUÍMICA chromate; ~ **férrico** *m* QUÍMICA iron chromate

cromatóforo *m* QUÍMICA plastid

cromatografía *f* LAB, QUÍMICA *análisis* chromatography; ~ **de capa delgada** *f* (*CCD*) LAB, QUÍMICA *análisis* thin-layer chromatography (*TLC*); ~ **de capa fina** *f* LAB *análisis*, QUÍMICA thin-layer chromatography (*TLC*); ~ **por filtración en gel** *f* LAB, QUÍMICA *análisis* gel permeation chromatography; ~ **gaseosa** *f* ALIMENT, QUÍMICA, TERMO gas chromatography; ~ **de gases** *f* ALIMENT, QUÍMICA, TERMO gas chromatography; ~ **de gases en fase líquida** *f* GAS, QUÍMICA, TERMO gas liquid chromatography, gas liquid partition chromatography; ~ **de iones** *f* LAB *análisis* ion chromatograph; ~ **líquida a alta presión** *f* (*CLAP*) ALIMENT, LAB, QUÍMICA high-pressure liquid chromatography (*HPLC*); ~ **de líquidos** *f* CONTAM *análisis-técnicas* liquid chromatography; ~ **en papel** *f* QUÍMICA paper chromatography; ~ **de partición** *f* QUÍMICA partition chromatography; ~ **de permeación de gel** *f* LAB *análisis* gel permeation chromatography; ~ **de permeación en gel** *f* QUÍMICA *análisis* gel permeation chromatography; ~ **de reparto** *f* QUÍMICA partition chromatography

cromatógrafo: ~ **de gases** *m* ALIMENT, GAS, LAB, QUÍMICA, TERMO gas chromatograph

cromeado *m* ELECTRÓN chrominance

crómico *adj* C&V, COLOR, QUÍMICA chromic

crominancia *f* TEC ESP chrominance, TV chroma, chrominance; ~ **retrasada** *f* TV lagging chrominance

cromita *f* C&V, MINERAL chromite

cromo *m* (*Cr*) METAL, QUÍMICA, REVEST chromium (*Cr*); ~ **limón** *m* COLOR lemon chrome

cromoanalizador *m* CINEMAT, FOTO, TV color analyzer (*AmE*), colour analyser (*BrE*)

cromodinámica: ~ **cuántica** *f* FÍS, FÍS PART quantum chromodynamics (*QCD*)

cromofórico *adj* QUÍMICA chromophoric

cromóforo *m* QUÍMICA chromogen

cromofotografía *f* FOTO color photography (*AmE*), colour photography (*BrE*)

cromógeno *m* QUÍMICA chromogen

cromometalografía *f* METAL color metallography (*AmE*), colour metallography (*BrE*)

cromopirómetro *m* FÍS RAD color pyrometer (*AmE*), colour pyrometer (*BrE*)

cromosfera *f* TEC ESP chromosphere

cromoso *adj* QUÍMICA chromous

cromotipia *f* FOTO, IMPR color-printing process (*AmE*), colour-printing process (*BrE*)

cromotrópico *adj* QUÍMICA *intermediario en pinturas* chromotropic

cronización *f* TELECOM timing

cronodesplazamiento *m* ELECTRÓN time shift

cronografía *f* INSTR, PROD time recording

cronógrafo *m* INSTR chronograph, PROD timer

cronograma *m* PROD timing diagram

cronointerruptor *m* ELEC, TELECOM time switch

cronología *f* TEC ESP *del lanzamiento* chronology; ~ **geológica** *f* AmL (*cf escala de tiempos geológicos Esp*) GEOL geological time scale

cronomedición *f* TELECOM timing

cronometración *f* TELECOM timing

cronometrado *m* ELECTRÓN clocking

cronometrador *m* ELECTRÓN time marker, PROD time-keeper

cronometraje *m* CINEMAT timing, PROD timekeeping, TELECOM timing

cronometrar *vt* CINEMAT clock, time, ELECTRÓN, INFORM&PD clock, PROD, TEC ESP time

cronometrista *m* PROD timekeeper

cronómetro *m Esp* (*cf medidor de intervalos de tiempo AmL*) CINEMAT timer, FÍS chronometer, stopclock, timer, LAB *medida de tiempo* chronometer, timer, PROD, TELECOM timer, TRANSP MAR chronometer; **~ para color** *m* CINEMAT color timer (*AmE*), colour timer (*BrE*); **~ regulador** *m* ING MECÁ regulator

cronstedtita *f* MINERAL cronstedtite

croquis *m* CONST sketch, ING MECÁ layout, outline, outline drawing, MECÁ layout; **~ acotado** *m* PROD dimensioned sketch; **~ de arboladura y de jarcia** *m* TRANSP MAR rigging drawing; **~ de fuegos** *m* D&A range card; **~ de montaje** *m* EMB layout; **~ de nivel** *m* MINAS *topografía* foresight

crossita *f* MINERAL crossite

crotílico *adj* QUÍMICA *ácido* crotylic

crotonaldehído *m* QUÍMICA butenal, crotonaldehyde

crotónico *adj* QUÍMICA *ácido* crotonic

cruce *m* CONST *tuberías* cross, junction, FERRO *equipo inamovible* crossover, TELECOM *telefonía* crosstalk, crossover, TEXTIL *vestimenta* traverse; **~ de caminos** *m* CONST crossroads; **~ de carreteras** *m* AUTO road junction, CONST crossroads, fork, interchange, TRANSP crossroads, road junction, VEH road junction; **~ de carreteras en T** *m* CONST, TRANSP T-junction; **~ por cero** *m* ING ELÉC zero crossing; **~ doble** *m* CONST, FERRO *equipo inamovible* double crossover; **~ del ecuador** *m* TEC ESP *navegación* equatorial crossing; **~ entre vías curvas** *m* FERRO *equipo inamovible* crossover between curved track; **~ de ferrocarril** *m* CONST fork; **~ de galerías** *m* MINAS *geología* back; **~ peatonal solamente** *m* TRANSP pedestrian-only crossing zone; **~ de vías con corazón giratorio** *m* TRANSP *ferrocarriles* swing nose crossing

crucero *m* D&A, TRANSP cruiser, TRANSP AÉR cruise, TRANSP MAR cruise, *buque de guerra* cruiser; **~ ascendente** *m* TRANSP AÉR climb cruise; **~ de combate** *m* TRANSP MAR battle cruiser; **~ a motor** *m* TRANSP motor cruiser; **~ nivelado** *m* TRANSP AÉR level cruise; **~ de pesca** *m* OCEAN fishing trip; **~ principal** *m* MINAS *geología* back; **~ de subida** *m* TRANSP AÉR climb cruise

cruceta *f* AGUA crosspiece, FÍS crosshead, ING MECÁ crosshead, universal joint, PROD crosshead, TRANSP AÉR crossbar, spider unit, TRANSP MAR *jarcias* crosstree, *mástil* jumper strut, VEH *junta universal* spider; **~ del diferencial** *f* VEH *transmisión* differential spider; **~ con guía de 4 barras** *f* ING MECÁ crosshead with 4-bar guide; **~ de palo** *f* TRANSP MAR crosstree

cruciferario *m* PROD *parrilla de hogar* cross bearer

crudo[1] *adj* C&V unfired, CARBÓN *petróleo bruto* crude, raw, QUÍMICA *petróleo*, TEC PETR crude, TERMO unbaked; **en ~** *adj* TEXTIL in the gray (*AmE*), in the grey (*BrE*)

crudo[2] *m* CONTAM, CONTAM MAR crude oil, PETROL, TEC PETR *petróleo* crude, crude oil; **~ denso** *m* TEC PETR *calidad de petróleo* heavy crude oil; **~ desasfaltado** *m* TEC PETR *calidad del petróleo* deasphalted oil (*DAO*); **~ ligero** *m* TEC PETR *calidad*

de petróleo light crude; **~ con mucho azufre** *m* TEC PETR *calidad de petróleo* sour crude; **~ pesado** *m* TEC PETR *calidad del petróleo* heavy crude; **~ con poco azufre** *m* TEC PETR sweet crude; **~ sintético** *m* TEC PETR *petroquímica* synthetic crude; **~ sulfuroso** *m* TEC PETR *calidad del petróleo* sour crude; **~ viscoso** *m* TEC PETR *calidad del petróleo* heavy crude

crujía[1]: **en ~** *adv* TRANSP MAR midship

crujía[2] *f* TRANSP MAR midship

crujido *m* MECÁ cracking

cruz *f* AGRIC *de un animal* withers, CINEMAT *del esmerilado*, INFORM&PD *cursor* cross hair, TRANSP MAR *de verga* sling, *del ancla* crown; **~ de las bitas** *f* TRANSP MAR crosspiece; **~ hendida para tornillos** *f* ING MECÁ cross-recess for screw; **~ hilo a hilo** *f* TEXTIL end-and-end lease; **~ de Malta** *f* MECÁ Maltese cross; **~ de producción** *f* TEC PETR Christmas tree; **~ de tubos** *f* ING MECÁ pipe cross

cruzamiento *m* FERRO *haz de vías paralelas* crossover, scissors crossing; **~ agudo** *m* FERRO *equipo inamovible* common crossing; **~ agudo en curva** *m* FERRO *equipo inamovible* curved common crossing; **~ agudo en recta** *m* FERRO straight common crossing; **~ oblicuo** *m* FERRO diamond crossing; **~ oblicuo no normalizado** *m* FERRO nonstandard diamond crossing; **~ oblicuo con superficie de deslizamiento** *m* FERRO single diamond crossing with slips; **~ con probador** *m* AGRIC *fitomejoramiento* top cross; **~ de vías** *m* FERRO rail crossover; **~ de vías de ventilación** *m* MINAS air bridge

cruzar[1] *vt* AGRIC steer, CONST bridge, cross, FERRO *puntos* cross over, GEOL cross, intersect; **~ mediante un puente** *vt* CONST *valle* bridge over

cruzar[2] *vi* TRANSP MAR cruise

Cs *abr* ELECTRÓN, ING ELÉC (*cesio*) Cs (*caesium BrE, cesium AmE*), METEO (*cirrostrato*) Cs (*cirrostratus*), QUÍMICA, TEC ESP (*cesio*) Cs (*caesium BrE, cesium AmE*)

CSS *abr* (*control del sistema de comunicaciones*) TEC ESP CSM (*communications system monitoring*)

CTCP *abr* (*conjunto de giro y transferencia de potencia*) TEC ESP BAPTA (*bearing and power-transfer assembly*)

Cu *abr* ELEC, METAL (*cobre*) Cu (*copper*), METEO (*cumulus*) Cu (*cumulus*), QUÍMICA (*cobre*) Cu (*copper*)

CU *abr AmL* (*computadora universal AmL*) INFORM&PD GP computer (*general-purpose computer*), universal computer

cuadalímetro *m* CONST flowmeter

cuaderna *f* CONST frame, TRANSP AÉR ring frame, ring fuselage, TRANSP MAR *construcción de barcos* frame; **~ maestra** *f* TRANSP MAR midship frame

cuadernal *m* ING MECÁ pulley, pulley block, tackle, MECÁ tackle, PROD hoisting tackle, tackle fall, TRANSP MAR block, pulley, pulley block; **~ de cabina** *m* PROD gin block, gin pulley; **~ de cabria** *m* PROD gin block, gin pulley; **~ de engranaje recto** *m* ING MECÁ spur-geared pulley block; **~ fijo** *m* TRANSP MAR *equipamientos* standing block; **~ de gancho** *m* ING MECÁ hook block, PROD hoisting block, tackle with hook block; **~ giratorio** *m* PROD monkey block; **~ móvil** *m* ING MECÁ runner, PROD *grúas* hoisting block; **~ de piezas** *m* ING MECÁ made block; **~ de polea** *m* ING MECÁ sheave pulley block

cuadernas: **~ transversales** *f pl* TRANSP MAR *cons-*

trucción naval transverse framing; **~ de trazado** *f pl* TRANSP MAR *plano de formas* station

cuadernillo *m* IMPR section, signature

cuaderno: **~ de bitácora** *m* TRANSP logbook; **~ de campo** *m* CONST *topografía* field book; **~ diario de máquinas** *m* TRANSP MAR engine-room log; **~ de máquinas** *m* TRANSP MAR engine-room log

cuadra *f* AmL (*cf manzana Esp*) CONST *de casas* block

cuadradillo *m* ALIMENT lump sugar

cuadrado¹ *adj* CONST, GEOM, IMPR, ING MECÁ, MATEMÁT, PROD square, squared

cuadrado² *m* CONST, GEOM square, IMPR *tipografía* quad, ING MECÁ, MATEMÁT square; **~ perfecto** *m* MATEMÁT perfect square; **~ en T** *m* ING MECÁ T-square

cuadrado³: **ser ~** *vi* PROD be square

cuadrafonía *f* ACÚST quadraphony

cuadrangular *adj* GEOM quadrangular

cuadrángulo *m* GEOM quadrangle

cuadrante *m* ELEC *de instrumento* dial, GEOM, ING MECÁ quadrant, TEC PETR *perforación* kelly; **~ de agrimensor** *m* CONST surveyor's dial; **~ de alcances** *m* D&A range dial; **~ atenuador** *m* CINEMAT dimmer quadrant; **~ calibrado** *m* PROD calibrated dial; **~ graduado** *m* CINEMAT scale dial, ING MECÁ dial plate, divided dial, INSTR graduated dial; **~ instrumental** *m* INSTR instrument dial; **~ de la palanca de control** *m* TRANSP AÉR control-lever quadrant

cuadrático¹ *adj* MATEMÁT quadratic

cuadrático² *m* ING ELÉC, MATEMÁT quadrac

cuadratín *m* IMPR em quad, *doce puntos* em

cuadratura¹: **en ~** *adj* ELECTRÓN, FÍS in quadrature

cuadratura² *f* GEN quadrature, CONST squareness; **~ del círculo** *f* GEOM squaring the circle; **~ de fase** *f* ELECTRÓN phase quadrature; **~ de fase sincronizada** *f* ELECTRÓN *en señales* locked in-phase quadrature; **~ gaussiana** *f* INFORM&PD Gaussian quadrature

cuadrete *m* ELECTRÓN, ING ELÉC quad; **~ estrella** *m* ING ELÉC star quad

cuadrícula *f* CINEMAT, ELECTRÓN, INSTR, METR, ÓPT graticule, TEXTIL grid, TV graticule; **~ externa** *f* ELECTRÓN external graticule; **~ de imagen** *f* ELECTRÓN frame grid; **~ interna** *f* ELECTRÓN internal graticule; **~ óptica** *f* METR optical graticule

cuadriculado *m* PROD checkering (*AmE*), chequering (*BrE*)

cuadrilateral *adj* GEOM quadrilateral

cuadrilátero¹ *adj* GEOM four-sided

cuadrilátero² *m* GEOM quadrangle

cuadrilla *f* AGRIC, TEC PETR *perforación* crew; **~ de perforación** *f* TEC PETR drilling crew; **~ de sondeo** *f* TEC PETR *perforación* drilling crew; **~ en voladizo** *f* CONST flying squad

cuadrípolo *m* ELEC *red de distribución*, FÍS, FÍS RAD quadripole, quadrupole, ING ELÉC quadripole, four-terminal network, quadrupole; **~ activo** *m* ING ELÉC active quadrupole; **~ eléctrico** *m* ELEC, FÍS RAD, ING ELÉC electric quadrupole; **~ lineal** *m* ELEC linear four-terminal network; **~ pasivo** *m* ING ELÉC passive quadripole

cuadrivalencia *f* QUÍMICA quadrivalence

cuadrivalente *adj* QUÍMICA quadrivalent

cuadro *m* CARBÓN table, CINEMAT frame, CONST *de cantidades, precios* table, ELECTRÓN board, ENERG

RENOV housing, FÍS chart, FOTO, TEC ESP, TELECOM frame; **~ de actividad presente** *m* TRANSP *tráfico* presence loop; **~ aditivo** *m* ELECTRÓN add-in board; **~ aislado** *m* ELEC insulating board; **~ de alineación** *m* PETROL alignment frame; **~ de alineamiento** *m* TEC ESP *comunicaciones* frame alignment; **~ analógico** *m* ELECTRÓN analog board; **~ anterior** *m* FOTO front frame; **~ de aviso general** *m* TRANSP AÉR general warning panel; **~ de balance térmico** *m* TERMO heat balance chart; **~ en blanco** *m* CINEMAT blank frame; **~ bloqueado** *m* CINEMAT, FOTO, TV freeze-frame; **~ de calibración** *m* INFORM&PD calibration chart; **~ de campo** *m* AUTO field frame; **~ central de distribución de acumuladores** *m* TELECOM central battery switchboard; **~ cero** *m* CINEMAT zero frame; **~ del circuito** *m* ELEC, ELECTRÓN circuit board; **~ de conexiones** *m* ING ELÉC pinboard; **~ congelado** *m* CINEMAT, FOTO, TV freeze-frame, frozen frame; **~ de conmutación** *m* TELECOM switchboard; **~ de conmutación con clavija y cordón** *m* TELECOM plug-and-cord switchboard; **~ de conmutación telefónica** *m* TELECOM switchboard, telephone switchboard; **~ conmutador** *m* TELECOM switchboard, telephone switchboard; **~ conmutador central** *m* TELECOM common battery switchboard; **~ conmutador con clavija y cordón** *m* TELECOM plug-and-cord switchboard; **~ conmutador sin dicordios** *m* TELECOM cordless switchboard; **~ conmutador manual** *m* TELECOM manual switchboard; **~ conmutador múltiple** *m* TELECOM multiple switchboard; **~ conmutador para servicio interurbano** *m* TELECOM toll switch; **~ de conmutadores magnético** *m* TELECOM magneto switchboard; **~ de contactos** *m* METR *electrónica* level; **~ de contactos enchufables** *m* ING ELÉC plugboard; **~ de contactos motorizado** *m* INSTR *telefonía automática* motor-driven level; **~ de control** *m* CALIDAD control chart, TRANSP MAR *de instalación eléctrica* electrical panel, *de instrumentos* control panel; **~ de control eléctrico** *m* ELEC, NUCL electrical control board; **~ cromático** *m* TV color field (*AmE*), colour field (*BrE*); **~ a cuadro** *m* CINEMAT, TV frame by frame; **~ DCME** *m* TELECOM DCME frame; **~ de distribución** *m* ELEC distributing board, ING ELÉC distribution board, switchboard, TELECOM distribution board, TRANSP MAR *circuito eléctrico* switch panel, VEH *sistema de encendido* distributing board; **~ de distribución sin hilos conductores** *m* TELECOM cordless switchboard; **~ eléctrico impreso en resina de epoxia** *m* ELECTRÓN glass epox-printed circuit board; **~ de esquema** *m* ELEC mimic board; **~ fijo** *m* CINEMAT, FOTO, TV freeze-frame, frozen frame; **~ de filiación** *m* ELECTRÓN daughter board; **~ de flujo calorífico** *m* TERMO heat flow chart; **~ de fuerzas** *m* PROD force table; **~ de fusibles** *m* ELECTRÓN, ING ELÉC, TV fuse holder; **~ de fusibles de distribución** *m* ELEC distribution fuse board; **~ graduado** *m* INSTR trial frame; **~ indicador** *m* ING ELÉC annunciator, PROD annunciator panel; **~ individual** *m* CINEMAT single frame; **~ instantáneo** *m* CINEMAT, TV flash frame; **~ de instrumentos** *m* AUTO dashboard, INSTR instrument board, VEH dashboard; **~ interurbano** *m* TELECOM toll switch; **~ de lubricación** *m* MECÁ lubrication chart; **~ luminoso** *m* FERRO *equipo inamovible* reflectorized board; **~ de mandos** *m* TEX-

TIL control panel; **~ de maniobras** *m* PETROL draw works; **~ manual** *m* TELECOM manual switchboard; **~ de modem** *m* ELECTRÓN modem board; **~ móvil** *m* ELEC *de galvanómetro*, ING ELÉC moving coil; **~ múltiple** *m* TELECOM multiple switchboard; **~ de operadora** *m* TELECOM telephone switchboard; **~-patrón de imágenes** *m* TV definition test pattern; **~ portador** *m* MINAS *galerías, pozos* curb, kerb; **~ portátil** *m* MINAS *armazón de sustentación* crib; **~ de pozo** *m* CARBÓN *minas* sinking trestle; **~ de pulsadores** *m* ELEC *aislación* pressboard; **~ receptor** *m* ELECTRÓN receiver board; **~ reflectante** *m* FERRO *equipo inamovible* reflectorized board; **~ de sincronización** *m* TEC ESP *comunicaciones* frame synchronization; **~ de suma acumulada** *m* CALIDAD *estadística* cusum chart; **~ de tajo** *m* MINAS *galerías* gate end box; **~ de tamaño medio** *m* ELECTRÓN half-sized board; **~ de terminales** *m* ING ELÉC terminal block

cuadros *m pl* TEXTIL checks; **~ tipo cazadora canadiense** *m pl* TEXTIL lumberjack checks

cuádruple: **~ amplificador operacional** *m* ELECTRÓN quad operational amplifier; **~ de pares retorcidos** *m* ING ELÉC multiple twin quad

cuadruplete *m* ING ELÉC quadding

cuadruplexo *adj* TV quadruplex

cuajaleches *m* ALIMENT lady's bedstraw

cuajo *m* AGRIC rennet, ALIMENT rennet, *industria quesera* curd

cualidad: **~ del borde** *f* ELECTRÓN edge steepness; **~ de trabajable** *f* CONST workability

cualidades: **~ técnicas** *f pl* ING MECÁ performance properties

cualificación: **~ espacial** *f* TEC ESP space qualification

cualímetro *m* LAB *medida del poder de penetración de los rayos X* penetrometer

cuántico *adj* GEN quantum

cuantificación *f* GEN quantification, quantization; **~ de bloques** *f* TELECOM block quantization; **~ de riesgos** *f* CALIDAD, SEG risk quantification; **~ de una señal** *f* TELECOM signal quantization; **~ de señal de entrada** *f* ELECTRÓN input signal quantization

cuantificador *m* CONST quantifier, ELECTRÓN, FÍS PART, INFORM&PD quantizer, MATEMÁT quantifier, TEC ESP, TELECOM, TV quantizer

cuantificar *vt* CONST quantify, ELECTRÓN, FÍS, FÍS PART, INFORM&PD quantize, MATEMÁT quantify, TEC ESP, TELECOM, TV quantize

cuantización: **~ espacial** *f* FÍS spatial quantization

cuanto *m* GEN quantum; **~ de radiaciones Roentgen** *m* NUCL X-ray quantum; **~ de rayos gamma** *m* FÍS PART, FÍS RAD gamma-ray quantum

cuarcita *f* CONST, GEOL, PETROL quartzite

cuarcítico *adj Esp* (*cf cuarzoso AmL*) GEOL quartzose

cuarenta: **~ rugientes** *m pl* OCEAN *Atlántico austral* roaring forties

cuarentena *f* AGRIC, SEG, TRANSP quarantine

cuark *m* FÍS, FÍS PART, NUCL quark

cuarta: **~ generación** *f* INFORM&PD fourth generation; **~ nota de la escala diatónica** *f* ACÚST subdominant; **~ perfecta** *f* ACÚST perfect fourth

cuartas *f pl* TRANSP MAR points of the compass

cuarteado *m* P&C, REVEST, TRANSP AÉR alligatoring

cuarteamiento *m* MINAS *de la roca por explosiones de cargas* springing, P&C, REVEST *defecto* alligatoring; **~ de rocas** *m* MINAS *canteras* plugging

cuartear: **~ la aguja** *vt* TRANSP MAR box the compass

cuartearse *v refl* MINAS split

cuartel *m* CARBÓN *corta forestal* compartment; **~ de bomberos** *m* TERMO fire station; **~ de rescate** *m* SEG rescue station

cuarteo: **~ por malla** *m* C&V sieve fraction

cuarteto *m* FÍS quartet, INFORM&PD nibble; **~ inferior** *m* PROD lower nibble

cuartillo *m* METR *unidad británica (=1,136 litros) y estadounidense (=1,101 litros) de capacidad de líquidos* quart

cuarto *m* CONST room, METR *peso* quarter; **~ bocel** *m* CONST, ING MECÁ quarter round; **~ de círculo** *m* ING MECÁ quadrant; **~ de crisoles** *m* C&V pot room; **~ de derrota** *m* TRANSP MAR *construcción naval* chartroom; **~ frío** *m* ING MECÁ, REFRIG, TERMO coldroom; **~ limpio** *m* LAB, SEG clean room; **~ del motor de extracción** *m* MINAS hoist room; **~ oscuro** *m* CINEMAT darkroom, loading room, FOTO darkroom; **~ de radiotelegrafía** *m* TRANSP MAR radio room; **~ trasero** *m* AGRIC *de una res* hindquarter

cuartón *m* CONST scantling

cuarzo *m* ELECTRÓN, FÍS, MINERAL quartz; **~ ahumado** *m* MINERAL smoky quartz; **~ amarillo** *m* MINERAL citrine; **~ amarillo con crocidolita** *m* MINERAL tiger's eye; **~ aventurina** *m* MINERAL aventurine quartz; **~ azul** *m* MINERAL blue quartz, sapphire quartz; **~ citrino** *m* MINERAL citrine quartz; **~ fundido** *m* ÓPT, TELECOM fused quartz; **~ ojo de gato** *m* MINERAL cat's eye quartz; **~ verde** *m* MINERAL prase

cuarzoarenita *f* GEOL quartzarenite

cuarzocitrino *m* MINERAL false topaz

cuarzodiorita *f* PETROL quartzdiorite

cuarzoso *adj AmL* (*cf cuarcítico Esp*) GEOL quartzose

cuásar *m* TEC ESP *astronomía* quasar

cuasi: **~-partícula** *f* FÍS quasi-particle

cuasicolisión *f* TRANSP AÉR near collision, near miss

cuasiestático *adj* TEC ESP quasi-statical

cuasina *f* QUÍMICA quassin

cuasireticuloide *m* CARBÓN near mesh

cuaternario[1] *adj* GEOL, NUCL, QUÍMICA, TEC PETR *geología* quaternary

cuaternario[2] *m* GEOL, NUCL, QUÍMICA, TEC PETR *geología* quaternary

cuaternio *m* MATEMÁT quaternion

cuatricromía *f* IMPR printing with four colors (*AmE*), printing with four colours (*BrE*)

cuatro: **de ~ espesores** *adj* PROD four-ply; **de ~ polos** *adj* ELEC *generador* four-pole

CUB *abr* (*conmutador unipolar bidireccional*) ELEC, ING ELÉC DPSTS (*double-pole single-throw switch*)

cuba *f AmL* (*cf jaula Esp*) ALIMENT vat, C&V glass tank, CONTAM tank, LAB *material de vidrio* trough, glass tank, MINAS kieve, kibble, cage, PAPEL *máquina de cartón* vat, PROD vat, *de alto horno* tunnel, tub, fire room, mantle, *transformador eléctrico* tank, trough, bowl, QUÍMICA vat; **~ de amalgamación** *f* CARBÓN pan, PROD pan, pan mill, grinding pan; **~ del carburador** *f* AUTO, VEH carburetor float chamber (*AmE*), carburettor float chamber (*BrE*), carburetor barrel (*AmE*), carburettor barrel (*BrE*); **~ colectora** *f Esp* (*cf artesa colectora Esp*) MINAS collecting vat; **~ de combustible** *f* TEC ESP *vehículos* fuel tank; **~ conductimétrica** *f* ELECTRÓN, LAB *análisis, electroquímica* conductance cell; **~ criógenica** *f* TEC ESP *motores cohete* cryogenic tank; **~ de cromatografía** *f*

LAB *análisis*, QUÍMICA chromatography tank; ~ **de decantación** *f* CARBÓN settling tank; ~ **de electrochapado** *f* ELEC, ING ELÉC, REVEST electroplating vat; ~ **electroforética** *f* ELEC, LAB, QUÍMICA electrophoresis cell; ~ **de electrolítica** *f* ELEC, ING ELÉC, REVEST electroplating vat; ~ **electrolítica** *f* ELEC, FÍS, ING ELÉC electrolytic cell; ~ **empleada para aplicación de revestimientos** *f* REVEST coat trough, coating trough; ~ **de enfriamiento** *f* PROD, REFRIG, TERMO cooling basin; ~ **para enfriar** *f* C&V *en el temple* bosh; ~ **exterior** *f* REFRIG *de mueble frigorífico* cabinet shell; ~ **de extracción** *f* PROD hoisting bucket; ~ **de filtración** *f* AGUA filtration vat; ~ **del flotador** *f* AUTO, VEH float chamber; ~ **de hidrógeno** *f* TEC ESP hydrogen tank; ~ **inclinable** *f* TRANSP tilting skip; ~ **interior** *f* REFRIG *de un refrigerador* liner; ~ **de lavado de Baum** *f* CARBÓN Baum washbox; ~ **de lavado de diafragma** *f* CARBÓN diaphragm-type washbox; ~ **de maceración** *f* ALIMENT *fermentación* mash tun, masher; ~ **de mezclado** *f* PROC QUÍ mixing vessel; ~ **de nivel constante** *f* AUTO *flotador*, VEH *carburador* float chamber; ~ **para petróleo** *f* ING ELÉC oil tank; ~ **de precipitación** *f* PROC QUÍ precipitating tank, precipitation tank; ~ **primaria** *f* AUTO primary barrel; ~ **de recogida** *f* CONTAM MAR collection basin; ~ **de sedimentación** *f* CARBÓN settling tank, PROC QUÍ sedimentation tank, settling tub, settling vat, REVEST coater trough, coating trough; ~ **de sinterización** *f* PROC QUÍ sintering pan; ~ **de trituración** *f* PROD grinding pan

cubanita *f* MINERAL cubanite

cubebina *f* QUÍMICA cubebin

cubertada *f* TRANSP MAR deck cargo, deck load

cubeta *f* CARBÓN *geología* basin, container, pan, FOTO dish, GEOL basin, sub-basin, INFORM&PD *memoria* bucket, ING MECÁ *barómetro* bulb, LAB *contenedor* Buchner funnel, *análisis* cell, *de barómetro* cup, MINAS *para el lavado de minerales* trunk, *recipiente* tub, P&C *pinturas, instrumento* cup, PROD tank; ~ **abisal** *f* OCEAN deep-sea trough; ~ **de bitácora** *f* FÍS compass bowl; ~ **de centrífuga** *f* LAB *separación* centrifuge bucket; ~ **de desescarcha** *f* REFRIG *para fluido del deshielo* drip tray; ~ **draga** *f* MINAS grab; ~ **plástica** *f* FOTO plastic dish; ~ **de revelado regulada termostaticamente** *f* FOTO thermostatically-controlled developing dish; ~ **sacalodos** *f* MINAS sludger

cubicación *f* CARBÓN cubage, CONST measurement of quantities, METR cubing, cubature, cubic measurement, *determinación del contenido cúbico* cubage; ~ **de obra hecha por horas** *f* PROD taking-off sheets per hours; ~ **de obra hecha materiales** *f* PROD taking-off sheets materials

cubicar *vt* METR *sólidos* measure, PROD *toneles* gage (*AmE*), gauge (*BrE*)

cubicita *f* MINERAL cubicite

cúbico[1] *adj* CONST, CRISTAL, GEOM, MATEMÁT, METR cubic

cúbico[2]: ~ **de caras centradas** *m* (*CCC*) CRISTAL, METAL, QUÍMICA *cristalografía* face-centered cubic (*AmE*) (*FCC*), face-centred cubic (*BrE*) (*FCC*); ~ **de cuerpo centrado** *m* QUÍMICA body-centered cubic (*AmE*), body-centred cubic (*BrE*)

cubierta[1]: **bajo** ~ *adj* TRANSP MAR *buque* below deck; **con** ~ *adj* TRANSP MAR *construcción naval* decked; **sin** ~ *adj* TRANSP MAR open; **con** ~ **de dos capas de algodón** *adj* ELEC *conductor* double cotton-covered (*DCC*); **con** ~ **de goma pura doble** *adj* ELEC double-pure-rubber-covered (*DPRC*)

cubierta[2] *f* ALIMENT *para embutidos*, C&V casing, CONST covering, roofing, deck, IMPR cover, ING MECÁ cover, hood, INSTR cover, LAB *de contenedor* lid, MECÁ cover, case, lid, MINAS *de mina, pozo* deck, PROD jacket, SEG cover, TEC ESP shroud, cover, *protección de vehículos* casing, TRANSP AÉR canopy, TRANSP MAR *construcción naval* deck; ~ **de abrigo** *f* TRANSP MAR shelter deck; ~ **a un agua** *f* CONST shed roof; ~ **aislante** *f* PROD insulating lagging; ~ **de alojamientos** *f* TRANSP MAR accommodation deck; ~ **de banco** *f* C&V bench cloth; ~ **para bandeja** *f* Esp (*cf cubierta para pálet AmL*) EMB pallet hood; ~ **del bastidor** *f* ING MECÁ frame cap; ~ **de botes** *f* TRANSP MAR *en un buque* boat deck; ~ **con brusca** *f* TRANSP MAR *construcción naval* cambered deck; ~ **del cabezal del portaobjetivos** *f* INSTR turret cap; ~ **del cable** *f* ELEC, ING ELÉC, P&C cable covering (*AmE*), cable sheathing (*BrE*); ~ **del castillo** *f* TRANSP MAR foredeck, *para faena de anclas* anchor deck; ~ **corrida** *f* TRANSP MAR *construcción naval* flush deck; ~ **a cuatro aguas** *f* CONST hip roof; ~ **con curvatura transversal** *f* TRANSP MAR *construcción naval* cambered deck; ~ **de desembarco** *f* TRANSP landing deck; ~ **de despegue** *f* TRANSP MAR *portaaviones* flat top; ~ **a dos aguas** *f* CONST span roof, *tejado* ridge roof; ~ **con falso tirante** *f* CONST collar roof; ~ **forrada** *f* TRANSP MAR *construcción naval* sheathed deck; ~ **del fuelle** *f* FOTO bellows covering; ~ **de fusible** *f* ELEC fuse cover; ~ **guardapolvos** *f* ING MECÁ dust cover; ~ **inferior** *f* TRANSP MAR lower deck, *entrepuente* orlop deck; ~ **interior** *f* ELEC *aislamiento de cable* inner covering; ~ **con lámina plástica** *f* REVEST plastic-laminate covering; ~ **de la mansarda** *f* CONST *tejado* gambrel roof (*AmE*), mansard roof (*BrE*); ~ **de nivelación** *f* CONST grading envelope; ~ **de la noria** *f* C&V cover tile (*BrE*), spout cover (*AmE*); ~ **para pálet** *f* AmL (*cf cubierta para bandeja Esp*) EMB pallet hood; ~ **para paleta** *f* Esp EMB pallet hood; ~ **del paracaídas de frenado** *f* TRANSP AÉR drag chute cover; ~ **de paseo** *f* TRANSP MAR shelter deck, promenade deck; ~ **plástica** *f* REVEST plastic covering; ~ **con pliegues diagonales** *f* P&C diagonal ply tire (*AmE*), diagonal ply tyre (*BrE*); ~ **con pliegues radiales** *f* AUTO, P&C, VEH radial-ply tire (*AmE*), radial-ply tyre (*BrE*); ~ **con pliegues transversales** *f* P&C crossply tire (*AmE*), crossply tyre (*BrE*); ~ **para polvos** *f* SEG dust cover; ~ **de popa** *f* TRANSP MAR afterdeck; ~ **posterior interna** *f* IMPR inside back cover; ~ **para prevenir accidentes** *f* SEG cover to prevent accidents; ~ **principal** *f* TRANSP MAR main deck; ~ **protectora** *f* EMB protective cover, ING MECÁ housing, protective cap, NUCL cover plate, PROD covering plate, REVEST, SEG protective cover; ~ **protectora de tubos catódicos** *f* ING MECÁ cover plate; ~ **puente** *f* TRANSP MAR bridge deck; ~ **resistente al ácido** *f* SEG acid-resisting covering; ~ **con revestimiento** *f* TRANSP MAR *construcción naval* sheathed deck; ~ **de rueda** *f* VEH wheel cover; ~ **de la salida de la unidad de enfriamiento de aire** *f* TRANSP AÉR blanking cover for air-cooling unit outlet; ~ **superior** *f* TRANSP MAR upper deck;

~ **térmica** *f* TEC ESP heat shroud; ~ **de toldilla** *f* TRANSP MAR *construcción naval* poop deck; ~ **de unión** *f* TRANSP AÉR coupling cover; ~ **vegetal** *f* AGRIC vegetable cover; ~ **del ventilador** *f* AUTO fan shroud; ~ **de vuelo** *f* TRANSP AÉR, TRANSP MAR *de buque portaaviones* flight deck

cubierto *adj* ALIMENT coated, TERMO *fuego de horno* banked up; ~ **de esmalte** *adj* REVEST enamel-covered; ~ **con hoja de cobre** *adj* ELEC *cable*, REVEST copper-clad

cubilete *m* PAPEL beaker

cubilote *m* C&V, INSTAL TERM cupola, PROD cupola furnace, cupola; ~ **con antecristol** *m* PROD cupola with receiver

cubo *m* CONST *recipiente* bucket, pail, GEOM cube, INFORM&PD hub, bucket, LAB *contenedor* Buchner funnel, TEC ESP *de germanio, de silicio* chip, TRANSP skip, TRANSP AÉR hub; ~ **autobasculante** *m* CONST self-dumping bucket; ~ **basculante** *m* CONST dump bucket (*AmE*), tipping bucket (*BrE*); ~ **de clasificación** *m* AGRIC holding bin; ~ **del cojinete de desenganche** *m* AUTO release-bearing hub; ~ **de descarga** *m* CONST, TRANSP dumping bucket; ~ **de desechos** *m* TEC ESP disposal tank; ~ **de elección** *m* INFORM&PD hub polling; ~ **con fondo móvil** *m* CONST bucket with drop-bottom; ~ **de la hélice** *m* TRANSP AÉR propeller hub, TRANSP MAR propeller boss, propeller hub; ~ **para mineral** *m* MINAS ore dump; ~ **de rotor** *m* TRANSP AÉR rotor hub; ~ **del rotor principal** *m* TRANSP AÉR main-rotor hub

cuboflash *m* FOTO flashcube

cuboide *m* GEOM cuboid

cubre: ~ **testero** *m* PAPEL *bobina* roll head

cubrecadena *f* ING MECÁ chain guard

cubreculata *m* D&A *cañones* breech cover

cubrehierbas *m* AGRIC trash spring

cubrejunta *f* FERRO *vía* fishplate block, MECÁ butt plate, PROD butt strip; ~ **de angular** *m* FERRO splice bar

cubrejunta *f* CONST butt joint, PROD butt strap, junction plate, welt

cubremuelas *m* ING MECÁ, SEG *máquinas* wheel guard

cubreobjeto *m* C&V cover glass, cover slip, INSTR *microscopio* cover slip, LAB *microscopía* cover slip, cover glass, *microscopio* microscope slide cover slip; ~ **con escala** *m* C&V, LAB cover-glass gage (*AmE*), cover-glass gauge (*BrE*)

cubretasa *f* C&V tuckstone

cubriente *m* REVEST *barniz* opaque

cubrimiento: ~ **parcial** *m* TRANSP MAR *pintura* fleet

cubrir *vt* ALIMENT coat, CONST *tejados* cap, GEOL drape, MINAS *con carbón* land; ~ **de nuevo** *vt* CARBÓN recover; ~ **con planchas** *vt* TRANSP MAR *construcción naval* plate; ~ **con vegetación** *vt* AGRIC outgrow

cuchara *f* C&V *para el rodillo en vidrio rolado* dish, CARBÓN *sondeos* auger, CONST *excavadoras, maquinaria* bucketful, ladle, bucket, CONTAM MAR scoop, MINAS sludger, *AmL* (*cf cuchara limpiadora Esp*) *pozo de sondeo* cleaner, PETROL *perforación de sondeos* bailer, PROD ladle, TRANSP MAR *dragado* bailer, grab; ~ **de albañil** *f AmL* CONST, ING MECÁ (*cf llana de albañil Esp, cf llana para enlucir Esp*) bricklayer's trowel, plastering trowel; ~ **de arena** *f* MINAS *sondeos* sludge pump; ~ **de arrastre** *f* CARBÓN *minas* scraper, MINAS scraper, *pega* hole scraper, *sondeos* slusher; ~ **bivalva** *f* CONST clam shell bucket, grab bucket;

~ **de correa** *f* MINAS belt cleaner; ~ **espumadora** *f* PROD skimming ladle; ~ **excavadora** *f* CARBÓN earth grab, TRANSP digging bucket; ~ **de exterior** *f* MINAS surface auger; ~ **de horquilla** *f* PROD *fundería* shank, shank ladle; ~ **limpiadora** *f* Esp (*cf cuchara AmL*), MINAS *pozo de sondeo* cleaner; ~ **de mano** *f* PROD *fundería* shank, shank ladle; ~ **sacabarro** *f* CARBÓN *minería*, MINAS scraper

cuchareado *m* C&V ladling

cucharilla *f* MINAS sludger

cucharón *m* C&V, CONST ladle, CONTAM *maquinaria industrial*, CONTAM MAR *draga*, LAB *para manipular líquidos* scoop, PROD ladle, TRANSP skip; ~ **bivalvo** *m* TRANSP grapple; ~ **de descarga** *m* TRANSP dumping bucket; ~ **de fundir** *m* PROD *fundería* casting ladle; ~ **de grúa** *m* PROD *fundería* crane ladle; ~ **de mandíbulas** *m* CONST grab bucket; ~ **sobre vagoneta** *m* PROD *fundición* car ladle

cucharros: ~ **de popa** *m pl* TRANSP MAR *construcción* buttocks

cuchilla *f* AGRIC knife cutter, CARBÓN *herramientas* iron, CONST blade, ING MECÁ cutter, cutter bar, blade, MECÁ blade, PAPEL blade, doctor; ~ **de aire** *f* PAPEL air blade; ~ **a chorro de aire** *f* P&C *equipo para revestir o pintar* air knife; ~ **circular** *f* AGRIC roller mill; ~ **de corte** *f* ING MECÁ shear blade; ~ **desbastadora** *f* CONST drawknife; ~ **desgotadora** *f* PAPEL drainage foil; ~ **de doble mango** *f* CONST *carpintería* spokeshave; ~ **de dos mangos** *f* CONST drawknife; ~ **de encuadernador** *f* IMPR skiver; ~ **giratoria** *f* IMPR rotary cutter; ~ **lateral** *f* IMPR edge doctor; ~ **niveladora de aplanadora** *f* TRANSP grader leveling blade (*AmE*), grader levelling blade (*BrE*); ~ **oscilante** *f* IMPR reciprocating blade, PAPEL oscillating doctor; ~ **de parada** *f* CONST stopping knife; ~ **de perforación** *f* IMPR perforation blade; ~ **para pizarra** *f* CONST slate knife; ~ **de platina** *f* PAPEL bed knife; ~ **de plegado en ángulo recto** *f* IMPR chopper-fold blade; ~ **plegadora** *f* IMPR tucking blade; ~ **recta de arado** *f* AGRIC standing knife cutter; ~ **recta de arado de doble filo** *f* AGRIC quincy cutter; ~ **del refino** *f* PAPEL refiner bar; ~ **tangente** *f* IMPR doctor blade; ~ **tangente de borde regulable** *f* IMPR adjustable-edge doctor blade; ~ **de tijera rotatoria** *f* ING MECÁ rotary-shear blade; ~ **de trocear** *f* ING MECÁ parting blade, parting-off blade; ~ **de vidriero** *f* CONST putty knife

cuchillas *f pl* C&V shears; ~ **de cepilladora** *f pl* ING MECÁ plane iron; ~ **cobresoldadas** *f pl* ING MECÁ brazed-on tips; ~ **de herramientas descartables** *f pl* ING MECÁ throwaway tip; ~ **de herramientas desechables** *f pl* ING MECÁ throwaway tip; ~ **de la troceadora** *f pl* PAPEL chipper knife

cuchillo: ~ **de corte** *m* CONST hacking knife; ~ **de espátula** *m AmL* (*cf espátula Esp*) LAB *pinturas* palette knife; ~ **de hoja retráctil** *m* ING MECÁ retractable-blade knife; ~ **para sacos** *m* EMB sack knife; ~ **para uso general** *m* ING MECÁ utility knife

cuele *m* Esp (*cf huida AmL*) MINAS cut, *de túneles* driving; ~ **canadiense** *m* MINAS *voladuras* burn cut; ~ **convergente** *m* MINAS *voladuras* V-cut

cuello *m* AUTO, C&V neck, CONST neck, throat, ING MECÁ journal, collar, LAB *de retorta* neck, MECÁ gland, neck, collar, PETROL collar, PROD neck, TEC PETR collar; ~ **acampanado** *m* C&V flared neck; ~ **de**

boca ancha *m* EMB wide-mouth neck; ~ **caído** *m* C&V slug, slug in neck; ~ **de cisne** *m* CINEMAT *brazo acodado* gooseneck, swanneck, CONST swanneck, MECÁ gooseneck; ~ **de eje** *m* CONST *mecánica* gudgeon; ~ **hueco** *m* C&V hollow neck; ~ **mal soplado** *m* C&V hollow neck; ~ **en papel** *m* TEXTIL paper collar; ~ **del puerto** *m* C&V port neck; ~ **del radiador** *m* AUTO, VEH radiator filler neck; ~ **del tanque** *m* C&V tank neck; ~ **del tubo** *m* TV tube neck

cuenca *f* AGUA basin, *ríos* drainage area, CARBÓN, GEOL, HIDROL basin; ~ **de aguas freáticas** *f* AGRIC, AGUA, CARBÓN, HIDROL ground-water basin; ~ **de aguas subterráneas** *f* AGUA groundwater catchment; ~ **de arco volcánico** *f* GEOL back-arc basin; ~ **asociada a fallas direccionales** *f* GEOL *geología estructural* pull-apart basin; ~ **de captación** *f* AGUA catchment basin, drainage basin, watershed, HIDROL drainage area; ~ **carbonífera** *f* CARBÓN, GEOL coal basin, coal field, MINAS coal field; ~ **colectora** *f* CONST *para drenaje, para embalse* catchment area; ~ **de concentración** *f* OCEAN concentration basin; ~ **de drenaje** *f* CONST drainage area; ~ **experimental** *f* AGUA experimental basin; ~ **fallada** *f* TEC PETR *geología* fault basin; ~ **fluvial** *f* AGRIC, AGUA river basin, HIDROL river basin, drainage basin, TRANSP MAR river basin; ~ **geológica** *f* GEOL, TEC PETR basin; ~ **hidrográfica** *f* AGUA catchment basin, drainage basin, intake, watershed, GEOL hydrographic basin, HIDROL watershed, drainage basin, intake, *impluviométrica* catchment area; ~ **hidrográfica calibrada** *f* CONTAM calibrated watershed; ~ **hidrográfica menor** *f* AGRIC minor watershed; ~ **hundida** *f* TEC PETR *geología* subsident basin; ~ **de hundimiento tectónico** *f* GEOL fault basin; ~ **imbrífera** *f* HIDROL catchment area; ~ **intermontañosa** *f* GEOL *tectónica* intermontane basin; ~ **de lavado** *f* OCEAN fishing basin; ~ **lechera** *f* AGRIC dairy belt; ~ **marginal interna** *f* OCEAN back-arc basin; ~ **oceánica** *f* GEOL, OCEAN ocean basin, oceanic basin; ~ **petrolífera** *f* TEC PETR petroleum basin; ~ **de recepción** *f* AGUA *hidrología* catch basin; ~ **del río** *f* AGUA, HIDROL river bed; ~ **de sedimentación** *f* RECICL sedimentary deposit, sedimentation basin; ~ **sedimentaria** *f* CARBÓN basin, GAS sedimentary basin, GEOL basin, sedimentary basin, TEC PETR sedimentary basin; ~ **subsidiente** *f* TEC PETR *geología* subsident basin; ~ **vertiente** *f* AGUA drainage basin

cuenco *m* AGUA chamber, lock chamber, TRANSP MAR *canal, puerto* lock chamber; ~ **de cristalización** *m* CRISTAL, PROC QUÍ crystallization basin; ~ **del embalse** *m* AGUA reservoir basin; ~ **de esclusa** *m* AGUA, TRANSP MAR lock chamber

cuenta *f* CONST *de vidrio* bead, INFORM&PD bead, tally; ~ **atrás** *f* ELECTRÓN, TEC ESP *lanzamiento de misiles, naves espaciales y satélites* countdown; ~ **de defectos** *f* CALIDAD defect counting; ~ **regresiva** *f* ELECTRÓN, TEC ESP *lanzamiento de misiles, naves espaciales y satélites* countdown; ~ **de vidrio** *f* C&V glass bead

cuentaburbujas *m* C&V bubbler

cuentafotogramas *m* AmL (*cf contador de fotogramas Esp*) CINEMAT, FOTO frame counter

cuentagotas *m* EMB drop counter, LAB dropper

cuentahilos *m* IMPR linen tester, PAPEL line tester, thread counter

cuentakilómetros *m* AUTO mileage recorder (*BrE*), odometer (*AmE*), VEH *instrumento* odometer (*AmE*), trip counter (*BrE*); ~ **parcial** *m* VEH *instrumento* odometer (*AmE*), trip counter (*BrE*)

cuentapasos *m* CONST odometer

cuentarrevoluciones *m* AUTO, CINEMAT, FÍS tachometer, ING MECÁ revolution counter, INSTR rev counter, revolution counter, tachometer, TV tachometer, VEH rev counter, tachometer

cuentavueltas *m* AUTO, CINEMAT, FÍS tachometer, ING MECÁ revolution counter, INSTR revolution counter, tachometer, TV, VEH tachometer; ~ **de órbitas** *m* TEC ESP orbit counter

cuerda *f* ACÚST chord, C&V cord, CONST line, ELEC cord, ENERG RENOV, GEOM chord, ING MECÁ cord, PAPEL *la medida de madera para pasta* cord, rope, SEG, TEXTIL rope, TRANSP AÉR chord, TRANSP MAR chord, *construcción* tie plate; ~ **aerodinámica media** *f* TRANSP AÉR mean aerodynamic chord (*MAC*); ~ **de alambre de hierro** *f* PROD iron wire rope; ~ **alar** *f* TRANSP AÉR aerofoil chord (*BrE*), airfoil chord (*AmE*); ~ **detonante** *f* MINAS detonating cord, detonating fuse, primacord; ~ **del enturbado** *f* CONST string of casing; ~ **de escape** *f* TRANSP AÉR escape rope; ~ **de extensión** *f* TRANSP AÉR extension cord; ~ **de fibra textil** *f* ING MECÁ fiber rope (*AmE*), fibre rope (*BrE*); ~ **freno** *f* MINAS gland; ~ **de leña** *f* CONST cordwood, METR cord of wood; ~ **de maniobra** *f* PROD *aparejos* hand rope; ~ **media geométrica** *f* GEOM mean geometric chord; ~ **media de la superficie de control** *f* TRANSP AÉR mean chord of the control surface; ~ **de nylon** *f* ING MECÁ nylon rope; ~ **ondulada** *f* C&V wavy cord; ~ **de la pala** *f* TRANSP AÉR blade chord; ~ **para pasar la banda de papel en la sequería** *f* PAPEL rope carrier; ~ **de piano** *f* ING MECÁ piano wire; ~ **de sisal** *f* ING MECÁ sisal rope; ~ **de sostén** *f* CONST supporting rope; ~ **de la superficie aerodinámica** *f* TRANSP AÉR aerofoil chord (*BrE*), airfoil chord (*AmE*); ~ **de tracción** *f* ING MECÁ traction rope; ~ **vibratoria** *f* FÍS vibrating string

cuerina *f* AmL (*cf cuero artificial Esp*) P&C *plásticos* leatherette

cuernecillo *m* AGRIC *malezas* birdsfoot trefoil

cuerno *m* TRANSP AÉR horn

cuernos: ~ **polares** *m pl* TV pole tips

cuero: ~ **artificial** *m* Esp (*cf cuerina AmL*) P&C *plásticos* leatherette, PAPEL artificial leather; ~ **en bruto** *m* PROD green hide; ~ **embutido en forma de U** *m* PROD cup leather; ~ **lavable** *m* CONST wash leather; ~ **de montaña** *m* MINERAL *especie de amianto* mountain leather; ~ **de res** *m* AGRIC cattle hide; ~ **verde** *m* PROD green hide

cuerpo[1]: **con** ~ *adj* IMPR *tinta* full-bodied

cuerpo[2] *m* CONST pipe, IMPR *de tinta, de tipos* body, ING MECÁ shank, INSTAL HIDRÁUL *de biela de bomba, de cilindro, de estator de turbina de vapor* shaft, barrel, cylinder, LAB *de jeringa* barrel, MECÁ cylinder, chamber, barrel, PAPEL bulk, PROD cylinder, TELECOM, TEXTIL body, TRANSP MAR *de tanque* shell, *del buque* body; ~ **absorbente de neutrones incidentes** *m* CINEMAT, FÍS, FÍS RAD, IMPR, TEC ESP, TV black body; ~ **de acoplamiento** *m* ING MECÁ cone union body; ~ **de aguja** *m* PROD switch body; ~ **de arado apto para rastrojo** *m* AGRIC stubble bottom; ~ **de arado múltiple** *m* AGRIC semi-digger bottom;

~ **basculante** *m* TRANSP tilting body; ~ **de biela** *m* AUTO, ING MECÁ, VEH connecting-rod shank; ~ **de bomba** *m* AGUA pump barrel; ~ **de la bomba de agua** *m* VEH *sistema de refrigeración* water pump housing; ~ **de la bujía** *m* AUTO, ELEC, VEH spark plug body; ~ **del cable** *m* ELEC, ING ELÉC cable shaft; ~ **del cabrestante** *m* CONST capstan screw; ~ **de caldera** *m* INSTAL HIDRÁUL boiler shell; ~ **de la cámara** *m* CINEMAT, FOTO, TV camera body; ~ **celeste** *m* TEC ESP celestial body; ~ **central** *m* TRANSP MAR *del buque* midship body; ~ **cilíndrico** *m* ING MECÁ barrel, PROD *de una caldera* shell; ~ **del cilindro** *m* ING MECÁ, INSTAL HIDRÁUL, PROD, VEH *motor* cylinder barrel; ~ **de compresor centrífugo** *m* REFRIG casing; ~ **de la cruceta** *m* ING MECÁ crosshead body; ~ **cuasiestelar** *m* (*QSO*) TEC ESP quasistellar object (*QSO*); ~ **emisor** *m* ELEC, ELECTRÓN, ING ELÉC emitter; ~ **falso** *m* C&V, IMPR false body; ~ **fundido a presión** *m* PROD die-cast body; ~ **gris** *m* TV gray body (*AmE*), grey body (*BrE*); ~ **impresor** *m* IMPR unit; ~ **impresor de retiración** *m* IMPR perfecting unit; ~ **del inducido** *m* ING ELÉC armature spider; ~ **del microscopio** *m* INSTR microscope body; ~ **de motor** *m* TEC ESP *vehículos* engine body; ~ **en movimiento** *m* MECÁ body in motion; ~ **muerto** *m* CONTAM MAR, PETROL, TRANSP MAR mooring buoy; ~ **negro** *m* CINEMAT, FÍS, FÍS RAD, IMPR, TEC ESP, TV black body; ~ **del papel** *m* IMPR, PAPEL specific index, bulking index; ~ **del pistón** *m* AUTO piston body, ING MECÁ piston head; ~ **de la polea** *m* ING MECÁ pulley shell; ~ **de polea** *m* ING MECÁ shell; ~ **de popa** *m* TEC ESP afterbody; ~ **del programa** *m* INFORM&PD program body; ~ **radiactivo** *m* FÍS, FÍS RAD radioactive body; ~ **del radiador** *m* AUTO, VEH radiator frame; ~ **en reposo** *m* MECÁ body at rest; ~ **romo** *m* TRANSP AÉR blunt body; ~ **de la rueda de radios** *m* VEH spoke wheel center (*AmE*), spoke wheel centre (*BrE*); ~ **de sifón** *m* ENERG RENOV siphon barrel; ~ **de unión del cono** *m* ING MECÁ cone union body; ~ **de la válvula** *m* AUTO valve body, INSTAL HIDRÁUL valve chest; ~ **de válvula** *m* ING MECÁ valve body

cuerpos¹: **de ~ centrados** *adj* METAL space-centred

cuerpos²: ~ **intercambiables** *m pl* FERRO *vehículos* swap bodies; ~ **rotatorios** *m pl* FÍS FLUID spinning bodies

cuesco *m* C&V *hornos* bear, PROD bear, *metalurgia* sow

cuesta: ~ **arriba** *adj* CONST uphill

cuidado *m* SEG, TEXTIL *del tejido o de las prendas* care; ~ **del tejido** *m* TEXTIL fabric care

culata *f* AUTO cylinder head, D&A breech, butt, FÍS *de un imán* yoke, ING MECÁ cylinder head, head, INSTAL HIDRÁUL cylinder cover, cylinder head, VEH *motor* cylinder head; ~ **abovedada** *f* PROD dome head; ~ **de balancines** *f* AUTO, VEH rocker box; ~ **del cilindro** *f* ING MECÁ head, PROD cylinder head; ~ **del cilindro posterior** *f* INSTAL HIDRÁUL *motor alternativo* back-cylinder cover; ~ **del fusil** *f* D&A rifle butt

culicida *m* QUÍMICA mosquitocide

cullet: ~ **ecológico** *m* C&V ecology cullet; ~ **enfriado en agua** *m* AmL (*cf fundente enfriado en agua Esp*) C&V quenched cullet (*BrE*), shredded cullet (*AmE*); ~ **foráneo** *m* C&V foreign cullet; ~ **propio** *m* C&V factory cullet

culm *m* CARBÓN culm

culminación: ~ **axial** *f* GEOL axial culmination

culombímetro *m* FÍS, METEO coulomb gage (*AmE*), coulomb gauge (*BrE*)

culombio *m* (*C*) ELEC, ELECTRÓN, FÍS, ING ELÉC, METR coulomb (*C*)

cultivador *m* AGRIC tiller; ~ **con armazón balanceado** *m* AGRIC balanced-frame cultivator; ~ **de asiento** *m* AGRIC ridging cultivator; ~ **en hileras** *m* AGRIC row-crop cultivator; ~ **de mariscos** *m* OCEAN shellfish farmer; ~ **de matorrales** *m* AGRIC scrub cultivator; ~ **rotativo** *m* AGRIC rotary tiller; ~ **subsuperficial** *m* AGRIC field cultivator; ~ **tipo compactador giratorio** *m* AGRIC revolving-packer-type tiller

cultivadora *f* CONST cultivator; ~ **de mariscos** *f* OCEAN shellfish farmer

cultivo *m* AGRIC crop, farming, culture; ~ **de abono en verde** *m* AGRIC green manure crop; ~ **abusivo** *m* AGRIC overcropping; ~ **al aire libre** *m* AGRIC outdoor crop; ~ **de alevinos** *m* OCEAN stocking; ~ **asociado** *m* AGRIC companion crop; ~ **de cereal encamado** *m* AGRIC laid grain; ~ **de cobertura** *m* AGRIC cover crop; ~ **comercial** *m* AGRIC cash crop; ~ **por curvas de nivel** *m* AGRIC contour farming; ~ **de decrecida** *m* AGRIC flood-recession crop; ~ **que se destina a forraje** *m* AGRIC feed crop; ~ **destinado a producción de heno** *m* AGRIC hay crop; ~ **de emergencia** *m* AGRIC emergency crop; ~ **de esponjas** *m* OCEAN sponge culture; ~ **extensivo** *m* AGRIC field crop; ~ **para forraje** *m* AGRIC forage crop; ~ **en franjas** *m* AGRIC alley cropping, strip cropping; ~ **en franjas contraviento** *m* AGRIC wind stripping cropping; ~ **en franjas paralelas** *m* AGRIC field strip cropping; ~ **de hortalizas que se comercializan frescas** *m* AGRIC truck crop; ~ **industrial** *m* AGRIC industrial crop; ~ **inicial** *m* AGRIC culture starter; ~ **intercalado** *m* AGRIC intercropping; ~ **en invernadero** *m* AGRIC greenhouse crop; ~ **en líneas separadas** *m* AGRIC row crop; ~ **de maduración tardía** *m* AGRIC late-maturing crop; ~ **de mariscos** *m* OCEAN shellfish culture; ~ **de montaña** *m* AGRIC hill agriculture; ~ **del nácar** *m* OCEAN *ostras* mother-of-pearl culture; ~ **de peces** *m* OCEAN fish farming; ~ **de pecten** *m* OCEAN scallop culture; ~ **en pendiente** *m* AGRIC hillside farming; ~ **de raíces** *m* AGRIC rose chafer; ~ **de secano** *m* AGRIC dry farming; ~ **en surco único** *m* AGRIC single-crop farming

cultivos: ~ **de bacterias** *m pl* RECICL bacteria beds; ~ **básicos** *m pl* AGRIC basic crops; ~ **comestibles** *m pl* AGRIC edible crops; ~ **dobles** *m pl* AGRIC double cropping; ~ **extensivos** *m pl* AGRIC extensive crops; ~ **de oleaginosas** *m pl* AGRIC oil crops; ~ **en pie** *m pl* AGRIC growing crops; ~ **que producen muchos residuos** *m pl* AGRIC high-residue crops

cumaldehído *m* QUÍMICA cumaldehyde

cumálico *adj* QUÍMICA *ácido* coumalic

cumalina *f* QUÍMICA coumalin, coumaline

cumarana *f* QUÍMICA coumaran, coumarane

cumárico *adj* QUÍMICA *ácido* coumaric

cumarina *f* QUÍMICA coumarine, cumarin, *perfumería, síntesis dicumarol* coumarin

cumbre *f* CONST, MINAS brow; ~ **del descenso** *f* TRANSP AÉR top of descent

cumbrera *f* AmL (*cf corona Esp*) AGUA cap still, capping, CONST *tejado* hip, *tejados* ridgepiece, *arquitectura* ridge, ridge line, MINAS *entibación de galerías* crown piece, crown, *entibación* bonnet, cap piece, headpiece, *geología* cap

cumengeíta f MINERAL cumengeite
cumeno m QUÍMICA isopropylbenzene, *aditivo en motores* cumene
cúmico *adj* QUÍMICA *ácido* cumic
cumil m QUÍMICA cumyl
cummingtonita f MINERAL cummingtonite
cumplido: ~ **de cable** m TRANSP MAR cable's length
cumplimentación: ~ **de pedidos de clientes** f PROD customer order servicing
cumplimiento m ING MECÁ performance
cumplir: ~ **con** vt CALIDAD, SEG *las normas* meet
cúmulo: ~ **volcánico** m GEOL cumulo volcano
cumulonimbo m METEO cumulonimbus (*Cb*)
cumulonimbus m METEO cumulonimbus (*Cb*)
cumulus m (*Cu*) METEO cumulus (*Cu*); ~ **congestus** m METEO cumulus congestus; ~ **humilis** m METEO cumulus humilis
cuna f CONST cradle, MECÁ gimbal, MINAS cradle rocker; ~ **de hierro** f CONST cradle iron; ~ **de madera** f PROD wooden cradle; ~ **del misil** f D&A missile cradle; ~ **del montante** f PROD *parrilla* bearer cradle
cuña f Esp (*cf chaveta AmL*) CINEMAT *de cámara o grúa* wedge, CONST wedge, plug, spile, *carpintería* feather, FERRO *junta de carril* shim, IMPR wedge, ING MECÁ wedge, feather, liner, gib, quoin, taper key, keeper, block, MECÁ wedge, shim, key, MINAS *entibación* gad, headboard, wedge, chuck, PROD feather, TEC PETR *perforación* slip, TRANSP MAR *meteorología* ridge; ~ **de agua de mar** f OCEAN estuarios saltwater wedge; ~ **de apriete** f ING MECÁ keying wedge, tightening wedge; ~ **de bloqueo** f ING MECÁ scotch; ~ **de calzo** f CONST plug; ~ **de cantera** f CONST stone wedge; ~ **cilíndrica** f ING MECÁ *de piedra* chip; ~ **de control** f CINEMAT *de película* control wedge; ~ **de cuarzo** f CRISTAL quartz wedge; ~ **de desapriete** f ING MECÁ loosening wedge; ~ **de desviación** f TEC PETR whipstock; ~ **desviadora** f TEC PETR *perforación* whipstock; ~ **escalonada** f CINEMAT *de película*, FOTO step wedge; ~ **escalonada sensitométrica** f CINEMAT *de película* sensitometric step wedge; ~ **de falla inversa** f CONST upthrusted wedge; ~ **de madera** f MINAS lid; ~ **de madera para separar rocas fracturadas por voladuras** f AmL MINAS jack; ~ **de metal que queda adherida al filo de la cuchilla** f ING MECÁ *corte de metales dúctiles* built-up edge; ~ **múltiple** f MINAS compound wedge, multiple wedge; ~ **neutra** f CINEMAT *de película* neutral wedge; ~ **para partir madera** f CONST timber-splitting wedge~ **con tacón** f ING MECÁ taper key with gib head; ~ **sin tacón** f ING MECÁ taper key without gib head; ~ **de tensión** f ING MECÁ tightening wedge; ~ **de tierra** f CINEMAT film horse, MINAS *filones* horse; ~ **variable** f CINEMAT *en inclinación* fox wedge
cuñero m ING MECÁ keyway, PETROL roughneck, TEC PETR *perforación* floorman, roughneck; ~ **de perforación** m PETROL, TEC PETR *perforador* roughneck
cuneta f CARBÓN ditch, MECÁ gutter, PETROL ditch; ~ **de guardia** f AGUA berm ditch; ~ **paralela** f CONST parallel gutter
cunicultura f AGRIC rabbit breeding
cuota: ~ **de pesca** f OCEAN catch quota
cupla f PETROL, TEC PETR collar
cuprato m QUÍMICA cuprate
cúprico *adj* QUÍMICA cupric

cúprido m QUÍMICA cupride
cuprífero *adj* MINAS copper-bearing
cuprita f MINERAL, QUÍMICA cuprite, red copper ore
cuproamonio m QUÍMICA, TEXTIL cuprammonium
cuproapatito m MINERAL, QUÍMICA cuproapatite
cuprodescloizita f MINERAL, QUÍMICA cuprodescloizite
cupromagnesita f MINERAL, QUÍMICA cupromagnesite
cupromanganeso m MINERAL, QUÍMICA cupromanganese
cuproplumbito m MINERAL, QUÍMICA cuproplumbite
cuproscheelita f MINERAL, QUÍMICA cuproscheelite
cuproso *adj* QUÍMICA cupreous, cuprous
cuprotungstita f MINERAL, QUÍMICA cuprotungstite
cúpula f C&V dome, CARBÓN *bóveda* cupola, CONST *arquitectura*, GAS dome, PETROL cap, TEC ESP astrodrome, TRANSP AÉR canopy; ~ **desprendible** f TRANSP AÉR jettisonable canopy; ~ **gasífera** f PETROL gas cap; ~ **lanzable** f TRANSP AÉR jettisonable canopy; ~ **de radar** f FÍS RAD, TEC ESP, TELECOM, TRANSP AÉR, TRANSP MAR radome; ~ **de la torre** f Esp (*cf corona de la torre AmL*) TEC PETR *perforación* derrick crown, crown block; ~ **de vapor** f INSTAL HIDRÁUL steam dome
cura f CONST *madera* seasoning; ~ **por reposo** f C&V *del cristal, termómetro* ageing
curación f MECÁ curing
curado¹: **no** ~ *adj* P&C uncured; ~ **artificialmente** *adj* TERMO artificially aged
curado² m ALIMENT ageing, CONST cure period, curing, MECÁ curing, OCEAN curing, *tratamiento para conservar pescado* bloating, REFRIG *de la carne* ageing, TEXTIL *polimerización* curing; ~ **al aire** m P&C air cure; ~ **artificial** m METAL, TERMO *metales* artificial ageing; ~ **en frío** m P&C air cure; ~ **por haz electrónico** m IMPR *tintas, barnices*, NUCL *de lacas o barnices* electron-beam curing (*EBC*); ~ **rápido** m P&C fast curing, set
curar vt ALIMENT, TEXTIL *polimerización* cure; ~ **artificialmente** vt TERMO artificially age
curbadura f TRANSP AÉR camber
cúrcuma f QUÍMICA turmeric
curcumina f QUÍMICA *indicador* curcumine
cureña f D&A carriage; ~ **de cañón** f D&A gun carriage
curie m (*Ci*) FÍS, FÍS RAD curie (*Ci*)
curio m (*Cm*) FÍS RAD, QUÍMICA curium (*Cm*)
curiosear vt INFORM&PD browse
curratura: ~ **del material** f NUCL material buckling
currentímetro m OCEAN current meter
curricán m OCEAN troll
curricanero m OCEAN troller
cursa f TEXTIL *mecánica* throw
cursivas f pl IMPR italics
curso m Esp (*cf ley AmL*) HIDROL *de un río* course, ING MECÁ travel, run, MINAS *de minerales* tenor, TEC ESP course, TELECOM *tráfico* routing; ~ **de agua** m HIDROL streamway; ~ **del agua** m CONTAM watercourse; ~ **de agua aislado** m HIDROL perched watercourse; ~ **de agua anastomizante** m PETROL braided stream; ~ **de agua entrante** m HIDROL influent stream; ~ **de agua subterráneo** m HIDROL subsurface flow; ~ **de enlace** m INFORM&PD linkage path; ~ **normal** m TEC PETR normal trend; ~ **superior** m HIDROL *de un río* upper part
cursor¹: **por** ~ *adj* PROD cursor hit
cursor² m ELEC moveable contact, IMPR, INFORM&PD

cursor, LAB *de instrumento* slide, PROD *cuba de amalgamación* moveable contact, spider; ~ **invisible** *m* PROD invisible cursor; ~ **de potenciómetro** *m* ELEC *resistor* potentiometer slider; ~ **de reóstato** *m* ELEC *resistencia*, ING ELÉC rheostat slider; ~ **de rumbos electrónico** *m* TRANSP MAR *radar* electronic bearing-cursor

curtido *m* QUÍMICA tannage

curva *f* C&V bend, CONST *deformación* bulge, curve, GEOM curve, HIDROL *de un río* sweep, ING MECÁ knee, MATEMÁT curve, TRANSP MAR *madera* knee; ~ **de acordamiento** *f* ING MECÁ fillet; ~ **de acuerdo** *f* ING MECÁ fillet, radius; ~ **adiabática** *f* FÍS, TERMO adiabatic curve; ~ **alabeada** *f* GEOM skewed curve; ~ **de brillo** *f* AmL (*cf curva de luminosidad Esp*) TV brightness curve; ~ **de calentamiento** *f* TERMO heating curve; ~ **de calentamiento** *f* P&C *prueba* heating curve, TERMO heating curve, heating-up curve; ~ **característica** *f* ACÚST, CINEMAT, ELEC, ELECTRÓN, FÍS, FOTO, TEC ESP characteristic curve; ~ **característica voltaje-corriente** *f* ING ELÉC current-voltage characteristic, voltage-current characteristic; ~ **de carga** *f* ELEC *red de distribución* load curve; ~ **de cargas-deformaciones** *f* TRANSP MAR *arquitectura naval* stress-strain diagram; ~ **cáustica** *f* FÍS caustic curve; ~ **cerrada** *f* GEOM closed curve, sharp curve; ~ **cerrada no simple** *f* GEOM nonsimple closed curve; ~ **de clasificación** *f* ACÚST rating curve; ~ **de compensación** *f* ELECTRÓN equalization curve; ~ **de compresibilidad** *f* TERMO compressibility curve; ~ **de compresión** *f* CARBÓN compression curve; ~ **de contacto** *f* ING MECÁ curve of contact; ~ **de copo de nieve** *f* GEOM snowflake curve; ~ **de la corriente** *f* PROD current ramp; ~ **de densidad** *f* IMPR density curve; ~ **densimétrica** *f* CARBÓN densimetric curve, specific gravity curve; ~ **de desintegración** *f* FÍS RAD decay curve; ~ **de disparo** *f* PROD trip curve; ~ **de distribución de frecuencia** *f* MATEMÁT frequency distribution curve; ~ **de distribución normal** *f* MATEMÁT normal distribution curve; ~ **de duración de la carga** *f* ELEC *red de distribución* load duration curve; ~ **duración-carga** *f* PROD life-load curve; ~ **de empujes** *f* TRANSP MAR buoyancy curve; ~ **de energía de enlace** *f* NUCL binding-energy curve; ~ **de enlace** *f* TRANSP AÉR fillet; ~ **envolvente** *f* MECÁ envelope curve; ~ **de error** *f* TELECOM error pattern; ~ **de esfuerzo y deformación** *f* P&C *propiedad física, prueba* stress-strain curve; ~ **espacial** *f* GEOM space curve; ~ **en el espacio** *f* GEOM space curve; ~ **de estabilidad** *f* TRANSP MAR *diseño naval* stability curve; ~ **de evolución** *f* TRANSP MAR *atracamiento de barcos* turning circle; ~ **de experiencia** *f* TEXTIL experience curve; ~ **exponencial** *f* ELEC, MATEMÁT exponential curve; ~ **del filtro** *f* TEC ESP *comunicaciones* filter template; ~ **de forma de campana** *f* GEOM bell-shaped curve; ~ **de frecuencia** *f* MATEMÁT frequency curve; ~ **de ganancia** *f* ELECTRÓN gain curve; ~ **de Gauss** *f* GEOM Gaussian curve; ~ **granulométrica** *f* C&V particle-size curve; ~ **hipsográfica** *f* GEOL hypsographic curve; ~ **de histéresis** *f* ELEC *magnetismo* B/H loop, hysteresis curve, hysteresis loop, P&C *prueba* hysteresis loop; ~ **de igual sonoridad** *f* ACÚST *audición* equal-loudness contour; ~ **de imanación** *f* ING ELÉC magnetization curve; ~ **de imanación inicial** *f* FÍS

initial magnetization curve; ~ **de isopeso** *f* TRANSP AÉR isoweight curve; ~ **isotérmica** *f* FÍS isothermal curve; ~ **de luminosidad** *f* Esp (*cf curva de brillo AmL*) TV brightness curve; ~ **de magnetización** *f* FÍS, ING ELÉC magnetization curve; ~ **de magnetización inicial** *f* FÍS initial magnetization curve; ~ **magnitud-frecuencia** *f* ACÚST magnitude-frequency curve; ~ **de nivel** *f* AGRIC contour, CONST *topografía*, GEOL *isolínea* contour line; ~ **de nivel de ruido** *f* ACÚST noise contour; ~ **de nivelado** *f* CARBÓN grading curve; ~ **oblicua** *f* GEOM skewed curve; ~ **osculante** *f* GEOM osculating curve; ~ **piezométrica** *f* INSTAL HIDRÁUL pressure curve; ~ **plana** *f* GEOM flat curve; ~ **de ponderación** *f* ACÚST weighting curve; ~ **de ponderación A** *f* ACÚST A-weighting curve; ~ **de ponderación B** *f* ACÚST B-weighting curve; ~ **de ponderación C** *f* ACÚST C-weighting curve; ~ **de ponderación D** *f* ACÚST D-weighting curve; ~ **de presiones** *f* INSTAL HIDRÁUL pressure curve; ~ **de probabilidad** *f* MATEMÁT probability curve; ~ **de profundidad** *f* OCEAN depth curve; ~ **del punto de fusión** *f* TERMO melting point curve; ~ **radial larga** *f* GEOM long-radius curve; ~ **de referencia de aislamiento** *f* ACÚST insulation reference curve; ~ **de rendimiento térmico** *f* TERMO heat rate curve; ~ **de rendimiento de tiempo-luz** *f* FOTO time-light output curve; ~ **de rendimiento del ventilador** *f* TERMOTEC fan efficiency curve; ~ **de resonancia** *f* TV resonance curve; ~ **de respuesta** *f* ELECTRÓN response curve; ~ **de respuesta de amplitud** *f* ELECTRÓN amplitude-amplitude response curve; ~ **de respuesta de amplitud de frecuencia** *f* ELECTRÓN amplitude-frequency response curve; ~ **de respuesta elíptica** *f* ELECTRÓN elliptic response curve; ~ **de respuesta fase-frecuencia** *f* ACÚST phase-frequency response curve; ~ **de respuesta de frecuencia** *f* ELECTRÓN frequency-response curve; ~ **de respuesta en frecuencia** *f* ELECTRÓN frequency-response curve; ~ **de respuesta total** *f* ACÚST overall response curve; ~ **de retroceso** *f* AGUA, HIDROL recession curve; ~ **de salida de pista de alta velocidad** *f* TRANSP AÉR high-speed exit taxiway; ~ **sensitométrica** *f* CINEMAT sensitometric curve; ~ **subsuperficial** *f* GEOL subsurface contour; ~ **de temperatura de caldeo** *f* TERMO heating temperature curve; ~ **de temperatura crítica** *f* TERMO critical-temperature curve; ~ **de temperaturas** *f* TERMO temperature curve; ~ **de tensión-deformación** *f* P&C *propiedad física, prueba* stress-strain curve; ~ **de tiempo y temperatura** *f* CINEMAT time temperature curve; ~ **de la trama** *f* IMPR screen curve; ~ **de transición** *f* CONST *topografía* transition curve; ~ **de unión** *f* ING MECÁ radius; ~ **velocidad-profundidad** *f* TEC PETR *geofísica* velocity-depth curve; ~ **de virado** *f* IMPR tone curve; ~ **de volumen por velocidad** *f* TRANSP speed-volume curve; ~ **de volúmenes acumulados en el desagüe** *f* HIDROL discharge mass curve

curvado[1] *adj* GEOM, MATEMÁT curved

curvado[2] *m* REVEST camber

curvadora: ~ **de chapas** *f* ING MECÁ plate-bending rolls; ~ **de rieles** *f* CONST jim crow; ~ **de tubos** *f* CONST pipe bender

curvar *vt* CONST bend, ING MECÁ deflect, MATEMÁT curve, MINAS *construcción* arch, PROD crook, bow, curve, bend

curvarse *v refl* CONST camber, MECÁ buckle; **~ hacia abajo** *v refl* TRANSP MAR *quilla* sag

curvas *f pl* TEC PETR bends; **~ cruzadas de estabilidad** *f pl* TRANSP MAR *diseño de buques* crosscurves; **~ de desviación** *f pl* PETROL departure curves; **~ hidrostáticas** *f pl* TRANSP MAR *diseño de barcos* hydrostatic curves; **~ de tubos atornilladas y enchufadas** *f pl* ING MECÁ pipe bends screwed and socketed

curvatubos *m* ING MECÁ pipe bend

curvatura *f* ACÚST camber, C&V *vidrio hueco* bending, CONST bow, FÍS, GEOM, MATEMÁT curvature, MECÁ bend, braking, METAL deflection, ÓPT *de lente*, TEC ESP *trajes espaciales, vehículos* bending; **~ de campo** *f* CINEMAT, FÍS curvature of field; **~ de la cinta** *f* TV tape curvature; **~ de Gauss** *f* GEOM Gaussian curvature; **~ de Gauss negativa** *f* GEOM negative Gauss curvature; **~ de Gauss positiva** *f* GEOM positive Gauss curvature; **~ geométrica** *f* NUCL geometric buckling; **~ intrínsica** *f* GEOM intrinsic curvature; **~ negativa** *f* GEOM negative curvature; **~ negativa de Gauss** *f* GEOM negative Gauss curvature; **~ de pandeo** *f* TEC PETR sag bend; **~ de plano H** *f* ING ELÉC H plane bend; **~ positiva** *f* GEOM positive curvature; **~ positiva de Gauss** *f* GEOM positive Gauss curvature; **~ principal** *f* C&V principal curvature; **~ seccional** *f* GEOM sectional curvature; **~ secundaria** *f* C&V secondary curvature; **~ de superficies** *f* GEOM curvature of surfaces; **~ terrestre** *f* TEC ESP earth curvature; **~ total** *f* GEOM Gaussian curvature

curvilíneo *adj* FÍS, GEOM curvilinear

cuscuta *f* AGRIC dodder

cusenusoide *f* GEOM cosine wave

cúspide *f* MINAS *filones* apex; **~ del limbo** *f* INSTR limb top

custodia *f* CONTAM safekeeping

cúter *m* TRANSP MAR *embarcación* cutter; **~ de crucero** *m* TRANSP MAR cruising cutter

CV *abr* (*caballo de vapor*) AUTO, ING MECÁ hp (*horsepower*)

CVA *abr* (*control automático de volumen*) ING ELÉC AVC (*automatic volume control*)

D

D *abr* (*deuterio*) FÍS, NUCL, QUÍMICA D (*deuterium*)

da: **que ~ al mar** *adj* TRANSP MAR *navegación* seaward; **que ~ al oeste** *adj* TRANSP MAR westward

D-a-A *abr* (*digital a analógico*) CINEMAT, TV D-to-A (*digital-to-analog*)

DAC *abr* ELECTRÓN, FÍS, INFORM&PD, ING ELÉC, TELECOM, TV (*conversión digital-analógica*) DAC (*digital-to-analog conversion*), (*conversor digital-analógico, convertidor digital-analógico*) DAC (*digital-to-analog converter*), (*dispositivo acoplado por carga*) CCD (*charge-coupled device*)

dacita *f* MINERAL, PETROL dacite

Dacrón® *m* TEXTIL, TRANSP MAR Dacron®

dado *m* C&V, PROD die; **~ de terraje automática** *m* ING MECÁ machine die plate

dador *m* FÍS, FÍS PART, QUÍMICA *átomo* donor

dafnetina *f* QUÍMICA daphnetin

dafnina *f* QUÍMICA daphnin

dafnita *f* MINERAL daphnite

daltón *m* FÍS RAD dalton

dama *f* PROD *horno alto* dam

damajuana *f* C&V demijohn, carboy

damourita *f* MINERAL damourite

danaíta *f* MINERAL danaite

danalita *f* MINERAL danalite

danburita *f* MINERAL danburite

Daniense *adj* GEOL Danian

dañino *adj* SEG, TEC PETR harmful; **~ a los ojos** *adj* SEG harmful to the eyes

dannemorita *f* MINERAL dannemorite

daño: **~ foliar** *m* AGRIC leaf damage; **~ por frío** *m* REFRIG *alimentos* chilling injury, cold injury; **~ de insectos** *m* AGRIC insect damage

daños: **~ causados por la mar** *m pl* TRANSP MAR sea damage; **~ por fuego** *m pl* CALIDAD, SEG fire damage; **~ por el incendio** *m pl* CALIDAD, SEG fire damage; **~ por irradiación** *m pl* FÍS RAD radiation damage

dar¹: **~ de alta** *vt* TRANSP MAR *para el servicio* put into commission; **~ de baja** *vt* MECÁ scrap; **~ un baño de metal a** *vt* TRANSP MAR *construcción naval* plate; **~ brillo a** *vt* METAL, PROD gild; **~ un buen resguardo a** *vt* TRANSP MAR *atracamiento* give a wide berth to; **~ una capa de pintura a** *vt* P&C coat; **~ concavidad a** *vt* PROD cross; **~ corriente** *vt* ING ELÉC switch on; **~ cuerda a** *vt* ING MECÁ wind up; **~ energía** *vt* ELEC, ING ELÉC, PROD energize; **~ juego a** *vt* ING MECÁ give clearance to; **~ una mano de pintura a** *vt* P&C coat; **~ tolerancia a** *vt* ING MECÁ give clearance to; **~ vuelta a** *vt* TRANSP MAR *cabo* belay; **~ vueltas a** *vt* ING MECÁ swing

dar²: *vi* ING MECÁ yield; **~ la alarma** *vi* SEG give the alarm; **~ un bandazo** *vi* TRANSP MAR *barco* lurch; **~ bandazos** *vi* TEC ESP *automóviles* yaw; **~ un bordo** *vi* TRANSP MAR *navegación* make a tack; **~ cabezadas** *vi* TRANSP MAR pitch; **~ contramarcha** *vi* ING MECÁ reverse, reverse the motion; **~ una espía** *vi* TRANSP MAR *en tiempo inestable* stream a warp; **~ exceso de exposición** *vi* FOTO overexpose; **~ fondo** *vi* TRANSP MAR *terreno panta-*

noso bring up; **~ forma** *vi* ING MECÁ, PROD shape; **~ formato** *vi* INFORM&PD format; **~ golpecitos** *vi* FÍS tap; **~ guiñadas** *vi* ENERG RENOV yaw; **~ lustre** *vi* MECÁ polish; **~ máquina atrás** *vi* ING MECÁ reverse, reverse the motion; **~ marcha atrás** *vi* FERRO *vehículos* back up, ING MECÁ reverse, reverse the motion; **~ nombre** *vi* ELECTRÓN initialize; **~ órdenes al timón** *vi* TRANSP MAR helm; **~ paso** *vi* CONST unstop; **~ peralte** *vi* CONST camber; **~ una señal** *vi* TRANSP MAR signal; **~ sustento** *vi* ING MECÁ support; **~ timbre** *vi* AmL (*cf sonar Esp*) TELECOM ring; **~ la vela** *vi* TRANSP MAR *barco* make sail; **~ vueltas** *vi* TEC ESP spin, turn, TRANSP AÉR spin

darcy *m* HIDROL darcy

dardo: **~ de llama** *m* TERMO *quemador* darting flame

dársena *f* TRANSP MAR dock; **~ de construcción** *f* TRANSP MAR fitting-out berth; **~ de limpieza** *f* OCEAN scouring basin; **~ de maniobra** *f* TRANSP MAR *puerto* turning basin; **~ de marea** *f* ENERG RENOV, OCEAN tidal basin, TRANSP MAR tidal dock

darta *f* PROD *moldería* scab

datación *f* FÍS, GEOL dating; **~ por carbono** *f* FÍS RAD carbon dating; **~ por carbono-14** *f* FÍS RAD, GEOL carbon-14 dating; **~ de la época** *f* GEOL age dating; **~ K-Ar** *f* GEOL K-Ar dating; **~ mediante trazas de fisión** *f* FÍS fission track dating; **~ radiactiva** *f* FÍS, FÍS RAD radioactive dating; **~ por trazos** *f* GEOL fission track dating

datáfono® *m* ELECTRÓN, INFORM&PD, SEG, TELECOM dataphone®

datagrama *m* INFORM&PD, TELECOM datagram

datarom® *f* ÓPT datarom®

datiscina *f* QUÍMICA datiscin

dato *m* CONST datum, INFORM&PD item, TEC ESP datum

datolita *f* MINERAL datholite, datolite

datos *m pl* GEN data; **~ de abordos** *m pl* TEC ESP housekeeping; **~ analógicos** *m pl* ELECTRÓN analog data; **~ sobre la calidad de servicio** *m pl* TELECOM performance data; **~ sobre las características de funcionamiento** *m pl* TELECOM performance data; **~ condición** *m pl* TELECOM status data; **~ conectados preparados para línea** *m pl* TELECOM connect data set to line; **~ de control** *m pl* INFORM&PD control data; **~ corregidos** *m pl* INFORM&PD corrected data; **~ digitales** *m pl* ELECTRÓN digital data; **~ digitalizados** *m pl* ELECTRÓN, INFORM&PD, TELECOM digitized data; **~ de emisión** *m pl* CONTAM emission data; **~ de enmascaramiento** *m pl* PROD mask data; **~ de entrada** *m pl* INFORM&PD input data; **~ estadísticos** *m pl* INFORM&PD statistical data; **~ de estado** *m pl* TELECOM status data; **~ de filtración** *m pl* PROD mask data; **~ geométricos** *m pl* CONST, GEOM geometric data; **~ inmediatos** *m pl* INFORM&PD immediate data; **~ de instrucciones** *m pl* PROD step data; **~ legibles por computadora** *m pl* AmL (*cf datos legibles por ordenador Esp*) INFORM&PD machine-readable data; **~ legibles por ordenador** *m pl* Esp (*cf datos legibles por computadora AmL*) INFORM&PD machine-readable data;

~ **de mantenimiento N para el usuario** *m pl* TELE-COM N-user data; ~ **meteorológicos** *m pl* CONTAM, METEO, TEC ESP meteorological data; ~ **de motor instantáneo** *m pl* PROD instantaneous motor data; ~ **no analizados** *m pl* ELECTRÓN, INFORM&PD raw data; ~ **del plan de vuelo** *m pl* TRANSP AÉR flight-plan data; ~ **procesados** *m* INFORM&PD processed data; ~ **sin procesar** *m pl* ELECTRÓN, INFORM&PD raw data; ~ **de prueba** *m pl* INFORM&PD test data; ~ **recibidos** *m pl* INFORM&PD, TELECOM received data; ~ **de salida** *m pl* INFORM&PD output data; ~ **sin sentido** *m pl* INFORM&PD meaningless data; ~ **situación** *m pl* TELECOM status data; ~ **status** *m pl* TELECOM status data; ~ **técnicos** *m pl* PROD engineering data; ~ **transmitidos** *m pl* TELECOM transmitted data; ~ **de vuelo** *m pl* TRANSP AÉR flight data

datum: ~ **de la carta** *m* TRANSP MAR *navegación* chart datum

davyna *f* MINERAL davyne

dB *abr* (*decibelio*) GEN dB (*decibel*)

dBA *abr* (*decibelio A*) GEN dBA (*decibel A*)

dBB *abr* (*decibelio B*) GEN dBB (*decibel B*)

dBC *abr* (*decibelio C*) GEN dBC (*decibel C*)

dBD *abr* (*decibelio D*) GEN dBD (*decibel D*)

DBO *abr* (*demanda bioquímica de oxígeno*) CONTAM, HIDROL, TEC PETR BOD (*biochemical oxygen demand*)

DCB *abr* ELECTRÓN, IMPR, INFORM&PD (*decimal codificado en binario*) BCD (*binary-coded decimal*)

DCL: ~ **transflexivo** *m* ELECTRÓN transflective LCD

DCME *abr* (*equipo de multiplicación del circuito digital*) TELECOM DCME (*digital circuit multiplication equipment*)

DDE *abr* (*entrada directa de datos*) ELECTRÓN, INFORM&PD DDE (*direct data entry*)

DEA *abr* (*dietanolamina*) QUÍMICA DEA (*diethanolamine*)

debate *m* CARBÓN contest

debilitamiento *m* CARBÓN reduction, TEC ESP *comunicaciones* de-emphasis, fading; ~ **de la señal** *m* TELECOM signal weakening

debituse *m* C&V debiteuse

década *f* ELECTRÓN, NUCL, TRANSP AÉR decade

decaédrico *adj* GEOM decahedral

decaedro *m* GEOM decahedron

decágono *m* GEOM decagon

decagramo *m* METR decagram

decahidronaftaleno *m* QUÍMICA decalin, *solvente* dec-ahydronaphthalene

decaimiento *m* NUCL decay; ~ **orbital** *m* TEC ESP orbital decay

decalaje *m* ING ELÉC displacement; ~ **circular** *m* INFORM&PD circular shift; ~ **de fase** *m* FÍS ONDAS *electricidad* phase angle

decalar *vt* ING ELÉC displace

decalina® *f* QUÍMICA decalin®

decalitro *m* METR decaliter (*AmE*), decalitre (*BrE*)

decámetro *m* METR decameter (*AmE*), decametre (*BrE*)

decano *m* QUÍMICA decane

decanol *m* QUÍMICA decanol

decantación *f* ALIMENT decantation, decanting, CARBÓN thickening, CONST settling, CONTAM, PROC QUÍ decantation, decanting, QUÍMICA decantation, *proceso* decanting, TEC ESP separation, TEC PETR

decantation, decanting; ~ **garantizada** *f* ENERG RENOV guaranteed draw off

decantador *m* ALIMENT, CONTAM, PROC QUÍ, QUÍMICA, TEC PETR decanter

decantar *vt* ALIMENT, CONTAM, PROC QUÍ, QUÍMICA, TEC PETR decant

decapado *m* PROD *metales* scouring, QUÍMICA, TRANSP MAR *mantenimiento de buques* pickling; ~ **por baño ácido** *m* AmL (*cf desoxidación por baño ácido Esp*) NUCL corrosion pickling; ~ **preliminar con ácido** *m* Esp (*cf decapado previo con ácido AmL*) NUCL acid prepickling; ~ **previo con ácido** *m* AmL (*cf decapado preliminar con ácido Esp*) NUCL acid prepickling

decapaje *m* QUÍMICA pickling

decapante *m* MECÁ flux, QUÍMICA etching; ~ **de pintura** *m* CONST paint stripper

decapar *vt* MECÁ descale, PROD *metales* scour

decatizado *m* TEXTIL decatizing

decatizadora *f* TEXTIL decatizing machine

decatizar *vt* TEXTIL decatize

decelerar *vt* FÍS, MECÁ decelerate

dechenita *f* MINERAL dechenite

decibelio *m* (*dB*) GEN decibel (*dB*); ~ **A** *m* (*dBA*) GEN decibel A (*dBA*); ~ **B** *m* (*dBB*) GEN decibel B (*dBB*); ~ **C** *m* (*dBC*) GEN decibel C (*dBC*); ~ **D** *m* (*dBD*) GEN decibel D (*dBD*)

decígramo *m* (*dg*) METR decigram (*dg*)

decilo *m* QUÍMICA decyl

decimal: ~ **codificado binario** *m* (*DCB*) IMPR, INFORM&PD binary-coded decimal (*BCD*); ~ **codificado en binario** *m* (*DCB*) ELECTRÓN, INFORM&PD binary-coded decimal (*BCD*); ~ **empaquetado** *m* INFORM&PD packed decimal; ~ **infinito** *m* MATEMÁT nonterminating decimal; ~ **no finalizador** *m* MATE-MÁT nonterminating decimal; ~ **periódico** *m* MATEMÁT recurring decimal, repeating decimal; ~ **recurrente** *m* MATEMÁT repeating decimal

decímetro *m* METR decimeter (*AmE*), decimetre (*BrE*); ~ **cuadrado** *m* METR square decimeter (*AmE*), square decimetre (*BrE*); ~ **cúbico** *m* METR cubic decimeter (*AmE*), cubic decimetre (*BrE*)

décimo *m* IMPR tenthmo

decineperio *m* (*dN*) ACÚST, ELECTRÓN, METR, TELECOM, TV decineper (*dN*)

decinormal *adj* QUÍMICA *solución* decinormal

decisión *f* INFORM&PD decision, ING MECÁ ruling

declaración *f* PROD statement; ~ **de carga** *f* TRANSP, TRANSP AÉR, TRANSP MAR cargo manifest; ~ **compuesta** *f* INFORM&PD compound statement; ~ **de conformidad** *f* PROD conformance statement; ~ **de conformidad en el cumplimiento del protocolo** *f* TELECOM protocol-implementation conformance statement (*PICS*); ~ **del contenido** *f* EMB *aduanas* contents declaration; ~ **de datos** *f* INFORM&PD data declaration; ~ **ejecutable** *f* INFORM&PD *programa* executable statement; ~ **de impacto ambiental** *f* CONTAM, NUCL environmental-impact statement; ~ **indefinida** *f* INFORM&PD undefined statement; ~ **de lenguaje** *f* INFORM&PD language statement

declarativo *adj* INFORM&PD declarative

declinación *f* FÍS, PETROL, TRANSP AÉR declination, TRANSP MAR variation, *astronavegación* declination; ~ **magnética** *f* FÍS, GEOFÍS magnetic declination, TRANSP MAR magnetic declination, magnetic variation

declinómetro *m* FÍS *brújula de declinación* declinometer

declive *m* CONST falling gradient, gradient, incline, pitch, slope, GEOM gradient, slope, ING ELÉC *navegación* pitch, MINAS *geología* dip; **~ del arrecife** *m* GEOL, OCEAN reef glacis; **~ continental** *m* GEOFÍS, GEOL, OCEAN, TEC PETR *geología* continental slope; **~ lateral de la carretera** *m* CONST curb (*AmE*), kerb (*BrE*)

decocción *f AmL* ALIMENT decoction

decodificación *f* ELECTRÓN, INFORM&PD, TEC ESP, TELECOM, TV decoding; **~ de la decisión blanda** *f* TELECOM soft decision decoding; **~ de decisión dura** *f* TELECOM hard decision decoding; **~ secuencial** *f* TELECOM sequential decoding; **~ Viterbi** *f* TELECOM Viterbi decoding

decodificador *m* ELECTRÓN, INFORM&PD, TEC ESP, TELECOM decoder; **~ ADPCM** *m* TELECOM ADPCM decoder; **~ de color** *m* CINEMAT, TV color decoder (*AmE*), colour decoder (*BrE*); **~ de instrucciones** *m* INFORM&PD instruction decoder

decodificar *vt* ELECTRÓN decipher, decode, INFORM&PD, TEC ESP, TELECOM, TV decode; **~-codificar** *vt* TV decode-encode

decoloración *f* C&V discoloration (*AmE*), discolouration (*BrE*), DETERG bleaching, FÍS fading, P&C *defecto* color removal (*AmE*), colour removal (*BrE*), QUÍMICA bleaching; **~ acromática** *f* IMPR achromatic color removal (*AmE*), achromatic colour removal (*BrE*)

decolorante *m* C&V decolorizer (*AmE*), decolourizer (*BrE*)

decolorar *vt* COLOR stain, QUÍMICA bleach, TEXTIL stain

decolorarse *v refl* COLOR, TEXTIL fade

decompilador *m* INFORM&PD decompiler

decompresor: **~ de gas** *m* GAS, INSTAL TERM, TERMO gas-pressure reducing valve

decorado *m* CINEMAT, TV set; **~ de porcelana** *m* C&V porcelain decoration; **~ en proceso** *m* C&V decoration in the make

decorador *m* C&V earthenware decorator; **~ de alfarería** *m* C&V pottery decorator; **~ de mayólica** *m* C&V majolica painter; **~ de porcelana** *m* C&V porcelain gilder

decrecida *f* CONTAM, GEOL, HIDROL *corrientes subterráneas* subsidence

decreciente *adj* ING MECÁ dying

decrecimiento: **~ adiabático** *m* TERMO adiabatic lapse rate; **~ logarítmico** *m* FÍS logarithmic decrement

decrementador *m* INFORM&PD *microprocesador* decrementer

decrementar *vt* INFORM&PD decrement

decremento[1] *adj* INFORM&PD decoupled

decremento[2] *m* FÍS decrement; **~ logarítmico medio** *m* NUCL *de energía* average logarithmic energy decrement; **~ de reactividad** *m* NUCL decrement in reactivity

dedal *m* C&V, PROD thimble, SEG finger stall; **~ oviforme** *m* PROD egg-shaped thimble

dedicado *adj* INFORM&PD, TELECOM dedicated

dedo: **~ de arrastre** *m* ING MECÁ catch pin, drive pin; **~ del distribuidor** *m* AUTO distributor finger; **~ de enganche** *m* ING MECÁ keeper, pawl; **~ de retención** *m* ING MECÁ finger

D₂O *abr* (*óxido de deuterio*) FÍS, NUCL, QUÍMICA D_2O (*deuterium oxide*)

deducción: **~ por agua dulce** *f* TRANSP MAR *franco-bordo* freshwater allowance

deducciones: **~ en el franco bordo** *f pl* TRANSP MAR freeboard allowances

defasado *adj* ELEC *corriente*, ELECTRÓN out of phase

defasaje *m* ELEC phase difference; **~ de nivel** *m* ELECTRÓN level shifting

defecación: **~ de cal** *f* ALIMENT lime defecation

defecto *m* CALIDAD, CRISTAL defect, flaw, ELEC, ELECTRÓN fault, IMPR, INFORM&PD bug, ING ELÉC defect, fault, ING MECÁ vice (*BrE*), vise (*AmE*), MECÁ fault, flaw, METAL defect, NUCL flaw, *en un material* defect, PAPEL *papel, cartón* flaw, PROD fault, QUÍMICA defect, TEXTIL fault; **~ de alineación** *m* ING MECÁ misalignment; **~ de alineamiento** *m* ING MECÁ misalignment; **~ de aroma** *m* ALIMENT off-flavor (*AmE*), off-flavour (*BrE*); **~ de colimación** *m* CINEMAT, FOTO collimating fault; **~ de construcción** *m* CONST constructional defect; **~ de contacto** *m* ING ELÉC contact fault; **~ crítico** *m* CALIDAD critical defect; **~ por desgaste** *m* NUCL wear-out defect; **~ de difusión** *m* ELECTRÓN diffusion defect; **~ de diseño** *m* NUCL *de un elemento de combustible* design-related defect; **~ de distribución** *m* C&V defect in distribution; **~ de fabricación en el combustible** *m* NUCL fabrication-related fuel defect; **~ Frenkel** *m* CRISTAL Frenkel defect; **~ de horneado** *m* ALIMENT baking fault; **~ inherente a la madera** *m* CALIDAD wood defect; **~ lineal** *m* METAL linear defect; **~ de masa** *m* FÍS mass defect; **~ no revelado** *m* CALIDAD unrevealed failure; **~ en el original** *m* IMPR blemish on copy; **~ puntual** *m* CRISTAL point defect; **~ de la red cristalina** *m* ELECTRÓN *semiconductores* lattice defect; **~ relacionado con la operación** *m* NUCL operations-related defect; **~ reticular** *m* METAL lattice defect; **~ de Schottky** *m* CRISTAL Schottky defect

defectos[1]: **sin ~** *adj* PAPEL *papel, cartón* flawless, TEXTIL faultless

defectos[2] *m pl* PAPEL specks; **~ en el cable** *m pl* ELEC, ING ELÉC cable defects; **~ del material** *m pl* ING MECÁ material defects; **~ superficiales producidos por los cilindros** *m pl* ING MECÁ roll marking

defectoscopio: **~ de aislamiento** *m* ELEC, ING ELÉC earth detector (*BrE*), ground detector (*AmE*)

defectuoso *adj* CALIDAD, ELEC defective, faulty, EMB, ING MECÁ faulty, MECÁ defective, faulty, PROD faulty, TEXTIL defective, faulty

defensa *f* CONTAM MAR fender, HIDROL screen, INSTAL HIDRÁUL breasting parapet, PROD, SEG guard, TRANSP MAR fender; **~ aérea** *f* D&A air defence (*BrE*), air defense (*AmE*); **~ antiaérea pasiva** *f* D&A passive air defence (*BrE*), passive air defense (*AmE*); **~ antisubmarina** *f* D&A, TRANSP MAR antisubmarine defence (*BrE*), antisubmarine defense (*AmE*); **~ civil** *f* D&A civil defence (*BrE*), civil defense (*AmE*); **~ contra las inundaciones** *f* AGRIC flood control; **~ de correa** *f* ING MECÁ, SEG belt guard; **~ de costa** *f* D&A, HIDROL, TRANSP MAR coastal defence (*BrE*), coastal defense (*AmE*); **~ costera** *f* D&A, HIDROL, TRANSP MAR *naval* coastal defence (*BrE*), coastal defense (*AmE*); **~ flotante de muelle** *f* TRANSP MAR camel; **~ por láser** *f* ELECTRÓN laser weapon; **~ de proa** *f* TRANSP MAR *equipamiento de cubierta* bow fender, noseband; ~ **de taludes** *f* CONST slope protection; ~ **de tela metálica** *f* PROD wire guard

defensas *f pl* SEG *edificio* safeguarding

deficiencia: ~ **de agua** *f* AGUA water deficiency; ~ **de nitrógeno** *f* AGRIC nitrogen deficiency

déficit *m* PROD shortage; ~ **en oxígeno** *m* (*DO*) HIDROL oxygen deficit (*OD*); ~ **de reactividad** *m* NUCL deficit reactivity; ~ **de saturación** *m* METEO saturation deficit

definición *f* CINEMAT, ELECTRÓN, FOTO, INFORM&PD definition, ING MECÁ sharpness, TV resolution; ~ **de ambigüedad** *f* TEC ESP ambiguity resolution; ~ **de la celda** *f* IMPR cell definition; ~ **del conjunto de datos** *f* INFORM&PD data set definition; ~ **de datos** *f* INFORM&PD data definition; ~ **de la fase de ambigüedad** *f* TEC ESP phase-ambiguity resolution; ~ **horizontal** *f* TV horizontal resolution; ~ **de la imagen** *f* TELECOM picture definition; ~ **de márgenes** *f* IMPR margin settings; ~ **de problemas** *f* INFORM&PD problem definition; ~ **de tareas** *f* INFORM&PD job control language (*JCL*), job definition

definido: ~ **por el usuario** *adj* INFORM&PD user-defined

deflagración *f* MINAS *explosivo*, QUÍMICA deflagration

deflagrar[1] *vt* SEG *arder con flama* deflagrate

deflagrar[2] *vi* MINAS *explosivo*, QUÍMICA deflagrate

deflector *m* ING MECÁ, MINAS baffle, PAPEL deflector, PROC QUÍ baffle, deflector, TELECOM, TRANSP AÉR deflector, VEH *silenciador de escape* baffle; ~ **de aceite** *m* ING MECÁ oil slinger; ~ **aerodinámico** *m* TRANSP, TRANSP AÉR spoiler; ~ **del chorro** *m* TEC ESP jet deflector; ~ **contra la agitación del aceite** *m* TRANSP AÉR anti-surge baffle; ~ **parabólico** *m* TV dishpan

deflegmador *m* NUCL, QUÍMICA dephlegmator

deflegmar *vt* NUCL, QUÍMICA dephlegmate

deflexión *f* ELEC, FÍS, FÍS RAD, ING MECÁ deflection, MECÁ braking, deflection, PETROL foldover, TRANSP AÉR, TV deflection; ~ **del alerón** *f* TRANSP AÉR aileron deflection; ~ **asimétrica** *f* TV asymmetrical deflection; ~ **por campos eléctricos** *f* FÍS RAD deflection by electric fields; ~ **por campos magnéticos** *f* FÍS RAD deflection by magnetic fields; ~ **de flujo hacia arriba** *f* TRANSP AÉR upwash; ~ **horizontal** *f* ELECTRÓN, TV horizontal deflection; ~ **del izador** *f* TRANSP AÉR elevator deflection; ~ **por ruido de la base de tiempos** *f* NUCL grass; ~ **vertical** *f* ELECTRÓN, TV vertical deflection

defluviación *f* OCEAN *descargas de los ríos* plume; ~ **formando penacho** *f* OCEAN *descargas de los ríos* plume

defoliante *m* AGRIC, QUÍMICA defoliant

deforestación *f* AGRIC deforestation

deformación *f* C&V buckle, CARBÓN deformation, FERRO *de la rueda* warping, FÍS *de núcleos atómicos* deformation, FOTO buckling, GEOL strain, deformation, ING MECÁ strain, set, offset, distortion, MECÁ deflection, buckling, strain, deformation, METAL kink, deformation, NUCL *de las varillas, vainas del combustible* buckling, *del núcleo* deformation, P&C *propiedad física, prueba* deformation, distortion, strain, PROD *piezas fundidas* warping, QUÍMICA deformation, TRANSP MAR strain; ~ **acelerada** *f* METAL accelerated creep; ~ **por agotamiento** *f* METAL *fatiga* exhaustion creep; ~ **por cizallamiento**

f ING MECÁ shearing strain; ~ **por compresión** *f* MECÁ compressive strain; ~ **compresiva** *f* GEOL compressive strain; ~ **cóncava del centro del buque** *f* TRANSP MAR *construcción naval* sagging; ~ **del contorno del grano** *f* METAL sliding; ~ **de corte** *f* ING MECÁ shearing strain; ~ **debida al esfuerzo cortante** *f* ING MECÁ shear; ~ **dependiente del tiempo** *f* METAL creep strain; ~ **elástica** *f* EMB elastic deformation, ING MECÁ lag, temporary set; ~ **por esfuerzo cortante** *f* ING MECÁ shearing strain; ~ **homogénea** *f* GEOL homogeneous deformation; ~ **de la inflorescencia** *f* AGRIC crazy top; ~ **lateral** *f* NUCL lateral yielding; ~ **local** *f* METAL local yielding; ~ **nuclear** *f* NUCL nuclear deformation; ~ **de la pala** *f* TRANSP AÉR blade distortion; ~ **permanente** *f* ING MECÁ permanent set, METAL permanent deformation, yielding, NUCL yielding; ~ **permanente por esfuerzos que sobrepasan el límite elástico** *f* PROD overstressing; ~ **plástica** *f* CONST, CRISTAL, FÍS plastic deformation, FÍS FLUID creeping motion, METAL plastic deformation, NUCL plastic yield, P&C *propiedad física, prueba* plastic deformation, plastic flow, plastic yield; ~ **progresiva** *f* MECÁ creep; ~ **progresiva logarítmica** *f* METAL logarithmic creep; ~ **de los rayos acústicos** *f* OCEAN *acústica submarina* sound ray curve; ~ **remanente** *f* ING MECÁ permanent set; ~ **reticular** *f* METAL lattice deformation; ~ **por sacudidas** *f* METAL jerky flow; ~ **de señal** *f* TELECOM signal distortion; ~ **por tensión** *f* ING MECÁ strain; ~ **torsional** *f* ING MECÁ torsional strain; ~ **por tracción** *f* ING MECÁ, METAL tensile strain; ~ **volumétrica** *f* ING MECÁ volumetric strain

deformado *adj* C&V out-of-shape, ELECTRÓN *señal* distorted, MECÁ warped

deformándose: ~ **continuamente** *adj* GEOL creeping

deformar *vt* CONST warp, ING MECÁ buckle, strain, MECÁ buckle, deform

deformarse *v refl* CONST bend, GEOL creep, ING MECÁ warp, MECÁ buckle, warp, TRANSP MAR warp

deformímetro *m* LAB *medida de presiones, fuerzas, desplazamientos y flujos* strain gage (*AmE*), strain gauge (*BrE*)

degasificador: ~ **de agua pesada** *m* NUCL heavy water degasifier

degaussado *m* FÍS, INFORM&PD degaussing

degaussar *vt* FÍS, INFORM&PD degauss

degeneración *f* CRISTAL, ELECTRÓN, FÍS, FÍS RAD degeneracy, ING MECÁ degradation, METAL, QUÍMICA degeneracy

degenerado *adj* ELECTRÓN, FÍS, METAL, QUÍMICA degenerate

degradable: no ~ *adj* EMB, RECICL nondegradable

degradación *f* CALIDAD degradation, CINEMAT grad, gradding, FÍS, INFORM&PD, ING MECÁ degradation, NUCL degradation, thindown, P&C *defecto* degradation, shading, TEC ESP *de equipos*, TV degradation; ~ **biológica** *f* CONTAM, CONTAM MAR, DETERG, EMB, HIDROL, RECICL biodegradation; ~ **de calidad** *f* TV quality degradation; ~ **energética** *f* NUCL energy degradation; ~ **de la energía** *f* ELEC, ING ELÉC dissipation; ~ **con garbo** *f* INFORM&PD graceful degradation; ~ **por irradiación** *f* FÍS RAD radiation degradation; ~ **parcial** *f* INFORM&PD graceful degradation; ~ **del suelo** *f* CONTAM land degradation; ~ **del terreno** *f* CONTAM land degradation

dejar¹: ~ **inactivo** *vt* PROD lay off; ~ **reproducir** *vt* PROD *fuego de un horno alto* allow to breed

dejar²: ~ **escapar el agua** *vi* CONST leak

DEL *abr* GEN (*diodo electroluminiscente, diodo emisor de luz*) LED (*light-emitting diode*)

delaminación *f* C&V, P&C *de plásticos, adhesivos, defecto* delamination

delantal *m* PAPEL, SEG apron; ~ **de cuero** *m* SEG leather apron; ~ **protector** *m* SEG protective apron

delantero *adj* AUTO, ELECTRÓN, FOTO, INFORM&PD, INSTR front-end, TEC ESP ahead, TRANSP AÉR forward, VEH front-end

delegado: ~ **de seguridad** *m* SEG safety representative

delessita *f* MINERAL delessite

delfinidina *f* MINERAL, QUÍMICA *antocianina* delphinin

delfinina *f* MINERAL, QUÍMICA *alcaloide* delphinine

delfinita *f* MINERAL, QUÍMICA delphinite, epidote

delga *f* ELEC, ING ELÉC *colector dinamico* commutator bar; ~ **de colector** *f* ELEC, ING ELÉC commutator bar, commutator segment

delicuescencia *f* ALIMENT, QUÍMICA *conversión en fluidos* deliquescence

delicuescente *adj* ALIMENT, QUÍMICA deliquescent

delimitación: ~ **de un terreno** *f* CONST *topografía* surface demarcation

delimitador *m* IMPR, INFORM&PD delimiter; ~ **de campo** *m* INFORM&PD field delimiter; ~ **de palabra** *m* INFORM&PD word delimiter

delineante *m* ING MECÁ, PROD, TRANSP MAR *diseño naval* draftsman (*AmE*), draughtsman (*BrE*); ~ **mecánico** *m Esp* (*cf dibujante mecánico AmL*) PROD mechanical draftsman (*AmE*), mechanical draughtsman (*BrE*); ~ **proyectista** *m* ING MECÁ design draftsman (*AmE*), design draughtsman (*BrE*)

delinear *vt* CONST line out, GEOM plot, IMPR outline, rule

delorencita *f* METAL, MINERAL, NUCL delorenzite

delta *m* GEOL, HIDROL, OCEAN, TEC PETR *geología* delta; ~ **de marea** *m* OCEAN tidal delta; ~ **de onda** *m* PETROL wave delta

deltaico *adj* GEOL, HIDROL, OCEAN, TEC PETR *geología* deltaic

demanda: ~ **en barras de central** *f* NUCL demand on plant buses; ~ **bioquímica de oxígeno** *f* (*DBO*) CONTAM *caracterización de aguas*, HIDROL *aguas residuales*, TEC PETR biochemical oxygen demand (*BOD*); ~ **de calor** *f* TERMOTEC heat demand; ~ **calorífica** *f* TERMO heat demand; ~ **cerrada** *f* PROD closed demand; ~ **de cloro** *f* HIDROL chlorine demand; ~ **de corriente** *f* PROD current requirement; ~ **corta** *f* PROD lumpy demand; ~ **de energía** *f* TERMO energy demand; ~ **de energía de alimentación del circuito** *f* ELEC circuit power requirement; ~ **de energía del circuito** *f* ELEC, PROD circuit power requirement; ~ **de energía eléctrica del circuito** *f* ELEC circuit power requirement; ~ **estacional** *f* GAS seasonal demand; ~ **interna** *f* PROD internal demand; ~ **máxima** *f* ELEC *fuente de alimentación*, ING ELÉC maximum demand; ~ **máxima de agua** *f* AGUA peak water demand; ~ **máxima de potencia** *f* PROD maximum power requirement; ~ **de nutrientes** *f* ALIMENT nutrient requirements; ~ **pico** *f* GAS peak demand; ~ **química de oxígeno** *f* (*DQO*) CONTAM *caracterización de aguas*, HIDROL *aguas residuales*, QUÍMICA chemical oxygen demand (*COD*); ~ **de señal de mando para varios puntos** *f* TELECOM

multipoint command token claim (*MCT*); ~ **por tráfico** *f* TRANSP traffic demand

demandas: ~ **competidoras** *f pl* PROD competing demands

demantoide *f* MINERAL demantoid

demarcación: ~ **de tráfico** *m* TRANSP traffic control

demasiada: ~ **semilla** *f* C&V heavy seed

demodulación *f* ELECTRÓN, FÍS, INFORM&PD, TEC ESP, TELECOM, TV demodulation; ~ **de la amplitud** *f* ELECTRÓN, INFORM&PD, TELECOM, TV amplitude demodulation; ~ **de banda estrecha** *f* ELECTRÓN narrow-band demodulation; ~ **de frecuencia** *f* ELECTRÓN, INFORM&PD, TELECOM, TV frequency demodulation; ~ **de subportador de cromeado** *f* ELECTRÓN chrominance subcarrier demodulation

demodulado *adj* ELECTRÓN, FÍS, INFORM&PD, TEC ESP, TELECOM, TV *señal* demodulated

demodulador *m* GEN demodulator; ~ **de crominancia** *m* TV chrominance demodulator; ~ **de cuadratura** *m* ELEC, ELECTRÓN quadrature demodulator; ~ **de estimación de frecuencia instantánea** *m* TEC ESP instantaneous frequency estimation demodulator; ~ **de frecuencia** *m* ELECTRÓN, INFORM&PD, TELECOM, TV frequency demodulator; ~ **I** *m* TV I demodulator; ~ **de Nyquist** *m* TV Nyquist demodulator; ~ **oscilador sincronizado por inyección** *m* TEC ESP injection-locked oscillator demodulator; ~ **Q** *m* ELECTRÓN Q-demodulator; ~ **de realimentación por compresión de frecuencia** *m* TEC ESP *comunicaciones* frequency compressive feedback demodulator; ~ **de subportador de cromeado** *m* ELECTRÓN chrominance subcarrier demodulator

demodular *vt* ELECTRÓN, FÍS, INFORM&PD, TEC ESP, TELECOM, TV demodulate

demolición: ~ **de macizos** *f* CARBÓN *minas* pillar drawing; ~ **de pilares** *f* CARBÓN *minas* pillar working

demora *f* ELEC, ELECTRÓN, INFORM&PD, ING ELÉC, TEC ESP, TRANSP, TRANSP AÉR turnaround, TRANSP MAR *navegación* bearing; ~ **de aguja** *f* TRANSP MAR *navegación* compass bearing; ~ **de entrega** *f* ING MECÁ lead time; ~ **en la entrega** *f* EMB delivery delay; ~ **para establecer la comunicación** *f* TELECOM call-setup delay; ~ **del fondeadero** *f* TRANSP MAR anchor bearing; ~ **de propagación** *f* TEC ESP propagation delay; ~ **de red** *f* INFORM&PD network delay; ~ **rotacional** *f* INFORM&PD, ÓPT rotational delay; ~ **de sincronizado controlada** *f* TV controlled delay lock; ~ **en la transmisión de grupo** *f* TELECOM group transmission delay

demorar *vt* ELEC *en relé, conmutador*, ING ELÉC delay

demostración: ~ **agrícola** *f* AGRIC farming demonstration; ~ **de vuelo** *f* TRANSP AÉR flight display

demostrar *vt* ING MECÁ support

demultiplexación *f* ELECTRÓN, FÍS, INFORM&PD, TELECOM demultiplexing

demultiplexado *m* ELECTRÓN, FÍS, INFORM&PD, TELECOM, TV demultiplexing

demultiplexador *m* ELECTRÓN, FÍS, INFORM&PD, TELECOM, TV demultiplexer

demultiplexor *m* ELECTRÓN, FÍS, INFORM&PD, TELECOM, TV demultiplexer

dendrita *f* CRISTAL, GEOL, METAL, MINERAL dendrite

dendrítico *adj* CRISTAL, GEOL, METAL, QUÍMICA dendritic

denier *m* TEXTIL *hilado* denier; ~ **del filamento** *m*

TEXTIL filament denier; ~ **total** *m* TEXTIL *hilatura* total denier

denitrificación *f* HIDROL, QUÍMICA denitrification

denominador *m* MATEMÁT denominator

densidad *f* CARBÓN, CONST, FÍS, GAS, GEOL density, HIDROL, IMPR body, INFORM&PD, MECÁ, P&C density, PAPEL consistency, PROD *fluido*, QUÍMICA density, TEC PETR *magnitud física* density, *propiedad física, unidades de medida* specific gravity, TEXTIL density; ~ **de almacenamiento** *f* INFORM&PD storage density, packing density; ~ **de un amperio** *f* ELEC ampere density; ~ **aparente** *f* C&V bulk density, CARBÓN apparent density, bulk density, P&C *propiedad física, prueba* apparent density, apparent powder density, bulk density, PAPEL *del papel y cartón* apparent density, TEC PETR *petrofísica, evaluación de la formación* bulk density; ~ **aparente de compactación** *f* PROC QUÍ compacted apparent density; ~ **aparente de los polvos de moldeo** *f* P&C apparent powder density; ~ **aparente de sedimentación** *f* PROC QUÍ apparent sedimentation density; ~ **API** *f* TEC PETR *unidad de medida* API gravity; ~ **atómica** *f* FÍS RAD atomic density; ~ **de bits** *f* INFORM&PD bit density; ~ **calorífica** *f* TERMO heat density; ~ **calorífica de salida** *f* NUCL heat output density; ~ **de carga** *f* ELEC, FÍS, FÍS RAD charge density; ~ **de carga lineal** *f* ELEC, FÍS, FÍS RAD linear charge density; ~ **de carga superficial** *f* ELEC, FÍS, FÍS RAD surface charge density; ~ **de carga volumétrica** *f* ELEC, FÍS, FÍS RAD volume charge density; ~ **de circulación** *f* TRANSP traffic density; ~ **de circulación equivalente** *f* *(ECD)* TEC PETR *perforación* equivalent circulating density *(ECD)*; ~ **de colisión** *f* FÍS RAD collision density; ~ **del componente** *f* ELECTRÓN component density; ~ **de corriente** *f* ELEC, FÍS, GEOFÍS, ING ELÉC, METAL current density; ~ **crítica** *f* TRANSP *tráfico* critical density; ~ **defectuosa** *f* ELECTRÓN defect density; ~ **de dislocación** *f* METAL dislocation density; ~ **de drenaje** *f* HIDROL drainage density; ~ **efectiva de partículas** *f* NUCL effective particle density; ~ **de los electrones** *f* CRISTAL, ELECTRÓN, FÍS, FÍS PART, QUÍMICA electron density; ~ **de electrones libres** *f* ELEC, ELECTRÓN, FÍS, FÍS PART, FÍS RAD free-electron density; ~ **electrónica** *f* CRISTAL, ELECTRÓN, FÍS, FÍS PART, QUÍMICA electron density; ~ **de empaquetado** *f* CRISTAL, TEXTIL packing density; ~ **de energía** *f* ENERG RENOV power density, TEC ESP energy density; ~ **de energía cinética** *f* ACÚST, FÍS kinetic energy density; ~ **de energía radiante** *f* FÍS, FÍS RAD, TELECOM radiant-energy density; ~ **de energía sónica** *f* ACÚST, FÍS sound energy density; ~ **de energía total** *f* ACÚST, FÍS total energy density; ~ **de equilibrio** *f* TEC PETR equilibrium density; ~ **equivalente** *f* TEC PETR *perforación* equivalent density; ~ **de errores** *f* TELECOM error density; ~ **espectral** *f* ACÚST, ELECTRÓN, FÍS spectral density; ~ **espectral de energía** *f* TEC ESP power spectral density; ~ **del espectro de energía** *f* TEC ESP power spectral density; ~ **especular** *f* ACÚST specular density; ~ **en estado seco** *f* CARBÓN, CONST dry density; ~ **de flujo** *f* ELEC *magnetismo*, FÍS flux density; ~ **del flujo electrostático** *f* ELEC electrostatic flux density; ~ **de flujo de energía** *f* TEC ESP power-flux density; ~ **de flujo magnético** *f* ELEC, FÍS, GEOFÍS magnetic flux density; ~ **de flujo de potencia** *f* ÓPT, TEC ESP power-flux density; ~ **del flujo de radiación** *f* FÍS RAD, NUCL radiation-flux density; ~ **de flujo radiante** *f* FÍS, FÍS RAD, ÓPT, TELECOM radiant flux density; ~ **de flujo remanente** *f* ELEC, FÍS, ING ELÉC remanent flux density, residual flux density; ~ **de frenado** *f* FÍS, NUCL slowing-down density; ~ **de grabación** *f* INFORM&PD recording density, storage density, TV packing density; ~ **de inducción magnética** *f* ELEC magnetic induction density; ~ **de integración** *f* ELECTRÓN integration density; ~ **del lodo** *f* PETROL mud density, TEC PETR *perforación* mud weight; ~ **de lodo equivalente** *f* *(EMW)* TEC PETR equivalent mud weight *(EMW)*; ~ **luminosa difusa** *f* ÓPT diffuse light density; ~ **de lutita** *f* TEC PETR shale density; ~ **en masa** *f* C&V, CARBÓN, P&C, TEC PETR bulk density; ~ **de masa comprimida** *f* P&C *propiedad física, prueba* packing density; ~ **neutral equivalente** *f* IMPR equivalent neutral density; ~ **óptica** *f* ÓPT optical density; ~ **de partición** *f* CARBÓN partition density; ~ **de paso** *f* ELECTRÓN gate density; ~ **de pelo** *f* TEXTIL *tejidos* density of pile; ~ **de la población de átomos excitados** *f* FÍS RAD population density of excited atoms; ~ **de potencia** *f* TELECOM power density; ~ **de potencia del haz** *f* ELECTRÓN beam-power density; ~ **de probabilidad** *f* FÍS probability density; ~-**promedio** *f* TRANSP average density; ~ **de radiación** *f* FÍS, FÍS RAD, TELECOM radiation density; ~ **real** *f* CARBÓN relative density; ~ **de la red de avenamiento** *f* HIDROL drainage density; ~ **de reflectancia** *f* ÓPT reflectance density; ~ **de registro** *f* ELECTRÓN packing density, INFORM&PD recording density; ~ **relativa** *f* CARBÓN real density, relative density, FÍS relative density; ~ **de relleno** *f* P&C *propiedad física, prueba* packing density; ~ **seca** *f* CARBÓN, CONST dry density; ~ **de la sección eficaz** *f* NUCL cross-section density; ~ **de separación** *f* CARBÓN separation density; ~ **de siembra** *f* AGRIC plant density, planting rate, seeding rate; ~ **de sinterización** *f* PROC QUÍ sintering density; ~ **del soporte** *f* CINEMAT base density; ~ **superficial** *f* ING ELÉC surface density; ~ **superficial de carga** *f* ELEC *electroestática*, FÍS, FÍS RAD surface charge density; ~ **de la tinta en las masas** *f* IMPR solid ink density; ~ **total** *f* C&V, CARBÓN, P&C, TEC PETR *petrofísica, evaluación de la formación* bulk density; ~ **de tráfico** *f* TRANSP traffic density; ~ **de la trama** *f* TEXTIL weft density; ~ **de transmitancia** *f* ÓPT transmittance density; ~ **de traza** *f* ÓPT *de emulsión fotográfica nuclear* track density; ~ **del trépano** *f* PETROL, TEC PETR bit density; ~ **por unidad de volumen** *f* C&V, CARBÓN, P&C, TEC PETR bulk density; ~ **de vapor** *f* FÍS, QUÍMICA, TERMO vapor density *(AmE)*, vapour density *(BrE)*; ~ **verdadera** *f* NUCL true density; ~ **volumétrica aparente** *f* C&V, CARBÓN, P&C, TEC PETR *petrofísica, evaluación de la formación* bulk density

densidades: ~ **energéticas de la radiación** *f pl* FÍS RAD energy densities of radiation

densimetría *f* ALIMENT densimetry

densímetro *m AmL* (*cf hidrómetro Esp*) GEN *aparato de medida* densimeter, hydrometer, water gage *(AmE)*, water gauge *(BrE)*; ~ **de ácidos** *m* AUTO acid hydrometer; ~ **de flotación** *m* ELEC *acumulador* aerometer

densitometría *f* ACÚST, CINEMAT, FÍS, FOTO, ÓPT

densitometry; **~ gamma de borde K** *f* NUCL K-edge gamma densitometry

densitómetro *m* ACÚST, CINEMAT, FÍS, FOTO, ÓPT densitometer; **~ de color** *m* PAPEL color densitometer (*AmE*), colour densitometer (*BrE*); **~ comparador** *m* CINEMAT comparator densitometer; **~ de cuña** *m* FOTO wedge densitometer

denso *adj* MECÁ heavy, QUÍMICA *solución* thick

dentado[1] *adj* AUTO cogged, ING MECÁ gearing, notched, stepped, cogged, MECÁ cogged, jagged, SEG jagged, VEH cogged

dentado[2] *m* ING MECÁ toothing; **~ en espiral** *m* ING MECÁ involute serration; **~ de involuta** *m* ING MECÁ involute serration

dentar *vt* CONST indent

dentellado *adj* ING MECÁ notched, pinked

dentellar *vt* ING MECÁ notch, pink

dentellón *m* CONST *muro* toothing

dentro: ~ del chip *adj* ELECTRÓN on chip

denudación *f* GEOL, METAL denudation, PETROL weathering

denuncia: ~ por infracción de tráfico *f* TRANSP traffic demand

denunciar *vt* TRANSP MAR *falta* log

departamento *m* ING MECÁ compartment; **~ de corte** *m* C&V cutter's bay; **~ de diseño** *m* PROD design department; **~ de diseño e ingeniería** *m* ING MECÁ engineering and design department; **~ de efectos especiales** *m* CINEMAT, TV special effects department; **~ de equipaje** *m* TRANSP baggage compartment (*AmE*), baggage room (*AmE*), luggage compartment (*BrE*); **~ de ingeniería** *m* ING MECÁ, MECÁ engineering department; **~ de proyectos** *m* TRANSP MAR *diseño de barcos* design department; **~ de publicidad** *m* IMPR advertising department; **~ de recepción** *m* PROD reception department; **~ receptor** *m* PROD receiving department; **~ técnico** *m* ING MECÁ, MECÁ engineering office; **~ de vía y obras** *m* FERRO way and works department (*BrE*)

dependiente *adj* PROD dependent; **~ de la computadora** *adj AmL* (*cf dependiente del ordenador Esp*) INFORM&PD *machine-dependent;* **~ de la máquina** *adj* INFORM&PD *machine-dependent;* **~ del ordenador** *adj Esp* (*cf dependiente de la computadora AmL*) INFORM&PD *machine-dependent*

deplexión *f* GEOL depletion

desplome *m* CONST subsidence

depocentro *m AmL* (*cf depósito central Esp*) GEOL depocenter (*AmE*), depocentre (*BrE*)

depolarización *f* TEC ESP *comunicaciones* depolarization; **~ magnética de la radiación de resonancia** *f* FÍS RAD magnetic depolarization of resonance radiation

deposición: CONTAM *depuración* sedimentation, deposition, ELECTRÓN *fabricación de circuito impreso*, METAL deposition; **~ axial** *f* C&V axial deposition; **~ axial en fase vapor** *f* ÓPT, TELECOM vapor phase axial deposition (*AmE*) (*VAD*), vapour phase axial deposition (*BrE*) (*VAD*); **~ axial por fase de vapor** *f* ÓPT, TELECOM vapor phase axial deposition (*AmE*), vapour phase axial deposition (*BrE*); **~ axial de plasma** *f* TELECOM axial plasma deposition; **~ catódica** *f* ING ELÉC sputtering; **~ electrolítica** *f* CONST, ELEC, ING ELÉC, REVEST electrodeposition; **~ electrónica** *f* ING ELÉC sputter-

ing; **~ epitaxial** *f* CRISTAL, ELECTRÓN epitaxial deposition; **~ del estratificador** *f* ELECTRÓN layer deposition; **~ en fase vapor** *f* ELECTRÓN, TERMO vapor deposition (*AmE*), vapour deposition (*BrE*); **~ húmeda** *f* CONTAM *depuración* wet deposition; **~ lateral de plasma** *f* TELECOM lateral plasma deposition; **~ de limo** *f* AGUA silt storage space; **~ química en fase de vapor** *f* ÓPT, TELECOM vapor phase chemical deposition (*AmE*), vapour phase chemical deposition (*BrE*); **~ radiactiva** *f AmL* (*cf precipitación radiactiva Esp*) AGRIC radioactive fallout, CONTAM, FÍS, FÍS RAD, NUCL fallout, radioactive fallout; **~ regulada** *f* CONTAM regulated deposition; **~ total** *f* CONTAM total deposition; **~ en vacío** *f* ELECTRÓN vacuum deposition; **~ en vapor** *f* ELECTRÓN, TERMO vapor deposition (*AmE*), vapour deposition (*BrE*); **~ de vapor químico** *f* (*DVQ*) ELECTRÓN, ÓPT, TELECOM chemical vapor deposition (*AmE*) (*CVD*), chemical vapour deposition (*BrE*) (*CVD*); **~ de vapor químico de plasma activado** *f* (*DVQPA*) ELECTRÓN, ÓPT *proceso*, TELECOM plasma-activated chemical vapor deposition (*AmE*) (*PCVD*), plasma-activated chemical vapour deposition (*BrE*) (*PCVD*)

depositado: ~ en tela *m* C&V laying on cloth; **~ en yeso** *m* C&V laying on plaster

depositar *vt* CONST set; **~ sedimentos en** *vt AmL* (*cf rellenar Esp*) GEOL *ríos* aggrade

depositarse *v refl* CARBÓN *un sedimento* settle

depósito *m* AGUA deposit, CARBÓN container, deposit, pan, settling, CINEMAT vault, CONST deposit, *combustible* bunker, CONTAM storage tank, repository, tank, deposition, storage facility, EMB deposit, FERRO depot, FÍS FLUID *de fluidos* reservoir, GAS pool, GEOL deposit, HIDROL reservoir, settling, basin, lens, INFORM&PD repository, ING MECÁ receiver, pit, MECÁ bin, tank, NUCL repository, PAPEL tank, PETROL reservoir, *petróleo* pool, PROD tank, warehouse, bin, storage tank, storage, QUÍMICA *asentamientos* deposit, REFRIG, TEC ESP tank, TEC PETR reservoir, TRANSP container, VEH *gasolina, gas-oil* tank, *aceite, carburante, líquido* reservoir; **~ de aceite** *m* AUTO, MECÁ oil pan, MINAS *yacimiento petrolífero* oil reservoir, PROD oil box; **~ ácido** *m* CONTAM acid deposit; **~ ácido seco** *m* CONTAM *contaminación terrestre* dry acid deposit; **~ adicional** *m* TEC ESP additional tank; **~ de agua** *m* AGUA water tank, CONST water tanker, FERRO *vehículos* water tank, MINAS water reservoir; **~ de agua caliente** *m* INSTAL TERM hot water tank; **~ de agua caliente presurizada** *m* INSTAL TERM pressurized hot-water tank; **~ de agua elevado** *m* HIDROL overhead water tank; **~ de agua para limpiar** *m* AGUA flush box; **~ de agua purificada** *m* AGUA, HIDROL purified-water reservoir, purified-water tank; **~ de aire** *AmL m* (*cf caldecín Esp*) ING MECÁ air receiver, MINAS air reservoir, TEC PETR *perforación, producción* air tank; **~ de aire de arranque** *m* TRANSP MAR *buques* air tank; **~ de aire comprimido** *m* ING MECÁ air receiver; **~ de aireación** *m* AGUA aeration basin; **~ de alimentación** *m* CONST, EMB feed hopper, ING MECÁ feed magazine, feed hopper, MECÁ, PROD feed hopper; **~ alimentador** *m* ING MECÁ magazine; **~ de alivio del presionador** *m* NUCL pressurizer relief-tank; **~ almacén** *m* MINAS *productos petrolíferos* depot; **~ para almacenamiento** *m* TEC ESP storage tank;

~ **de almacenamiento de agua de recarga** m NUCL refueling water storage tank (*AmE*), refuelling water storage tank (*BrE*); ~ **aluvial** m AGUA aggradational deposit, alluvial deposit, PETROL alluvial deposit; ~ **de arenas** m HIDROL *aguas residuales* sand trap; ~ **de aspiración** m INSTAL HIDRÁUL suction tank; ~ **de bomba** m MINAS standage; ~ **de botellas** m EMB bottle deposit; ~ **de cadáveres** m REFRIG mortuary; ~ **calcáreo** m AGUA, ALIMENT, HIDROL limescale; ~ **calorifugado** m TERMO heat-insulated container; ~ **de captación** m AGUA impounding reservoir; ~ **de captación de aguas** m HIDROL infiltration basin; ~ **de carbón conocido** m CARBÓN known coal deposit; ~ **de carbones** m CONTAM *buques* bunker tank; ~ **de carga** m AGUA *tuberías forzadas* head bay, CONTAM cargo tank, TRANSP freight station (*AmE*), goods depot (*BrE*), goods station (*BrE*), freight depot (*AmE*); ~ **para cargamento** m *Esp* (*cf receptáculo para cargamento AmL*) MINAS loading pocket; ~ **de cargas rodadas** m TRANSP, TRANSP MAR roll-on/roll-off depot (*ro-ro depot*); ~ **de cartuchos** m D&A cartridge depot; ~ **de cenizas** m PROD ash dump; ~ **central** m *Esp* (*cf depocentro AmL*) GEOL depocenter (*AmE*), depocentre (*BrE*); ~ **cilíndrico** m PROD *para mezcla, lavado* drum; --**cisterna** m TRANSP *carretera* tanker; ~ **de cloración** m HIDROL chlorination tank, chlorination vessel; ~ **de coches** m TRANSP car pool; ~ **de combustible** m AUTO fuel tank, D&A fuel dump, MECÁ, TRANSP AÉR, VEH fuel tank; ~ **de compensación** m HIDROL compensation basin, compensation reservoir; ~ **criogénico** m TEC ESP *aviones* cryogenic tank; ~ **de cromo** m REVEST chromium deposit; ~ **debajo de la bancada** m ING MECÁ tray; ~ **de decantación** m HIDROL clarifier, settling tank; ~ **de decantación final** m AGUA final-settling tank; ~ **de desagüe** m HIDROL catchment basin, ING MECÁ, REFRIG drain pan; ~ **de desengrasado** m ALIMENT degreasing tank; ~ **detrítico** m OCEAN silt; ~ **de dinamita** m *AmL* (*cf polvorín Esp*) MINAS *explosivos* dynamite store; ~ **de dragados** m *AmL* (*cf depósito de ripios Esp*) MINAS spoil bank; ~ **de drenaje** m ING MECÁ, REFRIG drain pan; ~ **de drenajes del refrigerante del reactor** m NUCL, REFRIG reactor-coolant drain tank; ~ **electrolítico** m CONST, ELEC, ING ELÉC, REVEST electrodeposit; ~ **elevado de agua** m HIDROL water tower; ~ **para embotellado** m ALIMENT bottling tank; ~ **de escarcha** m REFRIG frost deposit; ~ **de escorias** m PROD slag pot; ~ **esférico** m TEC ESP spherical tank; ~ **estructural** m ING MECÁ integral tank; ~ **de evaporación** m HIDROL evaporation tank; ~ **de evaporación de clase A** m HIDROL class A evaporating pan; ~ **de expansión** m AUTO, INSTAL HIDRÁUL, MECÁ, TERMOTEC expansion tank; ~ **de fangos** m AGUA *alcantarillado* catch basin, HIDROL *aguas residuales* sand trap; ~ **flexible** m TRANSP AÉR bladder tank; ~ **de fluido hidráulico** m ING MECÁ hydraulic fluid reservoir; ~ **de flujo radial** m HIDROL *aguas residuales* radial-flow tank; ~ **de fondo** m AGUA bottom deposit; ~ **franco** m TRANSP MAR bonded warehouse; ~ **de frente abierto** m PROD open-front bin; ~ **galvánico** m REVEST galvanic deposition; ~ **de gas** m GAS, PROD gas holder; ~ **de gas natural** m GAS natural gas deposit; ~ **de gas a presión** m GAS, ING MECÁ, MECÁ, NUCL, PROD, TERMO gas cylinder; ~ **de**

gasolina m *Esp* (*cf depósito de nafta AmL*) AUTO, VEH *carburante* gas tank (*AmE*), gasoline tank (*AmE*), petrol tank (*BrE*); ~ **de grisú** m MINAS gas feeder; ~ **hidráulico** m INSTAL HIDRÁUL hydraulic reservoir; ~ **para hielo** m REFRIG *en arcón refrigerado* ice bunker; ~ **inferior** m AUTO lower tank; ~ **por inmersión** m REVEST immersion plating; ~ **integral** m ING MECÁ integral tank; ~ **de latas** m RECICL can bank; ~ **de latas de aluminio** m RECICL aluminium can bank (*BrE*), aluminum can bank (*AmE*); ~ **de limpia** m AGUA flush pond, flush tank; ~ **del líquido de frenos** m AUTO brake-fluid reservoir, ING MECÁ brake-fluid tank; ~ **de líquido hidráulico** m ING MECÁ hydraulic fluid reservoir; ~ **de locomotoras** m FERRO *vehículos* engine shed (*BrE*), locomotive depot (*AmE*); ~ **de máquinas** m FERRO *vehículos* engine shed (*BrE*), locomotive depot (*AmE*); ~ **de matrices** m IMPR matrix magazine; ~ **menero** m MINAS ore deposit; ~ **de mercancías** m TRANSP freight shed (*AmE*), goods shed (*BrE*); ~ **mezclador** m HIDROL mixing basin; ~ **de mineral** m MINAS mineral deposit, *minería exterior* deposit of ore; ~ **para mineral** m MINAS ore bin; ~ **para minerales** m *AmL* MINAS ore hopper; ~ **de municiones** m D&A ammunition depot, ammunition dump, MINAS *explosivos* magazine; ~ **de nafta** m *AmL* (*cf depósito de gasolina Esp*) AUTO, VEH *carburante* gas tank (*AmE*), gasoline tank (*AmE*), petrol tank (*BrE*); ~ **de nieve** m HIDROL snow storage; ~ **de paquetes** m TRANSP parcels depot; ~ **de petróleo** m CONTAM bunker tank, ING MECÁ oilmeter, MINAS *yacimiento petrolífero* oil reservoir; ~ **portátil de agua** m D&A portable water-storage tank; ~ **portátil para gasolina** m D&A jerrican, jerry can, TRANSP jerry can; ~ **de presión sin costura** m ING MECÁ seamless pressure vessel; ~ **para la propagación de bacterias** m ALIMENT bacteria propagation tank; ~ **provisional de gasolina** m *Esp* (*cf depósito provisional de nafta AmL*) D&A gas dump (*AmE*), gasoline dump (*AmE*), petrol dump (*BrE*); ~ **provisional de nafta** m *AmL* (*cf depósito provisional de gasolina Esp*) D&A gas dump (*AmE*), gasoline dump (*AmE*), petrol dump (*BrE*); ~ **de reextracción** m NUCL backwash tank; ~ **regulador de humedad** m CONST wet standpipe; ~ **de retención** m AGUA detention basin, detention reservoir, HIDROL detention reservoir; ~ **de retención de efluentes activos** m NUCL active effluent holdup tank; ~ **de retorno de salmuera** m REFRIG brine return-tank; ~ **de ripios** m *Esp* (*cf depósito de dragados AmL*) MINAS spoil bank; ~ **de sal** m GAS salt deposit; ~ **de salmuera** m REFRIG brine tank; ~ **sapropélico** m GEOL sapropel deposit; ~ **en seco** m CONST dry standpipe; ~ **seco** m CONTAM *depuración* dry deposition; ~ **secundario** m PROD branch warehouse; ~ **de sedimentación** m AGUA, HIDROL settling basin, RECICL sedimentation tank, TEC PETR *perforación, producción* settling tank; ~ **sedimentario** m RECICL sedimentary deposit; ~ **de sedimentos radiales** m HIDROL *aguas residuales* radial-sludge tank; ~ **de seguridad** m SEG safety storage tank; ~ **de suministro** m AUTO, TRANSP AÉR supply tank; ~ **de suministro de emergencia** m TRANSP AÉR emergency supply tank; ~ **superior del radiador** m AUTO, VEH radiator header; ~ **de toba volcánica** m GEOL tuff deposit; ~ **de vagones** m FERRO *equipo inamovible* truck shed (*AmE*), wagon shed (*BrE*); ~ **de vapor** *m* INSTAL HIDRÁUL steam chest

depósitos: ~ **basales de un canal** *m pl* GEOL *lecho del río* channel lag; ~ **clasificados por grupos** *m pl* GEOL coset deposits; ~ **de combustión** *m pl* TERMO combustion deposits; ~ **de crecidas** *m pl* HIDROL flood deposits; ~ **cubiertos para equipos móviles** *m pl* FERRO stabling; ~ **fosfáticos** *m pl* GEOL phosphatic deposits; ~ **intermareales** *m pl* GEOL intertidal deposits; ~ **en superficie** *m pl* HIDROL *agua en lagos o embalses* surface storage; ~ **supramareales** *m pl* GEOL supratidal deposits

depreciación *f* ING MECÁ degradation, PROD depreciation

depresión *f* CONST hollow, GEOL sink, trough, ING MECÁ depression, METEO cyclone, depression, MINAS *galería* sink; ~ **angosta** *f* OCEAN *geología submarina* trough; ~ **aparente** *f* TRANSP MAR *sextante* dip; ~ **axial** *f* GEOL axial depression; ~ **baja** *f* METEO low depression, shallow depression; ~ **molecular del punto de congelamiento** *f* TERMO molecular depression of freezing point; ~ **orográfica** *f* METEO lee depression, orographic depression; ~ **del pH** *f* CONTAM *tratamiento químico* pH depression; ~ **poco profunda** *f* METEO low depression, shallow depression; ~ **del potencial hidrógeno** *f* CONTAM *tratamiento químico* pH depression; ~ **profunda** *f* METEO deep depression; ~ **a sotavento** *f* METEO lee depression

depuración *f* AGRIC roguing, AGUA cleansing, purification, CONTAM *tratamiento de gases* scrubbing, GAS, HIDROL purification, INFORM&PD debugging, PAPEL cleaning, PROC QUÍ purification, scrubbing, QUÍMICA scavenging, purification, RECICL purification, TEC ESP regeneration, TEC PETR *refino* purification, purging, *producción* scrubbing, TERMO *de gases* scrubbing, purification; ~ **adicional** *f* CONTAM supplementary purification; ~ **del agua** *f* AGUA water purification; ~ **de las aguas** *f* AGUA, HIDROL water purification; ~ **de aguas residuales** *f* AGUA wastewater purification, RECICL wastewater; ~ **del aire** *f* CARBÓN air cleaning; ~ **atmosférica** *f* CONTAM *tratamiento de gases* atmospheric scrubbing; ~ **en caja** *f* CARBÓN jigging; ~ **del lodo** *f* RECICL sludge processing; ~ **primaria** *f* CARBÓN prescreening; ~ **química del carbón** *f* CARBÓN, CONTAM, QUÍMICA chemical coal cleaning

depurador *m Esp* (*cf separador por lavado AmL*) CARBÓN scrubber, INFORM&PD debugger, MINAS scrubber, PAPEL cleaner, PROC QUÍ depurator, PROD strainer, QUÍMICA purifier, TEC ESP scrubber; ~ **de aire** *m* CARBÓN, ING MECÁ air cleaner, MINAS air scrubber; ~ **de entrada de aire** *m* PROD air inlet purifier; ~ **de gas** *m* GAS, PROD gas purifier; ~ **de gases** *m* NUCL, PROC QUÍ gas scrubber; ~ **de nudos** *m* PAPEL knotter screen; ~ **de tamiz** *m* PAPEL screener

depurar[1]: **sin** ~ *adj* AGUA, HIDROL raw

depurar[2] *vt* AGUA cleanse, purify, GAS, HIDROL purify, INFORM&PD debug, PAPEL clean, PROC QUÍ, QUÍMICA, RECICL, TERMO purify

derecha: **hacia la** ~ *adv* MECÁ right-hand

derecho[1]: **al** ~ *adj* IMPR positive-reading

derecho[2]: ~ **de inscripción** *m* AGRIC registration feed; ~ **marítimo** *m* TRANSP MAR maritime law; ~ **de paso** *m* AGRIC, CONST right of way

derechos: ~ **de alijo** *m pl* TRANSP MAR lighterage

charges; ~ **de almacenaje** *m pl* PROD warehousing charges; ~ **de depósito en aduana** *m pl* PROD warehousing charges; ~ **de esclusa** *m pl* AGUA lockage, TRANSP MAR *canalón de trancanil interior* lock dues; ~ **para explotar yacimientos minerales** *m pl* TEC PETR *licencias de exploración* mineral rights; ~ **de faros y balizas** *m pl* TRANSP MAR *navegación* light dues; ~ **de licencia de explotación** *m pl* TEC PETR *licencias, concesiones* royalty; ~ **de no emisión** *m pl* TELECOM nonbroadcast rights; ~ **de pesca** *m pl* TRANSP MAR fishing rights; ~ **portuarios** *m pl* TRANSP MAR port charges; ~ **de puerto** *m pl* TRANSP MAR harbor dues (*AmE*), harbour dues (*BrE*); ~ **de radiodifusión** *m pl* TV broadcasting rights; ~ **de remolque** *m pl* TRANSP MAR towage; ~ **de televisión** *m pl* TELECOM, TV television rights

deriva[1]: **a la** ~ *adv* TEC ESP off-course, TRANSP MAR *navegación* adrift

deriva[2] *f* GEN *navegación* drift; ~ **ascendente de subida de crucero** *f* TRANSP AÉR cruise-climb drift up; ~ **del Atlántico Norte** *f* GEOL, OCEAN North Atlantic Drift; ~ **continental** *f* GEOL, OCEAN *tectónica* continental drift; ~ **de corriente** *f* ELEC, ELECTRÓN, TELECOM frequency drift; ~ **de la corriente** *f* ELEC, ELECTRÓN, TELECOM current drift; ~ **de frecuencia** *f* ELEC, ELECTRÓN, TELECOM frequency drift; ~ **magnética** *f* TRANSP magnetic drag; ~ **del oscilador** *f* ELEC, ELECTRÓN, TELECOM oscillator drift; ~ **del Pacífico Norte** *f* GEOL, OCEAN North Pacific Drift; ~ **del servomecanismo** *f* TEC ESP servo system drift

derivación *f* AGUA diversion, CARBÓN bypass, CONST branch, bypass, ELEC *de red de distribución* branch, *circuito* bypass, *conexión* tapping, *devanado* tap, FÍS branching, GAS bypass, ING ELÉC bias, shunt, branch, ING MECÁ, INSTAL HIDRÁUL bypass, PETROL bypassing, PROD bypass, TEC ESP spin-off, switching, TEC PETR *tuberías, conductos*, TRANSP bypass; ~ **Ayrton** *f* ELEC *resistencia* universal shunt; ~ **de capacidad** *f* PROD output biasing; ~ **central** *f* ELEC, ING ELÉC *transformador* center tap (*AmE*), centre tap (*BrE*), common branch; ~ **central secundaria** *f* ING ELÉC secondary center tap (*AmE*), secondary centre tap (*BrE*); ~ **de circuitos** *f* ELEC *conexión* tapping; ~ **común** *f* ELEC *fuente de alimentación*, ING ELÉC common branch; ~ **electródica** *f* ELEC, ING ELÉC electrode bias; ~ **del filtro** *f* PROD filter bypass; ~ **de fuga de energía** *f* ING ELÉC drain bias; ~ **de galvanómetro** *f* ELEC, ING ELÉC *instrumento, resistencia* galvanometer shunt; ~ **manual** *f* PROD manual bypass; ~ **negativa** *f* ELEC minus tapping; ~ **positiva** *f* ELEC plus tapping; ~ **de red** *f* ING ELÉC network spur; ~ **subterránea** *f* TRANSP buried loop; ~ **a tierra** *f* ELEC, ING ELÉC earth leakage (*BrE*), ground leakage (*AmE*)

derivaciones: **sin** ~ *adj* ELEC *bobina* untapped

derivada: ~ **temporal** *f* FÍS time derivative

derivado[1] *adj* FÍS bypass, ING ELÉC shunt

derivado[2] *m* CARBÓN, CONTAM by-product, ELECTRÓN derivative, ING MECÁ by-product, MATEMÁT derivative, PROD by-product, QUÍMICA by-product, derivative, TEC PETR by-product

derivador *m* ELEC resistor shunt; ~ **Ayrton** *m* ELEC *resistencia* universal shunt; ~ **del embrague** *m* TRANSP AÉR clutch pick-off; ~ **universal** *m* ELEC *resistencia* universal shunt

derivar[1] *vt* CARBÓN *electricidad* tap, CONST branch, divert, FÍS bias, ING ELÉC shunt, bias, PROD tap, QUÍMICA derive

derivar[2] *vi* METR *instrumento de medida*, TRANSP MAR drift

derivómetro *m* TRANSP AÉR driftmeter

dernbachita *f* MINERAL dernbachite

derogación: **~ de salida** *f* PROD output override

derrabe *m* CARBÓN slope failure

derramamiento *m* AGUA, CONTAM, CONTAM MAR, GAS, TEC PETR spillage

derramarse *v refl* AGUA run out, ING MECÁ run

derrame *m* AGUA overflow, runoff, spillage, spill, CONST *hidráulica*, CONTAM *procesos industriales* runoff, CONTAM MAR runoff, spillage, *producción* spill, EMB pour spout, ENERG RENOV runoff, GAS spillage, HIDROL runoff, ING MECÁ overflow, PROD break, QUÍMICA overflow, TEC PETR spillage; **~ ácido** *m* CONTAM acid runoff; **~ de hidrocarburos** *m* CONTAM MAR oil spill; **~ de petróleo en el mar** *m* CONTAM, CONTAM MAR black tide, chocolate mousse; **~ petrolero** *m* CONTAM oil spill

derrapar *vi* TEC ESP, TRANSP AÉR yaw, VEH skid

derrape *m* TEC ESP *aviones*, TRANSP AÉR yaw

derretir *vt* ALIMENT thaw, ING MECÁ *metales* cast, METEO melt, thaw, TERMO melt

derretirse *v refl* ALIMENT thaw, ING MECÁ run, METEO, TERMO thaw

derribar *vt* CARBÓN blast, TRANSP AÉR shut down; **~ con calor** *vt* CARBÓN blast by heating

derribo *m* MECÁ breaking-in

derribos *m pl* CARBÓN, GEOL debris

derroche *m* ELEC, ING ELÉC dissipation; **~ de energía** *m* GAS energy waste

derrota *f* OCEAN *navegación* ship course, track, TRANSP MAR *navegación* course to steer, *comercial, mercantil* route; **~ indicada por la estela** *f* TRANSP MAR *navegación* leeway track, wake track; **~ marítima** *f* TRANSP MAR shipping route, *navegación* sea route; **~ meteorológica** *f* TRANSP MAR weather route; **~ óptima** *f* TRANSP MAR *recomendada por los servicios meteorológicos* weather routing; **~ ortodrómica** *f* TRANSP MAR great-circle course; **~ de la pala** *f* TRANSP AÉR blade tracking; **~ perimétrica** *f* TRANSP AÉR perimeter track

derrotero *m* TRANSP MAR sailing directions; **~ por radio** *m* TRANSP route guidance by radio

derrubiar *vt* AGUA, ENERG RENOV scour

derrubio *m* ENERG RENOV *presas*, HIDROL scour

derrubios *m pl* AGUA, ENERG RENOV scouring

derrumbamiento *m* CONST earth fall, fall of rock, MINAS *galería* caving-in, PETROL, TEC PETR *perforación* caving; **~ gravitacional** *m* TEC ESP gravitation collapse; **~ de tierras** *m* MINAS *taludes* run of ground

derrumbar *vt* CONST batter

derrumbarse *v refl* CONST crumble, fall in, give way, GEOL collapse, MINAS *terreno, mina* run

derrumbe *m* CONST collapse, *muros* batter, ING MECÁ collapsing, MINAS breaking down, *de minas* nip, PETROL, TEC PETR *perforación* caving; **~ de galería** *m* CARBÓN slope failure; **~ de tierra** *m* MINAS *taludes* breaking ground

desabobinado *m* ING MECÁ unwinding

desaceleración *f* FÍS, MECÁ, TEC ESP deceleration; **~ de frenado** *f* TRANSP braking deceleration

desacelerómetro *m* INSTR, METR, TEC ESP decelerometer

desacentuación *f* ACÚST, ELECTRÓN, TEC ESP *señales sonoras* de-emphasis

desacidificación *f* HIDROL deacidification

desacolchar *vt* TRANSP *contenedor* unstuff

desacoplado *adj* ELEC, INFORM&PD, TELECOM decoupled

desacoplamiento *m* ELEC *en circuito de CA*, ING ELÉC decoupling, NUCL uncoupling, QUÍMICA decoupling, TEC ESP uncoupling, TELECOM decoupling; **~ de señales** *m* TEC ESP *comunicaciones* decoupling

desacoplar *vt* GAS release, PROD lay off

desacoplo *m* ELEC *en circuito de CA*, ING ELÉC, QUÍMICA, TEC ESP *alimentación de dispositivos* decoupling; **~ cuasiestelar** *m* TEC ESP quasi-stellar decoupling

desactivación *f* CARBÓN deactivation, ING ELÉC disabling, turn-off, ING MECÁ deactivation, NUCL decay, QUÍMICA deactivation; **~ del estado excitado** *f* FÍS RAD excited-state deactivation

desactivar *vt* CARBÓN deactivate, D&A *mina* disarm, INFORM&PD disable, ING ELÉC disable, turn off, ING MECÁ deactivate, PROD *electrónica* turn off, QUÍMICA deactivate

desadaptación *f* ING MECÁ mismatch; **~ de impedancias** *f* ING ELÉC impedance mismatch

desafilado *adj* MECÁ dull

desagrupación *f* ELECTRÓN underbunching

desaguadero *m* AGRIC drain

desaguador *m* MINAS dewaterer

desaguar *vt* AGUA, HIDROL drain, MINAS *mina* dewater, *pozos* drain, RECICL drain

desagüe[1]: **de ~ automático** *adj* TRANSP MAR self-draining

desagüe[2] *m* AGUA outflow, outfall, pumping, CONST runoff, drain, dewatering, HIDROL *desembocadura* outlet, catchment, ING MECÁ discharge, outlet, drainage, overflow, LAB sump, MECÁ overflow, MINAS drainage, pumping, dewatering, drainage, PROD delivery, outflow, outlet, drain, RECICL drain, drainage, outfall, sewerage, TEC PETR drain, pumping, TRANSP MAR drain; **~ de aguas pluviales** *m* CONST, RECICL storm drain; **~ de la bañera** *m* TRANSP MAR cockpit drainage; **~ principal** *m* CONST main drain; **~ vertical** *m* CARBÓN vertical drainage

desahogo *m* ING MECÁ, MECÁ relief

desaireador *m* TEC PETR *perforación, producción*, TRANSP AÉR deaerator

desajustado *adj* FÍS mismatched, ING MECÁ displaced

desajuste *m* CRISTAL *defecto del cristal* misfit, ING ELÉC, ING MECÁ mismatch, MECÁ backlash, METAL misfit

desalabear *vt* PROD straighten

desalación *f* AGUA desalination, HIDROL desalination, leach, PROC QUÍ desalination, desalting, TEC PETR desalting

desalar *vt* AGUA, HIDROL, PROC QUÍ desalt, desalinate, QUÍMICA *jabón* salt out, TEC PETR desalt, TRANSP MAR desalinate

desalcalinización *f* C&V, QUÍMICA dealkalization

desalineación *f* IMPR misalignment, ING MECÁ misalignment, drift

desalineado *adj* ING MECÁ out-of-true, TRANSP AÉR out-of-track

desalineamiento *m* ING MECÁ misalignment, TEC PETR skew

desalinización *f* AGUA, HIDROL, OCEAN desalination, PROC QUÍ desalination, desalting, TEC PETR desalting

desalinizar *vt* AGUA, HIDROL desalinate, PROC QUÍ desalinate, desalt, TEC PETR, TRANSP MAR desalinate

desamarrar *vti* TRANSP MAR *barco* unmoor

desamarre *m* TRANSP MAR unmooring

desaminasa *f* QUÍMICA deaminase

desapareado *adj* FÍS, FÍS PART *electrones* unpaired

desapareamiento *m* QUÍMICA decoupling

desaparejar *vt* TRANSP MAR *barco* lay up, *mástil, abanico* unrig

desaparejo *m* TRANSP MAR *barco, mástil, abanico* laying-up, unrigging

desaparición *f* INFORM&PD drop-out; **~ gradual** *f* CINEMAT lap dissolve; **~ gradual de la imagen** *f* CINEMAT, TV fade-out

desapilar *vt* EMB, PROD destack

desapriete *m* PROD *tuerca* looseness; **~ de frenos** *m* FERRO *vehículos* brakes off

desarbolar *vt* TRANSP MAR dismast

desarbole *m* TRANSP MAR dismasting

desarenación: **~ al tonel** *f* PROD rattling

desarenado *m* PROD *piezas fundidas* cleaning, dressing; **~ en el tambor giratorio** *m* PROD *fundición* tumbling

desarenador *m* AGUA sand trap, PETROL desander, PROD *fundición* cleaner, TEC PETR *perforación* desander

desarenar *vt* PROD *fundición* clean

desarmable *adj* ING MECÁ, PROD collapsible

desarmado *m* ING MECÁ dismantling

desarmar[1] *vt* CONST break down, break up, D&A disarm, dismantle, ING MECÁ disassemble, dismantle, TRANSP MAR *buque de guerra* lay up

desarmar[2] *vti* INFORM&PD disarm

desarmar[3] *vi* ING MECÁ come apart; **~ la escenografía** *vi* CINEMAT kill a set

desarmarse *v refl* CONST break up

desarme *m* INFORM&PD, ING MECÁ disassembly; **~ manual** *m* TEC ESP manual disarming

desarreglo *m* ING MECÁ mismatch

desarrollador: **~ de urdimbre** *m* TEXTIL let-off motion

desarrollar *vt* CONST, GEOM, ING MECÁ develop

desarrollo *m* CONST development, FÍS course, evolution, ING MECÁ, TEC PETR, TERMO development; **~ activo** *m* PROD active development; **~ de un key seat** *m* TEC PETR *perforación* key seating; **~ múltiple** *m* ENERG RENOV multiple development; **~ de un ojo de cerradura** *m* TEC PETR *perforación* key seating; **~ y producción subsiguiente** *m* ING MECÁ development and subsequent manufacture; **~ de programas** *m* INFORM&PD program development; **~ de prueba** *m* METAL competition growth; **~ de software** *m* INFORM&PD software development; **~ de soporte lógico** *m* INFORM&PD software development

desarticulación *f* ING MECÁ disjointing

desarticulado *m* ING MECÁ disjointing

desarticular *vt* ING MECÁ disjoint

desatornillado *m* ING MECÁ unscrewing, TEC PETR *tuberías* screwing-out

desatracamiento *m* MINAS *de barrenos* unramming

desatracar *vt* TRANSP MAR fend off

desatracarse *v refl* TRANSP MAR *navegación a vela* bear away

desaturación *f* OCEAN desaturation

desaturar *vt* TV desaturate

desbalanceado *adj* ING MECÁ out-of-balance

desbarbado *m* ING MECÁ *madera, metal*, MECÁ deburring, PROD chipping

desbarbador *m* AGRIC awner, PROD sprue cutter; **~ de punta de diamante** *m* PROD chipping chisel

desbarbadora *f* ING MECÁ grinding machine, trimming machine, MECÁ grinding machine

desbarbar *vt* ING MECÁ, MECÁ debur, PAPEL trim, PROD *fundición* dress, *rebabas de la fundición* remove

desbastación *f* PROD *a buril* chipping

desbastado *m* CARBÓN roughing, roughing-down, ING MECÁ stock removal, PROD *con máquinas-herramientas* roughing, *trabajo* deburring, roughing-down; **~ por láser** *m* ELECTRÓN laser trimming; **~ de orillas** *m* C&V cutting off the edges; **~ con plantilla** *m* SEG jig routing

desbastador *m* PROD *fundición* dresser

desbastadora *f* ING MECÁ trimmer, surface planer

desbastar *vt* CARBÓN *una piedra* hew, CONST trim, ING MECÁ surface, MECÁ cut, plane, PROD *lima* clean, *soportes, dentaduras de lima* cut; **~ en caliente** *vt* MINAS *minerales* break down

desbaste *m* AGRIC *ganado* shrinkage, CARBÓN slab, *metales* roughing, CONST *piedra, madera* rough dressing, spalling, MINAS ragging, PROD *trabajo* roughing; **~ a la muela** *m* CARBÓN, MINAS coarse grinding; **~ plano** *m* PROD slabbing

desbituminación *f* CARBÓN debituminization

desbloqueado *adj* INFORM&PD deblocking

desbloqueador: **~ de bebedero** *m* ING MECÁ gate accentuator

desbloquear *vt* CINEMAT unlock, INFORM&PD deblock, ING MECÁ, PROD *control* back off

desbloqueo *m* INFORM&PD deblocking, unlocking

desbobinador *m* PROD decoiler

desbobinadora *f* EMB unwinding machine

desbobinar *vt* CINEMAT unwind

desbordamiento *m* AGUA inundation, flooding, overflow, CARBÓN overflow, ELECTRÓN flooding, HIDROL flooding, overflow, INFORM&PD, ING MECÁ, MECÁ, TELECOM overflow; **~ acordado** *m* TV agreed spillover; **~ de capacidad** *m* PROD overrange

desbordarse *v refl* HIDROL *río* flood, overflow; **~ bruscamente** *v refl* HIDROL *río* burst its banks

desborde *m* INFORM&PD overflow; **~ de página** *m* IMPR page bursting; **~ transgresivo** *m* GEOL onlap

desbroce: **~ de la cantera** *m* MINAS *preparación* quarry stripping

desbrote *m* AGRIC disbudding

descafeinado *adj* ALIMENT decaffeinated

descalce *m* AmL (*cf desmonte Esp*) MINAS holing, cut

descalcificación *f* HIDROL, QUÍMICA decalcification

descalcificar *vt* HIDROL, QUÍMICA decalcify

descamación *f* PROD *exfoliación* scaling

descanso: **~ de rodamiento** *m* ING MECÁ roller bearing

descantear *vt* CONST *carpintería* splay, ING MECÁ remove

descantilladura: **~ de silicio** *f* ELEC, INFORM&PD silicon chip

descapotable *m* VEH convertible

descarbonatación *f* CARBÓN, GEOL *reacción* decarbonation

descarbonatar *vt* CARBÓN, GEOL decarbonate

descarbonización *f* CARBÓN *culatas de los motores*

decarbonization, decarbonizing, QUÍMICA, TERMO decarbonizing, decarbonization

descarbonizar *vt* CARBÓN *culatas de motores* decarb (*AmE*), decarbonize, decarbonate, QUÍMICA, TERMO decarbonize

descarboxilación *f* QUÍMICA decarboxylation

descarboxilasa *f* QUÍMICA decarboxylase

descarburación *f* CARBÓN *aceros* decarbonization, decarbonizing

descarburar *vt* CARBÓN *aceros* decarb (*AmE*), decarbonize

descarga *f* AGRIC dump, AGUA discharge, outflow, outlet, C&V chute, CARBÓN *mercancías* landing, CONTAM *procesos industriales* dumping, discharge, CONTAM MAR discharge, D&A *de arma automática* burst, salvo, round, *de conjunto de armas* volley, ELEC *de condensador, arco* discharge, FÍS discharge, unloading, HIDROL *alcantarillado* outfall, discharge, ING ELÉC discharge, ING MECÁ unloading, discharge, outlet, INSTAL HIDRÁUL, METAL unloading, MINAS unloading, dumping, landing, NUCL *de bomba* delivery, *de bomba, compresor* discharge, *en un tubo de gas* breakdown, PAPEL *de lejiadora, tina* dumping, PROD outlet, *tuberías, agua, gas* outflow, QUÍMICA unloading, RECICL discharge, dumping, REFRIG *de compresor*, TELECOM discharge, TRANSP break bulk; **~ en abanico** *f* ELEC *en máquina*, FÍS, ING ELÉC brush discharge; **~ accidental** *f* CONTAM *procesos industriales*, CONTAM MAR accidental discharge; **~ de acumulador** *f* ELEC accumulator discharge; **~ de agua** *f* AGUA scouring; **~ de agua a alta presión** *f* CONTAM MAR high-pressure flushing; **~ de agua del primario** *f* NUCL primary letdown; **~ de aguas bajas** *f* AGUA low-water discharge; **~ de aguas residuales** *f* AGUA wastewater outfall; **~ de alta frecuencia** *f* GAS high-frequency discharge; **~ de arco** *f* GAS arc discharge; **~ del arco** *f* ELEC, ING ELÉC arc discharge; **~ en arco** *f* ELEC, ING ELÉC arc discharge; **~ automática** *f* ELEC, ING ELÉC self-discharge; **~ autosostenida** *f* ELEC, ING ELÉC self-sustained discharge; **~ en avalancha** *f* ELECTRÓN avalanche breakdown; **~ azulada** *f* ING ELÉC brush; **~ de bandejas** *f Esp* (*cf descarga de pálets AmL*) EMB pallet unloading; **~ de batería** *f* ELEC, TEC ESP battery discharge; **~ de CA** *f* ING ELÉC AC discharge; **~ de calor** *f* RECICL heat discharge, thermal discharge; **~ calórica** *f* RECICL heat discharge; **~ calorífica** *f* RECICL heat discharge; **~ de chispas** *f* FÍS spark discharge, **~ en circuito abierto** *f* ING ELÉC, TEC ESP self-discharge; **~ de combustible** *f* TEC ESP fuel dumping; **~ de condensador** *f* ING ELÉC capacitor discharge; **~ de contenedores** *f* TRANSP container unpacking; **~ de los contenedores** *f* TRANSP container stripping; **~ controlada** *f* CONTAM *procesos industriales*, RECICL controlled dumping; **~ en corona** *f* ELEC *alta tensión*, FÍS, GAS, ING ELÉC, P&C, TEC ESP corona discharge; **~ deliberada** *f* CONTAM MAR intentional discharge; **~ destructiva** *f* ING ELÉC destructive breakdown; **~ disruptiva** *f* ELEC *de dieléctrico* breakdown, *de electrodo* flash-over, ING ELÉC disruptive discharge, electrical breakdown, sparking, flash-over, P&C *prueba eléctrica* breakdown, TERMO flash-over; **~ disruptiva de escobilla** *f* ELEC *en máquina* brush sparking; **~ por efecto corona** *f* ELEC, FÍS, GAS, ING ELÉC, P&C, TEC ESP corona discharge; **~ eléctrica** *f* CONST breakdown,

ELEC, ELECTRÓN, ING ELÉC electric discharge, lightning discharge; **~ espontánea** *f* AUTO, ING ELÉC, TEC ESP self-discharge; **~ exterior** *f* ELEC arc-over; **~ por el extremo** *f* CONST end dump; **~ de gas** *f* ELECTRÓN, GAS, NUCL gas discharge; **~ incondensable** *f* ING ELÉC noncondensed discharge; **~ industrial** *f* AGUA industrial discharge; **~ por ionización acumulativa** *f* ELECTRÓN avalanche breakdown; **~ luminiscente** *f* ELEC, ELECTRÓN, FÍS, FÍS RAD glow discharge, ING ELÉC glow discharge; **~ al mar** *f* CONTAM, CONTAM MAR discharge at sea; **~ en el mar** *f* CONTAM, CONTAM MAR discharge at sea; **~ de la memoria principal** *f* INFORM&PD *programas* roll-out; **~ de microondas** *f* GAS microwave discharge; **~ del motor** *f* TRANSP AÉR exhaust fumes; **~ no autónoma** *f* CONST non-self-sustained discharge; **~ de paletas** *f Esp* (*cf descarga de pálets AmL*) EMB pallet unloading; **~ de pálets** *f AmL* (*cf descarga de bandejas Esp, descarga de paletas Esp*) EMB pallet unloading; **~ parcial** *f* ELEC *condensador* partial discharge; **~ radiante** *f* ELEC *en máquina*, FÍS, ING ELÉC brush discharge; **~ ramificada** *f* IMPR treeing; **~ del rayo** *f* ELEC, ING ELÉC lightning discharge; **~ de rayos canales** *f* FÍS, NUCL canal-ray discharge; **~ de referencia** *f* TEC ESP reference burst; **~ residual** *f* ELEC, ING ELÉC residual discharge; **~ sedimentaria** *f* AGUA sediment discharge; **~ de sedimento** *f* AGUA silt discharge; **~ superficial** *f* ELEC *falla del aislador* tracking; **~ térmica** *f* CONTAM *procesos industriales* thermal load, RECICL, TERMO heat discharge, thermal discharge; **~ a tierra** *f* METEO thunderbolt; **~ Townsend** *f* ELECTRÓN Townsend discharge; **~ a través de un aislante** *f* FÍS, ING ELÉC breakdown; **~ al vacío** *f* ELEC, ING ELÉC vacuum discharge

descargadero: **~ de crecidas** *m* AGUA *presas* weir

descargado *adj* AGUA discharged, CINEMAT unloaded, CONTAM MAR unladen, ELEC *batería* discharged, FOTO unloaded

descargador *m* CONTAM MAR unloader, ING ELÉC spark gap, REFRIG unloader; **~ de chispa micrométrico** *m* ELEC micrometric spark discharger; **~ de chispa protector** *m* ELEC protective spark discharger; **~ de fondo** *m* AGUA *presa* sluiceway; **~ giratorio** *m* ELEC *chispa*, ING ELÉC rotary discharger; **~ con pistón hidráulico** *m* CONST hydraulic piston discharger; **~ de silo** *m* AGRIC silo unloader; **~ de sobretensiones** *m* ELEC *línea de alimentación* surge diverter

descargamento *m* PROD *navíos* off-load

descargar *vt* CARBÓN dump, CONST *vertedero* shoot, CONTAM *procesos industriales* discharge, CONTAM MAR unload, dump, discharge, D&A, INFORM&PD unload, ING MECÁ relieve, INSTAL HIDRÁUL *válvula* relieve, balance, MECÁ relieve, MINAS land, PROD relieve, RECICL discharge, dump, TRANSP MAR *mercancías* land, *buque* discharge, *cargamento* unload; **~ de la memoria principal** *vt* INFORM&PD roll out

descarnador *m* CONST, ING MECÁ, MECÁ scraper

descarrilamiento *m* FERRO derailment, TRANSP AÉR runoff

descarrilar[1] *vt* FERRO derail

descarrilar[2] *vi* TRANSP AÉR run off

descartar *vt* MECÁ, PROD scrap

descascarado *adj* ALIMENT shelled

descascarador *m* ALIMENT decorticator

descascaradora *f* AGRIC huller

descascarar *vt* AGRIC, ALIMENT *granos, etc* husk

descascarillado *m* CONST scaling; ~ **del barniz** *m* REVEST varnish tear; ~ **del borde** *m* C&V edge peeling

descascarillar *vt* MECÁ descale

descascarillarse *v refl* PROD *pinturas, soldeo* peel off

descebado *m* ING ELÉC *del tubo fluorescente* misfire, *magnetismo* de-energization

descebar *vt* AGUA drain, INSTAL HIDRÁUL *bombas* dewater, drain, dry up, MECÁ drain

descebarse *v refl* MECÁ, PROD *de inyectores, bombas* fail

descebo *m* PROD *de inyectores, bombas* failure

descendente *adj* FÍS FLUID, HIDROL downstream

descender[1] *vt* AGUA draw down, IMPR drop; ~ **el fondo** *vt* MINAS bring down the face

descender[2] *vi* AGUA, HIDROL *agua* recede; ~ **en barrena** *vi* TEC ESP *aviación*, TRANSP AÉR spin; ~ **con motor desembragado** *vi* VEH coast

descendiente[1] *adj* FÍS FLUID, HIDROL downstream

descendiente[2] *m* INFORM&PD *árbol* descendant, NUCL decay product; ~ **radiactivo** *m* FÍS, FÍS RAD daughter product

descenso *m* CONST going, CONTAM subsidence, FÍS *de temperatura* lowering, GEOL subsidence, drop, HIDROL *del nivel de agua* subsidence, PROD *de husillo perforador, portaherramientas* fall, TEC ESP *vehículos espaciales* descent, *vuelo espacial* drop, chute, TEC PETR *geología* subsidence, TRANSP AÉR let-down; ~ **de crucero** *m* TRANSP AÉR cruise descent; ~ **de emergencia** *m* TEC ESP *accidentes en vuelo* forced landing, TRANSP AÉR emergency descent, forced landing; ~ **forzoso** *m* TEC ESP *vehículos*, TRANSP AÉR forced landing; ~ **manual** *m* ING MECÁ hand downfeed; ~ **del nivel** *m* AGUA *embalse* draw-down; ~ **del nivel de agua acuífera** *m* ENERG RENOV draw-down of water in aquifer; ~ **del nivel hidrostático** *m* GEOFÍS, HIDROL falling water table; ~ **del pH** *m* CONTAM *tratamiento* pH drop; ~ **poco pronunciado** *m* TRANSP AÉR shallow descent; ~ **de temperatura** *m* GAS temperature drop; ~ **térmico** *m* PROD thermal relief

descentrado[1] *adj* ING MECÁ off-center (*AmE*), offcentre (*BrE*), displaced

descentrado[2] *m* ING MECÁ setting over

descentralizado *adj* INFORM&PD, TELECOM decentralized

descentramiento *m* CINEMAT offset, ING MECÁ offsetting, MECÁ eccentricity, offset, PROD off-setting, *ejes* runout

descentrar *vt* ING MECÁ set over, throw off center (*AmE*), throw off centre (*BrE*)

deschaladora: ~ **de maíz** *f* AmL (*cf deshojadora de maíz Esp*) AGRIC corn husker (*AmE*), maize husker (*BrE*)

descifrado *m* ELECTRÓN, INFORM&PD decoding, TEC ESP, TELECOM, TV descrambling, decoding

descifrador *m* ELECTRÓN, INFORM&PD decoder, SEG, TEC ESP, TELECOM, TV descrambler, decoder

descifrar *vt* ELECTRÓN, INFORM & PD decipher, decode, decrypt, SEG decode, descramble, TEC ESP, TELECOM, TV unscramble, decipher, decode

descifre *m* INFORM&PD decryption

desclasificador *m* CARBÓN outsize

declinación *f* METAL disclination

descloizita *f* MINERAL descloizite

descloración *f* HIDROL dechlorination

descodificación *f* ELECTRÓN, TELECOM deciphering

descodificar *vt* TELECOM decipher

descohesión *f* METAL decohesion

descolgado[1] *adj* TELECOM off the hook

descolgado[2] *m* P&C *pintura, defecto* flop

descoloración *f* COLOR staining, IMPR color removing (*AmE*), colour removing (*BrE*), QUÍMICA, TEXTIL staining

descolorar *vt* COLOR, QUÍMICA, TEXTIL stain

descolorido *adj* CINEMAT washed out

descompensado *adj* TRANSP AÉR out of trim

descomponer[1] *vt* NUCL, QUÍMICA decompose

descomponer[2] *vi* ALIMENT putrefy

descomponerse *v refl* SEG break down

descomposición *f* ALIMENT decomposition, C&V *química* breaking down, CONST breakdown, EMB, HIDROL, INFORM&PD, METAL decomposition, MINAS *filones* floor heave, *minerales* breakdown, NUCL decomposition, P&C *física, química, procesamiento* breakdown, QUÍMICA, TEC PETR, TERMO decomposition; ~ **en ácido** *f* NUCL wet ashing; ~ **del agua por irradiación** *f* NUCL water decomposition under irradiation; ~ **anaeróbica** *f* HIDROL *aguas residuales*, RECICL anaerobic decomposition; ~ **espinodal** *f* METAL spinodal decomposition; ~ **funcional** *f* INFORM&PD functional decomposition; ~ **del haz** *f* ELECTRÓN beam splitting; ~ **térmica** *f* P&C *propiedad, prueba, procesamiento*, TERMO thermal decomposition; ~ **en tomas** *f* CINEMAT shot breakdown

descompresión *f* ING MECÁ, MECÁ relief, OCEAN, PETROL decompression, TEC ESP, TRANSP AÉR *aviones* depressurization; ~ **explosiva** *f* OCEAN explosive decompression, *buceo* blow-up ascent, TERMO explosive decompression

descomprimir *vt* CINEMAT unsqueeze, OCEAN, TEC ESP, TRANSP AÉR depressurize

descompuesto *adj* CINEMAT out of order, ELEC *fallo*, ING MECÁ out-of-order; **no** ~ *adj* QUÍMICA undecomposed

desconchado *m* CARBÓN *crisoles* scalping

desconectable *adj* ING MECÁ detachable

desconectado *adj* INFORM&PD, TELECOM off-line

desconectador *m* ING ELÉC disconnect switch, isolating switch, ING MECÁ tripping device, PROD disconnect switch, multigage (*AmE*), multigauge (*BrE*), tripping device, VEH *cambio de velocidades automático* kick down; ~ **accionado por varillas** *m* PROD rod-operated disconnect switch; ~ **múltiple** *m* PROD multigage isolator (*AmE*), multigauge isolator (*BrE*)

desconectar[1] *vt* ELEC *conexión* unplug, ELECTRÓN turn off, INFORM&PD disconnect, ING ELÉC turn off, disconnect, switch off, ING MECÁ cut out, disconnect, disjoint, release, MECÁ release, PROD release, *cinta* lay off, QUÍMICA disconnect, TELECOM disconnect, unplug, TV switch off; ~ **correa** *vt* ING MECÁ fork a belt off; ~ **rápidamente** *vt* ELEC, PROD snap

desconectar[2] *vi* ING MECÁ switch off, trip

desconectarse: ~ **del sistema** *v refl* INFORM&PD log off, log out

desconexión *f* ELEC, ING ELÉC disconnection, release, ING MECÁ off, trip, tripping, PROD cutoff, TELECOM disconnection, prefix; ~ **circuital de acción retardada** *f* ING ELÉC delayed action circuit-breaking; ~ **de la corriente residual** _f_ ING ELÉC zero current turn off; ~ **cuasiestelar** _f_ TEC ESP quasistellar decoupling; ~ **en derivación** _f_ PROD shunt trip; ~ **por falta de corriente** _f_ ING ELÉC no-volt release; ~ **de interrogación** _f_ PROD examine-off; ~ **láser automática** _f_ TELECOM automatic laser shutdown; ~ **de la potencia principal** _f_ PROD mainpower disconnect; ~ **rápida** _f_ INSTAL HIDRÁUL, TRANSP AÉR quick disconnect; ~ **de señal mecánica** _f_ FERRO off aspect; ~ **del sistema** _f_ INFORM&PD log-off, log-out; ~ **total** _f_ ING ELÉC _corriente_ insulation

descongelación _f_ ALIMENT, ING MECÁ defrosting, INSTAL TERM de-icing, REFRIG thawing, _alimentos_ defrosting, TRANSP AÉR de-icing; ~ **por alta frecuencia** _f_ REFRIG high-frequency thawing; ~ **dieléctrica** _f_ REFRIG dielectric thawing; ~ **por microondas** _f_ REFRIG microwave thawing

descongelador _m_ INSTAL TERM, TEC ESP, VEH _accesorio_ de-icer

descongelar _vt_ ALIMENT, REFRIG thaw, _alimentos_ defrost

desconmutación _f_ ELECTRÓN decommutation

desconmutador _m_ ELECTRÓN decommutator

descontaminación _f_ GEN decontamination

descontaminado _adj_ GEN decontaminated, depolluted

desconvolución _f_ ELECTRÓN, GEOL, TELECOM deconvolution

descorchadora _f_ EMB uncorking machine

descornador: ~ **para animales de uno a dos años** _m_ AGRIC Leavitt clippers

descortezado[1] _adj_ ALIMENT shelled

descortezado[2] _m_ PAPEL debarking

descortezador _m_ ALIMENT decorticator

descortezadora _f_ PAPEL debarker; ~ **de tambor** _f_ PAPEL drum debarker

descortezarse _v refl_ PROD _soldadura_ peel off

descrestado: ~ **de ondas** _m_ ELECTRÓN, TELECOM, TV peak clipping

descripción: ~ **de archivos** _f_ INFORM&PD file description; ~ **de la célula de mantenimiento** _f_ TELECOM maintenance cell description (_MCD_); ~ **de datos** _f_ INFORM&PD data description; ~ **estadística del movimiento turbulento** _f_ FÍS FLUID statistical description of turbulent motion; ~ **de ficheros** _f_ INFORM&PD file description; ~ **general de la acción** _f_ CINEMAT action outline; ~ **de problemas** _f_ INFORM&PD problem description; ~ **de ruta** _f_ TRANSP, TRANSP AÉR, TRANSP MAR route description

descripciones: ~ **para la elección entre variantes de productos** _f pl_ PROD product-variant option descriptions

descriptación _f_ INFORM&PD, TELECOM decryption

descriptor _m_ INFORM&PD descriptor

descrudado _m_ TEXTIL _acabado_ scouring

descrudecedor _m_ DETERG scouring liquid

descuadrado _adj_ IMPR, ING MECÁ out-of-square

descubierto[1]: al ~ _adj_ ING ELÉC bare

descubierto[2]: al ~ _adv_ MINAS above-ground

descubrimiento _m_ GEOL _de intrusión_ deroofing, TEC PETR discovery; ~ **del núcleo** _m_ NUCL core uncovery; ~ **de petróleo** _m_ PETROL, TEC PETR oil discovery

desdoblamiento _m_ CRISTAL, NUCL splitting; ~ **múltiple** _m_ NUCL multiple splitting

desecación _f_ ALIMENT desiccation, CARBÓN drying, CONST _maderas_ seasoning, EMB desiccation, MINAS _galerías_ drainage, draining, PROC QUÍ, QUÍMICA, REFRIG, TEC PETR, TERMO desiccation; ~ **por aspiración** _f_ TEXTIL suction dewatering; ~ **dentro de paquetes** _f_ EMB, REFRIG in-package desiccation

desecado[1] _adj_ GEN desiccated

desecado[2] _m_ TERMO drying-out

desecador _m_ ALIMENT _maquinaria para proceso de alimentos_ desiccator, CARBÓN dryer, INSTAL TERM, LAB, PROC QUÍ desiccator, QUÍMICA desiccator, dryer, TERMO dryer; ~ **al vacío** _m_ LAB _material de vidrio, secado_ vacuum desiccator; ~ **por vacío** _m_ INSTAL TERM vacuum desiccator, vacuum dryer

desecante[1] _adj_ EMB, INSTAL TERM, QUÍMICA, REFRIG, TERMO desiccant

desecante[2] _m_ EMB _gel de sílice_, PROC QUÍ, QUÍMICA desiccant, TERMO drying agent, _sustancia higroscópica_ desiccant

desecar _vt_ ALIMENT, EMB desiccate, INSTAL HIDRÁUL dewater, drain, dry up, INSTAL TERM desiccate, dry, PROC QUÍ, QUÍMICA, REFRIG, TEC PETR desiccate, TERMO desiccate, dry, dry out; ~ **completamente** _vt_ HIDROL dry up

desechable _adj_ INFORM&PD nonreusable, TEC ESP _equipos_ expendable

desechar _vt_ CALIDAD, CINEMAT reject, CONST throw back to waste, ING MECÁ discard, reject, PROD reject, scrap, TEXTIL discard, reject

desecho _m_ CALIDAD, CARBÓN, CINEMAT reject, CONTAM residue, MECÁ scrap, PROD reject, RECICL waste product, TEXTIL reject

desechos _m pl_ AmL (_cf ripios Esp_) AGUA refuse, CARBÓN refuse, debris, scum, tailings, CONST spoil to waste, tailings, EMB garbage (_AmE_), rubbish (_BrE_), refuse, MINAS deads, attle, PROD garbage (_AmE_), rubbish (_BrE_), waste, refuse, scrap, RECICL garbage (_AmE_), rubbish (_BrE_), waste, refuse; ~ **combustibles semisólidos** _m pl_ CONTAM semisolid combustible waste; ~ **de cribado** _m pl_ CARBÓN, MINAS _mineral quebrantado_ oversize; ~ **de fundición** _m_ PROD _funderías_ scrap; ~ **a gran escala** _m pl_ CONTAM _contaminantes_ macrowaste; ~ **indefinidos** _m pl_ CONTAM, RECICL indeterminate waste; ~ **industriales** _m pl_ CONTAM, RECICL industrial waste; ~ **líquidos** _m pl_ CONTAM, RECICL liquid waste; ~ **municipales** _m pl_ AGUA, CONTAM, RECICL municipal waste; ~ **patógenos** _m pl_ RECICL pathogenic waste, pathological waste; ~ **secos** _m pl_ PAPEL dry broke; ~ **sólidos** _m pl_ CONTAM solid waste

desembarcadero _m_ MINAS _ríos_, TEC ESP landing stage, TRANSP MAR _amarre_ landing place, wharf

desembarcar[1] _vt_ D&A _tropas_ discharge, TRANSP MAR land, _carga, cargamento_ unload, _tripulantes_ pay off, unship

desembarcar[2] _vi_ FERRO _pasajeros_ detrain

desembarco _m_ OCEAN landing

desembarque _m_ TRANSP MAR landing

desembarrancar _vt_ TRANSP MAR _barco perdido_ float off

desembocadura _f_ AGUA outfall, HIDROL, MINAS _río_ mouth, OCEAN outlet, RECICL _río_ outfall, TRANSP MAR _río_ mouth; ~ **del río** _f_ AGUA, HIDROL, MINAS, OCEAN, TRANSP MAR river mouth

desembocar: ~ **en** _vi_ HIDROL _río_ flow into; ~ **en el mar** _vi_ HIDROL discharge into the sea, outfall to sea

desembragado _adj_ ING MECÁ out of gear

desembragar[1] *vt* AUTO unclutch, ING MECÁ disengage, unclutch, MECÁ disengage, release, PROD release

desembragar[2] *vi* AUTO declutch, ING MECÁ declutch, release, MECÁ, VEH declutch; ~ **correa** *vi* ING MECÁ fork a belt off

desembrague *m* AUTO clutch throwout, disengage, FERRO *vehículos* releasing, ING MECÁ stop motion, tripping, VEH clutch throwout; ~ **automático** *m* ING MECÁ automatic release; ~ **rápido** *m* ING MECÁ, PROD, TEC ESP quick release; ~ **de transmisión** *m* AUTO, ING MECÁ unclutch

desempaquetar *vt* INFORM&PD unpack

desemulsificante *m* AGUA, CONTAM, CONTAM MAR, QUÍMICA emulsion-breaker

desemulsionador *m* CONTAM MAR de-emulsifier

desemulsionar *vt* AGUA, CONTAM, QUÍMICA break an emulsion

diseñado: ~ **expresamente para un fin determinado** *adj* EMB purpose-designed

desencadenado *adj* TELECOM break-free

desencallar *vt* TRANSP MAR *barco* refloat

desencasquetado *m* ING MECÁ uncapping

desencasquillado *m* ING MECÁ uncapping

desencepar *vt* TRANSP MAR *ancla* clear

desenchavetarse *v refl* ING MECÁ work loose

desenchufar *vt* ELEC, TELECOM, TV unplug

desencolado *m* TEXTIL *tejidos* desizing

desencolar *vt* TEXTIL *tejidos* desize

desencriptado *m* SEG, TELECOM descrambling

desencriptador *m* SEG, TELECOM descrambler

desencriptar *vt* SEG, TELECOM descramble

desenergizarse *v refl* FÍS de-energize

desenfocado *adj* CINEMAT, FOTO out-of-focus

desenfocar[1] *vt* ELECTRÓN defocus

desenfocar[2] *vti* CINEMAT defocus

desenganchado *m* NUCL unlatching

desenganchador *m* ING MECÁ tripper

desenganchar[1] *vt* FERRO uncouple, ING MECÁ, MECÁ disengage, PROD release; ~ **bruscamente** *vt* PROD snap

desenganchar[2] *vi* ING MECÁ trip

desenganche *m* FERRO uncoupling, ING ELÉC dropout, ING MECÁ trip, tripping, unhooking, unscrewing, MINAS *varillas enganchadas* unmaking, TEC ESP uncoupling; ~ **de acción rápida** *m* ING MECÁ fast-acting trip; ~ **instantáneo** *m* ING MECÁ instantaneous release; ~ **del piloto automático** *m* TRANSP AÉR autopilot disengagement; ~ **rápido** *m* ELEC *conmutador* instantaneous release; ~ **de la tapa posterior** *m* FOTO back-cover release; ~ **del tren de aterrizaje** *m* TRANSP AÉR landing gear unlocking; ~ **ultra rápido** *m* ING MECÁ instantaneous release

desengastar *vt* TRANSP MAR *mecha de un palo* unstep

desengranar *vt* ING MECÁ release, throw out of gear, MECÁ release

desengrasado *m* ALIMENT, C&V, CONST, ELEC, ING MECÁ, MECÁ degreasing

desengrasador *m* ALIMENT, C&V, CONST, ELEC, ING MECÁ, MECÁ degreaser

desengrasante *m* ALIMENT, C&V, CONST, ELEC, ING MECÁ, MECÁ degreasing

desengrasar *vt* ALIMENT, C&V, CONST, ELEC, ING MECÁ, MECÁ degrease

desengrase *m* ALIMENT, C&V, CONST, ING MECÁ, MECÁ degreasing; ~ **alcalino** *m* DETERG alkaline cleaning

desenhebrar *vt* CINEMAT unthread

desenlace *m* AGRIC *leguminosas* tripping

desenlodador *m* CARBÓN desliming screen

desenlodadura *f* CARBÓN desliming

desenlodar *vt* CARBÓN, METAL, MINAS deslime

desenmangar: ~ **por presión** *vt* ING MECÁ, PROD force off

desenredado *m* ING MECÁ unwinding, MINAS *cables de extracción* clearing

desenredar *vt* MINAS *cable de extracción* clear, OCEAN unravel

desenrolar *vt* TRANSP MAR *de tripulación* pay off

desenrole *m* TRANSP MAR *de tripulación* paying off

desenrollable *adj* TEC ESP *antena* unfurlable

desenrollado *m* PAPEL reeling-off, winding off

desenrollador *m* TEXTIL *de tejido* unwinder

desenrolladora: ~ **de película plástica** *f* AGRIC mulcher-transplanter

desenrollamiento *m* ING MECÁ unwinding

desenroscado *m* ING MECÁ unscrewing, TEC PETR *perforación* breakout

desensamblador *m* INFORM&PD disassembler

desensamblar *vt* INFORM&PD disassemble

desensibilización *f* FOTO, TELECOM desensitization

desensibilizador *m* FOTO, TELECOM desensitizer

desensibilizar *vt* FOTO, TELECOM desensitize

desentrañar *vt* OCEAN gut

desenvainado *m* NUCL decanning; ~ **mecánico** *m* NUCL mechanical decanning, mechanical decladding; ~ **por medios químicos** *m* NUCL, PROC QUÍ, QUÍMICA chemical decanning, chemical decladding

desenvenenamiento *m* NUCL depoisoning

desenvolvente *adj* GEOL deconvolved

desequilibrado *adj* ELEC unbalanced, ING MECÁ out-of-balance, METR out of balance, TEC ESP unbalanced, TRANSP AÉR out of trim

desequilibrio *m* ING MECÁ, MECÁ unbalance, METAL misfit; ~ **de la compactación** *m* GEOL, TEC PETR compaction disequilibrium; ~ **de la distribución del tráfico** *m* TELECOM traffic distribution imbalance; ~ **de fases** *m* PROD phase unbalance; ~ **giroscópico** *m* TRANSP AÉR gyro unbalance; ~ **de impedancias** *m* ING ELÉC impedance mismatch; ~ **de la intensidad del tráfico** *m* TELECOM traffic load imbalance; ~ **residual admisible** *m* ING MECÁ permissible residual unbalance; ~ **residual permitido** *m* ING MECÁ permissible residual unbalance; ~ **térmico** *m* TERMO thermal imbalance

desescamado *m* DETERG, MECÁ descaling, PROD scaling

desescamar *vt* DETERG, MECÁ descale, PROD scale

desescarchar *vt* REFRIG *tuberías* defrost

desescarche *m* REFRIG *tuberías* defrosting; ~ **por aspersión** *m* REFRIG water defrosting; ~ **automático** *m* REFRIG autodefrost; ~ **desde el exterior** *m* REFRIG external defrosting; ~ **desde el interior** *m* REFRIG internal defrosting; ~ **eléctrico** *m* ELEC, REFRIG electric defrosting; ~ **por gas caliente** *m* REFRIG hot gas defrosting; ~ **por inversión de ciclo** *m* REFRIG, TERMO reverse-cycle defrosting; ~ **por lluvia** *m* REFRIG water defrosting; ~ **manual** *m* REFRIG manual defrost; ~ **natural cíclico** *m* REFRIG off-cycle defrosting; ~ **temporizado** *m* REFRIG timed defrosting

desescoriado *m* MINAS *hornos* flushing, PROD *fundería* skinning, *fundición* skimming

desescoriador *m* CONTAM, PROD *colada de alto horno*

skimmer; **~ por ciclón** *m* CONTAM cyclone recovery skimmer; **~ de cinta** *m* CONTAM belt skimmer; **~ de cinta absorbente** *m* CONTAM absorbent belt skimmer; **~ de cinta transportadora** *m* CONTAM conveyor-belt skimmer; **~ de disco** *m* CONTAM disc skimmer (*BrE*), disk skimmer (*AmE*); **~ de vertedero** *m* CONTAM weir skimmer

desescoriar *vt* CARBÓN deslag, CONTAM, PROD *alto horno* skim off

desespumante *m* ALIMENT, CARBÓN, EMB, P&C, PAPEL defoamer, PROC QUÍ foam breaker, QUÍMICA, TEC PETR defoamer

desetiquetado: ~ de la lata *m* EMB can delabeling (*AmE*), tin delabelling (*BrE*)

desexcitación *f* ELEC, FÍS *electroimán* de-energization, GEOFÍS de-excitation, ING ELÉC *electroimán* de-energization

desexcitar *vt* ING ELÉC, PROD de-energize

desfasado[1] *adj* ELEC *corriente alterna* out-of-phase, *corriente* dephased, ELECTRÓN phase-shifted, TELECOM *corriente alterna*, TV out-of-phase, VEH *motor* offset

desfasado[2] *m* ELEC *corriente alterna* phase displacement

desfasado[3]**: estar ~** *vi* ELEC, FÍS *electricidad* lag

desfasador *m* ELEC *corriente alterna*, ELECTRÓN, FÍS ONDAS, TELECOM phase shifter, phase changer; **~ de guía de ondas** *m* ELECTRÓN, FÍS ONDAS, ING ELÉC, NUCL, TELECOM waveguide phase shifter; **~ de guíaondas** *m* ING ELÉC, NUCL waveguide phase shifter; **~ de microondas** *m* ELECTRÓN, FÍS ONDAS, ING ELÉC, TELECOM microwave phase shifter; **~ múltiple** *m* ELEC *corriente alterna* phase splitter

desfasaje *m* ELEC *corriente alterna* phase angle, phase displacement, phase difference, phase shift, FÍS ONDAS phase difference, phase-out, TELECOM phase difference, TV phase shift

desfasamiento *m* ELEC *transformador* phase difference, *corriente alterna* phase displacement, FÍS ONDAS, TELECOM phase difference

desfasar *vt* ELEC, ELECTRÓN, FÍS ONDAS, ING ELÉC, TELECOM phase-shift

desfase *m* ELEC *corriente alterna* phase displacement, lag, phase shift, *transformador* phase difference, ELECTRÓN phase difference, lag, level shifting, FÍS ONDAS phase difference, ING ELÉC lag, PETROL phase difference; **~ de nivel** *m* NUCL level shift

desfibrado *m* PAPEL defibering (*AmE*), defibring (*BrE*), *madera* grinding, *pasta, papel* pulping

desfibrador *m* PAPEL *pasta papel, papelote* pulper; **~ de cadena** *m* PAPEL chain grinder; **~ de cadenas** *m* PAPEL caterpillar grinder; **~ de cajones** *m* PAPEL pocket grinder; **~ continuo** *m* PAPEL continuous grinder; **~ de muela** *m* PAPEL *pasta mecánica* grinder; **~ transversal** *m* PAPEL cross grinding

desfibradora *f* PAPEL *pasta mecánica* grinding machine, TEXTIL chafer

desfiguración *f* CARBÓN deformation

desfiladero *m* CONST, GEOL defile; **~ abisal** *m* *Esp* OCEAN abyssal pass

desfloculación *f* HIDROL, P&C, QUÍMICA deflocculation

desfloculador *m* HIDROL, P&C, QUÍMICA deflocculating agent

desfloculante *m* HIDROL, P&C, QUÍMICA deflocculant

desfondadora: ~ hidráulica *f* MINAS impact ripper; **~ de terrenos** *f* *AmL* (*cf escariador de terrenos Esp*) MINAS ripper

desfondamiento *m* MECÁ breaking-in

desfondeadora: ~ montada detrás *f* AUTO, TRANSP, VEH rear-mounted ripper

desforradora: ~ de cables *f* ELEC, ING ELÉC, ING MECÁ, PROD cable-sheath stripper

desforramiento *m* ELEC, ING ELÉC, ING MECÁ, PROD *de cables* stripping

desfunción *f* ING MECÁ malfunction

desgarrador *m* MINAS *tubería, pozos, petróleo* ripper

desgarramiento *m* ING MECÁ pulling

desgarro *m* CONST ripping, ING MECÁ gash, PAPEL tear, tearing; **~ de imagen** *m* TV picture break-up

desgarros: ~ de imagen claros *m pl* TV highlight tearing

desgasado *m* PROC QUÍ degassing, QUÍMICA outgassing

desgasar *vt* PROC QUÍ degas, QUÍMICA outgas

desgaseador *m* GEOL, MINAS *persona* getter

desgasificación *f* ELECTRÓN degasification, *de lámparas eléctricas* degassing, HIDROL, MINAS degasification, PROC QUÍ degasifying, degasification; **~ atmosférica** *f* HIDROL atmospheric degassing

desgasificador *m* PROC QUÍ, TERMOTEC degasser

desgasificar *vt* ELECTRÓN *lámparas eléctricas*, HIDROL degas

desgastado *adj* MECÁ worn, TEC PETR *perforación* dull, TEXTIL worn-out; **~ por las inclemencias** *adj* AGRIC weathered

desgastar *vt* CONST, GEOL, HIDROL erode, ING MECÁ wear down, wear off, METEO erode, TEXTIL wear out; **~ en caliente** *vt* MINAS *minerales, lingotes* break down; **~ por rozamiento** *vt* MECÁ abrade

desgastarse: ~ a la intemperie *v refl* CONST weather

desgaste *m* CARBÓN, CONST attrition, CONTAM MAR wear, GAS deterioration, HIDROL ablation, ING MECÁ attrition, wear, wearing, MECÁ float, fretting, wear, PAPEL wear, RECICL wastage, TRANSP MAR *velas* chafing; **~ por abrasión** *m* CONST, P&C *propiedad física, prueba* abrasive wear; **~ abrasivo** *m* P&C *propiedad física, prueba* abrasive wear; **~ por acción química** *m* PETROL chemical weathering; **~ de la barrena** *m* TEC PETR *perforación* bit wear; **~ del corte** *m* ING MECÁ wire erosion; **~ del filo de corte** *m* ING MECÁ wire erosion; **~ por frotamiento de ajuste** *m* MECÁ abrasion fretting corrosion; **~ de una herramienta** *m* ING MECÁ float; **~ mecánico** *m* CONST, PROD mechanical wear; **~ natural** *m* ING MECÁ wear and tear; **~ normal** *m* ING MECÁ wear and tear; **~ por ovalización** *m* AUTO *del cilindro* out-of-round wear; **~ químico** *m* NUCL wastage; **~ por rozamiento** *m* C&V attrition, MECÁ abrasion, fretting, NUCL fretting, P&C abrasion, PROD fretting; **~ del trépano** *m* TEC PETR *perforación* bit wear; **~ por el uso** *m* ING MECÁ, TEXTIL wear and tear

desgatado *adj* IMPR *tipo* blunt

desgausar *vt* ING MECÁ, PROD degauss

desglosador *m* INFORM&PD decollator

desglosar *vt* INFORM&PD decollate

desglose *m* CONST, TEXTIL *del mercado* breakdown

desgomado *m* ALIMENT, TEXTIL *de seda* degumming

desgomar *vt* ALIMENT, TEXTIL *seda* degum

desgotador *m* PAPEL drainer

desgotar *vt* PAPEL drain

desgote *m* PAPEL drainage

desgranadora *f* AGRIC sheller; **~ de maíz** *f* AGRIC corn sheller (*AmE*), maize sheller (*BrE*); ~ **de maíz a resorte** *f* AGRIC spring corn sheller (*AmE*), spring maize sheller (*BrE*)

desgrane *m* AGRIC shattering

desgrasador *m* GEN degreaser, scouring liquid

desgrasar *vt* GEN degrease, scour

desguace: ~ **de buques** *m* TRANSP MAR shipbreaking

desguarnecer *vt* TRANSP *contenedor* unstuff

desguarnir *vt* TRANSP MAR *mástil, abanico* unrig

desguazar *vt* PROD, TRANSP MAR *buques* scrap

deshabilitar *vt* ING ELÉC disable

deshechos *m pl* METAL debris

deshelador *m* AUTO, INSTAL TERM, REFRIG, TRANSP AÉR de-icer

deshelar[1] *vt* INSTAL TERM de-ice

deshelar[2] *vti* AUTO, REFRIG, TRANSP AÉR de-ice

deshelarse *v refl* METEO thaw

deshidratación *f* GEN *ingeniería de lodos* de-watering, *producción* dehydration; ~ **por aspersión** *f* DETERG spray-drying; ~ **por congelación** *f* ALIMENT, EMB, PROC QUÍ, REFRIG, TERMO freeze-drying; ~-**congelación** *f* REFRIG dehydrofreezing

deshidratado *adj* ALIMENT dehydrated, DETERG waterless; ~ **por aspersión** *adj* DETERG spray-dried; ~ **por congelación** *adj* ALIMENT, EMB, PROC QUÍ, REFRIG, TERMO freeze-dried

deshidratador *m* MINAS dewaterer, PROC QUÍ, REFRIG dehydrator; ~ **por congelación** *m* ALIMENT, PROC QUÍ, REFRIG freeze-dryer; ~ **con filtro** *m* REFRIG filter-drier

deshidratante *m* REFRIG desiccant

deshidratar *vti* AGRIC, GEOL dehydrate, INSTAL HIDRÁUL dewater, drain, dry up, QUÍMICA, REFRIG, TEC PETR dehydrate

deshidrogenación *f* ALIMENT, QUÍMICA dehydrogenation

deshidrogenado *adj* ALIMENT, QUÍMICA dehydrogenated

deshidrogenar *vt* ALIMENT, QUÍMICA dehydrogenate

deshidrogenasa *f* ALIMENT, DETERG, QUÍMICA dehydrogenase

deshidroluminosterol *m* QUÍMICA dehydroluminosterol

deshielo *m* HIDROL *ríos helados* breaking up, ING MECÁ defrosting, METEO thaw, TRANSP AÉR de-icing; ~ **del motor** *m* TRANSP AÉR engine de-icing; ~ **de la nieve** *m* AGUA snow melt (*AmE*), snow water (*BrE*); ~ **de la superficie aerodinámica** *m* TRANSP AÉR aerofoil de-icing (*BrE*), airfoil de-icing (*AmE*)

deshojado: ~ **en el suelo** *m* AGRIC in-place topping

deshojador *m* AGRIC leaf stripper

deshojadora: ~ **de maíz** *f* *Esp* (*cf deschaladora de maíz AmL*) AGRIC corn husker (*AmE*), maize husker (*BrE*)

deshollejar *vt* ALIMENT husk

deshollinar *vt* CONST *chimeneas* sweep

deshornadora *f* CARBÓN *horno de coque* ram pump

deshornar *vt* PROD draw the charge

deshulladora *f* CARBÓN, MINAS coal-mining machine

deshumidificación *f* INSTAL TERM, NUCL, REFRIG, SEG dehumidification

deshumidificador *m* INSTAL TERM, NUCL, REFRIG, SEG dehumidifier; ~ **de aire** *m* SEG air dehumidifier; ~ **de superficie** *m* REFRIG surface dehumidifier

deshumidificar *vt* INSTAL TERM, NUCL, REFRIG dehumidify

deshumificador *m* CONST moisture absorber

designación: ~ **de byte** *f* INFORM&PD, PROD, TELECOM byte designation; ~ **E/S** *f* PROD I/O designation; ~ **por láser** *f* ELECTRÓN laser designation; ~ **de objetivos por láser** *f* D&A laser designation system; ~ **de terminales** *f* PROD terminal designation

designador: ~ **de objetivo por láser** *m* D&A laser target designator (*LTD*); ~ **de pista** *m* TRANSP AÉR runway designator

designar *vt* INFORM&PD assign, TELECOM designate

desigualdad *f* INFORM&PD inequality

desiltor *m* AmL (*cf separador de limolita Esp*) TEC PETR *perforación* desilter

desimanación *f* GEN demagnetization

desimanador *m* GEN demagnetizer

desimanar *vt* GEN demagnetize

desimanarse *v refl* GEN demagnetize

desimantación *f* CARBÓN, ELEC, FÍS, ING MECÁ, MECÁ, TV demagnetization

desimantado *m* FÍS, INFORM&PD degaussing

desimantar *vt* CARBÓN, ELEC, FÍS demagnetize, INFORM&PD degauss, ING MECÁ, MECÁ, TV demagnetize

desimantarse *v refl* CARBÓN, ELEC, FÍS demagnetize, INFORM&PD degauss, ING MECÁ, MECÁ, TV demagnetize

desincronizado *adj* PROD gated off, TV out-of-sync

desincrustación *f* DETERG descaling, PROD fur, *calderas* descaling

desincrustador *m* MINAS scaling bar

desincrustante *m* PROD *calderas* disincrustant

desincrustar *vt* DETERG, PROD *calderas* descale

desincrustrador: ~ **de hielo** *m* TEC ESP, TRANSP AÉR de-icer

desinfección *f* AGUA, HIDROL, SEG disinfection

desinfectante *m* AGUA, HIDROL, SEG disinfectant

desinfectar *vt* AGUA, HIDROL, SEG disinfect

desinfestar *vt* SEG disinfest

desintegración *f* ELECTRÓN, FÍS RAD decay, MECÁ breaking up, P&C *física, química, procesamiento* breakdown, PAPEL *de la pasta, papel* slushing, TEC PETR *ingienería de lodos* decomposition; ~ **alfa** *f* FÍS, FÍS PART, FÍS RAD alpha decay; ~ **beta** *f* FÍS, FÍS PART, FÍS RAD beta decay; ~ **catalítica** *f* QUÍMICA *de aceite pesado* catalytic cracking, TEC PETR *refino* catalytic cracking, cracking; ~ **catódica** *f* ELECTRÓN, ING ELÉC cathode disintegration; ~ **espontánea** *f* FÍS RAD spontaneous decay; ~ **por estallido** *f* METAL crack branching; ~ **gamma** *f* FÍS, FÍS PART, FÍS RAD gamma decay; ~ **en polvo** *f* P&C *operación* grinding; ~ **radiactiva** *f* FÍS PART, FÍS RAD radioactive decay

desintegrador *m* ALIMENT disintegrator, PAPEL *pasta, papelote* pulper; ~ **de balas** *m* PAPEL bale pulper; ~ **de bolas** *m* AmL (*cf molino de bolas Esp*) LAB ball mill; ~ **catalítico** *m* TEC PETR *refino* cat cracker; ~ **discontinuo** *m* PAPEL batch pulper; ~ **de nudos** *m* PAPEL knot breaker; ~ **de la pila** *m* PAPEL beaterbreaker; ~ **planetario** *m* LAB *desintegración* planetary mill

desintegrar *vt* QUÍMICA disintegrate

desintegrarse *v refl* ELECTRÓN, FÍS decay, FÍS RAD decay, disintegrate

desintercalado *m* ELECTRÓN de-interleaving

desintercalar *vt* ELECTRÓN de-interleave

desintonización *f* ELECTRÓN, TELECOM, TV detuning

desintonizar *vt* ELECTRÓN, TELECOM, TV detune

desionizador *m* LAB *de agua* de-ionizer

deslastraje: ~ del agua de lastre *m* CONTAM, CONTAM MAR, TRANSP MAR *petroleros* deballasting water

deslastrar *vt* CONTAM, TRANSP MAR *cargamento, lastre* discharge

deslinde *m* CONST setting, setting-out

deslingotado *m* CARBÓN *lingotes*, PROD *fundería* stripping

deslingotar *vt* CARBÓN *lingotes*, PROD *fundería* strip

deslizadera *f* CARBÓN chute, CONST skid track, MECÁ, MINAS chute, TRANSP guideway; ~ en curso *f* TRANSP guideway at grade; ~ para frenado *f* TEC ESP brake chute

deslizador *m* IMPR slide; ~ de escape de emergencia *m* SEG, TRANSP AÉR emergency slide

deslizamiento *m* CONST slippage, CRISTAL glide, gliding, slip, ELEC *máquina* slip, GAS creep, ING ELÉC slip, ING MECÁ shear, MECÁ slide, METAL glide, P&C *defecto de la pintura* running, PROD *cilindros de laminador* slipping, QUÍMICA glide, TEC ESP *geología* drift, TEC PETR creep, TRANSP AÉR skidding; ~ de aterrizaje *m* TRANSP AÉR landing skid; ~ basal *m* METAL basal slip; ~ de la capa de revestimiento *m* REVEST coating slip; ~ de la cinta *m* TV tape slippage; ~ conjugado *m* METAL conjugate slip; ~ cruzado *m* CRISTAL cross slip; ~ cuasiconstante *m* NUCL quasiconstant slip; ~ discontinuo *m* METAL discontinuous glide; ~ Doppler *m* PETROL Doppler shift; ~ del embrague *m* AUTO, VEH clutch slip; ~ fácil *m* CRISTAL easy glide; ~ por gravedad *m* GEOL gravity slide; ~ gravitacional *m* GEOL gravity gliding; ~ de la grieta *m* GEOL break thrust; ~ del haz *m* METAL *deformación* pencil glide; ~ de la hélice *m* TRANSP MAR slip; ~ de la imagen *m* CINEMAT frame slip, picture slip, TV frame slip, picture drift; ~ lateral *m* FÍS sideslip; ~ de línea *m* TV line slip; ~ de la muestra obtenida de acero líquido *m* METAL *deformación* pencil glide; ~ del pantógrafo *m* FERRO, ING MECÁ pantograph slippage; ~ piramidal *m* METAL pyramidal slip; ~ de planos en una deformación *m* METAL pencil glide; ~ de polos *m* ELEC *máquina* pole slip; ~ prismático *m* CRISTAL prismatic slip; ~ del registro *m* CINEMAT, TV registration drift; ~ relativo *m* ELEC *máquina* relative slip; ~ de roca *m* GEOL rock slip; ~ de rocas *m* GEOL rockslide; ~ de tierras *m* FERRO *equipo inamovible*, GEOL *geomorfología* landslide; ~ de la tinta *m* IMPR ink creep; ~ transversal *m* CRISTAL, METAL cross slip

deslumbramiento *m* IMPR glare, SEG dazzle

deslustrado *adj* TEXTIL lusterless (*AmE*), lustreless (*BrE*)

deslustrar *vt* TEXTIL *acabado de telas* deluster (*AmE*), delustre (*BrE*)

deslustre *m* P&C dullness

desmagnetización *f* CARBÓN, ELEC demagnetization, FÍS degaussing, demagnetization, INFORM&PD degaussing, ING MECÁ, MECÁ demagnetization, TV degaussing, demagnetization; ~ adiabática *f* FÍS adiabatic demagnetization

desmagnetizado *m* FÍS, INFORM&PD, TV degaussing

desmagnetizador *m* FÍS, INFORM&PD degausser, ING MECÁ demagnetizer, TV degausser

desmagnetizante *m* CARBÓN, FÍS, INFORM&PD, ING MECÁ demagnetizer, TV degausser, demagnetizer

desmagnetizar *vt* CARBÓN, ELEC demagnetize, FÍS degauss, demagnetize, INFORM&PD degauss, ING MECÁ, MECÁ demagnetize, TV degauss, demagnetize

desmagnetizarse *v refl* CARBÓN, ELEC, FÍS, ING MECÁ, MECÁ, TV demagnetize

desmalezado: ~ por fuego *m* AGRIC flake

desmantelamiento *m* GEOL *de intrusión* deroofing, NUCL *de una instalación nuclear* dismantling

desmantelar *vt* TRANSP MAR *barco* lay up

desmembramiento *m* CONST breaking up

desmenuzable *adj* P&C friable

desmenuzadora *f* AGRIC chopper, shredder; ~ en seco *f* PAPEL shredder

desmenuzar *vt* CARBÓN mill

desmetilación *f* QUÍMICA demethylation

desmetilar *vt* QUÍMICA demethylate

desmina *f* MINERAL desmine

desmineralización *f* HIDROL, PROC QUÍ, TERMOTEC demineralization

desmineralizar *vt* HIDROL, PROC QUÍ, TERMOTEC demineralize

desmochadora *f* AGRIC topper

desmochar *vt* AGRIC dehorn

desmodulación *f* ELECTRÓN, FÍS, INFORM&PD, TELECOM, TV demodulation; ~ de fase *f* ELECTRÓN, TELECOM phase demodulation

desmodulador *m* ELECTRÓN, FÍS, INFORM&PD, TEC ESP, TELECOM, TV demodulator; ~ de extensión umbral *m* TELECOM threshold extension demodulator; ~ de fase *m* ELECTRÓN phase demodulator; ~ de filtro rastreador *m* TEC ESP tracking filter demodulator; ~ de umbral extendido *m* TEC ESP threshold extension demodulator; ~ de umbral mejorado *m* TEC ESP threshold extension demodulator

desmodular *vt* ELECTRÓN, FÍS, INFORM&PD, TEC ESP, TELECOM, TV demodulate

desmoldar *vt* C&V *fabricación del vidrio* break off

desmoldeado *m* C&V, P&C *procesamiento* demolding (*AmE*), demoulding (*BrE*)

desmoldeadora *f* PROD *fundición* stripping machine; ~ de palanca *f* PROD *fundición* lever draft machine (*AmE*), lever draught machine (*BrE*)

desmoldear *vt* PROD *fundería* strip, *metal en cable* draw

desmoldeo *m* C&V demold (*AmE*), demould (*BrE*), demolding (*AmE*), demoulding (*BrE*), ING MECÁ draft (*AmE*), draught (*BrE*), P&C demolding (*AmE*), demoulding (*BrE*), PROD *fundería* stripping, draught (*BrE*), draft (*AmE*)

desmontable *adj* ELEC removable, FOTO detachable, removable, INFORM&PD removable, ING MECÁ collapsible, detachable, removable, sectional, ÓPT removable, PROD made in sections, removable, collapsible, TRANSP AÉR removable

desmontado *m* ING MECÁ dismantling

desmontaje *m* D&A dismantlement, INFORM&PD disassembly, ING MECÁ disassembly, dismantling, dismount, removal, dismantlement, MINAS *del entibado* removal, NUCL *de un elemento de combustible* dismantling, PROD stripping, *máquinas* teardown; ~ de los contenedores *m* TRANSP container stripping; ~ de las varillas de la locomotora *m* FERRO *vehículos* removal of locomotive rods

desmontar[1] *vt* COLOR strip, CONST break, break down, D&A dismantle, IMPR strip, ING MECÁ disassemble, dismantle, dismount, remove, unclamp, unmake, MECÁ scrap, TELECOM *línea* recover

desmontar[2] *vi* ING MECÁ come apart

desmonte *m* Esp (*cf descalce AmL*) AGRIC clearing, CARBÓN *canteras, minas* stripping, CONST *movimiento de tierras* sidehill cut, GEOL dump, MINAS cut; **~ de la montera** *m* MINAS *yacimiento* stripping; **~ de terrenos** *m* AGRIC land clearing

desmoronamiento *m* CONST caving, OCEAN *litoral* nip, slumping, PETROL, TEC PETR caving

desmoronarse *v refl* CONST *muros, paredes, estructuras* crumble

desmotropía *f* QUÍMICA *equilibrio isomérico* desmotropy

demuestre *m* ALIMENT sampling, MINAS ore sampling

desmulsificante *m* ALIMENT de-emulsifier

desmultiplexado *m* ELECTRÓN, INFORM&PD, TELECOM demultiplexing

desmultiplexaje: ~ en el reparto del tiempo *m* ELECTRÓN time-division demultiplexing

desmultiplexar *vt* ELECTRÓN, TELECOM demultiplex

desmultiplicación *f* ING MECÁ gear

desmultiplicador: ~ de impulsos *m* ELECTRÓN scaler

desnatado *m Esp* (*cf desnate AmL*) ALIMENT, C&V, P&C creaming

desnatador *m* C&V, CONTAM *tratamiento de líquidos* skimmer; **~ de banda** *m* CONTAM belt skimmer; **~ de banda transportadora** *m* CONTAM conveyor-belt skimmer; **~ de cinta absorbente** *m* CONTAM absorbent belt skimmer; **~ de disco** *m* CONTAM disc skimmer (*BrE*), disk skimmer (*AmE*); **~ de vertedero** *m* CONTAM weir skimmer

desnatadora: ~ por ciclón *f* CONTAM *tratamiento de líquidos* cyclone recovery skimmer

desnatar *vt* CONTAM *tratamiento de líquidos* skim off

desnate *m AmL* (*cf desnatado Esp*) ALIMENT, C&V, P&C creaming

desnitrificación *f* CONTAM *tratamiento de aguas y gases*, HIDROL denitrification

desnivel *m* CARBÓN drop, CONST dislevelment, GEOL drop, ING MECÁ head, offset, INSTAL HIDRÁUL *diferencia de nivel* elevation head, fall; **~ de la carga hidráulica** *m* HIDROL fall in hydraulic head; **~ efectivo** *m* INSTAL HIDRÁUL effective head; **~ del terreno** *m* CONST fall of earth, fall of ground

desnivelado *adj* ING MECÁ off-center (*AmE*), off-centre (*BrE*)

desnudar *vt* TRANSP MAR *palos, vergas* unrig

desnutrido *adj* AGRIC underfed

desobediencia: ~ de las reglas de seguridad *f* SEG breach of the safety rules

desobrecalentador *m* TERMOTEC desuperheater

desobrecalentar *vt* TERMOTEC desuperheat

desocupado *adj* PROD idle

desocupar *vt* MECÁ, SEG evacuate

desodorización *f* HIDROL *aguas residuales* deodorizing

desoldado *m* ING MECÁ unsoldering

desorber *vt* CARBÓN, QUÍMICA desorb

desorción *f* CARBÓN, QUÍMICA desorption

desorden: en ~ *adj* ELEC out-of-order

desordenado *adj* ELEC *fallo, error* out-of-order

desorientación *f* METAL disorientation

desorientado *adj* GEOL *textura* unoriented

desoxidación *f* ING MECÁ derusting, PROD stripping, *metales* scouring, QUÍMICA deoxidization; **~ por baño ácido** *f Esp* (*cf decapado por baño ácido AmL*) NUCL corrosion pickling; **~ por frote** *f* PROD rubbing-off of rust

desoxidante *m* ING MECÁ rust remover, QUÍMICA deoxidizer

desoxidar *vt* PROD scale, *metales* scour, QUÍMICA bleach; **~ por frote** *vt* PROD rub the rust off

desoxigenación *f* MINAS *del aire* deoxidation

desoxigenar *vt* CARBÓN *alimentación de calderas* deactivate

despachador *m* INFORM&PD dispatcher; **~ de carga** *m* ELEC *red de distribución* load dispatcher

despachar *vt* PROD, TRANSP dispatch, TRANSP MAR *buque de aduanas* clear, *mercancías* ship

despacho *m* TELECOM *tráfico* routing, TRANSP MAR *aduanas* clearance; **~ de aduanas** *m* TRANSP MAR customs clearance; **~ de cargas** *m* NUCL load dispatch office; **~ del distribuidor de vagones** *m* FERRO car distributor office (*AmE*), wagon distributor office (*BrE*)

despaletizado *adj* EMB depalletization

despanojado *m* AGRIC detasseling (*AmE*), detasselling (*BrE*)

despanojar *vt* AGRIC detassel

despaquetificador *m* TELECOM depacketizer

desparramado: ~ de bronce fundido sobre las superficies *m* ING MECÁ *soldadura* wetting

despastaje *m* AGRIC *forestación* scalping

despatillador *m* PAPEL deflaker

despegamiento: ~ de molusco *m* OCEAN mollusc detaching

despegue *m* TEC ESP *naves espaciales, proyectiles balísticos* blast-off, liftoff, *motores* kickoff, TRANSP AÉR takeoff; **~ abortado** *m* TRANSP AÉR aborted takeoff; **~ asistido por cohete** *m* TRANSP AÉR jet-assisted takeoff (*AmE*) (*JATO*), rocket-assisted takeoff (*BrE*) (*RATO*); **~ y aterrizaje convencional** *m* D&A, TRANSP AÉR conventional takeoff and landing (*CTOL*); **~ de avión remolcado** *m* TRANSP AÉR aeroplane tow launch (*BrE*), airplane tow launch (*AmE*); **~ con ayuda de reactores** *m* TRANSP AÉR jet-assisted takeoff (*AmE*) (*JATO*), rocket-assisted takeoff (*BrE*) (*RATO*); **~ con cohetes auxiliares** *m* TRANSP AÉR jet-assisted takeoff (*AmE*) (*JATO*), rocket-assisted takeoff (*BrE*) (*RATO*); **~ hacia adelante** *m* TRANSP AÉR forward takeoff; **~ hacia atrás** *m* TRANSP AÉR backward takeoff, rearward takeoff; **~ interrumpido** *m* TRANSP AÉR aborted takeoff; **~ propulsado** *m* TRANSP AÉR power takeoff; **~ de salto** *m* TRANSP AÉR jump takeoff; **~ vertical** *m* CONST liftoff, D&A, TRANSP AÉR vertical takeoff

despejar *vt* CONST clear out, D&A unmask, INFORM&PD, MINAS *frente de avance* clear

despeje *m* ING MECÁ, MECÁ clearance; **~ lateral** *m* ING MECÁ side clearance

despellejado *adj* AGRIC, ALIMENT husked

despellejar *vt* AGRIC, ALIMENT husk, OCEAN *pescado* skin

despepitado *m* AGRIC ginning

despepitar *vt* AGRIC gin

desperdicio *m* ALIMENT, CONTAM, RECICL kitchen waste; **~ de aceite** *m* CONTAM oil waste; **~ combustible** *m* CONTAM combustible waste; **~ combustible semisólido** *m* CONTAM semisolid combustible waste; **~ mayor** *m* CONTAM macrowaste; **~ voluminoso** *m* CONTAM, RECICL bulky waste

desperdicios *m pl* AGUA, CARBÓN refuse, EMB garbage (*AmE*), refuse, rubbish (*BrE*), PROD garbage (*AmE*), refuse, rubbish (*BrE*), skim, waste, RECICL garbage (*AmE*), refuse, rubbish (*BrE*); **~ de algodón** *m pl* PROD, TEXTIL cotton waste; **~ de bajo nivel** *m pl*

CONTAM, RECICL low-level waste; ~ **de carbón de leña** *m pl* CARBÓN charcoal duff; ~ **de carbón vegetal** *m pl* CARBÓN charcoal duff; ~ **de colada** *m pl* PROD *fundería* returns; ~ **de criba** *m pl* QUÍMICA screening; ~ **del cribado** *m pl* CARBÓN riddlings, screenings, HIDROL screenings, PROD riddlings; ~ **domésticos** *m pl* CONTAM, RECICL household waste; ~ **indefinidos** *m pl* CONTAM, RECICL indeterminate waste; ~ **industriales** *m* AGUA, CONTAM, RECICL industrial waste; ~ **de molienda** *m pl* AGRIC milling waste; ~ **no determinados** *m pl* CONTAM, RECICL indeterminate waste; ~ **residuales** *m pl* CONTAM, RECICL sewage waste; ~ **sólidos** *m pl* CONTAM, RECICL solid waste

despezonado *m* AGRIC removal of stalk from fruit

despiece *m* MECÁ exploded view

despilaramiento *m* CARBÓN, MINAS pillar drawing, pillar working

despilfarro *m* PROD wastage

despimpollar *vt* AGRIC remove the side shoots of

despintar: ~ **con soplete** *vt* C&V, TERMO burn off

despinzado *m* TEXTIL *lana de tenería* plucking

desplazable *adj* MECÁ floating

desplazado *adj* ING MECÁ displaced, off-center (*AmE*), off-centre (*BrE*), NUCL *átomo* displaced

desplazamiento *m* ACÚST, AGUA, AUTO displacement, CONST offset, FÍS, GEOFÍS displacement, GEOL displacement, drift, landslip, shift, HIDROL drift, transfer, IMPR scrolling, INFORM&PD *direcciones* displacement, shift, ING MECÁ movement, displacement, drift, translation, travel, INSTAL HIDRÁUL *de la válvula de corredera* stroke, MECÁ motion, METAL sliding, MINAS offset, NUCL travel, OCEAN transfer, drift, PETROL drift, TEC ESP drift, transfer, TRANSP AÉR drift, TRANSP MAR *diseño naval* displacement, TV shift; ~ **por absorción de laguna reticular** *m* METAL vacancy-absorbing jog; ~ **angular** *m* ELEC *corriente alterna* angular displacement; ~ **aritmético** *m* INFORM&PD arithmetic shift; ~ **atómico** *m* METAL atomic displacement; ~ **axial** *m* TV axial displacement; ~ **base** *m* INFORM&PD base displacement; ~ **de una capa** *m* AmL (*cf falla Esp*) MINAS *geología* floor heave; ~ **en carga** *m* TRANSP MAR *de barco* loaded displacement, *diseño naval* load displacement; ~ **de la carga de combustible líquido** *m* TEC ESP *misiles* sloshing; ~ **cero** *m* C&V zero displacement; ~ **cíclico** *m* INFORM&PD cyclic shift; ~ **circular** *m* INFORM&PD circular shift; ~ **continuo** *m* ELECTRÓN follow range; ~ **de corta duración** *m* ELECTRÓN short-term drift; ~ **por cuadratura** *m* TV quadrature displacement; ~ **a la derecha** *m* INFORM&PD right shift; ~ **direccional** *m* Esp (*cf desplazamiento de rumbo AmL*) GEOL *geología estructural* strike slip; ~ **Doppler** *m* TEC ESP *vuelo espacial, astronomía* Doppler shift; ~ **eléctrico** *m* ELEC, FÍS, ING ELÉC electric displacement; ~ **de los electrones** *m* ELECTRÓN, NUCL electron drift; ~ **por emisión de lagunas reticulares** *m* METAL vacancy-emitting jog; ~ **de fase** *m* ELEC *corriente alterna* phase shift, phase difference, phase displacement, ELECTRÓN phase shift, PETROL phase difference; ~ **de frecuencia** *m* ELECTRÓN frequency hopping, frequency shift, INFORM&PD, TELECOM, TV frequency shift; ~ **en frecuencia** *m* ELECTRÓN, TELECOM frequency displacement; ~ **de frecuencia por efecto Doppler-Fizeau** *m* ACÚST Doppler-Fizeau displacement shift

~ **de frecuencia de la onda portadora** *m* TELECOM carrier-frequency offset; ~ **horizontal** *m* GEOFÍS, GEOL horizontal displacement, offset, MECÁ offset; ~ **de imagen** *m* TV picture shift; ~ **de inclinación** *m* GEOL dip slip; ~ **por inducción** *m* ELEC *corriente alterna* induction displacement; ~ **a la izquierda** *m* INFORM&PD left shift; ~ **de Lamb** *m* FÍS Lamb shift; ~ **lateral** *m* GEOL lateral shift; ~ **de letras** *m* INFORM&PD letters shift; ~ **leve** *m* MECÁ float; ~ **ligero** *m* TRANSP MAR light displacement; ~ **lógico** *m* INFORM&PD end-around shift, logical shift; ~ **de nivel** *m* NUCL level displacement; ~ **por pantalla** *m* INFORM&PD scrolling; ~ **pequeño** *m* ING MECÁ float; ~ **permanente del umbral** *m* ACÚST permanent threshold shift; ~ **pesado** *m* TRANSP MAR *de casco* heavy displacement; ~ **de petróleo por gas** *m* GAS gas drive; ~ **de la punta de la grieta** *m* NUCL crack opening displacement; ~ **en rosca** *m* TRANSP MAR *buque vacío* light displacement; ~ **rotativo** *m* HIDROL spindrift; ~ **sobre ruedas en la pista** *m* TEC ESP *aviones* roll; ~ **de rumbo** *m* AmL (*cf desplazamiento direccional Esp*) GEOL *geología estructural* strike slip; ~ **por sacudidas** *m* METAL jog; ~ **de la subportadora** *m* TV subcarrier offset; ~ **temporal del umbral** *m* ACÚST temporary threshold shift; ~ **térmico** *m* TERMO heat displacement; ~ **de trazado** *m* TRANSP MAR *diseño naval* molded displacement (*AmE*), moulded displacement (*BrE*); ~ **de válvula** *m* INSTAL HIDRÁUL valve motion

desplazamientos: ~ **geométricos** *m pl* GEOM, ING MECÁ geometric displacements

desplazar *vt* INFORM&PD push, shift, ING ELÉC displace, PROD, TEXTIL shift, TRANSP remove, transport; ~ **por inercia hasta parar** *vt* PROD coast-to-stop; ~ **por la pantalla** *vt* INFORM&PD scroll

desplazarse *v refl* ING MECÁ run, MECÁ slide, PROD shift

desplegable *adj* TEC ESP, TELECOM *antena* unfurlable

desplegador: ~ **acabador** *m* TEXTIL finisher scutcher

desplegar *vt* INFORM&PD pop

desplomado *adj* CONST off plumb, out-of-plumb, GEOL, MINAS caved

desplomarse *v refl* GEOL collapse

desplome *m* CONST *arquitectura* batter, FERRO *equipo inamovible* subsidence, GEOL collapse; ~ **que obstruye una galería** *m* MINAS horse; ~ **del terreno** *m* MINAS *galerías* run of ground

despojos *m pl* AGUA refuse, CARBÓN refuse, debris, EMB, PROD, RECICL garbage (*AmE*), refuse, rubbish (*BrE*)

despolarización *f* FÍS, ING ELÉC, QUÍMICA depolarization

despolarizador *m* FÍS, ING ELÉC, QUÍMICA depolarizer

despolarizar *vt* FÍS, ING ELÉC, QUÍMICA depolarize

despolimerización *f* P&C, QUÍMICA depolymerization

despolvoreo *m* PROD dust removal

despostillada *f* C&V chip

desprender[1] *vt* CONTAM *procesos industriales* discharge, FÍS give off, *energía* release, ING MECÁ throw off, disengage, MECÁ disengage, QUÍMICA unfasten, *gas* evolve, TERMO give off, *vapor* emit

desprender[2]: ~ **el modelo** *vi* PROD *fundería* rap; ~ **rayos** *vi* GEN radiate

desprenderse *v refl* CONST break, ING MECÁ come away, come off, part

desprendible *adj* TEC ESP jettisonable

desprendimiento *m* CARBÓN *de gases* discharge, CONST *de piedra* spalling, landslide, *de piedras* earth fall, ING MECÁ detachment; ~ **cristalográfico** *m* CRISTAL, METAL crystallographic slip; ~ **de gas** *m* GAS, PROC QUÍ, QUÍMICA gas leaking; ~ **por gravedad** *m* GEOL gravity slide; ~ **del óxido** *m* TV oxide shedding; ~ **de partículas de tinta ya impresa** *m* IMPR chalking; ~ **de polvillo** *m* PAPEL dusting; ~ **del revestimiento** *m* MINAS *pared* walling scaffold; ~ **de tierras** *m* GEOL landslip, MINAS run of ground; ~ **de vapor** *m* CONTAM steam emission; ~ **violento de vapor** *m* CARBÓN scalping; ~ **de virutas** *m* ING MECÁ chip removal

despresionización *f* PROD depressurization

despresionizar *vt* PROD depressurize

despresurización *f* GEN depressurization

despresurizar *vt* GEN depressurize

desproporción *f* IMPR imbalance

despumación *f* PROD skimming

despumar *vt* PROD skim, skim off

despuntado: ~ **plástico** *m* METAL plastic blunting

despurinación *f* QUÍMICA *reacción* depurination

desrebabar *vt* PROD trim, trim off the burr from, trim off the rough edges; ~ **en el tonel** *vt* PROD *fundería* rattle

destacarse *v refl* TRANSP MAR *buque* swing away

destajo: **a** ~ *adv* PROD on contract

destalonado *m* ING MECÁ backing off, clearance, relief, MECÁ *herramientas* relief

destalonar *vt* ING MECÁ, INSTAL HIDRÁUL, MECÁ, PROD relieve

destape *m* MINAS baring

destarcador *m* *Esp* (*cf polvorero AmL*) MINAS *persona* picker

destellar *vi* TERMO blaze up, flare up

destello *m* TRANSP AÉR flare; ~ **solar** *m* GEOFÍS solar flare

destemplar *vt* MECÁ, METAL anneal

desteñible: **no** ~ *adj* COLOR, QUÍMICA, TEXTIL fast

desteñido: ~ **por gases** *m* COLOR gas fading

desteñir *vi* COLOR bleed off, fade, TEXTIL fade

desteñirse *v refl* PROD *colores en telas* run

destilación *f* CONST filtration, QUÍMICA *producto*, TEC PETR *refino*, TERMO distillation; ~ **del agua salada** *f* AGUA saline water conversion; ~ **por arrastre de vapor** *f* PROC QUÍ steam distillation; ~ **por ascensión** *f* PROC QUÍ distillation by ascent; ~ **azeotrópica** *f* ALIMENT azeotropic distillation; ~ **por bajada de temperatura** *f* TERMO distillation by descent; ~ **por caída** *f* PROC QUÍ distillation by descent; ~ **por caída de temperatura** *f* TERMO distillation by descent; ~ **en corriente de vapor** *f* AGUA, ALIMENT, QUÍMICA, TERMO flash distillation; ~ **por elevación de temperatura** *f* TERMO distillation by ascent; ~ **de equilibrio** *f* AGUA, ALIMENT, QUÍMICA, TERMO flash distillation; ~ **por etapas** *f* NUCL batch distillation; ~ **por expansión brusca** *f* AGUA, ALIMENT, QUÍMICA, TERMO flash distillation; ~ **extractiva** *f* ALIMENT extractive distillation; ~ **fraccionada** *f* PROC QUÍ, TEC PETR *refino* fractional distillation; ~ **instantánea** *f* AGUA, ALIMENT, QUÍMICA, TERMO flash distillation; ~ **por lotes** *f* NUCL batch distillation; ~ **de muestras** *f* TEC PETR *refino* fractional distillation; ~ **seca** *f* PROC QUÍ degasifying, degassing, QUÍMICA dry distillation; ~ **solar** *f* ENERG RENOV solar distillation; ~ **por subida de temperatura** *f* TERMO distillation by ascent; ~ **al vacío** *f* PROC QUÍ,

QUÍMICA vacuum distillation; ~ **en vacío** *f* ALIMENT, FÍS vacuum distillation; ~ **al vapor** *f* QUÍMICA steam distillation; ~ **con vapor** *f* QUÍMICA steam distillation

destilado[1] *adj* HIDROL, PROC QUÍ, QUÍMICA, TERMO, VEH distilled

destilado[2] *m* PROC QUÍ distillate, distillation, QUÍMICA, TEC PETR, TERMO distillate; ~ **medio** *m* QUÍMICA, TEC PETR *refino* middle distillate

destilador *m* PROC QUÍ, QUÍMICA distiller

destilados: ~ **ligeros** *m pl* QUÍMICA, TEC PETR *refino* light distillates; ~ **medios** *m pl* TEC PETR *refino* medium distillates

destilar *vt* HIDROL, PROC QUÍ, QUÍMICA, TERMO, VEH distil (*BrE*), distil off (*BrE*), distill (*AmE*), distill off (*AmE*); ~ **repetidas veces** *vt* QUÍMICA cohobate; ~ **al vapor** *vt* QUÍMICA steam distil (*BrE*), steam distill (*AmE*)

destinado: **estar** ~ **en un buque** *vi* TRANSP MAR be posted to a ship; **ser** ~ **a un buque** *vi* TRANSP MAR be posted to a ship

destinar *vt* INFORM&PD assign

destinatario *m* INFORM&PD addressee

destino: **al** ~ *adv* ING MECÁ home

destintado *m* PAPEL *papelote* de-inking

destornillado *m* ING MECÁ unbolting

destornillador: ~ **de afinación** *m* ING MECÁ tuning screwdriver; ~ **de berbiquí** *m* ING MECÁ screwdriver bit; ~ **de comprobación de corriente** *m* ING MECÁ voltage tester screwdriver; ~ **eléctrico de comprobación de fuerza** *m* ING MECÁ voltage tester screwdriver; ~ **eléctrico a pilas** *m* ING MECÁ cordless power screwdriver; ~ **para ensayos de tensión** *m* ING MECÁ voltage tester screwdriver; ~ **de golpe** *m* ING MECÁ impact screwdriver; ~ **de impacto** *m* ING MECÁ impact screwdriver; ~ **de par prefijado** *m* ING MECÁ torque screwdriver; ~ **probador de tensión** *m* ING MECÁ voltage tester screwdriver; ~ **de reglaje** *m* ING MECÁ tuning screwdriver; ~ **de sintonización** *m* ING MECÁ tuning screwdriver; ~ **de tirafondos** *m* FERRO sleeper screwdriver (*BrE*), tie screwdriver (*AmE*); ~ **para tornillos de cabeza perdida** *m* ING MECÁ, PROD flat-head screwdriver; ~ **torsiométrico** *m* ING MECÁ torque screwdriver; ~ **de traviesas** *m* FERRO railroad tie screwdriver (*AmE*), railway sleeper screwdriver (*BrE*), sleeper screwdriver (*BrE*), tie screwdriver (*AmE*); ~ **para tuercas** *m* ING MECÁ nut spinner

destornillar *vt* NUCL unscrew

destroza *f* *Esp* (*cf banco AmL*) MINAS *galería* stope; ~ **de cabeza** *f* *Esp* (*cf rebaje de cabeza AmL*) MINAS overhand stope

destrozar *vt* MINAS *mineral* stope

destrucción *f* FÍS RAD destruction; ~ **algácea** *f* HIDROL algal destruction; ~ **de la basura** *f* RECICL waste disposal

destructor *m* D&A destroyer, TERMO destructor, TRANSP MAR destroyer

destruir *vt* CONST batter; ~ **por fuego** *vt* TERMO destroy by fire

desulfonación *f* QUÍMICA desulfonation (*AmE*), desulphonation (*BrE*)

desulfuración *f* CARBÓN *aceros, gases de combustión*, CONTAM *depuración-purificación*, PROD, QUÍMICA, TEC PETR *producción, refino* desulfurization (*AmE*), desulphurization (*BrE*)

desulfuramiento *m* CARBÓN, CONTAM, PROD, QUÍMICA,

TEC PETR *producción, refino* desulfurization (*AmE*), desulphurization (*BrE*)

desulfurar *vt* CARBÓN, CONTAM, PROD, QUÍMICA, TEC PETR desulfurize (*AmE*), desulphurize (*BrE*)

desulfurización *f* CARBÓN, CONTAM, PROD, TEC PETR *producción, refino* desulfurization (*AmE*), desulphurization (*BrE*)

desunión *f* TEC ESP separation

desunir *vt* ING MECÁ, MECÁ disengage

desvainado *adj* AGRIC, ALIMENT husked, shelled

desvalvar *vt* OCEAN *moluscos valvados* shuck

desvanecer *vt* CINEMAT fade, PROC QUÍ dissolve out, TV fade

desvanecimiento *m* CINEMAT fade, fading, FOTO fade, IMPR *color, imagen* fading, TV fade; **~ a blanco** *m* CINEMAT fade-to-white, FOTO fade, TV fade-to-white; **~ cruzado** *m* CINEMAT cross fade; **~ gradual** *m* CINEMAT lap dissolve; **~ de la imagen** *m* CINEMAT picture-fading; **~ intenso** *m* ELECTRÓN *señales* deep fading; **~ por iris** *m* CINEMAT iris fade; **~ por multitrayectoria** *m* TELECOM multipath fading; **~ a negro** *m* CINEMAT, TV fade-to-black; **~ de la película** *m* CINEMAT film fading; **~ polaroid** *m* CINEMAT Polaroid fade; **~ de Rayleigh** *m* TELECOM Rayleigh fading; **~ selectivo** *m* TELECOM selective fading; **~ de la señal** *m* TEC ESP, TELECOM fading; **~ por trayectoria de propagación múltiple** *m* TELECOM multipath fading

desvarar *vt* TRANSP MAR *barco* refloat, *barco perdido* float off

desvastado *adj* ELEC *corriente alterna* wattless

desvatiado *adj* ELEC *corriente alterna* wattless

desviación *f* AGUA diversion, diverting, CARBÓN bypass, CONST baffle, bypass, deflection, ELEC *circuito* bypass, ELECTRÓN deflection, FERRO switching, FÍS deviation, GAS deviation, bypass, GEOFÍS deflection, INFORM&PD deviation, deflection, bias, ING ELÉC switch, bend, ING MECÁ off-setting, bypass, deflection, deviation, INSTAL HIDRÁUL bypass, MATEMÁT deviation, MECÁ offset, deflection, MINAS *pozo* offset, NUCL shift, deflection, PETROL drift, PROD deviation, off-setting, bypass, TEC ESP deviation, *navegación* drift, *de naves espaciales* deflection, TEC PETR *perforación* bypass, sidetracking, TELECOM deviation, TRANSP *de tráfico* diversion, bypass, TRANSP AÉR drift, TV deviation; **~ E** *f* FÍS ONDAS, ING ELÉC *microondas* E bend; **~ angular** *f* ING ELÉC angular deviation; **~ característica** *f* ELEC *término básico*, MATEMÁT root-mean-square deviation (*RMS deviation*), PROD standard deviation; **~ de condiciones de criticidad** *f* NUCL *de un reactor nuclear* deviation from criticality; **~ del consumo** *f* PROD consumption deviation; **~ electromagnética** *f* ING ELÉC electromagnetic deflection; **~ estándar** *f* INFORM&PD, MATEMÁT standard deviation; **~ de fase** *f* NUCL phase shift; **~ de frecuencia** *f* ELEC *de onda* frequency deviation, ELECTRÓN frequency departure, frequency deviation; **~ de frecuencia eficaz** *f* TEC ESP root-mean-square frequency deviation (*RMS frequency deviation*); **~ de frecuencia fraccionaria** *f* ELECTRÓN fractional-frequency deviation; **~ de frecuencia máxima** *f* TEC ESP, TELECOM peak frequency deviation; **~ horizontal** *f* ELECTRÓN horizontal deflection, TV x-deflection; **~ del izador** *f* TRANSP AÉR elevator deflection; **~ lateral** *f* ING MECÁ drift; **~ de la línea base** *f* PETROL baseline shift;

~ magnética *f* FÍS PART magnetic deflection, GEOFÍS magnetic deflection, magnetic deviation, NUCL magnetic deflection; **~ máxima admisible** *f* ING MECÁ maximum permissible deviation; **~ media** *f* ELEC *término general*, MATEMÁT average deviation, mean deviation; **~ media absoluta** *f* PROD mean absolute deviation; **~ media cuadrática** *f* ELEC, MATEMÁT *término básico* root-mean-square deviation (*RMS deviation*); **~ mínima** *f* FÍS minimum deviation; **~ normal** *f* ELEC, MATEMÁT *término básico* root-mean-square deviation (*RMS deviation*); **~ normalizada** *f* ELECTRÓN standard deviation, MATEMÁT standard division; **~ ortogonal** *f* GEOL offset; **~ en el plano E** *f* ING ELÉC E-plane bend; **~ del plazo de entrega** *f* PROD lead time deviation; **~ promedio** *f* ELEC average deviation; **~ radial** *f* NUCL radial shift; **~ selectiva** *f* TRANSP *de tráfico* selective diversion; **~ señalizada** *f* TRANSP advisory diversion; **~ sistemática** *f* CALIDAD systematic variation; **~ stándard** *f* FÍS standard deviation; **~ térmica** *f* TERMO heat shunt; **~ típica** *f* ELEC *término básico* root-mean-square deviation (*RMS deviation*), INFORM&PD standard deviation, typical deviation, MATEMÁT root-mean-square deviation (*RMS deviation*), standard deviation, PROD *estadística* standard deviation; **~ total** *f* TRANSP complete diversion; **~ de la vertical** *f* MINAS *sondeos* drift; **~ vertical** *f* ELECTRÓN vertical deflection, TV y-deflection

desviado *adj* FÍS bypass, shunt, ING MECÁ off-center (*AmE*), off-centre (*BrE*), offset, out-of-true; **~ negativamente** *adj* GEOM negatively skewed

desviador *m* PROC QUÍ deflector, TEC PETR *perforación* diverter, TRANSP AÉR deflector; **~ de flujo** *m* TEC PETR *perforación* diverter; **~ de llamadas** *m* TELECOM call diverter

desviar *vt* AGUA deflect, CONST divert, FÍS RAD deflect, ING MECÁ deflect, throw off center (*AmE*), throw off centre (*BrE*), PROC QUÍ, TV deflect

desviejar *vt* AGRIC *producción animal* cull

desvío *m* Esp (*cf reencaminamiento AmL*) AGUA diverting, CARBÓN bypass, CONST bypass, turnout, ELEC *circuito* bypass, FERRO siding, FÍS yaw, shunt, yawing, GAS, ING MECÁ, INSTAL HIDRÁUL bypass, PETROL bypassing, PROD bypass, TEC ESP shift, TEC PETR bypass, TELECOM *del tráfico* rerouting, TRANSP bypass, TRANSP MAR *de la brújula* deviation, compass error; **~ auxiliar** *m* PETROL loop; **~ de llamadas** *m* TELECOM call diversion; **~ provisional** *m* TRANSP bypass road; **~ suplementario** *m* PETROL loop; **~ tangencial de la pala** *m* TRANSP AÉR blade sweep

desviscosificador *m* TEC PETR *producción* visbreaker

desvitrificación *f* C&V, GEOL devitrification

desvitrificar *vt* C&V, GEOL devitrify

deswatado *adj* ELEC *corriente alterna* wattless

desyerbador: **~-acolchador** *m* AGRIC weeder-mulcher

detallar *vt* TEXTIL itemize

detalle: **~ del ángulo** *m* TV corner detail; **~ de la puntada** *m* TEXTIL stitch detail; **~ de las sombras** *m* FOTO shadow detail

detallista *mf* PROD retailer

detección *f* ACÚST, ELECTRÓN, FÍS RAD detection, INFORM&PD sensing, TEC ESP *comunicaciones* detection; **~ de averías** *f* ELEC, GAS, INFORM&PD, TELECOM fault detection; **~ de capacitancia** *f* ÓPT capacitance sensing; **~ coherente** *f* ELECTRÓN coherent detection; **~ de colisiones** *f* INFORM&PD, TELECOM collision detection (*CD*); ~ **y corrección de errores** *f* INFORM&PD error report; ~ **de corriente** *f* ING ELÉC current sensing; ~ **a distancia** *f* CONTAM MAR, GEOL, INFORM&PD, TEC ESP, TELECOM, TV remote sensing; ~ **por eco** *f* OCEAN echo detection; ~ **de electrones** *f* ELECTRÓN, FÍS PART electron detection; ~ **de errores** *f* ELECTRÓN, INFORM&PD, TELECOM, TV error detection; ~ **de escapes** *f* EMB leakage detection, GAS leak detection; ~ **de fallas** *f AmL* (*cf detección de fallo Esp*) ELEC, GAS fault detection, INFORM&PD error detection, TELECOM fault detection; ~ **de fallo** *f Esp* (*cf detección de fallas AmL*) ELEC, GAS, TELECOM fault detection; ~ **de fugas** *f* GAS leak detection; ~ **de fugas por medio de helio** *f* NUCL helium leak detection; ~ **de fugas por recogida y análisis de los gases de escape** *f* NUCL sipping; ~ **del grupo de señales único** *f* TEC ESP unique word detection; ~ **de hadrones** *f* FÍS PART hadron detection; ~ **heterodina** *f* TELECOM heterodyne detection; ~ **lineal** *f* ELECTRÓN linear detection; ~ **por manipulación** *f* TELECOM manipulation detection; ~ **de marcas** *f* INFORM&PD mark detection; ~ **mediante lápiz fotosensible** *f* INFORM&PD light pen detection; ~ **de objetos por el radar** *f* TRANSP, TRANSP AÉR, TRANSP MAR radar detection; ~ **de la onda portadora** *f* ELECTRÓN, TELECOM carrier detection; ~ **óptica** *f* ELECTRÓN, ÓPT, TELECOM optical detection; ~ **de pérdidas** *f* GAS leak detection; ~ **de portadora** *f* ELECTRÓN, TELECOM carrier detection; ~ **de posición de rotación** *f* INFORM&PD rotation position sensing; ~ **radárica** *f* TRANSP, TRANSP AÉR, TRANSP MAR radar detection; ~ **de radiación** *f* FÍS RAD, TEC ESP radiation monitoring; ~ **de referencias** *f* INFORM&PD mark sensing; ~ **remota** *f* CONTAM MAR, GEOL, INFORM&PD, TEC ESP, TELECOM, TV remote sensing; ~ **de señales** *f* TEC ESP demodulation, TELECOM signal detection; ~ **síncrona** *f* ELECTRÓN synchronous detection; ~ **sobrecarrera** *f* PROD overtravel detection; ~ **de tierra** *f* PROD ground sensing; ~ **vocal** *f* ELECTRÓN, TEC ESP, TELECOM speech detection

detectividad *f* ÓPT, TELECOM detectivity; ~ **específica** *f* ÓPT, TELECOM specific detectivity; ~ **normalizada** *f* ÓPT, TELECOM normalized detectivity

detector *m* ELEC *desmodulación* detector, ELECTRÓN sensor, *desmodulación* detector, FÍS detector, INFORM&PD sensor, ING ELÉC barretter, detector, ING MECÁ detector, INSTR monitor, ÓPT detector, TELECOM sensor, *desmodulación* detector, TRANSP MAR *meteorología* sensor; ~ **de accidentes** *m* SEG, TRANSP accident detector; ~ **de aceleración** *m* TRANSP AÉR acceleration detector; ~ **activo** *m* TEC ESP active sensor; ~ **de análisis del tráfico** *m* TRANSP traffic analysis detector; ~ **bidireccional** *m* ING ELÉC bidirectional coupler; ~ **del bucle de inducción** *m* TRANSP induction loop detector; ~ **de bucle magnético** *m* TRANSP magnetic loop detector; ~ **de burbujas** *m* OCEAN *buceo* bubble detector; ~ **de cable** *m* ELEC, ING ELÉC *instrumento*, INSTR cable detector; ~ **de captura electrónica** *m* CONTAM *análisis técnicas* electron-capture detector; ~ **de características de vehículos** *m* TRANSP vehicle characteristic detector; ~ **de centelleo** *m* FÍS RAD scintillation counter; ~ **Cerenkov** *m* FÍS PART, FÍS RAD *partículas subatómicas* Cerenkov detector; ~ **de clasificación** *m* TRANSP classification detector; ~ **climático** *m* TRANSP climatic detector; ~ **cons-**

tante de ondas de radar *m* FÍS ONDAS, TRANSP, TRANSP MAR continuous-wave radar detector; ~ **de contacto** *m* TRANSP contact detector; ~ **de contador** *m* PROD set counter sensor; ~ **cuadrático** *m* ING ELÉC square-law detector; ~ **de datos de la banda de frecuencias telefónicas** *m* TELECOM voice-band data detector; ~ **de datos de la banda de frecuencias vocales** *m* TELECOM voice-band data detector; ~ **de defectos** *m* PAPEL void detector; ~ **de defectos de puesta a masa** *m* ELEC, ING ELÉC earth detector (*BrE*), ground detector (*AmE*); ~ **DELPHI** *m* FÍS PART DELPHI detector; ~ **de destellos** *m* FÍS RAD scintillation counter; ~ **de desviación** *m* TRANSP AÉR deviation detector; ~ **direccional** *m* TRANSP directional detector; ~ **de doble hoja** *m* IMPR double sheet detector; ~ **de enganche de fase** *m* TEC ESP phase-locked demodulator; ~ **de errores** *m* ELECTRÓN, INFORM&PD, TELECOM error detector; ~ **de estado sólido** *m* FÍS RAD solid-state detector; ~ **de fallas** *m* ELEC *cable* fault detector; ~ **de fallo en la portadora de datos** *m* TELECOM data carrier failure detector; ~ **de fallos por sensación artificial** *m* TRANSP AÉR artificial feel failure detector; ~ **de la falta de cápsula** *m* EMB *botellas* missing cap detector; ~ **de fase** *m* ELECTRÓN, TELECOM, TV phase detector; ~ **de filtrado y muestreo** *m* TELECOM filter-and-sample detector; ~ **fotoeléctrico** *m* TRANSP *control de tráfico* photoelectric detector; ~ **fotoeléctrico de haz reflejado** *m* FÍS, ÓPT, TRANSP reflected-beam photoelectric detector; ~ **de fotones** *m* FÍS PART photon detector; ~ **de frecuencia** *m* ELECTRÓN frequency detector; ~ **de las frecuencias vocales** *m* ELECTRÓN, TELECOM speech detector; ~ **de fuga de gas** *m* GAS, SEG, TERMO gas-leak detector; ~ **de fuga a tierra** *m* ELEC, ING ELÉC earth-leakage detector (*BrE*), ground-leakage detector (*AmE*); ~ **de fugas** *m* EMB, INSTAL TERM, LAB, SEG leak detector; ~ **de fugas de las botellas** *m* EMB bottle-leak detector; ~ **de gases** *m* D&A, GAS, INSTR, LAB, MINAS gas detector; ~ **de Geiger** *m* FÍS, FÍS PART, FÍS RAD, NUCL Geiger tube; ~ **de Geiger-Müller** *m* FÍS RAD Geiger-Müller tube; ~ **de grietas** *m* ING MECÁ, NUCL crack detector; ~ **hidráulico** *m* TRANSP hydraulic detector; ~ **de hielo** *m* TRANSP AÉR ice detector; ~ **de humos** *m* SEG smoke detector; ~ **de incendios** *m* SEG fire detector; ~ **de infrarroja activo** *m* ELECTRÓN, FÍS, TRANSP active infrared detector; ~ **de infrarrojos** *m* ELECTRÓN, FÍS, TRANSP infrared detector; ~ **de infrarrojos activos** *m* ELECTRÓN, FÍS, TRANSP active infrared detector; ~ **de ionización** *m* FÍS RAD ionization detector; ~ **Jodel** *m* NUCL Jodel detector; ~ **LEP** *m* FÍS PART LEP detector; ~ **lineal** *m* ELECTRÓN linear detector; ~ **de llamas** *m* INSTAL TERM flame detector; ~ **lógico** *m* INFORM&PD logical sensor; ~ **luminoso** *m* TELECOM light detector; ~ **de luz** *m* PROD light sensor; ~ **magnético** *m* TRANSP magnetic detector; ~ **magnético para grietas** *m* ING MECÁ magnetic crack detector; ~ **de metales** *m* CONST, EMB metal detector; ~ **de metano** *m Esp* (*cf grisuscopio AmL*) GAS, MINAS gas detector, methane detector; ~ **de minas** *m* D&A mine detector; ~ **de movimiento dinámico** *m* TRANSP dynamic movement detector; ~ **neumático** *m* TRANSP *control de tráfico* pneumatic detector; ~ **de neutrones de resonancia** *m* ELECTRÓN, FÍS RAD resonance neutron detector; ~ **de nieve** *m* TRANSP snow detector; ~ **del núcleo de inducción del** circuito resonante *m* TRANSP *tráfico* resonant-circuit induction loop detector; ~ **de núcleo de inducción de desfase** *m* TRANSP *control de tráfico* phase-displacement induction-loop detector; ~ **del objetivo** *m* TRANSP *tráfico* objective detector; ~ **de ocupación** *m* TRANSP *tráfico* occupancy detector; ~ **OPAL** *m* FÍS PART OPAL detector; ~ **óptico** *m* ELECTRÓN, ÓPT, TELECOM optical detector; ~ **óptico de exceso de velocidad** *m* TRANSP optical speed trap detector; ~ **de la palabra** *m* TELECOM speech detector; ~ **de partículas neutrales** *m* NUCL neutral particle detector; ~ **de partículas neutras** *m* NUCL neutral particle detector; ~ **pasivo de infrarrojos** *m* FÍS, GAS, TRANSP passive infrared detector; ~ **de peces** *m* OCEAN fish detector; ~ **de picos a diodos** *m* ELECTRÓN diode peak detector; ~ **piezoeléctrico** *m* TRANSP *tráfico* piezoelectric detector; ~ **piezo-sensible** *m* TRANSP *tráfico* pressure-sensitive detector; ~ **de porcentaje de aumento** *m* TERMO rate-of-rise detector; ~ **de portadora de datos** *m* TELECOM data carrier detector (*DCD*); ~ **de posición** *m* TEC ESP attitude sensor; ~ **de presencia constante** *m* TRANSP continuous-presence detector; ~ **de presencia dinámica** *m* TRANSP dynamic presence detector; ~ **de presencia de intrusos** *m* TELECOM intruder presence detector; ~ **de presencia limitada** *m* TRANSP limited-presence detector; ~ **de presión** *m* GAS pressure detector; ~ **principal** *m* ELECTRÓN first detector; ~ **de puntos delgados** *m* C&V thin spot detector; ~ **de radar** *m* D&A, FÍS RAD, TRANSP radar sensor, TRANSPAÉR observation, radar sensor, TRANSP MAR radar sensor; ~ **radárico de impulsos** *m* TRANSP *control de tráfico* pulsed radar detector; ~ **de radiación** *m* FÍS RAD, LAB *instrumento* radiation detector; ~ **de radiaciones de transición** *m* FÍS RAD transition radiation detector; ~ **de rayos X** *m* INSTR X-ray gate; ~ **de recalentamiento de la caja de grasa** *m* FERRO *equipo inamovible* hot box detector; ~ **de retenciones** *m* TRANSP queue detector; ~ **de rotura de la banda** *m* IMPR web break detector; ~ **de rotura del filamento** *m* C&V strand break detector; ~ **selectivo de vehículos** *m* TRANSP selective vehicle detector; ~ **de señales** *m* TELECOM signal detector; ~ **de señalización** *m* TELECOM signaling detector (*AmE*), signalling detector (*BrE*); ~ **de separación** *m* TRANSP gap detector; ~ **de silicio** *m* ELECTRÓN, FÍS RAD silicon detector; ~ **sónico** *m* TRANSP sonic detector; ~ **sónico de profundidad** *m* D&A, OCEAN, TRANSP MAR sonic depth finder; ~ **de temperatura** *m* PROD, TERMO temperature detector; ~ **térmico** *m* TERMO heat detector, heat sensor; ~ **terrestre** *m* GEOFÍS earth sensor; ~ **de tiempo de vuelo** *m* FÍS PART, FÍS RAD time-of-flight detector (*TOFD*); ~ **de tipo banda** *m* TEC ESP strip-type detector; ~ **de tonalidad del impulso** *m* ELECTRÓN beat-note detector; ~ **de tráfico** *m* TRANSP traffic detector; ~ **de tránsito** *m* TRANSP passage detector; ~ **del tránsito de estrellas** *m* TEC ESP star transit detector; ~ **ultrasónico** *m* FERRO ultrasonic probe, TRANSP ultrasonic detector; ~ **ultrasónico de impulsos** *m* TRANSP pulsed ultrasonic detector; ~ **ultrasónico de ondas continuas** *m* FÍS ONDAS, TRANSP continuous-wave ultrasonic detector; ~ **de la velocidad** *m* TRANSP speed detector; ~ **de velocidad axial** *m* TEC ESP axial velocity sensor; ~ **de vibraciones y choques** *m* ING MECÁ vibration and shock pick-up;

~ **de viento** *m* TRANSP MAR wind sensor; ~ **vocal** *m* TEC ESP voice detector

detención *f* INFORM&PD halt, ING MECÁ stop, stopping, VEH stop; ~ **por corrientes de Foucault** *f* ING ELÉC eddy-current braking; ~ **momentánea periódica** *f* ING ELÉC periodic damping

detener *vt* INFORM&PD halt, stop, ING MECÁ hold up

detenerse *v refl* INFORM&PD halt

detenido: ~ **por el mal tiempo** *adj* TRANSP MAR weather-bound

detergencia *f* DETERG, QUÍMICA detergency

detergente *m* DETERG detergent, scouring liquid, QUÍMICA, TEC PETR, TEXTIL *acabado* detergent; ~ **aniónico** *m* DETERG anionic detergent; ~ **no jabonoso** *m* DETERG nonsoapy detergent; ~ **sintético** *m* DETERG synthetic detergent

deteriorado *adj* ING MECÁ, MECÁ worn, TELECOM damaged, TEXTIL worn

deterioro *m* GAS deterioration, ING MECÁ wear, wearing, MECÁ wear, P&C *término general* deterioration, TEXTIL wear; ~ **por acción de la intemperie** *m* CONST weathering; ~ **auditivo producido por ruidos** *m* SEG noise-induced hearing impairment; ~ **de la calidad de transmisión debido al ruido** *m* TELECOM noise transmission impairment (*NTI*); ~ **por congelación** *m* REFRIG frostbite; ~ **debido a la compresión** *m* EMB compression damage; ~ **por exposición a la intemperie artificial** *m* P&C *prueba* accelerated weathering, artificial weathering; ~ **por irradiación** *m* FÍS RAD radiation damage; ~ **por uso** *m* ING MECÁ wear and tear

determinación: ~ **de contaminación** *f* ALIMENT, CALIDAD, CONTAM contaminant determination; ~ **de la dimensión de un pedido** *f* PROD lot sizing; ~ **de distancia** *f* TEC ESP ranging; ~ **de edades por medio del radiocarbono** *f* FÍS, FÍS RAD, GEOFÍS radiocarbon dating; ~ **de la humedad** *f* EMB moisture determination; ~ **de órbita** *f* TEC ESP orbit determination; ~ **de la posición** *f* D&A position-finding; ~ **de posición por radio** *f* TELECOM, TRANSP AÉR, TRANSP MAR radio fix; ~ **radioeléctrica de la posición** *f* TRANSP radio position fixing; ~ **radiométrica de la edad** *f* FÍS RAD, GEOFÍS, GEOL radiometric age determination; ~ **del sentido de una indicación** *f* NUCL sensing; ~ **de la trayectoria** *f* TEC ESP, TELECOM *satélite* tracking

determinante *m* INFORM&PD, MATEMÁT *de matriz* determinant

detonable *adj* MINAS *explosivo* detonable, TERMO detonatable

detonación *f* AUTO knocking, pinging (*AmE*), pinking (*BrE*), ING MECÁ knocking, MINAS *explosivo* detonation, report, TEC ESP report, TERMO detonation, VEH knocking, pinging (*AmE*), pinking (*BrE*); ~ **por simpatía** *f* MINAS *explosivo* sympathetic detonation

detonador *m Esp* (*cf cebo AmL*) D&A detonator, MINAS *explosivo* exploder, squib, detonator, blasting-cap, cap, *voladuras* fuse, PETROL *explosivos* cap, TEC ESP, TERMO detonator; ~ **de acción retardada** *m* MINAS delayed-action detonator; ~ **de alta tensión** *m* MINAS *voladuras* high-tension detonator; ~ **por descarga del condensador** *m* MINAS condenser discharge exploder; ~ **a distancia** *m* INSTR dew cap; ~ **eléctrico** *m* MINAS electric detonator; ~ **eléctrico de bajo voltaje** *m* MINAS low-tension detonator; ~ **eléctrico con retardo** *m* MINAS short-delay electric detonator; ~ **instantáneo** m MINAS *voladuras* instantaneous detonator; ~ **progresivo** m MINAS ordinary detonator; ~ **con retardo** m MINAS *explosivo* short-delay detonator; ~ **de retardo** m MINAS delay cap, *voladuras* delay detonator

detonante: ~ **de gas** m GAS, PETROL gas exploder

detonar vt D&A, MINAS, TERMO detonate

detorsión f ING MECÁ untwisting

detrítico adj GEOL clastic, PETROL clastic, detrital, RECICL clastic

detritos m pl AGUA refuse, CARBÓN *de rocas* debris, *basura* refuse, CONTAM refuse, debris, EMB refuse, GEOL *de rocas* clast, debris, rubble, PETROL clast, PROD refuse, RECICL *de rocas* clast, refuse; ~ **de la perforación** m pl CARBÓN drilling debris; ~ **de perforación** m pl GAS, TEC PETR drill cuttings; ~ **de sondeos** m pl CARBÓN, GAS, TEC PETR *perforación* drill cuttings

detritus m AGUA refuse, CARBÓN *de rocas* debris, *basura* refuse, EMB refuse, GEOL *de rocas* rubble, debris, clast, PETROL clast, PROD refuse, RECICL *de rocas* clast, refuse; ~ **de la perforación** m CARBÓN drilling debris; ~ **de perforación** m GAS, TEC PETR drill cuttings; ~ **de sondeos** m CARBÓN, GAS, TEC PETR drill cuttings

deuterio m (*D*) FÍS, NUCL, QUÍMICA deuterium (*D*)

deuterón m FÍS, FÍS PART, QUÍMICA deuteron

deuteroproteosa f QUÍMICA deuteroproteose

deutón m FÍS, FÍS PART, QUÍMICA deuton

Deutsche: ~ **Industrienorm** m (*DIN*) CALIDAD, FOTO, IMPR, ING MECÁ, PROD Deutsche Industrienorm (*DIN*)

devanadera f ING ELÉC, PROD bobbin, reel, spool, TEXTIL reel, spool

devanado m ELEC *transformador* winding, FERRO *cable* unwinding, ING ELÉC winding, reeling, P&C reeling, VEH *generador* winding; ~ **abierto** m ELEC *transformador, reactor* open winding; ~ **de alimentación** m ELEC fed-in winding; ~ **alternado** m ELEC sandwich winding; ~ **de alto voltaje** m ELEC *transformador* high-voltage winding; ~ **en anillo** m ELEC *bobina* ring winding; ~ **de baja tensión** m ELEC low-voltage winding; ~ **de bajo voltaje** m ELEC low-voltage winding; ~ **de barras** m ELEC bar winding; ~ **en barras** m ING ELÉC squirrel-cage winding; ~ **bifilar** m ELEC *de bobina* double-layer winding, *inductor* bifilar winding; ~ **bipolar** m ELEC bipolar winding; ~ **de bobina** m ELEC coil winding; ~ **de bobinas concéntricas** m ELEC concentric winding; ~ **en cadena** m ING ELÉC chain winding; ~ **de campo** m ELEC, ING ELÉC field winding; ~ **por capas** m ING ELÉC layer winding; ~ **de capas múltiples** m ELEC bank winding, *bobina* banked winding; ~ **casual** m ING ELÉC random winding; ~ **en circuito abierto** m ELEC open-circuit winding; ~ **compensador** m ING ELÉC compensation winding; ~ **compound** m ELEC, ING ELÉC compound winding; ~ **conformado** m ELEC form-wound coil; ~ **de confusión** m ELEC mush winding; ~ **de conmutación** m ELEC commutating winding; ~ **en cruce** m ING ELÉC frogleg winding; ~ **de derivación** m ELEC, ING ELÉC bias winding; ~ **en derivación** m ELEC, ING ELÉC shunt winding; ~ **diametral** m ING ELÉC full-pitch winding; ~ **diferencial** m ING ELÉC differential winding; ~ **en disco** m ELEC, ING ELÉC disc winding (*BrE*), disk winding (*AmE*); ~ **en disco doble** m ELEC double

disc winding (*BrE*), double disk winding (*AmE*); ~ **doble** m ELEC duplex lap winding, *corriente continua* compound winding, ING ELÉC compound winding; ~ **en doble jaula de ardilla** m ING ELÉC double squirrel cage winding; ~ **de dos circuitos** m ELEC *máquina eléctrica* duplex lap winding; ~ **de drenaje** m ELEC *regulación* bleeder winding; ~ **dúplex** m ELEC *máquina eléctrica* duplex lap winding; ~ **entre espiras** m ELEC turn-to-turn winding; ~ **espiral en forma de disco plano** m ELEC, ING ELÉC pancake coil; ~ **estabilizador** m ELEC stabilizing winding; ~ **del estátor del arrancador** m AUTO starter field winding; ~ **del estator bifásico** m ING ELÉC two-phase stator winding; ~ **del estator trifásico** m ING ELÉC three-phase stator winding; ~ **estatórico** m ING ELÉC stator winding; ~ **de excitación** m ELEC *de máquina* field winding, *electromagnetismo* excitation winding, ING ELÉC field winding; ~ **de fase** m ELEC phase winding; ~ **Gramme** m ELEC *bobina* Gramme winding; ~ **de hilos sacados** m ELEC pull-through winding; ~ **imbricado** m ELEC *máquina eléctrica*, ING ELÉC lap winding; ~ **imbricado de circuito único** m ELEC simplex lap winding; ~ **imbricado simple** m ELEC simplex lap winding; ~ **inclinado** m ELEC edge winding; ~ **del inducido** m ELEC, ING ELÉC armature winding, rotor winding; ~ **inductor** m ELEC *máquina* field winding, *transformador* primary winding, *electromagnetismo* excitation winding, ING ELÉC field winding, primary winding; ~ **del inductor del arrancador** m AUTO starter field winding; ~ **de interferencia** m ELEC mush winding; ~ **mixto** m ELEC compounding, ING ELÉC compound winding, compounding; ~ **monofásico** m ING ELÉC single-phase winding; ~ **del motor** m PROD motor winding; ~ **múltiple** m ELEC, ING ELÉC multiple winding; ~ **de panal** m ING ELÉC honeycomb winding; ~ **paralelo múltiple** m ELEC multiplex lap winding; ~ **de paso entero** m ELEC full-pitch winding; ~ **de paso fraccionario** m ELEC, ING ELÉC fractional pitch winding; ~ **de paso largo** m ELEC, ING ELÉC long-pitch winding; ~ **de paso parcial** m ELEC, ING ELÉC fractional pitch winding; ~ **en pi** m ING ELÉC pi winding; ~ **primario** m ELEC *transformador* primary winding; ~ **primario derivado** m ING ELÉC tapped primary winding; ~ **con puntos intermedios** m ING ELÉC tapped winding; ~ **de ranura fraccional** m ELEC, ING ELÉC fractional slot winding; ~ **de realimentación** m ING ELÉC feedback winding; ~ **en rombo** m ELEC diamond winding; ~ **del rotor bifásico** m ING ELÉC two-phase rotor winding; ~ **del rotor trifásico** m ING ELÉC three-phase rotor winding; ~ **rotórico** m ELEC, ING ELÉC rotor winding; ~ **de salida** m ELEC output winding; ~ **secundario** m AUTO, ELEC, ING ELÉC secondary winding; ~ **secundario derivado** m ING ELÉC tapped secondary winding; ~ **en serie** m ELEC, ING ELÉC series winding; ~ **en serie y derivación** m ELEC, ING ELÉC compound winding; ~ **en shunt** m ELEC shunt winding; ~ **de tambor** m ING ELÉC drum winding; ~ **de tensión intermedia** m ELEC intermediate voltage winding; ~ **terciario** m ELEC stabilizing winding, tertiary winding; ~ **toroidal** m ELEC *bobina* ring winding; ~ **de tres ranuras** m ELEC three-slot winding; ~ **uniforme** m ING ELÉC diamond winding; ~ **de varias capas** m ELEC, ING ELÉC multiple winding

devanador: ~ **de cable** *m* PETROL cable reel

devanadora *f* ING ELÉC, PROD bobbin, reel, spool, TEXTIL reel

devenado: ~ **mixto** *m* ELEC, ING ELÉC compound winding

devitrificación *f* C&V, GEOL devitrification

Devónico *adj* GEOL Devonian

deweylita *f* MINERAL deweylite

dextral *adj* GEOL dextral

dextrano *m* QUÍMICA *polisacárido* dextran

dextrina *f* ALIMENT, P&C, QUÍMICA dextrin, dextrine

dextrógiro *adj* ALIMENT, FÍS, QUÍMICA dextrorotatory

dextrosa *f* AGRIC, ALIMENT, QUÍMICA dextrose

dextroso *adj* MECÁ right-hand

deyecciones: ~ **volcánicas** *f pl* GEOL ejectamenta

DFA *abr* MATEMÁT (*distribución de frecuencias acumulativas*) CDF (*cumulative distribution function*), QUÍMICA (*difenilamina*) DFA (*diphenylamine*)

dg *abr* (*decígramo*) METR dg (*decigram*)

DHA *abr* (*dihidroxiacetona*) QUÍMICA DHA (*dihydroxyacetone*)

día: ~ **de estadías** *m* TEC PETR *comercio, transporte* lay day; ~ **de producción** *m* PROD production day; ~ **sideral** *m* FÍS, TEC ESP sidereal day; ~ **del vencimiento** *m* PROD due date

diabasa *f* GEOL, PETROL diabase (*AmE*), dolerite (*BrE*); ~ **esferoidal** *f* GEOL pillow lava

diablo *m* PAPEL rag thrasher, PETROL, TEC PETR *perforación* pig

diábolo *m* OCEAN net roller

diacetilacetona *f* QUÍMICA diacetylacetone

diacetileno *m* QUÍMICA diacetylene

diacetilmorfina *f* QUÍMICA diacetylmorphine, heroin

diacetilo *m* QUÍMICA diacetyl

diácido *m* QUÍMICA diacid

diaclasa *f* CARBÓN joint, pile joint, CONST crack, GEOL joint, MINAS *rocas* seam; ~ **longitudinal** *f* CARBÓN, GEOL, MINAS back; ~ **de tensión** *f* GEOL tension gash

diaclasado *m* GEOL, MINAS jointing; ~ **columnar** *m* GEOL columnar jointing

diacrítico *adj* INFORM&PD diacritical

diácrono *adj* GEOL diachronous

diádico *adj* INFORM&PD dyadic

diadoquia *f* GEOL diadochy

diadoquita *f* MINERAL diadochite

diaesquístico *adj* GEOL, PETROL, TEC PETR diaschistic

diafonía *f* ACÚST crosstalk, flutter, FÍS, INFORM&PD crosstalk, TELECOM flutter; ~ **posicional** *f* TV positional crosstalk

diáfono *m* TRANSP MAR *bocina* diaphone

diaforita *f* MINERAL diaphorite

diafragma *m* ACÚST diaphragm, CINEMAT aperture plate, diaphragm, FOTO diaphragm, stop, ING ELÉC septum, ING MECÁ diaphragm, MECÁ diaphragm, *arandela* shim; ~ **de abertura** *m* METAL aperture diaphragm; ~ **ajustable** *m* CINEMAT adjustable stop, FÍS adjustable aperture; ~ **de campo** *m* METAL field diaphragm; ~ **cerrado** *m* ACÚST closed diaphragm; ~ **de filtración** *m* PROC QUÍ sieve diaphragm; ~ **frontal** *m* FOTO front diaphragm; ~ **del iris** *m* INSTR iris diaphragm; ~ **de montaje** *m* METAL field diaphragm; ~ **del objetivo** *m* INSTR objective aperture; ~ **protector** *m* MECÁ bursting disc (*BrE*), bursting disk (*AmE*); ~ **de rendija** *m* FOTO slit diaphragm; ~ **totalmente automático** *m* FOTO fully-automatic diaphragm

diafragmar *vt* CINEMAT, FOTO *disminuir el diafragma* stop down

diagénesis *f* GEOL, PETROL, TEC PETR diagenesis

diagenético *adj* GEOL, PETROL, TEC PETR *yacimientos* diagenetic

diagnosis: ~ **de averías** *f* INFORM&PD fault diagnosis; ~ **de fallos** *f* INFORM&PD fault diagnosis

diagnóstico[1] *adj* INFORM&PD diagnostic

diagnóstico[2]: ~ **de compilador** *m* INFORM&PD compiler diagnostic; ~ **de error** *m* INFORM&PD, TELECOM error diagnosis; ~ **de errores** *m* INFORM&PD, TELECOM error diagnostics; ~ **por rayos X** *m* INSTR X-ray diagnostics

diagonal[1] *adj* GEOM diagonal

diagonal[2] *f AmL* (*cf jabalcón Esp*) CARBÓN bias, NUCL diagonal; ~ **de cruz** *f* INSTR star diagonal

diagrafía *f* ENERG RENOV, GAS, GEOL, PETROL, TEC PETR *petrofísica, evaluación de la formación* log; ~ **acústica** *f* TEC PETR acoustic log; ~ **del autopotencial** *f* TEC PETR self-potential log; ~ **eléctrica especial múltiple** *f* TEC PETR multiple special electrical logging; ~ **gamma-gamma** *f* TEC PETR gamma-gamma log; ~ **neutrón** *f* ENERG RENOV, GAS, PETROL, TEC PETR neutron log, neutron logging; ~ **de neutrón impulsado** *f* TEC PETR pulsed-neutron log; ~ **de neutrón-gamma** *f* TEC PETR neutron-gamma log; ~ **de neutrón-neutrón** *f* TEC PETR neutron-neutron log; ~ **nuclear** *f* TEC PETR nuclear log; ~ **de la porosidad** *f* TEC PETR porosity log; ~ **potencial espontáneo** *f* TEC PETR spontaneous potential log; ~ **del pozo** *f* TEC PETR well logging; ~ **de proximidad** *f* TEC PETR proximity log; ~ **pseudosónica** *f* TEC PETR pseudo-sonic log; ~ **radioactiva** *f* TEC PETR radioactive log; ~ **de resonancia magnética nuclear** *f* TEC PETR NMR log, nuclear magnetic resonance log; ~ **sónica** *f* TEC PETR sonic log; ~ **de la temperatura del pozo** *f* TEC PETR temperature well logging; ~ **de temperaturas** *f* TEC PETR temperature logging

diagrafiado: ~ **acústico de pozos** *m* TEC PETR acoustic well logging

diagrama *m* ELEC, ELECTRÓN, FÍS, INFORM&PD diagram, ING MECÁ outline, scheme, MATEMÁT, PROD diagram, TEC ESP *antenas* pattern, TELECOM, TV diagram; ~ **de ajuste de fases** *m* TRANSP phasing diagram; ~ **de análisis del lodo** *m* TEC PETR *ingeniería de lodos, perforación* mud analysis log; ~ **de balance térmico** *m* TERMO heat balance diagram, heat balance chart; ~ **de barras** *f Esp* (*cf tabla de barras AmL*) INFORM&PD bar chart ~ **de bloques** *m* INFORM&PD, ING ELÉC block diagram; ~ **de cableado** *m* ELEC, ELECTRÓN, ING ELÉC, TELECOM wiring diagram; ~ **de campo lejano** *m* ÓPT, TELECOM far-field pattern; ~ **de campo próximo** *m* ÓPT, TELECOM near-field pattern; ~ **de carga** *m* ELEC *red de distribución* load curve, ING ELÉC, MECÁ load diagram; ~ **del ciclo** *m* VEH *motor* pressure-indicator card; ~ **cinemático** *m* ING MECÁ kinematic diagram; ~ **de circuito** *m* ELEC, ELECTRÓN, ING ELÉC, TV circuit diagram; ~ **del circuito** *m* TV circuit diagram; ~ **del circuito de combustible** *m* TRANSP MAR fuel-system diagram; ~ **del circuito eléctrico** *m* ING ELÉC, TRANSP MAR electrical wiring diagram; ~ **de circulación de energía** *m* TERMO energy-flow chart; ~ **de la circulación del tráfico** *m* TRANSP traffic flow diagram; ~ **climático** *m* METEO climagram, climatic graph; ~ **de colocación** *m* ING MECÁ, MECÁ layout; ~ **de conexiones** *m* AUTO, ELEC wiring diagram, ING ELÉC connection diagram, TELECOM wiring diagram; ~ **de conexiones de entrada** *m* PROD input connection diagram; ~ **de conjuntos** *m* TELECOM block diagram; ~ **de constitución** *m* METAL *aleaciones* phase diagram; ~ **del controlador** *m* PROD controller layout; ~ **de copolarización** *m* TEC ESP copolar pattern; ~ **de cromaticidad** *m* TV chromaticity diagram; ~ **de difracción** *m* CRISTAL diffraction pattern; ~ **de difracción de campo lejano** *m* ÓPT, TELECOM far-field diffraction pattern; ~ **de difracción de campo próximo** *m* ÓPT, TELECOM near-field diffraction pattern; ~ **de difracción de electrones** *m* CRISTAL electron-diffraction pattern; ~ **de difracción de Fraunhofer** *m* FÍS, ÓPT, TELECOM Fraunhofer-diffraction pattern; ~ **de difracción de Fresnel** *m* ACÚST, FÍS, ÓPT, TELECOM Fresnel diffraction pattern; ~ **de difracción de rayos X** *m* CRISTAL X-ray diffraction pattern; ~ **direccional de radiación** *m* TV beam pattern; ~ **de distribución** *m* HIDROL, ING MECÁ distribution diagram, MECÁ layout drawing; ~ **de distribución acumulativa** *m* HIDROL cumulative distribution diagram; ~ **eléctrico** *m* AUTO, ELEC wiring diagram; ~ **de ensamblaje** *m* ING MECÁ block diagram; ~ **de equilibrio líquido-vapor** *m* TERMO liquid vapor equilibrium diagram (*AmE*), liquid vapour equilibrium diagram (*BrE*); ~ **de escalones** *m* PROD ladder diagram; ~ **espacio-tiempo** *m* TRANSP time-space diagram; ~ **esquemático del cableado** *m* NUCL schematic wiring diagram; ~ **de estado** *m* INFORM&PD state diagram; ~ **de exploración por radar** *m* D&A, FÍS RAD, TRANSP, TRANSP AÉR, TRANSP MAR radar scan pattern; ~ **de la fase de color** *m* TV color-phase diagram (*AmE*), colour-phase diagram (*BrE*); ~ **de fases** *m* NUCL, QUÍMICA, TRANSP *control de tráfico* phase diagram; ~ **de Feynman** *m* FÍS Feynman diagram; ~ **de flujo** *m* *Esp* (*cf flujograma AmL*) ELECTRÓN signal flow-graph, FÍS, INFORM&PD flowchart; ~ **de flujo calorífico** *m* TERMO heat flow chart, heat flow diagram; ~ **de flujo energético** *m* TERMO energy-flow chart; ~ **de flujo térmico** *m* TERMO heat flow chart, heat flow diagram; ~ **del flujo de la velocidad** *m* TRANSP speed flow diagram; ~ **de flujos** *m* CARBÓN *ingeniería* flow sheet; ~ **funcional** *m* INFORM&PD functional diagram, ING ELÉC block diagram, ING MECÁ functional diagram; ~ **de funcionamiento de circuito** *m* ING MECÁ functional diagram; ~ **general** *m* ING MECÁ combined diagram; ~ **indicador** *m* FÍS indicator diagram; ~ **de la instalación de tuberías** *m* TRANSP MAR *construcción* piping plan; ~ **isócrono** *m* GEOL isochrone diagram; ~ **de Laue** *m* CRISTAL, FÍS RAD Laue diagram, Laue pattern; ~ **del lodo** *m* TEC PETR *evaluación de la formación, perforación* mud log; ~ **lógico** *m* ELECTRÓN, INFORM&PD logic diagram; ~ **meteorológico** *m* TEC ESP weather pattern; ~ **muaré** *m* ACÚST moiré pattern; ~ **del ojo** *m* TELECOM eye diagram; ~ **de onda sísmica** *m* GEOFÍS, PETROL seismic wave trace; ~ **oxígeno-salinidad** *m* OCEAN oxygen-salinity diagram; ~ **de la permeabilidad** *m* TEC PETR *evaluación de la formación, petrofísica* permeability logging; ~ **PERT** *m* INFORM&PD PERT chart; ~ **polar** *m* FÍS polar diagram; ~ **de polarización cruzada** *m* TEC ESP, TELECOM cross-polar pattern; ~ **de polvo** *m* CRISTAL powder pattern; ~ **de posición electrónica** *m* ELECTRÓN Applegate diagram; ~ **de presiones** *m* INSTR pressure graph; ~ **del proceso operativo del sistema** *m* ELEC *red de distribución* system operational diagram; ~ **del promedio de distribución** *m* HIDROL average distribution diagram; ~ **de proximidad** *m* TEC PETR proximity log; ~ **de puntos** *m* TV dot grating; ~ **de radiación** *m* FÍS, FÍS RAD, ÓPT, TEC ESP, TELECOM, TV radiation pattern; ~ **de radiación de campo lejano** *m* ÓPT, TELECOM far-field radiation pattern; ~ **de radiación de campo próximo** *m* TELECOM near-field radiation pattern; ~ **de radiación de equilibrio** *m* ÓPT, TELECOM equilibrium radiation pattern; ~ **de rayos X** *m* INSTR X-ray diagram; ~ **de refracción de las olas** *m* TRANSP MAR *diseño de barcos* wave-refraction diagram; ~ **de refracción de ondas** *m* OCEAN wave-refraction diagram; ~ **de la relación temperatura-salinidad** *m* OCEAN potential temperature-salinity diagram; ~ **de Rieke** *m* ELECTRÓN Rieke diagram; ~ **de sectores** *m* MATEMÁT pie chart; ~ **de sincronización** *m* TRANSP phasing diagram; ~ **de sincronización de válvulas** *m* AUTO valve-timing diagram; ~ **del sistema** *m* ELEC *red de distribución* system diagram; ~ **de temperatura potencial-salinidad** *m* OCEAN potential temperature-salinity diagram; ~ **de temperatura-salinidad** *m* OCEAN T-S diagram, temperature-salinity diagram; ~ **de temporización** *m* INFORM&PD timing diagram; ~ **totalizado** *m* ING MECÁ combined diagram; ~ **de transición de estados** *m* TELECOM state transition diagram; ~ **de tratamiento térmico** *m* TERMO heat treatment diagram; ~ **unilineal** *m* ELEC *red de distribución* single-line diagram; ~ **de válvulas de Corliss** *m* PROD diagram of Corliss valve gear; ~ **de Veitch** *m* INFORM&PD Veitch diagram; ~ **de velocidad** *m* ENERG RENOV, ING MECÁ velocity diagram; ~ **de Venn** *m* INFORM&PD, MATEMÁT Venn diagram; ~ **de vías** *m* FERRO *equipo inamovible* track diagram; ~ **V-N de ráfagas** *m* TEC ESP, TRANSP AÉR, TRANSP MAR gust envelope, gust V-n diagram

diaklasita *f* MINERAL diaclasite

diálaga *f* MINERAL diallage

dialisado *m* QUÍMICA dialysate

dialítico *adj* QUÍMICA dialytic

dialogita *f* MINERAL dialogite, oligonite

diálogo *m* INFORM&PD dialog (*AmE*), dialogue (*BrE*); ~ **comprimido** *m* ELECTRÓN, TELECOM compressed speech

dialqueno *m* MINERAL, QUÍMICA, TEC PETR *refino* dialkene

diamagnética *f* FÍS, FÍS RAD diamagnetics

diamagnético *adj* ELEC nonmagnetic, diamagnetic, FÍS, FÍS RAD, PETROL, QUÍMICA diamagnetic

diamagnetismo *m* ELEC, FÍS, FÍS RAD, ING ELÉC, PETROL, QUÍMICA diamagnetism

diamante: ~ **cortavidrios** *m* C&V, CONST glass cutter; ~ **industrial** *m* MINAS *herramienta* industrial diamond; ~ **negro** *m* MINERAL black diamond; ~ **reavivador** *m* ING MECÁ dresser cutter; ~ **sin tallar** *m* MINAS stone; ~ **de tornear** *m* ING MECÁ turning diamond; ~ **de vidriero** *m* C&V, CONST glazier's diamond

diamantífero *adj* MINAS diamond-bearing, diamondiferous

diametral *adj* ELEC, GEOM diametric, diametrical

diametralmente: ~ **opuesto** *adv* ELEC, GEOM diametrically opposed

diamétrico *adj* ELEC, GEOM diametric, diametrical

diámetro *m* GEOM diameter; ~ **adimensional** *m* ENERG RENOV nondimensional diameter; ~ **admisible sobre la bancada** *m* ING MECÁ swing of the bed, swing over bed, swing over saddle; ~ **admisible sobre la torreta** *m* ING MECÁ swing of the rest; ~ **angular** *m* TEC ESP angular diameter; ~ **de apriete** *m* EMB *torno, abonzaderas*, PROD *plato de torno* holding capacity; ~ **através de planos** *m* ING MECÁ diameter across flats; ~ **del cabeceo** *m* TRANSP AÉR pitch diameter; ~ **de campo de modos** *m* ÓPT, TELECOM mode field diameter; ~ **del centro del cabeceo** *m* TRANSP AÉR pitch center diameter (*AmE*), pitch centre diameter (*BrE*); ~ **del centro del paso** *m* TRANSP AÉR pitch center diameter (*AmE*), pitch centre diameter (*BrE*); ~ **de chorro** *m* ENERG RENOV jet diameter; ~ **del círculo de cabeceo** *m* TRANSP AÉR pitch-circle diameter; ~ **del círculo del paso** *m* TRANSP AÉR pitch-circle diameter; ~ **del círculo primitivo** *m* ING MECÁ pitch diameter; ~ **del cuerpo del remache** *m* ING MECÁ rivet shank diameter; ~ **de ecualización** *m* ACÚST diameter equalization; ~ **exterior** *m* ING MECÁ, MECÁ outside diameter (*OD*); ~ **del haz** *m* ÓPT beam diameter; ~ **del haz guiado** *m* ÓPT guided beam diameter; ~ **interior** *m* AUTO bore, ING MECÁ bore, inner diameter (*ID*), inside diameter (*ID*), internal diameter, MECÁ caliper (*AmE*), calliper (*BrE*), inner diameter (*ID*), inside diameter (*ID*), PROD internal diameter, VEH *motor, cilindro* bore; ~ **interior del cilindro** *m* ING MECÁ cylinder bore; ~ **interior de cilindros** *m* C&V *botellas* bore (*BrE*), corkage (*AmE*); ~ **interno** *m* ING MECÁ, MECÁ inner diameter (*ID*), inside diameter (*ID*); ~ **interno nominal** *m* NUCL nominal bore (*NB*); ~ **máximo admisible** *m* ING MECÁ swing; ~ **máximo admisible sobre la escotadura** *m* ING MECÁ swing in gap; ~ **máximo de apriete** *m* ING MECÁ holding capacity; ~ **máximo torneable** *m* ING MECÁ swing; ~ **medio del alma** *m* TELECOM average core diameter; ~ **medio de chapeado** *m* TELECOM average cladding diameter; ~ **menor** *m* ING MECÁ minor diameter; ~ **nominal** *m* ING MECÁ nominal diameter; ~ **normal** *m* TELECOM stock diameter; ~ **del núcleo** *m* ÓPT, TELECOM core diameter; ~ **del paso** *m* TRANSP AÉR pitch diameter; ~ **primitivo** *m* ING MECÁ pitch diameter; ~ **promedio del núcleo** *m* ÓPT *fibras ópticas* average core diameter; ~ **promedio de revestimiento** *m* ÓPT *fibras ópticas* average cladding diameter; ~ **del revestimiento** *m* ÓPT *fibras ópticas* cladding diameter; ~ **del rodillo** *m* TEXTIL roll diameter; ~ **del rotor** *m* ENERG RENOV rotor diameter; ~ **de la superficie de referencia** *m* ÓPT, TELECOM reference-surface diameter

diamida *f* QUÍMICA diamide

diamina *f* QUÍMICA diamine

diaminodifenilmetano *m* P&C, QUÍMICA diaminodiphenylmethane

diapirismo: ~ **esquistoso** *m* GEOL, TEC PETR shale diapirism; ~ **salino** *m* GEOL, TEC PETR salt diapirism

diapiro *m* GAS dome, GEOL, TEC PETR *geología* diapir

diapositiva *f* CINEMAT slide, FOTO diapositive, slide, transparency, IMPR slide, transparency, transparent positive; ~ **negra** *f* IMPR skeleton black; ~ **proy-**ectable *f* CINEMAT lantern slide; ~ **de referencia** *f* CINEMAT line-up slide

diarilos *m pl* QUÍMICA diaryls

diario: ~ **de a bordo** *m* TRANSP AÉR flight log; ~ **de navegación** *m* TRANSP AÉR flight log, TRANSP MAR ship's log; ~ **de operaciones** *m* TRANSP logbook

diarrea: ~ **del ganado** *f* AGRIC scours

diásporo *m* MINERAL diaspore

diastasa *f* ALIMENT, QUÍMICA diastase

diastema *m* GEOL, PETROL diastem

diastereoisómero *m* QUÍMICA *estereoquímica* diastereo isomer

diastereómero *m* QUÍMICA diastereomer

diatérmano *adj* TERMO diathermanous

diatérmico *adj* FÍS diathermal, TERMO diathermic, diathermal

diatomea *f* PETROL diatom, diatomite

diatómico *adj* QUÍMICA diatomic

diatomita *f* GEOL, TERMO, TERMOTEC kieselguhr, diatomite

diatónica *adj* ACÚST *gama* diatonic

diatrema *f* GEOL *chimenea volcánica* diatreme

diazo *m* IMPR, QUÍMICA diazo

diazoacético *adj* QUÍMICA diazoacetic

diazobenceno *m* QUÍMICA diazobenzene

diazoimida *f* QUÍMICA diazoimide

diazol *m* QUÍMICA diazole

diazomina *f* QUÍMICA diazomine

diazonio *m* QUÍMICA diazonium

diazotar *vt* QUÍMICA diazotize

diazotizar *vt* QUÍMICA diazotize

dibásico *adj* QUÍMICA dibasic

dibencilamina *f* QUÍMICA dibenzylamine

dibenzoantraceno *m* QUÍMICA dibenzanthracene

dibenzoil *adj* QUÍMICA dibenzoyl

dibenzopirrol *m* QUÍMICA dibenzopyrrole

dibromhidrina *f* QUÍMICA dibromohydrin

dibromobenceno *m* QUÍMICA dibromobenzene

dibujante *m* MECÁ, PROD draftsman (*AmE*), draughtsman (*BrE*); ~ **mecánico** *m* AmL (*cf delineante mecánico Esp*) PROD mechanical draftsman (*AmE*), mechanical draughtsman (*BrE*); ~**proyectista** *m* ING MECÁ, MECÁ design draftsman (*AmE*), design draughtsman (*BrE*), designer draftsman (*AmE*), designer draughtsman (*BrE*)

dibujar *vt* Esp (*cf graficar AmL*) MATEMÁT *gráfica* graph, plot

dibujo: ~ **acotado** *m* ING MECÁ outline drawing, PROD dimensioned sketch; ~ **animado** *m* CINEMAT animated cartoon, cartoon, TV cartoon; ~ **de la banda de rodadura** *m* VEH *neumático* tread design; ~ **de color** *m* AmL (*cf gráfico en color Esp*) INFORM&PD color graphic (*AmE*), colour graphic (*BrE*); ~ **de contorno aproximado y sin detalles** *m* ING MECÁ outline drawing; ~ **a escala** *m* GEOM scale drawing; ~ **a escala mitad** *m* PROD half-size drawing; ~ **al lavado** *m* ING MECÁ wash (*BrE*); ~ **lineal** *m* GEOM line drawing; ~ **con líneas** *m* INFORM&PD raster graphics; ~ **mecánico** *m* PROD mechanical drawing; ~ **a mitad de tamaño natural** *m* PROD half-size drawing; ~ **de montaje** *m* ING MECÁ, MECÁ assembly drawing; ~ **de producción** *m* Esp (*cf plano de producción AmL*) PROD production drawing; ~ **técnico** *m* ING MECÁ engineering drawing; ~ **tramado de fondo** *m* IMPR background pattern; ~ **a trazos** *m* ING MECÁ outline; ~ **en tres dimensiones** *m* INFORM&PD three-dimensional graphic; ~ **tridimensional** *m* INFORM&PD three-dimensional graphic

dibutirina *f* QUÍMICA dibutyrin

dicarburo: ~ **de uranio** *m* NUCL, QUÍMICA uranium dicarbide

diccionario: ~ **de datos** *m* INFORM&PD data dictionary; ~ **de sinónimos** *m* INFORM&PD *léxico* thesaurus

dickinsonita *f* MINERAL dickinsonite

dicloroacetona *f* QUÍMICA dichloroacetone

diclorobenceno *m* QUÍMICA dichlorobenzene

diclorodifluorometano *m* QUÍMICA, REFRIG dichlorodifluoromethane, freon

dicloroetano *m* QUÍMICA dichloroethane

diclorofluorometano *m* PROD freon

diclorohidrina *f* QUÍMICA dichlorohydrin

dicloruro *m* QUÍMICA dichloride

dicordio *m* TELECOM cord

dicordios *m pl* TELECOM cord circuit

dicotiledón *m* AGRIC dicotyledon

dicroico *adj* CINEMAT, ELECTRÓN, FÍS, FOTO, ÓPT, TELECOM dichroic

dicroísmo *m* CRISTAL, FÍS, QUÍMICA dichroism

dicroíta *f* MINERAL cordierite, dichrote, iolite

dicromato *m* QUÍMICA dichromate

didimio *m* QUÍMICA didymium

dieciseisavo *m* IMPR sixteenmo

diédrico *adj* GEOM dihedral

diedro[1] *adj* GEOM dihedral

diedro[2] *m* TRANSP AÉR dihedral

dieldrina *f* ALIMENT, QUÍMICA dieldrin

dieléctrico[1] *adj* ELEC, FÍS, ING ELÉC, QUÍMICA, TELECOM dielectric

dieléctrico[2]: ~ **absorbente** *m* ELEC absorptive dielectric; ~ **de absorción** *m* ELEC absorptive dielectric; ~ **de aire** *m* ELEC air dielectric; ~ **disipativo** *m* ELEC lossy dielectric; ~ **ideal** *m* ELEC perfect dielectric; ~ **imperfecto** *m* ELEC imperfect dielectric; ~ **no polar** *m* FÍS nonpolar dielectric; ~ **de papel impregnado en aceite** *m* ING ELÉC solid dielectric; ~ **de pequeñas pérdidas** *m* ELEC low-loss dielectric; ~ **con pérdida** *m* ELEC imperfect dielectric; ~ **perfecto** *m* ELEC perfect dielectric; ~ **polar** *m* FÍS polar dielectric; ~ **de puerta** *m* ING ELÉC gate dielectric

dieno *m* QUÍMICA *compuesto* diene

diente *m* CONST, ING MECÁ, MECÁ cog, prong (*BrE*), tine (*AmE*), tooth, trip catch, PROD tooth; ~ **de aguas arriba** *m* AGUA cutoff wall; ~ **angular** *m* ING MECÁ herringbone tooth; ~ **bihelicoidal** *m* ING MECÁ herringbone tooth; ~ **con bisel** *m* ING MECÁ fleam tooth; ~ **biselado** *m* ING MECÁ fleam tooth; ~ **de bloqueo del engranaje planetario** *m* AUTO sun gear lock-out tooth; ~ **común** *m* ING MECÁ peg tooth; ~ **contorneado** *m* PROD crosscut tooth; ~ **de encaje** *m* ING MECÁ pawl; ~ **de engranaje** *m* ING MECÁ gear tooth; ~ **de engrane** *m* CINEMAT sprocket tooth; ~ **de gancho** *m* ING MECÁ hook tooth, hooked tooth; ~ **guía** *m* ING MECÁ guide tooth; ~ **del inducido** *m* ELEC armature tooth; ~ **de porcelana** *m* C&V porcelain tooth; ~ **de rastra** *m* AGRIC harrow tooth; ~ **recto** *m* ING MECÁ peg tooth; ~ **de retención** *m* ING MECÁ stud

dientes[1]: con ~ *adj* ING MECÁ notched

dientes[2] *m pl* ING MECÁ teeth; ~ **cortantes** *m pl* CONST cutting teeth; ~ **destalonados** *m pl* ING MECÁ relieved teeth, *engranaje, cremallera* backed-off teeth; ~ **de engranaje** *m pl* ING MECÁ wheel tooth

diéresis *f* IMPR diaeresis (*BrE*), dieresis (*AmE*)

dietanolamina *f* (*DEA*) DETERG, QUÍMICA diethanolamine (*DEA*)

dietileno *m* DETERG diethylene

difásico *adj* TRANSP diaphasic

difenilamina *f* (*DFA*) QUÍMICA diphenylamine (*DFA*)

difenilcarbinol *m* QUÍMICA diphenylcarbinol

difenilo *m* QUÍMICA biphenyl, diphenyl

diferencia: ~ **del ángulo de fase** *f* ING ELÉC angle of phase difference; ~ **de caminos** *f* FÍS path difference; ~ **de color** *f* TV color difference (*AmE*), colour difference (*BrE*); ~ **común** *f* MATEMÁT common difference; ~ **de diámetro entre el perno y el agujero correspondiente** *f* ING MECÁ drift; ~ **entre el máximo y el mínimo** *f* FÍS peak-to-peak value; ~ **de fase** *f* ACÚST, ELEC phase difference, ELECTRÓN differential phase, phase difference, FÍS, FÍS ONDAS, PETROL, TELECOM phase difference, TV differential phase, phase difference; ~ **de ganancia** *f* ELECTRÓN, TV differential gain; ~ **del índice de refracción** *f* ÓPT ESI refractive-index difference; ~ **de índice refractivo del perfil de índice escalonado** *f* TELECOM ESI refractive-index difference; ~ **de intervalo** *f* ACÚST interval difference; ~ **media de temperatura** *f* INSTAL TERM, REFRIG mean temperature difference; ~ **de nivel** *f* ACÚST level difference; ~ **de niveles** *f* HIDROL *esclusas de canal* lift; ~ **de potencial** *f* ELEC, FÍS, ING ELÉC, TELECOM potential difference (*pd*); ~ **de presión** *f* INSTAL HIDRÁUL differential pressure, pressure difference, TEC PETR *perforación* differential pressure; ~ **de temperatura** *f* FÍS, TERMO temperature difference; ~ **de las temperaturas medias** *f* INSTAL TERM, REFRIG mean temperature difference; ~ **de tensión** *f* ELEC, FÍS, ING ELÉC, TELECOM voltage difference; ~ **en el umbral de audición** *f* ACÚST hearing threshold difference

diferenciación *f* GEOL, MATEMÁT differentiation; ~ **implícita** *f* MATEMÁT implicit differentiation; ~ **magmática** *f* GEOL magmatic differentiation; ~ **metamórfica** *f* GEOL metamorphic differentiation

diferenciado *adj* ELECTRÓN *señal*, GEOL *roca*, MATEMÁT differentiated

diferencial[1] *adj* MATEMÁT differential

diferencial[2] *m* AUTO, C&V, ING MECÁ, MATEMÁT, MECÁ, VEH *transmisión* differential; ~ **autoblocante** *m* AUTO nonslip differential; ~ **de deslizamiento controlado** *m* MECÁ, VEH controlled slip differential; ~ **de deslizamiento limitado** *m* AUTO, VEH *transmisión* limited-slip differential; ~ **de desplazamiento limitado** *m* AUTO, VEH *transmisión* limited-slip differential; ~ **de engranaje planetario** *m* AUTO, VEH planetary-gear differential; ~ **de tracción** *m* AUTO, VEH traction differential

diferencias: sin ~ **de calados** *fra* TRANSP MAR on even keel

difracción *f* GEN diffraction; ~ **acústica** *f* ACÚST acoustic diffraction; ~ **de electrones** *f* CRISTAL, ELECTRÓN electron diffraction, FÍS PART *técnica* high-energy electron diffraction (*HEED*), FÍS RAD electron diffraction; ~ **de espectroscopia electrónica** *f* FÍS, FÍS RAD electron-spectroscopic diffraction; ~ **de Fraunhofer** *f* FÍS, ÓPT, TELECOM Fraunhofer diffraction; ~ **de Fresnel** *f* ACÚST, FÍS, ÓPT, TELECOM Fresnel diffraction; ~ **del haz atómico** *f* NUCL atomic beam diffraction; ~ **múltiple** *f* TELECOM multiple diffraction; ~ **de neutrones** *f* CRISTAL neutron diffraction; **~ de onda** *f* TELECOM wave diffraction; **~ de rayos X** *f* CRISTAL, ELECTRÓN, FÍS ONDAS, METAL, TEC ESP X-ray diffraction

difractometría *f* CRISTAL, GEOL, TEC PETR diffractometry

difractómetro *m* CRISTAL, FÍS RAD, INSTR, TEC PETR diffractometer; **~ de cuarto círculos** *m* CRISTAL, INSTR four-circle diffractometer; **~ de polvo** *m* CRISTAL, INSTR powder diffractometer; **~ de rayos X** *m* CRISTAL, FÍS RAD, INSTR X-ray diffractometer

difuminación *f* P&C scumble

difundir *vt* GEN diffuse

difusibilidad: ~ hidráulica *f* HIDROL hydraulic diffusivity

difusiómetro *m* OCEAN scattering meter, scatterometer

difusión *f* ACÚST, CARBÓN, CONST, CONTAM diffusion, CRISTAL scattering, ELECTRÓN scattering, diffusion, FÍS diffusion, ÓPT scattering, QUÍMICA *de fluidos* diffusion, TEC ESP, TELECOM scattering, TV diffusion; **~ a bajo ángulo** *f* CRISTAL small-angle scattering; **~ de base** *f* ELECTRÓN *transistor* base diffusion; **~ capilar** *f* HIDROL capillary diffusion; **~ conjunta** *f* METAL bulk diffusion; **~ en directo por satélite** *f* TELECOM direct broadcasting by satellite (*DBS*); **~ doble** *f* ELECTRÓN double diffusion; **~ de impurezas** *f* ELECTRÓN impurity diffusion; **~ de lagunas reticulares** *f* METAL vacancy diffusion; **~ lateral** *f* ELECTRÓN lateral diffusion; **~ de líneas** *f* TV line diffusion; **~ planar** *f* ELECTRÓN planar diffusion; **~ del plano de exfoliación** *f* METAL grain-boundary diffusion; **~ de Raman** *f* FÍS, FÍS RAD Raman scattering; **~ de Rayleigh** *f* FÍS RAD, ÓPT, TELECOM Rayleigh scattering; **~ de Rutherford** *f* FÍS Rutherford scattering; **~ selectiva** *f* ELECTRÓN selective diffusion; **~ sonora** *f* ACÚST sound diffusion; **~ superior-inferior** *f* ELECTRÓN *de impurezas epitaxiales* top-bottom diffusion; **~ térmica** *f* FÍS, NUCL, TERMO thermal diffusion; **~ de Thomson** *f* FÍS Thomson scattering; **~ tipo p** *f* ELECTRÓN p-type diffusion; **~ tubular** *f* METAL pipe diffusion; **~ turbulenta** *f* NUCL eddy diffusion, turbulent diffusion; **~ vertical** *f* CONTAM vertical dispersion; **~ de volumen** *f* METAL bulk diffusion, volume diffusion; **~ de la vorticidad** *f* FÍS FLUID *conductividad térmica* vorticity diffusion

difusividad *f* FÍS diffusivity; **~ térmica** *f* FÍS, TERMO, TERMOTEC thermal diffusivity

difusómetro *m* INSTR diffusion apparatus, OCEAN diffusiometer, PROC QUÍ diffusion apparatus

difusor[1] *adj* C&V *vidrio* diffusing

difusor[2] *m* AGUA *ventilador centrífugo* volute chamber, CARBÓN *carburador* mixer, CINEMAT, ENERG RENOV, FOTO, PROC QUÍ, REFRIG diffuser, TEC ESP *carburador* mixer, TELECOM diffuser, VEH *carburador* venturi; **~ de aire** *m* REFRIG air diffuser; **~ lineal** *m* REFRIG slot diffuser; **~ de paja** *m* AGRIC straw spreader; **~ perfecto reflectante** *m* PAPEL perfect reflecting diffuser; **~ de techo** *m* REFRIG ceiling diffuser

digestibilidad *f* AGRIC digestibility; **~ real** *f* AGRIC true digestibility

digestión *f* PAPEL boiling, cooking, QUÍMICA, RECICL digestion; **~ aeróbica** *f* RECICL aerobic digestion; **~ aeróbica de fangos** *f* AGUA aerobic sludge digestion; **~ anaeróbica** *f* RECICL anaerobic digestion; **~ de lodos** *f* HIDROL *alcantarillado*, RECICL sludge digestion

digestor *m* HIDROL, PAPEL, PROD, RECICL, TERMO digester; **~ aeróbico** *m* RECICL aerobic digester; **~ vertical** *m* PROD vertical digester

digitación *f* GEOL *de yacimientos, estratos*, PETROL fingering

digital[1] *adj* ELECTRÓN, INFORM&PD digital; **~ a analógico** *adj* (*D-a-A*) INFORM&PD, TV digital-to-analog

digital[2] *m* ALIMENT digitalis

digitalina *f* QUÍMICA digitalin

digitalización *f* ELECTRÓN, FÍS, INFORM&PD digitization, ING MECÁ digitalization, NUCL, TELECOM, TV digitization; **~ de imagen** *f* ELECTRÓN, TELECOM, TV image digitization; **~ de la voz** *f* ELECTRÓN, TELECOM, TV voice digitization

digitalizado *adj* ELEC, ELECTRÓN, FÍS, FOTO, INFORM&PD, NUCL digitized, TEC ESP *instrumentos ópticos* digitalized, TELECOM digitized

digitalizador *m* ELECTRÓN, INFORM&PD, TELECOM digitizer; **~ de imagen** *m* ELECTRÓN image digitizer

digitalizar *vt* GEN digitize

digitización: ~ de imagen *f* ELECTRÓN, TELECOM, TV image digitization; **~ de señal** *f* ELECTRÓN, TELECOM, TV signal digitization; **~ de la voz** *f* ELECTRÓN, TELECOM, TV voice digitization

digitizador: ~ de señal *m* ELECTRÓN, TELECOM, TV signal digitizer

dígito: ~ de acarreo *m* INFORM&PD carry digit; **~ de arrastre** *m* INFORM&PD carry digit; **~ binario** *m* (*bit*) ELECTRÓN, IMPR, INFORM&PD, TELECOM binary digit (*bit*); **~ de control** *m* INFORM&PD check digit; **~ más significativo** *m* (*DMS*) INFORM&PD, MATEMÁT, PROD most significant digit (*MSD*); **~ de menor peso** *m* INFORM&PD, MATEMÁT, PROD least significant digit (*LSD*); **~ menos significativo** *m* (*DMS*) INFORM&PD, MATEMÁT, PROD least significant digit (*LSD*); **~ de paridad** *m* INFORM&PD parity digit; **~ de relleno** *m* TELECOM stuffing digit; **~ significativo** *m* MATEMÁT significant figure; **~ de signo** *m* INFORM&PD sign digit

dígitos: ~ idénticos consecutivos *m pl* TELECOM consecutive identical digits (*CID*); **~ de identificación nacional** *m pl* TELECOM national identification digits (*NID*)

diglicidiléter *m* P&C diglycidyl ether

diguanida *f* QUÍMICA diguanide

dihidroacridina *f* QUÍMICA dihydroacridine

dihidrobenceno *m* QUÍMICA dihydrobenzene

dihidrocarveol *m* QUÍMICA dihydrocarveol

dihidrocarvona *f* QUÍMICA dihydrocarvone

dihidroergotamina *f* QUÍMICA dihydroergotamine

dihidroestreptomicina *f* QUÍMICA dihydrostreptomycin

dihidronaftaleno *m* QUÍMICA dihydronaphthalene

dihidrotaquisterol *m* QUÍMICA dihydrotachysterol

dihidroxiacetona *f* (*DHA*) QUÍMICA dihydroxyacetone (*DHA*)

diisocianato: ~ de difenilmetano *m* (*MDI*) P&C diphenylmethane diisocyanate (*MDI*); **~ de hexametileno** *m* P&C hexamethylene diisocyanate; **~ de tolueno** *m* (*TDI*) P&C toluene diisocyanate (*TDI*)

dilatable *adj* CARBÓN dilatable

dilatación *f* ELECTRÓN, FÍS expansion, FOTO extension, GAS dilatation, GEOL extension, GEOM dilation, ING MECÁ extension, stretch, INSTAL HIDRÁUL dilatation, expansion, INSTR enlargement, MATEMÁT, MECÁ expansion, METAL expansion, growth, P&C expansion, swelling, QUÍMICA, REFRIG, TEC ESP, TERMO expansion; **~ del cristal** *f* METAL crystal growth; **~ facetada** *f* C&V faceted expansion, METAL faceted expansion, faceted growth; **~ del papel** *f* PAPEL paper swelling; **~ temporal** *f* FÍS time dilation; **~ térmica** *f* TEC PETR *termodinámica*, TELECOM thermal expansion, TERMO heat expansion, heat dilatation, thermal expansion

dilatador[1] *adj* CARBÓN dilatant

dilatador[2] *m* GEOM dilator

dilatancia *f* CARBÓN, P&C *viscosidad, prueba* dilatancy

dilatante *adj* CARBÓN dilatant

dilatarse *v refl* INSTAL HIDRÁUL *vapor, gases* expand

dilatómetro *m* FÍS dilatometer

dilución *f* PROC QUÍ dilution, PROD *de pinturas* thining, thinning down, QUÍMICA dilution, thinning, TEC PETR *ingeniería de lodos* dilution; **~ de espuma** *f* PROC QUÍ foam dilution

diluente *m* COLOR extender, P&C *revestimientos, pinturas, adhesivos* diluent

diluido *adj* METAL *aleación, solución* dilute; **no ~** *adj* QUÍMICA undiluted

diluir *vt* CARBÓN dilute, COLOR distemper, METAL dilute, P&C *revestimientos, pinturas, adhesivos* dilute, *viscosidad* cut, PROC QUÍ dilute, PROD *pinturas* thin, QUÍMICA dilute; **~ con agua** *vt* QUÍMICA water

diluvio *m* METEO deluge

diluyente *m* COLOR distemper, thinner, P&C *revestimientos, pinturas, adhesivos* diluent, PETROL thinner, QUÍMICA thinner, diluent; **~ al aceite** *m* COLOR oil-bound distemper; **~ de pintura** *m* COLOR, REVEST paint thinner

dimensión *f* METR measure

dimensionamiento *m* CARBÓN sizing, INFORM&PD dimensioning, sizing, ING MECÁ, PROD dimensioning; **~ horizontal y vertical** *m* PROD horizontal and vertical dimensioning

dimensiones: **~ de conexión** *f pl* ING MECÁ connecting dimensions; **~ de conexión del manguito portaherramientas** *f pl* ING MECÁ connecting dimensions of chucks; **~ de conexión del plato** *f pl* ING MECÁ connecting dimensions of chucks; **~ de conexión del portabrocas** *f pl* ING MECÁ connecting dimensions of chucks; **~ exteriores** *f pl* EMB outside dimensions, overall dimensions, ING MECÁ overall dimensions, outside dimensions; **~ extremas** *f pl* ING MECÁ overall dimensions; **~ de pantalla** *f pl* CINEMAT screen ratio; **~ superiores** *f pl* GEOM higher dimensions; **~ totales** *f pl* EMB, ING MECÁ overall dimensions

dimérico *adj* QUÍMICA dimeric

dímero *m* QUÍMICA dimer

dímeros: **~ alquil-ceténicos** *m pl* PAPEL alkyl ketone dimers (*AKD*)

dimetilacético *adj* QUÍMICA dimethylacetic

dimetilamina *f* (*DMA*) QUÍMICA dimethylamine (*DMA*)

dimetilanilina *f* QUÍMICA xylidine, *intermediario* dimethylaniline

dimetilarsina *f* QUÍMICA dimethylarsine

dimetilbenceno *m* QUÍMICA xylene, dimethylbenzene

dimetilfenol *m* QUÍMICA xylenol

dimetilformamida *f* (*DMF*) QUÍMICA dimethylformamide (*DMF*)

dimetilglioxima *f* QUÍMICA dimethylglyoxime

dimetilo *m* QUÍMICA dimethyl

dimetilsulfona *f* QUÍMICA dimethyl sulfone (*AmE*), dimethyl sulphone (*BrE*)

dimetoximetano *m* QUÍMICA dimethoxymethane

dimorfismo *m* CRISTAL, QUÍMICA *cristales* dimorphism

DIN *abr* (*Deutsches Institut für Normung*) PROD DIN (*German Institute for Normalization*)

dina *f* METR *unidad de fuerza* dyne

dinaftilo *m* QUÍMICA dinaphthyl

dinámica *f* GEN dynamics; **~ estructural** *f* ING MECÁ structural dynamics; **~ de los fluidos** *f* FÍS FLUID fluid dynamics; **~ de fluidos computacional** *f* (*CFD*) FÍS FLUID, TRANSP AÉR computational fluid dynamics (*CFD*); **~ de las masas en movimiento** *f* NUCL particle dynamics; **~ oceánica** *f* TRANSP MAR ocean dynamics; **~ de partículas** *f* FÍS PART, NUCL particle dynamics; **~ solar** *f* ENERG RENOV solar dynamics

dinámico *adj* FÍS FLUID *proceso*, METAL dynamical

dinamita *f* MINAS dynamite; **~ amoniacal** *f* MINAS ammonia dynamite; **~ de base explosiva** *f* MINAS extradynamite; **~ de gelatina** *f* MINAS gelatine dynamite; **~ goma** *f* MINAS blasting gelatine; **~ semigelatinosa** *f* MINAS semigelatin dynamite, slurry; **~ de temperatura de congelación baja** *f* MINAS low-freezing dynamite

dinamitero *m* CARBÓN, CONST, MINAS blaster, shotfirer

dinamoeléctrico *adj* ELEC, ING ELÉC dynamo-electric

dinamo, dínamo *f* ELEC, ING ELÉC, TRANSP MAR, VEH dynamo; **~ acílica** *f* ELEC, ING ELÉC acyclic dynamo; **~ de anillo plano** *f* ELEC, ING ELÉC flat-ring dynamo; **~ de baliza** *f* ELEC, ING ELÉC beacon generator; **~ para carga de baterías** *f* ELEC, ING ELÉC battery charger; **~ de compensación** *f* ELEC, ING ELÉC balancer, buffer dynamo; **~ de control** *f* ELEC, ING ELÉC control dynamo; **~ de corriente constante** *f* ELEC, ING ELÉC constant-current dynamo; **~ de cuna** *f* ELEC, ING ELÉC cradle dynamo; **~ dinamométrica** *f* ELEC dynamometric dynamo; **~ elevadora de tensión** *f* ELEC, ING ELÉC booster dynamo; **~ excitada en derivación** *f* ELEC, ING ELÉC shunt dynamo, shunt-wound dynamo; **~ excitada independientemente** *f* ELEC, ING ELÉC separate excited dynamo; **~ excitada en serie** *f* ELEC, ING ELÉC series dynamo; **~ excitadora** *f* ELEC, ING ELÉC exciter; **~ excitatriz** *f* ELEC, ING ELÉC exciting dynamo; **~-freno** *f* ELEC, ING ELÉC brake dynamo; **~ de polos interiores** *f* ELEC, ING ELÉC internal pole dynamo; **~ reguladora** *f* ELEC, ING ELÉC regulating dynamo; **~ serie** *f* ELEC, ING ELÉC series-wound dynamo; **~ de voltaje constante** *f* ELEC, ING ELÉC constant-current dynamo, constant-voltage dynamo

dinamógrafo *m* MECÁ dynamograph

dinamómetro *m* INSTR, LAB, MECÁ dynamometer, OCEAN dynamic tide; **~ de torsión** *m* LAB *resistencia de materiales* torsion meter; **~ de tracción** *m* PAPEL tensile strength tester

dinamotor *m* ELEC, ING ELÉC dynamotor

dinitrobenceno *m* QUÍMICA dinitrobenzene

dinitrocresol *m* (*DNC*) QUÍMICA dinitrocresol (*DNC*)

dinitrofenol *m* QUÍMICA dinitrophenol

dinitronaftaleno *m* QUÍMICA dinitronaphthalene

dinitrotolueno *m* (*DNT*) QUÍMICA dinitrotoluene (*DNT*)

dinky *m* CINEMAT dinky

dino: **~ magnético** *m* ELEC *automotor* mag-dyno

dínodo m ELECTRÓN electron mirror, FÍS dynode, INSTR electron mirror

dintel m CONST beam, lintel, MINAS *galería* cap piece; ~ **de madera** m CONST bressummer (*BrE*), brestsummer (*AmE*); ~ **del puerto** m C&V port sill

dinucleótido m QUÍMICA dinucleotide

diodo m GEN diode; ~ **activador** m ELECTRÓN trigger diode; ~ **de aislamiento** m ELECTRÓN isolation diode; ~ **con aislante Schottky** m ELECTRÓN Schottky-barrier diode; ~ **de aleación** m ELECTRÓN alloy diode; ~ **de almacenamiento de carga** m ELECTRÓN charge-storage diode; ~ **de amplificador paramétrico** m ELECTRÓN parametric amplifier diode; ~ **de amplificador paramétrico de arseniuro de galio** m ELECTRÓN gallium arsenide parametric amplifier diode; ~ **de arseniuro de galio** m ELECTRÓN, FÍS, ÓPT gallium arsenide diode; ~ **atenuador** m ELECTRÓN attenuator diode; ~ **atenuador NIP** m ELECTRÓN PIN attenuator diode; ~ **de avalancha** m ELECTRÓN, FÍS, TELECOM avalanche diode; ~ **de avalancha de silicio** m ELECTRÓN silicon avalanche diode; ~ **de baja dispersión** m ELECTRÓN low-leakage diode; ~ **de baja potencia** m ELECTRÓN low-power diode; ~ **de banda S** m ELECTRÓN S-band diode; ~ **de barrera intrínseca** m ELECTRÓN intrinsic barrier diode; ~ **de base doble** m ELECTRÓN double base diode; ~ **bipolar** m ELECTRÓN bipolar diode; ~ **de bloqueo** m ELECTRÓN, FÍS clamping diode; ~ **de bloqueo Schottky** m ELECTRÓN Schottky clamping diode; ~ **de Burrus** m ÓPT, TELECOM Burrus diode; ~ **coaxial** m ELECTRÓN coaxial diode; ~ **de conmutación** m ELECTRÓN, TELECOM, TV diode switch, switching diode; ~ **conmutador** m TELECOM diode switch, switching diode; ~ **de contacto de punta** m ELECTRÓN point-contact diode; ~ **de corriente elevada** m ELECTRÓN high-current diode; ~ **de cristal** m ELECTRÓN crystal diode; ~ **de cuatro capas** m ELECTRÓN four-layer diode; ~ **con desbloqueo** m ELECTRÓN gated diode; ~ **de descarga** m ELECTRÓN avalanche diode; ~ **detector** m ELECTRÓN detector diode; ~ **detector con aislante Schottky** m ELECTRÓN Schottky-barrier detector diode; ~ **detector de contacto de punta** m ELECTRÓN point-contact detector diode; ~ **de detector de silicio** m ELECTRÓN silicon detector diode; ~ **dopado con oro** m ELECTRÓN gold-doped diode; ~ **de efecto Gunn** m ELECTRÓN, FÍS Gunn-effect diode; ~ **electroluminiscente** m (*DEL*) GEN light-emitting diode (*LED*); ~ **electroluminiscente de emisión superficial** m ELECTRÓN ÓPT, TELECOM surface-emitting electroluminescent diode, surface-emitting light-emitting diode; ~ **electromagnético** m ELECTRÓN magnetodiode; ~ **de emisión de borde** m ELECTRÓN, ÓPT edge-emitting diode; ~ **emisivo** m ELECTRÓN emissive diode; ~ **emisor** m ELECTRÓN, ING ELÉC emitter diode; ~ **emisor de luz** m (*DEL*) ELEC, ELECTRÓN, FÍS, FÍS RAD, INFORM&PD, ÓPT, TELECOM, TV light-emitting diode (*LED*); ~ **de enclavamiento por corriente continua** m ELECTRÓN DC clamp diode; ~ **de enfoque** m ING ELÉC, INSTR, TV focusing diode; ~ **epitaxial planar** m ELECTRÓN planar epitaxial diode; ~ **Esaki** m ELECTRÓN Esaki diode, tunnel diode, FÍS Esaki diode; ~ **de fijación** m ELECTRÓN, FÍS clamping diode; ~ **de fijación por corriente continua** m ELECTRÓN, TV DC clamp diode; ~ **fotodetector** m ELECTRÓN, TELECOM diode photodetector; ~ **fotoemisor de emisión marginal** m TELECOM edge-emitting light-emitting diode (*ELED*); ~ **de ganancia** m ELECTRÓN *televisión* efficiency diode; ~ **de gas** m ELECTRÓN, GAS gas diode; ~ **de germanio** m ELEC, ELECTRÓN germanium diode; ~ **de gran capacidad de conmutación** m ELECTRÓN high-speed switching diode; ~ **de gran conductancia** m ELECTRÓN high-conductance diode; ~ **de Gunn** m ELECTRÓN, FÍS Gunn diode; ~ **de homounión p-n** m ELECTRÓN p-n homojunction diode; ~ **inverso** m ELECTRÓN backward diode; ~ **láser** m ELECTRÓN, FÍS, ÓPT, TELECOM laser diode; ~ **láser de baja potencia** m TELECOM low-power laser diode; ~ **de láser de funcionamiento continuo** m ELECTRÓN CW laser diode; ~ **láser de inyección** m ELECTRÓN, ÓPT, TELECOM injection laser diode (*ILD*); ~ **de láser en monohetero-unión** m ELECTRÓN single-heterojunction laser diode; ~ **limitador** m ELECTRÓN, TV clipper diode, diode limiter, limiter diode; ~ **luminiscente** m ELECTRÓN luminescent diode; ~ **mesa** m ELECTRÓN mesa diode; ~ **mezclador** m ELECTRÓN diode mixer, mixer diode; ~ **mezclador con aislante Schottky** m ELECTRÓN Schottky-barrier mixer diode; ~ **mezclador de contacto de punta** m ELECTRÓN point-contact mixer diode; ~ **mezclador de silicio** m ELECTRÓN silicon mixer diode; ~ **de microondas** m ELECTRÓN microwave diode; ~ **modulador** m ELECTRÓN modulator diode; ~ **NIP** m ELECTRÓN PIN diode; ~ **parásito** m ELECTRÓN parasitic diode; ~ **PIN** m ELECTRÓN, TELECOM PIN diode; ~ **planar** m ELECTRÓN planar diode; ~ **de poca potencia** m ELECTRÓN low-power diode; ~ **portador con gran actividad** m ELECTRÓN hot carrier diode; ~ **de potencia** m ELECTRÓN power diode; ~ **de protección** m ELECTRÓN protection diode; ~ **receptor** m ELECTRÓN receiver diode; ~ **rectificador** m ELECTRÓN, ING ELÉC rectifier diode; ~ **rectificador con aislante Schottky** m ELECTRÓN Schottky-barrier rectifier diode; ~ **rectificador de gas** m ELEC, ING ELÉC gas-filled rectifier diode; ~ **de recuperación** m ELECTRÓN efficiency diode; ~ **de recuperación escalonada** m ELECTRÓN step recovery diode; ~ **de recuperación rápida** m ELECTRÓN fast-recovery diode; ~ **de referencia de voltaje** m ELECTRÓN voltage reference diode; ~ **regulador de voltaje** m ELECTRÓN voltage regulator diode; ~ **de resistencia negativa** m ELECTRÓN negative resistance diode; ~ **de rueda libre** m ELECTRÓN, PROD freewheeling diode; ~ **de ruidos** m ELECTRÓN noise diode; ~ **de ruptura brusca** m ELECTRÓN snap-off diode; ~ **de Schottky** m ELECTRÓN, FÍS Schottky diode; ~ **semiconductor** m ELECTRÓN, INFORM&PD semiconductor diode; ~ **Shockley** m ELECTRÓN Shockley diode; ~ **de silicio** m ELEC, ELECTRÓN silicon diode; ~ **de silicio de contacto de punta** m ELECTRÓN point-contact silicon diode; ~ **superluminiscente** m ÓPT, TELECOM superluminescent diode (*SLD*); ~ **superradiante** m ÓPT, TELECOM superradiant diode (*SRD*); ~ **térmico** m ELECTRÓN thermal diode; ~ **de tiempo de tránsito** m ELECTRÓN transit-time diode; ~ **de tiempo de tránsito por avalancha** m FÍS avalanche transit-time diode; ~ **de tiempo de tránsito por avalancha con ionización por choque** m ELECTRÓN, FÍS impact ionization avalanche transit-time diode (*IMPATT diode*); ~ **de tiempo de tránsito por inyección en barrera** m FÍS

barrier injection transit-time diode (*BARITT diode*); ~ **de transferencia de electrones** *m* ELECTRÓN transferred-electron diode; ~ **de transferencia electrónica** *m* FÍS electron-transfer diode; ~ **TRAPATT** *m* ELECTRÓN, FÍS TRAPATT diode (*trapped plasma avalanche transit-time diode*); ~ **de TTAICH** *m* ELECTRÓN, FÍS IMPATT diode (*impact ionization avalanche transit-time diode*); ~ **túnel** *m* ELECTRÓN, FÍS tunnel diode; ~ **de unión** *m* ELEC *semiconductor*, ELECTRÓN junction diode; ~ **de unión NP** *m* ELEC *semiconductor* PN junction diode; ~ **de unión PN** *m* ELECTRÓN PN junction diode; ~ **de unión de silicio** *m* ELECTRÓN silicon junction diode; ~ **de vacío** *m* ELECTRÓN vacuum diode; ~ **varactor** *m* ELECTRÓN, FÍS, TV varactor diode; ~ **varactor hiperabrupto** *m* ELECTRÓN, FÍS hyperabrupt varactor diode; ~ **de vector mayoritario** *m* ELECTRÓN majority-carrier diode; ~ **Zener** *m* ELECTRÓN, FÍS, TELECOM Zener diode; ~ **Zener de silicio** *m* ELECTRÓN, FÍS, TV silicon Zener diode; ~ **Zener con temperatura compensada** *m* ELECTRÓN temperature-compensated Zener diode
diodos: ~ **coincidentes** *m pl* ELECTRÓN matched diodes
dioico *adj* AGRIC *planta* dioecious
diol *m* QUÍMICA diol
diolefina *f* TEC PETR *refino* diolefin
diópsido *m* MINERAL diopside
dioptasa *f* MINERAL dioptase, emerald copper
dioptría *f* FÍS, FOTO, ÓPT, TEC ESP diopter (*AmE*), dioptre (*BrE*); ~ **positiva** *f* CINEMAT plus diopter (*AmE*), plus dioptre (*BrE*)
diorita *f* PETROL diorite
dióxido *m* QUÍMICA dioxide; ~ **de azufre** *m* CONTAM, PROC QUÍ, QUÍMICA sulfur dioxide (*AmE*), sulphur dioxide (*BrE*); ~ **de carbono** *m* CARBÓN, ING MECÁ, QUÍMICA carbon dioxide; ~ **de carbono sólido** *m* QUÍMICA dry ice; ~ **de estaño** *m* QUÍMICA stannic oxide; ~ **de manganeso** *m* ING ELÉC manganese dioxide; ~ **de nitrógeno** *m* CONTAM, QUÍMICA, SEG nitrogen dioxide; ~ **de silicio** *m* ELECTRÓN silicon dioxide, EMB silica gel; ~ **sulfúrico** *m* CONTAM sulfur dioxide (*AmE*), sulphur dioxide (*BrE*); ~ **de titanio** *m* P&C *pigmento*, QUÍMICA titanium dioxide
dioxina *f* QUÍMICA dioxin
dioxitartárico *adj* QUÍMICA dioxytartaric
dioxopurina *f* QUÍMICA xanthine
DIP *abr* (*empaque doble en línea, paquete en línea doble*) INFORM&PD, ING ELÉC DIP (*dual-in-line package*)
dipalmitina *f* QUÍMICA dipalmitin
diparaclorobencilo *m* QUÍMICA diparachlorobenzyl
dipiro *m* MINERAL dipyre
diplacusia *f* ACÚST *audición* diplacusis
diplexor *m* TELECOM *antena* diplexer; ~ **de polarización** *m* TEC ESP, TELECOM polarization diplexer
diploacusia *f* ACÚST *audición* diplacusis
diploide *adj* AGRIC diploid
diploma: ~ **de vuelo por instrumentos** *m* TRANSP AÉR instrument rating
dipolar *adj* QUÍMICA *molécula* dipolar
dipolo *m* ACÚST, ELEC, FÍS, ING ELÉC, METAL, TELECOM dipole; ~ **activo** *m* FÍS, ING ELÉC, TELECOM active dipole; ~ **eléctrico** *m* ELEC, FÍS, ING ELÉC electric dipole; ~ **hertziano** *m* FÍS, TEC ESP, TELECOM Hertzian dipole; ~ **magnético** *m* FÍS magnetic dipole;

~ **pasivo** *m* FÍS, ING ELÉC, TELECOM passive dipole; ~ **de semionda** *m* FÍS, ING ELÉC, TELECOM half-wave dipole
dique *m* AmL (*cf terraplén Esp*) AGUA dam, retaining wall, dike, breakwater, CARBÓN *geología* rib, embankment, CONST *almacenamiento de agua* breakwater, retaining wall, groin (*AmE*), dike (*AmE*), batardeau, dyke (*BrE*), embankment, groyne (*BrE*), ENERG RENOV dike (*AmE*), dyke (*BrE*), FERRO embankment, GEOL levee, hog back, HIDROL *terraplenar, abrir zanjas* dike (*AmE*), dyke (*BrE*), MINAS embankment, OCEAN breakwater, groin (*AmE*), groyne (*BrE*), TRANSP MAR breakwater, *construcción* dock; ~ **de aguas profundas** *m* CONST *puertos* deep-water dock; ~ **avertedero** *m* AGUA spillway dam; ~ **de carenas** *m* CONST, TRANSP MAR dry dock; ~ **sin central** *m* ENERG RENOV dead dike (*AmE*), dead dyke (*BrE*); ~ **de construcción** *m* TRANSP MAR fitting-out berth; ~ **de desvío** *m* AGUA diversion dam; ~ **de embalse** *m* AGUA storage dam; ~ **de escollera** *m* AGUA rock-fill dam; ~ **flotante** *m* CONST floating dock, TRANSP pontoon dock, TRANSP MAR floating dock, wet dock; ~ **de mar** *m* AGUA sea dike (*AmE*), sea dyke (*BrE*); ~ **de marea** *m* TRANSP MAR wet dock; ~ **mixto** *m* GEOL *intrusión ígnea* composite dike, composite dyke; ~ **de presa** *m* AGUA, CONST, HIDROL cofferdam, TRANSP MAR cofferdam, daggerboard; ~ **para regeneración de costas** *m* OCEAN spur dike (*AmE*), spur dyke (*BrE*); ~ **rompeolas** *m* AGUA, CONST mole; ~ **seco** *m* CONST dry dock, TRANSP MAR dry dock, graving dock; ~ **de tierra** *m* AGUA, HIDROL earth dam
dirección[1]: **en** ~ *adv* MINAS *filones* along the strike
dirección[2] *f* AUTO, D&A steering, ELECTRÓN *señales* directing, GEOL *de vector* sense, HIDROL *de corriente* drift, INFORM&PD address, *memoria* location, ING MECÁ guiding, MINAS *filón, veta* trend, *filones* bearing, *de filón* course, TEC ESP vector, pointing, course, TELECOM, TRANSP, VEH steering; ~ **absoluta** *f* INFORM&PD absolute address; ~ **asistida** *f* AUTO, VEH power steering, power-assisted steering; ~ **de buzamiento** *f* GEOL, MINAS strike; ~ **calculada** *f* INFORM&PD generated address; ~ **de calidad** *f* CALIDAD, CONST, ING MECÁ, PROD quality management; ~ **del campo inductor** *f* ELEC, ING ELÉC field direction; ~ **de compactación** *f* GEOL, TEC PETR *geología* compaction trend; ~ **compartida** *f* TELECOM party address; ~ **del contador** *f* PROD counter address; ~ **de la corriente** *f* ELEC current direction, HIDROL *río* direction of flow, ING ELÉC current direction, TRANSP MAR current set; ~ **del crucero más resistente** *f* MINAS *canteras* head; ~ **del curso** *f* HIDROL *río* direction of flow; ~ **directa** *f* INFORM&PD direct address; ~ **a distancia** *f* AUTO, CINEMAT, CONST, D&A, ELEC, FOTO, ING MECÁ, MECÁ, TEC ESP, TELECOM, TRANSP, TV, VEH remote control; ~ **efectiva** *f* Esp (*cf direccional efectivo AmL*) INFORM&PD effective address; ~ **electrónica** *f* INFORM&PD e-mail address; ~ **de encaminamiento** *f* TELECOM routing address; ~ **de encaminamiento inverso** *f* TELECOM reverse-routing address; ~ **explícita** *f* Esp (*cf direccional explícito AmL*) INFORM&PD explicit address; ~ **de fabricación** *f* PAPEL machine direction; ~ **de fin de instrucción** *f* PROD end-statement address; ~ **de flujo** *f* INFORM&PD flow direction; ~ **en función del temporizador** *f* PROD timer address; ~ **generada** *f* Esp (*cf direccional*

generado AmL) INFORM&PD generated address; ~ **del haz electrónico** *f* ELECTRÓN electronic-beam steering; ~ **inmediata** *f* INFORM&PD immediate address; ~ **inversa** *f* INFORM&PD reverse direction, ING ELÉC inverse direction, reverse direction; ~ **inversa de flujo** *f* ELEC, INFORM&PD reverse-direction flow; ~ **invertida** *f* ING ELÉC inverse direction; ~ **lateral** *f* TRANSP lateral guidance; ~ **longitudinal según el lado más corto** *f* IMPR, PAPEL short grain; ~ **magnética** *f* GEOFÍS magnetic bearing, TRANSP AÉR magnetic heading; ~ **de memoria** *f* INFORM&PD memory location, storage location; ~ **en memoria** *f* PROD location address; ~ **normal** *f* GEOL normal trend; ~ **de operaciones** *f* NUCL operative management; ~ **original** *f* INFORM&PD source address; ~ **con palabra de la tabla de datos** *f* PROD data table word address; ~ **paleocorriente** *f* GEOL palaeocurrent direction (*BrE*), paleocurrent direction (*AmE*); ~ **del paso del cableado** *f* ELEC *de componente de cable* direction of lay; ~ **de piñón y cremallera** *f* AUTO, VEH rack-and-pinion steering; ~ **de propagación de las olas** *f* OCEAN wave propagation direction; ~ **pulsátil** *f* PROD pulsating address; ~ **por radio** *f* TRANSP radio steering; ~ **real** *f* INFORM&PD real address; ~ **receptora** *f* PROD receiving address; ~ **de retenida** *f* PROD latch address; ~ **de retorno** *f* INFORM&PD return address; ~ **de la rueda del morro** *f* TRANSP AÉR nose-wheel steering; ~ **simple** *f* INFORM&PD single address; ~ **terrestre** *f* TELECOM ground address; ~ **total de la calidad** *f* CALIDAD total quality management (*TQM*); ~ **del tráfico** *f* TRANSP traffic management; ~ **transversal** *f* PAPEL cross direction; ~ **del tren de aterrizaje del morro** *f* TRANSP AÉR nose gear steering; ~ **de la ubicación de terminales de E/S** *f* PROD I/O terminal location address; ~ **única** *f* INFORM&PD single address; ~ **de vergencia** *f* GEOL facing direction; ~ **del viento** *f* METEO, TRANSP AÉR, TRANSP MAR wind direction

direccionable *adj* INFORM&PD addressable

direccional: ~ **directo** *m* INFORM&PD direct address; ~ **efectivo** *m AmL* (*cf dirección efectiva Esp*) INFORM&PD effective address; ~ **explícito** *m AmL* (*cf dirección explícita Esp*) INFORM&PD explicit address; ~ **de la fuente** *m* INFORM&PD source address; ~ **generado** *m AmL* (*cf dirección generada Esp*) INFORM&PD generated address; ~ **inmediato** *m* INFORM&PD immediate address, immediate addressing; ~ **real** *m* INFORM&PD real address; ~ **de retorno** *m* INFORM&PD return address; ~ **único** *m* INFORM&PD single address

direccionamiento *m* INFORM&PD, TELECOM addressing; ~ **adaptable** *m* INFORM&PD adaptive routing; ~ **asociativo** *m* INFORM&PD associative addressing; ~ **de base** *m* INFORM&PD base address; ~ **centralizado** *m* INFORM&PD centralized routing; ~ **complementario de E/S** *m* PROD complementary I/O addressing; ~ **diferido** *m* INFORM&PD deferred addressing; ~ **directo** *m* INFORM&PD direct addressing; ~ **duplicado de entrada y salida** *m* PROD duplicate I/O addressing; ~ **extendido** *m* INFORM&PD extended addressing; ~ **extensible** *m* INFORM&PD extensible addressing; ~ **implícito** *m* INFORM&PD implied addressing; ~ **indexado** *m* INFORM&PD indexed addressing; ~ **indicado** *m* INFORM&PD indexed addressing; ~ **indirecto** *m* INFORM&PD indirect addressing; ~ **inherente**

m INFORM&PD inherent addressing; ~ **inmediato** *m* INFORM&PD immediate addressing; ~ **lógico** *m* INFORM&PD logical addressing; ~ **múltiple** *m* TELECOM multiaddressing; ~ **relativo** *m* INFORM&PD relative addressing; ~ **simbólico** *m* INFORM&PD symbolic addressing; ~ **único de E/S** *m* INFORM&PD, PROD unique I/O location addressing

direccionar *vt* INFORM&PD address

directiva: ~ **de compilador** *f* INFORM&PD compiler directive

directividad *f* TEC ESP, TELECOM, TV directivity; ~ **de la antena** *f* TELECOM, TV aerial directivity (*BrE*), antenna directivity (*AmE*)

directo *adj* AGUA, ELEC, INFORM&PD, TELECOM, TRANSP, TRANSP AÉR direct

director *m* ACÚST master, CINEMAT, TEC ESP *dispositivos de guía* director; ~ **de la academia de TV** *m* TV TV academy leader; ~ **de aeropuerto** *m* TRANSP AÉR airport manager; ~ **de animación** *m* CINEMAT animation director; ~ **cinematográfico** *m* CINEMAT film director; ~ **del estudio** *m* TV studio manager; ~ **de fotografía** *m* CINEMAT director of photography (*DP*); ~ **de proyecto** *m* CONST, PROD project manager; ~ **de vuelo** *m* TRANSP AÉR flight director

directorio *m* INFORM&PD, TELECOM directory; ~ **de archivos** *m* INFORM&PD file directory; ~ **de ayuda** *m* INFORM&PD, PROD help directory; ~ **de consulta principal** *m* PROD master help directory; ~ **de contenido** *m* INFORM&PD contents directory; ~ **electrónico** *m* TELECOM electronic directory; ~ **de ficheros** *m* INFORM&PD file directory; ~ **de impresión** *m* IMPR, INFORM&PD printer directory; ~ **telefónico** *m AmL* (*cf guía telefónica Esp*) TELECOM telephone directory; ~ **de teléfonos** *m AmL* (*cf guía telefónica Esp*) TELECOM telephone directory

directriz *f* INFORM&PD directive, ING MECÁ guiding, guideline, INSTAL HIDRÁUL *turbina* guide, PROD guideline; ~ **de compilación** *f* INFORM&PD compiler directive

dirigido: ~ **por láser** *adj* ELECTRÓN laser-guided; ~ **por radio** *adj* FÍS, TELECOM, TRANSP radio-controlled

dirigir *vt* CINEMAT direct, PROD *negocio* operate, run, TELECOM direct; ~ **por radio** *vt* TELECOM, TRANSP radioguide

disacárido *m* QUÍMICA disaccharide

disanalita *f* MINERAL dysanalyte

disarmónico *adj* GEOL disharmonic

discado: ~ **de corto código** *m* TELECOM short code dialing (*AmE*), short code dialling (*BrE*); ~ **de dígito único** *m* TELECOM single digit dialing (*AmE*), single digit dialling (*BrE*); ~ **de marcación con pulsador** *m* TELECOM push-button dial

disclasita *f* MINERAL dysclasite

disclinación: ~ **por alabeo** *f* METAL twist disclination; ~ **en cuña** *f* METAL wedge disclination

disco *m* ACÚST record, ELECTRÓN wafer, EMB disc (*BrE*), disk (*AmE*), INFORM&PD platter, disk, ING MECÁ disc (*BrE*), disk (*AmE*), TELECOM dial, TRANSP AÉR, TV disc (*BrE*), disk (*AmE*); ~ **abrasivo** *m* ING MECÁ abrasive disc (*BrE*), abrasive disk (*AmE*); ~ **accionador** *m* TRANSP AÉR actuator disc (*BrE*), actuator disk (*AmE*); ~ **de Airy** *m* FÍS Airy disc (*BrE*), Airy disk (*AmE*); ~ **cambiable** *m* INFORM&PD exchangeable disk; ~ **de capacitancia** *m* ÓPT capacitance disc (*BrE*), capacitance disk (*AmE*); ~ **cilíndrico de amianto** *m* C&V asbestos roll disc

(*BrE*), asbestos roll disk (*AmE*); ~ **compacto** *m* (*CD*) INFORM&PD compact disk (*CD*), ÓPT compact disc (*CD*) (*BrE*), compact disk (*CD*) (*AmE*); ~ **compacto de audio** *m* ÓPT audio compact disc (*BrE*), audio compact disk (*AmE*), CD audio disc (*BrE*), CD audio disk (*AmE*); ~ **compacto interactivo** *m* ÓPT interactive compact disc (*BrE*), interactive compact disk (*AmE*); ~ **compacto de memoria de sólo lectura** *m* (*CD-ROM*) ÓPT compact disc read only memory (*BrE*) (*CD-ROM*), compact disk read only memory (*AmE*) (*CD-ROM*); ~ **compacto de memoria de sólo lectura programable** *m* ÓPT CD-PROM; ~ **compacto de música** *m* ÓPT compact music disc (*BrE*), compact music disk (*AmE*); ~ **conducido** *m* AUTO *embrague* driven plate, ING MECÁ driven disc (*BrE*), driven disk (*AmE*), VEH *embrague* driven plate; ~ **conductor** *m* ING MECÁ driving disc (*BrE*), driving disk (*AmE*); ~ **de corindón** *m* PROD corundum wheel; ~ **cortante** *m* ING MECÁ revolving cutter; ~ **de datos borrable** *m* INFORM&PD erasable data disk; ~ **de datos para escribir una sola vez** *m* ÓPT write-once data disc (*BrE*), write-once data disk (*AmE*), write-once disc (*BrE*), write-once disk (*AmE*); ~ **de datos no borrable** *m* ÓPT nonerasable data disc (*BrE*), nonerasable data disk (*AmE*); ~ **de datos ópticos** *m* INFORM&PD optical data disk, ÓPT optical data disc (*BrE*), optical data disk (*AmE*); ~ **directo** *m* ÓPT direct disc (*BrE*), direct disk (*AmE*); ~ **dividido en sectores por software** *m* INFORM&PD soft-sectored disk; ~ **divisor** *m* ING MECÁ division plate; ~ **de doble cara** *m* INFORM&PD double-sided disk, double-sided floppy diskette; ~ **de dos caras** *m* ÓPT two-sided disc (*BrE*), two-sided disk (*AmE*); ~ **duro** *m* INFORM&PD, TELECOM hard disk, Winchester disk; ~ **elaborado** *m* ACÚST processed disc (*BrE*), processed disk (*AmE*); ~ **electrónico de capacitancia** *m* ÓPT capacitance electronic disc (*BrE*) (*CED*), capacitance electronic disk (*AmE*) (*CED*); ~ **de embrague** *m* AUTO, VEH clutch disc (*BrE*), clutch disk (*AmE*), clutch plate; ~ **para engranaje** *m* ING MECÁ gear blank; ~ **de esfuerzo** *m* C&V strain disk (*AmE*), strain disc (*BrE*); ~ **de esmeril horizontal** *m* C&V horizontal grinding disc (*BrE*), horizontal grinding disk (*AmE*); ~ **de excéntrica** *m* ING MECÁ eccentric disc (*BrE*), eccentric disk (*AmE*), eccentric sheave, sheave; ~ **excéntrico** *m* ING MECÁ eccentric disc (*BrE*), eccentric disk (*AmE*); ~ **extraíble** *m* INFORM&PD removable disk; ~ **de Faraday** *m* FÍS Faraday's disc (*BrE*), Faraday's disk (*AmE*); ~ **de fibra vulcanizada** *m* ING MECÁ vulcanized fiber disk (*AmE*), vulcanized fibre disc (*BrE*); ~ **fijo** *m* INFORM&PD fixed disk; ~ **flexible** *m* IMPR, INFORM&PD, TELECOM floppy disk; ~ **flexible de doble lado** *m* INFORM&PD double-sided disk, double-sided floppy diskette; ~ **flotante** *m* MINAS *gasógenos sin agua* piston; ~ **de freno** *m* FERRO *vehículos* brake flange, MECÁ, VEH brake disc (*BrE*), brake disk (*AmE*); ~ **del freno** *m* VEH brake plate; ~ **de fricción** *m* ING MECÁ friction disc (*BrE*), friction disk (*AmE*); ~ **graduado** *m* ING MECÁ graduated dial; ~ **de gramófono** *m* ACÚST gramophone record; ~ **Hager** *m* C&V Hager disc (*BrE*), Hager disk (*AmE*); ~ **iluminado** *m* FOTO illuminated dial; ~ **de impresión** *m* IMPR printing disc (*BrE*), printing disk (*AmE*); ~ **indicador** *m* ING MECÁ locating disc (*BrE*),

locating disk (*AmE*); ~ **interactivo** *m* ÓPT interactive disc (*BrE*), interactive disk (*AmE*); ~ **intercambiable** *m* INFORM&PD exchangeable disk; ~ **de laca** *m* ACÚST lacquer disc (*BrE*), lacquer disk (*AmE*); ~ **de lagunas reticulares** *m* METAL vacancy disc (*BrE*), vacancy disk (*AmE*); ~ **láser** *m* ÓPT, TV laser disc (*BrE*), laser disk (*AmE*), laservision disc (*BrE*) (*LV disc*), laservision disk (*AmE*) (*LV disk*); ~ **de lectura multiespiral** *m* TV multispiral scanning disc (*BrE*), multispiral scanning disk (*AmE*); ~ **lineal** *m* ÓPT linear disc (*BrE*), linear disk (*AmE*); ~ **maestro** *m* INFORM&PD master disk; ~ **magnético** *m* ELEC magnetic disc (*BrE*), magnetic disk (*AmE*), INFORM&PD magnetic disk, ING ELÉC magnetic disc (*BrE*), magnetic disk (*AmE*); ~ **magnetoóptico** *m* (*disco m-o*) INFORM&PD magneto-optic disk (*m-o disk*), ÓPT magneto-optical disc (*BrE*) (*m-o disc*), magneto-optical disk (*AmE*) (*m-o disk*); ~ **matriz** *m* ACÚST mother; ~ **m-o** *m* (*disco magnetoóptico*) INFORM&PD m-o disk (*magneto-optic disk, magneto-optical disk*), ÓPT m-o disc (*BrE*) (*magneto-optical disc*), m-o disk (*AmE*) (*magneto-optic disk, magneto-optical disk*); ~ **de movimiento lento** *m* TV slow-motion disc (*BrE*), slow-motion disk (*AmE*); ~ **del núcleo** *m* ING MECÁ core plate; ~ **óptico** *m* INFORM&PD optical disk, ÓPT optical disc (*BrE*), optical disk (*AmE*); ~ **óptico borrable** *m* ÓPT erasable optical disc (*BrE*), erasable optical disk (*AmE*); ~ **óptico digital** *m* ÓPT digital optical disc (*BrE*), digital optical disk (*AmE*); ~ **óptico digital de datos** *m* INFORM&PD optical digital data disk, ÓPT optical digital data disc (*BrE*), optical digital data disk (*AmE*); ~ **óptico láser** *m* INFORM&PD laser-optic disk, ÓPT laser-optic disc (*BrE*), laser-optic disk (*AmE*); ~ **óptico registrable** *m* ÓPT writable optical disc (*BrE*), writable optical disk (*AmE*); ~ **óptico reutilizable** *m* ÓPT reusable optical disc (*BrE*), reusable optical disk (*AmE*); ~ **original** *m* ACÚST original disc (*BrE*), original disk (*AmE*), COMP&DP master disk, ÓPT master disc (*BrE*), master disk (*AmE*); ~ **posterior del rodete** *m* REFRIG *ventilador* impeller backplate; ~ **de presión** *m* ING MECÁ bearing plate; ~ **principal** *m* INFORM&PD master disk; ~ **de pulir** *m* ING MECÁ, METAL polishing wheel, PROD buff wheel, polishing wheel; ~ **ranurador** *m* C&V slitting disc (*BrE*), slitting disk (*AmE*); ~ **de Rayleigh** *m* ACÚST Rayleigh disc (*BrE*), Rayleigh disk (*AmE*); ~ **reflector** *m* ÓPT reflective disc (*BrE*), reflective disk (*AmE*); ~ **registrable** *m* ÓPT writable disc (*BrE*), writable disk (*AmE*); ~ **del rotor** *m* TRANSP AÉR rotor disc (*BrE*), rotor disk (*AmE*); ~ **de sectores blandos** *m* INFORM&PD soft-sectored disk; ~ **con secuencia de arranque** *m* INFORM&PD bootable disk, PROD bootable disc (*BrE*), bootable disk (*AmE*); ~ **separable** *m* INFORM&PD removable disk; ~ **de silicio** *m* ELECTRÓN silicon wafer; ~ **del sistema** *m* INFORM&PD system disk; ~ **de una sola cara** *m* INFORM&PD, ÓPT single-sided disc (*BrE*), single-sided disk (*AmE*); ~ **de una sola escritura** *m* INFORM&PD write-once disk; ~ **sólido experimental** *m* ING MECÁ experimental solid-disc flywheel (*BrE*), experimental solid-disk flywheel (*AmE*); ~ **de un solo lado** *m* INFORM&PD single-sided disk; ~ **de sólo lectura** *m* ÓPT read-only disc (*BrE*), read-only disk (*AmE*); ~ **de taladrar abovedado** *m* ING MECÁ dome pad; ~ **transmisivo** *m* ÓPT transmissive disc

(*BrE*), transmissive disk (*AmE*); ~ **transparente** *m* ÓPT transparent disc (*BrE*), transparent disk (*AmE*); ~ **de trapo para pulir** *m* PROD calico mop, rag wheel; ~ **de válvula** *m* AUTO valve disc (*BrE*), valve disk (*AmE*); ~ **de velocidad angular constante** *m* ÓPT CAV disc (*BrE*), CAV disk (*AmE*), constant angular velocity disc (*BrE*), constant angular velocity disk (*AmE*) (*CAV disk*); ~ **de velocidad lineal constante** *m* ÓPT CVL disc (*BrE*), CVL disk (*AmE*), constant linear velocity disc (*BrE*), constant linear velocity disk (*AmE*); ~ **de visión láser** *m* ÓPT laservision disc (*BrE*), laservision disk (*AmE*), LV disc (*BrE*), LV disk (*AmE*)

disconformidad *f* GEOL disconformity; ~ **angular** *f* GEOL angular disconformity

discontinuidad *f* GEOL unconformity, PROD breakdown; ~ **de Gutenberg** *f* GEOL Gutenberg discontinuity; ~ **hidráulica** *f* HIDROL hydraulic discontinuity; ~ **magnética** *f* GEOFÍS magnetic discontinuity; ~ **de Moho** *f* GEOL Moho discontinuity, Mohorovicic discontinuity

discontinuo *adj* ELECTRÓN, INFORM&PD discrete, ING MECÁ stepped, TELECOM discrete

discordancia *f* ACÚST discordance, GEOL disconformity, unconformity, discordance, TEC PETR unconformity; ~ **angular** *f* GEOL angular unconformity; ~ **erosional** *f* GEOL erosional unconformity; ~ **del kimeridgiense** *f* GEOL, TEC PETR Cimmerian unconformity; ~ **progresiva** *f* GEOL progressive unconformity

discos: ~ **del núcleo** *m pl* ING ELÉC core laminations

discrasita *f* MINERAL dyscrasite

discrepancia *f* FÍS, ING MECÁ, MATEMÁT discrepancy

discreto *adj* ELECTRÓN, INFORM&PD, TELECOM discrete

discriminación *f* ELECTRÓN discrimination; ~ **del filtro** *f* ELECTRÓN filter discrimination; ~ **selectiva cromática** *f* ELECTRÓN sampling

discriminador *m* ELECTRÓN, ING MECÁ, ÓPT, TELECOM, TV discriminator; ~ **cromático** *m* TELECOM, TV sampler; ~ **de fase** *m* ELECTRÓN phase discriminator; ~ **de protocolo** *m* TELECOM protocol discriminator (*PD*)

discriminante *m* ING MECÁ discriminant

disección *f* GEOM, LAB dissection

disector: ~ **de imagen** *m* ELECTRÓN image dissector

diseminación *f* TEC ESP scattering

diseminado *m* QUÍMICA spreading

diseminadora: ~ **de piedra** *f* TRANSP stone spreader

diseño: ~ **antisísmico** *m* CONST seismic design; ~ **asistido por computadora** *m* AmL (*CAD, cf diseño asistido por ordenador Esp*) ELEC, IMPR, INFORM&PD, MECÁ, PROD, TELECOM computer-aided design (*CAD*); ~ **asistido por ordenador** *m* Esp (*CAD, cf diseño asistido por computadora AmL*) ELEC, IMPR, INFORM&PD, MECÁ, PROD, TELECOM computer-aided design (*CAD*); ~ **base** *m* NUCL base design; ~ **calcado** *m* IMPR traced design; ~ **del chip** *m* ELECTRÓN chip design; ~ **del circuito** *m* ELECTRÓN, ING ELÉC circuit design; ~ **de circuito analógico** *m* ELECTRÓN, ING ELÉC, TELECOM analog circuit design; ~ **de circuito digital** *m* ELECTRÓN, ING ELÉC, TELECOM digital circuit design; ~ **de circuito integrado** *m* ELECTRÓN integrated circuit design; ~ **para esfuerzo de pandeo** *m* TEC ESP *nave espacial* design to buckling strength; ~ **para esfuerzos máximos** *m* TEC ESP *vehículos* design to ultimate strength;

~ **estructurado** *m* INFORM&PD structured design; ~ **y fabricación asistidos por computadora** *m* AmL (*cf diseño y fabricación asistidos por ordenador Esp*) INFORM&PD, PROD computer-aided design and manufacture; ~ **y fabricación asistidos por ordenador** *m* Esp (*cf diseño y fabricación asistidos por computadora AmL*) INFORM&PD, PROD computer-aided design and manufacture; ~ **factorial** *m* INFORM&PD, MATEMÁT factorial design; ~ **de fallo a prueba de fallo** *m* NUCL fail-safe design; ~ **de fallo sin riesgo de fallo** *m* NUCL fail-safe design; ~ **funcional** *m* INFORM&PD functional design; ~ **jerárquico orientado al objeto** *m* (*HOOD*) INFORM&PD hierarchical object-oriented design (*HOOD*); ~ **lógico** *m* ELECTRÓN, INFORM&PD logic design; ~ **del manguito** *m* INSTAL HIDRÁUL sleeve pattern; ~ **de máquina herramienta** *m* ING MECÁ machine tool design; ~ **para obtener resistencia máxima** *m* TEC ESP design to ultimate strength; ~ **orientado al objeto** *m* INFORM&PD object-oriented design (*OOD*); ~ **de pavimento** *m* CONST pavement design; ~ **perfeccionado** *m* PROD improved pattern; ~ **de programas** *m* INFORM&PD program design; ~ **para punto de deformación** *m* TEC ESP design to yield point; ~ **a punto de rotura** *m* TEC ESP *vehículos* design to yield point; ~ **de rayado cruzado** *m* CINEMAT, ING MECÁ cross-hatch pattern; ~ **del reactor** *m* NUCL reactor design; ~ **de reactores** *m* NUCL reactor art, reactor design; ~ **del software** *m* INFORM&PD software design; ~ **de soporte lógico** *m* INFORM&PD software design

disfunción *f* CALIDAD, INFORM&PD, METR, SEG, TEC ESP malfunction

disgregación *f* CONST *madera* breaking down, ELECTRÓN disjunction, GEOL *mineral, roca* breakdown; ~ **térmica** *f* ELECTRÓN, TERMO thermal breakdown

disgregador *m* PROD disintegrator

disgregadora: ~ **montada detrás** *f* AUTO, TRANSP, VEH rear-mounted ripper

disilano *m* QUÍMICA disilane

disimetría *f* GEOM dissymmetry, ING MECÁ unbalance, MATEMÁT, QUÍMICA dissymmetry

disimétrico *adj* GEOM, MATEMÁT, QUÍMICA dissymmetric, dissymmetrical

disipación *f* FÍS RAD *de calor, energía* dissipation; ~ **de calor por aletas** *f* NUCL fin cooling; ~ **del calor por recipiente** *f* TEC ESP heat sink; ~ **por fuga** *f* ELEC *corriente* leakage loss; ~ **térmica** *f* TERMO heat dissipation

disipador *m* ING ELÉC sink; ~ **de calor** *m* ELEC, TERMO heat sink; ~ **de energía** *m* CARBÓN pie de presas splitter, HIDROL *presas* energy dissipator; ~ **de niebla** *m* TRANSP AÉR fog dispersal; ~ **de sobretensiones** *m* ELEC *línea de alimentación* surge diverter, *transmisión* surge arrester; ~ **de sobrevoltajes** *m* ING ELÉC surge arrester; ~ **térmico** *m* ELEC, ING ELÉC, PROD, TERMO heat sink

disipativo *adj* ELECTRÓN, ING ELÉC, TELECOM lossy

disjunto *adj* MATEMÁT disjoint

dislocación *f* CRISTAL, METAL, QUÍMICA dislocation, glide; ~ **andada por obstáculos discretos** *f* CRISTAL pinning; ~ **articulada en escalera** *f* METAL stair rod dislocation; ~ **en cuña** *f* METAL edge dislocation; ~ **en desarrollo** *f* METAL grown-in dislocation; ~ **por desplazamiento** *f* METAL glissile dislocation; ~ **epitaxial** *f* METAL epitaxial dislocation; ~ **de falla**

f GEOL, TEC PETR fault trap; **~ fija** *f* CRISTAL sessile dislocation, METAL immobile dislocation; **~ helicoidal** *f* CRISTAL helical dislocation, METAL helical dislocation, screw dislocation; **~ de imagen** *f* METAL image dislocation; **~ inmóvil** *f* METAL immobile dislocation; **~ mixta** *f* CRISTAL, METAL mixed dislocation; **~ no decorada** *f* METAL undecorated dislocation; **~ parcial** *f* CRISTAL, METAL partial dislocation; **~ perfecta** *f* CRISTAL perfect dislocation; **~ del piso** *f* AmL (*cf descomposición Esp*) MINAS *filones* floor heave; **~ poligonal** *f* METAL polygonal dislocation; **~ Shockley** *f* METAL Shockley dislocation; **~ de Taylor-Orowan** *f* CRISTAL edge dislocation, Taylor-Orowan dislocation; **~ tipo cuña** *f* CRISTAL edge dislocation; **~ de tornillo** *f* CRISTAL screw dislocation, METAL helical dislocation; **~ por tornillo desplazado** *f* METAL jogged screw dislocation; **~ Volterra** *f* METAL Volterra dislocation; **~ en zigzag** *f* METAL zigzag dislocation

disluíta *f* MINERAL dysluite

disminución *f* ELECTRÓN decrement, droop, HIDROL *de la superficie piezométrica* decline, IMPR abatement, INFORM&PD decrement, METEO lessening, PROD thinning down, thinning out; **~ de amplitud de onda** *f* OCEAN *olas* wave decay; **~ de la definición** *f* FOTO decrease in definition; **~ del empuje** *f* TEC ESP thrust decay; **~ del espesor** *f* C&V attenuation; **~ de frecuencia** *f* ELEC frequency fall-off; **~ logarítmica** *f* ELECTRÓN logarithmic decrement; **~ de órbita** *f* TEC ESP orbital decay; **~ del pH** *f* CONTAM *tratamiento químico* pH drop, pH depression; **~ del potencial hidrógeno** *f* CONTAM *tratamiento químico* pH depression; **~ de la presión** *f* INSTAL HIDRÁUL loss of pressure, pressure loss

disminuir[1] *vt* PROD thin, thin out

disminuir[2] *vi* CARBÓN reduce

disminuirse *v refl* PROD thin out

dismutación *f* QUÍMICA *reacción* dismutation

disnea *f* OCEAN dyspnea, dyspnoea

disociable *adj* CARBÓN, CRISTAL, GAS, QUÍMICA dissociable

disociación *f* CARBÓN, CRISTAL, QUÍMICA dissociation; **~ de gas** *f* GAS gas dissociation; **~ térmica** *f* TERMO thermal dissociation

disociado *adj* CARBÓN, CRISTAL, GAS, QUÍMICA dissociated; **no ~** *adj* METAL *dislocación* undissociated

disociar *vi* CARBÓN, CRISTAL, GAS, QUÍMICA *compuestos* dissociate

disodilo *m* MINERAL dysodile

disolubilidad *f* PROC QUÍ, QUÍMICA dissolubility

disolución *f* PROC QUÍ dissolution, QUÍMICA dissolution, dissolving; **~ alcalina** *f* DETERG alkaline solution; **~ de ganga** *f* CARBÓN barren solution

disoluto *adj* PROC QUÍ, QUÍMICA dissoluble

disolvedor *m* PAPEL *recuperación lejías negras* dissolving tank

disolvencia *f* CINEMAT dissolve

disolvente[1] *adj* CONTAM *características químicas*, QUÍMICA solvent

disolvente[2] *m* ALIMENT, CARBÓN, CINEMAT solvent, COLOR thinner, CONTAM *características químicas* solvent, dispersant, DETERG, EMB, METAL solvent, P&C solvent, diluent, PROC QUÍ dissolvent, dispersant, dissolver, QUÍMICA dissolver, thinner, dissolvent, solvent, TEC PETR *ingeniería de lodos* dispersant, TEXTIL solvent; **~ activo** *m* EMB active solvent;

~ alifático *m* QUÍMICA, TEC PETR *refino, petroquímica* aliphatic solvent; **~ aprótico** *m* QUÍMICA aprotic solvent; **~ para los fabricantes de barnices** *m* REVEST varnish maker's naphtha; **~ hidrocarbúrico halogenado** *m* PROD halogenated hydrocarbon solvent; **~ de incrustaciones** *m* FERRO, INSTAL TERM *calderas* scale solvent; **~ del óxido** *m* ING MECÁ rust remover; **~ selectivo** *m* TEC PETR *refino, petroquímica* selective solvent

disolver *vt* CINEMAT, GAS dissolve, GEOL *geoquímica* leach, P&C cut, PROC QUÍ dissolve away, QUÍMICA dissolve

disonancia *f* ACÚST, FÍS ONDAS *del sonido* dissonance

disparador *m* CINEMAT *cámara* trigger, *obturador* shutter release, D&A trigger, ELEC cortocircuito trip gear, FOTO trigger, shutter release, INFORM&PD trigger, ING MECÁ release, tripper, trip gear, trigger, MINAS shooter; **~ de acción retardada** *m* FOTO delayed action release; **~ automático** *m* FOTO automatic timer; **~ de azul** *m* TV blue gun; **~ de barrenos** *m* Esp (*cf polvorero AmL*) MINAS blaster, picker, shooter; **~ de botón** *m* FOTO finger release; **~ de cable** *m* FOTO cable release; **~ de dos cables** *m* FOTO double cable release; **~ electromagnético** *m* ELEC, FOTO electromagnetic shutter-release; **~ de interruptor** *m* AUTO breaker triggering; **~ sin interruptor** *m* AUTO breakerless triggering; **~ de inundación** *m* ELECTRÓN flooding gun; **~ neumático** *m* FOTO pneumatic release

disparar[1] *vt* CINEMAT release, D&A *arma* discharge, *cartucha* fire, ELECTRÓN, FÍS, INFORM&PD trigger, PROD release, SEG, TELECOM *alarma* trigger

disparar[2] *vi* ING MECÁ trigger, trip

disparo *m* D&A, GEOFÍS shot, INFORM&PD triggering, ING ELÉC *relé* drop-out, ING MECÁ fire, MINAS shot, *voladuras* firing, TEC PETR *perforación* shot; **~ accidental** *m* NUCL squitter; **~ automático** *m* AUTO self-firing; **~ de cañón** *m* D&A gunfire; **~ completo** *m* D&A round, round of ammunition, MINAS *cañones* round; **~ eléctrico** *m* ELEC electrical firing, MINAS *voladuras* electrical firing, electrical shot-firing; **~ experimental** *m* TEC ESP test firing; **~ falso** *m* ELEC *relé* false tripping; **~ de flash** *m* Esp (*cf centelleo AmL*) CINEMAT, FOTO, ING ELÉC flashing; **~ de fuente de energía** *m* GEOFÍS reflection shooting; **~ instantáneo** *m* MINAS *voladuras* instantaneous firing; **~ por medios eléctricos** *m* AmL (*cf explosor Esp*) MINAS *explosivos* electric firing; **~ parcial** *m* NUCL partial trip; **~ por perdida a tierra** *m* ELEC, PROD ground fault trip (*AmE*), earth fault trip (*BrE*); **~ de rayos X** *m* INSTR X-ray flash; **~ del reactor** *m* NUCL reactor trip; **~ de reflexión** *m* GEOFÍS reflection shooting; **~ de refracción** *m* GEOFÍS refraction shooting; **~ de verificación** *m* GEOFÍS, TEC PETR checkshot

disparos: **~ simultáneos** *m pl* MINAS simultaneous firing

dispensadora *f* EMB dispensing machine

dispersador: **~ de bruma** *m* TRANSP AÉR fog dispersal

dispersante *m* CONTAM *características químicas* dispersant, P&C *de pigmentos, polímeros* dispersing agent, PROC QUÍ, TEC PETR *ingeniería de lodos* dispersant

dispersión *f* AmL (*cf grieta Esp*) ACÚST dispersion, scattering, CARBÓN *de rayos* diffusion, scattering, dispersion, CINEMAT scattering, CONTAM *tratamiento* *de gases* diffusion, CONTAM MAR *de los contaminantes* spreading, CRISTAL dispersion, scattering, ELEC *de corriente, carga* leakage, ELECTRÓN backscattering, debunching, scattering, ENERG RENOV scattering, FÍS dispersion, leakage, FÍS ONDAS dispersion, FÍS RAD *de energía* dispersion, *de partículas* scattering, INFORM&PD spread, *variación* dispersion, ING ELÉC leakage, spill, INSTR drift, METAL dispersion, MINAS *de un filón* split, NUCL scattering, ÓPT, P&C, PROC QUÍ, QUÍMICA dispersion, TEC ESP *balística* pattern, scattering, scatter, TEC PETR scattering, TELECOM scattering, dispersion; **~ acústica** *f* ACÚST acoustic dispersion; **~ aleatoria** *f* NUCL random scattering; **~ anómala** *f* CRISTAL anomalous scattering, FÍS anomalous dispersion; **~ atómica** *f* NUCL atomic scattering; **~ de carga** *f* ELEC *capacitor* charge leakage; **~ de Compton** *f* FÍS, FÍS RAD Compton scattering; **~ de un contaminante** *f* MINAS *desde un puerto o estuario* flushing; **~ cromática** *f* ÓPT chromatic dispersion; **~ elástica** *f* FÍS, FÍS PART, FÍS RAD elastic scattering; **~ elástica de electrones de alta energía** *f* FÍS, FÍS PART elastic scattering of high-energy electrons; **~ de energía** *f* TEC ESP energy dispersal; **~ eólica** *f* CONTAM, GAS wind dispersion; **~ de fase no lineal** *f* TEC ESP phase non-linear distortion; **~ por fibra** *f* TELECOM fiber scattering (*AmE*), fibre scattering (*BrE*); **~ de flujo** *f* ING ELÉC flux leakage; **~ de frecuencias** *f* ELECTRÓN frequency offset; **~ a gran ángulo** *f* NUCL large-angle scattering; **~ de gran ángulo** *f* NUCL wide-angle scattering; **~ de guía de ondas** *f* ÓPT, TELECOM waveguide dispersion; **~ de guíaondas** *f* ÓPT, TELECOM waveguide dispersion; **~ de impulsos** *f* ÓPT pulse spreading, TELECOM pulse dispersion, pulse spreading; **~ de impurezas** *f* ELECTRÓN impurity scattering; **~ inelástica** *f* FÍS inelastic scattering, NUCL inscattering; **~ inelástica de neutrones** *f* FÍS inelastic neutron scattering; **~ inversa** *f* TEC ESP backscattering; **~ de junta** *f* ING ELÉC junction leakage; **~ lineal** *f* FÍS RAD linear dispersion; **~ por lluvia** *f* TEC ESP rain scatter; **~ macromolecular** *f* ALIMENT macromolecular dispersion; **~ magnética** *f* ING ELÉC magnetic dispersion; **~ del material** *f* ÓPT material dispersion, TELECOM material dispersion, material scattering; **~ modal** *f* ÓPT modal dispersion; **~ de neutrones** *f* FÍS PART neutron scattering; **~ no lineal** *f* TELECOM nonlinear scattering; **~ de la onda** *f* TELECOM wave dispersion; **~ del perfil** *f* ÓPT, TELECOM profile dispersion; **~ de pulsos** *f* ÓPT pulse dispersion; **~ por pulsos** *f* ÓPT impulse dispersion; **~ de la radiación** *f* ELECTRÓN scattering; **~ Raman coherente anti-Stokes** *f* FÍS coherent anti-Stokes Raman scattering (*CARS*); **~ de Rayleigh** *f* FÍS, FÍS RAD, ÓPT, TELECOM Rayleigh scattering; **~ de rayos X** *f* CRISTAL X-ray scattering; **~ sectorial** *f* TRANSP *control de tráfico* platoon dispersion; **~ a tierra** *f* ELEC, ING ELÉC earth leakage (*BrE*), ground leakage (*AmE*); **~ troposférica** *f* TELECOM tropospheric scatter; **~ de vapor** *f* GAS vapor dispersion (*AmE*), vapour dispersion (*BrE*); **~ vertical** *f* CONTAM *tratamiento de gases* vertical dispersion

dispersividad *f* PROC QUÍ dispersivity

disperso *adj* FÍS stray, QUÍMICA, TELECOM dispersed

dispersoide *m* PROC QUÍ dispersoid

dispersor: **~ de energía** *m* TEC ESP scrambler

display: **~ electroluminiscente** *m* AmL (*cf pantalla electroluminiscente Esp*) ELEC, ELECTRÓN, INFORM&PD electroluminescent display; **~ matricial** *m* TELECOM matrix display

disponer *vt* INFORM&PD arm; **~ en capas** *vt* TELECOM lay

disponibilidad: **~ de circuitos** *f* TELECOM circuit availability; **~ factible** *f* PROD achievable availability; **~ inherente** *f* TRANSP AÉR inherent availability; **~ operacional** *f* PROD operational availability; **~ de suministro** *f* ELEC, ING ELÉC feedstock

disposición *f* AGRIC regulation, INFORM&PD *configuración* layout, ING ELÉC arrangement, ING MECÁ layout, *de taller* arrangement, MECÁ layout, PROD layout, setup, TEC ESP inclination; **~ antiparalelo** *f* ING ELÉC antiparallel arrangement; **~ balanceada** *f* ING ELÉC symmetrical arrangement; **~ de basuras** *f* AGUA waste disposal; **~ en capas** *f* TELECOM laying; **~ de cienos** *f* AGUA sludge disposal; **~ de los contactos** *f* ING ELÉC contact arrangement; **~ equilibrada** *f* IMPR balance; **~ de los explosivos** *f* MINAS blasting-pattern; **~ general** *f* CONST general arrangement, ING MECÁ general assembly, general arrangement, MECÁ general arrangement; **~ en oposición** *f* ING ELÉC back-to-back arrangement; **~ en paralelo** *f* ING ELÉC parallel arrangement, PROD parallel condition; **~ en paralelo de las fibras** *f* TEXTIL parallelization of fibers (*AmE*), parallelization of fibres (*BrE*); **~ de polos** *f* PROD pole arrangement; **~ de sensores desplazados** *f* GEOFÍS *prospección sísmica* offset; **~ de simbología polar** *f* ELECTRÓN *amplificador* totem pole arrangement; **~ subterránea de basuras** *f* AGUA underground waste disposal; **~ trifásica equilibrada** *f* PROD three-phase balanced condition; **~ vertical de páginas en el pliego** *f* IMPR sheetwise

dispositivo *m* CONTAM MAR device, INFORM&PD drive, device, feature, ING MECÁ mechanism, appliance, fixture, gear, contrivance, device, TELECOM device; **~ para absorción de ruidos** *m* ACÚST, ING MECÁ noise absorption device; **~ de acceso aleatorio** *m* INFORM&PD random-access device; **~ de acceso a la red** *m* TELECOM network gateway; **~ de acceso en serie** *m* INFORM&PD serial access device; **~ accionado hidráulicamente** *m* ING MECÁ hydraulically-operated device; **~ de accionamiento por tiempos programados** *m* LAB, INSTR program timer (*AmE*), programme timer (*BrE*); **~ acoplado por carga** *m* (*DAC*) ELECTRÓN, FÍS, INFORM&PD, ING ELÉC, TELECOM, TV charge-coupled device (*CCD*); **~ de actuación activo** *m* ELECTRÓN *transistor* active pullup device; **~ de actuación rápida** *m* TELECOM bucket-brigade device (*BBD*); **~ de alimentación mediante mesa giratoria** *m* EMB turntable feed; **~ de almacenamiento** *m* INFORM&PD storage device; **~ para almacenamiento de energía** *m* ING ELÉC, TEC ESP, TERMO energy-storage device; **~ analizador de barrido** *m* FÍS ONDAS, FÍS RAD scanner; **~ analógico** *m* TELECOM analog device; **~ antibalance** *m* TRANSP MAR antirolling device; **~ antienclavamiento** *m* MECÁ antilocking system; **~ antillamarada** *m* SEG, VEH *motor* flame trap; **~ antipar** *m* TRANSP AÉR antitorque device; **~ antitorsión** *m* TRANSP AÉR antitorque device; **~ de apertura-cierre** *m* ELEC *conmutador* make-and-break device; **~ de apoyo** *m* ING ELÉC lever; **~ de arranque** *m* ELEC *motor* starting device; **~ de**

arranque contactor *m* AUTO, ELEC, ING ELÉC, VEH contactor starter; **~ de arranque en frío** *m* AUTO, VEH *motores, carburadores* cold-start device; **~ atornillador** *m* NUCL screwing device; **~ automático** *m* TELECOM automatic device; **~ de aviso del arranque** *m* TRANSP headway warning device; **~ de aviso de entrada en pérdida** *m* TRANSP AÉR stall warning device; **~ de ayuda de navegación** *m* TRANSP AÉR navigational aid;

~ b **~ de bajo nivel** *m* ELECTRÓN low-level device; **~ de barrido** *m* TV scanning device; **~ basculante** *m* NUCL tilting device; **~ de bifurcación** *m* ÓPT branching device; **~ de bloqueo** *m* ING ELÉC blocking device, PROD locking attachment; **~ de bloqueo de máximo de corriente** *m* ELEC overcurrent blocking device; **~ de bloqueo de sobrecorriente** *m* ELEC overcurrent blocking device;

~ c **~ del cable de elevación** *m* ING ELÉC *grúas* cable harness; **~ de cálculo** *m* INFORM&PD computing device; **~ de calefacción** *m* ING MECÁ heating device; **~ de canal p** *m* ELECTRÓN p-channel device; **~ de capa delgada** *m* ELECTRÓN thin-film device; **~ de capa gruesa** *m* ELECTRÓN thick-film device; **~ captador** *m* ING MECÁ pick-up gear; **~ de captura electrostática** *m* NUCL electrostatic collector; **~ de cierre** *m* ELEC, ING MECÁ locking device, INSTAL HIDRÁUL cutoff device, SEG locking device; **~ de cinta para copia de seguridad** *m* INFORM&PD streaming tape drive; **~ para circular la cola** *m* TEXTIL size circulator unit; **~ de codificación** *m* TRANSP coding device; **~ de colación** *m* INFORM&PD collator; **~ colocado en la parte baja** *m* ING MECÁ heel block; **~ de compresor axial** *m* ING ELÉC bucket-brigade device (*BBD*); **~ de comprobación** *m* ING MECÁ tryout facility; **~ de cómputo** *m* INFORM&PD computing device; **~ concentrador** *m* ENERG RENOV concentrator; **~ concentrador solar** *m* ENERG RENOV solar concentrator; **~ de conducción del haz** *m* ELECTRÓN beam lead device; **~ de conducción de tráfico** *m* TELECOM traffic-carrying device; **~ del conducto de debate** *m* INSTR discussion tube arrangement; **~ para conectar dos sistemas** *m* ELECTRÓN interface; **~ de conmutación** *m* ING ELÉC, TELECOM switching device; **~ de conmutación óptico mecánico** *m* AmL (*cf conmutador óptico mecánico Esp*) ÓPT, TELECOM mechanical optical switch; **~ contra chispas** *m* ELEC *relé* spark arrester; **~ contra el derrapaje** *m* AUTO, VEH *frenos* antiskid device; **~ contra el deslizamiento** *m* AUTO, SEG, VEH *frenos* antiskid device; **~ de control** *m* ELEC, ING ELÉC control unit; **~ de control de la conexión de señalización** *m* TELECOM signaling connection control part (*AmE*), signalling connection control part (*BrE*); **~ de control de errores** *m* INFORM&PD, TELECOM error-control device (*ECD*); **~ de control de navegación** *m* TRANSP cruise control device; **~ de control en paralelo** *m* ELEC *cambiador de toma* parallel control device; **~ de control de señalización** *m* TELECOM signaling control part (*AmE*) (*ISCP*), signalling control part (*BrE*) (*ISCP*); **~ de control de tiempo** *m* TELECOM timer; **~ controlado por corriente** *m* ING ELÉC current-controlled device; **~ copiador** *m* ING MECÁ copying attachment, copying unit; **~ de corriente alterna** *m* ING ELÉC alternating-current machine; **~ corta-corriente** *m* ING ELÉC interlock;

~ de cortocircuito *m* ELEC, ING ELÉC short circuiting device; **~ de cuantificación** *m* ELECTRÓN, FÍS PART, INFORM&PD, TEC ESP, TELECOM, TV quantizer;

~ d **~ en derivación** *m* ELEC *fuente de alimentación* shunting device; **~ descodificador** *m* ELECTRÓN, INFORM&PD, TEC ESP, TELECOM, TV decoding device; **~ desconector** *m* ING MECÁ tripping device; **~ de desconexión automática** *m* ELEC *seguridad* automatic cutout switch; **~ para desenroscar** *m* ING MECÁ rack; **~ desviador** *m* ENERG RENOV deflector; **~ detector** *m* CONTAM MAR sensing device; **~ digital** *m* ELECTRÓN digital device; **~ de disparo** *m* ING MECÁ, PROD, SEG trip, tripping device; **~ de disparo del lanzamiento** *m* TEC ESP launcher release gear; **~ de distribución extraíble** *m* ELEC draw-out switchgear; **~ distribuidor** *m* ING MECÁ dividing attachment; **~ de disyunción mecánica** *m* ELEC *cortocircuito* mechanical tripping device; **~ divisor** *m* ING MECÁ dividing attachment;

~ e **~ electrónico** *m* ELECTRÓN electronic device, electron device, ING ELÉC electron device, electronic device, METR electronic gage (*AmE*), electronic gauge (*BrE*); **~ eléctronico de control de tiempo** *m* TELECOM timer; **~ de elevación vertical** *m* ENERG RENOV lift-type device; **~ elevador de escobillas** *m* ELEC *máquina* brush-lifting device; **~ de encendido** *m* ING ELÉC, TEC ESP igniter, TERMOTEC ignition device; **~ enchufable** *m* ELEC *conexión* plug pin; **~ de enclavamiento** *m* PROD locking attachment; **~ para encolar** *m* EMB gluing device; **~ de enganche** *m* FERRO *vehículos* coupling hook; **~ de entrada** *m* INFORM&PD, ING ELÉC input device; **~ de entrada/salida** *m* INFORM&PD input/output device; **~ de escudriñado de líneas** *m* INFORM&PD raster-scan device; **~ de estabilización** *m* TRANSP stabilization device; **~ de estado sólido** *m* ELEC *componente*, ELECTRÓN, TELECOM solid-state device; **~ de estangulación** *m* REFRIG throttling device; **~ evolvente paralelo** *m* ING MECÁ parallel involute gear; **~ de exploración** *m* ELEC *conmutador*, TELECOM scanner; **~ de exploración aleatoria** *m* INFORM&PD random-scan device; **~ de exploración por puntos** *m* INFORM&PD raster-scan device; **~ de exploración de tramas** *m* INFORM&PD raster-scan device; **~ explorador** *m* ELECTRÓN scanner; **~ externo** *m* INFORM&PD external device;

~ f **~ de fijación** *m* ELEC, SEG locking device, ING MECÁ holding device, holding fixture; **~ de fijación automático** *m* ING MECÁ robot gripping device; **~ de fijación electromagnética** *m* ING MECÁ electromagnetic fixing-device; **~ fluídico** *m* ING MECÁ fluidic device; **~ fotoeléctrico** *m* ELECTRÓN photoelectric device;

~ g **~ para girar la bobina** *m* IMPR web turning device;

~ h **~ hipersustentador** *m* TRANSP AÉR high lift device;

~ i **~ de imagen** *m* INFORM&PD imager; **~ de impregnación** *m* TEXTIL quetch unit; **~ de impresión suplementaria** *m* IMPR imprinting unit; **~ impresor** *m* IMPR output device; **~ de indicación digital** *m* ELEC, INSTR digital read-out; **~ de indicación numérica** *m* ELEC, INSTR digital read-out; **~ indicador de continua** *m* ELEC direct read-out instrument; **~ indicador de llamada** *m* TELECOM call-indicating device; **~ de inferencia** *m* INFORM&PD

inference engine; ~ **informático** *m* INFORM&PD computing device;~ **de inmovilización** *m* PROD locking attachment; ~ **inmovilizador** *m* ING MECÁ locking device; ~ **inmovilizador suave** *m* MINAS spark-out stop; ~ **para insertar hojas** *m* EMB leaflet insertor; ~ **de interbloqueo** *m* CINEMAT interlock device; ~ **interruptor** *m* ING MECÁ, PROD, SEG trip, tripping device; ~ **de introducción de datos** *m* INFORM&PD input device;

■**l** ~ **de lanzamiento** *m* TEC ESP launcher; ~ **de lavado** *m* VEH *ventanas* washer; ~ **de lectura** *m* ELEC, INSTR read-out; ~ **de lectura de continua** *m* ELEC direct read-out instrument; ~ **de limitación binaria** *m* ELECTRÓN binary subtractor; ~ **de llamada** *m* TELECOM ringer; ~ **de llenado** *m* EMB filling device; ~ **localizador** *m* ING MECÁ locating device; ~ **de lógica** *m* ELECTRÓN, INFORM&PD logic device; ~ **lógico** *m* ELECTRÓN, INFORM&PD logic device;

■**m** ~ **de marcación de cranes** *m* ING MECÁ locating device; ~ **de mediación** *m* TELECOM message dropping (*MD*), mediation device; ~ **de medida** *m* CONST, ELÉC, ING ELÉC measuring device, ING MECÁ gage (*AmE*), gauge (*BrE*), INSTR gage (*AmE*), gauge (*BrE*), measuring device, LAB gage (*AmE*), gauge (*BrE*), METR measuring device; ~ **de memoria** *m* INFORM&PD storage device; ~ **modular** *m* ELEC *equipo* module; ~ **monitor** *m* ELEC *instrumento* monitor; ~ **de montaje** *m* ING MECÁ fitting device; ~ **de muestreo** *m* LAB sampling device;

■**n** ~ **nominal telefónico** *m* TELECOM telephony-rated device;

■**o** ~ **de OAS** *m* (*dispositivo de onda acústica de superficie*) ELECTRÓN, TELECOM SAW device (*surface acoustic wave device*); ~ **olfateador** *m* GAS sniffer device; ~ **de onda acústica de superficie** *m* (*dispositivo de OAS*) ELECTRÓN, TELECOM surface acoustic wave device (*SAW device*); ~ **óptico de búsqueda** *m* TV optical scanning device (*OSD*); ~ **óptico-electrónico** *m* ING ELÉC, ÓPT optoelectronic device; ~ **optoelectrónico** *m* ING ELÉC, ÓPT optoelectronic device;

■**p** ~ **de parada mecánico** *m* ING MECÁ mechanical stop unit; ~ **en paralelo** *m* ELEC *fuente de alimentación* shunting device; ~ **periférico** *m* INFORM&PD, ING ELÉC, TELECOM peripheral device; ~ **p-n-p-n** *m* ELECTRÓN p-n-p-n device; ~ **portaherramientas** *m* ING MECÁ tool-holding fixture; ~ **posicionador** *m* ING MECÁ locating device; ~ **programable** *m* INFORM&PD, TELECOM programmable device; ~ **programable de campo** *m* INFORM&PD field-programmable device; ~ **programable por usuario** *m* INFORM&PD user programmable device; ~ **de protección** *m* ELEC, SEG guard; ~ **de protección contra sobrecarga** *m* ING ELÉC, SEG overload protection device; ~ **de protección contra la sobretensión** *m* ING ELÉC, SEG overvoltage protection device; ~ **de protección con espacio de aire** *m* ELEC, SEG air gap protector; ~ **de protección de máximo de corriente** *m* ELEC, SEG overcurrent protection; ~ **de protección de máximo de tensión** *m* ELEC, SEG overvoltage protection; ~ **protector** *m* ING MECÁ, MECÁ, SEG guard; ~ **protector de enclavamiento** *m* SEG interlocking guard; ~ **protector con espacio de aire** *m* ELEC, SEG air gap protector; ~ **protector fotoeléctrico** *m* SEG photo-electric guard; ~ **protector telescópico** *m* SEG

máquinas cepilladoras telescopic guard; ~ **de prueba para el anclaje del batán** *m* TEXTIL batt-anchorage testing device; ~ **para pruebas** *m* ING MECÁ test rig; ~ **con puerta metálica** *m* ELECTRÓN metal gate; ~ **de puesta en cola** *m* TELECOM queueing device; ~ **de puesta en fase del obturador** *m* CINEMAT shutter phasing device;

■**q** ~ **Q** *m* NUCL Q-device;

■**r** ~ **de ramificación** *m* ÓPT branching device; ~ **reavivador** *m* PROD *muelas abrasivas* dresser; ~ **de recogida** *m* CONTAM, CONTAM MAR collection device; ~ **de recuperación** *m* CONTAM, CONTAM MAR recovery device; ~ **recuperador** *m* CONTAM, CONTAM MAR recovery device; ~ **de regulación** *m* REFRIG *caudal de fluido* refrigerant metering device; ~ **de regulación de temperatura** *m* REFRIG temperature controller; ~ **de relleno** *m* TELECOM stuffing device; ~ **de retención** *m* ING MECÁ holding device, holding fixture; ~ **de retención de líquido** *m* REFRIG liquid trap; ~ **de retenida** *m* MINAS retainer; ~ **de ruptura** *m* REFRIG rupture member;

■**s** ~ **de salida** *m* INFORM&PD, PROD output device; ~ **Schottky** *m* ELECTRÓN Schottky device; ~ **de secuencia** *m* INFORM&PD, ING ELÉC, PROD, TELECOM sequencer; ~ **de seguridad** *m* CONST safety lock, ING MECÁ, MINAS, SEG safety appliance, safety device; ~ **de seguridad antienganche de la película** *m* CINEMAT buckle trip; ~ **de seguridad y control** *m* INSTAL TERM, SEG control and safety device; ~ **de seguridad por corte** *m* REFRIG, SEG safety cutout; ~ **semiconductor** *m* ELECTRÓN, INFORM&PD semiconductor device; ~ **semiconductor discreto** *m* ELECTRÓN discrete semiconductor device; ~ **sensible a la temperatura** *m* PROD temperature sending device; ~ **de separación** *m* ING MECÁ dividing attachment; ~ **de separación del tráfico** *m* TRANSP MAR *navegación* traffic separation scheme; ~ **silenciador** *m* TELECOM muting device; ~ **de silicio** *m* ING ELÉC silicon device; ~ **SIMD** *m* INFORM&PD SIMD machine; ~ **SISD** *m* INFORM&PD SISD machine; ~ **de soporte de la hoja** *m* EMB foil backing machine; ~ **de soporte lógico** *m* INFORM&PD software tool; ~ **de sujeción** *m* CARBÓN attachment, ING MECÁ fitting device, INSTR attachment, TEXTIL pinning-up device; ~ **de sujeción automático** *m* ING MECÁ robot gripping device; ~ **de sujeción electromagnética** *m* ING MECÁ electromagnetic fixing-device; ~ **sujetador** *m* ING MECÁ locking device; ~ **de supresión** *m* PROD suppression device; ~ **de suspensión** *m* MINAS suspension gear;

■**t** ~ **de tensión** *m* TEXTIL tension device; ~ **terminal** *m* INFORM&PD terminal device; ~ **terminal de fibra óptica** *m* ÓPT, TELECOM fiber-optic terminal device (*AmE*), fibre-optic terminal device (*BrE*); ~ **terminal de fibra óptica de transmisión** *m* TELECOM transmit fiber optic terminal device (*AmE*), transmit fibre optic terminal device (*BrE*); ~ **terminal de fibroóptica** *m* ÓPT fiber-optic terminal device (*AmE*), fibre-optic terminal device (*BrE*); ~ **terminal de recepción de fibra óptica** *m* ELECTRÓN, ÓPT, TELECOM receive fiber optic terminal device (*AmE*), receive fibre optic terminal device (*BrE*); ~ **de termosellado** *m* EMB heat-sealing device; ~ **de tiempo de tránsito** *m* ING ELÉC transit-time device; ~ **de tolva y cono** *m* PROD *horno alto* bell-and-hopper; ~ **trabador** *m* ELEC *seguridad*, ING MECÁ, SEG

locking device; ~ **de transferencia de carga** *m* (*DTC*) FÍS, ING ELÉC, TEC ESP, TELECOM charge-transfer device (*CTD*); ~ **de transmisión secreta** *m* TELECOM scrambler; ~ **de transmisión secreta de modos** *m* TELECOM mode scrambler; ~ **transportador** *m* ING MECÁ propelling gear; ~ **de traqueo** *m* PAPEL Fourdrinier shake;
▶▶ ~ **de variación de potencia** *m* REFRIG capacity controller; ~ **de vigilancia** *m* FERRO *equipo inamovible* vigilance device; ~ **de visualización** *m* INFORM&PD display device; ~ **visualizado de niveles múltiples** *m* PROD multiple rung display; ~ **visualizador** *m* TELECOM alphageometric display; ~ **visualizador del número llamado** *m* TELECOM called number display

dispositivos: ~ **para levantar grandes pesos** *m pl* MINAS lift; ~ **de la mudada** *m pl* TEXTIL doffing devices; ~ **de salvamento** *m pl* SEG, TRANSP AÉR, TRANSP MAR rescue apparatus; ~ **de seguridad** *m pl* SEG *edificios* safety fittings; ~ **de unión** *m pl* ING MECÁ fixtures

disprosio *m* (*Dy*) CARBÓN, METAL, QUÍMICA *elemento metálico trivalente* dysprosium (*Dy*)

disquete *m* IMPR, INFORM&PD, TELECOM diskette, floppy disk; ~ **de doble cara** *m* INFORM&PD double-sided disk, double-sided floppy diskette

disrupción *f* ELEC *dieléctrico* breakdown, NUCL *reacción en cadena* disruption; ~ **en avalancha** *f* ELEC *diodo* avalanche breakdown

disruptivo *adj* ING ELÉC disruptive

distancia[1]: **a** ~ *adj* INFORM&PD, TELECOM, TV remote

distancia[2] *f* CONST distance piece, FÍS distance, ING MECÁ gap, distance, MATEMÁT distance, MECÁ gap, METR distance, TRANSP *tráfico* gap, TRANSP AÉR spacing, clearance, TRANSP MAR *cable, navegación, equipamiento eléctrico* range; ~ **aceleración-parada** *f* TRANSP AÉR accelerate-stop distance; ~ **aceleración-parada disponible** *f* TRANSP AÉR accelerate-stop distance available; ~ **aceleración-parada requerida** *f* TRANSP AÉR accelerate-stop distance required; ~ **de advertencia** *f* FERRO *equipo fijado* warning distance; ~ **de agrupación** *f* ELECTRÓN drift space; ~ **de aislamiento** *f* ING ELÉC insulation distance; ~ **entre aislamientos** *f* ING ELÉC insulation distance; ~ **angular de fase** *f* ELEC *corriente alterna* phase angle; ~ **anterior a la señalización** *f* FERRO *equipo fijado* pre-signaling distance (*AmE*), pre-signalling distance (*BrE*); ~ **de apriete** *f* EMB *torno, abonzaderas* holding capacity; ~ **de aproximación máxima** *f* TRANSP MAR *navegación* closest approach distance; ~ **de aterrizaje** *f* TRANSP AÉR landing distance; ~ **del borde al centro del remache** *f* ING MECÁ edge distance; ~ **de la broca a la columna** *f* ING MECÁ distance of drill from column; ~ **cenital** *f* TEC ESP zenith distance; ~ **entre centros** *f* ING MECÁ center distance (*AmE*), centre distance (*BrE*), distance between centers (*AmE*), distance between centres (*BrE*); ~ **entre centros de chumacera** *f* ING MECÁ distance between centers of journals (*AmE*), distance between centres of journals (*BrE*); ~ **entre centros de cojinete** *f* ING MECÁ distance between centers of journals (*AmE*), distance between centres of journals (*BrE*); ~ **entre centros de muñones** *f* ING MECÁ distance between centers of journals (*AmE*), distance between centres of journals (*BrE*); ~ **entre los contornos de nivel** *f* GEOL

contour interval; ~ **de corte** *f* ING MECÁ cutting stroke; ~ **crítica** *f* PETROL, TEC PETR critical distance; ~ **al diagnóstico** *f* TELECOM diagnostic aid; ~ **disponible de aterrizaje** *f* TRANSP AÉR landing distance available; ~ **disponible de despegue** *f* TRANSP AÉR takeoff distance available; ~ **disruptiva** *f* ING ELÉC break distance; ~ **efectiva** *f* METAL working distance; ~ **entre ejes** *f* VEH wheelbase; ~ **entre electrodos** *f* ELEC, ING ELÉC spark gap; ~ **focal** *f* CINEMAT, ELECTRÓN, FÍS, FÍS RAD, FOTO, TV focal length; ~ **focal variable** *f* FOTO variable focal length; ~ **de formación de ráfaga** *f* TEC ESP, TRANSP AÉR, TRANSP MAR gust gradient distance; ~ **de frenado** *f* AUTO, VEH braking distance; ~ **de la graduación** *f* IMPR shingle distance; ~ **de Hamming** *f* TELECOM Hamming distance; ~ **interelectródica** *f* ELEC parallel spark gap, *chispa* coordinating gap, *electrodos* spark gap; ~ **interplanar** *f* CRISTAL, ÓPT spacing; ~ **de interrupción** *f* ING ELÉC break distance; ~ **lejana** *f* TEC ESP far range; ~ **máxima chispeante** *f* ING ELÉC sparking distance; ~ **en millas** *f* TRANSP mileage; ~ **mínima de enfoque** *f* CINEMAT minimum focusing distance; ~ **nominal de detección** *f* PROD rated sensing distance; ~ **del objetivo a la placa** *f* FOTO camera extension; ~ **permitida** *f* TRANSP *entre vehículos* tolerable gap; ~ **entre pistas** *f* ÓPT, TV track pitch; ~ **plana** *f* ING MECÁ plain length; ~ **de planeo** *f* TRANSP AÉR gliding distance; ~ **de proyección** *f* CINEMAT throw; ~ **de puntería** *f* D&A sighting range; ~ **entre puntos** *f* ING MECÁ distance between centers (*AmE*), distance between centres (*BrE*); ~ **entre raíces** *f* CONST, FERRO gap; ~ **recorrida** *f* TRANSP MAR *navegación* run distance; ~ **registrada por la corredera** *f* TRANSP MAR logged distance; ~ **requerida de despegue** *f* TRANSP AÉR takeoff distance required; ~ **de salto** *f* ELEC *chispa* coordinating gap, FÍS ONDAS *onda radioeléctrica reflejada* skip distance; ~ **de seguridad entre convoyes** *f* FERRO *equipo inamovible*, SEG block headway; ~ **entre señales** *f* TRANSP signaling distance (*AmE*), signalling distance (*BrE*); ~ **de la señalización** *f* TRANSP signaling distance (*AmE*), signalling distance (*BrE*); ~ **de separación** *f* ING MECÁ distance apart; ~ **de toma de la fotografía** *f* FOTO shooting distance; ~ **útil** *f* ING MECÁ cutting stroke, METAL working distance; ~ **de visibilidad** *f* CONST sight distance, TRANSP visibility distance; ~ **de visibilidad en adelantamiento** *f* TRANSP *tráfico* passing sight distance; ~ **de visibilidad en parada** *f* TRANSP *tráfico* stopping sight distance; ~ **visual** *f* TRANSP *tráfico* sight distance

distanciador *m* PROD spacer; ~ **tubular** *m* PROD thimble

distanciamiento: ~ **entre las plantas** *m* AGRIC plant spacing

distanciómetro *m* ELEC *medición*, ING MECÁ telemeter

distena *f* MINERAL cyanite, disthene, kyanite

distintivo *m* PROD tag; ~ **de llamada** *m* TELECOM, TRANSP MAR call sign

distorsión[1]: **sin** ~ *adj* ELECTRÓN undistorted

distorsión[2] *f* ACÚST, ELEC, ELECTRÓN, FÍS, TELECOM, TV distortion; ~ **de abertura** *f* ELECTRÓN aperture distortion; ~ **del agujero taladrado** *f* AmL (*cf apiñamiento Esp*) MINAS squeezing; ~ **alineal** *f* ELECTRÓN, TELECOM nonlinear distortion; ~ **amortiguada** *f* CINEMAT cushion distortion; ~ **de amplitud** *f* ACÚST amplitude distortion, ELECTRÓN,

FÍS, FÍS ONDAS, TELECOM amplitude distortion, amplitude-amplitude distortion; ~ **de amplitud de frecuencia** *f* ELECTRÓN, FÍS, FÍS ONDAS, TELECOM amplitude-frequency distortion; ~ **angular** *f* CINEMAT angular distortion; ~ **armónica** *f* ACÚST, ELECTRÓN, TELECOM harmonic distortion; ~ **de atenuación** *f* ELECTRÓN, FÍS attenuation distortion; ~ **azimutal** *f* TV azimuth distortion; ~ **en barril** *f* CINEMAT positive distortion; ~ **de campo** *f* NUCL field flutter; ~ **por capa adyacente** *f* TV print-through; ~ **cóncava** *f* CINEMAT pillow distortion; ~ **de contacto** *f* ACÚST tracing distortion; ~ **por corriente continua** *f* NUCL direct current distortion; ~ **en corsé** *f* CINEMAT, FÍS, FOTO pincushion distortion; ~ **cromática** *f* ÓPT chromatic distortion; ~ **cruzada** *f* TELECOM crossover distortion; ~ **de cuantificación** *f* ELECTRÓN quantization distortion; ~ **cúbica** *f* METAL cubical distortion; ~ **debida al retardo** *f* TV delay distortion; ~ **dinámica** *f* ACÚST dynamic distortion; ~ **del eco** *f* ELECTRÓN echo distortion; ~ **ecoica** *f* ELECTRÓN echo distortion; ~ **esférica** *f* CINEMAT spherical distortion; ~ **esferoide** *f* CINEMAT, FÍS, FOTO barrel distortion; ~ **de fase** *f* ACÚST, ELECTRÓN, FÍS, TELECOM, TV phase distortion; ~ **final** *f* ELECTRÓN end distortion; ~ **de frecuencia** *f* ELECTRÓN, FÍS, FÍS ONDAS, TELECOM frequency distortion; ~ **de la imagen** *f* TV tilt; ~ **intermodal** *f* TELECOM intermodal distortion; ~ **de intermodulación** *f* ELECTRÓN intermodulation distortion; ~ **intramodal** *f* ÓPT, TELECOM intramodal distortion; ~ **lineal** *f* CINEMAT, TELECOM linear distortion; ~ **lineal de retardo de grupo** *f* ING ELÉC, TEC ESP group-delay linear distortion; ~ **máxima** *f* TV peak distortion; ~ **modal** *f* TELECOM modal distortion; ~ **de modos** *f* TELECOM mode distortion; ~ **multimodal** *f* ÓPT, TELECOM multimode distortion; ~ **multimodo** *f* ÓPT, TELECOM multimode distortion; ~ **negativa** *f* CINEMAT, FÍS, FOTO barrel distortion; ~ **no lineal** *f* ELECTRÓN, TELECOM nonlinear distortion; ~ **oblícua** *f* TV skew; ~ **óptica** *f* C&V optical distortion; ~ **de parche verde** *f* C&V green patch distortion; ~ **de pastilla** *f* ELECTRÓN *semiconductores* wafer distortion; ~ **positiva** *f* CINEMAT positive distortion; ~ **de retardo de la envolvente** *f* TEC ESP *comunicaciones* envelope delay distortion; ~ **S** *f* TV S distortion; ~ **del segundo armónico** *f* ACÚST, ELECTRÓN second harmonic distortion; ~ **de señal** *f* ELECTRÓN, TELECOM signal distortion; ~ **por sobremodulación** *f* TV overshoot distortion, underswing; ~ **del tercer armónico** *f* ACÚST, ELECTRÓN third harmonic distortion; ~ **termoelástica** *f* TEC ESP thermoelastic distorsion; ~ **del tiempo de retardo** *f* ELECTRÓN time-delay distortion; ~ **trapezoidal** *f* CINEMAT keystone distortion, ELECTRÓN keystone distortion, trapezoidal distortion

distorsionado *m* TV scrambling
distorsionador *m* TEC ESP, TV scrambler; ~ **de modos** *m* ÓPT mode scrambler
distorsionante *adj* C&V *espejo* distorting
distorsionar *vt* TEC ESP, TV scramble
distribución *f* C&V, ELEC distribution, EMB dispensing, GAS spread, distribution, INFORM&PD *placa de circuito impreso* layout, ING MECÁ distribution, release, *de tipo corredera* valve motion, PROD distribution, dispatching, TEC ESP wiring, TEXTIL dispensing, TRANSP dispatching, VEH *sistema de encendido* distribution; ~ **de agua** *f* AGUA, CONST water distribution; ~ **de amplitud de ruido** *f* TELECOM noise amplitude distribution (*NAD*); ~ **bifilar** *f* ELEC *red* two-wire system; ~ **binomial** *f* FÍS, MATEMÁT binomial distribution; ~ **de Bose-Einstein** *f* FÍS Bose-Einstein distribution; ~ **de canales** *f* TV channel allocation; ~ **canónica** *f* FÍS canonical distribution; ~ **de carga** *f* CONST, TRANSP AÉR load distribution; ~ **combinada** *f* TELECOM combined distribution frame (*CDF*); ~ **por conductos** *f* REFRIG duct distribution; ~ **por corredera** *f* ING MECÁ link gear, link motion, link valve motion; ~ **de corriente** *f* ELEC current distribution; ~ **del cuadro posterior** *f* PROD back panel layout; ~ **de la densidad electrónica** *f* CRISTAL, ELECTRÓN, FÍS, FÍS PART, QUÍMICA electron-density distribution; ~ **eléctrica** *f* ELEC *conexión* wiring; ~ **de electricidad** *f* ELEC distribution of electricity; ~ **de energías espectrales** *f* ENERG RENOV spectral energy distribution; ~ **de error** *f* MATEMÁT error distribution; ~ **espacial** *f* CONTAM spatial distribution; ~ **espacial de la carga eléctrica** *f* FÍS RAD charge cloud; ~ **en el espacio** *f* CONTAM spatial distribution; ~ **de espacios entre letras** *f* IMPR letterfit; ~ **de estado cuasiconstante** *f* METAL quasi-steady-state distribution; ~ **en estrella** *f* TELECOM star distribution; ~ **exponencial** *f* INFORM&PD exponential distribution; ~ **de fase** *f* METAL phase distribution; ~ **de Fermi-Dirac** *f* FÍS Fermi-Dirac distribution; ~ **de fertilizantes** *f* AGRIC fertilizer distribution; ~ **de filtros programables** *f* ELECTRÓN mask-programmable array; ~ **de flujos no viscosos** *f* FÍS FLUID inviscid flow distribution; ~ **fortuita** *f* METAL random arrangement, random distribution; ~ **de frecuencias** *f* INFORM&PD, MATEMÁT, METAL frequency distribution, TELECOM, TV frequency allocation; ~ **de frecuencias acumulativas** *f* (*DFA*) MATEMÁT cumulative distribution function (*CDF*); ~ **de fuerza** *f* PROD power distribution; ~ **de Gauss** *f* FÍS, INFORM&PD, MATEMÁT Gaussian distribution; ~ **gaussiana** *f* FÍS, INFORM&PD, MATEMÁT Gaussian distribution; ~ **geométrica** *f* GEOM, MATEMÁT geometric distribution; ~ **granulométrica** *f* GEOL grain-size distribution; ~ **hipergeométrica** *f* MATEMÁT hypergeometric distribution; ~ **de intensidad** *f* CRISTAL, FÍS RAD intensity distribution; ~ **interior de la cabina** *f* TRANSP AÉR cabin layout; ~ **intrínseca** *f* FÍS RAD intrinsic distribution; ~ **logarítmica normal** *f* TELECOM log-normal shadowing; ~ **logarítmico-normal** *f* MATEMÁT log-normal distribution; ~ **de Maxwell** *f* FÍS Maxwell distribution; ~ **por mecanismo de palanca** *f* INSTAL HIDRÁUL hook gear valve motion; ~ **modal de no equilibrio** *f* ÓPT, TELECOM nonequilibrium mode distribution; ~ **en modo equilibrio** *f* ÓPT, TELECOM equilibrium-mode distribution; ~ **en modo no equilibrado** *f* ÓPT, TELECOM nonequilibrium mode distribution; ~ **normal** *f* FÍS, INFORM&PD, MATEMÁT normal curve distribution, normal distribution; ~ **en planta** *f* PROD *diseño* layout; ~ **de Poisson** *f* FÍS, INFORM&PD, MATEMÁT Poisson distribution; ~ **polinomial** *f* MATEMÁT multinomial distribution; ~ **de la potencia** *f* NUCL power shape; ~ **a priori** *f* MATEMÁT prior distribution; ~ **de probabilidad** *f* INFORM&PD probability distribution; ~ **de probabilidad de amplitud** *f* TELECOM amplitude probability distribution; ~ **radial de potencia** *f*

NUCL radial power distribution; **~ ramificada** *f* TELE-COM tree distribution; **~ de residuos** *f* AGRIC slash disposal; **~ por sector Allan** *f* ING MECÁ Allan's link motion; **~ por sector Stephenson** *f* ING MECÁ link motion; **~ de superficies cultivadas** *f* AGRIC crop ratio; **~ t** *f* MATEMÁT t-distribution; **~ del tamaño de los granos** *f* C&V granule size distribution; **~ del tamaño de partículas** *f* QUÍMICA, CONST particle-size distribution; **~ por tamaños** *f* CARBÓN size distribution; **~ de tiempo** *f* TELECOM time sharing; **~ toda en marcha adelante** *f* ING MECÁ full-forward gear; **~ toda en marcha hacia adelante** *f* ING MECÁ full gear forward; **~ transversal de la energía** *f* FÍS RAD transverse energy distribution; **~ por válvulas** *f* INSTAL HIDRÁUL *máquina de vapor* poppet valve gear, VEH *motor* valve gear

distribuidor *m* AGRIC *sembradora* feed, *semillas* haulier, CONST valve, ELEC distributor, EMB distributor, dispenser, INFORM&PD circulator, dispatcher, supplier, ING MECÁ *automóviles* ignition distributor, distributor, INSTAL HIDRÁUL *máquina de vapor* slide, LAB *material de vidrio* manifold, MECÁ distributor, *tuberías* manifold, NUCL, PROD manifold, TEC ESP valve, circulator, VEH distributor, *válvula de corredera* slide valve; **~ de aceite** *m* ING MECÁ oil distributor; **~ de agua** *m* CONTAM water supplier; **~ automático de etiquetas** *m* EMB label dispenser; **~ automático de llamadas** *m* TELECOM automatic call distributor (*ACD*); **~ de bebedero caliente** *m* ING MECÁ hot runner manifold; **~ de bebidas** *m* REFRIG beverage dispenser; **~ binario** *m* ELECTRÓN binary divider; **~ de cables** *m* ELEC, ING ELÉC cable distributor; **~ de canal de colada caliente** *m* ING MECÁ hot runner manifold; **~ de carga** *m* ELEC *red de distribución* load dispatcher; **~ del caudal** *m* PAPEL flow distributor; **~ central de impulsos** *m* TELECOM central pulse distributor; **~ de concha** *m* INSTAL HIDRÁUL *máquina de vapor* D-valve, plain slide valve, slide valve; **~ de descarga** *m* INSTAL HIDRÁUL draft box (*AmE*), draught box (*BrE*); **~ de encendido** *m* PROD *motores* timer; **~ del encendido** *m* *Esp* AUTO, ING MECÁ ignition distributor; **~ de entrada** *m* TRANSP AÉR inlet manifold (*BrE*), intake manifold (*AmE*); **~ de exploración** *m* TELECOM scanner distributor; **~ de fertilizantes** *m* AGRIC fertilizer spreader; **~ de fluido frigorífico** *m* REFRIG refrigerant distributor; **~ de fondo giratorio** *m* AGRIC revolving-bottom spreader; **~ de fuerza del freno** *m* AUTO brake-power distributor; **~ de ignición** *m* AUTO ignition distributor; **~ de líneas urbanas** *m* TELECOM local line concentrator; **~ de llamadas** *m* TELECOM call distributor; **~ de lumbreras múltiples** *m* PROD multiported valve; **~ de muestras** *m* LAB sample divider; **~ oscilante** *m* PROD swinging valve; **~ de piquera caliente** *m* ING MECÁ hot runner manifold; **~ de potencia del freno** *m* AUTO, VEH brake-power distributor; **~ de purín** *m* AGRIC liquid-manure spreader; **~ con recubrimiento** *m* INSTAL HIDRÁUL *máquina de vapor* lap valve; **~ sin recubrimiento** *m* INSTAL HIDRÁUL lapless valve; **~ rotativo** *m* INSTAL HIDRÁUL, MECÁ cock; **~ rotativo de válvula** *m* INSTAL HIDRÁUL valve cock; **~ rotativo de vapor** *m* INSTAL HIDRÁUL pit cock; **~ de semilla** *m* AGRIC drill spout; **~ de señales** *m* TELECOM signal distributor (*SD*); **~ de teja de expansión** *m* INSTAL HIDRÁUL riding cut-off valve, PROD independent

cutoff valve; **~ de vaciadero caliente** *m* ING MECÁ hot runner manifold; **~ de válvula** *m* INSTAL HIDRÁUL *distribución de vapor*, PROD *máquina de vapor* poppet valve; **~ del vapor** *m* INSTAL HIDRÁUL steam valve; **~ de velas** *m* C&V *vidrio* gob distributor

distrito: **~ minero** *m* MINAS mining area; **~ rural** *m* ING ELÉC rural district

disuelto *adj* CONTAM dissolved, METAL *aleación, solución* dilute, QUÍMICA dissolved

disulfuro *m* QUÍMICA disulfide (*AmE*), disulphide (*BrE*); **~ de carbono** *m* ALIMENT carbon disulfide (*AmE*), carbon disulphide (*BrE*), QUÍMICA carbon bisulfide (*AmE*), carbon bisulphide (*BrE*)

disyunción *f* GEOL fracture, INFORM&PD *operación OR* disjunction, ING ELÉC switching, ING MECÁ tripping; **~ exclusiva** *f* INFORM&PD exjunction; **~ prismática** *f* GEOL prismatic jointing

disyuntor *m* CINEMAT, CONST circuit breaker, ELEC *cortocircuito* switch, release switch, *conmutador* circuit breaker, FERRO disconnector, ING ELÉC release switch, cutout, circuit breaker, ING MECÁ trip, tripping device, PROD tripping device, TRANSP MAR contact breaker; **~ de acción rápida** *m* ING MECÁ fast-acting trip; **~ de aceite** *m* ELEC, ING ELÉC oil switch; **~ en aceite** *m* ELEC, ING ELÉC oil circuit breaker; **~ al aire** *m* ELEC *conmutador* air breaker, air circuit breaker; **~ de aire comprimido** *m* INSTAL HIDRÁUL, PROD pressure switch; **~ automático** *m* ELEC, ING ELÉC, MECÁ, PROD limit switch, SEG automatic cutout switch; **~ de baño de aceite** *m* ELEC bulk-oil circuit-breaker; **~ en baño de aceite** *m* ELEC, ING ELÉC, PROD oil switch; **~ de campo** *m* NUCL field-breaking switch; **~ a chorro de aire** *m* ELEC *cortocircuito* air blast breaker; **~ de chorro de aire** *m* ING ELÉC air blast circuit breaker; **~ de contracorriente** *m* ELEC, ING ELÉC reverse-current circuit breaking; **~ de gravedad** *m* ING ELÉC gravity switch; **~ de inyección magnética** *m* ELEC magnetic blowout circuit breaker; **~ magistral** *m* ELEC master switch; **~ de máxima** *m* ING ELÉC maximum cutout, overcurrent trip; **~ mecánico** *m* ELEC *cortocircuito* mechanical tripping device; **~ de múltiples desconexiones** *m* ELEC multibreak circuit breaker; **~ de seguridad** *m* ELEC, ING ELÉC, MECÁ, PROD, SEG limit switch; **~ de sobrecorriente** *m* ELEC overcurrent circuit breaker, overcurrent switch; **~ de sobreintensidad** *m* ELEC, ING ELÉC overcurrent circuit breaker; **~ de soplado magnético** *m* ELEC, ING ELÉC magnetic blowout circuit breaker; **~ telefónico** *m* ING ELÉC, TELECOM telephone switchgear; **~ térmico** *m* ELEC, ING ELÉC, TERMO thermal circuit breaker

ditionato *m* QUÍMICA dithionate

diurno *adj* GEOFÍS, METEO diurnal

divalencia *f* QUÍMICA divalence

divalente *adj* QUÍMICA divalent

divergencia *f* FÍS, MATEMÁT, NUCL, ÓPT, TEC ESP, TRANSP AÉR divergence, VEH *ruedas delanteras* toe-out; **~ del haz** *f* ÓPT, TELECOM beam divergence; **~ lateral** *f* TRANSP AÉR lateral divergence; **~ longitudinal** *f* TRANSP AÉR longitudinal divergence; **~ de salida** *f* TELECOM output divergence

divergente[1] *adj* FÍS, MATEMÁT, NUCL, ÓPT, TEC ESP, TRANSP AÉR divergent

divergente[2] *m* TRANSP diverging

diversidad *f* MECÁ manifold; **~ biológica** *f* AGRIC

biodiversity; **~ de frecuencias** *f* TELECOM frequency diversity

dividendo *m* INFORM&PD, MATEMÁT dividend

dividido *adj* INFORM&PD partitioned

dividador: **~ de potencia** *m* ELEC, FÍS potential divider

dividir *vt* CINEMAT *negativo* break down, INFORM&PD, MATEMÁT divide

dividirse *v refl* CONST branch off

divinilo *m* QUÍMICA divinyl

divisa *f* D&A stripe

divisible *adj* ING MECÁ sectional, MATEMÁT divisible

división *f* CONST division, split, GEOL division, IMPR split, ING MECÁ compartment, parting, segment, MATEMÁT division; **~ de las aguas** *f* HIDROL parting of the waters; **~ de ambiente** *f* INFORM&PD environment division; **~ de la amplitud** *f* FÍS amplitude division; **~ binaria** *f* ELECTRÓN binary division; **~ en compartimientos** *f* MINAS *pozo* dividing into compartments; **~ cronostratigráfica** *f* GEOL chronostratigraphic division; **~ de datos** *f* INFORM&PD data division; **~ de escala** *f* ELEC *instrumentos* scale division; **~ de fase** *f* ELEC *corriente alterna* phase splitting; **~ de identificación** *f* INFORM&PD identification division; **~ de impreso por zonas** *f* IMPR zoning; **~ de lote** *f* PROD lot splitting; **~ naval con atribuciones especiales** *f* D&A, TRANSP MAR navy task force; **~ de procedimientos** *f* INFORM&PD procedure division; **~ de tiempo** *f* ELECTRÓN time division; **~ del tiempo** *f* ELECTRÓN time slicing

divisional *adj* ING MECÁ sectional

divisor *m* ELECTRÓN divider, INFORM&PD divider, divisor, ING MECÁ divider, separator, MATEMÁT divisor, TEC PETR *refino* catchpit, TELECOM divider; **~ de arco** *m* AGRIC *cosechadora* loop divider; **~ de corriente** *m* ELEC *circuito* current divider; **~ de energía** *m* ING ELÉC power divider; **~ de fase** *m* ELEC *corriente alterna* phase splitter; **~ de flujo** *m* PROD flow divisor; **~ de la frecuencia de la imagen** *m* TV field divider; **~ de frecuencia para obtener la señal de línea** *m* TV line divider; **~ de frecuencias** *m* ELEC, TELECOM frequency divider; **~ del haz** *m* CINEMAT, ÓPT, TV beam splitter; **~ del haz con prisma** *m* CINEMAT prism-beam splitter; **~ de la luz** *m* CINEMAT beam splitter; **~ de muestra** *m* CARBÓN riffle sampler; **~ de potencia** *m* ELEC *fuente de alimentación*, TELECOM power divider; **~ de potencial** *m* ELEC, TELECOM voltage divider; **~ de potencial inductivo** *m* ELEC *autotransformador* inductive potential divider; **~ primo** *m* MATEMÁT prime factor; **~ de señal** *m* TV signal splitter; **~ de tensión** *m* ELEC voltage divider, ING ELÉC bleeder, voltage divider, TELECOM, TV voltage divider; **~ de tensión ajustable** *m* ELEC adjustable voltage divider; **~ de tensión por resistencias** *m* ING ELÉC resistor voltage divider; **~ de tensión resistivo** *m* ELEC resistive voltage divider; **~ de voltaje** *m* ELEC, FÍS voltage divider, ING ELÉC bleeder, TELECOM voltage divider; **~ de voltaje ajustable** *m* ELEC adjustable voltage divider; **~ de voltaje capacitivo** *m* ING ELÉC capacitive voltage divider

divisores *m pl* GEOM dividers, ING MECÁ beam dividers

divisoria: **~ de aguas** *f* AGUA water parting, HIDROL water parting, watershed

diyodometano *m* QUÍMICA diiodomethane

DLE *abr* (*escape de enlace de datos*) INFORM&PD DLE (*data link escape*)

DMA *abr* ELECTRÓN, INFORM&PD (*acceso directo a la memoria*) DMA (*direct memory access*), QUÍMICA (*dimetilamina*) DMA (*dimethylamine*)

DMF *abr* (*dimetilformamida*) QUÍMICA DMF (*dimethylformamide*)

DML *abr* (*lenguaje de manipulación de datos*) INFORM&PD DML (*data manipulation language*)

DMS *abr* INFORM&PD, MATEMÁT, PROD (*dígito más significativo*) MSD (*most significant digit*); (*dígito menos significativo*) LSD (*least significant digit*)

dN *abr* (*decineperio*) ACÚST, ELECTRÓN, FÍS, METR, TELECOM, TV dN (*decineper*)

DNC *abr* (*dinitrocresol*) AGRIC, QUÍMICA DNC (*dinitrocresol*)

DNT *abr* (*dinitrotolueno*) QUÍMICA *colorantes y explosivos* DNT (*dinitrotoluene*)

DO *abr* (*déficit en oxígeno*) HIDROL OD (*oxygen deficit*)

dobladillo *m* TEXTIL hem

doblado *m* MECÁ bending; **~ en frío** *m* MECÁ cold bending; **~ de hilo a dos cabos** *m* TEXTIL two plies of a two-ply yarn

doblador: **~ de frecuencia** *m* ELEC, ELECTRÓN, FÍS, TELECOM frequency doubler; **~ de ingenio** *m* IMPR *encuadernación* plough folder (*BrE*), plow folder (*AmE*); **~ de tensión** *m* ELEC *circuito rectificador* voltage doubler; **~ de voltaje** *m* ELEC *circuito rectificador* voltage doubler, TELECOM doubler

dobladora *f* IMPR bender; **~ de chapas** *f* MECÁ *herramientas* brake; **~ a embudo** *f* IMPR former folder

doblaje *m* CINEMAT, TV dubbing; **~ de cinta de video** *AmL*, **~ de cinta de vídeo** *m* *Esp* TV videotape dubbing; **~ de RF** *m* TV RF dub

doblar *vt* CINEMAT *sonido* dub, MINAS crimp, TRANSP MAR *amarras* double, *cabo* round, TV dub

doble[1] *adj* GEOM dual; **de ~ acción** *adj* AUTO, MECÁ double-acting; **de ~ alimentación** *adj* ELEC *motor* double-fed; **de ~ efecto** *adj* ENERG RENOV *servomotor*, MECÁ, REFRIG *compresor* double-acting; **~ mando** *adj* ELEC *potenciómetro* dual-ganged; **de ~ seguridad** *adj* MECÁ, SEG fail-safe; **de ~ torsión** *adj* CONST double-threaded

doble[2]: **~ árbol de levas en cabeza** *m* VEH *motor* double overhead camshaft; **~ arrollamiento** *m* ELEC *de bobina* double-layer winding; **~ bisel** *m* C&V double bevel; **~ carril** *m* ELECTRÓN double rail; **~ casco** *m* TRANSP MAR *construcción naval* double hull; **~ cruzamiento oblicuo con deslizamientos** *m* FERRO *configuración de la vía* double diamond crossing with slips; **~ dúo** *m* PROD four-high; **~ eje** *m* VEH *de camión* tandem axle; **~ enlace** *m* QUÍMICA double bond; **~ fondo** *m* TRANSP MAR *buque* double bottom; **~ fondo celular** *m* TRANSP MAR *construcción naval* cellular bottom; **~ fondo de construcción celular** *m* TRANSP MAR *construcción naval* cellular bottom; **~ forro** *m* TRANSP MAR *construcción naval* double skin; **~ piso** *m* CONST double floor; **~ puente** *m* ELEC *instrumento* Kelvin bridge; **~ sostenido** *m* ACÚST *música* double sharp; **~ teñido** *m* TEXTIL double dyeing; **~ vidrio** *m* CONST, TERMOTEC double glazing

doble[3]: **~ aislación** *f* ING ELÉC double insulation; **~ alimentación** *f* ING ELÉC dual power supply; **~ bemol** *f* ACÚST *música* double flat; **~ cadena** *f* AUTO, VEH *transmisión* duplex chain; **~ capa** *f* REVEST double layer; **~ densidad** *f* INFORM&PD double density; ~ **difusión** f ELECTRÓN double diffusion; ~ **exploración entrelazada** f TV twin interlaced scanning; ~ **exposición** f CINEMAT, FOTO double exposure; ~ **ligadura** f QUÍMICA *alqueno insaturación* double bond; ~ **modulación** f ELECTRÓN double modulation; ~ **pase** f PROD *soldadura* backing pass; ~ **precisión** f INFORM&PD double precision; ~ **ranuración** f PROD dual slot; con ~ **rosca** f CONST double-threaded; ~ **tracción con unidad tractora en cola** f FERRO *vehículos* banking, pusher operation; ~ **vía** f ELECTRÓN double rail

doblemente: ~ **acoplado** adj ELEC *potenciómetro* dual-ganged

doblete m FÍS, IMPR, QUÍMICA *espectroscopia* doublet; ~ **acromático** m FÍS achromatic doublet

doblez m C&V lap, CRISTAL kink, ING MECÁ ply, MECÁ bending, PETROL dogleg, TEC ESP *trajes espaciales* bending, TEXTIL *tejido* crease; ~ **del fondo** m EMB bottom fold

doceavo m IMPR duodecimo

dócil adj CONST *hormigón* workable

docilidad f CONST *de hormigón* workability

dócima f CARBÓN *muestreo de minerales* test

docosanoico adj QUÍMICA behenic

documentación f VEH data sheet; ~ **del buque** f TRANSP MAR ship's papers; ~ **de la fabricación** f PROD manufacturing documents, manufacturing papers; ~ **técnica del producto** f PROD, SEG technical product documentation; ~ **de vuelo** f TRANSP AÉR flight documentation

documental m CINEMAT, TV documentary

documento m INFORM&PD document; ~ **de carga y descarga** m PROD turnaround document; ~ **fuente** m INFORM&PD source document; ~ **nítido** m IMPR high key document; ~ **original** m INFORM&PD source document; ~ **de respuesta** m PROD turnaround directive, turnaround document

documentos: ~ **de embarque** m pl TRANSP MAR shipping documents; ~ **de fabricación** m pl PROD manufacturing documents, manufacturing papers; ~ **de taller** m pl PROD shop papers

dodecaédrico adj GEOM dodecahedral

dodecaedro m GEOM dodecahedron

dodecágono m GEOM dodecahedron

dodecano m DETERG, QUÍMICA dodecane

dodecilbenceno m DETERG, QUÍMICA dodecyl benzene

doéglico adj QUÍMICA doeglic

dolerita f GEOL, MINERAL, PETROL dolerite

dolerofana f MINERAL dolerophane

dolerofanita f MINERAL dolerophanite

dolina f AGUA doline, GEOL, MINAS sinkhole

dolly f CINEMAT, TV billyboy dolly, dolly; ~ **para el boom** f CINEMAT, TV boom dolly; ~ **para travelling** f CINEMAT, TV traveling dolly (*AmE*), travelling dolly (*BrE*)

dolomía f GEOL dolostone, MINERAL, PETROL, QUÍMICA, TEC PETR dolomite

dolomita f C&V, GEOL, MINERAL, PETROL, QUÍMICA, TEC PETR dolomite; ~ **alveolar** f GEOL *formación* alveolar dolomite; ~ **calcárea** f GEOL calcareous dolomite; ~ **calcítica** f GEOL calcitic dolomite; ~ **celular** f GEOL *formación alpina* cellular dolomite; ~ **férrica** f GEOL *roca* ferroan dolomite

dolomitización f GEOL *diagénesis* intrusion

domeykita f MINERAL domeykite

dominante adj ACÚST dominant

dominio m FÍS, MATEMÁT *función* domain; ~ **entero** m MATEMÁT *integral* domain of integrity; ~ **espacial** m ELECTRÓN spatial domain; ~ **de frecuencia** m ELECTRÓN frequency domain; ~ **de gestión** m TELECOM management domain (*MD*); ~ **de gestión privada** m TELECOM private-management domain (*PRMD*); ~ **de integridad** m MATEMÁT *integral* domain of integrity; ~ **magnético** m ING ELÉC magnetic domain; ~ **minero** m MINAS mining area; ~ **temporal** m ELECTRÓN time domain; ~ **Tetiense** m GEOL Tethyan realm

dominios: ~ **de Weiss** m pl FÍS Weiss domains

domo m GAS, GEOL, TEC PETR dome; ~ **esquistoso** m GEOL, TEC PETR shale dome; ~ **salífero** m GEOL, TEC PETR salt dome; ~ **salino** m GEOL, TEC PETR *geología* salt dome; ~ **toloide** m GEOL *estructura volcánica* tholoid dome; ~ **volcánico** m GEOL cumulo volcano, *característica volcánica* cumulo dome

donador m ELECTRÓN, INFORM&PD *chip* donor; ~ **de red** m CONTAM net donator

donante m INFORM&PD *chip*, QUÍMICA *átomo* donor

donut m GEOM doughnut, torus

dopado[1] adj ELECTRÓN, INFORM&PD doped

dopado[2]: ~ **de base** m ELECTRÓN base doping; ~ **por difusión** m ELECTRÓN diffusion doping; ~ **con oro** m ELECTRÓN *semiconductor* gold doping; ~ **del semiconductor** m ELECTRÓN semiconductor doping

dopaje m TELECOM doping

dopamina f QUÍMICA *catecolamina* dopamine

dopante m ELECTRÓN, INFORM&PD, ÓPT, TELECOM dopant

dopar vt ELECTRÓN, INFORM&PD, ÓPT, TELECOM dope

doplerita f MINERAL dopplerite

dorado[1] adj QUÍMICA *espécimen* gold-plated

dorado[2] m IMPR gilding, METAL gilding, gold plating, PROD gilding, TEC ESP gold plating

dorador m METAL, PROD *persona* gilder

doradura f METAL, PROD gilding

dorar vt METAL, PROD gild

dorsal f TRANSP MAR *meteorología* ridge; ~ **anticiclónica** f TRANSP MAR ridge; ~ **oceánica** f GEOL midocean ridge, OCEAN oceanic ridge; ~ **submarina** f OCEAN ridge

dorso[1]: ~ **contra dorso** adj CONST back-to-back

dorso[2] m IMPR reverse

dos: **de** ~ **bocas** adj PROD *llave* double-ended; **de** ~ **hilos** adj ELEC *devanado*, FÍS bifilar; **con** ~ **máquinas** adj ING MECÁ, TRANSP AÉR twin-engined; **de** ~ **motores** adj ING MECÁ, TRANSP AÉR twin-engined; **de** ~ **pisos** adj EMB *paleta de carga* double-decked

DOS abr (*sistema operativo de discos*) INFORM&PD DOS (*disk operating system*)

dosaje m CARBÓN, METR dosage

dosel: ~ **vegetal** m AGRIC canopy

dosificación f CARBÓN dosage, titration, EMB dosing, IMPR metering, METR apportionment, batching, dosage, gradation, percentage determination, QUÍMICA dosing; ~ **de isótopos** f CARBÓN isotope dosage, isotope measurement; ~ **manual** f EMB hand dosing

dosificado: ~ **por volumen** m EMB volume dosing

dosificador m ALIMENT dispenser, EMB dosing apparatus

dosificadora f EMB *máquina* dosing machine

dosificar vt CARBÓN titrate

dosimetría f FÍS, FÍS RAD dosimetry; **~ de alto nivel** f FÍS RAD high level dosimetry; **~ de alto nivel energético** f FÍS RAD high level dosimetry; **~ personal** f FÍS RAD personal dosimetry; **~ química** f FÍS RAD, QUÍMICA chemical dosimetry; **~ de radiación** f FÍS RAD radiation dosimetry

dosímetro m ACÚST dose-meter, AGRIC fertilizer proportioner, FÍS, FÍS RAD dosimeter; **~ de alarma acústica** m FÍS RAD sound alarm radiation dosimeter; **~ de radiación** m FÍS RAD radiation dosimeter; **~ de rayos X** m INSTR X-ray dosimeter; **~ termoluminiscente** m FÍS RAD thermoluminescent dosimeter

dosis f FÍS RAD, METR, QUÍMICA dose; **~ absorbida** f FÍS, FÍS RAD absorbed dose; **~ acumulada del personal** f FÍS RAD dose accumulated by workers; **~ acumulativa** f ELECTRÓN implant dose; **~ de cemento** f CARBÓN *hormigones* settlement gage (*AmE*), settlement gauge (*BrE*); **~ colectiva de la subpoblación** f CONTAM *toxicología* subpopulation collective dose; **~ equivalente** f FÍS, FÍS RAD dose equivalent; **~ equivalente efectiva colectiva** f NUCL collective effective dose equivalent, SE; **~ equivalente efectiva integrada** f NUCL committed effective dose equivalent; **~ equivalente en renguenios en el hombre** f (*rem*) FÍS RAD roentgen equivalent for man (*rem*); **~ de exposición** f FÍS RAD exposure dose; **~ de irradiación** f CONTAM, FÍS RAD, TEC ESP radiation dose; **~ letal** f CONTAM *toxicología*, FÍS lethal dose; **~ letal del 50%** f FÍS, NUCL mean lethal dose (*LD50*); **~ letal mediana** f FÍS, NUCL mean lethal dose (*LD50*); **~ máxima admisible** f FÍS RAD maximum permissible dose; **~ media letal** f CONTAM *toxicología*, FÍS RAD median lethal dose; **~ mortal** f CONTAM *toxicología*, FÍS lethal dose; **~ en profundidad** f FÍS RAD depth dose; **~ de radiación** f CONTAM *legislación*, FÍS RAD, TEC ESP radiation dose; **~ registrada** f FÍS RAD recorded dose; **~ de respuesta** f CONTAM dose response; **~ universal** f CONTAM *legislación* population dose

dotación f TRANSP MAR manning, *bote, buque* crew; **~ lógica** f PETROL software; **~ de oxígeno** f SEG oxygen supply

dotar vt TRANSP MAR *buque de marinería, tripulación* man

dotes: **~ de mando** f pl D&A leadership

dovela f CONST arch stone, *arquitectura* voussoir; **~ superior** f CONST keying up

DPC abr (*centro de proceso de datos*) TELECOM DPC (*data processing center AmE, data processing centre BrE*)

DQO abr (*demanda química de oxígeno*) CONTAM, HIDROL, QUÍMICA COD (*chemical oxygen demand*)

dracma m METR dram

dracone m CONTAM MAR dracone

draga f AGRIC dredge, CONST *maquinaria* bucket, dredge, CONTAM MAR dredger, MECÁ drag, MINAS dredge, drag, dredger, PROD drag, TRANSP MAR dredge; **~ de almeja** f TRANSP MAR dipper dredger, grab dredger; **~ aspirante** f AGUA suction dredge, hydraulic dredge; **~ bivalva** f CONST grab dredge; **~ de bombeo** f CONST, TRANSP MAR pump dredge, pump dredger; **~ de cangilones** f AGUA, CONST, TRANSP MAR bucket dredger; **~ en catamarán** f AGUA catamaran dredge; **~ de cuchara** f AGUA spoon dredger, spoondredge, TRANSP MAR dipper dredger;

~ de cuchara bivalva f CONST clam shell bucket dredger; **~ disgregadora** f TRANSP MAR cutter dredger; **~ excavadora** f CONST *maquinaria, obras públicas* dragline excavator; **~ excavadora sobre orugas** f CONST crawler dragline excavator; **~ fluvial** f AGRIC, AGUA, HIDROL river dredge; **~ gánguil** f TRANSP MAR hopper barge, hopper dredger; **~ hidrogéfica** f OCEAN floating hydrographic dredge; **~ mixta** f TRANSP MAR dual-purpose dredger; **~ de prospección** f MINAS prospecting dredge; **~ romperrocas** f TRANSP MAR rock-cutting dredger; **~ de rosario** f TRANSP MAR ladder dredge; **~ de succión** f AGUA suction dredge, CONST pump dredger, OCEAN suction dredge, TRANSP MAR pump dredger, suction dredger; **~ de succión con cántara** f TRANSP MAR suction hopper dredger

dragado m CONTAM MAR, HIDROL, MINAS, TRANSP MAR dredging; **~ de aluviones** m MINAS dredge mining; **~ de arena** m CONST sand dredging; **~ circular** m AGUA swinging across the face; **~ hidrográfico** m OCEAN wire dragging, wire dredging; **~ de minas** m D&A, TRANSP MAR minesweeping; **~ a remolque** m TRANSP MAR trail dredging

dragadora: **~ bivalva** f CONST grab dredger

dragalina f Esp (*cf cuchara sacabarro AmL*) CONST dragline excavator, MINAS scraper, *barco* dredge boat; **~ móvil** f CONST *maquinaria obras públicas* walking dragline excavator; **~ sobre orugas** f CONST crawler dragline excavator

dragaminas m D&A, TRANSP MAR minesweeper

dragar vt CONST clear away, CONTAM MAR, TRANSP MAR *canal* dredge

dragón m CONTAM MAR dracone

DRAM abr (*memoria dinámica de acceso aleatorio*) COMP&DP DRAM (*dynamic random access memory*)

drama: **~ documental** m CINEMAT, TV docudrama

drao m TRANSP MAR *colisión* ram

drapeado m TEXTIL drape

dren m CONST drain

drenaje m GEN drain, *de fluidos* drainage; **~ de aguas pluviales** m CONST storm drain; **~ anular** m AGUA annular drainage; **~ arterial** m AGUA arterial drainage; **~ de espuma** m PROC QUÍ foam drainage; **~ por expansión de gas** m GAS, TEC PETR *recuperación de petróleo* gas cap drive; **~ por gravedad** m PETROL gravity draining; **~ mediante zanjas** m HIDROL *aguas residuales* ditch drainage; **~ natural** m AGUA natural drainage; **~ químico** m NUCL chemical drains; **~ del radiador** m AUTO, VEH radiator draining; **~ subterráneo** m AGUA underground drainage; **~ topero** m AGRIC, CONST mole drainage; **~ vertical** m CARBÓN vertical drainage

drenar vt AGUA, CONST, HIDROL, MECÁ, QUÍMICA, RECICL drain

driza f TRANSP MAR *jarcia de labor* halyard; **~ de fuera** f TRANSP MAR *puntal de carga* outhaul

drusa f GEOL druse

drusado adj GEOL, TEC PETR vuggy

drúsico adj GEOL drusy

DST abr (*prueba de producción con tubería de perforación*) TEC PETR DST (*drill stem test*)

DTC abr (*dispositivo de transferencia de carga*) FÍS, ING ELÉC, TEC ESP, TELECOM CTD (*charge-transfer device*)

DTE abr INFORM&PD, TELECOM, TRANSP AÉR (*conjunto terminal de datos, equipo terminal de datos*) DTE (*data terminal equipment*)

DTL abr (circuito a base de diodos y transistores)
ELECTRÓN, INFORM&PD DTL (diode-transistor logic)
D₂O abr (óxido de deuterio) FÍS D_2O (deuterium oxide)
dual adj GEOM dual
dualidad: ~ **onda-corpúsculo** f FÍS, FÍS ONDAS, FÍS
PART wave-particle duality; ~ **onda-partícula** f FÍS,
FÍS ONDAS, FÍS PART wave-particle duality
dúctil adj CRISTAL ductile, ING MECÁ tensile, METAL
ductile
ductilidad f CRISTAL, ING MECÁ, METAL ductility,
tensility
ducto m ING MECÁ, MECÁ duct; ~ **de aire** m ING MECÁ
air duct; ~ **para cables** m AmL (cf conducto para
cables Esp) PROD wiring duct; ~ **de descarga** m
AGUA delivery main
duela f PROD barril stave
duendecillo m INFORM&PD sprite
dufrenita f MINERAL dufrenite
dufrenoysita f MINERAL dufrenoysite
dulcina f ALIMENT, QUÍMICA dulcin
dulcita f QUÍMICA dulcine
dumortierita f MINERAL dumortierite
dumper m CONST, TRANSP dumper (AmE), tipper (BrE)
duna f GEOL dune
dunita f GEOL, MINERAL, PETROL dunite
dupe m CINEMAT dupe; ~ **reversible** m CINEMAT
reversal dupe; ~ **de segunda generación** m CINE-
MAT second generation dupe
dúplex[1] adj INFORM&PD, TELECOM duplex
dúplex[2] m CONST maisonette, INFORM&PD, TELECOM
duplex; ~ **todo completo** m INFORM&PD operation
duplex
duplexor m INFORM&PD, TELECOM duplexer
duplicación f ACÚST duplication; ~ **de alta velocidad** f
AmL (cf tiraje de alta velocidad Esp) TV high-speed
duplication
duplicado: ~ **de diapositivas** m FOTO slide duplication
duplicador m ING MECÁ doubler; ~ **de frecuencia** m
ELEC, ELECTRÓN, FÍS, TELECOM frequency doubler
duplicar vt INFORM&PD, TELECOM duplicate
duplo adj ING ELÉC duplex
duque: ~ **de alba** m TRANSP MAR mooring pile, puerto
de amarre, radiofaro dolphin
durabilidad f GEN durability; ~ **del concreto** f AmL
(cf durabilidad del hormigón Esp) CONST concrete
durability; ~ **del hormigón** f Esp (cf durabilidad del
concreto AmL) CONST concrete durability; ~ **química**
f C&V, QUÍMICA chemical durability
duración[1]: **de ~ limitada** adj MECÁ expendable
duración[2] f CINEMAT película running time, screen
time, screening time, ING ELÉC lifetime, ING MECÁ
duration, life, INSTAL HIDRÁUL de compresión del
vapor duration, METR durability, PROD duration, TEC
ESP range, TEXTIL durability; ~ **de almacenado** f
CINEMAT, IMPR, P&C shelf life; ~ **de almacenaje** f ING
ELÉC storage time; ~ **en almacenaje** f IMPR, P&C shelf
life; ~ **de almacenaje útil** f ING ELÉC storage time;
~ **de apertura/cierre** f ELEC relé make-break time;
~ **del arco** f ELEC polo, fusible, interruptor arcing
time; ~ **de cierre** f ELEC relé make time; ~ **completa
a mitad de altura** f ÓPT full-duration half maximum
(FDHM); ~ **de disipación** f ING ELÉC storage time;
~ **del establecimiento del frente del impulso** f TV
leading-edge pulse time; ~ **de la expansión** f INSTAL
HIDRÁUL del vapor expansion duration; ~ **de la
expansión del vapor** f INSTAL HIDRÁUL duration of

expansion of steam; ~ **de exploración de línea** f TV
trace interval; ~ **de la exposición** f CINEMAT, FOTO
exposure duration; ~ **del flash** f FOTO flash duration;
~ **de funcionamiento** f TRANSP running time;
~ **hasta la rotura** f METAL time to rupture; ~ **de la
herramienta** f ING MECÁ tool life; ~ **del impulso** f
ELECTRÓN pulse length, TELECOM pulse length, pulse
width; ~ **de integración** f ELECTRÓN integration
time; ~ **de liberación** f ING ELÉC release time; ~ **de
llamada** f TELECOM call duration; ~ **de la
luminosidad** f FÍS RAD luminosity lifetime; ~ **de la
marea** f OCEAN tide duration; ~ **de una marea
completa** f OCEAN full-tide duration; ~ **mecánica** f
ING ELÉC, PROD mechanical life; ~ **media** f CALIDAD
mean life; ~ **media antes de fallas** f CALIDAD mean
time to failure; ~ **de memoria** f ELECTRÓN canal
subterráneo, transistores bulk lifetime; ~ **de parada** f
NUCL outage time; ~ **prevista** f NUCL de materiales
serviceability; ~ **del pulso** f ELECTRÓN pulse length;
~ **de servicio** f NUCL service life; ~ **del servicio** f
CARBÓN, CONST, EMB, ING ELÉC, TEC ESP service life;
~ **total mitad del máximo** f TELECOM full-duration
half maximum (FDHM); ~ **de la travesía** f TRANSP
MAR crossing time; ~ **útil en almacenaje** f CINEMAT,
IMPR, P&C shelf life; ~ **de la utilización** f CARBÓN,
CONST, EMB, ING ELÉC, NUCL, TEC ESP service life;
~ **de vida** f FÍS nucleónica, TELECOM lifetime; ~ **de
vida media** f CALIDAD mean life; ~ **de vuelo** f TRANSP
AÉR flying time
duradero adj CALIDAD durable
duraluminio m MECÁ duralumin
durangita f GEOL, MINERAL durangite
duraznillo: ~ **negro** m AGRIC black nightshade
dureno m QUÍMICA durene
dureza f CONST, CRISTAL hardness, ING MECÁ roughness, stiffness, MECÁ hardness, roughness, METAL
propiedad física, prueba, P&C propiedad física hardness, PROD muelas abrasivas grade, QUÍMICA de agua,
sustancia hardness; ~ **del agua** f AGUA water hardness; ~ **Brinell** f ING MECÁ, MECÁ Brinell hardness;
~ **de carbonatos** f HIDROL carbonate hardness; ~ **de
corte** f CARBÓN cutoff grade; ~ **a la melladura** f P&C
propiedad física, prueba indentation hardness; ~ **de
miga** f ALIMENT crumb firmness; ~ **de péndulo** f P&C
prueba, revestimientos pendulum hardness; ~ **de la
radiación** f FÍS RAD, ING ELÉC, NUCL, TELECOM
radiation hardness; ~ **Shore** f P&C propiedad física,
prueba Shore hardness; ~ **de la superficie** f ING
MECÁ surface hardness
durmiente m AmL (cf solera Esp) CONST mudsill, sill
plate, groundsill, MINAS construcción sole, sill,
groundsill, TRANSP MAR construcción naval sleeper,
deadwood, rock; ~ **de concreto** m AmL (cf durmiente de hormigón Esp) CONST, FERRO concrete
sleeper (BrE), concrete tie (AmE); ~ **de hormigón**
m Esp (cf durmiente de concreto AmL) CONST, FERRO
concrete sleeper (BrE), concrete tie (AmE); ~ **de
madera** m AmL (cf traviesa de madera Esp) CONST,
FERRO wooden sleeper (BrE), wooden tie (AmE)
durmientes: ~ **de madera** m pl PROD wooden dunnage
duro: ~ **al timón** adj TRANSP MAR buque hard on the
helm
durómetro m GEN durometer, hardness tester;
~ **Vickers** m ING MECÁ Vickers hardness-testing
machine
durra f AGRIC sorgo durra

DVQ *abr* (*deposición de vapor químico*) ELECTRÓN, ÓPT, TELECOM CVD (*chemical vapor deposition AmE, chemical vapour deposition BrE*)

DVQPA *abr* (*deposición de vapor químico de plasma activado*) ELECTRÓN, ÓPT, TELECOM PCVD (*plasma-activated chemical vapor deposition AmE, plasma-activated chemical vapour deposition BrE*)

Dy *abr* (*disprosio*) CARBÓN, METAL, QUÍMICA Dy (*dysprosium*)

E

é *abr* (*electrón*) GEN e (*electron*)

E *abr* (*exa-*) METR E (*exa-*)

ebanista *m* CONST cabinetmaker, joiner

ebanistería *f* CONST cabinetmaking, joinery

EBCDIC *abr* (*código ampliado de caracteres decimales codificados en binario*) INFORM&PD EBCDIC (*extended binary-coded decimal-interchange code*)

ebonita *f* ELEC *aislador* ebonite, P&C *polímero, vulcanización* ebonite, vulcanite

EBR *abr* (*eficacia biológica relativa*) NUCL RBE (*relative biological effectiveness*)

ebullición[1]: **de ~** *adj* FÍS, TERMO boiling; **en ~** *adj* FÍS, TERMO boiling

ebullición[2] *f* GEN boiling; **~ en condiciones de saturación** *f* NUCL saturated boiling; **~ laminar** *f* NUCL sheet boiling; **~ libre** *f* REFRIG pool boiling; **~ nucleada** *f* REFRIG nucleate boiling; **~ en película** *f* REFRIG film boiling; **~ pelicular** *f* NUCL film boiling; **~ pelicular anular** *f* NUCL annular film boiling; **~ por reflujo** *f* PROC QUÍ reflux boiling; **~ en rollo** *f* TEXTIL roll boiling; **~ subenfriada** *f* PROD subcooled boiling; **~ violenta** *f* NUCL violent boiling

ebullir *vi* QUÍMICA simmer

ECCS *abr* NUCL railroad container (*AmE*), railway container (*BrE*)

ECD *abr* (*densidad de circulación equivalente*) TEC PETR ECD (*equivalent circulating density*)

ecdisona *f* QUÍMICA ecdysone

ecgonina *f* QUÍMICA ecgonine

echar[1]: **~ al mar** *vt* OCEAN, TRANSP MAR *mercancías* jettison

echar[2]: **~ el ancla** *vi* TRANSP MAR cast anchor; **~ a andar** *vi* ING MECÁ start; **~ el cerrojo** *vi* CONST bolt; **~ por delante** *vi* TRANSP MAR *velas* lay aback; **~ a perder** *vi* ALIMENT spoil

echazón *f* CONTAM MAR, OCEAN jetsam, jettison, TRANSP MAR *desechos* jetsam; **~ flotante** *f* TRANSP MAR flotsam; **~ hundida** *f* TRANSP MAR jetsam

eclímetro *m* GEN clinometer

eclínometro *m* MINAS clinometer

eclipse *m* TEC ESP eclipse

eclisa *f* FERRO joint bar (*AmE*), splice bar, fishplate (*BrE*), *equipo inamovible* applying joint bar (*AmE*), fishplating (*BrE*), PROD butt strip; **~ aislante** *f* FERRO insulating fishplate (*BrE*), insulating joint bar (*AmE*); **~ central horizontal estabilizadora** *f* TRANSP AÉR horizontal-stabilizer center joint bar (*AmE*), horizontal-stabilizer centre fishplate (*BrE*)

eclisaje *m* FERRO *equipo inamovible* applying joint bar (*AmE*), fishplating (*BrE*); **~ de emergencia** *m* FERRO emergency fishplating

eclogita *f* PETROL eclogite

eclosión: **~ de huevos** *f* AGRIC egg hatch

eco *m* ACÚST echo, reverberation, ELECTRÓN echo, FÍS, FÍS ONDAS *sonido reflejado* echo, reverberation, GEOFÍS reverberation, INFORM&PD, OCEAN, TEC ESP, TV echo; **~ falso** *m* OCEAN false echo, TEC ESP *radar* ghost echo; **~ fantasma** *m* TEC ESP ghost echo; **~ fluctuante** *m* ACÚST flutter echo; **~ de impresión** *m* IMPR printing echo; **~ múltiple** *m* ACÚST, OCEAN multiple echo, TV flutter echo; **~ negativo** *m* TV negative echo; **~ parásito por reflexión marina** *m* TEC ESP *radar* sea clutter; **~ permanente** *m* TEC ESP permanent echo; **~ pulsante** *m* ACÚST flutter echo; **~ radárico** *m* TRANSP radar echo, TRANSP AÉR radar blip, radar echo, TRANSP MAR radar echo; **~ de segundo impulso** *m* TRANSP MAR *radar* second trace echo

ecodetector *m* D&A sound detector

ecógrafo *m* FÍS echograph

ecolocación *f* ACÚST, TRANSP MAR echolocation

ecología *f* CONTAM ecology

ecometría *f* GAS echometry

economía: **~ agrícola comercial** *f* AGRIC cash farming economy; **~ de la calidad** *f* CALIDAD quality economics; **~ de combustible** *f* TERMO fuel economy; **~ de la recirculación del agua** *f* AGUA, HIDROL recirculating water economy

económico *adj* ING MECÁ cost-effective

economizador *m* INSTAL HIDRÁUL boiler economizer, economizer, INSTAL TERM economizer; **~ de agua** *m* AGUA, HIDROL water-saving; **~ de calor** *m* TERMO heat economizer

ecoplex *m* INFORM&PD echoplex

ecos: **~ de mar** *m pl* TEC ESP, TELECOM *radar*, TRANSP MAR sea clutter; **~ parásitos del mar** *m pl* TELECOM *radar*, TRANSP MAR sea clutter; **~ parásitos de pantalla** *m pl* ELECTRÓN, TEC ESP, TELECOM, TRANSP MAR clutter; **~ perturbadores de radar** *m pl* ELECTRÓN, TEC ESP, TELECOM, TRANSP MAR clutter

ecosistema *m* CONTAM ecosystem; **~ hidrotérmico** *m* OCEAN hydrothermal ecosystem

ecosonda *f* FÍS ONDAS echo sounder, echo sounding, OCEAN echo sounding, TRANSP MAR *equipamiento electrónico* echo sounder, *navegación* echo sounding; **~ de canal múltiple** *f* OCEAN, TRANSP MAR multibeam echo sounder; **~ de haces múltiples** *f* OCEAN, TRANSP MAR multibeam echo sounder

ecosondador *m* TRANSP MAR depth sounder, *navegación* depth finder

ectoparásito *m* AGRIC external parasite

ecuación *f* ELEC, FÍS, MATEMÁT, MECÁ, QUÍMICA, TERMO equation; **~ de Arrhenius** *f* QUÍMICA Arrhenius equation; **~ de Bernoulli** *f* FÍS FLUID Bernoulli's equation; **~ Bethe-Goldstone** *f* NUCL Bethe-Goldstone equation; **~ de Boltzmann** *f* FÍS RAD Boltzmann's equation; **~ característica** *f* TEC ESP characteristic equation; **~ de Clapeyron** *f* FÍS Clapeyron's equation; **~ de continuidad** *f* ELEC, FÍS equation of continuity; **~ diferencial** *f* MATEMÁT differential equation; **~ dimensional** *f* FÍS dimensional equation; **~ de dispersión** *f* FÍS dispersion equation; **~ de enésimo grado** *f* MATEMÁT equation of nth degree; **~ de equilibrio** *f* MECÁ equation of equilibrium; **~ de estado** *f* MECÁ equation of state; **~ de estado del núcleo** *f* FÍS RAD nuclear equation of state; **~ de estado térmico** *f* TERMO equation of thermal state; **~ de estado termodinámico** *f* TERMO thermodynamic equation of state; ~ **para el flujo viscoso** *f* FÍS FLUID viscous flow equation; ~ **fotoeléctrica de Einstein** *f* ING ELÉC Einstein photoelectric equation; ~ **de Fresnel** *f* ACÚST Fresnel's equation; ~ **de gas** *f* GAS, MECÁ gas equation; ~ **general del círculo** *f* GEOM general equation of the circle; ~ **de Hamilton-Jacobi** *f* FÍS Hamilton-Jacobi equation; ~ **de Klein-Gordon** *f* FÍS Klein-Gordon equation; ~ **de Laplace** *f* FÍS Laplace's equation; ~ **de Navier-Stokes** *f* FÍS Navier-Stokes equation; ~ **nuclear** *f* FÍS RAD nuclear equation of state; ~ **de onda** *f* ELEC, FÍS, FÍS RAD wave equation; ~ **de origen y destino** *f* TRANSP *tráfico* origin and destination equation (*O-D equation*); ~ **de Poisson** *f* ELEC *problemas de potencial,* FÍS Poisson's equation; ~ **de propagación** *f* FÍS propagation equation; ~ **Schlueter del movimiento** *f* NUCL Schlueter equation of motion; ~ **de Schrödinger** *f* FÍS Schrödinger equation; ~ **termodinámica de estado** *f* TERMO thermodynamic equation of state; ~ **de la transferencia radiativa** *f* FÍS RAD equation of radiative transfer; ~ **de Van der Waals** *f* FÍS Van der Waals equation; ~ **de vorticidad** *f* FÍS FLUID vorticity equation

ecuaciones: ~ **canónicas** *f pl* FÍS canonical equations; ~ **para deformación plástica de fluencia** *f pl* FÍS FLUID *flujo viscoso* equations of creeping motion; ~ **de Ehrenfest** *f pl* FÍS Ehrenfest's equations; ~ **de Hamilton** *f pl* FÍS Hamilton's equations; ~ **de Lagrange** *f pl* FÍS Lagrange's equations; ~ **de Maxwell** *f pl* ELEC *electromagnetismo,* FÍS Maxwell's equations; ~ **de las ondas electromagnéticas** *f pl* ELEC, FÍS, FÍS ONDAS, FÍS RAD electromagnetic-wave equations; ~ **simultáneas** *f pl* MATEMÁT simultaneous equations

ecuador *m* GEOFÍS, METEO, TEC ESP, TRANSP MAR equator; ~ **geomagnético** *m* GEOFÍS geomagnetic equator; ~ **magnético** *m* FÍS magnetic equator, GEOFÍS dip equator, magnetic equator

ecualización *f* GEN equalization; ~ **de adaptamiento** *f* ING ELÉC adaptive equalization; ~ **automática** *f* TV auto-equalization

ecualizador *m* GEN equalizer; ~ **de amplitud** *m* TELECOM amplitude equalizer; ~ **de fase** *m* ELEC *corriente alterna* phase equalizer

edad: ~ **del agua freática** *f* AGUA, HIDROL age of groundwater; ~ **de emplazamiento** *f* GEOL emplacement age

edáfico *adj* AGRIC edaphic

edafología *f* AGRIC, CARBÓN *ciencia del suelo* edaphology, soil science

edenita *f* MINERAL edenite

edestina *f* QUÍMICA edestin

edición *f* INFORM&PD editing; ~ **abreviada** *f* IMPR abridged edition; ~ **automática** *f* TV auto-editing, automatic editing; ~ **del código de tiempo** *f* TV time code editing; ~ **por cuadruplex física** *f* TV physical-quadruplex editing; ~ **electrónica** *f* CINEMAT electronic editing (*EE*), ELECTRÓN, INFORM&PD electronic publishing, TV electronic editing (*EE*); ~ **de enlace** *f* INFORM&PD link editing; ~ **extraordinaria** *f* IMPR special edition; ~ **físicohelicoidal** *f* TV physical-helical editing; ~ **fuera de línea** *f* TV off-line editing; ~ **gráfica** *f* INFORM&PD graphic software package, graphical editing; ~ **de lujo** *f* IMPR de luxe edition; ~ **manual** *f* TV manual editing; ~ **mecánica** *f* TV mechanical editing; ~ **original** *f* IMPR original edition; ~ **revisada** *f* IMPR revised edition; ~ **en tablero de ajedrez** *f* CINEMAT checkerboarding (*AmE*), chequerboarding (*BrE*); ~ **de textos** *f* TELECOM text editing; ~ **de tipo aditivo** *f* TV add-on edit

edificación *f* CONST building, construction; ~ **antisísmica** *f* CONST earthquake-proof construction; ~ **moderna** *f* CONST modern construction

edificar *vt* CONST build, build up, put up

edificio *m* CONST, MINAS, NUCL building, construction; ~ **administrativo** *m* CONST, MINAS *oficinas* administration building; ~ **anexo** *m* ING MECÁ annex block (*AmE*), annexe block (*BrE*); ~ **para el ganado** *m* AGRIC animal building; ~ **de turbinas** *m* NUCL turbine building

edingtonita *f* MINERAL edingtonite

editar *vt* INFORM&PD edit

editor *m* INFORM&PD editor; ~ **de enlace** *m* INFORM&PD link editor, linkage editor; ~ **gráfico** *m* INFORM&PD graphic software package; ~ **de pantalla llena** *m* INFORM&PD full-screen editor; ~ **en pantalla de página completa** *m* INFORM&PD full-screen editor; ~ **de textos** *m* INFORM&PD, TELECOM text editor

EDP *abr* (*procesamiento electrónico de datos*) INFORM&PD EDP (*electronic data processing*)

edredón *m* TEXTIL eiderdown

EDTA *abr* (*ácido etilendiamino tetra-acético*) DETERG, QUÍMICA EDTA (*ethylenediamin tetra-acetic acid*)

educación: ~ **para la seguridad** *f* SEG safety education

eductor *m* ING MECÁ eductor

edulcorado *adj* ALIMENT sweetened

edulcorante *m* ALIMENT sweetener

EEA *abr* (*espectroscopía de electrones Auger*) FÍS, QUÍMICA AES (*Auger electron spectroscopy*)

EEII *abr* (*potencia equivalente radiada isotrópica*) TEC ESP EIRP (*equivalent isotropically-radiated power*)

EEMTEC *abr* (*especificación de equipos mezclados en transistores de efecto de campo*) ELECTRÓN MESFET (*metal semiconductor field-effect transistor*)

EEROM *abr* (*ROM borrable eléctricamente*) INFORM&PD EEROM (*electrically-erasable ROM*)

EFA *abr* (*ácido graso esencial*) QUÍMICA, ALIMENT EFA (*essential fatty acid*)

efectividad: ~ **de aleta** *f* REFRIG fin efficiency; ~ **del avión** *f* TRANSP AÉR aircraft effectivity

efecto[1]: **de ~ doble** *adj* ING MECÁ double-acting

efecto[2]:

▄ a ▄ ~ **de abanderamiento** *m* TRANSP AÉR feathering effect; ~ **acustoóptico** *m* ÓPT acousto-optic effect; ~ **agudo** *m* CONTAM acute effect; ~**-almacenamiento** *m* ING ELÉC storage effect; ~ **de apantallamiento** *m* ELEC *campo electromagnético, jaula de Faraday* shielding effect; ~ **de apiaramiento** *m* CONTAM MAR herder effect; ~ **arrecife** *m* TEC PETR *geología,* GEO reef effect; ~ **de atracción** *m* TRANSP attractive effect; ~ **Auger** *m* FÍS, FÍS RAD Auger effect; ~ **autocatalíptico** *m* METAL autocatalytic effect;

▄ b ▄ ~ **Barkhausen** *m* FÍS Barkhausen effect; ~ **Barnett** *m* FÍS Barnett effect; ~ **Becquerel** *m* ELEC *celda electrolítica* Becquerel effect; ~ **de blindaje** *m* ELEC screening effect, *campo electromagnético* shielding effect, TELECOM screen effect; ~ **de**

borde m FÍS, ING ELÉC *condensador* edge effect; ~ **brillante** m COLOR gloss effect;

~ c ~ **calorífico** m TERMO heat effect; ~ **en cámara** m CINEMAT in-camera effect; ~ **de campo** m ELECTRÓN *transistores*, FÍS, INFORM&PD, ÓPT, TEC ESP, TELECOM field effect; ~ **del campo de gravitación sobre la nave** m TEC ESP swing-by effect; ~ **de captura** m ELECTRÓN, TELECOM capture effect; ~ **de carena líquida** m TRANSP MAR *diseño naval* free-surface effect; ~ **de catapulta del campo gravitatorio** m TEC ESP swing-by effect; ~ **de caudal intrínseco** m TEC ESP *dinámica de fluidos* base flow effect; ~ **chimenea** m TERMOTEC chimney effect; ~ **cinetoisotópico** m NUCL kinetic isotope effect; ~ **de cizallamiento** m PAPEL shearing effect; ~ **Coanda** m FÍS FLUID Coanda effect; ~ **de coincidencia** m ACÚST coincidence effect; ~ **de Compton** m FÍS, FÍS RAD Compton effect; ~ **Compton inverso** m FÍS inverse Compton effect; ~ **de contorno** m ACÚST contour effect; ~ **de contraste** m CINEMAT, FÍS, FOTO, TV contrast effect; ~ **de Coriolis** m FÍS FLUID Coriolis effect; ~ **corona** m ELEC *alta tensión* corona discharge, *descarga* corona effect, FÍS corona discharge, GAS, ING ELÉC corona effect, corona discharge, P&C corona effect, corona discharge, TEC ESP corona discharge, corona effect; ~ **Cotton-Mouton** m FÍS Cotton-Mouton effect; ~ **crónico** m CONTAM chronic effect;

~ d ~ **desagradable** m CONTAM chronic effect; ~ **de desenfoque** m CINEMAT defocus effect; ~ **detergente** m DETERG detergent effect; ~ **diferencial** m TRANSP AÉR differential effect; ~ **dinamo** m TEC ESP dynamo effect; ~ **Doppler** m ELECTRÓN, FÍS, FÍS ONDAS, FÍS RAD, NUCL, PETROL Doppler effect; ~ **Doppler-Fizeau** m ACÚST Doppler-Fizeau effect;

~ e ~ **Early** m ELECTRÓN Early effect; ~ **Einstein-de-Haas** m FÍS Einstein-de-Haas effect; ~ **electroacústico** m ACÚST, ELEC, TELECOM acousto-electric effect; ~ **electrodérmico** m ACÚST, ELEC electrodermal effect; ~ **electromérico** m ELEC, QUÍMICA electromeric effect; ~ **electroóptico** m ELEC, ÓPT, TELECOM electro-optic effect; ~ **electroóptico de Kerr** m ELEC, FÍS, ÓPT Kerr electro-optical effect; ~ **enmascarador** m ACÚST masking effect; ~ **de envenenamiento por xenón** m NUCL xenon poisoning effect; ~ **de estado sólido** m NUCL solid-state effect; ~ **de estricción neoclásica** m NUCL neoclassical-pinch effect; ~ **de estricción toroidal** m NUCL toroidal pinch effect; ~ **estructural** m CONST structural effect;

~ f ~ **Faraday** m ELEC, ELECTRÓN, FÍS, ING ELÉC Faraday effect; ~ **fotoeléctrico** m ELECTRÓN, FÍS, FÍS RAD, ÓPT, TELECOM photoelectric effect; ~ **fotoeléctrico externo** m ELEC, FOTO, ÓPT, TELECOM external photoelectric effect; ~ **fotoeléctrico extrenor** m ELEC, FOTO, ÓPT, TELECOM external photoelectric effect; ~ **fotoeléctrico interno** m ELECTRÓN, ÓPT, TELECOM internal photoelectric effect; ~ **fotoeléctrico inverso** m ELECTRÓN inverse photoelectric effect; ~ **fotoemisivo** m FÍS RAD, ÓPT, TELECOM photoemissive effect; ~ **fotonuclear directo** m NUCL direct photonuclear effect; ~ **fotoquímico** m ENERG RENOV photochemical effect; ~ **fotovoltaico** m ENERG RENOV, FÍS, ING ELÉC, ÓPT, TEC ESP, TELECOM photovoltaic effect; ~ **fugoide** m TRANSP AÉR phugoid effect;

~ g ~ **Gunn** m ELECTRÓN, FÍS Gunn effect;

~ h ~ **Hall** m ELEC *electromagnetismo*, FÍS, FÍS RAD, TEC ESP Hall effect; ~ **Hall cuántico** m FÍS quantum Hall effect; ~ **Hanle** m FÍS RAD Hanle effect; ~ **de las heladas** m GEOL frost action; ~ **de Holden** m NUCL Holden effect;

~ i ~ **de iluminación** m CINEMAT lighting effect; ~ **de impresión magnética** m ACÚST magnetic printing effect; ~ **insoportable** m CONTAM chronic effect; ~ **de intervalo** m TV gap effect; ~ **invernadero** m AmL (*cf efecto invernadero por dióxido de carbono Esp*) CONTAM carbon dioxide greenhouse effect, ENERG RENOV, FÍS, GEOFÍS, METEO, REFRIG greenhouse effect; ~ **invernadero por dióxido de carbono** m Esp (*cf efecto invernadero AmL*) CONTAM carbon dioxide greenhouse effect, ENERG RENOV, FÍS, GEOFÍS, METEO, REFRIG greenhouse effect;

~ j ~ **jaula** m C&V bird cage; ~ **de Josephson** m ELECTRÓN, FÍS, NUCL Josephson effect; ~ **Josephson de CA** m ELECTRÓN AC Josephson effect; ~ **Josephson de CC** m FÍS, ING ELÉC DC Josephson effect; ~ **Joule** m ELEC *calor, corriente*, FÍS, ING ELÉC Joule effect, NUCL Joule heating;

~ k ~ **de Kelvin** m ELEC, ING ELÉC, TERMO, TRANSP AÉR Kelvin effect *electricidad* skin effect; ~ **Knudsen** m NUCL *flujo de moléculas térmicas* Knudsen effect;

~ l ~ **de Langmuir** m METEO Langmuir effect; ~ **larsen** m CINEMAT larsen effect; ~ **lateral** m INFORM&PD side effect; ~ **de latitud magnética** m GEOFÍS magnetic latitude effect; ~ **de lecho adyacente** m PETROL adjacent bed effect; ~ **letal** m CONTAM *toxicología* lethal effect; ~ **de linea larga** m ELECTRÓN long-line effect; ~ **local** m TELECOM *telefonía* sidetone;

~ m ~ **de magnetización cruzada** m ELEC cross-magnetizing effect; ~ **magnetoóptico** m FÍS, ÓPT magneto-optical effect; ~ **magnetoóptico de Kerr** m FÍS, ÓPT Kerr magneto-optical effect; ~ **Magnus** m FÍS Magnus effect; ~ **marginal** m TRANSP AÉR fringe effect; ~ **de masa** m NUCL packing effect; ~ **mecanotérmico** m TERMO mechanothermal effect; ~ **Meissner** m ELECTRÓN *superconductores*, FÍS Meissner effect; ~~**memoria** m ING ELÉC storage effect; ~ **microfónico coclear** m ACÚST cochlear microphonic effect; ~ **mortal** m CONTAM *toxicología* lethal effect; ~ **Mösbauer** m FÍS Mössbauer effect; ~ **muaré** m TV moiré effect; ~ **de muesca** m ING MECÁ notch effect; ~ **de Mullin** m P&C *prueba* Mullin's effect;

~ n ~ **nido de pájaro** m C&V bird's nest; ~ **nocturno** m CINEMAT night effect;

~ o ~ **de obturación** m TRANSP AÉR blanking effect; ~ **ojo de gato** m MINERAL chatoyant; ~ **ondulatorio** m TV flutter effect, fringe effect; ~ **óptico** m CINEMAT, ÓPT optical effect;

~ p ~ **de palanca** m ING MECÁ leverage; ~ **de pantalla** m ELEC screening effect, *campo electromagnético* shielding effect, TELECOM screen effect; ~ **Paschen-Back** m FÍS Paschen-Back effect; ~ **de paso cercano** m TEC ESP *vehículos* fly-by effect; ~ **peculiar** m ING ELÉC fringe effect; ~ **pelicular** m ING ELÉC, TEC PETR skin effect; ~ **Peltier** m ELEC, FÍS Peltier effect; ~ **de perforación** m PETROL borehole effect; ~ **de persiana** m TV Venetian blind effect;

~ piezoeléctrico *m* FÍS, ING ELÉC piezoelectric effect; **~ piezoeléctrico directo** *m* FÍS, ING ELÉC direct piezoelectric effect; **~ piezoeléctrico inverso** *m* FÍS, ING ELÉC inverse piezoelectric effect; **~ POGO** *m* TEC ESP POGO effect; **~ posterior** *m* FÍS RAD after-effect; **~ de presión dinámica** *m* TRANSP AÉR ram effect; **~ de proximidad** *m* ING ELÉC proximity effect; **~ de punto de contacto** *m* ELECTRÓN pinch-off effect;

~ r **~ de Raman** *m* FÍS, FÍS RAD Raman effect; **~ reactivo** *m* MINAS backlash; **~ de reciprocidad** *m* CINEMAT, FOTO, TV reciprocity effect; **~ relativista** *m* FÍS, TEC ESP relativity effect; **~ de remanso** *m* ENERG RENOV backwater effect; **~ de retrodispersión** *m* NUCL backscatter effect; **~ retrógrado del yugo de deflexión** *m* TV deflection yoke pullback;

~ s **~ Sabattier** *m* CINEMAT Sabattier effect; **~ sacudida** *m* TRANSP whiplash effect; **~ del samario** *m* NUCL samarium effect; **~ Schottky** *m* ELECTRÓN Schottky effect; **~ secundario** *m* FÍS RAD after-effect; **~ Seebeck** *m* ELEC *termoeléctrico* Seebeck effect, *de circuito, temperatura* thermoelectric effect, FÍS, ING ELÉC Seebeck effect; **~ de separación** *m* PROC QUÍ separation effect; **~ de separación elemental** *m* NUCL elementary separation effect; **~ sinergético** *m* CONTAM synergetic effect, P&C synergistic effect; **~ de sinergia** *m* P&C synergism effect; **~ sinérgico** *m* P&C synergistic effect; **~ de sinergismo** *m* P&C synergism effect; **~ sistemático** *m* CALIDAD systematic effect; **~ sobrecarga** *m* GEOL, TEC PETR *geología* overburden effect; **~ Stark** *m* FÍS Stark effect; **~ Stark lineal** *m* FÍS linear Stark effect; **~ Stark no lineal** *m* FÍS nonlinear Stark effect; **~ subletal** *m* CONTAM *toxicología* sublethal effect; **~ superficial** *m* FÍS, PETROL skin effect; **~ de superficie** *m* TRANSP AÉR skin effect; **~ de superficie libre** *m* TRANSP MAR *granos* free-surface effect; **~ de sustentación** *m* TRANSP MAR *arquitectura naval* lift effect;

~ t **~ de la tasa de dosis** *m* FÍS RAD dose rate effect; **~ térmico** *m* GAS thermal effect, thermic effect, TERMO heat effect; **~ termoeléctrico** *m* ELEC *de circuito, temperatura* thermoelectric effect; **~ termomecánico** *m* TERMO thermomechanical effect; **~ de Thomson** *m* FÍS Thomson effect; **~ de tierra** *m* TRANSP AÉR ground effect; **~ de torbellino** *m* GAS vortex effect; **~ tornasolado** *m* MINERAL chatoyant; **~ tóxico** *m* CONTAM *toxicología* toxic effect; **~ tóxico acumulativo** *m* CONTAM *toxicología* cumulative toxic effect; **~ de triodo** *m* ELECTRÓN triode action; **~ túnel** *m* ELECTRÓN tunnel effect, tunneling (*AmE*), tunnelling (*BrE*), FÍS, TRANSP tunnel effect; **~ Tyndall** *m* FÍS Tyndall effect;

~ v **~ Venturi** *m* METEO Venturi effect;

~ w **~ Wigner** *m* FÍS Wigner effect;

~ x **~ del xenón** *m* NUCL xenon effect;

~ z **~ Zeeman** *m* FÍS, NUCL Zeeman effect; **~ Zeeman anómalo** *m* FÍS anomalous Zeeman effect; **~ Zeeman normal** *m* FÍS normal Zeeman effect; **~ Zener** *m* ELECTRÓN Zener effect

efectos *m pl* CINEMAT, TV effects (*FX*); **~ biológicos** *m pl* FÍS RAD biological effects; **~ de blocaje** *m pl* TELECOM blockage effects; **~ de compresibilidad** *m pl* TRANSP AÉR compressibility effects; **digitales de video** *AmL*, **~ digitales de vídeo** *m pl Esp* TV digital video effects (*DVE*); **~ elásticos** *m pl* FÍS FLUID elastic after-effects; **~ especiales** *m pl* CINEMAT, TV effects (*FX*), special effects; **~ fisiológicos** *m pl* SEG *riesgos ocupacionales* physiological effects; **~ giromagnéticos** *m pl* FÍS gyromagnetic effects; **~ no estocásticos de la radiación** *m pl* NUCL nonstochastic radiation effects; **~ ópticos efectuados con la cámara** *m pl* CINEMAT, TV camera opticals; **~ de pantalla de la onda** *m pl* TELECOM wave shadowing effects; **~ de sonido** *m pl* TV sound effects; **~ sonoros** *m pl* CINEMAT, TV effects (*FX*), sound effects

efectuar[1]: **~ en registro intermedio** *vt* INFORM&PD buffer

efectuar[2]: **~ descargas de** *vi* CONTAM MAR, TRANSP MAR *aguas sucias* discharge; **~ el predoblaje de** *vi* CINEMAT predub; **~ una prueba de** *vi* INFORM&PD test; **~ un reconocimiento de** *vi* CONTAM MAR, MINAS survey

efectuar[3]: **~ una toma con dolly** *fra* CINEMAT dolly; **~ una toma con dolly ampliando la imagen** *fra* CINEMAT dolly back; **~ una toma con dolly reduciendo la imagen** *fra* CINEMAT dolly in

efedrina *f* QUÍMICA ephedrine

efemérides *f* TEC ESP *navegación* ephemerides

efervescencia *f* ALIMENT effervescence, QUÍMICA slugging, effervescence

efervescente *adj* ALIMENT effervescent, FÍS bubbling, QUÍMICA effervescent, REVEST rimmed

eficacia *f* GEN efficiency, CONTAM MAR effectiveness; **~ del acoplamiento** *f* ÓPT, TELECOM coupling efficiency; **~ de la antena** *f* TV aerial efficiency (*BrE*), antenna efficiency (*AmE*); **~ del aumento** *f* INSTR magnification effectiveness; **~ axial** *f* ACÚST axial efficiency; **~ biológica relativa** *f* (*EBR*) NUCL relative biological effectiveness (*RBE*); **~ de la cámara** *f* CINEMAT, TV gun efficiency; **~ característica** *f* ACÚST characteristic efficiency; **~ de combustión** *f* TEC ESP combustion efficiency; **~ eléctrica** *f* ELEC electrical efficiency; **~ específica** *f* ACÚST specific efficiency; **~ del filtrado** *f* CARBÓN screening efficiency; **~ de funcionamiento** *f* PROD operating force, performance efficiency; **~ de la iluminación** *f* TRANSP AÉR lighting efficiency; **~ luminosa** *f* FÍS luminous efficacy; **~ parafónica** *f* ACÚST close-talking efficiency; **~ del proceso de reproducción** *f* NUCL breeding-process efficiency; **~ de la rectificación** *f* ELEC, ING ELÉC rectification efficiency; **~ relativa** *f* ACÚST relative efficiency; **~ de tensión de campo libre** *f* ACÚST free-field tension efficiency; **~ térmica** *f* AUTO, TERMO thermal efficiency; **~ de la termoaislación** *f* TERMO heat insulation effectiveness; **~ de tracción** *f* ACÚST tension efficiency; **~ transductiva** *f* ACÚST transducer efficiency; **~ volumétrica** *f* AUTO volumetric efficiency

eficiencia *f* GEN efficiency; **~ de acoplamiento** *f* ÓPT, TELECOM coupling efficiency; **~ adiabática** *f* FÍS, TERMO adiabatic efficiency; **~ aerodinámica** *f* TRANSP AÉR aerodynamic efficiency; **~ de la caldera** *f* INSTAL TERM boiler efficiency; **~ del canal** *f* TELECOM channel efficiency; **~ de carga reducida** *f* ING MECÁ part-load efficiency; **~ de combustión** *f* TERMO combustion efficiency; **~ de cuadro** *f* TEC ESP frame efficiency; **~ cuántica** *f* ELECTRÓN, FÍS, FÍS PART, FÍS RAD, NUCL, ÓPT, TELECOM quantum efficiency; **~ cuántica diferencial** *f* FÍS, ÓPT differential quantum efficiency; **~ del depósito de difusión** *f* TEC ESP nozzle efficiency; **~ eléctrica** *f* ELEC electrical efficiency; ~ **de la fuente de alimentación** *f* TELE-COM source power efficiency; ~ **de la fuente de energía** *f* TELECOM source power efficiency; ~ **hidráulica** *f* HIDROL hydraulic efficiency; ~ **luminosa** *f* FÍS luminous efficiency; ~ **luminosa espectral** *f* FÍS spectral luminous efficiency; ~ **mecánica** *f* ENERG RENOV, ING MECÁ, MECÁ mechanical efficiency; ~ **pesquera** *f* OCEAN fishing efficiency; ~ **de potencia de la fuente** *f* ÓPT source power efficiency; ~ **radiante** *f* FÍS, FÍS RAD radiant efficiency; ~ **de rotor** *f* TRANSP AÉR rotor efficiency; ~ **térmica** *f* FÍS thermal efficiency, GAS thermal efficiency, thermic performance, TERMO thermal efficiency; ~ **de transformador** *f* ELEC transformer efficiency; ~ **de la unión soldada** *f* CONST *soldadura* joint efficiency; ~ **en la utilización del agua** *f* HIDROL water-use efficiency

eflorescencia *f* ALIMENT bloom, efflorescence, C&V, QUÍMICA, REVEST efflorescence; ~ **de cuarzo** *f* C&V batch melting line

eflorescente *adj* ALIMENT, C&V, QUÍMICA, REVEST efflorescent

eflorescer *vi* ALIMENT, C&V, QUÍMICA, REVEST effloresce

efluente *m* AGUA effluent, outflow, GAS, HIDROL, NUCL effluent, PROD waste, *agua, gas, tuberías* outflow, RECICL, TEC PETR effluent; ~ **acuoso** *m* CONTAM *procesos industriales* aqueous effluent; ~ **de aguas residuales** *m* RECICL sewage effluent; ~ **de alcantarillas** *m* AGUA sewage effluent; ~ **acuífero** *m* CONTAM *procesos industriales* aqueous effluent; ~ **ensilado** *m* HIDROL silage effluent; ~ **gaseoso** *m* GAS gaseous effluent; ~ **industrial** *m* AGUA, HIDROL industrial effluent; ~ **líquido** *m* GAS liquid effluent; ~ **standard** *m* RECICL effluent standard; ~ **tóxico** *m* RECICL toxic effluent

efluvio *m* ELEC *ionización del gas*, ELECTRÓN, FÍS, FÍS RAD, ING ELÉC glow discharge, QUÍMICA effluvium; ~ **eléctrico** *m* ELEC *en máquina*, FÍS, ING ELÉC brush discharge

EFTS *abr* (*sistema de transferencia electrónica de fondos*) INFORM&PD EFTS (*electronic funds transfer system*)

efusión *f* CONST outpouring, ELECTRÓN, FÍS, NUCL effusion

EGA *abr* (*arreglo gráfico extendido*) INFORM&PD EGA (*extended graphic arrangement*)

egipcio *m* IMPR *estilo de tipo* Egyptian

egirina *f* MINERAL aegirine

egirita *f* MINERAL aegirite, aegyrite

EHC *abr* (*empaquetado hexagonal compacto*) CRISTAL HCP (*hexagonal close-packed structure*)

eicosano *m* QUÍMICA eicosane

eigenfrecuencia *f* GEN eigenfrequency

einstenio *m* (*Es*) FÍS, FÍS RAD, QUÍMICA einsteinium (*Es*)

eje *m* AUTO sleeve, CONST axle, CRISTAL, FÍS axis, GEOM axial line, axis, ING MECÁ line of centers (*AmE*), line of centres (*BrE*), mandrel, mandril, shaft, MECÁ axis, shaft, METEO axis, PROD spindle, TRANSP MAR *de circulación* track, *hélice, motor* shaft, VEH *transmisión* axle;

~ a ~ **de abscisas** *m* CONST, GEOM, MATEMÁT x-axis; ~ **de accionamiento** *m* TRANSP MAR *motor* driving shaft; ~ **acodado** *m* ING MECÁ crank; ~ **de acoplamiento de plato** *m* ING MECÁ coupling spindle;

~ **aerodinámico** *m* TRANSP MAR *diseños de barcos* aerodynamic axis; ~ **de afiladora** *m* ING MECÁ grinding spindle; ~ **del anticiclón** *m* METEO axis of an anticyclone; ~ **de arrastre** *m* TRANSP AÉR drag axis; ~ **de articulación** *m* ING MECÁ fulcrum pin, hinge pin; ~ **del avión** *m* TRANSP AÉR aircraft axis;

~ b ~ **del balancín** *m* AUTO, VEH rocker-arm shaft; ~ **de balancines** *m* AUTO, VEH rocker shaft, rocker-arm shaft; ~ **de banda ancha** *m* TV wideband axis; ~ **binario de inversión** *m* CRISTAL twofold screw axis; ~ **binario de rotación** *m* CRISTAL twofold rotation axis; ~ **de bisagra** *m* ING MECÁ hinge shaft; ~ **de la bomba del aceite** *m* AUTO oil pump spindle; ~ **B-Y** *m* TV B-Y axis;

~ c ~ **del cabeceo** *m* TEC ESP, TRANSP AÉR pitch axis; ~ **del cable** *m* P&C cable core; ~ **de cambio** *m* ING MECÁ lifting shaft; ~ **del cambio de marcha** *m* ING MECÁ tumbling shaft; ~ **de cambio de paso** *m* TRANSP AÉR pitch-change axis; ~ **de campana** *m* ING MECÁ hub-type spindle; ~ **cardan** *m* AUTO cardan shaft; ~ **cardánico** *m* MECÁ cardan shaft; ~ **central de broca espiral** *m* PROD grinding line; ~ **del cigüeñal** *m* AUTO, ING MECÁ, MECÁ, TRANSP MAR, VEH crankshaft; ~ **cilíndrico hueco** *m* ING MECÁ hollow circular shaft; ~ **codificador** *m* PROD encoder shaft; ~ **de cola** *m* TRANSP MAR *construcción naval* propeller shaft; ~ **del conducto** *m* PROD conduit hub; ~ **conductor** *m* ING MECÁ drive shaft (*AmE*), driving shaft, propeller shaft (*BrE*); ~ **conductor de la caja de cambio de velocidades** *m* ING MECÁ gearbox input shaft; ~ **conductor flexible** *m* ING MECÁ flexible drive shaft; ~ **conductor inclinado** *m* TRANSP AÉR inclined drive shaft; ~ **conductor principal** *m* TRANSP AÉR main drive shaft; ~ **de conexión** *m* TRANSP AÉR connecting shaft; ~ **de conmutación** *m* ELEC *máquina* axis of commutation; ~ **corto acampanado de anilla achaflanada** *m* TRANSP AÉR bevel ring flared stub shaft; ~ **corto acampanado de anilla cónica** *m* TRANSP AÉR bevel ring flared stub shaft; ~ **de crujía** *m* TRANSP MAR center line (*AmE*), centre line (*BrE*), fore-and-aft line; ~ **en cuadratura** *m* ELEC *máquina* quadrature axis; ~ **de cubo** *m* ING MECÁ hub-type spindle;

~ d ~ **de declinación** *m* INSTR declination axis; ~ **delantero** *m* VEH *ruedas* front axle; ~ **dentado de accionamiento** *m* NUCL toothed rack; ~ **de la depresión** *m* METEO axis; ~ **de derrape** *m* TRANSP AÉR yaw axis; ~ **de derrota** *m* TRANSP MAR *de circulación* track; ~ **descentrado** *m* ING MECÁ eccentric shaft; ~ **desplazable** *m* ING MECÁ floating shaft; ~ **del diferencial** *m* AUTO differential shaft, ING MECÁ, VEH *transmisión* live axle; ~ **de Dion** *m* AUTO Dion axle; ~ **de la dirección** *m* VEH *ruedas* steering axle; ~ **de dirección de derrumbe axial** *m* TRANSP axially-collapsing steering column; ~ **de dislocación en tornillo** *m* CRISTAL screw axis; ~ **de distribución** *m* AUTO, ELEC, MECÁ, TRANSP MAR, VEH camshaft; ~ **del distribuidor** *m* AUTO, VEH *sistema de encendido* distributor shaft;

~ e ~ **de elevación** *m* TRANSP AÉR lift shaft; ~ **del embrague** *m* AUTO, VEH clutch shaft; ~ **de empuje** *m* TEC ESP thrust axis; ~ **de engranajes** *m* AUTO, ING MECÁ, MECÁ gear shaft; ~ **de entrada** *m* AUTO, ING MECÁ, VEH *caja de cambio de velocidades, embrague* input shaft; ~ **de entrada de piñones cónicos** *m* TRANSP AÉR input bevel pinion shaft; ~ **excéntrico** *m*

ING MECÁ eccentric shaft; ~ **de expansión** *m* ING MECÁ extension shaft; ~ **de extensión** *m* ING MECÁ extension shaft;

~f ~ **de fibra** *m* ÓPT, TELECOM fiber axis (*AmE*), fibre axis (*BrE*); ~ **de la fibra** *m* ÓPT, TELECOM fiber axis (*AmE*), fibre axis (*BrE*); ~ **fijo** *m* AUTO, ING MECÁ, VEH dead axle; ~ **flexible** *m* ING MECÁ flexible shaft; ~ **flotante** *m* ING MECÁ floating shaft, floating spindle; ~ **de freno** *m* AUTO, VEH brake shaft; ~ **de fresa** *m* ING MECÁ milling cutting arbor; ~ **de la fresadora** *m* PROD milling-machine arbor;

~g ~ **giratorio** *m* ING MECÁ, VEH *transmisión* live axle; ~ **de giro** *m* TEC ESP spin axis; ~ **de giro de la mangueta de la rueda** *m* MECÁ kingpin; ~ **de giro de la rueda auxiliar** *m* VEH *semirremolque* fifth-wheel kingpin axis; ~ **graduable** *m* FERRO adjustable axle; ~ **de grúa** *m* CONST jib post; ~ **de guiñada** *m* TEC ESP, TRANSP AÉR yaw axis;

~h ~ **de la hélice** *m* TRANSP AÉR propeller shaft; ~ **horizontal** *m* CONST horizontal axis, MATEMÁT horizontal axis, x-axis, TV x-axis; ~ **hueco** *m* ING MECÁ hollow shaft;

~i ~ **I** *m* TV I axis; ~ **del impulsor** *m* ING MECÁ input shaft; ~ **impulsor** *m* AUTO drive shaft (*AmE*), propeller shaft (*BrE*), ING MECÁ live axle, driving shaft, VEH *transmisión* live axle; ~ **del inducido** *m* ELEC *máquina* armature shaft; ~ **de inercia** *m* MECÁ axis of inertia; ~ **inferior** *m* AUTO lower shaft, VEH *transmisión* dropped axle; ~ **intermedio** *m* AUTO countershaft, ING MECÁ countershaft, intermediate shaft, jackshaft, lay shaft, PROD jackshaft; ~ **intermedio autónomo** *m* ING MECÁ self-contained countershaft; ~ **intermedio inversor** *m* AUTO, VEH *de marchas* reverse idler shaft; ~ **intermedio de marcha atrás** *m* AUTO, VEH reverse idler shaft; ~ **de inversión** *m* CRISTAL inversion axis; ~ **de inversión de marcha** *m* ING MECÁ tumbling shaft;

~j ~ **con juntas universales** *m* MECÁ cardan shaft;

~l ~ **lateral** *m* TRANSP AÉR lateral axis; ~ **de levas** *m* AUTO, ELEC camshaft, ING MECÁ camshaft, tumbling shaft, wiper shaft, MECÁ, TRANSP MAR, VEH camshaft; ~ **de levas en cabeza** *m* AUTO overhead camshaft (*OHC*); ~ **de levas en cabeza indirecto** *m* AUTO indirect overhead camshaft; ~ **libre** *m* ING MECÁ free shaft; ~ **de longitud de la pala** *m* TRANSP AÉR blade-span axis; ~ **longitudinal** *m* ING MECÁ center line (*AmE*), centre line (*BrE*), longitudinal axis, MECÁ, PROD center line (*AmE*), centre line (*BrE*), TEC ESP longitudinal axis, *aeronaves, transbordadores* center line (*AmE*), centre line (*BrE*), TRANSP AÉR longitudinal axis, TRANSP MAR *diseño de barcos* fore-and-aft line; ~ **longitudinal del buque** *m* TRANSP MAR *diseño de barcos, movimientos* center line (*AmE*), centre line (*BrE*);

~m ~ **del macho** *m* PROD *fundería* core arbor; ~ **magnético** *m* GEOFÍS, TEC ESP magnetic axis; ~ **de mando** *m* VEH *transmisión* driving axle; ~ **de maniobra** *m* PROD operating shaft; ~ **de marcha** *m* ING MECÁ lifting shaft; ~ **mayor** *m* GEOM major axis; ~ **de la mecha del timón** *m* TRANSP MAR *diseño de barcos* center line of rudderstock (*AmE*), centre line of rudderstock (*BrE*); ~ **menor** *m* GEOM minor axis; ~ **de mordazas** *m* ING MECÁ collet-type spindle; ~ **del motor** *m* ING MECÁ motor shaft; ~ **motor** *m* ING MECÁ live axle, main shaft, drive shaft (*AmE*), propeller shaft (*BrE*), driving shaft, transmission

shaft, MECÁ main shaft, VEH *transmisión* live axle; ~ **motor de la caja de cambio de velocidades** *m* ING MECÁ gearbox input shaft; ~ **motor principal** *m* ING MECÁ driving shaft; ~ **motriz** *m* AUTO, VEH drive shaft (*AmE*), driving axle, propeller shaft (*BrE*); ~ **motriz flexible** *m* ING MECÁ flexible drive shaft; ~ **motriz moleteado** *m* PROD knurled operating shaft; ~ **de movimiento planetario** *m* ING MECÁ planet spindle, planet-action spindle; ~ **muerto** *m* AUTO dead axle, ING MECÁ dead spindle, dead axle;

~n ~ **neutral** *m* CONST neutral axis; ~ **neutro** *m* CONST neutral axis; ~ **normal** *m* TRANSP AÉR normal axis;

~o ~ **óptico** *m* CRISTAL, FÍS optic axis, FOTO optical axis, ÓPT, TELECOM optic axis, optical axis; ~ **de ordenadas** *m* CONST, GEOM y-axis, MATEMÁT vertical axis, y-axis; ~ **de oscilación** *m* FÍS axis of oscillation; ~ **oscilante** *m* AUTO swing axle;

~p ~ **de paso** *m* TRANSP AÉR pitch axis; ~ **perforador equilibrado** *m* ING MECÁ counterbalanced drilling spindle; ~ **de pie de biela** *m* VEH *motor* piston pin; ~ **del piñón de ataque** *m* AUTO, VEH *árbol de mando* drive pinion shaft; ~ **de pinzas** *m* ING MECÁ collet-type spindle; ~ **de la pista** *m* TRANSP AÉR runway centerline (*AmE*), runway centreline (*BrE*); ~ **de pista extendido** *m* TRANSP AÉR extended runway centerline (*AmE*), extended runway centreline (*BrE*); ~ **de pivote** *m* VEH *de remolque* pivot axle; ~ **del pivote de quinta rueda** *m* VEH *semirremolque* fifth-wheel kingpin axis; ~ **del pivote de la rueda auxiliar** *m* VEH *semirremolque* fifth-wheel kingpin axis; ~ **con plato** *m* ING MECÁ flanged shaft; ~ **del plegador** *m* TEXTIL beam shaft; ~ **del pliegue** *m* GEOL, PETROL fold axis; ~ **polar** *m* INSTR polar axis; ~ **portabalancines** *m* AUTO, VEH *motor, válvulas* rocker box; ~ **portabrocas** *m* ING MECÁ, MECÁ boring spindle; ~ **portacuchillas** *m* ING MECÁ cutter spindle; ~ **portador** *m* AUTO, FERRO, ING MECÁ, MECÁ, VEH carrying axle; ~ **portafresas** *m* ING MECÁ cutter arbor, PROD milling-machine arbor, milling-machine cutter arbor; ~ **portafresas de una fresadora** *m* PROD milling-machine cutter arbor; ~ **portahélice** *m* TRANSP MAR *construcción naval* propeller shaft; ~ **portaherramienta** *m* CARBÓN cutter arm, ING MECÁ *máquinas herramientas* arbor; ~ **portaherramientas** *m* ING MECÁ boring spindle, cutter spindle, MECÁ boring spindle; ~ **portamuelas** *m* ING MECÁ wheel spindle; ~ **portarrodillo** *m* PROD roller pin; ~ **posicionador** *m* ING MECÁ locating arbor; ~ **posterior** *m* AUTO, MECÁ, VEH rear axle; ~ **primario** *m* AUTO primary shaft, ING MECÁ input shaft, VEH *caja de cambio de velocidades, embrague* input shaft, main shaft; ~ **primario de la caja de cambio de velocidades** *m* ING MECÁ gearbox input shaft; ~ **principal** *m* AUTO main shaft, CRISTAL *rotación*, FÍS *de un sólido rígido* principal axis, ING MECÁ, MECÁ, TRANSP AÉR, VEH main shaft; ~ **principal de la caja de cambio de velocidades** *m* ING MECÁ gearbox input shaft; ~ **propulsor** *m* AUTO output shaft, shaft drive, VEH *transmisión* drive shaft (*AmE*), propeller shaft (*BrE*), shaft drive; ~ **propulsor de dos piezas** *m* AUTO two-piece propeller shaft; ~ **de proyección** *m* CINEMAT axis of projection, projection axis; ~ **del punto fijo** *m* ING MECÁ live spindle;

~r ~ **de rectificadora** *m* ING MECÁ grinding spindle;

~ **de referencia** *m* ACÚST, TV reference axis; ~ **de resistencia aerodinámica** *m* TRANSP AÉR drag axis; ~ **de revolución** *m* MECÁ axis of revolution; ~ **rígido** *m* AUTO rigid axle, ING MECÁ droop-restraining shaft, VEH *suspensión* rigid axle; ~ **de rodillos** *m* AUTO roller shaft; ~ **de rotación** *m* CRISTAL rotation axis, ING MECÁ pivot, MECÁ axis of rotation, pivot, TEC ESP spin axis; ~ **de rotación cuaternario** *m* CRISTAL fourfold rotation axis; ~ **de rotación senario** *m* CRISTAL six-fold rotation axis; ~ **de rotación ternario** *m* CRISTAL three-fold rotation axis; ~ **rotativo** *m* VEH *transmisión* drive; ~ **de rotor** *m* ELEC *generador* rotor shaft; ~ **del rotor principal** *m* TRANSP AÉR main-rotor shaft; ~ **R-Y** *m* TV red color difference axis (*AmE*) (*R-Y axis*), red colour difference axis (*BrE*) (*R-Y axis*);

~ **s** ~ **de salida** *m* AUTO output shaft, ING MECÁ input shaft, VEH *caja de cambios* output shaft, *suspensión de las ruedas* trailing arm; ~ **de sector** *m* AUTO sector shaft; ~ **secundario** *m* AUTO countershaft, ING MECÁ countershaft, intermediate shaft, jackshaft, VEH *caja de cambios* output shaft; ~ **secundario completo** *m* ING MECÁ self-contained countershaft; ~ **secundario independiente** *m* ING MECÁ self-contained countershaft; ~ **de segundo movimiento** *m* ING MECÁ second motion shaft; ~ **semiflotante** *m* AUTO, VEH semifloating axle; ~ **semiportante** *m* VEH *posterior* semifloating axle; ~ **de la sierra** *m* ING MECÁ saw arbor; ~ **de sierra circular** *m* ING MECÁ saw arbor; ~ **de simetría** *m* CRISTAL, GEOM, ING MECÁ axis of symmetry; ~ **de simetría cuaternario** *m* CRISTAL fourfold rotation axis; ~ **sinclinal** *m* ING MECÁ synclinal axis;

~ **t** ~ **de la taladradora** *m* ING MECÁ drilling spindle; ~ **de tambor** *m* ING MECÁ drum shaft; ~ **de tijeras** *m* ING MECÁ scissor joint; ~ **del timón** *m* TRANSP MAR *construcción naval* tiller axle; ~ **totalmente flotante** *m* AUTO full-floating axle; ~ **de tracción** *m* TEC ESP thrust axis; ~ **de la transmisión** *m* AUTO, VEH drive shaft (*AmE*), propeller shaft (*BrE*); ~ **de transmisión** *m* ING MECÁ transmission shaft; ~ **de transmisión de dos piezas** *m* AUTO, VEH two-piece drive shaft; ~ **transmisor** *m* ING MECÁ transmission shaft; ~ **transmisor flexible** *m* ING MECÁ flexible drive shaft; ~ **transversal** *m* ING MECÁ jackshaft, VEH *suspensión de las ruedas* transverse control arm; ~ **trasero** *m* AUTO rear-axle shaft, rear axle, MECÁ rear axle, rear-axle shaft, VEH rear-axle shaft, *transmisión* rear axle; ~ **trasero portador** *m* VEH *de camión* trailing axle; ~ **del tren de aterrizaje** *m* TRANSP AÉR landing gear shaft;

~ **v** ~ **de válvulas** *m* AUTO valve shaft; ~ **vertical** *m* EMB center (*AmE*), centre (*BrE*), ING MECÁ upright shaft, MECÁ vertical axis, y-axis, TEC ESP yaw axis, TV y-axis; ~ **V-Y** *m* TV G-Y axis;

~ **x** ~ **de las x** *m* CONST, GEOM, MATEMÁT x-axis; ~ **x** *m* FÍS, GEOM, MATEMÁT x-axis;

~ **y** ~ **de las y** *m* CONST, GEOM, MATEMÁT y-axis; ~ **y** *m* FÍS, GEOM, MATEMÁT y-axis;

~ **z** ~ **z** *m* FÍS z-axis; ~ **de zona** *m* CRISTAL zone axis; ~ **zonal** *m* CRISTAL zone axis

ejecución *f* CALIDAD performance, INFORM&PD running, execution, *proceso* run, ING MECÁ performance; ~ **concurrente** *f* INFORM&PD concurrent execution; ~ **de instrucciones** *f* INFORM&PD instruction execution; ~ **de programas** *f* INFORM&PD program execution; ~ **del proyecto** *f* PROD project production; ~ **de prueba** *f* INFORM&PD test run; ~ **de un realce** *f* MINAS raising; ~ **simultánea** *f* INFORM&PD concurrent execution

ejecutar *vt* INFORM&PD execute, PROD handle, run, make

ejecutivo *m* INFORM&PD executive

ejecutor *m* INFORM&PD executor

ejemplar *m* IMPR copy, MECÁ specimen, QUÍMICA sample

ejemplo *m* INFORM&PD instance, sample

ejercer *vt* FÍS exert

ejercicio: ~ **de botes** *m* TRANSP MAR boat drill; ~ **de emergencia** *m* SEG, TRANSP MAR emergency drill; ~ **de evacuación para casos de emergencia** *m* SEG, TRANSP MAR emergency drill

ejercicios: ~ **de contraincendios** *m* PROD, SEG, TERMO fire drill; ~ **de salvamento de incendios** *m* PROD, SEG, TERMO fire drill

ejes: ~ **de cabeceo, balanceo y guiñada** *m pl* TRANSP AÉR pitch, roll and yaw axes; ~ **de cabeceo, inclinación y guiñada** *m pl* TEC ESP pitch, roll and yaw axes; ~ **circulares** *m pl* ING MECÁ circular shafts; ~ **de coordenadas** *m pl* GEOM coordinate axes; ~ **cristalinos** *m pl* CRISTAL crystal axes; ~ **oblicuos** *m pl* MATEMÁT oblique axes; ~ **plegables** *m pl* TRANSP AÉR folding axes

ekebergita *f* MINERAL ekebergite

elaboración *f* C&V, CONST manufacture, manufacturing, CARBÓN process, ING MECÁ, PROD manufacture, manufacturing; ~ **cervecera** *f* ALIMENT brewing; ~ **por electroerosión** *f* ING MECÁ, PROD electrodischarge machining (*EDM*); ~ **de redes** *f* OCEAN net making; ~ **de un semiconductor** *f* ELECTRÓN semiconductor fabrication

elaborar *vt* C&V manufacture, CONST make, manufacture, ING MECÁ manufacture, work, PROD *herramienta* make, manufacture

elágico *adj* QUÍMICA ellagic

elagitanina *f* QUÍMICA ellagitannin

elaídico *adj* QUÍMICA elaidic

elaidina *f* QUÍMICA elaidin

elasmosina *f* MINERAL elasmosine

elasticidad *f* ACÚST compliance, elasticity, FÍS elasticity, ING MECÁ resilience, MECÁ, METAL, P&C, TRANSP MAR elasticity; ~ **acústica** *f* ACÚST acoustic compliance, acoustic elasticity; ~ **de miga** *f* ALIMENT crumb elasticity; ~ **de la pastilla** *f* ELECTRÓN wafer yield; ~ **de torsión** *f* ING MECÁ rotational elasticity, torsional elasticity; ~ **torsional** *f* ING MECÁ torsional elasticity

elástico *adj* TEXTIL stretchy

elastómero *m* P&C, PROD, QUÍMICA, REFRIG, TEC ESP, TEC PETR elastomer; ~ **sintético** *m* TEC PETR synthetic elastomer; ~ **termoplástico** *m* P&C thermoplastic rubber

elaterita *f* MINERAL elaterite

elección *f* CARBÓN culling, sorting; ~ **por pasado de lista** *f* INFORM&PD roll call polling

electreto *m* FÍS electret

electricidad *f* GEN electricity; ~ **atmosférica** *f* GEOFÍS, METEO atmospheric electricity; ~ **estática** *f* CONST static electricity, ING ELÉC static, TEXTIL static electricity; ~ **estática engendrada por frotamiento** *f* ELEC frictional electricity; ~ **estática sobre la película** *f* FOTO static on film; ~ **generada** *f* ELEC,

ING ELÉC, NUCL electricity generated; ~ **solar** *f* ING ELÉC solar electricity

eléctrico[1] *adj* GEN electric

eléctrico[2]: **~-óptico** *m* ELEC, ÓPT, PROD, TELECOM electrical-optical (*E-O*)

electrificación *f* ELEC *fuente de alimentación*, FÍS, ING ELÉC electrification

electrificar *vt* GEN electrify

electrizar *vt* GEN electrify

electro: **~-elongámetro** *m* ING ELÉC strain gage (*AmE*), strain gauge (*BrE*)

electroaccionado *adj* ELEC, FOTO, ING ELÉC, MECÁ electrically-driven

electroacústica *f* ACÚST, ING ELÉC electroacoustics

electroaislamiento *m* ING ELÉC insulation

electroanálisis *m* QUÍMICA electroanalysis

electrobús *m* ELEC, TRANSP electrobus

electrocaldera *f* ELEC, ING MECÁ electric steam-boiler

electrocalefactor *m* ELEC, INSTAL TERM, TERMO electric heater

electrocalentador *m* ELEC, INSTAL TERM, TERMO electric hot-plate

electrocapilar *adj* QUÍMICA electrocapillary

electrocapilaridad *f* QUÍMICA electrocapillarity

electrochapado[1] *adj* ELEC, ING ELÉC, REVEST electroplated

electrochapado[2] *m* ELEC, ING ELÉC, REVEST electroplating

electrochapar *vt* ELEC, ING ELÉC, REVEST electroplate

electrochoque *m* ELEC, ING ELÉC, SEG electric shock

electrocinado *adj* REVEST *galvanización por baño* zinc-coated

electrocinética *f* ELEC, FÍS, ING ELÉC, QUÍMICA electrokinetics

electrocinético *adj* ELEC, FÍS, ING ELÉC, QUÍMICA electrokinetic

electrococleografía *f* ACÚST electrocochleography

electrodeposición *f* CONTAM *depuración*, METAL deposition, REVEST brush plating; **~ regulada** *f* CONTAM regulated deposition; **~ seca** *f* CONTAM dry deposition; **~ total** *f* CONTAM total deposition

electrodepositado *adj* ELEC, ING ELÉC, REVEST electroplated

electrodepositar *vt* ELEC, ING ELÉC, REVEST electroplate

electrodepósito *m* CONST, ELEC, ING ELÉC, REVEST electrodeposit

electrodiálisis *f* ELEC, QUÍMICA electrodialysis

electrodinámica *f* ACÚST, ELEC, FÍS electrodynamics; **~ covariante** *f* ELEC covariant electrodynamics; **~ cuántica** *f* ELEC, ELECTRÓN, FÍS, FÍS PART quantum electrodynamics (*QED*)

electrodinamómetro *m* ELEC, FÍS, INSTR electrodynamometer

electrodo *m* GEN electrode, ELEC *pila* plate, *electroquímica* electrode; **~ acelerador** *m* NUCL accelerating electrode; **~ anticátodo** *m* ING ELÉC target electrode; **~ del arranque** *m* ING ELÉC starter electrode; **~ auxiliar** *m* C&V auxiliary electrode; **~ base** *m* ING ELÉC base electrode; **~ de batería** *m* ELEC battery plate; **~ bimetálico** *m* ELEC bimetallic electrode; **~ bipolar** *m* ELEC bipolar electrode; **~ del blanco** *m* TV target electrode; **~ de la bujía** *m* AUTO, ELEC, VEH spark plug electrode; **~ de calomelanos** *m* ELEC calomel electrode; **~ captador** *m* ING ELÉC collector electrode; **~ de carbón** *m* ELEC, ING ELÉC *pila* carbon

electrode; **~ del casquillo** *m* ING ELÉC base electrode; **~ cebador** *m* ING ELÉC keep-alive electrode; **~ de conexión a tierra** *m* AUTO, ELEC, FÍS, ING ELÉC, VEH earth electrode (*BrE*), ground electrode (*AmE*); **~ consumible** *m* CONST consumable electrode; **~ de corte** *m* TV splitting electrode; **~ deflector** *m* NUCL deflecting electrode; **~ de deflexión electromagnética** *m* ING ELÉC deflection electrode; **~ de desviación radial** *m* TV radial-deflecting electrode; **~ de emisión secundaria** *m* ING ELÉC, ÓPT reflecting electrode; **~ emisor** *m* ELECTRÓN, ING ELÉC emitter electrode; **~ de encendido** *m* ING ELÉC starter electrode; **~ de enfoque** *m* ING ELÉC, INSTR, TV focusing electrode; **~ de entrada** *m* ING ELÉC input electrode; **~ de escobilla** *m* ING ELÉC brush rod; **~ excitador** *m* ING ELÉC keep-alive electrode; **~ explorador** *m* GAS probe electrode; **~ en forma de peine** *m* TEC ESP comb-shaped electrode; **~ fungible** *m* CONST consumable electrode; **~ hueco semicilíndrico** *m* FÍS dee; **~ de mando** *m* ING ELÉC control electrode; **~ de masa** *m* AUTO, ELEC, FÍS, ING ELÉC, VEH earth electrode (*BrE*), ground electrode (*AmE*); **~ mecánico** *m* CONTAM mechanical collector; **~ de modulación** *m* TV modulation electrode; **~ negativo** *m* ELEC *pila*, ING ELÉC negative electrode; **~ no consumible** *m* CONST, ING ELÉC nonconsumable electrode; **~ de una pila** *m* ELEC battery plate; **~ de plata** *m* LAB silver electrode; **~ de posaceleración** *m* TV intensifier electrode; **~ positivo compacto** *m* ING ELÉC heavy anode; **~ positivo secundario** *m* ING ELÉC second anode; **~ primario** *m* ING ELÉC initiating electrode; **~ receptor** *m* CONTAM precipitation collector; **~ recubierto** *m* MECÁ coated electrode; **~ de referencia** *m* ELEC, ELECTRÓN, LAB reference electrode; **~ de reflexión** *m* ING ELÉC reflecting electrode, reflector electrode, ÓPT reflecting electrode, TV reflector electrode; **~ regulador** *m* ING ELÉC control electrode; **~ revestido** *m* CONST covered electrode, MECÁ coated electrode; **~ de salida** *m* ING ELÉC output electrode; **~ selectivo de iones** *m* (*ESI*) LAB ion selective electrode (*ISE*); **~ de señal** *m* ELECTRÓN signal electrode; **~ sensor** *m* ING ELÉC sensing electrode; **~ de soldadura por arco** *m* ELEC arc welding electrode; **~ de tierra** *m* AUTO, ELEC, FÍS, ING ELÉC, VEH earth electrode (*BrE*), ground electrode (*AmE*); **~ de la toma de corriente** *m* ING ELÉC collector electrode; **~ de vidrio** *m* LAB glass electrode

electrodofiltros: **~ de doble etapa** *m pl* CONTAM double bucket collector

electrodoméstico *m* ELEC electrical household appliance

electrodos: **~ colectores de doble etapa** *m pl* CONTAM double bucket collector; **~ colectores de dos cámaras** *m pl* CONTAM double bucket collector

electroducto *m* ELEC, ING ELÉC electrical transmission line

electroestricción *f* FÍS electrostriction

electroextracción *f* QUÍMICA electro-extraction

electrofílico *adj* QUÍMICA electrophilic

electrofiltro *m* CONTAM collecting electrode, receptor

electrofluorescencia *f* ELEC, NUCL electrofluorescence

electroforesis *f* ELEC, LAB, QUÍMICA electrophoresis

electroformación *f* TERMO cold casting

electróforo *m* ING ELÉC electrophorus

electrofotografía *f* IMPR, INFORM&PD electrophotography

electrofusión *f* ELEC, GAS, ING ELÉC electric smelting, electrofusion

electrogenerador *m* ELEC, ING ELÉC, NUCL, TV electric generator; **~ eólico** *m* ING ELÉC wind-driven generator

electrohorno *m* CARBÓN, ELEC, ING ELÉC, TERMO electric furnace, electric oven; **~ de resistencia** *m* FÍS, ING ELÉC electric-resistance furnace

electroimán *m* GEN electromagnet, electric magnet; **~ de arrollamientos superconductores** *m* TRANSP superconducting magnet; **~ del campo** *m* ING ELÉC field magnet; **~ de concentración** *m* CINEMAT, FOTO, ING ELÉC, TEC ESP, TV focusing magnet; **~ de elevación** *m* ELEC, ING MECÁ lifting magnet; **~ elevador** *m* ELEC, ING MECÁ lifting magnet; **~ de enganche** *m* ELEC latching electromagnet; **~ de foco** *m* CINEMAT, FOTO, ING ELÉC, TEC ESP, TV focusing magnet; **~ en herradura** *m* ELEC horseshoe magnet; **~ levantador** *m* ELEC, ING MECÁ lifting magnet; **~ portador** *m* ELEC, ING MECÁ lifting magnet; **~ de relé** *m* ELEC, ING ELÉC relay magnet; **~ de soplado** *m* ING ELÉC blowout coil; **~ de suspensión** *m* ELEC, ING MECÁ lifting magnet

electroimpulsado *adj* GEN electrically-driven

electrólisis *f* GEN electrolysis

electrolítico *adj* GEN electrolytic

electrólito *m* GEN electrolyte; **~ seco** *m* ING ELÉC dry electrolyte; **~ sólido para alta temperatura** *m* TRANSP high-temperature solid electrolyte cell

electrolizador *m* GEN electrolyzer

electrolizar *vt* GEN electrolyze

electrolocomotora: ~ con rectificador *f* FERRO rectifier locomotive

electroluminiscencia *f* GEN electroglow, electroluminescence; **~ por capa delgada** *f* ELECTRÓN thin-film electroluminescence (*TFEL*)

electromagnético *adj* GEN electromagnetic; **~ transversal** *adj* (*TEM*) ING ELÉC transverse electromagnetic (*TEM*)

electromagnetismo *m* GEN electromagnetism

electromecánica *f* GEN electromechanics

electrometeoro *m* GEN electrometeor

electrometría *f* GEN electrometry

electrómetro *m* GEN electrometer; **~ bifilar** *m* ELEC, INSTR bifilar electrometer; **~ de cuadrantes** *m* ING ELÉC, INSTR quadrant electrometer; **~ de dos hilos** *m* ELEC bifilar electrometer; **~ de Hoffman** *m* ELEC, INSTR Hoffman electrometer; **~ de torsión** *m* GEOFÍS torsion electrometer

electromotor *m* GEN electric motor, electromotor

electromóvil *m* TRANSP electromobile

electrón *m* (*é*) GEN electron (*e*); **~ de alta energía** *m* ELECTRÓN high-energy electron; **~ antienlazante** *m* FÍS RAD antibonding electron; **~ Auger** *m* FÍS, FÍS RAD Auger electron; **~ de la capa Q** *m* ELECTRÓN, NUCL Q-shell electron; **~ de capas internas del núcleo** *m* FÍS RAD inner shell electron; **~ de conversión** *m* ELECTRÓN, FÍS RAD conversion electron; **~ deslocalizado** *m* ELECTRÓN, FÍS delocalized electron; **~ enlazado** *m* FÍS PART bound electron; **~ fugitivo térmico** *m* ELECTRÓN thermal runaway electron; **~ de gran potencia** *m* ELECTRÓN high-energy electron; **~ interno** *m* FÍS RAD inner electron; **~ libre** *m* ELEC, ELECTRÓN, FÍS, FÍS PART, FÍS RAD free electron; **~ ligado** *m* FÍS PART bound electron; **~ negativo** *m* ELECTRÓN, FÍS PART negatron; **~ no**

enlazante *m* NUCL nonbonding electron; **~ óptico** *m* FÍS RAD optical electron; **~ orbital** *m* FÍS RAD orbital electron; **~ oscilatorio** *m* NUCL oscillating electron; **~ pesado** *m* QUÍMICA meson; **~ Q** *m* ELECTRÓN, NUCL Q-electron; **~ de retroceso** *m* ELECTRÓN, FÍS, FÍS PART, FÍS RAD recoil electron; **~ secundario** *m* ELECTRÓN, FÍS RAD secondary electron; **~ de valencia** *m* FÍS, METAL valence electron

electronegativo *adj* FÍS RAD, QUÍMICA electronegative

electrones: ~ pareados *m pl* FÍS RAD *superconductividad* paired electrons

electrónica: ~ aeronáutica *f* TRANSP AÉR avionics; **~ para el consumidor** *f* ELECTRÓN consumer electronics; **~ del estado sólido** *f* ELECTRÓN solid-state electronics; **~ industrial** *f* ING ELÉC industrial electronics; **~ molecular** *f* ELECTRÓN molecular electronics; **~ remota** *f* ELECTRÓN, INSTR remote electronics

electrónicamente: ~ controlado *adj* ELECTRÓN electronically controlled

electrónico *adj* ELEC, ELECTRÓN electronic; **~ a electrónico** *adj* ELECTRÓN, TV electronic-to-electronic (*E-E, E-to-E*)

electronvoltio *m* (*eV*) ELECTRÓN, FÍS, FÍS PART, ING ELÉC, NUCL electron-volt (*eV*)

electroóptico[1] *adj* ELEC, ELECTRÓN, ÓPT, TELECOM electro-optic, electro-optical

electroóptico[2] *m* ELEC, ELECTRÓN, ÓPT, TELECOM electrical-optical (*E-O*)

electroósmosis *f* QUÍMICA electro-osmosis

electroosmótico *adj* QUÍMICA electro-osmotic

electroplastia *f* CONST, ELEC, ING ELÉC, PROD, REVEST electrodeposition, electroplating

electropositivo *adj* FÍS RAD, QUÍMICA electropositive

electropulido *m* METAL electropolishing

electroquímica *f* ELEC, ING ELÉC, QUÍMICA electrochemistry

electroquímico *adj* ELEC, ING ELÉC, QUÍMICA electrochemical

electrorrevestido *adj* ELEC, ING ELÉC, REVEST electroplated

electroscopio *m* ELEC, FÍS, ING ELÉC, INSTR electroscope; **~ de láminas de oro** *m* FÍS, ING ELÉC, INSTR gold leaf electroscope

electroshock *m* ELEC, ING ELÉC electric shock

electrosíntesis *f* QUÍMICA electrosynthesis

electrostática *f* ELEC, ELECTRÓN, FÍS, ING ELÉC, TELECOM electrostatics

electrostático *adj* ELEC, ELECTRÓN, FÍS, ING ELÉC, TELECOM electrostatic

electrotecnia *f* ELEC electrotechnics, electrical engineering, ING ELÉC electrical engineering

electrotecnología *f* ELEC electrotechnology

electrotérmico *adj* ELEC, TERMO electrothermal

electrotipo *m* IMPR electro, electrotype

electroválvula *f* ELEC, PROD electrically-operated valve

electroventilador *m* AUTO power fan

elemento *m* CARBÓN component, CINEMAT *de batería* cell, CONST utensil, ELEC *circuito* element, *parte* component, couple, ELECTRÓN, FÍS, FÍS PART element, INFORM&PD item, element, ING ELÉC cell, primary cell, couple, ING MECÁ element, item, part, INSTAL HIDRÁUL *de caldera dividida* element, unit, MATEMÁT element, MECÁ component, NUCL, ÓPT element, PROD *de caldera* section, element, QUÍMICA element, TEC ESP *electricidad* cell; **~ absorbente** *m* NUCL absorber member; **~ absorbente de acero borado** *m* NUCL boronated-steel absorber; ~ **de acceso a la red** *m* TELECOM gateway network element; ~ **activo** *m* ING ELÉC active element; ~ **de acumulador** *m* ING ELÉC secondary cell; ~ **de acumulador Edison** *m* ELEC Edison cell; ~ **de acumuladores alcalinos** *m* ELECTRÓN, ING ELÉC alkaline storage cell; ~ **alcalino** *m* ING ELÉC alkaline cell; ~ **de almacenamiento** *m* INFORM&PD storage element; ~ **anterior** *m* FOTO front element; ~ **aprovechable** *m* NUCL valuable element; ~ **de arranque** *m* INFORM&PD start element; ~ **atenuador** *m* ELECTRÓN attenuating element; ~ **de batería** *m* ELEC *pilas y baterías eléctricas* cell, *de acumulador* battery cell, ING ELÉC battery cell, VEH accumulator; ~ **de la batería** *m* AUTO battery cell; ~ **de una batería** *m* TEC ESP battery cell; ~ **bimetálico** *m* ELEC *en termómetros, termostatos* bimetallic strip, REFRIG bimetallic element; ~ **de Bragg** *m* ING ELÉC Bragg cell; ~ **de calefacción** *m* TERMOTEC heating element; ~ **calefactor** *m* ELEC heating element, PROD heater element, TRANSP AÉR heating element; ~ **calentador** *m* FOTO heater element; ~ **de la capa de detención** *m* ING ELÉC barrier-layer cell; ~ **captador** *m* NUCL sensing element; ~ **de carbón-zinc** *m* ING ELÉC zinc-carbon cell; ~ **central de soporte de carga** *m* ÓPT central load-bearing element; ~ **circuital** *m* ELECTRÓN, ING ELÉC circuit element; ~ **de circuito** *m* ELECTRÓN, ING ELÉC circuit element; ~ **de circuito integrado** *m* ELECTRÓN integrated circuit element; ~ **de circuito lineal** *m* ELECTRÓN linear circuit element; ~ **clorídico de plata y magnesio** *m* ING ELÉC magnesium silver chloride cell; ~ **de cloruro de zinc** *m* ING ELÉC zinc chloride cell; ~ **de combustible** *m* NUCL fuel assembly; ~ **de combustible de acumulador** *m* TRANSP secondary fuel cell; ~ **de combustible anular** *m* NUCL annular fuel element; ~ **de combustible dañado** *m* NUCL damaged fuel assembly; ~ **de combustible desenvainado** *m* NUCL uncanned fuel element; ~ **de combustible con fugas** *m* NUCL leaking fuel assembly; ~ **de combustible sin instrumentación** *m* NUCL uninstrumented fuel assembly; ~ **de combustible limitador** *m* NUCL limiting fuel assembly; ~ **de combustible nuevo** *m* NUCL new-fuel assembly, new-fuel element; ~ **de combustible periférico** *m* NUCL peripheral fuel assembly; ~ **de combustible revestido de grafito** *m* NUCL graphite-clad fuel element; ~ **de combustible semihomogéneo** *m* NUCL semihomogeneous fuel element; ~ **de combustible de uranio** *m* NUCL uranium fuel element; ~ **de combustible sin vaina** *m* NUCL canless fuel assembly, uncanned fuel element; ~ **de combustible venteado** *m* NUCL vented fuel assembly; ~ **de compensación** *m* NUCL shim element, shim member; ~ **componente** *m* MECÁ component; ~ **de comunicación en tránsito** *m* TELECOM transit connection element (*TCE*); ~ **concentrado** *m* ING ELÉC lumped element; ~ **conducido** *m* ING MECÁ driven element; ~ **conductor** *m* ING MECÁ driving element; ~ **de conexión de acceso** *m* TELECOM access connection element; ~ **de conexión en tránsito** *m* TELECOM transit connection element (*TCE*); ~ **de conexiones** *m* TELECOM connection element (*CE*); ~ **de conmutación digital** *m* TELECOM digital switching element (*DSE*); ~ **de contacto en vacío** *m* PROD vacuum contact element; ~ **para el control de**

asociación *m* TELECOM association control service element (*ACSE*); ~ **de control de la asociación de señalización** *m* TELECOM signalling association control service element (*AmE*) (*SACSE*), signalling association control service element (*BrE*) (*SACSE*); ~ **de control fino** *m* NUCL fine-control member; ~ **de control periférico** *m* NUCL edge-control element, peripheral control element; ~ **de corriente** *m* FÍS current element; ~ **de datos** *m* INFORM&PD, TELECOM data element; ~ **destructivo** *m* SEG destructive element; ~ **detector** *m* ELECTRÓN sensing element; ~ **de detenida** *m* INFORM&PD stop element; ~ **de disipación térmica** *m* ELEC, TERMO heat sink; ~ **Edison** *m* ELEC Edison cell; ~ **enchufable** *m* ELEC plug pin, ING ELÉC plug-in unit; ~ **de enganche** *m* ING MECÁ attachment fitting; ~ **estándar** *m* ING ELÉC standard cell; ~ **estándar Weston** *m* ING ELÉC Weston standard cell; ~ **de ferroníquel** *m* ING ELÉC ferronickel cell; ~ **de fijación** *m* ING MECÁ attaching part; ~ **del filtro** *m* PROD filter element; ~ **finito** *m* ING MECÁ, MECÁ finite element; ~ **de fusible** *m* ELECTRÓN, ING ELÉC fuse element; ~ **fusible de cartucho** *m* ING ELÉC fuse link; ~ **gelatinizado** *m* ING ELÉC gel cell; ~ **giratorio** *m* ING MECÁ rotating part; ~ **H** *m* ING ELÉC H cell; ~ **identidad** *m* MATEMÁT identity element; ~ **de imagen** *m* TELECOM TV picture dot; ~ **incompatible** *m* GEOL incompatible element; ~ **de información** *m* ELECTRÓN data element; ~ **infrarrojo** *m* TERMOTEC infrared element; ~ **inicial** *m* INFORM&PD start element; ~ **intercalable de PVC** *m* EMB PVC insert-fitment; ~ **intranuclear de control de la potencia** *m* NUCL in-core power manipulator; ~ **Leclanch** *m* ING ELÉC Leclanch cell; ~ **limitado** *m* ING MECÁ, MECÁ finite element; ~ **litófilo** *m* GEOL lithophile element; ~ **lógico** *m* ELECTRÓN, INFORM&PD logic element; ~ **de memoria** *m* INFORM&PD storage element, ING ELÉC storage cell, storage element; ~ **micronutritivo** *m* HIDROL micronutrient; ~ **modular** *m* ELEC *equipo* module; ~ **móvil** *m* CONTAM mobile component; ~ **no lineal** *m* CONST nonlinear element; ~ **nutritivo** *m* CONTAM MAR nutrient; ~ **de origen natural** *m* NUCL naturally-occurring element; ~ **de parada** *m* INFORM&PD stop element; ~ **parásito** *m* FÍS parasitic element; ~ **pasivo** *m* ELECTRÓN passive element, TELECOM passive component; ~ **pentavalente** *m* QUÍMICA pentavalent element; ~ **piezoeléctrico** *m* ING ELÉC piezoelectric element; ~ **de pila** *m* ELEC *de acumulador*, ING ELÉC battery cell; ~ **en plata** *m* ING ELÉC silver cell; ~ **principal** *m* CONST *estructuras* principal member; ~ **programable** *m* CONST programmable box; ~ **que se encuentra en una concentración pequeña** *m* CONTAM trace element; ~ **radiactivo** *m* FÍS RAD radioactive element; ~ **del radiador** *m* AUTO, VEH radiator core, radiator element; ~ **reactivo** *m* ELEC, ING ELÉC reactive element; ~ **del recalentador** *m* PROD superheater element; ~ **de red** *m* TELECOM network element (*NE*); ~ **de una red de antena** *m* TEC ESP bay; ~ **de red sin jerarquía digital sincrónica** *m* TELECOM non-SDH network element (*NNE*); ~ **de red del servicio de aplicación de gestión** *m* TELECOM network management application service element (*NM-ASE*); ~ **reflectante** *m* CONST *carreteras* cat's eye; ~ **de reserva** *m* ING ELÉC backup; ~ **resistente** *m* ELEC, ING ELÉC, TELECOM resistive element; ~ **secundario**

m QUÍMICA secondary element; ~ **de seguridad** *m* TRANSP MAR safety factor; ~ **de sembrado** *m* NUCL seed assembly, seed element; ~ **sensible** *m* METEO, TRANSP MAR meteorological sensor; ~ **sensor** *m* NUCL sensing element; ~ **del servicio de aplicación** *m* TELECOM application service element (*ASE*); ~ **de servicio de la aplicación para la gestión del sistema** *m* TELECOM system management application service element; ~ **de servicio de información de gestión común** *m* TELECOM common management information service element (*CMISE*); ~ **del servicio de operación remota** *m* TELECOM remote-operation service element (*ROSE*); ~ **de silicio** *m* ING ELÉC silicon cell; ~ **sobrecalentador** *m* PROD superheater unit; ~ **de sobrerreactividad** *m* NUCL booster element; ~ **solar desconcentrador** *m* ING ELÉC nonconcentrator solar cell; ~ **de sujeción de cabeza chata ranurada** *m* ING MECÁ cheese-head fastener; ~ **de suspensión elástica** *m* TRANSP sandow; ~ **tensor** *m* CONTAM MAR tension member; ~ **terminal** *m* ING ELÉC terminating element; ~ **termostático** *m* REFRIG thermostatic element; ~ **de textura** *m* GEOL fabric element; ~ **de tipo aditivo** *m* TEC ESP added-on component; ~ **de transición** *m* QUÍMICA transition element; ~ **transplutónico** *m* NUCL transplutonium element; ~ **traza** *m* GEOL *geoquímica*, QUÍMICA trace element; ~ **trazador** *m* QUÍMICA tracer element; ~ **de valor** *m* NUCL valuable element; ~ **de variación de fase** *m* ELECTRÓN phase-shifting element; ~ **de visualización digital** *m* ELECTRÓN digital display; ~ **voltaico** *m* ELEC voltaic cell

elementos: ~ **de calibración múltiple estándar** *m pl* METR standard multigaging elements (*AmE*), standard multigauging elements (*BrE*); ~ **electropositivos** *m pl* FÍS RAD, QUÍMICA electropositive elements; ~ **finitos para ingeniería mecánica** *m pl* ING MECÁ finite elements for mechanical engineering; ~ **de impresión** *m pl* IMPR printing implements; ~ **limitados para ingeniería mecánica** *m pl* ING MECÁ finite elements for mechanical engineering; ~ **de programación** *m pl* ELEC *informática*, INFORM&PD software; ~ **protectores hechos de plástico** *m pl* SEG plastic protective elements; ~ **de simetría** *m* CRISTAL symmetry elements; ~ **de sujeción de tubos** *m pl* ING MECÁ pipe-clamping elements

eleolita *f* MINERAL elaeolite

elevación *f* CONST elevation, heightening, lift, raising, FÍS pitch, *de temperatura* lift, raising, HIDROL *río* raising, ING MECÁ lifting, height, holding up, MECÁ elevation, MINAS lifting, OCEAN heave, TEC ESP altitude, TRANSP lifting, TRANSP AÉR lift, TRANSP MAR *navegación* elevation; ~ **aerodinámica** *f* TRANSP aerodynamical levitation; ~ **con amortiguación por aire** *f* TRANSP air cushion levitation; ~ **continental** *f* GEOL *submarina* continental rise; ~ **electrodinámica** *f* ELEC, TRANSP electrodynamic levitation; ~ **frontal** *f* ING MECÁ front elevation; ~ **inicial** *f* TRANSP AÉR initial climb-out; ~ **mediante colchón de aire** *f* TRANSP air cushion levitation; ~ **molecular del punto de ebullición** *f* TERMO molecular elevation of boiling point; ~ **del negro** *f* TV black lift; ~ **sobre el nivel del mar** *f* METEO elevation above sea level; ~ **del penacho de humos** *f* CONTAM plume rise; ~ **a una potencia** *f* MATEMÁT involution

elevadizo *m* TRANSP turntable

elevado *adj* ING MECÁ high

elevador *m* CARBÓN follower, CONST lift, lifter, elevator (*AmE*), ING MECÁ lifter, LAB *soporte* jack, MECÁ lift (*BrE*), elevator (*AmE*), MINAS lifter, PETROL riser, PROD elevator, TEC PETR riser, *perforación* elevator, TRANSP AÉR hoist; ~ **ácido** *m* PROC QUÍ acid elevator; ~ **aerodinámico** *m* TRANSP aerodynamic lift; ~~**arrancador** *m* AGRIC *remolachas* lifter; ~ **de arrastre** *m* TRANSP drag lift; ~ **de barcas** *m* TRANSP, TRANSP MAR boat elevator; ~ **de barcos** *m* TRANSP, TRANSP MAR boat elevator, boat lift; ~ **de bobinas** *m* EMB reel lifter; ~ **de cadena** *m* ING MECÁ chain elevator (*AmE*), chain lift (*BrE*); ~ **de cangilones** *m* AGRIC, CONST bucket elevator; ~ **de cinta** *m* AGRIC belt elevator; ~ **en cola** *m* AGUA tail elevator; ~ **de correa** *m* AGRIC belt elevator; ~ **de cubos basculantes** *m* TRANSP tilt bucket elevator; ~ **eléctrico** *m* ELEC, MECÁ electric hoist; ~ **hidrodinámico** *m* TRANSP hydrodynamic lift; ~ **en horizontal** *m* TRANSP horizontal elevator; ~ **de horquilla** *m* AGRIC fork lift; ~ **del izador** *m* TRANSP AÉR elevator hoist; ~ **del motor** *m* TRANSP AÉR engine hoist; ~ **neumático** *m* TRANSP airlift; ~~**noria** *m* TRANSP scoop wheel elevator; ~~**reductor** *m* ING ELÉC reversible booster; ~ **para residuos** *m* CONST tailings elevator; ~ **de tensión** *m* ELEC *transformador*, ELECTRÓN voltage booster; ~ **para tubería de pozo** *m* CONST casing elevator (*AmE*), casing lift (*BrE*); ~ **de vehículos pesados** *m* TRANSP heavy vehicle elevator (*AmE*), heavy vehicle lift (*BrE*); ~ **de voltaje** *m* ELEC *transformador*, ELECTRÓN, ING ELÉC, TEC ESP, TV voltage booster

elevar *vt* MECÁ lift, MINAS raise, PROD hoist; ~ **al cuadrado** *vt* MATEMÁT square; ~ **al cubo** *vt* MATEMÁT cube; ~ **a una potencia** *vt* MATEMÁT raise to a power; ~ **la potencia de** *vt* TEC ESP *sistemas de alimentación energética* boost

elevón *m* TRANSP elevon

eliminación *f* CONST removal, removing, CONTAM *tratamiento de aguas y gases* removal, CONTAM MAR disposal, ELEC expulsion, ELECTRÓN blanking, ING ELÉC, TEC PETR expulsion, ING ELÉC, TEC PETR *producción* water knockout; ~ **de aguas cloacales** *f* RECICL sewage disposal; ~ **de aguas residuales** *f* RECICL wastewater disposal; ~ **de banda lateral** *f* ELECTRÓN sideband suppression; ~ **de la basura** *f* RECICL waste disposal; ~ **de basuras** *f* AGUA waste disposal; ~ **de la capa de metal electrodepositado** *f* PROD stripping; ~ **de la capa superior del suelo** *f* CONST topsoil stripping; ~ **de defectos superficiales** *f* CARBÓN *aceros* air conditioning; ~ **de desechos** *f* RECICL waste disposal; ~ **de diodo** *f* ELECTRÓN diode suppression; ~ **ecológica** *f* RECICL ecology cullet; ~ **de ecos** *f* ELECTRÓN echo cancellation; ~ **de espiras** *f* ELECTRÓN, GEOL, TELECOM deconvolution; ~ **de fangos** *f* HIDROL *alcantarillado* sludge dewatering; ~ **de hierro** *f* C&V removal of iron; ~ **de interferencias** *f* ELECTRÓN interference rejection, TELECOM interference suppression; ~ **de lodos por chorro de agua y aspiración** *f* NUCL sludge lancing; ~ **marina** *f* RECICL marine disposal; ~ **de obstrucciones** *f* CONST cleaning-off; ~ **de parásitos** *f* ING ELÉC noise suppression; ~ **de partículas contaminantes del aire** *f* CONTAM particulate removal of air pollutants; ~ **de polvo seco** *f*

SEG dry dust removal; ~ **de la portadora** *f* TV zero carrier; ~ **de residuos** *f* AGUA, RECICL waste disposal; ~ **de ruido de fritura** *f* PROD contact noise suppression; ~ **de silencio** *f* TELECOM silence elimination; ~ **de la vaina** *f* NUCL decanning, decladding; ~ **de venenos del reactor** *f* CONTAM MAR, NUCL, SEG nuclear poison removal; ~ **de venenos del reactor nuclear** *f* CONTAM, NUCL, SEG nuclear reactor poison removal

eliminado: ~ **del agua del crudo** *m* TEC PETR *producción, refino* oil scrubbing; ~ **en montaje** *m* CINEMAT, TV edit-out

eliminador *m* PROD suppressor; ~ **de cuerdas** *m* PAPEL ragger; ~ **de diodos** *m* ELECTRÓN diode suppressor; ~ **de ecos** *m* ELECTRÓN echo canceller; ~ **de efectos locales** *m* TELECOM antisidetone; ~ **de electricidad estática** *m* TEXTIL static eliminator; ~ **de impurezas gruesas** *m* PAPEL junk remover; ~ **de interferencias** *m* TV interference eliminator; ~ **de ruidos parásitos** *m* ELECTRÓN parasitic suppressor; ~ **de salvado** *m* ALIMENT bran finisher; ~ **de sobrevoltaje** *m* PROD surge suppressor

eliminar *vt* C&V *hierro* de-iron, INFORM&PD clear *errores* debug, PROC QUÍ wash away, SEG *polvo* remove, TRANSP *aeropuerto*, TRANSP AÉR push back; ~ **el anhídrido carbónico de** *vt* CARBÓN *cemento* decarbonate; ~ **el gas de** *vt* TERMO free from gas; ~ **por lavado** *vt* HIDROL wash away; ~ **lodos** *vt* CARBÓN deslime; ~ **obstáculos de** *vt* CONST clear away; ~ **quemando** *vt* TERMO burn off

elipse *f* GEOM ellipse; ~ **de inercia** *f* MECÁ ellipse of inertia

elipsógrafo *m* ING MECÁ trammel

elipsoide[1] *adj* FÍS, GEOL, GEOM, ING MECÁ ellipsoid

elipsoide[2] *m* FÍS, GEOL, GEOM, ING MECÁ ellipsoid; ~ **achatado** *m* GEOM oblate ellipsoid; ~ **atachado** *m* GEOM spheroid; ~ **de deformación** *m* GEOL strain ellipsoid; ~ **por deformación** *m* GEOL deformation ellipsoid; ~ **oblato** *m* FÍS oblate ellipsoid; ~ **prolate** *m* FÍS prolate ellipsoid; ~ **de revolución** *m* GEOM spheroid; ~ **de tensión** *m* ING MECÁ strain ellipsoid

elipsómetro *m* FÍS ellipsometer

elipticidad *f* CRISTAL axial ratio, TEC ESP elliptical polarization

elíptico *adj* ELECTRÓN, GEOM elliptic, elliptical

eliptona *f* QUÍMICA elliptone

elongación *f* ACÚST displacement, C&V, FÍS, ING MECÁ, METAL, P&C elongation; ~ **del cilindro** *f* C&V elongation of the cylinder

elución *f* ALIMENT, NUCL, PROC QUÍ, QUÍMICA elution

eluido *m* ALIMENT, NUCL, PROC QUÍ, QUÍMICA eluate

eluir *vt* ALIMENT, NUCL, PROC QUÍ, QUÍMICA elute

elutriación *f* C&V, PROC QUÍ, QUÍMICA elutriation

elutriar *vt* C&V, PROC QUÍ, QUÍMICA elutriate

eluyente *m* ALIMENT, NUCL, PROC QUÍ, QUÍMICA eluent

elvan *m* GEOL elvan

elvanita *f* PETROL elvanite

EM *abr* (*fin del soporte*) INFORM&PD EM (*end of medium*)

emanación *f* AGUA effluent, FÍS, FÍS RAD emanation, GAS effluent, seepage, HIDROL effluent, NUCL effluent, emanation, QUÍMICA emanation, RECICL, TEC PETR *refino* effluent; ~ **activa** *f* NUCL active emanation; ~ **rádica** *f* FÍS RAD, QUÍMICA radium emanation; ~ **de radio** *f* FÍS RAD, QUÍMICA radium emanation;

~ **radioactiva** *f* FÍS RAD radioactive emanation; ~ **tóxica** *f* RECICL toxic effluent

emanaciones: ~ **de barniz** *f pl* SEG lacquer fumes; ~ **de pintura** *f pl* SEG paint fumes

emasculación *f* AGRIC emasculation

embadurnamiento: ~ **con aceite** *m* PROD *metales* oiling

embadurnar *vt* PROD daub

embalado[1]: ~ **para enviar por correo** *adj* EMB mail order-packed

embalado[2] *m* ING MECÁ racing, PROD wrapping

embalador *m* ING MECÁ packer

embaladora *f* AGRIC baler; ~ **de porciones simples** *f* EMB single portion packaging machine

embalaje *m* EMB packaging, ING MECÁ, OCEAN packing, PAPEL, PROD packaging, TEC ESP package, TEXTIL packaging; ~ **del aerosol** *m* EMB aerosol packing; ~ **de alimentos** *m* ALIMENT, EMB food packaging; ~ **aséptico** *m* EMB aseptic packaging; ~ **barrera** *m* EMB barrier packaging; ~ **caducado** *m* EMB CA packaging; ~ **para calentar en hornos de microondas** *m* EMB microwavable packaging; ~ **para climas tropicales** *m* EMB tropical packaging; ~ **combinado** *m* EMB combined packaging; ~ **defectuoso** *m* EMB defective packaging; ~ **dosificador** *m* EMB dosing packing; ~ **eficaz** *m* EMB efficient packaging; ~ **embutido** *m* EMB deep-drawn packaging; ~ **de exportación** *m* EMB export packaging; ~ **expositor** *m* EMB display packaging; ~ **industrial** *m* EMB industrial packing; ~ **interior** *m* EMB interior packaging; ~ **a medida** *m* EMB contract packaging; ~ **de poliestireno expandido** *m* EMB expanded-polythene packaging; ~ **para productos ultracongelados** *m* EMB, REFRIG deep-freeze packaging; ~ **programado** *m* EMB carded packaging; ~ **de protección** *m* EMB barrier packaging; ~ **a prueba de niños** *m* EMB, SEG child-resistant packaging; ~ **que evita la oxidación** *m* EMB rust-preventive packaging; ~ **relleno de espuma solidificable y almohadillado** *m* EMB foam packaging and cushioning; ~ **reticulado** *m* C&V basketweave packing; ~ **retornable** *m* EMB, RECICL returnable packaging; ~ **reutilizable** *m* EMB, RECICL reusable packaging; ~ **transparente** *m* EMB see-through packaging; ~ **para el transporte marítimo** *m* EMB overseas packaging, seaworthy packaging; ~ **con ventana** *m* EMB *para inspección* window packaging

embalamiento *m* ING MECÁ, VEH overspeed

embalar *vt* CONST *transporte* case, VEH *el motor* rev up

embaldosado *m* CONST flagging, tile flooring

embalsar *vt* AGUA dam, CONST bank up, MINAS *agua* dam

embalse *m* AGUA reservoir, CARBÓN pond, HIDROL storage basin, PAPEL pond, PROD storage reservoir; ~ **de almacenamiento** *m* ENERG RENOV storage basin; ~ **de almacenamiento inferior** *m* ENERG RENOV lower-storage basin; ~ **de almacenamiento superior** *m* ENERG RENOV upper storage basin; ~ **de detención** *m* AGUA detention basin; ~ **eutrófico** *m* RECICL eutrophic lake; ~ **de filtrado** *m* RECICL filter basin; ~ **de retención** *m* AGUA detention reservoir, impounding reservoir, flood-control reservoir, HIDROL detention reservoir, flood-control reservoir; ~ **secundario** *m* AGUA, HIDROL compensating reservoir; ~ **de sedimentación** *m* RECICL sedimentation basin; ~ **de tierra** *m* AGUA earth reservoir; ~ **útil** *m* AGUA live storage

embarcación *f* CONST boat, CONTAM MAR, OCEAN boat, vessel, TEC ESP craft, TRANSP MAR boat, craft, ship; ~ **de abastecimiento** *f* TEC PETR supply boat; ~ **sobre aletas hidrodinámicas sustentadoras VEE** *f* TRANSP VEE foil craft; ~ **de apoyo** *f* TEC PETR standby boat, TRANSP MAR support boat; ~ **auxiliar** *f* TRANSP MAR auxiliary vessel, tender; ~ **de un buque** *f* TRANSP MAR ship's boat; ~ **camaronera** *f* OCEAN shrimper; ~ **contraincendios** *f* SEG, TRANSP MAR fireboat; ~ **dedicada a la pesca de ostras con rastro** *f* TRANSP MAR oyster dredger; ~ **dedicada a la recuperación de hidrocarburos** *f* CONTAM MAR oil-recovery vessel; ~ **dedicada a la remoción de hidrocarburos** *f* CONTAM MAR oil-clearance vessel; ~ **para la defensa de radas y puertos** *f* TRANSP MAR seaward defence boat; ~ **de descarga** *f* CONTAM MAR lightering vessel; ~ **de desembarco** *f* TRANSP MAR landing craft; ~ **fluvial** *f* TRANSP MAR river boat; ~ **a moto** *f* TRANSP MAR motorboat; ~ **a motor** *f* ELECTRÓN motorboating; ~ **para navegación en canales** *f* TRANSP MAR canal boat; ~ **neumática** *f* TRANSP hovering craft; ~ **patrullera** *f* D&A patrol craft, patrol vessel, TRANSP MAR patrol boat; ~ **patrullera del servicio de aduanas** *f* TRANSP MAR customs patrol boat; ~ **de práctico** *f* TRANSP MAR pilot boat; ~ **de salvamento** *f* SEG, TRANSP AÉR, TRANSP MAR rescue boat

embarcadero *m* AGRIC livestock loading ramp, TRANSP boathouse, loading dock, TRANSP MAR landing place, pier; ~ **flotante** *m* MINAS landing stage

embarcar[1] *vt* TRANSP MAR *mercancías* ship

embarcar[2] *vi* TRANSP MAR board; ~ **agua** *vi* TRANSP MAR ship water

embarcarse *v refl* TRANSP MAR *barco* board

embargo: ~ **preventivo** *m* TRANSP MAR *barco* arrest

embarque *m* MINAS loading, PROD shipment, shipping, TEC ESP handling, TRANSP MAR boarding; ~ **de mercancías** *m* TRANSP MAR shiploading; ~ **por propulsión propia** *m* TRANSP, TRANSP MAR roll-on/roll-off (ro-ro)

embarrancado *adj* OCEAN, TRANSP MAR aground

embarrancar *vi* OCEAN, TRANSP MAR run aground, run ashore

embaste *m* MINAS *pared* pre-split basting

embate *m* TEC ESP buffet; ~ **ascendente** *m* OCEAN uprush, *de la ola al romper contra estructuras* swash, TRANSP MAR swash

embeber *vt* TERMO moisten

embellecedor: ~ **del tubo de escape** *m* AUTO tailpipe extension

embestir: ~ **deliberadamente** *vt* TRANSP MAR *embarcación* run down

embisagrado *adj* ING MECÁ hinged

embobinador *m* C&V winder

embobinar *vt* TV spool

embocadura *f* CONST *túnel* portal, MINAS *hidrografía* mouth, PROD *de tubo* funnel, TEC ESP nozzle

embolada *f* ING MECÁ piston stroke

embolia *f* OCEAN gas embolism; ~ **gaseosa** *f* OCEAN aeroembolism, air embolism; ~ **de nitrógeno** *f* OCEAN decompression sickness

embolita *f* MINERAL embolite

émbolo *m* ING MECÁ, INSTAL HIDRÁUL piston, sucker, MECÁ piston, plunger, MINAS *bomba* piston, PROD,

VEH *bombín de freno, collarín de desembrague* plunger; ~ **accionado por vapor** *m* INSTAL HIDRÁUL steam piston; ~ **del cilindro de aire** *m* ING MECÁ air piston; ~ **de guía de ondas** *m* ING ELÉC waveguide plunger; ~ **de guíaondas** *m* ING ELÉC waveguide plunger; ~ **macizo** *m* ING MECÁ plunger; ~ **tubular** *m* INSTAL HIDRÁUL *motor sin cruceta* solid piston

embolsado *m* *Esp* EMB (*cf ensacado AmL*) bag packaging, bagging, OCEAN (*cf cierre de la bolsa AmL*) purse-fastener, pursing; ~ **de la captura** *m* OCEAN pursing; ~ **de comida diversa** *m* EMB multi-snack bagging; ~ **manual** *m* EMB handbagging

embolsador *m* AGRIC sacker

embolsadora *f* *AmL* (*cf ensacadora Esp*) AGRIC bag-filling machine, EMB bagging machine

embolsamiento: ~ **de aire** *m* FÍS FLUID *tuberías* airlock

embolsar *vt* ALIMENT bag

emborronamiento *m* CINEMAT *de imagen* blurring; ~ **debido al mar** *m* TELECOM *pantalla del radar* sea clutter; ~ **por el obturador** *m* CINEMAT shutter blur; ~ **de pantalla** *m* ELECTRÓN, TEC ESP, TELECOM, TRANSP MAR clutter

embotado *adj* ING MECÁ blunt, MECÁ dull

embotamiento *m* PROD gumming, gumming-up; ~ **de plomo** *m* REVEST lead glazing

embotar *vt* ING MECÁ, MECÁ blunt

embotarse *v refl* PROD gum up

embotellado[1] *adj* AGRIC, ALIMENT, EMB bottled

embotellado[2] *m* AGRIC, ALIMENT racking

embotellador *m* AGRIC, ALIMENT, EMB bottle filler

embotelladora *f* ALIMENT, EMB bottling machine

embotellamiento *m* ALIMENT, TRANSP bottleneck; ~ **de tráfico** *m* TRANSP traffic jam

embotellar *vt* AGRIC, ALIMENT, EMB bottle

embragado *m* ING MECÁ engaging

embragador *m* AUTO clutch fork, ING MECÁ clutch fork, coupler, VEH clutch fork

embragar[1] *vt* ING MECÁ, MECÁ engage, *correa* fork a belt on

embragar[2] *vi* ING MECÁ, MECÁ engage

embrague *m* AUTO clutch, ING MECÁ clutch, engagement, MECÁ clutch, engagement, engaging, TRANSP MAR, VEH clutch; ~ **por aceite** *m* VEH viscous clutch; ~ **de agarre** *m* AUTO grabbing clutch; ~ **automático** *m* AUTO self-adjusting clutch, VEH automatic clutch; ~ **autorregulable** *m* AUTO self-adjusting clutch; ~ **de canal en V** *m* ING MECÁ V-groove clutch; ~ **centrífugo** *m* AUTO, VEH centrifugal clutch; ~ **de cinta** *m* ING MECÁ band clutch, strap clutch; ~ **cónico** *m* AUTO cone clutch, ING MECÁ cone clutch, scroll clutch, V-groove clutch, *engranajes* spiral clutch, VEH cone clutch; ~ **de cono** *m* AUTO, ING MECÁ, VEH cone clutch; ~ **de diafragma** *m* AUTO diaphragm clutch; ~ **por dientes** *m* VEH dog clutch; ~ **de disco doble en seco** *m* AUTO double plate dry clutch; ~ **de disco único en seco** *m* AUTO single dry plate clutch; ~ **de discos** *m* ING MECÁ plate clutch, MECÁ, VEH disc clutch (*BrE*), disk clutch (*AmE*); ~ **de discos múltiples** *m* AUTO multiple disc clutch (*BrE*), multiple disk clutch (*AmE*), ING MECÁ multiplate clutch, multiple disc clutch (*BrE*), multiple disk clutch (*AmE*), multiple plate clutch; ~ **de doble disco** *m* AUTO two-disc clutch (*BrE*), two-disk clutch (*AmE*); ~ **de dos platos** *m* AUTO two-plate clutch; ~ **electromagnético** *m* ELEC magnetic clutch, electromagnetic clutch, ING ELÉC, ING MECÁ magnetic clutch, MECÁ, TRANSP electromagnetic clutch, VEH magnetic clutch; **~ de espiral** *m* ING MECÁ coil clutch; **~ de fricción** *m* ING MECÁ friction clutch; **~ con funcionamiento en seco** *m* VEH dry clutch; **~ de garras** *m* AUTO dog clutch, grabbing clutch, ING MECÁ claw clutch, dog clutch, VEH dog clutch; **~ hidráulico** *m* INSTAL HIDRÁUL hydraulic clutch, VEH viscous clutch; **~ hidromecánico** *m* ING MECÁ hydromechanical clutch; **~ magnético** *m* ELEC, ING ELÉC magnetic clutch, ING MECÁ magnetic clutch, magnetic face, VEH magnetic clutch; **~ de mandíbulas** *m* AUTO jaw clutching; **~ de manguito** *m* ING MECÁ muff coupling; **~ monodisco** *m* MECÁ, VEH disc clutch (*BrE*), disk clutch (*AmE*); **~ de mordazas** *m* AUTO jaw clutching, ING MECÁ jaw clutch; **~ neumático** *m* ING MECÁ air clutch, pneumatic clutch; **~ pluridisco** *m* AUTO multiple disc clutch (*BrE*), multiple disk clutch (*AmE*), ING MECÁ multiplate clutch, multiple disc clutch (*BrE*), multiple disk clutch (*AmE*), multiple plate clutch; **~ polidisco** *m* AUTO multiple disc clutch (*BrE*), multiple disk clutch (*AmE*), ING MECÁ multiplate clutch, multiple disc clutch (*BrE*), multiple disk clutch (*AmE*), multiple plate clutch; **~ de ranura en V** *m* ING MECÁ V-groove clutch; **~ de resorte helicoidal** *m* AUTO, ING MECÁ, MECÁ, VEH coil-spring clutch; **~ de rodillos** *m* AUTO roller clutch; **~ de rotación libre** *m* AUTO overrunning clutch; **~ de rueda libre** *m* AUTO, TRANSP AÉR overrunning clutch; **~ en seco** *m* ING MECÁ metal-to-metal clutch; **~ de sobremarcha** *m* TRANSP AÉR overrunning clutch; **~ de uñas** *m* AUTO, ING MECÁ dog clutch

embriaguez: **~ de las profundidades** *f* OCEAN *buzos* rapture of the deep

embridado *adj* GEN flanged

embridar[1] *vt* ING MECÁ flange, MECÁ *conductos, tuberías* clamp

embridar[2] *vi* TEC PETR *conductos, tuberías* flange up

embrión *m* AGRIC embryo

embrollo *m* SEG entanglement

embudo *m* ALIMENT funnel, C&V cone, funnel, CARBÓN *geología* bowl, CONST funnel, ING MECÁ hopper, LAB funnel, MECÁ hopper, funnel, MINAS crater, PROC QUÍ funnel, PROD filler, funnel, QUÍMICA funnel, TEC ESP *vuelos* drogue; **~ antirebase** *m* AmL (*cf embudo de seguridad Esp*) LAB safety funnel; **~ de aproximación** *m* TRANSP AÉR approach funnel; **~ de Buchner** *m* LAB Buchner funnel; **~ para el combustible** *m* INSTAL TERM fuel funnel; **~ de contracción** *m* MINAS *lingotes* sinkhole; **~ de decantación** *m* CARBÓN cone separator, MINAS *química*, QUÍMICA separator; **~ de desagüe** *m* PROD drain cup; **~ de despegue** *m* TRANSP AÉR takeoff funnel; **~ de extracción** *m* PROC QUÍ separating funnel; **~ de filtración** *m* LAB, QUÍMICA filter funnel; **~ de filtro de vidrio sinterizado** *m* LAB sintered glass filter funnel; **~ de guía** *m* PETROL guide funnel; **~ de levigación** *m* PROC QUÍ, QUÍMICA elutriating funnel; **~ con llave** *m* LAB tap funnel, thistle funnel; **~ de porcelana** *m* C&V porcelain funnel; **~ de seguridad** *m* Esp (*cf embudo antirebase AmL*) LAB safety funnel; **~ de separación** *m* LAB separating funnel, separation funnel, QUÍMICA separation funnel; **~ separador** *m* QUÍMICA separator funnel

embutición *f* MECÁ, P&C forming, PROD swaging,

fundición drawing, forming, *estampado* shaping, REVEST forming; **~ profunda** *f* MECÁ deep drawing; **~ del tapón corona** *f* EMB crown cup

embutido[1] *adj* CONST encased, embedded, ING MECÁ embedded, MECÁ dished; **~ en la pared** *adj* CONST sunk into the wall

embutido[2] *m* PROD *metal en chapas* stamp; **~ en caliente** *m* EMB hot stamping; **~ de hilo** *m* ING MECÁ thread insert; **~ superficial** *m* ING MECÁ dimpled hole, METAL dimple

embutidor *m* CONST rivet snap, ING MECÁ, MECÁ punch, PROD *obrero de chapistería* shaper; **~ de clavos** *m* CONST nail set

embutidora *f* PROD shaper

embutir *vt* CONST, ING MECÁ embed, PROD dish; **~ en la prensa** *vt* PROD press

EMC *abr* (*error medio cuadrático*) INFORM&PD, MATEMÁT MSE (*mean square error*)

emendido: **~ sin platinos** *m* VEH contactless ignition (*BrE*), pointless ignition (*AmE*)

emergencia *f* AGRIC, GEOL *del fondo del mar* emergence

emergente *adj* AGRIC, GEOL emergent

emergido *adj* AGRIC, GEOL emerged

emerylita *f* MINERAL emerylite

emetina *f* QUÍMICA emetin

emisión *f* AGUA effluent, CONTAM emission, ENERG RENOV emittance, FÍS ONDAS *radar* pulse, FÍS PART, FÍS RAD emission, GAS, HIDROL effluent, INFORM&PD, INSTAL HIDRÁUL release, NUCL issue, effluent, RECICL effluent, TEC ESP vent, TEC PETR effluent, TELECOM *de ondas* launching, transmission, TRANSP AÉR emission, TV broadcast (*BC*), broadcasting, VEH emission; **~ acústica** *f* NUCL acoustic emission; **~ beta** *f* FÍS RAD beta emission; **~ del calor** *f* TERMO heat emission; **~ de campo** *f* ELECTRÓN, FÍS field emission; **~ cargada de vapor** *f* CONTAM steam-laden emission; **~ en color** *f* CINEMAT, FOTO color cast (*AmE*), colour cast (*BrE*); **~ comunitaria** *f* TV community broadcasting; **~ de contaminante atmosférico** *f* CONTAM air pollution emission; **~ cooperativa** *f* METAL cooperative emission; **~ deseada** *f* TELECOM wanted emission; **~ en diferido** *f* TV delayed broadcast; **~ en directo** *f* Esp (*cf emisión en vivo AmL*) TV live broadcast; **~ doméstica** *f* CONTAM domestic emission; **~ por efecto de campo** *f* ELECTRÓN, FÍS field emission; **~ de electrones** *f* ELECTRÓN, FÍS PART electron emission; **~ específica** *f* ING ELÉC specific emission; **~ espontánea** *f* ELECTRÓN, FÍS, FÍS RAD, TELECOM spontaneous emission; **~ espontánea de radiación a velocidad constante** *f* FÍS RAD spontaneous emission of radiation at a constant rate; **~ espuria conducida** *f* TELECOM conducted spurious emission; **~ estándar** *f* RECICL effluent standard, TV broadcast standard; **~ estimulada** *f* ELECTRÓN, FÍS, FÍS RAD, ÓPT, TELECOM stimulated emission; **~ fotoeléctrica** *f* ELECTRÓN photoelectric emission; **~ fría** *f* ING ELÉC cold emission; **~ gamma** *f* FÍS RAD gamma emission; **~ de gases de escape** *f* AUTO, MECÁ, TRANSP exhaust-gas emission; **~ interferente** *f* D&A *de radio* jammer; **~ inversa** *f* ELECTRÓN reverse emission; **~ iónica secundaria** *f* TELECOM secondary ionic emission; **~ de láser** *f* ELECTRÓN laser emission; **~ local** *f* TRANSP area broadcasting; **~ no deseada** *f* TELECOM unwanted emission; **~ de ondas con longitud** *f* FÍS RAD millimeter-wavelength emission

(*AmE*), millimetre-wavelength emission (*BrE*); ~ **de ondas milimétricas** *f* FÍS RAD millimeter-wavelength emission (*AmE*), millimetre-wavelength emission (*BrE*); ~ **de la orden de fabricación** *f* PROD production-order issue; ~ **parásita** *f* TELECOM unwanted emission; ~ **parásita conducida** *f* TELECOM conducted spurious emission; ~ **perturbadora** *f* D&A *de radio* jammer; ~ **primaria** *f* ELECTRÓN primary emission; ~ **de ráfagas** *f* VEH headlamp flasher; ~**-recepción** *f* (*ER*) ELECTRÓN transmit-receive (*TR*); ~ **de rejilla secundaria** *f* ELECTRÓN secondary grid emission; ~ **retardada** *f* TV delayed broadcast; ~ **secundaria** *f* ELECTRÓN, FÍS secondary emission; ~ **sincrotrónica** *f* FÍS, FÍS RAD synchrotron emission; ~ **standard** *f* RECICL effluent standard; ~ **térmica** *f* FÍS RAD thermal emission; ~ **termoiónica** *f* FÍS, FÍS RAD, ING ELÉC thermionic emission; ~ **de vapor** *f* CONTAM steam emission; ~ **por video Umatic** *AmL*, ~ **por vídeo Umatic** *f Esp* TV broadcast video Umatic (*BVU*); ~ **de videográficos** *f* TV broadcast videographics; ~ **en vivo** *f AmL* (*cf emisión en directo Esp*) TV live broadcast

emisiones: ~ **del motor del avión** *f pl* TRANSP AÉR aircraft engine emissions

emisividad *f* FÍS, FÍS RAD, ÓPT, TELECOM, TERMOTEC, TV emissivity; ~ **espectral** *f* FÍS spectral emissivity; ~ **luminosa** *f* FÍS luminous emissivity; ~ **térmica** *f* TERMO thermal emissivity

emisor[1] *adj* FÍS, FÍS RAD, ÓPT, TELECOM, TV radiating

emisor[2] *m* ELEC, ELECTRÓN, FÍS, FÍS RAD, INFORM&PD, ING ELÉC emitter, sender, transmitter; ~ **alfa** *m* FÍS, FÍS PART, FÍS RAD alpha emitter; ~ **beta** *m* FÍS, FÍS PART, FÍS RAD beta emitter; ~ **de contaminantes** *m* CONTAM pollution emitter, pollution source; ~ **de electrones** *m* ELECTRÓN, FÍS PART electron emitter; ~ **de gas** *m* ELECTRÓN, GAS gas maser; ~ **del intercambio** *m* TELECOM interchange sender; ~ **K** *m* NUCL K-emitter; ~ **de llamada** *m* TELECOM call sender; ~ **óptico** *m* ÓPT optical emitter; ~ **perturbador** *m* D&A jammer; ~ **de radiación** *m* ELECTRÓN, FÍS RAD, NUCL radiation source; ~ **de radiación de cuatro niveles** *m* ELECTRÓN four-level maser; ~ **de radiodifusión de sonido** *m* TELECOM sound broadcast transmitter; ~ **de rayos x** *m* NUCL X-emitter; ~**-receptor** *m* INFORM&PD emitter-receiver, sender-receiver, TELECOM transmitter-receiver; ~**-receptor multifrecuencia** *m* TELECOM MF sender-receiver, multiple frequency sender-receiver, multifrequency sender-receiver; ~ **de televisión** *m* ELECTRÓN, FÍS, ING ELÉC, TELECOM, TV television transmitter; ~ **de transistor** *m* ELECTRÓN transistor emitter; ~ **del transistor** *m* ELECTRÓN transistor emitter; ~ **de transmisión** *m* TELECOM transmission sender

emisora *f* TRANSP station; ~ **auxiliar de televisión** *f* TV booster station; ~ **de difusión local** *f* TRANSP area broadcasting station; ~ **de radio móvil** *f* TELECOM mobile radio station; ~ **de radiodifusión local** *f* TRANSP *radio*, TV area broadcasting station; ~ **de TV de refuerzo** *f* TV booster station

emitancia: ~ **de radiación** *f* ÓPT, TELECOM *en punto de superficie* radiant emittance; ~ **radiante** *f* ÓPT, TELECOM radiant emittance

emitir[1] *vt* ELECTRÓN beam, FÍS ONDAS broadcast, GAS release, INFORM&PD emit, TELECOM broadcast, TERMO emit, TRANSP MAR transmit, TV broadcast

emitir[2]: ~ **impulsos** *vi* PROD pulse off; ~ **rayos** *vi* GEN radiate

emódico *adj* QUÍMICA emodic

emodina *f* QUÍMICA emodin

empacado *m* PROD packaging; ~ **automático en cinta** *m* ING ELÉC tape-automated bonding (*TAB*)

empacador *m* ING MECÁ packer

empacadora *f* AGRIC baler; ~ **de heno** *f* AGRIC hay baler; ~ **de pacas cilíndricas** *f* AGRIC roll baler

empaletado *m* MECÁ blading

empalizada *f* CONST enclosure, palisade, barricade, INSTAL HIDRÁUL stockade

empalmado *m* PAPEL splicing

empalmador *m* CONST, PROD jointer, TELECOM jointer, splicer; ~ **de cable** *m* ING ELÉC cable connector; ~ **de cables** *m* TELECOM cable jointer; ~ **volante** *m* IMPR *en las rotativas* flying paster

empalmadora *f* CINEMAT splicer, CONST jointer, IMPR autopaster, PAPEL paster, PROD, TELECOM jointer; ~ **por cemento** *f* CINEMAT cement splicer; ~ **de cinta** *f* CINEMAT tape splicer; ~ **guillotina** *f* CINEMAT, FOTO, TV guillotine splicer; ~ **en seco** *f* CINEMAT dry splicer; ~ **a tope** *f* CINEMAT butt-end splicer

empalmar *vt* CINEMAT splice, CONST assemble, bind, branch, joggle, join on, match, ELEC connect, ING ELÉC connect, patch, MINAS *sondas* couple, PAPEL, TELECOM, TEXTIL splice; ~ **con junta de enchufe** *vt* CONST *tuberías* socket

empalme *m* CINEMAT splice, CONST lap joint, connection, junction, union, *carpintería* scarf, ELEC junction, joint, FÍS coupling, IMPR splice, joint, ING ELÉC connector, coupling, ING MECÁ lengthening piece, coupling, couple, MECÁ bonding, coupling, NUCL jointing, ÓPT, PAPEL splice, PROD junction, jointing, TEC ESP bending, TELECOM splice, joint, jointing, TEXTIL splicing, TV splice; ~ **de acoplamiento** *m* CINEMAT patch splice; ~ **de aleación** *m* ELECTRÓN alloy junction; ~ **de bastidor** *m* CINEMAT patch bay; ~ **biselado** *m* CONST scarf jointing, splayed miter joint (*AmE*), splayed mitre joint (*BrE*); ~ **de cable eléctrico** *m* FERRO electric-cable joint; ~ **de cableado** *m* ELEC, ING ELÉC cable junction; ~ **de cables** *m* ELEC cable joint, cable fitting, cable splicing, ING ELÉC cable fitting, cable joint, cable splicing, TELECOM cable joint; ~ **por cemento** *m* CINEMAT cement splice; ~ **con cinta** *m* CINEMAT tape splice; ~ **de cola de milano** *m* CONST dovetail joint, *carpintería* fantail joint; ~ **de cordón** *m* CONST spigot joint; ~ **cruzado continuo** *m* ELEC continuous cross-bonding, continuous shield-bonding; ~ **con cubrejuntas** *m* PROD welted joint; ~ **defectuoso** *m* CINEMAT blooping notch; ~ **diagonal** *m* CINEMAT diagonal joiner, diagonal splice; ~ **de enchufeo y cordón** *m* CONST spigot-and-faucet joint; ~ **ferroviario** *m* FERRO rail junction, railroad junction (*AmE*), railway junction (*BrE*); ~ **ferroviario principal** *m* FERRO major railway junction; ~ **de fibra óptica** *m* ÓPT, TELECOM fiber-optic splice (*AmE*), fibre-optic splice (*BrE*), optical fiber splice (*AmE*), optical fibre splice (*BrE*); ~ **por fusión** *m* ÓPT, TELECOM fusion splice; ~ **a inglete** *m* CONST splayed miter joint (*AmE*), splayed mitre joint (*BrE*); ~ **de Josephson** *m* FÍS Josephson junction; ~ **de llave** *m* ING MECÁ key joint; ~ **con manguito y pasadores** *m* ING MECÁ pin coupling; ~ **mecánico** *m*

ÓPT, TELECOM, TV mechanical splice; **~ a media madera** *m* CONST halving; **~ a media madera con cola de milano** *m* CONST dovetail halved joint; **~ óptico** *m* ÓPT, TELECOM optical splice; **~ recto** *m* ELEC straight joint; **~ de retención** *m* ELEC stop joint; **~ roscado** *m* ING MECÁ threaded fitting; **~ por secciones** *m* ELEC sectionalizing joint; **~ en T** *m* ELEC T-joint; **~ por torcedura** *m* PROD *de alambres* twist joint; **~ de tuberías** *m* CONST pipe connection; **~ de tubos flexibles** *m* MINAS *sondeos* flexible-hose coupling; **~ de las varillas de perforación** *m* MINAS *sondeos* drill pipe coupling

empañamiento *m* C&V *por la formación de sulfato durante el recocido* bloom

empapar *vt* ALIMENT, C&V steep

empaparse *v refl* TEXTIL soak

empapelado *m* REVEST paper covering, paper wall-covering

empaque *m* ING MECÁ packing, QUÍMICA wadding; **~ doble en línea** *m* (*DIP*) INFORM&PD, ING ELÉC dual-in-line package (*DIP*); **~ con papel sencillo** *m* EMB smooth plain packing; **~ sencillo** *m* C&V straight packing

empaquetada *f* REVEST package lacquer

empaquetado¹: **~ para enviar por correo** *adj* EMB mail order-packed

empaquetado² *m* ALIMENT packaging, CARBÓN piling, EMB bundle, PROD packing, piling, stuffing, wrapping, TEXTIL packing; **~ con un agente secante** *m* EMB packing with siccative; **~ hexagonal compacto** *m* (*EHC*) CRISTAL hexagonal close-packed structure (*HCP*); **~ manual** *m* EMB handpacking; **~ con un material con acolchado de burbujas** *m* EMB bubble pack

empaquetador *m* IMPR bundle tier, ING MECÁ, TEXTIL *persona* packer; **~-desempaquetador** *m* TELECOM packetizer-depacketizer

empaquetadora *f* EMB bundling machine, parcelling machine; **~ de botellas** *f* EMB bottle-packing machine; **~ de laberinto** *f* PROD *turbinas* labyrinth packing; **~ de latas** *f* EMB can-packing machine (*AmE*), tin-packing machine (*BrE*); **~ con un material estirable totalmente automática** *f* EMB fully-automatic stretch wrapper

empaquetadura *f* AUTO gasket, MECÁ packing, PROD hemp gasket, packaging, gland, stuffing, TRANSP MAR gasket; **~ de cáñamo** *f* PROD hemp gasket, hemp packing; **~ de cuero** *f* PROD leather gasket; **~ de la mesa** *f* PROD body gasket; **~ metálica** *f* MECÁ gasket; **~ del obturador del pistón** *f* ING MECÁ piston-seal housings; **~ del pistón** *f* ING MECÁ piston packing; **~ al vacío sin apoyo** *f* EMB unsupported shrink wrapping

empaquetamiento: **~ a cantidad** *m* EMB, PROD quantity packing; **~ compacto** *m* CRISTAL close packing; **~ estanco a los fluidos** *m* PROD fluid-tight packing; **~ de prensaestopas** *m* TEC PETR gland

empaquetar *vt* INFORM&PD, MECÁ pack, PAPEL bundle, pack, PROD pile

emparejado *m* ELEC, ING ELÉC pairing

emparejar *vt* IMPR jog, MECÁ level, PROD match

emparrillado *m* PROD fire grate, grating, grid; **~ de barrotes** *m* HIDROL *aguas residuales* bar screen; **~ basculante** *m* PROD drop grate, *horno* dump grate; **~ móvil** *m* PROD shaking grate

empastar *vt* REVEST coat

empaste *m* AGRIC bloat; **~ de la plancha por exceso de tinta** *m* IMPR flooding

empate: **~ de tornillo** *m* ING MECÁ screw coupling

empedrado *m* CONST pavement (*BrE*), sidewalk (*AmE*)

empedrador *m* CONST paver, pavior (*AmE*), paviour (*BrE*)

empenaje *m* TRANSP AÉR aircraft tail unit, empennage, tail unit

empernado *m* CONST, ING MECÁ bolting

empernar *vt* CONST dowel

empezar¹ *vt* C&V *el corte* start

empezar²: **~ a tomar salida** *vi* TRANSP MAR gather way

emplazamiento¹: **en el ~** *adv* CONST on site

emplazamiento² *m Esp* (*cf anchurón AmL*) CARBÓN stall, room, INFORM&PD *memoria* location, MECÁ seat, MINAS *galerías* stall; **~ de almacenamiento** *m* INFORM&PD storage location; **~ de atrapado** *m* ING ELÉC trapping site; **~ de carga** *m* C&V batch house; **~ exclusivo** *m* TRANSP exclusive site; **~ de lanzamiento** *m* TEC ESP launch site, launching site; **~ para módulos de E/S** *m* PROD I/O module placement; **~ de perforación** *m* MINAS drill site, PETROL drilling site; **~ protegido** *m* INFORM&PD protected location; **~ variado** *m* TEC ESP site diversity

emplear: **~ de nuevo** *vt* RECICL reuse

emplectita *f* MINERAL emplectite

empleo *m* MECÁ application

emplomadura *f* CONST, PROD leading

emplomar *vt* CONST, PROD lead, REVEST coat

empobrecer *vt* AGUA, ELECTRÓN, NUCL, QUÍMICA deplete

empobrecimiento *m* AGRIC impoverishment, CONTAM poisoning, ELECTRÓN depletion, METAL downgrading, NUCL depletion, impoverishment, PROD *de aleaciones* downgrading; **~ del aire en oxígeno** *m* MINAS oxygen depletion of the air

empolvado: **~ cuádruple** *m* IMPR four-way powdering

empotrado *adj* CONST clamped, encased, embedded, ELEC, EMB clamped, ING MECÁ built-in, clamped, embedded, MECÁ built-in, clamped, PROD clamped; **~ en la pared** *adj* CONST sunk into the wall

empotramiento *m* CONST abutment, embedment, housing, retreat, ING MECÁ fixing

empotrar *vt* CONST abut, embed, encase, *carpintería* house, ING MECÁ embed, MINAS insulate, PROD clamp together, clamp

empresa *f* PROD, TEXTIL firm; **~ certificada** *f* CALIDAD certified company; **~ constructora** *f* CALIDAD, CONST contractor; **~ eléctrica** *f* ELEC, ING ELÉC electric utility; **~ electrocomercial** *f* ELEC, ING ELÉC electric utility; **~ operativa de Bell** *f* TELECOM Bell operating company (*BOC*); **~ petrolífera** *f* PROD petroleum company; **~ de servicio público** *f* PROD, TRANSP utility

empujador *m* AUTO rocker arm, SEG push stick, TRANSP MAR pusher tug, VEH rocker arm, valve lifter; **~ de leva** *m* ING MECÁ, MECÁ, TRANSP AÉR cam follower; **~ de loza** *m* C&V ware pusher; **~ del pedal de embrague** *m* AUTO, VEH clutch-pedal push rod; **~ de válvula** *m* INSTAL HIDRÁUL valve tappet

empujadora: **~ niveladora** *f AmL* (*cf motoniveladora Esp*) MINAS bulldozer

empujar *vt* ING MECÁ drive into, MECÁ boost

empujaválvulas *m* AUTO tappet, valve tappet; **~ hidráulico** *m* AUTO hydraulic valve lifter; **~ plano** *m* AUTO flat-bottom tappet

empuje *m* GEN thrust; **~ de agua** *m* TEC PETR water drive; **~ por agua** *m* TEC PETR water drive; **~ axial máximo** *m* ENERG RENOV maximum axial thrust; **~ bruto** *m* TRANSP AÉR gross thrust; **~ estático** *m* D&A, TRANSP AÉR static thrust; **~ garantizado** *m* TRANSP AÉR guaranteed thrust; **~ por gas en disolución** *m* TEC PETR solution gas drive; **~ del gas en solución** *m* TEC PETR solution gas drive; **~ por gas en solución** *m* GAS gas drive; **~ gasífero** *m* GAS gas drive; **~ hacía arriba** *m* FÍS upthrust; **~ de la hélice** *m* TRANSP propeller thrust; **~ hidráulico** *m* ENERG RENOV hydraulic thrust, TRANSP MAR *colisión* ram; **~ hidrodinámico** *m* HIDROL hydrodynamic thrust; **~ hidrostático** *m* PETROL water drive; **~ horizontal** *m* ENERG RENOV side thrust, MINAS *bóvedas* drift; **~ invertido** *m* TRANSP AÉR reverse thrust; **~ iónico** *m* TEC ESP ion thruster; **~ lateral** *m* ACÚST, ING MECÁ side thrust; **~ de leva** *m* VEH *motor* cam following; **~ por montera de gas** *m* GAS, TEC PETR gas cap drive; **~ orientable** *m* TEC ESP vectored thrust; **~ pasivo** *m* CONST *geotecnia* passive earth pressure; **~ del rotor** *m* TRANSP AÉR rotor thrust; **~ de tierras** *m* CARBÓN earth pressure; **~ de tierras pasivo** *m* CARBÓN passive earth pressure; **~ en vacío** *m* TEC ESP vacuum thrust

empujón *m* C&V push-up

empuñacubos *m* ING MECÁ hub grip

empuñadura *f* CINEMAT, FOTO handgrip, ING MECÁ grip, handgrip, handle, MECÁ, MINAS handle, PROD hilt, TEC ESP handle, TRANSP AÉR hold, VEH handle; **~ de cubos** *f* ING MECÁ hub grip; **~ con gatillo** *f* CINEMAT trigger grip; **~ con interruptor** *f* CINEMAT handgrip with shutter release; **~ para pilas** *f* FOTO battery grip; **~ de pistola con disparador** *f* FOTO pistol-grip with shutter release; **~ de sujeción** *f* ING MECÁ clamping handle; **tipo fusil** *f* CINEMAT, FOTO rifle grip

empuñar *vt* PROD grip

EMS *abr* (*especificación de memoria expandida*) INFORM&PD EMS (*expanded-memory specification*)

Emsiense *adj* GEOL Emsian

emulación *f* ELECTRÓN, INFORM&PD, TELECOM emulation

emulador *m* ELECTRÓN, INFORM&PD, TELECOM emulator

emular *vt* ELECTRÓN, INFORM&PD, TELECOM emulate

emulgente[1] *adj* GEN emulsifying

emulgente[2] *m* GEN emulsifier, emulsion binder

emulsificable *adj* GEN emulsifiable

emulsificación *f* GEN emulsification

emulsificado *adj* GEN emulsified

emulsificador *m* GEN emulsifier

emulsificante *m* GEN emulsifier

emulsificar *vt* AmL (*cf emulsionar Esp*) GEN emulsify

emulsión[1] *f* GEN emulsion; **~ de asfáltica** *f* CONST bitumen emulsion; **~ de bitumen** *f* ING MECÁ bitumen emulsion; **~ para blocs** *f* IMPR padding emulsion; **~ de cloruro de plata** *f* FOTO silver chloride emulsion; **~ despegable** *f* IMPR emulsion stripping; **~ directa** *f* IMPR direct emulsion; **~ de haluro de plata** *f* FOTO silver halide emulsion; **~ para imágenes directas** *f* FOTO printing-out emulsion; **~ infrarroja** *f* FOTO infrared emulsion; **~ ortocromática** *f* CINEMAT, FOTO orthochromatic emulsion; **~ pancromática** *f* CINEMAT, FOTO panchromatic emulsion; **~ de petróleo bruto en agua**

de mar *f* CONTAM, CONTAM MAR seawater-in-crude oil emulsion; **~ sensible al infrarrojo** *f* FOTO infrared-sensitive emulsion

emulsión[2]: **con la ~ hacia adentro** *fra* CINEMAT emulsion in; **con la ~ hacia afuera** *fra* CINEMAT emulsion out

emulsionabilidad *f* GEN emulsifiability

emulsionado *adj* GEN emulsified

emulsionante[1] *adj* GEN emulsifying

emulsionante[2] *m* GEN emulsifier, emulsion binder

emulsionar *vt* Esp (*cf emulsificar AmL*) GEN emulsify

emulsor *m* GEN emulsifier

EMW *abr* (*peso de lodo equivalente*) TEC PETR EMW (*equivalent mud weight*)

enanismo *m* AGRIC dwarfism; **~ amarillo de la cebada** *m* AGRIC *virus* barley yellow-dwarf; **~ del maíz** *m* AGRIC corn stunt (*AmE*), maize stunt (*BrE*)

enantal *m* QUÍMICA enanthal

enántico *adj* QUÍMICA enanthic

enantilato *m* QUÍMICA enanthylate

enantílico *adj* QUÍMICA enanthylic

enantina *f* QUÍMICA enanthin

enantiómero *m* CRISTAL enantiomer, optical antipode, optical isomer, QUÍMICA enantiomer

enantiomórfico *adj* CRISTAL, QUÍMICA enantiomorphic, enantiomorphous

enantiomorfismo *m* CRISTAL, QUÍMICA enantiomorphism

enantiomorfo *m* CRISTAL, QUÍMICA enantiomer, enantiomorph

enantiotrópico *adj* CRISTAL, QUÍMICA enantiotropic

enantol *m* QUÍMICA enanthol

enarbolar *vt* TRANSP MAR *bandera* fly

enarenado *m* PROD sanding

enarenadora *f* TRANSP grit spreader

enarenar *vt* PROD sand

enargita *f* MINERAL enargite

encabezamiento *m* IMPR head, INFORM&PD header, heading, ING MECÁ heading

encabritamiento *m* TRANSP AÉR pitch-up

encadenamiento *m* ING MECÁ, MECÁ linkage; **~ de datos** *m* INFORM&PD data chaining

encajado *adj* CONST encased, MATEMÁT nested; **~ a presión** *adj* PROD snap-on

encajadura: **~ de ladrillos** *f* CONST nogging

encajar *vt* CARBÓN table, CONST encase, house, MATEMÁT nest

encajarse *v refl* GEOL wedge out, GEOM fit together

encaje *m* AGRIC nick, C&V lacing, PROD slot, TEXTIL lace

encajonar *vt* CONST box

encalado *m* AGRIC liming, COLOR whitewash

encalamiento: **~ de lagos** *m* CONTAM lake liming

encalladero *m* OCEAN, TRANSP MAR sandbank, shallows, shoal,

encallado *adj* OCEAN, TRANSP MAR aground

encallar *vi* OCEAN, TRANSP MAR go aground, ground, run aground, run ashore

encamado *m* AGRIC lodging

encaminador *m* INFORM&PD, TELECOM router

encaminamiento *m* INFORM&PD, TELECOM routing; **~ adaptable** *m* INFORM&PD adaptive routing; **~ centralizado** *m* INFORM&PD centralized routing; **~ inverso** *m* TELECOM reverse routing; **~ de mensajes** *m* INFORM&PD message routing

encamisado *m* NUCL cladding, jacketing, TEC ESP *del combustible nuclear* cladding

encanillar *vt* ING MECÁ wind

encanto *m* FÍS PART, NUCL charm

encapsulación *f* NUCL, P&C encapsulation, PROD, TEC ESP potting

encapsulado[1]: ~ **al vacío** *adj* ING ELÉC vacuum-encapsulated

encapsulado[2] *m* EMB capping, PROD, TEC ESP potting; ~ **acústico** *m* ACÚST acoustic enclosure; ~ **con una sola línea de conexiones** *m* ELECTRÓN single in-line package (*SIP*)

encapsulador: ~ **automático** *m* EMB automatic capper

encapsuladora *f* EMB capper

encarado *m* GEOL facing

encarar *vt* TEXTIL *un tejido* face

encargada: ~ **del coche cama** *f* FERRO sleeping car attendant; ~ **del envío y recepción de mercancías** *f* TRANSP MAR shipping clerk; ~ **del guión** *f* CINEMAT script person; ~ **de seguridad** *f* SEG security officer

encargado[1]: ~ **a medida** *adj* TEXTIL custom-ordered

encargado[2] *m* CONST foreman, PROD charge hand; ~ **del coche cama** *m* FERRO sleeping car attendant; ~ **del envío y recepción de mercancías** *m* TRANSP MAR shipping clerk; ~ **del guión** *m* CINEMAT script person; ~ **de mantenimiento** *m* CONST maintainer; ~ **de seguridad** *m* SEG security officer; ~ **de seguridad en el trabajo** *m* SEG safety officer

encartado *adj* IMPR tipped in, ING MECÁ enclosed

encarte *m* IMPR tip; ~ **publicitario** *m* IMPR art insert

encartonadora *f* EMB cartoner

encasquillado *adj* AUTO, D&A, ING MECÁ, TEC ESP jammed; ~ **con bronce de cañón** *m* ING MECÁ gunmetal bushing

encasquillamiento *m* AUTO, D&A, ING MECÁ, TEC ESP jamming

encasquillarse *v refl* AUTO, D&A, ING MECÁ, TEC ESP jam

encastrado *adj* ING MECÁ inserted

encastrar *vt* CONST abut, ING MECÁ insert, MINAS insulate

encastre *m* CONST retreat, ING MECÁ insert; ~ **de aire a presión** *m* ING MECÁ compressed-air socket; ~ **a presión** *m* PROD press fit; ~ **receptor del cerrojo** *m* ING MECÁ keeper

encauzador *m* INSTAL HIDRÁUL guide vane, stationary vane, MECÁ guide vane

encauzamiento *m* AGUA, ENERG RENOV, GEOL channeling (*AmE*), channelling (*BrE*), INFORM&PD pipelining, PROD channeling (*AmE*), channelling (*BrE*)

encauzar[1] *vt* AGUA canalize, ENERG RENOV, GEOL, PROD channel

encauzar[2]: ~ **hacia la salida** *vi* PROD channel out

encefalopatía: ~ **espongiforme bovina** *f* (*BSE*) AGRIC bovine spongiform encephalopathy (*BSE*), mad cow disease

encendedor *m* ELEC, ING MECÁ, TEC ESP igniter, TERMOTEC firelighter, lighter, VEH cigar lighter; ~ **eléctrico** *m* ELEC electric lighter; ~ **de gas** *m* GAS, PROD, TERMOTEC gas lighter

encender[1] *vt* ALIMENT *calentar horno*, C&V fire, CINEMAT burn in, ING ELÉC turn on, switch on, PROD burn, *luz* turn on, *horno alto* blow in, QUÍMICA,

TEC ESP ignite, TERMO warm up, set ablaze, set fire to, burn

encender[2] *vi* NUCL ignite

encenderse *v refl* QUÍMICA ignite, TERMO glow

encendido[1] *adj* QUÍMICA glowing, TERMO ablaze, glowing; **de ~ y apagado** *adj* ING MECÁ on-off

encendido[2] *m* AUTO ignition, CINEMAT ignition, turn-on, ELEC ignition, ING ELÉC firing, turn-on, MECÁ ignition, PROD sparkling, QUÍMICA ignition, TEC ESP firing, TRANSP AÉR, VEH ignition; ~ **por acumuladores** *m* AUTO battery ignition; ~ **por apertura y cierre** *m* AUTO, VEH make-and-break ignition; ~ **de arco** *m* ELEC arc ignition, arc striking; ~ **del arco** *m* CINEMAT, ING ELÉC arc ignition; ~ **automático** *m* AUTO self-ignition, GAS automatic ignition, TRANSP AÉR autoignition; ~ **por chispa** *m* ELEC spark ignition; ~ **electrónico** *m* AUTO, ELEC, ELECTRÓN, VEH electronic ignition; ~ **por magneto** *m* AUTO, VEH magneto ignition; ~ **mecánico** *m* ING MECÁ mechanical firing; ~ **sin platinos** *m* AUTO contactless ignition (*BrE*), pointless ignition (*AmE*); ~ **de posición** *m* TEC ESP position light; ~ **prematuro** *m* AUTO premature ignition, VEH backfire, preignition, premature ignition

encenegar *vt* HIDROL silt up

encenegarse *v refl* HIDROL silt up

encentar *vt* TRANSP MAR *fardo* broach

encepado: ~ **de cabezas de pilotes** *m* CARBÓN pile cap

enceparse *v refl* TRANSP MAR *ancla, cuerda, motor* foul

encerado *m* CINEMAT waxing, P&C, TEXTIL oilskin, TRANSP MAR tarpaulin

enceradora *f* CINEMAT waxing machine

encerar *vt* CINEMAT wax, MECÁ polish, REVEST wax

encerrado *adj* ING MECÁ enclosed

encerramiento: ~ **de una sola nave** *m* PROD single-bay enclosure

encerrar *vt* CONTAM MAR entrap, ING MECÁ encompass; ~ **entre corchetes** *vt* IMPR bracket

enchapado: ~ **en oro** *adj* QUÍMICA gold-plated

enchapar *vt* MECÁ plate

enchaquetado *adj* MECÁ jacketed

enchaquetar *vt* PROD jacket

encharcamiento *m* CONTAM lagooning, NUCL water-logging

enchavetado[1] *adj* CONST, ING MECÁ, MECÁ, PROD keyed

enchavetado[2] *m* CONST keying, ING MECÁ forelocking, keying, wedging, MECÁ, PROD keying; ~ **interior** *m* CONST *carpintería* keying in; ~ **plano** *m* ING MECÁ key on flat

enchavetar *vt* CONST, ING MECÁ, MECÁ, PROD key

enchufable *adj* ING ELÉC pluggable, PROD plug-in

enchufado[1] *adj* ING ELÉC jacked in; ~**-desenchufado** *adj* ING MECÁ on-off; ~ **a la red** *adj* ELEC, ING ELÉC current-operated (*AmE*), mains-operated (*BrE*)

enchufado[2] *m* TEC PETR stabbing

enchufar *vt* CINEMAT, ELEC, ING ELÉC, TELECOM, TV plug in, switch on

enchufe *m* CINEMAT, CONST plug, ELECTRÓN, FÍS socket, plug, MECÁ socket, TELECOM outlet, socket outlet; ~ **acoplador y zócalo conector** *m* ELEC coupler plug y socket connection; ~ **de aire a presión** *m* ING MECÁ compressed-air socket; ~ **de bayoneta** *m* ELEC bayonet socket; ~ **de cierre por torsión** *m* CINEMAT twist lock plug; ~ **compatible** *m*

ING ELÉC compatible plug; ~ **conector para remolques** *m* AUTO, ELEC connector socket for trailer; ~ **de conexión a la red** *m* ELEC, ING ELÉC current plug (*AmE*), mains plug (*BrE*); ~ **de corriente** *m* ELEC, ING ELÉC current socket (*AmE*), mains socket (*BrE*); ~ **de la corriente** *m* ELEC, ING ELÉC current plug (*AmE*), mains plug (*BrE*); ~ **de cuatro direcciones** *m* ELEC, ING ELÉC four-way extension socket; ~ **de encendido y apagado** *m* CINEMAT on-off switch; ~ **con energía de alimentación** *m* ELEC power outlet; ~ **de extensión** *m* ING MECÁ extension socket; ~ **hembra** *m* CINEMAT, ELEC, ING ELÉC, LAB, PROD, TELECOM electric socket, female connector, jack, jack socket, socket; ~ **hembra de acción inmediata** *m* ING ELÉC snap-in socket; ~ **hembra de montaje exterior** *m* ELEC surface-mounted socket; ~ **hembra múltiple** *m* ELEC multiple socket, ING ELÉC jack; ~ **hembra a prueba de choques eléctricos** *m* ELEC, SEG shockproof socket; ~ **hembra a prueba de sacudidas eléctricas** *m* ELEC, SEG shockproof socket; ~ **hembra a rosca** *m* ING ELÉC socket; ~ **hembra de seguridad** *m* CINEMAT safety jack; ~ **hembra sobresaliente** *m* ELEC surface-mounted socket; ~ **hembra superficial** *m* ELEC surface socket; ~ **hembra de tres alfileres** *m* ELEC three-pin socket; ~ **hembra de tres clavijas** *m* ELEC three-pin socket; ~ **hermético** *m* ELEC watertight socket outlet; ~ **hermético al agua** *m* ELEC watertight socket outlet; ~ **impermeable** *m* ELEC watertight socket outlet; ~ **impermeable al agua** *m* ELEC watertight socket outlet; ~ **macho** *m* CINEMAT jack plug, ELEC electric plug, plug, ING ELÉC electric plug, male connector, plug, LAB electric plug, PROD plug, TELECOM male plug; ~ **macho con contactos blindados** *m* ELEC socket with shrouded contacts; ~ **del martillo** *m* CONST hammer plug; ~ **del micrófono** *m* CINEMAT mike tap; ~ **para módulo de memoria** *m* PROD memory module socket; ~ **múltiple** *m* ELEC adaptor; ~ **mural** *m* ELEC wall outlet, wall socket; ~ **mural de luz** *m* ELEC light wall socket; ~ **de pared** *m* ELEC *conexión* wall socket; ~ **de pesca** *m* TEC PETR *perforación* overshot; ~ **polarizado** *m* ING ELÉC polarized plug; ~ **portátil** *m* ELEC portable-socket outlet, ING MECÁ extension socket; ~ **reglamentario** *m* ELEC Home Office socket; ~ **supletorio** *m* ING MECÁ extension socket; ~ **tomacorriente** *m* ELEC *conexión*, ING ELÉC current plug (*AmE*), mains plug (*BrE*), PROD plug-in switch

encierro *m* AGRIC confinement

encintado *m* IMPR stripping; ~ **de cerrojo** *m* CONST keeper

enclaustrado *adj* ING MECÁ enclosed

enclavado *adj* PROD interlocked

enclavamiento *m* CONST locking, INFORM&PD latch, ING MECÁ locking, interlocking, MECÁ interlocking, NUCL, PROD interlock, TRANSP AÉR interlock control, TV interlock; ~ **condicional** *m* FERRO permissive block; ~ **eléctrico** *m* ELEC, PROD electric interlock; ~ **de estado sólido** *m* FERRO Solid State Interlocking (*BrE*) (*SSI*); ~ **de fase** *m* TELECOM phase locking; ~ **del itinerario** *m* FERRO route locking; ~ **mecánico** *m* ING MECÁ mechanical locking, PROD mechanical interlock; ~ **mediante relés controlados por computadora** *m* AmL (*cf enclavamiento mediante relés controlados por ordenador Esp*) FERRO, INFORM&PD computer-controlled all-relay interlock-ing; ~ **mediante relés controlados por ordenador** *m Esp* (*cf enclaviamiento mediante relés controlados por computadora AmL*) FERRO, INFORM&PD computer-controlled all-relay interlocking; ~ **de seguridad** *m* ING MECÁ safety interlock; ~ **vertical** *m* TV vertical lock

enclavar *vt* FERRO lock, INFORM&PD latch, ING ELÉC interlock, ING MECÁ lock, interlock, MECÁ lock

enclave *m* AGUA offset, PETROL enclave

enclavijamiento *m* CONST, PROD pegging

enclavijar *vt* CONST, PROD dowel, peg

encofrado *m* CONST falsework, formwork, *hormigón* shuttering, MECÁ casing; ~ **deslizante** *m* CONST sliding formwork; ~ **de madera** *m* PROD crib; ~ **metálico** *m* CONST steel form; ~ **móvil** *m* CONST moving formwork; ~ **de tablas** *m* CONST strip form-work

encoger *vti* C&V, TERMO, TEXTIL shrink

encogible *adj* C&V, TERMO, TEXTIL shrinkable

encogimiento *m* C&V, CINEMAT shrinkage, CONST shrinking, FÍS shrinkage, GEOL shortening, METAL constriction, shrinkage, P&C, PAPEL shrinkage, PROD, TELECOM shrinkage, shrinking, TEXTIL shrinkage; ~ **diferencial** *m* TEXTIL differential shrinkage; ~ **de la orilla** *m* C&V edge creep; ~ **residual** *m* TEXTIL residual shrinkage

encolado *m* EMB, IMPR gluing, PAPEL gluing, sizing, PROD *soldadura* gluing, shutting together, shutting up, TEXTIL sizing; ~ **en caliente** *m* EMB hot gluing; ~ **interno** *m* PAPEL beater sizing; ~ **en masa** *m* PAPEL beater sizing, stock sizing, stuff sizing; ~ **de plegador a plegador** *m* TEXTIL beam-to-beam sizing; ~ **de un solo cabo** *m* TEXTIL single-end sizing; ~ **superficial** *m* PAPEL surface sizing; ~ **de la urdimbre en plegador múltiple** *m* TEXTIL multiple beam sizing, multiple beam slashing; ~ **de urdimbre seccional** *m* TEXTIL sectional warp sizing, sectional warp slashing

encoladora *f* CINEMAT joiner, EMB binding machine, gluing machine, gumming machine, IMPR gluing machine, TEXTIL sizing machine; ~ **por aire caliente** *f* TEXTIL hot air sizing machine

encoladura *f* EMB, IMPR, PAPEL, PROD gluing

encolar[1]: **sin ~** *adj* PAPEL unsized

encolar[2] *vt* EMB, P&C, PAPEL glue, PROD gum, *soldadura* shut up, QUÍMICA glue, TEXTIL glue, size

encomendar *vt* CONST commission

encontrar[1] *vt* MINAS *un filón* strike

encontrar[2]: ~ **temporal** *vi* TRANSP MAR *barco* make heavy weather

encorchadora: ~ **de botellas** *f* EMB bottle-corking machine

encorchetar *vt* ING MECÁ clasp, hook

encorvado *adj* C&V bent

encorvar *vt* PROD curve

encribado *m* PROD cribwork

encrinita *f* GEOL encrinite

encriptación *f* ELECTRÓN, INFORM&PD, TEC ESP encryption, TELECOM encryption, scrambling, TV encryption; ~ **de datos** *f* INFORM&PD data encryption

encriptador *m* TELECOM scrambler; ~ **de voz digital lineal** *m* TELECOM linear digital-voice scrambler

encriptar *vt* D&A scramble, ELECTRÓN, INFORM&PD encrypt, TELECOM encrypt, scramble, TV encrypt

encristalado *m* PROD glazing

encristalar *vt* PROD glaze

encuadernación *f* IMPR bookbinding; ~ **americana** *f* IMPR perfect binding, unsewn binding; ~ **con bordes cerrados** *f* IMPR yapp binding; ~ **en cartoné** *f* IMPR inboard binding; ~ **sin costura** *f* IMPR perfect binding, unsewn binding; ~ **manual** *f* IMPR handbinding; ~ **a media piel** *f* IMPR halfbinding, quarterbinding; ~ **en piel de becerro** *f* IMPR calf binding; ~ **en rústica** *f* IMPR paper binding; ~ **de tapas blandas** *f* IMPR soft-cover binding; ~ **de tapas flexibles** *f* IMPR limp binding

encuadernado: ~ **sin costura** *adj* IMPR perfect-bound; ~ **flexible** *adj* IMPR limp-bound; ~ **a media pasta** *adj* IMPR half-bound; ~ **a media piel** *adj* IMPR half-bound, quarter-bound; ~ **piel** *adj* IMPR full-bound; ~ **con tapas blandas** *adj* IMPR soft-bound; ~ **de tapas duras** *adj* IMPR hard-bound; ~ **a toda piel** *adj* IMPR full-bound

encuadernador *m* EMB binder, IMPR bookbinder

encuadernadora *f* EMB binding machine; ~ **sin costura** *f* IMPR perfect binder

encuadernar *vt* EMB, IMPR bind

encuadrar *vt* CINEMAT, FOTO frame, ING MECÁ align, TV frame

encuadre *m* CONST mapping; ~ **ajustado** *m* CINEMAT tight framing

encuelladero *m* TEC PETR derrick monkey board, footboard (*AmE*), monkey board (*BrE*), stabbing board

encuellador *m* TEC PETR derrick man

encuentro *m* MINAS *de un filón, de una veta* strike; ~ **del electrón-positrón** *m* FÍS PART electron-positron interaction; ~ **espacial** *m* TEC ESP space rendezvous; ~ **en órbita terrestre** *m* TEC ESP earth-orbit rendezvous

encuñarse *vi* GEOL wedge out

endentamiento *m* ING MECÁ toothing

endentar *vt* ING MECÁ engage, mesh, ratch, MECÁ engage

enderezador *m* ING MECÁ straightener; ~ **de alambre** *m* CONST wire stretcher; ~ **pantógrafo** *m* FERRO, ING MECÁ pantograph dresser

enderezadora *f* ING MECÁ straightening machine

enderezamiento *m* ING MECÁ rectification, PROD *tubos curvados* straightening, TRANSP AÉR flare-out

enderezar *vt* CONST straighten, ING MECÁ true, PROD *varilla* straighten

enderezatubos *m* PROD swage

endoatmosférico *adj* TEC ESP endoatmospheric

endoenzima *f* QUÍMICA endoenzyme

endogenético *adj* PETROL endogenetic

endógeno *adj* PETROL endogenic

endolinfa *f* ACÚST endolymph

endomorfo *m* MINERAL endomorph

endoscopia *f* FÍS, INSTR, NUCL, TEC ESP endoscopy

endoscopio *m* FÍS, INSTR, NUCL, TEC ESP endoscope

endosperma *f* AGRIC, ALIMENT endosperm

endotérmico *adj* C&V, QUÍMICA, TEC ESP, TEC PETR endothermic, TERMO heat-absorbing

endulzado *adj* ALIMENT sweetened

endulzante *m* ALIMENT sweetener

endurecedor *m* CINEMAT, FOTO, P&C *aditivo* hardener; ~ **de amidas** *m* P&C *adhesivos, revestimientos* amide hardener; ~ **de anhídrido** *m* P&C *adhesivos, resinas, revestimientos* anhydride hardener; ~ **rápido** *m* REFRIG *para helados* rapid hardener

endurecer *vt* PROD, TERMO harden;

~ **estructuralmente** *vt* TERMO precipitation harden; ~ **en frío** *vt* PROD, TERMO cold-harden; ~ **por precipitación** *vt* TERMO precipitation harden; ~ **por solubilización** *vt* TERMO *aleaciones* precipitation harden; ~ **superficialmente** *vt* PROD chill harden

endurecerse *v refl* PROD harden; ~ **por envejecimiento** *v refl* METAL age

endurecido[1] *adj* ING MECÁ hard-faced; **no** ~ *adj* P&C uncured; ~ **por envejecimiento** *adj* TERMO aged

endurecido[2]: ~ **por zonas** *m* C&V zone toughening

endurecimiento *m* CARBÓN hydrogenation, *hormigón, resinas* setting, CINEMAT tanning, CONST *hormigón*, CRISTAL hardening, GEOL induration, MECÁ curing, METAL hardening, P&C *pinturas, revestimientos* hardening, *polímeros* set, REFRIG *de helados* hardening, TRANSP MAR *material* curing; ~ **por acritud** *m* CRISTAL, MECÁ, METAL work hardening; ~ **por acritud lineal** *m* METAL linear work hardening; ~ **por compresión** *m* P&C *polímeros* compression set; ~ **por contacto** *m* REFRIG *de los helados*, TERMO contact hardening; ~ **por deformación** *m* CRISTAL, MECÁ, METAL work hardening; ~ **por deformación en frío** *m* METAL *acritud* strain hardening; ~ **por dispersión** *m* METAL dispersion hardening; ~ **por envejecimiento** *m* METAL, P&C age hardening; ~ **del esmalte** *m* C&V hardening on the glazing; ~ **estructural** *m* CRISTAL age hardening, precipitation hardening, PROD precipitation hardening; ~ **por fatiga** *m* METAL fatigue hardening; ~ **en frío** *m* P&C *adhesivos, polímeros* cold setting; ~ **por haz de electrones** *m* IMPR *tintas, barnices*, NUCL electron-beam curing (*EBC*); ~ **por haz electrónico** *m* IMPR *tintas, barnices*, NUCL electron-beam curing (*EBC*); ~ **por inducción** *m* ING ELÉC, METAL induction hardening; ~ **por irradiación** *m* METAL irradiation hardening; ~ **por medios mecánicos** *m* CRISTAL, MECÁ, METAL work hardening; ~ **por medios mecánicos lineales** *m* METAL linear work hardening; ~ **con nitruro** *m* MECÁ nitride hardening; ~ **por nitruro de titanio** *m* ING MECÁ titanium nitride hardening; ~ **por obstaculización** *m* METAL obstacle hardening; ~ **por precipitación** *m* TERMO precipitation hardening; ~ **por puesta en orden** *m* METAL *aleaciones* order hardening; ~ **químico** *m* METAL, QUÍMICA chemical hardening; ~ **por radiación** *m* FÍS RAD, ING ELÉC radiation hardening, METAL irradiation hardening, NUCL radiation hardening; ~ **rápido** *m* P&C fast curing; ~ **residual** *m* P&C residual set; ~ **retardado** *m* TERMO delayed hardening; ~ **por saturación** *m* METAL saturation hardening; ~ **por solubilización de un componente** *m* METAL precipitation hardening; ~ **superficial** *m* MECÁ hard facing, METAL surface hardening, PROD *funderías* chill hardening; ~ **de la superficie** *m* CONST set; ~ **por temple** *m* METAL, NUCL quench hardening

enengrecido *adj* PAPEL blackened

energía *f* GEN energy, power; ~ **absorbida** *f* FÍS RAD, METAL absorbed energy; ~ **activa** *f* ELEC *de sistema* active energy, ING ELÉC active power; ~ **de activación** *f* CRISTAL, FÍS RAD, METAL, NUCL activation energy; ~ **acumulada** *f* METAL stored energy; ~ **acústica** *f* ACÚST, ING ELÉC acoustic energy, sound energy; ~ **acústica instantánea por unidad de volumen** *f* ACÚST instantaneous acoustic energy per unit volume; ~ **de adherencia** *f* METAL bond energy; ~ **aerodinámica** *f* ENERG RENOV aerodynamic power;

~ **almacenada** *f* FÍS, FÍS RAD, METAL stored energy, PROD stored-up energy; ~ **aparente** *f* ELEC apparent energy, FÍS, ING ELÉC apparent power; ~ **calorífica** *f* TERMO heat energy; ~ **de cambio** *f* METAL exchange energy; ~ **cinética** *f* FÍS, GEOFÍS, ING MECÁ, MECÁ, TEC ESP kinetic energy; ~ **cinética acústica instantánea por unidad de volumen** *f* ACÚST instantaneous acoustic kinetic energy per unit volume; ~ **cinética angular** *f* ING MECÁ angular kinetic energy; ~ **cinética lineal** *f* ING MECÁ linear kinetic energy; ~ **cinética de onda por metro de cresta** *f* ENERG RENOV wave momentum per metre of crest; ~ **cinética de rotación** *f* ING MECÁ angular kinetic energy; ~ **de cohesión** *f* METAL cohesive energy; ~ **de colisión** *f* FÍS PART collision energy; ~ **de combustión** *f* TERMO combustion energy; ~ **compleja** *f* ELEC complex power; ~ **de contracción** *f* METAL constriction energy, constriction power; ~ **convergente** *f* FÍS ONDAS converging energy; ~ **crítica** *f* METAL critical energy, threshold energy; ~ **en cuadratura** *f* ELEC quadrature power; ~ **culombiana** *f* FÍS RAD coulomb energy; ~ **derivada** *f* TEC PETR derived energy; ~ **de desarrollo** *f* METAL formation energy; ~ **de desintegración** *f* FÍS, FÍS PART, FÍS RAD disintegration energy; ~ **de desintegración alfa** *f* FÍS, FÍS PART, FÍS RAD alpha disintegration energy; ~ **de desintegración beta** *f* FÍS, FÍS PART, FÍS RAD beta disintegration energy; ~ **disipada** *f* ELEC *pérdidas* dissipated power; ~ **disipada sin efectuar trabajo útil** *f* ING MECÁ loss; ~ **de disociación** *f* GAS dissociation energy; ~ **disponible** *f* ENERG RENOV available power; ~ **eficaz** *f* ELEC effective power; ~ **eléctrica** *f* GEN electric energy, electrical energy, electric power, electrical power; ~ **electrocinética** *f* ELEC, FÍS, ING ELÉC, QUÍMICA electrokinetic energy; ~ **electromagnética** *f* ELEC, FÍS, ING ELÉC electromagnetic energy; ~ **electroquímica** *f* ELEC, ING ELÉC, QUÍMICA electrochemical energy; ~ **de enlace** *f* CRISTAL, FÍS PART, FÍS RAD binding energy (*BE*), METAL bond energy, QUÍMICA binding energy (*BE*); ~ **de enlace ficticia** *f* NUCL fictitious binding energy; ~ **de enlace media** *f* NUCL mean bond energy; ~ **de entrada** *f* ELECTRÓN inputting; ~ **de entrada óptica** *f* ING ELÉC optical input power; ~ **eólica** *f* ENERG RENOV, FÍS wind energy, METEO wind power; ~ **de escisión del campo cristalino** *f* CRISTAL, FÍS RAD crystal-field splitting energy; ~ **específica** *f* GAS specific energy, TEC ESP specific energy, specific power; ~ **específica interna** *f* FÍS specific internal energy; ~ **de excitación** *f* FÍS RAD excitation energy; ~ **de Fermi** *f* FÍS Fermi energy; ~ **de formación** *f* METAL formation energy; ~ **de fotones** *f* FÍS PART photon energy; ~ **fotónica** *f* FÍS PART photon energy; ~ **geotérmica** *f* ENERG RENOV geothermal power, FÍS, GEOFÍS geothermal energy; ~ **por haz láser** *f* ELECTRÓN laser-beam energy; ~ **hidráulica** *f* AGUA, ENERG RENOV water power, INSTAL HIDRÁUL hydraulic power, water power, TERMO, TEXTIL water power; ~ **hidroeléctrica** *f* ELEC, ENERG RENOV hydroelectric power, TERMO water power; ~ **de impacto** *f* METAL impact energy; ~ **de interacción** *f* METAL interaction energy; ~ **de interfase** *f* METAL interface energy; ~ **interna** *f* FÍS, FÍS RAD *termodinámica* internal energy; ~ **interna molar** *f* FÍS molar internal energy; ~ **de ionización** *f* FÍS, FÍS RAD *de un átomo*, GAS

ionization energy; ~ **liberada** *f* CARBÓN yield; ~ **libre** *f* FÍS, METAL, QUÍMICA, TERMO free energy; ~ **libre de Gibbs** *f* FÍS Gibbs' free energy; ~ **libre de Helmholtz** *f* FÍS Helmholtz free energy; ~ **luminosa** *f* FÍS ONDAS luminous energy; ~ **de la luz** *f* FÍS RAD light energy; ~ **magnética** *f* ELEC, FÍS, ING ELÉC magnetic energy; ~ **de la marea** *f* ENERG RENOV tidal power, OCEAN tidal energy; ~ **de mareas** *f* FÍS tidal energy; ~ **máxima disponible** *f* ING MECÁ available power; ~ **mecánica** *f* ING MECÁ mechanical energy; ~ **metabolizable** *f* AGRIC metabolizable energy (*ME*); ~ **necesaria del circuito** *f* ELEC circuit power requirement; ~ **neta** *f* AGRIC net energy (*NE*); ~ **nuclear** *f* ELEC nuclear energy, FÍS nuclear energy, nuclear power, NUCL nuclear power; ~ **de las olas** *f* OCEAN wave energy; ~ **de onda** *f* ENERG RENOV wave power, FÍS ONDAS wave energy; ~ **potencial** *f* FÍS potential energy; ~ **potencial acústica instantánea por unidad de volumen** *f* ACÚST instantaneous acoustic potential energy per unit volume; ~ **potencial electrostática** *f* GEOFÍS electrostatic potential energy; ~ **producida por la unidad** *f* NUCL unit output; ~ **del proyectil** *f* D&A projectile energy; ~ **de punto cero** *f* FÍS zero point energy; ~ **del punto cero** *f* FÍS RAD zero point energy; ~ **radiada** *f* FÍS RAD radiated energy; ~ **radiante** *f* FÍS, FÍS RAD, TELECOM radiant energy; ~ **reactiva** *f* ELEC *sistema de CA*, ING ELÉC reactive energy; ~ **de referencia** *f* ACÚST reference energy; ~ **residual** *f* INSTAL TERM residual energy, TERMO waste energy; ~ **de retroceso** *f* D&A recoil energy; ~ **rotacional** *f* PROD rotational energy; ~ **de rozamiento** *f* METAL friction force; ~ **de Rydberg** *f* FÍS Rydberg energy; ~ **sensible** *f* GAS sensitive energy; ~ **de simetría nuclear** *f* FÍS RAD, NUCL nuclear symmetry energy; ~ **solar** *f* ENERG RENOV, FÍS, ING ELÉC solar energy; ~ **sonora** *f* ACÚST sound energy; ~ **suministrada** *f* PROD output; ~ **superficial** *f* FÍS surface energy; ~ **térmica** *f* TERMO heat energy, thermal energy, thermal power, TERMOTEC thermal energy; ~ **térmica oceánica** *f* OCEAN ocean thermal energy; ~ **térmica producida** *f* TERMO heat output; ~ **termoeléctrica** *f* FÍS, ING ELÉC thermoelectric power; ~ **de toma** *f* ELEC *de devanado* tapping power; ~ **por unidad de superficie** *f* TEC ESP power per unit area; ~ **de unificación electrodébil** *f* FÍS, FÍS PART electroweak unification energy; ~ **de unión** *f* CRISTAL, FÍS PART, FÍS RAD binding energy (*BE*), METAL bond energy, QUÍMICA binding energy (*BE*); ~ **vibracional** *f* NUCL vibrational energy; ~ **del viento** *f* ENERG RENOV wind power; ~ **del yacimiento** *f* PETROL, TEC PETR reservoir energy

energización *f* ELEC, FÍS RAD, ING ELÉC, PROD energization

energizar *vt* ELEC, FÍS RAD, ING ELÉC, PROD power up

enfajadora *f* EMB paper-banding machine

enfajillado *m* EMB *periódicos, revistas* banding

enfajilladora *f* EMB banding machine

enfaldillado *m* PROD flanging

enfaldilladora *f* PROD flanger

enfaldillar *vt* PROD flange

enfardador *m* ING MECÁ packer

enfardadora *f* AGRIC baler, EMB baling press; ~ **de fardos cilíndricos** *f* AGRIC roll baler; ~ **de heno** *f* AGRIC hay baler

enfardonadora *f* EMB baling press

enfermedad: ~ **del buzo** *f* OCEAN, TRANSP MAR bends;

~ de los cajones *f* OCEAN, TRANSP MAR *buceo* caisson disease; **~ de las cámaras de sumersión** *f* OCEAN, TRANSP MAR *buceo* caisson disease; **~ del césped** *f* AGRIC lawn disease; **~ de las plantas** *f* AGRIC plant disease; **~ del trebol de olor** *f* AGRIC sweet clover disease; **~ de la vaca loca** *f* AGRIC mad cow disease; **~ viral** *f* AGRIC virus disease

enfermería: **~ de primeros auxilios** *f* SEG first-aid treatment room

enfilación *f* TRANSP MAR *cable, equipamiento eléctrico, navegación* range, *navegación* leading line, transit, *navegación costera* alignment

enfilar *vt* PROD thread

enfocado[1] *adj* CINEMAT, ELECTRÓN, FÍS, FÍS ONDAS, TV, FOTO focused, in focus

enfocado[2] *adv* CINEMAT, ÓPT, TV in focus

enfocador *m* FOTO, INSTR finder

enfocar[1] *vt* GEN adjust focus, focus; **~ constantemente** *vt* CINEMAT pull focus; **~ con cremallera** *vt* CINEMAT rack focus; **~ al infinito** *vt* FOTO focus for infinity; **~ en movimiento** *vt* CINEMAT follow focus

enfocar[2]: **~ en movimiento** *vi* TV follow focus

enfoque *m* CINEMAT, ELECTRÓN, FOTO, TV focus, focusing; **~ automático** *m* CINEMAT autofocus, FOTO autofocus, automatic focusing; **~ continuo** *m* ING ELÉC searchlight; **~ por control remoto** *m* TV remote-control focusing; **~ diferencial** *m* CINEMAT differential focusing; **~ electromagnético** *m* ING ELÉC electromagnetic focusing; **~ electrostático** *m* ING ELÉC electrostatic focusing; **~ estático** *m* TV static focus; **~ del haz** *m* ELECTRÓN beam focusing; **~ de haz de electrones** *m* ELECTRÓN electron-beam focusing; **~ magnético** *m* ING ELÉC magnetic focusing; **~ por mando a distancia** *m* TV remote-control focusing; **~ en película** *m* CINEMAT, TV focus on film; **~ a través del objetivo** *m* CINEMAT through-the-lens focusing; **~ ultrafino** *m* NUCL ultrafine focus

enfriado[1] *adj* ALIMENT, REFRIG, TERMO chilled, cooled; **~ con aceite** *adj* TERMO oil-cooled; **~ por agua** *adj* GAS, TERMO water-cooled; **~ al aire** *adj* TERMO air-cooled; **~ por aire** *adj* ELEC, ING MECÁ, PAPEL air-cooled; **~ por chorro de aire** *adj* ING MECÁ air-cooled; **~ con gas** *adj* GAS, REFRIG, TERMO gas-cooled; **~ con hielo** *adj* TERMO ice-cooled; **~ con líquido** *adj* TERMO liquid-cooled

enfriado[2]: **~ por aire forzado** *m* ING MECÁ, PROD, REFRIG forced-air cooling; **~ por aire a presión** *m* ING MECÁ, PROD, REFRIG forced-air cooling; **~ a presión** *m* ING MECÁ, PROD, REFRIG forced cooling

enfriador[1] *adj* ALIMENT, ING MECÁ, REFRIG refrigerating

enfriador[2] *m* ING MECÁ coolant, cooler, NUCL *de los sistemas auxiliares*, PROD, REFRIG, TEC PETR, TERMO, TERMOTEC cooler; **~ de aceite** *m* ELEC *transformador*, REFRIG, TERMO oil cooler; **~ con acumulación de hielo** *m* REFRIG ice bank cooler; **~ de agua** *m* REFRIG water chiller; **~ de agua potable** *m* REFRIG drinking-water cooler; **~ de aire** *m* NUCL, REFRIG air cooler; **~ de aire de calor sensible** *m* REFRIG sensitive heat air cooler; **~ de aire con relleno húmedo** *m* REFRIG wetted-pad evaporative cooler; **~ con aletas dobles** *m* REFRIG two-way finned cooler; **~ del anillo batidor** *m* REFRIG sparge ring cooler; **~ de botella** *m* REFRIG bottle-type liquid cooler; **~ de botellas** *m* REFRIG bottle cooler; **~ de cerveza** *m* REFRIG beer cooler; **~ de chimenea** *m* AmL (*cf torre de enfriado Esp*); MINAS cooling tower; **~ de convección** *m* REFRIG, TERMO convection cooler; **~ de cortina** *m* REFRIG irrigation cooler; **~ del fluido** *m* REFRIG fluid cooler; **~ de gases** *m* GAS, REFRIG gas cooler; **~ de horquilla** *m* C&V hairpin cooler; **~ húmedo de aire** *m* REFRIG wet-type cooler; **~ intermedio** *m* REFRIG interstage cooler; **~ intermedio de expansión** *m* REFRIG flashstage cooler; **~ de leche** *m* AGRIC, REFRIG milk cooler; **~ de leche con anillo de aspersión** *m* AGRIC, REFRIG sparge ring type milk cooler; **~ de leche por aspersión** *m* AGRIC, REFRIG cascade milk cooler; **~ de leche en cántaros** *m* AGRIC, REFRIG churn milk cooler; **~ de leche por inmersión** *m* AGRIC, REFRIG immersion milk cooler; **~ de leche por turbina** *m* AGRIC, REFRIG turbine milk cooler; **~ de líquido** *m* ING MECÁ liquid cooler, REFRIG liquid chiller; **~ de mosto** *m* REFRIG *de cerveza* wort cooler; **~ por pulverización** *m* REFRIG spray-type cooler; **~ radiante** *m* REFRIG, TERMO radiant cooler; **~ de salmuera** *m* REFRIG brine cooler; **~ de tambor** *m* REFRIG drum cooler; **~ de tambor-agitador** *m* REFRIG spin chiller; **~ de tipo seco** *m* REFRIG dry-type cooler; **~ de tubos de aletas** *m* REFRIG, TERMO finned cooler

enfriamiento[1]: **con ~ por aire** *adj* ING MECÁ air-cooled; **de ~ por aire** *adj* ELEC air-cooled

enfriamiento[2] *m* ALIMENT refrigeration, C&V, GAS, ING MECÁ cooling, PROD *hornos altos* slacking, REFRIG, TERMO chilling, cooling, refrigeration, TEXTIL cooling; **~ por ablación** *m* METEO ablation; **~ ablacionante** *m* TEC ESP ablative cooling; **~ del aceite** *m* ELEC oil cooling; **~ en aceite** *m* TERMO oil quenching; **~ por aceite** *m* ING MECÁ oil cooling; **~ de agregados** *m* REFRIG aggregate cooling; **~ por agua** *m* METAL water quenching; **~ por agua a presión** *m* ING MECÁ, REFRIG forced-water cooling; **~ por aire** *m* ING ELÉC, ING MECÁ air cooling; **~ por aire o aceite** *m* TERMO air-or-oil cooling; **~ aleatorio** *m* FÍS PART stochastic cooling; **~ de ambientes** *m* REFRIG environment cooling; **~ artificial** *m* ALIMENT, REFRIG, TERMO refrigeration; **~ por aspersión** *m* REFRIG spray cooling; **~ bifásico** *m* NUCL two-phase cooling; **~ de la camisa** *m* NUCL jacket cooling; **~ de la chispa** *m* ING ELÉC spark quenching; **~ de choque** *m* REFRIG shock chilling; **~ por chorro de aire frio** *m* REFRIG jet cooling; **~ por circulación forzada** *m* REFRIG forced-draft cooling (*AmE*), forced-draught cooling (*BrE*); **~ por condensación superficial** *m* REFRIG, TEC ESP sweat cooling; **~ por convección** *m* REFRIG, TERMO convection cooling; **~ por convección natural** *m* NUCL, REFRIG natural-convection cooling; **~ por corriente de aire** *m* REFRIG air-blast cooling; **~ criogénico** *m* REFRIG cryocooling; **~ en el crisol** *m* C&V pot cooling; **~ de la cubierta** *m* TRANSP film cooling; **~ por descarga** *m* TEC ESP dump cooling; **~ de electrones** *m* ELECTRÓN, FÍS PART electron cooling; **~ estocástico** *m* FÍS PART stochastic cooling; **~ por evaporación** *m* REFRIG, TERMO evaporative cooling; **~ evaporativo** *m* PROC QUÍ evaporation cooling; **~ con hielo** *m* REFRIG icing; **~ por inmersión** *m* REFRIG flood-type cooling; **~ de la lámina** *m* TEC ESP film cooling; **~ lento** *m* METAL, REFRIG slow quenching; **~ localizado** *m* REFRIG spot cooling; **~ magnético** *m* REFRIG magnetic cooling; **~ mediante superficie fría** *m* REFRIG surface cooling; **~ natural al aire libre** *m* TERMO natural cooling; **~ nuclear** *m* NUCL, REFRIG nuclear cooling; **~ por porosidad del metal** *m* TEC ESP sweat cooling; **~ radiante** *m* REFRIG, TERMO radiant cooling; **~ rápido** *m* METAL, REFRIG quenching, TERMO rapid cooling; **~ rápido por aire** *m* REFRIG, TERMO rapid air cooling; **~ por recuperación** *m* REFRIG, TERMO regenerative cooling; **~ por rociado** *m* C&V splat cooling; **~ a una temperatura baja** *m* REFRIG, TERMO cooling to low temperature; **~ termoeléctrico** *m* REFRIG thermoelectric cooling; **~ con tiro natural** *m* TERMO natural-draft cooling (*AmE*), natural-draught cooling (*BrE*); **~ por transpiración** *m* GAS, REFRIG transpiration cooling, TEC ESP sweat cooling; **~ de vacío** *m* REFRIG vacuum cooling; **~ por ventilador** *m* REFRIG, TERMO fan cooling

enfriar[1] *vt* ALIMENT, C&V, ING MECÁ, REFRIG, TERMO chill, refrigerate; **~ en aceite** *vt* TERMO oil quench; **~ con agua** *vt* QUÍMICA water-cool; **~ al aire** *vt* TERMO air-cool; **~ repentinamente** *vt* MECÁ quench; **~ por ventilador** *vt* REFRIG, TERMO fan cool

enfriar[2] *vti* MECÁ chill

enfurtido *m* TEXTIL *operación de acabado* milling

engalanar *vt* TRANSP MAR dress ship overall

engalgar *vt* TRANSP MAR *ancla* back

enganchado *m* PAPEL threading of paper

enganchador *m* MINAS dog, hoisting dog, *obrero* onsetter (*BrE*), platman (*AmE*); **~ de sondas** *m* MINAS grab; **~ de vagones** *m* FERRO shunter

enganchar *vt* CINEMAT rack, FERRO couple, ING MECÁ attach, clasp, engage, hook, lock, trip in, MECÁ engage, hitch, MINAS hitch, *en la cámara de enganche* cut a plat, cut a station, TEC PETR *perforación* latch

enganche *m* AUTO insert, FERRO coupling, ING ELÉC, ING ELÉC coupling, joint, *electricidad* crawling, ING MECÁ trip catch, coupling link, joint, couple, link, MECÁ hitch, shackle, MINAS *Esp* (*cf muesca en la roca AmL, filón para sostener una apea AmL*) hitch, onsetting, TRANSP coupling, TRANSP AÉR hold, lock-on, TV crawling, VEH hitch, coupling; **~ acústico** *m* TELECOM squealing; **~ articulado** *m* TRANSP articulated coupling; **~ automático** *m* FERRO automatic coupling; **~ de barra** *m* MINAS *sondeos* rod joint; **~ de barras** *m* MINAS *sondeos* rod coupling; **~ de cadena** *m* ING MECÁ chain coupling; **~ de la calle** *m* CARBÓN *pozos de minas* processing, MINAS *pozos* bank, pit eye, pit mouth; **~ del capó** *m* AUTO, VEH bonnet catch (*BrE*), hood catch (*AmE*); **~ de capota del techo** *m* VEH top catch; **~ de carga** *m* TRANSP AÉR load hook-up; **~ de la compuerta del tren de aterrizaje** *m* TRANSP AÉR landing gear door latch; **~ contra ráfagas** *m* TEC ESP, TRANSP AÉR, TRANSP MAR gust lock; **~ D** *m* FERRO D-link; **~ de fase** *m* TELECOM phase locking; **~ de frecuencia** *m* TEC ESP frequency tracking; **~ inferior** *m* MINAS *pozo* plat; **~ superior** *m* CARBÓN *minas* landing, MINAS apex *jaula de extracción* banking, *jaula de mina* landing stage, pit bank; **~ del tren de aterrizaje** *m* TRANSP AÉR landing gear uplock; **~ de vagoneta Baum** *m* CARBÓN Baum jig; **~ de vagonetas** *m* CARBÓN, MINAS jig

enganchón *m* TEXTIL *malla enganchada* snag

engarce *m* ING MECÁ crimping bush

engarzador: **~ de cartuchos** *m* ING MECÁ crimping tool; **~ terminal** *m* ING MECÁ terminal crimper

engarzar *vt* ING MECÁ set

engastar *vt* ING MECÁ set

engaste *m* MECÁ insert

engomado[1] *adj* P&C, REVEST rubberized

engomado[2] *m* PAPEL gumming; **~ en caliente** *m* EMB hot gluing; **~ en franjas** *m* IMPR strip gumming

engomador: **~ de márgenes** *m* EMB margin gluer

engomadora *f* EMB gumming machine; **~ de bordes** *f* EMB edge-gumming machine; **~ por puntos múltiples** *f* EMB multipoint gluing machine

engomar *vt* P&C rubberize, PROD gum, REVEST coat, rubberize

engordar *vt* AGRIC fatten

engorde *m* AGRIC fattening

engoznado *adj* ING MECÁ hinged

engranado[1] *adj* ING MECÁ geared, gearing, MECÁ geared

engranado[2] *m* ING MECÁ engaging; **~ angular** *m* AUTO, ING MECÁ bevel gearing

engranaje[1]: **de ~** *adj* ING MECÁ, MECÁ geared

engranaje[2] *m* AUTO gear, CINEMAT sprocket, ING MECÁ gear, gear wheel, nest, cogwheel, engagement, pitching, MECÁ engagement, engaging, gear, gear wheel, TRANSP AÉR, TRANSP MAR, VEH gear; **~ acampanado** *m* ING MECÁ bell gear; **~ de accesorio** *m* TRANSP AÉR accessory gearbox; **~ de alta multiplicación** *m* ING MECÁ high gear; **~ alternativo** *m* ING MECÁ reciprocal gear; **~ de ángulo** *m* ING MECÁ, MECÁ bevel gear; **~ de anillo** *m* AUTO, VEH ring gear; **~ anular** *m* AUTO, VEH ring gear; **~ anular fijo** *m* TRANSP AÉR fixed-ring gear; **~ anular del volante** *m* ING MECÁ, VEH flywheel ring gear; **~ de aparcamiento** *m* AUTO parking gear; **~ de arco** *m* ING MECÁ sector gear; **~ del arrancador** *m* AUTO starter gear; **~ de arranque** *m* ING MECÁ starter gear, starting gear; **~ de arranque y paro** *m* ING MECÁ stopping-and-starting gear; **~ de avance** *m* PROD feed gear; **~ bihelicoidal** *m* ING MECÁ herringbone gear, V-gear, MECÁ herringbone gear; **~ de bloqueo para estacionamiento** *m* AUTO parking lock-gear; **~ de cadena** *m* ING MECÁ chain gear; **~ de cambio** *m* ING MECÁ change gear; **~ de cambio de velocidad** *m* ING MECÁ change gear; **~ de cebado** *m* ING MECÁ strike gear; **~ cebador** *m* ING MECÁ striking gear; **~ central** *m* AUTO sun gear; **~ de cheurón** *m* MECÁ herringbone gear; **~ cicloidal** *m* MECÁ cycloidal gear; **~ del cigüeñal** *m* AUTO, VEH crankshaft gear; **~ cilíndrico** *m* ING MECÁ circular gear, circular gearing; **~ cilíndrico de dientes rectos** *m* ING MECÁ spur gear; **~ compensador** *m* ING MECÁ equalizing gear; **~ conducido** *m* ING MECÁ following gear, driven wheel; **~ conductor** *m* ING MECÁ driving gear; **~ cónico** *m* AUTO bevel gearing, ING MECÁ bevel gear, bevel gearing, conical gear, MECÁ bevel gear, VEH bevel-gear set; **~ cónico de dentadura hipoide** *m* MECÁ hypoid bevel gear; **~ cónico del diferencial** *m* VEH differential bevel gear; **~ cónico espiral** *m* AUTO spiral bevel gearing; **~ de corona y piñón** *m* AUTO, VEH ring-and-pinion gearing; **~ de cremallera** *m* AUTO rack-and-pinion steering, FERRO, ING MECÁ, INSTR, PROD, TRANSP rack and pinion, VEH rack-and-pinion steering; **~ de cruz de malta** *m* ING MECÁ Geneva wheel; **~ cruzado** *m* ING MECÁ stepped gear; **~ de dentadura frontal** *m* ING MECÁ face gear; **~ de dentadura helicoidal** *m* MECÁ heavy-duty gear; **~ de desembrague delantero** *m* ING MECÁ disengaging gear in front; **~ desmultiplicador** *m* ING MECÁ reducing gear; ~ **desplazable** *m* VEH sliding gear; ~ **de dientes angulares** *m* ING MECÁ, MECÁ herringbone gear; ~ **de dientes cruzados** *m* ING MECÁ step tooth gear; ~ **de dientes de flancos rectos** *m* ING MECÁ flank gear, straight-flank gear; ~ **de dientes helicoidales** *m* ING MECÁ enveloping-tooth wheel, VEH helical gear; ~ **de dientes interiores** *m* ING MECÁ annular gear; ~ **de dientes rectos** *m* AUTO straight-tooth meshing gear; ~ **diferencial** *m* ING MECÁ balance gear, compensating gear; ~ **de la dirección** *m* AUTO steering gear; ~ **de la dirección de bola de recirculación** *m* AUTO recirculating-ball steering-gear; ~ **de distribución** *m* AUTO, VEH *válvula* timing gear; ~ **doble helicoidal** *m* MECÁ herringbone gear; ~ **del eje de levas** *m* ING MECÁ camshaft gear; ~ **de eje secundario** *m* AUTO countershaft gear; ~ **elíptico** *m* ING MECÁ elliptical gear; ~ **del embrague** *m* AUTO, ING MECÁ, VEH clutch gear; ~ **encerrado** *m* ING MECÁ shrouding gear; ~ **epicíclico** *m* MECÁ, TRANSP MAR epicyclic gear; ~ **epicicloidal** *m* ING MECÁ, MECÁ epicycloidal gear; ~ **escalonado** *m* ING MECÁ stepped gear; ~ **de espina de pescado** *m* ING MECÁ, MECÁ herringbone gear; ~ **espiral** *m* ING MECÁ spiral; ~ **de estacionamiento** *m* AUTO parking gear; ~ **exterior** *m* ING MECÁ outside gear; ~ **fijo** *m* ING MECÁ fixed wheel; ~ **de fresadoras** *m* ING MECÁ mill gearing; ~ **globoide** *m* ING MECÁ globoid gear, PROD hourglass screw gear; ~ **de gran multiplicación** *m* ING MECÁ high gear; ~ **de hélice** *m* ING MECÁ screw gearing; ~ **helicoidal** *m* AUTO helical gear, ING MECÁ helical gear, curved worm gear, screw gear, spiral, worm gear, worm wheel, MECÁ worm gear; ~ **helicoidal doble** *m* ING MECÁ double helical gear; ~ **helicoidal y tornillo sin fin** *m* ING MECÁ screw gearing; ~ **hipoidal** *m* ING MECÁ hypoid gear; ~ **hipoide** *m* AUTO, ING MECÁ, VEH hypoid gearing; ~ **impulsado** *m* ING MECÁ follower; ~ **inferior** *m* CINEMAT bottom sprocket; ~ **de inglete** *m* CONST miter gear (*AmE*), mitre gear (*BrE*); ~ **interior** *m* AUTO, ING MECÁ inside gear, internal gear; ~ **intermedio** *m* ING MECÁ intermediate gear, VEH idle gear; ~ **intermedio de contramarcha** *m* AUTO, VEH reverse idler gear; ~ **intermedio inversor de marchas** *m* AUTO, VEH reverse idler gear; ~ **intermedio de retroceso** *m* AUTO, VEH reverse idler gear; ~ **intermedio de transmisión** *m* ING MECÁ transmission gear; ~ **interno** *m* AUTO, ING MECÁ inside gear, internal gear; ~ **de inversión** *m* CINEMAT reversing gear; ~ **de linterna** *m* ING MECÁ lantern gear, lantern gearing, cog and round; ~ **de mando principal** *m* AUTO main drive gear; ~ **de mando del tacómetro** *m* AUTO, VEH speedometer drive gear; ~ **de marcha atrás** *m* AUTO, VEH reverse gear; ~ **montado sobre muelles** *m* VEH sprung gear; ~ **motor** *m* ING MECÁ driving gear; ~ **motriz** *m* ING MECÁ driving gear; ~ **multiplicador** *m* ING MECÁ multiplying gear, multiplying gearing, multiplying wheel; ~ **parcialmente dentado** *m* ING MECÁ mutilated gear; ~ **de perfil de evolvente en círculo** *m* ING MECÁ involute gear; ~ **de piñón** *m* AUTO pinion gear; ~ **planetario** *m* AUTO planet gear, side gear, sun gear, ING MECÁ planet gear, planet gearing, TRANSP AÉR, VEH planet gear; ~ **planetario del diferencial** *m* AUTO differential side gear; ~ **planetario de primera etapa** *m* TRANSP AÉR first-stage planet gear; ~ **principal** *m* ING MECÁ driving gear; ~ **protegido** *m* ING MECÁ flanged gear;

~ **receptor** *m* CINEMAT take-up sprocket; ~ **reductor** *m* ING MECÁ gear ratio, VEH reduction gear; ~ **reductor de velocidad** *m* ING MECÁ, VEH reducing gear; ~ **de rueda dentada y tornillo sin fin** *m* ING MECÁ screw gear; ~ **satélite** *m* ING MECÁ tap with metric thread; ~ **secundario** *m* ING MECÁ follower; ~ **de segmento** *m* ING MECÁ segment gear; ~ **de sincronización** *m* AUTO timing gear; ~ **sinfín** *m* ING MECÁ worm gear; ~ **solar de primera etapa** *m* TRANSP AÉR first-stage sun gear; ~ **de tornillo sinfín** *m* ING MECÁ, MECÁ screw gear, worm gear; ~ **de tornillo sinfín de precisión** *m* ING MECÁ fine worm drive; ~ **tosco de fundición** *m* PROD rough cast gear; ~ **de transmisión** *m* ING MECÁ driving gear, transmission gear; ~ **de transporte de la película** *m* CINEMAT, FOTO film-transport sprocket

engranajes[1] *m pl* AUTO, PROD, ING MECÁ, MECÁ, TRANSP, TRANSP AÉR gears; ~ **a cheurones** *m pl* ING MECÁ chevron cut gears; ~ **cilíndricos** *m pl* ING MECÁ cylindrical gears; ~ **cilíndricos para ingeniería pesada** *m pl* ING MECÁ cylindrical gears for heavy engineering; ~ **a conexión continua** *m pl* AUTO constant-mesh gears; ~ **cónicos con sistema de dientes rectos y helicoidales** *m pl* ING MECÁ bevel gears with straight and spiral tooth system; ~ **encartados** *m pl* ING MECÁ guarded gears; ~ **enclaustrados** *m pl* ING MECÁ enclosed gears; ~ **flotantes** *m pl* TRANSP AÉR floating gears; ~ **fresados** *m pl* ING MECÁ cut gears; ~ **de inversión de marcha** *m pl* ING MECÁ screw reversing gears; ~ **planetarios** *m pl* AUTO planetary gears; ~ **protegidos** *m pl* ING MECÁ guarded gears; ~ **para trabajos pesados** *m pl* ING MECÁ heavy-duty gear

engranajes[2]: **sin** ~ *fra* ING MECÁ, MECÁ gearless

engranar[1] *vt* ING MECÁ mesh, gear, engage, trip in, pitch, throw into gear, MECÁ engage, lock; ~ **con** *vt* ING MECÁ *ruedas dentadas* pitch

engranar[2] *vi* ING MECÁ, MECÁ engage

engranarse *v refl* ING MECÁ mesh

engrane *m* ING MECÁ gearing, mesh, meshing, pitching, MECÁ gear, meshing, PROD, TEC ESP mesh; ~ **angular** *m* MECÁ angular meshing

engranes: ~ **resguardados** *m pl* SEG guarded gears

engrapado *m* CONST joggle; ~ **talonario** *m* IMPR stab stitching

engrapadora: ~ **eléctrica automática** *f* ELEC, PROD electric stapling-machine

engrasador *m* ING ELÉC, ING MECÁ lubricator, MINAS *herramienta* oiler, PROD lubricator, *persona* greaser, oiler, TEXTIL *persona* oiler, VEH grease cap; ~ **de aire comprimido** *m* CONST, ING MECÁ compressed-air lubricator; ~ **de aire a presión** *m* CONST, ING MECÁ compressed-air lubricator; ~ **de casco** *m* PROD helmet cap lubricator; ~ **a compresión** *m* PROD spring grease lubricator; ~ **de copa** *m* ING MECÁ oil cup, PROD grease cup; ~ **de resorte** *m* PROD spring grease lubricator

engrasar *vt* AUTO, ING MECÁ, MECÁ, PROD, VEH grease, oil

engrase *m* AUTO lubricación, PROD greasing, oiling, REFRIG lubrification, TEXTIL *de las máquinas* oiling; ~ **por alimentación forzada** *m* AUTO forced-feed lubrication; ~ **por barboteo** *m* REFRIG splash lubrification; ~ **forzado** *m* REFRIG forced lubrification; ~ **a presión** *m* REFRIG forced lubrification

engravillado *m* CONST gritting, sanding

engrilletar *vt* TRANSP MAR shackle
engrosamiento *m* CONST bulge, METAL *coalescencia* coarsening
engruado *m* ING MECÁ cranage
engrudo *m* TEC ESP paste
engrumación *f* ALIMENT, QUÍMICA clotting
enguatar *vt* TEXTIL wad
enhebrado: ~ **automático de la película** *m* CINEMAT automatic film-threading
enhebrador: ~ **de cinta magnética** *m* INFORM&PD tape leader
enhebramiento: ~ **manual** *m* CINEMAT manual threading; ~ **de la película** *m* CINEMAT film threading
enhebrar *vt* CINEMAT, PROD, TEXTIL thread
enhédrico *adj* GEOL automorphic
enhidrita *f* MINERAL enhydrite
enhielamiento *m* METEO, TRANSP MAR icing
enjalbegado *m* COLOR whitewash
enjambre *m* TEC ESP *astronomía* cluster
enjarciar *vt* TRANSP MAR *palos* rig
enjaretado *m* TRANSP MAR *cabina* grating
enjaular *vt* MINAS *vagones* cage
enjuagadora: ~ **de botellas** *f* EMB bottle-rinsing machine
enjugador *m* TEXTIL *rodillo de goma* squeegee
enlace *m* CONST, CRISTAL, EMB bond, INFORM&PD *comunicaciones* trunk (*AmE*), link (*BrE*), linkage, ING ELÉC crawling, trunking, ING MECÁ link, linkage, tie, INSTR attachment, MECÁ bonding, link, linkage, P&C *químico* bond, linking, linkage, PROD junction, QUÍMICA *entre átomos* bond, linkage, linking, TEC ESP, TEC PETR bonding, TELECOM link, linking, trunking; ~ **de acceso** *m* TELECOM private line (*PL*), *telefonía* access link; ~ **alimentador** *m* TEC ESP *comunicaciones* feeder link; ~ **ascendente** *m* TEC ESP, TELECOM uplink; ~ **básico** *m* INFORM&PD basic coding, basic linkage; ~ **común** *m* TELECOM trunk; ~ **de comunicación** *m* INFORM&PD communication link; ~ **de comunicaciones** *m* INFORM&PD bus, highway, trunk; ~ **de control de la llave de cierre del combustible** *m* TRANSP AÉR fuel shut-off cock control link; ~ **de coordinación** *m* QUÍMICA coordination bond; ~ **coordinado** *m* METAL coordinate linkage; ~ **covalente** *m* CRISTAL, METAL covalent bond, QUÍMICA covalent bond, sigma bond; ~ **cruzado** *m* QUÍMICA cross link; ~ **de datos** *m* ELECTRÓN, INFORM&PD, TEC ESP data link; ~ **delta** *m* QUÍMICA delta bonding; ~ **descendente** *m* TEC ESP, TELECOM down link; ~ **de desplazamiento** *m* ING MECÁ shifting link; ~ **de desviación** *m* ING MECÁ shifting link; ~ **de desvío** *m* ING MECÁ shifting link; ~ **directo** *m* TEC ESP forward link; ~ **dirigido** *m* CRISTAL directed bond; ~ **entre procesadores** *m* TELECOM interprocessor link; ~ **entre satélites** *m* TEC ESP intersatellite link; ~ **etérico** *m* DETERG ether linkage; ~ **eterlénico** *m* DETERG etherlene linkage; ~ **de fibra óptica** *m* ÓPT, TELECOM optical fiber link (*AmE*), optical fibre link (*BrE*); ~ **de flujo magnético** *m* ENERG RENOV magnetic flux linkage; ~ **en frío** *m* TERMO cold bond, cold bonding; ~ **de fuerza** *m* TRANSP AÉR force link; ~ **heterodino** *m* TELECOM heterojunction; ~ **de hidrógeno** *m* QUÍMICA hydrogen bond; ~ **homogéneo** *m* TELECOM homojunction; ~ **intermedio** *m* TELECOM intermediate trunk (*IT*); ~ **iónico** *m* CRISTAL, FÍS RAD, QUÍMICA ionic bond; ~ **por ionización meteórica** *m* TEC ESP

meteor burst link; ~ **marítimo** *m* TRANSP MAR sea link; ~ **mediante modem** *m* PROD modem link; ~ **de microondas** *m* TELECOM radio link; ~ **por microondas** *m* TV microwave link; ~ **de modulación digital** *m* TELECOM digital modulation link; ~ **multipunto** *m* INFORM&PD multidrop link, multipoint link; ~ **multiterminal mediante módem** *m* PROD multipoint modem link; ~ **de oferta** *m* TELECOM trunk offer (*TKO*); ~ **oficina central** *m* TELECOM central office trunk; ~ **óptico** *m* ÓPT optical link; ~ **óptico de datos** *m* ÓPT optical data bus; ~ **de par** *m* TRANSP AÉR torque link; ~ **de paro** *m* ING MECÁ knock-off link; ~ **peptídico** *m* QUÍMICA peptide link; ~ **pi** *m* QUÍMICA pi bond; ~ **con el prestador de servicios** *m* TELECOM service provider link (*SPL*); ~ **primario interactivo** *m* TELECOM interactive primary link (*IPL*); ~ **principal de distribución** *m* TELECOM distribution primary link (*DPL*); ~ **de programas** *m* INFORM&PD program linking; ~ **de puente de hidrógeno** *m* CRISTAL hydrogen bond; ~ **de punto a punto** *m* ING ELÉC point-to-point link; ~ **químico** *m* CRISTAL, QUÍMICA, TEC PETR chemical bond; ~ **de radio** *m* FÍS RAD radio beam; ~ **radioeléctrico** *m* TRANSP MAR radio link; ~ **radiofónico** *m* TRANSP AÉR radio link; ~ **radiotelefónico** *m* TELECOM, TRANSP, TRANSP MAR radiotelephone link; ~ **por rayos infrarrojos** *m* TV infrared link; ~ **de recepción** *m* TRANSP receiving trunk; ~ **por satélite** *m* TEC ESP, TV satellite link; ~ **de señal de control** *m* TEC ESP command link; ~ **sigma** *m* QUÍMICA covalent bond, sigma bond; ~ **térmico** *m* TERMO thermal link; ~ **terminal** *m* QUÍMICA endlink; ~ **tipo margarita** *m* INFORM&PD multidrop link; ~ **para la transmisión de datos** *m* TELECOM data link connection (*DLC*); ~ **de Van der Waals** *m* CRISTAL Van der Waals bond; ~ **con varios puntos mediante módem** *m* PROD multipoint modem link
enlaces: ~ **extremos del inducido** *m pl* ELEC armature end connection
enladrillado *m* CONST brick pavement
enlatado[1] *adj* ALIMENT, EMB canned (*AmE*), tinned (*BrE*)
enlatado[2] *m* ALIMENT, EMB canning (*AmE*), tinning (*BrE*)
enlatar *vt* ALIMENT, EMB can (*AmE*), tin (*BrE*)
enlazador *m* INFORM&PD linker
enlazamiento *m* INFORM&PD linkage
enlazar *vt* CONST bind, ELEC, INFORM&PD, ING ELÉC connect, TELECOM link up, link together
enlechado *m* CONST, NUCL grouting
enlistonado *m* CONST lathing, PROD lagging
enlomado *m* IMPR lining-up
enlosado *m* CONST flagging, tile floor
enlucido *m* CONST plastering, plasterwork, *albañilería* rendering; ~ **acústico** *m* REVEST acoustic plaster; ~ **antisonoro** *m* REVEST acoustic plaster
enlucir *vt* CONST plaster
enmaderado *m* CONST boxing
enmangado: ~ **en caliente** *m* PROD shrinking-on
enmangamiento *m* PROD *herramientas* handling
enmangar: ~ **en caliente** *vt* PROD *herramientas* shrink on; ~ **por presión** *vt* ING MECÁ, PROD force on
enmanguitado: ~ **en caliente** *m* PROD shrinking-on
enmanguitar: ~ **en caliente** *vt* PROD shrink on
enmarañar *vt* TEXTIL snarl

enmascarado *m* IMPR masking

enmascaramiento *m* ACÚST, ELECTRÓN, INFORM&PD masking; ~ **de ruido** *m* TELECOM noise masking

enmascarar *vt* TEC ESP jam

enmasillar *vt* CONST putty

enmendar *vt* AGRIC, ING MECÁ amend, TRANSP MAR *rumbo* change, cant

enmienda *f* AGRIC *suelos*, ING MECÁ amendment; ~ **del suelo** *f* AGRIC soil amendment

ENN *abr* (*extracto no nitrogenado*) AGRIC NFE (*nitrogen-free extract*)

ennegrecer *vt* C&V, CARBÓN, PAPEL, TERMO *por calor* blacken

ennegrecido *adj* C&V, CARBÓN, PAPEL, TERMO blackened

ennegrecimiento *m* C&V, CARBÓN, PAPEL, TERMO *defecto del papel* blackening; ~ **de moldes** *m* CARBÓN *fundición* dust

enol *m* QUÍMICA enol

enolasa *f* QUÍMICA enolase

enólico *adj* QUÍMICA enolic

enolización *f* QUÍMICA enolization

ENQ *abr* (*fin de consulta*) INFORM&PD ENQ (*end of query*)

enquilladura *f* TRANSP MAR *botes* skeg

enrarecerse *v refl* PROD thin out

enrarecimiento *m* ACÚST, FÍS, FÍS ONDAS rarefaction, QUÍMICA *gas* depletion

enrasado *m* CONST furring

enrasamiento *m* MECÁ, PROD leveling (*AmE*), levelling (*BrE*)

enrasar *vt* PROD flush

enrase *m* CARBÓN liquid limit device, MECÁ, PROD leveling (*AmE*), levelling (*BrE*)

enredadera: ~ **anual** *f* AGRIC *maleza del trigo* black bindweed; ~ **de campanillas** *f* AGRIC morning glory; ~ **europea** *f* AGRIC bindweed

enredado *adj* ING MECÁ intricate, SEG, TEXTIL entangled

enredar[1] *m* OCEAN netting

enredar[2] *vt* ING MECÁ foul, SEG, TEXTIL entangle

enredarse *v refl* TRANSP MAR *ancle, cabo, motor* foul

enredo *m* SEG entanglement

enrejado *m* C&V grillage, jalousie, CONST lattice, trellis, trelliswork; ~ **de alambre** *m* CONST wire netting; ~ **con cable con forma de diamante** *m* CONST diamond wire lattice; ~ **de control** *m* ING ELÉC control grid; ~ **para el ganado** *m* AGRIC, CONST cattle grid

enriado *m* TEXTIL *fibras* retting; ~ **por rociado** *m* TEXTIL *fibras* dew retting

enriaje: ~ **de cáñamo** *m* AGRIC retting

enriquecer *vt* ALIMENT, FÍS RAD, NUCL enrich

enriquecido *adj* ALIMENT fortified; **no** ~ *adj* NUCL unenriched

enriquecimiento *m* ALIMENT enrichment, fortification, CARBÓN upgrading, FÍS, NUCL enrichment; ~ **gaseoso** *m* GAS, TERMO gas enrichment; ~ **isotópico** *m* FÍS PART isotopic enrichment; ~ **con oxígeno** *m* C&V oxygen enrichment

enrocamiento *m* AGUA rock fill

enrojecido *adj* TERMO, TERMOTEC red-hot

enrolar *vt* TRANSP MAR *tripulantes* take on

enrollado[1]: ~ **en espiral** *adj* GEOM involute; ~ **sobre el plegador** *adj* TEXTIL wound onto the beam

enrollado[2] *m* ING ELÉC reeling, ING MECÁ winding, P&C reeling, PAPEL reeling, winding; ~ **en espiral** *m* ING ELÉC *conductores* pigtail; ~ **de sacos** *m* PROD bag rolling, bail

enrollador *m* PAPEL winder; ~ **de la hoja** *m* AGRIC leaf roller; ~ **de película** *m* CINEMAT, FOTO, TV film winder

enrolladora *f* ING ELÉC, P&C, PAPEL reeling; ~ **pope** *f* PAPEL *final de la máquina* pope reel

enrollamiento *m* INFORM&PD scrolling, ING MECÁ convolution, winding motion, PROD coiling; ~ **yuxtapuesto** *m* ELEC *bobina* bank winding, banked winding, ING ELÉC banked winding

enrollar *vt* CINEMAT spool, CONTAM MAR reel in, ELEC coil, INFORM&PD scroll, ING MECÁ wind, wrap, PROD *cables* coil, TEXTIL *tejido* wind

enromar *vt* ING MECÁ blunt

enroscado *m* ING MECÁ screwing, winding

enroscadura *f* ING MECÁ convolution

enroscamiento *m* TEC PETR screwing, *tuberías* screwing-in

enroscar *vt* ING MECÁ screw

enrosque *m* ING MECÁ threading, TEC PETR *tuberías, conductos* screwing-in

enrutamiento *m* INFORM&PD routing; ~ **alternativo** *m* TELECOM alternative routing; ~ **centralizado** *m* INFORM&PD centralized routing; ~ **del tráfico enlace por enlace** *m* TELECOM link-by-link traffic routing

ensacado *m* AmL (*cf embolsado Esp*) EMB bag packaging

ensacador *m* AGRIC sacker

ensacadora *f* Esp (*cf embolsadora AmL*) AGRIC bag-filling machine, EMB bagging machine

ensamblado *m* ING MECÁ assembly

ensamblador *m* CONST joiner, INFORM&PD assembler, assembly; ~ **cruzado** *m* INFORM&PD cross assembler; ~**-desensamblador de paquetes** *m* (*PAD*) INFORM&PD, TELECOM packet assembler-disassembler (*PAD*); ~ **de macros** *m* INFORM&PD macro assembler

ensambladora *f* CONST, PROD, TELECOM jointer

ensambladura: ~ **en cola de milano** *f* MECÁ dovetail joint; ~ **empotrada** *f* CONST housed joint

ensamblaje *m* CONST framing, joggle, ING MECÁ assembling, TEC ESP assembly building; ~ **cruzado** *m* INFORM&PD cross assembly

ensamblajes *m pl* PROD work assembly

ensamblar *vt* CARBÓN *carpintería* table, CONST brace, joggle, INFORM&PD assemble, ING MECÁ *herramientas* gang; ~ **con cola de milano** *vt* CONST dovetail; ~ **con espigas** *vt* CONST dowel

ensample *m* ING MECÁ couple

ensanchado *adj* MECÁ, TEXTIL flared

ensanchador *m* ING MECÁ counterbore, reamer, MECÁ drift, reamer, TEXTIL stretcher; ~ **de bebedero** *m* ING MECÁ gate accentuator; ~ **de contramarcos** *m* CONST casing expander

ensanchamiento *m* CONTAM MAR spreading, ING MECÁ bell mouth, counterboring, INSTR enlarging, PROD flaring, counterboring, TEC ESP bell mouth; ~ **de curva** *m* CONST curve widening; ~ **debido a una resonancia** *m* FÍS RAD resonance broadening; ~ **Doppler** *m* CRISTAL Doppler broadening, line broadening, FÍS, FÍS RAD Doppler broadening; ~ **del fondo** *m* MINAS *barrenos* chambering; ~ **del fondo por explosión de una carga** *m* MINAS *pozos* springing; ~ **de impulso** *m* TELECOM pulse widening;

~ **inhomogéneo** *m* FÍS RAD inhomogeneous broading; ~ **de una línea espectral** *m* CRISTAL Doppler broadening, line broadening, FÍS, FÍS RAD Doppler broadening; ~ **de las líneas espectrales debido a la presión** *m* FÍS RAD pressure broadening; ~ **polar** *m* ING ELÉC, ING MECÁ pole piece; ~ **por presión** *m* FÍS pressure broadening; ~ **de pulsos** *m* ÓPT, TELECOM pulse broadening; ~ **de taladros con mandril** *m* ING MECÁ drifting; ~ **por trépano** *m* MINAS *pozos* underreaming; ~ **de la vía** *m* FERRO *vehículos* gage widening (*AmE*), gauge widening (*BrE*)

ensanchar *vt* CONTAM MAR spread, ING MECÁ counterbore, INSTR enlarge, PETROL ream, PROD counterbore, TEC ESP spread

ensanche *m* ING MECÁ, TEC ESP bell mouth; ~ **de banda** *m* FÍS RAD bandspread; ~ **de base** *m* ELECTRÓN base widening

ensayado: ~ **in situ** *adj* MECÁ field-tested; ~ **sobre el terreno** *adj* MECÁ field-tested

ensayador: ~ **de capas** *m* PETROL formation tester

ensayar *vt* QUÍMICA assay, test, TELECOM test

ensayo *m* AGRIC trial, AGUA analysis, CALIDAD testing, CARBÓN assay, test, trial, CONST test, INFORM&PD testing, ING MECÁ test run, testing, MECÁ, METAL, METR test, PROD dry run, proof, QUÍMICA assay, test, testing, REFRIG probe, test, TEC ESP, TELECOM, TEXTIL test; ~ **de abrasión de Los Angeles** *m* CONST Los Angeles abrasion test; ~ **de absorción de tinta** *m* IMPR *papel* absorption ink test; ~ **de aceptación** *m* CARBÓN acceptance test; ~ **de adaptación ambiental** *m* TEC ESP test of environmental stress; ~ **de adhesión** *m* EMB bonding test; ~ **de aguante** *m* CALIDAD endurance test; ~ **de aguja de contacto** *m* ING MECÁ touch needle test; ~ **de almacenado acelerado** *m* EMB accelerated storage test; ~ **ambiental** *m* TEC ESP environmental test; ~ **de aprobación** *m* CALIDAD approval testing, TEC ESP approval test; ~ **de asentamiento** *m* CONST slump test; ~ **atómico** *m* D&A atomic test; ~ **de bombeo** *m* AGUA, CARBÓN, INSTAL HIDRÁUL pumping test; ~ **de caducidad** *m* EMB shelf life test; ~ **de caída** *m* EMB drop test; ~ **de campo** *m* AGRIC, TELECOM field trial; ~ **censurado** *m* TEC ESP censured test; ~ **al choque mecánico** *m* METR mechanical shock test; ~ **climático** *m* EMB, TEC ESP climatic test; ~ **de colisión** *m* TRANSP collision test; ~ **de compresión** *m* CARBÓN, EMB, MECÁ, METAL compression test; ~ **en condiciones ambientales controladas** *m* MECÁ environmental testing; ~ **en conducto** *m* REFRIG in-duct method; ~ **de consolidación** *m* CARBÓN, PETROL, TEC PETR consolidation test; ~ **de control** *m* CARBÓN, QUÍMICA control assay; ~ **de corte en bloque** *m* ING MECÁ block-shear test; ~ **de densidad** *m* CONST density test; ~ **de desarrollo** *m* TEC ESP *equipos* development test; ~ **de desgaste por abrasión** *m* IMPR *papel* abrasion test; ~ **de deterioro por el calor** *m* P&C heat ageing test; ~ **de doblado** *m* EMB folding strength test; ~ **de duración** *m* EMB durability test, ING ELÉC life test; ~ **de dureza** *m* FÍS hardness test, MECÁ hardness tester; ~ **de dureza Brinell** *m* ING MECÁ, MECÁ Brinell ball test; ~ **de dureza Rockwell** *m* ING MECÁ, MECÁ Rockwell hardness test; ~ **de enfriamiento** *m* ALIMENT, REFRIG, TERMO refrigeration test; ~ **de envejecimiento** *m* EMB ageing test; ~ **de estampación** *m* PROD *metales* stamping test; ~ **estático** *m* TEC ESP static test; ~ **de**

fractura *m* ING MECÁ fracture test; ~ **al freno** *m* ING MECÁ brake test; ~ **frigorífico** *m* ALIMENT, REFRIG, TERMO refrigeration test; ~ **de fugas** *m* EMB leakage test; ~ **de fugas por medio de helio** *m* NUCL helium leak test; ~ **de funcionamiento** *m* ING ELÉC reliability testing; ~ **de fusión** *m* TERMO melting test; ~ **gamamétrico de minerales** *m* NUCL gammametric ore assay; ~ **granulométrico** *m* CONST grading analysis; ~ **de impacto** *m* EMB, REVEST, TRANSP impact test; ~ **de impacto en probeta entallada** *m* NUCL notch impact test; ~ **de levigación** *m* CONST elutriation test; ~ **por líquidos densos** *m* CARBÓN heavy liquid test; ~ **con líquidos penetrantes** *m* MECÁ dye penetrant test; ~ **manométrico** *m* INSTAL HIDRÁUL pressure test; ~ **Marshall** *m* CONST Marshall test; ~ **de Martens** *m* P&C Martens test; ~ **de materiales** *m* ING MECÁ material testing; ~ **mecánico** *m* ING MECÁ mechanical testing; ~ **de minerales** *m* NUCL ore assaying; ~ **con un modelo** *m* TRANSP MAR model test; ~ **con modelos** *m* TRANSP MAR *arquitectura naval* model test; ~ **no destructivo** *m* FÍS, ING MECÁ, TEC ESP nondestructive test, nondestructive testing (*NDT*); ~ **no destructivo de materiales** *m* NUCL nondestructive materials testing; ~ **de oxidación** *m* ING MECÁ rust-proofing; ~ **del paquete** *m* EMB package test; ~ **de penetración** *m* CARBÓN penetration test; ~ **de penetración fluorescente** *m* MECÁ fluorescent penetration test; ~ **piloto** *m* CARBÓN pilot test, INFORM&PD beta test; ~ **de plegado** *m* EMB folding test; ~ **a presión hidráulica** *m* MECÁ hydrotest; ~ **Proctor** *m* CONST *geotecnia* Proctor test; ~ **de producción** *m* CARBÓN testing, PETROL production test; ~ **de programa** *m* ING MECÁ test run; ~ **de puesta en régimen** *m* REFRIG pull-down test; ~ **por rayos X** *m* NUCL X-ray testing; ~ **de resistencia al calor y a los golpes** *m* TERMO heat shock test; ~ **de resistencia al doblado** *m* IMPR bending test; ~ **de rotura** *m* ING MECÁ fracture test; ~ **de sedimentación** *m* CONST *geotecnica* sedimentation test; ~ **de selección** *m* CALIDAD screening test; ~ **de separación por líquidos de densidad intermedia** *m* CARBÓN float-and-sink analysis; ~ **subjetivo** *m* TELECOM subjective test; ~ **de tensión longitudinal** *m* ING MECÁ longitudinal-traction test; ~ **de la tinta al frotamiento** *m* IMPR ink rub resistance test; ~ **torsional** *m* ING MECÁ torsional test; ~ **de tracción** *m* ING MECÁ tensile test; ~ **triaxial** *m* CONST *geotecnia* triaxial test; ~ **del túnel aerodinámico** *m* TEC ESP wind tunnel testing; ~ **ultrasónico** *m* ING MECÁ, NUCL ultrasonic testing; ~ **ultrasónico no destructivo** *m* ING MECÁ, NUCL nondestructive ultrasonic testing; ~ **por ultrasonidos** *m* NUCL ultrasonic testing; ~ **de unión** *m* PROD bonding test; ~ **al uso** *m* TEXTIL wearer trial ~ **de verificación** *m* CAL; IDAD verification testing, QUÍMICA check assay; ~ **por vía húmeda** *m* CARBÓN chemical analysis, QUÍMICA wet assay; ~ **de vibración** *m* EMB jarring test

ensayos: ~ **de recepción** *m pl* MECÁ acceptance testing; ~ **de resistencia al ambiente** *m pl* MECÁ environmental testing

ensenada *f* HIDROL inlet, OCEAN cove

enseñanza *f* ING MECÁ training; ~ **asistida por computadora** *f* AmL (*cf enseñanza asistida por ordenador Esp*) INFORM&PD computer-assisted instruction (*CAI*); ~ **asistida por ordenador** *f* Esp

(*cf enseñanza asistida por computadora AmL*)
INFORM&PD computer-assisted instruction (*CAI*);
~ **auxiliada por computadora** *f AmL* (*cf enseñanza
auxiliada por ordenador Esp*) INFORM&PD computer-
assisted learning (*CAL*); ~ **auxiliada por ordenador**
f Esp (*cf enseñanza auxiliada por computadora AmL*)
INFORM&PD computer-assisted learning (*CAL*)

enseñar: ~ **la quilla al sol** *vi* TRANSP MAR turn turtle

enser: ~ **doméstico** *m* ELEC home appliance;
~ **doméstico eléctrico** *m* ELEC electrical household
appliance; ~ **portátil** *m* ELEC portable appliance

enseres: ~ **y accesorios** *m pl* CONST fixtures and
fittings

ensilado *m* AGRIC silo filling; ~ **de maíz** *m* AGRIC corn
silage (*AmE*), maize silage (*BrE*)

ensilaje *m* AGRIC silage; ~ **con condiciones de baja
humedad** *m* AGRIC haylage; ~ **hecho con materiales no curados** *m* AGRIC hay crop silage

ensombrecimiento *m* TELECOM shadowing

ensortijamiento *m* PROD kink

ensortijar *vt* TEXTIL snarl

enstatita *f* MINERAL enstatite

ensuciamiento *m* PROD gumming, gumming-up

ensuciar *vt* CONTAM MAR foul

ensuciarse *v refl* PROD gum up, TRANSP MAR *los fondos
del buque* foul

entablado *m* CONST boarding

entablamiento *m* PROD entablature; ~ **de lona** *m*
MINAS canvas table

entablillar *vt* CONST batten

entablonado *m AmL* (*cf forro Esp*) MINAS *cimbras*
lagging

entalcadora *f* PROD talking unit

entalingadura *f* TRANSP MAR cable clinch

entalingar *vt* TRANSP MAR *el cable al arganco del ancla*
bend, *la cadena del ancla* clinch

entalla *f* AGUA notch, CARBÓN kerf, MECÁ, METAL
notch, PROD notch, notching; ~ **en V** *f* NUCL V-
shaped notch

entallado *adj* ING MECÁ notched

entalladura *f* AGUA notch, CARBÓN kerf, CONST
cerrajería notch, ING MECÁ gash, MECÁ, METAL notch,
PROD notch, notching

entallar *vt* AGUA, CONST, ING MECÁ, MECÁ, METAL,
PROD notch

entalpia *f* GEN *transferencia de calor* enthalpy;
~ **específica** *f* FÍS specific enthalpy; ~ **de
formación** *f* TERMO enthalpy of formation; ~ **de
transición** *f* NUCL transition enthalpy; ~ **de
vaporización** *f* PROC QUÍ, TERMO enthalpy of vaporization, evaporation enthalpy

entaponado *m* C&V *botellas* bore (*BrE*), corkage
(*AmE*)

entarimado *m* C&V palleting, CINEMAT batten, CONST
boarding, planking, *solado* flooring, SEG flooring

entarimar *vt* CONST board, floor

entenalla *f* ING MECÁ, MECÁ hand vice (*BrE*), hand vise
(*AmE*)

enteramina *f* QUÍMICA enteramine

entero *m* INFORM&PD, MATEMÁT integer; ~ **negativo** *m*
MATEMÁT negative integer

enterrado *adj* GEOL concealed

enterramiento *m* GEOL, TEC PETR burial

enterrar: ~ **con arado** *vt* AGRIC plough in (*BrE*),
plough under (*BrE*), plow in (*AmE*), plow under
(*AmE*)

entibación *f* CONST propping, timber framing, timbering, MINAS lining, piling; ~ **adosada** *f* MINAS skin-to-
skin timbering; ~ **con agujas** *f AmL* (*cf posteo Esp*)
MINAS forepoling; ~ **apoyada en los pilotes** *f* CAR-
BÓN interpile sheeting; ~ **armada** *f AmL* (*cf
entibación de muestra Esp*) MINAS reinforced
timbering, *túneles* herringbone timbering;
~ **autoavanzante** *f AmL* (*cf taco autoavanzante
Esp*) MINAS self-advancing chock; ~ **de la galería
principal** *f* CARBÓN entry timbering; ~ **hincada** *f*
MINAS forepoling; ~ **de muestra** *f Esp* (*cf entibación
armada AmL*) MINAS herringbone timbering, *túneles*
reinforced timbering

entibado *m* CONST timbering, MINAS, PROD crib;
~ **adosado** *m AmL* (*cf entibado ajustado Esp*) MINAS
close timbering, *galerías* skin-to-skin timbering;
~ **ajustado** *m Esp* (*cf entibado adosado AmL*) MINAS
close timbering, skin-to-skin timbering; ~ **de la
galería transversal** *m* MINAS blocking board;
~ **hasta el frente de ataque** *m* MINAS *galerías* skin-
to-skin timbering; ~ **reforzado** *m* MINAS *galerías*
reinforced timbering

entibar[1] *vt* CONST brace together, timber

entibar[2] *vi* PROD *remaches* hold up

entibo: ~ **provisional** *m* MINAS forepoling, lagging

entidad: ~ **de aplicación para la gestión del sistema**
f TELECOM system management application entity
(*SMAE*); ~ **para aplicaciones ISCP** *f* TELECOM
ISCP application entity (*ISCP AE*); ~ **funcional** *f*
TELECOM functional entity (*FE*); ~ **para la gestión
de capas** *f* TELECOM layer management entity
(*LME*); ~ **N** *f* TELECOM N-entity

entidades: ~ **de mantenimiento instalaciones de
abonados** *f pl* TELECOM customer installation maintenance entities (*CIME*); ~ **nobles** *f pl* TELECOM peer
entities

entintar *vt* COLOR ink, IMPR *forma* roll-up, *imagen,
plancha* king; ~ **con tampón** *vt* IMPR dab

entonación *f* ACÚST intonation

entorno *m* GEN environment; ~ **espacial** *m* TEC ESP
space environment; ~ **físico-químico** *m* CONTAM
physico-chemical environment; ~ **hostil** *m* FÍS RAD
hostile environment; ~ **ionizado** *m* TEC ESP ionized
environment; ~ **del lanzamiento** *m* TEC ESP launch
environment; ~ **marino** *m* CONTAM marine environment; ~ **de mensajería por intercambio de datos
electrónicos** *m* TELECOM EDI-messaging environment (*EDIME*); ~ **operativo del usuario** *m*
INFORM&PD user operating environment;
~ **peligroso** *m* FÍS RAD, SEG dangerous environment;
~ **vibratorio** *m* TEC ESP vibratory environment

entrada *f* AGUA, AUTO inlet, CONST gate, ingress,
mouth, throat, ELEC induction, *de corriente, tensión*
input, FÍS, GEOL input, INFORM&PD entry, ING MECÁ
intake, ear, inlet, INSTAL HIDRÁUL *máquinas* intake,
induction, MECÁ inlet, MINAS *pozos* mouth,
inlet, PROD inlet, ingate, gate, TEC ESP threshold,
TELECOM inlet, TRANSP AÉR *de aire* intake, TV input;
~ **de acción** *f* INFORM&PD *cuadro de medidas* action
entry; ~ **de aceite** *f* AUTO oil inlet; ~ **acústica** *f* ING
ELÉC acoustic admittance; ~ **de agua** *f* NUCL water
ingress, REFRIG water inlet; ~ **de agua fría** *f* INSTR
cool water inlet; ~ **de aire** *f* C&V air inlet, ING MECÁ
soplador air inlet, air intake, TRANSP, TRANSP AÉR
variable geometry inlet; ~ **de aire del motor** *f*
TRANSP AÉR engine air-intake; ~ **al alimentador** *f*

C&V forehearth entrance; ~ **de almacenamiento** *f* INFORM&PD storage entry; ~ **de amplificador vertical** *f* ELECTRÓN vertical amplifier input; ~ **de la aplicación para la gestión del sistema** *f* TELECOM system management application entry; ~ **de aplicaciones** *f* TELECOM application entry (*AE*); ~ **de bucle** *f* PROD tieback input; ~ **de bucle cerrado** *f* CONST noninverting input; ~ **de CA** *f* ING ELÉC AC input; ~ **de cargamento** *f* FERRO freight inwards (*AmE*), goods inwards (*BrE*); ~ **de CC** *f* ING ELÉC DC input; ~ **al chorreador** *f* C&V forehearth entrance; ~ **en circuito** *f* ING ELÉC switching-in; ~ **en comunicación** *f* INFORM&PD handshake; ~ **del conducto** *f* PROD conduit entry; ~ **de corriente** *f* ELEC, ELECTRÓN electrical input, ING ELÉC current input, electrical input; ~ **de corriente externa** *f* ELEC, ELECTRÓN external input; ~ **de corriente invertida** *f* ELEC, ELECTRÓN *de amplificador diferencial* inverting input; ~ **de corriente en el TEC** *f* ELECTRÓN FET input; ~ **corta** *f* TRANSP AÉR undershoot; ~ **de datos** *f* ELECTRÓN, INFORM&PD data entry, input; ~ **de datos de voz** *f* INFORM&PD voice data entry; ~ **diferencial** *f* ELECTRÓN differential input; ~ **digital** *f* ELECTRÓN digital input; ~ **directa** *f* ELECTRÓN direct input; ~ **directa de datos** *f* (*DDE*) ELECTRÓN, INFORM&PD direct data entry (*DDE*); ~ **de distribución** *f* ING ELÉC partition gate; ~ **doble tipo transformador** *f* PROD transformer-type dual input; ~ **de energía controlada por el voltaje** *f* ING ELÉC voltage-controlled input; ~ **de energía eléctrica** *f* ELEC, ELECTRÓN, ING ELÉC electrical input; ~ **de energía óptica** *f* ING ELÉC optical input; ~ **equilibrada** *f* ING ELÉC balanced input; ~ **en espera** *f* INFORM&PD input queue; ~ **de existencias** *f* PROD entry in stock; ~ **de fluido** *f* ING MECÁ fluid inlet; ~ **de flujo** *f* TRANSP AÉR inflow; ~ **de flujo del rotor** *f* TRANSP AÉR rotor inflow; ~ **de hombre** *f* ING MECÁ manhole; ~ **de información de la giroscópica** *f* TRANSP MAR compass input; ~ **inmediata** *f* PROD immediate input; ~ **larga** *f* TRANSP AÉR overshoot, overshooting; ~ **de línea** *f* TV line-in; ~ **manual** *f* INFORM&PD manual input; ~ **máxima** *f* INFORM&PD fan-in; ~ **de memoria** *f* INFORM&PD storage entry; ~ **de mercancías** *f* FERRO, PROD freight inwards (*AmE*), goods inwards (*BrE*); ~ **no regulada** *f* PROD unregulated input; ~ **de obra muerta** *f* TRANSP MAR tumblehome; ~ **en órbita** *f* TEC ESP entry into orbit; ~ **de par termoeléctrico** *f* PROD thermocouple input; ~ **en pérdida de la pala en retroceso** *f* TRANSP AÉR retreating blade stall; ~ **de potencia máxima** *f* ING ELÉC maximum power input; ~ **de presión** *f* TRANSP AÉR pressure inlet; ~ **principal** *f* CARBÓN, MINAS main gate; ~ **protegida** *f* ING ELÉC guarded input; ~ **sin referencia a tierra** *f* ING ELÉC floating input; ~ **RGB** *f* INFORM&PD, TV RGB input; ~**-salida diferida** *f* TV spooling; ~ **y salida discontínuas** *f pl* PROD discrete input-output; ~ **y salida discretas** *f pl* PROD discrete input-output; ~**-salida mediante registro intermedio** *f* INFORM&PD buffered input-output; ~**-salida secuencial** *f* (*SIO*) INFORM&PD sequential input-output (*SIO*), series input-output (*SIO*); ~**-salida en serie** *f* (*SIO*) INFORM&PD sequential input-output (*SIO*), series input-output (*SIO*); ~ **y salida de la tubería de revestimiento** *f* PETROL, TEC PETR round trip; ~ **de secuenciador** *f* PROD sequencer input;

~ **de señal de referencia** *f* ELECTRÓN reference-signal input; ~ **de sincronización** *f* TV sync input; ~ **en el sistema** *f* INFORM&PD log-in, log-on; ~ **de termopar** *f* PROD thermocouple input; ~ **en tiempo real** *f* ELECTRÓN, INFORM&PD, TELECOM real-time input; ~ **de trabajos a distancia** *f* INFORM&PD remote job entry (*RJE*); ~ **de ventilación** *f* MINAS ventilation door

entrada/salida *f* (*E/S*) ELEC *de corriente, tensión,* INFORM&PD, PROD input/output (*I/O*); ~ **en paralelo** *f* INFORM&PD parallel input-output

entramado *m* CONST frame; ~ **doble** *m* CONST double floor; ~ **eléctrico** *m* ING ELÉC carcass; ~ **de madera** *m* CONST half-timbering

entrampado *m* SEG trapping

entrante[1] *adj* GEOM re-entrant

entrante[2]: ~ **de pared** *m* CONST bay, recess

entrar[1]: ~ **en cuadro por arriba** *vt* CINEMAT bump up

entrar[2] *vi* TRANSP MAR *marea* flood; ~ **en barrena** *vi* TEC ESP, TRANSP AÉR spin; ~ **en circuito** *vi* TELECOM go into circuit; ~ **en dársena** *vi* TRANSP MAR *buque* dock; ~ **en dique seco** *vi* TRANSP MAR *buque* dry-dock; ~ **en marcha** *vi* ING MECÁ come into gear; ~ **en el sistema** *vi* INFORM&PD, PROD log in, log on; ~ **en el sistema usuario** *vi* INFORM&PD, PROD log in, log on~ **a través del circuito** *vi* TELECOM go via the circuit; ~ **en vigencia** *vi* ING MECÁ take effect

entrecruzado *m* ING MECÁ interlocking

entrecruzamiento *m* P&C, QUÍMICA cross link, cross linking

entreejes *adj* ING MECÁ center to center (*AmE*), centre to centre (*BrE*)

entrega *f* ALIMENT, C&V, IMPR delivery, NUCL *de una central nuclear nueva* handover, PROD delivery; ~ **en almacén** *f* PROD warehouse delivery; ~ **como pago parcial** *f* PROD trade-in; ~ **a cuenta** *f* PROD trade-in; ~ **directa** *f* PROD direct delivery; ~ **a domicilio** *f* FERRO door-to-door delivery; ~ **física** *f* TELECOM physical delivery; ~ **posterior** *f* PROD subsequent delivery; ~ **de presión** *f* INSTAL HIDRÁUL pressure delivery; ~ **secuenciada** *f* PROD time-phased delivery; ~ **única** *f* PROD single delivery

entregado: ~ **en obra** *adj* CONST site-delivered

entrehierro *m* ELEC *máquinas,* FÍS air gap, INFORM&PD head gap, air gap, ING ELÉC air gap, ING MECÁ gap; ~ **de la cabeza** *m* TV head gap; ~ **de cabeza magnética** *m* TV magnetic-head gap; ~ **del cabezal** *m* INFORM&PD, TV head gap

entrelazado *m* INFORM&PD interleaving, ING MECÁ, MECÁ interlocking, TELECOM interworking, TV interlace; ~ **de frecuencias** *m* TV frequency interlace; ~ **progresivo** *m* TV progressive interlace; ~ **secuencial** *m* TV sequential interlace

entrelazamiento *m* TEXTIL interlacing; ~ **y rizado** *m* TEXTIL interlacing and crimping

entrelazar *vt* CONST *carpintería* weave, ELECTRÓN interlace, INFORM&PD interleave, interlace, TEXTIL interlace

entremezclar *vt* ELECTRÓN interlace

entrenador: ~ **avanzado** *m* TRANSP AÉR advanced trainer; ~ **básico** *m* TRANSP AÉR basic trainer; ~ **básico de vuelo de instrumentos** *m* TRANSP AÉR basic instrument flight trainer

entrenamiento *m* ING MECÁ training; ~ **para la campaña "La Seguridad es Primero"** *m* SEG "safety first" training

entrepuente *m* TRANSP MAR between decks, *construcción* tweendeck

entrerrosca *f* CONST, ING MECÁ nipple

entresacar *vt* MINAS sort

entretejido *adj* PROD, TEXTIL interwoven

entretela *f* TEXTIL interlining, *vestimenta* canvas

entretelar *vt* TEXTIL interline

entrevía *f* FERRO rail gage (*AmE*), rail gauge (*BrE*)

entropía *f* GEN entropy; ~ **de activación** *f* METAL activation entropy; ~ **configuracional** *f* METAL configurational entropy; ~ **específica** *f* FÍS specific entropy; ~ **de fusión** *f* TERMO entropy of fusion; ~ **de vaporización** *f* TERMO entropy of vaporization; ~ **vibracional** *f* METAL vibrational entropy

entubación *f* MINAS *pozos* tubbing, TEC PETR *conductos, tuberías* piping; ~ **por contracción** *f* PROD shrink tubing

entubado *m* MINAS *sondeos* tubbing, TEC PETR *perforación* casing; ~ **por contracción** *m* PROD shrink tubing; ~ **flexible** *m* ING MECÁ flexible tubing; ~ **de junta de inserción** *m* PROD inserted-joint casing; ~ **de mampostería** *m* MINAS stone tubbing; ~ **del pozo** *m* CARBÓN well casing

entubar *vt* CONST case, MINAS *pozo* line, *sondeos* tub

enturbamiento: ~ **de aceites por el frío** *m* REFRIG winterization

enumeración *f* MATEMÁT enumeration

enumerar *vt* INFORM&PD list

enunciado: ~ **de Clausius** *m* FÍS Clausius statement; ~ **inicial** *m* PROD start statement; ~ **de Kelvin** *m* FÍS Kelvin statement

envainado *m* NUCL jacketing, TEC ESP cladding

envainar *vt* AGRIC, ELEC, ING ELÉC, MECÁ, ÓPT, PROD sheathe

envasado[1]: ~ **bajo atmósfera controlada** *adj* ALIMENT *carne*, EMB controlled-atmosphere packed; ~ **con gas** *adj* ALIMENT gas-packed; ~ **al vacío** *adj* ALIMENT vacuum-packed

envasado[2] *m* ALIMENT, EMB packaging, ING MECÁ, MECÁ packing, PROD packaging; ~ **bajo ambiente controlado** *m* EMB controlled-atmosphere packed; ~ **en bandejas** *m* EMB tray packaging; ~ **en bolsa** *m* EMB bag packaging

envasadora: ~ **en bandejas** *f* EMB tray-packing machine

envase *m* ALIMENT flask, CARBÓN container, EMB package, packaging, ING MECÁ, MECÁ packing, PAPEL, PROD packaging, TEC ESP, TEXTIL package; ~ **de aerosol** *m* ING MECÁ aerosol container; ~ **para alimentos** *m* ALIMENT, EMB food packaging; ~ **biodegradable** *m* EMB, RECICL biodegradable packaging; ~ **"bolsa en caja"** *m* EMB bag-in-a-box packaging; ~ **"bolsa en lata"** *m* EMB bag-in-a-can packaging; ~ **de cartón a prueba de líquidos** *m* EMB liquid-proof carton; ~ **de cartulina** *m* EMB, PAPEL cardboard packaging; ~ **de espuma plástica endurecida** *m* EMB plastic-foam packaging; ~ **en lámina al vacío** *m* EMB blister card, blister pack; ~ **no utilizable para alimentos** *m* EMB nonfood packaging; ~ **prefabricado** *m* EMB prefabricated package; ~ **de tres paquetes** *m* EMB triple pack; ~ **para la venta al por menor** *m* EMB retail package

envases: ~ **vacíos** *m pl* EMB empties

envejecer: ~ **artificialmente** *vt* TERMO artificially age

envejecido: ~ **artificialmente** *adj* TERMO artificially aged

envejecimiento[1]: **de** ~ **natural** *adj* TERMO naturally aged

envejecimiento[2] *m* ALIMENT, CINEMAT, CONST ageing, CRISTAL age hardening, ageing, ELEC, EMB, IMPR, P&C, PAPEL ageing, PROD, TEC ESP burn-in, TELECOM ageing; ~ **artificial** *m* METAL, TERMO *metales* artificial ageing; ~ **en caliente** *m* P&C heat ageing; ~ **por calor** *m* P&C heat ageing; ~ **en el horno** *m* P&C oven ageing; ~ **interrumpido** *m* P&C interrupted ageing; ~ **natural** *m* METAL, P&C, TERMO natural ageing; ~ **progresivo** *m* P&C *revestimientos* progressive ageing; ~ **por temple** *m* METAL *tratamiento maraging*, NUCL quench ageing

envenenamiento *m* QUÍMICA poisoning; ~ **directo** *m* CONTAM direct poisoning

enverdecer *vt* COLOR green

envergadura *f* FÍS *de una ala* span, TEC ESP spread, TRANSP AÉR span; ~ **alar** *f* TRANSP AÉR wing span

envergar *vt* TRANSP MAR *velas* bend

envés: ~ **contra envés** *adj* TEXTIL *tejido* back-to-back

enviar *vt* INFORM&PD send, TRANSP MAR *mercancías* ship; ~ **por fax** *vt* IMPR, INFORM&PD, TELECOM fax; ~ **muestras** *vt* TELECOM sample; ~ **una señal** *vt* ELECTRÓN, TRANSP MAR signal

envío *m* INFORM&PD sending; ~ **de la carga por avión** *m* EMB air freight; ~ **de carga parcial** *m* TRANSP partload consignment; ~ **por correo** *m* PROD posting; ~ **directo** *m* PROD drop shipment; ~ **de intercambio de datos electrónicos** *m* TELECOM EDI forwarding, electronic data interchange forwarding; ~ **de transmisión** *m* TELECOM transmission sending; **~-recepción automáticos** *m* (*ASR*) INFORM&PD automatic send-receive (*ASR*)

envoltorio: ~ **de amianto** *m* AmL (*cf envoltorio de asbesto Esp*) CONST asbestos-plaited packing; ~ **de asbesto** *m* Esp (*cf envoltorio de amianto AmL*) CONST asbestos-plaited packing; ~ **encogible** *m* TERMO shrink film (*BrE*), shrink wrap (*AmE*); ~ **termosellado** *m* TERMO heat-sealed wrapping

envoltura *f* ALIMENT casing, CONST housing, ELEC *cable* jacket (*AmE*), sheath (*BrE*), EMB wraparound, ING MECÁ wrapping, jacket, NUCL jacket, PROD *cables eléctricos* sheath, jacket, sheathing, TEC ESP *protección de vehículos* casing, TELECOM envelope, TEXTIL wrapping, TV envelope; ~ **de alambre** *f* ING ELÉC wire wrapping; ~ **de la bandeja con un material retráctil** *f* Esp (*cf envoltura de la pálet con un material retráctil AmL*) EMB pallet shrink-wrapping; ~ **cinemática** *f* FERRO *vehículos* kinematic envelope; ~ **conformada por estiramiento** *f* EMB stretch wrapping; ~ **de contracción** *f* ALIMENT, EMB shrink film (*BrE*), shrink wrap (*AmE*); ~ **de fruta** *f* ALIMENT, EMB fruit wrapper; ~ **interior** *f* EMB interior wrapping; ~ **metálica** *f* ELEC metallic sheath; ~ **del motor** *f* TEC ESP motor case; ~ **de la pálet con un material retráctil** *f* AmL (*cf envoltura de la bandeja con un material retráctil Esp, envoltura de la paleta con un material retráctil Esp*) EMB pallet shrink-wrapping; ~ **de la paleta con un material retráctil** *f* Esp (*cf envoltura de la pálet con un material retráctil AmL*) EMB pallet shrink-wrapping; ~ **de papel** *f* EMB paper wrapping; ~ **con película** *f* EMB film wrap; ~ **de plomo** *f* ELEC *cable* lead sheath, lead sheathing; ~ **protectora** *f* EMB protective wrapper, REVEST protective covering; ~ **repelable** *f* EMB peel-off wrapping; ~ **retráctil** *f* EMB shrink wrapping

envolvedora: **~ con manguitos** *f* EMB wraparound sleeving machine

envolvente *m* ING ELÉC envelope, envelopment; **~ de cilindros** *m* PROD roll shell; **~ de la señal** *m* TELECOM signal envelope; **~ de vuelo** *m* TRANSP AÉR flight envelope

envolver *vt* ING MECÁ wrap, PROD jacket, TEXTIL wrap

envuelta *f* INSTAL HIDRÁUL *turbina* card wire, clothing, enclosed casing, PROD, REVST deading, jacket, TEC ESP skin, TERMO jacket; **~ de agua** *f* TERMO water jacket; **~ aisladora** *f* PROD insulating jacketing; **~ aislante** *f* PROD insulating jacketing, insulating lagging; **~ de caracol** *f* PROD *bomba centrífuga, turbina, ventilador* spiral casing; **~ de chapa aislando la caldera** *f* INSTAL HIDRÁUL boiler jacketing; **~ exterior calefactora** *f* TERMO heating jacket; **~ termoaislante** *f* INSTAL HIDRÁUL lagging; **~ de turbina** *f* TRANSP MAR turbine casing

enyesado: **~ de perlita** *m* TERMOTEC perlite plaster

enzima *f* ALIMENT, DETERG, QUÍMICA, RECICL enzyme; **~ digestiva** *f* RECICL digestive enzyme

EOB *abr* (*fin de bloque*) INFORM&PD EOB (*end of block*)

Eoceno *adj* GEOL Eocene

EOD *abr* (*fin de datos*) INFORM&PD EOD (*end of data*)

EOF *abr* (*fin de fichero*) INFORM&PD EOF (*end of file*)

EOJ *abr* (*fin de trabajo*) INFORM&PD EOJ (*end of job*)

eólico *adj* GEOL, PETROL aeolian

EOM *abr* (*fin de mensaje*) INFORM&PD, TELECOM EOM (*end of message*)

eosina *f* QUÍMICA eosin

EOT *abr* INFORM&PD, TELECOM (*fin de transmisión*) EOT (*end of transmission*), TELECOM (*fin de movimiento*) EOT (*end of transaction*)

EP *abr* (*equilibrio potencial*) GEOFÍS EP (*equilibrium potential*)

EPA *abr* (*equipo de pruebas automático*) INFORM&PD ATE (*automatic test equipment*)

epicentro *m* FÍS, GEOFÍS, GEOL epicenter (*AmE*), epicentre (*BrE*)

epicíclico *adj* ING MECÁ, MECÁ, TRANSP MAR epicyclic

epicicloidal *adj* ING MECÁ, MECÁ, TRANSP MAR epicycloidal

epicicloide *f* GEOM epicycloid

epicinconina *f* QUÍMICA epicinchonine

epiclástico *adj* PETROL epiclastic

epiclorhidrina *f* QUÍMICA epichlorhydrin

epicontinental *adj* GEOL, PETROL epicontinental

epicoprostanol *m* QUÍMICA epicoprostanol

epidehidroandrosterona *f* QUÍMICA epidehydroandrosterone

epidiascopio *m* FOTO, INSTR, ÓPT epidiascope

epidiorita *f* PETROL epidiorite

epídota *f* MINERAL, QUÍMICA delphinite, epidote

epigenético *adj* GEOL epigenetic

epigénico *adj* GEOL epigenic

epimerización *f* QUÍMICA epimerization

epímero *m* QUÍMICA epimer

epirogénesis *f* GEOL epeirogenesis

episodio: **~ de contaminación atmosférica** *m* CONTAM air pollution incident, air pollution problem

epistilbita *f* MINERAL epistilbite

epitaxia *f* CRISTAL, ELECTRÓN, FÍS RAD, METAL, PROC QUÍ, TELECOM epitaxy; **~ en fase líquida** *f* ELECTRÓN liquid-phase epitaxy; **~ en fase vapor** *f* PROC QUÍ vapor phase epitaxy (*AmE*), vapour phase epitaxy

(*BrE*); **~ por haz molecular** *f* ELECTRÓN, FÍS RAD molecular-beam epitaxy (*MBE*)

epitaxial *adj* CRISTAL, ELECTRÓN, FÍS RAD, METAL, PROC QUÍ, TELECOM epitaxial

epitermal *adj* GEOL, MINERAL epithermal

epizona *f* GEOL epizone

epizootia *f* AGRIC *ganado* epidemic disease

EPNL *abr* (*nivel efectivo de ruido percibido*) ACÚST EPNL (*effective perceived-noise level*)

época *f* AGRIC, GEOL, PETROL age, date, era, period, season; **~ de emplazamiento** *f* GEOL emplacement age; **~ glacial** *f* GEOL glacial stage; **~ húmeda** *f* CONTAM wet period; **~ de parición** *f* AGRIC calving season; **~ de pastoreo** *f* AGRIC grazing season; **~ de polaridad** *f* GEOL polarity epoch; **~ de recolección** *f* AGRIC picking season; **~ de siembra** *f* AGRIC planting season; **~ solsticial** *f* TEC ESP solstitial period

epoxia: **~ en oro** *f* ELECTRÓN gold epoxy

epóxido *m* P&C, QUÍMICA epoxide

EPROM *abr* (*PROM borrable*) INFORM&PD EPROM (*erasable PROM*)

epsomita *f* MINERAL epsomite

eptano *m* TEC PETR heptane

equiangular *adj* GEOM equiangular

equidimensional *adj* GEOL equant

equidistante *adj* CONST, GEOM, ING MECÁ equally-spaced, equidistant

equigranular *adj* GEOL equigranular

equilateral *adj* GEOM equilateral

equilátero *adj* GEOM equilateral

equilenina *f* QUÍMICA equilenin

equilibrado[1] *adj* TRANSP push-pull, TRANSP AÉR balanced; **~ con respecto tierra** *adj* ING ELÉC balanced to earth

equilibrado[2] *m* AUTO balancing, METR *de una balanza* adjustment, NUCL, TELECOM balancing, TRANSP AÉR balancing, trim, TRANSP MAR counterflooding; **~ de impedancias** *m* FÍS, ING ELÉC impedance matching; **~ lateral** *m* TRANSP AÉR lateral trim; **~ de la pala** *m* TRANSP AÉR blade balance; **~ del rotor flexible** *m* ING MECÁ flexible-rotor balance

equilibrador[1] *adj* ELEC, TELECOM *electricidad* push-pull

equilibrador[2] *m* GEN equalizer; **~ de agua** *m* HIDROL *capa de agua* water balance; **~ de CC** *m* ELEC *generador* DC balancer; **~ de corriente** *m* ELEC, FÍS, INSTR current balance; **~ de corriente alterna** *m* ELEC alternating-current balancer; **~ estático** *m* ING ELÉC static balancer

equilibrar *vt* ING MECÁ, MECÁ counterbalance, TEC ESP trim, *dirigibles* top up

equilibrio[1] *m* GEN balance, equalization, equilibrium; **~ aerodinámico** *m* TRANSP AÉR aerodynamic balance; **~ de amperios** *m* ELEC, INSTR ampere balance; **~ del anhídrido carbónico** *m* HIDROL carbonic acid equilibrium; **~ del avión** *m* TRANSP AÉR aircraft balance; **~ de colores** *m* CINEMAT, FOTO color balance (*AmE*), colour balance (*BrE*); **~ de la columna** *m* CONST pillar balance; **~ de la corteza terrestre** *m* ENERG RENOV isostasy; **~ cromático** *m* COLOR color match (*AmE*), color matching (*AmE*), colour match (*BrE*), colour matching (*BrE*), TV chromatic balance; **~ dinámico** *m* ING MECÁ dynamic balance, TRANSP AÉR dynamic balancing; **~ energético** *m* GAS energy equilibrium; **~ energético del suelo** *m* AGRIC, HIDROL energy balance of

the soil; ~ **de energía en turbulencia** *m* FÍS FLUID energy balance in turbulence; ~ **estable** *m* FÍS stable equilibrium; ~ **estático** *m* ING MECÁ static balance; ~ **estático de las muelas** *m* ING MECÁ static balance of grinding wheels; ~ **de fases** *m* TERMO phase equilibrium; ~ **de grises** *m* IMPR gray balance (*AmE*), grey balance (*BrE*); ~ **hidrófilo** *m* DETERG hydrophile balance; ~ **hidrológico** *m* HIDROL hydrologic balance; ~ **hidrostático** *m* FÍS FLUID, HIDROL hydrostatic balance, TERMO hydrostatic equilibrium; ~ **inestable** *m* FÍS unstable equilibrium; ~ **metaestable** *m* FÍS metastable equilibrium; ~ **de la onda portadora** *m* TV carrier balance; ~ **óptico** *m* IMPR optical balance; ~ **potencial** *m* (*EP*) GEOFÍS equilibrium potential (*EP*); ~ **radiactivo** *m* FÍS RAD radioactive equilibrium; ~ **secular** *m* FÍS secular equilibrium; ~ **de tensiones** *m* ELEC voltage balance; ~ **térmico** *m* TERMO heat balance, temperature balance, thermal equilibrium, TERMOTEC thermal equilibrium; ~ **termodinámico** *m* TEC ESP thermodynamic equilibrium; ~ **de torsión** *m* FÍS torsion balance; ~ **transiente** *m* FÍS transient equilibrium; ~ **del túnel aerodinámico** *m* TRANSP AÉR wind tunnel balance; ~ **en vuelo rectilíneo** *m* TEC ESP *aviones* trim
equilibrio²: **no en ~** *fra* FÍS not in equilibrium
equimolecular *adj* QUÍMICA equimolecular
equinoccio: ~ **de primavera** *m* TEC ESP vernal equinox; ~ **vernal** *m* TEC ESP vernal equinox
equinocromos *m pl* QUÍMICA echinochromes
equinodermos *m pl* GEOL, OCEAN echinoderms
equipado: ~ **con** *adj* TELECOM fitted with; ~ **con batería** *adj* ING ELÉC battery-powered
equipaje: ~ **de cabina** *m* TRANSP, TRANSP AÉR cabin baggage; ~ **de mano** *m* TRANSP AÉR hand baggage
equipamiento *m* CONST equipment, fitting out, ING MECÁ outfit, equipment, PROD equipment; ~ **electrónico industrial** *m* ING ELÉC industrial electronic equipment; ~ **de E/S** *m* PROD I/O hardware; ~ **de montaje** *m* PROD mounting hardware set; ~ **opcional** *m* PROD optional hardware; ~ **de plataforma** *m* TEC PETR *operaciones costa-fuera* platform equipment
equipar *vt* ING MECÁ fit out, MINAS equip, PROD *fábricas, talleres* equip, rig out, rig up, rig, TELECOM, TRANSP MAR *buque* equip; ~ **con** *vt* ING MECÁ fit with
equipartición *f* FÍS equipartition
equipo *m* C&V *grupo de trabajadores* shop, CINEMAT *de personas* crew, CONST furnishing, equipment, tackle, FOTO apparatus, outfit, ING MECÁ outfit, equipment, kit, gear, PROD *fábricas, talleres* equipment, crew, *de personas* team, TEC ESP team, VEH unit;
~ a ~ **del abonado** *m* TELECOM customer equipment (*CEQ*); ~ **de acabado en línea** *m* EMB in-line finishing equipment; ~ **de acceso externo** *m* TELECOM external access equipment (*EA*); ~ **accesorio** *m* MECÁ ancillary equipment; ~ **agitador** *m* PROC QUÍ stirring equipement; ~ **de aire acondicionado con compresor** *m* CONST compressed air-conditioning unit; ~ **de aire comprimido** *m* CONST, ING MECÁ compressed-air equipment; ~ **aislador de ruidos** *m* SEG noise-insulating equipment; ~ **de alimentación en cadena** *m* EMB line-feeding equipment; ~ **para almacenamiento de datos** *m* INFORM&PD data-switching exchange (*DSE*); ~ **de alta tensión** *m* ELEC high-voltage equipment; ~ **de análisis de agua** *m* LAB water analysis kit; ~ **anamorfizador** *m*

CINEMAT anamorphoser; ~ **de apisonar** *m* MINAS tamping material; ~ **autoelevable** *m* TEC PETR *perforación* jack-up rig; ~ **auxiliar** *m* MECÁ ancillary equipment;
~ b ~ **del barco** *m* OCEAN vessel equipment; ~ **de bombeo** *m* AGUA pumping equipment; ~ **de a bordo** *m* TEC ESP on-board equipment; ~ **de a bordo de ambulancia** *m* TRANSP AÉR ambulance installation; ~ **de a bordo del avión** *m* TRANSP AÉR aircraft equipment; ~ **de buceo** *m* OCEAN diving gear, PETROL diving equipment;
~ c ~ **de cabezal de pozo** *m* PETROL wellhead equipment; ~ **de calibrado en pozos de perforación** *m* GEOFÍS *sondeos* borehole logging-equipment; ~ **calibrador** *m* METR gage stand (*AmE*), gauge stand (*BrE*); ~ **de cámara** *m* CINEMAT camera equipment, TV camera equipment, *referido al personal* camera crew; ~ **de camuflaje antiaéreo** *m* D&A anti-air camouflage equipment; ~ **de canales** *m* TELECOM channel equipment; ~ **para captación de polvo** *m* CONST dedusting unit; ~ **de carga** *m* TELECOM charging equipment; ~ **del casco** *m* TRANSP AÉR headgear; ~ **central** *m* IMPR front-end equipment; ~ **para cerrar la tapa** *m* EMB cap-sealing equipment; ~ **del circuito de terminación de datos** *m* TELECOM data circuit-terminating equipment; ~ **del cliente** *m* TELECOM customer equipment (*CEQ*); ~ **para combate de incendios** *m* SEG, TERMO firefighting equipment; ~ **para combate de incendios y rescate** *m* SEG, TERMO firefighting and rescue equipment; ~ **de combinador** *m* ELEC control gear; ~ **compacto** *m* ELECTRÓN solid-state device; ~ **del compartimento de carga** *m* TRANSP AÉR cargo compartment equipment; ~ **compensador de campo** *m* ING MECÁ field-balancing equipment; ~ **completo** *m* INSTR stand; ~ **común** *m* TELECOM common equipment; ~ **de conmutación** *m* INFORM&PD switching equipment; ~ **de conmutación por la voz** *m* TELECOM voice-switching equipment; ~ **conmutador digital** *m* ING ELÉC digital switching equipment; ~ **de control** *m* CALIDAD inspection equipment, ELEC *instrumento* control gear, monitor, TELECOM control equipment; ~ **de control automático** *m* TEC PETR automatic control equipment; ~ **de control común** *m* TELECOM common control equipment; ~ **de control reglamentario** *m* PROD standard control equipment; ~ **de conversión de doble ingreso** *m* PROD dual-input conversion kit; ~ **convertidor de señal** *m* ELECTRÓN, TELECOM signal conversion equipment; ~ **para copiado de diapositivas** *m* FOTO slide-copying device; ~ **de craqueo con catalizador fluidizado** *m* PROC QUÍ cracking equipment with fluidized catalyst; ~ **de cubierta** *m* TRANSP MAR deck fittings, deck gear;
~ d ~ **de desactivación de explosivos** *m* D&A bomb-disposal team; ~ **desanamorfizador** *m* CINEMAT deanamorphoser; ~ **de desarenado** *m* ING MECÁ dressing equipment; ~ **de desimanación** *m* PROD demagnetizing equipment; ~ **de desmontaje** *m* TEC ESP strapdown equipment; ~ **de distribución** *m* ELEC switchgear; ~ **dosificador** *m* EMB metering equipment; ~ **de drenaje del freno** *m* AUTO VEH brake bleeder unit;
~ e ~ **eléctrico** *m* ELEC, ING ELÉC electrical equipment; ~ **eléctrico de protección** *m* ELEC, SEG

electrical protection equipment; ~ **electrónico** *m* ELECTRÓN electronic equipment; ~ **electrónico aeronáutico de a bordo** *m* TRANSP AÉR avionics; ~ **electrónico doméstico** *m* ING ELÉC domestic electronic equipment; ~ **electrónico del usuario** *m* ING ELÉC consumer electronic equipment; ~ **para elevar** *m* ING MECÁ lifting equipment; ~ **para embalar y manejar bobinas** *m* EMB reel-wrapping-and-handling equipment; ~ **de emergencia** *m* SEG, TRANSP AÉR emergency equipment; ~ **emulsionante** *m* PROC QUÍ emulsifying machine; ~ **de engranaje planetario** *m* AUTO planetary gear set; ~ **de engranajes** *m* ING MECÁ gear assembly; ~ **para enjuagar contenedores** *m* EMB container-rinsing equipment; ~ **de enrollado-desenrollado** *m* EMB wind-unwind equipment; ~ **de ensayo** *m* ING MECÁ tryout facility; ~ **equilibrador de campo** *m* ING MECÁ field-balancing equipment; ~ **de estabilización automática** *m* TRANSP automatic stabilization equipment (*ASE*); ~ **de excavaciones** *m* ING MECÁ earthmoving machinery; ~ **de exploración** *m* TEC PETR *perforación* exploration rig; ~ **de extracción** *m* MINAS *pozo* hoisting system; ~ **de extracción Soxhlet** *m* LAB Soxhlet extraction equipment;

~f ~ **de fibra óptica** *m* LAB, ÓPT, TELECOM fiber-optics equipment (*AmE*), fibre-optics equipment (*BrE*); ~ **fibroóptico** *m* LAB, ÓPT, TELECOM fiber-optics equipment (*AmE*), fibre-optics equipment (*BrE*); ~ **fijo** *m* CONST fixed equipment; ~ **físico** *m* INFORM&PD hardware; ~ **para flejar** *m* EMB strapping equipment; ~ **de flotación** *m* PROC QUÍ flotation equipment; ~ **de flotación de emergencia** *m* TRANSP AÉR emergency flotation gear; ~ **de fondeo** *m* TRANSP MAR *amarres* ground tackle; ~ **de fondeo y amarre** *m* TRANSP MAR mooring gear; ~ **de formación** *m* TEC PETR *perforación* rig; ~ **para formación de perfiles estampados** *m* PROD roll-forming equipment; ~ **fotográfico** *m* FOTO photographic apparatus; ~ **de fracción** *m* PROC QUÍ fractionating apparatus;

~g ~ **generador de fuerza eléctrica auxiliar** *m* TRANSP AÉR auxiliary power unit; ~ **giroscópico del piloto automático** *m* TRANSP AÉR autopilot gyro unit; ~ **de grabación a distancia** *m* INSTR telerecording equipment; ~ **de grapado** *m* EMB stapling equipment;

~h ~ **de herramientas** *m* ING MECÁ outfit, set of tools; ~ **de herramientas de carburo al tungsteno** *m* ING MECÁ tungsten-carbide tooling; ~ **de herramientas para jardín** *m* CONST lawn rake kit; ~ **hidráulico** *m* ING MECÁ hydraulic equipment; ~ **hidroestático** *m* ING MECÁ hydrostatic equipment; ~ **de humidificación** *m* EMB moistening equipment;

~i ~ **de iluminación** *m* FOTO lighting equipment; ~ **impelente** *m* MINAS plunger set; ~ **inamovible** *m* PROD fixed plant; ~ **inhabilitado** *m* TELECOM disabled equipment; ~ **para insertar una señal de división** *m* EMB division-inserting equipment; ~ **de inspección** *m* CALIDAD, EMB inspection equipment; ~ **de intercepción** *m* TELECOM interception equipment; ~ **de interconexión digital** *m* TELECOM digital cross-connect equipment (*DXC*); ~ **de interdicción de llamadas** *m* TELECOM call-barring equipment; ~ **para inyecciones** *m* CONST grouting equipment; ~ **de irradiación por cobalto 60** *m* NUCL cobalt 60 irradiation plant;

~l ~ **de levigación** *m* PROC QUÍ, QUÍMICA elutriating machine; ~ **limpiador de gases** *m* SEG cleaning equipment; ~ **de limpieza y de comprobación de desagües** *m* CONST drain testing and cleaning equipment; ~ **en locales del abonado** *m* TELECOM customer premises equipment; ~ **de lubricación** *m* CONST lubricating unit; ~ **de lubricación manual** *m* ING MECÁ manual lubricating equipment;

~m ~ **de manejo manual** *m* ING MECÁ manual handling equipment; ~ **de manejo mecánico continuo** *m* ING MECÁ continuous mechanical handling equipment; ~ **de maniobra** *m* ING ELÉC switchgear, PROD control equipment; ~ **de manipulado** *m* EMB handling equipment; ~ **de manipulado de paquetes** *m* EMB pack-handling equipment; ~ **manipulador mecánico continuo** *m* ING MECÁ continuous mechanical handling equipment; ~ **manipulador mecánico continuo para materiales a granel** *m* ING MECÁ continuous mechanical handling equipment for loose bulk materials; ~ **de manutención y llenado** *m* EMB handling-and-filling equipment; ~ **del maquinista** *m* FERRO driving crew; ~ **para marcar** *m* EMB marking equipment; ~ **de máxima tensión** *m* ELEC highest voltage for equipment; ~ **para la medición de la distancia de visibilidad** *m* TRANSP visibility distance-measuring equipment; ~ **de medición de distancias** *m* METR, TRANSP AÉR distance-measuring equipment (*DME*); ~ **de medición excéntrica** *m* METR cam measuring equipment; ~ **de medida del contorno** *m* METR contour-measuring equipment; ~ **para medir ruidos y vibraciones** *m* INSTR, SEG noise-and-vibration measuring equipment; ~ **de mezcla y combinación** *m* EMB mixing-and-blending equipment; ~ **motor** *m* CONTAM MAR power pack; ~ **de movimiento de tierra** *m* ING MECÁ earthmoving machinery; ~ **de multiplicación del circuito digital** *m* (*DCME*) TELECOM digital circuit multiplication equipment (*DCME*);

~n ~ **neumático** *m* ING MECÁ pneumatic equipment; ~ **de nivelación** *m* CONST leveling instrument (*AmE*), levelling instrument (*BrE*); ~ **de numeración automática** *m* TELECOM automatic numbering equipment;

~o ~ **para obtener películas por extrusión** *m* EMB film extrusion equipment; ~ **opcional** *m* ING MECÁ optional equipment; ~ **optativo** *m* ING MECÁ optional equipment;

~p ~ **de perforación** *m* MINAS *sondeos* drilling rig, PETROL drilling platform, TEC PETR drilling rig, drilling platform, rig; ~ **de perforación flotante** *m* TEC PETR *perforación costa-fuera* floating rig; ~ **de perforación geotérmico** *m* ENERG RENOV geothermal drilling equipment; ~ **perforador de percusión** *m* MINAS *sondeos* percussion rig; ~ **perforador tipo barco** *m* PETROL ship-type rig; ~ **para perforar** *m* MINAS *sondeos* drilling equipment; ~ **pesado** *m* ING MECÁ heavy-duty gear; ~ **de potencia auxiliar** *m* TRANSP AÉR auxiliary power unit; ~ **de potencia hidráulica** *m* TEC ESP hydraulic power pack; ~ **de presión** *m* ING MECÁ, TEC PETR pressure equipment; ~ **de presurizado** *m* ING MECÁ, TEC PETR pressure equipment; ~ **procesador de mensajes** *m* TRANSP MAR *comunicación por satélite* message processing equipment; ~ **de protección al aparato respiratorio** *m* SEG respiratory protective equipment; ~ **de pro-**

tección individual *m* D&A individual protection equipment (*IPE*); ~ protector eléctrico *m* ELEC, SEG electrical protection equipment; ~ de prueba *m* NUCL test assembly; ~ de pruebas automático *m* (*EPA*) INFORM&PD automatic test equipment (*ATE*); ~ de pruebas servohidráulico *m* ING MECÁ servohydraulic test equipment; ~ de pulverización *m* PROC QUÍ pulverizing equipment; ~ de purga del freno *m* AUTO brake bleeder unit; ~ para purificar y desodorizar aire *m* SEG air-purification-and-deodorization equipment;

■ r ~ de radar *m* TRANSP radar equipment; ~ de radio *m* TRANSP radio equipment; ~ de reavivado *m* ING MECÁ dressing equipment; ~ de reavivar muelas abrasivas *m* ING MECÁ, PROD grinding-wheel dressing equipment; ~ de rebarbado *m* ING MECÁ dressing equipment; ~ de recogida de datos *m* TELECOM data collection equipment; ~ para recolección de datos *m* TEC ESP data collection platform (*DCP*); ~ de rectificar ruedas abrasivas *m* ING MECÁ, PROD grinding-wheel dressing equipment; ~ recuperable *m* TEC ESP recovery package; ~ de refrigeración *m* ING MECÁ, REFRIG, TEXTIL cooling equipment; ~ refrigerante *m* ING MECÁ, REFRIG, TEXTIL cooling equipment; ~ de relleno *m* MINAS stowing equipment; ~ de reparación de averías *m* CONST breakdown gang; ~ de rescate *m* SEG, TRANSP AÉR, TRANSP MAR rescue equipment; ~ de rescate aeromarítimo *m* D&A, SEG, TRANSP air-sea rescue equipment; ~ de reserva *m* TEC PETR *perforación* backup, casing; ~ respiratorio *m* OCEAN respirator; ~ respiratorio autónomo *m* OCEAN aqualung; ~ para resucitación *m* SEG resuscitation equipment; ~ de revelado *m* CINEMAT processing equipment; ~ para el revelado continuo *m* FOTO continuous processing machine;

■ s ~ salvavidas *m* SEG, TRANSP MAR life-saving apparatus; ~ de sangrado del freno *m* AUTO brake bleeder unit; ~ de seguridad *m* SEG safety equipment; ~ de señalización por computadora *m* *AmL* (*cf equipo de señalización por ordenador Esp*) FERRO computerized signaling equipment (*AmE*), computerized signalling equipment (*BrE*); ~ de señalización de líneas *m* TELECOM line signaling equipment (*AmE*), line signalling equipment (*BrE*); ~ de señalización por ordenador *m Esp* (*cf equipo de señalización por computadora AmL*) FERRO computerized signaling equipment (*AmE*), computerized signalling equipment (*BrE*); ~ de servicio de tierra *m* TRANSP AÉR ground service equipment; ~ de sobrevivencia *m* TEC ESP survival kit; ~ de soldadura por alta frecuencia *m* EMB high-frequency welding equipment; ~ de soldadura por resistencia *m* CONST, ELEC, ING ELÉC, ING MECÁ, TERMO resistance-welding equipment; ~ de soldadura para varios operarios *m* PROD multiple-operator welding set; ~ de soldeo por resistencia *m* CONST, ELEC, ING ELÉC, ING MECÁ, TERMO resistance-welding equipment; ~ de sondeo *m* MINAS drilling equipment; ~ de sondeo marítimo *m* CONST marine drilling rig; ~ de superficie *m* MINAS *sondeos* surface equipment; ~ de supervivencia *m* D&A, SEG, TRANSP AÉR survival kit;

■ t ~ telefónico con teclado *m* TELECOM key telephone set; ~ de terminación de línea *m* (*ETL*) INFORM&PD line termination equipment (*LTE*),

TELECOM line-terminating equipment; ~ terminal *m* TELECOM terminal equipment; ~ terminal B de la RDSI *m* TELECOM B-ISDN terminal equipment (*TE-LB*); ~ terminal del circuito de datos *m* TELECOM data circuit terminal equipment; ~ terminal de datos *m* (*DTE*) INFORM&PD, TELECOM, TRANSP AÉR data terminal equipment (*DTE*); ~ terminal móvil de procesamiento de datos *m* TELECOM mobile data-processing terminal equipment; ~ terminal para el proceso de datos *m* TELECOM data-processing terminal equipment; ~ terminal de la red *m* TELECOM network terminal equipment (*NTE*); ~ para el termosellado *m* EMB heat-sealing equipment; ~ de trabajo *m* PROD team; ~ para transfusión de sangre *m* SEG blood transfusion equipment; ~ de transmisión plesiosincrónico *m* TELECOM plesiosynchronous transmission equipment; ~ de tubos *m* ELECTRÓN multigun tube;

■ v ~ de vacío *m* ING MECÁ vacuum equipment;

■ z ~ para el zunchado con alambre *m* EMB wire-strapping equipment

equipos: ~ centrales de carga *m pl* TELECOM central charging equipment; ~ de conmutación *m pl* NUCL switchgear; ~ de conmutación y distribución *m pl* ELEC switchgear; ~ de producción *m pl* PROD production facilities; ~ de soldadura oxiacetilénica *m pl* PROD oxyacetylene welding equipment; ~ de tarificación de llamadas *m pl* TELECOM call-charging equipment; ~ de tierra *m pl* TEC ESP ground facilities

equipotencia: ~ del campo *f* ELEC, ING ELÉC field strength

equipotencial[1] *adj* ELEC, FÍS, GEOFÍS, ING ELÉC, TEC ESP equipotential

equipotencial[2] *m* GEN equipotential

equivalencia *f* INFORM&PD, MATEMÁT, NUCL, TRANSP equivalence; ~ de masa y energía *f* FÍS mass-energy equivalence

equivalente[1]: ~ a tejido *adj* NUCL tissue equivalent

equivalente[2]: ~ de agua *m* FÍS water equivalent; ~ de arena *m* CONST *geotecnica* sand equivalent; ~ a un barril de crudo *m* TEC PETR barrel oil equivalent (*BOE*); ~ a coche de pasajeros *m* TRANSP passenger car equivalent; ~ demográfica *m* AGUA population equivalent; ~ de Joule *m* ELEC, MECÁ, TERMO Joule's equivalent; ~ mecánico del calor *m* MECÁ, TERMO mechanical equivalent of heat; ~ por millón *m* CONTAM *unidad de concentración* equivalent per million; ~ de poblaciones *m* AGUA population equivalent; ~ térmico *m* TERMO heat equivalent, thermal equivalent

ER *abr* (*emisión-recepción*) ELECTRÓN TR (*transmit-receive*)

Er *abr* (*erbio*) QUÍMICA Er (*erbium*)

era *f* GEOL date, era, PETROL era, *estratigrafía* age; ~ de colada *f* METAL, PROD pig bed; ~ Cuaternaria *f* GEOL, TEC PETR Quaternary era; ~ espacial *f* TEC ESP space age; ~ Terciaria *f* GEOL, TEC PETR Tertiary era

eratema *m* GEOL erathem

erbio *m* (*Er*) QUÍMICA erbium (*Er*)

erección *f* ING MECÁ setting

eremeyevita *f* MINERAL eremeyevite

ergio *m* METR erg

ergodisipación *f* PROD power dissipation

ergol *m* TEC ESP, TERMO ergol

ergolimitador *m* ING MECÁ power limiter

ergómetro *m* ING ELÉC energy meter, INSTR, METR ergmeter
ergonomía *f* EMB, INFORM&PD, ING MECÁ, PROD, SEG, TEC ESP ergonomics
ergonómico *adj* EMB, INFORM&PD, ING MECÁ, PROD, SEG, TEC ESP ergonomic
ergonomista *m* EMB, INFORM&PD, ING MECÁ, PROD, SEG, TEC ESP ergonomist
ergotinina *f* QUÍMICA ergotinine
erguir *vt* ING MECÁ hold up
erial¹ *adj* AGRIC barren
erial² *m* AGRIC waste land
erigir *vt* CONST set
erinita *f* MINERAL erinite
eritrina *f* MINERAL cobalt bloom, QUÍMICA erythrin
eritrita *f* MINERAL erythrite
eritritol *m* QUÍMICA erythritol
eritropsina *f* QUÍMICA erythropsin
eritrosa *f* QUÍMICA erythrose
eritrosina *f* QUÍMICA eritrosine, erythrosine
eritrulosa *f* QUÍMICA erythrulose
erosión *f* CARBÓN erosion, *por gases, llamas* cutting, CONST, ENERG RENOV, FERRO, GEOL, HIDROL, METEO erosion, PETROL weathering, PROD *metal por chorro de oxígeno* cutting, QUÍMICA erosion, TEC PETR scouring; **~ por el agua** *f* HIDROL erosion by water; **~ por agua de superficie** *f* CARBÓN surface water erosion; **~ asurcada** *f* AGRIC rill erosion; **~ de una capa uniforme de suelo** *f* AGRIC sheet erosion; **~ costera** *f* HIDROL coastal erosion; **~ debido al agua** *f* HIDROL water erosion; **~ eólica** *f* CARBÓN weathering, CONTAM blowout, GEOL wind erosion; **~ fluvial** *f* HIDROL river erosion; **~ interna** *f* CARBÓN internal erosion; **~ por la lluvia** *f* GEOFÍS, GEOL, METEO rain erosion; **~ mecánica** *f* PETROL mechanical weathering; **~ natural** *f* AGRIC natural erosion; **~ subsuperficial** *f* AGUA subsurface erosion; **~ superficial** *f* HIDROL surface erosion
erosionado: ~ por el agua *adj* HIDROL water-eroded; **~ por la lluvia** *adj* GEOFÍS, GEOL, METEO rain-eroded; **~ por el viento** *adj* METEO wind-eroded
erosionar *vt* CONST, GEOL, HIDROL, METEO erode
errata *f* IMPR misprint
error *m* ELECTRÓN, FÍS, INFORM&PD error, ING MECÁ deviation, MECÁ fault; **~ absoluto** *m* INFORM&PD absolute error; **~ accidental** *m* CARBÓN, METR random error; **~ de ajuste** *m* TRANSP MAR *navegación* compass error; **~ aleatorio** *m* CARBÓN *estadística*, FÍS, METR, TELECOM random error; **~ de alineación** *m* TELECOM alignment fault; **~ analítico** *m* CARBÓN analysis error; **~ de angulación** *m* IMPR angling error; **~ angular del registro lateral** *m* ACÚST *grabación sonora* lateral-tracking angle error; **~ de ángulo** *m* METR angle error; **~ de ángulo de seguimiento vertical** *m* ACÚST *grabación sonora* vertical-tracking angle error; **~ de aproximación** *m* TELECOM approximation error; **~ de apuntamiento** *m* TEC ESP pointing error; **~ de base de tiempo** *m* TV timebase error; **~ de bloqueo distante** *m* TELECOM far-end block error (*FEBE*); **~ de cableado** *m* PROD wiring error; **~ casual** *m* CARBÓN, METR random error; **~ de codificación** *m* ELECTRÓN, INFORM&PD coding error; **~ de color** *m* TV color error (*AmE*), colour error (*BrE*); **~ compensado** *m* INFORM&PD balanced error; **~ de concentricidad núcleo-revestimiento** *m* ÓPT, TELECOM core-cladding concentricity error;

~ de concentricidad en el revestimiento del alma *m* ÓPT, TELECOM core-cladding concentricity error; **~ de concentricidad de la superficie de referencia del núcleo** *m* ÓPT, TELECOM core reference surface concentricity error; **~ de contrapresión** *m* TRANSP AÉR backlash error; **~ de convergencia** *m* TV convergence error; **~ cuadrantal** *m* TRANSP AÉR quadrantal error; **~ por cuadratura** *m* TV quadrature error; **~ de cuantificación** *m* ELECTRÓN, TELECOM quantization error; **~ de deriva** *m* TRANSP AÉR drift error; **~ de desincronización** *m* TELECOM out-of-sync error, out-of-synchronization error; **~ de desplazamiento** *m* TRANSP AÉR displacement error, drift error; **~ de diámetro menor** *m* METR minor diameter error; **~ en el diámetro del paso** *m* METR pitch-diameter error; **~ digital** *m* TELECOM digital error; **~ de distorsión oblícua** *m* TV skew error; **~ de empuje** *m* TEC ESP thrust misalignment; **~ de equivalencia de designación** *m* PETROL aliasing error; **~ de escritura** *m* INFORM&PD write error; **~ estándar** *m* MATEMÁT standard error; **~ estático** *m* TV static error; **~ por exceso** *m* ING ELÉC drop-in; **~ de exploración** *m* TV scanning error; **~ de explotación** *m* CONTAM *procesos industriales* operational error; **~ de fase** *m* ELECTRÓN, TV phase error; **~ de fase aleatoria** *m* TV random-phase error; **~ de fase casual** *m* TV random-phase error; **~ fatal** *m* INFORM&PD fatal error; **~ fijo en el radioaltímetro** *m* TRANSP AÉR fixed error on radio altimeter; **~ de fluctuación** *m* TV fluctuating error; **~ de frecuencia** *m* ELEC *de onda*, ELECTRÓN frequency deviation; **~ de fuera de sincronismo** *m* TELECOM out-of-sync error, out-of-synchronization error; **~ de fuera de sincronización** *m* TELECOM out-of-sync error, out-of-synchronization error; **~ gamma** *m* TV gamma error; **~ geométrico** *m* TV geometric error; **~ de hardware** *m* INFORM&PD hardware error; **~ de hechura** *m* METR *de la pieza de trabajo* form error; **~ de histéresis** *m* TRANSP AÉR hysteresis error; **~ humano** *m* CALIDAD, SEG human error; **~ indefinido** *m* INFORM&PD undefined error; **~ informático** *m* INFORM&PD machine error; **~ de instalación** *m* PROD installation error; **~ instrumental** *m* ING MECÁ, INSTR, TRANSP MAR instrument error; **~ de instrumento** *m* ING MECÁ, INSTR, TRANSP MAR instrument error; **~ intermitente** *m* ELEC intermittent fault; **~ en la intersección de trayectorias** *m* TRANSP MAR cross-track error; **~ intrínseco** *m* INSTR, METR intrinsic error; **~ irrecuperable** *m* INFORM&PD irrecoverable error; **~ del juego de los engranajes** *m* TRANSP AÉR backlash error; **~ de lectura** *m* INFORM&PD read error; **~ de linealidad** *m* TV linearity error; **~ de mantenimiento de altura** *m* TRANSP AÉR height-keeping error; **~ de máquina** *m* INFORM&PD machine error; **~ de marcación** *m* TELECOM dialing error (*AmE*), dialling error (*BrE*); **~ máximo** *m* ING MECÁ tolerance; **~ máximo de intervalo** *m* TELECOM maximum time interval error (*MTIE*); **~ máximo de intervalo relativo** *m* TELECOM maximum relative time interval error (*MRTIE*); **~ máximo permisible** *m* METR maximum permissible error; **~ mecánico** *m* TV mechanical error; **~ de la medida** *m* METR uncertainty of measurement; **~ de medida** *m* METR measuring error; **~ medio** *m* ELEC mean error; **~ medio cuadrático** *m* (*EMC*) INFORM&PD, MATE-

MÁT mean square error (*MSE*); ~ **de mensaje** *m* PROD message slip; ~ **de noventa grados** *m* TV ninety degree error; ~ **por omisión** *m* CALIDAD error of omission; ~ **de operación** *m* CARBÓN operating error, CONTAM *procesos industriales* operational error; ~ **operativo** *m* CONTAM *procesos industriales* operational error; ~ **de paralelismo** *m* GEOFÍS *debido a desigualdad de altura* tilt error; ~ **de paridad** *m* INFORM&PD parity error; ~ **de paridad de la memoria** *m* INFORM&PD memory parity error; ~ **de período de ejecución** *m* INFORM&PD, PROD run time error; ~ **permanente** *m* INFORM&PD permanent error; ~ **de procesado** *m* IMPR, INFORM&PD processing error; ~ **de puntería** *m* TEC ESP pointing error; ~ **recuperable** *m* INFORM&PD recoverable error; ~ **de redondeo** *m* INFORM&PD, MATEMÁT, PROD rounding error; ~ **de referencia** *m* TV guide error; ~ **relativo** *m* INFORM&PD relative error; ~ **de restricción de presentación de acontecimientos** *m* TELECOM event-presentation restriction error; ~ **de ruptura** *m* INFORM&PD truncation error; ~ **de seguimiento** *m* METR, TELECOM tracking error; ~ **de seguimiento instantáneo** *m* TEC ESP instantaneous-tracking error; ~ **de seguimiento lateral** *m* TEC ESP along-track error; ~ **de seguimiento oblicuo** *m* TEC ESP across-track error; ~ **de seguimiento transversal** *m* TEC ESP across-track error; ~ **sintáctico** *m* INFORM&PD syntax error; ~ **de sintaxis** *m* INFORM&PD syntax error; ~ **sistemático** *m* FÍS, METR systematic error; ~ **de software** *m* INFORM&PD software error; ~ **de tecleado** *m* INFORM&PD, TELECOM keying error; ~ **técnico de vuelo** *m* TRANSP AÉR flight technical error; ~ **tipográfico** *m* IMPR literal, literal error, typo; ~ **transitorio** *m* INFORM&PD transient error; ~ **de transmisión** *m* TELECOM transmission error; ~ **trivial** *m* METAL banal slip, trivial slip; ~ **de truncamiento** *m* INFORM&PD truncation error; ~ **de velocidad** *m* TV velocity error

errores: ~ **de forma de onda** *m pl* FÍS ONDAS waveform errors

erúcico *adj* QUÍMICA erucic

erupción *f* CARBÓN *pozo petróleo* weathering, FÍS FLUID eruption, inrush, GAS eruption, PETROL, TEC PETR *perforación* blowout; ~ **cromosférica** *f* TEC ESP solar flare; ~ **solar** *f* TEC ESP solar flare

E/S¹ *abr* (*entrada/salida*) ELEC, INFORM&PD, PROD I/O (*input/output*)

E/S² : ~ **directa** *f* PROD immediate I/O

Es *abr* (*einstenio*) FÍS, FÍS RAD, QUÍMICA Es (*einsteinium*)

ESA *abr* (*Agencia Espacial Europea*) TEC ESP ESA (*European Space Agency*)

esbeltez *f* ING MECÁ fineness ratio

esbozo *m* CONST sketch, TEXTIL scratching

ESC *abr* (*tecla de salida*) INFORM&PD ESC (*escape key*)

ESCA *abr* (*Espectroscopía Electrónica para Análisis Químico*) QUÍMICA ESCA (*Electron Spectroscopy for Chemical Analysis*)

escabeche *m* ALIMENT brine, OCEAN marinade

escafandra: ~ **articulada** *f* Esp (*cf traje de inmersión articulado AmL*) OCEAN articulated diving suit; ~ **autónoma** *f* OCEAN aqualung

escala *f* ACÚST, CONST, ELEC, GEOM scale, IMPR scale, step, ING MECÁ window, LAB graduation, METR scale, MINAS *dragas de rosario* digging ladder, dredge ladder, TEC ESP range, TRANSP AÉR stopover, TRANSP

MAR *de rastra* ladder; ~ **uno a uno** *f* ING MECÁ full scale; ~ **de la abertura** *f* FOTO aperture scale; ~ **absoluta de Kelvin** *f* CONST Kelvin scale; ~ **absoluta de temperatura** *f* QUÍMICA Kelvin scale; ~ **de alturas** *f* METR height gage (*AmE*), height gauge (*BrE*); ~ **Aristoxene-Zarlin** *f* ACÚST Aristoxene-Zarlin scale; ~ **de armónicos menores** *f* ACÚST harmonic minor scale; ~ **de aumentos** *f* PROD *salarios* sliding scale; ~ **de bajada a la cámara** *f* TRANSP MAR companionway ladder; ~ **de Baumé** *f* ALIMENT, FÍS Baumé scale; ~ **de Beaufort** *f* METEO, TRANSP MAR Beaufort scale; ~ **de Brix** *f* ALIMENT Brix scale; ~ **de calados** *f* TRANSP MAR *diseño naval* draft marks (*AmE*), draught marks (*BrE*); ~ **de cámara** *f* TRANSP MAR companionway ladder; ~ **de cargas** *f* CONST scale of charges; ~ **completa** *f* ING MECÁ full scale; ~ **cromática** *f* ACÚST chromatic scale; ~ **defectuosa** *f* ACÚST defective scale; ~ **Delezenne** *f* ACÚST Delezenne scale; ~ **de disparo** *f* PROD trip scale; ~ **de distancias** *f* CINEMAT, FOTO distance scale; ~ **de draga** *f* MINAS dredge bucket ladder; ~ **de dureza** *f* MECÁ hardness scale; ~ **de enfoque** *f* CINEMAT, FOTO, TV focusing range; ~ **eptavalente** *f* ACÚST heptatonic scale; ~ **exavalente** *f* ACÚST hexatonic scale; ~ **expandida** *f* METR expanded scale; ~ **de exposiciones** *f* CINEMAT, FOTO exposure scale; ~ **Fahrenheit** *f* FÍS, METEO Fahrenheit scale; ~ **del fiel** *f* METR beam scales; ~ **del fotómetro** *f* FOTO light meter scale; ~ **de frecuencia** *f* ELECTRÓN frequency scale; ~ **de funciones de onda radiales** *f* FÍS RAD scale of radial wavefunctions; ~ **del gas perfecto** *f* FÍS perfect-gas scale; ~ **gradual** *f* PROD sliding scale; ~ **de grises** *f* CINEMAT, FOTO, IMPR, INFORM&PD, TV gray scale (*AmE*), grey scale (*BrE*); ~ **de grises lineal** *f* CINEMAT linear gray scale (*AmE*), linear grey scale (*BrE*); ~ **hedónica** *f* ALIMENT hedonic scale; ~ **híbrida** *f* PETROL hybrid scale; ~ **hidrométrica** *f* OCEAN depth gage (*AmE*), depth gauge (*BrE*); ~ **hidrométrica de Baumé** *f* ALIMENT, FÍS Baumé scale; ~ **de imagen** *f* CINEMAT image scale; ~ **de la imagen** *f* FOTO scale of image; ~ **instrumental** *f* INSTR instrument range; ~ **de justificación** *f* IMPR justifying scale; ~ **de Kelvin** *f* CINEMAT, QUÍMICA, TEC ESP Kelvin scale; ~ **lineal** *f* ELEC, ELECTRÓN, INSTR, METR linear scale; ~ **logarítmica** *f* ELEC, ELECTRÓN, INSTR, METR logarithmic scale; ~ **de marea** *f* OCEAN tide scale; ~ **de mareas** *f* OCEAN tidal scale, tide gage (*AmE*), tide gauge (*BrE*); ~ **mayor** *f* ACÚST major scale; ~ **mayor de igual temperamento** *f* ACÚST major scale of equal temperament; ~ **Mercator-Holder** *f* ACÚST Mercator-Holder scale; ~ **de mina** *f* MINAS ladder way; ~ **móvil** *f* PROD *salarios* sliding scale; ~ **musical** *f* ACÚST musical scale, gamut; ~ **no lineal** *f* CONST, METR nonlinear scale; ~ **del nonio** *f* ING MECÁ vernier scale; ~ **pantonal** *f* ACÚST pantonal scale; ~ **pentatónica** *f* ACÚST pentatonic scale; ~ **de peso muerto** *f* TRANSP MAR *diseño naval* dead weight scale; ~ **de Pitágoras** *f* ACÚST Pythagorean scale; ~ **de la positivadora** *f* CINEMAT printer scale; ~ **preeptavalente** *f* ACÚST preheptatonic scale; ~ **profundidad de campo** *f* CINEMAT, FOTO depth of field scale; ~ **de profundidad de foco** *f* CINEMAT, FOTO depth of focus scale; ~ **de proporciones** *f* ING MECÁ diagonal scale; ~ **de Rameau-Bach** *f* ACÚST Rameau-Bach scale; ~ **real** *f* TRANSP MAR accommodation ladder; ~ **real menor** *f* ACÚST real minor scale; ~ **de resistencias** *f* ING ELÉC resistor ladder; ~ **de Richter** *f* CONST, GEOFÍS, GEOL Richter scale; ~ **del rosario de la draga** *f* MINAS dredge bucket ladder; ~ **de salvamento** *f* CONST, SEG, TERMO *de los bomberos* fire-escape; ~ **suspendida** *f* ING MECÁ cradle; ~ **de temperatura** *f* TERMO temperature scale; ~ **de tiempos geológicos** *f Esp* (*cf cronología geológica AmL*) GEOL geological time scale; ~ **de vernier** *f* ING MECÁ vernier scale; ~ **volante** *f* ING MECÁ cradle

escalado *adj* INFORM&PD scalar

escalamera *f* TRANSP MAR oarlock (*AmE*), rowlock (*BrE*)

escalar[1] *m* FÍS, MATEMÁT scalar

escalar[2] *vt* INFORM&PD scale

escalas: ~ **de sacos** *f pl* EMB sack scales

escaldado *adj* ALIMENT, SEG, TERMO scalding

escaldadura *f* ALIMENT, SEG, TERMO scald

escaldar *vt* ALIMENT blanch, scald, SEG, TERMO scald

escalera *f* CONST ladder, stair, SEG ladder, TV staircase; ~ **abierta** *f* CONST open wall string; ~ **de acceso al compartimento de vuelo** *f* TRANSP AÉR flight-compartment access stairway; ~ **con armazón** *f* CONST staircase; ~ **de cadena** *f* CONST chain ladder; ~ **de cangilones** *f* CONST bucket ladder; ~ **de caracol** *f* CONST spiral stairs; ~ **para caso de incendio** *f* CONST, SEG, TERMO *en edificios* fire-escape; ~ **cerrada** *f* CONST dogleg stairs; ~ **corrediza** *f* CONST traveling ladderway (*AmE*), traveling staircase (*AmE*), travelling ladderway (*BrE*), travelling staircase (*BrE*); ~ **de eje abierto** *f* CONST open-newel stair; ~ **en espiral** *f* CONST spiral stairs; ~ **extensible** *f* CONST extension ladder; ~ **helicoidal** *f* CONST corkscrew stairs; ~ **de incendios** *f* CONST, SEG fire ladder; ~ **de mano** *f* CONST stepladder; ~ **mecánica** *f* CONST escalator, moving stairway (*BrE*), elevator, moving staircase (*BrE*), MECÁ elevator, TRANSP escalator, moving stairway (*BrE*); ~ **móvil** *f* CONST, TRANSP escalator, moving staircase (*BrE*), moving stairway (*BrE*), traveling ladderway (*AmE*), travelling ladderway (*BrE*); ~ **de ochos** *f* CONST figure-of-eight stairs; ~ **de peces** *f* ENERG RENOV fish pass; ~ **de peces tipo vertedero** *f* ENERG RENOV overfall type fish pass; ~ **de pilares macizos** *f* CONST solid-newel stair; ~ **plegable** *f* CONST extension ladder; ~ **portátil** *f* SEG portable ladder; ~ **de pozo** *f Esp* (*cf pozo de las escalas AmL*) MINAS ladder shaft, ladder road; ~ **de seguridad** *f* SEG safety ladder; ~ **de seguridad de metal ligero** *f* SEG fall-safe light metal ladder; ~ **triple** *f* CONST triple ladder; ~ **voladiza** *f* CONST hanging stairs

escaleras *f pl* CONST stairs; ~ **del avión** *f pl* TRANSP AÉR air stairs; ~ **con descansillos** *f pl* CONST stairs interrupted by landings; ~ **de ida y vuelta** *f pl* CONST half turn stairs

escalón *m* CARBÓN berm, CONST step, *escaleras* flyer, *escalera de mano* rung, FERRO step, ING MECÁ lift, MINAS stope, lift, stage, *pozo* stave, PROD rung, *pozo de mina* stave; ~ **de cielo** *m* MINAS overhand stope; ~ **compensado** *m* CONST balance step; ~ **de compresión** *m* ING MECÁ compression stage; ~ **de crecimiento** *m* CRISTAL step; ~ **curvo** *m* CONST curved step; ~ **de diagrama** *m* PROD ladder diagram rung; ~ **de escalera de curva** *m* CONST winder; ~ **incompleto** *m* PROD incomplete rung; ~ **normal** *m*

CONST flier; ~ **oscilante** *m* CONST dancing step; ~ **de presión** *m* INSTAL HIDRÁUL *de turbina, compresor* pressure stage; ~ **recto** *m* CONST flier; ~ **del temporizador** *m* PROD timer rung; ~ **de testero** *m* MINAS overhand stope; ~ **de tomas** *m* ELEC tapping step; ~ **de voltaje** *m* FÍS voltage step

escalonado[1] *adj* ING MECÁ stepped

escalonado[2]: ~ **vertical** *m* EMB stacking-up

escalonamiento *m* TV staggering

escalonar *vt* IMPR *composición* skew, *las hojas* steep

escalones[1]: **en** ~ *adj* ING MECÁ stepped

escalones[2] *m pl* CONST fliers; ~ **en voladizo** *m* CONST hanging steps

escalpelo *m* LAB scalpel

escama *f* DETERG scale, GEOL flake, PROD *exfoliación* scale; ~ **de caldera** *f* DETERG boiler scale

escamación *f* PROD scaling

escamoteamiento *m* D&A, TEC ESP *radio* scrambler

escamotear *vt* D&A, TEC ESP radio scramble

escamoteo *m* D&A, TEC ESP radio scrambling

escampavía: ~ **pesquera** *f* OCEAN patrol boat

escandallada *f* OCEAN, TRANSP MAR sounding

escandallar[1] *vt* OCEAN *navegación* sound

escandallar[2] *vi* TRANSP MAR *navegación* sound

escandallo *m* OCEAN, TRANSP MAR sounding lead

escandio *m* (*Sc*) QUÍMICA scandium (*Sc*)

escanear *vt* IMPR, INFORM&PD scan

escaneo *m* IMPR, INFORM&PD scan, scanning; ~ **de marcas** *m* INFORM&PD mark scanning; ~ **de referencias** *m* INFORM&PD mark scanning

escáner *m* IMPR, INFORM&PD scanner; ~ **para código de barras** *m* INFORM&PD, PROD bar-code scanner; ~ **holográfico** *m* INFORM&PD holographic scanner; ~ **óptico** *m* INFORM&PD optical scanner; ~ **de pilas** *m* IMPR pile scanner; ~ **plano** *m* INFORM&PD flat-bed scanner; ~ **de tambor** *m* IMPR drum scanner

escantilladora *f* ING MECÁ beveling machine (*AmE*), bevelling machine (*BrE*)

escantillón *m* CONST *madera* scantling, ELEC *fabricación, medición* template, INSTR bevelling (*BrE*), beveling (*AmE*), TRANSP MAR *construcción naval* scantling

escapar *vt* MINAS *grisú* baffle, TEC ESP *gases* blow

escaparate *m Esp* (*cf vidriera AmL*) CONST glazing, TEXTIL display; ~ **frigorífico** *m* REFRIG refrigerated window

escaparse *v refl* AGUA run off

escape *m Esp* (*cf vaciado AmL*) AGUA leakage, runoff, AUTO exhaust, CONTAM *en tuberías* washout, ELEC *de carga, corriente*, GAS leakage, leak, ING ELÉC leakage, ING MECÁ outlet, INSTAL HIDRÁUL eduction, release, INSTAL TERM *del vapor* exhaust, eduction, MINAS *del aire* exhausting, PROD *vapor* release, TEC ESP *máquina de vapor* stray radiation, chugging, TRANSP AÉR, VEH exhaust; ~ **de acción instantánea** *m* ELEC *conmutador* instantaneous release; ~ **de aire** *m* CONTAM *procesos industriales* blowout; ~ **de áncora** *m* ING MECÁ lever escapement; ~ **anticipado** *m* PROD *válvula de corredera* early release; ~ **de carga** *m* ELEC *capacitor* charge leakage; ~ **delantero** *m* ING MECÁ crank-end release; ~ **de enlace de datos** *m* (*DLE*) INFORM&PD data link escape (*DLE*); ~ **de escoria** *m* PROD *horno alto* breakout; ~ **de gas** *m* CONTAM waste gas, GAS, TERMO gas leak; ~ **libre** *m* ELEC *cortocircuito* trip free release; ~ **libre regulador** *m* AUTO, ELEC, ING ELÉC regulator cutout; ~ **magnético** *m* ING

ELÉC magnetic leakage; ~ **de metal fundido** *m* PROD *del horno, molde* runout; ~ **de metal líquido** *m* PROD *del molde* breakout

escapular *vt* TRANSP MAR *punta* clear

escarabajo *m* AGRIC beetle, PROD *vaciado* sand hole; ~ **escoriador** *m* AGRIC chafer; ~ **de harina** *m* AGRIC flour beetle; ~ **de savia del maíz** *m* AGRIC corn sap beetle (*AmE*), maize sap beetle (*BrE*)

escarbador *m* ING MECÁ scraper

escarceo *m* OCEAN cross ripple, TRANSP MAR race

escarcha *f* METEO hoar frost, white frost, REFRIG frost; ~ **cristalina** *f* REFRIG hoar frost; ~ **opaca** *f* REFRIG rime

escarchado[1] *adj* ALIMENT, C&V, METR, REFRIG frosted

escarchado[2] *m* ALIMENT, C&V, METR frosting

escardador *m* AGRIC weeder

escardillo *m* AGRIC weeding hoe

escariado *m* C&V reaming, CONST reaming, reaming-out, ING MECÁ reaming, reaming-out, counterboring, MECÁ, MINAS reaming, reaming-out, PROD counter-boring, reaming-out, reaming, TEC PETR reaming, reaming-out, VEH reaming

escariador *m* C&V reamer, ING MECÁ, MECÁ counter-bore, octagonal reamer, opening bit, reamer, scraper MINAS, VEH reamer; ~ **de acabado** *m* ING MECÁ finishing reamer; ~ **de acabado para conos morse** *m* ING MECÁ finishing reamer for Morse tapers; ~ **ajustable** *m* ING MECÁ adjustable reamer; ~ **auto-alimentador** *m* ING MECÁ self-feeding reamer; ~ **de concreto** *m* AmL (*cf escariador de hormigón Esp*) TRANSP concrete scraper; ~ **cónico** *m* ING MECÁ cone bit; ~ **de cuchillas ajustables** *m* ING MECÁ adjustable blade reamer; ~ **desbarbador** *m* ING MECÁ *tubos* burring reamer; ~ **de diámetro regulable** *m* PROD expansion reamer; ~ **estructural** *m* ING MECÁ bridge reamer; ~ **helicoidal de mano** *m* ING MECÁ twist hand reamer; ~ **de hormigón** *m* Esp (*cf escariador de concreto AmL*) TRANSP concrete scra-per; ~ **hueco** *m* ING MECÁ shell reamer; ~ **hueco con taladro interior cónico** *m* ING MECÁ shell reamer with taper bore; ~ **de mandril** *m* ING MECÁ chucking reamer; ~ **de mano de cuatro ranuras** *m* ING MECÁ four-flute twist hand reamer; ~ **para máquina** *m* ING MECÁ chucking reamer; ~ **mecánico de ranuras profundas** *m* ING MECÁ long-fluted machine reamer; ~ **con piloto desmontable** *m* ING MECÁ counterbore with detachable pilot; ~ **con piloto sólido** *m* ING MECÁ counterbore with solid pilot; ~ **quitarrebabas** *m* ING MECÁ burring reamer; ~ **de ranuras finas** *m* ING MECÁ fine-fluted reamer; ~ **de terrenos** *m* Esp (*cf desfondadora de terrenos AmL*) MINAS ripper; ~ **de tracción** *m* ING MECÁ *herramienta* broach

escariadora: ~ **de áridos** *f* TRANSP aggregate scraper

escariar *vt* ING MECÁ, MECÁ, PETROL, PROD counter-bore, ream

escarificación *f* CONST *obras públicas* scarification; ~ **de la cinta** *f* TV tape cupping

escarificador *m* Esp (*cf aparato disgregador de suelos AmL*) AGRIC rooter, scarifier, CONST ripper, scarifier, MINAS ripper; ~ **trasero de la cargadora** *m* CONST loader back hoe

escarificadora *f* AGRIC ripper, TRANSP scarifier; ~ **para carreteras** *f* CONST, TRANSP road ripper

escarificar *vt* CONST *obras públicas* scarify

escarpa *f* OCEAN scarp; ~ **submarina** *f* OCEAN sea scarp

escarpe *m* OCEAN escarpment, PROD *soldadura* scarf

escarpia *f* CONST tenterhook, OCEAN escarpment; ~ **roscada** *f* FERRO *equipo inamovible* screw spike

escarpiador *m* AmL (*cf perno Esp*) MINAS spike

escarvar: ~ **con pico** *vti* CONST break with a pick

escasez *f* PROD shortage; ~ **de existencias** *f* PROD stock shortage

escatol *m* QUÍMICA skatole

escatolcarboxílico *adj* QUÍMICA skatolecarboxylic

escatoxisulfato *m* QUÍMICA skatoxysulfate (*AmE*), skatoxysulphate (*BrE*)

escatoxisulfúrico *adj* QUÍMICA skatoxysulfuric (*AmE*), skatoxysulphuric (*BrE*)

escena: ~ **retrospectiva** *f* CINEMAT flashback; ~ **de transición** *f* CINEMAT connecting scene

escenario *m* CINEMAT stage

eschinita *f* MINERAL eschinite

escindible *adj* FÍS, FÍS PART, NUCL fissionable

escintilación *m* TEC ESP scintillation

escintilador *m* FÍS RAD scintillator

escisión: ~ **de haces prismática** *f* INSTR beam-splitting prism; ~ **del hidrocarburo por vapor** *f* TEC PETR *refino* steam cracking; ~ **múltiple** *f* NUCL multiple splitting

escleroclasa *f* MINERAL scleroclase

esclerómetro *m* INSTR, LAB, MECÁ, METR hardness tester

escleroproteína *f* ALIMENT scleroprotein, QUÍMICA albuminoid

escleroscopio *m* INSTR, LAB Shore hardness tester

esclusa *f* AGUA canal lock, lock, sluice, CONST lock, HIDROL single lock, weir, lock, MINAS sluice, PROD *fundería* gate shutter, RECICL sluice, TRANSP MAR *canalón de francanil, puertos* lock, *de un canal navegable* sluice; ~ **de aire** *f* TEC ESP airlock; ~ **de barras** *f* AGUA bar weir; ~ **calibrada** *f* HIDROL calibrated weir; ~ **contra la marea** *f* AGUA tide lock; ~ **de cuenco** *f* AGUA, HIDROL, TRANSP MAR navigation lock; ~ **de descarga** *f* AGUA discharge sluice, tailsluice; ~ **de desviación** *f* AGUA diversion sluice; ~ **elevadora** *f* AGUA lift lock; ~ **de emergencia** *f* NUCL emergency airlock; ~ **de entrada mareal** *f* TRANSP MAR tide gate; ~ **a gran altura** *f* HIDROL high lift lock; ~ **de navegación** *f* AGUA, HIDROL, TRANSP MAR navigation lock; ~ **neumática** *f* AGUA, MINAS, TRANSP AÉR airlock; ~ **de salida** *f* AGUA tail lock; ~ **de toma** *f* AGUA intake sluice; ~ **de travesaños corredizos** *f* AGUA stop log weir

esclusada *f* AGUA feed, locking

esclusaje *m* AGUA lockage, run of sluices, sluicing

esclusas: ~ **escalonadas** *f pl* HIDROL, TRANSP MAR flight of locks, staggered locks; ~ **superpuestas** *f pl* HIDROL, TRANSP MAR flight of locks

esclusero *m* HIDROL, TRANSP MAR lock keeper

escoba *f* CONST broom; ~ **mecánica** *f* CONST mechan-ical broom

escobén *m* TRANSP MAR *buque* hawse

escobilla *f* AUTO, CONST, ELEC, MECÁ, PROD brush, VEH *accesorio* wiper; ~ **del arrancador** *f* AUTO, VEH starter brush; ~ **de carbón** *f* ALIMENT, ELEC, ING ELÉC carbon brush; ~ **de colector** *f* ELEC, ELECTRÓN commutator brush; ~ **del generador** *f* ELEC, VEH generator brush; ~ **de grafito** *f* ELEC graphite brush; ~ **laminar** *f* ELEC, ING ELÉC laminated brush; ~ **de láminas** *f* ELEC, ING ELÉC laminated brush; ~ **del**

limpiaparabrisas *f* VEH wiper blade; ~ **neumática** *f* CINEMAT air squeegee; ~ **para la toma de corriente** *f* ING ELÉC current-collecting brush

escobillón *m* LAB *para limpiar tubos* brush, tube brush; ~ **para cañones** *m* D&A gun swab brush

escofina *f* CONST *carpintería* rasp, rasp cut, MECÁ file; ~ **encorvada** *f* ING MECÁ riffler; ~ **de picadura sencilla** *f* MECÁ float; ~ **de talla simple** *f* MECÁ float

escofinado *m* CONST rasping

escoger *vt* MINAS pick, select, sort; ~ **a mano** *vt* CARBÓN cull by hand, *minería* cobb (*AmE*), cob (*BrE*), MINAS *minerales* cobb (*AmE*), cob (*BrE*)

escogido: ~ **a mano** *m* CARBÓN hand selection; ~ **manual** *m* CARBÓN hand selection

escogimiento *m* CARBÓN, MINAS pick

escolecita *f* MINERAL mesotype, scolecite

escollera *f* AGUA breakwater, rock fill, CONST riprap, rock fill, breakwater, OCEAN breakwater, TRANSP MAR breakwater, *dique, rompeolas* pier

escolleras *f pl* OCEAN *modificaciones artificiales del litoral* spur dike

escollo *m* TRANSP MAR *geografía* rock

escoltar *vt* D&A, TRANSP MAR convoy

escombrera *f* AGUA refuse dump, CARBÓN dump, CONST spoil heap, disposal site, waste heap, dump skip, CONTAM landfill site, MINAS waste heap, dump, waste dump, spoil bank, waste, RECICL dump, refuse dump, landfill site, tip (*BrE*), waste heap, dump site, dumping ground; ~ **al aire libre** *f* RECICL open dump; ~ **para escorias** *f* PROD slag dump, slag tip; ~ **municipal** *f* CONTAM, RECICL municipal dump

escombro *m* CONST rubble, spoil, FERRO *equipo fijado* spoil

escombros *m pl* CARBÓN dust, MINAS deads, mullock, RECICL garbage (*AmE*), rubbish (*BrE*); ~ **metalúrgicos** *m pl* CARBÓN, METAL metallurgical waste (*BrE*), metalurgical waste (*AmE*)

escondrijo *m* INFORM&PD cache

escopeta *f* D&A gun

escopladura: ~ **por llama** *f* CONST flame-gouging

escopleadora *f* PROD mortising machine

escoplear *vt* IMPR mortise

escoplo *m* ING MECÁ, MECÁ, MINAS chisel, cutting tool; ~ **plano** *m* MINAS flat chisel; ~ **para tornear** *m* ING MECÁ turning gouge; ~ **triangular** *m* ING MECÁ parting tool

escora *f* TRANSP MAR beaching leg, *buque, navegación* heel, *construcción de buques, geografía* shore; ~ **estética** *f* TRANSP MAR list

escorado: ~ **del núcleo** *m* NUCL flux tilting

escorar[1] *vt* TRANSP MAR *buques* shore up

escorar[2] *vi* TRANSP MAR *buque* cant, heel; ~ **a una banda** *vi* TRANSP MAR *navegación* cant over

escorbuto *m* AGRIC scurvy

escoria *f* C&V scum, slag glass, CARBÓN *de metal* dross, slag, CONST clinker, slag, GEOL slag, scoria, PROC QUÍ clinker, PROD skim, TEC PETR slag, TERMO ash, TERMOTEC slag; ~ **de fragua** *f* PROD hammer slag; ~ **hiperácida** *f* PROD highly acid slag; ~ **hiperbásica** *f* PROD highly basic slag; ~ **de horno alto** *f* PROD blast-furnace slag; ~ **del martillado** *f* PROD hammer slag; ~ **volcánica** *f* GEOL scoria, volcanic cinder

escorial *m* CARBÓN *minas* dump, MINAS dump, ore dump, waste dump, slag heap, pit heap, PROD cinder bank, slag dump, slag heap

escorias *f pl* CARBÓN scum, PROD clinker, *batiduras de forja* cinders

escorodita *f* MINERAL scorodite

escorrentía *f* AGRIC, AGUA runoff; ~ **anual** *f* AGUA annual runoff; ~ **laminar** *f* AmL (*cf flujo laminar Esp*) MINAS *hidrogeología* flow sheet; ~ **media por hora** *f* HIDROL mean hourly runoff

escota *f* TRANSP MAR *cuerda, metal, vela* sheet; ~ **de la mayor** *f* TRANSP MAR mainsheet

escotadura *f* CONST recess, ING MECÁ, TEC ESP *máquinas* gap

escotero *m* TRANSP MAR *equipamiento de cubierta* main-sheet track

escotilla *f* CONST flap door, TEC ESP, TRANSP AÉR, TRANSP MAR hatch; ~ **de acceso a un tanque** *f* TRANSP MAR *petrolero* tank hatch; ~ **de bajada** *f* TRANSP MAR *buque* companionway hatch; ~ **de carga** *f* TEC ESP cargo hatch, TRANSP cargo hatch, cargo hatchway, TRANSP MAR cargo hatch; ~ **de inspección** *f* TEC ESP inspection hatch; ~ **de salvamiento** *f* TEC ESP *vehículos* escape hatch; ~ **de tambucho** *f* TRANSP MAR *buques* companionway hatch; ~ **trasera** *f* TRANSP hatchback

escotín *m* TRANSP MAR *cuerda, metal, vela* sheet

escribir *vti* INFORM&PD write; ~ **en el disco** *vti* INFORM&PD save; ~ **con el ordenador** *vti* INFORM&PD type; ~ **con el teclado** *vti* INFORM&PD type

escritura *f* INFORM&PD write; ~ **agrupada** *f* INFORM&PD gather write; ~ **directa de haz de electrones** *f* ELECTRÓN electron-beam direct writing; ~ **de transferencia en bloques** *f* PROD block-transfer write; ~ **única lectura múltiple** *f* ÓPT write-once read many times (*WORM*)

escrutación *f* TEC PETR scanning

escrutinio *m* INFORM&PD polling

escuadra *f* CONST angle bracket, squaring, *topografía* cross, GEOM set square, ING MECÁ square, scale, try square, set square, MECÁ angle bracket, METR *instrumento de dibujo y graduación* square, MINAS end bracket; ~ **de ángulo** *f* METR angle plate; ~ **en ángulo** *f* CONST angle tie; ~ **de apoyo** *f* ING MECÁ, METR angle plate; ~ **calibradora** *f* ING MECÁ caliper square (*AmE*), calliper square (*BrE*); ~ **para la canaleta** *f* CONST gutter bracket; ~ **de comprobación** *f* ING MECÁ try square; ~ **de dibujo sin centro** *f* ING MECÁ framed set square; ~ **de ebanista** *f* CONST joiner's bevel; ~ **en T** *f* ING MECÁ T-square; ~-**transportador** *f* METR bevel protractor; ~ **para trazar** *f* ING MECÁ angle plate; ~ **para verificar** *f* ING MECÁ try square

escuadrado *m* PAPEL squaring

escuadrar *vt* PROD square

escuadrilla *f* TRANSP MAR flotilla

escualano *m* QUÍMICA squalane

escualeno *m* QUÍMICA squalene

escucha[1] *f* PROD monitoring, TELECOM listening in

escucha[2]**: a la** ~ *fra* TELECOM standby

escudete *m* CONST *cerrajería* escutcheon (*AmE*), scutcheon (*BrE*)

escudo *m* CONST *túneles*, GEOL shield, ING MECÁ *carro de torno* apron of a lathe, INSTAL HIDRÁUL breasting parapet, TEC ESP shield; ~ **ablativo** *m* TEC ESP ablation shield; ~ **de cerradura** *m* CONST *cerrajería* escutcheon (*AmE*), key plate, scutcheon (*BrE*); ~ **de la cerradura** *m* CONST lock cup; ~ **continental** *m*

GEOFÍS, GEOL continental shield; **~ contra la radiación** *m* FÍS RAD, SEG radiation shield; **~ para cuello** *m* SEG neck shield; **~ para frenado** *m* TEC ESP braking shield; **~ de lava** *m* GEOL lava shield; **~ de mano** *m* SEG hand shield; **~ térmico** *m* TEC ESP, TERMO heat shield

escudriñado: ~ de líneas *m* INFORM&PD raster scan

escuela: ~ técnica *f* PROD technical school

esculina *f* QUÍMICA esculin

escurridera *f* LAB draining rack

escurridor *m* ALIMENT draining rack, CINEMAT, FOTO squeegee; **~ de caucho** *m* CINEMAT, FOTO rubber squeegee

escurridora *f* TEXTIL wringer

escurrimiento *m* AGUA draining, runoff, CONTAM, CONTAM MAR, ENERG RENOV runoff, HIDROL runoff, trickling, QUÍMICA seepage; **~ directo** *m* HIDROL direct runoff; **~ máximo diario** *m* HIDROL maximum daily runoff; **~ máximo horario** *m* HIDROL maximum hourly runoff; **~ mínimo diario** *m* HIDROL minimum daily runoff; **~ mínimo horario** *m* HIDROL minimum hourly runoff

escurrir[1] *vt* CINEMAT, FOTO squeegee, TEXTIL wring

escurrir[2] *vi* CONTAM MAR, HIDROL, MINAS leach

escurrirse *v refl* AGUA run out, CONST leak

escúter: ~ submarino *m* OCEAN minisubmersible

esencia: ~ de capoc *f* QUÍMICA kapok oil; **~ de kapoc** *f* QUÍMICA kapok oil; **~ de trementina** *f* P&C rosin, QUÍMICA spirit of turpentine; **~ de verbena** *f* QUÍMICA verbena oil; **~ de verveína** *f* QUÍMICA vervein oil

esencial *adj* MECÁ overriding

eserina *f* QUÍMICA eserine

esfalerita *f* MINAS *mineral* sphalerite, MINERAL, QUÍMICA blende, sphalerite, zinc blende

esfena *f* MINERAL sphene

esfera *f* ELEC *de instrumento* dial, FÍS, GEOM sphere, OCEAN net roller, VEH dial; **~ de actividad** *f* TEC ESP range; **~ celeste náutica** *f* OCEAN nautical celestial globe; **~ de Fermi** *f* FÍS Fermi sphere; **~ hidrotermal** *f* OCEAN warm-water sphere; **~ magnética** *f* GEOFÍS, TEC ESP magnetosphere; **~ de reflexión** *f* CRISTAL sphere of reflection; **~ del reloj** *f* ING MECÁ dial

esferas *f pl* METEO sferics (*AmE*), spherics (*BrE*)

esfericidad *f* GEOM sphericity

esférico *adj* GEN spherical

esféricos *m pl* METEO sferics (*AmE*), spherics (*BrE*)

esferoide *m* FÍS, GEOM spheroid; **~ achatado** *m* GEOM oblate spheroid; **~ alargado** *m* GEOM prolate spheroid

esferómetro *m* FÍS spherometer

esferosiderita *f* MINERAL spherosiderite

esferulita *f* GEOL, MINERAL, TEC PETR spherulite

esfingosina *f* QUÍMICA sphingosine

esfoliación: ~ por fractura *f* GEOL fracture cleavage

esfragidita *f* MINERAL sphragidite

esfuerzo *m* C&V strain, CARBÓN, CONST, CONTAM MAR, ING MECÁ, METAL, P&C stress, TEC ESP spurt, TEC PETR, TRANSP MAR *construcción naval* stress; **~ ácido** *m* CONTAM acid stress; **~ alternativo** *m* METAL cyclical stress, fluctuating stress; **~ aplicado** *m* METAL applied stress; **~ cíclico** *m* METAL cyclical stress, fluctuating stress; **~ de cizalla** *m* P&C shear, shear stress; **~ de cizallamiento** *m* ING MECÁ shearing stress; **~ de compresión** *m* GEOL compressive stress, ING MECÁ positive stress; **~ de correa** *m* ING MECÁ, METAL belt stress; **~ cortante** *m* CONST *resistencia de*

materiales shear force, shearing stress, ING MECÁ *metal*, P&C shear, shear stress; **~ cortante crítico** *m* METAL critical shear strain; **~ cortante descompuesto** *m* METAL resolved shear stress; **~ cortante sencillo** *m* METAL simple shear stress; **~ de corte** *m* FÍS shear strain, ING MECÁ shearing stress, MECÁ shear strength; **~ crítico** *m* METAL critical stress; **~ crítico de la fractura** *m* METAL critical fracture stress; **~ debido al cambio de temperatura** *m* ING MECÁ stress due to temperature change; **~ de deformación** *m* ING MECÁ strain; **~ de dilatación** *m* INSTAL HIDRÁUL expansion stress; **~ efectivo** *m* CARBÓN, GEOL effective stress, METAL working stress, TEC PETR effective stress; **~ de flexión** *m* MECÁ, P&C bending stress; **~ de frenado** *m* TRANSP brake effort; **~ de inflexión circunferencial** *m* TRANSP MAR bending circumferential stress; **~ interno** *m* METAL internal stress; **~ local** *m* METAL local stress; **~ longitudinal** *m* ING MECÁ direct stress, METAL longitudinal stress; **~ medio** *m* METAL mean stress; **~ mínimo** *m* METAL minimum stress; **~ pesquero** *m* OCEAN fishing effort; **~ real** *m* METAL applied stress; **~ repetitivo** *m* METAL stress; **~ residual** *m* C&V residual stress; **~ de Reynolds** *m* FÍS FLUID Reynolds stress; **~ de rotura** *m* ING MECÁ, MECÁ breaking stress, TRANSP MAR *cuerda* breaking strain; **~ de rotura máximo** *m* ING MECÁ ultimate strength; **~ de rotura uniforme** *m* METAL true fracture stress; **~ de separación** *m* NUCL separative effort; **~ tangencial** *m* ING MECÁ tangential stress; **~ temporal** *m* C&V temporary stress; **~ térmico** *m* TERMO thermal stress; **~ torsional** *m* ING MECÁ torsional stress; **~ de tracción** *m* ING MECÁ pull, MECÁ, METAL, NUCL tensile stress; **~ de tracción máximo** *m* NUCL ultimate tensile stress; **~ útil de tensión** *m* ING MECÁ effective tension; **~ útil de tracción** *m* ING MECÁ effective tension; **~ viscoso** *m* FÍS FLUID viscous stress

esfumado *adj* COLOR rainbow-effect

esfumar *vt* IMPR tone down

ESI *abr* (*electrodo selectivo de iones*) LAB *electroquímica* ISE (*ion selective electrode*)

eslabón *m* ING MECÁ *cadena*, MECÁ, TRANSP MAR link; **~ alimentador** *m* TEC ESP feeder link; **~ de arrastre** *m* TRANSP AÉR drag link; **~ de cadena** *m* AUTO, VEH *motocicletas* chain link; **~ de comunicación** *m* INFORM&PD link (*BrE*), trunk (*AmE*); **~ de contador irreversible** *m* PROD up counter rung; **~ desbloqueado** *m* PROD unlatch rung; **~ giratorio** *m* ING MECÁ swivel; **~ inicial** *m* PROD start rung; **~ para reparar cadenas** *m* ING MECÁ repair link; **~ de unión** *m* ING MECÁ connecting link

eslabonamiento *m* ING MECÁ, MECÁ linkage

eslinga *f* CONST *tejados, vigas* span, span piece, CONTAM MAR, PROD, TRANSP MAR sling; **~ de cable de acero** *f* SEG wire rope sling; **~ de carga** *f* TRANSP AÉR cargo sling; **~ de cuerda** *f* SEG rope-type sling; **~ de gancho de arrastre** *f* ING MECÁ dog hook sling; **~ del izador** *f* TRANSP AÉR hoisting sling; **~ de tipo cinto** *f* SEG belt-type sling; **~ tipo fibra** *f* SEG fiber-type sling (*AmE*), fibre-type sling (*BrE*)

eslingada *f* TRANSP MAR hoist, *carga* sling

eslora *f* TRANSP MAR *construcción de barcos* deck girder, girder, ship girder; **~ entre perpendiculares** *f* TRANSP MAR *diseño de buques* length between perpendiculars; ~ **total** *f* TRANSP MAR *construcción naval* length overall

esmaltación: ~ **defectuosa** *f* ELECTRÓN defect annealing; ~ **isócrona** *f* METAL isochronal annealing; ~ **isotérmica** *f* METAL isothermal annealing; ~ **por láser** *f* ELECTRÓN laser annealing

esmaltado[1] *adj* COLOR, INSTAL TERM, PROD, REVEST enameled (*AmE*), enamelled (*BrE*)

esmaltado[2] *m* COLOR, INSTAL TERM, PROD, REVEST enameling (*AmE*), enamelling (*BrE*), glazing; ~ **por calor** *m* INSTAL TERM bright annealing; ~ **electroforético** *m* COLOR electrophoretic enameling (*AmE*), electrophoretic enamelling (*BrE*); ~ **electrostático** *m* COLOR, ELEC electrostatic enameling (*AmE*), electrostatic enamelling (*BrE*); ~ **a prueba de intemperie** *m* C&V antistorm glazing

esmaltador *m* REVEST enameller

esmaltar *vt* COLOR enamel, FOTO, PROD glaze, REVEST enamel; ~ **al horno** *vt* COLOR stove enamel

esmalte *m* C&V, COLOR, CONST, P&C, PROD, REVEST enamel; ~ **al agua** *m* COLOR water enamel; ~ **de alambre** *m* REVEST wire enamel; ~ **antiácido** *m* REVEST acid-proof enamel; ~ **de concreto** *m* *AmL* (*cf esmalte de hormigón Esp*) COLOR concrete enamel; ~ **con efecto martillado** *m* COLOR hammer enamel; ~ **de hormigón** *m* *Esp* (*cf esmalte de concreto AmL*) COLOR concrete enamel; ~ **de hornear** *m* PROD baking enamel; ~ **para horno** *m* COLOR baking enamel, stove enamel; ~ **mate** *m* REVEST mat enameling (*AmE*), mat enamelling (*BrE*); ~ **metálico** *m* ELECTRÓN metal glaze; ~ **polimerizado** *m* ING MECÁ polymerized enamel; ~ **de porcelana** *m* REVEST porcelain enamel; ~ **a prueba de ácidos** *m* COLOR acid-proof enamel; ~ **recocido** *m* COLOR baked enamel; ~ **resistente al ácido** *m* REVEST acid-proof enamel; ~ **secado en estufa** *m* REVEST stove enamel; ~ **secado al horno** *m* REVEST baked enamel; ~ **sintético** *m* COLOR synthetic enamel; ~ **transparente** *m* C&V transparent enamel; ~ **vítreo** *m* C&V, REVEST vitreous enamel

esmaltina *f* MINERAL smaltite

esmectita *f* CARBÓN, GEOL, QUÍMICA, TEC PETR smectite

esmeralda *f* MINERAL emerald

esmeraldita *f* MINERAL smaragdite

esmeril *m* C&V grinding unit, *vidrio plano pulido* grinder, ING MECÁ, MECÁ, METAL, PROD emery; ~ **de disco** *m* C&V disc grinder (*BrE*), disk grinder (*AmE*); ~ **doble** *m* C&V twin grinder; ~ **de dos filos** *m* C&V double-edge grinder; ~ **universal** *m* C&V universal grinder

esmerilación *f* TEXTIL emerizing

esmerilado *m* C&V, MECÁ grinding; ~ **burdo** *m* C&V rough grinding; ~ **con chorro de arena** *m* C&V plain sandblast; ~ **con disco** *m* C&V, PROD disc grinding (*BrE*), disk grinding (*AmE*); ~ **fino** *m* C&V smooth grinding; ~ **fino de bordes** *m* C&V edge fine-grinding; ~ **y pulido** *m* C&V grinding and polishing; ~ **de superficies planas** *m* ING MECÁ surface grinding; ~ **de un tapón** *m* C&V grinding in of a stopper; ~ **universal** *m* PROD universal grinding

esmerilador: ~ **neumático de válvulas** *m* AUTO air-operated valve grinder; ~ **de válvulas por aire** *m* AUTO air-operated valve grinder

esmeriladora *f* ING MECÁ surface grinder, MECÁ grinder, PROD emery grinder

esmerilar *vt* CARBÓN *el vidrio* polish with emery, ING MECÁ, MECÁ, PROD grind

esmitsonita *f* MINERAL smithsonite

esmog *m* CONTAM, METEO smog; ~ **fotoquímico** *m* CONTAM photochemical smog

esnorkel *m* CINEMAT snorkle

esonita *f* MINERAL essonite

espaciado[1] *adj* TELECOM spaced-out; ~ **fino** *adj* IMPR one-point letter-spaced; ~ **óptico** *adj* IMPR optically letter-spaced

espaciado[2] *m* *Esp* (*cf alzado AmL*) CINEMAT spacing, ING MECÁ pitch, MINAS *de los cables de extracción* clearing; ~ **de canales** *m* INFORM&PD, TV channel spacing; ~ **entre letras** *m* IMPR letter spacing; ~ **horizontal** *m* PROD horizontal spacing; ~ **de línea** *m* IMPR line spacing; ~ **en negro** *m* CINEMAT black spacing; ~ **reticular** *m* METAL lattice spacing; ~ **del retículo** *m* NUCL lattice-pitch spacing; ~ **del tren** *m* FERRO train spacing; ~ **vertical** *m* PROD vertical spacing

espaciador *m* ING MECÁ packing piece, distance piece, distance piece spacer, PROD, TEC ESP spacer; ~ **acopado** *m* ING MECÁ dishpan spacer; ~ **cónico** *m* ING MECÁ conical spacer; ~ **de precisión** *m* ING MECÁ close tolerance spacer

espaciamiento *m* ÓPT, TRANSP spacing; ~ **reticular** *m* METAL lattice spacing

espacio *m* CONST room, FÍS, GEOM, INFORM&PD space, ING MECÁ gap, INSTAL HIDRÁUL *entre pistón y tapa del cilindro* space, MECÁ blank, MINAS room, TEC ESP space; ~ **aéreo** *m* TEC ESP aerospace, TRANSP AÉR airspace; ~ **aéreo controlado** *m* TRANSP AÉR controlled airspace; ~ **de aire** *m* ALIMENT head space, ELEC *capacitor, transformador* air gap; ~ **de almacenado** *m* EMB shelf space; ~ **de almacenaje** *m* EMB storage space; ~ **anular** *m* GAS annular space, PETROL annulus, TEC PETR *perforación* annular space; ~ **anular de bajada** *m* NUCL *de un reactor PWR* downcomer; ~ **aséptico** *m* REFRIG clean space; ~ **atonal** *m* ACÚST atonal space; ~ **en blanco** *m* IMPR, INFORM&PD blank; ~ **de cabeza** *m* ALIMENT head space; ~ **para cables agrupados** *m* ELECTRÓN bunching space; ~ **de captación** *m* ELECTRÓN catcher space; ~ **de carga** *m* TRANSP, TRANSP AÉR, TRANSP MAR, VEH cargo space; ~ **cerrado** *m* SEG confined space; ~ **de cifra** *m* IMPR figure space; ~ **de configuración** *m* FÍS, GEOM configuration space; ~ **cósmico** *m* TEC ESP cosmic space, outer space; ~ **de direcciones** *m* INFORM&PD address space; ~ **en disco** *m* INFORM&PD disk space; ~ **distante** *m* TEC ESP far range; ~ **elíptico** *m* GEOM elliptical space; ~ **de entrada** *m* ELECTRÓN input gap; ~ **entre alas adyacentes** *m* TEC ESP, TRANSP AÉR *biplanos, triplanos* gap; ~ **entre bloques** *m* INFORM&PD interblock gap (*IBG*); ~ **entre los electrodos** *m* ELEC *capacitor, transformador* air gap; ~ **euclídeo** *m* FÍS, GEOM Euclidean space; ~ **euclídico** *m* FÍS, GEOM Euclidean space; ~ **de fase** *m* FÍS phase space; ~ **fino** *m* IMPR thin space; ~ **para gases de fisión** *m* NUCL *del elemento de combustible* fission-gas plenum; ~ **grueso** *m* IMPR thick space, three-to-em space; ~ **hiperbólico** *m* GEOM hyperbolic space; ~ **de interacción** *m* ELECTRÓN *tubos de campo cruzado* interaction space; ~ **interelectródico principal** *m* ING ELÉC main gap; ~ **interestelar** *m* TEC ESP interstellar space;

~ **interlineal** *m* INFORM&PD line feed (*LF*);
~ **interlobular** *m* TEC ESP *antena, biplanos, radar*
gap; ~ **intermedio** *m* MECÁ gap; ~ **interno del**
átomo *m* NUCL atomic interspace; ~ **lateral** *m* ING
MECÁ side clearance; ~ **lejano** *m* TEC ESP *cosmología*
far range; ~ **de la letra** *m* IMPR letter spacing; ~ **libre**
m MECÁ clearance, TRANSP nonreserved space; ~ **libre**
lateral *m* TRANSP lateral clearance; ~ **de línea** *m* IMPR
line space; ~ **de máquinas** *m* TRANSP MAR engine
room; ~ **de medio cuadratín** *m* IMPR en space; ~ **de**
Minkowski *m* FÍS Minkowski space; ~ **de**
modulación *m* ELECTRÓN buncher space;
~ **modulado** *m* ACÚST modulated space; ~ **muerto**
m ING MECÁ clearance; ~ **negro del cátodo** *m*
ELECTRÓN, ING ELÉC cathode dark space; ~ **nocivo**
m ING MECÁ piston clearance; ~ **de ocupación** *m*
TRANSP box; ~ **oscuro de Crookes** *m* ELECTRÓN, FÍS
Crookes dark space; ~ **oscuro de Faraday** *m*
ELECTRÓN, FÍS Faraday dark space; ~ **oscuro**
Hittorf *m* ELECTRÓN Hittorf dark space;
~ **perjudicial** *m* ING MECÁ piston clearance, INSTAL
HIDRÁUL *cilindro de vapor* clearance, clearance space,
waste space; ~ **para las piernas** *m* AUTO, TRANSP,
VEH legroom; ~ **poroso** *m* GEOL pore space;
~ **profundo** *m* TEC ESP *de cosmos* deep space; ~ **de**
protección *m* TELECOM guard space; ~ **publicitario**
m TV advertising slot; ~ **que ocupa la máquina** *m*
EMB floor space; ~ **recíproco** *m* CRISTAL reciprocal
space; ~ **reflector** *m* ELECTRÓN reflector space; ~ **sin**
reservar *m* TRANSP nonreserved space; ~ **de**
retroceso *m* IMPR, INFORM&PD backspace (*BS*);
~ **de sedimentación** *m* AUTO sediment space; ~ **de**
tres cuadratines *m* IMPR three-to-em space; ~ **vacío**
m CARBÓN porous layer, GEOL *en una roca* pore space,
MINAS *en agujero de explosión* airspace, TEC ESP
cisterna, TEC PETR ullage; ~ **variable** *m* IMPR variable
space
espadilla *f* CONST beating pick, TRANSP MAR scull
espalda: ~ **contra espalda** *adj* TEXTIL *confección*
back-to-back; ~ **con espalda** *adj* PROD back-to-back
espaldón *m* CONST haunch
esparavel *m* CONST *albañilería* hawk
esparcedora *f* REVEST spreading machine
esparciador: ~ **de estiércol** *m* AGRIC manure spreader
esparcido *m* C&V guniting, gunning, QUÍMICA spreading
esparcidor *m* AGRIC, AGUA spreader
esparcidora *f* MINAS *de arena, grava*, TRANSP *cons-*
trucción de carreteras spreader; ~ **de arena** *f* TRANSP
grit spreader; ~ **de grava** *f* TRANSP grit spreader;
~ **macadán** *f* TRANSP macadam spreader
esparcimiento *m* ÓPT scattering, TEC ESP scatter; ~ **en**
la fibra *m* ÓPT fiber scattering (*AmE*), fibre scattering
(*BrE*); ~ **material** *m* ÓPT material scattering; ~ **no**
lineal *m* ÓPT nonlinear scattering; ~ **de Rayleigh** *m*
FÍS RAD, ÓPT, TELECOM Rayleigh scattering
esparcir *vt* AGUA sprinkle
esparcirse *v refl* CONST branch off
esparita *f* PETROL esparite
espárrago *m* AUTO, ING MECÁ, PROD, VEH stud; ~ **de la**
culata del cilindro *m* ING MECÁ cylinder-head stud
espartalita *f* MINERAL spartalite
esparteína *f* QUÍMICA spartein
esparto *m* MINERAL, TEC PETR spar
espata *f* AGRIC, ALIMENT husk; ~ **de maíz** *f* AGRIC
corn husk (*AmE*), maize husk (*BrE*)

espatillar *vt* ING MECÁ *chapas* chamfer
espato: ~ **azul** *m* MINERAL blue-spar, lazulite; ~ **bruno**
m MINERAL brown spar; ~ **clorítico** *m* MINERAL
chlorite spar; ~ **ferrífero** *m* MINERAL chalybite;
~ **flúor** *m* C&V, MINERAL, QUÍMICA fluorspar; ~ **de**
Groenlandia *m* QUÍMICA cryolite; ~ **de itrio** *m*
MINERAL xenotime; ~ **lustroso** *m* MINERAL satin
spar; ~ **manganoso** *m* MINERAL dialogite;
~ **pesado** *m* MINERAL heavy spar; ~ **tornasolado** *m*
MINERAL schiller spar
espátula *f Esp* (*cf cuchillo de espátula AmL*) CONST
putty knife, ING MECÁ palette knife, LAB spatula,
pinturas palette knife; ~ **de ganchos** *f* PROD *molder-*
ías lifter; ~ **para la tinta** *f* IMPR ink knife
especial: ~ **para pantalla** *adj* INFORM&PD screen-
based
especialista *m* ING MECÁ troubleshooter; ~ **en efectos**
especiales *m* CINEMAT, TV special effects person;
~ **en redes** *m* INFORM&PD network specialist
especialización *f* ING MECÁ training
especializado *adj* INFORM&PD, TELECOM dedicated
especificación *f* PROD, TEXTIL, TRANSP MAR specification; ~ **de aceptación** *f* CALIDAD acceptance
specification; ~ **de las adquisiciones** *f* TEC ESP
procurement specification; ~ **de British Standards**
f ING MECÁ British Standards Specification (*BSS*);
~ **de control** *f* CALIDAD inspection specification;
~ **del equipo** *f* CALIDAD equipment specification;
~ **de equipos mezclados en transistores de efect** *f*
(*EEMTEC*) ELECTRÓN MESFET; ~ **estándar** *f* PROD
standard specification; ~ **experimental** *f* TEC ESP test
specification; ~ **de materias** *f* CALIDAD materials
specification; ~ **de memoria expandida** *f* (*EMS*)
INFORM&PD expanded-memory specification (*EMS*);
~ **normal** *f* PROD standard specification; ~ **normali-**
zada de colores *f* IMPR standard color specification
(*AmE*), standard colour specification (*BrE*); ~ **de**
proceso *f* CALIDAD process specification; ~ **del**
producto *f* CALIDAD product specification; ~ **de**
programas *f* INFORM&PD program specification;
~ **reglamentaria** *f* PROD standard specification;
~ **técnica** *f* CONST technical specification
especificaciones *f pl* CONST, ING MECÁ specification;
~ **estándar** *f pl* ING MECÁ standard specification;
~ **de funcionamiento** *f pl* ING MECÁ performance
specification; ~ **legales** *f pl* ING MECÁ standard
specification; ~ **normales** *f pl* ING MECÁ standard
specification; ~ **reglamentarias** *f pl* ING MECÁ standard specification; ~ **y requisitos del diseño** *f pl*
CALIDAD, PROD design specification and requirements; ~ **de seguridad** *f pl* TRANSP safety
specifications
especificador: ~ **de la sección de tabla de datos** *m*
PROD data table section specifier
especificar *vt* INFORM&PD address
específico *adj* FÍS specific; ~ **para máquina** *adj*
INFORM&PD machine-oriented
espécimen *m* GEN specimen; ~ **de ensayo** *m* P&C test
specimen
espectador *m* TELECOM, TV viewer; ~ **de televisión** *m*
TELECOM, TV television viewer
espectometría: ~ **de ultravioleta** *f* LAB ultraviolet
spectrophotometry
espectro *m* GEN spectrum; ~ **de absorción** *m* FÍS, FÍS
RAD, TEC ESP absorption spectrum; ~ **de absorción**
de rayos X *m* FÍS, NUCL X-ray absorption spectrum;

~ **acústico** *m* ACÚST acoustical spectrum; ~ **de AF** *m* ELECTRÓN HF spectrum; ~ **ajeno** *m* ELECTRÓN aliased spectrum; ~ **de alta frecuencia** *m* ELECTRÓN high-frequency spectrum; ~ **de alta resolución** *m* FÍS RAD high-resolution scan; ~ **de amplitud** *m* FÍS ONDAS, PETROL amplitude spectrum; ~ **del arco** *m* FÍS arc spectrum; ~ **atómico** *m* FÍS, NUCL atomic spectrum; ~ **de banda** *m* FÍS, FÍS RAD band spectrum; ~ **beta** *m* FÍS, FÍS RAD beta-ray spectrum; ~ **de Brocken** *m* METEO phantom horizon; ~ **calorífico** *m* TERMO heat spectrum; ~ **característico de rayos X** *m* FÍS RAD characteristic X-ray spectrum; ~ **de chispas** *m* FÍS RAD spark spectrum; ~ **continuo** *m* ACÚST, ELECTRÓN, FÍS, FÍS RAD, TEC ESP continuous spectrum; ~ **continuo electrónico** *m* ELECTRÓN, FÍS PART, NUCL electron continuum; ~ **de corriente** *m* AGUA flow pattern; ~ **cromático** *m* FOTO chromatic spectrum; ~ **de difracción** *m* FÍS, FÍS RAD, NUCL, ÓPT diffraction spectrum; ~ **discontinuo** *m* FÍS RAD discontinuous spectrum; ~ **electromagnético** *m* ELEC, ELECTRÓN, FÍS RAD, ING ELÉC electromagnetic spectrum; ~ **de emisión** *m* FÍS, FÍS RAD, TEC ESP emission spectrum; ~ **energético** *m* ING ELÉC power spectrum; ~ **de energía** *m* TEC ESP energy spectrum; ~ **del estado de la mar** *m* OCEAN sea state spectrum; ~ **de excitación de fluorescencia** *m* FÍS RAD fluorescence excitation spectrum; ~ **de fisión de Watt** *m* NUCL Watt's fission spectrum; ~ **fotoelectrónico de rayos X** *m* NUCL X-ray photoelectron spectrum; ~ **de frecuencias** *m* ACÚST, ELECTRÓN, FÍS ONDAS, INFORM&PD frequency spectrum; ~ **gamma** *m* FÍS RAD gamma-ray spectrum; ~ **infrarrojo** *m* FÍS, FÍS RAD infrared spectrum; ~ **de interferencia** *m* FÍS ONDAS interference pattern; ~ **iónico** *m* FÍS RAD ion spectrum; ~ **lineal** *m* ACÚST, FÍS, FÍS RAD, ÓPT, TELECOM line spectrum; ~ **de líneas** *m* ACÚST, FÍS, FÍS RAD, ÓPT, TELECOM line spectrum; ~ **de líneas luminosas** *m* TEC ESP bright line spectrum; ~ **de líneas oscuras** *m* TEC ESP dark line spectrum; ~ **de llama** *m* FÍS, TERMO flame spectrum; ~ **de luz** *m* FÍS ONDAS light spectrum; ~ **de masa** *m* FÍS, FÍS RAD, QUÍMICA mass spectrum; ~ **de mesones ligeros** *m* FÍS RAD light meson spectrum; ~ **de microondas** *m* FÍS, FÍS RAD microwave spectrum; ~ **molecular** *m* FÍS RAD molecular spectrum; ~ **de ondas** *m* FÍS ONDAS, OCEAN wave spectrum; ~ **de ondas de radiocomunicación** *m* ELECTRÓN, FÍS RAD radio spectrum; ~ **óptico** *m* ELECTRÓN, FÍS RAD, ÓPT, TELECOM optical spectrum; ~ **permitido** *m* NUCL allowed spectrum; ~ **de la presión acústica** *m* Esp (*cf espectro de la presión sonora AmL*) ACÚST, CONTAM sound pressure spectrum; ~ **de la presión sonora** *m AmL* (*cf espectro de la presión acústica Esp*) ACÚST, CONTAM sound pressure spectrum; ~ **prismático** *m* FÍS RAD prismatic spectrum; ~ **puro** *m* FÍS RAD pure spectrum; ~ **de radiación** *m* FÍS RAD, GEOFÍS radiation spectrum; ~ **de radiaciones nucleares** *m* FÍS RAD, NUCL, TEC ESP nuclear radiation spectrum; ~ **radioeléctrico** *m* ELECTRÓN, FÍS RAD radio spectrum; ~ **de rayas** *m* ACÚST, FÍS, FÍS RAD, ÓPT, TELECOM line spectrum; ~ **de rayas del hidrógeno** *m* FÍS RAD hydrogen spectral line; ~ **de rayas de láser** *m* FÍS RAD laser spectral line; ~ **de rayos X** *m* FÍS RAD X-ray spectrum; ~ **de resolución temporal** *m* FÍS RAD time-resolved spectrum; ~ **de resonancia** *m* FÍS ONDAS resonance spectrum; ~ **de respuesta de banda estrecha** *m* NUCL narrow-band response spectrum; ~ **resuelto en función del tiempo** *m* FÍS RAD time-resolved spectrum; ~ **de rotación** *m* FÍS RAD rotation spectrum; ~ **rotacional** *m* FÍS rotational spectrum; ~ **térmico** *m* TERMO thermal spectrum; ~ **de transmisión** *m* NUCL transmission spectrum; ~ **de turbulencia** *m* FÍS FLUID spectrum of turbulence; ~ **de vibración y rotación** *m* FÍS vibration-rotation spectrum; ~ **vibracional** *m* FÍS vibrational spectrum; ~ **visible** *m* FÍS visible spectrum; ~ **de vuelo** *m* TRANSP AÉR flight spectrum

espectroanálisis: ~ **de absorción** *m* FÍS RAD absorption spectroanalysis; ~ **molecular** *m* FÍS RAD molecular spectroanalysis

espectrofotometría *f* GEN spectrophotometry; ~ **cinética** *f* FÍS RAD kinetic spectrophotometry; ~ **de infrarrojo** *f* LAB, QUÍMICA infrared spectrophotometry; ~ **de ultravioleta** *f* LAB, QUÍMICA ultraviolet spectrophotometry

espectrofotómetro *m* GEN, spectrophotometer; ~ **de absorción** *m* FÍS RAD absorption spectrophotometer; ~ **de absorción atómica** *m* FÍS RAD atomic absorption spectrophotometer; ~ **de haz simple** *m* FÍS RAD single-beam spectrophotometer; ~ **de infrarrojo** *m* LAB, QUÍMICA infrared spectrophotometer; ~ **de ultravioleta visible** *m* LAB, QIMICA ultraviolet visible spectrophotometer

espectrografía: ~ **de masas por sonda lasérica** *f* NUCL laser probe mass spectrography; ~ **por rayos X** *f* NUCL X-ray spectrography; ~ **reticular en vacío** *f* FÍS ONDAS vacuum-grating spectrograph

espectrógrafo *m* GEN spectrograph; ~ **acústico** *m* ACÚST sound spectrograph; ~ **beta semicircular** *m* NUCL semicircular beta spectrograph; ~ **de difracción** *m* FÍS, FÍS RAD, NUCL, ÓPT diffraction spectrograph; ~ **difractivo** *m* FÍS, FÍS RAD, NUCL, ÓPT diffraction spectrograph; ~ **magnético** *m* GEOFÍS magnetic spectrograph; ~ **de masas** *m* FÍS, FÍS RAD mass spectrograph; ~ **prismático** *m* FÍS RAD prismatic spectrograph; ~ **de rayos X** *m* FÍS RAD, INSTR X-ray spectrograph; ~ **de retículo** *m* NUCL grating spectrograph

espectrometría *f* GEN spectrometry; ~ **de absorción** *f* TELECOM absorption spectrometry; ~ **alfa** *f* FÍS RAD alpha-ray spectrometry; ~ **de emisiones** *f* FÍS RAD emission spectral analysis; ~ **de masas** *f* FÍS, FÍS RAD, QUÍMICA mass spectrometry; ~ **de masas de iones-secundarios** *f* FÍS secondary-ion mass spectrometry (*SIMS*); ~ **de rayos X** *f* NUCL X-ray spectrometry

espectrómetro *m* GEN spectrometer; ~ **de absorción** *m* INSTR, LAB, TELECOM absorption spectrometer; ~ **de absorción atómica** *m* INSTR atomic absorption, LAB atomic absorption spectrometer; ~ **de barrido** *m* FÍS RAD scanning spectrometer; ~ **de centelleo** *m* FÍS RAD scintillation spectrometer, NUCL scintillation coincidence spectrometer; ~ **de coincidencia de haz triple** *m* NUCL triple-beam coincidence spectrometer; ~ **de cristal** *m* CRISTAL, FÍS RAD crystal spectrometer; ~ **de fluorescencia** *m* FÍS, FÍS RAD, INSTR fluorescence yield spectrometer; ~ **de masas** *m* FÍS, FÍS RAD, LAB mass spectrometer; ~ **de rayos gamma** *m* FÍS ONDAS, FÍS RAD, INSTR gamma-ray spectrometer; ~ **de rayos infrarrojos** *m* FÍS RAD infrared spectrometer; ~ **de rayos X** *m* FÍS RAD X-ray spectrometer; ~ **de rayos X fluorescente** *m* ELEC, FÍS, FÍS RAD, INSTR fluorescent X-ray spectrometer;

~ **Slatis-Siegbahn** *m* NUCL Slatis-Siegbahn spectrometer; ~ **trocoidal de masas** *m* NUCL trochoidal mass spectrometer

espectroquímico *adj* QUÍMICA spectrochemical

espectroradiómetro *m* ENERG RENOV spectroradiometer

espectros: ~ **de energía y disipación en turbulencia** *m pl* FÍS FLUID energy and dissipation spectra in turbulence

espectroscopía *f* GEN, spectroscopy; ~ **de absorción** *f* FÍS, FÍS RAD absorption spectroscopy; ~ **de absorción atómica** *f* FÍS, QUÍMICA atomic absorption spectroscopy; ~ **atómica** *f* FÍS RAD atomic spectroscopy; ~ **de electrones Auger** *f* (*EEA*) FÍS, QUÍMICA Auger electron spectroscopy (*AES*); ~ **electrónica** *f* FÍS, FÍS RAD electron spectroscopy; ~ **de emisión por llama** *f* FÍS, TERMO flame-emission spectroscopy; ~ **de foto-electrones de rayos X** *f* FÍS X-ray photoelectron spectroscopy; ~ **fotoelectrónica de rayos X** *f* QUÍMICA X-ray photoelectron spectroscopy XPS; ~ **infrarroja** *f* FÍS, QUÍMICA infrared spectroscopy; ~ **láser** *f* FÍS RAD laser spectroscopy; ~ **a la llama** *f* FÍS, TERMO flame spectroscopy; ~ **de microondas** *f* FÍS, FÍS RAD microwave spectroscopy; ~ **por microondas** *f* FÍS, FÍS RAD microwave spectroscopy; ~ **por pérdida electroenergética** *f* FÍS RAD electron-energy-loss spectroscopy; ~ **de Raman** *f* FÍS, FÍS RAD Raman spectroscopy; ~ **de rayos X** *f* NUCL X-ray spectroscopy; ~ **de resonancia magnética** *f* FÍS RAD magnetic resonance spectroscopy; ~ **por transformada de Fourier** *f* ELECTRÓN, FÍS Fourier transform spectroscopy

Espectroscopía: ~ **Electrónica para Análisis Químico** *f* (*ESCA*) QUÍMICA Electron Spectroscopy for Chemical Analysis (*ESCA*)

espectroscopio *m* GEN spectroscope

espejo *m* ACÚST, AUTO, CINEMAT, ELECTRÓN reflector, FÍS mirror, reflector, FOTO, ING MECÁ, INSTR, ÓPT, TEC ESP reflector, TRANSP MAR *botes* transom, VEH mirror; ~ **auxiliar** *m* FOTO auxiliary mirror; ~ **de bruja** *m* C&V witch mirror; ~ **de calor** *m* ENERG RENOV heat mirror; ~ **de cavidad** *m* INSTR cavity mirror; ~ **cóncavo** *m* FÍS, INSTR concave mirror; ~ **de concentración** *m* INSTR concentrating mirror; ~ **convexo** *m* FÍS, INSTR convex mirror; ~ **de cortesía** *m* VEH vanity mirror; ~ **de desviación** *m* INSTR deviation mirror; ~ **dicroico** *m* CINEMAT, ÓPT, TELECOM dichroic mirror; ~ **del distribuidor** *m* INSTAL HIDRÁUL portface, valve seat; ~ **electrónico** *m* ELECTRÓN, INSTR electron mirror; ~ **elíptico** *m* FÍS, INSTR elliptical mirror; ~ **esférico** *m* FÍS, INSTR, LAB, ÓPT, TELECOM circular mirror, spherical mirror; ~ **de falla** *m* GEOL slickenside; ~ **falso** *m* C&V see-through mirror (*AmE*), two-way mirror (*BrE*); ~ **giratorio** *m* CINEMAT, FÍS rotating mirror, FOTO swing-up mirror, INSTR rotating mirror; ~ **de iluminación** *m* INSTR illuminating mirror, illumination mirror; ~ **de iluminación para círculo vertical** *m* INSTR illumination mirror for vertical circle; ~ **interior día-noche** *m* VEH day-night mirror; ~ **inversor** *m* INSTR inverting mirror; ~ **de Lloyd** *m* FÍS Lloyd's mirror; ~ **de luz reflejada** *m* INSTR, ÓPT reflected-light mirror; ~ **magnético** *m* FÍS magnetic mirror; ~ **de mano** *m* INSTR hand glass, hand mirror; ~ **metálico** *m* INSTR metallic mirror; ~ **parabólico** *m* FÍS parabolic mirror; ~ **plano** *m* FÍS plane mirror, INSTR flat mirror, plane mirror; ~ **plateado** *m* INSTR silvered mirror; ~ **poligonal** *m* INSTR, ÓPT polygonal mirror; ~ **primario** *m* INSTR primary mirror; ~ **principal** *m* INSTR main mirror; ~ **reflex** *m* FOTO reflex mirror; ~ **de reposición** *m* INSTR make-up mirror; ~ **reticulado** *m* TEC ESP reticulated mirror; ~ **retrovisor** *m* AUTO, C&V, INSTR, TRANSP, VEH driving mirror, rear-view mirror; ~ **retrovisor regulable** *m* AUTO, TRANSP, VEH adjustable rear-view mirror; ~ **rotatorio** *m* CINEMAT, FÍS, INSTR rotating mirror; ~ **secundario** *m* INSTR secondary mirror, TEC ESP secondary reflector; ~ **semiplateado** *m* INSTR semisilvered mirror; ~ **semireflector** *m* INSTR semireflecting mirror; ~ **superficial** *m* INSTR surface mirror; ~ **de superficie secundaria** *m* (*SSM*) TEC ESP second surface mirror (*SSM*); ~ **ultravioleta** *m* FÍS RAD ultraviolet mirror; ~ **ustorio** *m* INSTR burning-mirror, burning reflector; ~ **visual** *m* INSTR sighting mirror

espejos: ~ **de Fresnel** *m pl* FÍS Fresnel mirrors

espeque *m* MECÁ brake, handspike, TRANSP MAR *marcas de navegación* perch

espera[1]: **en** ~ *adj* INFORM&PD, TELECOM standby

espera[2]: **en** ~ *adv* TELECOM on hold

espera[3] *f* TRANSP AÉR holding; ~ **de autorización para aterrizar** *f* TEC ESP hold; ~ **de instrucciones** *f* TRANSP AÉR hold; ~ **por mal tiempo** *f* (*WOW*) TEC PETR *perforación* waiting on weather (*WOW*); ~ **de permiso** *f* TRANSP AÉR hold

espera[4]: **a la** ~ *fra* CONTAM MAR, D&A on standby

esperanza: ~ **de vida** *f* CONST life expectancy, TELECOM life expectancy, lifetime expectancy

espermidina *f* QUÍMICA spermidine

espermina *f* QUÍMICA spermin

esperrilita *f* MINERAL sperrylite

esperssartina *f* MINERAL spessartine

espesado *m* ALIMENT, C&V thickening

espesador *m* CARBÓN, P&C thickener, PAPEL decker, thickener; ~ **de filtro** *m* CARBÓN filter thickener

espesamiento *m* CARBÓN thickening; ~ **de cienos** *m* AGUA sludge thickening; ~ **del lodo** *m* HIDROL sludge thickening; ~ **de la tinta** *m* IMPR piling

espesante *m* ALIMENT thickener, thickening, CARBÓN thickener, DETERG thickening agent, IMPR bodying agent, PROC QUÍ, QUÍMICA thickener

espesar *vt* PROC QUÍ thicken; ~ **por ebullición** *vt* PROC QUÍ thicken by boiling

espesarse *v refl* PROC QUÍ thicken

espesímetro *m* ING MECÁ, MECÁ thickness gage (*AmE*), thickness gauge (*BrE*)

espesor *m* C&V thickness, IMPR caliper (*AmE*), calliper (*BrE*), MECÁ, P&C thickness, PAPEL thickness, calliper (*BrE*), caliper (*AmE*), QUÍMICA thickness; ~ **de arco** *m* ING MECÁ circular thickness; ~ **de la camisa** *m* ING MECÁ thickness of lining; ~ **de la capa de revestimiento** *m* P&C *pinturas, adhesivos*, REVEST coating thickness; ~ **circular** *m* ING MECÁ circular thickness; ~ **compactado** *m* CONST compacted thickness; ~ **curvilíneo** *m* ING MECÁ *dientes engranajes* circular thickness; ~ **de desplazamiento** *m* ING MECÁ displacement thickness; ~ **de la hoja** *m* SEG width of blade; ~ **de líneas** *m* METR *en una graduación* thickness of lines; ~ **de la mano** *m* P&C *adhesivos, pinturas*, REVEST coating thickness; ~ **medio** *m* FÍS half thickness; ~ **medio de una hoja** *m* PAPEL bulking thickness; ~ **mínimo teórico** *m*

TRANSP MAR minimum theoretical thickness; ~ **nominal** *m* CALIDAD nominal thickness; ~ **óptico** *m* ÓPT, TEC ESP, TELECOM optical thickness; ~ **de la película** *m* P&C *adhesivos, pinturas* film thickness; ~ **del revestimiento** *m* ING MECÁ thickness of lining; ~ **de semirreducción** *m* FÍS half-value thickness, NUCL half value layer (*HVL*); ~ **de una sola hoja** *m* PAPEL single-sheet thickness; ~ **trabajado** *m* CARBÓN worked thickness

espí *m* TRANSP MAR spinnaker

espiar *vt* TRANSP MAR *barco* warp, *embarcación* kedge

espiarse *v refl* TRANSP MAR kedge

espiche *m* TRANSP MAR *construcción de barcos* molding frame (*AmE*), moulding frame (*BrE*)

espiga *f* AGRIC ear, spike, CONST dowel, dowel pin, peg, tenon, spigot, *carpintería* trunnel, treenail, *remache* shank, *cerradura* stem, ING MECÁ collar, fang, *de válvula* pin, pivot, tongue, shank, tang, MECÁ peg, collar, pivot, *de válvula* pin, PETROL stinger, PROD *de válvula* pin, TEC PETR stinger; ~ **achatada** *f* ING MECÁ flatted parallel shank tool; ~ **de acoplamiento directo** *f* ING MECÁ direct-fitting shank; ~ **de acoplamiento con rosca** *f* ING MECÁ coupling-spigot with thread; ~ **de apoyo de la maza** *f* PROD tup pallet; ~ **de apoyo superior** *f* PROD top pallet; ~ **de apoyo para el yunque** *f* PROD pallet for anvil; ~ **de cara visible** *f* CONST barefaced tenon; ~ **conductora** *f* ING MECÁ driving tenon; ~ **de cono Morse** *f* ING MECÁ Morse taper shank; ~ **cuadrada** *f* ING MECÁ flatted parallel shank tool; ~ **de dos caras** *f* ING MECÁ flatted parallel shank tool; ~ **estática** *f* D&A static pin; ~ **de guía** *f* ING MECÁ, MECÁ guide pin; ~ **de horquilla** *f* ING MECÁ, MECÁ clevis pin; ~ **invisible** *f* CONST stub tenon; ~ **de madera** *f* PROD dowel pin; ~ **de la pala** *f* TRANSP AÉR blade shank; ~ **pasante** *f* CONST *carpintería* through-tenon; ~ **posicionadora** *f* ING MECÁ locating pin, locating spigot, location spigot, MECÁ locating pin; ~ **con rosca para extracción** *f* ING MECÁ dowel pin with extracting thread; ~ **con rosca extractora** *f* ING MECÁ dowel pin with extracting thread; ~ **roscada** *f* ING MECÁ notched stem; ~ **térmica** *f* NUCL thermal spike; ~ **de tope** *f* CONST shouldered tenon; ~ **sin trillar** *f* AGRIC unthreshed head

espigadora *f* ING MECÁ tenoning machine; ~**-deschaladora de maíz** *f* AGRIC corn picker-husker (*AmE*), maize picker-husker (*BrE*); ~**-desgranadora de maíz** *f* AGRIC corn picker-sheller (*AmE*), maize picker-sheller (*BrE*); ~ **de maíz** *f* AGRIC corn snapper (*AmE*), maize snapper (*BrE*)

espigar *vt* AGRIC head

espigón *m* CONST *ingeniería civil* groyne (*BrE*), jetty, groin (*AmE*), GEOL hinge, ING MECÁ newel, TRANSP MAR *protección de costas* groin (*AmE*), groyne (*BrE*), pier

espigueta *f* MINAS shooting needle, *aguja de polvorero* aiguille

espiguilla *f* AGRIC spikelet

espilita *f* TEC PETR spilite

espilosita *f* TEC PETR spilosite

espín *m* (*s*) FÍS PART, FÍS RAD spin (*s*); ~ **controlado** *m* TRANSP AÉR controlled spin; ~ **del electrón** *m* ELECTRÓN, FÍS, FÍS PART, NUCL, QUÍMICA electron spin; ~ **entero** *m* FÍS integral spin; ~ **impar-impar** *m* NUCL odd-odd spin; ~ **isobárico** *m* FÍS isobaric spin; ~ **isotópico** *m* FÍS isotopic spin; ~ **nuclear** *m* FÍS, FÍS

RAD nuclear spin; ~ **semientero** *m* FÍS half-integral spin

espinasterol *m* QUÍMICA spinasterol

espinel *m* MINERAL, QUÍMICA spinel

espinela *f* MINERAL, QUÍMICA spinel; ~ **rojo-amarillenta** *f* MINERAL rubicelle; ~ **rosa** *f* MINERAL balas ruby

espira *f* AGRIC whorl, ELEC *devanado* turn, ELECTRÓN convolution, ING ELÉC turn, ING MECÁ spiral, coil, spire, helix; ~ **cortocircuitada** *f* ING ELÉC shorted turn; ~ **de cortocircuito** *f* ING ELÉC slug; ~ **inactiva** *f* ING MECÁ dead end

espiral¹ *adj* GEOM spiral; **en** ~ *adj* ING MECÁ spiral

espiral² *f* CALIDAD, FOTO, GEOM spiral, ING ELÉC slug, spiral, ING MECÁ helix, spiral, METAL, ÓPT spiral; ~ **de calidad** *f* CALIDAD quality spiral; ~ **de crecimiento** *f* METAL growth spiral; ~ **de Ekman** *m* OCEAN Ekman spiral; ~ **de fibra** *f* ÓPT fiber pigtail (*AmE*), fibre pigtail (*BrE*); ~ **hacia la derecha** *f* ING MECÁ right-handed spiral; ~ **hiperbólica** *f* ING MECÁ reciprocal spiral; ~ **de relevado** *f* FOTO developing spiral; ~ **del tanque de unidades múltiples** *m* FOTO multiunit tank spiral

espiralado *adj* ING MECÁ spiral

espirano *m* QUÍMICA spirane

espiras: ~ **del bobinado flojas** *f pl* ING ELÉC loosely-wound turns

espíritu *m* QUÍMICA spirit

espiroidal *adj* ING MECÁ spiral

espita *f* CONST bib, faucet, PROD spout; ~ **de drenaje del radiador** *f* AUTO, VEH radiator drain cock

espitar *vt* TRANSP MAR *tonel* broach

espodumena *f* MINERAL spodumene

espoiler *m* TRANSP, TRANSP AÉR spoiler

espoleta *f* D&A *en artillería*, MINAS fuse; ~ **cohética** *f* D&A rocket pistol; ~ **de mando** *f* MINAS master fuse; ~ **de la mecha detonante** *f* TEC ESP primacord fuse; ~ **de percusión** *f* MINAS percussion fuse; ~ **de percusión y retardo** *f* D&A retard-and-impact fuse; ~ **progresiva** *f* MINAS ordinary fuse; ~ **de seguridad** *f* MINAS safety fuse

espolón *m* CONST cutwater, *puertos* jetty, OCEAN groin (*AmE*), groyne (*BrE*), PETROL, TEC PETR stinger; ~ **articulado** *m* PETROL articulated stinger; ~ **coralífero** *m* GEOL, OCEAN reef spur; ~ **fijo** *m* PETROL fixed stinger; ~ **rígido** *m* PETROL rigid stinger

espolvoreado: ~ **en cuatro direcciones** *m* IMPR four-way dusting

espolvoreador *m* AGRIC duster; ~ **de mochila** *m* AGRIC knapsack duster

espolvoreadora *f* IMPR dusting unit

espolvorear *vt* AGRIC, ALIMENT, IMPR dust

espondeo *m* ACÚST spondee

espongina *f* QUÍMICA spongine

esponja: ~ **de platino** *f* QUÍMICA platinum sponge

esponjamiento *m* CONST *tierras* bulking

espreado *m* C&V gunning, *de una mezcla a presión sobre una superficie* guniting

esprín *m* OCEAN *amarra* spring line, TRANSP MAR spring; ~ **de proa** *m* TRANSP MAR bow spring

esprines: ~ **de proa y de popa** *m pl* TRANSP MAR *amarres* cross springs

espuerta *m Esp* (*cf caldero AmL*) MINAS *minerales o carbón* kibble

espuma *f* ALIMENT, C&V foam, scum, CARBÓN *metalurgia* dross, foam, froth, DETERG, P&C, PAPEL, PROC

QUÍ foam, QUÍMICA foam, scum, TERMO, TEXTIL foam; **~ para cerrar compartimientos** *f* EMB close-cell foam; **~ extintora** *f* SEG, TERMO, TRANSP AÉR extinguishing foam; **~ de flotación** *f* PROC QUÍ flotation froth; **~ de látex** *f* P&C latex foam; **~ de mar** *f* MINERAL meerschaum, OCEAN sea foam; **~ de oro** *f* MINAS floating gold; **~ de poliéster** *f* P&C polyester foam, polyether foam; **~ de poliuretano** *f* P&C, TRANSP MAR polyurethane foam; **~ de silice** *f* C&V silica scum; **~ de vidrio** *f* C&V foam glass

espumación *f* ALIMENT, C&V, DETERG, PROC QUÍ, PROD, TEC PETR foaming

espumadera *f* CONTAM MAR, PROD skimmer

espumado[1] *adj* TEC PETR *ingeniería de lodos* foaming

espumado[2] *m* ALIMENT, C&V, DETERG, PROC QUÍ foaming, PROD skimming, foaming

espumador *m* CONTAM skimmer; **~ por ciclón** *m* CONTAM cyclone recovery skimmer; **~ de cinta** *m* CONTAM belt skimmer; **~ de cinta absorbente** *m* CONTAM absorbent belt skimmer; **~ de cinta transportadora** *m* CONTAM conveyor-belt skimmer; **~ de disco** *m* CONTAM disc skimmer (*BrE*), disk skimmer (*AmE*); **~ de vertedero** *m* CONTAM weir skimmer

espumante *m* C&V, DETERG, PROC QUÍ, TEC PETR foamer

espumar *vt* CONTAM *tratamiento de líquidos* skim off, PROC QUÍ foam, PROD skim off, skim, TEXTIL foam

espumoso *adj* C&V, DETERG, P&C, PROC QUÍ, PROD foamy, TERMO gassy, TEXTIL foamy

espúreo *adj* INFORM&PD *gráficos* aliasing

esqueje *m* CARBÓN *de un árbol* cutting

esqueleto *m* CONST framing

esquema *m* INFORM&PD layout, scheme, ING MECÁ outline, scheme, PROD diagram; **~ del chip** *m* ELECTRÓN chip layout; **~ de conexiones** *m* AUTO wiring diagram, ELEC circuit diagram, wiring diagram, ELECTRÓN, ING ELÉC circuit diagram, TELECOM wiring diagram; **~ de distribución de la cabina** *m* TRANSP AÉR cabin layout; **~ eléctrico** *m* ELEC circuit diagram, wiring diagram, ELECTRÓN, ING ELÉC circuit diagram, TELECOM wiring diagram; **~ eléctrico del circuito** *m* ELEC, TELECOM wiring diagram; **~ lógico** *m* PROD logic diagram; **~ de montaje** *m* ING ELÉC connection diagram, ING MECÁ, MECÁ layout, PROD mounting layout; **~ de puesta a punto** *m* AUTO timing diagram; **~ quark-leptón** *m* FÍS, FÍS PART quark-lepton scheme; **~ de reglaje** *m* AUTO timing diagram; **~ de sincronismo** *m* INFORM&PD timing diagram, timing analysis; **~ de sincronización de válvulas** *m* AUTO valve-timing diagram; **~ unifilar** *m* ELEC *red de distribución* single-line diagram; **~ unilineal** *m* ELEC *red de distribución* single-line diagram; **~ de velocidad** *m* ENERG RENOV velocity diagram

esquila: **~ mecánica** *f* AGRIC mechanical shearing

esquiladora *f* AGRIC, TEXTIL *para ovejas* shearing machine; **~ mecánica** *f* CARBÓN shearing machine

esquilar *vt* AGRIC clip, TEXTIL shear

esquina *f* CONST angle tie, cant, FOTO corner mount, ING MECÁ angle; **~ despostillada** *f* C&V chipped corner

esquinal *m* CONST corner post

esquisto *m* TEC PETR shale; **~ azul** *m* GEOL blueschist; **~ bituminoso** *m* GEOL, PETROL, TEC PETR oil shale; **~ calcáreo** *m* GEOL calc-schist; **~ carbonoso** *m*

CARBÓN bone coal; **~ graptolítico** *m* GEOL graptolitic shale; **~ lustroso** *m* GEOL lustrous schist; **~ de sericita** *m* GEOL *roca metamórfica* sericite schist; **~ verde** *m* GEOL greenschist

esquistosidad *f* GEOL foliation, schistosity; **~ de cizalla** *f* GEOL slip cleavage; **~ de crenulación** *f* GEOL slip cleavage; **~ superpuesta** *f* GEOL slip cleavage

esquistoso[1] *adj* CARBÓN, GEOL shaly, PETROL aschistic

esquistoso[2]: **~ carbonoso** *m* MINAS bone coal

estabilidad *f* CARBÓN, INFORM&PD stability, ING MECÁ set, MINAS *terrenos*, QUÍMICA, TELECOM, TRANSP MAR *diseño de barcos* stability; **~ absoluta** *f* TELECOM *telefonía* absolute stability; **~ sin avería** *f* TRANSP MAR *diseño de barcos* intact stability; **~ al calor** *f* P&C heat stability; **~ de la capa límite** *f* FÍS FLUID *líquidos* boundary-layer stability; **~ de compensación** *f* TRANSP AÉR trim stability; **~ contra la guiñada** *f* TEC ESP, TRANSP directional stability; **~ dimensional** *f* CINEMAT, EMB, IMPR, NUCL *de un elemento de combustible*, PAPEL dimensional stability; **~ dimensional frente al agua** *f* IMPR, PAPEL hygrostability; **~ dinámica** *f* TELECOM, TRANSP AÉR dynamic stability; **~ durante el almacenado** *f* EMB shelf stability; **~ estática** *f* TRANSP AÉR static stability; **~ de fase** *f* ELEC *corriente alterna*, TELECOM phase stability; **~ de frecuencia de corta duración** *f* ELECTRÓN short-term frequency stability; **~ frente al calor** *f* EMB heat stability; **~ de la galería** *f* CARBÓN slope stability; **~ de la imagen** *f* CINEMAT image steadiness; **~ inicial** *f* TRANSP MAR *arquitectura naval* initial stability; **~ intrínseca** *f* FÍS FLUID intrinsic stability; **~ de larga duración** *f* ELECTRÓN long-term stability; **~ lateral** *f* TRANSP lateral stability; **~ longitudinal** *f* TRANSP longitudinal stability; **~ mecánica** *f* P&C mechanical stability; **~ con la palanca de mandos fijada** *f* TRANSP AÉR fixed-stick stability; **~ química** *f* CARBÓN chemical stability; **~ de ruta** *f* TEC ESP, TRANSP, TRANSP AÉR directional stability; **~ térmica** *f* P&C, TELECOM, TERMO heat stability, thermal stability; **~ transversal** *f* TRANSP AÉR rolling stability, TRANSP MAR *diseño de barcos* transverse stability; **~ vertical** *f* TRANSP vertical stability

estabilización *f* GEN stabilization, FÍS settling; **~ en azimut** *f* TRANSP MAR azimuth stabilization; **~ por balanceo** *f* TRANSP sway stabilization; **~ con cal** *f* CONST lime stabilization; **~ con cemento** *f* CONST cement stabilization; **~ por centro inferior de gravedad** *f* TRANSP stabilization by low center of gravity (*AmE*), stabilization by low centre of gravity (*BrE*); **~ de la deriva** *f* TV drift lock; **~ por doble rotación** *f* TEC ESP dual-spin stabilization; **~ por frío** *f* REFRIG chill-proofing, *del vino* cold stabilization; **~ del giro** *f* CINEMAT, TRANSP, TRANSP AÉR gyro stabilization; **~ por giro** *f* TEC ESP spin stabilization; **~ giroscópica** *f* CINEMAT, TRANSP, TRANSP AÉR gyro stabilization; **~ por gradiente de gravedad** *f* TEC ESP gravity-gradient stabilization; **~ por gradiente de gravitación** *f* TEC ESP gravity-gradient stabilization; **~ a gran profundidad** *f* CARBÓN deep stabilization; **~ mediante centro de gravedad bajo** *f* TRANSP stabilization by low center of gravity (*AmE*), stabilization by low centre of gravity (*BrE*); **~ de muestra** *f* CALIDAD sample stabilization; **~ de la producción** *f* PROD production smoothing; **~ química** *f* HIDROL *aguas residuales* chemical stabilization; **~ por**

realimentación *f* ELECTRÓN feedback stabilization; ~ **de la rotación** *f* TEC ESP stabilization of rotation; ~ **de sedimentos** *f* HIDROL *aguas residuales* sludge stabilization; ~ **del suelo** *f* CONST soil stabilization; ~ **superficial** *f* CARBÓN shallow stabilization; ~ **de tipos de cambio en contratos** *f* PROD contract pegging; ~ **triaxial** *f* TEC ESP three-axis stabilization

estabilizado[1] *adj* QUÍMICA stabilized, TEC ESP despun, TERMO *aleaciones* aged

estabilizado[2] *m* TRANSP AÉR trim

estabilizador *m* GEN stabilizer; ~ **compensador** *m* TEC ESP bucking regulator; ~ **de frecuencia** *m* ELEC frequency stabilizer; ~ **giroscópico** *m* CINEMAT, TRANSP, TRANSP AÉR gyro stabilizer; ~ **horizontal** *m* TRANSP AÉR horizontal stabilizer, tailplane; ~ **del par** *m* AUTO, VEH torque stabilizer; ~ **de tensión** *m* ELEC voltage regulator; ~ **vertical** *m* TRANSP AÉR tail fin; ~ **de voltaje** *m* ING ELÉC voltage stabilizer

estabilizar *vt* GEN stabilize, TRANSP AÉR trim; ~ **por reposo** *vt* TERMO *una aleación* age

estabilizarse *v refl* CARBÓN *tiempo* settle

estable *adj* QUÍMICA stable; ~ **a la luz** *adj* EMB stable-to-light

establecer *vt* ING MECÁ set, TELECOM *comunicación* set up; ~ **el emplazamiento de** *vt* PROD locate

establecido *adj* ING MECÁ set

establecimiento *m* CRISTAL setting-up; ~ **de una comunicación de señal en el aire** *m* TELECOM off-air call setup; ~ **ganadero** *m* AGRIC ranch; ~ **de llamadas** *m* TELECOM call setup

estaca *f* CONST boning-stick, peg, picket, stake; ~ **de rasante** *f* FERRO grade stake

estacada *f* AGUA groin (*AmE*), groyne (*BrE*), CONST paling, TRANSP MAR groin (*AmE*), groyne (*BrE*); ~ **de pilotes** *f* AGUA pile groin (*AmE*), pile groyne (*BrE*)

estacado *m* CONST *topografía* staking

estacha *f* TRANSP MAR hawser, mooring line; ~ **de amarre** *f* TRANSP MAR hawser, mooring line; ~ **de remolque** *f* CONTAM MAR, TRANSP MAR towline

estación *f* TEC ESP, TRANSP *transmisiones* station; ~ **aeronáutica móvil** *f* TEC ESP mobile aeronautical station; ~ **de aforo** *f* AGUA gaging station (*AmE*), gauging station (*BrE*); ~ **almacén** *f* TRANSP station; ~ **de ambulancias** *f* SEG ambulance station; ~ **de auto-servicio** *f* TRANSP self-service station; ~ **auxiliar** *f* PETROL booster station; ~ **base** *f* D&A base station, FERRO home station, TELECOM base station, TV key station; ~ **bloqueada** *f* TELECOM station barred; ~ **de bombas** *f* AGUA pump station; ~ **de bombeo** *f* AGUA pumping plant, pumping-station, CONST, MINERAL, PETROL, TEC PETR pumping-station; ~ **de calefacción centralizada** *f* TERMO district heating station; ~ **de cambio de ancho de vía** *f* FERRO change-of-gage station (*AmE*), change-of-gauge station (*BrE*); ~ **de carga** *f* MINAS onsetting station, PROD *teleféricos* load terminal; ~ **central** *f* GAS central station; ~ **de clasificación** *f* FERRO, TRANSP classification yard (*AmE*), marshalling yard (*BrE*), shunting yard (*BrE*), switching yard (*AmE*); ~ **de comunicaciones a bordo** *f* TELECOM on-board communication station; ~ **conjunta de campo** *f* PETROL field joint station; ~ **de consulta** *f* INFORM&PD inquiry station; ~ **de contenedores** *f* TRANSP container station; ~ **de control de la red** *f* ELEC net control station; ~ **de conversión** *f* ELEC, ING ELÉC converting station; ~ **de coordinación de**

redes *f* TEC ESP network coordination station (*NCS*); ~ **costera** *f* TELECOM coastal station; ~ **de datos** *f* INFORM&PD data station; ~~**depósito** *f* TRANSP station; ~ **depuradora de aguas** *f* MINAS water-cleansing plant; ~ **depuradora de aguas cloacales** *f* AGUA sewage disposal plant; ~ **de desembarco** *f* MINAS landing station; ~ **de desembarque** *f* MINAS landing station; ~ **de dirección de la red** *f* ELEC net control station; ~ **directa** *f* PROD forward station; ~ **de distribución** *f* ELEC *suministro* switching station, MINAS *productos petrolíferos* depot; ~ **de distribución de energía** *f* ELEC *suministro* switching station; ~ **eléctrica** *f* ELEC *red de distribución* electric-power station; ~ **elevadora** *f* ELEC *transformador* step-up station; ~ **elevadora de presión** *f* PETROL booster station; ~ **de empalme** *f* FERRO railroad junction (*AmE*), railway junction (*BrE*); ~ **de enganche** *f* MINAS onsetting station; ~ **de enlace** *f* FERRO junction station (*BrE*), tie station (*AmE*); ~ **de envío de información** *f* TELECOM information-sending station; ~ **esclava** *f* INFORM&PD slave station; ~ **espacial** *f* TEC ESP space station; ~ **experimental** *f* AGRIC experiment station; ~ **de ferrocarril** *f* FERRO railroad center (*AmE*), railway centre (*BrE*); ~ **de ferrocarril principal** *f* FERRO depot; ~ **fija aeronáutica** *f* TRANSP AÉR aeronautical fixed station; ~ **de filtración** *f* AGUA filter plant; ~ **fluviométrica** *f* AGUA gaging station (*AmE*), gauging station (*BrE*); ~ **de gasolina** *f* Esp (*cf estación de nafta AmL*) AUTO, VEH gas station (*AmE*), gasoline station (*AmE*), petrol station (*BrE*), road gas station (*AmE*), road gasoline station (*AmE*); ~ **generadora** *f* FÍS, ING MECÁ, MINAS, TELECOM power station; ~ **geodésica** *f* TRANSP MAR geodesic station; ~ **inferior del pozo** *f* CARBÓN, MINAS bottom station; ~ **intermedia** *f* TRANSP *cablevías, funiculares* valley station; ~ **de lanzamiento** *f* D&A *de cohetes* launch station; ~ **de lanzamiento de globos** *f* METEO, TEC ESP balloon-release station; ~ **de lluvias** *f* AGRIC, AGUA, GEOL, METEO rainy season; ~ **maestra** *f* INFORM&PD master station; ~ **de maniobras** *f* FERRO switching station (*BrE*), switching yard (*AmE*); ~ **mareográfica** *f* OCEAN tide station; ~ **marítima** *f* TRANSP MAR harbor station (*AmE*), harbour station (*BrE*); ~ **marítima móvil** *f* TEC ESP mobile maritime station; ~ **de mercancías** *f* TRANSP freight depot (*AmE*), freight station (*AmE*), goods depot (*BrE*), goods station (*BrE*); ~ **meteorológica** *f* METEO, TEC ESP, TRANSP AER, TRANSP MAR meteorological station, weather station; ~ **meteorológica aeronáutica** *f* METEO, TRANSP AÉR aeronautical meteorological station; ~ **meteorológica automática** *f* METEO automatic weather station; ~ **meteorológica de buque** *f* TRANSP MAR weather station cabinet; ~ **meteorológica oceánica** *f* METEO weather ship; ~ **móvil** *f* (*MS*) IMPR, TEC ESP, TELECOM mobile station (*MS*); ~ **móvil radioeléctrica** *f* TELECOM mobile radio station; ~ **móvil de radionavegación** *f* TELECOM radio-navigation mobile station; ~ **móvil terrestre** *f* TELECOM land mobile station (*LMS*); ~ **de nafta** *f* AmL (*cf estación de gasolina Esp*) AUTO, VEH gas station (*AmE*), gasoline station (*AmE*), petrol station (*BrE*), road gas station (*AmE*), road gasoline station (*AmE*); ~ **de observación meteorológica** *f* METEO meteorological station; ~ **orbital** *f* TEC ESP orbital station; ~ **de petición** *f* INFORM&PD inquiry station;

~ **pluviométrica** *f* AGRIC, METEO rainfall station; ~ **principal** *f* INFORM&PD master station; ~ **de radio** *f* TELECOM radio station; ~ **de radio móvil** *f* TELECOM mobile radio station; ~ **de radio de tierra** *f* TRANSP AÉR ground radio station; ~ **de radiocomunicación móvil** *f* TELECOM mobile radio station; ~ **radiogoniométrica** *f* TELECOM, TRANSP, TRANSP AÉR, TRANSP MAR radio direction-finding station; ~ **de rastreo** *f* TEC ESP tracking station; ~ **de rayos X** *f* PETROL X-ray station; ~ **receptora de información** *f* CONTAM *automatización del control ambiental*, TELECOM information receiver station (*IRS*), information-receiving station (*IRS*); ~ **para recubrir tubería** *f* PETROL dope station; ~ **de recuento** *f* TRANSP counting station; ~ **de recuento continuo** *f* TRANSP continuous counting station; ~ **de red** *f* INFORM&PD network station; ~ **reductora** *f* ELEC *transformador* step-down station; ~ **de referencia** *f* TEC ESP reference station; ~ **de refuerzo** *f* HIDROL boosting station, PETROL booster station; ~ **relé de radar** *f* D&A, TRANSP, TRANSP AÉR, TRANSP MAR radar relay station; ~ **repetidora** *f* TEC ESP relay, TELECOM relay station, TV relay, relay station; ~ **repetidora de la cadena** *f* TV network broadcast repeater station; ~ **repetidora para enlace móvil-base** *f* TELECOM mobile-to-base relay; ~ **repetidora orbital** *f* TEC ESP active satellite; ~ **repetidora de radar** *f* TRANSP radar picket station; ~ **de rescate** *f* SEG rescue station; ~ **retransmisora** *f* TELECOM relay; ~ **seca** *f* AGRIC, METEO dry season; ~ **de seguimiento** *f* TEC ESP tracking station; ~ **de seguimiento con radar** *f* D&A, TRANSP, TRANSP AÉR, TRANSP MAR radar tracking station; ~ **de señales** *f* TRANSP MAR signal station; ~ **de servicio** *f* AUTO, PROD, TRANSP, VEH gas station (*AmE*), gasoline station (*AmE*), petrol station (*BrE*), road gas station (*AmE*), road gasoline station (*AmE*); ~ **del servicio de salvamento de náufragos** *f* SEG, TRANSP MAR *en tierra* lifeboat station; ~ **sísmica** *f* GEOFÍS seismic station; ~ **situada en la trayectoria** *f* TEC ESP *misiles balísticos, misiones espaciales* downrange station; ~ **de soldado** *f* PETROL welding station; ~ **subordinada** *f* INFORM&PD slave station; ~ **superior** *f* TRANSP *cablevías, funiculares* top station; ~ **térmica y eléctrica combinada** *f* TERMO combined heat and power station; ~ **terminal** *f* TRANSP dead end station, terminal station, through station; ~ **terrena** *f* FÍS earth station, TEC ESP ground station, earth station, terrestrial station, TELECOM, TV earth station; ~ **terrena emisora de la señal de control** *f* TEC ESP command earth station; ~ **terrena de mando** *f* TEC ESP command earth station; ~ **terrestre** *f* FÍS earth station, TEC ESP ground station, earth station, terrestrial station, TELECOM, TV earth station; ~ **terrestre para aeronaves** *f* TEC ESP airborne earth station; ~ **terrestre aerotransportada** *f* TEC ESP airborne earth station; ~ **terrestre de buque** *f* TRANSP MAR ship earth station; ~ **terrestre costera** *f* TRANSP MAR coast earth station; ~ **terrestre embarcada** *f* TEC ESP shipborne earth station; ~ **terrestre móvil** *f* TEC ESP mobile land station; ~ **terrestre receptora** *f* TEC ESP, TELECOM, TV receiving earth station; ~ **terrestre transportable** *f* TEC ESP transportable earth station; ~ **testigo** *f* TEC ESP reference station; ~ **de trabajo** *f* INFORM&PD, PROD work station; ~ **de transbordo** *f*

FERRO junction station, MINAS transfer station; ~ **de translación** *f* TV translator station; ~ **de la tripulación** *f* TRANSP AÉR crew station

estacionado *adj* TELECOM, TV parked
estacionalidad *f* PROD seasonality
estacionamiento *m* CONST lay-by
estacionar *vt* TELECOM, TV park
estadía *f* AmL (*cf permanencia Esp*) CONST stay, TEC PETR, TRANSP lay day
estadio *m* METR furlong
estadiómetro *m* CONST stadiometer
estadística *f* MATEMÁT statistic; ~ **de Bose-Einstein** *f* FÍS Bose-Einstein statistics; ~ **cuántica** *f* FÍS quantum statistics; ~ **de Fermi-Dirac** *f* FÍS Fermi-Dirac statistics
estadísticas *f pl* MATEMÁT statistics
estadístico *adj* CALIDAD, MATEMÁT statistical
estado[1]: **de ~ sólido** *adj* TV solid-state
estado[2]: **en ~ de conducción** *adv* ELEC conducting; **de ~ sólido** *adv* INFORM&PD solid-state; **en ~ tosco de forjado** *adv* PROD as forged; **en ~ tosco de fundición** *adv* PROD as cast; **en ~ tosco de laminación** *adv* PROD as rolled
estado[3] *m* FÍS state, INFORM&PD state, status, TRANSP MAR state; ~ **UNO** *m red Olivetti* ELECTRÓN ONE state; ~ **activado** *m* METAL activated state; ~ **de activado** *m* ING ELÉC on state; ~ **actual de la tecnología** *m* PROD state of the art, TELECOM current state of the art; ~ **del acumulador** *m* TELECOM battery condition; ~ **aislante** *m* ING ELÉC nonconducting state; ~ **de alta impedancia** *m* ING ELÉC high impedance state; ~ **atómico** *m* NUCL atomic state; ~ **cuántico** *m* FÍS, FÍS PART, TEC ESP quantum state; ~ **cuasiestacionario** *m* FÍS quasi-steady state; ~ **de cuentas** *m* PROD statement; ~ **de desactivado** *m* ING ELÉC off state; ~ **despresionizado** *m* PROD depressurized condition; ~ **dieléctrico** *m* ING ELÉC nonconducting state; ~ **dispositivo de E/S** *m* PROD I/O device status; ~ **de equilibrio** *m* TERMO state of equilibrium; ~ **de espera** *m* INFORM&PD wait state; ~ **de espigado** *m* AGRIC heading stage; ~ **estacionario** *m* ELEC steady state, FÍS, FÍS RAD stationary state, TELECOM steady state, TERMO stationary state; ~ **de excitación** *m* FÍS, FÍS PART, FÍS RAD, METAL, QUÍMICA excited state; ~ **excitado** *m* FÍS, FÍS PART, FÍS RAD, METAL, QUÍMICA excited state; ~ **final** *m* PROD last state; ~ **del fondo del pozo** *m* TEC PETR *perforación* bottom-hole conditions; ~ **fundamental** *m* FÍS, FÍS PART, FÍS RAD, QUÍMICA ground state; ~ **gaseoso** *m* GAS gaseous state; ~ **de gran impedancia** *m* ING ELÉC high impedance state; ~ **hidrométrico** *m* HIDROL hydrometric state; ~ **higrométrico** *m* AGUA moisture content; ~ **inicial** *m* INFORM&PD initial state; ~ **ionizado** *m* FÍS PART ionized state; ~ **K** *m* NUCL K-state; ~ **libre** *m* TELECOM idle state; ~ **lógico** *m* ELECTRÓN logic state; ~ **de la máquina** *m* ING MECÁ machine status; ~ **de la mar** *m* TRANSP MAR sea conditions, sea state; ~ **metaestable** *m* FÍS, METAL metastable state; ~ **microscópico** *m* FÍS RAD microscopic state; ~ **neutro** *m* FÍS PART neutral state; ~ **de ocupación** *m* TELECOM busy state, busy status; ~ **permanente** *m* ELEC steady state; ~ **sin pesadez** *m* TEC ESP weightlessness; ~ **de plataforma** *m* TRANSP AÉR ramp status; ~ **de preparación para emergencias** *m* NUCL emergency preparedness; ~ **de procesos** *m*

INFORM&PD process state; ~ **real** *m* NUCL actual state; ~ **regular** *m* FÍS steady state; ~**-resistencia** *m* ING ELÉC on resistance; ~ **de salida de existencias** *m* PROD stock issue status; ~ **sólido** *m* ELEC, FÍS PART, FÍS RAD, INFORM&PD solid state; ~ **transitorio** *m* ELEC transient state; ~ **triaxial de esfuerzo** *m* METAL triaxial state of stress; ~ **de valencia** *m* NUCL valence state; ~ **vítreo** *m* C&V glassy state, vitreous state

estado[4]: **en ~ bruto** *fra* PROD in the crude state; **en ~ de servicio** *fra* ING MECÁ in full working order

estado[5]: **estar en ~ de reposo** *vi* INFORM&PD quiesce

estalagnómetro *m* FÍS stalagmometer

estallar *vi* CONST, MECÁ burst, PROD snap, TEC ESP blow up

estallido *m* CARBÓN *pozo petróleo* weathering, crack, METAL crack, MINAS report, TEC ESP boom, blowout, report, TELECOM burst, TERMO outbreak, TEXTIL *tejido* burst; ~ **alternante** *m* TV alternating burst

estampa *f* ING MECÁ die, die-casting die, press tool, MECÁ die, dolly, PROD swage, *forja* swage block, *mandril para tubos* drift, *roscado fileteado* die plate; ~ **de corte** *f* ING MECÁ shearing die; ~ **de doblar** *f* ING MECÁ bending die; ~ **ede forjado en frío** *f* TERMO cold-forging die; ~ **para embutir** *f* ING MECÁ die for pressing; ~ **para estampar** *f* ING MECÁ die for pressing; ~ **de forja** *f* ING MECÁ shaping machine, PROD swage block; ~ **de forjado en caliente** *f* ING MECÁ hot forging die; ~ **de forjado en frío** *f* ING MECÁ, PROD cold-forging die; ~ **formadora** *f* C&V blank mold (*AmE*), blank mould (*BrE*); ~ **hembra** *f* PROD *martillo pilón, martinete* bottom die; ~ **honda de embutir** *f* ING MECÁ deep drawing die; ~ **para indentar hidráulica** *f* ING MECÁ hydraulic-bulging die; ~ **inferior** *f* PROD pallet for anvil, *fragua, terja* bottom swage, *martillo pilón* tup pallet, *martillo pilón, martinete* bottom die, bottom pallet; ~ **de mano** *f* PROD hand stamp; ~ **para paneles de automóviles** *f* ING MECÁ die for motor body panels; ~ **para paneles de coches** *f* ING MECÁ die for motor body panels; ~ **partida** *f* ING MECÁ split die; ~ **perforadora** *f* ING MECÁ piercing-die; ~ **para polvos metálicos** *f* ING MECÁ die for metallic powders; ~ **progresiva** *f* ING MECÁ progression die; ~ **redonda superior** *f* ING MECÁ top rounding tool; ~ **superior** *f* ING MECÁ top swage, PROD *martinete* top die, top pallet; ~ **superior para hierros redondos** *f* ING MECÁ top rounding tool

estampación *f* IMPR stamping, PROD *marcado* stamping, TEXTIL *del tejido* printing; ~ **automática en pantalla** *f* TEXTIL automatic screen printing; ~ **en caliente** *f* PROD drop forging; ~ **con molde** *f* TEXTIL block printing; ~ **con rodillos** *f* TEXTIL roller printing; ~ **en tamiz de seda** *f* TEXTIL silkscreen printing

estampado[1] *adj* EMB, IMPR, ING MECÁ, PAPEL embossed; ~ **en caliente** *adj* IMPR hot-stamped, PROD drop-forged; **según** ~ *adj* PROD as stamped; ~ **en tosco** *adj* PROD rough-stamped

estampado[2] *m* EMB embossing, IMPR blocking, embossing, stamping, PAPEL embossing, PROD swaging, *marcado* stamping, *trabajos con láminas de metal* stamp, TEXTIL *del tejido* printing; ~ **por caída** *m* MECÁ drop forging; ~ **en caliente** *m* IMPR hot stamp imprint; ~ **por corrosión** *m* TEXTIL discharge printing; ~ **en dos piezas** *m* PROD die in two halves; ~ **en frío** *m* ING MECÁ cold hobbing; ~ **a la lionesa** *m* TEXTIL hand screen printing; ~ **a mano con molde** *m*

TEXTIL hand block printing; ~ **a máquina** *m* IMPR roller printing; ~ **en relieve** *m* P&C embossing; ~ **en una sola pieza** *m* PROD *redondos para pernería, tornillería* die in one piece

estampador *m* PROD *obrero* stamper

estampadora *f* PROD stamping machine, press; ~ **de cabeza horizontal** *f* ING MECÁ traversing head shaping machine; ~ **diseñada para cintería** *f* TEXTIL printer designed for narrow fabric

estampadura *f* PROD tooling

estampar *vt* IMPR *forma* set a forme, PROD press, stamp, TEXTIL print; ~ **en caliente** *vt* IMPR hot stamping, MECÁ swage, PROD drop-forge; ~ **en frío** *vt* PROD, TERMO cold-draw; ~ **en relieve** *vt* IMPR die-stamping

estampido *m* CARBÓN *del cañón* crack; ~ **sónico** *m* ACÚST, FÍS, SEG, TRANSP AÉR sonic boom

estampillar *vt* PROD stamp

estamux *m* INFORM&PD statmux

estañado[1] *adj* C&V, CONST, ELEC, QUÍMICA tinned

estañado[2] *m* C&V, CONST, ELEC, QUÍMICA tinning

estañador *m* PROD tin plate worker

estañadura *f* ELEC *de conductor del cable* tinning, PROD tin plate working

estanato *m* QUÍMICA stannate

estancamiento *m* *Esp* (*cf encharcamiento AmL*) CONTAM lagooning

estancar *vt* MINAS insulate

estancia *f* *Esp* (*cf estadía AmL*) AGRIC cattle ranch, CONST stay

estanco *adj* ALIMENT airtight, CONST watertight, MECÁ leak-tight, REFRIG leak-free, TEC PETR airtight, TRANSP MAR *buque* watertight; ~ **al aceite** *adj* PROD oiltight; ~ **al agua** *adj* EMB, TERMOTEC, TEXTIL watertight; ~ **a gases** *adj* GAS, ING MECÁ, MECÁ, PROC QUÍ, PROD, QUÍMICA, TERMO gas-proof, gastight; ~ **al petróleo** *adj* PROD oiltight; ~ **al vapor** *adj* TERMOTEC steamtight;

estand *m* INSTR *exposiciones* stand; ~ **de columna** *m* INSTR pillar stand

estándar *m* GEN standard; ~ **de aspereza superficial** *m* METR surface roughness standard; ~ **de calidad** *m* METR standard of quality; ~ **de emisión** *m* CONTAM *legislación* emission standard; ~ **francés** *m* ING MECÁ French standard; ~ **de referencia de dureza** *m* ING MECÁ hardness reference standards

estándares: ~ **de ingeniería** *m pl* ING MECÁ engineering standards

estandarización *f* GAS standardization, INFORM&PD normalization, standardization, QUÍMICA, TELECOM standardization

estandarizar *vt* GEN standardize

estánico *adj* QUÍMICA stannic

estannina *f* MINERAL stannite, tin pyrite

estannoso *adj* QUÍMICA stannous

estaño *m* (*Sn*) METAL, MINAS, MINERAL, PROD, QUÍMICA tin (*Sn*); ~ **leñoso** *m* MINERAL wood tin; ~ **en lingotes** *m* PROD block tin; ~ **nativo** *m* MINAS mine tin; ~ **de roca** *m* MINAS lode tin, mine tin; ~ **vidirioso** *m* MINERAL tinstone; ~ **xiloide** *m* MINERAL wood tin

estañoso *adj* QUÍMICA stannous

estanque *m* AGUA basin, reservoir, CARBÓN pond, PROD tank; ~ **de amortiguación** *m* HIDROL stilling basin; ~ **de clarificación** *m* TEC PETR clarifying basin; ~ **colector** *m* CARBÓN tailing pond; ~ **de decantación** *m* CARBÓN clear pond, HIDROL *aguas* *residuales* settling basin; **~ decantador** *m* AGUA settling basin, CARBÓN settling basin, slurry pond, HIDROL, MINAS *sondeos* settling basin; **~ de deposición** *m* PROC QUÍ settling pool, settling reservoir; **~ de derrame de gravedad** *m* TRANSP AÉR gravity spillway dam; **~ de desenlodamiento** *m* HIDROL desilting basin; **~ de enfriamiento** *m* AGUA cooling pond; **~ de engorde** *m* OCEAN oyster-fattening pond; **~ de maduración** *m* AGUA maturation pond; **~ de petróleo** *m* TEC PETR petroleum basin; **~ de prueba** *m* AGUA test basin; **~ de sedimentación** *m* PROC QUÍ settling pool, settling reservoir; **~ subsidente** *m* GEOL, TEC PETR subsident basin

estanqueidad *f* CONST sealing, watertightness, FERRO *vehículos* sealing, ING MECÁ tightness, NUCL leaktightness, PROD *de junta* imperviousness; **~ ultrasónica** *f* EMB, TERMO ultrasonic sealing, ultrasonic welding

estante *m* CONST, LAB shelf, QUÍMICA stand, TEC ESP bay

estantería *f* EMB compartment case; **~ de compuerta** *f* REFRIG *en frigorífico doméstico* door rack

estaqueado *m* CONST, PROD pegging

estaquidrina *f* QUÍMICA stachydrine

estaquillado *m* CONST staking, *topografía* picking

estaquillar *vt* CONST mark out

estaquiosa *f* QUÍMICA stachyose

estarcido *m* C&V, PROD stencil

estática[1]: **contra la ~** *adj* ELEC *materiales*, ELECTRÓN antistatic

estática[2] *f* FÍS, ING ELÉC statics; **~ en la pantalla** *f* INFORM&PD hash

estático *adj* ELEC quiescent, ELECTRÓN passive, INFORM&PD static, ING ELÉC quiescent, static

estatitrón *m* ELEC, FÍS Van de Graaff generator

estativo *m* CINEMAT chestpod

estator *m* AUTO stator, CARBÓN *turbina de vapor* shaft, ELEC *máquina*, FÍS stator, ING ELÉC frame, stator, INSTAL HIDRÁUL *de ventilador* casing, TRANSP AÉR *máquina* stator; **~ bifásico** *m* ING ELÉC two-phase stator; **~ del compresor** *m* TRANSP AÉR compressor stator; **~ devanado** *m* ING ELÉC wound stator; **~ de polos salientes** *m* ING ELÉC salient pole stator; **~ trifásico** *m* ING ELÉC three-phase stator

estatorreactor *m* TERMO ramjet engine, TRANSP AÉR athodyd, ramjet

estatua: **~ de terracota** *f* C&V terracotta statue

estaurolita *f* MINERAL staurolite

estaurotida *f* MINERAL staurotide

estay *m* ING MECÁ, TRANSP MAR stay; **~ de caldera** *m* ING MECÁ boiler stay; **~ mayor** *m* TRANSP MAR *jarcia* mainstay; **~ de mesana** *m* TRANSP MAR *jarcia firme* aft stay; **~ del trinquete** *m* TRANSP MAR forestay

este: **hacia el ~** *adv* CARBÓN open east

estearato *m* QUÍMICA stearate

esteárico *adj* QUÍMICA stearic

estearilo *m* QUÍMICA stearyl

estearina *f* ALIMENT, QUÍMICA stearin, glyceryl tristearate

esteatita *f* MINERAL steatite

estefanita *f* MINERAL stephanite

estela *f* CONTAM MAR, FÍS, FÍS FLUID, OCEAN, TRANSP MAR wake; **~ de un cilindro** *f* FÍS FLUID wake of a cylinder; **~ del cohete** *f* D&A rocket plume; **~ de condensación** *f* TRANSP AÉR condensation trail;

~ del escape del motor *f* TRANSP AÉR exhaust trail; **~ formada aguas abajo** *f* FÍS FLUID downstream wake; **~ formada aguas arriba** *f* FÍS FLUID upstream wake

estelar *adj* TEC ESP stellar

estemple[1] *adj* MINAS, PROD *minería* chock block

estemple[2] *m* AmL (*cf puntal Esp*) CARBÓN *minas* prop, MINAS pit wood, prop, pit prop

estenopetografía *f* FOTO pinhole photography

estenoscopio *m* FOTO pinhole camera

estequiometría *f* QUÍMICA stoichiometry

estequiométrico *adj* QUÍMICA stoichiometric; **no ~** *adj* QUÍMICA nonstoichiometric

éster *m* DETERG, P&C, QUÍMICA ester; **~ adípico** *m* P&C, QUÍMICA adipic ester; **~ fosfático** *m* DETERG phosphate ester; **~ maleico** *m* DETERG maleic ester; **~ metílico** *m* DETERG methyl ester

esterar *vt* TEXTIL mat

estercolar *vt* AGRIC dig in manure, manure

estercorita *f* MINERAL stercorite

estereoespecífico *adj* P&C, QUÍMICA stereospecific

estereofonía *f* ACÚST stereophony

estereofónico *adj* ACÚST binaural, stereophonic

estereoisómero *m* QUÍMICA stereoisomer

estereomicroscopio *m* INSTR, LAB stereomicroscope; **~ de distancia focal regulable** *m* INSTR ZOOM stereomicroscope; **~ con zoom** *m* INSTR ZOOM stereomicroscope

estereoquímica *f* QUÍMICA stereochemistry

estereorradián *m* GEOM steradian

estereoscopía *f* FOTO stereoscopy

estereoscopio *m* FOTO stereoscope

estereovisión *f* TV stereovision

estérico *adj* QUÍMICA steric

esterificación *f* ALIMENT, QUÍMICA esterification

esterificar *vt* ALIMENT, QUÍMICA esterify

estéril *adj* CARBÓN barren, MINAS hungry

estériles *m pl* MINAS deads, leavings, mullock, waste; **~ de flotación** *m pl* CARBÓN flotation tailings

esterilización *f* INFORM&PD sanitization

esterilla: **~ antiestática** *f* INFORM&PD antistatic mat; **~ de asbesto** *f* TEXTIL asbestos mat; **~ de fibra troceada** *f* TRANSP MAR chopped strand mat

estero: **~ mareal** *m* OCEAN tide flat

esteroide *m* QUÍMICA steroid

esterol *m* QUÍMICA sterol

esteroradian *m* ELECTRÓN, FÍS steradian

esteva *f* INSTR *arado* arm

estiaje *m* ENERG RENOV, HIDROL *en ríos o lagos* low water; **~ mínimo** *m* AGUA minimum low water

estiba *f* AGRIC stower, C&V *de vidrio plano* stack, TRANSP MAR stowage; **~ atravesada** *f* TRANSP MAR aburton stowage

estibador *m* C&V stacker, TRANSP MAR docker (*BrE*), longshoreman (*AmE*), stevedore

estibaje *m* TEC ESP, TRANSP MAR stowage

estibar *vt* TEC ESP stow, trim, TRANSP MAR stow, *el ancla* ship

estibiconita *f* MINERAL stibiconite

estibilita *f* MINERAL stibilite

estibina *f* MINERAL stibnite

estibioso *adj* QUÍMICA stibious

estibnita *f* MINERAL, QUÍMICA stibnite

estiércol *m* AGRIC manure

estifnato *m* QUÍMICA styphnate

estigmasterol *m* QUÍMICA stigmasterol

estilbeno *m* QUÍMICA stilbene

estilbita *f* MINERAL stilbite

estilete: ~ **arrancamaterial** *m* ACÚST cutting stylus; ~ **para repostado en vuelo** *m* TRANSP in-flight refueling probe (*AmE*), in-flight refuelling probe (*BrE*)

estilite *m* ACÚST stylus instrument

estilización *f* TEXTIL styling

estilo[1]: ~ **red** *adj* TRANSP AÉR network-like

estilo[2] *m* CONST *arquitectura* style, IMPR *del tipo* face, style, TEXTIL style; ~ **de impresión** *m* IMPR printstyle; ~ **de letra de imprenta** *m* IMPR, INFORM&PD typeface; ~ **de línea** *m* INFORM&PD line style; ~ **moderno** *m* IMPR modern face; ~ **propio** *m* IMPR house style; ~ **de tipo para libros** *m* IMPR bookface

estilolito *m* GEOL stylolite

estilpnomelana *f* MINERAL stilpnomelane

estima *f* OCEAN, TRANSP MAR *navegación* dead reckoning

estimación *f* MATEMÁT estimation, MINAS *de los daños* extent, TRANSP AÉR dead reckoning; ~ **del coste de amortización** *f* PROD replacement cost valuation; ~ **del coste de reposición** *f* PROD replacement cost valuation; ~ **de las reservas de petróleo** *f* CALIDAD gaging (*AmE*), gauging (*BrE*)

estimador *m* CONST quantity surveyor, MATEMÁT estimator

estimar *vt* METR estimate, measure

estimulación *f* PETROL, TEC PETR stimulation; ~ **petrolífera** *f* PETROL oil stimulation

estimulador: ~ **de la entrada** *m* PROD input stimulator

estimular *vt* FÍS RAD, PROD stimulate

estímulo *m* ACÚST, FÍS RAD stimulus, ING MECÁ impulse; ~ **homogéneo** *m* FÍS RAD homogeneous stimulus

estípula *f* CARBÓN prop

estiracitol *m* QUÍMICA styracitol

estirado[1] *adj* C&V, ING MECÁ drawn, PROD drawn, draw-down, REVEST drawn; ~ **en caliente** *adj* TERMO hot-drawn; ~ **en frío** *adj* PROD, TERMO cold-drawn

estirado[2] *m* CONST draw casting; ~ **en caliente** *m* TERMO hot drawing; ~ **de hilos** *m* PROD wire drawing; ~ **horizontal** *m* C&V horizontal drawing-process

estiraje *m* CONST stretching, PROD draw-down, drawing, *fundición* drag, TEXTIL drawing, *hilado* draft (*AmE*), draught (*BrE*); ~ **trefilado** *m* METAL drawing

estiramato *m* QUÍMICA styramate

estiramiento *m* CONST stretching, ING MECÁ strain, stretch, MINAS creep, P&C stretch; ~ **acelerado por termofluencia** *m* METAL accelerated creep; ~ **continuo y lento** *m* MECÁ creep; ~ **en frío** *m* MECÁ cold drawing; ~ **nulo** *m* METAL zero creep; ~ **parabólico** *m* METAL parabolic creep; ~ **de recuperación** *m* METAL recovery creep; ~ **repentino de la sección transversal** *m* INSTAL HIDRÁUL sudden enlargement of cross section

estirar *vt* ING MECÁ strain, tighten, TEXTIL stretch, *hilado* draw; ~ **en caliente** *vt* TERMO hot-draw; ~ **en frío** *vt* TERMO cold-draw

estireno *m* QUÍMICA styrene, vinylbenzene, TEC PETR styrene

estiroleno *m* QUÍMICA styrolene

estirón *m* C&V pull, ING MECÁ, MECÁ, PROD lug

estoa *f* OCEAN slack tide, slack water, stand of tide, tidal stand, TRANSP MAR *marea* stand of tide, *navegación* slack water

estocástico *adj* INFORM&PD, MATEMÁT stochastic

estofar *vt* ALIMENT braise

estolcita *f* MINERAL stolzite

estopa *f* PROD *fibra de cáñamo o lino* tow, TRANSP MAR *cuerdas* oakum; ~ **de filtración** *f* PROC QUÍ filter stuff

estopín *m* AmL (*cf mecha Esp*) D&A fuse, MINAS fuse, primer, TEC ESP *artillería* tube; ~ **de fricción** *m* MINAS *cañones* friction fuse; ~ **de mando** *m* MINAS master fuse; ~ **de percusión** *m* MINAS percussion fuse; ~ **progresivo** *m* MINAS ordinary fuse; ~ **de seguridad** *m* MINAS safety fuse

estopor *m* TRANSP MAR *equipamiento de cuerda* bow stopper

estorbo *m* SEG hindrance

estradiol *m* QUÍMICA estradiol

estrangulador *m* AUTO butterfly, choke, C&V choke, PETROL choke, *perforación de sondeos* bean, PROD restrictor, TEC PETR *conductos, tuberías*, VEH *carburador* choke; ~ **de aire** *m* AUTO choke; ~ **automático** *m* AUTO, VEH automatic choke; ~ **de emergencias** *m* TEC PETR *perforación, producción* storm choke

estrangulamiento *m* AUTO choking, FERRO *vehículos de vapor*, ING MECÁ throttling, TRANSP AÉR choking

estrategia: ~ **para el enrutamiento del tráfico** *f* TELECOM traffic-routing strategy; ~ **de reglamentación** *f* TRANSP regulation strategy; ~ **para el trazado de rutas de tráfico** *f* TELECOM traffic-routing strategy

estratificación[1]: ~ **gruesa** *adj* GEOL thick-bedded; ~ **potente** *adj* GEOL thick-bedded

estratificación[2] *f* CARBÓN, ENERG RENOV stratification, GEOL, bedding, layering, stratification, ING MECÁ banding, METAL layering, stratification, TEC PETR bedding, stratification; ~ **de la carga** *f* TRANSP charge stratification; ~ **contorsionada** *f* GEOL contorted bedding; ~ **cruzada** *f* GEOL cross bedding; ~ **cruzada en festón** *f* GEOL festoon cross-bedding; ~ **cruzada en surco** *f* GEOL trough cross-bedding, festoon cross-bedding; ~ **entrecruzada** *f* GEOL *acuosa o eólica* current bedding; ~ **gradada** *f* GEOL graded bedding; ~ **regresiva** *f* GEOL regressive overlap

estratificado[1] *adj* CARBÓN, GEOL, ING MECÁ, METAL, TEC PETR stratified

estratificado[2] *m* MECÁ laminate; ~ **por difusión** *m* ELECTRÓN diffused layer

estratificador *m* ELECTRÓN layer; ~ **activo** *m* ELECTRÓN active layer; ~ **amorfo** *m* ELECTRÓN amorphous layer; ~ **añadido tipo p** *m* ELECTRÓN p-type implanted layer; ~ **depositado** *m* ELECTRÓN deposited layer; ~ **de dióxido de silicio** *m* ELECTRÓN silicon dioxide layer; ~ **epitaxial en fase de vapor crecida** *m* ELECTRÓN vapor-phase-grown epitaxial layer (*AmE*), vapour-phase-grown epitaxial layer (*BrE*); ~ **epitaxial de tipo n** *m* ELECTRÓN n-type epitaxial layer; ~ **epitaxial tipo p** *m* ELECTRÓN p-type epitaxial layer; ~ **epitáxico de silicio** *m* ELECTRÓN silicon epitaxial layer; ~ **evaporado** *m* ELECTRÓN evaporated layer; ~ **fotoemisivo** *m* ELECTRÓN photoemissive layer; ~ **de interconexión** *m* ELECTRÓN interconnection layer; ~ **de inversión** *m* ELECTRÓN inversion layer; ~ **magnético epitaxial** *m* ELECTRÓN magnetic epitaxial layer; ~ **de metalización** *m* ELECTRÓN metalization layer (*AmE*), metallization layer (*BrE*); ~ **de pasivación** *m* ELECTRÓN passivation layer; ~ **de polisilicio** *m* ELECTRÓN polysilicon layer;

~ **de semiconductor** *m* ELECTRÓN semiconductor layer

estratificar *vt* CARBÓN stratify

estratificarse *v refl* CARBÓN stratify

estratigrafía *f* CARBÓN stratigraphy

estrato *m* AGUA layer, CARBÓN stratum, CONST, GAS, GEOL bed, PETROL layer, TEC PETR horizon; ~ **acuífero** *m* AGUA, HIDROL water-bearing stratum; ~ **con concreciones de siderita** *m* MINAS ball vein; ~ **empinado** *m* CARBÓN, MINAS edge seam; ~ **índice** *m* NUCL key bed; ~ **permeable** *m* GAS permeable layer; ~ **poroso** *m* GAS porous layer; ~ **rocoso** *m* AGUA, CARBÓN, GEOL rock layer; ~ **sedimentario** *m* GAS sediment layer; ~ **yesífero** *m* CONST chalk stratum; ~ **de yeso** *m* CONST chalk stratum

estratocúmulo *m* (*Sc*) METEO stratocumulus (*Sc*)

estratos: ~ **del lecho** *m pl* HIDROL bed strata

estratosfera *f* METEO stratosphere

estratotipo *m* GEOL type section

estrechamiento *m* FÍS, HIDROL, ING MECÁ contraction, MINAS *de filones* contraction, nip, squeeze, TEXTIL contraction; ~ **de la calzada** *m* TRANSP bottleneck

estrechar *vt* C&V tighten, ING MECÁ tighten, *filones* pinch

estrechez *f* C&V crimp, ING MECÁ tightness

estrecho *m* OCEAN *accidentes geográficos* sound, *geomorfología* gut, TRANSP MAR *navegación* sound

estrella *f* GEN star; ~ **activa** *f* INFORM&PD *red* active star; ~ **antigiratoria** *f* TRANSP AÉR nonrotating star; ~ **binaria** *f* TEC ESP apastron, periastron; ~ **D** *f* ÓPT, TELECOM D-star; ~ **enana** *f* TEC ESP dwarf star; ~ **fija** *f* TRANSP AÉR nonrotating star; ~ **fugaz** *f* METEO, TEC ESP meteor; ~ **del inducido** *f* ELEC armature spider; ~ **con intensidad luminosa pulsante** *f* TEC ESP quasar; ~ **de neutrones** *f* TEC ESP neutron star; ~ **pasiva** *f* INFORM&PD *red* passive star; ~ **radioeléctrica** *f* FÍS RAD, TEC ESP radio star

estrellada: ~ **por presión** *f* C&V pressure check; ~ **en la vela** *f* C&V parison check

estrépito *m* TEC ESP noise

estrés: ~ **permanente** *m* C&V permanent stress

estría *f* C&V score, CONST flute, IMPR crease, score, ING MECÁ groove, spire, MECÁ cable, flute, spline, ÓPT groove, P&C scratch, PROD groove, ridge, TEC ESP spline; ~ **bacteriana** *f* AGRIC bacterial stripe; ~ **de pared recta** *f* ING MECÁ straight-sided spline; ~ **principal** *f* ING MECÁ master spline

estriación *f* ING MECÁ serration, METAL striation

estriado[1] *adj* CONST ribbed, ING ELÉC fluted, ING MECÁ corrugated, ribbed, fluted, MECÁ fluted, knurled, ribbed, PAPEL striped, PROD grooved

estriado[2] *m* CONST fluting, IMPR scoring, ING MECÁ knurling, PROD grooving; ~ **en espiral** *m* ING MECÁ involute serration; ~ **de involuta** *m* ING MECÁ involute serration

estriadora *f* ING MECÁ knurling tool

estriar *vt* C&V, IMPR score

estrías[1]: **sin** ~ *adj* ING MECÁ, MECÁ fluteless

estrías[2] *f pl* MECÁ lining

estribación *f* GEOL offset

estribo *m* ACÚST stapes, stirrup, CONST *monumento* step, *hormigón armado* stirrup, *arquitectura, obras hidráulicas* abutment, *dique, puente* pier, ING MECÁ stirrup; ~ **de apoyo** *m* CONST cradle stirrup; ~ **de ballestas** *m* ING MECÁ spring washer

estribor *m* TEC ESP, TRANSP MAR starboard

estricción *f* METAL necking

estriol *m* QUÍMICA estriol

estrobo *m* C&V sling; ~ **de cable de acero** *m* SEG wire rope sling; ~ **de cadena** *m* SEG chain sling; ~ **de cuerda** *m* SEG rope-type sling; ~ **tipo fibra** *m* SEG fiber-type sling (*AmE*), fibre-type sling (*BrE*)

estroboscopio *m* FÍS, FÍS ONDAS stroboscope

estrofantina *f* QUÍMICA strophanthin

estromatolito *m* GEOL stromatolite

estrona *f* QUÍMICA estrone

estróncico *adj* QUÍMICA strontic

estroncio[1]: **de** ~ *adj* QUÍMICA strontic

estroncio[2] *m* (*Sr*) QUÍMICA strontium (*Sr*); ~ **radioactivo** *m* FÍS RAD, QUÍMICA radiostrontium

estropeado *adj* ELEC *fallo* out-of-order; **muy** ~ *adj* TEXTIL *específico de las prendas de vestir* worn-out; ~ **por el tiempo** *adj* TRANSP MAR *buque* weather-beaten

estropear *vt* ALIMENT spoil

estructura *f* CONST frame, structure, ELECTRÓN pattern, FERRO, FÍS structure, GEOL *petrografía* fabric, ING ELÉC, MECÁ frame, PAPEL formation; ~ **en A** *f* CONST A frame; ~ **acanalada** *f* METAL banded structure; ~ **de acero** *f* CONST steel construction; ~ **aerodinámica** *f* VEH aerodynamic shape; ~ **alveolada** *f* CONST honeycomb structure; ~ **alveolar** *f* GEOL honeycomb structure; ~ **amorfa** *f* P&C amorphous structure; ~ **anormal** *f* METAL abnormal structure; ~ **en árbol** *f* INFORM&PD, TELECOM tree structure; ~ **de archivos** *f* INFORM&PD file structure; ~ **atómica** *f* FÍS PART atomic structure, NUCL, TEC ESP atomistic structure; ~ **bayoneta** *f* PROD girder-type frame; ~ **de bloques** *f* INFORM&PD block structure; ~ **de cables flotantes** *f* ÓPT, TELECOM loose-cable structure; ~ **de cables sueltos** *f* ÓPT, TELECOM loose-cable structure; ~ **en capas** *f* CRISTAL layer structure; ~ **de capas de aplicación** *f* TELECOM application layer structure; ~ **en capas de aplicación extendida** *f* TELECOM extended application layer structure (*XALS*); ~ **en carrusel** *f* TEC ESP *comunicaciones* carousel structure; ~ **de celosía** *f* CONST lattice truss; ~ **celular** *f* TELECOM cellular structure; ~ **columelar** *f* GEOL mullion structure; ~ **de concreto** *f* AmL (*cf estructura de hormigón Esp*) CONST concrete structure; ~ **de crecimiento** *f* METAL *cristalografía* growth pattern; ~ **cristalina** *f* CARBÓN, CRISTAL, METAL crystal structure; ~ **de cubierta** *f* TRANSP MAR *construcción naval* deck structure; ~ **de datos** *f* INFORM&PD data structure; ~ **de datos de las instrucciones del secuenciador** *f* PROD sequencer instruction data form; ~ **de defectos** *f* METAL defect structure; ~ **de deformación** *f* GEOL boudinage; ~ **direccional** *f* GEOL directional structure; ~ **de dominios** *f* FÍS, METAL domain structure; ~ **de drenaje** *f* GEOL, HIDROL drainage pattern; ~ **de drenaje dendrítico** *f* HIDROL dendritic drainage pattern; ~ **electrónica** *f* NUCL electronic structure; ~ **de empaquetamiento compacto** *f* CRISTAL close-packed structure; ~ **de enmascaramiento de un solo nivel** *f* ELECTRÓN single-level masking structure; ~ **de la escalera** *f* CONST staircase; ~ **estratificada** *f* GEOFÍS layered structure; ~ **en estrella** *f* TELECOM star structure; ~ **para la extracción de agua** *f* CONST water extraction structure; ~ **fabricada** *f* TEC ESP *nave* fabricated structure; ~ **de falla inclinada** *f*

GEOL basin-and-range structure; **~ de ficheros** *f* INFORM&PD file structure; **~ fija** *f* CONST permanent structure; **~ del filtro** *f* ELECTRÓN filter shaping; **~ fina** *f* FÍS, FÍS RAD, NUCL fine structure; **~ fluidal** *f* GEOL fluidal structure; **~ de forja** *f* PROD stamp guide; **~ formada por redes cúbicas de caras centradas** *f* CRISTAL, QUÍMICA FCC-based structure; **~ del fuelle** *f* FOTO bellows frame; **~ de garno** *f* CARBÓN grain structure; **~ de grafito** *f* NUCL graphite structure; **~ grumosa** *f* ING MECÁ crumb structure; **~ hiperfina** *f* FÍS, FÍS RAD hyperfine structure; **~ de la hoja** *f* PAPEL formation; **~ de hormigón** *f* *Esp* (*cf estructura de concreto AmL*) CONST concrete structure; **~ de hundimiento por gravedad** *f* GEOL gravity-collapse structure; **~ imbricada** *f* GEOL *fallas superpuestas, escamosas* imbricated structure; **~ inferior** *f* TRANSP AÉR bottom structure; **~ inflada** *f* CONST inflated structure; **~ interna** *f* FÍS PART internal structure; **~ laminar** *f* GEOL laminar structure, METAL lamellar structure; **~ de laminilla** *f* METAL platelet structure; **~ lateral** *f* ELECTRÓN *circuitos integrados monolíticos* lateral structure; **~ de listado** *f* INFORM&PD list structure; **~ longitudinal** *f* TRANSP MAR *construcción naval* longitudinal framing; **~ de madera** *f* CONST timber frame; **~ metálica** *f* CONST metalic structure (*AmE*), metallic structure (*BrE*); **~ modulada** *f* METAL modulated structure; **~ de molde de carga** *f* GEOL *sedimentaria* load cast; **~ monocasco** *f* TEC ESP monocoque structure; **~ nervada** *f* CONST *arquitectura* ribbed frame; **~ de nido de abejas** *f* TRANSP AÉR honeycomb structure; **~ de onda lenta** *f* FÍS, ING ELÉC slow-wave structure; **~ de la partícula** *f* CARBÓN grain structure; **~ de pilotaje** *f* CONST piling frame; **~ primaria** *f* TRANSP AÉR primary structure; **~ de programa** *f* INFORM&PD program structure; **~ del puente** *f* CONST bridge truss; **~ de retención** *f* AGUA retaining structure; **~ reticular del núcleo** *f* NUCL core grid structure; **~ secundaria** *f* TEC ESP secondary structure; **~ sedimentaria** *f* GEOL sedimentary structure; **~ del tejado** *f* CONST roof frame; **~ tentativa** *f* CRISTAL trial structure; **~ tipo caja** *f* TRANSP AÉR box-type structure; **~ tipo panal de bajo peso** *f* EMB lightweight honeycomb structure; **~ de toma** *f* AGUA intake structure; **~ de tramas digitales** *f* TELECOM digital frame structure; **~ de tubos sueltos** *f* TELECOM loose-tube structure; **~ widmanstatten** *f* METAL widmanstatten structure; **~ zunchada** *f* METAL banded structure

estructuración *f* ELECTRÓN patterning, GEN implementation, setup

estructural *adj* CONST structural

estructuras: ~ de descarga *f pl* AGUA outlet works; **~ de partes soldadas** *f pl* TEC ESP *vehículos* fabric

estruendo *m* TEC ESP ring

estrujadora *f* AGRIC *uvas* crusher

estrujar *vt* CONTAM MAR wring, TEXTIL crush

estuarino *adj* GEOL, OCEAN, TRANSP MAR estuarine

estuario *m* ENERG RENOV, GEOL, HIDROL, OCEAN, TRANSP MAR estuary

estucado *m* PAPEL, REVEST coat, coating; **~ de alto brillo** *m* PAPEL, REVEST cast coating; **~ con cepillo** *m* PAPEL, REVEST brush coating; **~ a cuchilla** *m* PAPEL, REVEST blade coating; **~ por cuchilla oscilante** *m* PAPEL, REVEST trailing-blade coating; **~ fuera de máquina** *m* PAPEL, REVEST off-machine coating;

~ con labio soplante *m* PAPEL, REVEST air knife coating; **~ en máquina** *m* PAPEL, REVEST on-machine coating; **~ en prensa encoladora** *m* PAPEL, REVEST size-press coating; **~ con rodillo** *m* PAPEL, REVEST roller coating; **~ con rodillo grabado** *m* PAPEL, REVEST gravure coating; **~ con rodillos alisadores** *m* PAPEL, REVEST smoothing-roll coating; **baño de ~** *m* PAPEL, REVEST coating color (*AmE*), coating colour (*BrE*)

estucadora *f* PAPEL, REVEST coater, coating machine; **~ de cepillos** *f* PAPEL, REVEST brush coater; **~ de cortina** *f* PAPEL, REVEST curtain coater; **~ de cuchilla** *f* PAPEL, REVEST blade coater, knife coater; **~ fuera de máquina** *f* PAPEL, REVEST off-machine coater; **~ de labio soplante** *f* PAPEL, REVEST air knife coater; **~ offset** *f* PAPEL, REVEST offset coater; **~ por rociado** *f* PAPEL, REVEST spray coater; **~ de rodillo escurridor** *f* PAPEL, REVEST squeeze-roll coater; **~ de rodillo igualador** *f* PAPEL, REVEST metering-roll coater; **~ de rodillo invertido** *f* PAPEL, REVEST reverse-roll coater; **~ con tobera de aire** *f* PAPEL, REVEST air jet coater; **~ de varilla igualadora** *f* PAPEL, REVEST metering rod coater

estucar *vt* PAPEL, REVEST coat

estuche *m* MECÁ case, sheath; **~ de ampollas** *m* EMB ampoule box (*BrE*), ampule box (*AmE*); **~ de cuero** *m* FOTO leather case; **~ de llaves** *m* AUTO, VEH case of box spanners (*BrE*), case of box wrenches (*AmE*); **~ de llaves de tubo** *m* AUTO, VEH case of box spanners (*BrE*), case of box wrenches (*AmE*), case packing; **~ de llaves de vaso** *m* AUTO, VEH case of box spanners (*BrE*), case of box wrenches (*AmE*), case packing; **~ portaobjetivos** *m* FOTO lens case

estudio *m* CONST survey, ING MECÁ study; **~ aeromagnético** *m* GEOL, PETROL aeromagnetic survey; **~ de alta resolución de contornos lineales** *m* FÍS RAD high-resolution study of line profiles; **~ de alta resolución de contornos de rayas** *m* FÍS RAD high-resolution study of line profiles; **~ sobre el aprovechamiento de las aguas** *m* HIDROL *valoración* water resources study; **~ en campo** *m* AGRIC field study; **~ cinematográfico** *m* CINEMAT studio; **~ de doblaje** *m* CINEMAT dubbing theater (*AmE*), dubbing studio (*BrE*); **~ del envejecimiento** *m* TEC ESP *resistencia de materiales* ageing study; **~ de factibilidad** *m* CONST, INFORM&PD feasibility study; **~ sobre frecuencias de tiempo** *m* TRANSP *control de tráfico* time-lapse survey; **~ hidrológico** *m* CONST *hidráulica* hydrologic study; **~ de impacto** *m* AGUA, GAS impact study; **~ magnético** *m* GEOFÍS magnetic survey; **~ sobre origen y destino** *m* TRANSP *tráfico* origin and destination survey (*O-D survey*); **~ sísmico** *m* CONST, GEOFÍS, TEC PETR seismic survey; **~ topográfico** *m* CONST topographical survey; **~ del tráfico** *m* TRANSP traffic survey; **~ de viabilidad** *m* INFORM&PD feasibility study

estufa *f* GEN oven, PROD calcining kiln, stove, TERMO calcining kiln; **~ bacteriológica** *f* LAB bacteriological oven; **~ con convección forzada** *f* LAB oven with forced convection; **~ con convección natural** *f* LAB oven with natural convection; **~ de Cowper** *f* C&V Cowper stove; **~ de desecación** *f* CARBÓN drying oven; **~ eléctrica** *f* ELEC, INSTAL TERM, TERMO electric convector, electric heater; **~ de esmaltado** *f* REVEST enameling stove (*AmE*), enamelling stove (*BrE*); **~ de machos** *f* PROD core stove, *fundición* core oven; ~ **para machos** *f* PROD *funderías* kettle; ~ **de secado** *f* PROD, TEXTIL drying stove; ~ **de secado de rayos infrarrojos** *f* PROD infrared oven; ~ **de secar moldes** *f* PROD *fundición* mold drier (*AmE*), mould drier (*BrE*); ~ **de temperatura constante** *f* LAB, TERMO constant temperature oven; ~ **al vacío** *f* LAB vacuum oven

estufación *f* PROD, REVEST stoving

estufado *m* PROD, REVEST stoving

estufar *vt* PROD kiln

estupefaciente[1] *adj* QUÍMICA stupefacient

estupefaciente[2] *m* QUÍMICA stupefacient

etal *m* QUÍMICA ethal

etalón *m* FÍS etalon

etamín *m* TEXTIL bolting fabric

etanal *m* QUÍMICA ethanal

etano *m* QUÍMICA, TEC PETR ethane

etanol *m* ALIMENT, FOTO, QUÍMICA, TEC PETR ethanol

etanolamina *f* DETERG, QUÍMICA ethanolamine

etanólisis *f* QUÍMICA ethanolysis

etanotiol *m* QUÍMICA ethanethiol

etapa[1]: **de ~ tardía** *adj* GEOL late-stage

etapa[2] *f* GEOL, PAPEL stage, PROD stage, step; ~ **de ascenso** *f* TEC ESP ascent stage; ~ **de aterrizaje** *f* TEC ESP lander stage; ~ **de compresión** *f* ING MECÁ compression stage; ~ **de concentración** *f* TELECOM concentration stage; ~ **conductora** *f* AUTO driver stage; ~ **de conmutación** *f* TELECOM switching stage; ~ **de control cíclico** *f* TRANSPAÉR cyclic-control step; ~ **criogénica** *f* TEC ESP cryogenic stage; ~ **de descenso** *f* TEC ESP descent stage; ~ **de despegue** *f* TEC ESP kick stage; ~ **de distribución** *f* TELECOM distribution stage; ~ **de escape de la tierra** *f* TEC ESP earth escape stage; ~ **espacial** *f* TELECOM space stage; ~ **excitadora** *f* TEC ESP *electrónica* driver stage; ~ **de expansión** *f* TELECOM expansion stage; ~ **de FI** *f* TELECOM IF stage; ~ **ficticia** *f* TEC ESP *comunicaciones* dummy stage; ~ **de frecuencia intermedia** *f* TELECOM intermediate frequency stage; ~ **de impresión** *f* FOTO printing stage; ~ **interglacial** *f* GEOL interglacial stage; ~ **justa** *f* TEC ESP just stage; ~ **mecánica** *f* FOTO, INSTR mechanical stage; ~ **mezcladora** *f* ELECTRÓN mixer stage; ~ **de muestreo** *f* INSTR specimen stage; ~ **orbital** *f* TEC ESP orbiter stage; ~ **en perigeo** *f* TEC ESP perigee stage; ~ **del prototipo** *f* FÍS RAD prototype stage; ~ **de reposición** *f* PROD reset rung; ~ **de reposo** *f* AGRIC resting stage; ~ **de retardo** *f* ELEC, ING ELÉC on-delay phase; ~ **S** *f* TELECOM S-stage; ~ **de salida** *f* NUCL outlet edge; ~ **de selección** *f* TELECOM selection stage; ~ **simulada** *f* TEC ESP dummy stage; ~ **de teleconmutación** *f* TELECOM, TV remote-switching stage; ~ **de temporización** *f* TELECOM time stage (*T-stage*); ~ **temporizadora** *f* TELECOM time stage (*T-stage*); ~ **terminal** *f* TELECOM terminating stage; ~ **de transbordo** *f* TEC ESP transfer stage

etapas: por ~ *adv* INFORM&PD staging

eteno *m* ALIMENT, DETERG, GAS, P&C, QUÍMICA, TEC PETR ethene, ethylene

éter *m* QUÍMICA ether; ~ **butílico** *m* ALIMENT butyl ether; ~ **dietílico** *m* DETERG diethyl ether; ~ **diglicidílico** *m* P&C, QUÍMICA diglycidyl ether; ~ **glicólico** *m* DETERG, QUÍMICA glycol ether; ~ **metílico terciario-butílico** *m* QUÍMICA, TEC PETR methyl tertiary-butyl ether; ~ **óxido** *m* DETERG ether oxide; ~ **polivinílico** *m* P&C polyvinyl ether

etéreo *adj* QUÍMICA ethereal

etil: ~ **mercaptano** *m* QUÍMICA ethanethiol

etilación *f* QUÍMICA ethylation

etilamina *f* QUÍMICA ethylamine

etilanilina *f* QUÍMICA ethylaniline

etilar *vt* QUÍMICA ethylate

etilato *m* QUÍMICA ethylate

etilenglicol *m* DETERG, QUÍMICA, REFRIG ethylene glycol, glycol

etilénico *adj* QUÍMICA ethylenic

etileno *m* ALIMENT, DETERG, GAS, P&C, QUÍMICA, TEC PETR ethene, ethylene; ~ **acetato de vinilo** *m* (*EVA*) P&C, QUÍMICA ethylene-vinyl acetate (*EVA*)

etílico *adj* QUÍMICA ethylic

etilideno *m* QUÍMICA ethylidene

etilmorfina *f* QUÍMICA ethylmorphine

etilo *m* QUÍMICA, TEC PETR ethyl

etilsulfúrico *adj* QUÍMICA ethylsulfuric (*AmE*), ethylsulphuric (*BrE*)

etino *m* QUÍMICA ethyne

etiónico *adj* QUÍMICA ethionic

etiqueta *f* ELEC *de artefacto* nameplate, EMB label, INFORM&PD label, tag, flag, TEXTIL tag, label; ~ **de advertencia** *f* EMB caution label, SEG warning label, caution label; ~ **de archivo** *f* INFORM&PD file label; ~ **autoadhesiva** *f* EMB self-adhesive label, band label; ~ **autoadhesiva fácil de quitar** *f* EMB easy-peel-off self-adhesive label; ~ **en blanco** *f* EMB blank ticket; ~ **de calidad** *f* CALIDAD, EMB quality label; ~ **de cinta** *f* INFORM&PD tape label; ~ **de código de colores** *f* PROD color code letter (*AmE*), colour code letter (*BrE*); ~ **de cola** *f* INFORM&PD trailer label; ~ **colgante** *f* EMB hangtag; ~ **de control** *f* EMB control tag; ~ **cosida** *f* TEXTIL sewn-in label; ~ **de la declaración** *f* INFORM&PD statement label; ~ **del destinatario** *f* EMB address label; ~ **de encabezamiento** *f* INFORM&PD header label; ~ **encolada** *f* EMB gummed label; ~ **engomada** *f* EMB glued tab, gummed label; ~ **de la envoltura** *f* EMB wraparound label; ~ **fácil de quitar** *f* EMB easy-opening tag; ~ **de fichero** *f* INFORM&PD file label; ~ **gofrada** *f* EMB embossed label; ~ **de identificación para eslingas** *f* SEG sling identification tag; ~ **de instrucción** *f* INFORM&PD statement label; ~ **de marcado** *f* EMB marking label; ~ **para material móvil** *f* CONST, FERRO rolling-stock label; ~ **de oreja** *f* AGRIC ear tag; ~ **de polipropileno orientada** *f* EMB orientated polypropylene label; ~ **posterior** *f* EMB back label; ~ **preimpresa** *f* EMB preprinted label; ~ **retornable** *f* EMB return label; ~ **de seguridad** *f* TELECOM *de transmisiones* security label; ~ **separable formando una banda** *f* EMB band label; ~ **del soporte** *f* EMB header label; ~ **termoactivada** *f* EMB heat-activated label; ~ **termosellable** *f* EMB heat seal label; ~ **de termotransferencia** *f* EMB heat transfer label; ~ **termotransferible** *f* EMB hot transfer label; ~ **de título** *f* INFORM&PD header label; ~ **para la utilización de aplicaciones** *f* TELECOM application appliance label; ~ **de vidrio** *f* C&V vitreous enamel label; ~ **de volumen** *f* INFORM&PD volume label

etiquetado *m* EMB, INFORM&PD, PROD labeling (*AmE*), labelling (*BrE*), TEXTIL labeling (*AmE*), labelling (*BrE*), tagging; ~ **por aire a presión** *m* EMB air blast labeling (*AmE*), air blast labelling (*BrE*); ~ **de conservación** *m* TEXTIL care labeling (*AmE*), care labelling (*BrE*); ~ **de cuidado** *m* TEXTIL care labeling

(*AmE*), care labelling (*BrE*); ~ **de mantenimiento** *m* TEXTIL care labeling (*AmE*), care labelling (*BrE*); ~ **textil** *m* TEXTIL textile labeling (*AmE*), textile labelling (*BrE*)

etiquetador *m* EMB labeler (*AmE*), labeller (*BrE*); ~ **manual** *m* EMB hand labeler (*AmE*), hand labeller (*BrE*); ~ **de la parte delantera del paquete** *m* EMB front of pack labeler (*AmE*), front of pack labeller (*BrE*)

etiquetadora: ~ **con etiquetas autoadhesivas** *f* EMB pressure-sensitive labeler (*AmE*), pressure-sensitive labeller (*BrE*); ~ **mecánica** *f* EMB labeling machine (*AmE*), labelling machine (*BrE*)

etiquetaje *m* PROD labeling (*AmE*), labelling (*BrE*)

etiquetar *vt* INFORM&PD flag, label, tag, TEXTIL label, tag

ETL *abr* (*equipo de terminación de línea*) INFORM&PD LTE (*line termination equipment*)

etoxilación *f* DETERG ethoxylation

Eu *abr* (*europio*) QUÍMICA Eu (*europium*)

eucairita *f* MINERAL eucairite, eukairite

eucaliptol *m* QUÍMICA cineol, eucalyptol

euclasa *f* MINERAL euclase

euclídeo *adj* GEOM Euclidean

euclídico *adj* GEOM Euclidean

eucolita *f* MINERAL eucolite

eucroíta *f* MINERAL euchroite

eudialita *f* MINERAL eudialite, eudialyte

eudiometría *f* INSTR, QUÍMICA eudiometry

eudiómetro *m* INSTR, QUÍMICA eudiometer

eudnofita *f* MINERAL eudnophite

eugenol *m* QUÍMICA eugenol

eugeosinclinal *f* GEOL eugeosyncline

euhédrico *adj* GEOL euhedral, idioblastic

eulitina *f* MINERAL eulytine

euosmita *f* MINERAL euosmite

európico *adj* QUÍMICA europic

europio *m* (*Eu*) QUÍMICA europium (*Eu*)

europoso *adj* QUÍMICA europous

eusinquita *f* MINERAL eusynchite

eustático *adj* ENERG RENOV, GEOFÍS, GEOL eustatic

eutéctico[1] *adj* C&V, METAL, QUÍMICA eutectic

eutéctico[2] *m* C&V, METAL, QUÍMICA eutectic

eutectoide *m* METAL eutectoid

eutexia *f* QUÍMICA eutexia

eutrofía *f* DETERG, RECICL eutrophy

eutrofiar *vt* DETERG, RECICL eutrophy

eutrófico *adj* DETERG, RECICL eutrophic

eutrofización *f* DETERG, RECICL eutrophication

euxenita *f* MINERAL euxenite

euxínico *adj* GEOL euxinic

eV *abr* (*electronvoltio*) ELECTRÓN, FÍS, FÍS PART, ING ELÉC, NUCL eV (*electron-volt*)

EVA *abr* (*copolímero etileno, etileno-acetato de vinilo*) P&C EVA (*ethylene-vinyl acetate*)

evacuación *f* CONTAM *tratamiento de aguas y gases* removal, MECÁ ejection; ~ **de aguas residuales** *f* RECICL sewage disposal, wastewater disposal; ~ **de desechos** *f* CARBÓN reject disposal; ~ **de elementos nutritivos** *f* HIDROL *del agua* nutrient removal; ~ **de emergencia de edificios** *f* SEG emergency evacuation of buildings; ~ **de recipiente** *f* REFRIG pumpdown; ~ **de residuos** *f* RECICL sewage disposal, waste disposal; ~ **de residuos por transmutación nuclear** *f* NUCL waste disposal by nuclear transmutation

evacuador: ~ **de crecidas** *m* AGUA *presas*, HIDROL flood spillway; ~ **de vapor** *m* INSTAL HIDRÁUL steam outlet

evacuar *vt* AGUA blow off, ELECTRÓN *tubo*, FÍS evacuate, HIDROL, MECÁ drain, evacuate, MINAS *grisú*, RECICL drain

evaluación *f* GEN assessment; ~ **de los daños** *f* TRANSP MAR damage assessment; ~ **de las existencias** *f* PROD stock valuation; ~ **de formación** *f* GEOL, PETROL, TEC PETR formation evaluation; ~ **del impacto medioambiental** *f* CONTAM, NUCL environmental-impact assessment; ~ **de prestaciones** *f* INFORM&PD benchmarking; ~ **prudente** *f* PROD conservative estimate; ~ **del rendimiento** *f* INFORM&PD benchmarking; ~ **del riesgo** *f* CALIDAD, CONTAM, CONTAM MAR, SEG risk assessment, risk evaluation

evaluar[1] *vt* METR measure, PROD take stock of, TELECOM test, TEXTIL assess; ~ **el rendimiento de** *vt* INFORM&PD benchmark

evaluar[2]: ~ **la distribución** *vi* ING MECÁ *de carburos en herramientas de acero* assess the distribution; ~ **prestaciones** *vi* INFORM&PD benchmark

evaporable *adj* INSTAL TERM evaporable

evaporación *f* GEN evaporation; ~ **anual** *f* AGUA annual evaporation; ~ **efectiva** *f* AGUA effective evaporation; ~ **instantánea** *f* ALIMENT, NUCL, QUÍMICA flash evaporation; ~ **latente** *f* HIDROL latent evaporation; ~ **relativa** *f* HIDROL relative evaporation; ~ **súbita** *f* ALIMENT, NUCL, QUÍMICA flash evaporation; ~ **térmica** *f* METAL, TERMO thermal evaporation; ~ **total** *f* HIDROL total evaporation

evaporador *m* GEN evaporator, PROC QUÍ, QUÍMICA evaporator, vaporizer; ~ **acumulador de hielo** *m* REFRIG ice bank evaporator; ~ **alimentado por bomba** *m* REFRIG pump-fed evaporator; ~ **de baño de agua** *m* NUCL water bath evaporator; ~ **centrífugo** *m* INSTR, LAB centrifugal evaporator; ~ **de efecto único** *m* ALIMENT *maquinaria* single effect evaporator; ~ **envolvente** *m* REFRIG wraparound evaporator; ~ **inundado** *m* REFRIG flooded evaporator; ~ **de lluvia** *m* REFRIG spray-type evaporator; ~ **multitubular de envolvente** *m* REFRIG shell-and-tube evaporator; ~ **de porcelana** *m* C&V porcelain evaporating-basin; ~ **en régimen seco** *m* REFRIG dry expansion evaporator; ~ **de residuos** *m* NUCL waste evaporator; ~ **de residuos radiactivos** *m* CONTAM, FÍS, FÍS RAD, NUCL radioactive-waste evaporator; ~ **de tipo vertical** *m* REFRIG vertical-type evaporator; ~ **de vacío** *m* ALIMENT vacuum evaporator

evaporar *vt* GEN evaporate; ~ **a sequedad** *vt* PROC QUÍ, QUÍMICA evaporate dry

evaporarse[1] *vi* FOTO, PROC QUÍ evaporate

evaporarse[2] *v refl* ALIMENT boil away, INSTAL TERM, QUÍMICA evaporate, TERMO boil away

evaporímetro *m* HIDROL evaporimeter, PROC QUÍ evaporating pan, evaporation meter, evaporimeter

evaporita *f* ENERG RENOV, GEOL, PETROL, TEC PETR evaporite

evapotranspiración *f* AGUA, HIDROL evapotranspiration; ~ **potencial** *f* AGUA, HIDROL potential evapotranspiration; ~ **real** *f* AGUA, HIDROL actual evapotranspiration; ~ **relativa** *f* AGUA, HIDROL relative evapotranspiration

evasé *adj* MECÁ, TEXTIL flared

evento *m* GEN event; ~ **crítico** *m* PROD critical event;

~ justificativo del puntero *m* TELECOM pointer justification-event (*PJE*)

evolución *f* FÍS, QUÍMICA evolution, TERMO development; **~ de emergencia** *f* TRANSP MAR emergency turn; **~ espacial** *f* CONTAM spatial trend; **~ de la playa** *f* OCEAN beach growth; **~ temporal** *f* CONTAM temporal fluctuation, temporal variation, time trend

evolucionar *vi* TRANSP MAR *navegación* turn

evoluta *f* GEOM, MATEMÁT evolute

exa- *pref* (*E*) METR exa- (*E*)

exactitud *f* FÍS, ING MECÁ accuracy, METR accuracy, *de una balanza* correctness, PROD exactitude; **~ de medida** *f* METR accuracy of measurement; **~ de pendiente** *f* TEC ESP tracking accuracy; **~ de rastreo** *f* TEC ESP tracking accuracy; **~ del registro** *f* CINEMAT, TV registration accuracy; **~ de la situación del buque** *f* TRANSP MAR *navegación* accuracy of ship's position

examen *m* MECÁ test; **~ de bit** *m* PROD bit examining; **~ médico de admisión** *m* SEG pre-employment health screening; **~ minucioso** *m* CINEMAT very close-up, PROD minute examination, TEC ESP close-up, scanning; **~ por partículas magnéticas** *m* MECÁ magnetic particle examination; **~ ultrasónico** *m* FÍS ONDAS ultrasonic inspection, MECÁ ultrasonic examination

examinado: ~ sobre la investigación de la calidad *m* CALIDAD auditee

examinar *vt* INFORM&PD browse, MINAS develop, prove, PROD scan, TEC ESP inspect, TRANSP scan

excavabilidad *f* CARBÓN, MINAS excavatability

excavación *f* CARBÓN cutting, digging, excavating, excavation, CONST cutting, digging, excavating, excavation, pit, MINAS heading, *de la veta, del filón, de la capa de carbón* excavation, *de una cantera* opening, *del pozo* excavating, *perforación* cutting; **~ de aireación** *f* MINAS fan cut; **~ a cielo abierto** *f* MINAS open digging, open pit, open-pit mine; **~ en cuña** *f* MINAS wedge cut; **~ magmática** *f* GEOL magmatic stoping; **~ paralela** *f* MINAS parallel cut; **~ y revestimiento** *f* CONST *túneles* cut and cover; **~ con trépano de granalla de acero** *f* MINAS opening shot

excavaciones *f pl* MINAS diggings

excavado *m* CARBÓN concave

excavador *m* CARBÓN stripper

excavadora *f* AGRIC earthmover, CARBÓN digger, excavator, CONST digging machine, excavator, scraper, ING MECÁ earthmoving machinery, earthmover, MINAS dintheader, excavator, TRANSP backhoe loader, excavator; **~ de arrastre** *f* MINAS slusher; **~ de arrastre concatenada** *f* HIDROL *aguas residuales* chain scraper; **~ de avance** *f* MINAS heading machine; **~ bivalva** *f* CONST grab crane; **~ de cable** *f* MINAS scraper; **~ de cangilones** *f AmL* (*cf rotopala Esp*) CONST bucket excavator, MINAS bucket-wheel excavator; **~ de cucharón** *f* MINAS bucket-chain excavator; **~ desplazable** *f* MINAS *perforación* collar grab; **~ de galerías** *f AmL* (*cf topo Esp*) MINAS roadheader; **~ de hormigón** *f* TRANSP concrete scraper; **~ mecánica de operación manual** *f* ING MECÁ hand-operated power shovel; **~ mecánica de zanjas** *f* PETROL mechanical trencher; **~ con púas** *f* CONST pronged shovel; **~ de rosario** *f* MINAS dredger, bucket-chain excavator; **~ con rueda de rosario** *f AmL* (*cf rotopala Esp*) MINAS bucket-wheel excava-

tor; **~-transportadora de origen** *f* TRANSP caterpillar hauling scraper

excavar *vt* CONST hollow out, MINAS cut a shaft, hole, *en escalones* stope; **~ hacia arriba** *vt* MINAS *chimeneas* raise

excedente: ~ de agua *m* AGUA surplus water; **~ de capacidad de rotación** *m* ENERG RENOV *motor eléctrico* spinning reserves

exceder[1] *vt* MECÁ overshoot

exceder[2]: **~ el tiempo asignado** *vi* PROD time out

excederse *v refl* MECÁ overshoot, TV overrun

excéntrica *f* AUTO cam, tappet, CINEMAT cam, ING MECÁ cam, eccentric, wiper, MECÁ cam, eccentric, TEXTIL eccentric, TRANSP AÉR, VEH cam; **~ de cilindro acanalada** *f* ING MECÁ grooved cylinder cam; **~ del freno** *f* AUTO brake cam; **~ lobulada** *f* ING MECÁ chambered eccentric; **~ de paro** *f* ING MECÁ knock-off cam; **~ de válvula** *f* INSTAL HIDRÁUL valve eccentric

excentricidad *f* ACÚST eccentricity, ING MECÁ throw, MECÁ eccentricity, TEC ESP eccentric anomaly

excéntrico *adj* ING MECÁ, TEC ESP off-center (*AmE*), off-centre (*BrE*)

exceso *m* INFORM&PD, ING MECÁ overflow, TRANSP AÉR overrun; **~ de atenuación** *m* TELECOM excess attenuation; **~ de curado** *m* P&C overcure; **~ de energía** *m* TERMO excess energy; **~ de equipaje** *m* TRANSP AÉR excess baggage; **~ de excavación** *m* MINAS overbreak; **~ de gasolina** *m* TEC ESP *carburador, motores* loading; **~ de machos en un rodeo** *m* AGRIC overmating; **~ de madurez** *m* AGRIC overripeness; **~ de masa** *m* FÍS mass excess; **~ de neutrones** *m* FÍS neutron excess; **~ de peso** *m* EMB excess weight; **~ de pintura en el borde** *m* P&C *defecto en la pintura* fat edge; **~ de presión** *m* REFRIG excess pressure; **~ de refrigerante** *m* REFRIG overcharge; **~ de revelado** *m* CINEMAT, FOTO overdevelopment; **~ de sobrecarga operativa** *m* PROD excess of operating overload; **~ de tensión** *m* METAL, PROD overstressing; **~ de tinta** *m* IMPR flood; **~ de velocidad** *m* ING MECÁ, MECÁ, PROD overspeed

excipiente: ~ líquido *m* TEC ESP vehicle

excitación[1]: **con ~ insuficiente** *adj* ING MECÁ underdriven

excitación[2] *f* GEN excitation; **~ aleatoria** *f* INFORM&PD, TELECOM random excitation; **~ de base** *f* ELECTRÓN base drive; **~ de CA** *f* ING ELÉC AC excitation; **~ por choque** *f* TELECOM shock excitation; **~ colectiva** *f* FÍS RAD collective excitation; **~ en derivación** *f* ING ELÉC shunt excitation; **~ espontánea** *f* NUCL spontaneous excitation; **~ por impacto** *f* FÍS RAD impact excitation; **~ por impulsos** *f* TELECOM impulse excitation; **~ independiente** *f* ING ELÉC independent excitation, separate excitation; **~ del inductor** *f* ELEC field excitation; **~ por láser** *f* FÍS RAD laser excitation; **~ parásita** *f* TELECOM shock excitation; **~ por pulsos** *f* FÍS RAD impulse excitation; **~ de radiación** *f* ELECTRÓN, FÍS RAD radiation excitation; **~ en serie** *f* ING ELÉC series excitation

excitado *adj* ELEC energized, FÍS, FÍS PART, FÍS RAD, GAS excited, ING ELÉC, PROD energized, QUÍMICA excited

excitador *m* ELEC *generador*, ENERG RENOV *turbinas* exciter; **~ electrostático** *m* ACÚST electrostatic exciter; **~ de semiconductor de óxido de metal** *m* (*excitador de SOM*) ELECTRÓN metal-oxide semiconductor driver

excitancia *f* FÍS RAD excitance

excitar *vt* ELEC, ING ELÉC, PROD energize

excitatriz *f* ELEC exciting dynamo

excitón *m* FÍS exciton

excitrón *m* ING ELÉC excitron

exclusión *f* CALIDAD, ELECTRÓN, INFORM&PD rejection

excrementos *m pl* HIDROL *aguas negras*, RECICL faeces (*BrE*), feces (*AmE*)

exductor *m* TRANSP AÉR exducer

exento: **~ de carbono** *adj* METAL carbon-free; **~ de carga** *adj* C&V batch-free; **~ de ceniza** *adj* CONTAM *caracterización de combustibles* ash-free; **~ de escorias** *adj* TERMO free from slag; **~ de gas** *adj* TERMO free from gas; **~ de materia mineral** *adj* CONTAM mineral-matter-free; **~ de pérdidas** *adj* ELEC *dieléctrico* loss-free; **~ de visión** *adj* TRANSP AÉR visual-exempted

exfoliación *f* CRISTAL cleavage, FÍS exfoliation, GEOL cleavage, exfoliation, foliation, METAL cleavage, P&C *defecto* delamination, PROD scaling, QUÍMICA peeling-off; **~ por deformación** *f* GEOL strain slip cleavage; **~ menor** *f* GEOL *minería* false cleavage; **~ pizarrosa** *f* GEOL slaty cleavage, flow cleavage; **~ del plano axial** *f* GEOL axial plane cleavage, axial plane foliation

exfoliar *vt* P&C *adhesivos, defecto* delaminate

exfoliarse *v refl* PROD *soldadura* peel off, QUÍMICA peel off in flakes

exhaustor *m* OCEAN *extracción de nódulos polimetálicos* air exhauster

exhibición *f* TEXTIL display

exhibir *vt* ING MECÁ hold up, TEXTIL display

exhudar *vt* ALIMENT sweat

exigencias: **~ alimenticias** *f pl* ALIMENT food requirements

existencia *f* ING MECÁ life, PROD store; **~ mecánica** *f* ING ELÉC, PROD mechanical life; **~ de pedido** *f* PROD picking stock

existencias *f pl AmL* (*cf reservas Esp*) CARBÓN, MINAS *de carbón, minerales* offtake, PROD, TEC PETR stock; **~ de almacén** *f pl* PROD inventory; **~ asignadas** *f pl* PROD allocated stock; **~ bajo inspección** *f pl* PROD stock at inspection; **~ disponibles** *f pl* PROD available stock, stock on hand; **~ disponibles según planificación y previsión** *f pl* PROD planned-projected available stock; **~ iniciales** *f pl* PROD opening stock; **~ en obra** *f pl* PROD floor stock; **~ en pedido** *f pl* PROD stock on order; **~ previstas disponibles** *f* PROD projected available stock; **~ de seguridad** *f pl* PROD safety stock; **~ del sistema** *f pl* PROD system stock

exitación: **~ energética** *f* ÓPT energy exitance

exitancia: **~ radiante** *f* FÍS, FÍS RAD, ÓPT, TELECOM *en punto de superficie* radiant exitance

éxito *m* INFORM&PD hit

exoatmosférico *adj* TEC ESP *cosmología* exoatmospheric

exosfera *f* TEC ESP exosphere

exotérmico *adj* QUÍMICA, TEC ESP, TEC PETR, TERMO exothermal, exothermic

expandido *adj* INFORM&PD unpacked; **~ por vapor** *adj* PROC QUÍ vapor-expanded (*AmE*), vapour-expanded (*BrE*)

expandir *vt* INFORM&PD unpack

expandirse *v refl* GEOM expand, IMPR *la tinta* feather, INSTAL HIDRÁUL *vapor, gases* expand

expansibilidad *f* TERMO expansibility

expansión *f* ELECTRÓN, FÍS expansion, INSTAL HIDRÁUL dilatation, expansion, INSTR enlargement, *turbina* stage, MATEMÁT, MECÁ expansion, METAL expansion, growth, P&C, QUÍMICA, REFRIG, TEC ESP, TERMO expansion; **~ adiabática** *f* FÍS, TERMO adiabatic expansion; **~ de ancho de banda** *f* ELECTRÓN bandwidth expansion; **~ por calor** *f* TERMO expansion due to heat; **~ del cristal** *f* METAL crystal growth; **~ cúbica** *f* METAL cubical expansion; **~ directa** *f* REFRIG direct expansion; **~ facetada** *f* C&V, METAL faceted expansion, faceted growth; **~ del fondo marino** *f* GEOFÍS sea floor spreading; **~ del fondo oceánico** *f* GEOL ocean floor spreading; **~ de los fondos oceánicos** *f* OCEAN *tectónica de placas* seafloor spreading; **~ de fuelle** *f* MECÁ expansion bellows; **~ isotérmica** *f* FÍS isothermal expansion; **~ Joule** *f* FÍS Joule expansion; **~ de Joule-Kelvin** *f* FÍS Joule-Kelvin expansion; **~ de Joule-Thomson** *f* FÍS Joule-Thomson expansion; **~ polar** *f* ING ELÉC *generadores* pole shoe; **~ de la señal** *f* TELECOM signal expansion; **~ por tensiones** *f* FERRO stress expansion; **~ térmica** *f* P&C thermal expansion, TERMO heat expansion

expansividad: **~ cúbica** *f* FÍS cubic expansivity

expansor: **~ de termocupla** *m AmL* (*cf expansor de termopar Esp*) PROD thermocouple expander; **~ de termopar** *m Esp* (*cf expansor de termocupla AmL*) PROD thermocouple expander

expedición *f* TRANSP MAR shipment; **~ de materiales** *f* PROD material issue

expedidor *m* EMB deliverer, PROD expeditor, TELECOM originator, TRANSP dispatcher, TRANSP MAR shipper; **~ aduanero** *m* TRANSP forwarding agent (*BrE*), freight agent (*AmE*); **~ de agrupamiento de mercancías** *m* FERRO *vehículos* groupage traffic forwarder

expedidora *f* TRANSP MAR shipper

expediente *m* PROD file

expedir *vt* TRANSP MAR ship

expendedora: **~ de billetes** *f* TRANSP ticket slot machine

experiencia: **~ de estabilidad** *f* TRANSP MAR *arquitectura naval* inclining test

experimental *adj* GEN experimental

experimento *m* CARBÓN test, trial, experiment, FÍS, FÍS PART, FÍS RAD experiment, MECÁ test, experiment, NUCL experiment, PROD proof, QUÍMICA, TEC ESP, TERMO experiment; **~ del anticátodo fijo** *m* FÍS PART fixed-target experiment; **~ en caliente** *m* NUCL in-pile experiment, in-reactor experiment; **~ de Cavendish** *m* FÍS Cavendish experiment; **~ de colisión** *m* FÍS PART collision experiment; **~ de criticidad inicial** *m* NUCL first critical experiment; **~ de estricción triaxial** *m* NUCL triaxial pinch experiment; **~ de Franck-Hertz** *m* FÍS Franck-Hertz experiment; **~ de fusión** *m* TERMO melting test; **~ con haz de alta energía** *m* FÍS RAD high-energy beam experiment; **~ con haz rápido** *m* FÍS RAD fast-beam experiment; **~ de ignición** *m* NUCL ignition experiment; **~ en el interior del reactor** *m* NUCL in-pile experiment; **~ de Melde** *m* FÍS Melde's experiment; **~ de Michelson y Morley** *m* FÍS Michelson-Morley experiment; **~ de Millikan** *m* FÍS Millikan's experiment; **~ radiativo** *m* FÍS RAD radiative experiment; **~ con retroflector lasérico** *m* FÍS RAD laser retroflector experiment; ~ **de Rowland** *m* FÍS Rowland's experiment; ~ **de Stern-Gerlach** *m* FÍS Stern-Gerlach experiment; ~ **sobre el terreno** *m* TEC ESP field trial; ~ **de tiempo de vuelo en un haz térmico** *m* FÍS RAD thermal beam time-of-flight experiment

experto: ~ **en fitomejoramiento** *m* AGRIC plant breeder; ~ **en pesca** *m* OCEAN fishing expert

explanación *f* CONST grading, leveling (*AmE*), levelling (*BrE*), FERRO *vía férrea* subgrade, *infraestructura* road bed, MECÁ, PROD leveling (*AmE*), levelling (*BrE*)

explanada *f* ING MECÁ platform

explanadora *f* CONST levelling machine (*BrE*), grader, leveling machine (*AmE*), *maquinaria obras públicas* bulldozer, CONTAM MAR, TRANSP grader; ~ **de carros** *f* D&A tank bulldozer, tank dozer; ~ **de cuchilla basculante** *f* TRANSP tiltdozer; ~ **de hoja oblicua** *f* CONST *maquinaria de obras públicas* angledozer; ~ **pendular** *f* TRANSP tilting dozer

exploración *f* ELECTRÓN scanning, sweep, FÍS scanning, GAS exploration, probe, INFORM&PD scan, scanning, MINAS cutting, prospecting, TEC ESP scanning, TEC PETR scanning, exploration, prospecting, TRANSP AÉR scan, TRANSP MAR *radar* scan, scanning, TV scan, scanning, sweep; ~ **aleatoria** *f* INFORM&PD random scan; ~ **de alta velocidad** *f* TV high-velocity scanning; ~ **ampliada** *f* ELECTRÓN expanded sweep; ~ **aproximada** *f* TV coarse scanning; ~ **de barrido** *f* INFORM&PD raster scan; ~ **circular** *f* ELECTRÓN circular scan; ~ **por contacto** *f* INFORM&PD contact scanning; ~ **continua de prueba** *f* PROD test continuous scan; ~ **en contrafase** *f* ELEC, ELECTRÓN, ÓPT push-pull scanning; ~ **cuádruple** *f* TV quadruple scanning; ~ **con draga hidrográfica** *f* OCEAN wire-drag survey; ~ **electrónica** *f* IMPR electronic scanning; ~ **entrelazada** *f* ELECTRÓN, TV interlaced scanning; ~ **de entrelazado de líneas** *f* TV line-interlaced scanning; ~ **de frecuencias** *f* TELECOM frequency scanning; ~ **frontal** *f* TV front scanning; ~ **geofísica** *f* GEOFÍS geophysical exploration; ~ **a gran distancia** *f* TEC ESP, TRANSP AÉR ranging; ~ **del haz** *f* ELECTRÓN beam scanning; ~ **por haz de electrones** *f* ELECTRÓN electron-beam scanning; ~ **helicoidal** *f* ELECTRÓN, INFORM&PD, TV helical scan, helical scanning; ~ **horizontal** *f* TV horizontal scanning; ~ **de la imagen** *f* ELECTRÓN raster scanning; ~ **inversa** *f* TV reverse scan; ~ **de línea** *f* ELECTRÓN, INFORM&PD line scanning; ~ **de la línea del sonido** *f* TV slit scanning; ~ **lineal** *f* ELECTRÓN linear scan; ~ **por líneas** *f* FÍS, FÍS ONDAS rectilinear scanning, TELECOM line scanning, TV rectilinear scanning; ~ **por líneas sucesivas** *f* TV sequential scanning; ~ **magnetométrica** *f* GEOFÍS, TEC PETR magnetometer survey; ~ **de marcas** *f* INFORM&PD mark scanning; ~ **ortogonal** *f* TV orthogonal scanning; ~ **oscilatoria** *f* ELECTRÓN, TELECOM, TV oscillatory scanning; ~ **petrolífera** *f* TEC PETR oil exploration; ~ **de programa** *f* PROD program scanning (*AmE*), programme scanning (*BrE*); ~ **por puntos** *f* INFORM&PD raster scan; ~ **por puntos sucesivos** *f* TV dot interlace scanning; ~ **por radar** *f* D&A, FÍS RAD, TRANSP, TRANSP AÉR, TRANSP MAR radar scan; ~ **radárica** *f* D&A, FÍS RAD, TRANSP, TRANSP AÉR, TRANSP MAR radar scan; ~ **rápida** *f* ELECTRÓN fast sweep; ~ **retardada** *f* ELECTRÓN delayed sweep, delaying sweep, TV delayed scanning; ~ **por sectores** *f* D&A sector scan; ~ **segmentada** *f*

TV segmented scanning; ~ **simple de prueba** *f* PROD test single scan; ~ **sísmica** *f* CARBÓN, GEOFÍS, TEC PETR seismic exploration; ~ **subterránea** *f* GAS underground exploration; ~ **del suelo** *f* CARBÓN soil exploration; ~ **de telecine** *f* TV telecine scan; ~ **total de imagen** *f* PROD raster scanning; ~ **de trama** *f* INFORM&PD, TV raster scan, raster scanning; ~ **ultrarrápida** *f* TV high-velocity scanning; ~ **vectorial** *f* ELECTRÓN vector scanning; ~ **de velocidad variable** *f* TV variable-speed scanning; ~ **vertical** *f* TV vertical scanning

explorador *m* ACÚST driver, ELEC *conmutador*, TELECOM scanner; ~ **de base plana** *m* INFORM&PD flat-bed scanner; ~ **de frecuencias** *m* TELECOM frequency scanner; ~ **de hélice** *m* TRANSP MAR helical scanner; ~ **holográfico** *m* INFORM&PD holographic scanner; ~ **de imagen** *m* TV image scanner; ~ **indirecto móvil** *m* TEC ESP flying-spot scanner; ~ **óptico** *m* INFORM&PD optical scanner; ~ **plano** *m* INFORM&PD flat-bed scanner; ~ **de punto deslizante** *m* TV flying-spot scanner; ~ **de punto deslizante de tubo** *m* TV flying-spot tube scanner; ~ **de títulos** *m* TV caption scanner; ~ **de transparencias** *m* TV slide scanner; ~ **a varilla** *m* INFORM&PD wand scanner

explorar *vt* FÍS ONDAS *radar*, IMPR, INFORM&PD scan, MINAS costean, PROD, TRANSP, TRANSP MAR, TV scan

explosímetro *m* INSTR, LAB explosimeter

explosión *f* CARBÓN blasting, blowout, CONST bursting, ING MECÁ explosion, MINAS *voladuras* report, PROD, SEG explosion, TEC ESP report, spurt, *despegue de un cohete* blast, TEC PETR *pozo, refinería, gas* explosion, TELECOM blast; ~ **de caldera** *f* INSTAL HIDRÁUL, SEG boiler explosion; ~ **de la demanda** *f* PROD requirement explosion; ~ **de gas** *f* CARBÓN, TEC PETR gas explosion; ~ **instantánea** *f* TERMO flash fire; ~ **por medios eléctricos** *f* MINAS electric firing; ~ **nuclear** *f* D&A, NUCL nuclear explosion; ~ **de polvos** *f* CARBÓN dust explosion; ~ **prematura** *f* CONST *motores, maquinaria* backfire; ~ **primaria** *f* MINAS primary blasting; ~ **primigenia** *f* FÍS, TEC ESP *cosmogénesis* big bang; ~ **del proyectil** *f* D&A shell burst; ~ **retardada** *f* MINAS delayed explosion; ~ **superficial** *f* D&A surface burst

explosionar *vti* GEN explode

explosiones: ~ **al carburador** *f pl* AUTO, VEH backfire; ~ **sucesivas** *f pl* MINAS multiple row blasting

explosivo[1] *adj* GEN explosive

explosivo[2] *m* GEN explosive; ~ **del acuífero** *m* AmL (*cf explosivo para aguas Esp*) MINAS watergel explosive; ~ **admisible** *m* MINAS permissible explosive (*AmE*), permitted explosive (*BrE*); ~ **para aguas** *m* Esp (*cf explosivo del acuífero AmL*) MINAS watergel explosive; ~ **autorizado** *m* MINAS permissible explosive (*AmE*), permitted explosive (*BrE*); ~ **a base de nitrato** *m* MINAS nitrate-based explosive; ~ **con base de nitroglicerina** *m* MINAS high explosive; ~ **de cebado** *m* MINAS priming explosive; ~ **cloratado** *m* MINAS chlorate explosive; ~ **compuesto de nitrato amónico y fueloil** *m* (*NAFO*) MINAS ammonium nitrate fuel oil (*ANFO*); ~ **deflagrante** *m* MINAS deflagrating explosive; ~ **detonante** *m* MINAS detonating explosive, high explosive; ~ **detonante secundario** *m* MINAS secondary high explosive; ~ **para la explotación de minas de carbón** *m* CARBÓN coal-mining explosive,

MINAS coal mining explosive; ~ **fluidizado** *m* MINAS slurry blasting agent (*SBA*), slurry explosive (*SE*); ~ **iniciador** *m* MINAS initiating explosive; ~ **lento** *m* MINAS low explosive; ~ **moldeable** *m* *AmL* (*cf explosivo plástico Esp*) D&A, MINAS plastic explosive (*PE*); ~ **no detonante** *m* MINAS low explosive; ~ **de oxígeno líquido** *m* MINAS liquid-oxygen explosive; ~ **de perclorato** *m* MINAS perchlorate explosive; ~ **plástico** *m* *Esp* (*cf explosivo moldeable AmL*) D&A, MINAS plastic explosive (*PE*); ~ **de pólvora** *m* MINAS powder explosive; ~ **primario** *m* MINAS primary explosive; ~ **rompedor** *m* D&A, MINAS high explosive; ~ **rompedor secundario** *m* MINAS secondary high explosive; ~ **secundario** *m* MINAS secondary explosive; ~ **de seguridad** *m* MINAS permissible explosive (*AmE*), permitted explosive (*BrE*), safety explosive; ~ **semigelatinoso** *m* MINAS semigelatin explosive; ~ **para trabajo en roca** *m* MINAS rock-work explosive

explosor *m* CARBÓN, CONST blaster, MINAS blasting machine, exploder, *explosivos* electric firing, *máquina* blaster, *pega* electric blasting-cap; ~ **de dínamo** *m* MINAS dynamo blaster, dynamo exploder; ~ **eléctrico** *m* MINAS electric blasting machine; ~ **de esferas** *m* ING ELÉC sphere gap; ~ **giratorio** *m* ELEC *chispa*, ING ELÉC rotary discharger; ~ **de pulsador** *m* MINAS push-down blasting machine

explotabilidad *f* MINAS mineability, workability

explotable *adj* MINAS mineable, workable

explotación[1]: **de ~ de reserva** *adj* TELECOM stand-by working

explotación[2]: ~ **agrícola** *f* AGRIC farm; ~ **agrícola colectiva** *f* AGRIC communal farm; ~ **agrícola económicamente viable** *f* AGRIC farm of economic size; ~ **de aluviones** *f* MINAS alluvial mining; ~ **aurífera** *f* MINAS placer workings; ~ **por bancos** *f* MINAS combined stoping, stoping; ~ **de canteras** *f* MINAS quarrying; ~ **de canteras de arcilla** *f* MINAS clay mining; ~ **a cielo abierto** *f* MINAS open digging, open pit, open-cast mining, open-cut mining, openwork; ~ **dedicada al cultivo de raíces y tubérculos** *f* AGRIC root-crop farm; ~ **en descubierto** *f* CARBÓN, MINAS, PROD *minería* stripping; ~ **por escalones** *f* MINAS combined stoping, stoping; ~ **por excavadoras** *f* CARBÓN stripping; ~ **forestal** *f* PROD *silvicultura* logging; ~ **por frentes invertidas** *f* *Esp* (*cf explotación por gradas al revés AmL*) MINAS overhand stoping; ~ **por galerías** *f* MINAS drift mining; ~ **ganadera** *f* *Esp* (*cf cabaña AmL*) AGRIC livestock farm; ~ **por gradas** *f* MINAS combined stoping, stoping; ~ **por gradas al revés** *f* *AmL* (*cf explotación por frentes invertidas Esp*) MINAS overhand stoping; ~ **por gradas invertidas** *f* MINAS overhand stoping; ~ **hidráulica** *f* CARBÓN, MINAS piping; ~ **hortícola** *f* AGRIC market garden; ~ **lechera** *f* AGRIC dairy farming; ~ **y mantenimiento** *f* TELECOM operations and maintenance (*OAM*); ~ **y mantenimiento de la capa física** *f* TELECOM physical-layer operations and maintenance (*PL-OAM*); ~ **manual** *f* TELECOM manual working; ~ **de minas** *f* *Esp* (*cf colocación de minas AmL*) MINAS mining; ~ **de minas de carbón** *f* CARBÓN, MINAS coal mining; ~ **minera profunda** *f* MINAS deep mine; ~ **de minerales** *f* CARBÓN, MINERAL mineral processing; ~ **mixta** *f* AGRIC mixed farming; ~ **pesquera racional** *f* OCEAN rational

fishery management; ~ **por pilares** *f* CARBÓN *minas* pillar working; ~ **de pláceres** *f* MINAS placer workings; ~ **preparatoria** *f* *AmL* (*cf labores preparatorias Esp*) MINAS dead workings; ~ **a roza abierta** *f* *AmL* (*cf explotación en trinchera Esp*) MINAS openwork; ~ **semiautomática de la línea principal** *f* TELECOM semiautomatic trunk working; ~ **por socavones** *f* MINAS drift mining; ~ **subterránea** *f* MINAS closed work; ~ **subterránea de yacimientos auríferos** *f* MINAS drift mining; ~ **por tajos largos** *f* MINAS longwall stoping; ~ **de tierras** *f* AGRIC land management; ~ **en trinchera** *f* *Esp* (*cf explotación a roza abierta AmL*) MINAS openwork

explotado: ~ **en una cantera** *adj* CONST, GEOL, MINAS quarried

explotar[1] *vt* CARBÓN blast, operate lut, work out, MINAS develop, *canteras* open up, *una mina* mine, operate, run, *filón* cut a lode, PROD *un negocio* operate, *fábricas, minas* run

explotar[2] *vi* MECÁ burst, TERMO explode

expoliador *m* TRANSP AÉR spoiler

exponencial[1] *adj* ELEC, ELECTRÓN, INFORM&PD exponential

exponencial[2] *m* ELEC, MATEMÁT exponential

exponente *m* ELEC, ELECTRÓN exponent, IMPR inferior character, inferior letter, superscript, INFORM&PD explicit address, superscript, exponent, ING MECÁ index, MATEMÁT exponent, power; ~ **polarizado** *m* INFORM&PD biased exponent; ~ **de transferencia entre imágenes** *m* ACÚST image transfer exponent

exponer *vt* CINEMAT, FOTO, IMPR, TV expose

exportación *f* INFORM&PD copy-out

exposición *f* CINEMAT, FÍS, FOTO exposure, IMPR exposure, insolation, TV exposure; ~ **automática** *f* FOTO automatic exposure; ~ **auxiliar** *f* IMPR flash exposure; ~ **doble** *f* CINEMAT, FOTO double exposure; ~ **fotográfica** *f* ACÚST photographic exposure; ~ **a grandes rasgos** *f* ING MECÁ outline; ~ **humana a vibraciones mecánicas** *f* SEG human exposure to mechanical vibrations; ~ **a humos** *f* SEG exposure to fumes; ~ **instantánea** *f* FOTO instantaneous exposure; ~ **a la intemperie** *f* CONTAM MAR weathering, P&C exposure to weather; ~ **intermedia** *f* IMPR intermediate exposure; ~ **luminosa** *f* FÍS light exposure; ~ **para luz día** *f* FOTO daylight exposure; ~ **a materiales radiactivos por emergencia** *f* NUCL emergency exposure to radioactive materials; ~ **de medio tono** *f* IMPR halftone exposure; ~ **múltiple** *f* CINEMAT, FOTO multiple exposure; ~ **parcial** *f* IMPR partial exposure; ~ **principal** *f* IMPR main exposure; ~ **a la radiación** *f* FÍS, FÍS RAD radiant exposure, NUCL, P&C, SEG exposure to radiation; ~ **a radiaciones externas por emergencia** *f* NUCL emergency exposure to external radiations; ~ **al ruido ocupacional** *f* SEG occupational noise exposure; ~ **en superficie** *f* OCEAN exposure at the surface; ~ **suplementaria** *f* IMPR bump exposure; ~ **visual en vehículo** *f* TRANSP in-vehicle visual display

exposímetro *m* CINEMAT exposure meter, exposure timer, FÍS exposure meter, FOTO exposure meter, exposure timer; ~ **acoplado** *m* FOTO coupled exposure meter; ~ **acústico** *m* SEG sound exposure meter; ~ **para ampliaciones** *m* FOTO, IMPR enlarging-meter; ~ **por coincidencia de agujas** *m* CINEMAT, FOTO exposure meter using needle-matching system; ~ **incorporado** *m* FOTO built-in exposure meter

expositor *m* EMB, TRANSP display; ~ **de múltiples niveles** *m* PROD multiple rung display; ~ **en el punto de venta** *m* EMB point-of-sale display

expresión *f* INFORM&PD expression; ~ **de búsqueda** *f* INFORM&PD search word; ~ **condicional** *f* INFORM&PD conditional expression

exprimidor *m* CONTAM *métodos de separación* wringer

exprimir *vt* PROD squeeze out

expulsado *adj* SEG ejection

expulsador: ~ **de balas** *m* AGRIC bale ejector; ~ **de fardos** *m* AGRIC bale ejector

expulsatestigos *m* MINAS *sondeos* core pusher

expulsión *f* ELEC, ING ELÉC expulsion, MECÁ ejection, scavenging, TEC PETR expulsion, scavenging; ~ **de gas** *f* OCEAN gas expulsion; ~ **periódica** *f* TEC ESP *reactor nuclear* chugging

expulsor *m* ING MECÁ, MECÁ, P&C ejector

extendedura: ~ **de carbón** *f Esp* (*cf apilador de carbón AmL*) CARBÓN, MINAS trimmer

extender *vt* IMPR *la composición* keep out; ~ **capa de base sobre** *vt* CONST *carreteras* metal

extenderse *v refl* CONTAM MAR spread, GEOM expand

extendible: ~ **por el usuario** *adj* PROD field-expandable

extendido *m* QUÍMICA spreading; ~ **de capa de rodadura** *m* CONST metaling (*AmE*), metalling (*BrE*)

extensibilidad *f* INFORM&PD, ING MECÁ, P&C extensibility; ~ **del nudo** *f* TEXTIL knot extensibility

extensible *adj* ING MECÁ tensile, PAPEL stretchable

extensímetro *m* MECÁ extensometer, OCEAN strain gage (*AmE*), strain gauge (*BrE*); ~ **de resistencia eléctrica** *m* ING ELÉC, MECÁ strain gage (*AmE*), strain gauge (*BrE*)

extensión *f* C&V spread, ELECTRÓN range, expansion, FÍS expansion, FOTO, GEOL extension, GEOM, INFORM&PD range, ING MECÁ extension, scope, stretch, INSTAL HIDRÁUL, MATEMÁT, MECÁ, METAL expansion, MINAS extent, P&C, QUÍMICA, REFRIG, TEC ESP, TERMO expansion; ~ **de agua** *f* OCEAN sheet of water; ~ **de cámara única** *f* FOTO single-camera extension; ~ **de código** *f* INFORM&PD code extension; ~ **crítica de la grieta** *f* METAL critical crack-length; ~ **de difusión** *f* ELECTRÓN diffusion length; ~ **para dos cámaras** *f* FOTO double-camera extension; ~ **de la entrada de aire del motor** *f* TRANSP AÉR engine air-intake extension; ~ **de fuelle** *f* FOTO bellows extension; ~ **interna** *f* TELECOM internal extension; ~ **del período de estancia del combustible en el reactor** *f* NUCL stretch-out; ~ **de la señal** *f* TELECOM signal extension; ~ **telefónica** *f* TELECOM telephone extension; ~ **de tierra con pastizales naturales** *f* AGRIC range; ~ **de la toma de aire del motor** *f* TRANSP AÉR engine air-intake extension; ~ **del tren de aterrizaje** *f* TRANSP AÉR landing gear extension; ~ **del trípode** *f* FOTO tripod extension; ~ **de la vida de la central** *f* NUCL plant-lifetime extension

extensómetro *m* INSTR strain gage (*AmE*), strain gauge (*BrE*), METR extensometer, P&C strain gage (*AmE*), strain gauge (*BrE*)

extensor *m* ING MECÁ stretcher, PROD expander; ~ **de línea** *m* FÍS line stretcher; ~ **de tubos** *m* CONST *tubería* tube expander

exterior[1]: *adj* ING MECÁ outboard, outer, overall, TRANSP MAR outboard; **del** ~ *adj* MINAS surface; **en el** ~ *adj* MINAS above-ground; **hacia el** ~ *adj* ING MECÁ, TRANSP MAR outboard

exterior[2]: **al** ~ *adv* TRANSP MAR overboard

exterior[3] *m* CINEMAT exterior, MINAS grass; ~ **fuera del estudio** *m* CINEMAT location

externo *adj* ING MECÁ outer

extinción *f* ELECTRÓN decay, FÍS quenching, FÍS RAD *de la luz*, ING ELÉC, ING MECÁ extinction, NUCL quenching, TEC ESP *motor de chorro* burnout; ~ **del arco** *f* ELEC arc extinction; ~ **del fuego por descarga de agua** *f* MINAS flushing; ~ **de incendios** *f* CONST, SEG, TERMO firefighting; ~ **de la llama** *f* TERMO flameout, *por falta de combustible* burnout; ~ **de la luminiscencia** *f* FÍS RAD quenching of luminescence; ~ **del motor** *f* TEC ESP, TRANSP AÉR, VEH engine flameout; ~ **primaria** *f* CRISTAL primary extinction; ~ **secundaria** *f* CRISTAL secondary extinction; ~ **selectiva** *f* FÍS RAD selective quenching

extinguido: **no** ~ *adj* QUÍMICA *cal* unslaked

extinguidor: ~ **de bióxido de carbono** *m* SEG carbon dioxide fire-extinguisher; ~ **de incendios** *m AmL* (*cf extintor de incendios Esp*) CONST, ING MECÁ, SEG, TERMO, TRANSP AÉR fire extinguisher

extinguirse *v refl* ING ELÉC go out

extintor: ~ **de bióxido de carbono** *m* SEG carbon dioxide fire-extinguisher; ~ **de espuma** *m* CONST, SEG, TRANSP AÉR foam extinguisher; ~ **de fuego a base de agua** *m* SEG water fire-extinguisher; ~ **de halón** *m* SEG halon fire extinguisher; ~ **de incendios** *m Esp* (*cf extinguidor de incendios AmL*) CONST, ING MECÁ, SEG, TERMO, TRANSP AÉR fire extinguisher; ~ **de incendios a base de espuma** *m* CONST, SEG, TRANSP AÉR foam fire-extinguisher; ~ **de incendios fijo** *m* SEG fixed fire-extinguisher; ~ **de incendios interconstruido** *m* SEG fixed fire extinguisher; ~ **de incendios móvil** *m* SEG mobile fire-extinguisher; ~ **de incendios de polvo** *m* SEG dry-powder fire extinguisher, powder fire extinguisher; ~ **de incendios de polvo químico seco** *m* SEG dry-powder fire extinguisher; ~ **de incendios portátil** *m* SEG portable fire-extinguisher; ~ **de líquidos no inflamables** *m* SEG nonflammable-liquid extinguisher; ~ **de soda-ácido** *m* SEG soda acid fire-extinguisher

extirpación *f* NUCL ablation

extracción *f* AGUA draw-off, drawing, *embalse* drawdown, ALIMENT *molienda, panadería, repostería* dockage, CARBÓN *minas* output, extraction, CONST, GAS extraction, INFORM&PD abstraction, fetch, ING MECÁ extraction, pulling out, MINAS *de mineral* drawing, hoisting, output, NUCL *de una barra de control de reactor PWR* lift, *en la separación isotópica* output, P&C extraction, PROD *minas* output, hoisting, QUÍMICA, TEC PETR extraction, TRANSP MAR *del vaciado del molde* removal; ~ **de agua** *f* AGUA, MINAS, TEC PETR pumping, pumping out, TRANSP MAR pumping out; ~ **del agua** *f* TEXTIL dewatering; ~ **del agua bajo carga** *f* TEXTIL dewatering under load; ~ **de aire** *f* CONST taking out of wind; ~ **de barro para alfarería** *f* C&V potter's clay extraction; ~ **de una capa espesa de carbón** *f* CARBÓN thick-seam winning; ~ **de características** *f* INFORM&PD feature extraction; ~ **del carbón** *f* CARBÓN, MINAS coal extraction; ~ **en cubos** *f Esp* (*cf extracción del mineral AmL*) MINAS hub; ~ **de datos** *f* ELECTRÓN data extraction, INFORM&PD data abstraction, data extraction; ~ **de los depósitos solubles** *f* PROC QUÍ soluble deposit extraction, washing out; ~ **del diamante** *f* MINAS diamond mining; ~ **disolvente** *f*

ALIMENT solvent extraction; ~ **con disolventes** *f* CARBÓN solvent leaching; ~ **por etapas** *f* NUCL batch extraction; ~ **de finos por métodos húmedos** *f* CARBÓN deslurrying; ~ **de gas** *f* OCEAN gas extraction; ~ **de gas sulfídrico de los hidrocarburos** *f* TEC PETR *refino, producción* sweetening; ~ **de la herrumbre** *f* ING MECÁ derusting; ~ **de instrucciones** *f* INFORM&PD instruction-fetching; ~ **intermitente** *f* CARBÓN skip extraction; ~ **por lotes** *f* NUCL batch extraction; ~ **a mano** *f* CARBÓN waling; ~ **mediante lectura** *f* INFORM&PD read-out; ~ **de menas** *f* MINAS ore extraction; ~ **de mercaptanes de gasolina** *f* TEC PETR *refino, producción* sweetening; ~ **del mineral** *f* AmL (*cf extracción en cubos Esp*) MINAS hub; ~ **de minerales** *f* MINAS ore extraction, ore mining; ~ **del modelo** *f* PROD delivery of pattern from mold (*AmE*), delivery of pattern from mould (*BrE*); ~ **no programada** *f* NUCL unscheduled withdrawal; ~ **del óxido** *f* ING MECÁ derusting; ~ **de piedra** *f* MINAS quarrying; ~ **de piedras del carbón** *f* CARBÓN waling; ~ **de los pilares** *f* CARBÓN, MINAS pillar extraction; ~ **por presión** *f* MINAS squeezing; ~ **de una raíz** *f* MATEMÁT evolution; ~ **de sólidos** *f* CARBÓN filtration; ~ **por tajos largos** *f* MINAS longwall extraction; ~ **de testigos** *f* Esp (*cf corte núcleos AmL*) PETROL, TEC PETR *perforación* coring; ~ **de testigos de pared** *f* TEC PETR sidewall coring; ~ **de tritio** *f* NUCL tritium extraction; ~ **con vapor** *f* ALIMENT steam extraction; ~ **de vapor** *f* MECÁ bleeding; ~ **del viento** *f* CONST taking out of wind

extractar *vt* INFORM&PD abstract

extracto *m* CARBÓN, CONST, GAS, QUÍMICA extract; ~ **de acetona** *m* P&C acetone extract; ~ **de levadura** *m* ALIMENT yeast extract; ~ **de malta** *m* ALIMENT malt extract; ~ **no nitrogenado** *m* AGRIC nitrogen-free extract (*NFE*)

extractor *m* ALIMENT exhauster, CARBÓN extractor, *prensa troqueladora* stripper, CONST, GAS extractor, ING MECÁ extractor, exhaust fan, sucker, tackle, LAB extraction fan, MECÁ exhaust fan, P&C *moldeo* extractor, PAPEL exhaust fan, PROC QUÍ, TEC PETR extractor, TERMOTEC exhaust fan; ~ **de aire** *m* ING MECÁ, MECÁ, PAPEL, TERMOTEC exhaust fan, exhauster; ~ **centrífugo** *m* NUCL, PROC QUÍ centrifugal exhauster, centrifugal extractor; ~ **de clavos** *m* CONST nail extractor, EMB nail puller; ~ **de cubos** *m* ING MECÁ hub extractor; ~ **de dos brazos** *m* ING MECÁ two-leg puller; ~ **de engranajes** *m* VEH gear puller; ~ **de gas** *m* GAS, NUCL gas stripper; ~ **de grampas** *m* AmL (*cf extractor de grapas Esp*) ING MECÁ staple remover; ~ **de grapas** *m* Esp (*cf extractor de grampas AmL*) ING MECÁ staple remover; ~ **de humos** *m* PROC QUÍ fume extractor; ~ **de pernos** *m* ING MECÁ stud extractor; ~ **de polvo** *m* CARBÓN, CONTAM *tratamiento de gases* cyclone; ~ **de sedimentos** *m* OCEAN sediment probe; ~ **de testigos** *m* MINAS core breaker, core catcher, TEC PETR *perforación* corer; ~ **de tirafondos** *m* FERRO sleeper-screw extractor (*BrE*), spike puller (*AmE*); ~ **de vahos** *m* ING MECÁ, MECÁ, PAPEL, TERMOTEC exhaust fan

extradós *m* CONST extrados

extraer[1] *vt* AGUA draw down, pump out, CARBÓN extract, CONST extract, pump out, pull out, INFORM&PD Esp (*cf traer AmL*) fetch, ING MECÁ pump out, MINAS hoist, develop, tap, *mineral* mine,

PROD *agua, gas* tap, pump out, QUÍMICA extract, separate out, TRANSP MAR *el vaciado del molde* remove; ~ **el agua de** *vt* TEXTIL dewater; ~ **carbón** *vt* CARBÓN break coal; ~ **por destilación** *vt* QUÍMICA, TERMO distil off (*BrE*), distill off (*AmE*); ~ **gas de** *vt* MINAS degas; ~ **por lectura** *vt* INFORM&PD read out; ~ **mediante lectura** *vt* INFORM&PD read out; ~ **por salado** *vt* QUÍMICA salt out; ~ **a la superficie** *vt* MINAS hoist to the surface

extraer[2]: ~ **un filón** *vi* MINAS cut a lode; ~ **finos por métodos húmedos** *vi* CARBÓN deslurry; ~ **la jaula** *vi* MINAS cut a landing; ~ **mineral** *vi* MINAS raise ore; ~ **minerales** *vi* MINAS hoist ore

extraerse *v refl* QUÍMICA separate out

extrafuerte *adj* DETERG heavy-duty

extragaláctico *adj* TEC ESP extragalactic

extragrande *adj* ING MECÁ oversize

extrapolación *f* CALIDAD extrapolation

extrapolar *vt* CALIDAD extrapolate

extremidad: ~ **del cable** *f* ELEC, ING ELÉC cable end; ~ **del pilote** *f* CARBÓN pile tip

extremo[1] *adj* ING MECÁ outer; **de ~ a extremo** *adj* INFORM&PD end-to-end, ING MECÁ overall

extremo[2] *m* INFORM&PD tail, OCEAN *redes planctónicas* bottom; ~ **del ala** *m* OCEAN *tecnología de pesca* wing end, TRANSP AÉR wing tip; ~ **alado** *m* ING MECÁ flanged edge; ~ **alámbrico** *m* ING ELÉC wire end; ~ **de biela** *m* ING MECÁ end of connecting-rod; ~ **de cabio** *m* CONST *arquitectura* tail; ~ **del cable** *m* ELEC, ING ELÉC, MINAS cable end; ~ **del cojinete** *m* ING MECÁ bearing end; ~ **cónico del árbol** *m* AUTO tapered axle end; ~ **cónico del eje** *m* ING MECÁ conical shaft end; ~ **corto** *m* CINEMAT short end; ~ **delantero** *m* ING MECÁ crank end; ~ **delantero del cigüeñal** *m* AUTO, VEH crankshaft front end; ~ **embutido** *m* NUCL *de la pastilla de combustible* dishing shallow depression; ~ **de la enrolladora** *m* PAPEL reeling end; ~ **ensanchado** *m* MECÁ *de tubo* faucet; ~ **frontal del TEC** *m* ELECTRÓN FET front end; ~ **de la grieta** *m* METAL crack tip; ~ **húmedo** *m* CINEMAT wet end; ~ **inicial** *m* TEXTIL *tejido* head end; ~ **inicial de la banda** *m* IMPR web lead; ~ **libre** *m* MINAS free end; ~ **de la pala** *m* TRANSP AÉR blade tip; ~ **posterior** *m* ING MECÁ crank end, TRANSP, VEH rear end; ~ **de red de cerco** *m* OCEAN wing end; ~ **de salida** *m* PROD delivery; ~ **del segmento** *m* AUTO ring end; ~ **superior** *m* AUTO rear end; ~ **a tierra** *m* ING ELÉC shore end; ~ **de varilla de conexión** *m* ING MECÁ end of connecting-rod

extremos: ~ **fríos** *m pl* ELEC *termopar*, ING ELÉC cold junction; ~ **libres** *m pl* ELEC *termopar*, ING ELÉC cold junction

extrudibilidad *f* GEN extrudability

extrudir *vi* GEN extrude

extrusión *f* GEN extrusion; ~ **de película** *f* P&C *procesamiento* film extrusion; ~ **en rosario** *f* ING MECÁ beaded extrusion

extrusionadora *f* EMB, P&C extrusion machine, PAPEL extruder, extrusion machine

extrusor: ~ **de caucho** *m* ING MECÁ, P&C rubber extruder

extrusora *f* P&C extruder, extruding machine; ~ **de tornillo** *f* ALIMENT screw extruder

exudación *f* CARBÓN *del lingote* cobbles, GAS seepage, P&C, QUÍMICA exudation

exudado *m* REFRIG drip

exudar *vt* P&C exude, *plastificación* bleed, QUÍMICA exude

exyunción *f* INFORM&PD exjunction

eyección *f* MECÁ ejection

eyecciones *f pl* GEOL ejecta; **~ volcánicas** *f pl* GEOL ejecta, volcanic ejecta

eyectable *adj* TRANSP AÉR jettisonable

eyector *m* ING MECÁ eductor, ejector, kicker actuator, MECÁ, P&C ejector; **~ de resortes** *m* ING MECÁ spring ejector; **~ de vapor** *m* INSTAL HIDRÁUL steam ejector

F

f *abr* (*femto*) METR f (*femto*)

F *abr* GEN (*farad, faradio*) F (*farad, faraday*)

fábrica[1] *f* C&V shop, CONST plant, ING MECÁ, MECÁ factory, PAPEL mill, PROD factory, plant, works, SEG factory; **~ de cartuchos** *f* D&A cartridge factory; **~ de cemento** *f* C&V cement works; **~ de esmaltado metálico** *f* REVEST metal enameling works (*AmE*), metal enamelling works (*BrE*); **~ de gas** *f* Esp (*cf factoría de gas industrial AmL*) GAS gas works, MINAS gas-filled workings, PROD, TERMOTEC gas works; **~ de ladrillo** *f* CONST brickworks; **~ de laminación** *f* ING MECÁ, PROD rolling mill; **~ de láminas de aluminio integrales** *f* PROD integral aluminium foil forming plant (*BrE*), integral aluminum foil forming plant (*AmE*); **~ de láminas metálicas delgadas** *f* PROD foil-forming plant; **~ de malta** *f* ALIMENT malt house (*AmE*), maltings (*BrE*); **~ de mosaicos, azulejos y tejas** *f* C&V tile factory, tilery; **~ de papel** *f* PAPEL, RECICL paper mill; **~ de pasta** *f* PAPEL, RECICL pulp mill; **~ de pólvoras** *f* PROD powder mill, powder works; **~ de productos químicos** *f* PROC QUÍ chemical plant; **~ siderúrgica** *f* PROD ironworks; **~ de tubos** *f* PROD tube works

fábrica[2]: **en ~** *fra* PROD ex-works

fabricación *f* C&V, CONST, ING MECÁ manufacture, manufacturing, output, MINAS output, PROD make, manufacture, manufacturing; **~ asistida por computadora** *f* AmL (*CAM, cf fabricación asistida por ordenador Esp*) ELEC, INFORM&PD, PROD, TELECOM computer-aided manufacturing (*CAM*); **~ asistida por ordenador** *f* Esp (*CAM, cf fabricación asistida por computadora AmL*) ELEC, INFORM&PD, PROD, TELECOM computer-aided manufacturing (*CAM*); **~ de chapas** *f* PROD *hierro* plating; **~ de circuito integrado** *f* ELECTRÓN integrated circuit fabrication; **~ de clavos** *f* PROD nailmaking; **~ de clisés** *f* IMPR block making; **~ por descarga eléctrica** *f* ING MECÁ, PROD electro-discharge machining (*EDM*); **~ discontinua** *f* PROD discontinued manufacturing; **~ discreta** *f* PROD discrete manufacturing; **~ por electroerosión** *f* ING MECÁ, PROD electro-discharge machining (*EDM*); **~ de encargo** *f* PROD customization; **~ de espejos** *f* C&V mirror-making; **~ informatizada** *f* Esp (*cf fabricación integrada por computadora AmL*) INFORM&PD computer-integrated manufacturing (*CIM*); **~ integrada por computadora** *f* AmL (*cf fabricación informatizada Esp*) INFORM&PD computer-integrated manufacturing (*CIM*); **~ de machos** *f* PROD core making; **~ a máquina** *f* ING MECÁ, MECÁ machining; **~ en medio alcalino** *f* PAPEL alkaline papermaking; **~ de modelos** *f* PROD pattern making; **~ de papel** *f* PAPEL paper making; **~ de pastillas** *f* ELECTRÓN wafer fabrication; **~ sobre pedido** *f* PROD custom production; **~ de película por soplado** *f* P&C film blowing; **~ de piezas en tosco** *f* ING MECÁ blanking operation; **~ de plantillas** *f* CARBÓN jigging; **~ de primordios** *f* ING MECÁ blanking operation; **~ en**

serie *f* PROD gang work, mass production, TRANSP MAR *construcción naval* series production

fabricante *m* GEN maker, manufacturer; **~ de aparejos y pertrechos** *m* TRANSP MAR equipment manufacturer; **~ de azulejos** *m* C&V tile maker, tiler; **~ de barniz** *m* REVEST varnish maker; **~ de bolsas** *m* EMB bagmaker; **~ de calderas** *m* INSTAL HIDRÁUL, MECÁ boilermaker; **~ de cemento** *m* CONST cement manufacturer; **~ de los componentes** *m* PROD component manufacturer; **~ de género de punto por metros** *m* TEXTIL *artículos en pieza* yard good knitter; **~ de herramientas** *m* MECÁ, PROD toolmaker; **~ de máquinas** *m* ING MECÁ engine builder; **~ de modelos** *m* PROD pattern maker; **~ de motor** *m* ING MECÁ, TEC ESP engine manufacturer; **~ de motores** *m* ING MECÁ engine builder; **~ de pernos** *m* ING MECÁ bolt cutter; **~ de porcelana** *m* C&V porcelain-maker; **~ de tinta** *m* IMPR ink maker

fabricar *vt* GEN make, manufacture

faceta: **~ de despegue** *f* METAL cleavage facet; **~ de exfoliación** *f* METAL cleavage facet; **~ plana** *f* C&V flat facet; **~ de superficie de bruñido** *f* ACÚST surface burnishing facet; **~ superior** *f* CARBÓN *diamante* table

facetado *adj* C&V, METAL, NUCL faceted

facha: **en ~** *adv* TRANSP MAR aback

fachada *f* CONST elevation, facade, front, PROD *caldera, horno* front; **~ marítima** *f* OCEAN seafront

fachear[1] *vt* TRANSP MAR *velas* back

fachear[2] *vi* TRANSP MAR *velas* lay aback

facia *f* CONST fascia

facies *f pl* GEOL *metamórficas y sedimentarias*, PETROL facies; **~ metamórfica** *f* GEOL metamorphic facies

fácil: **~ de manejar** *adj* INFORM&PD user-friendly; **~ para el usuario** *adj* INFORM&PD user-friendly

facilidad: **~ de acceso** *f* MECÁ ease of access; **~ de manejo** *f* INFORM&PD user-friendliness, MECÁ ease of operation; **~ de mantenimiento** *f* CALIDAD, MECÁ maintainability, NUCL serviceability, PROD, TEC ESP maintainability; **~ de maquinado** *f* MECÁ ease of machining, machinability; **~ para el temple** *f* METAL hardenability

facolita *f* GEOL, MINERAL phacolite

facolito *m* GEOL, MINERAL phacolith

facsímil *m* IMPR, INFORM&PD, TELECOM facsimile; **~ de alta velocidad** *m* TELECOM high-speed facsimile; **~ tipo A** *m* IMPR A-type facsimile

factor *m* TRANSP freight agent (*AmE*), goods agent (*BrE*); **~ de abrasión** *m* MECÁ abrasion factor; **~ de absorción** *m* ACÚST, TELECOM absorption factor; **~ de acoplamiento electromecánico** *m* ACÚST, ELEC electromechanical-coupling factor; **~ de actividad** *m* TEC ESP *comunicaciones* activity factor; **~ de actividad vocal** *m* TELECOM speech activity factor; **~ de acumulación** *m* NUCL buildup factor; **~ aerodinámico** *m* TRANSP AÉR aerodynamic factor; **~ de aislación térmica** *m* TERMO heat insulation factor; **~ de amortiguación** *m* ELEC *oscilación*, FÍS damping factor; **~ de amortiguamiento**

hidrodinámico *m* ENERG RENOV hydrodynamic damping factor; ~ **de amplificación** *m* CINEMAT, ELECTRÓN, INSTR amplification factor, magnification factor; ~ **de amplitud** *m* ING ELÉC crest factor; ~ **de anisoelasticidad** *m* TEC ESP *giróscopos* anisoelasticity factor; ~ **aplanador** *m* PROD smoothing factor; ~ **de atenuación de ráfagas** *m* TEC ESP, TRANSP AÉR, TRANSP MAR gust alleviation factor; ~ **de aumento** *m* CINEMAT, FOTO, INSTR, ÓPT magnification factor; ~ **de calidad** *m* ACÚST, CALIDAD, CONTAM, FÍS quality factor, Q-factor; ~ **de calor sensible** *m* REFRIG *enfriador de aire* sensitive heat ratio; ~ **de camión** *m* TRANSP lorry factor (*BrE*), truck factor (*AmE*); ~ **de capacidad** *m* TEC PETR capacity factor; ~ **de capacidad anual** *m* ENERG RENOV annual capacity factor; ~ **de carga** *m* MECÁ, REFRIG load factor, TEC PETR capacity factor (*AmE*), load factor (*BrE*), loading factor, TRANSP AÉR load factor, loading factor; ~ **de carga límite** *m* TRANSP AÉR limit load factor; ~ **de carga de ráfaga** *m* TEC ESP, TRANSP AÉR, TRANSP MAR gust load factor; ~ **de cargamento** *m* TEC ESP loading factor; ~ **de cementación** *m* PETROL cementation factor; ~ **de compresibilidad** *m* PETROL, TEC PETR, TERMO compressibility factor; ~ **de corrección** *m* METR, TRANSP AÉR correction factor; ~ **de corrección para la resistencia aerodinámica** *m* TRANSP AÉR correction factor for induced drag; ~ **de corrección para retardo inducido** *m* TRANSP AÉR correction factor for induced drag; ~ **de cresta** *m* ING ELÉC *corriente alterna* crest factor; ~ **de curva** *m* ENERG RENOV curve factor; ~ **de Debye-Waller** *m* CRISTAL Debye-Waller factor; ~ **de descontaminación** *m* FÍS RAD decontamination factor; ~ **de desguace** *m* PROD scrap factor, scrapping factor; ~ **de desmultiplicación de impulsos** *m* ELECTRÓN scaling factor; ~ **de desviación** *m* ELECTRÓN deflection factor; ~ **de difusión térmica** *m* FÍS, NUCL, TERMO thermal diffusion factor; ~ **direccionador** *m* *Esp* (*cf factor directivo AmL*) ENERG RENOV directionality factor; ~ **de directividad** *m* ACÚST directivity factor; ~ **directivo** *m* *AmL* (*cf factor direccionador Esp*) ENERG RENOV directionality factor; ~ **de disipación** *m* ELEC, FÍS dissipation factor, *energía* loss factor; ~ **de disminución** *m* ING ELÉC decay factor; ~ **de dispersión** *m* CRISTAL scattering factor; ~ **de dispersión atómica** *m* CRISTAL atomic scattering factor; ~ **de dispersión de perfil** *m* ÓPT profile-dispersion factor; ~ **de disponibilidad de energía** *m* ING ELÉC energy-availability factor; ~ **ecológico** *m* CONTAM ecological factor; ~ **de eficacia de la pala** *m* TRANSP AÉR blade efficiency factor; ~ **energético de conversión** *m* FÍS RAD energy-conversion factor; ~ **de energía** *m* TEC ESP activity factor; ~ **de energía de reflectancia** *m* ÓPT reflectance energy factor; ~ **de enriquecimiento elemental** *m* NUCL elementary enrichment factor; ~ **de entrada de elementos** *m* ING ELÉC fan-in factor; ~ **de entrecruzamiento entre vehículos** *m* TRANSP weaving factor; ~ **de escala** *m* INFORM&PD scale factor; ~ **específico de volumen** *m* METAL volume size factor; ~ **estacional** *m* PROD seasonal factor; ~ **de estallido** *m* PAPEL burst factor; ~ **de estructura** *m* CRISTAL structure factor; ~ **eta** *m* FÍS eta factor, NUCL effective neutron number; ~ **de exposición** *m* CINEMAT, FOTO exposure factor; ~ **de filtrado** *m* PROD smoothing factor; ~ **del filtro** *m*

CINEMAT, FOTO, TV filter factor; ~ **de fisión rápida** *m* FÍS, FÍS PART, NUCL fast-fission factor; ~ **de flexibilidad** *m* ING MECÁ flexibility factor; ~ **de forma** *m* PETROL geometric factor, TERMOTEC *aislamiento* shape factor; ~ **de forma de partículas** *m* PROC QUÍ particle-shape factor; ~ **de formación** *m* PETROL, TEC PETR formation factor; ~ **de frecuencia de pulsos** *m* TV pulse-rate factor; ~ **g** *m* FÍS g-factor; ~ **de gas** *m* GAS gas factor; ~ **hiperglicémicoglicogenolítico** *m* QUÍMICA glucagon; ~ **hora punta** *m* FERRO, TRANSP peak-hour factor (*PHF*), rush-hour factor; ~ **de incrustaciones** *m* REFRIG fouling factor; ~ **de irradiación óptima** *m* NUCL advantage factor; ~ **de Landé** *m* FÍS Landé factor; ~ **de luminosidad** *m* FÍS RAD luminosity factor; ~ **de lutita** *m* TEC PETR shale factor; ~ **de máxima** *m* ELECTRÓN, TEC ESP peak factor; ~ **de mejora de la preacentuación** *m* TEC ESP pre-emphasis improvement factor; ~ **de mitigación** *m* TRANSP AÉR alleviation factor; ~ **de modelo de energía** *m* ENERG RENOV energy-pattern factor; ~ **de modulación** *m* ELECTRÓN, FÍS modulation factor; ~ **de multiplicación de gas** *m* ELECTRÓN gas multiplication factor; ~ **de nivelación familiar** *m* PROD smoothing factor; ~ **de ondulación** *m* CONST ripple factor; ~ **de operación empírico** *m* ING MECÁ empirical operation factor; ~ **de orientación** *m* METAL orientation factor; ~ **de pantalla** *m* NUCL screen factor; ~ **de pérdida del extremo de la pala** *m* TRANSP AÉR blade-tip loss factor; ~ **de pérdida transductiva** *m* ACÚST transducer loss-factor; ~ **de pérdidas** *m* ELEC *energía* loss factor; ~ **de pesada** *m* TEC ESP weighting factor; ~ **de ponderación** *m* TEC ESP weighting factor; ~ **de ponderación de ganancia** *m* ELECTRÓN gain-weighting factor; ~ **ponderado sofométrico** *m* TEC ESP psophometric-weighting factor; ~ **de potencia** *m* ELEC, FÍS, ING ELÉC, P&C, PROD power factor; ~ **de potencia absorbida** *m* PROD input power factor; ~ **de potencia sellado** *m* PROD sealed power factor; ~ **primo** *m* MATEMÁT prime factor; ~ **principal** *m* PROD chief factor; ~ **de proteína animal** *m* (*FPA*) AGRIC, ALIMENT animal protein factor (*APF*); ~ **de pulimentación familiar** *m* PROD smoothing factor; ~ **de punta** *m* ELECTRÓN, TEC ESP peak factor; ~ **R** *m* CRISTAL R factor; ~ **de recuperación** *m* PETROL, TEC PETR *producción, yacimientos* recovery factor; ~ **de recuperación de energía** *m* ENERG RENOV, RECICL, TERMO energy-recovery factor; ~ **de reducción** *m* CARBÓN reduction ratio; ~ **de reducción de interferencia** *m* TEC ESP interference reduction factor; ~ **de reducción del tiempo de tránsito** *m* PROD transit-time reduction factor; ~ **de reducción del volumen de los residuos** *m* NUCL waste volume reduction factor; ~ **de reflectancia** *m* FÍS RAD, ÓPT, PAPEL reflectance factor; ~ **de reflectancia en el azul** *m* PAPEL blue reflectance factor; ~ **de reflectancia difusa en el azul** *m* PAPEL diffuse blue reflectance factor; ~ **de reflectancia direccional en el azul** *m* PAPEL directional blue reflectance factor; ~ **de reflexión** *m* ELECTRÓN, FÍS, ÓPT reflection factor; ~ **de retorno** *m* TEC ESP return factor; ~ **de retrodifusión** *m* FÍS PART, TEC ESP backscattering factor; ~ **de retrodispersión** *m* TEC ESP, TELECOM backscattering factor; ~ **de ruido** *m* ELECTRÓN, FÍS, IMPR, TEC ESP noise factor; ~ **de rumbo sinuoso** *m* TRANSP, TRANSP AÉR weaving factor; ~ **de seguridad** *m* CARBÓN, ELEC, SEG, TRANSP MAR safety factor; **~ de sobrecarga** *m* ELEC *fuente de alimentación* overload factor; **~ de supresión** *m* TEC ESP suppression factor; **~ de temperatura** *m* CRISTAL temperature factor; **~ temporal** *m* PROD seasonal factor; **~ térmico de fisión** *m* FÍS thermal fission factor; **~ de termoaislación** *m* TERMO heat insulation factor; **~ de toma** *m* ELEC tapping factor; **~ de trabajo** *m* MECÁ, PROD duty cycle; **~ de transferencia** *m* TV transfer ratio; **~ de transmisión** *m* FÍS ONDAS transmittance; **~ de utilización** *m* MECÁ load factor; **~ de utilización térmica** *m* FÍS thermal utilization factor; **~ de ventaja** *m* NUCL advantage factor; **~ vibratorio** *m* ACÚST flutter factor; **~ de volumen de la formación** *m* PETROL, TEC PETR formation volume factor; **~ de volumen de la formación de gas** *m* PETROL gas formation volume factor; **~ de volumen de la formación petrolífera** *m* PETROL oil formation volume factor

factoría: **~ comercial y naval** *f* TRANSP MAR shipyard; **~ de gas industrial** *f* AmL (*cf fábrica de gas Esp*) GAS, PROD, TERMOTEC gas works

factorial[1] *adj* INFORM&PD, MATEMÁT factorial

factorial[2] *m* INFORM&PD, MATEMÁT factorial

factorización *f* MATEMÁT factorization

factorizar *vt* MATEMÁT factorize

factor-Q: **~ de la bobina** *m* ELEC, FÍS coil Q-factor

factura *f* IMPR, PROD, TEXTIL invoice

facturación *f* TELECOM billing; **~ detallada** *f* TELECOM itemized billing

faena *f* MINAS labor (*AmE*), labour (*BrE*)

fahlunita *f* MINERAL fahlunite

faja *f* INFORM&PD band; **~ sin arar** *f* AGRIC border; **~ de corte** *f* ING MECÁ land; **~ de rizos** *f* OCEAN *veleros* reef band; **~ de seguridad** *f* AmL (*cf cinturón de seguridad Esp*) AUTO, SEG, TEC PETR, TRANSP, TRANSP AÉR seat belt

falaceado *m* TRANSP MAR *cuerdas* whipping

falca *f* TRANSP MAR *tabla amovible, volante* gunwale, washboard

falda *f* CONST *colina, montaña* slope, CONTAM MAR, D&A skirt, ING MECÁ flap; **~ del casco de frenado** *f* TRANSP AÉR drag cup skirt; **~ del émbolo** *f* AUTO, VEH piston skirt; **~ del pistón** *f* AUTO, VEH piston skirt; **~ de pistón acanalada** *f* AUTO, VEH split-piston skirt; **~ de pistón dividida** *f* AUTO, VEH split-piston skirt

faldilla *f* CONST flap, CONTAM MAR skirt, ING MECÁ flap, PROD *de polea ranurada* flange; **~ de la vasija** *f* NUCL vessel support skirt

faldón *m* C&V skirt, CONST hip, *edificación* skirt, ING MECÁ flap, TEXTIL *vestimenta* lap; **~ conector** *m* TEC ESP connecting skirt

falla *f* CALIDAD failure, CARBÓN *minas* break, *geología* fault, CONST *arco* thrust, ELEC *equipo* failure, fault, ELECTRÓN fault, FÍS breakdown, GAS failure, GEOFÍS, GEOL, ING ELÉC fault, ING MECÁ breakdown, MECÁ fault, flaw, MINAS floor heave, fault, break, OCEAN heave, PETROL, PROD, TEC PETR fault; **~ antitética** *f* GEOL antithetic fault; **~ en el cable** *f* ELEC, ING ELÉC cable defects; **~ concéntrica** *f* GEOL concentric dike; **~ contemporánea** *f* GEOL contemporaneous fault; **~ de crecimiento** *f* GEOL growth fault; **~ deposicional** *f* GEOL depositional fault; **~ de desgarramiento** *f* GEOL tear fault; **~ en dirección**

dextral *f* GEOL dextral fault; **~ en dirección siniestral** *f* GEOL sinistral fault; **~ distensiva** *f* GEOL distensional fault; **~ distributiva** *f* GEOL distributive fault; **~ eléctrica** *f* CALIDAD electrical fault; **~ de Griffith** *f* C&V Griffith flaw; **~ hidráulica** *f* SEG hydraulic failure; **~ horizontal** *f* GEOL horizontal displacement; **~ humana** *f* CALIDAD human failure; **~ intermitente** *f* ELEC intermittent fault; **~ inversa** *f* GEOL reverse fault; **~ en la línea** *f* ING ELÉC line fault; **~ lístrica** *f* GEOL listric fault; **~ no revelada** *f* CALIDAD unrevealed failure; **~ normal** *f* GEOL normal fault; **~ ortogonal** *f* GEOL dip fault; **~ revelada** *f* CALIDAD revealed failure; **~ de señal** *f* AmL (*cf fallo de señal Esp*) TELECOM signal fail (*SF*); **~ sincrónica** *f* GEOL contemporaneous fault; **~ siniestra** *f* TEC PETR sinistral fault; **~ sinsedimentaria** *f* GEOL synsedimentary fault; **~ sintética** *f* GEOL synthetic fault; **~ transcurrente** *f* GEOL transcurrent fault; **~ transformante** *f* GEOL transform fault; **~ transversal** *f* PETROL dip fault; **~ de vaciado** *f* C&V casting scar; **~ vertical de desgarre** *f* GEOL wrench fault

fallar *vi* ELECTRÓN malfunction, INFORM&PD crash, ING ELÉC, TELECOM malfunction

falleba *f* ING MECÁ bolt, catch; **~ de emergencia** *f* SEG panic bolts; **~ de ventana** *f* CONST sash fastener

fallo[1]: **~ leve** *adj* TEC ESP fail-soft

fallo[2] *m* CALIDAD failure, malfunction, CONST breakdown, failure, INFORM&PD malfunction, failure, fault, ING ELÉC failure, ING MECÁ malfunction, failure, MECÁ, METAL failure, METR, SEG malfunction, TEC ESP malfunction, burnout, failure, report, TELECOM failure; **~ de bajo nivel** *m* PROD low-level fault; **~ de base** *m* CONST base failure; **~ básico** *m* TRANSP AÉR basic failure; **~ del bucle** *m* PROD loop fault; **~ catastrófico** *m* ING ELÉC catastrophic failure; **~ de los componentes** *m* PROD component failure; **~ en la continuidad** *m* TELECOM continuity fault; **~ de encendido** *m* ING ELÉC misfire, VEH mismatch; **~ del equipo procesador** *m* PROD processor hardware fault; **~ de fatiga incipiente** *m* TRANSP AÉR incipient fatigue failure; **~ fortuito** *m* ING ELÉC random failure; **~ de frenos** *m* TRANSP brake failure; **~ inducido** *m* INFORM&PD induced failure; **~ de las instalaciones** *m* TRANSP AÉR facility failure; **~ interno de circuito** *m* PROD internal circuit fault; **~ de línea** *m* TELECOM line fault; **~ del motor** *m* AUTO, VEH engine breakdown; **~ de potencia** *m* ING ELÉC power failure; **~ prematuro** *m* ING ELÉC early failure; **~ principal** *m* TEC ESP mean anomaly; **~ de reciprocidad** *m* CINEMAT, FOTO, TV reciprocity failure; **~ de la señal** *m* IMPR drop-out; **~ de señal** *m* Esp (*cf falla de señal AmL*) TELECOM signal fail (*SF*); **~ súbito** *m* NUCL fast burst; **~ de la tabla de datos** *m* PROD data table failure; **~ técnico** *m* TELECOM technical breakdown; **~ de tensión** *m* TELECOM power fail; **~ en el tiro** *m* AmL (*cf bocazo Esp*) MINAS misfire; **~ del transmisor** *m* TV transmitter failure

falsa: **~ escuadra** *f* CONST miter square (*AmE*), mitre square (*BrE*), METR bevel square; **~ filigrana** *f* PAPEL simulated watermark; **~ imitación** *f* ELECTRÓN imitative deception; **~ de madera** *f* PROD *moldeo* match plate; **~ ventana** *f* CONST blank window

falso[1] *adj* INFORM&PD false, TEXTIL mock

falso[2]: **~ codaste** *m* TRANSP MAR *yates* skeg; **~ contacto** *m* ING ELÉC bad contact; **~ embalaje** *m*

EMB dummy pack; ~ **espejo** *m* C&V see-through mirror (*AmE*), two-way mirror (*BrE*); ~ **estay** *m* TRANSP MAR jumper stay; ~ **flete** *m* TEC PETR dead freight; ~ **liso** *m* TEXTIL *tejidos* false plain; ~ **pilote** *m* CARBÓN pile block; ~ **suelo** *m* REFRIG false floor; ~ **techo** *m* REFRIG false ceiling; ~ **tirante** *m* ING MECÁ collar beam, collar tie; ~ **topacio** *m* MINERAL false topaz

falta[1] *f* ACÚST absence, ING MECÁ deviation, MECÁ, PROD *de peso* fault; ~ **de adaptación** *f* ING MECÁ mismatch; ~ **de alimentación** *f* IMPR starvation; ~ **de contraste** *f* IMPR flatness; ~ **de nitidez debido al movimiento** *f* CINEMAT motion unsharpness; ~ **de papel** *f* INFORM&PD paper low; ~ **de peso** *f* EMB short weight; ~ **de plasticidad** *f* MECÁ, METAL blue brittleness, brittleness; ~ **de registro** *f* IMPR misregistration; ~ **de tensión de CA** *f* ING ELÉC AC power failure

falta[2]: ~ **temporaria de operador** *fra AmL* (*cf operador temporalmente fuera de servicio Esp*) TELECOM operator temporarily unavailable

falto *adj* TRANSP MAR *amarre, sistemas de carga* broken; ~ **de calidad** *adj* TEXTIL off-quality; ~ **de revelado** *adj* CINEMAT underdeveloped

famatinita *f* MINERAL famatinite

Fameniese *adj* GEOL Famennian

familia: ~ **de circuito lógico de transistor a transistor** *f* ELECTRÓN transistor-transistor logic family; ~ **de CLTT** *f* ELECTRÓN TTL logic family; ~ **de computadoras** *f AmL* (*cf familia de ordenadores Esp*) INFORM&PD computer family; ~ **de desintegración radiactiva** *f* FÍS PART, FÍS RAD radioactive-decay series; ~ **de elementos** *f* FÍS RAD family of elements; ~ **lógica** *f* ELECTRÓN, INFORM&PD logic family; ~ **de ordenadores** *f Esp* (*cf familia de computadoras AmL*) INFORM&PD computer family; ~ **de partículas** *f* FÍS PART family of particles; ~ **de productos** *f* PROD product family; ~ **de tipos** *f* IMPR type family

fan: ~ **anticondensación** *m* TRANSP AÉR defogging fan

fanerítico *adj* GEOL, PETROL faneritic

Fanerozoico *adj* GEOL Phanerozoic

fango *m* AGUA loam, mud, *ductos agua* slime, CARBÓN dirt, silt, slime, GEOL *sedimentología* ooze, HIDROL slush, sludge, MINAS slush, OCEAN mud, PROD, REFRIG sludge, TEC PETR mud; ~ **activado** *m* CONTAM, HIDROL, RECICL activated sludge; ~ **activo** *m* CONTAM, HIDROL, RECICL activated sludge; ~ **de aguas cloacales** *m* RECICL sludge; ~ **de alcantarilla** *m* AGUA sewage sludge; ~ **almacenado** *m* CARBÓN stowing dirt; ~ **cloacal** *m* AGUA, HIDROL sewage sludge; ~ **concentrado** *m* CARBÓN concentrated sludge; ~ **digerido** *m* HIDROL *aguas residuales* digested sludge; ~ **glaciárico** *m* HIDROL glacier silt; ~ **de lavado** *m* CARBÓN, DETERG, MINAS *concentración* slurry; ~ **líquido** *m* HIDROL *aguas negras* sludge liquor; ~ **margoso** *m* GEOL marly loam; ~ **de muela** *m* PROD *muelas abrasivas* wheel swarf; ~ **no tratado** *m* AGUA, CONTAM, RECICL raw sludge; ~ **pelágico rico en oozoos** *m* GEOL oozue; ~ **de retorno activado** *m* HIDROL activated return sludge; ~ **de sondeo** *m* CARBÓN drillings; ~ **sin tratar** *m* AGUA, CONTAM, RECICL raw sludge; ~ **Venturi** *m* CARBÓN Venturi sludge

fangolita *f* GEOL mudstone

fangos *m pl* CARBÓN stowing material, *metalurgia*

slimes; ~ **de alcantarilla** *m pl* AGUA sewage sludge; ~ **de carbón** *m pl* CARBÓN, MINAS coal sludge; ~ **espesos** *m pl* CARBÓN thickened slime; ~ **de lavado colables** *m pl* MINAS pourable slurry; ~ **del polvo de lavado** *m pl* DETERG washing powder slurry; ~ **de sondeo** *m pl* MINAS drillings

fanotrón *m* ELECTRÓN, GAS gas diode

fantoma *m* NUCL phantom

FAQ *abr* (*preguntas más frecuentes*) INFORM&PD FAQ (*frequently asked questions*)

farad *m* (*F*) ELEC, ELECTRÓN, FÍS, ING ELÉC, METR, NUCL farad (*F*), faraday (*F*)

faradio *m* (*F*) ELEC, ELECTRÓN, FÍS, ING ELÉC, METR, NUCL farad (*F*), faraday (*F*)

faradización *f* ELECTRÓN, TEC ESP *electrónica* screening

faradizar *vt* CARBÓN faredice, ELECTRÓN, TEC ESP screen

farallón *m* OCEAN *ríos* bluff

fardo *m* AGRIC bale, EMB bale, bundle, PAPEL, TEXTIL bale, TRANSP parcel; ~ **cilíndrico** *m* AGRIC round bale; ~ **de lana** *m* TEXTIL woolpack

farelita *f* MINERAL farelite

farero *m* TRANSP MAR lighthouse keeper

farináceo *adj* ALIMENT farinaceous

farinógrafo *m* ALIMENT farinograph

farmacolita *f* MINERAL pharmacolite

farmacosiderita *f* MINERAL pharmacosiderite

farnesol *m* QUÍMICA farnesol

faro *m* AUTO headlamp, CONST beacon, FERRO indicator lamp, TRANSP MAR lighthouse, VEH headlamp; ~ **de charnela del tren de aterrizaje** *m* TRANSP AÉR landing gear hinge beam; ~ **delantero** *m* AUTO headlamp; ~ **orientable** *m* AUTO, VEH spotlight; ~ **de radar** *m* D&A, FÍS, FÍS RAD, TELECOM, TRANSP, TRANSP AÉR, TRANSP MAR radar beacon; ~ **testigo** *m* FERRO indicator lamp, telltale lamp

farol *m* ING ELÉC lamp; ~ **de intemperie** *m* CONST hurricane lamp

faros: ~ **basculantes** *m pl* AUTO, VEH dimmed beams (*AmE*), dipped headlights (*BrE*); ~ **inclinables** *m pl* AUTO, VEH headlights (*BrE*), low beams (*AmE*)

FAS *abr* (*franco al costado del buque*) TRANSP MAR FAS (*free alongside ship*)

fasaje *m* ELEC *corriente alterna*, ELECTRÓN, TRANSP, TV phasing

fase[1]: **en** ~ *adj* ELEC in-phase, ELECTRÓN in-phase, in phase, FÍS, FÍS ONDAS in-phase, TV in-phase, in phase, phased

fase[2]: **en** ~ *adv* ELEC, ELECTRÓN, FÍS, FÍS ONDAS, TV in phase

fase[3] *f* ACÚST, ELEC, ELECTRÓN, FÍS, FÍS ONDAS phase (*ph*), INSTR *compresor* stage, METAL, PETROL phase (*ph*), PROD stage, step, QUÍMICA, TELECOM, TERMO, TRANSP phase (*ph*); ~ **de acabado** *f* TEXTIL finishing stage; ~ **acuosa** *f* CARBÓN aqueous phase; ~ **de amplificación** *f* ELECTRÓN amplifier stage; ~ **de aprendizaje** *f* ELECTRÓN learning phase; ~ **de aproximación** *f* TRANSP AÉR approach phase; ~ **de ascenso** *f* TRANSP AÉR climb phase; ~ **de aterrizaje no pilotado** *f* TEC ESP unmanned landing stage; ~ **del color** *f* TV color phase (*AmE*), colour phase (*BrE*); ~ **de compresión** *f* ING MECÁ compression stage; ~ **constante** *f* ACÚST, ELECTRÓN, FÍS, ÓPT, TELECOM, TV phase constant; ~ **de crecimiento** *f* METAL growth step; ~ **de crominancia** *f* TV chrominance phase; ~ **en cuadratura** *f* ELECTRÓN quadrature phase; ~ **de**

datos *f* TELECOM data phase; ~ **de deformación** *f* GEOL deformational phase; ~ **de desarrollo** *f* TEC PETR development phase; ~ **de despegue** *f* TRANSP AÉR takeoff phase; ~ **de desplazamiento** *f* ELEC *corriente alterna* displacement phase; ~ **diferencial** *f* ELECTRÓN, TV differential phase; ~ **de dilatación** *f* METAL growth step; ~ **de ejecución** *f* INFORM&PD execute phase; ~ **de entrada** *f* ELECTRÓN input stage; ~ **de escape** *f* AUTO, VEH exhaust stroke; ~ **escrita** *f* ELECTRÓN *tubos de almacenamiento* written state; ~ **esméctica** *f* CRISTAL smectic phase; ~ **para establecer una comunicación** *f* TELECOM call-setup phase; ~ **estacionaria** *f* TELECOM stationary phase; ~ **de exploración** *f* TEC PETR exploration phase; ~ **de FI** *f* ELECTRÓN IF stage; ~ **de fluctuación** *f* TEC ESP phase jitter; ~ **de frecuencia intermedia** *f* ELECTRÓN intermediate frequency stage; ~ **gaseosa** *f* TERMO gaseous phase; ~ **generatriz** *f* METAL parent phase; ~ **de giro a la izquierda** *f* TRANSP left-turn phase; ~ **grabada** *f* ELECTRÓN *tubos de almacenamiento* written state; ~ **de la imagen negativa** *f* TV negative-picture phase; ~ **de imagen positiva** *f* TV positive picture phase; ~ **de impulso** *f* ELECTRÓN pulse phase; ~ **inicial** *f* METAL initial stage; ~ **de instalación** *f* PROD setup phase; ~ **integral** *f* TEC ESP phase integral; ~ **interglacial** *f* GEOL interglacial phase; ~ **lógica** *f* ELECTRÓN logic state; ~ **de luz roja** *f* TRANSP *semáforos* red phase; ~ **de luz verde** *f* TRANSP *semáforos* green phase; ~ **de la marea** *f* OCEAN tide phase; ~ **mesomórfica** *f* CRISTAL mesomorphic phase; ~ **de montaje** *f* PROD setup phase; ~ **nemática** *f* CRISTAL nematic phase; ~ **nocturna** *f* TEC ESP nocturnal phase; ~ **opuesta** *f* ELEC *corriente alterna* opposite phase; ~ **de organización** *f* PROD setup phase; ~ **orogénica** *f* GEOL orogenic phase; ~ **peatonal** *f* TRANSP pedestrian phase; ~ **de preparación** *f* PROD setup phase; ~ **de producción** *f* TEC PETR production phase; ~ **de purga** *f* NUCL *pérdida de refrigerante* blowdown; ~ **de recuperación de la central** *f* NUCL plant-recovery phase; ~ **de referencia** *f* TV reference phase; ~ **de RF** *f* ELECTRÓN, ING ELÉC RF stage; ~ **de salida vertical** *f* ELECTRÓN vertical output stage; ~ **sedimentaria** *f* GEOL sedimentary phase; ~ **de señal** *f* ELECTRÓN signal phase; ~ **de señal de referencia** *f* ELECTRÓN reference-signal phase; ~ **de sincronización cromática** *f* TV burst phase; ~ **sólida** *f* TERMO solid phase; ~ **de sólo gas** *f* TERMO gas-only phase; ~ **sólo líquida** *f* TERMO liquid-only phase; ~ **de subida** *f* TRANSP AÉR climb phase; ~ **de la subportadora** *f* TV subcarrier phase; ~ **a tierra** *f* PROD phase to ground; ~ **de transbordo** *f* TEC ESP transfer stage; ~ **transitoria** *f* METAL transient phase; ~ **de transmisión del enlace ascendente** *f* TELECOM uplink transmission phase; ~ **vapor** *f* PROC QUÍ, TERMO vapor phase (*AmE*), vapour phase (*BrE*); ~ **de variaciones cíclicas** *f* TEC ESP phase jitter

fase[4]: **estar en ~** *vi* ELEC *fuente de alimentación de CA* be in phase

faseolina *f* QUÍMICA phaseolin

faseolunatina *f* QUÍMICA phaseolunatin

fases: ~ **entre acoplamiento** *f pl* ELECTRÓN coupling between stages

fasor *m* ELEC *forma de onda* phasor

fassaita *f* MINERAL fassaite, pyrgom

fatiga *f* CRISTAL fatigue, GEOL stress, ING MECÁ stress, strain, MECÁ, METAL, P&C fatigue, stress, TRANSP AÉR fatigue; ~ **acelerada por rozamiento** *f* NUCL chafing; ~ **acústica** *f* METAL sonic fatigue; ~ **ambiental** *f* TEC ESP environmental stress; ~ **auditiva** *f* ACÚST hearing fatigue; ~ **por contacto** *f* METAL fretting fatigue; ~ **por corrosión** *f* METAL corrosion fatigue; ~ **de flujo** *f* ING MECÁ flow fatigue; ~ **del metal** *f* CRISTAL metal fatigue; ~ **metálica** *f* CRISTAL metal fatigue; ~ **con pocos ciclos** *f* CRISTAL low-cycle fatigue; ~ **de rotura** *f* ING MECÁ ultimate strength; ~ **por rozamiento** *f* METAL fretting fatigue; ~ **tangencial** *f* ING MECÁ tangential stress; ~ **térmica** *f* TERMO thermal fatigue

fatigado: ~ **con ácido** *adj* CONTAM *procesos industriales* acid-stressed

faujasita *f* MINERAL faujasite

fauserita *f* MINERAL fauserite

favor: **a ~ del viento** *adv* CONTAM, TRANSP AÉR downwind

faxear *vt* IMPR, INFORM&PD, TELECOM fax

fayalita *f* MINERAL fayalite

FBT *abr* (*fin del bloque de transmisión*) INFORM&PD ETB (*end of transmission block*)

FDP *abr* (*función de densidad de probabilidad*) MATEMÁT PDF (*probability density function*)

FDT *abr* (*fin de texto*) INFORM&PD ETX (*end of text*)

FDX *abr* (*sistema de transmisión birideccional*) INFORM&PD FDX (*full duplex*)

Fe *abr* (*hierro*) METAL, QUÍMICA Fe (*iron*)

FEB *abr* (*frecuencia extremadamente baja*) ELEC, ELECTRÓN, TELECOM ELF (*extremely low frequency*)

fecal *adj* HIDROL *aguas residuales*, RECICL faecal (*BrE*), fecal (*AmE*)

fecha: ~ **de caducidad** *f* ALIMENT best before, expiry date, EMB best before, use-by date, FÍS PART, FÍS RAD, NUCL best before; ~ **de caída** *f* TEC ESP *vehículos en órbita* decay date; ~ **de entrada inmediata** *f* PROD immediate input date; ~ **de entrega** *f* EMB, PROD delivery date; ~ **de fabricación** *f* EMB date of manufacture; ~ **y hora de transmisión** *f* TELECOM date and time of transmission; ~ **de intercambio** *f* TELECOM interchange date; ~ **de llegada al taller** *f* PROD shop arrival date; ~ **de llegada del flete** *f* PROD date freight inward; ~ **de producción** *f* PROD production date; ~ **de salida al aire** *f* TV air date; ~ **de salida inmediata** *f* PROD immediate output date; ~ **de taller** *f* PROD shop date; ~ **de vencimiento** *f* CINEMAT expiry date, PROD due date

fechador *m* INFORM&PD dater

feculoso *adj* QUÍMICA starchy

fecundar *vt* AGRIC fertilize; ~ **por fertilización** *vt* AGRIC cross-fertilize

felandreno *m* QUÍMICA phellandrene

feldespato *m* C&V, MINERAL, QUÍMICA feldspar, felspar; ~ **alcalino** *m* TEC PETR alkali feldspar; ~ **aventurina** *m* MINERAL aventurine feldspar; ~ **verde** *m* MINERAL green feldspar; ~ **vidrioso** *m* C&V glassy feldspar

felpa: ~ **obturada** *f* C&V clogged felt

felsita *f* PETROL petrosilex, felsite

FEM[1] *abr* (*fuerza electromotriz*) ELEC, FÍS, ING ELÉC EMF (*electromotive force*)

FEM[2]: ~ **de contacto** *f* ELEC *entre dos metales*, FÍS contact EMF; ~ **del transformador** *f* ELEC transformer EMF

femto *pref* (*f*) METR femto (*f*)

fenacetina *f* QUÍMICA phenacetin

fenacetúrico *adj* QUÍMICA phenaceturic
fenacil *m* QUÍMICA phenacyl
fenacilo *m* QUÍMICA phenacyl
fenacina *f* QUÍMICA phenazine
fenadona *f* QUÍMICA phenadone
fenantraquinona *f* QUÍMICA phenanthraquinone
fenantrazina *f* QUÍMICA phenanthrazine
fenantridina *f* QUÍMICA phenanthridine
fenantridona *f* QUÍMICA phenanthridone
fenantril *m* QUÍMICA phenanthryl
fenantrilo *m* QUÍMICA phenanthryl
fenantrol *m* QUÍMICA phenanthrol
fenantrolina *f* QUÍMICA phenanthroline
fenaquita *f* MINERAL phenakite
fenato *m* QUÍMICA phenate
fenazocina *f* QUÍMICA phenazocine
fenazona *f* QUÍMICA phenazone
fencona *f* QUÍMICA fenchone
fenetedina *f* QUÍMICA phenetidine
fenetol *m* QUÍMICA phenetole
fenicita *f* MINERAL phoenicite
fénico *adj* QUÍMICA carbolic
fenicrocoíta *f* MINERAL phoenicochroite
fenil *m* QUÍMICA phenyl
fenilacetamida *f* QUÍMICA phenylacetamide
fenilacétivo *adj* QUÍMICA phenylacetic
fenilado *adj* QUÍMICA phenylated
fenilalanina *f* QUÍMICA phenylalanine
fenilamina *f* QUÍMICA phenylamine
fenilendiamina *f* QUÍMICA phenylenediamine
feniletilamina *f* QUÍMICA phenylethylamine
feniletileno *m* QUÍMICA phenylethylene
fenilglicina *f* QUÍMICA phenylglycine
fenilglicol *m* QUÍMICA phenylglycol
fenilglicólico *adj* QUÍMICA phenylglycolic
fenilhidrazina *f* QUÍMICA phenylhydrazine
fenilhidrazona *m* QUÍMICA phenylhydrazone
fenilhidroxiacético *adj* QUÍMICA phenylhydroxyacetic
fenilhidroxilamina *f* QUÍMICA phenylhydroxylamine
fenílico *adj* QUÍMICA phenylic
fenilmetano *m* QUÍMICA phenylmethane
fenilmetilcetona *f* QUÍMICA acetophenone
fenilo *m* QUÍMICA phenyl
fenilpirazol *m* QUÍMICA phenylpyrazole
fenilpropiólico *adj* QUÍMICA phenylpropiolic
fenilurea *f* QUÍMICA phenylurea
feniramina *f* QUÍMICA pheniramine
fenita *f* PETROL phanite
fenoclasto *m* GEOL, PETROL phenoclast
fenocristal *m* GEOL, PETROL phenocryst
fenocristalino *adj* GEOL, PETROL phenocrystalline
fenol *m* HIDROL, QUÍMICA phenol
fenolato *m* QUÍMICA phenolate
fenolftaleína *f* QUÍMICA phenolphthalein
fenólico *adj* QUÍMICA phenolic
fenolsulfónico *adj* QUÍMICA phenolsulfonic (*AmE*), phenolsulphonic (*BrE*)
fenómeno: ~ **atmosférico** *m* CONTAM atmospheric phenomenon; ~ **cooperativo** *m* FÍS cooperative phenomenon; ~ **metereológico** *m* CONTAM atmospheric phenomenon; ~ **transitorio** *m* TELECOM transient phenomenon
fenómenos: ~ **de inestabilidad** *m pl* FÍS FLUID instability phenomena; ~ **sísmicos** *m pl* GEOFÍS seismic phenomena
fenosafranina *f* QUÍMICA phenosafranine

fenotiazina *f* QUÍMICA phenothiazine
fenotipo *m* AGRIC phenotype
fenoxazina *f* QUÍMICA phenoxazine
fenoxibenceno *m* QUÍMICA phenoxybenzene
fenóxido *m* QUÍMICA phenoxide
fenqueno *m* QUÍMICA fenchene
fenquilo *m* QUÍMICA fenchyl
fermentación *f* AGRIC, ALIMENT, HIDROL, OCEAN, PROC QUÍ, QUÍMICA fermentation; ~ **acética** *f* ALIMENT acetic fermentation; ~ **aeróbica** *f* ALIMENT aerobic fermentation; ~ **alcohólica** *f* ALIMENT alcoholic fermentation; ~ **por ebullición** *f* PROC QUÍ boiling fermentation; ~ **láctica** *f* AGRIC lactic acid fermentation; ~ **de masa panaria** *f* ALIMENT *repostería, panadería* panary fermentation; ~ **metánica** *f* HIDROL *aguas residuales* methane fermentation; ~ **de metano** *f* QUÍMICA methane fermentation; ~ **en salmuera** *f* OCEAN brine fermentation
fermentador *m* AGRIC, ALIMENT, HIDROL, OCEAN, PROC QUÍ, QUÍMICA fermenter
fermentar *vi* ALIMENT, HIDROL, QUÍMICA ferment
fermio *m* (*Fm*) FÍS RAD, QUÍMICA fermium (*Fm*)
fermión *m* FÍS, FÍS PART fermion
ferrallista *m* CONST fabricator
ferrato *m* QUÍMICA ferrate
ferredoxina *f* QUÍMICA ferredoxin
férreo *m* CARBÓN iron
ferricianógeno *m* QUÍMICA ferricyanogen
ferricianuro *m* QUÍMICA ferricyanide
férrico *adj* PROD, QUÍMICA ferric; **no** ~ *adj* QUÍMICA nonferric
ferrimagnético *adj* FÍS ferrimagnetic
ferrimagnetismo *m* FÍS ferrimagnetism
ferrioxálico *adj* QUÍMICA ferrioxalic
ferrita *f* ELEC, FÍS, ING ELÉC, QUÍMICA ferrite; ~ **bainítica** *f* METAL bainitic ferrite
ferrítico *adj* MECÁ, METAL, TEC ESP ferritic
ferritina *f* QUÍMICA ferritin
ferrocarril: ~ **de circunvalación** *m* FERRO loop line; ~ **de cremallera** *m* FERRO, TRANSP rack railroad (*AmE*), rack railway (*BrE*); ~ **de cremallera y piñón** *m* AUTO, FERRO rack-and-pinion railroad (*AmE*), rack-and-pinion railway (*BrE*); ~ **metropolitano urbano y de cercanías** *m* TRANSP urban and regional metropolitan railroad (*AmE*), urban and regional metropolitan railway (*BrE*); ~ **modelo** *m* FERRO model railroad (*AmE*), model railway (*BrE*); ~ **de montaña** *m* FERRO mountain railroad (*AmE*), mountain railway (*BrE*); ~ **subterráneo** *m* FERRO subway (*AmE*), underground (*BrE*), underground railway (*BrE*); ~ **de tránsito rápido** *m* FERRO, NUCL, TRANSP rapid-transit railroad (*AmE*), rapid-transit railway (*BrE*); ~ **de vía ancha** *m* FERRO broad-gage railroad (*AmE*), broad-gauge railway (*BrE*); ~ **de vía estrecha** *m* FERRO *equipo inamovible* narrow-gauge railway (*BrE*), narrow-gage railroad (*AmE*), *vehículos* light railway (*BrE*), light railroad (*AmE*)
ferrocianato *m* QUÍMICA ferrocyanate
ferrocianógeno *m* QUÍMICA ferrocyanogen
ferrocianuro *m* QUÍMICA ferrocyanide
ferroelectricidad *f* ELEC, FÍS, ING ELÉC ferroelectricity
ferroeléctrico *adj* ELEC, FÍS, ING ELÉC ferroelectric
ferromagnético *adj* ELEC, ELECTRÓN, FÍS, ING ELÉC ferromagnetic
ferromagnetismo *m* ELEC, ELECTRÓN, FÍS, ING ELÉC ferromagnetism

ferromolibdeno *m* METAL ferromolybdenum
ferroníquel *m* METAL ferronickel
ferroprusiato *m* QUÍMICA ferroprussiate
ferrorresonancia *f* ING ELÉC ferroresonance
ferroso: no ~ *adj* QUÍMICA nonferrous
ferrosoférrico *adj* QUÍMICA ferrosoferric
ferruginoso *adj* GEOL ferruginous
férula *f* ING MECÁ, MECÁ, ÓPT, TELECOM ferrule
ferúlico *adj* QUÍMICA ferulic
festón *m* GEOL festoon
festoneado *m* ING MECÁ pinging (*AmE*), pinking (*BrE*), TV scallop, scalloping
festonear *vt* ING MECÁ pink (*BrE*), ping (*AmE*)
FI *abr* ELECTRÓN (*fidelidad*) FI (*fidelity*), ELECTRÓN, TELECOM, TV (*frecuencia intermedia*) IF (*intermediate frequency*)
fiabilidad *f* CALIDAD dependability, reliability, CONTAM MAR, CRISTAL, ELEC, INFORM&PD, ING ELÉC reliability, ING MECÁ dependability, reliability, METR, TEC ESP, TELECOM, TV reliability; ~ **del hardware** *f* INFORM&PD hardware reliability; ~ **humana** *f* CALIDAD human reliability; ~ **de las instalaciones** *f* TRANSP AÉR facility reliability; ~ **prevista** *f* TEC ESP predicted reliability; ~ **de la señalización** *f* TELECOM signaling reliability (*AmE*), signalling reliability (*BrE*)
fiador *m* CONST gutter bracket, ING MECÁ grip, keeper, catch, click, safety catch, trip, trip catch, detent, dog, pawl, fastener, MECÁ catch, dog, pawl, stop, fastener, PROD grip, fastener; ~ **de acero inoxidable anticorrosivo** *m* ING MECÁ corrosion-resistant stainless steel fastener; ~ **giratorio** *m* ING MECÁ pawl; ~ **de la mesa portapiezas** *m* ING MECÁ workholding pallet; ~ **de perno de pistón** *m* ING MECÁ, MECÁ circlip; ~ **roscado** *m* MECÁ threaded fastener; ~ **de rueda** *m* ING MECÁ pawl; ~ **de la rueda del tren de aterrizaje** *m* TRANSP AÉR landing gear boot retainer; ~ **de tuerca** *m* ING MECÁ nut lock
fibra *f* GEN fiber (*AmE*), fibre (*BrE*); **contra** ~ *f* IMPR *sentido transversal del papel, cartón* cross grain; ~ **abierta** *f* TEXTIL open fiber (*AmE*), open fibre (*BrE*); ~ **ahusada** *f* TELECOM tapered fiber (*AmE*), tapered fibre (*BrE*); ~ **aislante** *f* ING ELÉC fish paper; ~ **de alta tenacidad** *f* TEXTIL high-tenacity fiber (*AmE*), high-tenacity fibre (*BrE*); ~ **de alto módulo** *f* TEXTIL high-tenacity fiber (*AmE*), high-tenacity fibre (*BrE*); ~ **de amianto** *f* AmL (*cf fibra de asbesto Esp*) CONST asbestos string; ~ **artificial** *f* EMB, TERMOTEC, TEXTIL man-made fiber (*AmE*), man-made fibre (*BrE*); ~ **de asbesto** *f* Esp (*cf fibra de amianto AmL*) CONST asbestos string; ~ **de baja atenuación** *f* ÓPT, TELECOM low-loss fiber (*AmE*), low-loss fibre (*BrE*); ~ **básica** *f* C&V basic fiber (*AmE*), basic fibre (*BrE*); ~ **bruta** *f* ALIMENT crude fiber (*AmE*), crude fibre (*BrE*); ~ **buffer** *f* TELECOM fiber buffer (*AmE*), fibre buffer (*BrE*); ~ **de carbono** *f* P&C, TEC ESP, VEH carbon fiber (*AmE*), carbon fibre (*BrE*); ~ **cerámica** *f* C&V, INSTAL TERM ceramic fiber (*AmE*), ceramic fibre (*BrE*); ~ **cerámica refractaria formada por monocristales** *f* METAL whisker; ~ **de coco** *f* TEXTIL, TRANSP MAR *cuerdas* coir; ~ **cónica** *f* TELECOM tapered fiber (*AmE*), tapered fibre (*BrE*); ~ **continua** *f* METAL continuous fiber (*AmE*), continuous fibre (*BrE*), P&C roving; ~ **cortada** *f* C&V chopped strand, TEXTIL staple fiber (*AmE*), staple fibre (*BrE*); ~ **cruda** *f* AGRIC crude fiber (*AmE*),

crude fibre (*BrE*); ~ **dañada** *f* C&V damaged yarn; ~ **dietética** *f* ALIMENT dietary fiber (*AmE*), dietary fibre (*BrE*); ~ **digestible** *f* AGRIC digestible fiber (*AmE*), digestible fibre (*BrE*); ~ **dopada** *f* ÓPT doped fiber (*AmE*), doped fibre (*BrE*); ~ **emisora** *f* ÓPT, TELECOM launching fiber (*AmE*), launching fibre (*BrE*); ~ **estirada por calor** *f* TEXTIL heat-stretched fiber (*AmE*), heat-stretched fibre (*BrE*); ~ **ficticia** *f* ÓPT, TELECOM dummy fiber (*AmE*), dummy fibre (*BrE*); ~ **de guía débil** *f* ÓPT, TELECOM weakly-guiding fiber (*AmE*), weakly-guiding fibre (*BrE*); ~ **hasta la acera** *f* TELECOM fiber to the curb (*AmE*) (*FTTC*), fibre to the kerb (*BrE*) (*FTTC*); ~ **hasta el domicilio** *f* TELECOM fiber to the home (*AmE*) (*FTTH*), fibre to the home (*BrE*) (*FTTH*); ~ **hasta el edificio** *f* TELECOM fiber to the building (*AmE*) (*FTTB*), fibre to the building (*BrE*) (*FTTB*); ~ **hasta la oficina** *f* TELECOM fiber to the office (*AmE*) (*FTTO*), fibre to the office (*BrE*) (*FTTO*); ~ **helicoidal** *f* ÓPT fiber helix (*AmE*), fibre helix (*BrE*); ~ **de índice enjaretada** *f* FÍS, TRANSP MAR graded-index fiber (*AmE*), graded-index fibre (*BrE*); ~ **de índice escalonado** *f* ÓPT, TELECOM step index fiber (*AmE*), step index fibre (*BrE*); ~ **de índice gradual** *f* ÓPT, TELECOM graded-index fiber (*AmE*), graded-index fibre (*BrE*); ~ **de índice parabólico** *f* FÍS, ÓPT, TELECOM parabolic index fiber (*AmE*), parabolic index fibre (*BrE*); ~ **con índice de pasos** *f* ÓPT, TELECOM step index fiber (*AmE*), step index fibre (*BrE*); ~ **de índice uniforme** *f* ÓPT, TELECOM uniform-index fiber (*AmE*), uniform-index fibre (*BrE*); ~ **de lanzamiento** *f* ÓPT, TELECOM launching fiber (*AmE*), launching fibre (*BrE*); ~ **mineral** *f* TERMOTEC mineral fiber (*AmE*), mineral fibre (*BrE*); ~ **monomodo** *f* ÓPT, TELECOM monomode fiber (*AmE*), monomode fibre (*BrE*), single-mode fiber (*AmE*), single-mode fibre (*BrE*); ~ **multimodo** *f* FÍS, ÓPT, TELECOM multimode fiber (*AmE*), multimode fibre (*BrE*); ~ **natural** *f* TEXTIL natural fiber (*AmE*), natural fibre (*BrE*); ~ **neutra** *f* CONST neutral axis; ~ **óptica** *f* C&V, FÍS, INFORM&PD, ING ELÉC, ÓPT, TEC ESP fiber-optic (*AmE*), fibre-optic (*BrE*), ÓPT, TELECOM optical fiber (*AmE*), optical fibre (*BrE*); ~ **óptica monomodo** *f* ÓPT, TELECOM monomode optical fiber (*AmE*), monomode optical fibre (*BrE*), single-mode optical fiber (*AmE*), single-mode optical fibre (*BrE*); ~ **óptica multimodal** *f* ÓPT, TELECOM multimode optical fiber (*AmE*), multimode optical fibre (*BrE*); ~ **óptica de plástico** *f* ÓPT, TELECOM all-plastic optical fiber (*AmE*), all-plastic optical fibre (*BrE*); ~ **óptica terminal** *f* ÓPT optical fiber pigtail (*AmE*), optical fibre pigtail (*BrE*); ~ **óptica totalmente de plástico** *f* ÓPT, TELECOM all-plastic optical fiber (*AmE*), all-plastic optical fibre (*BrE*); ~ **óptica totalmente de vidrio** *f* ÓPT, TELECOM all-glass optical fiber (*AmE*), all-glass optical fibre (*BrE*); ~ **óptica de tránsito** *f* ÓPT, TELECOM transit fiber optic (*AmE*), transit fibre optic (*BrE*); ~ **óptica unimodal** *f* ING ELÉC single-mode optical fiber (*AmE*), single-mode optical fibre (*BrE*); ~ **óptica de vidrio** *f* ÓPT, TELECOM all-glass optical fiber (*AmE*), all-glass optical fibre (*BrE*); ~ **PCS** *f* (*fibra de sílice con revestimiento plástico*) ÓPT, TELECOM PCS fiber (*AmE*) (*plastic-clad silica fiber*), PCS fibre (*BrE*) (*plastic-clad silica fibre*); ~ **de pequeñas pérdidas** *f* ÓPT, TELECOM low-loss fiber (*AmE*), low-loss fibre (*BrE*); ~ **de perfil de**

índice potencial f ÓPT, TELECOM power-law index fiber (*AmE*), power-law index fibre (*BrE*); **~ de perfil de índice uniforme** f ÓPT, TELECOM uniform-index profile fiber (*AmE*), uniform-index profile fibre (*BrE*); **~ plástica** f ÓPT, TELECOM plastic fiber (*AmE*), plastic fibre (*BrE*); **~ de plástico** f ÓPT, TELECOM all-plastic fiber (*AmE*), all-plastic fibre (*BrE*), plastic fiber (*AmE*), plastic fibre (*BrE*); **~ preformada** f ÓPT, TELECOM pre-formed fiber (*AmE*), pre-formed fibre (*BrE*); **~ primaria** f C&V primary fiber (*AmE*), primary fibre (*BrE*); **~ de primera ventana** f TELECOM first-window fiber (*AmE*), first-window fibre (*BrE*); **~ pristina** f C&V pristine fiber (*AmE*), pristine fibre (*BrE*); **~ protegida** f ÓPT, TELECOM buffered fiber (*AmE*), buffered fibre (*BrE*); **~ para recubrir** f TEXTIL wrapper fiber (*AmE*), wrapper fibre (*BrE*); **~ relajada** f TEXTIL relaxed fiber (*AmE*), relaxed fibre (*BrE*); **~ rizada** f TEXTIL crimped fiber (*AmE*), crimped fibre (*BrE*); **~ de sección decreciente** f ÓPT, TELECOM tapered fiber (*AmE*), tapered fibre (*BrE*); **~ de segunda ventana** f ÓPT, TELECOM second window fiber (*AmE*), second window fibre (*BrE*); **~ de silicato de aluminio** f TERMOTEC aluminium silicate fibre (*BrE*), aluminum silicate fiber (*AmE*); **~ de sílice** f ÓPT, TELECOM silica fiber (*AmE*), silica fibre (*BrE*), all-silica fiber (*AmE*), all-silica fibre (*BrE*), TERMOTEC silica fiber (*AmE*), silica fibre (*BrE*); **~ de sílice dopada** f ÓPT, TELECOM doped silica fiber (*AmE*), doped silica fibre (*BrE*); **~ de sílice con revestimiento plástico** f (*fibra PCS*) ÓPT, TELECOM plastic-clad silica fiber (*AmE*) (*PCS fiber*), plastic-clad silica fibre (*BrE*) (*PCS fibre*); **~ sintética** f TERMOTEC, TEXTIL synthetic fiber (*AmE*), synthetic fibre (*BrE*); **~ soplada** f TELECOM blown fiber (*AmE*), blown fibre (*BrE*); **~ terminal** f ÓPT fiber pigtail (*AmE*), fibre pigtail (*BrE*); **~ de termoadhesión** f TEXTIL thermobonding fiber (*AmE*), thermobonding fibre (*BrE*); **~ texturizada** f C&V textured yarn; **~ de torsión** f ING ELÉC torsion string; **~ unimodo** f ÓPT, TELECOM monomode fiber (*AmE*), monomode fibre (*BrE*); **~ de variación parabólica de índice** f FÍS, ÓPT, TELECOM parabolic index fiber (*AmE*), parabolic index fibre (*BrE*); **~ de ventana doble** f ÓPT double window fiber (*AmE*), double window fibre (*BrE*), dual-window fiber (*AmE*), dual-window fibre (*BrE*), TELECOM double window fiber (*AmE*), double window fibre (*BrE*); **~ de vidrio** f C&V, CONST, ELEC, EMB, INFORM&PD, ING ELÉC, ÓPT, P&C, PROD, REFRIG, TELECOM, TERMOTEC, TEXTIL, TRANSP MAR, VEH fiberglass (*AmE*), fibreglass (*BrE*), glass-fiber (*AmE*), glass-fibre (*BrE*); **~ de vidrio compuesta** f ÓPT compound glass fiber (*AmE*), compound glass fibre (*BrE*); **~ de vidrio compuesto** f ÓPT compound glass fiber (*AmE*), compound glass fibre (*BrE*); **~ de vidrio de multicomponentes** f ÓPT multicomponent glass fiber (*AmE*), multicomponent glass fibre (*BrE*); **~ de vidrio para núcleo grande** f ING ELÉC large-core glass fiber (*AmE*), large-core glass fibre (*BrE*); **~ de vidrio con revestimiento plástico** f ÓPT plastic-clad glass fiber (*AmE*), plastic-clad glass fibre (*BrE*); **~ de vidrio textil** f C&V, TEXTIL textile glass fiber (*AmE*), textile glass fibre (*BrE*); **~ virgen** f PAPEL virgin fiber (*AmE*), virgin fibre (*BrE*); **~ de viscosa** f PROD viscose fiber (*AmE*), viscose fibre (*BrE*)

fibras: ~ de algodón f pl AGRIC lint; **~ muy cortas** f pl PAPEL fines; **~ proteínicas** f pl TEXTIL protein fibers (*AmE*), protein fibres (*BrE*); **~ de resistencia** f pl ING ELÉC resistor string; **~ secundarias** f pl PAPEL *papelote* secondary fibers (*AmE*), secondary fibres (*BrE*)

fibravidrio f GEN fiberglass (*AmE*), fibreglass (*BrE*), glass-fiber (*AmE*), glass-fibre (*BrE*)

fibrilación f PAPEL fibrillating

fibrocemento m CONST asbestos cement

fibroína f QUÍMICA fibroin

fibroóptica f GEN fiber-optics (*AmE*), fibre-optics (*BrE*), optical fiber (*AmE*), optical fibre (*BrE*)

ficha: ~ de existencias f PROD stock record card; **~ Hollerith** f INFORM&PD Hollerith card; **~ de inspección** f PROD inspection card; **~ de mantenimiento** f PROD maintenance data card; **~ de ocho columnas** f INFORM&PD eight-column card; **~ perforada** f INFORM&PD punch card (*AmE*), punched card (*BrE*); **~ de procesamiento** f TELECOM processing card; **~ técnica** f VEH data sheet, specifications sheet; **~ telefónica** f TELECOM telephone token

fichero m IMPR, INFORM&PD, PROD, TELECOM file; **~ de datos** m INFORM&PD data file; **~ físico** m INFORM&PD physical file; **~ de imagen** m ELECTRÓN image file; **~ de impresión** m IMPR print file; **~ indexado** m INFORM&PD indexed file; **~ lógico** m INFORM&PD logical file; **~ maestro** m INFORM&PD master file; **~ memorizado** m PROD unload file; **~ de programa** m INFORM&PD program file; **~ de salida** m IMPR, INFORM&PD output file; **~ secuencial** m INFORM&PD sequential file; **~ secuencial indexado** m INFORM&PD indexed sequential file; **~ de transacciones** m INFORM&PD movement file

ficina f QUÍMICA ficin

ficología f OCEAN phycology

fidelidad f (*FI*) ELECTRÓN fidelity (*FI*)

fiebre: ~ aftosa f AGRIC aphthous fever, foot-and-mouth disease; **~ de embarque** f AGRIC shipping fever; **~ de la leche** f AGRIC milk fever; **~ de Malta** f AGRIC Bang's disease, brucellosis, undulant fever

fiel m ING MECÁ tongue, METR *aguja* pointer; **~ divisor** m ING MECÁ index

fieltro m PAPEL, TEXTIL felt; **~ abrillantador** m PAPEL calendering felt; **~ agujado** m PAPEL needled felt; **~ alquitranado** m CONST tarred felt; **~ arrancador** m PAPEL lick-up; **~ de cartón** m PAPEL board felt; **~ del cilindro aspirante** m PAPEL suction roll felt; **~ de fibra de carbono** m TEC ESP carbon fiber felt (*AmE*), carbon fibre felt (*BrE*); **~ de fibra continua** m P&C continuous strand mat; **~ de fibra recortada** m P&C chopped strand mat; **~ húmedo** m PAPEL press felt, wet felt; **~ impermeable** m CONST roofing felt; **~ marcador** m PAPEL marking felt; **~ de prensa** m PAPEL press felt, *material sintético* press fabric; **~ de la prensa invertida** m PAPEL reverse-press felt; **~ secador** m PAPEL dryer felt; **~ separador superior** m PAPEL lick-up overfelt

fierro m *AmL* (*cf hierro Esp*) CARBÓN, GEOL, METAL iron

figura: ~ alargada f GEOM oblong, rectangle; **~ de corrosión** f CRISTAL, METAL etch pit; **~ de difracción** f CRISTAL, METAL, ÓPT diffraction pattern; **~ de difracción de campo lejano** f ÓPT, TELECOM far-field diffraction pattern; **~ de difracción de campo**

próximo *f* ÓPT, TELECOM near-field diffraction pattern; ~ **de interferencia** *f* CRISTAL, ÓPT interference figure; ~ **de Lichtenberg** *f* FÍS, FÍS ONDAS Lichtenberg figure; ~ **ocho** *f* GEOM pretzel; ~ **polar** *f* CRISTAL pole figure

figuras: ~ **de Bitter** *f pl* FÍS Bitter pattern; ~ **de Lissajous** *f pl* ELECTRÓN, FÍS Lissajous figures; ~ **planas** *f pl* GEOM plane figures; ~ **semejantes** *f pl* GEOM similar figures

fijación *f* CARBÓN martingale, CONST lock, *armadura de hormigón* fastening, ELECTRÓN clamping, ING MECÁ fixing, clamping, trunnion, locking, fastening, MECÁ fastening, clamping, METAL *mecanización* pinning, binding, locking, PROD fastening, clamping, TRANSP AÉR holding, setting, TV clamping; ~ **del altímetro** *f* FÍS, GEOFÍS, TEC ESP, TRANSP AÉR altimeter setting; ~ **de apertura del tren de aterrizaje** *f* TRANSP AÉR landing gear downlock; ~ **de aproximación** *f* TRANSP AÉR approach fix; ~ **por calor** *f* TEXTIL *sintéticos, en tintura* heat setting; ~ **de carteles o anuncios** *f* PROD posting; ~ **del color** *f* TV color lock (*AmE*), colour lock (*BrE*); ~ **de encadenados** *f* TV slavelock; ~ **de estructura** *f* CONST frame fixing; ~ **de fase** *f* TV phase locking; ~ **de función** *f* ELECTRÓN mode locking; ~ **hidráulica** *f* ING MECÁ hydraulic clamping; ~ **de imagen** *f* TV picture lock; ~ **de la imagen** *f* CINEMAT image steadiness, TELECOM picture lock; ~ **de imágenes** *f* TV pixlock; ~ **de inclinación** *f* TV tip engagement, tip penetration; ~ **de intervalo** *f* TV gap setting; ~ **lateral** *f* TV side lock; ~ **del número de piezas de una serie** *f* PROD lot sizing; ~ **a perno simple** *f* PROD single-bolt clamping; ~ **de la posición** *f* D&A plotting, position-fixing; ~ **del punto medio** *f* FERRO *infraestructura* midpoint anchor; ~ **a rosca** *f* ING MECÁ screw fixing; ~ **del umbral** *f* TEC ESP thresholding; ~ **del umbral posterior** *f* TV back-porch clamping

fijador: ~ **de botones** *m* PROD button-fastener; ~ **de cabeza moleteada** *m* ING MECÁ knurled-head fastener; ~ **de rosca externa** *m* ING MECÁ external-threaded fastener; ~ **de rosca interior** *m* ING MECÁ internal thread fastener; ~ **roscado** *m* ING MECÁ threaded fastener; ~ **roscado para madera** *m* ING MECÁ coach bolt

fijar *vt* CONST fix, secure, set, FÍS ONDAS focus, INFORM&PD *variable* set, ING MECÁ fix, lock, tighten, set, MECÁ fasten, apply, clamp, PROD clamp, grip, TELECOM *plazos* schedule, TRANSP MAR fit; ~ **por calor** *vt* C&V, MECÁ *esmaltes, colores* anneal; ~ **con calor en seco** *vt* TEXTIL dry-heat-set; ~ **al muro** *vt* MINAS wall in; ~ **por presión** *vt* MECÁ crimp; ~ **con vapor** *vt* TEXTIL steam set

fijo *adj* ING MECÁ, MECÁ fixed

fil: **a ~ de popa** *adv* TRANSP MAR dead astern; **a ~ de roda** *adv* TRANSP MAR dead ahead, fine on the bow

fila *f* IMPR file, INFORM&PD row, ING MECÁ line, MATEMÁT *matriz*, TEC ESP array, TRANSP line (*AmE*), queue (*BrE*); ~ **de bastidores** *f* TELECOM suite; ~ **de coches** *f* AUTO, TRANSP, VEH line of cars (*AmE*), line of traffic (*AmE*), queue of cars (*BrE*), queue of traffic (*BrE*); ~ **de entrada** *f* INFORM&PD entry queue; ~ **en espera** *f* INFORM&PD queue; ~ **de espera del dispositivo** *f* INFORM&PD device queue; ~ **de ficha** *f* INFORM&PD bit position; ~ **de paneles**

conmutadores *f* TELECOM suite of switchboards; ~ **reticular** *f* CRISTAL lattice row

filamento *m* C&V strand, yarn, CONST, ELEC filament, FÍS filament, strand, ING ELÉC *tubo electrónico* heater, filament, INSTR filament, MINAS *mineral* thread, P&C *de tejidos, refuerzo* filament, PROD thread, TEC ESP strand, TEXTIL filament; ~ **de amianto** *m AmL* (*cf filamento de asbesto Esp*) CONST asbestos thread; ~ **de asbesto** *m Esp* (*cf filamento de amianto AmL*) CONST asbestos thread; ~ **continuo** *m* C&V continuous filament; ~ **continuo de vidrio** *m* C&V glass continuous filament yarn; ~ **en doble espiral** *m* ELEC *bombilla* coiled coil filament; ~ **doblemente arrollado** *m* ELEC *bombilla* coiled coil filament; ~ **fino** *m* ELECTRÓN *diodos de contacto de punta* whisker; ~ **metálico** *m* ING ELÉC metal filament; ~ **de tubo electrónico** *m* ELECTRÓN electron-tube heater; ~ **de tungsteno** *m* ELEC tungsten filament; ~ **de tungsteno toriado** *m* ING ELÉC thoriated tungsten filament; ~ **de wolframio incandescente** *m* FÍS RAD glowing tungsten filament

filar[1]: ~ **cadena de ancla** *fra* TRANSP MAR slip one's cable

filar[2] *vt* TRANSP MAR *cables* pay out, *cadenas, cuerdas* slip; ~ **poco a poco** *vt* TRANSP MAR *cadena* check

fileta *f* TEXTIL *tejido* creel; ~ **de almacenamiento** *f* TEXTIL *de canillas* magazine creel; ~ **del urdidor** *f* TEXTIL warping creel

filetado *adj* ING MECÁ chasing

filete *m* CONST *arquitectura* fillet, FÍS filament, IMPR rule, ING MECÁ screw, thread, worm; ~ **adornado** *m* IMPR ornamental rule; ~ **de alisar** *m* PROD *funderías* fillet slick; ~ **cuadrado** *m* ING MECÁ square thread; ~ **externo** *m* ING MECÁ external thread; ~ **extrafino** *m* IMPR hairline, hairline space; ~ **extrafino doble** *m* IMPR double hairline; ~ **del husillo aplicado** *m* C&V applied thread; ~ **puntillado** *m* IMPR leaders; ~ **de roscar** *m* ING MECÁ screw thread; ~ **de tornillo de una pulgada** *m* ING MECÁ inch screw thread

fileteado *m* ING MECÁ screw cutting, screwing, screwing-up

fileteadora *f* OCEAN filleting machine

filetear *vt* ING MECÁ chase

filetes: ~ **exteriores** *m pl* ING MECÁ external threads

filícico *adj* QUÍMICA filicic

filigrana *f* CINEMAT watermark, PAPEL rubbermark, watermark

filipsita *f* MINERAL phillipsite

filita *f* GEOL, PETROL phyllite

filler *m* CONST *granulometría de carretera* filler metal

filmación: ~ **fotograma a fotograma** *f* CINEMAT single-frame filming

filmador *m* CINEMAT, TV cameraman

filo[1]: **sin ~** *adj* MECÁ dull; **de ~ cortante** *adj* ING MECÁ sharp-edged

filo[2] *m* C&V arris, CONST *corte* bit, bite, ING MECÁ edge, lip, sharpness, wire edge, METEO *del viento* eye; ~ **cortante** *m* CONST cutting edge, keen edge, ING MECÁ, MECÁ cutting edge; ~ **normal al eje** *m* MECÁ *herramientas* flat; ~ **recto** *m* ING MECÁ straight edge; ~ **de la reja** *m* AGRIC share throat

filocho *m* PAPEL breaker

filón[1] *m* CARBÓN bed, *minas* coal seam, layer, GEOL layer, MINAS lode, country rock, seam, ore feeder, coal seam, *mineral* lead; ~ **agotado** *m* MINAS exhausted vein; ~ **alargado horizontalmente** *m*

MINAS course of ore; ~ **atravesado** *m* MINAS cross vein; ~ **de carbón** *m* CARBÓN, GEOL, MINAS coal measure; ~ **crucero** *m* AmL (*cf filón transversal Esp*) GEOL, MINAS contralode, counter, counterlode, cross lode, cross vein; ~ **de cuarzo** *m* GEOL, MINAS, MINERAL quartz vein; ~ **de cuarzo aurífero** *m* MINAS gold reef, reef; ~ **en escalones** *m* MINAS lob; ~ **explotado** *m* MINAS worked-out lode; ~ **de fuerte pendiente** *m* MINAS steep vein; ~ **gaseoso** *m* GAS gaseous vein; ~ **madre** *m* MINAS mother lode; ~ **de minerales** *m* CARBÓN stratum; ~ **principal** *m* MINAS main lode, master lode; ~ **de roca** *m* MINAS jack leg; ~ **para sostener una apea** *m* MINAS hitch; ~ **transversal** *m* Esp (*cf filón crucero AmL*) GEOL, MINAS contralode, counter, counterlode, cross lode, cross vein; ~ **tubular** *m* MINAS gold reef, reef

filón² : **estar en el** ~ *vi* MINAS *mineral* be in ore

filonita *f* GEOL phyllonite

filtrabilidad *f* CARBÓN filterability

filtración *f* AGUA filtration, seepage, CARBÓN filtration, CONST filtering, filtration, leak, CONTAM leaching, ELECTRÓN, GEOL filtering, HIDROL leach, INFORM&PD, PROC QUÍ filtering, PROD filtration, QUÍMICA straining, filtering, filtration, seepage, percolation, RECICL, TEC PETR filtration, TELECOM filtering; ~ **acelerada** *f* ALIMENT, PROC QUÍ accelerated filtration; ~ **por adaptiva** *f* TELECOM adaptive filtering; ~ **por aspiración** *f* PROC QUÍ suction filtration; ~ **de banda estrecha** *f* ELECTRÓN narrow-band filtering; ~ **centrífuga** *f* CARBÓN, ING MECÁ, PROC QUÍ centrifugal filtration; ~ **cronovariable** *f* GEOFÍS *procesado sísmico* time-variable filtering; ~ **de datos discontinuos** *f* ELECTRÓN sampled-data filtering; ~ **equilibrada** *f* ELECTRÓN matched filtering; ~ **fuera de banda** *f* ELECTRÓN out-of-band filtering; ~ **de interferencia electromagnética** *f* ELEC, ELECTRÓN, ING ELÉC electromagnetic interference filtering; ~ **Kalman** *f* ELECTRÓN Kalman filtering; ~ **lenta de arena** *f* AGUA slow sand filtration; ~ **de línea de retorno** *f* PROD return-line filtration; ~ **multicapa** *f* AGUA multilayer filtration; ~ **de paso bajo** *f* ACÚST, ELEC, ELECTRÓN, FÍS, INFORM&PD, PETROL, TELECOM low-pass filtering; ~ **de paso de banda** *f* ACÚST, ELEC, ELECTRÓN band-pass filtering, low-pass filtering, FÍS band-pass filtering, INFORM&PD, PETROL, TELECOM band-pass filtering, low-pass filtering; ~ **primaria** *f* CARBÓN coarse filtration; ~ **por proyección** *f* CARBÓN spraying screen; ~ **al vacío** *f* ALIMENT, LAB, PROC QUÍ vacuum filtration; ~ **variable en el tiempo** *f* GEOFÍS *procesado sísmico* time-variable filtering

filtrado *m* CARBÓN filtrate, straining work, CONST filtering, ELECTRÓN masking, filtering, GEOL, INFORM&PD filtering, PROC QUÍ filtering, filtrate, QUÍMICA filtrate, filtering, RECICL filtrate, TEC ESP screening, TELECOM filtering; ~ **activo** *m* ELECTRÓN active filtering; ~ **adaptivo** *m* INFORM&PD adaptive filtering; ~ **alineal** *m* TELECOM nonlinear filtering; ~ **analógico** *m* ELECTRÓN analog filtering; ~ **anti-ajeno** *m* ELECTRÓN anti-aliasing filtering; ~ **de banda ancha** *m* ELECTRÓN, TELECOM wideband filtering; ~ **de capacidad total** *m* ELECTRÓN all-capacitor filtering; ~ **digital** *m* ELECTRÓN, INFORM&PD, TELECOM digital filtering; ~ **Doppler** *m* ELECTRÓN Doppler filtering; ~ **de eliminación de banda** *m* ELECTRÓN band-stop filtering; ~ **de entrada** *m* ELECTRÓN, TELECOM input filtering; ~ **en espira** *m* ELECTRÓN convolutional filtering; ~ **de ganancia permanente** *m* ELECTRÓN fixed-gain filtering; ~ **lineal** *m* ELECTRÓN, TELECOM linear filtering; ~ **de lodo** *m* PETROL mud filtrate; ~ **multidimensional** *m* TELECOM multidimensional filtering; ~ **no lineal** *m* ELECTRÓN, TELECOM nonlinear filtering; ~ **de OAS** *m* ELECTRÓN SAW filtering; ~ **de pasabanda** *m* ELECTRÓN, FÍS, PETROL, TELECOM band-pass filtering; ~ **de paso alto** *m* ELECTRÓN, TELECOM high-pass filtering; ~ **de paso de banda** *m* ELECTRÓN, FÍS, PETROL, TELECOM band-pass filtering; ~ **de paso banda estrecha** *m* ELECTRÓN, TELECOM narrow-band low-pass filtering; ~ **de peine** *m* ELECTRÓN, TELECOM comb filtering; ~ **primario** *m* CARBÓN prescreening; ~ **recursivo** *m* ELECTRÓN, INFORM&PD, TELECOM recursive filtering; ~ **transversal** *m* ELECTRÓN transversal filtering; ~ **al vacío** *m* CARBÓN vacuum filtration; ~ **en vacío** *m* ALIMENT, LAB, PROC QUÍ vacuum filtration

filtraje *m* ELEC, ING ELÉC smoothing; ~ **multidimensional** *m* TELECOM multidimensional filtering

filtrar¹ *vt* GEN filter, PROC QUÍ filter, filter out, QUÍMICA separate out, strain, filter, RECICL drain, filter

filtrar² *vti* CONST leak, HIDROL leach

filtrarse *v refl* HIDROL leach, QUÍMICA seep, separate out

filtro *m* GEN filter, CARBÓN filter, screen, GAS filter, strainer, GEOL, HIDROL screen, MECÁ filter, screen, TEC ESP screen, TEC PETR *refino, sísmica* filter, strainer; **▸a** ~ **absorbente de calor** *m* FOTO heat-absorbing filter; ~ **de absorción** *m* CINEMAT, INSTR absorption filter; ~ **abstractivo de rayos infrarrojos** *m* INSTR red-abstracting filter; ~ **de aceite** *m* AUTO, ING MECÁ, MECÁ, VEH oil filter; ~ **de aceite de capacidad total** *m* VEH full-flow oil filter; ~ **de aceite en derivación** *m* AUTO bypass-oil cleaner; ~ **activo** *m* ELECTRÓN, TELECOM active filter; ~ **activo de tercer orden** *m* ELECTRÓN third-order active filter; ~ **acústico** *m* ACÚST, ELECTRÓN, ING MECÁ acoustic filter; ~ **adaptable digital** *m* ELECTRÓN digital matched filter; ~ **adaptado** *m* ELECTRÓN matched filter; ~ **adaptivo** *m* ELECTRÓN adaptive filter; ~ **de agua** *m* AGUA, TRANSP MAR water filter; ~ **para agua calcárea** *m* AGUA hard water filter; ~ **de aire** *m* AUTO air cleaner, CARBÓN air cleaner, air filter, ING MECÁ air cleaner, MECÁ air filter, MINAS air scrubber, PROC QUÍ, TERMOTEC air filter, VEH air cleaner, air filter; ~ **de aire autolimpiable** *m* TERMOTEC self-cleaning air filter; ~ **de aire en baño de aceite** *m* AUTO oil bath air cleaner, oil bath air filter; ~ **de aire electrostático** *m* ELEC, SEG, TERMOTEC electrostatic air filter; ~ **de aire húmedo** *m* TERMOTEC wet-air filter; ~ **de aire limpiable** *m* CONTAM, TERMOTEC cleanable air filter; ~ **de aire para partículas submicrónicas** *m* SEG submicron particulate air filter; ~ **de aire seco** *m* TERMOTEC dry-air filter; ~ **de aislamiento** *m* ELECTRÓN isolation filter; ~ **ajustado** *m* ELECTRÓN tuned filter; ~ **de alimentación de energía** *m* ELECTRÓN power-supply filter; ~ **de alineación** *m* ELECTRÓN tracking filter; ~ **de alta precisión** *m* ING MECÁ high-stop filter; ~ **de alto filtrado** *m* ING MECÁ high-stop filter; ~ **analógico** *m* ELECTRÓN analog filter; ~ **de ancho de banda**

ajustable *m* ACÚST adjustable bandwidth filter; ~ **anti-ajeno** *m* ELECTRÓN anti-aliasing filter; ~ **anti-ajeno de tres ceros** *m* ELECTRÓN three-zeros anti-aliasing filter; ~ **antiparasitario** *m* ELECTRÓN line filter; ~ **antipolvo** *m* GAS antidust filter; ~ **de arena** *m* AGUA, CARBÓN, HIDROL sand filter; ~ **de arena lento** *m* AGUA, HIDROL slow sand filter; ~ **de armónicos** *m* ELECTRÓN, TEC ESP, TELECOM harmonic filter; ~ **de aspiración** *m* AGUA suction filter, REFRIG suction strainer; ~ **por aspiración** *m* PROD suction filter;

~ b ~ **de baja presión** *m* PROD low-pressure filter; ~ **de banda ancha** *m* ELECTRÓN, TELECOM wideband filter; ~ **de banda eliminada** *m* ELECTRÓN, INFORM&PD, TELECOM notch filter; ~ **de banda estrecha** *m* ELECTRÓN, TELECOM narrow-band filter; ~ **de banda lateral inferior** *m* ELECTRÓN, TELECOM lower-sideband filter; ~ **de banda lateral residual** *m* ELECTRÓN, TELECOM vestigial sideband filter; ~ **de banda de octava** *m* ELECTRÓN octave band filter; ~ **de banda pasante** *m* ELECTRÓN, TELECOM bandpass filter; ~ **de barra libre** *m* ELECTRÓN free bar filter; ~ **de bifurcación** *m* ELECTRÓN *en programa* branching filter; ~ **biológico** *m* CONTAM biological filter, HIDROL, PROC QUÍ biofilter, biological filter; ~ **Butterworth** *m* ELECTRÓN, FÍS Butterworth filter;

~ c ~ **de canal** *m* ELECTRÓN channel filter; ~ **de canal de ondas** *m* ELECTRÓN waveguide filter; ~ **de cañamazo** *m* *AmL* (*cf filtro de lona Esp*) MINAS *amalgama* canvas filter; ~ **de carbón** *m* LAB charcoal filter; ~ **de carbón activado** *m* HIDROL, NUCL activated carbon filter; ~ **de carbón bruto** *m* CARBÓN, MINAS raw-coal screen; ~ **de cartucho** *m* ING MECÁ, VEH *lubricación* cartridge filter; ~ **de Cauer** *m* FÍS Cauer filter; ~ **celular de cristal** *m* CRISTAL, ELECTRÓN crystal ladder filter; ~ **centrífugo** *m* CARBÓN, ING MECÁ, PROC QUÍ centrifugal filter; ~ **centrípeto** *m* ING MECÁ centripetal filter; ~ **de Chebyshev** *m* ELECTRÓN, FÍS Chebyshev filter; ~ **Chebyshev en octava posición** *m* ELECTRÓN, FÍS eighth-order Chebyshev filter; ~ **de CL** *m* ELECTRÓN LC filter; ~ **coaxial** *m* ELECTRÓN coaxial filter; ~ **del combustible** *m* AUTO, TRANSP AÉR fuel filter; ~ **de combustible de baja presión** *m* TRANSP AÉR low-pressure fuel filter; ~ **de combustible de dos etapas** *m* AUTO two-stage fuel filter; ~ **de compensación de colores** *m* CINEMAT, TV color compensating filter (*AmE*), colour compensating filter (*BrE*); ~ **compensador** *m* FOTO compensating filter; ~ **compensador de color** *m* CINEMAT, TV color compensating filter (*AmE*), colour compensating filter (*BrE*); ~ **de compresión** *m* ELECTRÓN compression filter; ~ **de compresión de OAS** *m* ELECTRÓN SAW compression filter; ~ **de comunicaciones** *m* ELECTRÓN communications filter; ~ **de conformación de la señal** *m* TELECOM signal-shaping filter; ~ **constante de corta duración** *m* PROD short time constant filter; ~ **de conversión** *m* CINEMAT conversion filter; ~ **de conversión de color** *m* CINEMAT, TV color-conversion filter (*AmE*), colour-conversion filter (*BrE*); ~ **de correa móvil** *m* NUCL traveling belt filter (*AmE*) (*TBF*), travelling belt filter (*BrE*) (*TBF*); ~ **de corrección** *m* FOTO correction filter; ~ **de corrección del color** *m* CINEMAT, ELECTRÓN, TV color-correction filter (*AmE*), colour-correction filter (*BrE*); ~ **corrector de color** *m* CINEMAT, ELECTRÓN,

TV color-correction filter (*AmE*), colour-correction filter (*BrE*); ~ **de corte rápido** *m* ELECTRÓN sharpcutoff filter; ~ **de cristal** *m* CRISTAL, ELECTRÓN, TELECOM crystal filter; ~ **de cristal de cuarzo** *m* ELEC, ELECTRÓN quartz-crystal filter; ~ **de cristal gris** *m* INSTR gray glass filter (*AmE*), grey glass filter (*BrE*);

~ d ~ **DAC** *m* ELECTRÓN CCD filter; ~ **decapolar** *m* ELECTRÓN ten-pole filter; ~ **de densidad neutra** *m* CINEMAT, FOTO, IMPR neutral-density filter; ~ **dentro del objetivo** *m* CINEMAT between-the-lens filter; ~ **de depulpación** *m* CARBÓN depulping screen; ~ **de desacoplamiento** *m* ELECTRÓN decoupling filter; ~ **desecador** *m* PROC QUÍ dryer filter; ~ **desenlodador** *m* CARBÓN desliming screen; ~ **de detención de banda** *m* ELECTRÓN, TELECOM bandstop filter; ~ **dicroico** *m* CINEMAT, ELECTRÓN, ÓPT, TELECOM dichroic filter; ~ **de diez polos** *m* ELECTRÓN ten-pole filter; ~ **difusor** *m* CINEMAT frost; ~ **digital** *m* ELECTRÓN, TELECOM digital filter; ~ **digital de respuesta a impulsos infinitos** *m* ELECTRÓN, TELECOM IIR digital filter, infinite-impulse-response digital filter; ~ **direccional** *m* ELECTRÓN directional filter; ~ **discreto** *m* ELECTRÓN discrete filter; ~ **distribuidor** *m* CARBÓN disc filter (*BrE*), disk filter (*AmE*); ~ **doble** *m* PROD duplex filter; ~ **de doble sintonización** *m* ELECTRÓN double-tuned filter; ~ **Doppler** *m* ELECTRÓN Doppler filter;

~ e ~ **de efecto de difusión** *m* CINEMAT diffusion effect filter; ~ **de efectos** *m* CINEMAT effect filter; ~ **eléctrico** *m* ELEC electrical filter, ELECTRÓN electrical filter, *nivelación* smoothing filter, ING ELÉC electrical filter; ~ **electromecánico** *m* ELEC, ELECTRÓN, ING ELÉC electromechanical filter; ~ **electrostático** *m* ELEC, ELECTRÓN electrostatic filter; ~ **de eliminación de banda** *m* ELECTRÓN, INFORM&PD, TELECOM band-stop filter; ~ **de eliminación de banda estrecha** *m* ELECTRÓN narrowband rejection filter; ~ **eliminador de banda** *m* ELECTRÓN, INFORM&PD, TELECOM band-stop filter; ~ **eliminador de banda activo** *m* ELECTRÓN active band-stop filter; ~ **eliminador de banda pasivo** *m* ELECTRÓN passive band-stop filter; ~ **eliminador de banda de segundo orden** *m* ELECTRÓN second-order band-stop filter; ~ **eliminador de interferencias** *m* ELEC, ELECTRÓN, FÍS, ÓPT, TELECOM interference filter; ~ **elíptico** *m* ELECTRÓN elliptic filter; ~ **de energía electrónica** *m* ELECTRÓN, FÍS PART, FÍS RAD electronenergy filter; ~ **enroscable** *m* CINEMAT screw in filter; ~ **de entalla** *m* ELECTRÓN, INFORM&PD, TELECOM notch filter; ~ **entallado activo** *m* ELECTRÓN active notch filter; ~ **de entrada** *m* ELECTRÓN input filter, PROD inlet strainer, TEC ESP input filter; ~ **de entrada inductiva** *m* ELEC, ELECTRÓN, ING ELÉC choke input filter; ~ **entre lentes** *m* CINEMAT between-the-lens filter; ~ **de escalera** *m* ELECTRÓN ladder filter; ~ **en escalera RC** *m* ELECTRÓN, ING ELÉC RC ladder filter; ~ **de escurrimiento** *m* CONTAM *tratamiento de líquidos* trickling filter; ~ **de espejo en cuadratura** *m* ELECTRÓN, TELECOM quadrature mirror filter; ~ **en espira** *m* ELECTRÓN convolutional filter; ~ **estático** *m* ELECTRÓN, TELECOM passive filter, passive filtering; ~ **de estrella** *m* CINEMAT star filter; ~ **de expansión** *m* ELECTRÓN expansion filter; ~ **de expansión de OAS** *m* ELECTRÓN SAW expansion filter;

~ f ~ **de FI** *m* ELECTRÓN, TELECOM IF filter; ~ **para**

fluctuaciones *m* CONST, FÍS RAD ripple filter; **~ de formación de señal** *m* ELECTRÓN, TELECOM signal-shaping filter; **~ fotoequilibrante** *m* CINEMAT light-balancing filter; **~ de frecuencia** *m* ACÚST, ELECTRÓN, FÍS, INFORM&PD, ING ELÉC, TEC ESP, TELECOM band-pass filter (*BPF*); **~ de frecuencia intermedia** *m* ELECTRÓN, TELECOM intermediate frequency filter; **~ de función** *m* ELECTRÓN mode filter;

~ g **~ de gamella** *m* CARBÓN pan filter; **~ de ganancia constante** *m* ELECTRÓN fixed-gain filter; **~ de ganancia permanente** *m* ELECTRÓN fixed-gain filter; **~ de gasolina** *m Esp* (*cf filtro de nafta AmL*) AUTO, VEH gas filter (*AmE*), gasoline filter (*AmE*), petrol filter (*BrE*); **~ de gelatina** *m* CINEMAT gelatine filter; **~ GFI** *m* ELECTRÓN YIG filter; **~ giratorio** *m* CARBÓN drum filter, FOTO swing-in filter, HIDROL drum filter; **~ de goteo** *m* CONTAM trickling filter; **~ graduado** *m* CINEMAT, FOTO, TV graduated filter; **~ de gran capacidad** *m* ELECTRÓN high-order filter; **~ de grava** *m* CONST *ingeniería civil* ripple filter;

~ h **~ de hendidura** *m* TELECOM notch filter;

~ i **~ ideal** *m* ELECTRÓN ideal filter; **~ de imagen** *m* TELECOM picture filter; **~ de impulsos** *m* CONST, FÍS RAD ripple filter; **~ de impurezas** *m* GAS dross filter; **~ incorporado** *m* CINEMAT built-in filter; **~ de inductancia-capacitancia** *m* ELECTRÓN inductance-capacitance filter; **~ de información de muestra de paso bajo** *m* ELECTRÓN low-pass sampled data filter; **~ de infrarrojos** *m* FÍS RAD infrared filter; **~ integrado** *m* ELECTRÓN integrated filter; **~ de interferencia** *m* ELEC *de circuito*, ELECTRÓN, FÍS, ÓPT, TELECOM interference filter; **~ de interferencia electromagnético** *m* ELEC, ELECTRÓN, ING ELÉC electromagnetic interference filter;

~ k **~ Kalman** *m* ELECTRÓN Kalman filter;

~ l **~ de lana de vidrio** *m* C&V glass-wool filter; **~ de lente frontal** *m* INSTR, ÓPT front-lens filter; **~ de línea** *m* ELECTRÓN line filter; **~ de línea de abastecimiento** *m* P&C supply line filter; **~ lineal** *m* ELECTRÓN, TELECOM linear filter; **~ de lona** *m Esp* (*cf filtro de cañamazo AmL*) MINAS *amalgama* canvas filter; **~ longitudinal** *m* ELECTRÓN longitudinal filter; **~ de luz** *m* IMPR, INSTR light filter; **~ de luz difusa** *m* INSTR stray-light filter;

~ m **~ magnético** *m* ING MECÁ magnetic filter; **~ de malla** *m* P&C *pintura* filter screen; **~ de malla ancha** *m* AGUA coarse filter; **~ de malla de cristal** *m* CRISTAL, ELECTRÓN crystal-lattice filter; **~ de MCI** *m* ELECTRÓN PCM filter; **~ mecánico** *m* ELECTRÓN mechanical filter; **~ mecánico de aire** *m* TERMOTEC mechanical air filter; **~ de membrana** *m* LAB, PROC QUÍ membrane filter; **~ de microondas** *m* ELECTRÓN microwave filter; **~ de modos** *m* ELECTRÓN, ÓPT, TELECOM mode filter; **~ monolítico** *m* ELECTRÓN, TELECOM monolithic filter; **~ para monóxido de carbono** *m* SEG carbon monoxide filter; **~ montado** *m* FOTO mounted filter; **~ multibanda** *m* ELECTRÓN, TELECOM multiband filter; **~ multicanal** *m* ELECTRÓN, TELECOM multichannel filter; **~ multipolar** *m* ELECTRÓN multipole filter; **~ multiseccional** *m* ELECTRÓN multisection filter;

~ n **~ de nafta** *m AmL* (*cf filtro de gasolina Esp*) AUTO, VEH gas filter (*AmE*), gasoline filter (*AmE*), petrol filter (*BrE*); **~ para neblina** *m* CINEMAT haze filter; **~ neumático** *m* PROC QUÍ pneumatic filter;

~ neutro *m* FOTO, IMPR neutral-density filter; **~ de nivelación** *m* ELECTRÓN, ING ELÉC smoothing filter; **~ no recurrente** *m* ELECTRÓN, TELECOM nonrecursive filter; **~ no recursivo** *m* ELECTRÓN, TELECOM non-recursive filter;

~ o **~ de OAS** *m* ELECTRÓN SAW filter; **~ ocular** *m* SEG eye filter; **~ de onda acústica** *m* ACÚST, ELECTRÓN acoustic wave filter; **~ de ondas** *m* FÍS ONDAS wave filter; **~ de ondulación** *m* CONST, FÍS RAD ripple filter; **~ óptico** *m* ELECTRÓN, IMPR, ÓPT, TELECOM optical filter; **~ de orden impar** *m* ELECTRÓN odd-order filter; **~ en orden par** *m* ELECTRÓN even-order filter; **~ oscuro** *m* INSTR dark filter;

~ p **~ de papel sin ceniza** *m* ALIMENT ashless filter paper; **~ para partículas coloidales** *m* QUÍMICA ultrafilter; **~ pasa-alto** *m AmL* (*cf filtro de paso alto Esp*) ACÚST, ELECTRÓN, FÍS, INFORM&PD, PETROL, TEC ESP, TELECOM, TV high-pass filter; **~ pasivo** *m* ELECTRÓN, TELECOM passive filter, passive filtering; **~ de paso alto** *m Esp* (*cf filtro pasa-alto AmL*) ACÚST, ELECTRÓN, FÍS, INFORM&PD, PETROL, TEC ESP, TELECOM, TV high-pass filter; **~ de paso alto de banda ancha** *m* ELECTRÓN wideband high-pass filter; **~ de paso alto de segundo orden** *m* ELECTRÓN second-order high-pass filter; **~ de paso bajo** *m* ACÚST, ELEC low-pass filter, ELECTRÓN low-order filter, low-pass filter, FÍS, INFORM&PD, PETROL, TELECOM low-pass filter; **~ de paso bajo de banda ancha** *m* ELECTRÓN wide-band low-pass filter; **~ de paso bajo de segundo orden** *m* ELECTRÓN second-order low-pass filter; **~ de paso de banda** *m* ACÚST, ELEC, ELECTRÓN, FÍS RAD, INFORM&PD, PETROL, TELECOM band-pass filter (*BPF*); **~ paso banda** *m* ACÚST, ELEC, ELECTRÓN, FÍS RAD, INFORM&PD, PETROL, TELECOM band-pass filter (*BPF*); **~ de paso de banda activo** *m* ELECTRÓN active band-pass filter; **~ de paso banda de autorreglaje** *m* ELECTRÓN self-tracking band-pass filter; **~ de paso de banda constante** *m* ELECTRÓN flat-band pass filter; **~ de paso de banda estático** *m* ELECTRÓN passive band-pass filter; **~ de paso de banda estrecha** *m* ELECTRÓN narrow-band low-pass filter; **~ de paso de banda de GFI** *m* ELECTRÓN YIG band-pass filter; **~ paso banda de microondas** *m* ELECTRÓN microwave band-pass filter; **~ de paso de banda pasivo** *m* ELECTRÓN passive band-pass filter; **~ de paso de banda de segundo orden** *m* ELECTRÓN second-order bandpass filter; **~ del patín** *m* INSTR *pantógrafo* pan filter; **~ de peine** *m* ELECTRÓN, TELECOM comb filter; **~ percolador** *m AmL* CONTAM *tratamiento de líquidos* trickling filter; **~ perfecto** *m* ELECTRÓN ideal filter; **~ de placas** *m* CARBÓN roll screen; **~ plano óptico** *m* CINEMAT optical flat filter; **~ plano de porcelana** *m* C&V porcelain filter-plate; **~ de plantilla** *m* TEC ESP filter jig, filter template; **~ de plomo** *m* TRANSP lead filter; **~ de polarización** *m* CINEMAT, FOTO, INSTR polarization filter; **~ polarizador** *m* CINEMAT, FOTO, INSTR polarizing filter; **~ polinómico** *m* ELECTRÓN polynomial filter; **~ de polvo** *m* CARBÓN dust filter; **~ prensa** *m* ALIMENT, C&V, CARBÓN, LAB, PAPEL, PROC QUÍ filter press; **~ de presión** *m* AGUA, CARBÓN, PROC QUÍ pressure-filter; **~ primario** *m* AGUA primary filter; **~ de primer orden** *m* ELECTRÓN first-order filter; **~ de protección ocular** *m* SEG eye-protection filter; **~ pseudoelíptico** *m* TEC ESP pseudo-elliptic filter;

~ de pulsaciones *m* CONST, FÍS RAD ripple filter; **~ de purificación de agua** *m* AGUA, PROC QUÍ water purification filter;

~r **~ de realimentación inversa** *m* ELECTRÓN inverse feedback filter; **~ de recepción** *m* ELECTRÓN, TELECOM receive filter; **~ de recubrimiento programable** *m* ELECTRÓN mask-programmable filter; **~ recursivo** *m* ELECTRÓN, INFORM&PD, TELECOM recursive filter; **~ de red** *m* ELECTRÓN lattice filter; **~ de reglaje** *m* ELECTRÓN tracking filter; **~ resonador dieléctico** *m* ELECTRÓN, TEC ESP dielectric resonator filter; **~ resonador de disco** *m* ELECTRÓN disc resonator filter (*BrE*), disk resonator filter (*AmE*); **~ de resonancia** *m* ELECTRÓN, FÍS RAD resonance filter; **~ para respirador** *m* SEG respiratory filter; **~ de respuesta de impulso finito** *m* ELECTRÓN, TELECOM finite-impulse response filter (*FIR filter*); **~ de respuesta de impulso infinita** *m* ELECTRÓN IIR filter, infinite-impulse-response filter; **~ de retención de banda de microondas** *m* ELECTRÓN microwave band-stop filter; **~ de retorno** *m* PROD return filter; **~ retraíble** *m* FOTO retractable filter; **~ rotativo por succión** *m* PROD suction strainer; **~ rotatorio** *m* AGUA, CARBÓN rotary filter;

~s **~ de sacos** *m* CARBÓN bag filter; **~ secador** *m* PAPEL dryer felt; **~ de seda metálica** *m* P&C *pintura* filter screen; **~ de segundo orden** *m* ELECTRÓN second-order filter; **~ selectivo de frecuencia** *m* ELECTRÓN frequency-selective filter; **~ de señales parásitas** *m* ELECTRÓN *radar* clutter filter; **~ de separación** *m* ELECTRÓN, TV separation filter; **~ de separación de colores** *m* CINEMAT, FOTO, TV color-separation filter (*AmE*), colour-separation filter (*BrE*); **~ sintonizable de microondas** *m* ELECTRÓN microwave tunable filter; **~ sintonizable de varias octavas** *m* ELECTRÓN multioctave tunable filter; **~ de sintonización electrónica** *m* ELECTRÓN, TELECOM electronically-tuned filter; **~ de sintonización rápida** *m* ELECTRÓN fast-tuned filter; **~ de una sola banda lateral** *m* ELECTRÓN single-sideband filter; **~ de una sola sección** *m* ELECTRÓN single-section filter; **~ de sonido** *m* ELECTRÓN audio filter; **~ de succión** *m* AGUA suction filter; **~ por succión** *m* PROD suction filter; **~ de suministro de energía** *m* ELECTRÓN power-supply filter; **~ de supresión** *m* ELECTRÓN rejection filter; **~ de supresión de banda** *m* FÍS RAD band-rejection filter; **~ supresor de banda escalonada** *m* GEOFÍS *procesado sísmico* notch filter;

~t **~ de tamiz fino** *m* ING MECÁ high-stop filter; **~ de la tapa del doble fondo** *m* PROD tank top filter; **~ de tela metálica** *m* P&C, PROD filter screen; **~ de tercer orden** *m* ELECTRÓN third-order filter; **~ térmico** *m* CINEMAT heat filter; **~ tetrapolar** *m* ELEC, ELECTRÓN four-pole filter; **~ de tipo múltiple** *m* ELECTRÓN multiple filter; **~ de tipo nth** *m* ELECTRÓN nth-order filter; **~ de todo paso** *m* ELECTRÓN, TELECOM all-pass filter; **~ de tope inferior** *m* ING MECÁ low-stop filter; **~ transversal** *m* ELECTRÓN, TELECOM transversal filter; **~ de tres ceros** *m* ELECTRÓN three-zeros filter;

~u **~ ultravioleta** *m* CINEMAT, FOTO, SEG ultraviolet filter;

~v **~ de vacío** *m* CARBÓN vacuum filter; **~ variable con el tiempo** *m* ELECTRÓN time-varying filter;

~ para varios usos *m* HIDROL multimedia filter; **~ de ventana** *m* ELECTRÓN window filter; **~ de volumen de datos discontinuos** *m* ELECTRÓN sampled-data size filter;

~w **~ Wratten** *m* CINEMAT Wratten filter;

~z **~ de zumbido** *m* CONST, FÍS RAD ripple filter

fin: **~ anticipado** *m* TV early finish; **~ anticipado de audio** *m* TELECOM early-finish audio; **~ anticipado de video** *AmL*, **~ anticipado de vídeo** *m* *Esp* TV early-finish video; **~ de archivo** *m* (*EOF*) INFORM&PD end of file (*EOF*); **~ de bifurcación** *m* PROD branch end; **~ de bloque** *m* (*EOB*) INFORM&PD end of block (*EOB*); **~ del bloque de transmisión** *m* (*FBT*) INFORM&PD end of transmission block (*ETB*); **~ del carrete** *m* INFORM&PD end of reel; **~ de consulta** *m* (*ENQ*) INFORM&PD end of query (*ENQ*); **~ de datos** *m* (*EOD*) INFORM&PD end of data (*EOD*), DP end of data; **~ de dirección** *m* INFORM&PD end of address; **~ direccional** *m* INFORM&PD end of address; **~ de la emisión** *m* INSTAL HIDRÁUL cutoff; **~ de fichero** *m* (*EOF*) INFORM&PD end of file (*EOF*); **~ de inyección** *m* INSTAL HIDRÁUL cutoff; **~ del medio** *m* (*EM*) INFORM&PD end of medium (*EM*); **~ de mensaje** *m* (*EOM*) INFORM&PD, TELECOM end of message (*EOM*); **~ de movimiento** *m* (*EOT*) TELECOM end of transaction (*EOT*); **~ de plazo** *m* TV deadline; **~ de la sesión** *m* INFORM&PD log-off, log-out; **~ del soporte** *m* (*EM*) INFORM&PD end of medium (*EM*); **~ de tabulación** *m* IMPR tab stop; **~ de tarea** *m* (*EOJ*) INFORM&PD end of job (*EOJ*); **~ de texto** *m* (*FDT*) INFORM&PD end of text (*ETX*); **~ de trabajo** *m* (*EOJ*) INFORM&PD end of job (*EOJ*); **~ de tramo** *m* CONST *escaleras* nosing line; **~ de transmisión** *m* (*EOT*) INFORM&PD, TELECOM end of transmission (*EOT*); **~ del vuelo** *m* TEC ESP, TRANSP AÉR end of flight

final¹: **del ~** *adj* TEXTIL downstream

final² *m* MINAS *marina* free end; **~ del cable** *m* MINAS *marina* cable end; **~ de carrera** *m* ING MECÁ end of stroke, end of travel; **~ del carrete** *m* CINEMAT, TV reel end; **~ de pasada** *m* ING MECÁ end of travel; **~ del templador** *m* C&V end of lehr

finalización: **~ anormal** *f* INFORM&PD abnormal termination

finamente: **~ estratificado** *adj* GEOL thin-bedded, thinly-bedded

fineza *f* ENERG RENOV lift-to-drag ratio

finos *m pl* CARBÓN *decantación* overflow, *minería* tails, *polvo* undersize, *metalurgia* slimes, *pulvimetalurgia* fines, CONST *áridos* fines, MINAS *minerales* fines, *de carbón* slack, PAPEL limits; **~ de antracita** *m pl* CARBÓN duff; **~ de carbón** *m pl* CARBÓN, MINAS coal slake; **~ de coque** *m pl* CARBÓN coking duff; **~ de proa** *m pl* TRANSP MAR *diseño de barco* bow entrance

finura *f* CONST *de arena*, TRANSP MAR *construcción de barcos* fineness

fírico¹ *adj* GEOL phyric

fírico² *m* PETROL phyric

firma: **~ digital** *f* TELECOM digital signature; **~ electrónica** *f* INFORM&PD electronic signature

firme: **~ rígido** *m* CONST *carreteras* rigid construction

fisalita *f* MINERAL physalite

fisetina *f* QUÍMICA fisetin

física: **~ de alta energía** *f* FÍS high-energy physics;

~ de altas energías *f* FÍS RAD high-energy physics; **~ atómica** *f* FÍS, FÍS PART atomic physics; **~ cuántica** *f* FÍS, FÍS PART quantum physics; **~ cuaternaria** *f* FÍS, FÍS PART, NUCL quaternary physics; **~ estadística** *f* FÍS statistical physics; **~ del estado sólido** *f* FÍS PART solid-state physics; **~ experimental** *f* FÍS, FÍS PART experimental physics; **~ de los fluidos** *f* FÍS FLUID fluid physics; **~ matemática** *f* FÍS mathematical physics; **~ nuclear** *f* FÍS, FÍS PART, NUCL nuclear physics; **~ nuclear de baja energía** *f* FÍS PART, FÍS RAD, NUCL low-energy nuclear physics; **~ nuclear de energías intermedias** *f* NUCL medium-energy nuclear physics; **~ de partículas** *f* FÍS, FÍS PART particle physics; **~ de radiación** *f* FÍS RAD, NUCL radiation physics; **~ radiológica** *f* FÍS RAD, NUCL health physics; **~ sanitaria** *f* FÍS RAD, NUCL health physics

físicamente: ~ conectado *adj* INFORM&PD hard-wired
físico: ~ nuclear *m* FÍS, FÍS PART, NUCL nuclear physicist; **~ térmico** *m* FÍS, TERMO heat physicist
físil *adj* FÍS, FÍS PART, NUCL fissile
fisio *m* NUCL fissium
fisiólogo: ~ vegetal *m* AGRIC plant physiologist
fisión: ~ cuaternaria *f* FÍS, FÍS PART, NUCL quaternary fission; **~ espontánea** *f* FÍS, FÍS RAD, NUCL spontaneous fission; **~ por neutrones térmicos** *f* FÍS, FÍS PART, NUCL thermal neutron fission; **~ nuclear** *f* FÍS, FÍS PART, NUCL nuclear fission
fisionable *adj* FÍS, FÍS PART, NUCL fissile, fissionable
fisionado: no ~ *adj* FÍS, FÍS PART, NUCL unfissioned
fisioquímica *f* QUÍMICA physiochemistry
fisostigmina *f* QUÍMICA physostigmine
fisura *f* CARBÓN crack, *rocas* crevice, seam, CONST *geología* fissure, crack, *hormigón* cracking, CRISTAL cleavage, GEOFÍS *materiales* fissure, GEOL cleat, joint, MECÁ crack, MINAS *rocas* seam, OCEAN fissure, PROD *chapas de calderas* groove, *moldes, fundición* surface crack; **~ por fatiga** *f* TEC ESP fatigue crack; **~ fina** *f* ING MECÁ fine crack; **~ incipiente** *f* TRANSP AÉR incipient crack; **~ interna** *f* METAL internal crack; **~ del terreno relleno de hielo** *f* CARBÓN ice lens
fisuración *f* GEOL jointing, PROD grooving; **~ conjugada** *f* GEOL conjugate jointing; **~ en cuña** *f* METAL wedge crack; **~ por fragmentación** *f* METAL cleavage crack; **~ hidráulica** *f* ENERG RENOV hydraulic fracturing
fitasa *f* ALIMENT phytase
fitina *f* ALIMENT *repostería*, QUÍMICA phytin
fitocídico *adj* AGRIC, QUÍMICA phytocidal
fitofisiólogo *m* AGRIC plant physiologist
fitopatólogo *m* AGRIC phytopathologist, plant pathologist
fitotécnia *f* AGRIC plant production
fitotóxico *adj* AGRIC, QUÍMICA phytotoxic
flameado *m* CONST, GAS, MECÁ, PROD torching
flameo *m* TRANSP AÉR flutter
flamómetro *m* AmL (*cf fotómetro de llama Esp*) INSTR, LAB flame photometer
flanco *m* CONST flank, ELECTRÓN edge, GEOL limb, ING MECÁ flank, INSTR *de pliegues* side, MECÁ flap, flank, PROD side, TEC ESP flank; **~ anterior del impulso** *m* TELECOM pulse leading-edge; **~ de bajada** *m* ELECTRÓN falling edge, ING ELÉC trailing edge; **~ invertido** *m* GEOL overturned limb; **~ del pliegue** *m* GEOL fold limb; **~ posterior del impulso** *m* TELECOM pulse trailing-edge; **~ posterior de la pista de video** *AmL*, **~ posterior de la pista de vídeo** *m Esp* TV trailing-edge video track; **~ posterior de un pulso** *m* TV trailing edge; **~ de subida** *m* ELECTRÓN low-to-high transition, falling rising

flanja *f* ING MECÁ flange; **~ de extrusión** *f* ING MECÁ extrusion flange

flap *m* TEC ESP, TRANSP, TRANSP AÉR flap; **~ alar** *m* TRANSP AÉR wing flap; **~ de aterrizaje** *m* TRANSP, TRANSP AÉR landing flap; **~-borde de ataque** *m* TRANSP, TRANSP AÉR leading-edge flap; **~ del borde de salida** *m* TRANSP, TRANSP AÉR trailing-edge flap; **~ desacelerador** *m* TRANSP, TRANSP AÉR speed brake; **~ doble** *m* TRANSP, TRANSP AÉR split flap; **~ de inclinación** *m* TRANSP, TRANSP AÉR droop flap; **~ de intradós** *m* TRANSP, TRANSP AÉR split flap; **~ ranurado** *m* TRANSP, TRANSP AÉR slot flap; **~ de refrigeración** *m* TRANSP, TRANSP AÉR cooling flap; **~ de resistencia al arrastre** *m* TRANSP, TRANSP AÉR drag brake; **~ soplado** *m* TRANSP, TRANSP AÉR blown flap

flash *m* FOTO, IMPR flash; **~ anular** *m* FOTO ring flash; **~ de cubo** *m* FOTO flashcube; **~ de rayos X** *m* INSTR X-ray flash; **~ de relleno** *m* CINEMAT, FOTO fill-in flash
flauta *f* TEC ESP flute
flavana *f* QUÍMICA flavan
flavanona *f* QUÍMICA flavanone
flavina *f* QUÍMICA flavin
flavona *f* ALIMENT, QUÍMICA flavone
flavonoide *m* ALIMENT, QUÍMICA flavonoid
flavonol *m* QUÍMICA flavonol
flavoproteína *f* ALIMENT, QUÍMICA flavoprotein
flavopurpurina *f* QUÍMICA flavopurpurin
flecha *f* CONST dip, ING MECÁ set, shaft, NUCL *cable* sag; **~ abisal** *f* OCEAN abyssal spit; **~ litoral** *f* OCEAN spit; **~ del viento** *f* METEO wind arrow, wind shaft
flejado *m* EMB hooping; **~ de la bandeja** *m Esp* (*cf flejado de la pálet AmL*) EMB pallet strapping material; **~ de la pálet** *m AmL* (*cf flejado de la bandeja Esp, flejado de la paleta Esp*) EMB pallet strapping material; **~ de la paleta** *m Esp* (*cf flejado de la pálet AmL*) EMB pallet strapping material
flejadora *f* EMB strapping machine
fleje *m* CONST, EMB, ING MECÁ, PROD band, hoop, strap, strip; **~ de acero** *m* CONST, EMB strapping steel; **~ de fardo** *m* EMB bale hoop; **~ de hierro** *m* CONST, EMB hoop iron
fletador *m* TEC PETR, TRANSP AÉR, TRANSP MAR charterer
fletamento *m* TEC PETR, TRANSP AÉR, TRANSP MAR charter, charterage; **~ a plazo** *m* TRANSP MAR time charter; **~ a término** *m* TRANSP MAR time charter; **~ por tiempo** *m* TRANSP MAR time charter; **~ por viaje** *m* TRANSP MAR voyage charter
fletamentos *m pl* TEC PETR, TRANSP AÉR, TRANSP MAR chartering
flete *m* TEC ESP, TEC PETR, TRANSP MAR cargo, freight (*AmE*), goods (*BrE*); **~ aéreo** *m* EMB air freight; **~ por embarcación** *m* TEC PETR bare-boat charter; **~ de embarcación descubierta** *m* TEC PETR, TRANSP MAR bare-boat charter; **~ fijo** *m* TEC PETR, TRANSP MAR lump sum freight; **~ de llegada** *m* PROD incoming freight; **~ de retorno** *m* TRANSP MAR return cargo; **~ temporal** *m* TEC PETR time charter; **~ terrestre** *m* AGRIC land freight; **~ por viaje** *m* TEC PETR voyage charter; **~ de vuelta** *m* TRANSP MAR back freight, home freight

fletero *m AmL* (*cf buque de carga Esp*) TRANSP MAR freighter

flexión: ~ lateral *f* ING MECÁ *columnas* buckling; **~ periférica** *f* ING ELÉC edgewise bend

flexografía *f* IMPR aniline printing, aniline rubber-plate printing, flexography

flobafeno *m* QUÍMICA phlobaphene

floculación *f* ALIMENT, CARBÓN flocculation, GAS flocking, P&C *caucho*, PROC QUÍ, QUÍMICA, TEC PETR flocculation

floculador *m* ALIMENT, CARBÓN, P&C, PROC QUÍ *máquina*, QUÍMICA flocculant, flocculator

floculante[1] *adj* ALIMENT, CARBÓN, P&C, PROC QUÍ, QUÍMICA flocculent

floculante[2] *m* ALIMENT, CARBÓN, P&C, PROC QUÍ, QUÍMICA flocculant

flocular *vi* ALIMENT, CARBÓN, P&C, PROC QUÍ, QUÍMICA flocculate

floculencia *f* ALIMENT, CARBÓN, P&C, PROC QUÍ, QUÍMICA flocculence

flóculo *m* ALIMENT, CARBÓN, P&C, PROC QUÍ, QUÍMICA flocculate

flogopita *f* MINERAL phlogopite

flojo *adj* PROD loose

flor[1]**: a ~ de agua** *adj* TRANSP MAR awash

flor[2]**: ~ morada** *f* AGRIC *malezas* blue devil; **~ de saúco** *f* ALIMENT elderflower

flor[3]**: a ~ del agua** *fra* AGUA on a level with the water

floración *f* AGRIC blooming; **~ algácea** *f* OCEAN algal bloom; **~ femenina del maíz** *f* AGRIC silking; **~ térmica** *f* ELECTRÓN thermal blooming

florecido *m* REVEST blooming

florecimiento: ~ algáceo *m* AGUA, OCEAN algal bloom

flores: ~ de cobalto *f pl* MINERAL cobalt bloom; **~ de níquel** *f* MINERAL nickel bloom

florescencia: ~ algológica *f* AGUA, OCEAN algal bloom; **~ del fitoplancton** *f* OCEAN phytoplankton bloom; **~ fitoplanctónica** *f* OCEAN phytoplankton bloom

florético *adj* QUÍMICA phloretic

floretina *f* QUÍMICA phloretin

floricultura *f* AGRIC floriculture

floridizina *f* QUÍMICA phloridzin

florizina *f* QUÍMICA phlorhizin, phlorrhizin

florol *m* QUÍMICA phlorol

florón *m* IMPR fleuron, rosette

flos: ~ ferri *m* MINERAL flos ferri

flota: ~ mercante *f* TRANSP MAR merchant fleet; **~ pesquera** *f* TRANSP MAR fishing fleet; **~ de transporte fluvial** *f* TRANSP MAR river fleet

flotabilidad *f* FÍS, FÍS FLUID, HIDROL, INSTAL HIDRÁUL, TRANSP AÉR, TRANSP MAR buoyancy; **~ inducida térmicamente** *f* CONTAM, HIDROL, INSTAL HIDRÁUL thermally-induced buoyancy

flotación *f* CONTAM MAR *buque* water line, MINAS *minería*, PROC QUÍ, QUÍMICA, TEC PETR flotation, TRANSP MAR flotation, *arquitectura naval* waterplane, *barco* water line; **~ por aire** *f* CONTAM, HIDROL, INSTAL HIDRÁUL *tratamiento de líquidos* air flotation; **~ sin carga** *f* TRANSP MAR light waterline; **~ con carga completa** *f* TRANSP MAR *equipamiento de buques* load waterline; **~ colectiva** *f* CARBÓN *minerales* bulk flotation; **~ por espuma** *f* CARBÓN flotation, froth flotation; **~ inducida cinéticamente** *f* CONTAM *tratamiento de aguas y sólidos en suspensión* kinetically-induced buoyancy;

~ inducida térmicamente *f* CONTAM, HIDROL, INSTAL HIDRÁUL thermally-induced buoyancy; **~ de máxima carga** *f* TRANSP MAR deep-water line; **~ de proyecto** *f* TRANSP MAR *diseño de barcos* design waterline; **~ simultánea** *f* CARBÓN *minerales* bulk flotation; **~ única** *f* CARBÓN single flotation; **~ de verano** *f* TRANSP MAR *proyecto del buque* summer load waterline

flotador *m* AGUA, AUTO float, C&V floater, CONST ball, float, OCEAN surface float, PETROL float, PROD floater, TRANSP MAR, VEH float; **~ de alarma en la caldera** *m* INSTAL HIDRÁUL boiler emergency float, boiler float; **~ del carburador** *m* AUTO, VEH carburetor float (*AmE*), carburettor float (*BrE*); **~ de deriva** *m* OCEAN drifting float; **~ de péndulo** *m* C&V pendulum floater

fluato *m* QUÍMICA fluate

fluctuación[1]**: sin ~** *adj* ELECTRÓN jitter-free

fluctuación[2] *f* ACÚST flutter, ripple, ELEC fluctuation, oscillation, ELECTRÓN oscillation, INFORM&PD jitter, ING MECÁ oscillation, REFRIG hunting, TEC ESP ripple, shift, weft, TRANSP AÉR flutter, TV wow; **~ de la capa freática** *f* HIDROL water table fluctuation; **~ de la carga** *f* ELEC load fluctuation; **~ de corriente** *f* ELEC current fluctuation; **~ de imagen** *f* TV picture flutter; **~ de impulsos** *f* TELECOM pulse jitter; **~ de la longitud de onda** *f* ÓPT wavelength fluctuation; **~ del punto cero** *f* FÍS RAD zero point fluctuation; **~ de la temperatura** *f* REFRIG temperature fluctuation; **~ temporal** *f* CONTAM temporal fluctuation, temporal variation, time trend; **~ de tensión** *f* ELEC *fuente de alimentación* voltage fluctuation; **~ transitoria** *f* TELECOM transient fluctuation; **~ de velocidad** *f* FÍS FLUID velocity fluctuation; **~ de voltaje** *f* ELEC *fuente de alimentación* voltage fluctuation

fluctuante *adj* ELEC, ELECTRÓN, FÍS ONDAS oscillating

fluctuar *vi* ELEC fluctuate, ELECTRÓN flicker, oscillate

fluellita *f* MINERAL fluellite

fluencia *f* CARBÓN *geología* flow, *magnetismo* creep, FÍS fluence, GEOL flow, MECÁ creep; **~ a alta temperatura** *f* CRISTAL creep; **~ brusca** *f* METAL jerky flow; **~ en caliente** *f* TERMO *circuitos* hot creep; **~ energética** *f* FÍS energy fluence; **~ en frío** *f* TERMO *circuitos* cold creep; **~ parabólica** *f* METAL parabolic creep; **~ de partículas** *f* FÍS, FÍS PART particle fluence; **~ plástica** *f* METAL plastic flow; **~ primaria invertida** *f* METAL inverse primary creep; **~ terciaria** *f* METAL tertiary creep; **~ de la tinta** *f* IMPR ink flow; **~ transitoria** *f* METAL transient creep

fluídica *f* FÍS FLUID fluidics

fluidificación *f* CARBÓN, FÍS FLUID, PETROL, TEC PETR fluidization

fluidificador *m* PAPEL flow box

fluidímetro *m* GEN flowmeter; **~ instantáneo** *m* AGUA instant flowmeter

fluidización *f* CARBÓN, FÍS FLUID, PETROL, TEC PETR fluidization

fluido: ~ ambiente *m* FÍS FLUID ambient fluid; **~ criogénico** *m* REFRIG cryogenic fluid; **~ de formación** *m* GEOL, TEC PETR formation fluid; **~ para frenos hidráulicos** *m* ING MECÁ brake fluid; **~ frigorífico** *m* REFRIG, TERMO coolant, cooling medium; **~ frigorígeno** *m* REFRIG primary refrigerant; **~ monoatómico** *m* GAS monoatomic fluid;

~ **perfecto** *m* FÍS perfect fluid; ~ **silicónico** *m* ELEC *aislador* silicone fluid

fluidos: ~ **rotatorios** *m pl* FÍS FLUID rotating fluids

flujo: ~ **de aceite** *m* PROD oil flow; ~ **ahogado** *m* FÍS FLUID drowned flow; ~ **de aire** *m* CONST, TRANSP AÉR airflow; ~ **de aire aspirado** *m* ING MECÁ indraft of air (*AmE*), indraught of air (*BrE*); ~ **alterno** *m* ELEC, FÍS alternating flux; ~ **de arcilla** *m* GEOL clay flowage; ~ **ascendente** *m* NUCL upflow, upward flow; ~ **axial** *m* TRANSP AÉR, VEH axial flow; ~ **bidireccional** *m* INFORM&PD bidirectional flow; ~ **bifásico** *m* GAS biphasic flow, two-phase flow; ~ **de bit** *m* INFORM&PD bit stream; ~ **de calor** *m* TERMOTEC heat flow; ~ **calorífico** *m* GEOL, TEC ESP, TERMO heat flow; ~ **calorífico crítico** *m* PROD, TERMOTEC critical heat flux; ~ **en canal abierto** *m* FÍS FLUID open channel flow; ~ **característico** *m* HIDROL characteristic flow; ~ **compresible** *m* FÍS FLUID compressible flow; ~ **constante** *m* FÍS steady flow; ~ **continuo** *m* AGUA, HIDROL continuous flood, continuous high flow; ~ **de control** *m* INFORM&PD control flow; ~ **de convección libre** *m* FÍS FLUID free convection flow; ~ **convectivo** *m* FÍS FLUID convective flow; ~ **de la corriente** *m* OCEAN current stream; ~ **de cortadura** *m* FÍS FLUID shear flow; ~ **en cortocircuito** *m* ACÚST, TV short-circuit flux; ~ **de Couette** *m* FÍS FLUID Couette flow; ~ **crítico de calor** *m* NUCL critical heat flow; ~ **de datos** *m* INFORM&PD, ING ELÉC, TELECOM data stream; ~ **de desplazamiento** *m* ELEC, FÍS, ING ELÉC electric flux; ~ **en dirección inversa** *m* ELEC, INFORM&PD reverse-direction flow; ~ **de dispersión** *m* FÍS, ING ELÉC leakage flux; ~ **de Ekman** *m* HIDROL Ekman flow; ~ **eléctrico** *m* ELEC, FÍS, ING ELÉC electric flux; ~ **de electrones** *m* FÍS PART, TV electron stream; ~ **electrostático** *m* ELEC electrostatic flux; ~ **elevado continuo** *m* AGUA, HIDROL continuous flood, continuous high flow; ~ **energético** *m* FÍS radiant flux, FÍS RAD, ÓPT, TELECOM energy flux, radiant flux, TERMO energy flux; ~ **de energía sónica** *m* FÍS sound energy flux; ~ **entrante** *m* FÍS inward flux; ~ **entrópico** *m* TERMO entropic flux; ~ **equilibrado** *m* REFRIG balanced flow; ~ **estable** *m* TRANSP *tráfico* stable flow, TRANSP AÉR steady flow; ~ **estacionario** *m* FÍS FLUID steady flow; ~ **de fluidos** *m* FÍS FLUID fluid flow; ~ **freático** *m* HIDROL subsurface flow; ~ **friccional** *m* NUCL frictional flow; ~ **en frío** *m* MECÁ cold flow; ~ **de gas** *m* FÍS FLUID, GAS gas flow; ~ **gaseoso** *m* FÍS FLUID, GAS gas flow; ~ **por gravedad** *m* NUCL gravity flow; ~ **de la hélice** *m* TRANSP AÉR propeller wash, slipstream; ~ **hipersónico** *m* FÍS FLUID hypersonic flow; ~ **incompresible** *m* FÍS, FÍS FLUID incompressible flow; ~ **de inducción** *m* ING ELÉC induction flux; ~ **de inducción magnética** *m* ING ELÉC magnetic induction flux; ~ **inestable** *m* FÍS FLUID unstable flow, NUCL unstable flow, unsteady flow, TRANSP *tráfico* unstable flow; ~ **de información** *m* INFORM&PD information flow; ~ **de instrucciones** *m* INFORM&PD instruction stream; ~ **de instrucciones múltiple-flujo de datos múltiple** *m* INFORM&PD multiple-instruction multiple-data (*MIMD*); ~ **de instrucciones múltiple-flujo de datos único** *m* (*MISD*) INFORM&PD multiple-instruction single-data (*MISD*); ~ **irrotacional** *m* FÍS, FÍS FLUID irrotational flow; ~ **lácteo** *m* AGRIC rate of milking;

~ **laminar** *m* Esp (*cf escorrentía laminar AmL*) ACÚST, FÍS, FÍS FLUID, MECÁ, METAL laminar flow, viscous flow, MINAS flow sheet, NUCL, REFRIG, TRANSP AÉR laminar flow, viscous flow; ~ **laminar en tuberías** *m* FÍS FLUID laminar pipe flow; ~ **de lava** *m* GEOL lava flow, lava stream; ~ **de líquidos** *m* FÍS FLUID liquid flow; ~ **de llamadas** *m* TELECOM call flow; ~ **de lodo** *m* GEOL mud flow; ~ **luminoso** *m* FÍS, ING ELÉC luminous flux; ~ **magnético** *m* ELEC, ENERG RENOV, FÍS, FÍS RAD, ING ELÉC, TV magnetic flux; ~ **de la marea** *m* HIDROL tidal flow, TRANSP MAR flood stream; ~ **mareal** *m* AGUA tidal flow; ~ **de la masa** *m* TRANSP AÉR mass flow; ~ **de masa** *m* FÍS mass rate, FÍS FLUID *a través de una tubería* mass flux; ~ **de la masa de aire** *m* TRANSP AÉR mass airflow; ~ **másico** *m* NUCL mass flow; ~ **masivo** *m* GEOL mass flow; ~ **del motor** *m* TRANSP AÉR jet wash; ~ **neutrónico radial** *m* NUCL radial neutron flux; ~ **no estacionario** *m* FÍS FLUID unsteady flow; ~ **óptico** *m* ÓPT optical flux; ~ **de partículas beta** *m* ELEC, FÍS, FÍS PART, NUCL beta ray; ~ **de Poiseuille** *m* FÍS FLUID Poiseuille flow; ~ **previsto** *m* HIDROL design flood, design flow; ~ **de producción** *m* TEXTIL production flow; ~ **pulsátil** *m* FÍS FLUID, REFRIG pulsating flow; ~ **de purga** *m* PROD bleed flow; ~ **de radiación** *m* FÍS RAD, RAD flux of radiation; ~ **radiante** *m* FÍS, FÍS RAD, ÓPT, TELECOM radiant flux; ~ **en remolino** *m* GAS swirling flow; ~ **de retorno** *m* INSTAL HIDRÁUL *agua de caldera* back flow; ~ **rodado en hora punta** *m* AUTO vehicular flow at peak hour; ~ **rotatorio de Couette** *m* FÍS FLUID rotating Couette flow; ~ **del rotor** *m* TRANSP AÉR rotor stream; ~ **saliente** *m* FÍS outward flux; ~ **de tareas** *m* INFORM&PD job stream; ~ **térmico** *m* TEC ESP heat flux, TERMO, TERMOTEC flow of heat; ~ **térmico máximo** *m* NUCL peak heat flux; ~ **de tráfico** *m* TELECOM traffic flow; ~ **transversal** *m* VEH crossflow; ~ **en tuberías** *m* FÍS FLUID pipe flow; ~ **turbulento** *m* CARBÓN eddy flow, FÍS FLUID turbulent flow, REFRIG eddy flow; ~ **uniforme** *m* FÍS FLUID, TRANSP AÉR steady flow; ~ **viscoso** *m* ALIMENT, FÍS, METAL viscous flow; ~ **viscoso incompresible** *m* FÍS FLUID viscous incompressible flow

flujograma *m* AmL (*cf diagrama de flujo Esp, organigrama Esp*), FÍS, INFORM&PD flowchart

flujómetro *m* AGUA, CARBÓN flowmeter, CONST flowmeter, fluxmeter, ELEC flowmeter, *magnetismo* fluxmeter, FÍS fluxmeter, flowmeter, HIDROL flowmeter, ING ELÉC fluxmeter, flowmeter, INSTAL HIDRÁUL, LAB, PAPEL, PROC QUÍ, TEC PETR, TERMO flowmeter; ~ **de imán permanente** *m* NUCL permanent-magnet flowmeter

fluoanteno *m* QUÍMICA fluoanthene

fluoantreno *m* QUÍMICA fluoanthrene

fluoborato *m* QUÍMICA fluoroborate

fluocerina *f* MINERAL fluocerine

fluocerita *f* MINERAL fluocerite

fluoformo *m* QUÍMICA fluoroform

fluofosfato *m* QUÍMICA fluorophosphate

flúor *m* (*F*) C&V, CARBÓN, QUÍMICA fluorine (*F*)

fluoración *f* QUÍMICA fluoridation

fluoreno *m* QUÍMICA fluorene

fluorenona *f* QUÍMICA fluorenone

fluoresceína *f* QUÍMICA fluorescein

fluorescencia *f* GEN fluorescence; ~ **de impacto** *f* TV impact fluorescence; ~ **de rayos X** *f* FÍS RAD X-ray

fluorescence; ~ **sensibilizada** f FÍS RAD sensitized fluorescence; ~ **X** f FÍS RAD X-ray fluorescence
fluorhidrato m QUÍMICA hydrofluoride
fluorhídrico adj QUÍMICA hydrofluoric
fluorita f C&V fluorspar, MINERAL chlorophane, fluorite, fluorspar, QUÍMICA fluorspar
fluorítico adj QUÍMICA fluoritic
fluorografía f FÍS RAD fluorography
fluoroscopia f ING ELÉC fluoroscopy
fluoruro m QUÍMICA fluoride; ~ **de plata** m QUÍMICA silver fluoride; ~ **de polivinilo** m P&C polyvinyl fluoride; ~ **de uranio** m NUCL uranic fluoride; ~ **de vinilo** m P&C polyvinyl fluoride
fluosilicato m QUÍMICA silicofluoride, fluorosilicate
fluosilícico adj QUÍMICA fluorosilicic
fluosulfónico adj QUÍMICA fluosulfonic (AmE), fluosulphonic (BrE)
fluroaluminato m QUÍMICA fluroaluminate
fluviatil adj GEOL fluviatile
fluvioglaciar adj GEOL fluvio-glacial
fluviomarino adj GEOL fluvio-marine
fluviómetro m AGUA current meter
fluxímetro m CONST, ELEC, FÍS, ING ELÉC fluxmeter; ~ **de burbuja** m LAB bubble flowmeter; ~ **capilar** m LAB capillary flowmeter
fluxómetro m CONST, ELEC, FÍS, ING ELÉC fluxmeter; ~ **de integración** m TRANSP AÉR integrating flowmeter
fluyente adj AGUA, HIDROL curso de agua flowing
flysch m GEOL, PETROL flysch
Fm abr (fermio) FÍS RAD, QUÍMICA Fm (fermium)
fmm abr (fuerza magnetomotriz) ELEC, FÍS mmf (magnetomotive force)
FOB abr (franco a bordo) TEC PETR, TRANSP MAR FOB (free on board)
focalización: ~ **dinámica** f TV dynamic focusing
foco m C&V lamp bulb; ~ **de calor** m REFRIG, TERMO heat reservoir; ~ **de emisión puntal** m CONTAM emission point; ~ **fijo** m CINEMAT, FOTO, TV fixed-focus; ~ **del fondo** m CINEMAT back focus; ~ **lineal** m CONTAM line source, TV line focus; ~ **de mano** m ELEC, ING ELÉC flashlight; ~ **posterior** m FOTO rear focus; ~ **puntual** m CONTAM radioactive-point source, FÍS RAD point source, radioactive-point source, NUCL radioactive-point source; ~ **sísmico** m GEOFÍS seismic focus
focómetro m INSTR focimeter
foehn m METEO foehn, foehn wind
fogón m CONST, GAS hearth, INSTAL TERM firebox, PROD horno stoke hole
fogonadura f TRANSP MAR mast tabernacle
fogonazo: ~ **térmico** m NUCL thermal flash
fogonero m PROD persona stoker
föhn m METEO föhn, föhn wind
foliación f AGRIC árboles blooming
foliado adj GEOL foliated
foliar vt IMPR folio, paginate
folículo m AGRIC follicle
folínico adj QUÍMICA folinic
folíolo m AGRIC leaflet
fonación f ACÚST phonation
fondeadero m OCEAN roads, roadstead, TEC ESP marina berthing, TRANSP MAR roads, roadstead; ~ **de amarre** m TRANSP MAR mooring berth; ~ **limpio** m TRANSP MAR amarradero safe ground

fondeado: **estar ~ a la gira** adj TRANSP MAR ride at anchor
fondear[1] vt CONTAM MAR, METEO, TRANSP MAR moor
fondear[2] vi TRANSP MAR drop anchor, barco anchor; ~ **en la rada** vi TRANSP MAR anchor in the roads
fondeo m TRANSP MAR anchorage; ~ **de minas** m D&A, TRANSP MAR misión naval minelaying
fondo: **bajo ~** m OCEAN shallow, PROD shoal, TRANSP MAR geología shallows; ~ m C&V del cuello base, CARBÓN bottom, CONST head, FOTO background, HIDROL sill, de estanque, depósito bottom, IMPR background, ING MECÁ head, MINAS bottom, filones sill; ~ **ajedrezado** m C&V checker pattern; ~ **borroso** m FOTO background blur; ~ **del buque** m TRANSP MAR quickwork; ~ **caído** m C&V slugged bottom; ~ **de cilindro** m ING MECÁ back-cylinder cover; ~ **de cilindro posterior** m ING MECÁ back-cylinder cover; ~ **de color** m IMPR tint; ~ **de cuchara** m PROD fundición scull, skull; ~ **de fango** m MINAS mud bottom; ~ **del hueco** m TEC PETR perforación bottom hole; ~ **interior del buque** m TRANSP MAR quickwork; ~ **de mal tenedero** m TRANSP MAR foul bottom; ~ **del mar** m CONST, ENERG RENOV sea floor, OCEAN bed, sea bottom, sea floor, TEC PETR mud line; ~ **móvil** m CINEMAT moving background; ~ **con pestaña** m C&V flanged bottom; ~ **del piso** m MINAS pit bottom; ~ **del pozo** m MINAS sump shaft, TEC PETR perforación bottom hole, downhole; ~ **profundo marino** m GEOFÍS, OCEAN deep-sea floor; ~ **proyectado** m CINEMAT projected background; ~ **del río** m AGUA, HIDROL river bed; ~ **rocoso** m GEOFÍS, OCEAN, TEC PETR bedrock; ~ **de saco** m MINAS blind level; ~ **sucio** m TRANSP MAR foul bottom; ~ **del surco** m AGRIC sole; ~ **del tamiz** m PROC QUÍ sieve bottom; ~ **tramado** m IMPR screen tint; ~ **uniforme** m IMPR flat tint; ~ **del valle** m CONST bottom
fondos m pl TEC PETR perforación bottoms, TRANSP MAR quickwork; ~ **aplacerados** m pl TRANSP MAR soundings; ~ **marinos** m pl OCEAN bed
fonio m ACÚST, FÍS phon; ~ **Stevens** m ACÚST, FÍS Stevens's phon; ~ **Zwicker** m ACÚST, FÍS Zwicker's phon
fonocaptor: ~ **filtrador** m INSTR filter pick-up; ~ **láser** m ÓPT laser pick-up head; ~ **monofónico** m ACÚST monophonic pick-up; ~ **piezoeléctrico** m ING MECÁ piezoelectric pick-up
fonogoniómetro m ACÚST, FÍS ONDAS sound locator
fonolita f GEOL, PETROL, PROD clinkstone, phonolite, phonolyte
fonolocalización f ACÚST, FÍS ONDAS sound ranging
fonolocalizador m ACÚST, FÍS ONDAS sound locator
fonón m FÍS phonon
fonotelemetría f ACÚST, FÍS ONDAS sound ranging
fonovisión f ÓPT phonovision
fontanería f Esp (cf plomería AmL) CONST plumbing
fontanero m Esp (cf plomero AmL) CONST plumber
fontura: ~ **de agujas** f TEXTIL needle bed
foque m TRANSP MAR vela jib
foración: ~ **por fusión** f CARBÓN fusion drilling
foraminífero[1] adj GEOL, PETROL foraminiferal
foraminífero[2] m GEOL, PETROL foraminifera
forca: ~ **de charnela** f TRANSP AÉR hinge fork
forja f CONST, ING MECÁ, MECÁ lugar smithy, forge, procesamiento forging, METAL forge, PROD lugar forge, smithy, procesamiento forging; ~ **en caliente**

f ING MECÁ, PROD, TERMO hot forging; **~ con estampa** *f* ING MECÁ, PROD die forging; **~ con estampa abierta** *f* ING MECÁ, PROD open-die forging; **~ de estampado en caliente** *f* ING MECÁ, PROD drop forging; **~ en frío** *f* ING MECÁ, PROD, TERMO cold forging; **~ de herrero** *f* ING MECÁ, PROD blacksmith's forge; **~ a martinete** *f* ING MECÁ, PROD drop forging; **~ en matriz** *f* ING MECÁ, PROD die forging; **~ de titanio** *f* TEC ESP titanium forging; **~ en tosco** *f* ING MECÁ, PROD roughed forging; **~ con troquel** *f* ING MECÁ, PROD drop forging; **~ en troquel** *f* ING MECÁ, PROD die forging

forjado[1] *adj* CONST, ING MECÁ, MECÁ, METAL, PROD forged, wrought; **~ en caliente** *adj* ING MECÁ, PROD, TERMO hot-forged; **~ en frío** *adj* ING MECÁ, PROD, TERMO cold-forged; **~ en tosco** *adj* PROD rough-forged

forjado[2] *m* ING MECÁ, MECÁ, PROD die

forjador *m* CONST, ING MECÁ, MECÁ, PROD blacksmith; **~ de calderas** *m* INSTAL HIDRÁUL boilersmith

forjadora *f* ING MECÁ, MECÁ, PROD shaping machine

forjadura *f* ING MECÁ, MECÁ, PROD *procesamiento* forging; **~ en caliente** *f* ING MECÁ, PROD, TERMO hot forging; **~ en frío** *f* ING MECÁ, PROD, TERMO cold forging

forjar: **~ en caliente** *vt* ING MECÁ, PROD, TERMO hot-forge; **~ con estampa** *vt* MECÁ swage; **~ en frío** *vt* ING MECÁ, PROD, TERMO cold-forge

forma[1]: **en ~ de anillo** *adj* GEOM, ING MECÁ, MECÁ annular, ring-shaped; **en ~ de arco** *adj* CONST arched; **en ~ de lámina** *adj* ELEC laminated; **de ~ ovalada** *adj* GEOM oval-shaped; **en ~ de raspa de pescado** *adj* IMPR herringboned; **en ~ de serie** *adj* ING ELÉC serial form; **en ~ de T** *adj* GEOM, ING MECÁ T-shaped; **en ~ de V** *adj* GEOM, ING MECÁ V-shaped

forma[2] *f* IMPR chase, form (*AmE*), forme (*BrE*), INFORM&PD form, PAPEL *para papel a mano* mold (*AmE*), mould (*BrE*), TEXTIL shape; **~ aerodinámica** *f* VEH aerodynamic form; **~ para bobina** *f* ELEC, ING ELÉC coil form; **~ de bolsillo sobreimpresa** *f* IMPR patch-pocket form; **~ de cables** *f* ELEC, ING ELÉC cable form; **~ característica** *f* ING MECÁ characteristic shape; **~ de la celda** *f* IMPR cell shape; **~ cetónica** *f* QUÍMICA keto-form; **~ cuadrada** *f* CONST squareness; **~ dendrítica** *f* GEOL dendridic form; **~ dominante** *f* METAL habit; **~ de energía** *f* GAS energy form; **~ estándar** *f* MATEMÁT standard form; **~ impresora** *f* IMPR printing form; **~ del impulso** *f* ELECTRÓN pulse shape; **~ interior** *f* IMPR inside form; **~ de leva** *f* MECÁ cam shape; **~ de manejar el transportador** *f* EMB conveyor handling system; **~ normal** *f* INFORM&PD normal form, MATEMÁT standard form; **~ normal de Backus** *f* INFORM&PD Backus normal form; **~ de onda** *f* ACÚST, ELEC, ELECTRÓN, FÍS, FÍS ONDAS, INFORM&PD, ING ELÉC waveform; **~ de onda compleja** *f* ACÚST, ELEC, ELECTRÓN, FÍS, FÍS ONDAS, INFORM&PD, ING ELÉC complex waveform; **~ de onda en diente de sierra** *f* ACÚST, ELEC, ELECTRÓN, FÍS, FÍS ONDAS, ING ELÉC ramp waveform, sawtooth waveform; **~ de onda visualizada** *f* ELECTRÓN *osciloscopio* displayed waveform; **~ en paralelo** *f* GEOM, ING ELÉC parallel form; **~ de las partículas** *f* CARBÓN grain shape; **~ de pestaña** *f* IMPR tab form; **~ plana** *f* IMPR *estucado* flat form; **~ redonda** *f* PAPEL cylinder mold (*AmE*), cylinder mould (*BrE*); **~ de silla de montar simétrica** *f* GEOM symmetric saddle shape; **~ del surco** *f* ACÚST groove shape; **~ de la trama** *f* IMPR screen shape; **~ trans** *f* QUÍMICA trans form

formación: **~ absorbente** *f* PETROL thief formation; **~ de ampollas** *f* IMPR, PAPEL *defecto* blistering; **~ del arco eléctrico** *f* ING ELÉC, PROD arcing; **~ de arco en el magnetrón** *f* ING ELÉC magnetron arcing; **~ de arco de tensión** *f* ELEC spark-over; **~ de arrugas** *f* P&C *defecto* crawling; **~ de una barra de arena** *f* CONST gritting, sanding; **~ basada en computadoras** *f* *AmL* (*cf formación basada en ordenadores Esp*) INFORM&PD, TELECOM computer-based training; **~ basada en ordenadores** *f* *Esp* (*cf formación basada en computadoras AmL*) INFORM&PD, TELECOM computer-based training; **~ de bolitas** *f* P&C pelletizing; **~ de burbujas** *f* IMPR, PAPEL blistering; **~ en cadena** *f* QUÍMICA chain formation; **~ de la capa límite** *f* FÍS FLUID boundary-layer formation; **~ carbonífera** *f* CARBÓN, GEOL, MINAS coal formation; **~ de carburos metálicos** *f* METAL carbide formation; **~ de colchoneta** *f* C&V mat formation; **~ compacta** *f* PETROL tight formation; **~ de cristales en la superficie** *f* CARBÓN, QUÍMICA creep; **~ de escarcha** *f* REFRIG frost formation; **~ de escarcha en la aspiración** *f* REFRIG frost back; **~ de las espigas** *f* AGRIC ear-setting; **~ del frente de arranque del carbón** *f* MINAS coalface system; **~ de un gel** *f* P&C, PROC QUÍ, QUÍMICA gelation; **~ de gotas** *f* FÍS FLUID drop formation; **~ de granza** *f* P&C pelletizing; **~ de grietas** *f* METAL crack formation; **~ de grietas irregulares** *f* P&C alligatoring; **~ de grietas pequeñas** *f* C&V, CONST crazing, ING MECÁ cracking, P&C crazing; **~ en grupo** *f* METAL cluster formation; **~ de halo** *f* FOTO halation; **~ del haz** *f* ELECTRÓN beam shaping; **~ de haz electrónico** *f* ELECTRÓN electronic-beam forming; **~ de hendiduras** *f* METAL crack formation; **~ de hielo** *f* TRANSP MAR icing; **~ de un hueco** *f* CONST hollowing out; **~ de imágenes** *f* ELECTRÓN imaging; **~ de imágenes por espectroscopía electrónica** *f* FÍS, FÍS RAD electron-spectroscopic imaging; **~ de imágenes falsas** *f* CINEMAT streaking; **~ de imágenes térmicas** *f* D&A, FÍS RAD, TERMO thermal imaging (*TI*); **~ de impulsos** *f* INFORM&PD pulse shaping; **~ de incrustaciones** *f* AGUA scale formation; **~ por inyección de aire** *f* P&C blast forming; **~ de lagunas** *f* METAL void formation; **~ de líneas** *f* ELECTRÓN raster; **~ de maclas** *f* CRISTAL twin formation, METAL twinning; **~ metalífera** *f* GEOL, MINAS ore formation; **~ de miga** *f* ALIMENT crumb formation; **~ de minerales** *f* GEOL, MINAS ore formation; **~ de las nubes** *f* METEO cloud formation; **~ del papel** *f* PAPEL sheet formation; **~ de pares** *f* ELEC, ING ELÉC cable pairing; **~ petrolífera** *f* PETROL pay zone; **~ de picos de yodo** *f* NUCL iodine spiking; **~ plana** *f* ELEC, ING ELÉC *configuración de cable* flat formation; **~ productora** *f* GEOL, TEC PETR pay zone; **~ de quelatos** *f* NUCL chelate formation; **~ de referencia** *f* GEOL marker bed; **~ de la señal** *f* ELECTRÓN signal shaping; **~ sísmica** *f* GEOL seismic array; **~ de trébol** *f* ELEC *configuración de cable* trefoil formation; **~ de trifolio** *f* ELEC *configuración de cable* trefoil formation; **~ de umbrales** *f* ELECTRÓN, TEC ESP thresholding; **~ de vapor en el primario** *f* NUCL voiding; **~ de velo** *f* IMPR scumming; **~ de veneros**

interiores f CARBÓN *presas* piping; ~ **de vuelo** f TRANSP AÉR flight formation; ~ **de zonas** f METAL zone formation

formado[1]: ~ **por capas superpuestas** *adj* P&C, PROD laminated; ~ **de chapas** *adj* P&C, PROD laminated

formado[2]: ~ **automático** m C&V automatic forming

formador[1]: ~ **de aceite** *adj* QUÍMICA oil-forming; ~ **de orto** *adj* QUÍMICA orthoforming

formador[2] m PAPEL former; ~ **de bucles** m CINEMAT loop former; ~ **de redes** m C&V network former

formadora: ~ **de vellón** f TEXTIL sheeter box

formaldehído m ALIMENT, P&C, QUÍMICA, TEXTIL formaldehyde, methanal

formalina f QUÍMICA formalin

formamida f QUÍMICA formamide

formante m ACÚST formant

formar[1]: ~ **con un macho** *vt* PROD core up, *moldería* core, core out

formar[2]: ~ **cauce** *vi* HIDROL scour

formarse: ~ **un juicio de** *v refl* PROD take stock of

formatador m AmL (*cf formateador Esp*) INFORM&PD formatter; ~ **de texto** m AmL (*cf formeatador de texto Esp*) INFORM&PD text formatter

formateado m INFORM&PD formatting

formateador m Esp (*cf formatador AmL*) INFORM&PD formatter; ~ **de texto** m Esp (*cf formatador de texto AmL*) INFORM&PD text formatter

formatear *vt* INFORM&PD format

formato m IMPR, INFORM&PD format, QUÍMICA formate, TV format; ~ **alargado** m IMPR oblong size; ~ **apaisado** m IMPR, INFORM&PD landscape format; ~ **bruto** m IMPR, PAPEL untrimmed size; ~ **cortado** m IMPR, PAPEL *tamaño normal* trimmed size; ~ **de datos** m INFORM&PD data format; ~ **de diagrama de escalones** m PROD ladder diagram format; ~ **de dirección** m INFORM&PD address format; ~ **fijo** m INFORM&PD fixed format; ~ **de grabación** m INFORM&PD record format; ~ **de la imagen** m CINEMAT picture ratio; ~ **de impresión** m INFORM&PD print format; ~ **de instrucción** m INFORM&PD instruction format; ~ **libre** m INFORM&PD free format; ~ **neto** m IMPR, PAPEL trimmed size; ~ **permanente en sectores por hardware** m INFORM&PD hard sectoring; ~ **de presentación visual** m NUCL display format; ~ **de rodaje** m CINEMAT shooting ratio; ~ **vertical** m INFORM&PD portrait format, vertical format; ~ **de visualización** m PROD display format

formazilo m QUÍMICA formazyl

fórmico *adj* QUÍMICA formic

formilo m QUÍMICA formyl

formio m AGRIC New Zealand flax

formón m CONST, ING MECÁ chisel, chisel-and-point pick, framing chisel, paring chisel, paring machine; ~ **de achaflanado** m ING MECÁ beveled chisel (*AmE*), bevelled chisel (*BrE*); ~ **de auto-centrado** m CONST self-coring chisel; ~ **en bisel** m ING MECÁ beveled chisel (*AmE*), bevelled chisel (*BrE*); ~ **de bisel oblicuo** m CONST *carpintería* side chisel; ~ **de doble biselado** m CONST double-beveled chisel (*AmE*), double-bevelled chisel (*BrE*), double-beveled turning chisel (*AmE*), double-bevelled turning chisel (*BrE*); ~ **de mortaja** m CONST *carpintería* mortise chisel; ~ **de mortaja de cerradura** m CONST lock-mortise chisel; ~ **de perno** m CONST heading chisel; ~ **de punta redonda** m ING MECÁ round-nose chisel; ~ **de**

tornero con doble biselado m CONST double-beveled turning chisel (*AmE*), double-bevelled turning chisel (*BrE*)

fórmula f GEN formula; ~ **de Archie** f PETROL Archie's formula; ~ **de Balmer** f FÍS Balmer's formula; ~ **de Bazin** f INSTAL HIDRÁUL Bazin's formula; ~ **de carga** f C&V batch formula; ~ **de Chezy** f HIDROL Chezy's formula; ~ **de Clausius-Mosotti** f FÍS Clausius-Mosotti formula; ~ **de constitución** f QUÍMICA structural formula; ~ **empírica** f QUÍMICA empirical formula; ~ **estructural** f QUÍMICA structural formula; ~ **de Euler** f GEOM Euler's formula; ~ **para la hincadura de los pilotes** f CARBÓN pile-driving formula; ~ **de Kutter** f HIDROL Kutter's formula; ~ **de Lacey** f HIDROL Lacey's formula; ~ **de Lorentz-Lorenz** f FÍS Lorentz-Lorenz formula; ~ **de Manning** f HIDROL Manning's formula; ~ **de Planck** f FÍS Planck's formula; ~ **de Plank para la radiación** f FÍS RAD Planck's radiation formula; ~ **de Rayleigh-Jeans** f FÍS Rayleigh-Jeans formula; ~ **de revelado** f CINEMAT, FOTO developer formula; ~ **de Strickler** f HIDROL Strickler's formula; ~ **suave** f PROD bland formula

formulación: ~ **del color** f COLOR color formulation (*AmE*), colour formulation (*BrE*); ~ **de estucado** f PAPEL coating color (*AmE*), coating colour (*BrE*)

formulario: ~ **continuo** m IMPR, INFORM&PD, PAPEL continuous feed paper (*AmE*), continuous form (*AmE*), continuous stationery (*BrE*); ~ **de datos** m PROD mask data; ~ **estándar** m MATEMÁT standard form; ~ **normal** m MATEMÁT standard form

formularios: ~ **continuos comerciales** m pl PAPEL business forms; ~ **para copias múltiples** m pl PAPEL multicopy business forms; ~ **de papel autocopiativo** m pl PAPEL carbonless copy paper forms; ~ **con papeles carbonados** m pl PAPEL carbonized forms

fórmulas: ~ **de Fresnel** f pl FÍS Fresnel's formulae

fornitura f TEXTIL trimming

fornituras: ~ **para mecánicos** f pl PROD engineers' stores

foro: ~ **de interés** m INFORM&PD newsgroup

forona f QUÍMICA phorone

forrado[1] *adj* ELEC, ING ELÉC, REVEST covered; ~ **de algodón** *adj* ELEC, ING ELÉC, REVEST cotton-covered; ~ **con una capa de algodón** *adj* ELEC single-cotton-covered (*SCC*); ~ **con una capa de caucho** *adj* ELEC single-rubber-covered (*SRC*); ~ **con una capa de papel** *adj* ELEC single-paper-covered (*SPC*); ~ **de caucho** *adj* P&C, REVEST rubber-lined

forrado[2] m CONST *edificación* siding, PAPEL lining; ~ **interior** m EMB interior lining; ~ **de tubos** m TERMO, TERMOTEC flue lining

forraje m AGRIC fodder, forage; ~ **basto** m AGRIC roughage; ~ **suculento** m AGRIC succulage

forrar *vt* CONST sheet, PROD jacket, *cojinete con metal antifricción* line, REVEST line, coat; ~ **de nuevo** *vt* CARBÓN recover; ~ **con planchas** *vt* TRANSP MAR *construcción naval* plate

forro[1]: **de** ~ **liso** *adj* TRANSP MAR carvel-built

forro[2] m AUTO lining, pad, ELEC cable sheath (*BrE*), jacket (*AmE*), serving, FÍS jacket, IMPR lining, ING MECÁ backing, MECÁ liner, lining, MINAS *cimbras* lagging, PETROL liner, PROD *funda, estuche, cubierta* sheath, lining, REVEST lining, sheathing, TEC ESP coat, TRANSP AÉR liner, TRANSP MAR *construcción naval* skin, *de planchas* plating, VEH pad; ~ **del cable** m

ELEC, ING ELÉC, P&C cable covering (*AmE*), cable sheathing (*BrE*); ~ **calorífugo** *m* PROD, TERMO, TERMOTEC lagging; ~ **calorífugo del cilindro** *m* PROD, TERMO, TERMOTEC cylinder lagging; ~ **de chimenea** *m* TERMO, TERMOTEC flue lining; ~ **de cubierta** *m* TRANSP MAR *yates* deckhead; ~ **del embalse** *m* AGUA reservoir lining; ~ **del embrague** *m* AUTO, ING MECÁ, MECÁ, VEH clutch lining; ~ **enrasado** *m* TRANSP MAR *con cubrejuntas interiores* flush plating; ~ **exterior** *m* TRANSP MAR *construcción naval* outer hull, outer skin, shell; ~ **de freno** *m* AUTO, C&V, ING MECÁ, MECÁ, VEH brake lining; ~ **del freno de disco** *m* AUTO, MECÁ, VEH disc brake pad (*BrE*), disk brake pad (*AmE*); ~ **inhibidor** *m* TRANSP AÉR liner; ~ **interior** *m* TRANSP MAR *del casco* inner lining, inner skin; ~ **del lomo** *m* IMPR back lining; ~ **de madera** *m* CARBÓN planking; ~ **del pantoque** *m* TRANSP MAR *construcción naval* bilge plating; ~ **de plomo** *m* ELEC *cable* lead sheath, lead sheathing; ~ **protector** *m* REVEST protective sheathing; ~ **refractario** *m* INSTAL TERM, REVEST, TERMO refractory lining; ~ **de tablas** *m* MINAS lagging; ~ **de tablas entre pilotes** *m* CARBÓN interpile sheeting; ~ **de la zapata del freno** *m* AUTO, VEH brake pad

forsterita *f* MINERAL forsterite
fortalecimiento: ~ **del haz de electrones** *m* ELECTRÓN electron-beam annealing
fortificación *f* MINAS *en pozos* curb ring
fortín *m* D&A blockhouse
forzador *m* MINAS forcing
forzamiento *m* CONST *cerraduras*, MINAS forcing; ~ **de cemento** *m* TEC PETR *perforación* squeeze; ~ **descendente** *m* MINAS forcing down
forzar *vt* CINEMAT push, CONST *cerradura* pick, *con palanca* prize, ING MECÁ batter, overturn, strain, MINAS *tiro de la chimenea* force
forzoso *adj* TELECOM mandatory
fosa: ~ **bajo la tela** *f* PAPEL wire pit; ~ **de barro** *f* MINAS clay pit; ~ **colectora** *f* CONST collecting pit; ~ **de la desfibradora** *f* PAPEL defiberer pit (*AmE*), defibrer pit (*BrE*); ~ **de la muela desfibradora** *f* PAPEL grinder pit; ~ **noruega** *f* GEOFÍS, GEOL, TEC PETR Norwegian trench; ~ **oceánica** *f* GEOFÍS, GEOL, OCEAN ocean trench; ~ **periférica** *f* GEOFÍS, GEOL, OCEAN marginal trench; ~ **profunda del fondo marino** *f* GEOFÍS, GEOL, OCEAN deep-sea trench; ~ **profunda del mar** *f* GEOFÍS, GEOL, OCEAN deep-sea trench; ~ **de remoldeo** *f* PROD *fundería* setting-up pit; ~ **séptica** *f* AGUA, HIDROL, RECICL septic tank; ~ **séptica de oxidación** *f* AGUA, HIDROL, RECICL oxidation pond; ~ **submarina** *f* GEOFÍS, GEOL, OCEAN trench, TRANSP MAR ocean deeps
fosfamo *m* QUÍMICA phospham
fosfatación *f* QUÍMICA phosphatization, REVEST phosphate coating
fosfatado[1] *adj* QUÍMICA, REVEST phosphated
fosfatado[2] *m* QUÍMICA, REVEST phosphatization
fosfatasa *f* QUÍMICA phosphatase
fosfático *adj* AGRIC, GEOL, QUÍMICA phosphatic
fosfatización *f* DETERG phosphation
fosfato *m* ALIMENT, DETERG, HIDROL, QUÍMICA phosphate; ~ **ácido** *m* QUÍMICA superphosphate; ~ **de cal** *m* QUÍMICA phosphate of lime; ~ **cálcico** *m* ALIMENT, QUÍMICA calcium phosphate; ~ **trisódico** *m* (*TSP*) DETERG trisodium phosphate (*TSP*)
fosfina *f* QUÍMICA phosphine

fosfito *m* QUÍMICA phosphite
fosfocatálisis *f* QUÍMICA phosphocatalysis
fosfoglicérico *adj* QUÍMICA phosphoglyceric
fosfolípido *m* ALIMENT, QUÍMICA phospholipid
fosfomolíbdico *adj* QUÍMICA phosphomolybdic
fosfonio *m* QUÍMICA phosphonium
fosforado *adj* QUÍMICA phosphorated, phosphorized
fosforescencia *f* ELECTRÓN, FÍS, FÍS RAD, INFORM&PD, TEC ESP phosphorescence, TERMO afterglow, TRANSP MAR phosphorescence
fosfórico *adj* QUÍMICA phosphoric
fosforil *m* QUÍMICA phosphoryl
fosforilado *adj* QUÍMICA phosphorylated
fosforilasa *f* QUÍMICA phosphorylase
fosforita *f* MINERAL phosphorite
fosforizar *vt* QUÍMICA phosphorize
fósforo *m* (*P*) ELECTRÓN, FÍS, METAL, QUÍMICA phosphorus (*P*)
fosforocalcita *f* MINERAL phosphochalcite, phosphorochalcite
fosforogénico *adj* QUÍMICA phosphorogenic
fosfotungstato *m* QUÍMICA phosphotungstate
fosfuro *m* QUÍMICA phosphide
fosgenita *f* MINERAL, QUÍMICA horn lead, phosgenite
fosgeno *m* D&A, QUÍMICA phosgene
fósil: ~ **característico** *m* GEOL index fossil; ~ **derivado** *m* GEOL derived fossil; ~ **guía** *m* GEOL index fossil; ~ **indicador** *m* GEOL index fossil; ~ **traza** *m* GEOL trace fossil
fósiles: **sin** ~ *adj* CARBÓN, GEOL, MINAS barren
fosilífero *adj* CARBÓN, GEOL, MINAS, PETROL fossiliferous
foso *m* AGUA ditch, pit, CARBÓN ditch, trench, CONST ditch, pit, HIDROL ditch, MINAS trench, PROD engine pit, TEC PETR trench; ~ **de bio-oxidación** *m* AGUA, HIDROL *aguas negras* bio-oxidation ditch; ~ **de cateo** *m* AmL (*cf foso de exploración Esp*) MINAS exploration trench; ~ **de colada** *m* PROD *estructuras de acero* casting pit; ~ **decantador** *m* CARBÓN slurry trench wall; ~ **descubierto** *m* MINAS open pit; ~ **de exploración** *m* Esp (*cf foso de cateo AmL*) MINAS exploration trench; ~ **de inspección** *m* AUTO, PROD, VEH inspection pit; ~ **de limpieza** *m* PROD cleaning pit; ~ **de moldear** *m* PROD molding hole (*AmE*), moulding hole (*BrE*); ~ **de moldeo** *m* PROD molding hole (*AmE*), moulding hole (*BrE*); ~ **de reparaciones** *m* PROD inspection pit; ~ **de sedimentación** *m* PROC QUÍ sedimentation pit, settling pit; ~ **submarino** *m* OCEAN depth, sea deep, sea moat
foticón *m* ELECTRÓN photicon
foto: ~ **fija de acción** *f* CINEMAT action still
fotocátodo *m* ELEC, FÍS, ING ELÉC, TV photocathode; ~ **semi-transparente** *m* ELEC, FÍS, ING ELÉC, TV semitransparent photocathode
fotocelda *f* GEN photocell
fotocélula *f* GEN photocell; ~ **alcalina** *f* FOTO alkaline photocell; ~ **multiplicadora** *f* INSTR multiplier phototube; ~ **de silicio** *f* ELECTRÓN silicon photocell; ~ **de sulfuro de cadmio** *f* ELEC, ING ELÉC cadmium sulfide cell (*AmE*), cadmium sulphide cell (*BrE*)
fotoclinómetro *m* INSTR, PETROL photoclinometer
fotocomponedora *f* CINEMAT, FOTO, IMPR filmsetter
fotocomposición *f* IMPR, INFORM&PD photo-typesetting; ~ **asistida por computadora** *f* AmL (*cf fotocomposición asistida por ordenador Esp*) IMPR,

INFORM&PD computer-assisted typesetting (*CAT*); ~ **asistida por ordenador** *f Esp* (*cf fotocomposición asistida por computadora AmL*), IMPR, INFORM&PD computer-assisted typesetting (*CAT*)

fotocompositora *f* IMPR filmset (*BrE*), photocomposer (*AmE*)

fotoconductividad *f* ELEC, ELECTRÓN, FÍS, FOTO, ÓPT, TELECOM photoconductivity; ~ **extrínseca** *f* ELEC extrinsic photoconductivity; ~ **indirecta** *f* ELEC indirect photoconductivity; ~ **intrínseca** *f* ELEC intrinsic photoconductivity

fotoconductivo *adj* ELEC, ELECTRÓN, FÍS, FOTO, ÓPT, TELECOM photoconductive

fotoconductor *adj* ELEC, ELECTRÓN, FÍS, FOTO, ÓPT, TELECOM light-positive, photoconductive

fotocopia: ~ **en negativo** *f* IMPR white photocopy

fotocorriente *f* ELEC, ELECTRÓN, FÍS, FOTO, ÓPT photocurrent

fotodesintegración *f* FÍS, ÓPT photodisintegration

fotodesprendimiento *m* NUCL photodetachment

fotodetección *f* ELEC, ELECTRÓN, ÓPT photodetection

fotodetector *m* ELEC, ELECTRÓN photodetector, FÍS RAD light detector, ÓPT, TELECOM photodetector; ~ **de diodos** *m* ELECTRÓN diode photodetector; ~ **de semiconductor** *m* ELECTRÓN semiconductor photodetector

fotodiodo *m* ELEC, ELECTRÓN, FÍS, FOTO, ÓPT, TELECOM photodiode; ~ **de avalancha** *m* ELECTRÓN, ÓPT avalanche photodiode (*APD*); ~ **de avalancha de silicio** *m* ELECTRÓN silicon avalanche photodiode; ~ **de descarga** *m* ELECTRÓN, ÓPT avalanche photodiode (*APD*); ~ **de difusión** *m* ELECTRÓN diffused photodiode; ~ **de empobrecimiento** *m* ELECTRÓN depletion layer photodiode; ~ **de germanio de avalancha** *m* ELEC, ELECTRÓN germanium avalanche photodiode; ~ **PIN** *m* ELECTRÓN PIN photodiode; ~ **positivo-intrínseco-negativo** *m* ELECTRÓN PIN photodiode; ~ **de silicio** *m* ELECTRÓN silicon photodiode

fotodisociación *f* QUÍMICA photodissociation

fotoeléctrico *adj* ELEC, ELECTRÓN photoelectric

fotoelectrón *m* ELEC, ELECTRÓN photoelectron

fotoelectrónico *adj* ELEC, ELECTRÓN photoelectronic

fotoemisión *f* ELEC, ELECTRÓN, FOTO, ÓPT, TV photoemission

fotoemisor *adj* ELEC, ELECTRÓN, FOTO, ÓPT, TV photoemissive

fotoestable *adj* IMPR, ÓPT fast-to-light

fotoflood *f* FOTO photoflood bulb

fotofluorografía *f* FÍS RAD, NUCL, TEC ESP photofluorography

fotogenerador *m* ELECTRÓN, ING ELÉC photogenerator

fotograbado *m* ELECTRÓN, IMPR photoengraving; ~ **directo** *m* IMPR direct gravure; ~ **de línea** *m* IMPR line engraving

fotograbador *m* ELECTRÓN, IMPR process engraver

fotograbar *vt* ELECTRÓN, IMPR photoengrave

fotógrafa *f* FOTO photographer

fotografía *f* CINEMAT, FOTO, TV *retrato* photograph, *proceso* photography; ~ **de acción** *f* CINEMAT action still; ~ **aérea** *f Esp* (*cf aerofotografía AmL*) FOTO aerial photography; ~ **de archivo** *f* FOTO file picture; ~ **a baja velocidad** *f* CINEMAT, FOTO low-speed photography; ~ **con cadencia temporizada de toma de imágenes** *f* CINEMAT, FOTO time-lapse photography; ~ **en color** *f* FOTO *proceso* color photography (*AmE*), colour photography (*BrE*), *retrato* color picture (*AmE*), colour picture (*BrE*); ~ **coloreada** *f* FOTO *proceso* color photography (*AmE*), colour photography (*BrE*); ~ **corporativa** *f* FOTO commercial photography; ~ **infrarroja** *f* FOTO infrared photography; ~ **con luz artificial** *f* FOTO artificial-light photography; ~ **por luz ultravioleta** *f* FÍS RAD ultraviolet photography; ~ **de naturaleza muerta** *f* FOTO still-life photography; ~ **sin objetivo** *f* FOTO pinhole photography; ~ **panorámica** *f* FOTO panoramic photograph; ~ **de polvo** *f* CRISTAL powder photograph; ~ **principal** *f* CINEMAT, FOTO principal photography; ~ **de producción** *f* CINEMAT, FOTO production still; ~ **publicitaria** *f* FOTO advertising photography; ~ **por rayos X** *f* FÍS RAD, FOTO, INSTR X-ray photograph, *proceso* X-ray photography; ~ **de rotación** *f* CINEMAT, CRISTAL, FOTO rotation photography; ~ **de Schlieren** *f* FÍS, FOTO Schlieren photography; ~ **silueteada** *f* FOTO *con el fondo desvanecido* cutout photograph; ~ **submarina** *f* FOTO underwater photograph, *proceso* underwater photography; ~ **en triconomía** *f* FOTO three-color photography (*AmE*), three-colour photography (*BrE*); ~ **tricroma** *f* FOTO three-color photography (*AmE*), three-colour photography (*BrE*); ~ **ultrarrápida** *f* CINEMAT, FOTO ultrahigh-speed photography

fotografiar: ~ **con trípode** *vt* FOTO photograph with a tripod

fotógrafo *m* FOTO photographer; ~ **de paisajes** *m* FOTO landscape photographer

fotograma[1]: ~ **a fotograma** *adv* CINEMAT, FOTO, TV frame by frame

fotograma[2] *m* CINEMAT, FOTO, TV frame; ~ **en blanco** *m* CINEMAT, TV blank frame; ~ **completo** *m* CINEMAT, FOTO, TV full-frame; ~ **congelado** *m* CINEMAT, FOTO, TV frozen frame; ~ **instantáneo** *m* CINEMAT, TV flash frame; ~ **único** *m* CINEMAT, TV single frame

fotogrametría *f* CONST, TEC ESP photogrammetry

fotohaluros *m pl* FOTO photohalides

fotoiniciador *m* FÍS, P&C photoinitiator

fotoionización *f* FÍS, P&C photoionization

fotólisis *f* ENERG RENOV, QUÍMICA photolysis; ~ **instantánea** *f* QUÍMICA flash photolysis

fotolítico *adj* QUÍMICA photolytic

fotolitografía *f* ELECTRÓN, IMPR photolithography

fotoluminiscencia *f* FÍS RAD, QUÍMICA photoluminescence

fotomáscara *f* ELECTRÓN photomask

fotomecánico *adj* IMPR photomechanical

fotometría *f* FÍS photometry

fotómetro *m* CINEMAT, FÍS, FOTO exposure meter, light meter, photometer, INSTR, P&C photometer; ~ **de contraste** *m* FOTO contrast photometer, match photometer; ~ **de destellos** *m* FÍS flicker photometer; ~ **incorporado** *m* CINEMAT, FOTO built-in light meter; ~ **de llama** *m Esp* (*cf flamómetro AmL*) INSTR, LAB flame photometer; ~ **de Lummer-Brodhun** *m* FÍS, LAB Lummer-Brodhun photometer; ~ **microscópico** *m* INSTR microscopic photometer; ~ **a través del objetivo** *m* CINEMAT through-the-lens light meter (*TTL light meter*)

fotomicrografía *f* METAL photomicrograph

fotomicrograma *m* METAL photomicrogram

fotomicroscopio *m* INSTR, METAL photomicroscope

fotomontaje *m* FOTO photomounting

fotomultiplicador *m* ELECTRÓN, FÍS, FÍS PART, FÍS RAD photomultiplier

fotón *m* FÍS, FÍS PART, FÍS RAD, ÓPT photon; **~ de aniquiliación** *m* FÍS RAD annihilation photon; **~ gamma** *m* FÍS RAD gamma-ray photon; **~ de rayos X** *m* FÍS PART, FÍS RAD, NUCL X-ray photon; **~ ultravioleta** *m* FÍS PART ultraviolet photon

fotopolimerización *f* FÍS, IMPR, QUÍMICA photopolymerization

fotopolímero *m* FÍS, IMPR, QUÍMICA photopolymer

fotoprotector *m* ELECTRÓN photoresist; **~ negativo** *m* ELECTRÓN negative photoresist; **~ positivo** *m* ELECTRÓN positive photoresist

fotoquímica *f* CONTAM, FOTO, QUÍMICA photochemistry

fotoquímico *adj* CONTAM, FOTO, QUÍMICA photochemical

fotoreacción *f* FOTO, QUÍMICA photoreaction

fotoresistor *m* FÍS RAD, FOTO, METAL photoresistor

fotorreflector *m* CINEMAT photoflood

fotorresistente *adj* FOTO light-negative

fotosensibilidad *f* CINEMAT, ELECTRÓN, FÍS, FOTO, IMPR photosensitivity

fotosensible *adj* GEN light-sensitive, photosensitive

fotosensor *m* ELECTRÓN, FOTO, ÓPT photosensor

fotosfera *f* FÍS RAD, TEC ESP photosphere

fotosíntesis *f* ENERG RENOV, HIDROL, QUÍMICA photosynthesis

fototeca *f* FOTO picture library

fototelegrafía *f* TELECOM phototelegraphy

fototeodolito *m* INSTR phototheodolite

fototransistor *m* ELECTRÓN, FÍS RAD, INFORM&PD phototransistor; **~ de silicio** *m* ELECTRÓN silicon phototransistor

fototrazador *m* ELECTRÓN, INFORM&PD photoplotter

fototubo *m* ELECTRÓN, INSTR phototube; **~ de cesio** *m* ELECTRÓN, INSTR caesium phototube (*BrE*), cesium phototube (*AmE*); **~ de gas** *m* ELECTRÓN, INSTR gas phototube; **~ multiplicador** *m* ELECTRÓN, INSTR multiplier phototube; **~ multiplicador electrónico** *m* ELECTRÓN, FÍS RAD, INSTR electron-multiplier phototube; **~ de vacío** *m* ELECTRÓN, INSTR vacuum phototube

fotovaristor *m* ELECTRÓN photovaristor

fotovoltaico *adj* ELECTRÓN, FÍS, TEC ESP photovoltaic

foulardado *m* TEXTIL padding

Fourdrinier *m* PAPEL Fourdrinier

fowlerita *f* MINERAL fowlerite

FPA *abr* (*factor de proteína animal*) AGRIC, ALIMENT APF (*animal protein factor*)

FPLA *abr* (*arreglo lógico programable de campo*) INFORM&PD FPLA (*field-programmable logic array*)

Fr *abr* (*francio*) FÍS RAD, METAL, NUCL, QUÍMICA Fr (*francium*)

fracaso: **~ de la ley de reciprocidad** *m* CINEMAT, FOTO, TV reciprocity-law failure

fracción *f* MATEMÁT, TEC PETR *refino* fraction; **~ de bloqueo de conexión** *f* TEC ESP, TELECOM freeze-out fraction (*FOF*); **~ de bloqueo momentáneo** *f* TEC ESP, TELECOM freeze-out fraction (*FOF*); **~ compleja** *f* MATEMÁT complex fraction; **~ compuesta** *f* MATEMÁT complex fraction; **~ común** *f* MATEMÁT simple fraction; **~ densimétrica** *f* CARBÓN specific gravity fraction; **~ efectiva de neutrones diferidos** *f* NUCL effective delayed-neutron fraction; **~ de empaquetado** *f* FÍS, ÓPT, TELECOM packing fraction; **~ irreducible** *f* MATEMÁT reduced fraction; **~ molar** *f* REFRIG mole fraction; **~ molecular** *f* METAL, QUÍMICA mole fraction; **~ en peso** *f* NUCL weight fraction; **~ de precursores** *f* NUCL parent fraction; **~ seca** *f* TERMOTEC *vapor* dry fraction; **~ simple** *f* MATEMÁT simple fraction; **~ de tiempo** *f* INFORM&PD time slice; **~ de volumen** *f* METAL volume fraction

fraccionado: **~ por tamaños** *m* CARBÓN grain fraction, size fraction

fraccionamiento *m* QUÍMICA fractionation; **~ del tiempo** *m* INFORM&PD time slicing

fraccionario *adj* MATEMÁT fractional

fracciones: **~ densas** *f pl* TEC PETR heavy fractions; **~ de hidrocarburo pesado** *f pl* TEC PETR heavy hydrocarbon fractions; **~ de hidrocarburos ligeras** *f pl* TEC PETR light hydrocarbon fractions; **~ ligeras** *f pl* TEC PETR light fractions; **~ pesadas** *f pl* TEC PETR heavy fractions

fractal *m* INFORM&PD, MATEMÁT fractal

fractura *f* C&V fracture, CONST *minería, construcción* breaking, CRISTAL fracture, failure, GEOL joint plane, fracturing, ING MECÁ, MECÁ, METAL, NUCL, PETROL fracture, PROC QUÍ breaking, TEC PETR *perforación* fracture, fracturing; **~ de apilamiento** *f* CRISTAL stacking fault; **~ astillosa** *f* CARBÓN splintery fracture, CRISTAL ductile fracture, fibrous fracture; **~ de borde** *f* C&V edge fracture; **brecha por ~** *f* GEOL crush breccia; **~ concoidal** *f* C&V, CRISTAL conchoidal fracture; **~ concoide** *f* C&V, CRISTAL conchoidal fracture; **~ de cono y copa** *f* METAL *ensayo de tracción* cup and cone fracture; **~ de cono y embudo** *f* METAL *ensayo de tracción* cup and cone fracture; **~ cristalina** *f* CRISTAL, METAL crystalline fracture; **~ por deslizamiento** *f* NUCL sliding fracture; **~ deslizante** *f* NUCL gliding fracture; **~ dúctil** *f* CRISTAL, METAL ductile fracture; **~ por esfuerzo cortante** *f* METAL shear fracture; **~ en espiral** *f* C&V spiral fracture; **~ en estrella** *f* C&V star fracture; **~ por fatiga** *f* CRISTAL fatigue fracture, METAL fatigue failure; **~ fibrosa** *f* CRISTAL, METAL, NUCL fibrous fracture; **~ en forma de cuña** *f* METAL wedge-type fracture; **~ por fragilidad** *f* CRISTAL, MECÁ, METAL brittle fracture; **~ por fragilización** *f* NUCL brittle failure; **~ granular** *f* NUCL granular fracture; **~ hertziana** *f* C&V Hertzian fracture; **~ por impacto** *f* NUCL impact fracture; **~ inestable** *f* METAL unstable fracture; **~ plana** *f* GEOFÍS, GEOL, TEC PETR fracture plane; **~ retardada** *f* METAL delayed fracture; **~ de roca** *f* GEOL rock fracture; **~ semiquebradiza** *f* CRISTAL, MECÁ, METAL semibrittle fracture; **~ sesgada** *f* METAL slant fracture; **~ transversal** *f* C&V tranverse fracture

fracturación *f* GEOL, PETROL, TEC PETR fracturing; **~ en bloques** *f* GEOL block faulting; **~ hidráulica** *f* GEOL, PETROL, TEC PETR hydraulic fracturing

fracturar *vt* CONST break

fractus: **~ stratus** *m* METEO stratus fractus

fragata *f* D&A, TRANSP MAR frigate

frágil *adj* C&V, CRISTAL, GEOL, MECÁ, METAL, P&C brittle

fragilidad *f* C&V, CRISTAL, GEOL, MECÁ, METAL, P&C brittleness; **~ azul** *f* METAL blue brittleness; **~ por ductibilidad** *f* METAL temper brittleness; **~ en frío** *f* MECÁ, REFRIG, TERMO cold brittleness; **~ por revenido** *f* METAL temper brittleness

fragilización *f* GEN embrittlement

fragmentación f GEOL cataclasm, MECÁ breaking up, METAL fragmentation; ~ **de burbujas** f C&V *atrapadas en la masa del vidrio fundido* broken seed; ~ **de memoria** f INFORM&PD storage fragmentation

fragmento m ING MECÁ part, PROD scrap

fragmentos m pl CARBÓN debris; ~ **de fibras** m pl PAPEL dusting; ~ **de fisión** m pl FÍS, FÍS PART, NUCL fission fragments

fragua f CONST smithy, MINAS, PROD forge

fraguado adj CONST, ING MECÁ, MECÁ, METAL, PROD forged, wrought

fraguar vi CONST *cemento, hormigón* set

framboide m GEOL, MATEMÁT framboid

francio m (Fr) FÍS RAD, METAL, NUCL, QUÍMICA francium (Fr)

franco adj AGRIC loamy; ~ **a bordo** adj (FOB) TEC PETR, TRANSP MAR free on board (FOB); ~ **al costado del buque** adj (FAS) TRANSP MAR free alongside ship (FAS); ~ **en muelle** adj TRANSP MAR free on quay (FOQ)

francoarenoso adj AGRIC sandy loam

francobordo m TRANSP MAR *construcción naval* freeboard

frangulina f QUÍMICA frangulin

franja: ~-**almacén** f CARBÓN *minas* shrinkage limit; ~ **arenosa** f HIDROL river bar; ~ **brillante** f FÍS bright fringe; ~ **del camino** f CONST track strip; ~ **capilar** f HIDROL capillary fringe; ~ **continental** f OCEAN continental fringe; ~ **de control** f CINEMAT control strip; ~ **gamma** f CINEMAT gamma strip; ~ **límite** f ING MECÁ limit strip; ~ **de luz blanca** f FÍS white-light fringe; ~ **microscópica** f FÍS microstrip; ~ **oscura** f FÍS dark fringe; ~ **de tiempo** f TEC ESP time slot

franjas: ~ **de absorción** f pl FÍS, FÍS RAD absorption fringes; ~ **acromáticas** f pl FÍS achromatic fringes; ~ **de contorno** f pl FÍS contour fringes; ~ **de Fizeau** f pl FÍS Fizeau fringes; ~ **de Haidinger** f pl FÍS Haidinger fringes; ~ **en el impreso debidas a los engranajes** f pl IMPR gear streaks; ~ **de interferencia** f pl FÍS, FÍS ONDAS interference fringes; ~ **localizadas** f pl FÍS localized fringes; ~ **de Moiré** f pl FÍS Moiré fringes; ~ **no localizadas** f pl FÍS nonlocalized fringes

franklinita f MINERAL franklinite

franquear vt GEOL cross

franqueo m ING MECÁ unkeying; ~ **en forma de cono** m MINAS pyramid cut; ~ **vertical** m AmL (cf *separación vertical Esp*) MINAS clearance

franquía[1]: **en** ~ adj TRANSP MAR in the offing

franquía[2]: **en** ~ adv TRANSP MAR in the offing

franquía[3] f TRANSP MAR offing

frasco: ~ **de boca ancha** m LAB wide-mouth bottle; ~ **de cuello angosto** m LAB narrow-necked bottle; ~ **Dewar** m FÍS, LAB, PROC QUÍ Dewar flask; ~ **de filtración** m LAB filter tank, filtering tank, PROC QUÍ filter flask, filtering flask; ~ **para liofilización** m C&V lyophilization flask; ~ **para penicilina** m C&V penicillin phial (BrE), penicillin vial (AmE); ~ **pequeño** m LAB, QUÍMICA phial (BrE), vial (AmE); ~ **de reactivo** m LAB reagent bottle; ~ **con tapón** m LAB *recipiente para reactivos* stoppered bottle; ~ **de yodo** m LAB iodine flask

Frasniense adj GEOL Frasnian

frasqueta f IMPR frisket

frecuencia f GEN frequency, ELECTRÓN frequency, periodicity, INFORM&PD rate, frequency;

~ a ~ **acústica** f ACÚST, ELEC beat frequency, ELECTRÓN acoustic frequency (VF), audio frequency (AF), FÍS beat frequency, TELECOM voice frequency (VF); ~ **ajena** f ELECTRÓN aliased frequency; ~ **del alabeo angular** f TRANSP AÉR angular roll rate; ~ **de alimentación eléctrica** f ELEC, ING ELÉC current frequency (AmE), mains frequency (BrE); ~ **angular** f (W) ACÚST, ELEC, FÍS, FÍS ONDAS, FÍS RAD, ING ELÉC angular frequency (W); ~ **ascendente** f TEC ESP uplink frequency; ~ **audible** f ACÚST, ELECTRÓN, TELECOM audio frequency (AF); ~ **auxiliar** f CINEMAT pilot frequency;

~ b ~ **en banda intermedia** f ELECTRÓN midband frequency; ~ **de banda L** f TRANSP MAR L-band frequency; ~ **de banda lateral** f ELECTRÓN, TELECOM sideband frequency; ~ **de barrido** f ELECTRÓN sweep frequency, TRANSP AÉR scanning rate, TV sweep frequency; ~ **de base** f ELECTRÓN basic frequency, clock frequency, INFORM&PD clock frequency, clock rate, TV basic frequency; ~ **de la base de tiempo** f ELEC time base frequency; ~ **básica** f ELECTRÓN, TV basic frequency; ~ **de bombeo** f TEC ESP pump frequency;

~ c ~ **del cabeceo angular** f TRANSP AÉR angular pitch rate; ~ **característica** f FÍS RAD characteristic frequency; ~ **de centelleo** f CINEMAT, TV flicker frequency; ~ **de centelleo crítica** f CINEMAT, TV critical flicker frequency; ~ **central del canal** f TELECOM channel-center frequency (AmE) (CCF), channel-centre frequency (BrE) (CCF); ~ **del ciclotrón de electrones** f ELECTRÓN, FÍS, FÍS PART, NUCL electron-cyclotron frequency; ~ **ciclotrónica** f ELECTRÓN, FÍS, FÍS PART, NUCL cyclotron frequency; ~ **de cierre** f ACÚST, ELEC, ELECTRÓN, FÍS, TELECOM cutoff frequency; ~ **de coincidencia** f ACÚST coincidence frequency; ~ **de conmutación** f PROD switching frequency; ~ **de conmutación de colores** f TV color-sampling rate (AmE), colour-sampling rate (BrE); ~ **controlada** f ELECTRÓN controlled frequency; ~ **de conversión** f ELECTRÓN conversion frequency; ~ **de la corriente** f ELEC, ING ELÉC current frequency (AmE), mains frequency (BrE); ~ **de corte** f ACÚST, ELEC, ELECTRÓN, FÍS, TELECOM cutoff frequency; ~ **de corte teórica** f ACÚST, ELEC, ELECTRÓN, FÍS, TELECOM theoretical cutoff frequency; ~ **de corte del transistor** f ACÚST, ELEC, ELECTRÓN, FÍS, TELECOM transistor cutout frequency; ~ **por cristal** f CRISTAL, ELECTRÓN crystal frequency; ~ **crítica** f ELECTRÓN, FÍS ONDAS threshold frequency, TEC ESP *comunicaciones* critical frequency, TELECOM threshold frequency; ~ **de cruce** f ELECTRÓN, TELECOM crossover frequency;

~ d ~ **de Debye** f FÍS Debye frequency; ~ **dedicada** f TELECOM dedicated frequency; ~ **designada** f TELECOM designated frequency; ~ **desintonizada** f ELECTRÓN off-tune frequency; ~ **de deslizamiento** f TV sliding frequency; ~ **diferencial** f ELECTRÓN difference frequency; ~ **discreta** f ELECTRÓN discrete frequency; ~ **distribuida** f TV allocated frequency; ~ **Doppler** f ELECTRÓN Doppler frequency;

~ e ~ **emitida por las cuerdas vocales** f FÍS RAD voice frequency (VF); ~ **de empleo** f ING MECÁ operating frequency; ~ **de energía comercial** f ELEC, ING ELÉC commercial power frequency; ~ **de error** f ELECTRÓN, INFORM&PD, TELECOM error rate; ~ **de errores por manipulación** f INFORM&PD keying error rate; ~ **espacial** *f* FÍS, ING ELÉC spatial frequency; ~ **del estado fundamental** *f* FÍS, FÍS PART, FÍS RAD, QUÍMICA ground-state frequency; ~ **excluida** *f* ELECTRÓN rejected frequency; ~ **de exploración** *f* INFORM&PD scan rate; ~ **de exploración horizontal** *f* TV horizontal scanning frequency; ~ **extremadamente baja** *f* (*FEB*) ELEC, ELECTRÓN, TELECOM extremely low frequency (*ELF*);

~ f ~ **de fallos** *f* INFORM&PD, ING ELÉC, ING MECÁ, TEC ESP failure rate; ~ **de filtrado** *f* ELECTRÓN, FÍS RAD filter frequency; ~ **fundamental** *f* ACÚST, ELECTRÓN, TELECOM fundamental frequency;

~ g ~ **de la guiñada angular** *f* TRANSP AÉR angular yaw rate;

~ i ~ **de imagen** *f* CINEMAT frame rate, ELECTRÓN, TELECOM image frequency, video frequency, TV field frequency, frame rate; ~ **de imágenes** *f* CINEMAT, TV frame frequency; ~ **de impulsos** *f* ELECTRÓN pulse frequency; ~ **infrasónica** *f* ACÚST, FÍS infrasonic frequency; ~ **instantánea** *f* ELECTRÓN, TEC ESP instantaneous frequency; ~ **intermedia** *f* (*FI*) ELECTRÓN, TELECOM, TV intermediate frequency (*IF*); ~ **intermedia secundaria** *f* ELECTRÓN, TELECOM, TV second intermediate frequency;

~ l ~ **de Larmor** *f* FÍS Larmor frequency; ~ **lateral** *f* ELECTRÓN, TELECOM side frequency; ~ **lenta** *f* ELECTRÓN idler frequency; ~ **límite superior** *f* ING ELÉC maximum usable frequency (*MUF*); ~ **de línea** *f* TV line frequency; ~ **local del oscilador** *f* ELECTRÓN local oscillator frequency; ~ **localizada** *f* TV allocated frequency;

~ m ~ **de marcha continua** *f* ELECTRÓN free-running frequency; ~ **máxima de consumo** *f* ELÉC, ING ELÉC, TELECOM maximum usable frequency (*MUF*); ~ **máxima utilizable** *f* ELÉC, ING ELÉC, TELECOM maximum usable frequency (*MUF*); ~ **media** *f* ING ELÉC, INSTAL TERM medium frequency; ~ **de microondas** *f* ELECTRÓN, FÍS microwave frequency; ~ **de modulación** *f* ELEC, ELECTRÓN, TELECOM modulation frequency; ~ **modulada** *f* ELEC, ELECTRÓN, FÍS, FÍS ONDAS, FÍS PART, INFORM&PD, TELECOM, TV frequency modulation; ~ **de muestreo** *f* ELECTRÓN, INFORM&PD, PETROL, TELECOM sampling frequency, sampling rate; ~ **de muestreo cromático** *f* FÍS RAD color-sampling frequency (*AmE*), colour-sampling frequency (*BrE*); ~ **muy alta** *f* (*FMA*) ELECTRÓN, FÍS ONDAS, TEC ESP, TELECOM very high frequency (*VHF*); ~ **muy elevada** *f* (*FME*) ELECTRÓN, FÍS ONDAS, TEC ESP, TELECOM very high frequency (*VHF*);

~ n ~ **natural** *f* ACÚST, ELEC, ELECTRÓN, FÍS, FÍS ONDAS, FÍS RAD, ING MECÁ, PETROL, TEC ESP natural frequency; ~ **necesaria** *f* ENERG RENOV required frequency; ~ **de nivel de blanco** *f* TV white-level frequency; ~ **de nivel del negro** *f* TV black-level frequency; ~ **nominal** *f* ELEC rated frequency, ELECTRÓN center frequency (*AmE*), centre frequency (*BrE*); ~ **normalizada** *f* ÓPT, TELECOM normalized frequency; ~ **de Nyquist** *f* PETROL Nyquist frequency;

~ o ~ **de onda** *f* CONST ripple frequency, FÍS RAD ripple frequency, wave frequency; ~ **de la onda portadora** *f* ELECTRÓN, FÍS, TELECOM, TV carrier frequency; ~ **de ondulación** *f* CONST, FÍS RAD ripple frequency; ~ **óptica** *f* ELECTRÓN, ÓPT optical frequency; ~ **de oscilación** *f* ELECTRÓN, TELECOM oscillation frequency;

~ p ~ **de parpadeo** *f* CINEMAT, TV flicker frequency; ~ **de parpadeo de imagen** *f* CINEMAT, TV field rate flicker; ~ **perturbada** *f* FÍS RAD perturbed frequency; ~ **piloto** *f* TV control frequency; ~ **de plegamiento** *f* GEOL, PETROL folding frequency; ~ **de la portadora** *f* ELECTRÓN, INFORM&PD, TELECOM, TV carrier frequency; ~ **de la portadora de imagen** *f* TELECOM, TV picture-carrier frequency; ~ **de preselección** *f* ELECTRÓN preset frequency; ~ **principal** *f* ELECTRÓN master frequency; ~ **propia** *f* ACÚST, ELEC, ELECTRÓN, FÍS, FÍS ONDAS, FÍS RAD, ING MECÁ, PETROL, TEC ESP eigenfrequency, natural frequency; ~ **de prueba** *f* ELECTRÓN test frequency; ~ **de pulsación** *f* ACÚST, ELEC, FÍS beat frequency; ~ **de pulsos de sincronización** *f* TV sync tip frequency;

~ r ~ **del radar náutico** *f* TRANSP MAR marine-radar frequency; ~ **de radiación** *f* FÍS RAD radiation frequency; ~ **de radio** *f* (*RF*) GEN radio frequency (*RF*); ~ **radioeléctrica** *f* ACÚST, ELEC, ELECTRÓN, FÍS RAD, ING ELÉC, NUCL, TEC ESP, TELECOM, TRANSP, TRANSP AÉR, TV radio frequency (*RF*); ~ **de recepción** *f* ELECTRÓN, TELECOM, TV reception frequency; ~ **rechazada** *f* ELECTRÓN rejected frequency; ~ **rectilínea** *f* ELEC, ING ELÉC straight-line frequency; ~ **de red** *f* ELEC, ING ELÉC current frequency (*AmE*), mains frequency (*BrE*), power frequency; ~ **de la red eléctrica** *f* ELEC, ING ELÉC current frequency (*AmE*), mains frequency (*BrE*), power frequency; ~ **de referencia** *f* TELECOM reference frequency; ~ **de régimen** *f* ELEC rated frequency; ~ **relativa** *f* INFORM&PD relative frequency; ~ **de reloj** *f* ELECTRÓN clock frequency, INFORM&PD clock rate, clock frequency; ~ **de repetición** *f* ELECTRÓN repetition rate; ~ **de repetición de impulsos** *f* ELECTRÓN, FÍS, INFORM&PD, TELECOM pulse-repetition frequency (*PRF*); ~ **resonante** *f* ACÚST, ELECTRÓN, FÍS, FÍS ONDAS, FÍS RAD, TELECOM resonance frequency, resonant frequency;

~ s ~ **de segundo canal** *f* ELECTRÓN second-channel frequency; ~ **de señal** *f* ELECTRÓN signal frequency; ~ **de servicio** *f* ELEC rated frequency, ING MECÁ operating frequency; ~ **de sintonización normalizada** *f* ACÚST standard tuning frequency; ~ **de sonido** *f* ELECTRÓN audio frequency, sound frequency; ~ **de la subportadora** *f* ELECTRÓN, TELECOM, TV subcarrier frequency; ~ **subsónica** *f* ACÚST, ELECTRÓN, FÍS RAD subsonic frequency; ~ **superalta** *f* (*FSA*) ELECTRÓN, TEC ESP, TELECOM super-high frequency (*SHF*); ~ **supersónica** *f* FÍS RAD supersonic frequency;

~ t ~ **telefónica** *f* TELECOM voice frequency (*VF*); ~ **temporal** *f* ELECTRÓN time frequency; ~ **de trabajo** *f* ING MECÁ, TELECOM operating frequency; ~ **de transición** *f* ELECTRÓN, TELECOM crossover frequency; ~ **de transmisión** *f* ELECTRÓN, TELECOM, TV transmitting frequency; ~ **de transmisión simultánea** *f* ELECTRÓN multiplexing frequency; ~ **de turno** *f* PROD shift clock rate;

~ u ~ **ultra-alta** *f* (*FUA*) ELECTRÓN, TELECOM, TV ultrahigh frequency (*UHF*); ~ **ultraelevada** *f* ELECTRÓN, FÍS ONDAS, TELECOM, TV hyperfrequency, ultrahigh frequency (*UHF*); ~ **ultrasónica** *f* ACÚST, FÍS, FÍS RAD ultrasonic frequency; ~ **umbral** *f*

ELECTRÓN, FÍS, FÍS ONDAS threshold frequency; ~ **de utilización** *f* ING MECÁ operating frequency;

~ v ~ **de variación de velocidad** *f* ACÚST speed variation frequency; ~ **de video** *AmL,* ~ **de vídeo** *f Esp* ELECTRÓN, TELECOM, TV video frequency; ~ **vocal** *f (FV)* FÍS RAD, TELECOM voice frequency *(VF);* ~ **de voz** *f* FÍS RAD, TELECOM voice frequency *(VF);*

~ z ~ **de zumbido** *f* CONST, FÍS RAD ripple frequency

frecuencias: ~ **conversacionales** *f pl* ACÚST conversational frequencies; ~ **medias** *f pl* TELECOM *radiodifusión* middle frequencies; ~ **preferentes** *f pl* ACÚST preferred frequencies

frecuencímetro *m* ELECTRÓN, ING ELÉC, INSTR, TELECOM frequency meter

freibergita *f* MINERAL freibergite

freidora *f* ALIMENT deep-fryer

freieslebenita *f* MINERAL freieslebenite

frenada *f* TEC ESP *vehículos* deceleration

frenado *m* FERRO braking, FÍS drag, MECÁ braking, TEC ESP *descenso de aeronaves* drag; ~ **accidental** *m* FERRO, VEH accidental braking; ~ **aerodinámico** *m* TEC ESP aerodynamic braking, TRANSP AÉR dumping; ~ **combinado** *m* FERRO, VEH combined braking; ~ **de la corriente de Foucault** *m* ELEC *motor,* ING ELÉC eddy-current braking; ~ **de corrientes parásitas** *m* ELEC, ING ELÉC eddy-current braking; ~ **diferencial** *m* TRANSP AÉR differential braking; ~ **dinámico** *m* FERRO, TRANSP, VEH dynamic braking; ~ **eléctrico** *m* ELEC *motor, generador* electric braking; ~ **espontáneo** *m* FERRO, VEH spontaneous brake application; ~ **en función de la carga** *m* TRANSP load-sensitive braking; ~ **graduado** *m* FERRO, VEH graduated braking; ~ **inesperado** *m* FERRO unexpected braking; ~ **por inversión de las conexiones** *m* ING ELÉC plugging; ~ **por inversión de la secuencia de fases** *m* ING ELÉC plugging; ~ **inverso** *m* TRANSP reverse braking; ~ **normal** *m* FERRO normal brake application; ~ **regenerativo** *m* FERRO regenerative braking; ~ **reostático** *m* FERRO rheostatic braking, TRANSP dynamic braking, VEH rheostatic braking; ~ **suave** *m* FERRO, VEH smooth braking

freno *m* AUTO, ING MECÁ brake, MECÁ drag, VEH brake; ~ **de aceite** *m* FERRO, ING MECÁ oil dashpot; ~ **aerodinámico** *m* ING MECÁ, MECÁ, TRANSP, TRANSP AÉR air brake; ~ **aerodinámico de picado** *m* TRANSP AÉR diving brake; ~ **de aire** *m* ING MECÁ, VEH *carburador* air brake; ~ **de aire comprimido** *m* AUTO air pressure brake, ING MECÁ, MECÁ air brake, atmospheric brake; ~ **de ajuste automático** *m* AUTO, VEH self-adjusting brake; ~ **de aparcamiento** *m* TRANSP AÉR parking brake; ~ **automático** *m* AUTO self-acting brake, VEH self-acting brake, *sistema de frenos* automatic brake; ~ **autoregulable** *m* AUTO, VEH self-adjusting brake; ~ **de banda** *m* ING MECÁ, MECÁ band brake, ribbon brake; ~ **sobre el carril** *m* FERRO rail brake; ~ **de cinta** *m* ING MECÁ, MECÁ band brake, strap brake; ~ **compensador** *m* FERRO *infraestructura* compensating buffer; ~ **por compresión de aire en el motor** *m* AUTO, VEH exhaust brake; ~ **de corrientes parásitas** *m* FERRO *vehículos,* ING ELÉC, MECÁ, VEH eddy-current brake *(AmE),* eddy brake *(BrE);* ~ **de cuña** *m* ING MECÁ, MECÁ wedge brake; ~ **de disco** *m* AUTO, ING MECÁ, MECÁ, VEH disc brake *(BrE),* disk brake *(AmE);* ~ **de doble circuito** *m* TRANSP double-circuit brake, dual-circuit brake; ~ **de dos zapatas opuestas entre sí** *m* TRANSP clasp brake; ~ **electromagnético** *m* AUTO, FERRO, TRANSP, VEH electromagnetic brake; ~ **electroneumático** *m* FERRO, TRANSP, VEH electropneumatic brake; ~ **de emergencia** *m* TRANSP AÉR, VEH emergency brake; ~ **de estacionamiento** *m* AUTO, FERRO, MECÁ, VEH parking brake; ~ **de expansión interno** *m* ING MECÁ internal expanding brake; ~ **de fricción** *m* VEH friction brake; ~ **hidráulico** *m* VEH hydraulic brake; ~ **hidrocinético** *m* TRANSP hydrokinetic brake; ~ **hidroneumático** *m* TRANSP hydropneumatic brake; ~ **de husillo** *m* FERRO, VEH screw brake; ~ **de husillo con manivela** *m* FERRO, VEH screw brake with crank handle; ~ **de jaula** *m Esp (cf mecanismo moderador de velocidad al final de la carrera AmL)* MINAS overwinder; ~ **de línea doble** *m* AUTO twin-line brake; ~ **magnético** *m* FERRO, ING ELÉC, MECÁ eddy brake *(BrE),* eddy-current brake *(AmE);* ~ **de mano** *m* AUTO, ING MECÁ, MECÁ, VEH brake band, handbrake; ~ **de máquina herramienta** *m* ING MECÁ machine-tool brake; ~ **mecánico de vacío** *m* AUTO vacuum-assisted power brake; ~ **del molinete** *m* ING MECÁ pawl rim; ~ **del motor** *m* AUTO, VEH engine brake; ~ **neumático** *m* AUTO air pressure brake, ING MECÁ air brake, vacuum brake, VEH pneumatic brake; ~ **de palanca** *m* ING MECÁ lever brake; ~ **de parada** *m* FERRO stop brake, MECÁ parking brake; ~ **de picado** *m* TRANSP AÉR diving brake; ~ **de pie** *m* AUTO, VEH footbrake; ~ **de Prony** *m* AUTO, VEH Prony brake; ~ **del remolque** *m* VEH trailer brake; ~ **reostático** *m* TRANSP rheostatic brake; ~ **de servicio** *m* VEH service brake; ~ **sinterizado** *m* FERRO sintered brake; ~ **para sobrevelocidad** *m* ING MECÁ overspeed brake; ~ **de tambor** *m* AUTO drum brake; ~ **de tornillo** *m* ING MECÁ screw brake; ~ **para tuerca** *m* ING MECÁ nut lock; ~ **de vacío** *m* FERRO, ING MECÁ, VEH vacuum brake; ~ **de vacío automático** *m* FERRO, ING MECÁ, VEH automatic vacuum brake; ~ **de vacío automático doble** *m* FERRO, ING MECÁ, VEH duplex automatic vacuum brake; ~ **de vapor** *m* INSTAL HIDRÁUL steam brake; ~ **del ventilador** *m* ING MECÁ fan brake; ~ **de zapata** *m* VEH shoe brake

frenos: ~ **de zapatas** *m pl* FERRO, VEH disc brake calliper *(BrE),* disk brake caliper *(AmE)*

frente[1]: **de ~** *adj* SEG *colisión* head-on

frente[2] *m* CARBÓN *filón* breast, CONST heading, front, face, METEO front; ~ **anabático** *m* METEO anabatic front; ~ **de arranque** *m AmL (cf frente de trabajo Esp)* CARBÓN, MINAS bank, face, stope face, working face; ~ **de arranque del carbón** *m* CARBÓN, MINAS coal wall, coalface; ~ **de ataque** *m* CARBÓN advancing face, MINAS breast face; ~ **de avance** *m* MINAS heading face; ~ **de caldera** *m* INSTAL HIDRÁUL boiler front, foreboiler; ~ **caliente** *m* METEO warm front; ~ **del camino en construcción** *m* CARBÓN road head; ~ **de cantera** *m* MINAS quarry head; ~ **casi estacionario** *m* METEO stationary front; ~ **catabático** *m* METEO katabatic front; ~ **de choque** *m* GEOFÍS *tormenta magnética* bow shock; ~ **de choque de la magnetosfera** *m* TEC ESP magnetosphere bow shock; ~ **de dragado** *m* MINAS dredging face; ~ **estacionario** *m* METEO stationary front; ~ **de excavación** *m* MINAS digging face; ~ **de extracción** *m* MINAS crossheading; ~ **frío** *m* METEO cold front; ~ **del impulso** *m* TELECOM pulse leading-edge; ~ **inclinado** *m* MINAS overhanging face; ~ **no**

ocluido *m* METEO nonoccluded front; ~ **ocluido** *m* METEO occluded front; ~ **de onda** *m* ACÚST, FÍS, FÍS ONDAS, ING ELÉC, ÓPT, PETROL, TELECOM wavefront; ~ **de pliegue** *m* AmL (cf borde Esp) MINAS geología brow; ~ **polar** *m* METEO polar front; ~ **de presión** *m* D&A pressure front; ~ **de la señal** *m* ELECTRÓN signal edge; ~ **de tajo largo** *m* MINAS longwall face; ~ **de tajo largo diagonal** *m* MINAS diagonal ram longwall face; ~ **de tajo largo por elección neumática vertical** *m* MINAS vertical pneumatic pick longwall face; ~ **de tajo largo y horizontal** *m* MINAS horizontal cut longwall face; ~ **de tajo largo y laboreo horizontal** *m* MINAS horizontal ploughed long-wall face (BrE), horizontal plowed long-wall face (AmE); ~ **de trabajo** *m* Esp (cf cara de trabajo AmL, frente de arranque AmL) CARBÓN, MINAS bank, face, working face, stope face

freón *m* PROD, QUÍMICA, REFRIG freon

fresa *f* ING MECÁ milling cutter, cutter, MECÁ, PROD, SEG milling machine; ~ **para acanaladuras en T** *f* ING MECÁ T-slot cutter; ~ **con agujero roscado** *f* ING MECÁ cutter with tapped hole; ~ **angular** *f* MECÁ angular milling cutter; ~ **de ángulo** *f* ING MECÁ bevel cutter; ~ **de aplanar** *f* MECÁ hob; ~ **con buje** *f* ING MECÁ cutter with hub; ~ **de cantos redondeados** *f* ING MECÁ round-edge milling cutter; ~ **de cepillar** *f* ING MECÁ plain milling cutter; ~ **cilíndrica** *f* ING MECÁ plain milling cutter; ~ **de colas de milano** *f* ING MECÁ dovetail cutter; ~ **de colas de milano invertida** *f* ING MECÁ inverse dovetail cutter; ~ **cónica** *f* ING MECÁ single-angle cutter, single-angular cutter; ~ **cónica de ángulo** *f* ING MECÁ, MECÁ angle cutter; ~ **cónica de ángulo doble** *f* ING MECÁ, MECÁ double equal angle cutter; ~ **cónica para escariar** *f* ING MECÁ, MECÁ, PROD reamer; ~ **de cortar metales** *f* ING MECÁ metal-slitting saw; ~ **de un corte** *f* ING MECÁ plain milling cutter; ~ **de cuarto bocel** *f* CONST, ING MECÁ quarter-round milling cutter; ~ **de cuchillas insertadas** *f* ING MECÁ inserted blade milling cutter; ~ **de cuchillas postizas** *f* ING MECÁ inserted blade milling cutter; ~ **a derechas** *f* ING MECÁ right-hand milling cutter; ~ **de desbastar chaveteros** *f* ING MECÁ roughing slot mill; ~ **diagonal de corte doble** *f* ING MECÁ double equal angle cutter; ~ **de dientes destalonados** *f* ING MECÁ backed-off cutter, relieved milling cutter; ~ **de dientes en espiral** *f* ING MECÁ milling cutter with spiral teeth; ~ **de dientes helicoidales** *f* ING MECÁ spiral; ~ **de dientes insertados** *f* ING MECÁ inserted-tooth milling cutter; ~ **de dientes interrumpidos** *f* ING MECÁ nicked-tooth milling cutter; ~ **de dientes postizos** *f* ING MECÁ inserted-tooth milling cutter; ~ **con dientes rectos** *f* ING MECÁ milling cutter with straight teeth; ~ **con eje** *f* ING MECÁ cutter with shank; ~ **para engranajes** *f* ING MECÁ, MECÁ gear cutter; ~ **entrecruzada** *f* ING MECÁ interlocking milling cutter; ~ **entrelazada** *f* ING MECÁ interlocking milling cutter; ~ **para escariador** *f* ING MECÁ, MECÁ, METAL, PROD reamer cutter; ~ **estrecha** *f* PROD edge mill; ~ **con estrías espirales** *f* ING MECÁ, PROD reamer with spiral flutes; ~ **de forma** *f* ING MECÁ formed cutter, formed milling cutter; ~ **frontal** *f* ING MECÁ face milling cutter; ~ **frontal plana** *f* ING MECÁ face milling cutter; ~ **generadora** *f* ING MECÁ generating cutter; ~ **generatriz** *f* ING MECÁ hob;

~ **generatriz para tallar engranajes** *f* ING MECÁ gear hob; ~ **helicoidal** *f* ING MECÁ, MECÁ hob; ~ **horizontal** *f* ING MECÁ horizontal milling cutter; ~ **hueca** *f* ING MECÁ running-down cutter, shell mill; ~ **de machiembrar invertida** *f* ING MECÁ inverse dovetail cutter; ~ **matriz** *f* ING MECÁ, MECÁ gear hob, hob; ~ **de módulo** *f* ING MECÁ, MECÁ gear cutter; ~ **paralela** *f* ING MECÁ parallel milling cutter; ~ **para peines** *f* ING MECÁ hob; ~ **de perfil constante** *f* ING MECÁ backed-off cutter; ~ **de perfil invariable** *f* ING MECÁ backed-off cutter; ~ **perfilada** *f* ING MECÁ, MECÁ, PROD formed cutter, form milling, formed milling cutter; ~ **de perfilar** *f* ING MECÁ, PROD form-milling cutter; ~ **de perfilar con perfil fijo** *f* ING MECÁ, PROD form-milling cutter with constant profile; ~ **de perfiles** *f* ING MECÁ, PROD form-milling cutter; ~ **de planear** *f* ING MECÁ plain milling cutter; ~ **portamoleta** *f* PROD milling tool; ~ **radial** *f* ING MECÁ side lash; ~ **radial hueca** *f* ING MECÁ shell end mill; ~ **radial con insertos graduables** *f* ING MECÁ end mill with indexable inserts; ~ **rectilínea de cremallera** *f* ING MECÁ broach; ~ **de refrentar** *f* ING MECÁ face milling cutter, face cutter, plain milling cutter, spotface cutter; ~ **de roscar** *f* ING MECÁ thread-milling cutter; ~ **sierra** *f* PROD edge mill; ~ **universal con adaptador graduable** *f* ING MECÁ end mill with indexable inserts; ~ **universal de espiga** *f* ING MECÁ end mill; ~ **universal frontal** *f* ING MECÁ end mill; ~ **universal hueca** *f* ING MECÁ shell end mill; ~ **universal radial** *f* ING MECÁ end mill

fresado *m* ING MECÁ, MECÁ machine cutting, machining, MINAS machine cutting, milling, crushing, PROC QUÍ, PROD milling; ~ **convencional** *m* ING MECÁ conventional milling; ~ **de copia** *m* ING MECÁ copy milling; ~ **en dirección contraria al avance** *m* ING MECÁ conventional milling; ~ **con dos fresas acopladas** *m* ING MECÁ straddling; ~ **de engranajes de evolvente en círculos** *m* ING MECÁ involute gearing; ~ **de engranajes de perfil de evolvente** *m* ING MECÁ involute gearing; ~ **exterior** *m* ING MECÁ out-milling; ~ **en frío** *m* ING MECÁ cold hobbing; ~ **paralelo** *m* ING MECÁ conventional milling, straddling; ~ **de perfiles** *m* ING MECÁ, PROD form milling; ~ **en plano** *m* ING MECÁ plane milling; ~ **por plantilla** *m* ING MECÁ jig milling; ~ **químico** *m* PROD chemical milling; ~ **de reproducción** *m* ING MECÁ copy milling; ~ **en sentido contrario al avance** *m* ING MECÁ conventional milling; ~ **en serie** *m* PROD multiple milling; ~ **de superficies** *m* ING MECÁ surface milling

fresador: ~ **de avance automático** *m* ING MECÁ self-feeding reamer

fresadora *f* AGRIC, ALIMENT miller, ING MECÁ, MECÁ, PROD, SEG miller, milling machine; ~ **de cepillo** *f* CONST plane bit; ~**-cepillo** *f* PROD para metal slabber; ~ **de colas de milano** *f* ING MECÁ dovetail cutter; ~ **copiadora** *f* ING MECÁ copy-milling machine; ~ **copiadora hidráulica** *f* ING MECÁ hydraulic copy mill; ~ **de corte descendente** *f* ING MECÁ down cut milling; ~ **con cremallera** *f* ING MECÁ broaching machine; ~ **de cremalleras tipo superficie vertical** *f* ING MECÁ vertical surface-type broaching machine; ~ **para desbastar** *f* CARBÓN, PROD rougher; ~ **de engranajes** *f* ING MECÁ, MECÁ gear cutter, gear cutting, gear-cutting machine; ~ **de estampas** *f* ING MECÁ die-sinking machine; ~ **horizontal** *f* PROD para *metal* slabber, slabbing miller; ~ **plana** *f* ING MECÁ surface milling machine; ~ **en plano** *f* ING MECÁ plane milling machine; ~ **de precisión** *f* ING MECÁ precision milling; ~ **de reproducción** *f* ING MECÁ copy-milling machine; ~ **de roscar** *f* ING MECÁ thread-milling cutter; ~ **de rótula** *f* ING MECÁ knee-type milling machine; ~ **de tallar engranajes por disco** *f* ING MECÁ gear-milling machine; ~ **de tallar engranajes por fresa** *f* ING MECÁ gear-milling machine; ~ **para tallar engranajes por fresa matriz** *f* ING MECÁ gear-hobbing machine; ~ **de tres cortes con meseta** *f* ING MECÁ side-and-face milling cutter with plain; ~ **universal** *f* ING MECÁ universal milling machine

fresar *vt* MECÁ, PROD mill

fresas: ~ diagonales *f pl* ING MECÁ diagonal cutters

fricción[1]: **sin ~** *adj* ING MECÁ, MECÁ frictionless

fricción[2] *f* GEN friction; ~ **aérea** *f* TEC ESP air friction; ~ **de arranque** *f* ING MECÁ starting friction; ~ **de correa** *f* ING MECÁ belt friction; ~ **dinámica** *f* FÍS dynamic friction; ~ **de entrada** *f* IMPR metering friction; ~ **estática** *f* ING MECÁ static friction; ~ **interior** *f* C&V, ING MECÁ, METAL internal friction; ~ **interna** *f* C&V, ING MECÁ, METAL internal friction

friccional *f* OCEAN depth of frictional influence

friedelita *f* MINERAL friedelite

frigorífico *m* ALIMENT, ELEC, ING MECÁ, REFRIG, TERMO refrigerator; ~ **de absorción** *m* REFRIG, TERMO absorption refrigerator; ~ **para biberones** *m* REFRIG, TERMO nursery refrigerator; ~ **doméstico** *m* REFRIG, TERMO domestic refrigerator, household refrigerator; ~ **a gas** *m* GAS, ING MECÁ, REFRIG, TERMO gas refrigerator

frigorista *m* REFRIG, TERMO refrigeration engineer

frimartin *f* AGRIC freemartin

frimartinismo *m* AGRIC freemartin

frío: ~ artificial *m* ALIMENT, REFRIG, TERMO refrigeration

frisa *f* TRANSP MAR *construcción naval* seal, *cuerdas* eyelet hole, grommet, *para aferrar velas* gasket

frisar *vt* TRANSP MAR *construcción naval* seal

friseíta *f* MINERAL friesete

frita *f* C&V frit, glass frit; ~ **clara** *f* C&V clear frit

frontal *adj* AUTO, ELECTRÓN, FOTO, INFORM&PD, INSTR front-end, TEC ESP, TRANSP head-on, VEH front-end

frotador *m* ING ELÉC sliding contact

frotamiento *m* GEN friction

frote *m* ING MECÁ attrition, scotching

fructana *f* QUÍMICA fructan

fructosa *f* AGRIC grape sugar, ALIMENT, QUÍMICA fructose, laevulose (*BrE*), levulose (*AmE*)

fructosana *f* QUÍMICA fructosan

frunce *m* TEXTIL gathering

fruncido[1] *adj* TEXTIL gathered

fruncido[2] *m* TEXTIL gathering

fruncir *vt* TEXTIL crimp, gather

frusemida *f* QUÍMICA frusemide

frusto *m* GEOM frustum

fruta: ~ de pepitas *f* ALIMENT pomaceous fruit

frutas: ~ secas *f pl* AGRIC mummies

fruticultor *m* AGRIC fruit grower

fruticultura *f* AGRIC pomology

fruto: ~ pomoideo *m* ALIMENT pome fruit

FSA *abr* (*frecuencia superalta*) ELECTRÓN, TEC ESP, TELECOM SHF (*super-high frequency*)

FSO *abr* (*función del sistema operativo*) TELECOM OSF (*operating system function*)

ftalamida *f* QUÍMICA phthalamide

ftalato *m* QUÍMICA phthalate; ~ **de butilo** *m* P&C butyl phthalate; ~ **de dibutilo** *m* P&C dibutyl phthalate; ~ **de dioctilo** *m* P&C dioctyl phthalate

ftaleína *f* QUÍMICA phthalein

ftálico *adj* QUÍMICA phthalic

ftálido *m* QUÍMICA phthalide

ftalina *f* QUÍMICA phthalin

ftalocianina *f* P&C phthalocyanine

FTP *abr* (*protocolo de transferencia de archivos*) INFORM&PD FTP (*file transfer protocol*)

fucosa *f* QUÍMICA fucose

fucosterol *m* QUÍMICA fucosterol

fucoxantina *f* QUÍMICA fucoxanthin

fucsina *f* QUÍMICA fuchsin

fucsita *f* MINERAL avalite, fuchsite

fucsona *f* QUÍMICA fuchsone

fuego: ~ apagado *m* PROD blown fire; ~ **de cañón** *m* D&A gunfire; ~ **cruzado** *m* D&A *artillería* crossfire; ~ **eléctrico** *m* ELEC *calefacción*, INSTAL TERM electric fire; ~ **latente** *m* SEG, TERMO smoldering fire (*AmE*), smouldering fire (*BrE*); ~ **sin llama** *m* SEG, TERMO smoldering fire (*AmE*), smouldering fire (*BrE*); ~ **secundario** *m* AGRIC *patología vegetal* fire spot; ~ **suave** *m* C&V soft fire

fuel: ~ derivado *m* TEC PETR derived fuel; ~-**oil** *m* TEC PETR domestic fuel oil

fuelle *m* GEN bellows, blower; ~ **axial** *m* SEG axial blower; ~ **circular de ráfaga única** *m* ING MECÁ single-blast circular bellows; ~ **de comunicación** *m* FERRO, VEH gangway; ~ **de cuero** *m* FOTO leather bellows; ~ **extensible** *m* CINEMAT, FOTO extension bellows; ~ **de extensión** *m* CINEMAT, FOTO extension bellows; ~ **de fragua** *m* PROD forge bellows; ~ **de herrero** *m* ING MECÁ blacksmith's bellows; ~ **de mano** *m* C&V hand bellows; ~ **de parasol** *m* CINEMAT lens hood bellows

fueloil *m* AmL (*cf fuelóleo Esp*) CARBÓN, CONST, QUÍMICA fuel oil; ~ **para calderas** *m* TEC PETR *combustibles* residual fuel oil

fuelóleo *m* Esp (*cf fueloil AmL*) CARBÓN, CONST, QUÍMICA fuel oil; ~ **para calderas** *m* TEC PETR *combustibles* residual fuel oil

fuente *f* AGUA source, spring, ELECTRÓN, FÍS source, HIDROL spring, *de suministro de agua* source, IMPR, INFORM&PD font, ING ELÉC, INSTAL HIDRÁUL source, TEC PETR *de fluidos* head; ~ **de acoplamiento** *f* ELECTRÓN companion source; ~ **acústica** *f* ACÚST, CONTAM sound source; ~ **acústica de precisión** *f* ACÚST pinpoint acoustic source; ~ **acústica real** *f* ACÚST real sound source; ~ **acústica única** *f* ACÚST single acoustic source; ~ **de agua potable** *f* AGUA, HIDROL drinking water supply; ~ **de alimentación** *f* ELEC, ELECTRÓN, ING ELÉC, TELECOM, TV power supply; ~ **de alimentación de CC** *f* ELEC, ING ELÉC DC supply; ~ **de alimentación de corriente** *f* ELEC, ING ELÉC generator; ~ **de alimentación de corriente alterna** *f* ELEC, ELECTRÓN, ING ELÉC alternating-current supply; ~ **de alimentación doble** *f* ELEC, ING ELÉC duplicate supply; ~ **de alimentación eléctrica** *f* ELEC, ING ELÉC, PROD power-supply unit (*PSU*); ~ **de alimentación regulada** *f* ELEC, ING ELÉC regulated power supply; ~ **de alta tensión** *f* ELEC, ING ELÉC HT supply; ~ **de CA** *f* ELEC, ING ELÉC AC source; ~ **de**

calor *f* REFRIG, TERMO heat reservoir, heat source;
~ **conmutada de salida única** *f* ELEC, ING ELÉC
single-output switching power supply; ~ **de
contaminación** *f* CONTAM pollution source; ~ **de la
corriente** *f* FÍS current source; ~ **de corriente
continua** *f* ELEC, ING ELÉC direct current voltage
source; ~ **de datos** *f* ING ELÉC data source; ~ **DCE** *f*
TELECOM DCE source; ~ **difusa** *f* CINEMAT soft
source; ~ **distante** *f* CONTAM distant source; ~ **de
electrones** *f* ELECTRÓN, FÍS RAD electron source;
~ **de emisión** *f* CONTAM emission source; ~ **de
emisión constante** *f* CONTAM stationary emission
source; ~ **de emisión local** *f* CONTAM local emission
source; ~ **de emisión urbana** *f* CONTAM local emission source; ~ **emisora** *f* CONTAM emission source;
~ **encapsulada** *f* NUCL encapsulated source;
~ **energética** *f* ING MECÁ prime mover; ~ **de
energía** *f* ELEC, ELECTRÓN, power supply, ING ELÉC
power pack, power supply, ING MECÁ power producer, TEC ESP energy source, TELECOM power supply;
~ **de energía primaria** *f* ING MECÁ prime mover; ~ **de
energía renovable** *f* ENERG RENOV, FÍS renewable
source of energy; ~ **estacionaria de emisión** *f* CON-
TAM stationary emission source; ~ **de excitación** *f*
FÍS RAD excitation source; ~ **externa** *f* ELECTRÓN
external source; ~ **de frecuencia** *f* ELECTRÓN frequency source; ~ **de gas** *f* GAS, ING MECÁ gas spring;
~ **de iluminación difusa** *f* CINEMAT soft source;
~ **iluminada** *f* FÍS RAD illuminated source; ~ **de
información** *f* INFORM&PD information source;
~ **de iones** *f* FÍS ion source; ~ **de iones de cátodo
hueco** *f* FÍS, NUCL hollow cathode ion source; ~ **de
iones de metal líquido** *f* FÍS RAD liquid-metal ion
source; ~ **iónica** *f* PROD ion gun; ~ **de Lambert** *f* ÓPT,
TELECOM lambertian source; ~ **lambertiana** *f* ÓPT,
TELECOM lambertian source; ~ **libre** *f* NUCL free
source; ~ **lineal** *f* CONTAM line source; ~ **luminosa** *f*
FÍS, FÍS ONDAS, FÍS RAD, FOTO, LAB, ÓPT light source,
luminous source; ~ **luminosa avanzada** *f* (*ALS*) FÍS
PART advanced light source (*ALS*); ~ **de luz** *f* ELEC-
TRÓN light source; ~ **de luz fotométrica** *f* INSTR
photometric-light source; ~ **mayor** *f* CONTAM major
source; ~ **natural de energía** *f* ING MECÁ prime
mover; ~ **negativa** *f* ING ELÉC negative power supply;
~ **no blindada** *f* NUCL unshielded source; ~ **no
uniforme de radiación** *f* FÍS RAD nonuniform source
of radiation; ~ **de onda milimétrica** *f* ELECTRÓN
millimeter-wave source (*AmE*), millimetre-wave
source (*BrE*); ~ **perenne** *f* AGUA, HIDROL perennial
spring; ~ **de polarización** *f* ING ELÉC bias source;
~ **de potencia** *f* TELECOM power supply; ~ **de
presión hidráulica** *f* INSTAL HIDRÁUL hydraulic
pressure source; ~ **principal** *f* CONTAM major source;
~ **puntual** *f* CINEMAT *iluminacion*, CONTAM, FÍS, FÍS
RAD, FOTO, NUCL point source, radioactive-point
source; ~ **puntual colimada** *f* NUCL collimated point
source; ~ **de radiación** *f* ELECTRÓN, FÍS RAD, NUCL
radiation source; ~ **de radiación de frenado** *f* FÍS
RAD, NUCL bremsstrahlung source; ~ **de radiación
lineal** *f* CONTAM, FÍS RAD line source; ~ **de radiación
puntal** *f* FÍS RAD line source; ~ **de radiación puntual**
f CONTAM, FÍS RAD, NUCL radioactive-point source;
~ **radiactiva delgada** *f* FÍS RAD, NUCL thin source;
~ **radiactiva gruesa** *f* FÍS RAD, NUCL thick source;
~ **radioeléctrica** *f* FÍS RAD, TEC ESP radio source;
~ **de rayos X** *f* NUCL X-ray source; ~ **renovable** *f*

ENERG RENOV, FÍS renewable source, renewable source
of energy; ~ **de ruido blanco** *f* ELECTRÓN white noise
source; ~ **de ruidos** *f* CONTAM, ELECTRÓN noise
source; ~ **de ruidos complejos** *f* ELECTRÓN, FÍS,
TEC ESP, TELECOM random-noise source;
~ **secundaria** *f* FÍS, ÓPT secondary source;
~ **sellada** *f* NUCL sealed source; ~ **de señal
diferencial** *f* ELECTRÓN differential signal source;
~ **de señal externa** *f* ELECTRÓN, TELECOM external
signal source; ~ **de señal de microondas** *f* ELEC-
TRÓN microwave signal source; ~ **de señales de
sincronismo del multiplexor** *f* INFORM&PD, TELE-
COM multiplexer timing source (*MTS*); ~ **de sonido** *f*
ACÚST, CONTAM sound source; ~ **sonora** *f* ACÚST,
CONTAM sound source; ~ **sonora virtual** *f* ACÚST,
CONTAM virtual sound source; ~ **de suministro
electrónica** *f* ELEC, ELECTRÓN, ING ELÉC electronic
power supply; ~ **de tensión constante** *f* ELEC
constant-voltage source; ~ **termal** *f* ENERG RENOV,
GEOFÍS, HIDROL, TERMO, TERMOTEC geyser;
~ **térmica** *f* HIDROL, TERMO thermal spring; ~ **de
transporte** *f* CONTAM transportation source; ~ **de
voltaje** *f* FÍS voltage source

fuera[1]: ~ **del aire** *adj* TV off-air; ~ **de alcance** *adj* D&A
out-of-range; ~ **de bordo** *adj* ING MECÁ, TRANSP MAR
jarcias, motor outboard; ~ **del campo visual de la
cámara** *adj* CINEMAT off-camera; ~ **de centro** *adj*
ING MECÁ off-center (*AmE*), off-centre (*BrE*); ~ **del
chip** *adj* ELECTRÓN off chip; ~ **costa** *adj* GEOL,
OCEAN, TEC PETR *operaciones marinas* offshore;
~ **del eje** *adj* ING MECÁ off-center (*AmE*), off-centre
(*BrE*); ~ **de equilibrio** *adj* ING MECÁ out-of-balance;
~ **de escuadra** *adj* IMPR, ING MECÁ out-of-square;
~ **de fase** *adj* ELEC out-of-phase, ING MECÁ out-of-
parallel; ~ **de foco** *adj* CINEMAT, FOTO, TV out-of-
focus; ~ **de guía** *adj* TRANSP AÉR out-of-track; ~ **de
línea** *adj* INFORM&PD, TELECOM off-line; ~ **del marco**
adj IMPR out-of-frame; ~ **de paso** *adj* TRANSP AÉR
out-of-pitch; ~ **de la placa** *adj* ELECTRÓN off-board;
~ **de plomada** *adj* CONST out-of-plumb; ~ **de puntas**
adj TRANSP MAR *navegación* in the offing; ~ **de punto**
adj TV out-point; ~ **de servicio** *adj* ELEC, ING MECÁ,
PROD out-of-action, out-of-order, out-of-repair; ~ **de
su lugar** *adj* ING MECÁ off-center (*AmE*), off-centre
(*BrE*); ~ **de tolerancias** *adj* ING MECÁ out-of-toler-
ance; ~ **de tono** *adj* PAPEL *respecto al color de
referencia* off-shade; ~ **de trama de información**
adj TELECOM out-of-frame; ~ **de uso** *adj* ING MECÁ
out-of-use

fuera[2]: ~ **de línea** *adv* INFORM&PD, TELECOM off-line;
~ **de puntas** *adv* TRANSP MAR in the offing; ~ **de
trayectoría** *adv* TEC ESP off-course

fueraborda: ~ **hinchable** *m* TRANSP, TRANSP MAR
outboard inflatable

fuerza *f* AUTO power, CARBÓN strength, ELEC *campo
electromagnético* intensity, FÍS force, GAS *física*
strength, ING MECÁ, MECÁ impetus, power, stress,
ÓPT finesse, QUÍMICA *de ácido*, TEXTIL strength;
~ **aceleradora** *f* FÍS, TEC ESP accelerating force;
~ **de aceleración** *f* FÍS, TEC ESP accelerating force;
~ **ácida** *f* QUÍMICA acid strength; ~ **activa** *f* ING ELÉC
active power; ~ **de adherencia** *f* P&C bond strength;
~ **de adherencia de las capas** *f* P&C plybond
strength; ~ **de adherencia de los pliegues** *f* P&C
plybond strength; ~ **adhesiva** *f* P&C adhesive
strength; ~ **de la aguja** *f* ACÚST stylus force; ~ **de**

apriete *f* EMB clamping force; **~ ascensional** *f* FÍS lifting force, *temperaturas* lift, TRANSP lifting power; **~ de atracción** *f* ELEC, FÍS attractive force; **~ atractiva** *f* ELEC, FÍS attractive force; **~ de Bartlett** *f* NUCL Bartlett force; **~ sin carga** *f* NUCL no-load force; **~ central** *f* FÍS central force; **~ centrífuga** *f* CONTAM, FÍS, FÍS FLUID, PROC QUÍ, TEC ESP, TRANSP AÉR centrifugal force; **~ centrípeta** *f* FÍS centripetal force; **~ de cisión** *f* ING MECÁ shear force; **~ de cizallamiento** *f* ING MECÁ shear force; **~ coercitiva** *f* ELEC, FÍS, ING ELÉC, METAL coercive force; **~ de cohesión** *f* FÍS cohesive force; **~ compresiva** *f* ING MECÁ compressive force; **~ conservativa** *f* FÍS conservative force; **~ contraelectromotriz** *f* ELEC, FERRO, FÍS, ING ELÉC, VEH back electromotive force (*back emf*), counterelectromotive force; **~ contraria** *f* CONST, ELEC, FÍS, FÍS FLUID, GAS, ING ELÉC, ING MECÁ, MINAS, PAPEL, TEC PETR, TELECOM, TERMO, TEXTIL resistance; **~ de Coriolis** *f* FÍS, FÍS FLUID, METEO Coriolis force; **~ de corte** *f* ING MECÁ shear force; **~ debida a la viscosidad** *f* FÍS FLUID viscous force; **~ débil** *f* FÍS PART weak force, weak interaction; **~ directa** *f* ING MECÁ direct force; **~ de Ekman** *f* OCEAN Ekman forcing; **~ eléctrica** *f* ELEC, ING ELÉC electric force; **~ electromagnética** *f* ELEC, FÍS, FÍS PART, ING ELÉC electromagnetic force; **~ electromagnética aplicada** *f* FÍS, FÍS PART applied electromagnetic force; **~ electromotriz** *f* (*FEM*) ELEC, FÍS, ING ELÉC electromotive force (*EMF*); **~ electromotriz alterna** *f* ELEC, FÍS, ING ELÉC alternating electromotive force; **~ electromotriz aplicada** *f* ELEC, FÍS, ING ELÉC applied electromotive force; **~ electromotriz de CA** *f* ELEC, FÍS, ING ELÉC AC electromotive force; **~ electromotriz eficaz** *f* ELEC, FÍS, ING ELÉC effective electromotive force; **~ electromotriz estática** *f* ELEC, FÍS, ING ELÉC transformer electromotive force; **~ electromotriz inducida** *f* ELEC, FÍS, ING ELÉC induced electromotive force; **~ electromotriz subtransitoria longitudinal** *f* ELEC, FÍS, ING ELÉC direct axis subtransient electromotive force; **~ electromotriz transitoria longitudinal** *f* ELEC, FÍS, ING ELÉC direct axis transient electromotive force; **~ electrostática** *f* ELEC, ING ELÉC electrostatic force; **~ elevadora** *f* ING MECÁ, TRANSP lifting power; **~ de empuje** *f* FÍS FLUID buoyancy force; **~ de enlace** *f* CRISTAL bond strength; **~ de expulsión** *f* TEC ESP ejection force; **~ externa** *f* METAL external force; **~ de frenado** *f* ING MECÁ brake force, brake power; **~ de fricción** *f* FÍS, ING MECÁ force of friction, frictional force; **~ friccional** *f* FÍS, ING MECÁ friction force; **~ fuerte** *f* FÍS PART strong nuclear force; **~ G** *f* FÍS, TEC ESP, TRANSP AÉR g-force; **~ del gas** *f* MECÁ gas bearing; **~ giroscópica** *f* ENERG RENOV gyroscopic force; **~ de la gravedad** *f* FÍS, TEC ESP, TRANSP AÉR g-force; **~ hidroeléctrica** *f* ENERG RENOV hydroelectric power; **~ impulsora** *f* ING ELÉC active power; **~ de inercia** *f* FÍS, MECÁ inertial force; **~ de intercambio espacial** *f* NUCL position exchange force; **~ internuclear** *f* FÍS PART strong interaction, strong nuclear force; **~ iónica** *f* FÍS RAD ionic strength; **~ lateral** *f* ING MECÁ side thrust, TRANSP lateral force; **~ límite de cizallamiento** *f* ING MECÁ ultimate shear strength; **~ del líquido hidráulico** *f* ING MECÁ hydraulic fluid power; **~ de Lorentz** *f* ELEC, FÍS, ING ELÉC Lorentz force; **~ magnética** *f* ELEC, GEOFÍS,

PETROL magnetic force; **~ magnetizante** *f* ELEC magnetizing force; **~ magnetomotriz** *f* (*fmm*) ELEC, FÍS magnetomotive force (*mmf*); **~ de Majorana** *f* NUCL Majorana force; **~ máxima a la velocidad nominal del viento** *f* ENERG RENOV maximum power at rated wind speed; **~ motriz** *f* ING MECÁ drive power, mover, propelling force, propulsive force, MECÁ power, METAL driving force, TRANSP MAR *diseño de barcos* motive force; **~ motriz de arrastre** *f* ING MECÁ driver; **~ neutralizadora** *f* ING MECÁ counteracting force; **~ nuclear** *f* FÍS PART, NUCL nuclear force; **~ nuclear central** *f* FÍS PART, NUCL central nuclear force; **~ nuclear débil** *f* FÍS PART, NUCL weak nuclear force; **~ opuesta** *f* ING MECÁ counteracting force; **~ de par del motor** *f* AUTO, TRANSP AÉR, VEH engine torque; **~ de percusión** *f* ING MECÁ percussive force; **~ polar** *f* FÍS pole strength; **~ portante** *f* TEC ESP *relés* pull; **~ de producción** *f* FÍS yield stress; **~ del propulsante** *f* D&A propellant force; **~ de propulsión** *f* ING MECÁ drive power, propelling force, propulsive force; **~ de ráfaga** *f* TEC ESP, TRANSP AÉR, TRANSP MAR gust intensity; **~ repositora** *f* FÍS, MECÁ restoring force; **~ de repulsión** *f* ELEC, FÍS, ING ELÉC repulsive force; **~ repulsiva** *f* ELEC, FÍS, ING ELÉC repulsive force; **~ residual** *f* CARBÓN residual strength; **~ de resorte máxima** *f* TRANSP AÉR maximum spring back load; **~ de restauración** *f* FÍS, MECÁ restoring force; **~ de retroceso por muelle** *f* PROD spring return force; **~ de rozamiento** *f* FÍS, METAL friction force; **~ de separación** *f* FÍS tensile stress; **~ solenoide** *f* ING ELÉC solenoid actuation; **~ de sostén** *f* ING MECÁ lifting power; **~ de sujeción** *f* EMB clamping force; **~ sustentadora** *f* FÍS lifting force; **~ tangencial** *f* CARBÓN, ING MECÁ shear force, tangential force; **~ de tensión** *f* FÍS, ING MECÁ tensile force; **~ térmica** *f* ING MECÁ hot strength; **~ termoelectromotriz** *f* ING ELÉC thermoelectromotive force; **~ de trabajo** *f* PROD manpower; **~ de tracción** *f* ING MECÁ tensile force, tensile strength, METR tractive force; **~ de transmisión** *f* ING MECÁ drive power, TEXTIL driving force; **~ del viento** *f* METEO wind force
Fuerza: ~ Aérea Naval *f* D&A, TRANSP MAR Fleet Air Arm (*BrE*), Naval Air Service (*AmE*)
fuerzas: ~ acorazadas *f pl* D&A armor (*AmE*), armour (*BrE*); **~ asimétricas** *f pl* ING MECÁ out-of-balance forces; **~ de atracción de muy corto alcance** *f pl* FÍS PART attractive forces of very short range; **~ coplanares** *f pl* FÍS coplanar forces; **~ desbalanceadas** *f pl* ING MECÁ out-of-balance forces; **~ interatómicas** *f pl* FÍS RAD interatomic forces

fuga *f* AGUA leakage, C&V, CONST leak, CONTAM *en tuberías* washout, CONTAM MAR, ELEC, ENERG RENOV, FÍS, GAS leakage, leak, HIDROL oozing, ING ELÉC, SEG, TELECOM leakage; **~ de aire** *f* AUTO air leak, CONTAM *procesos industriales* blowout; **~ de carga** *f* ELEC charge leakage; **~ de corriente del condensador** *f* ING ELÉC capacitor leakage current; **~ de gas** *f* CONTAM waste gas, GAS, TERMO gas leak; **~ por la junta** *f* ING ELÉC junction leakage; **~ de partículas** *f* NUCL particle leakage; **~ a tierra** *f* ELEC, ING ELÉC, PROD earth fault (*BrE*), earth leakage (*BrE*), ground fault (*AmE*), ground leakage (*AmE*)
fugacidad *f* GAS, QUÍMICA, TERMO fugacity
fugas[1]**: sin ~** *adj* ING ELÉC lossless, MECÁ leak-tight

fugas[2]: ~ **de fundición** *f pl* PROD spillings; ~ **primario-secundarias** *f pl* NUCL primary-secondary leakage
fugoide *m* TRANSP AÉR phugoid
fulard: ~ **de aprestar** *m* TEXTIL *acabados* pad mangle
fulardar *vt* TEXTIL *tintura* pad
fulcro *m* FÍS fulcrum
fulguración: ~ **solar** *f* GEOFÍS solar flare
fulgurita *f* GEOL fulgurite
fulgurito *m* PROD fulgurite
fulmicotón *m* AmL (*cf nitroalgodón Esp*) MINAS *explosivos* guncotton
fulminación *f* QUÍMICA fulmination
fulminante *m* MINAS blasting-cap, PETROL cap, TERMO detonator
fulminar *vi* QUÍMICA fulminate
fulminato *m* QUÍMICA fulminate; ~ **de mercurio** *m* QUÍMICA mercury fulminate
fulveno *m* QUÍMICA fulvene
fumárico *adj* QUÍMICA fumaric
fumarola *f* GEOL fumarole
fumivoridad *f* CARBÓN, QUÍMICA consumption of smoke
fumívoro *adj* CARBÓN, QUÍMICA fumivorous
función *f* ELECTRÓN mode; ~ **de adquisición de órbita** *f* TEC ESP station acquisition function; ~ **de alarma** *f* TELECOM alarm function; ~ **de aplicaciones de gestión** *f* TELECOM management applications function (*MAF*); ~ **aritmética** *f* MATEMÁT, PROD arithmetic function; ~ **armónica** *f* ELECTRÓN harmonic function; ~ **de autocorrelación** *f* PETROL autocorrelation function; ~ **avanzada de comunicaciones** *f* (*ACF*) INFORM&PD advanced communications function (*ACF*); ~ **de barrido** *f* ELECTRÓN sweep mode; ~ **de la capa superior** *f* TELECOM higher-layer function (*HLF*); ~ **característica del filtro** *f* ELECTRÓN filter characteristic-function; ~ **completa** *f* ELECTRÓN integrated function; ~ **de comunicación de mensajes** *f* TELECOM message communication function (*MCF*); ~ **de control** *f* TELECOM control function; ~ **de convergencia dependiente de la red secundaria** *f* TELECOM subnetwork dependent convergence function (*SNDCF*); ~ **de correlación** *f* ELECTRÓN correlation function; ~ **de correlación cruzada** *f* ELECTRÓN, PETROL cross-correlation function; ~ **de creación de códigos de comprobación** *f* INFORM&PD hash function; ~ **crítica de seguridad** *f* NUCL, SEG critical safety function; ~ **DCME** *f* TELECOM DCME function; ~ **delta** *f* PETROL delta function; ~ **de densidad de probabilidad** *f* (*FDP*) MATEMÁT probability density function (*PDF*); ~ **de Dirac** *f* PETROL Dirac function; ~ **discriminante** *f* MATEMÁT discriminant function; ~ **de distribución** *f* MATEMÁT distribution function; ~ **de la distribución radial** *f* FÍS, FÍS RAD radial-distribution function; ~ **del elemento de red** *f* TELECOM network element function (*NERF*); ~ **de encaminamiento y de relevo** *f* TELECOM relaying and routing function; ~ **entrelazada de facsímil** *f* TELECOM facsimile interworking function (*FAXFIF*); ~ **escalonada** *f* ELECTRÓN, INFORM&PD step function; ~ **específica de Gibbs** *f* FÍS specific Gibbs function; ~ **excedente** *f* METAL excess function; ~ **de excitación** *f* FÍS RAD excitation function; ~ **de ganancia** *f* ELECTRÓN gain function; ~ **de gestión del equipo sincrónico** *f* TELECOM synchronous equipment management function (*SEMF*); ~ **Helmholtz específica** *f* FÍS specific Helmholtz function; ~ **implícita** *f* MATEMÁT implicit function; ~ **de impulsos** *f* ELECTRÓN impulse function, pulsed mode; ~ **incorporada** *f* INFORM&PD built-in function; ~ **de interrogación** *f* ELECTRÓN interrogation mode; ~ **Jost** *f* NUCL Jost function; ~ **de linealización** *f* PROD linearization function; ~ **lógica** *f* ELECTRÓN logical function; ~ **de mantenimiento** *f* TELECOM maintenance function; ~ **de mediación** *f* TELECOM mediation function; ~ **de nivel superior** *f* TELECOM higher-layer function (*HLF*); ~ **de onda** *f* FÍS, FÍS ONDAS wave function; ~ **de onda antisimétrica** *f* FÍS, FÍS ONDAS antisymmetric wave function; ~ **de onda simétrica** *f* FÍS, FÍS ONDAS symmetric wave function; ~ **de oscilación** *f* ELECTRÓN oscillation mode; ~ **de partición** *f* FÍS partition function; ~ **pasiva** *f* ELECTRÓN passive mode; ~ **por pasos** *f* INFORM&PD step function; ~ **periódica** *f* ELECTRÓN periodic function; ~ **de potencial** *f* FÍS potential function; ~ **de propagación de modulación** *f* ELECTRÓN modulation transfer function; ~ **propia** *f* ACÚST, FÍS eigenfunction; ~ **recursiva** *f* INFORM&PD recursive function; ~ **de regeneración** *f* ELECTRÓN, INFORM&PD refresh mode; ~ **relacionada con la comunicación de canal virtual** *f* TELECOM virtual channel connection-related function; ~ **relacionada con la conexión** *f* TELECOM connection-related function (*CRF*); ~ **de relleno** *f* METAL loading function; ~ **de respuesta de la banda base** *f* ÓPT, TELECOM baseband response function; ~ **saturada** *f* ELECTRÓN saturated mode; ~ **de servicio sin conexiones** *f* TELECOM connectionless service function (*CLSF*); ~ **sinusoidal** *f* ING ELÉC, MATEMÁT sinusoidal function; ~ **del sistema operativo** *f* (*FSO*) TELECOM operating system function (*OSF*); ~ **termodinámica** *f* TERMO thermodynamic function; ~ **de trabajo** *f* FÍS, ING ELÉC work function; ~ **de transferencia** *f* ÓPT, TELECOM transfer function; ~ **de transferencia de banda base** *f* ÓPT, TELECOM baseband transfer function; ~ **de transferencia en lazo abierto** *f* NUCL open-loop transfer function; ~ **de transmisión** *f* ELECTRÓN transmission mode, NUCL transmission function; ~ **trigonométrica** *f* GEOM trigonometrical function; ~ **para verificación de errores** *f* INFORM&PD hash function

funcionalidad: ~ **fiable** *f* TELECOM trusted functionality
funcionamiento[1]: **de** ~ *adj* ING MECÁ operating; **en** ~ *adj* ING MECÁ on
funcionamiento[2] *m* CALIDAD, INFORM&PD performance, ING MECÁ action, performance, run, running, working, MECÁ action, PAPEL, TEXTIL running, TRANSP AÉR run up area; ~ **accidental** *m* NUCL squitter; ~ **asíncrono** *m* ELEC asynchronous operation, asynchronous running, TELECOM hands-off operation; ~ **automático** *m* CONST automatic operation, INFORM&PD hands-off operation, unattended operation; ~ **automático y control de frenado** *m* TRANSP automatic running and braking control; ~ **en banda limitada** *m* TELECOM bandwidth-limited operation; ~ **con batería** *m* ING ELÉC, TELECOM battery operation; ~ **en bucle cerrado** *m* ELEC, ELECTRÓN closed-loop control; ~ **con carga** *m* ING MECÁ running under load; ~ **cerrado** *m* ELEC *relé* closing operation; ~ **en circuito abierto** *m* ELEC

open-circuit operation; ~ **defectuoso** *m* CALIDAD, INFORM&PD, ING MECÁ, METR, SEG, TEC ESP, TRANSP MAR malfunction; ~ **a dos niveles** *m* ING ELÉC bilevel operation; ~ **dúplex** *m* INFORM&PD duplex operation, operation duplex; ~ **estático** *m* ING ELÉC static operation; ~ **fuera de línea** *m* TELECOM off-line working; ~ **incorrecto** *m* CALIDAD, INFORM&PD, ING MECÁ, METR, SEG, TEC ESP malfunction; ~ **intermitente** *m* ELEC intermittent duty; ~ **de interredes** *m* INFORM&PD internetting; ~ **irregular** *m* TRANSP AÉR hunting; ~ **por línea independiente** *m* TELECOM off-line working; ~ **lineal** *m* ELECTRÓN linear behavior (*AmE*), linear behaviour (*BrE*); ~ **manual** *m* CONST off-line working, INFORM&PD, TELECOM manual operation; ~ **de la máquina** *m* TRANSP MAR engine operation; ~ **con operador** *m* CONST, INFORM&PD, TELECOM attended operation; ~ **con par motor pequeño** *m* PROD low-torque operation; ~ **a presión constante** *m* TEC ESP constant-pressure operation; ~ **a prueba de fallos** *m* INFORM&PD fail-safe operation; ~ **de la red** *m* INFORM&PD, TELECOM network performance (*NP*); ~ **en régimen húmedo** *m* REFRIG wet compression; ~ **en régimen seco** *m* REFRIG dry compression; ~ **remoto** *m* INFORM&PD, TELECOM, TV remote operation; ~ **en serie** *m* INFORM&PD serial operation; ~ **silencioso** *m* AUTO, ING MECÁ, VEH quiet running; ~ **simplex** *m* INFORM&PD simplex operation; ~ **en sobrecarga** *m* ING MECÁ running on overload; ~ **suave** *m* AUTO, ING MECÁ, VEH quiet running; ~ **en vacío** *m* ING ELÉC no-load operation, ING MECÁ running on no load; ~ **con varios canales** *m* ELECTRÓN multiplex operation; ~ **en vuelo** *m* TEC ESP in-flight operation

funcionar *vi* ING MECÁ operate; ~ **en carga** *vi* ING MECÁ run under load; ~ **mal** *vi* ELECTRÓN, ING ELÉC, ING MECÁ, MECÁ, TELECOM malfunction; ~ **con marcha lenta** *vi* ING MECÁ idle; ~ **en mínima** *vi* ING MECÁ idle; ~ **en vacío** *vi* ING MECÁ run on no load, idle

funda *f* AUTO boot (*BrE*), trunk (*AmE*), CINEMAT *de protección de cámara* cover, casing, *de cámara para insonorizar* barney, D&A scabbard, ELEC *cable* sheath (*BrE*), jacket (*AmE*), ELECTRÓN *de tubo electrónico* envelope, EMB sleeve, envelopment, FOTO covering, ING ELÉC sheath, INSTAL HIDRÁUL cleading, MECÁ sheath, TEC ESP *vehículos* case, TEXTIL cover; ~ **antihielo** *f* TRANSP AÉR boot; ~ **de bobina** *f* TEXTIL package sleeve; ~ **de botella** *f* EMB bottle jacket; ~ **de cable** *f* ELEC, ING ELÉC, ING MECÁ cable sheath; ~ **de colchón** *f* TEXTIL mattress cover; ~ **de empaquetamiento** *f* PROD packing slip; ~ **para envíos postales** *f* EMB mailing sleeve; ~ **metálica** *f* ELEC metallic sheath; ~ **protectora** *f* FOTO slip-on sleeve; ~ **de PVC** *f* ING ELÉC PVC sheath; ~ **de tapicería** *f* VEH *habitáculo* headlining; ~ **de tubo electrónico** *f* ELECTRÓN electron-tube envelope

fundación *f* AmL (*cf cimientos Esp*) CONST foundation; ~ **del mortero** *f* CONST, PROD *bocarte* mortar bed

fundamento: ~ **de la norma** *m* INFORM&PD rule base

fundamentos: ~ **informáticos** *m pl* INFORM&PD computer literacy

fundente *m* GEN flux; ~ **ecológico** *m* C&V ecology cullet; ~ **enfriado en agua** *m* Esp (*cf cullet enfriado en agua AmL*) C&V quenched cullet (*BrE*), shredded cullet (*AmE*); ~ **foráneo** *m* C&V foreign cullet;

~ **incoloro** *m* C&V colorless flux (*AmE*), colourless flux (*BrE*); ~ **propio** *m* C&V factory cullet; ~ **de soldadura** *m* CONST brazing flux; ~ **para soldar** *m* ELEC *conexiones* soldering flux; ~ **soluble en agua** *m* CONST water-soluble flux

fundería *f* CARBÓN, ING MECÁ, INSTAL TERM, PROD foundry, ironworks; ~ **de cobre** *f* PROD copper works; ~ **de hierro** *f* PROD iron foundry

fundible *adj* C&V, TERMO fusible

fundición *f* CARBÓN iron, smelting, foundry, CONST cast iron, IMPR casting, ING MECÁ cast iron, casting, foundry, INSTAL TERM foundry, MECÁ cast iron, casting, foundry, METAL cast iron, PROD founding, foundry; ~ **de afino** *f* PROD converter pig; ~ **de baja presión** *f* ING MECÁ low pressure casting; ~ **Bessemer** *f* PROD Bessemer iron, Bessemer pig; ~ **de bronce** *f* CONST brass foundry; ~ **bruta de primera fusión** *f* METAL, PROD pig iron; ~ **en cajas** *f* ING MECÁ box casting; ~ **centrífuga** *f* PROD centrifugal casting; ~ **de cobre** *f* PROD copper works; ~ **de cobre de berilio** *f* ING MECÁ beryllium copper casting; ~ **colada a presión por gravedad** *f* ING MECÁ gravity casting; ~ **de concha** *f* PROD chilled cast iron; ~ **en concha** *f* PROD chill casting, chilling; ~ **de coquilla** *f* PROD chill casting; ~ **en coquilla** *f* PROD chilled-iron casting; ~ **de grafito laminar** *f* MECÁ lamellar graphite cast iron; ~ **gris** *f* MECÁ gray cast iron (*AmE*), grey cast iron (*BrE*); ~ **de hierro** *f* PROD *procesamiento* iron founding, iron foundry, ironworks; ~ **inyectada por gravedad** *f* ING MECÁ gravity casting; ~ **de líneas de tipo controlado por cinta** *f* IMPR tape-controlled linecasting; ~ **con macho** *f* PROD core casting, cored casting; ~ **maleable** *f* MECÁ malleable cast iron; ~ **mecanizada** *f* PROD machined casting; ~ **de metal** *f* PROD metal founding; ~ **de plomo** *f* PROD lead-smelting works; ~ **de primera fusión** *f* PROD direct casting; ~ **en vacío** *f* NUCL vacuum casting

fundido[1] *adj* C&V, MECÁ, TERMO molten; ~ **a presión** *adj* MECÁ die-cast; ~ **en una sola pieza** *adj* MECÁ integrally cast; ~ **a troquel** *adj* MECÁ die-cast

fundido[2]: ~ **a blanco** *m* CINEMAT, TV fade-to-white; ~ **de bordes** *m* C&V edge melting; ~ **de cojinetes** *m* TEC ESP burnout; ~ **en coquilla** *m* PROD die-cast; ~ **cruzado** *m* CINEMAT, TV cross fade; ~ **de imagen** *m* CINEMAT, TV picture-fading; ~ **de matrices** *m* PROD die casting; ~ **a negro** *m* CINEMAT, TV fade-to-black; ~ **a presión** *m* PROD die-cast

fundidor *m* PROD *persona* caster, founder, pourer; ~ **por calentamiento** *m* CONST heating melter; ~ **de calor** *m* CONST heating melter

fundidora: ~ **de la composición** *f* IMPR composition caster; ~ **de silicio** *f* ELECTRÓN silicon foundry

fundir *vt* C&V fuse, ING MECÁ cast, INSTAL TERM melt, smelt, P&C fuse, PROD melt, *lingotes* cast, oil, QUÍMICA melt, TEC ESP *fusibles* blow, TERMO fuse, melt, melt down, TEXTIL melt; ~ **al descubierto** *vt* PROD cast in open sand; ~ **a presión** *vt* ING MECÁ die-cast

fundirse *vi* Esp (*cf combinarse AmL*) QUÍMICA coalesce

fungicida *m* AGRIC, P&C, QUÍMICA fungicide

fungistático *m* ALIMENT fungistat

funicular *m* FERRO, ING MECÁ, TRANSP cable railway, cableway, funicular; ~ **de pasajeros** *m* TRANSP passenger ropeway

funivía *f* TRANSP cableway

furfural *m* P&C furfural

furfuraldehído *m* P&C furfuraldehyde
furfuril *adj* QUÍMICA furfuryl
furgón *m* TRANSP truck car (*AmE*), truck wagon (*BrE*); ~ **de cola** *m* FERRO, VEH caboose (*AmE*), guard's van (*BrE*); ~ **de equipajes** *m* FERRO, VEH baggage car (*AmE*), luggage van (*BrE*); ~ **de mercancías** *m* TRANSP freight van (*AmE*), goods van (*BrE*); ~ **con techo corredizo** *m* TRANSP bogie wagon with swivelling roof (*BrE*), truck car with swiveling roof (*AmE*)
furgoneta *f* AUTO, VEH estate car (*BrE*), station wagon (*AmE*), pick-up lorry (*BrE*), pick-up truck (*AmE*); ~ **de correo** *f* AUTO, VEH mail van; ~ **eléctrica** *f* TRANSP, VEH electrovan; ~ **con equipo de TV de exteriores** *f* TRANSP shooting brake; ~ **de mercancías** *f* TRANSP freight van (*AmE*), goods van (*BrE*); ~ **con techo alzado** *f* VEH raised-roof van
furílico *adj* QUÍMICA furilic
furilo *m* QUÍMICA furile
furona *f* QUÍMICA furon
fusariosis *f* AGRIC fusarium wilt
fuselado¹ *adj* VEH fairing
fuselado² *m* TEC ESP, VEH fairing
fuselaje *m* TEC ESP, TRANSP, TRANSP AÉR fuselage; ~ **a chorro** *m* TEC ESP jet body
fusible *m* ELEC, ELECTRÓN, ING ELÉC, TEC ESP, TV fuse, limiter; ~ **de acción diferida** *m* ING ELÉC delayed-action fuse; ~ **de acción retardada** *m* CINEMAT, ING ELÉC, PROD slow-blow fuse; ~ **de alambre** *m* ELEC, ING MECÁ wire fuse; ~ **con alarma** *m* ELEC, ING MECÁ alarm fuse; ~ **apagachispas** *m* ING ELÉC blowout fuse; ~ **de cartucho** *m* ELEC, ING ELÉC cartridge fuse; ~ **de cinta** *m* ELEC, ING ELÉC fuse link, fuse strip, link fuse; ~ **de control** *m* ING MECÁ, PROD control fuse; ~ **dentro** *m* ELEC, ING ELÉC enclosed fuse; ~ **encapsulado** *m* ELEC, ING ELÉC cartridge fuse; ~ **encerrado** *m* ELEC, ING ELÉC, ING MECÁ enclosed fuse; ~ **de entrada de suministro de energía** *m* ING MECÁ, PROD power-supply input fuse; ~ **de**

expulsión *m* ELEC, ING ELÉC expulsion fuse; ~ **fundido** *m* ELEC, ING ELÉC blown fuse; ~ **de hilo descubierto** *m* ELEC, ING MECÁ link fuse; ~ **con indicador** *m* ELEC alarm fuse; ~ **con indicador de alarma** *m* ING MECÁ alarm fuse; ~ **del interruptor** *m* ELEC, ING ELÉC switch fuse; ~ **de línea de entrada** *m* ING MECÁ, PROD incoming line fuse; ~ **no restaurable** *m* ELEC, ING MECÁ nonrenewable fuse; ~ **de potencia absorbida** *m* PROD input power fuse; ~ **de protección contra rayos** *m* GEOFÍS, ING MECÁ lightning protection fuse; ~ **quemado** *m* ELEC, ING ELÉC blown fuse; ~ **de seguridad** *m* MINAS, SEG safety fuse; ~ **de tipo descubierto** *m* ELEC open fuse
fusil *m* D&A rifle; ~ **de asalto** *m* D&A assault rifle
fusión *f* C&V fusing, fusion, FERRO *vehículos* blending, FÍS fusion, melting, GAS fusion, HIDROL *glaciares* ablation, INFORM&PD merge, merging, INSTAL TERM melting, smelting, METEO melting, MINAS fusion, P&C *proceso de moldeo* casting, fusing, *operación* fusion, PROD melting, QUÍMICA fusion, TERMO melting, fusion, fusing, TEXTIL melting, TRANSP merging; ~ **de borde** *f* C&V edge fusion; ~ **por haz electrónico** *f* ELECTRÓN, NUCL electron-beam melting; ~ **por haz lasérico** *f* ELECTRÓN, NUCL laser fusion; ~ **hiperenergética** *f* NUCL high-energy fusion; ~ **inducida por haz lasérico** *f* ELECTRÓN, NUCL laser-driven fusion; ~ **de iones pesados** *f* FÍS PART, NUCL heavy-ion fusion; ~ **de lagunas** *f* METAL void coalescence; ~ **con láser** *f* ELECTRÓN laser melting; ~ **de nieves** *f* AGUA snow melt (*AmE*), snow water (*BrE*); ~ **nuclear** *f* FÍS PART, NUCL nuclear fusion; ~ **de pista de hielo** *f* REFRIG de-icing; ~ **por zonas** *f* METAL, NUCL zone melting
fusionar *vt* INFORM&PD merge
fusiosoldeo *m* CONST, PROD, TERMO fusion welding
fuste *m* CONST pipe, *columna* shaft, shank
FV *abr* (*frecuencia acústica, frecuencia vocal, frecuencia de voz*) FÍS RAD, TELECOM VF (*voice frequency*)

G

G *abr* (*giga-*) METR G (*giga-*)
Ga *abr* (*galio*) ELECTRÓN, FÍS, METAL, ÓPT, QUÍMICA Ga (*gallium*)
gabarra *f* PETROL, TRANSP, TRANSP MAR barge, lighter; **~ algibe** *f* TRANSP tank barge; **~ para conducción submarina** *f* TEC PETR *producción costa-fuera* bury barge; **~ EBCS** *f* TRANSP EBCS lighter; **~ grúa** *f* TEC PETR *operaciones marinas* crane barge; **~ de limpieza con rasera** *f* CONTAM MAR skimming barge; **~ de perforación** *f* TEC PETR *costa-fuera* drill barge; **~ con propulsión propia** *f* TRANSP MAR self-propelled barge; **~ sin propulsión propia** *f* TRANSP MAR dumb barge; **~ de reparaciones** *f* TEC PETR *producción costa-fuera* workover barge; **~ para tender oleoductos submarinos** *f* PETROL, TEC PETR *producción costa-fuera* lay barge; **~ de tendido** *f* TEC PETR *operaciones marinas* laying barge; **~ para tendido de tuberías** *f* TEC PETR *producción costa-fuera* pipelaying barge
gabarrero *m* TRANSP MAR bargee (*BrE*), bargeman (*AmE*), lighterman
gabinete *m* AmL (*cf armario Esp*) LAB *mobiliario* cupboard, MECÁ, P&C *endurecimiento de plásticos y pinturas* cabinet; **~ de exhibición refrigerado** *m* REFRIG refrigerated display cabinet; **~ de montaje mural** *m* LAB *mobiliario* wall cupboard; **~ de primeros auxilios** *m* SEG first-aid cabinet (*AmE*), first-aid cupboard (*BrE*); **~ de secado** *m* P&C *pintura* drying cabinet; **~ del televisor** *m* TELECOM, TV television cabinet; **~ vacuosecador** *m* ING MECÁ vacuum-drying cabinet
gabro *m* PETROL gabbro
gadolinio *m* (*Gd*) QUÍMICA gadolinium (*Gd*)
gadolinita *f* MINERAL gadolinite
gafas *f pl Esp* (*cf anteojos AmL*) INSTR, ÓPT glasses; **~ de alto rendimiento para visión nocturna** *f pl Esp* (*cf anteojos de alto rendimiento para visión AmL*) D&A high-performance night vision goggles; **~ antideslumbrantes** *f pl Esp* (*cf anteojos antideslumbrantes AmL*) INSTR antidazzle glasses, antiglare glasses; **~ industriales** *f pl Esp* (*cf anteojos industriales AmL*) LAB, PROD, SEG industrial eye protectors; **~ de lectura** *f pl Esp* (*cf anteojos de protección AmL*) INSTR, ÓPT reading glasses, reading spectacles; **~ multifocales** *f pl Esp* (*cf anteojos multifocales AmL*) INSTR, ÓPT multifocal glasses; **~ polarizantes** *f pl Esp* (*cf anteojos polarizantes AmL*) FOTO polarizing spectacles; **~ protectores** *f pl Esp* (*cf anteojos de protección AmL*) ING MECÁ, LAB, ÓPT, PROD, SEG glasses, goggles, protective glasses, protective spectacles; **~ protectores de radiaciones X** *Esp* (*cf anteojos protectores de radiaciones X AmL*) *f pl* INSTR, ÓPT, SEG X-ray protective glasses; **~ protectores de rayos X** *f pl Esp* (*cf anteojos protectores de rayos AmL*) SEG X-ray protective glasses; **~ de seguridad** *f pl Esp* (*cf anteojos de seguridad AmL*) C&V, INSTR, LAB, ÓPT, PROD, SEG eye-protection glasses, safety glasses, safety goggles, safety spectacles; **~ de sol** *f pl Esp*
(*cf anteojos de sol AmL*) INSTR, ÓPT sun spectacles; **~ de soldador** *f pl Esp* (*cf anteojos de soldador AmL*) INSTR, SEG welder's goggles; **~ para soldadores** *f pl Esp* (*cf anteojos para soldadores AmL*) INSTR, SEG welding goggles; **~ para soldar** *f pl Esp* (*cf anteojos para soldar AmL*) SEG welding visor (*BrE*), welding vizor (*AmE*); **~ de visión nocturna** *f pl Esp* (*cf anteojos de visión nocturna AmL*) D&A, ÓPT night vision goggles (*NVG*)
galactónico *adj* QUÍMICA galactonic
galactosa *f* ALIMENT, QUÍMICA galactose
galactosamina *f* QUÍMICA galactosamine
galápago *m* CARBÓN ingot, MINAS headboard, PROD ingot, *fundería, fundición* pig, TRANSP MAR *construcción* chock; **~ de plomo** *m* PROD *fundición* pig of lead
galato *m* QUÍMICA gallate
galaxia: **~ cuasiestelar** *f* (*QSG*) TEC ESP quasi-stellar galaxy (*QSG*)
galeína *f* QUÍMICA gallein
galena *f Esp* (*cf mena de plomo AmL*) MINAS lead ore, MINERAL lead glance, QUÍMICA galena; **~ de uranio** *f* NUCL uranium galena
galenita *f* MINERAL galena
galenobismutita *f* MINERAL galenobismutite
galerada *f* IMPR galley
galería *f* CARBÓN gallery, CONST drift, GEOL gallery, MINAS passageway, roadway, drive, *pozo de minas* shaft, NUCL gallery; **~ de acceso** *f* CARBÓN adit; **~ aerífera** *f* MINAS air passageway; **~ de aire caliente** *f* TRANSP AÉR hot air gallery; **~ de aireación** *f* MINAS dumb drift; **~ de arrastre** *f* AmL (*cf galería de transporte Esp*) CARBÓN, MINAS, PROD cross entry, gateway, haulage road, haulage way; **~ atascada** *f* Esp (*cf galería de tostación AmL*) MINAS stall; **~ de avance** *f* Esp (*cf galería del frente de ataque AmL*) ING MECÁ heading, MINAS airhead, airheading, driving level, head, stall road; **~ de base** *f* Esp (*cf galería inferior AmL*) MINAS bottom road; **~ de cabeza** *f* Esp (*cf galería superior AmL*) MINAS top road; **~ de cables** *f* ELEC, ING ELÉC *fuente de alimentación* cable manhole; **~ en capa** *f* MINAS level in the seam; **~ de captación** *f* AGUA infiltration gallery; **~ de carbón** *f* CARBÓN, MINAS coal drift, coal road; **~ de carga** *f* MINAS lodge; **~ de cateo** *f* AmL (*cf galería de exploración Esp*) MINAS exploration drive, exploration level, prospecting level; **~ ciega** *f* AmL (*ver fondo de saco*) MINAS blind level; **~ a cielo abierto** *f* MINAS surface drive; **~ de circulación del personal** *f* MINAS manway; **~ conductora** *f* CARBÓN conveyor road; **~ de desagüe** *f* AGUA water adit, *minas* offtake, MINAS drainage level, lodge, offtake, adit; **~ de dirección** *f* MINAS drifting level, level; **~ en dirección** *f* MINAS drive; **~ de drenaje** *f* HIDROL *presas* drainage gallery; **~ entibada adosada** *f* AmL (*cf galería entibada ajustada Esp*) MINAS close-timbered level; **~ entibada ajustada** *f Esp* (*cf galería entibada adosada AmL*) MINAS close-timbered level; **~ de entrada de aire** *f* CARBÓN, MINAS intake airway; **~ de entrada de aire principal** *f* CARBÓN, MINAS

main intake airway; **~ estéril** *f* CARBÓN, MINAS metal drift; **~ de estériles** *f Esp* (*cf galería de relleno AmL*) MINAS waste drive; **~ de exploración** *f Esp* (*cf galería de reconocimiento AmL*) MINAS exploration drive, exploration level, prospecting level; **~ de extracción** *f* AGUA offtake, CARBÓN gangway, MINAS offtake; **~ falsa** *f Esp* (*cf piso intermedio AmL*) MINAS blind level; **~ filoniana** *f* MINAS reef drive; **~ filtrante** *f* AGUA filter gallery; **~ de fondo** *f* MINAS bottom level, bottom road; **~ forestal** *f* AGRIC forest gallery; **~ del frente de ataque** *f AmL* (*cf galería de avance Esp*) ING MECÁ heading, MINAS airhead, airheading, driving level, head, stall road; **~ horizontal de avance** *f* MINAS drifting level; **~ inclinada** *f Esp* (*cf chiflón AmL*) CARBÓN dip, MINAS inclined hole, slope, incline hole, dip; **~ inferior** *f AmL* (*cf galería de base Esp*) MINAS bottom road; **~ inferior de desagüe** *f* AGUA deep adit; **~ de infiltración** *f* AGUA infiltration gallery; **~ de inspección** *f AmL* (*cf galería de visita Esp*) HIDROL, MINAS inspection gallery, manhole; **~ intermedia** *f* MINAS counter, counter gangway, counterlevel, subdrift, sublevel; **~ de labor atravesada** *f AmL* (*cf transversal Esp*) MINAS offset drive; **~ maestra** *f* CARBÓN, MINAS gangway; **~ de mina** *f* MINAS mine level; **~ minera** *f AmL* (*cf túnel Esp*) MINAS tunnel; **~ principal** *f* CARBÓN entry, main roadway, MINAS cross entry, level road, mainway; **~ principal lateral** *f* MINAS side entry; **~ principal paralela al frente** *f* MINAS lateral; **~ principal de retorno del aire** *f* MINAS main return airway; **~ de prospección** *f* MINAS prospecting level; **~ de reconocimiento** *f AmL* (*cf galería de exploración Esp*) MINAS exploration drive, exploration level, prospecting level; **~ de recorte** *f* MINAS offset drive; **~ de relleno** *f AmL* (*cf galería de estériles Esp*) MINAS waste drive; **~ de retorno del aire** *f* MINAS return airway; **~ en roca** *f* CARBÓN metal drift, MINAS gallery in dead ground, rock drift, stone drift; **~ en rocas** *f* CARBÓN, MINAS stone drift; **~ de servicio** *f* CONST service tunnel, MINAS passageway; **~ de sifón** *f* *AmL* (*cf galería falsa Esp*) MINAS blind level; **~ superior** *f AmL* (*cf galería de cabeza Esp*) MINAS top road; **~ de tostación** *f AmL* (*cf galería atascada Esp*) MINAS stall; **~ de transporte** *f Esp* (*cf galería de arrastre AmL*) CARBÓN, MINAS, PROD gateway, cross entry, haulage road, haulage way; **~ de trazado** *f* development heading, forewinning heading; **~ de ventilación** *f* CARBÓN, MINAS airhead, airheading, cross entry, fan drift, ventilation drive, airway, crossheading; **~ de ventilación lateral** *f* MINAS side entry; **~ de visita** *f Esp* (*cf galería de inspección AmL*) HIDROL, MINAS inspection gallery, manhole

galerías: ~ gemelas *f pl* MINAS double entries

galga *f* ING MECÁ gage (*AmE*), gauge (*BrE*), template, templet, INSTR, LAB gage (*AmE*), gauge (*BrE*), MECÁ, MINAS drag, PROD *alambres*, TEXTIL *punto*, VEH gage (*AmE*), gauge (*BrE*); **~ de alambres** *f* METR wire gage (*AmE*), wire gauge (*BrE*); **~ para la alineación de cojinetes del cigüeñal** *f* ING MECÁ crankshaft alignment gage (*AmE*), crankshaft alignment gauge (*BrE*); **~ de alturas** *f* METR height gage (*AmE*), height gauge (*BrE*); **~ de ángulos de corte de brocas** *f* ING MECÁ drill gage (*AmE*), drill gauge (*BrE*); **~ para ángulos de tuercas octogonales** *f* ING MECÁ octagonal nut angle gage (*AmE*), octagonal nut angle gauge (*BrE*); **~ de bloques** *f* METR slip gage (*AmE*), slip gauge (*BrE*); **~ de bloques de cerámica** *f* ING MECÁ ceramic slip gage (*AmE*), ceramic slip gauge (*BrE*); **~ de brocas** *f* ING MECÁ drill gage (*AmE*), drill gauge (*BrE*); **~ de carpintero** *f* CONST carpenters' gage (*AmE*), carpenters' gauge (*BrE*); **~ de chapa** *f* ING MECÁ sheet gage (*AmE*), sheet gauge (*BrE*); **~ de chapa de hierro** *f* ING MECÁ sheet iron gage (*AmE*), sheet iron gauge (*BrE*); **~ de clavija** *f* ING MECÁ, METR plug gage (*AmE*), plug gauge (*BrE*); **~ de comprobación del ángulo de la pala** *f* TRANSP AÉR blade-angle check-gage (*AmE*), blade-angle check-gauge (*BrE*); **~ de cuadrante** *f* ING MECÁ dial gage (*AmE*), dial gauge (*BrE*); **~ para dimensiones exteriores** *f* METR gap gage (*AmE*), gap gauge (*BrE*); **~ eléctrica de deformación** *f* ING ELÉC, MECÁ strain gage (*AmE*), strain gauge (*BrE*); **~ de espesor de tubos** *f* NUCL tube thickness gage (*AmE*), tube thickness gauge (*BrE*); **~ de espesores** *f* ING MECÁ, MECÁ thickness gage (*AmE*), thickness gauge (*BrE*), METR, NUCL, VEH feeler gage (*AmE*), feeler gauge (*BrE*); **~ para espesores de chapas** *f* ING MECÁ, MECÁ plate gage (*AmE*), plate gauge (*BrE*); **~ fina** *f* TEXTIL *género de punto* fine gage (*AmE*), fine gauge (*BrE*); **~ gruesa** *f* TEXTIL *género de punto* heavy gage (*AmE*), heavy gauge (*BrE*); **~ indicadora de cuadrante** *f* METR dial-indicating gage (*AmE*), dial-indicating gauge (*BrE*); **~ de láminas** *f* ING MECÁ sheet gage (*AmE*), sheet gauge (*BrE*); **~ para límites** *f* ING MECÁ, METR limit gage (*AmE*), limit gauge (*BrE*); **~ para medir la caída del cigüeñal** *f* ING MECÁ crankshaft alignment gage (*AmE*), crankshaft alignment gauge (*BrE*); **~ para medir la profundidad del dibujo del neumático** *f* VEH tread depth gage (*AmE*), tread depth gauge (*BrE*); **~ para medir profundidades** *f* METR depth gage (*AmE*), depth gauge (*BrE*); **~ de nonio de avance** *f* ING MECÁ sliding caliper gage (*AmE*), sliding calliper gauge (*BrE*); **~ de palastro** *f* ING MECÁ sheet iron gage (*AmE*), sheet iron gauge (*BrE*); **~ de paso de rosca** *f* ING MECÁ thread pitch gage (*AmE*), thread pitch gauge (*BrE*); **~ patrón** *f* ING MECÁ standard gage (*AmE*), standard gauge (*BrE*); **~ de perfil** *f* ING MECÁ form shim; **~ de plancha** *f* ING MECÁ sheet gage (*AmE*), sheet gauge (*BrE*); **~ de plancha de hierro** *f* ING MECÁ sheet iron gage (*AmE*), sheet iron gauge (*BrE*); **~ receptora** *f* METR receiver gage (*AmE*), receiver gauge (*BrE*); **~ de roscas** *f* ING MECÁ, METR screw gage (*AmE*), screw gauge (*BrE*), screw thread gage (*AmE*), screw thread gauge (*BrE*), thread gage (*AmE*), thread gauge (*BrE*); **~ simple** *f* ING MECÁ plain gage (*AmE*), plain gauge (*BrE*); **~ simple para roscas de tubos** *f* ING MECÁ plain gage for pipe threads (*AmE*), plain gauge for pipe threads (*BrE*); **~ de tolerancia** *f* ING MECÁ, METR limit gage (*AmE*), limit gauge (*BrE*); **~ de tubo** *f* ING MECÁ tube gage (*AmE*), tube gauge (*BrE*); **~ de la vía** *f* FERRO *equipo fijo* rail gage (*AmE*), rail gauge (*BrE*)

galgas: ~ de precisión *f pl* ING MECÁ close tolerance spacer

galibador *m* TRANSP MAR loftsman

galibar *vi* TRANSP MAR *arquitectura naval* lay down the lines

gálibo *m* CONST gage bar (*AmE*), gauge bar (*BrE*), ELEC *medición, fabricación* template, ING MECÁ former, PROD *moldeo en arcilla* template, TRANSP MAR *construcción naval* template, *construcción de madera, acero* mold (*AmE*), mould (*BrE*); **~ de carga** *m* FERRO *equipo inamovible* gage (*AmE*), gauge (*BrE*), *vehículos* loading gage (*AmE*), loading gauge (*BrE*); **~ de tolerancia** *m* ING MECÁ, METR limit gage (*AmE*), limit gauge (*BrE*)

gálico *adj* QUÍMICA gallic

galio *m* (*Ga*) ELECTRÓN, FÍS, METAL, ÓPT, QUÍMICA gallium (*Ga*)

gálivo *m* PROD former

gallardete *m* C&V burgee, TRANSP MAR pennant; **~ característico** *m* TRANSP MAR answering pennant; **~ de inteligencia** *m* TRANSP MAR answering pennant

galleta *f* CARBÓN cobbles, TRANSP MAR *del mástil* truck (*AmE*)

gallinaza *f* AGRIC poultry manure

gallinero *m* AGRIC coop

galón *m* C&V gallon jug, D&A stripe, METR gallon, TEXTIL braid, *en telas* gimp, TRANSP MAR *construcción naval* molding (*AmE*), moulding (*BrE*); **~ de algodón** *m* TEXTIL cotton braid

galpón *m* CONST shed

galvánico *adj* GEN galvanic

galvanización *f* ELEC, ING ELÉC, PROD, QUÍMICA, REVEST galvanization, zincking

galvanizado[1] *adj* ELEC, ING ELÉC, MECÁ, P&C, QUÍMICA, REVEST galvanized

galvanizado[2] *m* P&C *pintura* zinc coating, REVEST hot dip galvanizing; **~ en caliente** *m* REVEST hot dip galvanised coating

galvanizar *vt* CONST, ELEC, ING ELÉC, PROD, QUÍMICA, REVEST galvanize, zinc

galvano *m* ÓPT *de disco fonográfico* master

galvanómetro *m* GEN galvanometer; **~ de aguja** *m* ELEC, INSTR needle galvanometer; **~ aperiódico** *m* ELEC, INSTR deadbeat galvanometer; **~ de Arsonval** *m* ING ELÉC d'Arsonval galvanometer; **~ astático** *m* FÍS, ING ELÉC astatic galvanometer; **~ balístico** *m* ELEC, FÍS ballistic galvanometer; **~ de bobina móvil** *m* ELEC, FÍS, ING ELÉC moving-coil galvanometer; **~ de bucle** *m* ELEC, ING ELÉC loop galvanometer; **~ de cero** *m* ELEC, ING ELÉC, INSTR null galvanometer; **~ de cuadro** *m* ELEC, ING ELÉC loop galvanometer; **~ de cuadro móvil** *m* ELEC, FÍS, ING ELÉC moving-coil galvanometer; **~ de cuerda de torsión** *m* ELEC, ING ELÉC torsion-string galvanometer; **~ diferencial** *m* ELEC, ING ELÉC, INSTR differential galvanometer; **~ de Einthoven** *m* ELEC, ING ELÉC Einthoven galvanometer; **~ de espejo** *m* ELEC, ING ELÉC, INSTR mirror galvanometer; **~ de espejo cóncavo** *m* ELEC, ING ELÉC, INSTR reflecting-mirror galvanometer; **~ de espejo ustorio** *m* ELEC, ING ELÉC, INSTR reflecting-mirror galvanometer; **~ de fibras de torsión** *m* ELEC, ING ELÉC torsion-string galvanometer; **~ de haz luminoso** *m* FÍS RAD light beam galvanometer; **~ de Helmholtz** *m* ELEC Helmholtz galvanometer; **~ de imán móvil** *m* ELEC, ING ELÉC, INSTR moving-magnet galvanometer; **~ de punto explorador** *m* INSTR light spot galvanometer; **~ de punto luminoso** *m* INSTR light spot galvanometer; **~ de punto móvil** *m* INSTR light spot galvanometer; **~ de resonancia** *m* ELEC vibration galvanometer; **~ de señal nula** *m* ELEC, INSTR null galvanometer; **~ sinusoidal** *m* FÍS sine galvanometer; **~ tangente** *m* FÍS tangent galvanometer; **~ de tangentes** *m* ELEC, ING ELÉC tangent galvanometer;

~ de vibración *m* ELEC, FÍS, ING ELÉC vibration galvanometer

galvanoplacas *f pl* FÍS galvanoplates

galvanoplastia *f* CONST, ELEC, ING ELÉC, PROD, REVEST electrodeposition, electroforming, electroplate, electroplating, galvanoplastics

galvanoplastiado *adj* CONST, ELEC, ING ELÉC, PROD, REVEST electroplated

galvanoplastiar *vt* CONST, ELEC, ING ELÉC, PROD, REVEST electroplate

galvanotipia *f* IMPR electrotype

galvonoplastia *f* REVEST galvanic plating

gama *f* ELECTRÓN *de emisor*, GEOM, INFORM&PD range, ING MECÁ scope, *de velocidades, temperaturas* range, MECÁ, TEC ESP, TELECOM, TEXTIL range; **~ actual** *f* PROD current rating; **~ de alimentación** *f* ING MECÁ feed range; **~ de avances automáticos** *f* PROD automatic crossfeed range; **~ de avances transversales** *f* PROD automatic crossfeed range; **~ de corriente** *f* PROD current rating; **~ de densidad** *f* IMPR density range; **~ de destilación** *f* TEC PETR *refino* boiling range; **~ dinámica** *f* ACÚST, ELECTRÓN, FÍS RAD, PETROL, TV dynamic range; **~ dinámica de amplificador vertical** *f* ELECTRÓN vertical amplifier dynamic range; **~ de ebullición** *f* TEC PETR *refino* boiling range; **~ de energía** *f* FÍS RAD energy range; **~ equilibrada de tintas** *f* IMPR balanced process inks; **~ focal** *f* CINEMAT, FOTO, TV focal range; **~ de frecuencias** *f* ELEC, ELECTRÓN, TV frequency range; **~ de frecuencias acústicas** *f* ACÚST, FÍS RAD audio range; **~ de funciones** *f* ELECTRÓN moding; **~ de intensidad luminosa** *f* FOTO brightness range; **~ de longitud de olas** *f* TRANSP MAR waveband; **~ de longitud de ondas** *f* FÍS RAD waveband; **~ de modos** *f* ELECTRÓN moding; **~ de períodos de semidesintegración** *f* FÍS RAD range of half-life; **~ de productos** *f* PROD product range; **~ de señales de sincronización** *f* ELECTRÓN *oscilador* lock-in range; **~ de sintonización electrónica** *f* ELECTRÓN, TELECOM electronic-tuning range; **~ de temperatura crítica** *f* TERMO critical-temperature range; **~ de temperatura efectiva** *f* TERMO effective-temperature range; **~ de temperaturas** *f* TERMO temperature range; **~ de temperaturas de fusión** *f* TERMO melting range; **~ de tomas** *f* ELEC tapping range; **~ de zoom** *f* CINEMAT zoom range

gambir *m* QUÍMICA gambier

gamella *f* CARBÓN, MINAS pan, *lavado de minerales* buddle, cradle, trough

gameto *m* AGRIC gamete

gamexano *m* QUÍMICA *marca* gammexane

gamma *f* CINEMAT, NUCL, PETROL, TV gamma

gammagrafía *f* TEC PETR gamma-ray log; **~ de pozo** *f* TEC PETR gamma-ray well logging

gamuza *f* PROD chamois

ganadería *f* AGRIC animal husbandry, livestock husbandry

ganadero *m* AGRIC cattle dealer, cattle raiser

ganado: **~ de abasto** *m* AGRIC beef cattle; **~ básico** *m* AGRIC nucleus breeding herd; **~ de distinto pelaje** *m* AGRIC rainbow cattle; **~ de doble finalidad** *m* AGRIC dual-purpose cattle; **~ de engorde** *m* AGRIC beef herd, feeder; **~ gordo** *m* AGRIC fat cattle; **~ lechero** *m* AGRIC dairy cattle, milk cattle; **~ mayor** *m* AGRIC heavy livestock; **~ original de la explotación** *m* AGRIC foundation stock; **~ en pie** *m* AGRIC cattle on the hoof; ~ **de pura sangre** *m* AGRIC blooded stock; ~ **vacuno** *m* AGRIC beef cattle, cattle; ~ **vacuno de leche** *m* AGRIC dairy cattle, milk cattle; ~ **vacuno seleccionado** *m* AGRIC grade cattle

ganancia *f* ELEC, INFORM&PD gain, PROD return, yield, TEC ESP *comunicaciones*, TELECOM, TRANSP MAR, TV gain; ~ **absoluta** *f* TEC ESP absolute gain; ~ **de amplificación** *f* ELECTRÓN, FÍS amplifier gain; ~ **de antena** *f* FÍS, TEC ESP, TV aerial gain (*BrE*), antenna gain (*AmE*); ~ **asociada** *f* ELECTRÓN associated gain; ~ **en banda intermedia** *f* ELECTRÓN midband gain; ~ **bruta** *f* PROD gross profit; ~ **en bucle cerrado** *f* ELEC, ELECTRÓN, ING ELÉC closed-loop gain; ~ **de conversión** *f* ELECTRÓN conversion gain; ~ **de corriente** *f* ELEC, ELECTRÓN, FÍS current gain; ~ **debida a la codificación** *f* TELECOM transcoding gain (*TG*); ~ **de desviación angular relativa** *f* ACÚST relative angular-deviation gain; ~ **diferencial** *f* ELECTRÓN, TV differential gain; ~ **en fase de entrada** *f* ELECTRÓN input-stage gain; ~ **fotoconductiva** *f* ELECTRÓN photoconductive gain; ~ **impeditiva** *f* ELECTRÓN obstacle gain; ~ **por inserción** *f* FÍS insertion gain; ~ **de integración** *f* ELECTRÓN integration gain; ~ **interna** *f* ELECTRÓN internal gain; ~ **de interpolación** *f* TELECOM interpolation gain (*IG*); ~ **inversa** *f* ELECTRÓN inverse gain; ~ **isotrópica** *f* TEC ESP isotropic gain; ~ **en modo común** *f* ELECTRÓN, TELECOM common-mode gain; ~ **de multiplicación del circuito digital** *f* TELECOM digital circuit multiplication gain (*DCMG*); ~ **óptica** *f* ELECTRÓN optical gain; ~ **de peso** *f* AGRIC gaining; ~ **de potencia** *f* ELECTRÓN, FÍS power gain; ~ **en proceso de descarga** *f* ELECTRÓN avalanche gain; ~ **receptora** *f* ELECTRÓN receiver gain; ~ **relativa** *f* TEC ESP relative gain; ~ **de tensión** *f* ING ELÉC voltage gain; ~ **de tensión de conversión** *f* ING ELÉC conversion voltage gain; ~ **de transcodificación** *f* TELECOM transcoding gain (*TG*); ~ **de transmisión** *f* ACÚST, TELECOM transmission gain; ~ **unidad** *f* ELECTRÓN unity gain

gancho *m* C&V hook, CONST *clavo* claw, crow, hook, ING MECÁ hook, catch, INSTAL HIDRÁUL gab, MECÁ hook, dog, MINAS *sondeos* pole hook, NUCL grab, PETROL hook, PROD *cazos de colada* bail; ~ **de acero forjado** *m* ING MECÁ forged shackle; ~ **de acoplamiento** *m* ING MECÁ coupling hook; ~ **de agarre de dedo** *m* TRANSP AÉR finger-grip clip; ~ **de aguja** *m* TEXTIL *de género de punto* hook; ~ **de alza** *m* CONST, ING MECÁ lifting hook; ~ **en ancla** *m* ING MECÁ clip hook; ~ **en ángulo** *m* CONST hook and hinge; ~ **de apriete** *m* ING MECÁ dog hook; ~ **de arrastre** *m* ING MECÁ dog hook; ~ **basculante** *m* ING MECÁ swivel hook; ~ **para bloques de piedras** *m* CONST stone hook; ~ **de botavara** *m* TRANSP MAR *armamento del puntal de carga* gooseneck; ~ **de carga** *m* TRANSP MAR cargo hook; ~ **de carne** *m* ALIMENT meat hook; ~ **cerrado** *m* PROD eyehook; ~ **cerrado y dedal** *m* PROD eyehook and thimble; ~ **para clavar** *m* CONST hook to drive; ~ **de cola de cerdo** *m* PROD pigtail hook; ~ **de cuadernal** *m* ING MECÁ pulley-block hook, PROD tackle hook; ~ **de deslizamiento** *m* MINAS slip box; ~ **de desprendimiento** *m* MINAS detaching hook; ~ **disparador** *m* ING MECÁ trip dog, TRANSP MAR *equipamiento de cubierta* senhouse slip; ~ **de elevación** *m* MINAS *sondeos* lifting dog; ~ **de**

enlace *m* ING MECÁ clip hook; ~ **excéntrico** *m* ING MECÁ eccentric hook; ~ **en forma de S** *m* ING MECÁ S-shaped hook; ~ **de frenado** *m* TRANSP AÉR arresting hook; ~ **giratorio** *m* ING MECÁ, TRANSP MAR swivel hook; ~ **de grúa** *m* CONST crane hook; ~ **de grúa de acero forjado** *m* ING MECÁ forged-steel lifting hook; ~ **de hierro forjado** *m* ING MECÁ forged shackle; ~ **de izar** *m* CONST, ING MECÁ lifting hook; ~ **para levantar** *m* CONST, ING MECÁ lifting hook; ~ **de liberación** *m* CONST releasing hook; ~ **de mano** *m* MINAS *perforación* hand dog; ~ **mecánico** *m* CONST mechanical grab; ~ **de molde** *m* PROD gagger; ~ **con ojal** *m* ING MECÁ hook with eye; ~ **y ojo** *m* ING MECÁ hook and eye; ~ **del plegador** *m* PROD *fundería* beam hook; ~ **de pudelador** *m* PROD puddler's rabble; ~ **de pudelar** *m* PROD staff; ~ **roscado** *m* CONST screw hook; ~ **en S** *m* ING MECÁ S hook; ~ **de seguridad** *m* ING MECÁ clip hook, safety hook, spring governor; ~ **de seguridad doble** *m* ING MECÁ clip hook; ~ **de seguridad magnético** *m* ING MECÁ magnetic holdfast; ~ **soportatubos** *m* CONST pipe hook; ~ **de suelta de la carga** *m* TRANSP AÉR cargo release hook; ~ **de superficie** *m* CONST hook on flap; ~ **de suspensión** *m* CONST lifting hook, ING MECÁ lifting hook, suspension hook, PROD hanger; ~ **de tracción** *m* FERRO *enganche* draw hook bar, *vehículos* draw hook, PROD *moldura* draw hook, VEH *remolque* tow hook; ~ **de unión** *m* PROD dog iron

ganchos: ~ **gemelos** *m pl* CONST match hooks

ganga *f* CARBÓN gangue mineral, *mineral* tailings, *minerales* scalping, GEOL gangue, MINAS minestuff, attle, dead ground, stone; ~ **improductiva** *f* CARBÓN barren gangue

gánguil *m* AmL (*cf dragalina Esp*) MINAS *barco de arrastre* dredge boat, TRANSP MAR *bote* dredger; ~ **con cántara** *m* TRANSP MAR hopper; ~ **con tolvas** *m* TRANSP MAR hopper barge

gap: ~ **no deposicional** *m* GEOL nondepositional gap

garantía *f* GEN security, warranty; ~ **de calidad** *f* GEN quality assurance (*QA*), quality guarantee; ~ **de calidad en control y pruebas definitivas** *f* CALIDAD quality-assurance in final inspection and test

garantizado: ~ **contra falsificaciones** *adj* EMB tamperproof

garantizar *vt* GEN warrant

garfear *vt* ING MECÁ hook

garfio *m* CINEMAT claw, register peg, *del movimiento de la cámara* register pin, CONST, ING MECÁ hook, MINAS *jaula, accesorio de extracción* catch; ~ **de alimentación** *m* CINEMAT feed claw; ~ **de alimentación de la película** *m* CINEMAT film feeder pin; ~ **de arpeo** *m* TRANSP MAR grappling hook; ~ **de arrastre** *m* CINEMAT drive pin; ~ **auxiliar** *m* CINEMAT pilot claw; ~ **central** *m* CINEMAT, ING MECÁ central claw; ~ **doble** *m* CINEMAT double claw, ING MECÁ change hook; ~ **intermitente** *m* CINEMAT intermittent claw; ~ **lateral simple** *m* CINEMAT single-side claw; ~ **de tracción** *m* CINEMAT pull-down claw

garganta: **con** ~ *adj* PROD grooved

garganta *f* ACÚST, C&V throat, CARBÓN *poleas* jaw, CONST *chimenea* throat, ING MECÁ groove, jaw, METAL neck, PROD *tornillo* groove, TEC ESP *tobera de la nave espacial* throat; ~ **de entrada** *f* TRANSP AÉR inlet throat; ~ **recta** *f* C&V straight throat; ~ **de la tobera** *f* TEC ESP nozzle throat

garita *f* CARBÓN *vagón de ferrocarril* cupola, D&A

watchtower, FERRO cupola, TRANSP cab; ~ **del guardafrenos** *f* FERRO brakeman's cabin

garlito *m* OCEAN pound net

garlopa *f* CONST jack plane, jointing plane

garlopín *m* CONST fillister

garnierita *f* MINERAL garnierite

garra *f Esp* (*cf perro AmL*) CONST *clavo* claw, ING MECÁ finger, grip, dog, jaw, hook, prong (*BrE*), MECÁ catch, clutch, PROD *para argolla, anillas, asa de cajas* dog; ~ **de acero forjada en caliente** *f* PROD *de horno* drop-forged steel dog; ~ **de fijación** *f* ING MECÁ yoke; ~ **del plato** *f* ING MECÁ faceplate dog

garrafón *m* C&V, LAB *material de vidrio* carboy

garrapata *f* AGRIC cattle tick, tick

garrear: ~ **el ancla** *fra* TRANSP MAR drag anchor

garreo *m* TRANSP MAR *ancla, boya amarrada* dragging

garrucha *f AmL* (*cf carrucha Esp*) ING MECÁ cone sheave, MINAS pulley; ~ **diferencial de cadena** *f* ING MECÁ chain block

garrucho *m* TRANSP MAR *cabo* hank, hook; ~ **de cabo** *m* TRANSP MAR *vela* cringle; ~ **de la faja de rizos** *m* TRANSP MAR *veleros* reef cringle

garuar *vi* METEO drizzle

gas *m* GEN gas; ~ **ácido** *m* GAS, TEC PETR sour gas; ~ **agente nervioso** *m* D&A nerve agent; ~ **de agua** *m* GAS water gas; ~ **de aire** *m* GAS air; ~ **de alto horno** *m* GAS top gas, PROD blast-furnace gas, furnace gas; ~ **de alumbrado** *m* CARBÓN carbon monoxide; ~ **amoníaco** *m* DETERG, GAS, HIDROL ammonia gas; ~ **de amortiguación** *m* GAS cushion gas; ~ **asociado** *m* GAS, TEC PETR associated gas; ~ **atrapado** *m* GAS entrapped gas; ~ **con azufre** *m* GAS, PETROL, TEC PETR sour gas; ~ **de background** *m* GAS, TEC PETR *perforación* background gas (*BG*); ~ **biatómico** *m* GAS biatomic gas; ~ **en botellones** *m* GAS, TERMO bottled gas; ~ **para calefacción** *m* GAS, TERMO heating gas; ~ **caliente ionizado** *m* GAS, TEC ESP plasma; ~ **de campamento** *m* GAS, TEC PETR camping gas; ~ **de carbón** *m* CARBÓN, GAS coal gas; ~ **carbónico** *m* MINAS stythe; ~ **de chimenea** *m* GAS flue gas; ~ **ciudad** *m* GAS, TEC PETR town gas; ~ **cloacal** *m* GAS, RECICL sewage gas; ~ **combustible** *m* GAS combustible gas, fuel gas; ~ **de combustión** *m* GAS, PROD, TEC ESP, TERMO, TERMOTEC combustion gas, flue gas; ~ **comercializable** *m* GAS marketable gas; ~ **común** *m* GAS conventional gas; ~ **condensable** *m* GAS condensable gas; ~ **de condensado** *m* GAS condensate gas; ~ **de conexión** *m* GAS, TEC PETR *perforación* connection gas (*CG*); ~ **para consumo urbano** *m* GAS city gas, town gas; ~ **convertidor** *m* DETERG, GAS converter gas; ~ **corrosivo** *m* GAS, TEC PETR sour gas; ~ **de cortes** *m* GAS, TEC PETR *perforación* cuttings gas; ~ **crudo** *m* GAS raw gas; ~ **de desecho** *m* GAS waste gas; ~ **de desperdicio** *m* GAS waste gas; ~ **de destilación** *m* GAS, PROC QUÍ, QUÍMICA, TERMO distillation gas; ~ **de diapiro** *m* GAS dome gas; ~ **diatómico** *m* FÍS, GAS diatomic gas; ~ **diesel** *m* GAS diesel gas; ~ **de digestión de fangos cloacales** *m* GAS, RECICL sludge gas; ~ **de digestión de lodos** *m* GAS, HIDROL sludge digestion gas; ~ **disuelto** *m* GAS dissolved gas; ~ **de domo** *m* GAS dome gas; ~ **dulce** *m* GAS, TEC PETR sweet gas; ~ **de electrones** *m* ELECTRÓN, FÍS, GAS electron gas; ~ **de electrones degenerados** *m* ELECTRÓN, FÍS, GAS degenerate electron gas; ~ **embotellado** *m* TERMO bottled gas;

~ **energético** *m* GAS power gas; ~ **de entrada** *m* GAS input gas; ~ **envasado** *m* GAS bottled gas; ~ **de exhaustación** *m* PROD *hornos* waste gas; ~ **de gasógeno** *m* GAS, PROD generator gas; ~ **hidrógeno** *m* GAS, QUÍMICA hydrogen gas; ~ **de hulla** *m* GAS coal gas; ~ **húmedo** *m* FÍS, GAS, PETROL wet gas; ~ **ideal** *m* GAS ideal gas, TERMO perfect gas; ~ **in situ** *m* GAS, PETROL gas in place; ~ **inerte** *m* GAS, NUCL inert gas; ~ **de inyección** *m* GAS, PETROL input gas; ~ **inyectado** *m* GAS, PETROL injected gas; ~ **lacrimógeno** *m* D&A tear gas; ~ **licuado** *m* GAS liquefied gas, TERMO liquid gas; ~ **licuado de petróleo** *m* (*GLP, cf supergás*) AUTO, GAS, TEC PETR, TERMO, TERMOTEC, TRANSP, VEH liquefied petroleum gas, liquid petroleum gas (*LPG*); ~ **de lignito** *m* GAS brown-coal gas; ~ **limpio** *m* GAS cleaned gas; ~ **líquido** *m* GAS liquid gas; ~ **de maniobra** *m* GAS, TEC PETR *perforación* trip gas; ~ **para el mercado interno** *m* GAS domestic gas; ~ **metano** *m* GAS, QUÍMICA, SEG methane gas; ~ **mezclado** *m* GAS mixed gas; ~ **minero tóxico** *m* MINAS foul mine-gas; ~ **monoatómico** *m* FÍS, GAS, QUÍMICA mona-tomic gas; ~ **mostaza** *m* GAS, QUÍMICA mustard gas; ~ **nacional** *m* GAS domestic gas; ~ **natural** *m* CONTAM, GAS, TEC PETR, TERMO, TERMOTEC natural gas; ~ **natural dulce** *m* GAS sweet natural gas; ~ **natural húmedo** *m* GAS, TEC PETR wet natural gas; ~ **natural licuado** *m* (*GNL*) GAS, TEC PETR, TERMO liquefied natural gas (*LNG*); ~ **natural presurizado** *m* GAS, TRANSP pressurized natural gas; ~ **natural seco** *m* GAS, TEC PETR dry natural gas; ~ **natural sintético** *m* (*GNS*) GAS, TEC PETR synthetic natural gas (*SNG*); ~ **neutro** *m* TEC ESP neutral gas; ~ **de nitrógeno** *m* (N_2) GAS nitrogen gas (N_2); ~ **no asociado** *m* GAS, TEC PETR nonassociated gas; ~ **noble** *m* GAS, QUÍMICA noble gas; ~**-oil** *m* AUTO, PETROL, TEC PETR, TERMO, TRANSP, VEH diesel fuel, gas oil; ~ **originariamente in situ** *m* PETROL gas originally in place; ~ **de los pantanos** *m* MINAS marsh gas; ~ **de pantanos** *m* GAS, RECICL sludge gas; ~ **perfecto** *m* FÍS, GAS, TERMO ideal gas, perfect gas; ~ **de petróleo** *m* GAS, PETROL, TEC PETR petroleum gas; ~ **de petróleo licuado** *m* AUTO, GAS, TEC PETR, TERMO, TERMOTEC liquid petroleum gas (*LPG*); ~ **de petróleo licuado envasado** *m* GAS bottled liquefied petroleum gas; ~ **de plasmagén** *m* GAS plasmagene gas; ~ **pobre** *m* GAS, PETROL, PROD generator gas, lean gas; ~ **pobre de gasógeno** *m* GAS, PROD generator gas; ~ **portador** *m* CONTAM, GAS, NUCL carrier gas; ~ **de presurización** *m* TEC ESP pressurizing gas; ~ **protector** *m* GAS, NUCL blanket gas, cover gas; ~ **de prueba** *m* MINAS testing flame; ~ **purificado** *m* GAS purified gas; ~ **reductor** *m* GAS, TERMO, TERMOTEC reducing gas; ~ **de refinería petrolera** *m* GAS, TEC PETR, TERMO refinery gas; ~ **de reposición** *m* GAS make-up gas; ~ **residual** *m* CONTAM *procesos industriales* residual gas, waste gas, ELECTRÓN residual gas, GAS off-gas, PROD, QUÍMICA waste gas, RECICL sewage gas; ~ **rico** *m* GAS, PETROL rich gas; ~ **de ripios** *m* GAS, TEC PETR *perforación* cuttings gas; ~ **seco** *m* GAS, PETROL dry gas; ~ **seco natural** *m* GAS, PETROL natural dry gas; ~ **de sellado** *m* GAS, NUCL seal gas; ~ **sintético** *m* GAS manufac-tured gas, synthetic gas; ~ **sobrante** *m* CONTAM, GAS *procesos industriales* residual gas, waste gas; ~ **sucio** *m* GAS foul gas; ~ **de tanque digestor** *m* GAS,

HIDROL, TERMO digester gas; ~ **tóxico** *m* GAS, SEG toxic gas; ~ **transportado mediante conductos** *m* GAS ducting gas; ~ **para uso doméstico** *m* GAS domestic gas; ~ **de ventilación** *m* D&A sweep gas; ~ **de viaje** *m* GAS, TEC PETR *perforación* trip gas

gasa *f* CINEMAT scrim; ~ **difusora** *f* CINEMAT diffuser scrim; ~ **para encuadernar** *f* IMPR crash, super (*AmE*)

gaseado *m* TEXTIL singeing

gaseoso *adj* FÍS, GAS, MINAS, QUÍMICA, TERMO gaseous, gassy

gasero *m* TRANSP MAR gas carrier

gases: ~ **de descarga** *m pl* NUCL off-gas; ~ **de descarga por la chimenea** *m pl* NUCL stack gas; ~ **de escape** *m pl* AUTO, GAS, MECÁ, TERMO, TRANSP, TRANSP AÉR exhaust gases; ~ **y humos** *m pl* GAS, SEG, TRANSP AÉR gases and fumes; ~ **tóxicos** *m pl* GAS, MINAS *en túnel de ventilación* afterdamp

gasificación *f* GEN gasification; ~ **del carbón** *f* CARBÓN, GAS, TEC PETR coal gasification; ~ **del crudo** *f* TEC PETR *refino* oil gasification; ~ **en lecho fluidizado** *f* PROC QUÍ fluidized-bed gasification; ~ **subterránea** *f* GAS, TERMO underground gasification

gasificar *vt* GEN gasify, ALIMENT aerate

gasista *m* GAS, PROD gas fitter

gasoducto *m* GEN gas pipeline, gas pipe

gasógeno *m* GAS gas generator, PROD gas generator, gas producer, gasogene, gazogene, TEC ESP, TERMO gas generator; ~ **de acetileno** *m* PROC QUÍ acetylene generator; ~ **de succión** *m* PROD suction gas producer

gasoil: ~ **para motores marinos** *m* PETROL, TEC PETR, TERMO gas oil, TRANSP MAR marine diesel oil, VEH diesel fuel

gasóleo *m* AUTO diesel fuel, INSTAL TERM fuel oil, PETROL gas oil, TEC PETR gas oil, diesel fuel, TERMO gas oil, TRANSP diesel oil, VEH diesel fuel

gasolina *f* Esp (*cf nafta AmL*) AUTO, P&C, PETROL, TEC PETR, TERMO, VEH gas (*AmE*), gasoline (*AmE*), petrol (*BrE*); ~ **sin aditivo** *f* Esp (*cf nafta sin aditivo AmL*) AUTO, PETROL, VEH regular gas (*AmE*), regular gasoline (*AmE*), regular petrol (*BrE*); ~ **de alta gravedad** *f* Esp (*cf nafta de alta gravedad AmL*) AUTO, PETROL, VEH high-gravity gasoline (*AmE*), high-gravity petrol (*BrE*); ~ **de alto octanaje** *f* Esp (*cf nafta de alto octanaje AmL*) AUTO high-test gasoline (*AmE*), high-test petrol (*BrE*), premium fuel (*BrE*), premium gas (*AmE*), premium gasoline (*AmE*), premium petrol (*BrE*), VEH premium-grade gas (*AmE*), premium-grade gasoline (*AmE*), premium-grade petrol (*BrE*); ~ **de aviación** *f* Esp (*cf nafta de aviación AmL*) TEC PETR aviation gasoline (*AmE*), aviation petrol (*BrE*); ~ **de baja gravedad** *f* Esp (*cf nafta de baja gravedad AmL*) AUTO, VEH low-gravity gasoline (*AmE*), low-gravity petrol (*BrE*); ~ **baja en plomo** *f* Esp (*cf nafta de baja en plomo AmL*) AUTO, PETROL, VEH low lead gasoline (*AmE*), low lead petrol (*BrE*); ~ **de bajo octanaje** *f* Esp (*cf nafta de bajo octanaje AmL*) AUTO, PETROL, VEH low test gasoline (*AmE*), low test petrol (*BrE*); ~ **condensada de gas natural** *f* Esp (*cf nafta de gas natural AmL*) TEC PETR casing head gasoline (*AmE*), casing head petrol (*BrE*); ~ **libre de plomo** *f* Esp (*cf nafta libre de plomo AmL*) CONTAM, TEC PETR lead-free gasoline (*AmE*), lead-free petrol (*BrE*); ~ **ligera** *f* Esp (*cf nafta ligera AmL*) AUTO, PETROL,

VEH light gasoline (*AmE*), light petrol (*BrE*); ~ **para motores** *f* Esp (*cf nafta para motores AmL*) TEC PETR motor spirit; ~ **normal** *f* Esp (*cf nafta normal AmL*) AUTO, PETROL, VEH regular gas (*AmE*), regular gasoline (*AmE*), regular petrol (*BrE*); ~ **obtenida en el separador** *f* Esp (*cf nafta obtineda en el separador AmL*) TEC PETR casing head gasoline (*AmE*), casing head petrol (*BrE*); ~ **sin plomo** *f* Esp (*cf nafta sin plomo AmL*) AUTO, CONTAM, PETROL, TEC PETR, VEH unleaded gas (*AmE*), unleaded gasoline (*AmE*), unleaded petrol (*BrE*); ~ **regular** *f* Esp (*cf nafta regular AmL*) AUTO, PETROL, VEH regular gas (*AmE*), regular gasoline (*AmE*), regular petrol (*BrE*); ~ **sintética** *f* Esp (*cf nafta sintética AmL*) AUTO, PETROL, VEH synthetic gasoline (*AmE*), synthetic petrol (*BrE*); ~ **super** *f* Esp (*cf nafta super AmL*) AUTO, PETROL, TEC PETR, VEH four-star petrol (*BrE*), premium-grade gas (*AmE*), premium-grade gasoline (*AmE*), premium-grade petrol (*BrE*)

gasolinera *f* AUTO, TRANSP, VEH gas station (*AmE*), gasoline station (*AmE*), petrol station (*BrE*), road gas station (*AmE*), road gasoline station (*AmE*); ~ **auto-servicio** *f* TRANSP self-service station

gasometría *f* GAS, PROC QUÍ, PROD, QUÍMICA, TERMO gasometry

gasométrico *adj* GAS, PROC QUÍ, PROD, QUÍMICA, TERMO gasometric

gasómetro *m* GAS, INSTR, PROC QUÍ, PROD, QUÍMICA, TERMO gas holder, gas tank, gasometer

gasoscopio *m* GAS, INSTR gasometer, MINAS gas detector, gasometer, pit-gas indicator, PROC QUÍ, QUÍMICA gasometer

gastado *adj* ING ELÉC bare, QUÍMICA spent

gastar[1] *vt* ING MECÁ wear, wear down, RECICL waste

gastar[2] *vi* ING MECÁ wear out

gasterópodo *m* AGRIC, GEOL gastropod

gasto *m* AGUA outflow, AUTO consumption, CARBÓN *fluidos* rate of flow, flow rate, MECÁ consumption, PROD *tuberías, agua, gas* outflow, VEH *carburante* consumption; ~ **acumulativo** *m* ENERG RENOV cumulative discharge; ~ **calorífico** *m* TERMO heat input; ~ **del combustible** *m* TEC ESP, TRANSP AÉR fuel cost

gastos: ~ **de acarreo** *m pl* ING MECÁ carriage; ~ **de datos en línea** *m pl* PROD on-line data charge; ~ **de demora** *m pl* TEC PETR demurrage; ~ **de desembarque** *m pl* TRANSP MAR landing charges; ~ **diferidos** *m pl* PROD deferred charges; ~ **de embarque** *m pl* TRANSP MAR shipping charges; ~ **de expedición** *m pl* TRANSP MAR shipping charges; ~ **fijos de producción** *m pl* CINEMAT production overhead; ~ **de grúa** *m pl* ING MECÁ cranage; ~ **de licencia de explotación** *m pl* TEC PETR royalty; ~ **de producción** *m pl* CINEMAT production expenses; ~ **de transporte** *m pl* ING MECÁ carriage

gata *f* TEXTIL slub

gatillar *vi* ING MECÁ trigger

gatillo *m* D&A trigger, ING MECÁ cramp, ratchet, trigger, trip, trip catch, MECÁ ratchet; ~ **de acción rápida** *m* ING MECÁ fast-acting trip; ~ **de desconexión de color** *m* TV color gate (*AmE*), colour gate (*BrE*); ~ **disparador** *m* ING MECÁ trip dog

gato *m* CONST jack, ING MECÁ lifting jack, lifting screw, handscrew, jack, MECÁ, MINAS jack, PROD hoisting jack, jack, TRANSP MAR, VEH jack; ~ **de carrillo** *m* ING MECÁ traversing jack; ~ **de cremallera y piñón** *m* ING MECÁ rack-and-pinion jack; ~ **elevador** *m* MINAS

jack; ~ **de empuje** f PROD pushing jack; ~ **del flap** m
TRANSP AÉR flap jack; ~ **en forma de botella** m ING
MECÁ bottle jack; ~ **de gusano** m ING MECÁ screw
jack, screw lifting jack; ~ **hidráulico** m CONST, ING
MECÁ, INSTAL HIDRÁUL, MECÁ hydraulic jack,
hydraulic lifting jack; ~ **a husillo** m ING MECÁ jack-
off screw; ~ **de husillo** m ING MECÁ screw jack, screw
lifting jack; ~ **longitudinal** m ING MECÁ traversing
jack; ~ **mecánico** m ING MECÁ lifting jack; ~ **de**
palanca m ING MECÁ lifting jack, PROD lever jack;
~ **de pared** m CONST wall holdfast; ~ **para**
posicionar m ING MECÁ positioning screw jack;
~ **de rosca** m ING MECÁ lifting jack, screw jack,
screw lifting jack; ~ **telescópico** m ING MECÁ telescopic jack; ~ **de tornillo** m ING MECÁ jackscrew,
lifting screw, screw jack, screw lifting jack;
~ **universal** m AUTO universal jack

GATT *abr* (*Acuerdo General sobre Aranceles y Comercio*
Aduaneros) AGRIC GATT (*General Agreement on*
Tariffs and Trade)

gausio m ELEC, GEOL gauss

gausiómetro m ELEC, INSTR gaussmeter

gauss m ELEC gauss

Gauss: de ~ *adj* ACÚST, FÍS, INFORM&PD, ÓPT,
TELECOM, TV Gaussian

gaussiano *adj* ACÚST, FÍS, INFORM&PD, ÓPT, TELECOM,
TV Gaussian

gavión m CONST gabion

gaylussita f MINERAL gaylussite

gaza f TRANSP MAR eye

Gd *abr* (*gadolinio*) QUÍMICA Gd (*gadolinium*)

GE *abr* (*guerra electrónica*) D&A, ELECTRÓN EW
(*electronic warfare*)

Ge *abr* (*germanio*) ING ELÉC, METAL, QUÍMICA Ge
(*germanium*)

Gediniense *adj* GEOL Gedinnian

gedrita f MINERAL gedrite

gehlenita f MINERAL gehlenite

géiser m ENERG RENOV, GEOFÍS, HIDROL, TERMO, TER-
MOTEC geyser

geiserita f MINERAL geyserite

gel m GEN gel; ~ **de sílice** m ALIMENT, EMB, QUÍMICA
silica gel; ~ **de sílice azul** m EMB, QUÍMICA blue silica-
gel

gelación f P&C, PROC QUÍ, QUÍMICA gelation

gelatina f ALIMENT gelatine, COLOR animal size, EMB,
MINAS, QUÍMICA gelatine; ~ **detonante** f MINAS
blasting gelatine; ~ **explosiva** f MINAS blasting
gelatine; ~ **de pectina** f ALIMENT pectin jelly

gelatinas f pl ALIMENT finings

gélido *adj* ALIMENT, REFRIG, TERMO frozen

gelificación f ALIMENT, PROC QUÍ jellification

gelificante m ALIMENT, CONTAM MAR, PROC QUÍ gelling
agent

gelificar *vi* QUÍMICA gel

gelignita f MINAS, QUÍMICA gelignite

gelisuelo m CARBÓN, TEC PETR permafrost

gelosa f QUÍMICA gelose

gelsemina f QUÍMICA gelsemine

gema: ~ bruta f MINAS stone

gemela f VEH *ballesta* shackle

gemelo m C&V twin

gemelos m pl INSTR spy glass; ~ **de campaña** m pl
INSTR, ÓPT field glasses

gencianina f QUÍMICA gentianin

gene m AGRIC gene

generación f ELEC, ELECTRÓN, ENERG RENOV, ING
ELÉC, NUCL, PROD generating, generation;
~ **armónica** f ELECTRÓN harmonic generation; ~ **de**
CA f ING ELÉC AC generation; ~ **de CC** f ING ELÉC
DC generation; ~ **de copias** f TV generation copy;
~ **de dirección** f INFORM&PD address generation;
~ **eléctrica termonuclear** f ELEC, NUCL thermonuclear power generation; ~ **de electricidad** f ELEC,
ENERG RENOV, NUCL electricity generation;
~ **estructural** f ELECTRÓN pattern generation; ~ **de**
estructuras directa f ELECTRÓN direct pattern generation; ~ **de impulsos** f ELECTRÓN pulse generation;
~ **de informes** f INFORM&PD report generation;
~ **menguante** f ENERG RENOV, HIDROL, TRANSP MAR
ebb generation; ~ **de números aleatorios** f
INFORM&PD, MATEMÁT, TELECOM random-number
generation; ~ **de onda cuadrada** f ELECTRÓN
square-wave generation; ~ **de onda rectangular** f
ELECTRÓN square-wave generation; ~ **de ondas** f
TELECOM wave generation; ~ **de originales** f ÓPT
mastering; ~ **de la palabra** f TELECOM speech generation; ~ **de palabras** f ELECTRÓN word generation;
~ **de señal** f ELECTRÓN, FÍS, TELECOM signal generation; ~ **del sistema** f INFORM&PD system generation;
~ **de tiempo de retardo** f ELECTRÓN time-delay
generation; ~ **de vapor** f INSTAL HIDRÁUL getting
up steam, raising steam, steam raising, PROD getting
up steam, TRANSP MAR steam generation

generado ~ sobre el chip *adj* ELECTRÓN generated on
chip

generador[1] *adj* ELEC, ELECTRÓN, ING ELÉC, PROD
generating

generador[2] m CONST, ELEC generator, ELECTRÓN
source, FÍS, INFORM&PD generator, ING ELÉC power
source, generator, INSTAL HIDRÁUL, NUCL, PROD,
TELECOM, TRANSP MAR generator, VEH *equipo eléctrico* alternator, generator, *instalación eléctrica*
dynamo;

■ **a** ~ **de acetileno** m CONST, GAS, PROC QUÍ acetylene
generator; ~ **acíclico** m ELEC acyclic generator;
~ **acústico** m ING MECÁ acoustic generator; ~ **de**
agua dulce m PROC QUÍ soft-water generator; ~ **de**
alta frecuencia m ELEC high-frequency generator;
~ **de ampollas** m ELEC bulb generator; ~ **de**
armarios m TELECOM frequency multiplier;
~ **armónico** m ELECTRÓN harmonic generator; ~ **de**
armónicos m ELEC, FÍS harmonic generator;
~ **asíncrono** m ELEC asynchronous generator; ~ **de**
audiofrecuencia m ELECTRÓN audio-frequency signal generator, TELECOM tone generator;
~ **autorregulador** m ELEC booster generator;
~ **auxiliar** m ELEC, ELECTRÓN voltage booster;

■ **b** ~ **de baja frecuencia** m ELEC, ELECTRÓN low-
frequency generator; ~ **de baliza** m TEC ESP beacon
generator; ~ **de barras** m TV bar generator; ~ **de**
barras de color m TV color-bar generator (*AmE*),
colour-bar generator (*BrE*); ~ **de base de tiempo** m
ELECTRÓN time-base generator, TV time base generator; ~ **de la base de tiempos** m ELECTRÓN *TRC*
signal converter; ~ **de bombillas** m ELEC bulb
generator;

■ **c** ~ **de CA** m ING ELÉC AC generator, AC source;
~ **de caracteres** m ELECTRÓN, INFORM&PD,
TELECOM, TV character generator; ~ **de CC** m ELEC,
FÍS, ING ELÉC DC generator; ~ **de clasificación** m
INFORM&PD sort generator; ~ **de código de tiempo**

m CINEMAT, TV time-code generator; **~ de colores de fondo** *m* CINEMAT, TV color-background generator (*AmE*), colour-background generator (*BrE*); **~ de corriente** *m* ELEC, FÍS, ING ELÉC current generator, current source; **~ de corriente alterna** *m* ELEC, FÍS alternative-current generator, ING ELÉC alternating-current source, alternator; **~ de corriente continua** *m* ELEC direct-current generator; **~ de corriente de llamada** *m* TELECOM ringing machine; **~ de corriente de RF** *m* ELECTRÓN, ING ELÉC, TELECOM RF current source; **~ de cuatro polos** *m* ELEC, ELECTRÓN four-pole generator;

~ d **~ de datos** *m* INFORM&PD, ING ELÉC data source; **~ de datos de prueba** *m* INFORM&PD test data generator; **~ de datos de rumbo** *m* TRANSP AÉR course-data generator; **~ de datos triaxial** *m* (*TADG*) TEC ESP three-axis data generator (*TADG*); **~ de dientes de sierra** *m* TV sawtooth generator; **~ diesel de emergencia** *m* NUCL emergency diesel-generator; **~ diesel de reserva** *m* NUCL diesel generator standby power plant; **~ dirigible** *m* TEC ESP pointable generator; **~ de dos arrollamientos** *m* ELEC double-wound generator;

~ e **~ de efectos** *m* CINEMAT, TV effects generator; **~ de efectos especiales** *m* CINEMAT, TV special-effects generator; **~ eléctrico** *m* ELEC, ING ELÉC, NUCL, TV electric generator; **~ eléctrico de vapor** *m* ELEC, ING ELÉC steam-electric generator; **~ electrógeno** *m* CINEMAT genny; **~ electrónico de puntos** *m* IMPR electronic dot generator (*EDG*); **~ electroquímico secundario** *m* TRANSP secondary electrochemical generator; **~ electrostático** *m* ELEC, ING ELÉC electrostatic generator; **~ de encendido** *m* TRANSP AÉR ignition generator; **~ de energía óptica** *m* ELEC, ING ELÉC optical power source; **~ eólico** *m* ELEC wind-powered generator; **~ sin escobillas** *m* ELEC, ING ELÉC brushless generator; **~ de excitación mixta diferencial** *m* ELEC, ING ELÉC differentially-excited compound generator; **~ excitado por separado** *m* ELEC, ING ELÉC separate excited generator; **~ de exploración** *m* ELECTRÓN tracking generator;

~ f **~ de fase** *m* ELECTRÓN phase generator; **~ de filtraciones** *m* ELECTRÓN mask generation; **~ fotovoltaico** *m* ING ELÉC photovoltaic generator; **~ de fuerza** *m* ING ELÉC source; **~ de función** *m* ELECTRÓN function generator; **~ de función escalonada** *m* ELECTRÓN step function generator;

~ g **~ de gas** *m* GAS, PROD, TEC ESP, TERMO gas generator; **~ de gas inerte** *m* GAS inert gas generator; **~ de gases de combustión** *m* GAS, PROD, TEC ESP, TERMO gas generator;

~ h **~ de Hall** *m* AUTO Hall generator; **~ helioeléctrico** *m* ING ELÉC, TEC ESP solar generator; **~ hidráulico** *m* INSTAL HIDRÁUL, TEC ESP hydraulic generator; **~ hidroeléctrico** *m* ELEC, ING ELÉC hydro-electric generator; **~ homopolar** *m* ELEC, ING ELÉC homopolar generator;

~ i **~ de imagen de prueba** *m* TV pattern generator; **~ de imán permanente** *m* ENERG RENOV, ING ELÉC permanent-magnet generator; **~ de impedancia nula** *m* ING ELÉC zero impedance source; **~ de impulsos** *m* ELEC pulse generator, surge generator, *equipo de prueba* impulse generator, ELECTRÓN pulse generator, pulser, INFORM&PD pulse generator, ING ELÉC impulse generator, NUCL surge generator, TELECOM pulse

generator; **~ de impulsos de alta tensión** *m* ELEC high-voltage impulse generator, surge generator; **~ de impulsos de bloqueo** *m* CINEMAT, TV clamp pulse generator; **~ de impulsos de corriente** *m* ELEC surge generator; **~ de impulsos de sincronización** *m* CINEMAT, TV sync pulse generator; **~ de impulsos de supresión** *m* ELECTRÓN blanking generator; **~ inamovible** *m* TEC ESP fixed generator; **~ de inducción** *m* ELEC, ING ELÉC induction generator; **~ inductor** *m* ELEC, ING ELÉC inductor generator; **~ de información de dirección** *m* TRANSP AÉR heading data generator; **~ de interferencias** *m* ELECTRÓN interference generator; **~ isotópico** *m* TEC ESP isotopic generator;

~ l **~ de llamada** *m* TELECOM ringing machine; **~ de luz estándar** *m* ING ELÉC standard light source;

~ m **~ magnetoeléctrico** *m* ELEC, ING ELÉC magneto-electric generator; **~ magnetohidrodinámico** *m* ELEC, ING ELÉC magnetohydrodynamic generator; **~ Marx** *m* ING ELÉC impulse generator, NUCL Marx generator; **~ de medición de tiempo** *m* ELECTRÓN timing generator; **~ MHD** *m* ING ELÉC MHD generator; **~ de microondas** *m* ELECTRÓN microwave generator; **~ modulado por la imagen** *m* TELECOM picture-modulated generator; **~ multifrecuencia** *m* TELECOM multiple-frequency generator, multifrequency generator;

~ n **~ de niebla** *m* CINEMAT, TV fog gun; **~ de números aleatorios** *m* INFORM&PD, MATEMÁT, TELECOM random-number generator;

~ o **~ de olas** *m* TRANSP MAR *diseño naval* wave maker; **~ de onda cuadrada** *m* ELECTRÓN, TV square-wave generator; **~ de onda portadora** *m* FÍS ONDAS carrier-wave generator; **~ de onda rectangular** *m* ELECTRÓN square-wave generator; **~ de onda transmisora** *m* FÍS ONDAS carrier-wave generator; **~ de ondas** *m* ELEC surge generator, FÍS ONDAS wave generator; **~ de ondas acústicas** *m* ING MECÁ acoustic generator; **~ de ondas de choque** *m* ELEC surge generator; **~ de oxígeno** *m* CONST *soldadura* oxygen generator;

~ p **~ de palabras** *m* ELECTRÓN word generator; **~ patrón** *m* *AmL* TV pattern generator; **~ de picos de alta tensión** *m* ELEC surge generator; **~ de polarización** *m* ING ELÉC bias source; **~ polifásico** *m* ING ELÉC polyphase generator; **~ polimórfico** *m* ING ELÉC multiple current generator; **~ de polos exteriores** *m* ELEC exterior-pole generator, external-pole generator; **~ de polos interiores** *m* ELEC internal-pole generator; **~ de polos salientes** *m* ELEC salient pole generator; **~ de portadora** *m* ELECTRÓN carrier generation; **~ de programas de informes** *m* INFORM&PD report-program generator; **~ de pulsos** *m* FÍS, FÍS RAD pulse generator; **~ de puntos** *m* ELECTRÓN dot generator;

~ r **~ de radiofrecuencia** *m* ING ELÉC radio-frequency generator; **~ de rampa** *m* ING ELÉC ramp generator; **~ regulador** *m* ING ELÉC reversible booster; **~ de retardo** *m* ELECTRÓN delay generator; **~ de RF** *m* ELECTRÓN, ING ELÉC, TELECOM RF generator; **~ de ruido** *m* ELECTRÓN, TELECOM noise generator; **~ de ruido blanco** *m* ELECTRÓN white noise generator; **~ de ruido térmico** *m* ELECTRÓN thermal noise generator; **~ de ruidos complejos** *m* ELECTRÓN, FÍS, TEC ESP, TELECOM random-noise generator;

~ s ~ **de sección de RF** *m* ELEC, ELECTRÓN, ING ELÉC RF section-generator; ~ **de seguimiento** *m* TV tracking generator; ~ **de señal de AF** *m* ELECTRÓN HF signal generator; ~ **de señal analógica** *m* ELECTRÓN analog signal generator; ~ **de señal de FMA** *m* ELECTRÓN VHF signal generator; ~ **de señal de frecuencia audio** *m* ELECTRÓN audio-frequency signal generator; ~ **de señal de FSA** *m* ELECTRÓN, TEC ESP, TELECOM SHF signal generator; ~ **de la señal de identificación** *m* CINEMAT pilot-tone generator; ~ **de señal de microondas** *m* ELECTRÓN microwave-signal generator; ~ **de señal programable** *m* ELECTRÓN programmable-signal generator; ~ **de señal de prueba** *m* ELECTRÓN test signal generator; ~ **de señal sintetizada** *m* ELECTRÓN synthesized-signal generator; ~ **de señal de UHF** *m* ELECTRÓN UHF signal generator; ~ **de señales** *m* ELECTRÓN, FÍS, TELECOM signal generator; ~ **de señales acústicas** *m* ING MECÁ acoustic signal generator; ~ **de señales de sincronización de la emisora** *m* ELECTRÓN, TV station sync generator; ~ **de señales sinusoidales** *m* ELECTRÓN sinusoidal-signal generator; ~ **de señalización** *m* ING ELÉC signaling generator (*AmE*), signalling generator (*BrE*); ~ **sincrónico** *m* ELEC, TV sync generator, synchronous generator; ~ **de sincronismo de múltiplex** *m* TELECOM multiplex timing generator; ~ **de sincronismo sonoro** *m* CINEMAT, TV sound-sync generator; ~ **de sincronización** *m* CINEMAT, TV sync generator; ~ **de sincronización del regenerador** *m* TELECOM *de impulsos* regenerator timing generator; ~ **de sobrecorrientes** *m* ELEC surge generator; ~ **solar** *m* TEC ESP solar generator; ~ **solar principal** *m* TEC ESP main solar generator; **~ t** ~ **tacométrico** *m* ING MECÁ tachogenerator; ~ **en tándem** *m* FÍS tandem generator; ~ **de temporización del regenerador** *m* TELECOM *radiografía* regenerator test method; ~ **de tensión de corriente continua** *m* ING ELÉC direct current voltage source; ~ **termiónico** *m* ING ELÉC thermionic generator; ~ **termoeléctrico** *m* ING ELÉC thermoelectric generator; ~ **tipo turbina eólica** *m* ENERG RENOV wind-turbine generator; ~ **de tonalidad** *m* ELECTRÓN tone generator; ~ **de tono** *m* TELECOM tone generator; ~ **de torbellinos** *m* TRANSP vortex generator; ~ **de tramas** *m* INFORM&PD raster generator, TELECOM frame generator, TV raster generator; ~ **de transmisión de señales** *m* ING ELÉC signaling generator (*AmE*), signalling generator (*BrE*); ~ **de tres hilos** *m* ELEC three-wire generator; ~ **trifásico** *m* ELEC three-phase generator; ~ **trifilar** *m* ELEC three-wire generator; **~ u** ~ **ultrasónico** *m* FÍS RAD ultrasonic generator, LAB ultrasound generator; ~ **de ultrasonidos** *m* FÍS RAD ultrasonic generator, LAB ultrasound generator; **~ v** ~ **Van de Graaff** *m* ELEC, FÍS, ING ELÉC Van de Graaff generator; ~ **de vapor** *m* ING MECÁ boiler, INSTAL HIDRÁUL power boiler, steam boiler, steamer, INSTAL TERM steam generator, NUCL vapor generator (*AmE*), vapour generator (*BrE*); ~ **de vapor macromodular** *m* NUCL macromodular steam generator; ~ **de vapor de un solo paso** *m* NUCL once-through steam generator (*OTSG*); ~ **de vibraciones** *m* TEC ESP vibration generator; ~ **de voltaje** *m* FÍS voltage generator, ING ELÉC voltage source; ~ **de voltaje de CC** *m* ING ELÉC DC voltage source; ~ **de**

voltaje constante *m* ELEC, ING ELÉC constant-voltage source; ~ **de voltaje exterior** *m* ELEC, ING ELÉC external-voltage source; ~ **de voltaje de polarización** *m* ING ELÉC bias generator; ~ **de voltaje regulable** *m* ELEC, ING ELÉC variable-voltage generator;
~ z ~ **en zona de seguridad** *m* TV safe-area generator

general *adj* CONST overall, INFORM&PD global, ING MECÁ overall

generar[1] *vt* ELEC generate, ELECTRÓN *impulsos* beat, FÍS RAD, GEOM, NUCL generate, PROD *vapor* raise

generar[2]: ~ **un dupe** *vi* CINEMAT dupe

generatriz[1] *adj* ELEC, ING ELÉC, PROD generating

generatriz[2]: ~ **activa** *f* ING ELÉC active element; ~ **multifásica** *f* ING ELÉC polyphase generator; ~ **de tensión variable** *f* ELEC variable-voltage generator

género *m* GEOM genus, TEXTIL fabric; ~ **para entretelas** *m* TEXTIL interlining material; ~ **de punto por urdimbre** *m* TEXTIL warp knitting; ~ **para secadora** *m* TEXTIL dryer fabric; ~ **para vestidos** *m* TEXTIL dress material

géneros: ~ **en pieza** *m pl* TEXTIL piece goods; ~ **raschel estructurados direccionalmente** *m pl* TEXTIL directionally-structured raschel goods

genético *adj* AGRIC, FÍS RAD genetic

genetista *m* AGRIC *de especies vegetales* breeder

genisteína *f* QUÍMICA genistein

gentiobiosa *f* QUÍMICA gentiobiose

gentiopicrina *f* QUÍMICA gentiopicrin

gentisato *m* QUÍMICA gentisate

gentísico *adj* QUÍMICA *ácido* gentisic

gentisina *f* QUÍMICA gentisin

gentisínico *adj* QUÍMICA gentisic

gentita *f* MINERAL genthite

geobarómetro *m* GEOL, INSTR geobarometer

geocronología *f* GEOL, PETROL geochronology

geoda *f* GEOL geode

geodesia *f* GEOL geodesy

geodímetro *m* CONST, INSTR geodimeter

geodinámica *f* GEOL, TEC ESP geodynamics

geofísica *f* ACÚST, CARBÓN, FÍS, GEOFÍS, GEOL, TEC PETR geophysics

geófono *m* GEN geophone

geohidrología *f* CARBÓN geohydrology

geoide *m* GEOL, OCEAN geoid

geología *f* CARBÓN, GEOL, TEC PETR geology; ~ **isotópica** *f* GEOL isotope geology; ~ **del petróleo** *f* GEOL, TEC PETR petroleum geology

geológico *adj* GEOL, TEC PETR geological

geólogo: ~ **del subsuelo** *m* GEOL, PETROL, TEC PETR *geología* subsurface geologist

geomagnético *adj* FÍS, GEOFÍS, GEOL, TEC ESP geomagnetic

geomagnetismo *m* FÍS, GEOFÍS, GEOL, TEC ESP geomagnetism

geomecánica *f* CARBÓN soil mechanics

geómetra *m* GEOM geometer, geometrician

geometría *f* AUTO, FÍS RAD, GEOM, MATEMÁT, METAL, NUCL geometry; ~ **de la absorción** *f* FÍS RAD, GEOM geometry of absorption; ~ **afín** *f* GEOM affine geometry; ~ **algebraica** *f* GEOM algebraic geometry; ~ **analítica** *f* GEOM analytic geometry, coordinate geometry, analytical geometry; ~ **Cartesiana** *f* GEOM Cartesian geometry; ~ **de conjunto de puntos** *f* GEOM point-set geometry; ~ **de deslizamiento** *f*

GEOM, METAL geometry of glide; ~ **diferencial** *f* GEOM differential geometry; ~ **de la dirección** *f* AUTO steering geometry; ~ **elíptica** *f* GEOM elliptical geometry; ~ **esférica** *f* GEOM spherical geometry; ~ **euclídica** *f* GEOM Euclidean geometry; ~ **gráfica** *f* GEOM graphic geometry; ~ **hiperbólica** *f* GEOM hyperbolic geometry; ~ **de la irradiación** *f* GEOM, NUCL geometry of irradiation; ~ **localmente homogénea** *f* GEOM locally-homogeneous geometry; ~ **métrica** *f* GEOM metrical geometry; ~ **no euclídea** *f* GEOM non-Euclidean geometry; ~ **plana** *f* GEOM plane geometry; ~ **del plano** *f* GEOM plane geometry; ~ **proyectiva** *f* GEOM projective geometry; ~ **de puntos** *f* GEOM point-set geometry; ~ **de Riemann** *f* GEOM, MATEMÁT Riemann geometry, Riemannian geometry; ~ **riemanniana** *f* GEOM, MATEMÁT Riemann geometry, Riemannian geometry; ~ **sólida** *f* GEOM solid geometry; ~ **de sólidos** *f* GEOM solid geometry

geométrico *adj* GEOM, MATEMÁT geometric, geometrical

geomorfología *f* GEOL geomorphology

geón *m* GEOFÍS geon

geopetal *adj* PETROL geopetal

geopotencial *m* GEOL, GEOFÍS geopotential

geopresión *f* CARBÓN soil pressure

geoquímica *f* GEOL, PETROL, QUÍMICA, TEC PETR geochemistry

geosinclinal[1] *adj* GEOFÍS, GEOL, TEC PETR geosynclinal

geosinclinal[2] *m* GEOFÍS, GEOL, TEC PETR geosyncline

geotecnia *f* AGUA soil science, CARBÓN, CONST, GEOL soil mechanics

geotécnica *f* CARBÓN, GEOFÍS, GEOL geotechnics

geotectoclino *m* GEOFÍS geotectocline

geotectónico[1] *adj* CARBÓN, GEOFÍS, GEOL geotectonic

geotectónico[2] *m* CARBÓN, GEOFÍS, GEOL geotectonic

geotérmica *f* GEN geothermics

geotérmico *adj* ENERG RENOV, FÍS, GEOFÍS, GEOL, TEC PETR, TERMO geothermal

geotermómetro *m* GEN geothermometer

geranial *m* QUÍMICA citral

geranilo *m* QUÍMICA geranyl

geraniol *m* QUÍMICA geraniol

gerente: ~ **de diseño de proyectos** *m* ING MECÁ project-design manager; ~ **del estudio** *m* TV studio manager; ~ **de métodos** *m* ING MECÁ methods manager

germanio *m* (*Ge*) ING ELÉC, METAL, QUÍMICA germanium (*Ge*)

germen: ~ **cristalino** *m* CRISTAL seed crystal

germicida *m* AGRIC germ killer

germinación *f* CRISTAL nucleation, METAL *solidificación* nucleation, *deformación* twinning, METEO, NUCL nucleation; ~ **compuesta** *f* METAL compound twinning; ~ **de costado** *f* METAL edgewise growth; ~ **en la espiga** *f* AGRIC *cereales* outgrowth; ~ **mecánica continua** *f* METAL continual mechanical twinning

gersdorfita *f* MINERAL gersdorffite

gestión: ~ **de archivos** *f* INFORM&PD file management; ~ **de la base de datos** *f* NFORM&PD, TELECOM database managent (*DBM*); ~ **de colas** *f* INFORM&PD, TELECOM queue management; ~ **de configuración** *f* INFORM&PD configuration management (*CM*); ~ **de datos** *f* INFORM&PD, TELECOM data management (*DM*); ~ **de datos técnicos** *f* INFORM&PD, PROD technical data management; ~ **de la demanda** *f* PROD demand management; ~ **de emplazamiento de posición** *f* PROD location management; ~ **de emplazamiento de ubicación** *f* PROD location management; ~ **de errores** *f* INFORM&PD error management; ~ **de existencias** *f* PROD inventory; ~ **de la memoria** *f* INFORM&PD memory management; ~ **periférica** *f* TELECOM peripheral management; ~ **racional de los recursos pesqueros** *f* OCEAN rational fish stock management; ~ **de residuos radiactivos** *f* CONTAM, FÍS RAD, NUCL radioactive-waste management; ~ **de sistemas** *f* TELECOM systems management; ~ **y supervisión de la red** *f* TELECOM network supervision and management; ~ **de tareas** *f* INFORM&PD job control; ~ **por teclado** *f* TELECOM key management

gestor: ~ **de excepciones** *m* INFORM&PD exception handler; ~ **de memoria expandida** *m* INFORM&PD expanded-memory manager

getter *m* ELEC, TELECOM getter

GFI *abr* (*granate férrico de itrio*) ELECTRÓN YIG (*yttrium iron garnet*)

gibbsita *f* MINERAL gibbsite

gieseckita *f* MINERAL gieseckite

giga- *pref* (*G*) METR giga- (*G*)

gigabyte *m* INFORM&PD, ÓPT, TELECOM gigabyte

gigadisco *m* ÓPT gigadisk

gigantolita *f* MINERAL gigantolite

gigaocteto *m* INFORM&PD, ÓPT, TELECOM gigabyte

gilbertio *m* ING ELÉC gilbert

gilbertita *f* MINERAL gilbertite

gill *m* METR gill

gimnita *f* MINERAL gymnite

ginocárdico *adj* QUÍMICA gynocardic

giobertita *f* MINERAL giobertite

gira: **que ~ en el sentido de las agujas del reloj** *fra* MECÁ clockwise-rotating; **que ~ en contra de las agujas del reloj** *fra* MECÁ anticlockwise-rotating

girador *m* FÍS gyrator

giramachos *m* ING MECÁ tap holder, tap wrench

girar[1] *vt* ING MECÁ run, swing

girar[2] *vti* INFORM&PD rotate

girar[3] *vi* ING MECÁ swing, TEC ESP spin, turn, TRANSP AÉR spin; ~ **lentamente** *vi* VEH *motor* idle; ~ **por redondo** *vi* TRANSP MAR veer; ~ **en el sentido de las agujas del reloj** *vi* MECÁ rotate clockwise, TRANSP MAR *viento* veer

giratoria *f* *Esp* (*cf cabezal de sonda AmL*) MINAS *sondeos* swivel; ~ **de perforación** *f* *Esp* (*cf perforadora AmL*) MINAS swivel

giro *m* CONST slewing round, *grúa* swinging round, GEOM revolution, ING MECÁ revolution, turn, TEC ESP revolution, spin, rotation; ~ **excéntrico** *m* TEC ESP wobble; ~ **en exceso** *m* AUTO, VEH oversteer; ~ **hacia abajo** *m* TEC ESP spin-down; ~ **hacia arriba** *m* TEC ESP spin-up; ~ **patrón de aterrizaje** *m* TRANSP AÉR landing pattern turn; ~ **en vertical** *m* TRANSP vertical gyro

girobús *m* *Esp* TRANSP, VEH (*cf ómnibus, bus giroscópico AmL*) gyrobus

giroclinómetro *m* TEC ESP gyroclinometer

girocompás *m* TRANSP MAR gyrocompass

girodino *adj* TRANSP AÉR gyrodyne

giromeridiano *m* TEC ESP meridian gyro

girómetro *m* TEC ESP gyrometer; ~ **de fibra óptica** *m* TEC ESP optical fiber gyrometer (*AmE*), optical fibre gyrometer (*BrE*)

giropiloto *m* TRANSP MAR gyropilot
giroplano *m* TRANSP AÉR gyroplane
giroscópico *adj* GEN gyroscopic
giroscopio *m* CINEMAT, ENERG RENOV, FÍS, MECÁ, TEC ESP, TRANSP AÉR gyroscope; **~ de cabeceo** *m* TEC ESP pitch gyro; **~ direccional** *m* TRANSP directional gyro; **~ láser** *m* TEC ESP laser gyro; **~ que mide la velocidad angular de balanceo** *m* TEC ESP roll-rate gyro; **~ vertical** *m* TEC ESP vertical gyro
giróscopo: **~ y amplificador de control** *m* TEC ESP control gyro and amplifier; **~ azimutal** *m* TEC ESP azimuth gyro; **~ de control** *m* TEC ESP, TRANSP AÉR control gyro; **~ direccional** *m* TRANSP AÉR directional gyro; **~ libre** *m* TRANSP AÉR free gyro; **~ de régimen de cabeceo** *m* TRANSP AÉR pitch-rate gyro
girostático *adj* FÍS, QUÍMICA gyrostatic
giróstato *m* FÍS, QUÍMICA gyrostat
girotrón *m* TELECOM gyrotron
gis *m* C&V chalk
gismondina *f* MINERAL gismondine
Givetiense *adj* GEOL Givetian
glabro *adj* AGRIC glabrous
glacial *adj* GEOL glacial, TERMO ice-coded
glacis: **~ continental** *m* OCEAN continental rise
glairina *f* QUÍMICA glairin
glándula: **~ nectarífera** *f* AGRIC nectar gland
glaseado *m* ALIMENT, PROD glazing, gloss, glossing, TEXTIL glazing
glasear *vt* ALIMENT, PROD, TEXTIL glaze
glaserita *f* MINERAL glaserite
glasto *m* COLOR woad
glauberita *f* MINERAL glauberite
glaucodot *m* MINERAL glaucodote
glaucofana *f* MINERAL glaucophane
glaucolita *f* MINERAL glaucolite
glauconita *f* GEOL, MINERAL, PETROL glauconite
glauconítico *adj* GEOL, MINERAL, PETROL, TEC PETR glauconitic
gleba *f* AGRIC furrow slice; **~ del surco muerto** *f* AGRIC sole furrow
gliceraldehído *m* QUÍMICA glyceraldehyde
glicérico *adj* QUÍMICA glyceric
glicérido *m* ALIMENT, QUÍMICA glyceride; **~ de ácido graso** *m* ALIMENT, QUÍMICA fatty-acid glyceride
glicerilo *m* QUÍMICA glyceryl
glicerina *f* QUÍMICA glycerine
glicerofosfato *m* QUÍMICA glycerophosphate
glicerofosfórico *adj* QUÍMICA glycerophosphoric
glicerol *m* QUÍMICA glycerol
glicídico *adj* QUÍMICA glycidic
glicilglicina *f* QUÍMICA glycylglycine
glicina *f* ALIMENT, QUÍMICA glycine
glicirricina *f* QUÍMICA glycyrrhizine
glicógeno *m* ALIMENT, QUÍMICA glycogen
glicol *m* DETERG, QUÍMICA, REFRIG glycol
glicólico *adj* DETERG, QUÍMICA, REFRIG glycolic
glicolina *f* QUÍMICA glycoline
glicolípido *m* QUÍMICA glycolipid
glicólisis *f* ALIMENT, QUÍMICA glycolysis
glicósido *m* QUÍMICA glycocide
glicurónico *adj* QUÍMICA glycuronic
glioxal *m* QUÍMICA glyoxal
glioxalidina *f* QUÍMICA glyoxalidine
glioxalina *f* QUÍMICA glyoxaline
glioxílico *adj* QUÍMICA glyoxylic
glioxima *f* QUÍMICA glyoxime

global *adj* INFORM&PD global, ING MECÁ overall, TEC ESP global
globalmente *adv* CARBÓN in bulk
globo: **~ aerostático de transmisión de datos** *m* D&A data-transmission balloon; **~ aerostático de vigilancia** *m* D&A surveillance balloon; **~ piloto** *m* TRANSP AÉR pilot balloon; **~ sonda** *m* METEO sounding balloon, weather balloon, TEC ESP sounding balloon, TRANSP AÉR pilot balloon
globular *adj* METAL, QUÍMICA globular
glóbulo *m* QUÍMICA globule; **~ transitorio de plasma quark gluón** *m* FÍS PART transient globule of quark-gluon plasma
glonoina *f* AmL (*cf nitroglicerina Esp*) MINAS nitroglycerine
glorieta *f* AUTO, CONST, TRANSP, VEH rotary (*AmE*), roundabout (*BrE*), traffic circle (*AmE*)
glosopeda *f* AGRIC aphthous fever, foot-and-mouth disease
GLP *abr* (*gas licuado de petróleo*) GEN LPG (*liquefied petroleum gas, liquid petroleum gas*)
glucagón *m* QUÍMICA glucagon
glucamina *f* QUÍMICA glucamine
glucarónico *adj* QUÍMICA glucaronic
glúcido *m* QUÍMICA saccharide, glucide, biose
glucónico *adj* QUÍMICA gluconic
glucopiranosa *f* QUÍMICA glucopyranose
glucoproteína *f* QUÍMICA glucoprotein
glucosa *f* ALIMENT grape sugar, QUÍMICA glucose
glucosamina *f* QUÍMICA glucosamine
glucosana *f* QUÍMICA glucosan
glucósido *m* ALIMENT, QUÍMICA glucoside, glycoside
gluma *f* AGRIC glume
gluón *m* FÍS, FÍS PART gluon
glutacónico *adj* QUÍMICA glutaconic
glutamato *m* QUÍMICA glutamate; **~ monosódico** *m* (*GMS*) ALIMENT, QUÍMICA monosodium glutamate (*MSG*)
glutámico *adj* QUÍMICA glutamic
glutamina *f* QUÍMICA glutamine
glutaraldehído *m* QUÍMICA glutaraldehyde
glutárico *adj* QUÍMICA glutaric
glutatión *m* QUÍMICA glutathione
gluten[1]: **sin ~** *adj* AGRIC, ALIMENT gluten-free
gluten[2] *m* AGRIC, ALIMENT gluten
gmelinita *f* MINERAL gmelinite
GMT *abbr* (*hora media de Greenwich, hora del meridiano de Greenwich, hora solar media*) FÍS, TEC ESP, TRANSP AÉR GMT (*Greenwich Mean Time*)
gneis *m* CONST, GEOL, PETROL gneiss; **~ granítico** *m* GEOL granite-gneiss; **~ de inyección** *m* GEOL injection gneiss
gnéisico *adj* CONST, GEOL, PETROL gneissic
GNL *abr* (*gas natural licuado*) GAS, TEC PETR, TERMO LNG (*liquefied natural gas*)
GNS *abr* (*gas natural sintético*) GAS, TEC PETR SNG (*synthetic natural gas*)
gobernador *m* TRANSP MAR buque conn; **~ de centrifugación y vacío** *m* TRANSP AÉR centrifugal and vacuum governor; **~ de hélice** *m* TRANSP AÉR propeller governor; **~ de resorte** *m* ING MECÁ spring governor
gobernar *vt* TEC ESP, TELECOM control, TRANSP barcos, misiles steer, TRANSP MAR barco run, barco de vela steer, helm, *embarcación* navigate; **~ apartándose de** *vt* TRANSP MAR steer clear of; **~ hacia** *vt* TRANSP MAR

stand for; ~ **poniéndose en franquía de** *vt* TRANSP MAR steer clear of

gobierno *m* TEC ESP, TRANSP, TRANSP MAR *barco* control; ~ **a distancia** *m* GEN remote control

goetita *f* MINERAL goethite

gofrado[1] *adj* EMB, IMPR, ING MECÁ, PAPEL embossed

gofrado[2] *m* EMB embossing, IMPR embossing, *estampado en seco* blind embossing, PAPEL embossing

gofrar *vt* EMB, IMPR, ING MECÁ, PAPEL emboss

gola *f* MINAS *hidrografía* mouth, OCEAN *de red* net mouth

golfo *m* TRANSP MAR gulf

gollete *m* AUTO neck; ~ **de boca rebordeada** *m* C&V *en botellas* bead; ~ **encorvado** *m* C&V bent neck; ~ **de entrada** *m* TRANSP AÉR inlet throat

golpe[1] *m* ALIMENT bruise, FÍS stroke, ING MECÁ impingement, stroke, TEC ESP shock; ~ **de agua** *m* AGUA water inflow; ~ **de ariete** *m* FÍS FLUID *tuberías* water hammer, NUCL *en tubería*, REFRIG hammering, TEC PETR *tuberías* water hammer; ~ **lateral** *m* ING MECÁ side lash; ~ **de líquido** *m* REFRIG slugging; ~ **de mar** *m* OCEAN breaker; ~ **del pistón** *m* AUTO piston slap; ~ **de retorno** *m* FÍS return stroke; ~ **sónico** *m* SEG sonic bang

golpe[2]: **a** ~ **de aleta** *fra* EMB flap snap

golpeador *m* ING MECÁ plunger

golpeadora *f* PAPEL banger

golpear[1] *vt* CARBÓN hew, CONST batter, OCEAN *tecnología pesquera* strike; ~ **con el batán** *vt* TEXTIL *alfombras* beat; ~ **contra** *vt* ING MECÁ, MECÁ impinge on; ~ **ligeramente** *vt* PROD tap; ~ **con un martillo** *vt* CONST sledge

golpear[2] *vi* ING MECÁ knock

golpeo *m* CONTAM rapping; ~ **por autoencendido** *m* AUTO, VEH pinging (*AmE*), pinking (*BrE*); ~ **de la pala** *m* TRANSP AÉR blade slap; ~ **del pistón** *m* AUTO, VEH piston knock

golpetear *vi* ING MECÁ knock

golpeteo *m* AUTO knocking, FÍS tapping, ING MECÁ knocking, PROD *cojinetes* hammering, TEC ESP *motor del cohete* chugging, VEH knocking; ~ **del motor** *m* AUTO, VEH rumble; ~ **del pistón** *m* AUTO, VEH piston slap

goma *f* ALIMENT tragacanth; ~ **de acetato** *f* EMB acetate glue; ~ **acrílica** *f* EMB acrylic rubber; ~ **de acrilonitrilo** *f* MECÁ, P&C acrylonitrile rubber; ~ **de almidón** *f* ALIMENT, P&C, QUÍMICA dextrin; ~ **arábiga** *f* ALIMENT gum arabic, P&C *adhesivo* glue; ~ **de caseína** *f* ALIMENT casein glue; ~ **de caseinato** *f* ALIMENT caseinate gum; ~ **celular** *f* P&C, REFRIG, TERMOTEC cellular rubber; ~ **elástica** *f* P&C india rubber, oilcloth, natural rubber, rubber, QUÍMICA, TEXTIL oilcloth; ~ **espuma** *f* P&C *caucho* foam rubber, *plástico celular* foam; ~ **éster** *f* P&C ester gum; ~ **de guayaco** *f* ALIMENT guaiac gum; ~ **karaya** *f* ALIMENT, QUÍMICA karaya gum; ~ **laca** *f* COLOR shellac, shellack, ELEC *aislación* shellac, PROD lac; ~ **silicónica** *f* ELEC *aislador* silicone rubber; ~ **tragacanto** *f* ALIMENT gum tragacanth; ~ **vegetal** *f* PROD vegetable glue

góndola *f* TRANSP AÉR nacelle; ~ **del motor** *f* TRANSP AÉR engine nacelle, engine pod

Gondwana: de ~ *adj* GEOL Gondwanan

gong *m* TRANSP MAR gong

goniometría *f* TELECOM, TRANSP AÉR direction finding

(*DF*); ~ **automática** *f* (*ADF*) TELECOM, TRANSP AÉR automatic direction finding (*ADF*)

goniómetro *m* CRISTAL goniometer, D&A dial sight, FÍS RAD, GEOM goniometer, INSTR goniometer, sight, METR angle meter, angulometer, protractor, OCEAN angulometer, goniometer; ~ **de rayos catódicos** *m* ELECTRÓN, TELECOM cathode-ray direction finder; ~ **de reflexión** *m* GEOM reflecting goniometer; ~ **de tránsito** *m* CONST surveyor's transit

gopher *m* INFORM&PD gopher

GOR *abr* (*relación gas-petróleo*) TEC PETR GOR (*gas-to-oil ratio*)

gorgojeo *m* AmL TELECOM chirping

gorgojo *m* AGRIC grub

gorjeo *m* AmL (*cf chirrido Esp*) INFORM&PD chirp

gorrón *m* AUTO journal, stud, CONST *mecánica* gudgeon, ING MECÁ trunnion, pivot, pillow, MECÁ journal, pivot, PROD *cilindro laminador* neck; ~ **del castillete** *m* Esp (*cf polea del castillete de extracción AmL*) MINAS gudgeon

goslarita *f* MINERAL goslarite

gota *f* ALIMENT, C&V, CARBÓN, CONST drop; ~ **de aceite** *f* FÍS oil drop; ~ **de madreperla** *f* C&V mother-of-pearl bead; ~ **de nube** *f* METEO droplet; ~ **de vidrio** *f* C&V glass bead

gotear *vi* ALIMENT drip, CARBÓN seep, ENERG RENOV leak, HIDROL trickle, TRANSP MAR leak

goteo *m* ALIMENT drip, CARBÓN seepage, ENERG RENOV leakage, HIDROL trickling

gotero *m* LAB dropper, dropping bottle

gotícula *f* CONTAM *atmosférica* droplet; ~ **de líquido** *f* CONTAM liquid droplet; ~ **de líquido en suspensión** *f* CONTAM *tratamiento de gases* suspended liquid droplet

gotita *f* CONTAM *atmosférica*, METEO droplet; ~ **de líquido** *f* CONTAM liquid droplet

gozne *m* ING MECÁ hinge, joint, pin, INSTR hinge, MECÁ hinge, pin, PROD pin; ~ **en forma de H** *m* CONST H hinge

gozo: ~ **de índice** *m* TELECOM index dip

GPS *abr* (*sistema de posicionamiento global*) TRANSP AÉR, TRANSP MAR GPS (*global positioning system*)

GPU *abr* (*unidad de potencia de tierra*) TRANSP AÉR GPU (*ground power unit*)

grabación *f* ACÚST recording, CINEMAT recording, sound recording, ELEC recording, ELECTRÓN etching, GEOFÍS, GEOL, INFORM&PD, TELECOM, TV recording; ~ **de acanalado en V** *f* ELECTRÓN V-groove etching; ~ **de amplitud variable** *f* ACÚST variable-amplitude recording; ~ **de área variable** *f* TV variable-area recording; ~ **de audio** *f* TV audio record; ~ **de cintas de video** *AmL*, ~ **de cintas de vídeo** *f* Esp TV videotaping; ~ **de datos** *f* INFORM&PD, TELECOM data recording; ~ **de densidad variable** *f* ACÚST variable-density recording; ~ **de doble densidad** *f* INFORM&PD double-density recording; ~ **estereofónica** *f* ACÚST stereophonic recording; ~ **por haz láser** *f* ELECTRÓN, TV laser-beam recording; ~ **helicoidal** *f* TV helical-recording; ~ **instantánea** *f* ACÚST instantaneous recording; ~ **por láser** *f* TV laser-beam recording; ~ **laseróptica** *f* ÓPT laser-optic recording; ~ **magnética** *f* ACÚST, INFORM&PD magnetic recording; ~ **magnética longitudinal** *f* ACÚST longitudinal magnetic recording; ~ **magnética perpendicular** *f* ACÚST perpendicular magnetic recording; ~ **magné-**

tica transversal *f* ACÚST transverse magnetic recording; ~ **mecánica** *f* ACÚST mechanical recording; ~ **mediante láser** *f* INFORM&PD laser-beam recording; ~ **por el método ablativo** *f* ÓPT ablative-method recording; ~ **de modulación lateral** *f* ACÚST lateral recording; ~ **monofónica** *f* ACÚST monophonic recording; ~ **monopista** *f* ACÚST one-track recording; ~ **multipistas** *f* ACÚST multitrack recording; ~ **no transmitida** *f* TV off air recording; ~ **NRZ** *f* INFORM&PD NRZ recording; ~ **óptica** *f* ACÚST, ÓPT optical recording; ~ **pirata** *f* ACÚST, TV pirate recording; ~ **en profundidad** *f* ACÚST vertical recording; ~ **de prueba** *f* ACÚST test record; ~ **sin retorno a cero** *f* INFORM&PD nonreturn-to-zero recording; ~ **segmentada** *f* TV segmented recording; ~ **de sonido** *f* CINEMAT, TELECOM, TV sound recording; ~ **tetrafónica** *f* ACÚST tetraphonic recording; ~ **transversal** *f* TV transverse recording; ~ **de video** *AmL* ~ **de vídeo** *f* *Esp* TELECOM video recording; ~ **en videodisco** *f* TV videodisc recording (*BrE*), videodisk recording (*AmE*)

grabado[1]: ~ **en obra** *adj* PROD field-engraved

grabado[2] *m* ACÚST engraving, C&V engraving, figuring, intaglio, CINEMAT tracing, ELECTRÓN engraving, IMPR block, engraving, etching, ING MECÁ tracing, METAL etching; ~ **con ácido** *m* C&V clear etching; ~ **al ácido** *m* C&V acid badging, acid etching; ~ **al ácido brillante** *m* C&V bright etching; ~ **con agujas** *m* C&V needle etching; ~ **al buril** *m* IMPR line engraving; ~ **sin contraste** *m* IMPR flat etching; ~ **de corrosión de fondo plano** *m* METAL flat-bottomed etch pit; ~ **directo de trama gruesa** *m* IMPR quarter tone; ~ **de doble composición** *m* IMPR double offset gravure; ~ **electrónico** *m* IMPR electronic engraving; ~ **engomado** *m* C&V glue etching; ~ **en fondo** *m* C&V lettering on bottom; ~ **francés** *m* C&V French embossing; ~ **ilegible** *m* IMPR blind blocking; ~ **de línea** *m* IMPR line block, line work; ~ **en linóleo** *m* IMPR linocut; ~ **de matrices en hueco** *m* ING MECÁ die sinking; ~ **con molde** *m* ING MECÁ mold engraving (*AmE*), mould engraving (*BrE*); ~ **profundo** *m* C&V deep etching; ~ **con punta de diamante** *m* C&V diamond-point engraving; ~ **químico por plasma** *m* ELECTRÓN plasma etching; ~ **en relieve** *m* C&V engraving in relief, P&C embossing; ~ **en relieve al ácido** *m* C&V acid embossing; ~ **satinado** *m* C&V satin etch; ~ **para tricromía** *m* IMPR three-color black (*AmE*), three-colour black (*BrE*)

grabador *m* C&V, IMPR engraver; ~ **de acontecimientos** *m* INSTR event recorder; ~ **de cinta de video** *m* *AmL*, ~ **de cinta de vídeo** *m* *Esp* TV video recorder; ~ **de cinta de video digital** *m* *AmL*, ~ **de cinta de vídeo digital** *m* *Esp* TV digital videotape recorder (*DVTR*); ~ **de cintas de vídeo encadenado** *m* *Esp* (*cf grabador de cintas de video esclavo AmL*) TV slave video cassette recorder; ~ **de cintas de video esclavo** *m* *AmL* (*cf grabador de cintas de vídeo encadenado Esp*) TV slave video cassette recorder; ~ **de discos** *m* ACÚST, ELEC gramófono disc recorder (*BrE*), disk recorder (*AmE*); ~ **de doce puntos** *m* INSTR twelve-point recorder; ~ **duplicador** *m* INSTR duplicating recorder; ~ **de exploración transversal** *m* TV transverse-scanning recorder; ~ **gráfico** *m* INSTR, LAB chart recorder; ~ **de gráficos continuo** *m* INSTR, LAB continuous chart recorder; ~ **por impedancia** *m*

INSTR impedance recorder; ~ **de luz difusa** *m* INSTR diffuse-light recorder; ~ **magnético** *m* ACÚST, INFORM&PD magnetic recorder; ~ **de microfilme** *m* INFORM&PD microfilm recorder; ~ **de micropelícula** *m* INFORM&PD microfilm recorder; ~ **de multipistas** *m* INSTR multitrack recorder; ~ **de multipuntos** *m* INSTR multipoint recorder; ~ **de ocho pistas** *m* INSTR eight-channel recorder; ~ **de ocho puntos** *m* INSTR eight-point recorder; ~ **óptico por láser** *m* ELECTRÓN laser optical recorder; ~ **del perfil sísmico** *m* FÍS ONDAS, INSTR seismic-profile recorder; ~ **puntual** *m* INSTR point recorder; ~ **reproductor de video** *m* *AmL*, ~ **reproductor de vídeo** *m* *Esp* TV playback video-tape recorder; ~ **de salida** *m* INSTR output recorder; ~ **de sonidos** *m* INSTR sound recorder; ~ **de la trayectoria de vuelo** *m* INSTR, TRANSP AÉR flight-path recorder; ~ **de variables múltiples** *m* INSTR multivariable recorder; ~ **de vibraciones** *m* INSTR vibration recorder; ~ **de video digital** *m* *AmL*, ~ **de vídeo digital** *m* *Esp* TV digital videotape recorder (*DVTR*); ~ **de video en estéreo** *m* *AmL*, ~ **de vídeo en estéreo** *m* *Esp* TV stereo video cassette recorder; ~ **de video por exploración helicoidal** *m* *AmL*, ~ **de vídeo por exploración helicoidal** *m* *Esp* TV helical-scan videotape recorder; ~ **de video en formato B** *m* *AmL*, ~ **de vídeo en formato B** *m* *Esp* TV B-format video recorder; ~ **de video en formato C** *m* *AmL*, ~ **de vídeo en formato C** *m* *Esp* TV C-format videotape recorder; ~ **de videocasette** *m* TV videocassette recorder (*VCR*); ~ **de voz** *m* TEC ESP voice recorder

grabadora *f* ACÚST, ELEC, INFORM&PD, INSTR, TELECOM, TV recorder; ~ **de audio** *f* TV audio tape machine; ~ **de audio a cinta** *f* TV audio tape machine; ~ **de cinta de video** *f* *AmL*, ~ **de cinta de vídeo** *f* *Esp* TV video recorder, videotape recorder (*VTR*); ~ **de datos** *f* INFORM&PD data recorder; ~ **de datos de vuelo** *f* TRANSP AÉR flight-data recorder; ~ **digital de datos de vuelo** *f* TRANSP AÉR digital flight-data recorder; ~ **de pruebas de vuelo** *f* TEC ESP, TRANSP AÉR flight-test recorder; ~ **de sonidos y voces de la cabina de mando** *f* TRANSP AÉR cockpit voice recorder; ~ **de voz** *f* TEC ESP voice recorder; ~ **de vuelo** *f* TRANSP AÉR flight recorder

grabar *vt* ACÚST record, prerecord, C&V engrave, CRISTAL etch, ELECTRÓN engrave, etch, FÍS, GEOFÍS record, IMPR engrave, etch, INFORM&PD record, METAL, QUÍMICA etch, TV record

grada *f* AGRIC harrow, CONST *monumento* step, MINAS stope, TRANSP MAR *construcción naval* slip, *reparaciones* slipway; ~ **al revés** *f* MINAS overhand stope; ~ **de alambre** *f* CONST wire staple; ~ **de dientes** *f* AGRIC section harrow; ~ **de dientes para caballones** *f* AGRIC saddle-back harrow; ~ **de dientes rígidos** *f* AGRIC spike-tooth harrow; ~ **de discos** *f* AGRIC disc harrow (*BrE*), disc tiller (*BrE*), disk harrow (*AmE*), disk tiller (*AmE*); ~ **de discos de doble acción** *f* AGRIC tandem disc harrow (*BrE*), tandem disk harrow (*AmE*); ~ **de discos excéntricos** *f* AGRIC offset disc harrow (*BrE*), offset disk harrow (*AmE*); ~ **invertida** *f* MINAS overhand stope; ~ **niveladora** *f* AGRIC leveling harrow (*AmE*), levelling harrow (*BrE*); ~ **de púas** *f* AGRIC knife-tooth harrow; ~ **rodillo** *f* AGRIC soil surgeon

gradación *f* CINEMAT grading, key, COLOR, IMPR gradation; ~ **de color de una escena a la otra** *f*

CINEMAT scene-to-scene color grading (*AmE*), scene-to-scene colour grading (*BrE*); **~ de matiz** *f* COLOR nuance; **~ de tonalidad** *f* IMPR tone scale

gradar *vt* GEN grade

gradería *f AmL* (*cf banco ascendente Esp*) MINAS stoping

gradiente *m* GEN gradient; **~ adiabático** *m* TERMO adiabatic lapse rate; **~ adiabático seco** *m* CONTAM, TERMO *atmosférica* dry adiabatic lapse rate; **~ adiabático de temperatura** *m* FÍS FLUID adiabatic temperature gradient; **~ del ascenso** *m* TRANSP AÉR climb gradient; **~ de campo eléctrico** *m* ELEC, FÍS, ING ELÉC electric-field gradient; **~ del campo magnético** *m* ELEC, FÍS, ING ELÉC magnetic-field gradient; **~ de esfuerzo de Reynolds** *m* FÍS FLUID gradient of Reynolds stress; **~ de fractura** *m* GEOL, TEC PETR fracture gradient; **~ geotérmico** *m* ENERG RENOV, GEOFÍS, GEOL, TEC PETR, TERMO geothermal gradient; **~ hidráulico** *m* HIDROL hydraulic gradient; **~ nulo de presión** *m* FÍS FLUID zero pressure gradient; **~ de potencial** *m* ELEC potential gradient, *campo eléctrico* voltage gradient, FÍS potential gradient; **~ de presión** *m* FÍS FLUID, TEC PETR pressure gradient; **~ de presión del agua** *m* TEXTIL water gradient pressure; **~ de presión de formación** *m* GEOL, TEC PETR formation-pressure gradient; **~ de presión impuesto** *m* FÍS FLUID imposed pressure gradient; **~ de recubrimiento** *m* GEOL overburden gradient; **~ de sobrecarga** *m* TEC PETR overburden gradient; **~ de temperatura** *m* INSTAL TERM, TERMO temperature gradient; **~ de tensión** *m* ELEC voltage gradient; **~ térmico** *m* TERMO thermal gradient, heat gradient; **~ térmico vertical** *m* METEO, TERMO vertical temperature gradient; **~ vertical de temperatura** *m* METEO vertical temperature gradient

gradilla *f* LAB draining rack, *para tubos de ensayo* test-tube rack

gradiomanómetro *m* PETROL gradiomanometer

grado *m* CONST grade, FÍS degree, GEOM degree, grade, gradient, METR degree, MINAS extent, PROD grade, QUÍMICA degree, TERMOTEC degree, grade, TRANSP *de servicio* level; **~ de acidez** *m* CONTAM acidity level; **~ de acoplamiento** *m* ING MECÁ interface level; **~ de calidad ambiental** *m* CONTAM *legislación* ambient quality standard; **~ de calor** *m* TERMO degree of heat; **~ de capacidad-volumen** *m* TRANSP volume-capacity ratio; **~ Celsius** *m* METR degree Celsius; **~ centesimal** *m* FÍS grade; **~ de coherencia** *m* ÓPT, TELECOM degree of coherence; **~ del combustible** *m* TRANSP AÉR fuel grade; **~ de compactación** *m* CARBÓN degree of compaction; **~ de compresión** *m* ING MECÁ compression ratio; **~ de concentración** *m* CONTAM *soluciones* level; **~ de conformidad** *m* TRANSP degree of compliance; **~ de consolidación** *m* CARBÓN degree of consolidation, CONST degree of compaction; **~ de contaminación** *m* CONTAM, CONTAM MAR degree of pollution; **~ de contraste** *m* CARBÓN assay grade; **~ de conversión** *m* CONTAM transformation rate, TRANSP conversion degree; **~ de cumplimiento** *m* TRANSP *control de tráfico* obedience level; **~ de curado** *m* CONST rate of curing; **~ de depuración** *m* AGUA, HIDROL degree of purification; **~ de dureza blando** *m* PROD *muelas* soft grade; **~ de dureza duro** *m* PROD *muelas* hard grade; **~ de dureza media** *m* PROD *muelas* medium grade; **~ de dureza no carbonatada** *m* HIDROL noncarbonate hardness;

~ de elevación *m* ING ELÉC rate of rise, TEC ESP pitch; **~ de elevación de corriente** *m* ING ELÉC rate of current rise; **~ de elevación del voltaje** *m* ING ELÉC rate of voltage rise; **~ de expulsión** *m* TEC PETR expulsion rate; **~ del flujo de aire** *m* INSTAL TERM airflow rate; **~ hidrométrico** *m* METEO hydrometric degree; **~ de incombustibilidad** *m* PAPEL degree of incombustibility; **~ de ininflamabilidad** *m* PAPEL degree of non-flammability; **~ de interfaz** *m* ING MECÁ interface level; **~ Kelvin** *m* METR degree Kelvin; **~ de libertad** *m* ACÚST degree of freedom; **~ medio suave** *m* PROD *muelas* medium-soft grade; **~ metamórfico** *m* GEOL metamorphic grade; **~ de ocupación** *m* TRANSP *tráfico* occupancy rate; **~ del papel** *m* FOTO paper grade; **~ de polimerización** *m* P&C degree of polymerization; **~ de protección** *m* PROD degree of protection; **~ de quemado** *m* NUCL burn-up fraction; **~ de quemado de diseño** *m* NUCL design burn-up; **~ de quemado final del combustible** *m* NUCL final fuel burn-up; **~ de quemado potencial** *m* NUCL achievable burn-up; **~ de refino** *m* PAPEL freeness value; **~ de refrigeración** *m* REFRIG, TERMO cooling rate; **~ de rizado** *m* TEXTIL degree of crimp; **~ de saturación** *m* CARBÓN, TRANSP degree of saturation; **~ de secado** *m* CONST degree of drying; **~ de separación entre filas** *m* INFORM&PD row pitch; **~ de servicio** *m* TELECOM grade of service; **~ de temperatura** *m* TERMO degree of temperature; **~ de transformación** *m* CONTAM transformation rate; **~ de utilización** *m* TRANSP degree of utilization; **~ de utilización de la máquina** *m* PROD machine utilization degree; **~ de viscosidad** *m* VEH *del aceite* viscosity index

graduable *adj* ING MECÁ adjusting

graduación *f* CONST grading, IMPR *de tonos de color* graduation, ING MECÁ scales, LAB graduation, PROD *máquinas* staging, TRANSP AÉR setting; **~ del altímetro** *f* FÍS, GEOFÍS, TEC ESP, TRANSP AÉR altimeter setting; **~ de aumento** *f* INSTR magnification scale; **~ de los márgenes** *f* IMPR shingling; **~ octánica** *f* VEH octane rating; **~ de regulación en carga** *f* ELEC *transformador* on-load tap changing; **~ tonal** *f* FOTO tonal gradation

graduado: no ~ *adj* LAB, QUÍMICA *vaso* ungraduated

graduador *m* GEOM protractor, ING MECÁ index; **~ circular** *m* GEOM circular protractor; **~ óptico** *m* ELECTRÓN optical stepper; **~ de regulación en carga** *m* ELEC on-load tap changer

gradual *adj* PROD step

graduar *vt* IMPR adjust, INFORM&PD scale, MECÁ adjust, PROD gage (*AmE*), gauge (*BrE*), TEC ESP index

gráfica *f* MATEMÁT graph; **~ en banda de papel** *f* ELEC *registro* chart strip; **~ de flujo** *f* INFORM&PD flowgraph; **~ de humos** *f* SEG smoke chart; **~ PERT** *f* INFORM&PD PERT chart; **~ de rollo** *f* ELEC *registro* chart strip

graficar *vt AmL* (*cf dibujar una gráfica Esp*) MATEMÁT graph, plot

gráfico *m* IMPR chart, graphic, INFORM&PD graph, graphics, MATEMÁT graph, PROD *cálculo de muelles helicoidales* diagram; **~ de barras** *m* FÍS, INFORM&PD, MATEMÁT, TELECOM bar chart, histogram; **~ cartesiano** *m* INFORM&PD line graph; **~ en color** *m Esp* (*cf dibujo de color AmL*) INFORM&PD color graphic (*AmE*), colour graphic (*BrE*); **~ del flujo** *m*

TEC PETR flow sheet; ~ **de flujo térmico** *m* TERMO heat flow chart; ~ **de Gantt** *m* PROD Gantt chart; ~ **lineal** *m* INFORM&PD line graph; ~ **presión-profundidad** *m* TEC PETR pressure-vs-depth plot; ~ **de producción** *m* PROD diagram of output; ~ **de sectores** *m* INFORM&PD, MATEMÁT pie chart; ~ **de utilización** *m* ENERG RENOV utilization curve

gráficos: ~ **de barrido** *m pl* INFORM&PD raster graphics; ~ **comerciales** *m pl* IMPR, INFORM&PD business graphics; ~ **de computadora** *m pl AmL* (*cf gráficos de ordenador Esp*) INFORM&PD, TV computer graphics; ~ **coordinados con etiquetas de botella** *m pl* EMB graphics coordinated with bottle labels; ~ **de energía** *m pl* ENERG RENOV power curves; ~ **formados por puntos** *m pl* INFORM&PD raster graphics; ~ **de fuerza** *m pl* ENERG RENOV power curves; ~ **interactivos** *m pl* INFORM&PD interactive graphics; ~ **de ordenador** *m pl Esp* (*cf gráficos de computadora AmL*) INFORM&PD, TV computer graphics; ~ **de presentación** *m pl* INFORM&PD presentation graphics; ~ **vectoriales** *m pl* INFORM&PD vector graphics

grafilado *m* C&V knurling

grafitización *f* METAL graphitization

grafito *m* COLOR black lead, MINERAL graphite, QUÍMICA graphite, plumbago

grafo *m* MATEMÁT lattice

gragea *f* CARBÓN pellet

grama *f* AGRIC agropiron repens, creeping wheatgrass, quackgrass; ~ **rastrera** *f* AGRIC knot grass, spayed heifer

gramaje *m* IMPR, PAPEL basic weight, grammage, substance; ~ **de la capa del estucado** *m* IMPR, REVEST coating weight; ~ **real** *m* IMPR actual weight

gramil *m AmL* (*cf ancho de vía Esp*) CONST joiner's gauge (*BrE*), scratch gage (*AmE*), scratch gauge (*BrE*), scriber, *carpintería* scribing gage (*AmE*), scribing gauge (*BrE*), GEOL marking gage (*AmE*), marking gauge (*BrE*), ING MECÁ marking gauge (*BrE*), scriber, marking gage (*AmE*), PROD marking gage (*AmE*), marking gauge (*BrE*), TRANSP MAR *remachados* landing, marking gage (*AmE*), marking gauge (*BrE*); ~ **de ajustador** *m* ING MECÁ combination surface gage (*AmE*), combination surface gauge (*BrE*); ~ **de cuchilla** *m* CONST cutting gage (*AmE*), cutting gauge (*BrE*); ~ **de escuadra** *m* ING MECÁ scribing block; ~ **de mortaja** *m* CONST *carpintería* mortise gage (*AmE*), mortise gauge (*BrE*); ~ **de prisma combinado** *m* ING MECÁ combination surface gage (*AmE*), combination surface gauge (*BrE*); ~ **de trazador** *m* ING MECÁ combination surface gage (*AmE*), combination surface gauge (*BrE*); ~ **de trazar** *m* ING MECÁ scribing block; ~ **de trazar combinado** *m* ING MECÁ combination surface gage (*AmE*), combination surface gauge (*BrE*)

grammatita *f* MINERAL grammatite

gramo *m* FÍS, METR, QUÍMICA gram, gramme; ~ **equivalente** *m* QUÍMICA gram equivalent, gramme equivalent; **~-masa** *m* METR gram in mass, gramme in mass

gramófono *m* ACÚST, ING ELÉC record player

grampa *f* MECÁ clamp, clip, TRANSP, VEH clamp; ~ **para armaduras** *f AmL* (*cf grapa para armaduras Esp*) ELEC *cable* armor clamp (*AmE*), armour clamp (*BrE*); ~ **para cable** *f AmL* (*cf grapa para cable Esp*)

MECÁ cable clamp; ~ **a cadena** *f* ING MECÁ chain vice (*BrE*), chain vise (*AmE*)

gran[1]: **de** ~ **alcance** *adj* D&A long-range; **de** ~ **capacidad** *adj* ING MECÁ, MECÁ heavy-duty; **de** ~ **consumo de energía** *adj* TERMO energy-intensive; **con** ~ **contraste** *adj* CINEMAT contrasty; **de** ~ **duración** *adj* PROD long-life; **a** ~ **escala** *adj* GEOL large-scale, megascale; **de** ~ **par** *adj* MECÁ high-torque; **de** ~ **potencia** *adj* ING MECÁ, MECÁ heavy-duty; **de** ~ **resistencia a la tracción** *adj* MECÁ high-tensile

gran[2]: ~ **deshielo** *m* OCEAN debacle; ~ **ordenador** *m Esp* (*cf gran computadora AmL*) INFORM&PD main-frame; ~ **pedazo de hielo** *m* TEC ESP ice chunk

gran[3]: ~ **afluencia** *f* AGUA, HIDROL large inflow; ~ **cámara de reacción electrón-positrón** *f* (*LEP*) FÍS PART large electron-positron collider (*LEP*); ~ **cámara de reacción para hadrones** *f* (*LHC*) FÍS PART large hadron collider (*LHC*); ~ **computadora** *f AmL* (*cf gran ordenador Esp*) INFORM&PD mainframe; ~ **energía de unificación** *f* FÍS PART grand unification energy; ~ **explosión** *f* FÍS, TEC ESP Big Bang; ~ **extensión de hielo a la deriva** *f* TRANSP MAR ice pack, iceberg; ~ **limitación** *f* ELECTRÓN hard limiting; ~ **pendiente** *f* CONST high gradient; ~ **resistencia** *f* FÍS high resistance

granada *f* D&A grenade; ~ **anticarro** *f* D&A antitank grenade; ~ **sin carga** *f* D&A dummy grenade; ~ **de efecto retardado** *f* D&A time grenade; ~ **fumígena** *f* D&A smoke grenade; ~ **del fusil** *f* D&A rifle grenade; ~ **de instrucción** *f* D&A dummy grenade; ~ **lacrimógena** *f* D&A tear grenade; ~ **de mano** *f* D&A hand-grenade; ~ **de metralla** *f* D&A shrapnel

granalla: ~ **de acero al carbono** *f* CARBÓN carbon steel dust; ~ **de acero templado** *f* PROD chilled shot; ~ **de carbón** *f* CARBÓN, MINAS coal powder; ~ **en frío** *f* PROD *fundición* cold shot; ~ **de plomo** *f* PROD lead shot, QUÍMICA *copelación* lead button

granallar *vt* PROD *gránulos* shot

granate *m* CONST, MINERAL garnet; ~ **alumínico** *m* MINERAL aluminium garnet (*BrE*), aluminum garnet (*AmE*); ~ **alumínico de itrio** *m* ELECTRÓN yttrium aluminium garnet (*BrE*) (*YAG*), yttrium aluminum garnet (*AmE*) (*YAG*); ~ **férrico** *m* MINERAL iron garnet; ~ **férrico de itrio** *m* (*GFI*) ELECTRÓN yttrium iron garnet (*YIG*)

granazón *m* AGRIC granulation

grancilla *f* CARBÓN pea coal

grandes[1]: ~ **remolinos** *m pl* FÍS FLUID large eddies

grandes[2]: ~ **piezas de fundición** *f pl* PROD heavy castings; ~ **profundidades oceánicas** *f pl* TRANSP MAR *geografía* ocean depths

graneado *m* CARBÓN *pólvoras* granulation

granelero *m* TRANSP MAR bulk carrier

granero *m* ALIMENT granary

granetazo *m* ING MECÁ, MECÁ, PROD center pop (*AmE*), centre pop (*BrE*), center punch mark (*AmE*), centre punch mark (*BrE*)

granete *m* ING MECÁ, MECÁ center punch (*AmE*), centre punch (*BrE*), punch; ~ **punzón de marca** *m* PROD center punch mark (*AmE*), centre punch mark (*BrE*)

granitela *f* PETROL granitell

granitita *f* PETROL granitite

granitización *f* GEOL granitization

granito *m* CONST, GEOL, PETROL granite

granitoide *adj* GEOL granitoid

granizada *f* FÍS, METEO hail, hailstorm

granizar *vi* FÍS, METEO hail

granizo *m* FÍS, METEO hail, hailstone

granja *f Esp* (*cf tambo AmL*) AGRIC farm; ~ **avícola** *f* AGRIC poultry farm; ~ **lechera** *f* AGRIC dairy farm

granjero *m* AGRIC farmer

grano[1]: **sin ~** *adj* FOTO grainless; **de ~ fino** *adj* CINEMAT, FOTO, GEOL, IMPR, METAL, PAPEL fine-grained; **de ~ grueso** *adj* CINEMAT, FOTO, GEOL, IMPR, METAL, PAPEL coarse-grained; **de ~ medio** *adj* CINEMAT, FOTO, GEOL, IMPR, METAL, PAPEL medium-grained

grano[2] *m* GEN grain, AGRIC grain, kernel; ~ **alargado** *m* METAL elongated grain; ~ **para elaboración cervecera** *m* ALIMENT brewer's grain; ~ **equiaxial** *m* METAL equiaxed grain; ~ **equidimensional** *m* METAL equiaxed grain; ~ **fino** *m* CINEMAT, FOTO, METAL, MINAS, PROD fine grain; ~ **fino óptico** *m* CINEMAT optical fine grain; ~ **de la fotográfia** *m* FOTO, IMPR photographic grain; ~ **grueso** *m* CINEMAT, FOTO, GEOL, IMPR, METAL, MINAS, PAPEL coarse grain; ~ **de maíz entero** *m* AGRIC shelled corn (*AmE*), shelled maize (*BrE*); ~ **medio** *m* CINEMAT, FOTO, GEOL, IMPR, METAL, MINAS, PAPEL medium grain; ~ **de muela abrasiva** *m* MECÁ grit; ~ **panadero** *m* ALIMENT bread grain; ~ **del papel** *m* PAPEL paper grain; ~ **propulsor** *m* TEC ESP propellant grain; ~ **de sorgo** *m* AGRIC milo grain; ~ **triturado** *m* AGRIC crushed grain

granodiorita *f* PETROL granodiorite

granofiro *m* PETROL granophyre

granos *m pl* AGRIC breadstuff; ~ **alimenticios** *m pl* AGRIC, ALIMENT food grains; ~ **aloquímicos** *m pl* GEOL allochem; ~ **bien clasificados** *m pl* GEOL well-sorted grains; ~ **bien seleccionados** *m pl* GEOL well-sorted grains; ~ **rotos** *m pl* AGRIC screenings; ~ **visibles** *m pl* GEOL visible grains

granulación *f* C&V granulation, CINEMAT, FOTO, IMPR graininess, P&C *operación* pelletizing, PROC QUÍ granulation

granulado[1] *adj* CONST granular, EMB, GEOL granulated; ~ **de áspero** *adj* CONST rough-grained

granulado[2]: ~ **de carbón** *m* CARBÓN coal bean

granulador *m* C&V, PROC QUÍ granulator; ~ **de lecho fluidizado** *m* NUCL, PROC QUÍ fluid-bed granulator

granuladora *f* EMB granulating machine, PROC QUÍ granulating crusher, granulating machine

granulados *m pl* PROC QUÍ granulates

granular[1] *adj* CONST, GEOL granular

granular[2] *vt* EMB, GEOL, PROC QUÍ granulate, PROD grain

granularidad *f* FOTO, INFORM&PD granularity

granularse *v refl* PROD grain

granulita *f* PETROL granulite

gránulo *m* CARBÓN pellet, EMB granule, P&C pellet

granulometría *f* C&V screen analysis (*AmE*), granulometry, CARBÓN *arenas, gravas* grading, granulometry, sizing, CONST grading, PROC QUÍ granulometry

granulométrico *adj* C&V, CARBÓN, CONST, NUCL, PROC QUÍ granulometric

granza *f* AGRIC chaff, P&C pellet

granzas *f pl* CARBÓN screenings, PROD siftings

grapa *f* CONST cramp, staple, ELEC *cable* clamp, EMB staple, ING MECÁ clip, clamp, cramp, yoke, MECÁ clip, dog, clamp, PROD holdfast; ~ **de alineación** *f* PETROL line-up clamp; ~ **para armaduras** *f Esp* (*cf grampa para armaduras AmL, mordaza para armaduras AmL*) ELEC *cable* armor clamp (*AmE*), armour clamp (*BrE*); ~ **de bastidor** *f* CONST *cerrajería* sash cramp; ~ **de cable** *f* ELEC, ING ELÉC *fijación*, VEH *sistema eléctrico, controles* cable clip; ~ **para cable** *f Esp* (*cf grampa para cable AmL*) MECÁ cable clamp; ~ **de la cerradura** *f* CONST lock staple; ~ **para cerrar bolsas** *f* EMB bag staple; ~ **circular** *f* ING MECÁ, MECÁ circlip; ~ **de conexión** *f* CONST joint cramp; ~ **a cuñas** *f* TEC PETR spider; ~ **G acanalada** *f* ING MECÁ ribbed G-cramp; ~ **G estriada** *f* ING MECÁ ribbed G-cramp; ~ **de hierro** *f* CONST cramp iron; ~ **de intercambio** *f* PROD *paquete de baterías* exchange clip; ~ **magnética** *f* ING MECÁ magnetic holdfast; ~ **de manguera** *f* ING MECÁ hose clip; ~ **de seguridad** *f* SEG *para latas* safety clamp; ~ **de sondeo** *f* MINAS spear rod; ~ **en T** *f* CONST T-cramp; ~ **para tubos** *f* MINAS pipe clamps; ~ **de unión** *f* CONST joiner's cramp; ~ **para ventana** *f* CONST sash clamp

grapado[1] *adj* CONST, ELEC, EMB, ING MECÁ, MECÁ, PROD clamped

grapado[2]: ~ **del ángulo** *m* EMB, IMPR corner stapling; ~ **lateral** *m* EMB lateral stapling; ~ **de la tapa del fondo** *m* EMB top and bottom stapling

grapadora *f* EMB stapling machine

grapar *vt* CONST, ELEC clamp, EMB staple, ING MECÁ, MECÁ clamp

grapas: ~ **para marco** *f pl* CONST casing clamps; ~ **de reloj de arena** *f pl* ING MECÁ hourglass calipers (*AmE*), hourglass callipers (*BrE*)

grapón *m* ING MECÁ hook, cramp

grasa[1]: **sin ~** *adj* ALIMENT nonfat

grasa[2] *f* ALIMENT fat, AUTO, ING ELÉC, ING MECÁ, MECÁ, PROD, VEH grease; ~ **para altas temperaturas** *f* MECÁ high-temperature grease; ~ **animal** *f* ALIMENT animal fat; ~ **en bloque** *f* ING MECÁ block grease; ~ **de cable** *f* ING ELÉC cable grease; ~ **comestible fabricada** *f* ALIMENT manufactured edible fat; ~ **derretida** *f* ALIMENT rendered fat; ~ **de fundición** *f* PROD clinker; ~ **fundida** *f* ALIMENT rendered fat; ~ **grafitada** *f* MECÁ graphite grease; ~ **hidrogenada** *f* ALIMENT hydrogenated fat; ~ **de leche** *f* ALIMENT milk fat; ~ **para rosca** *f* PETROL dope; ~ **solidificada** *f* ALIMENT solidified fat; ~ **vegetal** *f* ALIMENT vegetable fat

grasas: **bajo en ~** *adj* ALIMENT low-fat

gratícula: ~ **reflectora** *f* FÍS, FÍS ONDAS, ÓPT reflection grating

grátil *m* TRANSP MAR *vela* head; ~ **alto** *m* TRANSP MAR *vela triangular* head

grauwaka *f* GEOL, PETROL graywacke (*AmE*), greywacke (*BrE*); ~ **de bajo grado** *f* GEOL, PETROL low-rank graywacke (*AmE*), low-rank greywacke (*BrE*)

grava *f* AGUA *piedra partida* break stone, CONST ballasting, gravel, rolling, roadstones, GEOL, PETROL gravel; ~ **adherente** *f* CONST tight gravel; ~ **de filtración** *f* AGUA, PROC QUÍ filter gravel; ~ **filtradora** *f* AGUA, PROC QUÍ filter gravel; ~ **filtrante** *f* AGUA, PROC QUÍ filter gravel; ~ **fina** *f* CONST fine gravel; ~ **hiperaurífera** *f* MINAS highly auriferous gravel

gravas: ~ **auríferas** *f pl* MINAS placer gold

gravedad[1]: **sin ~** *adj* FÍS, GEOFÍS gravity-free

gravedad[2] *f* FÍS, GEOFÍS, ING MECÁ, NUCL, TEC ESP, TEC PETR *física* gravity; ~ **API** *f* PETROL API gravity; ~ **específica** *f* FÍS, GEOFÍS, TEC PETR, TEXTIL specific gravity; ~ **nula** f FÍS, GEOFÍS, TEC ESP zero gravity (*zero-g*)

gravera f AGUA, CONST gravel pit, MINAS quarry bar

gravilla f CARBÓN chippings, CONST coarse aggregate, PETROL grit

gravimetría f GEN gravimetry

gravimétrico *adj* CARBÓN, FÍS, GEOFÍS, INSTR, TEC PETR gravimetric

gravímetro m CARBÓN, FÍS, GEOFÍS, INSTR, TEC PETR gravimeter, gravity meter

gravitación f FÍS, TEC ESP gravitation

gravitón m FÍS graviton

gray m (*Gy*) GEN gray (*Gy*)

greda f AGUA loam

greenockita f MINERAL greenockite

greenovita f MINERAL greenovite

greisenización f GEOL greisening, greisenization

grieta f *Esp* (*cf dispersión AmL*) C&V bruise, CARBÓN crack, crevice, *rocas* seam, CONST *hormigón* crack, fissure, cracking, CRISTAL crack, GEOFÍS fissure, GEOL joint, cleat, MECÁ crack, cracking, hot tear, METAL crack, MINAS seam, split, OCEAN fissure, PROD slit, split, TEC ESP slit; ~ **capilar** f CONST hairline crack, NUCL capillary crack; ~ **capilar interna** f MECÁ hairline crack; ~ **de contracción** f GEOL shrinkage crack, PROD *fundería* shrinkage crack, *fundición* contraction crack; ~ **por contracción** f CONST shrinkage crack; ~ **debajo del cordón de soldadura** f NUCL underbead crack; ~ **de desecación** f GEOL desiccation crack, mud crack; ~ **dúctil** f METAL ductile crack; ~ **por fatiga** f METAL fatigue crack; ~ **por fatiga bajo corrosión** f METAL, NUCL corrosion-fatigue crack; ~ **fina** f ING MECÁ fine crack, METAL fine slip; ~ **en forma de monedas** f METAL penny-shaped crack; ~ **por fragilización** f NUCL brittle crack; ~ **gasífera** f MINAS gas fissure; ~ **incipiente** f NUCL incipient crack; ~ **incipiente por fatiga** f NUCL fatigue precrack; ~ **interna** f METAL internal crack; ~ **microscópica** f METAL, ÓPT microcrack; ~ **de orilla** f C&V edge crack; ~ **en la parte inferior de la pared** f CONST foot-wall seam; ~ **pasante** f NUCL through-wall crack; ~ **principal** f METAL main crack; ~ **de refracción** f PROD contraction crack; ~ **semiinfinita** f NUCL semi-infinite crack; ~ **de termotratamiento** f TERMO heat-treatment crack; ~ **de ventilación** f MINAS split

grifería: ~ **con accesorios** f CONST cocks and fittings

grifo m CONST cock, faucet, tap, HIDROL valve, MECÁ cock, faucet, *tuberías* tap, PROD valve, TRANSP MAR cock; ~ **de aceite** m ING MECÁ oil cock; ~ **del agua** m CONST water cock; ~ **de aire** m ING MECÁ air tap; ~ **de aspersión** m AGUA flood cock, jet cock; ~ **de base** m CONST butt cock; ~ **de boca curva** m ING MECÁ bib-cock; ~ **de boca recta** m *Esp* (*ver canilla recta AmL*) CONST straight-nose cock; ~ **de boca roscada** m CONST bib; ~ **de botón** m CONST push-button faucet (*AmE*), push-button tap (*BrE*); ~ **de bronce** m CONST cock brass; ~ **calibrador** m AGUA trial cock; ~ **cebador** m AGUA priming cock; ~ **de cierre automático** m CONST *fontanería* self-closing cock (*BrE*), self-closing faucet (*AmE*); ~ **cilíndrico** m CONST cylinder cock; ~ **de comprobación** m AGUA try cock; ~ **de cuatro vías** m AGUA four-way cock; ~ **con cuello de cisne** m CONST swanneck cock; ~ **curvo** m CONST bib nozzle; ~ **de desagüe** m CONST waste cock, ING MECÁ bib tap, drain cock, PROD drain

cock; ~ **de descompresión** m INSTAL HIDRÁUL compression cock, pet cock; ~ **de dos caras** m AGUA twin cock; ~ **de dos vías** m AGUA two-way cock, INSTAL HIDRÁUL tapping; ~ **de drenaje** m AGUA water drain cock, ING MECÁ, PROD drain cock; ~ **del enfriador** m PROD cooler cock; ~ **engrasador** m PROD grease cock; ~ **de extracción** m ING MECÁ draw-off tap; ~ **de flotador** m ING MECÁ ball cock; ~ **del fuelle** m PROD blower-cock; ~ **de gas** m GAS, PROD gas cock, gas tap; ~ **de inyección** m INSTAL HIDRÁUL injection cock; ~ **con llave de horquilla** m CONST tap with crutch key; ~ **maestro** m ING MECÁ main tap; ~ **de manguera** m CONST bib-cock; ~ **de metal** m CONST cock metal; ~ **de prueba** m AGUA gage cock (*AmE*), gauge cock (*BrE*), test cock; ~ **de purga** m AGUA bleeding-cock, purge cock, blow-off cock, drain cock, ING MECÁ bleed valve, INSTAL HIDRÁUL cock, pit cock, safety-relief valve, PROD drain cock, TEC PETR *tuberías, bombas* bleed valve, VEH *radiador* drain cock; ~ **recto** m CONST straight-nose cock, ING MECÁ globe tap; ~ **rotativo de vapor** m INSTAL HIDRÁUL steam cock; ~ **de vaciado** m VEH *radiador* drain cock; ~ **de vaciamiento** m AGUA purging cock, PROD mud cock; ~ **del ventilador impelente** m PROD blower-cock

grilla: ~ **de modulación** f TV modulation grid

grillete m AUTO, CONST, CONTAM MAR shackle, ING MECÁ clevis, shackle, TRANSP MAR *pieza de cadena* shackle; ~ **de acero forjado** m ING MECÁ forged shackle; ~ **de ancla** m TRANSP MAR anchor shackle; ~ **automático** m TRANSP MAR *guarnición* snap shackle; ~ **en forma de D** m TRANSP MAR *guarnición* D-shackle; ~ **giratorio** m TRANSP MAR *guarnición* swivel; ~ **de hierro forjado** m ING MECÁ forged shackle; ~ **de unión** m ING MECÁ connecting link

grímpola f TRANSP MAR pennant

griota f GEOL millstone grit

gripado m AUTO jamming, VEH *cojinete, pistón* seizing

gris: ~ **neutro** m IMPR neutral gray (*AmE*), neutral grey (*BrE*); ~ **Oxford** m COLOR Oxford gray (*AmE*), Oxford grey (*BrE*)

grisú m CARBÓN, MINAS, TERMO firedamp, gas, pit gas

grisuómetro m GAS, INSTR, MINAS, PROC QUÍ, QUÍMICA gas detector, gasometer, methanometer, pit-gas detector, pit-gas indicator

grisuoso *adj* GAS, MINAS, TERMO gaseous, gassy

grisuscopio m *AmL* (*cf detector de metano Esp*) GAS, INSTR, MINAS, PROC QUÍ, QUÍMICA gas detector, gasometer, pit-gas detector

groera f TRANSP MAR *construcción naval* limber hole

groroilita f MINERAL groroilite

grosor m FÍS bulk, MECÁ, P&C, QUÍMICA thickness; ~ **de la cuerda** m TRANSP AÉR chordal thickness; ~ **óptico** m TEC ESP optical thickness; ~ **del refuerzo** m ING MECÁ thickness of lining

grosularia f MINERAL grossular

grossularita f MINERAL grossularite

GRP *abr* GEN (*plástico reforzado con fibra de vidrio, plástico reforzado con vidrio, poliéster reforzado con fibra de vidrio*) GRP (*glass-reinforced plastic, glass-reinforced polyester*)

grúa f CINEMAT crane, jenny, CONST crane, derrick, ENERG RENOV derrick, MECÁ crane, derrick, hoist, PROD, TRANSP AÉR hoist, TRANSP MAR crane; ~ **aérea** f CONST overhead crane; ~ **de auxilio** f FERRO *equipo*

inamovible wrecking crane, TRANSP wrecker; **~ para la botadura de embarcaciones** *f* TRANSP boat launching crane; **~ de botes** *f* MECÁ davit; **~ con brazo** *f* MINAS jumbo including boom; **~ de brazo** *f* CONST jib crane; **~ de brazo amantillable** *f* CONST luffing crane; **~ de brazo móvil** *f* TRANSP AÉR derrick; **~ de cadena** *f* ING MECÁ chain hoist; **~ de carga** *f* FERRO *infraestructura* loading crane, TRANSP MAR *construcción naval* cargo crane; **~ sobre carriles** *f* TRANSP rail crane; **~ corrediza** *f* MECÁ traveling crane (*AmE*), travelling crane (*BrE*); **~ de cubierta** *f* TRANSP MAR deck crane; **~ de electroimán** *f* PROD magnet crane; **~ de electroimán portador** *f* PROD magnet crane; **~ de extracción** *f* MINAS engine winding-house; **~ fija** *f* CONST derrick crane; **~ flotante** *f* TRANSP MAR floating crane, pontoon crane; **~ giratoria** *f* CONST rotary crane, swing crane; **~ giratoria de 360 grados** *f* CONST all-around swing crane (*AmE*), all-round swing crane (*BrE*); **~ goliath** *f* CONST goliath crane; **~ independiente** *f* CONST independent crane; **~ jumbo** *f* MINAS jumbo including boom; **~ locomóvil** *f* PROD traveling crane (*AmE*), traveling gantry crane (*AmE*), travelling crane (*BrE*), travelling gantry crane (*BrE*); **~ de mástil** *f* TRANSP MAR mast crane; **~ para mover materiales** *f* EMB material-handling crane; **~ móvil** *f* CINEMAT overhead crane, CONST mobile crane, portable crane; **~ de muelle** *f* TRANSP quay crane; **~ neumática** *f* ING MECÁ, PROD air hoist; **~ de pared** *f* CONST wall crane; **~ de pared rotatoria** *f* CONST rotary wall crane; **~ pesada de auxilio** *f* FERRO heavy breakdown crane; **~ de pescante** *f* NUCL jib crane; **~ sobre pontona** *f* CONST floating derrick; **~ portador** *f* PROD magnet crane; **~ pórtica** *f* CONST portal crane; **~ pórtico** *f* CONST, FERRO, NUCL, TRANSP MAR gantry crane; **~ pórtico con mecanismo de izado** *f* NUCL gantry with hoist; **~ de pórtico móvil** *f* TRANSP MAR traveling gantry (*AmE*), travelling gantry (*BrE*), traveling gantry crane (*AmE*), travelling gantry crane (*BrE*); **~ pórtico-castillete** *f* TRANSP MAR *puntal de carga* portal mast; **~ puente** *f* EMB overhead traveling crane (*AmE*), overhead travelling crane (*BrE*); **~ de puntal** *f* CONST postcrane; **~ de salvamento** *f* TRANSP salvage crane; **~ semicaballete** *f* CONST semigantry crane; **~ de taller** *f* PROD shop crane; **~ torre** *f* CONST tower crane; **~ para vaciar** *f* C&V casting crane; **~ para vehículos pesados** *f* TRANSP heavy vehicle elevator (*AmE*), heavy vehicle lift (*BrE*); **~ Vinten** *f* CINEMAT Vinten crane; **~-cesta de salvamento** *f* TRANSP salvage crane
gruaje *m* ING MECÁ cranage
gruesa *f* METR gross
grueso¹ *adj* AGRIC, GEOL *grano*, ING MECÁ coarse, MECÁ heavy
grueso²: **~ del libro sin tapas** *m* PAPEL bulk
gruesos *m pl* CARBÓN *decantación* underflow, oversize, MINAS oversize, *preparación minerales* coarse sands
grumete *m* TRANSP MAR ship's boy
grumo *m* PAPEL *de fibras* lump; **~ de resina depositable** *m* PAPEL pitch
grunerita *f* MINERAL grunerite
grupo *m* AGUA set, GEOL group, *de capas* set, INFORM&PD group, ING MECÁ cluster, MATEMÁT group, NUCL unit, PROD *de personas, máquinas* number, QUÍMICA, TELECOM group; **~ de**

acumuladores *m* TEC ESP battery; **~ de apartaderos** *m* FERRO group of sidings; **~ de apoyo especializado** *m* FÍS RAD specialized support group; **~ auxiliar** *m* ING ELÉC standby set; **~ básico** *m* TELECOM basic group; **~ bidireccional** *m* TELECOM both-way group; **~ bidireccional de módulos de E/S** *m* PROD bidirectional I/O module group; **~ de bifurcación** *m* PROD branch group; **~ cerrado de usuarios** *m* INFORM&PD, TELECOM closed user group (*CUG*); **~ de circuitos** *m* TELECOM circuit group, high-usage circuit group; **~ de circuitos de última alternativa** *m* TELECOM last-choice circuit group; **~ compacto** *m* ELECTRÓN all-solid state; **~ compresor-condensador** *m* REFRIG condensing unit; **~ compresor-condensador comercial** *m* REFRIG commercial condensing unit; **~ de condensadores** *m* ING ELÉC bank of capacitors, TEC ESP battery; **~ convertidor** *m* ING ELÉC, ING MECÁ motor-generator set; **~ convertidor de corriente alterna a continua** *m* FERRO, VEH AC/DC motor converter set; **~ coordinador** *m* METAL ligand; **~ de cremalleras** *m* CINEMAT rack line; **~ de datos de generación** *m* INFORM&PD generation data set; **~ de discusión** *m* INFORM&PD newsgroup; **~ doble de calabrote** *m* TRANSP MAR *nudo* carrick bend; **~ de documentos** *m* INFORM&PD, TEC ESP batch; **~ electrogenerador** *m* ELEC *fuente de alimentación* generating set; **~ electrógeno** *m* ING ELÉC generating set, generator set, motor-generator set, ING MECÁ motor-generator set, power unit, PROD power unit, TRANSP MAR *motores* power plant; **~ electrógeno accionado por motor diesel** *m* ING ELÉC diesel-driven generating set; **~ electrógeno de emergencia** *m* MECÁ emergency power-generator; **~ electrógeno de pista** *m* (*GPU*) TRANSP AÉR ground power unit (*GPU*); **~ electrógeno radioisótopo** *m* TEC ESP radioisotope power generator; **~ de engranajes conductores** *m* VEH *transmisión* drive train; **~ entrante** *m* TELECOM incoming group; **~ espacial** *m* CRISTAL space group; **~ frigorífico** *m* REFRIG refrigerating unit; **~ frigorífico amovible** *m* REFRIG clip-on refrigerating unit; **~ frigorífico en caballete** *m* REFRIG saddle unit; **~ frigorífico de pequeña potencia** *m* REFRIG fractional low-power condensing unit; **~ funcional** *m* QUÍMICA functional group; **~ de hombres para un lanzamiento** *m* TRANSP cluster; **~ impelente** *m* MINAS plunger set; **~ de inducido equilibrado** *m* ELEC *motor* balanced-armature unit; **~ de interés** *m* INFORM&PD newsgroup; **~ de líneas** *m* TELECOM line group; **~ de líneas equilibradas** *m* TELECOM balanced grading group; **~ litoestratigráfico** *m* GEOL lithostratigraphic unit; **~ maestro** *m* TELECOM master group; **~ de mercancías** *m* EMB commodity group, group of commodities; **~ modular** *m* PROD module group; **~ módulo de E/S** *m* PROD I/O module group; **~ motobomba** *m* CONTAM MAR pumping unit; **~ motopropulsor** *m* TRANSP AÉR power plant, VEH power unit; **~ motopropulsor de tierra** *m* (*GPU*) TRANSP AÉR ground power unit (*GPU*); **~ motor** *m* ING MECÁ, PROD power unit, TRANSP AÉR power plant, VEH power unit; **~ motor crítico** *m* TRANSP AÉR critical power unit; **~ motor-generador** *m* ING ELÉC, ING MECÁ motor-generator set; **~ de onda** *m* FÍS, FÍS ONDAS wave group; **~ pendiente** *m* QUÍMICA pendant group; **~ de personas con intereses comunes** *m*

TELECOM clique; ~ **pesado** *m* NUCL heavy group;
~ **de pilas** *m* TEC ESP battery; ~ **de pilotes** *m* CARBÓN
pile group; ~ **de potencia crítico** *m* TRANSP AÉR
critical power unit; ~ **de pozos** *m* AGUA set of wells;
~ **preferente** *m* TELECOM first-choice group; ~ **de
productos** *m* PROD product group; ~ **de programas**
m INFORM&PD, TEC ESP batch; ~ **puntual** *m* CRISTAL
point group; ~ **de purga** *m* REFRIG purge recovery
system; ~ **de relés** *m* TELECOM relay set; ~ **de
relevadores** *m* TELECOM relay set; ~ **de rescate** *m*
SEG rescue party; ~ **de reserva** *m* ING ELÉC standby
set; ~ **saliente** *m* TELECOM outgoing group;
~ **secundario** *m* TELECOM supergroup; ~ **de señales
único** *m* TEC ESP unique word (*UW*); ~ **terminal** *m*
QUÍMICA end group; ~ **transformador** *m* ING ELÉC
converter set; ~ **de última alternativa** *m* TELECOM
last-choice group; ~ **de unidad administrativa** *m*
TELECOM administrative unit group; ~ **de unidad
tributaria** *m* TELECOM tributary-unit group (*TUG*);
~ **de usuarios** *m* INFORM&PD user group;
~ **voluminoso** *m* QUÍMICA bulky group

G-T *abr* (*relación ganancia temperatura de ruido*) TEC
ESP G-T (*gain-to-noise-temperature ratio*)

guadaña *f* AGRIC scythe; ~ **para cereales** *f* AGRIC
reaping scythe

guaiacol *m* QUÍMICA guaiacol

guaiacónico *adj* QUÍMICA guaiaconic

guaiarético *adj* QUÍMICA guaiaretic

gualdera: ~ **de litera** *f* TRANSP MAR *construcción naval*
check, leeboard

guánguil: ~ **para minerales** *m AmL* (*cf silo para
minerales Esp*) MINAS ore hopper

guanidina *f* QUÍMICA guanidine

guanilo *m* QUÍMICA guanyl

guanina *f* QUÍMICA guanine

guano *m* QUÍMICA guano

guanosina *f* QUÍMICA guanosine

guante: ~ **del borde de ataque** *m* TRANSP AÉR leading-
edge glove

guanteletas: ~ **de cuero** *f pl* SEG leather gauntlets

guantera *f* VEH glove box

guantes: ~ **aislantes** *m pl* SEG insulating gloves; ~ **de
caucho** *m pl* P&C, SEG rubber gloves; ~ **de goma** *m pl*
P&C, SEG rubber gloves; ~ **de hule** *m pl* P&C, SEG
rubber gloves; ~ **industriales** *m pl* SEG industrial
gloves; ~ **protectores** *m pl* SEG protective gloves;
~ **protectores a prueba de aceite** *m pl* SEG oilproof
protective gloves; ~ **protectores a prueba de ácidos**
m pl SEG acid-proof protective gloves; ~ **resistentes
al calor** *m pl* LAB, SEG, TERMO heat-resistant gloves;
~ **termorresistentes** *m pl* LAB, SEG, TERMO heat-
resistant gloves; ~ **de trabajo** *m pl* SEG working
gloves; ~ **de trabajo hechos de caucho artificial** *m
pl* P&C, SEG working gloves made of artificial rubber;
~ **de trabajo hechos de neopreno** *m pl* P&C, SEG
working gloves made of artificial rubber

guarda *f* CONST ward, *cerrajería* stub, IMPR,
INFORM&PD, PROD guard; ~ **contra escombros** *f*
ING MECÁ guard against debris; ~ **para rajar** *f* SEG *en
sierras circulares* riving knife; ~ **tipo jaula** *f* SEG
guard

guardaaguas *m* CONST skirt

guardaarenas *m* AGUA sand trap

guardabalasto *m* FERRO *equipo inamovible* ballast
retainer

guardabanda *m* SEG *en maquinaria* belt guard

guardabarros *m* AUTO, TRANSP, VEH fender (*AmE*),
mudguard (*BrE*); ~ **delantero** *m* AUTO, TRANSP, VEH
front fender (*AmE*), front wing (*BrE*); ~ **trasero** *m*
AUTO, TRANSP, VEH rear fender (*AmE*), rear wing
(*BrE*)

guardacabo: ~ **oviforme** *m* PROD *para cable, cuerda*
egg-shaped thimble

guardacabos *m* TRANSP MAR *guarnición* thimble

guardacadena *f* ING MECÁ chain guard, SEG chain case

guardacantón *m* PROD edgestone

guardacorrea *m* ING MECÁ belt guard

guardacostas *m* TRANSP MAR *aduanas* revenue cutter,
cúter coastguard cutter

guardafrenos *m* FERRO brakeman

guardafuegos *m* TERMO fireguard

guardaagujas *m* FERRO, TRANSP switcher

guardahielos *m* TRANSP AÉR ice guard

guardamáquinas *m* SEG machine guard

guardamonte *m* D&A *de arma de fuego* trigger guard

guardapescas *m* TRANSP MAR fishery protection vessel

guardapolvo *m* CARBÓN dust guard, TEXTIL *prenda de
vestir* overall, VEH *de la palanca de cambio* dust boot;
~ **de protección** *m* TRANSP AÉR boot

guardapolvos *m* AUTO dust seal, CARBÓN dust guard

guardar *vt* INFORM&PD save, PROD hold, TEC ESP stow

guardarraíl *m* CONST guardrail

guardarropa *m* C&V closet

guardasalpicaduras *m* PROD splash guard, splash
wing

guardatuberías *m* TEC PETR pipe rack

guardavivos *m* CONST corner band

guarde: ~ **de contenido direccionable** *m AmL*
INFORM&PD content-addressable storage (*AmE*), content-addressable store (*BrE*)

guardia[1]: **de** ~ *adj* TRANSP MAR *navegación* on watch

guardia[2]: ~ **marina** *m* TRANSP MAR midshipman;
~ **nocturna** *m* D&A night watch

guardia[3]: ~ **de babor** *f* TRANSP MAR port watch; ~ **de
puerto** *f* D&A anchor watch

guardia[4]: **estar de** ~ *vi* D&A, TRANSP MAR keep watch

guardín *m* TRANSP MAR transmission wire, *construcción
naval* steering wire, tiller rope, steering chain

guarnecer *vt* CONST trim, PROD jacket

guarnecido: ~ **de hierro** *adj* PROD ironbound

guarnecimiento *m* PROD *funderías* daubing

guarnición *f* PROD *material* stuffing, *encofrado* sheathing, TRANSP AÉR chafing strip; ~ **de aislamiento de
la caldera** *f* INSTAL HIDRÁUL boiler jacketing; ~ **del
armazón** *f* ING MECÁ frame edging; ~ **del borde** *f* ING
MECÁ edging; ~ **de caldera** *f* INSTAL HIDRÁUL boiler
jacket; ~ **de cáñamo** *f* PROD hemp packing; ~ **de
carda** *f* INSTAL HIDRÁUL clothing, TEXTIL card
clothing; ~ **de cuero** *f* PROD leather packer; ~ **de
estopa** *f* PROD junk packing; ~ **de freno** *f AmL* (*cf
zapata de freno Esp*) AUTO, C&V, ING MECÁ, MECÁ,
VEH brake lining; ~ **metálica** *f* ING MECÁ, MECÁ, ÓPT,
TELECOM ferrule; ~ **de vapor** *f* INSTAL HIDRÁUL steam
packing

guarnimiento: ~ **del tope de roda** *m* TRANSP MAR
yates, accesorios de cubierta stem fitting

guata *f* EMB wadding, ING MECÁ padding, PAPEL
wadding, TEXTIL wad, *capa de fibras* batt; ~ **de
algodón hidrófilo** *f* TEXTIL wad of cotton wool;
~ **de celulosa** *f* PAPEL cellulose wadding; ~ **de fibra
de vidrio** *f* EMB glass wadding

guateado *m* TEXTIL wadding, wad; **~ de algodón hidrófilo** *m* TEXTIL wad of cotton wool

guaya *f AmL (cf cable de perforación Esp, línea de perforación Esp)* PETROL, TEC PETR drilling line

guayacán *m* MECÁ, TRANSP MAR lignum vitae

gubia *f* CONST gouge, ING MECÁ bead, chisel; **~ de maceta** *f* ING MECÁ firmer gouge; **~ para molduras convexas** *f* PROD bead tool; **~ punzón** *f* ING MECÁ firmer gouge

güejarita *f* MINERAL chalcostibite

guerra: ~ bacteriológica *f* D&A bacteriological warfare; **~ electrónica** *f (GE)* D&A, ELECTRÓN electronic warfare *(EW)*

guía *f* ACÚST cueing, C&V edge guide, guide, CINEMAT leader, cueing, IMPR key, lay, ING MECÁ guide, guiding, track, MINAS guide, ore feeder, *de filones* chase, PROD *para colocar el módulo en el chasis* track, TEXTIL guide, TRANSP MAR chock, *accesorio de cubierta* fairlead, *cabos* messenger line, heaving line, TV cueing; **~ de abertura** *f* CINEMAT aperture guide; **~ académica** *f* CINEMAT academy leader; **~ de acercamiento** *f* TEC ESP approach guidance; **~ acimutal inversa** *f* TRANSP AÉR back azimuth guidance; **~ activa** *f* TEC ESP active guidance; **~ alada** *f* ING MECÁ flanged guide; **~ de alambre** *f* C&V wire guide; **~ de alisadora** *f* ING MECÁ honing guide; **~ de amura** *f* TRANSP MAR *construcción* bow chock; **~ de aproximación** *f* TEC ESP approach guidance; **~ de la banda** *f* IMPR web guide; **~ de la barra de tracción** *f* FERRO *vehículos* drawbar guide; **~ blanca** *f* CINEMAT white leader; **~ de bruñidora** *f* ING MECÁ honing guide; **~ de cable** *f* VEH *controles* cable guide; **~ de cadena** *f* ING MECÁ, VEH *motor* chain guide; **~ de cinta** *f* TV tape guide; **~ de cojinete de bolas** *f* ING MECÁ ball-bearing guideway; **~ de cojinete falso** *f* ING MECÁ dummy bearing race; **~ para cojinetes** *f* ING MECÁ guide; **~ de cola** *f* CINEMAT tail leader; **~ de corte** *f* ING MECÁ cutting-off slide; **~ de la cruceta** *f* ING MECÁ crosshead guide; **~ delantera** *f* IMPR front guide; **~ de deslizamiento** *f* MECÁ runner, slide; **~ de elevación** *f* TRANSP AÉR elevation guidance; **~ de elevación de aproximación** *f* TRANSP AÉR approach elevation guidance; **~ de entrada** *f* CINEMAT head leader, IMPR guide pin, TRANSP AÉR fairlead; **~ de entrada de la cinta** *f* TV tape input guide; **~ estándar** *f* CINEMAT, TV standard leader; **~ de exposiciones** *f* FOTO exposure-calculating chart; **~ de extensión** *f* CINEMAT extension lead; **~ de extracción** *f* MINAS *sondeos* hoisting plug; **~ de fin de cinta** *f* CINEMAT run-out leader; **~ final** *f* CINEMAT, TV end leader; **~ frontal** *f* IMPR front lay; **~ hembra** *f* TV female guide; **~ de hilatura de porcelana** *f* C&V porcelain thread-guide; **~ del husillo eyector del casquillo de colada** *f* ING MECÁ sprue bush; **~ inercial** *f* TEC ESP inertial guidance; **~ iterativa** *f* TEC ESP iterative guidance; **~ lateral** *f* IMPR lay edge, TRANSP lateral guidance; **~ en línea** *f* ING MECÁ straight fence; **~ de la línea del cuadro** *f* CINEMAT, TV frame-line leader; **~ de lubricación** *f* MECÁ lubrication chart; **~ luminosa** *f* FÍS light guide; **~ de máquina** *f* CINEMAT, TV machine leader; **~ marginal** *f* CINEMAT edge guide; **~ de montaje** *f* PROD mounting bezel; **~ negra** *f* CINEMAT, TV black leader; **~ obturadora** *f* FERRO *infraestructura* link block guide; **~ oficial de horarios** *f* FERRO official timetable; **~ de onda** *f* FÍS ONDAS wavefront; **~ de**

onda regresiva *f* TELECOM backward-wave guide; **~ de ondas** *f* FÍS, FÍS ONDAS, ING ELÉC, TELECOM, TRANSP MAR *radar* waveguide; **~ de ondas adaptadas** *f* ELEC, ING ELÉC matched waveguide; **~ de ondas de alimentación** *f* ELEC, FÍS ONDAS, ING ELÉC feed waveguide; **~ de ondas calibrada** *f* ELEC, FÍS ONDAS, ING ELÉC slotted waveguide; **~ de ondas circular** *f* FÍS, FÍS ONDAS, ING ELÉC circular waveguide; **~ de ondas coplanar** *f* FÍS, FÍS ONDAS coplanar waveguide, ING ELÉC clopanar waveguide; **~ de ondas elástica** *f* FÍS, FÍS ONDAS, ING ELÉC, TEC ESP flexible waveguide; **~ de ondas en espiral** *f* TELECOM spiral waveguide; **~ de ondas flexible** *f* FÍS ONDAS, ING ELÉC, TEC ESP flexible waveguide; **~ de ondas no recíproca** *f* TELECOM nonreciprocal wave guide; **~ de ondas óptica** *f* ÓPT, TELECOM optical waveguide; **~ de ondas óptica de película delgada** *f* ÓPT thin-film optical waveguide; **~ de ondas paralelepipédica** *f* FÍS, FÍS ONDAS, ING ELÉC rectangular waveguide; **~ de ondas de película delgada** *f* TELECOM thin-film waveguide; **~ de ondas plana** *f* ING ELÉC planar waveguide; **~ de ondas progresivas** *f* TELECOM traveling waveguide *(AmE)*, travelling waveguide *(BrE)*; **~ de ondas rectangular** *f* FÍS, FÍS ONDAS, ING ELÉC rectangular waveguide; **~ de ondas con resalte** *f* ING ELÉC ridge waveguide; **~ de ondas tabicada** *f* ING ELÉC ridge waveguide; **~ opaca** *f* CINEMAT, TV opaque leader; **~ de orden** *f* INFORM&PD command language; **~ pasiva** *f* TRANSP AÉR homing passive guidance; **~ con patín** *f* ING MECÁ flanged guide; **~ de la película** *f* CINEMAT film channel, INSTR *cámara* film camera; **~ de película** *f* CINEMAT, FOTO film leader; **~ de pilón** *f* PROD stamp guide; **~ de presión** *f* CINEMAT pressure guide; **~ para el protector del objetivo** *f* FOTO lens cover slide; **~ de proyección SMPTE** *f* CINEMAT SMPTE projection leader; **~ rectilínea** *f* ING MECÁ crosshead guide, PROD *sierras* parallel fence; **~ de registro** *f* IMPR register gage *(AmE)*, register gauge *(BrE)*; **~ de resorte** *f* PROD spring guide; **~ retenida** *f AmL (cf plácer Esp)* MINAS *explotación de aluviones* placer; **~ de revelado** *f* FOTO developing reel *(AmE)*, developing spiral *(BrE)*; **~ de rolete** *f* TRANSP MAR *accesorio de cubierta* roller fairlead; **~ de salida de cintas** *f* TELECOM tape output guide; **~ semiactiva** *f* TRANSP AÉR homing semi-active guidance; **~ de sierra** *f* ING MECÁ saw guide; **~ de tela** *f* PAPEL wire guide; **~ telefónica** *f Esp (cf directorio telefónico AmL)* TELECOM telephone directory; **~ transparente** *f* CINEMAT, TV clear leader; **~ universal** *f* CINEMAT, TV universal leader; **~ del usuario** *f* TELECOM user guide; **~ de válvula** *f* AUTO, INSTAL HIDRÁUL, VEH *motor* valve guide; **~ de válvula de aguja** *f* AUTO, VEH needle valve guide; **~ para varillas** *f* PROD *sondeos* rod guide; **~ vectorial por radar** *f* D&A, TRANSP, TRANSP AÉR, TRANSP MAR radar vectoring

Guía: ~ de Explotación del Sistema de Satélites *f (SSOG)* TEC ESP Satellite System Operation Guide *(SSOG)*

guiacabos *m* TRANSP MAR *equipamiento de cubierta* fairlead

guiacinta: ~ de alineamiento *f* TV tape alignment guide

guiacojinetes: ~ de roscar *m* PROD *material de tornillos* die guide

guiaderas: ~ **de la jaula de extracción** *f pl* MINAS *pozos* cage guides

guiado *m* TEC ESP active guidance; ~ **astronómico** *m* TEC ESP celestial guidance; ~ **estelar** *m* TEC ESP stellar guidance; ~ **por inercia** *m* TEC ESP inertial guidance; ~ **por radar** *m* TRANSP, TRANSP AÉR, TRANSP MAR radar homing; ~ **por referencia celeste** *m* TEC ESP celestial guidance

guiador *m* ING MECÁ guide, guiding, MINAS *de la jaula* guide

guiante *m* ING MECÁ guiding

guíaondas *m* FÍS, FÍS ONDAS, ING ELÉC, TELECOM, TRANSP MAR *radar* waveguide; ~ **cilíndrico** *m* ING ELÉC wave duct; ~ **de cresta doble** *m* ING ELÉC double ridge waveguide; ~ **en espiral** *m* TELECOM spiral waveguide; ~ **por hélice** *m* ING ELÉC helix waveguide; ~ **monoconductor** *m* ING ELÉC uniconductor waveguide; ~ **óptico** *m* ING ELÉC, ÓPT, TELECOM optical waveguide; ~ **de película delgada** *m* TELECOM thin-film waveguide; ~ **revirado** *m* ING ELÉC twisted waveguide

guías: ~ **aéreas** *f pl* ING MECÁ aeroslides; ~ **de colores** *f pl* COLOR, IMPR color guides (*AmE*), colour guides (*BrE*); ~ **de entrada** *f pl* IMPR feed guides; ~ **de la jaula** *f pl* MINAS cage slides, *pozo* pit guides; ~ **del martinete** *f pl* CONST guidepoles

guijarro *m* GEOL cobble, PETROL pebble

guijarros *m pl* CARBÓN cobbles, TRANSP MAR *playa, orilla* shingle bank

guillame *m* CONST rabbet plane, *carpintería* fillister

guillotina *f* IMPR cutter, paper cutter, print-cutter, SEG guillotine; ~ **de dos cuchillas** *f* PAPEL dual-knife cutter

guillotinado *m* PAPEL guillotining

guiñada *f* ENERG RENOV, TEC ESP, TRANSP AÉR yaw

guiñar[1] *vt* ENERG RENOV, TEC ESP, TRANSP AÉR yaw

guiñar[2] *vi* ENERG RENOV, TEC ESP, TRANSP AÉR yaw

guinche *m* CONST windlass, ING MECÁ hoisting gear, PROD hoist, TRANSP MAR *accesorios de cubierta* winch; ~ **de manivela** *m* PROD hand winch; ~ **de mano** *m* PROD hand winch

guinda *f* TRANSP MAR *diseño naval* air draft (*AmE*), air draught (*BrE*)

guindaleza *f* ING MECÁ, TRANSP MAR hawser; ~ **del virador** *f* TRANSP MAR messenger line

guindar *vt* PROD hoist

guindola *f* TRANSP MAR boatswain's chair

guinea: ~ **de gallina** *f* AGRIC chickencorn

guión *m* CINEMAT scenario, IMPR dash, hyphen, INFORM&PD script; ~ **cinematográfico** *m* CINEMAT screenplay, script; ~ **completo** *m* CINEMAT *para el rodaje* shooting script; ~ **definitivo de rodaje** *m* CINEMAT, TV final-shooting script

guiones: ~ **conductores** *m pl* IMPR hyphen ladders

guionista *m* CINEMAT, TV script writer

guitarra: ~ **del cigüeñal** *f* AUTO, ING MECÁ, VEH crankweb

gulónico *adj* QUÍMICA gulonic

gulosa *f* QUÍMICA gulose

gummita *f* MINERAL gummite

gunita *f* CONST gunite

gunitado *m* CONST guniting

gusanillo: ~ **de rosca** *m* ING MECÁ *barrenas* auger bit

gusano: ~ **cortador** *m* AGRIC army cutworm, cutworm; ~ **del ejército** *m* AGRIC army worm; ~ **de la miasis** *m* AGRIC screw worm; ~ **militar** *m* AGRIC army worm; ~ **de las yemas** *m* AGRIC bud worm

gutapercha *f* P&C, QUÍMICA gutta-percha

guyot *m* GEOL, OCEAN guyot

Gy *abr* (*gray*) GEN Gy (*gray*)

H

h *abr* FÍS (*constante de Planck*) h (*Planck's constant*), METR (*hecto*) h (*hecto*)

H *abr* GEN (*hidrógeno*) H (*hydrogen*)

ha *abr* (*hectárea*) METR ha (*hectare*)

Ha *abr* (*hahnio*) FÍS RAD, QUÍMICA Ha (*hahnium*)

habilidad: **~ para aumentar de peso** *f* AGRIC gain ability

habilitación *f* CONST fitting out, ELECTRÓN enabling; **~ de vuelo por instrumentos para piloto único** *f* TRANSP AÉR single pilot instrument rating

habilitar *vt* ELECTRÓN, INFORM&PD, PROD *fuerzas, energías* enable

habitación *f* CONST room; **~ insonorizada** *f* ACÚST dead room; **~ de inspección de temperatura regulada** *f* ING MECÁ temperature-controlled inspection room

habitáculo *m* AUTO, VEH *del coche* interior; **~ subacuático** *m* OCEAN underwater habitat; **~ submarino** *m* OCEAN undersea habitat

hábitat *m* PETROL habitat

hábito: **~ cristalino** *m* CRISTAL habit

hacer[1]: **~ una bola de** *vt* CARBÓN pile; **~ una copia de reserva de** *vt* INFORM&PD back up; **~ entrega de** *vt* TRANSP MAR put into commission; **~ estanco** *vt* PROD insulate, TRANSP MAR *construcción naval* seal; **~ una evaluación de** *vt* SEG make a valuation of; **~ firme** *vt* TRANSP MAR *botavara* make fast; **~ flotar** *vt* FÍS float; **~ fluir** *vt* ING MECÁ make flush; **~ funcionar** *vt* MECÁ drive, PROD operate, *máquinas* run, VEH *motor* run in; **~ hermético** *vt* TRANSP MAR *construcción naval* seal; **~ listados de** *vt* INFORM&PD list; **~ a medida** *vt* TEXTIL customize; **~ pasar por un motón** *vt* TRANSP MAR *cabo* reeve; **~ salir** *vt* ING MECÁ make flush; **~ una vista detallada de** *vt* FOTO close up; **~ volcar** *vi* TRANSP MAR *embarcación* capsize

hacer[2]: **~ explosión** *vti* TERMO explode

hacer[3]: **~ agua** *vi* TRANSP MAR make water; **~ aguada** *vi* TRANSP MAR *barco* take on water; **~ un agujero** *vi* MINAS *sondeo* hole; **~ asientos nuevos** *vi* VEH *válvulas* reseat; **~ asientos nuevos** *vi* AUTO reseat; **~ barricadas** *vi* CONST barricade; **~ blanco** *vi* D&A score a hit; **~ un browse** *vi* INFORM&PD browse; **~ coincidir exactamente** *vi* ING MECÁ register; **~ comba** *vi* ING MECÁ bulge; **~ converger** *vi* FÍS ONDAS *haz de luz* focus; **~ copias** *vi* CINEMAT print-up; **~ corresponder** *vi* ING MECÁ register; **~ un crucero** *vi* TRANSP MAR cruise; **~ efecto** *vi* ING MECÁ take effect; **~ espuma** *vi* PROC QUÍ make foam; **~ excavaciones** *vi* AmL (*cf barrenar Esp*) MINAS *perforación* bore; **~ foco** *vi* CINEMAT adjust focus; **~ un hoyo** *vi* MINAS hole; **~ huecos en forma de panel alveolar** *vi* CONST honeycomb; **~ interfase** *vi* INFORM&PD interface; **~ una investigación** *vi* CONST make a survey; **~ una junta hermética** *vi* PROD make a tight joint; **~ un levantamiento topográfico** *vi* CONST make a survey; **~ una lista** *vi* INFORM&PD list; **~ una maniobra** *vi* TRANSP MAR maneuver (*AmE*), manoeuvre (*BrE*); **~ maniobras** *vi* AmL (*cf perforar Esp*) MINAS bore,

drill; **~ muescas** *vi* CINEMAT notch; **~ panorámicas** *vi* CINEMAT pan; **~ panorámicas hacia abajo** *vi* CINEMAT pan down; **~ un recorrido** *vi* TRANSP MAR *navegación* overhaul; **~ ruido** *vi* ING MECÁ knock; **~ saltar un arco** *vi* ELEC, TERMO flash over; **~ una señal** *vi* TRANSP MAR signal; **~ sondeos** *vi* CONST bore for water, MINAS costean; **~ sondeos para encontrar agua** *vi* AGUA bore for water; **~ una toma panorámica** *vi* CINEMAT pan; **~ trabajos de mina** *vi* MINAS mine; **~ vela** *vi* TRANSP MAR make sail; **~ un viaje** *vi* TEC PETR *perforación* make a round trip; **~ una vista detallada** *vi* CINEMAT close up

hacerse: **~ crítico** *v refl* NUCL go critical; **~ líquido** *v refl* QUÍMICA deliquesce; **~ a la mar** *v refl* TRANSP MAR put to sea; **~ a la vela** *v refl* TRANSP MAR set sail

haces: **~ de banda ancha** *m pl* FÍS RAD wideband beams; **~ de electrones y positrones de alta energía** *m pl* FÍS PART, NUCL high-energy electron-positron beams; **~ electrón-positrón hiperenergéticos** *m pl* FÍS PART, NUCL high-energy electron-positron beams; **~ iónicos convergentes** *m pl* FÍS RAD focused ion beams; **~ iónicos de elevada intensidad** *m pl* FÍS RAD high-intensity ion beams; **~ de kaones de alta energía** *m pl* FÍS PART high-energy kaon beams; **~ mesón K hiperenergéticos** *m pl* FÍS PART high-energy kaon beams; **~ de mesones K de alta energía** *m pl* FÍS PART high-energy kaon beams; **~ muón hiperenergéticos** *m pl* FÍS PART high-energy muon beams; **~ de muones de alta energía** *m pl* FÍS PART high-energy muon beams

hacha *f* CONST cleaver; **~ para cortar pizarra** *f* CONST slate ax (*AmE*), slate axe (*BrE*); **~ hendedora** *f* CONST splitting ax (*AmE*), splitting axe (*BrE*); **~ de mano** *f* PROD hatchet

hacienda *f* AGRIC cattle ranch

hadrón *m* FÍS, FÍS PART, FÍS RAD hadron

hafnio *m* (*Hf*) METAL, QUÍMICA hafnium (*Hf*)

hahnio *m* (*Ha, Hn*) FÍS RAD, QUÍMICA hahnium (*Ha, Hn*)

halar[1] *vt* TRANSP MAR *cabo, red* haul; **~ a bordo** *vt* OCEAN haul on board

halar[2]: **~ de las cuerdas de amarre** *vi* TRANSP MAR heave in the mooring ropes

halita *f* MINERAL, QUÍMICA halite, rock salt

hallazgo *m* AmL (*cf rumbo del filón Esp*) MINAS *de filón subterráneo* strike

halloysita *f* MINERAL halloysite

halmirolisis *f* GEOL halmyrolysis

halo *m* ELECTRÓN *tubo catódico* halation, halo, IMPR, TEC ESP halo; **~ de dispersión** *m* NUCL halo of dispersion; **~ eléctrico** *m* ELEC, TEC ESP *fenómenos de ionización* electroglow

halogenación *f* QUÍMICA halogenation

halógeno[1] *adj* QUÍMICA halogenous

halógeno[2] *m* QUÍMICA halogen

halogenuro *m* QUÍMICA halide

halografía *f* QUÍMICA halography

haloide *m* QUÍMICA haloid

haloideo *adj* QUÍMICA haloid

halón *m* TEC ESP halon
haloquinesis *f* GEOL, TEC PETR halokinesis
halotecnia *f* QUÍMICA halotechny
halotriquita *f* MINERAL halotrichite
haluro *m* QUÍMICA halide; ~ **alcalino** *m* CRISTAL alkali halide; ~ **de plata** *m* FOTO silver halide
hamiltoniano *adj* FÍS Hamiltonian
hangar *m* TRANSP AÉR hangar
haploide *adj* AGRIC haploid
hardware *m* INFORM&PD, PETROL, TELECOM hardware
harina: ~ **de arroz** *f* AGRIC, ALIMENT rice flour; ~ **blanqueada** *f* ALIMENT bleached flour; ~ **entera** *f* ALIMENT wheatmeal; ~ **fósil** *f* HIDROL diatomaceous earth; ~ **de germen de maíz** *f* AGRIC maize germ meal; ~ **de gluten** *f* AGRIC gluten meal; ~ **de grano grueso** *f* AGRIC coarse meal; ~ **de huesos y carne** *f* AGRIC meat-and-bone meal; ~ **de madera** *f* P&C wood flour; ~ **de maíz** *f* AGRIC corn meal (*AmE*), Indian meal (*BrE*); ~ **patentada** *f* ALIMENT patent flour; ~ **de plumas** *f* AGRIC feather meal; ~ **de sangre** *f* AGRIC blood meal; ~ **de trigo sin cerner** *f* AGRIC graham flour; ~ **de trigo sin refinar** *f* ALIMENT wheatmeal
harinoso *adj* ALIMENT farinaceous
harmalina *f* QUÍMICA harmaline
harmina *f* QUÍMICA harmine
harmotoma *m* GEOL, MINERAL harmotome; ~ **cálcico** *m* GEOL, MINERAL lime harmotome
harnero *m* CARBÓN cribble
hartita *f* MINERAL hartite
hastial *m* AmL (*cf costero Esp*) CONST gable, MINAS stone, side, side wall, *minería del carbón* cheeks
hatchetina *f* MINERAL hatchetine, mountain tallow
hato *m* TEXTIL cluster
hauerita *f* MINERAL hauerite
hausmannita *f* MINERAL hausmannite
Hauteriviense *m* GEOFÍS Hauterivian
hauyna *f* MINERAL hauyne, hauynite
hayada *f* GEOL, MINAS sinkhole
hayesina *f* MINERAL hayesine
haz *m* GEN beam, C&V *de fibras de vidrio* bundle, METAL jet; ~ **en abanico** *m* ELECTRÓN fan beam; ~ **de acumulación** *m* ELECTRÓN holding beam; ~ **atómico** *m* ELECTRÓN, NUCL atomic beam; ~ **azul** *m* ELECTRÓN, TV blue beam; ~ **de baja potencia** *m* ELECTRÓN low-energy beam; ~ **de cables** *m* ELEC, ELECTRÓN, ING ELÉC bunched cable (*BrE*), bundled cable (*AmE*), cable bundle; ~ **catódico** *m* ELECTRÓN, ING ELÉC, TV cathode-ray; ~ **de cátodos** *m* ELECTRÓN, ING ELÉC, TV cathode beam; ~ **de coherencia temporal** *m* ELECTRÓN temporally-coherent beam; ~ **de conductores** *m* ELEC, ELECTRÓN, ING ELÉC bunched conductor (*BrE*), bundled conductor (*AmE*); ~ **continuo** *m* ELECTRÓN continuous beam; ~ **convergente** *m* ELECTRÓN, FÍS ONDAS focused beam, ÓPT convergent beam; ~ **difractado** *m* CRISTAL diffracted beam; ~ **direccional** *m* ELECTRÓN, TELECOM, TV directional beam; ~ **dirigido** *m* FÍS ONDAS focused beam; ~ **divergente** *m* TV deflected beam; ~ **de electrones** *m* ELEC, ELECTRÓN, FÍS ONDAS, FÍS PART, INFORM&PD, METAL, NUCL, QUÍMICA, TELECOM, TV electron beam; ~ **electrónico** *m* ELEC, ELECTRÓN, FÍS ONDAS, FÍS PART, INFORM&PD, METAL, NUCL, QUÍMICA, TELECOM, TV electron beam; ~ **electrónico estrecho** *m* ELECTRÓN pencil beam; ~ **electrónico**

de exploración *m* ELECTRÓN scanning electron beam; ~ **eliminado** *m* ELECTRÓN blanked beam; ~ **enfocado** *m* FÍS ONDAS focused beam; ~ **de exploración de imagen** *m* ELECTRÓN raster-scanned beam; ~ **de exploración vectorial** *m* ELECTRÓN vector-scanned beam; ~ **explorador** *m* ELECTRÓN, TV scanning beam; ~ **explorador de electrones** *m* ELECTRÓN, TV electron scanning beam; ~ **explorador de retorno** *m* ELECTRÓN, TV return scanning-beam; ~ **extraído** *m Esp* (*cf haz eyectado AmL*) NUCL ejected beam; ~ **eyectado** *m AmL* (*cf haz extraído Esp*) NUCL ejected beam; ~ **de fibras** *m* ÓPT, TELECOM fiber bundle (*AmE*), fibre bundle (*BrE*); ~ **de fibras fundido** *m* C&V fused bundle; ~ **filiforme** *m* FÍS RAD pencil beam; ~ **gaussiano** *m* ÓPT, TELECOM Gaussian beam; ~ **global** *m* TEC ESP global beam; ~ **de gran potencia** *m* ELECTRÓN high-energy beam; ~ **guía** *m* TRANSP guide beam; ~ **hertziano** *m* ELECTRÓN, TV Hertzian beam; ~ **homocéntrico** *m* FÍS homocentric beam; ~ **ILS** *m* TEC ESP ILS beam; ~ **incidente** *m* CRISTAL, FÍS incident beam; ~ **de inundación** *m* CINEMAT flooded beam; ~ **de iones** *m* ELECTRÓN, FÍS RAD ion beam; ~ **iónico** *m* ELECTRÓN, FÍS RAD ion beam; ~ **iónico convergente de baja energía** *m* ELECTRÓN, FÍS RAD low-energy focused ion beam; ~ **de láser** *m* ELECTRÓN, FÍS ONDAS, FÍS RAD, NUCL, TELECOM laser beam; ~ **de láser de alta radiación** *m* ELECTRÓN high-irradiance laser beam; ~ **de láser continuo** *m* ELECTRÓN continuous-laser beam; ~ **láser de exploración** *m* ELECTRÓN, ÓPT scanning laser beam; ~ **de láser de funcionamiento continuo** *m* ELECTRÓN CW laser beam; ~ **lasérico** *m* ELECTRÓN, FÍS ONDAS, FÍS RAD, NUCL, TELECOM laser beam; ~ **lasérico de exploración** *m* ELECTRÓN, ÓPT scanning laser beam; ~ **de lectura** *m* ELECTRÓN, INFORM&PD, ÓPT, TELECOM read beam; ~ **localizador** *m* TRANSP AÉR localizer beam; ~ **luminoso** *m* CINEMAT, ELECTRÓN, FÍS, FOTO, ÓPT light beam, TEC ESP spot beam, TELECOM light beam; ~ **de luz** *m* CONST beam; ~ **de luz colimado** *m* ELECTRÓN collimated beam; ~ **de luz láser** *m* FÍS RAD laser light beam; ~ **de microondas** *m* TELECOM microwave beam; ~ **modulado** *m* ELECTRÓN modulated beam; ~ **molecular** *m* TELECOM molecule beam; ~ **monocromático coherente** *m* FÍS RAD coherent monochromatic beam; ~ **de neutrones** *m* FÍS PART neutron beam; ~ **paralelo** *m* ELECTRÓN, FÍS parallel beam; ~ **de partículas** *m* ELECTRÓN, FÍS PART particle beam; ~ **de partículas cargado** *m* ELECTRÓN, FÍS PART charged-particle beam; ~ **perfilado** *m* ELECTRÓN shaped beam; ~ **principal** *m* TEC ESP main beam; ~ **de protones** *m* ELECTRÓN, FÍS PART proton beam; ~ **pulsado de electrones** *m* ELECTRÓN, FÍS RAD pulsed-electron gun; ~ **de radar** *m* FÍS, FÍS RAD, TRANSP, TRANSP AÉR, TRANSP MAR radar beam; ~ **radárico** *m* FÍS, FÍS RAD, TRANSP, TRANSP AÉR, TRANSP MAR radar beam; ~ **radioeléctrico fino** *m* TEC ESP spot beam; ~ **de rayos** *m* TELECOM beam; ~ **de rayos gamma** *m* FÍS PART, FÍS RAD gamma-ray beam; ~ **de rayos X** *m* ELECTRÓN X-ray beam; ~ **reflejado** *m* FÍS, ÓPT, TRANSP reflected beam; ~ **rojo** *m* ELECTRÓN, TV red beam; ~ **de salida** *m* TRANSP AÉR outbound beam; ~ **transmitido** *m* FÍS AÉR transmitted beam; ~ **transversal de baja energía** *m* FÍS RAD low-energy transverse jet; ~ **de trayectoria de planeo** *m* TRANSP, TRANSP AÉR glide-path beam;

~ tubular *m* NUCL tube bundle; **~ de urdimbre** *m* C&V beamed yarn; **~ de velocidad modulada** *m* ELECTRÓN velocity-modulated beam; **~ verde** *m* ELECTRÓN, TV green beam

HDX *abr* (*semi-dúplex*) INFORM&PD HDX (*half-duplex*)

He *abr* (*helio*) FÍS RAD, NUCL, QUÍMICA, REFRIG, TEC PETR He (*helium*)

hebilla *f* ING MECÁ, PETROL buckle

hebra *f* C&V strand, thread, ÓPT strand, P&C roving, PROD thread, TEXTIL filament

hebronita *f* MINERAL hebronite

heces *f pl* HIDROL, RECICL faeces (*BrE*), feces (*AmE*)

hecho *adj* ING MECÁ, PROD shaped; **~ de encargo** *adj* EMB custom-designed, custom-made; **~ a medida** *adj* TEXTIL made-to-measure

hectárea *f* (*ha*) METR hectare (*ha*)

hecto *m* (*h*) METR hecto (*h*)

hectogramo *m* (*hg*) METR hectogram (*hg*)

hectólitro *m* (*hl*) METR hectoliter (*AmE*) (*hl*), hectolitre (*BrE*) (*hl*)

hectómetro *m* (*hm*) METR hectometer (*AmE*) (*hm*), hectometre (*BrE*) (*hm*)

hectovatio *m* (*hW*) ELEC, ING ELÉC, METR hectowatt (*hW*)

hedenbergita *f* MINERAL hedenbergite

hedifana *f* MINERAL hedyphane

helada *f* GEOL, METEO frost, silver frost

heladera *f* ALIMENT, ELEC, ING MECÁ, REFRIG, TERMO refrigerator

helado: **~ de leche** *m* REFRIG ice milk

heliantina *f* QUÍMICA helianthin, helianthine

hélice *f* CARBÓN auger, CONTAM MAR propeller, GEOM helix, ING MECÁ helix, screw, spiral, MECÁ screw, ÓPT QUÍMICA helix, TRANSP AÉR airscrew, propeller, TRANSP MAR propeller; **~ abierta** *f* TRANSP open propeller; **~ de aire de paso variable** *f* TRANSP variable-pitch air propeller; **~ amortajada** *f* TRANSP, TRANSP AÉR shrouded propeller; **~ antitorsión de cola** *f* TRANSP AÉR antitorque propeller; **~ canalizada** *f* TRANSP ducted propeller; **~ carenada** *f* TRANSP carinated propeller; **~ coaxial** *f* TRANSP AÉR coaxial propeller; **~ de cola** *f* TRANSP tail propeller; **~ contrarrotatoria** *f* TRANSP counter-rotating propeller; **~ elevadora** *f* TRANSP lift fan; **~ de la fibra** *f* ÓPT fiber helix (*AmE*), fibre helix (*BrE*); **~ impulsora** *f* TRANSP driving propeller, TRANSP AÉR pusher propeller; **~ loca** *f* TRANSP AÉR windmilling propeller; **~ marina** *f* TRANSP marine propeller; **~ mezcladora** *f* PROC QUÍ mixing propeller; **~ en molinete** *f* TRANSP AÉR windmilling propeller; **~ de palas orientables** *f* TRANSP MAR controllable pitch propeller; **~ de paso controlable** *f* TRANSP MAR controllable pitch propeller; **~ de paso fijo** *f* TRANSP AÉR fixed-pitch propeller; **~ con paso nulo** *f* CONTAM MAR zero-pitch propeller; **~ de paso regulable** *f* CONTAM MAR, TRANSP AÉR adjustable-pitch propeller; **~ de paso reversible** *f* TRANSP AÉR, TRANSP MAR reversible-pitch propeller; **~ de paso variable** *f* CONTAM MAR, TRANSP AÉR, TRANSP MAR variable-pitch propeller; **~ plegable** *f* TRANSP MAR folding propeller; **~ de propulsión directa** *f* TRANSP AÉR direct drive propeller; **~ Schottel** *f* TRANSP Schottel propeller; **~ supercavitante** *f* TRANSP supercavitating propeller; **~ en tobera fija** *f* TRANSP ducted propeller; **~ tractora** *f* TRANSP AÉR tractor propeller; **~ trasera** *f* TRANSP, TRANSP MAR rear propeller; **~ de velocidad**

constante *f* TRANSP AÉR constant-speed propeller; **~ ventilada** *f* TRANSP ventilated propeller

helicina *f* QUÍMICA helicin

helicoidal *adj* GEN helical, helicoidal

helicoide[1] *adj* ING MECÁ helicoid

helicoide[2] *m* GEOM helicoid

helicoideo *adj* ING MECÁ helicoid

helicóptero *m* TRANSP AÉR helicopter; **~ antisubmarino** *m* TRANSP AÉR antisubmarine helicopter; **~ antitanque** *m* TRANSP AÉR antitank helicopter; **~ de aplicación general** *m* TRANSP AÉR multipurpose helicopter; **~ armado** *m* D&A helicopter gunship; **~ de ataque** *m* D&A attack helicopter; **~ con base en portaviones** *m* TRANSP AÉR carrier-ship-borne helicopter; **~ con base en portahelicópteros** *m* TRANSP AÉR carrier-ship-borne helicopter; **~ de carga** *m* TRANSP AÉR cargo helicopter; **~ coaxial** *m* TRANSP AÉR coaxial helicopter; **~ de combate** *m* D&A, TRANSP AÉR combat helicopter; **~ compuesto** *m* TRANSP AÉR compound helicopter; **~ de doble rotor** *m* TRANSP AÉR dual-rotor helicopter, twin-rotor helicopter; **~ de ejecutivo** *m* TRANSP AÉR executive helicopter; **~ experimental** *m* TRANSP AÉR experimental helicopter; **~ grúa** *m* TRANSP AÉR crane helicopter; **~ ligero multimisión** *m* TRANSP AÉR light multi-role helicopter; **~ multimotor** *m* TRANSP AÉR multiengine helicopter; **~ de observación ligero** *m* TRANSP AÉR light observation helicopter; **~ pesado** *m* TRANSP AÉR heavy-lift helicopter; **~ polimotor** *m* TRANSP AÉR multiengine helicopter; **~ polivalente** *m* TRANSP AÉR multipurpose helicopter; **~ radiodirigido** *m* TRANSP AÉR drone helicopter; **~ a reacción** *m* TRANSP AÉR jet helicopter; **~ de rescate** *m* D&A, SEG, TRANSP AÉR rescue helicopter; **~ de rotores en tándem** *m* TRANSP, TRANSP AÉR tandem rotor helicopter; **~ de salvamento** *m* D&A, SEG, TRANSP AÉR rescue helicopter; **~ teledirigido** *m* TRANSP AÉR drone helicopter; **~ todo tiempo** *m* TRANSP AÉR all-weather helicopter; **~ todo uso** *m* TRANSP AÉR multipurpose helicopter; **~ de transporte** *m* TRANSP AÉR transport helicopter; **~ de transporte y rescate** *m* TRANSP, TRANSP AÉR transport and rescue helicopter; **~ tripulado** *m* TRANSP AÉR manned helicopter

helicotrema *m* ACÚST helicotrema

helimagnetismo *m* FÍS helimagnetism

helio *m* (*He*) FÍS RAD, NUCL, QUÍMICA, REFRIG, TEC PETR helium (*He*); **~ líquido** *m* TERMO liquid helium

heliodoro *m* MINERAL heliodor

helión *m* ELEC alpha particle

heliopila *f* ELEC, TEC ESP, TELECOM solar cell

helioscopio: **~ de polarización** *m* INSTR polarizing sun prism

heliotrópico *adj* ENERG RENOV heliotropic

heliotropina *f* QUÍMICA heliotropin

heliotropo *m* MINERAL bloodstone, heliotrope

helipuerto *m* CONTAM MAR, TEC PETR, TRANSP AÉR heliport; **~ de azotea** *m* TRANSP AÉR rooftop heliport

helitransporte *m* TRANSP AÉR heli-lifting

helvina *f* MINERAL helvine

hemateína *f* QUÍMICA haematein (*BrE*), hematein (*AmE*)

hemático *adj* QUÍMICA haematic (*BrE*), hematic (*AmE*)

hematina *f* QUÍMICA haematin (*BrE*), hematin (*AmE*)

hematites *f* MINERAL micaceous iron ore, QUÍMICA

haematite (*BrE*), hematite (*AmE*); ~ **parda** *f* QUÍMICA brown haematite (*BrE*), brown hematite (*AmE*)

hematolita *f* MINERAL haematolite (*BrE*), hematolite (*AmE*)

hematolito *m* MINERAL haematolite (*BrE*), hematolite (*AmE*)

hematoporfirina *f* QUÍMICA haematoporphyrin (*BrE*), hematoporphyrin (*AmE*)

hematoxilina *f* QUÍMICA haematoxylin (*BrE*), hematoxylin (*AmE*)

hembra: ~ **del cerrojo** *f* CONST box staple, ING MECÁ keeper; ~ **del enchufe** *f* ING ELÉC connector socket; ~ **de terraje hexagonal** *f* ING MECÁ hexagonal die nut; ~ **del timón** *f* TRANSP AÉR, TRANSP MAR rudder brace; ~ **para tomacorriente** *f* CINEMAT plug socket

hembrilla: ~ **del portaherramientas** *f* ING MECÁ tool rest

hemicelulosa *f* QUÍMICA hemicellulose

hemiciclo *m* NUCL half cycle

hemimetálico *adj* QUÍMICA hemimetallic

hemimorfita *f* MINERAL hemimorphite

hemipínico *adj* QUÍMICA hemipinic

hemisferio *m* GEOM hemisphere

hemoglobina *f* QUÍMICA haemoglobin (*BrE*), hemoglobin (*AmE*)

hemolisina *f* QUÍMICA haemolysin (*BrE*), hemolysin (*AmE*)

hemólisis *f* QUÍMICA haemolysis (*BrE*), hemolysis (*AmE*)

hemolítico *adj* QUÍMICA haemolytic (*BrE*), hemolytic (*AmE*)

hemopirrol *m* QUÍMICA haemopyrrole (*BrE*), hemopyrrole (*AmE*)

hemosiderina *f* QUÍMICA haemosiderin (*BrE*), hemosiderin (*AmE*)

hemotoxina *f* QUÍMICA haemotoxin (*BrE*), hemotoxin (*AmE*)

hendedor *m* IMPR creaser

hendido *m* PAPEL slitting

hendidura *f* AGUA notch, C&V split, CARBÓN crack, crevice, CONST split, splitting, MECÁ crack, notch, METAL cleavage, notch, crack, PROD notch, groove, slit, split, splitting, TEC ESP slit, TELECOM slot; ~ **anular** *f* C&V *en la cocción de botellas* annular crack; ~ **arriñonada** *f* ING MECÁ kidney-shaped slot; ~ **fina** *f* ING MECÁ fine crack; ~ **de la pizarra** *f* CONST slate splitting; ~ **transversal** *f* TELECOM transverse slot

henificación *f* AGRIC haymaking

henificadora *m* AGRIC haymaker

henil *m* AGRIC hayloft

heno: ~ **de cereales** *m* AGRIC grain hay; ~ **picado** *m* AGRIC ground hay

henrímetro *m* ELEC, INSTR inductance meter

henrio *m* (*H*) ELEC, FÍS, ING ELÉC, METR henry (*H*)

heparina *f* QUÍMICA heparin

hepatita *f* MINERAL hepatite

héptada *f* QUÍMICA heptad

heptaedro *m* GEOM heptahedron

heptagonal *adj* GEOM heptagonal

heptágono *m* GEOM heptagon

heptano *m* QUÍMICA heptane

heptavalente *adj* QUÍMICA heptavalent

hepteno *m* QUÍMICA heptene

heptileno *m* QUÍMICA heptylene

heptílico *adj* QUÍMICA heptylic

heptilo *m* QUÍMICA heptyl

heptino *m* QUÍMICA heptyne

heptodo *m* ELECTRÓN heptode

heptosa *f* QUÍMICA heptose

herbazal: ~ **de marisma** *m* OCEAN sea grass bed

herbicida *m* AGRIC herbicide

herciano *adj* C&V, ELEC, TEC ESP, TV Hertzian

hercinita *f* MINERAL hercynite

hercio *m* (*Hz*) GEN hertz (*Hz*)

herencia *f* INFORM&PD inheritance

herida *f* SEG wound; ~ **por cortadora** *f* SEG cutter wound

herméticamente: ~ **cerrado** *adj* ING ELÉC, MECÁ hermetically-sealed; ~ **obturado** *adj* ING ELÉC, MECÁ hermetically-sealed; ~ **sellado** *adj* ING ELÉC, MECÁ hermetically-sealed

hermeticidad *f* ING MECÁ, MECÁ tightness, PROD imperviousness; ~ **ultrasónica** *f* EMB ultrasonic sealing, TERMO ultrasonic welding

hermético[1] *adj* FÍS watertight, MECÁ leak-tight; ~ **al gas** *adj* GAS, ING MECÁ, MECÁ, PROC QUÍ, PROD, QUÍMICA, TERMO gas-proof, gastight; ~ **a la luz** *adj* CINEMAT light-tight; ~ **al polvo** *adj* ELEC, EMB, ING MECÁ, PROD dust-tight, dustproof

hermético[2]: ~ **contra fuegos** *m* TRANSP AÉR fireseal

hermetizado *adj* ALIMENT, TEC PETR airtight

heroína *f* QUÍMICA heroin

herrado *adj* PROD ironshod

herrajes *m pl* CONST ironwork; ~ **y accesorios** *m pl* CONST furniture and fittings; ~ **de seguridad** *m pl* SEG safety fittings

herramental *m* ING MECÁ outfit, PROD tooling; ~ **aislado** *m* SEG insulated tooling

herramentista *m* PROD toolmaker

herramienta *f* INFORM&PD tool, ING MECÁ bent tool, implement, tool, MECÁ, MINAS tool; ~ **de acabado acodada** *f* ING MECÁ cranked finishing tool; ~ **de acero al carbono** *f* ING MECÁ carbon steel tool; ~ **acodada** *f* ING MECÁ cranked tool; ~ **afilada** *f* ING MECÁ sharp-edged tool; ~ **de afino** *f* INSTR fining lap; ~ **de aire comprimido** *f* ING MECÁ pneumatic tool; ~ **para alisar** *f* PROD *moldería* slick; ~ **de ángulo fijo** *f* ING MECÁ single-angle cutter; ~ **de aristas interiores** *f* ING MECÁ inside corner tool; ~ **biseladora para barras** *f* ING MECÁ bar chamfering tool; ~ **para brochar** *f* ING MECÁ, MECÁ broaching tool; ~ **cónica para tuberías** *f* CONST pipe-coning tool; ~ **de copiar** *f* ING MECÁ forming tool; ~ **cortadora giratoria** *f* GAS rotary cutting tool; ~ **cortadora rotatoria** *f* GAS rotary cutting tool; ~ **de corte** *f* ING MECÁ knife tool, cutter, cutting tool, cutting-off tool, MECÁ cutting tool; ~ **de corte acodada** *f* ING MECÁ cranked knife tool; ~ **de corte acodada para torneado de reproducción** *f* ING MECÁ cranked knife tool for copy turning; ~ **de corte a derechas** *f* ING MECÁ right-hand knife-tool; ~ **de corte de ranuras en T** *f* ING MECÁ T-slot cutter; ~ **para dar forma** *f* C&V shaping tool; ~ **de debastar** *f* ING MECÁ roughing tool; ~ **de desbarbar** *f* ING MECÁ deburring tool; ~ **eléctrica de mano** *f* ELEC, ING MECÁ, SEG hand-held power tool; ~ **eléctrica portátil** *f* ELEC, ING MECÁ hand-held power tool, portable power tool; ~ **de encargo** *f* ING MECÁ one-off tooling; ~ **engarzadora** *f* ING MECÁ crimping tool; ~ **de ensamblar neumática** *f* ING MECÁ pneumatic assembly-tool; ~ **para ensanchar** *f* TEC PETR reamer;

~ **para entallar** *f* PROD carving tool; ~ **escantilladora** *f* INSTR beveling tool (*AmE*), bevelling tool (*BrE*); ~ **para escariar** *f* C&V, ING MECÁ, MECÁ, PROD reamer; ~ **de espigar** *f* ING MECÁ tenon drive; ~ **para estibar** *f* C&V stowing tool; ~ **de fibra de carbón** *f* ING MECÁ carbon fiber tool (*AmE*), carbon fibre tool (*BrE*); ~ **de fibra de carbono** *f* ING MECÁ carbon fiber tool (*AmE*), carbon fibre tool (*BrE*); ~ **para formar colas de milano** *f* ING MECÁ dovetail-form tool; ~ **de fresado** *f* PROD *moleta* milling tool; ~ **de gancho** *f* PROD *tornos* hanging tool; ~ **de giro** *f Esp* (*cf mango de maniobra AmL*) MINAS *sondeos* brace head, brace key, rod-turning tool, tiller; ~ **gruesa** *f* ING MECÁ roughing tool; ~ **para interiores** *f* ING MECÁ cutter bar, MECÁ inside tool; ~ **para lapear** *f* PROD lapping tool; ~ **magnética** *f* CONST magnetic tool rack; ~ **con mango brillante** *f* ING MECÁ tool with bright handle; ~ **de mano** *f* MECÁ, PROD hand tool; ~ **manual** *f* MECÁ, PROD hand tool; ~ **mecánica** *f* AGRIC, MECÁ power tool; ~ **motorizada** *f* MECÁ power tool; ~ **múltiple** *f* ING MECÁ gang tool; ~ **neumática** *f* ING MECÁ pneumatic tool; ~ **no en serie** *f* ING MECÁ one-off tooling; ~ **óptica** *f* C&V, ÓPT optical tool; ~ **óptica cóncava** *f* C&V, ÓPT concave optical tool; ~ **óptica convexa** *f* C&V, ÓPT convex optical tool; ~ **óptica plana** *f* C&V, ÓPT flat optical tool; ~ **de perfilar** *f* ING MECÁ forming tool; ~ **de pesca magnética** *f* TEC PETR *perforación* magnetic fishing tool; ~ **portaocho** *f* PROD staff; ~ **de programación automática** *f* (*APT*) INFORM&PD automatic programming tool (*APT*); ~ **de pulir** *f* ING MECÁ planishing tool; ~ **de punta** *f* ING MECÁ point tool; ~ **con punta adiamantada** *f* CONST, ING MECÁ, MECÁ diamond tool; ~ **con punta de diamante** *f* CONST, ING MECÁ, MECÁ diamond tool; ~ **de punta redonda** *f* ING MECÁ round-nose tool; ~ **con puntas de carburo** *f* ING MECÁ, MECÁ carbide-tipped tool; ~ **de ranurar** *f* ING MECÁ splining tool; ~ **para rebajar** *f* C&V chipping tool; ~ **de refrentar** *f* ING MECÁ facing tool, spotface cutter; ~ **de refrentar acodada** *f* ING MECÁ cranked facing tool; ~ **de refrentar barras** *f* ING MECÁ bar facing-tool; ~ **de refrentar para desbastar** *f* ING MECÁ facing tool for roughing; ~ **de roscar** *f* ING MECÁ thread chaser, threading tool; ~ **de roscar exteriormente** *f* ING MECÁ outside-threading tool; ~ **de roscar interiores** *f* ING MECÁ inside-threading tool; ~ **sacatestigos** *f Esp* (*cf cortanúcleos AmL*) TEC PETR coring tool; ~ **de software** *f* INFORM&PD software tool; ~ **de sondeo** *f* ING MECÁ, MECÁ, MINAS boring tool; ~ **de sujeción** *f* NUCL *de elementos de combustible* gripper tool; ~ **de taladrar en acabado** *f* ING MECÁ, MECÁ, MINAS boring tool; ~ **de taladrar en desbaste** *f* ING MECÁ, MECÁ, MINAS boring tool; ~ **de tornear** *f* ING MECÁ turning tool; ~ **de tornear con placa de acero al carbono** *f* ING MECÁ turning tool with carbide tip; ~ **de tornear y planear de placa de carburo al tungsteno** *f* ING MECÁ tungsten-carbide-tipped turning and planing tool; ~ **de tornear y refrentar acodada** *f* ING MECÁ cranked turning and facing tool; ~ **de torno** *f* ING MECÁ lathe tool; ~ **de torno copiador** *f* ING MECÁ copying lathe tool; ~ **de trocear** *f* ING MECÁ cutting-off tool, parting tool

herramientas: ~ **aisladas** *f pl* SEG insulated tools; ~ **de arrastre** *f pl* MINAS scraping tools; ~ **del bastidor de conductores** *f pl* ING MECÁ lead-frame tooling; ~ **de bicicleta** *f pl* TRANSP bicycle tools; ~ **de bronce** *f pl* PROD bronzeworking tools; ~ **de cepilladora** *f pl* ING MECÁ shaper tools; ~ **de cobre** *f pl* PROD copperworking tools; ~ **de combinación** *f pl* ING MECÁ combination tools; ~ **de construcción** *f pl* CONST builders' hardware; ~ **de copiar** *f pl* ING MECÁ forming tools; ~ **cortantes** *f pl* PROD edge tools; ~ **de corte y embutido** *f pl* ING MECÁ combination tools; ~ **con cuchillas de carburo** *f* ING MECÁ carbide tools; ~ **de desarenado** *f pl* MINAS cleaning tools; ~ **de desatornillar de impresión múltiple** *f pl* ING MECÁ multi-impression unscrewing tools; ~ **para desbastar** *f pl* C&V knob tools; ~ **de embotar** *f pl* ING MECÁ foil tooling; ~ **del freno de disco** *f pl* AUTO, VEH disc brake tools (*BrE*), disk brake tools (*AmE*); ~ **del freno de tambor** *f pl* AUTO, VEH drum brake tools; ~ **de hoja plástica** *f pl* ING MECÁ plastic-foil tooling; ~ **de latón** *f pl* PROD brassworking tools; ~ **de lavado** *f pl Esp* MINAS flushing tools; ~ **de limadora** *f pl* ING MECÁ shaper tools; ~ **de mango paralelo** *f pl* ING MECÁ parallel-shank tools; ~ **de mantenimiento in situ** *f pl* CONST on-site maintenance tools; ~ **de montaje** *f pl* ING MECÁ assembly tools; ~ **múltiples** *f pl* ING MECÁ combination tools; ~ **del panel de la carrocería del automóvil** *f pl* ING MECÁ motor-body panels tooling; ~ **pequeñas de acero de alta velocidad** *f pl* ING MECÁ high-speed steel small tools (*HSS small tools*); ~ **pequeñas de acero de corte rápido** *f pl* ING MECÁ high-speed steel small tools (*HSS small tools*); ~ **pequeñas de alta velocidad** *f pl* ING MECÁ high-speed small tools; ~ **de perfilar** *f pl* ING MECÁ forming tools; ~ **para pinchar** *f pl* C&V pinching tools; ~ **pre-producción** *f pl* ING MECÁ preproduction tooling; ~ **de progresión de varias etapas** *f pl* ING MECÁ multistage progression tooling; ~ **prototipo** *f pl* ING MECÁ prototype tooling; ~ **de rascado** *f pl* MINAS scraping tools; ~ **de reparación común** *f pl* ING MECÁ common-repair tools; ~ **de sangrar** *f pl* ING MECÁ parting tools; ~ **de servicio** *f pl* ING MECÁ service tools; ~ **de tornear** *f pl* ING MECÁ slide-rest tools; ~ **para torno de óxido cerámico** *f pl* ING MECÁ oxide-ceramic lathe tools; ~ **de vidriero** *f pl* C&V glassmaker's tools

herrar *vt* PROD tip

herrería *f* CONST, ING MECÁ, PROD blacksmith's shop, ironworking, smithery

herrero *m* CONST, ING MECÁ blacksmith, smith, PROD blacksmith, chainsmith

herrete *m* PROD tag

herrumbre *f* ALIMENT, AUTO, MECÁ, QUÍMICA, TRANSP MAR, VEH rust

herrumbrosidad *f* ALIMENT, AUTO, MECÁ, QUÍMICA, TRANSP MAR, VEH rustiness

hertz *m* (*Hz*) GEN hertz (*Hz*)

hertziano *adj* C&V, ELEC, TEC ESP, TV Hertzian

hervido *m* ALIMENT, REFRIG, TERMO boiling

hervidor *m* ALIMENT, REFRIG, TERMO boiler

hervir *vti* ALIMENT, FÍS, TERMO boil; ~ **a borbotones** *vti* TERMO boil fast; ~ **a fuego lento** *vti* ALIMENT, TERMO simmer; ~ **sin sacar de la bolsa** *vti* ALIMENT, EMB boil in bag

hervor *m* QUÍMICA slugging

hesperetina *f* QUÍMICA hesperetin

hesperidina *f* ALIMENT, QUÍMICA hesperidin
hesperitina *f* QUÍMICA hesperitin
hessita *f* MINERAL hessite
hessonita *f* MINERAL hessonite
Hetangiense *m* GEOFÍS Hettangian
heteroatómico *adj* QUÍMICA heteroatomic
heteroátomo *m* QUÍMICA heteroatom
heteroauxina *f* QUÍMICA heteroauxin
heterocíclico *adj* QUÍMICA heterocyclic
heterodino *m* TV *dispositivo* heterodyne, *proceso* heterodyning
heterogeneidad *f* ING MECÁ, METAL coring
heterogéneo *adj* METAL, NUCL heterogeneous
heterogenita *f* MINERAL heterogenite
heteromarcado *m* NUCL heterolabeling (*AmE*), heterolabelling (*BrE*)
heterométrico *adj* GEOL heterometric
heteromorfita *f* MINERAL heteromorphite
heteropolar *adj* QUÍMICA heteropolar
heterósido *m* QUÍMICA heteroside
heterosita *f* MINERAL heterosite
heterounión *f* ELECTRÓN, ÓPT heterojunction
heteroxantina *f* QUÍMICA heteroxanthine
heulandita *f* MINERAL heulandite
heurístico *adj* INFORM&PD heuristic
hexacloruro: ~ **de benceno** *m* ALIMENT, QUÍMICA benzene hexachloride
hexacontano *m* QUÍMICA hexacontane
hexacosano *m* QUÍMICA hexacosane
héxada *f* QUÍMICA hexad
hexadecano *m* QUÍMICA hexadecane
hexadecimal *m* INFORM&PD hexadecimal (*hex*)
hexádico *adj* QUÍMICA hexadic
hexadieno *m* QUÍMICA hexadiene
hexaedral *adj* GEOM hexahedral
hexaédrico *adj* GEOM hexahedral
hexaedro *m* GEOM hexahedron
hexafluoruro: ~ **de uranio** *m* NUCL, QUÍMICA uranium hexafluoride
hexahidrobenceno *m* QUÍMICA hexahydrobenzene
hexahidrobenzoico *adj* QUÍMICA hexahydrobenzoic
hexahidrofenol *m* QUÍMICA hexahydrophenol
hexahidropiridina *f* QUÍMICA hexahydropyridine
hexametilenotetramina *f* QUÍMICA hexamethylenetetramine
hexanitrito: ~ **de manitol** *m* QUÍMICA mannitol hexanitrite, nitromannitol
hexano *m* QUÍMICA, TEC PETR hexane
hexanoato: ~ **de dialquilo** *m* P&C, QUÍMICA adipic ester
hexanol *m* QUÍMICA hexanol, hexyl alcohol
hexavalente *adj* QUÍMICA hexavalent
hexeno *m* QUÍMICA hexene
hexileno *m* QUÍMICA hexylene
hexílico *adj* QUÍMICA hexylic
hexilo *m* QUÍMICA hexyl
hexino *m* QUÍMICA hexyne
hexodo *m* ELECTRÓN hexode
hexogen *m* QUÍMICA hexogen
hexosa *f* QUÍMICA hexose
hexosana *f* QUÍMICA hexosan
Hf *abr* (*hafnio*) METAL, QUÍMICA Hf (*hafnium*)
HF *abr* (*hiperfrecuencia*) GEN UHF (*ultrahigh frequency*)
hg *abr* (*hectograma*) METR hg (*hectogram*)

Hg *abr* (*mercurio*) ING ELÉC, METAL, QUÍMICA Hg (*mercury*)
hialita *f* MINERAL hyalite
hialoclastita *f* GEOL hyaloclastite
hialófana *f* MINERAL hyalophane
hialógeno *m* QUÍMICA hyalogen
hiato *m* GEOL *ausencia de sedimentos* gap, *grieta* hiatus, TEC PETR hiatus; ~ **tectónico** *m* GEOL lag fault
hibridación *f* QUÍMICA hybridization
híbrido: ~ **de injerto** *m* AGRIC graft hybrid; ~ **línea variedad** *m* AGRIC inbred variety cross
hidantoina *f* QUÍMICA hydantoin
hiddenita *f* MINERAL hiddenite
hidoscopio *m* PROD moisture detector, water detector
hidrácido *m* QUÍMICA hydracid
hidracina *f* NUCL, QUÍMICA, TEC ESP *propulsión* hydrazine
hidrante *m* AGUA hydrant, plug; ~ **de columna** *m* AGUA, ING MECÁ pillar hydrant; ~ **contra incendio tipo pilar** *m* SEG pillar fire-hydrant; ~ **para incendios** *m* AGUA, CONST, SEG, TERMO, TRANSP MAR fire hydrant (*BrE*), fireplug (*AmE*); ~ **para repostaje** *m* SEG, TEC PETR *refino* hydrant system
hidrargilita *f* MINERAL hydrargillite
hidrástico *adj* QUÍMICA hydrastic
hidrastina *f* QUÍMICA hydrastine
hidratación *f* CONST, GEOL, QUÍMICA hydration
hidratado *adj* GEOL hydrated, *minerales* hydrous, QUÍMICA hydrated
hidratar *vt* GEOL, QUÍMICA hydrate
hidrato *m* QUÍMICA hydrate; ~ **de cloral** *m* QUÍMICA chloralhydrate
hidratrópico *adj* QUÍMICA hydratropic
hidráulica *f* AGUA, INSTAL HIDRÁUL, MECÁ hydraulics; ~ **fluvial** *f* AGUA fluvial hydraulics
hidráulico *adj* GEN hydraulic
hidrazida *f* QUÍMICA hydrazide
hidrazina *f* QUÍMICA, TEC ESP hydrazine
hidrazoato *m* QUÍMICA hydrazoate
hídrico *adj* QUÍMICA hydric
hidrindeno *m* QUÍMICA hydrindene, indane
hidroala *m* TRANSP MAR hydrofoil; ~ **a reacción** *m* TRANSP MAR jetfoil
hidroaromático *adj* QUÍMICA hydroaromatic
hidroavión *m* TRANSP floatplane, naviplane, TRANSP AÉR flying boat, floatplane, seaplane
hidrobarómetro *m* MECÁ, METR depth gage (*AmE*), depth gauge (*BrE*)
hidrobilirrubina *f* QUÍMICA hydrobilirubin
hidrocarbonato *m* QUÍMICA hydrocarbonate
hidrocarbónico *adj* QUÍMICA hydrocarbonic
hidrocarburo *m* CONTAM MAR, GEOL, QUÍMICA, TEC PETR, VEH *combustible* hydrocarbon; ~ **alifático** *m* P&C, QUÍMICA aliphatic hydrocarbon; ~ **aromático** *m* P&C, QUÍMICA aromatic hydrocarbon; ~ **aromático policíclico** *m* CONTAM, QUíMICA polycyclic aromatic hydrocarbon; ~ **gaseoso** *m* GAS, QUÍMICA gas hydrocarbon; ~ **saturado** *m* QUÍMICA, TEC PETR saturated hydrocarbon
hidrocarburos: ~ **alterados por exposición a la intemperie** *m pl* CONTAM MAR weathered oil; ~ **blancos** *m pl* CONTAM MAR white product; ~ **intemperizados** *m pl* CONTAM MAR weathered oil; ~ **líquidos** *m pl* QUÍMICA liquid hydrocarbons; ~ **sólidos** *m pl* QUÍMICA solid hydrocarbons
hidrocelulosa *f* QUÍMICA hydrocellulose

hidrocianita *f* MINERAL hydrocyanite
hidrociclón *m* CARBÓN, PROC QUÍ, TEC PETR *perforación* hydrocyclone
hidrociencia *f* AGUA hydroscience
hidrocinámico *adj* QUÍMICA hydrocinnamic
hidrocincita *f* MINERAL hydrozincite
hidrocortisona *f* QUÍMICA hydrocortisone
hidrocotarnina *f* QUÍMICA hydrocotarnine
hidrocraking *m* TEC PETR hydrocracking
hidrocraqueador *m* TEC PETR hydrocracker
hidrodensímetro *m* AUTO acid hydrometer
hidrodeslizador *m* TRANSP, TRANSP MAR gliding boat
hidrodeslizamiento *m* AUTO, TRANSP AÉR aquaplaning
hidrodinámica *f* GEN hydrodynamics
hidrodinámico *adj* GEN hydrodynamic
hidroelectricidad *f* ELEC hydroelectricity
hidroextracción: ~ **por aspiración** *f* TEXTIL suction hydroextraction
hidroextractor *m* CONTAM wringer; ~ **centrífugo** *m* PROC QUÍ centrifugal hydroextractor
hidrófana *f* MINERAL hydrophane
hidrofílico *adj* CARBÓN hydrophilic
hidrófilo *adj* AGRIC, QUÍMICA hydrophilic, moisture-retaining
hidrofóbico *adj* CARBÓN, QUÍMICA hydrophobic
hidrófobo *adj* CARBÓN, QUÍMICA hydrophobic
hidrófono *m* ACÚST, OCEAN, TEC PETR, TELECOM hydrophone
hidrófugo *adj* REVEST, TEXTIL water-repellent
hidrogenación *f* ALIMENT, CARBÓN, DETERG, QUÍMICA hydrogenation, TEC ESP *aceites* hardening
hidrogenado *adj* ALIMENT, CARBÓN, DETERG hydrogenated, GEOL hydrogenous, QUÍMICA, TEC PETR hydrogenated
hidrogenador *m* ALIMENT hydrogenator
hidrogenar *vt* ALIMENT, CARBÓN, DETERG hydrogenate, PROD *aceites* harden, QUÍMICA, TEC PETR hydrogenate
hidrogénico *adj* GEOL hydrogenous
hidrogenizado *m* TEC PETR hydrogenation
hidrógeno *m* (*H*) GEN hydrogen (*H*); ~ **líquido** *m* QUÍMICA, TEC ESP, TERMO hydrogen liquid, liquid hydrogen; ~ **pesado** *m* FÍS, NUCL, QUÍMICA heavy hydrogen; ~ **sulfurado** *m* PROC QUÍ, QUÍMICA hydrogen sulfide (*AmE*), hydrogen sulphide (*BrE*)
hidrogeología *f* AGUA, GEOL, HIDROL hydrogeology
hidrogeoquímica *f* GEOL, HIDROL, QUÍMICA hydrogeochemistry
hidrografía *f* ENERG RENOV, GEOL, HIDROL hydrography, TRANSP MAR hydrography, *barco, costa* surveying
hidrografiar *vt* TRANSP MAR *barco, costa* survey
hidrográfico *adj* GEOL, HIDROL, TRANSP MAR hydrographic
hidrógrafo *m* AGUA hydrograph
hidrograma *m* AGUA hydrograph
hidrolisado: ~ **de caseína** *m* ALIMENT, QUÍMICA casein hydrolysate
hidrólisis *f* ALIMENT, GEOL, P&C, QUÍMICA hydrolysis; ~ **de llama** *f* ÓPT flame hydrolysis
hidrología *f* AGUA, CARBÓN, GEOL, HIDROL hydrology; ~ **kárstica** *f* AGUA karst hydrology
hidrólogo *m* CONST hydrologist
hidrolomita *f* MINERAL hydrodolomite
hidromagnesita *f* MINERAL hydromagnesite
hidromecánica *f* FÍS FLUID hydromechanics
hidrometalurgia *f* CARBÓN, METAL hydrometallurgy
hidrometría *f* AGUA, FÍS, HIDROL, QUÍMICA hydrometry

hidrómetro *m* *Esp* (*cf densímetro AmL*) GEN hydrometer, water gage (*AmE*), water gauge (*BrE*)
hidroniveladora *f* TRANSP hydroskimmer
hidroplano *m* OCEAN, TRANSP MAR *submarinos* hydroplane
hidroponía *f* AGRIC hydroponics
hidroquinona *f* QUÍMICA quinol, *fotografía antioxidante* hydroquinone
hidrorretención *f* HIDROL water retention
hidroscópico *adj* MECÁ hydroscopic
hidróscopo *m* PROD moisture detector, water detector
hidrosfera *f* HIDROL hydrosphere
hidrosiliación *f* QUÍMICA hydrosilyation
hidrosilicato *m* QUÍMICA hydrosilicate
hidrosol *m* QUÍMICA hydrosol
hidrosopolina *f* QUÍMICA hydrosopoline
hidrostática *f* GEN hydrostatics
hidrostático *adj* GEN hydrostatic
hidrotermal *adj* ENERG RENOV, GEOL hydrothermal
hidrotérmico *adj* ENERG RENOV, GEOL hydrothermal
hidrotermoterapia *f* AGRIC hot-water treatment
hidroturbina *f* TERMO water turbine
hidroxidecapado *adj* QUÍMICA hydroxy-capped
hidróxido: ~ **de aluminio** *m* HIDROL, P&C, QUÍMICA aluminium hydroxide (*BrE*), aluminum hydroxide (*AmE*); ~ **amónico** *m* QUÍMICA, TEC PETR ammonium hydroxide; ~ **de amonio** *m* AGRIC aqua ammonia; ~ **de níquel** *m* QUÍMICA, TEC ESP nickel hydroxide; ~ **potásico** *m* QUÍMICA potassium hydroxide; ~ **de potasio** *m* QUÍMICA potassium hydroxide; ~ **de sodio** *m* QUÍMICA sodium hydroxide
hidroxietilcelulosa *f* QUÍMICA hydroxyethylcellulose
hidroxilado *adj* ALIMENT, QUÍMICA hydroxylated
hidróxilo *m* ALIMENT, QUÍMICA hydroxyl
hidroxitiramina *f* QUÍMICA dopamine
hidruro *m* METAL, QUÍMICA hydride; ~ **pesado** *m* QUÍMICA deuteride
hielo: ~ **de agua de mar** *m* REFRIG sea-water ice; ~ **calibrado** *m* REFRIG sized ice; ~ **en cintas** *m* REFRIG ribbon ice; ~ **cristalino** *m* REFRIG crystal ice; ~ **a la deriva** *m* TRANSP MAR drift ice, pack ice; ~ **evaporativo** *m* PROC QUÍ, REFRIG evaporative ice; ~ **de fondo** *m* OCEAN anchor ice, bottom ice; ~ **fraccionado** *m* REFRIG processed ice; ~ **en laminillas** *m* REFRIG chipped ice; ~ **liso** *m* METEO glazed frost; ~ **marino** *m* REFRIG, TRANSP MAR sea ice; ~ **en montículos** *m* OCEAN hummocked ice, hummocky ice; ~ **nieve** *m* REFRIG snow ice; ~ **nieve humedecido** *m* REFRIG slush ice; ~ **opaco** *m* REFRIG white ice; ~ **oscuro** *m* METEO, REFRIG, TRANSP black ice; ~ **en placas** *m* REFRIG plate ice; ~ **de plataforma** *m* OCEAN shelf ice; ~ **seco** *m* ALIMENT, QUÍMICA, REFRIG dry ice; ~ **transparente** *m* REFRIG clear ice; ~ **triturado** *m* REFRIG crushed ice; ~ **troceado** *m* REFRIG broken ice; ~ **en tubos** *m* REFRIG shell ice
hierba: ~ **bahía** *f* AGRIC Bahia grass; ~ **mora** *f* AGRIC black nightshade
hierro *m* CARBÓN, METAL, QUÍMICA *Esp* (*Fe, cf fierro AmL*) iron (*Fe*); ~ **blando** *m* FÍS, METAL soft iron; ~ **de bobina móvil** *m* ING ELÉC, METAL soft iron; ~ **de calafate** *m* TRANSP MAR caulking iron; ~ **en una capa** *m* CONST single plane iron; ~ **en chapas** *m* METAL, PROD iron plate, plate iron; ~ **colado** *m* CONST, ING MECÁ, MECÁ, METAL cast iron; ~ **en corte transversal** *m* PROD cross-section iron; ~ **dulce** *m* ING ELÉC, METAL soft iron; ~ **escariador** *m* CONST

reaming iron; ~ **esmaltado** *m* METAL, PROD enameled iron (*AmE*), enamelled iron (*BrE*); ~ **estructural** *m* CONST structural iron; ~ **forjado** *m* CONST, METAL wrought iron; ~ **fraguado** *m* CONST, METAL wrought iron; ~ **fundido** *m* CONST, ING MECÁ, MECÁ, METAL cast iron; ~ **en H** *m* CONST H-iron; ~ **del inducido** *m* ING ELÉC armature iron; ~ **en láminas** *m* METAL, PROD iron plate, plate iron; ~ **en lingotes** *m* METAL, PROD pig iron; ~ **machihembrado** *m* CONST tonguing iron; ~ **de marcar en caliente** *m* EMB branding iron; ~ **en moldes** *m* PROD chilled iron; ~ **negro** *m* REVEST iron black; ~ **en planchas** *m* METAL, PROD iron plate, plate iron; ~ **ranurado** *m* CONST grooving iron; ~ **de rebajo** *m* CONST rabbet iron; ~ **de relleno** *m* CONST back iron; ~ **de sección cruciforme** *m* PROD *perfiles laminados* cross-section iron; ~ **de soldar** *m* CONST ELEC soldering iron; ~ **en U** *m* CONST channel iron

hierros: ~ **de cepillo** *m pl* ING MECÁ plane iron; ~ **machihembrados** *m pl* CONST tonguing-and-grooving irons

higiene: ~ **ambiental** *f* AGRIC, SEG environmental health; ~ **industrial** *f* SEG industrial hygiene; ~ **pública** *f* AGUA, SEG public health

higienización *f* AGUA sanitation, INFORM&PD sanitization

higrometría *f* GEN hygrometry, TRANSP MAR humidity measurement

higrómetro *m* GEN hygrometer; ~ **de cabello** *m* FÍS, LAB, REFRIG hair hygrometer; ~ **de condensación** *m* PROC QUÍ condensation hygrometer; ~ **eléctrico** *m* ELEC, REFRIG electrical hygrometer; ~ **electrolítico** *m* ELEC electrolytic hydrometer, REFRIG electrolytic hygrometer; ~ **de honda** *m* REFRIG sling hygrometer; ~ **orgánico** *m* REFRIG organic hygrometer; ~ **de punto de rocío** *m* PROC QUÍ condensation hygrometer, REFRIG dew-point hygrometer

higrorresistente *adj* TERMO humidity-resistant

higroscopio *m* AGUA, CONST, FÍS, HIDROL, QUÍMICA hygroscope

higrostato *m* REFRIG, TERMO humidistat

higrotermógrafo *m* AGRIC hygrothermograph

hijo *m* AGRIC tiller, NUCL daughter

hijuelo *m* AGRIC offset

hilada *f* CONST *albañilería* course; ~ **de tizones** *f* CONST *albañilería* header course, heading course

hilado[1] *adj* P&C *producto* spun; ~ **de fibra cortada** *adj* TEXTIL cut-staple spun

hilado[2] *m* C&V spinning, TEXTIL yarn; ~ **para alfombra** *m* TEXTIL carpet yarn; ~ **en bucle** *m* TEXTIL looped yarn; ~ **por centrifugación** *m* TEXTIL boxspun yarn; ~ **de continua de anillos** *m* TEXTIL ring-spun yarn; ~ **desgastado** *m* TEXTIL abraded yarn; ~ **de fantasía** *m* TEXTIL fancy yarn; ~ **de gran volumen** *m* TEXTIL high-bulk spun yarn; ~ **tangleado** *m* TEXTIL intermingled yarn

hilandera *f* C&V spinner

hilar *vt* TEXTIL spin

hilatura *f* TEXTIL spinning; ~ **del algodón** *f* TEXTIL cotton spinning; ~ **del algodón por condensador** *f* TEXTIL cotton condenser spinning; ~ **de anillos** *f* TEXTIL ring spinning; ~ **de estambre** *f* TEXTIL worsted spinning; ~ **de lana cardada** *f* TEXTIL woollen spinning; ~ **para rebajas** *f* TEXTIL sales yarn spinning; ~ **de la seda** *f* TEXTIL silk spinning; ~ **del yute** *f* TEXTIL jute spinning

hilaza: ~ **de amianto** *f Esp* (*cf lana de amianto AmL*) NUCL asbestos wool

hilera *f* AGRIC drill, swath, C&V string, INFORM&PD row, ING MECÁ screw plate, line, PETROL line-up, PROD *corte de tornillería, pernos* die head, *trefilado de alambre* die plate; ~ **de estirar** *f* ING MECÁ, PROD die; ~ **de gas** *f* GAS, PROD gas stock; ~ **de heno amontonada** *f* AGRIC windrow; ~ **de planchas** *f* TRANSP MAR *construcción naval* strake; ~ **para relojería** *f* PROD clock screw plate; ~ **tangencial** *f* ING MECÁ tangential diehead; ~ **transversal de perforaciones** *f* PETROL array; ~ **de trefilar** *f* ING MECÁ die; ~ **para tubos de rosca de gas** *f* GAS, ING MECÁ, PROD gas-thread pipe stock

hilerador *m* AGRIC rower

hileradora *f* AGRIC windrower

hilero *m* OCEAN, TRANSP MAR race; ~ **de marea** *m* OCEAN, TRANSP MAR *de estuario* tide race

hilo *m* C&V thread, IMPR *hilado* thread, *lino* linen, ING MECÁ screw thread, thread, worm, ÓPT strand, P&C yarn, filament, PAPEL thread, PROD *tornillo* fillet, thread, TELECOM wire, TEXTIL thread, yarn; ~ **A** *m* TELECOM *cable* tip, *de clavija* T-wire; ~ **aéreo** *m* PROD open wire; ~ **de algodón** *m* TEXTIL cotton yarn; ~ **alimentador** *m* TEXTIL feeder yarn; ~ **de amianto** *m AmL* (*cf hilo de asbesto Esp*) CONST asbestos twine; ~ **de asbesto** *m Esp* (*cf hilo de amianto AmL*) CONST asbestos twine; ~ **auxiliar** *m* ING ELÉC pilot wire; ~ **de blindaje** *m* ÓPT armor wire (*AmE*), armour wire (*BrE*), PROD shield drain wire; ~ **de cable** *m* TEXTIL rope yarn; ~ **cableado** *m* PROD stranded wire; ~ **de chenilla** *m* TEXTIL chenille yarn; ~ **de clavija** *m* ING ELÉC plug wire; ~ **de cobre** *m* CONST, ELEC, ING ELÉC, REVEST copper wire; ~ **colgante** *m* TEXTIL hanging thread; ~ **comercializado** *m* TEXTIL marked yarn; ~ **conductor** *m* ELEC, ING ELÉC conductor, conductor wire; ~ **conductor armado** *m* ELEC, ING ELÉC shielded wire; ~ **conductor entre el cebo eléctrico y el explosor** *m AmL* (*cf línea de tiro Esp*) MINAS *voladuras* leading wire; ~ **conductor flexible** *m* ELEC, ING ELÉC flexible wire; ~ **conductor del haz** *m* ELECTRÓN beam lead; ~ **conductor principal** *m* ING ELÉC lead-in wire; ~ **de conexión** *m* ELEC, ING ELÉC connecting wire; ~ **continuo** *m* TEXTIL continuous yarn; ~ **-cremallera** *m* ING ELÉC rack wiring; ~ **cuadrado** *m* ING MECÁ square thread; ~ **de desgarre** *m* PAPEL tearing wire; ~ **desnudo** *m* PROD open wire, TELECOM bare wire; ~ **doblado** *m* TEXTIL plied yarn; ~ **de drenaje** *m* ELEC drain wire; ~ **de drenaje blindado** *m* ELEC shield drain wire; ~ **de drenaje desnudo** *m* ELEC bare drain wire; ~ **electroaislado** *m* ELEC, ING ELÉC insulated wire; ~ **de encuadernar** *m* IMPR binding thread; ~ **enrollado en paralelo** *m* TEXTIL parallel-wound yarn; ~ **entre el cebo eléctrico y el explosor** *m* ELEC leading-out wire; ~ **esmaltado** *m* ELEC, ING ELÉC, REVEST enameled wire (*AmE*), enamelled wire (*BrE*); ~ **estándar de los Estados Unidos** *m* ING MECÁ US standard thread; ~ **exterior** *m* ING MECÁ external thread; ~ **de fibra discontinua** *m* TEXTIL staple fiber yarn (*AmE*), staple fibre yarn (*BrE*); ~ **fusible** *m* ELEC, ING ELÉC fuse wire; ~ **gaseado** *m* TEXTIL gassed yarn; ~ **grueso y delgado** *m* TEXTIL thick-and-thin yarn; ~ **de guardia** *m* ING ELÉC guard wire; ~ **hilado** *m* TEXTIL spinning yarn; ~ **de hilatura continua** *m* TEXTIL continuous-spun yarn;

~ **irregular** *m* TEXTIL irregular yarn; ~ **jaspeado** *m* TEXTIL jasp yarn; ~ **de los machos de roscar** *m* ING MECÁ tapping-screws thread; ~ **magnetofónico** *m* ACÚST magnetic wire; ~ **de masa** *m* ELEC, FÍS, ING ELÉC, PROD, TELECOM, TV earth wire (*BrE*), ground wire (*AmE*); ~ **metálico** *m* ING ELÉC, TELECOM wire; ~ **metalizado** *m* REVEST metal-coated thread; ~ **para neumáticos** *m* TEXTIL tire yarn (*AmE*), tyre yarn (*BrE*); ~ **neutro** *m* ELEC *circuito* neutral wire; ~ **de nylon** *m* P&C, TEXTIL nylon thread; ~ **ondulado** *m* TEXTIL crimped yarn; ~ **de paso ancho unificado** *m* (*UNC*) ING MECÁ unified coarse thread (*UNC*); ~ **de paso fino unificado** *m* (*UNF*) ING MECÁ unified fine thread (*UNF*); ~ **de paso largo** *m* TRANSPAÉR *hélices* coarse pitch thread; ~ **peinado** *m* TEXTIL combed yarn; ~ **plano** *m* TEXTIL flat yarn; ~ **de plomada** *m* CONST plumb line; ~ **de plomo conductor** *m* ING ELÉC lead wire; ~ **con poca altura** *m* ING MECÁ undercut; ~ **de puesta a masa** *m* ELEC, FÍS, ING ELÉC, PROD, TELECOM, TV earth lead (*BrE*), ground lead (*AmE*); ~ **de puesta a tierra** *m* ELEC, FÍS, ING ELÉC, PROD, TELECOM, TV earth wire (*BrE*), ground wire (*AmE*); ~ **de punta** *m* TELECOM T-wire, *cable* tip; ~ **de punta de electrodo** *m* ELEC electrode tip; ~ **recubierto** *m* TEXTIL covered yarn, wrapped yarn; ~ **de resistencia** *m* ELEC, ING ELÉC, METAL resistance wire; ~ **del retículo** *m* IMPR cross hair; ~ **de retorno** *m* ELEC, ING ELÉC return wire; ~ **de retorno por tierra** *m* ELEC, ING ELÉC shield drain wire, *cables para señales* drain wire; ~ **de retorno por tierra desnudo** *m* ELEC, ING ELÉC bare drain wire; ~ **de rosca de British Association** *m* ING MECÁ British Association screw thread (*BA screw thread*); ~ **de rosca cónico de British Standards** *m* ING MECÁ British Standard taper pipe thread (*BSPT*); ~ **de rosca en paralelo de British Standards** *m* ING MECÁ British Standard parallel pipe thread (*BSP*); ~ **de rosca de paso pequeño de British Standards** *m* ING MECÁ British Standard fine screw thread (*BSF screw thread*); ~ **de rosca trapezoidal asimétrico** *m* ING MECÁ asymmetrical trapezoidal-screw thread; ~ **de rosca trapezoidal métrico** *m* ING MECÁ metric trapezoidal-screw thread; ~ **de rosca Whitworth de British Standards** *m* ING MECÁ British Standard Whitworth thread (*BSW thread*); ~ **satinado** *m* TEXTIL glazed yarn; ~ **de seda** *m* TEXTIL silk yarn; ~ **sencillo** *m* ELEC *conductor* solid wire; ~ **telefónico** *m* ING ELÉC, TELECOM telephone wire; ~ **teñido** *m* TEXTIL dyed yarn; ~ **termoadherido** *m* TEXTIL bonded thread; ~ **de título grueso** *m* TEXTIL coarse yarn; ~ **de tornillo de una pulgada** *m* ING MECÁ inch screw thread; ~ **de tracción** *m* ELEC, ING ELÉC *instalaciones* pull-in wire; ~ **de trama** *m* PAPEL weft yarn; ~ **trapezoidal** *m* ING MECÁ trapezoidal thread; ~ **trapezoidal simétrico** *m* ING MECÁ symmetrical trapezoidal-screw thread; ~ **único** *m* ING ELÉC solid conductor; ~ **de urdimbre** *m* PAPEL warp yarn; ~ **de uso general** *m* ING MECÁ general-purpose screw thread; ~ **de vidrio** *m* ING ELÉC glass-fiber (*AmE*), glass-fibre (*BrE*); ~ **de vuelta** *m* ELEC, ING ELÉC return wire; ~ **Whitworth** *m* ING MECÁ Whitworth screw thread, Whitworth thread; ~ **de yute** *m* TEXTIL jute yarn; ~ **de zinc** *m* ING ELÉC zinc wire

hilos: ~ **externos** *m pl* ING MECÁ external threads; ~ **sobrantes** *m pl* TEXTIL *de urdimbre* spare ends; ~ **de títulos finos** *m pl* TEXTIL fine-counts yarns;

~ **totalmente estirados** *m pl* TEXTIL fully-drawn yarns; ~ **para trama** *m pl* TEXTIL weft

hinca *f* CARBÓN, MINAS *de pilotes* driving; ~ **de pilotes** *f* CARBÓN, MINAS pile-driving

hincado *m* CONST driving; ~ **de pilotes** *m* CONST pile-driving, spiling

hincadora: ~ **de pilotes** *f* CARBÓN pile-driver, CONST pile hammer (*AmE*), piling hammer (*BrE*); ~ **de postes** *f* CONST postdriver

hincar *vt* MINAS *pilotes* ram, PROD *pilotes* tamp

hinchamiento *m* METAL growth; ~ **dieléctrico** *m* NUCL dielectric swelling; ~ **hidráulico** *m* CARBÓN hydraulic bottom heave; ~ **del piso** *m* MINAS floor heave

hinchar *vt* AmL (*cf rellenar Esp*) MINAS fill

hincharse *vi* CONST belly out

hinchazón *f* P&C swelling, PAPEL bloating

hiocolánico *adj* QUÍMICA hyocholanic

hioscina *f* QUÍMICA hyoscine

hipautomorfo *adj* GEOL hypautomorphic

hiperafinea *f* OCEAN hyperapnea

hiperbalística *f* TEC ESP hyperballistics

hiperbalístico *adj* TEC ESP hyperballistic

hipérbola *f* GEOM hyperbola

hiperbólico *adj* GEOM, TEC ESP, TRANSP MAR hyperbolic

hiperboloide *m* GEOM hyperboloid

hipercarga *f* FÍS hypercharge

hiperconductividad *f* ELEC superconductivity

hiperconductor *m* ELEC superconductor

hipercubo *m* TRANSP superhigh cube (*SHC*)

hiperenvejecimiento *m* METAL overageing

hipereutéctico *adj* METAL *aceros* hypereutectic

hiperexcitación *f* ING ELÉC overcompounding

hiperfrecuencia *f* (*HF*) GEN ultrahigh frecuency (*UHF*)

hipergol *m* TEC ESP hypergol

hipergólico *adj* TEC ESP hypergolic

hipergrupo *m* TELECOM hypergroup

hiperón *m* FÍS, FÍS PART hyperon; ~ **sigma** *m* FÍS, FÍS PART sigma hyperon; ~ **xi** *m* FÍS, FÍS PART xi hyperon

hiperoxia *f* OCEAN hyperoxia, oxygen poisoning

hiperóxido *m* QUÍMICA hyperoxide

hiperpaginación *f* INFORM&PD *memoria* thrashing

hiperplano *m* GEOM hyperplane

hipersalinidad *f* OCEAN hypersalinity

hipersensibilizar *vt* CINEMAT hypersensitize

hipersónico *adj* FÍS, FÍS FLUID, TEC ESP hypersonic

hiperstena *f* MINERAL hypersthene

hipersuperficie *f* GEOM hypersurface

hipertexto *m* INFORM&PD hypertext

hipidiomorfo *adj* GEOL hypidiomorphic

hipoacúsia *m* ACÚST hypoacusis

hipocentro *m* GEOFÍS, GEOL earthquake focus, hypocenter (*AmE*), hypocentre (*BrE*)

hipoclorato *m* QUÍMICA hypochlorate

hipoclorito *m* QUÍMICA hypochlorite; ~ **cálcico** *m* DETERG chlorinated lime

hipocotilo *m* AGRIC hypocotyl

hipoeutéctico *adj* METAL *aceros* hypoeutectic

hipofosfato *m* QUÍMICA hypophosphate

hipofosfito *m* QUÍMICA hypophosphite

hipoide *f* MECÁ hypoid

hiposulfito *m* CINEMAT hypo; ~ **de sodio** *m* QUÍMICA sodium thiosulfate (*AmE*), sodium thiosulphate (*BrE*)

hipotenusa *f* GEOM hypotenuse

hipotermal *adj* GEOL hypothermal

hipotermia *m* TRANSP MAR hypothermia
hipotérmico *adj* GEOL hypothermal
hipótesis *f* QUÍMICA hypothesis
hipotónico *adj* QUÍMICA hypotonic
hipovoltaje *m* PROD undervoltage
hipsométrico *adj* COLOR FÍS, HIDROL, PETROL hypsometric
hipsómetro *m* FÍS hypsometer
hirviendo *adj* FÍS, TERMO boiling
hirviente *adj* FÍS, TERMO boiling
histamina *f* QUÍMICA histamine
histarazina *f* QUÍMICA hystarazin
histéresis *f* GEN hysteresis; **~ dieléctrica** *f* ELEC, ING
ELÉC dielectric hysteresis; **~ magnética** *f* ELEC, ING
ELÉC magnetic hysteresis, ING MECÁ magnetic lag
histograma *m* FÍS, INFORM&PD, MATEMÁT, TELECOM
bar chart, histogram; **~ de contactos** *m* ELEC, PROD
contact histogram
histona *f* QUÍMICA histone
historia: **~ de inventario** *f* PROD inventory profile
hito *m* CONST boundary, landmark, reference mark,
benchmark, boundary stone, guide post, *topografía*
monument, MECÁ milestone, PETROL Boundstone,
TRANSP milestone
hl *abr* (*hectólitro*) METR hl (*hectoliter AmE, hectolitre
BrE*)
hm *abr* (*hectómetro*) METR hm (*hectometer AmE,
hectometre BrE*)
Hn *abr* (*hahnio*) FÍS RAD, QUÍMICA Hn (*hahnium*)
Ho *abr* (*holmio*) QUÍMICA Ho (*holmium*)
hogar *m* CONST, GAS hearth, INSTAL HIDRÁUL *caldera de
vapor* furnace, fire chamber, INSTAL TERM *del horno*
hearth, PROD fire chamber, furnace, TERMO fire
chamber; **~ de aireación** *m* MINAS dumb furnace;
~ de caldera *m* INSTAL HIDRÁUL boiler flue, boiler
furnace, boiler house; **~ mecánico de alimentación
superior** *m* AGRIC, ING MECÁ, REFRIG overfeed stoker
hoguera *f* TERMO blaze
hoja *f* CARBÓN *de oro o plata* pan, CONST *de puerta* leaf,
GEOM nappe, MECÁ, NUCL, P&C, PAPEL sheet, PROD
defecto de forja shut, TEC ESP *armas* blade; **~ abrasiva
cilíndrica** *f* ING MECÁ cylindrical abrasive sheet;
~ abrasiva de cono truncado *f* ING MECÁ truncated-cone abrasive sheet; **~ aislante** *f* EMB
insulating sheet; **~ de alto brillo** *f* EMB high-gloss
foil; **~ de aluminio** *f* ALIMENT, EMB, TERMOTEC
aluminium foil (*BrE*), aluminum foil (*AmE*); **~ para
arreglos** *f* IMPR make-ready sheet; **~ de ballesta** *f*
VEH spring leaf; **~ bandera** *f* AGRIC flag leaf;
~ basculante *f* CONST pivot-hung sash; **~ de
cálculo** *f* INFORM&PD spreadsheet; **~ cambiable** *f*
IMPR loose-leaf; **~ de carga y centrado** *f* TRANSP AÉR
load-and-trim sheet; **~ de construcción** *f* TRANSP
MAR construction sheet; **~ de contacto** *f* ELEC *relé*
contact blade; **~ de continuidad** *f* CINEMAT, TV
continuity sheet; **~ corredera** *f* CONST *ventanas*
sliding sash; **~ cortadora** *f* C&V cutter blade;
~ cuadriculada *f* PROD grid sheet; **~ de datos de
laboratorio** *f* CINEMAT lab-data sheet; **~ delgada de
metal** *f* PROD foil; **~ de dos carillas** *f* IMPR pp;
~ electrónica *f* INFORM&PD spreadsheet; **~ para
embutir** *f* EMB deep drawing foil; **~ para embutir
en caliente** *f* EMB hot stamping foil; **~ de ensayo** *f*
PAPEL handsheet; **~ de entregas** *f* PROD delivery
sheet, delivery ticket; **~ de escalpelo** *f* LAB scalpel
blade; **~ esmaltadora** *f* FOTO glazing sheet; **~ de

especificaciones *f* CINEMAT, TV specification sheet;
~ final *f* IMPR end sheet; **~ de gradación** *f* CINEMAT,
TV grading sheet; **~ grande** *f* IMPR broadsheet; **~ de
guarda externa** *f* IMPR outer-end paper; **~ impresa** *f*
IMPR printed sheet; **~ de impresión** *f* IMPR printing
sheet; **~ intercalada** *f* IMPR interleaf; **~ de jornales** *f*
PROD time sheet; **~ de laboratorio** *f* PAPEL handsheet;
~ laminada *f* P&C laminated sheet; **~ maestra** *f* AUTO
de ballesta master leaf, ING MECÁ top plate; **~ de
mantenimiento básico** *f* PROD elementary servicing
sheet; **~ de material de revestimiento** *f* P&C surfacing sheet; **~ de metal** *f* EMB metal foil; **~ de
montaje** *f* IMPR goldenrod, PROD assembly sheet;
~ de muestra *f* PAPEL outturn sheet, specimen;
~ pandeada *f* C&V bow and warp (*AmE*), warped
sheet (*BrE*); **~ del portapiezas** *f* ING MECÁ work rest
blade; **~ de prueba** *f* IMPR advance sheet, press
proof, show sheet; **~ para pruebas de recepción** *f*
PROD acceptance test sheet; **~ para puntura** *f* IMPR
stabbing sheet; **~ de respaldo** *f* PAPEL backing; **~ de
ruta** *f* PROD route sheet, TRANSP MAR waybill; **~ de
ruta para el tráfico de carga partial** *f* FERRO routing
code for part-load traffic; **~ de segueta** *f* ING MECÁ
hacksaw blade; **~ de sierra** *f* ING MECÁ, PROD saw
blade; **~ de sierra alternativa** *f* ING MECÁ, PROD
reciprocating-saw blade; **~ de sierra alternativa
vertical** *f* ING MECÁ, PROD jigsaw blade; **~ de sierra
de cinta para metales** *f* ING MECÁ, PROD metal-
cutting band-saw blade; **~ de sierra circular para
madera** *f* ING MECÁ, PROD woodsawing circular saw
blade; **~ de sierra de contornear** *f* ING MECÁ, PROD
jigsaw blade; **~ de sierra para metales** *f* ING MECÁ,
PROD metal-cutting saw blade, hacksaw blade; **~ de
sierra de vaivén** *f* ING MECÁ, PROD jigsaw blade;
~ superior *f* IMPR upper sheet; **~ terminal** *f* AGRIC
flag leaf; **~ de tijeras** *f* C&V shear blade; **~ de trabajo**
f PROD operation ticket; **~ trapezoidal** *f* ING MECÁ
trapezoidal blade; **~ de ventana** *f* CONST sash;
~ vidriada *f* CONST glazed sash; **~ de vidrio** *f* CONST
pane
hojal: **~ del izador** *m* TRANSP AÉR hoisting eye
hojalata *f* METAL tin plate
hojalatero *m* PROD tinsmith
hojas: **~ de continuación** *f pl* IMPR continuation
sheets; **~ en parejas paralelas** *f pl* FOTO tramlines;
~ de protección *f pl* PAPEL outside; **~ sobrantes** *f pl*
IMPR overs; **~ sueltas** *f pl* IMPR *anuncios pequeños*
dodgers
hojear *vti* INFORM&PD browse
holgura *f* *Esp* (*cf margen de altura AmL*) IMPR
backlash, ING MECÁ clearance, lash, play, drift,
INSTAL HIDRÁUL clearance, clearance space, MECÁ
allowance, MINAS, PROD clearance, TEC ESP backlash;
~ del árbol de levas *f* AUTO, VEH camshaft clearance;
~ de la dirección *f* AUTO, VEH steering play; **~ entre
la pestaña y el carril** *f* FERRO flange-to-rail clearance; **~ del freno** *f* AUTO, VEH brake clearance; **~ del
gálibo** *f* FERRO *de carga* gage clearance (*AmE*), gauge
clearance (*BrE*); **~ longitudinal** *f* ING MECÁ, MECÁ
end play; **~ del pistón** *f* AUTO, VEH *motor* piston
clearance; **~ radial interna** *f* ING MECÁ radial internal
clearance; **~ de rueda** *f* FERRO *vehículos* wheel
clearance; **~ de taqués** *f* AUTO, VEH valve clearance;
~ de la válvula *f* AUTO, ING MECÁ, VEH valve
clearance
hollín *m* VEH *motor* carbon

Holoceno *m* GEOL Holocene

holografía *f* FÍS, FÍS ONDAS, FÍS RAD, INFORM&PD, TEC ESP holography

holograma *m* FÍS, FÍS ONDAS hologram

hombre: **~-año** *m* PROD man-year; **~-día** *m* PROD man-day; **~-hora** *m* PROD man-hour; **~-semana** *m* PROD man-week

hombrera *f* TEXTIL shoulder pad

hombro *m* C&V shoulder; **~ sucio** *m* C&V dirty shoulder

homilita *f* MINERAL homilite

homocíclico *adj* QUÍMICA homocyclic

homoclinal *m* GEOL homocline

homogeneización *f* INSTAL TERM, P&C homogenization

homogeneizante *m* INSTAL TERM, P&C homogenizing

homogéneo *adj* GEN homogeneous

homogenización *f* METAL homogenizing

homogenizador *m* LAB homogenizer

homografía *f* GEOM homograph

homográfico *adj* GEOM homographic

homógrafo *m* GEOM homograph

homologación *f* INFORM&PD certification, VEH *normativa* homologation

homólogo *adj* METAL, QUÍMICA, TEC PETR homologous

homométrico *adj* GEOL homometric

homopirrol *m* QUÍMICA homopyrrole

homopolimerización *f* P&C homopolymerization

homopolímero *m* P&C homopolymer

homotereftálico *adj* QUÍMICA homoterephthalic

homotético *adj* IMPR homothetical

homounión *f* ELECTRÓN, ÓPT homojunction

honda *f* TRANSP MAR *jarcia* sling

hondonada *f* OCEAN trough

hongo *m* AGRIC fungus, TRANSP MAR *ancla* mushroom anchor, *cubierta* mushroom ventilator; **~ de ventilación** *m* TRANSP MAR *accesorios de cubierta* mushroom ventilator

HOOD *abr* (*diseño jerárquico orientado al objeto*) INFORM&PD HOOD (*hierarchical object-oriented design*)

hopcalita *f* QUÍMICA hopcalite

hopeita *f* MINERAL hopeite

hora *f* FÍS hour; **~ de carga** *f* TELECOM busy hour; **~ cargada acorde al tiempo** *f* TELECOM time-consistent busy hour; **~ más cargada acorde al tiempo** *f* TELECOM time-consistent busy hour; **~ cargada máxima** *f* TELECOM peak busy hour; **~ cargada media** *f* TELECOM mean busy hour; **~ de cierre** *f* ELEC *relé* closing time; **~ de entrega** *f* PROD delivery time; **~ estimada de llegada** *f* TRANSP AÉR, TRANSP MAR estimated time of arrival (*ETA*); **~ estimada de salida** *f* TRANSP AÉR, TRANSP MAR estimated time of departure (*ETD*); **~-hombre** *f* CONST man-hour; **~-hombre indirecta** *f* ING MECÁ indirect man-hour; **~ de huso** *f* TRANSP MAR zone time; **~ de huso horario** *f* TRANSP MAR zone time; **~ legal** *f* PROD, TRANSP MAR standard time; **~ de llegada prevista** *f* TRANSP AÉR, TRANSP MAR estimated time of arrival (*ETA*); **~ de mano de obra** *f* CONST man-hour; **~ media de Greenwich** *f* (*TMG*) FÍS, TEC ESP, TRANSP AÉR Greenwich Mean Time (*GMT*); **~ media local** *f* TRANSP MAR local mean time; **~ media solar** *f* TEC ESP mean solar time; **~ del meridiano de Greenwich** *f* (*TMG*) FÍS, TEC ESP, TRANSP AÉR Greenwich Mean Time (*GMT*); **~ oficial** *f* PROD standard time; **~ pico** *f* TV peak time; **~ punta** *f*

AUTO rush hour, FERRO, TRANSP peak hour, rush hour; **~ punta cargada** *f* TELECOM peak busy hour; **~ de relevo** *f* TRANSP AÉR release time; **~ de salida prevista** *f* TRANSP AÉR, TRANSP MAR estimated time of departure (*ETD*); **~ sideral** *f* FÍS, TEC ESP sidereal time; **~ sidérea** *f* FÍS, TEC ESP sidereal time; **~ solar media** *f* FÍS, TEC ESP, TRANSP AÉR Greenwich Mean Time (*GMT*); **~ universal** *f* TEC ESP Universal Time; **~ universal coordinada** *f* TELECOM coordinated universal time (*UTC*)

horadar *vt* AmL (*cf colar Esp*) CARBÓN bore, tap, CONST, ING MECÁ bore, MINAS crossdrive, hole, bore, tap

horario *m* PROD, TRANSP schedule; **~ de circulación** *m* FERRO traffic schedule; **~ fijo de servicio** *m* TELECOM scheduled operating time; **~ oficial** *m* FERRO official timetable

horas: **~ extraordinarias** *f pl* PROD *de trabajo* overtime; **~ de funcionamiento** *f pl* PROD *de máquina* operating hours; **~ de radiodifusión** *f pl* TV broadcasting times; **~ de servicio** *f pl* PROD operating hours; **~ de trabajo** *f pl* PROD working hours; **~ de vuelo** *f pl* TRANSP AÉR flying hours

hordeína *f* QUÍMICA hordein

horizontal[1] *adj* GEOM horizontal

horizontal[2] *f* GEOM horizontal

horizontalidad *f* PROD levelness

horizonte *m* GEOL, TRANSP MAR horizon; **~ acuífero** *m* AGUA water-bearing stratum; **~ AP** *m* AGRIC AP horizon; **~ artificial** *m* TEC ESP, TRANSP AÉR artificial horizon; **~ fantasma** *m* GEOFÍS *prospección* phantom horizon; **~ giroscópico** *m* TEC ESP, TRANSP AÉR gyro horizon; **~ del giróscopo** *m* TEC ESP, TRANSP AÉR gyro horizon; **~ de lanzamiento** *m* PROD release horizon; **~ productivo** *m* GEOL, TEC PETR *yacimientos* producing horizon; **~ de pronóstico** *m* PROD forecast horizon; **~ de referencia** *m* GEOL datum horizon, *estratigrafía* key horizon; **~ visible** *m* TRANSP MAR visible horizon

horma *f* ING ELÉC former, ING MECÁ jig, TEXTIL stretcher

hormado *m* TEXTIL boarding

hormiga: **~ cortadora** *f* AGRIC leaf-cutting ant

hormigón *m* *Esp* (*cf concreto AmL*) C&V, CONST concrete; **~ con alta dosificación de cemento** *m* *Esp* (*cf concreto con alta dosificación de cemento AmL*) CONST fat concrete; **~ amasado en fábrica** *m* *Esp* (*cf concreto amasado en fábrica AmL*) CONST ready-mixed concrete; **~ armado** *m* *Esp* (*cf concreto armado AmL*) CONST, D&A reinforced concrete, TRANSP MAR *construcción naval* ferroconcrete; **~ asfáltico** *m* *Esp* (*cf concreto asfáltico AmL*) CONST asphalt concrete; **~ de blindaje** *m* *Esp* (*cf concreto de blindaje AmL*) CONST blinding concrete; **~ celular** *m* *Esp* (*cf concreto celular AmL*) CONST honeycombing; **~ centrifugado** *m* *Esp* (*cf concreto centrifugado AmL*) CONST spun concrete; **~ ciclópleo** *m* *Esp* (*cf concreto ciclópleo AmL*) CONST cyclopean concrete; **~ fabricado a pie de obra** *m* *Esp* (*cf concreto fabricado a pie de obra AmL*) CONST in situ concrete; **~ sin finos** *m* *Esp* (*cf concreto sin finos AmL*) CONST no-fines concrete; **~ fresco** *m* *Esp* (*cf concreto fresco AmL*) CONST green concrete; **~ in situ** *m* *Esp* (*cf concreto in situ AmL*) CONST in situ concrete; **~ ligero** *m* *Esp* (*cf concreto ligero AmL*) CONST air-entrained concrete, lightweight concrete; **~ en masa** *m* *Esp* (*cf* concreto en masa *AmL*) CONST mass concrete; ~ **mezclado a mano** *m Esp* (*cf concreto mezclado a mano AmL*) CONST hand-mixed concrete; ~ **mezclado en obra** *m Esp* (*cf concreto mezclado en obra AmL*) CONST site concrete; ~ **para pavimento** *m Esp* (*cf concreto para pavimento AmL*) CONST pavement-quality concrete (*PQC*); ~ **pobre** *m Esp* (*cf concreto pobre AmL*) CONST lean concrete; ~ **preamasado** *m Esp* (*cf concreto preamasado AmL*) CONST ready-mixed concrete; ~ **prefabricado** *m Esp* (*cf concreto prefabricado AmL*) CONST precast concrete; ~ **pretensado** *m Esp* (*cf concreto pretensado AmL*) CONST prestressed concrete; ~ **sumergido** *m Esp* (*cf concreto sumergido AmL*) CONST submerged concrete; ~ **vibrado** *m Esp* (*cf concreto vibrado AmL*) CONST vibrated concrete; ~ **zunchado** *m Esp* (*cf concreto zunchado AmL*) CONST hooped concrete, stirruped concrete

hormigonado *m* CONST concrete work, concreting

hormigonar *Esp* (*cf concretar AmL*) *vt* CONST concrete

hormigonera *f Esp* (*cf concretera AmL*) CARBÓN, CONST concrete mixer, mixer, ING MECÁ concrete mixer

hornada *f* C&V, P&C batch

hornblenda *f* MINERAL hornblende

hornblendita *f* PETROL hornblendite

horneado[1] *adj* C&V fired-on

horneado[2] *m* C&V firing on, kilning, ~ **del decorado** *m* C&V decoration firing; ~ **de esmalte** *m* C&V enamel-firing

hornear *vt* ALIMENT fire, C&V fire, kiln, PROD *machos, moldes* bake, kiln; ~ **al esmalte** *vt* COLOR enamelbake, stove enamel

hornillo *m AmL* MINAS *pega* chamber; ~ **de mina** *m AmL* (*cf perforación de producción Esp*) MINAS mine chamber, *voladuras* mining hole

horno *m* ALIMENT kiln, C&V tank, underfired furnace, CARBÓN roaster, CONST kiln, INSTAL TERM *en fábricas* furnace, LAB kiln, PROD, TERMO calcining kiln, furnace;

~ a ~ **de ablandamiento** *m* C&V softening furnace; ~ **de afinación** *m* INSTAL TERM, PROD, TERMO refining furnace; ~ **de afino** *m* INSTAL TERM, PROD, TERMO refining furnace; ~ **de aire** *m* INSTAL TERM air furnace; ~ **de aire forzado** *m* GAS, ING MECÁ forced-air furnace, forced-draft burner (*AmE*), forced-draught burner (*BrE*); ~ **de alfarería** *m* TERMO pottery kiln; ~ **de alta frecuencia** *m* ING ELÉC high-frequency furnace; ~ **alto** *m* PROD blast furnace; ~ **alto de coque** *m* CARBÓN, PROD, TERMO coke blast furnace; ~ **alto para coque** *m* PROD blast-furnace for coke; ~ **anular** *m* C&V annular kiln; ~ **de arco** *m* INSTAL TERM arc furnace; ~ **de arco directo** *m* INSTAL TERM direct arc furnace; ~ **de arco eléctrico** *m* CARBÓN, ELEC, ING ELÉC, ING MECÁ, INSTAL TERM, TERMO electric-arc furnace (*EAF*); ~ **de arco longitudinal** *m* INSTAL TERM longitudinal-arch kiln; ~ **de arco al vacío** *m* INSTAL TERM vacuum arc furnace; ~ **de arco voltaico** *m* CARBÓN, ELEC, ING ELÉC, ING MECÁ, INSTAL TERM, TERMO electric-arc furnace (*EAF*); ~ **de azulejos** *m* C&V tile kiln;

~ b ~ **bacteriológico** *m* LAB bacteriological oven; ~ **de baja frecuencia** *m* ING ELÉC low-frequency furnace; ~ **de baño salino** *m* INSTAL TERM salt-bath furnace; ~ **basculador** *m* PROD tilter; ~ **basculante** *m* INSTAL TERM, PROD, TERMO rolling furnace; ~ **de**

bombardeo por haz electrónico *m* ELECTRÓN, FÍS PART, NUCL electron-beam bombardment furnace;

~ c ~ **de caja** *m* INSTAL TERM box furnace; ~ **de cal** *m* INSTAL TERM, PROD lime kiln; ~ **de cal vertical** *m* INSTAL TERM, PROD vertical lime kiln; ~ **de calcinación** *m* C&V, CARBÓN, INSTAL TERM, PROD, TERMO calcining kiln, roaster, roasting furnace, roasting kiln, roasting oven; ~ **de calcinar** *m* PROD burning-house; ~ **calcinificador de porcelana** *m* C&V porcelain-calcining furnace; ~ **caldeado con gas** *m* GAS, INSTAL TERM, TERMO, TERMOTEC gas-fired furnace; ~ **de caldeo** *m* INSTAL TERM, TERMO, TERMOTEC heating furnace; ~ **de caldeo por rayos infrarrojos** *m* PROD infrared oven; ~ **de caldera** *m* INSTAL HIDRÁUL boiler flue, boiler furnace; ~ **calientarremaches** *m* CONST, PROD, TERMO rivet-heating furnace; ~ **de campana** *m* PROD bell furnace; ~ **de canal de inducción** *m* INSTAL TERM channel induction furnace; ~ **de carga por paquetes** *m* PROC QUÍ, TERMO batch furnace; ~ **en cascada** *m* INSTAL TERM, TERMO cascade furnace; ~ **castellano** *m* CARBÓN low-shaft furnace; ~ **de cementación** *m* ING MECÁ carburizing furnace, INSTAL TERM, TERMO cementation furnace; ~ **de cemento** *m* C&V cement kiln; ~ **para cerámica** *m* C&V ceramic kiln; ~ **de ciclón** *m* C&V cyclone furnace; ~ **circular intermitente** *m* C&V beehive kiln, PROD *coquificación* beehive oven; ~ **para cocer ladrillos** *m* CONST brick kiln; ~ **por convección** *m* INSTAL TERM, QUÍMICA, TERMO convection oven; ~ **con convección forzada** *m* LAB oven with forced convection; ~ **con convección natural** *m* LAB oven with natural convection; ~ **de copela** *m* PROD muffle furnace; ~ **de coque** *m* C&V, CARBÓN, INSTAL TERM, MINAS, PROD, TERMO coke oven; ~ **crematorio** *m* INSTAL TERM, TERMO cremator; ~ **de crisol** *m* INSTAL TERM, LAB, PROD, QUÍMICA, TERMO crucible furnace; ~ **de crisoles** *m* TERMO pot furnace; ~ **de cuba** *m* PROD calcining kiln, shaft furnace, TERMO calcining kiln; ~ **de cuba baja** *m* CARBÓN low-shaft furnace; ~ **de cubilote** *m* C&V cupola furnace;

~ d ~ **de decorado** *m* C&V decorating kiln; ~ **de desecar** *m* PROD drying furnace; ~ **de difusión** *m* ELECTRÓN, PROC QUÍ diffusion oven; ~ **discontinuo** *m* PROC QUÍ, TERMO batch furnace; ~ **de doble corona** *m* C&V double deck crown furnace;

~ e ~ **de efusión** *m* ELECTRÓN, FÍS, NUCL effusion oven; ~ **eléctrico** *m* CARBÓN, ELEC, INSTAL TERM, TERMO electric furnace, electric oven; ~ **de electroinducción** *m* ING ELÉC electric-induction furnace; ~ **elevado** *m* INSTAL TERM lift-up furnace; ~ **de ensayos** *m* INSTAL TERM assay furnace; ~ **de esmaltado** *m* REVEST enameling furnace (*AmE*), enamelling furnace (*BrE*), enameling kiln (*AmE*), enamelling kiln (*BrE*);

~ f ~ **de fabricar acero** *m* PROD steel furnace; ~ **de frecuencias medias** *m* ING ELÉC, INSTAL TERM medium-frequency furnace; ~ **de fuego cruzado** *m* C&V cross-fired furnace; ~ **de fusión** *m* C&V fusing oven, CARBÓN melting furnace, ING ELÉC smelting furnace, P&C fusing oven, TERMO fusing oven, smelting furnace;

~ g ~ **de galera** *m* PROD gallery furnace; ~ **de galerías** *m* PROD gallery furnace; ~ **de gas** *m* GAS, INSTAL TERM, PROD, TERMO, TERMOTEC gas furnace, gas-fired furnace; ~ **de gasóleo** *m* INSTAL TERM,

TERMOTEC oil-fired furnace; **~ giratorio** *m* C&V, CARBÓN, INSTAL TERM, PROD, TERMO rotary furnace, rotary kiln; **~ de gran vacío** *m* ING MECÁ high-vacuum furnace;

~ h **~ de hogar abierto** *m* INSTAL TERM, TERMO open-hearth furnace; **~ con hogar giratorio** *m* INSTAL TERM, TERMO rotary-hearth kiln; **~ con hogares múltiples** *m* INSTAL TERM, TERMO multiple hearth furnace;

~ i **~ con impulsor** *m* INSTAL TERM pusher furnace; **~ para incineración de basuras** *m* TERMO destructor; **~ de inducción** *m* FÍS, ING ELÉC, PROD induction furnace; **~ de inducción sin núcleo** *m* INSTAL TERM, TERMO coreless induction furnace; **~ industrial** *m* GAS industrial furnace, ING MECÁ industrial oven; **~ intermitente** *m* PROC QUÍ, TERMO batch furnace; **~ con inyección de aire caliente** *m* PROD hot blast furnace;

~ l **~ de ladrillo** *m* CONST brick arch; **~ de ladrillos** *m* CARBÓN cupola; **~ de lecho fluidizado** *m* NUCL, PROC QUÍ fluid-bed furnace, fluidized-bed kiln; **~ de lecho fluido** *m* NUCL, PROC QUÍ fluid-bed furnace; **~ lento** *m* CARBÓN low kiln; **~ de licuación** *m* PROD sweating furnace;

~ m **~ metalúrgico** *m* ING MECÁ, METAL metallurgical furnace (*BrE*), metalurgical furnace (*AmE*); **~ de microondas** *m* ALIMENT, ING ELÉC microwave oven; **~ de mufla** *m* INSTAL TERM, LAB muffle furnace, PROD muffle furnace, *para tostar minerales sin contacto con los productos de la combustión* blind roaster;

~ o **~ oscilante** *m* INSTAL TERM tilting furnace; **~ a oxígeno** *m* CARBÓN oxygen furnace;

~ p **~ para piritas** *m* PROD desulfurizing furnace (*AmE*), desulphurizing furnace (*BrE*); **~ de polimerización** *m* TEXTIL curing oven; **~ para el precaldeo de crisoles** *m* C&V arch; **~ de precalentado de crisoles** *m* C&V pot arch; **~ de precalentamiento** *m* P&C preheating oven; **~ de pruebas** *m* PROD test furnace; **~ con puertos traseros** *m* C&V horseshoe-fired furnace;

~ r **~ de recalentamiento** *m* C&V *para soplado de vidrio* glory hole; **~ de recalentar** *m* INSTAL TERM, TERMO heating furnace; **~ de recocer** *m* C&V annealing furnace; **~ de recocido** *m* C&V annealing kiln, annealing lehr, MECÁ, NUCL, TERMO annealing furnace; **~ de recocido con rodillos** *m* C&V annealing lehr with rollers; **~ recuperativo** *m* C&V, PROD, TERMOTEC recuperative furnace; **~ de reducción** *m* INSTAL TERM, PROD, TERMO reducing furnace, reduction furnace; **~ reductor** *m* INSTAL TERM, PROD, TERMO reducing furnace, reduction furnace; **~ de refinación** *m* INSTAL TERM, PROD, TERMO refining furnace; **~ refrigerado por agua** *m* INSTAL TERM, REFRIG, TERMO water-cooled furnace; **~ regenerativo** *m* C&V regenerative furnace; **~ de resistencia** *m* ING ELÉC, TERMO resistance furnace; **~ de resistencia eléctrica** *m* ELEC, FÍS, ING ELÉC, TERMO electric-resistance furnace; **~ de revenido con banda transportadora** *m* C&V conveyor-belt lehr; **~ de revenido de convección forzada** *m* C&V forced-convection lehr; **~ de revenido de recirculación continua** *m* C&V continuous recirculation lehr; **~ de reverbero** *m* PROD reverberatory, reverberatory furnace; **~ rotatorio** *m* C&V, CARBÓN, INSTAL TERM, TERMO rotary kiln;

~ s **~ de sales** *m* TERMO pot furnace; **~ con salida** **de escoria** *m* TERMOTEC slag tap furnace; **~ de secado** *m* ALIMENT drying kiln, drying oven, CARBÓN drying kiln, EMB drying oven, ING MECÁ drying furnace, INSTAL TERM kiln, PAPEL drying oven, PROD drying kiln, TEXTIL drying oven; **~ de secado continuo** *m* INSTAL TERM continuous kiln; **~ de secado giratorio** *m* INSTAL TERM bogie kiln; **~ de secado en vacío** *m* ALIMENT vacuum-drying oven; **~ secador** *m* ING MECÁ drying furnace; **~ de secar de caja** *m* INSTAL TERM box kiln; **~ para secar moldes** *m* PROD drying stove; **~ de sinterización** *m* C&V, INSTAL TERM fritting furnace, PROC QUÍ sintering furnace; **~ solar** *m* ENERG RENOV solar furnace; **~ de solera** *m* CARBÓN, PROD open-hearth furnace; **~ de solera básico** *m* C&V basic open-hearth furnace;

~ t **~ de tambor** *m* TERMO drum furnace, drum kiln; **~ de temperatura constante** *m* LAB, TERMO constant temperature oven; **~ de templado hermético** *m* INSTAL TERM, TERMO sealed-quench furnace; **~ de termodifusión** *m* AGUA, INSTAL TERM, TERMO soaking pit; **~ de tiraje forzado** *m* *AmL* (*cf horno de tiro forzado Esp*) ING MECÁ, PROD, TERMOTEC forced-draft furnace (*AmE*), forced-draught furnace (*BrE*); **~ de tiro forzado** *m* *Esp* (*cf horno de tiro forzado AmL*) ING MECÁ, PROD, TERMOTEC forced-draft furnace (*AmE*), forced-draught furnace (*BrE*); **~ de tiro natural** *m* P&C air oven; **~ de tostación** *m* C&V roasting kiln, CARBÓN calcining kiln, INSTAL TERM, PROD, TERMO roaster, roasting furnace, roasting kiln, roasting oven; **~ de túnel** *m* C&V, INSTAL TERM tunnel kiln; **~ de turbulencia** *m* INSTAL TERM cyclone furnace;

~ v **~ al vacío** *m* LAB vacuum oven; **~ de vacío** *m* ING MECÁ vacuum furnace; **~ vertical** *m* INSTAL TERM vertical shaft furnace; **~ de vidriado** *m* C&V glaze kiln

horquilla *f* AGRIC, C&V fork, CARBÓN *de biela* jaw, ING MECÁ clevis, fork, yoke, jaw, MECÁ clevis, TRANSP MAR *en la botavara* crutch, VEH *cambio, embrague* fork, *junta universal* yoke; **~ de acero forjado** *f* ING MECÁ forged shackle; **~ de arrastre** *f* VEH towing bracket; **~ de bisagra** *f* ING MECÁ hinge yoke; **~ para cargar** *f* AGRIC loading fork; **~ de cinta** *f* ING MECÁ strap fork; **~ de desembrague** *f* PROD *correa* belt fork, VEH clutch release fork; **~ doble** *f* ING MECÁ dual clevis; **~ doble de conexión** *f* TRANSP AÉR connecting twin-yoke; **~ del embrague** *f* AUTO, ING MECÁ, VEH clutch fork; **~ estabilizadora** *f* VEH *motocicleta* antidive fork; **~ para heno** *f* AGRIC pitchfork; **~ de hierro forjado** *f* ING MECÁ forged shackle; **~ roscada** *f* ING MECÁ, MECÁ clevis bolt; **~ selectora de velocidades** *f* VEH gearbox selector fork; **~ telescópica** *f* VEH *motocicleta* telescopic fork; **~ de válvula** *f* ING MECÁ valve yoke

horquillas: **~ del freno** *f pl* AUTO, VEH brake forks

hortaliza: **~ de hoja** *f* AGRIC potherb

hovercraft *m* TRANSP, TRANSP MAR hovercraft

hoya *f* AGUA pit, basin, drainage area, CONTAM MAR pit; **~ hidrológica** *f* AGUA drainage basin; **~ de reborde** *f* *AmL* (*cf hoyo de resalto Esp*) MINAS shoulder hole; **~ tributaria** *f* AGUA watershed

hoyada *f* *AmL* (*cf perforación descendente Esp*) MINAS sink

hoyo *m* AGUA pit; **~ de ataque químico** *m* CRISTAL etch pit; **~ de explosión** *m* MINAS shot hole; **~ de reborde** *m* *Esp* (*cf hoya de reborde AmL*) MINAS shoulder hole; ~ **de resalto** *m Esp* (*cf hoya de reborde AmL*) MINAS shoulder hole

hoyuelo *m* C&V dimple; ~ **superficial** *m* ING MECÁ dimpled hole

hubnerita *f* MINERAL hubnerite

HUC *abr* (*hora universal coordinada*) METEO, TELECOM UTC (*coordinated universal time, universal time coordinated*)

HUD *abr* (*imagen vertical*) INSTR HUD (*head-up display*)

hueco *m* C&V void, CARBÓN concave, void, CONST break, hollow, FÍS, ING ELÉC hole, ING MECÁ opening, hole, MECÁ blank, gap, METAL void, MINAS *geología* cavity, *arena, grava* airspace, NUCL *en semiconductor* hole, gap, P&C *defecto* void, QUÍMICA *de contenedor* ullage, TEC PETR *perforación* hole, well bore; ~ **de la concha** *m* INSTAL HIDRÁUL exhaust cavity; ~ **del cuadrante** *m* TEC PETR *perforación* rat hole; ~ **de diámetro pequeño** *m* TEC PETR *perforación* slim hole; ~ **de escotilla** *m* TRANSP MAR hatchway; ~ **facetado** *m* NUCL faceted bubble; ~ **de gas** *m* NUCL gas cavity; ~ **de gran diámetro** *m* TEC PETR *perforación* big hole; ~ **inestable** *m* TEC PETR *perforación* tight hole; ~ **del kelly** *m* TEC PETR *perforación* rat hole; ~ **metalizado** *m* ELEC metalized hole (*AmE*), metallized hole (*BrE*); ~ **perdido** *m* TEC PETR *perforación* lost hole; ~ **publicitario** *m* TV advertising slot; ~ **de la puerta** *m* CONST door opening; ~ **rata** *m Esp* (*cf vaina AmL*) TEC PETR *perforación* rat hole; ~ **ratón** *m Esp* (*cf vaina AmL*) ING ELÉC, TEC PETR *perforación* mouse hole; ~ **del tubo** *m* INSTR tube center (*AmE*), tube centre (*BrE*); ~ **de ventana** *m* CONST window opening

huecograbado *m* IMPR gravure, intaglio

huelgo *m* ING MECÁ lash, slack, clearance, gap, play, ply, drift, *entre piezas de máquinas* backlash, INSTAL HIDRÁUL *cilindro de vapor* clearance space, MECÁ gap, clearance, backlash, NUCL clearance, gap, PROD *engranajes* clearance; ~ **anular** *m* NUCL *entre combustible y vaina* annular air gap; ~ **axial** *m* ING MECÁ float, MECÁ axial clearance, float; ~ **entre dientes** *m* ING MECÁ side clearance; ~ **entre elementos de combustible** *m* NUCL water gap; ~ **en el fondo de los dientes** *m* ING MECÁ top and bottom clearance; ~ **interno radial** *m* ING MECÁ radial internal clearance; ~ **lateral** *m* ING MECÁ side clearance, side lash; ~ **vaina-pastilla** *m* NUCL *del elemento de combustible* clad fuel clearance

huella *f* CONST *escalón* tread, GAS, PETROL trace, PROD indentation

huérfano *m* IMPR orphan

huerta: ~ **casera** *f* AGRIC kitchen garden

huesecillo *m* ACÚST *audición* ossicle

hueso *m* AGRIC *fruticultura* core

huida *f AmL* (*cf cuele Esp*) MINAS cut, driving, TEC ESP *accidentes de vuelo* stray radiation

hule *m* P&C, TEXTIL oilcloth

hulla *f* CARBÓN bituminous coal, hard coal, pit coal; ~ **bituminosa** *f* CARBÓN bituminous coal, flaming coal; ~ **brillante** *f* CARBÓN anthracite; ~ **a granel** *f* CARBÓN, MINAS rough coal; ~ **grasa** *f* CARBÓN, MINAS fat coal, soft coal; ~ **grasa de llama larga** *f* CARBÓN, MINAS cherry coal, sintering coal; ~ **de llama larga** *f* CARBÓN, MINAS cannel coal, kennel coal; ~ **magra** *f* CARBÓN, TERMO lean coal; ~ **muerta** *f* CARBÓN, MINAS dull coal; ~ **pizarrosa** *f* CARBÓN, MINAS foliated coal, bone coal; ~ **seca** *f* CARBÓN, MINAS

cannel coal, kennel coal; ~ **semigrasa** *f* CARBÓN, MINAS cherry coal, sintering coal; ~ **en trozos gruesos** *f* CARBÓN, MINAS lump coal

hullera *f* CARBÓN, MINAS coal mine, coalworks, colliery

hullificarse *v refl* CARBÓN, MINAS carbonize

humboldita *f* MINERAL humboldtilite

humboldtina *f* MINERAL humboldtine

humbral: ~ **desplazado** *m* TRANSP AÉR displaced threshold

humear: ~ **sin llama** *vi* TERMO smolder (*AmE*), smoulder (*BrE*)

humectabilidad *f* P&C *adhesivos*, PETROL wettability

humectación *f* ING MECÁ wetting, QUÍMICA moistening, TERMO humidification

humectador *m* TERMO humidifier

humectante[1] *adj* CONTAM MAR, ING MECÁ, QUÍMICA wetting

humectante[2] *m* CONTAM MAR wetting agent, ING MECÁ wetting, QUÍMICA humectant

humectar *vt* CONTAM MAR, ING MECÁ, QUÍMICA, REFRIG, TERMO humidify, moisten, wet

humedad[1]: **sin** ~ *adj* CONTAM water-free

humedad[2] *f* GEN humidity, QUÍMICA moisture; ~ **absoluta** *f* FÍS, HIDROL, METEO, TERMOTEC absolute humidity; ~ **del aire** *f* CARBÓN air moisture, EMB, TERMO humidity of the air; ~ **atmosférica** *f* TERMO humidity of the air; ~ **específica** *f* CARBÓN moisture content, METEO specific humidity; ~ **relativa** *f* AGUA, FÍS, HIDROL, INSTAL TERM, METEO, P&C, PAPEL, REFRIG, TERMOTEC, TEXTIL relative humidity; ~ **residual** *f* EMB, REFRIG residual moisture

humedecedor *m* TERMOTEC damper, TRANSP AÉR humidifier; ~ **de barro** *m* C&V clay wetting

humedecer *vt* ALIMENT, FÍS, TERMO humidify, moisten, wet

humedecimiento *m* ING MECÁ wetting, METAL *deformaciones* dampening, PETROL damping

humedificador: ~ **de aire** *m* SEG air humidifier

húmedo: ~ **sobre húmedo** *fra* P&C *aplicación de pintura* wet on wet

humero *m* INSTAL TERM, PROD *de chimenea* flue; ~ **de gas** *m* GAS, INSTAL TERM, PROD gas flue; ~ **de retorno** *m* INSTAL TERM, PROD return flue

humidificación *f* CARBÓN air conditioning, ING MECÁ wetting, QUÍMICA hydration, TERMO humidification

humidificador *m* REFRIG, TERMO humidifier; ~ **de disco centrífugo** *m* REFRIG, TERMO spinning-disk humidifier; ~ **de disco giratorio** *m* REFRIG, TERMO spinning-disk humidifier; ~ **de vapor** *m* REFRIG, TERMO steam humidifier

humidificar *vt* REFRIG, TERMO humidify, moisten, wet

humidistato *m* REFRIG, TERMO humidistat

humita *f* MINERAL humite

humo: ~~**niebla** *f* CONTAM, METEO smog; ~~**niebla fotoquímica** *f* CONTAM photochemical smog

humos[1]: **sin** ~ *adj* ENERG RENOV smokeless

humos[2] *m pl* AUTO, QUÍMICA, SEG, TERMOTEC fumes; ~ **adhesivos** *m pl* SEG adhesive fumes; ~ **de escape** *m pl* AUTO, MECÁ, TERMO, TRANSP, TRANSP AÉR, VEH exhaust fumes, exhaust gases; ~ **de fábrica** *m pl* CONTAM, PROD, SEG factory fumes

humuleno *m* QUÍMICA humulene

humus *m* CARBÓN, CONST, MINAS, QUÍMICA humus

hundido *adj* CINEMAT sunk up; ~ **por una falla** *adj* GEOL downfaulted; ~ **de popa** *adj* TRANSP MAR down by the stern; **~ de proa** *adj* TRANSP MAR down by the head

hundimiento *m* CARBÓN settling, GEOL collapse, *de pliegues* plunge, MINAS *galería* gob, nip, sink, sinking; **~ por acción del agua** *m* CONTAM washout; **~ por la acción del agua** *m* FERRO *del terraplén* washout; **~ de mancha de aceite por corrientes marinas** *m* CONTAM oil slick sinking; **~ del terreno** *m* GEOL, TEC PETR subsidence; **~ de tierras** *m* MINAS run of ground

hundir *vt* CONST sink, MINAS pack, PROD, TEXTIL dip

huracán *m* METEO hurricane

hureaulita *f* MINERAL hureaulite

hurgón *m* PROD prick bar, *hornos* pricker

husillo *m* AUTO stud, C&V, CINEMAT spindle, CONST gin, ING MECÁ lead screw, mandrel, mandril, feed screw, guide screw, pin, screw, spindle, shaft, MECÁ lead screw, spindle, P&C mandrel, mandril, PROD *torno* spindle, *tornos* lead screw; **~ de avance** *m* ING MECÁ feed screw; **~ del carrete** *m* CINEMAT reel spindle; **~ circulante** *m* ING MECÁ circulating ball spindle; **~ circulante esférico** *m* ING MECÁ ball circulating lead screw; **~ del contrapunto** *m* ING MECÁ tail spindle; **~ del esmeril** *m* C&V grinder spindle; **~ eyector** *m* ING MECÁ sprue puller pin; **~ fijo** *m* ING MECÁ dead spindle; **~ de fresadora** *m* ING MECÁ milling-machine arbor; **~ de gato** *m* ING MECÁ jack-off screw; **~ giratorio** *m* ING MECÁ stud; **~ de la máquina herramienta** *m* ING MECÁ machine tool spindle; **~ de mordazas** *m* ING MECÁ collet-type spindle; **~ de la pala** *m* TRANSP AÉR blade spindle; **~ de pinzas** *m* ING MECÁ collet-type spindle; **~ principal del carro** *m* ING MECÁ feed shaft

huso *m* AUTO sleeve, PROD, TEXTIL spindle; **~ de la cabeza movible** *m* ING MECÁ live spindle; **~ flotante** *m* ING MECÁ floating spindle; **~ horario** *m* TRANSP MAR time zone; **~ de la muela de esmeril** *m* PROD emery-wheel spindle

hW *abr* (*hectovatio*) ELEC, ING ELÉC, METR hW (*hectowatt*)

hyperventilación *f* OCEAN hyperventilation

Hz *abr* (*hercio, hertz*) GEN Hz (*hertz*)

I

I *abr* (*iodo, yodo*) ELECTRÓN, FOTO, QUÍMICA, TEXTIL **I** (*iodine*)

IA *abr* AGRIC (*inseminación artificial*) **AI** (*artificial insemination*), INFORM&PD, TELECOM (*inteligencia artificial*) **AI** (*artificial intelligence*)

IAC *abr* (*identificación automática de coches*) TRANSP **ACI** (*automatic car identification AmE, automatic wagon identification BrE*)

IAV *abr* (*identificación automática de vehículos*) TRANSP **AVI** (*automatic vehicle identification*)

icnofósil *m* GEOL ichnofossil

icono *m* INFORM&PD icon

iconoscopio *m* ELECTRÓN iconoscope; **~ de imagen** *m* ELECTRÓN image iconoscope

icosaédrico *adj* GEOM icosahedral

icosaedro *m* GEOM icosahedron

ictiocola *f* ALIMENT isinglass

identidad *f* MATEMÁT *símbolo* identity; **~ del abonado llamado** *f* TELECOM called-line identity (*CDLI*); **~ de la estación barco** *f* TELECOM ship-station identity; **~ de estación costera** *f* TELECOM coastal-station identity; **~ de la estación llamada** *f* TELECOM called-station identity (*CSI*); **~ de grupo llamante** *f* TELECOM group call identity; **~ de línea conectada** *f* TELECOM connected-line identity (*COLI*); **~ de la línea de llamada** *f* TELECOM called-line identity (*CDLI*); **~ de llamada** *f* TELECOM call identity

identificación *f* ACÚST, INFORM&PD, ING MECÁ, SEG *de riesgos* identification; **~ automática de coches** *f* (*IAC*) TRANSP automatic car identification (*AmE*) (*ACI*), automatic wagon identification (*BrE*) (*ACI*); **~ automática de vagones** *f* TRANSP automatic car identification (*AmE*) (*ACI*), automatic wagon identification (*BrE*) (*ACI*); **~ automática de vehículos** *f* TRANSP automatic vehicle identification (*AVI*); **~ del avión** *f* TRANSP AÉR aircraft identification; **~ de la cadena** *f* TV network identification; **~ de contenido** *f* SEG *de contenedores de gases industriales* identification of contents; **~ del destinatario** *f* EMB address label; **~ del emisor** *f* TELECOM sender identification; **~ del emisor del intercambio** *f* TELECOM interchange sender identification; **~ de la emisora** *f* TV station identification; **~ de la línea de llamada** *f* TELECOM called-line identification (*CLI*); **~ de llamada** *f* TELECOM call identification; **~ mediante código de barras secreto** *f* EMB hidden bar-code identification; **~ de multiflexión** *f* TELECOM multiplexing identification; **~ del plan de numeración** *f* TELECOM numbering-plan identification (*NPI*); **~ radárica** *f* D&A, TRANSP, TRANSP AÉR, TRANSP MAR radar identification; **~ del receptor del intercambio** *f* TELECOM interchange receiver identification

identificador *m* INFORM&PD, TELECOM identifier (*ID*); **~ de archivos** *m* INFORM&PD file identifier; **~ de autorización y formato** *m* TELECOM authority and format identifier (*AFI*); **~ de canal virtual** *m* TELECOM virtual channel identifier (*VCI*); **~ de conexión para la transmisión de datos** *m* TELECOM data link connection identifier (*DLCI*); **~ de conformidad adjudicación** *m* TELECOM submission identifier; **~ de dominio inicial** *m* TELECOM initial-domain identifier (*IDI*); **~ del enlace para la transmisión de datos** *m* TELECOM data link connection identifier (*DLCI*); **~ del punto de acceso al servicio** *m* TELECOM service access point identifier (*SAPI*); **~ del punto final terminal** *m* TELECOM terminal end-point identifier (*TEI*); **~ sintáctico** *m* TELECOM syntax identifier; **~ de la vía de transmisión virtual** *m* TELECOM virtual path identifier (*VPI*)

identificar *vt* GEN identify

identificarse *v refl* PROD *comunicación* log in

ideograma *m* INFORM&PD ideogram

idioblástico *adj* GEOL idioblastic

idiomórfico *adj* GEOL automorphic

idiomorfo *adj* CRISTAL, METAL, PETROL, TEC PETR idiomorphic, idiomorphous

iditol *m* QUÍMICA iditol

idocrasa *f* MINERAL idocrase, vesuvianite

idónico *adj* QUÍMICA idonic

idosa *f* QUÍMICA idose

idosacárico *adj* QUÍMICA idosaccharic

idranal *m* QUÍMICA idranal

idrialina *f* MINERAL idrialite

IEM *abr* ELEC (*inducción electromagnética, interferencia electromagnética*) **EMI** (*electromagnetic induction, electromagnetic interference*), ELECTRÓN, INFORM&PD (*interferencia electromagnética*) **EMI** (*electromagnetic interference*), ING ELÉC, PROD (*inducción electromagnética, interferencia electromagnética*) **EMI** (*electromagnetic induction, electromagnetic interference*), TEC ESP, TELECOM (*interferencia electromagnética*) **EMI** (*electromagnetic interference*)

IFRB *abr* (*Comité Internacional de Registro de Frecuencias*) TEC ESP **IFRB** (*International Frequency Registration Board*)

IGE *abr* (*integración a gran escala*) ELECTRÓN, FÍS, INFORM&PD, TELECOM, TRANSP MAR **LSI** (*large-scale integration*)

ignición *f* GEN ignition; **~ de apertura y cierre** *f* AUTO, VEH make-and-break ignition; **~ de arco** *f* ELEC arc ignition; **~ automática** *f* GAS automatic ignition; **~ de avance** *f* AUTO, VEH advanced ignition; **~ por condensador** *f* AUTO, VEH capacitor ignition; **~ sin contactos** *f* AUTO, VEH contactless ignition (*BrE*), pointless ignition (*AmE*); **~ electromagnética** *f* AUTO, ELEC, VEH electromagnetic ignition; **~ electrónica** *f* AUTO, ELEC, ELECTRÓN, VEH electronic ignition; **~ espontánea** *f* TEC ESP spontaneous ignition; **~ de fricción** *f* MINAS frictional ignition; **~ hipergólica** *f* TEC ESP hypergolic ignition; **~ por magneto** *f* AUTO, VEH magneto ignition; **~ sin platinos** *f* AUTO, VEH contactless ignition (*BrE*), pointless ignition (*AmE*); **~ prematura** *f* AUTO, VEH premature ignition; **~ por rozamiento** *f* MINAS frictional ignition; **~ sincronizada** *f* TEC ESP phased ignition; **~ transistorizada sin platinos** *f* AUTO, VEH contactless transistorized ignition (*BrE*), pointless transistorized ignition (*AmE*)

ignifugación *f* GEN fireproofing, flameproofing

ignifugado *adj* GEN fireproofed, flameproofed

ignifugar *vt* GEN fireproof

ignífugo *adj* GEN fire-resistant, fire-resisting, fire-retarding, fireproof, flame-resistant, flameproof, nonflammable

ignimbrita *f* GEOL ignimbrite

ignitor *m* ING ELÉC, TEC ESP igniter

ignitrón *m* ELEC, ING ELÉC *tubo electrónico* ignitron

igual: de ~ a igual *adj* INFORM&PD peer-to-peer

igualación *f* GEN equalization, MECÁ, PROD equalization, leveling (*AmE*), levelling (*BrE*); **~ de colores** *f* CINEMAT color matching (*AmE*), colour matching (*BrE*); **~ de gruesos** *f* ING MECÁ thicknessing; **~ de presión** *f* OCEAN pressure equalization; **~ de temperatura** *f* TERMO temperature equalization, temperature equalizing

igualado *m* PAPEL guillotining trimming; **~ de la madera** *m* REVEST dubbing

igualador *m* GEN equalizer; **~ de corriente alterna** *m* ELEC alternating-current balancer; **~ de coseno** *m* TV cosine equalizer; **~ de fase** *m* ELEC *corriente alterna* phase equalizer; **~ de la pila de salida** *m* IMPR delivery jogger; **~ de pilas** *m* IMPR pile jogger

igualar *vt* CINEMAT match, IMPR *hojas* jog, ING MECÁ true up, MECÁ level, PAPEL *paquete de hojas* knock up, PROD *madera* flush, trim

igualdad *f* INFORM&PD equality

IL *abr* (*lógica integrada a inyección*) ELECTRÓN IL (*integrated injection logic*)

ilegalidad *f* TV bootleg

iliácico *adj* QUÍMICA iliac

ilita *f* CARBÓN, GEOL, MINERAL, TEC PETR illite

ilmenita *f* MINERAL ilmenite

ILS *abr* (*sistema de aterrizaje por instrumentos*) TRANSP AÉR ILS (*instrument landing system*)

ilsemannita *f* MINERAL ilsemannite

iluminación *f* CINEMAT illumination, lighting, ELECTRÓN lighting, ING ELÉC illumination, searchlight, METAL illumination; **~ de borde** *f* ING ELÉC edge lighting; **~ en campo claro** *f* FÍS bright-field illumination; **~ en campo oscuro** *f* FÍS dark-field illumination; **~ de contorno** *f* CINEMAT outline lighting; **~ a contraluz** *f* CINEMAT, FOTO, ING ELÉC, TV backlighting; **~ sin contraste** *f* CINEMAT, FOTO, TV flat lighting; **~ difusa** *f* CINEMAT, FOTO, TV diffused light, soft light; **~ direccional** *f* CINEMAT, FOTO, TV directional lighting; **~ disponible** *f* CINEMAT, FOTO, TV available light; **~ por efecto de campo** *f* INSTR field illumination; **~ para efectos** *f* CINEMAT, FOTO, TV effect lighting; **~ eléctrica** *f* ELEC *instalación* electric lighting; **~ de emergencia** *f* ELEC emergency lighting; **~ de la escala** *f* TRANSP MAR *radar* scale illumination; **~ especial** *f* CINEMAT *para establecer el tono emocional de una escena* mood lighting; **~ existente** *f* CINEMAT, FOTO, TV existing light; **~ fluorescente** *f* ELEC, FÍS, FÍS RAD, GAS, ING ELÉC fluorescent lighting; **~ del fondo** *f* CINEMAT, FOTO, TV background lighting; **~ sobre fondo oscuro** *f* FÍS dark-ground illumination; **~ frontal** *f* CINEMAT, FOTO, TV front lighting; **~ inactínica** *f* CINEMAT, FOTO darkroom lighting; **~ incidente** *f* INSTR incident illumination; **~ indirecta** *f* CINEMAT, FOTO indirect lighting, ING ELÉC indirect illumination;

~ inferior *f* CINEMAT, TV bottom lighting; **~ interior** *f* ELEC, ING ELÉC indoor lighting; **~ por láser** *f* ELECTRÓN laser illumination; **~ mediante proyectores** *f* ELEC floodlighting; **~ de mucho contraste** *f* CINEMAT high key; **~ oblicua** *f* METAL oblique illumination; **~ óptima del objeto** *f* FÍS RAD optimum object illumination; **~ plana** *f* CINEMAT, FOTO, TV flat lighting; **~ posterior** *f* CINEMAT backlight; **~ posterior para animación** *f* CINEMAT animation backlight; **~ proveniente de una fuente puntual** *f* FOTO point-source light; **~ proyectada** *f* ELEC floodlighting; **~ de realce** *f* CINEMAT, FOTO, TV modeling light (*AmE*), modelling light (*BrE*); **~ por rebote** *f* CINEMAT, FOTO, TV bounce light, bounce lighting; **~ para resaltar el cabello** *f* CINEMAT hair light; **~ de seguridad** *f* CINEMAT, FOTO darkroom lighting; **~ de tierra** *f* TRANSP AÉR ground lighting; **~ de tungsteno** *f* CINEMAT tungsten lighting; **~ uniforme** *f* CINEMAT, FOTO, TV flat lighting; **~ vertical** *f* INSTR vertical illumination

iluminado *adj* GEN illuminated; **~ con luz posterior** *adj* FOTO backlit

iluminador *m* CINEMAT, FOTO, TV lighting cameraman

iluminancia *f* FÍS illuminance

iluminar *vt* CONST beacon; **~ frontalmente** *vt* FOTO front-light; **~ lateralmente** *vt* FOTO sidelight; **~ con luz de arco voltaico** *vt* CINEMAT arc; **~ con luz posterior** *vt* FOTO backlight

ilustración *f* IMPR figure, illustration; **~ que abarca dos páginas** *f* IMPR gutter bleed; **~ que forma el fondo** *f* IMPR background art

ilustraciones *f pl* IMPR art, artwork

ilustrar *vt* IMPR illustrate

ilvaíta *f* MINERAL ilvaite, lieberenite

imagen *f* ACÚST, C&V image, CINEMAT field, frame, image, picture, ELECTRÓN frame, image, FÍS image, FOTO field, frame, image, picture, IMPR image, INFORM&PD frame, picture, ING ELÉC, INSTR, METAL, NUCL, PROD image, TEC ESP image, *televisión* frame, TV frame, image, picture, field; **~ aérea** *f* CINEMAT, FOTO, TV aerial image; **~ de alta frecuencia** *f* TEC ESP high-frequency image; **~ ampliada** *f* FÍS, FOTO enlarged image; **~ aumentada del visor** *f* FOTO magnified-viewfinder image; **~ de baja frecuencia** *f* TEC ESP low-frequency image; **~ binaria** *f* TEC ESP, TELECOM binary image; **~ borrosa** *f* FÍS blurred image, FOTO fuzzy image; **~ borrosa por movimiento** *f* FOTO motion blur; **~ de campo claro** *f* CRISTAL bright-field image; **~ de campo oscuro** *f* CRISTAL dark-field image; **~ cegada** *f* IMPR *planchas tipográficas* blind image; **~ de la claqueta** *f* CINEMAT picture clap; **~ a color** *f* FOTO color picture (*AmE*), colour picture (*BrE*); **~ congelada** *f* TELECOM freeze-picture; **~ de las costas por radar** *f* TRANSP, TRANSP MAR radar coast image; **~ cromática** *f* LAB, TEC ESP color display (*AmE*), colour display (*BrE*); **~ descomprimida** *f* CINEMAT unsqueezed image; **~ desenfocada** *f* CINEMAT, FOTO, TV out-of-focus image; **~ digital** *f* ELECTRÓN digital image; **~ digitalizada** *f* ELECTRÓN, INFORM&PD digitized image; **~ doble** *f* CINEMAT double image; **~ de un eco** *f* TELECOM pip; **~ eléctrica** *f* ELEC, ING ELÉC electric image; **~ electrónica** *f* ELECTRÓN electron image; **~ especular** *f* CINEMAT, QUÍMICA mirror image; **~ fantasma** *f* CINEMAT, FOTO ghost, ghost image, IMPR echo image, printing echo, TV ghost,

ghost image; ~ **fija** *f* CINEMAT, TV still frame; ~ **de grano fino** *f* CINEMAT, FOTO, TV fine-grain image; ~ **de grano grueso** *f* CINEMAT, FOTO, TV coarse-grain image; ~ **incolora en bajo relieve** *f* IMPR blind stamp; ~ **instantánea** *f* FOTO, INFORM&PD snapshot; ~ **intermedia** *f* FÍS intermediate image; ~ **inversa** *f* MATEMÁT inverse image; ~ **invertida** *f* CINEMAT, FÍS, FOTO, TV inverted image, reversed image; ~ **invertida lateralmente** *f* CINEMAT, FÍS, FOTO, TV laterally-inverted image; ~ **latente** *f* CINEMAT, FÍS, FOTO, IMPR, ÓPT, TV latent image; ~ **ligeramente difusa** *f* CINEMAT, FOTO, TV soft focus; ~ **de multisensores** *f* TEC ESP multisensor image; ~ **negativa** *f* FOTO negative image; ~ **no nítida** *f* FÍS blurred image; ~ **numérica** *f* ELECTRÓN, INFORM&PD digitized image; ~ **óptica** *f* ELECTRÓN, ÓPT optical image, optical pattern; ~ **de pantalla** *f* INFORM&PD screen image; ~ **de pantalla grande** *f* CINEMAT wide-screen picture; ~ **de paso alto** *f* TEC ESP high-pass image; ~ **de paso bajo** *f* TEC ESP low-pass image; ~ **patrón** *f* TV patterning; ~ **patrón de prueba** *f* ELECTRÓN test pattern; ~ **positiva** *f* FOTO positive image; ~ **posterior** *f* TV after image; ~ **por radar** *f* TEC ESP radar image; ~ **de rayos X** *f* INSTR, LAB X-ray image; ~ **real** *f* FÍS real image; ~ **regenerada** *f* ELECTRÓN, INFORM&PD refreshed image; ~ **revelada** *f* CINEMAT, FOTO developed image; ~ **secundaria** *f* CINEMAT afterimage; ~ **térmica** *f* D&A, FÍS RAD, TERMO thermal imaging *(TI)*; ~ **tridimensional** *f* TELECOM three-dimensional image; ~ **vertical** *f* (*HUD*) INSTR head-up display (*HUD*); ~ **virtual** *f* CINEMAT, FÍS, FÍS ONDAS, FOTO, TV virtual image

imágenes: ~ **eléctricas** *f pl* ELEC *problemas de potencial* electrical images; ~ **recogidas en la calle por cámaras móviles** *f pl* TV field pick-up; ~ **del sistema** *f pl* ELEC *red de distribución* system pattern

imán *m* ELEC, FÍS, LAB magnet; ~ **en U** *m* ING MECÁ U magnet; ~ **anular** *m* *Esp* (*cf imán con forma de anillo AmL*) FÍS, TV ring magnet, annular magnet; ~ **de Bitter** *m* FÍS Bitter magnet; ~ **de cavidad esférica** *m* GEOFÍS, TEC ESP magnet spheric cavity; ~ **de desviación** *m* TV deflection magnet; ~ **de dirección** *m* TRANSP guidance magnet; ~ **director** *m* TRANSP MAR *montado en la rosa* compass needle; ~ **elevador** *m* TRANSP lift magnet; ~ **de enfoque** *m* CINEMAT, FOTO, ING ELÉC, TEC ESP, TV focusing magnet; ~ **con forma de anillo** *m* *AmL* (*cf imán anular Esp*) FÍS, TV annular magnet, ring magnet; ~ **de haz azul** *m* TV blue-beam magnet; ~ **en herradura** *m* ELEC, FÍS, ING MECÁ horseshoe magnet; ~ **laminado** *m* ELEC laminated magnet, FÍS compound magnet; ~ **de maniobra** *m* PROD operating magnet; ~ **permanente** *m* ELEC, FÍS, ING ELÉC, ING MECÁ, TELECOM, TRANSP permanent magnet; ~ **posicionador haz** *m* TV beam-positioning magnet; ~ **de propulsión** *m* TRANSP propulsion magnet; ~ **de rayos rojos** *m* ELECTRÓN, TV red-beam magnet; ~ **recto** *m* FÍS bar magnet; ~ **superconductor** *m* FÍS PART superconducting magnet; ~ **toroidal** *m* FÍS toroidal magnet; ~ **de vías** *m* FERRO track magnet

imanación *f* GEN magnetization; ~ **espontánea** *f* FÍS RAD spontaneous magnetization; ~ **remanente** *f* ELEC, FÍS, ING ELÉC, PETROL remanence; ~ **residual** *f* ELEC, FÍS remanence, FÍS RAD residual magnetization, ING ELÉC, PETROL remanence

imanar *vt* ELEC, FÍS, ING ELÉC magnetize

imán: ~~**-guía** *m* TRANSP guidance magnet

imantación *f* GEN magnetization; ~ **perpendicular** *f* TV perpendicular magnetization; ~ **remanente** *f* FÍS RAD residual magnetization

imantar *vt* GEN magnetize

imbibición *f* HIDROL imbibition

imbornal *m* TRANSP MAR scupper, *construcción naval* limber hole

imbornales: ~ **de la bañera** *m pl* TRANSP MAR cockpit drainage

imbricado *adj* CONST, GEOL imbricate, imbricated

IME *abr* (*integración a media escala, integración de mediana escala*) ELECTRÓN, FÍS, INFORM&PD, TELECOM MSI (*medium-scale integration*)

imida *f* QUÍMICA *síntesis orgánica* imide

imidazol *m* QUÍMICA glyoxaline

imido *m* QUÍMICA imido

imidógeno *m* QUÍMICA imidogen

imina *f* QUÍMICA imine

imitación *f* PAPEL bogus

IML *abr* (*módulo de imágen por cristal líquido*) ELEC, ELECTRÓN, FÍS, INFORM&PD, TELECOM, TV LCD module (*liquid crystal display module*)

impactar *vt* INFORM&PD hit; ~ **contra** *vt* SEG crash into

impacto *m* ACÚST, CARBÓN, CONST, FÍS impact, INFORM&PD hit, ING MECÁ impingement, MECÁ impact, MINAS *proyectiles* shot, TEC ESP crash; ~ **ambiental** *m* CONTAM, CONTAM MAR, NUCL environmental impact; ~ **ecológico** *m* CONTAM, CONTAM MAR, NUCL environmental impact; ~ **elástico** *m* NUCL elastic impact; ~ **en el estante** *m* EMB *mercadotecnia* shelf impact; ~ **de haz luminoso** *m* TRANSP spot

impagado *adj* PROD outstanding

imparcial *adj* INFORM&PD unbiased

impartir *vt* ING MECÁ impart

impecable *adj* TEXTIL sleek

impedancia *f* ELEC, FÍS impedance, ING ELÉC impedance, impedor, TELECOM impedance; ~ **de acoplamiento** *f* ELEC coupling impedance; ~ **activa** *f* ELEC *componente* active impedor; ~ **acústica** *f* ACÚST, FÍS, ING ELÉC acoustic impedance; ~ **acústica específica** *f* ACÚST, FÍS, ING ELÉC specific acoustic impedance; ~ **acústica de transferencia** *f* ACÚST, FÍS, ING ELÉC transfer acoustic impedance; ~ **bloqueada** *f* ING ELÉC blocked impedance; ~ **característica** *f* ACÚST characteristic impedance, ELEC *cable* surge impedance, FÍS, ING ELÉC characteristic impedance; ~ **de carga** *f* FÍS, TELECOM load impedance; ~ **sin carga** *f* ING ELÉC blocked impedance; ~ **cargada** *f* ACÚST loaded impedance; ~ **en circuito abierto** *f* ELEC, ING ELÉC open-circuit impedance; ~ **compleja** *f* ELEC, FÍS, ING ELÉC complex impedance; ~ **en cortocircuito** *f* ELEC, ING ELÉC short-circuit impedance; ~ **del diodo** *f* ELECTRÓN diode impedance; ~ **directa del diodo** *f* ELECTRÓN diode forward impedance; ~ **elástica** *f* PETROL elastic impedance; ~ **eléctrica bloqueada mecánicamente** *f* ACÚST, ELEC, ING ELÉC mechanically-blocked electrical impedance; ~ **eléctrica mocional libre** *f* ACÚST, ELEC, ING ELÉC free electrical motional impedance; ~ **electrocinética** *f* ACÚST, ELEC electrical-kinetic impedance; ~ **electrovibratoria libre** *f* ACÚST free electrical vibration impedance; ~ **de entrada** *f* ELEC, FÍS, ING ELÉC, TELECOM, TV input impedance; ~ **equilibrada** *f* ING ELÉC matched impedance; ~ **del**

generador *f* ING ELÉC source impedance; ~ **homopolar** *f* ELEC *devanado polifásico* zero sequence impedance; ~ **imagen** *f* ACÚST, FÍS, ING ELÉC, TELECOM image impedance; ~ **intrínseca** *f* ELEC, ING ELÉC *electromagnetismo* intrinsic impedance; ~ **iterativa** *f* ACÚST, ELEC, FÍS, ING ELÉC iterative impedance; ~ **de línea** *f* ELEC, ING ELÉC line impedance; ~ **mecánica** *f* ACÚST, ELEC, ING ELÉC mechanical impedance; ~ **mecánica libre** *f* ACÚST, ELEC, ING ELÉC free mechanical impedance; ~ **mecánica de transferencia** *f* ACÚST, ELEC, ING ELÉC transfer mechanical impedance; ~ **mutua** *f* TELECOM mutual impedance; ~ **negativa** *f* ING ELÉC negative impedance; ~ **propia** *f* ELEC *cable* surge impedance; ~ **de radiación** *f* FÍS RAD, TELECOM radiation impedance; ~ **reflejada** *f* ING ELÉC, ÓPT reflected impedance; ~ **de salida** *f* ELEC, FÍS, ING ELÉC, TELECOM, TV output impedance; ~ **de terminación** *f* ELEC, ING ELÉC terminating impedance; ~ **de transferencia** *f* ELEC, TELECOM transfer impedance; ~ **de transición** *f* ELEC *de resistor, reactancia* transition impedance

impedancias: ~ **conjugadas** *f pl* ACÚST conjugate impedances

impedimento *m* ELECTRÓN *en circuitos lógicos* inhibition

impedor: ~ **activo** *m* ELEC *componente* active impedor

impeler: ~ **con fuerza** *vt* TRANSP MAR ram

impenetrable *adj* GEOL, MECÁ, TEC PETR impervious; ~ **al polvo** *adj* ELEC dustproof

imperativo *adj* INFORM&PD imperative

imperdible *adj* MECÁ captive

imperfección *f* CRISTAL inclusion, ING MECÁ vice (*BrE*), vise (*AmE*), MECÁ flaw; ~ **cristalina** *f* CALIDAD, CRISTAL, ÓPT crystal defect; ~ **de superficie** *f* CALIDAD surface defect

imperfecto *adj* GEN faulty

impermeabilidad *f* AGUA, CONST, EMB impermeability, ENERG RENOV imperviousness, GAS, GEOL, HIDROL, PETROL impermeability, REVEST water repellency, TELECOM impermeability; ~ **al agua** *f* CONST watertightness

impermeabilización[1]: **de** ~ *adj* PROD staunch

impermeabilización[2] *f* CONST waterproofing, PROD staunchness, REVEST, TEXTIL waterproofing; ~ **al moho** *f* REVEST mildew-proofing; ~ **asfáltica** *f* CONST asphalt tanking

impermeabilizar *vt* CONST make impermeable, MINAS insulate, REVEST, TEXTIL waterproof

impermeable *adj* GEN impermeable, impervious, waterproof, watertight; ~ **al agua** *adj* EMB watertight; ~ **al gas** *adj* GAS, ING MECÁ, MECÁ, PROC QUÍ, PROD, QUÍMICA, TERMO gas-proof, gastight

ímpetu *m* FÍS, ING MECÁ, MECÁ impetus, momentum

impide: ~~**erupciones** *m* TEC PETR blowout preventer (*BOP*); ~~**reventones** *m* TEC PETR blowout preventer (*BOP*)

implantación: ~ **de arsénico** *f* ELECTRÓN arsenic implantation; ~ **de ion radiactivo** *f* ELECTRÓN, FÍS PART, FÍS RAD radioactive-ion implantation; ~ **de iones** *f* ELECTRÓN, FÍS PART ion implantation; ~ **iónica** *f* ELECTRÓN, FÍS PART ion implantation

implantar *vt* GEN implant

implante *m* AGRIC implant

implementación *f* GEN implementation

implementar *vt* ING MECÁ, TELECOM implement

implemento *m* ING MECÁ implement; ~ **lister** *m* AGRIC lister

implosión *f* ELECTRÓN, NUCL, PROD, QUÍMICA implosion

implosionar *vi* ELECTRÓN, NUCL, PROD, QUÍMICA implode

impolarizado *adj* INFORM&PD unbiased

imponer *vt* IMPR impose

importación *f* INFORM&PD copy-in

importancia *f* MINAS *de los daños* extent

imposición *f* IMPR imposition; ~ **a blanco y voltereta** *f* IMPR work and whirl; ~ **e impresión a blanco y vuelta** *f* IMPR half sheetwork; ~ **para imprimir a blanco y retiración en dos formas distintas** *f* IMPR sheetwise make-up

imposta *f* CONST impost, springing line, *arquitectura* fascia

impregnación *f* CONST *tratamiento químico*, EMB, ING ELÉC impregnation, P&C impregnation, rubberizing, frictioning, PAPEL, QUÍMICA impregnation, REVEST rubberizing, impregnation, TERMO moistening, TEXTIL impregnation

impregnado *adj* GEN impregnated; ~ **de caucho** *adj* P&C, REVEST rubberized, TERMO moistened

impregnante *m* GEN impregnant

impregnar *vt* CONST, EMB, ING ELÉC, PAPEL, QUÍMICA, REVEST impregnate, TERMO moisten, TEXTIL impregnate; ~ **de caucho** *vt* P&C, REVEST rubberize

imprenta *f* IMPR print shop, printing works; ~ **fósil** *f* GEOL fossil imprint

impresión[1]: **en** ~ *adv* IMPR now printing

impresión[2] *f* C&V impression, printing, FOTO printing, IMPR presswork, printer output, printing, INFORM&PD print, printing, PROD *de moldeo* impression, print, TEXTIL impression; ~ **a la anilina** *f* EMB aniline printing; ~ **aterciopelada** *f* IMPR flock printing; ~ **bicolor** *f* IMPR two-color printing (*AmE*), two-colour printing (*BrE*); ~ **a blanco y retiración** *f* IMPR sheetwork, work and back; ~ **a blanco y voltereta** *f* IMPR tumble printing; ~ **a blanco y vuelta** *f* IMPR twelve ways back up; ~ **en bronce** *f* IMPR bronzing; ~ **en color** *f* IMPR colorwork (*AmE*), colourwork (*BrE*); ~ **en colores** *f* IMPR process printing; ~ **consecutiva** *f* IMPR crash printing; ~ **por contacto** *f* CINEMAT, FOTO, IMPR contact printing; ~ **continua** *f* IMPR endless printing; ~ **de copias definitivas** *f* CINEMAT, IMPR release printing; ~ **sin costura** *f* IMPR seamless printing; ~ **a cuatro colores** *f* IMPR four-color printing (*AmE*), four-colour printing (*BrE*); ~ **del dorso** *f* IMPR back printing, backing-up; ~ **con dos fotograbados superpuestos** *f* IMPR duotype; ~ **de envases** *f* IMPR packaging printing; ~ **de etiquetas formando bobinas** *f* EMB roll label printing; ~ **en flexografía** *f* IMPR aniline rubber-plate printing; ~ **flexográfica** *f* IMPR, P&C flexographic printing; ~ **de fondos** *f* IMPR underprinting; ~ **de formularios** *f* IMPR business-forms printing; ~ **fuera de máquina** *f* EMB, IMPR printing off-line; ~ **en huecograbado** *f* IMPR gravure printing, intaglio printing, rotogravure printing, P&C rotogravure printing; ~ **en hueco-offset** *f* IMPR offset deep printing; ~ **in situ** *f* EMB printing on site; ~ **invertida** *f* IMPR back printing; ~ **láser** *f* IMPR, INFORM&PD, ÓPT laser printing; ~ **limpia** *f* IMPR clean printing; ~ **litográfica** *f* IMPR offset; ~ **manchada** *f* IMPR smudging print; ~ **manual** *f* IMPR handprinting;

~ **en máquina** *f* EMB, IMPR printing on-line; ~ **por máquina** *f* IMPR printing by machine; ~ **mediante grabados de madera** *f* IMPR block print, block printing; ~ **por medios ópticos** *f* IMPR optical printing; ~ **metalgráfica** *f* IMPR tin printing; ~ **en minúscula** *f* IMPR lower printing; ~ **a molde sencillo** *f* IMPR one-up; ~ **multicolor** *f* IMPR multicolor printing (*AmE*), multicolour printing (*BrE*); ~ **en negativo** *f* CINEMAT, FOTO, IMPR reverse printing; ~ **en negro** *f* IMPR printing in black; ~ **en oro** *f* IMPR gold blocking; ~ **en papel carbro** *f* FOTO, IMPR carbro printing; ~ **con plancha** *f* IMPR plate printing; ~ **de un plano** *f* CONST drawing print; ~ **policroma** *f* IMPR process printing; ~ **principal** *f* TEXTIL over-riding impression; ~ **por proyección de haz de electrones** *f* ELECTRÓN, IMPR electron-beam projection printing; ~ **a prueba de alcohol** *f* IMPR alcohol-proof printing; ~ **realizada con poca presión** *f* IMPR kiss impression; ~ **por reflexión** *f* IMPR reflection print; ~ **a registro** *f* IMPR register; ~ **con registro exacto** *f* EMB accurate print registration; ~ **de remiendos** *f* IMPR jobbing; ~ **del reverso** *f* EMB, IMPR reverse-side printing; ~ **en rotativa** *f* IMPR rotary printing; ~ **de salida** *f* INFORM&PD printout; ~ **a sangre** *f* IMPR bleed printing; ~ **en seco** *f* IMPR blind blocking, blind printing; ~ **serigráfica rotativa** *f* IMPR rotary-screen printing; ~ **sobre superficie grabada** *f* IMPR etched-surface printing; ~ **tipográfica** *f* IMPR relief printing; ~ **tricromía** *f* IMPR three-color printing (*AmE*), three-colour printing (*BrE*); ~ **a varios colores** *f* IMPR process printing

impreso *m* INFORM&PD form, printout; ~ **comercial** *m* IMPR business form; ~ **personalizado** *m* IMPR customized form; ~ **satinado** *m* IMPR glossy print

impresor *m* IMPR print-maker, printer, TELECOM printer; ~ **analizador** *m* IMPR, INSTR scanner printer; ~ **a proyección** *m* CINEMAT projection printer; ~ **telegráfico** *m* TELECOM teletype (*TTY*)

impresora *f* IMPR, INFORM&PD, TELECOM printer, printer machine, printing machine; ~ **de acceso inmediato** *f Esp* (*cf impresora de impacto al vuelo AmL*) IMPR, INFORM&PD hit-on-the-fly printer; ~ **de alta velocidad** *f* IMPR high-speed printer; ~ **de banda** *f* IMPR, INFORM&PD band printer, belt printer; ~ **de cadena** *f* IMPR, INFORM&PD chain printer (*AmE*), train printer (*BrE*); ~ **de cajas en troquel** *f* EMB, IMPR sheet-fed carton printer; ~ **de caracteres** *f* IMPR, INFORM&PD character printer; ~ **a chicler de tinta** *f AmL* (*cf impresora de chorro de tinta Esp*, *impresora de inyección de tinta Esp*), IMPR, INFORM&PD bubble-jet printer, inkjet printer; ~ **de chorro de tinta** *f Esp* (*cf impresora a chicler de tinta AmL*) IMPR, INFORM&PD bubble-jet printer, inkjet printer; ~ **de cinta** *f* IMPR, INFORM&PD belt printer; ~ **en color** *f* FOTO, IMPR color-printing machine (*AmE*), colour-printing machine (*BrE*); ~ **por contacto** *f* CINEMAT, FOTO, IMPR contact printer; ~ **y copiadora láser** *f* IMPR, ÓPT laser printer-copier; ~ **electrofotográfica** *f* ELEC, IMPR, INFORM&PD electrophotographic printer; ~ **electrográfica** *f* ELEC, IMPR, INFORM&PD electrographic printer; ~ **electrosensible** *f* ELEC, IMPR, INFORM&PD electrosensitive printer; ~ **electrostática** *f* ELEC, IMPR, INFORM&PD electrostatic printer; ~ **electrotérmica** *f* ELEC, IMPR, INFORM&PD electrothermal printer; ~ **de etiquetas con el código de barras** *f* EMB bar-code label printer; ~ **de impacto** *f* IMPR, INFORM&PD impact printer; ~ **sin impacto** *f* IMPR, INFORM&PD nonimpact printer; ~ **de impacto al vuelo** *f AmL* (*cf impresora de acceso inmediato Esp*) IMPR, INFORM&PD hit-on-the-fly printer; ~ **de inyección de tinta** *f Esp* (*cf impresora a chicler de tinta AmL*) IMPR, INFORM&PD bubble-jet printer, inkjet printer; ~ **láser** *f* IMPR, INFORM&PD, ÓPT laser printer; ~ **de línea** *f* IMPR, INFORM&PD line printer; ~ **magnetográfica** *f* IMPR, INFORM&PD magnetographic printer; ~ **de margarita** *f Esp* (*cf impresora a rueda de mariposa AmL*) IMPR, INFORM&PD daisy-wheel printer; ~ **matricial** *f* IMPR, INFORM&PD matrix printer, dot matrix printer; ~ **de matriz de puntos** *f* IMPR, INFORM&PD dot matrix printer; ~ **offset de oficina** *f* IMPR office printing machine; ~ **de páginas** *f* IMPR, INFORM&PD page-printer; ~ **de puntos** *f* IMPR, INFORM&PD dot printer, dot matrix printer; ~ **ranuradora** *f* EMB *fabricación de cajas de cartón ondulado* printer-slotter; ~ **de rodillo** *f* IMPR, INFORM&PD barrel printer (*BrE*), drum printer (*AmE*); ~ **a rueda de mariposa** *f AmL* (*cf impresora de margarita Esp*, *impresora de tipo margarita Esp*), IMPR, INFORM&PD daisywheel printer; ~ **en serie** *f* IMPR, INFORM&PD serial printer; ~ **serigráfica** *f* EMB screen printing machine; ~ **de tambor** *f* IMPR, INFORM&PD barrel plotter (*BrE*), barrel printer (*BrE*), drum plotter (*AmE*), drum printer (*AmE*); ~ **térmica** *f* IMPR, INFORM&PD thermal printer; ~ **de tipo margarita** *f Esp* (*cf impresora a rueda de mariposa AmL*) IMPR, INFORM&PD daisywheel printer; ~ **tipográfica** *f* IMPR letter-press printing machine; ~ **de tren** *f* IMPR, INFORM&PD chain printer (*AmE*), train printer (*BrE*); ~ **para trucado** *f* CINEMAT trick printer

impresos *m pl* IMPR printed matter; ~ **comerciales** *m pl* IMPR commercial printing

imprimación *f* COLOR primer, PROD priming, REVEST primer, VEH *carrocería* undercoat (*AmE*), underseal (*BrE*); ~ **anticorrosiva** *f* REVEST anticorrosive primer; ~ **antioxidante** *f* REVEST antirusting primer; ~ **por auto-ataque químico** *f* REVEST self-etching primer; ~ **de revoque** *f* REVEST wash primer; ~ **en taller** *f* REVEST shop priming

imprimador *m* CONST *pintura*, P&C *adhesivos, pintura*, REVEST primer; ~ **de madera** *m* REVEST wood primer; ~ **de reacción** *m* REVEST reaction primer

imprimar *vt* PROD *pintura*, REVEST prime

imprimibilidad *f* IMPR, PAPEL printability

imprimir[1] *vt* IMPR, INFORM&PD print, TEXTIL impress; ~ **en blanco** *vt* IMPR print recto; ~ **el reverso del pliego de** *vt* IMPR back up; ~ **a sangre** *vt* IMPR bleed; ~ **por la segunda cara** *vt* IMPR perfect

imprimir[2]: ~ **el dorso** *vi* IMPR back

improcedente *adj* ELEC *fallo* out-of-order

improductivo *adj* CARBÓN, MINAS barren

impuesto: ~ **del aeropuerto** *m* TRANSP AÉR airport charge, airport tax

impuestos: ~ **de aterrizaje** *m pl* TRANSP AÉR landing charges (*BrE*), landing fees (*AmE*)

impulsado: ~ **por la energía del sol** *adj* ENERG RENOV solar-powered

impulsador: ~ **de espuma** *m* ALIMENT lather booster

impulsar *vt* ING MECÁ actuate, drive, power, MECÁ drive, PROD operate

impulsión *f* ING MECÁ momentum, impetus, impulse,

impulsion, discharge, propulsion, MECÁ drive, impulse, momentum, REFRIG *del compresor* discharge, TEC ESP actuation; ~ **por eje delantero** *f* MECÁ, VEH front-wheel drive; ~ **por engranajes** *f* ING MECÁ, MECÁ gear drive; ~ **de velocidad constante** *f* ING MECÁ constant-speed drive

impulsivo *adj* ING MECÁ impulsive

impulso[1]: ~ **modulado** *adj* ELECTRÓN pulse-modulated

impulso[2] *m* CINEMAT spike, ELEC *tensión* pulse, surge, ELECTRÓN pulse, FÍS impulse, FÍS ONDAS, INFORM&PD pulse, ING MECÁ movement, impetus, impulse, momentum, impulsion, propulsion, thrust, MECÁ pressure, drive, impulse, momentum, PETROL impulse, PROD thrust, TEC ESP impulse, pulse, thrust, TELECOM impulse, TEXTIL *transmisión de energía* drive; ~ **de activación** *m* INFORM&PD enable pulse; ~ **activador** *m* ELECTRÓN trigger pulse; ~ **activador de la línea** *m* TELECOM triggering lead pulse; ~ **actuante** *m* ING ELÉC firing pulse; ~ **acústico** *m* ACÚST, ELECTRÓN acoustic pulse, tone pulse; ~ **aleatorio** *m* ELECTRÓN, TELECOM random pulse; ~ **autorizado** *m* ELECTRÓN enable pulse; ~ **de azimut** *m* TEC ESP azimuth thrust; ~ **de borrado vertical** *m* TV vertical-blanking pulse; ~ **breve** *m* ELECTRÓN narrow pulse, short pulse, undershoot; ~ **de conexión** *m* ELECTRÓN turn-on pulse; ~ **de corriente** *m* ING ELÉC current pulse; ~ **corto** *m* ELECTRÓN narrow pulse, short pulse, undershoot, TRANSP MAR *radar* short pulse; ~ **descendente** *m* ELECTRÓN down pulse; ~ **de desconexión** *m* ELECTRÓN gating pulse; ~ **disparador de la línea** *m* TELECOM triggering lead pulse; ~ **de disparo de la línea** *m* TELECOM triggering lead pulse; ~ **eléctrico** *m* ELEC, ING ELÉC electric pulse; ~ **electromagnético** *m* ELEC, ING ELÉC, TELECOM electromagnetic pulse; ~ **electrónico de microsegundos** *m* ELECTRÓN, TEC ESP pulse; ~ **de elevación rápida** *m* ELECTRÓN fast-rise pulse; ~ **de energía electromagnética** *m* ELEC, ING ELÉC electromagnetic-energy pulse; ~ **de entrada** *m* ELECTRÓN input pulse; ~ **de escritura** *m* INFORM&PD write pulse; ~ **específico** *m* TEC ESP specific impulse; ~ **de extirpación** *m* NUCL ablating momentum; ~ **de fijación** *m* TELECOM *radar* strobe pulse; ~ **fortuito** *m* ELECTRÓN, TELECOM random pulse; ~ **fraccionado** *m* TV serrated pulse; ~ **de frecuencia** *m* ELECTRÓN frequency pushing; ~ **gaussiano** *m* ELECTRÓN, INFORM&PD, TELECOM Gaussian pulse; ~ **gigante** *m* ELECTRÓN giant pulse; ~ **de gran amplitud** *m* ELECTRÓN high-amplitude pulse; ~ **de hiperamplitud** *m* TV spike; ~ **inhibidor** *m* ELECTRÓN inhibiting pulse; ~ **iniciador de la línea** *m* TELECOM triggering lead pulse; ~ **de interrupción** *m* ELECTRÓN turn-off pulse; ~ **largo** *m* TRANSP MAR *radar* long pulse; ~ **por láser** *m* ELECTRÓN laser pulse; ~ **luminoso** *m* ELECTRÓN light pulse; ~ **de mando de la línea** *m* TELECOM triggering lead pulse; ~ **no cíclico** *m* ELECTRÓN nonrecurrent pulse; ~ **no disruptivo** *m* PROD withstand impulse; ~ **no recursivo** *m* ELECTRÓN nonrecursive pulse; ~ **de nube** *m* ING ELÉC cloud pulse; ~ **óptico** *m* ELECTRÓN, ÓPT optical pulse; ~ **parásito** *m* ELEC, ING ELÉC pulse spike; ~ **parásito de voltaje** *m* ELEC, ING ELÉC voltage spike; ~ **de pequeña duración** *m* ELECTRÓN short pulse; ~ **perfilado** *m* ELECTRÓN shaped pulse; ~ **periódico** *m* ELECTRÓN periodic

pulse; ~ **principal** *m* ELECTRÓN master pulse; ~ **del radar** *m* TRANSP MAR radar blip; ~ **rápido** *m* ELECTRÓN sharp pulse; ~ **con rayos láser** *m* ELECTRÓN laser burst; ~ **de rayos X** *m* ELECTRÓN X-ray pulse; ~ **rectangular** *m* ELECTRÓN rectangular pulse; ~ **regenerado** *m* ELECTRÓN regenerated pulse; ~ **de reloj** *m* ELECTRÓN, INFORM&PD clock pulse; ~ **de señal** *m* ELECTRÓN signal pulse; ~ **de sensibilización** *m* ELECTRÓN indicator gate; ~ **de sincronización** *m* CINEMAT sync pulse, ELECTRÓN clock pulse, timing pulse, INFORM&PD clock pulse, TEC ESP burst; ~ **sincronizador de imagen** *m* TV frame-sync pulse; ~ **sincronizador de trama** *m* TELECOM frame-sync pulse; ~ **de tensión** *m* ELEC voltage pulse; ~ **de tiempo** *m* ELECTRÓN, INFORM&PD clock pulse; ~ **de trabajo** *m* ING ELÉC make pulse; ~ **unitario** *m* TEC ESP unit thrust; ~ **variable** *m* ELECTRÓN shift pulse; ~ **violento** *m* ELECTRÓN hard pulse

impulsor[1] *adj* ING ELÉC active

impulsor[2] *m* AUTO, CARBÓN impeller, INFORM&PD driver, ING ELÉC actuator, ING MECÁ drive, driving, propeller, MECÁ actuator, NUCL, TRANSPAÉR impeller; ~ **analógico** *m* ING ELÉC analog actuator; ~ **anexo** *m* TEC ESP strap-on booster; ~ **de bombardero electrónico** *m* TEC ESP electron-bombardment thruster; ~ **de cinta** *m* INFORM&PD tape drive; ~ **de cinta a serpentina** *m* INFORM&PD streaming tape drive; ~ **de combustible sólido** *m* TEC ESP solid-fuel booster; ~ **de conos de fricción** *m* ING MECÁ cone drive; ~ **de dispositivo** *m* INFORM&PD device driver; ~ **por engranajes** *m* ING MECÁ, MECÁ gear drive; ~ **de giro** *m* TEC ESP spin thruster; ~ **lateral** *m* TRANSP MAR side thruster; ~ **lateral de popa** *m* TRANSP MAR stern thruster; ~ **de línea** *m* INFORM&PD line driver; ~ **de pala y ranura** *m* TRANSP AÉR blade-and-slot drive; ~ **de plasma** *m* TEC ESP plasma thruster; ~ **de polea cónica** *m* ING MECÁ cone-pulley drive; ~ **de popa** *m* TRANSP MAR stern thruster; ~ **recuperable** *m* TEC ESP recoverable orbiter, recoverable thruster; ~ **de visualización** *m* INFORM&PD display driver

impulsora *f* ING ELÉC drive coil

impulsos: ~ **de fijación** *m pl* TV clamping pulses; ~ **recurrentes** *m pl* ELECTRÓN recurrent pulses; ~ **por segundo** *m pl* TELECOM pulses per second (*pps*)

impureza *f* CALIDAD, ELECTRÓN, METAL, MINERAL, QUÍMICA impurity; ~ **donadora** *f* ELECTRÓN donor impurity; ~ **de tipo n** *f* ELECTRÓN n-type impurity; ~ **de tipo p** *f* ELECTRÓN p-type impurity

impurezas *f pl* PAPEL pulp-and-paper contraries; ~ **de base** *f pl* ELECTRÓN transistores base impurities; ~ **flotantes** *f pl* CONTAM MAR, QUÍMICA scum

impurificado *adj* FÍS doped

impurificador *m* FÍS dopant

impurificar *vt* FÍS dope, QUÍMICA impurify

in: ~ **situ** *adv* CONST, MECÁ in situ, on site

In *abr* (*indio*) METAL, QUÍMICA In (*indium*)

inactivación *f* QUÍMICA inactivation

inactivo[1] *adj* ELEC quiescent, ELECTRÓN, INFORM&PD inactive, ING ELÉC quiescent, PROD idle, QUÍMICA quiescent, *proceso químico* inactive, TRANSP MAR *buque* inactive

inactivo[2]: ~ **síncrono** *m* INFORM&PD synchronous idle (*SYN*)

inadaptación *f* ING MECÁ mismatch

inaglutinable *adj* CARBÓN noncaking

inalterado *adj* GEOL unaltered

inatacable *adj* MECÁ impervious

incandescencia *f* FÍS RAD, ING ELÉC, QUÍMICA incandescence; **~ residual** *f* ELECTRÓN, TEC ESP *tubos de descarga gaseosa* afterglow

incandescente *adj* FÍS RAD, ING ELÉC incandescent, QUÍMICA glowing, incandescent, TEC ESP glowing, TERMO glowing, white-hot

incapacidad: ~ permanente *f* SEG permanent disability

incapacitivo *adj* ELEC, ING ELÉC noncapacitive

incapaz: ~ de cumplir *adj* TELECOM unable to comply

incendiar[1] *vt* TERMO burn, set ablaze, set fire to

incendiar[2] *vi* QUÍMICA deflagrate

incendiario[1] *adj* D&A incendiary

incendiario[2] *m* TERMO arsonist

incendio *m* GAS, INSTAL TERM, SEG, TERMO fire; **~ doloso** *m* TERMO arson; **~ forestal** *m* AGRIC, TERMO forest fire; **~ de hidrocarburos** *m* SEG, TERMO hydrocarbon fire; **~ intencionado** *m* TERMO arson; **~ de rescoldo** *m* AGRIC smoldering fire (*AmE*), smouldering fire (*BrE*)

incidencia *f* ACÚST, AGRIC, FÍS incidence, ING MECÁ impingement, ÓPT, TRANSP AÉR incidence; **~ de Brewster** *f* FÍS Brewster incidence; **~ de plagas** *f* AGRIC pest incidence; **~ rasante** *f* FÍS, ÓPT grazing incidence

incidente *m* CONTAM, CONTAM MAR, ÓPT, TRANSP, TRANSP AÉR incident; **~ de base de diseño** *m* NUCL design basis event; **~ de contaminación atmosférica** *m* CONTAM air pollution incident, air pollution problem

incineración: ~ de cienos *f* AGUA sludge incineration

incinerado *adj* TERMO incinerated

incinerador *m* CONST, INSTAL TERM, TERMO incinerator; **~ de humos** *m* PROC QUÍ fume incinerator

incinerar *vt* INSTAL TERM incinerate, TEC ESP ignite, TERMO incinerate

incisión *f* MECÁ, METAL, PROD notch; **~ cortical** *f* AGRIC *forestación* girdling

incisiones *f pl* P&C *defecto de la pintura*, REVEST cissing

inclasificado *adj* GEOL nonsorted

inclinación *f* ACÚST *de aguja del gramófono*, CINEMAT rake, CONST *muros* batter, cant, battering, dip, inclination, slant, FÍS pitching, inclination, GEOL *de pliegues* pitch, hade, dip, GEOM inclination, grade, gradient, slant, IMPR slant, INFORM&PD skew, ING MECÁ cant, METR *de una balanza* bearings, MINAS *minería a cielo abierto* slope, *capa* dip, PETROL dip, PROD *descarga, basculamiento* tilting, TEC ESP inclination, TRANSP AÉR bank; **~ de acción automática** *f* ING MECÁ self-acting incline; **~ automática** *f* ING MECÁ self-acting incline; **~ y bies** *f* TEXTIL bow and bias; **~ de las caras laterales del molde** *f* PROD *fundición* draw taper; **~ de la columna de la dirección** *f* AUTO, VEH steering-axis inclination; **~ descendente** *f* GEOL downbending; **~ directa** *f* HIDROL direct runoff; **~ del eje en un coladero** *f* MINAS inclined shaft; **~ del eje de la dirección** *f* AUTO, VEH steering-axis inclination; **~ por gravedad** *f* CONST gravity incline; **~ del haz** *f* TV beam tilt; **~ inicial** *f* GEOL initial dip; **~ lateral** *f* MINAS bank; **~ magnética** *f* GEOFÍS, GEOL, PETROL, TEC PETR dip angle, magnetic dip, magnetic inclination; **~ negativa** *f* AUTO negative bank; **~ de órbita** *f* TEC ESP orbit inclination; **~ de la pala** *f* TRANSP AÉR blade tilt; **~ del pilote** *f* CARBÓN pile tip; **~ de pista** *f*

TRANSP AÉR runway gradient; **~ positiva** *f* AUTO positive bank; **~ de subida** *f* TEC ESP, TRANSP AÉR climb gradient; **~ del tejado** *f* CONST roof pitch; **~ transversal del pivote de la rueda** *f* AUTO kingpin inclination

inclinado *adj* CONST leaning, slanting, GEOL, GEOM inclined

inclinar[1] *vt* IMPR tilt

inclinar[2] *vi* CONST lean, slant, TEC ESP *cambios de órbita de astronaves* bank

inclinarse *v refl* CONST tip, tip up, GEOL pitch

inclinómetro *m* CARBÓN, FÍS inclinometer, GEOFÍS inclinometer, dip circle, INSTR inclinometer, MATEMÁT estimator, PETROL inclinometer; **~ de cable en tensión** *m* OCEAN *fondeo dinámico* taut-wire angle indicator; **~ magnético** *m* GEOFÍS magnetic inclinometer

incluido *adj* ING MECÁ embodied, included

inclusión *f* CALIDAD, CRISTAL, METAL, MINERAL, NUCL, PETROL inclusion; **~ estéril** *f* MINAS *filones* horse; **~ fluida** *f* GEOL fluid inclusion; **~ no metálica** *f* CALIDAD, METAL nonmetallic inclusion

incluye: que ~ todo *fra* ING MECÁ overall

incoherencia *f* FÍS, FÍS ONDAS, METAL, ÓPT, TELECOM incoherence

incoherente *adj* FÍS, FÍS ONDAS, METAL, ÓPT, TELECOM incoherent

incohesiva *adj* CARBÓN *roca* loose

incoloro *adj* ALIMENT, COLOR, QUÍMICA colorless (*AmE*), colourless (*BrE*)

incombustibilización *f* GEN fireproofing, flameproofing

incombustibilizado *adj* GEN fireproofed, flameproofed

incombustibilizar *vt* GEN fireproof

incombustible *adj* GEN fireproof, flame-resistant, flameproof, nonflammable

incomestible *adj* ALIMENT inedible

incomible *adj* ALIMENT inedible

incompactado *adj* GEOL, TEC PETR undercompacted

incompresibilidad *f* FÍS, FÍS FLUID, QUÍMICA, TRANSP AÉR incompressibility

incompresible *adj* FÍS, FÍS FLUID, QUÍMICA, TRANSP AÉR incompressible

incomunicado *adj* ING ELÉC cut-off

incondicional *adj* INFORM&PD, PROD unconditional

incondicionalmente *adv* INFORM&PD, PROD unconditionally

inconexo *adj* MATEMÁT disjoint

inconformidad *f* CALIDAD nonconformity

inconmensurable *adj* MATEMÁT incommensurable

incontrolado *adj* FERRO *vehículos* runaway

incorporado *adj* HIDROL incorporated, ING ELÉC built-in, ING MECÁ, MECÁ built-in, embodied

incorporar[1] *vt* HIDROL *arrastrar* incorporate into

incorporar[2]: **~ a una ruta principal** *vi* TRANSP join a traffic stream; **~ al tráfico** *vi* TRANSP join a traffic stream;

incorrosible *adj* METAL, QUÍMICA incorrodible

incremental *adj* ELEC, ELECTRÓN, INFORM&PD, MATEMÁT, TEXTIL incremental

incrementar *vt* ELEC, ELECTRÓN, INFORM&PD, MATEMÁT, TEXTIL increment

incremento *m* ELEC, ELECTRÓN, INFORM&PD, MATEMÁT, TEXTIL increment; **~ absoluto** *m* TEC ESP isotropic gain, absolute gain; **~ del cabeceo** *m*

TRANSP AÉR pitch increase; ~ **de carena** *m* TRANSP MAR *buque* sinkage; ~ **isotrópico** *m* TEC ESP absolute gain, isotropic gain; ~ **del paso** *m* TRANSP AÉR pitch increase; ~ **relativo** *m* TEC ESP relative gain; ~ **de situación anticiclónica** *m* METEO anticyclonic growth; ~ **súbito** *m* TEC ESP burst; ~ **de velocidad** *m* TEC ESP velocity increment

incrustación *f* CARBÓN deposit, DETERG scale, GEOL incrustation, INSTAL HIDRÁUL, INSTAL TERM, PROD *en calderas* scale, *de calderas* fur, QUÍMICA incrustation; ~ **sobre rocas** *f* GEOL *derivada de precipitación química* sinter

incrustado *adj* CONST, INFORM&PD, ING MECÁ embedded, INSTAL HIDRÁUL, INSTAL TERM, PROD *en calderas* furred

incrustar *vt* CONST, INFORM&PD, ING MECÁ embed, PROD, TEXTIL fur

incrustarse *v refl* INSTAL HIDRÁUL, INSTAL TERM, PROD *calderas* fur

incubador *m* LAB incubator; ~ **rápido** *m* AGRIC fast-breeder reactor; ~ **refrigerado** *m* LAB, REFRIG cooled incubator, refrigerated incubator

incubadora *f* INSTAL TERM incubator

incumplimiento: ~ **de especificación** *m* PROD non-conformance; ~ **de las normas de seguridad** *m* SEG breach of the safety rules

incunable *m* IMPR incunabulum

indamina *f* QUÍMICA indamine

indano *m* QUÍMICA indane

indantreno *m* QUÍMICA indanthrene

indantrona *f* QUÍMICA indanthrone

indazina *f* QUÍMICA indazine

indazol *m* QUÍMICA indazole

indehiscente *adj* AGRIC indehiscent

indemnización *f* SEG *por accidente* compensation; ~ **a trabajadores** *f* SEG *en caso de accidente industrial* workers' compensation

indemnizar: ~ **por lesiones sufridas en el trabajo** *vt* SEG compensate for injuries received at work

indeno *m* QUÍMICA indene

indentación *f* ING MECÁ, MECÁ, METAL, PROD dent, indentation

indentado *adj* ING MECÁ, MECÁ, METAL, PROD indented

independencia: ~ **de datos** *f* INFORM&PD data independence

independiente *adj* TEC ESP independent, self-contained; ~ **de la computadora** *adj* AmL (*cf independiente del ordenador Esp*) INFORM&PD machine-independent; ~ **del contexto** *adj* INFORM&PD context-free; ~ **de la máquina** *adj* AmL (*cf independiente del ordenador Esp*) INFORM&PD machine-independent; ~ **del ordenador** *adj* Esp (*cf independiente de la máquina AmL*) INFORM&PD machine-independent

indesplisable *adj* TEXTIL permanently-pleated

indexación *f* INFORM&PD indexing

indexar *vt* INFORM&PD index

indicación *f* ACÚST cueing, CINEMAT cue, cueing, ELEC *instrumento* read-out, INFORM&PD indexing, TELECOM indication (*Ind*), TV cueing; ~ **de alarma remota** *f* TELECOM remote-alarm indication (*RAI*); ~ **de cero con aguja** *f* TELECOM null pointer indication (*NPI*); ~ **de concatenación** *f* TELECOM concatenation indication (*CI*); ~ **de las cotas con sus tolerancias** *f* ING MECÁ, PROD *dibujos, planos* tolerancing; ~ **digital** *f* ELEC *instrumento* digital read-

out; ~ **de hora de señal verde** *f* TRANSP *control de tráfico* hour of green signal indication; ~ **de llamada en espera** *f* TELECOM call-waiting indication; ~ **lógica de anomalías** *f* PROD logical fault indication; ~ **lógica de averías** *f* PROD logical fault indication; ~ **lógica de defectos** *f* PROD logical fault indication; ~ **lógica de fallos** *f* PROD logical fault indication; ~ **numérica** *f* ELEC *instrumento* digital read-out; ~ **de polaridad** *f* PROD *pilas* terminal marking; ~ **de puntero nulo** *f* TELECOM null pointer indication (*NPI*); ~ **de rollo** *f* CINEMAT roll cue; ~ **de situación** *f* TELECOM status indication; ~ **de status** *f* TELECOM status indication; ~ **de vueltas** *f* ING MECÁ revolution indication

indicador *m* AGUA, AUTO indicator, ELEC *instrumento* read-out, IMPR telltale, INFORM&PD indicator, flag, ING MECÁ marker, telltale, indicator, INSTAL HIDRÁUL *cilindros* indicator, INSTR gage (*AmE*), gauge (*BrE*), LAB, MECÁ gauge (*BrE*), gage (*AmE*), PETROL tracer, TRANSP AÉR, VEH indicator;

~ a ~ **de abandono de red** *m* TELECOM network discard indicator; ~ **de agua** *m* AGUA water indicator; ~ **de aguja** *m* PROD *cambio de vía* switch indicator; ~ **de alarma en la caldera** *m* INSTAL HIDRÁUL *nivel de agua* boiler alarm; ~ **de alcohol** *m* ALIMENT spirit gauge; ~ **de altura** *m* GEOFÍS height gage (*AmE*), height gauge (*BrE*); ~ **del ángulo de ataque** *m* TRANSP AÉR angle of attack indicator; ~ **de ángulo del cable** *m* TRANSP MAR cable angle indicator; ~ **de apantallamiento** *m* TELECOM screening indicator (*SI*); ~ **audible de fin de cinta** *m* CINEMAT audible runout indicator; ~ **de audio activo** *m* TELECOM audio indicator active (*AIA*); ~ **de audio silencioso** *m* TELECOM audio indicator muted (*AIM*); ~ **automático visual** *m* TV autocue; ~ **de averías** *m* PROD fault indicator, TELECOM fault display; ~ **de aviso de proximidad** *m* TRANSP AÉR proximity-warning indicator; ~ **de aviso de proximidad de a bordo** *m* TRANSP AÉR airborne proximity-warning indicator;

~ b ~ **biológico** *m* CONTAM, HIDROL bio-indicator, biological indicator; ~ **de brújula girosincronizada** *m* TRANSP AÉR gyrosyn compass indicator;

~ c ~ **CA** *m* TRANSP MAR *radar* AC marker; ~ **de caída de imagen** *m* TV drop-frame indicator; ~ **de capacidad** *m* TELECOM facility indicator; ~ **de carga** *m* PROD *alto horno* stock indicator, TRANSP MAR *batería* charge indicator; ~ **de carga segura** *m* SEG safe load indicator; ~ **de caudal** *m* AGUA, CARBÓN, CONST, ELEC, FÍS, HIDROL, ING ELÉC, INSTAL HIDRÁUL, LAB, PAPEL, PROC QUÍ, TEC PETR, TERMO flowmeter; ~ **de combustible** *m* AUTO, TRANSP AÉR, VEH fuel gage (*AmE*), fuel gage indicator (*AmE*), fuel gauge (*BrE*), fuel gauge indicator (*BrE*), fuel indicator; ~ **de consumo de combustible** *m* TERMO, TRANSP AÉR fuel consumption meter; ~ **contestando categoría del abonado que llama** *m* TELECOM calling-party category response indicator; ~ **contestando dirección del abonado que llama** *m* TELECOM calling-party address response indicator; ~ **de control manual** *m* FOTO manual control indicator; ~ **de control del protocolo** *m* TELECOM protocol control indicator; ~ **de crédito** *m* TELECOM credit indicator; ~ **en el cuadro de instrumentos** *m* VEH dial;

~ d ~ **de datos de extremo a extremo** *m* TELECOM end-to-end information indicator; ~ **de defectos de**

aislamiento *m* ELEC, ING ELÉC earth indicator (*BrE*), ground indicator (*AmE*), leakage indicator; **~ de deposición de bloques** *m* TELECOM block-dropping indicator; **~ de deriva** *m* TRANSP AÉR drift indicator; **~ de desenganche** *m* CINEMAT trip indicator; **~ de desgaste del forro del freno** *m* AUTO, VEH brake-lining wear-indicator; **~ de desviación** *m* TRANSP AÉR deviation indicator; **~ de dirección** *m* AUTO, VEH direction indicator; **~ de dirección de aterrizaje** *m* TRANSP AÉR landing-direction indicator; **~ de disco abatible** *m* PROD *máquinas desmoldeadoras* drop; **~ de distancia** *m* AUTO, VEH trip mileage indicator; **~ de distancias** *m* AUTO range indicator, CINEMAT distance meter, D&A range dial, TRANSP MAR distance finder, VEH range indicator; **~ doble** *m* ING MECÁ dual indicator; **~ dual** *m* ING MECÁ dual indicator;

~ e **~ de encaminamiento** *m* INFORM&PD, TELECOM routing indicator (*RI*); **~ de ensayo** *m* TELECOM test indicator; **~ de E/S forzado** *m* PROD forced I/O indicator; **~ esférico** *m* METR ball gage (*AmE*), ball gauge (*BrE*); **~ de espesor Elcometer** *m* INSTR, P&C Elcometer thickness gage (*AmE*), Elcometer thickness gauge (*BrE*); **~ de estado** *m* PROD status indicator; **~ del estado del circuito** *m* TELECOM circuit state indicator; **~ del estado de E/S** *m* PROD I/O status indicator; **~ de estado de fusible fundido** *m* PROD fuse-blown status indicator; **~ para evitar abordajes** *m* TRANSP MAR anticollision marker;

~ f **~ de fallos** *m* CINEMAT failure indicator; **~ de finalización** *m* PROD end pointer; **~ de finura de grano Hegman** *m* P&C Hegman fineness-of-grind gage (*AmE*), Hegman fineness-of-grind gauge (*BrE*); **~ flotador** *m* METR ball gage (*AmE*), ball gauge (*BrE*); **~ de fondo** *m* TRANSP AÉR bottoming indicator; **~ de formación de hielo del avión** *m* TRANSP AÉR aircraft icing indicator; **~ de fugas** *m* ELEC, ING ELÉC leakage indicator; **~ de función forzada** *m* PROD forcing indicator; **~ de fusible fundido** *m* PROD blown-fuse indicator; **~ de fusible quemado** *m* PROD blown-fuse indicator;

~ g **~ de gas del escape** *m* AUTO, TRANSP, VEH exhaust-gas indicator;

~ h **~ de hielo del avión** *m* TRANSP AÉR aircraft icing indicator; **~ de humedad** *m* EMB humidity indicator;

~ i **~ de inclinación y cabeceo** *m* TRANSP AÉR bank-and-pitch indicator (*BrE*), turn-and-bank indicator (*AmE*); **~ de información solicitada** *m* TELECOM solicited-information indicator; **~ de interruptor apagado-encendido** *m* PROD on/off switch indicator;

~ k **~ de kilometraje** *m* AUTO, VEH trip mileage indicator;

~ l **~ de liberación** *m* TELECOM release indicator; **~ de línea activa** *m* ELEC, SEG, TRANSP AÉR live-line indicator; **~ de llamadas** *m* TELECOM calling indicator; **~ de longitud** *m* TELECOM length indicator (*LI*); **~ de longitud de la cinta** *m* TV tape-length indicator; **~ de Lorentz** *m* FÍS Lorentz gage (*AmE*), Lorentz gauge (*BrE*); **~ con luz verde** *m* ELECTRÓN green gun;

~ m **~ maestro** *m* ING MECÁ master indicator; **~ magnético** *m* ING MECÁ magnetic indicator; **~ de máquina de vapor** *m* INSTAL HIDRÁUL steam engine indicator; **~ de marcación omnidireccional** *m* TRANSP AÉR omnibearing indicator; **~ de**

marcaciones *m* TRANSP MAR *radar* bearing-marker; **~ de marcaciones electrónico** *m* TRANSP MAR *radar* electronic bearing-marker; **~ de marcha** *m* PROD run indicator; **~ de marea** *m* GEOFÍS tide pole; **~ de máxima** *m* ELECTRÓN peak indicator; **~ de metano** *m* MINAS methane indicator; **~ de método de extremo a extremo** *m* TELECOM end-to-end method indicator; **~ del método SCCP** *m* TELECOM SCCP method indicator; **~ de modificaciones** *m* TELECOM modification indicator; **~ múltiple** *m* ELEC, INSTR logger;

~ n **~ de neón** *m* ING ELÉC neon indicator; **~ de nivel** *m* AGUA, CARBÓN level indicator, HIDROL level gage (*AmE*), level gauge (*BrE*), ING MECÁ, NUCL level indicator; **~ de nivel de aceite** *m* REFRIG oil sight glass; **~ del nivel de agua** *m* AGUA, HIDROL water gage (*AmE*), water gauge (*BrE*), water-level indicator, PROD gage glass (*AmE*), gauge glass (*BrE*); **~ del nivel bajo del agua** *m* ING MECÁ *calderas* alarm gage (*AmE*), alarm gauge (*BrE*); **~ del nivel de escarcha** *m* REFRIG frost-level indicator; **~ del nivel de líquido** *m* EMB liquid-level indicator; **~ de nivel de marea** *m* GEOFÍS tide-level indicator; **~ del nivel de ruido medio** *m* ACÚST, CONTAM weighted noise-level indicator; **~ del nivel de ruido ponderado** *m* ACÚST, CONTAM weighted noise-level indicator; **~ de nivel de video** *AmL*, **~ de nivel de vídeo** *m Esp* TV video level indicator; **~ de nivel visible** *m* REFRIG gage glass (*AmE*), gauge glass (*BrE*); **~ de número incompleto del abonado que llama** *m* TELECOM calling-party number incomplete indicator;

~ o **~ de orientación de pista** *m* TRANSP AÉR runway-alignment indicator;

~ p **~ del paso del colectivo** *m* TRANSP AÉR collective-pitch indicator; **~ de paso de la pala** *m* TRANSP AÉR blade-pitch indicator; **~ de pendiente** *m* CARBÓN, CONST, GEOFÍS, GEOL, INSTR, METR, MINAS, TEC ESP clinometer; **~ de pieza** *m* ING MECÁ detail gage (*AmE*), detail gauge (*BrE*); **~ del plan de numeración** *m* TELECOM numbering-plan indicator; **~ de posible traslado de llamada** *m* TELECOM call-forwarding-may-occur indicator; **~ de posición** *m* ING MECÁ locator, PROD position indicator; **~ de posición del alerón** *m* TRANSP AÉR aileron-position indicator; **~ de posición de tomas** *m* ELEC tap-position indicator; **~ de posición del tren de aterrizaje** *m* TRANSP AÉR landing-gear position indicator; **~ de potencia** *m* PROD power indicator; **~ de presentación** *m* TELECOM presentation indicator (*PI*); **~ de presión** *m* AGRIC pressure gage (*AmE*), CONST pressure gauge (*AmE*), pressure gauge (*BrE*), pressure indicator, FÍS pressure gage (*AmE*), pressure gauge (*BrE*), ING MECÁ, INSTAL HIDRÁUL, INSTR, LAB pressure gage (*AmE*), pressure gauge (*BrE*), pressure indicator; **~ de presión de aceite** *m* AUTO, VEH oil-pressure gage (*AmE*), oil-pressure gauge (*BrE*); **~ de presión diferencial de cabina** *m* TRANSP AÉR cabin differential pressure gage (*AmE*), cabin differential pressure gauge (*BrE*); **~ de prueba** *m* TELECOM test indicator; **~ de punta** *m* PROD end pointer; **~ del punto de rocío** *m* TV dew indicator;

~ r **~ de radar** *m* INFORM&PD, TELECOM routing indicator (*RI*); **~ de radioactividad** *m* FÍS, FÍS RAD radioactive tracer; **~ de radiofaro omnidireccional** *m* TRANSP AÉR omnirange indicator; **~ del recipiente** *m* METR receiver gage (*AmE*), receiver gauge (*BrE*);

~ **de recomendación** *m* TELECOM recommendation indicator; ~ **de reexpedición** *m* TELECOM redirecting indicator; ~ **del régimen de ascenso** *m* TEC ESP, TRANSP AÉR rate-of-climb indicator; ~ **de regulación del caudal** *m* PROD flow-control indicator; ~ **remoto de rumbo** *m* TRANSP AÉR heading remote indicator; ~ **restringido de presentación de dirección** *m* TELECOM address presentation restricted indicator; ~ **de retrodispersión** *m* NUCL backscatter gage (*AmE*), backscatter gauge (*BrE*); ~ **de rumbo** *m* TEC ESP heading indicator, TRANSP AÉR, TRANSP MAR course indicator; ~ **de ruta** *m* INFORM&PD, TELECOM routing indicator (*RI*);

s ~ **de la sección de entrada** *m* PROD input section indicator; ~ **de servicio de conexión y desconexión** *m* PROD on/off service indicator; ~ **de silbato de alarma** *m* INSTAL HIDRÁUL boiler alarm; ~ **de sintonización** *m* ELECTRÓN tuning indicator; ~ **de sistema** *m* PROD system pointer; ~ **de situación horizontal** *m* TRANSP AÉR horizontal situation indicator; ~ **de sobrecarga** *m* ELEC overload indicator; ~ **solicitando dirección del abonado que llama** *m* TELECOM calling-party address request indicator, calling-party category request indicator; ~ **de solicitud de identidad de línea conectada** *m* TELECOM connected-line identity request indicator; ~ **sonoro del metraje** *m* CINEMAT click footage counter;

t ~ **de temperatura por acción remota** *m* INSTR, REFRIG, TERMOTEC remote temperature gage (*AmE*), remote temperature gauge (*BrE*); ~ **de la temperatura del aceite** *m* TRANSP AÉR oil-temperature indicator; ~ **de la temperatura del agua** *m* AUTO, VEH water temperature gage (*AmE*), water temperature gauge (*BrE*); ~ **de la temperatura del aire deshelador** *m* TRANSP AÉR de-icing air temperature indicator; ~ **de la temperatura del aire exterior** *m* TRANSP AÉR outside air temperature indicator; ~ **de la temperatura de la boquilla** *m* TRANSP AÉR nozzle temperature indicator; ~ **de la temperatura de la cabina** *m* TRANSP AÉR cabin temperature indicator; ~ **de la temperatura de la cabina de mando** *m* TRANSP AÉR cockpit temperature indicator; ~ **de la temperatura de la tobera** *m* TRANSP AÉR nozzle temperature indicator; ~ **de tensión** *m* ING ELÉC, METR strain gage (*AmE*), strain gauge (*BrE*); ~ **de tierra** *m* ELEC, ING ELÉC earth detector (*BrE*), earth indicator (*BrE*), ground detector (*AmE*), ground indicator (*AmE*); ~ **del tipo de película** *m* CINEMAT, FOTO, TV film-type indicator; ~ **del tiro** *m* INSTAL TERM, PROD draft gage (*AmE*), draught gauge (*BrE*); ~ **triaxial** *m* TEC ESP three-axis indicator;

u ~ **de usuario a usuario** *m* TELECOM user-to-user indicator;

v ~ **de vacío** *m* ING MECÁ vacuum gage (*AmE*), vacuum gauge (*BrE*); ~ **de vacío Pirani** *m* LAB Pirani vacuum gage (*AmE*), Pirani vacuum gauge (*BrE*); ~ **variable de distancia** *m* TRANSP MAR *radar* variable-range marker; ~ **de velocidad** *m* ING MECÁ motion indicator; ~ **de la velocidad del aire** *m* TEC ESP, TRANSP AÉR airspeed indicator; ~ **de la velocidad del ascenso** *m* TEC ESP, TRANSP AÉR rate-of-climb indicator; ~ **de la velocidad relativa** *m* TRANSP AÉR airspeed indicator; ~ **de la velocidad con respecto al aire** *m* TRANSP AÉR airspeed indicator; ~ **de la velocidad superficial** *m* ING MECÁ surface-speed indicator; ~ **de la velocidad vertical** *m* (*VSI*)

TEC ESP, TRANSP AÉR vertical speed indicator (*VSI*); ~ **verde de diodo emisor de luz del estado de potencia conectada** *m* PROD green LED status indicator DC power ON; ~ **de viraje e inclinación** *m* TRANSP AÉR bank-and-pitch indicator (*BrE*), turn-and-bank indicator (*AmE*); ~ **de voltaje** *m* ELEC, ING ELÉC voltage indicator

indicadores: ~ **cromáticos de dosis** *m pl* FÍS RAD color indicators of dose (*AmE*), colour indicators of dose (*BrE*)

indicán *m* QUÍMICA indican

indicar *vt* CINEMAT, TV cue

indicativo: ~ **de llamada del avión** *m* TRANSP AÉR aircraft call sign

indicatriz *f* CRISTAL indicatrix

índice *m* INFORM&PD, ING MECÁ, QUÍMICA, TEC ESP index; ~ **de acetato** *m* P&C acetyl value; ~ **de acidez** *m* P&C acid value, TEC PETR acid number; ~ **de aislamiento térmico** *m* FÍS, TERMO thermal insulation index; ~ **de anilina** *m* P&C aniline value; ~ **de aprovechamiento del forraje** *m* AGRIC feed conversion ratio; ~ **de aridez** *m* HIDROL index of aridity; ~ **de articulación** *m* (*AI*) ACÚST articulation index (*AI*); ~ **de aviones en miniatura** *m* TRANSP AÉR miniature-aircraft index; ~ **de calidad** *m* ACÚST, CALIDAD, CONTAM, FÍS quality factor; ~ **de capturas** *m* OCEAN catch index; ~ **de combustión** *m* TERMO combustion index; ~ **de compresión** *m* CARBÓN compression index, ING MECÁ compression ratio; ~ **de consistencia** *m* CARBÓN consistency index; ~ **de contaminación del aire por ácidos en estado gaseoso** *m* SEG gaseous-acid air pollution index; ~ **de descascaramiento** *m* CONST flakiness index; ~ **de desgote** *m* PAPEL freeness value; ~ **de desintegración** *m* ELECTRÓN decay rate; ~ **de directividad** *m* ACÚST directivity index; ~ **de erosión** *m* TEC ESP erosion rate; ~ **escalonado equivalente** *m* ÓPT, TELECOM equivalent step index (*ESI*); ~ **escalonado lateral** *m* IMPR side index; ~ **estacional** *m* PROD seasonal index; ~ **de estallido** *m* PAPEL burst index; ~ **de estructuras** *m* TEC ESP structure index; ~ **de éxito de llamadas** *m* TELECOM call success rate; ~ **de exposición** *m* CINEMAT, FOTO, TV exposure index; ~ **del factor de pérdida auditiva** *m* ACÚST hearing loss factor index; ~ **de fiabilidad** *m* CRISTAL reliability index; ~ **gradual** *m* ÓPT graded index; ~ **de grupo** *m* ÓPT, TELECOM group index; ~ **de hidrógeno** *m* PETROL hydrogen index; ~ **de horizonte** *m* ÓPT index dip; ~ **de humedad** *m* HIDROL index of humidity, METEO moisture index; ~ **de infiltración** *m* HIDROL index of infiltration; ~ **de inteligibilidad** *m* ACÚST intelligibility index; ~ **isosófico** *m* ACÚST isosophic index; ~ **de Langelier** *m* HIDROL Langelier's index; ~ **de liquidez** *m* CARBÓN liquidity index; ~ **de modulación** *m* ELECTRÓN modulation index; ~ **de nitidez** *m* TELECOM articulation index; ~ **de nucleación** *m* METAL, QUÍMICA nucleation rate; ~ **de octano** *m* AUTO, PETROL, VEH octane index, octane rating; ~ **de ocupación** *m* TRANSP occupancy rate; ~ **de opacidad** *m* C&V darkening index; ~ **de oxígeno** *m* TRANSP AÉR oxygen index, oxygen value; ~ **de pérdida de paso** *m* ACÚST transition loss index; ~ **de perfil parabólico** *m* ÓPT, TELECOM parabolic profile index; ~ **de plasticidad** *m* CARBÓN, CONST, GEOFÍS, P&C plasticity index; ~ **de porosidad**

secundaria *m* (*IPS*) TEC PETR secondary porosity index (*SPI*); ~ **principal** *m* INFORM&PD primary index; ~ **de productividad** *m* PETROL productivity index; ~ **de profundidad** *m* ÓPT index dip; ~ **de prolificidad** *m* AGRIC reproductive efficiency; ~ **de reducción de ruido** *m* ACÚST, SEG noise-reduction index; ~ **de reducción de sonido** *m* FÍS sound-reduction index; ~ **de referencia de la brújula** *m* TEC ESP, TRANSP AÉR, TRANSP MAR lubber line; ~ **de reflectancia en el azul** *m* PAPEL whiteness degree; ~ **de refracción** *m* CINEMAT, FÍS, FÍS ONDAS, FOTO, ÓPT, TEC ESP, TELECOM index of refraction, refractive index; ~ **de refracción absoluto** *m* FÍS RAD absolute refractive index; ~ **de refracción del aire** *m* FÍS ONDAS air refractive index; ~ **de refracción complejo** *m* FÍS complex refractive index; ~ **de rendimiento** *m* QUÍMICA performance index; ~ **de resistencia a la abrasión** *m* P&C, PAPEL abrasion resistance index; ~ **de resistencia del terreno** *m* (*CBR*) CONST California Bearing Ratio (*CBR*); ~ **de resistividad** *m* PETROL resistivity index; ~ **de ruido** *m* ACÚST, TELECOM noise index; ~ **de ruido y número de operaciones** *m* (*NNI*) ACÚST noise and number index (*NNI*); ~ **de ruido de tráfico** *m* (*TNI*) ACÚST traffic noise index (*TNI*); ~ **de saponificación** *m* ALIMENT saponification number; ~ **secundario** *m* INFORM&PD secondary index; ~ **de sedimentación** *m* RECICL sedimentation rate; ~ **de trastorno auditivo** *m* ACÚST impairment of hearing index; ~ **de viscosidad** *m* FÍS FLUID, TEC PETR, TERMO viscosity index; ~ **de volumen** *m* IMPR, PAPEL bulk, bulk index, bulking index (*BrE*); ~ **de yema de huevo** *m* ALIMENT egg-yolk index; ~ **de yodo** *m* ALIMENT iodine number, P&C iodine value

índices *m pl* CRISTAL indices; ~ **de Bragg** *m pl* CRISTAL Bragg indices; ~ **de Miller** *m* CRISTAL, METAL Miller indices

indicio *m* GEOL, TEC PETR *perforación* show; ~ **de gas** *m* GAS, TEC PETR *exploración* gas show; ~ **de petróleo** *m* TEC PETR *perforación, exploración* oil show

índico *adj* QUÍMICA indic

indicolita *f* MINERAL indicolite

indiferencia *f* QUÍMICA indifference

índigo *m* COLOR, QUÍMICA indigo

indigolita *f* MINERAL indigolite

indio *m* (*In*) METAL, QUÍMICA indium (*In*)

indirrubina *f* COLOR, QUÍMICA indirubin, indirubine

indisposición: ~ **espacial** *f* TEC ESP space sickness

indistinguibilidad *f* FÍS indistinguishability

indistorsionado *adj* ELECTRÓN undistorted

indofenina *f* COLOR, QUÍMICA indophenin

indofenol *m* QUÍMICA indophenol

indogenido *m* QUÍMICA indogenide

indógeno *m* QUÍMICA indogen

indol *m* QUÍMICA indole

indolacético *adj* QUÍMICA indoleacetic

indolina *f* QUÍMICA indolin, indoline

indona *f* QUÍMICA indone

indoxílico *adj* QUÍMICA indoxylic

indoxilo *m* QUÍMICA indoxyl

indoxilsulfúrico *adj* QUÍMICA indoxylsulfuric (*AmE*), indoxylsulphuric (*BrE*)

inducción *f* GEN induction; ~ **automática** *f* ING ELÉC self-induction; ~ **electromagnética** *f* (*IEM*) FÍS, PROD electromagnetic induction (*EMI*); ~ **electrostática** *f* ELEC, FÍS, ING ELÉC electrostatic induction; ~ **con entrehierro** *f* ELEC air-gap induction; ~ **con espacio de aire** *f* ELEC air-gap induction; ~ **H** *f* ELEC *máquina* H armature; ~ **del inducido** *f* ELEC armature induction; ~ **magnética** *f* ELEC, FÍS, ING ELÉC, PETROL, TELECOM magnetic induction; ~ **magnética máxima** *f* ELEC, ING ELÉC saturation induction; ~ **matemática** *f* INFORM&PD mathematical induction; ~ **mutua** *f* ELEC, FÍS, ING ELÉC mutual induction; ~ **con núcleo de aire** *f* ELEC aircore inductance; ~ **propia** *f* ELEC self-induction; ~ **remanente** *f* ELEC, FÍS, ING ELÉC remanent induction; ~ **del ruido de superficie** *f* TV surface induction; ~ **de saturación** *f* ELEC, ING ELÉC saturation induction

inducido *m* ELEC *circuito*, FÍS *electromotor*, ING ELÉC *circuito* armature; ~ **de anillo plano** *m* ELEC, ING ELÉC disc armature (*BrE*), disk armature (*AmE*); ~ **de arrollamiento bifilar** *m* ELEC, ING ELÉC double-winding armature, double-wound armature; ~ **articulado** *m* ELEC, ING ELÉC pivoted armature; ~ **axial** *m* ELEC, ING ELÉC axial armature; ~ **de barras** *m* ELEC *de máquina* bar armature; ~ **en circuito abierto** *m* ELEC, ING ELÉC open-coil armature; ~ **dentado cerrado** *m* ELEC *generador* closed-slot armature; ~ **de derivación** *m* ELEC, ING ELÉC bias winding; ~ **del devanado en bandas** *m* ELEC, ING ELÉC strip-wound armature; ~ **de devanado bifilar** *m* ELEC, ING ELÉC double-winding armature, double-wound armature; ~ **de devanado cerrado** *m* ELEC, ING ELÉC closed-coil armature; ~ **de disco** *m* ELEC, ING ELÉC disc armature (*BrE*), disk armature (*AmE*); ~ **de doble arrollamiento** *m* ELEC, ING ELÉC double-winding armature, double-wound armature; ~ **de doble devanado** *m* ELEC, ING ELÉC double-winding armature, double-wound armature; ~ **de dos arrollamientos** *m* ELEC, ING ELÉC double-winding armature, double-wound armature; ~ **fijo** *m* ELEC, ING ELÉC fixed armature; ~ **giratorio** *m* ELEC, ING ELÉC revolving armature; ~ **de jaula** *m* ELEC, ING ELÉC cage armature; ~ **sin núcleo** *m* ELEC, ING ELÉC coreless armature; ~ **de polos salientes** *m* ELEC, ING ELÉC salient pole rotor; ~ **provisto de eje** *m* ELEC, ING ELÉC axial armature; ~ **de tambor** *m* ELEC, ING ELÉC drum armature; ~ **de timbre** *m* ELEC *generador* bell-type armature

inducir *vt* ELEC, FÍS, ING ELÉC, QUÍMICA induce

inductancia *f* ELEC, FÍS, ING ELÉC, TELECOM inductance; ~ **ajustable** *f* ELEC adjustable inductance; ~ **externa** *f* ELEC external inductance; ~ **de filtraje** *f* ELEC *inductor* smoothing choke; ~ **incremental** *f* ELEC incremental inductance; ~ **mutua** *f* ELEC, FÍS, ING ELÉC mutual inductance; ~ **parásita** *f* ELEC, ING ELÉC parasitic inductance; ~ **primaria** *f* ELEC *transformador*, ING ELÉC primary inductance; ~ **del secundario** *f* ELEC, ING ELÉC secondary inductance; ~ **variable** *f* ELEC, ING ELÉC variable inductance

inductancímetro *m* ELEC, ING ELÉC inductance meter, inductometer

inductímetro *m* ELEC, ING ELÉC inductometer

inductivo: no ~ *adj* ELEC, ING ELÉC noninductive

inductómetro *m* ELEC, ING ELÉC, INSTR inductometer

inductor *m* ELEC *eléctrico, magnético* field, *electromagnetismo* inductance coil, *generador* exciter, ING ELÉC exciter, primary, TRANSP AÉR inducer; ~ **ajustable** *m* ELEC adjustable inductance coil; ~ **ajustable magnético** *m* GEOFÍS magnetic variometer; ~ **de corriente alterna** *m* ELEC alternating-current field; ~ **de filtro** *m* ELEC *red*, ING ELÉC filter choke; ~ **limitador de corriente** *m* ELEC, ING ELÉC current-limiting inductor; ~ **mutuo** *m* ELEC, FÍS, ING ELÉC mutual inductor; ~ **neutralizante** *m* TV field-neutralizing magnet; ~ **de potencia** *m* ING ELÉC power inductor; ~ **regulable** *m* ING ELÉC variometer; ~ **en serie** *m* ELEC series inductor; ~ **terrestre** *m* GEOFÍS earth inductor; ~ **de tierra** *m* GEOFÍS earth inductor

inductotermia *f* ING ELÉC, MECÁ, TERMO induction heating

indulina *f* QUÍMICA indulin

indumentaria: ~ **de mal tiempo** *f* TRANSP MAR foul-weather gear

industria: ~ **de automoción** *f* AUTO, PROD automotive industry; ~ **del automóvil** *f* AUTO, PROD automotive industry; ~ **azucarera** *f* ALIMENT sugar industry; ~ **del barro** *f* C&V clay industry; ~ **botellera** *f* C&V bottle industry; ~ **de la cerámica** *f* C&V ceramic industry; ~ **cervecera** *f* ALIMENT brewing industry; ~ **de la cocción de cal** *f* PROD lime-burning industry; ~ **conservera** *f* AGRIC, ALIMENT, EMB canning industry (*AmE*), preserving industry, tinning industry (*BrE*); ~ **de la conversión del papel** *f* EMB, PAPEL paper-converting industry; ~ **de la cristaleria y vidrios domésticos** *f* C&V tableware and domestic glass industry; ~ **gráfica** *f* IMPR printing trade; ~ **harinera** *f* ALIMENT milling, milling industry; ~ **lechera** *f* AGRIC, ALIMENT dairy industry; ~ **minera** *f* MINAS mining; ~ **minera del oro** *f* MINAS gold-mining industry; ~ **panadera** *f* ALIMENT bread-making; ~ **pesquera** *f* ALIMENT, OCEAN fish trade, industrial fishery; ~ **petrolífera costa-fuera** *f* TEC PETR offshore oil industry; ~ **petrolífera marina** *f* TEC PETR offshore oil industry; ~ **de pizarra** *f* MINAS slate industry; ~ **de la porcelana** *f* C&V porcelain industry; ~ **de transformación** *f* PROD process industry; ~ **vidriera** *f* C&V glazing industry

ineficiencia: ~ **de transferencia** *f* ING ELÉC transfer inefficiency

inelasticidad *f* FÍS, METAL anelasticity

ineluición *f* QUÍMICA nonelution

inencogibilidad *f* TEXTIL shrink resistance

inercia[1]: **de** ~ *adj* GEN inertial

inercia[2] *f* FÍS, ING MECÁ, MECÁ inertia; ~ **de rotación** *f* ING MECÁ rotational inertia; ~ **rotacional** *f* ING MECÁ rotational inertia; ~ **térmica** *f* GAS thermal inertia, thermic inertia, REFRIG thermal inertia, TERMO temperature lag, thermal lagging, thermal inertia

inercial *adj* GEN inertial

inertancia: ~ **acústica** *f* FÍS acoustic inertance

inerte *adj* GEN inert, CARBÓN, MINAS dead

inertización *f* NUCL, SEG, TEC PETR blanketing, inerting

inestabilidad *f* ELECTRÓN jitter, EMB, FERRO, ING ELÉC instability, ING MECÁ unbalance, QUÍMICA instability; ~ **de la anchura** *f* ELECTRÓN width jitter; ~ **de la base de tiempo** *f* ELECTRÓN time jitter; ~ **capilar** *f* FÍS FLUID capillary instability; ~ **de la carga** *f* ING ELÉC charge pump; ~ **en cáscara** *f* NUCL kink instability; ~ **del chorro** *f* FÍS FLUID jet instability; ~ **de la combustión** *f* TERMO combustion instability; ~ **condicional** *f* METEO conditional instability; ~ **de doble haz** *f* NUCL two-stream instability; ~ **elástica** *f* TRANSP AÉR elastic instability; ~ **del flujo** *f* FÍS FLUID flow instability; ~ **de un flujo rotatorio de Couette** *f*

FÍS FLUID instability of rotating Couette flow; ~ **de flujos de cortadura** *f* FÍS FLUID shear flow instability; ~ **del haz** *f* ELECTRÓN, TV beam jitter; ~ **helicoidal** *f* NUCL helical instability; ~ **hidrodinámica** *f* FÍS FLUID, GAS hydrodynamic instability; ~ **de la imagen** *f* TV jitter; ~ **lateral oscilatoria** *f* TRANSP AÉR Dutch roll; ~ **magnetohidrodinámica** *f* GEOFÍS magnetohydrodynamic instability; ~ **mecánica** *f* METAL mechanical instability; ~ **MHD** *f* GEOFÍS MHD instability; ~ **de modos** *f* ÓPT, TELECOM mode hopping; ~ **de oscilación** *f* ÓPT, TELECOM mode jumping; ~ **plástica** *f* METAL plastic instability; ~ **por retorcimiento** *f* NUCL kink instability; ~ **térmica** *f* P&C, TELECOM, TERMO thermal instability

inestable *adj* ELEC unstable, ELECTRÓN *oscilador* labile, FÍS, FÍS PART unstable, GEOL labile, METAL, METEO unstable, QUÍMICA labile, TEC ESP *comunicaciones* unbalanced, TRANSP unstable

inexactitud: ~ **de la medida** *f* METR inaccuracy of measurement

infantería *f* D&A infantry

inferencia *f* INFORM&PD inference

inferior *adj* GEOL lower

infiltración *f* AGUA infiltration, leakage, seepage, CARBÓN filtration, infiltration, seepage, CONTAM MAR seepage, ELEC *magnetismo* permeance, HIDROL infiltration, leaching, ooze, percolation, QUÍMICA seepage; ~ **de agua salada** *f* HIDROL salt water infiltration; ~ **de tubería** *f* AGUA piping seepage

infiltrar *vt* AGRIC percolate, HIDROL infiltrate, leach, ooze, percolate

infinitesimal *adj* GEOM infinitesimal

inflable *adj* CONTAM MAR, NUCL, REFRIG, TRANSP AÉR, TRANSP MAR inflatable

inflación *f* CONTAM MAR, PROD inflation

inflamabilidad *f* GEN flammability, inflammability

inflamable *adj* GEN flammable, inflammable; **muy** ~ *adj* GEN highly flammable; **no** ~ *adj* GEN nonflammable

inflamación *f* AUTO, ELEC, ING ELÉC, TRANSP AÉR, VEH *motor de combustión interna* ignition; ~ **de arco** *f* ELEC arc ignition

inflamado *adj* QUÍMICA, TERMO glowing

inflamar *vt* TEC ESP ignite

inflamarse *v refl* TERMO, QUÍMICA blaze up, glow

inflar[1] *vt* CONST, FÍS inflate

inflar[2] *vi* FÍS swell

inflexión *f* CONST dip, GEOM curvature

inflorescencia *f* AGRIC inflorescence, *de caña de azúcar* arrow

influencia *f* CONST, QUÍMICA influence

influirse: **sin** ~ **mutuamente** *fra* FÍS RAD without interacting

influjo *m* CARBÓN impact, influence; ~ **meteorítico** *m* TEC ESP meteorite influx

infopista *f* INFORM&PD information highway

información *f* CALIDAD data, CONST datum, CONTAM data, ELECTRÓN data, information, GEOL data, INFORM&PD *comunicaciones* data, information, METEO, TEC ESP data, TELECOM information, directory enquiries (*BrE*), enquiries service (*AmE*); ~ **acerca de la actividad naviera** *f* TRANSP MAR shipping intelligence; ~ **para administración de empresas** *f* TELECOM management information; ~ **de amplitud** *f* ELECTRÓN amplitude information; ~ **análoga** *f* ELECTRÓN analog information;

~ **analógica** *f* ELECTRÓN analog information; ~ **de autentificación** *f* TELECOM authentication information; ~ **de autorización** *f* TELECOM authorization information; ~ **del cabeceo** *f* TRANSP AÉR pitch information; ~ **de carga** *f* TELECOM charging information; ~ **de combate** *f* D&A combat intelligence; ~ **para el control del protocolo** *f* TELECOM protocol control information (*PCI*); ~ **de control de protocolos AAL** *f* TELECOM AAL protocol control information (*AAL-PCI*); ~ **sobre datos de productos** *f* PROD information product data; ~ **digital** *f* ELECTRÓN digital data; ~ **digitalizada** *f* ELECTRÓN, INFORM&PD, TELECOM digitized data; ~ **del directorio** *f* TELECOM directory assistance (*AmE*), directory enquiries (*BrE*), information; ~ **de encaminamiento** *f* INFORM&PD, TELECOM routing information; ~ **sobre enrutamiento** *f* INFORM&PD, TELECOM routing information; ~ **específica** *f* PROD proprietary information; ~ **de las existencias** *f* EMB off-the-shelf information; ~ **meteorológica** *f* CONTAM, METEO, TEC ESP, TRANSP AÉR meteorological data, weather report; ~ **no numerada** *f* TELECOM unnumbered information (*UI*); ~ **sin numerar** *f* TELECOM unnumbered information (*UI*); ~ **parásita** *f* INFORM&PD garbage, drop-in, ING ELÉC drop-in; ~ **del paso** *f* TRANSP AÉR pitch information; ~ **del precio de la llamada en curso** *f* TELECOM call-in-progress cost information; ~ **prevuelo** *f* TRANSP AÉR preflight information; ~ **recibida** *f* INFORM&PD, TELECOM received data; ~ **relativa a un embarque** *f* TRANSP MAR shipping intelligence; ~ **de rumbo** *f* TRANSP AÉR heading information; ~ **de salida** *f* ING MECÁ output; ~ **selectiva** *f* PROD selection information; ~ **de señalización** *f* TELECOM signaling information (*AmE*), signalling information (*BrE*); ~ **del tráfico** *f* TRANSP traffic information; ~ **del tráfico local** *f* TRANSP local traffic information; ~ **del tráfico de la zona** *f* TRANSP area traffic information; ~ **de última hora** *f* CINEMAT dope sheet; ~ **de usuario a usuario** *f* TELECOM user-to-user information (*UUI*); ~ **de vuelo** *f* TRANSP AÉR flight information

informática *f Esp* (*cf computación AmL*) INFORM&PD, TELECOM *teoría* computing, computing sciences, *procesamiento de datos* computation; ~ **comercial** *f* INFORM&PD commercial computing; ~ **distribuida** *f* INFORM&PD distributed data processing (*DDP*); ~ **integrada** *f* INFORM&PD integrated data processing

informatización *f* INFORM&PD process automation; ~ **de biblioteca** *f* INFORM&PD library automation; ~ **del diseño** *f* INFORM&PD design automation

informatizado *adj Esp* (*cf computadorizado AmL*) INFORM&PD, PROD computer-based, computerized

informe *m* INFORM&PD, METEO report, PROD brief, TRANSP AÉR report; ~ **de aceptación** *m* MECÁ acceptance report; ~ **de administración** *m* PROD management report; ~ **de aterrizaje** *m* METEO, TRANSP AÉR report for landing; ~ **de cámara** *m* CINEMAT, FOTO, TV camera report; ~ **de conclusiones** *m* PROD summary report; ~ **de datos de accidente** *m* TRANSP AÉR accident data reporting; ~ **de datos de incidente** *m* TRANSP AÉR incident data reporting; ~ **de despegue** *m* METEO, TRANSP AÉR report for take off; ~ **de diagnóstico** *m* INFORM&PD diagnostic report; ~ **de errores** *m* INFORM&PD error report; ~ **de existencias** *m* PROD inventory report-

ing; ~ **final de análisis de seguridad** *m* NUCL, SEG final safety analysis report (*FSAR*); ~ **de gerencia** *m AmL* (*cf informe de gestión Esp*) PROD management report; ~ **de gestión** *m Esp* (*cf informe de gerencia AmL*) PROD management report; ~ **de laboratorio** *m* CINEMAT lab report; ~ **meteorológico** *m* METEO, TEC ESP, TRANSP AÉR, TRANSP MAR weather report; ~ **de prueba** *m* INFORM&PD test report; ~ **de resumen** *m* PROD summary report; ~ **de sondeo** *m* CARBÓN sounding record; ~ **técnico** *m* TV technical report; ~ **del tiempo** *m* METEO, TEC ESP, TRANSP AÉR, TRANSP MAR weather report; ~ **de vuelo** *m* TEC ESP, TRANSP AÉR debriefing, flight report

infovía *f* INFORM&PD information highway

infraestructura *f* CONST substructure, ING MECÁ underframe; ~ **de vía angosta** *f* FERRO narrow-gage track system (*AmE*), narrow-gauge track system (*BrE*); ~ **de la vía férrea** *f* FERRO road bed

infraproteína *f* QUÍMICA infraprotein

infrarrojo[1] *adj* (*IR*) GEN infrared (*IR*)

infrarrojo[2] *m* (*IR*) GEN infrared (*IR*); ~ **cercano** *m* FÍS, FÍS RAD near infrared; ~ **intermedio** *m* FÍS, FÍS RAD middle infrared; ~ **lejano** *m* FÍS, FÍS RAD far infrared

infrasonido *m* ACÚST, FÍS infrasound

infusión *f* ALIMENT, C&V infusion

ingeniería: ~ **acústica** *f* ACÚST, ING MECÁ acoustic engineering; ~ **ambiental** *f* AGRIC environmental engineering; ~ **aséptica** *f* SEG aseptic engineering; ~ **de bajo nivel de ruido** *f* ACÚST, ING MECÁ low-noise engineering; ~ **civil** *f* CONST civil engineering; ~ **de combustión** *f* TERMO combustion engineering; ~ **de conocimiento** *f* INFORM&PD knowledge engineering; ~ **eléctrica** *f* ELEC, ING ELÉC electrical engineering; ~ **electrónica** *f* ELECTRÓN electronic engineering; ~ **espacial** *f* TEC ESP space engineering; ~ **de los fluidos** *f* FÍS FLUID fluid engineering; ~ **lumínica** *f* ING MECÁ light engineering; ~ **mecánica** *f* ING MECÁ, PROD mechanical engineering; ~ **minera** *f* MINAS mining engineering; ~ **naval** *f* TRANSP MAR marine engineering; ~ **pesada** *f* ING MECÁ heavy engineering; ~ **de precisión** *f* ING MECÁ precision engineering; ~ **de procesos** *f* INFORM&PD, TEC PETR process engineering; ~ **de reactores** *f* NUCL reactor engineering; ~ **de refrigeración** *f* ALIMENT, REFRIG, TERMO refrigeration engineering; ~ **sanitaria** *f* AGUA sanitary engineering; ~ **para seguridad** *f* SEG safety engineering; ~ **de seguridad y medio ambiente** *f* SEG environmental-and-safety engineering; ~ **sísmica** *f* GEOFÍS seismic engineering; ~ **de sistemas** *f* ING MECÁ systems engineering; ~ **de software** *f* INFORM&PD software engineering; ~ **de software asistida por ordenador** *f Esp* (*cf ingeniería de soporte lógico auxiliada por computadora AmL*) INFORM&PD computer-aided software engineering (*CASE*); ~ **solar** *f* ENERG RENOV solar engineering; ~ **de soporte lógico** *f* INFORM&PD software engineering; ~ **de soporte lógico auxiliada por computadora** *f AmL* (*cf ingeniería de software asistida por ordenador Esp*) INFORM&PD computer-aided software engineering (*CASE*); ~ **de termotransferencia** *f* ING MECÁ heat transfer engineering; ~ **de tráfico** *f* TRANSP traffic engineering; ~ **ultrasónica** *f* ING MECÁ ultrasonic engineering

ingeniero *m* FERRO, MECÁ engineer; ~ **asesor** *m* PROD consulting engineer; ~ **de a bordo** *m* TRANSP AÉR flight engineer; ~ **calefactor** *m* TERMO heating engineer; **~ de caminos, canales y puentes** *m* CONST civil engineer; **~ de campo** *m* MECÁ field engineer; **~ civil** *m* CONST civil engineer; **~ consultor** *m* PROD consulting engineer; **~ de desarrollo de investigación** *m* ING MECÁ research development engineer; **~ encargado del estudio de los métodos** *m* ING MECÁ methods engineer; **~ frigorista** *m* REFRIG, TERMO refrigeration engineer; **~ jefe** *m* CONST chief engineer; **~ jefe de montaje** *m* CONST chief erecting engineer; **~ jefe de obra** *m* CONST chief superintendent engineer; **~ de lodo** *m* TEC PETR mud engineer; **~ de mantenimiento de aviones** *m* TRANSP AÉR aircraft maintenance engineer; **~ mecánico** *m* ING MECÁ, PROD mechanical engineer; **~ minero** *m* MINAS mining engineer; **~ naval superior** *m* TRANSP MAR naval architect; **~ naval superior e ingeniero de máquinas** *m* TRANSP MAR marine architect and engineer; **~ de obra** *m* MECÁ field engineer; **~ de perforación** *m* PETROL, TEC PETR drilling engineer; **~ petrolero** *m* PETROL, TEC PETR petroleum engineer; **~ petrolífero** *m* PETROL, TEC PETR petroleum engineer; **~ técnico de obras públicas** *m* CONST civil engineer; **~ de termotransferencia** *m* TEC ESP, TERMO heat transfer engineer; **~ de vuelo** *m* TRANSP AÉR flight engineer

ingenio *m* IMPR *cortadora de cantos del libro* plough and press (*BrE*), plow and press (*AmE*)

ingletado *m* CONST beveling (*AmE*), bevelling (*BrE*), chamfering

ingletar *vt* CONST bevel, chamfer

inglete *m* C&V miter (*AmE*), mitre (*BrE*), CONST chamfer, miter (*AmE*), mitre (*BrE*), IMPR miter (*AmE*), mitre (*BrE*)

ingravidez *f* FÍS, TEC ESP weightlessness

ingrediente *m* ALIMENT ingredient, ING MECÁ material, P&C *aditivo* admixture, *término general* ingredient, QUÍMICA *formulación* ingredient

ingresar *vt* INFORM&PD access

ingreso *m* AmL (*cf acceso Esp*) CONST ingress, ELECTRÓN, INFORM&PD entry; **~ de agua** *m* PROD ingress of water; **~ de datos** *m* ELECTRÓN, INFORM&PD data entry; **~ dual** *m* PROD, TV dual input; **~ dual de tipo diodo** *m* PROD, TV diode-type dual input; **~ manual** *m* INFORM&PD manual input

ingresos: **~ agrícolas** *m pl* AGRIC farm income; **~ netos** *m pl* TEC PETR net income

inhabilitado *adj* ING ELÉC out of order, ING MECÁ out-of-order

inhibición *f* ELECTRÓN *en circuitos lógicos*, QUÍMICA inhibition

inhibidor *m* AGUA, ALIMENT, DETERG, P&C, QUÍMICA inhibitor; **~ anticorrosivo** *m* ALIMENT, AUTO, EMB, MECÁ, P&C, REVEST, TEC ESP, VEH corrosion inhibitor; **~ de catálisis** *m* ALIMENT anticatalyst; **~ de corrosión** *m* ALIMENT, AUTO, EMB, MECÁ, P&C, REVEST, TEC ESP, VEH corrosion inhibitor; **~ enzimático** *m* ALIMENT anti-enzyme; **~ de espuma** *m* PROC QUÍ foam inhibitor; **~ de la sinterización** *m* PROC QUÍ sintering inhibitor

inhibir *vt* GEN inhibit

inhomogeneidad *f* C&V inhomogeneity

iniciación *f* MINAS, QUÍMICA initiation; **~ del campo magnético** *f* FÍS RAD onset of magnetic field; **~ de la fractura** *f* NUCL initiation of fracture; **~ inversa** *f* MINAS inverse initiation; **~ de la mecha de seguridad** *f* MINAS, SEG safety fuse initiation; **~ de registro** *f* TELECOM log-in

iniciado: **~ por un acontecimiento** *adj* PROD event-driven

iniciador *m* Esp (*cf cebador AmL*) ALIMENT initiator, *fermentación, panadería, repostería* starter, MINAS *voladuras* primer, QUÍMICA initiator; **~ de asociaciones** *m* TELECOM association initiator; **~ de combustión** *m* TERMO, TRANSP AÉR combustion starter; **~ de la onda de choque** *m* TEC ESP shock wave initiator

inicial[1] *adj* GEOL *etapa* early

inicial[2] *f* IMPR initial; **~ que ocupa más de una línea de altura** *f* IMPR drop initial; **~ recuadrada** *f* IMPR factotum initial

iniciales: **~ decoradas** *f pl* IMPR versals

inicialización *f* ELECTRÓN, INFORM&PD initialization

inicializar *vt* ELECTRÓN, INFORM&PD initialize

iniciar *vt* INFORM&PD *sesión* sign on, log in, log on, ING MECÁ, PROD *agujero* start, TEC PETR *perforación* spud in

iniciarse *v refl* SEG *fuegos* break out

inicio *m* INFORM&PD start, ING MECÁ starting, QUÍMICA onset, TELECOM *de transmisión* start; **~ de arco** *m* ELEC arc striking; **~ del cable** *m* ELEC, ING ELÉC cable head; **~ de la cinta** *m* (*BOT*) INFORM&PD beginning of tape (*BOT*); **~ de encabezamiento** *m* INFORM&PD start of header (*SOH*); **~ de la entalla** *m* METAL crack initiation; **~ de la grieta** *m* METAL crack initiation; **~ de la hendidura** *m* METAL crack initiation; **~ indirecto** *m* MINAS indirect initiation; **~ de la perforación** *m* TEC PETR *de un pozo* spudding in; **~ repentino de tormenta** *m* TEC ESP sudden storm commencement (*SSC*); **~ de la sesión** *m* INFORM&PD log-in, log-on; **~ del sondeo** *m* TEC PETR *perforación* spudding in

ininflamable *adj* GEN fireproof, flameproof, nonflammable

injertar *vti* AGRIC bud

injerto: **~ de hendidura** *m* AGRIC veneer graft; **~ de incrustación** *m* AGRIC inlay graft; **~ a ojo dormido** *m* AGRIC grafting in the dormant bud; **~ de parche** *m* AGRIC patch budding; **~ de punta** *m* AGRIC scion grafting

inmantado *m* C&V magnetting

inmersión *f* C&V dip, CONST plunging, EMB, FÍS, GEOM immersion, IMPR dipping, ING ELÉC, ING MECÁ, LAB, METAL immersion, MINAS dip, OCEAN submergence, PETROL dip, PROD, QUÍMICA, REFRIG, REVEST immersion, TEC ESP dive; **~ en caliente** *f* EMB hot dipping; **~ de corriente** *f* ING ELÉC current sink; **~ electrolítica** *f* ELEC, FÍS, ING ELÉC electrolytic bath; **~ por saturación** *f* TEC PETR *operaciones costa-fuera* saturation diving

inmersiones: **~ convexas de superficies cerradas** *f pl* GEOM convex immersions of closed surfaces

inmiscible *adj* ALIMENT, FÍS FLUID, QUÍMICA, TEC PETR immiscible

inmodulado *adj* TEC ESP unmodulated

inmovilización *f* CONST fixing, ING MECÁ holding up, interlocking, MECÁ interlocking, TRANSP AÉR interlock control; **~ del retractor del tren de aterrizaje** *f* TRANSP AÉR landing-gear retraction lock

inmovilizador: **~ de ruedas** *m* AUTO, VEH wheel clamp; **~ de seguridad del tren de aterrizaje** *m* TRANSP AÉR landing-gear safety lock; **~ de tuerca** *m* ING MECÁ nut lock

inmovilizar *vt* CONST fix, ING MECÁ lock, hold up, MECÁ, TRANSP AÉR interlock, lock

inmunidad: ~ a los ruidos *f* INFORM&PD noise immunity

innavegable *adj* TRANSP MAR *buque* unseaworthy

inoculación *f* NUCL inoculation; **~ de legumbres** *f* AGRIC *fijación de nitrógeno* legume inoculation

inodoro *adj* EMB nonodorous, PROC QUÍ, QUÍMICA odorless (*AmE*), odourless (*BrE*)

inoperativo *adj* PROD nonoperating

inorgánico *adj* COLOR, ELECTRÓN, QUÍMICA inorganic

inosina *f* ALIMENT, QUÍMICA inosine

inositol *m* ALIMENT, QUÍMICA inositol

inoxidable *adj* FOTO *acero*, METAL stainless, QUÍMICA unoxidizable, inoxidizable

insaturable *adj* QUÍMICA unsaturable

insaturación *f* QUÍMICA insaturation

insaturado[1] *adj* ALIMENT, P&C, QUÍMICA, TEC PETR unsaturated

insaturado[2] *m* ALIMENT, P&C, QUÍMICA, TEC PETR unsaturate

inscripción *f* AGRIC registration, PROD posting

inscripto *adj* AGRIC *en el libro genealógico* registered

insecto *m* AGRIC bug, insect; **~ masticador** *m* AGRIC chewing insect; **~ minador** *m* AGRIC miner; **~ vector de la enfermedad** *m* AGRIC vector insect

inseguro *adj* SEG unsafe, TRANSP MAR *tiempo* unsettled

inseminación: ~ artificial *f* (*IA*) AGRIC artificial insemination (*AI*)

insensibilización *f* MINAS *explosivos* desensitization

insensibilizador *m* MINAS *explosivos* desensitizer

insensibilizar *vt* MINAS *explosivos* desensitize

insensible: ~ a fallos *adj* INFORM&PD fault-tolerant

inserción *f* AUTO, CINEMAT insert, IMPR insert, insertion, inset, offcut, INFORM&PD insert, PAPEL inserting, PROD insertion; **~ por corriente continua** *f* TV DC insertion; **~ a dos caras** *f* IMPR double-sided insert; **~ gráfica** *f* IMPR art insert; **~ independiente** *f* IMPR free-standing insert; **~ intercalada** *f* IMPR inset insert; **~ larga y fina** *f* IMPR knifecut; **~ de medianerías** *f* EMB partitioning insert; **~ rápida** *f* NUCL *de una barra de control* fast insertion; **~ del receptor** *f* TELECOM receiver inset; **~ volante** *f* IMPR flying insert

insertado[1] *adj* ING MECÁ inserted

insertado[2]: **~ de edición** *m* TV insert edit, insert editing; **~ de hilo** *m* ING MECÁ thread insert

insertador: ~ de pernos *m* ING MECÁ stud driver

insertar *vt* CINEMAT, INFORM&PD insert, ING MECÁ set, PAPEL, PROD insert

inserto *m* AUTO insert, ING MECÁ cutting-in, insert, PROD insert; **~ para cavidades** *m* ING MECÁ cavity insert; **~ de metal duro graduable** *m* ING MECÁ indexable hard metal insert; **~ de metal duro rotatorio** *m* ING MECÁ indexable hard metal insert; **~ de rosca** *m* ING MECÁ thread insert

inservible *adj* ELEC out-of-order, TRANSP MAR *buque* unseaworthy

insesgado *adj* INFORM&PD unbiased

insignia *f* EMB banderole, TRANSP MAR flag

insípido *adj* ALIMENT unflavored (*AmE*), unflavoured (*BrE*)

insolación *f* CINEMAT printing down, ENERG RENOV, FÍS, GEOFÍS insolation, IMPR *copiado de la plancha*

printing down, METEO, REFRIG *exposición a los rayos solares* insolation; **~ completa** *f* TEC ESP full sunlight

insoluble *adj* ALIMENT, QUÍMICA, TEC PETR insoluble; **~ en agua** *adj* ALIMENT, QUÍMICA, TEC PETR insoluble in water

insonorización *f* ACÚST, CINEMAT soundproofing, CONST deadening, SEG soundproofing, TRANSP AÉR, TRANSP MAR, VEH *carrocería* sound insulation

inspección *f* C&V, CALIDAD inspection, CONST *topografía* survey, MECÁ, PROD inspection, SEG *del lugar de trabajo* inspection, *de trabajadores* surveillance, TEC ESP inspection, TRANSP MAR survey; **~ de alimento** *f* ALIMENT, CALIDAD food inspection; **~ de los alrededores** *f* TEC ESP walkaround inspection; **~ de alta velocidad** *f* EMB high-speed inspection; **~ de calderas** *f* INSTAL TERM, SEG, TERMO boiler inspection; **~ por corrientes magnéticas** *f* FERRO, ING ELÉC, MECÁ eddy-current inspection; **~ por corrientes parásitas** *f* FERRO, ING ELÉC, MECÁ eddy-current inspection; **~ definitiva** *f* CALIDAD final inspection; **~ después de accidentes reportables** *f* SEG inspection following notifiable accidents; **~ de fábrica** *f* PROD, SEG factory inspection; **~ de fatiga** *f* TRANSP AÉR fatigue inspection; **~ ferroviaria** *f* FERRO rail inspection; **~ fitosanitaria** *f* AGRIC inspection for disease; **~ de garantía de calidad** *f* CALIDAD, CONST, ING MECÁ, MECÁ, NUCL, PROD, TEC ESP quality-assurance examination; **~ general** *f* TRANSP AÉR major inspection; **~ a gran velocidad** *f* EMB high-speed inspection; **~ de línea base** *f* MECÁ baseline inspection; **~ por líquidos penetrantes** *f* NUCL penetrant-liquids inspection; **~ normal** *f* CALIDAD normal inspection; **~ por partículas magnéticas** *f* ING MECÁ, NUCL magnetic particle inspection; **~ periódica** *f* ING MECÁ periodic inspection; **~ radiográfica** *f* FÍS, FÍS RAD, ING MECÁ radiographic examination; **~ con rayos X** *f* MECÁ X-ray examination; **~ por rayos X** *f* FÍS, FÍS RAD, NUCL X-ray inspection; **~ de recepción** *f* ING MECÁ acceptance inspection, PROD receiving inspection; **~ rutinaria** *f* TRANSP AÉR routine inspection; **~ sanitaria de la COSHH** *f* SEG health surveillance under COSHH; **~ del tren** *f* FERRO train supervision; **~ ultrasónica** *f* NUCL ultrasonic examination; **~ por ultrasonidos** *f* NUCL ultrasonic examination; **~ visual** *f* METR visual inspection; **~ volante** *f* CALIDAD patrol inspection

inspeccionar *vt* CALIDAD inspect, CONST survey, MECÁ inspect, METR survey, PROD oversee, SEG overlook, TEC ESP inspect

inspector *m* CALIDAD inspector, CONST surveyor, MECÁ inspector, METR *de cantidad*, TRANSP MAR *tasador de averías* surveyor; **~ de fábricas** *m* PROD, SEG factory inspector; **~ que controla las mediciones** *m* CONST quantity surveyor

inspirador *m* ING MECÁ inspirator

inspisación *f* QUÍMICA inspissation

instalación *f* CONST installation, *de equipos, instrumentos* setup, ELECTRÓN mounting, GAS assembly, INFORM&PD setup, installation, ING ELÉC wiring, ING MECÁ setting, PROD setup, *equipamiento* plant, installation, *para un fin determinado* layout, *de equipos, maquinaria* mounting, facility, TELECOM installation, TRANSP AÉR facility, TV setup; **~ de abastecimiento de agua** *f* AGUA waterworks; **~ de abrazaderas del tren de aterrizaje** *f* TRANSP AÉR landing-gear bracing installation; **~ de absorción** *f*

SEG, TEC PETR absorption plant; ~ **de alarma contra robo** *f* SEG theft-alarm installation; ~ **de alarmas contra humos y gases** *f* SEG smoke-and-gas-alarm installation; ~ **de almacenamiento** *f* TRANSP MAR storage facility; ~ **para apagar el coque** *f* CARBÓN, MINAS coke-quenching tower; ~ **asimétrica** *f* ING ELÉC unsymmetrical arrangement; ~ **automática de extinción por aspersor** *f* AGUA, SEG sprinkler; ~ **de baja tensión** *f* ING ELÉC low-voltage installation; ~ **bajo techo** *f* ELEC indoor installation; ~ **de bajo voltaje** *f* ING ELÉC low-voltage installation; ~ **de bombas** *f* AGUA, HIDROL, TRANSP MAR pumping plant; ~ **de bombeo** *f* AGUA, HIDROL, TRANSP MAR pumping-station; ~ **de bombeo de aguas residuales** *f* AGUA, HIDROL, RECICL sewage-pumping station; ~ **para buques** *f* TRANSP MAR *antenas, reflectores* ship-type rig; ~ **de cálculo** *f* INFORM&PD computing facility; ~ **de caldeado con petróleo** *f* ING MECÁ, INSTAL TERM, PETROL oil-fired installation; ~ **de calefacción** *f* INSTAL TERM, TERMO heating installation; ~ **de calefacción con petróleo** *f* ING MECÁ, INSTAL TERM, TERMO oil-fired installation; ~ **para cargamento** *f* MINAS loading system; ~ **para cintas de video** *f* AmL, ~ **para cintas de vídeo** *f* Esp TV videotape facility; ~ **de un circuito integrado** *f* ELECTRÓN integrated-circuit layout; ~ **de combate de incendios con halón, espuma y polvo químico** *f* SEG halon foam and powder fire-fighting installation; ~ **completa** *f* ING MECÁ turnkey installation; ~ **del componente** *f* ELECTRÓN component layout; ~ **de compresión** *f* GAS compression installation; ~ **para la comprobación visual de la bajada del tren de aterrizaje** *f* TRANSP AÉR landing-gear downlock visual check installation; ~ **de concentración de finos** *f* PROD *minerales* fine concentration mill; ~ **para concentrar minerales** *f* MINAS concentrator; ~ **de conductores** *f* ELEC wiring; ~ **de congelación** *f* ALIMENT, ING MECÁ, REFRIG freezing plant; ~ **para congelación rápida** *f* ALIMENT, ING MECÁ, REFRIG quick-freezing installation, quick-freezing plant; ~ **para congelado rápido** *f* ALIMENT, ING MECÁ, REFRIG quick-freezing installation, quick-freezing plant; ~ **contra rayos e instalación de tierras** *f* SEG lightning protection and earthing installation (*BrE*), lightning protection and grounding installation (*AmE*); ~ **de control del tráfico** *f* TRANSP traffic control installation; ~ **para cortar a la longitud deseada** *f* PROD cut-to-length line; ~ **de depuración** *f* AGUA, GAS, HIDROL, PROC QUÍ, QUÍMICA, RECICL, TEC PETR purification plant, purifying plant; ~ **de depuración de aguas residuales** *f* AGUA, RECICL sewage treatment plant; ~ **de desalinización de agua de mar** *f* AGUA desalination plant; ~ **de desempolvamiento** *f* PROD dust collector; ~ **de desintegración catalítica** *f* TEC PETR catalytic-cracking plant; ~ **para desulfurizar humos y gases** *f* CONTAM, SEG smoke gas desulfurization installation (*AmE*), smoke gas desulphurization installation (*BrE*); ~ **discrecional** *f* ELECTRÓN *pastillas* discretionary wiring; ~ **para distribuir el aire de ventilación** *f* MINAS air coursing, coursing; ~ **de dos B más D** *f* TELECOM two-B-plus-D arrangement; ~ **eléctrica** *f* ELEC electrical installation, *conexión* wiring, ING ELÉC, TRANSP MAR electrical installation; ~ **eléctrica antideflagrante** *f* ELEC, ING ELÉC, SEG, TERMO fireproof lighting installation, flameproof

lighting installation; ~ **eléctrica doméstica** *f* ELEC, ING ELÉC domestic electric installation, home electric installation; ~ **eléctrica por enlace directo** *f* TEXTIL direct link wiring; ~ **eléctrica incombustible** *f* ELEC, ING ELÉC, SEG, TERMO fireproof lighting installation, flameproof lighting installation; ~ **eléctrica a la intemperie** *f* ELEC, ING ELÉC outdoor electrical equipment, outdoor electrical installation; ~ **de emergencia** *f* TELECOM emergency installation; ~ **energética** *f* TRANSP MAR *motores* power plant; ~ **de energía** *f* ELEC *fuente de alimentación*, TELECOM power plant; ~ **de energía sin almacenamiento** *f* ENERG RENOV *energía hidroeléctrica*, INSTAL HIDRÁUL run-of-river station; ~ **de energía eléctrica a vapor** *f* ING ELÉC steam-electric power plant; ~ **de enfriamiento de baja temperatura** *f* ING MECÁ low-temperature cooling installation; ~ **para enfriar coque** *f* CARBÓN, MINAS coke-cooling tower; ~ **de envío de llamadas** *f* TELECOM call-forwarding installation; ~ **estacionaria para combate de incendios** *f* SEG stationary firefighting installation; ~ **experimental** *f* ING MECÁ, NUCL test rig; ~ **de extracción** *f* AmL (*cf instalación fija Esp*) MINAS close set, fixed set; ~ **para la extracción de humos y calor** *f* SEG smoke and heat exhaust installation; ~ **de extracción volante** *f* AmL MINAS suspended set; ~ **fija** *f* Esp (*cf instalación de extracción AmL*) MINAS close set, fixed set; ~ **filtradora** *f* AGUA filter plant; ~ **frigorífica** *f* ING MECÁ, REFRIG, TERMO refrigerating plant; ~ **frigorífica central** *f* ING MECÁ, REFRIG, TERMO central refrigerating plant; ~ **frigorífica marina** *f* REFRIG, TERMO, TRANSP MAR marine refrigeration plant; ~ **frigorífica múltiple** *f* REFRIG, TERMO multistage refrigerating plant; ~ **de fuerza** *f* TELECOM power plant; ~ **generatriz** *f* ELEC, ENERG RENOV, ING ELÉC, NUCL generating plant; ~ **geotérmica** *f* ELEC, ENERG RENOV, TERMO geothermal plant; ~ **hidráulica** *f* INSTAL HIDRÁUL, TRANSP MAR hydraulic system; ~ **indicadora** *f* ING MECÁ indicator plant; ~ **informática** *f* INFORM&PD computing facility; ~ **interior** *f* ELEC indoor installation; ~ **interna** *f* ING ELÉC internal installation; ~ **de lavado** *f* PROD washing plant; ~ **limpiadora de escape** *f* SEG exhaust-cleaning installation; ~ **para limpieza de gases de chimenea** *f* SEG flue-gas cleaning installation; ~ **llave en mano** *f* ING MECÁ turnkey installation; ~ **de matizado** *f* TEC PETR *ingeniería de lodos, perforación* blending plant; ~ **mezcladora** *f* PROD mixing plant; ~ **móvil** *f* ING MECÁ loose plant, portable plant, TELECOM mobile installation; ~ **N** *f* TELECOM N-facility; ~ **de ordeño por aspiración** *f* AGRIC releaser milking installation; ~ **en paralelo** *f* ING ELÉC parallel arrangement; ~ **de pararrayos y puesta a tierra** *f* SEG lightning protection and earthing installation (*BrE*), lightning protection and grounding installation (*AmE*); ~ **petroquímica** *f* QUÍMICA, TEC PETR petrochemical plant; ~ **de procesamiento** *f* GAS processing facility, processing installation; ~ **provisional** *f* Esp (*cf instalación provisoria AmL*) PROD, TEC ESP makeshift; ~ **provisoria** *f* AmL (*cf instalación provisional Esp*) PROD, TEC ESP makeshift; ~ **de prueba** *f* ING MECÁ tryout facility, PROD *motores, estructuras, equipamiento* rig; ~ **de pruebas** *f* ING MECÁ test facility, NUCL test rig; ~ **de purga** *f* TRANSP MAR drainage system; ~ **para purines** *f* AGRIC liquid-manure plant;

~ **de radar** *f* D&A, TRANSP, TRANSP AÉR, TRANSP MAR radar station; ~ **radioeléctrica** *f* TELECOM, TRANSP AÉR, TRANSP MAR radio facility; ~ **receptora de residuos en tierra** *f* CONTAM MAR shore reception facility; ~ **receptora en tierra** *f* CONTAM MAR shore reception facility; ~ **para reducir las emisiones de azufre** *f* CONTAM, SEG installation for reducing sulfur emissions (*AmE*), installation for reducing sulphur emissions (*BrE*); ~ **de refrigeración por absorción** *f* ING MECÁ, REFRIG absorption refrigerating installation; ~ **de refrigeración por aire** *f* ING MECÁ, REFRIG air-cooling installation; ~ **de regeneración** *f* PROD, RECICL reprocessing plant; ~ **para remoción de polvos en húmedo** *f* SEG wet dust removal installation; ~ **de rociadores de cubierta** *f* TRANSP MAR deck sprinkler system; ~ **de secado de gas** *f* GAS gas-drying plant; ~ **de señales** *f* TRANSP signal installation; ~ **en serie** *f* ING ELÉC series arrangement, tandem arrangement; ~ **solar** *f* ENERG RENOV solar farm; ~ **de sondeos** *f* AmL (*cf plataforma de sondeos Esp*) MINAS boring plant, drilling plant; ~ **de succión y filtros** *f* SEG *para polvos y rebabas* suction-and-filter installation; ~ **para el suministro de combustible** *f* MINAS bunker system; ~ **en superficie** *f* ELECTRÓN surface mounting; ~ **telegráfica** *f* TELECOM telegraph installation; ~ **de tierra** *f* TRANSP AÉR ground installation; ~ **de tratamiento de agua** *f* AGUA water treatment plant; ~ **para tratamiento de agua salada** *f* AGUA salt water plant; ~ **para el tratamiento del efluente** *f* AGUA, CONTAM, CONTAM MAR effluent treatment plant; ~ **de ventilación** *f* CONST fan station; ~ **de vías permanentes** *f* FERRO permanent-way installation; ~ **de vidrios en las ventanas** *f* C&V glazing installations *f pl* PROD, TRANSP AÉR, TV facilities; ~ **para acondicionamiento de aire** *f pl* SEG air-conditioning installations; ~ **complementarias de la central** *f pl* NUCL balance of plant (*BOP*); ~ **para el control de crecidas** *f pl* AGUA, HIDROL flood-control works; ~ **disponibles** *f pl* TRANSP AÉR facility availability; ~ **del estudio** *f pl* TV studio facilities; ~ **para extinguir el fuego por medio de aspersores y rocío** *f pl* SEG sprinkler and water spray fire-extinguishing installations; ~ **de lavado** *f pl* TEXTIL washing facilities; ~ **mecánicas para extracción del aire** *f pl* SEG mechanical exhaust air installations; ~ **portuarias** *f pl* TRANSP MAR port facilities; ~ **de producción** *f pl* PETROL, PROD, TV production facilities; ~ **de la ruta aérea** *f pl* TRANSP AÉR air-route facilities; ~ **técnicas** *f pl* MECÁ engineering facilities; ~ **en tierra** *f pl* TEC ESP ground facilities; ~ **ventiladoras de escape** *f pl* SEG exhaust vent installations

instalador *m* ING MECÁ, MECÁ erector, fitter; ~ **de carriles** *m* FERRO tracklayer; ~ **frigorista** *m* REFRIG refrigeration contractor; ~ **de gas** *m* GAS, PROD gas fitter; ~ **de tuberías** *m* CONST, ING MECÁ pipelayer

instalar *vt* CONST fix, lay, ELEC put in, ELECTRÓN mount, INFORM&PD install, set up, ING MECÁ set, OCEAN mount, PROD rig up, install, mount, *correa, cinta* lay on, *máquinas* rig, TRANSP MAR install, fit, TV set up

instancia *f* INFORM&PD instance

instantánea *f* FOTO shot, INFORM&PD snapshot

instantáneo *adj* ELECTRÓN fast-changing

instar *m* AGRIC instar

instituto: ~ **hidrográfico** *m* TRANSP MAR *marina española* hydrographic office; ~ **meteorológico** *m* METEO weather bureau

Instituto: ~ **Americano de Normalización Nacional** *m* (*ANSI*) PROD American National Standards Institute (*ANSI*); ~ **Británico de Normalización** *m* (*BSI*) PROD British Standards Institution (*BSI*); ~ **de Estudios Costeros** *m* CONTAM, CONTAM MAR Coastal Studies Institute (*CSI*); ~ **Europeo de Normas de Telecomunicaciones** *m* TELECOM European Telecommunication Standardization Institute (*ETSI*); ~ **de Operaciones Nucleares** *m* NUCL Institute of Nuclear Power Operations (*INPO*)

instrucción *f* EMB instruction, INFORM&PD instruction, *programación* statement, ING MECÁ training, instruction, PROD instruction; ~ **para abrir bifurcación** *f* PROD branch-open instruction; ~ **de acondicionamiento** *f* PROD conditioning instruction; ~ **de actualización de E/S directa** *f* PROD immediate I/O update instruction; ~ **de apertura** *f* PROD unlatch instruction; ~ **aritmética** *f* INFORM&PD arithmetic instruction; ~ **de asignación** *f* INFORM&PD assignment statement; ~ **automática** *f* INFORM&PD machine instruction; ~ **de bifurcación** *f* INFORM&PD branch instruction, jump instruction, PROD branching instruction; ~ **de búsqueda** *f* INFORM&PD fetch instruction; ~ **compuesta** *f* INFORM&PD compound statement; ~ **de condición** *f* PROD condition instruction; ~ **condicional** *f* INFORM&PD conditional instruction; ~ **controladora de bit** *f* PROD bit-controlling instruction; ~ **de detenida** *f* INFORM&PD stop instruction; ~ **de una dirección** *f* INFORM&PD one-address instruction; ~ **de dirección cero** *f* INFORM&PD zero-address instruction; ~ **de dirección única** *f* INFORM&PD single-address instruction; ~ **con direccional de dos más uno** *f* INFORM&PD two-plus-one address instruction; ~ **de direccionamiento múltiple** *f* INFORM&PD multiaddress instruction; ~ **de dos direccionales** *f* INFORM&PD two-address instruction; ~ **de dos direcciones** *f* INFORM&PD two-address instruction; ~ **de dos más una direcciones** *f* INFORM&PD two-plus-one address instruction; ~ **ejecutable** *f* INFORM&PD executable instruction; ~ **de entrada** *f* INFORM&PD input instruction, entry instruction; ~ **de entrada de secuenciador** *f* PROD sequencer input instruction; ~ **de entrada/salida** *f* INFORM&PD input/output instruction; ~ **de escritura** *f* INFORM&PD write instruction; ~ **examinadora de bit** *f* PROD bit-examining instruction; ~ **de extracción** *f* Esp (*cf instrucción de traída AmL*) INFORM&PD fetch instruction; ~ **ficticia** *f* INFORM&PD dummy instruction; ~ **final provisional para terminar temporalmente** *f* PROD temporary end instruction; ~ **de grabación** *f* INFORM&PD write instruction; ~ **gradual del secuenciador** *f* PROD sequencer step instruction; ~ **ilegal** *f* INFORM&PD illegal instruction; ~ **indefinida** *f* INFORM&PD undefined statement; ~ **inválida** *f* INFORM&PD illegal instruction; ~ **de llamada** *f* INFORM&PD call-by value; ~ **lógica** *f* INFORM&PD logic instruction; ~ **LOOP** *f* INFORM&PD LOOP statement; ~ **de una más una direcciones** *f* INFORM&PD one-plus-one address instruction; ~ **no operativa** *f* (*no op*) INFORM&PD no-operation instruction (*no op*); ~ **nula** *f* INFORM&PD do-nothing instruction, null instruction; ~ **del operador** *f* INFORM&PD operator

command; ~ **de parada** *f* INFORM&PD halt instruction, stop instruction; ~ **privilegiada** *f* INFORM&PD privileged instruction; ~ **de programa** *f* INFORM&PD program instruction; ~ **progresiva del secuenciador** *f* PROD sequencer step instruction; ~ **de ramificación** *f* PROD branching instruction; ~ **de retorno** *f* INFORM&PD return instruction; ~ **de salida del secuenciador** *f* PROD sequencer output instruction; ~ **de salto** *f* INFORM&PD jump instruction; ~ **de situación** *f* PROD condition instruction; ~ **del temporizador** *f* PROD timer instruction; ~ **para terminar bifurcación** *f* PROD branch-close instruction; ~ **de traída** *f* AmL (*cf instrucción de extracción Esp*) INFORM&PD fetch instruction; ~ **de tres direccionales** *f* INFORM&PD three-address instruction; ~ **de tres direcciones** *f* INFORM&PD three-address instruction
instrucciones *f pl* TEXTIL directions; ~ **de apertura** *f pl* EMB instructions for opening, opening instructions; ~ **dobles** *f pl* TRANSP AÉR dual instruction; ~ **de empleo** *f pl* ING MECÁ operating instructions, PROD instructions for use, TEXTIL directions; ~ **de lavado** *f pl* EMB, TEXTIL washing instructions; ~ **de manejo** *f pl* ING MECÁ operating instructions; ~ **de manutención e instalación** *f pl* EMB, ING MECÁ handling and installing instructions; ~ **náuticas** *f pl* TRANSP MAR *navegación* sailing directions; ~ **de operación** *f pl* ING MECÁ, NUCL operating instructions; ~ **de seguridad** *f pl* SEG safety instructions; ~ **técnicas** *f pl* FERRO, ING MECÁ technical instructions; ~ **de uso** *f pl* EMB directions for use, instructions for use; ~ **de utilización** *f pl* ING MECÁ operating instructions; ~ **de vuelo por instrumentos** *f pl* TRANSP AÉR instrument flight rules
instructor *m* ELECTRÓN, SEG, TRANSP, TRANSP AÉR, VEH instructor, trainer; ~ **de vuelo** *m* TRANSP AÉR flight instructor
instrumentación *f* ELECTRÓN, INSTR instrumentation
instrumental: ~ **de medida** *m* ELEC, METR measuring equipment; ~ **de pruebas** *m* ING MECÁ test equipment
instrumentar *vt* ING MECÁ implement
instrumentista *m* ING MECÁ, INSTR instrument maker
instrumento *m* AUTO, ELEC, ING ELÉC instrument, ING MECÁ appliance, tool, instrument, implement, INSTR instrument, gage (*AmE*), gauge (*BrE*), PROD, TEC ESP, TEC PETR, TELECOM, TRANSP, TRANSP AÉR, TRANSP MAR, VEH appliance, implement, instrument, tool; ~ **de ángulo amplio** *m* INSTR wide-angle instrument; ~ **Bourdon** *m* FÍS, TEC PETR Bourdon gage (*AmE*), Bourdon gauge (*BrE*); ~ **de calibración** *m* INSTR calibration instrument; ~ **de campo rotativo** *m* ELEC, FÍS, ING ELÉC, INSTR, TELECOM rotating-field instrument; ~ **de campo rotatorio** *m* ELEC, FÍS, ING ELÉC, INSTR, TELECOM rotating-field instrument; ~ **de cuadrante iluminado** *m* INSTR illuminated-dial instrument; ~ **digital** *m* ELEC, INSTR digital instrument; ~ **de estabilización** *m* TRANSP stabilization device; ~ **ferromagnético** *m* INSTR ferromagnetic instrument; ~ **de grabación por agujas** *m* INSTR stylus recording instrument; ~ **de horológico** *m* AmL (*cf instrumento de relojería Esp*) INSTR horological instrument; ~ **de imán móvil** *m* INSTR moving-iron instrument, moving-magnet instrument; ~ **indicador** *m* INSTR indicating instrument; ~ **de inducción** *m* ING ELÉC, INSTR induction instrument; ~ **de lectura digital de medidas** *m* METR digital read-out measuring instrument; ~ **de lectura directa** *m* ELEC, INSTR direct reading instrument; ~ **de medición** *m* CONST, ELEC, ELEC ENG, INSTR, METR measuring instrument; ~ **de medición analógico** *m* ELEC, INSTR analog measuring instrument; ~ **de medición análogo** *m* ELEC, INSTR analog measuring instrument; ~ **de medición de la base de las nubes** *m* INSTR, METEO, TRANSP AÉR cloudbase measuring instrument; ~ **de medición láser** *m* INSTR, METR laser measuring instrument; ~ **de medición de longitudes** *m* ING MECÁ, INSTR, METR length-measuring instrument; ~ **de medida** *m* CONST, ELEC, ELEC ENG, INSTR, METR measuring instrument; ~ **de medida diferencial** *m* INSTR differential measuring instrument; ~ **de medida de distancia electroóptica** *m* ELEC, INSTR, ÓPT, TELECOM electro-optical distance-measuring equipment, electro-optical distance-measuring instrument; ~ **de medida esférica** *m* INSTR, METR roundness-measuring instrument; ~ **de medida de lectura directa** *m* INSTR pointer instrument; ~ **de medida de rectilineidad** *m* INSTR, METR straightness-measuring instrument; ~ **de medida de superficie** *m* METR surface-measuring instrument; ~ **de medidas analógicas** *m* INSTR, METR analog measuring instrument; ~ **medidor de cable** *m* ELEC, ING ELÉC, INSTR cable gage (*AmE*), cable gauge (*BrE*); ~ **de mordazas** *m* ELEC, INSTR clip-on instrument; ~ **óptico** *m* FÍS, INSTR, ÓPT optical instrument; ~ **de precisión** *m* FÍS, INSTR precision instrument; ~ **de prueba con tenazas** *m* ELEC, INSTR tong-test instrument; ~ **para pruebas de tracción** *m* FÍS, ING MECÁ, P&C tensile tester; ~ **que se fija con presilla** *m* ELEC, INSTR clip-on instrument; ~ **reflector** *m* INSTR, ÓPT reflecting instrument; ~ **de relojería** *m* Esp (*cf instrumento de horológico AmL*) INSTR horological instrument; ~ **remolcado** *m* OCEAN towed instrument; ~ **para resquebrajar** *m* CONST *madera* cleaver; ~ **térmico** *m* INSTR, TERMO thermal instrument; ~ **termoeléctrico** *m* INSTR thermocouple instrument; ~ **de transposición** *m* ACÚST transposing instrument; ~ **de verificación** *m* CALIDAD, EMB checking apparatus
instrumentos: ~ **colocados en los satélites** *m pl* TEC ESP payload; ~ **giroscópicos** *m pl* TEC ESP, TRANSP AÉR gyro instruments; ~ **del giróscopo** *m pl* TEC ESP, TRANSP AÉR gyro instruments; ~ **de medidas dimensionales** *m pl* INSTR, METR dimensional measuring instruments; ~ **del motor** *m pl* TRANSP AÉR engine instruments; ~ **náuticos** *m pl* TRANSP MAR navigational instruments; ~ **ópticos para medidas dimensionales** *m pl* INSTR, METR, ÓPT optical instruments for dimensional measurement; ~ **de vuelo** *m pl* TRANSP AÉR flight instruments
insuficiencia *f* PROD *de peso* shortage, TEC ESP failure; ~ **de pasta** *f* CARBÓN, GEOL base failure
insuficientemente: ~ **expuesto** *adj* CINEMAT, FOTO, IMPR, TV underexposed
insulina *f* QUÍMICA insulin
integración *f* GEN integration; ~ **del circuito** *f* ELECTRÓN circuit integration; ~ **de circuito electrónico** *f* ELECTRÓN, TELECOM electronic circuit integration; ~ **digital** *f* ELECTRÓN digital integration; ~ **a escala de oblea de silicio** *f* INFORM&PD wafer scale integration; ~ **a escala de la pastilla** *f* ELECTRÓN wafer scale integration; ~ **a gran escala** *f* (*IGE*)

ELECTRÓN, FÍS, INFORM&PD, TELECOM, TRANSP MAR large-scale integration (*LSI*); ~ **a gran escala sobre pedido** *f* ELECTRÓN custom LSI; ~ **a media escala** *f* (*IME*) ELECTRÓN, FÍS, INFORM&PD, TELECOM medium-scale integration (*MSI*); ~ **de mediana escala** *f* (*IME*) ELECTRÓN, FÍS, INFORM&PD, TELECOM medium-scale integration (*MSI*); ~ **en muy gran escala** *f* ELECTRÓN, FÍS, INFORM&PD, TELECOM very large-scale integration (*VLSI*); ~ **a pequeña escala** *f* ELECTRÓN, FÍS, INFORM&PD, TELECOM small-scale integration (*SSI*); ~ **en red** *f* INFORM&PD networking; ~ **de la señal** *f* ELECTRÓN trace integration; ~ **tridimensional** *f* ELECTRÓN three-dimensional integration

integrado *adj* GEN integrated

integrador *m* ELECTRÓN integrator, ING ELÉC integrator, multiple contact switch, QUÍMICA, TRANSP AÉR integrator; ~ **activo** *m* ELECTRÓN active integrator; ~ **digital** *m* ELECTRÓN digital integrator; ~ **de error de rumbo** *m* TRANSP AÉR heading-error integrator; ~ **de etapa** *m* TEC ESP stage integrator; ~ **de paso lateral** *m* TRANSP AÉR lateral-path integrator

integral¹ *adj* ING MECÁ integral

integral² *f* MATEMÁT integral; ~ **de colisión** *f* FÍS, FÍS RAD collision integral; ~ **definida** *f* MATEMÁT definite integral; ~ **de Fourier** *f* ACÚST, ELECTRÓN, FÍS, MATEMÁT Fourier integral; ~ **indefinida** *f* MATEMÁT indefinite integral; ~ **de línea** *f* FÍS line integral; ~ **de Riemann** *f* GEOM, MATEMÁT Riemann integral; ~ **de superficie** *f* FÍS surface integral; ~ **de volumen** *f* FÍS volume integral

integrante *adj* ING MECÁ integral

integrar *vt* ELECTRÓN, ING MECÁ, MATEMÁT, TEC ESP integrate

integridad *f* TELECOM integrity; ~ **de datos** *f* TELECOM data integrity; ~ **del recinto de contención** *f* NUCL containment integrity

íntegro *adj* ING MECÁ integral

inteligencia: ~ **artificial** *f* (*IA*) INFORM&PD, TELECOM artificial intelligence (*AI*); ~ **electrónica** *f* ELECTRÓN electronic intelligence

inteligibilidad *f* ACÚST, TELECOM intelligibility

INTELSAT *abr* (*Satélite para Comunicaciones Internacionales*) TEC ESP INTELSAT (*International Telecommunications Satellite*)

intemperie *f* TEXTIL weathering

intemperización *f* CONTAM MAR weathering

intensidad *f* ELEC *campo electromagnético*, ELECTRÓN, GAS, ÓPT, QUÍMICA intensity, TRANSP MAR *de corriente* drift; ~ **en amperios** *f* ELEC *corriente* amperage; ~ **de campo** *f* ELEC, ING ELÉC field intensity; ~ **del campo** *f* ELEC, ING ELÉC field strength; ~ **de campo cercano** *f* TEC ESP near-field intensity; ~ **de campo coercitivo** *f* ELEC *magnetismo* coercive field strength; ~ **de campo distante** *f* TEC ESP, TELECOM far-field intensity; ~ **del campo eléctrico** *f* ELEC, FÍS, FÍS RAD, ING ELÉC, PETROL electric-field intensity, electric-field strength; ~ **de campo lejano** *f* TEC ESP, TELECOM far-field intensity; ~ **del campo magnético** *f* ELEC, FÍS, FÍS RAD, ING ELÉC, PETROL magnetic-field intensity, magnetic-field strength; ~ **coercitiva** *f* ELEC coercive intensity; ~ **del color** *f* IMPR chroma; ~ **de coloración** *f* IMPR tinting strength; ~ **constante** *f* ELEC, ING ELÉC constant current; ~ **de la corriente** *f* ELEC, ING ELÉC current intensity, current strength; ~ **de desexcitación** *f* ELEC, ING ELÉC drop-out; ~ **de**

la estela *f* FÍS FLUID wake intensity; ~ **de fondo** *f* CRISTAL background intensity; ~ **iónica** *f* FÍS RAD ionic strength; ~ **de irradiación** *f* FÍS RAD, TEC ESP radiation intensity; ~ **luminosa** *f* FÍS, FÍS RAD, ING ELÉC luminous intensity; ~ **de la luz** *f* FÍS RAD intensity of light; ~ **magnética** *f* GEOFÍS magnetic intensity; ~ **nominal de funcionamiento** *f* PROD rated operational current; ~ **de oscilación** *f* ELECTRÓN oscillating quantity; ~ **de polo magnético** *f* GEOFÍS magnetic pole strength; ~ **de la portadora de luminancia** *f* TV luminance carrier output; ~ **de la precipitación** *f* METEO rainfall intensity; ~ **de radiación** *f* FÍS RAD, TEC ESP radiation intensity; ~ **radiante** *f* FÍS, FÍS RAD, ÓPT, TELECOM *de fuente en dirección* radiant intensity; ~ **de ráfaga** *f* TEC ESP, TRANSP AÉR, TRANSP MAR gust intensity; ~ **de rompimiento instantáneo** *f* CALIDAD *de materia* instantaneous failure intensity; ~ **sísmica** *f* GEOFÍS, GEOL earthquake intensity; ~ **sónica** *f* ACÚST, FÍS sound intensity; ~ **de sonido** *f* ACÚST, FÍS loudness; ~ **sonora** *f* ACÚST, FÍS sound intensity; ~ **sonora de referencia** *f* ACÚST, FÍS reference sound intensity; ~ **del terremoto** *f* GEOFÍS, GEOL earthquake intensity; ~ **de tráfico** *f* TELECOM traffic load; ~ **de tráfico aleatorio equivalente** *f* TELECOM equivalent random-traffic intensity; ~ **de turbulencia** *f* TEC ESP, TRANSP AÉR, TRANSP MAR gust intensity; ~ **de la vibración** *f* SEG vibration severity

intensificación *f* ELECTRÓN, FOTO intensification; ~ **del anticiclón** *f* METEO anticyclonic growth; ~ **de la espumación** *f* DETERG foam boosting; ~ **de la imagen latente** *f* CINEMAT, FOTO latensification; ~ **de imágenes** *f* TEC ESP image enhancement; ~ **de línea** *f* ELECTRÓN trace intensification; ~ **con mercurio** *f* FOTO mercury intensification; ~ **de la presión** *f* PROD pressure intensification; ~ **química** *f* FOTO, QUÍMICA chemical intensification

intensificador: ~ **de cromo** *m* FOTO chrome intensifier; ~ **de empuje** *m* TRANSP AÉR thrust augmenter; ~ **de imagen** *m* ELECTRÓN, INSTR image intensifier; ~ **de imagen de microcanal** *m* ELECTRÓN microchannel image intensifier

intensificar *vt* ELECTRÓN, FOTO intensify, TEC ESP *motores de reacción* boost

intento: ~ **de llamada** *m* TELECOM call attempt; ~ **de llamada repetida** *m* TELECOM repeated-call attempt; ~ **de ocupación** *m* TELECOM bid

interacción *f* INFORM&PD, ING MECÁ, METAL, QUÍMICA *entre sustancias* interaction; ~ **de alcance finito** *f* NUCL finite-range interaction; ~ **débil** *f* FÍS weak interaction, FÍS PART, NUCL weak force, weak nuclear force, weak interaction; ~ **dinámica** *f* METAL dynamic interaction; ~ **dipolo-dipolo** *f* FÍS dipole-dipole interaction; ~ **electromagnética** *f* FÍS PART, FÍS RAD electromagnetic interaction; ~ **del electrón-positrón** *f* FÍS PART electron-positron interaction; ~ **fuerte** *f* FÍS strong interaction, FÍS PART, NUCL strong nuclear force; ~ **haz-plasma** *f* NUCL beam-plasma interaction; ~ **hombre-máquina** *f* (*MMI*) INFORM&PD man-machine interaction (*MMI*); ~ **mecánica pastilla-vaina** *f* NUCL pellet-clad mechanical interaction; ~ **nuclear** *f* FÍS PART, NUCL nuclear force; ~ **química pastilla-vaina** *f* NUCL pellet-clad chemical interaction; ~ **con el vecino más cercano** *f* CRISTAL nearest-neighbor interaction (*AmE*), nearest-neighbour interaction (*BrE*)

interaccionar: sin ~ adv FÍS RAD without interacting
interactivo adj INFORM&PD conversational, interactive, ÓPT interactive
interactuar vi FÍS RAD, QUÍMICA interact
interatómico adj CRISTAL, FÍS RAD interatomic
interbloquear vt ING ELÉC, ING MECÁ interlock
interbloqueo m CINEMAT, INFORM&PD, ING ELÉC interlock, ING MECÁ, MECÁ interlocking, PROD interlock
intercalación f CARBÓN filones break, ELEC intercalation, ELECTRÓN interleaving, IMPR sandwich, ING MECÁ cutting-in, MINAS break, TRANSP merging; ~ **de arcilla** f MINAS clay parting; ~ **clasificada** f INFORM&PD collation; ~ **de impulsos** f TELECOM pulse interleaving; ~ **de machos y hembras** f AGRIC maíz interplant; ~ **de roca** f CARBÓN filones stone band; ~ **de roca estéril** f CARBÓN, MINAS slurry basin; ~ **de vagones** f MINAS cage sheets
intercalado[1] adj ELEC intercalated, INFORM&PD embedded, ING MECÁ inserted
intercalado[2] m ING MECÁ insert, PAPEL interleaving; ~ **automático** m EMB automatic collation; ~ **de bandejas en movimiento** m EMB collating transit tray; ~ **en compartimentos** m EMB compartmented insert
intercalador adj GEOL, TEC PETR interlayer
intercalar vt ELECTRÓN interleave, IMPR collate, offcut, ING MECÁ, PROD insert
intercambiabilidad f FOTO, ING MECÁ, INSTR interchangeability
intercambiable adj FOTO, ING MECÁ, INSTR interchangeable
intercambiador m ING MECÁ exchanger; ~ **de aniones** m CARBÓN anion exchanger; ~ **aniónico** m DETERG anionic exchanger; ~ **de calor** m ENERG RENOV, ING MECÁ, NUCL, PROD, REFRIG, TEC PETR, TERMO, TERMOTEC, TRANSP MAR motor heat exchanger; ~ **de calor de aire a aire** m ING MECÁ air-to-air heat exchanger; ~ **de calor a contracorriente** m ALIMENT, TERMO counterflow heat exchanger; ~ **de calor por cruce de corrientes** m REFRIG, TERMO crossflow heat exchanger; ~ **de calor de flujos paralelos** m NUCL, REFRIG, TERMO, TERMOTEC parallel-flow heat exchanger; ~ **de calor con metal líquido** m TERMO, TERMOTEC liquid-metal heat exchanger; ~ **de calor multitubo** m TERMO, TERMOTEC multitube heat exchanger; ~ **de calor primario** m TRANSP AÉR primary heat exchanger; ~ **de calor refrigerante del combustible** m TRANSP AÉR fuel-coolant heat exchanger; ~ **de calor en separador de líquido** m REFRIG, TERMO heat exchanger suction accumulator; ~ **de calor de serpentín helicoidal** m NUCL helical coil-type heat exchanger; ~ **de calor de superficie extendida** m ING MECÁ extended-surface heat exchanger; ~ **de calor tubular** m ING MECÁ tube-type heat exchanger; ~ **de discos ópticos** m INFORM&PD, ÓPT optical disk exchanger; ~ **multitubular de envolvente** m REFRIG shell-and-tube heat exchanger; ~ **de presión criptofija** m TRANSP cryptosteady pressure exchanger; ~ **de segmento de tiempo** m TELECOM time slot interchanger (TSI); ~ **térmico** m ALIMENT heat exchanger
intercambiar vt INFORM&PD swap
intercambio m CONST carreteras, FERRO interchange, INFORM&PD exchange, swapping, ING MECÁ, PROD exchange; ~ **de autentificación** m TELECOM authentication exchange; ~ **calorífico** m FÍS, TERMO heat exchange; ~ **entre centrales** m TELECOM interexchange (BrE), interoffice (AmE); ~ **convectivo** m GAS convective exchange; ~ **de datos comerciales** m TELECOM trade data interchange (TDI); ~ **de datos electrónicos** m TELECOM electronic data interchange (EDI); ~ **de datos electrónicos para la administración, el comercio y el transporte** m TELECOM electronic data interchange for administration, commerce and transport; ~ **estérico** m DETERG ester interchange; ~ **de iones** m HIDROL, QUÍMICA ion exchange; ~ **por modem** m INFORM&PD, TELECOM modem interchange; ~ **de programas** m INFORM&PD roll in-roll out; ~ **térmico** m FÍS heat exchange, GAS thermal exchange, thermic exchange, TERMO heat exchange
intercepción f TEC ESP, TRANSP AÉR intercept; ~ **directa** f CONTAM direct interception; ~ **de haz** f TRANSP AÉR beam intercept; ~ **de llamada** f TELECOM call interception
interceptación f TEC ESP, TRANSP AÉR intercept
interceptador: ~ **de alcantarillado** m HIDROL interceptor sewer
interceptar vt ING MECÁ block, PROD metales fundidos shut off, TELECOM intercept
interceptor: ~ **de aceite** m AGUA oil trap
interconectado adj INFORM&PD attached, interfaced, TELECOM networked
interconectar vt INFORM&PD attach, interface, ING MECÁ interlock, TEC ESP interface
interconexión f ELEC sistemas interconnection, INFORM&PD interblock gap (IBG), attachment, interconnection, ING ELÉC interconnection, network, ING MECÁ interface, interlocking, MECÁ interlocking, TRANSP AÉR interconnection, interlocking; ~ **a bordo** f TEC ESP on-board switching; ~ **digital** f TELECOM digital interface; ~ **externa** f ELECTRÓN, TEC ESP, TELECOM external interface; ~ **de matrices** f INFORM&PD array interconnection; ~ **de redes** f INFORM&PD, TELECOM internetting, network interconnection; ~ **de sistemas abiertos** f (OSI) INFORM&PD, TELECOM open systems interconnection (OSI); ~ **universal S** f TELECOM red de transmisión digital de servicios integrados S-universal interface
interconexionado: ~ **a masa** m TEC ESP circuitos eléctricos bonding
intercrecimiento m GEOL textura mineral intergrowth
intercristalino adj METAL intergranular
interdicción: ~ **de llamadas** f TELECOM call barring
interdigitación f GEOL interdigitation, interfingering, INFORM&PD, ING ELÉC interdigitation
interduplicado m CINEMAT interdupe
interés: ~ **compuesto** m TEC PETR finanzas compound interest
interespacio m TEC ESP gap
interestelar adj TEC ESP interstellar
interfaces: ~ **de línea** f pl ING ELÉC line interfacing
interfaceta f ING MECÁ interface
interfase f INFORM&PD, ING MECÁ, METAL interface; ~ **accionada por órdenes** f INFORM&PD command-driven interface; ~ **de base común** f AmL (cf interfase de conductor común Esp) INFORM&PD bus interface; ~ **de conductor común** f Esp (cf interfase de base común AmL) INFORM&PD bus interface; ~ **de datos distribuidos por fibra** f TELECOM fiber-distributed data interface (AmE) (FDDI), fibre-distributed data interface (BrE) (FDDI); ~ **física**

para sincronismo del multiplexor *f* TELECOM multiplexer timing physical interface (*MTPI*); ~ **con el usuario** *f* INFORM&PD user interface

interfaz[1] *m* INFORM&PD, ING MECÁ, TELECOM interface; ~ **analógico** *m* TELECOM analog interface; ~ **de bus** *m* INFORM&PD bus interface; ~ **Centronics** *m* IMPR Centronics interface; ~ **del circuito de enlace** *m* TELECOM tie circuit interface; ~ **de computador** *m* AmL (*cf interfaz de ordenador Esp*) INFORM&PD, TELECOM computer interface; ~ **de computadora** *m* AmL (*cf interfaz de ordenador Esp*) INFORM&PD, TELECOM computer interface; ~ **computadora-PBX** *m* AmL (*cf interfaz ordenador-PBX Esp*) INFORM&PD, TELECOM computer-PBX interface (*CPI*); ~ **controlado por órdenes** *m* INFORM&PD, TELECOM command-driven interface; ~ **digital** *m* TELECOM digital cross-connect; ~ **digital directa** *m* TELECOM direct digital interface (*DDI*); ~ **de enlace digital** *m* TELECOM digital trunk interface (*DTI*); ~ **físico** *m* TELECOM *de circuitos* physical interface (*PI*); ~ **físico de jerarquía digital síncrona** *m* TELECOM synchronous digital hierarchy physical interface (*SDH physical interface*); ~ **hidrocarburos-agua** *m* CONTAM MAR oil-water interface; ~ **de modem** *m* INFORM&PD, TELECOM modem interface; ~ **de ordenador** *m* Esp (*cf interfaz de computadora AmL*) INFORM&PD, TELECOM computer interface; ~ **ordenador-PBX** *m* Esp INFORM&PD, TELECOM (*cf interfaz computador-PBX AmL, cf interfaz computadora-PBX AmL*) computer-PBX interface (*CPI*); ~ **en paralelo** *m* IMPR parallel interface; ~ **de la red del usuario** *m* INFORM&PD, TELECOM user-network interface (*UNI*); ~ **red-nodo** *m* TELECOM network-node interface; ~ **en serie** *m* IMPR, INFORM&PD, TELECOM serial interface

interfaz[2]: ~ **de línea** *f* ING ELÉC line interface; ~ **normalizada** *f* INFORM&PD, TELECOM standard interface; ~ **serie programable** *f* TELECOM serial programmable interface; ~ **de usuario** *f* TELECOM user interface; ~ **del usuario** *f* INFORM&PD user interface; ~ **usuario-red** *f* INFORM&PD, TELECOM user-network interface (*UNI*)

interferencia *f* GEN interference, ELECTRÓN interference, shot noise, INFORM&PD interference, spoofing, TELECOM interference, wave interference, squealing, TRANSP MAR radar interference; ~ **por agitación térmica** *f* TV shot noise; ~ **atmosférica** *f* GEOFÍS *estudios de campo magnético* atmospheric interference; ~ **axial** *f* ÓPT axial interference; ~ **de banda ancha** *f* ELECTRÓN wideband interference; ~ **de banda angosta** *f* ELECTRÓN, TELECOM narrow-band noise; ~ **de banda estrecha** *f* ELECTRÓN narrow-band interference; ~ **de banda lateral** *f* ELECTRÓN side-band interference, sideband interference; ~ **del canal propio** *f* TV co-channel interference; ~ **coherente** *f* GEOFÍS coherent noise; ~ **constructiva** *f* FÍS, FÍS ONDAS constructive interference; ~ **cruzada** *f* TV crosstalk; ~ **destructiva** *f* FÍS, FÍS ONDAS, ING ELÉC destructive interference; ~ **digital** *f* TELECOM digital interference; ~ **electromagnética** *f* (*IEM*) INFORM&PD, TEC ESP, TELECOM electromagnetic interference (*EMI*); ~ **entre canales** *f* ING ELÉC interchannel interference; ~ **entre los canales** *f* TV crosstalk; ~ **entre múltiples haces** *f* FÍS multiple beam interference; ~ **entre símbolos** *f* TELECOM intersymbol interference; ~ **de**

la frecuencia de imagen *f* ELECTRÓN image-frequency interference; ~ **de imagen por alimentación de la red** *f* TV hum bar; ~ **de imagen por variaciones de la red** *f* TV ripple; ~ **industrial** *f* ING ELÉC industrial interference; ~ **intencionada** *f* D&A *radio* jamming, ELECTRÓN jammer, jamming, TELECOM jamming; ~ **magnética** *f* GEOFÍS magnetic interference; ~ **de ondas** *f* TELECOM wave interference; ~ **óptica** *f* ÓPT, TELECOM optical interference; ~ **primaria** *f* FÍS RAD primary interference; ~ **radiada** *f* FÍS RAD, ING ELÉC radiated interference; ~ **radioeléctrica** *f* ELECTRÓN, FÍS ONDAS, FÍS RAD radio interference; ~ **de RF** *f* ELECTRÓN, TELECOM, TV RF interference; ~ **solar** *f* TEC ESP sun interference; ~ **superpuesta** *f* TV superimposed interference; ~ **televisiva** *f* ELECTRÓN, FÍS, ING ELÉC, TELECOM, TV television interference

interferencias[1]: **contra las** ~ *adj* ELEC *protección* anti-interference

interferencias[2] *f pl* FÍS ONDAS *radio* noise

interferir[1] *vt* D&A, ELECTRÓN, TEC ESP, TELECOM *radio* jam

interferir[2] *vi* ELECTRÓN, FÍS interfere

interferometría: ~ **axial** *f* ÓPT axial interferometry, TELECOM axial slab interferometry; ~ **de placas** *f* ÓPT slab interferometry; ~ **plana** *f* TELECOM slab interferometry; ~ **transversal** *f* ÓPT, TELECOM transverse interferometry

interferómetro *m* GEN interferometer; ~ **acústico** *m* ACÚST acoustic interferometer; ~ **de Fabry-Pérot** *m* FÍS, INSTR, TEC ESP Fabry-Pérot interferometer; ~ **láser** *m* ELECTRÓN, ING MECÁ laser interferometer; ~ **de Michelson** *m* FÍS Michelson interferometer; ~ **de radio** *m* FÍS RAD, GEOFÍS, INSTR radio interferometer; ~ **de Rayleigh** *m* FÍS Rayleigh interferometer

interfoliada *adj* GEOL *rocas metamórficas* interfoliated

interfoliar *vt* INFORM&PD interleave

interfolición *f* INFORM&PD interleaving

intergranular *adj* GEOL, METAL intergranular

interiónico *adj* QUÍMICA *distancia* interionic

interior[1] *adj* PAPEL *capa* middle, TRANSP AÉR inboard, TRANSP MAR inshore

interior[2] *m* CONST core; ~ **fuera del estudio** *m* CINEMAT location; ~ **planetario** *m* TEC ESP planetary interior

interisticio *m* INSTAL HIDRÁUL clearance space

interlínea *f* IMPR lead, INFORM&PD row pitch

interlineado *m* IMPR, INFORM&PD *tipografía* leading

interlinear *vti* IMPR, INFORM&PD lead

interlock: ~ **acanalado** *m* TEXTIL plain-rib interlock; ~ **de canalé plano** *m* TEXTIL plain-rib interlock

intermediario *m* CINEMAT in-betweener, TEC ESP spacer, TEC PETR *finanzas* broker

intermitente[1] *adj* ELEC *corriente alterna* periodic

intermitente[2] *m* AUTO, VEH blinker, *accesorio* direction indicator; ~ **delantero** *m* AUTO, VEH front flasher

intermitentes: ~ **de aviso** *m pl* AUTO, SEG, VEH flashing warning lights

intermodulación *f* ELECTRÓN, TEC ESP, TELECOM, TV intermodulation

intermolecular *adj* QUÍMICA intermolecular

internegativo *m* CINEMAT, IMPR internegative

Internet *m* INFORM&PD, TELECOM Internet

interno: ~ **superior del reactor** *m* NUCL upper core, upper internal

interoficinal *adj* TELECOM interexchange (*BrE*), interoffice (*AmE*)

interplanetario *adj* TEC ESP interplanetary

interplanta *f* PROD interplant

interpolación *f* ELEC, INFORM&PD, MATEMÁT, ÓPT, TELECOM interpolation; ~ **lineal** *f* TELECOM linear interpolation; ~ **no lineal** *f* TELECOM nonlinear interpolation; ~ **de señales vocales** *f* TELECOM interpolation of speech signals, speech interpolation (*SI*); ~ **vocal** *f* TEC ESP speech interpolation; ~ **vocal digital** *f* TELECOM digital speech interpolation (*DSI*)

interpolado *adj* ELECTRÓN *filtro* interpolating

interpolador *m* MATEMÁT, TELECOM interpolator

interpretación: ~ **de fotografía aérea** *f* CARBÓN airphoto interpretation; ~ **de la indicación** *f* ELEC, INSTR read-out

interpretar *vt* GEOM, INFORM&PD interpret

intérprete *m* INFORM&PD interpreter

interrevestimiento *m* TRANSP AÉR interlining

interrogación *f* INFORM&PD interrogation, query, TELECOM interrogation, polling; ~ **de la base de datos** *f* INFORM&PD database query; ~ **de búsqueda** *f* INFORM&PD search query; ~ **de datos** *f* INFORM&PD data query; ~ **de las estaciones de barco** *f* TELECOM ship polling; ~ **múltiple** *f* TELECOM multipolling; ~ **por el usuario** *f* INFORM&PD user query

interrogar *vt* INFORM&PD interrogate, query, TELECOM interrogate, poll

interrumpido *adj* ELEC *fallo* out-of-order

interrumpir[1] *vt* AGUA shut off, ELECTRÓN chop, turn off, INFORM&PD abort, break, halt, switch off, ING ELÉC interrupt, release, stick, ING MECÁ stop, trip, PROD break, stop

interrumpir[2]: ~ **el avance** *vi* ING MECÁ throw out of feed; ~ **la marcha** *vi* ING MECÁ throw out of action

interrupción *f* AGUA shutting-off, turning-off, ELEC release, *de equipo* failure, ELECTRÓN gating, GAS interruption, GEOL breakthrough, *brecha indeposicional breve* nonsequence, INFORM&PD time-out, break, ING ELÉC release, sticking, interruption, ING MECÁ stop, tripping, MINAS *de un filón* breakthrough, PROD stoppage, break, TEC ESP shutdown, switching, TELECOM interruption; ~ **de la comunicación** *f* INFORM&PD hand-up; ~ **destructiva** *f* ELECTRÓN destructive breakdown; ~ **de entrada/salida** *f* INFORM&PD input/output interrupt; ~ **externa** *f* INFORM&PD external interrupt; ~ **de fase** *f* TV phase failure; ~ **de frecuencia** *f* ELECTRÓN frequency cutoff; ~ **de función** *f* ELECTRÓN mode jump; ~ **por haz de electrones** *f* ELECTRÓN electron-beam cutting; ~ **de imagen** *f* TV picture failure; ~ **imprevista** *f* TELECOM unforeseen interruption; ~ **de la inyección** *f* ELECTRÓN injection locking; ~ **de la llama** *f* TERMO flame failure; ~ **de potencia** *f* PROD power outage; ~ **prioritaria** *f* INFORM&PD priority interrupt; ~ **en la sedimentación** *f* OCEAN sedimentation break; ~ **del servicio** *f* FÍS, ING ELÉC, TELECOM breakdown; ~ **en el servicio de red** *f* TELECOM network breakdown; ~ **en el servicio técnico** *f* TELECOM technical breakdown; ~ **del suministro eléctrico** *f* FÍS breakdown, INFORM&PD power-supply interrupt; ~ **técnica del servicio** *f* TELECOM transmission breakdown; ~ **térmica** *f* ELECTRÓN, TERMO thermal shutdown; ~ **vectorial** *f* Esp (*cf interrupción vectorizada AmL*) INFORM&PD vectored interrupt; ~ **vectorizada** *f* AmL (*cf interrupción vectorial Esp*) INFORM&PD vectored

interrupt; ~ **de la ventilación** *f* CARBÓN ventilation breakdown

interruptor *m* AUTO breaker, CINEMAT on-off switch, ELEC *conmutador* circuit breaker, switchgear, commutator, cut-out switch, bypass, on-off switch, release switch, switch, interrupter, ELECTRÓN chopper, FÍS, INFORM&PD switch, ING ELÉC release switch, on-off switch, interrupter, cut-out switch, circuit breaker, switch, ING MECÁ tripper, TELECOM interrupter, TV switch, switcher;

▪a ~ **de acción alternada** *m* ELEC, ING ELÉC alternate-action switch; ~ **de acción rápida** *m* ELEC, ING ELÉC snap-action switch; ~ **accionado neumáticamente** *m* ELEC pneumatically-operated switch; ~ **de accionamiento remoto** *m* TELECOM, TV remote switching; ~ **en aceite** *m* ELEC, ING ELÉC oil circuit breaker, oil switch; ~ **acoplado** *m* ELEC, ING ELÉC, PROD gang switch; ~ **de aire** *m* ELEC, ING ELÉC air-break switch; ~ **de alambre casero** *m* ELEC house-wiring switch; ~ **de alimentación** *m* ELEC, ING ELÉC on-off switch, power switch, ING MECÁ on-off switch; ~ **de arranque** *m* AUTO, VEH ignition starter switch; ~ **de arranque del motor** *m* AUTO, TRANSP AÉR, VEH crank switch, engine starter; ~ **de arranque Y-delta** *m* ELEC Y-delta starter, Y-delta starting switch; ~ **autocompensado** *m* ELEC, ELECTRÓN, ING ELÉC self-balancing switch; ~ **automático** *m* CONST circuit breaker, ELEC *relé* contactor, *cortocircuito* switch, ING ELÉC contactor, self-acting switch; ~ **automático por caída de presión** *m* ELEC, ING ELÉC pressure switch; ~ **automático miniatura** *m* ELEC, ING ELÉC miniature circuit breaker; ~ **automático de tipo fuelle** *m* PROD bellows-type pressure switch; ~ **auxiliar** *m* PROD auxiliary switch;

▪b ~ **en baño de aceite** *m* ELEC burk-oil circuit breaker, PROD oil break switch, oil interrupter, oil switch; ~ **barométrico** *m* TRANSP AÉR barometric switch; ~ **bidireccional** *m* ELEC, ING ELÉC bidirectional switch, double-throw switch; ~ **bipolar** *m* ELEC double-break switch, ING ELÉC two-pole switch, PROD double-break switch; ~ **de botón de presión** *m* ELEC, ING ELÉC, PROD push-button switch;

▪c ~ **de caja** *m* ELEC box switch; ~ **de caudal** *m* PROD flow switch; ~ **centrífugo** *m* ELEC, ING ELÉC, ING MECÁ centrifugal switch; ~ **a chorro de aire** *m* ELEC *cortocircuito* air blast breaker; ~ **de codillo** *m* ELEC, ELECTRÓN toggle switch; ~ **del colector** *m* ELEC, ELECTRÓN commutator switch; ~ **de la columna de dirección de autocancelación** *m* ELEC *automotor* self-canceling steering column switch (*AmE*), self-cancelling steering column switch (*BrE*); ~ **de la columna de mando de autocancelación** *m* ELEC *automotor* self-canceling steering column switch (*AmE*), self-cancelling steering column switch (*BrE*); ~ **para conectar y desconectar la alimentación** *m* ELEC on-off switch; ~ **de configuración simétrica** *m* ELEC, ELECTRÓN push-pull switch; ~ **de conmutador de pie** *m* AUTO foot dimmer (*AmE*), foot dipswitch (*BrE*); ~ **por contacto** *m* ELEC, ING ELÉC touch switch; ~ **de contacto de cuchilla** *m* ELEC, NUCL knife-edge switch; ~ **de contacto de dos sistemas** *m* AUTO, VEH two-system contact-breaker; ~ **de contacto manual** *m* ELEC touch-contact switch; ~ **de contacto momentáneo** *m* ELEC, ING ELÉC, PROD momentary-contact switch; ~ **de contacto montado en el piso** *m* ELEC floor contact switch;

~ **de contactos aislados** *m* CONST nonshorting switch; ~ **de continuidad** *m* ELEC continuity switch; ~ **de control colgante** *m* SEG pendant switch control; ~ **controlado por compuerta** *m* ELEC gate switch; ~ **controlado por silicio** *m* ELEC, ING ELÉC silicon-controlled switch; ~ **controlado por temperatura** *m* ELEC, ING ELÉC temperature-controlled switch; ~ **de cordón** *m* ELEC pull switch; ~ **de corriente** *m* ELEC, ING ELÉC current switch (*AmE*), mains switch (*BrE*), on-off switch; ~ **corta-corriente** *m* ELEC, ING ELÉC interlock switch; ~ **de corte rápido** *m* ELEC, ING ELÉC quick-break switch; ~ **cortocircuitantes** *m* ELEC, ING ELÉC shorting switch;

~ d ~ **de derivación** *m* ELEC shunting switch; ~ **derivante** *m* ELEC bypass switch; ~ **de descarga** *m* ELEC, ING ELÉC glow switch, ING MECÁ discharge button; ~ **deslizante** *m* ING ELÉC slide switch; ~ **detector** *m* ING ELÉC sensing switch; ~ **de disparo único** *m* FÍS single-throw switch; ~ **de disposición simétrica** *m* ELEC, ELECTRÓN push-pull switch; ~ **de doble ruptura** *m* ELEC, PROD double-break switch; ~ **de dos sistemas** *m* AUTO, VEH two-system contact-breaker;

~ e ~ **electrónico** *m* ELECTRÓN electronic chopper; ~ **embutido** *m* ELEC recessed switch; ~ **de empaquetado** *m* ELEC, TELECOM packet switch (*PS*); ~ **empotrado** *m* ELEC recessed switch; ~ **de encendido** *m* AUTO ignition starter switch, ING MECÁ on-off switch, VEH ignition switch; ~ **encendido-apagado** *m* ELEC, ING ELÉC on-off switch; ~ **enchufable** *m* PROD plug-in switch; ~ **sin enchufe** *m* PROD non-plug-in position switch; ~ **sin enchufe tomacorriente** *m* PROD non-plug-in switch; ~ **de entrada** *m* ELEC on-off switch; ~ **de escala** *m* ELEC scale switch; ~ **de esclusa** *m* ING MECÁ gate cutter; ~ **estrella-triángulo** *m* ING ELÉC star-delta switch; ~ **de excitación** *m* NUCL field-discharge switch; ~ **extensible** *m* ING MECÁ handle switch; ~ **de exterior** *m* ELEC outdoor switchgear;

~ f ~ **del faro delantero** *m* AUTO, VEH headlamp switch, headlight switch; ~ **de fin de carrera** *m* ELEC, ING ELÉC, MECÁ, PROD limit switch; ~ **de fin de carrera de sobrecarrera máxima** *m* PROD extreme-overtravel limit switch; ~ **de flotador** *m* ELEC, ING ELÉC, REFRIG float switch; ~ **de funcionamiento** *m* ELEC, ING ELÉC operating switch;

~ g ~ **giratorio de sectores sellado** *m* ING ELÉC sealed-wafer rotary switch; ~ **de graduación de la luz** *m* ELEC, ING ELÉC dimmer switch;

~ h ~ **horario** *m* ELEC, ING ELÉC switch clock, time switch, TELECOM time switch;

~ i ~ **ignífugo** *m* SEG, TERMO fireproof switch, flameproof switch; ~ **de imagen en negativo** *m* TV reverse-image switch; ~ **de impacto** *m* TRANSP AÉR crash switch; ~ **inmediato** *m* ING ELÉC snap-in switch; ~ **de inmersión** *m* PROD dip switch; ~ **instantáneo** *m* ELEC, ING ELÉC quick-break switch; ~ **de instrumento** *m* ING ELÉC instrument switch; ~ **intermedio-n** *m* ELEC n-way switch; ~ **intermitente** *m* AUTO, ELEC, VEH flasher, indicator; ~ **de inversión de fase** *m* ING ELÉC phase-reversal switch;

~ l ~ **de láminas** *m* ELEC, FÍS, ING ELÉC, TELECOM reed switch; ~ **de lengüeta** *m* ELEC, FÍS, ING ELÉC, TELECOM reed switch; ~ **limitador** *m* ELEC, ING ELÉC, MECÁ, PROD limit switch; ~ **de luz** *m* ELEC light

switch, ELECTRÓN light chopper; ~ **de luz de frenado** *m* AUTO, VEH brake light switch, stop light switch;

~ m ~ **maestro** *m* ELEC, ING ELÉC master switch; ~ **maestro de batería** *m* AUTO, VEH battery master switch; ~ **magnético** *m* AUTO, VEH *motor de arranque* solenoid; ~ **de mando** *m* ING ELÉC control switch; ~ **de mando de reglaje** *m* TRANSP AÉR trim control switch; ~ **manométrico** *m* ELEC manometric switch; ~ **de máxima** *m* ELEC overcurrent circuit breaker; ~ **de mercurio** *m* ELEC, ING ELÉC mercury interrupter, mercury switch; ~ **de mínima** *m* ING ELÉC minimum circuit breaker; ~ **del motor** *m* PROD motor switching; ~ **múltiple** *m* ELEC multiple switch;

~ n ~ **de n posiciones** *m* ELEC n-way switch; ~ **neumático de posición** *m* PROD air-operated position switch; ~ **de nivel** *m* PROD level switch; ~ **por nivel fluido** *m* PROD fluid-level switch; ~ **no enchufable** *m* PROD non-plug-in switch;

~ o ~ **on-off** *m* FÍS on-off switch; ~ **óptico** *m* ELEC, ING ELÉC, ÓPT optical switch; ~ **optoelectrónico** *m* ELEC, ING ELÉC, ÓPT optoelectronic switch; ~ **oscilante** *m* CINEMAT, ELEC, ING ELÉC rocker switch;

~ p ~ **de palanca** *m* ELEC knife switch, toggle switch, ING ELÉC lever switch; ~ **de palanca acodada** *m* ELEC toggle switch; ~ **de palanca acodillada** *m* ELEC toggle switch; ~ **de palanca de dos posiciones de contacto** *m* ELEC double-throw knife switch; ~ **de palanquita** *m* ELEC toggle switch; ~ **de pantalla** *m* ELEC, ING ELÉC deck switch; ~ **del paso del colectivo** *m* TRANSP AÉR collective-pitch switch; ~ **de paso de mando** *m* ELEC override switch; ~ **de pie** *m* AUTO, VEH foot dimmer, foot dipswitch; ~ **para las pilas** *m* FOTO battery switch; ~ **piloto** *m* ELEC pilot switch; ~ **de platinos dobles** *m* AUTO, VEH dual-point breaker; ~ **PLD** *m* ING ELÉC DIP switch; ~ **de posición** *m* ELEC, ING ELÉC, PROD position switch; ~ **de posición de codificador programable** *m* PROD programmable-encoder position switch; ~ **de posición metálica** *m* PROD metal-position switch; ~ **de posición no enchufable** *m* PROD non-plug-in position switch; ~ **de posición de palanca a resorte** *m* PROD wobble stick head; ~ **de posición del portaherramienta revólver** *m* PROD turret-head position switch; ~ **de posición termoplástico** *m* PROD thermoplastic position switch; ~ **de posición de tiempo retardado** *m* PROD time delay position switch; ~ **de posición de la torreta** *m* PROD turret-head position switch; ~ **de potencia** *m* PROD power switch; ~ **de la presión del aceite** *m* TRANSP AÉR oil-pressure switch; ~ **principal** *m* ELEC, ING ELÉC main switch, master switch, TV master switch; ~ **de proceso momentáneo** *m* ING ELÉC momentary action switch; ~ **de proximidad** *m* ELEC, ING ELÉC proximity switch; ~ **de proximidad inductivo** *m* ELEC, ING ELÉC inductive proximity switch; ~ **de pruebas** *m* ING ELÉC test switch; ~ **PST cuadripolar** *m* ING ELÉC 4 PST switch; ~ **de puerta** *m* ING MECÁ gate cutter; ~ **de puesta en marcha** *m* ING MECÁ on-off switch; ~ **de puesta a tierra** *m* ELEC, ING ELÉC earthing switch (*BrE*), grounding switch (*AmE*); ~ **de pulsador** *m* ELEC, ING ELÉC, PROD push-button switch; ~ **de puño** *m* ING MECÁ handle switch; ~ **de puntos dobles** *m* AUTO, VEH dual-point breaker;

~ r ~ **rápido** *m* ELEC, ING ELÉC quick-break switch; ~ **de red** *m* ELEC, ING ELÉC current switch (*AmE*), mains switch (*BrE*), on-off switch; ~ **del reductor de velocidad** *m* AUTO, VEH kick-down switch; ~ **del régimen de cambio de presión** *m* TRANSP AÉR pressure rate-of-change switch; ~ **de reloj** *m* ELEC, TELECOM time switch; ~ **de reserva** *m* PROD backup switch; ~ **de resorte** *m* ELEC spring commutator, spring switch; ~ **rotatorio de sectores** *m* ING ELÉC open-wafer rotary switch; ~ **de retorno de potencia** *m* PROD backup switch; ~ **de rótula** *m* ELEC toggle switch; ~ **de rueda moleteada** *m* PROD thumbwheel switch; ~ **de ruptura brusca** *m* ELEC, ING ELÉC quick-break switch; ~ **de ruptura del campo inductor** *m* NUCL field-discharge switch;

~ s ~ **seccionador** *m* ELEC *red de distribución* sectionalizing switch, ING ELÉC switch; ~ **de seccionamiento** *m* ELEC disconnecting switch; ~ **secundario-n** *m* ELEC n-way switch; ~ **de seguridad** *m* CONST RS flip-flop, ING ELÉC safety switch; ~ **selector del depósito de combustible** *m* TRANSP AÉR fuel-tank selector switch; ~ **selector de modo** *m* TRANSP AÉR mode selector switch; ~ **selector de ranura de monedas** *m* PROD coin-slot selector switch; ~ **de la señal de espira de autocancelación** *m* ELEC self-canceling turn signal switch (*AmE*), self-cancelling turn signal switch (*BrE*); ~ **de la señal de giro de autocancelación** *m* ELEC self-canceling turn signal switch (*AmE*), self-cancelling turn signal switch (*BrE*); ~ **de la señal de vuelta de autocancelación** *m* ELEC self-canceling turn signal switch (*AmE*), self-cancelling turn signal switch (*BrE*); ~ **simple** *m* ING ELÉC single-pole single-throw switch (*SPST switch*); ~ **simple cuadripolar** *m* ING ELÉC 4 PST switch; ~ **de sobrecorriente** *m* ELEC overcurrent switch; ~ **de sobrepaso** *m* ELEC bypass switch; ~ **subsidiario** *m* PROD backup switch;

~ t ~ **de tambor** *m* ELEC drum switch; ~ **de tambor inversor** *m* ELEC reversing-drum switch; ~ **telemandado** *m* TELECOM, TV remote switching; ~ **temporal** *m* ING ELÉC time switch; ~ **temporizado** *m* ELEC time switch, ING ELÉC stepping switch, TELECOM time switch; ~ **temporizador** *m* CINEMAT time switch; ~ **de tiempo** *m* ELEC, TELECOM time switch; ~ **de timbre** *m* CONST button; ~ **sin toma de corriente** *m* PROD non-plug-in position switch; ~ **de tono** *m* TEC ESP tone disabler, TRANSP AÉR beep switch; ~ **totalmente aislado** *m* ELEC all-insulated switch; ~ **con traba** *m* AmL (*cf cerrojo de bloqueo Esp*) TV switch lock; ~ **de transferencia de mando** *m* ELEC override switch; ~ **transisto de movimiento alternativo** *m* PROD up and lower transistor switch; ~ **de tres direcciones** *m* ELEC three-way switch;

~ u ~ **ultrarrápido** *m* ELEC, ING ELÉC quick-break switch; ~ **unipolar de bajo voltaje** *m* ING ELÉC tumbler switch; ~ **unipolar de doble vano** *m* ING ELÉC single pole double-throw switch (*SPDT switch*); ~ **unipolar y univanal** *m* ING ELÉC single-pole single-throw switch (*SPST switch*);

~ v ~ **de vacío** *m* ELEC *cortocircuito* vacuum switch; ~ **de válvula de estrangulación** *m* AUTO, VEH throttle valve switch; ~ **de varias posiciones** *m* ING ELÉC n-position switch; ~ **de la velocidad de transmisión de datos** *m* INFORM&PD, PROD data transmission rate switch; ~ **de vía única** *m* ING ELÉC single-throw switch; ~ **de volquete** *m* ELEC toggle switch; ~ **de voltaje de la red** *m* ELEC, ING ELÉC current switch (*AmE*), mains switch (*BrE*); ~ **de**

voltaje a tierra *m* ELEC, ING ELÉC earth arrester (*BrE*), ground arrester (*AmE*)
intersecar *vt* GEOM, ING MECÁ, MATEMÁT intersect
intersección *f* CONST *carreteras* intersection, GEOM meet, intersection, intercept, INFORM&PD intersection, MINAS *de un filón inclinado con un plano horizontal* strike drive, *galerías* drift, QUÍMICA *carreteras* intersection
intersectado[1] *adj* GEOM, ING MECÁ, MATEMÁT intersecting
intersectado[2] *m* GEOM, ING MECÁ, MATEMÁT intersecting
intersectar *vt* GEOM, ING MECÁ, MATEMÁT intersect
intersticio *m* CARBÓN, CONST, HIDROL interstice, ING MECÁ *turbinas* clearance, INSTAL HIDRÁUL, TEC ESP space; ~ **de aire** *m* TRANSP *de hovercraft* air gap
intervalo *m* ACÚST, GEOL interval, INFORM&PD gap, interval, PROD step, TEC ESP range, space, gap, TEC PETR interval, TRANSP gap, TV back gap, gap; ~ **de acidez** *m* CONTAM, QUÍMICA acidic area; ~ **de aire** *m* ELEC *capacitor, transformador*, TRANSP air gap; ~ **aislante residual** *m* ING ELÉC residual gap; ~ **aumentado** *m* ACÚST augmented interval; ~ **de borrado** *m* TELECOM, TV blanking interval; ~ **de borrado vertical** *m* TV vertical-blanking interval; ~ **de cabeza magnética** *m* TV magnetic-head gap; ~ **de caída** *m* ELECTRÓN fall time; ~ **de la carrera de ida** *m* TV forward-stroke interval; ~ **cerrado** *m* MATEMÁT closed interval; ~ **de cocción** *m* C&V *porcelanas* firing range; ~ **por conducción** *m* ING ELÉC on period; ~ **de confianza** *m* INFORM&PD, MATEMÁT confidence interval; ~ **cronometrado** *m* PROD timed interval; ~ **decreciente** *m* ACÚST diminished interval; ~ **de descarga gaseosa** *m* ELECTRÓN, GAS, NUCL gas-discharge gap; ~ **de energía** *m* FÍS RAD energy range; ~ **entre bloques** *m* INFORM&PD interblock gap (*IBG*); ~ **de la escala** *m* METR scale interval; ~ **de exploración** *m* TV scanning gap; ~ **de frecuencias** *m* ELEC *de corriente alterna*, ELECTRÓN, TV frequency range; ~ **de impulso** *m* ELECTRÓN pulse interval; ~ **de impulsos** *m* ELECTRÓN pulse spacing; ~ **de inactividad por avería** *m* INFORM&PD fault time; ~ **de interacción** *m* ELECTRÓN interaction gap; ~ **de luz verde** *m* TRANSP *control de tráfico* green time; ~ **magnético** *m* GEOL magnetic interval; ~ **de mareas** *m* ENERG RENOV, GEOFÍS, HIDROL, TRANSP MAR tidal range; ~ **de medida** *m* METR span; ~ **de miscibilidad** *m* METAL miscibility gap; ~ **de modulación de pulsos** *m* AmL TV pulse-interval modulation; ~ **de música** *m* ACÚST, FÍS ONDAS *entre dos notas* musical interval; ~ **negro** *m* TV black stretch; ~ **de órbita** *m* TEC ESP orbital period; ~ **de período** *m* NUCL period range; ~ **de planeamiento** *m* PROD planning interval; ~ **de posicionamiento** *m* ÓPT seek time; ~ **de pronóstico** *m* PROD forecast interval; ~ **de repetición de impulsos** *m* ELECTRÓN pulse-repetition interval; ~ **de resolución geométrica** *m* GEOM, NUCL geometric resolution length; ~ **de seguridad entre trenes y autobuses** *m* TRANSP safety headway; ~ **de supresión de la imagen** *m* TV horizontal blanking interval; ~ **de temperatura** *m* TERMO temperature range; ~ **de temperatura crítica** *m* TERMO critical-temperature range; ~ **de temperatura efectiva** *m* TERMO effective-temperature range; ~ **de temperatura de fusión** *m* TERMO melting range; ~ **de tiempo** *m* PROD time

interval; **~ de tiempo neto** *m* TRANSP net time interval; **~ de tiempo total** *m* TRANSP *control de tráfico* overall time interval; **~ de tolerancias** *m* CALIDAD tolerance interval; **~ de trazo** *m* ELECTRÓN scribing step, TV trace interval; **~ unitario** *m* TELECOM unit interval (*UI*); **~ vertical** *m* TV vertical interval

intervalos: **~ encajados** *m pl* MATEMÁT nested intervals; **~ principales en la escala diatónica** *m pl* ACÚST main intervals on the diatonic scale; **~ de salida** *m pl* FERRO, TRANSP *trenes, autobuses* time headway

intervención *f* INFORM&PD intervention, ING MECÁ cutting-in

intoxicación *f* ALIMENT food poisoning; **~ por exceso de oxígeno** *f* OCEAN oxygen poisoning

intraclasto *m* GEOL intraclast

intracratónico *adj* GEOL intracratonic

intradós *m* CONST soffit, *arco* intrados

intraesparita *f* TEC PETR intraesparite

intragranular *adj* GEOL, METAL intragranular

intramicrita *f* PETROL, TEC PETR intramicrite

intramolecular *adj* QUÍMICA intramolecular

intrarreacción *f* ING ELÉC inverse feedback

intrincado *adj* ING MECÁ intricate

intrínsecamente: **~ seguro** *adj* ING ELÉC intrinsically safe

intrínseco *adj* GEN intrinsic

introducción *f* INFORM&PD entry, input; **~ de datos** *f* ELECTRÓN, INFORM&PD data entry; **~ de datos en las casillas de un estado de obra por horas** *f* PROD taking-off sheets per hours; **~ de datos en las casillas de un estado de obra según materiales** *f* PROD taking-off sheets materials; **~ directa de la información** *f* ELECTRÓN direct writing; **~ de información directa por haz de electrones** *f* ELECTRÓN direct electron beam writing; **~ manual** *f* INFORM&PD manual input; **~ mediante el teclado** *f* INFORM&PD keyboard entry; **~ práctica** *f* PROD hands-on introduction

introducir *vt* ELECTRÓN input, *datos* enter, INFORM&PD enter, ING MECÁ drive in, drive into; **~ en** *vt* ING MECÁ fit in, insert, install; **~ gradualmente** *vt* CINEMAT, INFORM&PD key in; **~ en la jaula de extracción** *vt* MINAS *pozo* cage

introscopia *f* NUCL introscopy

intrusión *f* GEOL, PETROL, TELECOM, TRANSP AÉR intrusion; **~ de agua del mar** *f* AGUA sea water intrusion; **~ de roca ígnea** *f* CARBÓN, GEOL stock

intruso *adj* ENERG RENOV, GEOL intrusive

intumescencia: **~ salina** *f* GEOL, TEC PETR salt pillow

intumescente *adj* COLOR, P&C *pintura* intumescent

inulina *f* ALIMENT, QUÍMICA inulin

inundación *f* AGUA inundation, flood, flooding, CARBÓN overflow, ELECTRÓN flooding, HIDROL flood, torrent, inundation, flooding, overflow, ING MECÁ overflow, OCEAN flooding, TRANSP MAR *río, esclusas, marea* flood

inundado *adj* AGUA, HIDROL, OCEAN, TRANSP MAR *buque* flooded

inundar[1] *vt* TRANSP MAR *tanques* flood

inundar[2] *vi* HIDROL *río* flood

inundarse *v refl* HIDROL overflow

inutilizado *adj* ELEC, ING MECÁ out-of-action, out-of-order

invadeable *adj* HIDROL unfordable

invadido: **~ por las aguas** *adj* HIDROL waterlogged

inválido *adj* TELECOM invalid

invariable *adj* GEOM invariant, ING MECÁ set

invariante[1] *adj* GEOM invariant

invariante[2] *f* ELECTRÓN, FÍS, GEOM, MECÁ constant, invariant; **~ adiabática** *f* FÍS adiabatic invariant

invarianza: **~ del calibre** *f* FÍS gage invariance (*AmE*), gauge invariance (*BrE*)

invasión: **~ de agua salada** *f* HIDROL salt water invasion

invención *f* ING MECÁ contrivance

inventariar *vt* PROD schedule

inventario *m* EMB inventory, PROD inventory, tally, TEC PETR inventory; **~ de actividad** *m* NUCL activity inventory; **~ continuo** *m* PROD perpetual inventory; **~ disponible** *m* PROD available inventory; **~ durante la producción** *m* PROD in-process inventory; **~ de emisiones** *m* CONTAM emission inventory; **~ inmovilizado** *m* PROD inactive inventory; **~ de maquinaria** *m* PROD schedule of machinery; **~ de mercaderías** *m* PROD stocktaking; **~ periódico** *m* PROD cyclic inventory; **~ permanente** *m* PROD perpetual inventory

invernada *f* TRANSP MAR *embarcación* winter storage

invernadero *m* AGRIC greenhouse, hothouse; **~ multicuerpo** *m* AGRIC multispan greenhouse

invernar *vi* AGRIC overwinter

invernizar *vt* QUÍMICA *aceites comestibles y lubricantes* winterize

inversa *f* MATEMÁT inverse

inversión *f* CONTAM, CRISTAL inversion, ELECTRÓN inversion, reversing, FÍS inversion, ING MECÁ reversing, METR, PETROL inversion, PROD *fundería* rolling over, TEC ESP switching, TELECOM inversion, *señal de contestación* reversal, TRANSP *velocidad de hélices* reversal, TV inversion; **~ alternativa de señales** *f* TELECOM alternate mark inversion; **~ atmosférica** *f* CONTAM, METEO atmospheric inversion; **~ automática** *f* CINEMAT autoreverse; **~ del cabeceo** *f* TRANSP AÉR pitch reversing; **~ de cifras** *f Esp* (*cf movimiento de figuras AmL*) INFORM&PD figures shift; **~ de control** *f* TRANSP AÉR control reversal; **~ de corriente** *f* ING ELÉC reverse current; **~ de la esfera** *f* GEOM reversal of the sphere; **~ de fase** *f* CARBÓN, ELEC, PETROL, PROD phase reversal, phase inversion, TRANSP MAR *radar* backlash, TV phase reversal; **~ de fase de video** *f AmL*, **~ de fase de vídeo** *f Esp* TV video phase reversal; **~ de frecuencia** *f* TELECOM frequency inversion; **~ geomagnética** *f* GEOFÍS geomagnetic reversal; **~ ineficaz** *f* ELECTRÓN weak inversion; **~ lateral** *f* TV lateral inversion; **~ de líneas** *f* TELECOM line reversal; **~ magnética** *f* GEOL magnetic reversal; **~ de marcha** *f* FERRO *vehículos*, ING MECÁ, MINAS *corriente de aire* reversing; **~ meteorológica** *f* CONTAM, METEO meteorological inversion; **~ de microondas** *f* FÍS RAD microwave inversion; **~ óptica** *f* ÓPT, QUÍMICA optical inversion, Walden inversion; **~ del paso** *f* TRANSP, TRANSP AÉR pitch reversing; **~ del paso de las palas** *f* TRANSP, TRANSP AÉR blade-pitch reversal; **~ de población** *f* ELECTRÓN, FÍS, FÍS RAD population inversion; **~ de polaridad** *f* ELEC, ING ELÉC, TEC ESP polarity reversal; **~ de polaridad periódica** *f* TV periodic polarity inversion; **~ de polos** *f* ELEC, ING ELÉC, TEC ESP polarity reversal; **~ pronunciada** *f* ELECTRÓN strong inversion; **~ en régimen estacionario** *f* FÍS RAD steady-state inversion; **~ de la señal codificada** *f* TELECOM coded mark inversion (*CMI*); **~ de temperatura** *f* CONTAM, METEO, TERMO inversion layer, temperature inversion; **~ térmica** *f* CONTAM, METEO meteorological inversion; **~ de tonos** *f* IMPR tonal inversion; **~ de velocidad** *f* TEC PETR velocity inversion; **~ de Walden** *f* ÓPT, QUÍMICA optical inversion, Walden inversion

inverso *adj* CINEMAT reverse, ELECTRÓN, FÍS, GEOM inverse, INFORM&PD inverse, reverse, ING ELÉC, ING MECÁ, MATEMÁT, MINAS, PROD inverse

inversor *m* CINEMAT, ELEC inverter, ELECTRÓN reverser, FÍS, INFORM&PD, ING ELÉC, TELECOM inverter; **~ de arranque** *m* ELEC, ING ELÉC starting changeover switch; **~ de CC/CA** *m* ELECTRÓN AC/DC converter; **~ de corriente** *m* ELEC, ING ELÉC changeover switch, current reverser; **~ de corriente de arranque** *m* ELEC, ING ELÉC starting changeover switch; **~ de derecha a izquierda** *m* IMPR straight-line image reverser; **~ de empuje** *m* TRANSP, TRANSP AÉR thrust reverser; **~ estático** *m* TEC ESP static inverter; **~ de fase** *m* ELEC *circuito CA/CC* phase inverter; **~ de letras** *m* INFORM&PD letters shift; **~ de polaridad** *m* ELEC *conmutador* polarity reverser

invertasa *f* ALIMENT, QUÍMICA invertase, sucrase

invertido[1] *adj* CINEMAT inverted, ING MECÁ inverse, inverted

invertido[2] *m* ING MECÁ inverse

invertidor: **~ estático** *m* ING ELÉC static inverter

invertir *vt* CINEMAT, ELEC invert, ELECTRÓN reverse, FÍS, INFORM&PD, ING ELÉC invert *marcha* reverse, reverse the motion, ING MECÁ overturn, INSTAL HIDRÁUL *caudal de vapor* reverse, TELECOM invert

investigación *f* ING MECÁ investigation; **~ de accidente aéreo** *f* TRANSP AÉR air-disaster investigation; **~ animal** *f* AGRIC animal research; **~ aplicada** *f* CONTAM applied research; **~ atómica** *f* NUCL atomic research; **~ de la calidad** *f* CALIDAD audit; **~ con cámaras de reacción** *f* FÍS PART research with colliders; **~ de cobresoldabilidad** *f* ING MECÁ investigation of brazability; **~ de la contaminación** *f* CONTAM, CONTAM MAR pollution research; **~ detallada** *f* CONST *topografía* detailed survey; **~ de errores** *f* CALIDAD error retrieval; **~ espacial** *f* TEC ESP space research; **~ espacial tripulada** *f* TEC ESP manned space research; **~ geodésica** *f* CONST *topografía* geodesic survey; **~ geofísica** *f* CONST, GEOFÍS, GEOL, TEC PETR geophysical survey; **~ geológica** *f* CARBÓN, GEOL, TEC PETR geological survey; **~ hidrográfica** *f* OCEAN hydrographic survey; **~ hidrológica** *f* HIDROL hydrologic investigation; **~ de mercados** *f* CALIDAD, PROD market research; **~ microestructural por rayos X** *f* NUCL X-ray microstructure investigation; **~ nuclear** *f* FÍS RAD, NUCL nuclear research; **~ de operaciones** *f* (*OR*) INFORM&PD operational research (*OR*); **~ operativa** *f* INFORM&PD operations research; **~ pecuaria** *f* AGRIC animal research; **~ sobre sistemas de producción agrícola** *f* AGRIC farming systems research; **~ topográfica** *f* CONST topographical survey

investigar[1] *vt* INFORM&PD seek, SEG investigate

investigar[2] *vi* CONST make a survey

invierno *m* AGRIC, GEOL, METEO *trópicos* rainy season

invitación: **~ al envío** *f* INFORM&PD invitation to send; **~ a transmitir** *f* TELECOM invitation to transmit

involución *f* GEOM, MATEMÁT involution

involuta *f* GEOM involute; **~ de un círculo** *f* GEOM involute of a circle

involuto *adj* GEOM involute

inyección *f* GEN injection, PROD *maderas* impregnation, TEC ESP *de aire* blast; **~ de agua** *f* PETROL water flooding, PROD water feed, TEC PETR *recuperación secundaria de petróleo* water injection; **~ de aire** *f* CARBÓN air blast, TEC PETR air flooding; **~ de alto nivel** *f* ELECTRÓN *semiconductores* high-level injection; **~ de bajo nivel** *f* ELECTRÓN *semiconductores* low-level injection; **~ de combustible** *f* AUTO, TERMO, TERMOTEC, VEH fuel injection; **~ de combustible K Jetronic** *f* AUTO, TERMO, TERMOTEC, VEH K-Jetronic fuel injection; **~ directa** *f* AUTO, TRANSP direct injection; **~ electrónica** *f* AUTO, ELECTRÓN, VEH electronic injection; **~ de entrada** *f* ELECTRÓN first injection; **~ externa** *f* AUTO, VEH external injection; **~ forzada de cemento** *f* MINAS *pozos de petróleo* squeeze; **~ de gas** *f* GAS, TEC PETR gas injection; **~ del haz** *f* NUCL beam injection; **~ de haz atómico sin carga** *f* NUCL neutral-atom beam injection; **~ de haz atómico neutral** *f* NUCL neutral-atom beam injection; **~ integral** *f* AUTO, VEH integral injection; **~ de remolino** *f* GAS swirling injection; **~ del segundo armónico** *f* ELECTRÓN second harmonic injection; **~ de vapor** *f* NUCL, QUÍMICA sparging

inyecciones: **~ de enlechado** *f pl* CARBÓN grouting; **~ de lechada** *f pl* CONST grouting; **~ de mortero** *f pl* CONST grouting

inyectar *vt* CONST *lechada* grout, GAS, PROD inject, TEC ESP inject, *aire* blow; **~ aire a** *vt* PROD *fuegos, incendios*, TEC ESP *fuelles* blow

inyector *m* AUTO injector, CARBÓN *rueda Pelton, motor diesel* nozzle, ELECTRÓN injector, GAS injector, nozzle, ING MECÁ propeller, nozzle, INSTAL HIDRÁUL *calderas, motores*, MECÁ injector, PROD *rueda Pelton, motor diesel*, TEC ESP nozzle, TRANSP AÉR jet, VEH *de gasolina* injector; **~ de aire** *m* ING MECÁ air injector; **~ aspirador** *m* ING MECÁ inspirator, lifting injector; **~ aspirante** *m* ING MECÁ inspirator, lifting injector; **~ de autocebado** *m* ING MECÁ restarting injector; **~ automático** *m* ING MECÁ restarting injector, self-acting injector; **~ de combustible** *m* AUTO, ING MECÁ, TERMO, TERMOTEC, VEH fuel injector; **~ de combustible en los gases calientes de exhaustación** *m* ING MECÁ afterburner; **~ medio de contraste** *m* INSTR contrast medium injector; **~ de señal** *m* ELECTRÓN signal injector; **~ unido** *m* AUTO united injector; **~ de vapor** *m* INSTAL HIDRÁUL, TERMOTEC steam injector; **~ de vapor a presión** *m* INSTAL HIDRÁUL, TERMOTEC live-steam injector

iodo *m* (*I*) ELECTRÓN, FOTO, QUÍMICA, TEXTIL iodine (*I*); **~ radioactivo** *m* FÍS RAD, QUÍMICA radioiodine

ioduro: **~ de metilo** *m* QUÍMICA methyl iodide

iolita *f* MINERAL cordierite, dichrote, iolite

ión *m* ELEC *partícula cargada*, ELECTRÓN, FÍS, FÍS PART, GAS, QUÍMICA, TEC PETR excited atom, ion; **~ de alta energía** *m* ELECTRÓN, FÍS PART high-energy ion; **~ dipolar** *m* QUÍMICA dipolar ion; **~ extirpado** *m* NUCL ablated ion; **~ de gran potencia** *m* ELECTRÓN high-energy ion; **~ negativo** *m* ELEC, ELECTRÓN, FÍS, FÍS PART, FÍS RAD anion, negative ion; **~ positivo** *m* ELEC, ELECTRÓN, FÍS, FÍS PART, FÍS RAD, QUÍMICA cation, positive ion, positive-ionized atom

ion: ~ **negativo predominante** *m* CONTAM dominant anion; ~ **positivo predominante** *m* CONTAM dominant cation

iónico *adj* GEN ionic; **no** ~ *adj* GEN nonionic

ionio *m* QUÍMICA ionium

ionización *f* GEN ionization; ~ **acumulativa** *f* ING ELÉC avalanche; ~ **en cascada** *f* NUCL cumulative ionization; ~ **por choque** *f* FÍS, FÍS RAD collision ionization, impact ionization, ionization by collision; ~ **por colisión** *f* FÍS, FÍS RAD collision ionization, impact ionization, ionization by collision; ~ **instantánea** *f* FÍS RAD, TEC ESP burst; ~ **lineal** *f* FÍS, FÍS RAD linear ionization; ~ **primaria** *f* FÍS, FÍS RAD primary ionization; ~ **de radiación** *f* FÍS, FÍS RAD, GEOFÍS radiation ionization; ~ **secundaria** *f* FÍS, FÍS RAD secondary ionization; ~ **superficial por impacto láser** *f* NUCL laser impact surface ionization

ionizado *adj* GEN ionized

ionizadora *f* IMPR ionization unit

ionizar *vt* GEN *métodos radiométricos* ionize

ionográfico *adj* IMPR, INFORM&PD *impresora* ionographic

ionona *f* QUÍMICA *perfumería* ionone

ionosfera *f* FÍS, FÍS ONDAS, GEOFÍS E-layer, ionosphere; ~ **superior** *f* TELECOM upper ionosphere

ionotropía *f* QUÍMICA ionotropy

IPA *abr* (*ácido isopropílico*) QUÍMICA IPA (*isopropyl acid*)

ipecac *f* ALIMENT, QUÍMICA ipecac, ipecacuanha

ipecacuana *f* ALIMENT, QUÍMICA ipecac, ipecacuanha

ipecacuánico *adj* ALIMENT, QUÍMICA ipecacuanic

IPS *abr* (*índice de porosidad secundaria*) TEC PETR SPI (*secondary porosity index*)

ir[1]: ~ **más allá de** *vt* MECÁ overshoot

ir[2]: ~ **aguas arriba** *vi* HIDROL go upstream; ~ **a la deriva** *vi* TRANSP MAR drift; ~ **a la deriva arrastrando el ancla** *vi* TRANSP MAR club; ~ **escorado** *vi* TRANSP MAR list; ~ **marcha atrás** *vi* TRANSP MAR go astern

ir[3]: ~ **a la autoridad portuaria** *fra* TRANSP MAR report to the port authorities

IR *abr* (*infrarrojo*) GEN IR (*infrared*)

Ir *abr* (*iridio*) QUÍMICA Ir (*iridium*)

iraser *m* ELECTRÓN iraser

irídico *adj* QUÍMICA iridic

iridio *m* (*Ir*) QUÍMICA iridium (*Ir*)

iridiscencia *f* C&V, CONTAM MAR iridescence

iridito *m* QUÍMICA iridite

iridosmio *m* MINERAL iridosmine, iridium osmine

iris *m* CINEMAT, FÍS, TV iris; ~ **automático** *m* CINEMAT, FÍS, TV autoiris

irisación *f* C&V, CONTAM MAR iridescence

irona *f* QUÍMICA irone

irracional *adj* INFORM&PD, MATEMÁT irrational

irradiación *f* GEN irradiation, radiation; ~ **de alimentos** *f* ALIMENT, EMB, FÍS RAD irradiation of food; ~ **de los alimentos** *f* ALIMENT, FÍS RAD food irradiation; ~ **del ánodo** *f* FÍS PART target irradiation; ~ **del anticátodo** *f* FÍS PART target irradiation; ~ **del blanco** *f* NUCL target irradiation; ~ **espectral** *f* TELECOM spectral irradiance; ~ **excepcional concertada** *f* NUCL emergency exposure to external radiations; ~ **gamma por cobalto 60** *f* NUCL cobalt 60 gamma irradiation; ~ **iónica** *f* METAL ion bombardment; ~ **mínima** *f* FÍS RAD minimum irradiation; ~ **de protones** *f* TEC ESP proton irradiation; ~ **protónica** *f* TEC ESP proton irradiation; ~ **por rayos X** *f* NUCL X-ray irradiation

irradiado *adj* ALIMENT, FÍS, NUCL irradiated

irradiador: ~ **cilíndrico** *m* NUCL cylindrical irradiator; ~ **industrial** *m* NUCL industrial irradiator

irradiancia *f* FÍS, ÓPT irradiance; ~ **energética** *f* ÓPT energy irradiance; ~ **de energía espectral** *f* ÓPT spectral energy irradiance; ~ **espectral** *f* ÓPT spectral irradiance

irradiante *adj* FÍS RAD, ÓPT, TELECOM, TV radiating

irradiar *vt* GEN radiate

irrecobrable *adj* INFORM&PD *error* irrecoverable

irrecuperable *adj* INFORM&PD *error* irrecoverable

irreductible *adj* INFORM&PD *polinomio* irreductible

irregularidad *f* GEOL anomaly, ING MECÁ, MECÁ roughness, METAL disorder; ~ **real** *f* TEC ESP true anomaly; ~ **superficial** *f* ING MECÁ surface roughness; ~ **del surco** *f* ACÚST *discos gramofónicos* pinch effect

irregularidades: ~ **de la superficie** *f pl* ING MECÁ surface discontinuities

irreparable *adj* GEN beyond repair

irreproducible *adj* EMB nonreproductible

irreversible *adj* FÍS, QUÍMICA irreversible

irrigación *f* AGRIC, AGUA, HIDROL irrigation; ~ **por acequias** *f* AGRIC, AGUA, HIDROL ditch irrigation; ~ **con aguas cloacales** *f* AGRIC, AGUA, RECICL sewage farming; ~ **extensa** *f* AGRIC, AGUA, HIDROL broad irrigation; ~ **por inundación** *f* AGRIC, AGUA, HIDROL flood irrigation; ~ **por inundación superficial** *f* AGRIC, AGUA, HIDROL irrigation by surface flooding; ~ **por inundación de la superficie** *f* AGRIC, AGUA, HIDROL irrigation by surface flooding; ~ **subsuperficial** *f* AGRIC, AGUA, HIDROL subsurface irrigation

irrigar *vt* AGRIC, AGUA, HIDROL irrigate

irritante *adj* SEG irritant

irrotacional *adj* FÍS, FÍS FLUID, MATEMÁT irrotational

irrupción *f* AGUA irruption, FÍS FLUID eruption, inrush, HIDROL *de agua* inrush, MINAS *de agua* make, TEC ESP burst; ~ **de agua freática** *f* AGRIC, AGUA, CARBÓN, HIDROL ground-water inrush; ~ **glacial** *f* OCEAN glacial outburst

irse *v refl* TERMO *líquidos* boil over; ~ **al fondo** *v refl* TRANSP MAR *buque* go down; ~ **a pique** *v refl* TRANSP MAR *buque* founder, sink; ~ **a la ronza** *v refl* TRANSP MAR *buque* crab

isalobara *f* METEO isallobar

isático *adj* QUÍMICA isatic

isatina *f* QUÍMICA isatin

isatogénico *adj* QUÍMICA isatogenic

isatrópico *adj* QUÍMICA isatropic

isentrópico *adj* FÍS, TERMO isentropic

isetionato *m* QUÍMICA isethionate

isetiónico *adj* QUÍMICA isethionic

isla *f* HIDROL, OCEAN island; ~ **de hielo** *f* OCEAN ice island; ~ **pantanosa** *f* OCEAN marsh island; ~ **salobre** *f* OCEAN marsh island

isleo *m* TRANSP MAR *geografía* islet

isleta *f* CONST traffic island, HIDROL islet, TRANSP safety island, TRANSP MAR *geografía* islet

islote *m* HIDROL, TRANSP MAR *geografía* islet; ~ **de turbulencia** *m* FÍS FLUID turbulent plug

isoalilo *m* QUÍMICA isoallyl

isoamílico *adj* QUÍMICA isoamylic

isoamilo *m* QUÍMICA isoamyl

isoanomalía f GEOL isoanomaly
isoapiol m QUÍMICA isoapiol
isóbara, isobara f FÍS, METEO, PETROL, TRANSP MAR isobar
isóbata f GEOL, PETROL, TRANSP MAR isobath
isoborneol m QUÍMICA isoborneol
isobutano m PETROL, QUÍMICA, TEC PETR isobutane, methylpropane
isobuteno m QUÍMICA isobutene
isobutileno m QUÍMICA isobutylene
isobutílico adj QUÍMICA isobutylic
isobutilo m QUÍMICA isobutyl
isobutírico adj QUÍMICA isobutyric
isocianato m P&C, QUÍMICA isocyanate
isociánico adj QUÍMICA isocyanic
isocianuro m QUÍMICA isocyanide
isocíclico adj QUÍMICA isocyclic
isocincomerónico adj QUÍMICA isocinchomeronic
isóclino[1] adj GEOFÍS, GEOL, TEC PETR isoclinal
isóclino[2] m GEOFÍS, GEOL, TEC PETR isocline
isocora f FÍS, GEOL, QUÍMICA isochor, isochore
isocraqueo m TEC PETR isocracking
isócrona f GEOL isochrone; **~ mineral** f GEOL mineral isochrone; **~ de roca total** f GEOL whole rock isochrone
isocronismo m PETROL isochronism
isócrono adj INFORM&PD isochronous
isocrotónico adj QUÍMICA isocrotonic
isodulcital m QUÍMICA isodulcital
isoeléctrico m QUÍMICA punto, TRANSP isoelectric
isoentrópico adj FÍS, TERMO isentropic
isoespín m FÍS isospin
isoestrés m TRANSP AÉR isostress
isofacies f GEOL isofacies
isofencol m QUÍMICA isofenchol
isoflavona f QUÍMICA isoflavone
isoformato m QUÍMICA isoformate
isoforming m QUÍMICA isoforming
isogama f GEOL isogam
isogonal adj GEOFÍS, GEOM, MATEMÁT isogonal
isogónico adj GEOFÍS, GEOM, MATEMÁT isogonal
isograma m METEO, TEC PETR isogram
isohalina f GEOL isohaline
isohelia f METEO isohel
isohipsa f GEOL subsurface contour
isoipsa f GEOL, METEO isohypse
isoleucina f QUÍMICA aminoácido neutro isoleucine
isologo[1] adj QUÍMICA isologous
isologo[2] m QUÍMICA isolog (AmE), isologue (BrE)
isómera f QUÍMICA, TEC PETR isomer
isomería f QUÍMICA isomerism
isomérido m QUÍMICA isomeride
isomerismo m FÍS isomerism; **~ estructural** m QUÍMICA de compuesto structural isomerism; **~ nuclear** m FÍS, NUCL nuclear isomerism
isomerización f QUÍMICA, TEC PETR isomerization
isómero m FÍS, FÍS RAD, TEC PETR petroquímica isomer; **~ geométrico** m QUÍMICA geometric isomer; **~ óptico** m CRISTAL, QUÍMICA optical isomer
isométrico adj CRISTAL, MATEMÁT, QUÍMICA isometric
isomórfico adj CRISTAL, GEOL, MATEMÁT isomorphous
isomorfismo m CRISTAL, MATEMÁT, QUÍMICA isomorphism
isomorfo adj CRISTAL, GEOL, MATEMÁT isomorphous
isonicotínico adj QUÍMICA ácido isonicotinic
isonitrilo m QUÍMICA isonitrile

isooctano m QUÍMICA combustible de motores isooctane
isopaca f GEOL isopach
isoparafina f QUÍMICA isoparaffin
isopeletierina f QUÍMICA alcaloide isopelletierine
isopentano m QUÍMICA solventes combustibles isopentane
isopleta f GEOL, MATEMÁT, QUÍMICA nomografía isopleth
isopoliácido m QUÍMICA isopoly acid
isopreno m QUÍMICA isoprene
isoprenoide m QUÍMICA isoprenoid
isopropanol m DETERG, QUÍMICA isopropanol
isopropenilo m QUÍMICA isopropenyl
isopropilbenceno m QUÍMICA cumene, isopropylbenzene
isopropilcarbinol m QUÍMICA isopropylcarbinol
isopropílico adj ALIMENT, DETERG, QUÍMICA isopropyl
isopropilo m QUÍMICA isopropyl
isoquinolina f QUÍMICA isoquinoline
isosísmica adj GEOFÍS isoseismal
isostasía f GEOL isostasy
isostérico adj QUÍMICA isosteric
isosterismo m QUÍMICA propiedad de los isótopos isosterism
isotáctico adj QUÍMICA polímero isotactic
isoterma f GEN isotherm; **~ de adsorción** f NUCL adsorption isotherm; **~ crítica** f QUÍMICA critical isotherm; **~ de intercambio iónico** f FÍS RAD ion exchange isotherm
isotérmico adj GEN isothermal
isotermo[1] adj MECÁ isothermal
isotermo[2] m TRANSP AÉR isotherm
isótono m FÍS, NUCL isotone
isotopia f QUÍMICA isotopy
isótopo m FÍS, FÍS PART, GEOL, NUCL, QUÍMICA de un elemento isotope; **~ descendiente** m GEOL daughter isotope; **~ estable** m FÍS stable isotope; **~ fértil** m FÍS fertile isotope; **~ fisionable** m FÍS, FÍS PART, NUCL fissile isotope; **~ industrial** m FÍS, FÍS PART, NUCL industrial isotope; **~ radiactivo** m FÍS, FÍS PART, FÍS RAD radioactive isotope; **~ radiogénico** m FÍS, FÍS RAD, GEOL radiogenic isotope
isotropía f CRISTAL, GEOL, MATEMÁT isotropy
isotrópico adj ELECTRÓN, TEC ESP, TELECOM isotropic
isótropo[1] adj CRISTAL, GEOL, ÓPT isotropic
isótropo[2] m FÍS FLUID isotropic
isovainillina f QUÍMICA isovanilline
isovalerona f QUÍMICA isovalerone
isoxazolo m QUÍMICA isoxazole
istmo m TRANSP MAR geografía isthmus
itacónico adj QUÍMICA ácido itaconic
ítem m INFORM&PD item; **~ de datos** m INFORM&PD data item
iteración f CONTAM, MATEMÁT iteration
iterar vt INFORM&PD, MATEMÁT iterate
iterativo adj ACÚST, FÍS, INFORM&PD, ING ELÉC, TEC ESP, TELECOM iterative
iterbia f QUÍMICA ytterbia
iterbio m (Yb) METAL, QUÍMICA ytterbium (Yb)
itinerario m FERRO route, OCEAN survey traverse, TELECOM route, TRANSP MAR itinerary, comercial, mercantil route; **~ disponible para el tren** m FERRO available train path; **~ con enclavamiento de aproximación** m FERRO approach-locked route;

~ **de observación de mareas** *m* OCEAN survey traverse

itria *f* QUÍMICA yttria

itrio *m* (*Y*) METAL, QUÍMICA yttrium (*Y*)

itrita *f* MINERAL yttrite

itrocerita *f* MINERAL, QUÍMICA yttrocerite

itrotantalita *f* MINERAL yttrotantalite

itrotitanita *f* MINERAL yttrotitanite

ixora *f* AGRIC ixora

izada *f* ING MECÁ lifting, MINAS hoisting, lifting, *motor de extracción* hoist, PROD hoisting, TRANSP lifting

izado: ~ **por helicóptero** *adj* TRANSP AÉR, TRANSP MAR helicopter-lifted

izador *m* TRANSP AÉR elevator, hoist

izar[1] *vt* ING MECÁ hold up, MINAS raise, *pesos* hoist, PROD, TRANSP AÉR hoist, TRANSP MAR *buque* haul up, *vela, bandera, barco* hoist; ~ **arriba y claro** *vt* TRANSP MAR *ancla* clear; ~ **a bordo** *vt* TRANSP MAR *bote, con el chigre* hoist; ~ **a bordo del helicóptero** *vt* TRANSP MAR *rescate de aire-mar* winch into helicopter

izar[2]: ~ **la bandera nacional** *vi* TRANSP MAR hoist the colors (*AmE*), hoist the colours (*BrE*)

J

J *abr* (*julio*) ALIMENT, ELEC, FÍS, MECÁ, METR, TERMO J (*joule*)

jabalcón *m* CONST bow, brace, ING MECÁ collar beam, NUCL *Esp* (*cf diagonal AmL*) diagonal

jábega *f* OCEAN beach seine

jabón: ~ **desengrasador** *m* DETERG scouring soap; ~ **de fregado** *m* DETERG scrubbing soap; ~ **mineral** *m* MINERAL mountain soap; ~ **de montaña** *m* MINERAL mountain soap; ~ **de piedra pómez** *m* DETERG pumice soap; ~ **de tocador** *m* DETERG toilet soap

jaboncillo: ~ **de sastre** *m* MINERAL, TEXTIL French chalk, steatite

jácena *f* CONST girder, summer, summer beam; ~ **en doble T** *f* CONST I-girder

jacinto *m* MINERAL hyacinth, jacinth

jack: ~ **conmutador en contrafase** *m* ING MECÁ lift-pull-and-push jack and cramp; ~ **de enlace** *m* ING ELÉC spring jack

jadeíta *f* MINERAL jadeite

jadeo *m* OCEAN *medicina subacuática* panting, TEC PETR *movimientos de equipos flotantes* heave

jalápico *adj* QUÍMICA jalapic

jalapina *f* QUÍMICA jalapin

jalón *m* CONST *topografía* cross staff, poststaff, range pole, ranging pole, drop arrow, pole, rod

jalonamiento *m* CONST, PROD pegging, staking out

jalpaíta *f* MINERAL jalpaite

jamba *f* CONST jamb post, *ventana, puerta* post, jamb, ING MECÁ *de puerta* cheek; ~ **de piedra** *f* CONST jamb stone; ~ **de la puerta** *f* CONST doorpost

jambaje *m* CONST door framing

jamesonita *f* MINERAL jamesonite; ~ **sin hierro** *f* MINERAL feather ore

jangada *f* TRANSP MAR float, raft

japánico *adj* QUÍMICA japanic

jarabe: ~ **de maíz** *m* AGRIC corn syrup (*AmE*), maize syrup (*BrE*)

jarcia *f* OCEAN *veleros* rig, rigging, PROD tackle; ~ **firme** *f* TRANSP MAR standing rigging; ~ **firme y de labor** *f* OCEAN *veleros* rig; ~ **de labor** *f* TRANSP MAR running rigging; ~ **menuda** *f* TRANSP MAR *cuerdas* small stuff; ~ **móvil** *f* TRANSP MAR running rigging; ~ **muerta** *f* TRANSP MAR standing rigging

jardinería: ~ **ornamental** *f* AGRIC landscape gardening

jareta *f* OCEAN purse line

jargón *m* MINERAL jargon

jarosita *f* MINERAL jarosite

jarra: ~ **de barro** *f* C&V earthenware jar; ~ **de vidrio** *f* C&V, EMB glass jar

jarro: ~ **para conservas** *m* ALIMENT, C&V, EMB canning jar (*AmE*), preserving jar (*BrE*)

jaspe *m* GEOL, MINERAL jasper

jaspeado *m* IMPR, PAPEL marbling

jasperita *f* GEOL jasperite

jaspilita *f* GEOL jaspilite

jaula *f* AGRIC, EMB crate, ING MECÁ ball cage, race, MECÁ race, MINAS *Esp* (*cf cuba AmL*) cage, kibble, TRANSP *Esp* (*cf caja control AmL*) box; ~ **de ardilla** *f*

ING ELÉC squirrel cage; ~ **de articulación esférica** *f* ING MECÁ ball-joint cage; ~ **de bolas del rodamiento** *f* ING MECÁ ball-bearing cage; ~ **elevadora colgada** *f* MINAS suspended lift; ~ **elevadora suspendida** *f* MINAS suspended lift; ~ **de extracción** *f* MINAS bucket, mining bucket; ~ **de Faraday** *f* ELEC, FÍS, ING ELÉC Faraday cage; ~ **de ganado** *f* AGRIC cattle car (*AmE*), cattle wagon (*BrE*); ~ **de granade** *f* AGRIC cattle car (*AmE*), cattle wagon (*BrE*); ~ **para máquina** *f* SEG machine cage; ~ **neumática** *f* CARBÓN, MINAS pneumatic cell; ~ **protectora** *f* OCEAN *buceo* protection cage

javelización *f* QUÍMICA javellization

jefe *m* D&A leader; ~ **de la cámara hiperbárica** *m* OCEAN *buceo* caisson master; ~ **electricista** *m* CINE-MAT gaffer grip; ~ **eléctrico** *m* CINEMAT gaffer grip; ~ **de equipo** *m* TEC PETR *perforación* tool pusher; ~ **en lugar del siniestro** *m* CONTAM MAR on-scene commander; ~ **maquinista** *m* CINEMAT boss grip, FERRO head driver, TRANSP MAR *armada* marine engineer; ~ **de perforación** *m* PETROL tool pusher; ~ **de salto** *m* D&A dropmaster; ~ **de taladro** *m* TEC PETR *perforación* tool pusher; ~ **tramoyista** *m* CINE-MAT boss grip; ~ **de tren** *m* FERRO chief guard (*BrE*), conductor (*AmE*), guard (*BrE*)

jenkinsita *f* MINERAL jenkinsite

jerarquía *f* INFORM&PD hierarchy; ~ **de almacenamiento** *f* INFORM&PD storage hierarchy; ~ **de datos** *f* INFORM&PD data hierarchy; ~ **digital** *f* TELECOM digital hierarchy; ~ **digital plesiocrónica** *f* TELECOM plesiochronous digital hierarchy (*PDH*); ~ **digital síncrona** *f* TELECOM synchronous digital hierarchy (*SDH*); ~ **digital sincrónica** *f* TELECOM synchronous digital hierarchy (*SDH*); ~ **de la memoria** *f* INFORM&PD storage hierarchy, memory hierarchy

jerarquización *f* INFORM&PD nesting

jerarquizado *adj* INFORM&PD nested

jeremejevita *f* MINERAL jeremejevite

jeringa *f* LAB *equipo general* syringe; ~ **desechable** *f* LAB disposable syringe; ~ **de engrase** *f* VEH grease gun

jeroboam *f* C&V jeroboam

jervina *f* QUÍMICA jervine

JET *abr* (*Consejo Europeo de Investigación Nuclear*) FÍS RAD, NUCL JET (*Joint European Torus*)

jetfoil *m* TRANSP MAR jetfoil

jinetillo *m* AmL (*cf reiter Esp*) LAB *balanza* rider

Johannita *f* MINERAL, NUCL Johannite

jornada: ~ **laboral** *f* CONST workday

jornadas *f pl* CONST dayworks

jornales *m pl* PROD labor rate (*AmE*), labour rate (*BrE*)

joroba *f* MINAS *de un filón* hitch

joyero *m* MECÁ jeweler (*AmE*), jeweller (*BrE*)

joystick *m* Esp (*cf palanca de juego AmL*) INFORM&PD joystick

juego[1]: **de** ~ **completo** *adj* ING MECÁ, PROD complete-assembly

juego[2] *m* AGUA, CONST, INFORM&PD set, ING MECÁ lash,

set, cluster, clearance, play, slack, slackening, MECÁ backlash, clearance, PROD kit, TEC ESP backlash; **~ de agujas** *m* CONST, FERRO points (*BrE*), set of points (*BrE*), switch (*AmE*); **~ de alicates** *m* CONST plier-saw set; **~ del árbol de levas** *m* AUTO, VEH camshaft clearance; **~ axial** *m* ING MECÁ, MECÁ axial backlash, axial clearance, end play; **~ de caracteres** *m* INFORM&PD character set; **~ de cilindros** *m* PROD *laminadores* set of rolls; **~ de circuitos integrados para micrologicales** *m* PROD firmware chip set; **~ de conexiones** *m* PROD lug kit; **~ de contactos** *m* ELEC set of contacts; **~ de contactos del ruptor** *m* AUTO, VEH contact set; **~ de cuchillas** *m* PROD set of cutters; **~ de la dirección** *m* AUTO, VEH steering play; **~ del embrague** *m* AUTO, VEH clutch clearance; **~ de engranajes** *m* AUTO gear train, gears, set of gears, ING MECÁ set of change wheels, VEH gear train, gears, set of gears; **~ de engranajes de recambio** *m* ING MECÁ set of change wheels; **~ de estampas** *m* ING MECÁ die set; **~ de herramientas** *m* INFORM&PD toolkit, ING MECÁ kit, set of tools; **~ de instrucciones** *m* INFORM&PD instruction repertoire, PROD instruction set; **~ lateral** *m* ING MECÁ lateral play, side clearance, side lash; **~ de lengüetas** *m* PROD lug kit; **~ de lengüetas de conexión para terminales** *m* PROD terminal lug kit; **~ de llaves de tubo** *m* ING MECÁ set of box spanners; **~ longitudinal** *m* ING MECÁ, MECÁ end play; **~ de machos de roscar** *m* ING MECÁ, MECÁ combination tap assembly; **~ de maderas** *m* CONST set of timber; **~ de matrices circulares** *m* ING MECÁ circular die set; **~ de matrices rectangulares** *m* ING MECÁ rectangle die-set; **~ de matriz y troquel** *m* ING MECÁ die set; **~ de moldes** *m* ING MECÁ die set; **~ de objetivos** *m* CINEMAT lens set; **~ de palancas** *m* ING MECÁ leverage; **~ parcial de caracteres** *m* INFORM&PD subset, character subset; **~ del pedal del embrague** *m* AUTO, VEH clutch-pedal clearance; **~ de pesas** *m* LAB set of weights; **~ de piezas de recambio** *m* PROD set of spare parts; **~ de piezas de repuesto** *m* PROD set of spare parts; **~ del pistón** *m* AUTO, VEH piston clearance; **~ de platinos** *m* AUTO, VEH contact set; **~ de punzones** *m* TELECOM relay set; **~ de remaches** *m* CONST riveting set; **~ de rodillos** *m* PAPEL set of rolls; **~ de ruedas** *m* AUTO set of wheels, ING MECÁ set of change wheels, VEH set of wheels; **~ con separadores** *m* CONST set with stretcher piece; **~ de tamices** *m* PROC QUÍ sieve set; **~ de taqué** *m* AUTO, VEH valve clearance; **~ de tipos de caracteres** *m* IMPR, INFORM&PD font; **~ de tipos de letras** *m* IMPR, INFORM&PD font; **~ de troqueles** *m* ING MECÁ die set; **~ universal** *m* INFORM&PD universal set; **~ de la válvula** *m* AUTO, VEH valve clearance; **~ de velocidades** *m* ING MECÁ set of speeds; **~ Ward-Leonard** *m* ING ELÉC Ward-Leonard set

juglona *f* QUÍMICA juglone

jugo: **~ de caña** *m* AGRIC, ALIMENT cane juice

julio *m* (*J*) ALIMENT, ELEC, FÍS, MECÁ, METR, TERMO joule (*J*)

junquillo *m* CONST reed, *carpintería* rush

junta *f* AUTO gasket, CONST joint, union, ELECTRÓN clamping, FERRO *infraestructura* splice joint (*AmE*), ING ELÉC joint, ING MECÁ seam, clamping, couple, attach, joint, washer, coupling, MECÁ coupling, clamping, gasket, NUCL, PETROL joint, PROD jointing, *de molde* parting, clamping, *caja de moldeo, fundición* parting line, TELECOM splice, VEH *término genérico* gasket;

~ a **~ de abrazadera doble** *f* ING MECÁ double strap butt joint; **~ acanalada y embadurnada** *f* CONST grooved and feathered joint; **~ del acelerador** *f* TRANSP accelerator linkage; **~ acodada** *f* CONST elbow union, ING MECÁ cranked link; **~ aislante** *f* AUTO, ING ELÉC, VEH insulating joint; **~ con angular de hierro** *f* CONST angle iron joint; **~ angular machihembrada** *f* CONST angular grooved-and-tongued joint; **~ en ángulo** *f* CONST angle joint; **~ de anillo** *f* AUTO, VEH ring joint; **~ anular** *f* AUTO, VEH ring joint; **~ articulada** *f* CINEMAT joint, pair, ING MECÁ articulated joint, elbow joint, knuckle joint; **~ articulada de línea aérea** *f* FERRO *catenaria* overhead line knuckle; **~ de asbestos al cobre** *f* ING MECÁ copper asbestos gasket; **~ atornillada** *f* NUCL screwed joint;

~ b **~ con bisagra** *f* CONST hinge joint; **~ en bisel** *f* ING MECÁ bevel joint, chamfered joint, feathered joint; **~ biselada** *f* CONST splayed joint, *carpintería* scarf joint; **~ de la bomba de aceite** *f* REFRIG oil-pump gasket; **~ de brida** *f* ING MECÁ flanged fitting, stirrup joint; **~ de bridas** *f* ING MECÁ flange joint, flanged union; **~ de la bujía** *f* AUTO, ELEC, VEH spark plug gasket;

~ c **~ de cables** *f* ELEC, ING ELÉC *conexión*, TELECOM cable joint, cable splicing; **~ calafateada** *f* TRANSP MAR caulked joint; **~ cardan** *f* AUTO, ING MECÁ, MECÁ, VEH cardan joint, Hooke's joint, universal joint; **~ cardánica** *f* AUTO, ING MECÁ, MECÁ, VEH cardan joint, universal joint; **~ cardónica** *f* ING MECÁ, VEH gimbal joint; **~ de carpintero** *f* CONST carpenters' joint; **~ de carril** *f* FERRO *equipo inamovible* rail joint; **~ del cárter de aceite** *f* AUTO, VEH oil-pan gasket; **~ con chaflán** *f* ING MECÁ bevel joint; **~ de charnela** *f* CONST, ING MECÁ knuckle, knuckle joint, MECÁ knuckle joint; **~ cilíndrica** *f* REFRIG cylinder gasket; **~ de cobre y amianto** *f* AUTO, VEH copper asbestos gasket; **~ de cola de milano** *f* CONST dovetail joint, dovetailed joint; **~ de cola de milano a solape** *f* CONST dovetail lap joint; **~ de collarín** *f* ING MECÁ flange joint; **~ cónica, macho y hembra** *Esp* (*cf junta tipo Quickfit AmL, junta de cono y socket AmL*) LAB *material de vidrio* cone and socket joint; **~ cónica de vidrio esmerilado** *f* LAB *material de vidrio* conical ground glass point; **~ de cono y esfera** *f* PROD cup-and-ball joint; **~ de cono y socket** *f* AmL (*cf junta cónica, macho y hembra Esp*) LAB *material de vidrio* cone and socket joint; **~ de construcción** *f* CONST construction joint; **~ de contracción** *f* CONST contraction joint; **~ cuadrada** *f* CONST square joint; **~ de cuero** *f* PROD leather packer; **~ de culata** *f* AUTO, VEH head gasket, *motor* cylinder-head gasket;

~ d **~ desbloqueadora** *f* ING MECÁ gasket; **~ deslizante** *f* AUTO, VEH slip joint; **~ desmontable** *f* ING MECÁ detachable union; **~ de diente de sierra** *f* CONST scarf joint; **~ de dilatación** *f* CONST expansion joint, ELEC *conexión* dry joint, INSTAL HIDRÁUL expansion joint, expansion gap, slip joint, dry joint; **~ de dilatación térmica** *f* TERMO thermal expansion joint;

~ e **~ elástica** *f* ING MECÁ gasket; **~ elastomérica** *f* PROD elastomer seal; **~ embutida de mortaja y**

espiga *f* CONST *carpintería* mortise-and-tenon heel joint; ~ **empernada** *f* ING MECÁ bolted joint; ~ **encajada de espiga** *f* CONST tenoned and housed joint; ~ **encolada** *f* EMB glue joint, P&C joint line; ~ **engatillada** *f* PROD welted joint; ~ **ensanchadora** *f* ING MECÁ gasket; ~ **escalonada** *f* CONST step joint; ~ **esférica** *f* ING MECÁ, MECÁ ball-and-socket joint, ball joint, socket joint; ~ **de espiga** *f* CONST spur tenon joint, tenon joint, *carpintería* pegged tenon joint; ~ **de espiga dentada** *f* CONST tusk tenon joint; ~ **de espiga empernada** *f* CONST pinned tenon joint; ~ **de espiga y mortaja reforzada** *f* CONST *carpintería* haunched mortise-and-tenon joint; ~ **de la estampa formadora** *f* C&V blank mold seam (*AmE*), blank mould seam (*BrE*); ~ **estanca** *f* PROD close joint, TRANSP MAR *construcción naval* seal; ~ **estañosoldada** *f* PROD solder joint, soldering joint; ~ **de estanqueidad** *f* ING MECÁ, PROD gasket; ~ **de estanqueidad del asiento de la válvula** *f* REFRIG valve adaptor gasket; ~ **de estanqueidad elástica** *f* ING MECÁ flexible gasket; ~ **de estanqueidad flexible** *f* ING MECÁ flexible gasket; ~ **de estanqueidad de goma** *f* VEH *carrocería* rubber weather seal; ~ **de estanqueidad de la ventanilla** *f* VEH *carrocería* window seal; ~ **de estriba doble** *f* ING MECÁ double strap butt joint; ~ **de expansión** *f* C&V expansion space, CONST *carreteras* expansion joint; ~ **de expansión de fuelle** *f* ING MECÁ bellows expansion joint; ~ **de expansión ondulada** *f* NUCL corrugated expansion joint;

f ~ **de falsa espiga** *f* CONST feather joint; ~ **de fibra** *f* MECÁ fiber gasket (*AmE*), fiber joint (*AmE*), fiber washer (*AmE*), fibre gasket (*BrE*), fibre joint (*BrE*), fibre washer (*BrE*); ~ **de fibra óptica** *f* ÓPT, TELECOM optical fiber splice (*AmE*), optical fibre splice (*BrE*); ~ **flexible** *f* ING MECÁ flexible coupling, flexible joint; ~ **de fontanero** *f* *Esp* (*cf junta de plomero AmL*) CONST plumber's joint;

g ~ **de grano** *f* CRISTAL, METAL grain boundary;

h ~ **hermética** *f* ING MECÁ pressure-tight joint, seal, vacuum seal, LAB *equipo general*, TRANSP MAR *construcción naval* seal; ~ **homocinética** *f* VEH *eje de rueda* constant-velocity universal joint;

i ~ **a inglete** *f* CONST miter joint (*AmE*), mitre joint (*BrE*), splayed miter joint (*AmE*), splayed mitre joint (*BrE*); ~ **de inglete** *f* CONST chamfered joint; ~ **de inserción** *f* PROD inserted joint;

l ~ **de labios** *f* ING MECÁ lip seal, lip-type seal; ~ **con lengüeta postiza** *f* CONST slip-tongue joint; ~ **lisa** *f* ING MECÁ flush joint, straight joint, PROD flush joint;

m ~ **machihembra** *f* ING MECÁ, PROD flush joint; ~ **machihembrada** *f* CONST *carpintería* tongued-and-grooved joint; ~ **mecánica** *f* ÓPT, TELECOM, TV mechanical splice; ~ **a media fundición** *f* PROD halved joint, halving; ~ **a media madera** *f* CONST swallow-tail joint, *carpintería* halved joint; ~ **a medio hierro** *f* PROD halved joint, halving, push-through connection; ~ **metálica** *f* METAL *interfase* metal-to-metal joint, metallic bond; ~ **metálica de ventana** *f* CONST stanchion; ~ **de mortaja** *f* CONST *carpintería* mortise joint; ~ **de mortaja y espiga** *f* CONST *carpintería* mortise-and-tenon joint; ~ **de mortaja de espiga con cuña** *f* CONST *carpintería* wedged mortise-and-tenon joint; ~ **de mortaja estrecha** *f* CONST *carpintería* slit mortise joint; ~ **de**

mortaja única *f* CONST *carpintería* single-notch joint; ~ **muescada** *f* ING MECÁ key joint;

o ~ **obturadora** *f* ING MECÁ, MECÁ, PROD gasket; ~ **obturadora elástica** *f* ING MECÁ, MECÁ, PROD flexible gasket; ~ **obturadora flexible** *f* ING MECÁ, MECÁ, PROD flexible gasket; ~ **oxidativa** *f* GAS oxidative coupling;

p ~ **partida** *f* CONST break joint; ~ **de la pieza en tosco** *f* C&V blank seam; ~ **plana** *f* ING MECÁ *carpintería* butt joint, NUCL flat gasket, PROD butt joint; ~ **plana de cierre** *f* NUCL flat-packing gasket; ~ **plana doble** *f* ING MECÁ double strap butt joint; ~ **plana de empalme** *f* TEC PETR *tuberías* abutting joint; ~ **plana soldada por fusión** *f* ING MECÁ fusion-welded butt joint; ~ **de planeamiento** *f* PROD planning board; ~ **de plato** *f* ING MECÁ flanged fitting; ~ **de platos** *f* ING MECÁ flange joint; ~ **de plomero** *f* *AmL* (*cf junta de fontanero Esp*) CONST plumber's joint; ~ **de plomo** *f* CONST lead joint, lead jointing, lead packing; ~ **prominente** *f* C&V prominent joint;

q ~ **que no pierde** *f* ING MECÁ pressure-tight joint;

r ~ **con ranura** *f* CONST slot mortise joint; ~ **de ranura** *f* CONST notch joint; ~ **de ranura y lengüeta** *f* CONST *carpintería* joggle joint, tongue-and-groove joint; ~ **ranurada y con lengüeta** *f* CONST ploughed-and-feathered joint (*BrE*), plowed-and-feathered joint (*AmE*); ~ **ranurada y machihembrada** *f* CONST ploughed-and-tongued joint (*BrE*), plowed-and-tongued joint (*AmE*); ~ **de rebajo** *f* CONST, ING MECÁ rebated joint; ~ **por rebajo** *f* CONST rabbeted joint; ~ **rebordeada de mesa con llave** *f* CONST lipped table scarf with key; ~ **de recubrimiento** *f* CONST covering joint, ING MECÁ step joint; ~ **con recubrimiento de remache único** *f* CONST single-riveted lap joint; ~ **de remache** *f* CONST rivet joint; ~ **retacada** *f* TRANSP MAR caulked joint; ~ **de rosca y rótula** *f* ING MECÁ screw-and-socket joint; ~ **roscada** *f* ING MECÁ screw joint; ~ **roscada a izquierda y derecha** *f* ING MECÁ right-and-left screw link; ~ **rotativa** *f* ING ELÉC rotary joint; ~ **de rótula** *f* ING MECÁ globe joint, swivel joint, MECÁ socket joint, VEH *dirección* spherical joint, *mecanismo de mando de la dirección* ball joint; ~ **de rótula esférica** *f* ING MECÁ, MECÁ ball-and-socket joint; ~ **de rótula superior** *f* ING MECÁ, MECÁ upper ball joint;

s ~ **por secciones** *f* ELEC *cable* sectionalizing joint; ~ **de solapa remachada** *f* ING MECÁ riveted lap joint; ~ **de solape** *f* CONST lap, lap joint, overleap joint; ~ **soldada** *f* ING MECÁ wipe joint, wiped joint; ~ **de soldadura** *f* EMB welded body seam; ~ **de soldadura a tope con remache único** *f* CONST single-riveted butt joint; ~ **suspendida** *f* CONST suspended joint;

t ~ **de tablones** *f* CONST *carpintería* filleted joint; ~ **de tapón de espiga** *f* CONST plug-tenon joint; ~ **telescópica** *f* ING MECÁ telescopic joint; ~ **térmica** *f* TERMO heatsealing; ~ **tipo enchufe** *f* CONST faucet joint; ~ **tipo Quickfit** *f* *AmL* (*cf junta cónica, macho y hembra Esp*) LAB *material de vidrio* cone and socket joint; ~ **a tope** *f* CONST abutting joint, butt joint, *carpintería* heading joint, ING MECÁ straight joint, butt joint, PROD *soldaduras* butt joint, TRANSP MAR *a dos chapas* butt; ~ **al tope** *f* ING MECÁ, PROD flush joint; ~ **a tope con cubrejunta** *f* ING MECÁ *remachado*, PROD butt joint; ~ **a tope doble** *f* ING MECÁ double strap butt joint; ~ **a tope sin huelgo** *f* PROD closed butt joint; ~ **a tope soldada por fusión** *f*

ING MECÁ fusion-welded butt joint; ~ **tórica** f ING MECÁ, MECÁ, P&C, VEH *lubricación* O-ring; ~ **de transición** f ELEC *conexión de cable* transition joint; ~ **de transmisión** f AUTO universal joint, PROD transmission joint; ~ **de tuberías** f CONST, ING MECÁ, MINAS *sondeos* pipe joint;

~ u ~ **de unión** f CONST putty-joint; ~ **de unión para brocas** f MINAS drill joint; ~ **universal** f AUTO, ING MECÁ, VEH *transmisión* universal joint;

~ v ~ **vertical** f TEC PETR *tuberías* abutting joint; ~ **de vidrio esmerilado** f LAB ground-glass joint; ~ **de Viton** f PROD seal of Viton

juntar vt IMPR *tipografía* close up, TELECOM splice

juntera f PROD, TELECOM jointer

juntura f CONST *carpintería* jointer, jointing plane, INFORM&PD jump instruction, junction, PROD jointing; ~ **fría** f ING ELÉC cold junction; ~ **p-n** f FÍS p-n junction; ~ **rectificadora** f ING ELÉC rectifying junction

Jurásico adj GEOL *estratigrafía* Jurassic

justificación f IMPR *del texto* justification, *de columnas* tabbing, INFORM&PD, TELECOM justification; ~ **a la derecha** f IMPR, INFORM&PD right justification; ~ **a la izquierda** f IMPR, INFORM&PD left justification

justificado: ~ **a la derecha** adj IMPR, INFORM&PD flush right; ~ **a la izquierda** adj IMPR, INFORM&PD flush left

justificantes m pl CONTAM MAR *documentos* documentary proof

justificar[1]: **sin** ~ **por la derecha** adj IMPR rag-right (*AmE*), ragged-right (*BrE*); **sin** ~ **por la izquierda** adj IMPR rag-left (*AmE*), ragged-left (*BrE*)

justificar[2] vt IMPR, INFORM&PD justify; ~ **la composición** vt IMPR, INFORM&PD flush; ~ **a la derecha** vt IMPR, INFORM&PD right-justify; ~ **a la izquierda** vt IMPR, INFORM&PD left-justify

justo: ~ **a tiempo** fra EMB just-in-time (*JIT*)

juvenil adj GEOL juvenile

K

k *abr* (*kilo*) METR, FÍS k (*kilo*)
K *abr* FÍS, METR (*Kelvin*) K (*Kelvin*), QUÍMICA (*potasio*) K (*potassium*)
kaempferida *f* QUÍMICA kaempferide
kaempferol *m* QUÍMICA kaempferol
kali *m* QUÍMICA kali
kalinita *f* QUÍMICA potash alum
kaón *m* FÍS, FÍS PART kaon, K-meson, meson
kapnita *f* MINERAL kapnite
Karniense *adj* GEOL Carnian
Karst *m* GEOL *terrenos calizos* karst
Kazaniense *adj* GEOL *estratigrafía* Kazanian
Kcal *abr* (*kilocaloría*) ALIMENT Kcal (*kilocalorie*)
kCi *abr* (*kilocurie*) QUÍMICA kCi (*kilocurie*)
keilhauita *f* MINERAL keilhauite
kelly *m* TEC PETR *perforación* kelly
Kelvin *m* (*K*) FÍS, METR, TERMO Kelvin (*K*)
kemsoleno *m* QUÍMICA kemsolene
kerma *f* FÍS, NUCL kerma (*kinetic energy released in matter*)
kermanita *f* MINERAL kermanite
kernel *m* INFORM&PD kernel
kerogenita *f* GEOL kerogenite
kerógeno *m* GEOL, TEC PETR *compuestos del petróleo* kerogen
keroseno *m* PETROL, QUÍMICA, TEC PETR, TERMO, TRANSP, TRANSP AÉR kerosene; **~ de turbina de aviación** *m* TEC PETR *combustibles* aviation turbine kerosene (*ATK*)
kersantita *f* TEC PETR kersantite
keV *abr* (*kiloelectronvoltio*) QUÍMICA keV (*kilo-electronvolt*)
Kevlar *m* TRANSP MAR *material* Kevlar
key: **~ seating** *m* TEC PETR *perforación* key seating
kg *abr* (*kilogramo*) FÍS, METR kg (*kilogram*)
kHz *abr* (*kilohertz*) ELEC, TRANSP kHz (*kilohertz*)
kieselguhr *m* ALIMENT diatomaceous earth, infusorial earth
kieserita *f* MINERAL kieserite
kilo *m* (*k*) METR, FÍS kilo (*k*); **~-octeto** *m* INFORM&PD kilobyte (*kb*)
kilobyte *m* INFORM&PD kilobyte (*kb*)
kilocaloría *f* (*Kcal*) ALIMENT kilocalorie (*Kcal*)
kilocurie *m* (*kCi*) QUÍMICA kilocurie (*kCi*)
kiloelectronvoltio *m* (*keV*) QUÍMICA kilo-electronvolt (*keV*)
kilográmetro *m* METR kilogram meter (*AmE*), kilogram metre (*BrE*)
kilogramo *m* (*kg*) FÍS, METR kilogram (*kg*)

kilohertz *m* (*kHz*) ELEC, TRANSP kilohertz (*kHz*)
kilolitro (*kl*) *m* METR kiloliter (*AmE*) (*kl*), kilolitre (*BrE*) (*kl*)
kilometraje: **~ efectuado por el avión** *m* TRANSP AÉR aircraft-kilometer performed (*AmE*), aircraft-kilometre performed (*BrE*)
kilómetro *m* (*km*) METR kilometer (*AmE*) (*km*), kilometre (*BrE*) (*km*)
kilonem *m* (*kn*) QUÍMICA kilonem (*kn*)
kilovatio *m* (*kW*) ELEC, FÍS, ING ELÉC kilowatt (*kW*); **~-hora** *m* (*kWh*) ELEC, FÍS, ING ELÉC kilowatt-hour (*kWh*)
kilovoltio *m* (*kV*) ING ELÉC, MECÁ kilovolt (*kV*)
kilowatio *m* (*kW*) ELEC, FÍS, ING ELÉC kilowatt (*kW*)
kimberlita *f* GEOL, PETROL kimberlite
kimbombó *m* AGRIC okra
Kimeridgio *m* TEC PETR Kimmeridgian
Kimmeridgiense *adj* GEOL *estratigrafía*, TEC PETR Kimmeridgian
kit: **~ para cableado de fuerza** *m* PROD power-wiring kit kl *abr* (*kilolitro*) METR kl (*kiloleter*) AmE, kilolitre BrE
kl *abbr* (*kilolitro*) METR kl (*kiloliter AmE, kilolitre BrE*)
klistron *m* ELECTRÓN, FÍS, FÍS RAD, TEC ESP, TELECOM klystron; **~ de cavidad múltiple** *m* ELECTRÓN, FÍS, FÍS RAD multicavity klystron; **~ de reflexión** *m* ELECTRÓN, FÍS, FÍS RAD reflex klystron; **~ sintonizable** *m* TEC ESP tunable klystron; **~ de tres cavidades** *m* ELECTRÓN three-cavity klystron
km *abr* (*kilómetro*) METR km (*kilometer AmE, kilometre BrE*)
kmmererita *f* MINERAL kmmererite
kn *abr* (*kilonem*) QUÍMICA kn (*kilonem*)
knebelita *f* MINERAL knebelite
kotschubeyita *f* MINERAL kotschubeite
Kr *abr* (*criptón*) QUÍMICA Kr (*krypton*)
kremersita *f* MINERAL kremersite
krennerita *f* MINERAL krennerite
KSR *abr* (*transmisor-receptor a teclado*) INFORM&PD KSR (*keyboard send-receive*)
kV *abr* (*kilovoltio*) ELEC, ING ELÉC, MECÁ kV (*kilovolt*)
kW *abr* (*kilovatio, kilowatio*) ELEC, FÍS, ING ELÉC kW (*kilowatt*)
kWh *abr* (*kilovatio-hora*) ELEC, FÍS, ING ELÉC kWh (*kilowatt-hour*)
KWIC *abr* (*palabra clave en el contexto*) INFORM&PD KWIC (*keyword in context*)
KWOC *abr* (*palabra clave fuera de contexto*) INFORM&PD KWOC (*keyword out of context*)

L

l *abr* (*litro*) FÍS, METR l (*liter AmE, litre BrE*)
L *abr* (*luminosidad*) FÍS PART L (*luminosity*)
La *abr* (*lantano*) QUÍMICA La (*lanthanum*)
laberinto *m* ACÚST *audición*, MINAS labyrinth
lábil *adj* ELECTRÓN, GEOL, QUÍMICA labile
labio *m* C&V lip, GEOL limb; ~ del canal *m* C&V trough lip; ~ cortante *m* ING MECÁ, MECÁ cutting edge; ~ del delantal *m* PAPEL apron lip; ~ estrellado *m* C&V crizzled finish; ~ inferior *m* GEOL dropped side; ~ levantado *m* GEOL hanging wall; ~ de la ranura *m* C&V slot lip; ~ de la regleta *m* PAPEL slice lip; ~ soplante *m* PAPEL air knife
labor: ~ artística *f* IMPR artwork; ~ atravesada *f* MINAS offset; ~ bajo cubierta de rastrojo *f* AGRIC stubble mulch tilling; ~ a cielo *f* MINAS rising, top hole; ~ a cielo abierto *f* CARBÓN, MINAS rise workings; ~ complementaria *f* AGRIC secondary tillage; ~ de descubrimiento *f* MINAS discovery work; ~ escalonada *f* MINAS stope; ~ de extracción *f* MINAS pithead works; ~ de fondo *f* MINAS bottom; ~ de realce *f* MINAS overhand stoping
laboratorio *m* CINEMAT, FÍS, FOTO, LAB laboratory (*lab*), PROD charge chamber, *horno laboratory* (*lab*), reverberatory chamber, QUÍMICA laboratory (*lab*); ~ de análisis *m* QUÍMICA assay office; ~ analítico *m* CARBÓN analysis laboratory; ~ de biología marina *m* LAB, OCEAN marine biological laboratory; ~ de cierre *m* PROD locking lab; ~ de copias *m* CINEMAT printing lab; ~ externo *m* CINEMAT outside lab; ~ de física de partículas *m* (*CERN*) FÍS PART European laboratory for particle physics (*CERN*); ~ oceanográfico *m* OCEAN, TRANSP MAR oceanographic laboratory; ~ orbital tripulado *m* (*MOL*) TEC ESP manned orbiting laboratory (*MOL*); ~ orbitante *m* TEC ESP orbiting laboratory; ~ de pruebas *m* CALIDAD test laboratory; ~ de revelado *m* CINEMAT processing laboratory; ~ semicaliente *m* NUCL semihot laboratory, warm laboratory; ~ tibio *m* NUCL warm laboratory
laborear: ~ en un motón *vt* TRANSP MAR *cabo* reeve
laboreo: no ~ AGRIC no-tillage; ~ al aire libre *m* MINAS open-cast mining; ~ de arranque *m* MINAS stoped-out workings; ~ por arranque del filón entero *m* CARBÓN longwall system; ~ ascendente *m Esp* (*cf laboreo escalonado AmL*) MINAS raise stope; ~ por avalancha *m Esp* (*cf laboreo hidráulico AmL*) MINAS spatter work; ~ en capas *m* MINAS seam working; ~ sin dejar pilares *m* CARBÓN longwall system; ~ por derrumbe *m* MINAS caving; ~ escalonado *m AmL* (*cf laboreo ascendente Esp*) MINAS raise stope; ~ por escalones laterales *m* MINAS side stoping; ~ de filones *m* MINAS lode mining; ~ de fondo *m* MINAS bottom workings; ~ por grandes tajos *m* CARBÓN longwall system; ~ hidráulico *m AmL* (*cf laboreo por avalancha Esp*) MINAS spatter work; ~ por hundimiento *m* CARBÓN longwall system; ~ en manto *m* MINAS seam working; ~ por el método hidráulico *m* CARBÓN, MINAS piping; ~ de minas *m* MINAS mining; ~ de minerales *m* MINAS metal mining, ore mining; ~ mínimo *m* AGRIC minimum tillage; ~ de rocas *m* CARBÓN stone working; ~ subterráneo *m* MINAS underground workings
labores: ~ abandonadas *f pl* MINAS attle; ~ de exploración *f pl* MINAS exploration work; ~ mineras *f pl* MINAS diggings; ~ mineras abandonadas *f pl* MINAS abandoned workings; ~ mineras inferiores *f pl* MINAS lower workings; ~ preparatorias *f pl Esp* (*cf explotación preparatoria AmL*) MINAS dead work, dead workings
labrado[1] *adj* CONST *madera* carved, MECÁ, METAL wrought, TEXTIL *tela* embroidered, patterned; ~ áspero *adj* CONST rough-hewn
labrado[2] *m* CONST *piedra* dressing; ~ de fragmentos *m* MINAS *diamante* shatter cut
labrador: ~ de piedras *f* CONST stone dresser
labradorita *f* MINERAL labradorite
labranza: ~ conservacionista *f* AGRIC conservation tillage; ~ profunda *f* AGRIC deep ploughing (*BrE*), deep plowing (*AmE*)
labrar *vt* CONST *piedra* dress, *madera* carve, METAL work, PROD machine, TEXTIL *tela* embroider
laca *f* COLOR lacquer, lake, lac varnish, CONST lacquer, shellac, ELEC *aislación* shellac, MECÁ lacker, lacquer, PROD lacquer, REVEST lacquer, lake; ~ de aceite de linaza *f* REVEST linseed oil lacquer; ~ de acetilcelulosa *f* COLOR, REVEST acetyl cellulose lacquer; ~ de acetona *f* COLOR, REVEST acetone lacquer; ~ adhesiva *f* REVEST gum lac, gum lake; ~ al agua *f* COLOR, REVEST water lacquer; ~ aislante *f* REVEST insulating lacquer; ~ al alcohol *f* CONST spirit lacquer; ~ de alcohol *f* COLOR, REVEST spirit lacquer; ~ para aplicar sobre la impresión *f* IMPR overprint lacquer; ~ con capacidad de unirse *f* REVEST solderable lacquer; ~ de carmesí *f* COLOR, REVEST crimson lake; ~ de carmín *f* COLOR, REVEST carmine lacquer; ~ de celulosa *f* COLOR, REVEST cellulose lacquer; ~ china *f* COLOR, REVEST Chinese lacquer; ~ colorante *f* COLOR, REVEST stick lack; ~ coloreada *f* COLOR, REVEST colored lake (*AmE*), coloured lake (*BrE*); ~ conductora *f* REVEST conductive lacquer; ~ de enmascarar *f* COLOR, REVEST masking lacquer; ~ escarlata *f* COLOR, REVEST scarlet lake; ~ gomosa *f* REVEST gum lake; ~ de impresión *f* COLOR, IMPR printing lake; ~ indicadora de deformación *f* REVEST strain-indicating lacquer; ~ indicadora de temperatura *f* REVEST temperature-indicating lacquer; ~ japonesa *f* COLOR, REVEST japan, Japanese lacquer; ~ mate *f* COLOR, REVEST frosted lacquer, mat lacquer; ~ de nitrocelulosa *f* COLOR, CONST, REVEST nitrocellulose lacquer; ~ en paquete *f* REVEST package lacquer; ~ de piroxilina *f* COLOR, REVEST pyroxyline lacquer; ~ protectora *f* REVEST protecting lacquer, protective lacquer; ~ de resina gliptal *f* COLOR glyptal resin lacquer; ~ resinosa granulada *f* COLOR seed lac; ~ soldable *f* REVEST solderable lacquer; ~ termométrica *f* REVEST, TERMO temperature-indicating lacquer;

~ **transparente** *f* COLOR, REVEST transparent lacquer; ~ **uniforme** *f* REVEST isolac; ~ **vinílica** *f* CONST vinyl lacquer

lacado[1] *adj* COLOR, EMB, MECÁ, REVEST lacquered

lacado[2] *m* COLOR, EMB, REVEST lacquering; ~ **con barniz japonés** *m* COLOR, REVEST japan work; ~ **interior** *m* EMB, REVEST internal lacquering; ~ **japonés** *m* COLOR, REVEST japan work, japanning

lacaico *adj* QUÍMICA laccaic

lacar *vt* GEN lacquer, shellac

lacolito *m* GEOFÍS, GEOL, PETROL laccolith

lacrimógeno *m* QUÍMICA lachrymator

lactama *f* QUÍMICA lactam

lactamida *f* QUÍMICA lactamide

lactato *m* QUÍMICA lactate

lactenina *f* QUÍMICA lactenin

lácteo: no ~ *adj* ALIMENT nondairy

láctico *adj* QUÍMICA lactic

lactida *f* QUÍMICA lactide

lactobutirómetro *m* ALIMENT lactobutyrometer

lactómetro *m* ALIMENT lactometer

lactona *f* QUÍMICA lactone

lactónico *adj* QUÍMICA lactonic

lactonitrilo *m* QUÍMICA lactonitrile

lactonización *f* QUÍMICA lactonization

lactosa *f* QUÍMICA lactose

lacustre *adj* GEOL lacustrine

ladeado *adj* ING MECÁ out-of-true

ladear *vt* ING MECÁ deflect

ladera *f* CONST side, slope

lado[1]: **del** ~ **del mar** *adj* TRANSP MAR *navegación* seaward

lado[2] *m* CONST side, FÍS face, GEOM face, side, INSTR *cuerpos geométricos* edge, *de un hierro angular* arm, ÓPT side; ~ **del accionamiento** *m* PAPEL drive side, *máquina papel* backside; ~ **de admisión** *m* NUCL *de una turbina* inlet end; ~ **de barlovento** *m* METEO, TRANSP MAR windward side; ~ **caliente** *m* C&V working end; ~ **de carga** *m* C&V charging end; ~ **de la carne** *m* PROD *de una correa* flesh side; ~ **de la celda** *m* CINEMAT cell side; ~ **del cojinete** *m* ING MECÁ bearing end; ~ **de componentes** *m* TELECOM *placa de circuito impreso* component side; ~ **conductor** *m* PAPEL tending side; ~ **del conductor** *m* PAPEL front side, operating side; ~ **de descarga** *m* NUCL *de una bomba* delivery side; ~ **de la emulsión** *m* CINEMAT, TV emulsion side; ~ **esférico** *m* ING MECÁ ball end; ~ **exterior** *m* IMPR outer side; ~ **guía** *m* IMPR lay edge; ~ **idéntico del puente** *m* ELEC *circuito* equal-arm bridge; ~ **inferior** *m* IMPR lower side; ~ **del operador** *m* IMPR operator's side; ~ **del óxido** *m* CINEMAT, TV oxide side; ~ **secundario** *m* NUCL secondary side; ~ **de sotavento** *m* METEO, TRANSP MAR leeward side; ~ **tela** *m* IMPR, PAPEL wire side; ~ **de la transmisión** *m* MECÁ driving end, PAPEL drive side

lado[3]: **de** ~ **a lado** *fra* ING MECÁ edge-to-edge; **este** ~ **siempre hacia arriba** *fra* EMB this side up

lados: ~ **adyacentes** *m pl* GEOM adjacent sides; ~ **opuestos** *m pl* GEOM opposite sides

ladrillal *m* CONST brick field

ladrillo *m* C&V, CONST brick; ~ **abovedado** *m* CONST arch brick; ~ **aislante** *m* CONST baffle brick, TERMOTEC insulating brick; ~ **al hilo** *m* CONST stretcher; ~ **aplantillado** *m* CONST gage brick (*AmE*), gauge brick (*BrE*); ~ **de arcilla refractaria** *m* LAB *horno*

fireclay brick; ~ **de barro** *m* C&V clay brick; ~ **de canto de bisel** *m* CONST feather-edged brick; ~ **carbónico** *m* INSTAL TERM *material refractario* carbon brick; ~ **clinker** *m* C&V clinker brick; ~ **compacto** *m* CONST solid brick; ~ **crudo** *m* CONST cob, TERMO unburnt brick; ~ **de cuña** *m* CONST feather-edged brick; ~ **de diatomeas** *m* TERMOTEC diatomaceous brick; ~ **de dolomita** *m* CONST dolomite brick; ~ **macizo** *m* CONST solid brick; ~ **de madera** *m* CONST wood brick; ~ **de milpa** *m* C&V, CONST sun-dried brick; ~ **no cocido** *m* CONST, TERMO unburnt brick; ~ **perforado** *m* CONST perforated brick; ~ **quemado** *m* CONST burned brick; ~ **del quemador** *m* INSTAL TERM burner brick; ~ **refractario** *m* C&V firebrick, *para hornos de cemento* clinker brick, CONST firebrick, INSTAL TERM firebrick, refractory brick, LAB *aislamiento* refractory brick, TERMO refractory brick, firebrick; ~ **refractario para cámara de precalentamiento** *m* C&V checker brick; ~ **de refractario de magnesita** *m* CONST magnesite brick; ~ **secado al sol** *m* CONST sun-dried brick; ~ **de silicato de aluminio** *m* INSTAL TERM *revestimiento forro de hornos* aluminosilicate brick; ~ **ventilador** *m* CONST air brick; ~ **vidriado** *m* CONST glazed brick; ~ **de vidrio** *m* C&V, CONST glass brick; ~ **visto** *m* CONST facing brick

ladrillos: ~ **de respaldo** *m pl* PROD backing bricks

ladrón *m* ELEC, ING ELÉC adaptor, multiple plug, multiplug adaptor; ~ **de bombilla** *m* ELEC, ING ELÉC lamp holder

lago *m* AGUA, HIDROL lake; ~ **acidificado** *m* CONTAM acidified lake; ~ **ácido** *m* CONTAM acid lake; ~ **acidulado** *m* CONTAM acidified lake; ~ **de agua natural ácida** *m* CONTAM naturally-acid lake; ~ **artificial** *m* AGUA artificial lake, reservoir, HIDROL storage lake, storage lake dam; ~ **de enlace** *m* CONTAM transition lake; ~ **eutrófico** *m* RECICL eutrophic lake; ~ **de glaciar** *m* HIDROL glacier lake; ~ **no acídico** *m* CONTAM nonacidic lake; ~ **no ácido** *m* CONTAM nonacidic lake; ~ **de transición** *m* CONTAM transition lake; ~ **de unión** *m* CONTAM transition lake

lagrangiano *m* FÍS Lagrangian

laguna *f* AGUA, CONST, HIDROL lagoon, ING ELÉC hole, METAL void, *cristales* point defect; ~ **de cría** *f* AGRIC rearing pond; ~ **estratigráfica** *f* GEOL *grieta* hiatus; ~ **reticular** *f* METAL vacancy; ~ **solar** *f* ENERG RENOV solar pond

LAI *abr* (*límite anual de incorporación*) NUCL ALI (*annual limit on intake*)

lama *f* C&V slime

lámina *f* ELEC reed, GEOL lamina, slice, IMPR leaf, ING ELÉC reed, ING MECÁ plate, INSTAL HIDRÁUL *de agua entre dos placas de la caldera* film, MECÁ plate, sheet, NUCL sheet, P&C laminated sheet, film, sheet, PROD *hierro colado* plate, QUÍMICA lamina, TEC ESP *instrumentos de medición electrostática* blade, TELECOM reed; ~ **aislante** *f* EMB insulating sheet, ING ELÉC insulating plate; ~ **de aluminio** *f* EMB aluminium sheet (*BrE*), aluminum sheet (*AmE*); ~ **bimetálica** *f* ELEC *en termostatos, termómetros* bimetallic strip; ~ **caliente** *f* IMPR hot foil; ~ **de cobre** *f* CONST copper sheet; ~ **desgotadora** *f* PAPEL foil; ~ **de dispersión** *f* NUCL scattering foil; ~ **de espuma** *f* TEXTIL foam layer; ~ **foliar** *f* AGRIC leaf blade; ~ **fusible** *f* ELEC, ING ELÉC fuse link, fuse strip; ~ **de hierro** *f* PROD iron sheeting; ~ **intrusiva** *f* GEOL *elemento ígneo* intrusive sheet; ~ **metálica** *f* CRISTAL foil; ~ **metálica delgada** *f* EMB foil; ~ **ondulada** *f* PROD corrugated roll; ~ **de oro** *f* IMPR gold foil; ~ **de pestaña** *f* TRANSP AÉR edging panel; ~ **semireflectante** *f* FÍS semireflecting plate; ~ **soplada** *f* C&V blown sheet (*AmE*), cylinder glass (*BrE*); ~ **termoencogida** *f* REVEST heat-shrinking foil; ~ **vertiente** *f* HIDROL, INSTAL HIDRÁUL nappe

laminación *f* EMB, GEOL, ING ELÉC, METAL, P&C lamination, PROD *de metales* lamination, rolling, REVEST lamination; ~ **de acero de silicio** *f* ING ELÉC silicon steel lamination; ~ **convulta** *f* GEOL convolute lamination; ~ **cruzada bidireccional** *f* GEOL *estructura sedimentaria* herringbone cross lamination; ~ **en frío** *f* TERMO cold rolling; ~ **del inducido** *f* ING ELÉC rotor lamination; ~ **por sellado térmico** *f* REVEST heat seal laminating

laminado[1] *adj* ELEC *capacitor* laminated, FÍS FLUID laminar, GEOL foliated, laminated, P&C, PAPEL, PROD laminated, QUÍMICA laminar, TRANSP MAR laminated; ~ **en caliente** *adj* MECÁ, PROD, TERMO hot-rolled; ~ **en frío** *adj* MECÁ, PROD, TERMO cold-rolled; ~ **en tosco** *adj* PROD rough-rolled

laminado[2] *m* C&V laminating, ING ELÉC lamination, PAPEL pasting, REFRIG wire drawing; ~ **en caliente** *m* EMB heat lamination, TERMO hot roll, hot rolling; ~ **de cinco capas de película barrera** *m* EMB five-layer barrier film laminate; ~ **de circuito impreso** *m* ELECTRÓN printed-circuit laminate; ~ **de desbastes planos** *m* PROD slabbing; ~ **de fibra de vidrio** *m* C&V, EMB glass-fiber laminate (*AmE*), glass-fibre laminate (*BrE*); ~ **de películas** *m* EMB film lamination; ~ **plástico** *m* P&C plastic-based laminate

laminador *m* FERRO, ING MECÁ, PROD rolling mill; ~ **para alambres** *m* FERRO, PROD flatting mill; ~ **de canto** *m* P&C *equipo* edge mill; ~ **de carriles** *m* FERRO rail mill; ~ **de chapas** *m* ING MECÁ plate mill; ~ **para chapas finas** *m* PROD sheet mill; ~ **de desbaste** *m* ING MECÁ blooming mill; ~ **dúo** *m* PROD two-high mill, two-high roll, two-high train; ~ **de guías** *m* PROD *siderurgia* guide mill; ~ **pequeño** *m* PROD small bar mill; ~ **para perfiles** *m* ING MECÁ bar mill; ~ **para perfiles laminados** *m* PROD section mill; ~ **preliminar** *m* ING MECÁ blooming mill; ~ **para redondos** *m* CARBÓN, PROD rod mill; ~ **para redondos oscilantes** *m* CARBÓN, PROD vibrating rod mill; ~ **reversible** *m* PROD reversing mill; ~ **de tren trío** *m* ING MECÁ three-high mill, three-high rolls, three-high train; ~ **de tres cilindros superpuestos** *m* ING MECÁ three-high mill, three-high rolls, three-high train; ~ **de tres rodillos** *m* ING MECÁ three-high rolls; ~ **trío** *m* ING MECÁ three-high mill, three-high rolls; ~ **de vaivén** *m* PROD reciprocating mill, three-high mill, three-high train

laminadora *f* EMB calender unit, laminating machine, FERRO, PROD flatting mill

laminar[1] *adj* ELEC *capacitor* laminated, FÍS FLUID laminar, GEOL, MECÁ, MINERAL lamellar, QUÍMICA laminar

laminar[2] *vt* GEN laminate, PROD laminate, mill; ~ **en frío** *vt* TERMO cold-roll

laminarana *f* QUÍMICA laminarin

laminaria *f* ALIMENT kelp

laminarina *f* QUÍMICA laminarin

láminas: ~ **fijas** *f pl* ING ELÉC stator; ~ **de grisar** *f pl* IMPR mechanical tint

laminilla *f* METAL lamella; ~ **de centelleo** *f* CINEMAT, TV flicker blade; ~ **de macla** *f* METAL twin lamella; ~ **del obturador** *f* CINEMAT, FOTO shutter blade

laminoso *adj* GEOL, MECÁ, MINERAL lamellar

lampadita *f* MINERAL lampadite

lámpara *f* AUTO, CINEMAT lamp, ELECTRÓN valve, FOTO lamp, ING ELÉC bulb, lamp, VEH lamp; ~ **de acetileno** *f* CONST acetylene lamp; ~ **de advertencia** *f* ING ELÉC warning light; ~ **de alarma** *f* TELECOM alarm indication lamp; ~ **de alcohol** *f* LAB spirit lamp; ~ **de alta intensidad del proscenio** *f* ELEC apron floodlight; ~ **antiniebla** *f* AUTO, TRANSP, VEH fog lamp; ~ **de arco** *f* ELEC arc lamp; ~ **de arco eléctrico** *f* CINEMAT, ELEC brute; ~ **de arco con electrodos de carbón** *f* ELEC, ING ELÉC carbon arc lamp; ~ **de arco de mercurio** *f* ELEC, FÍS RAD, ING ELÉC mercury-arc lamp; ~ **de arco de xenón** *f* CINEMAT xenon arc lamp; ~ **de bolsillo** *f* ELEC, ING ELÉC flashlight, torch; ~ **de carbón** *f* CINEMAT carbon; ~ **de cebado en frío** *f* ELEC cold start lamp; ~ **para cobresoldadura** *f* PROD brazing lamp; ~ **colgante** *f* ELEC, ING ELÉC hanging lamp; ~ **condensadora** *f* CINEMAT, FOTO, TV condenser lamp; ~ **condensadora con recubrimiento especular** *f* CINEMAT mirror condenser lamp; ~ **con costados de vidrio para proteger la llama del viento** *f* CONST hurricane lamp; ~ **de decoración** *f* ELEC scenery lamp; ~ **de descarga** *f* ELEC discharge lamp; ~ **de descarga de alta intensidad** *f* CINEMAT high-intensity discharge lamp; ~ **de descarga luminiscente** *f* ELEC, ELECTRÓN, FÍS glow-discharge lamp; ~ **de descarga luminosa** *f* ELEC, GAS gas-discharge lamp, ING ELÉC discharge lamp; ~ **de descarga de vapor** *f* TERMO vapor discharge lamp (*AmE*), vapour discharge lamp (*BrE*); ~ **desnuda** *f* CINEMAT bare light; ~ **de destellos** *f* TEC ESP blinking light; ~ **de efluvios de neón** *f* ELEC neon glow lamp; ~ **de enfoque** *f* ELEC focusing lamp; ~ **estabilizadora de tensión** *f* ING ELÉC barretter; ~ **de estado** *f* TELECOM status lamp; ~ **excitadora** *f* CINEMAT exciter lamp; ~ **de fallas** *f* PROD fault light; ~ **de filamento de carbón** *f* ING ELÉC carbon filament lamp; ~ **de filamento doblemente arrollado** *f* ELEC *bombilla* coiled coil lamp; ~ **de flamear** *f* PROD fire lamp; ~ **de flash** *f* FOTO flash bulb; ~ **fluorescente** *f* ELEC, FÍS, FÍS RAD, GAS, ING ELÉC fluorescent lamp, fluorescent tube, gas-discharge lamp; ~ **fónica** *f* CINEMAT exciter lamp; ~ **giratoria de cabezal móvil** *f* CINEMAT capstan tach lamp; ~ **de gran amperaje del proscenio** *f* ELEC apron floodlight; ~ **halógena** *f* CINEMAT halogen, ELEC halogen lamp; ~ **halógena de cuarzo** *f* CINEMAT, ELEC quartz halogen lamp; ~ **de haz eléctrico filiforme** *f* ELEC, ING ELÉC pen light; ~ **HMI** *f* CINEMAT HMI lamp; ~ **de incandescencia** *f* ELEC, ING ELÉC glow lamp, incandescent lamp; ~ **incandescente** *f* ELEC, ING ELÉC incandescent lamp; ~ **indicadora** *f* CINEMAT pilot lamp, ELEC indicator lamp, GAS, ING ELÉC pilot lamp, repeater lamp, PROD indicator lamp; ~ **indicadora del detector de tierra** *f* ELEC, ING ELÉC, PROD earth detector light (*BrE*), ground detector light (*AmE*); ~ **indicadora del pulsador de pruebas** *f* PROD push-to-test indicating light; ~ **insonorizada** *f* CINEMAT sound film lamp; ~ **de inspección** *f* ING ELÉC inspection lamp; ~ **intermitente** *f* VEH *accesorio* flasher; ~ **de llamada** *f* TELECOM calling lamp; ~ **Lowell** *f* CINE-

MAT Lowell light; ~ **luminiscente de neón** f ELEC neon glow lamp, neon tube; ~ **luz día** f CINEMAT daylight lamp; ~ **marcadora del contorno** f VEH marker lamp; ~ **de neón** f ELEC, FÍS, ING ELÉC neon lamp, neon tube; ~ **de pared** f ELEC, ING ELÉC wall lamp; ~ **piloto** f ELEC, ING ELÉC, PROD, VEH pilot lamp (*AmE*), pilot light (*BrE*); ~ **piloto del pulsador de pruebas** f PROD push-to-test pilot light; ~ **de portafilamento de carbón** f ING ELÉC carbon-holder lamp; ~ **portátil** f CONST portable light, ELEC, ING ELÉC flashlight, portable lamp; ~ **posterior** f AUTO, VEH rear lamp; ~ **protectora aprobada** f CARBÓN approved safety lamp; ~ **protegida con metal** f CONST bonneted lamp; ~ **de proyección** f CINEMAT, FOTO projection lamp; ~ **del proyector** f CINEMAT, FOTO projector lamp; ~ **proyectora de haz concentrado** f ELEC spotlight; ~ **pulsatoria** f CINE-MAT pulsed lamp; ~ **quemadora de pintura** f CONST paint-burning lamp; ~ **para quemar pintura** f ING MECÁ blowlamp, blowtorch; ~ **de rayos ultravioleta** f LAB ultraviolet lamp; ~ **con recubrimiento especular** f FOTO mirror-coated lamp; ~ **con recubrimiento especular interno** f CINEMAT, FOTO, TV internal mirror lamp; ~ **con reflector** f CINEMAT, FOTO, TV reflector lamp; ~ **de refuerzo** f CINEMAT booster light; ~ **de repuesto** f FOTO spare bulb; ~ **de resistencia** f ING ELÉC barretter; ~ **de seguridad** f CARBÓN, MINAS, SEG Davy lamp, TRANSP MAR hurricane lamp; ~ **de señales** f FERRO *infraestructura* signal lamp; ~ **de señalización por destellos** f AUTO, VEH blinker; ~ **de señalización posterior** f AUTO, VEH rear lamp; ~ **de sodio** f ELEC sodium lamp; ~ **de soldar** f CONST blowtorch, ING MECÁ blowlamp, blowtorch; ~ **termiónica** f ELECTRÓN thermionic tube (*AmE*), thermionic valve (*BrE*), PROD, TEC ESP *radio* valve; ~ **testigo** f FERRO indicator lamp, telltale lamp, GAS, ING ELÉC pilot lamp, VEH pilot lamp (*AmE*), pilot light (*BrE*); ~ **de tubo fluorescente** f ING ELÉC discharge lamp; ~ **tubular** f ELEC tubular lamp; ~ **de tungsteno a luz de día** f CINEMAT tungsten-to-daylight; ~ **de vapor de mercurio** f ELEC, FÍS RAD, ING ELÉC mercury-arc lamp; ~ **de vapor de mercurio a alta presión** f ELEC, ING ELÉC high-pressure mercury lamp; ~ **de vapor de mercurio a poca presión** f ELEC, ING ELÉC low-pressure mercury lamp; ~ **de vapor de sodio** f ELEC, FÍS RAD, ING ELÉC sodium vapor lamp (*AmE*), sodium vapour lamp (*BrE*); ~ **de vapor de yodo con cubierta de cuarzo** f CINEMAT quartz iodine, quartz iodine lamp

lamparería f MINAS lamp room
lamparilla: ~ **de soldar** f TERMOTEC blowlamp
lampistería f CINEMAT, INSTR lamp house, MINAS lamp room
lamprófido m PETROL lamprophyre
lana f C&V, TEXTIL wool; ~ **de amianto** f AmL (*cf hilaza de amianto Esp*) NUCL asbestos wool; ~ **cardada** f AGRIC dressed wool; ~ **de escoria** f PROD cinder wool, TERMOTEC slag wool; ~ **de madera** f EMB wood wool; ~ **mineral** f REFRIG, TERMOTEC mineral wool; ~ **de papel** f PAPEL excelsior tissue; ~ **en rama** f AGRIC, TEXTIL raw wool; ~ **de roca** f TERMOTEC rock wool; ~ **suelta** f C&V loose wool; ~ **de vidrio** f C&V, EMB, P&C, REFRIG, TERMOTEC glass wool
laña f CONST cramp
lanarkita f MINERAL lanarkite

lance m OCEAN *de red, arte de pesca* setting
lancha f TRANSP MAR launch, longboat; ~ **de desembarco** f D&A landing craft; ~ **de motor** f TRANSP cruiser; ~ **patrullera** f CONTAM MAR, TRANSP MAR patrol boat; ~ **de práctico** f TRANSP MAR pilot cutter; ~ **de salvamento** f SEG, TRANSP AÉR, TRANSP MAR rescue boat
lanchada f TRANSP MAR *de pasajeros* boatload
lanchero m TRANSP MAR boatman
langita f MINERAL langite
lanolina f QUÍMICA lanolin, lanoline
lanoso adj TEXTIL woolly
lanosterol m QUÍMICA lanosterol
lantánido m QUÍMICA lanthanide
lantanita f MINERAL lanthanite
lantano m (*La*) QUÍMICA lanthanum (*La*)
lanza f ING MECÁ, PROD nozzle; ~ **de agua** f PROD nozzle; ~ **de manguera contra incendios** f CONST, SEG, TERMO fire-hose nozzle
lanzable adj TEC ESP jettisonable
lanzacabos m TRANSP MAR line thrower
lanzacohetes f D&A, TEC ESP rocket launcher
lanzadera f TEXTIL, TV shuttle; ~ **espacial** f TEC ESP, TELECOM space shuttle; ~ **de malla** f OCEAN *a máquina* netting needle; ~ **de redero** f OCEAN *labor manual* netting needle
lanzador m D&A launch vehicle, TEC ESP *de cohetes* launcher; ~ **de hielo** m REFRIG ice blower; ~ **de misiles colocado en un refugio subterráneo** m D&A, TEC ESP silo
lanzagranadas m D&A grenade launcher
lanzallamas m D&A flame-projector, flame-thrower
lanzamiento m INFORM&PD release, ING MECÁ launch, TEC ESP blast-off, TRANSP AÉR jettison; ~ **del codaste** m IMPR acknowledgement, TRANSP MAR aft rake; ~ **de combustible** m TRANSP AÉR fuel jettison; ~ **espacial** m TEC ESP space launch; ~ **en paracaídas** m TRANSP AÉR paradrop; ~ **de roda** m TRANSP MAR *diseño naval* stem rake
lanzar vt D&A *misil* launch, ING ELÉC ramp, OCEAN *red* shoot, TEC ESP *desde aviones* jettison; ~ **en chorro** vt FÍS FLUID jet; ~ **en paracaídas** vt TRANSP AÉR bail out, airdrop
lanzarse v refl TEC ESP *en emergencia* bail out; ~ **en paracaídas** v refl TRANSP AÉR bail out
lanzasondas m AmL (*cf punzón Esp*) MINAS pricker
lapaconitina f QUÍMICA lappaconitine
lapeado m PROD lapping
lapilli m pl GEOL lapilli
lapislázuli m QUÍMICA ultramarine
lápiz m GEOFÍS pen; ~ **de diamante** m C&V, CONST diamond pencil; ~ **especial fotosensible** m ING ELÉC pen light; ~ **fotosensible** m FÍS, INFORM&PD, TV light pen; ~ **graso** m CINEMAT grease pencil; ~ **lector** m INFORM&PD wand, wand scanner; ~ **lector de código de barras** m INFORM&PD bar-code pen; ~ **luminoso** m FÍS, INFORM&PD, TV light pen
laplaciano m FÍS, NUCL Laplacian
laqueado adj COLOR, MECÁ lacquered
laquear vt GEN lacquer
lardita f MINERAL lardite
larga: **de ~ duración** adj ALIMENT long-life; **de ~ vida** adj ALIMENT long-life
largar vt TRANSP MAR *cabos, cadenas* ease away, *amarre, bote* slip, *banderas de señales* fly, *estacha de amarre* cast off, *ancla* drop

largarse *v refl* TRANSP MAR get under way

largo: ~ **de popa** *m* TRANSP MAR *amarres* stern line; ~ **total** *m* ING MECÁ overall length

largometraje *m* CINEMAT feature

larguero *m* CARBÓN strut, CONST beam, girder, purlin post, bolster, sill, stringer, wall plate, batten, EMB side-frame, ING ELÉC boost, ING MECÁ stiffener, MINAS *Esp* (*cf reforzador AmL*) *entibación* stringer, TEC ESP *fuselajes* boom, spar, stringer, TRANSP AÉR spar; ~ **del bastidor** *m* VEH chassis member; ~ **de cerradura** *m* CONST shutting post; ~ **de entibación** *m* CONST *obra civil* waling; ~ **falso** *m* TRANSP AÉR false spar; ~ **múltiple** *m* ING MECÁ multistringer; ~ **de soporte de plano fijo** *m* TRANSP AÉR outrigger

laringófono *m* ACÚST throat microphone

lasca *f* CONST *piedra* chipping, ELEC chip, GEOL flake, ING MECÁ *de piedra* chip, TRANSP MAR figure-of-eight knot

lascar *vt* TRANSP MAR *cabo* check, *cabos* fleet

láser *m* ELECTRÓN, FÍS, FÍS ONDAS, FÍS RAD, IMPR, ÓPT laser; ~ **de alta potencia** *m* NUCL high-power laser; ~ **de anillo** *m* FÍS ring laser; ~ **antisatélite** *m* ELECTRÓN antisatellite laser; ~ **de argón** *m* ELECTRÓN argon laser; ~ **de argón ionizado** *m* ELECTRÓN ionized argon laser; ~ **de arseniuro de galio** *m* FÍS RAD gallium arsenide laser (*GaAs laser*); ~ **de baja potencia** *m* ELECTRÓN low-energy laser; ~ **bombeado eléctricamente** *m* ELECTRÓN electrically-pumped laser; ~ **bombeado ópticamente** *m* ELECTRÓN optically-pumped laser; ~ **de a bordo** *m* ELECTRÓN airborne laser; ~ **de cloruro xenónico** *m* ELECTRÓN xenon chloride laser; ~ **de color** *m* ELECTRÓN, FÍS dye laser; ~ **de colorante orgánico** *m* FÍS RAD organic dye laser; ~ **compacto** *m* ELECTRÓN solid-state laser; ~ **conectado en Q** *m* ELECTRÓN, NUCL Q-switched laser; ~ **continuo** *m* ELECTRÓN continuous laser; ~ **de cristal** *m* CRISTAL, ELECTRÓN crystal laser; ~ **de cristal de rubí** *m* ELECTRÓN, FÍS RAD ruby-crystal laser; ~ **de descarga eléctrica** *m* ELECTRÓN electric-discharge laser; ~ **dinámico de gas** *m* ELECTRÓN, ÓPT gas dynamic laser; ~ **de diodo** *m* ELECTRÓN, ÓPT diode laser; ~ **de dióxido de carbono** *m* ELECTRÓN carbon-dioxide laser; ~ **de elevada densidad energética** *m* NUCL high-power laser; ~ **de exploración** *m* ELECTRÓN, ÓPT read laser, scanning laser; ~ **de frecuencia única** *m* ELECTRÓN, TELECOM single-frequency laser; ~ **de función estabilizada** *m* ELECTRÓN mode-locked laser; ~ **de funcionamiento continuo** *m* ELECTRÓN CW laser; ~ **GAI** *m* ELECTRÓN YAG laser; ~ **de gas** *m* ELECTRÓN, FÍS RAD, GAS, ÓPT gas laser; ~ **de gas argón** *m* FÍS RAD, GAS argon gas laser; ~ **de gas atómico** *m* ELECTRÓN atomic gas laser; ~ **de gas y funcionamiento continuo** *m* ELECTRÓN CW gas laser; ~ **de gas molecular** *m* ELECTRÓN, GAS molecular gas laser; ~ **giroscópico** *m* ELECTRÓN, TEC ESP, TRANSP AÉR gyro laser; ~ **de gran potencia** *m* ELECTRÓN high-energy laser; ~ **de haz de electrones** *m* ELECTRÓN electron-beam laser; ~ **de haz verde** *m* ELECTRÓN, TV green-beam laser; ~ **de helio-neón** *m* (*láser de He-Ne*) FÍS, FÍS RAD helium-neon laser (*He-Ne laser*); ~ **de He-Ne** *m* (*láser de helio-neón*) FÍS, FÍS RAD He-Ne laser (*helium-neon laser*); ~ **de iluminación del blanco** *m* ELECTRÓN target-illuminating laser; ~ **de impulsos** *m* ELECTRÓN pulsed laser; ~ **de infrarrojos** *m* ELECTRÓN infrared laser; ~ **inyector** *m*

ELECTRÓN injection laser; ~ **iónico** *m* ELECTRÓN, ING MECÁ ion laser; ~ **de lectura** *m* ELECTRÓN, ÓPT read laser, scanning laser; ~ **líquido** *m* ELECTRÓN liquid laser; ~ **de líquido inorgánico** *m* ELECTRÓN inorganic liquid laser; ~ **de líquido orgánico** *m* ELECTRÓN organic liquid laser; ~ **de longitud de onda corta** *m* ELECTRÓN short-wavelength laser; ~ **de mercurio** *m* ELECTRÓN mercury laser; ~ **de mercurio y bromuro** *m* ELECTRÓN mercury bromide laser; ~ **molecular** *m* ELECTRÓN molecular laser; ~ **de monofrecuencia de baja energía** *m* ELECTRÓN, FÍS RAD low-energy single-frequency laser; ~ **de monóxido de carbono** *m* ELECTRÓN carbon monoxide laser; ~ **multimodo** *m* ELECTRÓN, ÓPT, TELECOM multimode laser; ~ **de neodimio** *m* ELECTRÓN neodymium laser; ~ **de onda continua** *m* ELECTRÓN, FÍS, FÍS ONDAS continuous-wave laser; ~ **paramétrico** *m* ELECTRÓN parametric laser; ~ **de pulsación corta** *m* FÍS ONDAS short-pulsed laser; ~ **pulsado** *m* FÍS RAD pulsed laser; ~ **de pulsos cortos** *m* FÍS ONDAS short-pulsed laser; ~ **Q conmutado** *m* ELECTRÓN, NUCL Q-switched laser; ~ **químico** *m* ELECTRÓN, QUÍMICA chemical laser; ~ **de rayos X** *m* ELECTRÓN, FÍS RAD, NUCL X-ray laser; ~ **rojo** *m* ELECTRÓN red laser; ~ **de rubí** *m* ELECTRÓN, FÍS RAD ruby laser; ~ **de semiconductor** *m* ELECTRÓN, FÍS RAD, ÓPT, TELECOM semiconductor laser; ~ **de señalización del blanco** *m* ELECTRÓN target designation laser; ~ **sincronizado por inyección** *m* ELECTRÓN, ÓPT, TELECOM injection-locked laser; ~ **térmicamente bombeado** *m* ELECTRÓN thermally-pumped laser; ~ **de tres niveles** *m* ELECTRÓN three-level laser; ~ **de vapor metálico** *m* ELECTRÓN metal vapor laser (*AmE*), metal vapour laser (*BrE*); ~ **verde-azul** *m* ELECTRÓN blue-green laser; ~ **para vidrio** *m* C&V, ELECTRÓN glass laser; ~ **de yodo** *m* ELECTRÓN iodine laser

LASER *abr* (*amplificación de la luz por estímulo en la emisión de radiaciones*) ELECTRÓN, FÍS ONDAS, FÍS RAD LASER (*light amplification by stimulated emission of radiation*)

Laserjet® *m* ÓPT Laserjet®

lastrabarrena *f* *Esp* (*cf portamechas AmL*) TEC PETR *perforación* drill collar

lastre[1]: **en** ~ *adj* TRANSP MAR *barco* in ballast

lastre[2] *m* CONST *náutica* ballast, TEC PETR dead weight tonnage, TRANSP AÉR, TRANSP MAR *construcción naval* ballast

lata *f* PROD can; ~ **de película** *f* CINEMAT can; ~ **de reserva de gasolina** *f* TRANSP jerry can

latencia *f* ACÚST latency, AGRIC dormancy, INFORM&PD latency

lateral[1] *adj* PETROL lateral

lateral[2]: ~ **adhesivo** *m* EMB adhesive side; ~ **encolado** *m* EMB adhesive side; ~ **soldado** *m* TELECOM soldered side

laterales: ~ **plegables** *m pl* EMB folding sides

laterita *f* CONST, PETROL laterite

látex *m* P&C latex; ~ **de caucho** *m* P&C rubber latex; ~ **centrifugado** *m* P&C centrifuged latex; ~ **desescoriado** *m* P&C skimmed latex; ~ **desnatado** *m* P&C creamed latex; ~ **estabilizado** *m* P&C stabilized latex; ~ **evaporado** *m* P&C evaporated latex; ~ **filtrado** *m* P&C skimmed latex; ~ **preservado** *m* P&C preserved latex; ~ **prevulcanizado** *m* P&C prevulcanized latex; ~ **sintético** *m* P&C synthetic latex

látice *m* MATEMÁT lattice

latiguillo *m* VEH brake line; ~ **de fibra óptica** *m* TELE-COM optical fiber pigtail (*AmE*), optical fibre pigtail (*BrE*)

latitud *f* TRANSP MAR latitude; ~ **de exposición** *f* CINEMAT, FOTO exposure latitude; ~ **geomagnética** *f* FÍS, GEOFÍS geomagnetic latitude; ~ **magnética** *f* FÍS, GEOFÍS magnetic latitude; ~ **norte** *f* TRANSP MAR northern latitude

latón *m* ING MECÁ, MECÁ brass; ~ **para cartuchería** *m* ING MECÁ, MECÁ cartridge brass; ~ **naval** *m* ING MECÁ, MECÁ naval brass

latonado *m* PROD, TERMO brazing

láudano *m* QUÍMICA laudanum

laudanosina *f* QUÍMICA laudanosine

lauegrama *m* FÍS RAD Laue diagram

laumontita *f* MINERAL laumonite, laumontite

laurencio *m* (*Lr*) FÍS RAD, QUÍMICA lawrencium (*Lr*)

láurico *adj* QUÍMICA lauric

laurionita *f* MINERAL laurionite

lava: ~ **almohadillada** *f* GEOL pillow lava; ~ **en bloques** *f* GEOL block lava

lavabilidad *f* TEXTIL washability

lavacepillos *m* TEXTIL *en operaciones de acabado* brush washer

lavadero *m* C&V washboard, MINAS washery, *minerales* jigging; ~ **de carbón** *m* CARBÓN coal washer, coal washery, coal-washing plant

lavado[1]: **no** ~ *adj* CARBÓN unwashed

lavado[2] *m* C&V washing, CARBÓN *minerales* enrichment, CONTAM scrubbing, PETROL washover, PROC QUÍ, PROD washing, QUÍMICA wash, TEXTIL laundering, washing; ~ **con agua caliente** *m* CONTAM MAR hot-water washing; ~ **a alta presión** *m* CONTAM MAR high-pressure washing; ~ **por arena** *m* TEC PETR sand-washing; ~ **atmosférico** *m* CONTAM atmospheric scrubbing; ~ **de bagazo** *m* QUÍMICA *industria cervecera* sparging; ~ **en batea** *m* QUÍMICA panning; ~ **en caja** *m* CARBÓN *minería* jigging; ~ **de caolín** *m* C&V china-clay washing; ~ **a contracorriente** *m* AGUA, HIDROL *agua tratada* back-washing; ~ **sobre cribas de sacudida** *m* CARBÓN *minería*, MINAS *preparación minerales* jigging; ~ **de crudo** *m* TEC PETR crude-oil washing (*COW*); ~ **de gas de combustión** *m* TERMO, TERMOTEC flue-gas scrubbing; ~ **interior con vapor** *m* PROC QUÍ washing out; ~ **con lija** *m* C&V emery washing; ~ **de mano** *m* PROD handwashing; ~ **de morteros** *m* CARBÓN *cal, cemento* filtration; ~ **en negro** *m* REVEST blackwash; ~ **de ojos** *m* SEG eye-wash; ~ **de oro** *m* MINAS gold cleanup; ~ **del oro** *m* MINAS flushing; ~ **previo** *m* MINAS ragging; ~ **en serie** *m* FOTO cascade washing

lavadoactivo *adj* DETERG washing-active

lavador *m* CARBÓN washer, *de mineral o de carbón* classifier, PROC QUÍ washer; ~ **de alfombras** *m* DETERG carpet shampoo; ~ **por aspersión preformada** *m* CONTAM pre-formed spray scrubber; ~ **de columna de platos** *m* CONTAM plate-column scrubber; ~ **a contracorriente** *m* CARBÓN *minerales* countercurrent classifier, counterflow classifier; ~ **del fieltro** *m* PAPEL felt washer; ~ **de gas** *m* GAS, PROD gas purifier; ~ **de gas de combustión** *m* GAS, TERMO, TERMOTEC flue-gas scrubber; ~ **de gases** *m* GAS, PROC QUÍ gas-scrubbing plant; ~ **de gases por vía húmeda** *m* CONTAM wet scrubber; ~ **de latas** *m* ALIMENT *maquinaria para proceso de alimentos* straight-through can washer; ~ **de lecho compacto** *m* CONTAM packed-bed scrubber; ~ **de lecho fijo** *m* CONTAM packed-bed scrubber; ~ **de minerales** *m* MINAS ore washer; ~ **de pantallas serigráficas** *m* IMPR screen washer; ~ **de placas** *m* CONTAM plate-column scrubber; ~ **por pulverización en contracorriente** *m* CONTAM pre-formed spray scrubber; ~ **rotativo** *m* PROD rotary washer; ~ **de sacudidas** *m* CARBÓN, MINAS jig; ~ **del sistema del blanqueo** *m* PAPEL bleaching washer; ~ **spray** *m* CONTAM pre-formed spray scrubber; ~ **de tromel** *m* CARBÓN trommel washer; ~ **Venturi** *m* CARBÓN, CONTAM Venturi scrubber

lavadora *f* PROD *máquina* washer, washing machine; ~ **de botellas** *f* ALIMENT, EMB bottle-washing machine; ~ **de cubetas** *f* MINAS trough washer; ~ **de gas de combustión** *f* GAS, TERMO, TERMOTEC flue-gas scrubber; ~ **rotatoria** *f* PROD rotary washing machine

lavamantillas *m* IMPR blanket washer

lavaojos *m* FOTO eye-cup, SEG eye-wash

lavaparabrisas *m* AUTO, VEH windscreen washer (*BrE*), windshield washer (*AmE*)

lavar *vt* CARBÓN clean, wash, enrich, MINAS, QUÍMICA wash, TEXTIL launder, wash; ~ **con agua caliente** *vt* CONTAM MAR wash with hot water; ~ **a la cuba** *vt* CARBÓN *minerales* dilute; ~ **con detergentes** *vt* DETERG scour; ~ **a máquina** *vt* ING MECÁ machine-flush; ~ **usando mangas de agua a baja presión** *vt* CONTAM MAR wash with low pressure hoses

lávica *adj* GEOL *textura* lava-like

lawsona *f* QUÍMICA lawsone

lawsonita *f* MINERAL lawsonite

laxmannita *f* MINERAL laxmannite

lazada *f* TEXTIL *vestimenta* loop

lazareto *m* TRANSP MAR *almacén* lazarette

lazo *m* ELEC *circuito*, ELECTRÓN, GEOM loop, ING MECÁ link, PETROL buckle; ~ **abierto** *m* ELEC *amplificador operacional* open loop; ~ **de auto-reposición** *m* INFORM&PD self-resetting loop; ~ **cerrado** *m* ELEC *circuito* closed loop, ELECTRÓN closed loop, feedback loop, ING ELÉC closed loop; ~ **de circulación de gas** *m* NUCL gas circulation loop; ~ **en condiciones de fallo** *m* NUCL *accidente* failed loop; ~ **de cortina** *m* TEXTIL curtain loop; ~ **Doppler-inercial** *m* TEC ESP Doppler-inertial loop; ~ **de enganche de fase** *m* TV phase-locked loop (*PLL*); ~ **de espera** *m* INFORM&PD wait loop; ~ **de evacuación de calor** *m* NUCL heat removal loop; ~ **infinito** *m* INFORM&PD infinite loop; ~ **intranuclear** *m* NUCL in pile loop; ~ **omega** *m* TV omega loop; ~ **del primario** *m* NUCL reactor loop; ~ **principal** *m* NUCL main leg; ~ **de prueba activo** *m* NUCL active test loop; ~ **de realimentación** *m* ELECTRÓN feedback loop; ~ **del refrigerante del reactor** *m* NUCL reactor loop; ~ **de video** *AmL*, ~ **de vídeo** *m Esp* TELECOM video loop

lazulita *f* MINERAL lazulite

lazurita *f* MINERAL lazurite

lb/pulg[2] *abr* (*libras por pulgada cuadrada*) METR, TEC PETR psi (*pounds per square inch*)

LBLOCA *abr* (*LOCA grande*) NUCL LBLOCA (*large break LOCA*)

LD50 *abr* (*dosis letal del 50%, dosis letal media*) FIS, NUCL LD50 (*mean lethal dose*)

LEA *abr* (*lógica de emisor acoplado*) INFORM&PD ECL (*emitter-coupled logic*)

leadhillita *f* MINERAL leadhillite

LEAR *abr* (*anillo antiprotón de baja energía*) FÍS PART LEAR (*low-energy antiproton ring*)

lechada *f* C&V, CARBÓN slurry, CONST limewashing, PAPEL slip, REVEST whitewash; **~ de almidón** *f* ALIMENT starch slurry; **~ de cal** *f* C&V lime slurry, CONST limewash; **~ de caolín** *f* PAPEL clay slip; **~ de cemento** *f* CONST grout, laitance, NUCL *ingeniería civil* grouting; **~ para inyección** *f* CONST grout

leche: **~ entera evaporada** *f* ALIMENT evaporated whole-milk; **~ de montaña** *f* MINERAL rock milk; **~ en polvo** *f* ALIMENT milk powder; **~ en polvo descremada** *f* ALIMENT skimmed milk powder; **~ TT** *f* ALIMENT TT milk

lechero *m* AGRIC dairyman

lecho *m* CARBÓN *ríos* bed, invert, CONST bed, ELEC *cable* bedding, GAS, GEOL bed, HIDROL bottom, *de un río* bed, ING MECÁ seating, TEC PETR bed; **~ aluvial** *m* AGUA alluvial bed; **~ arenoso** *m* HIDROL sandy bed; **~ bacterial** *m* HIDROL bacterial bed; **~ bacteriano** *m* HIDROL bacterial bed; **~ de bolas** *m* NUCL *reactor* pebble bed; **~ de cables** *m* ELEC, ING ELÉC *fuente de alimentación* cable trench; **~ de cantera** *m* MINAS quarry face; **~ de carbón activado** *m* NUCL activated charcoal bed; **~ carbonífero** *m* CARBÓN, GEOL, MINAS coal-bearing rock; **~ de cenizas** *m* PROD cinder bed; **~ de cimentación** *m* CARBÓN spot footing; **~ de coque** *m* CARBÓN coke bed; **~ de creciente** *m* HIDROL flood plain; **~ de filtración** *m* AGUA, NUCL, PROC QUÍ, RECICL filter bed; **~ filtrante** *m* AGUA, NUCL, PROC QUÍ, RECICL filter bed; **~ fluidificado** *m* INSTAL TERM, PROC QUÍ fluidized bed; **~ fluidizado** *m* INSTAL TERM, PROC QUÍ fluidized bed; **~ fluidizado particulado** *m* INSTAL TERM, PROC QUÍ particulate-fluidized bed; **~ fluvial** *m* HIDROL stream bed; **~ de grava** *m* CONST hardcore; **~ del mar** *m* OCEAN sea bed; **~ marino** *m* CONST sea bed; **~ mayor** *m* HIDROL flood bed; **~ de mortero** *m* CONST, PROD mortar bed; **~ normal** *m* HIDROL *de un río* normal bed; **~ oceánico** *m* OCEAN ocean floor; **~ percolador** *m* AGUA, NUCL, PROC QUÍ, RECICL filter bed; **~ de prueba** *m* INFORM&PD test bed; **~ para los rieles** *m* FERRO track bed; **~ de río** *m* CONST channel; **~ del río** *m* AGUA, HIDROL river bed; **~ de roca** *m* AGUA bed rock; **~ para el secado de lodos de alcantarilla** *m* HIDROL sludge-drying bed; **~ de siembra** *m* AGRIC seedbed; **~ para la traviesa** *m* FERRO sleeper bed; **~ de vía** *m* FERRO road bed

lecitina *f* ALIMENT lecithin

lector *m* IMPR, INFORM&PD, ÓPT read head, reader; **~ de caracteres en tinta magnética** *m* INFORM&PD magnetic ink character reader (*MICR*); **~ de cinta de papel** *m* INFORM&PD paper-tape reader; **~ de cinta perforada** *m* INFORM&PD, TELECOM punch-tape reader (*AmE*), punched-tape reader (*BrE*); **~ de cintas** *m* INFORM&PD tape reader; **~ de código de barras** *m* EMB bar-code reader, INFORM&PD bar-code scanner; **~ de códigos de barras y descifrador lógico** *m* EMB, INFORM&PD bar-code scanner and decoder-logic; **~ de discos flexibles** *m* IMPR, INFORM&PD, TELECOM floppy disk reader; **~ de discos ópticos** *m* INFORM&PD, ÓPT optical disk reader; **~ de documentos** *m* INFORM&PD document reader; **~ electromagnético** *m* ELEC *medición* pick-up; **~ de fichas perforadas** *m* INFORM&PD punch-card reader (*AmE*), punched-card reader (*BrE*); **~ de**

identificadores *m* INFORM&PD badge reader; **~ de marcas** *m* INFORM&PD mark reader; **~ de microfichas** *m* IMPR, INFORM&PD microfiche reader; **~ de microfilme** *m* INFORM&PD microfilm reader; **~ de micropelícula** *m* INFORM&PD microfilm reader; **~ óptico** *m* METR optical reader, TV optical scanning device (*OSD*); **~ óptico de caracteres** *m* (*OCR*) IMPR, INFORM&PD optical character reader (*OCR*); **~ óptico de marcas** *m* INFORM&PD optical mark reader; **~ de referencias** *m* INFORM&PD mark reader; **~ del sonido** *m* TV sound head; **~ de tarjetas** *m* TELECOM card reader; **~ de tarjetas inteligentes** *m* TELECOM smart-card reader; **~ de tarjetas magnéticas** *m* INFORM&PD magnetic-card reader; **~ de tarjetas perforadas** *m* INFORM&PD punch-card reader (*AmE*), punched-card reader (*BrE*); **~ de videodisco** *m* ÓPT videodisc player (*BrE*), videodisk player (*AmE*)

lectura[1]: **sólo de ~** *adj* INFORM&PD read only; **de ~ fija** *adj* INFORM&PD read only

lectura[2] *f* ACÚST reproducing, CONST reading, ELEC *instrumento* read-out, tracking, reading, ELECTRÓN reading, INFORM&PD read, reading, QUÍMICA read-out, reading, TV playback, scanning; **~ barométrica** *f* TRANSP MAR barometer reading; **~ de caracteres** *f* INFORM&PD character reader; **~ después de escritura** *f* INFORM&PD read-after-write; **~ destructiva** *f* INFORM&PD destructive read; **~ digital** *f* INFORM&PD, INSTR digital read-out; **~ directa** *f* ELEC, ING MECÁ direct reading; **~ directa después de escritura** *f* ÓPT direct read after write (*DRAW*); **~ dispersa** *f* INFORM&PD scatter read; **~ por dispersión** *f* INFORM&PD scatter read; **~ a distancia** *f* CONTAM MAR, GEOL, INFORM&PD, TEC ESP, TELECOM, TV remote sensing; **~ de documentos** *f* INFORM&PD document reading; **~ durante la escritura** *f* INFORM&PD read-while-write; **~-escritura** *f* INFORM&PD, ÓPT read-write; **~ de una hoja al azar** *f* IMPR random-sample reading; **~ labial** *f* ACÚST lip reading; **~ de luz reflejada** *f* CINEMAT, FOTO, ÓPT reflected-light reading; **~ de marcas** *f* INFORM&PD mark reading; **~ mecánica de amplitud constante** *f* ACÚST constant-amplitude mechanical reading; **~ mecánica de velocidad constante** *f* ACÚST constant-velocity mechanical reading; **~ de la mira** *f* CONST *topografía* staff reading; **~ negativa** *f* TV negative scanning; **~ no destructiva** *f* INFORM&PD nondestructive read; **~ óptica de marcas** *f* (*OMR*) INFORM&PD optical mark reading (*OMR*); **~ rápida** *f* TV fast playback; **~ de referencias** *f* INFORM&PD mark reading; **~ de salida** *f* INFORM&PD, TELECOM read out; **~ de salida digital** *f* INFORM&PD digital read-out; **~ de transferencia en bloques** *f* PROD block-transfer read; **~ tras escritura** *f* INFORM&PD read-after-write

LED: **~ indicador de bajo nivel de carga de acumulador** *m* PROD battery low LED; **~ superluminescente** *m* TELECOM superluminescent LED

ledeburita *f* METAL *constituyente* ledeburite

legible: **~ por computadora** *adj* AmL (*cf legible por ordenador Esp*) INFORM&PD machine-readable; **~ por ordenador** *adj* Esp (*cf legible por computadora AmL*) INFORM&PD machine-readable

legislación: **~ ambiental** *f* CONTAM environmental law; **~ básica** *f* SEG basic legislation

legumbre *f* ALIMENT legume, pulse
leguminosa: **~ de grano seco** *f* AGRIC grain legume
lejía *f* QUÍMICA lye; **~ de blanqueo** *f* PAPEL bleaching liquor; **~ de lavado** *f* DETERG scouring liquor; **~ negra** *f* PAPEL black liquor
lejiación *f* PAPEL cooking
lejiado *m* PAPEL boiling, digestion
lejiadora *f* PAPEL cooker, digester; **~ continua** *f* PAPEL continuous digester; **~ discontinua** *f* PAPEL batch digester; **~ vertical** *f* PAPEL vertical digester
leña *f* CONST cordwood
lencinita *f* MINERAL lenzinite
lengua *f* C&V tongue; **~ abisal** *f* OCEAN abyssal spit; **~ de arena** *f* OCEAN spit; **~ de soliflucción** *f* GEOFÍS *movimiento de materiales* solifluction tongue; **~ de vaca amarga** *f* AGRIC *malezas* bitter dock
lenguaje: **~ algorítmico** *m* INFORM&PD algorithmic language; **~ de alto nivel** *m* INFORM&PD high-level language; **~ ampliable** *m* INFORM&PD extensible language; **~ aplicativo** *m* INFORM&PD applicative language; **~ de bajo nivel** *m* INFORM&PD low-level language; **~ de comandos** *m* INFORM&PD command language; **~ comercial** *m* INFORM&PD commercial language; **~ de consulta** *m* (*QL*) INFORM&PD query language (*QL*); **~ de control de tareas** *m* INFORM&PD job control language (*JCL*); **~ de descripción de datos** *m* INFORM&PD data description language (*DDL*); **~ de descripción y especificación** *m* INFORM&PD, TELECOM specification and description language; **~ del diseño** *m* INFORM&PD design language; **~ ensamblador** *m* INFORM&PD assembly language; **~ de especificación** *m* INFORM&PD specification language; **~ estructurado por bloques** *m* INFORM&PD block-structured language; **~ extensible** *m* INFORM&PD extensible language; **~ formal** *m* INFORM&PD formal language; **~ fuente** *m* INFORM&PD source language; **~ funcional** *m* INFORM&PD functional language; **~ html** *m* INFORM&PD hypertext markup language (*html*); **~ interpretativo** *m* INFORM&PD interpretative language; **~ de invención conceptual** *m* INFORM&PD language construct; **~ de mandatos codificado en binario** *m* PROD binary command language; **~ de manipulación de datos** *m* (*DML*) INFORM&PD data manipulation language (*DML*); **~ de máquina** *m* INFORM&PD machine language; **~ de marcado generalizado estándar** *m* IMPR, INFORM&PD Standard Generalized Markup Language (*SGML*); **~ natural** *m* INFORM&PD natural language; **~ objeto** *m* INFORM&PD object language; **~ de órdenes** *m* INFORM&PD command language; **~ orientado al problema** *m* INFORM&PD problem-oriented language; **~ orientado a aplicaciones** *m* INFORM&PD application-oriented language; **~ orientado a los procedimientos** *m* INFORM&PD procedure-oriented language; **~ sin procedimientos** *m* INFORM&PD nonprocedural language; **~ de procesamiento de listas** *m* INFORM&PD list processing language (*LISP*); **~ de programación** *m* INFORM&PD programming language; **~ del proyecto** *m* INFORM&PD design language; **~ de representación de conocimiento** *m* INFORM&PD knowledge representation language (*KRL*); **~ SGML** *m* IMPR, INFORM&PD Standard Generalized Markup Language (*SGML*); **~ de simulación** *m* INFORM&PD simulation language; **~ en tiempo real** *m* CINEMAT, ELECTRÓN,

INFORM&PD, TELECOM, TV real-time language; **~ universal** *m* INFORM&PD general-purpose language
lengüeta *f* ACÚST molding (*AmE*), moulding (*BrE*), CONST feather tongue, feather, nosepiece, *carpintería* tongue, ELEC reed, EMB tab, ING ELÉC reed, ING MECÁ feather, MECÁ cog, PROD feather, tab, TEC ESP spline, TELECOM reed, TEXTIL latch, tongue; **~ de calibración de la pala** *f* TRANSP AÉR blade-trim tab; **~ de conexión a presión** *f* ING ELÉC crimp terminal lug; **~ de conexión terminal fruncida** *f* ING ELÉC crimp terminal lug; **~ de contacto** *f* ELEC *relé* contact blade; **~ postiza** *f* CONST loose tongue; **~ de puesta a tierra** *f* ELEC, ING ELÉC, PROD earth lug (*BrE*), ground lug (*AmE*)
leñoso *adj* PAPEL ligneous
lente *f* GEN lens; **~ acromática** *f* FÍS achromat, achromatic lens, FOTO single lens; **~ adicional** *f* INSTR supplementary lens; **~ de ampliación** *f* FOTO, INSTR close-up lens; **~ amplificadora** *f* INSTR multiplying glass; **~ anamórfica** *f* INSTR anamorphic lens; **~ anastigmática** *f* FÍS anastigmat lens; **~ de aproximación** *f* CINEMAT, FOTO, TV close-up attachment, diopter lens (*AmE*), dioptre lens (*BrE*); **~ de aproximación para primeros planos** *f* CINEMAT, FOTO, TV close-up attachment; **~ astigmática** *f* FÍS astigmatic lens; **~ de aumento** *f* FOTO viewing magnifier, INSTR lens magnification, magnifying lens; **~ auxiliar** *f* CINEMAT, FOTO, TV supplementary lens; **~ auxiliar para primeros planos** *f* CINEMAT, FOTO, TV auxiliary close-up lens; **~ bicóncava** *f* FÍS, INSTR biconcave lens; **~ biconvexa** *f* FÍS, INSTR biconvex lens; **~ bifocal** *f* C&V bifocal lens, CINEMAT split field lens, INSTR bifocal lens; **~ de brújula** *f* C&V compass lens; **~ de campo** *f* CINEMAT, FOTO, INSTR, ÓPT field lens; **~ de campo dividido** *f* CINEMAT, FOTO, ÓPT, TV split field lens; **~ sin cementar** *f* CINEMAT uncemented lens; **~ cilíndrica** *f* INSTR cylindrical lens; **~ colimada** *f* CINEMAT, FOTO, ÓPT collimated lens; **~ compuesta** *f* FOTO, ÓPT compound lens; **~ cóncava** *f* CINEMAT, FOTO, ING MECÁ, INSTR, ÓPT concave lens; **~ cóncava-convexa** *f* CINEMAT, FOTO, INSTR, ÓPT, TV concave-convex lens; **~ concéntrica de Fresnel** *f* INSTR, ÓPT concentric Fresnel lens; **~ condensadora** *f* CINEMAT, FOTO, INSTR, ÓPT condensing lens; **~ condensadora de luz** *f* CINEMAT, FOTO, INSTR, ÓPT condenser lens; **~ condensadora primaria** *f* CINEMAT first condenser lens, FOTO first condenser lamp, INSTR, ÓPT first condenser lens; **~ condensadora secundaria** *f* CINEMAT, FOTO second condenser lamp, INSTR, ÓPT second condenser lens; **~ de contacto** *f* C&V, ÓPT contact lens; **~ convergente** *f* C&V, CINEMAT, FÍS, FOTO, INSTR, ÓPT converging lens, crown-glass lens; **~ convexa** *f* FÍS, FOTO, INSTR, ÓPT convex lens; **~ con corrección de colores** *f* CINEMAT, FOTO, ÓPT color-corrected lens (*AmE*), colour-corrected lens (*BrE*); **~ correctora** *f* CINEMAT, FOTO, ÓPT correcting lens, correction lens; **~ sin corregir** *f* CINEMAT, FOTO, ÓPT uncorrected lens; **~ deformante** *f* INSTR, ÓPT distorting lens; **~ delgada** *f* FÍS thin lens; **~ de descentramiento** *f* CINEMAT offset lens; **~ de desdoblamiento** *f* INSTR, ÓPT folding lens; **~ de desdoblamiento iluminada** *f* INSTR, ÓPT illuminated folding lens; **~ difusora** *f* CINEMAT, FOTO soft-focus lens, INSTR diffuser lens; **~ dióptrica** *f* CINEMAT diopter lens (*AmE*), dioptre lens (*BrE*); **~ de distorsión** *f* CINEMAT, FOTO, ÓPT, TV

distortion lens, fish-eye lens; ~ **divergente** *f* CINEMAT, FÍS, FOTO, INSTR, ÓPT divergent lens, diverging lens, spreading lens; ~ **dividida** *f* INSTR split lens; ~ **dividida de Billet** *f* FÍS Billet split lens; ~ **doble** *f* FOTO doublet lens; ~ **electromagnética** *f* CINEMAT, ELEC, ING ELÉC, ÓPT, TV electromagnetic lens; ~ **electrónica** *f* CINEMAT, ELECTRÓN, FÍS, INSTR, ÓPT, TV electron lens; ~ **electrostática** *f* FÍS, FÍS RAD, ING ELÉC, ÓPT electrostatic lens; ~ **enfocadora** *f* CINEMAT, FOTO, ÓPT, TV viewing lens; ~ **de enfoque** *f* CINEMAT, FOTO, INSTR, ÓPT, TV focusing lens; ~ **equipotencial** *f* NUCL unipotential lens; ~ **escalonada** *f* CINEMAT step lens; ~ **de escalones** *f* INSTR, ÓPT echelon lens; ~ **de espejos** *f* FOTO mirror lens; ~ **estigmática** *f* FOTO stigmatic lens; ~ **del faro delantero** *f* AUTO, VEH headlamp lens; ~ **filtradora** *f* FOTO, INSTR, ÓPT filter lens; ~ **de flujo luminoso** *f* INSTR lumenized lens; ~ **de fluorita** *f* INSTR fluorite lens; ~ **de fluoruro de cuarzo acromática** *f* INSTR achromatic quartz fluoride lens; ~ **de foco fijo** *f* CINEMAT, FOTO, ÓPT, TV fixed-focus lens; ~ **de Fresnel** *f* CINEMAT, FÍS, FOTO, INSTR Fresnel lens; ~ **frontal** *f* INSTR, ÓPT front lens; ~ **de gran abertura** *f* CINEMAT wide aperture lens, FOTO large-aperture lens; ~ **granangular** *f* FOTO wide-angle lens; ~ **de imagen múltiple** *f* CINEMAT multi-image lens; ~ **de inmersión** *f* INSTR, LAB *microscopio* immersion lens; ~ **de inmersión en aceite** *f* INSTR, LAB *microscopio* oil immersion lens; ~ **intercambiable** *f* FOTO, INSTR, ÓPT interchangeable lens; ~ **interior** *f* FOTO, INSTR, ÓPT field lens; ~ **interior de gran densidad de energía** *f* INSTR high-power field glasses; ~ **intermedia** *f* INSTR intermediate lens; ~ **de inversión del haz** *f* TV beam-reversing lens; ~ **magnética** *f* FÍS, FÍS RAD, INSTR magnetic lens; ~ **de menisco** *f* FÍS, INSTR, ÓPT meniscus lens; ~ **de menisco negativo** *f* INSTR, ÓPT negative-meniscus lens; ~ **de menisco positivo** *f* INSTR, ÓPT positive-meniscus lens; ~ **monocromática** *f* INSTR, ÓPT monochromatic lens; ~ **montada a paño** *f* FOTO, ÓPT flush-mounted lens; ~ **negativa** *f* INSTR, ÓPT negative lens; ~ **del objetivo** *f* INSTR objective lens; ~ **objetivo** *f* FOTO taking lens; ~ **ocular** *f* CINEMAT, FÍS, FOTO, INSTR, LAB, ÓPT, PROD, TEC ESP eye-lens, eyepiece, eyepiece lens; ~ **óptica** *f* INSTR optical glass; ~ **partida** *f* INSTR split lens; ~ **con película antirreflectora** *f* FÍS bloomed lens; ~ **Petzval** *f* CINEMAT Petzval lens; ~ **plana** *f* INSTR flat lens; ~ **planocóncava** *f* FÍS, INSTR, ÓPT planoconcave lens; ~ **planoconvexa** *f* FÍS, INSTR, ÓPT planoconvex lens; ~ **positiva** *f* INSTR, ÓPT positive lens; ~ **posterior** *f* INSTR back lens; ~ **de proyección** *f* FOTO, INSTR projection lens; ~ **proyectora** *f* INSTR projector lens; ~ **radioeléctrica escalonada** *f* INSTR zoned lens; ~ **de rectificación** *f* INSTR, ÓPT rectification lens; ~ **con recubrimiento** *f* INSTR *de platino, plata*, REVEST coated lens; ~ **para retraso** *f* CINEMAT, TV close-up attachment; ~ **de retrato** *f* FOTO, INSTR portrait lens; ~ **para retratos** *f* CINEMAT, FOTO, ÓPT close-up attachment, portrait attachment; ~ **de retroenfoque** *f* CINEMAT, FOTO, ÓPT retrofocus lens; ~ **con revestimiento antirreflejante** *f* CINEMAT, FÍS, FOTO, ÓPT coated lens; ~ **seca** *f* LAB dry lens, immersion lens; ~ **semicircular** *f* INSTR semicircular

lens; ~ **teleconvertidora** *f* CINEMAT teleconverter lens; ~ **telefoto** *f* FÍS telephoto lens

lenteja *f* ING MECÁ *péndulo* ball

lentejones: ~ **de calcita fibrosa** *m pl* GEOL beef

lentes *f pl* INSTR, SEG glasses; ~ **bifocales** *f pl* INSTR, ÓPT bifocal glasses; ~ **cementadas** *f pl* C&V cemented lenses; ~ **esféricas** *f pl* FÍS spherical lens; ~ **de frenado** *f pl* FÍS stopped lens; ~ **de lectura** *f pl* INSTR, ÓPT reading glasses; ~ **multifocales** *f pl* INSTR, ÓPT multifocal glasses; ~ **de seguridad** *f pl Esp* (*cf anteojos de seguridad AmL*) C&V, INSTR, LAB, ÓPT, PROD, SEG eye-protection glasses, safety glasses, safety goggles, safety spectacles; ~ **telescópicas** *f pl* INSTR, ÓPT telescopic spectacles

lenticular *adj* FOTO lens-shaped, GEOL lensing, lenticular

lentilla: ~ **de contacto** *f* INSTR *corneal* contact lens

LEP *abr* (*gran cámara de reacción electrón-positrón*) FÍS PART LEP (*large electron-positron collider*)

lepidocroita *f* MINERAL goethite

lepidolita *f* MINERAL lepidolite

lepidomelana *f* MINERAL lepidomelane

leptoclorita *m* MINERAL leptochlorite

leptón *m* FÍS, FÍS PART, NUCL lepton; ~ **tau** *m* FÍS, FÍS PART, NUCL tau-lepton

L-escopolamina *f* QUÍMICA hyoscine

lesión *f* SEG injury; ~ **común en el área de trabajo** *f* SEG common injury in the workplace; ~ **industrial** *f* SEG industrial injury; ~ **ocular** *f* SEG eye injury; ~ **a los ojos** *f* SEG eye injury; ~ **producida por conservación en frío** *f* ALIMENT *empaquetado* cold storage injury

lesivo *adj* SEG injurious; ~ **a los ojos** *adj* SEG harmful to the eyes

letargia *f* FÍS RAD lethargy; ~ **neutrónica** *f* FÍS RAD lethargy

letra *f* IMPR, INFORM&PD letter; ~ **acentuada** *f* IMPR accented letter; ~ **de adorno** *f* IMPR swash letter; ~ **alta con trazo alto** *f* IMPR ascending letter; ~ **atada** *f* IMPR tied letter; ~ **de código de colores** *f* PROD color code letter (*AmE*), colour code letter (*BrE*); ~ **cursiva** *f* IMPR italic type, script; ~ **doble** *f* IMPR double letter; ~ **fundida en plomo** *f* IMPR lead printing letter; ~ **invertida** *f* IMPR turned letter; ~ **llena** *f* IMPR solid letter; ~ **negrita** *f* IMPR black letter; ~ **volada** *f* IMPR superscript

letras: ~ **enlazadas** *f pl* IMPR ligature

letrero *m* ELEC *artefacto* nameplate, ING MECÁ placard; ~ **de advertencia** *m* SEG warning sign; ~ **de emergencia** *m* SEG, TEC ESP emergency sign; ~ **obligatorio** *m* SEG mandatory sign; ~ **de peligro contra incendio** *m* SEG fire safety sign; ~ **para prevenir accidentes** *m* SEG accident prevention advertising sign; ~ **de prohibición** *m* SEG prohibition sign; ~ **de seguridad** *m* LAB safety placard, SEG safety sign; ~ **de seguridad fosforescente** *m* SEG phosphorescent safety-sign

letrina *f* AGUA cesspit

lettsomita *f* MINERAL lettsomite

leuchtenbergita *f* MINERAL leuchtenbergite

leucita *f* MINERAL, PETROL leucite

leuco *m* COLOR leuco

leucocrático *adj* GEOL leucocratic

leucófana *f* MINERAL leucophane

leucopirita *f* MINERAL leucopyrite

leucosoma *f* GEOL leucosome

leucoxeno *f* MINERAL leucoxene

leva *f* AUTO cam, tappet, CINEMAT cam, ING MECÁ lifter, cam, wiper, MECÁ cam, cog, TRANSP AÉR, VEH cam; **~ acodada** *f* ING MECÁ bellcrank; **~ de cilindro** *f* ING MECÁ, PROD barrel cam, cylinder cam; **~ de cilindro acanalada** *f* ING MECÁ grooved cylinder cam; **~ de dilatación** *f* INSTAL HIDRÁUL expansion cam; **~ del distribuidor** *f* VEH *sistema de encendido* distributor cam; **~ de espiral** *f* ING MECÁ involute cam; **~ de evolvente de círculo** *f* ING MECÁ involute cam; **~ excéntrica** *f* ING MECÁ eccentric cam; **~ de freno** *f* AUTO, MECÁ, VEH brake cam; **~ de huso** *f* PROD spindle cam; **~ impresora** *f* IMPR printing cam; **~ de paro** *f* ING MECÁ knock-off cam; **~ de seguridad** *f* ING MECÁ locking cam; **~ de tambor** *f* ING MECÁ barrel cam, PROD drum cam; **~ de tambor de lengüetas** *f* PROD strake drum cam; **~ triangular** *f* ING MECÁ triangular cam

levadura *f* ALIMENT *molienda, panadería* leaven; **~ baja** *f* ALIMENT bottom yeast; **~ de depósito** *f* ALIMENT bottom yeast; **~ de fondo** *f* ALIMENT bottom yeast; **~ química** *f* ALIMENT baking soda, *panadería, repostería* chemical leavening; **~ de solera** *f* ALIMENT bottom yeast

levana *f* QUÍMICA levan

levantacarriles *m* FERRO rail lifter

levantada *f* ING MECÁ take-up

levantamiento *m* CONST survey, GEOL *estructural* uplift, ING MECÁ lifting, METEO uplift, MINAS *del piso de la galería* lifting, NUCL upward heave, TRANSP lifting; **~ con la brújula** *m* MINAS *topografía* dialing (*AmE*), dialling (*BrE*); **~ catastral** *m* CONST cadastral survey; **~ por congelación** *m* CARBÓN *terrenos* frost heave; **~ de detalle** *m* CONST detailed survey; **~ gradual del piso por subpresiones** *m* CARBÓN, MINAS creep; **~ gravimétrico** *m* PETROL gravity survey; **~ por helada** *m* METEO frost heave; **~ magnético** *m* GEOFÍS *prospección* magnetic survey; **~ con método eléctrico** *m* PETROL electrical survey; **~ de planos** *m* CONST surveying; **~ de planos con brújula** *m* MINAS *topografía* compass survey; **~ del suelo** *m* MINAS *fallas* floor heave; **~ del suelo por congelación** *m* REFRIG frost heave; **~ topográfico** *m* CONST topographical survey

levantar[1] *vt* ING MECÁ hold up, MECÁ boost, lift, PROD hoist; **~ con el gato** *vt* AUTO, VEH jack; **~ incorrectamente** *vt* SEG lift incorrectly; **~ el morro** *vt* TRANSP AÉR pitch-up; **~ con palanca** *vt* ING MECÁ lever, lever up

levantar[2]: **~ presión** *vi* INSTAL HIDRÁUL, PROD get up steam

levantarse *v refl* TRANSP MAR *viento* rise

levantaválvulas *m* AUTO, VEH tappet, valve tappet; **~ hidráulico** *m* AUTO, VEH hydraulic valve lifter; **~ plano** *m* AUTO, VEH flat-bottom tappet; **~ de rodillos** *m* AUTO, VEH roller tappet

levante: **~ de la vía** *m* FERRO track raising

levar[1] *vt* TRANSP MAR *ancla* weigh

levar[2]: **~ anclas** *vi* TRANSP MAR cast off; **~ una de las dos anclas** *vi* TRANSP MAR unmoor

levigación *f* C&V, PROC QUÍ elutriation, PROD levigation, QUÍMICA elutriation

levigar *vt* C&V, PROC QUÍ, QUÍMICA elutriate

levitación *f* FÍS, TEC ESP, TRANSP levitation; **~ aerodinámica** *f* TRANSP aerodynamical levitation; **~ en altura** *f* TRANSP height hovering; **~ electrodinámica** *f* ELEC, TRANSP electrodynamic levitation; **~ por electroimán superconductor** *f* ELEC, FÍS, TRANSP superconducting magnet levitation; **~ electromagnética** *f* ELEC, FÍS, TRANSP electromagnetic levitation; **~ estática** *f* TRANSP static hovering; **~ por imanes permanentes** *f* FÍS, TRANSP levitation by permanent magnets; **~ magnética** *f* FÍS, TRANSP magnetic levitation; **~ mixta** *f* FÍS, TRANSP mixed levitation

levógiro *adj* FÍS, QUÍMICA laevorotatory (*BrE*), left-handed, levorotatory (*AmE*)

levulina *f* QUÍMICA levulin

levulínico *adj* QUÍMICA levulinic

levulosa *f* ALIMENT, QUÍMICA laevulose (*BrE*), levulose (*AmE*)

levyna *f* MINERAL levyne

levynita *f* MINERAL levynite

ley *f AmL* (*cf curso Esp*) METAL content, MINAS *mineral content, de minerales* tenor, *del oro o la plata* standard, SEG act; **~ de acción de masas** *f* FÍS, QUÍMICA law of mass action; **~ de altura cuadrantal** *f* TRANSP AÉR quadrantal height rule; **~ de Ampère** *f* ELEC, FÍS Ampère's law; **~ de las áreas de Kepler** *f* FÍS, TEC ESP Kepler's law of areas; **~ aurífera** *f* MINAS gold grade, gold tenor; **~ de Biot-Savart** *f* FÍS Biot-Savart's law; **~ de Boyle** *f* FÍS Boyle's law; **~ de Bragg** *f* CRISTAL Bragg's law; **~ cero** *f* FÍS, TERMO zeroth law; **~ de Charles** *f* FÍS Charles's law; **~ de conservación numérica de partículas** *f* NUCL particle number conservation law; **~ de conservación de la paridad** *f* FÍS RAD parity conservation law; **~ de los cosenos** *f* GEOM cosine rule; **~ de Coulomb** *f* ELEC, FÍS Coulomb's law; **~ del cuadrado** *f* FÍS square law; **~ de Curie** *f* FÍS, FÍS RAD, PETROL, TEC PETR Curie's law; **~ de Curie-Weiss** *f* FÍS, FÍS RAD Curie-Weiss law; **~ de Dalton** *f* FÍS Dalton's law; **~ de desintegración radiactiva** *f* FÍS, FÍS RAD, NUCL decay law, law of radioactive decay; **~ de desplazamiento de Wien** *f* FÍS Wien displacement law; **~ de la dilución de Ostwald** *f* QUÍMICA Ostwald's dilution law; **~ de Dulong y Petit** *f* FÍS Dulong and Petit's law; **~ de emisión del coseno** *f* ÓPT cosine emission law; **~ de estados correspondientes** *f* FÍS law of corresponding states; **~ de Fick** *f* FÍS Fick's law; **~ de Gauss** *f* FÍS Gauss's law; **~ de Gay-Lussac** *f* FÍS Gay-Lussac's law; **~ de Graham** *f* FÍS Graham's law; **~ de Hooke** *f* CONST, FÍS Hooke's law; **~ del inverso del cuadrado** *f* ACÚST, FÍS inverse square law; **~ del inverso del cuadrado de la distancia** *f* ACÚST, FÍS inverse square law; **~ de Joule** *f* FÍS Joule's law; **~ de Kirchhoff** *f* ELEC, FÍS, FÍS RAD, ING ELÉC Kirchhoff's law; **~ de Lambert** *f* FÍS, ÓPT Lambert's law; **~ de Lambert del coseno** *f* FÍS, ÓPT Lambert's cosine law; **~ de Lenz** *f* ELEC, FÍS Lenz's law; **~ de Malus** *f* FÍS Malus law; **~ de mineral** *Esp* (*cf contenido en mineral AmL*) *f* MINAS ore contents; **~ de Moseley** *f* MINAS Moseley's law; **~ de Ohm** *f* ELEC, FÍS, ING ELÉC Ohm's law; **~ de Paschen** *f* FÍS Paschen's law; **~ de Planck** *f* FÍS, FÍS RAD Planck's law; **~ de Poiseuille** *f* FÍS Poiseuille's law; **~ de Poisson** *f* FÍS, TEC ESP Poisson's law; **~ de reciprocidad** *f* CINEMAT, FÍS, FOTO, TV reciprocity law; **~ de salud y seguridad en el trabajo** *f* SEG Health and Safety at Work Act (*BrE*) (*HASAWA*); **~ de los senos** *f* GEOM sine rule; **~ de Snell** *f* FÍS Snell's law; **~ de Stefan** *f* FÍS, FÍS RAD Stefan's law; **~ de Stefan-Boltzmann** *f* FÍS, FÍS RAD Stefan-

Boltzmann law; ~ **de Steinmetz** *f* FÍS Steinmetz's law; ~ **de Stokes** *f* FÍS Stokes' law; ~ **de la termodinámica** *f* TERMO law of thermodynamics, thermodynamic law; ~ **de Weber-Fechner** *f* ACÚST Weber-Fechner law; ~ **de Wiedemann-Franz** *f* FÍS Wiedemann-Franz law; ~ **de Wien** *f* FÍS Wien law; ~ **de Young** *f* ACÚST Young's law

leyenda: ~ **doble** *f* PROD double legend

leyes: ~ **de Faraday** *f pl* ELEC, ELECTRÓN, FÍS, ING ELÉC, NUCL Faraday's laws; ~ **de Kepler** *f pl* FÍS Kepler's laws; ~ **sobre el medio ambiente** *f pl* CONTAM *legislación* environmental law; ~ **de la radiación** *f pl* FÍS RAD radiation laws; ~ **de la reflexión** *f pl* FÍS laws of reflection; ~ **de la refracción** *f pl* FÍS laws of refraction; ~ **de vibración de una cuerda fija** *f pl* FÍS ONDAS laws of vibration of a fixed string

lezna *f* CONST awl, bradawl, broach, ING MECÁ stabbing awl; ~ **de marcar** *f* CONST scratch awl; ~ **de trazado** *f* CONST scribe awl, *carpintería* scribing awl

LGN *abr* (*líquido de gas natural*) GAS, TEC PETR, TRANSP NGL (*natural gas liquid*)

LHC *abr* (*gran cámara de reacción para hadrones*) FÍS PART *en desarrollo en CERN* LHC (*large hadron collider*)

Li *abr* (*litio*) ING ELÉC, QUÍMICA Li (*lithium*)

liberación *f* ELEC, INFORM&PD, ING ELÉC, ING MECÁ release, QUÍMICA liberation, TELECOM release; ~ **de calor** *f* NUCL heat release; ~ **del mensaje** *f* TELECOM release message (*REL*)

liberador: ~ **de barrenas** *m* TEC PETR bit breaker

liberar *vt* FÍS liberate, GAS release

libethenita *f* MINERAL libethenite

libra *f* METR pound; ~ **avoirdupois** *f* METR pound avoirdupois

libras: ~ **por hora** *f pl* METR pounds per hour; ~ **por pulgada cuadrada** *f pl* METR, TEC PETR (*lb/pulg²*, *LPPC*) *unidades de presión* pounds per square inch (*psi*)

libre[1] *adj* ELEC *terminal*, ING MECÁ loose, MECÁ floating, QUÍMICA *no combinado* free, TEC ESP independent, TELECOM idle; ~ **de ácido** *adj* EMB acidless; ~ **de ácidos** *adj* PAPEL acid-free; ~ **de carbono** *adj* METAL carbon-free; ~ **de carga** *adj* TRANSP AÉR no-load; ~ **de ceniza** *adj* CONTAM ash-free; ~ **de contexto** *adj* INFORM&PD context-free; ~ **de gastos** *adj* TELECOM free of charge; ~ **de humedad** *adj* CONTAM water-free; ~ **de materia mineral** *adj* CONTAM mineral-matter-free; ~ **de pérdidas** *adj* ELEC *dieléctrico* loss-free; ~ **de semilla** *adj* C&V seed-free; ~ **de tierra** *adj* ELEC *circuito* earth-free (*BrE*), ground-free (*AmE*); ~ **de vacilación** *adj* INFORM&PD flicker-free

libre[2]: ~ **plática** *f* TRANSP MAR pratique, *comunicaciones* free pratique

librea *f* TRANSP AÉR livery

libreta: ~ **de borrador** *f* INFORM&PD scratch pad; ~ **de campo** *f* CONST *topografía* field book; ~ **de nivelación** *f* CONST *topografía* level book

libreto *m* INFORM&PD script; ~ **de instrucciones** *m* ING MECÁ instruction book

libro: ~ **de bolsillo** *m* IMPR pocketbook; ~ **de a bordo** *m* TRANSP AÉR journey logbook; ~ **en cartoné** *m* IMPR cased book; ~ **en cuarto** *m* IMPR quarto; ~ **encuadernado** *m* IMPR bound book; ~ **de faros** *m* TRANSP MAR list of lights; ~ **en folio** *m* IMPR folio;

~ **genealógico** *m* AGRIC heardbook; ~ **ilustrado** *m* IMPR picture book; ~ **impreso** *m* IMPR printed book; ~ **de instrucciones** *m* ING MECÁ instruction book; ~ **de navegación** *m* TRANSP AÉR journey logbook; ~ **registro** *m* PROD record book; ~ **de señales** *m* TRANSP MAR signal book; ~ **talonario** *m* IMPR stubook; ~ **de tapas blandas** *m* IMPR paperback; ~ **de vuelo** *m* TRANSP AÉR flight log, journey logbook

libros: ~ **y cuadernos de a bordo** *m pl* TRANSP MAR ship's books

licareol *m* QUÍMICA licareol

licencia *f* CALIDAD, TEC PETR licence (*BrE*), license (*AmE*); ~ **de construcción** *f* CONST planning permission; ~ **de edificación** *f* CONST planning permission; ~ **de exploración** *f* TEC PETR exploration licence (*BrE*), exploration license (*AmE*); ~ **de exportación** *f* TRANSP MAR export licence (*BrE*), export license (*AmE*); ~ **de importación** *f* TRANSP, TRANSP MAR import licence (*BrE*), import license (*AmE*); ~ **de producción** *f* TEC PETR production licence (*BrE*), production license (*AmE*); ~ **de prospección** *f* TEC PETR exploration licence (*BrE*), exploration license (*AmE*); ~ **de ruta** *f* TRANSP AÉR route licence (*BrE*), route license (*AmE*)

licenciamiento: ~ **de la planta** *m* NUCL plant licensing

licitación *f* MECÁ bid

licor *m* QUÍMICA liquor; ~ **de ajenjo** *m* QUÍMICA absinthe; ~ **de blanqueo** *m* PAPEL bleaching liquor; ~ **cervecero** *m* ALIMENT *fermentación* brewing liquor; ~ **de cienos** *m* AGUA sludge liquor; ~ **de lixiviación** *m* PROC QUÍ leach liquor; ~ **de maceración** *m* ALIMENT mash liquor; ~ **madre** *m* ALIMENT mother liquor; ~ **mezclado** *m* AGUA mixed liquor; ~ **salino** *m* NUCL salt liquor

licuación *f* FÍS, GAS, QUÍMICA, REFRIG, TERMO liquation

licuado *adj* FÍS, GAS, QUÍMICA, REFRIG, TERMO liquefied

licuar *vt* FÍS, GAS, QUÍMICA, REFRIG, TERMO liquefy

licuefacción *f* FÍS, GAS, QUÍMICA, REFRIG, TERMO liquefaction; ~ **del carbón** *f* CARBÓN, FÍS, GAS, MINAS, QUÍMICA, REFRIG, TEC PETR, TERMO coal liquefaction

licuefactor *m* FÍS, GAS, QUÍMICA, REFRIG, TERMO liquefier

licuescencia *f* FÍS, GAS, QUÍMICA, REFRIG, TERMO liquescence

licuescente *adj* FÍS, GAS, QUÍMICA, REFRIG, TERMO liquescent

lidar *n* ÓPT, TEC ESP lidar; ~ **sonda** *m* TEC ESP sounding lidar

lieberenita *f* MINERAL lieberenite

lievrita *f* MINERAL lievrite

LIFO *abr* (*último en entrar, primero en salir*) INFORM&PD, PROD LIFO (*last-in-first-out*)

ligada *f* TRANSP MAR *estibaje* lashing

ligado[1] *adj* C&V bonded

ligado[2]: ~ **mate** *m* C&V bonded mat

ligador *m* TV binder; ~ **transparente** *m* TV clear binder

ligadura *f* NUCL bond, QUÍMICA bond, linking; ~ **blindada** *f* ELEC cross bonding, shield-bonding; ~ **pi** *f* QUÍMICA pi bond; ~ **sólida** *f* ELEC solid bond; ~ **transversal** *f* P&C *polimerización, vulcanización*, QUÍMICA *plásticos termoestables, cauchos vulcanizados, proteínas* cross link

ligamento *m* CONST *albañilería* bonding, ING MECÁ

link, TEXTIL weave; ~ **de espiga** *m* TEXTIL herringbone weave; ~ **tafetán** *m* TEXTIL plain weave
ligamentos: ~ **en relieve** *m pl* TEXTIL relief weaves
ligamiento *m* TEC ESP *adhesivos sintéticos, recubrimientos ignífugos* bonding
ligando *m* QUÍMICA ligand
ligante *m* CONST *pintura* binder, bond, EMB *adhesivo* binder, PROD bonding; ~ **para estucados** *m* PAPEL, REVEST coating binder; ~ **de relleno** *m* CONST filled binder
ligar *vt* CONST bind, bond, ELECTRÓN alloy, INFORM&PD bind, METAL, PROD, QUÍMICA alloy, TRANSP MAR *cabos* seize
ligarse *v refl* CONST bind
ligazón *m* C&V bond, CARBÓN binding
ligera: ~ **apertura** *f* ING MECÁ *válvula* cracking open
ligero *adj* ING MECÁ light
lignana *f* QUÍMICA lignan
lignina *f* P&C, QUÍMICA lignin
lignito *m* CARBÓN, GEOL, MINERAL, PETROL, TERMO lignite; ~ **bituminoso** *m* CARBÓN pitch coal; ~ **en láminas delgadas** *m* CARBÓN paper coal; ~ **papiráceo** *m* CARBÓN papyraceous lignite; ~ **de pizarra esquistoso** *m* CARBÓN shaly brown coal, slate-foliated lignite; ~ **pizarroso** *m* CARBÓN slate-foliated lignite
ligroína *f* QUÍMICA ligroin
ligurita *f* MINERAL ligurite
lija *f* C&V emery; ~ **fina** *f* ING MECÁ fine emery cloth
lijado *m* C&V, CONST, MECÁ, P&C, PROD sanding, sandpapering
lijadora *f* ING MECÁ surface sander, PROD sandpapering machine; ~ **de disco** *f* CONST, PROD disc sandpapering machine (*BrE*), disk sandpapering machine (*AmE*); ~ **orbital** *f* ING MECÁ orbital sander; ~ **orbital rotatoria** *f* ING MECÁ rotary orbital sander; ~ **rotatoria a discos** *f* CONST, ING MECÁ disc sander (*BrE*), disk sander (*AmE*)
lijar *vt* CONST clean off, MECÁ sandpaper, PROD sand, sandpaper
lima *f* CONST *tejado* hip, ING MECÁ, MECÁ, PROD, VEH *utillaje* file; ~ **de acanalar** *f* ING MECÁ frame-saw file; ~ **achaflanada** *f* ING MECÁ screw-head file; ~ **de aguja** *f* ING MECÁ needle file; ~ **alméndrica** *f* ING MECÁ oval file; ~ **basta** *f* ING MECÁ rough file; ~ **bastarda** *f* MECÁ bastard file; ~ **de bordes paralelos** *f* ING MECÁ parallel file; ~ **de canto liso** *f* ING MECÁ safe-edge file; ~ **de cantos biselados** *f* PROD feather-edged file; ~ **de cantos redondeados** *f* ING MECÁ round-edge file; ~ **carrelera** *f* PROD hand file; ~ **de cerrajero** *f* ING MECÁ key file; ~ **cilíndrica** *f* ING MECÁ round file; ~ **de cola de milano** *f* PROD dovetail cutter; ~ **de cola de rata** *f* ING MECÁ rat-tail file, PROD gulleting saw file; ~ **cola de ratón** *f* ING MECÁ rat-tail file; ~ **para colas de milano** *f* ING MECÁ cant file; ~ **de corte** *f* PROD feather edge file; ~ **cuadrada** *f* ING MECÁ square file; ~ **cuadrada puntiaguda** *f* PROD entering file; ~ **de cuatro caras** *f* ING MECÁ square file; ~ **para dientes de sierra** *f* ING MECÁ saw file; ~ **doble** *f* PROD cross file; ~ **de dos mangos** *f* ING MECÁ two-tanged file; ~ **de embutición** *f* PROD drawing file; ~ **entrefina** *f* ING MECÁ second cut; ~ **espada** *f* ING MECÁ cant file; ~ **extradulce** *f* PROD extra-smooth file; ~ **de filo de cuchillo** *f* ING MECÁ knife-edge file, knife file; ~**-fresa** *f* PROD milling file; ~ **fresadora de mano** *f* ING MECÁ

circular cut file; ~ **gruesa** *f* ING MECÁ rough file; ~ **de guardas** *f* ING MECÁ key file; ~ **para guardas** *f* ING MECÁ warding file; ~ **de igualar** *f* PROD equalizing file; ~ **de lengua de pájaro** *f* PROD entering file; ~ **de media caña de doble picadura** *f* PROD double half round file; ~ **muza** *f* PROD crosscut file; ~ **de paletones** *f* ING MECÁ warding file; ~ **de picadura simple** *f* ING MECÁ float-cut file; ~ **plana de mano** *f* PROD hand file; ~ **de punto de contacto** *f* ING MECÁ contact-point file; ~ **ranuradora** *f* ING MECÁ cotter file; ~ **redonda** *f* ING MECÁ rat-tail file; ~ **para refrenar bridas** *f* PROD flange smoother; ~ **semidulce** *f* PROD middle cut; ~ **de sierra** *f* ING MECÁ saw file; ~ **tesa** *f* CONST hip rafter, *tejado* angle rafter; ~ **de tres bordes y extremos iguales** *f* PROD double-ended handsaw file; ~ **triangular** *f* ING MECÁ tri-square file, three-square file; ~ **triangular achatada** *f* ING MECÁ barrette file; ~ **triangular isósceles** *f* ING MECÁ cant file
limado *m* CONST rasping
limadora *f* ING MECÁ, MECÁ shaper, shaping machine, surface planer; ~ **accionada por biela** *f* ING MECÁ, MECÁ crank shaping machine; ~ **de cabeza horizontal** *f* ING MECÁ traversing head shaping machine; ~ **de cabezal doble** *f* ING MECÁ double-headed shaping machine; ~ **de carnero longitudinal** *f* ING MECÁ traversing head shaping machine; ~ **de carnero móvil** *f* ING MECÁ traverse shaper, traversing head shaping machine; ~ **de engranajes** *f* ING MECÁ gear-shaping machine; ~ **de engranajes blandos** *f* ING MECÁ gear-shaving machine; ~ **de engranajes sin templar** *f* ING MECÁ gear-shaving machine; ~ **de excéntrica** *f* ING MECÁ eccentric drive slotting machine; ~ **horizontal** *f* ING MECÁ traverse shaper; ~ **tipo banco** *f* ING MECÁ bench-type shaping machine; ~ **vertical de manivela** *f* ING MECÁ crank drive slotting machine
limaduras *f pl* CARBÓN stainless steel dust, steel alloy dust, CONST, MECÁ, PROD *de muelas abrasivas* grit; ~ **de hierro** *f pl* FÍS, QUÍMICA iron filings
limahoya *f* CONST valley
limatón *m* ING MECÁ round file
limbo *m* INSTR *del sol o la luna* edge, MATEMÁT limb, TRANSP MAR compass dial; ~ **graduado de la rosa** *m* TRANSP MAR compass dial
limen *m* TV limen
limitación *f* ELECTRÓN limiting, INFORM&PD constraint, MINAS, TEC PETR confinement; ~ **del quark** *f* FÍS, FÍS PART quark confinement; ~ **de ruidos** *f* ACÚST, TV noise modulation; ~ **de sobreintensidad** *f* ELEC, ING ELÉC current limiting; ~ **de tensión transitoria** *f* PROD transient voltage limitation; ~ **de velocidad** *f* FERRO *vehículos* speed restriction; ~ **de voltaje momentáneo** *f* PROD transient voltage limitation
limitaciones: ~ **de operación** *f pl* PROD operating limitations
limitado: ~ **a carretera** *adj* TRANSP road-bound; ~ **por entrada** *adj* INFORM&PD input-limited; ~ **por entrada/salida** *adj* INFORM&PD input/output limited; ~ **por los periféricos** *adj* INFORM&PD peripheral-limited; ~ **por el procesador** *adj* INFORM&PD processor-limited; ~ **en la salida** *adj* INFORM&PD output-limited
limitador *m* ELECTRÓN, ING ELÉC, ING MECÁ, TELECOM, TV limiter; ~ **de alimentación** *m* ING MECÁ feed limiter; ~ **de amplitud** *m* ELECTRÓN, TELECOM ampli-

tude limiter, clipper; ~ **automático de valor máximo** *m* TV automatic peak limiter; ~ **de avance** *m* ING MECÁ feed limiter; ~ **de blanco** *m* TV white limiter; ~ **del campo** *m* FÍS field stop; ~ **de corriente** *m* ELEC, ING ELÉC current limiter; ~ **de corriente de entrada** *m* ELEC, ING ELÉC inrush current limiter; ~ **de diodo** *m* ELECTRÓN diode limiter; ~ **de embalamiento** *m* ING MECÁ overspeed gear; ~ **de ferrita** *m* ING ELÉC ferrite limiter; ~ **horario** *m* ELEC, TELECOM time switch; ~ **inverso** *m* ELECTRÓN inverse limiter; ~ **de microondas** *m* ELECTRÓN microwave limiter; ~ **de potencia** *m* ING MECÁ power limiter; ~ **potente** *m* ELECTRÓN hard limiter; ~ **de pulsos** *m* TV pulse clipper; ~ **de señal** *m* ELECTRÓN signal limiter; ~ **de sobreintensidad** *m* ELEC, ING ELÉC current limiter; ~ **de tensión** *m* ELEC, ING ELÉC voltage limiter; ~ **del valor máximo** *m* CINEMAT, TV peak limiter; ~ **del valor de pico** *m* CINEMAT, TV peak limiter; ~ **de velocidad** *m* AUTO, VEH speed limiter; ~ **de voltaje** *m* ELEC, ING ELÉC voltage limiter; ~ **zonal** *m* TEC ESP limiter

limitar *vt* ELEC, ELECTRÓN, ING ELÉC limit

límite *m* CONST, GEOL boundary, ING MECÁ, MATEMÁT limit, PROD clearance, TRANSP *visibilidad* limit; ~ **de antifase** *m* CRISTAL antiphase boundary; ~ **anual de incorporación** *m* (*LAI*) NUCL annual limit on intake (*ALI*); ~ **aparente de elasticidad** *m* P&C *propiedad física, prueba* apparent modulus of elasticity; ~ **bien definido de fluencia** *m* METAL sharp yield point; ~ **de cada impresión** *m* IMPR cutoff; ~ **de calidad media después de control** *m* CALIDAD average outgoing quality limit (*AOQL*); ~ **de carga de agua** *m* INSTAL HIDRÁUL head limit; ~ **de carga de ráfaga** *m* TEC ESP, TRANSP AÉR, TRANSP MAR gust load limit; ~ **de la célula** *m* TELECOM cell boundary; ~ **de colorante** *m* COLOR color limit (*AmE*), colour limit (*BrE*); ~ **de congelación** *m* CARBÓN frost limit; ~ **de consistencia** *m* CARBÓN consistency limit, limit of consistency; ~ **de la construcción** *m* CONST building line; ~ **de control** *m* CALIDAD control limit; ~ **corriente de ciclo** *m* PROD run current limit; ~ **crítico de olores** *m* AGUA, HIDROL, RECICL odor threshold (*AmE*), odour threshold (*BrE*); ~ **de desconexión regulable** *m* PROD adjustable trip setting; ~ **de desgaste** *m* ING MECÁ wear limit; ~ **de dosis equivalente anual** *m* NUCL annual dose equivalent limit; ~ **de la ebullición nucleada** *m* NUCL critical heat flow; ~ **de elasticidad** *m* CONST, EMB, FÍS, MECÁ elastic limit, METAL lower yield point, yield point, yield stress, NUCL yield point, yield strength, P&C elastic limit; ~ **elástico** *m* CONST, EMB, FÍS, MECÁ, P&C elastic limit; ~ **elástico con deformación plástica** *m* NUCL plastic yield; ~ **de error** *m* METR *de instrumento de medida* limit of error; ~ **de especificación** *m* CALIDAD specification limit; ~ **de estirado cíclico** *m* METAL repeated yield limit; ~ **de exposición** *m* SEG exposure limit; ~ **de falla** *m* GEOL fault boundary; ~ **de fase** *m* METAL phase boundary; ~ **de fatiga** *m* CRISTAL, METAL endurance limit; ~ **de filtración** *m* PROC QUÍ filtering limit; ~ **final** *m* PROD end fence; ~ **de fluencia** *m* METAL creep limit, NUCL yield point; ~ **de flujo** *m* ING ELÉC flow limit; ~ **forestal** *m* AGRIC timberline; ~ **de grano** *m* CRISTAL, METAL grain boundary; ~ **inferior** *m* TELECOM lower limit; ~ **inferior de control** *m* CALIDAD lower control limit; ~ **inferior**

de detectabilidad *m* CONTAM lower limit of detectability; ~ **de interfase** *m* METAL interface boundary; ~ **justo de estirado** *m* METAL sharp yield point; ~ **de Kruskal** *m* NUCL Kruskal limit; ~ **líquido** *m* CONST *geotecnia* liquid limit; ~ **máximo de exposición** *m* SEG maximum exposure limit; ~ **mínimo de detección** *m* CONTAM lower limit of detectability; ~ **de la nieve** *m* METEO snow line; ~ **de placa** *m* GEOL plate boundary; ~ **de placas divergentes** *m* GEOL divergent plate boundary; ~ **de plasticidad** *m* CARBÓN, P&C plastic limit; ~ **plástico** *m* CONST, P&C plastic limit; ~ **de resistencia** *m* NUCL yield point; ~ **de rotura** *m* ING MECÁ ultimate strength, MECÁ breaking point; ~ **de seguridad** *m* SEG safe limit; ~ **superior** *m* TELECOM upper limit; ~ **superior de la capa de ozono** *m* METEO upper limit of ozone layer; ~ **superior de fiabilidad** *m* CALIDAD upper control limit; ~ **de temperatura de la vaina** *m* NUCL cladding temperature limit; ~ **de tolerancia** *m* CALIDAD, MECÁ, METR limit of tolerance, tolerance limit; ~ **de velocidad** *m* CONST speed limit, FERRO speed restriction, TRANSP speed limit

límites *m pl* CARBÓN *concesión minera*, CONST fines, GAS limits, MINAS fines; ~ **de temperatura en superficie** *m pl* SEG surface temperature limits

límnico *adj* GEOL limnic

limnímetro *m* HIDROL limnimeter, METR, OCEAN depth gage (*AmE*), depth gauge (*BrE*); ~ **de vernier** *m* METR vernier depth gage (*AmE*), vernier depth gauge (*BrE*)

limo *m* CARBÓN silt, CONST loam, GEOL loam, silt, OCEAN silt; ~ **fino** *m* GEOL fine silt; ~ **grueso** *m* GEOL coarse silt; ~ **en suspensión en el agua turbia** *m* MINAS floating dredge

limolita *f* GEOL siltstone

limonita *f* MINAS, MINERAL limonite, meadow ore

limpia: ~ **de cloacas** *f* AGUA, HIDROL, RECICL sewer cleaning

limpiado: ~ **con una corriente de gas** *adj* EMB gas-flushed

limpiador *m* AGUA cleaner, DETERG cleaning agent, PETROL pig, PROC QUÍ, QUÍMICA, SEG cleaner, TEC PETR *perforación* roustabout; ~ **de alfombras** *m* DETERG carpet cleaner; ~ **centrífugo** *m* PROC QUÍ centrifugal cleaner; ~ **de gas** *m* GAS gas cleaner, TEC PETR *producción* skimmer; ~ **industrial** *m* DETERG industrial cleaning material; ~ **sónico** *m* CINEMAT sonic cleaner; ~ **de superficie modular** *m* EMB modular surface cleaner; ~ **de tanques** *m* TEC ESP *petroleros* scrubber; ~ **de vidrio** *m* C&V, LAB glass washer

limpiadora *f* PROD *herramienta de moldeo* cleaner; ~ **de calles** *f* TRANSP street cleaner

limpiaparabrisas *m* VEH windscreen wiper (*BrE*), windshield wiper (*AmE*)

limpiar[1] *vt* CONST clear out, MINAS *chapa de oro* clean up, PROD *tubos de caldera, parrilla* clean out, clean, RECICL clean up; ~ **con chorro de arena** *vt* CONST, MECÁ sandblast; ~ **con chorro de granalla** *vt* PROD shot; ~ **por descarga de agua** *vt* AGUA, PROD *metales* scour; ~ **en seco** *vt* TEXTIL dry-clean; ~ **totalmente** *vt* CONST clean up; ~ **con vapor** *vt* TEXTIL steam

limpiar[2]: ~ **fondos** *vi* TRANSP MAR *mantenimiento de buques* grave, overhaul; ~ **los fondos con escobón** *vi* TRANSP MAR hog

limpiatubos *m* PROD tube-scraper, TEC PETR swab

limpidez *f* AGUA, HIDROL clearness

limpieza *f* AGUA cleaning-up, scouring, C&V swabbing, CONTAM scrubbing, CONTAM MAR clean-up, OCEAN scouring, PROD cleaning, clearing-away, RECICL cleaning-up; ~ **alcalina** *f* DETERG alkaline cleaning; ~ **de alcantarillas** *f* AGUA, HIDROL, RECICL sewer cleaning; ~ **ambiental** *f* SEG environmental cleanliness; ~ **ambiental en espacios cerrados** *f* SEG environmental cleanliness in enclosed spaces; ~ **con arena** *f* P&C sanding; ~ **atmosférica** *f* CONTAM *tratamiento de gases* atmospheric scrubbing; ~ **con chorro de arena** *f* CONTAM MAR, MECÁ, PROD, TRANSP MAR sandblasting; ~ **con chorro de perdigones** *f* CONST shot blasting; ~ **de cloacas** *f* AGUA sewer cleaning; ~ **de la costa desde la línea de pleamar** *f* CONTAM MAR shoreline clean-up; ~ **y desbroce** *f* CONST clearing and grubbing; ~ **con descarga de agua** *f* OCEAN scouring; ~ **por descarga de agua** *f* MECÁ, MINAS, PROD flushing; ~ **con descarga de agua a baja presión** *f* CONTAM MAR low-pressure flushing; ~ **con descarga a baja presión** *f* CONTAM MAR low-pressure flushing; ~ **de fachada** *f* CONST charring; ~ **del fondo** *f* PROD *hornos* rabbling; ~ **de gas** *f* GAS gas cleaning; ~ **de los gases de escape** *f* GAS, TERMO exhaust gas cleaning; ~ **del grisú acumulado por corriente de agua** *f* MINAS flushing; ~ **del litoral** *f* CONTAM MAR, OCEAN shoreline clean-up; ~ **por llama** *f* CONST flame cleaning; ~ **de metales** *f* CONST pickling; ~ **de la parrilla** *f* PROD *calderas* pricking; ~ **química del carbón** *f* CARBÓN, CONTAM, QUÍMICA chemical coal cleaning; ~ **de restricciones** *f* TEC PETR *tuberías* debottlenecking; ~ **de los rodillos** *f* IMPR, PAPEL washup; ~ **en seco** *f* DETERG dry-cleaning; ~ **en tambor giratorio** *f* PROD rattling; ~ **ultrasónica** *f* CINEMAT, TV ultrasonic cleaning; ~ **con vapor** *f* CONTAM MAR steam cleaning

limpión *m* ING MECÁ wipe
linalol *m* QUÍMICA linalool
linarita *f* MINERAL linarite
lindano *m* QUÍMICA lindane
lindero *m* CONST boundary
línea[1]: **en** ~ *adj* INFORM&PD on-line, ING MECÁ in-line, TELECOM on-line
línea[2] *f* GEN line, ELECTRÓN *en pantalla* trace, *de capa protectora* line, TELECOM route, routing, trunk, line;
▪ **a** ~ **abierta** *f* ING ELÉC open-wire line; ~ **de abonado** *f* TELECOM subscriber's line; ~ **de acceso** *f* INFORM&PD access line; ~ **de acción** *f* ING MECÁ line of action; ~ **aclínica** *f* GEOFÍS aclinic line; ~ **activa** *f* ELEC live line, ELECTRÓN, TV active line; ~ **adecuada** *f* TRANSP desire line; ~ **aérea** *f* CONST, ELEC, ING ELÉC, TELECOM overhead line, TRANSP, TRANSP AÉR airline; ~ **aérea regional** *f* TRANSP, TRANSP AÉR regional carrier; ~ **aérea de la sección** *f* TELECOM section overhead (*SOH*); ~ **aérea de la sección del regenerador** *f* TELECOM regenerator-section overhead; ~ **aérea secundaria** *f* TRANSP AÉR feeder airline; ~ **ágata** *f* IMPR *publicidad* agate line; ~ **agónica** *f* GEOFÍS agonic line; ~ **de agua** *f* CONTAM watercourse; ~ **de agua alta** *f* OCEAN high water mark; ~ **de aguante** *f* AmL (*cf cable de soporte Esp, cable de sujeción Esp*) TEC PETR *perforación* backup line; ~ **de aire comprimido** *f* CONST, ING MECÁ compressed-air line; ~ **con aislación de gas** *f* ELEC gas-insulated line; ~ **de alimentación** *f* ELEC power lead, ING ELÉC feeder, NUCL feedline, TEC ESP

comunicaciones feeder; ~ **de alimentación cruzada** *f* TRANSP AÉR crossfeed line; ~ **de alimentación resistente a los rayos** *f* ELEC lightning-resistant power line; ~ **alquilada** *f* INFORM&PD, TELECOM leased line; ~ **de alta frecuencia** *f* FÍS high-frequency line; ~ **alta habitual de la marea viva** *f* ENERG RENOV high-water ordinary spring tide (*HWOST*); ~ **de alta tensión** *f* PROD power line; ~ **de alta velocidad** *f* FERRO high-speed line; ~ **de alto voltaje** *f* ELEC, ING ELÉC power line; ~ **de alto voltaje de CA** *f* ELEC, ING ELÉC AC power line; ~ **de altura piezométrica hidráulica** *f* HIDROL hydraulic grade line; ~ **antinodal** *f* ACÚST antinodal line; ~ **de los ápsides** *f* TEC ESP line of apsides; ~ **de arranque** *f* CONST springing line; ~ **de arrufo** *f* TRANSP MAR *construcción naval* sheer line; ~ **artificial calibrada** *f* ING ELÉC slotted line; ~ **artificial mínima sin errores** *f* TELECOM minimum error-free pad (*MEFP*); ~ **atmosférica** *f* FÍS atmospheric line; ~ **auxiliar** *f* TELECOM *enlace privado* trunk; ~ **axial** *f* GEOM axial line;

▪ **b** ~ **de bajamar** *f* HIDROL, TRANSP MAR low-water mark; ~ **de banderines** *f* PETROL pennant line; ~ **de base** *f* CONST *agrimensura* base, IMPR, INSTR baseline, TEC ESP, TRANSP AÉR lubber line, TRANSP MAR lubber line, *diseño naval, navegación* baseline; ~ **base de lutita** *f* TEC PETR shale base line; ~ **bidireccional** *f* TELECOM both-way line; ~ **bidireccional ordinaria** *f* TELECOM ordinary both-way line; ~ **bifilar** *f* ELEC, ING ELÉC parallel-wire line; ~ **bipolar** *f* ELEC, ING ELÉC bipolar line; ~ **blanca discontinua** *f* CONST *carreteras* broken white line; ~ **de buzamiento** *f* GEOL line of dip;

▪ **c** ~ **de CA** *f* ING ELÉC AC line; ~ **del cabezal** *f* TEXTIL head line; ~ **de caída de tensión** *f* PROD drop line wiring; ~ **de cambio de fecha** *f* TRANSP MAR date line; ~ **de campo** *f* FÍS field line; ~ **de cangilones** *f* CONST *elevador, cinta transportadora* line of buckets; ~ **de caracteres** *f* IMPR character string; ~ **de carga** *f* TRANSP AÉR, TRANSP MAR load line; ~ **de carga de CA** *f* ING ELÉC AC load; ~ **de carga de verano** *f* TRANSP MAR *diseño naval* summer load waterline; ~ **central** *f* ING MECÁ, MECÁ, PROD, TEC ESP center line (*AmE*), centre line (*BrE*); ~ **de la central** *f* TELECOM exchange line; ~ **central de la pista** *f* TRANSP AÉR runway centerline (*AmE*), runway centreline (*BrE*); ~ **central de pista extendida** *f* TRANSP AÉR extended runway centerline (*AmE*), extended runway centreline (*BrE*); ~ **de centros** *f* ING MECÁ center line (*AmE*), centre line (*BrE*), line of centers (*AmE*), line of centres (*BrE*), MECÁ, PROD, TEC ESP center line (*AmE*), centre line (*BrE*); ~ **del chaflán** *f* TV bearding; ~ **de cinta** *f* ING ELÉC strip line; ~ **de clasificación** *f* FERRO sorting line; ~ **coaxial** *f* FÍS coaxial-line, ING ELÉC, TELECOM coaxial line; ~ **coaxial rígida** *f* FÍS, ING ELÉC, TELECOM rigid coaxial line; ~ **cocorriente** *f* OCEAN co-current line; ~ **colectiva** *f* ELEC bus line; ~ **de combustible** *f* AUTO, VEH fuel line; ~ **de comentarios** *f* PROD comment line; ~ **de comparación** *f* ING MECÁ, MECÁ, TEC ESP *cartas de navegación* datum line; ~ **de compensación del presionador** *f* NUCL pressurizer surge-line; ~ **compensada** *f* ELEC, ING ELÉC balanced-line; ~ **de comunicaciones** *f* ING ELÉC communications line; ~ **de conducción** *f* AUTO, VEH drive line; ~ **de** conducta *f* ING MECÁ, PROD guideline; ~ **de conductores paralelos** *f* ING ELÉC parallel-wire line; ~ **de conexión** *f* METAL tie line; ~ **de conmutación** *f* ELEC axis of commutation; ~ **de contacto** *f* IMPR, ING MECÁ line of contact, METAL tie line; ~ **de contacto ampliada** *f* PAPEL extended nip; ~ **de contacto blanda** *f* PAPEL soft nip; ~ **continua** *f* CONST *carreteras* continuous white line; ~ **de contorno** *f* GEOL subsurface contour; ~ **de control de fuego** *f* AGRIC fire-control line; ~ **conveniente** *f* TRANSP desire line; ~ **coronal de emisión** *f* FÍS RAD coronal emission line; ~ **de corriente** *f* HIDROL *hidrodinámica* stream line; ~ **corta** *f* IMPR catch line; ~ **corta de texto** *f* IMPR catch line; ~ **cortada** *f* GEOM broken line; ~ **de corte** *f* GEOL line of section, IMPR tear line, trim mark; ~ **de costa** *f* GEOL, HIDROL coastline, OCEAN, TRANSP MAR *hidrografía* coastline, shoreline; ~ **de costura** *f* C&V, TEXTIL seam line; ~ **de cota** *f* ING MECÁ dimension line; ~ **de crecida** *f* HIDROL high-water line; ~ **de cresta** *f* OCEAN crest line; ~ **del cuadro** *f* CINEMAT, TV frame line; ~ **de cuarto de onda** *f* FÍS quarter wave line; ~ **de cubierta** *f* TRANSP MAR deck line; ~ **de cuerda** *f* ENERG RENOV chord line; ~ **de curso oblicuo** *f* TRANSP AÉR slant course line; ~ **curva** *f* GEOM curved line;

■ **d** ~ **D del sodio** *f* FÍS sodium D-line; ~ **de datos** *f* TEC ESP *comunicaciones* datum line; ~ **dedicada** *f* INFORM&PD, TELECOM dedicated line; ~ **defectuosa** *f* ELEC, ING ELÉC faulty line; ~ **delgada** *f* IMPR fine line; ~ **de demora** *f* ELECTRÓN, FÍS, INFORM&PD, TV delay line; ~ **derivada** *f* ELEC branch line; ~ **de descarga del gas protector** *f* GAS, NUCL cover-gas discharge line; ~ **de descarga de gases** *f* GAS, NUCL gas vent; ~ **descendente** *f* FERRO down line; ~ **desequilibrada** *f* ELEC, ING ELÉC unbalanced line; ~ **de deslizamiento** *f* CRISTAL slip line; ~ **de desplazamiento** *f* TRANSP pipeline transportation; ~ **de desviación** *f* FERRO avoiding line; ~ **de desvío** *f* TRANSP bypass line; ~ **de dirección** *f* GEOL, MINAS *filones* line of strike; ~ **direccional** *f* ING MECÁ directing line; ~ **directa** *f* TELECOM throughline, direct line; ~ **de dislocación** *f* CRISTAL dislocation line; ~ **de distribución** *f* NUCL distributing pipe; ~ **divisoria** *f* CONST boundary line;

■ **e** ~ **ecuatorial** *f* TRANSP MAR line; ~ **de ejes** *f* ING MECÁ line of shafting; ~ **eléctrica** *f* CONST, ELEC overhead line; ~ **eléctrica resistente a los rayos** *f* ELEC lightning-resistant power line; ~ **eléctrica de retardo** *f* ELEC, ELECTRÓN electric delay line; ~ **electrizada** *f* ELEC live line; ~ **elevada** *f* TRANSP elevated line; ~ **de emisión** *f* FÍS emission line; ~ **de energía de CA** *f* ELEC, ING ELÉC AC power line; ~ **de energía eléctrica** *f* ELEC, ING ELÉC, TERMO electric power line; ~ **de enlace** *f* TELECOM tie line, trunk; ~ **para enroscar tubos** *f* TEC PETR *perforación* spinning line; ~ **entrante** *f* TELECOM incoming line; ~ **equilibrada** *f* ELEC, ING ELÉC balanced line; ~ **equipotencial** *f* ELEC, FÍS, ING ELÉC, TEC ESP equipotential line; ~ **E/S de CC de bajo nivel** *f* PROD low-level DC I/O line; ~ **de esmaltado** *f* REVEST enameling line *(AmE)*, enamelling line *(BrE)*; ~ **especializada** *f* INFORM&PD dedicated line; ~ **espectral** *f* FÍS, FÍS RAD, ÓPT, TELECOM spectral line; ~ **de espuma** *f* C&V foam line; ~ **de la espuma de sílice** *f* C&V silica scum line; ~ **de estación de**

clasificación *f* FERRO classification yardline; ~ **estacionada** *f* TELECOM parked line; ~ **de exploración** *f* ELECTRÓN scanning line, TEC ESP scan line, TV scanning line; ~ **de extensión** *f* ING MECÁ extension line;

■ **f** ~ **de fabricación** *f* TRANSP MAR *construcción naval* production line; ~ **de falla** *f* GEOL fault trace; ~ **de fase líquida** *f* METAL liquidus line; ~ **de fe** *f* TEC ESP, TRANSP AÉR, TRANSP MAR lubber line; ~ **de fe de la alidada** *f* TRANSP MAR *sextante* index bar; ~ **férrea de gravedad** *f* CONST gravity railroad; ~ **de fibra única** *f* ING ELÉC single-fiber line *(AmE)*, single-fibre line *(BrE)*; ~ **de flotación** *f* CONST, CONTAM MAR, TRANSP, TRANSP MAR water line; ~ **de flotación de construcción** *f* CONST, TRANSP, TRANSP MAR construction water line; ~ **de flujo** *f* INFORM&PD, PETROL, TEC PETR, TELECOM flow line; ~ **de flujo calorífico** *f* TERMO heat flow line; ~ **de flujo eléctrico** *f* ELEC, ING ELÉC line of flux; ~ **de flujo isodinámico** *f* GEOFÍS isodynamic flux line; ~ **focal sagital** *f* FÍS sagittal focal line; ~ **focal tangencial** *f* FÍS tangential focal line; ~ **de fondo de un valle** *f* GEOL talweg, thalweg; ~ **de fosos** *f* AGUA ditch line; ~ **de Fraunhofer** *f* FÍS Fraunhofer line; ~ **de fuerza** *f* ELEC *magnetismo, electrostática*, FÍS line of force, GEOFÍS, ING MECÁ force line; ~ **de fuerza de flujo** *f* ING ELÉC flux line; ~ **de fuerza magnética** *f* ING ELÉC magnetic line of force; ~ **de fuga** *f* ELEC leakage path, GEOM vanishing line, ING ELÉC leakage path; ~ **con fugas** *f* FÍS, ING ELÉC lossy line; ~ **de fundición** *f* C&V batch meltout line; ~ **de fusión de carga** *f* C&V batch melting line;

■ **g** ~ **generatriz** *f* ING MECÁ line of action; ~ **geodésica** *f* GEOM geodesic line; ~ **a gran distancia** *f* ING ELÉC long-distance line; ~ **de gran velocidad** *f* FERRO high-speed line; ~ **guía** *f* ING MECÁ leader line, PETROL guide line; ~ **guía de entrada** *f* TRANSP AÉR hold-short line *(AmE)*, lead-in line *(BrE)*; ~ **guía de salida** *f* TRANSP AÉR lead-out line *(BrE)*, hold-short line *(AmE)*;

■ **h** ~ **hiperconductora** *f* TELECOM superconductor line;

■ **i** ~ **de impresión** *f* IMPR line of impression; ~ **de inclinación** *f* GEOL line of dip; ~ **incompleta** *f* IMPR break line; ~ **inductora** *f* ING ELÉC field line; ~ **de influencia** *f* CONST influence line; ~ **de interconexión** *f* ELEC interconnecting line; ~ **interdigital** *f* FÍS interdigital line; ~ **internacional de cambio de fecha** *f* TRANSP MAR international date line; ~ **isobara** *f* TERMO isobaric line; ~ **isobárica** *f* OCEAN equal pressure; ~ **isóclina** *f* FÍS, GEOFÍS isoclinic line; ~ **isoclinal** *f* FÍS, GEOFÍS isoclinal line; ~ **isócora** *f* TERMO isochor, isochore; ~ **isodinámica** *f* GEOFÍS isodynamic line; ~ **isogala** *f* GEOL isogal; ~ **isógona** *f* GEOFÍS isogonic line; ~ **isograda** *f* GEOL *metamórfica* isograd; ~ **isosísmica** *f* GEOL isoseismic line; ~ **isoterma** *f* TERMO isothermal, TRANSP AÉR isotherm;

■ **k** ~ **Kikuchi** *f* NUCL Kikuchi line; ~ **Kossel** *f* NUCL Kossel line;

■ **l** ~ **laminar** *f* FÍS strip line; ~ **larga** *f* OCEAN long line; ~ **de larga distancia** *f* TELECOM long-distance line; ~ **libre** *f* TELECOM spare line; ~ **límite** *f* CRISTAL layer line, ING MECÁ clearance line; ~ **de liquidus** *f* METAL liquidus line; ~ **de litoral** *f* CONTAM MAR, TRANSP MAR shoreline; ~ **de llamada** *f* TELECOM calling line; ~ **de llamadas de salida bloqueadas** *f*

TELECOM outgoing-calls-barred line; ~ **de llaves** *f* TEC PETR *perforación* tong line; ~ **longitudinal** *f* FÍS pump line;

~ m ~ **de la marea alta** *f* HIDROL tidemark; ~ **de máxima carga** *f* TRANSP MAR plimsoll line; ~ **de máxima pleamar** *f* OCEAN *ordenación de costas* seamark; ~ **meridiana magnética** *f* INSTR compass meridian line; ~ **de mira** *f Esp* (*cf niveleta AmL*) MINAS lining sight; ~ **monofibra** *f* ELEC, ING ELÉC single-fiber line (*AmE*), single-fibre line (*BrE*); ~ **monopolar** *f* ELEC, ING ELÉC monopolar line; ~ **de montaje** *f* EMB, ING MECÁ, MECÁ, PROD, TRANSP MAR assembly line; ~ **de montaje en serie** *f* ING MECÁ, PROD production line; ~ **móvil** *f* CONST walking line;

~ n ~ **neumática** *f* ING MECÁ airline; ~ **de nivel** *f* C&V flux line; ~ **de nivel mínimo** *f* HIDROL *río*, OCEAN low-water mark; ~ **de nivelación** *f* CONST line of level; ~ **no presurizada** *f* NUCL unpressurized line; ~ **nodal** *f* ACÚST nodal line; ~ **de nodos** *f* TEC ESP line of nodes; ~ **de nódulos** *f* TEC ESP line of nodes;

~ o ~ **de onda media** *f* FÍS half wave line; ~ **de órdenes de ingeniería** *f* TELECOM engineering order wire (*EOW*); ~ **ordinaria** *f* TELECOM ordinary line; ~ **de orientación** *f* GEOL line of bearing;

~ p ~ **pendiente** *f* GEOL talweg, thalweg; ~ **con pérdidas** *f* FÍS, ING ELÉC lossy line; ~ **de perforación** *f Esp* (*cf guaya AmL*) PETROL, TEC PETR drilling line; ~ **de pleamar** *f* CONTAM MAR shoreline, TRANSP MAR shoreline, *navegación* seamark; ~ **de precisión** *f* ELECTRÓN *circuito integrado monolítico* fine line; ~ **de presión principal** *f* ING MECÁ pressure main line; ~ **principal** *f* ELEC trunk line, FERRO main line, trunk line, TELECOM main line; ~ **de principio de solidificación** *f* METAL liquidus line; ~ **privada** *f* TELECOM private line (*PL*); ~ **de proa** *f* TRANSP MAR heading; ~ **proa-popa** *f* TRANSP MAR fore-and-aft line; ~ **de producción** *f* EMB, ING MECÁ, MECÁ, PROD, TRANSP MAR assembly line; ~ **punto a punto** *f* INFORM&PD point-to-point line; ~ **de puntos** *f* IMPR bullet;

~ r ~ **ranurada** *f* ING ELÉC slotted line; ~ **real** *f* ING ELÉC side circuit; ~ **recta** *f* FÍS, GEOM straight line; ~ **de referencia** *f* CONST datum line, GEOM reference line, ING MECÁ, MECÁ datum line, TEC ESP *navegación* datum line, lubber line, TRANSP AÉR, TRANSP MAR lubber line; ~ **de referencia del fuselaje** *f* TRANSP AÉR fuselage datum line; ~ **de regreso del vapor** *f* AUTO vapor return line (*AmE*), vapour return line (*BrE*); ~ **de reserva** *f* TELECOM spare line; ~ **sin resonancia** *f* ELECTRÓN non-resonant line; ~ **resonante** *f* ELECTRÓN resonant line; ~ **restringida de llamadas entrantes** *f* TELECOM incoming calls barred line; ~ **de retardo** *f* ELECTRÓN, FÍS, INFORM&PD, TV delay line; ~ **de retardo acústico** *f* ACÚST, ELECTRÓN, INFORM&PD acoustic delay line; ~ **de retardo analógico** *f* TV analog delay line; ~ **de retardo binario** *f* TV binary delay line; ~ **de retardo cromático** *f AmL* (*cf retardo cromático Esp*) TV chroma delay; ~ **de retardo cuantificada** *f* TV quantized delay line; ~ **de retardo de cuarzo** *f* ELECTRÓN, TV quartz delay line; ~ **de retardo por derivación** *f* ELECTRÓN tapped delay line; ~ **de retardo dispersora** *f* ELECTRÓN dispersive delay line; ~ **de retardo de mercurio** *f* ELECTRÓN mercury delay line; ~ **de retardo de microondas** *f* ELECTRÓN

microwave delay line; ~ **de retardo de OAS** *f* ELECTRÓN SAW delay line; ~ **de retardo poligonal** *f* ELECTRÓN polygonal delay line; ~ **de retardo de semiconductor de óxido de metal** *f* ELECTRÓN metal-oxide semiconductor delay line; ~ **de retardo SOM** *f* ELECTRÓN MOS delay line; ~ **de retroceso del flujo** *f* INSTAL HIDRÁUL counterstream line; ~ **de rumbo** *f* TRANSP AÉR course line; ~ **de rumbos electrónica** *f* TRANSP MAR *radar* electronic bearing line;

~ s ~ **de salida** *f* TV outgoing line; ~ **saliente** *f* TELECOM outgoing line; ~ **de salmuera** *f* REFRIG brine line; ~ **seca** *f* PAPEL dry line; ~ **secundaria** *f* FERRO branch line, TRANSP secondary line; ~ **de señal** *f* ING ELÉC signal line; ~ **de separación de dos imágenes** *f* CINEMAT, TV frame line; ~ **de separación entre capas** *f* CARBÓN seam; ~ **de serie** *f* ING ELÉC serial line; ~ **de servicio** *f* ING ELÉC, TELECOM service line; ~ **de servicio compartido** *f* ING ELÉC, TELECOM shared service line; ~ **de sincronización** *f* PETROL timing line; ~ **de sonda** *f* OCEAN sounding line; ~ **de soporte** *f AmL* (*cf cable de soporte Esp, cable de sujeción Esp*) TEC PETR *perforación* backup line; ~ **de subida** *f* ELEC rising main; ~ **subterránea** *f* ING ELÉC, TELECOM underground line; ~ **de sujeción corta** *f* TRANSP AÉR hold-short line (*AmE*), lead-in line (*BrE*); ~ **de suministro** *f* ING ELÉC service line; ~ **de suministro cruzado** *f* TRANSP AÉR crossfeed line; ~ **superconductora** *f* TELECOM superconductor line;

~ t ~ **de tangencia** *f* PAPEL nip; ~ **de tangencia ampliada** *f* PAPEL extended nip; ~ **de tangencia de dos cilindros** *f* IMPR nip; ~ **telefónica** *f* ING ELÉC, TELECOM telephone line; ~ **telegráfica** *f* ING ELÉC telegraph line; ~ **de temporización constante** *f* ELECTRÓN constant delay line; ~ **terminada con una impedancia** *f* FÍS line terminated by an impedance; ~ **terrestre** *f* TELECOM land line; ~ **de tiempo isócrona** *f* GEOL isochronal time line; ~ **a tierra** *f* ELEC, ING ELÉC earth line (*BrE*), ground line (*AmE*); ~ **de tierra** *f* CONST baseline, TEC ESP *dibujo* datum line; ~ **de tiro** *f* MINAS *pega* shot-firing circuit, *voladuras* leading wire; ~ **de la trama** *f* IMPR screen ruling; ~ **de transferencia** *f* ING MECÁ transfer line; ~ **de transmisión** *f* CONST, ELEC, ELECTRÓN, FÍS transmission line, ING MECÁ line of shafting, TELECOM transmission line; ~ **de transmisión abierta** *f* ELEC, ING ELÉC open-wire transmission line; ~ **de transmisión acústica** *f* ELEC, ING ELÉC acoustic transmission line; ~ **de transmisión de alta tensión** *f* ELEC, ING ELÉC high-voltage transmission line; ~ **de transmisión armada** *f* ELEC, ING ELÉC shielded transmission line; ~ **de transmisión de corriente alterna** *f* ELEC, ING ELÉC alternating-current transmission line; ~ **de transmisión eléctrica** *f* ELEC, ING ELÉC electrical transmission line; ~ **de transmisión de gas** *f* GAS gas transmission line; ~ **de transmisión de media onda** *f* ELEC half-wave transmission line; ~ **de transmisión de microondas** *f* ING ELÉC microwave transmission line; ~ **de transmisión óptica** *f* ING ELÉC optical transmission line; ~ **de transmisión de placas paralelas** *f* D&A *radar* pillbox; ~ **de transporte de energía** *f* ING ELÉC power transmission line, transmission line, PROD power line, power wiring; ~ **troncal** *f* PETROL trunk line, TELECOM trunk; ~ **troncal principal** *f* TELECOM trunk;

~ **troquelada** *f* IMPR perforated line; ~ **de tubería** *f* CONST casing line;

▪ **u** ~ **uniforme** *f* ING ELÉC uniform line; ~ **de unión** *f* METAL tie line, TELECOM trunk;

▪ **v** ~ **de valle** *f* GEOL talweg, thalweg; ~ **de vía única** *f* FERRO single-track line; ~ **de visión** *f* TELECOM line of sight; ~ **visual** *f* MINAS lining sight

lineación *m* GEOL lineation

lineal *adj* FÍS linear; **no** ~ *adj* ELEC nonlinear

lineales *f pl* IMPR lineals

linealidad *f* ELEC, ELECTRÓN, TELECOM linearity

linear *vt* QUÍMICA *molécula*, TEC ESP linearize

linearizador *m* QUÍMICA, TEC ESP linearizer

linearizar *vt* QUÍMICA, TEC ESP linearize

líneas: ~ **de agua** *f pl* TRANSP MAR water lines; ~ **de Balmer** *f pl* FÍS PART Balmer series lines; ~ **de campo magnético** *f pl* FÍS RAD magnetic-field lines; ~ **concordantes** *f pl* GEOM concurrent lines; ~ **convergentes** *f pl* GEOM convergent lines; ~ **diagonales** *f pl* TRANSP MAR *arquitectura naval* diagonal lines; ~ **de fuerza** *f pl* GEOFÍS lines of force; ~ **de fuerza del campo magnético** *f pl* FÍS RAD magnetic-field lines; ~ **intersectantes** *f pl* GEOM intersecting lines; ~ **por minuto** *f pl* (*LPM*) IMPR, INFORM&PD lines per minute (*LPM*); ~ **paralelas** *f pl* GEOM parallel lines; ~ **de Paschen** *f pl* FÍS RAD Paschen series lines; ~ **perpendiculares** *f pl* GEOM perpendicular lines; ~ **que se cortan** *f pl* GEOM intersecting lines

liner *m* TEC PETR liner; ~ **liso** *m* TEC PETR blank liner; ~ **sin perforaciones** *m* TEC PETR blank liner

lingoide *adj* GEOL lingoid

lingote *m* CARBÓN ingot, IMPR clump, MECÁ ingot, MINAS *plata* bar, PROD ingot, *fundería, fundición* pig; ~ **de acero fundido** *m* PROD blank; ~ **de afino** *m* PROD conversion pig; ~ **de hierro** *m* CARBÓN iron, PROD pig of iron; ~ **de reducción directa** *m* CARBÓN ingot; ~ **de uranio** *m* NUCL uranium ingot

lingotera *f* ING MECÁ casting mold (*AmE*), casting mould (*BrE*), MECÁ, PROD ingot mold (*AmE*), ingot mould (*BrE*)

linguete *m* ING MECÁ click, finger, keeper, pawl; ~ **de retención** *m* ING MECÁ finger bar

lingüiforme *adj* GEOL linguiform

linneita *f* MINERAL linnaeite

lino: ~ **para cordel** *m* TEXTIL line flax

linoleato *m* QUÍMICA linoleate

linoleíco *adj* QUÍMICA linoleic

linoleína *f* QUÍMICA linoleine

linolenato *m* QUÍMICA linolenate

linolénico *adj* QUÍMICA linolenic

linotipia® *f* IMPR linotype®

linterna *f* ELEC, ING ELÉC flashlight, lamp, portable lamp, PROD lantern; ~ **de bolsillo** *f* ELEC, ING ELÉC flashlight; ~ **eléctrica** *f* ELEC, ING ELÉC flashlight

liofilia *f* QUÍMICA lyophily

liofílico *adj* QUÍMICA lyophilic

liofilización *f* AGRIC, ALIMENT, EMB, PROC QUÍ, REFRIG, TERMO freeze-drying, lyophilization; ~ **acelerada** *f* AGRIC, ALIMENT, EMB, REFRIG, TERMO accelerated freeze drying (*AFD*); ~ **continua** *f* AGRIC, ALIMENT, EMB, REFRIG, TERMO continuous freeze drying

liofilizado[1] *adj* AGRIC, ALIMENT, EMB, PROC QUÍ, REFRIG, TERMO freeze-dried

liofilizado[2] *m* AGRIC, ALIMENT, EMB, PROC QUÍ, REFRIG lyophilizate

liofilizador *m* AGRIC, ALIMENT, EMB, PROC QUÍ, REFRIG, TERMO freeze-dryer; ~ **de tambor** *m* REFRIG drum freeze dryer

liofilizar *vt* AGRIC, ALIMENT, EMB, PROC QUÍ, REFRIG, TERMO freeze-dry

liófobo *adj* QUÍMICA lyophobic

liogel *m* QUÍMICA lyogel

liosol *m* QUÍMICA lyosol

lipasa *f* QUÍMICA lipase

lípido *m* QUÍMICA lipid

lipofílico *adj* QUÍMICA lipophilic

lipófilo *m* QUÍMICA lipophile

lipoide *adj* QUÍMICA lipoid

lipopolisacárido *m* QUÍMICA lipopolysaccharide

lipositol *m* QUÍMICA lipositol

liposoluble *adj* QUÍMICA liposoluble

líquido[1] *adj* FÍS, QUÍMICA, REFRIG, TERMO liquid

líquido[2] *m* FÍS, QUÍMICA, REFRIG, TERMO liquid; ~ **asociado** *m* TEC PETR associated liquid; ~ **de coagulación** *m* PROC QUÍ coagulation liquid; ~ **colorante** *m* TEXTIL dye liquor; ~ **congelado** *m* QUÍMICA frozen liquid; ~ **congelante** *m* QUÍMICA freezing liquid; ~ **criogénico** *m* ING MECÁ cryogenic liquid; ~ **denso** *m* CARBÓN dense liquid; ~ **emulgente** *m* PROC QUÍ, SEG emulsifying liquid; ~ **emulsionante** *m* PROC QUÍ, SEG emulsifying liquid; ~ **de extracción** *m* PROC QUÍ extraction liquid; ~ **de flotación** *m* PROC QUÍ flotation liquid; ~ **de frenos** *m* AUTO, VEH brake fluid; ~ **de fresado** *m* PROC QUÍ milling liquid; ~ **de gas natural** *m* (*LGN*) GAS, TEC PETR, TRANSP natural gas liquid (*NGL*); ~ **incongelable** *m* REFRIG non freeze liquid; ~ **inflamable** *m* SEG flammable liquid; ~ **limpiador** *m* INSTR cleaning fluid; ~ **para maquinaria hidráulica** *m* INSTAL HIDRÁUL hydraulic fluid; ~ **muy inflamable** *m* SEG highly-flammable liquid; ~ **refrigerante** *m* AUTO, ING MECÁ, VEH coolant; ~ **residual** *m* HIDROL black liquor; ~ **de separación** *m* PROC QUÍ separation liquid; ~ **tóxico** *m* SEG toxic liquid

liquidus *m* QUÍMICA liquidus

lira *f* ING MECÁ swing frame, tangent plate; ~ **de dilatación** *f* REFRIG expansion bend

liroconita *f* MINERAL liroconite

lisa *f* EMB, PAPEL breaker stack, calender

lisérgico *adj* QUÍMICA lysergic

lisina *f* QUÍMICA lysine

liso *adj* PROD even, QUÍMICA unlined, TEXTIL plain

lisofosfatida *f* QUÍMICA lysophosphatide

lisolecitina *f* QUÍMICA lysolecithin

lista *f* INFORM&PD list; ~ **de acceso** *f* INFORM&PD access list; ~ **de aserrado** *f* PROD sawing list; ~ **de control** *f* ING MECÁ checklist; ~ **de desplazamiento ascendente** *f* INFORM&PD push-up list; ~ **de desplazamiento descendente** *f* INFORM&PD push-down list; ~ **de dibujos** *f* PROD drawing list; ~ **disponible** *f* INFORM&PD available list; ~ **de distribución** *f* TELECOM distribution list (*DL*); ~ **encadenada** *f* INFORM&PD chained list; ~ **de errores** *f* INFORM&PD error list; ~ **libre** *f* INFORM&PD free list; ~ **lineal** *f* INFORM&PD linear list; ~ **de materiales** *f* PROD bill of material; ~ **de materiales de compra** *f* PROD indented bill of material; ~ **de materiales para la fabricación** *f* PROD manufacturing bill of material; ~ **de materiales múltiples** *f* PROD multilevel bill of material; ~ **de montaje** *f* PROD assembly list; ~ **de**

números de teléfonos *f* TELECOM telephone number list; **~ de operaciones** *f* PROD routing list; **~ de pedido** *f* PROD pick list, picking list; **~ de planos** *f* PROD drawing list; **~ preferencial** *f* TEC ESP preferential list; **~ preferente** *f* TEC ESP preferential list; **~ de recambios** *f* PROD renewal parts list; **~ de refacciones** *f* AmL (*cf lista de repuestos Esp*) PROD renewal parts list; **~ de repuestos** *f Esp* (*cf lista de refacciones AmL*) PROD renewal parts list; **~ de tomas** *f* CINEMAT, TV shot list; **~ de tomas para montaje** *f* CINEMAT, TV editing shot list; **~ de trabajos** *f* PROD job sequence list; **~ de tripulación** *f* TRANSP MAR crew list

listado *m* CARBÓN table, INFORM&PD listing; **~ de programa** *m* INFORM&PD program listing

listo *adj* INFORM&PD, TELECOM ready; **~ para activar** *adj* INFORM&PD, TELECOM ready-to-activate; **~ para reproducir** *adj* IMPR camera-ready; **~ para ser transmitido** *adj* INFORM&PD, TELECOM ready-for-sending; **~ a transmitir** *adj* INFORM&PD, TELECOM ready-to-send

listón *m* C&V ribbon, CONST *madera* batten, baulk (*BrE*), lath, balk (*AmE*), cleat, GEOL lath, ING MECÁ batten, cleat; **~ de carga** *m* MINAS loading stick; **~ de defensa** *m* ING MECÁ wear strip; **~ para pizarra** *m* CONST slate lath; **~ tapajuntas** *m* ING MECÁ cover strip

listonaje *m* PROD lagging

listones: **~ de madera** *m pl* PROD wooden dunnage

lisura *f* PAPEL smoothness, PROD flatness

litarge *m* P&C, QUÍMICA litharge

litargirio *m* P&C, QUÍMICA litharge

litera *f* TRANSP MAR berth, bunk; **~ de bastidor tubular** *f* TRANSP MAR *plegable* pipe cot; **~ del patrón** *f* TRANSP MAR quarter berth

literal *m* INFORM&PD literal; **~ de la cadena** *m* INFORM&PD string literal; **~ numérico** *m* INFORM&PD numeric literal

litergol *m* TEC ESP lithergol

lithionita *f* MINERAL lithionite

lítico *adj* GEOL lithic

litio *m* (*Li*) ING ELÉC, QUÍMICA lithium (*Li*)

litoclasto *m* GEOL lithoclast

litofacies *f* GEOL lithofacies

litógeno *adj* GEOL lithogenous

litografía *f* ELECTRÓN, IMPR lithography; **~ de contacto** *f* ELECTRÓN, IMPR contact lithography; **~ descubierta** *f* ELECTRÓN, IMPR maskless lithography; **~ enmascarada** *f* ELECTRÓN, IMPR masked lithography, REVEST masking lithography; **~ por flujo de electrones** *f* ELECTRÓN electron-flood lithography; **~ de haz de electrones** *f* ELECTRÓN electron-beam lithography; **~ por haz electrónico para exploración de la imagen** *f* ELECTRÓN raster-scan electron-beam lithography; **~ por haz electrónico de exploración vectorial** *f* ELECTRÓN vector-scan electron-beam lithography; **~ por haz electrónico explorador** *f* ELECTRÓN scanning electron-beam lithography; **~ por haz de iones** *f* ELECTRÓN ion-beam lithography; **~ óptica** *f* ELECTRÓN optical lithography; **~ de proximidad** *f* ELECTRÓN proximity lithography; **~ por proyección** *f* ELECTRÓN projection lithography; **~ por proyección de haz de electrones** *f* ELECTRÓN electron-beam projection lithography, projection electron-beam lithography; **~ por rayos X** *f* ELECTRÓN X-ray lithography

litografiar *vt* IMPR lithograph

litológico *adj* GEOL lithologic, lithological

litomarga *f* MINERAL lithomarge

litopón *m* QUÍMICA lithopone

litoral[1] *adj* GEOL nearshore; **con ~** *adj* GEOL, HIDROL, TRANSP MAR *geografía* coastal; **en ~ de costa** *adj* TEC PETR onshore

litoral[2] *m* GEOL coast, shore, coastline, HIDROL coastline, OCEAN coastline, seaboard, TRANSP MAR *construcción naval* shore, *geografía* seaboard, coastline

litosfera *f* CONTAM, ENERG RENOV, GEOFÍS, GEOL lithosphere

litro *m* (*l*) FÍS, METR liter (*AmE*) (*l*), litre (*BrE*) (*l*)

liviano *adj* ING MECÁ light, MECÁ lightweight

lixiviabilidad *f* GEN leachability

lixiviación *f* AGRIC, ALIMENT, C&V leaching, CARBÓN filtration, CONTAM leaching, lixiviation, GAS leaching, QUÍMICA leaching, lixiviation, TELECOM leaching; **~ directa** *f* GAS direct leaching; **~ indirecta** *f* GAS indirect leaching; **~ primaria** *f* CARBÓN coarse filtration

lixiviador *m* NUCL leachant

lixiviar *vt* GEN leach

lixónico *adj* QUÍMICA lyxonic

lixosa *f* QUÍMICA lyxose

llama *f* QUÍMICA, SEG, TERMO, TERMOTEC blaze, flame, TRANSP AÉR flare; **~ auxiliar** *f* ING MECÁ pilot flame; **~ azul** *f* QUÍMICA oxidizing flame; **~ desoxidante** *f* TERMO, TERMOTEC reducing flame; **~ de encendido** *f* GAS pilot light, ING MECÁ pilot flame; **~ neutra** *f* CONST *soldadura* neutral flame; **~ oxidante** *f* CONST *soldadura*, QUÍMICA oxidizing flame; **~ piloto** *f* ING MECÁ pilot flame; **~ reductora** *f* TERMO, TERMOTEC reducing flame; **~ de reverbero** *f* PROD *horno* reverberatory flame

llamada *f* CONST, ELECTRÓN, ING ELÉC ringing, TELECOM ringing, call; **~ abreviada** *f* TELECOM abbreviated dialing (*AmE*), abbreviated dialling (*BrE*); **~ de alarma** *f* TELECOM alarm call; **~ automática** *f* TELECOM automatic recall; **~ del capitán** *f* TRANSP AÉR captain call; **~ cobrable de transferencia** *f* TELECOM transfer charge call; **~ a cobro revertido** *f Esp* (*cf llamada pagadera en destino AmL*) TELECOM collect call (*AmE*), reverse-charge call (*BrE*); **~ en cola de espera** *f* TELECOM waiting call; **~ de despegue** *f* METEO, TRANSP AÉR report for take off; **~ directa a distancia** *f* TELECOM direct distance dialing (*AmE*) (*DDD*), direct distance dialling (*BrE*) (*DDD*); **~ económica** *f* TELECOM cheap call; **~ de emergencia** *f* TELECOM, TRANSP emergency call; **~ entrante** *f* TELECOM incoming call; **~ errónea** *f* TELECOM faulty call; **~ en espera** *f* TELECOM call hold, call queued, call waiting; **~ falsa** *f* TELECOM false call; **~ con gastos de transferencia** *f* TELECOM transfer charge call; **~ global** *f* TELECOM global call; **~ gratuita** *f Esp* (*cf comunicación interurbana no tasada AmL*) TELECOM freephone call (*BrE*), toll-free call (*AmE*); **~ sin hilos** *f* TELECOM radio paging; **~ indecente** *f* TELECOM obscene call; **~ inefectiva** *f* TELECOM ineffective call; **~ con interferencia** *f* TELECOM nuisance call; **~ local** *f* TELECOM local call; **~ maliciosa** *f* TELECOM malicious call; **~ múltiple** *f* TELECOM meet-me conference call; **~ nacional** *f* TELECOM inland call; **~ no contestada** *f* TELECOM unanswered call; **~ por**

el nombre *f* INFORM&PD call by name; ~ **de la nota al pie** *f* IMPR footnote call out; ~ **obscena** *f* TELECOM obscene call; ~ **pagadera en destino** *f AmL* (*cf llamada a cobro revertido Esp*) TELECOM collect call (*AmE*), reverse-charge call (*BrE*); ~ **recordatoria** *f* TELECOM reminder call; ~ **por referencia** *f* INFORM&PD call by reference; ~ **retenida** *f* TELECOM booked call, call held; ~ **saliente** *f* TELECOM outgoing call; ~ **selectiva** *f* INFORM&PD, TELECOM polling; ~ **selectiva digital** *f* TELECOM digital selective calling (*DSC*); ~ **simplificada** *f* TELECOM abbreviated dialing (*AmE*), abbreviated dialling (*BrE*); ~ **de socorro** *f* TRANSP AÉR distress call; ~ **de subrutina** *f* INFORM&PD subroutine call; ~ **al supervisor** *f* INFORM&PD supervisor call (*SVC*); ~ **de tarifa local** *f* TELECOM local charge rate call; ~ **con tarjeta de crédito** *f* TELECOM, TV credit-card call; ~ **tasable de transferencia** *f* TELECOM transfer charge call; ~ **telefónica** *f* TELECOM telephone call; ~ **de timbre** *f* TELECOM ringing; ~ **de tres direcciones** *f* TELECOM three-way call; ~ **tridireccional** *f* TELECOM three-way call; ~ **por valor** *f* INFORM&PD call-by value; ~ **a la voz** *f* TRANSP MAR hail

llamadas: ~ **entrantes prohibidas** *f pl* TELECOM incoming calls barred (*ICB*); ~ **entrantes restringidas** *f pl* TELECOM incoming calls barred (*ICB*); ~ **por radio** *f pl* TELECOM radio paging; ~ **de salida bloqueadas** *f pl* TELECOM outgoing calls barred (*OCB*)

llamador *m* TELECOM ringer

llamar[1]: ~ **a la voz** *vt* TRANSP MAR hail

llamar[2] *vi* TELECOM ring

llamarada *f* TERMO blaze, TRANSP AÉR flare; ~ **de fuego** *f* CONST blazing fire; ~ **solar** *f* TEC ESP solar flare

llamarse: ~ **al norte** *v refl* TRANSP MAR *viento* veer northward

llameante *adj* TERMO ablaze

llamear *vi* TERMO blaze up

llana *f* CONST brick trowel, hawk; ~ **de albañil** *f Esp* (*cf cuchara de albañil AmL*) CONST, ING MECÁ bricklayer's trowel, plastering trowel; ~ **para enlucir** *f Esp* (*cf cuchara de albañil AmL*) CONST, ING MECÁ bricklayer's trowel, plastering trowel

llano[1] *adj* CONST flat, GEOM planar, PROD even

llano[2]: ~ **abisal** *m* GEOFÍS abyssal plain

llanta *f* FERRO *banda de rodamiento* tread, ING MECÁ hoop, flange, VEH *banda de metal* rim, *neumático* tire (*AmE*), tyre (*BrE*); ~ **de centrado de la rueda** *f* VEH drop center rim (*AmE*), drop centre rim (*BrE*); ~ **rigidizadora frontal** *f* TRANSP MAR faceplate; ~ **de rueda** *f* VEH wheel rim

llantén: ~ **lanceolado** *m* AGRIC rib grass

llanura: ~ **abisal** *f* GEOFÍS, GEOL, OCEAN abyssal plain; ~ **aluvial** *f* AGRIC first bottom, AGUA flood plain, GEOL alluvial plain, HIDROL flood plain; ~ **costera** *f* GEOL coastal plain; ~ **deltaica** *f* GEOL delta plain; ~ **de inundación** *f* GEOL flood plain

llave *f* CONST *de arco* key, keystone, *grifo* spigot, ING ELÉC key, switch, ING MECÁ key, LAB *material de vidrio, servicios* cock, faucet (*AmE*), stopcock (*BrE*), MECÁ key; ~ **de aceite** *f* ING MECÁ oil cock; ~ **de acero forjada en caliente** *f* PROD drop-forged steel spanner; ~ **acodada** *f* ING MECÁ bent spanner, cranked spanner; ~ **acodada de anillo para trabajos**

pesados *f* ING MECÁ heavy duty offset ring wrench; ~ **de admisión del aire** *f* ING MECÁ air inlet cock; ~ **del agua** *f Esp* (*cf canilla del agua AmL*) LAB *servicios* water tap; ~ **ajustable** *f* ING MECÁ screw wrench, MECÁ adjustable spanner, adjustable wrench; ~ **ajustable de rolleta** *f* ING MECÁ Clyburn wrench; ~ **ajustable de tuercas** *f* ING MECÁ screw wrench; ~ **Allen** *f* AUTO, ING MECÁ, VEH Allen key (*BrE*), Allen wrench (*AmE*); ~ **de anillo** *f* AUTO, ING MECÁ, VEH ring spanner (*BrE*), ring wrench (*AmE*); ~ **de anillo de media luna** *f* ING MECÁ half-moon ring wrench; ~ **de apretar tuercas en forma de S** *f* ING MECÁ S-shaped spanner; ~ **aprietatubos** *f* ING MECÁ pipe wrench; ~ **de apriete prefijado** *f* ING MECÁ torque spanner, torque wrench; ~ **de apriete de trinquete** *f* ING MECÁ ratchet spanner, ratchet wrench; ~ **articulada de boca de tubo para tuercas** *f* ING MECÁ hinged socket wrench; ~ **Berne** *f* FERRO Berne key; ~ **de boca** *f* ING MECÁ, MECÁ open-end wrench; ~ **de boca de perfil bajo** *f* ING MECÁ, MECÁ low-profile open-end wrench; ~ **de boca única** *f* ING MECÁ, MECÁ single-ended spanner; ~ **de cabeza hexagonal** *f* MECÁ hex head wrench; ~ **de cadena para tubos** *f* ING MECÁ chain pipe wrench; ~ **de cambio** *f* CINEMAT clutch; ~ **de la canilla** *f* CONST faucet key; ~ **de caños** *f* ING MECÁ alligator wrench; ~ **de casquillo abierto angular** *f* ING MECÁ angled open socket wrench; ~ **de cazoleta** *f* ING MECÁ, MECÁ box spanner, box wrench; ~ **de cerradura** *f* CONST lock key; ~ **de choque** *f* ING MECÁ impact wrench; ~ **ciega** *f* MECÁ blank; ~ **de cierre** *f* CONST cutoff cock; ~ **de cierre del combustible** *f* TRANSP AÉR fuel shut-off cock; ~ **de cinta** *f* ING MECÁ strap wrench; ~ **de combustible** *f* TRANSP AÉR fuel cock; ~ **de comprobación** *f* TV mono key; ~ **de conexión** *f* CONST union cock; ~ **de contacto** *f* ING ELÉC keyswitch; ~ **de contacto antirrobo** *f* AUTO antitheft ignition lock; ~ **de la corriente** *f* ELEC, ING ELÉC current switch (*AmE*), mains switch (*BrE*); ~ **de cremallera** *f* ING MECÁ monkey wrench, screw wrench; ~ **cruzada de tuercas de rueda** *f* ING MECÁ wheel nut cross brace; ~ **de cubo** *f* MECÁ box wrench, socket wrench; ~ **de cubo angulado** *f* ING MECÁ angled socket wrench; ~ **de cuello de cisne** *f* ING MECÁ gooseneck wrench; ~ **curva de tubería** *f* CONST parrot-nose pipe wrench; ~ **curvada** *f* ING MECÁ curved spanner; ~ **de derivación** *f* ELEC tap switch; ~ **de desagüe** *f* CONST cylinder cock, drain cock, drip cock; ~ **de dientes** *f* ING MECÁ hook and pin wrench, hook spanner; ~ **dinamométrica** *f* ING MECÁ, VEH *herramienta* torque spanner, torque wrench; ~ **de dos pasos** *f* LAB two-way tap; ~ **del eje** *f* ING MECÁ shaft key; ~ **de encendido** *f* AUTO ignition key; ~ **de enclavamiento** *f* ING MECÁ lock-grip wrench; ~ **de enroscar** *f* CONST screw-down cock; ~ **española** *f* ING MECÁ, MECÁ open-end wrench; ~ **española de perfil bajo** *f* ING MECÁ, MECÁ low-profile open-end wrench; ~ **de espiga** *f* CONST pin wrench; ~ **con espigas al frente** *f* ING MECÁ face spanner; ~ **de la espita** *f* CONST faucet key; ~ **extractora** *f* ING MECÁ shifting spanner; ~ **de flotador** *f* CONST, ING MECÁ ball cock; ~ **de gancho** *f* VEH *herramienta* spanner, wrench; ~ **de gancho ajustable** *f* VEH *herramienta* adjustable spanner, adjustable wrench; ~ **de gancho con espiga** *f* CONST pin spanner; ~ **de gancho y**

pasador *f* ING MECÁ hook and pin wrench, hook spanner; ~ **giramachos** *f* ING MECÁ tap wrench; ~ **de grasa** *f* ING MECÁ oil cock; ~ **del grifo** *f* CONST faucet key; ~ **de gusano** *f* ING MECÁ adjustable spanner, adjustable wrench; ~ **hexagonal** *f* ING MECÁ hexagonal key; ~ **de horquilla** *f* CONST crutch key; ~ **de impacto** *f* ING MECÁ impact wrench; ~ **inglesa** *f* ING MECÁ, VEH *herramientas* adjustable spanner, adjustable wrench, monkey wrench, screw wrench, turnscrew; ~ **inglesa acodada** *f* ING MECÁ offset ring wrench; ~ **inglesa ajustable** *f* ING MECÁ shifting spanner; ~ **inglesa de cremallera** *f* ING MECÁ monkey wrench; ~ **inglesa curvada en S** *f* ING MECÁ S wrench; ~ **inglesa de horquilla** *f* ING MECÁ fork wrench, gap spanner, single-ended spanner; ~ **inglesa torsiométrica** *f Esp* (*cf llave de par AmL*) ING MECÁ torque spanner, torque wrench; ~ **inglesa para tuercas** *f* MECÁ wrench; ~ **con interruptor de botón** *f* CONST push-button faucet (*AmE*), push-button tap (*BrE*); ~ **con limitador de par** *f* ING MECÁ torque spanner, torque wrench; ~ **de machos de terraja** *f* ING MECÁ tap wrench; ~ **de madera** *f* PROD crib; ~ **maestra** *f* CONST, PROD master key; ~ **de mandíbulas** *f* ING MECÁ alligator wrench; ~ **de mando** *f* ELEC control switch; ~ **de maniobra** *f* (*cf palanca de maniobra*) MINAS *sondeos* brace head, brace key, tiller; ~ **en mano** *f* CONST turnkey; ~ **de maquinista** *f* ING MECÁ C wrench, open-end wrench; ~ **de martillo para tuercas** *f* ING MECÁ screw hammer; ~ **de moleta** *f* ING MECÁ adjustable spanner, MECÁ adjustable wrench; ~ **de muletilla** *f* ING MECÁ, MECÁ box spanner, box wrench; ~ **de muletilla angular** *f* ING MECÁ angle box wrench; ~ **normal-reversa** *f* TV normal reverse switch; ~ **de par** *f AmL* (*cf llave inglesa torsiométrica Esp*) ING MECÁ torque spanner, torque wrench; ~ **de pasador** *f* CONST pinned key; ~ **de paso** *f* CONST bib-cock, cock key, master key, stopcock, *fontanería* switch cock, *tuberías* plug cock, LAB *control de fluido* faucet (*AmE*), stopcock (*BrE*), tap (*BrE*), PROD master key, TRANSP MAR cock; ~ **de pitones** *f* ING MECÁ pin wrench; ~ **principal** *f* CONST *cerrajería* barrel key (*AmE*), pipe key (*BrE*); ~ **de prueba** *f* AGUA gage cock (*AmE*), gauge cock (*BrE*); ~ **de purga** *f* ING MECÁ, PROD drain cock; ~ **regulable** *f* VEH adjustable spanner, adjustable wrench; ~ **de retenida** *f* MINAS *sondeos* rod support, supporting fork; ~ **con seguro** *f* ING MECÁ safety cock; ~ **selectora de canales** *f* TELECOM, TV channel-selector switch; ~ **sencilla** *f* ING MECÁ single-ended spanner; ~ **de suspensión** *f* MINAS *sondeos* lifting dog; ~ **de tetones** *f* ING MECÁ hook and pin wrench, hook spanner, VEH spanner, wrench; ~ **torsiométrica** *f* ING MECÁ torque spanner, torque wrench; ~ **de torsión** *f* ING MECÁ torque spanner, torque wrench; ~ **de tres puntos** *f* ELEC three-point switch, three-way switch; ~ **de tres vías** *f* CONST three-way cock; ~ **de tubista** *f* ING MECÁ pipe wrench; ~ **de tubo** *f* ING MECÁ tube wrench, box spanner, box wrench, MECÁ box spanner; ~ **de tubo angular** *f* ING MECÁ angle box wrench; ~ **de tubo con palanca en T** *f* ING MECÁ sliding tee socket wrench; ~ **de tubo en T** *f* ING MECÁ tee socket wrench; ~ **para tubos** *f* ING MECÁ alligator wrench, grip pipe wrench, pipe wrench, PROD cylinder wrench; ~ **tubular** *f* CONST piped key; ~ **de tuerca** *f* ING MECÁ screw key; ~ **de tuerca de boca tubular** *f* ING MECÁ angled open socket wrench; ~ **de tuercas** *f* ING MECÁ, MECÁ, VEH monkey wrench, screw wrench, spanner, wrench; ~ **de tuercas ajustable** *f* MECÁ adjustable spanner, adjustable wrench; ~ **para tuercas circulares con agujeros** *f* ING MECÁ hook and pin wrench, hook spanner; ~ **de tuercas de media luna de tetones** *f* ING MECÁ face spanner; ~ **para tuercas redondas** *f* ING MECÁ pin wrench; ~ **para tuercas salientes de trinquete** *f* ING MECÁ ratchet flare-nut wrench; ~ **universal** *f* ING MECÁ monkey wrench; ~ **de vagón** *f* FERRO carriage key; ~ **de vaso** *f* ING MECÁ, MECÁ box spanner, box wrench

llaves *f pl* TEC PETR *perforación* tongs; ~ **paralelas rectangulares** *f pl* ING MECÁ rectangular parallel keys; ~ **de tubería** *f pl* TEC PETR *perforación* pipe tongs

llavín *m* PROD latch key

llega: que ~ *adj* TRANSP MAR *tráfico portuario* inward-bound

llegada: ~ **de la jaula al exterior** *f* CARBÓN, MINAS landing; ~ **del viento** *f* ING MECÁ *alto horno* air inlet

llegar: ~ **a un puerto** *vi* TRANSP MAR make a port

llenado *m* ALIMENT refill, *conservas* filling; ~ **apisonado y pesado** *m* EMB weighing and punnet filling; ~ **de bolsas** *m* EMB bag filling; ~ **en caliente** *m* EMB hot filling; ~ **capilar** *m* ING MECÁ capillary filling; ~ **en condiciones asépticas** *m* EMB aseptic filling; ~ **a dos velocidades** *m* EMB two-speed filling; ~ **excesivo de un generador de vapor** *m* NUCL steam generator overfill; ~ **del fondo** *m* EMB bottom filling; ~ **de grano** *m* AGRIC seed filling; ~ **de horno** *m* C&V furnace-fill; ~ **por inyección** *m* EMB injection filling; ~ **de lingoteras** *m* PROD teeming; ~ **a presión** *m* EMB pressure-filling; ~ **en vacío** *m* ALIMENT vacuum-filling; ~ **de vagonetas** *m Esp* (*cf cargamento de vagonetas AmL*) MINAS filling; ~ **por volumen** *m* EMB volume filling

llenador *m* C&V filler

llenadora: ~ **de bolsas** *f* EMB bag-filling machine; ~ **de cajas** *f* EMB box-filling machine; ~ **por gravedad** *f* EMB gravity-filling machine; ~ **de latas** *f* EMB can-filling machine (*AmE*), tin-filling machine (*BrE*); ~ **de sacos** *f* EMB sack-filling machine

llenar *vt* C&V, EMB, MINAS, PROD fill; ~ **con agua** *vt* INSTAL HIDRÁUL prime; ~ **a tope** *vt* CONST fill up

lleno *adj* CONST filled; ~ **de gas** *adj* GAS, TERMO gas-filled; ~ **de gases** *adj* TERMO gassy

llevar[1]: ~ **a cabo** *vt* ING MECÁ implement; ~ **a cero** *vt* ING MECÁ set to zero, QUÍMICA zero; ~ **a hervor** *vt* TERMO bring to the boil

llevar[2]: ~ **escora** *vi* TRANSP MAR *buque* list

lloro *m* ACÚST flutter, wow, TV flutter; ~ **y trémolo** *m* ACÚST, TV wow and flutter

llovizna *f* METEO drizzle

lloviznar *vi* CONTAM mist, METEO drizzle

lluvia *f* CONST shower; ~ **ácida** *f* CONTAM acid precipitation, acid rain, GAS acid rain; ~ **de chispas** *f* CONST *soldadura* shower; ~ **cósmica** *f* TEC ESP *astronomía* cosmic shower; ~ **fuerte** *f* METEO heavy rain; ~ **limpia** *f* CONTAM clean rain; ~ **del monzón** *f* METEO monsoon rain; ~ **radiactiva ácida** *f* CONTAM acid fallout; ~ **radiactiva atmosférica** *f* CONTAM atmospheric fallout; ~ **radioactiva** *f* AGRIC, CONTAM, FÍS, FÍS RAD, NUCL radioactive fallout; ~ **útil** *f* METEO useful rain

LLV *abr* (*vehículo logístico lunar*) TEC ESP LLV (*lunar logistics vehicle*)

lm *abr* (*lumen*) METR lm (*lumen*)

LNP *abr* (*nivel de contaminación sonora*) ACÚST NPL (*noise pollution level*)

lo: ~ **que se ve es lo que se obtiene** *fra* (*WYSIWYG*) INFORM&PD what you see is what you get (*WYSIWYG*)

lobelina *f* QUÍMICA lobeline

lobinina *f* QUÍMICA lobinine

lobo *m* C&V bear, PROD *altos hornos* horse, *cuchara de colada* skull, bear, *metalurgia* sow

lóbulo *m* ING MECÁ, MECÁ, PROD lug, TEC ESP lobe; ~ **de descarga** *m* GEOL flow lobe; ~ **de lado alto** *m* ELECTRÓN high side lobe; ~ **lateral** *m* ELECTRÓN, FÍS, TEC ESP side lobe, sidelobe, TRANSP MAR sidelobe, *radar* side lobe; ~ **lateral alto** *m* ELECTRÓN high side lobe; ~ **de leva** *m* AUTO, TRANSP AÉR, VEH cam lobe; ~ **de solifluacción** *m* GEOFÍS *dinámica de fluidos* solifluction lobe; ~ **de solifluxión** *m* GEOL solifluction lobe

LOCA[1] *abr* (*accidente de pérdida de refrigerante*) NUCL, REFRIG, SEG LOCA (*loss of coolant accident*)

LOCA[2]: ~ **grande** *m* (*LBLOCA*) NUCL large break LOCA (*LBLOCA*); ~ **pequeño** *m* (*SBLOCA*) NUCL small break LOCA (*SBLOCA*); ~ **de rotura de la rama caliente** *m* NUCL hot leg LOCA; ~ **de rotura de la rama fría** *m* NUCL cold leg LOCA

local[1] *adj* INFORM&PD, TELECOM local

local[2]: ~ **para baterías** *m* PROD battery housing; ~ **de calderas** *m* INSTAL TERM boiler house

localización *f* CONST, ING ELÉC, ING MECÁ locating, location, METR traceability, PETROL, PROD, TELECOM, TRANSP MAR locating, location; ~ **acústica** *f* FÍS ONDAS sound ranging; ~ **automática de matrículas de buques** *f* TELECOM automatic location registration of ships; ~ **automática de vehículos** *f* TELECOM automatic vehicle location (*AVL*); ~ **de averías** *f* ING ELÉC fault-finding; ~ **de embarcaciones** *f* TRANSP MAR vessel location; ~ **general** *f* TELECOM general localization; ~ **inmediata de personal** *f* INFORM&PD *software* demand paging; ~ **de llamadas maliciosas** *f* TELECOM malicious-call tracing; ~ **de modalidad normal** *f* TEC ESP acquisition of normal mode; ~ **del objetivo** *f* TEC ESP acquisition; ~ **de órbita** *f* TEC ESP acquisition of orbit; ~ **de posición de vuelo** *f* TEC ESP acquisition of attitude; ~ **radiogoniométrica** *f* TELECOM, TRANSP AÉR, TRANSP MAR radio fix; ~ **de señal** *f* TEC ESP range finding

localizador *m* ACÚST, ING MECÁ locator, TRANSP localizer; ~ **acústico** *m* ACÚST sound locator; ~ **de brújula** *m* TRANSP AÉR compass locator; ~ **de cables** *m* ELEC, ING ELÉC, INSTR cable locator; ~ **de líneas** *m* TELECOM linefinder; ~ **de llamada** *m* TELECOM call trace; ~ **de señal** *m* CINEMAT, D&A, ELECTRÓN, FOTO, TV rangefinder; ~ **de taladro** *m* ING MECÁ drill locater; ~ **de trayectoria** *m* D&A pathfinder; ~ **de trayectoria de planeo** *m* TRANSP glide-path localizer

localizar *vt* CONST locate, INFORM&PD localize, ING ELÉC, PROD, TELECOM, TRANSP MAR locate

locis *m* AGRIC *genética* gene loci

loco *adj* ING MECÁ, PROD *rueda* loose

locomotora *f* FERRO locomotive, MINAS draft engine (*AmE*), draught engine (*BrE*); ~ **cisterna** *f* Esp (*cf locomotora ténder AmL*) MINAS tank engine, tank locomotive; ~ **controlada por tiristor** *f* TRANSP thyristor-controlled locomotive; ~ **de cremallera** *f* FERRO rack body truck, rack engine, TRANSP rack engine; ~ **diesel con transmisión hidráulica** *f* FERRO diesel-hydraulic locomotive; ~ **diesel de vía estrecha** *f* FERRO narrow-gage diesel locomotive (*AmE*), narrow-gauge diesel locomotive (*BrE*); ~ **diesel-eléctrica** *f* FERRO diesel-electric engine; ~ **diesel-eléctrica de maniobras** *f* FERRO diesel-electric shunting motor tractor; ~ **para dos voltajes** *f* ELEC, FERRO *vehículos* dual-current locomotive; ~ **eléctrica** *f* ELEC, FERRO electric locomotive; ~ **eléctrica de rectificador** *f* FERRO rectifier locomotive; ~ **empujadora** *f* FERRO banking locomotive (*BrE*), pusher locomotive (*AmE*); ~ **de ferrocarril de cremallera** *f* FERRO rack-rail locomotive; ~ **ligera** *f* FERRO light engine, light locomotive; ~ **de maniobra** *f* MINAS pug; ~ **de maniobras** *f* CONST *ferrocarriles* FERRO, TRANSP shunting engine (*BrE*), switch engine (*AmE*); ~ **para maniobras** *f* FERRO switcher, *vehículos* shunter, TRANSP switcher; ~ **para movimientos de vagones** *f* MINAS bulldozer; ~ **con rectificador de ignitrones** *f* FERRO ignitron locomotive; ~ **de refuerzo para subida de rampas** *f* FERRO banking locomotive (*BrE*), pusher locomotive (*AmE*); ~ **ténder** *f* AmL (*cf locomotora cisterna Esp*) MINAS tank engine, tank locomotive; ~ **de tracción** *f* FERRO traction locomotive; ~ **trasera** *f* FERRO banking locomotive (*BrE*), pusher locomotive (*AmE*); ~ **de tren de mercancías** *f* FERRO freight locomotive (*AmE*), freight porter (*AmE*), goods locomotive (*BrE*), goods porter (*BrE*)

locución: ~ **despejada** *f* TELECOM clear sixty-four service

lodazal: ~ **blando** *m* TRANSP MAR mudflat

lodo *m* AGUA sludge, mud, ALIMENT sludge, CARBÓN slime, *minería* slurry, CONST, CONTAM, HIDROL sludge, MINAS slurry, slush, OCEAN mud, PAPEL slime, PETROL, PROD, QUÍMICA, RECICL, TEC PETR sludge, mud; ~ **activado** *m* CONTAM, HIDROL, RECICL activated sludge; ~ **activo** *m* CONTAM, HIDROL, RECICL activated sludge; ~ **de agua dulce** *m* TEC PETR freshwater mud, freshwater sludge; ~ **de agua salada** *m* TEC PETR salt water mud, salt water sludge; ~ **de aguas cloacales** *m* AGUA sewage sludge; ~ **con aire** *m* TEC PETR aerated mud, aerated sludge; ~ **aireado** *m* TEC PETR aerated mud, aerated sludge; ~ **de alcantarillado** *m* HIDROL sewage sludge; ~ **con alto contenido en sólidos** *m* TEC PETR high-solid mud, high-solid sludge; ~ **con bajo contenido en sólidos** *m* TEC PETR low-solid mud, low-solid sludge; ~ **a base de agua** *m* TEC PETR water-based mud, water-based sludge; ~ **a base de arcilla** *m* TEC PETR clay-based mud, clay-based sludge; ~ **a base de petróleo** *m* PETROL oil-based mud; ~ **con base de petróleo** *m* PETROL oil-based mud; ~ **con base petróleo** *m* TEC PETR oil-based mud, oil-based sludge; ~ **con cal** *m* TEC PETR lime mud, lime sludge; ~ **colable** *m* MINAS pourable slurry; ~ **coloidal** *m* TEC PETR colloidal mud, colloidal sludge; ~ **concentrado** *m* CARBÓN concentrated sludge; ~ **contaminado** *m* TEC PETR contaminated mud, contaminated sludge; ~ **de emulsión** *m* PETROL emulsion mud; ~ **de glaciar** *m* HIDROL glacier mud; ~ **de globigerinas** *m* GEOL Globigerina ooze; ~ **glutinoso** *m* AGUA, CARBÓN silt; ~ **del hueco de superficie** *m* TEC PETR spud mud, spud sludge; ~ **de**

inyección *m* MINAS *sondeos* mud; **~ no tratado** *m* AGUA, CONTAM, RECICL raw sludge; **~ de perforación** *m* MINAS *sondeos* mud, mud bit, PETROL drilling mud, TEC PETR drilling mud, drilling sludge; **~ de perforación de agua dulce** *m* PETROL, TEC PETR freshwater drilling mud, freshwater drilling sludge; **~ de perforación a base de agua salada** *m* TEC PETR salt water drilling mud, salt water drilling sludge; **~ reciclado activado** *m* HIDROL *aguas de alcantarilla* activated recycled sludge; **~ residual** *m* AGUA, RECICL sewage sludge; **~ de retorno activado** *m* HIDROL activated return sludge; **~ seco** *m* TEC PETR dry mud, dry sludge; **~ surfactante** *m* TEC PETR surfactant mud, surfactant sludge; **~ tratado con cal** *m* TEC PETR lime-treated mud, lime-treated sludge; **~ sin tratar** *m* AGUA, CONTAM, RECICL raw sludge

lodos *m pl* CARBÓN stowing material; **~ de corrosión** *m pl* NUCL crud; **~ de decantación** *m pl* PAPEL dreg; **~ espesos** *m pl* CARBÓN thickened slurry

loess *m* AGRIC loess

lofina *f* QUÍMICA lophine

logarítmico *adj* INFORM&PD, MATEMÁT logarithmic

logaritmo *m* INFORM&PD, MATEMÁT log, logarithm, TELECOM logarithm; **~ de Briggs** *m* MATEMÁT common logarithm; **~ decimal** *m* MATEMÁT common logarithm; **~ natural** *m* MATEMÁT Naperian logarithm, natural logarithm; **~ neperiano** *m* MATEMÁT Naperian logarithm, natural logarithm

logátomo *m* ACÚST *palabra* logatom

logia *f* CONST *arquitectura* loggia

lógica *f* ELECTRÓN, INFORM&PD, TELECOM logic; **~ de acoplamiento directo** *f* TELECOM direct-coupled logic (*DCL*); **~ binaria** *f* INFORM&PD binary logic; **~ borrosa** *f* INFORM&PD fuzzy logic; **~ de cableado** *f* INFORM&PD hardwired logic; **~ celular** *f* ELECTRÓN cellular logic; **~ de computadoras** *f* AmL (*cf lógica de ordenadores Esp*) INFORM&PD computer logic; **~ de duda** *f* INFORM&PD fuzzy logic; **~ de emisor acoplado** *f* (*LEA*) INFORM&PD emitter-coupled logic (*ECL*); **~ físicamente conectada** *f* INFORM&PD hardwired logic; **~ formal** *f* INFORM&PD formal logic; **~ informática** *f* INFORM&PD computer logic; **~ integrada a inyección** *f* (*IL*) ELECTRÓN integrated injection logic (*IL*); **~ de matriz programable** *f* (*PAL*) INFORM&PD programmable array logic (*PAL*); **~ del modo de corriente** *f* ING ELÉC current-mode logic; **~ de nebulosa** *f* INFORM&PD fuzzy logic; **~ negativa** *f* ELECTRÓN negative logic; **~ de ordenadores** *f* Esp (*cf lógica de computadoras AmL*) INFORM&PD computer logic; **~ positiva** *f* ELECTRÓN positive logic; **~ de programación** *f* PROD programming logic; **~ de relé** *f* PROD relay logic; **~ de transistores acoplados** *f* ELECTRÓN transistor coupled logic; **~ transistor-transistor** *f* INFORM&PD transistor-transistor logic (*TTL*)

lógico *adj* INFORM&PD logical

logística *f* INFORM&PD logistics

logo *m* CONST, INFORM&PD, TELECOM logo

logotipo *m* CONST vibrating roller

lomo *m* IMPR shelfback, spine, ING MECÁ back, heel, MECÁ rib; **~ aserrado** *m* IMPR sawn-in back; **~ estucado** *m* IMPR, REVEST coated back; **~ falso** *m* IMPR false back; **~ plano** *m* IMPR flat back, square back; **~ redondeado** *m* IMPR rounded back; **~ unido** *m* IMPR tight back

lona *f* IMPR, TEXTIL, TRANSP MAR *tela* canvas; **~ antiescora** *f* TRANSP MAR *en yates pequeños* lee canvas; **~ del filtro prensa** *f* PROC QUÍ filter press cloth; **~ de salvamento** *f* SEG *en caso de incendios* jumping sheet; **~ para tabiques de ventilación** *f* MINAS brattice cloth; **~ de ventilación** *f* MINAS damp sheet

loneta *f* CONST duckboard

longevidad *f* AmL (*cf vida de la explotación Esp*) FÍS lifetime, METR durability, MINAS *de la mina* life, lifetime; **~ atmosférica** *f* CONTAM *radiactividad* atmospheric lifetime

longifoleno *m* QUÍMICA longifolene

longitud[1]: **de ~ normal** *adj* GEOFÍS *medida de ondas* long normal

longitud[2] *f* FÍS, GEOM, METR, PAPEL length, TEC ESP longitude, TELECOM length, TRANSP MAR longitude; **~ de aceite** *f* P&C *resina alquídica, barniz* oil length; **~ de aguja útil** *f* CARBÓN *minas* effective pile length; **~ de alargamiento** *f* PAPEL stretch length; **~ de ánima** *f* D&A *de un cañón* length of bore; **~ astronómica** *f* TEC ESP *astronomía* astronomic longitude; **~ básica de pista** *f* TRANSP AÉR runway basic length; **~ de bloque** *f* INFORM&PD block length; **~ cabeza comprendida** *f* ING MECÁ length overall; **~ cabeza no comprendida** *f* ING MECÁ length under head to point; **~ del cableado** *f* ELEC *componente del cable* length of lay; **~ de cadena** *f* P&C *polímeros* chain length; **~ de la cadena** *f* INFORM&PD string length; **~ de camino óptico** *f* ÓPT optical path length; **~ de campo equilibrada** *f* TRANSP AÉR balanced field length; **~ de canal** *f* ENERG RENOV length of channel; **~ celeste** *f* TEC ESP celestial longitude; **~ de coherencia** *f* FÍS, ÓPT, TELECOM coherence length; **~ completa** *f* TELECOM uncut length; **~ común** *f* ING MECÁ plain length; **~ del cuadro** *f* TEC ESP *comunicaciones* frame length; **~ de distribución en modo equilibrado** *f* ÓPT, TELECOM equilibrium-mode distribution length; **~ de una división de la escala** *f* METR scale spacing; **~ eclíptica** *f* TEC ESP celestial longitude; **~ de enlace** *f* CRISTAL bond length; **~ de equilibrio** *f* ÓPT, TELECOM equilibrium length; **~ de la escala** *f* METR scale length; **~ excedente de fibra** *f* ÓPT fiber excess length (*AmE*), fibre excess length (*BrE*); **~ de fibra** *f* TEXTIL staple length; **~ fija** *f* INFORM&PD fixed length; **~ de frenado** *f* FÍS slowing-down length; **~ de hilo sin nudos** *f* TEXTIL knotless yarn length; **~ de imagen** *f* TEC ESP frame length; **~ del impulso** *f* ELECTRÓN, TELECOM pulse length; **~ de instrucción** *f* INFORM&PD instruction length; **~ del intersticio** *f* ING ELÉC gap length; **~ del intervalo** *f* METR length of the interval; **~ de intervalos efectiva** *f* TV effective gap length; **~ libre** *f* ING MECÁ free length; **~ de línea** *f* IMPR argument; **~ máxima** *f* TEXTIL overall length; **~ media de vehículos** *f* TRANSP average vehicle length; **~ natural** *f* ING MECÁ free length; **~ de onda** *f* ACÚST, ELEC, ELECTRÓN, FÍS, FÍS ONDAS, METAL, ÓPT, PETROL, QUÍMICA, TELECOM wavelength; **~ de onda de Compton** *f* FÍS, FÍS RAD, NUCL Compton wavelength; **~ de onda de corte** *f* TELECOM cutoff wavelength; **~ de onda de la cresta intensidad** *f* ÓPT peak intensity wavelength; **~ de onda crítica** *f* FÍS, ÓPT cutoff wavelength; **~ de onda de entrada** *f* ELECTRÓN threshold wavelength; **~ de onda grabada** *f* ACÚST recorded wavelength; **~ de onda de guía** *f* FÍS guide wavelength; **~ de onda de intensidad de**

cresta *f* TELECOM peak intensity wavelength; **~ de onda de intensidad máxima** *f* TELECOM peak intensity wavelength; **~ de onda del pico de intensidad** *f* ÓPT peak intensity wavelength; **~ de la página** *f* IMPR length of page; **~ de página lógica** *f* IMPR logical page length; **~ de palabra** *f* INFORM&PD word length; **~ de palabra fija** *f* INFORM&PD fixed word length; **~ del paso** *f* CONST length of step; **~ del paso del pistón** *f* CONST length of piston stroke; **~ de la película** *f* CINEMAT film length; **~ del pilote** *f* CARBÓN pile length; **~ de registro** *f* INFORM&PD record length, register length; **~ del rodillo** *f* TEXTIL roll length; **~ de rotura** *f* PAPEL breaking length, tensile length, TEXTIL breaking length; **~ de rotura en húmedo** *f* PAPEL wet breaking length; **~ del saliente** *f* CONST projection length; **~ total** *f* ING MECÁ length overall, overall length, METR overall length; **~ de la trayectoria** *f* NUCL path length; **~ variable** *f* INFORM&PD variable length; **~ de la vía óptica** *f* TELECOM optical path length; **~ de la vía de transmisión óptica** *f* TELECOM optical path length

longitudinal[1] *adj* GEN longitudinal

longitudinal[2]: **~ de cubierta** *m* TRANSP MAR *diseño naval* deck longitudinal; **~ del fondo interior** *m* TRANSP MAR *diseño naval* inner bottom longitudinal

longitudinalmente *adv* ING MECÁ lengthways

lopolito *m* GEOL lopolith

losa *f* CARBÓN pan, slab, *de cimentación* raft, CONST slab, flag, paving stone; **~ continua** *f* CONST continuous slab; **~ continua de cimentación** *f* CARBÓN *hormigón* mat, raft; **~ del empino** *f* CONST *tejido* crown tile; **~ de fundación** *f* AGUA floor; **~ de hormigón** *f* CONST slab; **~ de piedra** *f* CONST flagstone; **~ radiante** *f* AmL (*cf calefacción empotrada en el suelo Esp*) TERMOTEC underfloor heating; **~-soporte** *f* TEC PETR *operaciones costa-fuera* template; **~ superior del reactor** *f* NUCL cover slab; **~ del techo** *f* CONST ceiling tile, roof plate

loseta *f* C&V flat tile, CONST, P&C, TEC ESP tile; **~ de barro** *f* C&V earthenware slab; **~ del bordillo** *f* CONST curbstone (*AmE*), kerbstone (*BrE*); **~ de pavimento** *f* CONST paving stone; **~ de PVC autoadhesivo para pisos y paredes** *f* P&C floor and wall self-adhesive PVC tile; **~ vítrea para pavimento** *f* C&V glass paving slab

lote *m* GEN batch, CONST set; **~ de control** *m* CALIDAD inspection lot; **~ de emulsión** *m* CINEMAT, FOTO emulsion batch; **~ de material** *m* TEXTIL, TRANSP batch; **~ de modificación** *m* PROD modification kit; **~ pequeño** *m* EMB short run; **~ de producción** *m* CALIDAD production batch; **~ de tareas** *m* INFORM&PD job batch

loweíta *f* MINERAL loeweite

lox *m* TEC ESP lox

loxodromía *f* TEC ESP loxodromics

loxodrómica *f* TRANSP MAR *navegación* rhumb line

loza *f* C&V slab; **~ de barro** *f* C&V pottery; **~ bizcochada no vidriada** *f* C&V biscuit ware; **~ de casa** *f* C&V household porcelain; **~ cocida** *f* C&V burnt earthenware; **~ cocida al bizcocho** *f* C&V biscuit-baked porcelain, biscuit-fired porcelain; **~ industrial** *f* OCEAN thick ware; **~ de mayólica** *f* C&V majolica ware; **~ prensada** *f* C&V pressed glass (*BrE*), press-ware (*AmE*); **~ a prueba de fuego** *f* C&V fireproof pottery, flameproof pottery; **~ vidriada** *f* C&V glazed earthenware

LPCS *abr* (*aspersión del núcleo a baja presión*) NUCL LPCS (*low pressure core spray*)

LPM *abr* (*líneas por minuto*) IMPR, INFORM&PD LPM (*lines per minute*)

LPPC *abr* (*libras por pulgada cuadrada*) METR, TEC PETR *unidades de presión* psi (*pounds per square inch*)

LQ *abr* IMPR, INFORM&PD (*calidad de letra*) LQ (*letter quality*)

Lr *abr* (*laurencio*) FÍS RAD, QUÍMICA Lr (*lawrencium*)

Lu *abr* (*lutecio*) METAL, QUÍMICA Lu (*lutetium*)

lubricación *f* AUTO, C&V, CONST, GAS, MECÁ lubrication, PROD lubrifaction, lubrification, lubrication, REFRIG lubrication, lubrification, TEC PETR, TEXTIL, VEH lubrication; **~ por alimentación forzada** *f* AUTO forced-feed lubrication; **~ con alimentación a presión** *f* AUTO pressure lubrication, ING MECÁ force-feed lubrication; **~ por barboteo** *f* AUTO, FERRO splash lubrication; **~ por cárter de aceite** *f* VEH *del motor* sump-type lubrication; **~ por cárter seco** *f* AUTO dry sump lubrication; **~ de goteo** *f* ING MECÁ drip feed lubrication; **~ periférica** *f* TRANSP AÉR boundary lubrication; **~ por salpicadura** *f* AUTO, FERRO splash lubrication

lubricado[1] *adj* ING MECÁ, PROD lubricated

lubricado[2]: **~ a presión** *m* ING MECÁ, REFRIG forced cooling

lubricador *m* PROD lubricator; **~ de aguja** *m* PROD needle lubricator; **~ automático** *m* ING MECÁ licker; **~ de bomba de aire** *m* ING MECÁ air pump lubricator; **~ de gota visible** *m* ING MECÁ sight-feed lubricator; **~ a gotas** *m* ING MECÁ drip feed lubricator; **~ de grasa** *m* PROD grease lubricator

lubricadora *f* ING MECÁ oil cup

lubricante *m* GEN lubricant; **~ conductor** *m* TEC ESP conductive grease; **~ para cuchillas** *m* ING MECÁ cutting oil, PROD cutting fluid; **~ de motor** *m* TEC PETR, VEH motor oil; **~ para motores de gran potencia** *m* TEC PETR heavy-duty oil (*HD*); **~ de la pestaña de la rueda** *m* FERRO wheel flange lubricant

lubricar *vt* GEN grease, lubricate

lubricación: **~ con alimentación a presión** *f* PROD force-feed lubrication

lubrificador: **~ de aire comprimido** *m* CONST, ING MECÁ compressed-air lubricator

lucero *m* TEC ESP *astronomía* star

luces *f pl* MINAS *de un diamante* flash; **~ de color de destellos** *f pl* TRANSP MAR alternating colored lights (*AmE*), alternating coloured lights (*BrE*); **~ del compartimiento de vuelo** *f pl* TRANSP AÉR flight-compartment lights; **~ cortas** *f pl* VEH dimmed beams (*AmE*), dipped headlights (*BrE*), low beams (*AmE*); **~ de distancia fija** *f pl* TRANSP AÉR fixed-distance lights; **~ de guía para vuelos de circunvalación** *f pl* TRANSP AÉR circling guidance light; **~ intermitentes** *f pl* SEG flashing lights; **~ de la zona de contacto con la pista** *f pl* TRANSP AÉR runway touchdown zone light

lucha *f* CARBÓN contest; **~ contra erosión** *f* AGRIC erosion control; **~ contra los insectos** *f* AGRIC insect control; **~ contra langosta** *f* AGRIC locust control; **~ contra las plagas** *f* AGRIC pest control; **~ contraincendios** *f* CONST, SEG, TERMO firefighting

Lucita® *f* P&C Lucite®

LUF *abr* (*mínima frecuencia empleada*) TEC ESP LUF (*lowest usable frequency*)

lugar[1]: **del ~** *adj* CONST on site

lugar²: en el ~ *adv* CONST on site
lugar³ *m* GEOM locus; **~ alternativo de descenso** *m* TEC ESP alternate landing site; **~ de construcción** *m* CONST construction site; **~ de corte** *m* GEOM *topología* cut locus; **~ en declive** *m* MINAS dip; **~ de desove** *m* OCEAN *peces y anfibios* spawning ground; **~ de fabricación** *m* PROD manufacturing location; **~ geométrico** *m* GEOM locus; **~ de lanzamiento** *m* TEC ESP launch site, launching site; **~ de paso** *m* ING MECÁ pitch-zone location; **~ propenso a incendios** *m* SEG firetrap; **~ de recogida por mando a distancia** *m* TV remote pickup point; **~ de refugio** *m* TRANSP MAR shelter; **~ de sondeo** *m* CONST boring site; **~ de trabajo** *m* EMB, ING MECÁ work station, SEG workplace
lugre: ~ cabotero *m* TRANSP MAR *pesca* coasting lugger
lumaquela *f* PETROL lumachelle
lumbre *f* GAS, INSTAL TERM, SEG, TERMO fire, firelight
lumbrera *f* ING MECÁ port, porthole, INSTAL HIDRÁUL, MECÁ port, TRANSP MAR deck light, deckhead light, *equipamiento de cubierta* skylight, VEH port; **~ de admisión** *f* AUTO inlet port, INSTAL HIDRÁUL *de cilindro de vapor* steam port, induction port, admission port, intake port, VEH inlet port, intake port; **~ de conexión** *f* ING MECÁ connection port; **~ de escape** *f* AUTO exhaust port, INSTAL HIDRÁUL *motores* exhaust port, outlet, vent hole, VEH exhaust port; **~ de transferencia** *f* VEH *motor de dos tiempos* transfer port
lumen *m* (*lm*) METR lumen (*lm*)
luminancia *f* GEN luminance; **~ cero** *f* TV zero luminance; **~ espectral** *f* FÍS spectral luminance
luminaria *f* ELEC *alumbrado* luminaire
luminiscencia *f* COLOR, FÍS, FÍS RAD, ING ELÉC, QUÍMICA, TV luminescence; **~ por rayos X** *f* NUCL X-ray luminescence; **~ remanente** *f* TERMO afterglow; **~ residual** *f* FÍS RAD, TERMO, TV afterglow
luminiscente *adj* COLOR, FÍS, FÍS RAD, ING ELÉC luminescent, QUÍMICA glowing, luminescent, TERMO glowing, TV luminescent
luminización *f* REVEST lumenizing
luminóforo *m* QUÍMICA phosphor
luminosidad *f* (*L*) ELECTRÓN brilliance, FÍS F-number, luminosity (*L*), FÍS PART luminosity (*L*), FOTO F-number (*L*), TV *Esp* (*cf brillo AmL*) brightness; **~ catódica** *f* FÍS cathode glow; **~ desarrollada** *f* FÍS RAD developed luminosity; **~ excesiva** *f* C&V *punto luminoso* blooming; **~ máxima** *f* FÍS RAD peak luminosity, TV peak brightness; **~ de pico** *f* FÍS RAD peak luminosity
luminoso *adj* CINEMAT, ÓPT, TV bright
luminotécnico *f* CINEMAT, ÓPT lighting engineer
lumnita *f* QUÍMICA lumnite
luna *f* C&V plate glass; **~ de cristal** *f* C&V crystal sheet glass
lunar *m* TEXTIL speckle
luneta *f* ING MECÁ centre rest (*BrE*), cat head, center rest (*AmE*), collar plate, steady, *torno* backrest; **~ de cojinetes** *f* ING MECÁ jaw steady rest; **~ fija** *f* ING MECÁ center rest (*AmE*), centre rest (*BrE*), fixed steady rest, steady rest; **~ fija de tres garras** *f* ING MECÁ three-jaw steady, three-jaw steady rest; **~ móvil** *f* ING MECÁ follow rest
luniforme *adj* GEOM crescent-shaped
lunnita *f* MINERAL lunnite

lupa *f* INSTR spectacle magnifier, *óptica* magnifier; **~ de aumento** *f* FÍS, LAB, MECÁ magnifying glass; **~ de campo** *f* C&V field glass magnifier; **~ de enfoque** *f* INSTR focusing magnifying glass; **~ de mano** *f* INSTR hand glass
lupia *f* PROD puddle ball
lupulina *f* QUÍMICA lupuline
lussatita *f* MINERAL lussatite
lustrado *m* PROD glossing, TEXTIL glazing
lustrar *vt* MECÁ, PROD polish, TEXTIL glaze
lustre *m* C&V, COLOR, P&C, PROD, TEXTIL glaze, gloss, glossiness, luster (*AmE*), lustre (*BrE*); **~ metálico** *m* QUÍMICA metaldehyde
lutación *f* PROD lutation
lutecio *m* (*Lu*) METAL, QUÍMICA lutetium (*Lu*)
luteína *f* QUÍMICA lutein
luten *m* PROD lute
luteocobáltico *adj* QUÍMICA luteocobaltic
luteol *m* QUÍMICA luteol
luteolina *f* QUÍMICA luteoline, *plantas* luteolin
lutidina *f* QUÍMICA lutidine
lutidínico *adj* QUÍMICA lutidinic
lutidona *f* QUÍMICA lutidone
lutita *f* GEOL lime mudrock, *sedimento de tipo arcilloso o barroso* lutite, TEC PETR *geología* shale; **~ de desprendimiento** *f* TEC PETR sloughing shale; **~ pegagosa** *f* PETROL, TEC PETR gumbo
lux *m* (*lx*) FÍS, FOTO, METR lux (*lx*)
luz *f* CONST span, FÍS, FOTO light, ING MECÁ opening, MECÁ clearance, ÓPT light, *de un faro* beacon, TRANSP MAR *de un faro* beacon, light, VEH *alumbrado* light, *de un faro* beacon; **~ actínica** *f* CINEMAT *de seguridad para manipulación de película*, FÍS RAD actinic light, IMPR *fotografía* light; **~ adicional** *f* CINEMAT kicker light; **~ de advertencia** *f* SEG warning light; **~ de advertencia de bajo nivel** *f* ING MECÁ low-level warning light; **~ de advertencia débil** *f* ING MECÁ low-level warning light; **~ de alcance** *f* TRANSP MAR *navegación* stern light; **~ alta** *f* TRANSP AÉR overhead light; **~ de alto** *f* VEH *alumbrado* stop lamp; **~ ambiental** *f* CINEMAT ambient light; **~ anticolisión** *f* TRANSP AÉR anticollision light; **~ de arco** *f* ELEC arc light; **~ de arco voltaico** *f* CINEMAT arc light; **~ de aterrizaje** *f* TRANSP AÉR landing light; **~ auxiliar orientable** *f* VEH *alumbrado* spotlight; **~ del avión** *f* TRANSP AÉR aircraft light; **~ de aviso** *f* CINEMAT cue light, SEG flashing warning light, TRANSP AÉR warning light, TV cue light, VEH flashing warning light; **~ de aviso del funcionamiento de los frenos** *f* VEH brake warning-light; **~ de aviso de la presión del aceite** *f* AUTO oil-pressure warning light; **~ de barra del ala** *f* TRANSP AÉR wing bar light; **~ blanca** *f* FÍS white light; **~ de la cabina de mando** *f* TRANSP AÉR cockpit light; **~ de carretera** *f* TEC ESP main beam; **~ cenicienta** *f* TEC ESP *astronomía* earthshine; **~ centelleante** *f* TEC ESP blinking light, TRANSP MAR quick-flashing light; **~ del cielo** *f* CINEMAT skylite; **~ de cobre** *f* C&V copper light; **~ coherente** *f* ELECTRÓN, FÍS ONDAS, ÓPT, TELECOM coherent light; **~ colimada** *f* *Esp* (*cf luz paralelada AmL*) ENERG RENOV collimated light; **~ combinada** *f* FOTO mixed light; **~ concentrada** *f* CINEMAT, ELEC, FOTO spotlight; **~ de control** *f* CINEMAT control light; **~ de cruce** *f* VEH passing light, *alumbrado exterior* dimmed beams (*AmE*), dipped headlights (*BrE*); **~ de cruce de pista** *f* TRANSP AÉR runway crossing

light; **~ de cuarzo** *f* CINEMAT quartz light; **~ de demarcación** *f* TRANSP AÉR boundary light; **~ para destacar** *f* CINEMAT accent light; **~ de destellos** *f* TRANSP MAR flashing light; **~ de detalle** *f* CINEMAT catch light; **~ difusa** *f* CINEMAT spill light; **~ directa** *f* FOTO direct light; **~ dirigida para seguimiento** *f* CINEMAT follow spot; **~ dispersada** *f* FÍS RAD scattered light; **~ del eje de la pista** *f* TRANSP AÉR runway-centerline light (*AmE*), runway-centreline light (*BrE*); **~ del eje de la pista de rodaje** *f* TRANSP AÉR taxiway centerline light (*AmE*), taxiway centreline light (*BrE*); **~ de emergencia** *f* ING ELÉC warning light, VEH hazard warning lamp; **~ de enfilación** *f* TRANSP MAR leading light; **~ estrecha** *f* CINEMAT strip light; **~ estroboscópica** *f* CINEMAT, TEC ESP strobe light; **~ fija** *f* TRANSP MAR fixed light; **~ de final de pista** *f* TRANSP AÉR runway-end light; **~ del flash** *f* FOTO flashlight; **~ de fondeo** *f* TRANSP MAR *amarres* anchor light, *navegación* riding light; **~ de fondo** *f* CINEMAT background lighting; **~ fría** *f* CINEMAT cold light; **~ de fuga** *f* CINEMAT leak light; **~ de funcionamiento** *f* FERRO *vehículos* running light; **~ de grupos de ocultaciones** *f* TRANSP MAR *navegación* group occulting light; **~ de identificación** *f* TRANSP AÉR identification light; **~ inactínica** *f* CINEMAT, FOTO safelight; **~ incidente** *f* CINEMAT, FÍS RAD, FOTO incident light; **~ incoherente** *f* FÍS ONDAS, TELECOM incoherent light; **~ indicadora** *f* CINEMAT marker light, tally light, ELEC indicator lamp, PROD, TRANSP AÉR indicator light, TV tally light; **~ indicadora de funcionamiento** *f* ING MECÁ running light; **~ indicadora de funcionamiento sin carga** *f* FERRO *vehículos* running light; **~ indicadora de grupo** *f* PROD cluster indicator light; **~ indicadora de inestabilidad** *f* ELECTRÓN pumping light; **~ infrarroja** *f* FÍS RAD infrared light; **~ intensa** *f* FÍS ONDAS intense light; **~ intermitente** *f* AUTO blinker, CINEMAT intermittent light, VEH blinker; **~ de inundación** *f* CINEMAT flood; **~ lateral** *f* VEH *de señalización* side marker light (*AmE*), sidelight (*BrE*); **~ libre** *f* FERRO *infraestructura* clear light; **~ de la línea central de la pista** *f* TRANSP AÉR runway-centerline light (*AmE*), runway-centreline light (*BrE*); **~ de magnesio** *f* FOTO flashlight; **~ marcadora** *f* CINEMAT marker light; **~ de marcha** *f* ING MECÁ running light; **~ de marcha atrás** *f* AUTO, VEH reverse light (*AmE*), reversing light (*BrE*); **~ de matrícula trasera** *f* AUTO rear license-plate lamp (*AmE*), rear marker-plate lamp (*AmE*), rear number-plate lamp (*BrE*); **~ mixta** *f* FOTO mixed light; **~ monocromática** *f* FÍS ONDAS monochromatic light; **~ natural** *f* CINEMAT, FOTO natural light; **~ de navegación** *f* TRANSP AÉR, TRANSP MAR navigation light; **~ de ocultaciones** *f* TRANSP MAR *navegación* occulting light; **~ de ojo de Houdini** *f* ELEC Houdini eye-light; **~ paralelada** *f* AmL (*cf luz colimada Esp*) ENERG RENOV collimated light; **~ parásita** *f* CINEMAT stray light; **~ parásita del objetivo** *f* FOTO lens flare; **~ de pare** *f* VEH *alumbrado* stop lamp; **~ piloto de encendido y apagado** *f* CINEMAT on-off pilot light; **~ de la pista de rodaje** *f* TRANSP AÉR taxiway light; **~ de pistola** *f* ELEC pistol light; **~ polarizada** *f* CINEMAT, FÍS, FÍS ONDAS, FÍS RAD, FOTO polarized light; **~ portátil** *f* CONST portable light; **~ de posición** *f* TEC ESP, VEH position light; **~ de positivado** *f* CINEMAT printer point; **~ de la positivadora** *f* CINEMAT printer light; **~ posterior** *f* CINEMAT backlight; **~ principal** *f* CINEMAT key lighting, keylight, FOTO key light; **~ reflejada** *f* CINEMAT, FOTO, INSTR, ÓPT reflected light; **~ de refuerzo** *f* CINEMAT booster light; **~ del regulador del timón de dirección** *f* TRANSP AÉR rudder-trim light; **~ de relleno** *f* CINEMAT, FOTO fill-in light; **~ de relleno sobre la cámara** *f* CINEMAT basher; **~ de resalte** *f* ING ELÉC floodlight; **~ roja** *f* TRANSP *control del tráfico* red phase; **~ de sector** *f* TRANSP MAR sector light; **~ de seguridad** *f* CINEMAT, FOTO safelight; **~ de situación** *f* TRANSP MAR, VEH position light; **~ solar** *f* TEC ESP sunlight; **~ de techo** *f* C&V roof light; **~ testigo** *f* ING ELÉC pilot light; **~ todo horizonte** *f* TRANSP MAR all-round light; **~ de tope** *f* TRANSP MAR steaming light, *navegación* masthead light; **~ transmitida** *f* FÍS RAD, FOTO transmitted light; **~ trasera** *f* AUTO, CONST rear light, VEH tail lamp, rear light; **~ ultravioleta** *f* FÍS RAD, P&C ultraviolet light; **~ en verde** *f* TRANSP *control del tráfico* green phase; **~ verde en sentido contrario** *f* TRANSP *control del tráfico* opposing green; **~ visible** *f* FÍS ONDAS, TELECOM visible light; **~ de zona de parada** *f* TRANSP AÉR stopway light

LV-Rom® *abr* (*memoria de solo lectura laservisión*) ÓPT LV-Rom® (*Laservision read-only memory*)

lx *abr* (*lux*) FÍS, FOTO, METR lx (*lux*)

M

m *abr* GEN m (*meter AmE, metre BrE*), METR (*mili-*) m (*milli-*)

M *abr* (*mega-*) METR M (*mega-*)

mA *abr* (*miliamperio*) ELEC, FÍS, ING ELÉC mA (*milli-ampere*)

maar *m* GEOL maar

MAC *abr* (*modulación de amplitud en cuadratura*) ELECTRÓN, INFORM&PD, TELECOM QAM (*quadrature amplitude modulation*)

macádam *m* CONST *carreteras* macadam, paving material (*AmE*), road metal (*BrE*)

macareo *m* OCEAN tidal bore, tidal wave, TRANSP MAR *de río* bore

macarrón *m* ELEC, ING ELÉC conductor; **~ del cable** *m* ELEC, ING ELÉC cable trough

maceración *f* ALIMENT, QUÍMICA maceration

macerado *m* ALIMENT, QUÍMICA mashing

macerador *m* ALIMENT, QUÍMICA macerator

maceral *m* GEOL maceral

macerar *vt* ALIMENT macerate, steep, QUÍMICA macerate

maceta *f* PROD *cantero* mallet

machacadora *f* CONST crusher, stone crusher, *obras públicas* stone breaker, ING MECÁ breaker, PROD *máquina trituradora* mill; **~ de Bradford** *f* ING MECÁ Bradford breaker; **~ de hielo** *f* ING MECÁ crushing machine for ice; **~ de impactos** *f* ING MECÁ impact breaker, impact crusher; **~ de mandíbulas** *f* ING MECÁ jaw breaker, jaw crusher, MINAS *preparación de minerales* alligator; **~ de mordazas** *f* CARBÓN, ING MECÁ jaw breaker, jaw crusher; **~ de quijadas** *f* ING MECÁ jaw breaker, jaw crusher

machacadura *f* ALIMENT bruise

machacamiento *m* PROD stamping

machacar *vt* ALIMENT crush, pound, CARBÓN, CONST crush, PROD stamp

machaqueo *m* CARBÓN *de piedra*, CONST, MINAS crushing

machero *m* PROD *fundería* pattern maker

machetazo *m* TRANSP MAR slamming

macheteo *m* TRANSP MAR slamming

machiembrado: **~ por frío** *m* REFRIG cold shrink-fitting

machihembrado *m* CONST bridle joint

machihembradora *f* CONST matching machine

macho *m* AUTO journal, C&V horse, ING MECÁ journal, plunger, tap, PROD core, core barrel, TRANSP MAR *del timón* pintle; **~ de acabar** *m* ING MECÁ plug tap; **~ accionado por resorte** *m* P&C *parte de equipo* spring-loaded core; **~ para agujeros de limpieza** *m* PROD washout plug; **~ de ahondar** *m* ING MECÁ bottoming tap; **~ de arcilla** *m* PROD *moldería* loam core; **~ aterrajado** *m* PROD *fundería* swept-up core; **~ de aterrajado preliminar** *m* PROD entering tap; **~ bombeado** *m* PROD chambered core; **~ de cabeza cuadrada** *m* ING MECÁ tap with square head; **~ cilíndrico** *m* ING MECÁ bottoming tap; **~ cónico** *m* PROD entering tap; **~ desmontable** *m* ING MECÁ collapsible core; **~ ensanchado** *m* PROD bellied core;

~ escariador *m* ING MECÁ, PROD reamer tap; **~ de fundería** *m* PROD baking core; **~ de fundición** *m* PROD core; **~ girado mecánicamente** *m* ING MECÁ machine tap; **~ y hembra** *m* MINAS *para varilla de perforación* box and pin; **~ intermedio para roscar** *m* ING MECÁ plug tap; **~ maestro para peines de roscar** *m* ING MECÁ hob; **~ maestro de roscar** *m* ING MECÁ master tap, hob, hob tap; **~ patrón de roscar** *m* ING MECÁ master tap; **~ de repasar roscas** *m* ING MECÁ tap-reseating tool; **~ de rosca métrica** *m* ING MECÁ tap with metric thread; **~ para rosca de tubería** *m* CONST pipe tap; **~ de rosca de 29 grados** *m* ING MECÁ Acme thread tap; **~ con rosca Whitworth** *m* ING MECÁ tap with Whitworth thread; **~ roscador a derechas** *m* ING MECÁ right-hand tap; **~ roscador de ranuras helicoidales** *m* ING MECÁ twist tap; **~ de roscar** *m* ING MECÁ tap; **~ de roscar de dentadura interrumpida** *m* ING MECÁ interrupted tooth tap; **~ de roscar estays de caldera** *m* ING MECÁ boiler-stay screwing-tap; **~ de roscar sin estrías** *m* ING MECÁ fluteless screwing tap; **~ de roscar de estrías rectas** *m* ING MECÁ straight tap; **~ de roscar final** *m* ING MECÁ bottoming tap, plug tap; **~ de roscar helicoidal** *m* ING MECÁ twist tap; **~ de roscar intermedio** *m* ING MECÁ second tap; **~ de roscar a mano** *m* PROD hand tap; **~ de roscar a máquina** *m* ING MECÁ machine tap; **~ de roscar con rosca rectificada** *m* ING MECÁ ground thread tap; **~ semicónico** *m* ING MECÁ plug tap; **~ de terraja** *m* ING MECÁ, MECÁ tap; **~ de tierra** *m* PROD *moldería* loam core

macizado *m* AmL (*cf rellenado Esp*) MINAS *muros de galerías* packing

macizar *vt* AmL (*cf rellenar Esp*) MINAS *muros de galerías* pack

macizo¹ *adj* CARBÓN, CONST, GEOL, MINAS solid

macizo² *m* CARBÓN mountain mass, CONST block; **~ de anclaje** *m* TEC PETR deadman; **~ autóctono** *m* GEOL terrane; **~ de fondo** *m* C&V *botellas* bottom block; **~ magmático** *m* CARBÓN, GEOL stock; **~ de mineral** *m* MINAS large body; **~ de protección** *m* MINAS boundary pillar; **~ de protección del pozo** *m* MINAS bottom pillar; **~ de relleno** *m* MINAS gob

macla *f* CRISTAL, METAL twin, MINERAL macle; **~ de compenetración** *f* CRISTAL interpenetration twin, penetration twin; **~ de contacto** *f* CRISTAL contact twin, juxtaposition twin; **~ de crecimiento** *f* METAL growth twin; **~ incoherente** *f* METAL incoherent twin; **~ lenticular** *f* METAL lenticular twin; **~ de yuxtaposición** *f* CRISTAL contact twin, juxtaposition twin

maclado¹ *adj* CRISTAL twinned

maclado² *m* CRISTAL twinning; **~ múltiple** *m* ING ELÉC multiple twin quad

maclaje *m* METAL twinning; **~ compuesto** *m* METAL compound twinning; **~ mecánico continuo** *m* METAL continual mechanical twinning

maclurina *f* QUÍMICA maclurin

macollo *m* AGRIC tiller

macro *m* INFORM&PD macro; **~-control** *m* TRANSP macrocontrol
macrocíclico *adj* QUÍMICA macrocyclic
macroclima *m* METEO macroclimate
macrocurva *f* ING ELÉC macrobend
macrodesecho *m* CONTAM macrowaste
macrodoblado *m* ÓPT macrobending
macrodureza *f* ING MECÁ macrohardness
macroelemento *m* AGRIC macroelement
macroensamblador *m* INFORM&PD macro assembler
macroestructura *f* QUÍMICA macrostructure
macroestructural *adj* QUÍMICA macrostructural
macroflexión *f* TELECOM macrobending
macroinstrucción *f* INFORM&PD macroinstruction
macromolécula *f* P&C, QUÍMICA macromolecule
macromolecular *adj* P&C, QUÍMICA macromolecular
macronutriente *m* AGRIC macronutrient
macropartícula *f* PROC QUÍ macroparticle
macropolímero *m* QUÍMICA macropolymer
macroprocesador *m* INFORM&PD macroprocessor
macroprograma *m* TRANSP macroprogram (*AmE*), macroprogramme (*BrE*)
macrorradiografía *f* NUCL macroradiography
macrosísmica *adj* GEOFÍS *vibración* macroseismic
mácula *f* TEC ESP speckle
madeja *f* TEXTIL hank, *de algodón, lana* rove; **~ enredada** *f* TEXTIL comingled yarn
madera *f* CONST, TRANSP MAR *de construcción* timber; **~ para ademas** *f* AmL (*cf madera para entibas Esp*) MINAS mining timber; **~ alabeada** *f* CONST warped timber; **~ para apeas** *f* MINAS mine timber, mining timber; **~ aserrada** *f* REVEST stuff; **~ para aserrar** *f* CONST saw timber; **~ blanda** *f* CONST softwood; **~ conglomerada** *f* Esp (*cf madera terciada pegada con adhesivo resinoso AmL*) P&C resin-bonded plywood; **~ de coníferas** *f* CONST, PAPEL, TRANSP MAR softwood; **~ para construcción** *f* CONST builders' timber, wood suitable for building; **~ contrachapada** *f* Esp (*cf madera terciada AmL*) CONST, EMB, TRANSP MAR plywood; **~ convertible en elementos de dimensiones comerciales** *f* CONST timber suitable for conversion into market forms; **~ cuadrada** *f* CONST square timber, squared timber; **~ curada** *f* CONST, TRANSP MAR *construcción naval* seasoned timber, seasoned wood; **~ dura** *f* CONST hardwood; **~ para entibación** *f* CONST propwood; **~ para entibas** *f* Esp (*cf madera para ademas AmL*) MINAS mining timber; **~ de frondosas** *f* PAPEL deciduous wood, hardwood; **~ lamelar** *f* MECÁ laminate; **~ laminada** *f* CONST, EMB plywood; **~ para listón** *f* CONST lath wood; **~ para minas** *f* MINAS mine timber, mining timber; **~ moldeada en frío** *f* TRANSP MAR *construcción naval* cold-molded wood (*AmE*), cold-moulded wood (*BrE*); **~ de montaña** *f* MINERAL mountain wood; **~ partida** *f* CONST broken-down timber; **~ para pasta** *f* PAPEL *materia prima* pulpwood; **~ en pie** *f* CONST standing timber; **~ de resinosas** *f* CONST, TRANSP MAR softwood; **~ revestida** *f* REVEST coated board; **~ de revestimiento** *f* CONST barked timber; **~ rústica** *f* CONST rough wood; **~ sazonada** *f* CONST, TRANSP MAR *construcción naval* seasoned timber, seasoned wood; **~ seca** *f* CONST, TRANSP MAR *construcción naval* seasoned timber, seasoned wood; **~ secada** *f* CONST, TRANSP MAR *construcción naval* seasoned timber, seasoned wood; **~ secada al aire** *f* PAPEL

yard lumber; **~ de teca** *f* CONST, TRANSP MAR teak; **~ terciada** *f* AmL (*cf madera contrachapada Esp*) CONST, EMB, TRANSP MAR plywood; **~ terciada pegada con adhesivo resinoso** *f* AmL (*cf madera conglomerada Esp*) P&C resin-bonded plywood; **~ de tres capas** *f* CONST, EMB three-ply wood
maderaje *m* CONST woodwork
maderista *m* MINAS deputy
madero *m* CONST, TRANSP MAR *construcción naval* plank
maderos *m pl* C&V woods, CARBÓN planking, CONST, TRANSP MAR *construcción naval* timber
madre *f* CARBÓN *de río* bed, TRANSP MAR capstan drum; **~ del río** *f* AGUA, HIDROL river bed
maduración *f* AGUA maturation, FOTO, IMPR ripening, REFRIG *de carnes* ageing, TEC PETR *generación de petróleo* maturation; **~ de cienos** *f* AGUA sludge ripening; **~ tardía** *f* AGRIC late maturity
madurar *vt* ALIMENT cure
maestra[1]: **en la** ~ *adv* TRANSP MAR *del buque* midship
maestra[2] *f* CONST *albañilería* screed heater, MINAS *albañilería* rod, TRANSP MAR *buques* midship, midship section
maestría: **~ de la calidad durante fabricación** *f* CALIDAD process quality control; **~ de procesos** *f* CALIDAD process control; **~ total de la calidad** *f* CALIDAD, PROD total quality control (*TQC*)
maestro *m* ACÚST master, mother, INFORM&PD master; **~ de mensajes** *m* TELECOM *Telecom Británica* Message-master®; **~ de obras** *m* CONST house builder, master builder; **~ velero** *m* TRANSP MAR sailmaker
máfico *adj* GEOL, PETROL mafic
magenta *f* COLOR, IMPR magenta
magma: **~ plutónico** *m* PETROL plutonic magma
magmático: **~ superior** *adj* GEOL late-stage magmatic; **~ tardío** *adj* GEOL late-stage magmatic
magnesia *f* P&C, QUÍMICA, TERMOTEC magnesia; **~ calcinada** *f* P&C calcined magnesia
magnesiano *adj* GEOL, QUÍMICA magnesian
magnésico *adj* GEOL magnesian, QUÍMICA magnesic
magnesio *m* (*Mg*) ING ELÉC, METAL, QUÍMICA magnesium (*Mg*)
magnesioferrita *f* MINERAL magnesioferrite
magnesita *f* MINERAL, P&C, QUÍMICA magnesite
magnesol *m* QUÍMICA magnesol
magnético *adj* GEN, magnetic; **no ~** *adj* GEN nonmagnetic; **~ transversal** *adj* (*TM*) ING ELÉC transverse magnetic (*TM*)
magnetismo *m* GEN magnetism; **~ interno** *m* TEC ESP internal magnetism; **~ remanente** *m* ELEC, ING ELÉC, PROD residual magnetism; **~ terrestre** *m* FÍS, GEOFÍS terrestrial magnetism
magnetita *f* MINERAL, NUCL magnetite
magnetización *f* GEN magnetization; **~ espontánea** *f* FÍS spontaneous magnetization; **~ remanente** *f* ELEC, ELECTRÓN, FÍS, PETROL remanent magnetization; **~ residual** *f* FÍS RAD residual magnetization; **~ de saturación** *f* ING ELÉC saturation magnetization
magnetizado *adj* GEN magnetized
magnetizar *vt* GEN magnetize
magneto *m* AUTO, ELEC *automotor*, ING ELÉC magneto, LAB magnet, TRANSP MAR, VEH *encendido* magneto; **~ de encendido** *m* AUTO, VEH ignition magneto
magnetoconductividad *f* TELECOM magnetoconductivity

magnetodinámica: ~ **de los gases** *f* NUCL magneto-gasdynamics (*MGD*)

magnetoelectricidad *f* ING ELÉC magnetoelectricity

magnetoestricción *f* ACÚST, FÍS, ING ELÉC magnetos-triction; ~ **negativa** *f* ING ELÉC negative magnetostriction; ~ **positiva** *f* ING ELÉC positive magnetostriction

magnetófono *m* ACÚST tape recorder, GEOFÍS magne-tophone, tape recorder; ~ **incremental** *m* INSTR incremental tape recorder

magnetogasodinámica *f* (*MGD*) NUCL magnetogas-dynamics (*MGD*)

magnetógrafo *m* GEOFÍS magnetograph

magnetógrama *m* GEOFÍS magnetogram

magnetohidrodinámica *f* FÍS, FÍS FLUID, GEOFÍS, ING ELÉC, TEC ESP magnetohydrodynamics (*MHD*)

magnetometría *f* GEN magnetometry

magnetómetro *m* GEN magnetometer; ~ **de apertura de flujo** *m* PETROL flux-gate magnetometer; ~ **de deflexión** *m* FÍS RAD deflection magnetometer; ~ **diferencial** *m* ELEC, INSTR differential magnet-ometer; ~ **de Hall** *m* FÍS Hall magnetometer; ~ **de resonancia del espín del electrón** *m* FÍS, FÍS PART, NUCL electron-spin resonance magnetometer; ~ **por resonancia protónica** *m* PETROL proton resonance magnetometer; ~ **de saturación** *m* ING ELÉC flux-gate magnetometer; ~ **vibratorio de muestreo** *m* NUCL vibrating sample magnetometer

magnetón *m* FÍS magneton; ~ **de Bohr** *m* FÍS, FÍS RAD Bohr magneton

magnetoplasma *m* NUCL magnetoplasma

magnetoresistencia *f* FÍS magnetoresistance

magnetoscopio *m* GEOFÍS, TELECOM magnetoscope, TV video recorder

magnetosfera *f* GEOFÍS, TEC ESP magnetosphere; ~ **externa** *f* GEOFÍS, TEC ESP external magnetosphere; ~ **interna** *f* GEOFÍS, TEC ESP internal magnetosphere

magnetostático *adj* FÍS, NUCL magnetostatic

magnetrón *m* ELECTRÓN, FÍS, ING MECÁ magnetron; ~ **de ánodo dividido** *m* ELECTRÓN split anode magnetron; ~ **ascendente** *m* CONST rising sun mag-netron; ~ **de cavidad** *m* ELECTRÓN, FÍS cavity magnetron; ~ **de cavidad múltiple** *m* ELECTRÓN, FÍS multicavity magnetron; ~ **coaxial** *m* ELECTRÓN coax-ial magnetron; ~ **de frecuencia constante** *m* ELECTRÓN fixed-frequency magnetron; ~ **de impulsos** *m* ELECTRÓN pulsed magnetron; ~ **industrial** *m* ELECTRÓN industrial magnetron; ~ **interdigital** *m* ELECTRÓN interdigital magnetron; ~ **modulado piezo-eléctricamente** *m* ING ELÉC piezoelectric-tuned magnetron; ~ **de onda milimétrica** *m* ELECTRÓN millimeter-wave magnetron (*AmE*), millimetre-wave magnetron (*BrE*); ~ **de onda progresiva** *m* ELECTRÓN traveling-wave mag-netron (*AmE*), travelling-wave magnetron (*BrE*); ~ **de ondas electrónicas** *m* ELECTRÓN, FÍS ONDAS electron-wave magnetron; ~ **de pequeño tamaño** *m* ELECTRÓN miniature magnetron; ~ **plurisectorial** *m* ELECTRÓN multisegment magnetron; ~ **sintonizable** *m* ELECTRÓN tunable magnetron; ~ **sintonizado mecánicamente** *m* ELECTRÓN mechanically-tuned magnetron

magnetrones: ~ **de banda X** *m pl* ELECTRÓN X-band magnetrons

magnificación *f* FÍS magnification; ~ **angular** *f* FÍS angular magnification; ~ **lateral** *f* FÍS, INSTR lateral magnification; ~ **longitudinal** *f* FÍS longitudinal magnification; ~ **transversal** *f* FÍS transverse magni-fication

magnificar *vt* CINEMAT, FÍS, INSTR magnify

magnitud *f* FÍS magnitude, REFRIG *del aire acondicio-nado* value, TEC ESP *de un desplazamiento* range; ~ **de ancho de banda** *f* ELECTRÓN power bandwidth; ~ **de flujo** *f* AGUA, TEC PETR flow rate; ~ **del impulso** *f* ELECTRÓN pulse width; ~ **periódica** *f* ACÚST periodic quantity; ~ **sometida a medición** *f* METR measurand

magnón *m* FÍS magnon

maimonete *m* TRANSP MAR crosspiece

maíz *m* AGRIC corn (*AmE*), maize (*BrE*); ~ **con alto porcentaje de humedad** *m* AGRIC high-moisture corn (*AmE*), high-moisture maize (*BrE*); ~ **blando** *m* AGRIC soft corn (*AmE*), soft maize (*BrE*); ~ **cose-chado a mano que permanece con la espata** *m* AGRIC snapped corn (*AmE*), snapped maize (*BrE*); ~ **dentado** *m* AGRIC dent corn (*AmE*), dent maize (*BrE*); ~ **para ensilaje** *m* AGRIC silage corn (*AmE*), silage maize (*BrE*); ~ **híbrido** *m* AGRIC hybrid corn (*AmE*), hybrid maize (*BrE*); ~ **húmedo** *m* AGRIC moist corn (*AmE*), moist maize (*BrE*); ~ **a medio moler** *m* AGRIC grits; ~ **tierno** *m* AGRIC green corn (*AmE*), green maize (*BrE*)

majadero *m* PROC QUÍ pestle

mejorar *vt* INFORM&PD upgrade

mal[1]: ~ **clasificado** *adj* GEOL nonsorted, *granulometría discontinua* nongraded; ~ **cortado** *adj* IMPR out-of-true; **en ~ estado** *adj* ELEC, ING MECÁ out-of-order

mal[2]: ~ **acoplamiento** *m* ING MECÁ mismatch; ~ **aislamiento** *m* ING ELÉC poor insulation; ~ **aislante** *m* ING ELÉC poor insulant; ~ **alineamiento** *m* ING MECÁ misalignment; ~ **conductor** *m* ELEC isolator, ING ELÉC poor con-ductor; ~ **fondeadero** *m* TRANSP MAR foul ground; ~ **recocido** *m* C&V bad annealing; ~ **tenedero** *m* TRANSP MAR foul bottom

mala: ~ **aislación** *f* ING ELÉC poor insulation; ~ **alineación** *f* ING MECÁ misalignment; ~ **clasificación** *f* ING MECÁ missorting; ~ **colocación** *f* ING MECÁ mismatch

malacate *m* MECÁ crab, hoist, MINAS hoist, PETROL *minería*, TEC PETR *perforación* draw works; ~ **neumático** *m* Esp (*cf winchy AmL*) ING MECÁ, TEC PETR air hoist

malacolita *f* MINERAL malacolite

malacón *m* MINERAL, NUCL malacon

malaquita *f* MINERAL malachite

malato *m* QUÍMICA malate

malaxación *f* ING MECÁ malaxage, P&C mastication

malaxador *m* ING MECÁ malaxator

maldonita *f* MINERAL maldonite

maleabilidad *f* MECÁ, METAL malleability

maleabilización *f* INSTAL TERM malleablizing

maleable *adj* MECÁ, METAL malleable

malecón *m* AmL (*cf terraplén Esp, dique Esp*) AGUA breakwater, dike, mole, CARBÓN embankment, CONST embankment, jetty, breakwater, mole, FERRO, MINAS embankment, OCEAN mole, breakwater, TRANSP MAR breakwater, mole, sea wall; ~ **de carga** *m* TRANSP loading dock

maleimida *f* QUÍMICA maleimide

malestar: ~ **espacial** *m* TEC ESP space sickness; ~ **producido por la radiación** *m* FÍS RAD radiation sickness

maletero *m* AUTO boot (*BrE*), trunk (*AmE*), FERRO, VEH *carrocería* luggage compartment

maletín: ~ **de herramientas** *m* ING MECÁ tool box

maleza: ~ **de hojas anchas** *f* AGRIC broadleaf weed

malfuncionamiento *m* CALIDAD malfunction, ELEC failure, INFORM&PD malfunction, ING MECÁ failure, malfunction, MECÁ, METAL failure, METR, SEG, TEC ESP malfunction; ~ **de control** *m* PROD control malfunction

málico *adj* QUÍMICA malic

malla *f* AGUA, ALIMENT, CARBÓN mesh, CONST netting, ELEC *de sistema* mesh, *cable* braid, network, ELECTRÓN *cristales* lattice, *de tubo de almacenamiento* mesh, FÍS *de un circuito eléctrico*, ING ELÉC, OCEAN, PAPEL, PROD mesh, TEXTIL *punto* mesh, stitch, *género de punto* loop, TRANSP MAR cable clinch; ~ **adiamantada** *f* METAL diamond lattice; ~ **de alambre** *f* LAB wire gauze; ~ **de almacenamiento** *f* ELECTRÓN storage mesh; ~ **de filamentos continua** *f* P&C continuous strand mat; ~ **de filamentos discontinua** *f* P&C chopped strand mat; ~ **de lizo** *f* TEXTIL heald; ~ **de memorización** *f* ELECTRÓN storage mesh; ~ **molecular** *f* QUÍMICA molecular sieve; ~ **suelta** *f* ING MECÁ mending link; ~ **de tamiz** *f* CARBÓN screening mesh; ~ **del tamiz** *f* CARBÓN mesh size, sieve mesh

mallardita *f* MINERAL mallardite

mallas *f pl* OCEAN webbing

mallazo *m* CARBÓN mesh, CONST mat reinforcement

malonamida *f* QUÍMICA malonamide

malonato *m* QUÍMICA malonate

malónico *adj* QUÍMICA malonic

malonitrilo *m* QUÍMICA malonitrile

malta *f* ALIMENT, QUÍMICA malt; ~ **empastada** *f* ALIMENT mash; ~ **remojada** *f* ALIMENT mash; ~ **tostada** *f* ALIMENT cured malt, kiln malt

maltacita *f* MINERAL malthacite

maltasa *f* ALIMENT, QUÍMICA maltase

maltear *vt* ALIMENT malt

maltenos *m pl* MINERAL maltha

maltería *f* ALIMENT malt house (*AmE*), maltings (*BrE*)

maltosa *f* QUÍMICA maltose

malva: ~ **india** *f* AGRIC passenger ship, velvet leaf

mamilar *adj* MINERAL *registro glacial* mammillary

mampara *f* CONST, SEG partition, screen; ~ **cortafuegos** *f* CONST, SEG, TEC PETR fire wall

mamparo *m* TEC ESP, TRANSP AÉR, TRANSP MAR *construcción naval* bulkhead; ~ **contraincendios** *m* SEG, TRANSP MAR *diseño de barcos* fire bulkhead; ~ **de crujía** *m* TRANSP MAR centerline bulkhead (*AmE*), centreline bulkhead (*BrE*); ~ **estanco** *m* ING MECÁ, TRANSP AÉR, TRANSP MAR pressure bulkhead; ~ **ignífugo** *m* TERMO, TRANSP MAR fire-resistant bulkhead, fire-resisting bulkhead; ~ **piroresistante** *m* TERMO, TRANSP MAR fire-resistant bulkhead; ~ **de presión** *m* TRANSP AÉR, TRANSP MÁR pressure bulkhead; ~ **transversal** *m* TRANSP MAR transverse bulkhead

mampirlán *m* CONST *escalera* nosing

mamposta *f* AmL (*cf columna Esp*) CARBÓN, MINAS *construcción de galerías* prop

mampostería *f* C&V, CONST masonry; ~ **de aglomerado** *f* CONST bonded masonry; ~ **ordinaria** *f* CONST rubble masonry; ~ **en seco** *f* CONST dry stone wall

manada *f* Esp (*cf rodeo AmL*) AGRIC *ganado bovino, ovino* herd

manantial *m* AGUA, HIDROL spring; ~ **de agua mineral** *m* AGUA, HIDROL mineral spring; ~ **de aguas termales** *m* AGUA, HIDROL thermal spring; ~ **de depresión** *m* AGUA depression spring; ~ **de desbordamiento** *m* AGUA contact spring; ~ **descendente** *m* AGUA depression spring; ~ **de falla** *m* AGUA fault spring; ~ **de fisura** *m* CARBÓN, GEOL fissure water; ~ **de Frank-Read** *m* CRISTAL Frank-Read source; ~ **intermitente** *m* AGUA intermittent spring; ~ **kárstico** *m* AGUA karstic spring; ~ **de ladera** *m* AGUA contact spring; ~ **perenne** *m* AGUA, HIDROL perennial spring; ~ **salino** *m* AGUA, HIDROL saline spring; ~ **sumergido** *m* AGUA submerged spring

manar *vi* AGUA, HIDROL flow along

mancha[1] *f* CALIDAD stain, CONTAM MAR slick, IMPR drag, smudge, OCEAN *de peces* shoal, PAPEL stain, TEC ESP speckle; ~ **de aceite flotando en el agua** *f* CONTAM, TEC PETR oil slick; ~ **de aceite flotante** *f* CONTAM, TEC PETR oil slick; ~ **bacteriana** *f* AGRIC bacterial spot; ~ **clara** *f* C&V clear spot; ~ **de difracción** *f* CRISTAL diffraction spot; ~ **foliar bacteriana** *f* AGRIC bacterial leaf-spot; ~ **de hidrocarburo** *f* TEC PETR hydrocarbon slick; ~ **de hidrocarburos** *f* CONTAM MAR slick, TEC PETR hydrocarbon slick; ~ **de hidrocarburos pequeña** *f* CONTAM MAR, TEC PETR patch; ~ **de hoja** *f* AGRIC leaf spot; ~ **iónica** *f* ELECTRÓN ion burn, TV ion spot; ~ **listada de cebada** *f* AGRIC barley stripe; ~ **luminosa** *f* ELECTRÓN light spot; ~ **por luz excesiva** *f* FOTO hot spot; ~ **negra** *f* C&V black stain; ~ **de papel** *f* C&V paper hum (*AmE*), paper stain (*BrE*); ~ **de resplandor** *f* CINEMAT flare spot; ~ **reticulada** *f* AGRIC net blotch; ~ **rojiza por pulido** *f* C&V *defecto del vidrio* burn mark; ~ **solar** *f* GEOFÍS sunspot

mancha[2]: **que no** ~ *adj* P&C nonstaining

manchado *m* C&V staining; ~ **con plata** *m* C&V silver staining

manchar *vt* CALIDAD, P&C, PAPEL stain

mancharse *v refl* QUÍMICA stain

manchas: ~ **negras** *f pl* C&V black staining; ~ **del papel** *f pl* PAPEL foxing

manchón *m* MINAS *de mineral* splash; ~ **de tela plástica encogible** *m* PAPEL fabric sleeve

mandarín: ~ **para ranuras** *m* PROD groove drift

mandato: ~ **ensamblador** *m* INFORM&PD assembler directive; ~ **de operador** *m* INFORM&PD operator command; ~ **único** *m* TEC ESP unique word (*UW*)

mandélico *adj* QUÍMICA mandelic

mandíbula *f* ING MECÁ jaw; ~ **del tornillo** *f* ING MECÁ vice jaw (*BrE*), vise jaw (*AmE*); ~ **trituradora** *f* PROC QUÍ crusher jaw

mando *m* ELEC control, ING MECÁ gearing, VEH control; ~ **del arrancador** *m* AUTO starter control; ~ **automático** *m* ELEC automatic control; ~ **automático del tren** *m* FERRO automatic train control (*ATC*); ~ **por conos de fricción** *m* ING MECÁ cone drive; ~ **de control** *m* CINEMAT control knob; ~ **de derogación de salida** *m* PROD output override instruction; ~ **directo** *m* MECÁ local control; ~ **a distancia** *m* GEN remote control; ~ **eléctrico** *m* ELEC, ING ELÉC electric control; ~ **de encuadre** *m* CINEMAT, TV framing knob; ~ **de enfoque** *m* CINEMAT, FOTO,

INSTR, TV focusing knob; ~ **por engranajes** *m* ING MECÁ, MECÁ gear drive; ~ **de foco** *m* CINEMAT, FOTO, TV follow focus; ~ **de frecuencia de la carga** *m* ELEC load frequency control; ~ **de maniobra del izador** *m* TRANSP AÉR elevator follow-up; ~ **manual** *m* ELEC manual control; ~ **manual del estrangulador** *m* AUTO manual choke control; ~ **manual de la mariposa** *m* AUTO hand throttle control, manual choke control; ~ **paso a paso** *m* ELEC step-by-step control; ~ **de puesta en marcha** *m* TEC ESP on-off control; ~ **del supresor de parásitos** *m* TRANSP MAR *radar* anticlutter control; ~ **por tornillo sin fin** *m* ING MECÁ worm gearing; ~ **de la válvula** *m* AUTO valve control

mandos: ~ **de estado sólido** *m pl* PROD solid-state controls; ~ **invertidos** *m pl* TRANSP AÉR reversed controls

mandril *m* C&V, CARBÓN mandrel, mandril, IMPR reelcore, spindle, ING MECÁ mandrel, chuck, arbor, mandril, broach, MINAS, P&C mandrel, mandril, PAPEL core; ~ **de acción rápida** *m* ING MECÁ quick-action chuck; ~ **autocentrador** *m* ING MECÁ self-centering chuck (*AmE*), self-centring chuck (*BrE*); ~ **de cambio rápido** *m* ING MECÁ quick-action chuck; ~ **de colada** *m* PROD runner stick; ~ **cortador** *m* ING MECÁ cutting drift; ~ **cuadrado** *m* MECÁ drift; ~ **para dar forma a la canalización de la caldera** *m* INSTAL HIDRÁUL boiler pipe shaping-mandrel; ~ **de diámetro regulable** *m* ING MECÁ expanding mandrel; ~ **de dientes insertados** *m* ING MECÁ inserted-tooth broach; ~ **ensanchador** *m* CONST turnpin; ~ **de ensanchar** *m* ING MECÁ cutting driftpin, driftpin; ~ **excéntrico** *m* ING MECÁ eccentric chuck; ~ **de expansión** *m* ING MECÁ expanding mandrel; ~ **extractor de tubos** *m* ING MECÁ tube-drawing mandrel; ~ **funcionando con cojinetes** *m* ING MECÁ mandrel running in bearings, mandril running in bearings; ~ **de papel** *m* EMB paper core; ~ **portafresas** *m* PROD milling-machine arbor, milling-machine cutter arbor; ~ **de tornillo** *m* ING MECÁ screw chuck; ~ **de tornillos** *m* ING MECÁ bell chuck; ~ **de torno** *m* MECÁ lathe chuck; ~ **de torno accionado mecánicamente** *m* ING MECÁ power-operated lathe chuck; ~ **de torno mecanoaccionado** *m* ING MECÁ power-operated lathe chuck; ~ **de torno motorizado** *m* ING MECÁ power-operated lathe chuck; ~ **para tubos de caldera** *m* INSTAL HIDRÁUL boiler-tube expander

mandrilado *m* ING MECÁ drifting

mandriladora *f* ING MECÁ broaching machine

mandrilar *vt* ING MECÁ broach, ream, MECÁ ream, PROD bore, bore out

mandrín *m* CARBÓN, CONST, ING MECÁ, MINAS, PROD mandrel, mandril

mandrinado *m* CARBÓN, CONST boring, ING MECÁ, MECÁ, MINAS *de rocas o piedras* boring, broaching, PROD boring

mandrinadora *f* ING MECÁ, MECÁ, MINAS boring cutter, boring machine; ~ **dúplex** *f* ING MECÁ, MECÁ, MINAS duplex boring machine

manecilla *f* ING MECÁ lever; ~ **desmontable** *f* ING MECÁ detachable handle; ~ **horaria** *f* METR *de reloj* hour hand; ~ **minutera** *f* METR *de reloj* minute hand; ~ **de la puerta** *f* VEH door handle; ~ **del zoom** *f* TV zoom lever

manejabilidad *f* ING MECÁ maneuverability (*AmE*), manoeuvrability (*BrE*)

manejar *vt* CONST, ING MECÁ run, PROD handle, operate, run, TRANSP MAR *embarcación* handle

manejo *m* ING MECÁ handling, maneuver (*AmE*), manoeuvre (*BrE*), PROD manipulating, manipulation, SEG *de materiales peligrosos* handling, TEC ESP control; ~ **de animales** *m* AGRIC cattle management; ~ **de bosque** *m* AGRIC forest management; ~ **del buque** *m* TRANSP MAR shiphandling; ~ **de capacidad del viento y del mar** *m* TRANSP wind and sea state capability handling; ~ **de la carga** *m* TRANSP AÉR cargo handling; ~ **de grúa** *m* CONST derricking crane; ~ **manual** *m* NUCL manual handling; ~ **de materiales** *m* NUCL materials handling; ~ **de pastizales naturales** *m* AGRIC range management; ~ **de riesgos** *m* CALIDAD, SEG risk management; ~ **de la seguridad** *m* CALIDAD safety management; ~ **de suelos** *m* AGRIC soil management

maneta *f* NUCL lever switch

manga *f* AGRIC chute, C&V sleeve, CONTAM *tratamiento de gases, filtración* bag, D&A sleeve, ING MECÁ journal, MINAS *termoiónica* baffle, TRANSP AÉR wind cone, TRANSP MAR *de barco* beam, breadth; ~ **para cargar animales** *f* AGRIC loading chute; ~ **para filtrar** *f* CARBÓN, PROC QUÍ filter bag; ~ **flexible de freno de aire comprimido** *f* AmL (*cf tubo flexible de freno de aire comprimido Esp*) ING MECÁ air-brake hose; ~ **para freno neumático** *f* AmL (*cf tubo para freno neumático Esp*) P&C air-brake hose; ~ **en el fuerte** *f* TRANSP MAR extreme breadth; ~ **indicadora del viento** *f* TRANSP AÉR wind cone, wind sock; ~ **máxima** *f* TRANSP MAR maximum beam; ~ **de mezclado** *f* PROC QUÍ mixing sieve; ~ **de la pala** *f* TRANSP AÉR blade sleeve; ~ **de trazado** *f* TRANSP MAR *diseño de barcos* molded breadth (*AmE*), moulded breadth (*BrE*); ~~**veleta** *f* TRANSP AÉR wind cone, wind sock; ~ **de ventilación** *f* TEC ESP vent

manganeso *m* (*Mn*) C&V, MECÁ, METAL, QUÍMICA manganese (*Mn*); ~ **gris** *m* MINERAL gray manganese ore (*AmE*), grey manganese ore (*BrE*)

mangánico *adj* QUÍMICA manganic

manganífero *adj* GEOL manganiferous

manganita *f* MINERAL, QUÍMICA manganite

manganocalcita *f* MINERAL manganocalcite

manganoso *adj* QUÍMICA manganous

mango *m* CINEMAT handgrip, CONST bolster, grab, D-handle, handle, ING MECÁ grip, handle, shaft, shank, MECÁ, MINAS handle, NUCL journal, PROD helver, hilt, *de martillo, pico* helve, handle, grip; ~ **de apriete** *m* ING MECÁ clamping handle; ~ **con disparador** *m* CINEMAT handgrip with shutter release; ~ **disparador** *m* CINEMAT trigger grip; ~ **de herramienta** *m* ING MECÁ tool holder; ~ **de herramienta con rodillos** *m* ING MECÁ roller tool chest; ~ **de maniobra** *m* AmL (*cf herramienta de giro Esp*) MINAS *sondeos* brace head, brace key, rod-turning tool, tiller; ~ **del martillo** *m* CONST hammer grab; ~ **de muletilla** *m* PROD crutch handle; ~ **de ojo** *m* PROD eyehandle; ~ **de la pala** *m* TRANSP AÉR blade shank; ~ **de palanca** *m* CONST lever handle; ~ **de pico** *m* PROD pick handle; ~ **de seguridad** *m* SEG safety handle

manguera *f* CONST, ING MECÁ hose, hosepipe, LAB hose, rubber tubing, MECÁ hose, P&C rubber tubing, PROD, TEC ESP hose; ~ **de acoplamiento** *f* FERRO *vehículos* coupling hose; **~ de acoplamiento de la cañería del freno** *f* FERRO main brake hose (*AmE*), main brake pipe (*BrE*); **~ de acoplamiento de la tubería del freno** *f* FERRO *vehículos* brake-hose coupling-head (*AmE*), brake-pipe coupling-head (*BrE*); **~ de agua** *f* MINAS water hose; **~ para agua a presión** *f* P&C hydraulic hose; **~ de aire** *f* CONST, MECÁ, MINAS air hose; **~ de aire comprimido** *f* CONST compressed-air line, ING MECÁ air hose, compressed-air line, MECÁ air hose; **~ de aspersión** *f* AGUA squirt hose; **~ blindada** *f* CONST armored hose (*AmE*), armoured hose (*BrE*); **~ de caucho para vapor** *f* ING MECÁ, P&C rubber hose for steam; **~ de descarga** *f* AGUA delivery hose; **~ flexible** *f* CONST flexible hose; **~ de gasolina** *f* Esp (*cf manguera de nafta AmL*) AUTO, P&C, VEH gas hose (*AmE*), gasoline hose (*AmE*), petrol hose (*BrE*); **~ de incendios** *f* CONST, SEG, TERMO fire hose; **~ de inyección de lodos** *f* PETROL mud hose; **~ de lodo** *f* TEC PETR *perforación* rotary hose; **~ metálica flexible** *f* ING MECÁ flexible metallic hose; **~ moldeada** *f* P&C molded hose (*AmE*), moulded hose (*BrE*); **~ de nafta** *f* AmL (*cf manguera de gasolina Esp*) AUTO, P&C, VEH gas hose (*AmE*), gasoline hose (*AmE*), petrol hose (*BrE*); **~ plana de lona** *f* SEG flat canvas hose; **~ plegable** *f* ING MECÁ collapsible water hose; **~ principal de inyección de aire** *f* FERRO *para equipos neumáticos* main air-supply hose, main air-supply pipe; **~ recubierta con tiras de metal flexible** *f* ING MECÁ strip-wound flexible metal hose; **~ reforzada con alambre** *f* ING MECÁ, P&C wire-reinforced hose; **~ reforzada con tela metálica** *f* ING MECÁ, P&C wire-reinforced hose; **~ con refuerzo textil** *f* ING MECÁ textile-reinforced hose; **~ resistente al aceite** *f* P&C oil-resisting hose; **~ resistente a la gasolina y al aceite** *f* Esp (*cf manguera resistente a la nafta y al aceite AmL*) P&C gas-and-oil-resistant hose (*AmE*), gasoline-and-oil-resistant hose (*AmE*), petrol-and-oil-resistant hose (*BrE*); **~ resistente a la nafta y al aceite** *f* AmL (*cf manguera resistente a la gasolina y al aceite Esp*) P&C gas-and-oil-resistant hose (*AmE*), gasoline-and-oil-resistant hose (*AmE*), petrol-and-oil-resistant hose (*BrE*); **~ semi-rígida de descarga** *f* SEG semirigid delivery hose; **~ de succión** *f* AGUA suction hose; **~ de succión de caucho** *f* AGUA, P&C rubber suction hose; **~ trenzada** *f* P&C braided hose; **~ de vacío** *f* ING MECÁ vacuum hose; **~ de vapor** *f* INSTAL HIDRÁUL steam hose; **~ de ventilación** *f* TRANSP MAR wind sail

manguerote *m* TRANSP MAR mushroom ventilator; **~ de ventilación** *m* ING MECÁ *barcos* air duct

mangueta *f* ING MECÁ grip, holdfast, journal, VEH *rueda* axle bushing, stub axle, *término genérico* journal; **~ de torsión** *f* VEH twist grip

manguito *m* AUTO sleeve, CONST *tuberías* boot, pipe joint, IMPR sleeve, ING MECÁ boot, collar, bushing, sleeve, muff, INSTAL HIDRÁUL shuttle, MECÁ socket, bushing, sleeve, collar, PROD sleeve, socket, VEH bush; **~ de acoplamiento** *m* ING MECÁ muff, muff coupling, coupling box, coupling sleeve, MECÁ coupling sleeve; **~ de acoplamiento de la canalización de frenos** *m* VEH brake hose; **~ aislador** *m* ING ELÉC, PROD insulating sleeve; **~ de aislamiento** *m* ING MECÁ insulation bush; **~ de ajuste** *m* AUTO adjusting sleeve; **~ de ajuste forzado** *m* ING MECÁ forced-fit bush; **~ de aleación de cobre** *m* ING MECÁ copper

alloy bush; **~ de alimentación** *m* ING MECÁ feed bush, feed sleeve; **~ del árbol de levas** *m* AUTO, VEH camshaft bushing; **~ de avance** *m* ING MECÁ feed sleeve; **~ de cable** *m* ELEC, ING ELÉC cable conduit; **~ de centrado** *m* ING MECÁ, MECÁ centering bush (*AmE*), centring bush (*BrE*); **~ de centrar** *m* ING MECÁ centering sleeve (*AmE*), centring sleeve (*BrE*); **~ del cilindro** *m* AUTO, VEH cylinder sleeve; **~ del cojinete de desembrague** *m* AUTO release-bearing sleeve, throw-out bearing sleeve, VEH throw-out bearing sleeve; **~ del cojinete de empuje** *m* AUTO, VEH clutch release bearing; **~ del cojinete principal** *m* AUTO main-bearing bushing; **~ para colada** *m* ING MECÁ pouring-sleeve; **~ cónico** *m* ING MECÁ self-holding taper, taper sleeve; **~ desconectable** *m* ING MECÁ loose sleeve; **~ deslizante** *m* AUTO sliding sleeve; **~ desplazable** *m* AUTO sliding sleeve; **~ de dilatación** *m* INSTAL HIDRÁUL expansion coupling; **~ de embrague** *m* MECÁ clutch; **~ del embrague** *m* AUTO, VEH clutch sleeve; **~ de empalme** *m* CONST, PROD, TELECOM jointer; **~ de enfriamiento** *m* PROD *toberas* jumbo; **~ expulsor** *m* ING MECÁ ejector sleeve; **~ del eyector** *m* ING MECÁ ejector sleeve; **~ flotante** *m* ING MECÁ floating bush; **~ guía** *m* ING MECÁ guide bush, NUCL guide bushing; **~ de guía de bronce** *m* ING MECÁ bronze guide bush; **~ de guía de eyección** *m* ING MECÁ ejection-guide bush; **~ de identificación** *m* ING MECÁ identification sleeve; **~ incandescente para gas** *m* TERMO gas mantle; **~ inferior** *m* VEH floor shift; **~ de inflación** *m* CONTAM MAR *de barrera flotante* inflation cuff; **~ de malla para la botella** *m* EMB bottle sleeve; **~ de ondulación** *m* ING MECÁ crimping bush; **~ de PCV** *m* PROD PCV sleeve; **~ de una pieza** *m* ING MECÁ unsplit bush; **~ piloto** *m* AUTO pilot bushing; **~ de pliegue** *m* ING MECÁ crimping bush; **~ portaherramienta** *m* ING MECÁ chuck; **~ portaherramienta neumático** *m* ING MECÁ air-operated chuck; **~ de presión** *m* ING MECÁ pressure sleeve; **~ de protección** *m* ING MECÁ protection sleeve; **~ del radiador** *m* AUTO, TRANSP AÉR, VEH radiator hose; **~ reductor** *m* ING MECÁ reduction sleeve; **~ de refuerzo** *m* ING MECÁ cat head; **~ roscado** *m* CONST *fontanería* nipple, ING MECÁ screw coupling, screw socket, *fontanería* nipple; **~ roscado apriete** *m* ING MECÁ right-and-left coupling; **~ secundario** *m* AUTO secondary sleeve; **~ para tubos** *m* CONST, ING MECÁ pipe coupling; **~ de unión** *m* ING MECÁ coupling sleeve; **~ de ventilación del radiador** *m* AUTO, VEH radiator-vent hose

manida *f* QUÍMICA mannide

manifiesto: **~ de carga** *m* TEC PETR *transporte* bill of lading, TRANSP, TRANSP AÉR cargo manifest, TRANSP MAR cargo manifest, bill of lading; **~ del contenido** *m* EMB *aduanas* contents declaration; **~ de embarque** *m* TRANSP MAR shipping bill

manifold *m* TEC PETR manifold

manigueta *f* ING MECÁ handle lever

manija *f* CONST *cerrajería* locking handle, VEH handle; **~ de enganche** *f* ING MECÁ attachment link

manilla *f* METR hand; **~ de apertura del maletero** *f* VEH boot handle (*BrE*), trunk handle (*AmE*); **~ de conexión del piloto automático** *f* TRANSP AÉR autopilot turn knob; **~ dentada** *f* ING MECÁ dented knob; **~ del elevalunas** *f* AUTO window winder; **~ de enganche del piloto automático** *f* TRANSP AÉR autopilot turn knob; ~ **de la ventanilla** *f* AUTO window winder

maniobra¹: **de** ~ **centrífuga** *adj* ENERG RENOV centrifugally operated

maniobra² *f* ING MECÁ maneuver (*AmE*), manoeuvre (*BrE*), handling, PROD manipulating, operating procedure, TEC ESP handling, TEC PETR *perforación* trip, TRANSP MAR maneuver (*AmE*), manoeuvre (*BrE*); ~ **de acoplamiento** *f* TRANSP rendezvous maneuver (*AmE*), rendezvous manoeuvre (*BrE*); ~ **de agrupamiento** *f* TRANSP rendezvous maneuver (*AmE*), rendezvous manoeuvre (*BrE*); ~ **de agujas** *f* FERRO switch gears; ~ **de amarre** *f* TRANSP MAR docking maneuver (*AmE*), docking manoeuvre (*BrE*); ~ **de apogeo** *f* TEC ESP apogee maneuver (*AmE*), apogee manoeuvre (*BrE*); ~ **de atraque** *f* TRANSP MAR docking maneuver (*AmE*), docking manoeuvre (*BrE*); ~ **de corrección** *f* TEC ESP correction maneuver (*AmE*), correction manoeuvre (*BrE*); ~ **a distancia** *f* AUTO, CINEMAT, CONST, D&A, ELEC, FOTO, ING MECÁ, MECÁ, TEC ESP, TELECOM, TRANSP, TV, VEH remote control; ~ **en rumbo sinuoso** *f* TRANSP weaving maneuver (*AmE*), weaving manoeuvre (*BrE*); ~ **de la señal de parada** *f* FERRO stop signal disc shunting (*BrE*), stop signal disk shunting (*AmE*); ~ **de separación** *f* TEC ESP separation maneuver (*AmE*), separation manoeuvre (*BrE*); ~ **de tierra** *f* TRANSP AÉR ground maneuver (*AmE*), ground manoeuvre (*BrE*); ~ **en vías horizontales** *f* FERRO, TRANSP shunting on level tracks; ~ **de vuelo** *f* TRANSP AÉR flight maneuver (*AmE*), flight manoeuvre (*BrE*); ~ **en zigzag** *f* TRANSP weaving maneuver (*AmE*), weaving manoeuvre (*BrE*)

maniobrabilidad *f* C&V workability, ING MECÁ, NUCL, TRANSP, TRANSP AÉR, TRANSP MAR maneuverability (*AmE*), manoeuvrability (*BrE*)

maniobrable *adj* TRANSP MAR maneuverable (*AmE*), manoeuvrable (*BrE*)

maniobrar¹ *vt* PROD operate, *palanca* shift

maniobrar² *vi* FERRO, ING MECÁ, PROD, TEC ESP, TRANSP, TRANSP MAR maneuver (*AmE*), manoeuvre (*BrE*)

maniobras *f pl* CARBÓN *de vagones* drilling, FERRO *vehículos* shunting

manipulación *f* CARBÓN process, ING MECÁ handling, PAPEL converting, handling, PROD handling, manipulating, manipulation, TELECOM handling, keying, TEXTIL handling; ~ **de cadenas** *f Esp* (*cf manipulación en series AmL*) INFORM&PD *texto* string manipulation; ~ **de camión a camión** *f* TRANSP truck-to-truck handling; ~ **de la carga** *f* TRANSP MAR cargo handling; ~ **por desplazamiento de frecuencia** *f* (*MDF*) ELECTRÓN, INFORM&PD, TELECOM, TV frequency-shift keying (*FSK*); ~ **por desplazamiento mínima** *f* TELECOM minimum-shift keying (*MSK*); ~ **por desplazamiento número** *f* TELECOM minimum-shift keying (*MSK*); ~ **por desviación de fase binaria** *f* TELECOM binary phase-shift keying (*BPSK*); ~ **por desviación de fase correlativa** *f* TELECOM correlative phase-shift keying; ~ **genética** *f* AGRIC genetic engineering; ~ **de mensajes** *f* TELECOM message handling (*MH*); ~ **en series** *f* INFORM&PD string manipulation; ~ **de sucesos** *f* INFORM&PD event handling; ~ **de variación de fase diferencial** *f* (*MVFD*) ELECTRÓN differential phase shift keying (*DPSK*)

manipulado *m* EMB handling

manipulador *m* ING MECÁ manipulating device, PAPEL converter, PROD *robot industrial* manipulating industrial robot, TV keyer; ~ **dactilográfico** *m* TELECOM keyboard sender; ~ **de excepciones** *m* INFORM&PD exception handler; ~ **de títulos** *m* TV title keyer; ~ **universal** *m* NUCL universal manipulator

manipular *vt* MECÁ handle, PAPEL convert, PROD handle, operate, run, TEXTIL handle

maniquí *m* NUCL phantom

manita *f* QUÍMICA mannitan, mannite

manitol *m* QUÍMICA mannitol

manivela *f* AUTO crank, FERRO crank handle, ING MECÁ handle, wheel, MECÁ crank, NUCL lever switch, TEXTIL handle, TRANSP AÉR, VEH crank; ~ **alzacristal** *f* VEH *carrocería* window crank (*AmE*), window regulator (*BrE*); ~ **de arranque** *f* VEH starting crank; ~ **de contrapeso** *f* ING MECÁ balance crank; ~ **elevalunas** *f* VEH *carrocería* window crank (*AmE*), window regulator (*BrE*); ~ **del freno** *f* ING MECÁ brake crank; ~ **del husillo de freno** *f* FERRO brake-screw handle; ~ **de platillo** *f* PROD disc crank (*BrE*), disk crank (*AmE*); ~ **de plato** *f* ING MECÁ wheel crank, PROD disc crank (*BrE*), disk crank (*AmE*); ~ **de rebobinado** *f* CINEMAT backwind handle; ~ **saliente** *f* ING MECÁ overhanging crank; ~ **del tornillo de freno** *f* FERRO brake-screw handle

mano¹: **a** ~ *adv* PROD by hand

mano² *f* CONST *pintura* coat, IMPR quire, P&C *de pintura, recubrimiento* coating, PAPEL *25 hojas de papel* quire; ~ **interior** *f* COLOR undercoat; ~ **de mortero** *f* PROC QUÍ, QUÍMICA pestle; ~ **de obra** *f* MINAS labor (*AmE*), labour (*BrE*), PROD workmanship, manual labor (*AmE*), manual labour (*BrE*), *fábricas* manpower; ~ **de obra complementaria** *f* PROD extra labor (*AmE*), extra labour (*BrE*); ~ **de obra mala** *f* PROD bad workmanship; ~ **de pintura** *f* PETROL coating, PROD, REVEST coat of paint, TEC PETR coating; ~ **superior de pintura** *f* P&C *recubrimiento* top coat of paint

manointensificación *f* PROD pressure intensification

manojo *m* TELECOM *de conductores, cables* bundle

manómetro *m* GEN manometer, pressure gauge (*AmE*), pressure gauge (*BrE*), pressure indicator; ~ **de aceite** *m* AUTO oil-pressure gage (*AmE*), oil-pressure gauge (*BrE*), PROD oil gage (*AmE*), oil gauge (*BrE*); ~ **del aceite** *m* ING MECÁ oil gage (*AmE*), oil gauge (*BrE*); ~ **de aire** *m* ING MECÁ, METR air gage (*AmE*), air gauge (*BrE*); ~ **de aire comprimido** *m* TEC PETR air pressure gage (*AmE*), air pressure gauge (*BrE*); ~ **de alarma** *m* ING MECÁ alarm gage (*AmE*), alarm gauge (*BrE*); ~ **de alta presión** *m* REFRIG high-pressure gage (*AmE*), high-pressure gauge (*BrE*); ~ **de aspiración** *m* REFRIG suction gage (*AmE*), suction gauge (*BrE*); ~ **de Bourdon** *m* FÍS, TEC PETR Bourdon gage (*AmE*), Bourdon gauge (*BrE*); ~ **compuesto** *m* REFRIG compound gage (*AmE*), compound gauge (*BrE*); ~ **de contrapeso** *m* FÍS dead weight pressure gage (*AmE*), dead weight pressure gauge (*BrE*); ~ **diferencial de cabina** *m* TRANSP AÉR cabin differential pressure gage (*AmE*), cabin differential pressure gauge (*BrE*); ~ **de ionización** *m* FÍS ionization gage (*AmE*), ionization gauge (*BrE*), REFRIG ionization vacuum gage (*AmE*), ionization vacuum gauge (*BrE*); ~ **metálico de esfera** *m* ING MECÁ dial-type metallic pressure gage (*AmE*), dial-type metallic pressure gauge (*BrE*); **~ metálico de presión tipo cuadrante** *m* ING MECÁ dial-type metallic pressure gage (*AmE*), dial-type metallic pressure gauge (*BrE*); **~ de montador** *m* REFRIG service gage (*AmE*), service gauge (*BrE*); **~ de muelle** *m* FÍS spring manometer; **~ de presión del aire** *m* TEC PETR air pressure gage (*AmE*), air pressure gauge (*BrE*); **~ de presión de doble combustible** *m* TRANSP AÉR dual-fuel pressure gage (*AmE*), dual-fuel pressure gauge (*BrE*); **~ de presión metálico tipo cuadrante** *m* ING MECÁ dial-type metallic pressure gage (*AmE*), dial-type metallic pressure gauge (*BrE*); **~ de profundidad** *m* MECÁ, METR depth gage (*AmE*), depth gauge (*BrE*); **~ de resistencia eléctrica** *m* METR wire gage (*AmE*), wire gauge (*BrE*); **~ térmico** *m* REFRIG thermal conductivity vacuum gage (*AmE*), thermal conductivity vacuum gauge (*BrE*); **~ de vacío** *m* ALIMENT, ING MECÁ, REFRIG vacuum gage (*AmE*), vacuum gauge (*BrE*); **~ de vapor** *m* INSTAL HIDRÁUL *indicador de presión* steam gage (*AmE*), steam gauge (*BrE*); **~ del viento** *m* PROD blast gage (*AmE*), blast gauge (*BrE*)

manónico *adj* QUÍMICA mannonic
manos: **~ libres** *adj* TELECOM hands-free
manosa *f* QUÍMICA mannose
manostato *m* QUÍMICA pressurestat
manovacuómetro *m* REFRIG compound gage (*AmE*), compound gauge (*BrE*)
manta *f* *Esp* (*cf cobija AmL*) TEC ESP, TEXTIL blanket; **~ contra incendios** *f* SEG fire blanket; **~ eléctrica** *f* ELEC electric blanket, TERMO electric heating-pad, TERMOTEC electric blanket; **~ hipotérmica** *f* REFRIG hypothermic blanket; **~ de rescate** *f* SEG rescue blanket
manteca *f* ALIMENT shortening, lard
mantenedor: **~ de llamas** *m* TRANSP AÉR flame holder; **~ de rumbo** *m* TRANSP AÉR heading hold
mantener[1] *vt* ING MECÁ support; **~ al ancho** *vt* TEXTIL *tela* keep at open width; **~ frío** *vt* EMB keep cool; **~ frío y seco** *vt* EMB keep cool and dry; **~ siempre de pie** *vt* EMB keep upright; **~ vertical** *vt* EMB keep upright
mantener[2]: **~ condiciones de zona aséptica** *vi* SEG maintain aseptic area conditions; **~ rumbo y velocidad** *vi* TRANSP MAR maintain course and speed; **~ un servicio de vigilancia** *vi* TRANSP MAR keep a lookout
mantener[3]: **~ el lugar de trabajo ordenado** *fra* SEG keep the work area tidy
mantenerse: **~ franco de otro buque** *v refl* TRANSP MAR *atraque, desatraque* fend off a collision; **~ sobre la máquina** *v refl* TRANSP MAR heave to; **~ a rumbo** *v refl* TRANSP MAR *red, cuerda, barco* haul
mantenimiento *m* INFORM&PD maintenance, ING ELÉC holding, MECÁ maintenance, PROD maintenance, upkeep, SEG, TEC ESP maintenance, TELECOM servicing, TEXTIL care, TV maintenance; **~ de archivos** *m* INFORM&PD file maintenance; **~ basado en condiciones** *m* CALIDAD condition-based maintenance; **~ de una conversación** *m* TELECOM hold; **~ correctivo** *m* INFORM&PD corrective maintenance; **~ directo** *m* NUCL *por medios manuales* direct maintenance; **~ económico** *m* ING MECÁ cost-effective maintenance; **~ de emergencia** *m* INFORM&PD emergency maintenance; **~ del equipo físico** *m* INFORM&PD hardware maintenance; **~ de ficheros**

m INFORM&PD file maintenance; **~ general** *m* NUCL general maintenance; **~ del hardware** *m* INFORM&PD hardware maintenance; **~ imprevisto** *m* CALIDAD unplanned maintenance; **~ local** *m* TELECOM local maintenance; **~ de las máquinas** *m* TRANSP MAR engine maintenance; **~ de nivel** *m* NUCL level holding; **~ de órbita** *m* TEC ESP station keeping; **~ en órbita** *m* TEC ESP station keeping; **~ pesado** *m* TRANSP AÉR heavy maintenance; **~ planificado** *m* CALIDAD planned maintenance; **~ en posición** *m* TEC ESP station keeping; **~ preventivo** *m* CONST, INFORM&PD, TELECOM preventive maintenance; **~ programado** *m* CALIDAD planned maintenance, INFORM&PD, NUCL scheduled maintenance; **~ de la red** *m* TELECOM network service (*NS*); **~ remediador** *m* INFORM&PD remedial maintenance; **~ rentable** *m* ING MECÁ cost-effective maintenance; **~ reparador** *m* INFORM&PD remedial maintenance; **~ del rumbo** *m* TRANSP AÉR, TRANSP MAR heading hold; **~ a terceros** *m* TRANSP AÉR third party maintenance; **~ de urgencia** *m* INFORM&PD emergency maintenance
mantequera *f* ALIMENT churn
mantequilla: **~ clarificada** *f* ALIMENT clarified butter; **~ cremosa** *f* ALIMENT creamery butter
mantilla[1]: **~ contra mantilla** *adj* IMPR blanket-to-blanket
mantilla[2] *f* IMPR blanket; **~ de caucho** *f* IMPR, P&C rubber blanket; **~ con depresiones** *f* IMPR engraved blanket; **~ de impresión** *f* IMPR impression blanket; **~ satinada** *f* IMPR glazed blanket; **~ superior** *f* IMPR top blanket
mantillo *m* CARBÓN duff, humus, MINAS humble hook, humus, QUÍMICA humus
mantisa *f* INFORM&PD, MATEMÁT mantissa
manto *m* GEOFÍS, GEOL mantle, MINAS *de terreno* seam, NUCL blanket; **~ aluvial** *m* HIDROL alluvial nappe; **~ de cabalgamiento** *m* GEOL overthrust nappe; **~ de calentamiento** *m* *Esp* (*cf canasta de calentamiento AmL*) LAB heating mantle; **~ compuesto** *m* GEOL composite sill; **~ freático** *m* AGUA, GEOL water-bearing stratum; **~ de hielo** *m* OCEAN ice sheet; **~ muy inclinado** *m* MINAS highly-inclined seam; **~ de recubrimiento** *m* GEOL fold nappe
manuable *adj* MECÁ handy
manual[1] *adj* CONST manual, MECÁ hand-operated, manual, PROD manual, SEG hand-held
manual[2] *m* ING MECÁ handbook, PROD handbook, manual; **~ autodidáctico** *m* PROD self-teaching manual; **~ de averías** *m* FERRO breakdown manual; **~ de calidad** *m* CALIDAD, TEC ESP quality manual; **~ del diseñador** *m* TELECOM designer handbook; **~ de instrucciones** *m* PROD instruction book, instruction manual; **~ de mantenimiento** *m* PROD, SEG, TRANSP AÉR maintenance manual; **~ de operación de la tripulación** *m* TRANSP AÉR crew operating manual; **~ de operaciones** *m* ING MECÁ operating manual, TRANSP MAR operations manual; **~ del usuario** *m* INFORM&PD user manual; **~ de vuelo** *m* TRANSP AÉR flight manual
manuar *m* TEXTIL drawframe
manubrio *m* MECÁ, TRANSP AÉR crank; **~ doble** *m* ING MECÁ duplex crank; **~ del motor** *m* ING MECÁ engine crank
manufactura *f* C&V, C&V, ING MECÁ manufacture, manufacturing, PROD manufacture, manufacturing, *textiles* make

manufacturar *vt* C&V, ING MECÁ, PROD work
manuscrito *m* IMPR script
manutención *f* EMB *transportes de materiales, objetos* handling, PROD upkeep
manzana *f Esp (cf cuadra AmL)* CONST block
mapa: ~ **de almacenamiento** *m* INFORM&PD storage map; ~ **de asignaciones** *m* TELECOM assignment map; ~ **de bits** *m* IMPR, INFORM&PD bit map; ~ **de carreteras** *m* AUTO, TRANSP, VEH road map; ~ **de colores** *m* INFORM&PD color map (*AmE*), colour map (*BrE*); ~ **de contorno de estructura** *m* GEOL structure contour map; ~ **de contorno de estructura de subsuperficie** *m* GEOL structure subsurface contour map; ~ **de direcciones** *m* INFORM&PD address mapping; ~ **espectral** *m* TEC ESP spectral map; ~ **estructural** *m* GEOL structural map; ~ **de flujo** *m* NUCL macroscopic flux variation, flux map; ~ **hipsométrico** *m* HIDROL hypsometric map; ~ **isobárico** *m* METEO isobaric map; ~ **de isópacas** *m* GEOL, TEC PETR isopach map; ~ **de Karnaugh** *m* INFORM&PD Karnaugh map; ~ **magnético** *m* GEOFÍS magnetic map; ~ **de memoria** *m* INFORM&PD memory map, storage map, PROD memory map ~ **potenciométrico** *m* TEC PETR *geología* potentiometric map; ~ **de presión de superficie** *m* METEO surface pressure chart; ~ **de red** *m* TELECOM network map; ~ **sinóptico** *m* METEO weather chart; ~ **sismológico** *m* FÍS, GEOFÍS seismological map; ~ **tectónico** *m* GEOL tectonic map; ~ **del tiempo** *m* METEO weather chart; ~ **topográfico** *m* D&A ordnance survey map; ~ **vial** *m* AUTO, TRANSP, VEH road map; ~ **de los vientos** *m* METEO wind chart
mapeado: ~ **de dirección** *m* INFORM&PD address mapping
maqueta *f* CINEMAT model set, PROD, TEC ESP, TRANSP MAR mock-up; ~ **blanda** *f* TEC ESP soft mock-up; ~ **a escala** *f* CINEMAT scale model; ~ **del libro** *f* IMPR mock-up
maquetista *m* CINEMAT model builder
máquina[1] *f* AUTO, C&V, ING ELÉC machine, ING MECÁ motor, appliance, contrivance, PROD machine, TEC ESP motor, TRANSP MAR engine; **~ a** ~ **para abrir galerías** *f* CARBÓN miner, MINAS inseam miner, miner; ~ **de acanalar** *f* ING MECÁ grooving machine, MINAS quarrying machine; ~ **de achaflanar** *f* ING MECÁ beveling machine (*AmE*), bevelling machine (*BrE*); ~ **de afilar** *f* ING MECÁ grinder, grinding machine, MECÁ grinding machine; ~ **de agotamiento** *f* AGUA, MINAS, TEC PETR pumping engine; ~ **alternativa** *f* ING MECÁ, TRANSP AÉR, TRANSP MAR, VEH reciprocating engine; ~ **alternativa de vapor** *f* INSTAL HIDRÁUL riding cut-off valve; ~ **alveolar** *f* AGRIC length grader; ~ **de amalgamación** *f* ING MECÁ amalgamator; ~ **amalgamadora** *f* CARBÓN amalgamator; ~ **amasadora de barro** *f* C&V clay-kneading machine; ~ **de amasar y mexclar** *f* ING MECÁ malaxator; ~ **de amolar** *f* MECÁ, ING MECÁ grinding machine; ~ **de amolar y cortar de disco abrasivo** *f* ING MECÁ foundry abrasive cutoff-and-grinding machine; ~ **de amolar herramientas** *f* ING MECÁ tool grinder; ~ **analítica** *f* INFORM&PD analytical engine; ~ **de anuncios registrados** *f* TELECOM recorded announcement machine; ~ **aplanadora de chapas** *f* FERRO plate-straightening machine; ~ **aplicadora de adhesivos** *f* EMB adhesive machine; ~ **aplicadora de**

fundas retráctiles *f* EMB shrink-sleeve wrapping machine; ~ **de aprendizaje** *f* ELECTRÓN learning machine; ~ **de aprestar cilíndrica** *f* TEXTIL cylinder-sizing machine; ~ **para arcosoldadura** *f* MECÁ arc welding machine; ~ **de arranque** *f AmL (cf máquina de mina Esp)* MINAS mining machine; ~ **de asentar** *f* MECÁ honing machine; ~ **de aserrar** *f* CONST sawing machine; ~ **para aserrar metales** *f* PROD metal-sawing machine; ~ **asfaltadora** *f* AUTO road-tarring machine, CONST, TRANSP asphalt-spreading machine (*AmE*), road-metal-spreading machine (*BrE*), road-tarring machine, VEH road-tarring machine; ~ **asíncrona** *f* ING ELÉC asynchronous machine; ~ **de aspersión** *f* C&V biscuit dipper; ~ **de aterrajar** *f* ING MECÁ tube-screwing machine; ~ **de Atwood** *f* FÍS Atwood's machine; ~ **de autoaprendizaje** *f* INFORM&PD self-learning machine; ~ **autodidáctica** *f* INFORM&PD self-learning machine; ~ **automática para aplicar etiquetas autoadhesivas** *f* EMB fully-automatic self-adhesive labeling machine (*AmE*), fully-automatic self-adhesive labelling machine (*BrE*); ~ **automática para tallar engranajes de dentadura recta** *f* ING MECÁ automatic spur gear cutting machine; ~ **auxiliar** *f* ING MECÁ auxiliary engine, TRANSP MAR auxiliary engine, auxiliary machinery; ~ **de avisos** *f* TELECOM announcement machine; ~ **azoladora de traviesa** *f* FERRO railroad tie-adzing machine (*AmE*), railway tie-adzing machine (*BrE*); **~ b** ~ **de balancín** *f* ING MECÁ beam engine; ~ **de barnizar** *f* PROD, REVEST varnishing machine; ~ **de barrenar** *f* ING MECÁ boring machine, MECÁ boring machine, drilling machine, MINAS boring machine; ~ **con biela conectada a una cruceta** *f* TRANSP MAR crosshead engine; ~ **bipolar** *f* ELEC bipolar machine; ~ **con brazo soporte** *f* ING MECÁ overarm machine; ~ **de brazo X** *f* C&V x-arm machine; ~ **de byte** *f* INFORM&PD, PROD, TELECOM byte machine; **~ c** ~ **de CA** *f* ING ELÉC AC machine; ~ **de cajera de polea** *f* ING MECÁ pulley key-seating machine; ~ **de calcular** *f* ELECTRÓN, FÍS, INFORM&PD, MATEMÁT calculating machine; ~ **de cantonear** *f* EMB banding machine; ~ **sin carga** *f* PROD machine running empty; ~ **de carga automática por el fondo y el lateral** *f* EMB automatic side and bottom loading machine; ~ **cargadora** *f* PROD charging machine; ~ **de cargar bolsas** *f* EMB bag-loading machine; ~ **de cartón** *f* PAPEL board machine; ~ **de cartón de doble tela** *f* PAPEL twin-wire board machine; ~ **de cartón de formas redondas** *f* PAPEL multivat board machine; ~ **de cartón húmedo** *f* PAPEL wet board machine; ~ **de CC** *f* ING ELÉC DC machine; ~ **de centrar** *f* ING MECÁ, MECÁ centering machine (*AmE*), centring machine (*BrE*); ~ **centrífuga** *f* FÍS, HIDROL, LAB, MECÁ, P&C, PROC QUÍ, QUÍMICA centrifuge; ~ **de cepillar y sacar a gruesos** *f* ING MECÁ planing and thicknessing machine; ~ **cerradora** *f* EMB closing machine; ~ **cerradora de latas** *f* EMB can-closing machine (*AmE*), tin-closing machine (*BrE*); ~ **cerradora al vacío** *f* EMB vacuum-closing machine; ~ **de cerrar** *f* EMB sealing machine; ~ **de cilindro invertido** *f* PROD inverted cylinder engine; ~ **con circulación de circuito abierto** *f* ING MECÁ machine with open-circuit ventilation; ~ **con circulación de circuito cerrado** *f* ING MECÁ machine with closed-circuit ventilation; ~ **de cizallar** *f* ING MECÁ shears; ~ **de**

cizallar y punzonar *f* ING MECÁ shearing and punching machine; ~ de clasificación e inspección de tabletas *f* EMB tablet sorting and inspection machine; ~ clasificadora *f* EMB sorting machine; ~ para clavar clavos *f* EMB nailing machine; ~ de CN *f* ING MECÁ NC machine; ~ colocadora de pernos *f Esp* MINAS roof-bolting drilling machine; ~ colocadora de redondos *f AmL* (*cf máquina colocadora de pernos Esp*) MINAS roof-bolting drilling machine; ~ para colocar y extraer objetos envasados en láminas al vacío *f* EMB case-packing and unpacking of ampoules machine; ~ para colocar fajas de papel *f* EMB paper-banding machine; ~ para colocar machos *f* PROD coring machine; ~ de colocar tapas *f* IMPR casing-in machine; ~ de colocar tapones *f* EMB corking machine; ~ para colocar los tapones roscados *f* EMB bottle-capping machine; ~ para colocar vía prefabricada *f* FERRO track panel laying machine; ~ combinada de mortajar, moldear, cepillar y acabado *f* ING MECÁ combined surfacing, planing, molding, and slot-mortising machine; ~ combinada de punzonar, cortar, forjar y estampar *f* ING MECÁ combined stamping, forging, shearing, and punching machine; ~ combinada de soldadura, contracción y doblado *f* ING MECÁ combined bending, shrinking, and welding machine; ~ compensadora *f* ING MECÁ balancing machine; ~ de comprobación *f* ING MECÁ trying-up machine; ~ de conformado y sellado en vertical y horizontal *f* EMB vertical and horizontal form-fill seal machine; ~ de contabilidad *f* INFORM&PD accounting machine; ~ contable *f* INFORM&PD accounting machine; ~ contadora de tapas *f* EMB countertop machine; ~ continua para envolver *f* EMB flow-wrapping machine; ~ de contrachapado *f* ING MECÁ veneer-splicing machine; ~ de control numérico *f* ING MECÁ numerical control machine; ~ controlada por láser *f* CONST laser-controlled machine; ~ coordinada de fresar y mandrinas *f* ING MECÁ coordinate boring-and-milling machine; ~ de corriente alterna *f* ELEC alternating-current machine; ~ cortacarriles *f* FERRO rail-cutting machine; ~ cortadora de formas *f* C&V shape-cutting machine; ~ cortadora de ladrillos y azulejos *f* C&V brick-and-tile machine; ~ de cortar *f* ING MECÁ cutter; ~ de cortar de disco abrasivo *f* ING MECÁ abrasive cutoff machine; ~ para cortar ladrillos y azulejos *f* C&V brick-and-tile machine; ~ de coser *f* IMPR, TEXTIL sewing machine; ~ de coser bolsas *f* EMB bag-stitching machine; ~ cribadora *f* CARBÓN, PROD riddler; ~ de cubitos *f* REFRIG ice maker; ~ para cubrir chocolate *f* ALIMENT chocolate-coating machine; ~ cultivadora *f* AGRIC rolling fork; ~ curvadora *f* EMB, MECÁ bending machine; ~ curvadora de carriles *f* FERRO rail-bending machine;

- d ~ de desagüe *f* AGUA, MINAS, TEC PETR pumping engine; ~ de desbarbar *f* ING MECÁ deburring machine; ~ de desbastar *f* PROD carving machine; ~ de desincrustar calderas *f* FERRO boiler-scaling appliance; ~ para desinfectar semillas contra tizón *f* AGRIC smutter; ~ de desmoldear *f* PROD *funderías* draft machine (*AmE*), draught machine (*BrE*); ~ de desmoldear de inversión *f* PROD roll-over drop machine, *funderías* roll-over draft machine (*AmE*), roll-over draught machine (*BrE*); ~ de desmoldear

de palanca *f* PROD *funderías* lever draught machine (*BrE*), lever draft machine (*AmE*); ~ para desnudar cables *f* ELEC, ING ELÉC, ING MECÁ cable-sheath stripper; ~ despanojadora *f* AGRIC detasseling machine (*AmE*), detasselling machine (*BrE*); ~ de despelusar *f* AGRIC scalper; ~ desterronadora *f* AGRIC scrubber; ~ para determinar la dureza por el método Brinell *f* ING MECÁ, MECÁ Brinell-hardness testing-machine; ~ de devanado *f* P&C, PAPEL reeling machine; ~ dinamoeléctrica excitada en serie *f* ELEC series-wound dynamo; ~ distribuidora *f* EMB dispensing machine; ~ dobladora *f* EMB, MECÁ bending machine; ~ dobladora y cosedora del fondo *f* EMB bottom-folding-and-seaming machine; ~ de dos mesas *f* C&V two-table machine;

- e ~ eléctrica *f* ELEC, ING ELÉC electric machine, electrical machine; ~ eléctrica de rotación *f* ELEC, ING ELÉC rotating electrical machine; ~ de electro-erosión por cables *f* ING ELÉC, ING MECÁ wire spark machine; ~ electroestática *f* ING ELÉC static electrical machine; ~ electrostática de influencia *f* ELEC, ING ELÉC electrostatic generator; ~ embaladora *f* IMPR bundle-type machine; ~ de embalaje exterior *f* EMB exterior-packaging machine; ~ embolsadora *f* EMB bagging machine; ~ embotelladora *f* ALIMENT, EMB bottling machine; ~ embutidora *f* EMB deep-drawing machine; ~ empalmadora *f* CINEMAT machine splicer; ~ empalmadora transversal de chapas de madera *f* ING MECÁ crosswise veneer-splicing machine; ~ de empaquetado al vacío *f* EMB, ING MECÁ vacuum-packaging tool; ~ empaquetadora al vacío *f* EMB vacuum-packaging machine; ~ para el empernado del techo *f* MINAS roof-bolting drilling machine; ~ de encabezar pernos *f* PROD bolt header; ~ encoladora *f* PAPEL pasting machine; ~ encoladora de bordes *f* EMB edge-gumming machine; ~ de enderezar y curvar chapas *f* ING MECÁ plate-flattening-and-bending machine; ~ engomadora *f* EMB glue-gumming machine; ~ enlatadora *f* EMB can-packing machine (*AmE*), tin-packing machine (*BrE*); ~ de ensamblar automática *f* ING MECÁ automatic assembly machine; ~ entalladora *f* CARBÓN nicking machine; ~ de entubación de vidrio verde *f* C&V bottle-casing machine; ~ para envasar en láminas al vacío *f* EMB blister-edge-and-foil machine, blister-packaging machine; ~ de envasar en porciones *f* EMB portioning machine; ~ de envolver *f* EMB enveloping machine, wrapping machine; ~ para envolver bandejas con un material estirable *f Esp* (*cf máquina para envolver pálets con un material estirable AmL*) EMB pallet stretch-wrapping machine; ~ de envolver en horizontal y vertical *f* EMB horizontal and vertical wrapping machine; ~ para envolver pálets con un material estirable *f AmL* (*cf máquina para envolver bandejas con un material estirable Esp*) EMB pallet stretch-wrapping machine; ~ de envolver con película *f* EMB film-wrapping machine; ~ escarbadora *f* CARBÓN scraper; ~ de escariar *f* PROD reamer; ~ escarificadora *f* CONST ripper; ~ de escribir *f* INFORM&PD typewriter; ~ esmaltadora *f* FOTO glazing machine; ~ de esmerilado plano *f* C&V flat-grinding machine; ~ esmeriladora *f* PROD emery-grinding machine, emery machine; ~ esmeriladora de ingletes *f* C&V miter grinding machine

(*AmE*), mitre grinding machine (*BrE*); ~ **de esmerilar y cortar de disco abrasivo** *f* ING MECÁ foundry abrasive cutoff-and-grinding machine; ~ **esparcidora de la cola** *f* EMB glue-spreading machine; ~ **de espigar** *f* ING MECÁ tenoning machine; ~ **estampadora** *f* ING MECÁ die-casting machine; ~ **de estampar** *f* TEXTIL printing machine; ~ **para estampar en relieve** *f* EMB, IMPR, PAPEL embossing machine; ~ **de estirado** *f* C&V drawing machine; ~ **de etiquetado semiautomática** *f* EMB semiautomatic labeling machine (*AmE*), semiautomatic labelling machine (*BrE*); ~ **etiquetadora** *f* EMB labeling machine (*AmE*), labelling machine (*BrE*), PROD label stamper; ~ **de excitación en serie** *f* ELEC series excited machine; ~ **excitada en serie** *f* ELEC series-wound machine; ~ **de expansión** *f* REVEST spreading machine; ~ **expendedora de billetes** *f* TRANSP ticket machine; ~ **explanadora** *f* CONTAM MAR, TRANSP grader; ~ **exploradora** *f* FERRO *locomotora* cow catcher, guard iron, pilot; ~ **de extracción** *f* MINAS hoist, hoisting engine, winding engine; ~ **de extracción sin cables** *f* MINAS ropeless hoisting apparatus; ~ **para extraer, llenar y tapar bandejas** *f* EMB tray denesting-filling-and-lidding machine; ~ **para extrudir** *f* EMB extrusion machine, P&C extruding machine, *equipo* extrusion machine, PAPEL extrusion machine; ~ **extrusionadora** *f* EMB, P&C, PAPEL extrusion machine;

■ **f** ~ **de fabricar bolsas** *f* EMB pouch-making machine; ~ **para fabricar bolsas** *f* EMB bag-making machine; ~ **para fabricar cajas de cartón** *f* EMB box-making machine, keyway-cutting machine; ~ **para fabricar envases flexibles** *f* EMB flexible packaging machine; ~ **para fabricar hojas de papel de estaño** *f* EMB tin-foiling machine; ~ **de fabricar tornillos** *f* MECÁ screw machine; ~ **de facsímil** *f* IMPR, INFORM&PD, TELECOM facsimile machine; ~ **de filetear** *f* ING MECÁ chasing machine, screw machine, screwing machine; ~ **de filetear tubos** *f* ING MECÁ tube-screwing machine; ~ **flejadora** *f* EMB strapper; ~ **flexográfica-dobladora-encoladora** *f* EMB flexo-folder-gluer; ~ **de flujo transversal** *f* TRANSP transverse-flux machine; ~ **para forjar pernos** *f* PROD bolt-forging machine; ~ **para formar, llenar y cerrar envases** *f* EMB form-fill-and-seal machine; ~ **de formas redondas** *f* PAPEL vat machine; ~ **de fresar chaveteros** *f* PROD keyway-cutting machine; ~ **frigorífica** *f* ALIMENT freezing machine, ING MECÁ, PROC QUÍ freezer, PROD freezing machine, REFRIG refrigerating machine, freezer, freezing machine, TERMO freezer; ~ **frigorífica de absorción** *f* REFRIG absorption refrigerating machine; ~ **frigorífica de aire** *f* REFRIG air-cycle refrigeration machine; ~ **fuente** *f* INFORM&PD source machine; ~ **con funcionamiento defectuoso** *f* MINAS crank-blasting machine; ~ **funcionando con carga** *f* ING MECÁ machine running under load; ~ **funcionando sin carga** *f* ING MECÁ machine running on no load; ~ **funcionando en vacío** *f* ING MECÁ machine running light, machine running on no load; ~ **de fundición y de composición simple** *f* IMPR single-type composing and casting machine; ~ **fundidora de tipos** *f* IMPR caster;

■ **g** ~ **de gas** *f* PROD, TERMO gas engine; ~ **generadora de energía** *f* ING MECÁ prime mover; ~ **para grabar** *f* IMPR etching machine; ~ **de grabar matrices en hueco** *f* ING MECÁ die-sinking machine; ~ **granuladora** *f* PROC QUÍ granulating machine; ~ **grapadora de ángulos** *f* EMB, IMPR corner-stapling machine;

■ **h** ~ **para hacer cajas** *f* EMB boxing machine; ~ **para hacer cerámica** *f* C&V ceramic machine; ~ **de hacer envases de cartón** *f* EMB cartoning machine; ~ **para hacer hielo** *f* REFRIG ice-making machine; ~ **de hacer juntas** *f* PROD jointing machine; ~ **de hacer pernos** *f* ING MECÁ bolt cutter; ~ **de hacer roscas a los tornillos** *f* ING MECÁ screw machine; ~ **para hacer las tapas** *f* IMPR case-making machine; ~ **de hacer tornillos** *f* ING MECÁ screw machine, tapping machine; ~ **herramienta** *f* ING MECÁ, MECÁ machine tool; ~ **herramienta de CN** *f* ING MECÁ NC machine tool; ~ **herramienta de control numérico** *f* ING MECÁ numerical control machine tool; ~ **herramienta para usos generales** *f* PROD general-purpose machine tool; ~ **homopolar** *f* ING ELÉC homopolar machine; ~ **horizontal de hacer envases de cartón** *f* EMB horizontal cartoning machine;

■ **i** ~ **de igualar tablillas** *f* ING MECÁ trying-up machine; ~ **impregnadora** *f* EMB impregnating machine; ~ **de imprimir** *f* IMPR printer machine, printing machine; ~ **inductora** *f* ING ELÉC inductor machine; ~ **de inferencia** *f* INFORM&PD inference engine; ~ **de influencia** *f* ELEC, ING ELÉC electrostatic generator; ~ **de inglete** *f* CONST miter machine (*AmE*), mitre machine (*BrE*); ~ **insonizadora** *f* SEG sound-absorbing machine; ~ **intermitente para cartón** *f* PAPEL intermittent board machine; ~ **de inyección** *f* SEG injection machine; ~ **IS** *f* C&V IS machine;

■ **l** ~ **lacadora** *f* EMB lacquering machine; ~ **ladrillera** *f* C&V brick-molding machine (*AmE*), brick-moulding machine (*BrE*); ~ **lapidadora** *f* MECÁ, PROD lapping machine; ~ **de lavar** *f* TEXTIL washing machine; ~ **para lavar minerales** *f* ING MECÁ patouillet; ~ **lijadora de disco** *f* CONST, PROD disc sandpapering machine (*BrE*), disk sandpapering machine (*AmE*); ~ **lijadora a discos** *f* CONST, ING MECÁ disc sanding machine (*BrE*), disk sanding machine (*AmE*); ~ **lijadora de doble disco** *f* PROD double disc sandpapering machine (*BrE*), double disk sandpapering machine (*AmE*); ~ **para limpiar** *f* PROD cleaning machine; ~ **de litografía de haz de electrones** *f* ELECTRÓN electron-beam lithography machine; ~ **de llamada** *f* TELECOM ringing machine; ~ **de llenado de cajas** *f* EMB carton filler; ~ **de llenado y cierre de tubos** *f* EMB tube-filling and closing machine; ~ **de llenado y precintado** *f* EMB fill-and-seal machine; ~ **de llenado al vacío** *f* EMB vacuum filling machine; ~ **llenadora** *f* EMB filling machine; ~ **llenadora y dosificadora** *f* EMB filling and dosing machine; ~ **llenadora y encapsuladora** *f* EMB filling and capping machine; ~ **llenadora de latas** *f* EMB can-filling machine (*AmE*), tin-filling machine (*BrE*); ~ **llenadora por peso** *f* EMB weight filling machine; ~ **para llenar automáticamente bolsas flexibles** *f* EMB automatic flexible bag-filling machine; ~ **para llenar bolsas** *f* EMB bag-filling machine; ~ **para llenar vasos con productos en polvo** *f* EMB in-cup powder-filling machine; ~ **lógica** *f* ELECTRÓN logical machine;

■ **m** ~ **de machihembrar** *f* CONST tonguing-and-grooving machine; ~ **magnetoeléctrica** *f* AUTO, ING ELÉC magneto; ~ **mandrinadora coordinada** *f* ING MECÁ coordinate boring-and-milling machine; ~ **de mandrinar** *f* MINAS borer; ~ **de mano** *f* SEG hand-guided machine; ~ **manual** *f* EMB hand-operated machine, SEG hand-held machine; ~ **manual de hojas** *f* IMPR hand sheet machine; ~ **marcadora** *f* EMB marking machine; ~ **marchando en vacío** *f* PROD empty machine; ~ **medidora automática** *f* AmL (*cf cajero automático Esp*) INFORM&PD automatic teller machine (*ATM*); ~ **de medir** *f* METR measuring machine; ~ **de mesa plana** *f* PAPEL Fourdrinier paper machine; ~ **metalgráfica** *f* IMPR metal decorating machine; ~ **mezcladora** *f* P&C blender; ~ **mezcladora de barro** *f* C&V clay-mixing machine; ~ **para mezclar la nieve y apisonarla** *f* PROD, TRANSP pulvimixer; ~ **de mina** *f* Esp (*cf máquina de arranque AmL*) MINAS mining machine; ~ **de modelar con torno** *f* CONST spindle-molding machine (*AmE*), spindle-moulding machine (*BrE*); ~ **moldeadora** *f* ING MECÁ die-casting machine; ~ **de moldear** *f* CONST, PROD molding machine (*AmE*), moulding machine (*BrE*); ~ **de moldear para banco** *f* PROD bench molding machine (*AmE*), bench moulding machine (*BrE*); ~ **de moldear por inyección** *f* SEG injection molding machine (*AmE*), injection moulding machine (*BrE*); ~ **de moldear ladrillos** *f* C&V brick-molding machine (*AmE*), brick-moulding machine (*BrE*); ~ **de moldeo por compresión** *f* EMB, ING MECÁ, P&C compression-molding machine (*AmE*), compression-moulding machine (*BrE*); ~ **de moldeo por inyección** *f* EMB injection blow molding machine (*AmE*), injection blow moulding machine (*BrE*); ~ **de moldurar** *f* PROD molding machine (*AmE*), moulding machine (*BrE*); ~ **monofásica** *f* ING ELÉC single-phase machine; ~ **montadora** *f* CINEMAT, TV editing machine; ~ **de montaje y cierre de cajas** *f* EMB carton erector and closer; ~ **para montar cajas plegables** *f* EMB folding-box setting machine; ~ **para montar, llenar y cerrar** *f* EMB case-erecting, filling and closing machine; ~ **de mortajar y taladrar** *f* PROD mortising-and-boring machine; ~ **motriz** *f* ING MECÁ prime mover; ~ **multifilamento** *f* TEXTIL multifilament machine; ~ **multivía** *f* EMB multilane machine;

■ **n** ~ **neumática** *f* ING MECÁ aspiring pump; ~ **de nieve** *f* CINEMAT snow machine; ~ **normadora de cajas** *f* EMB box-erecting machine; ~ **de numeración marginal** *f* CINEMAT edge-numbering machine;

■ **o** ~ **objeto** *f* INFORM&PD object machine; ~ **offset de bobina** *f* IMPR web offset press; ~ **de oxicorte** *f* CONST, MECÁ, TERMO flame-cutter; ~ **de oxicorte y refundición** *f* C&V burning-off and edge-melting machine;

■ **p** ~ **de papel** *f* PAPEL paper machine; ~ **de papel de doble tela** *f* PAPEL twin-wire paper machine; ~ **de papel de mesa invertida** *f* PAPEL harper machine; ~ **de papel tisú** *f* PAPEL tissue machine; ~ **parada** *f* ING MECÁ machine lying idle; ~ **paralizada** *f* ING MECÁ machine lying idle; ~ **de pasta en hojas plegadas** *f* PAPEL lap machine; ~ **perforadora** *f* MINAS drilling rig; ~ **perforadora ignífuga** *f* MINAS, SEG, TERMO fireproof drilling machine, flameproof drilling machine; ~ **de perforar** *f* ING MECÁ, MECÁ, MINAS drilling machine; ~ **pilón** *f* PROD overhead

engine; ~ **de pistón** *f* ING MECÁ piston engine; ~ **de pistón tubular** *f* TRANSP MAR trunk piston engine; ~ **de planchado final** *f* TEXTIL press-finishing machine; ~ **plegadora y cosedora** *f* EMB, PROD folding-and-seaming machine; ~ **plegadora y encoladora** *f* EMB creasing-and-gluing machine; ~ **plegadora y ranuradora** *f* EMB creasing-and-scoring machine; ~ **de plisar** *f* TEXTIL pleating machine; ~ **de poliexpansión** *f* PROD multiple expansion engine; ~ **de polo auxiliar** *f* ELEC interpole machine; ~ **de polo de conmutación** *f* ELEC interpole machine; ~ **para poner vidrios** *f* C&V glazing machine; ~ **portátil** *f* SEG portable machine; ~ **precintadora** *f* EMB sealing machine; ~ **de preparar cantos de chapa** *f* AmL (*cf máquina rebordeadora Esp*) INSTR para soldar edging machine; ~ **procesadora** *f* IMPR *fotografía* processing machine; ~ **propulsora** *f* TRANSP MAR propulsion engine; ~ **de protocolos de información de gestión común** *f* TELECOM common management information protocol machine (*CMIPM*); ~ **para prueba de dureza Rockwell** *f* ING MECÁ Rockwell hardness-testing machine; ~ **de prueba de dureza Vickers** *f* ING MECÁ Vickers hardness-testing machine; ~ **a prueba de grisú** *f* CARBÓN firedamp-proof machine; ~ **pulidora** *f* MECÁ, PROD lapping machine; ~ **pulidora circular** *f* CONST, ING MECÁ disc sanding machine (*BrE*), disk sanding machine (*AmE*); ~ **pulidora de discos** *f* CONST, ING MECÁ disc sanding machine (*BrE*), disk sanding machine (*AmE*); ~ **con pulsador** *f* ING MECÁ push-button machine; ~ **de punzonar y cizallar** *f* ING MECÁ punching-and-shearing machine;

■ **q** ~ **que corta inflorescencias masculinas** *f* AGRIC detasseling machine (*AmE*), detasselling machine (*BrE*); ~ **que imprime y coloca la etiqueta** *f* EMB print-and-apply labeling machine (*AmE*), print-and-apply labelling machine (*BrE*); ~ **quitanieves** *f* TRANSP snow blower;

■ **r** ~ **rafadora** *f* CARBÓN holing machine; ~ **rafadora de barra** *f* MINAS bar coal-cutting machine; ~ **rafadora de carbón** *f* CARBÓN, MINAS miner; ~ **rafadora para carbón** *f* CARBÓN coal cutter, coal-cutting machine, MINAS coal cutter, coal-cutting machine, in-seam miner; ~ **rafadora y rozadora** *f* CARBÓN holing-and-shearing machine; ~ **de ranurar** *f* CONST recessing machine, ING MECÁ grooving machine, PROD keyway-cutting machine; ~ **de ranurar portátil** *f* PROD keyway-cutting tool; ~ **Raschel de punto por urdimbre** *f* TEXTIL Raschel knitting machine; ~ **de rayado a pluma** *f* IMPR pen-ruling machine; ~ **de rebajar** *f* CONST recessing machine; ~ **rebordeadora** *f* Esp (*cf máquina de preparar cantos de chapa AmL*) INSTR *para soldar* edging machine; ~ **de recepción** *f* ELECTRÓN, TELECOM receive machine; ~ **de rectificación** *f* MECÁ, PROD lapping machine; ~ **rectificadora cilíndrica universal de control numérico** *f* ING MECÁ compact CNC universal cylindrical grinding machine; ~ **de rectificar** *f* ING MECÁ grinder, grinding machine, MECÁ grinding machine; ~ **de rectificar y bruñir** *f* MECÁ honing machine; ~ **de rectificar las correderas** *f* PROD link grinder; ~ **de rectificar muelas verticales** *f* PROD edge wheel grinding machine; ~ **recubridora de cortinas** *f* ING MECÁ, PAPEL, REVEST curtain-coating machine; ~ **para regar con alquitrán** *f* CONST tar sprayer; ~ **rellenadora** f PROD filler, TRANSP back filler; ~ **rellenadora por densidad** f EMB density filling machine; ~ **remachadora** f CONST riveting machine; ~ **de remoldear** f PROD *fundería* setting-up machine; ~ **de remolque** f TRANSP trailer-towing machine; ~ **de repicar** f PAPEL sharpening machine; ~ **de repique** f TELECOM ringing machine; ~ **de retiración** f IMPR perfecting machine; ~ **de retorno de la hoja** f EMB foil backing machine; ~ **retractiladora** f EMB film-applying lid and heat-sealing machine, film-wrapping machine; ~ **para revelado** f FOTO processing machine; ~ **revestidora de hilos** f PROD wire-coating machine; ~ **revestidora para perfiles laminados** f P&C bar coater; ~ **de revestimiento de cortinas** f ING MECÁ, PAPEL, REVEST curtain-coating machine; ~ **de revestir** f P&C, REVEST coating machine; ~ **de rodillos** f PAPEL cylinder machine; ~ **roscadora de interiores** f ING MECÁ tapping machine; ~ **roscadora de tubos** f ING MECÁ tube-screwing machine; ~ **de roscar** f ING MECÁ screw machine, screwing machine, threader; ~ **de roscar y aterrajar** f ING MECÁ screwing-and-tapping machine; ~ **rotativa de alta velocidad para comprimir tabletas** f EMB high-speed rotary tablet compression machine; ~ **rotatoria** f C&V rotating machine; ~ **de rotor cilíndrico** f ELEC cylindrical rotor machine;

■ **s** ~ **de secado de película** f CINEMAT film-drying machine; ~ **con secador yankee** f PAPEL MG machine; ~ **secadora** f P&C dryer; ~ **de secciones individuales** f C&V individual section machine; ~ **selladora de sacos o bolsas** f AGRIC, EMB bag-sealing machine; ~ **selladora al vacío** f EMB vacuum-sealing machine; ~ **semiautomática de flejado** f EMB semiautomatic strapping machine; ~ **serie** f ELEC series-wound machine; ~ **de simple distribución** f IMPR single-distributor machine; ~ **sincrónica** f ELEC synchronous machine; ~ **de sobreembalaje retráctil** f EMB shrink overwrapping machine; ~ **de sobreimpresión de etiquetas** f EMB label-overprinting machine; ~ **para soldadura a tope** f PROD butt-welding machine; ~ **para soldar, afilar y ajustar sierras de banda** f ING MECÁ band-saw brazing, sharpening and setting machine; ~ **para soldar cuchillas de sierras de banda** f ING MECÁ band-saw brazing apparatus; ~ **de soldar de mano** f PROD handle-welding machine; ~ **soplante** f PROD blowing-engine; ~ **soplante de pistón** f ING MECÁ piston-blower; ~ **de soplar** f PROD blower; ~ **de succión** f C&V suction machine; ~ **suministradora de cajas** f EMB carton-dosing machine;

■ **t** ~ **taladradora y fresadora coordinada** f ING MECÁ coordinate boring-and-milling machine; ~ **taladradora de traviesas** f FERRO tie-drilling machine; ~ **de taladrar** f ING MECÁ drilling machine, boring machine, MECÁ, MINAS boring machine; ~ **de taladrar con avance manual por palanca** f ING MECÁ lever-feed drilling machine; ~ **de taladrar carriles** f FERRO rail-drilling machine; ~ **de taladrar y roscar tuberías** f ING MECÁ tapping machine; ~ **de tallar engranajes** f ING MECÁ, MECÁ gear cutting, gear-cutting machine, shaper; ~ **de tallar engranajes con cuchilla** f ING MECÁ gear-shaping machine; ~ **de tallar engranajes por disco** f ING MECÁ gear-milling machine; ~ **de tallar engranajes por fresa** f ING MECÁ gear-milling machine; ~ **de tallar engranajes**

por fresa matriz f ING MECÁ gear-hobbing machine; ~ **de tallar piñones** f ING MECÁ pinion-cutting machine; ~ **de telecine** f TV film scanner, telecine machine; ~ **de teñir en plegador** f TEXTIL beam-dyeing machine; ~ **térmica** f ING MECÁ, MECÁ, TERMO heat engine; ~ **termoformadora al vacío** f EMB vacuum-thermoforming machine; ~ **de termosellado** f EMB heat-sealing machine; ~ **para termosellar y soldar** f EMB heat-sealing and welding machine; ~ **tijera de palanca con contrapeso** f PROD lever-shearing machine with counterweight; ~ **para trabajar el barro** f C&V clay-working machine; ~ **tractora** f ING MECÁ mover; ~ **de transferencia** f MECÁ transfer machine; ~ **de transmisión** f TELECOM *télex* transmit machine; ~ **transportadora** f *AmL* (*cf* cinta *transportadora Esp*) MINAS conveyor; ~ **tricilíndrica** f ING MECÁ triplex engine, three-cylinder engine; ~ **de tricotar por urdimbre** f TEXTIL warp-knitting machine; ~ **trifásica** f ING ELÉC three-phase machine; ~ **de triple expansión** f TRANSP MAR triple expansion reciprocating engine; ~ **troqueladora** f ING MECÁ die-casting machine; ~ **de tunelización** f MINAS tunneling machine (*AmE*), tunnelling machine (*BrE*); ~ **Turing** f INFORM&PD Turing machine;

■ **v** ~ **de vaciado continuo** f C&V ribbon machine; ~ **de vapor de agua** f INSTAL HIDRÁUL steam engine; ~ **de vapores combinados** f ING MECÁ binary engine, binary-heat engine; ~ **con ventilación de circuito abierto** f ING MECÁ machine with open-circuit ventilation; ~ **con ventilación de circuito cerrado** f ING MECÁ machine with closed-circuit ventilation; ~ **de verificación** f ING MECÁ testing machine; ~ **de viento** f CINEMAT wind machine; ~ **von Neumann** f INFORM&PD von Neumann machine;

■ **w** ~ **Wimshurst** f ING ELÉC Wimshurst machine
máquina[2]: **a toda** ~ *fra* ING MECÁ in full swing
maquinabilidad f ING MECÁ machinability
maquinación: ~ **láser** f MECÁ laser machining; ~ **por chispa eléctrica** m ING MECÁ electro-spark machining; ~ **por electrodescarga** ING MECÁ, PROD electro-discharge machining (*EDM*); ~ **electroerosivo** m ING MECÁ electro-spark machining; ~ **por fulguración** f ING MECÁ electro-spark machining;
maquinado[1] *adj* MECÁ machined; ~ **en basto** *adj* MECÁ rough-machined; ~ **por completo** *adj* MECÁ machined-all-over; ~ **con medidas aproximadas** *adj* MECÁ rough-machined
maquinado[2] m ING MECÁ, MECÁ machining, PROD tooling; ~ **en caliente** m METAL hot working; ~ **de cápsula roscada** m ING MECÁ screwcap tooling; ~ **con carburo al tungsteno** m ING MECÁ tungsten-carbide tooling; ~ **de casquillos roscados** m ING MECÁ screwcap tooling; ~ **por electrodescarga** m ING MECÁ, PROD electro-discharge machining (*EDM*); ~ **de engranajes de evolvente en círculos** m ING MECÁ involute gearing; ~ **de engranajes con fresa matriz** m ING MECÁ hob cutting; ~ **de engranajes de perfil evolvente** m ING MECÁ involute gearing; ~ **por fresa generatriz** m ING MECÁ, PROD hobbing; ~ **helicoidal** m ING MECÁ helicoidal machining; ~ **de precisión** m ING MECÁ precision machining; ~ **químico** m MECÁ chemical machining; ~ **de tapones de rosca** m ING MECÁ screwcap tooling; ~ **tridimensional** m ING MECÁ three-dimen-

sional machining; **~ ultrasónico** *m* FÍS RAD ultrasonic machining; **~ por vibración ultrasónica** *m* NUCL ultrasonic machining

maquinar *vt* MECÁ machine

maquinaria *f* ING MECÁ machinery, mechanism, tackle, PROD plant; **~ agrícola** *f* AGRIC, MECÁ agricultural machine, farm machinery; **~ de arranque** *f* MINAS drawing engine; **~ para cerrar bolsas** *f* EMB bag-sealing equipment; **~ para construcción de carreteras** *f* AUTO, CONST, TRANSP, VEH road-building machinery; **~ de conversión** *f* EMB conversion machinery; **~ para economizar el trabajo** *f* PROD labor-saving machinery (*AmE*), labour-saving machinery (*BrE*); **~ elevadora** *f* ING MECÁ elevating machinery; **~ para envolver alimentos** *f* EMB food-wrapping machinery; **~ de excavaciones** *f* ING MECÁ earthmoving machinery; **~ granuladora** *f* EMB granulating machine; **~ hidráulica** *f* ING MECÁ hydraulic machinery; **~ indicadora** *f* ING MECÁ indicator plant; **~ de mando** *f* ING ELÉC control gear; **~ de movimiento de tierra** *f* ING MECÁ earthmoving machinery; **~ para sobreenvolver** *f* EMB overwrapping machinery

máquinas *f pl* ING MECÁ machinery; **~ elevadoras** *f pl* ING MECÁ lifting machinery; **~ herramientas de precisión** *f pl* ING MECÁ precision machine-tools; **~ para tallar engranajes de evolvente en círculos** *f pl* ING MECÁ involute gear cutters; **~ para tallar engranajes de perfil evolvente** *f pl* ING MECÁ involute gear cutters

maquineta *f* TEXTIL dobby

maquinilla *f* CONTAM MAR winch, TEXTIL dobby, TRANSP MAR winch

maquinista *m* CINEMAT grip, CONST operator, FERRO *de tren* driver, engineer, ING MECÁ, MINAS operator, PROD machinist; **~ de apisonadora** *m* CONST roller operator; **~ de grúa** *m* CONST craneman; **~ naval** *m* TRANSP MAR engineer, marine engineer; **~ naval jefe** *m* TRANSP MAR chief engineer; **~ naval primero** *m* TRANSP MAR second engineer; **~ de rotativa** *m* IMPR rotary printer

mar[1]: **de ~** *adj* TRANSP MAR onshore; **~ adentro** *adj* TRANSP MAR offshore; **~ afuera** *adj* TRANSP MAR in the offing

mar[2]: **en el ~** *adv* TRANSP MAR at sea, overboard; **hacia el ~** *adv* TRANSP MAR *navegación* seaward; **por ~** *adv* TRANSP MAR by sea; **~ adentro** *adv* TRANSP MAR *navegación* seaward; **~ afuera** *adv* TRANSP MAR in the offing

mar[3] *m* HIDROL, OCEAN sea; **~ abierto** *m* GEOL offshore OCEAN offshore, open sea TRANSP MAR open sea **~ agitado** *m* HIDROL OCEAN wind stress sea; **~ calmo** *m* HIDROL, METEO, OCEAN, TRANSP MAR calm sea; **~ cerrado** *m* OCEAN closed sea; **~ confuso** *m* TRANSP MAR *tiempo atmosférico* confused sea; **~ encerrado** *m* OCEAN enclosed sea; **~ encrespado** *m* METEO heavy swell, OCEAN seaway, TRANSP MAR heavy swell; **~ epicontinental** *m* GEOL, PETROL epicontinental sea; **~ epírico** *m* GEOL epeiric sea; **~ de fondo** *m* METEO, OCEAN, OCEAN swell; **~ grueso** *m* METEO, OCEAN, TRANSP MAR rough sea; **~ interior** *m* HIDROL, METEO, OCEAN inland sea; **~ de leva** *m* METEO, OCEAN, TRANSP MAR swell; **~ libre** *m* OCEAN open seas, TRANSP MAR open water; **~ marginal** *m* OCEAN marginal sea; **~ muy grueso** *m* METEO, OCEAN, TRANSP MAR heavy seas; **~ patrimonial** *m* OCEAN

patrimonial sea; **~ picado** *m* METEO, OCEAN, TRANSP MAR choppy sea; **~ de proa** *m* TRANSP MAR head sea; **~ rizado** *m* METEO, TRANSP MAR catspaw; **~ tendido** *m* OCEAN, TRANS MAR ground swell; **~ Tethis** *m* GEOL Tethys Ocean; **~ de través** *m* TRANSP MAR beam sea; **~ verde** *m* OCEAN green sea; **~ de viento** *m* OCEAN wind stress sea

maraña *f* TEXTIL snarl

marbete *m* IMPR, PROD tag, TEXTIL label

marca *f* CONST marking, INFORM&PD flag, mark, tag, OCEAN *navegación* landmark, seamark, PROD mark, make, marking, TRANSP MAR mark, *navegación* seamark; **~ al agua** *f* PAPEL watermark; **~ de agua** *f* C&V watermark; **~ de alineación** *f* IMPR alignment mark; **~ angular** *f* IMPR angle mark; **~ de archivo** *f* INFORM&PD file mark; **~ de archivos** *f* INFORM&PD file mark; **~ de calidad** *f* CALIDAD, EMB quality mark; **~ de cinta** *f* INFORM&PD tape mark; **~ de contraste** *f* PROD *metales preciosos* hallmark; **~ de costilla** *f* C&V rib mark; **~ por desgaste** *f* TEXTIL scuff mark; **~ de la dirección del fieltro** *f* PAPEL felt-direction mark; **~ de encendido** *f* AUTO timing mark; **~ de enfilación** *f* TRANSP MAR *navegación* guiding mark, leading mark; **~ de enfriamiento** *f* C&V chill mark (*AmE*), chill wrinkle (*AmE*); **~ de erosión** *f* GEOL *sobre fondo no consolidado* flute mark; **~ de escaldado** *f* ALIMENT scald mark; **~ del estante** *f* C&V rack mark; **~ estática** *f* CINEMAT static mark; **~ de fábrica** *f* PROD trademark; **~ del fabricante** *f* PROD maker's mark; **~ falciforme** *f* GEOL *erosión glaciar* chatter mark; **~ de ficheros** *f* INFORM&PD file mark; **~ del fieltro** *f* PAPEL felt mark; **~ fija** *f* MECÁ benchmark; **~ final** *f* INFORM&PD end mark; **~ fósil** *f* GEOL fossil imprint; **~ fría** *f* C&V chill mark (*AmE*), chill wrinkle (*AmE*); **~ del ganado** *f* AGRIC cattle brand; **~ de gancho** *f* C&V hook mark; **~ de graduación** *f* LAB graduation mark; **~ graduada** *f* METR scale mark; **~ de grupo** *f* INFORM&PD group mark; **~ de identificación del componente** *f* PROD component identification marker; **~ indicadora** *f* CINEMAT, TV cue mark; **~ de inicio** *f* CINEMAT start mark; **~ de inicio de la positivadora** *f* CINEMAT printer start-mark; **~ de límite** *f* CONST boundary post; **~ de lubricante** *f* C&V dope mark, grease mark; **~ de marea** *f* GEOFÍS tidemark; **~ de marmol** *f* C&V marver mark; **~ de molde** *f* C&V mold mark (*AmE*), mould mark (*BrE*); **~ del nivel de aceite** *f* AUTO oil line, VEH *lubricación* oil-level mark; **~ de nivel inferior de agua** *f* CONST low-water mark; **~ de nivelación** *f* TRANSP AÉR leveling mark (*AmE*), levelling mark (*BrE*); **~ de patinaje de la rueda** *f* FERRO *vía* wheel slide mark; **~ de patinaje de la rueda en el carril** *f* FERRO *vía* wheel slip mark on rails; **~ de pelo** *f* C&V hackle (*AmE*), hackle mark (*BrE*); **~ de piel** *f* ALIMENT skin blemish; **~ de pleamar** *f* TRANSP MAR high-water mark; **~ de puesta a punto** *f* AUTO timing mark; **~ del pulidor** *f* C&V lap mark; **~ de ranura del molde** *f* C&V rib mark; **~ de referencia** *f* CONST *topografía* reference mark; **~ de referencia de tierra** *f* TRANSP AÉR landmark; **~ de rodillo** *f* C&V roll mark (*AmE*), roller mark (*BrE*); **~ de secado** *f* CINEMAT drying mark; **~ de sincronización** *f* AUTO timing mark, CINEMAT sync mark; **~ de sincronización de la imagen** *f* CINEMAT picture-cuing mark; **~ de la tela** *f* IMPR *papel* wire mark; **~ en tierra** *f* TRANSP MAR *navega-*

ción landmark; ~ **de las tijeras** *f* C&V *en el bombillo después de enfriar* shear mark; ~ **de tope** *f* TRANSP MAR *boyas* top mark

marcación *f* TELECOM marking, dialing (*AmE*), dialling (*BrE*), TRANSP MAR *navegación* bearing; ~ **automática** *f* TELECOM automatic dialing (*AmE*), automatic dialling (*BrE*); ~ **automática del último número** *f* TELECOM last-number redial; ~ **azimutal** *f* OCEAN azimuth bearing; ~ **de la brújula** *f* TRANSP AÉR, TRANSP MAR compass bearing; ~ **del centro de la pista** *f* TRANSP AÉR taxiway centerline marking (*AmE*), taxiway centreline marking (*BrE*); ~ **de cruce de la pista de rodaje** *f* TRANSP AÉR taxiway intersection marking; ~ **directa a distancia** *f* TELECOM direct distance dialing (*AmE*) (*DDD*), direct distance dialling (*BrE*) (*DDD*); ~ **directa entrante** *f* TELECOM direct inward dialing (*AmE*) (*DID*), direct inward dialling (*BrE*) (*DID*), direct dialing in (*AmE*), direct dialling in (*BrE*); ~ **directa internacional** *f* TELECOM International Direct Dialling (*BrE*) (*IDD*), International Direct Distance Dialing (*AmE*) (*IDDD*); ~ **directa saliente** *f* TELECOM direct outward dialing (*AmE*) (*DOD*), direct outward dialling (*BrE*) (*DOD*); ~ **de intercepción** *f* TRANSP AÉR intercept bearing; ~ **inversa** *f* TRANSP AÉR reciprocal bearing; ~ **lateral de la pista de rodaje** *f* TRANSP AÉR taxiway-edge marker; ~ **magnética** *f* MECÁ, TRANSP AÉR magnetic bearing; ~ **de pretoma** *f* TELECOM preseizure dialling (*BrE*), preseizure dialing (*AmE*); ~ **de radar** *f* TRANSP, TRANSP AÉR, TRANSP MAR radar bearing; ~ **radárica** *f* D&A, TRANSP, TRANSP AÉR, TRANSP MAR radar plotting; ~ **radiogoniométrica** *f* TELECOM, TRANSP AÉR, TRANSP MAR radio bearing; ~ **recíproca** *f* TRANSP AÉR, TRANSP MAR reciprocal bearing; ~ **relativa** *f* FÍS RAD relative bearing, OCEAN azimuth bearing, *navegación* relative bearing, TRANSP AÉR, TRANSP MAR relative bearing; ~ **en el reloj de entrada** *f* PROD timekeeping; ~ **en reposo** *f* AmL (*cf marcado en reposo Esp*) TELECOM on-hook dialing (*AmE*), on-hook dialling (*BrE*); ~ **de reserva** *f* TEC ESP backup bearing; ~ **de ruido** *f* ING MECÁ noise labeling (*AmE*), noise labelling (*BrE*); ~ **de toma previa** *f* TELECOM pre-seizure dialling (*BrE*), pre-seizure dialing (*AmE*)

Marcación: ~ **de Abonado Internacional** *f* TELECOM International Subscriber Dialing (*AmE*) (*ISD*), International Subscriber Dialling (*BrE*) (*ISD*)

marcada *f* TEXTIL marking

marcado[1] *adj* FÍS RAD labeling (*AmE*), labelling (*BrE*), NUCL, QUÍMICA labeled (*AmE*), labelled (*BrE*)

marcado[2] *m* CONST *topografía* monumenting, FÍS RAD labeling (*AmE*), labelling (*BrE*), NUCL labeled (*AmE*), labelled (*BrE*), PROD marking, QUÍMICA labeled (*AmE*), labelled (*BrE*); ~ **al ácido** *m* C&V acid mark; ~ **a cincel** *m* ING MECÁ chisel marking; ~ **en cuerda** *m* TEXTIL rope marking; ~ **electrolítico** *m* ING MECÁ electrolytic marking; ~ **por intercambio químico** *m* NUCL labeling by chemical exchange (*AmE*), labelling by chemical exchange (*BrE*); ~ **de la línea central de la pista** *m* TRANSP AÉR runway-centerline marking (*AmE*), runway-centreline marking (*BrE*); ~ **radioactivo** *m* QUÍMICA radio labeling (*AmE*), radio labelling (*BrE*); ~ **en reposo** *m* Esp (*cf marcación en reposo AmL*) TELECOM on-hook dialing (*AmE*), on-hook dialling (*BrE*)

marcador *m* INFORM&PD marker, timer, ING MECÁ, TELECOM marker; ~ **BOT** *m* INFORM&PD BOT marker; ~ **de fin de cinta** *m* INFORM&PD end of tape marker (*EOT marker*); ~ **flotante de radar** *m* TRANSP, TRANSP MAR radar marker float; ~ **de intervalo** *m* INFORM&PD interval timer; ~ **de juntas** *m* CONST, PROD, TELECOM jointer; ~ **de objetivos por láser** *m* D&A laser target marker; ~ **de perro guardián** *m* INFORM&PD watchdog timer; ~ **de precios** *m* EMB price marking; ~ **de principio de cinta** *m* INFORM&PD beginning of tape marker; ~ **de punta de diamante** *m* Esp (*cf pluma de punta de diamante AmL*) LAB diamond-tipped pen

marcadores: ~ **genéticos** *m pl* AGRIC genetic markers

marcaje *m* PROD *alambres* labeling (*AmE*), labelling (*BrE*); ~ **del eje de la pista** *m* TRANSP AÉR runway-centerline marking (*AmE*), runway-centreline marking (*BrE*)

marcar[1] *vt* CONST mark out, GEOL lay down, INFORM&PD flag, tag, PROD stamp, *con el gramil o el punzón* mark, TELECOM *un número* dial, TRANSP MAR *las velas* handle; ~ **con rayas** *vt* CONST line out, IMPR crimp; ~ **con tinta opaca** *vt* CINEMAT bloop

marcar[2]: ~ **el sincronismo** *vi* TELECOM time-tag; ~ **el tiempo** *vi* TELECOM time-tag

marcarse *v refl* TRANSP AÉR, TRANSP MAR *navegación* take a bearing

marcas: ~ **del cepillo** *f pl* C&V brushlines; ~ **de desgaste** *f pl* CINEMAT stress marks; ~ **de fuego** *f pl* C&V fire marks; ~ **de paro** *f pl* TEXTIL stopping marks; ~ **de las pinzas** *f pl* C&V tong marks; ~ **de revelado** *f pl* CINEMAT processing marks; ~ **superficiales producidas por los cilindros** *f pl* ING MECÁ roll marking; ~ **de viruelas** *f pl* C&V pockmarks

marcasita *f* MINERAL marcasite, spear pyrites, white iron pyrite

marcha[1]: **de** ~ **de máquina aislada** *adj* ING MECÁ light-running; **de** ~ **y parada** *adj* ING MECÁ on-off; **de** ~ **rápida** *adj* MECÁ high-speed; **de** ~ **en vacío** *adj* ING MECÁ light-running

marcha[2] *f* ING ELÉC turn, ING MECÁ movement, run, running, MECÁ action, motion, PROD working; ~ **activada** *f* METAL activated state; ~ **de aparcamiento** *f* AUTO parking gear; ~ **asincrónica** *f* ELEC asynchronous running; ~ **atrás** *f* FERRO *vehículos* backing, ING MECÁ backing, reversing motion, MECÁ reverse, TV reverse motion, VEH reverse; ~ **atrás de dos velocidades** *f* ING MECÁ two-speed counter motion; ~ **atrás del roscado** *f* ING MECÁ screw-cutting reverse; ~ **de bloqueo para estacionamiento** *f* AUTO parking lock-gear; ~ **fría** *f* PROD cold working; ~ **lenta** *f* ELEC *de motor* idling, PROD inching, VEH *de motor* idling; ~ **al ralentí** *f* VEH *de motor* idling; ~ **sin realizar ninguna función** *f* ELEC *de motor* idling; ~ **a seguir** *f* ING MECÁ procedure; ~ **silenciosa** *f* AUTO, ING MECÁ, VEH quiet running; ~ **con sobrecarga** *f* PROD *motores* overload running; ~ **en sobrecarga** *f* ING MECÁ running on overload; ~ **suave** *f* AUTO, ING MECÁ, VEH quiet running; ~ **en vacío** *f* ELEC *de motor* idling, ING MECÁ running on no load; ~ **a velocidad de crucero** *f* VEH cruise control

marchar: ~ **en carga** *vi* ING MECÁ run under load; ~ **por inercia** *vi* VEH coast; ~ **en vacío** *vi* ING MECÁ run light

marchitamiento *m* AGRIC fading, *patología vegetal* wilt

marchito *adj* AGRIC, ALIMENT withered

marco *m* AGUA frame, CONST door case, door casing, frame, *ventana* casement, *ventanas, puertas* casing, *carpintería* buck, IMPR frame, ING MECÁ case, *sierra* bow, MINAS *para galería* durn, *entibación* crown bearer, TELECOM standard; **~ adornado** *m* IMPR fancy frame; **~ de ajuste** *m Esp* (*cf marco de entibación AmL*) MINAS close set, timber set; **~ del codaste** *m* TRANSP MAR *pieza fundida, construcción metálica* stern frame; **~ de corte** *m* C&V cutting frame; **~ de cumbrera y dos pies** *m* MINAS ordinary timber set; **~ de diseño básico** *m* TEC ESP *nave* design basis case; **~ de entibación** *m AmL* (*cf marco de ajuste Esp*) MINAS close set, timber set; **~ de entibación de triple sección** *m* MINAS three-piece timber set; **~ falso** *m* TRANSP AÉR false frame; **~ de filtro** *m* PROC QUÍ filter frame; **~ lateral** *m* EMB side-frame; **~ de madera** *m* CONST, MINAS *entibación* timber frame; **~ de mina** *m* MINAS close set; **~ del muñón del estabilizador** *m* TRANSP AÉR fin stub frame; **~ neumático** *m* IMPR vacuum frame; **~ ordinario** *m* MINAS *entibación* double timbering; **~ de página** *m* INFORM&PD page frame; **~ portabolas** *m* ING MECÁ ball cage; **~ portabolas del cojinete** *m* ING MECÁ ball-bearing cage; **~ portador** *m AmL* (*cf marco soporte Esp*) MINAS *pozos* bearer set; **~ portaoriginales** *m* IMPR copy frame; **~ de pruebas** *m* INSTR trial frame; **~ de la puerta** *m* CONST doorframe; **~ de puerta** *m* VEH door casing; **~ punteado** *m* IMPR dotted frame; **~ del radiador** *m* AUTO, VEH radiator frame; **~ de referencia inercial** *m* TEC ESP inertial reference frame; **~ repartidor** *m* TELECOM distribution frame; **~ repartidor de dos caras** *m* TELECOM double-sided distribution frame; **~ con un solo pie derecho** *m* MINAS *entibación* half set; **~ soporte** *m Esp* (*cf marco portador AmL*) MINAS *pozos* bearer set; **~ de superficie** *m* MINAS *extracción en pozos* head frame, *pozo de mina* shaft collar; **~ tubular** *m* VEH *de motocicleta* tubular frame; **~ de ventana** *m* CONST, VEH *carrocería* window frame; **~ de ventana de dos hojas** *m* CONST French casement; **~ vidriado** *m* CONST glazed frame

marea *f* ENERG RENOV, OCEAN, TRANSP MAR tide; **~ alta** *f* AGUA, ENERG RENOV ebb tide, high tide, HIDROL ebb tide, high water, OCEAN ebb tide, high tide, TRANSP MAR ebb tide, high tide, high water; **~ ascendente** *f* ENERG RENOV rising tide, HIDROL flood tide, OCEAN, TRANSP MAR rising tide; **~ astronómica** *f* OCEAN astronomical tide; **~ atmosférica** *f* GEOFÍS atmospheric tide; **~ baja** *f* ENERG RENOV low tide, HIDROL, TRANSP MAR low tide, low water; **~ bajante** *f* OCEAN falling tide; **~ barométrica** *f* TRANSP MAR diurnal variation; **~ compuesta** *f* OCEAN compound tide; **~ creciente** *f* ENERG RENOV, OCEAN rising tide, TRANSP MAR rising tide, *navegación* flood tide; **~ descendente** *f* AGUA ebb tide, ENERG RENOV, HIDROL ebb tide, falling tide, OCEAN ebb tide, TRANSP MAR ebb tide, falling tide; **~ diurna** *f* TRANSP MAR diurnal tide, single day tide; **~ entrante** *f* AGUA flood, ENERG RENOV, OCEAN rising tide, TRANSP MAR rising tide, *de río, esclusa* flood; **~ de equilibrio** *f* OCEAN equilibrium tide; **~ equinoccial** *f* ENERG RENOV equinoctial tide, neap tide, HIDROL neap tide, spring tide, OCEAN neap tide, TRANSP MAR equinoctial tide, neap tide; **~ de equinoccio** *f* OCEAN spring tide; **~ ideal** *f* ENERG RENOV *de Newton* equilibrium tide; **~ menguante** *f* AGUA ebb tide, ENERG RENOV, HIDROL ebb tide, falling tide, OCEAN ebb tide, TRANSP MAR ebb tide, falling tide; **~ meteorológica** *f* METEO, OCEAN meteorological tide; **~ mixta** *f* OCEAN mixed tide; **~ muerta** *f* ENERG RENOV, HIDROL, OCEAN, TRANSP MAR neap tide, spring tide; **~ negra** *f* CONTAM, CONTAM MAR black tide, chocolate mousse; **~ roja** *f* OCEAN red tide; **~ semidiurna** *f* OCEAN, TRANSP MAR semidiurnal tide; **~ de sicigias** *f* OCEAN spring tide; **~ solar** *f* GEOFÍS solar tide; **~ terrenal** *f* OCEAN terrestrial tide; **~ terrestre** *f* OCEAN earth tide; **~ vaciante** *f* AGUA, ENERG RENOV, HIDROL, OCEAN, TRANSP MAR ebb tide; **~ viva** *f* ENERG RENOV spring tide

mareas: **~ mayores** *f pl* TRANSP MAR equinoctial tide
marejada *f* OCEAN ground swell
maremoto *m* GEOFÍS seaquake, submarine earthquake, OCEAN ground swell, seaquake
mareo: **~ espacial** *m* TEC ESP space sickness
mareógrafo *m* ENERG RENOV tide gage (*AmE*), tide gauge (*BrE*), GEOFÍS recording tide gage (*AmE*), tide gage (*AmE*), tide gauge (*BrE*), recording tide gauge (*BrE*), OCEAN marigraph, tide gage (*AmE*), tide gauge (*BrE*), tide recorder, recording tide gage (*AmE*), recording tide gauge (*BrE*)
mareograma *m* ENERG RENOV marigram, GEOFÍS cotidal line, OCEAN tide curve
marero *m* TRANSP MAR onshore wind
marga *f* AGUA loam, marl, CONST chalk marl, GEOL glauconite marl; **~ arcillosa** *f* AGUA clay marl; **~ arenosa** *f* GEOL sandy loam; **~ y caliza de aguas salobres** *f* GEOL brackish marl and limestone; **~ para cemento** *f* CONST cement marl; **~ de creta** *f* GEOL chalk marl; **~ silícea** *f* GEOL cherty marl
margarodita *f* MINERAL margarodite
margarato *m* QUÍMICA margarate
margárico *adj* QUÍMICA margaric
margarina *f* ALIMENT, QUÍMICA margarine
margarita *f* INFORM&PD *impresora* daisywheel, MINERAL margarite, TRANSP MAR *nudo* sheepshank
margas: **~ nodulares** *f pl* GEOL *pedregosa* nodular marls
margen *m* CONST bank, margin, ELECTRÓN range, GEOL margin, GEOM range, HIDROL, IMPR margin, INFORM&PD range, margin, ING MECÁ margin, offset, MECÁ allowance, clearance, TEC ESP *de esfuerzos*, TELECOM range, TRANSP AÉR clearance; **~ de absorción** *m* CRISTAL absorption edge; **~ de acabado** *m* PROD allowance for machining; **~ activo** *m* GEOL *cordillera o arcos insulares* active margin; **~ de altura** *m AmL* (*cf holgura Esp*) MINAS clearance; **~ de la banda de pista** *m* TV edge of track banding; **~ del borde posterior** *m* IMPR back-edge margin; **~ bruto** *m* PROD gross margin; **~ de cabeza** *m* IMPR head margin; **~ de capacitancia** *m* PROD capacitance tolerance; **~ de captura** *m* ELECTRÓN capture range; **~ continental** *m* OCEAN continental margin; **~ convergente activo** *m* GEOL *de capa activa* active plate margin, convergent margin; **~ por defecto** *m* ING MECÁ margin under; **~ derecho** *m* IMPR right margin; **~ dinámico** *m* ACÚST, FÍS RAD, PETROL, TV dynamic range; **~ de distancia** *m* ING MECÁ length margin; **~ de entrada** *m* IMPR grasping margin; **~ por exceso** *m* ING MECÁ margin over; **~ de fase** *m* ELECTRÓN phase margin; **~ de frecuencia** *m* ELECTRÓN frequency range; **~ de frecuencias** *m* ELEC *de corriente alterna*, TV frequency range; **~ de**

frecuencias audibles *m* ACÚST audible frequency range; ~ **inferior** *m* IMPR tail; ~ **interior** *m* IMPR inner margin; ~ **izquierdo** *m* IMPR left margin; ~ **del lomo** *m* IMPR binding edge; ~ **del medianil** *m* IMPR *entre dos páginas de un libro* gutter space; ~ **de parada** *m* NUCL shutdown margin; ~ **de pie** *m* IMPR tailband; ~ **de placa** *m* GEOL plate margin; ~ **de placa constructivo** *m* GEOL constructive plate margin; ~ **de placa destructivo** *m* GEOL destructive plate margin; ~ **de placa no-destructivo** *m* GEOL conservative plate margin; ~ **posterior** *m* IMPR back margin; ~ **próximo** *m* TEC ESP near range; ~ **de puntos** *m* GEOM range of points; ~ **de seguridad** *m* SEG safety margin, TRANSP AÉR clearance; ~ **suplementario** *m* IMPR apron

márgenes: ~ **de enlace** *m pl* TELECOM link margins; ~ **sobrecortados** *m pl* IMPR overcut margins

margoso *adj* GEOL marlaceous, marly

marialita *f* MINERAL marialite

marina[1]: **de** ~ *adj* TRANSP MAR marine

marina[2] *f* D&A, TRANSP MAR navy; ~ **civil** *f* TRANSP MAR merchant navy; ~ **de guerra** *f* D&A, TRANSP MAR navy; ~ **mercante** *f* TRANSP MAR mercantile marine, merchant marine; ~ **nacional** *f* D&A, TRANSP MAR national navy

marinero *m* TRANSP MAR sailor; ~ **de bote salvavidas** *m* SEG, TRANSP MAR lifeboatman; ~ **de embarcación de salvamento** *m* SEG, TRANSP MAR lifeboatman, sailor; ~ **novel** *m* TRANSP MAR sailor, ship's boy

marino *adj* AUTO, QUÍMICA, TRANSP MAR marine

mariposa *f* AUTO butterfly, choker plate, ING MECÁ thumbnut, wing, VEH *del carburador* choker plate, throttle; ~ **automática** *f* AUTO automatic choke; ~ **de válvula** *f* INSTAL HIDRÁUL throttle

mariquita: ~ **de San Antonio** *f* AGRIC ladybird

marisma *f* AGUA, HIDROL marsh, OCEAN salt marsh, TRANSP MAR mudflat

marisqueo *m* OCEAN mollusc harvesting

marítimo *adj* METEO maritime, TRANSP MAR marine, maritime, seaborne

marjal: ~ **marino** *m* OCEAN sea marsh

marlo *m* AmL (*cf mazorca Esp*) AGRIC ear, spike, *de cacao* cob, pod

marmita *f* QUÍMICA kettle

mármol *m* GEOL marble, MECÁ surface plate, MINAS marble; ~ **de ajustador** *m* ING MECÁ, METR surface plate, PROD engineer's surface plate; ~ **filítico** *m* GEOL phyllitic marble; ~ **modelado** *m* GEOL patterned marble; ~ **de sacador** *m* C&V marver; ~ **de trazado** *m* ING MECÁ, METR surface plate

marmoleado *m* AGRIC *de carne* marbling

marmolita *f* MINERAL marmolite

marquesina *f* CONST hood, umbrella roof, TRANSP *de estación* concourse; ~ **del andén** *f* FERRO platform awning

marquetería *f* C&V marquetry, CONST cabinetwork

marquito: ~ **portadiapositivas** *m* FOTO slide holder

marrano *m* PROD *metalurgia* sow, TEC PETR pig

marrón *m* COLOR, P&C maroon

martensita *f* CRISTAL martensite; ~ **termoelástica** *f* METAL thermoelastic martensite

martillado *m* PROD peening

martillaje *m* PROD malleation

martillazos *m pl* PROD hammering

martilleo *m* CONST, PROD hammering

martillo *m* ACÚST *audición* hammer, malleus, C&V hammer, CONST beater, ING MECÁ formwork hammer, MECÁ hammer, MINAS jar; ~ **de acuñar** *m* PROD keying hammer; ~ **de aire comprimido** *m* CONST pneumatic hammer-drill; ~ **de ajustador** *m* ING MECÁ fitter's hammer; ~ **aplanador** *m* PROD flatter; ~ **de aplanar** *m* ING MECÁ planisher; ~ **batidor** *m* ING MECÁ beating hammer; ~ **de boca cruzada** *m* CONST cross-peen hammer; ~ **de bola** *m* PROD pean hammer; ~ **con bolita** *m* ING MECÁ ball pane; ~ **de cabezas desmontables** *m* PROD metal-bound mallet; ~ **de calafate** *m* ING MECÁ caulking hammer; ~ **de chapista** *m* ING MECÁ bumping hammer, dinging hammer; ~ **de cincelar** *m* PROD chipping hammer; ~ **para clasificar** *m* MINAS picker; ~ **de cotillo revestido cuero** *m* ING MECÁ hide-faced mallet; ~ **desabollador** *m* ING MECÁ bumping hammer, dinging hammer; ~ **desincrustador** *m* PROD scaling hammer; ~ **de desincrustar** *m* PROD *calderas* pick; ~ **para desincrustar calderas** *m* INSTAL HIDRÁUL boiler-scaling hammer, furring hammer, pick; ~ **diesel** *m* CARBÓN diesel hammer; ~ **de doble cara** *m* CONST double-faced hammer; ~ **de dos manos** *m* CONST sledge, PROD sledgehammer; ~ **de dos manos de peña transversal** *m* CONST cross-peen sledgehammer; ~ **de fragua** *m* ING MECÁ drop hammer, PROD forge hammer, sledgehammer, drop hammer; ~ **de mano** *m* ING MECÁ blacksmith's hammer, PROD hand hammer; ~ **de mecánico** *m* ING MECÁ fitter's hammer; ~ **mecánico de forja** *m* ING MECÁ formwork hammer; ~ **de metal dúctil** *m* PROD soft metal hammer; ~ **neumático** *m* CONST pneumatic power-hammer, MECÁ air hammer, PETROL airgun; ~ **neumático para clavar** *m* FERRO spike driver; ~ **de orejas** *m* CONST claw hammer; ~ **de peña** *m* CONST cross-pane hammer, PROD pean hammer; ~ **perforador** *m* ING MECÁ hammer drill, MINAS jackleg drill, hammer drill, jackhammer, *sondeos* drilling machine; ~ **perforador de percusión** *m* ING MECÁ percussive drill, MINAS *sondeos* percussion drill; ~ **perforador de roca** *m* CARBÓN, CONST, MINAS rock drill; ~ **picador** *m* MINAS jackhammer, PROD pick hammer; ~ **picador de aire comprimido** *m* MINAS pneumatic pick; ~ **de picar calderas** *m* PROD scaling hammer; ~ **pilón** *m* CARBÓN drop hammer, pile hammer (*AmE*), piling hammer (*BrE*), CONST pile-driver, *teller* pile hammer (*AmE*), piling hammer (*BrE*), ING MECÁ drop hammer, power hammer, formwork hammer, MECÁ drop pile hammer, PROD drop hammer, power hammer; ~ **de pizarrero** *m* CONST slater's hammer; ~ **pulsante** *m* ING MECÁ beating hammer; ~ **de puño recto** *m* CONST straight-peen hammer; ~ **recto** *m* CONST straight-pane hammer; ~ **de remachar** *m* CONST riveting hammer, rivet knocking-off hammer, ING MECÁ rivet knocking-off hammer; ~ **de remache** *m* CONST, ING MECÁ rivet knocking-off hammer; ~ **de techador** *m* CONST roofer's hammer

martinete *m* CARBÓN *hinca de pilotes* pile ram, *electrotecnia* pile-driver, bell, CONST ram, ING MECÁ power hammer, MECÁ jack, PROD power hammer, rammer, stamp; ~ **de aire comprimido** *m* MINAS pneumatic hammer; ~ **de báscula** *m* PROD tilt hammer, tilting hammer; ~ **de caída libre** *m* CARBÓN, ING MECÁ, PROD drop hammer; ~ **de forja** *m* CARBÓN drop hammer, hammer, ING MECÁ formwork hammer; ~ **de forja de palanca accionado por**

leva *m* PROD tilt hammer; **~ de fricción** *m* PROD friction hammer; **~ hidráulico** *m* MECÁ hydraulic jack; **~ de hinca** *m* CONST pile-driver, MINAS ram; **~ de hinca pilotes** *m* CONST *obra civil* trip pile driver; **~ hueco** *m* ING MECÁ hollow ram; **~ de resorte** *m* PROD spring power hammer

martita *f* MINERAL martite

más¹: **~ significativo** *adj* INFORM&PD most significant

más²: **~ allá** *adv* TEC ESP ahead

masa¹: **de una ~ sobre un resorte vertical** *adj* ING MECÁ of a mass on a vertical spring

masa²: **en ~** *adv* CARBÓN in bulk

masa³ *f* ALIMENT dough, AUTO earth connection (*BrE*), ground connection (*AmE*), ELEC *conexión* ground (*AmE*), earth connection (*BrE*), ground connection (*AmE*), earth (*BrE*), FÍS ground (*AmE*), mass, earth (*BrE*), bulk, GEOFÍS mass, ING ELÉC earth (*BrE*), earth connection (*BrE*), ground (*AmE*), ground connection (*AmE*), OCEAN *de hielo flotante* floe, P&C dough, PROD ground (*AmE*), earth connection (*BrE*), ground connection (*AmE*), earth (*BrE*), QUÍMICA mass, TELECOM ground (*AmE*), earth (*BrE*), TEXTIL bulk, VEH *instalación eléctrica* earth (*BrE*), earth connection (*BrE*), ground (*AmE*), ground connection (*AmE*); **~ acústica** *f* ACÚST acoustic mass; **~ aerostática** *f* TEC ESP ballooning mass; **~ de aterrizaje permisible** *f* TRANSP AÉR allowable landing mass; **~ atómica** *f* (*A*) FÍS PART atomic mass (*A*), mass number (*A*), QUÍMICA atomic mass (*A*); **~ atómica relativa** *f* FÍS relative atomic mass; **~ barro** *f* C&V clay mass; **~ de cacao** *f* ALIMENT cocoa mass; **~ caliza con fragmentos de plantas fósiles** *f* CARBÓN seam nodules; **~ centrífuga** *f* NUCL centrifugal mass; **~ coalescida de polvo sin prensar** *f* PROD cake; **~ continental** *f* GEOFÍS, GEOL continental mass; **~ crítica** *f* FÍS critical mass, NUCL critical amount; **~ de cuarzo fundido** *f* METAL silica glass; **~ de despegue de diseño** *f* TRANSP AÉR design takeoff mass; **~ de despegue permisible** *f* TRANSP AÉR allowable takeoff mass; **~ efectiva** *f* ACÚST effective mass; **~ eléctrica** *f* ELEC, FÍS, PROD, TELECOM, VEH earth (*BrE*), ground (*AmE*); **~ del electrón** *f* ELECTRÓN, FÍS PART, QUÍMICA electron mass; **~ electrónica** *f* ELECTRÓN, FÍS PART, QUÍMICA electron mass; **~ flotante** *f* OCEAN *de hielo* floe, iceberg, raft; **~ giratoria** *f* NUCL gyrating mass; **~ gravitacional** *f* FÍS, GEOFÍS, GEOL gravitational mass; **~ inercial** *f* FÍS inertial mass; **~ molecular relativa** *f* FÍS relative molecular mass; **~ del neutrón** *f* FÍS RAD neutron mass; **~ no crítica** *f* NUCL off-critical amount; **~ polar** *f* ING ELÉC, ING MECÁ pole piece; **~ polar del condensador doble** *f* INSTR double-condenser pole piece; **~ polar del objetivo** *f* INSTR objective pole piece; **~ propulsora** *f* TEC ESP propellant mass; **~ reducida** *f* FÍS reduced mass; **~ en reposo** *f* FÍS, FÍS PART *del electrón* rest mass; **~ seca** *f* TEC ESP dry mass, dry weight; **~ seca sin combustible** *f* TEC ESP dry weight; **~ sísmica** *f* GEOFÍS seismic mass; **~ térmica** *f* TERMO thermal mass; **~ por unidad de longitud** *f* FÍS mass per unit length; **~ por unidad de volumen** *f* FÍS mass per unit volume

masas *f pl* IMPR solids; **~ llenas** *f pl* IMPR solids

máscara *f* CINEMAT, ELECTRÓN, FOTO, INFORM&PD mask, SEG full face-mask, *ropa protectora* mask, TELECOM mask; **~ de abertura** *f* ELECTRÓN aperture mask; **~ abierta por la cámara** *f* PROD train-through-the-lens mask; **~ académica** *f* CINEMAT academy mask; **~ antigás** *f* D&A gas mask; **~ de apertura** *f* TV aperture mask; **~ de buceo** *f* OCEAN diving mask; **~ de circuito integrado** *f* ELECTRÓN integrated circuit mask; **~ de humo** *f* TRANSP AÉR smoke mask; **~ de impresión** *f* FOTO printing mask; **~ de interrupción** *f* INFORM&PD interrupt mask; **~ inversa** *f* CINEMAT reverse mask; **~ litográfica** *f* FÍS RAD lithographic mask; **~ de metalización** *f* ELECTRÓN metalization mask (*AmE*), metallization mask (*BrE*); **~ óptica** *f* ELECTRÓN optical mask; **~ de oxígeno** *f* SEG, TEC ESP, TRANSP AÉR oxygen mask; **~ de pantalla** *f* CINEMAT screen mask; **~ de pastilla** *f* ELECTRÓN *semiconductores* wafer mask; **~ protectora para soldador** *f* SEG welder's shield; **~ de proximidad** *f* ELECTRÓN proximity mask; **~ de rayos X** *f* ELECTRÓN X-ray mask; **~ de reducción** *f* FOTO reduction mask; **~ de sombra** *f* ELECTRÓN, TV shadow mask; **~ de teclado** *f* PROD keytop overlay; **~ del teclado** *f* INFORM&PD keyboard mask; **~ de tipo bocallave** *f* CINEMAT keyhole mask

mascarilla *f* SEG half mask, quarter mask; **~ antipolvo** *f* SEG dust mask; **~ con filtro contra partículas** *f* SEG filtering facepiece; **~ de humo** *f* TRANSP AÉR smoke mask; **~ de oxígeno de colocación rápida** *f* TRANSP AÉR quick-donning oxygen mask; **~ de respiración** *f* SEG respirator

maser *m* FÍS, TELECOM maser; **~ de amoníaco** *m* ELECTRÓN ammonia maser; **~ compacto** *m* ELECTRÓN solid-state maser; **~ de hidrógeno atómico** *m* ELECTRÓN atomic hydrogen maser; **~ de impulsos** *m* ELECTRÓN pulsed maser; **~ de onda progresiva** *m* ELECTRÓN traveling-wave maser (*AmE*), travelling-wave maser (*BrE*); **~ óptico** *m* ELECTRÓN optical maser; **~ de tres niveles** *m* ELECTRÓN three-level maser

MASER *abr* (*amplificador de microondas por emisión estimulada de radiación*) ELECTRÓN, TEC ESP MASER (*microwave amplification by stimulated emission of radiation*)

masicote *m* MINERAL, QUÍMICA massicot

masilla *f* C&V, COLOR putty, CONST filler, putty, PETROL, TRANSP MAR mastic; **~ aislante** *f* REFRIG insulating mastic; **~ para carrocerías** *f* VEH body filler

masillado *m* CONST puttying

masivo *adj* GEOL *textura* massive

masonita *f* MINERAL masonite

masticación *f* P&C *operación* mastication

mástico: **~ aislante** *m* REFRIG insulating mastic

mástil *m* CONST, PETROL mast; **~ de antena** *m* TRANSP MAR, TV aerial mast (*BrE*), antenna mast (*AmE*); **~ con cruceta para señal** *m* FERRO *infraestructura* post bracket; **~ de electricidad** *m* CONST, ELEC electricity pylon; **~ de perforación** *m* PETROL drilling mast; **~ de radar** *m* TRANSP, TRANSP AÉR, TRANSP MAR radar mast; **~ de rotor** *m* TRANSP AÉR rotor mast; **~ de transmisión** *m* ELEC *red de distribución* transmission tower

mástique *m* PETROL mastic, PROD lute, TRANSP MAR mastic; **~ de fundición** *m* PROD fake; **~ para tapar** *m* PROD beaumontage

mastitis *f* AGRIC mastitis

mastoide *f* ACÚST *audición* mastoid; **~ artificial** *f* ACÚST artificial mastoid

MAT *abr* (*muy alta tensión*) TV EHT (*extra-high tension*)

mata *f* METAL *en bruto* matte; ~ **de plomo** *f* PROD lead matte

matadero *m* ALIMENT, REFRIG abattoir

matafuegos *m* CONST, ING MECÁ, SEG, TERMO, TRANSP AÉR extinguisher, fire extinguisher

matar *vt* TEC PETR *pozo* kill

mate *adj* COLOR, TEXTIL mat, matt

matemática *f* INFORM&PD, MATEMÁT mathematics

matemáticas *f pl* INFORM&PD, MATEMÁT mathematics

materia[1]: **sin ~ mineral** *adj* CONTAM mineral-matter-free

materia[2] *f* ING MECÁ material, PROD stuff; ~ **activa** *f* TEC ESP *acumuladores* paste; ~ **administrativa** *f* CONST administrative area; ~ **alóctona** *f* TELECOM allochthonous matter; ~ **autóctona** *f* CONTAM autochthonous matter; ~ **degenerada** *f* QUÍMICA degenerate matter; ~ **espumosa de aguas de cloaca** *f* HIDROL scum; ~ **estéril** *f* CARBÓN *minería* oxidic waste; ~ **extraña** *f* PROD foreign matter; ~ **fecal** *f* INSTAL HIDRÁUL, RECICL faecal matter (*BrE*), fecal matter (*AmE*); ~ **fértil** *f* TEC ESP *del reactor nuclear* blanket; ~ **fibrosa** *f* PAPEL bulk fiber (*AmE*), bulk fibre (*BrE*); ~ **granulosa** *f* CONTAM particulate materials, particulate matter; ~ **inerte** *f* AGRIC inert carrier; ~ **insoluble acuosa** *f* RECICL slurry; ~ **interestelar** *f* TEC ESP interstellar matter; ~ **nuclear comprimida** *f* NUCL compressed nuclear matter; ~ **orgánica** *f* AGRIC, CONTAM, GEOL, HIDROL, TEC PETR organic matter; ~ **orgánica digestible** *f* RECICL digestible organic matter (*DOM*); ~ **orgánica disuelta** *f* CONTAM dissolved organic matter; ~ **particulada** *f* CONTAM particulate materials, particulate matter; ~ **prima** *f* ALIMENT staple food, C&V, CARBÓN, P&C, PAPEL, PROD, TEC PETR, TEXTIL raw material; ~ **seca** *f* (*MS*) AGRIC dry matter (*DM*), PAPEL dry solids content; ~ **volátil** *f* CARBÓN volatile body

material[1]: **de ~ acrílico** *adj* CONST acrylic

material[2] *m* CONST material, materials, ING MECÁ material, PROD stuff, QUÍMICA material, SEG materials, TEXTIL materials, stuff; ~ **absorbente** *m* CONTAM MAR *para recoger hidrocarburos derramados* oil mop; ~ **absorbente poroso** *m* ACÚST porous absorber; ~ **de acolchado de burbujas de aire** *m* EMB air-bubble cushioning; ~ **adhesivo** *m* TEC ESP bonding agent; ~ **de aislamiento** *m* CONST insulating material; ~ **aislante** *m* CONST insulating material, ELEC *aislador* isolator, EMB, ING ELÉC, MECÁ insulating material; ~ **aislante acústico** *m* CONTAM acoustic insulating material; ~ **aislante cerámico** *m* C&V, ELEC, ING ELÉC ceramic insulating material; ~ **con ajuste índices** *m* ÓPT index-matching material; ~ **almohadillado** *m* EMB cushioning product; ~ **altamente radioactivo** *m* NUCL, RECICL high-level waste (*HLW*); ~ **amortiguador de sonidos de impacto** *m* ACÚST impact sound-reducing material; ~ **antiestático** *m* SEG antistatic material; ~ **de apisonar** *m* PROD tamping material; ~ **de apoyo** *m* PETROL propping agent; ~ **de armamento** *m* D&A ordnance material; ~ **de atraque** *m* MINAS stemming material; ~ **barrera** *m* EMB barrier material; ~ **bruto** *m* CARBÓN bulk material, run of mine; ~ **en bruto** *m* ING MECÁ stock, MECÁ blank; ~ **de capa delgada** *m* ELECTRÓN thin-film material; ~ **de capa gruesa** *m* ELECTRÓN thick-film material; ~ **científico** *m* TEC ESP payload; ~ **combustible** *m* CONTAM, SEG combustible material; ~ **compuesto** *m* METAL composite, TEC ESP *estructuras* composite material; ~ **conductor de descargas eléctricas** *m* SEG lightning conductor, lightning conductor material; ~ **de contacto** *m* PROD contact material; ~ **densificante** *m* PETROL weighting material; ~ **depositado** *m* CONTAM deposited matter; ~ **de desecho** *m* EMB scrap material, HIDROL night soil; ~ **detrítico** *m* GEOL rubble; ~ **diamagnético** *m* ELEC, FÍS diamagnetic material; ~ **dieléctrico** *m* FÍS, ING ELÉC dielectric material; ~ **dieléctrico, línea** lossy material; ~ **disipativo** *m* ELEC *dieléctrico, línea* lossy material; ~ **empleado para sellar la tapa** *m* EMB lid-sealing compound; ~ **enriquecido** *m* FÍS RAD enriched material; ~ **en espiras** *m* ELECTRÓN convolution product; ~ **de espuma** *m* CONST, EMB, PROD foam material; ~ **estabilizado** *m* CONST stabilized material; ~ **de estibaje** *m* CARBÓN stowing material; ~ **con estructura de panal** *m* EMB honeycomb material; ~ **extraído por excavación** *m* FERRO excavated material; ~ **extraído por extracción** *m* MINAS excavated material; ~ **ferromagnético** *m* ELEC, ING ELÉC, PETROL ferromagnetic material; ~ **fértil** *m* NUCL fertile material; ~ **fisionable** *m* FÍS, FÍS RAD fissile material, ING ELÉC active material, NUCL fissile material; ~ **fosforescente** *m* ELECTRÓN phosphorescent material; ~ **de gran permeabilidad y pequeña histéresis** *m* ING ELÉC soft magnetic material; ~ **de gran remanencia magnética** *m* FÍS hard magnetic material; ~ **granulado** *m* PROC QUÍ granuled material; ~ **granular** *m* CONST granular material; ~ **para imprimir** *m* IMPR stock; ~ **impuro** *m* HIDROL dry matter; ~ **indicador** *m* ING MECÁ indicator plant; ~ **inerte** *m* AGRIC inert matter; ~ **inflamable** *m* SEG flammable material; ~ **de insonorización** *m* ACÚST soundproofing material; ~ **laminar** *m* ELECTRÓN, MECÁ laminate; ~ **laminar reforzado con vidrio** *m* C&V, CONST, NUCL glass-reinforced laminate; ~ **ligador** *m* NUCL bonding material; ~ **de ligadura** *m* NUCL bonding material; ~ **limpiador** *m* DETERG cleaning material; ~ **de machaqueo** *m* CARBÓN, CONST crushed material; ~ **magnético** *m* ACÚST, ING ELÉC magnetic material; ~ **magnético blando** *m* FÍS soft magnetic material; ~ **magnético de gran remanencia** *m* MECÁ hard magnetic material; ~ **magnetoestrictivo** *m* ING ELÉC magnetostrictive material; ~ **de metal sinterizado** *m* ING MECÁ sintered metal material; ~ **de minas** *m* AmL (*cf accesorio de minas Esp*) MINAS mining appliance; ~ **móvil** *m* CONST, FERRO *vehículos* rolling stock; ~ **no refractario** *m* FÍS, ÓPT nonrefractory material; ~ **no sulfonado** *m* DETERG unsulfonated matter (*AmE*), unsulphonated matter (*BrE*); ~ **ópticamente activo** *m* ÓPT, TELECOM optically-active material; ~ **paramagnético** *m* PETROL paramagnetic material; ~ **para pavimentos** *m* CONST flooring, flooring material, SEG flooring, flooring material, paving; ~ **pendiente de clasificar** *m* MECÁ backlog; ~ **poroso** *m* ACÚST porous material; ~ **preimpregnado** *m* P&C preimpregnate, prepreg; ~ **de préstamo** *m* CONST *movimiento de tierras* borrow; ~ **de protección** *m* EMB barrier material; ~ **para protección contra golpes** *m* EMB cushioning product; ~ **protector contra el calor** *m* SEG heat-protective material; ~ **que disipa mucha energía** *m* ELEC *dieléctrico, línea* lossy material; ~ **con que se**

rellena algo *m* PROD stuffing; ~ **reciclado** *m* CONST, CONTAM recycled material, RECICL salvaged material, recycled material; ~ **recuperado** *m* RECICL salvaged material; ~ **reforzado para embalar** *m* EMB reinforced packaging material; ~ **refractario** *m* CARBÓN, GAS, INSTAL TERM, TERMO refractory material; ~ **reglamentario** *m* PROD standard item; ~ **para rellenar** *m* CONST back filler; ~ **de relleno** *m* CONST filler, GEOL *de una cuenca, filón* infilling, NUCL filling material, backfill material, TEC ESP potting compound; ~ **de relleno de cavidad** *m* ELECTRÓN cavity filler; ~ **de reproducción** *m* AGRIC *animales* breeding stock; ~ **de resistencia** *m* ELEC, ING ELÉC resistance material; ~ **resistente a los agentes químicos** *m* D&A chemical-agent-resisting material (*CARM*); ~ **de revestimiento** *m* REVEST coating material; ~ **rodante** *m* CONST, FERRO *vehículos* rolling stock; ~ **secante** *m* CONST blotter material; ~ **sedimentado** *m* CONTAM deposited matter; ~ **de sellado** *m* NUCL sealing material; ~ **de semiconductor** *m* ELECTRÓN semiconductor material; ~ **de soporte** *m* P&C substrate; ~ **para suelos** *m* CONST, SEG flooring; ~ **termosensible** *m* EMB heat-sensitive material; ~ **tóxico** *m* SEG toxic material; ~ **transportado** *m* GEOL allochthon; ~ **triturado** *m* CARBÓN, CONST crushed material; ~ **usado para el cerrado de la tapa** *m* EMB cap-sealing compound; ~ **usado para el sellado de la lata** *m* EMB can-sealing compound (*AmE*), tin-sealing compound (*BrE*); ~ **de la vaina** *m* NUCL cladding material; ~ **de vidrio** *m* C&V, LAB glassware

materiales: ~ **auxiliares** *m pl* PROD factory supplies; ~ **para cojinetes** *m pl* ING MECÁ bearing materials; ~ **de construcción** *m pl* CONST building materials; ~ **fundidos** *m pl* SEG molten materials; ~ **no isotrópicos** *m pl* ING MECÁ nonisotropic materials; ~ **ondulados** *m pl* EMB corrugated products; ~ **ortotrópicos** *m pl* ING MECÁ orthotropic materials; ~ **peligrosos** *m pl* QUÍMICA, SEG, TRANSP dangerous materials; ~ **plásticos** *m pl* SEG plastic materials; ~ **textiles impermeables al aire** *m pl* SEG air-impermeable clothing materials

materias *f pl* PROD stuff; ~ **estériles** *f pl* MINAS attle; ~ **fecales** *f pl* HIDROL, RECICL faecal matter (*BrE*), fecal matter (*AmE*); ~ **indeseables presentes en el papelote** *f pl* PAPEL waste paper contraries; ~ **de paleta** *f pl* ENERG RENOV blade materials; ~ **primas para alfarería** *f pl* C&V pottery raw materials; ~ **proyectadas** *f pl* GEOL ejectamenta, *volcanismo* ejecta; ~ **sólidas** *f pl* HIDROL, RECICL solid matter

matiz *m* COLOR tinge, tint, IMPR hue, tint, TEXTIL shade; ~ **fuera de color** *m* COLOR off-color shade (*AmE*), off-colour shade (*BrE*); ~ **igual** *m* COLOR self-shade

matizar *vt* COLOR, IMPR, TEXTIL tint

matlockita *f* MINERAL matlockite

matorral *m* CONST brush

matraca *f* PAPEL rag shredder; ~ **invertible** *f* ING MECÁ reversible ratchet

matraz *m* Esp (*cf redoma AmL*) LAB *material de vidrio* bolt-head flask (*BrE*), flask, matrass (*AmE*), QUÍMICA balloon; ~ **de absorción** *m* PROC QUÍ absorption vessel; ~ **aforado** *m* LAB long-necked flask; ~ **de Buchner** *m* LAB Buchner flask; ~ **de Claisen** *m* PROC QUÍ, QUÍMICA Claisen flask; ~ **cónico** *m* LAB conical flask; ~ **de cuello ancho** *m*

LAB wide-necked flask; ~ **de cuello angosto** *m* LAB narrow-necked flask; ~ **de cuello largo** *m* LAB long-necked flask; ~ **con cuello moldeado** *m* LAB flask with molded neck (*AmE*), flask with moulded neck (*BrE*); ~ **de decantación** *m* PROC QUÍ decantation vessel; ~ **de Delf** *m* HIDROL Delf flask; ~ **de destilación** *m* LAB distillation flask, PROC QUÍ, QUÍMICA distillation flask, distilling flask, TERMO distillation flask; ~ **Dewar** *m* FÍS, LAB, PROC QUÍ Dewar flask; ~ **de ebullición** *m* PROC QUÍ boiling flask; ~ **Erlenmeyer** *m* LAB conical flask, Erlenmeyer flask; ~ **de evaporación** *m* PROC QUÍ evaporating vessel; ~ **de filtración** *m* LAB filtration flask; ~ **de fondo plano** *m* LAB flat-bottomed flask; ~ **de fondo redondo** *m* LAB round-bottomed flask; ~ **graduado** *m* LAB graduated flask; ~ **de Kitasato** *m* PROC QUÍ Kitasato vessel; ~ **de lavado** *m* QUÍMICA wash bottle; ~ **con tapón** *m* LAB stoppered flask; ~ **de tres bocas** *m* LAB *material de vidrio* three-necked flask; ~ **de vacío** *m* LAB *material de vidrio* vacuum flask; ~ **volumétrico** *m* LAB flask, long-necked flask, volumetric flask

matrices *f pl* ING MECÁ die; ~ **de bombear hidráulicas** *f pl* ING MECÁ hydraulic-bulging die; ~ **para estampar** *f pl* ING MECÁ dies for stamping; ~ **de explosión** *f pl* ING MECÁ explosion die; ~ **de pandear hidráulicas** *f pl* ING MECÁ hydraulic-bulging die

matrícula *f* AUTO license plate (*AmE*), number plate (*BrE*), TRANSP AÉR registration; ~ **del buque** *f* TRANSP MAR ship's register

matriz *f* ACÚST mother, CONST form, matrix, ELEC template, GEOL matrix, IMPR mat, INFORM&PD array, matrix, ING MECÁ template, former, die-casting die, bolster, templet, INSTR former, MATEMÁT matrix, MECÁ die, METAL matrix, ÓPT stamper, P&C *parte de equipo* mold (*AmE*), mould (*BrE*), PROD die, former, matrix, mold (*AmE*), mould (*BrE*), TELECOM array, TV matrix; ~ **de acceso** *f* TELECOM access matrix; ~ **de autocentrado** *f* ING MECÁ self-centering dies (*AmE*), self-centring dies (*BrE*); ~ **en bajo relieve** *f* IMPR female die; ~ **de cadenas** *f* INFORM&PD *texto* string array; ~ **circular de filetear** *f* ING MECÁ circular screwing die; ~ **circular de filetear para roscas paralelas de tubería** *f* ING MECÁ circular screwing die for parallel pipe threads; ~ **de cobre de berilio** *f* ING MECÁ beryllium copper die; ~ **de colada a la cera perdida** *f* ING MECÁ investment-casting die; ~ **para colada por gravedad** *f* ING MECÁ gravity die-casting die; ~ **de conexión** *f* TELECOM connection matrix (*CM*); ~ **de conmutación** *f* TEC ESP, TELECOM, TV switching matrix; ~ **de conmutación digital** *f* TELECOM digital switching matrix; ~ **de conmutación de microondas** *f* TELECOM microwave switch matrix; ~ **de conmutación óptica** *f* TELECOM optical-switching matrix; ~ **de conmutación óptica integrada** *f* TELECOM integrated optical switching matrix; ~ **de conmutación optoelectrónica** *f* TELECOM optoelectronic switching matrix; ~ **de conmutación de video** *f* AmL, ~ **de conmutación de vídeo** *f* Esp TV video-switching matrix; ~ **de curvar** *f* ING MECÁ bending die; ~ **de datos** *f* INFORM&PD data array; ~ **descifradora** *f* TV decoding matrix; ~ **dispersa** *f* INFORM&PD sparse matrix; ~ **de doblar** *f* ING MECÁ bending die; ~ **para embutir** *f* ING MECÁ die for pressing; ~ **epóxica** *f* TEC ESP epoxy matrix; ~ **escasa** *f* INFORM&PD sparse matrix;

~ **para estampar** *f* ING MECÁ die for pressing; ~ **para estampar en caliente** *f* ING MECÁ swaging die; ~ **de extrusión** *f* ING MECÁ extrusion die; ~ **extrusora** *f* ING MECÁ extrusion die; ~ **de forjado en frío** *f* ING MECÁ, PROD, TERMO cold-forging die; ~ **de frente de onda** *f* INFORM&PD wavefront array; ~ **de fundición colada por gravedad** *f* ING MECÁ gravity die-casting die; ~ **para fundición por inyección a presión** *f* ING MECÁ pressure diecasting die; ~ **para fundición inyectada de presofusión** *f* ING MECÁ pressure diecasting die; ~ **honda de embutir** *f* ING MECÁ deep drawing die; ~ **lineal** *f* TV linear matrix; ~ **lógica programable** *f* INFORM&PD programmable logic array; ~ **lógica programada** *f* (*PLA*) INFORM&PD programmed logic array (*PLA*); ~ **para paneles de automóviles** *f* ING MECÁ die for motor body panels; ~ **para paneles de coches** *f* ING MECÁ die for motor body panels; ~ **para perforar** *f* ING MECÁ die for punching; ~ **para polvos metálicos** *f* ING MECÁ die for metallic powders; ~ **de puertas** *f* INFORM&PD gate array; ~ **pulvimetalúrgica** *f* ING MECÁ powder-metal die; ~ **de puntos** *f* ELECTRÓN, INFORM&PD dot matrix; ~ **para punzonar** *f* ING MECÁ blanking die; ~ **para punzonar** *f* ING MECÁ die for punching; ~ **de punzonar de precisión** *f* ING MECÁ fine blanking-die; ~ **rotativa** *f* ING MECÁ, PROD rotational mold (*AmE*), rotational mould (*BrE*); ~ **simétrica** *f* INFORM&PD symmetric matrix; ~ **sistólica** *f* INFORM&PD systolic array; ~ **de tiempo simétrica** *f* TELECOM symmetrical time matrix; ~ **de transferencia** *f* FÍS transfer matrix; ~ **de trefilar** *f* ING MECÁ, MECÁ die; ~ **de trefilar en caliente** *f* ING MECÁ hot wire-drawing die; ~ **triangular** *f* INFORM&PD triangular matrix
matrización: ~ **de discos** *f* PROD direct metal mastering (*DMM*)
matrizado *m* TV matrixing
matrizar *vt* IMPR die out
Matusalén *m* C&V Methuselah
mauveína *f* QUÍMICA mauveine
máxima[1]: ~ **y mínima** *adj* ING MECÁ go-and-not-go, go-no-go
máxima[2]: ~ **apertura numérica teórica** *f* TELECOM maximum theoretical numerical aperture; ~ **avenida** *f* AGUA high-water overflow; ~ **barométrica** *f* METEO barometric maximum; ~ **corriente** *f* ELEC maximum current; ~ **corriente con fluctuaciones** *f* PROD maximum ripple current; ~ **corriente permisible** *f* ELÉC, ING ELÉC maximum current rating; ~ **dosis corporal admisible para personas profesionales expuestas** *f* FÍS RAD, NUCL maximum permissible occupational whole-body dose; ~ **potencia** *f* TRANSP AÉR full throttle; ~ **potencia admisible** *f* TELECOM maximum admissible power; ~ **probabilidad** *f* MATEMÁT maximum likelihood; ~ **verosimilitud** *f* MATEMÁT maximum likelihood
maximización *f* TELECOM maximization
máximo[1]: *adj* ING MECÁ high; ~ **y mínimo** *adj* ING MECÁ go-and-not-go, go-no-go
máximo[2]: ~ **accidente creíble** *m* NUCL maximum credible accident; ~ **común divisor** *m* (*mcd*) MATE-MÁT greatest common divisor, highest common divisor, highest common factor (*HCF*); ~ **de corriente** *m* ELEC current peak; ~ **flujo de calor** *m*

NUCL maximum flux heat; ~ **rendimiento sostenible** *m* OCEAN maximal sustainable yield
máximos: ~ **principales** *m pl* FÍS principal maxima; ~ **secundarios** *m pl* FÍS secondary maxima
maxita *f* MINERAL maxite
maxvelio *m* (*Mx*) ELEC, ING ELÉC maxwell (*Mx*)
maxwell *m* (*Mx*) ELEC, ING ELÉC maxwell (*Mx*)
Mayday *f* TRANSP MAR Mayday
mayólica *f* C&V majolica
mayor[1]: **de ~ tamaño que el especificado** *adj* ING MECÁ oversize
mayor[2]: **al por ~** *adv* CARBÓN in bulk
mayor[3] *f* TRANSP MAR mainsail
mayordomo *m* C&V foreman
mayorista *m* PROD wholesale supplier
mayúscula *f* IMPR capital, capital letter; ~ **pequeña** *f* IMPR small cap
mayúsculas: ~ **de imprenta** *f pl* IMPR block capitals; ~ **rectas** *f pl* IMPR stand-up capitals
maza *f* CARBÓN *martillo pilón* ram pump, CONST monkey, beater, sledgehammer, pile hammer (*AmE*), piling hammer (*BrE*), drop hammer, MECÁ hammer, PROD pounder, *martillo pilón* tup; ~ **de hierro** *f* MECÁ mallet; ~ **hueca** *f* ING MECÁ hollow ram; ~ **de pilotes** *f* CONST maul; ~ **de solador** *f* CONST pavior's hammer (*AmE*), paviour's hammer (*BrE*)
mazarota *f* PROD head, riser, shrink head, *fundería* sinking head, sullage head, feeder, sullage piece
mazo *m* C&V mallet, CONST beetle, maul, ING MECÁ formwork hammer, ÓPT *de ondas, de fibras ópticas* bundle, PROD mallet; ~ **de cables** *m* TEC ESP wiring harness; ~ **de caída libre** *m* CONST free-falling stamp; ~ **de calafate** *m* PROD caulking hammer; ~ **de calafatear** *m* ING MECÁ, MECÁ, PROD caulking mallet; ~ **para estampar** *m* CONST *metales* bossing mallet; ~ **de fontanero** *m* Esp (*cf mazo de plomero AmL*) PROD lead dresser; ~ **de madera zunchado** *m* PROD ironbound mallet; ~ **de plomero** *m* AmL (*cf mazo de fontanero Esp*) PROD lead dresser; ~ **de romper piedras** *m* CARBÓN stone-splitting hammer
mazorca *f* Esp (*cf marlo AmL*) AGRIC ear, spike, *de cacao* cob, pod; ~ **de maíz** *f* AGRIC corn cob (*AmE*), maize cob (*BrE*); ~ **de maíz para asar** *f* AGRIC roaster ear; ~ **de maíz con grano** *f* AGRIC corn cob with grains (*AmE*), maize cob with grains (*BrE*)
Mb *abr* (*megabyte*) ELECTRÓN, INFORM&PD Mb (*megabyte*)
mcd *abr* (*máximo común divisor*) MATEMÁT HCF (*highest common factor*)
mcm *abr* (*mínimo común múltiple*) MATEMÁT LCM (*least common multiple*)
Md *abr* (*mendelevio*) FÍS RAD, NUCL, QUÍMICA Md (*mendelevium*)
MDCP *abr* (*modulación diferencial de código pulsado*) ELECTRÓN DPCM (*differential pulse code modulation*)
MDF *abr* ELECTRÓN (*manipulación por desplazamiento de frecuencia, multiplexación por division de frecuencias, multiplexado por división de frecuencias*) FSK (*frequency-shift keying*), FDM (*frequency-division multiplexing*), FÍS (*multiplexación por división de frecuencias, multiplexado por división de frecuencias*) FDM (*frequency-division multiplexing*), INFORM&PD, TELECOM (*multiplexación por división de frecuencias, multiplexado por división de frecuencias, manipulación por desplazamiento de frecuencia*) FDM (*frequency-*

division multiplexing), FSK (*frequency-shift keying*), TV (*manipulación por desplazamiento de frecuencia*) FSK (*frequency-shift keying*)

MDI *abr* (*diisocianato de difenilmetano*) P&C MDI (*diphenylmethane diisocyanate*)

meandro *m* HIDROL meander, *de un río* bend, loop; ~ **recortado** *m* AGUA cutoff

meato *m* ACÚST meatus

MEB *abr* (*microscopio electrónico de barrido*) ELEC, FÍS RAD, LAB SEM (*scanning electron microscope*)

mecánica *f* FÍS, ING MECÁ, MATEMÁT, MECÁ mechanics; ~ **analítica** *f* FÍS analytical mechanics; ~ **celeste** *f* TEC ESP celestial mechanics; ~ **cuántica** *f* FÍS, FÍS PART quantum mechanics; ~ **de los fluidos** *f* FÍS FLUID fluid mechanics; ~ **matricial** *f* FÍS, MECÁ matrix mechanics; ~ **naval** *f* TRANSP MAR marine engineering; ~ **Newtoniana** *f* FÍS, TEC ESP Newtonian mechanics; ~ **de ondas** *f* FÍS, FÍS ONDAS wave mechanics; ~ **ondulatoria** *f* FÍS ONDAS, ING ELÉC wave mechanics; ~ **racional** *f* ING MECÁ, MECÁ rational mechanics; ~ **relativista** *f* FÍS, TEC ESP relativistic mechanics; ~ **de rocas** *f* CARBÓN, GEOL rock mechanics; ~ **de rotura** *f* MECÁ fracture mechanics; ~ **del suelo** *f* CONST soil mechanics; ~ **de suelos** *f* CARBÓN, GEOL soil mechanics; ~ **de termotransferencia** *f* TEC ESP heat-transfer engineering

mecánico *m* ING MECÁ, MECÁ fitter, PROD mechanic; ~ **ajustador** *m* ING MECÁ artificer; ~ **de a bordo** *m* TRANSP AÉR flight engineer; ~ **formador** *m* ING MECÁ former; ~ **frigorista** *m* REFRIG, TERMO refrigeration engineer; ~ **de mantenimiento de aviones** *m* TRANSP AÉR aircraft-maintenance mechanic; ~ **de vuelo** *m* TRANSP AÉR flight engineer

mecanismo *m* INFORM&PD drive, ING ELÉC machine, ING MECÁ gearing, mechanism, appliance, gear, device, MECÁ gear, PROD device; ~ **de accionamiento** *m* CINEMAT drive mechanism; ~ **de accionamiento de las barras de control** *m* (*CRDM*) NUCL control rod drive mechanism (*CRDM*); ~ **de accionamiento del distribuidor** *m* VEH distributor drive; ~ **de accionamiento de frenos** *m* FERRO brake rigging; ~ **de accionamiento de frenos compensado** *m* FERRO compensated brake rigging; ~ **de accionamiento motorizado** *m* ELEC motor-drive mechanism; ~ **de aceleración** *m* TRANSP acceleration device; ~ **de advertencia para el intervalo de tiempo** *m* TRANSP headway warning device; ~ **de alimentación** *m* EMB feeding device, PROD feed gear; ~ **alternativo** *m* ING MECÁ reciprocal gear; ~ **de apertura** *m* EMB opening mechanism; ~ **de arranque** *m* TEC ESP kick-off mechanism; ~ **de arranque en serie** *m* ELEC *motores* series starter; ~ **de arrastre de la película** *m* CINEMAT film drive; ~ **automático de fuego** *m* D&A automatic firing unit; ~ **de avance** *m* VEH vacuum advance mechanism; ~ **de avance centrífugo** *m* AUTO, VEH centrifugal advance mechanism; ~ **de avance del garfio** *m* CINEMAT claw carriage; ~ **de avance de vacío** *m* AUTO vacuum advance mechanism; ~ **de báscula** *m* ELEC trip gear; ~ **basculador** *m* ELEC trip gear; ~ **basculante** *m* CINEMAT cradle, MINAS dump heap; ~ **bifásico** *m* ING ELÉC two-phase machine; ~ **de bloqueo** *m* CINEMAT locking mechanism, ING MECÁ clamping mechanism; ~ **de bloqueo de puerta** *m* VEH door-locking mechanism; ~ **de bobinado** *m* VEH winding mechanism; ~ **de la bomba** *m* AGUA, PROD pump gear; ~ **de**

cambio de marcha *m* ING MECÁ reversing gear; ~ **de cierre** *m* ING MECÁ locking mechanism; ~ **de cinta servocontrolada** *m* TV servo-controlled tape mechanism; ~ **de cinta servoregulada** *m* TV servo-controlled tape mechanism; ~ **de codificación** *m* TRANSP coding device; ~ **codificador** *m* TRANSP coding device; ~ **de compuerta** *m* INSTAL HIDRÁUL gate gear; ~ **de conos de fricción** *m* ING MECÁ cone gear; ~ **de contrarreajuste** *m* PROD antireset windup; ~ **de control** *m* CARBÓN control driving, ELEC switchgear; ~ **de control de crucero** *m* TRANSP cruise control device; ~ **de cruz de Malta** *m* CINEMAT Maltese cross assembly, ING MECÁ Maltese cross mechanism; ~ **de declinación** *m* INSTR declination gear; ~ **de descarga** *m* EMB dumping mechanism; ~ **de desciframiento** *m* ELECTRÓN, INFORM&PD, TEC ESP, TELECOM, TV decoding device; ~ **desconectador** *m* ELEC *cortocircuito* trip gear; ~ **de desembrague delantero** *m* ING MECÁ disengaging gear in front; ~ **de desenganche** *m* ELEC, ING MECÁ trip gear; ~ **de la dirección** *m* VEH *de motocicleta* steering head; ~ **de disparador** *m* ING MECÁ trip gear; ~ **de disparo** *m* ELEC trip gear, ING MECÁ release, trigger; ~ **de distribución tipo Joy** *m* INSTAL HIDRÁUL *distribución radial* Joy's valve-gear; ~ **de distribución por válvulas** *m* INSTAL HIDRÁUL valve gear; ~ **del distribuidor de expansión** *m* ING MECÁ expansion gear; ~ **del distribuidor de expansión automático** *m* ING MECÁ automatic expansion gear; ~ **electromecánico** *m* ELEC, ING ELÉC electromechanical device; ~ **electrónico** *m* ELECTRÓN, ING ELÉC electron device, electronic device; ~ **electrónico antibloqueo** *m* TRANSP electronic anti-locking device; ~ **elevador** *m* FERRO, ING MECÁ, MECÁ lifting gear; ~ **de eliminación alfa** *m* FÍS RAD alpha-elimination mechanism; ~ **de enganche** *m* ING ELÉC *relés* latching; ~ **de engranaje de cremallera y piñón** *m* ING MECÁ rack-and-pinion gear; ~ **de engranaje de las válvulas** *m* AUTO valve-gear mechanism; ~ **de engranajes** *m* ING MECÁ gear assembly; ~ **de enlace** *m* ING MECÁ link mechanism; ~ **de enrollar** *m* P&C *equipo*, PAPEL reeling machine; ~ **de escape** *m* ING MECÁ escapement mechanism; ~ **para evitar que se rebasen los enganches** *m* AmL (*cf topes Esp*) MINAS *jaula* overwind gear, overwinding gear; ~ **de excéntrica** *m* ING MECÁ eccentric gear; ~ **de expansión de termopar** *m* PROD thermocouple expander; ~ **extraíble** *m* TEXTIL take-away mechanism; ~ **de falla** *m* CALIDAD failure mechanism; ~ **de formación de imágenes** *m* ÓPT imaging mechanism; ~ **en funcionamiento** *m* ING MECÁ working mechanism; ~ **de gobierno** *m* ING ELÉC control gear, TRANSP MAR steering gear; ~ **grabador de cinta** *m* INFORM&PD tape deck; ~ **guiador** *m* MINAS *explotación de aluviones* placer; ~ **de humidificación** *m* EMB moistening device; ~ **impulsador de cinta** *m* TV tape drive; ~ **de interferencia** *m* ING ELÉC inference engine; ~ **interior** *m* AUTO internal gear; ~ **interno de sincronización** *m* PROD internal timing mechanism; ~ **de inversión** *m* ING MECÁ reversible gear, reversing gear; ~ **de inversión de marcha** *m* ING MECÁ change gear; ~ **inversor del avance** *m* ING MECÁ feed-reversing gear; ~ **de izada** *m* PROD hoisting gear; ~ **láser** *m* ÓPT laser mechanism; ~ **de mando** *m* ING MECÁ power unit; ~ **de mando de la dirección** *m* VEH steering gear; ~ **de mando del distribuidor** *m*

INSTAL HIDRÁUL valve gear; **~ de maniobra** *m* ING MECÁ rig, rigging; **~ de menguado** *m* TEXTIL fashioning mechanism; **~ moderador de velocidad al final de la carrera** *m* *AmL* (*cf freno de jaula Esp*) MINAS *jaula* overwinder; **~ moderador de velocidad y de sobrevelocidad** *m* MINAS overwinder and overspeeder; **~ móvil** *m* TRANSP MAR *motor* running gear; **~ de paro** *m* TEXTIL stop motion; **~ de paro en la fileta** *m* TEXTIL stop motion on creel; **~ del pistón** *m* C&V plunger-assist mechanism; **~ protector de circuito derivado** *m* PROD branch-circuit protective-device; **~ de puesta en marcha** *m* VEH starter; **~ de relleno** *m* MINAS pack system; **~ de relojería** *m* MECÁ clockwork; **~ de resorte** *m* MECÁ clockwork; **~ de resorte y fiador** *m* PROD spring and toggle mechanism; **~ de rueda libre** *m* ING MECÁ, MECÁ freewheel mechanism; **~ sacudidor de parrilla** *m* PROD grate-shaking rig; **~ de seguimiento** *m* ÓPT tracking mechanism; **~ de separación** *m* TEC ESP separation mechanism; **~ de servobloqueo del cabezal móvil** *m* CINEMAT capstan servolock; **~ de servomando** *m* TEC ESP feel mechanism; **~ de siega** *m* AGRIC mowing attachment; **~ sincrónico** *m* ING ELÉC synchronous machine; **~ de sondeo** *m* ING MECÁ rigging; **~ de suspensión** *m* MINAS suspension gear; **~ tornillado de cierre** *m* EMB screw-locking device; **~ de tornillo sin fin** *m* TRANSP MAR worm gear; **~ de trabajo** *m* ING MECÁ working mechanism; **~ de transmisión principal** *m* ING MECÁ main drive; **~ de transporte** *m* TV transport mechanism; **~ de trasmisión vertical** *m* ING MECÁ lift drive; **~ trifásico** *m* ING ELÉC three-phase machine; **~ de trinquete** *m* ING MECÁ ratchet motion, ratchet-and-pawl motion; **~ de trinquete de palanca** *m* ING MECÁ lever ratchet motion; **~ de unión** *m* ING MECÁ link mechanism; **~ de válvula de corredera** *m* INSTAL HIDRÁUL slide valve gear; **~ de válvula disparadora** *m* ING MECÁ trip valve gear; **~ de válvula invertida** *m* INSTAL HIDRÁUL drop valve gear

mecanismos *m pl* ING MECÁ machinery; **~ de inversión de la población** *m pl* FÍS RAD population-inversion mechanisms; **~ de población lasérica** *m pl* FÍS RAD laser population mechanisms

mecanización *f* PROD machining; **~ agrícola** *f* AGRIC, MECÁ farm mechanization; **~ de explotaciones agrícolas** *f* AGRIC mechanization of farms; **~ en serie** *f* PROD gang machining, multiple machining

mecanizado *m* PROD tooling; **~ por haz de electrones** *m* ELECTRÓN electron-beam machining

mecanizar *vt* PROD machine, tool

mecanoaccionamiento *m* ING MECÁ mechanical drive

mecate *m* SEG rope

mecedora: **~ de cubetas** *f* FOTO dish rocker

mecha *f* CONST bit, D&A fuse, ING MECÁ, MECÁ bit, MINAS (*cf estopín*) *explosivo* squib, *explosivos* gun-cotton, fuse, PETROL bit, TEC PETR (*cf barrena tricónica, cf barreno*) *perforación* three-cone bit, rock bit, bit, drill bit, tricone bit, rotary bit, TEXTIL sliver, *hilado* roving; **~ absorbente** *f* CONTAM MAR sorbent wick; **~ de arrastre** *f* TEC PETR *perforación* drag bit; **~ de Bickford** *f* MINAS blasting-fuse, common fuse, safety fuse, Bickford fuse; **~ centradora** *f* ING MECÁ center bit (*AmE*), center bit for bit stock (*AmE*), centre bit for bit stock (*BrE*); **~ para centrar** *f* ING MECÁ center drill (*AmE*), centre drill (*BrE*); **~ de cola de**

pescado *f* TEC PETR *perforación* roller bit; **~ común** *f Esp* MINAS blasting-fuse, common fuse, safety fuse, Bickford fuse; **~ cónica** *f AmL* (*cf barrena cónica Esp*) TEC PETR cone bit; **~ de corona** *f* TEC PETR crown bit; **~ de cuatro aletas** *f AmL* TEC PETR four-wing bit; **~ detonante** *f* MINAS detonating cord, detonating fuse, primacord; **~ eléctrica** *f* ELEC electric lighter; **~ para escariar** *f* TEC PETR reaming bit; **~ para formación dura** *f* TEC PETR hard of formation bit; **~ helicoidal** *f* TEC PETR spiral bit; **~ lenta** *f* MINAS blasting-fuse, common fuse, safety fuse, Bickford fuse; **~ de mando** *f* MINAS master fuse; **~ ordinaria** *f* MINAS common fuse; **~ piloto** *f* TEC PETR pilot bit; **~ de un pozo** *f* TEC PETR spudding bit; **~ rápida** *f* MINAS primacord; **~ sacatestigos** *f Esp* TEC PETR core bit; **~ de seguridad** *f* MINAS safety fuse; **~ tetracónica** *f AmL* (*cf barrena tetracónica Esp*) TEC PETR quadricone bit; **~ trenzada** *f* TEC ESP wick; **~ para vidrio** *f* C&V, ING MECÁ glass drill

mechera *f* TEXTIL roving frame

mechero: **~ de arranque** *m* C&V startup burner; **~ Bunsen** *m* LAB Bunsen burner; **~ de gas** *m* GAS, PROD, TERMO, TERMOTEC gas burner; **~-mariposa** *m* LAB flat-flame burner; **~ de Meker** *m* LAB Meker burner

mechinal *m* CONST *obra civil* weephole

mechón *m* PROD *moldería* swab

meconato *m* QUÍMICA meconate

mecónico *adj* QUÍMICA meconic

meconina *f* QUÍMICA meconin, meconine

medallón: **~ de vidrio** *m* C&V glass cameo

media[1]: **a ~ asta** *adv* TRANSP MAR *bandera* at half-mast; **a ~ eslora** *adv* TRANSP MAR midship

media[2] *f* INFORM&PD, MATEMÁT mean; **~ abrazadera** *f* ING MECÁ half cleat; **~ anchura** *f* FÍS half width; **~ aritmética** *f* FÍS average, INFORM&PD, MATEMÁT arithmetic mean; **~ armónica** *f* MATEMÁT harmonic mean; **~ de bits por muestra** *f* TELECOM average bits per sample; **~ brida** *f* ING MECÁ half clamp; **~ copa** *f* ING MECÁ half cup; **~ cuadrática** *f* CONST, ELEC, ELECTRÓN, MATEMÁT, TELECOM root-mean-square (*RMS*); **~ del error de trayectoria de planeo** *f* TRANSP AÉR mean glide path error; **~ flanja** *f* ING MECÁ half flange; **~ luna** *f* C&V feeder nose, GEOM meniscus; **~ marea** *f* TRANSP MAR half tide; **~ móvil autorregresiva** *f* (*ARMA*) INFORM&PD autoregressive moving average (*ARMA*); **~ onda** *f* ELEC *corriente alterna* half wave; **~ de opinión** *f AmL* (*cf puntuación media de opinión Esp*) TELECOM mean opinion score; **~ palabra** *f* INFORM&PD half word, nibble; **~ pestaña** *f* ING MECÁ half flange; **~ ponderada** *f* PROD weighted average; **~ vuelta** *f* ING MECÁ reversing, PROD half twist

mediacaña *f* IMPR double rule, double ruling

mediador *m* FÍS PART *de fuerza electromagnética* mediator

medialínea *f* GEOM half line

mediamarímetro *m* OCEAN mediamarimeter

mediana *f* GEOM, INFORM&PD, MATEMÁT median, TEC PETR median line

medianera *f* CONST *arquitectura* partition

medianería *f* CONST *arquitectura* partitioning

medianía *f* TRANSP MAR *geografía* middle ground

medianil *m* IMPR gutter; **~ posterior** *m* IMPR back gutter

mediante *m* ACÚST *música* mediant

medicamento: ~ sulfamídico *m* QUÍMICA sulfa drug (*AmE*), sulpha drug (*BrE*)

medicina: ~ aeroespacial *f* TEC ESP aerospace medicine

medicinal *adj* QUÍMICA medicinal

medición *f* ELECTRÓN, FÍS ONDAS, FÍS RAD measurement, GAS metering, ING MECÁ dimensioning, gaging (*AmE*), gauging (*BrE*), METR *tamaño, número* measure, gaging (*AmE*), measuring, gauging (*BrE*), measurement, QUÍMICA measuring, measurement, SEG measurement, measuring, TEC PETR *de producción, yacimiento* gaging (*AmE*), gauging (*BrE*), TELECOM metering; ~ de la activación neutrónica *f* NUCL neutron activation logging; ~ del alcance óptico *f* INSTR visual range meter; ~ del alcance visual *f* INSTR visual range meter; ~ de banda ancha *f* ELECTRÓN wideband measurement; ~ de cantidades *f* CONST measurement of quantities; ~ de carretera de acceso *f* TRANSP access road metering (*AmE*), slip road metering (*BrE*); ~ del caudal *f* HIDROL flow rate; ~ de la contaminación del aire *f* CONTAM, SEG measurement of air pollution; ~ electrónica de inyección de combustible *f* TRANSP electronic metering of fuel injection; ~ escalar *f* ING ELÉC scalar measurement; ~ y evaluación de la intensidad de las vibraciones *f* SEG measurement and evaluation of vibration severity; ~ de frecuencia instantánea *f* ELECTRÓN instantaneous frequency measurement (*IFM*); ~ de frecuencia instantánea digital *f* ELECTRÓN digital instantaneous frequency measurement; ~ geofísica *f* GAS, GEOFÍS geophysical measurement; ~ del grosor de paredes *f* NUCL wall thickness gaging (*AmE*), wall thickness gauging (*BrE*); ~ del grosor del revestimiento *f* ING MECÁ coated thickness measurement; ~ de grosores de capas *f* NUCL layer thickness gaging (*AmE*), layer thickness gauging (*BrE*); ~ de grosores de estratos *f* NUCL layer thickness gaging (*AmE*), layer thickness gauging (*BrE*); ~ de la humedad *f* METEO, TRANSP MAR humidity measurement; ~ en línea *f* NUCL on-line measurement; ~ de líquidos *f* METR liquid measure; ~ de partículas *f* PROC QUÍ particle sizing; ~ del punto de rocío *f* C&V dew point measurement; ~ de radiación *f* FÍS RAD radiation measurement; ~ de rampa *f* TRANSP ramp metering; ~ de rayos gamma *f* FÍS RAD, NUCL gamma-ray survey; ~ remota *f* NUCL remote metering; ~ repetible *f* METR repeatable measurement; ~ de la tasa de error *f* ELECTRÓN, INFORM&PD, TELECOM error-rate measurement; ~ del terreno *f* CONST, METR land measuring; ~ de la textura superficial *f* ING MECÁ surface texture measurement; ~ de tiempo *f* TELECOM timing; ~ de la vía de acceso *f* TRANSP access road metering (*AmE*), slip road metering (*BrE*)

mediciones: ~ de los contornos de rayas *f pl* FÍS RAD line profile measurements; ~ durante la perforación *f pl* (*MWD*) TEC PETR measurements while drilling (*MWD*)

medida *f* CARBÓN load, ELEC *medición, fabricación* gage (*AmE*), gauge (*BrE*), ELECTRÓN, FÍS ONDAS, FÍS RAD measurement, GEOM mensuration, METR measurement, measure, metering, QUÍMICA *cierta proporción, porcentaje* measure, measurement, measuring; ~ angular *f* METR angle measurement; ~ para áridos *f* AGRIC dry measure; ~ de carbón *f* CARBÓN

room; ~ del caudal *f* AGUA, FÍS, FÍS FLUID, HIDROL, MECÁ flow rate; ~ circular *f* GEOM circular measure; ~ correctiva *f* TELECOM corrective measure; ~ cuadrada *f* METR square measure; ~ cúbica *f* METR cubic measure; ~ del cuerpo *f* IMPR type size; ~ de la desviación *f* TELECOM deviation measurement; ~ de dispersión de un cristal *f* C&V Abb value; ~ estándar *f* METR measurement standard; ~ final *f* METR end measure; ~ del gasto *f* AGUA flow rate; ~ de humedad *f* EMB moisture test; ~ lineal *f* METR long measure; ~ de longitud *f* METR long measure; ~ para muros de ladrillo *f* MINAS rod; ~ no repetible *f* METR nonrepeatable measurement; ~ nominal *f* EMB, ING MECÁ nominal size; ~ para paja y heno *f* CARBÓN load; ~ de la permeabilidad mediante flujo lateral de aire *f* PAPEL paper side-air permeability measurement; ~ de la presión *f* TEC ESP pressure measurement; ~ preventiva *f* SEG precautionary measure; ~ real *f* ING MECÁ actual size; ~ de seguridad *f* SEG safety measure; ~ sólida *f* METR solid measure; ~ del tamaño de partículas *f* P&C particle-size measurement; ~ del terreno *f* CONST, METR land measure; ~ de tiempo *f* TELECOM timing

medidas: ~ de ajuste *f pl* ING MECÁ fitting dimensions; ~ para el control de crecidas *f pl* AGRIC, AGUA, HIDROL flood-control measures; ~ cuadradas *f pl* METR square measures; ~ de emergencia *f pl* SEG emergency measures; ~ estándar *f pl* METR standard measures; ~ del fondo del pozo *f pl* TEC PETR downhole measurements; ~ de instalación *f pl* ING MECÁ fitting dimensions; ~ para mitigar las crecidas *f pl* AGUA, HIDROL flood-mitigation measures; ~ de montaje *f pl* ING MECÁ fitting dimensions; ~ de protección contra las riadas *f pl* AGUA, HIDROL flood-relief measures; ~ de seguridad *f pl* SEG safety precautions; ~ seleccionadas *f pl* ING MECÁ selected sizes; ~ y tolerancias *f pl* ING MECÁ fits and clearances

medidor *m* CONST measuring apparatus, ELEC, measuring apparatus, meter, ELECTRÓN measurand, measuring apparatus, ING MECÁ, INSTR gage (*AmE*), gauge (*BrE*), measuring apparatus, LAB, MECÁ gage (*AmE*), gauge (*BrE*), METR measuring apparatus, TELECOM measuring apparatus, meter, timer; ~ acústico *m* SEG sound level meter; ~ acústico personal *m* SEG personal sound-exposure meter; ~ de aire *m* ING MECÁ air meter; ~ analógico *m* ING ELÉC analog meter; ~ de asentamiento *m* PROC QUÍ settlement meter; ~ audible de número de Mach *m* TRANSP AÉR audible machmeter; ~ de blandura *m* ALIMENT tenderometer; ~ Bourdon *m* FÍS, TEC PETR Bourdon gage (*AmE*), Bourdon gauge (*BrE*); ~ de CA *m* ELEC *instrumento* AC meter; ~ de cable *m* ELEC, ING ELÉC, INSTR cable gage (*AmE*), cable gauge (*BrE*); ~ de CC *m* ELEC *instrumento*, INSTR DC meter; ~ Charpy de resistencia a la rotura bajo impacto *m* P&C Charpy impact tester; ~ cinemático *m* FERRO kinematic gage (*AmE*), kinematic gauge (*BrE*); ~ del combustible *m* AUTO, TRANSP AÉR, VEH fuel gage indicator (*AmE*), fuel gauge indicator (*BrE*); ~ de la conductividad térmica *m* TERMO heat-conductivity meter; ~ de contaminación *m* CONTAM, D&A, NUCL, SEG contamination meter; ~ del contenido de berilio *m* NUCL beryllium content meter; ~ de corriente alterna *m* ELEC alternating-current meter; ~ de corrientes *m* AGUA,

INSTAL HIDRÁUL current meter; ~ **de deformación** *m* CONST, MECÁ, OCEAN strain gage (*AmE*), strain gauge (*BrE*); ~ **de la densidad beta** *m* NUCL beta-density gage (*AmE*), beta-density gauge (*BrE*); ~ **de diafragma** *m* GAS diaphragm meter; ~ **de diferencia de voltaje** *m* ING ELÉC potentiometer; ~ **de distancia** *m* ING MECÁ telemeter; ~ **de durezas** *m* MECÁ, METR, P&C hardness tester; ~ **eléctrico** *m* ELEC, SEG electrical measuring-apparatus; ~ **electromagnético** *m* ELEC moving-iron meter; ~ **electromagnético de caudal** *m* ELEC, ING ELÉC electromagnetic flowmeter; ~ **de energía** *m* FÍS watt-hour meter, ING ELÉC energy meter; ~ **de esfuerzos** *m* CONST strain gage (*AmE*), strain gauge (*BrE*); ~ **de espesores** *m* C&V, ING MECÁ, LAB thickness gage (*AmE*), thickness gauge (*BrE*); ~ **de estallido** *m* PAPEL burst tester; ~ **de la estructura** *m* PAPEL formation tester; ~ **de FEM** *m* ELEC counter EMF; ~ **de flexión** *m* P&C bending tester; ~ **de flujo** *m* GEN flowmeter; ~ **del flujo de corrientes de Foucault** *m* ELEC, NUCL eddy-current flow meter; ~ **del flujo de corrientes inducidas** *m* ELEC, NUCL eddy-current flow meter; ~ **de flujo de densidad magnética** *m* GEOFÍS magnetic flux density meter; ~ **de flujo luminoso** *m* FÍS lumenmeter; ~ **de flujo térmico** *m* TERMO heat flow meter; ~ **de frecuencia de lámina vibrante** *m* ING ELÉC vibrating-reed frequency meter; ~ **de frenada** *m* INSTR, METR, TEC ESP decelerometer; ~ **de fuga a tierra** *m* ELEC, ING ELÉC earth-leakage meter (*BrE*), ground-leakage meter (*AmE*); ~ **de gas** *m* GAS, LAB, TERMO gas meter; ~ **geopotencial** *m* GEOFÍS geopotential meter; ~ **de gravedad** *m* FÍS, GEOFÍS, INSTR gravity meter; ~ **de hierro móvil** *m* ELEC moving-iron meter; ~ **por hilo tenso** *m* OCEAN taut-wire indicator; ~ **de imán móvil** *m* INSTR moving-iron meter; ~ **de inclinación** *m* GEOFÍS tiltmeter; ~ **indicador de volumen** *m* TV volume indicator meter (*VI meter*); ~ **de integración** *m* ING ELÉC integrating meter; ~ **de intensidad de campo** *m* ELEC *instrumento*, ING ELÉC, INSTR field-strength meter; ~ **de intervalos de tiempo** *m* AmL (*cf cronómetro Esp*) LAB timer; ~ **de longitud de onda** *m* FÍS wavemeter; ~ **magnético de espesor** *m* LAB magnetic thickness gage (*AmE*), magnetic thickness gauge (*BrE*); ~ **de mareas** *m* GEOFÍS tide gage (*AmE*), tide gauge (*BrE*); ~ **de nivel** *m* CINEMAT, NUCL level meter; ~ **de núcleo móvil** *m* ING ELÉC moving-coil meter; ~ **de número de Mach** *m* TRANSP AÉR machmeter; ~ **de onda de frecuencia acústica** *m* FÍS ONDAS beat-frequency wavemeter; ~ **de ondas** *m* FÍS ONDAS, TRANSP MAR wavemeter; ~ **de ondas heterodinas** *m* FÍS ONDAS heterodyne wavemeter; ~ **de pérdida** *m* INSTR leakage meter; ~ **de perfil de contacto** *m* ING MECÁ contact profile meter; ~ **de pH** *m* CARBÓN, LAB pH meter; ~ **de picos de programas** *m* TV peak programme meter; ~ **de pieza** *m* ING MECÁ detail gage (*AmE*), detail gauge (*BrE*); ~ **de Piraní** *m* FÍS Pirani gage (*AmE*), Pirani gauge (*BrE*); ~ **de pistón giratorio** *m* GAS rotary-piston meter; ~ **de pistón rotatorio** *m* GAS rotary-piston meter; ~ **de porosidad mediante aspiración** *m* PAPEL suction porosity tester; ~ **de potencia activa** *m* ELEC active power meter; ~ **de potencia aparente** *m* ELEC apparent-power meter; ~ **de presión** *m* CARBÓN pressure gauge (*BrE*); ~ **de presiones** *m* CARBÓN crusher, pressure gage (*AmE*), pressure gauge (*BrE*), MECÁ crusher gage (*AmE*),

crusher gauge (*BrE*); ~ **de profundidad** *m* METEO bed gage (*AmE*), bed gauge (*BrE*), OCEAN depth meter; ~ **para pruebas de intensidad** *m* ING ELÉC current-testing meter; ~ **de radiación** *m* D&A radiation meter, FÍS ONDAS radiation counter, FÍS RAD, INSTR, NUCL radiation meter; ~ **de radiactividad** *m* FÍS RAD, INSTR, NUCL radioactivity meter; ~ **de resistencia** *m* ELEC, FÍS, ING ELÉC, INSTR, METR, MINAS, TELECOM resistance meter; ~ **de resistencia a la abrasión** *m* PAPEL abrasion tester; ~ **de resistencia de tierra** *m* ELEC, ING ELÉC earth-resistance meter (*BrE*), ground-resistance meter (*AmE*); ~ **de resistividad** *m* GEOFÍS, INSTR, TEC PETR resistivity meter; ~ **de retrodispersión beta** *m* NUCL beta-backscatter gage (*AmE*), beta-backscatter gauge (*BrE*); ~ **de roscas** *m* ING MECÁ thread gage (*AmE*), thread gauge (*BrE*); ~ **de sacarina** *m* FÍS, INSTR saccharimeter; ~ **sísmico** *m* FÍS ONDAS, INSTR seismic-profile recorder; ~ **de tensión** *m* FÍS strain gage (*AmE*), strain gauge (*BrE*); ~ **de tierra accidental** *m* ELEC, ING ELÉC earth-leakage meter (*BrE*), ground-leakage meter (*AmE*); ~ **de torbellino** *m* GAS vortex meter; ~ **de turbina** *m* GAS turbine meter; ~ **ultrasónico** *m* GAS ultrasonic meter; ~ **en V** *m* C&V V-gage (*AmE*), V-gauge (*BrE*); ~ **de vacío** *m* FÍS vacuum gage (*AmE*), vacuum gauge (*BrE*); ~ **de valor de pico** *m* TV peak meter; ~ **de vapor** *m* FÍS steam gage (*AmE*), steam gauge (*BrE*); ~ **de varias escalas** *m* ELEC, INSTR multirange meter; ~ **de Venturi** *m* FÍS venturi meter; ~ **de vibraciones** *m* INSTR vibration measurer; ~ **del volumen del tráfico** *m* TRANSP traffic-volume meter; ~ **VU** *m* ACÚST VU-meter

medio[1] *adj* ACÚST half, GEOL middle, ING MECÁ half; **del ~ ambiente** *adj* AGUA, CALIDAD, CONTAM, DETERG, EMB, GEOL, PROD environmental; ~ **tono** *adj* ACÚST halftone

medio[2] *m* AGUA, INFORM&PD medium, QUÍMICA vehicle; ~ **de acceso al videotex** *m* TELECOM videotex gateway; ~ **activo gaseoso** *m* ELECTRÓN gaseous active medium; ~ **activo de laser** *m* TELECOM active laser medium; ~ **de almacenaje magnético** *m* ING ELÉC magnetic storage medium; ~ **de almacenamiento** *m* INFORM&PD storage medium, NUCL storage environment; ~ **de almacenamiento de datos** *m* INFORM&PD data medium; ~ **de almacenamiento óptico** *m* ÓPT optical storage medium; ~ **de alta energía** *m* GEOL high-energy environment; ~ **ambiente** *m* CALIDAD, CONTAM, DETERG, EMB, GEOL, INFORM&PD, OCEAN, PROD environment; ~ **ambiente del baño** *m* C&V bath atmosphere; ~ **ambiente físico-químico** *m* CONTAM physico-chemical environment; ~ **ambiente marino** *m* CONTAM marine environment; ~ **ambiente natural** *m* AGUA, CONTAM natural environment; ~ **ambiente planificado** *m* CONTAM planned environment; ~ **de baja energía** *m* GEOL low-energy environment; ~ **bao** *m* TRANSP MAR half beam; ~ **birrefringente** *m* CRISTAL, FÍS, FÍS RAD, ÓPT birefringent medium; ~ **buje** *n* ING MECÁ half bushing; ~ **casquillo** *n* ING MECÁ half bushing; ~ **climático** *m* METEO climate environment; ~ **collarín** *m* ING MECÁ half flange; ~ **de comunicaciones** *m* INFORM&PD communication medium; ~ **de datos** *m* INFORM&PD data medium; ~ **denso** *m* CARBÓN dense medium, heavy medium; ~ **dieléctrico** *m* FÍS, ING ELÉC dielectric medium; ~ **disipativo** *m* FÍS dissipative medium; ~ **dispersante** *m* CARBÓN dispersing agent; ~ **de dispersión** *m* NUCL scattering medium; ~ **dispersivo** *m* FÍS, PROC QUÍ dispersive medium; ~ **fértil** *m* TEC ESP *exobiología* blanket; ~ **homogéneo** *m* FÍS homogeneous medium; ~ **de imán móvil** *m* ING ELÉC moving-magnet medium; ~ **de intercambio calorífico** *m* TERMO heat-exchanging medium; ~ **láser** *m* ÓPT, TELECOM laser medium; ~ **laser activo** *m* ÓPT active laser medium; ~ **manguito** *n* ING MECÁ half bushing; ~ **magnético** *m* ACÚST, INFORM&PD, ING ELÉC magnetic media; ~ **material** *m* FÍS medium; ~ **no dispersivo** *m* FÍS nondispersive medium; ~ **opaco** *m* FÍS opaque medium; ~ **óptico** *m* ÓPT optical medium; ~ **óptico para escribir una sola vez** *m* ÓPT write-once optical medium; ~ **de propagación** *m* ING ELÉC propagation medium; ~ **puente** *adj* ING MECÁ half bridge piece; ~ **reborde** *m* ING MECÁ half flange; ~ **de registro** *m* INFORM&PD recording medium; ~ **de salvamento** *m* SEG means of escape; ~ **de sólo lectura** *m* ÓPT read-only medium; ~ **termorrecuperador** *m* TERMO heat-exchanging medium; ~ **translúcido** *m* FÍS translucent medium; ~ **de transmisión** *m* INFORM&PD, ÓPT transmission medium; ~ **transparente** *m* FÍS transparent medium; ~ **vacío** *m* INFORM&PD empty medium; ~ **virgen** *m* INFORM&PD virgin medium

medioambientalmente: ~ **inocuo** *adj* CALIDAD, CONTAM, DETERG, EMB, PROD environment-friendly

mediomundo *m* *AmL* (*cf red barredera Esp*) OCEAN scoop net

medios *m pl* TV media; ~ **de carga y descarga** *m pl* TRANSP MAR cargo gear; ~ **de detección y supresión de explosiones** *m pl* SEG means of detecting and suppressing explosions; ~ **de impregnación impermeables** *m pl* REVEST water-repellent impregnation means; ~ **para impresión de alarma** *m pl* TELECOM alarm printout facility; ~ **de molturar** *m pl* CARBÓN grinding media; ~ **porosos** *m pl* GAS porous media

medir *vt* ELEC tape, METR, QUÍMICA measure; ~ **con cadena** *vt* CONST *topografía* chain; ~ **el tiempo de** *vt* PROD time

médula *f* AGRIC pith

megabit *m* ELECTRÓN megabit

megabyte *m* (*Mb*) ELECTRÓN, INFORM&PD megabyte (*Mb*)

megacantil *m* OCEAN very high cliff

megachip *m* ELECTRÓN, INFORM&PD megachip

megaciclo *m* ELEC megacycle

megadina *f* METR megadyne

megadoc *m* ÓPT megadoc

megahertz *m* ELEC, ELECTRÓN, PETROL, TV megahertz

megahertzio *m* ELEC, ELECTRÓN, PETROL, TV megahertz

megaóhmetro *m* ELEC, ELECTRÓN, GEOFÍS, ING ELÉC, INSTR Megger, megohmmeter

megaohmio *m* ELEC, ELECTRÓN, GEOFÍS, ING ELÉC megohm

megavatio *m* (*MW*) ELEC, ING ELÉC megawatt (*MW*)

Megger *m* ELEC, ELECTRÓN, GEOFÍS, ING ELÉC, INSTR Megger

megohm *m* ELEC, ELECTRÓN, GEOFÍS, ING ELÉC megohm

megóhmetro *m* ELEC, ELECTRÓN, GEOFÍS, ING ELÉC, INSTR Megger, megohmmeter

megohmio *m* ELEC, GEOFÍS, ING ELÉC *unidad* megohm

meionita *f* MINERAL meionite

mejillonero *m* OCEAN mussel bed

mejora *f* ALIMENT enrichment, CARBÓN advance, INFORM&PD upgrade, QUÍMICA refinement, TELECOM grading; ~ **genética de plantas** *f* AGRIC plant breeding; ~ **del hardware** *f* INFORM&PD hardware upgrade; ~ **de la imagen** *f* ELECTRÓN image enhancement; ~ **de pastizales** *f* AGRIC grass improvement

mejorador *m* ALIMENT, TEC PETR improver; ~ **de la corriente de grabación** *m* TV record-current optimizer; ~ **de pan** *m* ALIMENT bread improver; ~ **vegetal** *m* AGRIC plant breeder

mejoramiento *m* QUÍMICA upgrading; ~ **de la calidad** *m* CALIDAD quality improvement; ~ **estructural de suelo** *m* AGRIC improvement of soil structure; ~ **del ganado** *m* AGRIC cattle breeding; ~ **genético** *m* AGRIC breeding; ~ **genético de maíz** *m* AGRIC corn breeding (*AmE*), maize breeding (*BrE*); ~ **por mutación** *m* AGRIC mutation breeding

mejorante: ~ **del índice de viscosidad** *m* TEC PETR viscosity index improver

mejorar *vt* ALIMENT enhance

melaconita *f* MINERAL melaconite

melamina *f* QUÍMICA, TEXTIL melamine

melanina *f* QUÍMICA melanin

melanita *f* MINERAL melanite, topazolite

melanocrático *adj* GEOL melanocratic

melanosoma *m* GEOL melanosome

melanterita *f* MINERAL melanterite

melibiosa *f* QUÍMICA melibiose

melifana *f* MINERAL meliphane

melifanita *f* MINERAL meliphanite

melilita *f* MINERAL melilite

melinita *f* MINERAL melinite

melio *m* ACÚST mel

melita *f* MINERAL honey stone, mellite

melítico *adj* QUÍMICA mellitic

melitosa *f* QUÍMICA mellitose

mella *f* ING MECÁ indentation, dent, MECÁ, METAL dent, indentation

mellado *adj* ING MECÁ notched, MECÁ jagged

mellar *vt* ING MECÁ notch, MECÁ dent

melodía *f* ACÚST melody

membrana *f* ACÚST, CONST, FÍS membrane, MECÁ film, TEC ESP membrane; ~ **basilar** *f* ACÚST basilar membrane; ~ **bituminosa** *f* CONST bituminous membrane; ~ **de curado** *f* CONST curing membrane; ~ **de Reissner** *f* ACÚST Reissner's membrane; ~ **semipermeable** *f* FÍS semipermeable membrane; ~ **sintética** *f* CONST synthetic membrane; ~ **tectoria** *f* ACÚST tectorial membrane; ~ **timpánica** *f* ACÚST drum membrane, eardrum, tympanic membrane

membranas: ~ **para depuración** *f pl* CONTAM scrubber walls

membrecía *f* INFORM&PD membership

membrete: ~ **de carta** *m* IMPR letterhead

memoria *f* INFORM&PD memory, storage, storage memory, store memory (*BrE*), store (*BrE*), ING ELÉC memory, store, PROD statement, TV memory store; ~ **de acceso aleatorio** *f* (*RAM*) ELEC *computadoras*, IMPR, INFORM&PD, ING ELÉC random-access memory (*RAM*); ~ **de acceso aleatorio dinámica** *f* (*DRAM*) INFORM&PD dynamic RAM (*DRAM*); ~ **de acceso directo** *f* INFORM&PD direct access storage (*AmE*), direct access store (*BrE*), ING

ELÉC direct access memory; ~ **de acceso en serie** *f* INFORM&PD serial access storage; ~ **adicional** *f* INFORM&PD add-on memory; ~ **de añadidura** *f* INFORM&PD add-on memory; ~ **asociativa** *f* INFORM&PD associative memory, associative storage (*AmE*), associative store (*BrE*); ~ **auxiliar** *f* INFORM&PD auxiliary memory, auxiliary storage (*AmE*), auxiliary store (*BrE*), backing storage (*AmE*), backing store (*BrE*); ~ **borrable** *f* INFORM&PD erasable memory (*AmE*), erasable store (*BrE*); ~ **de burbuja** *f* INFORM&PD bubble memory; ~ **de burbuja magnética** *f* INFORM&PD, ING ELÉC magnetic bubble memory; ~ **de búsqueda en paralelo** *f* INFORM&PD parallel search storage (*AmE*), parallel search store (*BrE*); ~ **caché** *f* INFORM&PD cache memory; ~ **capacitiva** *f* INFORM&PD capacitor storage (*AmE*), capacitor store (*BrE*); ~ **central** *f* INFORM&PD main memory, main storage (*AmE*), main store (*BrE*), primary memory; ~ **de cilindro magnético** *f* INFORM&PD, ING ELÉC magnetic drum memory; ~ **compartida** *f* INFORM&PD shared memory; ~ **común** *f* TELECOM common store; ~ **de contenido direccionable** *f* INFORM&PD content-addressable memory (*CAM*), content-addressable storage (*AmE*), content-addressable store (*BrE*); ~ **de control** *f* TELECOM control memory; ~ **criogénica** *f* INFORM&PD cryogenic memory; ~ **de datos** *f* TELECOM data storage (*AmE*) (*DS*), data store (*BrE*) (*DS*); ~ **dinámica** *f* INFORM&PD dynamic memory, TELECOM buffer memory; ~ **de discos** *f* ÓPT jukebox; ~ **de doble acceso** *f* INFORM&PD dual-port memory; ~ **electrónica** *f* ELECTRÓN, ING ELÉC electronic memory; ~ **estable** *f* CONST nonvolatile memory; ~ **estática** *f* INFORM&PD, ING ELÉC static memory; ~ **externa** *f* INFORM&PD external memory, external storage (*AmE*), external store (*BrE*), ING ELÉC external memory; ~ **física** *f* ING ELÉC physical memory; ~ **de gran capacidad** *f* INFORM&PD mass storage (*AmE*), mass store (*BrE*), bulk memory; ~ **indeleble** *f* INFORM&PD nonerasable storage (*AmE*), nonerasable store (*BrE*); ~ **intermedia** *f* IMPR buffer, INFORM&PD buffer, cache memory, ING ELÉC buffer memory; ~ **intermedia de entrada** *f* INFORM&PD input buffer; ~ **intermedia de entrada-salida** *f* INFORM&PD buffered input-output; ~ **intermedia de impresión** *f* IMPR print buffer; ~ **intermedia de salida** *f* INFORM&PD output buffer; ~ **interna** *f* INFORM&PD, ING ELÉC internal memory, PROD internal storage (*AmE*), internal store (*BrE*); ~ **laseróptica** *f* ÓPT laser-optic memory (*LO-M*); ~ **de lectura-escritura** *f* INFORM&PD, ÓPT read-write memory; ~ **de línea de retardo** *f* PROD circulating shift register; ~ **de llamadas** *f* TELECOM call store; ~ **magnética de registro en paralelo** *f* ING ELÉC parallel storage; ~ **magnetoóptica** *f* ÓPT magneto-optic memory; ~ **de masa** *f* INFORM&PD mass storage (*AmE*), mass store (*BrE*); ~ **masiva** *f* INFORM&PD mass storage (*AmE*), mass store (*BrE*); ~ **de mensajes** *f* TELECOM message store (*MS*); ~ **no volátil** *f* INFORM&PD nonvolatile memory; ~ **de núcleo magnético** *f* ING ELÉC magnetic-core memory; ~ **de núcleos** *f* INFORM&PD core storage (*AmE*), core store (*BrE*); ~ **de núcleos de ferrita** *f* ING ELÉC ferrite core; ~ **óptica** *f* INFORM&PD optical storage (*AmE*), optical store (*BrE*), ING ELÉC, ÓPT optical memory; ~ **óptica borrable** *f* ÓPT erasable optical memory; ~ **óptica reprogramable** *f* ÓPT alterable optical memory; ~ **óptica de sólo lectura** *f* ÓPT optical read-only memory (*OROM*); ~ **de película delgada** *f* INFORM&PD thin-film memory; ~ **de película magnética fina** *f* INFORM&PD thin-film memory; ~ **permanente** *f* INFORM&PD, ING ELÉC permanent memory; ~ **principal** *f* INFORM&PD main storage (*AmE*), main store (*BrE*), primary memory, primary storage (*AmE*), primary store (*BrE*); ~ **principalmente de lectura** *f* INFORM&PD read-mostly memory; ~ **programable de solo lectura** *f* (*PROM*) INFORM&PD programmable read-only memory (*PROM*); ~ **real** *f* INFORM&PD real memory; ~ **remanente** *f* INFORM&PD nonvolatile memory; ~ **secundaria** *f* INFORM&PD secondary memory, backing storage (*AmE*), backing store (*BrE*); ~ **de semiconductores** *f* INFORM&PD semiconductor memory, ING ELÉC semiconductor memory, solid-state memory; ~ **de señales vocales** *f* TELECOM speech memory; ~ **en serie** *f* ING ELÉC serial memory; ~ **de sólo lectura** *f* (*ROM*) ELEC, INFORM&PD, ING ELÉC read-only memory (*ROM*); ~ **de sólo lectura del disco óptico** *f* INFORM&PD, ÓPT optical disk read-only memory; ~ **superconducente** *f* INFORM&PD superconducting memory; ~ **tampón de mandatos** *f* PROD command buffer; ~ **temporal** *f* INFORM&PD temporary storage (*AmE*), temporary store (*BrE*), ING ELÉC temporary memory; ~ **de trabajo** *f* INFORM&PD scratch pad memory, working storage (*AmE*), working store (*BrE*); ~ **de tránsito** *f* ING ELÉC bubble memory; ~ **de la vía de transmisión** *f* TELECOM path memory; ~ **virtual** *f* INFORM&PD virtual memory; ~ **volátil** *f* INFORM&PD, ING ELÉC volatile memory

memorización: ~ **de mensajes** *f* TELECOM message storing (*MS*)

memorizar *vt* ING ELÉC store

mena *f* CARBÓN ore, GEOL ore mineral; ~ **de plomo** *f* AmL (*cf galena Esp*) MINAS *mineral* lead ore; ~ **de zinc** *f* MINAS jack

mendelevio *m* (*Md*) FÍS RAD, NUCL, QUÍMICA mendelevium (*Md*)

mendipita *f* MINERAL mendipite

mendocita *f* MINERAL mendozite

meneado *m* C&V stirring

meneo *m* TEC ESP, TRANSP MAR buffeting

mengua *f* CONST *madera* wane

menguante *m* OCEAN ebb

menguar *vi* AGUA, ENERG RENOV, HIDROL, TRANSP MAR *la marea* ebb

menilita *f* MINERAL menilite

meniscal *m* TERMO meniscus

menisco *m* CONST, FÍS, TERMO meniscus; ~ **divergente** *m* FOTO divergent meniscus

menos: ~ **significativo** *adj* INFORM&PD least significant

mensaje *m* INFORM&PD message, prompt, TELECOM message; ~ **de advertencia** *m* TRANSP warning message; ~ **de ayuda** *m* INFORM&PD help message; ~ **de borrado y error** *m* TELECOM blank-and-burst message; ~ **en carretera** *m* AUTO, TRANSP, VEH road message; ~ **de confusión** *m* TELECOM confusion message; ~ **de consulta destinado a grupo de circuitos** *m* TELECOM circuit group query message (*CQM*); ~ **desconocido** *m* TELECOM unknown message; ~ **entrante** *m* INFORM&PD incoming message;

~ **de error** *m* INFORM&PD, TELECOM error message; ~ **de error de diagnóstico** *m* INFORM&PD diagnostic error message; ~ **de facsímil** *m* INFORM&PD, TELECOM facsimile message; ~ **de gestión consolidada de enlace entre capas** *m* TELECOM consolidated link-layer management message (*CLLM*); ~ **de información** *m* TELECOM INF information message (*INF*); ~ **de información de usuario a usuario** *m* TELECOM user-to-user information message (*USR*); ~ **por intercambio de datos electrónicos** *m* TELECOM EDI message (*EDIM*); ~ **de operador** *m* INFORM&PD operator message; ~ **de petición de información** *m* TELECOM information request message (*INR*); ~ **pregrabado** *m* TELECOM prerecorded message; ~ **prerregistrado** *m* TELECOM prerecorded message; ~ **de recomendación** *m* TRANSP advisory message; ~ **de regularidad de vuelo** *m* TRANSP AÉR flight-regularity message; ~ **de respuesta de consulta destinado a grupo de circuitos** *m* TELECOM circuit group query response message (*CQR*); ~ **de segmento único** *m* (*SSM*) TELECOM single-segment message (*SSM*); ~ **supervisor** *m* TELECOM supervisory message

mensajería *f* INFORM&PD messaging; ~ **electrónica** *f* ELECTRÓN, INFORM&PD, TELECOM electronic messaging; ~ **por intercambio de datos electrónicos** *f* TELECOM EDI messaging (*EDIMG*)

ménsula *f* CONST bracket, cantilever, cantilever beam, free beam, haunch, FERRO cantilever, ING MECÁ bracket, cantilever, MECÁ angle bracket, bracket, PAPEL cantilever; ~ **en escuadra** *f* CONST angle bracket; ~ **mural** *f* TELECOM wall bracket; ~ **de muro** *f* TELECOM wall bracket; ~ **del porta-macho** *f* PROD bearing bracket

mensuración *f* METR measuring
mensurar *vt* METR measure
mentano *m* QUÍMICA menthane
mentanodiamina *f* QUÍMICA menthanediamine
mentanol *m* QUÍMICA menthanol
mentanona *f* QUÍMICA menthanone
menteno *m* QUÍMICA menthene
mentenol *m* QUÍMICA menthenol
mentenona *f* QUÍMICA menthenone
mentilo *m* QUÍMICA menthyl
mentofurano *m* QUÍMICA menthofuran
mentol *m* QUÍMICA menthol
mentona *f* QUÍMICA menthone
menú *m* INFORM&PD menu; ~ **desplegable** *m* IMPR pull-down menu, INFORM&PD drop-down menu, pop-up menu, pull-down menu; ~ **drop down** *m* INFORM&PD drop-down menu, menu; ~ **del estado del sistema** *m* PROD system status menu; ~ **de pantalla** *m* INFORM&PD display menu; ~ **de visualización** *m* INFORM&PD display menu
menudo: ~ **de carbón** *m* CARBÓN coal slack, MINAS coal slack, small; ~ **de cok** *m* CARBÓN breeze; ~ **no lavado** *m* CARBÓN unwashed small; ~ **no tratado** *m* CARBÓN untreated small
menudos *m pl* CARBÓN fines, pea coal, smalls, CONST, MINAS fines; ~ **de carbón** *m pl* CARBÓN coal slake, duff, slack coal, MINAS coal slake; ~ **de coque** *m pl* CARBÓN coking duff; ~ **de criba** *m pl* CARBÓN undersize; ~ **lavados** *m pl* CARBÓN washed small
mepacrina *f* QUÍMICA mepacrine
meprobamato *m* QUÍMICA meprobamate
merbromina *f* QUÍMICA merbromin

mercado *m* PROD *comercial* market, outlet; ~ **para los tejidos de calada de fantasía** *m* TEXTIL fancy-woven-fabrics market; ~ **a término** *m* AGRIC futures market
mercancía: ~ **a granel** *f* EMB bulk goods; ~ **recibida** *f* EMB accepted stock
mercancías *f pl* EMB freight (*AmE*), goods (*BrE*); ~ **en cubierta** *f pl* TRANSP MAR deck cargo; ~ **en depósito** *f pl* TEC PETR bonded goods; ~ **empaquetadas** *f pl* EMB parcelled goods; ~ **en frío** *f pl* ALIMENT, EMB, REFRIG chilled goods; ~ **peligrosas** *f pl* EMB, SEG, TRANSP dangerous goods; ~ **refrigeradas** *f pl* ALIMENT, EMB, REFRIG chilled goods
mercaptal *m* QUÍMICA mercaptal
mercaptán *m* QUÍMICA thiol
mercaptano *m* CONTAM mercaptan, thiol, P&C, QUÍMICA, TEC PETR mercaptan
mercaptido *m* QUÍMICA mercaptide
mercaptoacético *adj* QUÍMICA mercaptoacetic
mercaptomerina *f* QUÍMICA mercaptomerin
mercerización *f* QUÍMICA mercerization
mercuración *f* QUÍMICA mercuration
mercurial *adj* QUÍMICA mercurial
mercúrico *adj* QUÍMICA mercuric
mercurificación *f* QUÍMICA mercurification
mercurio *m* (*Hg*) ING ELÉC mercury (*Hg*), quicksilver, METAL mercury (*Hg*), QUÍMICA mercury (*Hg*), quicksilver
meridiano[1] *adj* TELECOM meridional
meridiano[2] *m* TEC ESP meridian; ~ **geomagnético** *m* GEOFÍS geomagnetic meridian; ~ **magnético** *m* FÍS, GEOFÍS magnetic meridian
meridional *adj* TEC ESP meridional
meristema *m* AGRIC meristem
merma *f* ALIMENT, MECÁ ullage, PROD shrinkage, wastage, waste, TEC ESP *líquidos* ullage, TEXTIL waste; ~ **debida a la fricción** *f* ING MECÁ loss due to friction
merogénesis *f* GAS merogenesis
meroxeno *m* MINERAL meroxene
mesa *f* IMPR bank, ING MECÁ plate, platen, table, PROD *de yunque* body, face; ~ **de alimentación** *f* EMB feeding table, IMPR feed table, ING MECÁ traveling table (*AmE*), travelling table (*BrE*); ~ **para el alzado** *f* IMPR collating table, gathering table; ~ **de amalgamación** *f* CARBÓN amalgamating table; ~ **de animación** *f* CINEMAT animation bench; ~ **antivibratoria** *f* LAB antivibration table; ~ **ascendente** *f* ING MECÁ raising table; ~ **basculante** *f* ING MECÁ tilting table; ~ **de base en U** *f* ING MECÁ U-shaped base plate; ~ **con caras superior y laterales** *f* ING MECÁ table with top and side faces; ~ **circular de acción automática** *f* ING MECÁ self-acting circular table; ~ **circular automática** *f* ING MECÁ self-acting circular table; ~ **circular de movimiento automático** *f* ING MECÁ self-acting circular table; ~ **de clasificación** *f* MINAS picking table; ~ **de composición** *f* IMPR composition board; ~ **de concentración** *f* CARBÓN concentrating table; ~ **concentradora** *f* CARBÓN concentrating table; ~ **de control** *f* INSTR control desk; ~ **de control de iluminación** *f* CINEMAT light table; ~ **de control principal** *f* IMPR master control desk; ~ **del cortador** *f* C&V cutter's table; ~ **de costados abatibles** *f* ING MECÁ lift-up table; ~ **de cribado** *f* MINAS picking table; ~ **de derrota** *f* TRANSP MAR plotting table; ~ **para descargar el coke** *f* PROD

coking plate; ~ **de deslizamiento múltiple** *f* ING MECÁ table with compound slides; ~ **de distribución** *f* PROD spreading table; ~ **divisoria giratoria** *f* ING MECÁ rotary dividing table; ~ **divisoria rotatoria** *f* ING MECÁ rotary dividing table; ~ **elevadora** *f* MECÁ elevating table, PROD *tren laminador* lifting table; ~ **para empalmar** *f* IMPR splicing table; ~ **de examen angiográfico** *f* INSTR angiographic examination table; ~ **de examen para rayos X** *f* INSTR X-ray examination table; ~ **de fabricación** *f* PAPEL wire end; ~ **de faldón** *f* ING MECÁ lift-up table; ~ **fija** *f* C&V *manufactura de espejos* fixed table; ~ **fija portapiezas** *f* ING MECÁ work-holding fixed table; ~ **de formación en voladizo** *f* PAPEL roll-out Fourdrinier; ~ **giratoria** *f* MECÁ turntable; ~ **horizontal de montaje** *f* CINEMAT, TV flatbed editing table; ~ **del horno** *f* PROD dead plate; ~ **de imágenes de E/S** *f* PROD I/O image table; ~ **inclinable** *f* C&V tilt table, ING MECÁ tilting table; ~ **de inclinación multidireccional** *f* ING MECÁ table canting to any angle; ~ **inclinada de superficie plana** *f* MINAS plane table; ~ **de inversión** *f* PROD *fundería* roll-over table; ~ **de laboratorio** *f* LAB bench; ~ **de luz** *f* FOTO light box; ~ **de marear** *f* TRANSP MAR chart table; ~ **de marear las cartas** *f* TRANSP MAR chart table; ~ **mecedora** *f* C&V rocking table; ~ **de mediciones** *f* NUCL measuring desk; ~ **mezcladora** *f* TV mixing desk; ~ **de montaje** *f* CINEMAT table editing machine; ~ **de montaje Steenbeck** *f* CINEMAT Steenbeck; ~ **movible** *f* ING MECÁ traveling table (*AmE*), travelling table (*BrE*); ~ **móvil** *f* C&V moving table; ~ **de movimiento vertical** *f* ING MECÁ vertically-adjustable table; ~ **multidireccional** *f* ING MECÁ table with compound slides; ~ **neumática** *f* CARBÓN pneumatic table; ~ **de operadoras** *f* TELECOM telephone switchboard; ~ **ortogonal** *f* ING MECÁ compound table; ~ **oscilante** *f* CARBÓN oscillating table, ING MECÁ tilting table; ~ **de piezas en tosco** *f* C&V blank table; ~ **de la pila** *f* IMPR pile board; ~ **plana con viga voladiza** *f* PAPEL cantilevered Fourdrinier; ~ **portapiezas** *f* ING MECÁ work plate; ~ **refrigerada para pastelería** *f* ALIMENT, REFRIG refrigerated bakery slab; ~ **regulable vertical** *f* ING MECÁ table which can be raised or lowered; ~ **rotativa** *f* ING MECÁ revolving table, rotary table, METR, PETROL, PROD, TEC PETR rotary table; ~ **rotatoria** *f* ING MECÁ rotary table, table with compound slides, METR, PETROL, PROD, TEC PETR rotary table; ~ **de sacudidas** *f* CARBÓN shaking table, ING MECÁ concussion table, MINAS joggling table, PROD bumping tray, shaking table, bumping table; ~ **sacudidora** *f* CARBÓN shaking table; ~ **de secadoras** *f* PAPEL bank of dryers; ~ **de soldadura con aparatos de vacío** *f* SEG welding table with vacuum apparatus; ~ **de sonido** *f* FÍS ONDAS sounding board; ~ **de soplado** *f* C&V blow table; ~ **de taladrar** *f* CARBÓN drilling-machine table; ~ **trazadora** *f* INFORM&PD plotting board; ~ **trepidante** *f* PROD shaking table; ~ **de vaciado** *f* C&V casting table; ~ **de vaivén** *f* PROD bumping table; ~ **vibrante** *f* EMB jarring table, LAB *dispositivo electromecánico* shaker; ~ **vibratoria** *f* C&V vibrating table, CARBÓN shaking table, vibrating table; ~ **de yunque** *f* PROD crown
mesacónico *adj* QUÍMICA mesaconic
meseta *f* GEOL, OCEAN, TEC PETR plateau; ~ **de**

congelación *f* REFRIG freezing plateau; ~ **continental** *f* GEOFÍS, GEOL continental plate; ~ **de lava** *f* PETROL lava plateau; ~ **marginal** *f* OCEAN marginal plateau; ~ **submarina** *f* GEOL guyot, OCEAN submarine plateau, *geología* shelf; ~ **terrestre** *f* GEOFÍS earth plate
mesidina *f* QUÍMICA mesidine
mesitilénico *adj* QUÍMICA mesitylenic
mesitileno *m* QUÍMICA mesitylene
mesocarpio *m* AGRIC fruit flesh
mesocótilo *m* AGRIC mesocotyl
mesoestructura *f* CRISTAL mesostructure
mesofase *f* CRISTAL mesophase
mesolita *f* MINERAL mesolite, mesole
mesolitoral *m* OCEAN mesolittoral
mesomería *f* QUÍMICA mesomerism
mesomérico *adj* QUÍMICA mesomeric
mesón *m* FÍS, FÍS PART kaon, meson, NUCL meson; ~ **eta** *m* FÍS, FÍS PART eta-meson; ~ **fi** *m* FÍS PART phi-meson; ~ **K** *m* FÍS PART K-meson; ~ **negativo** *m* FÍS PART negative meson; ~ **neutro eta** *m* FÍS, FÍS PART eta neutral meson; ~ **pi** *m* FÍS, FÍS PART pi-meson, pion; ~ **positivo** *m* FÍS PART positive meson; ~ **ro** *m* FÍS PART rho-meson; ~ **tau** *m* FÍS PART tau-meson; ~ **zeta** *m* FÍS PART zheta-meson
mesorcinol *m* QUÍMICA mesorcinol
mesosfera *f* GEOFÍS, TEC ESP mesosphere
mesotartárico *adj* QUÍMICA mesotartaric
mesotipa: ~ **cálcica** *f* MINERAL mesotype
mesotorio *m* QUÍMICA mesothorium
mesoxálico *adj* QUÍMICA mesoxalic
mesozoico *m* GEOL, TEC PETR mesozoic
MET *abr* (*microscopía electrónica de transmisión*) FÍS RAD TEM (*transmission electron microscopy*)
metabasita *f* GEOL, PETROL metabasite
metabisulfito *m* QUÍMICA metabisulfite (*AmE*), metabisulphite (*BrE*)
metaborato *m* QUÍMICA metaborate
metabórico *adj* QUÍMICA metaboric
metacentro *m* FÍS, TRANSP MAR *arquitectura naval* metacenter (*AmE*), metacentre (*BrE*); ~ **longitudinal** *m* TRANSP MAR longitudinal metacenter (*AmE*), longitudinal metacentre (*BrE*); ~ **transversal** *m* TRANSP MAR transverse metacenter (*AmE*), transverse metacentre (*BrE*)
metacinabarita *f* MINERAL metacinnabarite
metaclorito *m* MINERAL metachlorite
metaclorotolueno *m* QUÍMICA metachlorotoluene
metacresol *m* QUÍMICA metacresol
metacrilato *m* P&C methacrylate, TRANSP MAR Perspex; ~ **de polimetilo** *m* (*PMMA*) P&C polymethyl methacrylate (*PMMA*)
metacrílico *adj* QUÍMICA metacrylic, methacrylic
metadino *m* ELECTRÓN metadyne
metadona *f* QUÍMICA methadone
metaestable *adj* FÍS RAD metastable
metaestánico *adj* QUÍMICA metastannic
metaformaldehído *m* QUÍMICA trioxane
metafosfato *m* QUÍMICA metaphosphate
metafosfórico *adj* QUÍMICA metaphosphoric
metal *m* C&V metal; ~ **alcalino** *m* METAL alkali metal; ~ **alcalinotérreo** *m* METAL alkaline-earth metal; ~ **antifricción** *m* ING MECÁ Babbitt's metal; ~ **de aportación** *m* CONST, MECÁ filler metal, PROD filler metal, *soldadura* filling metal; ~ **auxiliar** *m* CONST, MECÁ, PROD filler metal; ~ **Babbit** *m* ING MECÁ

Babbitt's metal; **~ blanco** *m* ING MECÁ Babbitt's metal; **~ de colada** *m* METAL, PROD pig metal; **~ compuesto** *m* METAL composite; **~ depositado** *m* CARBÓN deposit; **~ desnudo** *m* METAL bare metal; **~ duro** *m* ING MECÁ hard metal; **~ estable** *m* METAL noble metal; **~ fundido** *m* METAL, SEG molten metal; **~ líquido** *m* CARBÓN liquid metal; **~ madre** *m* PROD *fundería* sow; **~ de mazarota** *m* PROD head metal; **~ noble** *m* CONTAM, METAL noble metal; **~ pesado** *m* CARBÓN, FÍS RAD, RECICL heavy metal; **~ poroso** *m* MECÁ expanded metal; **~ puro** *m* METAL bare metal; **~ que queda en el bebedero** *m* PROD gate; **~ reflector** *m* INSTR metal reflector; **~ refractario** *m* METAL refractory metal; **~ de soldadura** *m* METAL weld metal; **~ de transición** *m* METAL transition metal

metalación *f* QUÍMICA metalation (*AmE*), metallation (*BrE*)

metaldehído *m* QUÍMICA metaldehyde

metalenguaje *m* INFORM&PD metalanguage

metales: ~ preciosos *m pl* METAL bullion

metálico[1] *adj* ELEC metallic; **no ~** *adj* QUÍMICA nonmetal

metálico[2] *m* METAL bullion

metalífero *adj* MINAS ore-bearing, QUÍMICA metalliferous

metalino *m* QUÍMICA metalline

metalización *f* CONST, ELECTRÓN, FÍS metalization (*AmE*), metallization (*BrE*), ING MECÁ metal coating, PROD, TEC ESP metalization (*AmE*), metallization (*BrE*); **~ por soplete** *f* NUCL flame-spraying

metalizado[1] *adj* ELEC metalized (*AmE*), metallized (*BrE*), ING MECÁ, MECÁ metal-clad, PROD metal-coated, REVEST metal-coated, metal-faced

metalizado[2] *m* C&V metalizing (*AmE*), metallizing (*BrE*), REVEST metal plating, TV metal coating; **~ al vacío** *m* C&V vacuum metallizing; **~ en vacío** *m* ELECTRÓN vacuum metallization

metalizar *vt* MECÁ plate, QUÍMICA sputter, REVEST metal-coat

metalografía *f* METAL metallography; **~ por rayos X** *f* NUCL X-ray metallography

metaloide *m* METAL, QUÍMICA metalloid

metaloideo *adj* METAL, QUÍMICA metalloidal

metaloídico *adj* METAL, QUÍMICA metalloidal

metamería *f* QUÍMICA metamerism

metamérico *adj* COLOR, IMPR, QUÍMICA metameric

metámero *m* QUÍMICA metamer

metamíctico *adj* GEOL metamict

metamorfismo *m* GEOL, PETROL metamorphism; **~ de alto grado** *m* GEOL high-grade metamorphism; **~ de bajo grado** *m* GEOL low-grade metamorphism; **~ de carga** *m* GEOL load metamorphism; **~ de contacto** *m* GEOL contact metamorphism; **~ dinámico** *m* ENERG RENOV, GEOL dynamic metamorphism; **~ estático** *m* GEOL burial metamorphism; **~ de grado intermedio** *m* GEOL medium-grade metamorphism; **~ isoquímico** *m* GEOL isochemical metamorphism; **~ mecánico** *m* GEOL dynamic metamorphism; **~ progresivo** *m* GEOL prograde metamorphism, progressive metamorphism; **~ regional** *m* ENERG RENOV regional metamorphism; **~ retrógrado** *m* ENERG RENOV retrograde metamorphism, GEOL retrograde metamorphism, retrogressive metamorphism

metanación *f* GAS methanation

metanal *m* ALIMENT, P&C, QUÍMICA, TEXTIL formaldehyde, methanal

metanero *m* TRANSP MAR liquefied-natural-gas carrier (*LNG carrier*), methane carrier

metanílico *adj* QUÍMICA methanilic

metanilo *m* QUÍMICA metanil

metano *m* AGRIC methane, CARBÓN firedamp, GAS methane, MINAS firedamp, marsh gas, QUÍMICA, SEG, TEC PETR methane, TERMO firedamp

metanobacteria *f* QUÍMICA methane bacterium

metanoico *adj* QUÍMICA methanoic

metanol *m* GAS methanol, P&C methyl alcohol, QUÍMICA methanol, methyl alcohol, TERMO wood alcohol

metarsenioso *adj* QUÍMICA metarsenious

metasedimento *m* GEOL metasediment

metasilicato *m* QUÍMICA metasilicate

metasilícico *adj* QUÍMICA metasilicic

metasomatismo *m* GEOL metasomatism; **~ de contacto** *m* GEOL contact metasomatism

metátesis *f* QUÍMICA metathesis

metavolcánico *m* GEOL metavolcanic

metaxita *f* MINERAL metaxite

meteorito *m* METEO, TEC ESP meteorite

meteorización *f* AGRIC, CARBÓN, GEOL, PETROL weathering; **~ por heladas** *f* GEOL frost work

meteoro *m* METEO, TEC ESP meteor

meteoroide *m* METEO, TEC ESP meteoroid

meteorología *f* METEO, TEC ESP, TRANSP MAR meteorology; **~ por observaciones de satélite** *f* TEC ESP satellite meteorology; **~ química** *f* CONTAM atmospheric chemistry

meter *vt* ING MECÁ drive in, drive into; **~ en** *vt* ING MECÁ fit in; **~ a bordo** *vt* TRANSP MAR put on board; **~ en dársena** *vt* TRANSP MAR *buque* dock; **~ dentro de** *vt* ING MECÁ fit into; **~ en dique seco** *vt* TRANSP MAR *buque* dry-dock; **~ en facha** *vt* TRANSP MAR *emisión* jam; **~ de orza** *vt* TRANSP MAR luff

metido: ~ en orza *adj* TRANSP MAR *timón* alee; **~ de popa** *adj* TRANSP MAR *buque* down by the stern; **~ de proa** *adj* TRANSP MAR *buque* down by the head

metil: ~ propano *m* QUÍMICA methylpropane

metilación *f* QUÍMICA methylation

metilal *m* QUÍMICA methylal

metilamina *f* QUÍMICA methylamine

metilanilina *f* QUÍMICA methylaniline

metilar *vt* QUÍMICA methylate

metilato *m* QUÍMICA methylate

metiletilcetona *f* QUÍMICA methyl ethyl ketone

metiletilcetoxima *f* P&C methyl ethyl ketoxime

metílico *adj* QUÍMICA methylic

metilmorfina *f* QUÍMICA codeine

metilnaftaleno *m* QUÍMICA methylnaphthalene

metilo *m* QUÍMICA methyl

metilpentosa *f* QUÍMICA methylpentose

metimazol *m* QUÍMICA methimazol

metiónico *adj* QUÍMICA methionic

método *m* CONST way, FÍS method, PROD mode, procedure, QUÍMICA method; **~ de acceso** *m* INFORM&PD access method; **~ aditivo** *m* CINEMAT, ELECTRÓN additive method; **~ de ajuste a cero** *m* ELEC null method; **~ de aleación** *m* ELECTRÓN alloying method; **~ de alimentar** *m* C&V method of feeding; **~ de alineación de haz de electrones** *m* ELECTRÓN electron-beam alignment method; **~ de alteración** *m* TELECOM cutback technique; **~ alter-**

nativo de pruebas *m* TELECOM alternative test method; ~ **por aproximaciones sucesivas** *m* CRISTAL trial and error method; ~ **del baño de sulfato de manganeso** *m* NUCL manganous sulfate bath method (*AmE*), manganous sulphate bath method (*BrE*); ~ **de cálculo de elementos finitos** *m* ING MECÁ, MECÁ finite-element calculation method; ~ **de calentamiento escalonado Ar-Ar** *m* GEOL Ar-Ar step heating method; ~ **de calentamiento gradual Ar-Ar** *m* GEOL Ar-Ar step heating method; ~ **del camino crítico** *m* INFORM&PD critical-path method (*CPM*); ~ **de campo próximo refractado** *m* ÓPT refracted near-field method; ~ **de cero** *m* ELEC null method; ~ **de cienos activos** *m* AGUA activated sludge process; ~ **de claves** *m* INFORM&PD hashing; ~ **para colocar las bolsas** *m* EMB bag-placing system; ~ **de control sanitario** *m* AGRIC disease control method; ~ **de crisol doble** *m* ÓPT double crucible method; ~ **del cristal giratorio** *m* FÍS RAD rotating-crystal method; ~ **del cristal oscilante** *m* FÍS RAD oscillating-crystal method; ~ **de dispersión difusa** *m* NUCL diffuse scattering method; ~ **de doble baño** *m* NUCL two-bath method; ~ **de engomado en frío** *m* EMB cold-gluing system; ~ **de ensayo alternativo** *m* (*ATM*) ÓPT alternative test method (*ATM*); ~ **de estimación del costo** *m* ING MECÁ method of costing; ~ **de excavación y revestimiento** *m* CONST cut and cover method; ~ **de expansión nodal** *m* NUCL nodal expansion method; ~ **de exploración por pistola de aire comprimido** *m* GEOFÍS air gun exploration method; ~ **de exploración sísmica** *m* GEOFÍS seismic exploration method; ~ **de fangos activados** *m* AGUA activated sludge process; ~ **de fijación de itinerarios** *m* FERRO method of routing; ~ **de flotación** *m* PETROL flotation method; ~ **de fusión por zonas flotantes** *m* NUCL floating-zone melting method; ~ **del haz luminoso retrodispersado** *m* ACÚST backscattered light beam method; ~ **de impresión reflex** *m* FOTO, IMPR reflex printing method; ~ **de integrales** *m* NUCL J-integral method; ~ **de interferencia** *m* ING MECÁ interference method; ~ **de inyección** *m* TRANSP MAR injection procedure; ~ **iterativo** *m* INFORM&PD iterative method; ~ **de Kjeldahl** *m* NUCL, QUÍMICA Kjeldahl method; ~ **de laboreo a lo largo de las minas** *m* MINAS flat-back stoping method; ~ **de Laue** *m* CRISTAL Laue method; ~ **de lectura cero** *m* ELEC null method; ~ **local refractado** *m* TELECOM refracted near-end method; ~ **magnético** *m* GEOFÍS magnetic method; ~ **de medición** *m* CONST method of measurement; ~ **micrográfico** *m* ING MECÁ, METAL micrographic method; ~ **de mínimos cuadrados** *m* CRISTAL, FÍS, MATEMÁT least squares method; ~ **Montecarlo** *m* INFORM&PD Monte Carlo method; ~ **de muestreo** *m* CALIDAD sampling method; ~ **nulo** *m* FÍS null method; ~ **de ondas distorsionadas** *m* NUCL distorted wave method; ~ **operatorio** *m* TEXTIL operating method; ~ **óptimo de operación** *m* GAS optimal operating method; ~ **de oxidación del mineral** *m* PROD ore process; ~ **de penetración** *m* NUCL penetration method; ~ **para planificar y controlar proyectos** *m* (*PERT*) TEC ESP program evaluation and review technique (*AmE*) (*PERT*), programme evaluation and review technique (*BrE*) (*PERT*); ~ **práctico de prueba** *m* TELECOM practi-

cal-test method; ~ **de prueba de referencia** *m* ÓPT, TELECOM reference test method (*RTM*); ~ **de prueba para sistemas de frenado** *m* ING MECÁ test procedure for braking systems; ~ **de pruebas prácticas** *m* ÓPT practical-test method; ~ **de rayos refractados** *m* FÍS, ÓPT, TELECOM refracted-ray method; ~ **de reciclaje** *m* CONST, CONTAM, PROD, RECICL recycling method; ~ **de reflexión** *m* NUCL reflection method; ~ **de reflexión de Fresnel** *m* ÓPT, TELECOM Fresnel reflection method; ~ **de reflexión de retroceso** *m* CRISTAL back-reflection method; ~ **de retrodispersión gamma** *m* NUCL gamma backscatter method; ~ **de señal desplazada** *m* TV offset signal method; ~ **de la sonda de oro** *m* NUCL gold probe method; ~ **substractivo** *m* ELECTRÓN subtractive method~ **al tanteo** *m* CRISTAL trial and error method; ~ **al tanteo** *m* CRISTAL trial and error method; ~ **de teñir con el mordiente después del colorante** *m* COLOR top chrome; ~ **de Verneuil en fase vapor** *m* ÓPT vapor phase verneuil method (*AmE*), vapour phase verneuil method (*BrE*)

metodología: ~ **descendente** *f* INFORM&PD top-down methodology; ~ **del fondo hacia arriba** *f* INFORM&PD bottom-up methodology; ~ **del software** *f* INFORM&PD software methodology; ~ **de soporte lógico** *f* INFORM&PD software methodology

métodos: ~ **de elementos finitos** *m pl* ING MECÁ finite elements methods; ~ **de elementos limitados** *m pl* ING MECÁ finite elements methods; ~ **de trabajo a prueba de averías** *m pl* PROD fail-safe work methods

metol *m* QUÍMICA methol, metol

metoxibenceno *m* QUÍMICA methoxybenzene

metóxido *m* QUÍMICA methoxide

metoxietanol *m* QUÍMICA methoxyethanol

metoxilo *m* QUÍMICA methoxyl

metraje *m* CINEMAT footage, TEXTIL yardage

metralla *f* CARBÓN *de bomba, de proyectil* pellet, scrap iron, D&A grapeshot

metralleta *f* D&A sub-machine gun

métrico *adj* METR metric, metrical

metro *m* (*m*) GEN meter (*AmE*) (*m*), metre (*BrE*) (*m*); ~ **agrimensor** *m* METR tape line; ~ **cuadrado** *m* METR square meter (*AmE*), square metre (*BrE*); ~ **cúbico** *m* MATEMÁT, METR cubic meter (*AmE*), cubic metre (*BrE*); ~ **cúbico normal** *m* PETROL normal cubic meter (*AmE*), normal cubic metre (*BrE*); ~ **de geometría de superficie** *m* METR surface geometry meter (*AmE*), surface geometry metre (*BrE*); ~ **longitudinal** *m* METR length meter (*AmE*), length metre (*BrE*); ~ **de madera** *m* TEXTIL yardstick; ~ **de medir** *m* METR *telas* measuring rod; ~ **patrón** *m* FÍS standard meter (*AmE*), standard metre (*BrE*); ~ **plegable** *m* CONST, METR folding rule (*AmE*), jointed rule (*BrE*), zigzag rule (*AmE*)

metrología *f* METR metrology

mezcla[1]: **sin ~** *adj* CONTAM MAR neat

mezcla[2] *f* ACÚST mixing, ALIMENT mix, mixture, C&V admix, mix, P&C blend, compound, mix, PROD mixing, QUÍMICA compound, mixture, TEC PETR *de fluidos* mixing, *química* mixture, TELECOM scrambling, TEXTIL admixture, blend; ~ **aceite-gasolina** *f* Esp (*cf mezcla aceite-nafta AmL*) AUTO, VEH gas-oil mixture (*AmE*), gasoline-oil mixture (*AmE*), petrol-oil mixture (*BrE*); ~ **aceite-nafta** *f* AmL (*cf aceite-gasolina Esp*) AUTO, VEH gas-oil mixture (*AmE*), gasoline-oil mixture (*AmE*), petrol-oil mixture (*BrE*); **~ de aceites minerales y vegetales** *f* PROD compound oil; **~ aceotrópica** *f* ALIMENT azeotropic mixture; **~ de adhesivos** *f* EMB mixed adhesive; **~ aditiva** *f* ELECTRÓN additive mixing; **~ de agua y aire** *f* TEC PETR *perforación, refino* mist; **~ de aire y gas** *f* GAS, TERMO gas-air mixture; **~ de amasado** *f* CONST batch mix; **~ antidetonante** *f* AUTO no-knock mixture; **~ de arrastre de aire** *f* CONST air entraining admixture; **~ atómica** *f* METAL atomic shuffling; **~ azeotrópica** *f* QUÍMICA constant boiling mixture, ALIMENT, REFRIG azeotropic mixture; **~ de bidón** *f* CONST drum mix; **~ en caliente** *f* CONST hot mix; **~ carburante** *f* TERMO fuel-air mixture; **~ de la carga** *f* C&V batch mixing; **~ de colores** *f* C&V, COLOR color mix (*AmE*), colour mix (*BrE*); **~ congeladora** *f* FÍS, REFRIG freezing mixture; **~ de desvanecido de imagen** *f* TV mix dissolve; **~ explosiva** *f* AUTO explosive mixture; **~ de fertilizantes** *f* AGRIC fertilizer mixture; **~ en frío** *f* CONST cold mix; **~ de fundiciones** *f* PROD blending; **~ de gasolina** *f* Esp (*cf mezcla de nafta AmL*) AUTO, VEH gas mixture (*AmE*), gasoline mixture (*AmE*), petrol mixture (*BrE*); **~ de granos molidos** *f* AGRIC mash; **~ de grava** *f* AUTO *para carreteras* gritter; **~ heliox** *f* OCEAN oxygen-helium mixture; **~ de hidrógeno-oxígeno** *f* OCEAN hydrogen-oxygen mixture; **~ insonorizante** *f* CONST pugging; **~ en el lugar** *f* CONST mix-in-place; **~ de nafta** *f* AmL (*cf mezcla de gasolina Esp*) AUTO, VEH gas mixture (*AmE*), gasoline mixture (*AmE*), petrol mixture (*BrE*); **~ en la obra** *f* CONST mix-in-place; **~ de papeles clasificados** *f* PAPEL mixed boards, mixed papers; **~ de papeles sin clasificar** *f* PAPEL mixed boards, mixed papers; **~ plastificante** *f* CONST plasticizer admixture, plastifying admixture; **~ pobre** *f* AUTO lean-burn, poor mixture, TRANSP AÉR, VEH lean mixture; **~ refrigerante** *f* FÍS freezing mixture, REFRIG cooling mixture, freezing mixture, TERMO cooling mixture; **~ de relleno** *f* CONST filler alloy; **~ respiratoria** *f* OCEAN breathing mixture; **~ rica** *f* AUTO, VEH *carburación* rich mixture; **~ de sincronismos** *f* AmL (*cf sincronismos mixtos Esp*) TV mixed syncs; **~ de sonidos** *f* ACÚST dubbing, mixing; **~ térmica** *f* NUCL thermal mixing; **~ variable** *f* AUTO variable mixture

mezclable *adj* QUÍMICA, TEC PETR miscible

mezclado *m* C&V mixing trough, CONST mixing, P&C composite, PROC QUÍ mixing, QUÍMICA stirring; **~ definitivo** *m* CINEMAT, TV final mix

mezclador *m* C&V *operador*, CARBÓN, CINEMAT mixer, CONST beater, ELECTRÓN *de radio, receptor de televisión*, FÍS mixer, ING MECÁ agitator, receiver, mixer, inspirator, LAB stirrer, MECÁ chamber, P&C mixer, TELECOM scrambler, TV mixer; **~ de ácido bórico** *m* NUCL boric acid blender; **~ activo** *m* ING MECÁ active mixer; **~ armónico** *m* ELECTRÓN harmonic mixer; **~ Banbury** *m* P&C Banbury mixer; **~ de barro** *m* C&V clay mixer; **~ de canal** *m* C&V trough mixer; **~ por cargas** *m* PROC QUÍ batch mixer; **~ compensado** *m* ELECTRÓN balanced mixer; **~ de cristal de silicio** *m* ELECTRÓN silicon crystal mixer; **~ de cristal de un solo frente** *m* ELECTRÓN single-ended crystal mixer; **~ discontinuo** *m* ALIMENT batch mixer; **~ de la distorsión de la imagen** *m* AmL (*cf corrector de distorsión de imagen Esp*) TV tilt mixer; **~ doble compensado** *m* ELECTRÓN double-balanced mixer; **~ equilibrado** *m* ING ELÉC balanced mixer; **~ de imagen** *m* TV vision mixer; **~ inicial** *m* ELECTRÓN first mixer; **~ de microondas** *m* ELECTRÓN microwave mixer; **~ de modos** *m* ÓPT, TELECOM mode mixer; **~ monocompensado** *m* ELECTRÓN single-balanced mixer; **~ de paletas** *m* INSTAL HIDRÁUL paddle mixer, pug mill, MINAS paddle mixer; **~ de pasta** *m* P&C *equipo* dough mixer; **~ de redes ATM de alta velocidad binaria** *m* TELECOM high bit-rate ATM network mixer; **~ de la señal de borrado y sincronismo** *m* TV blanking and sync signal mixer; **~ de soluciones** *m* P&C solution mixer; **~ de tinta** *m* IMPR ink mixer

mezcladora *f* C&V blender, CONST, ING MECÁ, P&C, PROC QUÍ, TEC ESP mixer; **~ Banbury** *f* P&C Banbury mixer; **~ de platos** *f* PROC QUÍ mixing pan mill; **~ del tipo Eirich** *f* C&V batch mixer

mezcladura *f* PROD mixing

mezclar *vt* CONST *mortero* mix, ING MECÁ blend, QUÍMICA mix, TELECOM scramble, TEXTIL blend, TV mix

mezclas: ~ alimentarias *f pl* AGRIC feed mashes

MF *abr* (*modulación de fase*) ELECTRÓN, FÍS, INFORM&PD, TELECOM, TV PM (*phase modulation*)

MFBE *abr* (*modulación de frecuencia de banda estrecha*) ELECTRÓN, TELECOM NBFM (*narrow-band frequency modulation*)

MFM *abr* (*modulación de frecuencia modificada*) ELECTRÓN MFM (*modified frequency modulation*)

MFWS *abr* (*sistema de agua de alimentación principal*) NUCL MFWS (*main feedwater system*)

mg *abr* (*miligramo*) METR mg (*milligram*)

Mg *abr* (*magnesio*) ING ELÉC, METAL, QUÍMICA Mg (*magnesium*)

MGD *abr* (*magnetogasodinámica*) NUCL MGD (*magnetogasdynamics*)

miargirita *f* MINERAL miargyrite

MIB *abr* (*microscopía iónica de barrido*) FÍS RAD SIM (*scanning ion microscopy*)

mica *f* GEN mica; **~ blanca** *f* P&C white mica; **~ ligada con vidrio aislante** *f* ING ELÉC glass-bonded mica; **~ moscovita** *f* P&C white mica; **~ rubí** *f* MINERAL goethite

micáceo *adj* GEOL micaceous

micaesquisto *m* GEOL, PETROL mica schist

micela *f* P&C, QUÍMICA micelle

micelio *m* AGRIC mycelium

micoproteína *f* ALIMENT mycoprotein

micotoxina *f* ALIMENT mycotoxin

micrita *f* GEOL, PETROL micrite

micrítica: ~ siderítica *f* GEOL sideritic mudstone

micro[1] *m* INFORM&PD micro; **~ control** *m* TRANSP microcontrol; **~-instrucción** *m* INFORM&PD firware; **~ ohmio** *m* ING ELÉC microhm; **~-relé** *m* ING ELÉC miniature relay

micro[2]: **~ curvatura** *f* ING ELÉC microbend

micro[3] *pref* METR micro-

microagujero *m* PAPEL *defecto* pinhole

microampere *m* ELEC, ING ELÉC microampere

microamperímetro *m* ELECTRÓN, ING ELÉC microammeter

microamperio *m* ELEC, ING ELÉC microampere

microanálisis *m* QUÍMICA microanalysis

microanalítico *adj* QUÍMICA microanalytic, microanalytical

microbanda *f* ELECTRÓN microstrip

microbomba *f* ING MECÁ micropump, TEC ESP microrocket

microburbuja *f* ÓPT microbubble

microbureta *f* QUÍMICA microburette

microcanal *m* ELECTRÓN microchannel

microchip *m* ELECTRÓN, FÍS RAD, INFORM&PD microchip

microcircuito *m* ELECTRÓN, INFORM&PD microcircuit; **~ híbrido** *m* ELECTRÓN hybrid microcircuit; **~ lógico** *m* ELECTRÓN logic microcircuit

microclima *m* METEO microclimate

microclina *f* MINERAL microcline, Amazon stone

microcódigo *m* INFORM&PD microcode

microcohete *m* TEC ESP microrocket

microcojinete: **~ de bolas** *m* ING MECÁ miniature ball bearing

microcomputadora *f* AmL (*cf microordenador Esp*) INFORM&PD desktop computer, microcomputer; **~ de mesa** *f* AmL (*cf microordenador de mesa Esp*) INFORM&PD desktop computer

microconmutador *m* ELEC, ING ELÉC microswitch

microcontrolador *m* INFORM&PD microcontroller

microcósmico *adj* QUÍMICA microcosmic

microcrespado *m* PAPEL microcreping

microcristalino *adj* QUÍMICA microcrystalline

microdeformación *f* METAL microstrain

microdestilación *f* QUÍMICA microdistillation

microdesvitrificación *f* C&V, ÓPT microdevitrification

microdetector: **~ de agentes químicos** *m* D&A miniature chemical-agent detector

microdisyuntor *m* ELEC, ING ELÉC miniature circuit breaker

microdoblado *m* ÓPT microbending

microdureza *f* METAL, P&C microhardness

microelectrónica *f* ELECTRÓN, INFORM&PD microelectronics

microelemento *m* AGRIC microelement, trace mineral, QUÍMICA microelement

microelementos *m pl* AGRIC minor elements

microempuje *m* TEC ESP microthruster

microesfera: **~ de vidrio** *f* C&V glass microsphere

microesferas *f pl* P&C *pigmentos, espuma* microspheres

microestructura *f* GEN microstructure; **~ fibrosa** *f* METAL fibrous microstructure

microfarad *m* ELEC, ING ELÉC microfarad

microfaradio *m* ELEC, ING ELÉC microfarad

microficha *f* IMPR, INFORM&PD microfiche

microfilm *m* FOTO, IMPR, INFORM&PD microfilm

microfilme *m* INFORM&PD microfilm

microfiltro *m* AGUA microfilter; **~ rotativo** *m* AGUA microstrainer

microfisura *f* METAL, ÓPT microcrack

microflexión *f* TELECOM microbending

microfluencia *f* METAL microcreep

microfónico: **~ coclear** *m* ACÚST cochlear microphonics

microfonista *m* CINEMAT poleman

micrófono *m* ACÚST, ELEC, FÍS, INFORM&PD microphone, PROD handset, TV microphone; **~ con amplificador** *m* TELECOM amplified handset; **~ bidireccional** *m* ACÚST bidirectional microphone; **~ de bobina móvil** *m* FÍS moving-coil microphone; **~ del boom** *m* CINEMAT, TV boom mike; **~ de carbón** *m* ACÚST carbon microphone; **~ cardioide** *m* ACÚST cardioid microphone; **~ de careta** *m* ACÚST mask microphone; **~ de cinta** *m* ACÚST ribbon microphone; **~ de condensador** *m* ACÚST condenser microphone; **~ de contacto** *m* ACÚST contact microphone; **~ de desviación de fase** *m* ACÚST phase-shift microphone; **~ diferencial** *m* ACÚST differential microphone; **~ direccional** *m* ACÚST directional microphone; **~ de electrete** *m* ACÚST electret-foil microphone; **~ electrodinámico** *m* ACÚST, ELEC electrodynamic microphone; **~ electromagnético** *m* ACÚST, ELEC electromagnetic microphone; **~ electrostático** *m* ACÚST electrostatic microphone; **~ de gradiente** *m* ACÚST gradient microphone; **~ hipercardiode** *m* ACÚST hypercardioid microphone; **~ inalámbrico** *m* D&A wireless microphone; **~ de intercomunicación** *m* CINEMAT, TV intercom microphone; **~ labial** *m* ACÚST lip microphone; **~ magnetodinámico** *m* ACÚST moving-coil microphone; **~ de magnetoestricción** *m* ACÚST magnetostriction microphone; **~ múltiple** *m* ACÚST multiple microphone; **~ omnidireccional** *m* ACÚST omnidirectional microphone; **~ patrón** *m* ACÚST standard microphone; **~ de la pértiga** *m* CINEMAT, TV boom mike; **~ piezoeléctrico** *m* ACÚST piezoelectric microphone, piezomicrophone; **~ de presión** *m* ACÚST pressure microphone; **~ semidireccional** *m* ACÚST semidirectional microphone; **~ de solapa** *m* ACÚST lapel microphone; **~ sonda** *m* ACÚST probe microphone; **~ térmico** *m* ACÚST hot-wire microphone; **~ de velocidad** *m* ACÚST velocity microphone

microfósil *m* GEOL microfossil

microfotografía *f* CINEMAT, FOTO microphotography

microfotómetro *m* FÍS RAD microphotometer

micrografía *f* CRISTAL micrograph; **~ electrónica** *f* FÍS RAD electron micrograph

microgranito *m* PETROL microgranite

microgranular *adj* GEOL microgranular

microgravedad *f* TEC ESP microgravity

microgravitación *f* TEC ESP microgravity

microhenrio *m* ELEC microhenry

microinstrucción *f* INFORM&PD microinstruction

microinterruptor *m* ELEC, ING ELÉC microswitch

microjeringa *f* LAB microsyringe

microlita *f* MINERAL microlite

micromacla *f* METAL microtwin

micromaquinado *m* FÍS RAD micromachining

micromaquinización *f* FÍS RAD micromachining

micrometría *f* ING MECÁ micrometry

micrométrico *adj* ING MECÁ micronic, METR micrometric

micrómetro *m* FÍS, ING MECÁ, LAB, MECÁ, METR micrometer (*AmE*), micrometre (*BrE*); **~ de exteriores** *m* ING MECÁ external micrometer; **~ indicador de cuadrante** *m* METR dial-indicating micrometer; **~ para interiores** *m* ING MECÁ internal micrometer; **~ de lectura digital** *m* METR digital read-out micrometer; **~ de lente ocular** *m* ING MECÁ, INSTR eyepiece micrometer; **~ de precisión** *m* ING MECÁ precision micrometer

micromicrón *m* METR micromicron

micromil *m* METR micromil

micromilímetro *m* METR micromillimeter (*AmE*), micromillimetre (*BrE*)

microminiaturización *f* ELECTRÓN microminiaturization

micrón *m* METR micron

micronizado *adj* P&C micronized

micronizar *vt* QUÍMICA micronize

microonda *f* ELEC, ELECTRÓN, FÍS, FÍS ONDAS, TEC ESP, TELECOM microwave

microordenador *m* *Esp* (*cf microcomputadora AmL*) INFORM&PD desktop computer, microcomputer; **~ de mesa** *m* *Esp* (*cf microcomputadora de mesa AmL*) INFORM&PD desktop computer

microorganismo *m* CONTAM, HIDROL microorganism

micropelícula *f* INFORM&PD microfilm

microperfil *m* PETROL microlog

micropertita *f* MINERAL microperthite

micropicadura *f* C&V, CALIDAD, METAL, ÓPT, P&C micropit

micropipeta *f* LAB micropipette

microplaca: **~ de acoplamiento** *f* ELECTRÓN companion chip; **~ artificial** *f* ELECTRÓN exotic chip; **~ diseñada sobre pedido** *f* ELECTRÓN custom-designed chip; **~ especial** *f* ELECTRÓN dedicated chip; **~ sobre pedido** *f* ELECTRÓN custom-designed chip

micropliegue *m* GEOL microfold

microprocesador *m* ELEC, ELECTRÓN, IMPR, INFORM&PD, ING MECÁ microprocessor, TEC ESP microprocessor, chip

microprograma *m* INFORM&PD microprogram

microprogramación *f* IMPR firmware, INFORM&PD microprogramming

microquímica *f* QUÍMICA microchemistry

microreología *f* C&V, METAL, P&C microrheology

microrruptor *m* ELEC, ING ELÉC microswitch

microscopía *f* FÍS, INSTR, LAB, ÓPT, QUÍMICA microscopy; **~ de barrido Auger** *f* NUCL scanning Auger microscopy; **~ electrónica** *f* ELECTRÓN, FÍS electron microscopy; **~ electrónica de transmisión** *f* (*MET*) FÍS RAD transmission electron microscopy (*TEM*); **~ de emisión** *f* METAL emission microscopy; **~ por interferencia axial** *f* ÓPT, TELECOM axial interference microscopy; **~ iónica de barrido** *f* (*MIB*) FÍS RAD scanning ion microscopy (*SIM*); **~ óptica** *f* METAL light microscopy

microscopio *m* FÍS, INSTR, LAB microscope, METR microscope, toolmaker's microscope, ÓPT, QUÍMICA microscope; **~ de asperezas superficiales** *m* INSTR surface-finish microscope; **~ de barrido** *m* TELECOM sweep microscope; **~ binocular** *m* INSTR, LAB binocular microscope; **~ de campo ancho universal** *m* INSTR universal wide-field microscope; **~ de campo electrónico** *m* INSTR field electron microscope; **~ compuesto** *m* FÍS, INSTR compound microscope; **~ de contraste de fase** *m* FÍS, LAB phase-contrast microscope; **~ corredizo** *m* LAB traveling microscope (*AmE*), travelling microscope (*BrE*); **~ de cuerpo articulado** *m* INSTR hinged body microscope; **~ de disección** *m* INSTR dissecting microscope; **~ de electrones** *m* FÍS scanning electron microscope (*SEM*); **~ de electrones de transmisión** *m* NUCL transmission electron microscope; **~ electrónico** *m* ELECTRÓN, FÍS, FÍS PART, INSTR, LAB, METAL, TELECOM electron microscope; **~ electrónico de alto voltaje** *m* INSTR high-voltage electron microscope; **~ electrónico de barrido** *m* (*MEB*) ELEC, FÍS RAD, LAB scanning electron microscope (*SEM*); **~ electrónico de difracción por transmisión** *m* LAB transmission electron microscope; **~ electrónico de emisión** *m* ELECTRÓN, INSTR emission electron microscope; **~ electrónico especular** *m* INSTR mirror electron

microscope; **~ electrónico de espejo exploratorio** *m* INSTR scanning-mirror electron microscope; **~ electrónico de exploración** *m* ELECTRÓN scanning electron microscope (*SEM*), INSTR scanning electron microscope (*SEM*), surface electron microscope (*SEM*); **~ electrónico de fotoemisión** *m* INSTR photoemission electron microscope; **~ electrónico de imán permanente** *m* INSTR permanent-magnet electron microscope; **~ electrónico de reflexión** *m* ELECTRÓN, INSTR, ÓPT reflection electron microscope; **~ electrónico selector de energía** *m* INSTR energy-selecting electron microscope; **~ electrónico de transmisión** *m* ELECTRÓN, FÍS, FÍS RAD, INSTR transmission electron microscope; **~ de elevada potencia** *m* INSTR high-power microscope; **~ de emisión** *m* INSTR emission microscope; **~ de emisión por efecto de campo** *m* ELECTRÓN, FÍS, INSTR field emission microscope; **~ de emisión iónica** *m* INSTR ion emission microscope; **~ de emisión termoiónica** *m* INSTR thermionic emission microscope; **~ estereoscópico** *m* LAB stereomicroscope; **~ estereoscópico de campo ancho** *m* INSTR wide-field stereo microscope; **~ de exploración electrónica** *m* INSTR surface electron microscope (*SEM*); **~ de fase de contraste** *m* INSTR contrasting phase microscope; **~ de fluorescencia** *m* FÍS, FÍS ONDAS, FÍS RAD, INSTR, QUÍMICA fluorescence microscope; **~ fotoeléctrico** *m* INSTR photoelectric microscope; **~ de interferencia** *m* FÍS, INSTR interference microscope; **~ interferencial** *m* METAL interference microscope; **~ invertido** *m* INSTR inverted microscope; **~ de iones de campo** *m* FÍS, INSTR field-ion microscope; **~ iónico de campo** *m* FÍS, INSTR field-ion microscope; **~ de laboratorio** *m* INSTR laboratory microscope; **~ de lectura** *m* INSTR reading microscope; **~ de luz reflejada** *m* INSTR, ÓPT reflected-light microscope; **~ de luz transmitida** *m* INSTR transmitted light microscope; **~ de luz transmitida con polarizador** *m* INSTR transmitted-light microscope with polarizer; **~ de luz transmitida con polarizador Zeiss sistema** *m* INSTR Zeiss system transmitted-light microscope with polarizer; **~ de luz ultravioleta** *m* FÍS RAD ultraviolet microscope; **~ de medidas** *m* INSTR measuring microscope; **~ metalográfico** *m* LAB, METAL metallographic microscope; **~ metalúrgico de campo ancho** *m* INSTR wide-field metallurgical microscope; **~ micrográfico** *m* INSTR micrographic microscope; **~ de plancton** *m* INSTR plankton microscope; **~ de polarización** *m* CRISTAL polarizing microscope; **~ polarizador** *m* INSTR polarizing microscope, LAB polarization microscope, polarizing microscope; **~ polarizante** *m* C&V, FÍS polarizing microscope; **~ con portaobjetos esférico** *m* INSTR ball-stage microscope; **~ protónico** *m* INSTR proton microscope; **~ de proyección** *m* INSTR projection microscope; **~ quirúrgico** *m* INSTR surgical microscope; **~ de rayos infrarrojos** *m* INSTR infrared microscope; **~ de rayos X** *m* FÍS RAD, NUCL X-ray microscope; **~ reflector** *m* INSTR, ÓPT reflecting microscope; **~ de rosca** *m* INSTR thread microscope; **~ de sección iluminada** *m* INSTR light section microscope; **~ de sombra** *m* INSTR shadow microscope; **~ de televisión** *m* INSTR, TELECOM, TV television microscope

microsecundo *m* INFORM&PD microsecond

microsegregación *f* METAL microsegregation
microsonda *f* FÍS RAD microprobe
microsurco *m* ACÚST microgroove
microtamiz *m* AGUA microstrainer
microteléfono *m* PROD, TELECOM handset
micrótomo *m* LAB microtome
microtubo: ~ **de propagación de ondas** *m* ELECTRÓN miniature traveling-wave tube (*AmE*), miniature travelling-wave tube (*BrE*)
microvolt *m* ELEC, FÍS, ING ELÉC microvolt
microvoltio *m* ELEC, FÍS, ING ELÉC microvolt
miembro *m* GEOL, INFORM&PD member, INSTR *de una ecuación* side, MATEMÁT member; ~ **de la caja lateral** *m* TRANSP AÉR edge-box member; ~ **central de resistencia** *m* ÓPT central strength member; ~ **cruzado de la suspensión delantera** *m* AUTO front suspension cross-member; ~ **de la cuerda** *m* TRANSP AÉR chord member; ~ **de la derecha** *m* FÍS *de una ecuación* right-hand side; ~ **del extremo de la montura** *m* EMB end-frame member; ~ **final** *m* GEOL end member; ~ **longitudinal** *m* TRANSP AÉR longitudinal member; ~ **transversal** *m* TRANSP MAR transverse member; ~ **de la tripulación** *m* TEC ESP crewmate
migmatita *f* GEOL, PETROL migmatite
migmatización *f* GEOL, PETROL migmatization
migración *f* COLOR migration (*AmE*), swealing (*BrE*), P&C, QUÍMICA, TEC PETR migration; ~ **electroforética** *f* NUCL electrophoretic migration; ~ **del plano de exfoliación** *f* METAL grain-boundary migration; ~ **del plastificante** *f* P&C plasticizer migration
migrar *vi* PETROL, QUÍMICA migrate
mijo *m* AGRIC millet; ~ **africano** *m* AGRIC finger millet
mil *m* MATEMÁT thousand; ~ **millones** *m pl* MATEMÁT a thousand million
milarita *f* MINERAL milarite
mildiú *m* AGRIC mildew; ~ **enanizante** *m* AGRIC downy mildew
miliámetro *m* ELEC, ING ELÉC, INSTR milliammeter
miliamperímetro *m* ELEC *instrumento*, ING ELÉC, INSTR milliammeter
miliamperio *m* (*mA*) ELEC, FÍS, ING ELÉC milliampere (*mA*)
milicrón *m* METR millicron
miligramo *m* (*mg*) METR milligram (*mg*)
milimétrico *adj* ELECTRÓN *onda* millimetric
milímetro *m* (*mm*) METR millimeter (*AmE*) (*mm*), millimetre (*BrE*) (*mm*)
milisegundo *m* (*ms*) INFORM&PD millisecond (*ms*)
milivatímetro *m* ELEC, ING ELÉC, INSTR milliwattmeter
milivatio *m* (*mW*) ELEC, ING ELÉC milliwatt (*mW*)
milivoltímetro *m* ELEC, INSTR millivoltmeter
milivoltio *m* (*mV*) ELEC, ING ELÉC millivolt (*mV*)
milla *f* METR, TRANSP mile; ~ **cuadrada** *f* METR square mile; ~ **marina** *f* METEO, OCEAN, TEC ESP, TRANSP MAR nautical mile; ~ **náutica** *f* METEO, OCEAN, TEC ESP, TRANSP MAR nautical mile
millas: ~ **por galón** *f pl* TRANSP miles per gallon (*mpg*)
millerita *f* MINERAL millerite
millones: ~ **de instrucciones por segundo** *m pl* (*MIPS*) IMPR, INFORM&PD millions of instructions per second (*MIPS*); ~ **de pies de gas cúbicos por día** *m pl* TEC PETR millions of cubic feet per day (*MCFD*)
milo: ~ **blanco precoz** *m* AGRIC early white milo

milonita *f* GEOL, PETROL mylonite; ~ **muy compacta** *f* GEOL flinty crush rock
milrayas *m* TEXTIL millerayes
mimetesita *f* MINERAL mimetite, mimetene
mina *f* CARBÓN, D&A, MINAS mine; ~ **en actividad** *f* MINAS active mine; ~ **anticarro** *f* D&A antitank mine; ~ **de arcilla** *f* MINAS clay pit; ~ **aurífera** *f* MINAS gold mine, placer mine; ~ **de caolín** *f* C&V china-clay quarry; ~ **de carbón** *f* CARBÓN coal mine, coal pit, collier, colliery, MINAS colliery, coal mine, coal pit, collier; ~ **de carbón a cielo abierto** *f* CARBÓN, MINAS daylight colliery; ~ **a cielo abierto** *f* CARBÓN opencast mine, MINAS daylight mine, open-pit mine, strip mine; ~ **de cobre** *f* MINAS copper mine; ~ **contra tropas** *f* D&A antipersonnel mine; ~ **de diamantes** *f* MINAS diamond mine; ~ **de dispersión de efectos** *f* D&A scatterable mine; ~ **en explotación** *f* MINAS active mine; ~ **explotada por pozos** *f* MINAS shaft mine; ~ **magnética** *f* D&A magnetic mine; ~ **operativa** *f* MINAS operational mine; ~ **de placeres** *f* MINAS placer mine; ~ **de plata** *f* MINAS silver mine; ~ **en servicio** *f* MINAS operational mine; ~ **subterránea** *f* MINAS mine, underground mine; ~ **terrestre** *f* MINAS mine
minabilidad *f* MINAS mineability
minable *adj* MINAS mineable
minado *adj* D&A mined
minador *m* D&A minelayer, MINAS roadheader, TRANSP MAR minelayer
minar *vt* CONST undermine, MINAS mine, undermine
minas *f pl* MINAS diggings
mineral[1] *adj* CARBÓN, GEOL, MINERAL, QUÍMICA mineral
mineral[2] *m* CARBÓN mineral, ore, MINAS metal, ore, mine, mine stone, stone, MINERAL, QUÍMICA mineral; ~ **agárico** *m* MINERAL agaric mineral; ~ **para alfarería** *m* C&V potter's ore; ~ **de alto grado** *m* CARBÓN high-grade ore; ~ **de arcilla** *m* CARBÓN, GEOL clay mineral; ~ **arcilloso** *m* GEOL clay mineral; ~ **aurífero** *m* MINAS gold ore; ~ **de azufre** *m* MINAS sulfur-ore (*AmE*), sulphur-ore (*BrE*); ~ **de baja ley** *m* MINAS low tenor of ore, poor grade; ~ **bocarteado** *m* MINAS milled ore; ~ **bruto** *m* CARBÓN run of mine; ~ **de calidad inferior** *m* MINAS dredge, low-grade ore; ~ **complejo** *m* CARBÓN complex ore; ~ **de cromo** *m* C&V, MINERAL chrome ore; ~ **cuyo metal es completamente amalgamable** *m* MINAS free-milling ore; ~ **en forma de cruz** *m* GEOL cross stone; ~ **guía** *m* GEOL index mineral; ~ **de hierro** *m* QUÍMICA iron ore; ~ **de hierro pardo** *m* QUÍMICA brown iron ore; ~ **índice** *m* GEOL index mineral; ~ **machacado** *m* MINAS broken ore, crushed ore; ~ **no fracturado** *m* MINAS unbroken ore; ~ **no refinado** *m* CARBÓN crude ore; ~ **normativo** *m* GEOL, MINERAL normative mineral; ~ **pobre** *m* MINAS low-grade ore, low tenor of ore; ~ **de poca ley** *m* MINAS low-grade ore; ~ **quebrantado** *m* MINAS broken ore, crushed ore; ~ **refractario** *m* MINAS stubborn ore; ~ **con sulfuros** *m* MINAS sulfide ore (*AmE*), sulphide ore (*BrE*); ~ **en la superficie** *m* MINAS ore at grass; ~ **en terrones** *m* MINAS lump ore; ~ **terroso** *m* MINAS glebe; ~ **triturado** *m* MINAS broken ore, crushed ore; ~ **triturado fino** *m* CARBÓN overflow; ~ **en trozos** *m* MINAS coarse ore, lump ore; ~ **uranífero** *m* NUCL uranium-bearing mineral; ~ **de valor** *m* MINAS color (*AmE*), colour (*BrE*)

mineralero *m* TRANSP MAR ore carrier; **~-petrolero** *m* TRANSP MAR ore-oil carrier

minerales: **~ escogidos** *m pl* MINAS picked ore; **~ finos que flotan al lavarlos** *m pl* CARBÓN float; **~ pesados** *m pl* GEOL heavy minerals; **~ seleccionados** *m pl* MINAS picked ore

mineralización *f* GEOL, MINERAL, QUÍMICA mineralization

mineralizador *m* GEOL, MINERAL, QUÍMICA mineralizer

mineralizante *adj* GEOL, MINERAL, QUÍMICA mineralizing

mineralogía *f* MINERAL, TEC PETR mineralogy

mineralógico *adj* GEOL, MINERAL mineralogic (*AmE*), mineralogical (*BrE*)

mineralogista *m* GEOL, MINERAL, TEC PETR mineralogist

minería *f* CARBÓN, HIDROL mining, MINAS mining, working mine; **~ del carbón** *f* CARBÓN, MINAS coal mining; **~ a cielo abierto** *f* CARBÓN, MINAS, NUCL open-pit mining; **~ metalífera** *f* MINAS ore mining; **~ de vanguardia** *f* MINAS advance mining

minero *m* CARBÓN, MINAS collier, miner; **~ que rescata madera de entibación** *m* CARBÓN stripper

minerogenético *adj* CARBÓN minerogenic

minerógeno *adj* CARBÓN minerogenous

mini[1] *m* INFORM&PD mini; **~-submarino** *m* OCEAN diving saucer, minisubmersible

mini[2]: **~-furgoneta** *f* TRANSP minivan

miniatura *f* CINEMAT miniature; **~ en primer plano** *f* CINEMAT foreground miniature

miniaturista *m* CINEMAT miniature maker

miniaturización *f* INFORM&PD miniaturization

minicomputadora *f* AmL (*cf* miniordenador *Esp*) INFORM&PD minicomputer

mínima: **~ frecuencia empleada** *f* (*LUF*) TEC ESP lowest usable frequency (*LUF*); **~ velocidad de emisión** *f* CONTAM lowest achievable emission rate

mínimo *m* GEOL low; **~ común denominador** *m* MATEMÁT least common denominator; **~ común múltiplo** *m* INFORM&PD, MATEMÁT least common multiple (*LCM*); **~ cotizable** *m* TELECOM minimum payable; **~ de tensión** *m* ELEC minimum voltage

minio *m* C&V red lead, MINERAL minium, red lead, P&C red lead, QUÍMICA red lead, minium

miniordenador *m* Esp (*cf* minicomputadora *AmL*) INFORM&PD minicomputer

minitractor: **~ submarino individual** *m* OCEAN minisubmersible

minúscula *f* IMPR lower case

minuto *m* CONST, FÍS, METR minute; **~ del arco** *m* FÍS arc minute; **~ degradado** *m* TELECOM degraded minute (*DM*)

miocina *f* QUÍMICA myocin

miogeosinclinal *m* GEOL miogeosyncline

mionita *f* MINERAL mionite

MIPS *abr* (*millones de instrucciones por segundo*) IMPR, INFORM&PD MIPS (*millions of instructions per second*)

mira[1] *f* CONST topografía staff, boning-rod, pole, poststaff, target, cross staff, MINAS hub, topografía rod; **~ de bombardeo** *f* D&A bomb sight; **~ de bombardeo aéreo** *f* D&A bomb sight; **~ de formación de imágenes térmicas** *f* TERMO thermal-imaging sight; **~ de nivelación** *f* CONST leveling pole (*AmE*), levelling pole (*BrE*), target-leveling rod (*AmE*), target-levelling rod (*BrE*), target-leveling

staff (*AmE*), target-levelling staff (*BrE*), TV hub; **~ óptica** *f* TEC ESP optical sight; **~ parlante** *f* CONST topografía self-reading staff, speaking rod; **~ telescópica del fusil** *f* D&A rifle telescope

mira[2]: **que ~ al este** *fra* TRANSP MAR eastward; **que ~ al oeste** *fra* TRANSP MAR westward

mirabilita *f* MINERAL mirabilite

mirador *m* CONST bay

miraguano *m* TERMOTEC kapok

mirando: **~ hacia atrás** *adj* PROD backward

mirbano *m* QUÍMICA mirbane

mirceno *m* QUÍMICA myrcene

miria *pref* METR myria-

miriágramo *m* METR myriagram

mirilla *f* INFORM&PD cross hair, INSTR viewing window, PROD peephole, TEC ESP inspection hole, peephole; **~ de observación** *f* C&V observation hole

mirístico *adj* QUÍMICA myristic

miristilo *adj* QUÍMICA myristyl

miristina *f* QUÍMICA myristin

mirónico *adj* QUÍMICA myronic

mirosina *f* QUÍMICA myrosin

miscible *adj* QUÍMICA, TEC PETR miscible

MISD *abr* (*multiflujo de instrucciones-monoflujo de datos*) INFORM&PD MISD (*multiple-instruction single-data*)

misil *m* D&A missile; **~ aerodinámico** *m* D&A aerodynamic missile; **~ aire-aire** *m* D&A air-to-air missile (*AAM*); **~ antiradiación** *m* D&A anti-radiation missile (*ARM*); **~ balístico** *m* D&A, ELECTRÓN, TEC ESP ballistic missile; **~ de crucero** *m* D&A, TRANSP MAR cruise missile; **~ desde la superficie a un blanco submarino** *m* D&A surface-to-underwater missile; **~ dirigido** *m* D&A guided missile; **~ estratégico** *m* D&A strategic missile; **~ de gran alcance** *m* D&A cruise missile; **~ de gran alcance** *m* TRANSP MAR cruise missile; **~ guiado** *m* D&A guided missile, guided weapon; **~ guiado anticarro** *m* D&A antitank guided missile (*ATGM*); **~ guiado de gran alcance** *m* D&A guided long-range missile; **~ guiado subsónico** *m* D&A subsonic guided missile; **~ radioguiado** *m* D&A guided missile; **~ superficie-superficie** *m* (*SSM*) TEC ESP surface-to-surface missile (*SSM*); **~ teledirigido** *m* D&A guided missile; **~ teleguiado** *m* D&A guided missile; **~ de tierra a aire** *m* D&A ground-to-air missile, surface-to-air missile (*SAM*); **~ de tierra a tierra** *m* D&A ground-to-ground missile; **~ tierra-aire** *m* D&A ground-to-air missile, surface-to-air missile (*SAM*); **~ de trayectoria rasante** *m* D&A skimmer missile

misión: **~ al espacio profundo** *f* TEC ESP deep space mission; **~ interplanetaria** *f* TEC ESP interplanetary mission; **~ de largo plazo** *f* TEC ESP long mission; **~ ofensiva sobre territorio enemigo** *f* TEC ESP sweep; **~ a un planeta exterior** *f* TEC ESP outer planet mission; **~ a un planeta interior** *f* TEC ESP inner planet mission; **~ terrestre en órbita** *f* TEC ESP earth-orbiting mission

mismo: **al ~ nivel** *adj* INFORM&PD peer-to-peer; **en el ~ sitio** *adj* CONST, MECÁ in situ

mispíquel *m* MINERAL mispickel

miticultura *f* OCEAN mussel culture

mitigación *f* CONTAM MAR mitigation

mitilotoxina *f* QUÍMICA mytilotoxine

mixtos *m pl* CARBÓN middlings, MINAS, MINERAL middles, middlings; **~ de arena gruesa** *m pl* MINAS,

MINERAL coarse sand middlings; ~ **puros** *m pl* CAR-BÓN true middlings; ~ **verdaderos** *m pl* CARBÓN true middlings

mixtura *f* ALIMENT mixture, P&C composite, QUÍMICA, TEC PETR mixture

mizzonita *f* MINERAL mizzonite

MLS *abr* (*sistema de aterrizaje de microondas*) TRANSP AÉR MLS (*microwave landing system*)

mm *abr* (*milímetro*) METR mm (*millimetre BrE, millimeter AmE*)

MMI *abr* (*interacción hombre-máquina*) INFORM&PD MMI (*man-machine interaction*)

MMS *abr* (*módulo de mando y servicio*) TEC ESP command and service module

Mn *abr* (*manganeso*) C&V, MECÁ, METAL, QUÍMICA Mn (*manganese*)

Mo *abr* (*molibdeno*) AGRIC, METAL, QUÍMICA Mo (*molybdenum*)

mobiliario *m* CONST furnishing; ~, **lámparas y accesorios** *m pl* CONST furniture, fixtures and fittings

mocional *adj* ACÚST motional

modacrílico *adj* TEXTIL modacrylic

modalidad[1]: ~ **de acceso** *f* INFORM&PD access mode; ~ **de explotación** *f* TELECOM mode of operation; ~ **de funcionamiento** *f* TELECOM mode of operation; ~ **de producción** *f* GAS production mode; ~ **de regeneración** *f* GAS regeneration mode; ~ **de utilización** *f* TELECOM mode of operation

modalidad[2]: **en ~ de espera** *fra* CONTAM MAR on standby

modelación *f* ELECTRÓN, PROD modeling (*AmE*), modelling (*BrE*); ~ **del haz** *f* ELECTRÓN beam forming; ~ **de la señal** *f* ELECTRÓN signal modeling (*AmE*), signal modelling (*BrE*)

modelado[1] *adj* ING MECÁ, PROD shaped

modelado[2] *m* CALIDAD, GEOL, INFORM&PD modeling (*AmE*), modelling (*BrE*), MECÁ, P&C forming, PROD forming, modeling (*AmE*), modelling (*BrE*), REVEST forming

modelador *m* PROD pattern maker

modelaje *m* PROD pattern making

modelar *vt* C&V barro throw, ING MECÁ, PROD shape

modelería *f* PROD pattern making

modelismo *m* INFORM&PD modeling (*AmE*), modelling (*BrE*)

modelista *m* PROD pattern maker, pattern molder (*AmE*), pattern moulder (*BrE*)

modelo *m* CARBÓN sample, ELEC template, INFORM&PD pattern, master, model, ING MECÁ standard, template, templet, MECÁ template, PROD mock-up, template, *plantilla* patron, pattern, TEC ESP model, pattern, mock-up, TELECOM model, standard, TRANSP MAR *arquitectura naval* model; ~ **de banda** *m* METAL band model; ~ **del bebedero** *m* PROD gate pin; ~ **de Bohr-Sommerfeld** *m* FÍS Bohr-Sommerfeld model; ~ **de Buchmann y Meyer** *m* ACÚST Buchmann and Meyer pattern; ~ **de burbujas** *m* METAL bubble model; ~ **de cableado** *m* TEC ESP wiring harness; ~ **de calibración** *m* CONTAM model calibration; ~ **de calificación** *m* TEC ESP qualification model; ~ **de capas** *m* FÍS shell model; ~ **del cliente** *m* ING MECÁ customer's model; ~ **de codificación** *m* ELECTRÓN coding scheme; ~ **colectivo** *m* FÍS collective model; ~ **compartimentado** *m* TELECOM compartmental model; ~ **de cuartetes** *m* NUCL quartet model;

~ **de datos** *m* INFORM&PD data model; ~ **de Debye** *m* FÍS Debye model; ~ **de difracción** *m* METAL diffraction pattern; ~ **de difracción de rayos X** *m* INSTR X-ray diffraction pattern; ~ **digital de fases múltiples** *m* GAS multiphase digital model; ~ **digital de simulación** *m* GAS digital simulation model; ~ **dinámico** *m* TEC ESP dynamic model; ~ **direccional** *m* ACÚST directional pattern; ~ **a escala** *m* GEOM, TRANSP MAR scale model; ~ **de esfera dura** *m* CRISTAL hard sphere model; ~ **espacial** *m* CONTAM spatial pattern; ~ **estocástico** *m* INFORM&PD stochastic model; ~ **estructural** *m* TEC ESP structural model; ~ **estructural de elemento finito** *m* TEC ESP finite-element structural model; ~ **experimental** *m* MECÁ experimental model; ~ **de fleje** *m* METAL band model; ~ **de fluctuación de carga** *m* TEC ESP load fluctuation pattern; ~ **de gas de fonones** *m* FÍS phonon-gas model; ~ **de la gota líquida** *m* FÍS liquid-drop model; ~ **hidrodinámico** *m* ENERG RENOV hydrodynamic model; ~ **de ingeniería** *m* TEC ESP engineering model; ~ **de integración** *m* TEC ESP *pruebas* integration model; ~ **de inundación** *m* GEOFÍS flooding pattern; ~ **de Jackson** *m* NUCL Jackson model; ~ **jerárquico** *m* INFORM&PD hierarchical model; ~ **matemático** *m* ELECTRÓN, GAS, INFORM&PD mathematical model; ~ **de momento variable de inercia** *m* NUCL variable moment of inertia model; ~ **nuclear** *m* FÍS nuclear model; ~ **de orden-desorden** *m* NUCL order-disorder model; ~ **de partículas independientes** *m* FÍS, NUCL independent particle model; ~ **de práctica de tráfico** *m* TRANSP traffic assignment model; ~ **de red** *m* INFORM&PD network model; ~ **reducido** *m* TELECOM reduced model; ~ **de referencia del protocolo** *m* TELECOM protocol-reference model (*PRM*); ~ **de referencia de siete capas** *m* INFORM&PD seven-layer reference model; ~ **de referencia de siete niveles** *m* INFORM&PD seven-layer reference model; ~ **relacional** *m* INFORM&PD relational model; ~ **de rotura** *m* MECÁ breaking pattern; ~ **sedimentario** *m* GAS sedimentary model; ~ **térmico** *m* TEC ESP thermal model; ~ **de transporte** *m* CONTAM transport model; ~ **vectorial** *m* FÍS vector model; ~ **de vuelo** *m* TEC ESP, TRANSP AÉR flight model; ~ **de Westcott** *m* NUCL Westcott model

módem *m* (*modulador-demodulador*) ELECTRÓN, IMPR, INFORM&PD, TELECOM modem (*modulator-demodulator*); ~ **de acción lenta** *m* ELECTRÓN low-speed modem; ~ **con acoplamiento acústico** *m* INFORM&PD acoustically-coupled modem; ~ **acústico** *m* INFORM&PD acoustic modem; ~ **de alta modulación** *m* ELECTRÓN high-speed modem; ~ **asíncrono** *m* ELECTRÓN asynchronous modem; ~ **de banda ancha** *m* ELECTRÓN wideband modem; ~ **de banda base** *m* INFORM&PD, TELECOM baseband modem; ~ **de fibra óptica** *m* ELECTRÓN, INFORM&PD, ÓPT fiber-optic modem (*AmE*), fibre-optic modem (*BrE*); ~ **integrado** *m* INFORM&PD integrated modem; ~ **listo para comunicar** *m* INFORM&PD data set ready (*DSR*); ~ **listo para funcionar** *m* INFORM&PD data set ready (*DSR*); ~ **de manipulación por desplazamiento de frecuencia** *m* ELECTRÓN frequency-shift keying modem; ~ **de MDF** *m* ELECTRÓN FSK modem; ~ **de MF** *m* ELECTRÓN FM modem (*frequency-modulation modem*); ~ **de modulación de frecuencia** *m* ELECTRÓN frequency-modulation modem (*FM modem*); ~ **sobre placa** *m* ELECTRÓN board-level modem; ~ **síncrono** *m* ELECTRÓN synchronous modem

moderación *f* NUCL slowing down

moderador *m* ELECTRÓN, FÍS, FÍS RAD moderator; ~ **orgánico** *m* NUCL organic moderator; ~ **de sobrevelocidad** *m* ING MECÁ overspeed gear, overspeeder; ~ **de velocidad de fin de carrera** *m* MINAS *jaula* overwind gear, overwinding gear

modernización: ~ **del equipo físico** *f* INFORM&PD hardware upgrade; ~ **del hardware** *f* INFORM&PD hardware upgrade

modernizar *vt* ING MECÁ update

modificación *f* CINEMAT, IMPR editing, ING MECÁ amendment, TV editing; ~ **de las características de la tinta mediante aditivos** *f* IMPR doping; ~ **de la concepción** *f* PROD engineering change; ~ **de ganancia** *f* ELECTRÓN gain change; ~ **de llamada entrante** *f* TELECOM in-call modification; ~ **de llamada mensaje concluído** *f* TELECOM call modification completed message (*CMC*); ~ **de llamada rechazar mensaje** *f* TELECOM call modification reject message (*CMRJ*); ~ **de llamada solicitar mensaje** *f* TELECOM call modification request message (*CMR*); ~ **de órbita** *f* TEC ESP orbit modification

modificador *m* CARBÓN modifier; ~ **de densidad** *m* ALIMENT density modifier; ~ **de fase** *m* ELEC phase advancer; ~ **de redes** *m* C&V network modifier; ~ **de la tensión interfacial** *m* CONTAM MAR surface-tension modifier

modificar *vt* ELECTRÓN, GEOM, INFORM&PD modify, ING MECÁ amend; ~ **por adición** *vt* COLOR dope

modo *m* ACÚST, INFORM&PD, PROD mode; ~ **de acceso** *m* INFORM&PD access mode; ~ **de activación** *m* ELECTRÓN enhancement mode; ~ **alterno** *m* ELECTRÓN alternate mode; ~ **de aporte y purga** *m* NUCL feed and bleed; ~ **armónico** *m* ELECTRÓN harmonic mode; ~ **autoservo** *m* TV autoservo mode; ~ **auxiliar** *m* TEC ESP standby mode; ~ **básico** *m* ING ELÉC fundamental mode; ~ **de caída de imagen** *m* TV drop frame mode; ~ **de captación normal** *m* TEC ESP normal mode acquisition; ~ **circuito para servicio portador** *m* TELECOM circuit mode bearer service; ~ **común** *m* ELECTRÓN, TELECOM common mode; ~ **conversacional** *m* INFORM&PD conversational mode; ~ **de corrección de posición** *m* TEC ESP station correction mode; ~ **dedicado** *m* INFORM&PD dedicated mode; ~ **desconectado normal** *m* TELECOM normal disconnected mode (*NDM*); ~ **diferencial** *m* ELECTRÓN differential mode; ~ **dominante** *m* FÍS, ING ELÉC dominant mode; ~ **E** *m* ÓPT, TELECOM E mode; ~ **edición** *m* CINEMAT, TV edit mode; ~ **de efecto túnel** *m* TELECOM tunneling mode (*AmE*), tunnelling mode (*BrE*); ~ **elástico** *m* TEC ESP *nave* elastic mode; ~ **eléctrico transversal** *m* (*modo TE*) ELEC, FÍS, ÓPT, TELECOM transverse electric mode (*TE mode*); ~ **electromagnético** *m* ÓPT electromagnetic mode; ~ **electromagnético transversal** *m* ÓPT, TELECOM transverse electromagnetic mode; ~ **de emergencia** *m* PROD backup mode, TEC ESP emergency mode; ~ **de empleo** *m* EMB instructions for use; ~ **de formación PIT** *m* ÓPT PIT-forming mode; ~ **con fugas** *m* ÓPT, TELECOM leaky mode; ~ **de funcionamiento** *m* ING ELÉC operating mode, TELECOM mode of operation; ~ **de**

funcionar *m* PROD modus operandi; ~ **fundamental** *m* FÍS, ING ELÉC, ÓPT, TEC ESP fundamental mode; ~ **fundamental de vibración** *m* TEC ESP fundamental vibration mode; ~ **de guía de ondas** *m* TELECOM waveguide mode; ~ **de guíaondas** *m* TELECOM waveguide mode; ~ **H** *m* TELECOM H mode, transverse electric mode (*TE mode*); ~ **híbrido** *m* ÓPT, TELECOM hybrid mode; ~ **HE11** *m* ÓPT HE11 mode; ~ **de impulsos** *m* ELECTRÓN pulsed mode; ~ **interactivo** *m* INFORM&PD conversational mode, interactive mode; ~ **interrumpido** *m* ELECTRÓN chopped mode; ~ **ligado** *m* ÓPT *de electrones*, TELECOM bound mode; ~ **longitudinal único** *m* TELECOM single-longitudinal mode (*SLM*); ~ **de lotes** *m* PROD batch mode; ~ **magnético transversal** *m* ÓPT, TELECOM transverse magnetic mode; ~ **de marcha** *m* PROD run mode; ~ **de marcha continua** *m* ELECTRÓN free-running mode; ~ **de mínima** *m* ELECTRÓN depletion mode; ~ **de múltiples destinos** *m* TELECOM multidestination mode; ~ **nativo** *m* INFORM&PD native mode; ~ **no inscrito** *m* PROD unpolled mode; ~ **no ligado** *m* ÓPT, TELECOM unbound mode; ~ **de no repetición de explotación en dos frecuencias** *m* TELECOM two-frequency operation non-repeater mode; ~ **normal** *m* ACÚST, FÍS normal mode; ~ **operatorio** *m* ING MECÁ procedure; ~ **pasivo** *m* ELECTRÓN passive mode; ~ **polarizado linealmente** *m* ÓPT, TELECOM linearly-polarized mode (*LP mode*); ~ **de procesamiento** *m* INFORM&PD processing mode; ~ **de propagación** *m* ÓPT propagation mode; ~ **de prueba por control a distancia** *m* PROD remote test mode; ~ **de pulsado** *m* ING ELÉC pulsed mode; ~ **que se propaga por el revestimiento** *m* ÓPT cladding mode; ~ **de radiación** *m* FÍS RAD, ÓPT, TELECOM radiation mode; ~ **de ráfaga** *m* INFORM&PD burst mode; ~ **de readquisición** *m* TEC ESP reacquisition mode; ~ **de regeneración** *m* ELECTRÓN, INFORM&PD refresh mode; ~ **remoto** *m* PROD remote mode; ~ **de reserva** *m* TEC ESP standby mode; ~ **restaurador** *m* TEC ESP restoration mode; ~ **silencioso** *m* IMPR quiet mode; ~ **TE** *m* (*modo eléctrico transversal*) ELEC, FÍS, ING ELÉC, ÓPT, TELECOM TE mode (*transverse electric mode*); ~ **de teleprogramación** *m* PROD remote program mode; ~ **TEM** *m* ÓPT, TELECOM TEM mode; ~ **de termoablación** *m* ÓPT heat ablation mode; ~ **TE/TM** *m* FÍS TE/TM mode; ~ **TM** *m* ÓPT, TELECOM TM mode; ~ **de transferencia asíncrona** *m* (*ATM*) TELECOM asynchronous transfer mode (*ATM*); ~ **de transferencia síncrona** *m* TELECOM synchronous transfer mode (*STM*); ~ **de transmisión** *m* ELECTRÓN transmission mode; ~ **transversal eléctrico** *m* ÓPT, TELECOM transverse electric mode (*TE mode*); ~ **de utilización** *m* TELECOM mode of operation; ~ **de varios destinos** *m* TELECOM multidestination mode; ~ **vertical** *m* TV vertical mode; ~ **de vibración fundamental** *m* ACÚST fundamental vibration mode; ~ **de vibración natural** *m* ACÚST, TEC ESP natural mode of vibration; ~ **de vibración propio** *m* ACÚST, TEC ESP natural mode of vibration

modos: ~ **acoplados** *m pl* ACÚST, ÓPT, TEC ESP, TELECOM coupled modes; ~ **de desintegración** *m pl* FÍS RAD decay modes; ~ **de desintegración prohibidos** *m pl* FÍS RAD forbidden decay modes; ~ **multilongitudinales** *m pl* TELECOM multilongitudinal modes (*MLM*); ~ **resonantes** *m pl* FÍS RAD resonant modes; ~ **de respuesta normal** *m pl* TELE-COM normal-response modes (*NRM*)

modulación *f* GEN modulation; ~ **acústico-óptica** *f* ELECTRÓN acousto-optic modulation; ~ **adaptable diferencial de impulsos en código** *f* (*ADPCM*) TELECOM adaptive differential pulse coded modulation (*ADPCM*); ~ **de alto nivel** *f* ELECTRÓN high-level modulation; ~ **de amplitud** *f* (*AM*) GEN amplitude modulation (*AM*); ~ **de amplitud constante** *f* ELECTRÓN constant amplitude modulation; ~ **de amplitud en cuadratura** *f* (*MAC*) ELECTRÓN, INFORM&PD, TELECOM quadrature amplitude modulation (*QAM*); ~ **de amplitud imprevista** *f* ELECTRÓN incidental amplitude modulation; ~ **de amplitud de impulso** *f* (*PAM*) ELECTRÓN, INFORM&PD, TEC ESP, TELECOM pulse-amplitude modulation (*PAM*); ~ **analógica** *f* ELECTRÓN, FÍS analog modulation; ~ **de la anchura del impulso** *f* (*PWM*) ELECTRÓN, TEC ESP pulse-width modulation (*PWM*); ~ **angular** *f* ELECTRÓN angle modulation; ~ **del ánodo** *f* ELECTRÓN anode modulation; ~ **de baja distorsión** *f* ELECTRÓN low-distortion modulation; ~ **de bajo nivel** *f* ELECTRÓN low-level modulation; ~ **de banda ancha** *f* ELECTRÓN wideband modulation; ~ **de banda lateral doble** *f* ELECTRÓN double sideband modulation; ~ **de banda lateral independiente** *f* ELECTRÓN independent sideband modulation; ~ **de base** *f* ELECTRÓN base modulation; ~ **binaria** *f* TELECOM binary modulation; ~ **de brillo** *f* ELECTRÓN brightness modulation; ~ **de canal** *f* ELECTRÓN channel modulation; ~ **de cátodo** *f* ELECTRÓN cathode modulation; ~ **por códigos de pulsos** *f* (*PCM*) ELECTRÓN, FÍS, FÍS RAD, INFORM&PD, TEC ESP, TELECOM, TV pulse-code modulation (*PCM*); ~ **del color** *f* TV color modulator (*AmE*), colour modulator (*BrE*); ~ **compuesta** *f* ELECTRÓN compound modulation; ~ **de conductividad** *f* ELECTRÓN conductivity modulation; ~ **continua de intensidad** *f* TV continuous beam modulation; ~ **controlada de la portadora** *f* ELECTRÓN controlled carrier modulation; ~ **de corriente** *f* ELECTRÓN current modulation; ~ **de corriente constante** *f* ELEC, ELECTRÓN constant-current modulation; ~ **cruzada** *f* ELECTRÓN cross modulation, TELECOM cross-modulation, TV cross modulation; ~ **de datos** *f* TELECOM data modulation; ~ **delta** *f* INFORM&PD, TEC ESP, TELECOM delta modulation; ~ **delta compandida** *f* TELECOM companded delta modulation (*CDM*); ~ **delta de pendiente variable** *f* TEC ESP variable-slope delta modulation; ~ **de densidad** *f* ELECTRÓN density modulation; ~ **descendente** *f* ELECTRÓN downward modulation; ~ **por desplazamiento de frecuencia** *f* TEC ESP frequency-shift signaling (*AmE*), frequency-shift signalling (*BrE*); ~ **por desviación de fase** *f* (*PSK*) INFORM&PD, TEC ESP phase-shift keying modulation (*PSK*); ~ **diferencial** *f* ELECTRÓN differential modulation; ~ **diferencial de código pulsado** *f* (*MDCP*) ELECTRÓN differential pulse code modulation (*DPCM*); ~ **digital** *f* ELECTRÓN, FÍS, TELECOM digital modulation; ~ **de diodo** *f* ELECTRÓN diode modulation; ~ **de diodo NIP** *f* ELECTRÓN PIN-diode modulation; ~ **directa** *f* ELECTRÓN direct modulation; ~ **por división de tiempo** *f* (*TDM*) FÍS, INFORM&PD, TEC ESP, TELECOM time-division modulation (*TDM*); ~ **doble** *f* ELECTRÓN double modulation; ~ **Doppler** *f* ELECTRÓN Doppler mod-

ulation; ~ **de la duración del impulso** *f* ELECTRÓN pulse-time modulation, pulse-duration modulation; ~ **por duración de los impulsos** *f* (*PDM*) TEC ESP pulse-duration modulation (*PDM*); ~ **de empuje** *f* TEC ESP thrust modulation; ~ **espacial** *f* ELECTRÓN spatial modulation; ~ **de espectro expandido** *f* TELECOM spread-spectrum modulation; ~ **externa** *f* ELECTRÓN external modulation; ~ **de fase** *f* (*MF*) ELECTRÓN, FÍS, INFORM&PD, TELECOM, TV phase modulation (*PM*); ~ **de fase digital** *f* TELECOM digital phase modulation; ~ **de fase de impulso** *f* ELECTRÓN pulse-phase modulation; ~ **del foco** *f* CINEMAT, FOTO, TV focus modulation; ~ **de frecuencia** *f* ELEC, ELECTRÓN, FÍS, FÍS ONDAS, FÍS PART, INFORM&PD, TELECOM, TV frequency modulation; ~ **de frecuencia de banda estrecha** *f* (*MFBE*) ELECTRÓN, TELECOM narrow-band frequency modulation (*NBFM*); ~ **de frecuencia directa** *f* ELECTRÓN direct frequency modulation; ~ **de frecuencia imprevista** *f* ELECTRÓN incidental frequency modulation; ~ **de frecuencia de impulsos** *f* ELECTRÓN, INFORM&PD pulse frequency modulation (*PFM*); ~ **de frecuencia indirecta** *f* ELECTRÓN indirect frequency modulation; ~ **de frecuencia modificada** *f* (*MFM*) ELECTRÓN modified frequency modulation (*MFM*); ~ **de frecuencia regularizadora** *f* TELECOM tamed frequency modulation; ~ **de frecuencia residual** *f* ELECTRÓN residual frequency modulation; ~ **de frecuencia de subportadora** *f* TELECOM subcarrier frequency modulation (*SFM*); ~ **de grilla** *f* AmL (*cf modulación de rejilla Esp*) TV grid modulation; ~ **por haz láser** *f* ELECTRÓN laser-beam modulation; ~ **imprevista** *f* ELECTRÓN incidental modulation; ~ **del impulso** *f* ELECTRÓN pulse modulation; ~ **de impulso cuantificado** *f* ELECTRÓN, TELECOM quantized-pulse modulation; ~ **de impulsos** *f* ELEC *regulación*, ELECTRÓN pulse modulation; ~ **de impulsos en amplitud** *f* (*PAM*) ELECTRÓN, INFORM&PD, TEC ESP, TELECOM pulse-amplitude modulation (*PAM*); ~ **por impulsos codificados** *f* (*PCM*) ELECTRÓN, FÍS, FÍS RAD, INFORM&PD, TEC ESP, TELECOM, TV pulse-code modulation (*PCM*); ~ **de impulsos cuantificados** *f* ELECTRÓN, TELECOM quantization pulse modulation; ~ **de impulsos en duración** *f* (*PDM*) TEC ESP pulse-duration modulation (*PDM*); ~ **por impulsos de duración variable** *f* TELECOM pulse-time modulation; ~ **por impulsos de frecuencia variable** *f* TEC ESP chirp modulation; ~ **de impulsos en posición** *f* TELECOM pulse-time modulation; ~ **de la intensidad** *f* ELECTRÓN intensity modulation; ~ **invertida** *f* ELECTRÓN inverse modulation; ~ **lineal** *f* ELECTRÓN linear modulation; ~ **de la luz** *f* ELECTRÓN light modulation; ~ **MCI-FM** *f* ELECTRÓN PCM-FM modulation; ~ **mecánica** *f* ELECTRÓN mechanical modulation; ~ **múltiple** *f* ELECTRÓN multiple modulation; ~ **negativa** *f* ELECTRÓN negative modulation; ~ **no coherente por barrido de frecuencia** *f* TELECOM noncoherent swept-tone modulation; ~ **de onda portadora** *f* FÍS ONDAS carrier-wave modulation; ~ **de onda sinusoidal** *f* ELECTRÓN sine wave modulation; ~ **de onda transmisora** *f* FÍS ONDAS carrier-wave modulation; ~ **óptica** *f* ELECTRÓN optical modulation; ~ **por la palabra** *f* TELECOM speech modulation, voice control; ~ **de portadora** *f* ELECTRÓN carrier modulation; ~ **por portadora**

flotante *f* ELECTRÓN floating-carrier modulation; ~ **de portadora variable** *f* ELECTRÓN variable-carrier modulation; ~ **de la posición del impulso** *f* ELECTRÓN pulse-position modulation; ~ **por posición de impulsos** *f* TEC ESP pulse-position modulation; ~ **positiva** *f* ELECTRÓN positive modulation; ~ **de la propagación del espectro** *f* ELECTRÓN spread-spectrum modulation; ~ **de pulsos en intervalo** *f* TV pulse-interval modulation; ~ **de rejilla** *f* (*cf modulación de grilla AmL*) ELECTRÓN, TV grid modulation; ~ **sin retorno** *f* TELECOM nonreturn modulation; ~ **de una sola banda lateral** *f* ELECTRÓN single-sideband modulation; ~ **sonora** *f* ACÚST sound modulation; ~ **de la subportadora** *f* TV subcarrier modulation; ~ **de surco compatible** *f* ACÚST compatible groove modulation; ~ **por tiempo de impulsos** *f* TELECOM time pulse modulation (*TPM*); ~ **por tiempos** *f* ELECTRÓN time modulation; ~ **triangular** *f* ELECTRÓN delta modulation; ~ **por variación de fase** *f* (*PSK*) IMFORM&PD, TEC ESP phase-shift keying modulation (*PSK*); ~ **a varios niveles** *f* ELECTRÓN multilevel modulation; ~ **de velocidad** *f* ELECTRÓN, FÍS, TV velocity modulation; ~ **vocal de banda angosta** *f* TELECOM narrow-band voice modulation (*NBVM*)

modulado: **no** ~ *adj* TEC ESP unmodulated; ~ **en amplitud** *adj* ELECTRÓN amplitude-modulated; ~ **en frecuencia** *adj* ELECTRÓN frequency-modulated

modulador *m* ELECTRÓN modulator, wobbulator, INFORM&PD, TELECOM, TV modulator; ~ **absorbente** *m* ELECTRÓN absorptive modulator; ~ **acústico-óptico** *m* ELECTRÓN, ÓPT acousto-optic modulator, acousto-optical modulator; ~ **de amplitud de la cuadratura** *m* ELECTRÓN, INFORM&PD, TELECOM quadrature amplitude modulator; ~ **de audio** *m* ELECTRÓN audio modulation; ~ **de banda lateral doble** *m* ELECTRÓN double sideband modulator; ~ **de color** *m* TV color modulator (*AmE*), colour modulator (*BrE*); ~ **compensado** *m* ELECTRÓN balanced modulator; ~**demodulador** *m* (*módem*) INFORM&PD, TELECOM modulator-demodulator (*modem*); ~ **de una dirección** *m* ELEC single-way modulator; ~ **electroóptico** *m* ELECTRÓN, ÓPT electro-optical modulator; ~ **de fase** *m* ELECTRÓN, FÍS, INFORM&PD, TELECOM phase modulator; ~ **final** *m* ELECTRÓN final modulator; ~ **de frecuencia** *m* ELEC, ELECTRÓN, FÍS, FÍS ONDAS, FÍS PART, INFORM&PD, TELECOM, TV frequency modulator; ~ **de impulsos** *m* ELECTRÓN pulse modulator; ~ **de lámpara de vacío** *m* ELECTRÓN vacuum tube modulator; ~ **lineal** *m* ELECTRÓN linear modulator; ~ **de luz** *m* ACÚST light modulator; ~ **de microondas** *m* TELECOM microwave modulator; ~ **óptico** *m* ELECTRÓN, TELECOM optical modulator; ~ **de la propagación del espectro** *m* ELECTRÓN spread-spectrum modulator; ~ **de serrodina** *m* ELECTRÓN serrodyne modulator; ~ **de una sola banda lateral** *m* ELECTRÓN single-sideband modulator; ~ **de sonido** *m* ELECTRÓN audio modulation; ~ **de subportador de cromeado** *m* ELECTRÓN chrominance subcarrier modulator; ~ **de transistor** *m* ELECTRÓN transistor modulator; ~ **de tubo de vacío** *m* ELECTRÓN vacuum tube modulator; ~ **de una vía** *m* ELEC single-way modulator

modular[1] *adj* INFORM&PD, MATEMÁT modular
modular[2] *vt* ELECTRÓN, FÍS, TELECOM, TV modulate

modularidad *f* CALIDAD, INFORM&PD modularity
módulo *m* GEN module, FÍS, MATEMÁT, PETROL modulus, TELECOM bay; ~ **de alargamiento** *m* P&C modulus of elongation; ~ **aparente de elasticidad** *m* P&C apparent modulus of elasticity; ~ **asistido** *m* TEC ESP manned module; ~ **aterrizador** *m* TEC ESP lander; ~ **aterrizador automático** *m* TEC ESP unmanned lander; ~ **aterrizador no pilotado** *m* TEC ESP unmanned lander; ~ **de aterrizaje** *m* TEC ESP lander; ~ **de calibración** *m* TRANSP AÉR calibration module; ~ **codificador** *m* PROD encoder module; ~ **de compresión** *m* P&C compression modulus; ~ **de control de línea** *m* PROD line drive connector; ~ **de corte** *m* FÍS, ING MECÁ, TEC PETR shear modulus; ~ **de deformación** *m* CARBÓN deformation modulus, CONST strain modulus; ~ **de descenso** *m* TEC ESP lander; ~ **de descenso automático** *m* TEC ESP unmanned lander; ~ **de descenso no pilotado** *m* TEC ESP unmanned lander; ~ **de dos ranuras** *m* PROD two-slot module; ~ **de elasticidad** *m* CARBÓN, CONST, FÍS, ING MECÁ, MECÁ, METAL, P&C modulus of elasticity, TEC PETR modulus of elasticity, stretch modulus, TRANSP AÉR, TRANSP MAR modulus of elasticity; ~ **de elasticidad cúbica** *m* FÍS bulk modulus; ~ **de elasticidad longitudinal** *m* METAL Young's modulus; ~ **de elasticidad volumétrico** *m* ING MECÁ bulk modulus of elasticity; ~ **electrolítico** *m* ELEC, FÍS, ING ELÉC electrolytic unit; ~ **electrónico** *m* ELECTRÓN electronic module, INFORM&PD board; ~ **de elongación** *m* P&C modulus of elongation; ~ **enchufable** *m* ING ELÉC plug-in module; ~ **de E/S** *m* PROD I/O module; ~ **para escape de la tripulación** *m* TEC ESP crew escape module; ~ **de estado sólido** *m* PROD solid-state module; ~ **para experimentación** *m* TEC ESP experiment module; ~ **de extensión** *m* INFORM&PD expansion card; ~ **de extensión del sistema** *m* PROD system-expander module; ~ **extensor** *m* PROD expander module; ~ **de habitación** *m* TEC ESP habitation module; ~ **de imagen por cristal líquido** *m* (*IML*) ELEC, ELECTRÓN, FÍS, INFORM&PD, TELECOM, TV liquid crystal display module (*LCD module*); ~ **inteligente** *m* INFORM&PD smart card; ~ **de interfaz** *m* ELECTRÓN, TEC ESP, TELECOM interface module (*IM*); ~ **de interfaz de líneas** *m* TELECOM line interface module; ~ **latente** *m* METAL latent modulus; ~ **de líneas** *m* TELECOM line module; ~ **de mando** *m* TEC ESP command module; ~ **de mando y servicio** *m* (*MMS*) TEC ESP command and service module; ~ **de memoria** *m* INFORM&PD, PROD memory module; ~ **de memoria RAM de CMOS** *m* PROD CMOS RAM memory module; ~ **de memoria de una sola fila alineada** *m* (*SIMM*) INFORM&PD single in-line memory module (*SIMM*); ~ **de microondas** *m* ELECTRÓN microwave module; ~ **objeto** *m* INFORM&PD object module; ~ **periférico** *m* TELECOM peripheral module; ~ **de portadora digital** *m* TELECOM digital carrier module; ~ **de potencia** *m* ING ELÉC, PROD power module; ~ **de ranura simple** *m* PROD single-slot module; ~ **de ranurado métrico evolvente** *m* ING MECÁ involute splinemetric module; ~ **de reacción del subsuelo** *m* CARBÓN subgrade reaction modulus; ~ **de regleta de terminales** *m* PROD terminal-strip module; ~ **del retículo** *m* NUCL lattice pitch; ~ **de rigidez** *m* CARBÓN rigidity modulus, FÍS modulus of rigidity, rigidity modulus;

~ **RTD remoto** *m* PROD remote RTD module; ~ **de servicio de la tripulación** *m* TEC ESP crew service module (*CSM*); ~ **simple** *m* PROD singled module; ~ **simple de memoria en línea** *m* (*SIMM*) INFORM&PD single in-line memory module (*SIMM*); ~ **de tarjeta** *m* ELECTRÓN card module; ~ **de transporte síncrono** *m* TELECOM synchronous transport module (*STM*); ~ **tripulado** *m* TEC ESP manned module; ~ **de VCL** *m* ELECTRÓN LCD module; ~ **de visualización por cristal líquido** *m* (*IML*) ELEC, ELECTRÓN, FÍS, INFORM&PD, TELECOM, TV liquid crystal display module (*LCD module*); ~ **de volumen** *m* PETROL bulk modulus; ~ **volumétrico** *m* FÍS bulk modulus; ~ **de voz** *m* PROD speech module; ~ **de Young** *m* CARBÓN, CONST, FÍS, ING MECÁ, MECÁ, METAL, P&C modulus of elasticity, Young's modulus, PETROL Young's modulus, TEC PETR, TRANSP AÉR modulus of elasticity, Young's modulus, TRANSP MAR modulus of elasticity; ~**-n de transporte síncrono** *m* TELECOM synchronous transport module-n (*STM-n*)

modus: ~ **operandi** *m* PROD modus operandi

mogotes *m pl* OCEAN hummocks

mofeta *f* CARBÓN firedamp, ENERG RENOV mofette, ING MECÁ gas spring, MINAS choke damp, firedamp, TERMO firedamp

mofle: ~ **catalítico** *m* AmL (*cf amortiguador de sonido catalítico Esp*) CONTAM catalytic muffler (*AmE*), catalytic silencer (*BrE*)

moho *m* AGRIC mold (*AmE*), mould (*BrE*), ALIMENT mildew, TRANSP MAR mold (*AmE*), mould (*BrE*); ~ **gris** *m* AGRIC gray mold (*AmE*), grey mould (*BrE*); ~ **de la hoja** *m* AGRIC leaf mold (*AmE*), leaf mould (*BrE*); ~ **negro** *m* AGRIC black mold (*AmE*), black mould (*BrE*); ~ **polvoriento** *m* ALIMENT powdery mildew; ~ **velloso** *m* ALIMENT downy mildew

mojado[1] *adj* TERMO wet

mojado[2]: ~ **de carga** *m* C&V batch wetting; ~ **del material seco** *m* TEXTIL wetting of dry material

mojadura *f* ING MECÁ wetting

mojar *vt* TERMO moisten, wet

mojón *m* CONST boundary, reference mark, boundary mark, *topografía* landmark, monument, MECÁ milestone, PETROL Boundstone, TRANSP milestone

mol *f* AGUA, FÍS, METR, QUÍMICA mole (*mol*)

MOL *abr* (*laboratorio orbital tripulado*) TEC ESP MOL (*manned orbiting laboratory*)

molal *adj* QUÍMICA molal

molalidad *f* QUÍMICA molality

molar *adj* QUÍMICA molar

molaridad *f* QUÍMICA molarity

molde *m* CONST form, ING MECÁ die, die-casting die, former, mold (*AmE*), mould (*BrE*), INSTR former, MECÁ frame, OCEAN mesh pin, P&C *parte de equipo* mold (*AmE*), mould (*BrE*), PROD *fundería* mold (*AmE*), mould (*BrE*), *encofrados de hormigón* form, former, mold (*AmE*), mould (*BrE*), TRANSP MAR *construcción naval* mold (*AmE*), mould (*BrE*); ~ **de acero** *m* CONST steel form; ~ **de aflojamiento rápido** *m* ING MECÁ, PROD quick-release die; ~ **alargador** *m* ING MECÁ stretch die; ~ **de arena** *m* PROD sand mold (*AmE*), sand mould (*BrE*); ~ **de arena al descubierto** *m* PROD open-sand molding (*AmE*), open-sand moulding (*BrE*); ~ **de autocentrado** *m* ING MECÁ self-centering dies (*AmE*), self-centring dies (*BrE*); ~ **de barro refractario** *m* C&V fireclay

mold (*AmE*), fireclay mould (*BrE*); ~ **caliente** *m* C&V hot mold (*AmE*), hot mould (*BrE*); ~ **de caucho** *m* ING MECÁ, P&C rubber mold (*AmE*), rubber mould (*BrE*); ~ **para caucho** *m* ING MECÁ mold for rubber (*AmE*), mould for rubber (*BrE*); ~ **a la cera perdida para colada** *m* ING MECÁ investment mold for casting (*AmE*), investment mould for casting (*BrE*); ~ **de cera perecedero** *m* ING MECÁ lost-wax mold for casting (*AmE*), lost-wax mould for casting (*BrE*); ~ **para chapas de figura** *m* PROD horse; ~ **de colada** *m* P&C casting mold (*AmE*), casting mould (*BrE*); ~ **de colada a la cera perdida** *m* ING MECÁ investment-casting die; ~ **de collarín** *m* CONST neck mold (*AmE*), neck molding (*AmE*), neck mould (*BrE*), neck moulding (*BrE*); ~ **de compresión** *m* ING MECÁ, P&C compression mold (*AmE*), compression mould (*BrE*); ~ **para cristalería** *m* ING MECÁ mold for glassware (*AmE*), mould for glassware (*BrE*); ~ **para curvar** *m* C&V bending mold (*AmE*), bending mould (*BrE*); ~ **deslizante** *m* CONST slip form; ~ **de doble cavidad** *m* C&V double cavity mold (*AmE*), double cavity mould (*BrE*); ~ **de dos fases** *m* ING MECÁ transfer mold (*AmE*), transfer mould (*BrE*); ~ **empastado** *m* C&V paste mold (*AmE*), paste mould (*BrE*); ~ **encapsulador** *m* ING MECÁ encapsulation mould; ~ **de endurecimiento térmico** *m* ING MECÁ thermosetting mold (*AmE*), thermosetting mould (*BrE*); ~ **de espuma estructural** *m* ING MECÁ structural foam mold (*AmE*), structural foam mould (*BrE*); ~ **para espuma estructural** *m* ING MECÁ mold for structural foam (*AmE*), mould for structural foam (*BrE*); ~ **de extrusión** *m* P&C extrusion die; ~ **de fijación por pernos** *m* ING MECÁ unscrewing core; ~ **de forjado en frío** *m* ING MECÁ, PROD, TERMO cold-forging die; ~ **en forma de C** *m* INSTR C-shaped frame; ~ **para formar por vacío** *m* ING MECÁ vacuum-forming mold (*AmE*), vacuum-forming mould (*BrE*); ~ **de fraguado térmico** *m* ING MECÁ thermosetting mold (*AmE*), thermosetting mould (*BrE*); ~ **frío** *m* C&V cold mold (*AmE*), cold mould (*BrE*); ~ **de fundición** *m* PROD box, TRANSP MAR mold (*AmE*), mould (*BrE*); ~ **para fundición** *m* ING MECÁ mold for casting (*AmE*), mould for casting (*BrE*); ~ **de fundición inyectada por gravedad** *m* ING MECÁ gravity die-casting die; ~ **para fundición por presión** *m* ING MECÁ pressure-mould for casting; ~ **de gravedad para fundición** *m* ING MECÁ gravity mold for casting (*AmE*), gravity mould for casting (*BrE*); ~ **para hielo** *m* REFRIG ice can; ~ **para hornear** *m* PROD baking mold (*AmE*), baking mould (*BrE*); ~ **de impresión múltiple** *m* ING MECÁ, P&C multi-impression mold (*AmE*), multi-impression mould (*BrE*); ~ **de impresión única** *m* ING MECÁ single impression mold (*AmE*), single impression mould (*BrE*); ~ **inclinable** *m* C&V tilting mold (*AmE*), tilting mould (*BrE*); ~ **por insuflación de aire comprimido** *m* ING MECÁ blow-mold (*AmE*), blow-mould (*BrE*); ~ **de inyección** *m* C&V injection mold (*AmE*), injection mould (*BrE*); ~ **de inyección para caucho** *m* ING MECÁ, P&C injection mold for rubber (*AmE*), injection mould for rubber (*BrE*); ~ **de inyección para goma** *m* ING MECÁ, P&C injection mold for rubber (*AmE*), injection mould for rubber (*BrE*); ~ **de inyección para resinas** *m* ING MECÁ injection mold for thermoplastics (*AmE*), injection mould for thermoplastics (*BrE*); ~ **de**

inyección para resinas termoestables *m* ING MECÁ injection mold for thermosetting resins (*AmE*), injection mould for thermosetting resins (*BrE*); **~ de inyección para resinas termofraguables** *m* ING MECÁ injection mold for thermosetting resins (*AmE*), injection mould for thermosetting resins (*BrE*); **~ de inyección para resinas termoindurantes** *m* ING MECÁ injection mold for thermosetting resins (*AmE*), injection mould for thermosetting resins (*BrE*); **~ de inyección para termoplásticos** *m* ING MECÁ injection mold for thermoplastics (*AmE*), injection mould for thermoplastics (*BrE*); **~ de laminación** *m* ING MECÁ, P&C lamination mold (*AmE*), lamination mould (*BrE*); **~ de lingote** *m* PROD pig mold (*AmE*), pig mould (*BrE*); **~ de madera** *m* TRANSP MAR wooden mold (*AmE*), wooden mould (*BrE*), wooden plug; **~ para materiales minerales** *m* ING MECÁ mold for mineral materials (*AmE*), mould for mineral materials (*BrE*); **~ para moldeo a la cera perdida** *m* ING MECÁ lost-wax mold for casting (*AmE*), lost-wax mould for casting (*BrE*); **~ de operación rápida** *m* P&C flash mold (*AmE*), flash mould (*BrE*); **~ de panadero** *m* ING MECÁ confectionery mold (*AmE*), confectionery mould (*BrE*); **~ de partes con pernos** *m* ING MECÁ unscrewing parts mold (*AmE*), unscrewing parts mould (*BrE*); **~ partido** *m* P&C *parte de equipo* split mold (*AmE*), split mould (*BrE*); **~ de la pasta** *m* C&V body mold (*AmE*), body mould (*BrE*); **~ de piezas con pernos** *m* ING MECÁ unscrewing parts mold (*AmE*), unscrewing parts mould (*BrE*); **~ para plásticos** *m* ING MECÁ mold for plastics (*AmE*), mould for plastics (*BrE*); **~ para plásticos endurecidos** *m* ING MECÁ mold for thermoset plastics (*AmE*), mould for thermoset plastics (*BrE*); **~ para plásticos termofraguados** *m* ING MECÁ mold for thermoset plastics (*AmE*), mould for thermoset plastics (*BrE*); **~ de poliester reforzado con fibra de vidrio** *m* C&V, ING MECÁ, MECÁ glass-reinforced polyester mold (*AmE*), glass-reinforced polyester mould (*BrE*); **~ para poliester reforzado con fibra de vidrio** *m* ING MECÁ mold for glass-reinforced polyester (*AmE*) (*mold for GRP*), mould for glass-reinforced polyester (*BrE*) (*mould for GRP*); **~ portátil** *m* P&C portable mold (*AmE*), portable mould (*BrE*); **~ positivo** *m* P&C positive mold (*AmE*), positive mould (*BrE*); **~ para productos alimenticios** *m* ING MECÁ mold for food products (*AmE*), mould for food products (*BrE*); **~ que pierde por las grietas** *m* PROD *fundería* running mould (*AmE*), running mould (*BrE*); **~ que produce más de una pieza por ciclo de moldeo** *m* ING MECÁ multi-impression hot runner mold (*AmE*), multi-impression hot runner mould (*BrE*); **~ quemado** *m* C&V burnt mold (*AmE*), burnt mould (*BrE*); **~ de ranura** *m* IMPR slot die; **~ rayado** *m* C&V scratched mold (*AmE*), scratched mould (*BrE*); **~ de rebaba** *m* P&C flash mold (*AmE*), flash mould (*BrE*); **~ rotativo** *m* ING MECÁ, PROD rotational mold (*AmE*), rotational mould (*BrE*); **~ semipositivo** *m* P&C semipositive mold (*AmE*), semipositive mould (*BrE*); **~ de succión** *m* C&V suction mold (*AmE*), suction mould (*BrE*); **~ termofraguante** *m* ING MECÁ thermosetting mold (*AmE*), thermosetting mould (*BrE*); **~ termoplástico** *m* ING MECÁ thermoplastic mold (*AmE*), thermoplastic mould (*BrE*); **~ para**

termoplásticos *m* ING MECÁ mold for thermoplastics (*AmE*), mould for thermoplastics (*BrE*); **~ de tierra** *m* PROD loam mold (*AmE*), loam mould (*BrE*); **~ de transferencia** *m* ING MECÁ transfer die, transfer mold (*AmE*), transfer mould (*BrE*); **~ de triple cavidad** *m* C&V triple-cavity mold (*AmE*), triple-cavity mould (*BrE*); **~ de vacío** *m* ING MECÁ vacuum mold (*AmE*), vacuum mould (*BrE*); **~ de varias partes** *m* C&V split mold (*AmE*), split mould (*BrE*)

moldeado[1]: **no ~** *adj* C&V not molded (*AmE*), not moulded (*BrE*)

moldeado[2] *m* ACÚST molding (*AmE*), moulding (*BrE*), MECÁ forming, P&C *procesamiento* forming, molding (*AmE*), moulding (*BrE*), PROD, REVEST forming; **~ de collarín** *m* CONST neck mold (*AmE*), neck molding (*AmE*), neck mould (*BrE*), neck moulding (*BrE*); **~ en dos fases** *m* P&C transfer molding (*AmE*), transfer moulding (*BrE*); **~ por estiramiento** *m* P&C *fibras* drawing; **~ por extrusión** *m* ING MECÁ extrusion die-casting; **~ en frío** *m* C&V cold molding (*AmE*), cold moulding (*BrE*), ING MECÁ cold hobbing, P&C cold molding (*AmE*), cold moulding (*BrE*); **~ en hueco** *m* P&C slush molding (*AmE*), slush moulding (*BrE*); **~ por inyección de aire** *m* P&C blast forming; **~ de inyección para goma** *m* ING MECÁ injection mold for rubber (*AmE*), injection mould for rubber (*BrE*); **~ de inyección para resinas** *m* ING MECÁ injection mold for thermoplastics (*AmE*), injection mould for thermoplastics (*BrE*); **~ de inyeccion para termoplásticos** *m* ING MECÁ injection mold for thermoplastics (*AmE*), injection mould for thermoplastics (*BrE*); **~ de película por colada** *m* P&C film casting; **~ con placa intermedia** *m* PROD plate molding (*AmE*), plate moulding (*BrE*); **~ a presión** *m* P&C pressure-forming; **~ rotativo de fibras y barnizado** *m* EMB rotary molding of fiber and rollercoat varnishing (*AmE*), rotary moulding of fibre and rollercoat varnishing (*BrE*); **~ por soplado y extrusión** *m* EMB extrusion-blow molding (*AmE*), extrusion-blow moulding (*BrE*); **~ por transferencia** *m* P&C transfer molding (*AmE*), transfer moulding (*BrE*); **~ por vacío** *m* P&C vacuum forming

moldeador *m* PROD molder (*AmE*), moulder (*BrE*); **~ con espuma de poliestireno y polietileno** *m* EMB foamed polystyrene and polyethylene molder (*AmE*), foamed polystyrene and polyethylene moulder (*BrE*)

moldeadora: **~ basculante de sacudidas** *f* PROD *fundería* roll-over draft machine (*AmE*), roll-over draught machine (*BrE*); **~ por inyección** *f* EMB injection blow molding machine (*AmE*), injection blow moulding machine (*BrE*); **~ mecánica** *f* PROD *fundería* squeezer

moldear *vt* C&V *barro* mould (*BrE*), mold (*AmE*), ING MECÁ *metales*, PROD cast

moldeo *m* MECÁ casting, P&C, PROD, TRANSP MAR molding (*AmE*), moulding (*BrE*); **~ en arena** *m* PROD sand casting; **~ en caja** *m* PROD flask molding (*AmE*), flask moulding (*BrE*); **~ en cajas** *m* PROD box molding (*AmE*), box moulding (*BrE*); **~ por capilaridad** *m* PROD capillary molding (*AmE*), capillary moulding (*BrE*); **~ a la cera perdida** *m* PROD lost-wax casting; **~ por compresión** *m* C&V, EMB *procesamiento* compression moulding (*BrE*), compression molding (*AmE*), ING MECÁ compression mould (*BrE*), compression mold (*AmE*), P&C compression mold (*AmE*), *procesamiento* compression moulding (*BrE*), compression mould (*BrE*), compression molding (*AmE*); ~ **por contacto** *m* C&V contact molding (*AmE*), contact moulding (*BrE*); ~ **en coquilla** *m* PROD chill mold (*AmE*), chill mould (*BrE*); ~ **por estirado** *m* P&C drawing; ~ **por insuflación de aire comprimido** *m* AmL (*cf moldeo por soplado Esp*) P&C blow-molding (*AmE*), blow-moulding (*BrE*); ~ **por inyección** *m* P&C, PROD injection molding (*AmE*), injection moulding (*BrE*); ~ **por inyección-soplado** *m* P&C injection blow molding (*AmE*), injection blow moulding (*BrE*); ~ **con machos** *m* ING MECÁ coring; ~ **de piezas batidas** *m* PROD false core molding (*AmE*), false core moulding (*BrE*); ~ **por prensado** *m* C&V, EMB, P&C compression moulding (*AmE*), compression moulding (*BrE*); ~ **rotacional** *m* P&C slush molding (*AmE*), slush moulding (*BrE*); ~ **sobre placa-modelo** *m* PROD card molding (*AmE*), card moulding (*BrE*); ~ **por soplado** *m* Esp (*cf moldeo por insuflación de aire comprimido AmL*) P&C blow-molding (*AmE*), blow-moulding (*BrE*); ~ **a la terraja** *m* PROD sweep molding (*AmE*), sweep moulding (*BrE*); ~ **a terraja** *m* PROD strickle molding (*AmE*), strickle moulding (*BrE*); ~ **por vacío** *m* P&C *procesamiento* vacuum forming

moldero *m* C&V mold maker (*AmE*), mould maker (*BrE*)

moldes: ~ **a la cera** *m pl* ING MECÁ wax investment molds (*AmE*), wax investment moulds (*BrE*); ~ **planos de grandes dimensiones** *m pl* ING MECÁ large plate molds (*AmE*), large plate moulds (*BrE*)

moldura *f* CONST bead, molding (*AmE*), moulding (*BrE*); ~ **de cromo** *f* VEH chrome strip; ~ **de plinto** *f* CONST *arquitectura* cap

molécula *f* GEN molecule; ~ **activada** *f* FÍS RAD activated molecule; ~ **diatómica** *f* FÍS, FÍS RAD, QUÍMICA diatomic molecule; ~ **polar** *f* P&C polar molecule; ~**-gramo** *f* METR gram molecule, gramme molecule

moler *vt* ALIMENT pound, CARBÓN grind, mill, ING MECÁ, MECÁ, PROC QUÍ, PROD grind, QUÍMICA pulverize, grind

molestia: ~ **progresiva** *f* OCEAN *buceo* niggles; ~ **sonora** *f* ACÚST noise nuisance

moleta *f* ING MECÁ knurl, knurling tool, cutter wheel, MECÁ, PAPEL knurling tool, PROD muller; ~ **cortante** *f* ING MECÁ cutting wheel; ~ **de repicar** *f* PAPEL burr; ~ **de tornero** *f* PROD nurling tool

moleteado[1] *adj* MECÁ knurled

moleteado[2] *m* C&V knurling, ING MECÁ knurl, knurling, PROD milling

moleteador *m* ING MECÁ knurl

moleteadora *f* ING MECÁ knurling tool

moletear *vt* MECÁ, PROD mill

molibdato *m* QUÍMICA molybdate

molibdenita *f* MINERAL molybdenite

molibdeno *m* (*Mo*) AGRIC, METAL, QUÍMICA molybdenum (*Mo*)

molibdita *f* MINERAL molybdic ocher (*AmE*), molybdic ochre (*BrE*), molybdite

molido *m* QUÍMICA grinding; ~ **óptimo** *m* CARBÓN optimum grind

molienda *f* AGRIC milling, ALIMENT milling, milling industry, CARBÓN crushing, milling, grinding, MINAS grinding, P&C crushing, PROC QUÍ milling, QUÍMICA grinding, milling; ~ **autógena** *f* CARBÓN autogenous milling; ~ **de bolas** *f* CARBÓN ball milling; ~ **en circuito cerrado** *f* CARBÓN closed-circuit grinding; ~ **excesiva** *f* CARBÓN overgrinding; ~ **fina** *f* CARBÓN fine grinding; ~ **húmeda** *f* AGRIC wet milling; ~ **en húmedo** *f* CARBÓN wet grinding; ~ **primaria** *f* CARBÓN primary crusher, primary grinding, MINAS primary crushing; ~ **en seco** *f* CARBÓN dry crushing; ~ **secundaria** *f* CARBÓN secondary crusher, secondary grinding; ~ **terciaria** *f* CARBÓN tertiary crushing

molinete *m* CONST gin, gin wheel, TRANSP MAR windlass; ~ **hidráulico** *m* HIDROL current meter; ~ **hidrométrico** *m* AGUA current meter; ~ **regulador de paletas** *m* PROD fly

molinillo *m* ALIMENT grinder

molino *m* GEN mill; ~ **abierto** *m* P&C open mill; ~ **aceitero** *m* AGRIC oil mill; ~ **de amalgamación** *m* PROD grinding pan; ~ **de amolar** *m* PROC QUÍ grinding mill; ~ **aplastador** *m* AGRIC roller-rusher, roller mill; ~ **de arroz** *m* AGRIC, ALIMENT rice mill; ~ **autógeno** *m* CARBÓN autogenous mill; ~ **de barras** *m* CARBÓN rod mill; ~ **de barro** *m* C&V clay mill; ~ **de bocartes** *m* CARBÓN stamp mill, PROD stamp, stamp mill, stamping mill; ~ **de bolas** *m* ALIMENT (*cf triturador de bolas*) ball mill, C&V (*cf triturador de bolas*) *para moler barro fino* clay mill, ball mill, CARBÓN (*cf triturador de bolas*), LAB (*cf desintegrador de bolas*, *cf pulverizador de bolas*), MINAS (*cf triturador de bolas*), P&C (*cf triturador de bolas*) ball mill, PROC QUÍ (*cf triturador de bolas*) ball mill, milling ball; ~ **de bolas vibratorias** *m* LAB vibrating-ball mill; ~ **de bolos** *m* MINAS, PROD pebble mill; ~ **de cascada** *m* CARBÓN cascade mill; ~ **de cemento** *m* CONST cement maker, cement mill; ~ **centrífugo** *m* PROC QUÍ centrifugal mill, centrifuge mill; ~ **de cilindros** *m* ING MECÁ rolls; ~ **coloidal** *m* ALIMENT colloid mill; ~ **de coque** *m* CARBÓN coke mill; ~ **desintegrador** *m* PROD disintegrator; ~ **de discos** *m* ALIMENT, CARBÓN disc mill (*BrE*), disk mill (*AmE*); ~ **de dos rodillos** *m* P&C open mill; ~ **de Fuller-Bonot** *m* CARBÓN Fuller-Bonot mill; ~ **de granos** *m* AGRIC grain mill; ~ **de guijarros** *m* CARBÓN, PROD pebble mill; ~ **para hacer pienso** *m* AGRIC feed mill; ~ **harinero** *m* AGRIC, ALIMENT flourmill; ~ **de machaqueo de piedras** *m* CONST stone mill; ~ **mareomotriz** *m* ENERG RENOV tide mill; ~ **de martillo** *m* AGRIC mill hammer, ALIMENT hammer mill; ~ **mezclador** *m* PROD mixing mill; ~ **de molienda en fino** *m* PROC QUÍ fine-crushing mill; ~ **de muelas** *m* AGRIC grinding mill, PAPEL edge runner, kollergang; ~ **de piedras** *m* CARBÓN pebble mill, PAPEL kollergang; ~ **de piedras de sílex** *m* CARBÓN pebble mill; ~ **para pintura** *m* C&V paint mill; ~ **de pisones** *m* CARBÓN stamp mill, PROD mill, stamp, stamp mill, stamping mill; ~ **de polvo** *m* ALIMENT dust mill; ~ **de pólvora** *m* PROD powder mill, powder works; ~ **de rodillos** *m* ALIMENT roller mill; ~ **en seco** *m* CARBÓN dryer mill; ~ **de tres bocartes** *m* MINAS three-stamp mill; ~ **de tres pisones** *m* MINAS three-stamp mill; ~ **tubular** *m* CARBÓN tube mill; ~ **de vidrio** *m* C&V cullet crusher; ~ **para vidrio** *m* C&V glaze grinder

molinos *m pl* CARBÓN grinding media; ~ **de compartimientos** *m pl* CARBÓN, MINERAL compartment mill

molón *m* PAPEL beater roll

molturación *f* CARBÓN, MINAS grinding, milling; ~ **autógena** *f* CARBÓN autogenous milling; ~ **de bolas** *f* CARBÓN ball milling; ~ **en circuito cerrado** *f* CARBÓN closed-circuit grinding; ~ **fina** *f* CARBÓN fine grinding

molturar *vt* CARBÓN mill, ING MECÁ, MECÁ, PROD *trigo* grind

molysita *f* MINERAL molysite

momento *m* GEN moment; ~ **de adrizamiento** *m* TRANSP MAR *arquitectura naval* righting moment; ~ **alrededor de un eje** *m* FÍS moment about an axis; ~ **de amortiguación** *m* TRANSP AÉR damping moment; ~ **angular** *m* FÍS, FÍS PART, MECÁ, TRANSP AÉR angular momentum; ~ **angular intrínseco** *m* FÍS PART intrinsic angular momentum; ~ **angular orbital** *m* FÍS, FÍS RAD orbital angular momentum; ~ **angular de spin** *m* FÍS spin angular momentum; ~ **angular total** *m* FÍS total angular momentum; ~ **anular** *m* TRANSP AÉR annular momentum; ~ **de balanceo** *m* TRANSP AÉR rolling moment; ~ **de batimiento** *m* TRANSP AÉR flapping moment; ~ **de cabeceo** *m* ENERG RENOV pitching moment; ~ **de charnela** *m* TRANSP AÉR hinge moment; ~ **de conexión** *m* ELECTRÓN turn-on time; ~ **cuadripolar** *m* FÍS quadrupole moment; ~ **cuadrupolar nuclear** *m* FÍS, NUCL nuclear quadrupole moment; ~ **dipolar** *m* ING ELÉC dipole moment, QUÍMICA dipolar moment; ~ **dipolar eléctrico** *m* ELEC, FÍS, ING ELÉC electric dipole moment; ~ **del dipolo eléctrico** *m* ELEC, FÍS, ING ELÉC electric dipole moment; ~ **eléctrico cuadripolar** *m* ELEC, FÍS, FÍS RAD, ING ELÉC quadrupole electric momentum; ~ **electromagnético** *m* ELEC electromagnetic moment, electromagnetic momentum, FÍS electromagnetic momentum; ~ **de empotramiento** *m* CONST fixed-end moment; ~ **de encendido** *m* VEH point; ~ **enderezador** *m* TRANSP AÉR restoring moment; ~ **de escora** *m* TRANSP MAR heeling moment; ~ **escorante** *m* TRANSP MAR heeling moment; ~ **de FI** *m* ELECTRÓN IF stage; ~ **flector** *m* CONST, PROD, TRANSP MAR bending moment; ~ **flector de arrufo** *m* TRANSP MAR sagging; ~ **de flexión** *m* FÍS bending moment; ~ **de una fuerza** *m* PROD torque; ~ **de giro** *m* CONST overturning moment; ~ **de guiñada** *m* ENERG RENOV, TRANSP AÉR yawing moment; ~ **de inercia** *m* CONST, FÍS, TRANSP MAR moment of inertia; ~ **de inercia de la pala** *m* TRANSP AÉR blade moment of inertia; ~ **instantáneo del dipolo eléctrico** *m* FÍS RAD instantaneous electric dipole momentum; ~ **lineal** *m* FÍS momentum; ~ **magnético** *m* ELEC, FÍS magnetic moment, FÍS PART magnetic momentum; ~ **magnético dipolar** *m* FÍS magnetic dipole moment; ~ **magnético del electrón** *m* ELECTRÓN, FÍS, NUCL electron magnetic moment; ~ **magnético del muón** *m* FÍS RAD muon magnetic momentum; ~ **de máxima** *m* ELECTRÓN peaking; ~ **máximo de flexión** *m* TEC ESP maximum bending moment; ~ **orbital** *m* FÍS RAD orbital momentum; ~ **de par** *m* ING MECÁ torque, MECÁ moment; ~ **de par de arranque** *m* ING MECÁ starting torque; ~ **de resistencia aerodinámica** *m* TRANSP AÉR drag moment; ~ **de rotura** *m* MECÁ failure moment; ~ **de torsión** *m* ING MECÁ torque, twisting moment, PROD torque; ~ **de torsión de la pala** *m* TRANSP AÉR blade-twisting moment; ~ **torsional** *m* ING MECÁ twisting moment; ~ **torsional en condiciones ambientales** *m* TEC ESP environmental torque; ~ **torsional constante del motor** *m* PROD constant motor torque; ~ **torsor** *m* ING MECÁ torque, twisting moment, MECÁ, TEC PETR torque

monacolato: ~ **de glicerilo** *m* ALIMENT, QUÍMICA glyceryl monacoleate

monádico *adj* INFORM&PD monadic

monazita *f* MINERAL monazite

monergol: ~ **líquido** *m* TERMO liquid monopropellant

monheimita *f* MINERAL monheimite

monitor *m* ELEC, FÍS RAD, INFORM&PD, INSTR, TELECOM, TV monitor; ~ **aéreo** *m* TV air monitor; ~ **de aire** *m* TV air monitor; ~ **de ambiente** *m* TV air monitor; ~ **de cámara** *m* CINEMAT camera monitor, TV camera monitor, preview monitor; ~ **de color** *m* TEC ESP color display (*AmE*), colour display (*BrE*); ~ **de doble columna** *m* TV dual-standard monitor; ~ **de efluentes** *m* AGUA, RECICL effluent monitor; ~ **de elementos de combustible defectuosos** *m* NUCL failed fuel-element monitor; ~ **de emisiones** *m* AGUA, RECICL effluent monitor; ~ **encadenado** *m* TV slave unit, slaving; ~ **del estudio** *m* TV studio monitor; ~ **de fondo** *m* TV background monitor; ~ **de forma de onda** *m* TV waveform monitor; ~ **de frecuencia** *m* TV frequency monitor; ~ **de imagen** *m* TV picture monitor; ~ **de imagen y configuración de onda** *m* TV image and waveform monitor; ~ **de imagen y forma de onda** *m* TV image and waveform monitor; ~ **de línea** *m* TV line monitor; ~ **maestro** *m* TV master monitor; ~ **de polvo** *m* NUCL dust monitor; ~ **principal** *m* TV master monitor; ~ **de programa** *m* TV program monitor (*AmE*), programme monitor (*BrE*); ~ **que no transmite** *m* TV off air monitor; ~ **de radiación** *m* FÍS RAD, NUCL, TEC ESP radiation monitor; ~ **de rango de fuente** *m* NUCL source range monitor; ~ **de rango de potencia** *m* NUCL power-range monitor; ~ **RGB** *m* INFORM&PD, TV RGB monitor; ~ **de rojo verde azul** *m* INFORM&PD, TV RGB monitor; ~ **de salida** *m* TV output monitor; ~ **de seis canales** *m* INSTR six-channel monitor; ~ **de seis pistas** *m* INSTR six-channel monitor; ~ **del sistema de satélites** *m* (*SSM*) TEC ESP satellite system monitor (*SSM*); ~ **en tiempo real** *m* ELECTRÓN, INFORM&PD, TELECOM real-time monitor; ~ **de voltaje de entrada** *m* PROD incoming voltage monitor

monitorado *m* ACÚST monitoring

monitorear *vt* INFORM&PD monitor

monitoreo *m* PROD, TV monitoring; ~ **de blanco-negro** *m* TV black-white monitoring

monitorización *f* TELECOM monitoring, monitoring and maintenance; ~ **a distancia** *f* INFORM&PD, TELECOM, TV remote monitoring; ~ **in situ** *f* NUCL in situ monitoring; ~ **de la radiación** *f* FÍS RAD, TEC ESP radiation monitoring; ~ **de radiación** *f* NUCL radiation monitoring

mono *m* SEG industrial overalls (*BrE*), coveralls (*AmE*); ~ **de trabajo** *m* *Esp* (*cf overoles AmL*) PROD overalls, SEG coveralls (*AmE*), industrial overalls (*BrE*), overalls (*BrE*)

monoacetina *f* QUÍMICA monoacetin

monoácido[1] *adj* QUÍMICA monoacidic

monoácido[2] *m* QUÍMICA monoacid

monoalcohólico *adj* QUÍMICA monoalcoholic

monoamida *f* QUÍMICA monoamide

monoamina *f* QUÍMICA monoamine

monoamino *adj* QUÍMICA monoamino

monoatómico *adj* QUÍMICA monoatomic
monoaural *adj* ACÚST monaural
monoauricular *adj* ACÚST monaural
monobaño *m* CINEMAT monobath
monobásico *adj* QUÍMICA monobasic
monobloque: ~ **en serie** *m* ELECTRÓN single in-line package (*SIP*)
monocapa *f* QUÍMICA monolayer
monocarril CONST, FERRO, MINAS, TRANSP monorail
monocasco *adj* TEC ESP monocoque, TRANSP MAR single-hull
monoclinal *m* GEOL monocline
monoclínico *adj* CRISTAL, QUÍMICA monoclinic
monocordio *m* ACÚST *música* monochord
monocristal *m* CRISTAL, ELECTRÓN single crystal; ~ **semiconductor** *m* ELECTRÓN semiconductor single crystal
monocromador *m* FÍS, ÓPT, TELECOM monochromator; ~ **de cuarzo** *m* NUCL quartz monochromator; ~ **mecánico de neutrones** *m* NUCL mechanical chopper
monocromático *adj* CINEMAT, FÍS, IMPR, ÓPT monochromatic
monocromía *f* CINEMAT monochrome
monocromo[1] *adj* IMPR, INFORM&PD monochrome
monocromo[2] *m* CINEMAT monochrome
monocultivo *m* AGRIC monoculture, one-crop farming
monoenergético *adj* FÍS monoenergetic
monoestable[1] *adj* ELECTRÓN, FÍS, INFORM&PD monostable
monoestable[2] *m* ELECTRÓN, FÍS monostable
monoestearina *f* QUÍMICA monostearin
monoestrato *m* QUÍMICA monolayer
monoetilénico *adj* QUÍMICA monoethylenic
monofásico *adj* ELEC *conductor* uniphase, *fuente de alimentación* monophase, *red de distribución* single-phase, ING ELÉC one-phase, single-phase, uniphase
monofilamento *m* P&C monofilament, TEXTIL *hilo* monofilament yarn
monogenético *adj* GEOL monogenetic
monohalogenado *adj* QUÍMICA monohalogenated
monohidratado *adj* QUÍMICA monohydrated
monohidrato *m* QUÍMICA monohydrate
monohídrico *adj* QUÍMICA monohydric
monoico *adj* AGRIC monoecious
monomérico *adj* P&C, QUÍMICA, TEC PETR monomeric
monómero *m* P&C, QUÍMICA, TEC PETR monomer
monomicta *adj* GEOL, MINERAL monomineralic
monomíctico *adj* GEOL monomict
monomineralógico *adj* GEOL, MINERAL monomineralic
monomio *m* MATEMÁT monomial
monopolar *adj* ELEC *terminal* single-pole, *fuente de alimentación* monopolar, ELECTRÓN one-pole, ING ELÉC unipolar
monopolo: ~ **magnético** *m* FÍS magnetic monopole
monopropulsante *m* QUÍMICA monopropellant; ~ **líquido** *m* TERMO liquid monopropellant
monopropulsor *m* QUÍMICA monopropellant
monorraíl *m* CONST, FERRO, MINAS, TRANSP monorail; ~**-alforja** *m* TRANSP saddlebag monorail
monorranura *f* PROD single slot
monosacárido *m* QUÍMICA monosaccharide
monosacarosas *f pl* QUÍMICA monosaccharoses
monosubstituido *adj* QUÍMICA monosubstituted
monosustituido *adj* QUÍMICA monosubstituted

monotrón *m* ELECTRÓN monotron
monotrópica *adj* METAL monotropic
monovalencia *f* QUÍMICA univalence, monovalence, monovalency
monovalente *adj* QUÍMICA univalent, monovalent
monóxido *m* QUÍMICA monoxide; ~ **de carbono** *m* CARBÓN, CONTAM, ING MECÁ, QUÍMICA, VEH *gases de escape* carbon monoxide
montabarcos *m* TRANSP boat lift
montabebedero *m* PROD *funderías* gate spool
montacarga: ~ **de cadena** *f* ING MECÁ chain block
montacargas *m* AGRIC, CONST, MECÁ elevator (*AmE*), lift (*BrE*), hoist, PROD goods lift, hoist, SEG, TRANSP MAR hoist; ~ **de cadena** *m* ING MECÁ chain hoist; ~ **eléctrico** *m* ELEC, MECÁ electric hoist; ~ **de freno de fricción** *m* PROD friction-brake hoist; ~ **mecánico de mano** *m* PROD hand-power warehouse goods lift; ~ **para servicio pesado** *m* CONST heavy-duty lift; ~ **para vagones** *m* FERRO car elevator (*AmE*), wagon hoist, wagon lift (*BrE*)
montado *adj* ING MECÁ, MECÁ fixed; ~ **sobre cojinetes sin rozamiento** *adj* ING MECÁ mounted on frictionless bearings; ~**-cremallera** *adj* ING ELÉC rack-mounted; ~ **en la fábrica** *adj* PROD factory-assembled; ~ **a mano** *adj* EMB hand-assembled; ~ **sobre pivote** *adj* MECÁ pivoted; ~ **sobre railes** *adj* TRANSP rail mounted; ~ **en tándem** *adj* PROD tandem-mounted
montador *m* CINEMAT editor, FERRO erector, ING MECÁ, MECÁ fitter, TV editor; ~ **de correa** *m* PROD belt mounter; ~ **de estructuras metálicas** *m* CONST steel fixer; ~ **de grandes cajas** *m* EMB large-case erector; ~ **de imágenes** *m* IMPR image setter; ~ **de tuberías** *m* CONST pipefitter, pipelayer
montaje[1]: **de** ~ *adj* CINEMAT, TV editing; **de** ~ **completo** *adj* ING MECÁ, PROD complete-assembly; **de** ~ **trasero** *adj* AGRIC rear-mounted
montaje[2] *m* ACÚST *cine* dubbing, CARBÓN setting, CINEMAT editing, CONST *estructuras* erection, setup, assembly, ELEC assembly, ELECTRÓN, FOTO mounting, GAS assembly, IMPR editing, assembly, ING ELÉC connection, ING MECÁ assembly, setting, fitment, fit, assembling, fitting, MECÁ fitting, assembly, PROD *de equipos y maquinaria* mounting, erection, setup, installation, TEC ESP assembly building, TV assemble edit, editing, VEH assembly; ~ **de aguja** *m* PROD switch assembly; ~ **al aire** *m* ING MECÁ chucking, faceplate mounting; ~ **del alojamiento del eje trasero** *m* AUTO, MECÁ, VEH rear-axle housing assembly; ~ **altazimutal** *m* INSTR altazimuth mounting; ~ **de amortiguador de pedal** *m* TRANSP AÉR pedal-damper assembly; ~ **antigolpes** *m* TEC ESP shock mount; ~ **del arrancador** *m* AUTO starter drive assembly; ~ **automático** *m* EMB automatic assembly work; ~ **axial de tipo inglés** *m* INSTR English-type axis mounting; ~ **azimut-elevación** *m* TEC ESP Az-El mount; ~ **azul** *m* IMPR blue key; ~ **de barra de puesta a tierra** *m* ELEC, ING ELÉC, PROD earth-bus mounting (*BrE*), ground-bus mounting (*AmE*); ~ **en batería** *m* PROD battery assembly; ~ **de bayoneta** *m* FOTO bayonet mount; ~ **del brazo basculante** *m* AUTO, VEH rocker-arm assembly; ~ **de bridas** *m* ING MECÁ flanged union; ~ **sobre bridas** *m* PROD flange mounting; ~ **para buques** *m* TRANSP MAR *antenas, reflectores* ship-type rig; ~ **de cable óptico** *m* TELECOM optical cable assembly; ~ **de cables** *m* ÓPT,

TELECOM cable assembly; ~ **con código de tiempo** *m* CINEMAT, TV time-code editing; ~ **combinado** *m* IMPR nesting form; ~ **de continuidad** *m* CINEMAT, TV continuity editing; ~ **cuádruple de trípode** *m* FOTO fourfold tripod stand; ~ **de la cuna** *m* D&A cradle mounting; ~ **declinación ángulo horario** *m* TEC ESP Ha-Dec mount; ~ **definitivo** *m* CINEMAT, TV final cut; ~ **en delta** *m* ING MECÁ delta fitting; ~ **de detalle** *m* CINEMAT, TV fine cut; ~ **del disco conducido** *m* AUTO driven plate assembly; ~ **de dos B más D** *m* TELECOM two-B-plus-D arrangement; ~ **del elemento frontal** *m* FOTO mount of front element; ~ **embutido** *m* ELEC *de conmutador* flush mounting, FOTO countersunk mount, countersunk setting, PROD flush mounting; ~ **empernado** *m* PROD bolt mounting; ~ **empotrado** *m* ELEC *de conmutador* flush mounting; ~ **encastrado** *m* FOTO sunk mount; ~ **de enfoque** *m* CINEMAT, TV focusing mount; ~ **entre puntos** *m* ING MECÁ chucking between centers (*AmE*), chucking between centres (*BrE*); ~ **en estrella** *m* ELEC *transformador, reactor* star connection, Y-connection; ~ **de estructuras metálicas** *m* CONST steel fixing; ~ **exterior** *m* ELEC surface mounting; ~ **externo** *m* PROD external set-up; ~ **final** *m* NUCL, PROD final assembly; ~ **flexible** *m* ING MECÁ flexible mounting; ~ **forzado** *m* ING MECÁ drive fit; ~ **a frotamiento dulce** *m* ING MECÁ push fit; ~ **a frotamiento duro** *m* ING MECÁ drive fit; ~ **de fuerza** *m* ING MECÁ, MECÁ force fit; ~ **de grupos de agujas** *m* PROD switch group assembly; ~ **Ha-Dec** *m* TEC ESP Ha-Dec mount; ~ **en herradura** *m* INSTR horseshoe mount, horseshoe mounting; ~ **en horquilla** *m* INSTR fork mounting; ~ **hundido** *m* FOTO sunk mount; ~ **intercalado** *m* CINEMAT, TV crosscutting; ~ **interior** *m* ING ELÉC indoor wiring; ~ **del izador** *m* TRANSP AÉR hoist fitting; ~ **a mano** *m* ING MECÁ handed assembly; ~ **manual** *m* FOTO manual cocking, ING MECÁ handed assembly; ~ **de máquinas con envase plegadizo** *m* EMB folding-box erecting machine; ~ **por medio de escenas de unión** *m* CINEMAT, TV intercutting; ~ **móvil** *m* INSTR mobile mounting; ~ **sobre original** *m* CINEMAT, TV editing on original; ~ **de originales** *m* IMPR paste-up; ~ **en panel** *m* ELEC, PROD panel mounting; ~ **a paño** *m* PROD flush mount; ~ **en paralelo** *m* ELEC parallel mounting, ING ELÉC paralleling; ~ **de película** *m* TELECOM assemble edit; ~ **sobre perno** *m* PROD stud mounting; ~ **de una pieza sobre el plato** *m* MECÁ chucking; ~ **de piso** *m* ING MECÁ floor-mounting; ~ **sobre polos** *m* PROD pole-mounting; ~ **preliminar** *m* CINEMAT, TV rough cut; ~ **de presión** *m* FOTO push-on mount; ~ **del propulsor del arrancador** *m* AUTO starter drive assembly; ~ **en puente** *m* ELEC bridge connection; ~ **del puente trasero** *m* AUTO, MECÁ, VEH rear-axle assembly; ~ **de punto a punto** *m* ING ELÉC point-to-point wiring; ~ **rápido** *m* CINEMAT, TV flash cutting (*AmE*), tight editing (*BrE*); ~ **de refractarios en un horno** *m* PROD setting; ~ **de refuerzo** *m* INSTR knee mounting; ~ **relacional** *m* CINEMAT, TV relational editing; ~ **de Rowland** *m* FÍS Rowland mounting; ~ **en serie** *m* ELEC *circuito* series connection, series mounting, ING ELÉC series arrangement, MINAS nest; ~ **del sistema** *m* PROD system setup; ~ **de sujeción** *m* ING MECÁ fixture; ~ **de superficie** *m* ELEC, INFORM&PD surface mounting; ~ **superior** *m* ENERG RENOV top assembly; ~ **sobre**

tacos elásticos *m* VEH rubber mounting; ~ **de termistancia** *m* ING ELÉC thermistor mount; ~ **de tipo germánico** *m* INSTR German-type mounting; ~ **de la torre de sondeo** *m* TEC PETR *perforación* rigging-up; ~ **del transportador de hilos** *m* TEXTIL yarn carrier assembly; ~ **en triángulo** *m* ELEC mesh connection, ING ELÉC delta connection; ~ **de verificación** *m* ING MECÁ checking fixture; ~ **vertical** *m* CONST crown post

montaje[3]: ~ **contra el desmontaje** *fra* TEC ESP strap-down-mounted

montaña: ~ **submarina** *f* OCEAN sea mount

montante *m* CARBÓN strut, CONST string board, jamb post, strut, transom, stile, *carpintería* sash bar, upright, *ventana, puerta* jamb, *puerta* post, ING MECÁ upright, cheek, housing, standard, PROD bearer, TRANSP AÉR rib, VEH *carrocería, motor* strut; ~ **ascendente** *m* CONST riser; ~ **del castillete de extracción** *m* AmL (*cf bastidor del castillete de extracción Esp*) MINAS head frame; ~ **de encuentro** *m* AGUA meeting post; ~ **de giro** *m* AGUA quoin post; ~ **de hierro** *m* CONST sash bar iron; ~ **de puerta** *m* VEH door pillar; ~ **de refuerzo** *m* ING MECÁ stiffener

montar[1] *vt* CINEMAT cock, edit, CONST assemble, join up, *particiones* set up, ELECTRÓN mount, FOTO *el obturador* cock, INFORM&PD set, mount, ING MECÁ set, fit, fit to, OCEAN *arte de pesca* set up, mount, PETROL rig up, PROD mount, rig, TRANSP MAR *puerta* clear, install, TV edit, set up; ~ **en** *vt* ING MECÁ fit in; ~ **con cabría** *vt* PROD rig; ~ **al extremo de** *vt* ING MECÁ fit on; ~ **a horcajadas** *vt* PROD straddle; ~ **inicialmente** *vt* CINEMAT, TV assemble edit; ~ **en seco** *vt* FOTO dry-mount; ~ **según pedido** *vt* PROD assemble to order

montar[2] ~ **originales** *vi* IMPR paste up; ~ **tuberías** *vi* CONST pipe

montavagones *m* FERRO wagon hoist, *vehículos* car elevator (*AmE*), wagon lift (*BrE*)

montea *f* TRANSP MAR lofting

monteador *m* TRANSP MAR loftsman

montebrasita *f* MINERAL montebrasite

montera *f* CARBÓN overburden, GEOL overplacement, LAB *de alambique* head, MINAS strippings, overburden, overplacement, *geología* cap; ~ **de gas** *f* TEC PETR *geología* gas cap; ~ **de oxidación** *f* GEOL gossan

monticellita *f* MINERAL monticellite

montículo *m* CONST barrow; ~ **alargado de grava y arena estratificado** *m* GEOL beaded esker

montículos *m pl* OCEAN hummocks

montmorillonita *f* CARBÓN, MINERAL, TEC PETR montmorillonite

montón *m* CARBÓN pile, CONST *de polvo* ridge, MINAS pillar, PROD hoisting tackle, TEXTIL pile; ~ **de abono vegetal** *m* AGRIC, RECICL compost heap; ~ **de almacenamiento** *m* INFORM&PD heap; ~ **exterior** *m* MINAS tip heap (*BrE*); ~ **fijo** *m* TRANSP MAR standing block; ~ **de finos** *m* MINAS slack heap; ~ **de tostación** *m* PROD open heap

montura *f* CINEMAT, FOTO mount, ING MECÁ fitment, holder; ~ **de bayoneta** *f* *Esp* (*cf casquillo de bayoneta AmL*) CINEMAT bayonet mount, ELEC *ampolla, bombilla* bayonet cap, FOTO bayonet base, bayonet socket, ING ELÉC bayonet base; ~ **C** *f* CINEMAT C-mount; ~ **de la cámara** *f* CINEMAT, FOTO, TV camera mount; ~ **de gafas** *f* INSTR spectacle frame; ~ **del objetivo** *f* CINEMAT mount, FOTO lens mount;

~ plástica *f* FOTO plastic mount; **~ registradora** *f* CINEMAT register mount; **~ de rosca** *f* CINEMAT screw mount; **~ por succión** *f* CINEMAT suction mount; **~ en superficie** *f* INFORM&PD surface mounting; **~ Tyler** *f* CINEMAT Tyler mount

monzón *m* METEO monsoon

monzonita *f* PETROL monzonite

moon: **~ pool** *m* TEC PETR *perforación costa-fuera* moon pool

morcilla *f* PROD horseshoe main

mordaza *f* CONST lion's claw, ING MECÁ chop, grip, holdfast, jaw, clip, vice cap (*BrE*), vise cap (*AmE*), MECÁ clamp, clip, jaw, NUCL jaw, PROD grip, TRANSP clamp, TRANSP MAR *equipamiento de barcos* cam cleat, VEH clamp; **~ de apriete** *f* PROD gripping jaws, gripping pad; **~ de cable** *f* MECÁ cable clamp; **~ de cable para armaduras** *f* AmL (*cf grapa para armaduras Esp*) ELEC armor clamp (*AmE*), armour clamp (*BrE*); **~ del cable de toma de corriente** *f* ELEC, ING ELÉC current lead cleat (*AmE*), mains lead cleat (*BrE*); **~ de cables metálicos** *f* PROD wire rope clamp; **~ de cocodrilo** *f* ING ELÉC, ING MECÁ crocodile clip; **~ de freno** *f* ING MECÁ brake jaw; **~ de guía** *f* ING MECÁ guide clamp; **~ del plato** *f* ING MECÁ faceplate dog, faceplate jaw; **~ revestida** *f* ING MECÁ lined clamp; **~ de sujeción** *f* PROD gripping jaws; **~ del tornillo** *f* ING MECÁ vice jaw (*BrE*), vise jaw (*AmE*); **~ de varillaje** *f* TEC PETR casing clamp

mordazas *f pl* ING MECÁ chaps, cheeks; **~ acanaladas** *f pl* MINAS *machacadora de piedra* corrugated jaws; **~ de bancada** *f pl* ING MECÁ bed bars; **~ desmontables** *f pl* ING MECÁ detachable jaws; **~ postizas** *f pl* ING MECÁ inserted jaws, detachable jaws; **~ de prensa** *f pl* IMPR vice jaws (*BrE*), vise jaws (*AmE*); **~ de quita y pon** *f pl* ING MECÁ inserted jaws; **~ de retención** *f pl* PROD nippers

mordedura *f* ING MECÁ bite, biting

morder *vt* ING MECÁ bite, enter, grip, bite into, *laminadoras* pinch, TRANSP MAR *ancla* grip

mordido: **~ húmedo** *f* IMPR dampening etch

mordiente *m* ING MECÁ biting, QUÍMICA mordant

mordisco *m* MINAS nip

morenosita *f* MINERAL morenosite

morfina *f* QUÍMICA morphine

morfolina *f* DETERG, QUÍMICA morpholine

morfométrico *adj* GEOL morphometric

morfotropía *f* QUÍMICA morphotropy

morfotrópico *adj* QUÍMICA morphotropic

morfotropismo *m* QUÍMICA morphotropism

morillo *m* INSTAL TERM firedog

morina *f* QUÍMICA morin

morindina *f* QUÍMICA morindin

morión *m* MINERAL morion

morrena *f* CARBÓN, GEOL moraine, MINAS drift

morro[1] *m* OCEAN *litoral* bluff, *de espigón, de escollera* pierhead, PROD *avión* nose, TRANSP MAR *de espigón* head; **~ de la escollera** *m* TRANSP MAR pierhead; **~ Krueger** *m* TRANSP AÉR droop nose; **~ de malecón** *m* OCEAN mole head

morro[2]: **con el ~ levantado** *fra* TRANSP AÉR in a nose-up attitude

morsa: **~ giratoria** *f* ING MECÁ swivel vice (*BrE*), swivel vise (*AmE*); **~ de máquina** *f* ING MECÁ machine vice (*BrE*), machine vise (*AmE*); **~ de máquina herramienta** *f* ING MECÁ machine vice (*BrE*), machine vise (*AmE*)

mortaja *f* CONST, PROD mortise; **~ ciega** *f* CONST stub mortise; **~ pasante** *f* CONST through-mortice

mortajado *m* ING MECÁ grooving

mortajadora *f* ING MECÁ, PROD mortising machine; **~ con acción de herramienta oscilante** *f* ING MECÁ mortising machine with oscillating tool action; **~ de excéntrica** *f* ING MECÁ eccentric drive slotting machine; **~ de manivela** *f* ING MECÁ crank drive slotting machine; **~ para muescas** *f* PROD keyway-cutting machine

mortero *m* CARBÓN slurry, CONST grout, mortar, D&A, LAB, MINAS, PROC QUÍ mortar, PROD *bocarte* mortar, mortar box, QUÍMICA mortar; **~ de ágata** *m* QUÍMICA agate mortar; **~ sin amalgamación interior** *m* AmL (*cf mortero concentrado Esp*) MINAS concentration mortar; **~ antisonoro** *m* REVEST acoustic plaster; **~ de asiento** *m* CONST bedding mortar; **~ de barro** *m* C&V clay mortar; **~ de cal** *m* PROD lime mortar; **~ de cal grasa** *m* CONST fat mortar; **~ concentrado** *m* Esp (*cf mortero sin amalgamación interior AmL*) MINAS concentration mortar; **~ de percusión** *m* LAB percussion mortar

morvenita *f* MINERAL morvenite

morvina *f* QUÍMICA morvin

mosaico *m* AGRIC *enfermedad*, ALIMENT *fitopatología* mosaic, C&V tessera, CONST *pavimentación* tile, ELECTRÓN mosaic; **~ de análisis de manchas** *m* LAB spotting tile; **~ cerámico para pavimentar** *m* C&V ceramic pavement slab; **~ insonizador** *m* SEG sound-proof tile; **~ de mayólica** *m* C&V majolica tile; **~ de vidrio** *m* C&V glass mosaic

mosandrita *f* MINERAL mosandrite

mosca: **~ de las cebollas** *f* AGRIC onion maggot; **~ de establo** *f* AGRIC stable fly; **~ minadora** *f* AGRIC leaf miner

moscovita *f* MINERAL muscovite

mosquete: **~ de rueda** *m* TERMO fire lobby, firelock

mosquetón *m* ING MECÁ spring governor; **~ con sacavueltas** *m* TRANSP MAR snap shackle

mosquita *f* AGRIC midge; **~ de agalla** *f* AGRIC gall midge

mosquitera *f* TEXTIL insect screen, mosquito net

mosquito *m* AGRIC gnat

mosquitocida *m* QUÍMICA mosquitocide

mostaza: **~ negra** *f* AGRIC rutabaga, *malezas* black mustard

mostrador *m* REFRIG display case; **~ de paquetes** *m* TRANSP parcels counter

mostrar *vt* INFORM&PD display, ING MECÁ hold up

mota *f* TEXTIL blotch

motas *f pl* PAPEL specks; **~ de corteza** *f pl* Esp (*cf puntos negros AmL*) PAPEL bark specks

moteado[1] *adj* IMPR mottled

moteado[2] *m* FOTO spotting

motear *vt* PAPEL, TEXTIL speckle

motivo *m* INFORM&PD pattern; **~ del estampado** *m* TEXTIL body of the print; **~ de falla humana** *m* AmL (*cf motivo de fallo humano Esp*) CALIDAD human failure cause; **~ de fallo humano** *m* Esp (*cf motivo de falla humana AmL*) CALIDAD human failure cause; **~ de retención** *m* PROD reason for hold; **~ de tramado** *m* INFORM&PD *gráficos* dithering

motocicleta: **~ de gran capacidad** *f* TRANSP large-capacity motorcycle; **~ de gran cilindrada** *f* AUTO heavyweight motorcycle; **~ ligera con arranque a pedal** *f* AUTO light motorcycle with kickstarter

motocrucero: ~ **con cámara** *m* TRANSP MAR cabin cruiser

motón *m* ING MECÁ clamp, pulley, TRANSP MAR block, pulley, *de roldana* pulley block; ~ **de briol** *m* TRANSP MAR fiddle block; ~ **giratorio** *m* ING MECÁ swivel block, PROD monkey block; ~ **móvil** *m* PETROL traveling block (*AmE*), travelling block (*BrE*); ~ **de quijada** *m* TRANSP MAR cheek block; ~ **de rabiza** *m* PROD tail block; ~ **de reenvío** *m* TRANSP MAR lead block, leading block; ~ **de retorno** *m* TRANSP MAR lead block, leading block

motonave *f* TRANSP MAR motor ship, motor vessel

motoniveladora *f* AUTO bulldozer, CONST leveling motor (*AmE*), levelling motor (*BrE*), motor grader, MINAS bulldozer, TRANSP motor grader; ~ **de carreteras** *f* AUTO road grader; ~ **de cuchilla basculante** *f* TRANSP tiltdozer

motor¹: ~ **de <1 HP** *abr* (*motor de potencia menor de un caballo*) ELEC FHP motor (*fractional horsepower motor*)

motor² *m* AUTO motor, engine, ELEC, ING ELÉC motor, ING MECÁ motor, mover, drive, driving, power unit, engine, MECÁ engine, motor, TEC ESP motor, TRANSP AÉR, TRANSP MAR, VEH engine;

~ a ~ **de accionamiento** *m* CINEMAT, FOTO drive motor; ~ **de aceite pesado** *m* AUTO heavy-oil engine; ~ **acorazado** *m* ELEC shell-type motor, totally-enclosed motor; ~ **aerobio** *m* TRANSP AÉR air-breathing engine; ~ **de aire** *m* TRANSP AÉR air motor; ~ **de aire caliente** *m* TRANSP hot-air engine; ~ **de aire caliente comprimido** *m* CONST, ING MECÁ compressed-air engine; ~ **de aire comprimido** *m* ING MECÁ air motor, pneumatic motor, MINAS compressed-air motor; ~ **de aire comprimido rotativo** *m* ING MECÁ rotary pneumatic engine; ~ **de alimentación universal** *m* ELEC all-current motor (*AmE*), all-mains motor (*BrE*); ~ **de alta tensión** *m* ELEC high-voltage motor; ~ **de alta velocidad** *m* ELEC high-speed motor; ~ **alternativo** *m* ING MECÁ, TRANSP AÉR, TRANSP MAR, VEH reciprocating engine; ~ **alternativo de combustión interna** *m* ING MECÁ, TRANSP, TRANSP AÉR, TRANSP MAR, VEH reciprocating internal-combustion engine; ~ **alternativo con turbina de gases de escape** *m* ING MECÁ compound engine; ~ **de anillo colector** *m* ELEC slip ring motor; ~ **de anillo rozante** *m* ELEC slip ring motor; ~ **antideflagrante** *m* ING ELÉC, TERMO fireproof motor, flameproof motor; ~ **de apogeo** *m* TEC ESP apogee motor; ~ **de apoyo giratorio** *m* ELEC swivel-bearing motor; ~ **de armadura de barras** *m* ELEC squirrel cage motor; ~ **de armadura de jaula** *m* ELEC squirrel cage motor; ~ **de armadura de jaula de ardilla** *m* ELEC squirrel cage motor; ~ **de arranque** *m* Esp (*cf burro de arranque AmL*) AUTO starter motor, starter, ELEC starting motor, ING ELÉC, ING MECÁ, VEH starter motor; ~ **de arranque por aire comprimido** *m* ING MECÁ air starter; ~ **de arranque y arrastre con capacitor** *m* ELEC, ING ELÉC capacitor start-and-run motor; ~ **de arranque con capacitor** *m* ELEC, ING ELÉC capacitor start motor; ~ **de arranque por condensador** *m* ELEC, ING ELÉC capacitor start motor; ~ **de arranque por engranajes deslizantes** *m* ELEC sliding-gear starting motor; ~ **de arranque por fase auxiliar** *m* ELEC split-phase motor; ~ **de arranque y funcionamiento con condensador** *m* ELEC, ING ELÉC capacitor start-and-run motor; ~ **de**

~ **arranque por rotor estatórico** *m* ELEC stator-rotor starter motor; ~ **de arrastre** *m* ELEC drive motor, ING MECÁ driver; ~ **arrollado en serie** *m* ELEC series motor; ~ **con arrollamientos en serie** *m* ELEC series motor; ~ **asincrónico** *m* ELEC asynchronous motor, induction motor, FERRO asynchronous motor; ~ **asincrónico compensado** *m* ELEC compensated induction motor; ~ **asincrónico sincronizado** *m* ELEC synchronous induction motor; ~ **asíncrono** *m* ELEC, FERRO, ING ELÉC asynchronous motor; ~ **asíncrono de histéresis** *m* ING ELÉC hysteresis motor; ~ **por autoencendido** *m* ING MECÁ compression ignition engine; ~ **autoexcitado** *m* ELEC self-excited motor; ~ **autosincrónico** *m* CINEMAT selsyn motor; ~ **auxiliar** *m* ELEC, ING MECÁ auxiliary motor, TRANSP MAR auxiliary engine; ~ **de avance por pasos** *m* INFORM&PD stepper motor;

~ b ~ **bajo el asiento** *m* AUTO underfloor engine; ~ **de balancín** *m* ING MECÁ beam engine; ~ **bicilíndrico** *m* ING MECÁ double-cylinder engine, duplex engine, duplex-cylinder engine, twin-cylinder engine, TRANSP twin engine; ~ **bifásico** *m* ELEC, ING ELÉC two-phase motor; ~ **bipolar** *m* ELEC two-pole motor; ~ **blindado** *m* ELEC sealed motor, totally-enclosed motor, ING ELÉC enclosed motor; ~ **de bobinado compuesto dúplex** *m* ING MECÁ duplex compound winding engine; ~ **de brida** *m* ELEC flange motor; ~ **de bulbo incandescente** *m* AUTO hot-bulb engine;

~ c ~ **de CA** *m* ELEC, FÍS, ING ELÉC AC motor; ~ **de cámara de precombustión** *m* TRANSP stratified charge engine; ~ **de campo desplazable** *m* TRANSP traveling-field motor (*AmE*), travelling-field motor (*BrE*); ~ **con carburador** *m* AUTO, VEH carburetor engine (*AmE*), carburettor engine (*BrE*); ~ **de Carnot** *m* FÍS Carnot engine; ~ **de carrera corta** *m* VEH short-stroke engine; ~ **de CC** *m* ELEC, FÍS, ING ELÉC DC motor; ~ **CC sin escobillas** *m* ELEC, ING ELÉC brushless DC motor; ~ **de CC por escobillas** *m* ING ELÉC brush-type DC motor; ~ **CC en serie** *m* ING ELÉC series DC motor; ~ **central** *m* AUTO center engine (*AmE*), centre engine (*BrE*), mid-engine, VEH center engine (*AmE*), centre engine (*BrE*); ~ **cerrado** *m* ELEC enclosed motor, totally-enclosed motor; ~ **a chorro** *m* TEC ESP jet engine; ~ **de chorro** *m* MECÁ, TERMO jet engine; ~ **de chorro al arco** *m* ING MECÁ arc jet engine; ~ **de ciclo Rankine** *m* INSTAL TERM, TERMO, TRANSP Rankine-cycle engine; ~ **de cilindro** *m* ING MECÁ drum drive; ~ **de un cilindro** *m* AUTO, ING MECÁ one-cylinder engine; ~ **de cilindro único** *m* AUTO one-cylinder engine, ING MECÁ one-cylinder engine, single-cylinder engine; ~ **de cilindros gemelos** *m* ING MECÁ twin-cylinder engine; ~ **de cilindros giratorios** *m* TRANSP revolving-cylinder engine; ~ **con los cilindros en línea** *m* TRANSP MAR in-line engine; ~ **de cilindros en línea** *m* ING MECÁ in-line cylinder engine, in-line engine; ~ **con cilindros en V** *m* TRANSP MAR V-engine; ~ **de cilindros en V** *m* ING MECÁ V-cylinder engine, VEH V-type engine; ~ **cohete** *m* ING MECÁ rocket engine; ~ **del cohete** *m* TEC ESP rocket engine; ~ **de cohete eléctrico** *m* ING MECÁ electric-rocket engine; ~ **de cohete de propulsión iónica** *m* ING MECÁ ion-rocket engine; ~ **de cohete químico** *m* ING MECÁ chemical-rocket engine; ~ **cohético** *m* D&A rocket motor, ING MECÁ rocket engine, TEC ESP rocket motor; ~ **de colector** *m* ELEC,

ING ELÉC, TRANSP commutator motor; ~ **de colector de CC** *m* (*motor de colector de corriente continua*) ING ELÉC commutator DC motor (*commutator direct current motor*); ~ **del colector de corriente** *m* ING ELÉC collector motor; ~ **de colector de corriente continua** *m* (*motor de colector de CC*) ING ELÉC commutator direct current motor (*commutator DC motor*); ~ **de combustible líquido** *m* TERMO liquid-fuel engine; ~ **de combustión** *m* ING MECÁ gas engine, motor, TERMO, TRANSP combustion engine; ~ **de combustión externa** *m* ING MECÁ, TRANSP external combustion engine; ~ **de combustión interna** *m* ING ELÉC, ING MECÁ, MECÁ, PROD, TEC PETR, VEH internal combustion engine; ~ **de combustión de petróleo** *m* ING MECÁ oil burner motor; ~ **compensado** *m* ELEC compensated motor; ~ **en compound** *m* ELEC, ING ELÉC compound motor; ~ **con condensador hendido de imán permanente** *m* ING ELÉC permanent-magnet split-capacitor motor; ~ **de conos de fricción** *m* ING MECÁ cone drive; ~ **controlado por el inducido** *m* ELEC armature-controlled motor; ~ **de corriente alterna** *m* ELEC, FÍS, ING ELÉC alternating-current motor; ~ **de corriente continua en serie** *m* ING ELÉC series direct current motor; ~ **crítico** *m* TRANSP AÉR critical engine; ~ **crítico inoperativo** *m* TRANSP AÉR critical engine inoperative; ~ **de cuatro cilindros** *m* TRANSP flat-four engine; ~ **de cuatro tiempos** *m* AUTO, ING MECÁ, MECÁ four-stroke engine, TRANSP *diesel* flat-four engine, TRANSP MAR, VEH four-stroke engine; ~ **de culata en F** *m* AUTO F-head engine; ~ **de culata en I** *m* AUTO I-head engine; ~ **de culata en L** *m* AUTO L-head engine; ~ **de culata en T** *m* AUTO T-head engine;

~ d ~ **delantero** *m* AUTO front engine, VEH front-mounted engine; ~ **de derivación** *m* TRANSP, TRANSP AÉR bypass engine; ~ **en derivación** *m* ELEC, ING ELÉC shunt motor; ~ **de descenso** *m* TEC ESP descent engine; ~ **descompensado** *m* ING ELÉC noncompensated motor; ~ **de desplazamiento constante** *m* PROD fixed-displacement motor; ~ **de desplazamiento variable** *m* PROD variable-displacement motor; ~ **de desviación** *m* TRANSP, TRANSP AÉR bypass engine; ~ **con devanado en espiral** *m* ING ELÉC pancake motor; ~ **de devanado mixto** *m* ELEC, ING ELÉC compound-winding motor; ~ **devanado en serie** *m* ELEC series motor; ~ **con devanados en cortocircuito** *m* ING ELÉC shaded-pole motor; ~ **diesel** *m* AUTO diesel engine, ING ELÉC internal combustion engine, ING MECÁ compression ignition engine, PROD, TRANSP MAR, VEH diesel engine; ~ **diesel con cámara de precombustión** *m* AUTO air-cell diesel engine; ~ **diesel de inyección indirecta** *m* TRANSP indirect-injection diesel engine; ~ **diesel lento** *m* TRANSP MAR low-speed diesel engine; ~ **diesel poco revolucionado** *m* TRANSP slow-running diesel engine; ~ **diesel preparado** *m* TRANSP improved diesel engine; ~ **diesel sobrealimentado** *m* TRANSP MAR turbocharged engine; ~ **diesel-hidráulico** *m* ING MECÁ diesel-hydraulic engine; ~ **doble** *m* ING MECÁ double engine; ~ **de doble árbol de levas en cabeza** *m* AUTO DOHC engine; ~ **de doble capacitor** *m* ELEC dual-capacitor motor; ~ **de doble combustible** *m* TRANSP dual-fuel engine; ~ **de doble condensador** *m* ELEC dual-capacitor motor; ~ **de doble jaula de ardilla** *m* ELEC double-squirrel cage motor; ~ **de doble velocidad** *m* ELEC double speed motor; ~ **de dos cilindros** *m* ING MECÁ twin-cylinder engine, TRANSP twin engine; ~ **de dos sentidos de rotación** *m* ING ELÉC, TRANSP reversible motor; ~ **de dos tiempos** *m* AUTO, ING MECÁ, MECÁ, TRANSP MAR, VEH two-stroke engine; ~ **de dos tiempos de tres partes** *m* AUTO, VEH three-part two-stroke engine;

~ e ~ **eléctrico** *m* AUTO, ELEC, ING ELÉC, TRANSP electric motor; ~ **eléctrico de baja velocidad** *m* ING ELÉC low-speed electric motor; ~ **eléctrico provisto de diafragma** *m* MECÁ canned motor; ~ **de émbolo** *m* ING MECÁ, TRANSP AÉR piston engine; ~ **de émbolos** *m* ING MECÁ, TRANSP AÉR, TRANSP MAR, VEH reciprocating engine; ~ **de empuje orientable** *m* TEC ESP vectored thrust engine; ~ **enfriado por aire** *m* ING MECÁ air-cooled motor; ~ **engranado** *m* MECÁ geared motor; ~ **de engranaje** *m* MECÁ geared motor; ~ **entre los lados de la línea** *m* ELEC across-the-line motor; ~ **de escape de lanzamiento** *m* TEC ESP launch escape motor; ~ **sin escobillas** *m* ELEC, ING ELÉC brushless motor; ~ **con estator devanado** *m* ING ELÉC wound stator motor; ~ **de estiraje** *m* MINAS drawing engine; ~ **de excitación compound** *m* ELEC, ING ELÉC compound-wound motor; ~ **de excitación compuesta** *m* ELEC, ING ELÉC compound motor, compound-wound motor; ~ **de excitación mixta** *m* ELEC, ING ELÉC compound motor, compound-wound motor; ~ **de excitación propia** *m* ELEC self-excited motor; ~ **de excitación separada** *m* ELEC separately excited motor; ~ **excitado en derivación** *m* ING ELÉC shunt-wound motor; ~ **excitado en serie** *m* ELEC series-wound motor, ING ELÉC series motor, series-wound motor; ~ **de expansión** *m* ING MECÁ expansion engine; ~ **de expansión única** *m* ING MECÁ single-expansion engine; ~ **de extracción** *m* MINAS hoist, hoisting engine;

~ f ~ **de fan canalizado** *m* TRANSP AÉR ducted-fan engine; ~ **de fase abierta** *m* ING ELÉC split-phase motor; ~ **de fase dividida** *m* ELEC split-phase motor; ~ **flotante** *m* AUTO floating engine; ~ **de freno** *m* ELEC brake motor; ~ **fueraborda** *m* TRANSP MAR outboard engine; ~ **de funcionamiento alternativo** *m* ING MECÁ, TRANSP AÉR, TRANSP MAR, VEH reciprocating engine;

~ g ~ **de gas** *m* AUTO liquid-petroleum-gas engine (*LPG engine*), GAS, ING MECÁ, PROD, TERMO gas engine; ~ **de gas industrial** *m* PROD gas motor; ~ **de gas licuado** *m* TRANSP liquid-gas engine; ~ **de gas licuado de petróleo** *m* AUTO liquid-petroleum-gas engine (*LPG engine*); ~ **de gas natural** *m* TRANSP, VEH natural-gas engine; ~ **a gasoil** *m* ING ELÉC internal combustion engine; ~ **de gasolina** *m* *Esp* (*cf motor de nafta AmL*) AUTO, TERMO, TRANSP MAR, VEH gas engine (*AmE*), gasoline engine (*AmE*), petrol engine (*BrE*); ~ **generatriz** *m* ING ELÉC motor generator; ~ **de GLP** *m* AUTO LPG engine (*liquid-petroleum-gas engine*); ~ **graduador** *m* ING MECÁ graduating engine; ~ **de gran par de arranque** *m* MECÁ high-torque motor; ~ **de gran velocidad** *m* TRANSP MAR high-speed engine;

~ h ~ **de hélice entubada** *m* ING MECÁ ducted-fan engine; ~ **híbrido** *m* TRANSP hybrid engine; ~ **hidráulico** *m* ING MECÁ hydraulic motor;

~ **horizontal** *m* AUTO horizontal engine, VEH horizontal engine, transverse engine;

i ~ **de ignición por compresión** *m* ING MECÁ compression ignition engine; ~ **de impacto de iones eléctronicos** *m* TEC ESP electron-impact ion engine; ~ **de impulsión** *m* ELEC drive motor; ~ **de impulsión en perigeo** *m* TEC ESP perigee kick motor; ~ **inclinado** *m* AUTO slanter engine; ~ **incombustible** *m* ING ELÉC, SEG, TERMO fireproof motor, flameproof motor; ~ **incorporado** *m* ING ELÉC built-in motor; ~ **de inducción** *m* ELEC, FÍS, ING ELÉC, PROD, TRANSP induction motor; ~ **de inducción por anillo deslizante** *m* ING ELÉC slip-ring induction motor; ~ **de inducción compensado** *m* ELEC compensated induction motor; ~ **de inducción lineal** *m* ELEC, TRANSP linear induction motor; ~ **de inducción lineal asíncrono** *m* ELEC, TRANSP asynchronous linear induction motor; ~ **de inducción monofásico** *m* ING ELÉC capacitor motor, single-phase induction motor; ~ **de inducción polifásico** *m* ING ELÉC polyphase induction-motor; ~ **de inducción de polos protegidos** *m* ELEC shaded-pole motor; ~ **de inducción por repulsión** *m* ELEC, ING ELÉC repulsion-induction motor; ~ **de inducción con rotor devanado** *m* ING ELÉC wound-rotor induction motor; ~ **de inducción trifásico** *m* TRANSP three-phase induction motor; ~ **inductor lineal simple** *m* ELEC, TRANSP single linear inductor motor (*SLIM*); ~ **de inversión rápida de marcha** *m* ELEC reversing motor; ~ **de inyección directa** *m* AUTO, TRANSP direct injection engine; ~ **iónico** *m* TEC ESP ion engine; ~ **irreversible** *m* ELEC nonreversible motor;

j ~ **de jaula de ardilla** *m* ELEC squirrel cage motor; ~ **de jaula de ardilla Boucherot** *m* ELEC double-squirrel cage motor;

l ~ **en línea** *m* AUTO in-line engine, ING MECÁ straight engine; ~ **en la línea** *m* ELEC across-the-line motor; ~ **lineal** *m* ELEC, TRANSP linear motor; ~ **lineal asincrónico** *m* ELEC, TRANSP linear induction motor; ~ **lineal de flujo transversal** *m* TRANSP transverse-flux linear motor; ~ **lineal de gran potencia** *m* TRANSP high-power linear motor; ~ **lineal de inducción** *m* ELEC, TRANSP linear induction motor; ~ **lineal primario simple** *m* TRANSP single primary-type linear motor; ~ **de líquido orgánico** *m* TRANSP organic fluid engine;

m ~ **medianamente revolucionado** *m* TRANSP MAR medium-speed engine; ~ **de <1 HP** *m* (*motor de potencia menor de un caballo*) ING ELÉC FHP motor (*fractional horsepower motor*); ~ **monocilíndrico** *m* AUTO one-cylinder engine, ING MECÁ one-cylinder engine, single-cylinder engine; ~ **monofásico** *m* ELEC, ING ELÉC single-phase motor; ~ **monofásico con condensador** *m* ING ELÉC capacitor motor; ~ **monofásico con devanado auxiliar de arranque** *m* ING ELÉC split-phase motor; ~ **monofásico de tres conductores** *m* ING ELÉC split-phase motor; ~ **de montaje trasero** *m* AUTO, VEH rear-mounted engine; ~ **multicarburante** *m* AUTO, TERMO multifuel engine; ~ **multicilíndrico** *m* AUTO multicylinder engine; ~ **multicombustible** *m* AUTO, TERMO multifuel engine;

n ~ **de nafta** *m* AmL (*cf motor de gasolina Esp*) AUTO, TERMO, TRANSP MAR, VEH gas engine (*AmE*), gasoline engine (*AmE*), petrol engine (*BrE*);

~ **neumático** *m* ING MECÁ air engine, air motor; ~ **nivelador** *m* CONST leveling motor (*AmE*), levelling motor (*BrE*); ~ **de no expansión** *m* ING MECÁ nonexpansion engine; ~ **no ventilado** *m* ELEC totally-enclosed motor;

o ~ **de ondas progresivas** *m* ELEC traveling-wave motor (*AmE*), travelling-wave motor (*BrE*);

p ~ **de par** *m* ELEC torque motor; ~ **de par constante** *m* ELEC torque motor; ~ **de par de torsión** *m* ELEC torque motor; ~ **paso a paso** *m* CINEMAT stepper motor, ELEC stepper motor, stepping motor, INFORM&PD stepper motor; ~ **paso a paso de reluctancia variable** *m* ING ELÉC variable-reluctance stepper motor; ~ **de paso solenoide** *m* ING ELÉC solenoid stepper motor; ~ **de pequeña carrera** *m* VEH short-stroke engine; ~ **con pequeño empuje** *m* TEC ESP low-thrust motor; ~ **de pistón doble** *m* AUTO twin-piston engine; ~ **de pistón giratorio** *m* TRANSP rotating-piston engine; ~ **de pistón rotativo** *m* AUTO rotary-piston engine; ~ **de pistones** *m* AUTO piston engine, ING MECÁ reciprocating engine, TRANSP AÉR piston engine, reciprocating engine, TRANSP MAR, VEH reciprocating engine; ~ **de pistones alternativos** *m* TRANSP AÉR reciprocating-piston engine; ~ **de pistones contrapuestos** *m* ING MECÁ opposed piston engine; ~ **plano** *m* AUTO, VEH flat engine; ~ **plano de dos pistones** *m* AUTO flat twin engine; ~ **de plasma** *m* TEC ESP plasma engine; ~ **de pluriexpansión** *m* ING MECÁ compound engine, compound expansion engine; ~ **poco revolucionado** *m* TRANSP MAR *diesel* low-speed engine; ~ **policombustible** *m* AUTO, TERMO multifuel engine; ~ **polifásico** *m* ELEC, ING ELÉC polyphase motor; ~ **de polo sombreado** *m* ELEC shaded-pole motor; ~ **de polos divididos** *m* ELEC split-pole motor; ~ **de polos protegidos** *m* ELEC shaded-pole motor; ~ **de potencia menor de un caballo** *m* (*motor de <1 HP*) ELEC, ING ELÉC fractional horsepower motor (*FHP motor*); ~ **primario en línea** *m* AUTO vehicle-mounted short primary linear motor; ~ **principal** *m* TRANSP AÉR master engine; ~ **productor de par** *m* ELEC torque motor; ~ **de propulsión a chorro** *m* TRANSP propjet engine; ~ **de propulsión para cohete electrónico** *m* TEC ESP electronic rocket-engine; ~ **propulsor** *m* TRANSP MAR propulsion engine; ~ **protegido contra el polvo** *m* ELEC dustproof motor; ~ **de puesta en marcha** *m* ELEC starting motor;

q ~ **de quemado pobre** *m* TERMO lean-burn engine;

r ~ **rápido** *m* MECÁ high-speed engine; ~ **a reacción** *m* TRANSP AÉR jet engine; ~ **de reacción** *m* ELEC reaction motor, MECÁ jet engine, TEC ESP jet; ~ **de reacción neumático** *m* ELEC air reactor; ~ **a reacción pulsátil** *m* TRANSP AÉR pulsating jet engine; ~ **a reacción pulsatorio** *m* TRANSP AÉR pulsating jet engine; ~ **rebajado** *m* AUTO, VEH flat engine; ~ **de refrigeración líquida** *m* AUTO liquid-cooled engine; ~ **refrigerado por agua** *m* AUTO, VEH water-cooled engine; ~ **refrigerado por aire** *m* VEH air-cooled engine; ~ **regulable** *m* CINEMAT wild motor; ~ **de reluctancia variable** *m* TRANSP variable-reluctance motor; ~ **de repulsión** *m* ELEC, ING ELÉC repulsion motor; ~ **de respuesta uniforme** *m* ELEC flat-response motor; ~ **reversible** *m* ING ELÉC, TRANSP reversible motor; ~ **rotativo** *m* AUTO rotary engine,

VEH rotary engine, Wankel engine; ~ **rotatorio** *m* ING MECÁ Wankel engine; ~ **con rotor bobinado** *m* ING ELÉC wound rotor motor; ~ **de rotor en cortocircuito** *m* ING ELÉC, PROD squirrel-cage motor; ~ **con rotor de jaula** *m* PROD squirrel-cage motor; ~ **con rotor en jaula de ardilla** *m* PROD squirrel-cage motor;

~ s ~ **seco** *m* TRANSP AÉR dry engine; ~ **semicerrado** *m* ELEC semienclosed motor; ~ **semidiesel** *m* AUTO, ING MECÁ semidiesel engine; ~ **de separación** *m* TEC ESP separation motor; ~ **sincrónico** *m* CINEMAT, ELEC, FÍS, ING ELÉC synchronous motor; ~ **sincrónico de inducción** *m* ELEC synchronous induction motor; ~ **sincronizado** *m* ING ELÉC synchronous induction motor; ~ **síncrono** *m* TRANSP synchronous motor; ~ **síncrono de arranque automático** *m* ING ELÉC self-starting synchronous motor; ~ **síncrono de imán permanente** *m* ING ELÉC permanent-magnet synchronous motor; ~ **síncrono polifásico** *m* ING ELÉC polyphase synchronous motor; ~ **síncrono de reluctancia** *m* ELEC, ING ELÉC, TRANSP reluctance motor; ~ **síncrono trifásico** *m* ING ELÉC three-phase synchronous motor; ~ **sin sobrealimentación** *m* AUTO unsupercharged engine; ~ **sobrealimentado** *m* TRANSP MAR *diesel* turbocharged engine; ~ **supercuadrado** *m* AUTO oversquare engine;

~ t ~ **de tambor** *m* ING MECÁ drum drive; ~ **térmico** *m* FÍS, ING MECÁ, MECÁ heat engine, TERMO heat engine, thermal engine; ~ **termopropulsor** *m* TRANSP AÉR thermal jet engine; ~ **tipo U** *m* AUTO U-type engine; ~ **tipo nonio** *m* TEC ESP vernier motor; ~ **tipo W** *m* AUTO W-type engine; ~ **tipo X** *m* AUTO X-type engine; ~ **para toda corriente** *m* ELEC all-current motor (*AmE*), all-mains motor (*BrE*); ~ **para toda potencia** *m* ELEC all-watt motor; ~ **para todo vatio** *m* ELEC all-watt motor; ~ **para todo watt** *m* ELEC all-watt motor; ~ **totalmente encerrado** *m* ELEC totally-enclosed motor; ~ **de tracción** *m* TRANSP traction engine; ~ **de tracción de corriente continua** *m* TRANSP direct current traction motor; ~ **de tracción y extracción** *m* MINAS hauling and winding engine; ~ **transversal** *m* VEH horizontal engine, transverse engine; ~ **de tranvía** *m* FERRO tramway motor unit (*BrE*); ~ **trasero** *m* AUTO, VEH rear engine, rear-mounted engine; ~ **de tres cilindros** *m* ING MECÁ three-cylinder engine; ~ **trifásico** *m* ELEC, ING ELÉC, PROD three-phase motor; ~ **trifásico de inducción** *m* ING ELÉC three-phase induction motor; ~ **tubular** *m* TRANSP tubular motor; ~ **de turbina** *m* AUTO, TRANSP AÉR turbine engine; ~ **de turbina a gas** *m* TRANSP AÉR gas-turbine engine; ~ **de turbo ventilador** *m* ING MECÁ turbofan engine; ~ **turborreactor** *m* TERMO turbojet engine; ~ **turboventilador** *m* TERMO turbo-fan engine;

~ u ~ **universal** *m* ELEC all-current motor (*AmE*), all-mains motor (*BrE*), universal motor, ING ELÉC universal motor;

~ v ~ **en V** *m* AUTO V-engine; ~ **en V de cuatro cilindros** *m* AUTO V4 engine (*V-four engine*); ~ **en V de ocho cilindros** *m* AUTO V8 engine (*V-eight engine*); ~ **en V de seis cilindros** *m* AUTO V6 engine (*V-six engine*); ~ **de válvula de camisa** *m* AUTO sleeve-valve engine; ~ **de válvula lateral** *m* AUTO side-valve engine; ~ **sin válvulas** *m* AUTO valveless engine; ~ **de válvulas en cabeza** *m* AUTO overhead-

valve engine (*OHV engine*); ~ **de válvulas en culata** *m* AUTO overhead-valve engine (*OHV engine*); ~ **de vapor** *m* AUTO, FÍS steam engine; ~ **de vapor con cilindro intermedio** *m* INSTAL HIDRÁUL intermediate-cylinder steam engine; ~ **de velocidad ajustable** *m* ELEC adjustable-speed motor; ~ **de velocidad constante** *m* ING MECÁ constant-speed drive; ~ **de velocidad gradual** *m* CINEMAT, ING ELÉC stepper motor; ~ **de velocidad gradual de imanes permanentes** *m* ING ELÉC permanent-magnet stepper motor; ~ **de velocidad gradual trifásico** *m* ING ELÉC three-phase stepper motor; ~ **de velocidad regulable** *m* ELEC adjustable-speed motor, adjustable varying speed motor, ING ELÉC variable-speed motor, MECÁ adjustable-speed motor; ~ **de velocidad variable** *m* ELEC adjustable varying speed motor; ~ **con ventilación forzada** *m* ELEC, ING ELÉC, ING MECÁ forced-ventilation motor; ~ **ventilado** *m* ELEC ventilated motor; ~ **de ventilador** *m* TRANSP fan engine; ~ **del ventilador** *m* REFRIG fan motor; ~ **de ventilador entubado** *m* ING MECÁ ducted-fan engine; ~ **vernier** *m* TEC ESP vernier motor; ~ **vertical** *m* AUTO, ING MECÁ vertical engine; ~ **vertical embridado** *m* MECÁ flange motor;

~ w ~ **Wankel** *m* AUTO rotary engine, Wankel engine, ING MECÁ, TERMO, TRANSP Wankel engine, VEH rotary engine, Wankel engine; ~ **Warren** *m* TRANSP Warren engine

motor³: **un ~ inoperativo** *fra* TRANSP AÉR one engine inoperative; **con ~ preparado** *fra* VEH hotted-up (*AmE*), souped-up (*BrE*)

motores¹: ~ **acoplados** *m pl* ING MECÁ coupled engines; ~ **gemelos** *m pl* ING MECÁ twin engines; ~ **a reacción de flujo doble** *m pl* TRANSP AÉR dual-flow jet engine

motores²: **todos los ~ en funcionamiento** *fra* TRANSP AÉR all engines operating; **todos los ~ en operación** *fra* TRANSP AÉR all engines operating

motorizado *adj* MECÁ motor-driven

mototractor: ~ **de vía estrecha** *m* FERRO light-rail motor tractor

motriz *adj* ING MECÁ drive, driving

mottramita *f* MINERAL mottramite

mover *vt* FOTO agitate, INSTAL HIDRÁUL *compuertas* gate, MECÁ drive; ~ **por chorro de agua** *vt* AGUA flush; ~ **con el gato** *vt* VEH jack; ~ **hacia adelante con cuidado** *vt* TRANSP AÉR ease forward; ~ **con pala** *vt* CONST shovel

moverse *v refl* PROD shift

móvil¹ *adj* ELEC moveable, removable, FOTO removable, GEOL creeping, INFORM&PD removable, ING MECÁ detachable, moveable, removable, ÓPT removable, PROD *carga* live, moveable, removable, TRANSP AÉR moveable, removable, TRANSP MAR moveable

móvil² *m* ING MECÁ mover

movilidad *f* FÍS, PETROL, QUÍMICA mobility; ~ **de Hall** *f* FÍS Hall mobility; ~ **iónica** *f* FÍS RAD ionic mobility; ~ **de llamada** *f* TELECOM call portability; ~ **del portador** *f* FÍS carrier mobility

movimiento¹: **en ~** *adj* TRANSP MAR *navegación* underway

movimiento² *m* ACÚST movement, AGUA, EMB, ENERG RENOV flow, FÍS motion, FÍS FLUID flow, GAS movement, HIDROL flow, ING MECÁ movement, action, run, working, INSTAL HIDRÁUL flow, MECÁ flow, motion, METAL, TERMO, TEXTIL flow; ~ **absoluto** *m* FÍS absolute motion, QUÍMICA absolute movement; ~ **acelerado** *m* MECÁ accelerated motion; ~ **alternativo** *m* ING MECÁ alternating motion, reciprocating motion, up-and-down motion, MECÁ reciprocating motion; ~ **alternativo vertical** *m* ING MECÁ up-and-down motion; ~ **armónico simple** *m* FÍS, FÍS ONDAS simple harmonic motion; ~ **de arrollado** *m* TEXTIL take-up motion; ~ **ascendente y descendente** *m* ING MECÁ up-and-down motion; ~ **de avance** *m* PROD feed motion; ~ **de avance principal** *m* ING MECÁ main feed motion; ~ **de avance y retroceso** *m* ING MECÁ back-and-forth motion, backward-and-forward motion; ~ **de avance transversal** *m* ING MECÁ surfacing motion; ~ **del avión** *m* TRANSP AÉR aircraft movement; ~ **de balanceo** *m* ING MECÁ seesaw motion; ~ **bascular** *m* ING MECÁ balancing motion; ~ **de bobinado** *m* ING MECÁ winding motion; ~ **de Brown** *m* FÍS, FÍS RAD Brownian motion, Brownian movement; ~ **browniano** *m* FÍS, FÍS RAD Brownian motion, Brownian movement; ~ **de cámara** *m* CINEMAT, FOTO, TV camera movement; ~ **circular** *m* FÍS circular motion; ~ **de las compuertas** *m* INSTAL HIDRÁUL gating; ~ **compuesto** *m* MECÁ compound motion; ~ **contrario** *m* ING MECÁ countermotion; ~ **de cruz de Malta** *m* CINEMAT Maltese cross movement; ~ **descendente** *m* PROD downward motion; ~ **de detención** *m* ING MECÁ stop motion; ~ **epirogénico** *m* GEOL epirogenic movement; ~ **eustático** *m* OCEAN eustatic movement; ~ **de figuras** *m* AmL (*cf inversión de cifras Esp*) INFORM&PD figures shift; ~ **de flexión** *m* CONST bending movement; ~ **de G** *m* TRANSP MAR shift of G; ~ **del garfio** *m* CINEMAT claw movement, pin movement; ~ **giratorio** *m* CINEMAT rotary movement, TRANSP turning movement; ~ **de giro** *m* CONST bending movement; ~ **helicoidal** *m* NUCL helicoidal motion; ~ **de impulsión completo** *m* ING MECÁ self-contained driving motion; ~ **de impulsión independiente** *m* ING MECÁ self-contained driving motion; ~ **inútil** *m* MECÁ backlash; ~ **lento variable** *m* TV variable slow motion; ~ **libre** *m* AUTO free travel; ~ **en línea recta** *m* FÍS, ING MECÁ, MECÁ motion in a straight line; ~ **de un líquido en el interior de un tanque** *m* TEC ESP sloshing; ~ **longitudinal** *m* ING MECÁ longitudinal traverse; ~ **a máquina** *m* PROD machine motion; ~ **marcha atrás** *m* FERRO *vehículos* backing movement; ~ **de la marea** *m* ENERG RENOV, TRANSP MAR tidal movement; ~ **de materiales** *m* NUCL materials handling; ~ **mixto** *m* MECÁ compound motion; ~ **no uniforme** *m* FÍS nonuniform motion; ~ **no viscoso** *m* FÍS FLUID inviscid motion; ~ **del objetivo** *m* FOTO lens movement; ~ **ondular** *m* FÍS ONDAS wave motion; ~ **opuesto** *m* ING MECÁ countermotion; ~ **de oscilación** *m* ING MECÁ seesaw motion, swinging movement; ~ **oscilante** *m* ING MECÁ swinging movement; ~ **oscilatorio** *m* ING MECÁ shaking motion; ~ **paulatino de terreno** *m* TEC PETR creep; ~ **pendular** *m* ING MECÁ pendulum motion; ~ **de péndulo** *m* ING MECÁ swinging movement; ~ **rápido** *m* TV fast motion; ~ **recíproco** *m* ING MECÁ, MECÁ reciprocating motion, TEC PETR reciprocal movement; ~ **rectilíneo** *m* FÍS motion in a straight line, rectilineal motion, rectilinear motion, FÍS ONDAS rectilineal motion, rectilinear motion, ING MECÁ motion in a straight line, rectilineal motion, recti-

linear motion, MECÁ rectilineal motion, rectilinear motion, motion in a straight line; ~ **relativo** *m* FÍS relative motion; ~ **retardado** *m* MECÁ lagging; ~ **de retardo** *m* ING MECÁ retarded motion; ~ **de retroceso** *m* FERRO *vehículos* backing movement, ING MECÁ backward motion; ~ **retrógrado** *m* ING MECÁ backward movement; ~ **rotativo** *m* HIDROL spindrift; ~ **de rueda dentada y trinquete** *m* ING MECÁ pawl-and-ratchet motion; ~ **en sentido de apertura** *m* NUCL *de una válvula* lift; ~ **sinusoidal** *m* FÍS sinusoidal motion; ~ **sísmico submarino** *m* GEOFÍS submarine earthquake; ~ **del suelo** *m* GAS ground movement; ~ **tectónico** *m* GEOFÍS tectonic quake; ~ **térmico aleatorio** *m* FÍS RAD, TERMO random thermal motion; ~ **de tierra** *m* GAS ground movement; ~ **de tierras** *m* CONST earthwork; ~ **de tierras con bulldozer** *m* CONST bulldozing; ~ **transversal** *m* ING MECÁ surfacing motion; ~ **de traslación** *m* ING MECÁ translatory motion; ~ **traslatorio** *m* ING MECÁ translatory motion; ~ **del tren** *m* FERRO train movement; ~ **del trinquete activador del puntal** *m* ING MECÁ strut-action pawl motion; ~ **uniforme** *m* FÍS uniform motion; ~ **universal** *m* FÍS FLUID universal motion; ~ **de vaivén** *m* ING MECÁ alternating motion, back-and-forth motion, reciprocating motion, seesaw motion, MECÁ reciprocating motion; ~ **de vaivén vertical** *m* ING MECÁ up-and-down motion; ~ **verdadero** *m* TRANSP MAR *radar* true motion; ~ **vertical** *m* ING MECÁ vertical traverse, up-and-down motion; ~ **vibratorio** *m* ING MECÁ shaking motion; ~ **volcánico** *m* GEOFÍS volcanic quake

movimientos: ~ **de desgarre horizontal** *m pl* GEOL *de placas* strike-slip movements; ~ **de las existencias** *m pl* PROD stock movements; ~ **de existencias planeados** *m pl* PROD planned stock movements

mozo *m* TRANSP MAR ship's boy

ms *abr* (*milisegundo*) INFORM&PD ms (*millisecond*)

MS *abr* AGRIC (*materia seca*) DM (*dry matter*), IMPR, TEC ESP, TELECOM (*estación móvil*) MS (*mobile station*)

MSIV *abr* NUCL MSIV (*mean steam isolation valve*)

MSLB *abr* (*rotura de la línea de vapor principal*) NUCL MSLB (*main steamline break*)

mSv: ~-**año** *m* FÍS RAD mSv-year

múcico *adj* QUÍMICA mucic

muciférico *adj* QUÍMICA mucipheric

muciforme *adj* QUÍMICA muciform

mucilaginoso *adj* QUÍMICA muciferous

mucina *f* QUÍMICA mucin

mucoitin: ~-**sulfúrico** *adj* QUÍMICA mucoitin-sulfuric (*AmE*), mucoitin-sulphuric (*BrE*)

mucónico *adj* QUÍMICA muconic

mucoproteína *f* QUÍMICA mucoprotein

mudada *f* TEXTIL *hilatura* doffing

mudar: ~ **el rumbo** *vi* TRANSP MAR sheer off

mudez *f* ACÚST muteness

mueble: ~ **de exposición a bajas temperaturas** *m* REFRIG low-temperature display case

muebles: ~ **de cocina** *m pl* TRANSP AÉR, TRANSP MAR galley furnishings

muela *f* ING MECÁ grinder, wheel; ~ **abrasiva** *f* C&V grinding wheel, ING MECÁ grinding wheel, wheel, PROD grinding wheel, grindstone; ~ **abrasiva de afilar** *f* PROD sharpening wheel; ~ **abrasiva de cubeta** *f* PROD dished wheel; ~ **abrasiva de**

desbastar *f* ING MECÁ rough grinding wheel, PROD reducing wheel, roughing wheel; ~ **abrasiva dura** *f* PROD hard wheel; ~ **abrasiva de núcleo entrante** *f* ING MECÁ tub wheel; ~ **abrasiva sucia** *f* PROD loaded wheel; ~ **de acero de blindaje** *f* PROD armor plate mill (*AmE*), armour plate mill (*BrE*); ~ **adiamantada** *f* ING MECÁ diamond grinding-wheel; ~ **de afilar** *f* ING MECÁ sharpener; ~ **de corte** *f* ING MECÁ cutting-off wheel; ~ **de cubeta** *f* PROD cup wheel; ~ **desbarbadora** *f* ING MECÁ trimming wheel; ~ **desfibradora** *f* PAPEL grindstone; ~ **dulce** *f* PROD soft wheel; ~ **de esmeril** *f* MECÁ emery wheel, PROD emery wheel, hone; ~ **de grano fino** *f* PROD fine-grained wheel; ~ **lapidaria** *f* PROD face wheel; ~ **de molino** *f* ALIMENT millstone; ~ **de plancha de blindaje** *f* PROD armor plate mill (*AmE*), armour plate mill (*BrE*); ~ **rebarbadora** *f* ING MECÁ trimming wheel; ~ **rectificadora** *f* ING MECÁ truing wheel; ~ **de rectificar** *f* ING MECÁ wheel; ~ **para rectificar interiores** *f* PROD internal grinding-wheel; ~ **de resina sintética** *f* ING MECÁ resin-bonded wheel; ~ **de satinado** *f* ING MECÁ satin-finishing wheel; ~ **de satinar** *f* ING MECÁ satin-finishing wheel; ~ **superior** *f* ALIMENT *de molienda* upper millstone; ~ **vertical** *f* PROD edge runner, edge-wheel

muellaje *m* TRANSP MAR dockage

muelle *m* ACÚST spring, AUTO quayside roadway, CONST *puertos* jetty, mole, FERRO *infraestructura* platform, ING MECÁ, MECÁ spring, OCEAN jetty, TRANSP MAR quay, VEH spring; ~ **de atraque** *m* TRANSP MAR quay; ~ **de atraque con tinglados** *m* TRANSP MAR wharf; ~ **para carbón** *m* CARBÓN, TRANSP coal wharf; ~ **de carga** *m* CONST, FERRO loading platform, TRANSP loading dock, loading platform; ~ **para carga y descarga** *m* TRANSP loading dock; ~ **para carga y descarga de contenedores** *m* TRANSP container berth; ~ **compensador** *m* ING MECÁ equalizer spring; ~ **de compresión** *m* INSTAL HIDRÁUL compression spring, open-coil spring, open-spiral; ~ **de contacto** *m* ELEC *relé* contact spring; ~ **de contenedores** *m* TRANSP container wharf; ~**-dársena independiente** *m* TRANSP off-line docking station; ~ **de desembarque** *m* TRANSP MAR landing pier; ~ **de disco** *m* ING MECÁ disk spring; ~ **elíptico** *m* ING MECÁ elliptic spring; ~ **de embarque** *m* TRANSP boarding platform; ~ **del embrague** *m* ING MECÁ, VEH clutch spring; ~ **de la escobilla** *m* ING ELÉC brush spring; ~ **en espiral** *m* ING MECÁ hairspring, close-coil spring; ~ **espiral** *m* AUTO, ING MECÁ, MECÁ, VEH coil spring, coiled spring; ~ **de extensión** *m* ING MECÁ extension spring; ~ **de flexión** *m* ING MECÁ flexion spring; ~ **de fricción** *m* ING MECÁ friction spring; ~ **de hélice cónica** *m* ING MECÁ volute spring; ~ **helicoidal** *m* AUTO coil spring, coiled spring, FÍS helical spring, ING MECÁ coil spring, coiled spring, spiral wheel, MECÁ coil spring, coiled spring, helical spring, VEH coil spring, coiled spring; ~ **helicoidal cerrado** *m* ING MECÁ close-coil spring, close-spiral spring; ~ **de hojas** *m* ING MECÁ plate spring; ~ **de inserción** *m* AUTO insert spring; ~ **de llamada** *m* ING MECÁ drawback spring; ~ **de mercancías** *m* TRANSP freight depot (*AmE*), freight station (*AmE*), goods depot (*BrE*), goods station (*BrE*); ~ **de palanca de desenganche** *m* AGRIC, AUTO release-lever spring; ~ **petrolero** *m* TRANSP oil pier; ~ **del plato de**

empuje *m* VEH pressure-plate spring; ~ **de recuperación** *m* AUTO retracting spring, return spring, ING MECÁ, PROD, VEH return spring; ~ **de recuperación del freno** *m* AUTO brake-release spring; ~ **recuperador** *m* AUTO, ING MECÁ, PROD, VEH return spring; ~ **de retenida** *m* ING MECÁ, MINAS retaining spring; ~ **de retorno** *m* VEH cushion spring; ~ **de retorno de horquilla** *m* VEH fork return spring; ~ **retráctil** *m* AUTO retracting spring; ~ **de retroceso del motor de arranque** *m* VEH starter slip ring; ~ **del ruptor** *m* AUTO breaker spring; ~ **semielíptico** *m* ING MECÁ half-elliptic spring; ~ **de somier** *m* PROD hourglass spring; ~ **de suspensión del péndulo** *m* GEOFÍS pendulum suspension-spring; ~ **con tinglados** *m* TRANSP MAR wharf; ~ **de torsión** *m* ING MECÁ torsion spring; ~ **transversal** *m* FERRO transverse spring; ~ **de válvula** *m* AUTO, VEH valve spring

muerte: ~ **bacteriana** *f* AGRIC bacterial blight; ~ **bacteriana del tallo** *f* AGRIC bacterial stem-blight; ~ **generalizada** *f* AGRIC blight; ~ **generalizada de la hoja** *f* AGRIC leaf blight; ~ **generalizada tardía** *f* AGRIC late blight; ~ **generalizada temprana** *f* AGRIC early blight; ~ **térmica** *f* FÍS heat death

muerto *m* TRANSP MAR mooring post; ~ **de amarre** *m* TRANSP MAR dolphin

muesca[1]: **con ~** *adj* ING MECÁ notched

muesca[2] *f* C&V notch, undercut, CINEMAT notch, CONST abutment, notch, indent, groove, IMPR mortise, ING ELÉC slot, ING MECÁ groove, indentation, cutout, MECÁ, METAL, PROD groove, indentation, nick, slot, notch, notching; ~ **efectuada con punzón** *f* METAL dimple; ~ **de expansión** *f* INSTAL HIDRÁUL expansion notch; ~ **marginal** *f* CINEMAT edge notch; ~ **de protección contra escritura** *f* INFORM&PD write-protect notch; ~ **de punto muerto** *f* ING MECÁ dead-center notch (*AmE*), dead-centre notch (*BrE*); ~ **en la roca** *f* AmL (*cf enganche Esp*) MINAS hitch

muescador *m* ING MECÁ nicker

muescadora *f* CINEMAT notcher

muescar *vt* C&V notch, undercut, CINEMAT notch, CONST abut, notch, indent, groove, IMPR mortise, ING ELÉC slot, ING MECÁ groove, indent, cutout, MECÁ, METAL, PROD groove, indent, nick, slot, notch

muescas: ~ **aceitadas** *f pl* ING MECÁ oil groove

muestra *f* AGUA, CALIDAD sample, CARBÓN sample, specimen, CRISTAL specimen, ELECTRÓN sample, FÍS sample, sampling, GEOL sample, IMPR swatching-out, INFORM&PD, MATEMÁT sample, MECÁ specimen, METAL sample, test piece, MINAS *de perforación* core, P&C, PAPEL sample, QUÍMICA specimen, sample, TEC ESP, TELECOM, TEXTIL sample; ~ **de aire** *f* SEG air sample; ~ **aleatoria** *f* AGRIC, ALIMENT, CALIDAD, CARBÓN, MATEMÁT, TELECOM random sample; ~ **para análisis** *f* CARBÓN analysis sample, analytical sample; ~ **al azar** *f* AGRIC, ALIMENT, CALIDAD, CARBÓN, MATEMÁT, TELECOM random sample; ~ **compuesta** *f* CALIDAD, CARBÓN composite sample; ~ **cónica** *f* CARBÓN cone sampler; ~ **de contraste** *f* CARBÓN, QUÍMICA check sample; ~ **de control** *f* CARBÓN, QUÍMICA check sample; ~ **de ensayo** *f* QUÍMICA assay sample; ~ **estadística de los datos de desintegración** *f* FÍS RAD statistical sample of decay data; ~ **estratificada** *f* CARBÓN stratified sampling; ~ **de formaciones** *f* TEC PETR cuttings;

~ de gas *f* TEC PETR gas show; **~ lateralmente rayada** *f* METAL sidegrooved specimen; **~ de material vegetal** *f* AGRIC plant sample; **~ mecánica** *f* CARBÓN mechanical sample; **~ no alterada** *f* CARBÓN undisturbed sample; **~ de prueba** *f* CALIDAD test portion; **~ de remoldeado** *f* CARBÓN remolded sample (*AmE*), remoulded sample (*BrE*); **~ de roza** *f* CARBÓN channel sample; **del sondeo** *f* CARBÓN, GAS, MINAS, TEC PETR core sample; **~ de tejido** *f* TEXTIL fabric sample; **~ de tela** *f* TEXTIL fabric sample; **~ testigo** *f* AGRIC control sample, CARBÓN, QUÍMICA check sample; **a todo el ancho** *f* TEXTIL full-width sample; **~ tomada en la bobina** *f* PAPEL reel sample

muestrador: **~ automático** *m* LAB autosampler

muestrario *m* TEXTIL swatch

muestras: **~ de formaciones** *f pl* CARBÓN, TEC PETR cuttings

muestreador *m* CARBÓN sampler, LAB sampling device, PROD, TELECOM sampler; **~ de agua** *m* LAB water sampler; **~ de pan de oro** *m* CARBÓN foil sampler; **~ de pan de plata** *m* CARBÓN foil sampler; **~ de pistón** *m* CARBÓN piston sampler

muestrear *vt* GEN sample

muestreo *m* AGUA, ELECTRÓN, PROD, TEC ESP, TELECOM sampling; **~ al azar** *m* AGRIC, ALIMENT, CALIDAD, CARBÓN, MATEMÁT, TELECOM random sampling; **~ aleatorio** *m* AGRIC, ALIMENT, CALIDAD, CARBÓN, MATEMÁT, TELECOM random sampling; **~ automático** *m* CALIDAD, PROD automatic sampling; **~ continuo** *m* CALIDAD continuous sampling; **~ discontinuo** *m* PROD batch process, batch sampling; **~ estratificado** *m* SEG stratified sampling; **~ con fines sanitarios** *m* SEG health-related sampling; **~ integral** *m* TELECOM integral sampling; **~ isocinético** *m* CALIDAD isokinetic sampling; **~ manual** *m* CARBÓN hand selection; **~ del mineral** *m* AmL (*cf desmuestre Esp*) MINAS ore sampling; **~ múltiple** *m* TELECOM multiple sampling; **~ óptimo** *m* TELECOM optimal sampling; **~ primario** *m* CARBÓN primary sampling; **~ proporcional** *m* CALIDAD proportional sampling; **~ sencillo** *m* CALIDAD discrete sampling; **~ sistemático** *m* INFORM&PD systematic sampling; **~ por tubo hincado** *m* MINAS drive sampler; **~ de volumen constante** *m* CONTAM, QUÍMICA constant volume sampling (*CVS*)

mufa: **~ para cerámica** *f* C&V ceramic kiln

mufla *f* C&V muffle, ELEC *aislamiento de cable* inner covering, INSTAL TERM *horno* crucible furnace, muffle, LAB *calentamiento* crucible furnace, PROD crucible furnace, tackle, muffle, tackle and fall, tackle block, QUÍMICA, TERMO crucible furnace; **~ de ensayos** *f* QUÍMICA assay furnace; **~ de inmersión** *f* C&V *horno de laboratorio* immersion muffle; **~ de sinterización** *f* PROC QUÍ sintering furnace

muletilla *f* PROD crutch key; **~ de válvula** *f* ING MECÁ valve yoke

muletón *m* TEXTIL molleton

mullita *f* MINERAL mullite

multicanal[1] *adj* TRANSP MAR *radio, comunicaciones por satélite* multichannel

multicanal[2] *m* ELECTRÓN multiplex channel

multicapa[1] *adj* QUÍMICA multilayer

multicapa[2]: **~ de película delgada** *f* ELECTRÓN multilayer thin films; **~ de película gruesa** *f* ELECTRÓN multilayer thick films

multicasco *m* TRANSP MAR multihull

multicavidad: **~ klistron** *f* ELECTRÓN, FÍS multicavity klystron

multicelular *adj* ACÚST *altavoz* multicellular

multicolor *adj* COLOR multicolored (*AmE*), multicoloured (*BrE*), varicolored (*AmE*), varicoloured (*BrE*)

multicomponente *adj* QUÍMICA multicomponent

multidentado *adj* QUÍMICA multidentate

multidireccionamiento *m* TELECOM multiaddressing

multiestandar *adj* TV multistandard

multifasado *adj* ING ELÉC polyphase

multifase *adj* GEOL polyphase

multifásico *adj* ELEC *motor*, ELECTRÓN, GEOL multiphase

multifilamento *m* TEXTIL *hilo* multifilament yarn

multihaz *m* TELECOM multiple beam

multimedia *adj* INFORM&PD multimedia

multímetro *m* ING ELÉC multimeter, voltameter (*BrE*), voltammeter (*AmE*), voltmeter, TV multimeter; **~ digital** *m* ING ELÉC digital multimeter

multinorma *adj* TV multistandard

multiperforadora *f* PROD gang punch

múltiple[1] *adj* GEN multiple, GEOM manifold

múltiple[2] *m* ING MECÁ inlet manifold (*BrE*), intake manifold (*AmE*), manifold, LAB manifold, METR, PETROL multiple; **~ de admisión** *m* ING MECÁ *automóviles* inlet manifold (*BrE*), intake manifold (*AmE*); **~ de agua de alimentación** *m* NUCL feedwater manifold; **~ de distribución** *m* TEC PETR manifold; **~ de escape** *m* TRANSP AÉR exhaust manifold; **~ de tubería** *m* TEC PETR pipe manifold; **~ de tuberías** *m* TEC PETR manifold

múltiples[1]: **de ~ hilos** *adj* ING MECÁ multiwire

múltiples[2]: **~ direcciones** *f pl* TELECOM multiaddressing; **~ frecuencias** *f pl* TELECOM multi frequency, multiple frequency (*MF*)

multiplete *m* FÍS, FÍS RAD multiplet

multipletes: **~ de carga** *m pl* FÍS RAD charge multiplets

multiplex *m* ELECTRÓN, INFORM&PD, TELECOM multiplex; **~ de datos** *m* TELECOM data multiplexer; **~ por distribución en el tiempo** *m* TELECOM time-division multiplex; **~ por división de tiempo** *m* TELECOM time-division multiplex; **~ óptico** *m* ELECTRÓN, TV optical multiplex; **~ de tiempo** *m* TV time multiplex; **~ de tiempo compartido** *m* TELECOM time-division multiplex

multiplexación *f* INFORM&PD, TELECOM multiplexing; **~ por división de frecuencias** *f* (*MDF*) FÍS, INFORM&PD, TELECOM frequency-division multiplexing (*FDM*); **~ por división de tiempo** *f* (*TDM*) FÍS, INFORM&PD, TEC ESP, TELECOM time-division modulation (*TDM*), time-division multiplexing (*TDM*); **~ estadística** *f* INFORM&PD statmux; **~ de longitud de onda** *f* ÓPT wavelength multiplexing

multiplexado: **~ digital** *m* ELECTRÓN digital multiplex; **~ por división de frecuencias** *m* (*MDF*) ELECTRÓN, FÍS, INFORM&PD, TELECOM frequency-division multiplexing (*FDM*); **~ de frecuencia** *m* TELECOM frequency multiplexing

multiplexar *vt* INFORM&PD, TELECOM multiplex

multiplexión *f* INFORM&PD, TELECOM multiplexing; **~ asíncrona por división en el tiempo** *f* TELECOM code-division multiplexing (*CDM*); **~ de conmutación** *f* TELECOM switching mux; **~ por división de códigos** *f* TELECOM code-division multiplexing (*CDM*); **~ por división de longitud de onda**

f TELECOM wavelength division multiplexing (*WDM*); **~ por división de tiempo** *f* TELECOM time-division multiplexing (*TDM*); **~ por división en tiempo asíncrono** *f* (*ATDM*) TELECOM asynchronous time division multiplexing (*ATDM*); **~ temporal** *f* FÍS time-division multiplexing; **~ en el tiempo** *f* TELECOM time multiplexing

multiplexor *m* ELECTRÓN, INFORM&PD, TELECOM, TV multiplexer (*mux*); **~ de canales de datos** *m* INFORM&PD data channel multiplexer; **~ de conmutación** *m* TELECOM switching multiplexer, switching mux; **~ de datos** *m* TRANSP data multiplexer; **~ estadístico** *m* INFORM&PD statistical multiplexer; **~ estadístico de conmutación** *m* TELECOM switching statistical multiplexer; **~ de frecuencias** *m* ELECTRÓN frequency multiplexer; **~ de MCI** *m* ELECTRÓN PCM multiplexer; **~ óptico** *m* ELECTRÓN, TV optical multiplexer; **~ síncrono** *m* TELECOM synchronous multiplexer (*SM*); **~ sumar-restar** *m* TELECOM add-drop multiplexer; **~ temporal** *m* ELECTRÓN time-division multiplexer

multiplicación *f* ING MECÁ gear, gear ratio, MATEMÁT multiplication; **~ binaria** *f* ELECTRÓN binary multiplication; **~ digital** *f* ELECTRÓN digital multiplication; **~ de frecuencia** *f* ELECTRÓN frequency multiplication; **~ con niebla artificial** *f* AGRIC mist propagation; **~ pequeña** *f* ING MECÁ low gear; **~ de semillas** *f* AGRIC seed production

multiplicador *m* ELECTRÓN, INFORM&PD multiplier, TELECOM multiplicator, multiplier; **~ binario** *m* ELECTRÓN, TELECOM binary multiplier; **~ de cuatro cuadrantes** *m* ELECTRÓN four-quadrant multiplier; **~ digital** *m* ELECTRÓN digital multiplier; **~ de diodo de frecuencia** *m* ELECTRÓN diode frequency multiplier; **~ de electrones** *m* ELECTRÓN, FÍS RAD electron multiplier; **~ electrónico** *m* ELECTRÓN, FÍS RAD electron multiplier; **~ de emisión secundaria** *m* TV secondary-emission multiplier; **~ de frecuencia** *m* ELECTRÓN frequency multiplier; **~ de frecuencia klistron** *m* ELECTRÓN klystron frequency multiplier; **~ de frecuencia-reactancia** *m* ELEC, ELECTRÓN, ING ELÉC reactance-frequency multiplier; **~ de frecuencias** *m* TELECOM harmonic generator; **~ klistron de frecuencia** *m* ELECTRÓN frequency-multiplier klystron; **~ de par** *m* ING MECÁ torque multiplier; **~ paralelo** *m* ELECTRÓN parallel multiplier; **~ de presión** *m* INSTAL HIDRÁUL intensifier, pressure transmitter; **~ de tensión** *m* ELEC *circuito rectificador* voltage multiplier; **~ de velocidad** *m* MECÁ overdrive; **~ de voltaje** *m* ELEC, ING ELÉC voltage multiplier

multiplicando *m* INFORM&PD, MATEMÁT multiplicand
multiplicar: ~ por *vt* MATEMÁT multiply by
multiplicativo *adj* MATEMÁT multiplicative
multiplicidad *f* FÍS multiplicity
multipolar *adj* ELEC multipolar
multipolo *m* FÍS multipole
multiposicionamiento *m* INFORM&PD multithreading
multiprocesador *m* INFORM&PD, TELECOM multiprocessor
multiprocesamiento *m* INFORM&PD multiprocessing
multiprogramación *f* INFORM&PD multiprogramming
multipuerto *adj* INFORM&PD multiport
multitarea *f* INFORM&PD multitasking
multivalencia *f* QUÍMICA multivalence
multivibrador *m* ELECTRÓN, FÍS, TELECOM multivibra-

tor; **~ astable** *m* ELEC free running oscillator, ELECTRÓN astable multivibrator, free running oscillator; **~ biestable** *m* ELECTRÓN bistable multivibrator; **~ de división** *m* ELECTRÓN dividing multivibrator; **~ monoestable** *m* ELECTRÓN monostable multivibrator, one-shot multivibrator; **~ de retardo** *m* ELECTRÓN delay multivibrator
mu-metal *m* FÍS mu-metal
mundial *adj* INFORM&PD, TEC ESP, TELECOM global, worldwide
muñeco *m* TEC PETR go devil
muñequera *f* SEG wrist protector
muñequilla *f* ING MECÁ crankpin; **~ de biela del cigüeñal** *f* VEH *motor* crankpin; **~ del cigüeñal** *f* AUTO crankpin; **~ de la cruceta** *f* ING MECÁ crosshead pin; **~ del pistón** *f* MECÁ, VEH gudgeon pin (*BrE*), wrist pin (*AmE*)
munición *f* D&A ammunition; **~ guiada** *f* D&A precision-guided munition (*PGM*)
municiones: ~ de agentes múltiples *f pl* D&A multiagent munitions
munidora *f* TEXTIL clipper dryerfelt
muñón *m* AUTO journal, CONST *mecánica* gudgeon, ING MECÁ gudgeon, journal, trunnion, crankpin, MECÁ journal, trunnion, NUCL trunnion; **~ del cigüeñal** *m* AUTO, VEH crank pin; **~ de colada** *m* ING MECÁ sprue pin; **~ de cruz** *m* MECÁ journal cross; **~ de la góndola del motor** *m* TRANSP AÉR engine-nacelle stub; **~ del pistón** *m* MECÁ gudgeon pin (*AmE*), wrist pin (*AmE*); **~ de resorte** *m* ING MECÁ spring washer; **~ roscado** *m* ING MECÁ teat screw; **~ de la rueda** *m* ING MECÁ wheel spindle
muón *m* FÍS, FÍS PART, NUCL muon
mura *f* TRANSP MAR tack
murchisonita *f* MINERAL murchisonite
murete: ~ de piedra para relleno *m* MINAS pack wall; **~ de piedra en seco para relleno** *m* MINAS pack wall
murexida *m* QUÍMICA murexide
muriatado *adj* QUÍMICA muriated
murmullo: ~ de la corriente *m* ELEC, ING ELÉC current hum (*AmE*), mains hum (*BrE*)
muro *m* CARBÓN bottom, floor, *de una capa de carbón* sidewalk (*AmE*), pavement (*BrE*), *minas* pit bottom, CONST *estructura* wall, GEOL *de filón, yacimiento* hog back, footwall, MINAS bottom, *de una mina* wall, *de filones* ledger wall, footwall, pit bottom, seat, ledger, *de sostenimiento* barrier, *de una capa de carbón* pavement (*BrE*), sidewalk (*AmE*), SEG fire stop; **~ base** *m* CONST base wall; **~ de base** *m* MINAS *filones* footwall; **~ de cabecera** *m* CONST head wall; **~ de caída** *m* AGUA lift wall; **~ de carga** *m* CONST load-bearing wall; **~ de cerramiento** *m* CONST enclosing wall; **~ ciego** *m* CONST blind wall; **~ con comba** *m* CONST bulged wall, bulging wall; **~ de contención** *m* CONST *almacenamiento de agua* retaining wall, ING MECÁ bulkhead; **~ de contención del río** *m* AGUA, CONST, HIDROL river wall; **~ de contención en voladizo** *m* CONST cantilever retaining wall; **~ cortafuegos** *m* CONST, SEG, TEC ESP, TEC PETR fire wall, TERMO fire stop; **~ cortina** *m* CONST curtain wall; **~ de cortina ajustable** *m* C&V adjustable curtain wall; **~ de defensa** *m* HIDROL *dique* apron; **~ de encauzamiento** *m* AGUA, CONST, HIDROL river wall; **~ de fondo** *m* CONST head wall; **~ de la galería de minas** *m* MINAS floor; **~ de guardia** *m* AGUA cutoff wall; **~ de impermeabilización** *m* HIDROL *de*

una presa core; ~ **interceptor** *m* AGUA cutoff wall; ~ **de ladrillos** *m* CONST brick wall; ~ **lateral** *m* C&V *del alimentador, del chorreador* breast wall, CONST flank wall, wingwall, MINAS side wall; ~ **líquido** *m* OCEAN waterwall; ~ **medianero** *m* MINAS midwall; ~ **de muelle** *m* TEC ESP bulkhead; ~ **del muelle** *m* CONST, TRANSP MAR *puertos* dock wall, sea wall; ~ **de protección** *m* MINAS barrier, TRANSP crash barrier; ~ **de retención** *m* CONST breast wall; ~ **de revestimiento** *m* CONST face wall; ~ **de soporte** *m* CONST retaining wall; ~ **de sostén** *m* MINAS pack; ~ **de sostenimiento** *m* C&V breast wall, PROD metering land; ~ **del sótano** *m* CONST basement wall; ~ **transversal** *m* CONST cross wall; ~ **transversal sumergido** *m* AmL (*cf solera Esp*) MINAS *canales* groundsill, sill, sole; ~ **yacente** *m* MINAS *filones* footwall

muscarina *f* QUÍMICA muscarine

musgo: ~ **marino de Irlanda** *m* CONTAM MAR Irish moss

música: ~ **electrónica** *f* ELECTRÓN electronic music; ~ **sintetizada** *f* ELECTRÓN synthesized music

mustio *adj* ALIMENT withered

mutación: ~ **inducida por radiación** *f* FÍS RAD, NUCL radiation-induced mutation

mutador *m* TEC ESP static converter

mutágeno *adj* AGRIC mutagenic

mutarotación *f* QUÍMICA mutarotation

mutismo *m* ACÚST muteness

mV *abr* (*milivoltio*) ELEC, ING ELÉC mV (*millivolt*)

MVFD *abr* (*manipulación de variación de fase diferencial*) ELECTRÓN DPSK (*differential phase shift keying*)

mW *abr* (*milivatio*) ELEC, ING ELÉC mW (*milliwatt*)

MW *abr* (*megavatio*) ELEC, ING ELÉC MW (*megawatt*)

MWD *abr* (*mediciones durante la perforación*) TEC PETR MWD (*measurements while drilling*)

Mx *abr* (*maxvelio, maxwell*) ELEC, ING ELÉC Mx (*maxwell*)

N

n *abr* FÍS PART (*neurón*) n (*neutron*), METR (*nano*) n (*nano*)

N *abr* ACÚST (*neperio*) N (*neper*), ELEC, FÍS (*newton*) N (*newton*), GAS (*nitrógeno*) N (*nitrogen*), QUÍMICA, TEC PETR (*nitrógeno*) N (*nitrogen*)

N₂O₅ *abr* QUÍMICA (*óxido de nitrógeno V*) N_2O_5 (*nitrogen pentoxide*)

Na *abr* (*sodio*) METAL, QUÍMICA Na (*sodium, natrium*)

nacer *vi* AGUA, HIDROL *fuente de un río* rise, *río* take its rise

naciente *m* AGUA, CONST, HIDROL head of water

nacimiento *m* AGUA, HIDROL source; **~ de un río** *m* CONST head of water

nacrita *f* MINERAL nacrite

NAFO *abr* (*explosivo compuesto de nitrato amónico y fueloil*) MINAS ANFO (*ammonium nitrate fuel oil*)

nafta *f* AmL (*cf gasolina Esp*) AUTO, P&C, PETROL gas (*AmE*), gasoline (*AmE*), petrol (*BrE*), QUÍMICA naphtha, TEC PETR *combustibles* gasoline (*AmE*), petrol (*BrE*), gas (*AmE*), TERMO naphtha, *refinería* petrol (*BrE*), gasoline (*AmE*), gas (*AmE*), VEH gas (*AmE*), gasoline (*AmE*), petrol (*BrE*)

naftaceno *m* P&C, QUÍMICA naphthacene

naftalendisulfónico *adj* QUÍMICA naphthalenedisulfonic (*AmE*), naphthalenedisulphonic (*BrE*)

naftalénico *adj* DETERG, P&C, QUÍMICA naphthalenic

naftaleno *m* DETERG, QUÍMICA naphthalene

naftano *m* QUÍMICA naphthane

naftenato *m* P&C naphthenate; **~ de calcio** *m* P&C *pinturas* calcium naphthenate; **~ cobáltico** *m* P&C *pinturas* cobalt naphthenate; **~ de cobalto** *m* P&C *pinturas* cobalt naphthenate; **~ de plomo** *m* P&C *pinturas* lead naphthenate

nafténico *adj* P&C, QUÍMICA naphthenic

nafteno *m* QUÍMICA naphthene

naftil *m* QUÍMICA naphthyl

naftilamina *f* QUÍMICA naphthylamine

naftileno *m* QUÍMICA naphthylene

naftílico *adj* QUÍMICA naphthylic

naftilo *m* QUÍMICA naphthyl

naftiónico *adj* P&C, QUÍMICA naphthionic

naftoico *adj* QUÍMICA naphthoic

naftoilo *m* QUÍMICA naphthoyl

naftol *m* QUÍMICA naphthol

naftolato *m* QUÍMICA naphtholate

naftolsulfónico *adj* QUÍMICA naphtholsulfonic (*AmE*), naphtholsulphonic (*BrE*)

naftoquinona *f* QUÍMICA naphthoquinone

nagatelita *f* NUCL nagatelite

nagyagita *f* MINERAL nagyagite

nailon¹: **~ reforzado** *adj* EMB nylon-reinforced

nailon² *m* P&C, QUÍMICA, TEXTIL nylon

nano *pref* (*n*) METR nano (*n*)

nanofosil *m* GEOL nannofossil

nanosegundo *m* FÍS, INFORM&PD, METR, TV nanosecond

nantocoíta *f* MINERAL nantokite

napa *f* TEXTIL *hilatura* lap; **~ de agua** *f* AGRIC water table; **~ ventilada** *f* AGUA ventilated nappe

narceína *f* QUÍMICA narceine

narcosis: **~ causada por nitrógeno** *f* OCEAN nitrogen narcosis; **~ por nitrógeno** *f* OCEAN nitrogen narcosis

narcótico¹ *adj* QUÍMICA narcotic

narcótico² *m* QUÍMICA narcotic

narcotina *f* QUÍMICA narcotine

narguile *m* OCEAN narghile, *buceo* surface air supply, surface demand lifeline

naringenina *f* QUÍMICA naringenin

naringina *f* QUÍMICA naringin

nariz *f* C&V nose, ING MECÁ nose, nozzle

narrativa *f* INFORM&PD narrative

nasa *f* OCEAN nasse, *pesca* pot, *trampa de pesca* fish basket, basket trap

NASA *abr* (*Administración Nacional de Aeronáutica y del Espacio*) TEC ESP NASA (*National Aeronautics and Space Administration*)

naso: **~ de papel rizado** *m* EMB crimp paper cup

nata *f* QUÍMICA scum; **~ de fundición** *f* PROD sullage; **~ de silice** *f* C&V silica scum

nativo *adj* GEOL, MINERAL native

natrocalcita *f* MINERAL gaylussite

natrolita *f* MINERAL natrolite

natrón *m* MINERAL, QUÍMICA natron

natural *adj* NUCL unenriched

naumannita *f* MINERAL naumannite

náutico *adj* OCEAN nautical, TRANSP, TRANSP AÉR navigational, TRANSP MAR nautical, navigational

nautófono *m* Esp (*cf bocina de boquilla AmL*) TRANSP MAR *ayuda a la navegación* reed

navaja: **~ difusora** *f* PROC QUÍ diffuser blade

naval *adj* D&A naval, TRANSP MAR naval, seaborne

nave *f* MECÁ bay, TEC ESP craft, TRANSP vessel, TRANSP MAR ship, vessel; **~ de abastecimiento** *f* TEC PETR supply vessel; **~ de apoyo** *f* TEC PETR, TRANSP MAR support vessel; **~ de cargamento** *f* TEC ESP cargo bay; **~ de colada** *f* PROD cast house; **~ de lanzamiento** *f* TEC ESP launching aircraft; **~ de suministro** *f* TEC PETR supply vessel; **~ de turbinas** *f* TRANSP MAR turbine vessel; **~ vacía** *f* EMB clean room

navecilla *f* LAB *análisis* boat; **~ tarada** *f* AmL (*cf pesasubstancias Esp*) LAB *equipo* weighing boat

navegabilidad *f* CONTAM MAR, TRANSP MAR *de un buque* navigability

navegable *adj* TRANSP MAR navigable; **no ~** *adj* HIDROL, TRANSP MAR unnavigable

navegación¹: **en ~** *adj* TRANSP MAR underway

navegación² *f* AGUA, HIDROL, OCEAN, TEC ESP, TRANSP, TRANSP AÉR navigation, TRANSP MAR shipping, *práctica* sailing; **~ de altura** *f* TRANSP MAR ocean navigation; **~ astronómica** *f* TEC ESP, TRANSP AÉR celestial navigation, TRANSP MAR astronomical navigation, celestial navigation; **~ de cabotaje** *f* HIDROL, OCEAN, TRANSP MAR *marina mercante* coastal navigation; **~ celestial** *f* TEC ESP, TRANSP AÉR, TRANSP MAR celestial navigation; **~ a ciegas** *f* OCEAN, TRANSP MAR dead reckoning; **~ costera** *f* HIDROL coastal shipping, *marítima* coastal navigation, OCEAN coastal navigation, TRANSP MAR coastal shipping, *marina mercante* coastal navigation; **~ dependiente** *f* TEC ESP dependent navigation; **~ por efecto Doppler** *f* PETROL, TEC ESP Doppler navigation; **~ estelar** *f* TEC ESP stellar navigation; **~ a la estima** *f* OCEAN, TRANSP MAR dead reckoning, navigation by dead reckoning; **~ geodésica** *f* TRANSP MAR geodesic navigation; **~ independiente** *f* TEC ESP independent navigation; **~ inercial** *f* TEC ESP, TRANSP AÉR inertial navigation; **~ por instrumentos** *f* TEC ESP *astronáutica* blind navigation; **~ interior** *f* TRANSP MAR inland navigation; **~ loxodrómica** *f* TRANSP MAR rhumb-line navigation; **~ oceánica** *f* TRANSP MAR deep-sea navigation, ocean navigation; **~ por un paso** *f* TRANSP MAR fairway navigation; **~ por radar** *f* FÍS RAD, TRANSP, TRANSP AÉR, TRANSP MAR radar navigation; **~ radiogoniométrica** *f* FÍS RAD, TRANSP AÉR, TRANSP MAR radio navigation; **~ por satélite** *f* TRANSP AÉR satellite navigation; **~ con la sonda en la mano** *f* OCEAN, TRANSP MAR *parajes desconocidos* navigation by sounding

navegante *m* TRANSP, TRANSP AÉR, TRANSP MAR navigator

navegar: **~ hacia el norte** *vi* TRANSP, TRANSP MAR stand to the north; **~ a orejas de mulo** *vi* TRANSP MAR boom out; **~ a vela** *vi* TRANSP MAR sail; **~ con el viento a un largo** *vi* TRANSP MAR run before the wind

naviera *f* TRANSP MAR shipowner

naviero *m* TRANSP MAR shipowner

navío *m* TRANSP MAR warship; **~ mercante** *m* TRANSP MAR merchant ship; **~ de repostaje** *m* TRANSP refueling craft (*AmE*), refuelling craft (*BrE*); **~ para repostar combustible** *m* TRANSP refueling craft (*AmE*), refuelling craft (*BrE*)

Nb *abr* (*niobio*) METAL, QUÍMICA Nb (*niobium*)

NCR *abr* (*autocopiativo no carbonado*) PAPEL NCR (*no carbon required*)

Nd *abr* (*neodimio*) ELECTRÓN, METAL, QUÍMICA Nd (*neodymium*)

N₂ *abr* (*gas de nitrógeno*) GAS N_2 (*nitrogen gas*)

Ne *abr* (*neón*) ING ELÉC, QUÍMICA Ne (*neon*)

neblina *f* CINEMAT aerial fog, air fog, CONTAM, METEO, TEC PETR *perforación* mist; **~ ácida** *f* CONTAM acid fog

nebula: **~ Veil** *f* TEC ESP Veil Nebula

nebulio *m* QUÍMICA nebulium

nebulizador *m* AGRIC mist sprayer, PROC QUÍ nebulizer, PROD spray, QUÍMICA nebulizer

nebulizar[1] *vt* AGRIC fog

nebulizar[2] *vi* CONTAM mist

nebulosa *f* TEC ESP nebula; **~ del Caballete** *f* TEC ESP Horsehead nebula; **~ del Cangrejo** *f* TEC ESP Crab nebula; **~ cometaria** *f* TEC ESP cometary nebula; **~ difusa** *f* TEC ESP diffuse nebula; **~ emisión** *f* TEC ESP emission nebula; **~ fría de baja densidad** *f* TEC ESP cold, low-density nebula; **~ oscura** *f* TEC ESP dark nebula; **~ planetaria** *f* TEC ESP planetary nebula; **~ radiante, brillante y caliente** *f* TEC ESP hot, bright and radiating nebula; **~ de tipo HII** *f* TEC ESP HII-type nebula

Nebulosa: **~ en Anillo** *f* TEC ESP Ring Nebula; **~ de Orión** *f* TEC ESP Orion Nebula

nebulosidad *f* GAS cloudiness

necesidad: **~ de nitrógeno** *f* AGRIC nitrogen requirement; **~ total** *f* PROD total requirement

necesidades: **~ totales** *f pl* PROD gross requirements

necrosado *adj* AGRIC necrotic

necrosis *f* AGRIC necrosis

NEF *abr* (*predicción de exposición al ruido*) ACÚST NEF (*noise exposure forecast*)

nefelina *f* MINERAL elaeolite, nepheline; **~ meteorizada** *f* MINERAL lieberenite

nefelinita *f* PETROL nephelinite, nephelinyte

nefelita *f* MINERAL nephelite

nefelometría *f* QUÍMICA nephelometry

nefelómetro *m* INSTR, LAB, QUÍMICA nephelometer, turbidity meter

nefoscópico *m* TRANSP AÉR ceilometer

nefrita *f* GEOL kidney stone, nephrite, MINERAL nephrite

negación *f* TELECOM denial

negador *m* INFORM&PD negator

negativización: **~ del disco** *f* ÓPT disc mastering (*BrE*), disk mastering (*AmE*)

negativo[1] *adj* ELEC *electrodo*, MATEMÁT negative

negativo[2] *m* ACÚST *grabación* peeling, CINEMAT, FÍS, FOTO, IMPR, MATEMÁT negative, ÓPT *de un disco fonográfico* master, PROD stencil plate; **~ de contacto** *m* FOTO contact negative; **~ contrastado** *m* CINEMAT, FOTO contrasted negative; **~ para copias** *m* CINEMAT release negative; **~ cortado** *m* CINEMAT, FOTO cut negative; **~ delgado** *m* CINEMAT thin negative; **~ deteriorado** *m* FOTO spoiled negative; **~ directo** *m* FOTO straight negative; **~ dupe** *m* CINEMAT dupe negative; **~ dupe a color** *m* CINEMAT color dupe neg (*AmE*), colour dupe neg (*BrE*); **~ duplicado** *m* CINEMAT dupe negative; **~ de fondo** *m* CINEMAT background negative; **~ para hacer discos** *m* ACÚST stamper; **~ intermedio** *m* CINEMAT intermediate negative; **~ óptico** *m* CINEMAT optical negative; **~ original** *m* CINEMAT master negative; **~ original sin cortar** *m* CINEMAT original uncut negative; **~ en papel** *m* FOTO paper negative; **~ reversible intermedio** *m* CINEMAT intermediate reversal negative; **~ para la selección de colores** *m* FOTO, TV color-separation negative (*AmE*), colour-separation negative (*BrE*); **~ con separación de colores** *m* CINEMAT separation negative; **~ para la separación de colores** *m* CINEMAT color-separation negative (*AmE*), colour-separation negative (*BrE*); **~ de sonido** *m* CINEMAT sound negative; **~ de sonido óptico** *m* CINEMAT optical sound negative; **~ de tres capas de emulsión** *m* CINEMAT three-strip negative

negatoscopio *m* Esp (*cf cajeta de iluminación AmL*) FOTO light box

negatrón *m* ELECTRÓN, FÍS PART negatron

negrita *f* IMPR bold, bold face

negro: **~ de carbón** *m* CARBÓN carbon black; **~ de espolvorear** *m* PROD *moldería* face dust; **~ de fundición** *m* PROD blackening, blacking, REVEST blackwash; **~ para fundición** *m* REVEST founder's black; **~ de humo** *m* CARBÓN, P&C *pigmento, carga* carbon black; **~ de humo activado** *m* P&C *pigmento, carga* activated carbon black; **~ de humo de gas natural** *m* TEC PETR *petroquímica* carbon black; **~ de humo de horno y módulo alto** *m* P&C high modulus furnace carbon black (*HMF carbon black*); **~ de humo de horno y partícula fina** *m* P&C fine furnace carbon black (*FF carbon black*); **~ de humo de horno y refuerzo medio** *m* P&C semireinforcing carbon black (*SRF carbon black*); **~ de imprenta** *m* COLOR printer's black, printing black; **~ líquido** *m*

PROD blackwash; ~ **de referencia** m TV reference black; ~ **de uranio** m NUCL uranium black

negrón m AGRIC black mold (AmE), black mould (BrE)

negrura: ~ **del estilo** f IMPR weight of face; ~ **del tipo** f IMPR weight of type

nematicida m AGRIC nematicide

nemátodo m AGRIC nematode; ~ **cecidio** m AGRIC cyst nematode

nemónico adj INFORM&PD mnemonic

neoabiético adj QUÍMICA neoabietic

neodimio m (Nd) ELECTRÓN, METAL, QUÍMICA neodymium (Nd)

neoergosterol m QUÍMICA neoergosterol

neón m (Ne) ING ELÉC, QUÍMICA neon (Ne)

neopreno m CONST, EMB, P&C, QUÍMICA neoprene

neotectónico adj GEOL neotectonic

neovolcánico adj GEOL neovolcanic

neper m ELECTRÓN, FÍS neper

neperio m (N) ACÚST neper (N)

neptunio m (Np) FÍS RAD, QUÍMICA neptunium (Np)

nerol m QUÍMICA nerol

nervado adj CONST, ING MECÁ, MECÁ ribbed

nervadura f CARBÓN, IMPR rib, ING MECÁ stiffening, PROD, TEC ESP rib, TRANSP AÉR rib, web; ~ **cruzada** f ING MECÁ cross girth; ~ **falsa** f TRANSP AÉR false rib; ~ **transversal** f ING MECÁ cross girth

nervaduras: **con** ~ adj CONST, ING MECÁ, MECÁ ribbed

nerviadura: ~ **central** f AGRIC midrib; ~ **foliar** f AGRIC leaf vein

nervio m CARBÓN botánica rib, ING MECÁ stiffener, MINAS geología horse; ~ **central** m METAL midrib; ~ **de fijación** m PROD locating rib; ~ **vivo del acero** m PROD sharp ridge of steel

nervión: ~ **de las guías del flap** m TRANSP AÉR flap track rib

Net: **la** ~ f INFORM&PD the Net

neto m TEC PETR net

netsonde m OCEAN netsonde

neumático[1] adj FÍS, VEH pneumatic

neumático[2] m AUTO, VEH pneumatic tire (AmE), pneumatic tyre (BrE), tire (AmE), tyre (BrE); ~ **de alta presión** m TRANSP high-pressure tire (AmE), high-pressure tyre (BrE); ~ **antihielo** m TRANSP AÉR de-icer boot; ~ **de borde** m TRANSP AÉR chine tire AmE, chine tyre BrE; ~ **con cámara** m AUTO, VEH balloon tire (AmE), balloon tyre (BrE); ~ **sin cámara** m AUTO, VEH tubeless tire (AmE), tubeless tyre (BrE); ~ **con capas de tejido cruzadas** m VEH bias-ply tire (AmE), cross-ply tyre (BrE); ~ **con capas de tejido sesgadas** m VEH bias-ply tire (AmE), cross-ply tyre (BrE); ~ **con clavos** m VEH studded tire (AmE), studded tyre (BrE); ~ **desgastado** m VEH bald tire (AmE), bald tyre (BrE); ~ **desinflado** m VEH flat tire (AmE), flat tyre (BrE); ~ **esculpido** m VEH studded tire (AmE), studded tyre (BrE); ~ **de faldilla** m TRANSP AÉR chine tire AmE, chine tyre BrE; ~ **para nieve** m VEH snow tire (AmE), snow tyre (BrE); ~ **con pliegues diagonales** m P&C diagonal ply tire (AmE), diagonal ply tyre (BrE); ~ **con pliegues radiales** m AUTO, P&C, VEH radial-ply tire (AmE), radial-ply tyre (BrE); ~ **con pliegues transversales** m P&C, VEH bias-ply tire (AmE), cross-ply tyre (BrE); ~ **con poco dibujo** m AUTO, VEH plain-tread tire (AmE), plain-tread tyre (BrE); ~ **radial** m AUTO, P&C, VEH radial tire (AmE),

radial tyre (BrE); ~ **de recambio** m AUTO, VEH spare tire (AmE), spare tyre (BrE), spare wheel (BrE); ~ **recauchutado** m AUTO, P&C, VEH retreaded tire (AmE), retreaded tyre (BrE); ~ **con relieve desgastado** m VEH plain-tread tire (AmE), plain-tread tyre (BrE); ~ **de repuesto** m AUTO, VEH spare tire (AmE), spare tyre (BrE), spare wheel (BrE); ~ **salvavidas** m TRANSP AÉR inflatable dinghy; ~ **sin talón** m VEH bald tire (AmE), bald tyre (BrE); ~ **con tejido en diagonal** m VEH bias-ply tire (AmE), cross-ply tyre (BrE); ~ **para todo terreno** m VEH all-terrain tire (AmE), all-terrain tyre (BrE), town-and-country tire (AmE), town-and-country tyre (BrE)

neumáticos m pl FÍS pneumatics

neumatóforo m CARBÓN float

neumatógeno m P&C blowing-agent

neumatolisis f GEOL pneumatolysis

neumoencefalitis: ~ **aviar** f AGRIC Newcastle disease

neumomotor m ING MECÁ pneumatic motor

neuramínico adj QUÍMICA neuraminic

neurodina f QUÍMICA neurodine

neurona f INFORM&PD neuron

neutralidad f ELEC, FÍS, QUÍMICA neutrality

neutralización f CARBÓN, DETERG, ELEC neutralization, ELECTRÓN, FÍS passivation, HIDROL de detergentes, ING ELÉC, P&C neutralization, QUÍMICA neutralization, passivation, TEC ESP passivation; ~ **de carga** f ELEC electroestática charge neutralization; ~ **de control** f TRANSP AÉR override control

neutralizador[1] adj QUÍMICA neutralizing

neutralizador[2] m QUÍMICA neutralizer, neutralizing, TEC ESP neutralizer; ~ **de ácidos** m CONTAM, QUíMICA acid neutralizing; ~ **acústico** m TEC ESP acoustic blanket; ~ **de tono** m TEC ESP tone disabler

neutralizante: ~ **de ácidos** m CONTAM, QUíMICA acid neutralizing

neutralizar vt ELECTRÓN, FÍS passivate, INFORM&PD disable, QUÍMICA, TEC ESP passivate, TEC PETR perforación, producción kill

neutreto: ~ **eta** m FÍS, FÍS PART eta neutral meson

neutrino m FÍS, FÍS PART neutrino; ~ **electrónico** m ELECTRÓN, FÍS electron neutrino; ~ **muón** m FÍS, FÍS PART muon neutrino; ~ **muónico** m FÍS, FÍS PART muon neutrino; ~ **tau** m FÍS, FÍS PART tau neutrino; ~ **tauón** m FÍS, FÍS PART tauon neutrino

neutro adj ELEC uncharged, FÍS, QUÍMICA neutral

neutrón m (n) ELEC, FÍS, FÍS PART, FÍS RAD, PETROL neutron (n); ~ **difuso** m NUCL scattered neutron; ~ **epicádmico** m NUCL epicadmium neutron; ~ **de fisión** m FÍS PART, FÍS RAD, NUCL fission neutron; ~ **de fisión no moderado** m FÍS PART, FÍS RAD, NUCL unmoderated fission neutron; ~ **inmediato** m FÍS RAD prompt neutron; ~ **lento** m FÍS, FÍS PART, NUCL slow neutron; ~ **naciente** m NUCL nascent neutron; ~ **que no ha entrado en colisión** m NUCL uncollided neutron; ~ **rápido** m FÍS, FÍS PART, NUCL fast neutron; ~ **retardado** m FÍS, FÍS RAD, NUCL delayed neutron; ~ **térmico** m FÍS, FÍS RAD, NUCL thermal neutron; ~ **ultrafrío** m NUCL ultracold neutron; ~ **virgen** m NUCL virgin neutron

nevera f ALIMENT, ELEC, ING MECÁ, REFRIG, TERMO refrigerator; ~ **de hielo** f REFRIG ice refrigerator

newjanskita f MINERAL nevyanskite

newsgroup m INFORM&PD newsgroup

newton m (N) ELEC unidad de fuerza, FÍS, METR newton (N)

newtoniano *adj* FÍS Newtonian

NH₃ *abr* (*amoniaco*) GEN NH₃ (*ammonia*)

Ni *abr* (*níquel*) ING ELÉC, METAL, MINERAL, QUÍMICA Ni (*nickel*)

nialamida *f* QUÍMICA nialamide

nicho *m* CONST *túnel* manhole, PROD bellhousing; ~ **ecológico** *m* AGRIC ecological niche; ~ **del sonar** *m* OCEAN *en casco del buque* sonar hole

nicoles: ~ **cruzados** *m pl* FÍS *prismas polarizados* crossed Nicols

nicolo *m* QUÍMICA niccolum

nicopirita *f* MINERAL nicopyrite

nicotina *f* QUÍMICA nicotine

nicotinamida *f* QUÍMICA niacinamide, nicotinamide

nicotínico *adj* QUÍMICA nicotinic

nicotirina *f* QUÍMICA nicotyrine

nidrazida *f* QUÍMICA nydrazid

niebla *f* CINEMAT aerial fog, air fog, METEO, REFRIG fog, mist; ~ **por advección** *f* METEO fog; ~ **marina** *f* METEO fog; ~ **precipitante** *f* METEO wet fog; ~ **que moja** *f* METEO damp fog, wet fog; ~ **salina** *f* P&C *prueba de pintura* salt spray

nieve *f* ELECTRÓN *televisión*, METEO, REFRIG, TV snow; ~ **ácida** *f* CONTAM acid snow; ~ **acumulada** *f* METEO snowdrift; ~ **blanda** *f* METEO soft snow; ~ **caída** *f* METEO snowfall; ~ **carbónica** *f* CARBÓN, REFRIG carbon dioxide snow; ~ **dura** *f* METEO hard snow; ~ **fundente** *f* REFRIG slush; ~ **de helero** *f* HIDROL glacier snow; ~ **limpia** *f* CONTAM clean snow

nilón *m* P&C *grupo de polímeros*, QUÍMICA nylon

nimbostratus *m* (*Ns*) METEO nimbostratus (*Ns*)

niobio *m* (*Nb*) METAL, QUÍMICA niobium (*Nb*)

niobita *f* MINERAL, QUÍMICA columbite, niobite

NIP *abr* (*número de identificación personal*) ELECTRÓN PIN (*personal identification number*)

niple *m* CONST, ING MECÁ nipple; ~ **de desagüe** *m* ING MECÁ drain nipple; ~ **de drenaje** *m* ING MECÁ drain nipple; ~ **de purga** *m* ING MECÁ drain nipple; ~ **tubular** *m* ING MECÁ barrel nipple

níquel *m* (*Ni*) ING ELÉC, METAL, MINERAL, QUÍMICA nickel (*Ni*); ~ **antimonial** *m* MINERAL breithauptite; ~ **arsenical** *m* MINERAL copper nickel, kupfernickel

niquelación *f* PROD nickeling

niquelado¹ *adj* QUÍMICA nickelic

niquelado² *m* ELEC *proceso* nickel plating, PROD nickel plating, nickelage; **~-plateado** *m* ING ELÉC nickel-silver

niqueladura *f* PROD nickeling

niquelar *vt* ELEC, PROD, REVEST nickel-plate

niquelina *f* MINERAL copper nickel, nickeline

niqueloceno *m* QUÍMICA nickelocene

niquelocre *m* MINERAL nickel-ocher (*AmE*), nickel-ochre (*BrE*)

niqueloso *adj* QUÍMICA nickelous

niqueluro *m* PROD nickelure

nistatina *f* QUÍMICA nystatin

nitidez *f* CINEMAT definition, sharpness, ELECTRÓN definition, FOTO definition, sharpness, IMPR sharpness, INFORM&PD definition; ~ **del contorno** *f* TV edge enhancement

nítido *adj* CINEMAT, FOTO, IMPR, PROD sharp

nitracina *f* QUÍMICA nitrazine

nitración *f* QUÍMICA nitration; ~ **en fase vapor** *f* PROC QUÍ vapor phase nitration (*AmE*), vapour phase nitration (*BrE*)

nitramina *f* QUÍMICA nitramine

nitrar *vt* QUÍMICA nitrate

nitratina *f* QUÍMICA nitratine

nitratita *f* MINERAL nitratite

nitrato *m* QUÍMICA nitrate; ~ **de celulosa** *m* CINEMAT, P&C cellulose nitrate, QUÍMICA nitrocellulose; ~ **de plata** *m* QUÍMICA silver nitrate; ~ **potásico** *m* ALIMENT, QUÍMICA potassium nitrate; ~ **de potasio** *m* C&V saltpeter (*AmE*), saltpetre (*BrE*), QUÍMICA potassium nitrate; ~ **sódico** *m* QUÍMICA sodium nitrate; ~ **de sodio** *m* C&V, QUÍMICA soda niter (*AmE*), soda nitre (*BrE*), sodium nitrate

nítrico *adj* QUÍMICA nitric

nitrificación *f* HIDROL, INSTAL HIDRÁUL, QUÍMICA nitrification

nitrificante *adj* HIDROL, INSTAL HIDRÁUL, QUÍMICA nitrifying

nitrificar *vt* HIDROL, INSTAL HIDRÁUL, QUÍMICA nitrify

nitrificarse *v refl* HIDROL, INSTAL HIDRÁUL, QUÍMICA nitrify

nitrilo *m* QUÍMICA nitrile, nitryl

nitrina *f* QUÍMICA nitrin

nitrito *m* QUÍMICA nitrite

nitritoide *adj* QUÍMICA nitritoid

nitro *m* QUÍMICA nitro; ~ **de Perú** *m* QUÍMICA nitratine, Peru saltpeter (*AmE*), Peru saltpetre (*BrE*); ~ **sódico** *m* MINERAL soda niter (*AmE*), soda nitre (*BrE*)

nitroalgodón *m* *Esp* (*cf fulmicotón AmL*) MINAS *explosivos* guncotton, nitrocellulose

nitroamina *f* QUÍMICA nitroamine

nitroanilina *f* QUÍMICA nitroaniline

nitrobenceno *m* QUÍMICA nitrobenzene

nitrocelulosa *f* QUÍMICA nitrocellulose, *fotografía, diálisis* collodion

nitrocelulósico *adj* QUÍMICA nitrocellulosic

nitrocloroformo *m* QUÍMICA nitrochloroform

nitroderivado *m* QUÍMICA nitro-compound

nitroetano *m* QUÍMICA nitroethane

nitrofenol *m* QUÍMICA nitrophenol

nitroformo *m* QUÍMICA nitroform

nitrogelatina *f* MINAS nitrogelatine

nitrógeno *m* (*N*) GAS, QUÍMICA, TEC PETR nitrogen (*N*); ~ **gaseoso** *m* QUÍMICA nitrogen gas; ~ **gaseoso protector** *m* NUCL nitrogen cover gas; ~ **líquido** *m* TEC ESP liquid nitrogen

nitrogenoso *adj* QUÍMICA nitrogenous

nitroglicerina *f* *Esp* (*cf glonoina AmL*) D&A, MINAS, QUÍMICA nitroglycerine

nitroglucosa *f* QUÍMICA nitroglucose

nitroindol *m* QUÍMICA nitroindole

nitrometano *m* QUÍMICA nitromethane

nitrómetro *m* QUÍMICA nitrometer

nitrón *m* QUÍMICA nitron

nitronaftaleno *m* QUÍMICA nitronaphthalene

nitronio *m* QUÍMICA nitronium

nitroparafina *f* QUÍMICA nitroparaffin

nitrosación *f* QUÍMICA nitrosation

nitrosato *m* QUÍMICA nitrosate

nitrosificante *adj* QUÍMICA nitrosifying

nitrosil *m* QUÍMICA nitrosyl

nitrosilo *m* QUÍMICA nitrosyl

nitrosito *m* QUÍMICA nitrosite

nitroso *adj* QUÍMICA nitrous

nitrosocloruro *m* QUÍMICA nitrosochloride

nitrosubstituido *adj* QUÍMICA nitrosubstituted

nitrosulfúrico *adj* QUÍMICA nitrosulfuric (*AmE*), nitrosulphuric (*BrE*)

nitrosustituido *adj* QUÍMICA nitrosubstituted
nitrotartárico *adj* QUÍMICA nitrotartaric
nitrotolueno *m* QUÍMICA nitrotoluene
nitrox *m* OCEAN nitrox
nitruración *f* ING MECÁ nitriding, QUÍMICA nitridation,
nitriding; **~ gaseosa** *f* GAS, TERMO gas nitriding
nitrurar *vt* ING MECÁ, QUÍMICA nitride
nitruro *m* QUÍMICA nitride; **~ de boro cúbico** *m* ING
MECÁ cubic boron nitride; **~ de boro sintetizado** *m*
QUÍMICA borazon; **~ de silicio** *m* ELECTRÓN silicon
nitride; **~ de telurio** *m* ING ELÉC tellurium nitride
nivación *f* QUÍMICA nivation
nivel[1]: **a ~** *adj* CONST flush; **a ~ del suelo** *adj* CONST
level with the ground
nivel[2]: **a ~** *adv* PROD flush
nivel[3] *m* GEN level, CARBÓN *aparatos topográficos*
bubble, INFORM&PD level, layer, MINAS counter, *de
maniobras* battery, PROD levelness;
▪ a **~ A** *m* CONST A level; **~ de abrasión agregado**
m CONST aggregate abrasion value (*AA*); **~ de aceite**
m AUTO oil line; **~ para aceite** *m* ING MECÁ oil gage
(*AmE*), oil gauge (*BrE*); **~ de aceleración sonora** *m*
ACÚST sound acceleration level; **~ de aceptor** *m*
ELECTRÓN acceptor level; **~ de acidez** *m* CONTAM
caracterización del medio acidity level; **~ de ácido** *m*
AUTO, CONST acid level; **~ de acreditación** *m*
INFORM&PD clearance level; **~ de agua** *m* QUÍMICA
water level; **~ de agua crítico** *m* AGUA critical water
level; **~ del agua de descarga** *m* ENERG RENOV
tailwater level; **~ de agua freática** *m* AGRIC, AGUA,
CARBÓN, CONST, HIDROL ground-water level; **~ de
agua subterránea** *m* AGRIC, AGUA, CARBÓN, CONST
ground-water level, HIDROL ground-water level, water table; **~ del agua de zona de transición** *m* AGUA
intermediate water level; **~ de aguas abajo** *m*
HIDROL downstream level, INSTAL HIDRÁUL tailwater,
waste water; **~ de aguas arriba** *m* INSTAL HIDRÁUL *en
el canal de toma de agua* headwater, headwater level;
~ de aguas colgadas *m* HIDROL perched water table;
~ de aire *m* CONST INSTRUM air level, spirit level;
~ de alimentación *m* AGRIC wintering level; **~ de
anteojo** *m* CONST *topografía* surveyor's level; **~ de
anteojo corto** *m* CONST dumpy level; **~ aparente del
agua** *m* AGUA apparent water table; **~ de aplicación**
m INFORM&PD application layer; **~ de autorización**
m INFORM&PD clearance level; **~ de avance** *m* CONST
rate of progress; **~ azul negro** *m* TV blue-black level;
▪ b **~ de bajamar** *m* AGUA low-water level; **~ bajo
de circuito lógico** *m* ELECTRÓN low logic level;
~ bajo de torsión *m* TEXTIL low level of twist; **~ de
base de las olas** *m* OCEAN wave base; **~ de blanco** *m*
TELECOM picture white, TV white level; **~ del blanco**
m TV bright level; **~ de blandura** *m* PROD soft grade;
~ de blandura intermedia *m* PROD medium-soft
grade; **~ de borrado** *m* TV blanking level; **~ de
burbuja de aire** *m* MECÁ bubble level, METR spirit
level;
▪ c **~ a caballo** *m* ING MECÁ striding level; **~ de
calidad aceptable** *m* CALIDAD acceptable quality
level (*AQL*); **~ de la carga** *m* PROD *alto horno* stock
line; **~ cero** *m* CINEMAT zero level; **~ del
componente** *m* ELECTRÓN component level; **~ de
conducción** *m* ELECTRÓN carrier level; **~ de
contaminación** *m* CONTAM degree of pollution;
~ de contaminación sonora *m* (*LNP*) ACÚST,
CONTAM, SEG noise pollution level (*NPL*); **~ de

control *m* CALIDAD inspection level, TRANSP AÉR
control stage; **~ de corriente continua** *m* TV DC
level; **~ de corte del pilote** *m* CARBÓN, CONST pile
cutoff level; **~ de cuantificación** *m* ELECTRÓN,
INFORM&PD quantization level; **~ cumbre de un
canal** *m* AGUA summit canal;
▪ d **~ de densidad de energía sonora** *m* ACÚST
sound energy density level; **~ de desgaste** *m* CONST
wear rate; **~ de disparo** *m* ELECTRÓN trigger level;
~ de distorsión por capa adyacente *m* TV print-
through level; **~ de distribución de datos técnicos**
m PROD engineering issue level; **~ donador** *m* ELEC-
TRÓN donor level; **~ de dopado** *m* ELECTRÓN doping
level;
▪ e **~ efectivo de ruido percibido** *m* (*EPNL*) ACÚST
effective perceived-noise level (*EPNL*); **~ de eficacia
de la máquina** *m* PROD machine utilization degree;
~ elevado de esfuerzo *m* PROD elevated stress level;
~ de emisión *m* INFORM&PD release level; **~ de
emisión espúreos** *m* TEC ESP spurious emission
level; **~ de emisión falso** *m* TEC ESP spurious
emission level; **~ por encima del umbral** *m* ACÚST
level above threshold; **~ energético** *m* FÍS, FÍS PART,
FÍS RAD, GEOFÍS, NUCL energy level; **~ energético
fundamental** *m* NUCL normal energy level; **~ energético normal** *m* NUCL normal energy level;
~ energético nulo *m* NUCL zero energy level; **~ de
energía** *m* FÍS, FÍS PART, FÍS RAD, GEOFÍS, NUCL,
QUÍMICA energy level, TEC ESP quantum state; **~ de
energía atómica** *m* NUCL atomic energy level; **~ de
energía vibracional de la molécula** *m* FÍS RAD
molecular vibrational energy level; **~ de enlace de
datos** *m* INFORM&PD data link layer; **~ de entrada** *m*
TELECOM, TV input level; **~ de entrada por debajo
de saturación** *m* TEC ESP *comunicaciones* input back-
off; **~ esférico** *m* INSTR *de burbuja de aire* circular
level; **~ de estiaje** *m* AGUA low-water level;
▪ f **~ de Fermi** *m* FÍS Fermi level; **~ físico** *m*
INFORM&PD *interconexión de sistemas abiertos* physical layer; **~ de fluido en reposo** *m* TEC PETR static
fluid level; **~ de fondo** *m* CONTAM background
concentration, background level, MINAS *plano horizontal* bottom level; **~ freático** *m* AGRIC, AGUA,
CARBÓN, CONST, HIDROL ground-water level,
ground-water table, phreatic level; **~ freático
aparente** *m* AGRIC, AGUA, GEOFÍS, HIDROL apparent
water table; **~ freático descendente** *m* AGRIC, AGUA,
GEOFÍS, HIDROL receding water table; **~ fundamental**
m NUCL normal level;
▪ g **~ de gastos de entretenimiento** *m* FERRO
maintenance level; **~ de gastos de mantenimiento**
m FERRO maintenance level;
▪ h **~ sin helar** *m* CARBÓN frost-free level;
~ hidrostático *m* OCEAN hydrostatic level; **~ de
horquetas** *m* CONST *topografía* Y-level;
▪ i **~ de impurezas** *m* ELECTRÓN impurity level;
~ de intensidad *m* ACÚST, FÍS intensity level, FÍS RAD
level of intensity, ING ELÉC intensity level; **~ de
intensidad de sonido** *m* ACÚST, FÍS sound intensity
level; **~ de intensidad sonora** *m* ACÚST, FÍS sound
intensity level; **~ de interfase** *m* ING MECÁ interface
level; **~ de interfaz** *m* ING MECÁ interface level; **~ de
interferencia admisible** *m* TEC ESP permissible level
of interference; **~ de interferencia con la palabra** *m*
(*SIL*) ACÚST speech interference level (*SIL*); **~ de

inyección *m* ELECTRÓN injection level; ~ **de irradiación de diseño** *m* NUCL design irradiation level;

~ l ~ **del lecho** *m* HIDROL bed level; ~ **libre de helada** *m* CARBÓN frost-free level; ~ **de limitación** *m* ELECTRÓN clipping level; ~ **límite** *m* ING ELÉC overload level; ~ **límite superficial** *m* METEO surface boundary level; ~ **de línea** *m* TV line level; ~ **de líquido** *m* EMB, TEC PETR liquid level; ~ **de llenado** *m* EMB fill level; ~ **de lógica** *m* ELECTRÓN, INFORM&PD logic level; ~ **lógico** *m* ELECTRÓN, INFORM&PD logic level; ~ **lógico alto** *m* ELECTRÓN logic high; ~ **lógico bajo** *m* ELECTRÓN logic low;

~ m ~ **del mar** *m* AGUA, CONST, ENERG RENOV, TRANSP MAR *navegación* sea level; ~ **de marea alta** *m* CONST high-water level; ~ **máximo de azul** *m* TV blue peak level; ~ **máximo de caudal** *m* HIDROL maximum flood level; ~ **máximo de rojo** *m* TV red peak level; ~ **máximo de verde** *m* TV green peak level; ~ **medio anual del agua** *m* AGUA annual mean water level; ~ **medio de dureza** *m* PROD medium grade; ~ **medio del mar** *m* OCEAN, TRANSP MAR *navegación* mean sea level; ~ **medio de marea** *m* OCEAN half tide level; ~ **de la meseta** *m* GEOL, TEC PETR plateau level; ~ **de modulación** *m* FÍS ONDAS modulation level;

~ n ~ **de negro** *m* TELECOM picture black, TV black level; ~ **nominal de aislación** *m* ELEC *tensiones de prueba* rated insulation level; ~ **normal** *m* NUCL normal level;

~ o ~ **del objetivo** *m* TV target layer; ~ **original del terreno** *m* CONST old ground level;

~ p ~ **de penetración** *m* CONST penetration grade; ~ **de penetración del TRC** *m* ELECTRÓN penetration CRT; ~ **percentil** *m* ACÚST percentile level; ~ **de potencia** *m* ACÚST power level, TELECOM output level; ~ **de potencia sónica** *m* FÍS sound power level; ~ **de potencia sonora** *m* ACÚST sound power level; ~ **potenciométrico** *m* TEC PETR *hidrogeología* potentiometric level; ~ **de presentación** *m* INFORM&PD presentation layer; ~ **de presión** *m* ACÚST pressure level; ~ **de presión acústica** *m* ACÚST, CONTAM, FÍS sound pressure level (*SPL*); ~ **de presión de sonido** *m* (*NPS*) ACÚST, CONTAM, FÍS sound pressure level (*SPL*); ~ **de presión sonora** *m* (*NPS*) ACÚST, CONTAM, FÍS sound pressure level (*SPL*); ~ **principal** *m* ING MECÁ striding level; ~ **de progreso** *m* CONST rate of progress; ~ **de protección** *m* PROD degree of protection; ~ **de puntería en elevación** *m* D&A range dial;

~ r ~ **de recepción** *m* ELECTRÓN reception level; ~ **de recorte** *m* TV clipping level; ~ **de red** *m* INFORM&PD network layer; ~ **de reducción de sondas** *m* OCEAN sounding datum level; ~ **de referencia** *m* CONST datum level, ELECTRÓN reference level; ~ **de referencia del audio** *m* ACÚST, TV reference audio level; ~ **de reproducción** *m* CINEMAT, TV reproduction level; ~ **rígido** *m* CONST *topografía* dumpy level; ~ **rojo-negro** *m* TV red-black level; ~ **de ruido** *m* CONTAM, FÍS ONDAS, TELECOM noise level;

~ s ~ **de salida** *m* TELECOM, TV output level; ~ **de salida por debajo de saturación** *m* TEC ESP output back-off; ~ **de señal** *m* ELECTRÓN, TELECOM signal level; ~ **de señal bajo** *m* ELECTRÓN low signal level; ~ **del seno de ola** *m* OCEAN wave trough level; ~ **de servicio** *m* PROD service level, TELECOM service layer;

~ **de sesión** *m* INFORM&PD session layer; ~ **de sobrecarga** *m* ING ELÉC overload level; ~ **de sonido** *m* ACÚST, FÍS loudness level; ~ **de sonoridad** *m* ACÚST, FÍS loudness level; ~ **sonoro** *m* ACÚST sound level; ~ **sonoro continuo equivalente** *m* ACÚST equivalent continuous-sound level; ~ **sonoro de impactos normalizado** *m* ACÚST normalized impact sound level; ~ **sonoro de pico** *m* ACÚST peak sound pressure; ~ **del suelo** *m* NUCL ground level; ~ **de supresión de la línea** *m* TV line blanking level;

~ t ~ **de tecleado** *m* TV key level; ~ **de tensión** *m* PROD voltage level; ~ **de tiro** *m* EMB *adhesivos* tack level; ~ **de transmisión de datos** *m* TELECOM data link layer; ~ **de transmisión del sonido de impacto** *m* ACÚST impact sound transmission level; ~ **de transporte** *m* PROD carry level; ~ **de Troughton** *m* CONST dumpy level;

~ u ~ **de umbral** *m* PROD threshold level; ~ **del umbral de audición** *m* ACÚST hearing threshold level;

~ v ~ **de velocidad acústica** *m* ACÚST acoustic velocity level; ~ **de ventilación** *m* CARBÓN coursing bubble; ~ **verde negro** *m* TV green black level; ~ **de versión** *m* INFORM&PD release level; ~ **de vía** *m* INSTR plate level; ~ **de video** *AmL*, ~ **de vídeo** *m* *Esp* TV video level; ~ **de video comprimido** *AmL*, ~ **de vídeo comprimido** *m* *Esp* TV compressed video level; geostrophic wind level; ~ **de viento geotrópico** *m* GEOFÍS, METEO geostrophic wind level; ~ **de visibilidad** *m* CONST sighted level; ~ **de vuelo** *m* TRANSP AÉR flight level

nivel[4]: **a ~ del agua** *fra* AGUA on a level with the water
nivelación *f* CARBÓN grading, CONST boning, grading, leveling (*AmE*), levelling (*BrE*), ELECTRÓN leveling (*AmE*), levelling (*BrE*), smoothing, ING MECÁ surfacing, MECÁ leveling (*AmE*), levelling (*BrE*), MINAS *topografía* foresight, PROD leveling (*AmE*), levelling (*BrE*); ~ **de la vía** *f* FERRO surfacing of the road (*BrE*), surfacing of the track (*AmE*)
nivelado[1] *adj* CONST leveled (*AmE*), levelled (*BrE*)
nivelado[2] *m* AUTO balancing
nivelado[3]: **estar ~ con** *vt* PROD be flush with
nivelador *m* TV clipper
niveladora *f* AGRIC buffalo grass, CONST grader; ~ **de motor** *f* CONST, TRANSP motor grader
nivelar *vt* CINEMAT level, CONST bone, grade, plumb, surveying, *topografía* level, ING MECÁ true, true up, MECÁ level, PROD level, flush, TRANSP AÉR level out; ~ **con nivel de burbuja** *vt* CONST bring the air bubble to the center of its run (*AmE*), bring the air bubble to the centre of its run (*BrE*)
niveleta *f AmL* (*cf línea de mira Esp*) MINAS lining sight
nivenita *f* NUCL nivenite
NLQ *abr* (*calidad cercana a la de carta*) INFORM&PD NLQ (*near-letter-quality*)
NNI *abr* (*índice de ruido y número de operaciones*) ACÚST NNI (*noise and number index*)
necesita: **que no ~ plancha** *adj* TEXTIL noniron
No *abr* (*nobelio*) FÍS RAD, QUÍMICA No (*nobelium*)
NO *abr* (*óxido nítrico*) CHEM, POLL NO (*nitric oxide*)
NO₂ *abr* QUÍMICA (*óxido de nitrogenó*) NO_2 (*nitrogen oxide*)
nobelio *m* (*No*) FÍS RAD, QUÍMICA nobelium (*No*)
noble *adj* QUÍMICA noble
noche: ~ **americana** *f* CINEMAT day for night

nocho: ~ **perecedero** m ING MECÁ collapsible core

nocivo adj QUÍMICA noxious, TEC PETR *sustancia* harmful

noctuido m AGRIC noctuid

nodal adj QUÍMICA nodal

nodo m GEN node; ~ **de acceso** m TELECOM access node; ~ **de acceso internacional** m TELECOM international gateway node (*IGN*); ~ **del árbol** m INFORM&PD leaf; ~ **ascendente** m TEC ESP ascending node; ~ **de conmutación del circuito virtual** m TELECOM virtual circuit switching node; ~ **de conmutación de paquetes** m INFORM&PD, TELECOM packet-switching node (*PSN*); ~ **descendente** m TEC ESP descending node; ~ **de entrada internacional** m TELECOM international gateway node (*IGN*); ~ **estrechado** m METAL constricted node; ~ **imperfecto** m ING ELÉC partial node; ~ **parcial** m ACÚST partial node; ~ **de transmisión** m TELECOM transmission node

nodular adj METAL globular

nódulo m C&V, CARBÓN, GEOL, NUCL, PETROL, QUÍMICA nodule; ~ **de corrosión** m NUCL corrosion nodule; ~ **de manganeso** m GEOL manganese nodule; ~ **de metal duro sinterizado** m ING MECÁ sintered hard metal pellet; ~ **de silex** m GEOL chert nodule

nogiratorio adj ING MECÁ nonswiveling (*AmE*), nonswivelling (*BrE*)

nombre m PROD *fábrica* make; ~ **de archivo** m INFORM&PD file name; ~ **cualificado** m INFORM&PD qualified name; ~ **de datos** m INFORM&PD data name; ~ **de fichero** m INFORM&PD file name; ~ **genérico** m INFORM&PD generic name; ~ **genérico de las avispas del género tentredínidos** m ALIMENT *fitopatología* sawfly; ~ **de procedimiento** m INFORM&PD procedure name; ~ **simbólico** m INFORM&PD symbolic name; ~ **del usuario** m INFORM&PD user name; ~ **de variable** m INFORM&PD variable name; ~ **vulgar** m QUÍMICA trivial name

nomenclatura f PROD parts list, QUÍMICA nomenclature

nómina f CINEMAT call sheet, report sheet

non: ~ **polar** adj P&C nonpolar

nonacosano m QUÍMICA nonacosane

nonano m QUÍMICA nonane

nonileno m QUÍMICA nonylene

nonílico adj QUÍMICA nonylic

nonio m ING MECÁ, MECÁ, METR vernier; ~ **de avance** m ING MECÁ sliding calipers (*AmE*), sliding callipers (*BrE*)

nonosa f QUÍMICA nonose

nontrorlita f MINERAL nontrorlite

noradrenalina f QUÍMICA noradrenalin, norepinephrina

noray m TRANSP MAR bollard, *amarre, radiofaro* post

norbergita f MINERAL norbergite

norbornadieno m QUÍMICA norbornadiene

norbornano m QUÍMICA norbornane

norbornileno m QUÍMICA norbornylene

nordestada f TRANSP MAR *viento* northeaster

nordestal adj TRANSP MAR *viento* northeast

nordeste[1]: **del** ~ adj TRANSP MAR northeasterly, *viento* northeast

nordeste[2] adv TRANSP MAR northeast; **del** ~ adv TRANSP MAR northeasterly; **en el** ~ adv TRANSP MAR northeast; **hacia el** ~ adv TRANSP MAR northeasterly; ~ **cuarta al este** adv TRANSP MAR *viento* northeast by

east; ~ **cuarta al norte** adv TRANSP MAR *viento* northeast by north

nordeste[3] m TRANSP MAR *viento* northeaster

norefedrina f QUÍMICA norephedrine

norepinefrina f QUÍMICA norepinephrina

noria f AGUA scoop water-wheel, scoop wheel, C&V spout; ~ **de agua** f TEXTIL water wheel

norita f PETROL norite

norma f ACÚST, CALIDAD standard, GEOL norm, ING MECÁ standard, SEG requirement, TELECOM regulation, standard, TV standard; ~ **de aprobación** f CALIDAD approval standard; ~ **ASA** f CINEMAT ASA standard; ~ **de banda baja** f TV low-band standard; ~ **de calidad ambiental** f CONTAM *legislación* ambient quality standard; ~ **de codificación** f TELECOM coding standard; ~ **de emisión** f CONTAM *legislación* emission standard; ~ **de emisiones ambientales** f CONTAM *legislación* ambient emission standard; ~ **de exploración** f TV scanning standard; ~ **industrial** f INFORM&PD industrial standard; ~ **para el intercambio de datos digitales** f IMPR digital data exchange standard (*DDES*); ~ **para el intercambio de datos numéricos** f IMPR digital data exchange standard (*DDES*); ~ **interna** f NUCL in-house standard; ~ **internacional** f CONST international standard; ~ **internacional de radiactividad** f CONTAM *legislación* international radioactivity standard; ~ **de laboratorio** f NUCL laboratory standard; ~ **obligatoria** f CALIDAD mandatory standard; ~ **de poner conductores eléctricos en espiral** f ELEC *electromagnetismo*, FÍS, ING ELÉC corkscrew rule; ~ **del sector** f INFORM&PD industrial standard; ~ **de seguridad** f SEG safety requirement; ~ **de seguridad basada en la identidad** f TELECOM identity-based security policy; ~ **técnica** f CONST technical regulation; ~ **de televisión** f TELECOM, TV television standard; ~ **de transmisión para filtros ultravioleta** f SEG transmission requirement for ultraviolet filter

Norma: ~ **Europea de Telecomunicaciones** f TELECOM European Telecommunication Standard (*ETS*)

normal[1] adj FÍS normal, INFORM&PD, ING MECÁ standard

normal[2] f GEOM, PETROL normal; ~ **común a los perfiles** f ING MECÁ line of action; ~ **de contacto** f ING MECÁ line of action, train of action

normalidad f CALIDAD, QUÍMICA normality

normalización f CALIDAD, GAS standardization, INSTAL TERM normalizing, MECÁ normalization, METAL normalization, *tratamiento* homogenizing, QUÍMICA, TELECOM, TV standardization; ~ **de envases** f EMB package for standardization; ~ **de métodos de prueba** f ING MECÁ standardization of test methods

normalizado adj INFORM&PD, ING MECÁ standard

normalizar vt CALIDAD standardize, INFORM&PD, MECÁ, METAL normalize, PROD, QUÍMICA standardize

normanita f MINAS normanite

normas f pl ING MECÁ, PROD, SEG guidelines, rules; ~ **básicas de seguridad** f pl SEG basic safety rules; ~ **y código de práctica locales** f pl PROD local standards and code of practice; ~ **de construcción para unidad motriz** f pl ING MECÁ design rules for drive power; ~ **dimensionales** f pl SEG dimensional requirements; ~ **sobre el esmerilado de metales** f pl SEG metal-grinding regulations; ~ **sobre esmeriles** f pl SEG abrasive-wheels regulations; ~ **sobre el lugar**

de trabajo *f pl* SEG workplace regulations; ~ **de seguridad** *f pl* SEG safety regulations, safety rules; ~ **de seguridad eléctrica** *f pl* ELEC, SEG electrical-safety requirements; ~ **de seguridad e higiene** *f pl* SEG health and safety requirements;

normativa *f* SEG regulation; ~ **de chapada** *f* REVEST pattern plating; ~ **sobre incendios** *f* SEG fire regulation; ~ **legal** *f* SEG statutory regulation; ~ **técnica** *f* CONST, TELECOM technical regulation

normativas: ~ **de seguridad aérea** *f pl* TRANSP AÉR air safety regulations

normorfina *f* QUÍMICA normorphine

nornarceína *f* QUÍMICA nornarceine

nornicotina *f* QUÍMICA nornicotine

nornordeste *m* TRANSP MAR north-northeast

nornoroeste *m* TRANSP MAR north-northwest

noroeste[1]: **del** ~ *adj* TRANSP MAR northwesterly

noroeste[2]: **hacia el** ~ *adv* TRANSP MAR northwesterly; ~ **cuarta al norte** *adv* TRANSP MAR *viento* northwest by north; ~ **cuarta al oeste** *adv* TRANSP MAR *viento* northwest by west

noroeste[3] *m* TRANSP MAR *viento* northwest wind, northwester

noropiánico *adj* QUÍMICA noropianic

nortazo *m* GEOL, METEO, TRANSP MAR *viento duro* north wind

norte[1]: **del** ~ *adj* TRANSP MAR northerly; ~ **arriba** *adj* TRANSP MAR *radar* north-up

norte[2]: **del** ~ *adv* TRANSP MAR northerly; **hacia el** ~ *adv* TRANSP MAR northerly; ~ **cuarta al nordeste** *adv* TRANSP MAR *viento* north by east; ~ **cuarta al noroeste** *adv* TRANSP MAR *viento* north by west

norte[3] *m* GEOL, METEO north wind, TRANSP MAR *punto de la brújula* north, *viento duro* north wind; ~ **verdadero** *m* TRANSP MAR *navegación* true north

Norte: ~ **magnético** *m* GEOFÍS magnetic North

norvalina *f* QUÍMICA norvaline

NOS *abr* (*sistema operativo de red*) INFORM&PD NOS (*network operating system*)

noscapina *f* QUÍMICA noscapine

noseana *f* MINERAL nosean

noselita *f* MINERAL noselite

nota *f* ACÚST note; ~ **de batido** *f* FÍS ONDAS beat note; ~ **de embarque** *f* TRANSP MAR shipping note; ~ **enarmónica** *f* ACÚST *música* enharmonic note; ~ **de expedición** *f* TRANSP MAR shipping note; ~ **marginal** *f* IMPR side note; ~ **modal** *f* ACÚST modal note; ~ **natural** *f* ACÚST natural; ~ **tonal** *f* ACÚST tonal note

notación *f* INFORM&PD, PROD, TELECOM notation; ~ **de base** *f* INFORM&PD radix notation; ~ **de base fija** *f* INFORM&PD fixed-base notation, fixed-radix notation; ~ **de base mixta** *f* INFORM&PD mixed-base notation, mixed-radix notation; ~ **binaria** *f* INFORM&PD binary notation; ~ **de coma fija** *f* INFORM&PD fixed-point notation; ~ **en coma flotante** *f* INFORM&PD, TELECOM floating-point notation; ~ **decimal** *f* INFORM&PD, MATEMÁT, TELECOM decimal notation; ~ **hexadecimal** *f* INFORM&PD hexadecimal notation; ~ **por infijos** *f* INFORM&PD infix notation; ~ **infix** *f* INFORM&PD infix notation; ~ **inversa por sufijos** *f* INFORM&PD reverse Polish notation; ~ **octal** *f* INFORM&PD octal notation; ~ **sin paréntesis** *f* INFORM&PD parenthesis-free notation; ~ **polaca inversa** *f* INFORM&PD reverse Polish notation; ~ **por prefijos** *f* INFORM&PD prefix notation;

~ **en punto flotante** *f* INFORM&PD, TELECOM floating-point notation; ~ **por sufijos** *f* INFORM&PD Polish notation, postfix notation, suffix notation

noticias: ~ **de carreteras** *f pl* AUTO, TELECOM, TRANSP, TV, VEH road news

noticiero: ~ **de televisión** *m* AmL (*cf programa de noticias Esp*) TV newscast

noticioso: ~ **de televisión** *m* AmL (*cf programa de noticias Esp*) TV newscast

notificación *f* CONTAM MAR notification; ~ **explícita de bloqueo regresivo** *f* TELECOM backward explicit congestion notification (*BECN*); ~ **explícita de congestión** *f* TELECOM explicit congestion notification (*ECN*); ~ **de falta de entrega** *f* TELECOM nondelivery notification (*NDN*); ~ **de intercambio de datos electrónicos** *f* TELECOM electronic data interchange notification (*EDIN*); ~ **de mejora** *f* SEG improvement notice; ~ **negativa** *f* TELECOM negative notification (*NN*); ~ **positiva** *f* TELECOM positive notification (*PN*); ~ **de prohibición** *f* SEG prohibition notice; ~ **de rechazo** *f* PROD rejection note; ~ **reenviada** *f* TELECOM forwarded notification

notificar: ~ **duración e importe** *vi* TELECOM advise duration and charge

noumeaíta *f* MINERAL noumeite

novilla *f* AGRIC heifer; ~ **castrada** *f* AGRIC spayed heifer; ~ **no cubierta** *f* AGRIC maiden heifer; ~ **que nunca ha parido** *f* AGRIC heiferette

novillo *m* AGRIC steer; ~ **destetado de engorde** *m* AGRIC feeder steer

novocaína *f* QUÍMICA novocaine

novolaca *f* P&C novolac

noy *m* ACÚST *unidad de ruido* noy

Np *abr* (*neptunio*) FÍS RAD, QUÍMICA Np (*neptunium*)

NPS *abr* (*nivel de presión sonora*) ACÚST SPL (*sound pressure level*)

Ns *abr* (*nimbostratus*) METEO Ns (*nimbostratus*)

NSSS *abr* (*sistema de producción nuclear de vapor*) NUCL NSSS (*nuclear steam supply system*)

NTSC *abr* (*Comité Nacional de Normas de Televisión*) TV NTSC (*National Television Standards Committee*)

nubarrón *m* METEO storm cloud

nube: ~ **ardiente** *f* GEOL *erupción volcánica* glowing cloud; ~ **caliente** *f* METEO warm cloud; ~ **de carga** *f* FÍS RAD charge cloud; ~ **de electrones** *f* ELECTRÓN, FÍS RAD, NUCL, QUÍMICA, TV electron cloud; ~ **electrónica** *f* ELECTRÓN, FÍS RAD, NUCL, QUÍMICA, TV electron cloud; ~ **galáctica** *f* TEC ESP galactic cloud; ~ **de gas** *f* GAS gas cloud; ~ **luminosa** *f* METEO luminous cloud; ~ **noctilucente** *f* METEO noctilucent cloud; ~ **de plasma** *f* GEOFÍS plasma cloud; ~ **subfundida** *f* METEO supercooled cloud; ~ **de tinta** *f* IMPR misting; ~ **de tormenta** *f* METEO thundercloud

nublado *adj* TV clouding

nubosidad *f* METEO *total o parcial* cloud amount, cloud cover, TRANSP MAR *tiempo atmosférico* cloud cover

nuboso *adj* METEO cloudy

nucleación *f* CRISTAL, METAL, METEO, NUCL nucleation; ~ **espontánea** *f* METAL spontaneous nucleation; ~ **de granos orientados** *f* METAL oriented nucleation; ~ **de la grieta** *f* METAL crack nucleation; ~ **de la hendidura** *f* METAL crack nucleation

nuclear *adj* D&A, ELEC, FÍS nuclear, NUCL nuclear, anular

nucleico *adj* QUÍMICA nucleic

nucleido *m* NUCL nuclide; **~ doblemente mágico** *m* NUCL twice magic nuclide; **~ hijo** *m* FÍS daughter product, FÍS RAD daughter nuclide, daughter product; **~ padre** *m* FÍS RAD, NUCL parent nuclide; **~ precursor** *m* NUCL parent nuclide

nucleína *f* QUÍMICA nuclein

núcleo *m* C&V nucleus, CINEMAT *para enrollar la película* CONST *presas, tornillos* ELEC, EMB core, FÍS, FÍS PART nucleus, core, GEOFÍS, GEOL core, INFORM&PD nucleus, kernel, core, MECÁ hub, NUCL nucleus, *del reactor* core, *de una ecuación integral* kernel, ÓPT core, PROD *fundería* newel, nucleus, TEC ESP, TEC PETR, TELECOM, TEXTIL core; **~ abierto** *m* ELEC *transformador* open core; **~ de acero silicioso** *m* ING ELÉC silicon steel core; **~ achatado** *m* NUCL oblate nucleus; **~ del aislador** *m* AUTO bead core; **~ alámbrico** *m* ING ELÉC wire core; **~ del ancla** *m* TRANSP MAR anchor boss; **~ anular** *m* NUCL annular core, ring core; **~ aterrajado** *m* PROD *fundería* struck-up core; **~ atómico** *m* FÍS PART, core, NUCL atomic core, atomic nucleus; **~ del azúcar** *m* ALIMENT sugar nucleus; **~ bien moderado** *m* NUCL well-moderated core; **~ de la bobina** *m* ELEC, ING ELÉC coil core; **~ bobinado** *m* ELEC *transformador* wire-wound core; **~ de cabeza magnética** *m* TV magnetic head core; **~ del cable** *m* P&C cable core; **~ central** *m* FÍS *fibra óptica* core, MINAS *de una roca para ser quitado* center core (*AmE*), centre core (*BrE*); **~ de chapas adosadas** *m* ING ELÉC laminated core; **~ de chapas en mazo** *m* ELEC *generador* laminated core; **~ del cometa** *m* TEC ESP comet core; **~ compensable** *m* ING ELÉC adjustable core; **~ compuesto** *m* FÍS, FÍS RAD, NUCL compound nucleus; **~ de condensación** *m* METEO, QUÍMICA condensation nucleus; **~ de la corriente** *m* OCEAN current core; **~ de difusión** *m* NUCL diffusion kernel; **~ de dislocación** *m* METAL dislocation core; **~ del electroimán** *m* ING ELÉC magnet core; **~ enrollado de cinta** *m* ING ELÉC tape-wound core; **~ espejo** *m* FÍS mirror nucleus; **~ de ferrita** *m* ELEC *de transformador* ferrite core, ELECTRÓN ferrite, INFORM&PD ferrite core, ING ELÉC ferrite core, *circuito* magnetic core, TV ferrite core, ferrite head; **~ ferromagnético** *m* ING ELÉC iron core; **~ de la fibra** *m* ÓPT fiber core (*AmE*), fibre core (*BrE*); **~ fundido** *m* TEC ESP molten core; **~ de la hélice** *m* TRANSP MAR propeller boss, propeller hub; **~ de hierro** *m* ELEC *transformador de máquina* core, ING ELÉC iron core; **~ de hierro fácilmente imanado por inducción** *m* ING ELÉC soft-iron core; **~ de hierro pulverizado** *m* ING ELÉC powdered-iron core; **~ en I** *m* ING ELÉC I core; **~ impar-impar** *m* FÍS, FÍS RAD odd-odd nucleus; **~ del inducido** *m* ELEC *generador*, ING ELÉC armature core; **~ inestable** *m* FÍS PART unstable nucleus; **~ interno** *m* GEOL *de la Tierra* inner core; **~ de laminación** *m* ELEC *generador* laminated core; **~ laminado** *m* ELEC *generador* laminated core; **~ lateral** *m* TEC PETR sidewall core; **~ licuado** *m* TEC ESP molten core; **~ magnético** *m* FÍS core, INFORM&PD, ING ELÉC *circuito* magnetic core; **~ de memoria** *m* ING ELÉC core; **~ de memoria magnética** *m* ING ELÉC magnetic core; **~ móvil** *m* ELEC *de transformador* moveable core, ING ELÉC plunger, MINAS core plunger; **~ oblato** *m* NUCL oblate nucleus; **~ de pared** *m* TEC PETR *evaluación de formación* sidewall core; **~ par-impar** *m* FÍS, FÍS RAD even-odd nucleus, odd-even nucleus; **~ par-par** *m* FÍS, FÍS RAD even-even nucleus; **~ del perfil de índices gradual** *m* ING ELÉC graded-index core; **~ primitivo** *m* INFORM&PD kernel; **~ primitivo de seguridad** *m* INFORM&PD security kernel; **~ del radiador** *m* AUTO, VEH *sistema de refrigeración* radiator core; **~ receptor plegable** *m* CINEMAT collapsible take-up core; **~ rectangular** *m* ING ELÉC square loop; **~ rectangular de ferrita** *m* ING ELÉC square-loop ferrite; **~ del relé** *m* ING ELÉC relay core; **~ de la resistencia** *m* ING ELÉC resistor core; **~ resistente** *m* CONST hardcore; **~ de retroceso** *m* FÍS recoil nucleus; **~ de ruido** *m* CONST noise bund; **~ de sal marina** *m* OCEAN sea salt nucleus; **~ saturado** *m* ING ELÉC saturated core; **~ de seguridad** *m* INFORM&PD security kernel; **~ del sistema operativo** *m* INFORM&PD operating system kernel; **~ superpesado** *m* FÍS superheavy nucleus; **~ T** *m* CINEMAT T-core; **~ terrestre** *m* ENERG RENOV, GEOFÍS, GEOL earth's core; **~ tórico** *m* ING ELÉC, NUCL toroidal core; **~ del transformador** *m* ELEC, ING ELÉC transformer core; **~ del uranio** *m* NUCL uranium nucleus

nucleofilia *f* QUÍMICA nucleophilicity

nucleofílico *adj* QUÍMICA nucleophilic

nucleohistona *f* QUÍMICA nucleohistone

nucleolina *f* QUÍMICA nucleolin

nucleón *m* FÍS, NUCL nucleon; **~ periférico** *m* FÍS, NUCL peripheral nucleon

nucleónica *f* FÍS, NUCL nucleonics

nucleopropulsado *adj* TEC ESP nuclear power

núclido *m* FÍS, FÍS PART nuclide

núclidos: **~ espejo** *m pl* FÍS mirror nuclides

nudillo *m* AUTO knuckle

nudo *m* AGRIC *caña* node, C&V knot, CRISTAL, ELEC node, ENERG RENOV *velocidad náutica* knot, FÍS ONDAS, ING ELÉC, METAL node, PAPEL knot, PROD *cable* kink, TEXTIL *hilatura*, TRANSP MAR knot; **~ de boza** *m* TRANSP MAR rolling hitch; **~ de carretera** *m* CONST interchange; **~ de comunicaciones bajo tierra** *m* TRANSP buried loop; **~ de conmutación** *m* TELECOM switching point; **~ de empalme** *m* ING ELÉC coupling loop; **~ estirado** *m* METAL extended node; **~ ferroviario** *m* FERRO railroad junction (*AmE*), railway junction (*BrE*); **~ fibroso** *m* C&V stringy knot; **~ lleno mal hecho** *m* TRANSP MAR granny knot; **~ prolongado** *m* METAL extended node; **~ de retenida** *m* TRANSP MAR stopper knot; **~ reticular** *m* CRISTAL lattice point; **~ de rizo** *m* TRANSP MAR reef knot; **~ simple** *m* TRANSP MAR overhand knot; **~ de tejedor** *m* TRANSP MAR sheet bend; **~ de unión** *m* ING MECÁ *tubería* butt joint

nudos: **sin ~** *adj* TEXTIL knotless

nudoso *adj* TEXTIL knotty

nueva: **~ conexión** *f* ING ELÉC, TELECOM reconnection; **~ extracción** *f* CARBÓN re-extraction; **~ línea transitoria** *f* IMPR soft return

nuevo[1] *adj* INFORM&PD, ING MECÁ, NUCL, QUÍMICA, TELECOM new

nuevo[2]: **~ levantamiento topográfico** *m* CONST resurvey; **~ método austríaco de perforación de túneles** *m* CONST new Austrian tunnelling method (*NATM*)

nuez: **~ del mandrín** *f* ING MECÁ mandrel nose, mandril nose

null *m* INFORM&PD null

nulo[1] *adj* QUÍMICA null

nulo[2] *m* INFORM&PD null

numeración *f* FOTO numbering, IMPR numbering, pagination; ~ **automática de páginas** *f* INFORM&PD automatic page numbering; ~ **continua** *f* IMPR crash numbering; ~ **de la escala** *f* METR scale numbering; ~ **marginal** *f* CINEMAT edge numbering; ~ **a pie de página** *f* IMPR drop folio; ~ **secuencial** *f* IMPR sequential numbering

numerado: ~ **automático de páginas** *m* INFORM&PD automatic page numbering

numerador *m* EMB numbering apparatus, MATEMÁT numerator

numeradora *f* IMPR numbering machine

numerar[1] *vt* PROD number

numerar[2]: ~ **páginas** *vi* IMPR paginate

numérico *adj* ELECTRÓN digital, MATEMÁT numerical

numerizar *vt* ING MECÁ digitize

número *m* GEN number, INFORM&PD count, number, TEXTIL *de pasadas de hilo* count; ~ **de Abbe** *m* FÍS, ÓPT Abb number; ~ **del abonado** *m* TELECOM subscriber number (*SN*); ~ **de abonado múltiple** *m* TELECOM multiple subscriber number (*MSN*); ~ **abreviado** *m* TELECOM abbreviated number; ~ **de acceso** *m* TELECOM access number; ~ **aleatorio** *m* INFORM&PD, MATEMÁT, TELECOM random number; ~ **de amasado** *m* CONST batch number; ~ **antiguo** *m* *Esp* (*cf número viejo AmL*) TELECOM old number; ~ **atómico** *m* (*Z*) FÍS, FÍS PART, QUÍMICA atomic number (*Z*); ~ **de barión** *m* FÍS, FÍS PART baryon number; ~ **bariónico** *m* FÍS, FÍS PART baryon number; ~ **de billete** *m* PROD ticket number; ~ **binario** *m* INFORM&PD binary number; ~ **de bloque** *m* TEC PETR block number; ~ **de boleto** *m* PROD ticket number; ~ **de calibre** *m* PROD *cable* gage number (*AmE*), gauge number (*BrE*); ~ **de caracteres por unidad de longitud** *m* IMPR character pitch; ~ **cardinal** *m* MATEMÁT cardinal number; ~ **de carreras** *m* TEXTIL run number; ~ **del catálogo** *m* PROD catalog number (*AmE*), catalogue number (*BrE*); ~ **de caucho** *m* CINEMAT rubber number; ~ **de Clarke** *m* GEOL, QIMICA Clarke number; ~ **de código** *m* CINEMAT code number; ~ **de coma flotante** *m* INFORM&PD, TELECOM floating-point number; ~ **combinatorio** *m* MATEMÁT binomial coefficient; ~ **complejo** *m* INFORM&PD, MATEMÁT complex number; ~ **compuesto** *m* MATEMÁT composite number; ~ **de coordinación** *m* CRISTAL, METAL, QUÍMICA coordination number; ~ **cuántico** *m* ELECTRÓN, FÍS, FÍS PART, NUCL, QUÍMICA quantum number; ~ **cuántico azimutal** *m* FÍS, FÍS RAD azimuthal quantum number; ~ **cuántico magnético** *m* FÍS magnetic quantum number; ~ **cuántico del momento angular orbital** *m* FÍS orbital angular momentum quantum number; ~ **cuántico del momento angular total** *m* FÍS total angular momentum quantum number; ~ **cuántico orbital** *m* FÍS, FÍS RAD orbital quantum number; ~ **cuántico principal** *m* FÍS principal quantum number, TRANSP AÉR main quantum number; ~ **cuántico rotacional** *m* FÍS rotational quantum-number; ~ **cuántico de spin** *m* FÍS spin quantum number; ~ **cuántico vibracional** *m* FÍS vibrational quantum number; ~ **de cursos** *m* TEXTIL *ligamento* number of repeats; ~ **desocupado** *m* TELECOM spare number; ~ **del directorio** *m* TELECOM directory number; ~ **disponible** *m* TELECOM spare number;

~ **de dureza de Brinell** *m* ING MECÁ, MECÁ Brinell hardness number; ~ **entero** *m* INFORM&PD integer, MATEMÁT natural whole number; ~ **entero natural** *m* GEOM, MATEMÁT natural whole number; ~ **equivocado** *m* TELECOM wrong number; ~ **de la estación de barco** *m* TELECOM ship station number; ~ **estérico** *m* DETERG ester number; ~ **de fabricación** *m* ING MECÁ serial number; ~ **de Froude** *m* FÍS Froude number; ~ **de generación** *m* INFORM&PD generation number; ~ **guía** *m* FOTO guide number, IMPR key number; ~ **de identificación personal** *m* (*NIP*) ELECTRÓN personal identification number (*PIN*); ~ **de identificación del tren** *m* FERRO train supervision number; ~ **imaginario** *m* MATEMÁT imaginary number; ~ **irracional** *m* INFORM&PD, MATEMÁT irrational number; ~ **isotópico** *m* FÍS, FÍS PART isotopic number; ~ **leptónico** *m* FÍS lepton number; ~ **limitador de la viscosidad** *m* FÍS FLUID limiting viscosity number; ~ **de línea** *m* INFORM&PD line number; ~ **de llamada gratuita** *m* TELECOM freephone number (*BrE*), toll-free number (*AmE*); ~ **llamado original** *m* TELECOM original called number; ~ **de lote de la emulsión** *m* CINEMAT, FOTO emulsion batch-number; ~ **de Mach** *m* ACÚST, FÍS, TRANSP AÉR Mach number; ~ **de Mach nunca a exceder** *m* TRANSP AÉR never-exceed Mach number; ~ **de Mach permisible máximo** *m* TRANSP AÉR maximum permissible Mach number; ~ **mágico** *m* FÍS magic number; ~ **marginal** *m* CINEMAT edge number; ~ **másico** *m* FÍS mass number; ~ **máximo de revoluciones** *m* VEH *del motor* peak revs; ~ **nacional** *m* TELECOM national number (*NN*); ~ **nacional significativo** *m* TELECOM national signifigant number (*NSN*); ~ **natural** *m* GEOM natural whole number, INFORM&PD, MATEMÁT natural number; ~ **de negativo** *m* CINEMAT negative number; ~ **de neutrones** *m* FÍS neutron number; ~ **no primo** *m* MATEMÁT composite number; ~ **nucleónico** *m* FÍS, NUCL nucleon number; ~ **de octano** *m* (*OCTN*) TEC PETR *calidad* , VEH *gasolina* octane number (*ON*); ~ **ocupado** *m* TELECOM busy number; ~ **de onda** *m* ACÚST, FÍS ONDAS, QUÍMICA wave number; ~ **de orden** *m* ING MECÁ serial number, TELECOM sequence number (*SN*); ~ **ordinal** *m* MATEMÁT ordinal number; ~ **de la página** *m* IMPR page number; ~ **de paletas** *m* ENERG RENOV blade quantity; ~ **de partículas** *m* METAL particle number; ~ **de pasadas** *m* CONST *carreteras* number of passes; ~ **de la pieza** *m* EMB part number; ~ **de pieza** *m* ING MECÁ item number; ~ **de piezas de la serie a fabricar** *m* PROD lot size; ~ **de pista** *m* TRANSP AÉR runway number; ~ **de porcinos nacidos en un año** *m* AGRIC hog run; ~ **primo** *m* MATEMÁT prime number; ~ **de protones** *m* FÍS PART *en el núcleo* proton number; ~ **pseudoaleatorio** *m* INFORM&PD pseudo-random number; ~ **racional** *m* INFORM&PD, MATEMÁT rational number; ~ **real** *m* INFORM&PD, MATEMÁT real number; ~ **de reexpedición** *m* TELECOM redirecting number, redirection number; ~ **de la remesa** *m* EMB batch number; **bajo** ~ **de Reynolds** *m* FÍS FLUID low Reynolds number; ~ **de Reynolds** *m* ENERG RENOV, FÍS, FÍS FLUID, FÍS RAD, TRANSP AÉR Reynolds number (*R*); ~ **de Reynolds 900** *m* (*Re900*) FÍS FLUID Reynolds number 900 (*Re900*); ~ **de rollo** *m* CINEMAT roll number; ~ **romano** *m* IMPR, MATEMÁT Roman numeral; ~ **de saponificación** *m*

ALIMENT saponification number; ~ **de Schmidt** *m* FÍS Schmidt number; ~ **secuencial** *m* TELECOM sequence number (*SN*); ~ **de señales que entran** *m* INFORM&PD fan-in; ~ **de señales que salen** *m* INFORM&PD fan-out; ~ **de serie** *m* FOTO, INFORM&PD, ING MECÁ serial number; ~ **de signatura** *m* IMPR signature number; ~ **Sommerfield** *m* ENERG RENOV Sommerfield number; ~ **de subritina** *m* PROD *programas* operation number; ~ **de teléfono gratuito** *m* TELECOM freephone number (*BrE*), toll-free number (*AmE*); ~ **de unidad** *m* ING MECÁ item number; ~ **universal** *m* TELECOM universal number; ~ **viejo** *m* *AmL* (*cf*

número antiguo Esp) TELECOM old number; ~ **de yodo** *m* ALIMENT iodine number

números: ~ **cuánticos** *m pl* FÍS PART spin (*s*); ~ **estilo moderno** *m pl* IMPR modern figures; ~ **inversamente proporcionales** *m pl* MATEMÁT inversely proportional numbers

nutación *f* FÍS, TEC ESP nutation

nutrición *f* ALIMENT nutrition

nutriente *m* ALIMENT, CONTAM MAR nutrient

nutritivo *adj* ALIMENT nutritious, nutritive

NVT *abr* (*terminal virtual en red*) INFORM&PD NVT (*network virtual terminal*)

nylon *m* P&C, QUÍMICA, TEXTIL nylon

O

O *abr* (*oxígeno*) GEN O (*oxygen*)
OA *abr* (*ofimática*) INFORM&PD OA (*office automation*)
oasis: **~ abisal** *m* OCEAN deep-sea oasis
obducción *f* GEOL obduction
obelisco *m* GEOM, IMPR obelisk
obencadura *f* TRANSP MAR shroud
obenque *m* TRANSP MAR shroud; **~ mayor** *m* TRANSP MAR main shroud; **~ del palo macho** *m* TRANSP MAR lower shroud
objetivo *m* CINEMAT lens, FÍS objective, FOTO lens, INSTR objective, LAB, METAL objective lens, TEC ESP lens, target, TV target; **~ acromático** *m* CINEMAT, FÍS, FOTO, IMPR, ÓPT achromatic lens, achromat; **~ anacromático** *m* CINEMAT, FÍS, FOTO, INSTR, ÓPT anachromatic lens, soft-focus lens; **~ anamórfico** *m* CINEMAT, FÍS, FOTO, ÓPT anamorphic lens; **~ anastigmático** *m* CINEMAT, FÍS, FOTO, ÓPT anastigmat, anastigmatic lens; **~ anastigmático simétrico** *m* CINEMAT, FOTO, ÓPT symmetrical anastigmat lens; **~ angular** *m* CINEMAT, FOTO narrow-angle lens; **~ de ángulo medio** *m* CINEMAT, FOTO medium-angle lens; **~ antirreflector** *m* CINEMAT, FOTO bloomed lens; **~ aplanático** *m* CINEMAT, FOTO aplanatic lens; **~ apocromático** *m* CINEMAT, FÍS, FOTO, ÓPT apochromat, apochromatic lens; **~ asférico** *m* CINEMAT aspherical lens; **~ astigmático** *m* CINEMAT, FÍS, FOTO, ÓPT astigmatic lens; **~ compuesto** *m* FOTO, RECICL compound lens; **~ condensador** *m* CINEMAT, FOTO, INSTR, ÓPT condensing lens; **~ con corrección de colores** *m* CINEMAT, FOTO, ÓPT color-corrected lens (*AmE*), colour-corrected lens (*BrE*); **~ de distancia focal regulable** *m* INSTR zoom lens; **~ doble** *m* CINEMAT, FÍS, FOTO, INSTR, ÓPT doubled lens, doublet, doublet lens; **~ de dosis absorbida total** *m* FÍS, FÍS RAD total absorption target; **~ de emisión secundaria** *m* ELECTRÓN secondary emission target; **~ esférico** *m* CINEMAT, FOTO, INSTR, ÓPT spherical lens; **~ de focal corta** *m* CINEMAT, FOTO, INSTR, ÓPT short-focus lens; **~ de foco largo** *m* CINEMAT, FOTO, INSTR, ÓPT long-focus lens; **~ de foco variable** *m* CINEMAT, FOTO, INSTR, ÓPT varifocal lens, zoom, zoom lens; **~ gran angular** *m* CINEMAT, FOTO, INSTR, ÓPT wide-angle lens; **~ de inmersión** *m* FÍS, INSTR, METAL immersion objective; **~ macro** *m* CINEMAT, FOTO, INSTR, ÓPT macro lens; **~ magnético** *m* ELECTRÓN, INSTR, NUCL magnetic objective; **~ menisco** *m* CINEMAT, FOTO, INSTR, ÓPT meniscus lens; **~ muy rápido** *m* CINEMAT, FOTO, ÓPT high-speed lens; **~ normal** *m* CINEMAT, FOTO, ÓPT standard lens; **~ panorámico** *m* CINEMAT, FOTO, ÓPT panoramic lens; **~ con preselector de abertura** *m* CINEMAT, FOTO, ÓPT lens with aperture preselector; **~ principal** *m* CINEMAT, FOTO, ÓPT prime lens; **~ de proyección** *m* CINEMAT, ÓPT, FOTO projection lens; **~ con revestimiento antirreflejante** *m* CINEMAT, FÍS, FOTO, INSTR, ÓPT, REVEST coated lens; **~ telescópico** *m* INSTR telescope objective; **~ triple** *m* CINEMAT, FOTO, ÓPT triplet lens; **~ zoom** *m* CINEMAT, FOTO, ÓPT

zoom, zoom lens; **~ zoom motorizado** *m* CINEMAT, FOTO, ÓPT motorized zoom lens
objetivos: **~ intercambiables** *m pl* INSTR, ÓPT interchangeable objectives
objeto: **~ en el espacio** *m* TEC ESP object in space; **~ de inmersión** *m* METAL immersion objective; **~ en órbita** *m* TEC ESP orbiting object; **~ sólido** *m* PROD solid object; **~ virtual** *m* FÍS virtual object; **~ volador no identificado** *m* (*OVNI*) TEC ESP, TRANSP AÉR unidentified flying object (*UFO*)
oblea *f* ELECTRÓN, INFORM&PD wafer; **~ de silicio** *f* ELECTRÓN, INFORM&PD silicon wafer
oblicuidad *f* CARBÓN bias, CONST, GEOM skewness, ING MECÁ cant
oblicuo *adj* CONST skew, slanting, GEOM oblique, MECÁ warped
obligación: **~ a permanecer en tierra** *f* TRANSP AÉR grounding
obligar: **~ a permanecer en tierra** *vt* TRANSP AÉR ground
obligatorio: **no ~** *adj* TELECOM nonmandatory
oblimak *m* NUCL oblimak
obliteración *f* IMPR, ING MECÁ overprinting
obra: **~ de albañilería** *f* CONST masonry work; **~ civil** *f* CONST civil engineering; **~ de concreto** *f* AmL (*cf obra de hormigón Esp*) CONST concrete work; **~ por contrata** *f* PROD contract work; **~ de fábrica de concreto** *f* AmL (*cf obra de fábrica de hormigón Esp*) CONST concrete masonry; **~ de fábrica de hormigón** *f* Esp (*cf obra de fábrica de concreto AmL*) CONST concrete masonry; **~ hecha con diversos trozos** *f* PROD patchwork; **~ hecha a mano** *f* PROD handwork; **~ de hierro** *f* CONST, METAL, PROD ironwork; **~ de hormigón** *f* Esp (*cf obra de concreto AmL*) CONST concrete work; **~ de mampostería** *f* CONST masonry work, stonework; **~ muerta** *f* TRANSP MAR deadworks, upper works, *construcción naval* topside; **~ muerta de proa** *f* TRANSP MAR prow; **~ realizada en cal** *f* REVEST limework; **~ viva** *f* TRANSP MAR *diseño de barcos* quickwork, ship's bottom, underwater hull
obrar: **~ entre sí** *vi* QUÍMICA interact
obras: **~ de blindaje** *f pl* CONST blind workings; **~ de cabecera** *f pl* ENERG RENOV *de presa* headworks; **~ fluviales** *f pl* AGUA river works; **~ públicas** *f pl* AGUA, CONST public works; **~ subterráneas** *f pl* CONST work underground
obrero *m* ING MECÁ, MINAS operator; **~ de cantera** *m* MINAS quarryman; **~ dragador** *m* MINAS, TRANSP MAR dredger; **~ metalúrgico** *m* METAL, PROD smelter; **~ portuario** *m* TRANSP MAR docker (*BrE*), longshoreman (*AmE*); **~ que trabaja a cielo abierto** *m* MINAS surface worker; **~ que trabaja en el exterior** *m* MINAS above-ground worker; **~ que trabaja en superficie** *m* MINAS surface worker; **~ de servicio de maniobras** *m* FERRO shunter
obscuridad: **~ atmosférica** *f* CONTAM atmospheric obscurity
observación *f* CONST observation, *topografía* sight,

PROD, TEC ESP monitoring, TRANSP MAR observation; ~ **astronómica** *f* TRANSP MAR sight; ~ **directa** *f* CONST fore observation; ~ **meteorológica** *f* METEO, TEC ESP weather watch; ~ **posterior** *f* CONST back observation; ~ **por sonda radárica** *f* FÍS RAD, TRANSP, TRANSP AÉR, TRANSP MAR radar sonde; ~ **del tiempo** *f* TEC ESP weather watch

observador *m* TEC ESP tracker, TRANSP AÉR observer; ~ **de artillería** *m* D&A artillery spotter

observar *vt* PROD, TEC ESP monitor

observatorio: ~ **astronómico orbitante** *m* TEC ESP orbiting astronomical observatory; ~ **contra incendios** *m* AGRIC fire danger station; ~ **espacial** *m* TEC ESP space observatory

obsidiana *f* C&V, MINERAL, PETROL obsidian

obstáculo *m* CONST obstruction, METAL obstacle, SEG hindrance; ~ **en un fluido estratificado** *m* FÍS FLUID obstacle in stratified fluid; ~ **en un fluido en rotación** *m* FÍS FLUID obstacle in rotating fluid

obstrucción *f* AUTO choking, C&V crimp, CARBÓN clogging, CONST obstruction, ING MECÁ jam, clogging, MECÁ bottleneck, MINAS *de la chimenea* clogging, stoping, PROD bottleneck, choking, clogging, *altos hornos* hang, stoppage, TRANSP AÉR choking; ~ **de la bocana** *f* TRANSP MAR *defensa de un puerto* boom; ~ **del cabezal** *f* TV head clogging

obstruido *adj* HIDROL silted up, ING MECÁ, MECÁ clogged

obstruir *vt* ING MECÁ foul, jam, clog, PAPEL clog, PROD choke, *altos hornos* clog, TRANSP MAR *puerto* block; ~ **con sedimentos** *vt* HIDROL silt up

obstruirse *v refl* HIDROL silt up, PROD hang

obtención *f* MINAS production; ~ **de componentes** *f* TEC ESP component procurement; ~ **de copos** *f* PAPEL fluff, fluffing; ~ **del envase prensado la hoja** *f* EMB press-through packaging sheet; ~ **de hojas de plástico** *f* EMB plastic sheeting; ~ **del promedio de señal** *f* ELECTRÓN signal averaging

obtener: ~ **acceso a** *vt* INFORM&PD access; ~ **muestras de** *vt* ELECTRÓN sample

obturación *f* AGUA blinding, ELEC *protección* seal, ING MECÁ packing, plugging, throttling, PROD *de cribas, tamices* blinding, blanking, TEC PETR *perforación* plugging, *de tuberías* seal, TRANSP MAR seal; ~ **de hoyo de disparo** *f* TEC PETR shot-hole plug

obturador *m* AUTO choke, choker plate, CINEMAT lens shutter, shutter, CONST casing packer, FOTO lens shutter, shutter, stop, GAS poker, ING MECÁ, INSTAL TERM obturator, INSTR shutter, MECÁ blank, gasket, plug, MINAS *Esp* (*cf portacebo AmL*) *cargas explosivas* shutter, OCEAN choke, PETROL stinger, PROD plug, shutter, obturator, QUÍMICA sealant, TEC ESP *Esp* (*cf portacebo AmL*) shutter, TEC PETR stinger, VEH choker plate; ~ **de abertura variable** *m* CINEMAT, FOTO variable-aperture shutter; ~ **de aceite** *m* ING MECÁ, MECÁ oil seal; ~ **de agujero de hombre** *m* MECÁ manhole gasket; ~ **ajustable** *m* CINEMAT, FOTO adjustable shutter; ~ **con ajuste B** *m* CINEMAT, FOTO shutter with B setting; ~ **anular** *m* PETROL packer; ~ **de Compur** *m* FOTO Compur shutter; ~ **de cortinilla** *m* CINEMAT, FOTO, TV focal plane shutter, roller-blind shutter; ~ **de desvanecimiento** *m* CINEMAT, FOTO, TV fade shutter; ~ **desviado** *m* C&V swung baffle; ~ **de disco** *m* CINEMAT, FOTO disc shutter (*BrE*), disk shutter (*AmE*); ~ **con disparador automático** *m* CINEMAT, FOTO self-cocking shutter;

~ **para ejes en rotación** *m* ING MECÁ lip seal, rotary shaft lip-type seal, rotary shaft seal; ~ **de espejo** *m* CINEMAT, FOTO mirror shutter; ~ **de filtraje** *m* ELEC *inductor* smoothing choke; ~ **de fuelle** *m* CINEMAT, FOTO bellows shutter; ~ **para fundidos** *m* CINEMAT, FOTO dissolving shutter; ~ **giratorio** *m* CINEMAT, FOTO, ÓPT, TV revolving shutter; ~ **de guillotina** *m* CINEMAT, FOTO, TV guillotine shutter; ~ **de hoja** *m* FOTO leaf shutter; ~ **laberíntico del eje alternativo** *m* ING MECÁ reciprocating-shaft seal; ~ **de laminilla** *m* CINEMAT, FOTO blade shutter; ~ **de laminilla simple** *m* CINEMAT, FOTO single blade shutter; ~ **longitudinal** *m* TEC ESP stringer; ~ **de nitrilo** *m* PROD nitrile seal; ~ **posterior** *m* CINEMAT, FOTO rear shutter; ~ **preajustado** *m* CINEMAT, FOTO preset shutter; ~ **protector** *m* CINEMAT, FOTO capping shutter; ~ **de proyector** *m* CINEMAT dowser; ~ **de pulsómetro** *m* INSTAL HIDRÁUL clapper valve; ~ **en ranura** *m* TV slit shutter; ~ **reflex** *m* CINEMAT, FOTO, TV reflex shutter; ~ **de rosca** *m* PROD stop screw; ~ **rotativo** *m* CINEMAT, FOTO, TV rotating shutter; ~ **de sector** *m* FOTO between-the-lens shutter; ~ **sincronizado XM** *m* FOTO XM synchronized shutter; ~ **Synchro-Compur** *m* FOTO Synchro-Compur shutter; ~ **de vacío** *m* ING MECÁ vacuum seal; ~ **variable** *m* CINEMAT, FOTO, TV variable shutter

obturar *vt* PROD blank off, seal up, insulate, stop, stop up, TRANSP MAR seal

obtusángulo *adj* GEOM obtuse-angular

obtusidad *f* GEOM obtuseness

obtuso *adj* GEOM obtuse, ING MECÁ *ángulo* blunt

obús *m* D&A howitzer

OC *abr* (*onda continua*) FÍS, FÍS ONDAS, ING ELÉC, TV CW (*continuous wave*)

oceanauta *m* OCEAN aquanaut

oceanografía *f* METEO, OCEAN, TRANSP MAR oceanography

oceanográfico *adj* METEO, OCEAN, TRANSP MAR oceanographical

oceanógrafo *m* METEO, OCEAN, TRANSP MAR oceanographer

oceanología *f* METEO, OCEAN, TRANSP MAR oceanology

ocluido *adj* METEO, MINAS, PAPEL, PROD, QUÍMICA occluded

ocluir *vt* MINAS insulate, PAPEL seal, PROD insulate, QUÍMICA occlude

oclusión *f* METEO, PAPEL occlusion, PROD stoppage, QUÍMICA occlusion; ~ **frontal** *f* METEO occluded front

OCR *abr* IMPR, INFORM&PD (*lector óptico de caracteres, reconocimiento óptico de caracteres*) OCR (*optical character reader, optical character recognition*)

ocre: ~ **cadmífero** *m* MINERAL cadmium ocher (*AmE*), cadmium ochre (*BrE*); ~ **de cobalto negro** *m* MINERAL black cobalt ocher (*AmE*), black cobalt ochre (*BrE*); ~ **de cromo** *m* MINERAL chrome ocher (*AmE*), chrome ochre (*BrE*); ~ **de molibdeno** *m* MINERAL molybdic ocher (*AmE*), molybdic ochre (*BrE*); ~ **negro** *m* MINERAL wad; ~ **de teluro** *m* MINERAL telluric ocher (*AmE*), telluric ochre (*BrE*)

o-cresol *m* QUÍMICA o-cresol

octacosano *m* QUÍMICA octacosane

octadecano *m* QUÍMICA octadecane

octaédrico *adj* CRISTAL, GEOM, QUÍMICA octahedral

octaedrita *f* MINERAL octahedrite

octaedro *m* CRISTAL, GEOM, QUÍMICA octahedron; ~ **cúbico** *m* GEOM cubic octahedron

octágono *m* GEOM octagon

octal[1] *adj* INFORM&PD octal

octal[2]: **~ codificado en binario** *m* INFORM&PD, PROD binary-coded octal

octámero *m* QUÍMICA octamer

octanaje *m* (*OCTN*) TEC PETR, VEH octane index, octane number (*ON*), octane rating

octanal *m* QUÍMICA, TEC PETR octanal, octyl aldehyde

octano *m* QUÍMICA, TEC PETR, VEH octane

octavalente *adj* QUÍMICA octavalent

octavo *m* IMPR eightvo (*8vo*)

octeno *m* QUÍMICA octene

octeto *m* ELECTRÓN eight-bit byte, INFORM&PD byte, eight-bit byte, octet, PROD byte, QUÍMICA octet, TELECOM byte

octilaldehído *m* QUÍMICA octanal, octyl aldehyde

octilo *m* QUÍMICA octyl

octina *f* QUÍMICA octyne

OCTN *abr* (*número de octano, octanaje*) TEC PETR *refino* ON (*octane number*)

octodo *m* ELECTRÓN octode

octógono *m* GEOM octagon

octosa *f* QUÍMICA octose

octovalente *adj* QUÍMICA octovalent

ocular *m* CINEMAT, FÍS eyepiece, FOTO eye-cup, eyepiece, INSTR, LAB, ÓPT, PROD, TEC ESP eyepiece; **~ ajustable** *m* CINEMAT, FOTO, ÓPT adjustable eyepiece; **~ con cruz y escala** *m* C&V centering lens with ruled cross (*AmE*), centring lens with ruled cross (*BrE*); **~ cuadriculado** *m* CINEMAT, ELECTRÓN, INSTR, METR, ÓPT, TV graticule; **~ erector** *m* INSTR *de telescopios*, ÓPT erecting lens; **~ de Huygens** *m* FÍS Huygens' eyepiece; **~ del microscopio** *m* INSTR, LAB microscope eyepiece; **~ de Ramsden** *m* FÍS Ramsden eyepiece; **~ regulable** *m* CINEMAT, FOTO adjustable eyepiece; **~ del visor** *m* CINEMAT, FOTO viewfinder eyepiece

ocultación: **~ del haz de electrones** *f* ELECTRÓN electron-beam mask; **~ de información** *f* INFORM&PD information hiding

ocultamiento: **~ de información** *m* INFORM&PD information hiding

oculto *adj* CONST encased, GEOL concealed, PROD shrouded

ocupación: **~ de circuitos** *f* TEC ESP, TELECOM hold

ocupancia: **~ del canal** *f* TELECOM channel occupancy; **~ espectral** *f* TELECOM spectral occupancy

ocurrencia: **~ de vuelo** *f* TEC ESP *nave espacial* flight occurrence

OD *abr* (*oxígeno disuelto*) CONTAM, QUÍMICA DO (*dissolved oxygen*)

odómetro *m* AUTO, CONST, TRANSP, VEH mileometer (*BrE*), odometer (*AmE*), trip counter (*BrE*), trip odometer (*AmE*)

odontógrafo *m* ING MECÁ odontograph

odontolita *f* MINERAL odontolite

odorante *m* PROC QUÍ odorizer, TEC PETR odorant

odorífero[1] *adj Esp* (*cf oloroso AmL*) CONTAM, PROC QUÍ, QUÍMICA odorous, odoriferous

odorífero[2] *m* QUÍMICA odoriphore

odorización *f* GAS odorization

O-E *abr* (*óptico-eléctrico*) TELECOM O-E (*optical-electrical*)

oersted *m* CONST, ELEC, FÍS oersted

oerstedio *m* CONST, ELEC, FÍS oersted

oeste *adj* TRANSP MAR west

oestes *m pl* OCEAN westerlies

oferta *f* MECÁ bid, TELECOM offer

offset *m* ELECTRÓN, IMPR lithography, offset; **~ seco** *m* IMPR dry offset

oficial: **~ de derrota** *m* D&A, TRANSP MAR navigation officer; **~ de máquinas** *m* TRANSP MAR engineer officer; **~ de la marina mercante con título de capitán** *m* TRANSP MAR master mariner; **~ de puente** *m* TRANSP MAR deck officer; **~ radiotelegrafista** *m* TRANSP MAR radio officer; **~ responsable de la carga** *m* TRANSP MAR cargo officer

oficina: **~ central manual** *f* TELECOM manual exchange; **~ de control de aproximación** *f* TRANSP AÉR approach control office; **~ de diseño** *f* CONST design office; **~ electrónica** *f* INFORM&PD electronic office; **~ de expediciones** *f* TRANSP forwarding office (*BrE*), freight office (*AmE*); **~ expedidora** *f* TRANSP forwarding office (*BrE*), freight office (*AmE*); **~ de fletes** *f* TRANSP forwarding office (*BrE*), freight office (*AmE*); **~ meteorológica** *f* METEO weather bureau; **~ para pago de sobreprecio en billetes** *f* FERRO excess-fare office; **~ de paquetería** *f* TRANSP parcels office; **~ de proyectos** *f* CONST design office, TRANSP MAR *diseño de barcos* design department, drawing office; **~ satélite** *f* TELECOM satellite exchange; **~ supervisora de la contratación de marinos** *f* TRANSP MAR shipping office; **~ técnica** *f* PROD engineering department

Oficina: **~ para la Investigación y Experimentos** *f* FERRO Office for Research and Experiments (*ORE*)

ofimática *f* (*OA*) INFORM&PD, PROD office automation (*OA*)

ofita *f* PETROL ophite

oftalmómetro *m* INSTR, ÓPT ophthalmometer

ohm *m* GEN ohm

óhmetro *m* GEN ohmmeter, resistance meter

óhmico *adj* GEN ohmic

ohmímetro *m* GEN ohmmeter, resistance meter

ohmio *m* GEN ohm; **~ patrón** *m* ELEC *unidad* standard ohm

oidio *m* AGRIC powdery mildew

oído *m* ING MECÁ inlet, ear, PROD *ventilador, bomba centrífuga* inlet; **~ artificial** *m* ACÚST artificial ear; **~ medio** *m* ACÚST middle ear

OIN *abr* (*Organización Internacional de Normalización*) ELEC, ING MECÁ, TELECOM ISO (*International Standards Organization*)

ojal *m* ING ELÉC *cable* grommet, ING MECÁ eye, PROD eyelet, thimble; **~ metálico** *m* EMB eyelet

ojete *m* IMPR eyelet, ING MECÁ, MECÁ clevis; **~ metálico** *m* PROD metal eyelet

ojiva *f* TEC ESP *proyectiles* windscreen (*BrE*), nosecone, windshield (*AmE*); **~ de puente** *m* CONST bay; **~ redondo de cabeza de biela** *m* ING MECÁ rod-end spherical eye; ~ **de tigre** m C&V, MINERAL tiger's eye; ~ **del tipo** m IMPR typeface

okenita f MINERAL okenite

okra f AGRIC, ALIMENT okra

ola: ~ **de aguas profundas** f FÍS, OCEAN, TRANSP MAR deep-water wave; ~ **de arena** f OCEAN sand wave; ~ **de calor** f METEO heat wave; ~ **capilar** f FÍS, OCEAN capillary wave; ~ **excepcional** f OCEAN phenomenal wave; ~ **fenomenal** f OCEAN phenomenal wave; ~ **fuerte** f OCEAN heavy sea; ~ **grande** f TRANSP MAR billow; ~ **de marea** f GEOFÍS, OCEAN, TRANSP MAR tidal wave; ~ **de proa** f TRANSP MAR bow wave; ~ **rompiente** f OCEAN breaking wave; ~ **seca** f OCEAN seiche; ~ **semidiurna** f OCEAN semidiurnal wave; ~ **significativa** f OCEAN significant wave; ~ **de superficie** f OCEAN surface wave; ~ **topográfica** f OCEAN topographic wave

olas: ~ **de viento** f pl OCEAN wind-driven sea

oleada f OCEAN en aguas someras billow, ground swell, heavy sea, TRANSP MAR del mar billow, en aguas someras ground swell

oleaginoso adj PETROL, TEC PETR oil-bearing

oleaje: ~ **de fondo** m OCEAN, TRANSP MAR en aguas someras ground swell; ~ **de temporal** m OCEAN storm surge

oleandrina f QUÍMICA oleandrin

oleato m QUÍMICA oleate

olefilo adj CONTAM MAR oleophilic

olefina f DETERG, PETROL, QUÍMICA, TEC PETR olefin

olefínico adj DETERG, PETROL, QUÍMICA, TEC PETR olefinic

oléico adj QUÍMICA oleic

oleína f QUÍMICA olein, triolein

oleoamortiguador m ING MECÁ oil dashpot

oleoducto m GEN oil pipeline, pipeline

oleoenfriamiento m ING MECÁ oil cooling

oleofílico adj QUÍMICA oleophilic

oleofosfórico adj QUÍMICA oleophosphoric

oleomargarina f QUÍMICA oleo oil

oleómetro m ING MECÁ oil gage (AmE), oil gauge (BrE), oilmeter, INSTR, QUÍMICA oleometer

oleoresina f QUÍMICA oleoresin

oleoresinoso adj QUÍMICA oleoresinous

oleovitamina f QUÍMICA oleovitamin

óleum m DETERG, QUÍMICA oleum

oligoclasa f MINERAL oligoclase, sunstone

oligoelemento m CONTAM características químicas trace element, HIDROL micronutrient, QUÍMICA trace element

oligomérico adj QUÍMICA oligomeric

oligomerización f QUÍMICA oligomerization

oligómero m QUÍMICA oligomer

oligomicina f QUÍMICA oligomycyn

oligonita f MINERAL oligonite

olistolito m GEOL olistolith

olivenita f MINERAL olivenite

olivicultor m AGRIC olive grower

olivino m MINERAL olivine, peridot

olla: ~ **de cierre hermético** f ALIMENT, ING MECÁ pressure cooker; ~ **exprés** f ALIMENT, ING MECÁ pressure cooker; ~ **a presión** f ALIMENT, ING MECÁ pressure cooker; ~ **de presión** f ALIMENT, ING MECÁ pressure cooker

ollao: ~ **con guardacabos metálico** m TRANSP MAR cuerdas eyelet hole, grommet

ollar: ~ **de empuje** m AUTO, VEH thrust collar

olor: **sin** ~ adj CONTAM, PROC QUÍ, QUÍMICA odorless (AmE), odourless (BrE)

oloroso adj AmL (cf odorífero Esp) CONTAM, PROC QUÍ, QUÍMICA odorous, odoriferous

OLRT abr (tiempo real en línea) INFORM&PD OLRT (on line real time)

omaso m AGRIC manyplies, psalterium, third stomach

OMI abr (Organización Marítima Internacional) TRANSP MAR IMO (International Maritime Organization)

omisión f IMPR out, INFORM&PD skip; ~ **en la cinta magnética** f INFORM&PD tape skip

omni: ~-**alcance de muy alta frecuencia** m TRANSP, TRANSP AÉR very high-frequency omnirange (VHFO)

ómnibus m AmL (cf autobús Esp, girobús Esp) TRANSP coach battery bus, bus on railroad car (AmE), bus on railroad tracks (AmE), bus on railway tracks (BrE), bus on railway wagon (BrE), dial-a-bus, gyrobus, hybrid bus, liquid-petroleum-gas bus, pressurized natural-gas bus; ~ **articulado** m AmL (cf autobús articulado Esp) AUTO, TRANSP bimodal bus; ~-**bimodal** m AmL (cf autobús bimodal Esp) AUTO bimodal bus, TRANSP bimodal bus, dual-mode bus; ~-**ferrobús** m AmL (cf autobús-ferrobús Esp) TRANSP, VEH road-rail bus; ~ **híbrido** m AmL (cf autobús híbrido Esp) TRANSP hybrid bus; ~ **oruga** m AmL (cf autobús oruga Esp) AUTO bimodal bus, TRANSP bimodal bus, dual-mode bus; ~ **de vapor** m AmL (cf autobús de vapor Esp) VEH steam bus

OMR abr (lectura óptica de marcas) INFORM&PD OMR (optical mark reading)

OMS abr (sistema de maniobra orbital) TEC ESP OMS (orbital maneuvering system AmE, orbital manoeuvring system BrE)

oncolito m GEOL oncolith

onda f GEN wave, CONST, HIDROL de agua ripple;

~ a ~ **acústica** f ACÚST, FÍS ONDAS, TELECOM sound-wave; ~ **acústica de superficie** f (OAS) ACÚST, ELECTRÓN, FÍS, FÍS ONDAS, TELECOM surface acoustic wave (SAW); ~ **acústica volumétrica** f ING ELÉC bulk acoustic-wave (BAW); ~ **aérea** f FÍS ONDAS airwave; ~ **aérea estacionaria** f FÍS ONDAS stationary aerial wave; ~ **de aire** f PETROL airwave; ~ **anormal** f OCEAN freak wave;

~ b ~ **de baja frecuencia** f ING ELÉC wave; ~ **de de Broglie** f FÍS, FÍS RAD, ING ELÉC de Broglie wave;

~ c ~ **calorífica** f TERMO heat wave; ~ **capilar** f FÍS, OCEAN capillary wave; ~ **celeste** f FÍS sky wave; ~ **centimétrica** f FÍS centimetre wave (BrE), centimeter wave (AmE); ~ **de choque** f FÍS, FÍS ONDAS, GEOFÍS shock wave, MINAS shock wave, explosiones collision wave; ~ **de choque adiabática** f TEC ESP adiabatic shock wave; ~ **de choque de cola** f TRANSP AÉR tail shock wave; ~ **de choque nuclear** f FÍS RAD, NUCL nuclear shock wave; ~ **cilíndrica** f ACÚST cylindrical wave; ~ **circular** f FÍS ONDAS, NUCL circular wave; ~ **de cizalladura** f ACÚST shear wave; ~ **coherente** f FÍS, FÍS ONDAS coherent wave; ~ **de combustión termonuclear** f NUCL thermonuclear combustion wave; ~ **compleja** f ELECTRÓN, TELECOM complex wave; ~ **de compresión** f ACÚST, FÍS ONDAS, GEOL compressional wave; ~ **compresional** f ACÚST, FÍS ONDAS, GEOL compressional wave; ~ **compuesta** f TELECOM composite wave; ~ **continua** f (OC) FÍS, FÍS ONDAS, ING ELÉC, TV continuous wave (CW); ~ **con-**

tinua modulada *f* TV modulated continuous wave; ~ **corta** *f* ELEC *radiación*, FÍS ONDAS *radio* short wave; ~ **de coseno** *f* GEOM cosine wave; ~ **de la crecida** *f* HIDROL flood wave; ~ **cuadrada** *f* ELECTRÓN, FÍS square wave, TELECOM square waveform; ~ **decamétrica** *f* TELECOM decametric wave; ~ **decimétrica** *f* FÍS ONDAS ultrahigh frequency (*UHF*); ~ **de detonación** *f* MINAS *planeada* detonation wave; ~ **en diente de sierra** *f* FÍS RAD sawtooth signal wave; ~ **difractada** *f* FÍS, FÍS ONDAS diffracted wave; ~ **de dilatación** *f* GEOL dilatational wave; ~ **dinámica** *f* MINAS, TEC PETR shock wave; ~ **directa** *f* GEOL direct wave; ~ **directa avanzada** *f* ELEC, FÍS ONDAS, ING ELÉC forward-traveling wave (*AmE*), forward-travelling wave (*BrE*); ~ **dirigida** *f* ÓPT guided wave, TEC ESP *radiocomunicaciones* beam, TELECOM guided wave; ~ **diurna** *f* OCEAN diurnal tide, diurnal wave;

~ **E** *f* FÍS ONDAS, ING ELÉC *guíaondas* E wave; ~ **de eco** *f* FÍS, FÍS ONDAS, ING ELÉC, OCEAN, ÓPT reflected wave; ~ **elástica** *f* FÍS, FÍS ONDAS, ING ELÉC elastic wave; ~ **eléctrica** *f* ELEC, FÍS ONDAS, ING ELÉC electric wave; ~ **eléctrica transversal** *f* (*onda TE*) ELEC, FÍS, ING ELÉC, TELECOM transverse electric wave (*TE wave*); ~ **electromagnética** *f* ELEC, ELECTRÓN, FÍS, FÍS ONDAS, FÍS RAD, ING ELÉC, TELECOM electromagnetic wave; ~ **electromagnética compuesta híbrida** *f* ELEC, FÍS ONDAS hybrid electromagnetic wave; ~ **electromagnética externa** *f* FÍS ONDAS, FÍS RAD external electromagnetic wave; ~ **electromagnética híbrida** *f* ELEC FÍS ONDAS hybrid electromagnetic wave; ~ **electromagnética transversal** *f* (*onda TEM*) ELEC, FÍS, ING ELÉC transverse electromagnetic wave (*TEM wave*); ~ **errante** *f* PROD transient current, TELECOM transient fluctuation; ~ **esférica** *f* ACÚST, FÍS, ING ELÉC spherical wave; ~ **estacionaria** *f* ACÚST standing wave, FÍS, FÍS ONDAS standing wave, stationary wave, ING ELÉC standing wave, TELECOM standing wave, stationary wave; ~ **estacionaria en la órbita del electrón** *f* FÍS ONDAS stationary waves in electron orbit; ~ **evanescente** *f* FÍS evanescent wave; ~ **de expansión** *f* TRANSP AÉR expansion wave; ~ **de explosión** *f* MINAS *voladuras* explosion wave; ~ **explosiva** *f* MINAS *planeada* detonation wave;

~ **en fluido** *f* FÍS FLUID fluid wave; ~ **fundamental** *f* OCEAN *mareas* ground wave, tidal component;

~ **de gravedad** *f* FÍS, FÍS ONDAS, FÍS RAD gravity wave, gravitational wave; ~ **gravitacional** *f* FÍS, FÍS ONDAS, FÍS RAD gravitational wave; ~ **guiada** *f* ÓPT, TELECOM guided wave; ~ **guiada ópticamente** *f* ING ELÉC, ÓPT, TELECOM optical guided wave;

~ **H** *f* ING ELÉC H wave; ~ **hacia adelante** *f* ING ELÉC forward wave; ~ **de hielo** *f* OCEAN ice ripple; ~ **hertziana** *f* ELEC, ELECTRÓN, FÍS, FÍS ONDAS, FÍS RAD, GEOFÍS, ING ELÉC, TELECOM electric wave, radio wave;

~ **incidente** *f* FÍS, FÍS ONDAS incident wave; ~ **interna** *f* OCEAN internal wave; ~ **irruptiva del fondo** *f* OCEAN bottom surge;

~ **larga** *f* ELEC *radiación*, FÍS ONDAS *radio* long wave; ~ **lenta** *f* ING ELÉC slow wave; ~ **longitudinal** *f* ACÚST, FÍS ONDAS longitudinal wave; ~ **longitudinal estacionaria** *f* ACÚST, FÍS ONDAS stationary longitudinal wave; ~ **luminosa** *f* FÍS ONDAS light wave; ~ **luminosa estacionaria** *f* FÍS ONDAS stationary light

wave; ~ **lunar** *f* OCEAN lunar wave; ~ **lunisolar** *f* OCEAN lunisolar wave; ~ **de luz** *f* ELECTRÓN light wave;

~ **magnética** *f* ING ELÉC magnetic wave; ~ **magnética transversal** *f* (*onda TM*) ELEC, FÍS, ING ELÉC transverse magnetic wave (*TM wave*); ~ **magnetohidrodinámica** *f* GEOFÍS magnetohydrodynamic wave; ~ **de marea** *f* OCEAN, TRANSP MAR tidal wave; ~ **mecánica** *f* ING ELÉC mechanical wave; ~ **media** *f* FÍS ONDAS medium wave; ~ **MHD** *f* GEOFÍS MHD wave; ~ **milimétrica** *f* FÍS millimeter wave (*AmE*), millimetre wave (*BrE*); ~ **de modulación** *f* ELECTRÓN modulating wave, modulation wave; ~ **modulada** *f* ELECTRÓN modulated wave; ~ **moduladora** *f* TV modulating wave; ~ **del morro** *f* TRANSP AÉR bow wave;

~ **nocturna** *f* TRANSP AÉR night wave;

~ **óptica** *f* ING ELÉC, ÓPT optical wave;

~ **P** *f* FÍS, GEOL P wave; ~ **periódica** *f* ING ELÉC periodic wave; ~ **periódica de sonido** *f* FÍS ONDAS periodic sound-wave; ~ **piloto** *f* TELECOM pilot, TV pilot tone; ~ **plana** *f* ACÚST, FÍS, ING ELÉC, ÓPT, TEC ESP, TELECOM plane wave; ~ **polarizada** *f* FÍS, FÍS ONDAS polarized wave; ~ **polarizada circularmente** *f* ACÚST, FÍS, FÍS ONDAS circularly-polarized wave; ~ **polarizada elípticamente** *f* ACÚST, FÍS, FÍS ONDAS elliptically-polarized wave; ~ **polarizada linealmente** *f* ACÚST, FÍS, FÍS ONDAS linearly-polarized wave; ~ **polarizada plana** *f* FÍS plane-polarized wave; ~ **de polarizada en el plano** *f* FÍS ONDAS plane-polarized wave; ~ **portadora** *f* FÍS, FÍS ONDAS carrier wave, TELECOM carrier, TRANSP MAR *radio*, TV carrier wave; ~ **portadora electrizada** *f* ING ELÉC charge carrier; ~ **portadora de imagen** *f* TV image carrier; ~ **portadora multicanal** *f* TEC ESP multichannel carrier; ~ **portadora piloto** *f* TELECOM pilot carrier; ~ **de presión** *f* TELECOM pressure wave; ~ **primaria** *f* GEOL P wave; ~ **principal** *f* OCEAN tidal component; ~ **progresiva** *f* ACÚST, ELECTRÓN, FÍS ONDAS, TELECOM traveling wave (*AmE*), travelling wave (*BrE*); ~ **progresiva inversa** *f* FÍS ONDAS, ING ELÉC reverse traveling wave (*AmE*), reverse travelling wave (*BrE*); ~ **prolongada** *f* AmL (*cf onda de choque Esp*) *explosiones* collision wave; ~ **de propagación hacia dentro** *f* TELECOM inward propagating wave; ~ **de propagación de salida** *f* TELECOM outward propagating wave;

~ **radárica** *f* FÍS, FÍS ONDAS, FÍS RAD radar wave; ~ **de radio** *f* ELEC, ELECTRÓN, FÍS, FÍS ONDAS, FÍS RAD, GEOFÍS, ING ELÉC, TELECOM radio wave; ~ **radioeléctrica** *f* ELEC, ELECTRÓN, FÍS, FÍS ONDAS, FÍS RAD, GEOFÍS, ING ELÉC, TELECOM radio wave; ~ **rápida** *f* ELECTRÓN, FÍS ONDAS, ING ELÉC fast wave; ~ **de Rayleigh** *f* FÍS, FÍS ONDAS, FÍS RAD, ÓPT, TELECOM Rayleigh wave; ~ **rectangular** *f* ELECTRÓN, FÍS ONDAS, ING ELÉC rectangular wave; ~ **reflejada** *f* FÍS, FÍS ONDAS, ING ELÉC, OCEAN, ÓPT reflected wave; ~ **de reflexión** *f* GEOFÍS reflection wave; ~ **de refracción** *f* FÍS ONDAS, GEOFÍS, PETROL refraction wave; ~ **regresiva** *f* ING ELÉC backward wave; ~ **de retorno** *f* ING ELÉC backward wave; ~ **de retroceso** *f* MINAS retonation wave; ~ **retrógada** *f* MINAS retonation wave; ~ **rotacional** *f* ACÚST rotational wave, FÍS shear wave;

~ **S** *f* FÍS S wave; ~ **secundaria** *f* FÍS secondary wave; ~ **senoidal** *f* ACÚST, ELEC, ELECTRÓN, FÍS, FÍS

ONDAS sine wave, sinusoidal wave; ~ **significativa** *f* OCEAN *equipos de sondeo* significant wave; ~ **sinusoidal** *f* ACÚST, ELEC, ELECTRÓN, FÍS, FÍS ONDAS sine wave, sinusoidal wave; ~ **sísmica** *f* FÍS ONDAS, GEOFÍS, GEOL seismic wave; ~ **de sobretensión** *f* TELECOM voltage surge; ~ **de sonido** *f* FÍS, FÍS ONDAS sound wave; ~ **de sonido adiabático** *f* FÍS, FÍS ONDAS adiabatic sound wave; ~ **sonora** *f* ACÚST, TELECOM soundwave; ~ **de spin** *f* FÍS spin wave; ~ **subterránea** *f* OCEAN ground wave; ~ **superficial** *f* ACÚST surface wave, FÍS wave surface, ÓPT surface wave; ~ **superficial acústica** *f* ING ELÉC acoustic wave; ~ **de superficie** *f* FÍS, GEOFÍS surface wave;

~ t ~ **TE** *f* (*onda eléctrica transversal*) ELEC, FÍS, FÍS ONDAS, ING ELÉC, TELECOM TE wave (*transverse electric wave*); ~ **TEM** *f* (*onda electromagnética transversal*) ELEC, FÍS, ING ELÉC TEM wave (*transverse electromagnetic wave*); ~ **de temporal** *f* OCEAN storm tide; ~ **terrestre** *f* TELECOM groundwave; ~ **de tierra** *f* FÍS RAD, OCEAN, TRANSP MAR ground wave; ~ **TM** *f* (*onda magnética transversal*) ELEC, FÍS, ING ELÉC TM wave (*transverse magnetic wave*); ~ **transitoria de tensión** *f* TELECOM voltage surge; ~ **transitoria de voltaje** *f* TELECOM voltage surge; ~ **transmisora** *f* FÍS, FÍS ONDAS, TRANSP MAR, TV carrier wave; ~ **transmitida** *f* FÍS transmitted wave; ~ **transversa** *f* FÍS ONDAS transverse wave; ~ **transversal** *f* ACÚST, FÍS, GEOL, ING ELÉC, TELECOM transverse wave; ~ **transversal estacionaria** *f* FÍS ONDAS stationary transverse wave;

~ u ~ **ultracorta** *f* FÍS ONDAS ultrashort wave; ~ **ultrasónica** *f* FÍS ONDAS ultrasonic wave;

~ v ~ **viajera** *f* FÍS traveling wave (*AmE*), travelling wave (*BrE*)

ondámetro *m* FÍS ONDAS wavemeter; ~ **de frecuencia acústica** *m* FÍS ONDAS beat-frequency wavemeter; ~ **heterodino** *m* FÍS ONDAS heterodyne wavemeter

ondas: ~ **armónicas** *f pl* FÍS ONDAS harmonic waves; ~ **planas paralelas** *f* FÍS ONDAS *de fuentes distantes* plane parallel waves; ~ **que se desplazan a la misma velocidad** *f pl* FÍS RAD waves travelling at the same speed; ~ **sinusoidales** *f pl* ELECTRÓN, PETROL *seismogram* singing; ~ **de volumen sin ser influidas por discontinuidades** *f pl* GEOL body waves

ondulación *f* ACÚST ripple, C&V *en el vidrio plano* bloach, CONST ripple, MECÁ flute, MINAS *geología* crimping, PAPEL corrugation; ~ **correspondiente a la vorticidad** *f* FÍS FLUID curl corresponding to vorticity; ~ **de la corriente** *f* ING ELÉC current ripple; ~ **superficial** *f* MECÁ buckling

ondulaciones: ~ **residuales del cauce** *f pl* HIDROL bottom ripples

ondulado[1] *adj* C&V wavy, ING MECÁ corrugated, PAPEL corrugated, fluted

ondulado[2] *m* PAPEL waviness, PROD crinkled finish

onduladora *f* PAPEL corrugator

ondulante *adj* TELECOM *superficie* undulating

ondular *vi* MINAS crimp

ondulatoria *adj* TELECOM undulating

ONE *abr* red Olivetti (*estado UNO*) ELECTRÓN ONE state

ónice *m* MINERAL onyx

ónix *m* MINERAL onyx; ~ **de Argelia** *m* MINERAL Algerian onyx, oriental alabaster; ~ **argelino** *m* MINERAL Algerian onyx, oriental alabaster; ~ **calcáreo** *m* MINERAL Algerian onyx, oriental alabaster

online *adj* INFORM&PD on-line

ontogenia *f* AGRIC ontogeny

onza *f* METR ounce; ~ **avoirdupois** *f* METR avoirdupois ounce; ~ **troy** *f* METR troy ounce

oolito *m* PETROL, TEC PETR oolith

oosparita *f* PETROL, TEC PETR oosparite

op: **no** ~ *abr* (*instrucción no operativa*) INFORM&PD no op (*no-operation instruction*)

opacado *m* C&V darkening

opacador *m* C&V opacifier

opacidad *f* C&V milkiness, FÍS, FÍS ONDAS opacity, GAS cloudiness, IMPR opacity, P&C *de superficies pintadas o revestidas* haze, QUÍMICA milkiness, opacity; ~ **de contraste** *f* PAPEL contrast ratio; ~ **sobre fondo blanco** *f* PAPEL opacity white backing; ~ **sobre fondo de papel** *f* PAPEL opacity paper backing; ~ **de impresión** *f* PAPEL printing opacity; ~ **de la pluma de humo** *f* C&V plume opacity

opacificante *adj* DETERG, QUÍMICA opacifying

opacímetro *m* INSTR, PAPEL, QUÍMICA opacimeter

opaco *adj* CINEMAT light-tight

opalescencia *f* HIDROL opalescence

opalino *m* C&V opaline

ópalo *m* MINERAL opal; ~ **de crisol** *m* C&V pot opal; ~ **de fuego** *m* MINERAL girasol, girasole; ~ **girasol** *m* MINERAL girasol, girasole; ~ **incoloro** *m* MINERAL hyalite

opción: ~ **según programa** *f* PROD scheduling option

OPEP *abr* (*Organización de Países Exportadores de Petróleo*) TEC PETR OPEC (*Organization of Petroleum-Exporting Countries*)

operación *f* CARBÓN process, INFORM&PD transaction, ING MECÁ, MATEMÁT working, TEC ESP actuation; ~ **O** *f* INFORM&PD OR operation; ~ **de aceleración-desaceleración** *f* PROD push-on-push-off operation; ~ **de amplitud de señal** *f* ELECTRÓN large-signal operation; ~ **AND** *f* INFORM&PD AND operation; ~ **aritmética** *f* INFORM&PD arithmetic operation; ~ **asíncrona** *f* ELEC *motor* asynchronous operation; ~ **asincrónica** *f* ELEC asynchronous operation; ~ **de atenuación limitada** *f* TELECOM attenuation-limited operation; ~ **automatizada** *f* INFORM&PD machine operation; ~ **de bifurcación del secuenciador** *f* PROD sequencer jump operation; ~ **binaria** *f* INFORM&PD binary operation; ~ **de buceo** *f* OCEAN diving operation; ~ **de CA** *f* ING ELÉC AC operation; ~ **de cadenas** *f* INFORM&PD *texto* string operation; ~ **de camión a camión** *f* TRANSP truck-to-truck operation; ~ **de la carga** *f* FÍS PART C operation, charge conjugation operation; ~ **centralizada** *f* INFORM&PD centralized operation; ~ **de coma flotante** *f* INFORM&PD, TELECOM floating-point operation; ~ **a conexión derivada** *f* PROD loop-through operation; ~ **controlada por la voz** *f* TELECOM voice-controlled operation; ~ **defectuosa** *f* ING MECÁ faulty operation; ~ **de descarga** *f* PROD download operation; ~ **desclasificada** *f* PROD out-of-sequence operation; ~ **discontinua** *f* ALIMENT, PROC QUÍ batchwise operation; ~ **de distorsión limitada** *f* TELECOM distortion-limited operation; ~ **duplex** *f* TELECOM operation duplex; ~ **en dúplex** *f* INFORM&PD operation duplex; ~ **de elevada intensidad** *f* FÍS RAD high-intensity operation; ~ **de**

equivalencia *f* INFORM&PD equivalence operation; ~ **errónea** *f* ING MECÁ faulty operation; ~ **O exclusiva** *f* INFORM&PD exclusive OR operation; ~ **EXOR** *f* INFORM&PD EXOR operation; ~ **de ficheros** *f* INFORM&PD, PROD file operation; ~ **en forma discontinua** *f* PROC QUÍ batchwise operation; ~ **en una frecuencia** *f* TELECOM single-frequency operation; ~ **para ganar experiencia práctica** *f* INFORM&PD hands-on operation; ~ **gradual de secuenciador reversible** *f* PROD reversing-sequencer step operation; ~ **ilegal** *f* INFORM&PD illegal operation; ~ **de impulsos** *f* ELECTRÓN pulsed operation; ~ **inautorizada** *f* PROD unauthorized operation; ~ **O inclusiva** *f* INFORM&PD inclusive OR operation; ~ **incorrecta** *f* ING MECÁ faulty operation; ~ **inicial** *f* INFORM&PD housekeeping operation; ~ **intermitente** *f* ALIMENT, PROC QUÍ batchwise operation; ~ **inválida** *f* INFORM&PD illegal operation; ~ **de inversión del tiempo** *f* FÍS PART T operation, time reversal operation; ~ **local** *f* TELECOM local operation; ~ **lógica** *f* ELECTRÓN, INFORM&PD logical operation, logic operation; ~ **sin manos** INFORM&PD hands-off operation; ~ **manual** *f* INFORM&PD manual operation; ~ **mecánica** *f* ING MECÁ mechanical operation; ~ **monádica** *f* INFORM&PD monadic operation; ~ **NAND** *f* INFORM&PD NAND operation; ~ **NO** *f* INFORM&PD NOR operation; ~ **de no equivalencia** *f* INFORM&PD non-equivalence operation; ~ **NOO** *f* INFORM&PD NOR operation; ~ **NOR** *f* INFORM&PD NOR operation; ~ **NOT** *f* INFORM&PD NOT operation; ~ **por número de línea constante** *f* TV constant line number operation; ~ **OR** *f* INFORM&PD OR operation; ~ **OR exclusiva** *f* INFORM&PD exclusive OR operation; ~ **OR inclusiva** *f* INFORM&PD inclusive OR operation; ~ **paralela** *f* IMPR parallel operation; ~ **sin personal** *f* INFORM&PD unattended operation; ~ **preparatoria** *f* INFORM&PD housekeeping operation; ~ **a presión constante** *f* TEC ESP constant-pressure operation; ~ **privilegiada** *f* INFORM&PD privileged operation; ~ **prolongada** *f* PROD stretch-out operation; ~ **de punto flotante** *f* INFORM&PD, TELECOM floating-point operation; ~ **de reciclado de aguas residuales** *f* NUCL wastewater recycling operation; ~ **remota** *f* INFORM&PD, TELECOM, TV remote operation; ~ **sin riesgo de fallo** *f* INFORM&PD fail-safe operation; ~ **de salvamento** *f* D&A rescue operation, SEG rescue operation, rescue work, TRANSP AÉR, TRANSP MAR rescue operation; ~ **por seguimiento de carga** *f* NUCL grid-following behavior (*AmE*), grid-following behaviour (*BrE*); ~ **en semidúplex** *f* INFORM&PD, TELECOM half duplex operation; ~ **semidúplex** *f* INFORM&PD, TELECOM half duplex operation; ~ **en serie** *f* INFORM&PD serial operation; ~ **en series** *f* INFORM&PD *texto* string operation; ~ **de servicio** *f* INFORM&PD housekeeping operation; ~ **de un solo paso** *f* INFORM&PD single-step operation; ~ **con terminación única** *f* PROD single-ended operation; ~ **terrena** *f* TEC ESP ground operation; ~ **terrestre** *f* TEC ESP ground operation; ~ **de tierra** *f* TRANSP AÉR ground operation; ~ **con torque pequeño** *f* PROD low-torque operation; ~ **de transformación de la paridad** *f* FÍS PART P operation, parity transformation operation; ~ **de tripulación doble** *f* TRANSP AÉR double crew operation; ~ **umbral** *f* INFORM&PD threshold operation; ~ **unaria** *f*

INFORM&PD unary operation; ~ **uniforme** *f* PROD smooth operation; ~ **en vacío** *f* TRANSP AÉR no-load operation; ~ **con valores lógicos** *f* INFORM&PD logical operation; ~ **vigilada** *f* INFORM&PD attended operation; ~ **XOR** *f* INFORM&PD XOR operation; ~ **Y** *f* INFORM&PD AND operation

operaciones: ~ **de dragado** *f pl* TRANSP MAR dredging operations; ~ **de limpieza** *f pl* CONST clearing operations; ~ **de manutención** *f pl* EMB work handling; ~ **de la perforación** *f pl* TEC PETR drilling operations; ~ **de servicio mayores y menores** *f pl* TRANSP AÉR minor and major servicing operations; ~ **de sondeo** *f pl* TEC PETR drilling operations; ~ **todo tiempo** *f pl* TRANSP AÉR all-weather operations

operado: ~ **a chorro** *m* PROD flush-operated

operador[1] *m* CONST, IMPR, INFORM&PD, ING MECÁ, TEC PETR operator, TELECOM operator, telephone operator, telephonist; ~ **de apisonadora** *m* CONST roller operator; ~ **aritmético** *m* INFORM&PD arithmetic operator; ~ **booleano** *m* INFORM&PD Boolean operator; ~ **del boom** *m* CINEMAT, TV boom operator; ~ **de cámara** *m* CINEMAT cameraman; ~ **de cámara de actualidades** *m* CINEMAT newsreel cameraman; ~ **del cuadro de conmutación** *m* TELECOM switchboard operator; ~ **del cuadro conmutador** *m* TELECOM switchboard operator; ~ **del dolly para el boom** *m* CINEMAT, TV boom dollyman; ~ **de grúa** *m* CINEMAT crane operator; ~ **del izador** *m* TRANSP AÉR hoist operator; ~ **jefe de reactor** *m* NUCL senior reactor operator; ~ **lógico** *m* ELECTRÓN logic operator, INFORM&PD logic operator, logical operator; ~ **de máquina tituladora** *m* CINEMAT caption roller; ~ **de palanca de pomo** *m* PROD knob lever operator; ~ **de pluma de grúa** *m* CINEMAT arm swinger; ~ **de puente móvil** *m* TRANSP MAR bridge-keeper; ~ **de radar** *m* D&A, TRANSP, TRANSP AÉR, TRANSP MAR radar operator; ~ **de reactor** *m* NUCL reactor operator; ~ **de reinserción de la portadora** *m* ELECTRÓN carrier reinsertion operator; ~ **de relación** *m* INFORM&PD relational operator; ~ **relacional** *m* INFORM&PD relational operator; ~ **del tablero de conmutación** *m* TELECOM switchboard operator; ~ **de telecomunicaciones** *m* TELECOM telecommunications operator; ~ **de templador** *m* C&V lehr attendant; ~ **temporalmente fuera de servicio** *fra Esp* (*cf falta temporaria de operador AmL*) TELECOM operator temporarily unavailable; ~ **de terminal de pantalla de visualización** *m* INFORM&PD, PROD visual display unit operator (*VDU operator*); ~ **de turno** *m* NUCL on-shift operator

operador[2]: **ningún** ~ **disponible** *fra* TELECOM no operator available

operadora: ~ **jefe de reactor** *f* NUCL senior reactor operator

operando *m* INFORM&PD operand

operar: ~ **simultáneamente** *vt* ING MECÁ, TEC PETR *herramientas* gang

operario *m* CONST, ING MECÁ, MINAS operator; ~ **de máquina herramienta** *m* PROD *excepto torno* machinist; ~ **no especializado** *m* TEC PETR roustabout

operatividad: ~ **sin carga** *f* ING ELÉC no-load operation

opiánico *adj* QUÍMICA opianic

opianilo *m* QUÍMICA opianyl

opianina *f* QUÍMICA opianine

opiato *f* QUÍMICA opiate

oposición[1]: **en ~ de fase** *adj* ELECTRÓN, FÍS in phase opposition
oposición[2] *f* CARBÓN contest, GEN resistance; **~ de fase** *f* ELECTRÓN, FÍS phase opposition
óptica: **~ correctora** *f* FOTO, ÓPT correcting optics; **~ electrónica** *f* ELECTRÓN, FÍS, INFORM&PD, ING ELÉC, ÓPT, TELECOM optoelectronics, electron optics; **~ de fibras** *f* C&V, CINEMAT, FÍS, INFORM&PD, ING ELÉC, ÓPT, TELECOM fiber-optics (*AmE*), fibre-optics (*BrE*); **~ física** *f* FÍS, ÓPT, TELECOM physical optics; **~ geométrica** *f* FÍS, ÓPT, TELECOM geometric optics; **~ de iluminación** *f* INSTR, ÓPT illumination optics; **~ junta** *f* ÓPT, TELECOM optical splice; **~ de la onda** *f* ÓPT, TELECOM wave optics; **~ ondulatoria** *f* ÓPT, TELECOM wave optics; **~ de profundidad** *f* TEC ESP optical depth; **~ de rayos** *f* ÓPT, TELECOM ray optics
ópticamente: **~ plano** *adj* FOTO optically flat
óptico: **~-eléctrico** *adj* (*O-E*) ELEC, ÓPT, TELECOM optical-electrical (*O-E*)
optimización *f* INFORM&PD optimization; **~ del peso** *f* TEC ESP weight optimization
optimizador: **~ del cabezal de video** *m* AmL, **~ del cabezal de vídeo** *m* Esp TV video head optimizer
optimizar *vt* INFORM&PD optimize
optoacoplador *m* ING ELÉC, TELECOM optocoupler
optoelectrónica *f* ELECTRÓN optronics, FÍS, INFORM&PD, ING ELÉC, ÓPT, TELECOM optoelectronics
optoelectrónico *adj* ELECTRÓN, FÍS, INFORM&PD, ÓPT, TELECOM optoelectronic
optómetro *m* INSTR optometer
oquedad *f* METAL, MINAS cavity
OR *abr* (*investigación de operaciones*) INFORM&PD OR (*operational research*)
orangita *f* MINERAL orangite
órbita[1]: **de ~** *adj* TEC ESP orbital; **de ~ sobre el polo** *adj* TEC ESP polar-orbiting; **en ~ terrestre** *adj* TEC ESP earth-orbiting
órbita[2] *f* TEC ESP orbit; **~ de aparcamiento** *f* D&A, TEC ESP *satélites* parking orbit; **~ de aparcamiento terrestre** *f* TEC ESP earth-parking orbit; **~ para aparcar sátelites fuera de servicio** *f* TEC ESP graveyard orbit; **~ arriñonada** *f* NUCL banana orbit; **~ de baja altitud** *f* TEC ESP low-altitude orbit; **~ circular** *f* TEC ESP circular orbit; **~ de corona** *f* TEC ESP halo orbit; **~ de deriva** *f* TEC ESP drift orbit; **~ de descenso** *f* TEC ESP descent orbit; **~ directa** *f* TEC ESP direct orbit; **~ ecuatorial** *f* TEC ESP equatorial orbit; **~ elíptica** *f* FÍS, TEC ESP elliptical orbit; **~ de espera** *f* TEC ESP parking orbit; **~ de estabilización** *f* TEC ESP parking orbit; **~ estacionaria** *f* TEC ESP stationary orbit; **~ geoestacionaria** *f* FÍS, GEOFÍS geostationary orbit, TEC ESP earth-synchronous orbit, geostationary orbit; **~ geosincrónica** *f* TEC ESP geosynchronous orbit; **~ hiperbólica** *f* TEC ESP hyperbolic orbit; **~ intermedia** *f* TEC ESP interim orbit; **~ de Kepler** *f* TEC ESP Keplerian orbit; **~ lunar** *f* TEC ESP lunar orbit; **~ no perturbada** *f* TEC ESP unperturbed orbit; **~ parabólica cercana** *f* TEC ESP near parabolic orbit; **~ polar** *f* TEC ESP polar orbit; **~ provisional** *f* TEC ESP interim orbit; **~ retrógrada** *f* TEC ESP retrograde orbit; **~ de satélite geoestacionario** *f* TEC ESP geostationary satellite orbit; **~ sincrónica con la Tierra** *f* TEC ESP earth-synchronous orbit; **~ síncronosolar** *f* TEC ESP sun-synchronous orbit; **~ solar síncrona** *f* TEC ESP sun-

synchronous orbit; **~ terrestre** *f* TEC ESP earth orbit; **~ de transferencia** *f* TEC ESP transfer orbit
orbital[1] *adj* FÍS, FÍS RAD, TEC ESP orbital
orbital[2] *m* FÍS, FÍS RAD, QUÍMICA orbital; **~ atómico** *m* FÍS, FÍS PART, FÍS RAD, QUÍMICA atomic orbital; **~ atómico antienlazante** *m* FÍS RAD antibonding atomic orbital; **~ híbrido** *m* QUÍMICA hybrid orbital; **~ molecular** *m* FÍS, FÍS RAD, QUÍMICA molecular orbital; **~ molecular antienlazante** *m* FÍS RAD antibonding molecular orbital
orbitar *vi* TEC ESP orbit
ORC *abr* (*osciloscopio de rayos catódicos*) ELECTRÓN, FÍS, FÍS RAD, INSTR, TV CRO (*cathode-ray oscilloscope, cathode ray*)
orceína *f* QUÍMICA orcein
orden[1]: **de ~ inferior** *adj* INFORM&PD low order; **de ~ superior** *adj* INFORM&PD high-order
orden[2] *m* ACÚST *de armónica* order, IMPR select, QUÍMICA order; **~ de aproximación** *m* TRANSP AÉR approach sequence; **~ armónico** *m* ELECTRÓN harmonic order; **~ de clasificación de caracteres** *m* INFORM&PD collating sequence; **~ de compra recomendado** *m* PROD recommended purchase order; **~ en consigna** *m* PROD held-up order; **~ de corto alcance** *m* CRISTAL, FÍS short-range order; **~-desorden** *m* METAL order-disorder; **~ de desviación** *m* CONST variation order; **~ de encendido** *m* VEH *del motor* firing order; **~ de filtrado** *m* ELECTRÓN filter order; **~ del filtro** *m* ELECTRÓN filter order; **~ de interferencia** *m* FÍS order of interference; **~ a larga distancia** *m* CRISTAL long-range order; **~ de largo alcance** *m* CRISTAL long-range order; **~ lexicográfico** *m* INFORM&PD lexicographic order; **~ magnético** *m* METAL magnetic order; **~ de magnitud** *m* FÍS, INFORM&PD order of magnitude; **~ a pequeña distancia** *m* CRISTAL short-range order; **~ de pequeño alcance** *m* METAL short-range order; **~ de prioridad** *m* INFORM&PD order of precedence, PROD priority order; **~ de producción** *m* PROD production order; **~ de producción recomendado** *m* PROD recommended production order; **~ de reacción** *m* METAL order of reaction; **~ de trabajo** *m* PROD *fabricación de piezas* work order; **~ de variación** *m* CONST variation order
orden[3]: **~ de almacén** *f* PROD warehouse order; **~ basada en tiempo finito** *f* PROD fine-time based instruction; **~ de batalla** *f* D&A order of battle; **~ de cambio** *f* MECÁ change order; **~ de compra a un precio concreto** *f* PROD open order; **~ de conexión de interrogación** *f* PROD examine-on instruction; **~ de desconexión de interrogación** *f* PROD examine-off instruction; **~ de desenganche** *f* PROD released order; **~ de expedición** *f* TRANSP MAR shipping order; **~ de inspección general** *f* PROD turnaround directive; **~ de lanzamiento de fabricación** *f* PROD order shop; **~ de reaprovisionamiento** *f* PROD replenishment order; **~ de retenida** *f* PROD latch instruction; **~ única** *f* TEC ESP unique word (*UW*); **~ urgente** *f* IMPR rush order
ordenación: **~ de territorio** *f* AGRIC regional-planning policy
ordenada *f* INFORM&PD, MATEMÁT ordinate
ordenador *m* Esp (*cf computadora AmL*) ELEC, INFORM&PD computer, machine; **~ anfitrión** *m* Esp (*cf computadora anfitrión AmL*) INFORM&PD host computer; **~ autónomo** *m* Esp (*cf computadora*

autónoma AmL) INFORM&PD off-line computer, stand-alone, NUCL (*cf computadora autónoma AmL*) off-line computer; **~ de a bordo** *m Esp* (*cf computadora de a bordo AmL*) TEC ESP on-board computer; **~ de casa** *m Esp* (*cf computadora de casa AmL*) INFORM&PD home computer; **~ central** *m Esp* (*cf computadora central AmL*) INFORM&PD host computer; **~ de conjunto de instrucciones complejas** *m Esp* (*cf computadora de conjunto de instrucciones complejas AmL*) INFORM&PD complex instruction set computer (*CISC*); **~ de cuarta generación** *m Esp* (*cf computadora de cuarta generación AmL*) INFORM&PD fourth-generation computer; **~ de datos de aire** *m Esp* (*cf computadora de datos de aire AmL*) TRANSP AÉR air data computer; **~ dedicado** *m Esp* (*cf computadora dedicada AmL*) ELECTRÓN, INFORM&PD digital computer; **~ de destino** *m Esp* (*cf computadora de destino AmL*) INFORM&PD target computer; **~ digital** *m Esp* (*cf computadora digital AmL*) ELECTRÓN, INFORM&PD digital computer; **~ digital de proceso** *m Esp* (*cf computadora digital de proceso AmL*) NUCL digital process computer system; **~ doméstico** *m Esp* (*cf computadora doméstica AmL*) INFORM&PD, TELECOM home computer, personal computer (*PC*); **~ especializado** *m Esp* (*cf computadora especializada AmL*) INFORM&PD dedicated computer, special-purpose computer; **~ de falda** *m Esp* (*cf computadora de falda AmL*) INFORM&PD laptop computer; **~ fuera de línea** *m Esp* (*cf computadora fuera de línea AmL*) INFORM&PD, NUCL off-line computer; **~ híbrido** *m Esp* (*cf computadora híbrida AmL*) INFORM&PD hub polling, hybrid computer; **~ con juego reducido de instrucciones** *m Esp* (*cf computadora con juego reducido de instrucciones AmL*) INFORM&PD reduced instruction set computer (*RISC*); **~ laptop** *m Esp* (*cf computadora laptop AmL*) INFORM&PD laptop computer; **~ de longitud de palabra fija** *m Esp* (*cf computadora de longitud de palabra fija AmL*) INFORM&PD fixed-wordlength computer; **~ de longitud de palabra variable** *m Esp* (*cf computadora de longitud de palabra variable AmL*) INFORM&PD variable-wordlength computer; **~ orientado a los caracteres** *m Esp* (*cf computadora orientada a los caracteres AmL*) INFORM&PD character-oriented machine; **~ en paralelo** *m Esp* (*cf computadora en paralelo AmL*) INFORM&PD parallel computer; **~ pasarela** *m Esp* (*cf computadora pasarela AmL*) INFORM&PD gateway computer; **~ personal** *m Esp* (*cf computadora personal AmL*) INFORM&PD, TELECOM personal computer (*PC*); **~ portátil** *m Esp* (*cf computadora portátil AmL*) INFORM&PD laptop computer; **~ de portilla** *m Esp* (*cf computadora de portilla AmL*) INFORM&PD gateway computer; **~ de primera generación** *m Esp* (*cf computadora de primera generación AmL*) INFORM&PD first-generation computer; **~ principal** *m Esp* (*cf computadora principal AmL*) INFORM&PD host computer; **~ con programa almacenado** *m Esp* (*cf computadora con programa almacenado AmL*) INFORM&PD stored program computer; **~ con programa almacenado en memoria** *m Esp* (*cf computadora con programa almacenado en memoria AmL*) INFORM&PD stored program computer; **~ puente entre redes** *m Esp* (*cf computadora puente entre redes AmL*) INFORM&PD gateway computer; **~ de quinta generación** *m Esp* (*cf computadora de*

quinta generación AmL) INFORM&PD fifth-generation computer; **~ satélite** *m Esp* (*cf computadora satélite AmL*) INFORM&PD satellite computer; **~ de secuencia** *m Esp* (*cf computadora de secuencia AmL*) INFORM&PD, ING ELÉC, PROD, TELECOM sequencer; **~ de segunda generación** *m Esp* (*cf computadora de segunda generación AmL*) INFORM&PD second-generation computer; **~ serie** *m Esp* (*cf computadora serie AmL*) INFORM&PD serial computer; **~ síncrono** *m Esp* (*cf computadora síncrona AmL*) INFORM&PD synchronous computer; **~ de sobremesa** *m Esp* (*cf computadora de sobremesa AmL*) INFORM&PD desktop computer; **~ solo** *m Esp* (*cf computadora sola AmL*) INFORM&PD stand-alone; **~ de tercera generación** *m Esp* (*cf computadora de tercera generación AmL*) INFORM&PD third-generation computer; **~ para tráfico** *m Esp* (*cf computadora para tráfico AmL*) TRANSP traffic computer; **~ universal** *m Esp* (*cf computadora universal AmL*) INFORM&PD general-purpose computer (*GP computer*), universal computer; **~ de uso general** *m Esp* (*cf computadora de uso general AmL*) INFORM&PD general-purpose computer (*GP computer*); **~ virtual** *m Esp* (*cf computadora virtual AmL*) INFORM&PD virtual machine; **~ de vuelo** *m Esp* (*cf computadora de vuelo AmL*) TRANSP AÉR flight computer

ordenamiento *m* FERRO, TRANSP marshaling (*AmE*), marshalling (*BrE*), MATEMÁT *de figuras, cifras,* TEC ESP array

ordenanza *f* TELECOM regulation

ordenanzas: ~ nacionales sobre el agua *f pl* AGUA water law

órdenes: ~ de impresión *f pl* IMPR printer commands

ordeño: ~ mecánico *m* AGRIC mechanical milking

ordinal *m* GEOM ordinal

ordinograma: ~ de programación *m* PROD process chart

oreador: ~ del heno *m* AGRIC tedder

oreja *f* ING MECÁ wing

orejeras *f pl* ACÚST, CONST, LAB, SEG ear protectors, earmuffs

orejeta *f* ING MECÁ wing, MECÁ lobe, TRANSP MAR *construcción naval* lug; **~ de cierre** *f* PROD board locking tab; **~ para izar** *f* CONST lifting lug, ING MECÁ handling lug, lifting lug

oreo *m* CARBÓN *minerales* weathering; **~ refrigerado** *m* ALIMENT, REFRIG carcass chilling

orgánico *adj* AGRIC, ALIMENT, QUÍMICA organic

organigrama *m Esp* (*cf flujograma AmL*) FÍS, INFORM&PD flowchart; **~ de datos** *m* INFORM&PD data flowchart; **~ de gerencia** *m AmL* (*cf organigrama de gestión Esp*) PROD management chart; **~ de gestión** *m Esp* (*cf organigrama de gerencia AmL*) PROD management chart

organismo: ~ de aprobación *m* CALIDAD certification body; **~ con competencias en urbanismos** *m* CONST chief building authority; **~ regulador** *m* NUCL regulatory body, TRANSP MAR regulatory agency

organización *f* INFORM&PD layout, PROD setup; **~ de aprobación** *f* CALIDAD certification organisation; **~ de archivos** *f* INFORM&PD file organization; **~ autorizada** *f* CALIDAD approved organisation; **~ de la eliminación de desperdicios** *f* RECICL waste management; **~ de ficheros** *f* INFORM&PD file organization; ~ **y mantenimiento** *f* TELECOM organization and maintenance (*OAM*); ~ **del tráfico marítimo** *f* TRANSP MAR routing system

Organización: ~ **Internacional de Normalización** *f* (*OIN*) ELEC, ING MECÁ, TELECOM International Standards Organization (*ISO*); ~ **Marítima Internacional** *f* (*OMI*) TRANSP MAR International Maritime Organization (*IMO*); ~ **de Paises Exportadores de Petróleo** *f* (*OPEP*) TEC PETR Organization of Petroleum-Exporting Countries (*OPEC*)

organizado: ~ **por funciones** *adj* PROD function-orientated

organizador *m* INFORM&PD scheduler; ~ **de tareas** *m* INFORM&PD job scheduler

órgano *m* ING MECÁ member; ~ **activo** *m* NUCL working part; ~ **de conmutación** *m* ING ELÉC switch gear; ~ **de Corti** *m* ACÚST *audición* organ of Corti; ~ **de entrada** *m* ELECTRÓN, INFORM&PD, ING ELÉC, PROD, TELECOM input device; ~ **de salida** *m* PROD, TELECOM output device

organoclorado[1] *adj* AGRIC, QUÍMICA, RECICL organochlorine

organoclorado[2] *m* RECICL organochlorine

organógeno *adj* CARBÓN organogenous

organomagnesio *m* QUÍMICA organomagnesium

organometálico *adj* QUÍMICA organometallic

órganos: ~ **de desprendimiento** *m pl* ELEC *cortocircuito*, ING ELÉC trip gear

organosol *m* QUÍMICA organosol

orientable *adj* TEC ESP vectored

orientación *f* CONST orientation, GEOL bearing, IMPR orientation, INFORM&PD prompt, TEC ESP heading; ~ **biaxial** *f* P&C *película plástica* biaxial orientation; ~ **del haz** *f* TEC ESP beam switching; ~ **de partículas magnéticas** *f* TV magnetic particle orientation; ~ **de pista** *f* TRANSP AÉR runway alignment; ~ **predominante** *f* CRISTAL, GEOL preferred orientation; ~ **preferente** *f* CRISTAL, GEOL preferred orientation; ~ **vertical** *f* IMPR portrait

orientado: ~ **hacia la computadora** *adj AmL* (*cf orientado hacia el ordenador Esp*) INFORM&PD machine-oriented; ~ **hacia el ordenador** *adj Esp* (*cf orientado hacia la computadora AmL*) INFORM&PD machine-oriented; ~ **a la palabra** *adj* INFORM&PD word-oriented; ~ **según la regla de la mano derecha** *adj* FÍS *de un sistema de coordenadas* right-handed; ~ **a la transmisión continua de bits** *adj* TELECOM continuous bit stream oriented (*CBO*)

orientar *vt* CONST orient

orificio *m* C&V feeder opening, orifice, neck ring, CONST hole, opening, ING MECÁ hole, opening, orifice, port, porthole, INSTAL HIDRÁUL port, *en pared gruesa* orifice, *en pared divisoria delgada* opening, MECÁ port, aperture, METAL aperture, NUCL hole, TEC ESP nozzle, TEC PETR orifice; ~ **abierto en la parte superior** *m* AGUA *vertederos de aforo* notch; ~ **de acceso** *m* NUCL access port; ~ **de admisión** *m* ING MECÁ inlet, INSTAL HIDRÁUL admission port, induction port, intake port; ~ **de admisión de vapor** *m* INSTAL HIDRÁUL steam inlet, steam port, steam way; ~ **de aligeramiento** *m* MECÁ, TRANSP AÉR lightening hole; ~ **de alivio** *m Esp* (*cf orificio de desahogo AmL*) MINAS easer hole; ~ **anegado** *m* INSTAL HIDRÁUL orifice, submerged orifice; ~ **de aristas vivas** *m* INSTAL HIDRÁUL sharp-edged orifice, standard ori-

fice; ~ **de aspiración** *m* PROD suction port; ~ **calibrado** *m* REFRIG orifice plate, TEC PETR gaged orifice (*AmE*), gauged orifice (*BrE*); ~ **ciego** *m* CONST blind hole; ~ **para clavija** *m* CONST dowel hole; ~ **de conexión** *m* ING MECÁ connection port; ~ **de desahogo** *m AmL* (*cf orificio de alivio Esp*) MINAS easer hole; ~ **de descarga** *m* AGUA, ING MECÁ, PROD *hidráulica* outlet; ~ **de despresurización** *m* TEC ESP vent hole; ~ **de ensanchamiento** *m* MINAS enlarging-hole; ~ **de entrada** *m* ING MECÁ, MECÁ, MINAS inlet; ~ **de entrada de agua** *m* AGUA water inlet; ~ **de entrada redondeada** *m* INSTAL HIDRÁUL rounded-approach orifice; ~ **de escape** *m* AUTO exhaust port, INSTAL HIDRÁUL eduction port, outlet, vent hole, exhaust port, VEH exhaust port; ~ **de evacuación** *m* INSTAL HIDRÁUL *de cilindro de vapor* eduction port, outlet, vent hole; ~ **del frente de arranque** *m* MINAS face hole; ~ **de grandes dimensiones** *m* HIDROL large orifice; ~ **guía** *m* ING MECÁ pilot hole; ~ **hundido** *m* PROD countersunk hole; ~ **de índice** *m* INFORM&PD *de disco magnético flexible* index hole; ~ **de inyección** *m* NUCL injection borehole; ~ **para inyectar** *m* P&C gate; ~ **de llegada** *m* AGUA infall; ~ **de llenado** *m* EMB fill hole; ~ **de llenado del molde** *m* C&V baffle hole; ~ **de medición** *m* ING MECÁ metering hole; ~ **de mina** *m AmL* (*cf barreno Esp*) MINAS drill hole; ~ **en pared delgada** *m* INSTAL HIDRÁUL sharp-edged orifice, standard orifice; ~ **de rebosadero** *m* CARBÓN drain; ~ **refractario** *m* C&V *salida del alimentador* bushing; ~ **respiradero** *m* TEC ESP vent hole; ~ **roscado** *m* NUCL threaded hole; ~ **de salida** *m* ING MECÁ, PROD outlet, TRANSP tailgate; ~ **de salida de agua** *m* AGUA water outlet port; ~ **de seguridad** *m* CONST refuge hole; ~ **de succión** *m* PROD suction port; ~ **sumergido** *m* INSTAL HIDRÁUL orifice, submerged orifice; ~ **con tapón** *m* CONST plug hole; ~ **de vaciado** *m* C&V tapping hole; ~ **variador** *m* CONTAM MAR *de un sumidero* gulley sucker; ~ **de ventilación** *m* C&V, CONST, P&C vent, vent hole, TEC PETR air vent; ~ **del visor del indicador del nivel de aceite** *m* REFRIG oil-sight o-ring

origen: ~ **de llamada** *m* TELECOM call origin; ~ **de un manantial** *m* AGUA, HIDROL spring head; ~ **del mensaje** *m* INFORM&PD message source; ~ **principal** *m* CONTAM major source

original *m* CINEMAT original, IMPR *manuscrito* manuscript, original, ÓPT master, TV original; ~ **de anuncio** *m* IMPR advertising copy; ~ **de bajo contraste** *m* CINEMAT low-contrast original; ~ **de cámara** *m* CINEMAT, TV camera original; ~ **ilegible** *m* IMPR bad copy; ~ **limpio** *m* IMPR clear copy; ~ **de línea** *m* IMPR line copy; ~ **listo para reproducir** *m* IMPR camera-ready copy

originales *m pl* IMPR art

orilla *f* AGUA riverbank, C&V shoreline, CONST bank, riverbank, HIDROL *marítima* coast, *río o lago* bank, margin, riverbank, OCEAN shore, TRANSP MAR bank, riverbank; ~ **del agua** *f* CONST water line; ~ **del bulbo** *f* C&V bulb edge; ~ **cónica reforzada** *f* C&V conical reinforced rim; ~ **de corte** *f* C&V *del diamante* cutting edge; ~ **despostillada** *f* C&V chipped edge; ~ **del río** *f* AGUA, CONST, HIDROL, TRANSP MAR riverbank; ~ **rizada** *f* C&V curled edge

orillas: ~ **burdas** *f pl* C&V edge as cut

orillo *m* CONST, METAL, P&C selvage, PAPEL selvage,

cortadoras, rebobinadoras trimming, TEXTIL selvage; **~ con tensión** *m* TEXTIL pulled-in selvage, pulled-in selvedge; **~ tirante** *m* TEXTIL tight selvage, tight selvedge

orinque *m* TRANSP MAR *para el ancla* anchor buoy rope, buoy rope, buoyline, tripping line

orla *f* IMPR border; **~ de adorno** *f* IMPR ornamental border; **~ ornamental** *f* C&V border; **~ de playa** *f* OCEAN beach cusp

orlón *m* QUÍMICA orlon

ornamentación *f* PROD fretting

ornamento: ~ floral *m* IMPR fleuron; **~ de porcelana** *m* C&V china ornamentation

ornitúrico *adj* QUÍMICA ornithuric

oro *m* (*Au*) METAL, QUÍMICA gold (*Au*); **~ de aluvión** *m* MINAS placer gold; **~ brillante** *m* C&V bright gold; **~ bruto** *m* C&V, METAL bullion; **~ corrido** *m* MINAS placer gold; **~ de espuma** *m* MINAS float gold; **~ flotante** *m* MINAS float gold, floating gold; **~ líquido** *m* C&V liquid gold; **~ molido** *m* C&V powdered gold; **~ en polvo** *m* C&V powdered gold; **~ que flota** *m* MINAS float gold

orogenia *f* GEOL, TEC PETR orogeny; **~ cimeriana** *f* GEOL, TEC PETR Cimmerian orogeny; **~ cimerianense** *f* GEOL, TEC PETR Cimmerian orogeny; **~ primaria** *f* GEOL internides; **~ secundaria** *f* GEOL externides

orogénico *adj* GEOL orogenic

orógeno *m* GEOL orogen, tectogene

oropel *m* PROD foil

oropimente *m* MINERAL orpiment

orotrón *m* ELECTRÓN orotron

orquilla: ~ para coger el coque *f* MINAS coke fork; **~ para escoger el coque** *f* CARBÓN coke fork

orsélico *adj* QUÍMICA orsellic

orselínico *adj* QUÍMICA orsellinic

orticonoscopio *m* ELECTRÓN, TV orthicon; **~ de imagen** *m* ELECTRÓN, TV image orthicon

ortiga *f* AGRIC hemp nettle

ortita *f* MINERAL cerine, orthite, NUCL orthite, QUÍMICA cerine

ortobásico *adj* QUÍMICA orthobasic

ortocarbónico *adj* QUÍMICA orthocarbonic

ortocentro *m* GEOM orthocenter (*AmE*), orthocentre (*BrE*)

ortoclasa *f* MINERAL moonstone, orthoclase

ortoclorito *m* MINERAL orthochlorite

ortoclorotolueno *m* QUÍMICA orthochlorotoluene

ortocromatismo *m* CINEMAT, FOTO orthochromatism

ortocromatización *f* CINEMAT, FOTO orthochromatization

ortocuarcita *f* GEOL orthoquartzite

ortodromía *f* TEC ESP, TRANSP MAR orthodromy

ortodromo *f* FÍS great circle

ortoescopico *adj* FÍS orthoscopic

ortofórmico *adj* QUÍMICA orthoformic

ortofosfato *m* QUÍMICA orthophosphate

ortofosfórico *adj* DETERG, QUÍMICA orthophosphoric

ortogonal *adj* GEOM orthogonal

ortohidrógeno *m* QUÍMICA orthohydrogen

ortosilicato *m* QUÍMICA orthosilicate

ortosilícico *adj* QUÍMICA orthosilicic

ortotropismo *m* AGRIC orthotropism

ortovanádico *adj* QUÍMICA orthovanadic

ortoxileno *m* QUÍMICA, TEC PETR orthoxylene

oruga *f* AGRIC caterpillar, CONST track strip, tread, D&A caterpillar; **~ de rodamiento** *f* ING MECÁ crawler track

orza *f* TRANSP MAR *construcción naval* centerboard (*AmE*), centreboard (*BrE*), fin keel; **~ de deriva** *f* TRANSP MAR *construcción naval* centerboard (*AmE*), centreboard (*BrE*), leeboard; **~ de deriva metálica** *f* TRANSP MAR *yates* centerplate (*AmE*), centreplate (*BrE*); **~ de eje** *f* TRANSP MAR *construcción naval* centerboard (*AmE*), centreboard (*BrE*); **~ metálica** *f* TRANSP MAR centerplate (*AmE*), centreplate (*BrE*)

orzar *vi* TRANSP MAR luff, *atraques* come to

Os *abr* (*osmio*) QUÍMICA Os (*osmium*)

osazona *f* QUÍMICA osazone

oscilación *f* CONST, ELEC, ELECTRÓN oscillation, FÍS beat, oscillation, FÍS ONDAS oscillation, ING ELÉC hunting, ING MECÁ swinging, seesawing, swaying, oscillation, shake, shaking, PAPEL swing, TEC ESP yaw, TELECOM oscillation, TRANSP AÉR hunting; **~ acústica** *f* ACÚST acoustic oscillation; **~ aeroelástica de hipofrecuencia** *f* TEC ESP, TRANSP AÉR judder; **~ amortiguada** *f* ELECTRÓN damped oscillation; **~ armónica** *f* FÍS harmonic oscillation; **~ de la carga** *f* TRANSP AÉR cargo swing; **~ cero** *f* TV zero beat; **~ constante** *f* ELECTRÓN sustained oscillation; **~ contínua** *f* ELECTRÓN continuous oscillation; **~ en diente de sierra** *f* ELEC saw tooth oscillation; **~ eléctrica** *f* ELEC electrical oscillation, ELECTRÓN electric oscillation, FÍS ONDAS electrical oscillation; **~ forzada** *f* ELECTRÓN, FÍS forced oscillation; **~ de frecuencia propia** *f* ELECTRÓN natural frequency oscillation; **~ fugoide** *f* TRANSP AÉR phugoid oscillation; **~ fundamental** *f* ING MECÁ natural oscillation; **~ de incidencia** *f* TRANSP AÉR incidence oscillation; **~ independiente** *f* ING ELÉC *circuito válvula triodo* overlapping; **~ iónica electrostática** *f* NUCL electrostatic ion-oscillation; **~ libre** *f* ELEC *galvanómetro*, FÍS, ING ELÉC free oscillation, ING MECÁ natural oscillation; **~ lineal** *f* ING MECÁ, TELECOM linear oscillation; **~ lineal libre** *f* ING MECÁ, TELECOM free linear oscillation; **~ longitudinal libre** *f* ING MECÁ, TELECOM free longitudinal oscillation; **~ mecánica** *f* ACÚST mechanical oscillation; **~ natural** *f* ING MECÁ natural oscillation; **~ del nivel de video** *AmL*, **~ del nivel de vídeo** *Esp f* TV fluttering video level; **~ no lineal** *f* ING MECÁ, TELECOM nonlinear oscillation; **~ parásita** *f* ELECTRÓN singing, FÍS, TELECOM parasitic oscillation; **~ de un péndulo** *f* ING MECÁ oscillation of a pendulum; **~ de pequeña amplitud** *f* ELECTRÓN dither; **~ periódica anormal** *f* REFRIG hunting; **~ periódica del espejo de agua** *f* ENERG RENOV seiche; **~ propia** *f* ING MECÁ natural oscillation; **~ de relajación** *f* ELECTRÓN, FÍS relaxation oscillation; **~ de un resorte** *f* ING MECÁ oscillation of a spring; **~ sinusoidal** *f* ELECTRÓN sinusoidal oscillation; **~ sostenida** *f* FÍS maintained oscillation; **~ torsional** *f* FÍS torsional oscillation; **~ transiente** *f* FÍS transient oscillation; **~ transitoria** *f* CONST, ELECTRÓN, ING ELÉC ringing

oscilaciones: ~ pendulares *f pl* ELEC *máquinas* hunting

oscilador *m* CINEMAT, ELEC oscillator, ELECTRÓN, FÍS carcinotron, oscillator, FÍS ONDAS oscillator, ING MECÁ, MECÁ rocker, TEC ESP oscillator, vibrator, TELECOM carcinotron, oscillator; **~ de acoplamiento electrónico** *m* ELECTRÓN electron-coupling oscillator; **~ ajustable de banda ancha** *m* ELECTRÓN wideband tunable oscillator; ~ **ajustado GFI** *m* (*oscilador ajustado de granate férrico de itrio*) ELECTRÓN YIG tuned oscillator (*yttrium iron garnet tuned oscillator*); ~ **armónico** *m* ELECTRÓN, FÍS harmonic oscillator; ~ **astable** *m* ELEC, ELECTRÓN free running oscillator; ~ **de audiofrecuencia** *m* ELECTRÓN audio frequency oscillator; ~ **autoexcitado** *m* ELECTRÓN self-excited oscillator; ~ **de baja derivación** *m* ELECTRÓN low-drift oscillator; ~ **de baja frecuencia** *m* ELECTRÓN low-frequency oscillator; ~ **de banda de octava** *m* ELECTRÓN octave band oscillator; ~ **de bifurcación** *m* ELECTRÓN fork oscillator; ~ **de bloqueo** *m* ELECTRÓN blocking oscillator; ~ **de búsqueda** *m* ELECTRÓN tracking oscillator; ~ **de cavidad** *m* ELECTRÓN cavity oscillator; ~ **de cierre de inyección** *m* ELECTRÓN injection-locked oscillator; ~ **con circuito de reacción abierto** *m* ELECTRÓN open-loop oscillator; ~ **coherente** *m* ELECTRÓN coherent oscillator; ~ **Colpitts** *m* ELECTRÓN Colpitts oscillator; ~ **compensador de rozamientos** *m* ELECTRÓN antistiction oscillator; ~ **en contrafase** *m* ELEC, ELECTRÓN, TELECOM push-pull oscillator; ~ **controlado** *m* ELECTRÓN controlled oscillator; ~ **controlado por cristal de cuarzo** *m* ELECTRÓN, TELECOM crystal-controlled oscillator (*CCO*); ~ **de conversión** *m* ELECTRÓN conversion oscillator; ~ **de corriente constante** *m* ELECTRÓN constant-current oscillator; ~ **de corriente controlada** *m* ELECTRÓN current-controlled oscillator; ~ **de cristal** *m* CRISTAL, ELEC, ELECTRÓN, FÍS RAD crystal oscillator; ~ **de cristal de aire libre** *m* ELEC, ELECTRÓN free-air crystal oscillator; ~ **de cristal completo** *m* ELECTRÓN simple-packaged crystal oscillator; ~ **de cristal de cuarzo** *m* CINEMAT, ELEC, ELECTRÓN, FÍS, TELECOM quartz-crystal oscillator; ~ **de cristal listo para funcionar** *m* ELEC, ELECTRÓN simple-packaged crystal oscillator; ~ **de cristal con temperatura compensada** *m* ELECTRÓN temperature-compensated crystal oscillator; ~ **de cristal con temperatura controlada** *m* ELECTRÓN temperature-controlled crystal oscillator; ~ **cristalino** *m* CRISTAL, ELEC, ELECTRÓN, FÍS RAD crystal oscillator; ~ **de cuarzo** *m* CINEMAT, ELEC, ELECTRÓN, FÍS, TELECOM quartz oscillator; ~ **de décadas** *m* ELECTRÓN decade oscillator; ~ **de desplazamiento de frecuencia** *m* ELECTRÓN frequency-hopping oscillator; ~ **de diodo** *m* ELECTRÓN diode oscillator; ~ **eléctrico** *m* ELEC, FÍS ONDAS electrical oscillator; ~ **estabilizado por cristal** *m* CRISTAL, ELEC, ELECTRÓN, FÍS RAD crystal oscillator; ~ **de excitación** *m* ELECTRÓN keep-alive oscillator; ~ **de frecuencia acústica** *m* FÍS RAD audio oscillator; ~ **de frecuencia de batido** *m* ELECTRÓN, FÍS, TELECOM, TRANSP AÉR, TV beat-frequency oscillator (*BFO*); ~ **de frecuencia constante** *m* ELECTRÓN fixed-frequency oscillator; ~ **de frecuencia de impulsos** *m* ELECTRÓN, FÍS, TELECOM, TRANSP AÉR, TV beat-frequency oscillator (*BFO*); ~ **de frecuencia portadora** *m* ELECTRÓN carrier-frequency oscillator; ~ **de frecuencia de pulsación** *m* ELECTRÓN, FÍS, TELECOM, TRANSP AÉR, TV beat-frequency oscillator (*BFO*); ~ **de frecuencia de sonido** *m* ELECTRÓN audio-frequency oscillator; ~ **de frecuencia variable** *m* ELECTRÓN variable-frequency oscillator; ~ **de gran estabilidad** *m* ELECTRÓN highly stable oscillator; ~ **Hartley** *m* ELECTRÓN Hartley oscillator; ~ **homodino** *m* ELECTRÓN homodyne oscillator; ~ **de impulsos** *m* ELECTRÓN pulsed oscillator; ~ **de interacción ampliado** *m* ELECTRÓN extended interaction oscillator; ~ **de interferencias** *m* ELECTRÓN jammer oscillator; ~ **invertidor** *m* ELECTRÓN inverter oscillator; ~ **klistron** *m* ELECTRÓN klystron oscillator; ~ **de lámpara de vacío** *m* ELECTRÓN vacuum tube oscillator; ~ **de línea estabilizada** *m* ELECTRÓN line-stabilized oscillator; ~ **de línea resonante** *m* ELECTRÓN resonant-line oscillator; ~ **local** *m* ELECTRÓN, FÍS, TEC ESP, TELECOM local oscillator (*LO*); ~ **local de exploración** *m* ELECTRÓN tracking local oscillator; ~ **local principal** *m* ELECTRÓN first local oscillator; ~ **local secundario** *m* ELECTRÓN second local oscillator; ~ **local sintetizado** *m* ELECTRÓN synthesized local oscillator; ~ **maestro** *m* ELECTRÓN, FÍS, TELECOM master oscillator (*MO*); ~ **maestro general** *m* TELECOM common master oscillator (*CMO*); ~ **de magnetrón** *m* ELECTRÓN magnetron oscillator; ~ **de microondas** *m* ELECTRÓN microwave oscillator; ~ **modulado** *m* ELECTRÓN modulated oscillator; ~ **de onda de medición de volumen** *m* ELECTRÓN bulk-wave oscillator; ~ **de onda regresiva** *m* ELECTRÓN, TELECOM backward-wave oscillator (*BWO*); ~ **de onda de retorno** *m* FÍS backward-wave oscillator (*BWO*); ~ **de onda de retorno con haz lineal** *m* ELECTRÓN linear-beam backward-wave oscillator; ~ **de onda sinusoidal** *m* ELECTRÓN sine wave oscillator; ~ **óptico** *m* ELECTRÓN optical oscillator; ~ **paramétrico** *m* TELECOM parametric oscillator; ~ **paramétrico óptico** *m* TELECOM optical parametric oscillator; ~ **patrón** *m* ELECTRÓN, FÍS master oscillator; ~ **de pequeña amplitud** *m* ELECTRÓN dither oscillator; ~ **Pierce** *m* ELECTRÓN Pierce oscillator; ~ **piezoeléctrico** *m* ELEC, ELECTRÓN, FÍS, ING ELÉC piezoelectric oscillator; ~ **de polarización** *m* ELECTRÓN bias oscillator; ~ **de potencia** *m* ELECTRÓN power oscillator; ~ **de potencia autoexcitado** *m* ELECTRÓN self-excited power oscillator; ~ **principal** *m* ELECTRÓN master oscillator, service oscillator, FÍS master oscillator; ~ **programable** *m* ELECTRÓN programmable oscillator; ~ **de pruebas** *m* ELECTRÓN test oscillator; ~ **de puente Wien** *m* ELECTRÓN Wien-bridge oscillator; ~ **RC** *m* (*oscilador resistencia-capacitancia*) ELECTRÓN, FÍS, ING ELÉC, TELECOM RC oscillator (*resistance-capitance oscillator*); ~ **de realimentación** *m* ELECTRÓN feedback oscillator; ~ **regulado por cristal** *m* CRISTAL, ELEC, ELECTRÓN, FÍS RAD crystal oscillator; ~ **de relajación** *m* ELECTRÓN relaxation oscillator, FÍS relaxation oscillator, *radio* blocking oscillator; ~ **de resistencia negativa** *m* ELECTRÓN, FÍS negative resistance oscillator; ~ **resistencia-capacitancia** *m* (*oscilador RC*) ELECTRÓN, FÍS, ING ELÉC, TELECOM RC oscillator (*resistance-capacitance oscillator*); ~ **de RF** *m* ELECTRÓN, TELECOM, TV RF oscillator; ~ **de servicio** *m* ELECTRÓN service oscillator; ~ **simétrico** *m* ELEC, ELECTRÓN, TELECOM push-pull oscillator; ~ **de sincronización cromática** *m* ELECTRÓN, TV burst-locked oscillator; ~ **de sincronización de fase** *m* ELECTRÓN phase-locked oscillator; ~ **sincronizado** *m* ELECTRÓN locked oscillator; ~ **sincronizado por inyección** *m* ELECTRÓN, TEC ESP injection-locked oscillator; ~ **sintetizado** *m* ELECTRÓN synthesized oscillator; ~ **sintonizable** *m* ELECTRÓN tunable oscillator; ~ **sintonizable**

continuamente *m* ELECTRÓN continuously-tunable oscillator; **~ sintonizable de varias octavas** *m* ELECTRÓN multioctave tunable oscillator; **~ de sintonización rápida** *m* ELECTRÓN fast-tuned oscillator; **~ sintonizado eléctricamente** *m* ELEC, ELECTRÓN, FÍS ONDAS electrically-tuned oscillator; **~ sintonizado electrónicamente** *m* ELECTRÓN, TELECOM electronically-tuned oscillator; **~ sintonizado mecánicamente** *m* ELECTRÓN mechanically-tuned oscillator; **~ de subportador de cromeado** *m* ELECTRÓN chrominance subcarrier oscillator; **~ de la subportadora** *m* ELECTRÓN, TV subcarrier oscillator; **~ de transistor** *m* ELECTRÓN transistor oscillator; **~ de transistor ajustado GFI** *m* ELECTRÓN YIG tuned transistor oscillator; **~ de triodo** *m* ELECTRÓN triode oscillator; **~ de TTAICH** *m* ELECTRÓN IMPATT oscillator; **~ de tubo electrónico** *m* ELECTRÓN electron tube oscillator; **~ de tubo de vacío** *m* ELECTRÓN vacuum tube oscillator; **~ de varactor ajustado** *m* ELECTRÓN varactor-tuned oscillator; **~ de variación de fase** *m* ELECTRÓN phase-shift oscillator; **~ de velocidad modulada** *m* ELECTRÓN velocity-modulated oscillator; **~ con voltaje controlado** *m* ELECTRÓN, TEC ESP, TELECOM voltage-controlled oscillator (*VCO*); **~ de voltaje controlado en servicio** *m* ELECTRÓN set-on voltage-controlled oscillator

osciladores: **~ acoplados** *m pl* ELECTRÓN, FÍS coupled oscillators

oscilante *adj* ELEC, ELECTRÓN, FÍS ONDAS oscillating, ING MECÁ reciprocating, MECÁ floating, pivoted, reciprocating, TELECOM, TV oscillatory

oscilar¹ *vt* CONST dance

oscilar² *vi* ELEC, ELECTRÓN, FÍS ONDAS oscillate; **~ verticalmente** *vi* TRANSP MAR *en la ola* heave

oscilatorio *adj* ELEC oscillating, ELECTRÓN oscillating, oscillatory, FÍS ONDAS oscillating, TELECOM, TV oscillatory

oscilógrafo *m* ELECTRÓN oscillograph, ING MECÁ recorder, QUÍMICA oscillograph; **~ registrador** *m* ACÚST recording oscillograph

oscilograma *m* TV oscillogram

osciloscopio *m* ELEC, ELECTRÓN, FÍS, FÍS ONDAS, INFORM&PD, INSTR, TV oscilloscope; **~ de almacenamiento** *m* FÍS storage oscilloscope; **~ medidor** *m* ELEC measuring oscilloscope; **~ de rayos catódicos** *m* (*ORC*) ELECTRÓN, FÍS, FÍS RAD, INSTR, TV cathode-ray oscilloscope (*CRO*)

osculador *adj* GEOM osculating

osculante *adj* GEOM osculating

osculatorio *adj* GEOM osculatory

oscurecer *vt* COLOR sadden

oscurecimiento *m* COLOR browning; **~ parcial** *m* ELECTRÓN, ING ELÉC, PROD brownout

oscuridad: **~ atmosférica** *f* CONTAM atmospheric obscurity

oseína *f* QUÍMICA ossein, TEXTIL bone glue

OSI *abr* (*interconexión de sistemas abiertos*) INFORM&PD, TELECOM OSI (*open systems interconnection*)

osículo *m* ACÚST ossicle

osmiato *m* QUÍMICA osmate

ósmico *adj* QUÍMICA osmic

osmio *m* (*Os*) QUÍMICA osmium (*Os*)

osmioso *adj* QUÍMICA osmious

osmiridio *m* MINERAL osmiridium

osmofórico *adj* QUÍMICA osmophoric

osmóforo *m* QUÍMICA osmophore

osmol *m* QUÍMICA osmol, osmole

osmolaridad *f* QUÍMICA osmolarity

ósmosis *f* GEN osmosis; **~ inversa** *f* CONST, HIDROL, PROC QUÍ reverse osmosis

osmótico *adj* GEN osmotic

osona *f* QUÍMICA osone

osotetrazina *f* QUÍMICA osotetrazine

osotriazol *m* QUÍMICA osotriazole

OSR *abr* (*reflector solar óptico*) TEC ESP OSR (*optical solar reflector*)

osta *f* TRANSP MAR boom guy

osteína *f* QUÍMICA ostein

osteófilo *m* NUCL bone-seeker

osteolita *f* MINERAL osteolite

ostrera *f* OCEAN *ostricultura* claire

ostricultor *m* OCEAN oyster farmer

ostricultura *f* OCEAN oyster culture

otosclerosis *m* ACÚST *audición* otosclerosis

otrelita *f* MINERAL ottrelite

OTV *abr* (*vehículo de transferencia entre órbitas*) TEC ESP OTV (*orbital transfer vehicle*)

ovalado *adj* AUTO out-of-round, GEOM egg-shaped, oval, IMPR, ING MECÁ, PETROL out-of-round

ovalización *f* AUTO, IMPR, ING MECÁ, PETROL out-of-roundness

ovalizado *adj* AUTO, IMPR, ING MECÁ, PETROL out-of-round

ovalizar *vt* ING MECÁ run out of true

óvalo *m* GEOM oval

overoles *m pl* AmL (*cf bata de trabajo Esp, cf mono de trabajo Esp*) PROD, SEG coveralls (*AmE*), industrial overalls (*BrE*), overalls (*BrE*)

overshot *m* TEC PETR *perforación* overshot

ovíparo *adj* AGRIC egg laying

oviscapto *m* AGRIC ovipositor

oviscopio *m* AGRIC egg candler

OVNI *abr* (*objeto volador no identificado*) TEC ESP, TRANSP AÉR UFO (*unidentified flying object*)

ovoglobulina *f* QUÍMICA ovoglobulin

ovoide *m* CARBÓN ovoid; **~ pequeño** *m* CARBÓN boulet

óvolo *m* CONST *arquitectura* ovolo

oxalatado *adj* QUÍMICA oxalated

oxalato *m* QUÍMICA oxalate

oxálico *adj* QUÍMICA oxalic

oxalilurea *f* QUÍMICA oxalylurea

oxaloacético *adj* QUÍMICA oxaloacetic

oxalúrico *adj* QUÍMICA oxaluric

oxámico *adj* QUÍMICA oxamic

oxamida *f* QUÍMICA oxamide

oxanílico *adj* QUÍMICA oxanilic

oxanilida *f* QUÍMICA oxanilide

oxazina *f* QUÍMICA oxazine

oxazol *m* QUÍMICA oxazole

oxetona *f* QUÍMICA oxetone

oxhídrico *adj* QUÍMICA oxyhydric, oxyhydrogen

oxiácido *m* QUÍMICA oxyacid

oxicelulosa *f* QUÍMICA oxycellulose

oxicianuro *m* QUÍMICA oxycyanide

oxicloruro *m* QUÍMICA oxychloride

oxicortadura *f* CONST, MECÁ, TERMO flame cutting

oxicorte *m* MECÁ oxycutting, PROD, TERMO gas cutting; **~ por arco** *m* PROD oxygen arc cutting

oxidabilidad *f* QUÍMICA oxidability, oxidizability

oxidable *adj* QUÍMICA oxidable, oxidizable

oxidación *f* ALIMENT, AUTO rust, CINEMAT, CONTAM, ELEC, HIDROL oxidation, MECÁ scale, rust, PROD scaling, QUÍMICA oxidation, rust, TRANSP MAR, VEH rust; ~ **biológica** *f* HIDROL biological oxidation; ~ **interior en fase de vapor** *f* TELECOM inside vapor phase oxidation (*AmE*), inside vapour phase oxidation (*BrE*) (*IVPO*); ~ **interna** *f* METAL internal oxidation; ~ **interna en fase de vapor** *f* TELECOM inside vapor phase oxidation (*AmE*), inside vapour phase oxidation (*BrE*) (*IVPO*); ~ **local** *f* ELECTRÓN local oxidation; ~**-reducción** *f* (*redox*) LAB, QUÍMICA oxidation-reduction, oxidoreduction (*redox*)
oxidado: **no ~** *adj* QUÍMICA unoxidized
oxidante[1] *adj* GEOL, QUÍMICA, SEG oxidizing; **no ~** *adj* GEOL, METAL, QUÍMICA, SEG nonoxidizing
oxidante[2] *m* CONTAM, QUÍMICA, TEC ESP, TERMO, TRANSP AÉR oxidizer, oxidizing agent
oxidar *vt* QUÍMICA oxidize
oxidarse *v refl* QUÍMICA oxidize
óxido *m* ALIMENT, AUTO, MECÁ rust, QUÍMICA oxide, rust, TRANSP MAR rust, TV oxide, VEH *carrocería* rust; ~ **de aluminio** *m* QUÍMICA aluminium oxide (*BrE*), aluminum oxide (*AmE*); ~ **de azufre** *m* CONTAM, QUÍMICA sulfur oxide (*AmE*), sulphur oxide (*BrE*); ~ **de azufre IV** *m* CONTAM, QUÍMICA sulfur dioxide (*AmE*), sulphur dioxide (*BrE*); ~ **de azufre VI** *m* CONTAM, QUÍMICA sulfur trioxide (*AmE*), sulfuric anhydride (*AmE*), sulphur trioxide (*BrE*), sulphuric anhydride (*BrE*); ~ **de boro** *m* C&V, QUÍMICA boric oxide; ~ **de campo** *m* ELECTRÓN field oxide; ~ **de carbono** *m* CARBÓN, CONTAM, ING MECÁ, QUÍMICA, VEH carbon monoxide; ~ **ceroso** *m* QUÍMICA cerous oxide; ~ **de circonio** *m* QUÍMICA zirconia; ~ **de cromo** *m* C&V, COLOR, QUÍMICA chromic oxide; ~ **de deuterio** *m* (D_2O) FÍS, NUCL, QUÍMICA deuterium oxide (D_2O); ~ **espeso** *m* ELECTRÓN thick oxide; ~ **estánico** *m* QUÍMICA stannic oxide; ~ **de etileno** *m* DETERG, QUÍMICA ethylene oxide; ~ **férrico** *m* PROD, QUÍMICA colcothar, ferric oxide; ~ **férrico finamente dividido** *m* PROD rouge; ~ **ferroso** *m* C&V, QUÍMICA ferrous oxide; ~ **de ferrotitanio** *m* MINERAL, QUÍMICA titanium iron oxide; ~ **de hierro** *m* P&C, QUÍMICA iron oxide; ~ **de hierro micáceo** *m* P&C, QUÍMICA micaceous iron oxide; ~ **de hierro negro** *m* COLOR, QUÍMICA black iron oxide; ~ **de iterbio** *m* C&V, QUÍMICA ytterbia, ytterbium oxide, yttrium oxide; ~ **de itrio** *m* QUÍMICA yttria; ~ **mineral** *m* QUÍMICA oxide ore; ~ **nítrico** *m* (*NO*) CONTAM, QUÍMICA nitric oxide (*NO*); ~ **de nitrógeno** *m* (NO_2) CONTAM, QUÍMICA nitrogen oxide (NO_2); ~ **de nitrógeno III** CONTAM, QUÍMICA nitrous oxide (N_2O); ~ **de nitrógeno V** *m* CONTAM, QUÍMICA (N_2O_5) nitrogen pentoxide (N_2O_5); ~ **nitroso** *m* CONTAM, QUÍMICA nitrous oxide (N_2O); **óxido-reducción** *m* (*redox*) LAB, QUÍMICA oxidation-reduction, oxidoreduction (*redox*); ~ **de plata** *m* ING ELÉC, QUÍMICA silver oxide; ~ **de plomo rojo** *m* C&V red lead, MINERAL minium, red lead, P&C red lead, QUÍMICA minium, red lead;

~ **propilénico** *m* DETERG, QUÍMICA propylene oxide; ~ **de silicio** *m* ELECTRÓN, QUÍMICA silicon oxide; ~ **de silicio vítreo** *m* TELECOM vitreous silica; ~ **superficial** *m* CONST surface rust; ~ **de tantalio** *m* ING ELÉC, QUÍMICA tantalum oxide; ~ **de titanio** *m* QUÍMICA titanium oxide; ~ **de torio** *m* QUÍMICA thoria; ~ **de uranio** *m* (UO_2) QUÍMICA uranium oxide (UO_2); ~ **de zinc** *m* ING ELÉC, QUÍMICA zinc oxide; ~ **de zinc activado** *m* P&C, QUÍMICA activated zinc oxide
oxidorreducción *f* (*redox*) LAB, QUÍMICA oxidation-reduction, oxidoreduction (*redox*)
oxifílico *adj* QUÍMICA oxyphilic
oxifiloso *adj* QUÍMICA oxyphilous
oxifluoruro *m* QUÍMICA oxyfluoride
oxifosfato *m* QUÍMICA oxyphosphate
oxigenable *adj* QUÍMICA oxygenizable
oxigenación *f* HIDROL, QUÍMICA oxygenation
oxigenado *adj* HIDROL, QUÍMICA oxygenated
oxigenar *vt* HIDROL, QUÍMICA oxygenate
oxigenasa *f* QUÍMICA oxygenase
oxigénico *adj* QUÍMICA oxygenic
oxígeno *m* (*O*) GEN oxygen (*O*); ~ **disuelto** *m* (*OD*) CONTAM, QUÍMICA dissolved oxygen (*DO*); ~ **líquido** *m* QUÍMICA, TEC ESP, TERMO liquid oxygen, lox
oxihemografía *f* QUÍMICA oxyhaemography (*BrE*), oxyhemography (*AmE*)
o-xileno *m* QUÍMICA o-xylene
oxima *f* QUÍMICA oxime
oximación *f* QUÍMICA oximation
oximetría *f* QUÍMICA oximetry
oximétrico *adj* QUÍMICA oximetric
oxímetro *m* QUÍMICA oximeter
oxisal *f* QUÍMICA oxysalt
oxisulfuro *m* QUÍMICA oxysulfide (*AmE*), oxysulphide (*BrE*)
oxitetraciclina *f* QUÍMICA oxytetracycline
oxitócico *adj* QUÍMICA oxytocic
oxo[1]: ~**-alcohol** *m* DETERG oxo alcohol
oxo[2]: ~**-síntesis** *f* DETERG oxo synthesis
oxoácido *m* QUÍMICA oxo acid, oxyacid
oxonio *m* QUÍMICA oxonium
oxozono *m* QUÍMICA oxozone
oyente: ~ **normal** *m* ACÚST *audición* normal listener
ozocerita *f* MINERAL ozocerite
ozonización *f* METEO, PROC QUÍ, QUÍMICA ozonization
ozonizado *adj* METEO, PROC QUÍ, QUÍMICA ozonized
ozonizador *m* METEO, PROC QUÍ, QUÍMICA ozonizer
ozonizar *vt* METEO, PROC QUÍ, QUÍMICA ozonize
ozono *m* CONTAM. METEO, PROC QUÍ, QUÍMICA, TEC ESP ozone
ozonólisis *f* PROC QUÍ, QUÍMICA ozonolysis
ozonoscópico *adj* QUÍMICA ozonoscopic
ozonoscopio *m* INSTR, QUÍMICA ozonoscope
ozonosfera *f* CONTAM, METEO, TEC ESP ozone layer, ozonosphere
ozonuro *m* QUÍMICA ozonide

P

p *abr* FÍS PART, FÍS RAD (*protón*) p (*proton*), METR (*peta, pico*) p (*peta, pico*), NUCL (*protón*) p (*proton*)

p⁺ *abr* (*positrón*) FÍS, FÍS PART p⁺ (*positron*)

P *abr* (*fósforo*) ELECTRÓN, FÍS, METAL, QUÍMICA P (*phosphorus*)

Pa *abr* ACÚST, FÍS (*pascal*), FÍS RAD (*pascal, protactinio*), METEO, METR (*pascal*), QUÍMICA (*pascal, protactinio*), TEC PETR (*pascal*) Pa (*pascal, protactinium*)

PA *abr* (*poliamida*) AGRIC, ELECTRÓN, P&C, QUÍMICA, TEXTIL PA (*polyamide*)

pabellón: **~ de conveniencia** *m* TRANSP MAR flag of convenience

PABX: **~ de control para programas almacenados** *m* TELECOM stored-program control PABX (*AmE*), stored-programme control PABX (*BrE*)

paca: **~ de heno muy compactada** *f* AGRIC wafer

pachnolita *f* MINERAL pachnolite

PAD *abr* EMB (*polietileno de alta densidad*) HDPE (*high-density polyethylene*), INFORM&PD, TELECOM (*ensamblador-desensamblador de paquetes*) PAD (*packet assembler-disassembler*)

padre *m* INFORM&PD *nódulo* parent; **~ donante** *m* AGRIC *mejora genética* nonrecurrent parent

pagar: **por ~** *adj* PROD outstanding

página *f* IMPR, INFORM&PD page; **~ accesoria** *f* IMPR oddments; **~ de anuncio opuesta a la de lectura** *f* IMPR facing matter; **~ de anuncios** *f* IMPR advertisement page; **~ apaisada** *f* IMPR broadside page, oblong page; **~ de composición perfecta** *f* IMPR perfectly-set page; **~ derecha** *f* IMPR odd page, recto, right-hand page, PAPEL recto; **~ frontal** *f* INFORM&PD home page; **~ gráfica** *f* IMPR photo page; **~ impar** *f* IMPR, PAPEL odd page, recto, right-hand page; **~ izquierda** *f* IMPR, PAPEL left-hand page, verso; **~ par** *f* IMPR, PAPEL left-hand page, verso; **~ principal** *f* INFORM&PD home page; **vertical** *f* IMPR deep page; **~ web** *f* INFORM&PD web page

paginación *f* IMPR pagination, INFORM&PD *presentación*, TELECOM paging; **~ discrecional** *f* INFORM&PD demand paging

paginador *m* TELECOM pager; **~ de mensajes** *m* TELECOM message pager; **~ de tono** *m* TELECOM tone pager

paginar *vt* IMPR folio, paginate, INFORM&PD page, paginate

páginas: **~ por minuto** *f pl* (*PPM*) INFORM&PD pages per minute (*PPM*)

pago: **~ de cargo directo** *m* TELECOM direct debit payment; **~ fácil** *m* TELECOM easy payment; **~ de peaje** *m* TRANSP toll payment; **~ de la renta** *m* TEC PETR *finanzas* rental payment

pairear *vi* TRANSP MAR *veleros* heave to

país: **~ productor de granos** *m* AGRIC grain-producing country

paisaje: **~ blanco** *m* METEO white-out

paisajismo *m* CONST landscaping

paja *f* ALIMENT *molienda* chaff, PAPEL straw

pajuela *f* CARBÓN *de oro* drop

pajuelas: **~ metálicas** *f pl* MINAS float mineral; **~ de oro** *f pl* CARBÓN float, MINAS float gold

PAL *abr* (*array lógico preformable, arreglo lógico programable, lógica de matriz programable*) INFORM&PD, TELECOM, TV, PAL (*programmable array logic*)

pala *f* CONST blade, spade, shovel, CONTAM MAR shovel, ING MECÁ vane, INSTAL HIDRÁUL blade, vane, LAB *para sustancias secas* scoop, MECÁ, TEC ESP, TRANSP AÉR, TRANSP MAR *de hélice* blade; **~ articulada** *f* TRANSP AÉR articulated blade; **~ de buey** *f* AGRIC rollover scraper; **~ cargadora** *f* CONST *maquinaria obras públicas* loading shovel, power loader, PROD *tractor* loader; **~ del compresor** *f* TRANSP AÉR compressor blade; **~ descentrada** *f* TRANSP AÉR offset blade; **~ de enhornar** *f* C&V battledore; **~ en estado de avance** *f* TRANSP AÉR advancing blade; **~ excavadora flotante** *f* TRANSP MAR *draga* dipper dredger; **~ excéntrica** *f* TRANSP AÉR offset blade; **~ fuera de paso** *f* TRANSP AÉR out-of-pitch blade; **~ de funcionamiento irregular** *f* TRANSP AÉR hunting blade; **~ mecánica** *f* CONST *maquinaria, obras públicas* power shovel, MECÁ crab, TRANSP MAR shovel dredger; **~ mecánica de ataque frontal** *f* CONTAM MAR front-end loader; **~ mecánica de operación manual** *f* ING MECÁ hand-operated power shovel; **~ mecánica trasera de tractor** *f* AGRIC back-hoe; **~ de mordaza** *f* CARBÓN jaw plate; **~ oscilante** *f* TRANSP AÉR hunting blade; **~ plegable** *f* TRANSP AÉR folding blade; **~ en retroceso** *f* TRANSP AÉR retreating blade; **~ retroexcavadora** *f* TRANSP backhoe loader; **~ de rotor** *f* TRANSP AÉR rotor blade; **~ de rotor movible** *f* TRANSP AÉR moveable rotor blade; **~ de rotor principal** *f* TRANSP AÉR main rotor blade; **~ del timón** *f* TRANSP MAR *construcción naval* rudder plane; **~ transportadora de oruga** *f* TRANSP caterpillar hauling scraper; **~ de turbina** *f* ING MECÁ turbine blade; **~ del ventilador** *f* AUTO fan blade

palabra *f* ACÚST speech, INFORM&PD word; **~ de búsqueda** *f* INFORM&PD search word; **~ cable** *f* INFORM&PD keyword; **~ del canal de cambio** *f* INFORM&PD channel address word (*CAW*); **~ de la categoría** *f* INFORM&PD status word; **~ de categoría de canales** *f* INFORM&PD channel status word; **~ clave en el contexto** *f* (*KWIC*) INFORM&PD keyword in context (*KWIC*); **~ clave fuera de contexto** *f* (*KWOC*) INFORM&PD keyword out of context (*KWOC*); **~ de comando en canal** *f* INFORM&PD channel command word; **~ de control** *f* INFORM&PD control word; **~ criptografiada** *f* TELECOM encrypted speech; **~ de datos** *f* INFORM&PD data word; **~ de dirección en canal** *f* INFORM&PD channel address word (*CAW*); **~ direccional del canal** *f* INFORM&PD channel address word (*CAW*); **~ de estado** *f* INFORM&PD status word; **~ de estado de canal** *f* INFORM&PD channel status word; **~ de estado del procesador** *f* INFORM&PD processor status word; **~ de estado de programa** *f* INFORM&PD program status word; **~ de mandato en canal** *f* INFORM&PD

channel command word; ~ **opcional** *f* INFORM&PD optional word; ~ **de orden del canal** *f* INFORM&PD channel command word; ~ **reservada** *f* INFORM&PD reserved word; ~ **de sincronización** *f* TELECOM syncword

palabras: ~ **por segundo** *f pl* IMPR words per second (*wps*)

palada *f* CONST shovelful

paládico *adj* METAL, QUÍMICA palladic

paladio *m* (*Pd*) METAL, QUÍMICA palladium (*Pd*)

paladioso *adj* METAL, QUÍMICA palladious

palagonita *f* PETROL palagonite

palanca *f* CONST crowbar, FÍS, ING ELÉC, ING MECÁ lever, INSTAL HIDRÁUL *posicionamiento de la válvula* gab hook, TRANSP AÉR, VEH lever; ~ **para accionar las agujas** *f* PROD switch-operating handle; ~ **acodada** *f* CARBÓN toggle, TRANSP AÉR bellcrank; ~ **acodada del colectivo** *f* TRANSP AÉR collective bellcrank; ~ **acodada en forma de lira** *f* TRANSP AÉR lyre-shaped bellcrank; ~ **acodada del mezclador** *f* TRANSP AÉR mixer bellcrank; ~ **de admisión de gases** *f* INSTAL HIDRÁUL throttle lever; ~ **de las agujas** *f* FERRO *infraestructura* points rod; ~ **de arranque** *f* AUTO, ING MECÁ starting handle, starting lever; ~ **de arrastre** *f* FOTO sliding leg, INSTR *engranajes* arm; ~ **articulada** *f* ING MECÁ pivoted lever; ~ **de bloqueo** *f* FERRO *equipo inamovible* locking bar; ~ **de cadena** *f* ING MECÁ chain lever; ~ **de cambio de marcha** *f* ING MECÁ reversing lever; ~ **de cambio de marchas** *f* AUTO, VEH gear change lever (*BrE*), gear shift lever (*AmE*); ~ **de cambio de velocidades** *f* AUTO, VEH gear change lever (*BrE*), gear shift lever (*AmE*); ~ **cíclica** *f* TRANSP AÉR cyclic stick; ~ **compuesta** *f* ING MECÁ compound lever; ~ **de contrapeso** *f* ING MECÁ balance lever; ~ **de control** *f* Esp (*cf palanca de juego AmL*) INFORM&PD joystick, TRANSP AÉR control lever; ~ **de control de la mariposa** *f* AUTO throttle control lever; ~ **de control del paso** *f* TRANSP AÉR pitch-control lever; ~ **de control de paso cíclico** *f* TRANSP AÉR cyclic-pitch control stick; ~ **corta** *f* MECÁ jemmy (*BrE*), jimmy (*AmE*); ~ **cruciforme** *f* ING MECÁ split-set collar; ~ **de desembrague** *f* ING MECÁ disengaging lever; ~ **de desembrague del plato de presión** *f* AUTO pressure-plate release lever; ~ **de desenganche** *f* AGRIC, AUTO release lever; ~ **de desenganche del pedal del embrague** *f* AUTO, VEH clutch-pedal release lever; ~ **desplazadora** *f* ING MECÁ shifting rod; ~ **enganchada** *f* ING MECÁ latched lever; ~ **equilibrada** *f* ING MECÁ counterbalanced lever; ~ **de freno** *f* FERRO, ING MECÁ brake lever; ~ **de freno de conexión** *f* ING MECÁ connecting-brake lever rod; ~ **de freno de estacionamiento** *f* VEH parking-brake lever; ~ **del freno de mano** *f* AUTO handbrake lever; ~ **de gases** *f* TRANSP AÉR throttle; ~ **de gatillo** *f* ING MECÁ ratchet lever; ~ **de inmovilización** *f* ING MECÁ clamping lever; ~ **del interruptor** *f* ING ELÉC switch lever; ~ **del izador** *f* TRANSP AÉR hoist lever; ~ **de juego** *f AmL* (*cf palanca de control Esp*) INFORM&PD joystick, TRANSP AÉR control lever; ~ **de lazo** *f* PROD loop lever; ~ **de mando** *f* CINEMAT joystick, ING MECÁ operating lever, MINAS control lever, TEC ESP handle, *aviones* joystick, VEH *dirección* drop arm; ~ **de maniobra** *f* ING MECÁ operating lever, MECÁ handspike, MINAS control lever, tiller; ~ **de maniobra de la aguja** *f* FERRO *equipo*

inamovible point-lock plunger; ~ **de maniobra de agujas** *f* FERRO pointer, switch lever; ~ **de mano** *f* ING MECÁ hand lever; ~ **de la mariposa** *f* INSTAL HIDRÁUL *válvulas* throttle lever; ~ **omnidireccional** *f* TEC ESP joystick; ~ **de paso cíclico** *f* TRANSP AÉR cyclic-pitch stick; ~ **del paso del colectivo** *f* TRANSP AÉR collective-pitch lever; ~ **del pedal de cambio de velocidades** *f* VEH *en motocicletas* foot-change lever; ~ **de pie de cabra** *f* CONST pick-and-claw crowbar, pinching bar; ~ **para piedra** *f* CONST rock lever; ~ **del plato de empuje** *f* VEH *embrague* pressure-plate release lever; ~ **de poleas** *f* ING MECÁ strap bar; ~ **de prensa** *f* ING MECÁ screw key; ~ **de primer orden** *f* ING MECÁ lever of the first kind; ~ **de primer tipo** *f* ING MECÁ lever of the first kind; ~ **de rebobinado** *f* CINEMAT, FOTO, TV rewind handle; ~ **para rebobinar la película** *f* CINEMAT, FOTO film-rewind handle; ~ **con recubrimiento de plomo** *f* INSTAL HIDRÁUL lap and lead lever; ~ **de resorte** *f* PROD spring return lever; ~ **de rodillos** *f* PROD roller lever; ~ **de segundo orden** *f* ING MECÁ lever of the second kind; ~ **de segundo tipo** *f* ING MECÁ lever of the second kind; ~ **de selector** *f* AUTO selector lever; ~ **selectora** *f* VEH selector lever; ~ **de suspensión de la rueda** *f* AUTO, VEH wheel-suspension lever; ~ **de tercer orden** *f* ING MECÁ lever of the third kind; ~ **de tercer tipo** *f* ING MECÁ lever of the third kind; ~ **de trinquete** *f* ING MECÁ ratchet lever; ~ **de varillas** *f* PROD rod lever; ~ **vertical** *f* PROD vertical lever; ~ **de volante** *f* ING MECÁ swing-bob lever; ~ **del zoom** *f* TV zoom lever

palangrero: ~ **atunero** *m* OCEAN tunny boat

palanquear *vt* ING MECÁ lever

palanqueo *m* OCEAN fish trade

palanqueta *f* ING MECÁ jemmy bar (*BrE*), jimmy bar (*AmE*), MECÁ handspike, jemmy (*BrE*), jimmy (*AmE*)

palanquita *f* TEC ESP horn; ~ **de indicación** *f* ING MECÁ pointer jack

palas: ~ **en cascada** *f pl* TRANSP AÉR cascade blades; ~ **extractoras de muestras** *f pl* CARBÓN spoon sampler

palatabilidad *f* AGRIC palatability

palear *vt* CONST shovel

paleoceanografía *f* GEOL, OCEAN palaeoceanography (*BrE*), paleoceanography (*AmE*)

paleomagnetismo *m* FÍS, GEOL palaeomagnetism (*BrE*), paleomagnetism (*AmE*)

paleopendiente *f* GEOL palaeoslope (*BrE*), paleoslope (*AmE*)

paleopresión *f* GEOL, TEC PETR palaeopressure (*BrE*), paleopressure (*AmE*)

paleozoico[1] *adj* GEOL, TEC PETR palaeozoic (*BrE*), paleozoic (*AmE*)

paleozoico[2] *m* GEOL, TEC PETR palaeozoic (*BrE*), paleozoic (*AmE*)

pálet *m AmL* (*cf bandeja Esp, paleta Esp*) C&V pan, tray, EMB pallet, tray, TRANSP pallet; ~ **apilable** *m AmL* TRANSP stacking pallet; ~ **de carga** *m AmL* TRANSP AÉR pallet; ~ **cilíndrico** *m AmL* TRANSP roller pallet; ~ **desechable** *m AmL* TRANSP expendable pallet; ~ **de dos accesos** *m AmL* TRANSP two-way pallet; ~ **estiba** *m AmL* TRANSP stevedore-type pallet; ~ **invertible** *m AmL* TRANSP reversible pallet; ~~**jaula** *m AmL* TRANSP crate pallet; ~ **no retornable** *m AmL* TRANSP one-way pallet; ~ **sencillo** *m AmL* EMB single-faced pallet, TRANSP single-decked pallet;

~ con separación desmontable *m AmL* TRANSP pallet with loose partition, single-faced pallet; **~ de una sola cara** *m AmL* TRANSP pallet with loose partition, single-faced pallet; **~ de un solo acceso** *m AmL* TRANSP one-way pallet; **~ de un solo tablero** *m AmL* TRANSP single-platform pallet; **~ de sujeción** *m AmL* TRANSP post pallet

paleta *f Esp* (*cf pálet AmL, bandeja Esp*) AGUA float, floatboard, C&V, CINEMAT paddle, CONST *albañilería* trowel, *de yesero, de turbina* bucket, vane, heart trowel, EMB pallet, ENERG RENOV *de turbina* blade, vane, ING MECÁ vane, blade, paddle, INSTAL HIDRÁUL *de turbina* vane, bucket, blade, MECÁ *de turbina* blade, bucket, vane, PROD *turbinas, bombas* vane, TRANSP pallet, TRANSP AÉR pallet, vane; **~ antivibratoria** *f* CINEMAT anti-flicker blade; **~ apilable** *f* TRANSP stacking pallet; **~ para apilar** *f* EMB stacking pallet; **~ batidora** *f* CONTAM MAR *para mezclar los dispersantes con el agua de mar* breaker board; **~ con caja alambrada** *f* EMB box pallet with mesh; **~ de carga para cajas** *f* EMB box pallet; **~ de carga de cuatro accesos** *f* EMB four-way pallet; **~ de carga extensible** *f* EMB expandable pallet; **~ en cascada** *f* TRANSP AÉR cascade vane; **~ cilíndrica** *f* TRANSP roller pallet; **~ de colores** *f* INFORM&PD color palette (*AmE*), colour palette (*BrE*); **~ curvada** *f* INSTAL HIDRÁUL curved vane; **~ desechable** *f* TRANSP expendable pallet; **~ directriz** *f* INSTAL HIDRÁUL wicket gate, *de turbina* guide blade, stationary vane, *turbina* stationary blade; **~ de doble plataforma** *f* EMB double platform pallet; **~ de dos accesos** *f* TRANSP two-way pallet; **~ con envoltura retráctil** *f* EMB shrink-wrapped pallet cover; **~ para esparcir** *f* P&C *aplicación de revestimientos* spreader-finisher; **~ estiba** *f* TRANSP stevedore-type pallet; **~ fija** *f* INSTAL HIDRÁUL stationary blade, stationary vane; **~ guiadora** *f* ENERG RENOV guide vane; **~-jaula** *f* TRANSP crate pallet; **~ de mezclador** *f* LAB *equipo general* stirrer blade; **~ móvil** *f* ENERG RENOV runner blade, INSTAL HIDRÁUL runner vane, *turbinas* runner blade; **~ no retornable** *f* TRANSP one-way pallet; **~ plana** *f* EMB flat pallet; **~ receptora** *f* ENERG RENOV, INSTAL HIDRÁUL runner blade; **~ de reloj** *f* ING MECÁ pawl; **~ sin retorno** *f* EMB nonreturnable pallet, one-way pallet; **~ reversible** *f* TRANSP reversible pallet; **~ del rodete** *f* ENERG RENOV *turbinas*, INSTAL HIDRÁUL runner blade; **~ de rotor** *f* ENERG RENOV, INSTAL HIDRÁUL runner blade; **~ de rueda hidráulica** *f* ING MECÁ paddle board; **~ sencilla** *f* EMB single-faced pallet, TRANSP single-decked pallet; **~ con separación desmontable** *f* TRANSP pallet with loose partition, single-faced pallet; **~ de una sola cara** *f* TRANSP pallet with loose partition, single-faced pallet; **~ de un solo acceso** *f* TRANSP one-way pallet; **~ de un solo tablero** *f* TRANSP single-platform pallet; **~ de sujeción** *f* TRANSP post pallet; **~ transportable de forma convencional** *f* EMB conventional transportable pallet; **~ de turbina** *f* ING MECÁ turbine blade; **~ del ventilador** *f* AUTO, MECÁ, REFRIG, TRANSP, VEH fan blade

paletaje *m* MECÁ blading

paletalizable *adj* EMB *material que es apto par ser colocado en paletas*, TRANSP palletizable

paletín *m* CONST heart trowel

paletización *f* EMB, TRANSP palletization

paletizado *m* EMB palletized board, TRANSP palletizing

paletizar *vt* EMB, TRANSP palletize

paletón *m* CONST *cerrajería* web; **~ de llave** *m* CONST key bit

palier *m* VEH *de rueda* spindle, *transmisión* axle shaft, halfshaft; **~ de magneto** *m* ING MECÁ magneto bearing

palizada *f* CONST paling

palmejar *m* TRANSP MAR *cubierta* stringer

palmitato *m* QUÍMICA palmitate; **~ de glicerilo** *m* QUÍMICA palmitin

palmítico *adj* QUÍMICA palmitic

palmitona *f* QUÍMICA palmitone

palo[1]: **a ~ seco** *adv* TRANSP MAR *navegar* ahull

palo[2] *m* CONST pole, stick, TRANSP MAR mast; **~ aerodinámico** *m* TRANSP MAR aerodynamic mast; **~ de carga** *m* TRANSP MAR mast crane; **~ macho** *m* TRANSP MAR *accesorios de cubierta* king post; **~ mayor** *m* TRANSP MAR mainmast, mizzen mast; **~ de mesana** *m* TRANSP MAR mizzen mast; **~ trinquete** *m* TRANSP MAR foremast

palomilla *f* CONST angle bracket, wall bracket, ING ELÉC bracket, ING MECÁ wing; **~ del canalón** *f* CONST gutter bracket; **~ de equilibrio** *f* TRANSP AÉR balance horn; **~ del estrangulador** *f* AUTO, VEH choker plate

palomillas: **~ de cables** *f pl* ELEC, ING ELÉC cable rack

palos: **~ y jarcia** *m pl* TRANSP MAR rigging

palpitación *f* OCEAN *medicina subacuática* panting

palustrín: **~ de hoja de laurel** *m* PROD leaf, leaf-shaped trowel

PAM *abr* ELECTRÓN, INFORM&PD, TEC ESP, TELECOM (*modulación de amplitud de impulso, modulación de impulsos en amplitud*) PAM (*pulse-amplitude modulation*)

pan *m* PROD *fundería, fundición* pig; **~ ázimo** *m* ALIMENT matzo

PAN *abr* (*peroxoacetilnitrato*) CONTAM, QUÍMICA PAN (*peroxoacetylnitrate*)

pana *f* TEXTIL corduroy, panne

panal *m* TEC ESP, TRANSP AÉR honeycomb; **~ de nido de abeja** *m* TEC ESP, TRANSP AÉR honeycomb

pancreatina *f* QUÍMICA pancreatin

pancromático *adj* CINEMAT, FOTO, IMPR, QUÍMICA panchromatic

pandear[1] *vt* MECÁ buckle

pandear[2] *vi* CONST belly, bulge, ING MECÁ bulge, *columnas* buckle, MECÁ buckle

pandeo *m* CONST *deformación* bulge, ING MECÁ bulge, *columnas* buckling, MECÁ, NUCL buckling, PETROL buckle, PROD buckling

pandermita *f* MINERAL pandermite

panel *m* CONST, ELEC, GAS, REFRIG panel; **~ de acceso** *m* TEC ESP, TRANSP AÉR access panel; **~ de acoplamiento** *m* INFORM&PD, TV jack panel (*BrE*), patch board (*BrE*), patch panel (*AmE*); **~ aislante** *m* TERMOTEC insulating board; **~ aislante para pared** *m* CONST insulating wall panel; **~ aislante tipo sandwich** *m* REFRIG sandwich-paned insulating panel; **~ alto** *m* TRANSP AÉR overhead panel; **~ en arco** *m* CONST arch panel; **~ articulado** *m* ING MECÁ hinged panel; **~ de calefacción** *m* ING MECÁ heating panel; **~ de calentamiento** *m* ING MECÁ heating panel; **~ de células solares** *m* ING ELÉC solar-cell panel; **~ central** *m* TRANSP AÉR center panel (*AmE*), centre panel (*BrE*); **~ de cierre** *m* ING MECÁ closing panel; **~ de conexiones** *m* INFORM&PD, TV patch board (*BrE*), patch panel (*AmE*); **~ de conmutación**

m CINEMAT jack field, TELECOM CB switchboard, TV switchboard; **~ de conmutación por conductos flexibles** *m* TELECOM cord switchboard; **~ de conmutación de datos** *m* TELECOM data-switching exchange (*DSE*); **~ de conmutación dicordio** *m* TELECOM double-cord switchboard; **~ de conmutación por dicordios** *m* TELECOM cord switchboard; **~ de conmutación sin dicordios** *m* TELECOM cordless switchboard; **~ conmutador automático-manual** *m* TELECOM automanual switchboard; **~ de control** *m* CONST control panel, ELEC control board, INFORM&PD, MECÁ, TEC ESP, TRANSP AÉR control panel; **~ de control por botón** *m* ELEC push-button control panel; **~ de control por botón de presión** *m* ELEC push-button control panel; **~ de control por botones de contacto** *m* ELEC push-button control panel; **~ de control de la centralita privada** *m* TELECOM PBX switchboard; **~ del control maestro** *m AmL* (*cf tablero de control principal Esp*) TV master control panel; **~ de control por pulsador** *m* ELEC push-button control panel; **~ de control del reactor** *m* NUCL reactor control board; **~ delgado** *m* C&V light panel; **~ de distribución** *m* ING ELÉC distribution board; **~ de enfriamiento** *m* REFRIG panel cooler; **~ de espuma amortiguador de ruidos** *m* SEG noise-abating foam panel; **~ de espuma amortiguador de sonidos** *m* SEG sound-absorbent foam panel; **~ experimental** *m* ING ELÉC breadboard model; **~ de extracción modular** *m* PROD module extraction pad; **~ del extremo** *m* EMB, ING MECÁ end panel; **~ extremo** *m* EMB, ING MECÁ end panel; **~ de fondo** *m* ELECTRÓN, INFORM&PD, PROD backplane; **~ frontal** *m* ING MECÁ front panel, TV faceplate; **~ de fusibles** *m* ELEC, ING ELÉC fuse board; **~ grueso** *m* C&V heavy panel; **~ informativo de dirección** *m* TRANSP *tráfico* directional aid; **~ de instrucciones** *m* PROD instruction panel; **~ de instrumentos** *m* AUTO, TRANSP AÉR, VEH *accesorios* instrument panel; **~ lateral** *m* EMB *de envases, cajas*, VEH *carrocería* side panel; **~ lenticular** *m* INSTR lens panel; **~ de luz** *m* C&V light panel; **~ de malla** *m* TEC ESP mesh sandwich; **~ de mandos** *m* TRANSP AÉR console; **~ de mensaje cambiante** *m* TRANSP *control de tráfico* variable-message sign; **~ multiuso** *m* ELECTRÓN general-purpose laminate; **~ de nido de abeja** *m* TEC ESP, TRANSP AÉR honeycomb; **~ del piso** *m* TRANSP AÉR floor panel; **~ prefabricado** *m* REFRIG prefabricated panel; **~ de pruebas y medidas** *m* ELEC test board; **~ de la puerta** *m* CONST door panel; **~ radiante a gas de alta temperatura** *m* INSTAL TERM high-temperature gas radiant panel; **~ radiante infrarrojo** *m* GAS infrared radiant panel; **~ de señalización eléctrica** *m* ING ELÉC annunciator; **~ solar** *m* CONST *edificación* solar collector, ENERG RENOV, FÍS, TEC ESP, TELECOM solar panel; **~ de VCL** *m* ELECTRÓN LCD panel; **~ de velocidad variable** *m* TRANSP *control de tráfico* variable-speed message sign; **~ de vidrio para calentar** *m* C&V glass-heating panel; **~ de vidrio macizo** *m* C&V glass-concrete panel; **~ de visualización de cristal líquido dicrico** *m* ELECTRÓN dichroic liquid visual display

panflavina *f* QUÍMICA panflavine
panícula: **~ de avena** *f* AGRIC oat panicle
panizo: **~ blanco** *m* AGRIC foxtail millet
paño[1]: **a ~** *adj* CONST flush
paño[2]: **a ~** *adv* PROD flush

paño[3]: **~ de ampliar** *m* C&V bolting cloth; **~ engomado** *m* P&C, REVEST rubberized cloth; **~ de presión** *m* FOTO pressure cloth; **~ de red** *m* OCEAN netting
panoja *f* AGRIC panicle, *maíz* tassel; **~ loca** *f* AGRIC crazy top
pañol: **~ de carbón** *m* CARBÓN, MINAS *buques* coal bunker; **~ del carbón** *m* CARBÓN *buques* bunker; **~ cerrado** *m* TRANSP MAR *alojamiento* closed locker; **~ de equipajes** *m* TRANSP *barcos* baggage room (*AmE*), luggage compartment (*BrE*); **~ para herramientas** *m* PROD *talleres* tool crib; **~ pequeño** *m* TRANSP MAR lazarette; **~ de señales** *m* TRANSP MAR signal locker; **~ de velas** *m* TRANSP MAR sail locker, sail loft
panorámica *f* CONST landscaping, INFORM&PD panning; **~ desde el cielo** *f* CINEMAT, TV skypan; **~ y exploración** *f* CINEMAT, TV pan and scan; **~ muy rápida** *f* CINEMAT, TV swish pan; **~ rápida** *f* CINEMAT flash pan, flick pan, whip pan, zip pan, TV flash pan, flick pan; **~ rápida de transición** *f* CINEMAT blur pan; **~ vertical** *f* CINEMAT tilt
paños *m pl* TRANSP MAR *de vela* sailcloth
pantalán *m* TRANSP MAR jetty
pantalla[1]: **en ~** *adj* IMPR screen-displayed
pantalla[2] *f* CARBÓN, CINEMAT screen, ELEC *cable* shield, *conductor del cable* screen, ELECTRÓN, FÍS screen, FÍS RAD shielding, IMPR *serigrafía* screen, INFORM&PD display, screen display, *visualización* screen, ING ELÉC shield, lampshade, MECÁ screen, PROD shield, TEC ESP *equipos* display, shield, screen, TV screen, shield, visual display unit (*VDU*); **~ acústica** *f* ACÚST acoustic barrier, acoustic screen, *altavoces* baffle; **~ de agua** *f* OCEAN waterwall; **~ de aislamiento** *f* ELEC *de conductor de cable* insulation screen; **~ aislante** *f* CONST baffle board; **~ del alma** *f* ELEC *conductor del cable* core screen; **~ de almacenamiento** *f* ELECTRÓN storage screen; **~ de aluminio** *f* ELECTRÓN aluminized screen; **~ alveolar** *f* CINEMAT beaded screen, glass-beaded screen, FOTO beaded screen; **~ amortiguadora de oscilación eléctrica** *f* TRANSP AÉR anti-surge baffle; **~ antideslumbrante** *f* TRANSP AÉR glare shield; **~ anti-ruido** *f* ACÚST noise shield; **~ antisobretensión eléctrica** *f* TRANSP AÉR anti-surge baffle; **~ antisobrevoltaje** *f* TRANSP AÉR anti-surge baffle; **~ automontable** *f* FOTO self-erecting screen; **~ de aviso** *f* CINEMAT, TV cue screen; **~ de ayuda** *f* INFORM&PD help screen; **~ de blindaje** *f AmL* (*cf pantalla de seguridad Esp*) FOTO safelight filter, LAB, SEG safety screen; **~ de color** *f* FOTO color screen (*AmE*), colour screen (*BrE*), TEC ESP, TRANSP AÉR color display (*AmE*), colour display (*BrE*); **~ en color** *f* FOTO, INFORM&PD, TEC ESP, TRANSP AÉR color display (*AmE*), colour display (*BrE*); **~ conductora** *f* ELEC, ING ELÉC conductor screen; **~ contra la radiación** *f* FÍS RAD, SEG radiation shielding; **~ de control** *f* INSTR control screen; **~ cortafuegos** *f* CONST, TEC PETR, SEG fire wall; **~ de cristal esmerilado** *f* CINEMAT, FOTO, INSTR ground-glass screen; **~ de cristal líquido** *f* ELEC, ELECTRÓN, FÍS, INFORM&PD, TELECOM, TV liquid crystal display (*LCD*); **~ deflectora** *f* NUCL *de la vasija del reactor* baffle; **~ desecadora** *f* PROC QUÍ desiccator screen; **~ digital** *f* INSTR digital display; **~ dividida** *f* INFORM&PD split screen; **~ eléctrica del cable** *f*

ELEC, ING ELÉC *conductor* cable screen; ~ **electroluminiscente** *f Esp* (*cf display electroluminiscente AmL*) ELEC, ELECTRÓN, INFORM&PD electroluminescent display; ~ **electromagnética** *f* ELEC, FÍS electromagnetic screen; ~ **electrostática** *f* ELEC, ELECTRÓN, FÍS, ING ELÉC, NUCL, VEH electrostatic screen, Faraday cage, Faraday shield; ~ **enderezadora de la corriente de aire** *f* TRANSP AÉR honeycomb; ~ **de enfoque** *f* CINEMAT, FOTO, INSTR, TV focusing screen; ~ **de enfoque intercambiable** *f* FOTO interchangeable focusing screen; ~ **estática** *f* ELEC static screen; ~ **de Faraday** *f* ELEC, FÍS, ING ELÉC, NUCL Faraday screen; ~ **de filtro** *f* TEC ESP filter mask; ~ **del filtro** *f* FOTO filter screen; ~ **final** *f* ELECTRÓN end screen; ~ **fluorescente** *f* ELEC *de osciloscopio*, ELECTRÓN, FÍS, FÍS RAD, INSTR fluorescent screen, TV phosphor screen; ~ **fraccionada** *f* CINEMAT split screen; ~ **ignífuga** *f* PROD fire screen; ~ **de imagen intermedia** *f* INSTR intermediate image screen; ~ **interna** *f* ELECTRÓN internal shield; ~ **de inyección** *f* CONST grout curtain; ~ **de larga persistencia** *f* TV long-persistence screen; ~ **luminosa** *f* TELECOM display screen; ~ **para la luz del día** *f* CINEMAT daylight screen; ~ **mate** *f* INSTR matt screen; ~ **de memorización** *f* ELECTRÓN storage screen; ~ **menú** *f* INFORM&PD menu screen; ~ **metalizada** *f* ELECTRÓN metalized screen (*AmE*), metallized screen (*BrE*); ~ **de mucho brillo** *f* ELECTRÓN high-brightness screen; ~ **múltiple** *f* TV multiscreen; ~ **ortocromática** *f* FOTO color screen (*AmE*), colour screen (*BrE*); ~ **partida** *f* CINEMAT split screen; ~ **perforada** *f* CONST punched-plate screen; ~ **perlada** *f* CINEMAT pearl screen; ~ **plana** *f* ELECTRÓN flat-panel display, INFORM&PD, TELECOM, TV flat screen; ~ **de plasma** *f* INFORM&PD plasma display; ~ **de protección termoaislante** *f* ING ELÉC, TRANSP AÉR heat shield; ~ **protectora** *f* PROD, SEG face shield, *para tubos de rayos catódicos* protective screen; ~ **de proyección** *f* CINEMAT, INSTR projection screen; ~ **de radar** *f* D&A, FÍS, FÍS RAD, TRANSP radar display, radar scope, TRANSP AÉR, TRANSP MAR radar display, radar scope, radar screen; ~ **de rayos catódicos** *f* TELECOM cathode-ray screen; ~ **reflectora** *f* ACÚST reflex baffle, CINEMAT, FOTO, ÓPT reflecting screen, TRANSP AÉR glare shield; ~ **rejilla azul** *f* TV blue screen grid; ~ **de relleno** *f* CINEMAT fill-in screen; ~ **de representación visual en vehículo** *f* TRANSP in-vehicle visual display; ~ **de resonancia** *f* CARBÓN resonance screen; ~ **de retroproyección** *f* CINEMAT rear-projection screen; ~ **de seda** *f* C&V silk screen; ~ **de seguridad** *f Esp* (*cf pantalla de blindaje AmL*) FOTO safelight filter, LAB, SEG safety screen; ~ **sensible al tacto** *f* INFORM&PD touch-sensitive screen; ~ **sensible al toque** *f* INFORM&PD touch-sensitive screen; ~ **térmica** *f* ING ELÉC heat shield, TEC ESP heat screen, thermal screen, thermal shield, TERMO, TRANSP AÉR heat shield; ~ **translúcida** *f* CINEMAT back-projection screen; ~ **para transparencias** *f* CINEMAT back-projection screen; ~ **de trazo oscuro** *f* ELECTRÓN dark-trace screen; ~ **de trenza de cobre** *f* ELEC *cable*, ELECTRÓN, PROD copper-braid shielding; ~ **tríptica** *f* CINEMAT triptych screen; ~ **de tubo de rayos catódicos** *f* TV oscilloscope, scope; ~ **vibratoria** *f* CARBÓN circular vibrating screen; ~ **visual** *f* TV video display unit

(*VDU*); ~ **de visualización** *f* ELECTRÓN display screen

pantano *m* AGUA marsh, reservoir, swamp, CARBÓN pond, HIDROL dam; ~ **salobre** *m* AGUA salt swamp

pantanoso *adj* CARBÓN boggy, GEOL, HIDROL marshy

pantelerita *f* PETROL pantellerite

pantocazo *m* TRANSP MAR *movimiento de barco* slamming

pantógrafo *m* ELEC *ferrocarril*, FERRO, ING MECÁ pantograph

pantotenato *m* QUÍMICA pantothenate; ~ **cálcico** *m* ALIMENT, QUÍMICA calcium pantothenate

panza *f* CONST, PROD belly

panzer *m* MINAS armored face conveyor (*AmE*), armoured face conveyor (*BrE*)

papaína *f* ALIMENT, QUÍMICA papain

papaveraldina *f* QUÍMICA papaveraldine

papaverina *f* QUÍMICA papaverine

papel *m* INFORM&PD form, PAPEL paper;

▸ **a** ~ **para ábacos** *m* PAPEL chart paper; ~ **abrasivo** *m* C&V glass paper, ELEC *limpieza* abrasive paper, ING MECÁ abrasive sheet, coated abrasive, MECÁ abrasive paper, PAPEL glass paper, REVEST coated abrasive; ~ **abrasivo con fibra de vidrio** *m* C&V, PAPEL glass paper; ~ **abrasivo de vidrio** *m* C&V glass passivation; ~ **abrillantado en máquina** *m* EMB, PAPEL machine-glazed paper; ~ **absorbente** *m* PAPEL blotting paper; ~ **sin acabado** *m* PAPEL paper without finish; ~ **aislador** *m* ELEC, PAPEL insulating paper; ~ **aislante** *m* ELEC, PAPEL insulating paper; ~ **aislante para cables eléctricos** *m* ELEC, PAPEL paper for conductor insulation; ~ **para aislantes laminados** *m* ELEC, PAPEL paper for laminated insulators; ~ **de alimentación continua** *m* IMPR, INFORM&PD, PAPEL continuous feed paper (*AmE*), continuous form (*AmE*), continuous stationery (*BrE*); ~ **alisado en húmedo** *m* PAPEL water-finished paper; ~ **alquitranado de dos hojas** *m* PAPEL union paper; ~ **alquitranado reforzado** *m Esp* (*cf papel embreado reforzado AmL*) PAPEL reinforced union paper; ~ **de aluminio** *m* PAPEL foil paper, TERMOTEC aluminium foil (*BrE*), aluminum foil (*AmE*); ~ **para ampliaciones** *m* FOTO, IMPR, PAPEL enlarging-paper; ~ **anti-adhesivo** *m* PAPEL anti-adhesive paper; ~ **anticongelante** *m* PAPEL anti-freeze paper; ~ **anticorrosivo** *m* EMB *para envolver metales*, PAPEL anticorrosion paper; ~ **antioxidante** *m* EMB, PAPEL antirust paper; ~ **antivaho** *m* PAPEL antitarnish paper; ~ **apergaminado** *m* PAPEL vegetable-sized paper; ~ **arte** *m* PAPEL art paper; ~ **asfaltado** *m* PAPEL asphalt paper; ~ **asfáltico** *m* CONST bituminized paper; ~ **para el asma** *m* PAPEL asthma paper; ~ **autocopiativo** *m* PAPEL carbonless copy paper; ~ **averiado** *m* PAPEL broke; ~ **de avión** *m Esp* (*cf papel para correo aéreo AmL*) PAPEL airmail paper;

▸ **b** ~ **de bajo gramaje** *m* PAPEL *menos de 40g/m^2* light weight paper; ~ **base** *m* PAPEL body paper; ~ **base para esténcil** *m* PAPEL duplicating stencil-base paper; ~ **biblia** *m* PAPEL bible paper; ~ **para billetes de banco** *m Esp* (*cf papel moneda AmL*) PAPEL banknote paper; ~ **bituminado** *m* EMB, PAPEL bitumen-coated paper, tar paper, tarred brown paper; ~ **para bolsas** *m* EMB, PAPEL bag paper; ~ **para bordear** *m* PAPEL edging paper; ~ **bromuro** *m* FOTO bromide paper; ~ **bromuro suave** *m* FOTO soft bromide paper;

~ c ~ **calandrado** *m* PAPEL calendered paper board; ~ **de calcar** *m* ING MECÁ, PAPEL tracing paper; ~ **de calco** *m* ING MECÁ, PAPEL tracing paper, translucent drawing paper; ~ **de calco aceitado** *m* PAPEL oil tracing paper; ~ **calibrado** *m* PAPEL even-thickness paper; ~ **con capa estucado de bajo gramaje** *m* PAPEL light-weight coated paper (*LWC paper*); ~ **para capas lisas interiores** *m* PAPEL *cartón ondulado* liner paper; ~ **para caras** *m* PAPEL liner; ~ **para caras con una capa de kraft** *m* PAPEL kraft face liner; ~ **carbón** *m* PAPEL carbon paper; ~ **para carteles** *m* PAPEL poster paper; ~ **cartográfico** *m* PAPEL map paper; ~ **de cartón** *m* PAPEL yellow strawboard, yellow strawpaper; ~ **para cartuchos** *m* PAPEL ammunition paper; ~ **cebolla** *m* PAPEL onionskin paper; ~ **sin cenizas** *m* PAPEL *para filtros de laboratorio* ashless paper; ~ **de china** *m* PAPEL Chinese paper; ~ **para ciclostilo** *m* PAPEL stencil duplicator copy paper; ~ **cloruro** *m* FOTO, PAPEL chloride paper; ~ **coloreado por ambas caras** *m* PAPEL two-side colored paper (*AmE*), two-side coloured paper (*BrE*); ~ **para condensadores** *m* PAPEL capacitor tissue paper; ~ **para conos de hilatura** *m* PAPEL paper for textile paper tubes; ~ **conteniendo pasta mecánica** *m* PAPEL mechanical woodpulp paper; ~ **continuo doblado** *m* *Esp* (*cf papel fanfold, papel membretado fanfold AmL*) INFORM&PD fanfold paper, fanfold stationery; ~ **contracolado** *m* PAPEL pasted paper; ~ **de copias** *m* PAPEL onionskin paper; ~ **para copias** *m* FOTO printing paper, PAPEL carbon copy paper; ~ **para correo aéreo** *m* *AmL* (*cf papel de avión Esp*) PAPEL airmail paper; ~ **para corrugar** *m* *AmL* (*cf papel para ondular Esp*) PAPEL corrugating medium, fluting corrugating paper, fluting medium; ~ **costero** *m* PAPEL outside defective paper; ~ **couché** *m* TRANSP baryta paper; ~ **crespado** *m* EMB, PAPEL crepe paper; ~ **crespado doble cara** *m* EMB, PAPEL double-face crepe paper; ~ **crispado** *m* PAPEL cockle-finished paper; ~ **cristal** *m* PAPEL glassine; ~ **para cubrir** *m* PAPEL tympan paper;

~ d ~ **de desecho** *m* RECICL waste paper; ~ **de desperdicio** *m* PAPEL retree; ~ **para diagramas** *m* PAPEL chart paper; ~ **para dibujo técnico** *m* CONST plotting paper; ~ **dieléctrico** *m* PAPEL electrical insulating paper; ~ **para documentos** *m* PAPEL archival paper, deed paper; ~ **para documentos de larga duración** *m* PAPEL paper for long storage documents; ~ **de dos capas** *m* PAPEL two-layer paper; ~ **duro de bromuro** *m* FOTO hard bromide paper;

~ e ~ **para edición** *m* PAPEL book paper; ~ **electrosensible** *m* IMPR, INFORM&PD electrosensitive paper; ~ **de embalaje** *m* EMB, PAPEL wrapping paper; ~ **embreado** *m* EMB, PAPEL bitumen-coated paper, tar paper, tarred brown paper; ~ **embreado reforzado** *m* *AmL* (*cf papel alquilvanado reforzado Esp*) PAPEL reinforced union paper; ~ **de empaquetadura** *m* ING MECÁ paper gasket; ~ **encerado** *m* ALIMENT, PAPEL greaseproof paper, parchment paper; ~ **encerado por ambas caras** *m* EMB double-face wax paper; ~ **encolado** *m* PAPEL *en masa* sized paper; ~ **con encolado superficial** *m* PAPEL surface-sized paper; ~ **sin encolar** *m* PAPEL waterleaf paper; ~ **engomado** *m* EMB gummed paper; ~ **entretelado en una cara** *m* PAPEL cloth-lined paper; ~ **para envases de aceite** *m* EMB, PAPEL oil-packing paper; ~ **de envolver** *m* EMB, PAPEL wrapping paper; ~ **esmaltado** *m* PAPEL, REVEST enameled paper (*AmE*), enamelled paper (*BrE*), glazed paper; ~ **de esmeril** *m* MECÁ, PAPEL, PROD emery paper; ~ **estucado** *m* EMB, PAPEL coated paper; ~ **estucado por una cara** *m* PAPEL single-coated paper; ~ **estucado con cepillo** *m* PAPEL brush-coated paper; ~ **estucado por cuchilla oscilante** *m* PAPEL trailing-blade-coated paper; ~ **con estucado esponjoso** *m* PAPEL bubble-coated paper; ~ **estucado para ilustraciones** *m* PAPEL halftone coated paper; ~ **estucado con labio soplante** *m* PAPEL air-knife-coated paper; ~ **estucado en prensa encoladora** *m* PAPEL size-press coated paper; ~ **estucado con rodillo grabado** *m* PAPEL gravure-coated paper; ~ **extrafuerte** *m* FOTO, PAPEL extra-hard paper; ~ **extrasuave** *m* FOTO, PAPEL extra-soft paper;

~ f ~ **fabricado en máquina de doble tela** *m* PAPEL twin-wire paper machine; ~ **fanfold** *m* *AmL* (*cf papel continuo doblado Esp*) INFORM&PD fanfold paper, fanfold stationery; ~ **para fichas perforadas** *m* *Esp* (*cf papel para tarjetas perforadas AmL*) PAPEL paper for punched cards; ~ **de filtro** *m* LAB, PAPEL, PROC QUÍ, QUÍMICA filter paper; ~ **de filtro sin cenizas** *m* LAB, PAPEL ashless filter paper; ~ **para formularios** *m* PAPEL autotype paper; ~ **para forrar cajas** *m* EMB, PAPEL case-lining paper; ~ **fotográfico** *m* FOTO, PAPEL photographic paper; ~ **fotográfico con capa fotosensible a la albúmina** *m* FOTO, PAPEL albumenized paper; ~ **fuertemente alisado** *m* PAPEL English finish paper;

~ g ~ **de gelatinobromuro** *m* FOTO, PAPEL bromide paper; ~ **gofrado** *m* EMB, IMPR, PAPEL embossed paper; ~ **de grabación** *m* INSTR recording paper; ~ **de gramaje elevado** *m* IMPR, PAPEL heavyweight paper; ~ **de guardas** *m* PAPEL end-leaf paper;

~ h ~ **para hectografía** *m* PAPEL spirit duplicator copy paper; ~ **para hilar** *m* PAPEL spinning paper;

~ i ~ **ignífugo** *m* PAPEL, TERMO fire-resistant paper; ~ **impregnado** *m* EMB, PAPEL impregnated paper; ~ **para imprimir por chorro de tinta** *m* PAPEL ink-jet printing paper; ~ **indicador** *m* FOTO indicator paper, QUÍMICA test paper;

~ j ~ **jacquard** *m* EMB, PAPEL jacquard paper; ~ **jaspeado** *m* PAPEL veined paper; ~ **para juntas** *m* PAPEL gasket paper;

~ k ~ **kraft** *m* EMB, PAPEL kraft paper; ~ **kraft para caras** *m* PAPEL kraft liner;

~ l ~ **para lectura óptica** *m* PAPEL OCR paper; ~ **libre de ácido** *m* EMB, PAPEL acid-free paper; ~ **libre de cloro** *m* PAPEL chlorine-free pulp; ~ **de lija** *m* C&V glass paper, ELEC *limpieza* abrasive paper, ING MECÁ coated abrasive, MECÁ sandpaper, PAPEL flint paper, glass paper, PROD sandpaper, REVEST coated abrasive; ~ **de lija impermeable** *m* ING MECÁ, PAPEL waterproof abrasive paper; ~ **de líneas** *m* GEOM lined paper;

~ m ~ **sin marca** *m* PAPEL wove paper; ~ **mate** *m* IMPR, PAPEL matt paper; ~ **membretado fanfold** *m* *AmL* (*cf papel plegado en acordeón Esp*) INFORM&PD fanfold paper, fanfold stationery; ~ **metalizado** *m* EMB, PAPEL metalized paper (*AmE*), metallized paper (*BrE*), PROD foil; ~ **milimetrado** *m* PAPEL scale paper; ~ **para mimiógrafo** *m* PAPEL stencil duplicator copy paper; ~ **moneda** *m* *AmL* (*cf papel para billetes de* *banco Esp*) PAPEL banknote paper; ~ **multicapas** *m* PAPEL multilayer paper;

~ n ~ **nitrado** *m* PAPEL nitrate paper; ~ **no empañante** *m* PAPEL non-tarnish paper; ~ **no oxidante** *m* PAPEL non-rust paper;

~ o ~ **offset** *m* PAPEL offset paper; ~ **ondulado** *m* EMB, PAPEL corrugated paper; ~ **ondulado sencillo** *m* PAPEL ripple paper; ~ **para ondular** *Esp* (*cf papel para corrugar AmL*) *m* PAPEL fluting corrugating paper, fluting medium, fluting paper, corrugating medium;

~ p ~ **de paja** *m* PAPEL yellow strawboard, yellow strawpaper; ~ **de paja mixto** *m* PAPEL mixed strawpaper; ~ **con pasta mecánica** *m* PAPEL wood-containing paper; ~ **sin pasta mecánica** *m* PAPEL freesheet paper; ~ **pegado** *m* PAPEL pasted paper; ~ **pergamino** *m* EMB, PAPEL parchment paper; ~ **de pergamino vegetal** *m* PAPEL vegetable parchment; ~ **pigmentado en prensa encoladora** *m* PAPEL size-press pigmented coated paper; ~ **pintado de madera** *m* REVEST woodchip wallpaper; ~ **pintado en relieve** *m* REVEST anaglyptic wall-paper; ~ **plegado en acordeón** *m* Esp (*cf papel fanfold, papel membretado fanfold AmL*) INFORM&PD fanfold paper, fanfold stationery; ~ **pluma** *m* PAPEL bulking paper; ~ **posterior protector** *m* FOTO backing paper; ~ **prensa** *m* PAPEL newsprint; ~ **prensado** *m* PAPEL presspaper; ~ **de prueba** *m* QUÍMICA test paper; ~ **a prueba de corrosión** *m* EMB, PAPEL corrosion-preventive paper; ~ **para pulir** *m* C&V, PAPEL polishing paper;

~ q ~ **quebrado** *m* PAPEL outside defective paper;

~ r ~ **reciclado** *m* CONTAM, PAPEL, RECICL recycled paper; ~ **recuperado** *m* PAPEL waste paper; ~ **reforzado** *m* PAPEL reinforced paper; ~ **resistente** *m* EMB, PAPEL grease-resistant paper; ~ **resistente a los álcalis** *m* EMB, PAPEL alkali-proof paper; ~ **resistente a las grasas** *m* EMB, PAPEL grease-resistant paper, greaseproof paper; ~ **resistente a la tracción en húmedo** *m* PAPEL wet-strength paper; ~ **revestido** *m* EMB, PAPEL, REVEST coated paper; ~ **revestido antiadherente** *m* PAPEL release-coated paper; ~ **con revestido barrera** *m* PAPEL barrier-coated paper; ~ **revestido por cortina** *m* PAPEL, REVEST curtain-coated paper; ~ **revestido por emulsión** *m* PAPEL emulsion-coated paper; ~ **revestido por extrusión** *m* PAPEL extrusion-coated paper; ~ **revestido con fusiones en caliente** *m* PAPEL hot-melt coated paper; ~ **revestido por inmersión** *m* PAPEL dip-coated paper; ~ **revestido por medio de disolventes** *m* PAPEL solvent-coated paper; ~ **revestido no adherente** *m* PAPEL release-coated paper; ~ **revestido de sulfato de bario** *m* TRANSP baryta paper; ~ **para rodillos de la calandra** *m* PAPEL calender bowl paper;

~ s ~ **satinado** *m* PAPEL glazed paper, supercalendered paper, glassine paper, REVEST glazed paper; ~ **satinado una cara** *m* PAPEL machine-glazed paper (*MG paper*); ~ **secante** *m* PAPEL blotting paper; ~ **seda** *m* PAPEL wrapping tissue; ~ **de seda para montaje en seco** *m* FOTO dry-mounting tissue; ~ **de seguridad** *m* PAPEL safety paper, security paper; ~ **sensibilizado** *m* FOTO, PAPEL sensitive paper; ~ **separador** *m* C&V pad; ~ **similsulfurizado** *m* EMB, PAPEL greaseproof paper; ~ **para sobres postales** *m* PAPEL envelope paper; ~ **soporte** *m* PAPEL base paper, body paper, body stock; ~ **soporte carbón** *m* PAPEL carbonizing base paper; ~ **soporte**

carbón de un solo uso *m* PAPEL one-time carbonizing base paper; ~ **soporte carbón de varios usos** *m* PAPEL multiple-use carbonizing base paper; ~ **soporte para ciclostilo** *m* PAPEL duplicating stencil-base paper; ~ **de soporte delgado** *m* FOTO single-weight paper; ~ **soporte para diazotipo** *m* PAPEL base paper for diazotype; ~ **soporte para estucar** *m* PAPEL, REVEST coating base paper; ~ **soporte fotográfico** *m* PAPEL photographic base-paper; ~ **de soporte grueso** *m* FOTO, PAPEL double weight paper; ~ **soporte para papeles pintados** *m* PAPEL wallpaper base; ~ **sulfurizado** *m* PAPEL vegetable parchment; ~ **superfino** *m* PAPEL onionskin paper; ~ **supersatinado** *m* PAPEL imitation art paper;

~ t ~ **de tamaño 41,9 x 53,3 cm** *m* IMPR large post paper; ~ **de tamaño 8,8 x 13,3 cm** *m* IMPR diamond paper; ~ **para tapizar paredes** *m* PAPEL wallpaper; ~ **para tarjetas perforadas** *m* AmL (*cf papel para fichas perforadas Esp*) PAPEL paper for punched cards; ~ **tela** *m* ING MECÁ tracing paper; ~ **con tela en el interior** *m* PAPEL fabric-pasted paper; ~ **teñido en calandra** *m* PAPEL padding; ~ **térmico** *m* INFORM&PD, PAPEL heat-sensitive paper; ~ **termoadherente** *m* PAPEL, REVEST heat-set adhesive paper; ~ **termográfico** *m* PAPEL thermographic paper; ~ **termosellable** *m* EMB, REVEST heat-sealable paper; ~ **de tina revestido** *m* PROD vat-lined board; ~ **tisú** *m* PAPEL tissue; ~ **tisú guata** *m* PAPEL soft tissue; ~ **para títulos** *m* PAPEL bond paper; ~ **tornasol** *m* QUÍMICA test paper, *indicador de pH* litmus paper; ~ **tornasol azul** *m* QUÍMICA blue litmus paper; ~ **tornasol rojo** *m* QUÍMICA red litmus paper; ~ **de trazado** *m* GEOM, PAPEL drawing paper; ~ **para tubos de hilatura** *m* PAPEL paper for textile paper tubes;

~ u ~ **usado** *m* RECICL waste paper;

~ v ~ **vegetal** *m* ING MECÁ, PAPEL tracing paper; ~ **velín** *m* PAPEL vellum paper; ~ **verjurado** *m* PAPEL laid paper; ~ **verjurado con acabado imitando a la goma** *m* EMB laid paper with rubber appearance; ~ **vitela** *m* PAPEL vellum paper

papelera *f* RECICL paper mill

papeles ~ **autocopiativos no carbonados** *m pl* PAPEL NCR papers; ~ **de cromatografía** *m pl* LAB, QUÍMICA chromatography papers; ~ **delgados** *m pl* PAPEL fine papers; ~ **e impresos leídos** *m pl* PAPEL mixed magazines, mixed papers; ~ **usados** *m pl* RECICL waste paper

papilla *f* ALIMENT slurry; ~ **de almidón** *f* ALIMENT starch slurry

paquete *m* EMB bundle, pack, package, FOTO pack, INFORM&PD package, packet, ING ELÉC package, PROD packet, *fabricación de hierro* pile, TEC PETR *perforación* skid, TELECOM, TEXTIL packet, TRANSP parcel; ~ **de añadidura por goteo** *m* ING ELÉC drop-in package; ~ **de aplicación** *m* INFORM&PD application package; ~ **aviónico del helicóptero** *m* TRANSP AÉR helicopter avionics package; ~ **de baterías** *m* PROD battery pack; ~ **de cigarrillos** *m* EMB cigarette pack (*AmE*), cigarette packet (*BrE*); ~ **conformado al vacío** *m* EMB vacuum-formed package; ~ **con contenido a la vista** *m* EMB visual pack; ~ **de datos** *m* TELECOM data packet; ~ **doble** *m* EMB twin pack; ~ **de una dosis** *m* EMB unit-dose pack (*AmE*), unit-dose packet (*BrE*); ~ **de edición gráfico** *m* INFORM&PD graphic software package; ~ **de**

energía *m* ELECTRÓN, FÍS, FÍS PART, INFORM&PD, NUCL, TEC ESP, TELECOM, TV quantum; ~ **para establecer la comunicación** *m* TELECOM call-setup packet; ~ **flexible** *m* EMB flexible package; ~ **en línea doble** *m* (*DIP*) INFORM&PD, ING ELÉC dual-in-line package (*DIP*); ~ **llenado al vacío** *m* EMB vacuum pack; ~ **matemático** *m* IMPR math package; ~ **mediante laminado** *m* EMB laminated pack; ~ **de ondas** *m* ELEC, FÍS wave packet; ~ **plano** *m* EMB flat pack; ~ **de programas** *m* INFORM&PD software package; ~ **que se puede volver a cerrar** *m* EMB reclosable pack; ~ **de ración** *m* EMB portion pack; ~ **sin retorno** *m* EMB one-way pack; ~ **retractilado** *m* EMB shrink pack; ~ **de software** *m* FÍS, INFORM&PD software package; ~ **de software gráfico** *m* INFORM&PD graphic software package; ~ **de solicitud de desconexión** *m* TELECOM clear-request packet; ~ **de soporte lógico** *m* INFORM&PD software package; ~ **de soporte lógico gráfico** *m* INFORM&PD graphic software package; ~ **de taller** *m* PROD shop packet; ~ **de tamaño económico** *m* EMB economy-size pack; ~ **termoconformado** *m* EMB skin pack; ~ **de trabajo** *m* TEC ESP work package; ~ **unitario** *m* EMB unit pack

par *m* CONST *arquitectura* principal, roof truss, principal rafter, ELEC, FÍS couple, ING ELÉC couple, pair, ING MECÁ moment; ~ **alámbrico** *m* ING ELÉC wire pair; ~ **antagonista** *m* MECÁ restoring torque, TRANSP AÉR antagonistic torque; ~ **apantallado** *m* ING ELÉC shielded pair; ~ **armónico** *m* ELECTRÓN, FÍS even harmonic; ~ **en autorrotación** *m* TRANSP AÉR windmill torque; ~ **bimetálico** *m* ELEC *en termostatos, termómetros* bimetallic strip; ~ **binario** *m* ELECTRÓN binary pair; ~ **blindado** *m* ING ELÉC shielded pair; ~ **de cable telefónico** *m* ING ELÉC, TELECOM telephone cable pair; ~ **de cables** *m* ELEC, ING ELÉC cable pair; ~ **coaxial** *m* ING ELÉC coaxial pair; ~ **complementario** *m* ELECTRÓN complementary pair; ~ **cónico** *m* VEH *transmisión* crown wheel and pinion; ~ **de electrones** *m* ELECTRÓN, FÍS PART, FÍS RAD electron pair; ~ **estereoscópico** *m* FOTO stereoscopic pair; ~ **del freno** *m* TRANSP AÉR brake torque; ~ **de fuerzas** *m* ING MECÁ torque; ~ **de fuerzas exterior** *m* ING MECÁ external torque; ~ **de fuerzas giroscópico** *m* TRANSP AÉR gyroscopic torque; ~ **galvánico** *m* ING ELÉC galvanic couple; ~ **del gradiente de gravedad** *m* TEC ESP gravity-gradient torque; ~ **de gran persistencia** *m* ELECTRÓN *transistores* long-tail pair; ~ **de la hélice** *m* TRANSP AÉR propeller torque; ~ **inicial de arranque** *m* ELEC *máquina eléctrica* locked-rotor torque; ~ **iónico** *m* FÍS RAD, QUÍMICA *de electrólito* ion pair; ~ **de llamada** *m* MECÁ restoring torque; ~ **motor** *m* AUTO *del motor*, MECÁ torque, PROD power-torquing, torque; ~ **del motor** *m* AUTO, TRANSP AÉR engine torque; ~ **motor al arranque** *m* PROD starting torque; ~ **motor de arranque** *m* PROD starting torque; ~ **motor de servicio** *m* PROD running torque; ~ **ordenado** *m* MATEMÁT ordered pair; ~ **de razamiento** *m* TEC ESP frictional torque; ~ **de reducción** *m* VEH reduction gear, *transmisión* crown and pinion; ~ **en régimen de molinete** *m* TRANSP AÉR windmill torque; ~ **de reposición** *m* MECÁ restoring torque; ~ **de resistencia** *m* AUTO resisting torque; ~ **del rotor** *m* TRANSP AÉR rotor torque; ~ **suelto** *m* QUÍMICA *de electrones* lone pair;

~ **térmico** *m* ING ELÉC, TEC PETR, TERMO thermocouple; ~ **termoeléctrico** *m* ING ELÉC, TERMO thermocouple; ~ **con tirantes** *m* CONST couple roof; ~ **de torsión** *m* ING ELÉC torque; ~ **torsor** *m* MECÁ, PROD torque; ~ **trasero** *m* AUTO, MECÁ, VEH rear-end torque; ~ **trenzado** *m* INFORM&PD, ING ELÉC twisted pair

parabánico *adj* QUÍMICA parabanic
parabenceno *m* QUÍMICA parabenzene
parábola *f* FOTO parabola
parabólica *f* TV satellite dish
parabrisas *m* AUTO, C&V, TEC ESP, TRANSP, TRANSP AÉR, VEH windscreen (*BrE*), windshield (*AmE*); ~ **laminado** *m* AUTO, C&V, TEC ESP, TRANSP, TRANSP AÉR, VEH laminated windscreen (*BrE*), laminated windshield (*AmE*); ~ **térmico** *m* AUTO, TEC ESP, TRANSP, TRANSP AÉR, VEH heated windscreen pane (*BrE*), heated windshield pane (*AmE*)
paracaídas *m* TEC ESP parachute, chute, TRANSP AÉR parachute; ~ **auxiliar** *m* D&A auxiliary parachute; ~ **de frenado** *m* TEC ESP drag chute, TRANSP brake parachute, deceleration parachute, TRANSP AÉR drag chute, drag parachute; ~ **frenante de cola** *m* TEC ESP parabrake; ~ **con piloto** *m* D&A parachute with pilot; ~ **retardador** *m* D&A retarder parachute; ~ **troncoidal** *m* TEC ESP *atmosférico* drogue chute
paracaidista *m* D&A parachutist
paracetaldehido *m* QUÍMICA paracetaldehyde
parachispas *m* ELEC *cortocircuito* arcing contacts, PROD fire screen, TERMO fireguard
parachoques *m* AUTO bumper, FERRO *vehículos* buffer-stop block, VEH *carrocería* bumper; ~ **delantero** *m* AUTO bullbar, front bumper; ~ **flexible** *m* VEH *carrocería* underrun bumper; ~ **trasero** *m* AUTO, VEH rear bumper; ~ **del tren de aterrizaje** *m* TRANSP AÉR landing gear bumper
paracianógeno *m* QUÍMICA paracyanogen
paracoro *m* QUÍMICA parachor
paracresol *m* QUÍMICA paracresol
paracusia *f* ACÚST paracusis
parada *f* CINEMAT click stop, CONST stop, stopping, ENERG RENOV stall, FÍS stop, FOTO click stop, INFORM&PD halt, stop, ING MECÁ holdup, jamming, knock-off, standstill, NUCL shutdown, PROD stoppage, TEC ESP shutdown, TEC PETR *perforación* stand of pipe, VEH stop; ~ **de arrastre** *f* TRANSP AÉR drag stop; ~ **caliente** *f* NUCL hot shutdown; ~ **del ciclo** *f* INFORM&PD cycle stealing; ~ **de emergencia** *f* INFORM&PD emergency shutdown, NUCL emergency shutdown, scram, SEG emergency stop; ~ **de emergencia del reactor** *f* NUCL emergency shutdown of the reactor, reactor trip; ~ **para escape de gases** *f* PROD dwell setting; ~ **por falta de papel** *f* IMPR, INFORM&PD form stop; ~ **forzosa del sistema** *f* INFORM&PD system crash; ~ **fría** *f* NUCL cold shutdown; ~ **por inercia de una bomba** *f* NUCL pump coast-down, pump run-down; ~ **manual** *f* NUCL manual shutdown; ~ **no programada** *f* NUCL non-scheduled outage; ~ **nominal** *f* PROD rated stop; ~ **de papel** *f* INFORM&PD form stop; ~ **periódica** *f* NUCL periodic shutdown; ~ **programada** *f* NUCL scheduled outage; ~ **remota** *f* NUCL remote shutdown; ~ **por reparaciones** *f* PROD stoppage for repairs; ~ **segura** *f* NUCL proper shutdown; ~ **técnica** *f* TRANSP AÉR technical stop; ~ **temporal** *f* PROD damping down;

~ **terminal** *f* INFORM&PD dead halt; ~ **del tranvía** *f* TRANSP streetcar stop (*AmE*), tram stop (*BrE*)
paradera *f AmL* (*cf barrera de seguridad Esp*) MINAS safety drag bar
paradiafonia *f* ING ELÉC near-end crosstalk
paradiclorobenceno *m* QUÍMICA paradichlorobenzene
paradigma *m* INFORM&PD paradigm
parado: estar ~ *vi* ING MECÁ lie idle
paradoja: ~ **de los gemelos** *f* FÍS twin paradox
parafenetolcarbamida *f* QUÍMICA dulcin
parafina *f* PETROL, QUÍMICA, TEC PETR, TERMO, TRANSP, TRANSP AÉR kerosene (*AmE*), paraffin (*BrE*); ~ **líquida** *f* QUÍMICA, TERMO liquid paraffin; ~ **sólida** *f* ELEC *aislador* kerosene wax (*AmE*), paraffin wax (*BrE*)
parafínico *adj* QUÍMICA paraffinic
paragénesis *f* GEOL paragenesis
paragonita *f* MINERAL paragonite
paraguas: ~ **de fijación con cavidad de metal** *m* CONST metal-cavity fixing umbrella
parahidrógeno *m* QUÍMICA parahydrogen
parahielos *m* TRANSP AÉR ice guard
parahilos *m* TEXTIL stop motion
paraisómero *m* QUÍMICA paraisomer
paraje: ~ **de aguas poco profundas** *m* TRANSP MAR, GEOL shallows; ~ **sondable con escandallo** *m* TRANSP MAR *fondo del mar* soundings
paral *m* ING MECÁ upright
paralaje *m* CINEMAT, FÍS, FOTO, ING MECÁ parallax
paraldehido *m* QUÍMICA paraldehyde
paralela: ~ **enclavijada** *f* ING MECÁ dowel-pin parallel
paralelamente *adv* GEOM parallel
paralelismo *m* GEOM, ING MECÁ, TEXTIL parallelism
paralelo[1]: *adj* ELEC *circuito*, GEOM, INFORM&PD, TEXTIL parallel; **en** ~ *adj* GEOM, TEXTIL in parallel; **no** ~ *adj* ING MECÁ off-center (*AmE*), off-centre (*BrE*), out-of-parallel
paralelo[2] *m* ING ELÉC bank
paralelogramo *m* GEOM, MECÁ parallelogram; ~ **de fuerzas** *m* FÍS, GEOM parallelogram of forces; ~ **de velocidades** *m* FÍS parallelogram of velocities
parálico *adj* GEOL paralic
parálisis: ~ **de los buzos** *f* OCEAN, TRANSP MAR caisson disease
paralización *f* MINAS lagging, PROD stoppage
paralizado: estar ~ *vi* ING MECÁ lie idle
parallamas *m* C&V *puerta de caldera* baffle plate, CONST, SEG, TEC ESP, TEC PETR, TERMO flame arrester, flame trap
paramagnético *adj* ELEC *material*, FÍS paramagnetic
paramagnetismo *m* ELEC, FÍS paramagnetism
paramento *m* CARBÓN *muros* face; ~ **de carga** *m* NUCL charge face
paramétrico *adj* INFORM&PD parametric
parametrización *f* INFORM&PD parameterization
parámetro *m* CONST *presas* face, facing, ELECTRÓN, FÍS, INFORM&PD parameter; ~ **de activación** *m* METAL activation parameter; ~ **de categoría fuera de rango** *m* TELECOM parameter out-of-range class; ~ **de la celda unidad** *m* CRISTAL cell parameter, unit-cell parameter; ~ **combinatorio** *m* INFORM&PD combinational setting; ~ **dinámico** *m* INFORM&PD dynamic parameter; ~ **de dispersión material** *m* ÓPT, TELECOM material dispersion parameter; ~ **de dispersión del perfil** *m* TELECOM profile-dispersion parameter; ~ **formal** *m* INFORM&PD formal para-

meter; ~ **híbrido** *m* ELECTRÓN *transistores* hybrid parameter; ~ **híbrido del transistor** *m* ELECTRÓN transistor hybrid parameter; ~ **de impacto** *m* FÍS impact parameter; ~ **de palabra clave** *m* INFORM&PD keyword parameter; ~ **de pequeña señal** *m* ELECTRÓN small-signal parameter; ~ **del perfil** *m* ÓPT profile parameter; ~ **de programa** *m* INFORM&PD program parameter; ~ **real** *m* INFORM&PD actual parameter; ~ **de la red** *m* CRISTAL lattice parameter; ~ **de red** *m* FÍS network parameter, METAL crystal-lattice parameter, lattice constant; ~ **de la rejilla** *m* ELECTRÓN grid parameter; ~ **reticular** *m* CRISTAL lattice parameter; ~ **de tráfico** *m* TRANSP traffic parameter; ~ **de visualización** *m* INFORM&PD display setting; ~ **Y** *m* ELECTRÓN Y-parameter; ~ **Z** *m* ELECTRÓN Z-parameter; ~ **Zener-Hollomon** *m* METAL Zener-Hollomon parameter
parantina *f* MINERAL paranthine
parapeto *m* CONST parapet
parapista *f* TEC ESP parafoil
parar[1] *vt* C&V stop, ING MECÁ hold up, switch off, TEC ESP shut down, TEXTIL stop, TRANSP AÉR shut down, VEH stall; ~ **una fuga en** *vt* AGUA stop a leak in
parar[2]: ~ **el avance** *vi* ING MECÁ throw out of feed; ~ **las máquinas** *vi* TRANSP MAR stop engines; ~ **la marcha** *vi* ING MECÁ throw out of action
parar[3] *vti* INFORM&PD stop, halt
pararenina *f* QUÍMICA parachymosin, pararennin
pararosanilina *f* QUÍMICA pararosaniline
pararrayos *m* ELEC lightning arrester, GEOFÍS lightning arrester, lightning surge arrester, ING ELÉC lightning arrester, SEG lightning rod, lightning protector, lightning arrester, TEC ESP, TRANSP MAR *accesorios de cubierta* lightning arrester; ~ **de alto voltaje** *m* SEG lightning arrester for high voltage; ~ **atmosférico** *m* GEOFÍS, METEO lightning conductor; ~ **de barra** *m* ELEC, SEG lightning rod; ~ **con fusible de expulsión** *m* ING ELÉC expulsion-type lightning arrester
pararse *v refl* ENERG RENOV stall, ING MECÁ come to a standstill, PROD give out, VEH stall
parásita *f* MINERAL parasite
parásito *adj* ELECTRÓN parasitic
parásitos *m pl* ELECTRÓN snow, TEC ESP interference; ~ **atmosféricos** *m pl* GEOFÍS atmospherics, METEO sferics (*AmE*), spherics (*BrE*)
parasol *m Esp* (*cf visera contra el sol AmL*) AUTO sun vizor (*AmE*), sun visor (*BrE*), CINEMAT umbrella, *de objetivos* lens hood, matte box, FOTO lens hood, VEH sun vizor (*AmE*), sun visor (*BrE*)
paratartárico *adj* QUÍMICA paratartaric
paratípico *adj* QUÍMICA paratypical
paratramas *m* TEXTIL weft stop motion
paraurdimbres *m* TEXTIL warp stop motion
paraxileno *m* QUÍMICA, TEC PETR paraxylene
parcela *f* AGRIC plot; ~ **demostrativa** *f* AGRIC demonstration plot; ~ **experimental** *f* AGRIC experimental plot; ~ **de mineral** *f* MINAS patch
parche *m* INFORM&PD patch
parchear *vt* INFORM&PD patch
parcheo *m* PROD patching, patching-up
parcial[1] *adj* ACÚST partial
parcial[2] *m* ACÚST partial
pareado *m* ELEC *cable*, ING ELÉC pairing
pared[1]: **de** ~ **delgada** *adj* ING MECÁ thin-walled
pared[2] *f Esp* (*cf respaldo AmL*) CARBÓN *fondo de horno*

breast, CONST *arquitectura*, FÍS FLUID wall, INSTR *de pozo* side, MINAS wall, *de filón* side, *filones* cheeks, PROD *de cilindro, caldera* wall; **~ acústica** *f* SEG sound-absorbent wall; **~ adiabática** *f* FÍS adiabatic wall; **~ amortiguadora de ruido** *f* SEG noise-abating wall; **~ de carga** *f* INSTAL HIDRÁUL pressure side; **~ sin carga** *f* CONST blank wall; **~ de chimenea** *f* CONST jamb; **~ ciega** *f* CONST dead wall; **~ del cilindro** *f* AUTO, VEH *motor* cylinder wall; **~ con comba** *f* CONST bulged wall, bulging wall; **~ delantera** *f* C&V front wall; **~ desviadora** *f* PAPEL baffle wall; **~ del dique** *f* TRANSP MAR dock wall; **~ de dislocación** *f* CRISTAL dislocation wall; **~ divisoria** *f* CONST *arquitectura* partition wall; **~ de filón** *f* GEOL vein wall; **~ frontal** *f* C&V front wall, CONST breast wall; **~ interior** *f* TELECOM inner wall; **~ de ladrillos** *f* CONST brick wall; **~ lateral** *f* AGUA, MINAS side wall; **~ lateral con aletas de ventilación** *f* TRANSP side wall with ventilation flaps; **~ lateral del dique** *f* TRANSP MAR dock wall; **~ lateral del puerto** *f* C&V port sidewall; **~ maestra** *f* CONST main wall; **~ medianera** *f* CONST *arquitectura* party wall; **~ por medio** *f* CONST wall plate; **~ metálica** *f* ÓPT *de una guía de ondas* sheath; **~ del pozo** *f* MINAS side wall; **~ que soporta la carga** *f* CONST load-bearing wall; **~ de sombra** *f* C&V *separa al fundidor del refinador* shadow wall; **~ de soporte** *f* CONST supporting wall; **~ del sótano** *f* CONST basement wall; **~ superior** *f* GEOL hanging wall; **~ termoaislante** *f* TRANSP AÉR heat-insulating wall; **~ transversal** *f* CONST cross wall; **~ trasera del puerto** *f* C&V port backwall (*BrE*), port endwall (*AmE*); **~ de la viga de la caja del fuselaje** *f* TRANSP AÉR fuselage-box beam wall; **~ en voladizo** *f* CONST overhanging wall; **~ de vuelta** *f* CONST *albañilería* return wall

paredes: de ~ de poco espesor *adj* ING MECÁ thin-walled

pareja *f* AmL (*cf triple Esp*) ING MECÁ couple, TEC PETR stand of pipe; **~ de engranajes cilíndricos** *f* AUTO cylindrical gear pair; **~ de transistores** *f* ELECTRÓN transistor pair

parentales *m pl* AGRIC breeding stock

paréntesis: ~ rectangulares *m pl* IMPR brackets

pares: ~ de Cooper *m pl* FÍS Cooper pairs; **~ de haces de elevada masa** *m pl* FÍS RAD pairs of high-mass jets

parfísico *m* ING ELÉC physical circuit

pargasita *f* MINERAL pargasite

parición *f* AGRIC calving; **~ invernal** *f* AGRIC winter calving; **~ de vacas en verano** *f* AGRIC summer calving

paridad *f* FÍS, FÍS RAD, INFORM&PD parity; **~ por bit intercalado** *f* TELECOM bit-interleaved parity (*BIP*); **~ horizontal** *f* INFORM&PD horizontal parity; **~ impar** *f* FÍS, INFORM&PD odd parity; **~ par** *f* FÍS, INFORM&PD even parity; **~ vertical** *f* INFORM&PD vertical parity

paridera *f* AGRIC calving, calving season; **~ de verano** *f* AGRIC summer calving

parilla: ~ eléctrica *f* ELEC, TERMO electric hot-plate

parir: ~ cerdos *vi* AGRIC farrow

parlante *m* ACÚST, ELEC, FÍS, TELECOM loudspeaker

paro *m* ING MECÁ stop, stop motion, TEXTIL stop; **~ de la máquina** *m* C&V machine stop; **~ de máquina** *m* PROD machine shutdown; **~ de la pala** *m* TRANSP AÉR blade stop; **~ de resorte** *m* ING MECÁ spring stop;

~ del sistema de la máquina *m* PROD machine-system shutdown

parpadeante *adj* INFORM&PD blinking

parpadear *vi* INFORM&PD blink

parpadeo[1]: sin ~ *adj* INFORM&PD *pantalla* flicker-free

parpadeo[2] *m* CINEMAT flicker, flickering, INFORM&PD *pantalla* blinking, flicker, TELECOM, TV flicker, flickering; **~ cromático** *m* TV chroma flutter, chromatic flicker; **~ de la imagen** *m* TV image flicker; **~ del nivel de blanco** *m* TV fluttering of brightness level

parque: ~ de arrabio *m* METAL, PROD pig-iron yard, pig yard; **~ de artillería** *m* D&A artillery park; **~ de carbón** *m* CARBÓN, MINAS coal yard; **~ de carbones** *m* CARBÓN, MINAS colliery; **~ de cultivo** *m* OCEAN *acuicultura* weir bed; **~ de cultivo de ostras** *m* OCEAN oyster farm; **~ de desguace** *m* PROD scrap-yard; **~ de intemperie** *m* NUCL plant switchyard; **~ de interruptores de la planta** *m* NUCL plant switchyard; **~ invertido** *m* ELECTRÓN inverted population; **~ de vagones** *m* FERRO *vehículos* fleet

párrafo *m* IMPR paragraph; **~ francés** *m* IMPR hanging indentation

parrilla *f* AUTO grille, INSTAL TERM grate, LAB *AmL* (*cf placa Esp*) hotplate, PROD grid, *del hogar* furnace grate, grate, fire grate; **~ articulada** *f* ING MECÁ chain grate; **~ de barrotes** *f* ING MECÁ bar screen; **~ a etapas** *f* PROD stepped grate; **~ fija** *f* INSTAL TERM fixed grate; **~ giratoria** *f* PROD rotary grate; **~ de tamiz** *f* PROC QUÍ sieve grate; **~ de vaivén** *f* PROD shaking grate

parsec *m* FÍS parsec

parte[1] *m* TEC ESP report, TELECOM message; **~ meteorológico** *m* TRANSP MAR weather report

parte[2] *f* ING MECÁ member, part, PROD section, TEXTIL part; **~ activa** *f* ING MECÁ working part; **~ del cable unido a la polea del castillete** *f* MINAS *pozo de extracción* lifting point; **~ de cámara** *f* CINEMAT, TV camera report; **~ central** *f* ING MECÁ, PROD *caja de moldeo* cheek, *de una caja de moldeo de tres partes* middle, middle part; **~ de coma fija** *f* INFORM&PD fixed-point part; **~ de control de la señalización** *f* TELECOM signaling control part (*AmE*) signalling control part (*BrE*) (*ISCP*); **~ delantera** *f* ING MECÁ head end; **~ delantera de la cámara** *f* FOTO camera front; **~ descubierta** *f* CONST *de una teja* bare, exposed part; **~ desgastada de una máquina** *f* ING MECÁ wearing part of a machine; **~ de desgaste** *f* ING MECÁ wear part; **~ de dominio inicial** *f* TELECOM initial-domain part (*IDP*); **~ específica del dominio** *f* TELECOM domain-specific part (*DSP*); **~ falsa** *f* ING MECÁ dummy part; **~ fija** *f* PROD stationary portion; **~ fraccionaria** *f* INFORM&PD fractional part; **~ frontal de la cámara** *f* FOTO camera front; **~ de funcionamiento** *f* PROD operation ticket; **~ horizontal del impulso** *f* ELECTRÓN pulse tilt; **~ húmeda** *f* PAPEL *máquina de papel o cartón* wet end; **~ inferior** *f* PROD *caja de moldeo* nowel; **~ inferior de una botella** *f* EMB base cup; **~ inicial** *f* IMPR *antes del prólogo* front matter; **~ intercambiable** *f* ING MECÁ interchangeable part; **~ interior** *f* HIDROL *alcantarillado* invert; **~ de una molécula** *f* QUÍMICA moiety; **~ móvil** *f* ING MECÁ running end, working part, NUCL *máquinas* working part; **~ móvil del medidor** *f* ING ELÉC meter movement; **~ plana** *f* ING MECÁ flat; **~ plana entre acanaladuras** *f* ING MECÁ land; **~ popel** *f* TRANSP MAR *de la quilla* heel; **~ posterior**

f FOTO back, TRANSP, VEH rear end; ~ **de programa** *f* INFORM&PD program part; ~ **protectora** *f* ING MECÁ wear part; ~ **radial** *f* FÍS RAD *de la función de ondas* radial part; ~ **recta entre dos curvas** *f* AGUA, HIDROL reach; ~ **redondeada** *f* CONST belly; ~ **reglamentaria** *f* ING MECÁ standard part; ~ **removible** *f* ELEC *circuito*, ING ELÉC removable part; ~ **roscada del husillo del cabezal** *f* ING MECÁ nose; ~ **seca** *f* PAPEL dry end; ~ **sujeta a desgaste** *f* CARBÓN wearing part; ~ **superior** *f* AmL (*cf cúspide Esp*) AUTO rear end, MINAS *filones* apex; ~ **superior del equipo receptor de transacción** *f* TELECOM transactional set-header; ~ **trasera** *f* AUTO rear, CONST back, TRANSP *de vagón* rear, VEH *de la carrocería* rear, tail; ~ **trasera deformable** *f* TRANSP deformable rear section; ~ **del usuario de la RDSI** *f* TELECOM ISDN user part; ~ **útil** *f* ING MECÁ working part

partes: ~ **por millón** *f* (*ppm*) QUÍMICA parts per million (*ppm*); ~ **oscilantes** *f pl* ING MECÁ, MECÁ reciprocating parts

partición *f* INFORM&PD partition, partitioning, METAL cleavage, NUCL partition, QUÍMICA partitioning; ~ **en ventanas** *f* INFORM&PD windowing

partícula *f* GEN particle, CARBÓN drop, *metales* grain; ~ **acídica** *f* CONTAM *caracterización de contaminantes* acid particle, acidic particle; ~ **alfa** *f* ELEC, FÍS, FÍS PART, FÍS RAD, NUCL alpha particle; ~ **de alta energía** *f* ELECTRÓN high-energy particle; ~ **de alta velocidad** *f* NUCL high-speed particle; ~ **beta** *f* ELEC, FÍS, FÍS PART, FÍS RAD, NUCL beta particle; ~ **de carga virtual** *f* FÍS PART virtual charged particle; ~ **cargada** *f* CONTAM, ELECTRÓN, FÍS PART, FÍS RAD, ING ELÉC, NUCL charged particle; ~ **coherente** *f* METAL coherent particle; ~ **de combustible no revestida** *f* NUCL uncoated fuel particle; ~ **de corto alcance** *f* NUCL short-range particle; ~ **de desintegración** *f* FÍS PART decay particle; ~ **desnuda** *f* NUCL bare particle; ~ **elemental** *f* FÍS, FÍS PART elemental particle, elementary particle; ~ **elemental de alta energía** *f* FÍS, FÍS PART high-energy elemental particle; ~ **de fluido** *f* FÍS FLUID fluid particle; ~ **en forma de aguja** *f* METAL needle-shaped particle; ~ **de fuga** *f* FÍS RAD particle; ~ **gamma** *f* ELEC, FÍS, FÍS PART, FÍS RAD, NUCL gamma particle; ~ **de gran potencia** *f* ELECTRÓN, FÍS PART, NUCL high-energy particle; ~ **de Higgs** *f* FÍS PART Higgs boson, Higgs particle; ~ **ilesa** *f* FÍS RAD particle; ~ **incidente** *f* NUCL impinging particle, incident particle; ~ **iniciadora** *f* NUCL initiating particle; ~ **ionizante** *f* FÍS RAD ionizing particle; ~ **J** *f* FÍS J particle; ~ **lambda** *f* FÍS lambda particle; ~ **magnética** *f* ING MECÁ, MECÁ, NUCL, TV magnetic particle; ~ **matemática** *f* NUCL mathematical particle; ~ **de metal adheridas** *f* METAL pinning; ~ **metálica** *f* PROD metal particle; ~ **neutra** *f* FÍS, FÍS PART, NUCL neutral particle; ~ **neutra virtual** *f* FÍS PART virtual neutral particle; ~ **omega menos** *f* FÍS omega minus particle; ~ **pesada** *f* GAS heavy particle; ~ **plasma** *f* GEOFÍS plasma particle; ~ **de polvo** *f* CONTAM dust particle; ~ **radiactiva** *f* FÍS PART radiating particle; ~ **rápida** *f* FÍS, FÍS PART, NUCL fast particle; ~ **relativista** *f* FÍS, FÍS PART, TEC ESP relativistic particle; ~ **retenida** *f* NUCL trapped particle; ~ **de retroceso de fisión** *f* NUCL fission recoil; ~ **saliente** *f* NUCL outcoming particle; ~ **secundaria** *f* NUCL secondary particle; ~ **sigma** *f* FÍS sigma particle; ~ **sólida** *f* CONTAM solid particle; ~ **subatómica** *f* FÍS subatomic particle; ~ **supersimétrica** *f* FÍS RAD supersymmetrical particle; ~ **suspendida** *f* CONTAM *caracterización de gases y líquidos* suspended particle; ~ **en suspensión** *f* CONTAM *caracterización de gases y líquidos* suspended particle; ~ **en suspensión en el aire** *f* METEO airborne particle; ~ **tau** *f* FÍS PART tau particle; ~ **de vida corta** *f* FÍS PART short-lifetime particle, short-lived particle; ~ **virtual** *f* FÍS PART virtual particle; ~ **W** *f* FÍS, FÍS PART W particle; ~ **xi** *f* FÍS, FÍS PART xi particle; ~ **Z** *f* FÍS, FÍS PART Z particle

particulado *m* PROC QUÍ particulate

partículas *f pl* CONTAM particulate matter; ~ **diseminadas** *f pl* METAL scattering; ~ **identificadas con spin 0, 1/2, 1, 3/2, 2** *f pl* FÍS PART particles identified as having spins 0, 1/2, 1, 3/2, 2; ~ **de tamaño menor de 76 milímetros** *f pl* CARBÓN dust; ~ **Z** *f pl* FÍS PART weak force

partida *f* MECÁ, TEXTIL, TRANSP batch; ~ **de material** *f* MECÁ, TEXTIL, TRANSP batch; ~ **subcontratada** *f* PROD subcontracting item

partido *adj* INFORM&PD partitioned

partidor *m* CARBÓN splitter, ING MECÁ separator; ~ **central** *m* ENERG RENOV *turbinas* central splitter edge; ~**-reductor del voltaje** *m* ING ELÉC voltage divider

partir[1] *vt* CONST split, INFORM&PD partition

partir[2] *vi* FERRO *equipo inamovible* split

partirse *v refl* ING MECÁ come away, come off, part

partón *m* FÍS parton

parvolina *f* QUÍMICA parvoline

pasa *f* OCEAN narrows

pasabajos *m* TRANSP AÉR low pass

pasabanda *m* FÍS pass band

pasada *f* INFORM&PD *secuencia* run, ING MECÁ cut, run, MECÁ cut, PROD *escofina* cut, *máquina herramienta* going over, pass, TEXTIL *género de punto* course, *tejeduría* pick; ~ **apretada** *f* TEXTIL tight pick; ~ **del barniz** *f* REVEST varnish run; ~ **en caliente** *f* PETROL hot pass; ~ **floja** *f* TEXTIL loose pick; ~ **del papel** *f* IMPR paper run; ~ **profunda** *f* ING MECÁ heavy cut; ~ **de prueba** *f* INFORM&PD test run; ~ **de puntadas del derecho** *f* TEXTIL course-of-face stitches; ~ **de puntadas del revés** *f* TEXTIL course-of-reverse stitches

pasadas: ~ **por minuto** *f pl* TEXTIL courses per minute; ~ **pares e impares** *f pl* TEXTIL even-and-odd courses; ~ **por pulgada** *f pl* TEXTIL picks per inch

pasadizo *m* AmL (*cf transversal Esp*) MINAS *túnel* passageway

pasado: ~ **de la cruz uno y uno** *m* TEXTIL one-and-one lease; ~ **de planchas** *m* IMPR plate-making

pasador[1]: **con** ~ *adj* MECÁ keyed

pasador[2] *m* AUTO stud, CONST barrel key (*AmE*), bolt, dowel pin, gate hook, key, latch catch, pin, ING MECÁ fastener, gudgeon, key, MECÁ catch, fastener, key, peg, pin, PROD fastener, pin, TRANSP MAR *cabos* marlinspike, VEH *motor, pistón* gudgeon pin (*BrE*), wrist pin (*AmE*); ~ **abierto** *m* ING MECÁ, MECÁ cotter pin; ~ **para abrir respiraderos** *m* PROD *fundería* riser pin, riser stick; ~ **acanalado** *m* ING MECÁ grooved pin; ~ **ahusado** *m* ING MECÁ taper pin; ~ **de aletas** *m* ING MECÁ cotter pin, split cotter pin, split pin, MECÁ cotter pin; ~ **angular** *m* ING MECÁ *matriz para fundición a presión* angle pin; ~ **de arrastre** *m* ING

MECÁ driver; ~ **de barra** *m* CONST latch pin; ~ **de bisagra** *m* CONST, ING MECÁ hinge pin; ~ **con cadena** *m* CONST pin chain; ~ **de cadena** *m* CONST pintle, ING MECÁ chain bolt, MINAS chain pin; ~ **de caja** *m* ING MECÁ box pin; ~ **de caja de moldear** *m* ING MECÁ flask pin; ~ **para casar agujeros** *m* ING MECÁ drift pin; ~ **cementado** *m* ING MECÁ hardened dowel pin; ~ **de centrar** *m* ING MECÁ centering pin (*AmE*), centring pin (*BrE*); ~ **de cerrojo** *m* CONST lock bolt, tower bolt; ~ **de la charnela de aleteo** *m* TRANSP AÉR flapping-hinge pin; ~ **de chaveta** *m* ING MECÁ, MECÁ, TRANSP MAR cotter pin; ~ **del cigüeñal** *m* AUTO crank pin; ~ **cilíndrico** *m* ING MECÁ cylindrical pin; ~ **conductor de corriente** *m* PROD current-conducting pin; ~ **cónico** *m* ING MECÁ drift pin, driftpin, drift punch, taper dowel, taper pin, MECÁ taper pin; ~ **cónico cuadrado** *m* ING MECÁ square driftpin; ~ **cónico de ensanchar** *m* ING MECÁ cutting driftpin; ~ **cónico con rosca exterior** *m* ING MECÁ taper pin with external thread; ~ **cónico con rosca extractora** *m* ING MECÁ taper dowel with extracting thread; ~ **cónico con rosca interior** *m* ING MECÁ taper pin with internal thread; ~ **para cono Morse** *m* ING MECÁ Morse taper pin; ~ **corto** *m* ING MECÁ, PROD stud; ~ **de la cruceta** *m* ING MECÁ crosshead pin; ~ **de culata del cilindro** *m* ING MECÁ cylinder-head stud; ~ **del émbolo** *m* AUTO piston pin; ~ **endurecido** *m* ING MECÁ hardened dowel pin; ~ **de enganche** *m* ING MECÁ coupling pin, PROD hitch pin; ~ **del expulsor** *m* MECÁ ejector pin; ~ **expulsor plano** *m* ING MECÁ flat ejector-pin; ~ **eyector** *m* ING MECÁ ejection pin; ~ **eyector plano** *m* ING MECÁ flat ejector-pin; ~ **de fijación** *m* ING MECÁ set pin, steady pin, PROD clamp pin, clamping pin, *moldería* steady pin; ~ **de horquilla** *m* ING MECÁ, MECÁ clevis pin; ~ **hueco** *m* ING MECÁ hollow pin; ~ **igualador de agujeros** *m* ING MECÁ driftpin; ~ **limitador** *m* FOTO stop pin; ~ **de llamada** *m* CONST draw bore pin; ~ **con llave de empernada** *m* CONST pinned-key lock; ~ **de machos** *m* PROD core pin; ~ **para madera** *m* ING MECÁ coach bolt; ~ **de manivela** *m* MECÁ crank pin; ~ **de muelle suspensor** *m* CONST *cerrajería* spring hanger pin; ~ **de palanca de desenganche** *m* AGRIC, AUTO release-lever pin; ~ **paralelo** *m* ING MECÁ parallel dowel, parallel pin; ~ **paralelo con rosca interna** *m* ING MECÁ parallel pin with internal thread; ~ **de película** *m* FOTO *de un trípode* sliding leg; ~ **del pistón** *m* AUTO piston pin; ~ **de pivote** *m* VEH *de remolque* pivot pin; ~ **de presión** *m* ING MECÁ pressure pin; ~ **prisionero** *m* ING MECÁ grub screw; ~ **de retención** *m* ING MECÁ locking stud; ~ **de retención del tren de aterrizaje** *m* TRANSP AÉR landing gear lock pin; ~ **de rosca externa** *m* ING MECÁ external-threaded fastener; ~ **con rosca extractora** *m* ING MECÁ dowel pin with extracting thread; ~ **roscado** *m* ING MECÁ grub screw, screw key; ~ **del sector de la excéntrica** *m* ING MECÁ link block pin; ~ **de seguridad** *m* CINEMAT safety pin, ING MECÁ locking pin, lockpin, TRANSP MAR *de grillete* safety pin; ~ **de seguridad de la espoleta** *m* D&A fuse safety-pin; ~ **de unión** *m* ING MECÁ coupling pin, driftpin; ~ **de ventana** *m* CONST window fastener; ~ **de ventosa de aire** *m* ING MECÁ air-vent pin

pasaje *m* TRANSP MAR *navegación* passage; ~ **archipelágico** *m* OCEAN archipelagic passage

pasajero *m* TEC ESP, TRANSP, TRANSP AÉR passenger; ~**-kilómetro** *m* TRANSP passenger kilometer (*AmE*), passenger kilometre (*BrE*); ~ **en tránsito** *m* TRANSP, TRANSP AÉR transit passenger

pasamanería *f* TEXTIL braid; ~ **de algodón** *f* TEXTIL *adorno* cotton braid

pasamanos *m* CONST handrailing, handrail, MECÁ handrail, NUCL handrailing, SEG *en escaleras* guard-rail, TRANSP MAR handrail, *construcción naval* bulwark rail; ~ **a la altura del codo** *m* CONST elbow-height handrail

pasar[1] *vt* FERRO pass; ~ **por** *vt* PROD thread; ~ **por alto** *vt* SEG overlook; ~ **por una esclusa** *vt* TRANSP MAR pass through a lock; ~ **de la marcha por engranajes a la marcha por correa** *vt* ING MECÁ 'change over from gear-drive to belt-drive; ~ **texto a la línea anterior** *vt* IMPR run back; ~ **a través de** *vt* PROD thread

pasar[2]: ~ **a la línea siguiente** *vi* IMPR run down; ~ **a mayúsculas** *vi* INFORM&PD shift

pasarela *f* CONST, FERRO footbridge, INFORM&PD gateway, MINAS passageway, TEC ESP walkway, TEC PETR manway; ~ **central** *f* TRANSP central gangway; ~ **de servicio** *f* MECÁ catwalk

pasarse: ~ **por ojo** *v refl* TRANSP MAR *buque* go down by the bows

pasavante *m* TRANSP MAR *documento* ship's passport

pascal *m* (*Pa*) GEN pascal (*Pa*)

pase: ~ **de programa** *m* PROD program rung (*AmE*), programme rung (*BrE*)

paseo: ~ **espacial** *m* TEC ESP spacewalk

pasillo *m* CONST, MECÁ, TEC ESP walkway, TRANSP lane, TRANSP MAR *construcción naval* alleyway; ~ **aéreo** *m* TRANSP AÉR air corridor; ~ **de ascensión** *m* TEC ESP, TRANSP AÉR climb corridor; ~ **mecánico** *m* TRANSP pedestrian conveyor, TRANSP AÉR moving sidewalk (*AmE*), passenger conveyor (*BrE*); ~ **rodante rápido para pasajeros** *m* TRANSP high-speed passenger conveyor; ~ **de servicio** *m* MECÁ catwalk; ~ **de trepada** *m* TEC ESP, TRANSP AÉR climb corridor

pasivación *f* ELECTRÓN, FÍS, QUÍMICA, TEC ESP passivation; ~ **de vidrio** *f* C&V, ELECTRÓN glass passivation

pasivado *adj* ELECTRÓN, FÍS, QUÍMICA, TEC ESP passivated

pasivar *vt* ELECTRÓN, FÍS, QUÍMICA, TEC ESP passivate

pasivo *adj* ELECTRÓN, INFORM&PD passive

paso[1] *m* C&V step, CARBÓN chute, CINEMAT pitch, film gage (*AmE*), gage (*AmE*), gauge (*BrE*), size, film gauge (*BrE*), CONST step, *escaleras* throat, thoroughfare, flyer, ELEC *bobina* pitch, *componente del cable* lay, ELECTRÓN increment, ENERG RENOV fairway, FOTO stop, pitch, HIDROL navigation channel, INFORM&PD pitch, pass, ING MECÁ pitch, thread, convolution, outlet, lead, MECÁ pitch, feedthrough, thread, aperture, METAL aperture, step, MINAS pass, *galerías* chute, OCEAN navigation channel, fairway, PROD step, stage, *tornillo* thread, TRANSP navigation channel, TRANSP AÉR pitch, TRANSP MAR *de la hélice* pitch, *de río* narrows, *navegación* fairway, *geografía* pass; ~ **abanderado** *m* TRANSP AÉR feathered pitch; ~ **acodado** *m* ING MECÁ crankpath; ~ **de agua** *m* PROD passage of water, water port; ~ **de aire** *m* ING MECÁ, MINAS air passage; ~ **de álabe** *m* ENERG RENOV, TRANSP AÉR *turbinas* blade pitch; ~ **de aletas** *m* REFRIG pitch of fins; ~ **alto** *m* TRANSP AÉR high pitch; ~ **amortiguador** *m* PROD buffer stage;

~ ancho *m* TRANSP AÉR coarse pitch; **~ aparente** *m* ING MECÁ divided pitch, *hélices* apparent pitch; **~ del arrollamiento** *m* ELEC *bobina* winding pitch; **~ banda** *m* ELECTRÓN *filtros* pass band, PETROL band pass; **~ de la banda sin problemas** *m* PAPEL *a través de la máquina, estucadoras* runnability; **~ de banda del receptor** *m* TV receiver band pass; **~ bastardo** *m* ING MECÁ bastard pitch, fractional pitch, *tornillos* odd pitch; **~ en la bobina** *m* ELEC coil pitch; **~ de cables a través de cubierta** *m* TRANSP MAR through-deck cable fitting; **~ cercano** *m* TEC ESP *cosmonave* fly-by; **~ cíclico** *m* TRANSP AÉR cyclic pitch; **~ cíclico lateral** *m* TRANSP AÉR lateral cyclic pitch; **~ cíclico longitudinal** *m* TRANSP AÉR fore-and-aft cyclic pitch; **~ cíclico de orden elevado** *m* TRANSP AÉR high-order cyclic pitch; **~ circular** *m* ING MECÁ circular pitch, *engranajes* arc pitch, MECÁ circular pitch; **~ circunferencial** *m* ING MECÁ *engranajes* arc pitch, circular pitch, MECÁ circular pitch; **~ del colectivo** *m* TRANSP AÉR collective pitch; **~ del conjunto** *m* NUCL assembly pitch; **~ constante** *m* ING MECÁ *tornillos* even pitch; **~ de la copia** *m* CINEMAT print pitch; **~ de la corriente por el aislante** *m* ING ELÉC insulation breakdown; **~ de corriente por aislante de sobrevoltaje** *m* ING ELÉC overvoltage breakdown; **~ de corte** *m* ING MECÁ pitch to be cut; **~ corto** *m* CINEMAT short pitch; **~ por la criba** *m* PROD screening; **~ de derivación** *m* HIDROL diversion cut; **~ de desbaste** *m* CARBÓN, PROD *por el laminador* roughing pass; **~ del devanado** *m* ELEC *bobina* winding pitch; **~ diametral del surco** *m* ACÚST diametral groove pitch; **~ de los dientes** *m* ING MECÁ pitch; **~ efectivo** *m* TRANSP AÉR effective pitch; **~ de elementos** *m* NUCL assembly pitch; **~ elevado** *m* FERRO overhead crossover, *infraestructura* overpass, TRANSP AÉR high pitch; **~ de enlace** *m* CONST, FERRO crossover; **~ excitador** *m* ACÚST driver, TEC ESP *electrónica* driver stage; **~ del filete** *m* ING MECÁ screw pitch; **~ fino** *m* TRANSP AÉR fine pitch; **~ geométrico** *m* ING MECÁ pitch, TRANSP AÉR geometric pitch; **~ grande** *m* CINEMAT long pitch, ING MECÁ coarse thread; **~ de la hélice** *m* TRANSP AÉR propeller pitch; **~ para los hombres** *m* MINAS manhole; **~ inferior** *m* CONST, FERRO underpass; **~ invertido** *m* TRANSP AÉR reverse pitch; **~ largo** *m* ING MECÁ coarse pitch; **~ libre** *m* ING MECÁ, FERRO clearance; **~ longitudinal** *m* INFORM&PD row pitch; **~ menor que el normal** *m* CINEMAT substandard gauge (*BrE*), substandard gage (*AmE*); **~ por el meridiano** *m* TEC ESP meridian transit; **~ a nivel** *m* CONST, FERRO *equipo inamovible* grade crossing (*AmE*), level crossing (*BrE*); **~ a nivel de ferrocarril** *m* AUTO, CONST, FERRO, TRANSP, VEH road over railway; **~ nominal** *m* TRANSP AÉR standard pitch; **~ de paleta** *m* ENERG RENOV, TRANSP AÉR *turbinas* blade pitch; **~ del papel sin problemas** *m* PAPEL *a través de la máquina, estucadoras* runnability; **~ de la perforación** *m* CINEMAT perforation pitch; **~ positivo** *m* CINEMAT positive pitch; **~ principal** *m* AmL (*cf coladero principal Esp*) MINAS main chute; **~ real** *m* ING MECÁ total pitch, true pitch; **~ de rosca** *m* ING MECÁ screw pitch, thread pitch, VEH thread pitch; **~ de rosca métrico estándar** *m* ING MECÁ ISO metric thread; **~ de rosca métrico normalizado** *m* ING MECÁ ISO metric thread; **~ de rosca métrico para tornillos de diámetros muy pequeños** *m* ING MECÁ ISO miniature metric thread; **~ subterráneo** *m* FERRO underpass; **~ subterráneo de peatones** *m* CONST pedestrian subway; **~ superior** *m* AUTO flyover, CONST flyover, overbridge, TRANSP, VEH flyover; **~ de tarea** *m* INFORM&PD job step; **~ de tornillo** *m* ING MECÁ screw pitch; **~ del tornillo de avance** *m* ING MECÁ pitch of lead screw; **~ del tornillo regulador** *m* ING MECÁ pitch of lead screw; **~ de torsión** *m* ING ELÉC lay; **~ de trabajos** *m* INFORM&PD job step; **~ variable** *m* REFRIG variable pitch

paso²: **~ a paso** *fra* CINEMAT, TV frame by frame

Paso: **~ del Ecuador** *m* TRANSP MAR crossing the line; **~ de la Línea** *m* TRANSP MAR crossing the line

pasta *f* ALIMENT dough, C&V body, *materias primas en la fabricación del vidrio* batch, CONST matrix, P&C dough, PAPEL bulk fibre (*BrE*), pulp, bulk fiber (*AmE*), TEC ESP paste; **~ abrasiva** *f* ING MECÁ, MECÁ, PROD grinding paste; **~ aceptada** *f* *Esp* (*cf pasta depurada AmL*) PAPEL accepted stock; **~ adiamantada** *f* METAL diamond paste; **~ de alto rendimiento** *f* PAPEL high-yield pulp; **~ amarilla de paja** *f* PAPEL yellow straw pulp; **~ de bambú** *f* PAPEL bamboo pulp; **~ blanqueada** *f* PAPEL bleached pulp; **~ de carbón** *f* REVEST coal prepared and blended; **~ de cemento** *f* CONST cement slurry; **~ concentrada** *f* PAPEL thick stock; **~ de coníferas** *f* PAPEL softwood pulp; **~ depurada** *f* *AmL* (*cf pasta aceptada Esp*) PAPEL accepted stock; **~ de descarga** *f* PAPEL brown stock; **~ destintada** *f* PAPEL de-inked paper stock; **~ detergente** *f* DETERG detergent paste; **~ para enmascarar** *f* COLOR masking paste; **~ de esparto** *f* PAPEL esparto pulp; **~ de frondosas** *f* PAPEL hardwood pulp; **~ para grabado profundo** *f* C&V deep etching paste; **~ para grabar** *f* C&V stamp etching paste; **~ grasa** *f* PAPEL wet stock; **~ kraft** *f* PAPEL kraft pulp; **~ libre de cloro** *f* PAPEL chlorine-free pulp; **~ de madera** *f* CARBÓN pulp, PAPEL wood pulp, PROD pulp, RECICL paper pulp; **~ magra** *f* PAPEL free stock, thin stock; **~ mecánica** *f* EMB mechanical pulp, PAPEL groundwood, mechanical pulp, mechanical woodpulp, wood pulp; **~ mecánica parda** *f* PAPEL brown mechanical pulp board; **~ mecánica de refino** *f* (*RMP*) PAPEL refiner mechanical pulp (*RMP*); **~ mecánica de refino con tratamiento en caliente** *f* PAPEL thermal refiner mechanical pulp; **~ de mercado** *f* PAPEL market pulp; **~ noble** *f* PAPEL dissolving pulp; **~ de paja** *f* PAPEL straw pulp; **~ de papel** *f* RECICL paper pulp; **~ poco refinada** *f* PAPEL fast pulp; **~ para pulir** *f* ING MECÁ, MECÁ, PROD grinding paste; **~ pulverizada** *f* ING MECÁ, MECÁ, PROD grinding paste; **~ que se factura** *f* PAPEL invoiced mass; **~ químico-mecánica** *f* PAPEL chemigroundwood; **~ químico-termomecánica** *f* PAPEL chemicothermo mechanical pulp (*CTMP*); **~ seca** *f* PAPEL dry pulp; **~ seca al aire** *f* PAPEL air-dry pulp; **~ de sellado** *f* CONST sealing compound; **~ semiblanqueada** *f* PAPEL semibleached pulp; **~ semiquímica** *f* PAPEL semichemical pulp; **~ a la sosa** *f* PAPEL soda pulp; **~ al sulfito fest** *f* PAPEL fest-sulfite cellulose (*AmE*), fest-sulphite cellulose (*BrE*); **~ en suspensión** *f* PAPEL stock, stuff; **~ termomecánica** *f* (*pasta TMP*) PAPEL thermomechanical pulp (*TMP*); **~ TMP** *f* (*pasta termomecánica*) PAPEL TMP (*thermomechanical pulp*); **~ de trapos** *f* PAPEL rag pulp; **~ para vender** *f* PAPEL market pulp pasteca *f* ING MECÁ return block; ~ **grande** *f* ING MECÁ bull block

pasteurización *f* ALIMENT, CALIDAD, QUÍMICA, TERMO pasteurization

pasteurizado *adj* ALIMENT, CALIDAD, QUÍMICA, TERMO pasteurized

pasteurizar *vt* ALIMENT, CALIDAD, QUÍMICA, TERMO pasteurize

pasticultura *f* AGRIC grass management

pastilla *f* AUTO pad, CARBÓN *Esp* (*cf pella AmL*) pellet, ELECTRÓN *semiconductores* wafer, PAPEL knot, *de fibras* lump, VEH *freno de disco* pad; ~ **abrasiva** *f* ING MECÁ honing stone; ~ **de circuito integrado** *f* ELECTRÓN integrated-circuit wafer; ~ **de combustible nuclear** *f* CARBÓN *reactor nuclear* pellet; ~ **cruda** *f* NUCL green pellet; ~ **con cuatro hileras de conexiones** *f* (*PCHC*) ELECTRÓN quad-in-line package (*QUIP*); ~ **desechable** *f* ING MECÁ throwaway tip; ~ **de dióxido de uranio** *f* NUCL uranium dioxide pellet; ~ **dispersora** *f* ELECTRÓN flip chip; ~ **epitaxial** *f* ELECTRÓN epitaxial wafer; ~ **de extracción** *f* ALIMENT expeller cake; ~ **de una fila de conexiones** *f* INFORM&PD single in-line package (*SIP*); ~ **de freno** *f* VEH brake pad; ~ **del freno de disco** *f* AUTO disc brake pad (*BrE*), disk brake pad (*AmE*); ~ **de fricción** *f* AUTO, VEH pad; ~ **de memoria** *f* ELECTRÓN bulk wafer; ~ **de presión** *f* AUTO pressure pad; ~ **con revestimiento de protección** *f* ELECTRÓN, REVEST resist-coated wafer; ~ **de semiconductor** *f* ELECTRÓN semiconductor wafer; ~ **de silicio** *f* ELECTRÓN silicon wafer; ~ **de UO₂** *f* NUCL UO₂ pellet; ~ **de UO₂-Gd₂-O₃** *f* (*pastilla de urania-gadolinia*) NUCL UO_2-Gd_2-O_3 pellet (*urania-gadolinia pellet*); ~ **de urania-gadolinia** *f* (*pastilla de UO_2-Gd_2-O_3*) NUCL urania-gadolinia pellet (*UO_2-Gd_2-O_3 pellet*)

pastillo: ~ **de invierno** *m* AGRIC annual bluegrass

pasto: ~ **azul de Kentucky** *m* AGRIC *malezas* bluegrass; ~ **colorado** *m* AGRIC barnyard grass; ~ **cuaresma** *m* AGRIC hairy crab grass; ~ **picado** *m* AGRIC minced grass; ~ **sudán** *m* AGRIC sudan grass; ~ **valcheta** *m* AGRIC barren brome; ~ **verde cortado fino** *m* AGRIC green chop

pastoreo *m* AGRIC grazing; ~ **en franjas** *m* AGRIC strip grazing; ~ **mixto** *m* AGRIC mixed grazing; ~ **sectorial** *m* AGRIC spot grazing

pastura *f* AGRIC pasture; ~ **nativa** *f* AGRIC native pasture; ~ **permanente** *f* AGRIC permanent pasture; ~ **renovada** *f* AGRIC renovated pasture; ~ **de rotación** *f* AGRIC *forraje* ley

pata *f* ING MECÁ leg, INSTR *de gafas* side; ~ **de conejo** *f* TRANSP MAR *cabo* back splice; ~ **de eslinga** *f* PROD sling dog; ~ **de gallo** *f* AGRIC barnyard grass, orchardgrass; ~ **de ganso** *f* TRANSP MAR *cabo* back splice; ~ **de perro** *f* TEC PETR *perforación* dogleg; ~ **de sujeción** *f* PROD *de carruaje de sierra para troncos* clamping dog; ~ **telescópica** *f* FOTO telescopic leg; ~ **del tren de aterrizaje** *f* TRANSP AÉR landing gear leg; ~ **del tren de aterrizaje del morro** *f* TRANSP AÉR nose gear leg; ~ **del trípode** *f* CINEMAT, FOTO, INSTR tripod leg

patada *f* TEC PETR *perforación* kick

patentado *adj* ING MECÁ, TEXTIL patented

patente *f* ING MECÁ, TEXTIL patent; ~ **de invención** *f* ING MECÁ patent; ~ **de navegación** *f* TRANSP MAR certificate of registry, ship's register; ~ **de perfeccionamiento** *f* ING MECÁ patent of improve-ment; ~ **de sanidad** *f* TRANSP MAR *documentación del barco* bill of health

patentes: ~ **solicitadas** *fra* TEXTIL patents applied for

patilla *f* ING ELÉC pin; ~ **de conexión a tierra** *f* ELEC, ELECTRÓN, PROD earth pin (*BrE*), ground pin (*AmE*); ~ **de gafas** *f* INSTR temple spectacles

patín *m* ELEC *conmutador* collector shoe, ING MECÁ pan, gib, runner, MECÁ skid, shoe, PROD *cuba de amalgamación* sole, *de chumacera* sole, TEC PETR *perforación* skid; ~ **de aterrizaje** *m* TRANSP AÉR landing skid; ~ **de chorro** *m* TEC PETR jet sled; ~ **de cola** *m* TRANSP AÉR tailskid; ~ **de cruceta** *m* ING MECÁ crosshead gib; ~ **de la cruceta** *m* ING MECÁ guide block, crosshead shoe, crosshead slipper; ~ **de freno** *m* AUTO, ING MECÁ, MECÁ, VEH brake pad, brake shoe; ~ **del pantógrafo** *m* FERRO, ING MECÁ pantograph slipper; ~ **de soporte** *m* TRANSP rest skid

pátina *f* QUÍMICA patina

patinaje: ~ **estacionario de la rueda** *m* AUTO wheel slip

patinamiento: ~ **del embrague** *m* AUTO, VEH clutch slip

patinar¹ *vt* TRANSP, TRANSP AÉR skid

patinar² *vi* VEH skid

patinazo *m* TRANSP, TRANSP AÉR, VEH skidding

patio *m* CONST yard; ~ **de cajas** *m* PROD box yard; ~ **de carga** *m* FERRO, VEH freight yard (*AmE*), goods yard (*BrE*); ~ **de mercancías** *m* FERRO, VEH freight yard (*AmE*), goods yard (*BrE*); ~ **de la mina** *m* MINAS pit bank; ~ **de tendido** *m* C&V laying yard

patógeno *adj* HIDROL pathogenic

patrón *m* AGRIC *injerto vegetal* rootstock, CARBÓN sample, ELEC *medición, fabricación* template, INFORM&PD template, pattern, ING MECÁ template, standard, MECÁ template, PROD *plantilla* patron, TEC ESP mock-up, pattern, TELECOM standard, TEXTIL *costura* pattern, TRANSP MAR *marina mercante* shipmaster; ~ **de ajuste de barras** *m* AmL (*cf carta de ajuste de barras Esp*) TV bar pattern; ~ **de calibración** *m* CONTAM *análisis instrumental* model calibration; ~ **de comprobación** *m* CONTAM *análisis instrumental* model calibration; ~ **de crecimiento** *m* CRISTAL, METAL growth pattern; ~ **de difracción** *m* METAL diffraction pattern; ~ **para duplicar** *m* PROD stencil; ~ **espacial** *m* CONTAM spatial pattern; ~ **estándar** *m* TV standard pattern; ~ **de la estructura del ojo** *m* TELECOM eye-shape pattern; ~ **de flujo** *m* FÍS FLUID flow pattern; ~ **de fractura** *m* C&V fracture pattern; ~ **de frecuencia** *m* ELECTRÓN frequency standard; ~ **de frecuencia del haz atómico** *m* NUCL atomic beam frequency standard; ~ **industrial** *m* PROD commercial standard; ~ **de interferencia** *m* FÍS ONDAS interference pattern; ~ **de Laue** *m* CRISTAL Laue pattern, FÍS RAD Laue pattern, Laue diagram; ~ **de medida** *m* ING MECÁ standard; ~ **de normas** *m* TV standard pattern; ~ **de ondas estacionarias** *m* FÍS ONDAS stationary wave pattern; ~ **original** *m* ELECTRÓN master pattern; ~ **de pesca** *m* OCEAN fish finder; ~ **de prueba** *m* CINEMAT test pattern; ~ **de prueba de barras de color** *m* AmL (*cf carta de prueba de barras de color Esp*) TV color-bar test pattern (*AmE*), colour-bar test pattern (*BrE*); ~ **de prueba electrónico** *m* ELECTRÓN electronic test-pattern; ~ **de radiación** *m* FÍS, FÍS RAD, ÓPT, TEC ESP, TELECOM, TV radiation pattern; ~ **radiactivo** *m* FÍS RAD radioactive standard; ~ **de referencia** *m* ING

MECÁ reference gage (*AmE*), reference gauge (*BrE*); ~ **de Segrè** *m* FÍS PART Segrè chart; ~ **suspendido** *m* IMPR stop pattern; ~ **de verificación** *m* CONTAM *análisis instrumental* model calibration

patrullera *f* TRANSP MAR *policía* patrol boat

pausa *f* ACÚST *música* rest, ING MECÁ stop, TV pause

pauta *f* TELECOM standard; ~ **meteorológica** *f* METEO, TEC ESP weather pattern; ~ **del tiempo** *f* PROD time key; ~ **del tráfico aéreo** *f* TRANSP AÉR air traffic pattern

pavimentación *f* CONST metaling (*AmE*), metalling (*BrE*), paving

pavimentadora *f* TRANSP pavement-spreading machine

pavimento *m* AmL (*cf adoquinado Esp*) CONST, TRANSP pavement (*BrE*), sidewalk (*AmE*); ~ **de ladrillo** *AmL* *m* CONST brick paving; ~ **móvil** *m* AmL TRANSP moving pavement (*BrE*), moving sidewalk (*AmE*); ~ **móvil articulado** *m* AmL TRANSP articulated-type moving pavement (*BrE*), articulated-type moving sidewalk (*AmE*); ~ **móvil cubierto** *m* AmL TRANSP cabin-type moving pavement (*BrE*), cabin-type moving sidewalk (*AmE*); ~ **de piedras** *m* AmL CONST pebble paving; ~ **rodante** *m* AmL TRANSP moving pavement (*BrE*), moving sidewalk (*AmE*); ~ **rodante articulado** *m* AmL TRANSP articulated-type moving pavement (*BrE*), articulated-type moving sidewalk (*AmE*); ~ **rodante cubierto** *m* AmL TRANSP cabin-type moving pavement (*BrE*), cabin-type moving sidewalk (*AmE*); ~ **rodante tipo cinta** *m* AmL TRANSP belt-type moving pavement (*BrE*), belt-type moving sidewalk (*AmE*); ~ **soporte de pista** *m* AmL REFRIG rink floor

Pb *abr* (*plomo*) METAL, MINERAL, QUÍMICA Pb (*lead*)

PB *abr* (*polibutileno*) P&C PB (*polybutylene*)

PBX *abr* (*central telefónica privada*) TELECOM PBX (*private branch exchange*)

PC[1] *abr* INFORM&PD, TELECOM (*computador personal AmL, computadora personal AmL, ordenador personal Esp*) PC (*personal computer*)

PC[2]: ~ **de entrada y salida local** *m* PROD local I/O PC

PCI *abr* (*placa de circuito impreso, tablero de circuito impreso, tarjeta de circuito impreso*) ELEC, ELECTRÓN, INFORM&PD, TELECOM, TV PCB (*printed-circuit board*)

PCM *abr* GEN (*modulación por códigos de pulsos, modulación por impulsos codificados*) PCM (*pulse-code modulation*)

p-cresol *m* QUÍMICA p-cresol

PD *abr* (*procesamiento de datos, proceso de datos*) INFORM&PD, TELECOM DP (*data processing*)

Pd *abr* (*paladio*) METAL, QUÍMICA Pd (*palladium*)

PDM *abr* (*modulación por duración de los impulsos*) TEC ESP PDM (*pulse-duration modulation*)

PDR *abr* (*revisión preliminar de diseño*) TEC ESP PDR (*preliminary-design review*)

PE *abr* (*codificación de fase*) INFORM&PD PE (*phase encoding*)

peana *f* PROD, TRANSP pedestal

pechblenda *f* GEOL, MINERAL, QUÍMICA pitchblende; ~ **variada** *f* NUCL varied pitchblende

pectasa *f* QUÍMICA pectase

pectato *m* QUÍMICA pectate

péctico *adj* QUÍMICA pectic

pectina *f* ALIMENT, QUÍMICA pectin

pectinicultura *f* OCEAN *acuicultura* scallop culture

pectinosa *f* QUÍMICA pectinose

pectizable *adj* QUÍMICA pectizable

pectización *f* QUÍMICA pectization

pectizar *vt* QUÍMICA pectize

pectolita *f* MINERAL pectolite

pectoral *m* ING MECÁ breastplate

pectosa *f* QUÍMICA pectose

pectoso *adj* QUÍMICA pectous

pedal *m* AUTO pedal, ING MECÁ treadle, VEH pedal; ~ **del acelerador** *m* AUTO, VEH accelerator pedal (*BrE*), gas pedal (*AmE*); ~ **autoregulado** *m* ING MECÁ pedal-operated control; ~ **de embrague** *m* AUTO, VEH clutch pedal; ~ **del freno** *m* AUTO, VEH brake pedal; ~ **del timón de dirección** *m* TRANSP AÉR, TRANSP MAR rudder pedal

pedazo *m* MINAS shatter

pedazos: ~ **seleccionados** *m pl* C&V selected chunks

pedernal *m* GEOL flint

pedestal *m* CONST tower, ING MECÁ base, bearing block, bracket, MECÁ bracket, pedestal, TV pedestal; ~ **de cojinete de bolas** *m* ING MECÁ ball-bearing plummet block; ~ **de control** *m* TRANSP AÉR control pedestal; ~ **para el lanzamiento** *m* TEC ESP launch pad; ~ **de oscilación** *m* MINAS *apoyo vigas* rocker; ~ **de roca** *m* CARBÓN rock shoe; ~ **de tornillo** *m* CONST screw shoe

pedido *m* INFORM&PD, TELECOM req, request; ~ **de almacén** *m* PROD warehouse order; ~ **de servicio** *m* TELECOM request for service

pedidos: ~ **pendientes** *m pl* PROD backlog of orders

pedimento *m*: ~ **del arrecife** *m* GEOL reef pediment; ~ **desértico** *m* GEOL desert pediment

pedir: ~ **socorro** *vi* SEG call for help

pedología *f* CARBÓN soil science, *medicina* edaphology

pedómetro *m* FÍS pedometer

pedraplén *m* AGUA rock fill, CONST riprap, MINAS *presa* rock work

pedregoso *m* CARBÓN *terrenos* gravel

pedrera *f* MINAS stone pit

pedrisco *m* METEO hailstone, hailstorm

pedruscos *m pl* Esp (*cf ganga AmL*) MINAS stone

pega *f* MINAS shot, blasting, *voladuras* round; ~ **del antepecho** *f* MINAS bench blasting; ~ **de banco** *f Esp* (*cf pega de la grada recta AmL*) MINAS bench blasting; ~ **de barrenos** *f* MINAS shot-firing, shotfiring, *voladuras* shot-by-shot firing; ~ **de barrenos eléctrica** *f* MINAS electric shot-firer, electric shot-firing; ~ **eléctrica** *f* ELEC electrical blasting, MINAS electrical blasting, *de barrenos* electric blasting; ~ **de la grada recta** *f* AmL (*cf pega de banco Esp*) MINAS bench blasting; ~ **del piso** *f* MINAS bench blasting; ~ **de pozos profundos** *f* MINAS long-hole blasting; ~ **por rotación** *f* MINAS rotation firing; ~ **secundaria** *f* MINAS secondary blasting; ~ **simultánea** *f* MINAS simultaneous shot firing; ~ **de varios barrenos en secuencia** *f* MINAS delay firing

pegado[1] *adj* IMPR pasted

pegado[2] *m* PAPEL pasting

pegador *m* CARBÓN shotfirer, MINAS shooter

pegamento *m* CONST cement, EMB *adhesivo* adhesive glue, P&C adhesive, TEC ESP cement; ~ **sin ácido** *m* FOTO acid-free glue; ~ **neutro** *m* FOTO acid-free glue; ~ **vegetal** *m* PROD vegetable glue

pegar *vt* CONST stick; ~ **con goma** *vt* PROD gum; ~ **en seco** *vt* FOTO dry-mount

pegarse *v refl* EMB adhere, PROD stick

peinado[1] *adj* TEXTIL combed

peinado[2] *m* ING MECÁ *roscado en el torno* chasing, TEXTIL *hilatura* top; ~ **mediante turbina** *m* TEXTIL turbo-top

peinadora *f* TEXTIL combing machine; ~ **circular** *f* TEXTIL circular combing machine; ~ **rectilínea** *f* TEXTIL rectilinear combing-machine

peinar *vt* TEXTIL comb

peinazo *m* CONST ledge; ~ **de cerradura** *m* CONST lock rail; ~ **superior** *m* CONST top rail

peine *m* ACÚST *de aguja del gramófono* rake, TEXTIL comb, *tejer* reed; ~ **de la cruz** *m* TEXTIL *dispositivo de tejeduría* leasing reed; ~ **desprendedor** *m* TEXTIL doffer comb; ~ **de roscar** *m* ING MECÁ screw tool, thread chaser; ~ **para roscar** *m* ING MECÁ chaser; ~ **del urdidor** *m* TEXTIL warping reed

pelado *adj* ALIMENT husked, peeled, shelled, ING ELÉC bare

peladura *f* TEXTIL peeling

pelar *vt* AGRIC husk, ALIMENT peel, shell, *granos* husk

pelargonato *m* QUÍMICA pelargonate

pelargónico *adj* QUÍMICA pelargonic

peldaño *m* CONST *de escalera* landing step, step, tread, *de escalera de mano* rung, round, PROD *de escalera* step, *escala* rung; ~ **de abanico** *m* CONST *escaleras* winder; ~ **de contador decreciente** *m* PROD down counter rung; ~ **extremo** *m* PROD end rung; ~ **de retenida** *m* PROD latch rung

peletierina *f* QUÍMICA pelletierine

peletización *f* P&C pelletizing

película *f* GEN film; ~ **de acetato** *f* EMB acetate film; ~ **de acetato de celulosa** *f* EMB cellulose acetate film; ~ **de acetato laminada** *f* EMB acetate laminate; ~ **con acolchado de burbujas** *f* EMB bubble film; ~ **adherente** *f* EMB, REVEST adhesive film, cling film; ~ **adhesiva** *f* EMB, REVEST adhesive film, cling film; ~ **aislante** *f* REVEST insulating film; ~ **animada** *f* CINEMAT animated film; ~ **antiácida** *f* REVEST anti-acid film; ~ **anti-corrosiva** *f* EMB anticorrosive film; ~ **anticuada** *f* CINEMAT outdated film; ~ **antivaho** *f* EMB antifog film; ~ **de archivo** *f* CINEMAT archival film; ~ **autoadhesiva** *f* EMB self-adhesive film; ~ **con banda incorporada** *f* CINEMAT striped film; ~ **con banda magnética** *f* CINEMAT magnetic-striped film; ~ **con banda sonora previamente colocada** *f* CINEMAT prestriped film; ~ **barrera** *f* EMB barrier film; ~ **base** *f* TV base film; ~ **en blanco y negro** *f* CINEMAT, FOTO black and white film; ~ **en bobina** *f* CINEMAT, FOTO, IMPR, TV roll film; ~ **calandrada** *f* P&C calendered film; ~ **de capa simple** *f* CINEMAT single-layer film; ~ **captadora** *f* CINEMAT, TV film pick-up; ~ **de carbón** *f* ING ELÉC carbon film; ~ **en carrete** *f* CINEMAT, FOTO, IMPR, TV roll film; ~ **de celofán** *f* EMB cellophane film; ~ **cinematográfica** *f* CINEMAT motion picture, movie; ~ **coextrusionada** *f* EMB co-extruded film; ~ **de cola** *f* EMB glue film; ~ **color para luz artificial** *f* CINEMAT artificial-light color film (*AmE*), artificial-light colour film (*BrE*); ~ **en color para luz artificial** *f* CINEMAT, FOTO artificial-light color film (*AmE*), artificial-light colour film (*BrE*); ~ **en color para la luz del día** *f* CINEMAT, FOTO daylight color film (*AmE*), daylight colour film (*BrE*); ~ **de color de múltiples capas** *f* IMPR multiple-layer color film (*AmE*), multiple-layer colour film (*BrE*); ~ **en color de tipo A** *f* CINEMAT A-type color film (*AmE*), A-type colour film (*BrE*);

~ **de condensador** *f* ELEC capacitor film; ~ **para conformar por estiramiento** *f* EMB stretch film; ~ **contraíble** *f* EMB, TERMO shrink film (*BrE*), shrink wrap (*AmE*); ~ **para copia** *f* CINEMAT print film; ~ **para copias duplicadas** *f* FOTO duplicating film; ~ **delgada de metal imperfectamente unida a la superficie** *f* CARBÓN *aceros* shell; ~ **delgada resistente** *f* ELECTRÓN resistive thin film; ~ **depositada en vacío** *f* ELECTRÓN vacuum-deposited film; ~ **descendente** *f* DETERG falling film; ~ **de 16 mm doble** *f* CINEMAT double 16 film (*AmE*), double 16 stock (*BrE*); ~ **para documentos** *f* FOTO document film; ~ **de dos capas** *f* CINEMAT bipack; ~ **dosimétrica** *f* FÍS RAD film dosimeter; ~ **para duplicación directa** *f* FOTO direct duplicating film; ~ **para duplicados** *f* CINEMAT printing stock; ~ **para embutir** *f* EMB deep drawing film; ~ **de emulsión triple** *f* CINEMAT three-emulsion film; ~ **encogible** *f* EMB, TERMO shrink film (*BrE*), shrink wrap (*AmE*); ~ **para estar en contacto con alimentos** *f* ALIMENT, EMB food-grade film; ~ **estereoscópica** *f* CINEMAT stereoscopic film; ~ **para etiquetas** *f* EMB label film; ~ **sin exponer** *f* CINEMAT, FOTO unexposed film; ~ **extruida** *f* P&C extruded film; ~ **de fondo** *f* CINEMAT background film; ~ **fundida** *f* P&C cast film; ~ **gamma** *f* FÍS RAD gamma film; ~ **gammamétrica** *f* FÍS RAD gamma film; ~ **de gran contraste** *f* CINEMAT high-contrast film; ~ **a granel** *f* AmL (*cf película en rollo Esp*) CINEMAT, FOTO, IMPR, TV bulk film, roll film; ~ **gruesa** *f* ELECTRÓN thick film; ~ **de hielo** *f* REFRIG glaze; ~ **infrarroja** *f* FOTO infrared film; ~ **ininflamable** *f* FOTO safety film; ~ **en lata** *f* FOTO bulk film; ~ **para luz del día** *f* CINEMAT, FOTO daylight film; ~ **magnética** *f* CINEMAT magnetic film, magnetic stock, REVEST magnetic film; ~ **magnética delgada** *f* ELECTRÓN magnetic thin film; ~ **membranosa** *f* P&C skin; ~ **metálica** *f* ELECTRÓN metal film; ~ **metálica vidriada** *f* REVEST metal glaze film; ~ **metálica vitrificada** *f* REVEST metal glaze film; ~ **metalizada** *f* EMB metalized film (*AmE*), metallized film (*BrE*); ~ **miniatura** *f* FOTO *de 35 mm* miniature film; ~ **moldeada por colada** *f* P&C cast film; ~ **muda** *f* CINEMAT mute film; ~ **multicapa en color** *f* CINEMAT multilayer color film (*AmE*), multilayer colour film (*BrE*); ~ **muy sensible** *f* CINEMAT high-speed film; ~ **de nitrato** *f* CINEMAT nitrate film; ~ **no inflamable** *f* IMPR safety film; ~ **de 8 mm doble** *f* CINEMAT double 8 film (*AmE*), double 8 stock (*BrE*); ~ **orientada biaxialmente** *f* P&C biaxially-oriented film; ~ **de óxido** *f* REVEST oxide film, rust film; ~ **pancromática** *f* CINEMAT, FOTO panchromatic film; ~ **de paso estrecho** *f* CINEMAT narrow-gage film (*AmE*), narrow-gauge film (*BrE*); ~ **de paso normal** *f* CINEMAT standard gage film (*AmE*), standard gauge film (*BrE*); ~ **de perforación doble** *f* CINEMAT double-perf stock; ~ **de perforaciones simples** *f* CINEMAT single perforation film; ~ **plana** *f* FOTO sheet film; ~ **de polietileno** *f* EMB polyethylene film (*PET film*); ~ **de polipropileno orientada** *f* EMB orientated polypropylene film; ~ **positiva** *f* CINEMAT positive film; ~ **protectora** *f* EMB barrier film, film coating, protective film, REVEST protective film; ~ **protectora de la superficie** *f* EMB surface protection film; ~ **en rama** *f* TV raw film; ~ **rápida** *f* CINEMAT fast film (*AmE*), fast stock (*BrE*); ~ **de rayos X** *f* INSTR X-ray film; ~ **retráctil** *f* EMB shrink film (*BrE*), shrink wrap (*AmE*); ~ **reversible** *f* CINE-MAT reversal film, FOTO reversal film; ~ **reversible a color** *f* CINEMAT, FOTO color reversal film (*AmE*), colour reversal film (*BrE*), reversal-type color film (*AmE*), reversal-type colour film (*BrE*); ~ **rígida** *f* FOTO cut film, TV cut slide; ~ **en rollo** *f Esp* (*cf película a granel AmL*) CINEMAT, FOTO, IMPR, TV bulk film, roll film; ~ **de seguridad** *f* FOTO safety film; ~ **de separación** *f* ING MECÁ peel shim; ~ **soplada** *f* P&C *producto semielaborado* blown film; ~ **de soporte de seguridad** *f* CINEMAT safety film; ~ **de televisión** *f* TELECOM, TV television film; ~ **para termoconformar** *f* EMB skin film; ~ **termocontraíble** *f* EMB, TERMO heat-shrinkable film (*BrE*), heat-shrinkable wrap (*AmE*); ~ **termoencogible** *f* EMB, TERMO heat-shrinkable film (*BrE*), heat-shrinkable wrap (*AmE*); ~ **tipo barrera de poliolefina** *f* EMB polyolefin barrier-film; ~ **transparente** *f* EMB transparent film; ~ **tripack** *f* FOTO tripack film; ~ **tripack integral** *f* CINEMAT integral tripack; ~ **de tungsteno** *f* CINEMAT tungsten film; ~ **velada** *f* CINEMAT, FOTO fogged film; ~ **virgen** *f* CINEMAT raw film, raw stock, stock, unexposed film, FOTO raw stock, unexposed film

peliculígeno *adj* P&C *aglutinante de pintura* film-forming

peligro[1]: **en** ~ *adj* TRANSP MAR *buque* in distress

peligro[2] *m* AUTO, CALIDAD hazard, SEG danger, TRANSP AÉR hazard; ~ **del ambiente** *m* CALIDAD environmental hazard; ~ **de asfixia** *m* SEG risk of suffocation; ~ **de belios** *m* C&V belshazzar; ~ **biológico** *m* SEG biological hazard; ~ **causado por ondas radioeléctricas** *m* SEG radio-wave hazard; ~ **causado por ultrasonidos** *m* SEG ultrasonic hazard; ~ **de choque con aves** *m* TRANSP AÉR bird-strike hazard; ~ **climatológico** *m* METEO, SEG climatic hazard; ~ **de corte con sierra tronzadera** *m* PROD, SEG crosscut-saw hazard; ~ **eléctrico** *m* ELEC, SEG electrical hazard; ~ **de incendio** *m* CONST, SEG, TERMO, TRANSP AÉR, TRANSP MAR fire hazard; ~ **de incendio eléctrico** *m* ELEC, SEG electrical fire risk; ~ **de mar** *m* TRANSP MAR maritime peril; ~ **mayor** *m* CALIDAD, SEG major hazard; ~ **mecánico** *m* MECÁ, SEG mechanical hazard; ~ **microbiológico** *m* SEG microbiological hazard; ~ **no mecánico** *m* SEG nonmechanical hazard; ~ **originado por la carpintería** *m* SEG woodworking machinery hazard; ~ **originado por la maquinaria** *m* SEG machinery hazard; ~ **originado por sierra de cinta** *m* SEG narrow-bandsaw hazard; ~ **originado por soldadura** *m* SEG welding hazard; ~ **originado por soldadura fuerte** *m* SEG brazing hazard; ~ **originado por vibración** *m* SEG vibration hazard; ~ **originado por vibraciones transmitidas manualmente** *m* SEG hand-transmitted vibration hazard; ~ **patógeno** *m* SEG pathogenic hazard; ~ **potencial** *m* SEG potential hazard; ~ **profesional** *m* SEG occupational hazard; ~ **químico** *m* QUÍMICA, SEG chemical hazard; ~ **de radiación** *m* NUCL, SEG, TELECOM radiation hazard; ~ **de radiación láser** *m* ELECTRÓN, FÍS RAD, SEG laser radiation hazard; ~ **sanitario** *m* SEG health hazard

peligro[3]: **ser un** ~ *vi* SEG be a safety risk

pelita *f* GEOL pelite

pelítico *adj* GEOL pelitic

pella *f AmL* (*cf pastilla Esp, píldora Esp*) CARBÓN pellet

pellet *m* CARBÓN pellet

pelo *m* C&V hackle (*AmE*), hackle mark (*BrE*), TEXTIL *del tejido* pile, *hebra* hair; ~ **sin cortar** *m* TEXTIL uncut pile; ~ **de óxido metálico** *m* METAL whisker

pelos: ~ **salientes** *m pl* TEXTIL projecting hairs

pelusa *f* ALIMENT bloom, IMPR *polvillo del papel* fuzz, lint, P&C *revestimientos* flock, PAPEL fuzz, *polvillo, fragmentos de fibras* dusting, fluff

peña *f* CARBÓN diamond, PROD *martillo* pane, peen; ~ **bombeada** *f* ING MECÁ ball peen, *martillo* ball pane

penacho *m* CONTAM *evacuación de gases* plume

penalización: ~ **por demora** *f* TEC PETR *comercio* demurrage

peñasco *m* CARBÓN diamond

pendiente[1] *adj* CONST slanting, PROD hanging, outstanding, pending

pendiente[2] *f* CARBÓN bias, slope, CONST cant, dip, downhill slope, drop, falling gradient, grade, pitch, slant, slope, upgrade, FERRO gradient, FÍS slope, GEOFÍS scree, GEOM slant, slope, MINAS dip, TEC ESP inclination; ~ **de arrecife** *f* OCEAN recifal slope; ~ **del arrecife** *f* GEOL, OCEAN reef slope; ~ **ascendente** *f* CONST rising gradient; ~ **de buzamiento** *f* GEOL dip slope; ~ **de la curva de sustentación** *f* TRANSP AÉR lift curve slope; ~ **descendente** *f* CONST downward gradient; ~ **empinada** *f* CONST steep gradient; ~ **de x por 1000** *f* CONST gradient of x in 1000; ~ **geotérmica** *f* ENERG RENOV, GEOFÍS, GEOL, TEC PETR geothermal gradient; ~ **inclinada** *f* GEOL dip slope; ~ **pequeña** *f* CONST low gradient; ~ **predeltaica** *f* OCEAN pre-delta slope; ~ **submarina** *f* OCEAN submarine slope; ~ **del tejado** *f* CONST pitch of roof, roof pitch; ~ **del terreno** *f* CONST fall of earth, fall of ground; ~ **transversal** *f* CONST *carreteras* crossfall; ~ **de x por ciento** *f* CONST gradient of x per cent

pendolón *m* CONST *estructura de la cubierta* king rod, ING MECÁ joggle post; ~ **de una cercha** *m* ING MECÁ joggle piece

pendulación *f* ING MECÁ swinging

pendular *adj* MECÁ pendular

penduleo *m* ELEC, TRANSP AÉR hunting

péndulo *m* FÍS, GEOFÍS, MECÁ pendulum; ~ **de compensación** *m* FÍS compound pendulum; ~ **compensado** *m* FÍS compensated pendulum; ~ **compuesto** *m* FÍS compound pendulum; ~ **de Foucault** *m* FÍS Foucault pendulum; ~ **de movimiento vertical** *m* FÍS pendulum bob; ~ **simple** *m* FÍS simple pendulum; ~ **torsional** *m* FÍS torsional pendulum

penecontemporáneo *adj* GEOL penecontemporaneous

penetración *f* CONST *herramientas* keenness, ELEC *magnetismo* permeance, ELECTRÓN, MECÁ penetration; ~ **del calor** *f* TERMO heating depth; ~ **de la pantalla** *f* ELECTRÓN screen penetration; ~ **del punzonado** *f* TEXTIL needling penetration; ~ **por revolución del trépano** *f* CARBÓN *sondeos* feed; ~ **del vapor** *f* TERMOTEC vapor permeance (*AmE*), vapour permeance (*BrE*); ~ **en la vasija del reactor** *f* NUCL vessel penetration

penetrar *vt* MECÁ bore

penetrómetro *m* CONST, LAB penetrometer, P&C penetration tester

península *f* HIDROL, TRANSP MAR peninsula

pennina *f* MINERAL pennine

penninita *f* MINERAL penninite

pennyweight *m* METR pennyweight
penol *m* TRANSP MAR yardarm
pentacloruro *m* QUÍMICA pentachloride
pentadieno *m* QUÍMICA pentadiene
pentaédrico *adj* GEOM pentahedral
pentaédro *m* GEOM pentahedron
pentagonal *adj* GEOM pentagonal
pentágono *m* GEOM pentagon
pentaédrico *adj* GEOM pentahedral
pentaedro *m* GEOM pentahedron
pentametilendiamina *f* QUÍMICA cadaverine, pentamethylendiamine
pentano *m* QUÍMICA, TEC PETR pentane
pentanoico *adj* QUÍMICA, TEC PETR pentanoic
pentanol *m* QUÍMICA pentanol
pentanona *f* QUÍMICA pentanone
pentaóxido: ~ **de dinitrógeno** *m* CONTAM nitric oxide (NO), nitrogen pentoxide (N_2O_5)
pentaprisma *m* FOTO pentaprism
pentasulfuro *m* QUÍMICA pentasulfide (*AmE*), pentasulphide (*BrE*)
pentationato *m* QUÍMICA pentathionate
pentatiónico *adj* QUÍMICA pentathionic
pentatómico *adj* QUÍMICA pentatomic
pentavalencia *f* QUÍMICA pentavalence
pentavalente *adj* QUÍMICA pentavalent
penteno *m* QUÍMICA pentene
pentilentetrazol *m* QUÍMICA pentylentetrazol
pentilo *m* QUÍMICA pentyl
pentiofeno *m* QUÍMICA penthiophene
pentita *f* QUÍMICA pentite
pentitol *m* QUÍMICA pentitol
pentlandita *f* MINERAL pentlandite
péntodo *m* ELECTRÓN, FÍS pentode
pentosa *f* QUÍMICA pentose
pentosan *m* QUÍMICA pentosan
pentosazona *f* QUÍMICA pentosazon
pentósido *m* QUÍMICA pentosid
pentosúrico *adj* QUÍMICA pentosuric
pentotal *m* QUÍMICA pentothal
pentóxido *m* QUÍMICA pentoxide
penumbra *f* FÍS penumbra
peón *m* AGRIC farm-hand; ~ **de carga** *m* FERRO freight porter (*AmE*), goods porter (*BrE*); ~ **de mercancías** *m* FERRO freight porter (*AmE*), goods porter (*BrE*); ~ **de patio** *m* TEC PETR roustabout; ~ **de perforación** *m* PETROL, TEC PETR roughneck
peonina *f* QUÍMICA peonin
pepena: ~ **a cielo abierto** *f* AmL (*cf basurero a cielo abierto Esp*) CONTAM below-cloud scavenging
pepita *f* AGRIC core, kernel, CARBÓN *de mineral* slug, MINAS *mineral* nugget, PETROL grapestone, nugget
pepitas: ~ **de mineral y combustible líquido** *f pl* MINAS prills-and-oil
pepsina *f* ALIMENT, QUÍMICA pepsin
pepsino *m* QUÍMICA pepsinum
pepsinógeno *m* QUÍMICA pepsinogen
peptizable *adj* ALIMENT, DETERG, P&C, QUÍMICA peptizable
peptización *f* ALIMENT, DETERG, P&C, QUÍMICA peptization
peptizador *m* ALIMENT, DETERG, P&C, QUÍMICA peptizer
peptizar *vt* ALIMENT, DETERG, P&C, QUÍMICA peptizate, peptize
peptólisis *f* QUÍMICA peptolysis

peptonizable *adj* QUÍMICA peptonizable
pequeña[1]: **a ~ escala** *adj* GEOL small-scale
pequeña[2]: ~ **apertura** *f* ING MECÁ *de una válvula* cracking open; ~ **botavara** *f* TRANSP MAR bumpkin; ~ **burbuja ocluida** *f* P&C *defecto* pinhole
pequeñísimo *adj* PROD ultrasmall
pequeño: ~ **espacio de publicidad en horas de máxima audiencia** *m* TV prime-time slot; ~ **recalentamiento** *m* REFRIG low superheat
pera *f* LAB *material de vidrio* pear-shaped vessel; ~ **de caucho** *f* Esp (*cf bulbo de caucho AmL*) LAB *pipeta*, P&C rubber bulb; ~ **de succión** *f* LAB *pipeta* pipetting bulb
peracético *adj* QUÍMICA peracetic
perácido *m* GEOL, QUÍMICA peracid
peralcalino *adj* GEOL, QUÍMICA peralkaline
peralte *m* CONST camber, cambering, FERRO superelevation of track, *infraestructura* cant, ING MECÁ *vía férrea* cant; ~ **del rail exterior** *m* FERRO, TRANSP superelevation of the outer rail
peraluminoso *adj* GEOL peraluminous
perborato *m* DETERG, QUÍMICA perborate
perbromuro *m* QUÍMICA perbromide
perca *f* ALIMENT, OCEAN *pez* bass
percarbonato *m* QUÍMICA percarbonate
percepción: ~ **de colisión** *f* INFORM&PD collision detection
perceptivo: **no ~** *adj* TELECOM nonmandatory
percha *f* METR *medida* perch, TEXTIL raising machine, TRANSP MAR *mástil, brazo de izado* spar, bumpkin
perchado *m* TEXTIL raising, *acabado* napping
perchar *vt* TEXTIL raise, *operación de acabado* nap
perclorado *adj* QUÍMICA perchlorinated
perclorato *m* QUÍMICA perchlorate; ~ **de amonio** *m* QUÍMICA, TEC ESP ammonium perchlorate
perclórico *adj* QUÍMICA perchloric
percloruro *m* QUÍMICA perchloride
percolación *f* AGUA seepage, ALIMENT percolation, CARBÓN *filtración* creep, seepage, CONST filtration, HIDROL percolation, infiltration, QUÍMICA leaching, seepage, percolation
percolar[1] *vt* AGUA, ALIMENT percolate, CARBÓN seep, CONST filter, seep, HIDROL percolate
percolar[2] *vi* QUÍMICA seep, leach, percolate
percromato *m* QUÍMICA perchromate
percrómico *adj* QUÍMICA perchromic
percusión *f* NUCL *electrón en capa atómica* knock-on; ~ **del extintor** *f* TRANSP AÉR extinguisher percussion; ~ **hidráulica por rotación** *f* MINAS hydraulic rotary percussion drilling
percusor *m* CARBÓN *de un arma* hammer
percutor *m* AmL (*cf martillo Esp*) D&A firing pin, percussion needle, ING MECÁ plunger, striker, MINAS jar; ~ **del sonar** *m* OCEAN *determinaciones por sonar* pinger
percylita *f* MINERAL percylite
perder *vi* MECÁ bleed, TRANSP MAR leak; ~ **el equilibrio** *vi* SEG overbalance
perderse *v refl* HIDROL *las aguas* run to waste; ~ **en un naufragio** *vr* TRANSP MAR be shipwrecked
pérdida *f* ELEC *de sincronismo* loss, *de corriente, carga* leakage, leak, ENERG RENOV *de calor* loss, GAS leak, leakage, ING ELÉC leakage, ING MECÁ, INSTAL HIDRÁUL, ÓPT loss, PROD wastage, waste, TEC PETR *perforaciones*, TELECOM loss, TRANSP AÉR stall; ~ **por absorción** *f* ING ELÉC, TV absorption loss; ~ **de**

acoplador *f* TELECOM coupler loss; **~ en el acoplamiento** *f* ÓPT, TELECOM coupling loss; **~ por acoplamiento** *f* ÓPT coupler loss, coupling loss, TELECOM coupling loss; **~ por acoplamiento de polarización** *f* TELECOM polarization-coupling loss; **~ de aire** *f* AUTO air leak; **~ de alineación de cuadros** *f* TELECOM loss of frame alignment (*LFA*); **~ angular** *f* TELECOM corner loss; **~ del aparato de medida** *f* ELEC meter loss; **~ por apartamento** *f* TELECOM pointing loss; **~ de apuntador** *f* TELECOM loss of pointer (*LOP*); **~ del apuntador** *f* TELECOM loss of pointer (*LOP*); **~ por apuntamiento de la antena** *f* TELECOM aerial-pointing loss (*BrE*), antenna-pointing loss (*AmE*); **~ de azimut** *f* AmL (*cf pérdida azimutal Esp*) TV azimuth loss; **~ azimutal** *f* Esp (*cf pérdida de azimut AmL*) TV azimuth loss; **~ básica del espacio libre** *f* TEC ESP, TELECOM free-space basic loss; **~ por bloqueo** *f* TEC ESP blocking loss; **~ en el cable** *f* ELEC, ING ELÉC cable loss; **~ de calidad** *f* CALIDAD degradation; **~ de la calidad** *f* AGRIC, CALIDAD quality loss; **~ de calor** *f* ELEC *resistor* heat loss, INSTAL HIDRÁUL abstracting of heat, loss of heat, REFRIG, TERMO, TERMOTEC heat loss; **~ de calor bruta** *f* TERMO, TERMOTEC gross heat loss; **~ de calor por efecto Joule** *f* ELEC *resistencia* TERMO, TERMOTEC Joule's heat loss; **~ de calor neta** *f* TERMO, TERMOTEC net heat loss; **~ calorífica** *f* FÍS heat loss; **~ de carga** *f* INSTAL HIDRÁUL drop of pressure, loss of pressure, pressure declination, pressure drop, pressure loss, TEC PETR *mecánica de fluidos* pressure drop; **~ de la carga** *f* ELEC load loss; **~ sin carga** *f* ELEC no-load loss; **~ causada por incendio** *f* SEG loss caused by fire; **~ de circulación** *f* TEC PETR lost circulation, *perforación* loss of returns; **~ en el cobre** *f* ELEC, FÍS, ING ELÉC copper loss; **~ por codo** *f* ING ELÉC bending loss; **~ de color** *f* C&V loss of color (*AmE*), loss of colour (*BrE*); **~ de la compacidad de la vía** *f* FERRO *equipo inamovible* loss of compactness of track; **~ del compresor** *f* TRANSP AÉR compressor stall; **~ por conmutación** *f* ING ELÉC switching loss; **~ de control** *f* INFORM&PD thrashing; **~ por copiado** *f* CINEMAT printing loss; **~ por corrientes parásitas** *f* ELEC, FÍS, TV eddy-current loss; **~ debida a la fricción** *f* ING MECÁ loss due to friction; **~ por desalineación** *f* ÓPT, TELECOM misalignment loss; **~ por desalineación angular** *f* ÓPT, TELECOM angular misalignment loss; **~ de desbordamiento** *f* TEC ESP, TELECOM spillover loss; **~ por desplazamiento lateral** *f* ÓPT, TELECOM lateral offset loss; **~ por desplazamiento longitudinal** *f* ÓPT, TELECOM longitudinal offset loss; **~ por desplazamiento transversal** *f* ÓPT, TELECOM transverse offset loss; **~ de desviación angular relativa** *f* ACÚST relative angular-deviation loss; **~ por desviación longitudinal** *f* ÓPT, TELECOM longitudinal offset loss; **~ dieléctrica** *f* ELEC, FÍS, ING ELÉC dielectric loss; **~ en el dieléctrico** *f* ELEC, FÍS, ING ELÉC dielectric loss; **~ directa** *f* HIDROL direct loss; **~ por disipación** *f* ELECTRÓN, FÍS, ING ELÉC dissipative loss; **~ por efecto Joule** *f* ELEC *calefacción* ohmic loss, *transformador* copper loss, ING ELÉC ohmic loss; **~ de eficacia de los frenos** *f* VEH *forros* brake fade; **~ eléctrica** *f* ELEC electric loss; **~ elevada** *f* TELECOM high loss; **~ del empalme** *f* TELECOM splice loss; **~ en el empalme extrínseco** *f* ÓPT extrinsic-joint loss, extrinsic-junction loss, TELECOM extrinsic-junction

loss; **~ de encendido** *f* CARBÓN ignition loss; **~ energética** *f* ELEC, FÍS, GAS energy loss; **~ de energía** *f* ELEC, FÍS, GAS energy loss, ING ELÉC drain; **~ de energía por efecto del viento** *f* ELEC *máquina* windage loss; **~ de energía electrónica** *f* FÍS RAD electron-energy loss; **~ en el entrehierro** *f* ACÚST, ÓPT gap loss; **~ por escurrimiento** *f* AGRIC seepage loss; **~ de espacio libre** *f* TEC ESP, TELECOM free-space loss; **~ de espacios** *f* TV spacing loss; **~ de espesor** *f* TV thickness loss; **~ por evaporación** *f* AGUA, PROC QUÍ evaporation loss; **~ de fase** *f* PROD phase loss; **~ fija** *f* ELEC *de máquina* fixed loss; **~ de filtrado** *f* PETROL filtrate loss; **~ del flujo de aire a los extremos** *f* ENERG RENOV tip loss; **~ de frecuencia** *f* TV frequency loss; **~ por fricción** *f* ENERG RENOV friction loss, ING MECÁ loss due to friction; **~ hidráulica** *f* NUCL hydraulic loss; **~ en el hierro** *f* ELEC *transformador*, FÍS core loss, iron loss; **~ por histéresis** *f* FÍS, ING ELÉC, P&C *propiedad física, prueba* hysteresis loss; **~ de horneado** *f* ALIMENT baking loss; **~ de humedad** *f* TERMO humidity loss; **~ de imagen fija** *f* TV loss of picture lock; **~ por infiltración de aire** *f* INSTAL TERM air-infiltration loss; **~ de información** *f* INFORM&PD drop-out; **~ por inserción** *f* ACÚST, FÍS, ÓPT, TELECOM insertion loss; **~ de intervalo** *f* TV gap loss; **~ intrínseca de empalme** *f* TELECOM intrinsic joint loss; **~ por inundación** *f* AGUA, ENERG RENOV flood loss; **~ por ionización** *f* FÍS PART, FÍS RAD ionization loss; **~ de Joule** *f* ELEC *resistencia* Joule's heat loss; **~ de lectura** *f* ACÚST reproducing loss; **~ de líneas** *f* ELEC *red de distribución* line loss; **~ de lodo** *f* TEC PETR *perforación* lost circulation, mud loss; **~ longitudinal en el entrehierro** *f* ÓPT longitudinal gap loss; **~ por macrodoblado** *f* ÓPT macrobend loss; **~ de macroflexión** *f* TELECOM macrobend loss; **~ magnética** *f* ING ELÉC magnetic leakage; **~ en el mar** *f* TRANSP MAR marine loss; **~ por microdoblado** *f* ÓPT microbend loss; **~ de microflexión** *f* TELECOM microbend loss; **~ de multitrama** *f* TELECOM loss of multiframe (*LOM*); **~ en el núcleo** *f* ELEC, FÍS core loss; **~ en el núcleo de hierro** *f* ELEC *de transformador* iron loss; **~ de nutriente** *f* ALIMENT nutrient loss; **~ óhmica** *f* ELEC *transformador* copper loss, *calefacción* ohmic loss, FÍS, ING ELÉC ohmic loss; **~ de la pala** *f* TRANSP AÉR blade stall; **~ por las paredes** *f* REFRIG *del calor* wall losses; **~ de potencia** *f* CINEMAT power drop, ELECTRÓN reflection loss, ING ELÉC power loss, ING MECÁ loss of power, PROD power-down; **~ de potencia durante la marcha** *f* PROD power loss ride-through; **~ de potencia del motor** *f* PROD motor power loss; **~ de presión** *f* ING MECÁ deflation, INSTAL HIDRÁUL loss of pressure, pressure decay, pressure loss, VEH deflation, *neumático* flattening; **~ por propagación** *f* FÍS propagation loss; **~ en la punta de la pala** *f* TRANSP AÉR blade-tip stall; **~ de puntería** *f* TEC ESP pointing loss; **~ de puntero** *f* TELECOM loss of pointer (*LOP*); **~ de puntos** *f* IMPR speckle; **~ por radiación** *f* FÍS PART, FÍS RAD radiation loss; **~ de reflexión** *f* ELECTRÓN reflection loss; **~ de reproducción** *f* CINEMAT reproduction loss, TV playback loss, reproduction loss; **~ de retorno óptica** *f* TELECOM optical return loss (*ORL*); **~ de retornos** *f* TEC PETR loss of returns, lost circulation; **~ por retrodifusión** *f* NUCL back diffusion loss; **~ por rozamiento con el aire** *f* ELEC *de máquina* windage loss; ~ **de señal** *f* TELECOM loss of signal (*LOS*); ~ **de separación** *f* TELECOM gap loss; ~ **de sincronización** *f* TV synchronization loss; ~ **sólida** *f* CARBÓN solid exchanger; ~ **de sumidero de calor** *f* NUCL loss of heat sink; ~ **del suministro del exterior** *f* NUCL loss of off-site power; ~ **de sustentación** *f* CARBÓN, TRANSP AER stall; ~ **de tensión** *f* ELEC voltage loss; ~ **térmica** *f* TERMO heat loss; ~ **a tierra** *f* ELEC, ING ELÉC earth fault (*BrE*), earth leakage (*BrE*), ground fault (*AmE*), ground leakage (*AmE*), PROD earth fault (*BrE*), ground fault (*AmE*); ~ **total** *f* HIDROL total loss; ~ **total de corriente alterna** *f* NUCL loss of off-site power, station blackout; ~ **de trama** *f* TELECOM loss of frame (*LOF*); ~ **del transformador** *f* ING ELÉC transformer loss; ~ **de transmisión** *f* TEC ESP, TELECOM transmission loss; ~ **por transmisión** *f* ACÚST, ÓPT transmission loss; ~ **de la unión** *f* TELECOM splice loss; ~ **uniones extrínsecas** *f* TELECOM extrinsic-joint loss; ~ **de velocidad** *f* NUCL velocity loss; ~ **de ventilación** *f* INSTAL TERM ventilation loss; ~ **de voltaje** *f* ELEC, ING ELÉC line drop

pérdidas[1]: **con** ~ *adj* ELEC, ING ELÉC lossy; **sin** ~ *adj* ELEC *dieléctrico* loss-free, ING ELÉC lossless

pérdidas[2]: ~ **acumuladas** *f pl* INFORM&PD walk down; ~ **por penetración** *f pl* TELECOM *radiopropagación* building-penetration loss; ~ **de retorno ópticas** *f pl* TELECOM optical return loss (*ORL*); ~ **en vacío** *f pl* ELEC no-load loss

perdigón *m* CARBÓN pellet

perecedero *adj* MECÁ expendable

pereirina *f* QUÍMICA pereirine

perenne *adj* AGRIC perennial

perfecto: **en** ~ **estado de funcionamiento** *fra* ING MECÁ in working order, in full working order

perfil *m* GEOL contour, log, IMPR, ING MECÁ outline, PETROL log, section, TEC PETR *evaluación de la formación* log, TRANSP MAR *construcción naval* section; ~ **de acero** *m* TRANSP MAR *construcción naval* steel section; ~ **de acero laminado** *m* TRANSP MAR *construcción naval* steel section; ~ **de activación** *m* TEC PETR *evaluación de la formación* activation log; ~ **acústico** *m* TEC PETR acoustic log; ~ **de adherencia del cemento** *m* (*CBL*) TEC PETR *perforación* cement bond log (*CBL*); ~ **aerodinámico** *m* VEH aerodynamic form; ~ **de ala** *m* ENERG RENOV aerofoil (*BrE*), airfoil (*AmE*); ~ **alfa** *m* ÓPT alpha profile; ~ **de alteración variable** *m* ELECTRÓN low-high-low doping profile; ~ **de aspereza** *m* ING MECÁ, MECÁ roughness profile; ~ **del autopotencial** *m* TEC PETR self-potential log; ~ **del bandaje** *m* FERRO tire profile (*AmE*), tyre profile (*BrE*); ~ **batimétrico** *m* OCEAN sounding profile, *levantamientos cartográficos con sonar* bottom profile; ~ **de calibración** *m* PETROL, TEC PETR caliper log (*AmE*), calliper log (*BrE*); ~ **de caracteres** *m* INFORM&PD character outline; ~ **de carril** *m* FERRO *vehículos* rail profile; ~ **de cemento nuclear** *m* PETROL nuclear cement log; ~ **de concentración de impurezas** *m* ELECTRÓN impurity concentration profile; ~ **de contacto** *m* TEC PETR contact log; ~ **convergente** *m* PETROL focused log; ~ **de la costa** *m* OCEAN *navegación* seascape; ~ **cuadrático** *m* ÓPT quadratic profile; ~ **de densidad** *m* TEC PETR density log; ~ **desguarnecido** *m* TEC PETR *sin lodo* strip log; ~ **de detalle** *m* PETROL detail log; ~ **del diente** *m* ING

MECÁ tooth profile; ~ **eléctrico** *m* PETROL electrical log, TEC PETR wireline log, electric log; ~ **de embalaje** *m* EMB packaging profile; ~ **ESI** *m* (*perfil de índice escalonado equivalente*) ÓPT, TELECOM ESI profile (*equivalent step-index profile*); ~ **del fondo** *m* OCEAN *topografía submarina* bottom profile; ~ **fotónico** *m* PETROL photon log; ~ **de fractura** *m* PETROL fracture log; ~ **de franja** *m* TEC PETR strip log; ~ **gamma** *m* ENERG RENOV, GAS, PETROL gamma log, neutron log, neutron logging, TEC PETR gamma log, neutron logging; ~ **gamma-gamma** *m* TEC PETR gamma-gamma log; ~ **de gas** *m* PETROL gas log; ~ **geológico** *m* GEOL geological column; ~ **geotérmico** *m* ENERG RENOV, TEC PETR geothermal log; ~ **gradado** *m* GEOL graded profile; ~ **hidráulico** *m* HIDROL hydraulic profile; ~ **hidrodinámico** *m* TRANSP AÉR hydrofoil; ~ **de impulso** *m* TELECOM pulse profile; ~ **de índice** *m* ÓPT, TELECOM index profile; ~ **de índice escalonado** *m* ÓPT, TELECOM step index profile; ~ **de índice escalonado equivalente** *m* (*perfil ESI*) ÓPT, TELECOM equivalent step-index profile (*ESI profile*); ~ **de índice gradual** *m* FÍS, ING ELÉC, ÓPT, TELECOM graded-index profile; ~ **con índice de pasos** *m* ÓPT, TELECOM step index profile; ~ **de índice potencial** *m* ÓPT, TELECOM power-law index profile; ~ **del índice de refracción** *m* CINEMAT, FÍS, FÍS ONDAS, FOTO, ÓPT, TEC ESP, TELECOM refractive-index profile; ~ **de inducción** *m* PETROL, TEC PETR induction log; ~ **de inducción doble** *m* PETROL dual-induction log; ~ **inverso** *m* GEOFÍS *sísmica* reversed profile; ~ **de inyección** *m* PETROL mud log, mud logging; ~ **laminado** *m* TRANSP MAR *construcción naval* rolled section; ~ **laminado en frío** *m* TERMO cold-rolled joist; ~ **de la leva** *m* VEH *motor* cam profile; ~ **longitudinal** *m* CONST *carreteras* longitudinal section, TRANSP MAR *diseño naval* buttock lines; ~ **por magnetismo nuclear** *m* PETROL nuclear magnetic log; ~ **mixto** *m* GEOL composite log; ~ **neutrón** *m* ENERG RENOV, GAS, PETROL, TEC PETR neutron log, neutron logging; ~ **neutrón impulsado** *m* TEC PETR pulsed-neutron log; ~ **de neutrones** *m* ENERG RENOV, GAS, PETROL, TEC PETR neutron log, neutron logging; ~ **neutrón-gamma** *m* TEC PETR neutron-gamma log; ~ **neutrónico de pared** *m* TEC PETR sidewall neutron log; ~ **neutrón-neutrón** *m* TEC PETR neutron-neutron log; ~ **de la pala** *m* TRANSP AÉR blade profile; ~ **parabólico** *m* ÓPT, TELECOM parabolic profile; ~ **del parámetro** *m* TELECOM parameter profile; ~ **de porosidad** *m* TEC PETR porosity log; ~ **de proximidad** *m* PETROL proximity log; ~ **pseudosónico** *m* TEC PETR pseudo-sonic log, pseudo-sonic profile; ~ **radiactivo** *m* TEC PETR nuclear log, radioactive log; ~ **radiométrico** *m* PETROL radiometric log; ~ **de rayos gamma** *m* TEC PETR gamma-ray log; ~ **de la regla** *m* CONST screed profile; ~ **de resistividad** *m* PETROL, TEC PETR proximity log, resistivity log; ~ **de resonancia magnética nuclear** *m* TEC PETR, nuclear magnetic resonance log (*NMR log*); ~ **sinergético** *m* TEC PETR synergetic log; ~ **sísmico vertical** *m* TEC PETR vertical seismic profile (*VSP*); ~ **sónico** *m* TEC PETR sonic log; ~ **de temperatura del pozo** *m* TEC PETR temperature well logging; ~ **de temperaturas** *m* TERMO temperature profile; ~ **térmico** *m* PETROL temperature log; ~ **transversal** *m* ING MECÁ cross section;

~ **tridimensional** *m* PETROL three-D log; ~ **de velocidad** *m* FÍS velocity profile

perfilado[1] *adj* ING MECÁ, PROD shaped

perfilado[2] *m* ING MECÁ profiling; ~ **eléctrico** *m* PETROL electrical survey; ~ **de resistividad** *m* PETROL, TEC PETR resistivity logging; ~ **del sondeo** *m* GEOL well logging

perfilador *m* ING MECÁ crusher; ~ **del subsuelo marino** *m* OCEAN *sedimentología marina* sub-bottom profiler

perfiladora *f* CONST finisher

perfilaje *m* TEC PETR *evaluación de la formación* logging; ~ **de activación** *m* TEC PETR activation logging; ~ **acústico de pozos** *m* TEC PETR acoustic well logging; ~ **eléctrico especial múltiple** *m* TEC PETR multiple special electrical logging; ~ **geotérmico** *m* ENERG RENOV, TEC PETR geothermal logging; ~ **neutrónico** *m* GAS neutron log, neutron logging; ~ **de permeabilidad** *m* TEC PETR permeability logging; ~ **del pozo** *m* TEC PETR well logging; ~ **de rayos gamma** *m* TEC PETR gamma-ray logging; ~ **de temperaturas** *m* TEC PETR temperature logging

perfilar *vt* IMPR outline

perfilómetro *m* ING MECÁ contour follower

perforabilidad *f* TEC PETR *perforación* drillability

perforable *adj* TEC PETR drillable

perforación *f* AGUA tapping, drilling, C&V perforation, CARBÓN boring, drilling, CINEMAT *de negativo* pinhole, punch, sprocket hole, CONST *carpintería* boring bit, hole, breakthrough, boring, ELEC *de dieléctrico* breakdown, ELECTRÓN punch-through, puncture, FOTO sprocket hole, perforation, GAS bore, drilling, borehole, GEOFÍS, GEOL, HIDROL borehole, INFORM&PD punching, ING ELÉC sink, ING MECÁ boring, MINAS cut, cross driving, cut hole, *de rocas o piedras* boring, *agujeros* borehole, driving, *sondeos* drilling, holing, PAPEL puncture, PETROL borehole, PROD drilling, perforation, boring; ~ **por aire comprimido** *f* TEC PETR air drilling; ~ **de amplio diámetro** *f* CARBÓN large-hole boring; ~ **de arrastre** *f* INFORM&PD sprocket hole; ~ **automática** *f* INFORM&PD *cinta* automatic punch; ~ **a baja temperatura** *f* TERMO low-temperature sinking; ~ **con barrena** *f* CONST *sondeos* boring with the bit; ~ **con barrena de diamantes** *f* TEC PETR diamond drilling; ~ **para barrenar** *f* TEC PETR shot drilling; ~ **con broca de diamantes** *f* TEC PETR diamond drilling; ~ **por cable** *f* PETROL cable drilling, TEC PETR churn drill, cable drilling; ~ **de caída libre** *f* CONST free-fall boring; ~ **de comprobación** *f* AmL (cf *perforación testiguera Esp*) MINAS testing drill; ~ **continua** *f* IMPR *formularios continuos* crash perforation; ~ **por corriente de agua** *f* TEC PETR wash boring; ~ **costa-fuera** *f* TEC PETR offshore drilling; ~ **descendente** *f* MINAS sink; ~ **con desviación** *f* CONST jump drilling; ~ **desviada** *f* TEC PETR deviated drilling, sidetrack drilling; ~ **al diamante** *f* MINAS diamond drilling; ~ **dieléctrica** *f* ELEC dielectric breakdown; ~ **direccional controlada** *f* TEC PETR directional drilling; ~ **dirigida** *f* TEC PETR directional drilling; ~ **doble** *f* CINEMAT double perforation; ~ **enfrentada** *f* MINAS *sondeos* offset; ~ **de evaluación** *f* TEC PETR appraisal drilling; ~ **exploratoria** *f* GAS exploration drilling, exploratory drilling, TEC PETR exploration drilling, exploratory drilling, wildcat drilling; ~ **para la**

extracción de metano *f* CARBÓN methane-draining boring; ~ **para facilitar el desgarro** *f* IMPR *formularios continuos* snap-out perforation; ~ **por fusión** *f* CARBÓN fusion drilling; ~ **por fusión de la roca mediante un soplete** *f* CARBÓN jet piercing; ~ **de galerías en dirección al filón** *f* MINAS drifting; ~ **de galerías transversales** *f* GEOL, MINAS crosscutting; ~ **por granalla** *f* CONST *sondeos* boring by shot drills; ~ **hacia arriba** *f* MINAS *barrenos* dry hole; ~ **de hoyo de disparo** *f* TEC PETR shot-hole drilling; ~ **en hueco no revestido** *f* TEC PETR open-hole drilling; ~ **inclinada** *f* MINAS angled drill, incline hole; ~ **del manto de recubrimiento** *f* CARBÓN overburden drill; ~ **marginal** *f* INFORM&PD sprocket hole; ~ **metalizada** *f* ELECTRÓN plated-through hole; ~ **del negativo** *f* CINEMAT negative perforation; ~ **neumática** *f* PETROL air drilling; ~ **por percusión** *f* CONST *sondeos* boring by percussion, TEC PETR percussion drilling, production drilling; ~ **por percusión con varillas** *f* CONST *sondeos* boring by percussion with rods; ~ **petrolífera no costera situada en la plataforma continental** *f* TEC PETR offshore well; ~ **con plantilla** *f* PETROL template drilling; ~ **positiva** *f* CINEMAT positive perforation; ~ **de pozo** *f* ENERG RENOV well sinking; ~ **a presión** *f* CARBÓN jet drilling; ~ **de producción** *f* Esp (cf *hornillo de mina AmL*) MINAS *voladuras* mining hole; ~ **recapitulativa** *f* INFORM&PD summary punching; ~ **redonda** *f* IMPR round-hole perforating; ~ **de registro de la película** *f* IMPR film register punch; ~ **de resúmenes** *f* INFORM&PD summary punching; ~ **por rotación** *f* CONST *sondeos* boring by rotation; ~ **rotativa** *f* CARBÓN, MINAS, PETROL, TEC PETR rotary drilling; ~ **con sacatestigos** *f* PETROL core drilling; ~ **térmica** *f* CARBÓN jet piercing; ~ **testiguera** *f* Esp (cf *perforación de comprobación AmL*) MINAS testing drill; ~ **con trépano de chorro** *f* TEC PETR jet-bit drilling; ~ **por tubos** *f* Esp (cf *punzado de la cañería AmL*) TEC PETR casing perforation; ~ **de túneles** *f* CONST, MINAS tunneling (*AmE*), tunneling work (*AmE*), tunnelling (*BrE*), tunnelling work (*BrE*); ~ **con turbina** *f* TEC PETR turbine drilling; ~ **Zener** *f* ELECTRÓN Zener breakdown

perforaciones: ~ **de registro** *f pl* CINEMAT, TV registration holes

perforado[1] *adj* MINAS holed; **no** ~ *adj* TELECOM *cinta* unperforated

perforado[2] *m* C&V bore (*BrE*), corkage (*AmE*); ~ **en cadena** *m* ING MECÁ lace punching; ~ **de profundidad** *m* ING MECÁ deep-hole boring

perforador[1] *adj* PROD perforating

perforador[2] *m* CONST auger, ING MECÁ drill, perforator, PETROL, TEC PETR driller; ~ **asistente** *m* PETROL, TEC PETR assistant driller; ~ **ayudante** *m* PETROL, TEC PETR assistant driller; ~ **de martillo** *m* ING MECÁ hammer drill; ~ **manual** *m* INFORM&PD handpunch; ~ **de porcelana** *m* C&V porcelain-borer, porcelain-driller, porcelain-piercer; ~ **del túnel con frente entero** *m* MINAS full-face tunnel borer

perforadora *f* AmL (cf *giratoria de perforación Esp*) CARBÓN drill, rock drill, CONST boring tool, rock drill, rock borer, EMB, IMPR perforating machine, INFORM&PD keypunch, perforator, punch, ING MECÁ, MECÁ boring cutter, boring machine, drilling machine, piercer, MINAS boring machine, rock drill,

drill, swivel, PROD perforator; **~ de aire comprimido** *f* CONST, ING MECÁ, MECÁ air drill, compressed-air drill; **~ de barrenado profundo** *f* MINAS shaft-sinking bar; **~ de cinta** *f* INFORM&PD tape punch; **~ de cinta de papel** *f* INFORM&PD paper-tape punch; **~ de columna** *f* ING MECÁ pillar-drilling machine, MINAS column; **~ de columna de banco** *f* ING MECÁ bench-pillar drilling machine; **~ de cuatro husillos** *f* PROD four-spindle drilling machine; **~ a diamante** *f* GAS diamond drill; **~ dúplex** *f* ING MECÁ duplex boring machine; **~ de mano** *f* INFORM&PD hand-punch, MECÁ hand drill; **~ de martillo** *f* CONST hammer drill; **~ mecánica** *f* MINAS machine drilling; **~ montada sobre carrillo** *f* AmL (*cf carro perforador Esp*) MINAS drill rig; **~ montada sobre embarcación** *f* TEC PETR *perforación costa-fuera* drill barge; **~ múltiple** *f* PROD gang drill; **~ neumática** *f* CONST, ING MECÁ compressed-air drill; **~ de percusión** *f* ING MECÁ hammer drill, percussive drill, plugger, MINAS hammer drill; **~ de percusión a mano** *f* ING MECÁ plug drill, plugger; **~ de percusión manual** *f* ING MECÁ plug drill; **~ portátil** *f* ING MECÁ portable drilling-machine; **~ de resúmenes** *f* INFORM&PD summary punch; **~ para roca** *f* CARBÓN, CONST, MINAS rock drill; **~ a rotación** *f* CARBÓN, CONST, MINAS, PETROL, TEC PETR rotary drill; **~ rotativa para carbón** *f* CARBÓN, MINAS coal drill; **~ sumaria** *f* INFORM&PD summary punch; **~ de tecla** *f* INFORM&PD keypunch; **~ de túneles** *f* CONST tunnel bar, tunnel-boring machine, tunneling machine (*AmE*), tunnelling machine (*BrE*); **~ con válvula de platillo** *f* CONST tappet valve drill

perforar[1]: **sin ~** *adj* TELECOM *cinta* unperforated
perforar[2] *vt* CARBÓN blast, bore, drill, perforate, CONST bite, bore, drill, hole, ING MECÁ bore, pink, MECÁ drill, MINAS crossdrive, hole, drill, bore, PROD pierce, bore; **~ en medio acuífero** *vt* MINAS bore against water
perforar[3]: **~ mar adentro** *vi* TRANSP MAR drill offshore
perforista *m* TEC PETR driller
pergamino: **~ vegetal** *m* PROD vegetable parchment
perhidrol *m* QUÍMICA perhydrol
perhidruro *f* QUÍMICA perhydride
periapsis *f* TEC ESP periapsis
periastro *m* TEC ESP periastron
periastrón *m* TEC ESP periastron
pericia: **~ marinera** *f* TRANSP MAR seamanship
periclasa *f* MINERAL periclase, periclasite
periclase *f* QUÍMICA periclase
periclina *f* GEOL, MINERAL pericline
peridotita *f* PETROL peridotite, peridotyte
peridoto *m* MINERAL peridot
periferia *f* CONST periphery
periférico[1] *adj* ING MECÁ outer
periférico[2] *m* INFORM&PD, TELECOM peripheral
perigeo *m* FÍS, TEC ESP perigee
perihelio *m* ENERG RENOV, FÍS, TEC ESP perihelion
perileno *m* QUÍMICA perylene
perilinfa *f* ACÚST *audición* perilymph
perilla *f* TRANSP MAR *del mástil* truck (*AmE*), lorry (*BrE*); **~ de ajuste de velocidad del obturador** *f* FOTO shutter speed setting knob; **~ con flecha** *f* ING MECÁ pointer knob; **~ con índice** *f* ING MECÁ pointer knob; **~ de mando** *f* ELEC control knob; **~ neumática** *f* LAB blow ball; **~ de trabazón** *f* INSTR locking knob; **~ del zoom** *f* TV zoom lever

perímetro *m* CONST contour, perimeter, GEOM perimeter; **~ de explosión** *m* MINAS perimeter blasting
periodicidad *f* CRISTAL, ELECTRÓN periodicity
periódico *adj* ACÚST, ELEC cyclic, periodic, ELECTRÓN, FÍS periodic, GEOL, INFORM&PD, ING ELÉC, ING MECÁ cyclic, MATEMÁT recurring, QUÍMICA, TEC PETR cyclic
periodo *ver período*
período *m* ACÚST period, ELEC cycle, ELECTRÓN period, periodic time, FÍS, GEOL period, ING ELÉC, ING MECÁ cycle, PETROL period, PROD *informática* time bucket, TEC ESP range; **~ de aceleración** *m* PROC QUÍ *de partículas* accelerating period; **~ de afino** *m* PROD fining period; **~ de almacenaje** *m* FOTO storage period; **~ de aprobación** *m* CALIDAD approval period; **~ de atenuación del eco** *m* TEC ESP hangover time; **~ de aumento del número de vehículos** *m* TRANSP *control de tráfico* vehicle extension period; **~ axial** *m* TEC ESP axial period; **~ de bloqueo** *m* ING ELÉC blocking period, off period, off-period; **~ de campo activo** *m* TV active field period; **~ de carga reducida** *m* ELEC off-peak period; **~ cargado** *m* TELECOM busy period; **~ de caudal mínimo** *m* HIDROL period of lowest flow; **~ de compresión** *m* INSTAL HIDRÁUL compression period; **~ de conducción** *m* ING ELÉC on period, on-air period; **~ de conservación** *m* ALIMENT, REFRIG keeping time; **~ de corte** *m* ING ELÉC off period, off-period; **~ Cretáceo** *m* GEOL, TEC PETR Cretaceous Period; **~ crítico de sequía** *m* AGRIC drought stress; **~ de curado** *m* CONST cure period; **~ de defectos rápido** *m* CALIDAD early-failure period; **~ de descanso** *m* TRANSP AÉR rest period; **~ de descarburación** *m* PROD boil period; **~ de ejecución** *m* INFORM&PD, PROD run time; **~ empleado** *m* ELECTRÓN clock period; **~ de enfriamiento** *m* C&V cooling-down period, GEOL cooling age, NUCL cooling-down period, REFRIG cooling-down period; **~ erróneo** *m* TELECOM erroneous period (*EP*); **~ de escape** *m* INSTAL HIDRÁUL *motores* release period; **~ de escorificación** *m* PROD *proceso Bessemer* slag formation period; **~ espacial** *m* ELECTRÓN spatial period; **~ de expansión** *m* INSTAL HIDRÁUL expansion period; **~ extenso** *m* OCEAN *derecho europeo* long period; **~ de falla a ritmo constante** *m* CALIDAD constant failure rate period; **~ de fallas rápido** *m* CALIDAD early-failure period; **~ de fallas por uso y desgaste** *m* CALIDAD wear-out failure period; **~ de flujo** *m* TEC PETR *pruebas de producción* draw-down; **~ de forjadura antes de recalentar de nuevo** *m* CARBÓN temperature; **~ fuera de puntas** *m* ELEC off-peak period; **~ de garantía** *m* PROD, TEC ESP guarantee period; **~ de hacer nuevos pedidos** *m* PROD reorder period; **~ húmedo** *m* CONTAM, METEO wet period; **~ de impulso** *m* ELECTRÓN period pulse; **~ inactivo** *m* ELECTRÓN idle period; **~ de indisponibilidad** *m* NUCL outage time; **~ de inducción** *m* METAL induction period; **~ de integración** *m* ING ELÉC integration period; **~ de interrupción** *m* CONST off-air period, GEOFÍS *métodos de prospección* time break, PROD interrupt period; **~ de interrupción del servicio** *m* MECÁ downtime; **~ Jurásico** *m* GEOL, TEC PETR Jurassic period; **~ de la llamada** *m* TELECOM ringing period; **~ sin lluvias** *m* AGRIC, CONTAM, METEO, TRANSP MAR rain-free period; **~ de mantenimiento** *m* CONST maintenance period; **~ de marcación** *m* TELECOM dialing period

(*AmE*), dialling period (*BrE*); **~ de marcha comprobatorio** *m* ING MECÁ test run; **~ de marcha continua** *m* ING MECÁ run; **~ de marcha a potencia reducida** *m* ELEC off-peak period; **~ de muestreo** *m* TEC PETR sample period; **~ de la ola** *m* TRANSP MAR *diseño de barcos* wave period; **~ orbital** *m* TEC ESP orbital period; **~ de oscilación** *m* ELEC *corriente alterna* period of oscillation, ELECTRÓN, GEOFÍS oscillation period; **~ de oscilación libre** *m* GEOFÍS *análisis de ondas* free oscillation period; **~ de paralización por avería** *m* TEC PETR *perforación* downtime; **~ Pérmico** *m* GEOL, TEC PETR Permian Period; **~ de persistencia** *m* TEC ESP hangover time; **~ de pronóstico** *m* PROD forecast period; **~ propio** *m* ELECTRÓN natural period; **~ de reciprocidad** *m* NUCL reciprocal period; **~ de reloj** *m* ELECTRÓN clock period; **~ de repetición de impulsos** *m* ELECTRÓN pulse-repetition period; **~ de reposo** *m* TRANSP AÉR rest period; **~ de semidesintegración** *m* FÍS, FÍS PART half-life, FÍS RAD radioactive half-life, GEOL half-life, NUCL half-life, radioactive half-life, QUÍMICA half-life; **~ de los solsticios** *m* TEC ESP solstitial period; **~ de tiempo** *m* ELECTRÓN time period, PROD time period, time phase, time phasing; **~ de tiempo calmado entre períodos tormentosos** *m* TEC PETR *operaciones costa-fuera* weather window; **~ de trabajo sin interrupción** *m* CARBÓN *alto horno* campaign; **~ Triásico** *m* GEOL, TEC PETR Triassic period; **~ de validez** *m* CONST validity period; **~ vegetativo** *m* AGRIC growing period; **~ verde** *m* TRANSP *control de tráfico* green period

perioduro *m* QUÍMICA periodide
periscopio *m* CINEMAT periscope, snorkel, FÍS, TEC ESP, TRANSP MAR *del submarino* periscope; **~ panorámico** *m* NUCL panorama periscope
peritación *f* CONST survey
peritectoide *m* METAL peritectoid
perito *m* METR, TRANSP MAR *tasador de averías* surveyor
perla *f* C&V, MECÁ bead; **~ de bórax** *f* CONST borax bead
perlas: **~ de coque** *f pl* CARBÓN globular coke
perlasa *f* DETERG pearl ash
perlita *f* PETROL pearlstone, perlite; **~ con cementita esferoidizada** *f* METAL divorced perlite; **~ expandida** *f* TERMOTEC expanded perlite; **~ globular** *f* METAL divorced perlite; **~ granular** *f* METAL divorced perlite; **~ laminar** *f* METAL lamellar perlite; **~ no laminar** *f* METAL nonlamellar perlite
perlítico *adj* PETROL perlic
perlón *m* QUÍMICA perlon
permahielo *m* CARBÓN, GEOL, TEC PETR permafrost
permalloy *m* ELEC, FÍS permalloy
permanencia *f Esp* (*cf estadía AmL*) CONST stay; **~ del color** *f* COLOR, P&C color fastness (*AmE*), colour fastness (*BrE*); **~ del combustible en el núcleo** *f* NUCL in-core fuel life; **~ del movimiento irrotacional** *f* FÍS FLUID permanence of irrotational motion
permanentemente: **~ conectado** *adj* ING ELÉC hardwired
permanganato *m* HIDROL, QUÍMICA permanganate; **~ de potasio** *m* QUÍMICA potassium permanganate
permangánico *adj* QUÍMICA permanganic
permeabilidad *f* GEN permeability, AGUA *de presas* seepage, ENERG RENOV perviousness; **~ absoluta** *f* ELEC, ING ELÉC, PETROL absolute permeability; **~ al**

agua *f* P&C water permeability; **~ al aire** *f* IMPR, TERMOTEC air permeability; **~ compleja** *f* ELEC *circuitos de corriente alterna* complex permeability; **~ efectiva** *f* PETROL effective permeability; **~ de espacio libre** *f* ING ELÉC permeability of free space; **~ específica** *f* ELEC, FÍS, ING ELÉC relative permeability; **~ a los fluidos** *f* FÍS FLUID, TERMOTEC fluid permeability; **~ del gas** *f* GAS, P&C, TERMO gas permeability; **~ intrínseca** *f* ELEC intrinsic permeability; **~ magnética** *f* ELEC, PETROL magnetic permeability; **~ relativa** *f* ELEC, FÍS, ING ELÉC relative permeability; **~ del vacío** *f* FÍS permeability of free space; **~ al vapor** *f* TERMO, TERMOTEC vapor permeability (*AmE*), vapour permeability (*BrE*); **~ de vapor** *f* TERMO, TERMOTEC vapor permeability (*AmE*), vapour permeability (*BrE*)
permeable *adj* GEN permeable; **~ a la grasa** *adj* EMB permeable to grease
permeación *f* ELEC, ING ELÉC permeance
permeámetro *m* CARBÓN permeameter
permeancia *f* ELEC, ING ELÉC permeance
permisividad: **~ eléctrica del vacío** *f* ING ELÉC electric constant
permiso *m* D&A leave, TEC PETR *licencias de explotación* block; **~ de aproximación** *m* TRANSP AÉR approach clearance; **~ de circulación** *m* AUTO, TRANSP, VEH road clearance; **~ de exportación** *m* TRANSP, TRANSP MAR export licence (*BrE*), export license (*AmE*); **~ de importación** *m* TRANSP, TRANSP MAR import licence (*BrE*), import license (*AmE*); **~ de operación** *m* TRANSP AÉR operating permit; **~ de tierra** *m* TRANSP MAR shore leave; **~ de vuelo** *m* TRANSP AÉR flight clearance; **~ de vuelo por instrumentos** *m* TRANSP AÉR instrument rating
permitido *adj* TRANSP AÉR cleared for take-off
permitividad *f* ELEC, FÍS, ING ELÉC, TEC ESP, TELECOM permittivity; **~ absoluta** *f* ELEC, FÍS, ING ELÉC absolute permittivity; **~ del aire** *f* ING ELÉC permittivity of air; **~ específica** *f* ING ELÉC dielectric constant; **~ relativa** *f* ELEC, FÍS, ING ELÉC relative permittivity; **~ del vacío** *f* ING ELÉC permittivity of free space
permonosulfúrico *adj* QUÍMICA permonosulfuric (*AmE*), permonosulphuric (*BrE*)
permuta *f* PROD trade-in
permutación *f* INFORM&PD swapping, permutation, MATEMÁT permutation; **~ a estado de espera** *f* TELECOM changeover to stand-by
permutador: **~ de cationes** *m* CARBÓN cation exchanger; **~ térmico primario** *m* TRANSP AÉR primary heat exchanger; **~ térmico refrigerador del combustible** *m* TRANSP AÉR fuel-coolant heat exchanger
permutar *vt* INFORM&PD swap
pernio *m* ING MECÁ hinge
pernitrato *m* QUÍMICA pernitrate
pernítrico *adj* QUÍMICA pernitric
perno *m* AUTO stud, CONST bolt, hinge, spike, D&A, FERRO bolt, ING MECÁ bolt, fastener, gudgeon, MECÁ bolt, MINAS *Esp* (*cf arrancatubos AmL*), jaula dog, spike, *del bocarte* boss, TRANSP, TRANSP AÉR, VEH bolt; **~ de anclaje** *m* CONST anchor bolt, foundation bolt, rag bolt, ING MECÁ anchor bolt, PROD holding-down bolt; **~ de anclaje lateral** *m* FERRO side-stay bolt; **~ para anclajes** *m* CONST anchoring bolt; **~ de anilla** *m* ING MECÁ, MECÁ eyebolt, PROD eye-headed bolt, VEH *embrague* eye bolt; **~ para apertura**

manual *m* D&A, TRANSP AÉR rip pin; ~ **con argolla** *m* TRANSP MAR *accesorios de cubierta* ring bolt; ~ **de argolla** *m* ING MECÁ, MECÁ eyebolt; ~ **arponado** *m* CONST bar bolt, bat bolt, bay bolt, fang bolt, jag bolt, jagged bolt, sprig bolt; ~ **de arrastre** *m* ING MECÁ lathe dog; ~ **banjo** *m* ING MECÁ banjo bolt; ~ **de barra de enganche** *m* VEH *remolque* drawbar bolt; ~ **de la bisagra** *m* REFRIG hinge spindle; ~ **de brida** *m* ING MECÁ stirrup bolt; ~ **con bridas** *m* ING MECÁ flanged bolt; ~ **sin cabeza** *m* ING MECÁ, MECÁ, PROD stud; ~ **de cabeza avellanada** *m* CONST, ING MECÁ, MECÁ countersunk-head bolt; ~ **sin cabeza de la culata del cilindro** *m* ING MECÁ cylinder-head stud; ~ **de cabeza embutida** *m* PROD flush bolt; ~ **con cabeza en forma de saliente cuadrado** *m* PROD feather-necked bolt; ~ **de cabeza hemisférica** *m* ING MECÁ round-head bolt, round-headed bolt; ~ **de cabeza hexagonal** *m* ING MECÁ hexagonal-head bolt, PROD hex-headed bolt; ~ **de cabeza hexagonal y tuerca hexagonal** *m* ING MECÁ hexagon-head bolt and hexagon nut; ~ **de cabeza de hongo** *m* ING MECÁ mushroom-head bolt; ~ **de cabeza moleteada** *m* ING MECÁ knurled-head fastener; ~ **con cabeza de muletilla** *m* ING MECÁ tee bolt (*T-bolt*); ~ **de cabeza plana** *m* ING MECÁ, PROD flat-head bolt; ~ **de cabeza redonda** *m* ING MECÁ mushroom-head bolt, cup-head bolt, round-head bolt, round-headed bolt; ~ **de cabeza redondeada** *m* ING MECÁ round-head bolt, round-headed bolt; ~ **de cabeza semiesférica** *m* CONST button-head bolt, ING MECÁ button-head bolt, cup-head bolt, round-head bolt, round-headed bolt; ~ **de cabeza en T** *m* ING MECÁ T-headed bolt; ~ **cabezal almenado** *m* ING MECÁ castellated-head fastener; ~ **de cadena** *m* ING MECÁ chain bolt; ~ **de caja de moldear** *m* ING MECÁ flask pin; ~ **con chaveta** *m* ING MECÁ cotter bolt; ~ **de chaveta** *m* ING MECÁ forelock bolt, key bolt; ~ **completamente roscado** *m* ING MECÁ tap bolt; ~ **común** *m* ING MECÁ machine bolt; ~ **de la correa** *m* PROD belt bolt; ~ **de la cruceta** *m* ING MECÁ gudgeon pin (*BrE*), wrist pin (*AmE*), crosshead pin; ~ **de la culata** *m* AUTO *cilindro* cylinder; ~ **de émbolo** *m* MECÁ gudgeon pin (*AmE*), wrist pin (*AmE*); ~ **del émbolo** *m* AUTO piston pin; ~ **de empotramiento** *m* ING MECÁ stone bolt; ~ **de enclavamiento** *m* CONST locking bolt; ~ **de enganche** *m* CONST *cerrajería* shackle; ~ **de entrada y salida** *m* ING MECÁ in-and-out bolt; ~ **estándarizado** *m* ING MECÁ unified bolt; ~ **de fijación** *m* ING MECÁ setscrew, tie bolt; ~ **para flejes** *m* CONST strap bolt; ~ **formando pivote** *m* ING MECÁ kingbolt, kingpin; ~ **de guía** *m* ING MECÁ, MECÁ guide pin; ~ **hecho a máquina** *m* ING MECÁ machine bolt; ~ **hueco** *m* ING MECÁ banjo bolt, hollow bolt; ~ **indicador de la carga** *m* ING MECÁ load-indicating bolt; ~ **para madera** *m* CONST *carpintería* timber dog, ING MECÁ coach bolt, coach screw; ~ **maestro** *m* ING MECÁ kingbolt, main pin; ~ **de montaje** *m* ING MECÁ fitting bolt; ~ **de ojo** *m* ING MECÁ, MECÁ eyebolt; ~ **de orejetas** *m* ING MECÁ wing bolt; ~ **de pala** *m* PROD spade lug; ~ **pasante** *m* ING MECÁ in-and-out bolt, through bolt; ~ **con pestañas** *m* ING MECÁ flanged bolt; ~ **pinzote** *m* ING MECÁ main pin, pintle; ~ **posicionador** *m* ING MECÁ locating screw; ~ **prisionero** *m* ING MECÁ stud bolt; ~ **de protección** *m* SEG security bolt; ~ **pulido** *m* ING MECÁ bright bolt; ~ **de puntal** *m* CONST stay

bolt; ~ **de reborde** *m* PROD lip bolt; ~ **regulable** *m* ING MECÁ expansion bolt; ~ **de resorte** *m* PROD spring bolt; ~ **para roca** *m* CONST rock bolt; ~ **de rosca externa** *m* ING MECÁ external-threaded fastener; ~ **roscado** *m* ING MECÁ screw bolt, MECÁ threaded fastener, VEH threaded bolt; ~ **de sujeción** *m* ING MECÁ fitting bolt, screw dog, tie bolt; ~ **sujetador** *m* ING MECÁ tie bolt; ~ **de suspensión** *m* ING MECÁ suspension stud; ~ **en T** *m* ING MECÁ tee bolt (*T-bolt*); ~ **tensador** *m* ING MECÁ straining screw; ~ **tensor** *m* ING MECÁ tension screw, tightening screw; ~ **torneado** *m* ING MECÁ machine bolt; ~ **de tornillo** *m* ING MECÁ screw bolt; ~ **de trabante** *m* ING MECÁ stud bolt; ~ **de transmisión** *m* PROD drive bolt; ~ **en U** *m* CONST, TRANSP MAR *construcción naval* U-bolt; ~ **de unión** *m* ING MECÁ assembling bolt, union screw

perol *m* ING MECÁ pan
peroxiácido *m* QUÍMICA peroxy acid
peroxidación *f* QUÍMICA peroxidation
peroxidar *vt* QUÍMICA peroxidize
peroxidisulfúrico *adj* QUÍMICA peroxydisulfuric (*AmE*), peroxydisulphuric (*BrE*)
peróxido *m* QUÍMICA peroxide; ~ **de benzoílo** *m* ALIMENT, QUÍMICA, P&C acetoxyl, benzoyl peroxide; ~ **de hidrógeno** *m* QUÍMICA, TEC ESP hydrogen peroxide; ~ **de nitrógeno** *m* CONTAM, QUÍMICA nitrogen peroxide
peroxoacetilnitrato *m* (*PAN*) CONTAM, QUÍMICA peroxoacetylnitrate (*PAN*)
peroxofosfato *m* QUÍMICA peroxophosphate
perpendicular[1] *adj* GEOM perpendicular
perpendicular[2] *f* GEOM perpendicular; ~ **corta** *f* TEC PETR short normal; ~ **media** *f* TRANSP MAR *diseño de barcos* perpendicular amidships; ~ **de popa** *f* TRANSP MAR *diseño de barcos* aft perpendicular; ~ **de proa** *f* TRANSP MAR *diseño de barcos* forward perpendicular
perpendiculares: ~ **entre sí** *adj* METR *ejes coordenados* perpendicular to each other
perpiaño *m* CONST bond stone, through-binder, *albañilería* perpend, through-stone
perrenato *m* QUÍMICA perrhenate
perrénico *adj* QUÍMICA perrhenic
perro *m* AmL (cf *garra* Esp) ING MECÁ, PROD dog; ~ **de arrastre** *m* ING MECÁ lathe carrier, dog, driver, driving dog, screw dog, lathe dog, turning carrier; ~ **de cojinete** *m* ING MECÁ jaw dog; ~ **de plato de torno** *m* ING MECÁ dog; ~ **de sujeción** *m* PROD gripping dog
persal *f* QUÍMICA persalt
perseulosa *f* QUÍMICA perseulose
persiana *f* Esp C&V louver (*AmE*), louvre (*BrE*), CONST shutter, MECÁ blind, louver (*AmE*), louvre (*BrE*); ~ **corredera** *f* Esp (cf *persiana corrediza* AmL) CONST sliding shutter; ~ **corrediza** *f* AmL (cf *persiana corredera* Esp) CONST sliding shutter
persistencia *f* ELEC, ELECTRÓN, INFORM&PD persistence; ~ **de la combustión durante el periodo de expansión** *f* TERMO afterburning; ~ **de la emulsión** *f* PROC QUÍ emulsion persistence; ~ **de espuma** *f* PROC QUÍ foam persistence; ~ **de la imagen** *f* AmL (cf *remanencia de la imagen* Esp) TV burn; ~ **luminosa** *f* PAPEL, TERMO, TRANSP MAR afterglow; ~ **luminica** *f* TERMO afterglow; ~ **variable** *f* ELECTRÓN variable persistence; ~ **de la visión** *f* CINEMAT persistence of vision

persona: ~ **encargada de extenser y recoger los geófonos** *f* GEOFÍS, TEC PETR jug hustler

personal *m* PROD *fábricas* manpower; ~ **de perforación** *m* TEC PETR drilling crew; ~ **de primeros auxilios** *m* SEG first-aid personnel; ~ **de tierra** *m* TRANSP AÉR ground staff; ~ **de topografía** *m* CONST leveling staff (*AmE*), levelling staff (*BrE*); ~ **del tren** *m* FERRO train crew

personalización *f* PROD customerization

personalizado *adj* IMPR customized

personalizar *vt* IMPR customize

perspectiva *f* GEOL, TEC PETR horizon; ~ **de planificación** *f* Esp (*cf planeación AmL*) PROD planning horizon

Perspex® *m* P&C Perspex®

persulfato *m* QUÍMICA persulfate (*AmE*), persulphate (*BrE*)

PERT *abr* (*técnica de revisión y evaluación de programas*) GEN PERT (*program evaluation and review technique AmE, programme evaluation and review technique BrE*)

pertenencia *f AmL* (*cf concesión Esp*) INFORM&PD membership, MINAS claim; ~ **a un grupo** *f* INFORM&PD membership; ~ **minera** *f* MINAS *legislación* mining area, mining claim

pértiga *f* CINEMAT, TV *para micrófonos* boom; ~ **Fisher** *f* CINEMAT, TV Fisher boom; ~ **sonda** *f* TRANSP MAR *equipamiento de cubierta* sounding pole

pertita *f* MINERAL, PETROL perthite

pertítico *adj* MINERAL, PETROL perthitic

perturbación *f* ELECTRÓN jammer, clutter, interference, jamming, *en el tubo de gas* breakdown, FÍS FLUID perturbation, METEO, PETROL disturbance, TEC ESP perturbation, clutter, TELECOM, TRANSP MAR *pantalla del radar* clutter; ~ **aerodinámica arrastrada** *f* TRANSP AÉR wake turbulence; ~ **atmosférica** *f* METEO atmospheric disturbance; ~ **de baja frecuencia** *f* TV glitch; ~ **eléctrica** *f* ELECTRÓN electrical noise; ~ **electromagnética** *f* TEC ESP, TELECOM electromagnetic disturbance; ~ **entre símbolos** *f* ELECTRÓN intersymbol interference; ~ **exterior** *f* ELECTRÓN external noise; ~ **externa** *f* ELEC external disturbance; ~ **por interferencia de ondas extrañas** *f* TELECOM jamming; ~ **por lluvia** *f* TRANSP MAR *radar* rain clutter; ~ **localizada** *f* FÍS FLUID localized disturbance; ~ **por mar** *f* TRANSP MAR *radar* sea clutter; ~ **de modulación** *f* ELECTRÓN modulation noise; ~ **momentánea** *f* PROD transient current; ~ **de órbita** *f* TEC ESP perturbation of orbit; ~ **oscilatoria** *f* TELECOM jitter, noise; ~ **por ruido** *f* ACÚST, CONTAM, SEG noise pollution; ~ **de señal por ruido** *f* ELECTRÓN signal buried in noise; ~ **silbante** *f* TEC ESP whistler; ~ **transitoria** *f* TELECOM transient fluctuation

perturbaciones: ~ **acústicas** *f pl* TELECOM acoustic noise; ~ **atmosféricas** *f pl* GEOFÍS atmospherics; ~ **transitorias de línea** *f pl* ELEC line transient

perturbador *m* ÓPT mode mixer; ~ **de conversación** *m* TEC ESP, TELECOM scrambler; ~ **de filetes de aire** *m* TRANSP AÉR spoiler

perturbar *vt* CONST disturb; ~ **con intención de interferir** *vt* TRANSP MAR *emisión* jam

perveancia *f* ING ELÉC, TELECOM perveance

peryodato *m* QUÍMICA periodate

pesada: ~ **de muy alta precisión** *f* EMB ultrahigh accuracy weighing

pesadez: ~ **de cola** *f* TRANSP AÉR tail heaviness; ~ **de morro** *f* TRANSP AÉR nose heaviness

pesado *adj* C&V *recipiente* heavy, ING MECÁ heavy-duty, MECÁ heavy

pesadora *f* PROD weighing machine

pesafiltro *m* LAB, QUÍMICA weighing bottle

pesar *vt* TEC ESP *dirigibles* top up

pesas *f pl* METR weights

pesasubstancias *m Esp* (*cf navecilla tarada AmL*) LAB weighing boat

pesca *f* PETROL, TEC PETR *perforación* fishing; ~ **de aguas profundas** *f* OCEAN deep-sea fishing; ~ **de altura** *f* OCEAN distant water fishing, offshore fishery, offshore fishing, TRANSP MAR deep-sea fishing; ~ **de arrastre** *f* OCEAN dragnet fishing; ~ **por arrastre** *f* CONTAM MAR, OCEAN trawl fishing, trawling, TRANSP MAR trawl fishing; ~ **de bajura** *f* OCEAN coastal fishery; ~ **a la cacea** *f* OCEAN trolling fishing; ~ **con carnada viva** *f* OCEAN live bait fishing; ~ **costera** *f* OCEAN coastal fishery, inshore fishery; ~ **al curricán** *f* OCEAN trolling fishing; ~ **desde la playa** *f* OCEAN beach fishing; ~ **en exceso** *f* OCEAN overfishing; ~ **de gran altura** *f* OCEAN offshore fishery, offshore fishing; ~ **liviana** *f* OCEAN light fishing; ~ **con nasa** *f* OCEAN creel fishing; ~ **subacuática** *f* OCEAN underwater fishing

pescado: ~ **desechado** *m* OCEAN trash fish

pescador: ~ **agricultor** *m* OCEAN coastal fisherman-farmer; ~ **con artes de playa** *m* OCEAN shore fisherman; ~ **de bajura** *m* OCEAN shore fisherman; ~ **campesino** *m* OCEAN coastal fisherman-farmer; ~ **de rastra** *m* MINAS, TRANSP MAR dredger

pescaherramientas *m* TEC PETR *perforación* fishing tool; ~ **abocinado** *m AmL* (*cf campana de pesca Esp*) MINAS *sondeos* horn socket, socket

pescante *m* MECÁ davit, NUCL jib, TEC PETR *perforación* fishing socket, fishing tool, TRANSP AÉR hoist arm, hoist boom, TRANSP MAR davit

pescar: ~ **con arte de arrastre** *vt* TRANSP MAR trawl; ~ **con red de arrastre** *vt* OCEAN trawl

pesebre *m* AGRIC rack, MINAS stall

pesidoto *m* MINERAL olivine

peso *m* GEN weight; ~ **aproximado** *m* EMB approximate weight; ~ **de aterrizaje** *m* TRANSP AÉR landing weight; ~ **de aterrizaje de diseño** *m* TRANSP AÉR design landing weight; ~ **atómico** *m* (*A*) FÍS atomic weight, QUÍMICA atomic mass (*A*), atomic weight (*A*); ~ **sobre la barrena** *m* TEC PETR weight on bit (*WOB*); ~ **sobre la broca** *m* TEC PETR *perforación* weight on bit (*WOB*); ~ **bruto** *m* EMB, METR, TEXTIL, TRANSP AÉR gross weight; ~ **bruto inicial** *m* TRANSP AÉR initial gross weight; ~ **del cable** *m* TRANSP AÉR cable weight; ~ **de calibración** *m* LAB calibration weight; ~ **en canal** *m* AGRIC carcass weight; ~ **en canal limpio** *m* AGRIC dressed carcass weight; ~ **de las cargas apiladas** *m* TRANSP MAR overstowage; ~ **de certificación** *m* TRANSP AÉR certification weight; ~ **sin combustible** *m* TEC ESP dry weight; ~ **comercial de una pasta** *m* PAPEL saleable mass; ~ **de compensación** *m* TRANSP AÉR balance weight; ~ **cursor** *m* METR *de balanza romana* bob; ~ **declarado de embarque** *m* TRANSP MAR shipping weight; ~ **deficiente** *m* METR short weight; ~ **de despegue** *m* TEC ESP liftoff weight; ~ **al destete** *m* AGRIC weaning weight; ~ **de diseño** *m* TRANSP AÉR design weight; ~ **de embarque** *m* AGRIC, PROD, TRANSP MAR shipping weight; ~ **de equilibrado de la pala** *m* TRANSP AÉR blade balance-weight; ~ **equilibrador** *m* ING MECÁ, MECÁ balance weight, balancing weight; ~ **específico** *m* P&C relative density, specific gravity, TRANSP MAR *del agua del mar* specific gravity; ~ **específico aparente** *m* PAPEL apparent specific gravity; ~ **en factura** *m* TEXTIL invoice weight; ~ **del fletado** *m* TRANSP AÉR fleet weight; ~ **funcional** *m* TEC ESP operating weight; ~ **garantizado** *m* TRANSP AÉR guaranteed weight; ~ **de la inyección** *m* PETROL *perforación* mud weight; ~ **del lodo** *m* TEC PETR *perforación* mud weight; ~ **de lodo equivalente** *m* TEC PETR *perforación* equivalent mud weight (*EMW*); ~ **máximo** *m* EMB maximum weight; ~ **sobre la mecha** *m* AmL (*cf peso sobre el trépano Esp*) (*PSM*) TEC PETR *perforación* weight on bit (*WOB*); ~ **mínimo** *m* EMB minimum weight; ~ **molecular** *m* TEC PETR molecular weight; ~ **muerto** *m* MECÁ, TEC PETR, TRANSP MAR dead weight tonnage, *diseño de barcos* dead weight; ~ **neto** *m* ALIMENT, METR, TEXTIL net weight; ~ **en orden de vuelo** *m* (*PMA*) TRANSP all-up weight (*AUW*); ~ **de la pasta seca al aire** *m* PAPEL air-dry mass; ~ **de pelo** *m* TEXTIL pile weight; ~ **en plataforma** *m* TRANSP AÉR ramp weight; ~ **de rodaje de diseño** *m* TRANSP AÉR design taxi weight; ~ **del rotor** *m* ENERG RENOV rotor weight; ~ **en seco** *m* EMB, TEC ESP, TRANSP AÉR dry weight; ~ **no suspendido** *m* VEH *ruedas, neumáticos, frenos* unsprung weight; ~ **del tejido** *m* TEXTIL fabric weight; ~ **total** *m* EMB, METR, TEXTIL, TRANSP, TRANSP AÉR gross weight; ~ **total autorizado** *m* VEH *normas* total permissible weight; ~ **total de despegue** *m* (*PMA*) TRANSP all-up weight (*AUW*); ~ **total máximo** *m* VEH maximum total weight; ~ **total del vehículo** *m* (*PTV*) VEH gross vehicle weight (*GVW*); ~ **del trépano** *m* PETROL bit weight; ~ **sobre el trépano** *m* Esp (*cf peso sobre la mecha AmL*) TEC PETR *perforación* weight on bit (*WOB*); ~ **troy** *m* METR troy weight; ~ **de las unidades** *m* METR avoirdupois weight; ~ **unitario** *m* CARBÓN unit weight; ~ **en vacío** *m* TRANSP AÉR empty weight; ~ **vivo** *m* AGRIC live weight; ~ **de vuelo de diseño** *m* TRANSP AÉR design flight weight

pesón *m* ING MECÁ spring balance

pesos: ~ **estándar** *m pl* METR standard weights; ~ **y medidas** *m pl* METR weights and measures

pesquería *f* ALIMENT, OCEAN, TRANSP MAR fishery; ~ **en aguas lejanas** *f* OCEAN offshore fishery; ~ **de aguas profundas** *f* OCEAN deep-sea fishery; ~ **de altura** *f* OCEAN deep-sea fishery; ~ **sedentaria** *f* OCEAN sedentary fishery

pestaña *f* GEN flange, TEXTIL *para cubrir una costura* flap; ~ **de asiento del cilindro** *f* VEH cylinder flange; ~ **del carril** *f* FERRO rail flange; ~ **del cubo** *f* AUTO, VEH hub flange; ~ **del índice** *f* IMPR index tab; ~ **del radiador** *f* AUTO, FERRO, VEH radiator flange; ~ **de la rueda** *f* AUTO, FERRO, VEH wheel flange

pestañadora *f* PROD flanger, flanging machine

peste: ~ **aviar** *f* AGRIC fowl pest; ~ **bovina** *f* AGRIC rinderpest

pesticida *m* AGRIC, QUÍMICA pesticide; ~ **organoclorado** *m* AGRIC organochlorine pesticide; ~ **organofosforado** *m* AGRIC organophosphorus pesticide

pestillo *m* CONST catch, hasp, latch, *cerradura* tumbler, INFORM&PD latch, ING MECÁ latch, bolt, catch, MECÁ latch; ~ **del capó** *m* AUTO, VEH bonnet catch (*BrE*), bonnet lock (*BrE*), hood lock (*AmE*) hood catch (*AmE*); ~ **incrustado** *m* CONST dormant lock; ~ **de la puerta** *m* CONST gate latch; ~ **de puerta** *m* VEH door catch; ~ **de resorte** *m* CONST latch bolt; ~ **vertical** *m* CONST lift latch

peta *pref* (*p*) METR peta (*p*)

petaca *f* PROD slabbing

petalita *f* MINERAL petalite

petardo *m* FERRO *de señales* detonator (*BrE*), torpedo (*AmE*)

petcita *f* MINERAL petzite

petición *f* INFORM&PD, TELECOM inquiry, req, request; ~ **de actualización rápida** *f* TELECOM fast-update request; ~ **de búsqueda** *f* INFORM&PD search query; ~ **de canal** *f* TEC ESP request channel; ~ **de compra** *f* PROD procurement request, procurement requisition; ~ **de comunicación** *f* TELECOM connection request (*CR*); ~ **de confirmación** *f* TELECOM acknowledgement request; ~ **de congelación de la imagen** *f* TELECOM freeze-picture request; ~ **de cotización** *f* PROD quotation request; ~ **de despeje** *f* INFORM&PD clear request; ~ **de herramientas** *f* PROD tool requisition; ~ **de imagen inmóvil de la videoseñal de mando** *f* AmL, ~ **de imagen inmóvil de la vídeoseñal de mando** *f* Esp TELECOM video command freeze-picture request (*VCF*); ~ **de liberación** *f* INFORM&PD clear request; ~ **de orden** *f* INFORM&PD command language; ~ **de piezas** *f* PROD parts requisition; ~ **de trabajo** *f* INFORM&PD job request

peto *m* TRANSP MAR *pieza fundida, construcción metálica* stern frame; ~ **de popa** *m* TRANSP MAR *buques de acero* transom

petrificación *f* GEOL lithification

petrolato *m* PETROL, QUÍMICA, TEC PETR mineral jelly, petrolatum, petroleum jelly, Vaseline ®

petroleno *m* QUÍMICA petrolene

petróleo[1]: **con** ~ *adj* PETROL, TEC PETR oil-bearing

petróleo[2] *m* AUTO mineral oil, CARBÓN mineral naphtha, mineral oil, rock oil, stone oil, P&C mineral oil, PETROL earth oil, oil, petroleum, rock oil, TEC PETR petroleum, oil; ~, **aceite y lubricantes** *m pl* D&A petroleum, oil and lubricants (*POL*); ~ **de alumbrado** *m* TEC PETR lamp oil; ~ **bruto obtenido por pirólisis** *m* CARBÓN coal oil; ~ **combustible** *m* TERMO oil fuel; ~ **crudo** *m* CONTAM MAR, PETROL, TEC PETR crude oil, TERMO live oil; ~ **emulsionado** *m* PETROL, TEC PETR cut oil; ~ **in situ** *m* PETROL, TEC PETR oil in place; ~ **lampante** *m* PETROL, QUÍMICA, TEC PETR kerosene, lamp oil, paraffin oil TERMO, TRANSP, TRANSP AÉR kerosene; ~ **de lámparas** *m* PETROL, TEC PETR lamp oil; ~ **ligero** *m* PETROL, TEC PETR light crude oil; ~ **sin movilidad** *m* PETROL, TEC PETR dead oil; ~ **muerto** *m* PETROL, TEC PETR dead oil; ~ **pesado** *m* PETROL, TEC PETR heavy-duty oil (*HD*)

petrolero[1] *adj* QUÍMICA petrolic

petrolero[2] *m* PETROL, TRANSP MAR *buque* oil tanker; ~ **para crudos** *m* CONTAM, CONTAM MAR, PETROL, TEC PETR, TRANSP MAR crude carrier; ~ **puente** *m* PETROL, TEC PETR, TRANSP shuttle tanker; ~ **para el transporte de crudos** *m* CONTAM, CONTAM MAR, PETROL, TEC PETR, TRANSP MAR crude carrier

petrolífero *adj* PETROL, TEC PETR oil-bearing

petrología *f* CARBÓN, PETROL, TEC PETR petrology

petroquímico[1] *adj* PETROL, QUÍMICA, TEC PETR petrochemical

petroquímico[2]: ~ **básico** *m* PETROL, QUÍMICA, TEC PETR basic petrochemical

petroquímicos *m pl* PETROL, QUÍMICA, TEC PETR petrochemicals

pez: ~ **mineral** *m* MINERAL asphalt

pezón *m* AUTO journal, INSTR *ejes* arm

pH *abr* (*potencial hidrógeno*) HIDROL, QUÍMICA pH (*potential of hydrogen*)

pheelgita *f* MINERAL pheelgite

PHMA *abr* (*polihexilmetacrilato*) P&C PHMA (*polyhexylmethacrylate*)

pHmetro *m* METR pH meter

PI *abr* ELECTRÓN, INFORM&PD (*procesamiento de información, tratamiento de la información*) IP (*information processing*)

PIA *abr* (*adaptador de interfaz de periféricos*) INFORM&PD PIA (*peripheral interface adaptor*)

piamontita *f* MINERAL piedmontite

piara *f* AGRIC herd

pica *f* IMPR pica

picada *f* TEXTIL *tejedura* picking

picado *m* CINEMAT tilt, TEC ESP *vehículos* dive, TEXTIL *vestimenta* padding, TRANSP AÉR dive; ~ **por autoencendido** *m* AUTO, VEH *motor* pinging (*AmE*), pinking (*BrE*); ~ **de las bielas** *m* AUTO, VEH rumble; ~ **de los cartones** *m* TEXTIL card cutting; ~ **en espiral** *m* TRANSP AÉR spiral dive; ~ **del ticket** *m* TRANSP ticket punch

picador *m* ALIMENT mincer, CARBÓN, MINAS *persona*; digger; ~ **de carbón** *m* CARBÓN *persona* coal cutter, MINAS dintheader, *persona* coal cutter

picadora *f* ALIMENT mincer; **~-ensiladora** *f* AGRIC silage cutter; ~ **de forrajes** *f* AGRIC forage chopper; ~ **de heno** *f* AGRIC hay chopper

picadura[1]: **de ~ cruzada** *adj* PROD *lima* double-cut

picadura[2] *f* CRISTAL etch pit, ING MECÁ, MECÁ cut, PROD cinder pit; ~ **basta** *f* ING MECÁ, PROD rough cut; ~ **bastarda** *f* ING MECÁ, PROD bastard cut; ~ **doble** *f* PROD crosscut; ~ **gruesa** *f* ING MECÁ, PROD rough cut; ~ **totalmente dulce** *f* PROD dead smooth cut

picafuegos *m* PROD prick bar, *hogar de locomotora* poker, pricker

picapedrero *m* CONST hewer

picaporte *m* CONST latch, hasp, *puertas* button, *cerrajería* thumb latch, ING MECÁ, MECÁ latch

picar[1] *vt* AGRIC chop, ALIMENT mince, CARBÓN hew, *un muro* polish with emery, CINEMAT tilt, CONST chip, pick, ING MECÁ *motor* pink, MINAS *carbón* cut, PROD *machacar con mazo* pound; ~ **cruzado** *vt* PROD crosscut; ~ **hacia abajo** *vt* CINEMAT tilt down

picar[2] *vi* TEC ESP, TRANSP AÉR dive

picazón: ~ **loca** *f* AGRIC *enfermedad animal* mad itch

picnita *f* MINERAL pycnite

picnómetro *m* FÍS, INSTR, LAB, TEC PETR pycnometer

pico[1]: ~ **a pico** *adv* TV peak-to-peak

pico[2] *m* C&V pick, spike, spout, CONST pickax (*AmE*), pickaxe (*BrE*), LAB *de recipiente* spout, NUCL peak, P&C *parte de equipo* nozzle, PROD spout, *herramienta* pick, *fuelle* nose, QUÍMICA peak, TV spike; ~ **de bigornia** *m* ING MECÁ beak iron; ~ **cangrejo** *m* TRANSP MAR gaff; ~ **para carbón** *m* CARBÓN, MINAS coal pick; ~ **de carga** *m* ELEC load peak, peak load; ~ **de corriente** *m* ELEC current peak; ~ **del crisol** *m* C&V pot spout; ~ **de descarga** *m* PROD *de crisol*,

recipiente lip; ~ **de fisión** *m* NUCL fission spike; ~ **de fugas** *m* FÍS RAD escape peak; ~ **de fugas de rayos gamma** *m* FÍS RAD gamma-ray escape peak; ~ **de fugas de rayos X** *m* FÍS RAD X-ray escape peak; ~ **de fugas simple** *m* FÍS RAD single escape peak; ~ **de loro** *m* TRANSP MAR anchor bill, *uñas de ancla* peak; ~ **con martillo** *m* CONST poll pick; ~ **de martillo** *m* PROD hammer head; ~ **másico de precursores** *m* NUCL parent mass peak; ~ **máximo de potencia** *m* TV peak power; ~ **de minero** *m* MINAS *herramienta* picker; ~ **de minero de dos puntas** *m* CARBÓN, MINAS mandrel, mandril; ~ **del negro** *m* AmL (*cf cresta del negro Esp*) TV black peak; ~ **de potencia de la envolvente** *m* TELECOM *de modulación* peak envelope power; ~ **de precursores** *m* NUCL parent peak; ~ **de pulverizadora** *m* AGRIC sprayer nozzle; ~ **de reactividad** *m* NUCL reactivity surge; ~ **de reflexión** *m* PETROL reflection peak; ~ **de resonancia** *m* ELECTRÓN, FÍS RAD resonance peak; ~ **de retrodispersión** *m* FÍS RAD backscatter peak; ~ **de salida** *m* AGUA spout; ~ **submarino** *m* GEOL, OCEAN seamount

pico[3] *pref* (*p*) METR pico (*p*)

picolina *f* QUÍMICA picoline

picos *m pl* TV peaks; ~ **de brillo** *m pl* AmL (*cf luminosidad máxima Esp*) TV peak brightness; ~ **de pares** *m pl* FÍS RAD pair peaks

picosegundo *m* FÍS, INFORM&PD, NUCL picosecond

picotita *f* MINERAL picotite

picrato *m* QUÍMICA picrate

pícrico *adj* QUÍMICA picric

picrita *f* PETROL picrite

picrol *m* QUÍMICA picrol

picrolita *f* MINERAL picrolite

picromerita *f* MINERAL picromerite

picrotina *f* QUÍMICA picrotine

pictograma *m* IMPR pictograph, SEG *de sustancias peligrosas* pictorial symbol

pie *m* CARBÓN *de muro* footing, CINEMAT *para un actor* cue, CONST foot, support, bottom, GEOM base, IMPR *de una ilustración* caption, tail, ING MECÁ leg, standard, METR *medida* foot, PROD standard, TRANSP MAR *de mastelero, de rollizo* foot; ~ **para accesorios** *m* FOTO accessory shoe; ~ **articulado** *m* FOTO *de un trípode* foldover leg; ~ **auxiliar** *m* FOTO *para accesorios* auxiliary shoe; ~ **de biela** *m* AUTO connecting-rod small end, small end, ING MECÁ connecting-rod small end, MECÁ crosshead, VEH connecting-rod small end, small end; ~ **del cabrestante** *m* ING MECÁ pawl head; ~ **de copa** *m* C&V stem; ~ **cuadrado** *m* METR square foot; ~ **cúbico** *m* METR cubic foot; ~ **derecho** *m* C&V uprighter; ~ **del flash** *m* FOTO flash shoe; ~ **en forma de herradura** *m* AmL (*cf pie de herradura Esp*) LAB *microscopio* horseshoe foot; ~ **de gallo** *m* TRANSP MAR *remolques, fondeo* bridle; ~ **de herradura** *m* Esp (*cf pie en forma de herradura AmL*) LAB *microscopio* horseshoe foot; **~-lambert** *m* CINEMAT *luminancia* foot lambert; ~ **de montaje** *m* FOTO mounting foot; ~ **negro** *m* AGRIC *patología vegetal* blackleg; ~ **de página** *m* IMPR bottom; ~ **de palo** *m* TRANSP MAR mast foot; ~ **de rey** *m* FÍS vernier caliper (*AmE*), vernier calliper (*BrE*), ING MECÁ caliper square (*AmE*), calliper square (*BrE*), INSTRUM, METR vernier caliper (*AmE*), vernier calliper (*BrE*); ~ **rígido** *m* FOTO rigid leg; ~ **de roda** *m* TRANSP MAR *construcción naval* forefoot; ~ **de rodadura** *m* METR running foot;

~ **del talud** *m* CARBÓN slope toe; ~ **del tipo de imprenta** *m* IMPR base; ~ **en "U"** *m* LAB *microscopio* horseshoe foot

piedra[1]: **de ~ dura** *adj* MECÁ jeweled (*AmE*), jewelled (*BrE*); **de ~ preciosa** *adj* MECÁ jeweled (*AmE*), jewelled (*BrE*)

piedra[2] *f* CARBÓN boulder, diamond, CONST stone, MINAS *canteras* grain, *geología* stone; ~ **de aceite** *f* ING MECÁ oilstone, PROD hone, honestone, oilstone; ~ **de aceite para repasar filos** *f* ING MECÁ honing stone; ~ **afiladora** *f* ING MECÁ oilstone; ~ **de afilar** *f* MECÁ honing stone, PROD hone, oilstone, grindstone, whetstone; ~ **de las Amazonas** *f* MINERAL Amazon stone; ~ **de amolar** *f* PROD whetstone; ~ **angular** *f* C&V corner block, PROD quoin; ~ **de árida para carretera** *f* CONST *carretera* chipping; ~ **de arranque de un arco** *f* CONST springer stone; ~ **azul** *f* QUÍMICA bluestone; ~ **para bordillo** *f* CONST curbstone (*AmE*), kerbstone (*BrE*); ~ **en bruto** *f* CONST rubble stone; ~ **caliza** *f* C&V, CONST, GEOL, PETROL limestone, QUÍMICA lime rock, limestone; ~ **de campana** *f* PETROL, PROD clinkstone; ~ **de carga** *f* C&V batch stone; ~ **de cemento** *f* CONST cement stone; ~ **para construcción** *f* CONST building stone; ~ **dentada** *f* CONST *albañilería* toothing stone; ~ **de devitrificación** *f* C&V devitrification stone; ~ **imán** *f* MINERAL lodestone; ~ **irregular** *f* CONST broken stone; ~ **jabón** *f* MINERAL bowlingite, saponite; ~ **labrada** *f* CONST cut stone, dressed stone; ~ **sin labrar** *f* CONST rubble stone; ~ **litográfica** *f* QUÍMICA lithographic stone; ~ **de la luna** *f* MINERAL moonstone; ~ **machacada** *f* CONST rubble; ~ **de machaqueo** *f* CONST rubble; ~ **moleña** *f* GEOL, MINERAL, PROD millstone grit; ~ **pasante** *f* CONST *albañilería* through-binder; ~ **de perpiaño** *f* CONST perpend stone; ~ **pómez** *f* C&V, GEOL pumice, MINERAL pumice-stone; ~ **de pulir de super acabado** *f* ING MECÁ superfinishing honing stone; ~ **de repasar filos** *f* ING MECÁ oilstone, PROD honestone; ~ **de sal** *f* AGRIC salt lick, licking block; ~ **sangre** *f* MINERAL bloodstone, heliotrope; ~ **de sol** *f* MINERAL sunstone; ~ **de superacabado** *f* ING MECÁ superfinishing stone; ~ **de la taza del horno** *f* C&V tank block; ~ **triturada para caminos** *f* CONST paving material (*AmE*), road metal (*BrE*); ~ **de unión** *f* CONST *albañilería* binding stone; ~ **de yeso** *f* CONST plaster stone

piedrecita *f* CONST, PETROL pebble

piel *f* CARBÓN skin, TEXTIL fur; ~ **sin curtir** *f* PROD green hide; ~ **ligera** *f* AGRIC kip

pielografía *f* INSTR pyelography

pienso: ~ **para animales** *m* AGRIC animal feed; ~ **completo** *m* AGRIC one-feed ration; ~ **compuesto** *m* AGRIC mixed feed; **concentrado** *m* AGRIC concentrate

pierna *f* INSTR *tenazas, tijeras, cizallas* arm

pies: ~ **por segundo** *m pl* CINEMAT feet per second

pieza[1]: **todo de una ~** *adj* PROD in one piece

pieza[2] *f* C&V piece, CARBÓN component, CONST member, room, ELEC *parte* component, INFORM&PD part, ING MECÁ member, part, section, item, MECÁ component, TEXTIL *de una prenda* panel; ~ **de acero fundido** *f* TRANSP MAR *construcción naval* steel casting; ~ **para acoplamiento** *f* TEC ESP docking piece; ~ **de apoyo en pared** *f* CONST *entibación* wall piece; ~ **aterrajada** *f* PROD strickled casting, *fundería* swept-up casting, *moldería* struck-up casting; ~ **atornillada** *f* ING MECÁ screw piece; ~ **batida** *f* PROD *funderías* false core; ~ **de brida** *f* ING MECÁ stirrup piece; ~ **bruta** *f* MECÁ blank; ~ **bruta de fundición** *f* PROD rough casting; ~ **cambiable** *f* ING MECÁ removable insert; ~ **de cierre** *f* PROD breeching piece, TEC ESP end piece; ~ **colada** *f* P&C casting; ~ **corta de conexión** *f* ELEC *en sistema* link; ~ **corta de conexión asincrónica** *f* ELEC *dos sistemas de CA* asynchronous link; ~ **cortada** *f* SEG cut piece; ~ **desgastable** *f* ING MECÁ wear part; ~ **desgastada de una máquina** *f* ING MECÁ wearing part of a machine; ~ **de desgaste** *f* GAS wearing part; ~ **desmontable** *f* ING MECÁ removable insert, PROD loose piece; ~ **para empaquetar** *f* C&V packing piece; ~ **de enganche** *f* CARBÓN, MINAS grip; ~ **de enrasar** *f* CONST furring piece; ~ **para ensayos de tracción** *f* ING MECÁ tensile test piece; ~ **entera** *f* TEXTIL *sin costuras* whole piece; ~ **estampada** *f* MECÁ forging; ~ **estampada en caliente** *f* TERMO hot forging; ~ **estampada en frío** *f* MECÁ die-stamper; ~ **fija** *f* ING MECÁ fixed member, fixture; ~ **forjada** *f* MECÁ, PROD forging; ~ **formada por elementos soldados** *f* MECÁ weldment; ~ **fundida** *f* ING MECÁ, P&C, PROD casting; ~ **fundida defectuosa** *f* PROD spoiled casting; ~ **giratoria** *f* ING MECÁ rotating part; ~ **intercalada** *f* MECÁ insert; ~ **maciza** *f* PROD solid blank; ~ **maciza de fundición** *f* PROD solid casting; ~ **a máquina** *f* MECÁ workpiece; ~ **matrizada** *f* PROD drop forging; ~ **media** *f* PROD average casting; ~ **metálica** *f* MECÁ hardware; ~ **moldeada** *f* ACÚST molding (*AmE*), moulding (*BrE*), PROD molded casting (*AmE*), moulded casting (*BrE*), molding (*AmE*), moulding (*BrE*); ~ **de mosaico** *f* C&V tessera; ~ **móvil** *f* ING MECÁ removable insert; ~ **múltiple** *f* ING MECÁ gang piece; ~ **normalizada** *f* ING MECÁ standard part; ~ **de obturación** *f* MECÁ bulkhead; ~ **ocular** *f* CINEMAT, FÍS, FOTO, INSTR, LAB, ÓPT, PROD, TEC ESP eyepiece; ~ **ocular de joyero** *f* INSTR jeweler's eyepiece (*AmE*), jeweller's eyepiece (*BrE*); ~ **de paro del eyector** *f* ING MECÁ ejector stop-piece; ~ **pequeña de fundición** *f* PROD fine castings; ~ **polar** *f* FÍS, ING ELÉC, ING MECÁ pole piece, TV pole shoe; ~ **polar del arrancador** *f* AUTO starter pole shoe; ~ **polar de la lente proyectora** *f* INSTR projector-lens pole piece; ~ **posterior** *f* CONST tailpiece; ~ **postiza** *f* MECÁ insert; ~ **de prueba** *f* FÍS test piece; ~ **para pruebas de tracción** *f* ING MECÁ tensile test piece; ~ **de recambio** *f* AUTO, VEH spare part; ~ **rechazada por defectuosa** *f* MECÁ discard; ~ **de refrentar** *f* ING MECÁ facing attachment; ~ **reglamentaria** *f* ING MECÁ standard part; ~ **removible** *f* ING MECÁ removable insert; ~ **de repuesto** *f* AUTO, ING MECÁ part, TRANSP MAR spare part, VEH part; ~ **roscada** *f* ING MECÁ screw piece; ~ **separable** *f* ING MECÁ removable insert; ~ **de separación** *f* EMB partition wall, ING MECÁ distance piece, distance piece spacer, PROD *de tubo* cutting-off piece; ~ **de separación redonda** *f* ING MECÁ round distance piece; ~ **soldada** *f* ING MECÁ welded fitting; ~ **suelta** *f* PROD loose piece; ~ **sujeta a desgaste** *f* ING MECÁ wear part; ~ **superior transformada** *f* TEXTIL *de una prenda* converted top; ~ **en T** *f* LAB *conector, adaptador* T-piece; ~ **de tensión** *f* ING MECÁ tension piece; ~ **sin terminar** *f* MECÁ blank; ~ **tosca de forja** *f* PROD blank; ~ **de trabajo** *f* ING MECÁ working part, MECÁ workpiece; **~ de unión** *f* ELECTRÓN *de semiconductor* bonding pad

piezas: ~ de recambio del automóvil *f pl* AUTO, VEH motorcar parts

piezoelectricidad *f* GEN piezoelectricity

piezoeléctrico *adj* GEN piezoelectric

piezómetro *m* CARBÓN, CONST, FÍS piezometer, LAB *presión de gas* pressure gage (*AmE*), pressure gauge (*BrE*)

piezostato *m* QUÍMICA pressurestat

pigeonita *f* MINERAL pigeonite

pigmentación *f* COLOR, TEXTIL pigmentation

pigmentado *adj* COLOR, TEXTIL pigmented

pigmentar *vt* COLOR, TEXTIL pigment

pigmento *m* C&V, COLOR, P&C pigment, QUÍMICA pigment, dye, TEXTIL pigment; **~ de alta pureza** *m* P&C *pintura* high purity pigment; **~ azul de hierro** *m* COLOR iron blue pigment; **~ azul satinado** *m* COLOR blue-glaze pigment; **~ azul ultramar** *m* COLOR ultramarine pigment; **~ benzol** *m* COLOR benzol dyestuff; **~ blanco** *m* COLOR white pigment; **~ de bronce** *m* COLOR bronze pigment; **~ calizo** *m* COLOR lime pigment; **~ colorante** *m* COLOR colored pigment (*AmE*), coloured pigment (*BrE*), coloring pigment (*AmE*), colouring pigment (*BrE*); **~ coloreado** *m* COLOR colored pigment (*AmE*), coloured pigment (*BrE*); **~ de cromato de bario** *m* COLOR barium chromate pigment; **~ de cromato de estroncio** *m* COLOR strontium chromate pigment; **~ de cromato de zinc** *m* COLOR zinc chromate pigment; **~ de cromo verde de plomo** *m* COLOR chrome-green lead pigment; **~ de cromo-ftalocianino azul de plomo** *m* COLOR chrome-phthalocyanine blue lead pigment; **~ esfumado** *m* COLOR shading-off pigment; **~ fotoresistente** *m* COLOR photoresistant pigment; **~ de hierro** *m* COLOR iron pigment; **~ de laca colorante** *m* COLOR lake pigment; **~ luminiscente** *m* COLOR luminescent pigment; **~ mate** *m* P&C *pintura* flatting pigment; **~ mineral** *m* COLOR mineral pigment; **~ nacarado** *m* COLOR, P&C *pintura* nacreous pigment; **~ opaco** *m* COLOR opaque pigment; **~ orgánico** *m* COLOR organic pigment; **~ órgano-metálico** *m* COLOR metallo-organic pigment; **~ de óxido crómico** *m* COLOR chromic oxide pigment; **~ de óxido de hierro** *m* COLOR iron oxide pigment; **~ de óxido de titanio** *m* COLOR titanium oxide pigment; **~ de polvo de zinc** *m* COLOR zinc dust pigment; **~ semitransparente** *m* COLOR water stain; **~ sólido** *m* P&C lasting pigment; **~ de tintura** *m* COLOR stainer pigment; **~ vegetal** *m* COLOR plant pigment

pila *f* CARBÓN pile, CONST *montón* pile, heap, ELEC *pilas y baterías* cell, INFORM&PD heap, battery, stack, INSTR cell, NUCL pile, TEC ESP battery cell; **~ de aire** *f* ING ELÉC air cell; **~ alcalina** *f* ING ELÉC alkaline battery; **~ atómica** *f* FÍS, NUCL atomic pile; **~ blanqueadora** *f* PAPEL bleacher; **~ de Bunsen** *f* ELEC Bunsen cell; **~ de combustible** *f* TEC ESP fuel cell; **~ desintegra-dora de trapos** *f* PAPEL breaker; **~ de desplazamiento ascendente** *f* INFORM&PD push-up stack; **~ de desplazamiento descendente** *f* INFORM&PD push-down stack; **~ de despolarización por aire** *f* ING ELÉC air cell; **~ de discos** *f* INFORM&PD disk pack; **~ de dos líquidos** *f* ELEC concentration cell; **~ eléctrica** *f* ELEC, ING ELÉC battery, electric cell, pile, TELECOM battery; **~ energética** *f* TEC ESP fuel

cell; **~ de equipo físico** *f* AmL (*cf pila de hardware Esp*) INFORM&PD hardware stack; **~ de espera** *f* TRANSP AÉR holding stack; **~ de estériles** *f* MINAS *escombrera* barrow; **~ filochadora** *f* PAPEL *para trapos* rag breaker; **~ flotante** *f* CARBÓN floating pile; **~ de gacetas** *f* C&V *hornos* bung of saggars; **~ de hardware** *f* Esp (*cf pila de equipo físico AmL*) INFORM&PD hardware stack; **~ hidroeléctrica** *f* ING ELÉC wet cell; **~ de hidrógeno con refrigeración directa** *f* TRANSP direct cold hydrogen cell; **~ de hojas** *f* IMPR pile; **~ holandesa** *f* PAPEL beater; **~ horizontal** *f* PETROL horizontal stack; **~ Leclanché** *f* LAB Leclanch cell; **~ de mercurio** *f* ING ELÉC mercury cell; **~ moldeada in situ** *f* CARBÓN cast-in-place pile; **~ de níquel y cadmio** *f* FOTO Nicad battery; **~ de óxido de plata** *f* ING ELÉC silver oxide cell; **~ patrón** *f* ELEC standard cell; **~ patrón Weston** *f* ELEC Weston standard cell; **~ de placas** *f* NUCL slab pile; **~ de plata y cadmio** *f* ING ELÉC silver-cadmium cell; **~ de plata y zinc** *f* ING ELÉC silver-zinc cell; **~ primaria de plata y zinc** *f* ING ELÉC silver-zinc primary cell; **~ recargable** *f* CINEMAT, FÍS, FOTO, ING ELÉC, TV rechargeable battery; **~ refinadora** *f* PAPEL beater; **~ seca** *f* ELEC, FÍS, ING ELÉC dry cell; **~ secundaria** *f* ELEC, ING ELÉC secondary cell; **~ de selenio** *f* ING ELÉC selenium cell; **~ sigma** *f* NUCL sigma pile; **~ solar** *f* ELEC, ING ELÉC, TEC ESP, TELE-COM solar cell; **~ solar de arseniuro de galio** *f* ELECTRÓN gallium arsenide solar cell; **~ solar de silicio** *f* ING ELÉC silicon solar cell; **~ de solicitudes** *f* INFORM&PD, TELECOM request stack; **~ de sulfuro de cadmio** *f* ELEC, ING ELÉC cadmium sulfide cell (*AmE*), cadmium sulphide cell (*BrE*); **~ de tareas** *f* INFORM&PD job stack; **~ termoeléctrica** *f* ING ELÉC thermocouple, thermopile, TEC ESP thermopile; **~ de trabajos** *f* INFORM&PD job stack; **~ voltaica** *f* ELEC voltaic cell, ING ELÉC voltaic pile; **~ Weston** *f* ELEC cadmium cell

pilar *m* C&V pillar, CARBÓN *horno cerámico* prop, CONST *puerta* post, pier, GEOFÍS pillar, ING MECÁ standard, TELECOM pillar; **~ de apoyo** *m* MINAS support pillar, supporting pillar, *construcción de galerías* support pier; **~ de bóveda** *m* MINAS *construcción de galerías* arch pillar; **~ de carbón** *m* CARBÓN broken rock, coal pillar, entry pillar, entry stump, MINAS coal pillar; **~ de erosión** *m* TEC PETR stack; **~ de esquina** *m* CONST corner pillar; **~ de galería** *m* CARBÓN entry pillar, entry stump; **~ de guía de eyección** *m* ING MECÁ ejection-guide pillar; **~ macizo** *m* AmL (*cf poste Esp*) MINAS *de mineral o carbón* post; **~ de madera** *m* AmL (*cf cuadro portátil Esp*) MINAS *armazón de sustentación* crib; **~ de mineral** *m* MINAS cranch, *sacado del tajo de arranque* broken ore; **~ de protección de galerías** *m* MINAS chain pillar; **~ del rellano** *m* CONST landing riser; **~ de seguridad** *m* CARBÓN rib, MINAS barrier, rib hole; **~ de sostén** *m* MINAS pack; **~ de sustentación** *m* MINAS support pier, support pillar, supporting pillar; **~ terraplenado** *m* CARBÓN embankment pile

pilarote *m* CONST *escaleras* newel, newel post

pildear *vt* TEXTIL *punto* pill

pildeo *m* TEXTIL *punto* pilling

píldora *f* AGRIC, CARBÓN Esp (*cf pella AmL*) pellet; **~ contra el mareo** *f* TRANSP MAR antiseasickness pill; **~ de efecto retardado** *f* AGRIC sustained release pellet (*SRP*)

pileta *f* AGUA basin, MINAS *colector de agua* sump; ~ **de mezcla** *f* AGUA mixing basin; ~ **de patio** *f* AGUA cesspool

pilocarpidina *f* QUÍMICA pilocarpidine

pilocarpina *f* QUÍMICA pilocarpine

pilón *m* CONST, ELEC, ING MECÁ pylon, PROD *cabeza de marinete* power hammer, standard, TRANSP AÉR pylon; ~ **de caída libre** *m* PROD gravity stamp; ~ **de forja** *m* ING MECÁ anvil block; ~ **triturador** *m* ING MECÁ tappet

pilot: ~ **chart** *m* TRANSP MAR *navegación* pilot chart

pilotaje *m* CONST pile work, piling, PROD, TEC PETR *operaciones costa-fuera* piling

pilotar *vt Esp* (*cf pilotear AmL*) TRANSP, TRANSP AÉR, TRANSP MAR pilot

pilote *m* CONST pier, pile, TRANSP MAR *amarres, radiofaro* post, *duques de alba* pile; ~ **de acero** *m* CARBÓN steel pile; ~ **de fricción** *m* CARBÓN, CONST friction pile; ~ **hincado por medio de gatos hidráulicos** *m* CARBÓN jacked pile; ~ **de hormigón armado** *m* CARBÓN concrete pile; ~ **inclinado** *m* C&V *muros* batter; ~ **de madera** *m* CARBÓN wooden pile; ~ **pretaladrado** *m* CARBÓN pre-bored pile; ~ **de rozamiento** *m* CARBÓN, CONST friction pile; ~ **segmentado** *m* CARBÓN segmented pile; ~ **de seguridad del pozo** *m* MINAS shaft safety pillar

pilotear *vt AmL* (*cf pilotar Esp*) TRANSP, TRANSP AÉR, TRANSP MAR pilot

piloto[1]: **sin** ~ *adj* TEC ESP unmanned

piloto[2] *m* ING ELÉC pilot lamp (*AmE*), pilot light (*BrE*), TELECOM, TRANSP AÉR, TRANSP MAR, TV pilot; ~ **automático** *m* TEC ESP, TRANSP AÉR automation; ~ **automático de autoridad limitada** *m* TRANSP AÉR limited-authority autopilot; ~ **cromático** *m* TV chroma pilot; ~ **de estado** *m* TELECOM status lamp; ~ **de indicación de alarma** *m* TELECOM alarm indication lamp; ~ **de primera clase** *m* TRANSP MAR *marina mercante* first mate; ~ **de pruebas** *m* TRANSP AÉR test pilot

pimárico *adj* QUÍMICA pimaric

pimélico *adj* QUÍMICA pimelic

pimelita *f* MINERAL pimelite

piña: ~ **de acollador** *f* TRANSP MAR single Matthew Walker, stopper knot

pinacol *m* QUÍMICA pinacol

pinacólico *adj* QUÍMICA pinacolic

pinacolina *f* QUÍMICA pinacoline

pinacolona *f* QUÍMICA pinacolone

pinacona *f* QUÍMICA pinacone

pináculo *m* CONST finial

pincel *m* P&C *pintura* paintbrush; ~ **para aplicar el diluente** *m* COLOR flat brush; ~ **luminoso** *m* INFORM&PD light gun; ~ **de rayos catódicos TRC** *m* ELEC CRT cathode-ray pencil; ~ **soplador** *m* CINEMAT, FOTO blower-brush

pincelada *f* REVEST brush coating

pincho: ~ **de pruebas** *m* ING ELÉC test prod

pineno *m* QUÍMICA terebenthene, pinene

pínico *adj* QUÍMICA pinic

pinita *f* MINERAL pinite

pino: ~ **negral** *m* CONST *carpintería*, TRANSP MAR *madera* pitch pine

piñón *m* AUTO pinion, ING MECÁ cogwheel, gear, gear wheel, pinion, MECÁ gear wheel, VEH *caja de cambio* pinion; ~ **de arrastre** *m* ING MECÁ driving pinion; ~ **de ataque** *m* AUTO drive pinion, ING MECÁ driving pinion, VEH drive pinion; ~ **de cadena** *m* ING MECÁ chain wheel, rag wheel; ~ **para cadena** *m* MECÁ sprocket wheel; ~ **de cadena de Galle** *m* ING MECÁ sprocket wheel; ~ **conductor** *m* AUTO drive pinion, ING MECÁ driving gear, driving pinion, VEH drive pinion; ~ **cónico del diferencial** *m* VEH *transmisión* differential bevel gear; ~ **de la cremallera** *m* ING MECÁ rack wheel; ~ **del diferencial** *m* AUTO, VEH differential pinion; ~ **diferencial** *m* VEH pinion gear; ~ **del eje intermediario** *m* VEH countershaft gear; ~ **de engranaje constante** *m* VEH constant mesh; ~ **fijo** *m* ING MECÁ fixed wheel; ~ **de linterna** *m* ING MECÁ lantern gear, lantern gearing; ~ **loco** *m* ING MECÁ idler; ~ **de mando** *m* AUTO, VEH drive pinion; ~ **motor** *m* ING MECÁ driving gear, driving pinion; ~ **del motor de arranque** *m* AUTO, VEH Bendix starter®; ~ **motriz** *m* AUTO drive pinion, ING MECÁ driving gear, VEH drive pinion; ~ **planetario** *m* AUTO planetary pinion, planet wheel, ING MECÁ planetary pinion, VEH planet wheel; ~ **planetario del diferencial** *m* VEH differential spider pinion; ~ **de puesta en marcha** *m* AUTO, VEH *del motor* starter motor pinion; ~ **satélite del diferencial** *m* VEH differential spider pinion; ~ **de la transmisión** *m* AUTO, VEH transmission pinion; ~ **de transmisión** *m* ING MECÁ idler

pinónico *adj* QUÍMICA pinonic

pinta *f* METR *medida* pint

pintado[1]: ~ **al óleo** *adj* COLOR oil-painted

pintado[2]: ~ **de la carretera** *m* CONST road painting; ~ **impermeable** *m* REVEST waterproof painting

pintar *vt* REVEST *con temple* coat

pintor: ~ **de pantalla** *m* INFORM&PD screen painter; ~ **de porcelana** *m* C&V china painter

pintura *f* CONST paint, paintwork, MECÁ, P&C, VEH paint; ~ **al aceite** *f* COLOR oil-bound paint; ~ **al aceite de secado rápido** *f* COLOR sharp oil paint; ~ **acrílica** *f* CONST acrylic paint; ~ **al agua** *f* CONST water-based paint; ~ **al agua al aceite** *f* COLOR oil-bound water paint; ~ **que no amarillea** *f* P&C, REVEST nonyellowing paint; ~ **antiácida** *f* CONST acid-resisting paint; ~ **antiacústica** *f* COLOR sound-deadening paint; ~ **anticorrosiva** *f* COLOR anticorrosive paint, P&C anticorrosive coating, REVEST anticorrosion paint, antirust paint; ~ **antideslizante** *f* REVEST, TRANSP MAR nonslip deck paint; ~ **antiincrustante** *f Esp* (*cf pintura antiséptica AmL*) P&C, REVEST, TRANSP MAR antifouling paint; ~ **antioxidante** *f* P&C antirust paint, REVEST rust-proofing paint; ~ **antipútrida** *f* P&C antifouling paint; ~ **antiséptica** *f Esp* (*cf pintura antiincrustante AmL*) P&C, REVEST, TRANS MAR antifouling paint; ~ **de apresto** *f* COLOR size paint; ~ **asfáltica** *f* P&C tar coating; ~ **al barniz** *f* COLOR, REVEST varnish paint; ~ **de barniz** *f* COLOR, REVEST painting varnish; ~ **bituminosa** *f* COLOR, CONST, P&C bituminous paint; ~ **brillante** *f* COLOR, CONST gloss paint; ~ **a la cal** *f* COLOR, CONST limewash, whitewash; ~ **de cal** *f* REVEST lime paint; ~ **de camuflaje** *f* COLOR camouflage paint; ~ **para casas** *f* P&C house paint; ~ **de caseína** *f* COLOR casein paint; ~ **a la celulosa en aerosol** *f* COLOR spray cellulose paint; ~ **celulósica** *f* P&C cellulose paint; ~ **de cemento** *f* COLOR cement paint; ~ **de cobertura** *f* COLOR masking paint, P&C, REVEST covering paint; ~ **contra incrustaciones** *f* COLOR, TRANSP MAR ship's bottom paint; ~ **de control**

térmico *f* COLOR thermal control paint; **~ para cristal recocido** *f* COLOR baked-glass painting; **~ Day-Glo®** *f* COLOR, P&C Day-Glo paint®; **~ de dispersión** *f* COLOR dispersion paint; **~ de emulsión** *f* COLOR, CONST, P&C, PROD emulsion paint; **~ para entorno húmedo** *f* COLOR sanitary paint; **~ al esmalte** *f* COLOR enamel paint, gloss paint, CONST gloss paint, REVEST enamel paint; **~ para exteriores** *f* COLOR, CONST outdoor paint; **~ externa** *f* COLOR outside paint; **~ de flotación** *f* TRANSP MAR boot topping; **~ fresca** *f* CONST wet paint; **~ fungicida** *f* CONST, PROD fungicide paint; **~ de grafito** *f* COLOR graphite paint; **~ ignífuga** *f* COLOR, CONST, SEG, TERMO fire-resistant paint, fireproof paint, fireproofing paint, flame-resistant paint, flameproof paint, flameproofing paint; **~ impermeable** *f* REVEST waterproof paint; **~ de imprimación** *f* COLOR, CONST priming paint, P&C, PROD primer; **~ de imprimación anticorrosiva** *f* P&C wash primer; **~ de imprimación permeable** *f* P&C permeable primer; **~ de imprimación con propiedades adhesivas y protectoras** *f* P&C wash primer; **~ incombustible** *f* P&C, REVEST, TERMO fireproof coating; **~ intumescente** *f* COLOR, REVEST intumescent paint; **~ luminiscente** *f* COLOR luminescent paint; **~ al látex** *f* COLOR latex paint; **~ luminosa** *f* COLOR luminous paint; **~ para madera** *f* COLOR, CONST wood paint; **~ metalizada** *f* COLOR, P&C metalized paint (*AmE*), metallized paint (*BrE*); **~ al óleo** *f* COLOR oil paint; **~ oleoresinosa** *f* CONST oleoresinous paint; **~ opaca** *f* COLOR, CONST flat paint; **~ para paredes** *f* COLOR wall paint; **~ penetrante** *f* COLOR penetrating paint; **~ de pigmento metálico** *f* COLOR metallic pigment paint; **~ a pistola** *f* COLOR, CONST spray paint; **~ plástica** *Esp*, **~ para plásticos** *f AmL* COLOR acetate color (*AmE*), acetate colour (*BrE*); **~ con plomo** *f* COLOR lead paint; **~ sin plomo** *f* COLOR, CONST lead-free paint; **~ al poliéster** *f* CONST polyester paint; **~ en porcelana** *f* C&V china painting, painting on porcelain; **~ previa al barnizado** *f* COLOR underglaze painting; **~ protectora** *f* C&V, REVEST protective paint; **~ protectora contra óxido** *f* COLOR rust-protective paint; **~ a prueba de fuego** *f* CONST, P&C, REVEST, TERMO fireproof coating; **~ por pulverización** *f* COLOR, CONST spray painting; **~ de pulverización electrostática** *f* COLOR, CONST electrostatic spray-painting; **~ rápida al óleo** *f* COLOR sharp oil paint; **~ de recubrimiento** *f* P&C, REVEST covering paint; **~ resistente al fuego** *f* COLOR, CONST, SEG, REVEST, TERMO fire-resistant paint, fireproof paint, flame-resistant paint, flameproof paint; **~ resistente a la intemperie** *f* COLOR weatherproof paint; **~ de rociar** *f* CONST, P&C spraying paint; **~ a rodillo** *f* COLOR, CONST roller painting; **~ de rodillo** *f* COLOR, CONST roller paint; **~ rugosa** *f* COLOR, P&C, REVEST wrinkle paint; **~ selladora** *f* COLOR, CONST sealing paint; **~ de silicato** *f* COLOR, REVEST silicate paint; **~ sombreadora** *f* COLOR shading paint; **~ para sombrear** *f* P&C shading paint; **~ al temple** *f* PROD distemper; **~ termosensible** *f* TERMO heat-sensitive paint; **~ con textura** *f* COLOR, CONST textured paint; **~ triestímulo** *f* P&C tristimulus paint; **~ para tubos** *f* COLOR tube paint; **~ vidriada** *f* COLOR, REVEST enamel paint; **~ en vidrio** *f* C&V painting on glass; **~ vítrea** *f* COLOR, REVEST enamel paint; **~ para viviendas** *f* COLOR, CONST house paint; **~ de zinc inorgánico** *f* COLOR inorganic zinc paint

pinza *f AmL* (*cf presilla AmL*) FOTO clamp, clip, ING MECÁ clamp, clip, LAB *para tubo de caucho* clamp, *soporte* clamp, MECÁ clip, TEXTIL clip, dart; **~ de amplitud por movimiento rápido** *f* INSTR latitude coarse-motion clamp; **~ para apretar tuberías** *f* CONST pipe twister; **~ de apriete** *f* ING MECÁ holding collet; **~ de cierre** *f* CINEMAT set clamp; **~ cocodrilo** *f* CINEMAT alligator clamp, ELEC, ING ELÉC alligator clip, crocodile clip, ING MECÁ crocodile clip; **~ para colgar el material revelado** *f* FOTO developing clip; **~ de conexión** *f* CINEMAT alligator clamp, ING ELÉC, ING MECÁ alligator clip; **~ de fijación** *f* ING MECÁ holding collet; **~ flotante** *f* AUTO floating tongs; **~ de freno fija** *f* AUTO fixed tongs; **~ de mandíbulas** *f* CINEMAT alligator clamp; **~ con mordazas** *f* LAB clamp with jaws; **~ para película** *f* FOTO film clip; **~ portapiezas** *f* ING MECÁ collet; **~ de retorta** *f* LAB retort clamp; **~ de talle** *f* TEXTIL waist dart; **~ terminal autodesprendible** *f* PROD self-lifting terminal clamp; **~ para tubos de goma** *f* ING MECÁ clamp

pinzas *f pl* C&V pinchers, ING MECÁ pliers, LAB tweezers; **~ ajustables** *f pl* ING MECÁ combination pliers; **~ del cortador** *f pl* C&V cutter's pliers; **~ de corte** *f pl* CONST, PROD cutting pliers; **~ para desatascar el agujero de colada** *f pl* PROD tapping bar; **~ de disección** *f pl* LAB *biología* dissecting scissors; **~ de gasista** *f pl* CONST gas pliers; **~ de junta de labios** *f pl* ING MECÁ lip-joint pliers; **~ de Mohr** *f pl Esp* (*cf presillas de Mohr AmL*) LAB Mohr's clips; **~ de mufla** *f pl* LAB crucible tongs; **~ portaobjetos** *f pl* INSTR stage clip; **~ para pruebas** *f pl* FOTO print tongs; **~ de punta de aguja** *f pl* ING MECÁ duckbill pliers, needle-nose pliers; **~ de puntas planas** *f pl* MECÁ, PROD flat-nose pliers, flat-nosed pliers; **~ rectas** *f pl* C&V straight pincers; **~ de retención** *f pl* PROD nippers; **~ para revelado** *f pl* CINEMAT, FOTO developing tongs; **~ de vidriero** *f pl* C&V glazier's pliers

piocha *f* PROD *herramienta* pick

piola *f* TRANSP MAR *construcción naval* lanyard

pión *m* FÍS, FÍS PART pi-meson, pion

pip *m* TELECOM *radar* pip

pipa: **~ del distribuidor** *f* AUTO, VEH *sistema de encendido* distributor arm

pipecolina *f* QUÍMICA pipecoline

piperacina *f* QUÍMICA piperazine

pipérico *adj* QUÍMICA piperic

piperidina *f* QUÍMICA piperidine

piperileno *m* QUÍMICA piperylene

piperonal *m* QUÍMICA heliotropin, piperonal

pipeta *f* LAB pipette, *análisis* burette, QUÍMICA *material de vidrio* burette; **~ de banda ancha** *f* QUÍMICA wideband pipette; **~ graduada** *f* LAB graduated pipette; **~ Pasteur** *f* LAB Pasteur pipette

pique: **~ de proa** *m* TRANSP MAR forepeak

piquera: **~ para la escoria** *f* PROD cinder notch

piqueta *f* CONST mattock, pick handle, pickax (*AmE*), pickaxe (*BrE*)

piquete *m* AGRIC trap, CONST *topografía* picket; **~ de anclaje** *m* CARBÓN anchor rod; **~ indicador de rasante** *m* FERRO *equipo inamovible* level indicator; **~ de toma de tierra** *m* FERRO earthing pole (*BrE*), earthing rod (*BrE*), grounding pole (*AmE*), grounding rod (*AmE*)

pirámide *f* GEOM pyramid; **~ ecológica** *f* CONTAM ecological pyramid; **~ truncada** *f* GEOM truncated pyramid

pirano *m* QUÍMICA pyran

piranosa *f* QUÍMICA pyranose

pirargirita *f* MINERAL pyrargyrite

pirazina *f* QUÍMICA pyrazine

pirazol *m* QUÍMICA pyrazol

pirazolina *f* QUÍMICA pyrazoline

pirazolona *f* QUÍMICA pyrazolone

pireno *m* QUÍMICA pyrene

piridazina *f* QUÍMICA pyridazin

piridina *f* QUÍMICA pyridine

piridona *f* QUÍMICA pyridone

pirimidina *f* QUÍMICA pyrimidine

pirita *f* GEOL, MINAS, MINERAL pyrite; **~ amarilla** *f* MINAS, MINERAL iron pyrite; **~ aurífera** *f* MINERAL auriferous pyrite; **~ blanca** *f* MINERAL white iron pyrite; **~ de cobre** *f* MINERAL copper pyrite; **~ cuprífera** *f* MINERAL copper pyrite; **~ de estaño** *f* MINERAL tin pyrite; **~ de hierro** *f* MINAS, MINERAL iron pyrite; **~ magnética** *f* MINERAL magnetic pyrite

pirítico *adj* MINAS, MINERAL pyritic

piroacético *adj* QUÍMICA pyroacetic

piroarsenato *m* QUÍMICA pyroarsenate

pirobórico *adj* QUÍMICA pyroboric

pirocatequina *f* QUÍMICA catechol, pyrocatechin

piroclástico *adj* GEOFÍS, GEOL, PETROL pyroclastic

pirocloro *m* MINERAL pyrochlore

pirocroíta *f* MINERAL pyrochroite

pirodesintegración *f* TEC PETR *refino* thermal cracking

piroelectricidad *f* CRISTAL, ELEC, TERMO pyroelectricity

piroeléctrico *m* CRISTAL, ELEC, TERMO pyroelectric

piroestato *m* FÍS pyrostat

pirofilita *f* MINERAL, TEC PETR pyrophyllite

pirofórico *adj* METAL, QUÍMICA pyrophoric

pirofosfato *m* QUÍMICA pyrophosphate

pirofosfórico *adj* QUÍMICA pyrophosphoric

pirofosforoso *adj* QUÍMICA pyrophosphorous

pirogálico *adj* FOTO, QUÍMICA pyrogallic

pirogalol *m* FOTO, QUÍMICA pyrogallol

pirogénico *adj* GEOFÍS pyrogenic

pirólisis *f* ALIMENT, GAS, QUÍMICA, TERMO pyrolysis; **~ de carburo metálico** *f* METAL carbide cracking; **~ catalítica a presión** *f* MECÁ cracking

pirolítico *adj* ALIMENT, GAS, QUÍMICA, TERMO pyrolytic

pirolusita *f* MINERAL pyrolusite

piromecónico *adj* QUÍMICA pyromeconic

piromelítico *adj* QUÍMICA pyromellitic

pirometría *f* GEN pyrometry

pirométrico *adj* GEN pyrometric

pirómetro *m* GEN pyrometer; **~ colorimétrico** *m* TERMO colorimetric pyrometer; **~ de filamento desvanecedor** *m* FÍS disappearing-filament pyrometer; **~ óptico** *m* FÍS, FÍS RAD, ÓPT optical pyrometer; **~ de radiación** *m* FÍS, FÍS RAD, GEOFÍS, INSTR, LAB radiation pyrometer; **~ de radiación total** *m* FÍS total radiation pyrometer

piromorfita *f* MINERAL pyromorphite

piromúcico *adj* QUÍMICA pyromucic

pirona *f* QUÍMICA pyrone

piropo *m* COLOR, MINERAL pyrope

pirorresistente *adj* GEN fire-resistant, fire-resisting, refractory

pirortita *f* MINERAL pyrorthite

pirosclerita *f* MINERAL pyrosclerite

piroscopio *m* FÍS pyroscope

pirostibita *f* MINERAL pyrostibite

pirosulfato *m* QUÍMICA pyrosulfate (*AmE*), pyrosulphate (*BrE*)

pirosulfito *m* QUÍMICA pyrosulfite (*AmE*), pyrosulphite (*BrE*)

pirosulfúrico *adj* QUÍMICA pyrosulfuric (*AmE*), pyrosulphuric (*BrE*)

pirosulfuril *m* QUÍMICA pyrosulfuryl (*AmE*), pyrosulphuryl (*BrE*)

pirotecnia *f* TEC ESP pyrotechnics

piroxenita *f* GEOL, MINERAL, PETROL pyroxenite

piroxeno *m* MINERAL pyroxene

piroxilina *f* AmL (*cf nitroalgodón Esp*) MINAS *explosivo* guncotton, QUÍMICA pyroxyline

pirrol *m* QUÍMICA pyrrole

pirrolidina *f* QUÍMICA pyrrolidine

pirrolidona *f* QUÍMICA pyrrolidone

pirrolina *f* QUÍMICA pyrroline

pirrotina *f* MINERAL iron pyrite, pyrrhotine

pirrotita *f* MINERAL pyrrhotite

piruvato *m* QUÍMICA pyruvate

pirúvico *adj* QUÍMICA pyruvic

pisahojas *m* IMPR hopper

piscicultor *m* OCEAN pisciculteur

piscicultura *f* OCEAN pisciculture; **~ marina** *f* OCEAN marine fish farming

piscina: **~ de combustible gastado** *f* Esp (*cf piscina de combustible quemado AmL*) NUCL discharge pond; **~ de combustible quemado** *f* AmL (*cf piscina de combustible gastado Esp*) NUCL discharge pond; **~ de desactivación** *f* AGUA cooling pond, NUCL decay cavity; **~ de enfriamiento** *f* NUCL cooling cavity; **~ de lodos** *f* PETROL *perforación* mud box; **~ de neutralización** *f* NUCL neutralization pond; **~ superior del recinto de contención** *f* NUCL upper containment pool; **~ de supresión de la presión** *f* NUCL pressure-suppression pool

piseta *f* LAB, QUÍMICA wash bottle

piso *m* CARBÓN sidewalk (*AmE*), bottom, floor, pavement (*BrE*), CINEMAT floor, CONST footway, causeway, pavement (*BrE*), sidewalk (*AmE*), floor, MINAS *de filones* footwall, bottom, level, ledger wall, *de la jaula de extracción* deck, sill *AmL*, *de galería* pavement, *de maniobras* battery, PROD floor; **~ de arado** *m* AGRIC ploughsole (*BrE*), plowsole (*AmE*); **~-base de trabajo** *m* TEC PETR rig floor; **~ de cargar** *m* MINAS plat; **~ de colada** *m* PROD *fundería* runner basin; **~ del equipo** *m* TEC PETR *perforación* derrick floor; **~ escalonado** *m* MINAS stope floor; **~ fijo** *m* MINAS fixed lift; **~ de galería** *m* MINAS drive; **~ de la galería de minas** *m* MINAS floor; **~ intermedio** *m* AmL (*cf galería falsa Esp*) MINAS blind level; **~ de loseta** *m* C&V terracotta floor; **~ de maniobra** *m* CARBÓN, MINAS *sobre la caña o comunicación de los pozos* brace; **~ de perforación** *m* PETROL drill floor; **~ de reposo** *m* MINAS stage; **~ del sótano** *m* PETROL, TEC PETR cellar deck; **~ superior** *m* C&V top floor; **~ del taladro** *m* TEC PETR *perforación* derrick floor, rig floor; **~ de terracota** *m* C&V terracotta floor; **~ de la torre** *m* TEC PETR *perforación* rig floor; **~ de trabajo** *m* TEC PETR *perforación* derrick floor

pisón *m* MINAS tamping rod, tamping stick, PROD stamp, *máquina* tamper; **~ de adoquinador** *m* CONST *pavimentación* rammer; **~ de aire**

comprimido *m* CARBÓN pneumatic ram, PROD *fundería* pneumatic rammer; ~ **de almadeneta** *m* PROD *bocartes* stamp head; ~ **chato** *m* PROD *fundería* flat rammer; ~ **de mano** *m* PROD hand rammer; ~ **de punta** *m* PROD *fundería* pegging peen, pegging rammer

pista *f* CONST track strip, INFORM&PD, ÓPT track, TEC ESP slot, TRANSP AÉR runway, TV track; ~ **en activo** *f* TRANSP AÉR active runway (*AmE*), runway in use (*BrE*); ~ **de aproximación por instrumentos** *f* TRANSP AÉR instrument approach runway; ~ **de aterrizaje** *f* TRANSP AÉR landing strip; ~ **de audio** *f* TV audio track; ~ **de audio de programas** *f* TV program audio track (*AmE*), programme audio-track (*BrE*); ~ **de auditoría** *f* INFORM&PD, PROD, TELECOM audit trail; ~ **avisadora** *f* CINEMAT, TV cue track; ~ **de aviso** *f* CINEMAT, TV cue track; ~ **de carreteo** *f* TRANSP AÉR taxiway; ~ **completa** *f* ACÚST full-track; ~ **de comprobación** *f* INFORM&PD audit trail; ~ **concéntrica** *f* ÓPT concentric track; ~ **de control** *f* ACÚST control track, TV control deck, control track; ~ **de control especial** *f* TV nonstandard control track; ~ **cronométrica** *f* INFORM&PD clock track; ~ **doble** *f* ACÚST dual track; ~ **espiral** *f* ÓPT spiral track; ~ **de grabación** *f* ACÚST, INFORM&PD recording track; ~ **de hielo** *f* REFRIG ice slab; ~ **húmeda** *f* IMPR wet track; ~ **sin instrumentos** *f* TRANSP AÉR noninstrument runway; ~ **libre** *f* INFORM&PD spare track; ~ **magnética compensadora** *f* TV balancing magnetic strip; ~ **de perforación** *f* INFORM&PD punching track; ~ **perimétrica** *f* TRANSP AÉR perimeter track; ~ **principal** *f* TRANSP AÉR primary runway; ~ **de pruebas** *f* TRANSP test track; ~ **de registro** *f* ACÚST, INFORM&PD recording track; ~ **de reloj** *f* INFORM&PD clock track; ~ **de reserva** *f* INFORM&PD spare track; ~ **de rodaje** *f* TRANSP AÉR apron taxiway, taxiway; ~ **de rodaje de salida rápida** *f* TRANSP, TRANSP AÉR, VEH rapid-exit taxiway; ~ **del rodamiento de bolas** *f* VEH ball-bearing race; ~ **semipreparada de aterrizaje** *f* TRANSP AÉR landing strip; ~ **en servicio** *f* TRANSP AÉR active runway (*AmE*), runway in use (*BrE*); ~ **de sonido** *f* ACÚST soundtrack, TV dialogue track (*BrE*), soundtrack; ~ **de sonido simétrica** *f* ACÚST, TV symmetrical soundtrack; ~ **sonora** *f* ACÚST, TV soundtrack; ~ **sonora de densidad variable** *f* TV variable-density track; ~ **unilateral** *f* ACÚST unilateral track; ~ **de video** *f* AmL, ~ **de vídeo** *f* Esp TV video track

pistas: ~ **por pulgada** *f pl* (*PPP*) IMPR, INFORM&PD tracks per inch (*TPI*)

pistilo *m* AGRIC pistil, LAB pestle

pistola *f* D&A gun; ~ **de aire** *f* AUTO airdraulic gun; ~ **de aire caliente** *f* ING MECÁ heat gun; ~ **de aire comprimido** *f* GEOFÍS airgun; ~ **para aplicar grapas** *f* EMB attaching gun; ~ **aspersora** *f* CONTAM MAR spray gun; ~ **de bengalas** *f* D&A flare pistol; ~ **de engrase** *f* AUTO, CONST, ING MECÁ, VEH grease gun; ~ **de extrusión** *f* ING MECÁ extrusion gun; ~ **de ligar** *f* ING MECÁ bonding gun; ~ **neumática** *f* AUTO airdraulic gun; ~ **de pintar** *f* EMB, P&C airbrush; ~ **reglamentaria** *f* D&A service pistol; ~ **de señales** *f* D&A flare pistol; ~ **de señales Very** *f* D&A, TRANSP MAR Very pistol; ~ **de soldar** *f* ELEC soldering gun

pistolete *m* TEC ESP *torpedos* igniter; ~ **de mina** *m* AmL (*cf barreno Esp*) CARBÓN jumper boring, MINAS

borer, drill, jumper, jumper drill, jumping drill, percussion drill, PROD jumper bar

pistón *m* AUTO piston, C&V plunger, CARBÓN *prensa hidráulica* ram pump, FÍS ram, piston, ING MECÁ piston, plunger, INSTAL HIDRÁUL piston, INSTR sucker, MECÁ piston, plunger, MINAS piston, P&C ram, PROD *bombas hidráulicas, prensas* plunger, TRANSP MAR, VEH *motor* piston; ~ **autotérmico** *m* AUTO autothermic piston; ~ **bimetálico** *m* AUTO bimetal piston; ~ **compensador** *m* INSTAL HIDRÁUL balance piston, balancing piston, dummy piston; ~ **con contravástago** *m* ING MECÁ piston with tailrod; ~ **de control** *m* ING MECÁ control piston; ~ **delantero** *m* AUTO front piston; ~ **elevador** *m* PROD lift piston, lifting piston; ~ **estriado** *m* AUTO ribbed piston; ~ **de falda estrecha** *m* AUTO full-slipper piston; ~ **de faldilla** *m* ING MECÁ plunger piston; ~ **inmovilizador** *m* ING MECÁ locking plunger; ~ **de media falda** *m* AUTO semislipper piston; ~ **de medio patín** *m* AUTO semislipper piston; ~ **neumático** *m* ING MECÁ air piston; ~ **de patín** *m* AUTO full-slipper piston; ~ **de plato** *m* INSTAL HIDRÁUL disc piston (*BrE*), disk piston (*AmE*); ~ **primario** *m* AUTO primary piston; ~ **ranurado** *m* AUTO ribbed piston; ~ **secundario** *m* AUTO secondary piston; ~ **tubular** *m* ING MECÁ plunger piston; ~ **unido a la biela** *m* AUTO, VEH piston locked to connecting rod; ~ **con válvula de charnela bola** *m* ING MECÁ piston with clack-valve; ~ **de la válvula de descarga de aceite** *m* AUTO oil relief valve plunger; ~ **de zapatilla** *m* AUTO full-slipper piston

pistonada *f* AUTO, ING MECÁ, VEH piston stroke

pistoneo *m* TEC PETR *perforación* swabbing

pistonófono *m* ACÚST, INSTAL HIDRÁUL pistonphone

pitada: ~ **corta** *f* TRANSP MAR *señales sonoras* short blast; ~ **larga** *f* TRANSP MAR *señales sonoras* long blast

pito: ~ **de alarma** *m* ING MECÁ alarm whistle

pitón *m* CONST dowel, screw eye, ING MECÁ nozzle

pittizita *f* MINERAL pitticite, pittizite

piválico *adj* QUÍMICA pivalic

pivote *m* ING MECÁ center point (*AmE*), centre point (*BrE*), gudgeon, pin, pivot, thrust bearing, INSTR, MECÁ, NUCL pivot, PROD *tambor del horno* gudgeon; ~ **de arrastre** *m* VEH *de remolque* bogie pivot (*BrE*), truck pivot (*AmE*); ~ **del bogie** *m* VEH *de remolque* bogie pin (*BrE*), bogie pivot (*BrE*), truck pin (*AmE*), truck pivot (*AmE*); ~ **de la bomba del aceite** *m* AUTO oil pump spindle; ~ **central** *m* ING MECÁ center pin (*AmE*), centre pin (*BrE*), kingbolt, kingpin, pintle; ~ **de la dirección** *m* ING MECÁ kingbolt, kingpin, VEH steering knuckle; ~ **de empuje** *m* ING MECÁ thrust block; ~ **del enganche** *m* VEH *de remolque* bogie pin (*BrE*), truck pin (*AmE*); ~ **eyector** *m* ING MECÁ ejection pin, ejector pin; ~ **de horquilla** *m* ING MECÁ, MECÁ clevis pin; ~ **de la horquilla de desembrague** *m* AUTO, VEH throw-out fork pivot; ~ **de orientación de la rueda** *m* ING MECÁ, MECÁ kingbolt, kingpin; ~ **de quinta rueda** *m* VEH fifth-wheel kingpin; ~ **de remolque** *m* VEH *de remolque* bogie pin (*BrE*), bogie pivot (*BrE*), truck pin (*AmE*), truck pivot (*AmE*); ~ **de la rótula de dirección** *m* AUTO, VEH steering knuckle pin; ~ **de sujeción de la boquilla** *m* AUTO nozzle holder spindle

pixel *m* IMPR, INFORM&PD pixel

pixeles: **~ por pulgada** *m pl* INFORM&PD pixels per inch

pizarra *f* C&V shale, CONST board, *piedra* slate, GEOL shale, slate, TEC PETR slate; **~ bituminosa** *f Esp* (*cf pizarra carbonosa candeloide AmL*) GEOL bituminous shale, MINAS jack, PETROL, TEC PETR oil shale; **~ carbonosa candeloide** *f AmL* (*cf pizarra bituminosa Esp*) GEOL bituminous shale, MINAS jack, PETROL, TEC PETR oil shale; **~ del itinerario real de vuelo** *f* TRANSP AÉR flight-progress board; **~ litográfica** *f* QUÍMICA lithographic slate; **~ moteada** *f* GEOL spotted slate; **~ negra** *f* GEOL black shale; **~ yesífera** *f* GEOL gypsiferous shale

pizarrosidad *f* GEOL flow cleavage, schistosity

pizarroso *adj* GEOL schistose

PLA *abr* (*matriz lógica programada*) INFORM&PD PLA (*programmed logic array*)

placa¹: **de ~ inferior** *adj* ELECTRÓN off-board; **de ~ interna** *adj* GEOL *composición tectónica* within-plate; **sobre la ~** *adj* ELECTRÓN on-board;

placa² *f* C&V pan, plate, CARBÓN *metalurgia* mat, CONST, ELEC plate, ELECTRÓN *circuitos impresos* board, GEOL plate, INFORM&PD, INFORM&PD *circuitos* board, platter, ING MECÁ plate, LAB *Esp* (*cf parrilla AmL*) hotplate, MECÁ, METAL, PROD plate, QUÍMICA tray;

■ a **~ absorbente** *f* ENERG RENOV, NUCL absorber plate; **~ de acero soldada** *f* CONST bonded steel plate; **~ de acumulador** *f* ELEC accumulator plate; **~ aislante** *f* ING ELÉC insulating plate; **~ alimentadora** *f* ING MECÁ feed plate; **~ de amalgamación** *f* CARBÓN amalgamation plate; **~ de amalgamación interna** *f* ING MECÁ inside amalgamation plate; **~ de análisis de manchas** *f* LAB spotting plate; **~ de anclaje** *f* CARBÓN anchor plate, CONST anchoring plate; **~ de antehogar** *f* PROD dumb plate; **~ antirozamiento** *f* TEC ESP antifret plate; **~ de apoyo** *f* CONST base plate, VEH *frenos* backing plate; **~ arenisca** *f* GEOL flag; **~ articulada** *f* ING MECÁ hinged panel; **~ de asiento** *f* CONST bedplate, wall plate, *de montaje* base plate, ING MECÁ base plate, MECÁ baseplate; **~ de aviso** *f* LAB safety placard;

■ b **~ base** *f* ELECTRÓN motherboard, INFORM&PD back plate, motherboard, MECÁ baseplate; **~ de base** *f* PROD *de chumacera* sole plate; **~ de base de cámara** *f* CINEMAT baseplate; **~ de base para troquelados finos** *f* ING MECÁ baseplate for fine blanking; **~ de batería** *f* AUTO, ELEC battery plate; **~ de la batería** *f* ING ELÉC battery plate; **~ de blindaje** *f* MECÁ armour plate (*AmE*), armour plate (*BrE*); **~ de borde** *f* CONST edge plate;

■ c **~ de caballete** *f* CONST *tejado* ridgeplate; **~ cabecera** *f* CONST head plate; **~ calentadora eléctrica** *f* ELEC, TERMO electric hot-plate; **~ calorífica** *f* ING MECÁ heating panel; **~ de características del motor** *f* PROD motor nameplate; **~ de centrado** *f* FERRO *infraestructura* center plate (*AmE*), centre plate (*BrE*); **~ de CI** *f* (*placa de circuito impreso*) ELEC, ELECTRÓN, INFORM&PD, TELECOM, TV PC board (*printed-circuit board*); **~ de cimentación** *f* CONST foundation plate, TEC ESP sole plate; **~ del circuito** *f* ELEC, ELECTRÓN circuit board; **~ de circuito impreso** *f* (*PCI, placa de CI*) ELEC, ELECTRÓN, INFORM&PD, TELECOM, TV printed-circuit board (*PC board, PCB*); **~ de circuito impreso de**

alta frecuencia *f* ELECTRÓN high-frequency printed-circuit board; **~ de circuito impreso a doble cara** *f* ELECTRÓN double-sided printed-circuit board; **~ para colocar un moldeo en el torno** *f* PROD main casting; **~ de colodión** *f* FOTO collodion plate; **~ con colodión de bromuro de plata** *f* FOTO silver bromide collodion plate; **~ de condensador** *f* ELEC, FÍS, ING ELÉC capacitor plate; **~ de conexión a tierra** *f* ELEC, ING ELÉC earth plate (*BrE*), ground plate (*AmE*); **~ del constructor** *f* ING MECÁ, MECÁ name-plate; **~ de contacto** *f* P&C feel plate; **~ de contactos** *f* ING ELÉC wafer; **~ de contención** *f* ING MECÁ retaining plate; **~ continental** *f* GEOFÍS *porción de litosfera terrestre*, GEOL continental plate; **~ continua** *f* ALIMENT continuous cooker; **~ correctora** *f* INSTR correcting plate; **~ correctora asférica** *f* TV aspheric corrector plate; **~ de criba** *f* PROC QUÍ sieve plate; **~ de cuarto de onda** *f* FÍS quarter-wave plate; **~ de cubierta** *f* ENERG RENOV *de colector solar* cover plate; **~ de cubierta de concreto** *f* AmL (*cf placa de cubierta de hormigón Esp*) CONST concrete roofing tile; **~ de cubierta de hormigón** *f* Esp (*cf placa de cubierta de concreto AmL*) CONST concrete roofing tile; **~ cubierta de terminales** *f* PROD terminal cover plate; **~ de cultivo** *f* AGRIC, LAB *bacteriología* culture plate, Petri dish;

■ d **~ de damas** *f* PROD *altos hornos* dam plate; **~ de defensa** *f* PROD safety guard plate; **~ deflectora** *f* CONST baffle plate, baffler, FÍS RAD deflection plate, PAPEL baffle, PROC QUÍ deflector plate, REFRIG *del aire* baffle plate, TEC ESP deflection plate, TRANSP AÉR deflector plate; **~ de deflexión vertical** *f* ELECTRÓN vertical deflection plate; **~ del delantal** *f* PAPEL apron board; **~ de deriva** *f* ING MECÁ drift plate; **~ de desgaste** *f* ING MECÁ wear plate, wearing plate; **~ de deslizamiento** *f* ING MECÁ drift plate; **~ de desviación** *f* ELECTRÓN deflection plate, ING MECÁ baffle, PROC QUÍ deflecting plate, TV deflection plate; **~ de desviación horizontal** *f* ELECTRÓN horizontal deflection plate, FÍS horizontal deflecting plate, TV horizontal deflection plate; **~ de desviación horizontal** *f* TV X plate; **~ desviadora** *f* ELECTRÓN, TV deflection plate; **~ directriz** *f* ING MECÁ, PROD guide plate; **~ divisora** *f* ING MECÁ index plate;

■ e **~ de ebullición** *f* PROC QUÍ boiling plate; **~ de endurecimiento de fondos** *f* REFRIG *para enfriar chocolate* bottomer slab; **~ de entintado** *f* C&V tint plate; **~ del estator** *f* ING ELÉC stator plate; **~ estrangulador** *f* AUTO, VEH choker plate; **~ experimental para componentes electrónicos** *f* INFORM&PD breadboard; **~ extrema del inducido** *f* ELEC *máquina* armature end plate; **~ del eyector** *f* P&C ejector plate; **~ eyectora** *f* ING MECÁ stripper plate;

■ f **~ de fijación del carro** *f* ING MECÁ core slide retaining plate; **~ de fondo** *f* MECÁ baseplate, PROD *de molde* bedplate, TRANSP MAR *tanques portátiles* head plate; **~ del fondo** *f* PROD bottom plate; **~ de fondo del inducido** *f* ELEC *máquina* armature end plate; **~ de forro** *f* TRANSP MAR *tanques portátiles* head plate; **~ fotográfica** *f* IMPR, INSTR photographic plate; **~ de fricción** *f* PROD friction plate; **~ frontal** *f* ELECTRÓN *de tubo catódico* faceplate, ING MECÁ faceplate, *torno* apron of a lathe; **~ de fundación** *f* ING MECÁ base plate, bedplate; **~ de fundición** *f* METAL, PROD iron plate, plate iron;

- g ~ **de galga** *f* PROD gaging plate (*AmE*), gauging plate (*BrE*); ~ **giratoria** *f* FERRO turntable, ING MECÁ, MECÁ, PROD swivel, TRANSP turntable; ~ **graduadora** *f* ING MECÁ index plate; ~ **gruesa** *f* ING MECÁ platen; ~ **de guarda** *f* CONST key drop, FERRO *vehículos* axle guide, PROD *vagones* guard plate; ~ **guía** *f* ING MECÁ, PROD guide plate, VEH *cambio de velocidades automático* gate; ~ **guía de garra** *f* ING MECÁ finger-guide plate;

- h ~ **de hornear** *f* ALIMENT baking sheet;

- i ~ **de identificación** *f* FERRO *infraestructura* signal identification plate, ING MECÁ, MECÁ nameplate; ~ **de identificación e instrucciones** *f* SEG, TEC ESP indicator plate; ~ **de identificación del vehículo** *f* AUTO vehicle tagging; ~ **identificadora del nombre** *f* ING MECÁ, MECÁ nameplate; ~ **de impacto** *f* CARBÓN *lámpara termoiónica* impact plate; ~ **inferior** *f* CONST bottom plate; ~ **de instrucciones** *f* PROD instruction plate; ~ **intensificadora de silicio** *f* ELECTRÓN silicon intensifier target;

- j ~ **de junta** *f* ING ELÉC junction plate;

- l ~ **lado posterior** *f* ING MECÁ back-end plate; ~ **limitadora** *f* FOTO stop plate; ~ **lógica** *f* ELECTRÓN, INFORM&PD logic card; ~ **de lógica mixta** *f* ELECTRÓN mixed-logic board;

- m ~ **de machos** *f* ING MECÁ, PROD core plate; ~ **del manguito** *f* INSTAL HIDRÁUL shuttle plate; ~ **de la máquina** *f* C&V machine tray; ~ **de matrícula** *f* ING MECÁ, VEH *carrocería* license plate (*AmE*), number plate (*BrE*); ~ **matriz** *f* ELECTRÓN motherboard, INFORM&PD back plate, motherboard; ~ **matriz sin bebedero** *f* ING MECÁ runnerless mold plate (*AmE*), runnerless mould plate (*BrE*); ~ **matriz sin canal de colada** *f* ING MECÁ runnerless mold plate (*AmE*), runnerless mould plate (*BrE*); ~ **metálica de viga** *f* CONST boom plate; ~ **metálica widmanstätten** *f* METAL widmanstätten plate; ~ **de microcanal** *f* ELECTRÓN microchannel plate; ~ **modelo de madera** *f* PROD *fundería* matchboard; ~ **para el montaje** *f* INSTR baseplate; ~ **de montaje instrumental** *f* INSTR instrument-mounting plate; ~ **de montura del objetivo** *f* FOTO lens mounting plate; ~ **móvil** *f* ING ELÉC, TRANSP *condensador variable* rotor plate;

- n ~ **negativa** *f* AUTO negative plate; ~ **negra** *f* FOTO dark slide;

- o ~ **de obturación** *f* TRANSP AÉR blanking plate; ~ **obturadora** *f* AUTO, VEH choker plate; ~ **de onda media** *f* FÍS half-wave plate; ~ **ondulada** *f* CONST corrugated iron, corrugated sheet iron;

- p ~ **de la palomilla** *f* AUTO throttle plate; ~ **pantalla** *f* NUCL screen plate; ~ **de pecho** *f* PROD conscience; ~ **perforada** *f* CONST punched plate, NUCL perforated plate; ~ **plana** *f* ENERG RENOV, FÍS FLUID flat plate; ~ **de porcelana** *f* C&V porcelain plate; ~ **portaestampa** *f* PRENSA bolster plate, cavity plate; ~ **portamodelo** *f* PROD carded pattern, *fundería* match plate, pattern plate; ~ **positiva** *f* AUTO positive plate; ~ **posterior** *f* ELECTRÓN *tubos de cámara* back plate, TV backplate; ~ **de presión** *f* AUTO, CINEMAT, FOTO pressure plate; ~ **de presión autoascensible** *f* PROD self-lifting pressure plate; ~ **de presión desmontable** *f* FOTO detachable pressure plate; ~ **de presión del embrague** *f* ING MECÁ clutch pressure plate; ~ **de**

presión montada sobre resortes *f* FOTO spring-mounted pressure plate; ~ **de presión con muelle** *f* CINEMAT spring pressure plate; ~ **protectora** *f* ING MECÁ wear plate, PROD covering plate; ~ **protectora del freno** *f* AUTO, VEH brake shield; ~ **protectora del techo** *f* NUCL roof-shielding plate; ~ **de pruebas** *f* C&V test plate; ~ **puente** *f* CONST bridge plate; ~ **de PVC** *f* ING ELÉC *tubo vacío* PVC sheath;

- q ~ **que cubre el cubo** *f* TRANSP AÉR hub cover plate;

- r ~ **de recubrimiento** *f* ING MECÁ cover, cover plate, PROD *fundería* top plate; ~ **de refuerzo** *f* ING MECÁ stiffening plate; ~ **de refuerzos de noyos** *f* ING MECÁ core plate; ~ **de registro** *f* CONST manhole plate; ~ **con remaches** *f* CONST riveted plate; ~ **de retención** *f* ING MECÁ retaining plate; ~ **de retención del carro** *f* ING MECÁ core slide retaining plate; ~ **de retención del expulsor** *f* ING MECÁ ejector retaining-plate; ~ **de retención del eyector** *f* ING MECÁ ejector retaining-plate; ~ **de retención de la maza** *f* ING MECÁ core slide retaining plate; ~ **de retención del portaestampas** *f* ING MECÁ core slide retaining plate; ~ **roblonada** *f* CONST riveted plate; ~ **de rolado del sacador** *f* C&V former roller; ~ **roscada de relojero** *f* ING MECÁ watch screw plate; ~ **de roscar** *f* ING MECÁ screw plate; ~ **de roscar en dos partes** *f* ING MECÁ two-part screw plate; ~ **de roscar en dos piezas** *f* ING MECÁ two-part screw plate; ~ **rotulada** *f* ING MECÁ, MECÁ nameplate;

- s ~ **de sangría** *f* PROD *altos hornos* guard plate; ~ **de seguridad** *f* ING MECÁ locking plate, lockplate; ~ **de señal** *f* ELECTRÓN signal plate; ~ **sensible al tacto** *f* INFORM&PD touchpad; ~ **soporte** *f* NUCL perforated plate; ~**-soporte** *f* TEC PETR *operaciones costa-fuera* bed plate, template; ~ **soporte central** *f* ING MECÁ center-bearing plate (*AmE*), centre-bearing plate (*BrE*); ~ **subordinada** *f* ELECTRÓN daughter board; ~ **de sujeción** *f* ING MECÁ locking plate, lockplate, bridge plate, NUCL tie plate; ~ **de sujeción superior** *f* NUCL upper tie plate; ~ **superior** *f* ING MECÁ top plate; ~ **superior del núcleo** *f* NUCL upper grid; ~ **de sustentación** *f* CARBÓN mat;

- t ~ **de tamiz** *f* PROC QUÍ sieve plate; ~ **terminal** *f* EMB, ING MECÁ end panel; ~ **terminal del inducido** *f* ELEC armature end plate; ~ **tricroma** *f* FOTO three-color plate (*AmE*), three-colour plate (*BrE*); ~ **tubular** *f* NUCL tube plate, PROD *de calderas* end plate, flue plate, flue sheet; ~ **tubular de caja de fuegos** *f* ING MECÁ back-tube sheet;

- u ~ **de unión** *f* PROD bonding strip; ~ **de unión interna** *f* ING MECÁ inside amalgamation plate;

- v ~ **del vértice** *f* TELECOM *de reflector de antena* vertex plate; ~ **de vidrio** *f* C&V, LAB glass plate

placas[1]: **entre** ~ *adj* GEOL *composición tectónica* intraplate

placas[2]: ~ **aislantes sin amianto** *f pl* SEG asbestos-free insulating plates; ~ **deflectantes verticales** *f pl* FÍS vertical deflecting plates

placer *m* CARBÓN gold diggings, gravel mine, MINAS placer, placer dirt

placeres: ~ **de oro** *m pl* MINAS placer gold

plaga *f* AGRIC pest

plagioclasa *f* MINERAL plagioclase

plan *m* CINEMAT schedule, INFORM&PD layout, ING MECÁ scheme, lay, PROD schedule, TRANSP MAR *navegación, arquitectura naval* design, plan; **~ de Babcock** *m* TELECOM Babcock plan; **~ de la cámara** *m* TRANSP MAR *construcción naval* cabin sole; **~ de canalización con dos frecuencias** *m* TELECOM two-frequency channeling plan (*AmE*), two-frequency channelling plan (*BrE*); **~ de construcción** *m* CONST construction schedule; **~ para contingencias** *m* CONTAM MAR, NUCL, SEG contingency plan; **~ de cultivo** *m* AGRIC cropping plan, planting scheme; **~ de doble nivel** *m* TELECOM dual-level plan; **~ de emergencia** *m* CONTAM MAR, SEG emergency plan; **~ de emergencia de la central** *m* NUCL site emergency plan; **~ de emergencia del emplazamiento** *m* NUCL site emergency plan; **~ de entregas** *m* PROD delivery schedule; **~ de explotación agrícola** *m* AGRIC farming plan; **~ general** *m* ING MECÁ outline; **~ de muestreo** *m* METR sampling plan; **~ de numeración privado** *m* TELECOM private-numbering plan (*PNP*); **~ operacional en vuelo** *m* TRANSP AÉR in-flight operational planning; **~ de pedidos** *m* PROD ordering policy; **~ de producción** *m* CINEMAT production schedule, PROD production plan, TV production schedule; **~ de protección contra incendios** *m* SEG, TERMO fire-protection plan; **~ reticular** *m* TELECOM lattice plan; **~ de seguridad** *m* TELECOM security policy; **~ de situación de pilotes** *m* CARBÓN pile situation plan; **~ de trabajo** *m* TEC ESP work package; **~ de trabajos** *m* CONST construction program (*AmE*), construction programme (*BrE*); **~ de trincaje** *m* TRANSP MAR *estibaje* lashing plan; **~ de vuelo** *m* TRANSP AÉR flight plan; **~ de vuelo archivado en vuelo** *m* TRANSP AÉR air-filed flight plan; **~ de vuelo repetitivo** *m* TRANSP AÉR repetitive flight plan

plana: **~ doble** *f* IMPR centre spread; **~ exterior** *f* IMPR outer form (*AmE*), outer forme (*BrE*); **~ interior** *f* IMPR inner form (*AmE*), inner forme (*BrE*)

planador *m* ING MECÁ planisher

planar *adj* GEOM planar

plancha *f* CARBÓN iron, CONST plate, slab, IMPR, ING MECÁ, MECÁ plate, P&C laminated sheet, TRANSP MAR *construcción naval* plate; **~ de ala** *f* CONST *estructuras metálicas* flange plate; **~ de aluminio anodizada** *f* IMPR anodized aluminium plate (*BrE*), anodized aluminum plate (*AmE*); **~ de arrastre de buceo** *f* OCEAN diving plate; **~ del azul** *f* IMPR blue printer; **~ bimetálica** *f* IMPR bimetal plate; **~ de blindaje** *f* MECÁ armor plate (*AmE*), armour plate (*BrE*); **~ de cerradura** *f* CONST *cerrajería* striking plate; **~ de cobre** *f* COLOR copperplate, CONST copper sheet, IMPR copperplate; **~ de cristal** *f* FOTO glass pressure plate; **~ de cubierta** *f* CONST flange plate; **~ para cubiertas** *f* TRANSP MAR deck plate; **~ deflector** *f* TRANSP AÉR deflector plate; **~ de desembarco** *f* TRANSP MAR gangway; **~ de desgaste** *f* ING MECÁ wearing plate; **~ electrónica** *f* ELEC, ELECTRÓN electronic plate; **~ enrollable al cilindro** *f* IMPR wrapround plate; **~ de estereotipia** *f* IMPR stereoplate, stereotype plate; **~ de fondo** *f* TRANSP MAR *tanques portátiles* head plate; **~ de forro** *f* TRANSP MAR *tanques portátiles* head plate; **~ para hacer rebotar la luz** *f* CINEMAT bounce board; **~ de impresión de matriz** *f* IMPR molded printing plate (*AmE*), moulded printing plate (*BrE*); **~ de impresión producida por láser** *f* IMPR laser-produced printing plate; **~ laminada** *f* P&C laminated sheet;

~ para libros *f* IMPR bookplate; **~ de línea** *f* IMPR line plate; **~ litográfica** *f* IMPR lithoplate; **~ para mamparas** *f* TRANSP MAR *construcción naval* bulkhead plate; **~ manual** *f* TEXTIL iron; **~ negativa** *f* IMPR negative plate, negative working plate; **~ del negro** *f* IMPR black printer; **~ de pantoque** *f* TRANSP MAR *construcción naval* bilge plate; **~ de papel** *f* IMPR paper plate; **~ pasada por rodillos conformados** *f* METAL sized rolled flat iron; **~ de paso entre buques** *f* TRANSP MAR *para embarco y desembarco* gangplank; **~ perforadora para cribas** *f* CARBÓN, FOTO screen plate; **~ del peto de popa** *f* TRANSP MAR *construcción naval* transom plate; **~ polimetálica** *f* IMPR multimetal plate; **~ sensible a la luz** *f* IMPR light-sensitive plate; **~ soporte** *f* MECÁ baseplate; **~ de tipos** *f* IMPR type plate; **~ trimetálica** *f* IMPR trimetallic plate; **~ de zinc** *f* IMPR zinc plate

planchada *f* TEC PETR *perforación* derrick floor, drilling floor, rig floor; **~ de perforación** *f* TEC PETR derrick floor

planchado *m* TEXTIL ironing, pressing

planchar *vt* TEXTIL iron

planchas: **~ del costado** *f pl* TRANSP MAR *construcción naval* side plating; **~ del fondo** *f pl* TRANSP MAR *construcción naval* bottom plating; **~ del forro exterior** *f pl* TRANSP MAR *construcción naval* shell plating

planchear *vt* TRANSP MAR *construcción naval* plate

planchero: **~ complementario de entrada** *m* PROD complementary input rack; **~ de machos** *m* PROD *fundición* core rack

plancheta *f* CONST *topografía* plane table, D&A *de artillería* range card

planeación *f* AmL (*cf perspectiva de planificación Esp*) PROD planning horizon; **~ ambiental** *f* AmL (*cf planificación ambiental Esp*) CONTAM environmental planning

planeador *m* D&A, TRANSP AÉR glider; **~ de órbita** *m* TEC ESP orbital glider; **~ orbital** *m* TEC ESP, TRANSP orbital glider; **~ de transporte** *m* D&A transport glider

planeamiento[1]: **de ~ en firme** *adj* PROD firm-planned

planeamiento[2]: **~ de listas de materiales** *m* PROD planning bill of material; **~ de las necesidades de producción** *m* PROD capacity requirement planning; **~ de los recursos para la fabricación** *m* PROD manufacturing resource planning; **~ de los requerimientos de materiales** *m* PROD material requirement planning

planear *vt* TELECOM *una serie de actividades* schedule

planeidad *f* ING MECÁ, PAPEL flatness, PROD levelness

planeo *m* TRANSP AÉR gliding flight; **~ en espiral** *m* TRANSP AÉR spiral glide

planero *m* TRANSP MAR hydrographic survey vessel

planeta *m* TEC ESP planet; **~ interior** *m* TEC ESP inner planet; **~ telúrico** *m* TEC ESP telluric planet

planetario *adj* MECÁ, TEC ESP planetary

planicie *f* CONST bottom; **~ abisal** *f* GEOFÍS, PETROL abyssal plain; **~ aluvial** *f* AGUA, HIDROL alluvial plain, flood plain; **~ de inundación** *f* AGUA, HIDROL flood plain

planificación *f* TELECOM planning; **~ ambiental** *f* Esp (*cf planeación ambiental AmL*) CONTAM environmental planning; **~ de la calidad** *f* CALIDAD quality planning; **~ de inserciones** *f* IMPR insertion schedule; **~ en períodos de tiempo** *f* PROD timed phased planning; **~ previa al vuelo** *f* TRANSP AÉR preflight planning; **~ de la producción** *f* CALIDAD, PROD production planning; **~ de programas** *f* INFORM&PD program scheduling; **~ de programas de producción** *f* PROD industrial engineering; **~ rural** *f* AGRIC regional planning

planificador *m* INFORM&PD scheduler; **~ de trabajo** *m* INFORM&PD job scheduler

planilla *f* INFORM&PD spread; **~ electrónica** *f* INFORM&PD spreadsheet

planimetría *f* GEOM planimetry

planímetro *m* CONST, GEOM planimeter

planisferio *m* TRANSP MAR star chart

plano¹ *adj* CONST flat, GEOL planar, GEOM planar

plano² *m* AmL (*cf trazado Esp*) CONST draft, FÍS plane, GEOM drawing, plane, ING MECÁ layout, draft, scheme, MECÁ layout, TRANSP MAR *arquitectura naval* drawing, plan, design; **~ aerodinámico** *m* TRANSP AÉR aerofoil (*BrE*), airfoil (*AmE*); **~ de alojamientos** *m* TRANSP MAR accommodation plan; **~ antideriva** *m* TRANSP MAR *construcción naval* fin keel; **~ de apoyo** *m* ING MECÁ *carretón, bogies* bolster; **~ automático** *m* ING MECÁ self-acting plane; **~ axial** *m* GEOL axial plane; **~ basal** *m* METAL basal plane; **~ básico del casco** *m* TRANSP MAR *construcción naval* hull drawing; **~ de Benioff** *m* GEOL Benioff plane; **~ bifilar** *m* TRANSP flat twin; **~ de capacidades** *m* TRANSP MAR *buque de carga* capacity plan; **~ de carga** *m* TRANSP MAR cargo plan; **~ de clivaje** *m* CRISTAL, QUÍMICA cleavage plane; **~ de cola** *m* TRANSP AÉR tail unit, tailplane; **~ de los conductos** *m* PROD duct layout; **~ conjugado** *m* FÍS, METAL conjugate plane; **~ de conjunto** *m* CONST, ING MECÁ general plan; **~ de construcción** *m* CONST construction plan, TRANSP MAR *máquinas* engineering drawing; **~ de construcción mecánica** *m* ING MECÁ engineering drawing; **~ constructivo** *m* CONST construction plan; **~ constructivo de mamparas** *m* TRANSP MAR *diseño de barcos* bulkhead plan; **~ de control** *m* TRANSP AÉR control plane; **~ de cortadura** *m* METAL plane of shear; **~ de corte** *m* METAL plane of shear; **~ cristalográfico** *m* CRISTAL, INSTR crystallographic plane; **~ de cuadernas** *m* TRANSP MAR *diseño de barcos* frame plan; **~ de la cubierta superior resistente** *m* TRANSP MAR *diseño de barcos* deck plan; **~ de deriva** *m* TRANSP MAR *construcción naval* fin; **~ de deriva vertical** *m* TRANSP AÉR tail fin; **~ de descompresión** *m* OCEAN decompression chart; **~ de deslizamiento** *m* CRISTAL glide plane, slip plane, GEOL slickenside, NUCL *de un cristal* glide plane; **~ de detalles** *m* CINEMAT very close-up, TV detail rendition; **~ de diaclasa** *m* GEOL joint plane; **~ de disposición general** *m* CONST, ING MECÁ, MECÁ, TRANSP MAR *construcción naval* general-arrangement drawing, general-arrangement plan; **~ del doble fondo** *m* TRANSP MAR *arquitectura naval* double-bottom plan; **~ E** *m* FÍS ONDAS, ING ELÉC *microondas* E plane; **~ de ensamblaje** *m* ING MECÁ assembly plan; **~ equipotencial** *m* ELEC, FÍS, GEOFÍS, ING ELÉC, TEC ESP equipotential surface; **~ espacial** *m* TEC ESP space plane; **~ de estiba** *m* TRANSP MAR cargo plan; **~ estructural** *m* METAL habit plane; **~ de fabricación** *m* PROD production drawing; **~ de falla** *m* GEOL fault plane; **~ de flotación** *m* TRANSP MAR *arquitectura naval* waterplane; **~ de flotación de proyecto** *m* TRANSP MAR *diseño de barcos* design

waterplane; **~ focal** *m* CINEMAT, ELECTRÓN, FÍS, FÍS RAD, FOTO, TV focal plane; **~ focal anterior** *m* CINEMAT, FOTO front focal plane; **~ focal posterior** *m* CINEMAT, FOTO rear focal plane; **~ de formas** *m* TRANSP MAR *diseño de barcos* body plan, lines drawing, lines plan; **~ fotográfico** *m* IMPR contour map; **~ general** *m* CINEMAT establishing shot, wide shot, CONST general plan, ING MECÁ general plan, combined diagram, NUCL general drawing, TV establishing shot; **~ del giróscopo** *m* TRANSP AÉR gyroplane; **~ H** *m* ING ELÉC H plane; **~ del hábito cristalino** *m* CRISTAL habit plane; **~ hidrodinámico** *m* TRANSP AÉR hydrofoil; **~ horizontal** *m* GEOM horizontal plane, TRANSP MAR *construcción naval* half-breadth plan; **~ horizontal de cola** *m* TRANSP AÉR tailplane; **~ de imagen** *m* CINEMAT image plane; **~ de incidencia** *m* FÍS plane of incidence; **~ inclinado** *m* AmL (*cf plano inclinado de criba Esp*) CARBÓN slope, CONST gravity plane, incline, inclined plane, FÍS inclined plane, GEOFÍS ramp, GEOM inclined plane, MINAS slope, *preparación de minerales* jig, PETROL, TEC PETR inclined ramp; **~ inclinado automotor** *m* ING MECÁ incline; **~ inclinado automotor para criba de minerales** *m* CARBÓN jig bed, MINAS jig; **~ inclinado automotor de simple efecto** *m* MINAS jig brow; **~ inclinado de criba** *m* Esp (*cf plano inclinado AmL*) CARBÓN slope, CONST gravity plane, incline, inclined plane, FÍS inclined plane, GEOFÍS ramp, GEOM inclined plane MINAS slope, *preparación de minerales* jig, PETROL, TEC PETR inclined ramp; **~ inclinado helicoidal** *m* MINAS spiral chute; **~ inclinado de vía única y contrapeso** *m* MINAS jig plane; **~ indicador de posiciones** *m* FÍS position-indicator plan; **~ de instalación** *m* TV set-up diagram; **~ de junta** *m* ING MECÁ parting; **~ lejano** *m* CINEMAT long shot; **~ longitudinal** *m* TRANSP MAR *construcción naval* sheer drawing; **~ de longitudinales** *m* TRANSP MAR *diseño de barcos* longitudinals plan, sheer draft; **~ de lucha contra incendios** *m* TRANSP MAR fire-control plan; **~ de luminancia cero** *m* TV zero-luminance plane; **~ de maclado** *m* METAL twinning plane; **~ de maclaje** *m* METAL twinning plane; **~ medio** *m* CINEMAT medium shot, mid-shot; **~ de montaje** *m* PROD *para maquinaria* erection plan; **~ de montaje general** *m* CONST general-layout drawing, ING MECÁ general assembly, general-layout drawing; **~ neutro de la cinta** *m* TV tape neutral plane; **~ de nivel** *m* ING MECÁ, MECÁ datum line; **~ nodal** *m* FÍS, FOTO nodal plane; **~ del núcleo** *m* ING ELÉC core plane; **~ óptico** *m* METR optical flat; **~ parcial** *m* ING MECÁ section drawing; **~ de la película** *m* CINEMAT film plane; **~ piramidal** *m* METAL pyramidal plane; **~ de polarización** *m* ÓPT plane of polarization; **~ posterior** *m* ELECTRÓN, INFORM&PD, PROD backplane; **~ principal** *m* TRANSP AÉR mainplane; **~ principal de crucero** *m* CARBÓN, MINERAL face cleat; **~ de producción** *m* AmL (*cf dibujo de producción Esp*) PROD production drawing; **~ de recorrido** *m* ELEC *circuito* layout; **~ de referencia** *m* (*PR*) CONST, GEOL datum plane (*DP*); **~ de referencia de la célula del aparato** *m* TRANSP AÉR airframe reference plane; **~ de reflexión** *m* CRISTAL mirror plane; **~ de refracción de las olas** *m* TRANSP MAR *diseño de barcos* wave-refraction diagram; **~ de restricción de nuevas construcciones** *m* AGUA, CONST zoning plan;

~ **reticular** *m* CRISTAL lattice plane; ~ **de la sección transversal** *m* TRANSP MAR *diseño de barcos* cross-sectional drawing; ~ **secundario de crucero** *m* CARBÓN, GEOL end cleat; ~ **de separación** *m* GEOL contact, ING MECÁ parting; ~ **de simetría** *m* CRISTAL mirror plane, ING MECÁ, METAL plane of symmetry, TRANSP MAR *diseño de barcos* sheer plan; ~ **de situación** *m* CINEMAT, TV establishing shot; ~ **de la superficie de los trabajos subterráneos** *m* MINAS plat; ~ **de tierra** *m* FÍS, GEOM earth plane (*BrE*), ground plane (*AmE*); ~ **topográfico** *m* CONST contour map, PAPEL map; ~ **total** *m* ING MECÁ combined diagram; ~ **de tracas de la cubierta superior resistente** *m* TRANSP MAR *diseño de barcos* deck plan; ~ **de traslación** *m* CRISTAL, METAL translation plane; ~ **de urbanismo** *m* CONST town planning map; ~ **de las válvulas de cubierta** *m* TEC PETR manifold; ~ **de velamen** *m* TRANSP MAR sail plan; ~ **de verificación** *m* ING MECÁ gage plane (*AmE*), gauge plane (*BrE*); ~ **vertical** *m* GEOM vertical plan

planos: ~ **intersectantes** *m pl* GEOM intersecting planes; ~ **principales** *m pl* FÍS principal planes; ~ **que se cortan** *m pl* GEOM intersecting planes

planta *f* GEN plant, CONST floor, plant; ~ **de agua pesada** *f* NUCL heavy-water plant; ~ **de aire acondicionado** *f* INSTAL TERM, REFRIG air-conditioning plant; ~ **de alquilación** *f* DETERG alkylation plant; ~ **de amasado** *f* CONST batch plant; ~ **de amolar** *f* PROC QUÍ grinding plant; ~ **anual** *f* AGRIC annual, winter annual; ~ **asfáltica** *f* CONST asphalt plant; ~ **aumentadora** *f* ENERG RENOV booster mill; ~ **de caldeo** *f* TERMO heating plant; ~ **de carbonización** *f* MINAS coking plant; ~ **central** *f* ELEC *fuente de alimentación* power plant; ~ **central de energía** *f* ING ELÉC central power plant; ~ **de clarificación** *f* AGUA clarification plant; ~ **clasificadora** *f* AGRIC grading plant; ~ **de compresión** *f* TERMO compression plant; ~ **de concreto** *f AmL* (*cf planta de hormigón Esp*) CONST concrete batching and mixing plant; ~ **de condensación** *f* PROC QUÍ condensing plant; ~ **de conversión** *f* ELEC *transformador*, ING ELÉC converting station; ~ **de conversión de uranio** *f* NUCL uranium-conversion plant; ~ **de cracking** *f* TEC PETR *refino* cracking plant; ~ **de cracking catalítico** *f* TEC PETR *refino* catalytic-cracking plant; ~ **de depuración** *f* GEN purification plant, purifying plant; ~ **de depuración de aguas residuales** *f* AGUA, HIDROL, RECICL sewage works, wastewater-treatment plant; ~ **de depuración de aguas residuales comunitarias** *f* AGUA community sewage works, sewage works, HIDROL, RECICL community sewage works; ~ **depuradora** *f* AGUA, HIDROL, RECICL purification plant; ~ **depuradora de aguas** *f* AGUA, HIDROL, RECICL water treatment plant; ~ **desalinizadora** *f* AGUA, CONST, INSTAL HIDRÁUL, PROC QUÍ desalination plant; ~ **desmineralizadora** *f* NUCL demineralizing plant; ~ **de difusión en contracorriente** *f* NUCL countercurrent-diffusion plant; ~ **de dragados** *f* MINAS dredge plant; ~ **eléctrica** *f* ELEC *fuente de alimentación* generating set, TELECOM power plant; ~ **de energía** *f* ELEC, TELECOM power plant; ~ **de energía hidroeléctrica** *f* ING ELÉC hydroelectric power plant; ~ **de energía solar fotovoltaica** *f* ENERG RENOV, ING ELÉC photovoltaic solar-power plant; ~ **de energía termo-**

eléctrica *f* ING ELÉC thermal-electric power plant; ~ **de enriquecimiento** *f* NUCL ore enrichment plant; ~ **de enriquecimiento de menas** *f* NUCL ore enrichment plant; ~ **de enriquecimiento por ultracentrifugado** *f* NUCL ultracentrifuge enrichment plant; ~ **extractora de grasa** *f* AGRIC rendering plant; ~ **de filtrado** *f* CARBÓN filtration plant, screening plant; ~ **filtradora** *f* AGUA, HIDROL filter plant; ~ **filtradora de rápidos** *f* AGUA, HIDROL rapid filter plant; ~ **de flotación** *f* PROC QUÍ flotation plant; ~ **de flotación por espuma** *f* MINAS froth flotation plant; ~ **frigorífica** *f* ING MECÁ, REFRIG, TERMO refrigerating plant; ~ **de fuerza** *f* TELECOM power plant; ~ **de generación** *f* ING ELÉC power plant; ~ **generadora** *f* ELEC, TELECOM power plant; ~ **generadora de vapor** *f* INSTAL TERM steam plant; ~ **geotérmica** *f* ELEC, ENERG RENOV geothermal plant; ~ **guacha** *f* AGRIC volunteer; ~ **para hacer hielo** *f* TERMO ice-making plant; ~ **de hormigón** *f Esp* (*cf planta de concreto AmL*) CONST concrete batching and mixing plant; ~ **huésped** *f* AGRIC host plant; ~ **de lavado** *f* PROD washing plant; ~ **de lixiviación** *f* CARBÓN leaching plant; ~ **de manejo del carbón** *f* CARBÓN, MINAS coal-handling plant; ~ **para la manipulación del combustible** *f* INSTAL TERM, TERMO fuel-handling plant; ~ **de mezclado** *f* TEC PETR *ingeniería de lodos, perforación* blending plant; ~ **motriz** *f* TRANSP AÉR power plant; ~ **de óleo-transformación** *f* RECICL oil regeneration plant; ~ **de ozonización** *f* HIDROL ozonization plant; ~ **petroquímica** *f* QUÍMICA, PETROL, TEC PETR petrochemical plant; ~ **piloto** *f* CARBÓN pilot plant; ~ **de potencia eléctrica solar** *f* ENERG RENOV, ING ELÉC solar electric power plant; ~ **de preparación de residuos para el almacenamiento** *f* NUCL storage head-end plant; ~ **de procesamiento** *f* GAS processing facility; ~ **de proceso de alimentos** *f* ALIMENT food processing plant; ~ **de producción de energía eléctrica** *f* ELEC, ENERG RENOV, ING ELÉC, NUCL generating plant; ~ **de purificación** *f* GEN purification plant, purifying plant; ~ **rastrera** *f* MINAS crawler base; ~ **de reacondicionamiento** *f* PROD, RECICL reprocessing plant; ~ **de recuperación de ácidos** *f* RECICL, TEC PETR acid recovery plant; ~ **de regeneración de aceite** *f* RECICL oil regeneration plant; ~ **de reprocesamiento químico** *f* NUCL, PROC QUÍ chemical reprocessing plant; ~ **para reproducción** *f* AGRIC stock plant; ~ **de secado de gas** *f* GAS gas-drying plant; ~ **de separación de isótopos de uranio** *f* NUCL uranium isotope separation plant; ~ **de sinterización** *f* PROC QUÍ sintering plant; ~ **de sondeo** *f* CONST *perforación* boreholing plant; ~ **textil** *f* AGRIC fiber crop (*AmE*), fibre crop (*BrE*); ~ **de transformación** *f AmL* (*cf planta de tratamiento Esp*) MINAS processing plant; ~ **de transformación de aceites** *f* RECICL oil regeneration plant; ~ **transformadora** *f* PROD, RECICL reprocessing plant; ~ **de tratamiento** *f Esp* (*cf planta de transformación AmL*) MINAS processing plant; ~ **de tratamiento de aguas** *f* CONST water treatment plant; ~ **de tratamiento de aguas residuales** *f* AGUA, HIDROL, RECICL wastewater-treatment plant; ~ **de tratamiento de basura** *f* RECICL waste treatment plant; ~ **de tratamiento completa** *f* AGUA comprehensive treatment plant; ~ **de tratamiento de desechos** *f* RECICL waste treatment plant; ~ **de**

tratamiento de efluentes *f* AGUA, HIDROL, RECICL effluent treatment plant; **~ de tratamiento de fangos activados** *f* HIDROL activated sludge treatment plant; **~ de tratamiento de residuos** *f* RECICL sewage treatment plant, waste treatment plant; **~ trituradora** *f* CARBÓN, PROC QUÍ crushing plant; **~ venenosa** *f* AGRIC poison plant; **~ vivaz** *f* AGRIC herbaceous perennial

plantación *f* AGRIC plantation, planting; **~ a raíz desnuda** *f* AGRIC bare-root method

plantel *m* *AmL* AGRIC nucleus breeding herd

plantilla *f* ELEC *medición, fabricación* template, IMPR *hoja de montaje* goldenrod, INFORM&PD template, ING MECÁ standard gauge (*BrE*), gauge (*BrE*), jig, gage (*AmE*), template, templet, standard gage (*AmE*), INSTR, LAB gage (*AmE*), gauge (*BrE*), MECÁ template, gage (*AmE*), gauge (*BrE*), MINAS jig, PROD pattern, template, patron, SEG *para máquinas cortadoras* jig, TEXTIL, TRANSP MAR, VEH *herramienta* template; **~ de cuatro círculos concéntricos de campo cercano** *f* ÓPT, TELECOM four-concentric-circle near-field template; **~ de devanado** *f* PROD former; **~ de fresado** *f* ING MECÁ milling jig; **~ de fundición** *f* MECÁ casting pattern; **~ guía** *f* ING MECÁ former; **~ de índice de refracción de cuatro círculos concéntricos** *f* ÓPT, TELECOM four-concentric-circle refractive-index template; **~ Johansson** *f* MECÁ Johansson gage (*AmE*), Johansson gauge (*BrE*); **~ litográfica** *f* FÍS RAD lithographic mask; **~ de montaje** *f* IMPR flat, ING MECÁ assembly jig; **~ posicionadora** *f* MECÁ jig; **~ de precisión** *f* ING MECÁ close tolerance spacer; **~ de recesos** *f* TRANSP AÉR jig pit; **~ sujetadora** *f* ING MECÁ fixture; **~ para taladrar** *f* CARBÓN jig, ING MECÁ drilling jig, drilling template, jig; **~ del teclado** *f* INFORM&PD keyboard template; **~ transparente** *f* CINEMAT transparent overlay

plantillero *m* PROD pattern maker

plántula *f* AGRIC seedling

plaqueado *m* MECÁ cladding, PROD plating; **~ galvánico** *m* REVEST galvanic plating

plaqueta *f* INFORM&PD chip, P&C *pigmento* platelet; **~ de encargo** *f* INFORM&PD custom chip; **~ iniciadora** *f* D&A, TEC ESP *proyectiles balísticos* detonating card; **~ de silice** *f* INFORM&PD silicon chip; **~ VLSI** *f* INFORM&PD VLSI chip; **~ de voz** *f* INFORM&PD speech chip

plaquita: ~ numeradora *f* ING MECÁ number plate

plasma *m* GEN plasma; **~ de hidrógeno** *m* GAS hydrogen plasma; **~ imantado** *m* NUCL magnetized plasma; **~ inductor** *m* GAS inductive plasma; **~ magnético** *m* NUCL magnetoplasma; **~ térmico** *m* GAS thermal plasma

plasmatrón *m* GAS plasmatron

plaste *m* CONST, C&V putty

plasticidad *f* C&V, CARBÓN, METAL, P&C plasticity; **~ del cristal** *f* CRISTAL, METAL crystal plasticity; **~ en frío** *f* P&C cold flow

plástico[1]**: no ~** *adj* CONST, P&C nonplastic

plástico[2] *m* GEN plastic; **~ acrílico** *m* EMB, P&C acrylic plastic; **~ armado** *m* C&V reinforced plastic, ELEC *aislador* laminated plastic, EMB reinforced plastic, P&C reinforced plastic, *producto semielaborado* laminated plastic; **~ armado con fibra de vidrio** *m* C&V, EMB, P&C, SEG glass-fiber reinforced plastic (*AmE*), glass-fibre reinforced plastic (*BrE*); **~ armado con fuerza de vidrio** *m* C&V, P&C glass-fiber reinforced

plastic (*AmE*), glass-fibre reinforced plastic (*BrE*); **~ biodegradable** *m* P&C, RECICL biodegradable plastic; **~ celular** *m* P&C, REFRIG cellular plastic, foamed plastic; **~ celular de celdillas abiertas** *m* P&C, REFRIG open-cell foamed plastic; **~ celular de celdillas cerradas** *m* P&C, REFRIG closed-cell foamed plastic; **~ celular de célula abierta** *m* P&C, REFRIG open-cell cellular plastic; **~ celular de célula cerrada** *m* P&C, REFRIG closed cell cellular plastic; **~ celular expandido** *m* P&C, REFRIG expanded cellular plastic; **~ celular extruido** *m* P&C, REFRIG extruded cellular plastic; **~ endurecido** *m* ING MECÁ thermoset plastic®; **~ esponjoso** *m* P&C foamed plastic; **~ espumado** *m* P&C foamed plastic; **~ expandido** *m* P&C, REFRIG expanded plastic; **~ fenólico** *m* P&C phenolic plastic; **~ formado por capas superpuestas** *m* ELEC, P&C laminated plastic; **~ lamelado** *m* ELEC, P&C laminated plastic; **~ laminado** *m* ELEC laminated plastic, P&C plastic-based laminate, TRANSP MAR laminated plastic; **~ moldeado** *m* P&C molded plastic (*AmE*), moulded plastic (*BrE*); **~ reforzado** *m* C&V, EMB, P&C reinforced plastic; **~ reforzado con fibra de carbono** *m* P&C carbon fiber reinforced plastic (*AmE*), carbon fibre reinforced plastic (*BrE*); **~ reforzado con fibra de vidrio** *m* (*GRP*) C&V, EMB, P&C, SEG glass-fiber reinforced plastic (*AmE*), glass-fibre reinforced plastic (*BrE*); **~ reforzado con vidrio** *m* (*GRP*) C&V, EMB, P&C, SEG glass-reinforced plastic (*GRP*); **~ rígido** *m* P&C rigid plastic; **~ termofraguado** *m* ING MECÁ thermoset plastic®

plastidio *m* QUÍMICA plastid

plastificado: no ~ *adj* P&C unplasticized; **~** *adj* REVEST plastic-coated

plastificante *m* CONST, P&C, QUÍMICA plasticizer; **~ externo** *m* P&C external plasticizer; **~ interno** *m* P&C internal plasticizer; **~ no migratorio** *m* P&C nonmigratory plasticizer

plastificar *vt* P&C plasticize

plastisol *m* P&C plastisol

plastómero *m* QUÍMICA, TEC PETR plastomer

plastómetro *m* IMPR, P&C plastimeter, plastometer

plata *f* (*Ag*) C&V, FOTO, ING ELÉC, METAL, QUÍMICA silver (*Ag*); **~ brillante** *f* C&V bright silver; **~ para bruñir** *f* C&V burnishing silver; **~ bruta** *f* C&V bullion; **~ residual** *f* FOTO, METAL residual silver

platabanda *f* CONST *estructuras metálicas* flange plate, platband, ING MECÁ cover strip

plataforma *f* *Esp* (*cf piso AmL*) CONST platform, bed, deck, *general* apron, FERRO *infraestructura*, ING MECÁ platform, INSTR stage, stand, MINAS *de la jaula de extracción* deck, PROD *de horno* landing, platform, lamp, TEC PETR *perforación* platform, drilling floor, TRANSP MAR *construcción naval* hull; **~ de abrasión** *f* OCEAN *erosión de costas* wave-cut bench; **~ de acero** *f* TEC PETR steel platform; **~ con alojamiento** *f* TEC PETR hotel platform; **~ de alojamiento** *f* TEC PETR flotel; **~ de apoyo** *f* TEC PETR booster platform; **~ arponera** *f* TRANSP MAR *pesca* pulpit; **~ astillero** *f* PETROL footboard (*AmE*), monkey board (*BrE*); **~ de aterrizaje de helicópteros** *f* TRANSP AÉR, TRANSP MAR helicopter landing deck; **~ de aterrizaje de un helipuerto** *f* TRANSP AÉR heliport deck; **~ de autobús** *f* AUTO bus bay; **~ autoelevable** *f* TEC PETR jack-up rig; **~ autolevadiza** *f* PETROL jack-up platform; **~ basculante** *f* TRANSP tipping platform; **~ en la**

boca del pozo *f* MINAS bracket; **~ de carbonatos** *f* GEOL carbonate platform; **~ de carga** *f* CONST, FERRO loading platform, MINAS landing stage, landing station, PROD *hornos* charging platform, TRANSP loading platform; **~ de carga sencilla** *f* EMB single-faced pallet; **~ de cemento** *f* TEC PETR concrete platform; **~ civil de helicópteros** *f* TRANSP AÉR helistop; **~ de concreto** *f* AmL (*cf plataforma de hormigón Esp*) TEC PETR concrete platform; **~ continental** *f* GEOFÍS, GEOL, OCEAN, TEC PETR *submarina* continental shelf, *tierra, suelo firme* continental platform; **~ continental insular** *f* GEOFÍS, GEOL, OCEAN insular shelf; **~ de corte de cosechadoras** *f* AGRIC auger table; **~ costa-fuera** *f* TEC PETR offshore platform; **~ deltaica** *f* GEOFÍS, GEOL, OCEAN deltaic platform; **~ de despegue y aterrizaje de helicópteros** *f* TRANSP AÉR helipad; **~ doble** *f* TRANSP AÉR dual platform; **~ de drenaje** *f* TRANSP AÉR drainage terrace; **~ de elevación** *f* TRANSP AÉR lifting platform; **~ de elevación portátil** *f* CONST portable hoisting-platform; **~ elevada** *f* FERRO elevated platform; **~ elevadora** *f* PROD platform lift; **~ de embarque** *f* TRANSP boarding platform; **~ de embolsado** *f* AGRIC bagging platform; **~ del encuellador** *f* TEC PETR *perforación* derrick monkey board, footboard (*AmE*), monkey board (*BrE*), stabbing board; **~ de enganche** *f* Esp (*cf anchurón de enganche AmL*) MINAS pit landing, platt; **~ de ensacado** *f* AGRIC bagging platform; **~ de entibado** *f* MINAS walling stage; **~ de espera** *f* TRANSP AÉR holding apron; **~ estabilizada** *f* TEC ESP stabilized platform; **~ estable** *f* ING MECÁ inertial platform; **~ de exploración** *f* TEC ESP scan platform, exploration platform; **~ fija metálica** *f* TEC PETR jacket platform; **~ flotante** *f* OCEAN floating platform; **~ flotante amarrada** *f* TEC PETR tethered buoyant platform (*TBP*); **~ giratoria** *f* FERRO, TRANSP AÉR turntable; **~ giro-estabilizada** *f* TEC ESP gyro-stabilised platform; **~ giroscópica** *f* TRANSP AÉR gyroscopic platform; **~ de gravedad** *f* TEC PETR gravity platform; **~ de guía** *f* ING MECÁ guide ramp; **~ para helicópteros** *f* CONTAM MAR helicopter pad; **~ de helipuerto** *f* TEC PETR helipad; **~ híbrida** *f* TEC PETR hybrid platform; **~ de hielo** *f* GEOFÍS, GEOL, OCEAN ice shelf; **~ de hormigón** *f* Esp HIDROL (*cf batiente de concreto AmL*) *presas* concrete apron, TEC PETR (*cf plataforma de concreto AmL*) concrete platform; **~ hotel** *f* TEC PETR accommodation platform, accommodation rig, hotel rig; **~ inercial** *f* ING MECÁ inertial platform; **~ inercial de desmontaje** *f* TEC ESP strapdown inertial platform; **~ inercial de navegación** *f* TEC ESP inertial navigation platform; **~ de inspección** *f* CONST inspection platform; **~ lanzacohetes** *f* D&A, TEC ESP rocket-launching site; **~ de lanzamiento** *f* D&A launch station, TEC ESP launch platform; **~ de lanzamiento de misiles teledirigidos** *f* D&A *submarinos* ramp; **~ levitante** *f* TRANSP hover pallet; **~ litoral** *f* GEOFÍS, GEOL, OCEAN coastal platform; **~ marina** *f* TEC PETR offshore platform; **~ para mercancías** *f* TRANSP AÉR pallet; **~ de montaje** *f* TRANSP AÉR jig pit; **~ móvil** *f* CONST moving platform, moving sidewalk (*AmE*), traveling platform (*AmE*), travelling platform (*BrE*), PROD *transporte de material*, TRANSP dolly; **~ del pararrayos** *f* ING MECÁ lightning brace; **~ de perforación** *f* PETROL, TEC PETR, TRANSP MAR *buque*

drilling platform; **~ plana** *f* OCEAN flat shelf; **~ de producción** *f* PETROL, TEC PETR production platform; **~ de producción metálica** *f* TEC PETR jacket; **~ protectora de pozo** *f* TEC PETR jacket; **~ de pruebas** *f* TEC ESP test stand; **~ de pruebas del motor** *f* TEC ESP engine test stand; **~ de pruebas oftálmicas** *f* INSTR ophthalmic test stand; **~ para recogida de datos** *f* METEO data collection platform; **~ para la recolección de datos** *f* TEC ESP *vehículos* data collection platform (*DCP*); **~ de refuerzo** *f* TEC PETR booster platform; **~ de las rotativas** *f* IMPR deck; **~ semisumergible** *f* TEC PETR semisubmersible platform, semisubmersible rig; **~ de sinterización** *f* PROC QUÍ sintering platform; **~ de sondeo** *f* TEC PETR rig; **~ de sondeos** *f* Esp (*cf instalación de sondeos AmL*) MINAS boring plant, drilling plant; **~ submarina** *f* TRANSP MAR *geografía* shelf; **~ sumergible** *f* TEC PETR submersible platform; **~ terminal** *f* TEC PETR terminal platform; **~ de la torre** *f* PETROL footboard (*AmE*), monkey board (*BrE*); **~ de trabajo** *f* C&V, CONST working platform; **~ de transporte** *f* CONST low boy trailer; **~ de tuberías** *f* PETROL pipe deck; **~ del tubo de rayos X** *f* INSTR X-ray tube stand; **~ universal** *f* INSTR universal stage

platea *f* AGUA floor

plateado *m* ACÚST, C&V, PROD silvering, silver plating; **~ de espejos** *m* C&V mirror-plating; **~ galvanoplástico** *m* ELEC, ING ELÉC, REVEST electrosilvering

platear *vt* ING ELÉC plate; **~ por electrólisis** *vt* ELEC, ING ELÉC, REVEST electrosilver

platería *f* ALIMENT silverware

platformación *f* TEC PETR *refino* platforming

platillo *m* ING MECÁ pan, disc (*BrE*), disk (*AmE*), LAB *balanza* pan, MECÁ plate, METR *de balanza o báscula* bowl; **~ de balanza** *m* ING MECÁ scale pan; **~ distribuidor de semilla** *m* AGRIC seed plate; **~ fijador** *m* ING MECÁ lockplate; **~ tarado** *m* LAB weighing dish

platina *f* CINEMAT platen, IMPR bed, imposing table, ING MECÁ platen, INSTR, LAB *microscopio* stage, MECÁ plate, P&C *parte de equipo* platen, PAPEL *de pila holandesa* beater plate, bed plate, TEXTIL *género de punto* sinker; **~ calentada** *f* P&C heated platen; **~ de Fedorov** *f* CRISTAL Fedorov stage, universal stage; **~ giratoria** *f* LAB *microscopio* revolving stage; **~ magnetofónica** *f* ACÚST tape deck; **~ microscópica** *f* INSTR *turbina de vapor* microscopic stage; **~ del microscopio** *f* INSTR microscope stage; **~ de la perforadora** *f* TEC PETR bed plate; **~ universal** *f* CRISTAL Fedorov stage, universal stage, U stage

platínico *adj* QUÍMICA platinic

platino *m* (*Pt*) ELEC, METAL, QUÍMICA platinum (*Pt*); **~ del ruptor** *m* AUTO, VEH contact breaker-point (*BrE*), points (*AmE*)

platinocloruro *m* QUÍMICA platinochloride

platinoiridio *m* QUÍMICA platiniridium, platinum-iridium

plató *m* CINEMAT lot, set

plato *m* INFORM&PD platter, ING MECÁ chuck, pan, plate, platen, QUÍMICA tray, dish, plate; **~ accionador** *m* TRANSP AÉR actuating plate; **~ de accionamiento del embrague** *m* AUTO, VEH clutch drive plate; **~ adaptador** *m* ING MECÁ adaptor plate; **~ de agujeros** *m* ING MECÁ chuck face; **~ de agujeros**

con cuatro garras *m* ING MECÁ chuck faceplate dogs; ~ **ahorquillado** *m* ING MECÁ fork chuck; ~ **con ajuste espiral** *m* ING MECÁ scroll chuck; ~ **al aire** *m* ING MECÁ face chuck; ~ **alimentador** *m* ING MECÁ feed plate; ~ **de arrastre** *m* ING MECÁ driver chuck, driver plate; ~ **autocentrador** *m* ING MECÁ self-centering chuck (*AmE*), self-centring chuck (*BrE*); ~ **autocentrante** *m* ING MECÁ scroll chuck; ~ **base en U** *m* ING MECÁ U-shaped base plate; ~ **de burbujeo** *m* PROC QUÍ bubble tray; ~ **de campana de cuatro tornillos** *m* ING MECÁ four-screw bell chuck; ~ **central de soporte** *m* ING MECÁ center-bearing plate (*AmE*), centre-bearing plate (*BrE*); ~ **combinado** *m* ING MECÁ combination chuck; ~ **para conectar las barrenas** *m* TEC PETR *perforación* bit breaker; ~ **para conectar las brocas** *m* TEC PETR *perforación* bit breaker; ~ **sin contrapuntos** *m* ING MECÁ face lathe; ~ **de control de engranaje planetario** *m* AUTO sun gear control plate; ~ **de copa** *m* ING MECÁ cup chuck; ~ **de cuatro garras independiente** *m* ING MECÁ four-jaw independent chuck; ~ **de cubierta del cubo** *m* TRANSP AÉR hub; ~ **de desgaste** *m* ING MECÁ wear plate, wearing plate; ~ **divisor** *m* ING MECÁ index dial, index plate, division plate; ~ **del eje del piñón** *m* AUTO pinion shaft flange; ~ **del eje del piñón de ataque** *m* VEH pinion shaft flange; ~ **electromagnético** *m* ELECTRÓN, ING MECÁ electromagnetic chuck; ~ **electromagnético rectangular para trabajos pesados** *m* ELECTRÓN, ING MECÁ heavy-duty rectangular magnetic chuck; ~ **de empuje** *m* ING MECÁ thrust plate, VEH *embrague* pressure plate; ~ **estacionario** *m* ING MECÁ stationary plate; ~ **excéntrico** *m* ING MECÁ eccentric chuck; ~ **eyector** *m* ING MECÁ ejector plate; ~ **de fijación** *m* ING MECÁ clamping plate; ~ **fijo** *m* ING MECÁ stationary plate; ~ **fijo del freno** *m* AUTO, VEH brake anchor-plate; ~ **de garras** *m* ING MECÁ jaw chuck, dog chuck, prong chuck; ~ **de garras independientes** *m* ING MECÁ combination chuck; ~ **graduador** *m* ING MECÁ index plate; ~ **de mordazas convergentes** *m* ING MECÁ draw-in chuck; ~ **de mordazas independientes** *m* ING MECÁ combination chuck; ~ **neumático** *m* IMPR air chuck, ING MECÁ *tornos* air chuck, air-operated chuck; ~ **perforado** *m* PROC QUÍ sieve tray; ~ **de pinzas** *m* ING MECÁ collet chuck; ~ **de porcelana** *m* C&V porcelain plate; ~ **portacojinete** *m* PROD *en el torno* die chuck; ~ **portafreno** *m* AUTO, ING MECÁ, MECÁ, VEH brake drum; ~ **portamachos** *m* ING MECÁ tapping chuck; ~ **portamandrín** *m* ING MECÁ catch plate; ~ **de prensa flotante** *m* ING MECÁ floating platen; ~ **de presión** *m* AUTO, ING MECÁ, PROD pressure plate; ~ **de presión del embrague** *m* ING MECÁ clutch pressure plate; ~ **de púas** *m* ING MECÁ prong chuck; ~ **de puntas** *m* ING MECÁ prong chuck; ~ **de quijadas convergentes** *m* ING MECÁ draw-in chuck; ~ **de resortes de mordazas convergentes** *m* ING MECÁ draw-in spring chuck; ~ **de resortes de quijadas convergentes** *m* ING MECÁ draw-in spring chuck; ~ **de retención** *m* ING MECÁ catch plate; ~ **de retención del expulsor** *m* ING MECÁ ejector retaining-plate; ~ **de retención del eyector** *m* ING MECÁ ejector retaining-plate; ~ **de roscar** *m* PROD *en el torno* die chuck; ~ **de rueda** *m* FERRO *vehículos* wheel web; ~ **de rueda plano** *m* ING MECÁ plain hub-flange; ~ **de sujeción concéntrico de tres garras** *m* ING

MECÁ three-jaw concentric gripping chuck; ~ **superior** *m* ING MECÁ top plate; ~ **de tetragarra independiente** *m* ING MECÁ four-jaw independent chuck; ~ **del tornillo** *m* ING MECÁ vice plate (*BrE*), vise plate (*AmE*); ~ **de tornillos** *m* ING MECÁ bell chuck; ~ **de tres garras** *m* ING MECÁ three-jaw chuck; ~ **de tres mordazas** *m* ING MECÁ spur chuck; ~ **de tres puntos** *m* ING MECÁ fork chuck, three-pronged chuck; ~ **de tulipa** *m* ING MECÁ fork chuck, spur chuck, three-pronged chuck; ~ **universal** *m* ELECTRÓN faceplate, ING MECÁ combination chuck, faceplate; ~ **universal al aire** *m* ING MECÁ faceplate chuck; ~ **de vidrio** *m* C&V glass dish

platos: ~ **hondos** *m pl* C&V hollow ware

plattnerita *f* MINERAL plattnerite

playa *f* GEOL, OCEAN beach, strand; ~ **baja** *f* GEOL, HIDROL foreshore; ~ **de clasificación** *f* TRANSP classification siding; ~ **elevada** *f* GEOFÍS raised beach; ~ **de grava** *f* OCEAN shingle beach; ~ **de guijarros** *f* OCEAN shingle beach

plaza: ~ **de aparcamiento** *f Esp* (*cf zona de estacionamiento AmL*) CONST parking area; ~ **de estacionamiento** *f AmL* (*cf zona de aparcamiento Esp*) CONST parking area

plazo *m* CINEMAT deadline; ~ **de construcción** *m* CONST construction time; ~ **de ejecución** *m* CONST completion time; ~ **entre la iniciación de una compra y el recibo del material** *m* PROD procurement lead time; ~ **de entrega** *m* ING MECÁ lead time, PROD delivery lead time, lead time; ~ **parcial a la fecha** *m* PROD partial time to date; ~ **de reaprovisionamiento** *m* PROD replenishment lead time; ~ **de vencimiento** *m* PROD due date

plazoleta *f* CONST piazzetta

pleamar *f* HIDROL flood tide, OCEAN flood tide, high tide, TRANSP MAR high tide, high water

plegable *adj* ING MECÁ, PROD collapsible, foldaway, folding

plegadizo *adj* ING MECÁ, PROD foldaway, folding

plegado[1] *adj* ING MECÁ, PROD folding; ~ **y clasificado** *adj* IMPR folded and collated

plegado[2] *m* ING ELÉC bend, PAPEL folding, PROD bending, folding; ~ **en acordeón** *m* IMPR accordion fold, concertina fold, over-and-back fold, zigzag fold, PAPEL accordion fold, PROD fanfold; ~ **en ángulo recto** *m* IMPR chopper fold; ~ **a bolsillo** *m* IMPR buckle-folder machine; ~ **en cuarto sin corte** *m* IMPR French fold; ~ **en cuatro sin cortes** *m* IMPR French folder; ~ **paralelo** *m* IMPR jaw fold; ~ **con plegadora de mordaza** *m* IMPR jaw fold; ~ **en zigzag** *m* IMPR accordion fold, concertina fold, over-and-back fold, zigzag fold

plegador[1] *adj* ING MECÁ folding

plegador[2] *m* C&V crimper, ELECTRÓN, PROD folding, TEXTIL beam; ~ **final** *m* ING MECÁ terminal crimper; ~ **medio seccional** *m* TEXTIL *para el tricotado por urdimbre* half sectional beam; ~ **posterior** *m* TEXTIL back beam; ~ **seccional** *m* TEXTIL *género de punto por urdimbre* sectional beam; ~ **de tejedor** *m* TEXTIL weaver's beam; ~ **de urdimbre** *m* TEXTIL warp beam

plegadora *f* EMB folder, IMPR folder, folder unit, ING MECÁ *para chapas delgadas* bending press, folding machine, PAPEL folding machine, PROD folder, folding machine; ~ **a bolsillo** *f* IMPR plate-folding machine, pocket folding-machine; ~ **de cartón** *f* ING MECÁ, PAPEL folding machine for cardboard; ~ **de cuchilla** *f*

IMPR blade folder; ~ **y curvadora de chapas** *f* ING MECÁ plate-folding-and-bending machine; **~-engomadora** *f* EMB folder-gluer; ~ **de lengüeta de encuadernador** *f* IMPR plough folder (*BrE*), plow folder (*AmE*); ~ **de martillo** *f* IMPR nip-and-tuck folder; ~ **mecánica** *f* ING MECÁ, PAPEL, PROD folding machine; ~ **de palastro** *f* MECÁ brake

plegamiento *m* GEOL fold, folding, PETROL fold; ~ **en bucles** *m* GEOL buckle folding, flexural slip folding; ~ **de cizalla** *m* GEOL shear folding; ~ **fluidal** *m* GEOL flow; ~ **similar** *m* PETROL similar folding

plegar[1] *vt* EMB crease, MECÁ, MINAS crimp, PAPEL fold, PROD fold back, TEC ESP stow

plegar[2]: ~ **hacia abajo** *vti* PROD fold down

plena[1]: a ~ **carga** *adj* TRANSP MAR *buque* loaded

plena[2]: ~ **abertura** *f* FOTO full aperture; ~ **carga** *f* ELEC *de generador*, PROD full load; ~ **luz solar** *f* TEC ESP *orbitografía* full sunlight; ~ **potencia** *f* TRANSP MAR *maquinaria* full power

plena[3]: **en ~ acción** *fra* ING MECÁ in full swing; **en ~ actividad** *fra* ING MECÁ in full swing; **en ~ producción** *fra* ING MECÁ in full swing

pleno: ~ **empuje** *m* TEC ESP full thrust; ~ **superior** *m* NUCL upper plenum, *de la vasija del reactor* top plenum; ~ **voltaje** *m* ELEC, ING ELÉC, PROD full voltage

pletina *f* CONST flange tile; ~ **biselada** *f* ING MECÁ bevel-edged flat

pliego *m* IMPR *de la tirada*, PAPEL sheet; ~ **anuncio de gran tamaño** *m* IMPR broadside; ~ **de condiciones** *m* PROD, TRANSP MAR *contracto de construcción* specification; ~ **doblado para formar seis hojas** *m* IMPR sixmo; ~ **doblado para formar 36 hojas** *m* IMPR thirty-sixmo; ~ **estropeado** *m* IMPR spoil; ~ **inserto de cuatro páginas** *m* IMPR wraparound

pliegos: ~ **para maculaturas** *m pl* IMPR spoils; ~ **sobrantes** *m pl* IMPR waste sheets

pliegue *m* AUTO ply, GEOL fold, IMPR gusset, ING MECÁ convolution, ply, MECÁ bend, bending, flute, MINAS crimping, crown, OCEAN *playa de arena* ripple mark, PETROL fold, REVEST lap, TEC ESP ply, *trajes espaciales* bending, TEXTIL *vestimenta* crease, pleat; ~ **en abanico** *m* IMPR fanfold; ~ **abierto** *m* PETROL open fold; ~ **de arrastre** *m* GEOL drag fold; ~ **en bucle** *m* GEOL buckling; ~ **cabrío** *m* GEOL zigzag fold; ~ **en caja** *m* GEOL box fold; ~ **central** *m* IMPR centerfold (*AmE*), centrefold (*BrE*); ~ **cobijante** *m* GEOL over-thrust fold; ~ **cosido** *m* TEXTIL *confección* knife pleat; ~ **cruzado** *m* GEOL cross fold; ~ **de dislocación** *m* METAL dislocation kink; **~-falla** *m* CARBÓN slide, GEOL break thrust; ~ **de flujo** *m* GEOL flow fold; ~ **geológico** *m* GEOL, TEC PETR overthrust; ~ **inclinado** *m* GEOL inclined fold; ~ **invertido** *m* GEOL inverted fold, overfold, overturned fold, TEXTIL inverted pleat; ~ **isoclinal** *m* PETROL isoclinal fold; ~ **monoclinal** *m* PETROL monocline fold; ~ **muy agudo y flanco plano** *m* GEOL kink fold; ~ **de la pala** *m* TRANSPAÉR blade folding; ~ **paralelo** *m* GEOL parallel fold; ~ **plano** *m* TEXTIL flat pleat; ~ **ptigmático** *m* GEOL ptygmatic fold; ~ **recumbente** *m* GEOL recumbent fold; ~ **de revestimiento** *m* GEOL sheath fold; ~ **en rodilla** *m* GEOL knee fold; ~ **simétrico** *m* GEOL upright fold; ~ **similar** *m* GEOL similar fold; ~ **sinclinal** *m* ING MECÁ synclinal flexure, synclinal fold; ~ **triangular** *m* IMPR delta fold; ~ **tumbado** *m* GEOL overfold,

recumbent fold; ~ **de velamen** *m* OCEAN *de barcos* cloth lap; ~ **en z** *m* IMPR z-fold; ~ **en zig-zag** *m* GEOL chevron fold

pliegues: ~ **con capas más potentes en flancos que en el eje** *m pl* GEOL drape folds; ~ **concéntricos** *m pl* GEOL concentric folds; ~ **en relevo** *m pl* GEOL en echelon folds

plinto *m* TRANSP, TRANSP MAR *aerodeslizador* skirt; ~ **de estabilidad** *m* TRANSP, TRANSP MAR *aerodeslizador* stability skirt; ~ **de geometría variable** *m* TRANSP variable geometry skirt; ~ **periférico** *m* TRANSP peripheral skirt; ~ **rígido** *m* TRANSP rigid skirt

pliolito *m* QUÍMICA pliolit

pliopelícula *f* QUÍMICA pliofilm

plisado[1] *adj* TEXTIL pleated

plisado[2] *m* TEXTIL pleating

plisadora *f* TEXTIL pleater

plisar *vt* TEXTIL pleat

plomada[1] *adj* CONST plumb

plomada[2] *f* CONST bob, plumb bob, plummet; ~ **óptica** *f* INSTR optical plummet

plombagina *f* AGRIC, QUÍMICA plumbago

plomería *f* AmL (*cf fontanería Esp*) CONST plumbing

plomero *m* AmL (*cf fontanero Esp*) CONST plumber

plomo *m* (*Pb*) METAL, MINERAL, QUÍMICA lead (*Pb*); ~ **amarillo** *m* MINERAL wulfenite; ~ **argentífero** *m* MINAS silver-lead ore; ~ **blanco** *m* MINERAL cerusite; ~ **córneo** *m* MINERAL horn lead, phosgenite; ~ **esponjoso** *m* AUTO sponge lead; ~ **en lincotes** *m* PROD pig lead; ~ **de obra** *m* MINAS silver-lead ore; ~ **en panes** *m* PROD pig lead; ~ **en planchas** *m* CONST sheet lead; ~ **rojo** *m* C&V, MINERAL, P&C, QUÍMICA red lead; ~ **tetraetilo** *m* QUÍMICA tetraethyl lead (*TEL*); ~ **verde** *m* MINERAL mimetite

plotter *m* ELEC, INFORM&PD plotter; ~ **digital** *m* INFORM&PD digital plotter; ~ **electrostático** *m* ELEC, INFORM&PD electrostatic plotter; ~ **gráfico** *m* INFORM&PD graph plotter, graphics plotter; ~ **incremental** *m* INFORM&PD incremental plotter; ~ **plano** *m* INFORM&PD flat-bed plotter; ~ **de tambor** *m* INFORM&PD barrel plotter (*BrE*), drum plotter (*AmE*)

pluma *f* C&V feather, CONST *grúa* boom, derrick, FÍS FLUID *flujos térmicos* plume, MECÁ boom, MINAS jumbo including boom, PETROL mast, TRANSP MAR *cargamento de la carga* derrick boom; ~ **auxiliar del salabardo** *f* TRANSP MAR *pesca* brailer boom; ~ **de carga** *f* TRANSP MAR *buque* derrick, *construcción naval* cargo derrick, *pesca* loading boom; ~ **de chimenea** *f* C&V stack plume; ~ **estilográfica de cartuchos** *f* IMPR cartridge pen; ~ **de grandes pesos** *f* TRANSP MAR *cargamento* heavy lift derrick; ~ **de grúa** *f* CONST derrick girt, ING MECÁ, MECÁ crane jib; ~ **mantélica** *f* GEOL mantle plume; ~ **metálica** *f* CONST *de grúa* boom plate; ~ **de punta de diamante** *f* AmL (*cf marcador de punta de diamante Esp*) LAB diamond-tipped pen; ~ **real** *f* TRANSP MAR *cargamento* heavy lift derrick; ~ **del salabardo** *f* TRANSP MAR *pesca* brailer boom

plumbato *m* QUÍMICA plumbate

plúmbico *adj* QUÍMICA plumbic

plumbicón *m* ELECTRÓN plumbicon

plumbífero *adj* MINAS lead-bearing

plumbita *f* QUÍMICA plumbite

plumboso *adj* QUÍMICA plumbous

plumosita *f* MINERAL feather ore
pluriaplicación *f* TEC ESP versatility
plurivalente *adj* QUÍMICA plurivalent
plutón *m* GEOL pluton
plutonio *m* (*Pu*) FÍS, FÍS RAD, GEOL, MINERAL, NUCL, QUÍMICA plutonium (*Pu*)
pluviógrafo *m* GEN rain gage (*AmE*), rain gauge (*BrE*)
pluviómetro *m* GEN rain gage (*AmE*), rain gauge (*BrE*)
pluviosidad *f* AGRIC, AGUA, ENERG RENOV, HIDROL, METEO rainfall
Pm *abr* (*promecio*) NUCL, QUÍMICA Pm (*promethium*)
PMA *abr* (*peso total de despegue*) TRANSP AUW (*all-up weight*)
PMI *abr* (*punto muerto inferior*) AUTO, MECÁ, VEH BDC (*bottom dead center AmE, bottom dead centre BrE*)
PMMA *abr* (*polimetacrilato de metilo*) P&C PMMA (*polymethyl methacrylate*)
PMS *abr* (*punto muerto superior*) AUTO, MECÁ, VEH TDC (*top dead center AmE, top dead centre BrE*)
p-n-p *abr* (*positivo-negativo-positivo*) ELECTRÓN, FÍS, FÍS RAD p-n-p (*positive-negative-positive*)
Po *abr* (*polonio*) FÍS RAD, QUÍMICA Po (*polonium*)
población: ~ **de electrones** *f* TEC ESP electron population
pobre *adj* MINAS *terrenos* hungry
pobreza *f* MINAS *del mineral* baseness
pocero *m* MINAS *obrero* digger, onsetter (*BrE*), platman (*AmE*)
poceta: ~ **para recoger el agua de un pozo** *f* MINAS curb; ~ **de recogida de aguas** *f* CARBÓN *pozos, minas* pocket, MINAS water ring
pocillo: ~ **de investigación** *m* Esp (*cf pozo de cateo AmL*) MINAS costean pit, exploration pit
poda: ~ **de extremos** *f* AGRIC heading back
podador *m* AGRIC, ING MECÁ pruner
podedumbre *f* CONST root
poder *m* GEN power; ~ **de aceleración** *m* AUTO, VEH accelerating power; ~ **calorífico** *m* FÍS, GEOL, TEC PETR calorific value, TERMO calorific power, calorific value, heating capacity, heating power, TERMOTEC calorific value; ~ **calorífico superior** *m* GAS gross calorific value; ~ **calorífugo** *m* TERMO heat insulation power; ~ **de cierre** *m* ELEC *de relé* making capacity; ~ **de cobertura** *m* COLOR, REVEST covering power; ~ **colorante** *m* COLOR coloring power (*AmE*), coloring value (*AmE*), colouring power (*BrE*), colouring value (*BrE*); ~ **cortante** *m* CONST *metales* bite; ~ **cubriente de la tinta** *m* IMPR ink coverage; ~ **de cubrimiento** *m* COLOR, P&C, REVEST covering power; ~ **detergente** *m* DETERG detergent power; ~ **dispersor** *m* FÍS dispersive power; ~ **endotérmico** *m* TERMO heat-absorbing power; ~ **de frenado** *m* FÍS slowing-down power, stopping power; ~ **de frenado másico total** *m* NUCL mass stopping power; ~ **gasificante** *m* GAS, TERMO gas-enriching value; ~ **inductor específico** *m* ELEC *capacitor* permittivity; ~ **luminoso** *m* FOTO candle power; ~ **luminoso eficaz** *m* FOTO effective candle-power; ~ **de magnificación** *m* FÍS magnifying power; ~ **óptico** *m* ÓPT optical power; ~ **de penetración** *m* FÍS RAD penetrating power; ~ **de penetración de la radiación** *m* FÍS, FÍS RAD, ING ELÉC, NUCL, TELECOM radiation hardness; ~ **portante** *m* ING MECÁ lifting power; ~ **de refracción** *m* FOTO refractive power; ~ **de resolución** *m* CINEMAT, ELECTRÓN, FÍS, FOTO, METAL resolving power; ~ **de resolución cromático** *m* FÍS chromatic resolving power; ~ **resolutivo** *m* CINEMAT, ELECTRÓN, FÍS, FOTO, METAL resolving power; ~ **de rotación** *m* FÍS rotatory power; ~ **de rozamiento** *m* FÍS, METAL friction force; ~ **de ruptura** *m* ELEC, ELECTRÓN *de interruptor, fusible*, ING ELÉC breaking capacity; ~ **de separación** *m* NUCL separating power; ~ **de separación elemental** *m* NUCL elementary separative power; ~ **solar** *m* ENERG RENOV, FÍS solar power; ~ **de termoaislación** *m* TERMO heat insulation power; ~ **termoaislante** *m* TERMO heat insulation power; ~ **de tinción** *m* COLOR, P&C color strength (*AmE*), colour strength (*BrE*); ~ **tintoreo** *m* COLOR, P&C color strength (*AmE*), colour strength (*BrE*); ~ **total de frenado atómico** *m* FÍS total atomic stopping power; ~ **total de frenado lineal** *m* FÍS total linear stopping power; ~ **total masivo de frenado** *m* FÍS total mass stopping power; ~ **de vaporización** *m* PROC QUÍ evaporative capacity
podocárpico *adj* QUÍMICA podocarpic
podofilina *f* QUÍMICA podophyllin
podredumbre *f* AGRIC, CONST, TRANSP MAR *en la madera* rot; ~ **negra** *f* AGRIC black rot; ~ **de plántula** *f* AGRIC damping off; ~ **seca** *f* AGRIC *patología vegetal*, CONST dry rot
podrido *adj* ALIMENT putrid
podrir *vi* ALIMENT putrefy
poise *m* METR poise; ~ **de alcohol** *m* ALIMENT *fermentación* spirit poise
polainas: ~ **protectoras** *f pl* SEG protective gaiters
polar[1] *adj* GEOM, QUÍMICA polar; **no** ~ *adj* FÍS, QUÍMICA nonpolar
polar[2] *m* METR pole
polaridad *f* ELEC, ELECTRÓN, FÍS, ING ELÉC, QUÍMICA, TV polarity; ~ **del impulso** *f* ELECTRÓN pulse polarity; ~ **del voltaje** *f* ING ELÉC voltage polarity
polarimetría *f* FÍS, FÍS RAD, LAB, QUÍMICA polarimetry
polarimétrico *adj* FÍS, FÍS RAD, LAB, QUÍMICA polarimetric
polarímetro *m* FÍS, FÍS RAD, LAB, QUÍMICA polarimeter
polariscopio *m* FÍS, FÍS RAD, LAB, QUÍMICA polariscope
polarizabilidad *f* FÍS, FÍS RAD, LAB, QUÍMICA polarizability
polarizable *adj* FÍS, FÍS RAD, LAB, QUÍMICA polarizable
polarización *f* ACÚST, ELEC, FÍS, FÍS RAD, FOTO polarization, INFORM&PD bias, ING ELÉC polarization, bias, TEC ESP, TELECOM polarization, TV bias, biasing; ~ **atómica** *f* FÍS RAD atomic polarization; ~ **automática de rejilla** *f* ING ELÉC self-bias; ~ **celular** *f* ING ELÉC cell polarization; ~ **circular** *f* FÍS, FÍS RAD, TELECOM circular polarization; ~ **circular a derechas** *f* TEC ESP right-hand circular polarization; ~ **circular dextrógira** *f* TEC ESP right-hand circular polarization; ~ **circular a izquierdas** *f* FÍS, TEC ESP left-hand circular polarization (*LHCP*); ~ **circular de la luz** *f* FÍS, FÍS RAD circular polarization of light; ~ **por corriente continua** *f* TV DC biasing; ~ **cruzada** *f* TEC ESP, TELECOM cross polarization; ~ **dieléctrica** *f* FÍS, ING ELÉC dielectric polarization; ~ **directa** *f* FÍS, ING ELÉC forward bias; ~ **eléctrica** *f* ELEC, FÍS electric polarization; ~ **del electrodo** *f* ING ELÉC electrode bias; ~ **electrónica** *f* ELECTRÓN, FÍS, FÍS RAD electronic polarization; ~ **elíptica** *f* FÍS, FÍS RAD elliptical polarization; ~ **horizontal** *f* FÍS, ING ELÉC, TELECOM horizontal polarization; ~ **inversa** *f* FÍS, ING ELÉC reverse bias; ~ **iónica** *f* FÍS ionic

polarization; ~ **lineal** *f* TELECOM linear polarization; ~ **magnética** *f* ACÚST magnetic polarization, ING ELÉC magnetic bias, magnetic polarization, magnetization, TV magnetic bias; ~ **magnética por CA** *f* TELECOM AC magnetic biasing; ~ **de montaje** *f* ING ELÉC mounting polarization; ~ **negativa** *f* ING ELÉC negative bias; ~ **negativa frontal** *f* FÍS, ING ELÉC forward bias; ~ **oblicua** *f* ING ELÉC slant polarization; ~ **de la onda** *f* TELECOM wave polarization; ~ **de la onda electromagnética** *f* ELEC, FÍS ONDAS, ING ELÉC electromagnetic-wave polarization; ~ **óptica** *f* TELECOM optical polarization; ~ **de orden** *f* INFORM&PD ordering bias; ~ **orientacional** *f* FÍS orientational polarization; ~ **ortogonal** *f* TELECOM orthogonal polarization; ~ **de la puerta** *f* ING ELÉC gate bias; ~ **de rejilla positiva** *f* ING ELÉC positive bias; ~ **del transistor** *f* ELECTRÓN transistor bias; ~ **en vacío** *f* NUCL vacuum polarization; ~ **vertical** *f* FÍS, ING ELÉC, TELECOM vertical polarization

polarizado *adj* ELECTRÓN, FÍS biased, LAB, METAL, ÓPT, TEC ESP, TELECOM polarized

polarizador *m* FÍS, LAB, METAL, ÓPT, TEC ESP, TELECOM polarizer

polarizar *vt* FÍS, ING ELÉC bias, LAB, METAL, ÓPT, TEC ESP, TELECOM polarize

polarografía *f* QUÍMICA polarography

polarográfico *adj* QUÍMICA polarographic

polarograma *m* QUÍMICA polarogram

polaroide *m* FÍS, FÍS RAD, FOTO Polaroid®

polarón *m* FÍS polaron

polea *f* CONST axle pulley, FÍS pulley, ING MECÁ pulley, sheave, stud, wheel, MECÁ stud, MINAS pulley, *sondeos* drive block, PROD sheaf, TRANSP MAR block, pulley; ~ **abombada** *f* PROD crowned pulley, crowning pulley; ~ **acanalada** *f* ING MECÁ sheave; ~ **acanalada cónica** *f* ING MECÁ cone sheave; ~ **acanalada escalonada** *f* ING MECÁ cone sheave; ~ **de acercamiento** *f* ING MECÁ nearing-up pulley; ~ **del aparejo** *f* ING MECÁ purchase block; ~ **de arrastre del ventilador** *f* AUTO, VEH *refrigeración* fan pulley; ~ **bombeada** *f* ING MECÁ round-faced pulley; ~ **de cadena** *f* ING MECÁ chain pulley, chain sheave, chain wheel; ~ **de cadena articulada** *f* ING MECÁ sheave; ~ **de cadena dentada** *f* ING MECÁ indented wheel; ~ **de cara combada** *f* PROD crown face pulley; ~ **del castillete de extracción** *f* AmL (*cf gorrón del castillete Esp*) MINAS hoisting pulley, pithead gear, pithead pulley, gudgeon; ~ **para cinta de tripa** *f* ING MECÁ pulley for gut band; ~ **de colocación** *f* ING MECÁ positioning block; ~ **conducida** *f* ING MECÁ follower; ~ **conductora** *f* ING MECÁ driving pulley, motorized driving pulley, PROD driving pulley, VEH *alternador* drive pulley; ~ **conductora sincrónica** *f* ING MECÁ synchronous belt drive; ~ **cónica** *f* ING MECÁ cone gear; ~ **cónica de cadena con cono** *f* PROD cupped chain sheave; ~ **cónica de tres diámetros** *f* ING MECÁ three-lift cone pulley, three-step cone pulley; ~ **cónica de tres velocidades** *f* ING MECÁ three-lift cone pulley; ~ **de cono** *f* ING MECÁ cone, cone pulley; ~**cono de dos escalones** *f* PROD two-step cone, two-step cone pulley; ~ **de contracono** *f* ING MECÁ overhead cone pulley; ~ **para correa** *f* ING MECÁ band pulley, band wheel, PROD belt pulley; ~ **de correa trapezoidal** *f* ING MECÁ V-belt drive; ~ **de cuadernal** *f* ING MECÁ sheave; ~ **dentada** *f* ING MECÁ cog belt;

~ **desmontable** *f* ING MECÁ parting pulley; ~ **de desviación** *f* ING MECÁ guide pulley; ~ **diferencial** *f* FÍS differential pulley, ING MECÁ chain block; ~ **directriz** *f* ING MECÁ guide pulley; ~ **de ensamblaje** *f* PROD made block; ~ **escalonada** *f* ING MECÁ cone pulley, step cone; ~ **de escalones** *f* ING MECÁ step cone; ~ **para escombros** *f* ING MECÁ rubbish pulley; ~ **fija** *f* ING MECÁ dead pulley, runner, standing block, MECÁ runner, PROD fast pulley, tight pulley; ~ **de fricción** *f* ING MECÁ friction wheel; ~ **de gancho** *f* ING MECÁ hook block, PROD tackle with hook block; ~ **de garganta** *f* ING MECÁ grooved pulley, grooved wheel, sheave; ~ **guía** *f* ING MECÁ leading-on pulley, guide pulley, idle pulley, runner, MECÁ idler, runner, PROD runner; ~ **guíafilme** *f* TV idler; ~ **impulsora** *f* ING MECÁ driving pulley, motorized driving pulley, PROD driving pulley; ~ **lateral** *f* CONST side pulley; ~ **de llanta plana** *f* ING MECÁ straight-faced pulley; ~ **loca** *f* AUTO idler pulley, CINEMAT idler, ING MECÁ idle, idle pulley, idler pulley, MECÁ idler, idler pulley; ~ **maestra** *f* PROD *sondeos* crown pulley; ~ **mandada** *f* CARBÓN, ING MECÁ follower; ~ **de mando** *f* VEH *alternador* drive pulley; ~ **de mortaja** *f* ING MECÁ, PROD mortise block; ~ **motorizada** *f* ING MECÁ motorized driving pulley; ~ **motriz** *f* ING MECÁ motorized driving pulley, driving pulley, PROD driving pulley; ~ **motriz de grabación** *f* TV record driver; ~ **móvil** *f* ING MECÁ running block, PETROL traveling block (*AmE*), travelling block (*BrE*), PROD traveling pulley block (*AmE*), travelling pulley block (*BrE*); ~ **muerta** *f* MECÁ idler; ~ **ovalada** *f* ING MECÁ oval pulley; ~ **partida** *f* ING MECÁ split pulley; ~ **con pestañas** *f* ING MECÁ flange pulley; ~ **portacuchillas** *f* ING MECÁ band-saw pulley; ~ **portasierra** *f* ING MECÁ band wheel, band-saw pulley; ~ **ranurada** *f* ING MECÁ scored pulley; ~ **de rebordes** *f* ING MECÁ flange pulley; ~ **redonda** *f* ING MECÁ rounding pulley; ~ **de retorno** *f* ING MECÁ return pulley, tail pulley, PROD tail sheave, turn pulley; ~ **de rosca** *f* CONST screw pulley; ~ **de sierra** *f* ING MECÁ saw pulley; ~ **síncrona** *f* ING MECÁ synchronous belt; ~ **soporte** *f* MINAS *vagonetas* hat roller; ~ **tensacorreas** *f* AUTO belt idler; ~ **de tensión** *f* ING MECÁ binding pulley, idle pulley, idler, MECÁ jockey pulley; ~ **tensora** *f* ING MECÁ idle pulley, jockey, jockey pulley, jockey-roller, jockey-wheel, tension pulley, MECÁ idler, jockey pulley; ~ **torneada y taladrada** *f* PROD turned-and-bored pulley; ~ **de tracción mecánica** *f* ING MECÁ motorized driving pulley; ~ **de transmisión** *f* TV idler; ~ **en V** *f* ING MECÁ V-pulley; ~ **de velocidad constante** *f* PROD constant-speed pulley; ~ **del ventilador** *f* AUTO, VEH fan pulley; ~ **volante** *f* ING MECÁ fly pulley

poleas: ~ **locas y fijas** *f pl* PROD fast and loose pulleys

polen *m* AGRIC pollen

polenta *f* AmL (*cf harina de maíz Esp*) AGRIC corn meal (*AmE*), Indian meal (*BrE*)

poliacrilamida *f* QUÍMICA, TEC PETR polyacrylamide

poliacrilato *m* P&C, QUÍMICA polyacrylate

poliacrilonitrilo *m* QUÍMICA polyacrylonitrile

polialcohol: ~ **vinílico** *m* P&C polyvinyl alcohol

poliamida *f* (*PA*) AGRIC, ELECTRÓN, P&C, QUÍMICA, TEXTIL polyamide (*PA*)

poliamina *f* P&C, QUÍMICA polyamine; ~ **alifática** *f* P&C, QUÍMICA aliphatic polyamine

polianita *f* MINERAL polianite

poliatómico *adj* GAS, QUÍMICA polyatomic
polibasita *f* MINERAL polybasite
polibutadieno *m* TEC ESP polybutadiene
polibutileno *m* (*PB*) P&C polybutylene (*PB*)
policarbonato *m* ELEC, P&C, QUÍMICA polycarbonate
policárpico *adj* AGRIC polycarpic
policíclico *adj* QUÍMICA polycyclic
policloropreno *m* QUÍMICA polychloroprene
policloruro: ~ **de vinilideno** *m* P&C polyvinylidene chloride (*PVDC*)
policondensación *f* QUÍMICA polycondensation
policrasa *f* MINERAL polycrase
policromático *adj* C&V polychromatic
policromía *f* IMPR color print (*AmE*), colour print (*BrE*)
policromo *adj* COLOR multicolored (*AmE*), multicoloured (*BrE*), varicolored (*AmE*), varicoloured (*BrE*), IMPR polychrome
polidentado *adj* QUÍMICA polydentate
polidimetilsiloxano *m* QUÍMICA polydimethylsiloxane
poliédrico *adj* CRISTAL, GEOM, METAL polyhedral
poliedro *m* CRISTAL, GEOM, METAL polyhedron; ~ **irregular** *m* GEOM irregular polyhedron
poliénico *adj* QUÍMICA polyenic
polieno *m* QUÍMICA polyene
poliéster *m* GEN polyester; ~ **no saturado** *m* TRANSP MAR polyester; ~ **reforzado con fibra de vidrio** *m* (*GRP*) C&V, CONST, ING MECÁ, MECÁ, P&C, PROD, TRANSP MAR glass-reinforced polyester (*GRP*); ~ **saturado** *m* P&C saturated polyester
poliesterificación *f* QUÍMICA polyesterification
poliestireno *m* (*PS*) EMB, P&C, QUÍMICA polystyrene (*PS*); ~ **expandido** *m* EMB expanded polystyrene; ~ **de impacto** *m* P&C impact polystyrene
polietilenglicol *m* DETERG, QUÍMICA polyethylene glycol
polietileno *m* GEN polyethylene; ~ **de alta densidad** *m* (*PAD*) EMB high-density polyethylene (*HDPE*); ~ **clorado** *m* P&C chlorinated polyethylene
polifásico *adj* ELEC *motor* multiphase, *motor de la red de CA* polyphase, ELECTRÓN, GEOL multiphase, ING ELÉC polyphase
polifenol *m* QUÍMICA polyphenol
polifosfato: ~ **sódico** *m* ALIMENT, QUÍMICA sodium polyphosphate
polifuncional *adj* QUÍMICA polyfunctional
poligonal *adj* FÍS, GEOM, METAL polygonal
poligonización *f* METAL polygonization
polígono *m* FÍS, GEOM, METAL polygon; ~ **cerdoso** *m* AGRIC *malezas* bristly lady's thumb; ~ **de lanzamiento** *m* D&A, TEC ESP *torpedo, misil* range; ~ **plano** *m* GEOM plane polygon; ~ **regular** *m* GEOM regular polygon; ~ **de tiro** *m* D&A shooting range; ~ **de tiro del fusil** *m* D&A rifle range
poligorskita *f* GEOL, TEC PETR polygorskite
polihalita *f* GEOL, MINERAL polyhalite
polihexilmetacrilato *m* (*PHMA*) P&C polyhexylmethacrylate (*PHMA*)
poliinsaturado *adj* ALIMENT, QUÍMICA polyunsaturated
poliisopreno *m* P&C natural rubber, polyisoprene, QUÍMICA polyisoprene
polilla *f* AGRIC moth; ~ **del grano** *f* AGRIC grain moth
polimería *f* QUÍMICA polymeria
polimérico *adj* QUÍMICA polymeric
polimerismo *m* QUÍMICA polymerism
polimerización *f* GEN polymerization; ~ **de adición** *f* P&C addition polymerization; ~ **en bloque** *f* DETERG, P&C block polymerization; ~ **por condensación** *f* P&C condensation polymerization; ~ **por emulsión** *f* P&C emulsion polymerization; ~ **por injerto de una cadena** *f* P&C graft polymerization; ~ **en solución** *f* P&C solution polymerization; ~ **en suspensión** *f* P&C suspension polymerization; ~ **a temperatura ambiente** *f* P&C *resinas sintéticas* air cure
polimerizador *m* TEXTIL *horno de curado* polymerizer
polimerizar *vt* GEN polymerize
polímero *m* GEN polymer; ~ **de adición** *m* P&C addition polymer; ~ **en bloque** *m* DETERG, P&C block polymer; ~ **por condensación** *m* P&C condensation polymer; ~ **en escalera** *m* P&C ladder polymer; ~ **injertado** *m* P&C graft polymer; ~ **lineal** *m* P&C linear polymer; ~ **obturador** *m* TEC PETR sealant polymer; ~ **ramificado** *m* P&C branched polymer; ~ **sellador** *m* TEC PETR sealant polymer
polimetacrilato *m* P&C polymethacrylate; ~ **de metilo** *m* (*PMMA*) P&C Perspex®, polymethyl methacrylate (*PMMA*)
polimetileno *m* QUÍMICA polymethylene
polímetro *m* ING ELÉC multimeter, voltameter (*BrE*), voltammeter (*AmE*), voltmeter, TV multimeter
polimignita *f* MINERAL polymignite
polimorfismo *m* CRISTAL, METAL, QUÍMICA allotropy, polymorphism
polín *m* TRANSP MAR engine seating; ~ **de la máquina principal** *m* TRANSP MAR engine frame; ~ **de la turbina** *m* TRANSP MAR turbine seating
polinización *f* AGRIC pollination; ~ **cruzada** *f* AGRIC cross pollination
polinomio *m* FÍS, INFORM&PD polynomial, MATEMÁT multinomial, polynomial; ~ **de Legendre** *m* FÍS Legendre polynomial; ~ **reducible** *m* INFORM&PD reducible polynomial
polinucleotido *m* QUÍMICA polynucleotide
poliol *m* P&C, QUÍMICA polyol
poliolefina *f* P&C, QUÍMICA, TEXTIL polyolefin
polióxido: ~ **de metileno** *m* (*POM*) P&C polyoxymethylene (*POM*)
polioxietileno *m* QUÍMICA polyoxyethylene
polioximetileno *m* (*POM*) QUÍMICA polyoxymethylene (*POM*)
polipasto *m* ING MECÁ block-and-tackle, MECÁ hoist, PROD *mecanismos de elevación* hoist, tackle, TEC PETR blocks, TRANSP MAR hoist; ~ **de cadena** *m* ING MECÁ chain pulley block; ~ **para chapas** *m* PROD plate hoist; ~ **móvil** *m* PROD traveling pulley block (*AmE*), travelling pulley block (*BrE*); ~ **con polea de gancho** *m* PROD tackle with hook block; ~ **simple** *m* PROD single tackle
polipéptido *m* QUÍMICA polypeptide
poliploide *adj* AGRIC polyploid
polipnea *f* OCEAN polypnea
polipropileno *m* GEN polypropylene
polisacárido *adj* AGRIC, QUÍMICA polysaccharide
polisilicio *m* ELECTRÓN polysilicon
polisiloxano *m* P&C, QUÍMICA polysiloxane
polisulfona *f* QUÍMICA polysulfone (*AmE*), polysulphone (*BrE*)
polisulfuro *m* P&C, QUÍMICA polysulfide (*AmE*), polysulphide (*BrE*)
politeno *m* ALIMENT, QUÍMICA, TEC PETR, TRANSP MAR polythene
politerpeno *m* QUÍMICA polyterpene

politetrafluoretileno *m* (*PTFE*) P&C, QUÍMICA polyte-trafluoroethylene (*PTFE*)

política: ~ **de la calidad** *f* CALIDAD quality policy

poliuretano *m* (*PUR*) CONST, P&C, QUÍMICA, TEXTIL polyurethane (*PUR*)

polivalencia *f* QUÍMICA polyvalence, polyvalency, TEC ESP versatility

polivalente *adj* QUÍMICA polyvalent

polivinilbenceno *m* QUÍMICA polyvinylbenzene

polivinilbutiral *m* P&C, QUÍMICA polyvinyl butyral

poliviniléter *m* P&C, QUÍMICA polyvinyl ether

polivinilo *m* QUÍMICA, TEXTIL polyvinyl

polivinilpirrolidona *f* (*PVP*) DETERG polyvinylpyrroli-done (*PVP*)

póliza: ~ **de fletamento** *f* TEC PETR, TRANSP MAR charter party; ~ **de fletamento por viaje** *f* TRANSP MAR voyage charter

polo¹: **de un solo** ~ *adj* ELEC *terminal* single-pole

polo² *m* GEN pole; ~ **antílogo** *m* CRISTAL antilogous pole; ~ **auxiliar** *m* ELEC commutating pole, *de motor de CC* interpole; ~ **de conmutación** *m* ELEC com-mutating pole, *de motor de CC* interpole; ~ **eléctrico** *m* ELEC, ING ELÉC electric pole; ~ **fijo** *m* ELEC, ING ELÉC fixed pole; ~ **del filtro** *m* ELECTRÓN filter pole; ~ **geomagnético** *m* GEOFÍS geomagnetic pole; ~ **inductor** *m* ING ELÉC field magnet, field pole; ~ **magnético** *m* ELEC, FÍS, GEOFÍS magnetic pole; ~ **magnético unitario** *m* TEC PETR unit magnetic pole; ~ **negativo** *m* ELEC, FÍS, ING ELÉC negative pole; ~ **no saliente** *m* ELEC nonsalient pole; ~ **norte** *m* FÍS north pole; ~ **norte magnético** *m* FÍS, GEOFÍS magnetic north pole; ~ **positivo** *m* ELEC, FÍS, ING ELÉC positive pole; ~ **principal** *m* ELEC *terminal* main pole; ~ **saliente** *m* ELEC, ING ELÉC salient pole; ~ **sur geomagnético** *m* GEOFÍS South geomagnetic Pole; ~ **sur magnético** *m* FÍS, GEOFÍS magnetic South Pole, magnetic south pole

polonio *m* (*Po*) FÍS RAD, QUÍMICA polonium (*Po*)

polos: ~ **del mismo nombre** *m pl* ELEC, FÍS like poles; ~ **de nombre contrario** *m pl* ELEC, FÍS unlike poles; ~ **opuestos** *m pl* ELEC, FÍS unlike poles; ~ **semejantes** *m pl* ELEC, FÍS like poles; ~ **de signo contrario** *m pl* ELEC, FÍS unlike poles

polución *f* GEN pollution; ~ **calorífica** *f* CONTAM, TERMO thermal pollution; ~ **térmica** *f* CONTAM, TERMO thermal pollution

polucionar *vt* GEN pollute

polucita *f* MINERAL pollucite

polutante: ~ **acuático** *m* AGUA aquatic pollutant

poluto *adj* CONTAM polluted

polvareda *f* CARBÓN dust

polvillo: ~ **incombustible** *m* CARBÓN, MINAS stone dust

polvo *m* C&V dirt, CARBÓN, CONST, CONTAM dust, P&C *material de carga, pigmentos,* QUÍMICA powder, SEG dust; ~ **abrasivo** *m* SEG abrasive dust; ~ **de amolado** *m* PROD *muelas abrasivas* wheel swarf; ~ **de antracita** *m* CARBÓN culm; ~ **de barreno** *m* MINAS *perforación* drillings; ~ **de barro** *m* C&V clay dust, clay powder; ~ **bituminoso** *m* CARBÓN flux powder; ~ **blanqueador** *m* ALIMENT bleaching powder; ~ **de blanqueo** *m* DETERG bleaching powder, chlorinated lime; ~ **de bronce** *m* COLOR bronze powder; ~ **de carbón** *m* CARBÓN coal dust, coal powder, culm, CONTAM *gases* coal dust, MINAS coal powder; ~ **de carga** *m* C&V batch dust; ~ **de cemento** *m* CONST

cement dust; ~ **de coque** *m* CARBÓN breeze; ~ **desengrasador** *m* DETERG scouring powder; ~ **de diamante** *m* MINAS diamond dust, diamond powder, PROD *para pulimentar* bort; ~ **de esmeril** *m* ING MECÁ, PROD emery powder; ~ **de lavado** *m* DETERG washing powder; ~ **de madera** *m* P&C *material de carga* wood flour; ~ **meteórico** *m* TEC ESP meteor dust; ~ **microscópico** *m* SEG microscopic dust; ~ **de muelas abrasivas** *m* MECÁ grit; ~ **pigmentado blanco** *m* COLOR white pigmented powder; ~ **de purpurina plateada** *m* COLOR silver bronze powder; ~ **radiactivo** *m* CONTAM, D&A, FÍS RAD, NUCL radio-active dust; ~ **radiactivo ácido y húmedo** *m* CONTAM wet acidic fallout; ~ **radiactivo ácido y seco** *m* CONTAM dry acidic fallout; ~ **radiactivo atmosférico** *m* CONTAM atmospheric fallout; ~ **de sílice** *m* SEG silica dust; ~ **de sílice cristalino** *m* CRISTAL, SEG crystalline silica dust; ~ **sinterizante** *m* PROC QUÍ sintering powder; ~ **en suspensión** *m* SEG airborne dust; ~ **de talla** *m* MINAS stone dust; ~ **del tragante** *m* CARBÓN flue dust; ~ **de vidrio** *m* C&V glass dust; ~ **de yeso** *m* CONST putty-powder

pólvora *f* D&A gunpowder

polvorero *m* CARBÓN, CONST blaster, MINAS *AmL* (*cf disparador de barrenos Esp, cf destarcador Esp*) blaster, shooter, *persona* picker

polvoriento *adj* CARBÓN dusty, PROC QUÍ pulverulent

polvorín *m* MINAS explosives magazine, powder store, *explosivos* dynamite store

polvorista *m* CARBÓN shotfirer

polvos *m pl* PROD grit; ~ **para moldeo** *m pl* P&C molding powder (*AmE*), moulding powder (*BrE*)

POM *abr* (*polióxido de metileno, polioximetileno*) P&C, QUÍMICA POM (*polyoxymethylene*)

pomo: ~ **oval** *m* CONST *cerradura* oval knob

ponderación *f* ACÚST, INFORM&PD weighting

poner¹ *vt* ING MECÁ set, *aro* collar, *correa* throw on, MECÁ apply, PROD locate; ~ **a uno** *vt* ELECTRÓN, INFORM&PD, PROD *bit* set; ~ **en acción** *vt* ING MECÁ throw into action, MECÁ actuate; ~ **a la altura** *vt* IMPR bring up; ~ **a cero** *vt* CINEMAT zero, ELECTRÓN reset, turn off, INFORM&PD *contador* reset, zeroize, ING MECÁ set to zero, PROD *aparatos* reset, QUÍMICA, TV zero; ~ **en circuito** *vt* ING ELÉC switch on; ~ **en comunicación** *vt* TELECOM put through; ~ **en condición inicial** *vt* PROD *aparatos* reset; ~ **en contenedores** *vt* EMB, TRANSP containerize; ~ **en cuarentena** *vt* AGRIC, SEG, TRANSP quarantine; ~ **a cubierto** *vt* CONST put under cover; ~ **derecho** *vt* PROD straighten; ~ **al día** *vt* ING MECÁ bring up to date, update; ~ **a escuadra** *vt* ING MECÁ true up; ~ **en espera** *vt* TELECOM put on hold; ~ **a flote** *vt* TRANSP MAR *embarcación* launch; ~ **en funcionamiento** *vt* CINEMAT run, PROD put into operation; ~ **a funcionar** *vt* ING MECÁ set to work; ~ **en un índice** *vt* INFORM&PD index; ~ **a son de mar** *vt* TRANSP MAR *velero* trim; ~ **en marcha** *vt* AUTO start, CONST commission, ING MECÁ restart, set in motion, start, throw into gear, PROD *máquinas* start, VEH run in, start; ~ **a masa** *vt* AUTO, ELEC, FÍS, ING ELÉC, PROD, TELECOM, VEH earth (*BrE*), ground (*AmE*); ~ **en movimiento** *vt* ING MECÁ set in motion, start up, PROD start; ~ **en negritas** *vt* IMPR, INFORM&PD embolden; ~ **de nuevo** *vt* ING MECÁ set up again; ~ **en órbita circular** *vt* TEC ESP circularize; ~ **en orden** *vt* CONST fix; ~ **en práctica** *vt* ING MECÁ implement; ~ **la proa a** *vt* TRANSP MAR head for; ~ **a punto** *vt* IMPR make ready; ~ **rumbo a** *vt* TRANSP MAR *navegación* steer for; ~ **en serie** *vt* INFORM&PD serialize; ~ **en servicio activo** *vt* TRANSP MAR put into commission; ~ **en situación de espera** *vt* CONTAM MAR *buque* put on stand-by; ~ **al socaire** *vt* CONST, D&A, TRANSP MAR shelter; ~ **tapas a** *vt* IMPR case; ~ **a tierra** *vt* AUTO, ELEC, FÍS, ING ELÉC, PROD, TELECOM, VEH earth (*BrE*), ground (*AmE*); ~ **a trabajar** *vt* ING MECÁ set to work; ~ **vidrios** *vt* C&V CONST glaze

poner[2]: ~ **bridas** *vi* ING MECÁ, PROD *tubos* flange; ~ **los cimientos** *vi* CONST lay the foundations; ~ **condiciones iniciales** *vi* ELECTRÓN reset; ~ **cristales** *vi* PROD *ventanas* glaze; ~ **forros nuevos** *vi* AUTO, VEH *frenos* reline; ~ **guardas** *vi* IMPR forward; ~ **hombreras** *vi* TEXTIL pad; ~ **mango** *vi* MECÁ handle; ~ **manguito** *vi* ING MECÁ collar; ~ **placa postiza a herramientas** *vi* PROD tip; ~ **la quilla a carenar** *vi* TRANSP MAR *para limpiar fondos de buques* careen

ponerse: ~ **a flote** *v refl* TRANSP MAR *de barco perdido* float off; ~ **en marcha** *v refl* ING MECÁ come into gear; ~ **a la vela** *v refl* TRANSP MAR get under way; ~ **vidrioso** *v refl* C&V, PROD glaze

poniente *adj* TRANSP MAR west

pontón *m* CONST pontoon, PETROL barge, pontoon, TRANSP MAR hulk, pontoon; ~ **de desembarco** *m* TRANSP landing pontoon

popa[1]: **hacia** ~ *adj* TRANSP MAR abaft, aft, after, stern

popa[2]: **a** ~ *adv* TEC ESP aft, TRANSP MAR abaft, astern; **hacia** ~ *adv* TRANSP MAR astern; **por la** ~ *adv* TRANSP MAR astern

popa[3] *f* TRANSP MAR *de barco* aft, stern; ~ **con bovedilla** *f* TRANSP MAR counterstern; ~ **de canoa** *f* TRANSP MAR *yates, veleros* canoe stern; ~ **de crucero** *f* TRANSP MAR cruiser stern; ~ **cuadra** *f* TRANSP MAR flat stern; ~ **cuadrada** *f* TRANSP MAR flat stern; ~ **elíptica** *f* TRANSP MAR elliptical stern; ~ **de espejo** *f* TRANSP MAR transom stern; ~ **llana** *f* TRANSP MAR square transom stern; ~ **redonda** *f* TRANSP MAR round stern

popel *m* TRANSP MAR aft

popelín *m* TEXTIL poplin

populina *f* QUÍMICA populin

porcelana *f* C&V, ELEC, ING ELÉC, TEXTIL porcelain; ~ **de alta tensión** *f* C&V, ELEC *aislación* high-voltage porcelain; ~ **artesanal** *f* C&V craft porcelain; ~ **artística** *f* C&V artistic porcelain; ~ **dura** *f* C&V hard porcelain; ~ **electrotécnica** *f* C&V, ING ELÉC electrotechnical porcelain; ~ **inglesa** *f* C&V English china; ~ **rechazada** *f* C&V porcelain reject; ~ **suave** *f* C&V soft porcelain; ~ **tipo china** *f* C&V china

porcelanas *f pl* C&V porcelain goods

porcentaje *m* METR rate, TEXTIL percentage; ~ **de animales criados** *m* AGRIC rearing proportion; ~ **de aplicación** *m* CONST application rate; ~ **de averías** *m* ING MECÁ failure rate; ~ **de bordes** *m* ELECTRÓN edge rate; ~ **característico de evacuación de aguas** *m* HIDROL characteristic floodwater flow rate; ~ **característico del nivel de estiaje** *m* HIDROL characteristic low-water flow rate; ~ **de carga** *m* METAL rate of loading; ~ **de carne en canal** *m* AGRIC, ALIMENT *carnicería* carcass dressing percentage; ~ **de caudal característico trimestral** *m* HIDROL three-month characteristic flow rate; ~ **de**

conversión *m* ELECTRÓN conversion rate; ~ **de la deformación del cuadro** *m* TV percentage tilt; ~ **de desviación** *m* ELECTRÓN deviation ratio; ~ **de disminución** *m* ELECTRÓN droop rate; ~ **empírico de utilización** *m* ING MECÁ empirical operation factor; ~ **de error** *m* ELECTRÓN, INFORM&PD, TELECOM error rate; ~ **de error de bit** *m* (*BER*) ELECTRÓN, INFORM&PD bit error rate (*BER*); ~ **en espera de reparación** *m* FERRO *vehículos* percentage awaiting repair; ~ **de extracción** *m* HIDROL *agua* extraction rate; ~ **de fallos** *m* TEC ESP *equipos* failure rate; ~ **de gas** *m* ELECTRÓN gas ratio; ~ **horizontal** *m* IMPR H-rate; ~ **de humedad** *m* CARBÓN moisture content; ~ **de impulsos** *m* ELECTRÓN pulse rate; ~ **inicial** *m* GEOL initial ratio; ~ **de mezcla** *m* TEXTIL blend ratio; ~ **de modulación** *m* ELECTRÓN percentage modulation; ~ **de movimiento lento** *m* PROD slow-moving percentage; ~ **de nucleación** *m* METAL nucleation rate; ~ **del peso de frenado** *m* FERRO *equipo inamovible* braked-weight percentage; ~ **de realimentación** *m* ELECTRÓN feedback ratio; ~ **de rechazo de modo común** *m* ELECTRÓN, TELECOM common-mode rejection ratio; ~ **de recombinación** *m* ELECTRÓN, FÍS recombination rate; ~ **de recuento** *m* ELECTRÓN counting rate; ~ **relativo de vapor seco** *m* HIDROL, INSTAL HIDRÁUL dryness fraction, relative dryness; ~ **de sincronización** *m* TV percentage synchronization; ~ **del tiempo de trabajo** *m* PROD working time percentage; ~ **de unidades inconformes** *m* CALIDAD percentage of nonconforming items; ~ **de vapor seco** *m* INSTAL HIDRÁUL dryness fraction

porche *m* AmL (*cf techo del porche Esp*) CONST porch roof

porción *f* ALIMENT, INFORM&PD slice, ING MECÁ part, TEC ESP range; ~ **individual** *f* ALIMENT single portion; ~ **de minerales** *f* Esp (*cf cantidad de minerales AmL*) MINAS parcel of ore; ~ **de original que compone un tipógrafo** *f* IMPR take

porfídico *adj* GEOL porphyraceous

porfidoblástico *adj* GEOL porphyroblastic

porfidoclasto *adj* GEOL porphyroclast

porfina *f* QUÍMICA porphin

porfirítico *adj* GEOL porphyritic

porfiropsina *f* QUÍMICA porphyropsin

porífero *adj* CARBÓN porous

poro *m* C&V leak, CARBÓN porous layer, GAS pore, P&C crater, PROD *fundición* blister

pororoca *f* OCEAN *corriente de marea* bore

porosidad *f* GEN porosity, CARBÓN *ingeniería* void ratio; ~ **aparente** *f* C&V apparent porosity; ~ **efectiva** *f* PETROL effective porosity; ~ **secundaria** *f* TEC PETR secondary porosity

porosímetro *m* PAPEL, PROC QUÍ porosimeter; ~ **de aspiración** *m* PAPEL suction porosimeter

poroso *adj* CARBÓN, ENERG RENOV porous; **no** ~ *adj* GAS nonporous

porpezita *f* MINERAL porpezite

porporciones: ~ **múltiples** *f pl* QUÍMICA multiple proportions

porta *f* TRANSP MAR port; ~ **de carga** *f* TRANSP MAR *construcción naval* cargo port; ~ **moleta** *f* ING MECÁ knurling

portabarras *m* ING MECÁ bar chuck

portabarrenas *m* ING MECÁ bit holder, drill

portaboquilla: ~ **de inyección** *f* AUTO injection nozzle holder

portabrocas *m* ING MECÁ bit holder, chuck, drill, drill chuck, point chuck, *laboreo de metales* drill holder, MECÁ pad, chuck, MINAS drill, drill holder; ~ **abovedado** *m* ING MECÁ dome pad; ~ **de cambio rápido** *m* ING MECÁ quick-change drill chuck; ~ **electromagnético** *m* ING MECÁ electromagnetic chuck; ~ **excéntrico** *m* ING MECÁ eccentric chuck; ~ **intercambiable** *m* ING MECÁ take-about chuck; ~ **de longitud exacta** *m* ING MECÁ dead-length-type chuck; ~ **de mango cuadrado** *m* ING MECÁ square bit drive; ~ **neumático** *m* ING MECÁ air-operated chuck; ~ **de tres mandíbulas** *m* ING MECÁ three-pronged chuck

portacabinas *m* TRANSP, TRANSP AÉR cabin conveyor

portacables *m* ELEC, ING ELÉC cable rack

portacadenas *m* CONST chainman

portacarros *m* ING MECÁ knee

portacebo *m* *AmL* (*cf obturador Esp*) MINAS, TEC ESP *cargas explosivas* shutter

portacojinetes *m* ING MECÁ pedestal, tower; ~ **radial** *m* ING MECÁ radial diehead; ~ **de roscar** *m* ING MECÁ screw box

portacuchillas *m* ING MECÁ blade holder, tool post, PAPEL knife holder

portada *f* IMPR title page, MINAS *entibación* anticline, arch; ~ **para machos** *f* PROD *moldería* core print

portadados *m* PROD die stock; ~ **circular** *m* ING MECÁ circular die stock

portadiapositivas *m* FOTO slide holder

portador *m* FÍS, INFORM&PD carrier, TEC ESP vector; ~ **de carga** *m* ELEC, TEC ESP charge carrier; ~ **de carga eléctrica** *m* ELEC, FÍS, FÍS RAD, ING ELÉC electric-current carrier; ~ **común** *m* TRANSP MAR *comunicación por satélite* common carrier; ~ **de corriente eléctrica** *m* ELEC, FÍS, FÍS RAD, ING ELÉC electric-current carrier; ~ **de datos** *m* INFORM&PD data carrier; ~ **del diferencial** *m* AUTO differential carrier; ~ **de emisión** *m* TELECOM transmission bearer; ~ **de MF** *m* ELECTRÓN FM carrier; ~ **minoritario** *m* ELECTRÓN minority carrier; ~ **modulado** *m* ELECTRÓN modulated carrier; ~ **óptico** *m* ELECTRÓN optical carrier; ~ **de pixeles** *m* IMPR, INFORM&PD pixel carrier; ~ **del planetario** *m* AUTO, VEH planet carrier; ~ **de RF** *m* ELECTRÓN, ING ELÉC, TELECOM RF carrier; ~ **de rodillos del flap** *m* TRANSP AÉR flap roller carriage; ~ **de transferencias fiable** *m* TELECOM reliable transfer server (*RTS*); ~ **de transmisión** *m* TELECOM transmission bearer

portadora *f* TV carrier; ~ **común** *f* TV common carrier; ~ **de datos** *f* TELECOM data carrier; ~ **de imagen** *f* ELECTRÓN, TELECOM picture carrier, TV picture carrier, vision carrier; ~ **de impulsos** *f* ELECTRÓN pulse carrier; ~ **modulada** *f* TV modulated carrier; ~ **modulada en amplitud** *f* ELECTRÓN amplitude-modulated carrier; ~ **monocanal** *f* (*SCPC*) TEC ESP, TELECOM single channel per carrier (*SCPC*); ~ **multicanálica** *f* TEC ESP multichannel carrier; ~ **a multidestinación** *f* TEC ESP multidestination carrier; ~ **piloto** *f* TELECOM pilot carrier; ~ **principal** *f* FÍS majority carrier; ~ **de pulsos** *f* TV pulse carrier; ~ **secundaria** *f* FÍS minority carrier; ~ **de sonido** *f* ELECTRÓN sound carrier; ~ **de video** *AmL*, ~ **de vídeo** *f Esp* TV video carrier

portaelectrodos *m* CONST, ELEC, ING ELÉC electrode holder

portaembudo *m* LAB funnel stand

portaempaquetadura *m* ING MECÁ packing gland

portaequipajes *m* AUTO, VEH boot (*BrE*), trunk (*AmE*)

portaescobillas *m* ELEC, ING ELÉC brush-holder; ~ **regulable** *m* ING ELÉC brush-rocker; ~ **tipo brazo** *m* ELEC arm-type brush holder; ~ **tipo palanca** *m* ELEC arm-type brush holder

portaespoleta *f* TEC ESP *cargas explosivas* shutter

portaestampa *m* ING MECÁ die set, die holder, PROD *para recortaduras* die holder

portafiltros *m* CINEMAT, TV filter holder

portafresas: ~ **de una máquina fresadora** *m* PROD milling-machine arbor

portafusibles *m* ELEC, ELECTRÓN, ING ELÉC fuse carrier, fuse holder, TV fuse holder

portaherramientas *m* ING MECÁ collet, cutter, cutter bar, holder, tool box, tool holder, tool post, MECÁ chuck; ~ **con agujero roscado** *m* ING MECÁ cutter with tapped hole; ~ **articulado** *m* ING MECÁ jointed tool-holder; ~ **basculante** *m* ING MECÁ swivel slide rest; ~ **de cambio rápido** *m* ING MECÁ rapid-change tool holder; ~ **compuesto** *m* ING MECÁ compound tool holder; ~ **conducido** *m* ING MECÁ driven tool holder; ~ **cónico** *m* ING MECÁ scroll chuck; ~ **de copiar** *m* ING MECÁ forming tool holder; ~ **cuadrado** *m* ING MECÁ square bit drive; ~ **electromagnético** *m* ING MECÁ electromagnetic chuck; ~ **con espiga** *m* ING MECÁ cutter with shank; ~ **de longitud exacta** *m* ING MECÁ dead-length-type chuck; ~ **con mango** *m* ING MECÁ cutter with shank; ~ **de mango cuadrado** *m* ING MECÁ square drive; ~ **múltiple** *m* ING MECÁ compound tool holder; ~ **oscilante** *m* ING MECÁ floating tool-holder; ~ **revólver** *m* ING MECÁ revolving tool holder, MECÁ turret, PROD *tornos* monitor; ~ **revólver de cuatro caras** *m* ING MECÁ four-tool tool post; ~ **en T** *m* ING MECÁ T-rest; ~ **de tornillo** *m* ING MECÁ screw chuck; ~ **del torno** *m* ING MECÁ lathe tool post; ~ **con vástago** *m* ING MECÁ cutter with shank

portainyector *m* AUTO nozzle holder

portal *m* CONST doorway, portal

portalámparas *m* CINEMAT lamp holder, FOTO lamp housing, ING ELÉC holder, lamp holder, MECÁ, PROD socket; ~ **aditivo** *m* CINEMAT additive lamphouse; ~ **de bayoneta** *m* ELEC bayonet lamp holder, bayonet socket; ~ **de llave** *m* TELECOM key-and-lamp unit; ~ **de vidrio** *m* C&V, ING ELÉC glass holder

portalentes *f* ÓPT lens turret

portalón *m* TRANSP MAR gangway

portaluneta *m* PROD die-head, diergol haed; ~ **revólver** *m* ING MECÁ revolving die-head

portamacho *m* PROD *fundición* bearing, seating, steady pin, *moldes* print

portamachos *m* ING MECÁ tap holder

portamaletas *m* AUTO boot (*BrE*), trunk (*AmE*)

portamatriz *m* ING MECÁ die set, die holder, PROD *para recortaduras* die holder

portamechas *m* *AmL* (*cf lastrabarrena Esp*) ING MECÁ drill chuck, drill collar

portamiras: ~ **operarios** *mpl* CONST staff holder

portamuestras *m* NUCL sample holder

portanegativos *m* FOTO negative carrier, negative sleeve

portaobjetivo *m* FOTO lens panel, INSTR, LAB *micro-*

scopio objective nose-piece; ~ **ajustable** *m* CINEMAT adjustable lens holder; ~ **revólver** *m* ING MECÁ, INSTR, LAB revolving nosepiece; ~ **rotativo** *m* ING MECÁ, INSTR, LAB *microscopio* revolving nosepiece

portaobjetivos *m* FOTO lens mount; ~ **giratorio** *m* IMPR, ÓPT lens turret; ~ **rotativo** *m* ÓPT lens turret

portaobjetos *m* C&V microscope slide, INSTR glass slide, object stage, stage base, LAB microscope slide, slide; ~ **cuadriculado** *m* INSTR *microscopio* finder

portaocular *m* INSTR eyepiece holder

portaorificio *m* C&V bushing assembly, neck ring holder

portaoriginales *m* CINEMAT, FOTO copy stand

portapelículas: ~ **con dispositivo de vacío** *f* IMPR vacuum film holder

portapiezas *m* ING MECÁ fixture, MECÁ jig; ~ **ranurado compuesto** *m* ING MECÁ compound slotted worktable

portapinza *m* AUTO, VEH *freno* caliper (*AmE*), calliper (*BrE*)

portapipetas *m* LAB pipette stand

portaplacas *m* FOTO plate-holder

portaplato *m* ING MECÁ chuck plate

portapoleas: ~ **de corona** *m* TEC PETR crown block

portaprobetas *m* NUCL specimen holder

portapunzón *m* ING MECÁ, MECÁ punch holder

portasección: ~ **de barra guadañadora** *f* AGRIC knife back

portasierra: ~ **de calar** *m* PROD fretsaw, fretsaw frame; ~ **de mano regulable** *m* ING MECÁ adjustable hacksaw frame; ~ **de marquetería** *m* PROD fretsaw, fretsaw frame

portaterrajas: ~ **de apertura automática** *m pl* ING MECÁ self-opening screwing head; ~ **y terraja** *m pl* ING MECÁ stocks and dies

portátil *adj* INFORM&PD portable

portatroquel *m* ING MECÁ, PROD *para recortaduras* die holder

portatubos *m* LAB test tube holder, tube holder

portavaso *m* LAB beaker holder

portaviones *m* D&A aircraft carrier, TRANSP MAR aircraft carrier, flat top

portazapata: ~ **de freno** *m* ING MECÁ brake block

porteador *m* TRANSP MAR sea carrier

portezuela *f* TRANSP AÉR flap, hatch; ~ **del capó** *f* TRANSP AÉR cowl flap; ~ **del piso** *f* TRANSP AÉR floor hatch; ~ **del tren de aterrizaje del morro** *f* TRANSP AÉR nose gear door; ~ **del tren de aterrizaje principal** *f* TRANSP AÉR main landing gear door

pórtico *m* AGUA, CONST, FERRO, TRANSP MAR gantry, TV porch; ~ **acarreador** *m* PROD traveling gantry crane (*AmE*), travelling gantry crane (*BrE*); ~ **de lanzamiento** *m* CONST launching gantry

portilla *f* INFORM&PD gateway, terminal port, TRANSP MAR *equipamiento de cubierta* porthole; ~ **de entrada/salida** *f* INFORM&PD input/output port; ~ **para terminal** *f* INFORM&PD terminal port

portillo *m* ING MECÁ port, MECÁ aperture, TRANSP MAR port, *equipamiento de cubierta* porthole; ~ **de guantes** *m* NUCL glove port

posarse *v refl* CARBÓN *líquidos* settle

poscalentamiento *m* PROD postheating

posenfriador *m* TERMOTEC aftercooler

posibilidad: ~ **de enganche** *f* FERRO *vehículos* couplability; ~ **de realización** *f* MINAS feasibility; ~ **de recorrido** *f* CONST rideability

posición[1]: ~ **de conectado** *adj* ING ELÉC on position

posición[2]: **a la ~ cero** *adv* ING MECÁ home; **a la ~ de salida** *adv* ING MECÁ home

posición[3] *f* GEN position, INFORM&PD *memoria* location, TEC ESP station; ~ **abierta** *f* ELEC *relé* open position; ~ **de abierto** *f* ING MECÁ off-position; ~ **de ajuste** *f* QUÍMICA set point; ~ **de aproximación** *f* TRANSP AÉR approach fix; ~ **de aproximación final** *f* TRANSP AÉR final-approach fix; ~ **de aproximación inicial** *f* TRANSP AÉR initial approach fix; ~ **de aproximación intermedia** *f* TRANSP AÉR intermediate approach fix; ~ **auxiliar de marcha** *f* TELECOM auxiliary service position; ~ **de balanceo** *f* TEC ESP roll attitude; ~ **de bit** *f* INFORM&PD bit position; ~ **de bloqueo** *f* PROD latching position; ~ **de celda binaria** *f* PROD memory bit location; ~ **de cero** *f* PROD home position; ~ **cerrada** *f* ELEC *relé* closed position; ~ **de cierre** *f* ING ELÉC on position; ~ **de conexión a masa** *f* ELEC *conmutador* earthing position (*BrE*), grounding position (*AmE*); ~ **de conexión a tierra** *f* ELEC *conmutador* earthing position (*BrE*), grounding position (*AmE*); ~ **de cortar** *f* ING MECÁ off-position; ~ **de desconexión** *f* ING MECÁ off-position; ~ **de desexcitación** *f* PROD de-energized position; ~ **de la emulsión** *f* CINEMAT emulsion position; ~ **de funcionamiento** *f* ING ELÉC on-position; ~ **de impresión** *f* INFORM&PD print position; ~ **de los impulsos** *f* ELECTRÓN pulse position; ~ **inicial** *f* PROD home position; ~ **en línea** *f* IMPR in-line position; ~ **de memoria** *f* INFORM&PD memory location; ~ **de memoria binaria** *f* PROD memory bit location; ~ **de montaje** *f* TRANSP AÉR rigging position; ~ **de morro contra el viento** *f* TRANSP AÉR nose-in positioning; ~ **morro de entrada** *f* TRANSP AÉR nose-in positioning; ~ **de morro a favor del viento** *f* TRANSP AÉR nose-out positioning; ~ **morro de salida** *f* TRANSP AÉR nose-out positioning; ~ **neutra** *f* AUTO, VEH *caja de cambio* neutral; ~ **"no"** *f* ING MECÁ off-position; ~ **oblicua** *f* CONST skewing; ~ **de orden inferior** *f* INFORM&PD low-order position; ~ **de parada** *f* ING MECÁ off-position; ~ **del paso de la pala** *f* TRANSP AÉR blade-pitch setting; ~ **protegida** *f* INFORM&PD protected location; ~ **de pruebas** *f* TELECOM test position; ~ **de pruebas y medidas** *f* TELECOM test position; ~ **en el reactor** *f* NUCL *de una barra de control* height position; ~ **de reposo** *f* ING MECÁ off-position, PROD home position; ~ **télex** *f* TELECOM telex position; ~ **de titulares** *f* TV caption stand; ~ **de vuelo** *f* TEC ESP attitude

posicionador *m* CINEMAT positioner, ING MECÁ jig, positioner; ~ **acústico** *m* OCEAN echo ranging; ~ **para montajes** *m* TRANSP MAR *construcción naval* erection jig; ~ **rotativo de la válvula** *m* VEH motor valve rotator

posicionamiento *m* IMPR *de la cinta* threading-in, ING MECÁ, TRANSP MAR positioning; ~ **dinámico** *m* PETROL, TEC PETR *operaciones costa-fuera* dynamic positioning; ~ **paralelo** *m* TRANSP AÉR parallel positioning; ~ **de la rejilla** *m* ELECTRÓN mask alignment

posicionar *vt* ING MECÁ fix

posincronización: ~ **del intervalo del campo de borrado** *f* TV postsync field-blanking interval

positivadora *f* CINEMAT, FOTO printer; ~ **aditiva** *f* CINEMAT additive printer; ~ **aditiva de color** *f* CINEMAT additive color printer (*AmE*), additive colour printer (*BrE*); ~ **automática** *f* FOTO automatic

printer; ~ **en cascada** *f* CINEMAT cascade printer;
~ **por contacto** *f* CINEMAT, FOTO, IMPR contact
printer; ~ **por contacto escalonado** *f* CINEMAT step
contact printer; ~ **intermitente por contacto** *f* CINE-
MAT intermittent contact printer; ~ **óptica** *f* CINEMAT
optical printer; ~ **óptica Oxberry** *f* CINEMAT Oxberry
optical printer; ~ **sustractiva a color** *f* CINEMAT
subtractive color printer (*AmE*), subtractive colour
printer (*BrE*); ~ **de ventanilla húmeda** *f* CINEMAT
wet gate printer
positivamente: ~ **inclinado** *adj* GEOM positively
skewed
positivar *vt* FOTO print
positivo[1] *adj* ELEC *electrodo* positive; ~**-negativo-
positivo** *adj* (*p-n-p*) ELECTRÓN, FÍS, FÍS RAD positive-
negative-positive (*p-n-p*)
positivo[2]: ~ **dupe** *m* CINEMAT dupe positive;
~ **duplicado** *m* CINEMAT dupe positive;
~ **intermedio** *m* CINEMAT intermediate positive,
FOTO lavender print
positrón *m* (*p*$^+$) FÍS, FÍS PART positron (*p*$^+$)
posluminiscencia *f* FÍS RAD, TEC ESP *tubos de descarga
gaseosa* afterglow
posmoldeado *adj* P&C *procesamiento* postforming
poso *m* AGUA sediment; ~ **radiactivo** *m* AGRIC fallout,
radioactive fallout, CONTAM radioactive fallout, fallout, FÍS, FÍS RAD, NUCL fallout, radioactive fallout
posquemador *m* NUCL afterburner
postal: **sobre** ~ *m* EMB correspondence pocket, PAPEL
banker shape, correspondence pocket
postámbulo *m* INFORM&PD postamble
postcinemático *adj* GEOL postkinematic
postcombustión: ~ **térmica** *f* TRANSP thermal postcombustion
postcosecha *f* AGRIC post-harvest
poste *m* Esp (*cf apoyo AmL*) AUTO post, CONST pillar,
pole, standard, strut, *topografía* poststaff, *vertical*
post, EMB strut, FERRO *infraestructura* stanchion, ING
MECÁ standard, MECÁ strut, MINAS *construcción*
upright, prop, *de mineral o carbón* post; ~ **de
anclaje** *m* CONST anchor post; ~ **de cambio de
rasante** *m* FERRO *equipo inamovible* gradient post;
~ **para descimbrado** *m* CONST striking post; ~ **de la
entrada** *m* CONST gatepost; ~ **extensible** *m* MINAS
jack leg; ~ **de grúa** *m* CONST derrick post; ~ **grúa** *m*
CONST gin; ~ **de guía** *m* PETROL guide post;
~ **indicador** *m* CONST finger post, guide post;
~ **indicador de la pendiente** *m* FERRO *equipo
inamovible* gradient post; ~ **de línea** *m* ELEC stay
pole; ~ **de quicio** *m* AGUA heel post, quoin post,
staple post; ~ **reflectante** *m* CONST *carreteras* reflecting stud; ~ **de retención** *m* TEC PETR *perforación*
backup post; ~ **de tabique** *m* ING MECÁ upright;
~ **del timón de dirección** *m* TRANSPAÉR rudder post;
~ **umbilical** *m* TEC ESP umbilical mast; ~ **de vaivén** *m*
CONST swinging post; ~ **del vano** *m* ELEC *red de
distribución* span pole
posteo *f* Esp (*cf entibación con agujas AmL*) MINAS
galerías forepoling
postes: ~ **acoplados** *m pl* FERRO twin pole, twin post
postizo *adj* ING MECÁ detachable
postprocesador *m* INFORM&PD postprocessor
postquemador *m* ING MECÁ, TERMO, TRANSP AÉR
afterburner
postsecador *m* PAPEL afterdryer
postsincronizar *vt* CINEMAT postsynchronize

postulado *m* GEOM, MATEMÁT postulate; ~ **del paralelo de Euclides** *m* GEOM Euclid's parallel postulate
potable *adj* AGUA potable, ALIMENT drinkable, fit to
drink
potasa *f* C&V, PETROL, QUÍMICA potash; ~ **cáustica** *f*
DETERG caustic potash; ~ **purificada** *f* DETERG pearl
ash
potásico *adj* GEOL potassic
potasio *m* (*K*) METAL, QUÍMICA potassium (*K*)
potea *f* PROD lute
potencia[1]: **poca** ~ *adj* ING ELÉC low power; **con**
~ **aumentada** *adj* VEH *motor, en sentido coloquial*
hotted-up (*AmE*), souped-up (*BrE*)
potencia[2] *f* AUTO power, CONST capacity, CRISTAL
energy, ELEC, FÍS, ING ELÉC energy, power, capacity,
ING MECÁ capacity, output, performance, power,
MATEMÁT power, MECÁ energy, power, torque,
MINAS *explosivos* strength, NUCL *de una central*
capacity, ÓPT *de lente* power, PROD *máquinas* torque,
strength, QUÍMICA energy, REFRIG capacity, TEC ESP
máquinas power, TERMO energy, VEH capacity;
~ **absorbida** *f* PROD input power; ~ **de aceleración**
f VEH accelerating power; ~ **activa** *f* ELEC *corriente
alterna*, FÍS, ING ELÉC active power, real power;
~ **acústica** *f* ING ELÉC acoustic energy; ~ **admisible
máxima** *f* TELECOM maximum admissible power;
~ **aparente** *f* ELEC, FÍS, ING ELÉC apparent power,
relative power; ~ **de aplastamiento** *f* SEG *de maquinaria* crushing power; ~ **base** *f* NUCL base power;
~ **bruta instalada** *f* NUCL gross installed capacity;
~ **bruta nominal** *f* ING MECÁ indicated horsepower;
~ **en caballos de vapor** *f* (*potencia en CV*) ING MECÁ
horsepower (*hp*); ~ **de caldeo** *f* TERMO heating
power; ~ **calórica** *f* TERMO caloric power;
~ **calorífica** *f* FÍS, GEOL, TEC PETR calorific value,
TERMO calorific value, *capacidad* calorific power,
heating power, caloric power, TERMOTEC calorific
value; ~ **de carga base** *f* NUCL off-peak power;
~ **compleja** *f* ELEC complex power; ~ **del
condensador** *f* REFRIG condenser heat; ~ **de
contracción** *f* METAL constriction energy, power
constriction; ~ **convergente** *f* FÍS ONDAS converging
energy, converging power; ~ **de cresta** *f* ING ELÉC
peak power; ~ **de crucero** *f* TRANSP AÉR cruising
power; ~ **en cuadratura** *f* ELEC quadrature power;
~ **en CV** *f* (*potencia en caballos de vapor*) ING MECÁ hp
(*horsepower*); ~ **de despegue homologada** *f* TRANSP
AÉR takeoff power rating; ~ **disipada** *f* ELEC dissipated power; ~ **disponible** *f* ELEC maximum
available power, ING ELÉC maximum available power,
power output, ING MECÁ, TELECOM available power;
~ **efectiva** *f* ING MECÁ brake power, effective horsepower, effective power, MECÁ actual horsepower,
actual power; ~ **efectiva en caballos de fuerza** *f*
ING MECÁ, PROD brake horsepower (*BHP*); ~ **al eje** *f*
ING MECÁ, PROD, TRANSP brake horsepower (*BHP*);
~ **de eje equivalente** *f* ING MECÁ equivalent shaft-
horsepower; ~ **eléctrica** *f* ELEC, ELECTRÓN, ING ELÉC
electrical power, electric power, NUCL *del reactor*,
TELECOM electrical power, TERMO electric power;
~ **electrocinética** *f* ELEC, FÍS, ING ELÉC, QUÍMICA
electrokinetic energy; ~ **de emanación** *f* FÍS RAD
emanating power; ~ **de emisión** *f* TV transmitter
power; ~ **de entrada** *f* ING ELÉC input power, power
input, PROD, TELECOM, TV input power; ~ **de entrada
a la línea** *f* PROD line input power; ~ **de entrada de**

video *AmL*, ~ **de entrada de vídeo** *f Esp* TV video input; ~ **eólica** *f* FÍS wind power; ~ **equivalente radiada isotrópica** *f* (*EEII*) TEC ESP *comunicaciones* equivalent isotropically-radiated power (*EIRP*); ~ **equivalente de ruido** *f* ÓPT, TELECOM noise equivalent power (*NEP*); ~ **específica** *f* TEC ESP specific power; ~ **de excitación de la rejilla** *f* ING ELÉC grid-driving power; ~ **explosiva** *f AmL* (*cf potencial enérgico Esp*) MINAS brisance; ~ **de fluido hidráulico** *f* ING MECÁ hydraulic fluid power; ~ **de fluido neumático** *f* ING MECÁ pneumatic fluid power; ~ **por flujo en peso** *f* ING MECÁ *hidráulica* head; ~ **fonética** *f* ACÚST phonetic power; ~ **de frenado** *f* TRANSP braking power; ~ **al freno** *f* ING MECÁ brake power, effective power, MECÁ actual horsepower; ~ **al freno en caballos** *f* ING MECÁ effective horsepower; ~ **al freno en caballos de fuerza** *f* (*BHP*) ING MECÁ, PROD, TRANSP brake horsepower (*BHP*); ~ **al freno en caballos hora** *f* PROD brake horsepower hour; ~ **de freno** *f* ING MECÁ, PROD brake horsepower (*BHP*); ~ **frigorífica** *f* REFRIG refrigerating capacity; ~ **frigorífica sensible** *f* REFRIG sensible cooling effect; ~ **frigorífica total** *f* REFRIG overall refrigerating effect, *enfriador de aire* total cooling effect; ~ **frigorífica útil** *f* REFRIG useful refrigerating effect; ~ **generada** *f* ELEC, ING ELÉC, ING MECÁ, PROD output; ~ **generada continua** *f* ELEC *generador* continuous output; ~ **del grupo** *f* NUCL unit capacity; ~ **del haz** *f* ELECTRÓN beam power; ~ **hidroeléctrica** *f* ELEC hydroelectric power; ~ **hora** *f* ING MECÁ horsepower hour; ~ **de impacto** *f* METAL impact energy, power energy; ~ **indicada** *f* ING MECÁ indicated horsepower; ~ **instalada** *f* ELEC installed capacity; ~ **instantánea máxima** *f* TEC ESP maximum instantaneous power; ~ **leudante** *f* ALIMENT *panadería, repostería y fermentación* raising power; ~ **luminosa** *f* FOTO light output; ~ **de mareas** *f* FÍS tidal power; ~ **másica** *f* ING MECÁ power-weight ratio; ~ **máxima** *f* ELEC maximum output, maximum power, ING ELÉC peak power, NUCL maximum capacity; ~ **máxima continua** *f* TRANSP AÉR maximum continuous power; ~ **máxima disponible** *f* ELEC, ING ELÉC maximum available power; ~ **máxima excepto al despegue** *f* TRANSP AÉR maximum except takeoff power (*METO power*); ~ **máxima instalada** *f* ELEC, ING ELÉC maximum demand; ~ **máxima posible** *f* NUCL maximum capacity; ~ **media** *f* TELECOM average power; ~ **media de la palabra** *f* ACÚST average speech power; ~ **de microondas** *f* ING ELÉC microwave power; ~ **del motor** *f* PROD motor rating; ~ **neta** *f* ING MECÁ net horsepower; ~ **nominal** *f* AUTO rated horsepower, ELEC rated power, ING MECÁ indicated horsepower, NUCL *del reactor* rated power capacity, VEH *motor* installed power, rated horsepower; ~ **normal** *f* ELEC rated power; ~ **de la onda** *f* FÍS ONDAS wave power; ~ **óptica** *f* ING ELÉC, ÓPT, TELECOM optical power; ~ **óptica de salida** *f* ÓPT, TELECOM optical power output; ~ **de pico de la envolvente** *f* TELECOM peak envelope power; ~ **de procesamiento** *f* TELECOM processing power; ~ **radiada** *f* ELEC radiated power, FÍS RAD radiated output, radiated power, NUCL, TELECOM, TV radiated power; ~ **radiada efectiva** *f* (*PRE*) TELECOM effective radiated-power (*ERP*); ~ **radiante** *f* FÍS, FÍS RAD, ÓPT, TELECOM radiant power; ~ **reactiva** *f* ELEC, FÍS, ING ELÉC reactive power; ~ **de red** *f* ING MECÁ net

horsepower; ~ **reflectora** *f* TEC ESP *superficies* albedo; ~ **reflejada** *f* ING ELÉC, ÓPT reflected power; ~ **refrigeradora neta** *f* REFRIG net refrigerating effect; ~ **de régimen** *f* ELEC rated power, MECÁ rating; ~ **requerida por máquinas herramientas** *f* ING MECÁ power required by machine tools; ~ **de salida** *f* ELEC output, output power, ING ELÉC output, output power, power rating, PROD output, TELECOM output power, TV output; ~ **de salida de CA** *f* ING ELÉC AC output, alternating-current output; ~ **de salida continua** *f* ELEC *generador* continuous output; ~ **de salida de diseño** *f* NUCL designed power required output; ~ **de salida del generador** *f* ELEC, ING ELÉC generator output power; ~ **de salida óptica** *f* ING ELÉC optical output power; ~ **de salida de ruido** *f* ELECTRÓN, NUCL noise power; ~ **de salida de saturación** *f* ELECTRÓN saturation output power; ~ **de salida de video** *AmL*, ~ **de salida de vídeo** *f Esp* TV video output; ~ **seca** *f* TRANSP AÉR dry power; ~ **de señal** *f* ELECTRÓN signal power; ~ **de señal de entrada** *f* ELECTRÓN input signal power; ~ **de servicio** *f* ELEC, ING ELÉC, MECÁ rating; ~ **sofométrica** *f* TELECOM psophometric power; ~ **sonora** *f* ACÚST sound power; ~ **sonora instantánea** *f* ACÚST instantaneous sound power; ~ **sonora instantánea por unidad de área** *f* ACÚST instantaneous sound power per unit area; ~ **sonora media por unidad de área** *f* ACÚST average sound power per unit area; ~ **sonora de referencia** *f* ACÚST reference sound power; ~ **de sostén** *f* PROD *plato magnético* holding power; ~ **suministrada** *f* ELEC, ING ELÉC, PROD output; ~ **suministrada continua** *f* ELEC *generador* continuous output; ~ **de superficie** *f* GAS surface power; ~ **térmica** *f* TERMO caloric power; ~ **termoeléctrica** *f* ING ELÉC, TERMO thermoelectric power; ~ **de toma** *f* ELEC *de devanado* tapping power; ~ **total de caballos** *f* ING MECÁ gross horsepower; ~ **útil** *f* ELEC *fuente de alimentación* maximum available power, *generador* output, output power, ING ELÉC *generador* output power, maximum available power, continuous output, output, ING MECÁ available power, output, PROD output; ~ **útil de CA** *f* ING ELÉC AC output; ~ **del viento** *f* METEO wind force; ~ **vocal instantánea** *f* ACÚST instantaneous speech power; ~ **vocal de pico** *f* ACÚST peak speech power; ~ **volúmica** *f* GAS volumic power

potenciador: ~ **de sabor** *m* ALIMENT flavor potentiator (*AmE*), flavour potentiator (*BrE*)

potencial *m* ELEC, FÍS, ING ELÉC, TEC ESP potential; ~ **absoluto** *m* ELEC absolute potential; ~ **activo** *m* ELEC active potential; ~ **de las alternancias en medio** *m* ENERG RENOV velocity potential; ~ **de campo** *m* PETROL field potential; ~ **cero** *m* ELEC zero potential; ~ **de contacto** *m* ELEC *termopar* contact EMF, contact potential, FÍS *efecto Volta* contact EMF; ~ **de crecimiento algáceo** *m* HIDROL algal growth potential; ~ **del cuadripolo** *m* NUCL quadrupole potential; ~ **eléctrico** *m* ELEC, FÍS, ING ELÉC electric potential; ~ **del electrodo** *m* ELEC, FÍS, ING ELÉC, QUÍMICA electrode potential; ~ **enérgico** *m Esp* (*cf potencia explosiva AmL*) MINAS brisance; ~ **escalar** *m* FÍS, ING ELÉC scalar potential; ~ **escalar magnético** *m* FÍS magnetic scalar potential; ~ **espontáneo** *m* GEOFÍS *fuentes eléctricas naturales* self-potential; ~ **evocado auditivo** *m* ACÚST hearing-evoked potential; ~ **de extinción** *m* ING ELÉC extinction potential; **~ flotante** *m* ING ELÉC floating potential; **~ de frenado** *m* FÍS stopping potential; **~ gravitacional** *m* FÍS, GEOFÍS, GEOL gravitational potential; **~ hidrógeno** *m* (*pH*) HIDROL, QUÍMICA potential of hydrogen (*pH*); **~ de ionización** *m* FÍS, FÍS RAD ionization potential; **~ lunisolar** *m* TEC ESP lunisolar potential; **~ magnético** *m* ING ELÉC magnetic potential; **~ magnético vectorial** *m* FÍS magnetic vector potential; **~ neto** *m* ELEC net output; **~ nuclear** *m* FÍS RAD, NUCL nuclear potential; **~ nuclear central** *m* NUCL central nuclear potential; **~ de núcleo** *m* NUCL nuclear potential; **~ del núcleo** *m* FÍS RAD nuclear potential; **~ nulo** *m* ELEC *tensión* zero potential; **~ de oxido-reducción** *m* CONTAM oxidation reduction potential; **~ del pozo** *m* PETROL well potential; **~ propio** *m* GEOFÍS *estudios de energía* self-potential; **~ químico** *m* FÍS, QUÍMICA chemical potential; **~ redox** *m* CONTAM oxidation reduction potential; **~ de reducción de oxidación** *m* CONTAM oxidation reduction potential; **~ de sedimentación** *m* PROC QUÍ sedimentation potential; **~ de segunda ionización** *m* FÍS second ionization potential; **~ Tabakin** *m* NUCL Tabakin potential; **~ termodinámico** *m* FÍS, TERMO thermodynamic potential; **~ terrestre** *m* ELEC, ING ELÉC, TEC ESP earth potential (*BrE*), ground potential (*AmE*); **~ de tierra** *m* ELEC, ING ELÉC, TEC ESP earth potential (*BrE*), ground potential (*AmE*); **~ vector** *m* FÍS, ING ELÉC vector potential; **~ vector magnético** *m* ELEC magnetic vector potential; **~ de Yukawa** *m* FÍS Yukawa potential

potenciales: **~ cocleares** *m pl* ACÚST cochlear potentials; **~ retardados** *m pl* FÍS retarded potentials

potenciómetro *m* CINEMAT, CONST, ELEC, FÍS, ING ELÉC, LAB potentiometer; **~ de ajuste** *m* ELEC *resistor* trimming potentiometer, TRANSP AÉR adjusting potentiometer; **~ de alambre bobinado** *m* ELEC wire-wound potentiometer; **~ de alambre devanado** *m* ELEC wire-wound potentiometer; **~ bobinado** *m* ELEC wire-wound potentiometer; **~ de CA** *m* ELEC AC potentiometer; **~ de CC** *m* ELEC *relé* DC potentiometer; **~ cifrado** *m* TEC ESP, TELECOM encoding potentiometer; **~ de control** *m* ING ELÉC control potentiometer; **~ del cursor** *m* ING ELÉC slide potentiometer; **~ giratorio** *m* ELEC, ING ELÉC, INSTR rotary potentiometer; **~ helicoidal** *m* ELEC multiturn potentiometer, *resistencia* helical potentiometer, ING ELÉC multiturn potentiometer; **~ de lectura** *m* ELEC read-out potentiometer; **~ magnetorresistor** *m* ELEC, INSTR magnetoresistor potentiometer; **~ multivuelta** *m* ELEC, ING ELÉC multiturn potentiometer; **~ no bobinado** *m* ELEC non-wirewound potentiometer; **~ no lineal** *m* CONST nonlinear potentiometer; **~ de regulación** *m* ELEC, ING ELÉC trimming potentiometer; **~ rotativo** *m* ELEC, ING ELÉC, INSTR rotary potentiometer; **~ de variación logarítmica** *m* ING ELÉC logarithmic potentiometer; **~ de velocidad** *m* PROD speed pot; **~ vernier** *m* ELEC vernier potentiometer

poza *f* HIDROL pothole

pozo *m* AGUA *en terreno acuífero* shaft in water-bearing ground, CARBÓN *minas* pit, drain, shaft, CONST hole, pothole, pit, well, CONTAM *instalaciones industriales* sink, HIDROL well, MINAS pit, shaft, PETROL, TEC PETR hole, well; **~ abierto** *m* PETROL open hole; **~ absorbente** *m* AGUA absorbing well, CARBÓN

drain; **~ de aireación** *m* MINAS air shaft; **~ de alivio** *m* AGUA relief well; **~ artesiano** *m* AGUA, HIDROL artesian well, PETROL flowing well; **~ de aspiración** *m* CARBÓN *minas* pumping pit; **~ de auxilio** *m* PETROL relief well; **~ barrenado** *m* AGUA, HIDROL bored well; **~ blindado** *m* MINAS ironclad shaft, metal-lined shaft; **~ de bombas** *m* AGUA pump shaft, pumping shaft; **~ bombeo** *m* CARBÓN pumping well; **~ de bombeo mecánico** *m* TEC PETR beam well; **~ para cabrestante** *m* CONST capstan pit; **~ del cañón** *m* D&A gun pit; **~ de captación** *m* HIDROL collector well; **~ de cateo** *m* *AmL* (*cf pozo de prospección Esp*) MINAS exploration shaft, costean pit, exploration pit, prospect hole, PETROL wildcat well; **~ cementado** *m* HIDROL cased well; **~ ciego** *m* CONST blind pit, MINAS staple; **~ de colada** *m* PROD molding hole (*AmE*), moulding hole (*BrE*); **~ colector** *m* AGUA collecting ditch, CONST collecting pit; **~ de compensación** *m* MINAS *sondeos* offset; **~ de comunicación** *m* *AmL* (*cf coladero Esp*) MINAS chute, mill hole, raise, winze; **~ de condensado** *m* GAS condensate well; **~ confidencial** *m* TEC PETR tight hole; **~ costa-fuera** *m* TEC PETR offshore well; **~ de delimitación** *m* TEC PETR outstep well, step-out well; **~ de desagüe** *m* MINAS drainage well, standage; **~ de desarrollo** *m* TEC PETR development well; **~ descubridor** *m* MINAS discovery shaft, TEC PETR discovery well; **~ desviado** *m* TEC PETR deviation well, deviated well; **~ de diámetro reducido** *m* PETROL rat hole, TEC PETR slim hole; **~ direccional** *m* TEC PETR directional well; **~ direccional controlado** *m* TEC PETR directional well; **~ donde no se encuentra ni gas ni petróleo** *m* MINAS dry hole; **~ de drenaje** *m* AGUA absorbing well, relief well; **~ drenante** *m* AGUA absorbing well; **~ enfundado** *m* HIDROL cased well; **~ de ensayo** *m* PETROL wildcat well; **~ entibado** *m* CONST timbered shaft; **~ de entrada de aire** *m* MINAS downcast, downcast shaft; **~ sin entubar** *m* PETROL open hole, TEC PETR uncased hole; **~ en erupción** *m* TEC PETR wild well; **~ de las escalas** *m* *AmL* (*cf escalera de pozo Esp*) MINAS ladder road, ladder shaft, ladder way, manway; **~ estéril** *m* *AmL* (*cf pozo seco Esp*) AGUA dry well, MINAS dry hole, NUCL dry well, PETROL, TEC PETR dry hole; **~ de evacuación** *m* MINAS drain shaft, drain well; **~ de evaluación** *m* *Esp* (*cf pozo excavado hacia arriba AmL*) PETROL, TEC PETR appraisal well, exploration well; **~ excavado hacia arriba** *m* *AmL* (*cf pozo de evaluación Esp*) PETROL, TEC PETR appraisal well, exploration well; **~ de exploración** *m* MINAS costean pit, discovery shaft, PETROL exploration well, wildcat well, TEC PETR exploration well; **~ exploratorio** *m* PETROL exploratory well, exploration well, TEC PETR exploration well, exploratory well; **~ de extensión** *m* TEC PETR extension well; **~ de extracción** *m* MINAS hoisting shaft, winding inset, winding shaft; **~ de filtración** *m* PROC QUÍ filtering well; **~ filtrante** *m* AGUA filter well; **~ de fondo arenoso** *m* AGUA absorbing well; **~ de franqueo** *m* *AmL* (*cf fondo del pozo Esp*) MINAS sump shaft; **~ gasífero** *m* PETROL gas well; **~ horadado** *m* CARBÓN sunk well; **~ improductivo** *m* MINAS dry hole, PETROL duster, dry hole, TEC PETR dry hole; **~ inclinado** *m* MINAS brow, incline hole, incline shaft, inclined hole, inclined shaft; **~ de inmersión** *m* PROD immersion well; **~ de** inspección *m* CONST, MINAS manhole; ~ **interior** *m* MINAS internal shaft; ~ **interior de comunicación** *m* MINAS staple pit, staple shaft; ~ **de inyección** *m* AGUA injection well, *sondeos* pressure wheel, PETROL, TEC PETR injection well; ~ **de inyección de aire** *m* TEC PETR air input well; ~ **maestro** *m* MINAS main shaft; ~ **marítimo** *m* TEC PETR offshore well; ~ **de mina** *m* MINAS mine shaft; ~ **negro** *m* AGUA cesspit; ~ **obligatorio** *m* TEC PETR *licencias, contratos* obligatory well; ~ **de observación** *m* AGUA, GAS observation well; ~ **en operación** *m* GAS operating well; ~ **de paso exterior** *m* TEC PETR outstep well; ~ **perforado** *m* HIDROL drilled well; ~ **petrolífero** *m* PETROL oil well; ~ **piloto** *m* PETROL rat hole; ~ **de potencial** *m* FÍS potential well; ~ **principal** *m* MINAS main shaft; ~ **de producción** *m* GAS, PETROL, TEC PETR production well; ~ **profundo** *m* AGUA, HIDROL deep well; ~ **de prospección** *m* *Esp* (*cf pozo de cateo AmL*) MINAS costean pit, exploration pit, exploration shaft, prospect hole; ~ **de quemado** *m* PETROL burn pit; ~ **de recarga** *m* AGUA *agua artesiana*, HIDROL recharge well; ~ **de restablecimiento** *m* AGUA *agua artesiana*, HIDROL recharge well; ~ **satélite** *m* TEC PETR satellite well; ~ **seco** *m* *Esp* (*cf pozo estéril AmL*) AGUA dry well, MINAS dry hole, NUCL dry well, PETROL dry hole, TEC PETR dry hole, dry well; ~ **de sedimentación** *m* PROC QUÍ settling sump; ~ **sísmico** *m* GEOFÍS seismic borehole; ~ **de sondeo** *m* AGUA bored well, CARBÓN borehole, test pit, GAS, GEOFÍS, GEOL borehole, HIDROL bored well, borehole, MINAS, PETROL borehole; ~ **de sondeo lateral** *m* MINAS side hole; ~ **de sondeo por tierra** *m* CARBÓN drop hole; ~ **submarino** *m* CONTAM subsea well; ~ **térmico** *m* TEC PETR thermowell; ~ **termostático** *m* PROD thermostat well; ~ **de toma** *m* MINAS intake shaft; ~ **torcido** *m* PETROL, TEC PETR crooked hole; ~ **vacío** *m* PETROL empty hole; ~ **para vástago de perforación** *m* PETROL rat hole; ~ **de ventilación** *m* *AmL* (*cf pozo ciego Esp*) MINAS airway, crossheading, downcast, downcast shaft, ventilation shaft, staple; ~ **vertedero** *m* *AmL* (*cf coladero Esp*) MINAS mill, mill hole, *mineral* coarse concentration mill; ~ **vertical** *m* MINAS vertical hole, vertical shaft; ~ **de visita** *m* ING MECÁ, MECÁ manhole; ~ **de Yukawa** *m* NUCL Yukawa well

ppm *abr* (*partes por millón*) QUÍMICA ppm (*parts per million*)

PPM *abr* (*páginas por minuto*) INFORM&PD PPM (*pages per minute*)

PPP *abr* (*pistas por pulgada*) IMPR, INFORM&PD TPI (*tracks per inch*)

PPS *abr* (*pulgadas por segundo*) INFORM&PD IPS (*inches per second*)

PR *abr* (*plano de referencia*) CONST, GEOL DP (*datum plane*)

Pr *abr* (*praseodimio*) QUÍMICA Pr (*praseodymium*)

practicable *m* CINEMAT scaffold

practicaje *m* OCEAN *reglamentaciones portuarias*, TRANSP MAR pilotage; ~ **en aguas costeras** *m* TRANSP MAR inshore pilotage

prácticas: ~ **de primeros auxilios** *f pl* SEG *clases de adiestramiento* first-aid work; ~ **para trabajar con seguridad** *f pl* SEG safe working practices

práctico *m* TRANSP MAR *navegación* pilot; ~ **de altura** *m* TRANSP MAR deep-sea pilot; ~ **de costa** *m* TRANSP MAR inshore pilot

pradera: ~ **cultivada** *f* AGRIC tame pasture; ~ **con pastos de alto porte** *f* AGRIC tall grass prairie; ~ **de pastos tiernos** *f* AGRIC sward; ~ **pobre** *f* AGRIC ranching

prado: ~ **en altoplanicie** *m* AGRIC upland meadow

pragma *m* INFORM&PD pragma

pragmática *f* INFORM&PD pragmatics

praseodimio *m* (*Pr*) QUÍMICA praseodymium (*Pr*)

praseolita *f* MINERAL praseolite

prasio *m* MINERAL prase

PRE *abr* (*potencia radiada efectiva*) TELECOM ERP (*effective radiated-power*)

preacentuación *f* ACÚST, TEC ESP, TV pre-emphasis; ~ **de video** *AmL*, ~ **de vídeo** *f Esp* TV video pre-emphasis

preadmisión *f* INSTAL HIDRÁUL *cilindro de motor* preadmission

preaireación *f* HIDROL preaeration

preajustar *vt* INFORM&PD preset

preajuste: ~ **del diafragma** *m* CINEMAT diaphragm presetting

prealmacenar *vt* INFORM&PD prestore

preámbulo *m* TEC ESP preamble

preamplificación *f* ACÚST pre-emphasis, ELECTRÓN preamplification, TEC ESP pre-emphasis

preamplificador *m* GEN preamplifier, TEC ESP head amplifier, preamplifier; ~ **con bajo nivel de ruidos** *m* ELECTRÓN, FÍS RAD low-noise preamplifier; ~ **mezclador** *m* ELECTRÓN mixer-preamplifier; ~ **de modulación** *m* ELECTRÓN modulator driver; ~ **SOM** *m* ELECTRÓN MOS driver

precalentador *m* GEN preheater; ~ **de aire** *m* ING MECÁ air preheater; ~ **de alto horno** *m* TERMO blast preheater

precalentamiento *m* GEN preheating; ~ **de crisoles** *m* C&V pot arching

precalentar *vt* GEN preheat

precámara *f* VEH prechamber

Precámbrico *m* GEOL Precambrian

precauciones *f pl* SEG precautions; ~ **contra incendios** *f pl* SEG, TERMO fire precautions; ~ **contra el polvo** *f pl* SEG precautions against dust; ~ **especiales** *f pl* SEG special precautions; ~ **que deben tomarse** *f pl* SEG precautions to be taken

precedencia *f* TEC ESP precession

preceptivo *adj* TELECOM mandatory

precesión *f* FÍS, TEC ESP precession; ~ **de Larmor** *f* FÍS Larmor precession

precintado[1] *adj* CONTAM MAR, EMB sealed

precintado[2]: ~ **por flejado** *m* EMB strapping seal; ~ **al vacío** *m* EMB chamber-type vacuum sealing

precintadora: ~ **de bandejas** *f* EMB tray sealer; ~ **de botellas** *f* EMB bottle-sealing machine; ~ **de cajas** *f* EMB case-sealing machine; ~ **de sacos** *f* EMB sack sealer

precintar *vt* CONTAM MAR seal, PROD insulate

precinto *m* EMB, ING MECÁ, MECÁ seal; ~ **adhesivo** *m* EMB adhesive tape; ~ **antiderrame** *m* EMB pour-spout seal; ~ **autoadhesivo** *m* EMB self-adhesive tape; ~ **engomado** *m* EMB glued seal; ~ **hermético** *m* EMB hermetic seal; ~ **intacto** *m* EMB freshness seal; ~ **laberinto** *m* EMB labyrinth seal; ~ **laminado autoadhesivo** *m* EMB self-adhesive laminated tape; ~ **de papel engomado** *m* EMB gummed paper tape; ~ **de plomo** *m* EMB lead seal; ~ **a prueba de**

falsificaciones _m_ EMB, SEG tamper-proof seal; ~ **a prueba de robos** _m_ EMB pilfer-proof seal
precio: ~ **de almacén** _m_ PROD warehouse price; ~ **APEX** _m_ (_precio de la excursión de reserva por adelantado_) TRANSP AÉR APEX fare (_advance purchase excursion fare_); ~ **base de acceso** _m_ TELECOM basic rate access (_BRA_); ~ **base de llamada** _m_ TELECOM basic call charge; ~ **base de servicio** _m_ TELECOM basic rate service; ~ **de coste** _m_ PROD cost price; ~ **de la excursión de reserva por adelantado** _m_ (_precio APEX_) TRANSP AÉR advance purchase excursion fare (_APEX fare_); ~ **de factura** _m_ PROD cost price; ~ **de llamada** _m_ TELECOM call charge; ~ **medio** _m_ TELECOM _tasa de conversación_ medium rate; ~ **de mercado** _m_ TEXTIL market price; ~ **a nivel de explotación agrícola** _m_ AGRIC farm gate price; ~ **al productor** _m_ AGRIC farm price; ~ **puesto en destino** _m_ TEC PETR _comercio_ landed price; ~ **de readquisición** _m_ TEC PETR _comercio_ buy-back price; ~ **de recompra** _m_ TEC PETR _comercio_ buy-back price; ~ **de reposición** _m_ PROD replacement price; ~ **según contrato** _m_ PROD contract price; ~ **de venta** _m_ PROD cost price
precipitabilidad _f_ QUÍMICA precipitability
precipitación _f_ AGRIC, AGUA rainfall, CONST precipitation, CONTAM _depuración_ deposition, precipitation event, sedimentation, ENERG RENOV, HIDROL rainfall, METAL precipitation, METEO precipitation, rainfall, QUÍMICA precipitation; ~ **ácida** _f_ CONTAM acid precipitation, acid rain, GAS acid rain; ~ **acídica húmeda** _f_ CONTAM wet acidic fallout; ~ **acídica seca** _f_ CONTAM dry acidic fallout; ~ **de contaminante** _f_ CONTAM pollutant deposition; ~ **continua** _f_ METAL, METEO continuous precipitation; ~ **efectiva** _f_ METEO effective precipitation; ~ **húmeda** _f_ CONTAM _depuración_ wet deposition, wet precipitation; ~ **idónea** _f_ METEO altitude of optimum rainfall; ~ **no interceptada** _f_ CONTAM throughfall; ~ **prevista** _f_ HIDROL design storm; ~ **primaria** _f_ HIDROL primary precipitation; ~ **química** _f_ HIDROL, QUÍMICA _aguas residuales_ chemical precipitation; ~ **radiactiva** _f_ _Esp_ (_cf deposición radiactiva AmL_) AGRIC radioactive fallout, CONTAM _en la atmósfera_ rainout, contamination fallout, radioactive fallout, fallout, FÍS, FÍS RAD, NUCL radioactive fallout, fallout; ~ **radiactiva ácida** _f_ CONTAM acid fallout; ~ **radiactiva ácida y húmeda** _f_ CONTAM wet acidic fallout; ~ **radiactiva ácida-seca** _f_ CONTAM dry acidic fallout; ~ **radiactiva atmosférica** _f_ CONTAM atmospheric fallout; ~ **recocida** _f_ PROC QUÍ precipitation annealing; ~ **regulada** _f_ CONTAM regulated deposition; ~ **seca** _f_ CONTAM _depuración_ dry deposition, _tratamiento de gases_ dry precipitation; ~ **total** _f_ CONTAM total deposition; ~ **por vía húmeda** _f_ CONTAM wet precipitation
precipitado[1] _adj_ QUÍMICA precipitated
precipitado[2] _m_ METAL, PROC QUÍ, QUÍMICA, TEC PETR _ingeniería de lodos_ precipitate; ~ **ácido** _m_ CONTAM, QUÍMICA acid deposit; ~ **electrolítico** _m_ CARBÓN deposit; ~ **de gas** _m_ METAL gas precipitate; ~ **de transición** _m_ METAL transitional precipitate
precipitador _m_ PROC QUÍ, QUÍMICA _tratamiento de gases_ precipitating agent, precipitator; ~ **electrostático** _m_ CONTAM _tratamiento de gases_ collecting electrode, electrostatic precipitator (_ESP_),

precipitation collector, GAS electrostatic precipitator (_ESP_); ~ **de polvos** _m_ CARBÓN baghouse
precipitante _m_ PROC QUÍ, QUÍMICA precipitating agent, precipitant
precipitar _vi_ PROC QUÍ, QUÍMICA precipitate
precisión _f_ FÍS accuracy, precision, GEOM sharpness, INFORM&PD, ING MECÁ, MECÁ, METR accuracy, precision, TRANSP MAR _de un cronómetro_ chronometer rate; ~ **de engranajes paralelos** _f_ ING MECÁ accuracy of parallel gears; ~ **de octeto** _f_ ELECTRÓN, INFORM&PD eight-bit accuracy; ~ **en la puntería** _f_ TEC ESP pointing accuracy; ~ **sencilla** _f_ INFORM&PD single precision; ~ **simple** _f_ INFORM&PD single precision; ~ **de la situación del buque** _f_ TRANSP MAR _navegación_ accuracy of ship's position; ~ **de tarado** _f_ NUCL set point accuracy
precocido _adj_ ALIMENT precooked
precombustión _f_ TERMO precombustion
precompresor _m_ REFRIG booster compressor
preconcentrado _m_ CARBÓN preconcentrate; ~ **de uranio** _m_ NUCL uranium preconcentrate
preconfigurar _vt_ INFORM&PD preset
precursor: ~ **ácido** _m_ CONTAM _características químicas_ acidic precursor
predeformación _f_ METAL prestrain
predeterminar _vt_ INFORM&PD preset
predicado _m_ INFORM&PD predicate
predicción _f_ AGUA forecasting, METEO forecast, forecasting; ~ **de exposición al ruido** _f_ (_NEF_) ACÚST noise exposure forecast (_NEF_); ~ **meteorológica** _f_ METEO forecasting, TEC ESP weather forecast; ~ **de órbita** _f_ TEC ESP orbit prediction; ~ **de ruido** _f_ ACÚST noise prediction; ~ **del tiempo** _f_ METEO forecasting
predisparo _m_ TEC ESP prefiring
predistorsión _f_ ACÚST predistortion
preeditar _vt_ CINEMAT pre-edit
preencendido _m_ VEH preignition
preendurecedor _m_ CINEMAT prehardener
pre-énfasis _m_ ACÚST, FÍS, TV pre-emphasis
preenfoque _m_ CINEMAT prefocusing
preenfriado _adj_ ALIMENT, REFRIG precooled
preenfriador _m_ ALIMENT, REFRIG precooler
preenfriamiento: ~ **de canales** _m_ ALIMENT, REFRIG carcass chilling
preenvejecimiento _m_ TELECOM preageing
preestucado _m_ PAPEL precoating
prefabricación _f_ CONST, TRANSP MAR _construcción naval_ prefabrication
prefabricado _adj_ CONST precast
prefabricados _m pl_ CONST precasting works
prefabricar _vt_ CONST, TRANSP MAR _construcción naval_ precast, prefabricate
preferencia[1]: ~ **de paso** _f_ TRANSP AÉR right of way
preferencia[2]: ~ **de paso a petición** _fra_ TRANSP demand right of way
preferencial _adj_ INFORM&PD foreground
preferente _adj_ INFORM&PD foreground
prefijación: ~ **de canales** _f_ _Esp_ (_cf presintonía de canales AmL_) TV presetting of channels
prefijado _adj_ ING MECÁ set
prefijo _m_ TELECOM prefix
prefiltrado _m_ ELECTRÓN prefiltering
prefiltro _m_ ELECTRÓN prefilter; ~ **multiseccional** _m_ ELECTRÓN multisection prefilter; ~ **de segundo orden** _m_ ELECTRÓN second-order prefilter
preflashing _m_ QUÍMICA preflashing

preforma *f* C&V, P&C *procesamiento* preform
preformar *vt* TELECOM preform
pre-fosa *f* GEOL fore deep
pregnano *m* QUÍMICA pregnane
pregnitano *m* QUÍMICA prehnitene
pregnítico *adj* QUÍMICA prehnitic
pregrabado *adj* ACÚST, INFORM&PD, TV prerecorded
pregrabar *vt* ACÚST, INFORM&PD, TV prerecord
pregunta *f* INFORM&PD inquiry
preguntas: ~ **más frecuentes** *fra* (*FAQ*) INFORM&PD frequently asked questions (*FAQ*)
preimpregnación *f* C&V preimpregnation
preimpresión *f* FOTO printing, IMPR preprint
preionización *f* FÍS RAD preionization
premaquinado *m* ING MECÁ premachined condition
premodelado *m* P&C preforming
premoldeado *adj* CONST precast
prendas *f pl* SEG clothing; ~ **acolchadas** *f pl* SEG padded clothing; ~ **antitérmicas** *f pl* SEG heat-proof clothing; ~ **asépticas** *f pl* SEG aseptic room clothing; ~ **estirilizadas** *f pl* SEG aseptic room clothing; ~ **exteriores** *f pl* TEXTIL outerwear; ~ **ignífugas** *f pl* SEG fireproof clothing, flameproof clothing; ~ **impermeables** *f pl* SEG weatherproof clothing; ~ **incombustibles** *f pl* SEG fireproof clothing, flameproof clothing; ~ **de malla de acero** *f pl* SEG chainmail garments; ~ **protectoras** *f pl* SEG protective clothing; ~ **protectoras sin amianto** *f pl* SEG asbestos-free protective clothing; ~ **protectoras antiestáticas** *f pl* SEG antistatic protective clothing; ~ **protectoras antitérmicas e ignífugas** *f pl* SEG protective clothing against heat and fire; ~ **protectoras de colores vivos** *f pl* SEG luminous and coloured protective clothing; ~ **protectoras desechables** *f pl* SEG disposable protective clothing; ~ **protectoras para soldadores** *f pl* SEG welders' protective clothing; ~ **protectoras termoaislantes** *f pl* SEG protective clothing for cold-storage work; ~ **protectoras de trabajo** *f pl* SEG workers' protective clothing; ~ **protectoras para trabajos de salvamento** *f pl* SEG protective clothing for rescue services; ~ **termoaislantes** *f pl* SEG heat-protective clothing; ~ **visibles a gran distancia** *f pl* SEG highly visible clothing
prenhita *f* MINERAL prehnite
prensa *f* C&V, EMB, IMPR press, ING MECÁ holdfast, shaping machine, PAPEL, PROD, TEXTIL press; ~ **de abanico** *f* IMPR printing fly press; ~ **de abertura múltiple** *f* P&C multiple daylight press; ~ **alimentada por bobinas** *f* IMPR *rotativa* reel-fed press; ~ **alisadora** *f* PAPEL smoothing press; ~ **de alta presión** *f* AGRIC high-density pick-up baler; ~ **para azulejos** *f* C&V tile press; ~ **de baja presión** *f* AGRIC low-density pick-up baler; ~ **de balancín** *f* ING MECÁ screw press, PROD fly press; ~ **de banco** *f* ING MECÁ vice (*BrE*), vise (*AmE*); ~ **de banda en línea** *f* IMPR in-line web press; ~ **para barro** *f* C&V clay press; ~ **a cadena** *f* ING MECÁ chain vice (*BrE*), chain vise (*AmE*); ~ **de cadena** *f* ING MECÁ chain vice (*BrE*), chain vise (*AmE*); ~ **de carrera ascendente** *f* P&C upstroke press; ~ **de carrera descendente** *f* P&C downstroke press; ~ **cilíndrica** *f* IMPR cylinder printing machine; ~ **de cilindro de parada** *f* IMPR stop-cylinder press; ~ **de cilindros** *f* PROC QUÍ, PROD rolling press; ~ **de comprobación** *f* ING MECÁ tryout press; ~ **de contactos** *f* CINEMAT, FOTO, IMPR con-tact-printing frame; ~ **de contrachapado** *f* ING MECÁ veneering press; ~ **de correa** *f* IMPR belt press; ~ **cortadora rotativa** *f* EMB roller and rotary cutting press; ~ **de cortezas** *f* PAPEL bark press; ~ **para curvar** *f* ING MECÁ bending press; ~ **desgotadora** *f* PAPEL dewatering press; ~ **de dos cilindros** *f* IMPR two-cylinder press; ~ **de dos colores** *f* IMPR two-color press (*AmE*), two-colour press (*BrE*); ~ **de dos cuerpos** *f* IMPR tandem-type press; ~ **de dos revoluciones** *f* IMPR two-revolution press; ~ **de embalar** *f* *Esp* (*cf prensa de enfardar AmL*) PAPEL baling press; ~ **de embutir** *f* ING MECÁ drawing press, shaping machine, MECÁ shaper, PROD flanger, flanging press; ~ **de embutir tipo banco** *f* ING MECÁ bench-type shaping machine; ~ **de empalmar** *f* CINE-MAT splicing block; ~ **de empaquetar** *f* PAPEL bundling press; ~ **de enchapar** *f* ING MECÁ veneering press; ~ **encoladora** *f* PAPEL size press; ~ **de encolar** *f* EMB glue press, ING MECÁ G-clamp; ~ **de enfardar** *f* *AmL* (*cf prensa de embalar Esp*) PAPEL baling press; ~ **de estampado en relieve** *f* IMPR stamping press; ~ **estampadora** *f* PROD stamping machine, stamping press; ~ **de estampar** *f* PROD stamping machine, stamping press; ~ **para estampar en seco** *f* EMB, IMPR, PAPEL embossing press; ~ **estiradora** *f* ING MECÁ drawing press; ~ **de estirar** *f* ING MECÁ drawing press; ~ **para estirar tubos** *f* PROD tube-drawing press; ~ **filtradora** *f* ALIMENT, C&V, CARBÓN, LAB, PAPEL, PROC QUÍ filter press; ~ **de forja de cuatro columnas** *f* PROD four-column forging press; ~ **de forjar** *f* PROD forging press; ~ **de fricción de cuello de cisne** *f* ING MECÁ swanneck screw press; ~ **gofradora** *f* EMB, IMPR, PAPEL embossing press; ~ **para hacer balas** *f* EMB baling press; ~ **para hacer blocs** *f* IMPR padding press; ~ **para hacer fardos** *f* EMB baling press; ~ **hidráulica** *f* LAB, P&C hydraulic press; ~ **de hojas** *f* IMPR sheet-fed machine; ~ **húmeda** *f* PAPEL couch press, wet press; ~ **a husillo** *f* ING MECÁ screw press; ~ **de husillo** *f* ING MECÁ, PROD arbor press, fly press; ~ **de husillo de cuello de cisne** *f* ING MECÁ swanneck fly press, swanneck screw press; ~ **de impresión de media anchura** *f* IMPR half width printing press; ~ **para impresión offset** *f* IMPR offset printing press; ~ **para impresión en plancha de cobre** *f* IMPR copperplate printing press; ~ **de imprimir** *f* IMPR printing press; ~ **industrial** *f* IMPR commercial press; ~ **inferior** *f* PAPEL bottom press; ~ **invertida** *f* PAPEL reversed press; ~ **con jaula con resortes** *f* C&V spring cage press; ~ **de línea de contacto ampliada** *f* PAPEL extended nip press; ~ **de lodos** *f* HIDROL *aguas de cloaca* sludge press; ~ **para loseta de barro** *f* C&V clay-plate press; ~ **manchón** *f* PAPEL couch press; ~ **a manchón aspirante** *f* PAPEL suction couch press; ~ **de mano** *f* ING MECÁ G-cramp; ~ **manual** *f* IMPR hand press; ~ **marcadora** *f* IMPR, PAPEL marking press; ~ **de matrices múltiples** *f* ING MECÁ gang press; ~ **mecánica** *f* SEG power press; ~ **mecánica de frente abierto** *f* ING MECÁ open-front mechanical power press; ~ **de moldeo por inyección** *f* P&C injection molding press (*AmE*), injection moulding press (*BrE*); ~ **para montaje en seco** *f* FOTO dry-mounting press; ~ **neumática** *f* IMPR printing frame; ~ **de palanca** *f* ING MECÁ lever press; ~ **de paso recto** *f* PAPEL straight-through press; ~ **plana** *f* IMPR flat-bed press; ~ **plana sacapruebas** *f* IMPR flat-bed

proofing press; ~ **planocilíndrica** ƒ IMPR flat-bed cylinder press; ~ **de platina** ƒ P&C platen press; ~ **para platos hondos** ƒ C&V hollow ware presser; ~ **de premoldeado** ƒ P&C preforming press; ~ **principal** ƒ PAPEL main press; ~ **punzonadora** ƒ ING MECÁ blanking press; ~ **punzonadora copiadora** ƒ ING MECÁ copy punch press; ~ **punzonadora de precisión** ƒ ING MECÁ fine blanking-press; ~ **de punzones múltiples** ƒ ING MECÁ gang press; ~ **ranurada** ƒ PAPEL grooved press, vented nip press; ~ **de rebordear** ƒ PROD flanging press; ~ **de recalcar** ƒ ING MECÁ upsetting-press; ~ **recortadora copiadora** ƒ ING MECÁ copy punch press; ~ **de recortar** ƒ ING MECÁ, PROD punching machine; ~ **de remachar** ƒ ING MECÁ rivet cold press; ~ **de rodillos** ƒ PROC QUÍ, PROD rolling press; ~ **rotativa** ƒ EMB, IMPR rotary printing press; ~ **de sujeción** ƒ ING MECÁ clamp; ~ **taladradora** ƒ ING MECÁ, MECÁ drill press; ~ **de talla dulce** ƒ IMPR plate-printing machine; ~ **de tornillo** ƒ ING MECÁ G-clamp, screw clamp, vice press (*BrE*), vise press (*AmE*), MECÁ vice (*BrE*), vise (*AmE*), PAPEL screw press; ~ **de torno** ƒ ING MECÁ standing block; ~ **transversal** ƒ ING MECÁ traversing head shaping machine; ~ **troqueladora** ƒ PROD stamping press; ~ **troqueladora copiadora** ƒ ING MECÁ copy punch press; ~ **troqueladora por estampación** ƒ PAPEL die-stamping press; ~ **de troquelar** ƒ ING MECÁ blanking press; ~ **de troquelar de precisión** ƒ ING MECÁ fine blanking-press; ~ **para tubos** ƒ CONST pipe vice (*BrE*), pipe vise (*AmE*)

prensado *m* C&V, EMB, TEXTIL pressing; ~ **automático** *m* C&V automatic pressing; ~ **en caliente** *m* EMB hot pressing; ~ **recto** *m* C&V straight pressing; ~ **semiautomático** *m* C&V semiautomatic pressing

prensador: ~ **de movimiento rápido acimutal** *m* INSTR azimuth coarse motion clamp

prensaestopas *m* ING MECÁ gland, packing gland, stuffing box, NUCL gland, PROD stuffing box, *máquina* packing gland, TEC ESP *máquina* packing gland; ~ **de grasa** *m* ING MECÁ grease-packing gland

prensar *vt* PAPEL, PROD press

prensas: ~ **superpuestas** ƒ *pl* PAPEL stacked presses

prensista *m* C&V presser

preparación ƒ ALIMENT pickling brine, FOTO making-up, INFORM&PD setup, ING MECÁ setting, setting up, MINAS *del manto, veta* opening up, OCEAN pickling brine; ~ **de apresto** ƒ COLOR sizing preparation; ~ **del archivo** ƒ INFORM&PD file preparation; ~ **de archivos** ƒ INFORM&PD file preparation; ~ **del canto escuadrado** ƒ CONST square-edge preparation; ~ **de los cantos** ƒ MECÁ edge preparation; ~ **del carbón** ƒ CARBÓN, MINAS coal preparation; ~ **de datos** ƒ INFORM&PD data preparation; ~ **de ficheros** ƒ INFORM&PD file preparation; ~ **mecánica** ƒ CARBÓN *de minerales* dressing; ~ **mecánica de carbones** ƒ CARBÓN, MINAS coal dressing; ~ **mecánica de minerales** ƒ MINAS ore dressing; ~ **de menas** ƒ MINAS ore dressing; ~ **del original en color** ƒ IMPR art color preparation (*AmE*), art colour preparation (*BrE*); ~ **de pastas** ƒ PAPEL stock preparation; ~ **previa** ƒ INFORM&PD, TEC ESP housekeeping; ~ **de la señal** ƒ PROD signal conditioning; ~ **de superficie** ƒ CONST surface preparation; ~ **de la superficie a soldar** ƒ CONST joint preparation; ~ **de la urdimbre** ƒ TEXTIL warp setting

preparado *adj* INFORM&PD, TELECOM ready; ~ **para enviar** *adj* INFORM&PD, TELECOM ready-to-send; ~ **para transmitir** *adj* INFORM&PD, TELECOM ready-to-send

preparador *m* CARBÓN dresser

preparar *vt* CINEMAT set up, INFORM&PD arm, set up

preplastificación ƒ P&C preplasticizing

pre-precio *m* EMB prepricing

prepreg *m* P&C preimpregnate, prepreg

preprocesado *adj* ELECTRÓN preprocessed

preprocesador *m* INFORM&PD preprocessor

preproducción ƒ PROD preproduction

preprogramado *adj* INFORM&PD preprogrammed

prerefrigeración ƒ FERRO precooling

pre-regulación: ~ **del RCS** ƒ ING ELÉC SCR preregulation

pre-regulador: ~ **del RCS** *m* ING ELÉC SCR preregulator

presa ƒ AGUA weir, *ríos* barrage, dam, retaining wall, CONST batardeau, dike, *almacenamiento de agua* retaining wall, ENERG RENOV dam, dike, HIDROL dike (*AmE*), dam, barrage,, MINAS dam, PROD grip; ~ **de agujas** ƒ AGUA needle dam; ~ **de aliviadero** ƒ TRANSP AÉR gravity spillway dam; ~ **para almacenamiento de residuos** ƒ CONST tailings dam; ~ **arco** ƒ AGUA, CONST arch dam; ~ **de arco simple** ƒ AGUA, CONST arch dam; ~ **bóveda** ƒ AGUA arch dam; ~~**bóveda** ƒ CONST arch dam; ~ **ciclópea** ƒ AGUA cyclopic barrage; ~ **con compuerta de rebosamiento** ƒ HIDROL overflow dam; ~ **de contención** ƒ OCEAN barrage; ~ **de contrafuertes** ƒ CONST, HIDROL buttress dam; ~ **de derivación** ƒ AGUA barrage, HIDROL diversion dam; ~ **de desagüe** ƒ HIDROL overflow dam; ~ **de descarga** ƒ AGUA discharge flume; ~ **de gravedad** ƒ HIDROL gravity dam; ~ **de gravedad de concreto ciclópeo** ƒ *AmL* (*cf presa de gravedad de hormigón ciclópeo Esp*) HIDROL cyclopean concrete gravity dam; ~ **de gravedad de hormigón ciclópeo** ƒ *Esp* (*cf presa de gravedad de concreto ciclópeo AmL*) HIDROL cyclopean concrete gravity dam; ~ **de gravedad hueca** ƒ HIDROL hollow gravity dam; ~ **de mampostería** ƒ HIDROL masonry dam; ~ **provisional** ƒ AGUA temporary dam; ~ **de rebose** ƒ AGUA overflow dam; ~ **de retención** ƒ AGUA retaining dam; ~ **de río** ƒ AGUA, HIDROL, TRANSP MAR river dam; ~ **sumergible** ƒ ENERG RENOV weir; ~ **de terraplén** ƒ AGUA, HIDROL earth dam; ~ **de tierra** ƒ AGUA, HIDROL earth dam; ~ **de tierra en mampostería** ƒ HIDROL masonry-earth dam; ~ **de vertedero** ƒ AGUA overflow dam

presbiacusia ƒ ACÚST presbyacusis (*BrE*), presbycusis (*AmE*)

presecador *m* PAPEL predryer

preseguimiento *m* PROD pre-expediting

preselección ƒ ELECTRÓN presetting, preset

preselector *m* ELECTRÓN, INSTAL TERM, VEH *caja de cambio* preselector; ~ **de la abertura** *m* FOTO aperture preselector

presentación ƒ ELEC, IMPR, INFORM&PD display, PROD statement, TELECOM *línea de enlace* offer, TRANSP MAR display; ~ **de la carta electrónica** ƒ TRANSP MAR electronic chart display; ~ **en diagrama de escalones** ƒ PROD ladder diagram display; ~ **de identificación de línea conectada** ƒ TELECOM connected-line identification presentation (*COLP*); ~ **de identificación de línea de llamada** ƒ TELECOM calling-line identification presentation; ~ **de líneas**

f INFORM&PD raster; ~ **de matrices** *f* TELECOM matrix display; ~ **en pantalla de la carta electrónica** *f* TRANSP MAR electronic chart display; ~ **visual** *f* ELEC display; ~ **visual de líneas** *f* INFORM&PD raster display

presentador: ~ **visual digital** *m* TELECOM digital readout

presentar: ~ **en pantalla** *vt* TEC ESP *instrumentos con TCR* display

presentarse: ~ **ante la autoridad portuaria** *v refl* TRANSP MAR report to the port authorities

preservación *f* TEC PETR conservation

preservar *vt* ALIMENT preserve

presiembra *f* AGRIC preplanting

presilenciador *m* AUTO premuffler

presilla *f* AmL (*cf pinza Esp*) CONST staple, ING ELÉC clip, ING MECÁ clasp, spring clip, LAB *para tubo de caucho* clip; ~ **boca de caimán** *f* ELEC *conexión* alligator clip; ~ **de la calidad** *f* CALIDAD quality loop; ~ **cocodrilo** *f* ELEC *conexión* alligator clip, crocodile clip; ~ **de conexión a masa** *f* ELEC, ING ELÉC earth clip (*BrE*), earthing clip (*BrE*), ground clip (*AmE*), grounding clip (*AmE*); ~ **de conexión a tierra** *f* ELEC, ING ELÉC earth clip (*BrE*), ground clip (*AmE*); ~ **de puesta a tierra** *f* ELEC, ING ELÉC earthing clip (*BrE*), grounding clip (*AmE*)

presillas: ~ **de Mohr** *f pl AmL* (*cf pinzas de Mohr Esp*) LAB Mohr's clips

presintonía: ~ **de canales** *f AmL* (*cf prefijación de canales Esp*) TV presetting of channels

presión[1]: **a** ~ *adj* PROD snap-on, *aire* live; **con** ~ **interior** *adj* MECÁ pressurized; **de** ~ **regulada** *adj* MECÁ pressurized

presión[2] *f* GEN pressure, CONST thrust, PROD feed; ~ **absoluta** *f* REFRIG absolute pressure; ~ **para actividad extravehicular** *f* TEC ESP extra-vehicular pressure garment; ~ **acústica** *f Esp* (*cf presión sonora AmL*) ACÚST, CONTAM sound pressure, TEC ESP acoustic pressure; ~ **acústica efectiva** *f* ACÚST, CONTAM effective sound pressure; ~ **acústica eficaz** *f* ACÚST, CONTAM effective sound pressure; ~ **acústica de referencia** *f* ACÚST, CONTAM reference sound pressure; ~ **aerodinámica** *f* TRANSP AÉR aerodynamic pressure; ~ **del agua** *f* AGUA, CONST, INSTAL TERM, TEXTIL water pressure; ~ **del agua embebida** *f* CARBÓN pore-water pressure; ~ **del agua intersticial** *f* CARBÓN pore-water pressure; ~ **de aire** *f* CARBÓN air pressure; ~ **del aire** *f* ING MECÁ air pressure; ~ **alta** *f* GAS high pressure; ~ **anormal** *f* GEOL, TEC PETR abnormal pressure; ~ **anormal positiva** *f* TEC PETR positive abnormal-pressure; ~ **ascensional** *f* TRANSP lifting pressure; ~ **de ataque** *f* ING MECÁ driving pressure; ~ **atmosférica** *f* FÍS, GAS atmospheric pressure, METEO atmospheric pressure, air pressure, TRANSP AÉR atmospheric pressure; ~ **baja** *f* FÍS, GAS low pressure; ~ **barométrica** *f* METEO barometric pressure; ~ **en boca de pozo** *f* ENERG RENOV wellhead pressure; ~ **de bomba** *f* TEC PETR pump pressure; ~ **en el cabezal del pozo** *f* PETROL wellhead pressure; ~ **de la cabina** *f* TEC ESP, TRANSP AÉR cabin pressure; ~ **de cálculo** *f* REFRIG design pressure; ~ **de calibración** *f* TRANSP AÉR calibration pressure; ~ **de la carga de agua** *f* INSTAL HIDRÁUL head of water pressure; ~ **de cierre** *f* TEC PETR *producción, yacimientos* shut-in pressure; ~ **del colector** *f* TRANSP AÉR manifold

pressure; ~ **de condensación** *f* REFRIG condensing pressure; ~ **controlada** *f* ING MECÁ controlled pressure; ~ **crítica** *f* FÍS, TERMOTEC critical pressure; ~ **de descarga** *f* AGUA delivery pressure, ING MECÁ, REFRIG discharge pressure; ~ **de detonación** *f* MINAS detonation pressure; ~ **diferencial** *f* INSTAL HIDRÁUL, TEC PETR *geología, producción, yacimientos* differential pressure; ~ **diferencial de cabina** *f* TRANSP AÉR cabin differential pressure; ~ **diferencial constante** *f* TRANSP AÉR constant differential pressure; ~ **dinámica** *f* METEO, TRANSP AÉR dynamic pressure; ~ **dinámica de la aleta de guía de entrada** *f* TRANSP AÉR intake guide vane ram; ~ **efectiva** *f* ING MECÁ active pressure, INSTAL HIDRÁUL actual pressure, effective pressure, rating pressure, working pressure, MECÁ active pressure; ~ **de empuje** *f* TEC ESP, TRANSP AÉR boost pressure; ~ **de endurecimiento** *f* METAL body force, press-hardening; ~ **de ensayo** *f* INSTAL HIDRÁUL test pressure; ~ **de entrada** *f* CARBÓN inlet pressure; ~ **de equilibrio** *f* OCEAN equipressure, REFRIG balance pressure; ~ **del escape** *f* TRANSP AÉR exhaust back-pressure; ~ **de estado constante** *f* ING MECÁ steady-state pressure; ~ **de estado estacionario** *f* ING MECÁ steady-state pressure; ~ **de estado permanente** *f* ING MECÁ steady-state pressure; ~ **de estallido** *f* C&V, EMB, PAPEL bursting pressure; ~ **estática** *f* ACÚST, REFRIG, TRANSP AÉR static pressure; ~ **de falla** *f* CARBÓN *geología* rock pressure; ~ **de fluidos** *f* FÍS FLUID fluid pressure; ~ **de flujo** *f* PETROL flowing pressure; ~ **de fondo** *f* PETROL bottom-hole pressure; ~ **de fracturación** *f* GEOFÍS, GEOL, TEC PETR *geología* fracture pressure; ~ **del freno** *f* AUTO brake pressure; ~ **de funcionamiento** *f* REFRIG operating pressure; ~ **de gas** *f* FÍS, GAS, INSTAL TERM, TERMO gas pressure; ~ **del gas embebido** *f* CARBÓN pore-gas pressure; ~ **del gas intersticial** *f* CARBÓN *presas* pore-gas pressure; ~ **geológica** *f* GEOFÍS, TEC PETR geopressure; ~ **geostática** *f* GEOFÍS, TEC PETR geostatic pressure; ~ **hermética** *f* PETROL shut-in pressure; ~ **hidráulica** *f* AGUA, HIDROL water pressure; ~ **hidrostática** *f* CARBÓN, FÍS FLUID, GAS, GEOL, HIDROL, OCEAN, REFRIG, TEC PETR hydrostatic pressure; ~ **de impacto** *f* TRANSP AÉR impact pressure; ~ **de impresión** *f* IMPR printing pressure; ~ **de impulsión** *f* AGUA delivery pressure; ~ **de inflación** *f* PROD inflation pressure; ~ **inicial del yacimiento** *f* PETROL initial reservoir pressure; ~ **insuficiente** *f* ING MECÁ underpressure; ~ **interior** *f* AGUA internal pressure; ~ **interna** *f* AGUA internal pressure; ~ **intersticial** *f* CARBÓN *presas*, GEOL, TEC PETR pore pressure; ~ **de leak-off** *f* TEC PETR *perforaciones* leak-off pressure; ~ **en la línea de tangencia** *f* PAPEL *rodillos, cilindros* nip pressure; ~ **lineal** *f* PAPEL linear pressure; ~ **litostática** *f* GEOL lithostatic pressure; ~ **manométrica** *f* REFRIG gage pressure (*AmE*), gauge pressure (*BrE*); ~ **del manto de recubrimiento** *f* CARBÓN overburden pressure; ~ **máxima admisible** *f* PROD maximum allowable pressure; ~ **máxima de servicio** *f* REFRIG design pressure; ~ **media efectiva del freno** *f* TRANSP AÉR brake mean-effective pressure; ~ **media en estacionamiento** *f* ING MECÁ average steady state pressure; ~ **del moldeo por inyección** *f* EMB injection molding pressure (*AmE*), injection moulding pressure (*BrE*); ~ **motriz** *f* ING MECÁ driving pressure;

~ negativa *f* ING MECÁ negative pressure; **~ normal** *f* GEOL, PETROL, TEC PETR standard pressure, normal pressure; **~ normal de la formación** *f* PETROL normal formation pressure; **~ osmótica** *f* FÍS, QUÍMICA, REFRIG, TEC PETR osmotic pressure; **~ con palanca** *f* ING MECÁ lever feed; **~ parcial** *f* FÍS, QUÍMICA partial pressure; **~ de percusión** *f* D&A percussion pressure; **~ sin pérdidas** *f* TEC PETR *perforación* leak-off pressure; **~ de poro** *f* TEC PETR pore pressure; **~ a pozo cerrado** *f* PETROL shut-in pressure; **~ de preconsolidación** *f* CARBÓN preconsolidation pressure; **~ promedio en estado de régimen** *f* ING MECÁ average steady state pressure; **~ de prueba** *f* ING MECÁ proof pressure, INSTAL HIDRÁUL, REFRIG proof pressure, test pressure, testing pressure; **~ de prueba hidráulica** *f* REFRIG hydraulic proof pressure; **~ pulsátil** *f* PROD pulsating pressure; **~ de punto de burbujeo** *f* PETROL bubble-point pressure; **~ de radiación** *f* ACÚST, FÍS, FÍS PART, FÍS RAD radiation pressure; **~ de la radiación solar** *f* TEC ESP solar radiation pressure; **~ de régimen permanente** *f* ING MECÁ steady-state pressure; **~ de régimen de trabajo** *f* INSTAL HIDRÁUL rating pressure, working pressure; **~ de regulación** *f* PROD set pressure; **~ de remanso** *f* FÍS FLUID stagnation pressure; **~ de reservorio** *f* GEOL, TEC PETR formation pressure; **~ de retroceso** *f* INSTAL HIDRÁUL *estator de turbina de vapor* back pressure; **~ de rotura por estallido** *f* C&V, EMB, PAPEL bursting pressure; **~ del ruido inherente** *f* ACÚST inherent noise pressure; **~ del ruido proprio** *f* ACÚST inherent noise pressure; **~ de ruptura** *f* C&V, EMB, PAPEL bursting pressure; **~ del silo** *f* CARBÓN silo pressure; **~ de soberempuje** *f* TRANSP AÉR boost pressure; **~ de sobrealimentación** *f* TRANSP AÉR boost pressure; **~ de sobrecarga** *f* GEOL, TEC PETR overburden pressure; **~ de sonido** *f* FÍS sound pressure; **~ sonora** *f* AmL (*cf presión acústica Esp*) ACÚST, CONTAM sound pressure, TEC ESP acoustic pressure; **~ sonora efectiva** *f* ACÚST, CONTAM effective sound pressure; **~ sonora eficaz** *f* ACÚST, CONTAM effective sound pressure; **~ sonora instantánea** *f* ACÚST instantaneous sound pressure; **~ sonora máxima** *f* ACÚST maximum sound pressure; **~ sonora de referencia** *f* ACÚST, CONTAM reference sound pressure; **~ subnormal** *f* GEOL, TEC PETR subnormal pressure; **~ de surgencia** *f* PETROL flowing pressure; **~ terrestre** *f* CARBÓN earth pressure; **~ terrestre activa** *f* CARBÓN active earth pressure; **~ terrestre en calma** *f* CARBÓN earth-pressure at rest; **~ de la toma de aire** *f* TRANSP AÉR air intake pressure; **~ de trabajo** *f* TEC ESP operating pressure; **~ de vapor** *f* FÍS vapor pressure (*AmE*), vapour pressure (*BrE*), INSTAL TERM steam pressure, vapor pressure (*AmE*), vapour pressure (*BrE*), REFRIG, TEC PETR vapor pressure (*AmE*), vapour pressure (*BrE*), TEXTIL steam pressure; **~ de vapor efectiva** *f* INSTAL HIDRÁUL, NUCL effective steam pressure; **~ de vapor saturado** *f* FÍS saturated vapor pressure (*AmE*), saturated vapour pressure (*BrE*); **~ del viento** *f* METEO wind pressure, OCEAN wind stress, PROD *horno alto* blast pressure; **~ del yacimiento** *f* PETROL, TEC PETR reservoir pressure

presión³: bajo ~ *fra* INSTAL HIDRÁUL *calderas* pressure-locked, under steam; **en ~** *fra* ING MECÁ under steam

presionar *vt* CARBÓN press, put pressure on

presionización *f* AmL (*cf presurización Esp*) TRANSP

AÉR pressurization; **~ de la cabina** *f* AmL TEC ESP, TRANSP AÉR cabin pressurization; **~ de caída** *f* AmL TEC ESP blow-down pressurization

presionizado *adj* MECÁ pressurized

presofundir *vt* ING MECÁ die-cast

presostato *m* INSTAL HIDRÁUL, PROD pressure switch, REFRIG pressostat, pressure-controller; **~ de aceite** *m* REFRIG oil pressostat; **~ de alta presión** *m* REFRIG high-pressure controller; **~ de baja presión** *m* REFRIG low-pressure controller; **~ combinado de alta y baja presión** *m* REFRIG dual-pressure controller; **~ diferencial** *m* REFRIG pressure differential cutout; **~ de seguridad de aceite** *m* REFRIG oil-pressure safety cutout; **~ de seguridad de alta presión** *m* REFRIG high-pressure safety cut-out; **~ de seguridad de baja presión** *m* REFRIG low-pressure safety cutout

prespan *m* PAPEL presspahn-transformer board

prestación *f* TELECOM *de servicios* provision

prestaciones: ~ en ascenso *f pl* TRANSP AÉR climb performance

prestador: ~ de servicios *m* TELECOM service provider (*SP*)

préstamo: ~ de tierras *m* FERRO *equipo fijado* spoil

presupuesto: ~ de agua *m* AGUA water budget; **~ de energía** *m* CONTAM *procesos industriales* energy budget; **~ preliminar** *m* CONST preliminary cost-estimate

presurización *f* Esp (*cf presionización AmL*) TRANSP AÉR pressurization; **~ de la cabina** *f* Esp TRANSP AÉR cabin pressurization; **~ de caída** *f* Esp TEC ESP blow-down pressurization

presurizar *vt* REFRIG pressurize

pretensado: ~ externo *m* CONST external pre-stressing

pretil: ~ del puerto *m* C&V port apron

pretratado *adj* ELECTRÓN preprocessed

pretratamiento *m* ELECTRÓN preprocessing, QUÍMICA pretreating

prevelar *vt* CINEMAT prefog

prevención *f* CALIDAD safety measure; **~ de accidentes** *f* SEG accident prevention; **~ de contaminación acústica** *f* Esp (*cf prevención de contaminación sonora AmL*) CONTAM prevention of noise pollution; **~ de contaminación del agua** *f* CONTAM prevention of water pollution; **~ de contaminación atmosférica** *f* CONTAM prevention of atmospheric pollution; **~ de contaminación sonora** *f* AmL (*cf prevención de contaminación aúcustica Esp*) CONTAM prevention of noise pollution; **~ contraincendios** *f* SEG, TERMO fire prevention; **~ de la corrosión** *f* EMB corrosion prevention; **~ de crecidas** *f* AGRIC, AGUA, HIDROL flood prevention; **~ de incendios** *f* SEG, TERMO fire prevention; **~ de pérdidas** *f* CALIDAD loss prevention; **~ de la propagación del fuego** *f* SEG, TERMO fire-spread prevention; **~ de riesgos** *f* CONST, SEG hazard prevention

previsión *f* AGUA, INFORM&PD forecasting; **~ de acarreo** *f* INFORM&PD carry lookahead, lookahead; **~ estadística** *f* PROD statistical forecasting; **~ intrínseca** *f* PROD intrinsic forecast; **~ del tiempo** *f* AGUA, HIDROL, TRANSP MAR weather forecast; **~ de tráfico** *f* FERRO traffic forecast, TRANSP traffic forecasting

priceíta *f* MINERAL priceite

prima: ~ de salvamento *f* TRANSP MAR *pago* salvage

primario *adj* GEOL *etapa* early, PROD *relés* trip-free

primer[1]: **en ~ plano** *adv* CINEMAT, FOTO, TV in the foreground; **en ~ término** *adv* CINEMAT, FOTO, TV in the foreground

primer[2]: **~ armónico** *m* ELECTRÓN *comprobación ortográfica* first harmonic; **~ cocido** *m* C&V first firing; **~ cocido oxidante** *m* C&V first oxidizing firing; **~ cuartillo** *m* D&A *guardia de mar* first dogwatch; **~ grupo de secadores** *m* PAPEL first dryer; **~ licor** *m* ALIMENT *fermentación* first runnings, fore-runnings; **~ miembro** *m* FÍS left-hand side; **~ montaje** *m* CINEMAT first assembly; **~ móvil** *m* ING MECÁ prime mover; **~ oficial** *m* TRANSP MAR first officer; **~ operador** *m* CINEMAT, TV chief cameraman; **~ par de rodillos** *m* C&V first pair of rollers; **~ plano** *m* CINEMAT close-up, head shot, FOTO, INFORM&PD foreground, TEC ESP *fotografía* close-up, TV foreground; **~ plano medio** *m* CINEMAT close medium shot, medium close-up; **~ potencial de ionización** *m* FÍS first ionization potential; **~ relevo** *m* MINAS forepoling; **~ término** *m* CINEMAT, FOTO, TV foreground

primera[1]: **de ~ prueba** *adj* TELECOM tested first

primera[2]: **~ barrena** *f* TEC PETR *perforación* spudding bit; **~ capa** *f* COLOR, REVEST first coat; **~ copia** *f* CINEMAT answer print, first answer print; **~ copia combinada de prueba** *f* CINEMAT first-trial composite; **~ corrección** *f* IMPR house corrections; **~ frecuencia intermedia** *f* ELEC, ELECTRÓN first intermediate frequency; **~ generación** *f* INFORM&PD first generation; **~ generación filial** *f* AGRIC F1 generation; **~ ley** *f* FÍS first law; **~ mano** *f* COLOR first coat, CONST undercoat, PROD priming, REVEST first coat; **~ parte del ciclo de combustible nuclear** *f* NUCL front end of nuclear fuel cycle; **~ prueba** *f* IMPR artist's proof, *capilla* advance sheet; **~ talla** *f* CARBÓN slab; **~ tinta** *f* IMPR first-down ink; **~ unidad** *f* CINEMAT first unit

primeras: **~ llegadas** *f pl* GEOL first arrivals

primerísimo: **~ plano** *m* CINEMAT, TV big close up (*BCU*), extreme close-up; **~ primer plano** *m* CINEMAT, TV tight closeup

primero[1] *m* IMPR first down

primero[2]: **~ en entrar, primero en salir** *fra* INFORM&PD first-in-first-out (*FIFO*)

primeros: **~ auxilios** *m pl* SEG first aid; **~ vecinos** *m pl* CRISTAL nearest neighbors (*AmE*), nearest neighbours (*BrE*)

primordial *adj* PROD overriding

primordio: **~ de engranaje** *m* ING MECÁ gear blank; **~ de rueda dentada** *m* ING MECÁ gear blank; **~ para ruedas de soldadura continua** *m* ING MECÁ seam welding wheel blank; **~ de tornillo sin roscar** *m* ING MECÁ, PROD screw blank

primulina *f* QUÍMICA primuline

principal[1] *adj* INFORM&PD host

principal[2]: **~ canalización neumática** *f* ING MECÁ air main; **~ ruta comercial** *f* TRANSP MAR main trading route

principales: **~ estándar** *m pl* FÍS primary standard

principio[1]: **al ~** *adv* TELECOM at the near end

principio[2] *m* GEN principle, ING MECÁ starting, TELECOM *de transmisión* start; **~ de acumulación y distribución** *m* ELECTRÓN *síntesis de frecuencia* add-and-divide principle; **~ de Arquímedes** *m* FÍS Archimedes' principle; **~ de Babinet** *m* FÍS Babinet's principle; **~ cero** *m* FÍS, TERMO zeroth law; **~ de combinación de Ritz** *m* FÍS Ritz combination principle; **~ de compresión** *m* INSTAL HIDRÁUL compression point; **~ de correspondencia** *m* FÍS correspondence principle; **~ de diseño ergonómico** *m* SEG ergonomic design principle; **~ de escape** *m* INSTAL HIDRÁUL release point; **~ de exclusión** *m* QUÍMICA exclusion principle; **~ de exclusión de Pauli** *m* FÍS, FÍS PART Pauli exclusion principle; **~ de la expansión** *m* INSTAL HIDRÁUL *diagramas* expansion point; **~ de Fermat** *m* FÍS Fermat's principle; **~ de Frank-Condon** *m* FÍS Frank-Condon principle; **~ de Huygens** *m* FÍS Huygens' principle; **~ de incertidumbre** *m* FÍS uncertainty principle; **~ de incertidumbre de Heisenberg** *m* FÍS PART Heisenberg uncertainty principle; **~ de Mach** *m* FÍS Mach principle; **~ de máxima entropía** *m* FÍS RAD maximum entropy principle, principle of maximum entropy; **~ de Pascal** *m* FÍS Pascal's principle; **~ de Pauli** *m* FÍS PART, QUÍMICA Pauli principle; **~ de Pauli generalizado** *m* FÍS PART generalized Pauli principle; **~ de que el contaminador paga** *m* CONTAM *legislación* polluter pays principle; **~ de superposición** *m* FÍS ONDAS principle of superposition; **~ de la termodinámica** *m* TERMO law of thermodynamics, thermodynamic law; **~ topológico abstracto** *m* GEOM abstract topological principle

prioridad *f* INFORM&PD precedence, priority; **~ interna** *f* PROD internal priority; **~ de interrupción** *f* INFORM&PD interrupt priority; **~ de mando** *f* TRANSP AÉR override control; **~ de operador** *f* INFORM&PD operator precedence; **~ de paso** *f* TRANSP AÉR, TRANSP MAR *navegación* right of way; **~ de pérdida de llamada** *f* TELECOM call loss priority (*CLP*)

prioritario *adj* INFORM&PD foreground, PROD overriding

prisionero[1] *adj* MECÁ captive

prisionero[2] *m* ING MECÁ screw, stud, PROD stud; **~ con cavidad hexagonal** *m* ING MECÁ Allen screw; **~ de chaveta** *m* ING MECÁ cotter stud bolt; **~ ciego** *m* ING MECÁ blind stud bolt; **~ guiador** *m* ING MECÁ locating screw; **~ posicionador** *m* ING MECÁ locating stud

prisma *m* GEN prism, AGUA tumbler; **~ accrecional** *m* GEOL accretionary prism; **~ de Amici** *m* FÍS Amici prism; **~ de ángulo pequeño** *m* FÍS small-angle prism; **~ de desviación** *m* INSTR, TELECOM deviation prism; **~ divisor del haz** *m* CINEMAT beam-splitting prism; **~ de Dove** *m* ÓPT Dove prism; **~ de enderezamiento de imagen** *m* INSTR image-erecting prism; **~ giratorio** *m* ÓPT rotating prism; **~ intermitente** *m* CINEMAT intermittent prism; **~ inversor** *m* INSTR inverting prism; **~ de marea** *m* ENERG RENOV tidal prism; **~ de Nicol** *m* FÍS Nicol prism; **~ polarizador** *m* INSTR polarizing prism; **~ Porro** *m* INSTR Porro prism; **~ reflector** *m* INSTR, ÓPT reflecting prism; **~ refringente** *m* FÍS, ÓPT refracting prism; **~ rotativo** *m* ÓPT rotating prism; **~ triangular** *m* FÍS roof prism; **~ de visión directa** *m* FÍS direct vision prism

privacidad *f* ACÚST, TELECOM privacy; **~ de la voz** *f* TELECOM voice privacy

privilegiado *adj* PROD overriding

proa[1]: **de ~** *adj* TRANSP MAR fore, forward

proa[2]: **a ~** *adv* TRANSP MAR fore, forward; **hacia ~** *adv* TRANSP MAR forward; **por la ~** *adv* TEC ESP, TRANSP MAR ahead; **~ arriba** *adv* TRANSP MAR course-up, head-up; **a ~ del través** *adv* TRANSP MAR forward of the beam

proa[3] *f* TRANSP AÉR nose, TRANSP MAR bow; **~ de un bajel** *f* TRANSP MAR prow; **~ de bulbo** *f* TRANSP MAR *construcción naval* bulbous bow; **~ curva entrante** *f* TRANSP MAR ram bow; **~ recta** *f* TRANSP MAR *construcción naval* straight stem; **~ redonda** *f* TRANSP MAR *construcción naval* spoon bow

probabilidad *f* INFORM&PD, MATEMÁT likelihood, probability; **~ de error** *f* TELECOM error probability; **~ de error humano** *f* CALIDAD human error probability; **~ de éxito del lanzamiento** *f* TEC ESP launch success probability; **~ de falsa alarma** *f* TELECOM false alarm probability; **~ de fisión espontánea** *f* FÍS, FÍS RAD, NUCL spontaneous fission probability; **~ matemática** *f* MATEMÁT mathematical chance, mathematical probability; **~ de permanencia de neutrones térmicos** *f* NUCL thermal neutron non-leakage probability; **~ de retraso excedente** *f* TELECOM probability of excess delay; **~ termodinámica** *f* FÍS, TERMO thermodynamic probability; **~ de transición** *f* FÍS transition probability

probabilidades *f pl* MATEMÁT odds

probador: **~ de aislamientos** *m* ELEC, INSTR growler; **~ de continuidad** *m* ELEC continuity tester; **~ de dureza** *m* ING MECÁ hardness tester; **~ de dureza de Shore** *m* LAB Shore hardness tester; **~ de hipervoltaje** *m* ELEC high-voltage tester; **~ de inducidos** *m* ELEC armature tester, growler, INSTR growler; **~ multifuncional** *m* ING ELÉC multifunction tester; **~ de rebabas de corte** *m* ING MECÁ flash tester; **~ de rebabas de forja** *m* ING MECÁ flash tester; **~ de rebabas de recalcar** *m* ING MECÁ flash tester; **~ de sobretensión** *m* ELEC high-voltage tester

probar[1] *vt* FÍS sample, INFORM&PD test, ING MECÁ support, MINAS prove, QUÍMICA, TELECOM test

probar[2] *vi* QUÍMICA assay; **~ la cuerda** *vi* ING MECÁ test the rope; **~ un proceso** *vi* QUÍMICA test a process; **~ la soga** *vi* ING MECÁ test the rope

probeta *f* ELEC probe, LAB measuring cylinder, METAL test piece, P&C specimen, PAPEL test piece, PROD test tube, QUÍMICA measuring tube; **~ cilíndrica** *f* CONST test cylinder, PROD *horno alto* test rod; **~ cúbica** *f* CONST *hormigón* test cube; **~ de ensayo** *f* METAL test specimen, specimen test, P&C test specimen; **~ para ensayos de tracción** *f* ING MECÁ tensile test piece; **~ con entalla en V para prueba de impacto Charpy** *f* NUCL Charpy V-notch impact specimen; **~ medidora electrónica** *f* ING MECÁ electronic gaging probe (*AmE*), electronic gauging probe (*BrE*); **~ normalizada** *f* PROD standard test piece; **~ con ranuras laterales** *f* METAL sidegrooved specimen; **~ con tapón** *f* LAB stoppered measuring cylinder; **~ tipo** *f* PROD standard test piece; **~ unificada** *f* PROD standard test piece

problema: **~ de contaminación atmosférica** *m* CONTAM air pollution problem; **~ cúbico** *m* GEOM cube problem; **~ de la duplicación del cubo** *m* GEOM cube problem; **~ del mapa de tres colores** *m* *Esp* (*cf problema del mapa tricromo AmL*) GEOM three-color map problem (*AmE*), three-colour map problem (*BrE*); **~ del mapa tricolor** *m* *Esp* (*cf problema del mapa tricromo AmL*) GEOM three-color map problem (*AmE*), three-colour map problem (*BrE*); **~ del mapa tricromo** *m* *AmL* (*cf problema del mapa de tres colores Esp, problema del mapa tricolor Esp*) GEOM three-color map problem (*AmE*), three-colour map problem (*BrE*); **~ pluripersonal** *m* TEC ESP many-body problem

procaína *f* QUÍMICA procaine

procedente: **~ del mar** *adj* TRANSP MAR seaborne; **~ del norte** *adj* TRANSP MAR northerly

procedimiento *m* CALIDAD procedure, CARBÓN, FOTO, IMPR, ING MECÁ process, PROD procedure, *artificial* process, QUÍMICA way; **~ de acercamiento** *m* TEC ESP approach procedure; **~ aditivo de color** *m* FOTO additive color process (*AmE*), additive colour process (*BrE*); **~ a la albúmina** *m* FOTO albumen process; **~ de apagado** *m* NUCL, TEC ESP *reactor* shutdown procedure; **~ de aparcamiento** *m* TRANSP AÉR docking procedure; **~ de aplicación de laser líquido** *m* ELECTRÓN liquid lasing medium; **~ de aproximación abortada** *m* TRANSP AÉR missed-approach procedure; **~ de aproximación frustrada** *m* TRANSP AÉR missed-approach procedure; **~ de aproximación por instrumentos** *m* TRANSP AÉR instrument approach procedure; **~ de aproximación de precisión** *m* TRANSP AÉR precision-approach procedure; **~ de aterrizaje** *m* TRANSP AÉR landing procedure; **~ de autentificación** *m* TELECOM authentication procedure; **~ de búsqueda binaria** *m* TELECOM binary-search procedure; **~ de cambio** *m* EMB changeover procedure; **~ de cierre** *m* NUCL, TEC ESP shutdown procedure; **~ al colodión** *m* IMPR collodion process; **~ al colodión húmedo** *m* IMPR wet-collodion process; **~ de control de transmisión síncrona** *m* INFORM&PD synchronous data-link control (*SDLC*); **~ para cortar los orillos** *m* TEXTIL selvage cutting process; **~ cuánticamente limitado** *m* ÓPT quantum-limited operation; **~ de despejo** *m* TELECOM clearing procedure; **~ de emergencia** *m* TRANSP AÉR emergency procedure; **~ para el encuentro espacial** *m* TEC ESP rendezvous procedure; **~ entrante** *m* TELECOM incoming procedure; **~ de espera** *m* TRANSP AÉR holding procedure; **~ de fabricación** *m* PROD make; **~ de flotación** *m* MINAS flotation process; **~ de funcionamiento parcial** *m* INFORM&PD fallback; **~ de fundición y mineral** *m* PROD pig-and-ore process; **~ de garantía de calidad para BS5750** *m* ING MECÁ quality-assurance procedure to BS5750; **~ limitado por el ancho de banda** *m* ÓPT bandwidth-limited operation; **~ limitado por atenuación** *m* ÓPT attenuation-limited operation; **~ limitado por ruido cuántico** *m* ÓPT quantum-noise-limited operation; **~ de limpieza** *m* CONTAM, CONTAM MAR clean-up technique; **~ de medida** *m* METR measurement process; **~ de operación de emergencia** *m* NUCL emergency operating procedure; **~ de permuta** *m* EMB changeover procedure; **~ de prueba para sistemas de frenado** *m* ING MECÁ test procedure for braking systems; **~ de relleno** *m* MINAS pack system; **~ saliente** *m* TELECOM outgoing procedure; **~ de soldadura** *m* CONST welding procedure; **~ soplado y resoplado** *m* C&V blow-and-blow process

procesado *m* CARBÓN process, TEC PETR *de líneas sísmicas* processing; **~ de campo** *m* PETROL field processing; **~ en discontinuo** *m* IMPR *revelado* batch processing; **~ de dos baños** *m* IMPR *revelado* two-bath processing; **~ de imagen digital** *m* ELECTRÓN digital image processing; **~ de planchas** *m* IMPR plate-processing; **~ serigráfico** *m* IMPR screen processing; ~ **de texto** *m* INFORM&PD text processing, word processing (*WP*); ~ **de transacciones** *m* INFORM&PD transaction processing (*TP*); ~ **vectorial** *m* INFORM&PD vector processing

procesador *m* IMPR, INFORM&PD, TELECOM processor; ~ **activo** *m* TELECOM active processor; ~ **acústico-óptico** *m* ELECTRÓN acousto-optic processor; ~ **administrativo** *m* TELECOM administrative processor; ~ **analógico de llamadas** *m* TELECOM analog call processor; ~ **de aplicaciones** *m* TELECOM applications processor; ~ **asociativo** *m* TELECOM associative processor; ~ **auxiliar** *m* TELECOM stand-by processor; ~ **central** *m* (*UCP*) INFORM&PD, TELECOM central processor; ~ **de coma flotante** *m* INFORM&PD, TELECOM floating-point processor (*FPP*); ~ **de comunicación** *m* INFORM&PD communication server; ~ **de comunicación hablada** *m* TELECOM voice message processor; ~ **de comunicaciones** *m* TELECOM communications processor; ~ **de comunicaciones inter-redes** *m* PROD gateway; ~ **de conducto** *m* INFORM&PD pipeline processor; ~ **de conmutación** *m* TELECOM switching processor; ~ **de conmutación de mensajes** *m* TELECOM message-switching processor; ~ **de conmutación de paquetes** *m* INFORM&PD, TELECOM packet-switching processor; ~ **delantero** *m* INFORM&PD front-end processor (*FEP*); ~ **de entrada/salida** *m* INFORM&PD input/output processor; ~ **esclavo** *m* INFORM&PD slave processor; ~ **de ficheros** *m* INFORM&PD file server; ~ **frontal** *m* INFORM&PD front-end processor (*FEP*); ~ **front-end** *m* TELECOM front-end processor (*FEP*); ~ **de interconexión** *m* INFORM&PD server; ~ **de llamadas** *m* TELECOM call processor; ~ **de llamadas híbrido** *m* TELECOM hybrid call processor; ~ **de macros** *m* INFORM&PD macro processor; ~ **maestro** *m* TELECOM master processor; ~ **de mantenimiento** *m* TELECOM maintenance processor; ~ **en máquina** *m* FOTO machine processor; ~ **de matrices** *m* INFORM&PD array processor; ~ **matricial distribuido** *m* INFORM&PD distributed array processor; ~ **de matriz distribuida** *m* INFORM&PD distributed array processor; ~ **de mensaje hablado** *m* TELECOM voice-message processor; ~ **de mensajes de interfaz** *m* TELECOM interface-message processor (*IMP*); ~ **de nodos** *m* INFORM&PD node processor; ~ **de pantalla** *m* INFORM&PD display processor; ~ **periférico** *m* INFORM&PD, TELECOM peripheral processor (*PP*); ~ **para planchas** *m* IMPR plate-processor; ~ **de punto flotante** *m* INFORM&PD, TELECOM floating-point processor (*FPP*); ~ **regional** *m* TELECOM regional processor (*RP*); ~ **de relación** *m* INFORM&PD relational processor; ~ **relacional** *m* INFORM&PD relational processor; ~ **de reserva** *m* TELECOM stand-by processor; ~ **de respaldo** *m* PROD backup processor; ~ **de señales** *m* ELECTRÓN, TELECOM, TV signal processor (*SP*); ~ **del sistema de conmutación** *m* TELECOM switching-system processor; ~ **subordinado** *m* INFORM&PD slave processor; ~ **del terminal** *m* INFORM&PD terminal server; ~ **de textos** *m* INFORM&PD word processing (*WP*), word processor (*WP*); ~ **de visualización** *m* INFORM&PD display processor

procesamiento *m* CARBÓN process, INFORM&PD, TELECOM processing; ~ **de archivos** *m* INFORM&PD file processing; ~ **automático de datos** *m* INFORM&PD automatic data processing; ~ **por batch** *m* INFORM&PD, PROC QUÍ batch processing; ~ **de consultas** *m* INFORM&PD inquiry processing, query processing; ~ **de datos** *m* (*PD*) INFORM&DP, TELECOM data processing (*DP*); ~ **de datos distribuidos** *m* INFORM&PD distributed data processing (*DDP*); ~ **digital distribuido** *m* NUCL distributed digital processing; ~ **en discontinuo** *m* PAPEL batch processing; ~ **de distribución abierta** *m* TELECOM open distribution processing (*ODP*); ~ **distribuido** *m* INFORM&PD distributed processing; ~ **electrónico de datos** *m* (*EDP*) INFORM&PD electronic data processing (*EDP*); ~ **de ficheros** *m* INFORM&PD file processing; ~ **de fondo** *m* INFORM&PD background processing; ~ **fuera de línea** *m* INFORM&PD off-line processing; ~ **de imagen** *m* INFORM&PD, TELECOM image processing; ~ **de imágenes** *m* INFORM&PD, TELECOM image processing, picture processing; ~ **de información** *m* (*PI*) ELECTRÓN, INFORM&PD information processing (*IP*); ~ **de interrogaciones** *m* INFORM&PD query processing; ~ **por irradiación** *m* FÍS RAD, NUCL radiation processing; ~ **en línea** *m* INFORM&PD in-line processing, on-line processing; ~ **de listados** *m* INFORM&PD list processing; ~ **de listas** *m* INFORM&PD list processing; ~ **por lotes** *m* INFORM&PD, PROC QUÍ batch processing; ~ **por lotes remoto** *m* INFORM&PD remote batch processing; ~ **múltiple** *m* INFORM&PD multiprocessing; ~ **óptico** *m* ÓPT, TELECOM optical processing; ~ **óptico de señales** *m* FÍS RAD, ÓPT optical signal processing; ~ **de la palabra** *m* TELECOM speech processing; ~ **en paralelo** *m* ELECTRÓN, INFORM&PD, TELECOM parallel processing; ~ **de película muy rápida** *m* CINEMAT high-speed film processing; ~ **de petición** *m* INFORM&PD inquiry processing; ~ **en primer plano** *m* INFORM&PD foreground processing; ~ **de prioridad subordinada** *m* INFORM&PD background processing; ~ **prioritario** *m* INFORM&PD foreground processing, priority processing; ~ **secuencial** *m* INFORM&PD in-line processing, sequential processing; ~ **de señal digital** *m* INFORM&PD digital signal processing; ~ **de señales** *m* INFORM&PD, TELECOM signal processing; ~ **de señales autoadaptivos** *m* ELECTRÓN adaptive signal processing; ~ **en serie** *m* INFORM&PD serial processing; ~ **simbólico** *m* INFORM&PD symbolic processing; ~ **simultáneo** *m* ELECTRÓN, INFORM&PD concurrent processing; ~ **subordinado** *m* INFORM&PD background processing; ~ **de texto** *m* INFORM&PD text processing, word processing (*WP*); ~ **en tiempo real** *m* CINEMAT, ELECTRÓN, INFORM&PD, TELECOM, TV real-time processing; ~ **de transacciones** *m* INFORM&PD transaction processing (*TP*); ~ **vectorial** *m* INFORM&PD vector processing; ~ **de la voz** *m* INFORM&PD speech processing

procesar *vt* ALIMENT, INFORM&PD, TELECOM process

proceso *m* GEN process, INFORM&PD running, ING MECÁ procedure, process; ~ **de abollado en caliente** *m* ING MECÁ hot-dimpling process; ~ **de abollado en frío** *m* ING MECÁ cold dimpling process; ~ **ácido** *m* PAPEL acid process; ~ **de acondicionamiento del agua** *m* NUCL water conditioning process; ~ **adaptivo** *m* INFORM&PD adaptive process; ~ **adiabático** *m* TERMO adiabatic process; ~ **aditivo** *m* ELECTRÓN additive process; ~ **aditivo de color** *m* CINEMAT additive color process (*AmE*), additive colour process (*BrE*); ~ **de agrupación** *m* PROD joining process; ~ **alfa triple** *m* NUCL triple alpha process; ~ **de alimentación de vela sencilla** *m* C&V single-gob feeding (*BrE*), single-gob process (*AmE*); ~ **anaglífico** *m* CINEMAT anaglyph process; ~ **de archivos** *m* INFORM&PD file processing; ~ **de atomización** *m* PROC QUÍ atomizing process; ~ **Bessemer** *m* PROD Bessemer process; ~ **de bobinado** *m* TEXTIL winding process; ~ **de las cámaras de plomo** *m* PROC QUÍ chamber process; ~ **de carga de crudo sobre residuos de limpieza de tanques** *m* CONTAM, TEC PETR load on top process; ~ **en cascada** *m* FÍS RAD cascade process; ~ **catalítico** *m* CONTAM catalytic process; ~ **catalítico fluidizado** *m* PROC QUÍ fluid-catalyst process; ~ **de cementación** *m* CONST cementation process; ~ **de cepillado** *m* CONST *carpintería* planing; ~ **por cilindro** *m* C&V cylinder process; ~ **de cilindro de vidrio plano** *m* C&V cylinder drawing process; ~ **de cloro-amoníaco** *m* HIDROL chlorine-ammonia process; ~ **de cocción** *m* PAPEL *fabricación de pasta* pulping; ~ **con colodión húmedo** *m* FOTO wet-collodion process; ~ **de concesión de la licencia de operación** *m* NUCL plant licensing; ~ **concurrente** *m* ELECTRÓN, INFORM&PD concurrent processing; ~ **de copiado de negativo** *m* IMPR negative-copying process; ~ **crown** *m* C&V crown process; ~ **de cuatricromía** *m* IMPR four-color process (*AmE*), four-colour process (*BrE*); ~ **de datos** *m* (*PD*) INFORM&PD, TELECOM data processing (*DP*); ~ **de depuración de los efluentes** *m* RECICL effluent-purification process; ~ **de depuración de las emanaciones** *m* RECICL effluent-purification process; ~ **de desencogimiento** *m* CINEMAT deshrinking process; ~ **de desulfuración por vía húmeda** *m* CONTAM *tratamiento de gases* wet desulfurization process (*AmE*), wet desulphurization process (*BrE*); ~ **de desulfuración por vía seca** *m* CONTAM *tratamiento de gases* dry desulfurization process (*AmE*), dry desulphurization process (*BrE*); ~ **de difusión térmica** *m* FÍS, NUCL, TERMO thermal diffusion process; ~ **de difusión única** *m* ELECTRÓN single-diffusion process; ~ **digital** *m* TELECOM digital processing; ~ **distribuido** *m* PROD distributed processing; ~ **endotérmico** *m* TEC PETR *refino*, TERMO endothermic process; ~ **de enlace en frío** *m* TERMO cold bonding; ~ **de entallar** *m* PROD notching process; ~ **de entrada a raudales** *m* TEXTIL flow-in process; ~ **de estirado continuo** *m* C&V continuous-drawing process; ~ **de estirado vertical** *m* C&V updraw process; ~ **estocástico** *m* FÍS stochastic process; ~ **exotérmico** *m* TEC PETR *refino*, TERMO exothermic process; ~ **de exploración** *m* TV scanning process; ~ **de extracción** *m* TELECOM drawing process; ~ **de fabricación** *m* CARBÓN processing, ING MECÁ production line, PROD make; ~ **Fingal** *m* NUCL *de vitrificación* Fingal process; ~ **de flotación** *m* CARBÓN flotation process; ~ **de formación catalítica** *m* TEC PETR *refino* platforming; ~ **de fundición con solidificación continua** *m* MECÁ continuous casting; ~ **de gelatino-bromuro** *m* FOTO gelatino-bromide process; ~ **de grabado al ácido** *m* IMPR photoetching; ~ **de Haber** *m* QUÍMICA Haber process; ~ **heliotérmico** *m* ENERG RENOV, TERMO heliothermal process; ~ **hidrotérmico** *m* ENERG RENOV, TERMO hydrothermal process; ~ **Imax** *m* CINEMAT Imax

process; ~ **de impresión** *m* C&V screen printing, IMPR printing process; ~ **de impresión en color** *m* FOTO, IMPR color printing process (*AmE*), colour printing process (*BrE*); ~ **indirecto** *m* IMPR *selección de colores* indirect process; ~ **de inspección** *m* METR inspection procedure; ~ **de inversión** *m* CINEMAT, FOTO, IMPR reversal process; ~ **de inyección de cemento** *m* CONST cement-injection process; ~ **isobárico** *m* TERMO isobaric process; ~ **isócoro** *m* TERMO isochoric process; ~ **isotérmico** *m* TERMO isothermic process; ~ **iterativo** *m* INFORM&PD iterative process; ~ **de Leblanc** *m* DETERG Leblanc process; ~ **limitado en la salida** *m* INFORM&PD output-limited process; ~ **litográfico** *m* ELECTRÓN lithographic process; ~ **de llenado en continuo** *m* EMB continuous motion weight filling; ~ **del mercaptol** *m* QUÍMICA mercaptol process; ~ **mesa** *m* ELECTRÓN mesa process; ~ **mixto** *m* ELECTRÓN mixed process; ~ **de muescar** *m* PROD notching process; ~ **osmótico** *m* PROC QUÍ osmotic process; ~ **oxo** *m* DETERG oxo process; ~ **ozalida** *m* IMPR ozalid process; ~ **con pantalla azul** *m* CINEMAT *para efectos especiales* blue-screen process; ~ **de placa de contacto en vacío** *m* ALIMENT vacuum contact plate process; ~ **de plancha humedecida** *m* IMPR wet-plate process; ~ **de polisilicio de un solo nivel** *m* ELECTRÓN single-level polysilicon process; ~ **positivo directo** *m* CINEMAT direct positive process; ~ **posterior** *m* ELECTRÓN postprocessing; ~ **preferencial** *m* INFORM&PD preferential process; ~ **de prensa con molde empastado** *m* C&V paste-mold press-and-blow process (*AmE*), paste-mould press-and-blow process (*BrE*); ~ **que sigue** *m* PROC QUÍ, PROD, TEXTIL downstream process; ~ **químico atmosférico** *m* CONTAM atmospheric chemical process; ~ **de radicales** *m* P&C radical process; ~ **de recombinación** *m* ELECTRÓN, FÍS recombination process; ~ **de recuperación** *m* PETROL, TEC PETR *producción, refino* recovery process; ~ **de refino del sobrante** *m* TEC PETR residue-refining process; ~ **de reformación catalítica** *m* TEC PETR *refino* platforming; ~ **de refrigeración** *m* REFRIG, TERMO chilling process; ~ **de reproducción** *m* NUCL breeding process; ~ **de retratamiento** *m* PROD, RECICL reprocessing; ~ **reversible a color** *m* CINEMAT, FOTO color reversal process (*AmE*), colour reversal process (*BrE*); ~ **Schfftan** *m* CINEMAT Schfftan process; ~ **de semitonos** *m* IMPR halftone process; ~ **de señal analógica** *m* ELECTRÓN analog signal processing; ~ **de señal avanzada** *m* ELECTRÓN advanced signal processing; ~ **de señal a bordo** *m* TEC ESP on-board processing; ~ **de señal coherente** *m* ELECTRÓN coherent signal processing; ~ **de señal DAC** *m* ELECTRÓN CCD signal processing; ~ **de señal digital** *m* (*PSD*) ELECTRÓN digital signal processing (*DSP*); ~ **de separación** *m* PROC QUÍ separation process, TEC PETR *refino* extraction process; ~ **de sinterización por reacción** *m* NUCL reaction-sintering process; ~ **de sobreflujo** *m* C&V overflow process; ~ **de soldadura** *m* CONST welding process; ~ **de Solvay** *m* DETERG Solvay's process; ~ **SOM con circuito de silicio con canal n** *m* ELECTRÓN n-channel silicon gate MOS process; ~ **de soplo con molde empastado** *m* C&V paste-mold press-and-blow process (*AmE*), paste-mould press-and-blow process (*BrE*); ~ **de sosa amoniacal de Solvay** *m* DETERG

Solvay's ammonia soda process; ~ **supervisor** *m*
PROD supervisory process; ~ **sustractivo** *m* CINEMAT
subtractive process; ~ **de tapado** *m* GEOL, TEC PETR
sealing; ~ **tectónico** *m* GEOL, TEC PETR tectonic
process; ~ **termodinámico** *m* TERMO thermodynamic
process; ~ **de termoendurecimiento** *m* IMPR *planchas presensibilizadas* baking process; ~ **de
transferencia de colorante** *m* CINEMAT, FOTO dye
transfer process; ~ **de transferencia de tintes** *m*
CINEMAT, FOTO dye transfer process; ~ **de
transformación** *m* GAS transformation process;
~ **de transformación a bordo** *m* TEC ESP on-board
processing; ~ **de trituración y lixiviación** *m* NUCL
combustibles de carburos grind-and-leach process;
~ **Umklapp** *m* NUCL Umklapp process; ~ **de vaciado
continuo** *m* C&V continuous-casting process; ~ **de
vacío y soplo** *m* C&V suck-and-blow process

prodigiosina *f* QUÍMICA prodigiosin

producción *f* ALIMENT *fuente* delivery, CARBÓN *extracción* yield, mine yield, production, get, *minas*
recovery, ELEC, ELECTRÓN, ENERG RENOV generation,
GAS yield, ING ELÉC generation, ING MECÁ output,
generation, performance, development, MINAS production, yield, output, PROD make, output, yield,
TEXTIL production; ~ **automática de informes** *f*
PROD automatic report generation; ~ **de burbujas
relativamente estables** *f* P&C *procesamiento* froth-
ing; ~ **del campo electrónico** *f* TV electronic-field
production; ~ **continua** *f* EMB, PROD continuous
production; ~ **de contornos** *f* PROD contour generation; ~ **del día** *f* C&V turnover; ~ **de electricidad** *f*
ELEC, ING ELÉC electricity generation; ~ **por encargo**
f PROD custom production; ~ **de energía** *f* ING ELÉC
power generation; ~ **de ensayo** *f* PETROL test production; ~ **excesiva** *f* PETROL excessive production;
~ **final** *f* TEC PETR ultimate recovery; ~ **frigorífica** *f*
REFRIG refrigeration output; ~ **global** *f* PROD aggregate output; ~ **a gran escala** *f* PROD mass
production; ~ **horaria** *f* EMB hourly output;
~ **intercalada** *f* IMPR insert production;
~ **intermitente** *f* PROD intermittent production;
~ **magnetohidrodinámica** *f* ING ELÉC magnetohydrodynamic generation (*MHD generation*);
~ **masiva** *f* PROD mass production; ~ **media diaria** *f*
PROD average daily output; ~ **MHD** *f* (*producción
magnetohidrodinámica*) ING ELÉC MHD generation
(*magnetohydrodynamic generation*); ~ **de neutrones** *f*
FÍS, FÍS PART neutron yield; ~ **de la palabra** *f* TELE-
COM speech production; ~ **de pares** *f* FÍS, FÍS PART
pair production; ~ **pecuaria** *f* AGRIC animal industry; ~ **porcina** *f* AGRIC hog crop; ~ **protegida** *f* ING
ELÉC guarded output; ~ **rectificada** *f* ELEC, ING ELÉC
rectified output; ~ **en serie** *f* PROD mass production,
TRANSP MAR *construcción naval* mass production,
series production; ~ **térmica** *f* ING MECÁ, TERMO
thermal output; ~ **térmica de imágenes** *f* D&A, FÍS
RAD, TERMO thermal imaging (*TI*); ~ **de la unidad** *f*
NUCL unit output; ~ **por unidad de área** *f* C&V
production per unit area; ~ **de vapor** *f* INSTAL
HIDRÁUL, PROD getting up steam, raising steam;
~ **vegetal** *f* AGRIC plant production; ~ **de voz** *f*
TELECOM speech production

producir[1] *vt* CINEMAT produce, ING MECÁ develop,
PROD raise, TV produce; ~ **en serie** *vt* ING ELÉC
serialize

producir[2]: ~ **vapor** *vi* INSTAL HIDRÁUL, PROD get up
steam

productividad *f* PROD productiveness, productivity;
~ **laboral** *f* PROD labor productivity (*AmE*), labour
productivity (*BrE*)

productivo *adj* TEXTIL yielding

producto: ~ **alimenticio** *m* ALIMENT foodstuff;
~ **autoadhesivo que se aplica fundido** *m* EMB
pressure-sensitive hot-melt adhesive; ~ **bruto
extraído de la mina** *m* CARBÓN run of mine;
~ **Cartesiano** *m* INFORM&PD Cartesian product;
~ **de cereales** *m* AGRIC grain product;
~ **congelado** *m* ALIMENT, EMB, PROD, REFRIG frozen
product; ~ **derivado** *m* QUÍMICA derived product;
~ **desechable** *m* ING MECÁ expendable item; ~ **de
desecho** *m* RECICL waste product;
~ **desemulsificante** *m* AGUA, CONTAM *tratamiento
de aguas*, CONTAM MAR, QUÍMICA demulsifying
product, emulsion-breaker; ~ **desemulsionante** *m*
AGUA, CONTAM *tratamiento de aguas*, CONTAM MAR,
QUÍMICA demulsifying product, emulsion-breaker;
~ **de destilación** *m* QUÍMICA distillation range, TEC
PETR *refino* straight-run product, TERMO distillation
range; ~ **con envoltura retráctil** *m* EMB shrink-
wrapped product; ~ **escalar** *m* FÍS scalar product;
~ **estanco** *m* PROD leak-free product; ~ **de
evaporación** *m* PROC QUÍ evaporation product;
~ **de intermodulación** *m* ELECTRÓN, TEC ESP, TELE-
COM intermodulation product; ~ **iónico** *m* QUÍMICA
ionic product; ~ **lácteo** *m* ALIMENT dairy product;
~ **de limpieza para calderas** *m* INSTAL TERM *tratamiento del agua* boiler-cleaning compound;
~ **liofilizado** *m* ALIMENT, EMB, REFRIG freeze-dried
product; ~ **molido** *m* CARBÓN, METAL pulp; ~ **nocivo**
m QUÍMICA toxicant; ~ **para obturar** *m* VEH *carrocería* sealant; ~ **petrolero** *m* CONTAM, PETROL
petroleum product; ~ **petrolífero** *m* CONTAM, PETROL
petroleum product; ~ **petroquímico básico** *m*
PETROL, QUÍMICA, TEC PETR basic petrochemical;
~ **a prueba de fugas** *m* PROD leak-free product;
~ **que evita que la tinta forme una película durante
su almacenado** *m* IMPR antiskinning agent; ~ **que
fluye fácilmente** *m* EMB free-flow product; ~ **que no
puede estar en contacto con alimentos** *m* EMB
non-food product; ~ **que retarda el crecimiento de
bacterias** *m* AGRIC bacteriostat; ~ **de refinación** *m*
CONTAM, CONTAM MAR refined product;
~ **refractario** *m* C&V, INSTAL TERM, PROD, QUÍMICA,
TERMO refractory; ~ **secundario** *m* CARBÓN,
CONTAM, ING MECÁ, PROD, QUÍMICA by-product, TEC
ESP spin-off, TEC PETR by-product; ~ **sinergético** *m*
DETERG synergist; ~ **de solubilidad** *m* QUÍMICA *de
electrólito* solubility product; ~ **todo uno** *m* MINAS
mine run; ~ **tóxico** *m* QUÍMICA toxicant; ~ **vectorial**
m FÍS vector product

productor *m* GAS producer; ~ **agrícola** *m* AGRIC
farmer; ~ **de energía** *m* ING MECÁ power producer;
~ **de granos** *m* AGRIC grain grower; ~ **de huevos** *m*
AGRIC egg producer; ~ **de maíz** *m* AGRIC corn grower
(*AmE*), maize grower (*BrE*)

productora *adj* ING ELÉC, PROD generating

productos: ~ **abrasivos aglomerados** *m pl* ING MECÁ
bonded abrasive products; ~ **acabados** *m pl* TEXTIL
finished goods; ~ **adsorbentes** *m pl* PAPEL adsorbent
pads; ~ **de ágata** *m pl* C&V agate ware; ~ **agrícolas
sin elaborar** *m pl* AGRIC raw agricultural products;

~ **de banda continua** *m pl* IMPR continuous web products; ~ **básicos** *m pl* AGRIC commodities; ~ **de celulosa moldeada** *m pl* PAPEL molded pulp products (*AmE*), moulded pulp products (*BrE*); ~ **cruzados** *m pl* ELECTRÓN cross products; ~ **lácteos** *m pl* AGRIC, ALIMENT dairy produce; ~ **lavados** *m pl* PROD wash; ~ **nuevos** *m pl* ING MECÁ new products; ~ **perecederos** *m pl* AGRIC perishable goods; ~ **petroquímicos** *m pl* QUÍMICA, PETROL, TEC PETR petrochemicals; ~ **de producción** *m pl* INFORM&PD, TELECOM software products; ~ **para el revelado** *m pl* CINEMAT, FOTO color-processing chemicals (*AmE*), colour-processing chemicals (*BrE*); ~ **para el revelado en color** *m pl* FOTO, QUÍMICA color-processing chemicals (*AmE*), colour-processing chemicals (*BrE*); ~ **software** *m pl* TELECOM software products

proel *adj* TRANSP MAR fore, forward

profesional *m* PROD practician

profesionalmente: ~ **expuesto** *adj* NUCL occupationally exposed

profundamente: ~ **arraigado** *adj* AGRIC deep-rooted

profundidad *f* AGUA, FÍS, MINAS, OCEAN depth; ~ **de agua** *f* ENERG RENOV, HIDROL, TRANSP MAR depth of water; ~ **del agua disponible** *f* OCEAN *barcos* water clearance; ~ **del agua freática** *f* AGRIC, AGUA, CARBÓN, HIDROL ground-water depth; ~ **de avance** *f* ING MECÁ infeed; ~ **bajo la circunferencia primitiva** *f* ING MECÁ depth below pitch line; ~ **bajo la curva primitiva** *f* ING MECÁ depth below pitch line; ~ **de buceo** *f* OCEAN diving depth; ~ **de caldeo** *f* TERMO heating depth; ~ **de cambio** *f* Esp (*cf campo de imagen eficaz AmL*) FOTO effective image-field; ~ **de campo** *f* CINEMAT depth of field, depth of focus, FOTO, METAL, ÓPT depth of field, TELECOM field depth, TV depth of field; ~ **cartografiada** *f* TRANSP MAR charted depth; ~ **de cementación** *f* ING MECÁ case depth; ~ **de compensación** *f* OCEAN compensation depth; ~ **de corte** *f* ING MECÁ, MECÁ, PROD *de sierras mecánicas* depth of cut; ~ **de la corteza** *f* FÍS skin depth; ~ **destructiva** *f* OCEAN *ensayos hiperbáricos* collapse depth; ~ **eficaz de operación** *f* OCEAN operational depth; ~ **del embalse** *f* PAPEL pond depth; ~ **de enfoque** *f* TV depth of focus; ~ **de entrehierro** *f* ACÚST gap depth; ~ **equivalente** *f* TEC PETR equivalent depth; ~ **focal** *f* GEOL focal depth; ~ **de foco** *f* FÍS, FOTO, ÓPT depth of focus; ~ **de grabado** *f* IMPR engraving depth; ~ **de la influencia** *f* OCEAN depth of frictional influence; ~ **de inmersión** *f* TRANSP depth of immersion; ~ **de intervalo** *f* TV gap depth; ~ **de investigación** *f* GEOFÍS, PETROL depth of investigation; ~ **de matiz** *f* TEXTIL *tintura* depth of shade; ~ **máxima** *f* OCEAN maximum depth; ~ **de modulación** *f* ELECTRÓN, FÍS modulation depth; ~ **óptica** *f* OCEAN *mediciones submarinas* optical depth; ~ **de la pala** *f* TRANSP AÉR blade depth; ~ **de pasada** *f* ING MECÁ infeed; ~ **de penetración** *f* CONST, ELECTRÓN penetration depth, NUCL depth of penetration; ~ **de penetración de la helada** *f* CARBÓN frost-penetration depth; ~ **de la rosca** *f* CONST depth of thread; ~ **de servicio** *f* OCEAN *operaciones de inmersión* operating depth; ~ **de siembra** *f* AGRIC depth of sowing; ~ **termométrica** *f* OCEAN thermometric depth; ~ **del vidrio** *f* C&V glass depth

profundización *f* MINAS sinking; ~ **a baja temperatura** *f* TERMO low-temperature sinking;

~ **de pozos** *f* MINAS shaft deepening; ~ **de pozos por cimentación del terreno** *f* CARBÓN grouting

profundizar *vt* MINAS *pozo de sondeo* put down

progesterona *f* QUÍMICA progesterone

progradación *f* OCEAN progradation

progradar *vi* GEOL prograde

programa *m* INFORM&PD program, ING MECÁ scheme, PROD program (*AmE*), programme (*BrE*), schedule, TELECOM, TV program (*AmE*), programme (*BrE*); ~ **de adiestramiento** *m* ING MECÁ training scheme; ~ **almacenado** *m* INFORM&PD stored program; ~ **de análisis sintáctico** *m* INFORM&PD parser; ~ **de aplicación** *m* INFORM&PD application program; ~ **de apoyo** *m* INFORM&PD support program; ~ **de área** *m* TRANSP area program (*AmE*), area programme (*BrE*), local program (*AmE*), local programme (*BrE*); ~ **de asignación de tráfico** *m* TRANSP traffic assignment program (*AmE*), traffic assignment programme (*BrE*); ~ **de ayuda** *m* INFORM&PD help program; ~ **de base** *m* INFORM&PD background program; ~ **de biblioteca** *m* INFORM&PD library program; ~ **bibliotecario** *m* INFORM&PD librarian program; ~ **de bombeo** *m* PETROL, TEC PETR pumping schedule; ~ **de bromeo** *m* INFORM&PD spoofing program; ~ **de capacitación** *m* ING MECÁ training scheme; ~ **de clasificación** *m* INFORM&PD sort program; ~ **de construcción** *m* CONST construction program (*AmE*), construction programme (*BrE*), construction schedule; ~ **de control** *m* TRANSP control program (*AmE*), control programme (*BrE*); ~ **de control para microprocesadores** *m* INFORM&PD critical-path method (*CPM*); ~ **para controlar el tráfico** *m* TRANSP traffic control program (*AmE*), traffic control programme (*BrE*); ~ **de diagnóstico** *m* INFORM&PD diagnostic program; ~ **de edición** *m* INFORM&PD editing; ~ **de edición publicitario** *m* INFORM&PD desktop publishing (*DTP*); ~ **emitido desde el estudio** *m* TV studio broadcast; ~ **de empuje** *m* TEC ESP thrust program (*AmE*), thrust programme (*BrE*); ~ **de encaminamiento del tráfico** *m* TRANSP traffic-routing program (*AmE*), traffic-routing programme (*BrE*); ~ **para encaminar el tráfico** *m* TRANSP traffic-routing program (*AmE*), traffic-routing programme (*BrE*); ~ **de ensamblaje** *m* INFORM&PD assembler; ~ **de enseñanza** *m* ING MECÁ training scheme; ~ **de entrenamiento** *m* ING MECÁ training scheme; ~ **de errores** *m* INFORM&PD error program; ~ **escrito del trabajo de cada parte** *m* CINEMAT, TV cue sheet; ~ **espacial** *m* TEC ESP space program (*AmE*), space programme (*BrE*); ~ **de evaluación de prestaciones** *m* INFORM&PD benchmark; ~ **de exploración** *m* PROD program scanning (*AmE*), programme scanning (*BrE*); ~ **de fabricación** *m* PROD production schedule; ~ **de fechado** *m* INFORM&PD dating program; ~ **de flujo** *m* PETROL flow schedule; ~ **de fondo** *m* INFORM&PD background program; ~ **fuente** *m* INFORM&PD source program; ~ **generador** *m* INFORM&PD generating program; ~ **guardado en memoria** *m* INFORM&PD stored program; ~ **hidrológico internacional** *m* AGUA, HIDROL international hydrological program (*AmE*), international hydrological programme (*BrE*); ~ **de intercalación** *m* INFORM&PD collator; ~ **de interferencia** *m* INFORM&PD spoofing program; ~ **jerárquico orientado a objetos** *m* INFORM&PD

hierarchical object-oriented design (*HOOD*); ~ **local** *m* TRANSP area program (*AmE*), area programme (*BrE*), local program (*AmE*), local programme (*BrE*); ~ **maestro** *m* TRANSP master program (*AmE*), master programme (*BrE*); ~ **de noticias** *m Esp* (*cf noticioso de televisión AmL*) TV newscast; ~ **objeto** *m* INFORM&PD object program; ~ **de obra** *m* CONST construction program (*AmE*), construction programme (*BrE*); ~ **de ordenación del tráfico** *m* TRANSP traffic-routing program (*AmE*), traffic-routing programme (*BrE*); ~ **de perforación** *m* TEC PETR drilling program (*AmE*), drilling programme (*BrE*); ~ **post mortem** *m* INFORM&PD postmortem program; ~ **preferencial** *m* INFORM&PD foreground program; ~ **de previsión del tráfico** *m* TRANSP traffic forecasting program (*AmE*), traffic forecasting programme (*BrE*); ~ **de primer plano** *m* INFORM&PD foreground program; ~ **principal de producción** *m* PROD master production schedule; ~ **para proceso por ordenador de los registros eléctricos** *m* TEC PETR program for computer-processing wire-line logs (*AmE*), programme for computer-processing wire-line logs (*BrE*); ~ **de producción** *m* PROD production schedule; ~ **de pronóstico del tráfico** *m* TRANSP traffic forecasting program (*AmE*), traffic forecasting programme (*BrE*); ~ **de prueba** *m* METR test program (*AmE*), test programme (*BrE*); ~ **de rastreo** *m* INFORM&PD trace program; ~ **de recocido** *m* C&V annealing schedule; ~ **de recorrido más corto** *m* TRANSP *tráfico* shortest-route program (*AmE*), shortest-route programme (*BrE*); ~ **residente** *m* INFORM&PD resident program; ~ **reubicable** *m* INFORM&PD relocatable program; ~ **de ruta más corta** *m* TRANSP *tráfico* shortest-route program (*AmE*), shortest-route programme (*BrE*); ~ **semafórico** *m* TRANSP semaphoric program (*AmE*), semaphoric programme (*BrE*); ~ **de señales de tráfico** *m* TRANSP traffic signals program (*AmE*), traffic signals programme (*BrE*); ~ **de señalización del tráfico** *m* TRANSP traffic signals program (*AmE*), traffic signals programme (*BrE*); ~ **de servicio** *m* INFORM&PD service program; ~ **de simulación** *m* INFORM&PD simulation program; ~ **de soldadura** *m* CONST welding program (*AmE*), welding programme (*BrE*); ~ **de soporte** *m* INFORM&PD support program; ~ **de televisión por satélite** *m* TV satellite telecast; ~ **de trabajo** *m* TEC ESP work package; ~ **de trayectoria más corta** *m* TRANSP shortest-path program (*AmE*), shortest-path programme (*BrE*); ~ **de trazado** *m* INFORM&PD trace program; ~ **de utilidades** *m* INFORM&PD utility program; ~ **utilitario** *m* INFORM&PD utility program; ~ **de utilitarios** *m* INFORM&PD utility program; ~ **de zona** *m* TRANSP area program (*AmE*), area programme (*BrE*), local program (*AmE*), local programme (*BrE*)

programable *adj* CONST, ELECTRÓN, INFORM&PD, PROD, TELECOM programmable; ~ **mediante máscaras** *adj* INFORM&PD mashing-programmable, mask programmable

programación *f* INFORM&PD programming, TELECOM planning, TV schedule, scheduling; ~ **de acceso mínimo** *f* INFORM&PD minimum-access programming; ~ **de base** *f* PROD fundamental programming; ~ **convexa** *f* INFORM&PD convex programming; ~ **dinámica** *f* INFORM&PD dynamic programming; ~ **directa** *f* PROD forward scheduling; ~ **educativa** *f* TV educational broadcasting; ~ **estructurada** *f* INFORM&PD structured programming; ~ **en línea** *f* PROD on-line programming; ~ **lineal** *f* ELECTRÓN, INFORM&PD linear programming; ~ **matemática** *f* INFORM&PD mathematical programming; ~ **modular** *f* INFORM&PD modular programming; ~ **no lineal** *f* INFORM&PD nonlinear programming; ~ **paso a paso de trayectoria posterior** *f* PROD trail-edge one-shot programming; ~ **de perforación** *f* TEC PETR drilling program (*AmE*), drilling programme (*BrE*); ~ **de la producción** *f* PROD production scheduling; ~ **en serie** *f* INFORM&PD serial programming; ~ **de sistemas** *f* INFORM&PD systems programming; ~ **en una sola operación** *f* PROD one-shot programming; ~ **de trayectoria más corta** *f* TRANSP shortest-path program (*AmE*), shortest-path programme (*BrE*); ~ **única de borde de ataque** *f* PROD leading-edge one-shot programming; ~ **y verificación de la producción** *f* PROD production scheduling and control

programado *adj* INFORM&PD, TRANSP programmed

programador *m* INFORM&PD, PROD programmer; ~ **portátil** *m* PROD hand-held programmer, handset; ~ **principal** *m* PROD master scheduler; ~ **de PROM** *m* INFORM&PD PROM programmer; ~ **de proveedores** *m* PROD vendor scheduler; ~ **de riego** *m* AGRIC irrigation controller

programar *vt* INFORM&PD program, PROD, TELECOM schedule, TV program (*AmE*), programme (*BrE*)

progresar *vt* ING MECÁ develop

progresión *f* ELEC *fase* advance, MATEMÁT progression

progreso *m* CARBÓN advance, PROD progress

prohíbe: se ~ la entrada *fra* SEG no admittance

prohibida: ~ la entrada *fra* SEG no admittance

prohibido: ~ el paso *fra* SEG no admittance

prolongación *f* C&V, FÍS, ING MECÁ, METAL, P&C elongation; ~ **de duración de impulso** *f* TELECOM pulse widening

PROM[1] *abr* (*memoria programable de solo lectura*) INFORM&PD PROM (*programmable read-only memory*)

PROM[2]: ~ **borrable** *f* (*EPROM*) INFORM&PD erasable PROM (*EPROM*)

promecio *m* (*Pm*) NUCL, QUÍMICA promethium (*Pm*)

promedio *m* ELEC average value, FÍS average, INFORM&PD mean, MATEMÁT average, mean; ~ **anual de tráfico diario** *m* TRANSP average annual daily traffic (*AADT*); ~ **aritmético** *m* MATEMÁT arithmetic mean; ~ **de caudal anual** *m* HIDROL average annual flow; ~ **de caudal característico** *m* HIDROL average characteristic flow; ~ **de caudal diario** *m* HIDROL average daily flow; ~ **del caudal mensual** *m* HIDROL average monthly flow; ~ **de conmutación de colores** *m* TV color-sampling rate (*AmE*), colour-sampling rate (*BrE*); ~ **de control anual** *m* HIDROL average yearly flow; ~ **del cuadrado de la velocidad** *m* FÍS mean square velocity; ~ **cuadrático del nivel de agua** *m* ENERG RENOV root-mean-square water level; ~ **de demora por vehículo** *m* TRANSP average delay per vehicle; ~ **global de la velocidad del viaje** *m* TRANSP average overall travel speed; ~ **de intervalo de tiempo** *m* TRANSP average time interval; ~ **de llamadas del abonado** *m* TELECOM subscriber calling rate; ~ **de muestreo mixto** *m* HIDROL mixed average sample; ~ **ponderado** *m* PROD weighted average; ~ **de retraso por vehículo** *m* TRANSP average delay per vehicle; ~ **tiempo por jornada** *m* TRANSP average journey time; ~ **de tiempo parado** *m* TRANSP average stopped time; ~ **tiempo por viaje** *m* TRANSP average journey time; ~ **de tráfico diario** *m* TRANSP average daily traffic (*ADT*)

prominencia *f* TEC ESP prominence

promoción *f* QUÍMICA promoting

promontorio *m* OCEAN foreland, headland, TRANSP MAR *geografía* bluff, promontory; ~ **submarino** *m* OCEAN *geología submarina* tablemount

promotor *m* QUÍMICA promoter; ~ **de adhesión** *m* P&C adhesion promoter

pronóstico *m* INFORM&PD forecasting; ~ **intrínseco** *m* PROD intrinsic forecast; ~ **del tiempo** *m* METEO forecasting, TEC ESP weather forecast; ~ **de tráfico** *m* TRANSP traffic forecasting

prontosil *m* QUÍMICA prontosil

propadieno *m* QUÍMICA propadiene

propagación *f* FÍS ONDAS, INFORM&PD, TELECOM propagation; ~ **por multitrayecto** *f* TELECOM multipath propagation; ~ **de la onda acústica** *f* ING ELÉC acoustic wave propagation; ~ **de ondas** *f* ENERG RENOV, FÍS ONDAS wave propagation; ~ **de ondas de radio** *f* FÍS ONDAS, FÍS RAD radio-wave propagation; ~ **rectilínea** *f* FÍS, FÍS ONDAS, TV rectilinear propagation; ~ **del sonido** *f* TELECOM sound transmission; ~ **submarina** *f* TELECOM underwater propagation; ~ **subterránea** *f* TELECOM subterranean propagation; ~ **por trayectoria múltiple** *f* TELECOM multipath propagation

propagado: ~ **por agua contaminada** *adj* HIDROL waterborne

propagador[1]: **no** ~ **de llama** *adj* INSTAL TERM, PROD, SEG, TERMO, TERMOTEC fire-resistant, fire-resisting

propagador[2]: ~ **de noticias** *m* TEC ESP circulator

propaganda *f* IMPR advertising, TV commercial

propagar *vt* FÍS ONDAS propagate

propagarse *v refl* ÓPT propagate

propano *m* QUÍMICA, TEC PETR propane

propanóico *adj* QUÍMICA propanoic

propanol *m* QUÍMICA propanol

propanona *f* P&C acetone, propanone, QUÍMICA propanone

propargílico *n* QUÍMICA propargyl

propedéutica *f* QUÍMICA propedeutics

propelente *m* CONTAM, QUÍMICA propellant; ~ **almacenable** *m* TEC ESP storable propellant; ~ **sólido** *m* TEC ESP solid propellant

propenílico *adj* QUÍMICA propenylic

propeno *m* QUÍMICA, TEC PETR propene

propenóico *adj* QUÍMICA propenoic

propensión *f* ELEC *reactancia* susceptibility

propenso: ~ **a sufrir accidentes** *adj* SEG accident-prone

propergol *m* CONTAM *combustible* propellant

propfan *m* TRANSP AÉR *hélice* propfan

propiedad *f* GAS, GEOM, QUÍMICA property; ~ **dieléctrica** *f* ELEC, FÍS dielectric property; ~ **dinámica** *f* P&C dynamic property; ~ **física** *f* FÍS, P&C physical property; ~ **hipergólica** *f* TEC ESP hypergolic property; ~ **del revestimiento** *f* REVEST coating property; ~ **de las rocas al ser perforadas** *f* TEC PETR drillability; ~ **térmica** *f* CONST, TERMO thermal property; ~ **topológica** *f* GEOM, INFORM&PD topological property

propiedades: ~ **aislantes** *f pl* EMB insulating properties; ~ **autocurativas** *f pl* TEXTIL self-healing properties; ~ **comunes de las ondas electromagnéticas** *f pl* FÍS RAD common properties of electromagnetic waves; ~ **de la deformación plástica** *f pl* P&C plastic-flow properties; ~ **de elasticidad** *f pl* FÍS FLUID elastic properties; ~ **eléctricas** *f pl* CONST, ELEC, ING ELÉC electrical properties; ~ **de fluencia** *f pl* MECÁ creep properties; ~ **geométricas** *f pl* GEOM geometric properties; ~ **geotécnicas** *f pl* CARBÓN, GEOL geotechnical properties; ~ **mecánicas** *f pl* CONST, FÍS FLUID, ING MECÁ, P&C mechanical properties; ~ **piezoeléctricas** *f pl* ING ELÉC piezoelectric properties; ~ **plásticas** *f pl* FÍS FLUID plastic properties; ~ **reológicas** *f pl* FÍS FLUID rheological properties

propietario: ~ **del buque** *m* TRANSP MAR shipowner

propilamina *f* QUÍMICA propylamine

propileno *m* QUÍMICA, TEC PETR propylene

propílico *adj* QUÍMICA propylic

propilo *m* QUÍMICA propyl

propino *m* QUÍMICA propyne

propinóico *adj* QUÍMICA propynoic

proporción[1] *f* FÍS, MATEMÁT, METR, REFRIG ratio; ~ **de aire** *f* AUTO air ratio; ~ **del ancho de la pala** *f* TRANSP AÉR blade-width ratio; ~ **de aprovechamiento** *f* METAL recovery rate; ~ **de arena a lutita** *f* TEC PETR sand-shale ratio; ~ **del aspecto** *f* TRANSP AÉR aspect ratio; ~ **del aspecto de la pala** *f* TRANSP AÉR blade-aspect ratio; ~ **aurea** *f* GEOM golden section; ~ **de brillos** *f* CINEMAT brightness ratio; ~ **carne-grasa** *f* AGRIC meat-fat ratio; ~ **de ceniza** *f* PROD ash content; ~ **de la conicidad de la pala** *f* TRANSP AÉR blade-taper ratio; ~ **de desvío** *f* TRANSP AÉR bypass ratio; ~ **del diámetro de avance** *f* TRANSP AÉR advance diameter ratio; ~ **dimensional** *f* MECÁ aspect ratio; ~ **de los engranajes de control** *f* TRANSP AÉR control gearing ratio; ~ **de entrada del flujo** *f* TRANSP AÉR inflow ratio; ~ **de esbeltez** *f* TRANSP AÉR fineness ratio; ~ **de estrechamiento de la pala** *f* TRANSP AÉR blade-taper ratio; ~ **grosor-cuerda** *f* TRANSP AÉR thickness-cord ratio; ~ **ideal de la mezcla** *f* AUTO ideal mixture ratio; ~ **del incremento del avance de las ruedas** *f* ING MECÁ wheel-feed increment range; ~ **inicial** *f* GEOL initial ratio; ~ **de llamadas falsas** *f* TELECOM false-calling rate; ~ **de la masa del aeroplano** *f* TRANSP AÉR aeroplane mass ratio (*BrE*), airplane mass ratio (*AmE*); ~ **de la mezcla** *f* AUTO mixture ratio, CONST mix proportion, mixing rate; ~ **de mezcla** *f* TEXTIL blend ratio; ~ **de mezcla del combustible oxidante** *f* TEC ESP fuel-oxidizer mixture ratio; ~ **de mezcla ideal** *f* AUTO maximum output mixture ratio; ~ **de pasadas** *f* TEXTIL pick rate; ~ **perfecta de la mezcla** *f* AUTO perfect mixture ratio; ~ **de planeo** *f* TRANSP AÉR glide ratio; ~ **de presión del motor** *f* TRANSP AÉR engine-pressure ratio; ~ **de transmisión** *f* AUTO transmission ratio; ~ **de unidades inconformes** *f* CALIDAD proportion of nonconforming items

proporción[2]: **en** ~ **inversa a** *fra* ING MECÁ in inverse proportion to

proporcional: **no** ~ *adj* ELEC *de variación* nonlinear

proporcionalidad *f* ELEC linearity

proporcionamiento *m* QUÍMICA proportioning

proporcionar: ~ **una fuente de emisión espontánea**

de luz *vi* FÍS RAD provide spontaneous emission of light

proporciones: ~ equivalentes *f pl* QUÍMICA equivalent proportions

propuesta *f* INFORM&PD, TELECOM bid; **~ de modificación** *f* PROD modification proposal; **~ de trabajo** *f* TEC ESP work package

propulsante *m* QUÍMICA propellant; **~ cohético** *m* D&A, TEC ESP rocket propellant; **~ criogénico** *m* TEC ESP cryogenic propellant; **~ líquido** *m* TEC ESP, TERMO liquid propellant

propulsar *vt* ING MECÁ drive

propulsión[1]**: de ~ mecánica** *adj* MECÁ motor-driven; **con ~ nuclear** *adj* TRANSP MAR *de barco* nuclear-powered; **con ~ propia** *adj* CONTAM MAR, TRANSP MAR *barcaza, draga* self-propelled

propulsión[2] *f* GEN propulsion; **~ por accionamiento espiral con paso variable** *f* TRANSP propulsion by spiral-drive with varying pitch; **~ por chorro** *f* D&A jet propulsion; **~ por chorro de agua** *f* TRANSP hydrojet propulsion, water jet propulsion; **~ a chorro hidráulica** *f* TRANSP hydraulic jet propulsion; **~ del cohete** *f* D&A, TEC ESP rocket propulsion; **~ coloidal** *f* TEC ESP colloid propulsion; **~ diesel-eléctrica** *f* TRANSP diesel-electric drive; **~ electrotérmica** *f* ELEC, TEC ESP electrothermal booster; **~ por haz láser** *f* TEC ESP laser propulsion; **~ por hélice** *f* TRANSP propeller drive; **~ híbrida** *f* TRANSP hybrid propulsion; **~ de hidracina** *f* TEC ESP hydrazine propulsion; **~ iónica** *f* TEC ESP ion propulsion, ionic propulsion; **~ con líquido bipropulsor** *f* TEC ESP liquid-bipropellant propulsion; **~ del motor a reacción** *f* TRANSP AÉR reaction jet propulsion; **~ nuclear** *f* TEC ESP nuclear power, nuclear propulsion; **~ por presión de aire** *f* TRANSP propulsion by air pressure; **~ por reacción** *f* D&A jet propulsion, TEC ESP propulsion by reaction; **~ del reactor** *f* TEC ESP, TRANSP AÉR jet propulsion; **~ sobre las ruedas delanteras** *f* MECÁ front-wheel drive; **~ por ruedas motrices estacionarias fijas** *f* TRANSP propulsion by stationary drive-wheels; **~ por vela solar** *f* TEC ESP solar-sail propulsion

propulsivo *adj* ING MECÁ impulsive

propulsor *m* AUTO, CARBÓN impeller, ING MECÁ drive, propeller, MECÁ impeller, MINAS booster, TEC ESP thruster, TRANSP AÉR impeller; **~ por aerosol** *m* TEC PETR aerosol propellant; **~ de aerosol de hidrocarburos** *m* TEC PETR hydrocarbon aerosol propellant; **~ almacenable** *m* TEC ESP storable propellant; **~ de chorro** *m* MECÁ jet engine; **~ de cohete** *m* D&A, TEC ESP rocket motor; **~ de hélice** *m* ING MECÁ propeller; **~ de hélice hidráulica** *m* TRANSP MAR water propeller (*BrE*), water screw (*AmE*); **~ híbrido** *m* TEC ESP hybrid propellant; **~ hidráulico** *m* *AmL* (*cf columna hidráulico Esp*) MINAS hydraulic prop; **~ iónico de Hall** *m* TEC ESP Hall ion-thruster; **~ líquido** *m* TEC ESP, TERMO liquid propellant; **~ de líquidos densos** *m* TEC ESP slush propellant; **~ mixto** *m* TEC ESP hybrid propellent; **~ monopropelante** *m* TEC ESP monopropellant thruster; **~ monopropulsante** *m* TEC ESP monopropellant thruster; **~ nominal** *m* TEC ESP *en el vacío, de la aeronave* nominal thrust; **~ Schottel** *m* TRANSP Schottel propeller; **~ sólido** *m* TEC ESP solid propellant; **~ de tiro** *m* TEC ESP pitch-thruster

prorratear *vt* PROD apportion

prosopita *f* MINERAL prosopite

prospección *f* GAS exploration, GEOL survey, MINAS prospecting, prospection, cutting, TEC PETR exploration, prospecting; **~ aeromagnética** *f* GEOL, PETROL aeromagnetic survey; **~ aerotransportada** *f* GEOL airborne survey; **~ efectuada desde un avión** *f* GEOL airborne survey; **~ fotográfica aérea** *f* GEOL aerial photographic survey; **~ geofísica** *f* CONST geophysical survey, GEOFÍS geophysical exploration, geophysical prospecting, geophysical survey, GEOL, TEC PETR geophysical survey; **~ gravimétrica** *f* GEOFÍS, GEOL, TEC PETR gravimetric survey; **~ magnética** *f* GEOFÍS, GEOL magnetic prospecting, magnetic survey; **~ magneto-telúrica** *f* GEOFÍS, GEOL magneto-telluric prospecting; **~ petrolífera** *f* GEOL, TEC PETR oil exploration; **~ sísmica** *f* GEOFÍS seismic prospecting, seismic prospection

protactinio *m* (*Pa*) FÍS RAD, QUÍMICA protactinium (*Pa*)

protágono *m* QUÍMICA protagon

proteasa *f* QUÍMICA protease

protección *f* CONST shield, ELECTRÓN resist, FÍS RAD shielding, ING ELÉC protection, ING MECÁ guard, LAB *seguridad* shield, MECÁ guard, ÓPT *fibras ópticas* buffering, PROD security, SEG guard, protection, guarding, TEC ESP protection, shielding, TELECOM security, security audit; **~ acústica** *f* CONST acoustic fencing; **~ de almacenamiento** *f* INFORM&PD storage protection; **~ ambiental** *f* CONTAM, PROD environmental protection; **~ antiestática** *f* ING MECÁ antistatic protection; **~ antihielo** *f* TEC ESP de-icer boot; **~ apretada** *f* ÓPT tight buffer, tight buffering; **~ de archivos** *f* INFORM&PD file protection; **~ auditiva** *f* ACÚST, CONST, SEG ear protection; **~ automática del tren** *f* FERRO *equipo inamovible* automatic train protection (*ATP*); **~ de bajos** *f* AUTO, VEH underbody protection; **~ de los bordes** *f* EMB edge protection; **~ breve** *f* ING ELÉC short-term protection; **~ de camuflaje** *f* D&A camouflage supports; **~ en caso de falla** *f* ELEC fail-safe; **~ catódica** *f* CARBÓN, HIDROL cathodic protection; **~ climática** *f* SEG climatic protection; **~ contra cortocircuitos** *f* ING ELÉC, PROD short-circuit protection; **~ contra descarga eléctrica** *f* GEOFÍS, ING ELÉC lightning protection; **~ contra escombros** *f* ING MECÁ guard against debris; **~ contra escritura** *f* INFORM&PD *disco flexible* write protection; **~ contra explosiones** *f* CARBÓN blast shelter; **~ contra la exposición a riesgos en el trabajo** *f* SEG protection against exposure at work; **~ contra fuga a tierra** *f* ELEC, ING ELÉC, PROD earth-fault protection (*BrE*), ground-fault protection (*AmE*); **~ contra gases** *f* D&A protection against gas; **~ contra goteo de agua** *f* PROD protection against dripping water; **~ contra incendios** *f* SEG, TERMO fire protection; **~ contra incendios de estructuras** *f* SEG structural fire protection; **~ contra el ingreso** *f* PROD ingress protection; **~ contra interrupción de una fase** *f* PROD phase-failure protection; **~ contra intrusos** *f* PROD antitamper cover; **~ contra mar gruesa** *f* PROD protection against heavy seas; **~ contra pérdida a tierra** *f* ELEC, ING ELÉC, PROD earth-fault protection (*BrE*), ground-fault protection (*AmE*); **~ contra el polvo** *f* EMB dust guard; **~ contra radiación** *f* CONTAM, D&A, FÍS RAD, ING ELÉC, SEG radiation protection; **~ contra rayos** *f* GEOFÍS, ING ELÉC lightning protection; **~ contra ruidos de compresores** *f*

SEG noise protection for compressors; ~ **contra ruidos y vibraciones** *f* SEG noise-and-vibration protection; ~ **contra sobrecarga** *f* ELEC, ING ELÉC overload protection; ~ **contra sobrecarga momentánea** *f* ELEC, PROD momentary-overload protection; ~ **contra sobrecorriente** *f* ELEC overcurrent protection; ~ **contra la sobretensión** *f* ELEC, ING ELÉC, SEG excess-voltage protection, overvoltage protection; ~ **contra el sobrevoltaje** *f* ELEC *seguridad*, SEG excess-voltage protection; ~ **contra sobrevoltaje momentáneo** *f* ING ELÉC surge protection; ~ **contra tierra accidental** *f* ELEC, ING ELÉC, PROD earth-fault protection (*BrE*), ground-fault protection (*AmE*); ~ **contra voltaje inverso** *f* ELEC, ING ELÉC, SEG reverse-voltage protection; ~ **por contraseña** *f* INFORM&PD password protection; ~ **de la corriente de entrada** *f* ING ELÉC inrush current protection; ~ **de corta duración** *f* ING ELÉC short-term protection; ~ **de costas** *f* AGUA shore protection; ~ **de datos** *f* INFORM&PD data protection, data security; ~ **diferencial de corriente** *f* ELEC current differential protection; ~ **elevada** *f* EMB *envases* high barrier; ~ **de error** *f* TELECOM error protection; ~ **de errores** *f* ELECTRÓN error protection; ~ **de la fibra** *f* ÓPT fiber buffer (*AmE*), fibre buffer (*BrE*); ~ **de ficheros** *f* INFORM&PD file protection, file security; ~ **del fiel** *f* METR *balanza delicada* beam support; ~ **fija** *f* SEG fixed guard; ~ **flexible** *f* VEH *carrocería* underrun guard; ~ **flotante** *f* ÓPT loose buffer; ~ **frente a la oxidación** *f* EMB rust protection; ~ **por fusibles** *f* ING ELÉC fuse protection; ~ **con gas inerte** *f* NUCL inert gas blanketing; ~ **del haz de electrones** *f* ELECTRÓN electron-beam resist; ~ **de inglete** *f* CONST miter fence (*AmE*), mitre fence (*BrE*); ~ **de línea** *f* ELEC, GEOFÍS line protection; ~ **de la mantilla** *f* IMPR blanket wrap; ~ **mediante disyuntor** *f* SEG trip guard; ~ **de la memoria** *f* INFORM&PD memory protection; ~ **de memoria** *f* INFORM&PD storage protection; ~ **del número de orden** *f* TELECOM sequence number protection (*SNP*); ~ **del número secuencial** *f* TELECOM sequence number protection (*SNP*); ~ **ocular** *f* SEG eye-protector; ~ **de ojos, cara y cuello** *f* SEG eye, face and neck protection; ~ **óptica** *f* ELECTRÓN optical resist; ~ **personal** *f* SEG personal protection; ~ **por plasma desarrollado** *f* ELECTRÓN plasma-developed resist; ~ **positiva** *f* ELECTRÓN positive resist; ~ **preventiva contra incendios** *f* SEG preventive fire protection; ~ **de punta de espolón en las ruedas locas** *f* ING MECÁ protection of pinch points on idlers; ~ **radiológica** *f* CONTAM, FÍS RAD, ING ELÉC radiation protection, NUCL health physics, SEG radiation protection; ~ **de rayos X** *f* ELECTRÓN X-ray resist; ~ **de la red** *f* ELEC network protection; ~ **de RF** *f* ELECTRÓN, TELECOM, TV RF shielding; ~ **de la sección de múltiplex** *f* TELECOM multiplex section protection (*MSP*); ~ **selectiva de campo** *f* TELECOM selective field protection; ~ **de sobrecorriente** *f* ING ELÉC overcurrent protection; ~ **sobrepotencial** *f* ING ELÉC overpotential protection; ~ **de la soldadura** *f* SEG welding protection; ~ **superficial** *f* REVEST surface protection; ~ **de la superficie** *f* REVEST surface protection; ~ **superior** *f* SEG *vehículos de gran sustentación* overhead guard; ~ **temporal** *f* PROD transient protection; ~ **térmica** *f* TEC ESP thermal protection; ~ **en el trabajo de**

soldadura *f* SEG welding protection; ~ **transitoria** *f* PROD transient protection; ~ **transitoria de la entrada** *f* PROD input transient protection

protectivo *m* TEC ESP shielding
protector[1] *adj* SEG, TEXTIL protective
protector[2] *m* CONST life preserver, ELEC, ING MECÁ, MECÁ, PROD guard; ~ **de ablación** *m* TEC ESP ablation shield; ~ **antihielo** *m* TRANSP AÉR ice guard; ~ **auditivo** *m* ACÚST, CONST, LAB, SEG earmuff, earplug; ~ **auditivo para el trabajo** *m* SEG industrial hearing protector; ~ **auricular** *m* ACÚST, CONST, LAB, SEG earplug; ~ **del cabello** *m* SEG hair protector; ~ **de cadena** *m* VEH *transmisión* chain guard; ~ **de la caja de cambio de velocidades** *m* ING MECÁ gearbox guard; ~ **del cárter** *m* VEH *lubricación del motor* sump guard; ~ **del catello** *m* VEH chain guard; ~ **contra escritura** *m* INFORM&PD *disco flexible* write protect; ~ **contra el viento** *m* METEO wind break; ~ **eléctrico** *m* ELEC, GEOFÍS line protector; ~ **eléctrico de cortacircuito** *m* ELEC, GEOFÍS line protector cutout; ~ **elevado** *m* SEG *vehículos de gran sustentación* overhead guard; ~ **de erupción** *m* TEC PETR blowout preventer (*BOP*); ~ **con espacio de aire** *m* ELEC, SEG air gap protector; ~ **de iluminación de 250 watt** *m* CINEMAT inky, inky-dink; ~ **de lente** *m* CINEMAT, FOTO lens cap; ~ **de manos para soldar** *m* CONST welding handshield; ~ **de motor programable** *m* PROD programmable-motor protector; ~ **de multicapas** *m* ELECTRÓN multilayer resist; ~ **negativo** *m* ELEC, ELECTRÓN, ING ELÉC negative resist; ~ **de objetivo** *m* CINEMAT, FOTO lens cap; ~ **ocular** *m* SEG eye-protector; ~ **ocular personal** *m* SEG personal eye-protector; ~ **ocular personal para soldar** *m* SEG personal eye-protector for welding; ~ **de oídos** *m* ACÚST, CONST, LAB, SEG ear-protector, earmuff; ~ **del plato** *m* ING MECÁ chuck guard; ~ **de sobretensión de carácter sólido** *m* ING ELÉC solid-state surge arrester; ~ **de sobrevelocidad** *m* ING MECÁ overspeed protection; ~ **de surco** *m* ACÚST groove guard; ~ **de terminal de tierra** *m* ELEC, ING ELÉC earth-terminal arrester (*BrE*), ground-terminal arrester (*AmE*); ~ **del ventilador** *m* REFRIG fan guard
protegemuelas *m* SEG *máquinas herramientas* wheel guard
proteger *vt* QUÍMICA screen; ~ **contra** *vt* SEG guard against
protegido *adj* SEG *maquinaria* guarded; ~ **contra las explosiones** *adj* EMB explosion-proof; ~ **contra goteo** *adj* ING MECÁ drip-proof, drip-proof screen-protected; ~ **contra goteras** *adj* ING MECÁ drip-proof, drip-proof screen-protected; ~ **contra la luz** *adj* EMB lightproof; ~ **contra el polvo** *adj* EMB, PROD dustproof; ~ **contra las radiaciones** *adj* FÍS RAD radiation-proof; ~ **contra salpicaduras** *adj* ING MECÁ drip-proof, drip-proof screen-protected; ~ **frente a la humedad** *adj* EMB moistureproof; ~ **del fuego** *adj* SEG, TERMO fireproofed, flameproofed; ~ **por fusibles** *adj* ING ELÉC fuse-protected
proteína *f* QUÍMICA protein; ~ **bruta** *f* ALIMENT crude protein; ~ **láctea** *f* ALIMENT milk protein; ~ **vegetal** *f* ALIMENT vegetable protein; ~ **vegetal texturizada** *f* (*PVT*) ALIMENT textured vegetable protein (*TVP*)
proteínico *adj* QUÍMICA proteinic
proterozoico *m* GEOL proterozoic
prótesis: ~ **auditiva** *f* ACÚST hearing aid
protesta: ~ **de mar** *f* TRANSP MAR *avería* ship's protest

prótido *m* QUÍMICA protid, protide
protobastita *f* MINERAL protobastite
protocolo *m* INFORM&PD, QUÍMICA, TEC ESP, TELECOM protocol; **~ de acceso a la línea** *m* TELECOM line access protocol (*LAP*); **~ de canal único** *m* INFORM&PD single-channel protocol; **~ de comunicación** *m* INFORM&PD link protocol; **~ de control de enlace de datos** *m* INFORM&PD data-link control protocol; **~ de convergencia sin conexiones** *m* TELECOM connectionless convergence protocol (*CLCP*); **~ de enlace** *m* INFORM&PD link protocol; **~ entre extremos** *m* INFORM&PD end-to-end protocol; **~ de entrelazado** *m* TELECOM inter-working protocol (*IWP*); **~ de extremo a extremo** *m* INFORM&PD end-to-end protocol; **~ de información de gestión común** *m* TELECOM common management information protocol (*CMIP*); **~ multicanal** *m* INFORM&PD multichannel protocol; **~ N** *m* TELECOM N-protocol; **~ del nivel físico** *m* INFORM&PD physical-layer protocol; **~ a nivel de red sin conexiones** *m* TELECOM connectionless network layer protocol (*CNLP*); **~ de pruebas** *m* COLOR testing record sheet; **~ de reserva rápida** *m* TELECOM fast-reservation protocol (*FRP*); **~ de señalización** *m* INFORM&PD signaling protocol (*AmE*), signalling protocol (*BrE*); **~ de sesión** *m* TELECOM session protocol (*SP*); **~ de transferencia de archivos** *m* INFORM&PD file transfer protocol (*FTP*); **~ de transporte** *m* INFORM&PD transport protocol
Protocolo: **~ de Operación para Mantenimiento** *m* TELECOM *Red de Área local* Maintenance Operation Protocol (*MOP*)
protólisis *f* QUÍMICA protolysis
protolito *m* GEOL protolith
protón *m* (*p*) FÍS, FÍS PART, FÍS RAD, ING ELÉC, NUCL, QUÍMICA proton (*p*); **~ de alta energía** *m* TEC ESP high-energy proton; **~ solar** *m* GEOFÍS solar proton
protoplasma *m* TEC ESP plasma
prototipo *m* ING MECÁ standard, TRANSP MAR *arquitectura naval* prototype
protóxido *m* QUÍMICA protoxide; **~ de plomo** *m* MINERAL, QUÍMICA massicot
protractor *m* METR protractor
protuberancia *f* PROD *de pieza de metal* burr, TEC ESP prominence; **~ del pistón** *f* AUTO piston boss
proustita *f* MINERAL proustite
proveedor *m* CALIDAD supplier; **~ de agua** *m* AGUA, HIDROL, CONTAM water supplier; **~ de efectos navales** *m* TRANSP MAR *provisiones* ship chandler; **~ al por mayor** *m* Esp (*cf proveedor al mayoreo AmL*) PROD wholesale supplier; **~ al mayoreo** *m* AmL (*cf proveedor al por mayor Esp*) PROD wholesale supplier; **~ del servicio de mantenimiento** *m* TELECOM maintenance service provider (*MSP*); **~ del sistema** *m* TELECOM system provider
proveer *vt* ING MECÁ fit out; **~ con** *vt* ING MECÁ fit with; **~ de víveres** *vt* TRANSP MAR *buque* supply with provisions
provincia: **~ metalogenética** *f* GEOL metallogenetic province; **~ paleogeográfica** *f* GEOL palaeogeographic province (*BrE*), paleogeographic province (*AmE*)
provisión: **~ de agua dulce** *f* AGUA freshwater stock; **~ de combustible** *f* TEC ESP fuel man; **~ y mantenimiento de la administración explotadora** *f* TELECOM operation administration maintenance

and provisioning (*OAMP*); **~ de la red abierta** *f* TELECOM open network provision (*ONP*)
provisional *adj* Esp (*cf provisorio AmL*) PROD, TEC ESP makeshift
provisorio *adj* AmL (*cf provisional Esp*) PROD, TEC ESP makeshift
provocar *vt* ELEC induce; **~ cortocircuito a** *vt* PROD short out
proximidad: **~ del mercado** *f* TEXTIL market closeness; **~ del plasma** *f* TEC ESP plasma environment
próximos: **~ vecinos** *m pl* CRISTAL nearest neighbors (*AmE*), nearest neighbours (*BrE*)
proyección *f* CONST, CRISTAL, FÍS, FOTO, GEOM projection, PROD tab; **~ centrográfica** *f* TRANSP MAR *navegación* gnomonic projection; **~ de chispas del arco** *f* ING ELÉC arcing; **~ cónica** *f* GEOM conical projection; **~ continuada de un haz luminoso** *f* ING ELÉC searchlight; **~ electrónica de imágenes** *f* ELECTRÓN electronic imaging; **~ estereográfica** *f* CRISTAL, METAL stereographic projection; **~ frontal** *f* CINEMAT, TV front projection; **~ gnomónica** *f* TRANSP MAR *navegación* gnomonic projection; **~ horizontal** *f* ING MECÁ plan view; **~ horizontal del plano de formas** *f* TRANSP MAR *construcción naval* half-breadth plan; **~ de hormigón** *f* Esp (*cf chorreado con granalla AmL*) MINAS shotting; **~ de la imagen** *f* ELECTRÓN image projection; **~ de imágenes electrónicas** *f* ELECTRÓN electron imaging; **~ de inclinación** *f* TV tip projection; **~ a lo largo del eje c** *f* CRISTAL projection along the c-axis; **~ de Mercator** *f* TEC ESP, TRANSP MAR Mercator projection; **~ mercatoriana** *f* TEC ESP, TRANSP MAR Mercator projection; **~ ortodrómica** *f* TEC ESP, TRANSP MAR orthodromic projection; **~ ortogonal** *f* GEOM orthogonal projection; **~ reflex** *f* CINEMAT, FOTO, TV reflex projection; **~ sobre superficie uniforme** *f* GEOL equal-area projection; **~ por transparencia** *f* CINEMAT rear projection
proyeccionista *m* CINEMAT projectionist
proyectar *vt* CINEMAT project, screen, CONST design, GEOM project, MECÁ design, PROD schedule; **~ corto** *vt* ELECTRÓN undershoot
proyectil *m* D&A missile, projectile, shell; **~ aerodinámico** *m* D&A aerodynamic missile; **~ balístico** *m* D&A, ELECTRÓN, TEC ESP ballistic missile; **~ cohético** *m* D&A rocket-assisted projectile; **~ de fragmentación** *m* D&A fragmenting shell; **~ de gran alcance** *m* D&A long-range weapon; **~ inteligente** *m* D&A smart shell; **~ por láser** *m* ELECTRÓN laser weapon; **~ nuclear** *m* D&A, NUCL nuclear shell; **~ que no ha estallado** *m* TEC ESP blind
proyectista *m* PROD *persona* designer, TRANSP MAR ship designer; **~ principal** *m* CONST chief designer
proyecto *m* CONST design, draft (*AmE*), draught (*BrE*), INFORM&PD design, ING MECÁ lay, scheme, TRANSP MAR *navegación, arquitectura naval* design, plan; **~ de almacenamiento** *m* ENERG RENOV *energía hidroeléctrica* storage scheme; **~ de almacenamiento bombeado** *m* ENERG RENOV *energía hidroeléctrica* pumped storage scheme; **~ para el aprovechamiento de las aguas** *m* HIDROL water resource development project; **~ de calidad** *m* CALIDAD quality plan; **~ de desarrollo** *m* TEC PETR development project; **~ económico** *m* TEC PETR economic project; **~ de energía sin almacenamiento** *m* ENERG RENOV *energía*

hidroeléctrica, INSTAL HIDRÁUL run-of-river scheme; ~ **experimental** *m* TEC PETR pilot project; ~ **y fabricación subsiguiente** *m* ING MECÁ development and subsequent manufacture; ~ **hidroéléctrico** *m* CONST hydroelectric project; ~ **llave en mano** *m* ING MECÁ turnkey project; ~ **de producción** *m* TEC PETR development project; ~ **de regadío** *m* HIDROL irrigation project

proyector *m* CINEMAT projection box, multigage (*AmE*), multigauge (*BrE*), projector, FOTO, INSTR projector; ~ **de acuidad** *m* INSTR acuity projector; ~ **de agudeza** *m* INSTR acuity projector; ~ **de alumbrado exterior** *m* ELEC, VEH headlamp; ~ **analítico** *m* CINEMAT analytic projector; ~ **analizador** *m* CINEMAT analysing projector (*BrE*), analyzing projector (*AmE*); ~ **del arco** *m* CINEMAT arc projector; ~ **de arco de xenón** *m* CINEMAT xenon arc projector; ~ **automático de diapositivas** *m* FOTO automatic slide changer; ~ **de banda doble** *m* CINEMAT double-band projector (*AmE*), double-headed projector (*BrE*); ~ **de bucle continuo** *m* CINEMAT continuous-loop projector; ~ **de cabezal doble** *m* CINEMAT double-band projector (*AmE*), double-headed projector (*BrE*); ~ **de cine** *m* CINEMAT, ELEC *alumbrado* cinema projector (*BrE*), movie projector (*AmE*); ~ **cinematográfico** *m* CINEMAT, ELEC *alumbrado* cinema projector (*BrE*), movie projector (*AmE*); ~ **convergente** *m* ELEC *alumbrado* intensive projector; ~ **de diapositivas** *m* FOTO slide projector; ~ **espectral** *m* FÍS RAD spectrum projector; ~ **de exploración** *m* D&A searchlight; ~ **de fondo** *m* CINEMAT backlight; ~ **de imágenes patrón** *m* CINEMAT pattern projector; ~ **intensificador** *m* ELEC cone light; ~ **de luz** *m* CINEMAT, FOTO spotlight; ~ **de luz abierto** *m* CINEMAT open-face spotlight; ~ **de luz estroboscópica** *m* TEC ESP strobe light projector; ~ **de medición óptico** *m* ING MECÁ optical measuring projector; ~ **multipaso** *m* CINEMAT multigage projector (*AmE*), multigauge projector (*BrE*); ~ **de nieve** *m* REFRIG snow gun; ~ **de opaco** *m* FOTO erect-image viewfinder; ~ **de paso múltiple** *m* CINEMAT multigage projector (*AmE*), multigauge projector (*BrE*); ~ **de película magnética** *m* CINEMAT magnetic film projector; ~ **de perfiles** *m* METR profile projector; ~ **del proscenio** *m* ELEC *alumbrado* apron floodlight; ~ **de video** *AmL*, ~ **de vídeo** *Esp m* TV video projector

proyectores: ~ **dobles** *m pl* CINEMAT twin projectors

prueba[1]: **a** ~ *adj* EMB antitheft; **a** ~ **de ácidos** *adj* PAPEL acid-proof; **a** ~ **de averías** *adj* GEN fail-safe; **a** ~ **de calor** *adj* SEG heat-proof; **a** ~ **de explosión** *adj* EMB, TRANSP AÉR explosion-proof; **a** ~ **de fallos** *adj* GEN fail-safe; **a** ~ **de fallos leves** *adj* INFORM&PD fail-safe; **a** ~ **de filtraciones** *adj* MECÁ leak-tight; **a** ~ **de fuego** *adj* GEN fireproof, flameproof; **a** ~ **de gases** *adj* GEN gas-proof, gastight; **a** ~ **de golpes** *adj* CINEMAT shockproof; **a** ~ **de grasa** *adj* EMB, PAPEL greaseproof; **a** ~ **de heladas** *adj* AGRIC, CARBÓN frost-resistant; **a** ~ **de incendios** *adj* GEN fireproof, flameproof; **a** ~ **de intemperie** *adj* PAPEL, SEG weatherproof; **a** ~ **de ladrones** *adj* SEG thief-proof; **a** ~ **de polvo** *adj* ELEC *equipo*, SEG dustproof; **a** ~ **de radiaciones** *adj* FÍS RAD radiation-proof; **a** ~ **de robos** *adj* EMB pilfer-proof, SEG *cerradura* burglar-proof; **a** ~ **de seguridad** *adj* GEN fail-safe; **a** ~ **de vapores** *adj* SEG vapor-proof (*AmE*), vapour-proof (*BrE*)

prueba[2] *f* CARBÓN test, experiment, CINEMAT check print, ELEC proof test, FÍS experiment, probe, test, FÍS PART, FÍS RAD experiment, FOTO proof, INFORM&PD test, testing, ING MECÁ test run, testing, MECÁ test, experiment, METAL test, NUCL experiment, QUÍMICA experiment, testing, assay, test, REFRIG test, TEC ESP test, experiment, TERMO experiment, TEXTIL trial; ⬛**a** ~ **de aceptación** *f* CARBÓN, INFORM&PD, TEC ESP acceptance test; ~ **de adherencia** *f* P&C adhesion test; ~ **de adherencia de ángulos** *f* P&C angle peeling test; ~ **de adhesión** *f* P&C adhesion test; ~ **de adhesión del recubrimiento del cable** *f* ING MECÁ cord-to-coating bond test; ~ **del agua salada** *f* C&V salt spray test; ~ **de aislamiento** *f* ING ELÉC flash test; ~ **del ambiente** *f* CALIDAD environmental testing; ~ **de amplitud variable** *f* METAL variable-amplitude test; ~ **de aplastamiento** *f* FÍS flattening test; ~ **de apriete** *f* GAS, MECÁ tightness test; ~ **de aprobación** *f* CALIDAD, TEC ESP approval test; ~ **de aptitud para conducir** *f* AUTO, VEH *normas* driving test; ~ **del árbol** *f* ENERG RENOV drill stem test (*DST*); ~ **de arranque a cero potencia** *f* NUCL startup zero power test; ~ **de articulación silábica** *f* TELECOM syllable articulation test; ~ **asistida por ordenador** *f Esp* (*cf prueba auxiliada por computador AmL, prueba auxiliada por computadora AmL*) INFORM&PD computer-aided testing (*CAT*); ~ **atómica** *f* D&A atomic test; ~ **auditiva** *f* ACÚST, SEG hearing test; ~ **de autoadrizamiento** *f* TRANSP MAR *bote salvavidas* self-righting test; ~ **automática** *f* TELECOM automatic attempt; ~ **auxiliada por computadora** *f AmL* (*cf prueba asistida por ordenador Esp*) INFORM&PD computer-aided testing (*CAT*);

⬛**b** ~ **de la bajada del tren de aterrizaje** *f* TRANSP AÉR landing gear drop test; ~ **de balance** *f* TRANSP MAR *diseño de barcos* roll test; ~ **beta** *f* INFORM&PD beta test; ~ **a la bola Brinell** *f* ING MECÁ, MECÁ Brinell ball test; ~ **de la bolsa de municiones** *f* C&V shot bag test; ~ **de bombeo** *f* AGUA *pozos*, CARBÓN, INSTAL HIDRÁUL *pozos* pumping test; ~ **de Brinell** *f* ING MECÁ, MECÁ Brinell test; ~ **del bucle** *f* ELEC loop test; ~ **en bucle** *f* ELEC loop test; ~ **en bucle de Allen** *f* ELEC Allen's loop test;

⬛**c** ~ **de calibración** *f* TRANSP AÉR calibration test; ~ **de calificación** *f* CALIDAD, NUCL, TEC ESP qualification test; ~ **de cámara** *f* CINEMAT, TV camera test; ~ **de carga** *f* CONST loading test, ELEC load test; ~ **sin carga** *f* ELEC no-load test; ~ **a cero potencia** *f* NUCL zero power test; ~ **de certificación** *f* GAS, TRANSP AÉR certification test; ~ **de Charpy** *f* FÍS Charpy test; ~ **al choque** *f* FÍS, METR impact test; ~ **de choque de calor** *f* TERMOTEC *circuitos impresos* heat shock test; ~ **cinematográfica** *f* CINEMAT screen test; ~ **en circuito abierto** *f* ELEC open-circuit test; ~ **de cizallamiento** *f* ING MECÁ shear test; ~ **de cizallamiento en bloque** *f* ING MECÁ block-shear test; ~ **climática** *f* EMB, TEC ESP climatic test; ~ **de cocido e irradiación UV** *f* C&V bake and UV-irradiation test; ~ **de colisión** *f* TRANSP collision test; ~ **combinada** *f* CINEMAT composite check print; ~ **comparativa** *f* FÍS comparative test; ~ **de comportamiento mecánico** *f* GAS mechanical behavior test (*AmE*), mechanical behaviour test (*BrE*); ~ **de compresión** *f* CARBÓN, EMB, MECÁ, METAL compression test; ~ **de conden-**

sación de vapor de agua *f* ING MECÁ water vapor condensation test (*AmE*), water vapour condensation test (*BrE*); **~ de confiabilidad** *f* ING ELÉC reliability test; **~ de conformidad** *f* CALIDAD compliance test, conformity test; **~ de consolidación** *f* CARBÓN, PETROL, TEC PETR consolidation test; **~ continua de cualificación** *f* NUCL ongoing qualification test; **~ de continuidad** *f* ELEC *circuito* continuity test; **~ de control** *f* QUÍMICA control assay; **~ por control a distancia** *f* PROD remote test; **~ de corta duración** *f* ELEC short-time test; **~ de corte** *f* ING MECÁ shear test; **~ de cortocircuito repentina** *f* ELEC sudden short-circuit test; **~ de crudo** *f* PETROL, TEC PETR crude assay;

~d **~ de decrepitación** *f* C&V decrepitation test; **~ de deformación estática** *f* ING MECÁ static strain test; **~ de densidad** *f* CONST density test; **~ de desarrollo** *f* TEC ESP development test; **~ de descarga disruptiva** *f* ELEC *equipo de alta tensión* puncture test; **~ de descascarillado** *f* PROD peeling test; **~ de descongelación** *f* ING MECÁ defrosting test; **~ de desgaste por abrasión** *f* ING MECÁ abrasion test; **~ de desgaste por frotamiento** *f* ING MECÁ attrition test; **~ de deshielo** *f* ING MECÁ defrosting test; **~ por destilación** *f* PROC QUÍ, QUÍMICA distillation test; **~ destructiva** *f* ING MECÁ destructive test; **~ para determinar contenido graso** *f* AGRIC fat test; **~ de diagnóstico** *f* INFORM&PD diagnostic test; **~ dinámica** *f* METAL dynamic test; **~ de dispersión** *f* GAS dispersion test, IMPR scatter proof; **~ disruptiva** *f* ING ELÉC flash test; **~ a distancia** *f* INFORM&PD remote test; **~ de duración** *f* ING MECÁ test run, PROD endurance test; **~ de duración de herramientas** *f* ING MECÁ tool-life testing; **~ de dureza Rockwell** *f* ING MECÁ Rockwell hardness test;

~e **~ de efectividad hidráulica** *f* ING MECÁ hydraulic performance test; **~ eléctrica** *f* ELEC, ING ELÉC electrical test; **~ electrónica** *f* IMPR soft proof; **~ de embalamiento** *f* ELEC *máquina* overspeed test; **~ de emulsión** *f* PROC QUÍ emulsion test; **~ de ensayo** *f* FOTO test print; **~ con entalla Charpy en V** *f* MECÁ Charpy V-notch test; **~ de envejecimiento acelerado** *f* P&C accelerated ageing test; **~ de escritorio** *f* INFORM&PD data-switching exchange (*DSE*); **~ de escurrimiento** *f* C&V creep test; **~ de espuma** *f* TRANSPAÉR foaming test; **~ de estabilidad** *f* CINEMAT steadiness test; **~ de estampación** *f* PROD *de metales* stamping test; **~ de exceso de velocidad** *f* ELEC *máquina* overspeed test; **~ por excitación de oscilaciones transitorias** *f* ELEC, ING ELÉC ringing test; **~ de exfoliación** *f* PROD peeling test; **~ de exposición** *f* CINEMAT, FOTO exposure test; **~ de extintor de espuma** *f* SEG, TRANSP AÉR foam extinguisher test;

~f **~ en factor de potencia cero** *f* ELEC zero power factor test; **~ en factor de potencia nulo** *f* ELEC zero power factor test; **~ de fatiga** *f* CRISTAL, MECÁ, METAL, P&C, TRANSPAÉR fatigue test; **~ de fatiga con carga axial** *f* ING MECÁ axial load fatigue testing; **~ de fatiga rápida** *f* METAL rapid-fatigue test; **~ de fiabilidad** *f* ING MECÁ test for accuracy; **~ de fijación** *f* CINEMAT, FOTO *de cámara* steadiness test; **~ de fil derecho** *f* INFORM&PD leapfrog test; **~ final** *f* IMPR final proof; **~ de la flexibilidad de la temperatura subambiente** *f* ING MECÁ subambient temperature

flexibility test; **~ de flexión al choque** *f* ING MECÁ blow-bending test; **~ de flexión y compresión** *f* ING MECÁ test for bending and for compression; **~ de flexión en probeta entallada** *f* ING MECÁ notch bending test; **~ de flexión rotativa** *f* METAL rotating bending test; **~ de floculación** *f* PROC QUÍ flocculation test; **~ de formación** *f* GEOL, PETROL, TEC PETR formation test; **~ de formación de lodo** *f* ELEC *aceite para transformadores* sludge formation test; **~ fotográfica** *f* IMPR photographic proof; **~ de fractura** *f* ING MECÁ fracture test; **~ de fragilidad** *f* METR impact test; **~ de frecuencia** *f* FÍS RAD frequency test; **~ de frenado** *f* CONST brake testing; **~ del freno** *f* ING MECÁ brake test; **~ de frío** *f* AGRIC cold test; **~ de fuga visual** *f* REFRIG visual leak test; **~ de fugas** *f* CONST, NUCL, REFRIG leak test, leak testing; **~ funcional** *f* INFORM&PD functional test, NUCL proving run, proving trial, trial run, trial, PROD operational test, TELECOM functional test; **~ de funcionamiento** *f* ELEC performance test, INFORM&PD functional test, ING MECÁ running test, test run, TELECOM functional test; **~ de fusión** *f* TERMO melting test;

~g **~ de galera** *f* IMPR string proof; **~ en galera** *f* IMPR slip proof; **~ de galerada** *f* IMPR galley proof; **~ de gama** *f* IMPR progressive proof; **~ del grano** *f* C&V grain test; **~ de grietas** *f* ING MECÁ crack test;

~h **~ del haz electrónico** *f* ELECTRÓN, FÍS RAD electron-beam test; **~ sobre hoja de acetato** *f* IMPR acetate proof; **~ de homologación** *f* GAS standard compliance test, TEC ESP type approval; **~ de homologación de tipo** *f* TRANSP AÉR type test; **~ de huésped** *f* AGRIC host indexing;

~i **~ de impacto** *f* METAL, TRANSP impact test; **~ de impacto Charpy** *f* MECÁ, METAL, P&C Charpy impact test; **~ de impacto con dardo** *f* C&V dart impact test; **~ de impacto de tracción** *f* METAL tensile impact test; **~ de imprenta** *f* IMPR proof; **~ de impulsión** *f* ING ELÉC impulse test; **~ de impulso hidráulico** *f* ING MECÁ hydraulic impulse test; **~ de incombustibilidad** *f* SEG noncombustibility test; **~ de inflamabilidad** *f* SEG fire test; **~ de inflamabilidad para muebles** *f* SEG fire test for furniture; **~ inicial de descarga parcial** *f* ELEC partial-discharge inception test; **~ de integridad de la formación** *f* TEC PETR leak-off test (*LOT*); **~ interna de diagnóstico** *f* PROD internal diagnostic test; **~ de intervalos de formación** *f* PETROL formation interval test (*FIT*); **~ de investigación** *f* TEC ESP investigation test; **~ isotérmica** *f* METAL isothermal test;

~j **~ de Jominy** *f* MECÁ Jominy test;

~l **~ del lápiz** *f* CINEMAT pencil test; **~ de lazo** *f* ELEC loop test; **~ de lazo de Allen** *f* ELEC Allen's loop test; **~ de leak-off** *f* TEC PETR *perforación* leak-off test (*LOT*); **~ de límite** *f* PROD limit test; **~ limpia** *f* IMPR clean proof; **~ por lote** *f* PROD batch test;

~m **~ manual** *f* TELECOM manual attempt; **~ de mar** *f* TRANSP MAR *construcción naval* sea trial; **~ en el mar** *f* CONTAM MAR field trial; **~ marginal** *f* INFORM&PD marginal test; **~ de materiales** *f* ING MECÁ material testing; **~ mecánica** *f* ING MECÁ mechanical testing; **~ mecanográfica** *f* IMPR typescript proof; **~ de mercancías de entrada** *f* PROD good inwards test; **~ por el método del anillo de Allen** *f* ELEC Allen's loop test; **~ de muestreo** *f* CALIDAD *agua* sample captor, sampling probe;

~ n ~ **nítida** *f* IMPR clean proof; ~ **no concluyente** *f* FÍS inconclusive test; ~ **no destructiva** *f* FÍS, ING MECÁ nondestructive test; ~ **del núcleo** *f* ELEC *máquina* core test;

~ o ~ **operacional** *f* METR, PROD operational test;

~ p ~ **de página** *f* IMPR page proof, page shoe; ~ **paramétrica** *f* TELECOM parametric test; ~ **de penetración** *f* CARBÓN penetration test; ~ **de penetración del pistón** *f* CARBÓN ram penetration test; ~ **de penetración ponderal** *f* CARBÓN weight penetration test; ~ **piloto** *f* CARBÓN pilot test; ~ **a pincel** *f* IMPR brush proof; ~ **de plegado** *f* PROD plating-out test; ~ **positiva** *f* FOTO positive, print; ~ **de precisión** *f* ING MECÁ accuracy test, test for accuracy; ~ **de prensa** *f* IMPR advance sheet, press proof, show sheet; ~ **de presión** *f* C&V pressure check, INSTAL HIDRÁUL pressure test; ~ **de producción** *f* CARBÓN *pozos de petróleo* testing, PETROL production test; ~ **de producción con tubería de perforación** *f* (*DST*) TEC PETR drill stem test (*DST*); ~ **de producto** *f* CALIDAD product test; ~ **de programa** *f* INFORM&PD program testing; ~ **de puesta en servicio** *f* TELECOM commissioning test;

~ r ~ **de rasgado** *f* P&C tearing test; ~ **de rebatimiento del collarín** *f* PROD *para tubos* flanging test; ~ **de recepción** *f* INFORM&PD, TEC ESP, TELECOM acceptance test, TRANSP AÉR acceptance trial, TRANSP MAR acceptance test, code checking, *de buques* acceptance trial; ~ **remota** *f* INFORM&PD remote test; ~ **de rendimiento** *f* ELEC performance test, INFORM&PD benchmark; ~ **de reproducción** *f* IMPR reproduction pull; ~ **para reproducción** *f* IMPR repro proof; ~ **de resiliencia Charpy** *f* MECÁ, METAL, P&C Charpy impact test; ~ **de resiliencia en probeta entallada** *f* MECÁ notched-bar impact test; ~ **de resistencia** *f* ING MECÁ endurance test, MECÁ breaking test, METR, TEC ESP *entrenamiento de astronautas* endurance test; ~ **de resistencia de adhesión** *f* ING MECÁ adhesion strength test; ~ **de resistencia al calor y a los golpes** *f* TERMO heat shock test; ~ **de resistencia a la compactación** *f* CARBÓN Proctor compaction test; ~ **de resistencia a esfuerzos alternos cíclicos** *f* PROD endurance test; ~ **de rigidez dieléctrica** *f* ING ELÉC dielectric rigidity test; ~ **de rotura** *f* ING MECÁ fracture test, MECÁ breaking test;

~ s ~ **secuencial** *f* TELECOM sequential test; ~ **de selección** *f* TEC ESP screening test; ~ **selectiva interna por saltos** *f* INFORM&PD leapfrog test; ~ **de separación del árido** *f* CONST aggregate stripping test; ~ **de sequedad** *f* REFRIG dryness test; ~ **de significación** *f* Esp (*cf prueba de significancia AmL*) INFORM&PD, MATEMÁT significance test; ~ **significancia** *f* AmL (*cf prueba de significación Esp*) INFORM&PD, MATEMÁT significance test; ~ **del sistema** *f* INFORM&PD system testing; ~ **de sobrecarga** *f* ELEC overload test; ~ **de sobrepresión** *f* PROD overpressure test; ~ **de sobretensión** *f* ELEC, PROD *electricidad* overpressure test; ~ **de sobretensión inducida** *f* ELEC *transformador* induced overvoltage test; ~ **de sobrevelocidad** *f* ELEC *máquina* overspeed test; ~ **subjetiva** *f* TELECOM subjective test;

~ t ~ **de tensión** *f* ELEC test voltage; ~ **de tensión disruptiva** *f* ELEC *equipo de alta tensión* puncture test; ~ **de tensión no disruptiva** *f* ELEC *instalación* withstand voltage test; ~ **sobre el terreno** *f* CONTAM MAR field trial; ~ **en tierra** *f* TEC ESP, TRANSP AÉR ground test; ~ **de tiro** *f* TEC ESP firing test; ~ **de tiro de recepción** *f* TEC ESP *artillería* acceptance firing test; ~ **de torsión** *f* ING MECÁ torsional test, METAL torsion test; ~ **torsional** *f* ING MECÁ torsional test; ~ **de tracción** *f* ING MECÁ, METAL tensile test; ~ **triangular** *f* ALIMENT triangle test, triangle testing; ~ **truncada** *f* TEC ESP truncated test;

~ u ~ **sin utilizar el proceso de impresión** *f* IMPR prepress proof;

~ v ~ **de vacío** *f* REFRIG vacuum test; ~ **en vacío** *f* ELEC no-load test; ~ **de validez** *f* TELECOM validity check; ~ **de velocidad excesiva** *f* ELEC *máquina* overspeed test; ~ **de vibración** *f* METR vibration test; ~ **de voltaje no disruptivo** *f* ELEC *instalación* withstand voltage test; ~ **de vuelo** *f* TEC ESP, TRANSP AÉR flight test; ~ **de vuelo libre** *f* TEC ESP free-flight test;

~ z ~ **de zona** *f* PROD zone test

prulaurasina *f* QUÍMICA prulaurasin

prusiato *m* QUÍMICA prussiate

PS *abr* (*poliestireno*) EMB, P&C, QUÍMICA PS (*polystyrene*)

psamita *f* PETROL psamite

PSD *abr* (*proceso de señal digital*) ELECTRÓN DSP (*digital signal processing*)

pseudo: ~ **rabia** *f* AGRIC mad itch

pseudoácido *m* QUÍMICA pseudo-acid

pseudoaleatorio *adj* TELECOM pseudo-random

pseudoamorfo *m* GEOL pseudomorph

pseudocódigo *m* INFORM&PD pseudocode

pseudoesfera *f* GEOM pseudosphere

pseudofijación *f* TV pseudo-lock

pseudoinstrucción *f* INFORM&PD pseudo-instruction

pseudoionona *f* QUÍMICA pseudo-ionone

pseudolenguaje *m* INFORM&PD pseudo-language

pseudomalaquita *f* MINERAL pseudomalachite

pseudomería *f* QUÍMICA pseudomerism

pseudomérico *adj* QUÍMICA pseudomeric

pseudómero *m* QUÍMICA pseudomer

pseudonitrol *m* QUÍMICA pseudonitrole

pseudoplástico *m* QUÍMICA pseudo-plastic

pseudorruido *m* TELECOM pseudonoise

psicometría *f* FÍS psychrometry

psicrómetro *m* LAB psychrometer; ~ **de aspiración** *m* REFRIG aspiration psychrometer; ~ **aspirado** *m* REFRIG aspirated psychrometer

psilomelana *f* MINERAL psilomelane

PSK *abr* INFORM&PD, TEC ESP (*modulación por desviación de fase, modulación por varación de fase*) PSK (*phase-shift keying modulation*)

PSM *abr* (*peso sobre la mecha AmL*) TEC PETR WOB (*weight on bit*)

Pt *abr* (*platino*) ELEC, METAL, QUÍMICA Pt (*platinum*)

PTC[1] *abr* INSTR (*transportador de cinta cóncava parabólico*) PTC (*parabolic trough conveyor*), TEC ESP (*control térmico pasivo*) PTC (*passive thermal control*)

PTC[2]: ~ **para enrollamiento** *m* PROD winding RTD

pterina *f* QUÍMICA pterin

PTFE *abr* (*politetrafluoretileno*) P&C, QUÍMICA PTFE (*polytetrafluoroethylene*)

ptomaína *f* QUÍMICA ptomaine

PTV *abr* (*peso total del vehículo*) VEH GVW (*gross vehicle weight*)

Pu *abr* (*plutonio*) FÍS, FÍS RAD, GEOL, MINERAL, NUCL, QUÍMICA Pu (*plutonium*)

púa *f* CONST prong (*BrE*), tine (*AmE*), ING MECÁ pin, prong (*BrE*), tine (*AmE*), tooth, MECÁ, PROD pin

publicación: ~ **electrónica** *f* ELECTRÓN, INFORM&PD electronic publishing; ~ **interna** *f* IMPR house organ; ~ **de mesa** *f* AmL (*cf autoedición Esp*) INFORM&PD desktop publishing (*DTP*); ~ **óptica** *f* ÓPT optical publishing

publicidad *f* IMPR advertising

pudelación: ~ **húmeda** *f* PROD pig boiling

pudelador: ~ **mecánico** *m* PROD puddler's rabble

pudelaje: ~ **caliente** *m* PROD pig boiling

pudinga *f* PETROL, TEC PETR conglomerate

pudrición *f* CONST rotting, *de madera* wet rot; ~ **seca** *f* TRANSP MAR *de madera* dry rot

pudrir *vi* ALIMENT putrefy

pudrirse *v refl* AGRIC, CONST, TRANSP MAR rot

puente[1]: **de** ~ *adj* PROD bridged

puente[2] *m* C&V *en el alimentador de un horno de vidrio* bridge, CONST *del andamio* ledger, ELEC *instrumento* bridge, *conexión* jumper, IMPR beam, INFORM&PD bridge, ING ELÉC jumper, ING MECÁ bridge, gap bridge, INSTR *conexión* bridge, PROD *hornos* bridge, flame bridge, fire bridge, TELECOM, TRANSP MAR *de mando* bridge; ~ **de acero** *m* CONST steel bridge; ~ **aéreo** *m* TRANSP AÉR air bridge, boarding bridge, intercity air service, shuttle; ~ **aéreo de pasajeros** *m* TRANSP AÉR passenger bridge; ~ **Anderson** *m* ELEC *circuito* Anderson bridge; ~ **con arcos disminuidos** *m* CONST bridge with diminished arches; ~ **de autopista** *m* CONST highway bridge; ~ **Bailey** *m* D&A Bailey bridge; ~ **bajo la línea férrea** *m* FERRO, TRANSP bridge under railroad (*AmE*), bridge under railway (*BrE*); ~ **de barcas** *m* D&A, TRANSP, TRANSP MAR floating bridge, pontoon bridge; ~ **báscula** *m* CONST weigh bridge; ~ **basculante** *m* CONST balance bridge, swing bridge; ~ **de caballetes** *m* CONST trestle bridge; ~ **de cadena** *m* CONST chain bridge; ~-**canal** *m* CONST eaves trough; ~ **de capacidad** *m* ELEC *circuito* capacity bridge; ~ **de capacidades** *m* ELEC, ING ELÉC, INSTR capacitance bridge; ~ **capacímetro** *m* ELEC, ING ELÉC, INSTR capacitance bridge; ~ **de capacitancias** *m* ELEC, ING ELÉC, INSTR capacitance bridge; ~ **de carga** *m* TRANSP loading bridge; ~ **de carretera** *m* AUTO, CONST, TRANSP, VEH road bridge; ~ **de CC** *m* ING ELÉC DC bridge; ~ **de celosía** *m* CONST frame bridge, lattice bridge, truss bridge; ~ **de centrar** *m* PROD centering bridge (*AmE*), centring bridge (*BrE*); ~ **colgante** *m* CONST cable-stayed bridge, suspension bridge, D&A suspension bridge; ~ **completo** *m* ELEC *de instrumento* full bridge; ~ **de conductancia** *m* ELEC, FÍS, ING ELÉC conductance bridge; ~ **conector** *m* ELEC jumper; ~ **de conexión** *m* ELEC jumper, *en sistema* link, ING MECÁ link, TELECOM jumper; ~ **de conexión asincrónico** *m* ELEC *dos sistemas de CA* asynchronous link; ~ **de conexión a tierra** *m* ING ELÉC bonding jumper; ~ **de conferencia** *m* TELECOM conference bridge; ~ **de contrapeso** *m* CONST counterpoise bridge; ~ **corredizo** *m* CONST roller bridge, rolling bridge; ~ **de corriente alterna** *m* ELEC, FÍS alternating-current bridge; ~ **de cuerdas** *m* CONST bowstring bridge; ~ **de cursor** *m* ING ELÉC slide bridge; ~ **doble** *m* INSTR double bridge; ~ **de encuentro** *m* TELECOM meet-me bridge; ~ **de las escobillas** *m* ING ELÉC brush-rocker; ~ **de estribo** *m* CONST running bridge; ~ **de extensímetro** *m* ING ELÉC strain gage bridge (*AmE*), strain gauge bridge (*BrE*); ~ **ferroviario** *m* CONST railroad bridge (*AmE*), railway bridge (*BrE*), FERRO railroad bridge (*AmE*), railway bridge (*BrE*), shuttle train; ~ **flotante** *m* D&A, TRANSP MAR floating bridge; ~ **giratorio** *m* CONST drawbridge, pivot bridge, swivel bridge, turn bridge, TRANSP MAR *esclusas, hidroductos internos* swing bridge; ~ **de gobierno** *m* TRANSP MAR *buques* wheel house; ~-**grúa** *m* CONST bridge, bridge crane, MECÁ, PROD overhead crane, traveling crane (*AmE*), travelling crane (*BrE*); ~ **grúa para la cuchara de colada** *m* PROD ladle crane; ~-**grúa de taller** *m* PROD shop traveler (*AmE*), shop traveller (*BrE*); ~ **de hierro** *m* CONST iron bridge; ~ **de hilo y cursor** *m* ING ELÉC slide wire bridge; ~ **del indicador de tensión** *m* ING ELÉC strain gage bridge (*AmE*), strain gauge bridge (*BrE*); ~ **de inductancias** *m* ELEC inductance bridge; ~ **inferior** *m* CONST undergrade bridge; ~ **de Kelvin** *m* ELEC, FÍS, ING ELÉC, INSTR Kelvin bridge; ~ **Kelvin doble** *m* ELEC, INSTR double Kelvin bridge; ~ **levadizo** *m* CONST bascule bridge, counterpoise bridge, drawbridge, lifting bridge, balance bridge, TRANSP lift bridge, TRANSP MAR *esclusas, hidroductos internos* lifting bridge; ~ **lleno** *m* ELEC *de instrumento* full bridge; ~ **con lomo** *m* CONST hog-backed bridge; ~ **de madera** *m* CONST timber bridge; ~ **de mando** *m* TRANSP AÉR flight deck; ~ **de maniobras** *m* CONST turning bridge; ~ **de medida** *m* ELEC, INSTR bridge; ~ **de medida Anderson** *m* ELEC *circuito* Anderson bridge; ~ **de medida universal** *m* ELEC universal bridge; ~ **de medida de varias tomas** *m* ELEC, INSTR padded bridge; ~ **de medidas** *m* ING MECÁ measuring bridge; ~ **para medidas de capacidad** *m* ELEC *circuito* capacity bridge; ~ **de medidas con corriente alterna** *m* ELEC alternating-current bridge; ~ **para medidas de inductancia** *m* ELEC, ING ELÉC inductance bridge; ~ **para medir impedancias** *m* ELEC, ING ELÉC impedance bridge; ~ **militar portátil** *m* CONST Bailey bridge; ~ **Miller** *m* ELEC *de circuito* Miller bridge; ~ **móvil** *m* CONST drawbridge, TRANSP MAR *esclusas, hidroductos internos* moveable bridge; ~ **de navegación** *m* TRANSP MAR navigating bridge; ~ **de Nernst** *m* ING ELÉC Nernst bridge; ~ **oblicuo** *m* CONST oblique bridge, skew bridge; ~ **Owen** *m* ELEC *circuito* Owen bridge; ~ **de peaje** *m* CONST toll bridge; ~ **para peatones** *m* CONST, FERRO footbridge; ~ **pivotante** *m* CONST pivot bridge, turn bridge; ~ **de pontones** *m* D&A, TRANSP, TRANSP MAR floating bridge, pontoon bridge; ~ **provisional** *m* CONST temporary bridge; ~ **rectificador** *m* ING ELÉC rectifier bridge; ~ **de resonancia** *m* ELEC resonance bridge; ~ **rodante de carga** *m* TRANSP transtainer, transtainer crane; ~ **de Schering** *m* FÍS, ING ELÉC Schering bridge; ~ **de tablero inferior** *m* CONST bottom-road bridge, through bridge; ~ **de tablero superior** *m* CONST deck bridge, top-road bridge; ~ **térmico** *m* TERMO, TERMOTEC heat bridge; ~ **de termistores** *m* ING ELÉC thermistor bridge; ~ **Thomson** *m* ELEC, INSTR Kelvin bridge, Thomson bridge; ~ **Thomson doble** *m* ELEC, INSTR double Kelvin bridge, double Thomson bridge; ~ **de tramos iguales** *m* CONST bridge with equal bays; ~ **de transmisión** *m* TELECOM transmission bridge; ~ **trasero** *m* AUTO, MECÁ, VEH rear-axle shaft, *transmisión* back axle; ~ **trasero de doble reducción** *m* AUTO, VEH double-reduction rear axle; ~ **trasero flotante** *m* VEH *transmisión* floating axle; ~ **de unión** *m* PROD bonding jumper; ~ **universal** *m* ELEC universal bridge; ~ **de viga cajón** *m* CONST box-girder bridge; ~ **de vigas** *m* CONST girder bridge; ~ **con vigas en arco** *m* CONST arched-beam bridge; ~ **de vigas en voladizo** *m* CONST cantilever bridge; ~ **volante** *m* CONST, TRANSP MAR flying bridge; ~ **Wheatstone** *m* ELEC, FÍS, ING ELÉC Wheatstone bridge; ~ **de Wien** *m* FÍS Wien bridge; ~ **Wien** *m* ELECTRÓN Wien bridge

puentear *vt* ING ELÉC short

puentes: **con** ~ *adj* PROD bridged

puerta *f* AGUA, CONST gate, CONTAM MAR trawl board, ELECTRÓN gate, INFORM&PD gate, gating, ING ELÉC gate, OCEAN trawl board, PROD *de esclusas* gate, *para observar el comportamiento de los mecanismos* door, TRANSP MAR *de esclusa* sluicegate, trawl board, VEH *carrocería* door; ~ **abatible** *f* CONST collapsible gate; ~ **de acceso** *f* CONST manhole door, TRANSP AÉR access door; ~ **para acoplamiento** *f* TEC ESP docking port; ~ **de aguas arriba** *f* AGUA head gate; ~ **de aluminio** *f* ELECTRÓN aluminium gate (*BrE*), aluminum gate (*AmE*); ~ **amortiguadora de ruidos** *f* SEG *láminas de acero* noise abatement door; ~ **amortiguadora de sonido** *f* SEG sound-absorbent door; ~ **analógica** *f* ELECTRÓN analog gate; ~ **AND** *f* INFORM&PD AND gate; ~ **arrolladiza** *f* CONST rolling door; ~ **del arte de arrastre** *f* OCEAN otter board; ~ **autoalineada** *f* ELECTRÓN self-aligned gate; ~ **de balcón** *f* CONST French doors (*AmE*), French windows (*BrE*); ~ **de carga** *f* C&V filling point, TRANSP AÉR loading door; ~ **de carga trasera** *f* TRANSP tail-loading gate; ~ **de cierre automático** *f* CONST self-closing door; ~ **de coladero** *f* MINAS chute door, chute gate; ~ **del compartimento de cargo** *f* TRANSP AÉR cargo compartment door; ~ **contra incendios** *f* SEG, TERMO fire door; ~ **contrafuegos** *f* TERMO fire door; ~ **de control de amplitud** *f* ELECTRÓN amplitude gate; ~ **corredera** *f* CONST sliding door; ~ **corrediza** *f* CONST sliding door; ~ **corrediza del engranaje principal** *f* TRANSP AÉR main-gear sliding door; ~ **cortafuga de frío** *f* REFRIG port door; ~ **cristalera** *f* CONST French doors (*AmE*), French windows (*BrE*); ~ **de cristales** *f* CONST French doors (*AmE*), French windows (*BrE*); ~ **cuantificada** *f* TV quantized gate; ~ **delantera** *f* AUTO front door; ~ **de descarga** *f* MECÁ tailgate; ~ **de diodo** *f* ELECTRÓN diode gate; ~ **electrónica habilitada** *f* ELECTRÓN enabled gate; ~ **elevadiza** *f* AUTO lift gate; ~ **de emergencia** *f* CONST emergency door; ~ **empernada** *f* PROD plug door; ~ **de entrada** *f* ELECTRÓN input gate; ~ **de equivalencia** *f* INFORM&PD equivalence gate; ~ **de esclusa** *f* AGUA, HIDROL, TRANSP MAR *de canal* lock gate; ~ **O exclusiva** *f* INFORM&PD exclusive OR gate; ~ **de exploración** *f* TV scanning gate; ~ **falsa** *f* CONST blind door; ~ **flotante** *f* ELECTRÓN floating gate; ~ **de fogón** *f* PROD stoking door; ~ **giratoria** *f* CONST swing door; ~ **de guillotina trasera** *f* C&V *en un horno de vidrio* back tweel; ~ **de hogar** *f* PROD stoking door; ~ **del hogar** *f* PROD, SEG, TERMO fire door; ~ **del horno** *f* SEG, TERMO fire door; ~ **ignífuga** *f* CONST, SEG, TERMO fire-resistant door; ~ **O inclusiva** *f* INFORM&PD inclusive OR gate; ~ **insonorizada** *f* SEG soundproof door; ~ **insonorizadora** *f* SEG sound-insulated door; ~ **levadiza** *f* CONST lift gate, VEH *carrocería* rollup door; ~ **de limpieza** *f* PROD

cleaning door; ~ **lógica** *f* ELECTRÓN, FÍS, INFORM&PD logic gate; ~ **O lógica** *f* ELECTRÓN inclusive OR gate; ~ **lógica integrada** *f* ELECTRÓN integrated logic gate; ~ **lógica mezcladora** *f* ELECTRÓN inclusive OR gate; ~ **lógica óptica** *f* ELECTRÓN optical logic gate; ~ **sin modular** *f* CONST blank door; ~ **NAND** *f* ELECTRÓN, FÍS, INFORM&PD NAND gate; ~ **NI exclusiva** *f* INFORM&PD exclusive NOR gate; ~ **NO** *f* INFORM&PD NOR gate; ~ **de no equivalencia** *f* INFORM&PD non-equivalence gate; ~ **NOO** *f* INFORM&PD NOR gate; ~ **NO-O** *f* ELECTRÓN NOR gate; ~ **NOR** *f* FÍS, INFORM&PD NOR gate; ~ **NOR exclusiva** *f* INFORM&PD exclusive NOR gate; ~ **NOT** *f* ELECTRÓN, FÍS, INFORM&PD NOT gate; ~ **NO-Y de tres entradas** *f* ELECTRÓN three-input NAND gate; ~ **O** *f* ELECTRÓN, FÍS, INFORM&PD OR gate; ~ **OR** *f* ELECTRÓN, FÍS, INFORM&PD OR gate; ~ **OR exclusiva** *f* INFORM&PD exclusive OR gate; ~ **OR inclusiva** *f* INFORM&PD inclusive OR gate; ~ **pirorresistente** *f* CONST, SEG, TERMO fire-resistant door; ~ **principal** *f* *Esp* (*cf alcancía principal AmL*) MINAS, TERMO main chute; ~ **de protección** *f* SEG security door; ~ **de protección contra humos** *f* SEG smoke protection doors; ~ **de protección contra incendios** *f* CONST, SEG, TERMO fire-protection door; ~ **de red** *f* OCEAN otter board; ~ **de regulación** *f* CARBÓN regulator door, MINAS *ventilación* regulating door; ~ **reguladora interior de la ventilación** *f* MINAS gage door (*AmE*), gauge door (*BrE*); ~ **de salida** *f* TELECOM output port; ~ **de seguridad** *f* CONST safety door, SEG security door; ~ **de silicio** *f* ELECTRÓN silicon gate; ~ **de la torre** *f* C&V tower door; ~ **trasera** *f* AUTO lift gate, rear door, C&V end door, TRANSP tailgate, VEH rear door; ~ **de tres entradas** *f* ELECTRÓN three-input gate; ~ **de tres estados** *f* ELECTRÓN three-state gate; ~ **de la tripulación** *f* TRANSP AÉR crew door; ~ **de vaivén** *f* CONST swinging door; ~ **de ventilación** *f* CONST *minería* flap door, ventilating door, MINAS ventilation door; ~ **vidriada** *f* CONST glass door; ~ **de vidrio** *f* C&V *en una terraza* patio door, CONST glazed door; ~ **de visita** *f* MECÁ, PROD inspection door; ~ **Y** *f* ELECTRÓN, INFORM&PD AND gate

puertas: ~ **plegables** *f pl* CONST folding doors

puerto[1] *m* GEN port, TEC ESP *de mar* loading dock, TRANSP MAR harbor (*AmE*), harbour (*BrE*); ~ **de acceso** *m* TELECOM access port; ~ **de aerodeslizadores** *m* TRANSP MAR hoverport; ~ **en aguas profundas** *m* TRANSP MAR deep-water harbor (*AmE*), deep-water harbour (*BrE*); ~ **de amarre** *m* TRANSP MAR mooring berth; ~ **artificial** *m* TRANSP MAR artificial port; ~ **asíncrono** *m* TELECOM asynchronous port; ~ **base** *m* TRANSP MAR *de compañia, armada* home port; ~ **con base naval** *m* TRANSP MAR home port; ~ **comercial** *m* TRANSP MAR commercial port, trading port; ~ **de datos** *m* TELECOM data port; ~ **dedicado** *m* TELECOM dedicated port; ~ **deportivo** *m* TRANSP MAR marina; ~ **para discado público** *m* TELECOM public dial-up port; ~ **de embarque** *m* TRANSP MAR shipping port; ~ **de entrada** *m* C&V inlet port; ~ **de entrada/salida** *m* INFORM&PD input/output port; ~ **de entrega de mando** *m* TRANSP MAR *de buque, armada* port of commissioning; ~ **de escala** *m* TRANSP MAR port of call; ~ **de expedición** *m* TRANSP MAR shipping port; ~ **exterior** *m* OCEAN, TRANSP MAR outer harbor (*AmE*), outer harbour (*BrE*), outer port, outport; ~ **fluvial** *m* TRANSP MAR river port;

~ **franco** *m* TRANSP MAR free port; ~ **de gran calado** *m* TRANSP MAR deep-water harbor (*AmE*), deep-water harbour (*BrE*); ~ **de impresora** *m* IMPR printer port; ~ **de inscripción** *m* TRANSP MAR port of registration; ~ **interior** *m* TRANSP MAR inner harbor (*AmE*), inner harbour (*BrE*), inner port; ~ **de mar** *m* TRANSP MAR seaport; ~ **de marcación privado** *m* TELECOM private dial-up port; ~ **para marcación pública** *m* TELECOM public dial-up port; ~ **de marea** *m* TRANSP MAR tidal harbor (*AmE*), tidal harbour (*BrE*), tidal port; ~ **de matrícula** *m* TRANSP MAR port of documentation (*AmE*), port of registry (*BrE*), *de compañía, armada* home port; ~ **natural** *m* TRANSP MAR natural harbor (*AmE*), natural harbour (*BrE*); ~ **de observación** *m* TEC ESP viewing port; ~ **de paquetes** *m* TELECOM packet port; ~ **paralelo** *m* INFORM&PD parallel port; ~ **de registro** *m* TRANSP MAR port of registration; ~ **de salida** *m* C&V outlet port; ~ **síncrono** *m* TELECOM synchronous port; ~ **de terminal** *m* INFORM&PD terminal port; ~ **de transbordo rodado** *m* TRANSP, TRANSP MAR roll-on/roll-off port (*ro-ro port*); ~ **de tránsito** *m* TRANSP MAR port of transit

puerto² : **estar en un** ~ *vi* TRANSP MAR make a port

puesta: ~ **en aplicación** *f* INFORM&PD implementation; ~ **en bandera** *f* TRANSP AÉR *de hélice* feathering; ~ **en bandera automática** *f* TRANSP AÉR *de hélice* autofeathering; ~ **a cero** *f* INFORM&PD clear to zero, ING MECÁ resetting, NUCL zero adjustment, zero setting, PROD resetting; ~ **a cero automática** *f* ELEC control automatic reset; ~ **en cola** *f* TELECOM queueing; ~ **en cola de espera** *f* TELECOM queueing; ~ **en derivación** *f* ELEC *conexión* tapping; ~ **al día** *f* TELECOM updating; ~ **en explotación** *f* AGUA harnessing; ~ **en fase** *f* ELEC *corriente alterna*, ELECTRÓN, TRANSP, TV phasing; ~ **a flote** *f* TRANSP MAR *de barco* launching; ~ **a grueso** *f* ING MECÁ thicknessing; ~ **de huevos** *f* AGRIC egg batch; ~ **en marcha** *f* CONST *instalaciones* commissioning, ING MECÁ coming into gear, running-in, starting; ~ **en marcha del arrollamiento de piezas** *f* ELEC *motor* part-winding starting; ~ **en marcha en caliente** *f* TRANSP AÉR hot start; ~ **en marcha directa** *f* ELEC *motor* direct starting; ~ **en marcha mediante pulsador** *f* ELEC *motor* push-button starter; ~ **en marcha del sistema** *f* PROD system startup; ~ **en marcha por tensión parcial** *f* ELEC *motor* partial-voltage starting; ~ **a masa** *f* TEC ESP earthing (*BrE*), grounding (*AmE*); ~ **en nivel** *f* INFORM&PD retrofit; ~ **en obra** *f* PROD planting; ~ **en órbita** *f* TEC ESP injection orbit, orbital injection, orbiting; ~ **en página** *f* PROD *tipografía* setup; ~ **en paralelo** *f* ING ELÉC paralleling; ~ **a punto** *f* AUTO timing, CONST *instalaciones* commissioning, ING MECÁ setting up, VEH *encendido, juego válvulas y balancines* timing; ~ **a punto de cámara** *f* CINEMAT, TV camera line-up; ~ **a punto del encendido** *f* AUTO timing of the ignition; ~ **a punto de la mantilla** *f* IMPR blanket make-ready; ~ **a punto previa** *f* IMPR premake-ready; ~ **de quilla** *f* TRANSP MAR keel laying; ~ **en servicio** *f* MECÁ commissioning, TELECOM launching; ~ **en temperatura** *f* REFRIG cooling down; ~ **a tierra** *f* GEN earth (*BrE*), earthing (*BrE*), ground (*AmE*), grounding (*AmE*); ~ **a tierra de protección múltiple** *f* ELEC protective multiple earthing (*BrE*), protective multiple grounding (*AmE*); ~ **a tierra del punto medio** *f* ELEC midpoint earthing (*BrE*), midpoint grounding (*AmE*)

puesto¹: ~ **en cola** *adj* TELECOM queueing; ~ **en cola de espera** *adj* TELECOM queueing; ~ **a tierra** *adj* ELEC, ING ELÉC, TELECOM earthed (*BrE*), grounded (*AmE*)

puesto² *m* INFORM&PD station; ~ **de atraque** *m* TRANSP MAR berth; ~ **de atraque en una marina** *m* TRANSP MAR *construcción naval, cuerdas* slip; ~ **de control** *m* TRANSP AÉR console; ~ **de embarque en los botes salvavidas** *m* SEG, TRANSP MAR *a bordo* lifeboat station; ~ **de marcación** *m* TELECOM dial-up port; ~ **de operador** *m* TELECOM operating position, operator position; ~ **de operador de tráfico** *m* TELECOM traffic operator position; ~ **de operadora** *m* TELECOM operating position, operator position; ~ **de pilotaje** *m* VEH *habitáculo del coche* cockpit; ~ **de primeros auxilios** *m* SEG, TRANSP MAR first-aid post; ~ **principal** *m* INFORM&PD master station; ~ **de reunión** *m* TRANSP MAR *para emergencia a bordo* assembly area; ~ **de señalización** *m* FERRO *equipo inamovible* signal box (*BrE*), signal tower (*AmE*); ~ **de servicio** *m* TRANSP service station; ~ **de trabajo** *m* INFORM&PD, ING MECÁ work station; ~ **de trabajo aséptico** *m* REFRIG clean-work station; ~ **de trabajo gráfico** *m* INFORM&PD graphics work station; ~ **de trabajo de ingeniería** *m* INFORM&PD engineering work-station; ~ **de trabajo para operador de tráfico** *m* TELECOM traffic operator position

pugna *f* CARBÓN contest

pujamen *m* TRANSP MAR *apoyo* foot

pulegona *f* QUÍMICA pulegone

pulgada *f* METR inch; ~ **cuadrada** *f* METR square inch; ~ **cúbica** *f* METR cubic inch

pulgadas: ~ **por segundo** *f pl* (*PPS*) INFORM&PD inches per second (*IPS*)

pulgón *m* AGRIC flea beetle, *patología vegetal* aphid; ~ **afidio** *m* AGRIC wingless aphid; ~ **de la hoja del maíz** *m* AGRIC corn leaf aphid; ~ **verde** *m* AGRIC greenbug

pulido¹: ~ **con fuego** *adj* C&V fire-polished; **todo** ~ *adj* ING MECÁ bright all over;

pulido² *m* C&V burnishing, *con papel estaño* burnish, IMPR burnishing, ING MECÁ planishing, MECÁ burnishing, METAL polishing, P&C *operación* sanding, polishing, PROD, REVEST polishing; ~ **al ácido** *m* C&V acid polishing; ~ **para carros** *m* AmL (*cf pulido para coches Esp*) REVEST car polish; ~ **con chorro de arena** *m* C&V sandblast obscuring, sandblasting; ~ **para coches** *m* Esp (*cf pulido para carros AmL*) REVEST car polish; ~ **con corcho** *m* C&V cork polishing; ~ **con disco** *m* C&V disc polishing (*BrE*), disk polishing (*AmE*); ~ **a fuego** *m* C&V fire-polishing; ~ **en húmedo** *m* PROD wet polishing; ~ **húmedo** *m* C&V wet polishing; ~ **por laminación en frío** *m* ING MECÁ planishing; ~ **mecánico** *m* ING MECÁ, METAL mechanical polishing; ~ **mojado y seco** *m* C&V wet and dry polishing; ~ **químico** *m* METAL, QUÍMICA chemical polishing; ~ **del vidrio** *m* C&V, REVEST glass glazing

pulidor *m* C&V lap, *instrumento para pulir vidrio* polisher, CONST finisher, PROD *persona* polisher; ~ **continuo** *m* C&V continuous polisher; ~ **doble** *m* C&V twin polisher; ~ **-esmerilador continuo** *m* C&V continuous grinder and polisher; ~ **de felpa** *m* C&V felt polisher; ~ **de papel** *m* C&V paper polisher; ~ **de**

porcelana *m* C&V porcelain polisher; ~ **de tela** *m* C&V cloth polisher

pulidora *f* C&V burnisher, ING MECÁ *rueda de pulir* buffer, polishing wheel, surface sander, METAL *máquina* polishing wheel, PROD polishing machine, *persona* polisher, REVEST glazing mill; ~ **de acabado** *f* ING MECÁ finishing sander; ~ **de alta velocidad** *f* ING MECÁ high-speed grinding machine; ~ **cilíndrica de control numérico computarizado** *f* ING MECÁ CNC cylindrical grinder; ~ **circular** *f* CONST, ING MECÁ disc sander (*BrE*), disk sander (*AmE*); ~ **de correa de esmeril** *f* PROD emery-belt polishing machine; ~ **de disco** *f* CONST, ING MECÁ disc sander (*BrE*), disk sander (*AmE*); ~ **a discos** *f* CONST, ING MECÁ disc sander (*BrE*), disk sander (*AmE*); ~ **interna de producción de control numérico computerizado** *f* ING MECÁ CNC production internal grinding machine; ~ **de superficies de control electrónico** *f* CONST, ELECTRÓN, ING MECÁ electronically-controlled surface grinder; ~ **de superficies exteriores de control numérico computerizado** *f* ING MECÁ CNC surface grinder; ~ **de trapo** *f* PROD *para pulir* mop-end brush

pulimentación *f* PROD polishing

pulimentado *m* PROD polishing

pulimentar *vt* ING MECÁ grind, MECÁ polish, grind, lap, PROD buff, grind, lap, polish, *con tela de esmeril* rub down

pulimento *m* CONST *ingeniería civil* gritting, sanding, ING MECÁ surfacing, PROD buffing; ~ **brillante** *m* PROD brilliant polish; ~ **doble** *m* C&V twin polishing

pulir *vt* C&V polish, CARBÓN finish, polish, smooth, DETERG scour, ING MECÁ grind, true up, surface, MECÁ grind, lap, PROD buff, grind, lap, polish; ~ **hasta secar** *vt* C&V polish till dry

pulpa *f* AGUA, PAPEL, PROD pulp; ~ **de alto rendimiento** *f* PAPEL high-yield pulp; ~ **amarilla de paja** *f* PAPEL yellow straw pulp; ~ **blanqueada** *f* PAPEL bleached pulp; ~ **sin blanquear** *f* PAPEL unbleached pulp; ~ **al cloro-sosa** *f* PAPEL soda-chlorine pulp; ~ **de coníferas** *f* PAPEL softwood pulp; ~ **de cuero** *f* PAPEL leather pulp; ~ **para disolver** *f* PAPEL dissolving pulp; ~ **de esparto** *f* PAPEL esparto pulp; ~ **extrablanqueada** *f* PAPEL fully-bleached pulp; ~ **de filtración** *f* PROC QUÍ filtering pulp; ~ **filtrada** *f* PROC QUÍ filtered pulp; ~ **de frondosas** *f* PAPEL hardwood pulp; ~ **húmeda** *f* PAPEL wet pulp; ~ **kraft** *f* PAPEL kraft pulp; ~ **de madera** *f* RECICL paper pulp; ~ **mecánica** *f* PAPEL groundwood pulp; ~ **mecánica parda** *f* PAPEL brown mechanical pulp; ~ **poco refinada** *f* PAPEL fast pulp; ~ **química** *f* PAPEL chemical pulp; ~ **seca** *f* PAPEL dry pulp; ~ **semiblanqueada** *f* PAPEL semibleached pulp; ~ **semiquímica** *f* PAPEL semichemical pulp; ~ **a la sosa** *f* PAPEL soda pulp; ~ **al sulfato** *f* PAPEL sulfate pulp (*AmE*), sulphate pulp (*BrE*); ~ **al sulfito** *f* PAPEL sulfite pulp (*AmE*), sulphite pulp (*BrE*); ~ **al sulfito neutro** *f* PAPEL neutral sulfite pulp (*AmE*), neutral sulphite pulp (*BrE*); ~ **en suspensión** *f* PAPEL stuff; ~ **de trapos** *f* PAPEL rag pulp; ~ **de viscosa** *f* PROD viscose pulp

púlpito *m* TRANSP MAR *equipamiento de cubierta* pulpit; ~ **de popa** *m* TRANSP MAR *yates* pushpit, stern pulpit

pulsación *f* ELEC *fuente de alimentación de CA* angular frequency, ELECTRÓN beating, beat, ING ELÉC pulsa-tion, TRANSP AÉR beat; ~ **angular** *f* ACÚST angular pulsing

pulsador *m* ING ELÉC button, push-button, ING MECÁ plunger, push-button, PETROL button, QUÍMICA push-button, TELECOM press-button, push-button; ~ **para abrir semáforo** *m* TRANSP *control de tráfico* pedestrian push-button; ~ **de arranque** *m* VEH *del motor* starter button; ~ **de cierre momentáneo** *m* PROD momentary-close push-button; ~ **del conmutador** *m* ING ELÉC switch handle; ~ **de contacto** *m* ING ELÉC contact button; ~ **de contacto momentáneo** *m* PROD momentary-contact push-button; ~ **de contacto sostenido** *m* PROD maintained-contact push-button; ~ **de desconexión** *m* CINEMAT, FOTO, TV release button; ~ **de detención** *m* TV stop key; ~ **de grabación** *m* TV record button; ~ **del haz** *m* NUCL beam pulser; ~ **iluminado** *m* PROD illuminated push-button; ~ **de interrogación** *m* TELECOM polling key; ~ **de parada** *m* PROD stopbutton switch; ~ **de reinicio** *m* INFORM&PD reset button; ~ **de reposición** *m* CINEMAT reset knob, PROD reset push, TV reset knob; ~ **de semáforo para peatones** *m* TRANSP pedestrian push-button; ~ **de toma de líneas** *m* TELECOM line seizure button

pulsancia *f* FÍS pulsatance

pulsar¹ *vt* INFORM&PD *teclas* hit, PROD pulse on, *tecla, botón* press

pulsar²: ~ **para hablar** *fra* TELECOM push-to-talk

pulso *m* ELECTRÓN, FÍS, FÍS ONDAS, ÓPT, TV pulse; ~ **activador** *m* TV triggering pulse; ~ **de aire** *m* PETROL air pulse; ~ **aislado** *m* ELECTRÓN single pulse; ~ **de borrado** *m* TV blanking pulse; ~ **de corta duración** *m* ELECTRÓN narrow pulse; ~ **de desconexión** *m* TV gating pulse; ~ **diente de sierra** *m* TV serrated pulse; ~ **de disparo** *m* ING ELÉC firing pulse; ~ **de excitación** *m* TV driving pulse; ~ **de habilitación** *m* ELECTRÓN enable pulse; ~ **de imagen** *m* CINEMAT, TV frame pulse; ~ **inicial** *m* CINEMAT, TV head position pulse; ~ **de montaje** *m* CINEMAT, TV edit pulse; ~ **en picosegundos** *m* FÍS PART picosecond pulse; ~ **de reloj** *m* CINEMAT, TV sync pulse; ~ **de RF** *m* ELECTRÓN, TELECOM, TV RF pulse; ~ **de la señal en curso** *m* TELECOM course blip pulse; ~ **de la señal de video** *AmL*, ~ **de la señal de vídeo** *m Esp* TV video signal pulse; ~ **de sincronización** *m* TV synchronization pulse, timing pulse; ~ **de sincronización de la subportadora de color** *m* TV color burst (*AmE*), colour burst (*BrE*); ~ **sincronizador vertical** *m* TV vertical sync pulse; ~ **de subida** *m* ELECTRÓN fast-rise pulse; ~ **tacométrico** *m* TV tach pulse

pulsómetro *m* ING MECÁ vacuum pump

pulsorreactor *m* TRANSP aeropulse, pulsojet, TRANSP AÉR pulsating jet engine, pulse jet

pulsos: ~ **de borrado** *m pl AmL* (*cf pulsos de permanencia Esp*) TV unblanking pulses; ~ **codificados** *m pl* TV encoded pulses; ~ **de ecualización** *m pl* TV equalizing pulses; ~ **largos separados por pausas cortas** *m pl* ELECTRÓN boxcar pulse; ~ **de permanencia** *m pl Esp* (*cf pulsos de borrado AmL*) TV unblanking pulses

pulverización *f* CARBÓN grinding, pulverization, HIDROL spraying, NUCL atomization, P&C *operación* grinding, PROD grinding, spray, QUÍMICA pulverization, *de líquido* atomization; ~ **catódica** *f* ELEC, ING ELÉC cathode sputtering; ~ **por falla** *f* GEOL clay gouge, fault gouge; ~ **de salmuera** *f* REFRIG brine spray

pulverizado *adj* P&C *pigmentos, material de carga, operación* micronized, PROC QUÍ pulverized

pulverizador *m* AGRIC duster, AGUA, CONST pulverizer, EMB aerosol, atomizer, PROC QUÍ atomizer, pulverizer, pulverizing equipment, PROD spray, spray producer, QUÍMICA sprayer; ~ **de aire** *m* PAPEL air shower; ~ **de aire comprimido** *m* AGRIC compressed-air sprayer, EMB airbrush; ~ **de bolas** *m* AmL (*cf molino de bolas Esp*) LAB ball mill; ~ **de mochila** *m* AGRIC knapsack sprayer; ~ **de pintura** *m* ING MECÁ paint spray

pulverizar *vt* AGRIC dust, AGUA sprinkle, CARBÓN grind, pulverize, COLOR spray, CONST pulverize, PROC QUÍ, QUÍMICA pulverize, grind

pulverulento *adj* GEOL powdery, PROC QUÍ pulverulent

pulvimetal *m* METAL metal powder

pumita *f* MINERAL pumice-stone

puncionar *vt* CARBÓN tap

puño *m* ING MECÁ grip, handle, MECÁ handle, PROD *de herramientas* handle, hilt, TRANSP MAR *en las velas* clew; ~ **de amura** *m* TRANSP MAR *de vela* tack; ~ **de escota** *m* TRANSP MAR *en las velas* clew; ~ **de la pala** *m* TRANSP AÉR blade cuff; ~ **de palanca de aletas** *m* PROD wing lever knob; ~ **de pico** *m* TRANSP MAR *vela cangreja* peak

punta[1]: **en ~** *adj* ING MECÁ sharp; **sin ~** *adj* MECÁ, TEC PETR *perforación* dull; **de ~ redonda** *adj* ING MECÁ round-nosed

punta[2] *f* CARBÓN *lingotes* sharp end, slope top, tip, CONST brad, point, nail claw, FERRO *de la aguja de cambio* tip, ING ELÉC *filtros eléctricos* overshoot, ING MECÁ point, nose, prong (*BrE*), tine (*AmE*), PROD *cortada de una barra de acero* crop end, TRANSP *aviones* spike, TRANSP MAR point; ~ **acampanada** *f* C&V flared end; ~ **del ala** *f* TRANSP AÉR wing tip; ~ **para alambre de espino** *f* CONST barbed-wire nail; ~ **de arena** *f* OCEAN *en playa festoneada* beach cusp; ~ **del cable** *f* ELEC, ING ELÉC, MINAS *atado a la jaula de extracción* cable end; ~ **de carburo** *f* ING MECÁ carbide tip; ~ **de carga** *f* ELEC *fuente de alimentación*, ING ELÉC maximum demand, peak load; ~ **de caucho** *f* CINEMAT, FOTO, P&C rubber tip; ~ **de contacto** *f* ING MECÁ prod; ~ **de descarga terrestre** *f* GEOFÍS earth spike; ~ **descartable** *f* ING MECÁ throwaway tip; ~ **del destornillador** *f* ING MECÁ turnscrew bit; ~ **de destornillador de trinquete** *f* ING MECÁ spiral ratchet screwdriver end; ~ **de diamante** *f* CONST, ING MECÁ diamond point; ~ **del electrodo de la bujía** *f* AUTO, ELEC spark plug point; ~ **de espolón** *f* ING MECÁ pinch point; ~ **de flecha** *f* CONST, EMB arrowhead; ~ **giratoria** *f* ING MECÁ live center (*AmE*), live centre (*BrE*); ~ **de jalón** *f* CONST *topografía* cross-staff head; ~ **de lanza** *f* EMB arrowhead; ~ **de la pala** *f* TRANSP AÉR blade tip; ~ **para pasar la banda** *f* PAPEL tail end; ~ **para pasar la hoja** *f* PAPEL leader; ~ **de prueba** *f* ING MECÁ prod; ~ **de red de cerco** *f* OCEAN wing end; ~ **rómbica** *f* CONST, ING MECÁ diamond point; ~ **de trazar** *f* ELECTRÓN, ING MECÁ scriber

puntada *f* TEXTIL stitch; ~ **de corte** *f* TEXTIL hacking stitch; ~ **engarzada** *f* TEXTIL meshed stitch

puntal *m* Esp (*cf estemple AmL*) AUTO post, strut, CARBÓN prop, CONST brace, raking shore, shore, stanchion, standard, strut, *grúa* post, prop, *vigas* spur, EMB strut, GEOFÍS pillar, ING MECÁ prop, MECÁ strut, MINAS bracket, tree, punch prop, prop, TRANSP MAR *construcción naval* pillar, samson post, *geografía* shore, stanchion, *del agua* depth; ~ **de amarre** *m* CONTAM MAR, TRANSP MAR *fondeo de embarcaciones* mooring bracket; ~ **de caballetes** *m* CONST trestle shore; ~ **de carga** *m* TRANSP MAR cargo boom; ~ **de entrepuente** *m* TRANSP MAR *construcción naval* deck pillar; ~ **de francobordo** *m* TRANSP MAR *diseño naval* depth for freeboard; ~ **grueso** *m* CARBÓN *minas, túneles* strut; ~ **horizontal** *m* TRANSP AÉR horizontal strut; ~ **de la horquilla de desembrague** *m* AUTO, VEH throw-out fork strut; ~ **inclinado** *m* CONST raker; ~ **MacPherson** *m* AUTO, VEH MacPherson strut; ~ **de madera** *m* MINAS pit wood; ~ **máximo** *m* TRANSP MAR *construcción naval* extreme depth; ~ **de mina** *m* MINAS pit prop; ~ **oblicuo del tren de aterrizaje principal** *m* TRANSP AÉR main landing gear brace strut; ~ **de pantoque** *m* TRANSP MAR *construcción de barcos* bilge shore; ~ **de registro** *m* TRANSP MAR *diseño de barcos* registered depth; ~ **de seguridad** *m* Esp (*cf tentemozo AmL*) MINAS safety drag bar; ~ **separador del ademado** *m* MINAS studdle; ~ **de trazado** *m* TRANSP MAR *arquitectura naval* molded depth (*AmE*), moulded depth (*BrE*); ~ **vertical** *m* CONST vertical shore

puntales *m pl* CONST set of shores; ~ **de la escotilla de bajada** *m pl* TRANSP MAR *construcción naval* companionway posts

punteadora: ~ **barrenadora** *f* ING MECÁ jig boring, jig-boring machine

puntear *vt* FOTO dodge, TRANSP MAR *el rumbo* plot

punteo *m* TRANSP MAR *navegación* plotting; ~ **de radar** *m* D&A, TRANSP, TRANSP AÉR, TRANSP MAR radar plotting

puntería *f* CONST, D&A sighting, TEC ESP pointing

puntero *m* CONST point, ELEC, IMPR, INFORM&PD, TELECOM pointer

punterola *f* Esp (*cf barra puntiaguda AmL*) MINAS *entibación* gad

puntiagudo *adj* ING MECÁ sharp

puntilla *f* TEXTIL lace

punto[1]: **de ~ menor** *adj* TRANSP MAR *carta de navegación* small-scale; ~ **a punto** *adj* ELEC, INFORM&PD, ING ELÉC, PROD, TRANSP point-to-point; **en ~** *adj* TV in-point;

punto[2]: ~ **por punto** *adv* IMPR dot-to-dot

punto[3] *m* ELECTRÓN dot, FÍS *de contacto* point, FOTO stop, GEOM point, IMPR point, dot, INFORM&PD pixel, dot, ING MECÁ center (*AmE*), centre (*BrE*), TEXTIL dot, stitch, TRANSP MAR point, fix, TV dot, VEH *encendido* point; ~ **a** ~ **de ablandamiento Littleton** *m* C&V Littleton softening point; ~ **acanalado** *m* TEXTIL ribbed stitch; ~ **de acceso** *m* INFORM&PD port; ~ **de acceso al mantenimiento de la red** *m* TELECOM network service access point (*NSAP*); ~ **de acceso al servicio** *m* TELECOM service access point (*SAP*); ~ **de acceso al servicio de transmisión** *m* TELECOM transport service access point (*TSAP*); ~ **de ajuste** *m* QUÍMICA set point; ~ **de alineación** *m* ELECTRÓN tie-down point; ~ **de apoyo** *m* CONST bearing, trig point, ELEC *de eje* bearing, FÍS fulcrum, ING MECÁ point of support, bearing point, MECÁ bearing; ~ **de aproximación** *m* TRANSP AÉR approach point; ~ **de aproximación final** *m* TRANSP AÉR final approach point; ~ **de aproximación inicial** *m* TRANSP AÉR initial approach point; **~ de aproximación máxima** *m* TRANSP MAR *navegación por radar* closest point of approach; **~ armónico** *m* GEOM harmonic point; **~ de arranque** *m* TV start mark; **~ de atrás** *m* ING MECÁ *telares* back center (*AmE*), back centre (*BrE*); **~ auxiliar de conmutación** *m* TELECOM auxiliary switching point; **~ azeotrópico** *m* REFRIG azeotropic point;

~b **~ de bifurcación** *m* ELECTRÓN *en el circuito*, INFORM&PD branch point;

~c **~ caliente** *m* C&V hot spot; **~ de cambio** *m* CONST turning point, TRANSP battery exchange point; **~ de carga** *m* ELEC charging station, INFORM&PD load point, ING ELÉC charging station, TRANSP charging point; **~ de carga de baterías** *m* TRANSP battery loading point; **~ de cebado** *m* ELECTRÓN *de circuito* singing point; **~ cenital** *m* TEC ESP zenith point; **~ central** *m* ING MECÁ center point (*AmE*), centre point (*BrE*); **~ de coincidencia** *m* GEOM point of concurrence; **~ de combustión** *m* REFRIG fire point; **~ de concurrencia** *m* GEOM point of concurrence; **~ de condensación** *m* FÍS FLUID drop point, PETROL, TERMOTEC dew point; **~ de congelación** *m* ALIMENT, C&V, EMB, FÍS freezing point, FÍS FLUID congealing point, ING MECÁ, METAL freezing point, P&C freezing point, melting point, PROC QUÍ, REFRIG, TERMO, TRANSP MAR freezing point; **~ de conmutación** *m* TELECOM switching point; **~ de cono Morse** *m* ING MECÁ Morse taper center (*AmE*), Morse taper centre (*BrE*); **~ de contacto** *m* ING MECÁ contact point, TEC ESP *aterrizaje* touchdown point, TRANSP AÉR contact point; **~ de contacto de los círculos primitivos** *m* ING MECÁ pitch point; **~ de contacto eléctrico** *m* TELECOM electrically-held crosspoint; **~ de contacto electrónico** *m* TELECOM electronic crosspoint; **~ de control** *m* INFORM&PD checkpoint; **~ de control de derrota** *m* TRANSP MAR *navegación* way point; **~ de corte** *m* CINEMAT cutting point, GEOM *topología* cut point; **~ de corte del flujo** *m* INSTAL HIDRÁUL cutoff point; **~ de crecimiento** *m* AGRIC growing point; **~ de cristalización** *m* CRISTAL, PROC QUÍ crystallization point; **~ crítico** *m* FÍS critical point; **~ de cruce** *m* TV crossover; **~ de cruce bifilar** *m* TELECOM two-wire crosspoint; **~ de cruce de conmutación óptica** *m* TELECOM optical-switching crosspoint; **~ de cruce de dos conductores** *m* TELECOM two-wire crosspoint; **~ de cruce de dos hilos** *m* TELECOM two-wire crosspoint; **~ de cruce de lengüeta rem** *m* TELECOM remreed crosspoint; **~ de cruce metálico** *m* TELECOM metallic crosspoint; **~ de cruce optoelectrónico** *m* TELECOM optoelectronic crosspoint; **~ de cruce RCS** *m* TELECOM SCR crosspoint; **~ de cruce del rectificador controlado por silicio** *m* TELECOM silicon-controlled rectifier crosspoint; **~ de cruce del relé de láminas** *m* ELEC, ING ELÉC, TELECOM reed-relay crosspoint; **~ de cruce de semiconductores** *m* TELECOM semiconductor crosspoint; **~ de Curie** *m* ELEC, FÍS, PETROL, REFRIG, TEC PETR Curie point;

~d **~ débil** *m* IMPR soft dot; **~ decimal** *m* INFORM&PD, MATEMÁT decimal point; **~ de deformación** *m* C&V deformation point; **~ de descongelación** *m* QUÍMICA thawing point; **~ de deshielo** *m* QUÍMICA thawing point; **~ deslizante** *m* TV flying spot; **~ de distribución de cables** *m* ELEC,

ING ELÉC cable distribution point; **~ de división** *m* ING MECÁ division point;

~e **~ de ebullición** *m* ALIMENT, FÍS, P&C, TERMO boiling-point; **~ de emisión** *m* CONTAM emission point; **~ de encendido** *m* MECÁ, TEC PETR ignition point; **~ de entrada** *m* INFORM&PD entry point, TV in-edit; **~ de enturbamiento** *m* REFRIG cloud point; **~ de equilibrio** *m* METAL saddle point, TRANSP AÉR balance point; **~ de equilibrio de un resorte** *m* FÍS spring balance; **~ de equivalencia** *m* QUÍMICA *de una reacción* end point; **~ de escarcha** *m* REFRIG frost point; **~ de escogimiento** *m* CALIDAD sampling point; **~ de espera** *m* TRANSP AÉR holding point; **~ de espera de rodaje** *m* TRANSP AÉR taxi holding position; **~ espigado** *m* ING MECÁ herringbone; **~ estacionario** *m* TEC ESP stationary point; **~ de estancamiento** *m* TRANSP AÉR stagnation point; **~ estático** *m* FÍS stagnation point; **~ eutéctico** *m* METAL, QUÍMICA eutectic point; **~ de evaporación** *m* INSTAL TERM, PROC QUÍ, QUÍMICA evaporating point; **~ de existencias de llegada** *m* PROD inbound stock point; **~ de exploración** *m* TV scanning dot, scanning spot; **~ de extracción del agua** *m* TEXTIL dewatering point;

~f **~ fijo** *m* CONST fixed point, ING MECÁ live center (*AmE*), live centre (*BrE*), dead spindle, MECÁ dead center (*AmE*), dead centre (*BrE*), PROD *tornos* fast headstock, TELECOM fixed point; **~ final** *m* IMPR full point, full stop; **~ final de la conexión de canal virtual** *m* TELECOM virtual channel connection endpoint (*VCCE*); **~ final de la conexión de la vía de transmisión virtual** *m* TELECOM virtual path connection endpoint (*VPCE*); **~ final de la vía de transmisión** *m* TELECOM transmission path endpoint (*TPE*); **~ de floculación** *m* PROC QUÍ flocculation point; **~ de fluencia** *m* IMPR *tintas* flow point, TEC PETR flow-line temperature; **~ de fluidez** *m* FÍS FLUID, TEC PETR, TERMO pour point; **~ de flujo** *m* FÍS FLUID flow point; **~ focal** *m* CINEMAT focal point, ELECTRÓN focal spot, FOTO, TV focal point; **~ fosforescente de fondo** *m* TV phosphor-dot faceplate; **~ de fricción** *m* ING MECÁ friction point; **~ de fuga** *m* GEOM vanishing point; **~ de funcionamiento** *m* ING ELÉC operating point; **~ de fusión** *m* FÍS, P&C, QUÍMICA melting point, REFRIG *grasa* drop point, melting point, TERMO, TEXTIL melting point;

~g **~ de gota** *m* METEO melting point; **~ de goteo** *m* ALIMENT dew point, REFRIG pour point;

~i **~ de ignición** *m* AUTO ignition point, TRANSP MAR *del combustible* flash-point; **~ de la imagen** *m* IMPR picture dot; **~ de impacto del haz** *m* TV beam impact point; **~ de inflamabilidad en cubeta cerrada** *m* P&C *prueba* closed-cup flash point; **~ de inflamación** *m* ALIMENT, AUTO, FÍS, P&C, QUÍMICA, REFRIG, TEC PETR flash point, TERMO flash point, kindling point, TERMOTEC flash point; **~ de inflexión** *m* ING MECÁ hinging; **~ de inicio** *m* MECÁ breaking up; **~ de intercepción** *m* TEC ESP intercept point; **~ intermedio de aproximación** *m* TRANSP AÉR intermediate approach point; **~ de interrupción** *m* INFORM&PD, PROD breakpoint; **~ de interrupción de contacto** *m* AUTO, VEH contact breaker-point (*BrE*), points (*AmE*); **~ de intersección** *m* CONST *topografía* intersection point, GEOM point of intersection, TEC ESP fix; **~ de intersección con la órbita**

de un astro *m* TEC ESP astro fix; **~ de introducción** *m* INFORM&PD entry point;

~ k **~ kilométrico** *m* FERRO milepost;

~ l **~ lambda** *m* FÍS lambda point; **~ de licuefacción** *m* TEC PETR pour point; **~ límite de retorno** *m* TRANSP AÉR point of no return; **~ de llenado del molde** *m* C&V baffle mark; **~ luminoso** *m* ELECTRÓN light spot;

~ m **~ más alto de pleamar** *m* OCEAN higher high water; **~ de máxima corriente** *m* ING ELÉC current antinode; **~ máximo de la carrera** *m* ING MECÁ top of stroke; **~ máximo de la curva del esfuerzo del aleteo** *m* TRANSP AÉR flapping stress peak; **~ máximo de la curva del esfuerzo del batimiento** *m* TRANSP AÉR flapping stress peak; **~ de medición de ruido lateral** *m* TRANSP AÉR lateral-noise measurement point; **~ de medida del ruido de aproximación** *m* TRANSP AÉR approach-noise measurement point; **~ de medida del ruido del paso de vuelo** *m* TRANSP AÉR flyover-noise measurement point; **~ de medida del ruido de referencia de aproximación** *m* TRANSP AÉR approach reference noise measurement point; **~ medio del recorrido** *m* ING MECÁ midtravel; **~ de mediodía** *m* TRANSP MAR *navegación celeste* noon sight; **~ del mínimo** *m* METAL saddle point; **~ de mira** *m* D&A foresight, front sight, MINAS *armas* foresight; **~ muerto** *m* INFORM&PD deadlock, MECÁ dead center (*AmE*), dead centre (*BrE*), VEH *caja de cambio* neutral, *motor, pistón* dead centre (*BrE*), dead center (*AmE*); **~ muerto delantero** *m* ING MECÁ crank-end dead center (*AmE*), crank-end dead centre (*BrE*); **~ muerto inferior** *m* (*PMI*) AUTO, MECÁ, VEH bottom dead center (*AmE*) (*BDC*), bottom dead centre (*BrE*) (*BDC*); **~ muerto superior** *m* (*PMS*) AUTO, MECÁ, VEH top dead center (*AmE*) (*TDC*), top dead centre (*BrE*) (*TDC*); **~ muerto trasero** *m* ING MECÁ crank-end dead center (*AmE*), crank-end dead centre (*BrE*); **~ de muestreo** *m* CALIDAD, CARBÓN sampling point;

~ n **~ de Néel** *m* FÍS, REFRIG Néel point; **~ negro** *m* C&V black speck; **~ negro de adorno** *m* IMPR bullet; **~ neutro** *m* FÍS neutral point, MECÁ neutral; **~ de niebla** *m* PETROL dew point; **~ de nivelación** *m* CONST leveling point (*AmE*), levelling point (*BrE*); **~ de no retorno** *m* TRANSP AÉR point of no return;

~ o **~ observado** *m* OCEAN observed position; **~ de obstrucción del filtro frío** *m* PROC QUÍ filtering limit; **~ de osculación** *m* GEOM point of osculation;

~ p **~ de parada** *m* CALIDAD hold point; **~ paramagnético de Curie** *m* FÍS RAD paramagnetic Curie point; **~ de paso** *m* FERRO passing point; **~ de paso sin personal** *m* FERRO unmanned passing point; **~ de pedido** *m* PROD *gestión de inventario* order point; **~ peligroso** *m* SEG danger point; **~ posterior** *m* ING MECÁ *telares* back center (*AmE*), back centre (*BrE*); **~ de producción** *m* FÍS yield point; **~ de profundidad** *m* GEOL, PETROL, TEC PETR depth point; **~ de profundidad común** *m* GEOL, PETROL, TEC PETR common depth point; **~ de prueba** *m* TELECOM test point;

~ q **~ de quiebre** *m* TEXTIL break-even point;

~ r **~ radiotransmisor** *m* TRANSP point-source radio transmitter; **~ de reanudación** *m* INFORM&PD restart point; **~ de reblandecimiento** *m* P&C, REFRIG softening point; **~ de recalada** *m* TRANSP MAR *navegación* landfall; **~ de referencia** *m* ACÚST refer-ence point, CONST datum level, datum point, mark point, GEOM reference point, QUÍMICA set point, TRANSP AÉR landmark, TV cue; **~ de referencia interna** *m* TELECOM internal reference point (*IRP*); **~ de referencia de la sedimentación** *m* PROC QUÍ settlement reference marker; **~ de rehacer pedidos** *m* PROD reorder point; **~ de reinicio** *m* INFORM&PD restart point; **~ de reja** *m* AGRIC share point; **~ de remanso** *m* FÍS FLUID stagnation point; **~ reticular** *m* METAL lattice point; **~ de reunión** *m* TRANSP assembly point; **~ de rocío** *m* FÍS, METEO, NUCL, PETROL, REFRIG, TERMO, TERMOTEC, TRANSP AÉR dew point; **~ de rocío equivalente** *m* REFRIG apparatus dew point; **~ de rotura** *m* CONST breaker point; **~ de ruptura** *m* CARBÓN breakthrough point;

~ s **~ de salida** *m* INFORM&PD exit point; **~ de saturación** *m* TEC ESP saturation point, TERMOTEC *temperatura* dew point; **~ de señalización** *m* TELECOM signaling point (*AmE*), signalling point (*BrE*) (*SP*); **~ de separación** *m* ING MECÁ division point; **~ de sobrevuelo** *m* TEC ESP *nave espacial* fly-by point; **~ de solidificación** *m* REFRIG solidification point; **~ solsticial** *m* TEC ESP solstitial point; **~ de suavizado** *m* TEXTIL softening point; **~ del subsatélite** *m* TEC ESP subsatellite point; **~ de suelta** *m* INSTAL HIDRÁUL release point;

~ t **~ de tangencia** *m* GEOM point of tangency; **~ de tarado** *m* NUCL limit setting; **~ de terminación del biselado** *m* CONST chamfer stop; **~ de tiempos iguales** *m* TRANSP AÉR equal-time point; **~ tipográfico** *m* IMPR typographic point; **~ de toma** *m* ING ELÉC tapping point; **~ del torno** *m* ING MECÁ lathe center (*AmE*), lathe centre (*BrE*); **~ de trabajo** *m* PROD work station; **~ de la trama** *m* IMPR halftone dot; **~ de transferencia de la señalización** *m* TELECOM signaling transfer point (*AmE*) (*STP*), signalling transfer point (*BrE*) (*STP*); **~ de transformación** *m* C&V transformation point; **~ de transición** *m* NUCL transition point; **~ transparente** *m* FOTO *de un negativo* pinhole; **~ de tres dientes** *m* ING MECÁ fork center (*AmE*), fork centre (*BrE*), prong center (*AmE*), prong centre (*BrE*); **~ de triangulación** *m* CONST *topografía* triangulation point; **~ triple** *m* FÍS, METAL, TERMO triple point; **~ de turbidez** *m* DETERG cloud point; **~ de turbulencia** *m* FÍS FLUID turbulent spot;

~ u **~ de unión** *m* ING ELÉC junction point;

~ v **~ de vaporización** *m* ALIMENT, AUTO, FÍS, P&C, QUÍMICA, REFRIG, TEC PETR, TERMO, TERMOTEC flash point; **~ de venta** *m* (*PV*) INFORM&PD point of sale (*POS*); **~ de venta electrónico** *m* INFORM&PD electronic point of sale (*EPOS*); **~ de verificación** *m* TEC ESP check point; **~ vernal** *m* TEC ESP vernal point; **~ de vista de cámara** *m* CINEMAT, TV camera viewpoint; **~ de vista técnico** *m* TELECOM technical viewpoint

punto[4]: estar a ~ de empezar el bocarteo *vi* MINAS be about to start crushing

puntos: ~ cardinales *m pl* FÍS, TRANSP MAR *brújula* cardinal points; **~ circulares en infinito** *m pl* GEOM circular points at infinity; **~ conductores** *m pl* IMPR dot leaders (*BrE*), leader dots (*AmE*); **~ conjugados** *m pl* FÍS conjugate points; **~ de enlace** *m pl* FERRO junction points; **~ faltantes** *m pl* IMPR, PAPEL missing dots; **~ negros** *m pl* AmL (cf *suciedad Esp, motas de corteza Esp*) PAPEL *pasta* bark specks; **~ nodales** *m pl*

FÍS nodal points; ~ **principales** *m pl* FÍS principal points; ~ **de referencia en la cinta transportadora** *m pl* C&V belt marks; ~ **suspensivos** *m pl* IMPR ellipsis

puntuación: ~ **media de opinión** *f* Esp (*cf media de opinión AmL*) TELECOM mean opinion score

puntual *adj* FÍS point

punturas *f pl* IMPR pins

punzado: ~ **de la cañería** *m* AmL (*cf perforación por tubos Esp*) TEC PETR casing perforation

punzón *m* Esp (*cf lanzasondas AmL*) CONST broach, nail punch, punch, *carpintería* scriber, ING MECÁ piercer, punch, stabbing awl, MECÁ punch, MINAS pricker, PROD *moldes* piercer; ~ **autocentrador** *m* ING MECÁ bell-centering punch (*AmE*), bell-centring punch (*BrE*); ~ **cilíndrico** *m* ING MECÁ key drift, drift; ~ **de clavo** *m* CONST pin punch; ~ **cónico cuadrado** *m* ING MECÁ square drift; ~ **cuadrado** *m* MECÁ drift; ~ **doble** *m* ING MECÁ duplex punching bear; ~ **embutidor** *m* ING MECÁ drawing punch; ~ **de estampar** *m* ING MECÁ top swage; ~ **expulsador** *m* ING MECÁ drift; ~ **eyector de perforación** *m* ING MECÁ perforating ejector-punch; ~ **de garantía** *m* PROD inspection stamp, *de ensayador* hallmark, mark; ~ **de garantía del fabricante** *m* PROD maker's mark; ~ **de garganta** *m* PROD groove punch; ~ **para grabar en disco virgen** *m* ACÚST cutting stylus; ~ **de guata** *m* ING MECÁ wad punch; ~ **de guía** *m* ING MECÁ, MECÁ guide pin; ~ **de mano** *m* ING MECÁ die, PROD handpunch; ~ **de marcar** *m* CONST scratch awl, ING MECÁ, MECÁ center punch (*AmE*), centre punch (*BrE*), prick punch, PROD marking awl; ~ **de mecánico** *m* ING MECÁ prick punch; ~ **de ojo** *m* PROD eyed punch; ~ **de palanca doble** *m* ING MECÁ duplex lever punch; ~ **de perforar** *m* ING MECÁ, MECÁ center punch (*AmE*), centre punch (*BrE*); ~ **de punto** *m* PROD finder point punch; ~ **redondo de cierre esférico** *m* ING MECÁ ball-lock round-punch; ~ **revólver** *m* ING MECÁ revolving head punch; ~ **sacabocados de precisión** *m* ING MECÁ fine blanking-die; ~ **de trazar** *m* ING MECÁ scriber

punzonado *m* ING MECÁ, PROD punching, TEXTIL needling

punzonadora *f* ING MECÁ, PROD punching machine; ~ **copiadora** *f* ING MECÁ copy punch press; ~ **doble** *f* ING MECÁ duplex punching bear; ~ **múltiple** *f* PROD multiple punching machine; ~ **de palanca** *f* PROD lever punching machine; ~ **a tornillo** *f* ING MECÁ screw punching bear; ~ **de tornillo** *f* ING MECÁ screw bear

punzonamiento *m* PROD *marcado* stamping

punzonar *vt* CONST, ING MECÁ, MECÁ, PROD *troqueles* punch

pupila: ~ **de entrada** *f* FÍS entrance pupil; ~ **de salida** *f* FÍS exit pupil

pupinización *f* ING ELÉC coil loading

pupitre: ~ **de control** *m* ELEC console; ~ **de mando** *m* ELEC console, control console, INSTR control desk; ~ **del maquinista** *m* FERRO *vehículos* driving desk

PUR *abr* (*poliuretano*) CONST, P&C, QUÍMICA, TEXTIL PUR (*polyurethane*)

pureza *f* CARBÓN, GAS purity; ~ **del aire** *f* CONTAM air purity; ~ **de polarización** *f* TEC ESP polarization purity; ~ **radiactiva** *f* FÍS RAD radioactive purity; ~ **radionucléidica** *f* FÍS RAD radioactive purity

purga[1]: **de** ~ **automática** *adj* TRANSP MAR self-draining

purga[2] *f* AGUA blowdown, CARBÓN blow off, draining, INFORM&PD purging, ING MECÁ bleeding, drainage, MECÁ flushing, NUCL bleeding, PROD drain, TEC PETR bleeding, TERMOTEC *de calderas* blow down, TRANSP MAR drain, VEH *frenos* bleeding; ~ **de aire** *f* ING MECÁ air valve, MECÁ vent, NUCL air drain; ~ **de aire del compresor** *f* ING MECÁ compressor air bleed; ~ **de batería** *f* TEC ESP battery drain; ~ **de gas** *f* REFRIG gas purging

purgado *m* TEC PETR purging; ~ **del gas** *m* REFRIG gas purging

purgador *m* FERRO *vehículos* drain valve, water strainer, ING MECÁ clearer, bleed plug, PROD *tuberías de vapor o aire comprimido* strainer; ~ **de aceite** *m* REFRIG oil drain; ~ **de agua de flotador** *m* INSTAL HIDRÁUL *máquina de vapor* float trap; ~ **de aire** *m* EMB air purger, ING MECÁ air valve, REFRIG gas purger; ~ **de arena y grava** *m* HIDROL sand and gravel trap; ~ **de expansión** *m* INSTAL HIDRÁUL expansion trap; ~ **de gases no condensables** *m* REFRIG gas purger; ~ **del retorno del líquido** *m* REFRIG dump trap liquid return; ~ **del vapor** *m* INSTAL HIDRÁUL steam trap

purgar *vt* AGUA blow down, blow off, CARBÓN drain, *gases* vent, ING MECÁ bleed, MECÁ bleed, drain, PETROL bleed down, bleed off, PROD drain, TERMOTEC *calderas* blow down

purificación *f* GEN purification, AGUA cleansing; ~ **adicional** *f* CONTAM supplementary purification; ~ **del agua** *f* AGUA, HIDROL, PROC QUÍ water purification; ~ **biológica** *f* AGUA, HIDROL *aguas residuales* biological purification; ~ **completa** *f* NUCL complete purification; ~ **por fusión de zonas** *f* METAL zone refining; ~ **química** *f* AGUA, HIDROL *aguas residuales*, PROC QUÍ chemical purification; ~ **suplementaria** *f* CONTAM supplementary purification; ~ **por zonas** *f* METAL zone refining

purificador *m* PROC QUÍ, QUÍMICA purifier, TEC ESP scrubber; ~ **del agua** *m* TEXTIL water purifier; ~ **de agua por intercambio iónico** *m* LAB ion-exchange water purifier; ~ **de aire** *m* CARBÓN, ING MECÁ, VEH *carburador* air cleaner; ~ **de vapor** *m* INSTAL HIDRÁUL steam separator

purificar *vt* AGUA cleanse, purify, CARBÓN clean, GAS, HIDROL, PROC QUÍ purify, QUÍMICA purify, bleach, RECICL, TEC PETR, TERMO purify

purina *f* QUÍMICA purine

puro *adj* CARBÓN clean, CONTAM MAR neat, PROD *oro, plata* solid, TEXTIL sheer

purpurar *vt* COLOR purple

purpurina *f* COLOR, QUÍMICA purpurin

purpuroxantina *f* QUÍMICA purpuroxanthin

pústula: ~ **bacteriana** *f* AGRIC bacterial pustule

putrefacción *f* RECICL digestion; ~ **aeróbica** *f* RECICL aerobic digestion; ~ **anaeróbica** *f* RECICL anaerobic digestion

putrefacto *adj* ALIMENT putrescent, putrid

puzolana *f* PETROL puzolan

PV *abr* (*punto de venta*) INFORM&PD POS (*point of sale*)

PVC[1] *abr* CONST, ELEC, EMB (*cloruro de polivinilo*) PVC (*polyvinyl chloride*), INFORM&PD (*circuito virtual permanente*, PVC (*permanent virtual circuit*), ING ELÉC, P&C (*cloruro de polivinilo*) PVC (*polyvinyl* *cholride*), TELECOM (*circuito virtual permanente*) PVC (*permanent virtual circuit*)

PVC^2: **~ no plastificado** *m* P&C unplasticized PVC; **~ rígido** *m* EMB rigid PVC

PVDC *abr* (*cloruro de polivinilideno*) P&C PVDC (*polyvinylidene chloride*)

PVP *abr* (*polivinilpirrolidona*) DETERG PVP (*polyvinyl-pyrrolidone*)

PVT *abr* (*proteína vegetal texturizada*) ALIMENT TVP (*textured vegetable protein*)

PWM *abr* ELECTRÓN, TEC ESP (*modulación de la anchura del impulso*) PWM (*pulse-width modulation*)

p-xileno *m* QUÍMICA p-xylene

pyrgom *m* MINERAL pyrgom

Q

QL *abr* (*lenguaje de consulta*) INFORM&PD QL (*query language*)

QSG *abr* (*galaxia cuasiestelar*) TEC ESP QSG (*quasi-stellar galaxy*)

QSO *abr* (*cuerpo cuasiestelar*) TEC ESP QSO (*quasi-stellar object*)

QSS *abr* (*radiofuente cuasiestelar*) TEC ESP QSS (*quasi-stellar radio source*)

QTOL *abr* (*avión de despegue y aterrizaje silencioso*) TRANSPAÉR QTOL aircraft (*quiet take-off and landing aircraft*)

quántum *m* GEN quantum

quark *m* FÍS, FÍS PART t-quark, quark, NUCL quark; **~ abajo** *m* FÍS, FÍS PART d-quark, down quark; **~ arriba** *m* FÍS, FÍS PART up quark; **~ azul** *m* FÍS, FÍS PART blue quark; **~ belleza** *m* (*b-quark*) FÍS, FÍS PART beauty quark (*b-quark*); **~ encanto** *m* FÍS, FÍS PART c-quark, charm quark; **~ extrañeza** *m* FÍS, FÍS PART s-quark; **~ rojo** *m* FÍS, FÍS PART red quark; **~ top** *m* FÍS, FÍS PART top quark; **~ up** *m* FÍS, FÍS PART up quark; **~ verdad** *m* FÍS, FÍS PART top quark, truth quark; **~ verde** *m* FÍS, FÍS PART green quark

quebradizo *adj* C&V, CRISTAL, GEOL, MECÁ, METAL, P&C brittle

quebrado: **~ complejo** *m* MATEMÁT complex fraction; **~ compuesto** *m* MATEMÁT complex fraction; **~ común** *m* MATEMÁT simple fraction

quebradora *f* PROC QUÍ breaker; **~ de quijadas** *f* LAB jaw crusher

quebrantado *adj* TRANSP MAR *buque* broken-backed, hogged

quebrantadora *f* CARBÓN coal cracker, ING MECÁ breaker, crusher, PROD *máquinas trituradoras* mill, TELECOM crusher; **~ de carbón** *f* CARBÓN, MINAS coal crusher; **~ de conos** *f* CARBÓN cone crusher; **~ de hielo** *f* ING MECÁ crushing machine for ice; **~ de mandíbulas** *f* CARBÓN, PROD jaw crusher; **~ de martillos** *f* CARBÓN hammer crusher, ING MECÁ impact breaker, impact crusher; **~ de mineral** *f* MINAS ore crusher; **~ de mortero** *f* PROD mortar mill

quebrantarrocas *m* CONST rock breaker

quebrantarse *v refl* TRANSP MAR *buque* hog

quebrarse *v refl* CONST break

queche *m* TRANSP MAR ketch

quehacer *m* MINAS labor (*AmE*), labour (*BrE*)

quelación *f* DETERG, HIDROL, NUCL, QUÍMICA chelation, chelating

quelatador *m* DETERG, HIDROL, NUCL, QUÍMICA chelating agent

quelatar *vt* DETERG, HIDROL, NUCL, QUÍMICA chelate

quelato *m* DETERG, HIDROL, NUCL, QUÍMICA chelate

quemado[1]: **de ~ escaso** *adj* TERMO lean-burn; **de ~ pobre** *adj* TERMO lean-burn

quemado[2] *m* CINEMAT burn, NUCL burn-up, TEC ESP burnout; **~ por antorcha** *m* TEC PETR *refino* flaring; **~ fácil** *m* C&V easy fire; **~ final** *m* NUCL ultimate burn-up; **~ medio del núcleo** *m* NUCL core average burn-up; **~ neutral** *m* NUCL neutral burn-out; **~ óptimo** *m* NUCL optimum burn-up; **~ por resonancia** *m* TERMO resonant burning

quemador *m* GAS, LAB, TELECOM, TERMO, TERMOTEC burner; **~ de aceite** *m* AGRIC oil burner; **~ de aire caliente** *m* AGRIC hot-air heater; **~ de anillo** *m* CONST ring burner; **~ atmosférico** *m* TERMO atmospheric burner; **~ atomizante** *m* TERMO atomizing burner; **~ automático de aceite** *m* TERMO, TERMOTEC automatic oil burner; **~ auxiliar** *m* VEH *motor* afterburner; **~ de azulejos** *m* C&V tile burner; **~ de chorro de aire** *m* ING MECÁ air-blast burner; **~ de corona** *m* GAS gas ring; **~ de corriente inducida** *m* ING MECÁ induced-draft burner (*AmE*), induced-draught burner (*BrE*); **~ de cortezas** *m* PAPEL bark burner; **~ de fueloil** *m* TERMO oil burner; **~ de gas** *m* GAS, PROD, TERMO, TERMOTEC gas burner; **~ de gas a baja presión** *m* TERMO, TERMOTEC low-pressure gas burner; **~ con ignición** *m* TERMO, TERMOTEC ignition burner; **~ infrarrojo** *m* GAS infrared burner; **~ interno** *m* C&V internal burner; **~ invertido** *m* PROD inverted burner; **~ de llama corta** *m* TERMOTEC short-flame burner; **~ nebulizante** *m* TERMO vaporizing burner; **~ de petróleo** *m* ING MECÁ oil burner; **~ principal** *m* TERMOTEC main burner; **~ de pulverización** *m* ING MECÁ pulverization burner, TERMO, TERMOTEC atomizing burner; **~ pulverizador de aceite** *m* TERMO, TERMOTEC atomizing oil burner; **~ pulverizante** *m* TERMO atomizing burner; **~ recto** *m* PROD erect burner, upright burner; **~ de tiraje forzado** *m AmL* (*cf quemador de tiro forzado Esp*) GAS, ING MECÁ forced-air furnace, forced-draft burner (*AmE*), forced-draught burner (*BrE*); **~ de tiro por aspiración** *m* GAS, ING MECÁ forced-air furnace, induced-draft burner (*AmE*), induced-draught burner (*BrE*); **~ de tiro forzado** *m Esp* (*cf quemador de tiraje forzado AmL*) GAS, ING MECÁ forced-air furnace, forced-draft burner (*AmE*), forced-draught burner (*BrE*); **~ trasero** *m* ING MECÁ afterburner; **~ con varias entradas y una salida** *m* TRANSP duplex burner; **~ con varias salidas** *m* CONST rose burner; **~ de ventilador** *m* ING MECÁ fan-powered burner

quemadura *f* SEG burn, TERMO burning; **~ por congelación** *f* REFRIG freezer burn; **~ de hoja** *f* AGRIC leaf scorch; **~ de primer grado** *f* SEG first-degree burn; **~ química** *f* QUÍMICA, SEG chemical burn; **~ por radiación** *f* FÍS RAD radiation burn

quemar *vt* MINAS combust, TERMO burn

quemarse *v refl* TERMO burn; **~ al rojo** *v refl* QUÍMICA *sin llamas* glow red

querargirita *f* MINERAL kerargyrite

queratina *f* QUÍMICA *tejidos epidérmicos* keratin

queratinización *f* QUÍMICA keratinization

queratinoso *adj* QUÍMICA keratinous

queratogenoso *adj* QUÍMICA keratogenous

quercetina *f* QUÍMICA quercetin

quercita *f* QUÍMICA quercite

quercitánico *adj* QUÍMICA quercitannic

quercitina *f* QUÍMICA quercitin

quermes: ~ mineral *m* MINERAL kermes mineral
quermesita *f* MINERAL kermesite
querógeno *m* PETROL, QUÍMICA kerogen
queroseno *m* PETROL, QUÍMICA, TEC PETR, TERMO, TRANSP, TRANSP AÉR kerosene
queso *m* TEXTIL *hilado* cheese; ~ blanco *m* ALIMENT fromage frais; ~ curado *m* ALIMENT curd cheese; ~ fresco *m* ALIMENT fromage frais
quetazina *f* QUÍMICA ketazine
quiastolita *f* MINERAL chiastolite
quicial *m* CONST hinge post, hinging post
quiebra *f* CARBÓN, MINAS break
quiescente *adj* INFORM&PD quiescing
quieto *adj* INFORM&PD quiescent
quijada *f* ING MECÁ, TRANSP MAR *de motón* cheek, jaw
quijo *m* GEOL *montera de oxidación* gossan
quilate *m* Esp (*cf carate AmL*) METR carat, carat fine; ~ métrico *Esp m* METR metric carat
quildrenita *f* MINERAL childrenite
quilla *f* TRANSP AÉR, TRANSP MAR keel; ~ de balance *f* TRANSP MAR bilge keel; ~ de lastre *f* TRANSP MAR *yates de vela* ballast keel; ~ a nivel *f* TRANSP MAR level keel; ~ retráctil *f* TRANSP MAR drop keel; ~ vertical *f* TRANSP MAR center girder (*AmE*), centre girder (*BrE*)
química *f* QUÍMICA chemistry; ~ de los alimentos *f* ALIMENT food chemistry; ~ aplicada *f* QUÍMICA applied chemistry; ~ atmosférica *f* CONTAM, QUÍMICA atmospheric chemistry; ~ inorgánica *f* QUÍMICA inorganic chemistry; ~ de minas *f* MINERAL, QUÍMICA mineral chemistry; ~ orgánica *f* QUÍMICA organic chemistry; ~ de las radiaciones *f* FÍS RAD, NUCL, QUÍMICA radiation chemistry; ~ teórica *f* QUÍMICA theoretical chemistry; ~ de tintes *f* COLOR teintochemistry; ~ de la tintura *f* COLOR teintochemistry; ~ de los volátiles *f* NUCL all-volatile chemistry (*AVT*)
químico: ~ intermedio *m* TEC PETR *refino* intermediate chemical
quimicoanalítico *adj* QUÍMICA chemicoanalytical
quimicofísico *adj* FÍS, QUÍMICA chemicophysical
quimicometalurgia *f* METAL, QUÍMICA chemicometallurgy
quimicometalúrgico *adj* METAL, QUÍMICA chemicometallurgical
quimicomineralógico *adj* MINERAL, QUÍMICA chemicomineralogical
químicos: ~ para el revelado en color *m pl* CINEMAT, FOTO, QUÍMICA color-processing chemicals (*AmE*), colour-processing chemicals (*BrE*)
quimioluminiscencia *f* FÍS, FÍS RAD chemiluminescence
quimioterapia *f* QUÍMICA chemotherapy
quina *f* QUÍMICA cinchona
quinacrina *f* QUÍMICA quinacrine

quináldico *adj* QUÍMICA quinaldic
quinaldina *f* QUÍMICA quinaldine
quinaldínico *adj* QUÍMICA quinaldinic
quinalizarina *f* QUÍMICA quinalizarin
quinamina *f* QUÍMICA quinamine
quincallería *f* MECÁ hardware
quincita *f* MINERAL quincite
quingombó *m AmL* (*cf kimbombó Esp*) AGRIC okra
quinhidrona *f* QUÍMICA quinhydrone
quinicina *f* QUÍMICA quinicine
quínico *adj* QUÍMICA quinic
quinidina *f* QUÍMICA quinidine
quinina *f* QUÍMICA quinine
quinínico *adj* QUÍMICA quininic
quinita *f* QUÍMICA quinite
quinitol *m* QUÍMICA quinitol
quinoa *f* AGRIC chenopodium album, lamb's quarter, QUÍMICA quinoa
quinoide *adj* QUÍMICA quinoid
quinol *m* QUÍMICA quinol
quinolina *f* CARBÓN, QUÍMICA quinoline
quinolínico *adj* CARBÓN, QUÍMICA quinolinic
quinona *f* QUÍMICA quinone
quinonoide *adj* QUÍMICA quinonoid
quinovina *f* QUÍMICA quinovin
quinoxalina *f* QUÍMICA quinoxalin, quinoxaline
quinta *f* ACÚST *música* fifth; ~ generación *f* INFORM&PD fifth generation; ~ perfecta *f* ACÚST perfect fifth; ~ rueda *f* VEH fifth wheel
quintal *m* CARBÓN coal basket, METR hundredweight (*BrE*), quintal, *peso de metal* centner; ~ métrico *m* METR metric centner, metric quintal
quinteto *m* FÍS *espectroscopia* quintet
quiolita *f* MINERAL chiolite
quiral *adj* QUÍMICA *estereoquímica* chiral
quiste *m* AGRIC cyst
quitafusibles *m* PROD fuse puller
quitamiedos *m* CONST guardrail
quitanieves *m* AGRIC snow plough (*BrE*), snow plow (*AmE*)
quitapiedras *m* FERRO *locomotora* cow catcher, guard iron, pilot
quitapinturas *m* COLOR paint remover
quitapolvo *m* FOTO duster
quitar[1] *vt* PROC QUÍ *lavar* wash away, SEG *polvo* remove, TRANSP MAR *velas* haul down
quitar[2]: ~ andar *vi* TRANSP MAR check; ~ la brida *vi* ING MECÁ unclamp; ~ los fangos *vi* CARBÓN, METAL, MINERAL deslime; ~ el filo *vi* ING MECÁ take the edge off, take the wire-edge off; ~ los pernos *vi* NUCL unbolt; ~ la rebaba *vi* ING MECÁ take the burr off; ~ el tapón *vi* PROD draw the plug
quitina *f* QUÍMICA chitin
quitosamina *f* QUÍMICA chitosamine

R

R *abr* (*número de Reynolds, renguenio, roentgen*) FÍS, FÍS RAD, **R** (*Reynolds number, roentgen*)

Ra *abr* (*radio*) FÍS RAD, QUÍMICA **Ra** (*radium*)

rabera *f* ING MECÁ tang

rabia *f* AGRIC rabies; **~ bovina** *f* AGRIC bovine rabies

rabitita *f* NUCL rabbittite

rabo *m* ING MECÁ tang

racemato *m* QUÍMICA racemate

racémico *adj* QUÍMICA racemic

racemización *f* QUÍMICA racemization

racemizado *adj* QUÍMICA racemizing

racemizar *vt* QUÍMICA racemize

racha *f* METEO, TEC ESP, TRANSP AÉR, TRANSP MAR *de viento* gust

racimo *m* AGRIC cluster

ración: ~ alimenticia recomendada *f* AGRIC recommended dietary allowance; **~ balanceada** *f* AGRIC balanced ration; **~ reforzada** *f* AGRIC enriched ration; **~ de terminación** *f* AGRIC hot ration

racleta *f* IMPR wiping

racón *m* TRANSP MAR racon

racor *m* VEH connector; **~ de carga** *m* REFRIG charging connection; **~ para tubería de aire** *m* ING MECÁ air connection

rada *f* OCEAN, TRANSP MAR *geomorfología litoral* roads, roadstead; **~ abierta** *f* OCEAN, TRANSP MAR open roadstead

radar *m* GEN radar; **~ acústico** *m* FÍS ONDAS sound ranging; **~ aéreo de exploración lateral** *m* (*RAEL*) CONTAM MAR, TRANSP AÉR sideways-looking airborne radar (*SLAR*); **~ aerotransportado** *m* TELECOM airborne radar; **~ de alerta previa** *m* TRANSP MAR early-warning radar; **~ con capacidad de recalada** *m* TRANSP MAR homing radar; **~ central** *m* D&A station radar; **~ de cita** *m* TEC ESP rendezvous radar; **~ de corto alcance** *m* D&A short-range radar; **~ de exploración** *m* TEC ESP tracking radar; **~ de exploración aérea** *m* TRANSP MAR air search radar; **~ de exploración en superficie** *m* TRANSP MAR surface search radar; **~ de gran alcance** *m* D&A long-range radar; **~ de impulsos sincronizados** *m* FÍS, FÍS RAD coherent-pulse radar, radar with coherent pulse, TRANSP, TRANSP AÉR, TRANSP MAR radar with coherent pulse; **~ de largo alcance** *m* D&A long-range radar; **~ náutico** *m* TRANSP MAR navigation radar; **~ de onda continua** *m* TRANSP MAR continuous-wave radar; **~ de ondas ultracortas** *m* D&A ultrashort wave radar; **~ con panel de retención de imagen** *m* TRANSP MAR radar imager; **~ de perseguimiento** *m* D&A pursuit radar; **~ por radiación óptica** *m* METEO, TEC ESP light detection and ranging; **~ de rastreo** *m* TEC ESP tracking radar; **~ de recalada** *m* TRANSP MAR homing radar; **~ de seguimiento** *m* TEC ESP tracking radar; **~ táctico** *m* TELECOM tactical radar (*TR*); **~ para tráfico** *m* AUTO, FÍS RAD, TELECOM, TRANSP, VEH road traffic radar; **~ ultrasónico** *m* FÍS RAD ultrasonic radar; **~ de vigilancia** *m* D&A surveillance radar

radarista *m* D&A, TRANSP, TRANSP AÉR, TRANSP MAR radar operator

radiación *f* GEN irradiation, radiation, ray, TERMO, TV radiance; **~ actínica** *f* FÍS RAD actinic radiation; **~ alfa** *f* ELEC alpha radiation, FÍS alpha ray, FÍS RAD alpha radiation; **~ de alta energía** *f* FÍS RAD high-energy radiation; **~ de aniquilación** *f* FÍS RAD annihilation radiation; **~ antigua** *f* GEOL, TEC ESP relict radiation; **~ atmosférica** *f* METEO atmospheric radiation; **~ atómica** *f* NUCL atomic radiation; **~ de baja energía** *f* FÍS RAD low-energy radiation; **~ beta** *f* ELEC beta radiation, FÍS, FÍS PART beta ray, FÍS RAD beta radiation; **~ del big bang** *f* FÍS, TEC ESP *cosmogénesis* big bang radiation; **~ blanca** *f* FÍS RAD, TEC ESP *rayos X* white radiation; **~ blanda** *f* FÍS RAD soft radiation; **~ de bremsstrahlung** *f* FÍS, FÍS RAD bremsstrahlung; **~ de Cerenkov** *f* FÍS PART, FÍS RAD Cerenkov radiation; **~ ciclotrónica** *f* FÍS, FÍS RAD cyclotron radiation; **~ coherente** *f* ÓPT, TELECOM coherent radiation; **~ cósmica** *f* FÍS, FÍS PART, FÍS RAD, TEC ESP cosmic radiation; **~ cósmica de fondo** *f* FÍS cosmic background radiation, cosmic ray background, FÍS PART cosmic background radiation, FÍS RAD cosmic ray background, cosmic background radiation, TEC ESP cosmic background radiation, cosmic ray background; **~ de cuerpo negro** *f* FÍS, FÍS RAD black-body radiation; **~ difusa** *f* ENERG RENOV, TEC ESP diffuse radiation; **~ directa** *f* ENERG RENOV direct radiation; **~ dispersada** *f* FÍS RAD scattered radiation; **~ dura** *f* FÍS RAD hard radiation; **~ electromagnética** *f* ELEC, ELECTRÓN, FÍS, FÍS ONDAS, FÍS RAD, ING ELÉC, METEO, ÓPT, PROD, TELECOM electromagnetic radiation; **~ de electrones** *f* TEC ESP electron irradiation; **~ electrónica** *f* ELECTRÓN electron ray; **~ emitida** *f* FÍS RAD emitted radiation; **~ de fondo** *f* FÍS background radiation, big bang radiation, FÍS RAD, NUCL background radiation, TEC ESP big bang radiation; **~ de fondo de microondas** *f* FÍS, FÍS RAD microwave background radiation; **~ fotónica de frenado electromagnético** *f* FÍS, FÍS RAD bremsstrahlung; **~ por fugas** *f* ING ELÉC leakage radiation; **~ gamma** *f* ELEC, FÍS, FÍS ONDAS, FÍS PART, FÍS RAD gamma radiation; **~ gamma instantánea** *f* FÍS RAD instant gamma radiation; **~ homogénea** *f* FÍS homogeneous radiation; **~ incoherente** *f* FÍS, TELECOM incoherent radiation; **~ infrarroja** *f* ENERG RENOV, FÍS, FÍS ONDAS, FÍS RAD, TEC ESP infrared radiation; **~ de ionización** *f* FÍS ONDAS ionizing radiation; **~ ionizante** *f* CONTAM, ELEC, FÍS, FÍS ONDAS ionizing radiation; **~ IR** *f* FÍS RAD IR radiation; **~ láser** *f* ELECTRÓN, FÍS RAD laser radiation; **~ mixta** *f* FÍS RAD mixed radiation; **~ monocromática** *f* FÍS RAD, ÓPT, TELECOM monochromatic radiation; **~ no ionizante** *f* FÍS RAD nonionizing radiation; **~ nuclear** *f* CONTAM, FÍS RAD, NUCL, SEG nuclear radiation; **~ óptica** *f* ÓPT, TELECOM optical radiation; **~ parásita** *f* TEC ESP *comunicaciones* stray radiation, TV parasitic radiation; **~ perdida** *f* FÍS RAD stray radiation; **~ de**

plasma *f* GAS plasma radiation; ~ **primigenia** *f* TEC ESP *cosmología* fossil radiation; ~ **primitiva** *f* GEOL relict radiation, TEC ESP fossil radiation, relict radiation; ~ **residual** *f* GEOL relict radiation, TEC ESP relict radiation, *cosmología* fossil radiation; ~ **resonante** *f* ELECTRÓN, FÍS RAD resonance radiation; ~ **secundaria** *f* FÍS RAD secondary radiation; ~ **sincrotrón** *f* FÍS RAD synchrotron radiation; ~ **solar** *f* CONTAM, ENERG RENOV, FÍS RAD, GEOFÍS solar radiation; ~ **térmica** *f* EMB heat radiation, FÍS, NUCL thermal radiation, TERMO heat radiation, thermal radiation; ~ **terrestre** *f* GEOFÍS earth radiation; ~ **ultravioleta** *f* CONTAM, ENERG RENOV, FÍS, FÍS RAD, GEOFÍS, ÓPT, TEC ESP ultraviolet radiation; ~ **ultravioleta cercana** *f* FÍS RAD near ultraviolet; ~ **ultravioleta profunda** *f* FÍS RAD, ING ELÉC deep ultraviolet radiation; ~ **visible** *f* ÓPT, TEC ESP, TELECOM visible radiation; ~ **X** *f* FÍS RAD X-ray radiation; ~ **X de fondo** *f* FÍS RAD X-ray background radiation

radiactinio *m* FÍS RAD, QUÍMICA, TRANSP MAR radioactinium

radiactividad *f* GEN radioactivity; ~ **ambiental** *f* FÍS RAD ambient radioactivity, environmental radioactivity; ~ **artificial** *f* FÍS, FÍS RAD artificial radioactivity; ~ **de fondo** *f* FÍS RAD ambient radioactivity, environmental radioactivity; ~ **inducida** *f* FÍS PART induced radioactivity; ~ **natural** *f* FÍS, FÍS RAD natural radioactivity; ~ **residual** *f* FÍS RAD residual radioactivity

radiactivo *adj* CONTAM radioactive, FÍS radioactive, secular, FÍS PART, FÍS RAD radioactive, GEOFÍS secular, NUCL, QUÍMICA, SEG, TEC PETR radioactive

radiador *m* AUTO radiator, MECÁ heater, radiator, TERMO *para calentamiento* radiator, TERMOTEC *intercambiador de calor* cooler, VEH *sistema de refrigeración* radiator; ~ **del aceite** *m* ELEC *transformador* oil cooler; ~ **de aletas** *m* AUTO finned radiator; ~ **de almacenamiento calorífico** *m* ING MECÁ storage heater; ~ **celular de bandas** *m* AUTO ribbon cellular radiator; ~ **celular de cintas** *m* AUTO ribbon cellular radiator; ~ **de cuerpo negro** *m* FÍS, FÍS RAD black-body radiator; ~ **derecho** *m* AUTO upright radiator; ~ **eléctrico** *m* ELEC *calefacción*, INSTAL TERM electric heater; ~ **de elementos** *m* INSTAL TERM ribbed radiator; ~ **de flujo cruzado** *m* AUTO, VEH crossflow radiator; ~ **de gas** *m* GAS, INSTAL TERM, TERMO gas fire; ~ **de Lambert** *m* ÓPT, TELECOM lambertian radiator; ~ **lambertiano** *m* ÓPT, TELECOM lambertian radiator; ~ **de panal** *m* AUTO honeycomb radiator; ~ **de Planck** *m* CINEMAT, FÍS, FÍS RAD, IMPR, TEC ESP, TV black body; ~ **plano** *m* INSTAL TERM flat radiator; ~ **de salida** *m* TERMOTEC aftercooler; ~ **total** *m* FÍS RAD total radiator; ~ **de tubos y aletas** *m* AUTO tube and fin radiator; ~ **de tubos con pestañas** *m* AUTO flanged tube radiator; ~ **tubular** *m* AUTO tubular radiator; ~ **vertical** *m* AUTO upright radiator

radial *adj* GEN radial

radián *m* ELEC *ángulo*, ELECTRÓN, FÍS, GEOM radian

radiancia *f* GEN radiance; ~ **espectral** *f* TELECOM spectral radiance

radiante *adj* ELEC radiative, FÍS radiant, FÍS RAD radiant, radiating, radiative, INSTAL TERM radiant, ÓPT radiant, radiating, QUÍMICA glowing, TEC ESP radiant, radiative, TELECOM radiant, radiating,

TERMO glowing, radiant, radiative, TERMOTEC radiant, TV radiant, radiating

radiar[1] *vt* GEN radiate, FÍS ONDAS, TELECOM broadcast

radiar[2] *vti* TERMO *calor* glow

radiativo *adj* ELEC, FÍS RAD, TEC ESP, TERMO radiative

radical *m* GAS, MATEMÁT, P&C, QUÍMICA radical; ~ **ácido** *m* QUÍMICA acid radical; ~ **libre** *m* ALIMENT, P&C, QUÍMICA free radical

radícula *f* AGRIC radicle

radiestesia *f* GEOFÍS, ING ELÉC geophysical prospecting

radio[1] *m* (*Ra*) CONST, FÍS radius, FÍS RAD radium (*Ra*), GEOM radius, INSTR *ruedas* arm, ÓPT radius, QUÍMICA radium (*Ra*); ~ **de acción** *m* D&A operating range, range, TEC ESP *aviones*, TRANSP AÉR range; ~ **de acción diario** *m* TRANSP AÉR day range; ~ **de acción nocturno** *m* TRANSP AÉR night range; ~ **de acción de operaciones** *m* TRANSP AÉR operating range; ~ **atómico** *m* NUCL, QUÍMICA atomic radius; ~ **axial** *m* ÓPT axial radius; ~ **de Bohr** *m* FÍS Bohr radius; ~ **del cabeceo** *m* TRANSP AÉR pitch radius; ~ **clásico** *m* FÍS, GEOM classical radius; ~ **de la curva vertical** *m* CONST vertical curve radius; ~ **de curvatura** *m* GEOM radius of curvature, MECÁ, PROD bending radius; ~ **de giro** *m* CONST, ING MECÁ radius of gyration, VEH steering circle; ~ **horizontal** *m* CONST horizontal radius; ~ **iónico** *m* CRISTAL, FÍS RAD ionic radius; ~ **medular** *m* AGRIC pith ray; ~ **de mezcla** *m* ING MECÁ blending radius; ~ **de operaciones** *m* TRANSP AÉR operating range; ~ **de oscilación** *m* PROD swing radius; ~ **de la pala** *m* TRANSP AÉR blade radius; ~ **del paso** *m* TRANSP AÉR pitch radius; ~ **de plegado** *m* PROD bend radius; ~ **primitivo** *m* ING MECÁ pitch radius; ~ **del rotor** *m* TRANSP AÉR rotor radius; ~ **de Schwarzschild** *m* FÍS Schwarzschild radius; ~ **táctico de acción** *m* D&A tactical radius of action; ~~**transmisor de carretera** *m* AUTO, FÍS RAD, TELECOM, TRANSP, VEH roadside radio transmitter; ~~**transmisor de tráfico** *m* TRANSP traffic radio transmitter; ~ **de Van der Waals** *m* FÍS Van der Waals radius

radio[2]: ~ **celular** *f* INFORM&PD cellular radio; ~ **de mochila** *f* D&A backpack radio, manpack radio; ~ **móvil celular** *f* TELECOM mobile radio cell; ~ **del paquete** *f* TELECOM packet radio; ~ **portátil** *f* CONST hand-held mobile radio; ~ **sonoboya** *f* TRANSP MAR radio sonobuoy

radioactividad *f* GEN radioactivity; ~ **de ambiente** *f* AGRIC radioactivity in the environment

radioactivo *adj* GEN radioactive

radioaficionado *m* TELECOM radio amateur

radioaltímetro *m* TRANSP AÉR radio altimeter

radioanálisis *m* FÍS RAD radioanalysis

radioastrometría *f* FÍS, TEC ESP radio astrometry

radioastronomía *f* FÍS, TEC ESP radio astronomy

radioaviso: ~ **de niebla** *m* METEO, TRANSP MAR *meteorología* fog warning

radiobaliza *f* D&A radio beacon, FERRO, TEC ESP beacon, TRANSP, TRANSP AÉR, TRANSP MAR radio beacon; ~ **de abanico** *f* TRANSP AÉR, TRANSP MAR fan marker beacon; ~ **aerotransportada** *f* D&A airborne marker balloon; ~ **automática** *f* TRANSP marker; ~ **de demarcación** *f* TRANSP AÉR marker beacon; ~ **de emergencia** *f* TEC ESP *accidentes de vuelo*, TRANSP AÉR emergency beacon; ~ **exterior** *f* TRANSP AÉR outer marker; ~ **de haz de abanico** *f* TRANSP

AÉR, TRANSP MAR fan marker beacon; ~ **de identificación** *f* TRANSP AÉR identification beacon; ~ **intermedia** *f* TRANSP AÉR middle marker; ~ **de localización de emergencia** *f* TRANSP AÉR emergency-location beacon; ~ **de localización de náufragos** *f* TRANSP MAR *navegación* personal-location beacon, *rescate por satélite* personal-locator beacon; ~ **de localización de siniestros** *f* (*RLS*) TRANSP MAR *para buques* emergency position-indicating radio beacon (*EPIRB*), *por satélite* distress beacon; ~ **marcadora de transmisión constante** *f* D&A, TRANSP, TRANSP AÉR, TRANSP MAR radar marker beacon; ~ **marcadora de transmisión secuencial** *f* D&A, TRANSP, TRANSP AÉR, TRANSP MAR radar marker beacon

radiobalizada: ~ **de peligro** *f* TRANSP AÉR hazard beacon

radiobrújula *f* TRANSP AÉR, TRANSP MAR radio compass

radiocarbono *m* FÍS, FÍS RAD, GEOFÍS radiocarbon

radiocesio *m* FÍS RAD radiocaesium (*BrE*), radiocesium (*AmE*)

radiocobalto *m* FÍS RAD, QUÍMICA radiocobalt

radiocompás *m* D&A homing device, TRANSP AÉR, TRANSP MAR *navegación* radio compass

radiocomunicación *f* TELECOM radiocommunication, TRANSP MAR radio sonobuoy

radioconducción: ~ **desde el punto de destino** *f* TELECOM, TRANSP radio homing

radiocromatografía *f* FÍS RAD, QUÍMICA radiochromatography

radiodemora *f* TELECOM, TRANSP AÉR, TRANSP MAR radio bearing

radiodiagnóstico *m* FÍS RAD radiodiagnosis

radiodifusión *f* TEC ESP *telecomunicaciones* broadcasting, TELECOM broadcast, radio broadcasting, broadcasting, bearer channel (*BC*); ~ **directa via satélite** *f* TELECOM direct broadcasting by satellite (*DBS*); ~ **en el exterior** *f* *AmL* (*cf radioemisión en el exterior Esp*) TV outside broadcast; ~ **por hilo** *f* TV wired broadcasting; ~ **local** *f* TRANSP area broadcasting; ~ **de paquetes** *f* TELECOM packet broadcasting; ~ **por paquetes** *f* TELECOM packet broadcasting; ~ **remota** *f* TV remote broadcast; ~ **por satélite directa** *f* TELECOM direct satellite broadcasting; ~ **simultánea** *f* TV simulcast broadcasting; ~ **sonora** *f* TELECOM sound broadcasting; ~ **telemandada** *f* TV remote broadcast; ~ **televisiva** *f* TELECOM television broadcasting; ~ **por UHF** *f* TV UHF broadcasting

radiodifusor *m* TEC ESP broadcasting

radioeco: ~ **meteórico** *m* TEC ESP meteor echo

radioemisión: ~ **en el exterior** *f* *Esp* (*cf radiodifusión en el exterior AmL*) TV outside broadcast

radioemisora *f* TV broadcasting station; ~ **de servicio móvil** *f* TELECOM mobile radio station

radioenlace: ~ **de microondas** *m* TELECOM radio link

radioestación: ~ **móvil** *f* TELECOM mobile radio station; ~ **repetidora** *f* ING ELÉC radio relay

radioestrella *f* FÍS RAD, TEC ESP radio star

radioestroncio *m* FÍS RAD, QUÍMICA radiostrontium

radiofaro *m* CONST beacon, D&A radio beacon, FERRO, TEC ESP beacon, TRANSP radio beacon, marker, TRANSP AÉR, TRANSP MAR *en tierra* radio beacon; ~ **contestador** *m* TRANSP AÉR responder beacon; ~ **direccional** *m* TRANSP, TRANSP AÉR, TRANSP MAR

radio homing beacon; ~ **indicador de posición de emergencia mediante satélite** *m* TELECOM satellite emergency position-indicating radio beacon; ~ **indicador de posición para emergencias** *m* TELECOM emergency position-indicating radio beacon (*EPIRB*); ~ **marítimo** *m* TRANSP MAR maritime radio beacon; ~ **omnidireccional** *m* TRANSP AÉR omnidirectional radiorange; ~ **omnidireccional de frecuencia muy alta** *m* (*RFMA*) TRANSP, TRANSP AÉR very high-frequency omnirange (*VHFO*), VHF omnidirectional radio range (*VOR*); ~ **de peligro** *m* TRANSP AÉR hazard beacon; ~ **para radar** *m* FÍS RAD, TRANSP ramark; ~ **de recalada** *m* TRANSP, TRANSP AÉR, TRANSP MAR homing beacon, radio homing beacon; ~ **respondedor** *m* TRANSP AÉR responder beacon; ~ **de rumbo** *m* TRANSP radio marker

radiofotografía *f* FÍS RAD fluorography, X-ray photograph, FOTO *medicina* X-radiograph

radiofrecuencia *f* (*RF*) ACÚST, ELEC, ELECTRÓN, TELECOM, TV, WATER TRANSP radio frequency (*RF*)

radiofuente *f* FÍS RAD, TEC ESP radio source; ~ **cuasiestelar** *f* (*QSS*) TEC ESP quasi-stellar radio source (*QSS*)

radiogénico *adj* FÍS RAD, GEOL radiogenic

radiogoniometría *f* FÍS radiogoniometry, radio direction finding, FÍS RAD radiogoniometry, TELECOM direction finding (*DF*), radio direction finding, TRANSP radio direction finding, TRANSP AÉR radio direction finding, radiogoniometry, TRANSP MAR radiogoniometry, radio direction finding

radiogoniómetro *m* CRISTAL radiogoniometer, FÍS, FÍS RAD, GEOM, INSTR , radio direction finder, radiogoniometer, OCEAN angulometer, direction finder, radiogoniometer, TRANSP radio direction finder, TRANSP AÉR direction finder, radio direction finder, radiogoniometer, TRANSP MAR radiogoniometer, *navegación* radio direction finder; ~ **automático** *m* TRANSP AÉR automatic direction finder; ~ **de navegación** *m* TRANSP AÉR, TRANSP MAR radio compass

radiografía *f* C&V shadowgraph, FÍS, FÍS RAD, INSTR radiography; ~ **electrónica** *f* ELECTRÓN, NUCL electron radiography; ~ **gamma** *f* FÍS RAD gamma radiography; ~ **por partículas cargadas** *f* FÍS PART, FÍS RAD, NUCL charged-particle radiography; ~ **por proximidad** *f* ELECTRÓN X-ray proximity printing; ~ **de resolución temporal** *f* NUCL time-resolved radiography; ~ **resuelta en función del tiempo** *f* NUCL time-resolved radiography

radiografiar *vt* FÍS, FÍS RAD radiograph, INSTR radiograph, X-ray

radiógrafo *m* FÍS radiograph, radiographer, FÍS RAD radiographer, radiograph, INSTR radiograph, radiographer

radiograma: ~ **en código** *m* D&A radio code message

radioguía *f* TELECOM, TRANSP radioguidance

radiointerferencia *f* ELECTRÓN, FÍS ONDAS, FÍS RAD radio interference

radioisómero *m* FÍS RAD, QUÍMICA radioisomer

radioisótopo *m* AGRIC, FÍS, FÍS PART, FÍS RAD, QUÍMICA radioisotope; ~ **de larga vida** *m* FÍS RAD long-lived radioisotope

radiólisis *f* FÍS RAD, QUÍMICA radiolysis

radiolita *f* MINERAL radiolite

radiolocalización *f* D&A, FÍS RAD, TRANSP, TRANSP AÉR, TRANSP MAR radiolocation; ~ **direccional** *f* FÍS,

TELECOM, TRANSP, TRANSP MAR radio direction finding

radiología *f* FÍS RAD radiology

radiólogo *m* FÍS, FÍS RAD, INSTR radiographer

radioluminiscencia *f* FÍS RAD radioluminescence

radiomando *m* D&A, TEC ESP *aviación* homing

radiomarcación *f* TELECOM, TRANSP AÉR, TRANSP MAR radio bearing

radiomecánico *m* D&A, FÍS RAD, MECÁ radiomechanic

radiometalografía *f* FÍS RAD X-ray metallography

radiometría *f* ENERG RENOV, FÍS RAD, GEOFÍS, GEOL, METEO, QUÍMICA radiometry

radiométrico *adj* ENERG RENOV, FÍS RAD, GEOFÍS, GEOL, QUÍMICA radiometric

radiómetro *m* ENERG RENOV, FÍS radiometer, FÍS ONDAS radiation counter, scintillation counter, GEOFÍS, INSTR, QUÍMICA radiometer; **~ acústico** *m* ACÚST acoustic radiometer; **~ espectral** *m* ENERG RENOV spectral pyranometer; **~ del sol** *m* ENERG RENOV pyrheliometer; **~ solar** *m* ENERG RENOV, INSTR pyranometer

radiomicrómetro *m* ING MECÁ radiomicrometer

radiomimético *adj* QUÍMICA radiomimetic

radionavegación *f* FÍS RAD, TRANSP AÉR, TRANSP MAR radio navigation

radionucleido *m* FÍS, FÍS RAD radionuclide; **~ natural** *m* FÍS, FÍS RAD natural radionuclide

radiooperador *m* TRANSP MAR radio operator

radiopatía *f* FÍS RAD radiation sickness

radioquímica *f* FÍS RAD, QUÍMICA radiochemistry

radioquímico *adj* FÍS RAD, QUÍMICA radiochemical

radiorreceptor *m* FÍS, FÍS RAD radio receiver

radioscopia *f* ING ELÉC fluoroscopy

radioseñal *f* ELECTRÓN radio signal, receiver signal, FÍS RAD receiver signal, ING ELÉC radio signal, receiver signal, TELECOM receiver signal

radiosensibilidad *f* FÍS RAD radio sensitivity

radiosonda *f* FÍS RAD, GEOFÍS, TELECOM radiosonde

radiotelefonear *vti* TELECOM, TRANSP, TRANSP MAR radiophone

radiotelefonía *f* (*RT*) TELECOM, TRANSP, TRANSP MAR radiotelephony (*RT*); **~ buque a costa** *f* TELECOM, TRANSP MAR ship-to-shore radio

radioteléfono *m* TELECOM, TRANSP, TRANSP MAR radiophone, radiotelephone; **~ de ondas métricas** *m* TRANSP MAR VHF radio telephone

radiotelescopio *m* FÍS, INSTR radio telescope

radioterapia *f* FÍS RAD, INSTR radiation treatment, radiotherapy

radiotomografía *f* ING MECÁ tomography

radiotoxicidad *f* FÍS RAD, INSTR radiotoxicity

radiotransmisor *m* FÍS, FÍS RAD, TELECOM radio transmitter

radiotrazador *m* FÍS, FÍS RAD radioactive tracer

radioyodo *m* FÍS RAD, QUÍMICA radioiodine

radomo *m* FÍS, TEC ESP, TELECOM, TRANSP AÉR radome, TRANSP MAR radar dome, radome; **~ húmedo** *m* TELECOM wet radome

radón *m* (*Rn*) FÍS, FÍS RAD, QUÍMICA radon (*Rn*)

raedera *f* AGRIC skimmer

RAEL *abr* (*radar aéreo de exploración lateral*) CONTAM MAR, TRANSP AÉR SLAR (*sideways-looking airborne radar*)

rafado[1] *adj* AmL (*cf rozado Esp*) MINAS holed

rafado[2] *m* MINAS kerving, kirving

rafadora *f* AmL (*cf acanaladora Esp*) CARBÓN *minas*

cutter arm, MINAS *canteras* channeler (*AmE*), channeller (*BrE*), channeling machine (*AmE*), channelling machine (*BrE*); **~ de afuste** *f* MINAS *explotación de canteras* bar channeler (*AmE*), bar channeller (*BrE*); **~ de cadena para carbón** *f* MINAS chain coal cutting machine; **~ de cantera** *f* MINAS quarrying machine; **~ para carbón** *f* CARBÓN, MINAS coal-mining machine; **~-cargadora** *f* MINAS shearer; **~ de deslizamiento** *f* MINAS track channeler (*AmE*), track channeller (*BrE*); **~ de franja pequeña** *f* CARBÓN short-wall coal-cutting machine

rafadura *f* MINAS kirving; **~ a cielo abierto** *f* MINAS surface cut; **~ paralela** *f* MINAS parallel cut

ráfaga *f* D&A *de arma de fuego*, INFORM&PD *de errores* burst, METEO *de viento* gust, TEC ESP gust, *pulsos electromagnéticos* burst, TRANSP AÉR, TRANSP MAR *de viento* gust, TV burst; **~ de aire** *f* TEC ESP weft; **~ en la capa sensible** *f* IMPR comet in coating; **~ de datos** *f* TEC ESP data burst; **~ descendente** *f* TRANSP AÉR *de aire* down gust; **~ de impulsos** *f* INFORM&PD pulse train; **~ de neutrones** *f* NUCL neutron burst; **~ de nieve** *f* METEO snow flurry

rafagosidad *f* TEC ESP, TRANSP AÉR, TRANSP MAR gust intensity

rafinosa *f* QUÍMICA raffinose

raíces: **~ adventicias** *f pl* AGRIC brace roots

raído *adj* ING ELÉC bare, SEG jagged, TEXTIL threadbare

rail *m* CONST, FERRO, ING ELÉC rail, ING MECÁ guiderail; **~ conductor** *m* ELEC, FERRO, ING ELÉC conductor rail; **~ de cremallera** *m* FERRO *equipo inamovible* rack rail; **~ dentado** *m* ING MECÁ cog rail; **~ elástico** *m* TRANSP resilient rail; **~ de estabilización** *m* TRANSP stabilization rail; **~ paramagnético** *m* TRANSP paramagnetic rail; **~ de soporte** *m* TRANSP bearing rail

raíz *f* C&V root, CARBÓN *dientes* pulp, INFORM&PD, MECÁ root; **~ del ala** *f* TRANSP AÉR wing root; **~ alar** *f* TRANSP AÉR wing root; **~ cuadrada** *f* MATEMÁT square root; **~ cuadrada y ángulos de costado** *f* METR *perfiles laminados* square root and edge angles; **~ cuadrada media** *f* CONST, ELEC, ELECTRÓN, MATEMÁT, TELECOM root-mean-square (*RMS*); **~ cuadrada del valor cuadrático medio** *f* FÍS root-mean-square value; **~ de la pala** *f* TRANSP AÉR blade root; **~ pivotante** *f* AGRIC taproot; **~ principal** *f* AGRIC main root

raja *f* CARBÓN crack, PROD slit, split, TEC ESP slit

rajado[1] *adj* ING MECÁ, MECÁ cracked

rajado[2]: **~ de pieza tosca** *m* C&V blank cracking

rajadura *f* MECÁ crack; **~ del diamante** *f* PROD diamond cleaving

rajamiento *m* PROD splitting

rajarse *v refl* CONST split

rajatuercas *m* ING MECÁ nut splitter

ralentizar *vi* VEH *motor* idle

raleo *m* AGRIC thinning

rallado *adj* SEG jagged

ralstonita *f* MINERAL ralstonite

RAM[1] *abr* (*memoria de acceso aleatorio*) ELEC, IMPR, INFORM&PD, ING ELÉC RAM (*random-access memory*)

RAM[2]: **~ dinámica** *f* INFORM&PD dynamic RAM; **~ estática** *f* INFORM&PD static RAM (*SRAM*)

rama *f* ELEC *de red de distribución* branch, IMPR *imposición de forma* chase, MINAS *de vagonetas* rake; **~ acústica** *f* FÍS acoustic branch; **~ B** *f* TELECOM Bleg; ~ **caliente** *f* NUCL hot leg; ~ **central** *f* ELEC *fuente de alimentación,* ING ELÉC common branch; ~ **común** *f* ELEC *fuente de alimentación,* ING ELÉC common branch; ~ **conjugada** *f* ING ELÉC conjugate branch; ~ **fría** *f* NUCL cold leg; ~ **idéntica del puente** *f* ELEC *circuito* equal-arm bridge; ~ **local** *f* TELECOM leg; ~ **negra** *f* AGRIC *malezas* bitterweed; ~ **principal lateral** *f* AGRIC main branch; ~ **sesgada** *f* IMPR bias chase

ramal *m* CONST y-branch, FERRO *equipo inamovible* branch line, feeder line, FÍS branch, GEOL *ígneo* feeder, MINAS ore feeder, NUCL *de una tubería* leg, PETROL *río* arm, PROD *ventilación minas* split; ~ **de clasificación** *m* FERRO *vehículos* sorting siding; ~ **cloacal** *m* AGUA branch sewer; ~ **conducido** *m* PROD *correas* idle side, slack side; ~ **libre** *m* MINAS *cuerdas,* PROD free end; ~ **primario** *m* ING ELÉC primary tap; ~ **tenso** *m* PROD *correas* driving side; ~ **de tubería** *m* AGUA, CONST *fontanería* branch pipe

ramark *m* D&A radar marker beacon, FÍS RAD ramark, TRANSP, TRANSP AÉR radar marker beacon, TRANSP MAR radar marker beacon, ramark

rame: ~ **de agujas** *m* TEXTIL pin stenter; ~ **de pinzas** *m* TEXTIL *acabado de tejidos* clip stenter

rameado *m* TEXTIL tentering

ramificación *f* AGUA branching, ELEC *conexión* tapping, ELECTRÓN *de circuito* branch, FERRO *equipo inamovible* feeder line, FÍS *de tuberías* branching, GEOL *ígnea* feeder, ING ELÉC tapping; ~ **axilar** *f* AGRIC axillary branching; ~ **de entrada** *f* ING ELÉC input tapping

ramificado *adj* QUÍMICA branched

ramificador: ~ **múltiple** *m* TRANSP AÉR manifold

ramificar *vi* CONST bend, branch

ramio *m* PAPEL china grass

ramnoxantina *f* QUÍMICA frangulin

ramoneo: ~ **bentónico** *m* OCEAN marine grazing

ramp *f* NUCL ramp

rampa *f* CONST *ascendente* rising gradient, slope, gradient, incline, *de una calzada* cambering, camber, FERRO *equipo inamovible* bank, ING ELÉC, MINAS, PETROL ramp, TEC ESP *minería* chute, TRANSP, TRANSP AÉR ramp, TRANSP MAR *de buque pesquero* slipway; ~ **de acceso** *f* CONST access ramp, FERRO *equipo inamovible* contact ramp, TRANSP access ramp; ~ **automática** *f* ING MECÁ self-acting plane; ~ **de carga** *f* TRANSP AÉR loading ramp; ~ **coclear** *f* ACÚST cochlear duct; ~ **descendente automática** *f* PROD automatic ramp down; ~ **deslizante** *f* SEG *aviones* toboggan; ~ **de engrase** *f* PROD lubricating rack, feed rack; ~ **de engrase de múltiples salidas** *f* PROD multiple feed rack; ~ **de escape** *f* TRANSP AÉR escape chute; ~ **de freno** *f* TEC ESP brake chute; ~ **para ganado** *f* Esp *(cf embarcadero AmL)* AGRIC livestock loading ramp; ~ **guía** *f* ING MECÁ guide ramp; ~ **de lanzamiento** *f* D&A *para misiles guiados,* TEC ESP launching ramp; ~ **de paso del personal** *f* TEC PETR manway; ~ **para sacos** *f* AGRIC bag chute; ~ **de salvamento** *f* SEG, TRANSP AÉR, TRANSP MAR rescue chute; ~ **timpánica** *f* ACÚST scala timpani; ~ **para vehículos** *f* AUTO vehicle ramp; ~ **vestibular** *f* ACÚST scala vestibuli

rampante *adj* CONST *arquitectura* rampant

rancho *m* AGRIC ranch

rango *m* D&A rank, ELECTRÓN, FÍS range, GAS limits, GEOM, INFORM&PD range, ING MECÁ class, *de traduc-*

tores range, MATEMÁT rank, MECÁ, TV range; ~ **de ablandamiento** *m* C&V softening range; ~ **de alimentación** *m* ING MECÁ feed range; ~ **del brillo** *m* TV brightness ratio; ~ **de energía** *m* FÍS RAD energy range; ~ **espectral** *m* FÍS spectral range; ~ **nominal** *m* NUCL rated range; ~ **de puntos** *m* GEOM range of points; ~ **de regulación** *m* ING ELÉC regulation range; ~ **de sintonización** *m* FÍS RAD tuning range; ~ **de temperatura** *m* TERMO temperature range; ~ **de temperatura crítica** *m* TERMO critical-temperature range; ~ **de temperatura efectiva** *m* TERMO effective-temperature range; ~ **de temperatura de fusión** *m* TERMO melting range; ~ **de trabajo** *m* C&V working range; ~ **de transformación** *m* C&V transformation range; ~ **de ultravioleta cercano** *m* FÍS RAD near ultraviolet

rangua *f* ELEC *contadores eléctricos* bearing, *máquina* brass, ING MECÁ footstep bearing, step box, pillow, footstep; ~ **del eje** *f* ING MECÁ shaft step

ranura *f* C&V slot, CARBÓN *tabla de diamantes* kerf, CONST rebate, notch, groove, EMB slit, FÍS slot line, FÍS ONDAS slit, ING ELÉC slot, ING MECÁ rebate, spire, spline, groove, MECÁ groove, channel, notch, spline, flute, METAL notch, NUCL *del rotor de una turbina* groove, slot, ÓPT groove, PROD slot, nick, notch, groove, notching, TEC ESP slit, slot, spline, TELECOM, TV slot; ~ **del ala** *f* TRANSP AÉR wing slot; ~ **de la aleta** *f* REFRIG fin slot; ~ **de alimentación** *f* INFORM&PD feed hole; ~ **para atornillador-destornillador** *f* ING ELÉC screwdriver slot; ~ **de carga** *f* AmL *(cf ranura para introducir la cinta de video Esp)* TV loading slot; ~ **cerrada** *f* AUTO lock groove; ~ **de chaveta** *f* ING MECÁ keyway, cotter slot, MECÁ keyway; ~ **para destornillador** *f* ING ELÉC screwdriver slot; ~ **de encaje** *f* CONST fillister; ~ **de engrase** *f* ING MECÁ oil-channel; ~ **de enhebrado** *f* CINEMAT threading slot; ~ **en espiral** *f* ING MECÁ involute spline; ~ **de extensión** *f* ELECTRÓN, INFORM&PD expansion slot; ~ **para filtros** *f* CINEMAT filter slot; ~ **con forma cuadrada** *f* CONST square-mouthed rabbet; ~ **con forma de riñón** *f* ING MECÁ kidney-shaped slot; ~ **helicoidal** *f* ING MECÁ helical groove; ~ **para introducir la cinta de video** *f* Esp *(cf ranura de carga AmL)* TV loading slot; ~ **de involuta** *f* ING MECÁ involute spline; ~ **longitudinal** *f* TELECOM longitudinal slot; ~ **de lubricación** *f* ING MECÁ oil groove; ~ **para módulo de chasis E/S** *f* PROD I/O chassis module slot; ~ **de mordaza** *f* ING MECÁ clamp slot; ~ **de pared recta** *f* ING MECÁ straight-sided spline; ~ **de pasador** *f* ING MECÁ cotter slot; ~ **principal** *f* ING MECÁ master spline; ~ **de punto muerto** *f* ING MECÁ dead-center notch *(AmE),* dead-centre notch *(BrE)*; ~ **de riñón** *f* ING MECÁ kidney-shaped slot; ~ **del segmento** *f* AUTO, VEH *motor* piston-ring groove; ~ **en T** *f* ING MECÁ tee slot *(T-slot)*; ~ **de tiempo** *f* TV time slot; ~ **transversal** *f* TELECOM transverse slot; ~ **en V** *f* ELECTRÓN *transistores,* ING MECÁ V-groove; ~ **visualizadora de tiempo** *f* ELECTRÓN time slot

ranurado[1] *adj* ING ELÉC, ING MECÁ fluted, MECÁ fluted, splined, PROD grooved, grooving, slitting

ranurado[2] *m* ING MECÁ grooving, PROD nicking, nicking machine

ranuradora *f* C&V slitter, ING MECÁ grooving machine, routing machine; ~ **de excéntrica** *f* ING MECÁ eccentric drive slotting machine; ~ **de manivela** f ING MECÁ crank drive slotting machine

ranurar vt CONST recess, ING MECÁ notch, PROD nick

ranuras[1]: **con ~** adj ING ELÉC, ING MECÁ, MECÁ fluted, PROD grooved; **sin ~** adj ING MECÁ, MECÁ fluteless

ranuras[2]: **~ y estriaciones** f pl ING MECÁ splines and serration

rapidez f FÍS speed; **~ de ascenso** f TEC ESP climb rate; **~ de conmutación** f ING ELÉC switching speed; **~ en el envío de la señal** f ELECTRÓN signal agility

rápido adj ING ELÉC fast-acting

raqueta: ~ subterránea f TRANSP buried loop

raquis m AGRIC rachis

raquitismo m AGRIC rickets

rarefacción f ACÚST, FÍS, FÍS ONDAS rarefaction

rarefactor m ELEC válvula, osciloscopio getter

raridad f PROD thinness

rarificación f ACÚST, FÍS, FÍS ONDAS rarefaction

ras[1]: **a ~** adj CONST, PROD flush

ras[2]: **estar a ~ de** vt PROD be flush with

rasante adj ACÚST, FÍS grazing

rascado m PROD pintura stripping

rascador m ING MECÁ, MECÁ scraper

rascadora: ~ de áridos f TRANSP aggregate scraper; **~ de concreto** f AmL (cf rascadora de hormigón Esp) TRANSP concrete scraper; **~ de hormigón** f Esp (cf rascadora de concreto AmL) TRANSP concrete scraper

rascadura f TEXTIL scratch

rascar vt CONST, CONTAM MAR scrape, MINAS, PROD scrap

rasera f CONTAM MAR belt skimmer, skimmer; **~ con aliviadero** f CONTAM MAR weir skimmer; **~ autopropulsada** f CONTAM MAR self-propelled skimmer; **~ centrífuga** f CONTAM MAR centrifugal skimmer; **~ de cinta** f CONTAM MAR belt skimmer; **~ con cinta olefila** f CONTAM MAR oleophilic belt skimmer; **~ de compuerta** f CONTAM MAR weir skimmer; **~ de disco** f CONTAM MAR disc skimmer (BrE), disk skimmer (AmE); **~ hidrodinámica** f CONTAM MAR hydrodynamic skimmer; **~ con maroma absorbente** f CONTAM MAR rope skimmer; **~ de maroma continua** f CONTAM MAR rope skimmer; **~ de succión directa** f CONTAM MAR direct-suction skimmer; **~ de tambor** f CONTAM MAR drum skimmer; **~ vortical** f CONTAM MAR vortex skimmer

rasgado[1] adj P&C prueba, falla torn

rasgado[2] m CONST ripping; **~ del fondo** m C&V bottom tear

rasgadura f C&V tear, tearing, TEXTIL crack; **~ de la pieza en tosco** f C&V blank tear

rasgar vt CONST rip

rasgo: ~ diagonal m IMPR scratch

rasgos: ~ ascendentes m pl IMPR ascenders

rasguño m TEXTIL scuffing

raso m TEXTIL satin

raspado m CONST chipping, rasping

raspador m ALIMENT scraper, C&V squeegee, CONST, ING MECÁ, MECÁ scraper, PROD tuberías pig, TEC PETR scraper; **~ giratorio** m AGRIC rotary scraper; **~ de tuberías** m TEC PETR producción pig

raspadura f C&V bruise, CONTAM MAR, MINAS, PROD scraping

raspar vt CONST scrape

raspatubos m PETROL pig

raspón m C&V scuff mark

rasqueta f CONST shave hook, CONTAM MAR scraper,

EMB, IMPR doctor blade, ING MECÁ, MECÁ scraper, PAPEL doctor blade; **~ del limpiaparabrisas** f VEH accesorio wiper blade; **~ para limpiar barrenos** f MINAS sludger; **~ de mariscador** f OCEAN mollusc detacher; **~ de la tinta** f IMPR ink blade; **~ del tintero** f IMPR fountain blade; **~ para tubos de caldera** f PROD tube-scraper

rasqueteado m CONTAM MAR, MINAS, PROD scraping

rastra f AGRIC smoothing harrow, MINAS pesca dredge, PROD de cinta transportadora flight, nowel, TRANSP MAR pesca dredge net; **~ autodescargante** f AGRIC self-dump rake; **~ lagrangiana** f OCEAN, TRANSP MAR investigación oceanografía Lagrangian drifter; **~ oceanográfica** f OCEAN oceanographic dredge

rastreador m TEC ESP tracker; **~ decodificador** m INSTR character tracer; **~ solar** m GEOFÍS astronomía sun tracker

rastrear vt INFORM&PD trace, TRANSP MAR fondo del mar drag

rastreo m INFORM&PD trace, TEC ESP, TELECOM tracking, TRANSP MAR del fondo del mar dragging; **~ automático** m TEC ESP autotracking; **~ de captación** m TRANSP captation drag; **~ de frecuencia** m TEC ESP comunicaciones frequency tracking; **~ de memoria** m PROD memory tracking; **~ de minas** m D&A, TRANSP MAR minesweeping

rastrero adj AGRIC rampant

rastrillado m AGRIC raking, TEXTIL hilatura hackling

rastrillador m CARBÓN, PROD rougher

rastrillar vt TEXTIL hilado hackle

rastrillo m ACÚST de aguja del gramófono, AGRIC, CONST rake; **~ de arrastre** m AGRIC drag rake; **~ henificador combinado** m AGRIC combined side-deliver rake and swath turner; **~ recolector-cargador** m AGRIC rake-bar loader

rastro m GAS, PETROL trace, TEC ESP slot; **~ de ganancia** m PETROL gain trace

rastrojo m AGRIC stubble; **~ del sorgo** m AGRIC sorghum stalk field

rata: ~ de agotamiento f AmL (cf tasa de agotamiento Esp) PETROL, TEC PETR depletion rate, pumping rate

raticida m AGRIC rodenticide

ratio: ~ contable m CONST accounting ratio; **~ de temperaturas** m TERMO temperature ratio

ratón m INFORM&PD mouse; **~ de dos botones** m INFORM&PD two-button mouse; **~ paralelo** m INFORM&PD parallel mouse; **~ de tres botones** m INFORM&PD three-button mouse

ratonera f ENERG RENOV equipo de perforación geotérmica mouse hole, PETROL rat hole

raudales m pl AGUA, HIDROL floods

raya f C&V, CINEMAT scratch, D&A de un cañón rifle, FÍS RAD line, INFORM&PD stripe, ING MECÁ line, P&C defecto scratch, PAPEL defecto streak, PROD espectro line, TEXTIL stripe; **~ de absorción** f FÍS absorption line; **~ espectral** f FÍS RAD, TELECOM spectral line; **~ longitudinal** f CINEMAT longitudinal scratch; **~ oscura** f FÍS absorption line; **~ de resonancia** f ELECTRÓN, FÍS RAD resonance line; **~ del torpedo** f D&A torpedo furrow; **~ transversal** f CINEMAT transversal scratch

rayado[1] adj GEOL banded; **no ~** adj QUÍMICA unlined

rayado[2] m C&V scoring, D&A de un arma rifling, ING MECÁ hatching, ruling; **~ cruzado** m ING MECÁ cross hatching

rayadura: ~ de la base f FOTO base scratch

rayar: ~ **en cruz** *vt* PROD cross-hatch
rayas: ~ **agrietadas verticales** *f pl* C&V brushlines, brushmarks; ~ **continuas** *f pl* IMPR feint rules; ~ **del revelador** *f pl* CINEMAT, FOTO developer streaks
Rayl *m* ACÚST Rayl
rayo *m* ELEC *fenómeno* lightning stroke, ray, *fenómeno meteorológico* lightning, ELECTRÓN, FÍS ray, FÍS PART *de partículas* beam, FÍS RAD beam, ray, GAS *luz* beam, GEOM half line, INFORM&PD, ING ELÉC ray, METEO lightning strike, thunderbolt, NUCL, ÓPT, TELECOM, TV ray; ~ **alfa** *m* FÍS alpha ray; ~ **anódico** *m* FÍS RAD anode ray; ~ **del ánodo** *m* ING ELÉC anode ray; ~ **axial** *m* ÓPT, TELECOM axial ray; ~ **beta** *m* ELEC, FÍS, FÍS PART, NUCL beta ray; ~ **canal** *m* FÍS, NUCL canal ray; ~ **catódico** *m* ELECTRÓN, FÍS, FÍS RAD, ING ELÉC, TV cathode ray (*CRO*); ~ **cósmico** *m* FÍS, FÍS RAD, TEC ESP cosmic ray; ~ **delta** *m* FÍS RAD, NUCL delta ray; ~ **de efecto túnel** *m* TELECOM tunneling ray (*AmE*), tunnelling ray (*BrE*); ~ **de electrones** *m* FÍS ONDAS electron beam; ~ **extraordinario** *m* FÍS extraordinary ray; ~ **con fugas** *m* ÓPT, TELECOM leaky ray; ~ **gamma** *m* ELECTRÓN, FÍS, FÍS ONDAS, FÍS PART, FÍS RAD, PETROL gamma ray; ~**guía** *m* TRANSP guide beam; ~ **incidente** *m* CRISTAL incident beam, FÍS, FÍS ONDAS incident ray; ~ **láser** *m* ELECTRÓN, FÍS ONDAS, FÍS RAD, NUCL, TELECOM laser beam; ~ **luminoso** *m* CINEMAT, FÍS, ÓPT light ray; ~ **meridiano** *m* ÓPT meridian ray; ~ **no meridiano** *m* ÓPT nonmeridian ray; ~ **no meridional** *m* ÓPT skew ray; ~ **oblicuo** *m* TELECOM skew ray; ~ **óptico** *m* ÓPT optical ray; ~ **ordinario** *m* FÍS ordinary ray; ~ **paraxial** *m* FÍS, ÓPT, TELECOM paraxial ray; ~ **positivo** *m* ELECTRÓN positive ray, FÍS canal ray; ~ **principal** *m* FÍS principal ray; ~ **reflejado** *m* FÍS, FÍS ONDAS, ÓPT reflected ray; ~ **refractado** *m* FÍS, ÓPT, TELECOM refracted ray; ~ **de Roentgen** *m* ELEC *radiación* X-ray; ~ **rojo** *m* ELECTRÓN, TV red beam; ~ **sesgado** *m* TELECOM skew ray; ~ **sísmico** *m* GEOL *geofísica* ray path; ~ **ultravioleta** *m* ÓPT ultraviolet ray; ~ **UV** *m* ÓPT UV ray; ~ **X** *m* ELEC, ELECTRÓN, FÍS, FÍS RAD, METAL X-ray; ~ **X blando** *m* FÍS RAD soft X-ray; ~ **X duro** *m* FÍS, FÍS RAD hard X-ray; ~ **X hiperenergético** *m* FÍS RAD hard X-ray; ~ **X secundario** *m* FÍS RAD secondary X-ray
rayón *m* C&V scratch, P&C *tipo de polímero*, TEXTIL rayon
razón *f* FÍS, MATEMÁT, METR, REFRIG *en matemáticas* ratio; ~ **de abundancias** *f* NUCL abundance ratio; ~ **adiabática de secado** *f* CONTAM *atmosférica* dry adiabatic lapse rate; ~ **armónica** *f* GEOM cross-ratio, harmonic ratio; ~ **de asentamiento** *f* C&V setting rate; ~ **atómica** *f* NUCL atomic ratio; ~ **áurea** *f* GEOM golden ratio; ~ **axial** *f* TEC ESP axial ratio; ~ **común** *f* MATEMÁT common ratio; ~ **de dosis** *f* FÍS RAD dose rate; ~ **electrónica-atómica** *f* FÍS PART, NUCL electron-to-atom ratio; ~ **de espiras** *f* ELEC *transformador* turn ratio, turns ratio; ~ **de flujo crítico** *f* NUCL critical flux ratio; ~ **del inventario fisionable** *f* NUCL fissile inventory ratio; ~ **inversa** *f* ING MECÁ inverse ratio, reciprocal ratio; ~ **invertida hombre-hora** *f* ING MECÁ indirect man-hour ratio; ~ **de ionización** *f* FÍS PART ionization rate; ~ **de luminancia** *f* FÍS RAD luminance ratio; ~ **máxima de fundido** *f* C&V maximum melting rate; ~ **de Poisson** *f* FÍS Poisson's ratio; ~ **de potencia de salida de portadora deseada a indeseada** *f* TELECOM wanted-to-unwanted carrier-power ratio; ~ **de rechazo del modo común** *f* ELECTRÓN, TELECOM common-mode rejection ratio; ~ **de reexpedición** *f* TELECOM redirecting reason; ~ **de reexpedición original** *f* TELECOM original redirection reason; ~ **de temperaturas** *f* TERMO temperature ratio; ~ **de transformación** *f* CONTAM *procesos químicos* transformation rate, ELEC *transformador* turn ratio, turns ratio, winding ratio; ~ **trigonométrica** *f* GEOM trigonometrical ratio
Rb *abr* (*rubidio*) QUÍMICA Rb (*rubidium*)
RBMA *abr* (*residuos de baja y mediana actividad*) NUCL L&ILW (*low and intermediate level waste*)
RC *abr* (*resistencia-capacitancia*) ELEC, ELECTRÓN, ING ELÉC RC (*resistance-capacitance*)
RCCP *abr* (*revisión de la configuración según la cláusula primera*) TEC ESP *normas de seguridad para vehículos espaciales* FACR (*first-article configuration review*)
RCD *abr* (*revisión crítica de diseño*) TEC ESP CDR (*critical design review*)
RCS *abr* ING ELÉC (*rectificador controlado de silicio*) SCR (*silicon rectifier, silicon-controlled rectifier*), NUCL (*sistema del refrigerante del reactor*) RCS (*reactor coolant system*), TELECOM (*rectificador controlado de silicio*) SCR (*silicon rectifier, silicon-controlled rectifier*)
RDP *abr* (*rompevientos de playa*) OCEAN BOP (*blow-out preventer*)
RDSI *abr* (*red digital de servicios integrados*) TELECOM ISDN (*integrated services digital network*)
RDSI-BA *abr* (*red digital de servicios integrados de banda ancha*) TELECOM B-ISDN (*broadband ISDN, broadband-integrated services digital network*)
Re *abr* (*renio*) QUÍMICA Re (*rhenium*)
reabastecimiento *m* ING MECÁ, TRANSP AÉR *de combustible* refueling (*AmE*), refuelling (*BrE*); ~ **en vuelo** *m* TRANSP AÉR *de combustible* in-flight refueling (*AmE*), in-flight refuelling (*BrE*)
reacción[1]: **de** ~ *adj* TEC ESP *electricidad* regenerative
reacción[2] *f* ELECTRÓN regenerative feedback, FÍS reaction, INFORM&PD feedback, retroaction, QUÍMICA reaction; ~ **absorbente de neutrones** *f* NUCL neutron-absorbing reaction; ~ **en cadena** *f* FÍS, FÍS RAD, ING MECÁ, METAL chain reaction; ~ **de Cannizzaro** *f* QUÍMICA Cannizzaro reaction; ~ **capacitiva** *f* ELEC, ING ELÉC capacitive feedback; ~ **catalítica** *f* GAS catalytic reaction; ~ **convergente** *f* NUCL convergent reaction; ~ **crítica** *f* FÍS RAD critical reaction; ~ **de deuterio catalizado** *f* NUCL catalyzed deuterium reaction; ~ **de Diels-Alder** *f* QUÍMICA Diels-Alder reaction; ~ **sin difusión** *f* METAL diffusionless reaction; ~ **endotérmica** *f* ENERG RENOV *energía solar*, GAS endothermic reaction; ~ **eutéctica** *f* METAL, QUÍMICA eutectic reaction; ~ **en fase de vapor** *f* ELECTRÓN vapor phase reaction (*AmE*), vapour phase reaction (*BrE*); ~ **del freno** *f* TRANSP brake reaction; ~ **de Friedel-Crafts** *f* QUÍMICA *orgánica* Friedel-Crafts reaction; ~ **de inducción** *f* ELEC *máquina* armature reaction; ~ **inducida por radiación** *f* FÍS RAD radiation-induced reaction; ~ **del inducido** *f* ELEC *máquina*, ING ELÉC armature reaction; ~ **inductiva** *f* ING ELÉC inductive feedback; ~ **de intercambio energético** *f* NUCL energy-exchange reaction; ~ **inversa** *f* NUCL reverse reaction; ~ **isotérmica** *f* METAL isothermal reaction; ~ **masiva** *f* METAL massive reaction; ~ **monofásica** *f* METAL monophase reaction; ~ **monomolecular** *f* QUÍMICA

monomolecular reaction; ~ **normal** *f* FÍS normal reaction; ~ **nuclear** *f* FÍS, FÍS RAD, NUCL, QUÍMICA nuclear reaction; ~ **nuclear artificial** *f* NUCL artificial nuclear reaction; ~ **nuclear inducida** *f* NUCL induced nuclear reaction; ~ **nuclear secundaria** *f* NUCL secondary nuclear reaction; ~ **peritéctica** *f* METAL peritectic reaction; ~ **pirogénica** *f* GEOFÍS, TERMO pyrogenic reaction; ~ **de primer orden** *f* METAL, QUÍMICA first-order reaction; ~ **de radicales libres** *f* ALIMENT, P&C, QUÍMICA free-radical reaction; ~ **secundaria** *f* QUÍMICA side reaction; ~ **subcrítica** *f* FÍS RAD *fisión nuclear* subcritical reaction; ~ **supercrítica** *f* FÍS RAD *fisión nuclear* supercritical reaction; ~ **tardía** *f* CINEMAT double take; ~ **termonuclear** *f* FÍS, NUCL thermonuclear reaction; ~ **de transferencia** *f* NUCL transfer reaction; ~ **de transferencia polinuclear** *f* NUCL *más de cuatro nucleones* many-nuclear transfer reaction; ~ **ultrarrápida** *f* QUÍMICA ultrarapid reaction; ~ **por vía húmeda** *f* QUÍMICA wet reaction

reacondicionadora *f* ING MECÁ dresser cutter

reacondicionamiento *m* ING MECÁ restyling, PROD, RECICL reprocessing, TEC ESP *nave* refurbishing, TRANSP restyling

reacondicionar *vt* ING MECÁ, TRANSP recondition, restyle

reacoplarse *v refl* TEC ESP redock

reactancia *f* CONST *electricidad* ballast, ELEC, ELECTRÓN, FÍS, ING ELÉC reactance; ~ **de acoplamiento** *f* ING ELÉC swinging choke; ~ **acústica** *f* ACÚST, FÍS, ING ELÉC acoustic reactance; ~ **auxiliar** *f* ING ELÉC ballast; ~ **bloqueada** *f* ING ELÉC blocked reactance; ~ **de capacidad** *f* ELEC, FÍS, ING ELÉC capacitive reactance; ~ **capacitiva** *f* ELEC, FÍS, ING ELÉC *circuito de corriente alterna* capacitive reactance; ~ **común** *f* ELEC *inductores* common reactance; ~ **del condensador** *f* ING ELÉC capacitor reactance; ~ **controlada** *f* ING ELÉC blocked reactance; ~ **de filtro** *f* ELEC, ING ELÉC filter choke; ~ **del inducido** *f* ELEC *máquina* armature reactance; ~ **inductiva** *f* ELEC, FÍS inductive reactance; ~ **mecánica** *f* ACÚST mechanical reactance; ~ **negativa** *f* ING ELÉC negative reactance; ~ **de puesta a tierra** *f* ELEC, ING ELÉC earthing reactor (*BrE*), grounding reactor (*AmE*); ~ **de la secuencia de fases positivas** *f* ELEC positive-phase sequence reactance; ~ **en serie** *f* ING ELÉC series reactance; ~ **variable** *f* ELECTRÓN varactor; ~ **variable de generador armónico** *f* ELECTRÓN harmonic generator varactor

reactivación *f* CARBÓN, GEOL, QUÍMICA reactivation, TEC ESP recovery, TEC PETR rejuvenation; ~ **por calor** *f* IMPR heat reactivation

reactivar *vt* CARBÓN, GEOL, QUÍMICA reactivate

reactividad *f* FÍS, NUCL, QUÍMICA reactivity; ~ **negativa** *f* NUCL negative reactivity; ~ **xenón** *f* NUCL xenon reactivity

reactivo[1] *adj* ELEC *corriente alterna* wattless, TEC ESP regenerative; **no** ~ *adj* QUÍMICA nonreactive

reactivo[2] *m* AGRIC reagent, CARBÓN *añadido a una pulpa para aumentar la adherencia* reagent, collecting agent, FÍS reactant, FOTO, P&C reagent, QUÍMICA reactant, reagent; ~ **captante** *m* CARBÓN collecting reagent; ~ **de Grignard** *m* QUÍMICA Grignard reagent

reactor *m* AUTO, ELEC, ELECTRÓN reactor, ING ELÉC impedance coil, reactor, MECÁ jet engine, NUCL reactor, QUÍMICA reactor, kettle, TERMO *chorro de gases* jet engine, TRANSP AÉR jet; ~ **de absorción** *m* PROC QUÍ absorption reactor; ~ **activado por acelerador lineal** *m* NUCL linear-accelerator-driven reactor (*LADR*); ~ **con agitación** *m* PROC QUÍ stirred reactor; ~ **de agua en ebullición** *m* FÍS, NUCL boiling-water reactor (*BWR*); ~ **de agua en ebullición de ciclo directo** *m* NUCL direct cycle boiling reactor; ~ **de agua en ebullición de circulación natural** *m* NUCL natural circulation boiling-water reactor; ~ **de agua ligera activado por acelerador** *m Esp* (*cf reactor de agua liviana activado por acelerador AmL*) NUCL accelerator driven light water reactor; ~ **de agua liviana activado por acelerador** *m AmL* (*cf reactor de agua ligera activado por acelerador Esp*) NUCL accelerator driven light water reactor; ~ **de agua pesada** *m* NUCL heavy water reactor; ~ **de agua presurizada** *m* FÍS, NUCL pressurized-water reactor; ~ **de aire comprimido** *m* ELEC air reactor; ~ **de alto flujo** *m* NUCL high-flux reactor; ~ **de autoinducción variable** *m* ING ELÉC swinging choke; ~ **blindado** *m* ELEC sealed reactor; ~ **catalítico** *m* AUTO, CONTAM catalytic converter, PROC QUÍ catalytic reactor, VEH catalytic converter; ~ **de ciclo de agotamiento** *m* NUCL one-cycle reactor; ~ **de ciclo doble** *m* NUCL dual-cycle reactor; ~ **de ciclo único** *m* NUCL one-cycle reactor; ~ **de cojinete de bolas** *m* ING MECÁ ball-bearing reactor; ~~**colector de escape térmico** *m* TRANSP thermal exhaust manifold reactor; ~ **con combustible de torio** *m* NUCL thorium-fueled reactor (*AmE*), thorium-fuelled reactor (*BrE*); ~ **comercial de construcción en serie** *m* NUCL series-produced power reactor; ~ **en condiciones próximas a la criticidad** *m* NUCL near-critical reactor; ~ **de conmutación** *m* ING ELÉC transductor; ~ **convertidor de calor** *m* NUCL thermal converter reactor; ~ **crítico con neutrones retardados** *m* NUCL delayed critical reactor; ~ **cuasicrítico** *m* NUCL near-critical reactor; ~ **cuasirreproductor** *m* NUCL quasi-breeder reactor; ~ **de decantación** *m* PROC QUÍ decantation reactor; ~ **de desalación** *m* NUCL desalination reactor; ~ **desnudo** *m* NUCL bare reactor; ~ **de doble flujo** *m* TERMO, TRANSP fan jet engine; ~ **emisor de neutrones** *m* NUCL neutron source reactor; ~ **enriquecido** *m* NUCL enriched reactor; ~ **de ensayo de materiales** *m* NUCL materials testing reactor (*MTR*); ~ **de entrenamiento** *m* NUCL training reactor; ~ **de epitaxia** *m* CRISTAL, ELECTRÓN, METAL, PROC QUÍ epitaxy reactor; ~ **de filtro** *m* ELEC red, ING ELÉC filter choke; ~ **de flujo doble** *m* TRANSP AÉR dual-flow jet engine; ~ **de flujo de sólidos** *m* NUCL flowable solids reactor (*FSR*); ~~**fuente** *m* NUCL source reactor; ~ **de fusión de espejos en tándem** *m* NUCL tandem mirror fusion reactor (*TMR*); ~ **de haz** *m* NUCL beam reactor; ~ **heterogéneo** *m* NUCL heterogeneous reactor; ~ **híbrido de agua ligera** *m Esp* (*cf reactor híbrido de agua liviana AmL*) NUCL light-water hybrid reactor (*LWHR*); ~ **híbrido de agua liviana** *m AmL* (*cf reactor híbrido de agua ligera Esp*) NUCL light-water hybrid reactor (*LWHR*); ~ **homogéneo** *m* TRANSP homogeneous reactor; ~ **imagen** *m* NUCL image reactor; ~ **imagen de flujo negativo** *m* NUCL negative-flux image reactor; ~ **intrínsecamente estable** *m* NUCL inherently stable reactor; ~ **de inyección de**

aire *m* TRANSP air injection reactor (*AIR*); **~ de laboratorio** *m* NUCL laboratory reactor; **~ limitador de corriente** *m* ELEC, ING ELÉC current-limiting reactor; **~ magnox** *m* NUCL magnox reactor; **~ moderado por agua pesada** *m* NUCL heavy water reactor; **~ moderado por agua pesada en ebullición** *m* NUCL boiling-heavy-water moderated reactor; **~ moderado por berilio** *m* NUCL beryllium-moderated reactor; **~ negativo** *m* NUCL negative reactor; **~ neumático** *m* ELEC air reactor; **~ neutro trifásico** *m* ELEC three-phase neutral reactor; **~ nuclear** *m* CONST, ELEC, FÍS, NUCL nuclear reactor; **~ nuclear natural** *m* NUCL natural nuclear reactor; **~ de núcleo gaseoso** *m* NUCL gaseous core reactor; **~ de núcleo magnético saturable** *m* ING ELÉC saturable reactor; **~ de núcleo sembrado** *m* NUCL seed-core reactor; **~ de placas** *m* NUCL slab reactor; **~ de potencia cero** *m* NUCL zero-energy reactor, zero-power reactor; **~ de potencia producido en serie** *m* NUCL series-produced power reactor; **~ prefabricado** *m* NUCL package reactor; **~ presurizado** *m* FÍS, NUCL reactor pressure vessel; **~ quimiconuclear** *m* NUCL, QUÍMICA chemonuclear fuel reactor; **~ rápido** *m* FÍS, FÍS PART, NUCL fast reactor; **~ refrigerado por agua** *m* NUCL water-cooled reactor; **~ refrigerado por agua ligera** *m* Esp (*cf reactor refrigerado por agua liviana AmL*) NUCL light-water-cooled reactor; **~ refrigerado por agua liviana** *m* AmL (*cf reactor refrigerado por agua ligera Esp*) NUCL light-water-cooled reactor; **~ refrigerado por dispersión** *m* NUCL dispersion-cooled reactor; **~ refrigerado por nitrógeno** *m* NUCL nitrogen-cooled reactor; **~ refrigerado por sodio** *m* NUCL sodium-cooled reactor; **~ reproductor** *m* FÍS, NUCL breeder reactor; **~ reproductor rápido** *m* FÍS, NUCL fast-breeder reactor; **~ reproductor refrigerado por gas** *m* NUCL gas-cooled breeder reactor (*GCBR*); **~ resistor** *m* TEC ESP resistojet; **~ saturable** *m* FÍS saturable reactor; **~ secundario** *m* NUCL secondary reactor; **~ sumergido** *m* NUCL underwater reactor; **~ de temperaturas muy elevadas** *m* NUCL ultrahigh temperature reactor; **~ térmico** *m* FÍS RAD thermal reactor; **~ a tierra** *m* ELEC, ING ELÉC earthing reactor (*BrE*), grounding reactor (*AmE*); **~ de tipo seco de devanado no encapsulado** *m* ELEC nonencapsulated-winding dry-type reactor; **~ de torio** *m* NUCL thorium-fueled reactor (*AmE*), thorium-fuelled reactor (*BrE*); **~ transportable** *m* NUCL transportable reactor; **~ de uranio** *m* NUCL uranium reactor; **~ de uranio moderado por agua pesada** *m* NUCL uranium heavy water reactor; **~ de vigilancia de los efectos de irradiación** *m* NUCL radiation effects reactor

readherencia *f* FÍS FLUID *de remolinos* reattachment; **~ de turbulencia** *f* FÍS FLUID turbulent reattachment

reafilado *m* CARBÓN regrinding, ING MECÁ resharpening, PROD dressing, regrinding

reafilar *vt* CARBÓN regrind, ING MECÁ, MECÁ reset, PROD regrind

reagrupación: ~ de vacíos *f* INFORM&PD garbage collection

reagrupamiento: ~ de Beckmann *m* QUÍMICA *de cetoxima* Beckmann rearrangement; **~ óptimo** *m* ELECTRÓN *de electrones* ideal bunching

reajustar *vt* CINEMAT, ING MECÁ, MECÁ, PROD *instrumento, mecanismo* reset

reajuste *m* CINEMAT, INFORM&PD reset, ING MECÁ resetting, PROD reset, resetting, TV reset

realambrado *m* ELEC, ING ELÉC rewiring

realce *m* ING MECÁ boss, embossment, MINAS *labor minera* top hole

realidad: ~ virtual *f* INFORM&PD virtual reality

realimentación *f* GEN feedback; **~ AGC** *f* TELECOM feedback AGC; **~ en circuito cerrado** *f* ELEC, ELECTRÓN, ING ELÉC, PROD closed-loop feedback; **~ en corriente** *f* ELEC, ELECTRÓN, ING ELÉC current feedback; **~ digital** *f* ELECTRÓN, ING ELÉC, TELECOM digital feedback; **~ inductiva** *f* ELECTRÓN, ING ELÉC inductive feedback; **~ intrínseca** *f* ELECTRÓN, ING ELÉC, NUCL inherent feedback; **~ inversa** *f* ELECTRÓN, ING ELÉC inverse feedback, negative feedback; **~ negativa** *f* ELECTRÓN negative feedback, ING ELÉC inverse feedback, negative feedback; **~ en paralelo** *f* ELECTRÓN, ING ELÉC shunt feedback; **~ de reactividad** *f* NUCL reactivity feedback; **~ selectiva** *f* ELECTRÓN, ING ELÉC selective feedback; **~ en serie** *f* ELECTRÓN, ING ELÉC series feedback; **~ síncrona** *f* TV sync feedback; **~ de voltaje** *f* ELECTRÓN, ING ELÉC voltage feedback

realización *f* GEOL, TEXTIL implementation; **~ de muestras** *f* TEXTIL patterning

realizador: ~ de documentales *m* CINEMAT, TV *producción* documentary film-maker

realizar: ~ una evolución *vi* TRANSP MAR *navegación* turn

realzar *vt* TEXTIL enhance

reamolar *vt* CARBÓN, PROD regrind

reanudación *f* INFORM&PD restart; **~ del sistema** *f* INFORM&PD warm start

reanudamiento *m* TRANSP AÉR restart

reanudar *vt* INFORM&PD, TRANSP AÉR restart

reaprovisionamiento *m* ING MECÁ *de combustible* refueling (*AmE*), refuelling (*BrE*); **~ por gravitación** *m* TEC ESP gravity refueling (*AmE*), gravity refuelling (*BrE*)

reapuntalar *vt* CONST repoint

reareación *f* AGUA reaeration

rearmar *vt* CONST reassemble, ING MECÁ, MECÁ, PROD *instrumento, mecanismo* reset

rearrancar *vt* ING MECÁ restart

rearranque *m* TRANSP AÉR restart; **~ en molinete** *m* TRANSP AÉR windmilling restart

rearreglo *m* QUÍMICA rearrangement

reavivado *m* PROD *rueda de esmeril* dressing

reavivador: ~ de muelas *m* ING MECÁ dresser cutter; **~ de muelas abrasivas** *m* ING MECÁ wheel dresser

reavivamuelas *m* ING MECÁ dresser cutter

rebaba *f* ING MECÁ burr, MECÁ burr, chip, scale, METAL scrap, PROD *de madera, metales* chip, *de pieza de metal* burr, *de pieza fundida* sprue, *en moldes* scale; **~ lateral** *f* P&C *procesamiento* flash; **~ de metal duro** *f* ING MECÁ hard metal burr

rebabar *vt* MECÁ trim, PROD *cortar* clip

rebabarse *v refl* P&C deflash

rebabeo *m* C&V deburring

rebada *f* ING MECÁ *fundición*, PROD *por rotura del molde* breakout

rebaja *f* ING MECÁ off; **~ de concentración** *f* QUÍMICA *mezcla* depletion

rebajado *m* ING MECÁ relief, *mecanizado* backing off, MECÁ relief

rebajador *m* CONST rabbet plane; **~ de Farmer** *m* FOTO

Farmer's reducer; ~ **de rayos** *m* CONST *carpintería* spokeshave

rebajar *vt* CARBÓN depress, CONST cope, lower, recess, ING MECÁ relieve, true up, INSTAL HIDRÁUL, MECÁ, PROD relieve

rebaje *m* CONST recessing, abutment, rebate, IMPR split, ING MECÁ offset, rebate, MECÁ, METAL, PROD notch; ~ **de cabeza** *m* AmL (*cf destroza de cabeza Esp*) MINAS overhand cut; ~ **cruzado para tornillos** *m* ING MECÁ cross-recess for screw; ~ **hasta su peso** *m* C&V chipping to the weight

rebajo *m* CONST rabbet, *carpintería* scarf, ING MECÁ relief, notch, necking, MECÁ relief, notch, METAL relief, notch, necking, PROD notch; ~ **superpuesto** *m* CONST lapped scarf

rebanadora: ~ **de testigos** *f* PETROL core slicer

rebaño *m* Esp (*cf rodeo AmL*) AGRIC *ganado caballar* herd

rebarba *f* IMPR matrix hairline, PROD *funderías* fin

rebasamiento: ~ **de costos** *m* TEC ESP cost overrun; ~ **de la velocidad límite** *m* VEH *motor* overrun

rebasar *vt* FERRO *estación, señal de stop*, TRANSP AÉR *pista* overshoot

rebase *m* TRANSP AÉR overrun; ~ **de los enganches** *m* MINAS *jaula de minas* overwinding

reblandecimiento: ~ **por acritud** *m* METAL work softening; ~ **por fatiga** *m* METAL fatigue softening

rebobinado *m* CINEMAT rewind, rewinding, ELEC rewinding, FOTO rewinding, rewind, spooling, INFORM&PD rewinding, rewind, PAPEL rewinding, rewind, TEXTIL rewind, rewinding, TV rewinding, rewind; ~ **apretado** *m* Esp (*cf bobinado apretado AmL*) FOTO tight spooling; ~ **automático** *m* Esp (*cf bobinado automático AmL*) CINEMAT, FOTO auto winding; ~ **B** *m* Esp (*cf bobinado B AmL*) CINEMAT, TV B-wind; ~ **a motor** *m* CINEMAT motor rewind

rebobinadora *f* CINEMAT, FOTO, INFORM&PD, PAPEL, TEXTIL, TV rewinder; ~ **automática** *f* FOTO automatic rewinder; ~ **de película** *f* CINEMAT, FOTO, TV film winder; ~ **positiva** *f* CINEMAT positive rewinder; ~ **de potencia** *f* CINEMAT power rewinder

rebobinar[1] *vt* CINEMAT, FOTO backwind, rewind, INFORM&PD, PAPEL, TEXTIL rewind

rebobinar[2] *vi* FOTO backwind

reborde *m* (*cf collarín Esp*) ACÚST *mecánica* flange, sill, bead, lip, edge, ledge, seam, rib, lap joint, ridge; ~ **del arrecife** *m* GEOL, OCEAN reef front; ~ **central de caída de la llanta** *m* AUTO drop center rim (*AmE*), drop centre rim (*BrE*); ~ **conector** *m* TEC ESP connecting flange; ~ **de extrusión** *m* ING MECÁ extrusion flange; ~ **de hierro** *m* CONST edging iron; ~ **de protección** *m* TRANSP AÉR coaming; ~ **del tapacubo** *m* AUTO hub flange

rebordeado[1] *adj* AUTO, ING ELÉC, ING MECÁ, MECÁ flanged, PROD beaded, flanged, TEC ESP, VEH flanged

rebordeado[2] *m* PROD crimping

rebordeadora *f* PROD flanger, flanging machine

rebordear *vt* C&V bead down, CONST bead over, MECÁ crimp

rebordes: **con ~** *adj* GEN flanged

rebosadero *m* INSTAL HIDRÁUL weir

rebosadura *f* ING MECÁ overflow

rebosamiento *m* CONTAM MAR, GAS spillage

rebosar *vi* AGUA run over, HIDROL *río* overtop its banks, PROD *de moldes* break out

rebosarse *v refl* TERMO boil over

rebose *m* AGUA, ING MECÁ, TELECOM overflow

rebotar *vi* AUTO, D&A, P&C rebound

rebote *m* AUTO rebound, D&A *bombardeo* rebound, ricochet, P&C *propiedad física, prueba* rebound; ~ **de contacto** *m* ELEC *relé*, PROD contact bounce

rebrote *m* AGRIC regrowth; ~ **lateral** *m* AGRIC offshoot

REC *abr* (*reconocimiento*) INFORM&PD ACK (*acknowledgement*)

recalada *f* D&A, TEC ESP *aviación* homing, TRANSP MAR *navegación* landfall; ~ **por radio** *f* TELECOM, TRANSP radio homing

recalar: ~ **a tierra** *vi* TRANSP MAR *navegación* stand inshore

recalcado *m* METAL jump, PROD swaging; ~ **de cabezas de pernos** *m* PROD jumping bolt heads

recalcador *m* PROD *sierras* swage

recalcar *vt* PROD *neumático, perno* jump

recalce *m* CONST underpinning

recalentado[1] *adj* VEH *motor, coloquial* hotted-up (*AmE*), souped-up (*BrE*)

recalentado[2] *m* C&V reheat (*AmE*), reheating (*BrE*), REFRIG, TERMOTEC, TRANSP reheating

recalentador *m* PROD chauffer, superheater, REFRIG, TERMOTEC reheater; ~ **de aire** *m* ING MECÁ air reheater

recalentamiento *m* C&V reheating (*BrE*), ING ELÉC, ING MECÁ overheating, METAL superheating, PROD overheating, superheating, QUÍMICA superheating, REFRIG reheating, superheat, reheat, SEG, TERMO overheating, TERMOTEC reheat (*AmE*), reheating, TRANSP reheat, reheating, VEH *del motor* overheating

recalentar *vt* REFRIG reheat, TERMO overheat, TERMOTEC reheat

recalentarse *v refl* TERMO overheat

recalescencia *f* FÍS recalescence

recalibración *f* FÍS RAD, NUCL recalibration

recalibrado[1] *adj* FÍS RAD, NUCL recalibrated

recalibrado[2] *m* FÍS RAD, NUCL recalibration

recalzo: ~ **con pilotes** *m* PROD piling; ~ **de la vía** *m* FERRO *infraestructura* track raising

recámara *f* D&A *cañones* breech, chamber; ~ **fotográfica** *f* INSTR photo-chamber; ~ **de muestras** *f* INSTR specimen chamber; ~ **de observación** *f* INSTR viewing chamber; ~ **de visión** *f* INSTR viewing chamber

recambio *m* TRANSP MAR, VEH spare part; ~ **del conjunto de la batería** *m* PROD replacement battery assembly

recambios: ~ **del automóvil** *m pl* VEH motorcar parts

recanteadora: ~ **de chapa** *f* ING MECÁ routing machine

recantear *vt* PROD mill

recardación *f* MINAS *de carbón, minerales* offtake

recarga *f* AGUA recharge, CINEMAT, FOTO reloading, HIDROL *agua de captación* recharge, INFORM&PD reloading, ING ELÉC charge build-up, PROD, TV reloading, ~ **acuífera** *f* HIDROL aquifer recharge; ~ **artificial** *f* HIDROL *acuífero* artificial recharge; ~ **de combustible en condiciones de parada** *f* NUCL off-load charging

recargable *adj* CINEMAT, ELEC, FÍS, FOTO, ING ELÉC, TV rechargeable

recargado *m* GEN recharging

recargar *vt* CINEMAT recharge, reload, ELEC, ELECTRÓN, FÍS recharge, FOTO recharge, reload, INFORM&PD reload, ING ELÉC recharge, PROD reload,

TV recharge, reload; ~ **a fondo** *vt* TEC ESP *acumuladores* top up

recebado: ~ **del arco eléctrico** *m* TEC ESP reignition

recebar *vt* ING MECÁ restart

recepción[1]: **sólo para ~** *adj* INFORM&PD, TELECOM receive-only

recepción[2] *f* INFORM&PD receipt, reception, ING MECÁ pick-up, PROD reception, TEC ESP acceptance; ~ **de diversidad temporal** *f* TELECOM time-diversity reception; ~ **programada** *f* PROD scheduled receipt

receptáculo *m* PROD receptacle, socket, QUÍMICA receptacle, RECICL bank; ~ **aerodinámico** *m* TEC ESP pod; ~ **de alimentación** *m* ELEC *conexión* power outlet; ~ **para una ametralladora** *m* TEC ESP *de avión* pod; ~ **para cargamento** *m* *AmL* (*cf depósito para cargamento Esp*) MINAS loading pocket; ~ **de clavija** *m* ELEC *conexión* plug receptacle; ~ **a prueba de incendios** *m* SEG *de aparato eléctrico* fireproof enclosure, flameproof enclosure; ~ **de suministro eléctrico** *m* ELEC *conexión* power outlet

receptor *m* CONTAM *electrofiltros* receptor, ELECTRÓN, FÍS receiver, INFORM&PD sink, ING ELÉC receiver, ING MECÁ receiver, load, PROD, QUÍMICA, TEC ESP, TEC PETR receiver, TELECOM receiver, recipient, TRANSP *radio, radar, satélite*, TV receiver; ~ **de alarma por láser** *m* ELECTRÓN laser warning receiver; ~ **de banda ancha** *m* TELECOM wideband receiver; ~ **de banda estrecha** *m* TELECOM narrow-band receiver; ~ **de fibra óptica** *m* ING ELÉC, ÓPT, TELECOM fiber-optic receiver (*AmE*), fibre-optic receiver (*BrE*); ~ **fotoeléctrico** *m* ÓPT photoelectric receiver; ~ **del goniómetro** *m* TRANSP MAR, direction-finding receiver (*DF receiver*); ~ **de guiado** *m* TEC ESP guidance receiver; ~ **impresor de tinta** *m* INSTR *telegrafía* ink writer; ~ **inalámbrico de corrección auditiva** *m* TELECOM wireless hearing-aid receiver; ~ **integrado PIN-FET** *m* ÓPT PIN-FET integrated receiver; ~ **del intercambio** *m* TELECOM interchange recipient; ~ **de mensajes** *m* INFORM&PD message sink; ~ **de modem** *m* ELECTRÓN modem receiver; ~ **monocromo** *m* TV monochrome receiver; ~ **multifrecuencia** *m* TELECOM multifrequency receiver; ~ **óptico** *m* ÓPT, TELECOM optical receiver; ~ **óptico regenerativo** *m* ÓPT optical regenerative receiver; ~ **optoelectrónico** *m* TELECOM optoelectronic receiver; ~ **portátil** *m* TELECOM portable receiver, hand-held receiver; ~ **de la señal de control** *m* TEC ESP command receiver; ~ **de señales** *m* TELECOM signal receiver; ~ **de televisión** *m* ELECTRÓN, FÍS, ING ELÉC television receiver, TELECOM, TV television receiver, television set; ~ **de televisión digital** *m* TV digital TV receiver; ~ **en tierra** *m* TEC ESP earth receiver; ~ **de transmisión** *m* *Esp* (*cf recipiente de transmisión AmL*) TELECOM transmission recipient; ~-**transmisor** *m* TELECOM transponder

recesión *f* AGUA *del agua subterránea* depletion

recesivo *adj* AGRIC recessive

rechazamiento *m* CALIDAD, ELECTRÓN, INFORM&PD rejection

rechazar *vt* CALIDAD discard, reject, CINEMAT reject, CONST throw back to waste, ING MECÁ, PROD, TEXTIL discard, reject

rechazo *m* CALIDAD nonacceptance, rejection, reject, CARBÓN, CINEMAT reject, ELEC repulsion, ELECTRÓN rejection, FÍS repulsion, INFORM&PD rejection, ING ELÉC repulsion, PROD *inspecciones* reject, TELECOM

repudiation, TEXTIL reject; ~ **de calor** *m* TERMOTEC heat rejection; ~ **de FI** *m* TELECOM IF rejection; ~ **horizontal** *m* GEOL heave; ~ **del modo común** *m* TELECOM common-mode rejection; ~ **vertical** *m* AmL MINAS *fallas* floor heave

rechazos *m pl* CARBÓN, MINAS *mineral quebrantado* oversize; ~ **de depuración** *m pl* PAPEL tailings, screening

rechupado *m* *AmL* (*cf revenido Esp*) MINAS lingotes drawing

rechupe *m* CARBÓN *metalurgia* piping, *lingote* void, *lingote metalúrgico* cavity, METAL shrinkage cavity, MINAS *lingote metalúrgico* cavity, *lingote* sinkhole, PROD *fundición* draw

recibido: ~ **para embarque** *m* TRANSP MAR waybill

recibidor *m* TELECOM recipient; ~ **de fluidos** *m* FÍS FLUID, ING MECÁ fluid receiver

recibir[1]: **sólo para ~** *adj* INFORM&PD, TELECOM receive-only

recibir[2] *vt* ELECTRÓN, INFORM&PD, TELECOM receive

recibo *m* PROD receipt ticket; ~ **de existencias** *m* PROD stock receipt; ~ **de inventario** *m* PROD inventory receipt; ~ **del piloto** *m* TRANSP MAR *marina mercante* mate's receipt; ~ **de productos** *m* CALIDAD receipt for goods; ~ **de taller** *m* PROD shop receipt

reciclable *adj* CONTAM, EMB, RECICL recyclable

reciclado *m* GEN recycling, RECICL salvaging; ~ **de basura** *m* RECICL waste recycling; ~ **de residuos** *m* RECICL waste recycling

reciclaje *m* GEN recycling; ~ **del escape** *m* TRANSP exhaust recycling

reciclar *vt* CARBÓN recirculate, CONST, CONTAM, EMB recycle, RECICL recycle, salvage

reciente *m* GEOL recent

recierre: ~ **automático de circuito** *m* ELEC *interruptor* automatic circuit-recloser

recinto *m* ACÚST, FÍS, NUCL enclosure; ~ **acústico** *m* ACÚST acoustic enclosure; ~ **campana** *m* C&V bell jar; ~ **de contención** *m* NUCL containment; ~ **frigorífico para carretillas** *m* REFRIG roll-in refrigerator; ~ **magnético** *m* GEOFÍS magnetic bay; ~ **de protección del reactor** *m* NUCL guard vessel

recio *adj* ING MECÁ coarse

recipiente *m* AGRIC *fertilizantes, plaguicidas* container, CINEMAT bin, EMB container, ING MECÁ pan, receiver, LAB vessel, QUÍMICA recipient, RECICL bank, TELECOM recipient, TEXTIL, TRANSP MAR vessel; ~ **de admisión de la muestra** *m* NUCL sample admission vessel; ~ **aerosol** *m* EMB, ING MECÁ aerosol container; ~ **aislado** *m* TERMO insulated container; ~ **de aspiración centrado** *m* INSTR centering suction holder (*AmE*), centring suction holder (*BrE*); ~ **calorifugado** *m* TERMO heat-insulated container; ~ **para ceniza** *m* INSTAL TERM ash box; ~ **cerrado** *m* LAB *material de vidrio* closed vessel; ~ **criogénico de presión** *m* ING MECÁ cryogenic pressure vessel; ~ **de cuello angosto** *m* C&V narrow-neck container; ~ **de descarga de sólidos pulverizados** *m* IMPR airlock; ~ **equilibrador** *m* REFRIG equalizer tank; ~ **de equilibrio** *m* REFRIG balance tank; ~ **evacuador de aceites** *m* CONTAM oil-clearance vessel; ~ **de expansión** *m* ING MECÁ expansion vessel; ~ **de fluido hidráulico** *m* ING MECÁ hydraulic fluid reservoir; ~ **de gas** *m* GAS, PROD, TERMO gas holder; ~ **de gas a presión** *m* GAS, NUCL, PROD, TERMO gas bottle; ~ **de hojalata** *m* CINEMAT *para un rollo de película* can;

~ **para incineración** *m* LAB *análisis*, TERMO combustion boat; ~ **intermedio** *m* ING MECÁ receiver-space; ~ **de líquido** *m* REFRIG *del condensador* liquid receiver; ~**-n virtual** *m* TELECOM virtual container-n (*VC-n*); ~ **del núcleo** *m* NUCL core catcher; ~ **de poliéster de fibra de vidrio** *m* C&V, PROD glass-polyester enclosure; ~ **a presión** *m* CONST, ING MECÁ, MECÁ, REFRIG pressure vessel; ~ **de presión chapado** *m* ING MECÁ clad pressure vessel; ~ **de presión revestido** *m* ING MECÁ clad pressure vessel; ~ **para recortes** *m* CINEMAT trim bin; ~ **para recubrir** *m* REVEST coater pan; ~ **refractario** *m* QUÍMICA melting pot; ~ **de seguridad** *m* LAB safety container, SEG *para almacenar líquidos inflamables* safety vessel; ~ **termoaislado** *m* TERMO heat-insulated container; ~ **de transmisión** *m* AmL (*cf receptor de transmisión Esp*) TELECOM transmission recipient; ~ **virtual** *m* TELECOM virtual container
reciprocante *adj* ING MECÁ, MECÁ reciprocating
reciprocidad *f* GEOM reciprocation
recíproco[1] *adj* GEN reciprocal
recíproco[2] *m* GEOM, INFORM&PD reciprocal, ING MECÁ inverse, MATEMÁT reciprocal
recirculación *f* C&V recycling, *de las corrientes del vidrio en el tanque u horno* recirculation, CARBÓN, CONST, CONTAM recycling, NUCL *de productos de fisión* recirculation, PAPEL, PROD, QUÍMICA, RECICL recycling; ~ **de gases de escape** *f* AUTO, MECÁ, TRANSP exhaust-gas recirculation; ~ **de gases de escape con inyección de aire** *f* AUTO, TRANSP exhaust-gas recirculation with air injection
recircular *vt* CARBÓN recirculate, CONST, CONTAM, EMB, RECICL recycle
reclamación: ~ **de tráfico** *f* TRANSP traffic demand
reclasificación: ~ **en un nivel inferior** *f* ING MECÁ, PROD downgrading
reclavado *m* CARBÓN *traviesa ferrocarril* redriving
reclutamiento *m* ACÚST *audición*, D&A recruitment
recobrar *vt* CARBÓN recover, TEC ESP *vehículos* capture
recocer *vt* C&V, INSTAL TERM, MECÁ, NUCL, TERMO anneal; ~ **en crisol** *vt* TERMO pot anneal; ~ **en horno** *vt* TERMO pot anneal
recocido[1] *adj* C&V, CRISTAL, INSTAL TERM, MECÁ, NUCL, TERMO annealed; ~ **con gas** *adj* TERMO annealed under gas
recocido[2] *m* C&V, CRISTAL, INSTAL TERM, MECÁ, NUCL, TERMO anneal, annealing; ~ **afinante del grano** *m* TERMO grain-refining anneal; ~ **blando** *m* METAL soft annealing, TERMO *metales* soft anneal; ~ **blando a fondo** *m* TERMO *metales* dead soft anneal; ~ **brillante** *m* INSTAL TERM bright annealing; ~ **por difusión** *m* PROC QUÍ diffusion annealing; ~ **ferritizante** *m* METAL ferritizing annealing; ~ **a fondo** *m* TERMO dead anneal; ~ **isócrono** *m* METAL isochronal annealing; ~ **isotérmico** *m* METAL isothermal annealing; ~ **por precipitación** *m* TERMO precipitation anneal; ~ **para relajamiento de esfuerzos interiores** *m* TERMO stress-relieving anneal; ~ **por solubilización** *m* METAL solution annealing; ~ **suave** *m* TERMO soft anneal; ~ **suave a fondo** *m* TERMO dead soft anneal
recodo *m* CARBÓN turning, HIDROL oxbow, MECÁ, TRANSP MAR *de río, canal, conducto* bend
recogedor: ~ **de aceite** *m* MECÁ oil pan; ~ **de aire** *m* TRANSP AÉR air scoop; ~ **de velas** *m* C&V parison gatherer; ~ **de vidrio sobrante** *m* C&V *al cortar artículos de vidrio* cullet catcher

recogedora: ~**-enfardadora** *f* AGRIC pick-up baler
recogepastas *m* PAPEL save-all
recoger *vt* C&V gather, RECICL clean up, TRANSP AÉR stow, TRANSP MAR *velas* haul down; ~ **a mano** *vt* CARBÓN handpick
recogida *f* CONTAM MAR *de hidrocarburos* collection; ~ **de basuras** *f* RECICL refuse collection, waste collection; ~ **de datos** *f* INFORM&PD, TELECOM data collection; ~ **de desperdicios** *f* RECICL waste collection; ~ **y entrega** *f* FERRO collection and delivery; ~ **selectiva** *f* RECICL selective collection; ~ **por separado** *f* RECICL separate collection
recogido *adj* TRANSP clustered
recolección *f* AGRIC picking; ~ **de basura** *f* RECICL waste collection; ~ **de basuras** *f* RECICL refuse collection; ~ **de datos** *f* INFORM&PD data collection, data gathering, TELECOM data collection; ~ **de desperdicios** *f* AmL RECICL waste collection; ~ **electrónica de noticias** *f* TV electronic news gathering (*ENG*); ~ **de mariscos** *f* OCEAN mollusc harvesting; ~ **selectiva** *f* RECICL selective collection; ~ **por separado** *f* RECICL separate collection
recolectora: ~ **de algodón** *f* AGRIC cotton stripper
recombinación *f* ELECTRÓN, FÍS recombination; ~ **radiativa** *f* ELECTRÓN, FÍS RAD radiative recombination
recomendación *f* CALIDAD, PROD, TEC ESP, TELECOM recommendation, TRANSP advisory message
recomendaciones: ~ **de seguridad** *f pl* TRANSP AÉR safety recommendations
recompensa: ~ **por el salvamento** *f* TRANSP MAR salvage award
recomposición *f* GEOL *geocronología* resetting, ING MECÁ, TRANSP restyling
recompresión *f* OCEAN recompression
reconectar *vt* ING ELÉC, TELECOM reconnect
reconexión *f* ING ELÉC, TELECOM *de no pagadores* reconnection; ~ **automática de circuito** *f* ELEC *interruptor* automatic circuit-recloser
reconfigurable *adj* INFORM&PD reconfigurable
reconfiguración *f* INFORM&PD reconfiguration
reconfigurar *vt* INFORM&PD reconfigure
recongelación *f* FÍS, REFRIG *al disminuir la presión* regelation
recongelar *vt* REFRIG, TERMO refreeze
reconocedor: ~ **de voz** *m* TELECOM voice recognizer
reconocer *vt* CONTAM MAR survey, INFORM&PD *caracteres* acknowledge, MINAS develop, prove, TRANSP MAR *buque, costa* survey
reconocimiento *m* (*REC*) ACÚST recognition, D&A reconnaissance, INFORM&PD acknowledgement (*ACK*), recognition, semantic analysis, MINAS development, PROD inspection, TRANSP MAR *buques, costa* surveying, TV reconnaissance; ~ **del acordonado** *m* TRANSP cordon-line survey; ~ **aéreo** *m* CONTAM MAR aerial reconnaissance, TRANSP AÉR aerial survey; ~ **de caracteres** *m* INFORM&PD character recognition; ~ **de caracteres en tinta magnética** *m* INFORM&PD magnetic ink character reader (*MICR*); ~ **por chimeneas de relleno** *m* MINAS development by rock chutes; ~ **de la configuración** *m* ELECTRÓN pattern recognition; ~ **de la estructura** *m* ELECTRÓN pattern recognition; ~ **fotográfico** *m* TRANSP AÉR photo reconnaissance; ~ **geológico** *m* CARBÓN, GEOL, TEC PETR geological survey; ~ **de la mina** *m* MINAS mine development; ~ **negativo** *m* INFORM&PD negative

acknowledgement (*NAK*); **~ óptico de caracteres** *m* (*OCR*) IMPR, INFORM&PD optical character recognition (*OCR*); **~ de patrones** *m* TELECOM pattern recognition; **~ positivo** *m* INFORM&PD positive acknowledgement; **~ con rayos X** *m* TEC ESP X-ray inspection; **~ de recibo** *m* TELECOM acknowledgement; **~ técnico** *m* TELECOM technical acknowledgement; **~ topográfico** *m* AGRIC land survey; **~ topográfico por fotografía aérea** *m* GEOL aerial photographic survey; **~ de tramas** *m* INFORM&PD pattern recognition; **~ de la voz** *m* INFORM&PD speech recognition, TELECOM voice recognition

reconstitución *f* TEC ESP regeneration

reconstrucción *f* CONST rebuilding, reconstruction, INFORM&PD reconstruction

reconstruir *vt* CONST rebuild, reconstruct

reconvertidor *m* ELECTRÓN reconverter

recopilador: **~ de delgas** *m* ING ELÉC commutator; **~ de potencia de arco de plasma** *m* TRANSP plasma arc-power collector

record *m* FÍS record; **~ de marea** *m* GEOFÍS *prospección* tide record

recorrer *vt* ING MECÁ run, TRANSP MAR *buque* overhaul, *de buque* refit

recorrida *f* TRANSP MAR *de buque* refit; **~ de la máquina** *f* TRANSP MAR engine overhaul; **~ del motor** *f* TRANSP MAR engine overhaul

recorrido *m* AGUA runoff, CONST ride, *carretera* course, CONTAM, CONTAM MAR runoff, D&A range, ENERG RENOV runoff, FERRO *de línea* route, HIDROL runoff, ING MECÁ travel, run, stroke, NUCL *de una carrera* length of stroke, stroke, *de una barra de control* range of movement, travel, TEC ESP range, path, course, TELECOM range, TEXTIL *mecánica* course; **~ del agua en superficie** *m* HIDROL running-off over the surface; **~ del arranque** *m* CONST *arquitectura* springing course; **~ de aspersión** *m* CONTAM MAR spraypath; **~ de aterrizaje** *m* TRANSP AÉR landing run; **~ automático** *m* TELECOM automatic roaming; **~ balístico** *m* TEC ESP ballistic path; **~ del cable** *m* ELEC, ING ELÉC *fuente de alimentación* cable run; **~ del carro** *m* TEXTIL *género de punto* traverse; **~ de la cinta** *m* TV tape run; **~ en ciudad** *m* VEH *conducción* urban cycle; **~ de despegue** *m* TRANSP AÉR takeoff run; **~ efectuado a velocidad de crucero** *m* VEH cruising range; **~ del electrón** *m* ELECTRÓN, FÍS PART, NUCL, TV electron path; **~ de estiramiento** *m* CONST stretching course; **~ de filtración** *m* ELECTRÓN mask run-out; **~ de frenado** *m* CONST friction course; **~ del freno** *m* TRANSP brake pitch; **~ ideal** *m* OCEAN *de rayo acústico* sound path; **~ libre medio** *m* ACÚST *partículas* mean free path, FÍS mean free path, diffusion mean free path, NUCL diffusion mean free path; **~ libre medio total** *m* NUCL total mean free path; **~ en millas** *m* TRANSP mileage; **~ óptico** *m* TELECOM optical path; **~ óptimo** *m* TELECOM optimal path; **~ del regenerado** *m* PROD rework routing; **~ de retorno** *m* INSTAL HIDRÁUL *del émbolo, de cilindros* back stroke; **~ sísmico** *m* GEOFÍS, TEC PETR seismic path; **~ del timón de dirección** *m* TRANSP AÉR, TRANSP MAR rudder travel; **~ de tornillos de carpintero** *m* CONST clamp course; **~ del torpedo** *m* D&A torpedo track; **~ de trabajo** *m* ING MECÁ length of stroke; **~ urbano** *m* VEH *conducción* urban cycle

recortado[1] *adj* PROD trimmed off

recortado[2] *m* IMPR cutout, INFORM&PD scissoring; **~ inverso** *m* INFORM&PD reverse clipping

recortador *m* TELECOM, TV clipper; **~ de pulsos** *m* TV pulse clipper

recortadora *f* ING MECÁ punching machine, trimmer, PROD punching machine; **~ de chapa** *f* ING MECÁ power nibbler (*BrE*), power nibbling machine (*AmE*); **~ de chapa de uña vibratoria** *f* ING MECÁ nibbler, nibbling machine

recortar *vt* CINEMAT crop, trim, CONST cope, IMPR crop, forward, pare, trim, ING MECÁ shear, MECÁ, PROD trim, TV crop

recorte *m* *Esp* (*cf arranque AmL*) CINEMAT cropping, trim, IMPR cropping, patch, *papel* trim, INFORM&PD chad, clipping, scissoring, MECÁ cut, MINAS offset, *de mineral* stoping, breakage, drawing, breaking, holing, PAPEL trim, TV cropping; **~ de blanco** *m* TV white clip; **~ de bridas** *m* PROD flange cutout; **~ inverso** *m* INFORM&PD reverse clipping; **~ por láser** *m* ELECTRÓN laser trimming; **~ de onda** *m* ELECTRÓN peak clipping; **~ de la plancha al ras** *m* IMPR flushing; **~ de prensa** *m* PROD *periódicos* press cutting; **~ de tiempo para uso múltiple** *m* INFORM&PD time slicing; **~ de ventana** *m* INFORM&PD window clipping

recortes *m pl* CINEMAT offcuts, PAPEL shavings, TEC PETR cuttings

recristalización *f* CRISTAL, METAL, QUÍMICA recrystallization; **~ secundaria** *f* METAL secondary recrystallization; **~ terciaria** *f* METAL tertiary recrystallization

recta *f* GEOM straight line, ING MECÁ line; **~ de carga del transistor** *f* ELECTRÓN transistor d-c load line

rectangular *adj* GEOM rectangular

rectangularidad *f* GEOM rectangularity

rectángulo *m* GEOM rectangle; **~ Berne** *m* FERRO Berne rectangle

rectas: **~ concordantes** *f pl* GEOM concurrent lines; **~ convergentes** *f pl* GEOM convergent lines; **~ paralelas** *f pl* GEOM parallel lines; **~ perpendiculares** *f pl* GEOM perpendicular lines

rectificación *f* CARBÓN regrinding, ELEC, ING ELÉC, ING MECÁ rectification, PROD regrinding, QUÍMICA *aparatos de medida* rectification; **~ de una alternancia** *f* ELEC single-wave rectification; **~ de media onda** *f* ELEC, ING ELÉC half-wave rectification; **~ monofásica** *f* ELEC single-wave rectification; **~ de una onda** *f* ELEC single-wave rectification; **~ de onda completa** *f* ELEC, FÍS, ING ELÉC, PROD full-wave rectification; **~ en semilongitud de onda** *f* ING ELÉC half-wave rectification; **~ termiónica** *f* ING ELÉC thermionic rectification

rectificado[1] *adj* ELEC, ING ELÉC, ING MECÁ, QUÍMICA rectified

rectificado[2] *m* ING MECÁ honing, MECÁ, METAL grinding, PROD grinding, lapping; **~ del carrete de la válvula** *m* PROD valve spool grinding; **~ cilíndrico** *m* ING MECÁ, MECÁ cylindrical grinding; **~ de cilindros** *m* AUTO, ING MECÁ, VEH reboring; **~ medio** *m* PROD *entre grueso, fino* medium grinding; **~ por muela de superficies curvas** *m* PROD contour grinding; **~ de perfiles** *m* ING MECÁ form grinding; **~ con plantillas** *m* ING MECÁ profile grinding; **~ de precisión** *m* PROD *máquinas herramientas* precision grinding; **~ de precisión sin puntos** *m* ING MECÁ, MECÁ centerless precision grinding (*AmE*), centreless

precision grinding (*BrE*); **~ de precisión en seco** *m* ING MECÁ dry precision grinding; **~ de punteadora** *m* ING MECÁ jig grinding; **~ sin puntos** *m* ING MECÁ, MECÁ centerless grinding (*AmE*), centreless grinding (*BrE*); **~ de roscas** *m* ING MECÁ thread grinding; **~ simultáneo interior y exterior** *m* ING MECÁ internal and external simultaneous grinding; **~ de la subportadora** *m* TV subcarrier rectification; **~ de superficies frontales** *m* PROD face grinding; **~ de superficies planas** *m* ING MECÁ surface grinding
rectificador *m* GEN rectifier; **~ de aceite** *m* REFRIG oil still; **~ de alimentación** *m* ELEC power rectifier; **~ de alta tensión** *m* ELEC high-voltage rectifier; **~ de alto voltaje** *m* ELEC high-voltage rectifier; **~ para aplicaciones de corrientes fuertes** *m* ELEC power rectifier; **~ para aplicaciones de potencia** *m* ELEC power rectifier; **~ del arco** *m* CINEMAT, ING ELÉC arc rectifier; **~ de arco mercurial** *m* ING ELÉC mercury-arc rectifier; **~ de arco de mercurio regulado por rejilla** *m* ING ELÉC grid-controlled mercury arc rectifier; **~ de carga** *m* ELEC *batería, acumulador* charging rectifier; **~ controlado de silicio** *m* (*RCS*) ING ELÉC, TELECOM silicon-controlled rectifier (*SCR*); **~ controlado de silicio activado a la luz** *m* ING ELÉC light-activated silicon-controlled rectifier; **~ de corriente** *m* ELEC, ING ELÉC current rectifier (*AmE*), mains rectifier (*BrE*); **~ a diodo** *m* ELECTRÓN diode rectifier; **~ de una dirección** *m* ELEC single-way rectifier; **~ de disco** *m* PROD disc grinder (*BrE*), disk grinder (*AmE*); **~ de disco seco** *m* ELEC metal rectifier; **~ de efluvio** *m* ING ELÉC glow-discharge rectifier; **~ electrolítico** *m* ELEC, FÍS, ING ELÉC electrolytic rectifier; **~ de estado sólido** *m* ELECTRÓN solid-state rectifier; **~ exafásico** *m* ING ELÉC six-phase rectifier; **~ excitador** *m* ELEC *transformador*, ELECTRÓN voltage booster; **~ de frecuencia de fases** *m* ELECTRÓN phase-sequence rectifier; **~ de gas** *m* ELEC, GAS, ING ELÉC gas-filled rectifier; **~ de germanio** *m* ELEC, ELECTRÓN germanium rectifier; **~ de gran potencia** *m* ING ELÉC high-power rectifier; **~ ideal** *m* ING ELÉC ideal rectifier; **~ de ignitrones** *m* ING ELÉC ignitron rectifier; **~ indirecto** *m* ELEC indirect rectifier; **~ inversor de CC/CA** *m* ELECTRÓN inverted rectifier; **~ de media onda** *m* ELEC, ING ELÉC half-wave rectifier; **~ de mercurio** *m* ING ELÉC mercury rectifier; **~ metálico** *m* ELEC, ING ELÉC metallic rectifier; **~ monofásico en puente** *m* ELEC single-phase bridge rectifier; **~ de muelas abrasivas** *m* ING MECÁ wheel dresser; **~ de onda completa** *m* ELEC, FÍS, ING ELÉC, PROD full-wave rectifier; **~ de óxido de cobre** *m* ING ELÉC copper-oxide rectifier; **~ de placas puntuales** *m* ELEC point-plate rectifier; **~ de placas secas** *m* ELEC metal rectifier; **~ polianódico** *m* ING ELÉC multianode rectifier; **~ de positivo-negativo** *m* ING ELÉC positive-negative rectifier, p-n rectifier; **~ de potencia** *m* ELEC power rectifier; **~ de potencia de conmutación rápida** *m* ING ELÉC fast-switching power rectifier; **~ de puente** *m* ELEC, ING ELÉC bridge rectifier; **~ de rápida recuperación** *m* ING ELÉC fast reverse-recovery rectifier; **~ de red** *m* ELEC, ING ELÉC current rectifier (*AmE*), mains rectifier (*BrE*); **~ seco** *m* ELEC metal rectifier; **~ de secuencia de fases** *m* ELEC *sistema trifásico* phase-sequence rectifier; **~ de selenio** *m* ELEC, FÍS, ING ELÉC selenium rectifier; **~ sellado** *m* ING ELÉC sealed rectifier; **~ de**

semiconductor *m* ELEC, ING ELÉC semiconductor rectifier; **~ de semilongitud de onda** *m* ING ELÉC half-wave rectifier; **~ de semionda** *m* FÍS half-wave rectifier; **~ de silicio** *m* ELEC, FÍS, ING ELÉC silicon rectifier (*SCR*); **~ de silicio de gran potencia** *m* ING ELÉC high-power SCR; **~ de superficies frontales** *m* ING MECÁ face grinding; **~ unianodal** *m* ING ELÉC single-anode rectifier; **~ de vapor de mercurio** *m* ING ELÉC mercury vapor rectifier (*AmE*), mercury vapour rectifier (*BrE*); **~ de una vía** *m* ELEC single-way rectifier; **~ de voltaje de la red** *m* ELEC, ING ELÉC current rectifier (*AmE*), mains rectifier (*BrE*)
rectificadora *f* ING MECÁ, MECÁ grinder, grinding machine; **~ de alta velocidad** *f* ING MECÁ high-speed grinding machine; **~ de banco** *f* ING MECÁ bench grinder; **~ y bruñidora de interiores** *f* ING MECÁ honing machine; **~ cilíndrica** *f* ING MECÁ, MECÁ cylindrical grinder; **~ cilíndrica de control numérico computarizado** *f* ING MECÁ CNC cylindrical grinder; **~ de eje de levas** *f* ING MECÁ camshaft grinding machine; **~ de fresas de refrentar** *f* ING MECÁ face-milling grinder; **~ de interiores** *f* ING MECÁ internal grinder; **~ óptica para perfiles** *f* ING MECÁ optical-profile grinder; **~ de plantillas** *f* ING MECÁ profile grinder; **~ sin puntos** *f* ING MECA, MECÁ centerless grinder (*AmE*), centreless grinder (*BrE*); **~ de superficies de control electrónico** *f* CONST, ELECTRÓN, ING MECÁ electronically-controlled surface grinder; **~ de superficies exteriores** *f* ING MECÁ surface-grinding machine; **~ de superficies frontales** *f* ING MECÁ, PROD face grinder, face grinding; **~ de superficies planas** *f* ING MECÁ plain-grinder, surface grinder; **~ tangencial de curvas de unión** *f* ING MECÁ tangent radius dresser; **~ vertical** *f* ING MECÁ vertical cylinder-grinding machine
rectificar *vt* AUTO rebore, CARBÓN *herramientas* regrind, CINEMAT rectify, CONST straighten, ELEC *corriente alterna*, ELECTRÓN, FÍS, ING ELÉC rectify, ING MECÁ *muelas abrasivas* grind, true, ream, *cilindros* rebore, MECÁ *con muela abrasiva* ream, hone, lap, grind, PETROL rectify, ream, PROD *tablas de datos* adjust, *con muela abrasiva* lap, *colada en piedra de esmeril* clean up, grind, regrind, *válvulas, cilindros, tapones de vidrio* straighten, QUÍMICA, TELECOM, TV rectify, VEH *un cilindro* rebore
rectilíneo *adj* FÍS, ING MECÁ, MECÁ, TELECOM, TV rectilinear
recuadro *m* CINEMAT framing mask, IMPR box, INFORM&PD frame, TV framing mask
recubierto[1] *adj* ELEC, ING ELÉC covered; **~ de caucho** *adj* P&C, REVEST rubber-covered; **~ de cinc** *adj* REVEST zinc-coated; **~ a máquina** *adj* EMB machine-coated
recubierto[2] *m* EMB covering, ING ELÉC, REVEST coating, covering; **~ de fotopolímero** *m* IMPR photopolymer coating; **~ de vidrio** *m* ING ELÉC glass cladding
recubriente *m* TEC PETR *seguridad* blanketing
recubrimiento[1]: **con ~ de cobre** *adj* ELEC *cable*, REVEST copper-clad
recubrimiento[2] *m* C&V overlaying (*AmE*), CARBÓN *geología* overburden, GEOL cover, INSTAL HIDRÁUL *de estator de turbina de vapor* cover, lap, encasing, MINAS *de terreno* seam, overburden, PETROL coating, PROD relining, lapping, lining, REVEST *geología* coating, relining, TEC ESP *sistemas de lanzamiento* shroud,

vehículos coat, TEC PETR coating; **~ a la admisión** *m* INSTAL HIDRÁUL outside lap; **~ de alfas** *m* TV alpha wrap; **~ anti-deslizante para suelos** *m* SEG antislip material floor covering; **~ antifricción** *m* ING MECÁ antifriction lining; **~ antirreflexión** *m* TEC ESP antireflection coating; **~ de aspiración** *m* REFRIG suction cover; **~ del borde de ataque** *m* TRANSP AÉR leading-edge glove; **~ de la caja aspirante** *m* PAPEL suction box cover; **~ de la carga con hielo triturado** *m* REFRIG top icing; **~ de concreto** *m AmL* (*cf recubrimiento de hormigón Esp*) COLOR, PETROL, REVEST concrete coating; **~ de conversión** *m* REVEST conversion coating; **~ electrolítico** *m* ELEC electroplate, ING ELÉC electroplate, plating, REVEST electroplate; **~ electrolítico de oro** *m* METAL, TEC ESP gold plating; **~ electrolítico de plomo** *m* ELEC, ING ELÉC, REVEST electroplated terne; **~ electrolítico en tambor** *m* PROD barrel plating; **~ electrostático con pintura en polvo** *m* COLOR, P&C, REVEST electrostatic powder coating; **~ de escape** *m* INSTAL HIDRÁUL *de válvula distribuidora de corredera* exhaust cover, inside cover, inside lap, *distribuidores de máquina de vapor alternativa* exhaust lap; **~ con espuma solidificable** *m* EMB flow-foam wrap; **~ exterior** *m* INSTAL HIDRÁUL *distribuidor máquina de vapor* outside lap, *válvula de corredera* steam lap; **~ exterior negativo** *m* INSTAL HIDRÁUL *distribuidor máquina de vapor* outside clearance; **~ en frío** *m* C&V cold end coating; **~ galvánico** *m* REVEST galvanic plating; **~ de gas argón** *m* NUCL argon gas blanket; **~ horizontal** *m* MINAS *geología* floor heave; **~ de hormigón** *m Esp* (*cf recubrimiento de concreto AmL*) COLOR, PETROL, REVEST concrete coating; **~ por inmersión** *m* PROD dip coating, REVEST dip coating, immersion plating; **~ interior** *m* INSTAL HIDRÁUL *distribuidor de máquina alternativa de vapor* inside cover, TRANSP AÉR inner shroud; **~ interior del silo** *m* EMB bin liner; **~ en el lado caliente** *m* C&V hot end coating; **~ de lecho fluidizado** *m* PROC QUÍ, REVEST fluidized-bed coating; **~ litográfico** *m* ELECTRÓN lithographic mask; **~ de madera** *m* REVEST board sheathing; **~ magnético** *m* CINEMAT magnetic coating; **~ metálico** *m* EMB, ING MECÁ, PROD, REVEST, TELECOM metal coating, metallic coating; **~ de molde** *m* C&V mold coating (*AmE*), mould coating (*BrE*); **~ no metálico** *m* REVEST nonmetallic coating; **~ de origen** *m* ELECTRÓN master mask; **~ de pintura impermeable** *m* REVEST waterproof painting; **~ con pintura en polvo** *m* P&C powder coating; **~ de plástico** *m* TELECOM plastic coating; **~ plástico** *m* ING ELÉC plastic cladding; **~ del pozo** *m* CARBÓN well casing; **~ con propiedades ópicas** *m* FÍS coated lens; **~ con propiedades ópticas** *m* CINEMAT, FOTO, INSTR, ÓPT, REVEST coated lens; **~ protector** *m* EMB protection coat, PAPEL protective coating; **~ protectora repelable** *m* EMB peelable protective coating; **~ pulvimetalúrgico** *m* REVEST coating powder, powder coating; **~ de rocas** *m* MINAS *geología* rock capping; **~ tipo barrera** *m* PAPEL protective coating; **~ vítreo** *m* C&V, REVEST glass glazing

recubrir *vt* CARBÓN recover, CONST *solado* floor, PROD *cojinete con metal antifricción* line, REVEST coat, TEXTIL surface

recuento *m* OCEAN *pesca* tally; **~ de carreteras de acceso** *m* TRANSP access road count (*AmE*), slip road count (*BrE*); **~ de ciclos** *m* PROD cycle count-

ing; **~ del tráfico** *m* TRANSP traffic count; **~ de vías de acceso** *m* TRANSP access road count (*AmE*), slip road count (*BrE*)

recuesto *m AmL* (*cf lugar en declive Esp*) MINAS dip

recultivo *m* AGRIC, CONTAM *recuperación del suelo* recultivation

recupera: ~-pastas *m* PAPEL save-all

recuperación *f* C&V recuperation, CARBÓN recovery, CONTAM *procesos industriales, subproductos* recovery, reclamation, CONTAM MAR recovery, INFORM&PD retrieval, METAL recovery, MINAS *del entibado* drawing, OCEAN, P&C, PETROL recovery, PROD, QUÍMICA recuperation, RECICL recovery, reclamation, salvaging, TEC ESP recovery, TERMOTEC recuperation, TEXTIL recovery, retrieval, TRANSP AÉR, TRANSP MAR *de cosas o personas en el mar* recovery; **~ del archivo** *f* INFORM&PD file recovery; **~ de archivos** *f* INFORM&PD file recovery; **~ de las arrugas** *f* TEXTIL *del tejido* crease recovery; **~ de averías** *f* PROD fault recovery; **~ de la barrera** *f* CONTAM MAR boom retrieval; **~ del calor** *f* C&V reheating (*BrE*), RECICL heat recovery, REFRIG reheat, reheating, TERMOTEC heat recovery, reheating, reheat (*AmE*), TRANSP reheat, reheating; **~ del calor de los gases de desechos** *f* TERMO waste gas heat recovery; **~ del calor de los gases residuales** *f* TERMO *recirculación* waste gas heat recovery; **~ del calor residual** *f* TERMO waste heat recovery; **~ de cargas dinámicas** *f* ING ELÉC recovery from dynamic loads; **~ de crudo por inyección de gas** *f* TEC PETR *recuperación* gas lift; **~ de datos** *f* INFORM&PD data retrieval, garbage collection; **~ de disolvente** *f* ALIMENT solvent recovery; **~ de documentos** *f* INFORM&PD document retrieval; **~ de energía** *f* ENERG RENOV, RECICL, TERMO energy recovery, energy recuperation; **~ de equipaje** *f* TRANSP baggage retrieval; **~ de errores** *f* INFORM&PD, TELECOM error recovery; **~ de espacio** *f* INFORM&PD garbage collection; **~ de fallos** *f* INFORM&PD failure recovery; **~ de ficheros** *f* INFORM&PD file recovery; **~ de herramientas perdidas** *f* PETROL, TEC PETR *perforación* fishing; **~ de información** *f* INFORM&PD information retrieval; **~ inmediata** *f* TV instant replay; **~ de llamada** *f* TELECOM call retrieval; **~ de lubricantes usados** *f* ING MECÁ oil reclaiming; **~ de materiales** *f* RECICL materials reclamation; **~ mejorada de petróleo** *f* TEC PETR *producción* enhanced oil recovery (*Eof*); **~ de mensajes** *f* INFORM&PD, TELECOM message retrieval; **~ del metal** *f* CARBÓN metal recovery; **~ de palabra clave** *f* INFORM&PD keyword retrieval; **~ de plata** *f* CINEMAT silver recovery; **~ del pliegue** *f* TEXTIL *de las prendas* crease recovery; **~ de presión dinámica** *f* TRANSP AÉR ram recovery; **~ primaria** *f* PETROL, TEC PETR *mecanismos de producción* primary recovery; **~ del punto de control** *f* INFORM&PD checkpoint recovery; **~ del reloj** *f* TEC ESP *comunicaciones* clock recovery; **~ secundaria** *f* TEC PETR *mecanismos de producción* secondary recovery; **~ terciaria** *f* TEC PETR *mecanismos de producción* tertiary recovery; **~ de tierras** *f* AGRIC land reclamation

recuperador *m* C&V recuperator, ELECTRÓN regenerator, INSTAL TERM, PROD, QUÍMICA recuperator; **~ de calor** *m* METAL, PROD checker (*AmE*), chequer (*BrE*); **~ de dirección electrónico** *m* TRANSP electronic direction-reverser

recuperar *vt* CARBÓN, CONTAM recover, INFORM&PD retrieve, PAPEL *fibras, cargas, energía* save, RECICL salvage, TELECOM recover, TEXTIL *extraer* retrieve
recurrente *adj* MATEMÁT recurring
recursión *f* ELEC, INFORM&PD, TELECOM recursion
recursivo *adj* ELEC, INFORM&PD, TELECOM recursive
recurso *m* CONTAM MAR, INFORM&PD, PROD resource; ~ crítico *m* INFORM&PD critical resource; ~ de interconexión de sistemas abiertos *m* TELECOM open systems interconnection resource (*OSI resource*)
recursos *m pl* TV facilities; ~ de agua subterránea *m pl* AGRIC, AGUA, CARBÓN, HIDROL ground-water resources; ~ compartidos *m pl* INFORM&PD resource sharing; ~ energéticos no utilizados *m pl* PROD lockout power sources; ~ de equipo físico *m pl* INFORM&PD hardware resources; ~ geotérmicos *m pl* ENERG RENOV geothermal resources; ~ de hardware *m pl* INFORM&PD hardware resources; ~ humanos *m pl* PROD manpower; ~ identificados *m pl* ENERG RENOV identified resources; ~ naturales *m pl* AGRIC natural resources; ~ de software *m pl* INFORM&PD software resources; ~ de soporte lógico *m pl* INFORM&PD software resources
red *f* CONST *cables* netting, network, ELEC *de distribución* mains supply (*BrE*), mains (*BrE*), current supply (*AmE*), *fuente de alimentación* network, ELECTRÓN *cristales* lattice, FÍS, INFORM&PD, ING ELÉC network, MATEMÁT lattice, OCEAN lint, P&C *polímero*, PROD *sistema de líneas cruzadas*, TEC ESP, TRANSP network; **■ a** ~ de abonados *f* TELECOM customer network (*CN*); ~ de acceso *f* TELECOM access network; ~ activa *f* ING ELÉC active network; ~ adiamantada *f* METAL diamond lattice; ~ de adición *f* ELEC *circuito* adding network; ~ aérea *f* TELECOM overhead network; ~ agallera *f* OCEAN tangle net; ~ de alcantarillado *f* HIDROL sewer network, sewerage system, RECICL sewerage system; ~ para alcanzar las fichas *f* INFORM&PD token-passing network; ~ en anillo *f* INFORM&PD ring network, token ring, TELECOM ring network; ~ de antenas *f* TEC ESP aerial array (*BrE*), antenna array (*AmE*), array aerial (*BrE*), array antenna (*AmE*); ~ de antenas parabólicas *f* TEC ESP parabolic mesh antenna; ~ antitorpedos *f* D&A para la marina torpedo net; ~ en árbol *f* INFORM&PD, TELECOM tree network; ~ de área local *f* INFORM&PD, TELECOM local area network (*LAN*); ~ de área metropolitana *f* TELECOM metropolitan area network (*MAN*); ~ de arenque *f* OCEAN herring net; ~ de arrastre *f* CONTAM MAR *pesca* trawl net, OCEAN dragnet, tow net, trawl net, TRANSP MAR *pesca* trawl net; ~ de arrastre de vara *f* OCEAN shrimping net; **■ b** ~ de baja tensión *f* ELEC *fuente de alimentación* low-voltage network; ~ barredera *f* (*cf mediomundo AmL*) OCEAN scoop net, TEXTIL *tejedura* trawl; ~ biconductora aislada en T *f* ING ELÉC twin-T network; ~ bidireccional *f* ING ELÉC bidirectional network; ~ bifásica *f* ELEC *fuente de alimentación* two-phase network; ~ bifilar *f* ELEC two-wire network; ~ de bloqueo *f* ING ELÉC, TELECOM blocking network; ~ en bucle *f* INFORM&PD loop network; ~ en bus *f* INFORM&PD bus network; **■ c** ~ por cable *f* TV cabled network; ~ de cables *f* ELEC, ING ELÉC cable network; ~ camaronera *f* OCEAN shrimping net; ~ cambiadora de fase *f* ELEC *corriente alterna* phase-shifting network;

~ de camuflaje de tipo hoja cortada *f* D&A incised-leaf-type camouflage net; ~ de canalización de gas *f* GAS, TEC PETR, TERMO gas grid; ~ de carga *f* TRANSP MAR net sling; ~ en cascada *f* ING ELÉC ladder network; ~ CC *f* (*red cúbica centrada*) CRISTAL BCC lattice; ~ de CC *f* ING ELÉC DC network; ~ CCC *f* CRISTAL, METAL, QUÍMICA FCC lattice; ~ celular *f* TELECOM cellular network; ~ centrada en la base *f* METAL base-centered lattice (*AmE*), base-centred lattice (*BrE*); ~ de cerco *f* OCEAN beach seine, *tecnología pesquera* seine, TRANSP MAR *pesca* seine net; ~ de cerco danesa *f* OCEAN *arte de pesca* Danish seine; ~ de cerco con jareta *f* OCEAN purse seine; ~ de circuitos conmutados *f* INFORM&PD circuit-switched network; ~ en círculo *f* INFORM&PD, TELECOM ring network; ~ cloacal *f* RECICL sewerage system; ~ de compartimiento de recursos *f* INFORM&PD resource-sharing network; ~ de compensación *f* ING ELÉC balancing network; ~ de compensación térmica *f* ING ELÉC temperature-compensating network; ~ de computadoras *f* AmL (*cf red de ordenadores Esp*) INFORM&PD computer network; ~ de comunicación *f* INFORM&PD communication network; ~ de comunicación colectiva *f* TV conference network; ~ de comunicación de datos *f* INFORM&PD data communication network; ~ de comunicaciones urbanas *f* TELECOM local communications network; ~ de concentración *f* TELECOM concentration network; ~ conductora de energía *f* ING ELÉC power-transmission network; ~ de conductores comunes *f* INFORM&PD bus network; ~ de conductos *f* REFRIG *del aire acondicionado* duct work; ~ conexa *f* ELEC, ING ELÉC connected network; ~ de conmutación *f* ING ELÉC, TELECOM switching network; ~ de conmutación de banda ancha *f* TELECOM broadband switching network, wideband switching network; ~ de conmutación de banda estrecha *f* TELECOM narrow-band switching network; ~ con conmutación de circuitos *f* INFORM&PD circuit-switched network; ~ de conmutación de circuitos *f* TELECOM circuit-switching network; ~ de conmutación digital *f* TELECOM digital switching network; ~ de conmutación con escala *f* TELECOM store-and-forward switching network; ~ con conmutación de mensajes *f* INFORM&PD message-switched network; ~ de conmutación óptica *f* TELECOM optical-switching network (*OSN*); ~ de conmutación de paquetes *f* ELECTRÓN, INFORM&PD, TELECOM packet-switching network; ~ de conmutación por paquetes *f* ELECTRÓN, INFORM&PD, TELECOM packet-switched network (*PSN*); ~ de conmutación de relés por láminas *f* ELEC, ING ELÉC, TELECOM reed-relay switching network; ~ de conmutación de tráfico *f* TELECOM message-switching network; ~ conmutada *f* INFORM&PD, ING ELÉC, TELECOM switched network; ~ conmutada canadiense *f* TELECOM Canadian switched network (*CSN*); ~ conmutada en estrella *f* TELECOM switched-star network; ~ de conversación *f* TELECOM voice network; ~ de coordenadas de Mercator *f* OCEAN Mercator plotting chart; ~ correctora *f* TV shaping network; ~ de corriente alterna *f* ELEC alternating-current network; ~ de corrimiento de fase *f* FÍS phase-shifting network; ~ cristalina *f* CRISTAL, ELECTRÓN crystal lattice, QUÍMICA lattice; ~ cristalina de Bravais *f*

CRISTAL Bravais lattice; ~ **cronizadora** *f* TELECOM time-division network; ~ **de cuadripolos** *f* FÍS, ING ELÉC ladder network; ~ **cúbica de caras centradas** *f* CRISTAL, METAL, QUÍMICA face-centered cubic lattice (*AmE*), face-centred cubic lattice (*BrE*); ~ **cúbica centrada** *f* (*red CC*) CRISTAL body-centered cubic lattice (*AmE*), body-centred cubic lattice (*BrE*); ~ **cúbica sencilla** *f* CRISTAL simple cubic lattice; ~ **de cuerpo centrado** *f* METAL body-centered lattice (*AmE*), body-centred lattice (*BrE*);

~ d ~ **de datos** *f* INFORM&PD, TELECOM data network; ~ **de datos internacional de conmutación de paquetes** *f* TELECOM international packet-switched data network; ~ **en delta** *f* ING ELÉC delta network; ~ **en delta bifilar** *f* ELEC *fuente de alimentación* two-wire delta network; ~ **de deriva** *f* TRANSP MAR fishing drift-net; ~ **desfasadora** *f* ELEC *corriente alterna* phase-shifting network; ~ **desplazadora de fase** *f* ELEC *corriente alterna* phase-shifting network; ~ **de destino** *f* TELECOM destination network (*DN*); ~ **de difracción** *f* FÍS, ÓPT diffraction grating; ~ **de difracción glaseada** *f* FÍS blazed grating; ~ **difractora** *f* FÍS ONDAS diffraction grating, grating; ~ **digital integrada** *f* TELECOM integrated digital network (*IDN*); ~ **digital de servicios integrados** *f* (*RDSI*) TELECOM integrated services digital network (*ISDN*); ~ **digital de servicios integrados de banda ancha** *f* (*RDSI-BA*) TELECOM broadband-integrated services digital network (*B-ISDN*); ~ **de distribución** *f* AGUA current, mains (*BrE*), distribution system, CONST distribution network, ELEC distribution network, distribution system, grid, high-voltage grid, supply network, GAS distribution system, ING ELÉC distribution network, distribution system, grid, TELECOM distribution network; ~ **de distribución del conductor auxiliar** *f* ELEC pilot network; ~ **de distribución eléctrica** *f* ELEC current, ING ELÉC current, mains (*BrE*); ~ **de distribución de gas** *f* GAS, TEC PETR, TERMO gas grid; ~ **de distribución local** *f* TELECOM local distribution network; ~ **de distribución de potencia** *f* TEC ESP power distribution network; ~ **de distribución trifásica** *f* ELEC three-phase supply network; ~ **distribuida** *f* INFORM&PD distributed network; ~ **distribuidora de energía** *f* TEC ESP power converter; ~ **doble T en paralelo** *f* ING ELÉC twin-T network; ~ **de dos accesos** *f* ING ELÉC two-port network; ~ **de dos terminales** *f* FÍS *cuadripolo* two-terminal network;

~ e ~ **eléctrica** *f* ELEC electric-power system, electrical network, *de distribución* mains supply (*BrE*), current supply (*AmE*), mains (*BrE*), ELECTRÓN, FÍS electrical network, ING ELÉC electric-power system, network, TERMO electric-power system, TV current (*AmE*), mains (*BrE*); ~ **eléctrica del avión** *f* TRANSP AÉR aircraft mains; ~ **eléctrica de cuatro bornes** *f* ING ELÉC four-terminal network; ~ **eléctrica de cuatro terminales** *f* ING ELÉC four-terminal network; ~ **eléctrica nacional** *f* ELEC, NUCL national grid; ~ **de energía eléctrica** *f* ELEC grid, *fuente de alimentación* high-voltage grid, ING ELÉC grid, power system; ~ **de enmalle** *f* OCEAN tangle net, TRANSP MAR fishing drift-net; ~ **de enmalle de deriva** *f* TRANSP MAR drift net; ~ **equilibrada** *f* ELEC *fuente de alimentación*, ING ELÉC balanced network; ~ **de equilibrado de impedancias** *f* FÍS impedance-matching network; ~ **de equilibrio** *f* ELEC *fuente de*

alimentación balanced network; ~ **de equilibro** *f* ING ELÉC balanced network; ~ **en escalera** *f* FÍS ladder network; ~ **espacial** *f* CRISTAL space lattice; ~ **espacio-tiempo-espacio** *f* TELECOM space-time-space network; ~ **estacada** *f* OCEAN stake net; ~ **en estrella** *f* INFORM&PD, TEC ESP, TELECOM star network; ~ **estrictamente antibloqueante** *f* TELECOM strictly nonblocking network; ~ **de expansión** *f* TELECOM expansion network;

~ f ~ **de ferrocarriles interceptada** *f* MINAS trapped rail system; ~ **de fibra óptica** *f* ING ELÉC, ÓPT, TELECOM fiber-optic network (*AmE*), fibre-optic network (*BrE*); ~ **fija aeronáutica** *f* TRANSP AÉR aeronautical fixed service (*BrE*), aeronautical fixed system (*AmE*); ~ **fija de telecomunicaciones aeronáuticas** *f* TRANSP AÉR aeronautical fixed telecommunication network;

~ g ~ **de gasoductos** *f* GAS, TEC PETR, TERMO gas grid; ~ **geodésica** *f* CARBÓN geodesic survey; ~ **de gestión de jerarquía digital síncrona** *f* TELECOM SDH management network (*SMN*); ~ **de gestión de telecomunicaciones** *f* TELECOM telecommunications management network (*TMN*); ~ **de gran amplitud** *f* IMPR wide-area network;

~ h ~ **hexagonal compacta** *f* NUCL hexagonal close-packed lattice; ~ **hidráulica** *f* HIDROL water system;

~ i ~ **de igualación de impedancias** *f* ING ELÉC impedance-matching network; ~ **por impedancia** *f* PROD impedance network; ~ **impulsora** *f* ING ELÉC active network; ~ **informática** *f* INFORM&PD, TELECOM computer network; ~ **de instalación del abonado** *f* TELECOM subscriber premises network (*SPN*); ~ **de integración** *f* TRANSP AÉR integrating network; ~ **interactiva** *f* TELECOM interactive network; ~ **de interconexión** *f* TELECOM interconnection network; ~ **interurbana** *f* TELECOM trunk network;

~ l ~ **L** *f* ING ELÉC L-network; ~ **con línea alquilada** *f* INFORM&PD leased line network; ~ **de la línea de transmisión** *f* ELEC transmission line network; ~ **lineal** *f* FÍS linear network; ~ **lineal de corriente** *f* ELEC linear current network; ~ **lineal de cuatro polos** *f* ELEC linear four-terminal network; ~ **de líneas arrendadas** *f* INFORM&PD leased line network;

~ m ~ **de malla centrada** *f* METAL body-centered lattice (*AmE*), body-centred lattice (*BrE*); ~ **de malla compacta** *f* METAL close-packed lattice; ~ **de malla pequeña** *f* METAL close-packed lattice; ~ **mallada** *f* ELEC *fuente de alimentación* mesh network; ~ **de mallas** *f* ELEC *fuente de alimentación*, INFORM&PD meshed network; ~ **metropolitana** *f* TELECOM metropolitan network; ~ **con modulación de amplitud de impulsos** *f* TELECOM pulse-amplitude modulation network (*PAM network*); ~ **de muestreo** *f* CALIDAD sampling network; ~ **multietapa** *f* TELECOM multistage network; ~ **mundial** *f* TEC ESP worldwide network;

~ n ~ **nacional de energía eléctrica** *f* ELEC, ING ELÉC grid; ~ **neural** *f* INFORM&PD neural network; ~ **neurológica** *f* INFORM&PD neural network; ~ **neuronal** *f* INFORM&PD neural network; ~ **no bloqueante de reordenamiento** *f* TELECOM rearrangeable non-blocking network; ~ **de no bloqueo** *f* TELECOM nonblocking network; ~ **no lineal** *f* CONST nonlinear network; ~ **no mallada** *f* TELECOM tree network;

~ o ~ **del objetivo** *f* TV target mesh; ~ **de ondas estacionarias** *f* FÍS ONDAS stationary wave pattern; ~ **óptica pasiva** *f* TELECOM passive optical network (*PON*); ~ **de orden superior** *f* TELECOM high-order network; ~ **de ordenadores** *f Esp* (*cf red de computadoras AmL*) INFORM&PD, TELECOM computer network; ~ **ordinaria** *f* TELECOM ordinary network;

~ p ~ **pasiva** *f* ELEC, ING ELÉC passive network; ~ **de paso de testigo** *f* INFORM&PD *red en anillo* token-passing network; ~ **pequeña unida a un mango largo** *f* OCEAN push net; ~ **de pesca** *f* OCEAN fishing net; ~ **Petri** *f* INFORM&PD Petri net; ~ **petrogenética** *f* GEOL *petrología metamórfica* petrogenetic grid; ~ **pi** *f* FÍS, ING ELÉC pi network; ~ **de picos** *f* TV peaking network; ~ **plegada** *f* TELECOM folded network; ~ **polifásica** *f* ELEC *fuente de alimentación* polyphase network; ~ **poligonal** *f* ELEC *fuente de alimentación*, TEC ESP mesh network; ~ **de ponderación** *f* ACÚST weighting network; ~ **privada** *f* TELECOM private network; ~ **protectora** *f* OCEAN netting; ~ **pública** *f* TELECOM public network; ~ **pública de datos** *f* (*RPD*) INFORM&PD public data network (*PDN*); ~ **pública móvil** *f* TELECOM public mobile network; ~ **en puente** *f* ING ELÉC lattice network; ~ **con puente en H** *f* ING ELÉC bridged-H network; ~ **con puente en T** *f* ING ELÉC bridged-T network; ~ **puenteada** *f* ING ELÉC lattice network;

~ r ~ **de radiodifusión** *f* TV broadcasting network; ~ **radioeléctrica** *f* TELECOM radio system; ~ **ramificada** *f* TELECOM tree and branch network, tree network; ~ **recíproca** *f* CRISTAL reciprocal lattice; ~ **de recursos compartidos** *f* INFORM&PD resource-sharing network; ~ **reflectora** *f* FÍS, FÍS ONDAS, ÓPT reflection grating; ~ **de refracción** *f* LAB, ÓPT refraction grating; ~ **repartida** *f* INFORM&PD distributed network; ~ **de resistencias** *f* ING ELÉC resistor network; ~ **reticular** *f* TEC ESP mesh network; ~ **rural** *f* TELECOM rural network;

~ s ~ **secundaria** *f* TELECOM subnetwork; ~ **de seguimiento espacial y adquisición de datos** *f* D&A space tracking and data acquisition network (*STADAN*); ~ **de seguridad** *f* SEG safety net; ~ **de seguridad para andamiajes** *f* SEG scaffolding protective net; ~ **de señalización** *f* TELECOM signaling network (*AmE*), signalling network (*BrE*); ~ **de servicio público** *f* TELECOM public network; ~ **de servicio público móvil** *f* TELECOM public mobile network; ~ **sincrónica** *f* INFORM&PD synchronous network; ~ **de sincronización** *f* TELECOM synchronization network; ~ **de sistema de espera** *f* TELECOM queueing network; ~ **sub-Clos** *f* TELECOM sub-Clos network; ~ **sumadora** *f* ELEC *circuito* adding network;

~ t ~ **en T** *f* FÍS, ING ELÉC, TELECOM T-network; ~ **del tamiz** *f* PROC QUÍ sieve netting; ~ **de telecomunicaciones** *f* INFORM&PD, TELECOM telecommunications network; ~ **telefónica** *f* TELECOM telephone network, voice network; ~ **telefónica conmutada** *f* TELECOM public switched telephone network (*PSTN*); ~ **telefónica de líneas conmutadas** *f* TELECOM switched telephone network; ~ **telefónica privada** *f* TELECOM private telephone network; ~ **telefónica pública** *f* TELECOM public telephone network; ~ **telefónica pública de**

líneas conmutadas *f* TELECOM public switched telephone network; ~ **telefónica tradicional** *f* TELECOM traditional telephone network; ~ **telefónica urbana de conexiones** *f* TELECOM trunk-switching exchange area; ~ **telemática** *f* INFORM&PD data communication network; ~ **de televisión por abono** *f AmL* (*cf red de televisión de pago Esp*) TELECOM, TV pay-television network; ~ **de televisión por cable** *f* TELECOM, TV cable television network; ~ **de televisión de pago** *f Esp* (*cf red de televisión por abono AmL*) TELECOM pay-television network; ~ **de televisoras** *f* TV network; ~ **térmica equivalente** *f* TERMO, TERMOTEC equivalent thermal network; ~ **tetrapolar lineal** *f* ELEC linear four-terminal network; ~ **de tierra** *f* TELECOM ground network; ~ **de tiro** *f* OCEAN tow net; ~ **token ring** *f* INFORM&PD token ring; ~ **de tracción** *f* ELEC traction network; ~ **de transmisión** *f* GAS, TELECOM transmission network; ~ **de transmisión de datos por paquetes** *f* TELECOM packet data-transmission network; ~ **de transmisión por paquetes** *f* ELECTRÓN, INFORM&PD, TELECOM packet-switched network (*PSN*); ~ **de transmisión de voz y datos** *f* TELECOM speech-data network; ~ **de transporte de datos** *f* INFORM&PD, TELECOM data-transport network; ~ **en triángulo** *f* ING ELÉC delta network; ~ **en triángulo bifilar** *f* ELEC *fuente de alimentación* two-wire delta network; ~ **trifilar** *f* ING ELÉC three-wire mains (*BrE*), three-wire supply network (*AmE*); ~ **TS** *f* TELECOM TS network; ~ **de tuberías** *f* CONST pipage, piping network, system of pipes; ~ **de tubos refrigerantes** *f* REFRIG *pista de patinaje sobre hielo* pipe-cooling grid;

~ u ~ **urbana** *f* TELECOM urban network;

~ v ~ **de valor agregado** *f AmL* INFORM&PD, TELECOM value-added network (*VAN*); ~ **de valor añadido** *f Esp* INFORM&PD, TELECOM value-added network (*VAN*);

~ y ~ **Y** *f* FÍS Y-network

Red: la ~ *f* INFORM&PD Internet, the Net

redada *f* TRANSP MAR *pesca* haul

redero *m* OCEAN net mender

redes *f pl* OCEAN fish traps

redestilación *f* QUÍMICA redistillation

redestilar *vt* QUÍMICA redistil (*BrE*), redistill (*AmE*)

rediente: ~ **del casco** *m* TRANSP AÉR hull step

redistribución *f* ING ELÉC redistribution; ~ **de los elementos de combustible en el núcleo** *f* NUCL core shuffling; ~ **radial del combustible** *f* NUCL radial shuffling

redoma *f AmL* (*cf matraz Esp*) QUÍMICA balloon

redondeamiento *m* GEOL roundness

redondear *vt* INFORM&PD, MATEMÁT, PROD round, round off; ~ **por defecto** *vt* INFORM&PD, MATEMÁT, PROD round down; ~ **por exceso** *vt* INFORM&PD, MATEMÁT, PROD round up

redondeo *m* IMPR, INFORM&PD, MATEMÁT, PROD rounding; ~ **y encuadernación** *m* IMPR rounding and binding; ~ **y enlomado** *m* IMPR rounding and backing; ~ **de esquina** *m* IMPR corner cut; ~ **del lomo** *m* IMPR back rounding

redondez *f* GEOL roundness

redondilla *f* IMPR Roman type

redondo: ~ **de saneado del techo** *m AmL* (*cf bulón de saneo Esp*) MINAS *galerías* scaling bar

redox[1] *abr* (*oxidación-reducción, óxido-reducción,*

oxidoreducción) LAB, QUÍMICA redox (*oxidation-reduction, oxidoreduction*)

redox[2]: ~ **cell** n CHEM, LAB célula de redox *f*, celda de redox *f*

reducción *f* AGRIC *de existencias*, CARBÓN *química* reduction, CINEMAT, CONST, FOTO, IMPR reducing, INFORM&PD zoom-out, ING MECÁ gear ratio, off, depression, PROD thinning out, thinning down, QUÍMICA reducing, reduction, TEXTIL cutback; ~ **de aire libre** *f* GEOFÍS, GEOL free-air reduction; ~ **por alejamiento de la lente** *f* INFORM&PD *efecto* zoom-out; ~ **catalítica selectiva** *f* CONTAM *procesos químicos* selective catalytic reduction; ~ **cenital** *f* TEC ESP zenith reduction; ~ **de color de fondo** *f* IMPR under-color removal (*AmE*), under-colour removal (*BrE*); ~ **de contenido de dióxido de azufre** *f* CONTAM sulfur dioxide reduction (*AmE*), sulphur dioxide reduction (*BrE*); ~ **de contraste** *f* CINEMAT, FÍS, FOTO, TV contrast reduction; ~ **de coste en el montaje mecánico** *f* ING MECÁ cost reduction in mechanical assembly; ~ **de crecidas** *f* AGUA flood abatement; ~ **de datos** *f* INFORM&PD data reduction; ~ **por dióxido de azufre** *f* CONTAM *procesos químicos, depuración* sulfur dioxide reduction (*AmE*), sulphur dioxide reduction (*BrE*); ~ **de engranajes** *f* ING MECÁ back speed; ~ **del giro** *f* TEC ESP spin-down; ~ **granulométrica** *f* PROC QUÍ particle-size reduction; ~ **en lecho fluidizado** *f* PROC QUÍ fluidized-bed reduction; ~ **de la llama** *f* C&V flame attenuation; ~ **de marcha** *f* AUTO kick down; ~ **de muelle** *f* METR scaling a spring; ~ **de oleaje** *f* OCEAN *medidas de protección para playas, barcos* swell abatement; ~~**oxidación** *f* LAB, QUÍMICA redox; ~ **de potencia** *f* ING ELÉC power-down; ~ **de presión** *f* FÍS pressure reduction; ~ **del punto** *f* IMPR dot etching; ~ **de ruidos** *f* ACÚST, INSTR, SEG *para transportadores oscilantes* noise reduction; ~ **de la sección de paso** *f* ING MECÁ throttling; ~ **de tamaño** *f* PROC QUÍ size reduction; ~ **en tamaño** *f* PROD draft (*AmE*), draught (*BrE*); ~ **del tamaño de partículas** *f* CARBÓN, PROC QUÍ particle-size reduction; ~ **de transmisión** *f* AUTO transmission reduction; ~ **de velocidad** *f* AUTO kick down; ~ **del volumen y aumento de densidad** *f* CARBÓN, PETROL, TEC PETR, TRANSP consolidation

reducir[1] *vt* ALIMENT boil down, CARBÓN reduce, CINEMAT *abertura del diagrama* close down, CONST reduce, FOTO reduce, close down, ING MECÁ, MATEMÁT reduce, PROD *chapa, placa* thin down, QUÍMICA *óxido* reduce, TERMO boil down; ~ **el ancho de** *vt* CONST reduce to width; ~ **a cenizas** *vt* QUÍMICA reduce to ashes; ~ **el diámetro de la punta de** *vt* PROD tag; ~ **a fragmentos** *vt* CARBÓN fragment; ~ **a la mínima expresión** *vt* MATEMÁT reduce to its lowest terms

reducir[2]: ~ **gases** *vi* TRANSP AÉR throttle back; ~ **la velocidad** *vi* ING MECÁ, TRANSP reduce speed

reducirse *v refl* PROD thin out

reductivo *adj* QUÍMICA reductive

reductor *m* CINEMAT, CONST reducer, ELECTRÓN *atenuador* pad, FOTO reducer, ING ELÉC negative booster, pad, ING MECÁ divider, QUÍMICA reducer, TELECOM restrictor; ~ **de alumbrado** *m* ELEC *control* dimmer; ~ **bidireccional** *m* PROD two-way restrictor; ~ **de iluminación** *m* ELEC *control* dimmer; ~ **de luz** *m* ELEC *control* dimmer; ~ **de potencia** *m* REFRIG capacity reducer; ~ **de presión** *m* ELEC *válvula,*

osciloscopio getter, TRANSP AÉR pressure reducer; ~ **proporcional** *m* FOTO proportional reducer; ~ **roscado** *m* ING MECÁ *tuberías*, MECÁ bushing; ~ **de ruidos** *m* TELECOM noise reducer; ~ **selectivo** *m* FOTO selective reducer; ~ **de tensión** *m* ING ELÉC adaptor; ~ **de tiro** *m* MINAS *chimenea* baffle; ~ **de tubería** *m* CONST pipe reducer; ~ **de voltaje** *m* ELEC voltage divider, ING ELÉC adaptor

redundancia *f* INFORM&PD, TEC ESP redundancy; ~ **de reserva** *f* TEC ESP standby redundancy

reeditar *vt* CINEMAT, TV re-edit

reejecución *f* INFORM&PD rerun

reejecutar *vt* INFORM&PD rerun

reelaboración *f* PROD, RECICL reprocessing; ~ **final** *f* NUCL tail-end process

reembarcar *vi* TRANSP MAR *pasajeros* re-embark

reembarque *m* TRANSP MAR *de cosas* reshipment

reemplazamiento: ~ **isomórfico** *m* CRISTAL isomorphous replacement

reemplazo *m* INFORM&PD *direcciones* displacement; ~ **de lámpara** *m* FOTO lamp replacement

reemplear *vt* RECICL reuse

reencaminamiento *m* AmL (*cf desvío Esp*) TELECOM *del tráfico* rerouting

reencender *vt* TRANSP AÉR restart

reencendido *m* TEC ESP reignition, TRANSP AÉR restart; ~ **en molinete** *m* TRANSP AÉR windmilling restart

reencuadrar *vt* CINEMAT reframe

reengaste *m* ING MECÁ, PROD resetting

reenrollar *vt* CINEMAT, FOTO, INFORM&PD, PAPEL, TEXTIL rewind

reentelar *vt* CARBÓN *alas de aviones* recover

reentrada: ~ **a través de la atmósfera** *f* TEC ESP atmospheric reentry

reentrante[1] *adj* GEOM, INFORM&PD re-entrant

reentrante[2] *m* GEOM, INFORM&PD re-entrant

reenvío *m* VEH *transmisión* backlash

reescribir *vt* INFORM&PD rewrite

reescritura *f* INFORM&PD rewrite

reestañado *m* PROD retinning

reestructuración *f* PROD restructuring, TEC PETR *puntos de venta de combustibles* revamping

reetiquetado: ~ **de latas** *m* EMB can relabeling (*AmE*), tin relabelling (*BrE*)

reexpedir *vt* TELECOM redirect

reextraer *vt* NUCL strip

referencia[1]: **de** ~ *adj* ING MECÁ standard

referencia[2] *f* ACÚST, CINEMAT cueing, CONST datum, *topografía* benchmark, INFORM&PD mark, ING MECÁ marker, guide, guiding, PROD *línea* guiding mark, mark, *topografía* marking, TEC ESP ranging, TV cueing; ~ **de aplicación** *f* TELECOM application reference; ~ **de blanco** *f* TV white reference; ~ **de la cara del reloj** *f* TRANSP AÉR clock-face reference; ~ **del control de intercambios** *f* TELECOM interchange control reference; ~ **de densidad óptica externa** *f* IMPR reflection density reference (*RDR*); ~ **estándar** *f* METR reference standard; ~ **de la estructura** *f* ELECTRÓN *litografía* pattern registration; ~ **externa** *f* INFORM&PD external reference; ~ **de fase** *f* TV phase reference; ~ **inercial** *f* TEC ESP *máquinas* inertial frame; ~ **de llamada** *f* TELECOM call reference; ~ **local** *f* TELECOM local reference; ~ **de perforación** *f* ING MECÁ, MECÁ punch mark; ~ **del receptor** *f* TELECOM recipient reference; ~ **del recibidor** *f* TELECOM recipient reference; ~ **de sub-**

portadora de crominancia f TV chrominance subcarrier reference; **~ de voltaje** f ING ELÉC voltage reference

referente m INFORM&PD marker

refinación f C&V refining, INFORM&PD refinement, PAPEL, PROD refining, QUÍMICA refining, upgrading, TEC PETR refining

refinado[1] adj C&V, CONTAM, CONTAM MAR refined, PAPEL beaten, refined; **no ~** adj ALIMENT unrefined, CARBÓN crude, raw

refinado[2] m C&V refining, CONTAM, CONTAM MAR refined product, PAPEL beating, refining, PROD, QUÍMICA, TEC PETR refining, processing; **~ de uranio** m NUCL uranium refining

refinador: **~ no presurizado** m PAPEL nonpressurized refiner

refinamiento m CRISTAL refinement; **~ por pasos** m INFORM&PD stepwise refinement; **~ por zonas** m CRISTAL zone refining

refinar vt C&V refine, CONST *carpintería, metales* try, try up, CONTAM, CONTAM MAR, PAPEL refine, QUÍMICA upgrade, TEC PETR purify

refinería f PROD *para hierro fundido*, TEC PETR refinery; **~ metalúrgica** f METAL, PROD metal refinery; **~ de sal** f PROD salt refinery, salt works

refino m C&V refining, PAPEL refining, *de pastas papeleras* freeness, PROD, QUÍMICA, TEC PETR refining; **~ cónico** m PAPEL conical refiner; **~ de discos** m PAPEL disc refiner (*BrE*), disk refiner (*AmE*); **~ con disolvente** m TEC PETR solvent refining; **~ de grano** m METAL grain refinement; **~ a presión atmosférica** m PAPEL atmospheric refiner; **~ de troceados** m PAPEL chip refining

refinómetro m PAPEL freeness tester

reflectancia f GEN reflectance, reflectivity; **~ espectral** f FÍS spectral reflectance

reflectante[1] adj ACÚST, ING ELÉC, ÓPT reflecting

reflectante[2]: **~ trasero** m AUTO, TRANSP, TRANSP MAR, VEH rear reflector

reflectividad f GEN reflectivity

reflectometría: **~ óptica en el dominio temporal** f ÓPT, TELECOM optical time domain reflectometry (*OTDR*)

reflectómetro m P&C *instrumento*, PAPEL reflectometer

reflector m AUTO reflector, CINEMAT reflector, scoop, ELECTRÓN, FÍS, FOTO reflector, ING ELÉC repeller, ING MECÁ, INSTR, ÓPT, TEC ESP reflector; **~ de concentración** m INSTR concentrating reflector; **~ conformado** m TEC ESP shaped reflector; **~ descentrado** m TEC ESP offset reflector; **~ difusor** m CINEMAT shadow light; **~ difusor principal** m CINEMAT shadow key; **~ de enfoque variable** m FOTO variable-focus reflector; **~ de espejo frío** m CINEMAT cold-mirror reflector; **~ flexible** m TEC ESP flexible reflector; **~ de Lambert** m ÓPT lambertian reflector; **~ lambertiano** m ÓPT lambertian reflector; **~ de lente escalonada** m CINEMAT spot; **~ de luz indirecta** m INSTR indirect light reflector; **~ con montaje giratorio** m FOTO swivel-mounted reflector; **~ de observación** m INSTR observation mirror; **~ octante** m INSTR octant mirror; **~ parabólico** m CINEMAT, D&A, FOTO parabolic reflector, TV satellite dish; **~ parabólico de radar** m TRANSP AÉR radar dish; **~ planar** m INSTR planar reflector; **~ plateado** m INSTR silvered reflector; **~ principal** m TEC ESP main reflector; **~ de radar** m D&A, SEG, TRANSP, TRANSP AÉR, TRANSP MAR radar reflector; **~ rígido** m TEC ESP rigid reflector; **~ de salida** m INSTR output mirror; **~ secundario** m TEC ESP secondary reflector; **~ solar óptico** m (*OSR*) TEC ESP optical solar reflector (*OSR*); **~ de sujeción** m FOTO clamping reflector

reflejado adj GEOM *ángulo* reflex

reflejar vti GEN reflect

reflejo m FÍS ONDAS *de dos ondas que se mezclan* interference pattern, GEOM reflex

reflex: **~ a través del objetivo** m CINEMAT through-the-lens reflex

reflexión f GEN reflection; **~ ausente** f CRISTAL absent reflection; **~ de Bragg** f CRISTAL Bragg reflection; **~ difusa** f ELECTRÓN scattering, FÍS RAD, ING ELÉC diffuse reflection; **~ especular** f CINEMAT, FÍS, TELECOM specular reflection; **~ de Fresnel** f ÓPT, TELECOM Fresnel reflection; **~ interna** f FÍS RAD internal reflection; **~ interna total** f C&V, FÍS ONDAS total internal reflection; **~ en la ionosfera** f FÍS ONDAS reflection from ionosphere; **~ por multitrayecto** f TELECOM *de ondas* multipath reflection; **~ por multitrayectoria** f TELECOM *de ondas* multipath reflection; **~ parcial** f FÍS ONDAS *de ondas luminosas* partial reflection; **~ de rayos X** f FÍS RAD X-ray reflection; **~ selectiva** f FÍS selective reflection; **~ total** f FÍS, ÓPT, TELECOM total reflection

reflexividad f ENERG RENOV, ÓPT, TEC ESP, TERMOTEC reflectivity

refluir vi TRANSP MAR *marea* set out

reflujo[1]: **sin ~** adj QUÍMICA no-reflux

reflujo[2] m ENERG RENOV, HIDROL falling tide, NUCL reflux, OCEAN ebb, tide flow, PROC QUÍ, QUÍMICA, TEC PETR reflux, TRANSP MAR falling tide, *de la marea* ebb

reforestación f AGRIC reafforestation, reforestation

reforma: **~ agraria** f AGRIC land reform

reformado m TEC PETR *refino* reforming; **~ catalítico** m TEC PETR catalytic reforming; **~ térmico** m TEC PETR thermal reforming; **~ por vapor** m TEC PETR steam reforming; **~ de la vaporización** m TEC PETR steam reforming

reformar vt ING MECÁ amend

reformateado m INFORM&PD reformatting

reforzado adj ING MECÁ, MECÁ heavy-duty; **~ con fibra de vidrio** adj EMB, PROD glass-fiber reinforced (*AmE*), glass-fibre reinforced (*BrE*); **~ con vitrofibra** adj EMB, PROD glass-fiber reinforced (*AmE*), glass-fibre reinforced (*BrE*)

reforzador m *AmL* (*cf larguero Esp*) ING ELÉC boost, ING MECÁ stiffener, MINAS stringer, TEC ESP stiffening; **~ de la imagen** m TV image flicker; **~ de sabor** m ALIMENT flavor enhancer (*AmE*), flavour enhancer (*BrE*)

reforzamiento m CONST reinforcement; **~ por dispersión** m PROC QUÍ dispersion strengthening; **~ de partículas** m METAL particle reinforcement

reforzar vt ING MECÁ *banco de torno*, MECÁ brace, PROD strengthen

refracción f ACÚST, FÍS, FÍS ONDAS, ÓPT, TELECOM, TRANSP MAR *sextante* refraction; **~ óptica** f TELECOM optical refraction

refractado adj ACÚST, FÍS, ÓPT, TELECOM refracted

refractar vt ACÚST, FÍS, ÓPT, TELECOM refract

refractario[1] adj GEN fireproof (*Am*), flameproof (*BrE*), heat-resistant, refractory

refractario[2] m C&V, INSTAL TERM, PROD, QUÍMICA,

TERMO refractory; ~ **básico** *m* C&V magnesite chrome refractory; ~ **de circona** *m* C&V zirconia refractory; ~ **de espinela** *m* C&V spinel refractory; ~ **sinterizado** *m* C&V sintered refractory; ~ **de zircón** *m* C&V zircon refractory

refractividad *f* GEN refractivity; ~ **molecular** *f* FÍS molecular refractivity

refractómetro *m* FÍS, FÍS RAD, INSTR, LAB refractometer; ~ **de Abbé** *m* FÍS Abbé refractometer; ~ **binocular** *m* INSTR binocular refractometer; ~ **del cénit** *m* INSTR vertex refractometer; ~ **de inmersión** *m* INSTR dipping refractometer; ~ **de Pulfrich** *m* FÍS Pulfrich refractometer; ~ **de Rayleigh** *m* FÍS Rayleigh refractometer

refractor *m* FÍS, GEOFÍS, INSTR, ÓPT refractor

refrentado: ~ **de bridas** *m* PROD flange face; ~ **posterior** *m* ING MECÁ back spotfacing

refrescadura: ~ **en salmuera** *f* ALIMENT brine cooling

refrescamiento: ~ **en salmuera** *m* ALIMENT brine cooling

refrescar *vt* ELECTRÓN *memorias dinámicas*, INFORM&PD refresh

refrigeración *f* ALIMENT refrigeration, C&V, GAS, ING MECÁ cooling, REFRIG, TERMO cooling, refrigeration, TEXTIL cooling; ~ **absorbente** *f* ENERG RENOV absorption cooling; ~ **del aceite** *f* ELEC *transformador* oil cooling; ~ **por aceite a presión** *f* ING MECÁ, REFRIG forced-oil cooling; ~ **por agua helada** *f* REFRIG hydro-cooling; ~ **por agua a presión** *f* ING MECÁ, REFRIG forced-water cooling; ~ **por aire** *f* AUTO, ING MECÁ air cooling; ~ **por aire o aceite** *f* TERMO *transformadores* air-or-oil cooling; ~ **a baja temperatura** *f* REFRIG, TERMO cooling to low temperature; ~ **centralizada** *f* REFRIG aggregate cooling; ~ **de combustible** *f* REFRIG, TERMO fuel cooling; ~ **confortable** *f* REFRIG, TERMO comfort cooling; ~ **de un contenedor de combustible nuclear** *f* NUCL jacket cooling; ~ **por convección** *f* REFRIG, TERMO convection cooling; ~ **crítica** *f* REFRIG superchilling; ~ **por evaporación** *f* REFRIG, TERMO evaporative cooling; ~ **inadecuada del núcleo** *f* NUCL inadequate core cooling; ~ **por inmersión** *f* REFRIG immersion cooling; ~ **límite** *f* REFRIG superchilling; ~ **por líquido** *f* TERMO liquid cooling; ~ **mecánica** *f* ING MECÁ, REFRIG mechanical refrigeration; ~ **del núcleo en condiciones de parada** *f* NUCL shutdown cooling; ~ **por presión de aire** *f* REFRIG pressure-cooling; ~ **rápida** *f* REFRIG, TERMO quick chilling, rapid chilling, rapid cooling; ~ **rápida por aire** *f* REFRIG, TERMO rapid air cooling; ~ **regenerativa** *f* REFRIG, TERMO regenerative cooling; ~ **con tiro natural** *f* TERMO natural-draft cooling (*AmE*), natural-draught cooling (*BrE*); ~ **por ventilación** *f* REFRIG, TERMO fan cooling

refrigerado[1] *adj* ALIMENT chilled, refrigerated, AUTO refrigerated, REFRIG, TERMO chilled, cooled, refrigerated, VEH refrigerated; ~ **con aceite** *adj* TERMO oil-cooled; ~ **por agua** *adj* PROD, TERMO, TEXTIL water-cooled; ~ **por aire** *adj* ELEC *máquina*, ING MECÁ, TERMO air-cooled; ~ **con gas** *adj* GAS, REFRIG, TERMO gas-cooled; ~ **por líquido** *adj* TERMO liquid-cooled; ~ **por metal líquido** *adj* TERMO liquid-metal-cooled

refrigerado[2]: ~ **por aire forzado** *m* ING MECÁ, PROD, REFRIG forced-air cooling; ~ **por aire a presión** *m* ING MECÁ, PROD, REFRIG forced-air cooling

refrigerador[1] *adj* ALIMENT, ING MECÁ, REFRIG refrigerating

refrigerador[2] *m* ALIMENT refrigerator, ELEC fridge, refrigerator, ING MECÁ cooler, freezer, refrigerator, PROC QUÍ freezer, REFRIG, TERMO cooler, freezer, refrigerator; ~ **de absorción** *m* REFRIG absorption refrigerator, TERMO absorption-type refrigerator; ~ **de aceite** *m* ELEC, TERMO oil cooler; ~ **de aletas** *m* TERMO finned cooler; ~ **de compresión** *m* ING MECÁ, REFRIG, TERMO compression refrigerator; ~ **de convección** *m* TERMO convection cooler; ~ **por convección** *m* REFRIG convection cooler; ~ **criogénico** *m* REFRIG cryogenic refrigerator; ~ **de dilución** *m* FÍS, REFRIG dilution refrigerator; ~ **por dilución de helio** *m* FÍS helium-dilution refrigerator; ~ **de doble temperatura** *m* REFRIG dual-temperature refrigerator; ~ **doméstico** *m* ING MECÁ domestic refrigerator; ~ **a gas** *m* GAS, ING MECÁ, REFRIG, TERMO gas refrigerator; ~ **de hielo** *m* REFRIG ice refrigerator; ~ **de leche** *m* AGRIC, REFRIG milk cooler; ~ **radiante** *m* REFRIG, TERMO radiant cooler; ~ **tubular** *m* ALIMENT tubular cooler

refrigerante[1] *adj* AUTO, CONTAM refrigerant, ING MECÁ chilling, cooling, refrigerant, PROC QUÍ refrigerant, QUÍMICA, REFRIG, TERMO chilling, cooling, refrigerant, TERMOTEC cooling, VEH refrigerant

refrigerante[2] *m* AUTO refrigerant, CINEMAT condenser, CONST coolant, CONTAM *procesos industriales* refrigerant, ELEC condenser, FÍS coolant, ING ELÉC condenser, ING MECÁ coolant, refrigerant, LAB *material de vidrio, refrigeración* condenser, PROC QUÍ, QUÍMICA, REFRIG condenser, refrigerant, TEC PETR *refino* condenser, TERMO refrigerant, TERMOTEC, TRANSP MAR condenser, VEH condenser, *sistema de refrigeración* refrigerant; ~ **de aceite** *m* VEH *lubricación* oil cooler; ~ **de emergencia del núcleo** *m* NUCL emergency core coolant; ~ **fluorocarbonado** *m* REFRIG fluorocarbon refrigerant; ~ **halocarbonado** *m* REFRIG halocarbon refrigerant; ~ **líquido** *m* ING MECÁ liquid cooler, TERMO liquid coolant; ~ **orgánico** *m* ING MECÁ organic refrigerant; ~ **primario** *m* REFRIG primary refrigerant; ~ **secundario** *m* REFRIG secondary refrigerant, cooling medium, TERMO cooling medium

refrigerar *vt* ALIMENT, ING MECÁ, REFRIG, TERMO refrigerate; ~ **con aire** *vt* TERMO air-cool; ~ **por compresión** *vt* ING MECÁ, REFRIG, TERMO refrigerate by compression; ~ **con ventilador** *vt* REFRIG, TERMO fan cool

refuerzo[1]: **con ~ de madera doble** *adj* EMB double-battened

refuerzo[2] *m* CINEMAT replenishment, CONST batten, reinforcement, stiffening, *hormigón* reinforcing, FOTO replenishment, ING MECÁ boss, stiffening, PROD welt, strengthening, TEC ESP shroud, stiffener, TELECOM enhancement, TEXTIL stiffener, *de libro* backing, TRANSP AÉR brace, TV replenishment; ~ **de alfombra** *m* TEXTIL underlayer; ~ **de cartón** *m* EMB, PAPEL cardboard backing; ~ **de corriente continua** *m* PROD DC boost; ~ **de eslabón** *m* ING MECÁ stud; ~ **de espuma** *m* TEXTIL foam backing; ~ **del expositor con cinta** *m* EMB tape-hanging display reinforcement; ~ **de fibra de vidrio** *m* C&V glass-fiber reinforcement (*AmE*), glass-fibre reinforcement (*BrE*); ~ **de la imagen** *m* TV image enhancement; ~ **de látex** *m* TEXTIL latex backing; ~ **longitudinal** *m*

CONST longitudinal reinforcement; ~ **con malla de alambre** *m* C&V wire mesh reinforcement; ~ **de mampara** *m* TRANSP MAR *construcción naval* bulkhead stiffener; ~ **para neumáticos** *m* TEXTIL tire reinforcement (*AmE*), tyre reinforcement (*BrE*); ~ **de una pieza** *m* C&V bait; ~ **de pilotes** *m* CARBÓN pile splice; ~ **del plan de la cámara** *m* TRANSP MAR *construcción naval* cabin sole reinforcement; ~ **principal** *m* CONST main reinforcement; ~ **con tela de alambre** *m* C&V wire mesh reinforcement; ~ **tipo caja** *m* TRANSP AÉR box-type stiffener; ~ **de voltaje** *m* AmL TV voltage booster

refuerzos: ~ **de madera** *m pl* TRANSP MAR *para equipamientos de cubierta* wood reinforcements

refugio *m* CONST manhole, shelter, *tráfico* traffic island; ~ **contra gases** *m* D&A anti-gas shelter; ~ **nuclear** *m* NUCL fallout shelter; ~ **de protección colectiva** *m* D&A collective protection shelter

refundir *vt* PROD recast

regadera *f* CONST *jardinería* sprinkler; ~ **asfáltica** *f* CONST tar sprinkler

regadío *m* AGUA *agricultura* irrigation

regala *f* TRANSP MAR *buques* gunnel, gunwale, *construcción naval* covering board

regar *vt* AGRIC, AGUA, CONST irrigate, water

regasificación *f* GAS regasification

regatón *m* ING MECÁ, MECÁ, ÓPT, TELECOM ferrule

regeneración *f* AGUA, CINEMAT regeneration, replenishment, ELECTRÓN refresh, regeneration, FÍS ONDAS feedback, FOTO replenishment, GAS regeneration, INFORM&PD refresh, regeneration, PROD *producción* reprocessing, QUÍMICA regeneration, RECICL reprocessing, TEC ESP regeneration, recovery, reconditioning, TEC PETR reclaiming, TELECOM regeneration, TV replenishment; ~ **digital** *f* ELECTRÓN digital regeneration; ~ **de energía** *f* TERMO energy regeneration; ~ **de imagen** *f* INFORM&PD image refreshing; ~ **de impulsos** *f* ELECTRÓN pulse regeneration; ~ **natural** *f* PROD restocking; ~ **de pulsos** *f* FÍS, TV pulse regeneration; ~ **de la señal** *f* ELECTRÓN signal regeneration

regenerador[1] *adj* TEC ESP regenerative

regenerador[2] *m* TEC ESP, TELECOM regenerator; ~ **digital** *m* ELECTRÓN digital regenerator; ~ **de impulsos** *m* ELECTRÓN, FÍS RAD impulse regenerator; ~ **de la señal** *m* ELECTRÓN signal regenerator

regenerar *vt* AGUA regenerate, CARBÓN *lanas, lubricantes* recover, CINEMAT regenerate, CONST, CONTAM recycle, ELECTRÓN regenerate, refresh, EMB recycle, INFORM&PD regenerate, refresh, QUÍMICA regenerate, RECICL recycle, TEC ESP, TELECOM regenerate

regenerativo *adj* TEC ESP regenerative

regido: ~ **por menús** *adj* INFORM&PD menu-driven

régimen[1]: **a** ~ **de vaporización** *adj* ING MECÁ *calderas* in steam

régimen[2] *m* ELEC *aparato* rating, *equipo* operating conditions, HIDROL *río* regime; ~ **de aprovechamiento de la tierra** *m* AGRIC land use pattern; ~ **de ascenso** *m* TEC ESP, TRANSP AÉR rate of climb (*RC*); ~ **de carga de aparatos** *m* TELECOM apparatus charge rate; ~ **continuo** *m* ELEC *generador* continuous rating; ~ **de control químico por fosfatos** *m* NUCL phosphate chemistry regime; ~ **de control químico por volátiles** *m* NUCL all-volatile chemistry regime; ~ **de corriente** *m* AGRIC rate of flow; ~ **de descarga** *m* AGRIC rate of flow, RECICL discharge

rate; ~ **de flujo** *m* P&C *propiedad física* rate of flow; ~ **de funcionamiento** *m* ELEC *equipo* operating conditions; ~ **de marcha** *m* PROD *máquinas* working conditions; ~ **de marcha normal** *m* ING MECÁ normal working conditions; ~ **máximo** *m* HIDROL peak rate; ~ **nominal** *m* PROD *contactos de relé* rating; ~ **normal** *m* PROD *contactos de relé* rating; ~ **de perforación normalizado** *m* TEC PETR *perforación* normalized drilling rate (*NDR*); ~ **permanente** *m* ELEC *generador* continuous rating, ING ELÉC, TELECOM steady state; ~ **pluvial** *m* AGRIC, METEO, TRANSP MAR rainfall pattern; ~ **de sedimentación** *m* RECICL sedimentation rate; ~ **de señalización de datos** *m* TELECOM data signaling rate (*AmE*), data signalling rate (*BrE*); ~ **de tempestividad** *m* PROD timeliness rating; ~ **de tenencia de tierras** *m* AGRIC land tenure; ~ **total** *m* HIDROL *de la capa de agua* total rate; ~ **de trabajo** *m* MECÁ duty cycle, rating, PROD *contactos de relé* rating; ~ **de transferencia de calor** *m* INSTAL TERM, TERMO rate of heat transfer; ~ **transitorio** *m* ELEC transient state; ~ **turbillonario de fondo** *m* FÍS FLUID background vorticity; ~ **de viraje** *m* TRANSP AÉR rate of turn; ~ **de voltaje inverso** *m* PROD inverse voltage rating

regímenes: ~ **degradados** *m pl* TEC ESP degraded operating conditions

región: ~ **activa** *f* FÍS *reactor nuclear* core; ~ **del alma** *f* TELECOM core area; ~ **de alta presión** *f* OCEAN *meteorología marina* sea high; ~ **de campo** *f* TEC ESP Fraunhofer region; ~ **de campo lejano** *f* ÓPT, TELECOM far-field region; ~ **de campo próximo** *f* ÓPT, TELECOM near-field region; ~ **de deslizamiento suave** *f* METAL easy-glide region; ~ **dopada positivamente** *f* FÍS positively-skewed doped region; ~ **con dopado negativo** *f* FÍS negatively doped region; ~ **de Fraunhofer** *f* TEC ESP Fraunhofer region; ~ **de Fresnel** *f* INFORM&PD, TELECOM Fresnel region; ~ **intermedia de la vasija del reactor** *f* NUCL reactor-vessel beltline region; ~ **del mar** *f* TRANSP MAR sea area; ~ **de movilidad de la corriente** *f* ELECTRÓN drift region; ~ **óptica** *f* FÍS optical branch; ~ **de origen** *f* GEOL *de material clástico o magmas* source region; ~ **petrolífera** *f* TEC PETR petroleum basin; ~ **p**$^+$ *f* ELECTRÓN p$^+$ region; ~ **p**$^-$ *f* ELECTRÓN p$^-$ region; ~ **productora de maíz** *f* AGRIC corn-growing area (*AmE*), maize-growing area (*BrE*); ~ **receptora** *f* CONTAM receptor region; ~ **de tráfico** *f* TRANSP traffic region; ~ **de transmisiones atmosféricas** *f* FÍS, FÍS RAD atmospheric window; ~ **de transmisiones por radio** *f* FÍS, FÍS RAD radio window

registrado: ~ **de rayos gamma** *m* TEC PETR *evaluación de la formación, petrofísica* gamma-ray well logging

registrador *m* ACÚST recorder, ELEC recorder, *mediciones* logger, INFORM&PD recorder, logger, ING MECÁ register, recorder, INSTR recorder, logger, OCEAN probe, TELECOM register, recorder, TV recorder; ~ **automático** *m* ELEC *mediciones*, INSTR logger; ~ **automático de infracciones** *m* TRANSP automatic infringement recorder; ~ **de banda** *m* ELEC strip chart recorder; ~ **de banda de papel** *m* ELEC strip chart recorder; ~ **de cantidades** *m* INSTR quantity recorder; ~ **de carta en rollo** *m* ELEC strip chart recorder; ~ **de cinta** *m* INSTR strip chart recorder; ~ **de cintas de vídeo a intervalos prefijados** *m* PROD time-lapse video tape recorder; ~ **de coordenadas rectangulares** *m* ELEC X-Y recorder; ~ **de**

corrimiento *m* TELECOM shift register; **~ de curvas XY** *m* ELEC X-Y recorder; **~ de datos** *m* ELEC *mediciones* logger, INFORM&PD data logger, data recorder, INSTR logger; **~ de datos numéricos incremental** *m* INSTR incremental digital recorder; **~ de datos de vuelo** *m* TEC ESP flight-data recorder; **~ de deformación del puente** *m* CONST *topografía* bridge-deflective recorder; **~ de desviación del puente** *m* CONST *topografía* bridge-deflective recorder; **~ digital** *m* NUCL, TELECOM, TV digital recorder; **~ digital de datos de vuelo** *m* TRANSP AÉR digital flight-data recorder; **~ de dimensiones** *m* INSTR quantity recorder; **~ de disco** *m* ELEC disc recorder (*BrE*), disk recorder (*AmE*); **~ de distancia** *m* AUTO trip odometer (*AmE*), trip recorder (*BrE*); **~ de dos coordenadas** *m* TEC ESP X-Y recorder; **~ de dos etapas** *m* TELECOM two-state register; **~ electrónico de las incidencias de vuelo** *m* INFORM&PD, INSTR, TRANSP AÉR black box; **~ del encendido** *m* INSTR spark recorder; **~ escribiente** *m* AmL (*cf registrador por tinta Esp*) INSTR ink recorder; **~ estilográfico** *m* AmL (*cf registrador gráfico Esp*) ELEC, INSTR, LAB *aparato* pen plotter (*AmE*), pen recorder (*BrE*); **~ de eventos** *m* ELEC event recorder; **~ de fenómenos** *m* ELEC event recorder; **~ gráfico** *m Esp* (*cf registrador estilográfico AmL*) ELEC, INSTR, LAB *aparato* pen plotter (*AmE*), pen recorder (*BrE*); **~ gráfico en banda de papel** *m* ELEC strip chart recorder; **~ gráfico en coordenadas cartesianas** *m* ELEC X-Y recorder; **~ de gráfico radial** *m* INSTR radial-chart recorder; **~ de horas de funcionamiento** *m* TRANSP MAR operating hours indicator; **~ de horas de funcionamiento de la máquina** *m* ING MECÁ, TRANSP MAR engine-hours indicator; **~ de horas de funcionamiento del motor** *m* ING MECÁ, TRANSP MAR engine-hours indicator; **~ de impulso** *m* PROD shift register; **~ indicador de volumen** *m* TV volume indicator meter (*VI meter*); **~ intermedio** *m* INFORM&PD buffer register; **~ ionosférico** *m* GEOFÍS ionospheric recorder; **~ de líneas** *m* INSTR line recorder; **~ mecánico** *m* ACÚST mechanical recorder; **~ de nivel** *m* ACÚST level recorder; **~ de olas** *m* OCEAN wave recorder; **~ potenciométrico** *m* INSTR potentiometric recorder; **~ de producción** *m* INSTR production recorder; **~ de profundidad** *m* OCEAN depth recorder; **~ por serie** *m* ING ELÉC serializer; **~ de sonidos** *m* INSTR sound recorder; **~ del tiempo** *m* INSTR time recorder; **~ por tinta** *m Esp* (*cf registrador escribiente AmL*) INSTR ink recorder; **~ traductor** *m* TELECOM register translator; **~ para el trazado de curvas en coordenadas cartesianas** *m* ELEC X-Y recorder; **~ de valor medio** *m* INSTR mean value recorder; **~ variable** *m* ELECTRÓN shift register; **~ de velocidad** *m* INSTR speed recorder; **~ del viento** *m* METEO wind recorder; **~ de voces de la cabina de mando** *m* TRANSP AÉR cockpit voice recorder; **~ de vuelo** *m* TRANSP AÉR flight recorder; **~ de XY** *m* ELEC X-Y recorder

registramiento: **~ doble** *m* INFORM&PD double buffering

registrar[1] *vt* ACÚST record, CINEMAT clock, log, ELECTRÓN clock, FÍS record, FOTO register, GEOFÍS record, INFORM&PD clock, enter, log, record, register, TEC PETR *evaluación de la formación, petrofísica* log, TELECOM register, TRANSP MAR *acaecimientos en el diario de navegación* log, TV record; **~ en memoria auxiliar** *vt* INFORM&PD roll out

registrar[2]: **~ la entrada** *vi* INFORM&PD log in, log on; **~ la salida** *vi* INFORM&PD log off, log out

registro *m* ACÚST record, recording, AGRIC registration, CINEMAT recording, register, registration, ELEC *fuente de alimentación* recording, manhole, inspection hatch, GEOFÍS recording, trace, GEOL log, recording, HIDROL *de alcantarillado* manhole, IMPR register, INFORM&PD registration, log, register, record, recording, ING ELÉC register, ING MECÁ manhole, regulator, MECÁ, MINAS manhole, PETROL record, log, PROD *de horno* register, *de recalentadores* damper, tally, QUÍMICA record, REFRIG *en conductos de aire acondicionado* damper, TEC PETR *evaluación de la formación* log, TELECOM logging, recording, TRANSP AÉR registration, TV recording, registration; **~ del abonado** *m* TELECOM subscriber's store; **~ de activación** *m* TEC PETR *petrofísica, evaluación de la formación* activation log; **~ de la activación neutrónica** *m* NUCL neutron activation logging; **~ acumulador** *m* INFORM&PD accumulator register; **~ acústico** *m* GEOFÍS, TEC PETR *petrofísica, evaluación de la formación* acoustic log; **~ acústico de pozos** *m* TEC PETR *evaluación de la formación* acoustic well logging; **~ de adherencia del cemento** *m* (*CBL*) TEC PETR *perforación* cement bond log (*CBL*); **~ aeronáutico** *m* TRANSP AÉR aeronautical register; **~ de análisis del lodo** *m* TEC PETR *ingeniería de lodos, perforación* mud analysis log; **~ analógico** *m* TELECOM analog recording; **~ del autopotencial** *m* TEC PETR *evaluación de la formación* self-potential log; **~ de averías** *m* INFORM&PD failure logging; **~ de banda baja** *m* TV low-band recording; **~ de bolas** *m* ING MECÁ ball register; **~ de borrado** *m* INFORM&PD deletion record; **~ de botonera** *m* IMPR pin register; **~ de cables** *m* ELEC, ING ELÉC *fuente de alimentación* cable manhole; **~ de cálculos de ingeniería** *m* ING MECÁ engineering calculations record; **~ de calibración** *m* TEC PETR *petrofísica* caliper log (*AmE*), calliper log (*BrE*); **~ de la calidad** *m* CALIDAD quality record; **~ de cámara** *m* CINEMAT, TV camera log; **~ de cambios** *m* INFORM&PD change record; **~ de categoría** *m* INFORM&PD status register; **~ de código agrupado** *m* INFORM&PD group code recording (*GCR*); **~ de cola** *m* INFORM&PD trailer record; **~ de conexiones** *m* ELEC, ING ELÉC cable box; **~ de contacto** *m* TEC PETR *evaluación de la formación* contact log; **~ de continuidad** *m* CINEMAT, TV continuity log; **~ de control** *m* INFORM&PD control register; **~ de control de secuencia** *m* INFORM&PD sequence control register; **~ de control secuencial** *m* INFORM&PD sequence control register; **~ cronométrico** *m* INFORM&PD clock register; **~ de datos** *m* ELEC data recording, INFORM&PD data logging, data record, data recording, NUCL, TELECOM data logging; **~ de datos de la memoria** *m* INFORM&PD memory data register (*MDR*); **~ de decalaje de bit** *m* PROD bit-shift register; **~ defectuoso** *m* TV misregistration; **~ de la densidad** *m* GEOFÍS *de pozos de perforación*, TEC PETR *evaluación de la formación, petrofísica* density log; **~ derecho de desplazamiento** *m* PROD shift right register; **~ de desplazamiento** *m* INFORM&PD,

TELECOM shift register; **~ de desplazamiento analógico** *m* INFORM&PD analog shift register; **~ de desvíos** *m* PROD shift register; **~ diagonal** *m* IMPR diagonal register; **~ digital** *m* TELECOM, TV digital recording; **~ de dirección** *m* INFORM&PD address register; **~ de direccionamiento de base** *m* INFORM&PD base address register; **~ del diseño gráfico** *m* CINEMAT, TV registration of artwork; **~ de doble densidad** *m* INFORM&PD double-density recording;

~ e **~ eléctrico** *m* PETROL electrical log, TEC PETR electric log, *evaluación de la formación, petrofísica* wireline log; **~ eléctrico especial múltiple** *m* TEC PETR *evaluación de la formación, petrofísica* multiple special electrical logging; **~ de enmiendas** *m* INFORM&PD amendment record; **~ de entrada** *m* INFORM&PD log-in, log-on, input record; **~ de entrada/salida** *m* INFORM&PD input/output register; **~ entrante** *m* TELECOM incoming register; **~ de equilibrio** *m* REFRIG equalizing damper; **~ espontáneo** *m* ENERG RENOV spontaneous log; **~ de estado** *m* INFORM&PD status register; **~ de etiqueta** *m* INFORM&PD label record; **~ exacto** *m* IMPR hairline register; **~ de existencias** *m* PROD stock record; **~ de exploración** *m* TV scan registration;

~ f **~ de fallos** *m* INFORM&PD failure logging; **~ de fango** *m* GEOFÍS *pozos de perforación* mud log; **~ físico** *m* INFORM&PD physical record;

~ g **~ gamma-gamma** *m* TEC PETR *evaluación de la formación, petrofísica* gamma-gamma log; **~ genealógico** *m* AGRIC herd book; **~ geofísico** *m* GEOFÍS, GEOL, TEC PETR geophysical log; **~ geotérmico** *m* ENERG RENOV, TEC PETR *evaluación de la formación, petrofísica* geothermal log; **~ gráfico** *m* INSTR, LAB chart recorder; **~ de guardia** *m* PETROL guard log; **~ de guillotina** *m* REFRIG slide damper;

~ h **~ por haz de láser** *m* INFORM&PD laser-beam recording; **~ de hincadura de los pilotes** *m* CARBÓN pile-driving record;

~ i **~ de índice** *m* INFORM&PD index register; **~ de inspección** *m* METR inspection record; **~ de instrucción** *m* INFORM&PD instruction register; **~ intermedio** *m* INFORM&PD *ordenador* buffer; **~ intermedio de entrada** *m* INFORM&PD input buffer; **~ intermedio de entrada/salida** *m* INFORM&PD input/output buffer; **~ izquierdo de desplazamiento** *m* PROD shift left register;

~ l **~ lateral** *m* IMPR side register; **~ de Laue** *m* CRISTAL, FÍS RAD Laue pattern; **~ de limpieza** *m* PROD *calderas* washout hole; **~ lineal** *m* IMPR linear register; **~ de llamada automática** *m* IMPR automatic call recording; **~ de llamadas** *m* TELECOM call logging, message log, message register; **~ de localizaciones móviles** *m* TELECOM mobile location registration; **~ con lodo** *m* GEOFÍS *métodos de sondeo* mud log; **~ del lodo** *m* TEC PETR *evaluación de la formación, perforación* mud log; **~ lógico** *m* INFORM&PD logical record; **~ de longitud fija** *m* INFORM&PD fixed-length record; **~ de longitud variable** *m* INFORM&PD variable-length record; **~ longitudinal** *m* TV longitudinal recording;

~ m **~ maestro** *m* INFORM&PD master record; **~ magnético** *m* TELECOM, TV magnetic recording; **~ de mantenimiento** *m* TRANSP AÉR maintenance record; **~ de marea** *m* GEOFÍS *prospección geofísica*

tide record; **~ de mariposa** *m* REFRIG butterfly-damper; **~ en memoria auxiliar** *m* INFORM&PD roll-out; **~ del mensaje** *m* TELECOM call logging, message log, message register; **~ de microresistividad** *m* GEOFÍS microresistivity log; **~ monitor** *m* PETROL monitor record;

~ n **~ neto** *m* TRANSP MAR *tonelaje* net register; **~ neutrón** *m* ENERG RENOV, GAS, PETROL, TEC PETR *evaluación de la formación* neutron log, neutron logging; **~ neutrón impulsado** *m* TEC PETR *evaluación de la formación* pulsed-neutron log; **~ de neutrones** *m* TEC PETR *evaluación de la formación, petrofísica* neutron logging; **~ neutrón-gamma** *m* TEC PETR *evaluación de la formación* neutron-gamma log; **~ neutrónico** *m* ENERG RENOV, GAS, PETROL, TEC PETR neutron log, neutron logging; **~ neutrón-neutrón** *m* TEC PETR *evaluación de la formación* neutron-neutron log;

~ o **~ de operación** *m* INFORM&PD operation register; **~ de operaciones** *m* INFORM&PD transaction record, TELECOM log; **~ óptico** *m* TELECOM optical recording; **~ de origen** *m* TELECOM originating register;

~ p **~ de perforación** *m* ENERG RENOV well logging; **~ periférico** *m* IMPR circumferential register; **~ de permeabilidad** *m* TEC PETR *evaluación de la formación* permeability logging; **~ de persiana** *m* REFRIG single-leaf damper; **~ de persianas** *m* REFRIG multi-leaf damper; **~ de la porosidad** *m* TEC PETR *evaluación de la formación* porosity log; **~ de potencial espontáneo** *m* GEOFÍS *método de sondeo*, TEC PETR *evaluación de la formación* spontaneous potential log; **~ de potencial propio** *m* GEOFÍS *método de sondeo* self-potential log; **~ del pozo** *m* TEC PETR *petrofísica, evaluación de la formación* well logging; **~ en pozo de perforación** *m* GEOFÍS *métodos de sondeo* well log; **~ principal** *m* ACÚST master; **~ de programas de televisión** *m* CINEMAT telerecording; **~ provisorio** *m* INFORM&PD temporary register; **~ pseudosónico** *m* TEC PETR *evaluación de la formación* pseudo-sonic log, pseudo-sonic profile;

~ r **~ radiactivo** *m* TEC PETR *evaluación de la formación* nuclear log, radioactive log; **~ de rayos gamma** *m* ENERG RENOV, GEOFÍS, TEC PETR *evaluación de la formación* gamma-ray log; **~ regulador del tiro** *m* PROD *chimenea* draft regulator *(AmE)*, draught regulator *(BrE)*; **~ de reloj** *m* INFORM&PD clock rate; **~ de resistencia** *m* ENERG RENOV, GEOFÍS, PETROL, TEC PETR resistivity log; **~ de resistividad** *m* ENERG RENOV, GEOFÍS, PETROL, TEC PETR resistivity log;

~ s **~ de salida** *m* INFORM&PD log-off, log-out, output record; **~ de seguridad** *m* CONST safety record; **~ del sobrecalentador** *m* PROD superheater damper; **~ de sondeos** *m* GEOFÍS well logging; **~ en sondeos** *m* GEOFÍS *propiedades de materiales* borehole logging; **~ sónico** *m* TEC PETR *evaluación de la formación* sonic log; **~ de sonido** *m* TELECOM sound recording; **~ de sonido óptico** *m* CINEMAT optical sound recording; **~ sonoro** *m* ACÚST *cine* record, CINEMAT, TELECOM sound recording; **~ sonoro defectuoso** *m* CINEMAT blooping patch; **~ de supresión** *m* INFORM&PD deletion record;

~ t **~ de temperatura** *m* ENERG RENOV temperature log, TEC PETR *evaluación de la formación* temperature logging, TEXTIL temperature recorder; **~ de tempe-**

ratura del pozo *m* TEC PETR *evaluación de la formación* temperature well logging; ~ **temporal** *m* INFORM&PD temporary register; ~ **de tiempo** *m* TELECOM timing; ~ **totalizador** *m* INFORM&PD accumulator register; ~ **de transacciones** *m* INFORM&PD transaction record; ~ **de la tubería del viento** *m* PROD *horno alto* blast gate;

■ **V** ~ **de velocidad** *m* TEXTIL speed recorder; ~ **de la velocidad de sonido** *m* ENERG RENOV acoustic velocity log; ~ **de video** *AmL*, ~ **de vídeo** *Esp m* TELECOM video recording

Registro: ~ **Internacional de Compuestos Químicos Potencialmente Tóxicos** *m* CONTAM, SEG International Register of Potentially Toxic Chemicals (*IRPTC*)

registros *m pl* IMPR lay marks

regla *f* C&V straight edge, GEOM ruler, METR straight edge; ~ **de acero** *f* METR steel straight edge; ~ **de Ampère** *f* ELEC *electromagnetismo* Ampère's rule; ~ **de las áreas** *f* TRANSP AÉR area rule; ~ **de cálculo** *f* INFORM&PD slide rule; ~ **del cortador** *f* C&V cutter's straight edge; ~ **del coseno** *f* GEOM cosine rule; ~ **deslizante** *f* INFORM&PD slide rule; ~ **de fase de Gibbs** *f* NUCL, QUÍMICA Gibbs' phase rule; ~ **graduada** *f* ING MECÁ, INSTR rule; ~ **maestra** *f* CONST *albañilería* screed heater; ~ **de la mano derecha** *f* ELEC, FÍS, ING ELÉC right-hand rule; ~ **de la mano izquierda** *f* ELEC, FÍS left-hand rule; ~ **de máquina herramienta dividida a máquina** *f* ING MECÁ machine-divided machine-tool scale; ~ **de la mesa para cortar** *f* C&V cutter's table ruler; ~ **de modelista** *f* PROD *fundería* shrink rule; ~ **paralela** *f* ING MECÁ parallel rule; ~ **de paralelas** *f* ING MECÁ, TRANSP MAR *navegación* parallel ruler; ~ **patrón graduada** *f* METR line-graduated master scales; ~ **plegable** *f* CONST folding rule (*AmE*), jointed rule (*BrE*), zigzag rule (*AmE*); ~ **precisa** *f* METR knife-edge rule; ~ **de producción** *f* INFORM&PD production rule; ~ **que considera la contracción de los metales** *f* PROD *moldeo* contraction rule; ~ **del sacacorchos** *f* ELEC, FÍS, ING ELÉC corkscrew rule; ~ **de selección** *f* FÍS, NUCL selection rule; ~ **del seno** *f* GEOM sine rule; ~ **de senos** *f* METR sine bar; ~ **de sumas de Kuhn-Thomas-Reich** *f* NUCL Kuhn-Thomas-Reich sum rule; ~ **T** *f* ING MECÁ T-square; ~ **del tirabuzón** *f* ELEC, FÍS, ING ELÉC corkscrew rule, right-hand rule; ~ **para trazar paralelas** *f* ING MECÁ, TRANSP MAR *navegación* parallel ruler; ~ **de los tres dedos** *f* ELEC *electromagnetismo*, FÍS, ING ELÉC right-hand rule

reglaje *m* ING ELÉC regulation, ING MECÁ adjustment, governing, TRANSP AÉR setting, trim, trimming, VEH *de encendido, juego, válvulas y balancines* timing; ~ **en alcance** *m* TEC ESP ranging; ~ **del altímetro normal** *m* TRANSP AÉR standard altimeter setting; ~ **del alza** *m* PROD *artillería* range adjustment; ~ **de la chispa** *m* AUTO spark timing; ~ **del encendido** *m* VEH ignition timing; ~ **exacto** *m* TEC ESP tracking; ~ **del giróscopo** *m* TRANSP AÉR gyro resetting; ~ **de ignición** *m* AUTO ignition setting; ~ **interno** *m* PROD internal setup; ~ **de machos** *m* PROD *fundería* setting; ~ **de la pala** *m* TRANSP AÉR blade setting; ~ **del paso de la pala** *m* TRANSP AÉR blade-pitch setting; ~ **de precisión** *m* ELECTRÓN, INSTR, NUCL, PROD, TV fine adjustment; ~ **del rodillo** *m* TEXTIL roller setting; ~ **silencioso** *m* TELECOM squelch; ~ **de subida** *m*

TRANSP AÉR climb setting; ~ **de taqué** *m* AUTO valve setting; ~ **del timón de dirección** *m* TRANSP AÉR, TRANSP MAR rudder trim; ~ **del timón de profundidad** *m* TRANSP AÉR elevator trim; ~ **de válvula** *m* AUTO valve setting; ~ **de válvulas variables** *m* TRANSP variable-valve timing

reglamentación: ~ **de métodos de prueba** *f* ING MECÁ standardization of test methods

reglamentario *adj* ING MECÁ standard

reglamento *m* ACÚST, SEG, TELECOM regulation; ~ **ferroviario** *m* FERRO railroad regulations (*AmE*), railway regulations (*BrE*); ~ **para metales no ferrosos** *m* SEG nonferrous metals regulations

reglamentos: ~ **del aire** *m pl* TRANSP AÉR rules of the air; ~ **especiales** *m pl* SEG special regulations

reglas: ~ **de construcción para potencia de conducción** *f pl* ING MECÁ design rules for drive power; ~ **de diseño para potencia motriz** *f pl* ING MECÁ design rules for drive power; ~ **de Fleming** *f pl* ELEC Fleming's rules; ~ **de Hund** *f pl* FÍS Hund rules; ~ **de vuelo por instrumentos** *f pl* TRANSP AÉR instrument flight rules

regleta *f* PAPEL deckle board; ~ **de bornas** *f* PROD terminal strip; ~ **de bornes** *f* ELEC *conexión* terminal block; ~ **de clavijas** *f* ING ELÉC jack strip; ~ **de clavijas bipolares hembra** *f* ING ELÉC socket board; ~ **de clavijas enchufables** *f* ING ELÉC plugboard; ~ **de componer** *f* IMPR composing stick; ~ **de conexión** *f* TELECOM connection strip; ~ **de conexiones** *f* ING ELÉC terminal barrier strip, PROD terminal barrier strip, terminal strip; ~ **de fusibles** *f* ELEC fuse strip; ~ **de goma** *f* IMPR squeegee; ~ **de nivel** *f* PAPEL slice; ~ **de terminales** *f* ELEC *conexión* terminal block; ~ **de terminales de entrada** *f* PROD input-terminal strip

regolito *m* GEOL regolith

regrabación *f* ACÚST rerecording

regrabado *m* IMPR re-etching

regresar *vi* INFORM&PD return

regresión *f* GEOL *marina*, INFORM&PD, MATEMÁT regression, TEC ESP *alas de los aviones* sweep; ~ **de la capa freática** *f* GEOFÍS, HIDROL falling water table; ~ **múltiple** *f* AGRIC multiple regression

regreso: ~ **inactivo** *m* ING MECÁ idle return; ~ **a la superficie** *m* OCEAN ascent; ~ **a la superficie con ayuda del globo de ascensión** *m* OCEAN balloon surfacing

regruesamiento *m* ING MECÁ thicknessing

reguarnecer *vt* ING MECÁ repack

reguera *f* PROD *fundería* main gate, runner, sow

regulable *adj* ING MECÁ adjustable, adjusting, MECÁ adjustable; **no ~** *adj* ING MECÁ, MECÁ fixed

regulación *f* ELEC control, IMPR metering, ING ELÉC regulation, ING MECÁ adjustment, regulating, regulation, setting, *fluidos* throttling, MINAS, NUCL regulating, PROD, TEC ESP monitoring, TELECOM regulation, TRANSP AÉR trimming; ~ **de ajuste de imagen** *f* TV framing control; ~ **del brillo** *f AmL* (*cf regulación de la luminosidad Esp*) TV brightness control; ~ **de cabeceo** *f* TRANSP AÉR pitch trim; ~ **del caudal** *f* PROD flow control; ~ **del cero mecánico** *f* ELEC *de instrumento*, INSTR mechanical zero adjustment; ~ **a cero voltios** *f* TV offset, zero volt adjustment; ~ **de la circulación** *f* TRANSP traffic control; ~ **de la combustión** *f* TRANSP MAR *motores* combustion control; ~ **continua** *f* ELEC stepless control; ~ **por contrarreacción** *f* ING ELÉC feedback control; ~ **del control de temperatura** *f* TEC ESP temperature control regulation; ~ **de corriente** *f* ELEC, ING ELÉC current regulation; ~ **derivada** *f* ELEC derivative control; ~ **discontinua** *f* TEC ESP on-off control; ~ **a distancia** *f* AUTO, CINEMAT, CONST, D&A, ELEC, FOTO, ING MECÁ, MECÁ, TEC ESP, TELECOM, TRANSP, TV, VEH remote control; ~ **de fase** *f* TELECOM phase regulation; ~ **de frecuencia** *f* TELECOM frequency regulation; ~ **de frecuencia de la carga** *f* ELEC *máquina eléctrica* load frequency control; ~ **frente a la reposición** *f* PROD set versus reset; ~ **de fuerza centrípeta** *f* ACÚST *grabación* antiskating; ~ **de ganancia** *f* ELECTRÓN gain setting, gain trimming; ~ **lineal** *f* ELEC linear control; ~ **de la luminosidad** *f* *Esp* (*cf regulación del brillo AmL*) TV brightness control; ~ **del nivel del agua embalsada** *f* AGUA storage level regulation; ~ **de pérdida de potencia** *f* ELEC power loss ride-through; ~ **de potencia frigorífica** *f* REFRIG capacity control; ~ **de la presión** *f* PROD pressure-monitoring; ~ **de puesta a tierra** *f* PROD earthing regulation (*BrE*), grounding regulation (*AmE*); ~ **del RCS** *f* ING ELÉC SCR regulation; ~ **de recorte** *f* ING ELÉC switching regulation; ~ **del régimen de cambio de presión** *f* TRANSP AÉR pressure rate-of-change regulating; ~ **de ríos** *f* AGRIC, AGUA, HIDROL river training; ~ **de serie** *f* ING ELÉC series regulation; ~ **de tensión** *f* ELEC voltage control; ~ **de tensión continua** *f* ELEC *control* direct voltage regulation; ~ **en el tiempo** *f* TELECOM timing; ~ **por todo o nada** *f* ELEC on-off control; ~ **vertical de alineamiento** *f* TV vertical-linearity control; ~ **de voltaje** *f* ING ELÉC voltage regulation

regulador *m* AUTO governor, regulator, ELEC regulator, ENERG RENOV governor, ING ELÉC, ING MECÁ controller, governor, regulator, INSTAL HIDRÁUL throttle, INSTR monitor, LAB regulator, MECÁ governor, buffer, P&C *control de pH* buffer, PROD *locomotoras* throttle valve, TEC ESP *dispositivos automáticos* controller, TEC PETR *tuberías, conductos* choke, TELECOM regulator, TRANSP MAR *motor*, VEH *limitador de la velocidad del motor* governor; ~ **de aire** *m* ING MECÁ air regulator; ~ **del árbol de levas** *m* ELEC *conmutador* camshaft controller; ~ **del arco** *m* ELEC arc regulator; ~ **automático** *m* ENERG RENOV automatic governor, ING MECÁ self-acting regulator; ~ **automático de corriente** *m* ELEC *relé* automatic current controller; ~ **automático del flujo de aguas residuales** *m* HIDROL automatic sewage flow regulator; ~ **de bolas** *m* ING MECÁ ball governor, PROD fly-ball governor; ~ **de bomba de aire** *m* ING MECÁ air pump throttle; ~ **del campo inductor** *m* ELEC, ING ELÉC field regulator; ~ **de carga** *m* TEC ESP charging regulator; ~ **centrífugo** *m* PROD fly-ball governor; ~ **de consistencia** *m* PAPEL consistency regulator; ~ **de corriente** *m* ELEC *control*, ING ELÉC current regulator; ~ **de corriente de aire** *m* INSTAL TERM draft regulator (*AmE*), draught regulator (*BrE*); ~ **del cuadro** *m* CINEMAT frame adjuster; ~ **en derivación** *m* ING ELÉC shunt regulator; ~ **de derivación de gases calientes** *m* REFRIG hot gas bypass regulator; ~ **de descarga** *m* ENERG RENOV, TEC ESP discharge regulator; ~ **del distribuidor** *m* INSTAL HIDRÁUL slide-throttle valve; ~ **de dos contactos** *m* AUTO two-contact regulator; ~ **del eje de levas** *m*

ELEC *conmutador* camshaft controller; ~ **de excitación** *m* ELEC *de motor*, ING ELÉC field regulator; ~ **de frenado** *m* TRANSP braking governor, VEH brake compensator; ~ **de fuerza centrífuga** *m* ACÚST advance ball; ~ **de la fuerza de frenado** *m* ING MECÁ brake-pressure regulator; ~ **de gases** *m* TRANSP AÉR throttle; ~ **de gases totalmente abierto** *m* TRANSP AÉR full-open throttle; ~ **de gasto** *m* AGUA flow-regulating valve; ~ **de hélice** *m* TRANSP AÉR propeller governor; ~ **hidráulico** *m* AGUA, INSTAL HIDRÁUL cataract; ~ **hidromecánico** *m* ENERG RENOV hydromechanical governor; ~ **de impulsión** *m* TEC ESP boosting regulator; ~ **de inducción** *m* ELEC *transformador* induction regulator; ~ **de inercia** *m* ING MECÁ inertia governor; ~ **manorreductor** *m* ING ELÉC pressure regulator, INSTAL HIDRÁUL pressure reducer; ~ **manual** *m* TERMOTEC *tiro de chimeneas* manual damper; ~ **de la mariposa** *m* VEH *del carburador* throttle slide; ~ **de mariposa** *m* CARBÓN *chimeneas*, PROD *tiro de chimeneas* throttle valve; ~ **de mesa central** *m* ING MECÁ loaded governor; ~ **de oxígeno** *m* TEC ESP oxygen regulator; ~ **del paso del colectivo** *m* TRANSP AÉR collective-pitch follow-up; ~ **del pedal** *m* VEH *freno, embrague* pedal adjuster; ~ **de posición** *m* CINEMAT positioner; ~ **de potencia** *m* REFRIG capacity controller; ~ **de presión** *m* CONST, INSTAL HIDRÁUL pressure regulator; ~ **de la presión del gas** *m* FÍS, GAS, INSTAL TERM, TERMO gas-pressure regulator; ~ **de presión en tuberías neumáticas** *m* ING MECÁ airline pressure regulator; ~ **de profundidad** *m* METR *arados* depth gage (*AmE*), depth gauge (*BrE*); ~ **del RCS** *m* ING ELÉC SCR regulator; ~ **recargado** *m* ING MECÁ loaded governor; ~ **de recorte** *m* ING ELÉC switching regulator; ~ **reductor-elevador** *m* TEC ESP *motores de aeronaves* buck-boost regulator; ~ **de resorte** *m* INSTAL HIDRÁUL spring-loaded valve; ~ **del resorte** *m* ING MECÁ spring governor; ~ **en serie** *m* ING ELÉC series regulator; ~ **de temperatura** *m* INSTAL TERM, PROD temperature regulator, REFRIG temperature controller; ~ **de tensión** *m* AUTO, ELEC, INFORM&PD, VEH *instalación eléctrica* voltage regulator; ~ **de tiempo** *m* PROD timer; ~ **tipo succión** *m* AUTO suction-type governor; ~ **del tiro** *m* PROD draft regulator (*AmE*), draught regulator (*BrE*); ~ **todo o nada** *m* TRANSP AÉR hit-or-miss governor; ~ **de tono** *m* TRANSP AÉR beeper trim; ~ **transistorizado** *m* AUTO transistorized regulator; ~ **de vagonetas** *m* MINAS tub controller; ~ **de vapor** *m* INSTAL HIDRÁUL *máquina de vapor* steam governor; ~ **de velocidad** *m* ACÚST advance ball, ELEC *máquina* speed control, ENERG RENOV speed control device; ~ **por velocidad excesiva** *m* ENERG RENOV overspeed control; ~ **de voltaje** *m* AUTO, ELEC, INFORM&PD, ING ELÉC, PROD, TV, VEH *instalación eléctrica* voltage regulator; ~ **del voltaje de inducción** *m* ING ELÉC induction voltage regulator; ~ **de Watt** *m* ING MECÁ ball governor, PROD fly-ball governor

reguladora: ~ **de mariposa** *f* *AmL* (*ver válvula de mariposa Esp*) MINAS *para el tiro de chimenea* throttle valve

regular[1] *adj* FÍS steady-state, GEOL even

regular[2] *vt* CINEMAT *volumen sonoro* monitor, CONTAM regulate, INFORM&PD normalize, ING ELÉC regulate, ING MECÁ set, MECÁ normalize, adjust, METAL normalize, PROD *tiro de un soplador* regulate, *controlar* adjust, QUÍMICA buffer, TEC ESP time, regulate

regularidad *f* PROD levelness; **~ de superficie** *f* PROD levelness

regularización *f* AGUA regularization, CARBÓN *de ríos, de pendientes* grading, ING MECÁ regulating, regulation, TELECOM standardization

rehabilitación *f* CONTAM *contaminación del suelo* rehabilitation, TEC PETR *perforación* workover

rehacimiento *m* PROD *de juntas* remaking

rehincado *m* CARBÓN *pilotes* redriving

reignición *f* TEC ESP reignition

reimpresión *f* IMPR rerun

reimpresiones: ~ separadas *f pl* IMPR off-prints

reimprimir *vt* CINEMAT, FOTO, IMPR reprint

reinfestación *f* AGRIC reinfestation

reinicialización *f* CINEMAT, INFORM&PD, PROD, TV reset

reinicializar *vt* CINEMAT, INFORM&PD, TV reset

reiniciar *vt* INFORM&PD, TELECOM *sistema* restart

reinicio *m* INFORM&PD, TELECOM restart; **~ tras fallo en alimentación** *m* INFORM&PD power-fail restart (*PFR*)

reintegración *f* QUÍMICA reintegration

reintento *m* INFORM&PD retry

reinyección *f* TEC PETR *producción* reinjection

reiter *m* *Esp* (*cf jinetillo AmL*) LAB rider

reja *f* AGRIC *que puede separarse de la costanera* slip share, CONST screen, window bar, trellis, PROD grating; **~ de arado** *f* AGRIC colter (*AmE*), coulter (*BrE*), plough share (*BrE*), plow share (*AmE*), share; **~ de barrotes** *f* ING MECÁ bar screen; **~ circular** *f* AGRIC disc coulter (*BrE*), disk colter (*AmE*); **~ común americana** *f* AGRIC regular share; **~ de irrigación** *f* AGRIC irrigation shovel

rejalgar *m* MINERAL, QUÍMICA realgar

rejilla *f* CARBÓN, ELEC grid, ELECTRÓN mask, HIDROL grid, MECÁ rack, PROD grating, grid, RECICL screen, TEXTIL *tejido* cannage; **~ de abertura** *f* ELECTRÓN aperture grill (*AmE*), aperture grille (*BrE*); **~ aislante** *f* ELECTRÓN barrier grid; **~ de la caldera** *f* INSTAL TERM boiler grate; **~ cóncava** *f* FÍS, TEC ESP concave grating; **~ de control** *f* ING ELÉC *tubo vacío* control grid; **~ de desionización** *f* ELECTRÓN deionizing grid; **~ de diamante** *f* AGUA diamond riffle; **~ de difracción** *f* FÍS, FÍS ONDAS, TELECOM *óptica* diffraction grating, grating; **~ de dispersión** *f* PROC QUÍ dispersion grating; **~ en escalones** *f* FÍS echelette grating, echelon grating; **~ O exclusiva** *f* ELECTRÓN exclusive OR gate; **~ de fase** *f* ELEC *corriente alterna* phase grid; **~ generadora** *f* FÍS FLUID generating grid; **~ generadora de turbulencia** *f* FÍS FLUID turbulence-generating grid; **~ de hilo metálico** *f* P&C *parte de equipo* wire mesh; **~ de inyección** *f* ELECTRÓN, TRANSP AÉR injection grid; **~ libre** *f* ING ELÉC floating grid; **~ de modulación** *f* TV modulation grid; **~ móvil** *f* ELECTRÓN free grid, INSTAL TERM moving grate; **~ NO-O exclusiva** *f* ELECTRÓN exclusive NOR gate; **~ de pantalla** *f* ELECTRÓN screen grid; **~ pantalla verde** *f* TV green screen grid; **~ de polarización** *f* TEC ESP polarization grid; **~ de potencia** *f* ING ELÉC power grid; **~ de protección** *f* ELEC *seguridad* guard, PROD wire guard; **~ protectora** *f* SEG *para tubo de rayos catódicos* protective screen; **~ del radiador** *f* AUTO, VEH *sistema de refrigeración* radiator blind, radiator grille; **~ reflectora** *f* FÍS, FÍS ONDAS, ÓPT reflection grating;

~ de refracción *f* LAB *óptica*, ÓPT refraction grating; **~ resonante** *f* ELECTRÓN resonator grid; **~ de seguridad soldada a presión** *f* SEG pressure-welded safety grating; **~ de soporte del núcleo** *f* NUCL grid-support plate; **~ supresora** *f* ELECTRÓN suppressor grid, ING ELÉC suppression grid; **~ de transmisión** *f* FÍS transmission grating; **~ de tubo electrónico** *f* ELECTRÓN electron-tube grid; **~ de ventilación** *f* ING MECÁ louver (*AmE*), louvre (*BrE*)

rejuntado *m* CONST pointing, PROD *muros* jointing

rejuvenecimiento *m* GEOL rejuvenation

relación *f* FÍS ratio, INFORM&PD relation, MATEMÁT, METR, REFRIG ratio; **~ de abundancia** *f* FÍS PART *de los isótopos* abundance ratio; **~ de abundancias** *f* NUCL abundance ratio; **~ de actividad de fichero** *f* INFORM&PD file-activity ratio; **~ agua-cemento** *f* CONST water-cement ratio; **~ agua-petróleo** *f* PETROL water-oil ratio (*WOR*); **~ del aire al combustible** *f* TRANSP air-fuel ratio; **~ de amplitud de ondas estacionarias** *f* TELECOM standing-wave ratio (*SWR*); **~ del ancho a la altura de la imagen** *f* CINEMAT aspect ratio; **~ de aspecto** *f* INFORM&PD aspect ratio; **~ axial** *f* CRISTAL, TEC ESP axial ratio; **~ del baño** *f* TEXTIL *tintura* liquor-to-goods ratio; **~ de cableado** *f* ELEC *componente del cable* lay ratio; **~ capacidad-volumen** *f* TRANSP V-C ratio, volume-capacity ratio; **~ de combustión** *f* TERMO combustion index; **~ de compresión** *f* ELECTRÓN, ING MECÁ, REFRIG, TERMO, VEH *motores* compression ratio; **~ de compresión de la lente** *f* CINEMAT lens squeeze ratio; **~ común** *f* MATEMÁT common ratio; **~ de concentración** *f* ENERG RENOV concentration ratio; **~ contable** *f* CONST accounting ratio; **~ de contraste** *f* CINEMAT, TV contrast ratio; **~ de controles de acceso** *f* TELECOM access control list; **~ de densidad onda portadora-ruido** *f* TELECOM carrier-to-noise density ratio; **~ densidad onda portadora-ruido intermodulación** *f* TELECOM carrier-to-intermodulation noise density ratio; **~ de densidades de sinterizado** *f* PROC QUÍ sintered density ratio; **~ densidad-volumen** *f* TRANSP volume-density relationship; **~ de desecación** *f* REFRIG *en alimentos liofilizados* desiccation ratio; **~ de desmultiplicación** *f* VEH *montaje del eje trasero* axle ratio; **~ desmultiplicadora** *f* CARBÓN reduction ratio; **~ día-hora cargada** *f* TELECOM day-to-busy-hour ratio; **~ del diámetro del cabeceo** *f* TRANSP AÉR pitch-diameter ratio; **~ del diámetro del paso** *f* TRANSP AÉR pitch-diameter ratio; **~ del diferencial** *f* AUTO differential ratio; **~ de las dimensiones de la imagen** *f* CINEMAT picture ratio; **~ de dispersión** *f* FÍS dispersion relation; **~ de la dosis de respuesta** *f* CONTAM *contaminación radiactiva* dose-response relationship; **~ de electrón a átomo** *f* FÍS PART, NUCL electron-to-atom ratio; **~ de engranajes** *f* ING MECÁ, MECÁ, VEH *caja de cambio* gear ratio; **~ entre el área de garganta y la abertura** *f* TEC ESP port-to-throat area ratio; **~ entre el diámetro del alma y el espesor del aislamiento** *f* ÓPT, TELECOM core-cladding ratio; **~ entre dimensiones** *f* FÍS *de objetos*, ING ELÉC aspect ratio; **~ entre dos presiones** *f* TEC ESP pressure ratio; **~ entre empuje y masa** *f* TEC ESP thrust-to-mass ratio; **~ entre empuje y peso** *f* TEC ESP thrust-to-weight ratio; **~ entre el número de espiras del secundario y del primario** *f* ING ELÉC turns ratio; **~ entre la resisten-**

cia en húmedo y en seco *f* PAPEL wet-strength retention; ~ **de espiras** *f* AUTO turns ratio; ~ **de la fineza** *f* TRANSPAÉR fineness ratio; ~ **del flujo de aire** *f* INSTAL TERM airflow rate; ~ **ganancia temperatura de ruido** *f* TEC ESP *comunicaciones* gain-to-noise-temperature ratio (*G-T*); ~ **gas-petróleo** *f* TEC PETR *producción* gas-to-oil ratio (*GOR*); ~ **giromagnética** *f* FÍS gyromagnetic ratio; ~ **hombre-máquina** *f* PROD man-machine ratio, TELECOM man-machine relationship; ~ **de iluminación** *f* CINEMAT lighting ratio; ~ **de impedancia** *f* PROD impedance ratio; ~ **de impulsividad** *f* TELECOM impulsiveness ratio; ~ **inversa** *f* MATEMÁT inverse ratio; ~ **del límite de fatiga a la resistencia de rotura por tracción** *f* NUCL endurance ratio; ~ **de mezcla** *f* ING MECÁ blending radius; ~ **de movilidad** *f* PETROL mobility ratio; ~ **de multiplicación** *f* ING MECÁ, MECÁ, VEH gear ratio; ~ **no restringida de datos digitales** *f* TELECOM unrestricted digital data ratio; ~ **de onda estacionaria** *f* (*ROE*) FÍS, TEC ESP, TELECOM standing-wave ratio (*SWR*); ~ **de ondas estacionarias de tensión** *f* TEC ESP voltage standing wave ratio (*VSWR*); ~ **petróleo-gas en solución** *f* TEC PETR solution gas-oil ratio; ~ **de potencia máxima-media** *f* NUCL peak-to-average power ratio; ~ **potencia-masa** *f* ING MECÁ power-weight ratio; ~ **puntal-calado** *f* TRANSP MAR *diseño de barcos* depth-to-draft ratio (*AmE*), depth-to-draught ratio (*BrE*); ~ **de ramificación** *f* FÍS *elementos radiactivos* branching ratio; ~ **de rechazo de canal adyacente** *f* TELECOM adjacent channel rejection ratio; ~ **recíproca** *f* ING MECÁ inverse ratio; ~ **de reducción** *f* ING MECÁ, MECÁ, VEH gear ratio; ~ **de rodaje** *f* CINEMAT shooting ratio; ~ **señal-ecos parásitos** *f* TELECOM signal-to-clutter ratio; ~ **señal-ruido** *f* ACÚST, ELECTRÓN, FÍS, INFORM&PD, TEC PETR, TELECOM, TV signal-to-noise ratio (*SNR*); ~ **señal-ruido en el circuito de entrada** *f* ELECTRÓN input signal-to-noise ratio; ~ **señal-ruido y distorsión** *f* (*relación SINAD*) TELECOM signal-to-noise and distortion ratio (*SINAD ratio*); ~ **de sequedad** *f* REFRIG dryness ratio; ~ **SINAD** *f* (*relación señal-ruido y distorsión*) TELECOM SINAD ratio (*signal-to-noise and distortion ratio*); ~ **de temperaturas** *f* TERMO temperature ratio; ~ **de tensión nominal** *f* ELEC *transformador*, ING ELÉC, PROD rated-voltage ratio; ~ **de transformación** *f* AUTO turns ratio, ELEC *transformador* turn ratio, turns ratio, winding ratio, ING ELÉC turns ratio; ~ **de transformación de corriente** *f* ELEC *transformador* current-transformation ratio; ~ **de transmisión** *f* ING MECÁ, VEH gear ratio; ~ **de vacío** *f* PROD void ratio; ~ **de la velocidad periférica** *f* ENERG RENOV tip-speed ratio; ~ **velocidad-densidad** *f* TRANSP speed-density relationship; ~ **velocidad-flujo** *f* TRANSP speed-flow relationship; ~ **volumétrica** *f* ING MECÁ compression ratio; ~ **de vueltas** *f* AUTO turns ratio, ELEC *transformador* turn ratio, turns ratio; ~ **de Wylie** *f* PETROL Wylie relationship

relacional *adj* CINEMAT, INFORM&PD, TELECOM relational

relajación *f* ELEC *oscilación*, METAL, P&C, TEXTIL relaxation; ~ **de esfuerzos interiores** *f* TERMO stress relief; ~ **térmica** *f* C&V *del cristal, termómetro* ageing

relajar *vt* ELEC, METAL, P&C, TEXTIL relax

relámpago *m* ELEC *fenómeno*, METEO lightning; ~ **difuso** *m* METEO sheet lightning

relampagueo *m* ELEC *fenómeno*, ING ELÉC lightning

relantizador *m* ING ELÉC slug

relatividad *f* FÍS, TEC ESP relativity

relativista *adj* TEC ESP relativistic

relativo: ~ **al globo terráqueo** *adj* TEC ESP global

relave *m* CARBÓN *minería* tailings, tails

relé *m* GEN relay, ING MECÁ trip;

~ a ~ **de acción diferida** *m* ELEC slow-acting relay, time-delay relay, time-locking relay; ~ **de acción instantánea** *m* ELEC instantaneous relay; ~ **de acción lenta** *m* ELEC slow-acting relay; ~ **de acción rápida** *m* ELEC high-speed relay; ~ **de acción retardada** *m* ELEC slow-acting relay; ~ **de acción unipolar** *m* ING ELÉC single-pole single-throw relay (*SPST relay*); ~ **accionado fotoeléctricamente** *m* ING ELÉC photoelectrically-operated relay; ~ **accionado por fuerza centrífuga** *m* ELEC centrifugal relay; ~ **accionado por la voz** *m* TELECOM voice-operated relay; ~ **accionador** *m* ELEC trip relay, tripping relay; ~ **de aceleración** *m* ELEC acceleration relay; ~ **de acoplamiento** *m* ELEC interlocking relay; ~ **actuado por la voz** *m* TELECOM voice-operated relay; ~ **con alarma** *m* ELEC alarm relay; ~ **de alcancía de pago previo** *m* ELEC coin-box relay; ~ **de aleta** *m* ELEC vane-type relay; ~ **de apertura** *m* ELEC release relay; ~ **de apertura sin tensión** *m* ELEC no-voltage release relay; ~ **de autorreposición** *m* ELEC self-resetting relay; ~ **de avance de tres pasos** *m* ELEC three-step relay;

~ b ~ **de baja carga** *m* ELEC underload relay; ~ **de baja corriente** *m* ELEC undercurrent relay; ~ **base-móvil** *m* TELECOM base-to-mobile relay; ~ **biestable** *m* ING ELÉC bistable relay; ~ **bimetálico térmico de capa** *m* PROD thermal-bimetallic overlay relay; ~ **bimetálico térmico de sobrecarga** *m* PROD thermal-bimetallic overload relay;

~ c ~ **de caja** *m* ELEC box relay; ~ **de caja de monedas** *m* ELEC coin-box relay; ~ **de camarín** *m* ING ELÉC cage relay; ~ **de capa de red** *m* TELECOM network layer relay (*NLR*); ~ **capacitivo** *m* ELEC capacitance relay; ~ **de carga mínima** *m* ELEC underload relay; ~ **de cascada** *m* ELEC stepping switch; ~ **de CBB** *m* ING ELÉC DPDT relay; ~ **de CC** *m* ING ELÉC DC relay; ~ **centrífugo** *m* ELEC centrifugal relay; ~ **de cierre** *m* ELEC make relay; ~ **de cierre con llave** *m* ING ELÉC lock-up relay; ~ **de cierre-apertura** *m* ELEC all-or-nothing relay; ~ **de conexión magnética** *m* ING ELÉC magnetic latching relay; ~ **de conmutación** *m* ELEC changeover relay, switch relay, ING ELÉC changeover relay; ~ **conmutador de batería** *m* AUTO battery changeover relay; ~ **de contacto de láminas flexibles** *m* ELEC, ING ELÉC, PROD reed-contact relay; ~ **para contador** *m* ING ELÉC meter-type relay; ~ **de control** *m* ELEC control relay; ~ **de control principal** *m* PROD master control relay; ~ **de corriente** *m* ELEC, ING ELÉC current relay; ~ **de corriente activa** *m* ELEC working current relay; ~ **de corriente alterna** *m* ELEC alternating-current relay; ~ **de corriente inversa** *m* ELEC, ING ELÉC reverse-current relay; ~ **de corriente de régimen** *m* ELEC working current relay; ~ **de corriente de trabajo** *m* ELEC working current relay; ~ **de corte** *m* ELEC cutoff relay;

~ d ~ **dependiente del tiempo** *m* ELEC time-dependent relay; ~ **desconectador** *m* ELEC trip relay, tripping relay; ~ **de desconexión** *m* ING ELÉC disconnect relay; ~ **de desenganche** *m* ELEC trip relay, tripping relay; ~ **de desexcitación** *m* ELEC release relay; ~ **de desexcitación sin tensión** *m* ELEC no-voltage release relay; ~ **detector** *m* ING ELÉC sensing relay; ~ **detector de hielo** *m* TRANSP AÉR ice detector relay; ~ **detonante** *m* MINAS *explosivos* detonating relay; ~ **diferenciador** *m* ELEC discriminating relay; ~ **diferencial** *m* ELEC, ING ELÉC differential relay; ~ **de dínamo de CA** *m* ING ELÉC AC armature relay, AC capacitor; ~ **direccional** *m* ELEC, ING ELÉC directional relay; ~ **discriminador** *m* ELEC discriminating relay; ~ **disparador** *m* ELEC trip relay, tripping relay; ~ **de distancia** *m* ELEC distance relay; ~ **de doble armadura** *m* ELEC double-armature relay; ~ **de dos elementos** *m* ELEC two-element relay; ~ **de dos etapas** *m* ELEC two-stage relay;

~ **e** ~ **eléctrico** *m* ELEC, ING ELÉC electric relay; ~ **electrodinámico** *m* ELEC electrodynamic relay; ~ **electromagnético** *m* ELEC, ING ELÉC electromagnetic relay; ~ **electromagnético de sobrecarga** *m* ELEC, ING ELÉC electromagnetic overload relay, magnetic overload relay; ~ **electromecánico** *m* ELEC, ING ELÉC electromechanical relay; ~ **electrónico** *m* ELEC, ELECTRÓN electronic relay; ~ **electrostático** *m* ELEC electrostatic relay; ~ **enchufable** *m* ING ELÉC plug-in relay; ~ **de enclavamiento** *m* ELEC interlock relay, interlocking relay, latching relay, ING ELÉC interlock relay; ~ **de enclavamiento de potencia** *m* PROD power-interlock relay; ~ **enganchador** *m* ELEC latching relay; ~ **de enganche a bobina doble** *m* ING ELÉC dual-coil latching relay; ~ **de enlace unipolar** *m* ING ELÉC single-coil latching relay; ~ **de equilibrado de corriente** *m* ELEC current balance relay; ~ **equilibrador** *m* ELEC balancing relay; ~ **de equilibrio** *m* ELEC balance relay; ~ **de equilibrio de fases** *m* ELEC phase-balance relay; ~ **esclavo rápido** *m* TRANSP AÉR fast-slaving relay; ~ **estable en la posición central** *m* ELEC center-stable relay (*AmE*), centre-stable relay (*BrE*); ~ **de estado sólido** *m* ING ELÉC solid-state relay; ~ **de estado sólido fotoacoplado** *m* ING ELÉC photo-coupled solid-state relay; ~ **estático** *m* ELEC, TELECOM static relay; ~ **de estrecho margen funcional** *m* ELEC marginal relay; ~ **excitado** *m* PROD energized relay; ~ **extrarrápido** *m* ELEC, ING ELÉC instantaneous relay;

~ **f** ~ **fotoeléctrico** *m* ELEC, ING ELÉC photoelectric relay; ~ **de frecuencia** *m* ING ELÉC frequency relay; ~ **de funcionamiento** *m* ING ELÉC operate relay;

~ **g** ~ **galvanométrico** *m* ELEC moving-coil relay; ~ **de gas** *m* ELEC, ELECTRÓN, GAS, ING ELÉC gas-filled relay;

~ **h** ~ **de hilo caliente** *m* ELEC hot-wire relay; ~ **de hipocorriente** *m* ELEC undercurrent relay;

~ **i** ~ **de imán permanente** *m* ELEC permanent-magnet relay; ~ **de impedancia** *m* ELEC impedance relay; ~ **impolarizado** *m* ELEC unbiased polarized relay; ~ **de impulsión** *m* ING ELÉC impulse relay; ~ **con indicador** *m* ELEC alarm relay; ~ **de inducción** *m* ELEC induction relay; ~ **del inducido** *m* ING ELÉC armature relay; ~ **de inducido de CA** *m* ING ELÉC AC armature relay; ~ **inductivo** *m* ELEC, ING ELÉC induction relay; ~ **instrumental** *m* ING ELÉC instru-

ment-type relay; ~ **de intensidad** *m* ELEC, ING ELÉC current relay; ~ **interruptor** *m* ELEC *conmutador*, ING ELÉC relay interrupter; ~ **de inversión de corriente** *m* ELEC, ING ELÉC reverse-current relay; ~ **de inversión de fases** *m* ELEC reverse-phase relay; ~ **inversor** *m* ELEC, ING ELÉC changeover relay;

~ **j** ~ **de jaula** *m* ING ELÉC cage relay;

~ **l** ~ **de láminas** *m* ELEC, ING ELÉC, PROD, TELECOM reed relay; ~ **de láminas flexibles** *m* ELEC, ING ELÉC, PROD, TELECOM reed relay; ~ **de láminas magnéticas** *m* ELEC, ING ELÉC, PROD, TELECOM reed relay; ~ **de láminas resonantes** *m* ING ELÉC resonant-reed relay; ~ **lento** *m* ELEC slow-acting relay; ~ **de línea** *m* ELEC line relay; ~ **lleno de gas** *m* ELEC, ELECTRÓN, GAS, ING ELÉC gas-filled relay;

~ **m** ~ **de máxima** *m* ELEC maximum-voltage relay, overcurrent relay, overload relay, surge relay, *protección* overvoltage relay, PROD overload relay; ~ **de máxima y mínima** *m* ELEC over-and-under current relay; ~ **de máximo y mínimo de corriente** *m* ELEC over-and-under current relay; ~ **de máximo de tensión** *m* ELEC maximum-voltage relay, *protección* overvoltage relay; ~ **de mediciones** *m* ING ELÉC measuring relay; ~ **de mercurio** *m* ELEC mercury relay; ~ **de mínima** *m* ELEC minimum current relay, minimum-power relay, underload relay; ~ **de mínimo de carga** *m* ELEC underload relay; ~ **de mínimo de corriente** *m* ELEC minimum current relay; ~ **de mínimo de potencia** *m* ELEC minimum-power relay; ~ **de múltiples posiciones** *m* ELEC multi-position relay, stepping switch;

~ **n** ~ **N** *m* TELECOM N-relay; ~ **neutral** *m* ELEC neutral relay; ~ **neutro** *m* CONST nonpolarized relay, ELEC neutral relay, nonpolarized relay, ING ELÉC neutral relay; ~ **no polarizado** *m* CONST nonpolarized relay, ELEC neutral relay, nonpolarized relay, unbiased polarized relay; ~ **de núcleo buzo** *m* ELEC solenoid relay; ~ **de núcleo móvil** *m* ING ELÉC plunger relay;

~ **o** ~ **óptico** *m* CINEMAT light valve, ING ELÉC optical relay;

~ **p** ~ **de paquete en línea doble** *m* ING ELÉC dual-in-line package relay; ~ **plano** *m* ELEC slave relay, ING ELÉC flat relay; ~ **PLD** *m* ING ELÉC DIP relay; ~ **polar neutral** *m* ING ELÉC neutral polar relay; ~ **polarizado** *m* ELEC, ING ELÉC biased relay, polarized relay; ~ **sin polarizar** *m* ELEC unbiased polarized relay; ~ **de potencia** *m* ING ELÉC, PROD power relay; ~ **de potencia activa** *m* ELEC active power relay; ~ **de protección diferencial** *m* ELEC differential protection relay; ~ **de protección de distancia** *m* ELEC distance relay; ~ **protector** *m* ING ELÉC protective relay;

~ **r** ~ **rápido** *m* ELEC high-speed relay; ~ **de reactancia** *m* ELECTRÓN, ING ELÉC reactance relay; ~ **reed con contactos de mercurio** *m* ING ELÉC mercury-wetted reed relay; ~ **de reposición automática** *m* ELEC self-resetting relay; ~ **residual** *m* ELEC, ING ELÉC residual relay; ~ **de resonancia** *m* ELEC tuned relay; ~ **retardado** *m* ELEC time-delay relay, time-lag relay, ING ELÉC slow-operate relay; ~ **de retardo** *m* ELEC delay relay, time-delay relay, on-delay relay, ING ELÉC on-delay relay, TELECOM delay-mode relay; ~ **de retardo dependiente** *m* ELEC inverse time relay; ~ **de retardo en desconexión** *m* ING ELÉC *temporizador* off-delay relay; ~ **de retardo**

independiente *m* ELEC definitive time relay; **~ de retardo de tiempo** *m* ELEC time-delay relay; **~ de retención** *m* ELEC, ING ELÉC biased relay; **~ de retenida** *m* ING ELÉC latching relay; **■ s** **~ secundario** *m* ELEC secondary relay; **~ secundario teleaccionado** *m* ING ELÉC sequence relay; **~ de seguridad** *m* ING ELÉC guarding relay; **~ selector** *m* ELEC discriminating relay, ING ELÉC selector relay; **~ de semiconductor** *m* ING ELÉC semiconductor relay; **~ para sentido de fuerza** *m* ELEC power-directional relay (*PDR*); **~ de simetría** *m* ELEC balance relay; **~ sincronizador** *m* ELEC synchronizing relay; **~ de sobrecarga** *m* ELEC, ING ELÉC, PROD overload relay; **~ de sobrecarga de aleación eutéctica** *m* PROD eutectic-alloy overload relay; **~ de sobrecorriente** *m* ELEC overcurrent relay; **~ de sobreintensidad** *m* ELEC surge relay; **~ de sobretensión** *m* ELEC surge relay, *protección* overvoltage relay, ING ELÉC overvoltage relay; **~ solenoide** *m* ING ELÉC solenoid relay; **~ sólido acoplado ópticamente** *m* ING ELÉC optically-coupled solid-state relay; **~ subminiatura** *m* ING ELÉC subminiature relay; **■ t** **~ telefónico** *m* ING ELÉC, TELECOM telephone relay; **~ temporizado** *m* ELEC time-locking relay, ING ELÉC delay relay, time-delay relay; **~ temporizador** *m* PROD timer; **~ temporizador del filtro de entrada** *m* PROD input-filter time delay; **~ de tensión** *m* ING ELÉC voltage relay; **~ térmico** *m* ELEC, ING ELÉC thermal relay; **~ para todas las aplicaciones** *m* ING ELÉC general-purpose relay; **~ de todo-nada** *m* ELEC, ING ELÉC all-or-nothing relay; **~ de tubo fotoeléctrico** *m* ING ELÉC phototube relay; **■ u** **~ UB** *m* ING ELÉC DPST relay; **~ unipolar** *m* ING ELÉC single-throw relay; **~ unipolar bidireccional** *m* ING ELÉC double-pole single-throw relay; **~ unipolar con contacto de cambio** *m* ING ELÉC single-pole single-throw relay (*SPST relay*); **~ unipolar de doble vano** *m* ING ELÉC single-pole single-throw relay (*SPST relay*); **~ unipolar y univanal** *m* ING ELÉC single-pole single-throw relay (*SPST relay*); **■ v** **~ variable en función del tiempo** *m* ELEC time-dependent relay; **~ de vuelta al reposo** *m* ELEC release relay

relevador *m* ING MECÁ relay, release, TEC ESP, TELECOM, TV relay; **~ con acción de retardo** *m* ELEC time-delay relay; **~ para corriente alterna** *m* ELEC alternating-current relay; **~ de corriente inversa** *m* ELEC, ING ELÉC reverse-current relay; **~ diferencial** *m* ELEC, ING ELÉC differential relay; **~ de disparo** *m* FOTO trigger relay; **~ electromagnético de sobrecarga** *m* ELEC, ING ELÉC electromagnetic overload relay, magnetic overload relay; **~ electrónico** *m* ELEC, ELECTRÓN electronic relay; **~ electrostático** *m* ELEC electrostatic relay; **~ estable en la posición central** *m* ELEC center-stable relay (*AmE*), centre-stable relay (*BrE*); **~ de intensidad** *m* ELEC, ING ELÉC current relay; **~ multiposición** *m* ELEC multiposition relay; **~ neutral** *m* ELEC neutral relay; **~ neutro** *m* ELEC neutral relay; **~ polarizado** *m* ELEC polarized relay; **~ retardado** *m* ELEC time-delay relay; **~ de retardo** *m* ELEC time-delay relay; **~ de sincronización** *m* ELEC synchronizing relay

relevo *m* PROD *de obreros* shift; **~ automático** *m* TELECOM automatic changeover

relictual *adj* GEOL *de mineral o estructura* relictual

relieve[1]: **en ~** *adj* EMB, IMPR, ING MECÁ, PAPEL embossed

relieve[2] *m* ING MECÁ embossment; **~ submarino** *m* OCEAN submarine relief

relinga *f* TRANSP MAR *velas* bolt rope

rellamada: **~ al último número** *f* TELECOM last-number recall

rellano *m* CONST landing, *escalera* half space

rellenado[1] *adj* EMB, MECÁ padded; **~ con polvo** *adj* ING ELÉC powder filled

rellenado[2] *m* MINAS *de galerías* packing, NUCL backfill; **~ aséptico** *m* ALIMENT *maquinaria para proceso de alimentos* aseptic filling

rellenador: **~ de calas** *m* AmL (*cf rellenazanjas Esp*) TRANSP back filler

rellenar *vt* *Esp* CONST backfill, bank up, FERRO refill, GEOL (*cf depositar sedimentos en ríos AmL*) aggrade, IMPR build, pad, fill, INFORM&PD build, MINAS (*cf hinchar AmL, macizar AmL, verter AmL*) pack, fill, *minas* stow, TEXTIL pad; **~ con ceros** *vt* INFORM&PD zero fill; **~ para formar un talud** *vt* CONST bank up with earth; **~ con masilla** *vt* CONST putty

rellenazanjas *m* *Esp* (*cf rellenador de calas AmL*) TRANSP back filler

relleno[1] *adj* CONST filled; **de ~** *adj* CONST filled

relleno[2] *m* C&V fill, CONST *albañilería* filling, backfill, CONTAM *instalaciones industriales* backfill, ELEC *cable* bedding, filler, EMB padding, GAS filling, IMPR plug, INFORM&PD filler, filling, padding, ING ELÉC filler, ING MECÁ padding, packing, MINAS *de galerías* packing, *murete de piedra* building, stowing, NUCL filling, P&C *ingrediente* filler, PROD stuffing, filling-up, REVEST backing, TELECOM stuffing, TEXTIL pad, TRANSP *contenedores* stuffing; **~ aséptico** *m* ALIMENT *maquinarias para proceso de alimentos* aseptic filling; **~ de bits** *m* INFORM&PD bit stuffing; **~ de canal** *m* GEOL *depósito sedimentario* channel fill; **~ con desechos** *m* MINAS goaf; **~ digital** *m* TELECOM digital filling, digital pad; **~ filoniano** *m* MINAS lode filling; **~ de galerías** *m* MINAS pack; **~ geotrópico** *m* GEOL geotropic filling; **~ por gravedad** *m* TEC ESP gravity filling; **~ hidráulico** *m* MINAS slush; **~ hidráulico con fangos** *m* PROD *minas* silting; **~ de minas** *m* MINAS pithead building; **~ natural de aguas freáticas** *m* AGUA natural groundwater recharge; **~ seleccionado** *m* CONST *carreteras* selected fill; **~ de tierra** *m* CARBÓN, CONTAM, RECICL landfill; **~ del tráfico** *m* TELECOM traffic padding; **~ de zanjas** *m* CARBÓN backfill

reloj *m* ELECTRÓN clock, FOTO timer, METR clock, REFRIG *temporizador* timer; **~ de arena** *m* C&V hour glass; **~ atómico** *m* FÍS, TELECOM atomic clock; **~ de cesio** *m* TEC ESP caesium clock (*BrE*), cesium clock (*AmE*); **~ de control** *m* SEG works recording clock; **~ de cristal de cuarzo** *m* ELEC, ELECTRÓN quartz-crystal clock; **~ para cuarto oscuro** *m* CINEMAT, FOTO darkroom timer; **~ desconectador** *m* ING MECÁ cutout clock; **~ eléctrico sincrónico** *m* ELEC synchronous electric clock; **~ electrónico** *m* ELECTRÓN electronic clock, FOTO electronic timer, TELECOM electronic clock; **~ fechador** *m* TELECOM time stamp; **~ para fichar** *m* PROD time recorder; **~ interno** *m* TELECOM internal clock; **~ maestro** *m* INFORM&PD, TELECOM master clock; **~ parlante** *m* TELECOM *información telefónica de la hora* speaking clock; **~ patrón** *m* INFORM&PD, TELECOM master clock; ~ **principal** _m_ INFORM&PD, TELECOM master clock; ~ **de referencia** _m_ TELECOM reference clock; ~ **registrador** _m_ PROD time recorder; ~ **telefónico** _m_ TELECOM speaking clock

relojeras _f pl_ ING MECÁ watch casings

reluctancia _f_ ELEC _magnetismo_ magnetic resistance, reluctance, FÍS reluctance, ING ELÉC magnetic resistance, reluctance; ~ **específica** _f_ ELEC _magnetismo_, FÍS, ING ELÉC reluctivity

reluctividad _f_ ELEC _magnetismo_, FÍS, ING ELÉC reluctivity

rem _m_ (_dosis equivalente en renguenios en el hombre_) FÍS RAD rem (_roentgen equivalent for man_)

remachado _m_ CONST riveting, _de clavos_ driving-in, ING MECÁ, MECÁ, PROD riveting; ~ **embutido** _m_ PROD flush riveting; ~ **de empalme** _m_ PROD joint riveting; ~ **en frío** _m_ CONST cold riveting; ~ **de junta** _m_ PROD joint riveting; ~ **de juntas transversales** _m_ PROD butt riveting; ~ **al martillo** _m_ PROD hammer riveting; ~ **de martillo** _m_ PROD hammer pick; ~ **a paño** _m_ PROD flush riveting; ~ **de recubrimiento** _m_ PROD lap riveting; ~ **solapado** _m_ EMB lap riveting; ~ **al tresbolillo** _m_ PROD cross riveting

remachadora _f_ CONST, ING MECÁ, MECÁ, PROD riveter; ~ **de aire comprimido** _f_ PROD pneumatic riveter; ~ **hidráulica de bóvedas** _f_ INSTAL HIDRÁUL dome riveter; ~ **hidráulica fija** _f_ PROD stationary hydraulic riveter

remachar _vt_ PROD clinch; ~ **en caliente** _vt_ ING MECÁ hot-rivet

remache _m_ CONST rivet, riveting, ING MECÁ, MECÁ, PROD rivet, riveting, fastener; ~ **de arrastre** _m_ ING MECÁ driver; ~ **avellanado** _m_ CONST, ING MECÁ, MECÁ countersunk riveting; ~ **bifurcado** _m_ ING MECÁ, PROD bifurcated rivet; ~ **de cabeza avellanada elevada** _m_ ING MECÁ raised countersunk-rivet; ~ **de cabeza cónica** _m_ CONST cone-head rivet, steeple-head rivet; ~ **de cabeza de cono truncado** _m_ CONST cone-head rivet; ~ **de cabeza embutida** _m_ CONST countersunk-head rivet, ING MECÁ countersunk-head rivet, flush-head rivet, MECÁ countersunk-head rivet, PROD flush-head rivet; ~ **de cabeza esférica** _m_ CONST button-head rivet; ~ **de cabeza ovalada** _m_ ING MECÁ oval countersunk rivet; ~ **de cabeza perdida** _m_ PROD flat countersunk rivet; ~ **de cabeza plana** _m_ CONST cheese-head rivet; ~ **de cabeza redonda** _m_ CONST bullhead rivet, bullheaded rivet, cup-head rivet; ~ **de cabeza redonda embutida** _m_ CONST countersunk buttonhead rivet; ~ **de cabeza troncocónica** _m_ CONST, ING MECÁ, PROD pan-head rivet; ~ **de caldera** _m_ INSTAL HIDRÁUL boiler rivet; ~ **de caña hendida** _m_ ING MECÁ, PROD bifurcated rivet; ~ **ciego** _m_ ING MECÁ blind rivet; ~ **de cobre** _m_ CONST copper rivet; ~ **para correa** _m_ PROD belt rivet; ~ **de cuerpo largo** _m_ ING MECÁ long-shank top; ~ **explosivo** _m_ MECÁ explosive-type rivet; ~ **hermético** _m_ CONST tight riveting; ~ **hueco** _m_ ING MECÁ hollow rivet, TEC ESP Dzus fastener; ~ **de ojal** _m_ PROD eyed rivet-snap; ~ **percusivo** _m_ ING MECÁ percussion rivet; ~ **ranurado** _m_ CONST slotted clinch rivet, slotted rivet; ~ **de redoblar** _m_ CONST slotted clinch rivet, ING MECÁ clinch rivet; ~ **superior de caldera de cobre** _m_ PROD brazier head rivet; ~ **tubular de forma especial** _m_ PROD pop rivet

remanencia _f_ ELEC, FÍS, ING ELÉC, PETROL remanence; ~ **de la imagen** _f_ Esp (_cf persistencia de la imagen AmL_) TV burn

remanente _m_ INFORM&PD, MATEMÁT remainder

remanso _m_ AGUA damming, HIDROL backwash water, OCEAN, TRANSP MAR _navegación_ slack water

remar _vi_ TRANSP MAR row

rematar _vt_ CONST _tejidos_ cap

remate _m_ CONST _arquitectura_ crowning, _tejado_ finial; ~ **ciego** _m_ TEC PETR blind auction

remendar _vt_ TEXTIL _prendas_ mend, _ropa_ darn

remesa _f_ MECÁ batch, PROD consignment

remetido _m_ TEXTIL _tejidos_ drafting

remiendo _m_ ING ELÉC patching, PROD patching, patching-up, patchwork, TEXTIL _ropa_ mending; ~ **de redes** _m_ OCEAN net mending

remitente _m_ TELECOM originator

remoción _f_ C&V stirring, _de pintura o capas_ stripping, CONTAM _tratamiento de aguas y gases_, CONTAM MAR removal, MINAS _de rocas después de una voladura_ lashing; ~ **del caldo** _f_ PROD _hornos_ rabbling; ~ **de la inflorescencia masculina del maíz** _f_ AGRIC detasseling (_AmE_), detasselling (_BrE_); ~ **de metal** _f_ ING MECÁ metal removal; ~ **de partículas contaminantes del aire** _f_ CONTAM _tratamiento de gases_ particulate removal of air pollutants

remodelación _f_ TEC PETR _de punto de venta de combustibles_ revamping

remodulación _f_ ELECTRÓN remodulation

remojar _vt_ ALIMENT, C&V, TEXTIL soak, steep

remojo _m_ ALIMENT, C&V, TEXTIL soaking

remolcador _m_ CONTAM MAR tugboat, TRANSP MAR tug; ~ **de altura** _m_ TRANSP MAR sea tug; ~ **de blancos** _m_ D&A target tug; ~ **empujador** _m_ TRANSP MAR pusher tug; ~ **espacial** _m_ TEC ESP space tug; ~ **del planeador** _m_ D&A glider tug; ~ **de salvamento** _m_ TRANSP MAR salvage tug

remolcar _vt_ CONTAM MAR, TRANSP MAR tow; ~ **en flecha** _vt_ TRANSP MAR tow astern

remoldeo _m_ PROD remolding (_AmE_), remoulding (_BrE_)

remoler _vt_ CARBÓN, PROD regrind

remolino _m_ ACÚST eddy, CARBÓN vortex, FÍS eddy, HIDROL backwater, eddy, vortex, whirlpool, MECÁ vortex, METEO eddy wind, OCEAN eddy, TRANSP MAR _agua, viento_ eddy, _del viento contra la vela_ eddy wind; ~ **de retorno** _m_ HIDROL back eddy; ~ **vertical** _m_ CARBÓN _ríos_ boil; ~ **de viento** _m_ METEO, TRANSP MAR whirlwind

remolque _m_ AGRIC loading wagon, CONST bogie (_BrE_), low boy trailer, trailer (_AmE_), truck (_AmE_), CONTAM MAR boom-towing, FERRO, ING MECÁ, MECÁ bogie (_BrE_), trailer (_AmE_), truck (_AmE_), PETROL tug, TRANSP caravan (_BrE_), trailer (_AmE_), TRANSP AÉR towing, TRANSP MAR tow, towing, warp, VEH bogie (_BrE_), caravan (_BrE_), _tipo de vehículo_ trailer (_AmE_), truck (_AmE_); ~ **articulado** _m_ AmL (_cf trailer Esp_) CONST, FERRO, ING MECÁ, MECÁ, VEH bogie (_BrE_), trailer (_AmE_), truck (_AmE_); ~ **de barca** _m_ TRANSP boat trailer; ~ **basculante trasero** _m_ TRANSP rear-tipping trailer; ~ **de cabina** _m_ VEH house bogie (_BrE_), house trailer (_AmE_); ~ **de carretera** _m_ AUTO, VEH road trailer; ~ **por cremallera** _m_ FERRO, TRANSP rack-railroad trailer (_AmE_), rack-railway trailer (_BrE_); ~ **de dos ruedas** _m_ TRANSP _vehículos_ semi-trailer; ~ **empujado** _m_ TRANSP push towing; ~ **por empuje** _m_ TRANSP push tow; ~ **mediante empuje** _m_

TRANSP push tug; ~ **de plataforma baja** *m* CONST low-bed trailer; ~ **portaequipajes** *m* TRANSP barrow; ~ **de uso múltiple** *m* TRANSP all-purpose trailer; ~ **sobre vagón plataforma** *m* TRANSP trailer on flatcar (*TOFC*)

remonte *m* OCEAN *de pescado, para el desove* run

remosqueo *m* IMPR slur, smearing

remoto *adj* INFORM&PD, TELECOM, TV remote

remover *vt* ING MECÁ remove; ~ **la capa de óxido de** *vt* MECÁ descale

removible *adj* GEN removable

removido: ~ **neumático** *m* CARBÓN *de vagones* pneumatic handling

rendija *f* FÍS, FOTO slit, MECÁ aperture, PROD slit, split, TEC ESP slit, slot, TELECOM slot; ~ **transversal** *f* TELECOM transverse slot

rendijas: ~ **de Young** *f pl* FÍS Young's slits

rendimiento *m* AGRIC yield, AGUA delivery, CARBÓN recovery, yield, ELEC *equipo* efficiency, *en una producción* output, ELECTRÓN efficiency, yield, ENERG RENOV yield, FÍS efficiency, GAS yield, INFORM&PD performance, yield, ING MECÁ output, ING MECÁ efficiency, output, performance, MECÁ efficiency, MINAS *de trabajo*, NUCL *de un sistema*, PROD efficiency, output, return, yield, QUÍMICA yield, TEC ESP *telecomunicaciones* efficiency, *comunicaciones* gain, TEC PETR, TEXTIL yield; ~ **acumulado de fisión** *m* NUCL cumulative fission yield; ~ **adiabático** *m* TERMO adiabatic efficiency; ~ **aerodinámico** *m* ENERG RENOV lift-to-drag ratio, FÍS lift-drag ratio; ~ **de aislación térmica** *m* TERMO heat insulation effectiveness; ~ **Auger** *m* FÍS Auger yield; ~ **del bocarte** *m* PROD mill result; ~ **calorífico** *m* TERMO calorific output, heat efficiency, heat throughput; ~ **calorífugo** *m* TERMO heat insulation effectiveness; ~ **en canal** *m* ALIMENT carcass yield; ~ **en carne de vaca** *m* AGRIC beef yield; ~ **de colector solar** *m* ENERG RENOV collector efficiency; ~ **del combustible** *m* TEC ESP fuel efficiency; ~ **de combustión** *m* TERMO combustion efficiency; ~ **constante** *m* ELEC *equipo* constant duty; ~ **cuántico** *m* ELECTRÓN, FÍS, FÍS PART, FÍS RAD, NUCL, ÓPT, TELECOM quantum efficiency; ~ **cuántico de conversión** *m* NUCL conversion quantum efficiency; ~ **cuántico diferencial** *m* TELECOM differential quantum efficiency; ~ **cuántico de la luminiscencia** *m* FÍS RAD luminescence quantum yield; ~ **diario promedio** *m* ENERG RENOV average daily output; ~ **eléctrico** *m* ELEC electrical efficiency, ING ELÉC electric efficiency, NUCL *del reactor* electrical output; ~ **energético** *m* GAS power output, TERMO energy efficiency; ~ **de energía** *m* ENERG RENOV energy extraction, power output; ~ **de la estampa formadora** *m* C&V blank mold turnover (*AmE*), blank mould turnover (*BrE*); ~ **de fabricación** *m* ELECTRÓN fabrication yield; ~ **de fase** *m* ELECTRÓN stage efficiency; ~ **global** *m* INFORM&PD throughput, PROD aggregate output; ~ **hidráulico** *m* ENERG RENOV hydraulic efficiency; ~ **hidrodinámico** *m* TRANSP MAR *arquitectura naval* lift-drag ratio; ~ **horario** *m* EMB hourly output; ~ **iónico** *m* FÍS RAD ionic yield; ~ **de la leche** *m* AGRIC milk yield; ~ **máximo** *m* ELEC *de generador* maximum output; ~ **máximo del peso** *m* TEC ESP weight optimization; ~ **mecánico** *m* ENERG RENOV, ING MECÁ mechanical efficiency; ~ **del motor** *m* VEH engine capacity; ~ **en neutrones** *m* FÍS, FÍS PART

neutron yield; ~ **de neutrones térmicos** *m* NUCL thermal neutron yield; ~ **en pares de iones** *m* FÍS RAD ionic yield; ~ **de perforación** *m* CONST drilling rate; ~ **promedio** *m* TERMO average output; ~ **térmico** *m* GAS thermal efficiency, thermic performance, REFRIG calorific output, TERMO calorific output, heat efficiency, heat rate, heat throughput, thermal output, thermal efficiency, TERMOTEC thermal efficiency; ~ **térmico del generador de vapor** *m* NUCL thermal steam generator output; ~ **de la tinta** *m* IMPR hold-out; ~ **de toma** *m* ELEC tapping duty; ~ **total** *m* ENERG RENOV overall efficiency, PROD aggregate output, throughput, TEC PETR throughput; ~ **de la trituradora** *m* CARBÓN crushing efficiency; ~ **de turbina** *m* ENERG RENOV turbine efficiency, turbine output; ~ **del ventilador** *m* TERMOTEC fan efficiency; ~ **volumétrico** *m* AUTO, ENERG RENOV, ING ELÉC, TEC ESP, VEH *motor* volumetric efficiency

rendimientos: ~ **y reducciones** *m pl* PROD returns and allowances

rendir *vt* TEXTIL yield

renguenio *m* (*R*) FÍS, FÍS RAD roentgen (*R*)

renguenografía *f* FÍS RAD X-ray photograph

renguenoluminiscencia *f* FÍS, FÍS RAD, NUCL roentgenoluminescence

renguenometalografía *f* FÍS roentgenometallography, FÍS RAD X-ray metallography, roentgenometallography, METAL, NUCL roentgenometallography

rénico *adj* QUÍMICA rhenic

renina *f* AGRIC *caseína*, ALIMENT rennin

renio *m* (*Re*) QUÍMICA rhenium (*Re*)

renivelación *f* CONST *topografía* releveling (*AmE*), relevelling (*BrE*)

renovación *f* AGRIC renovation, PROD renovation, restoration; ~ **de aire** *f* TRANSP AÉR air renewal; ~ **de existencias** *f* PROD restocking; ~ **de los fondos oceánicos** *f* OCEAN ocean floor spreading, seafloor renewal; ~ **de vías y balasto** *f* FERRO *infraestructura* track and ballast renewal

Re900 *abr* (*número de Reynolds 900*) FÍS FLUID Re900 (*Reynolds number 900*)

renta: ~ **neta** *f* TEC PETR *finanzas* net income

renvalso *m* CONST *carpintería* rabbet; ~ **de ventana** *m* CONST window rabbet

reología *f* CARBÓN, FÍS, FÍS FLUID, GAS, METAL, P&C rheology

reóstato *m* AUTO, ELEC, FÍS rheostat, ING ELÉC rheostat, variable resistance, LAB rheostat; ~ **del arco** *m* ELEC arc rheostat; ~ **de arranque** *m* ELEC *resistencia*, ING ELÉC starting rheostat; ~ **de arranque del motor** *m* ING ELÉC motor starter; ~ **de campo** *m* ELEC *de máquina*, ING ELÉC field rheostat; ~ **de campo tipo crisol** *m* ELEC *resistencia* pot-type field rheostat; ~ **de cilindro** *m* ELEC *resistencia* cylinder rheostat; ~ **de cursor** *m* ELEC *resistencia* slide rheostat; ~ **de excitación** *m* ELEC *de máquina* field regulator, field rheostat, ING ELÉC field regulator; ~ **de excitación tipo crisol** *m* ELEC *resistencia* pot-type field rheostat; ~ **potenciométrico** *m* ELEC *resistor* potentiometer rheostat; ~ **en puente** *m* ELEC *resistor* potentiometer rheostat; ~ **regulador de excitación** *m* ING ELÉC field rheostat; ~ **de resistencia líquida** *m* ELEC *resistencia* liquid rheostat; ~ **de resistencia líquida de arranque** *m* ELEC *resistor* liquid starter resistance

reoxidación *f* CONTAM *procesos químicos*, QUÍMICA reoxidation

reparación[1]: **bajo ~** *adj* GEN under repair; **sin ~ posible** *adj* GEN beyond repair

reparación[2] *f* ELECTRÓN repairing, INFORM&PD retrieval, ING MECÁ repair, repairing, PROD mending, repair, repairing, TEC ESP reconditioning, TEC PETR *perforación* workover, TEXTIL *resarcimiento* retrieval, TRANSP MAR *de buque* refit, repair, repairing; **~ automática** *f* TEXTIL automatic repair; **~ de avería** *f* TELECOM fault clearance, fault maintenance; **~ de redes** *f* OCEAN net mending

reparaciones[1]: **en ~** *adj* GEN under repair

reparaciones[2]: **~ pequeñas** *f pl* FERRO, PROD minor repairs

reparador: ~ de redes *m* OCEAN net repairer

reparar *vt* CONST fix, ELECTRÓN repair, PROD *desgaste* make good, repair, TRANSP MAR *buque* refit, repair; **~, inspeccionar y pintar** *vt* FERRO repair, inspect and paint

repartición *f* INFORM&PD allocation; **~ cromática** *f* AmL TV chromatic splitting; **~ de tiempo** *f* TELECOM time sharing

repartido *adj* INFORM&PD, PROD apportioned, distributed

repartidor *m* EMB distributor, TRANSP *carga, grúas* dispatcher, spreader; **~ de la carga** *m* ELEC *red de distribución* load dispatcher; **~ de conexiones** *m* TELECOM junction distribution frame (*JDF*); **~ digital** *m* TELECOM digital distribution frame; **~ de enlace** *m* TELECOM junction distribution frame (*JDF*); **~ principal** *m* TELECOM main distribution frame (*MDF*), trunk distribution frame (*TDF*); **~ de una sola cara** *m* TELECOM single-sided distribution frame

repartir *vt* INFORM&PD distribute, PROD apportion, distribute

reparto *m* ING ELÉC, QUÍMICA partitioning

repasado *m* TEXTIL mending

repasadora: ~ de aristas *f* ING MECÁ edge-trimmer; **~ de roscas** *f* ING MECÁ rethreading file, thread restorer

repasar *vt* ING MECÁ chase, TEXTIL *acabado* mend, TRANSP MAR *jarcia* overhaul

repaso *m* ING MECÁ rectification, PROD *de una pieza* cleaning-up, QUÍMICA review

repavimentación *f* CONST *carreteras* remetalling (*BrE*), repaving (*AmE*), resurfacing (*BrE*)

repavimentar *vt* CONST remetal (*BrE*), repave (*AmE*), resurface (*BrE*)

repelado *m* IMPR picking, PAPEL fluff, fluffing, linting, peeling

repelar *vt* PAPEL pick

repelente: ~ de la humedad *adj* EMB moisture-repellent

repercusión *f* CARBÓN impact, ING MECÁ, MINAS *de explosión* backlash

reperfilado *m* CONST reprofiling

repertorio *m* INFORM&PD repertoire; **~ de instrucciones** *m* INFORM&PD instruction repertoire

repetibilidad *f* AGRIC, METR repeatability; **~ de posicionamiento** *f* ING MECÁ positioning repeatability

repetición *f* CINEMAT *de una escena ya filmada* retake, CONTAM iteration, IMPR repeat, INFORM&PD repeat, retry, TV repeat, replay, retake; **~ de llamada** *f* TELECOM recall

repetidor *m* ELECTRÓN repeater, transponder, FÍS repeater, IMPR transponder, INFORM&PD repeater, TEC ESP relay, repeater, TELECOM relay, repeater, transponder, TRANSP MAR repeater, TV relay, repeater; **~ de cable** *m* ELEC, ING ELÉC cable repeater; **~ de cuatro hilos** *m* ELECTRÓN four-wire repeater; **~ de líneas** *m* TELECOM line repeater; **~ no regenerativo** *m* ELECTRÓN nonregenerative repeater; **~ óptico** *m* ÓPT, TELECOM optical repeater; **~ óptico sumergido** *m* TELECOM submerged repeater; **~ de la portadora** *m* ELECTRÓN carrier repeater; **~ portátil** *m* TV portable relay; **~ de programas** *m* TV program repeater (*AmE*), programme repeater (*BrE*); **~ reactivo** *m* ELECTRÓN regenerative repeater; **~ regenerador óptico** *m* TELECOM optical regenerative repeater; **~ regenerativo óptico** *m* TELECOM optical regenerative repeater; **~ de rumbo** *m* TRANSP AÉR heading repeater; **~ sumergido** *m* TELECOM submerged optical repeater; **~ unidireccional** *m* ELECTRÓN one-way repeater

repicado *m* PAPEL *de muela desfibradora* dressing

repintado *adj* IMPR offset

repintar *vt* IMPR offset

repinte *m* IMPR set-off

repisa *f* ING MECÁ, MECÁ bracket; **~ de almacenaje** *f* C&V storage rack; **~ para clavija** *f* ING MECÁ plug seat

replantar *vt* AGRIC replant

replanteo *m* TEC ESP ranging

repliegue *m* GEOL refolding

repoblación: ~ forestal *f* AGRIC reafforestation, reforestation; **~ íctica** *f* OCEAN stocking

reponer *vt* INFORM&PD *contador*, PROD reset

reposabrazo *m* VEH armrest

reposabrazos *m* INSTR handrest

reposacabeza *m* VEH headrest

reposapiés *m* VEH *en motocicletas* footrest

reposar *vi* INFORM&PD quiesce

reposición *f* AGUA make-up, CINEMAT, INFORM&PD reset, ING MECÁ resetting, PROD reset, resetting, TEC ESP *telefonía automática* homing, TV reset; **~ automática** *f* ELEC *control* automatic reset, TEC PETR automatic feed; **~ baja** *f* PROD low reset; **~ de control principal** *f* PROD master control reset (*MCR*); **~ de la manada** *f* *Esp* (*cf reposición del rodeo AmL*) AGRIC *ganado caballar* herd replacement; **~ a mano** *f* ELEC *controles* manual reset; **~ manual** *f* ELEC *controles* manual reset; **~ del rebaño** *f* *Esp* (*cf reposición del rodeo AmL*) AGRIC *ganado bovino, ovino* herd replacement; **~ remota** *f* PROD remote reset; **~ del rodeo** *f* *AmL* (*cf reposición del rebaño Esp*) AGRIC herd replacement; **~ del temporizador** *f* PROD timer reset

reposicionamiento *m* ING MECÁ, PROD resetting

repositorio *m* CONTAM, INFORM&PD, NUCL repository

reposo[1]: **en ~** *adj* INFORM&PD quiescent, quiescing

reposo[2]: **~ periódico** *m* ELECTRÓN periodic refresh; **~ vegetal** *m* AGRIC dormant period

repostado *m* TEC PETR fueling (*AmE*), fuelling (*BrE*), TRANSP AÉR fueling (*AmE*), fuelling (*BrE*), refueling (*AmE*), refuelling (*BrE*); **~ por gravedad** *m* TRANSP AÉR gravity refueling (*AmE*), gravity refuelling (*BrE*); **~ en vuelo** *m* TRANSP AÉR in-flight refueling (*AmE*), in-flight refuelling (*BrE*)

repostador *m* TRANSP AÉR, VEH refueler (*AmE*), refueller (*BrE*)

repostaje: ~ **por gravitación** *m* TEC ESP gravity refueling (*AmE*), gravity refuelling (*BrE*); ~ **en vuelo** *m* D&A air-to-air refueling (*AmE*), air-to-air refuelling (*BrE*), TEC ESP receiver

repostar *vt* D&A, TRANSP, TRANSP AÉR, VEH refuel

represa *f* AGUA, PETROL, TEC PETR reservoir; ~ **térmica** *f* AUTO heat dam

represamiento *m* AGUA, MINAS damming

represar *vt* AGUA, MINAS dam

representación *f* ELEC *término general* display, INFORM&PD representation, TEC ESP pattern; ~ **binaria** *f* INFORM&PD binary representation; ~ **cabeza abajo** *f* TEC ESP head-down display; ~ **cabeza arriba** *f* TEC ESP head-up display (*HUD*); ~ **de color equivocada** *f* IMPR wrong color rendering (*AmE*), wrong colour rendering (*BrE*); ~ **de coma variable** *f* INFORM&PD variable-point representation; ~ **de conocimiento** *f* INFORM&PD knowledge engineering; ~ **de datos** *f* INFORM&PD data representation; ~ **digital** *f* ELECTRÓN, INFORM&PD digital representation; ~ **geométrica** *f* GEOM geometric representation; ~ **gráfica** *f* INFORM&PD graphical representation, TRANSP AÉR plot; ~ **invertida** *f* TEC ESP head-down display; ~ **de magnitudes con signo** *f* INFORM&PD signed magnitude representation; ~ **numérica** *f* INFORM&PD number representation, numeric representation; ~ **simplificada de agujeros de centro** *f* ING MECÁ simplified representation of center holes (*AmE*), simplified representation of centre holes (*BrE*); ~ **en trama de alambre** *f* INFORM&PD wireframe representation; ~ **vectorial** *f* ING ELÉC phasor representation; ~ **visual** *f* TEC ESP *radar, informática* display; ~ **visual cromática** *f* TEC ESP color display (*AmE*), colour display (*BrE*)

representante: ~ **local del propietario de una mina** *m* MINAS deputy

representar *vt* TEC ESP *instrumentos con TRC* display; ~ **por imágenes** *vt* INFORM&PD image

represurización: ~ **de aire** *f* TEC PETR air repressuring

reprise *m* AUTO pick-up

reprocesado *m* PROD, RECICL reprocessing

reprocesamiento *m* PROD, RECICL reprocessing

reprocesar *vt* PROD, RECICL reprocess

reproceso *m* PROD, RECICL reprocessing

reproducción *f* ACÚST *de algo grabado, registrado* playback, reproducing, reproduction, FOTO duplicate; ~ **de audio** *f* TV audio playback; ~ **bicolor** *f* IMPR duotone; ~ **de los colores** *f* CINEMAT color rendition (*AmE*), colour rendition (*BrE*); ~ **fiel** *f* FOTO accurate reproduction; ~ **fuera de la época** *f* AGRIC out-of-season breeding; ~ **de imágenes por espectroscopía electrónica** *f* FÍS, FÍS RAD electron-spectroscopic imaging; ~ **del mejillón** *f* OCEAN *cría de mejillones* mussel breeding; ~ **múltiple** *f* ACÚST multiplay; ~ **en semitonos** *f* IMPR halftone reproduction; ~ **de tonos** *f* IMPR tone reproduction; ~ **de valores tonales** *f* FOTO reproduction of tonal values

reproducibilidad *f* CARBÓN, METR reproducibility

reproducir *vt* AGRIC breed, FOTO duplicate, IMPR copy, print, TV play

reproducirse *v refl* AGRIC, PROD breed

reproductibilidad *f* CARBÓN reproducibility, METR *de las medidas* repeatability, reproducibility

reproductividad: ~ **neta** *f* NUCL breeding-process efficiency

reproductor *m* AGRIC *ganadería* breeder; ~ **de CD de audio** *m* ÓPT audio CD player; ~ **de CD-ROM** *m* ÓPT CD-ROM player; ~ **de cinta perforada** *m* INFORM&PD, TELECOM punch-tape reproducer (*AmE*), punched-tape reproducer (*BrE*); ~ **de cintas** *m* INFORM&PD tape reproducer; ~ **de cintas de video** *AmL*, ~ **de cintas de vídeo** *Esp m* TV videotape player; ~ **de discos** *m* ÓPT disc player (*BrE*), disk player (*AmE*); ~ **de discos compactos de audio** *m* ÓPT audio compact disc player, CD audio player; ~ **de discos ópticos** *m* INFORM&PD optical disk player, ÓPT optical disc player (*BrE*), optical disk player (*AmE*); ~ **de fichas perforadas** *m* INFORM&PD punch-card reproducer (*AmE*), punched-card reproducer (*BrE*); ~ **magnético** *m* TV magnetic reproducer; ~ **de sonido óptico** *m* CINEMAT optical sound reproducer; ~ **de tarjetas perforadas** *m* INFORM&PD punch-card reproducer (*AmE*), punched-card reproducer (*BrE*); ~ **de videodisco** *m* ÓPT videodisc player (*BrE*), videodisk player (*AmE*); ~ **de visión láser** *m* ÓPT laservision player (*LV player*)

reproductores *m pl* AGRIC breeding stock

reprografía *f* INFORM&PD reprographics

reprogramación *f* PROD rescheduling

reptante *adj* AGRIC creeping

repuesto *m* AUTO, TRANSP MAR spare part

repulsado: ~ **de aumento de fluencia plástica** *m* ING MECÁ flow spinning

repulsión *f* ELEC, FÍS, ING ELÉC repulsion; ~ **culombiana** *f* FÍS coulomb repulsion; ~ **magnética** *f* ING ELÉC magnetic repulsion

repuntar: ~ **para corriente** *vi* TRANSP MAR *marea* set in; ~ **para menguante** *vi* TRANSP MAR *marea* set out

repunte *m* OCEAN *marea* turn, TRANSP MAR *marea* turn, *navegación* slack water; ~ **de la marea** *m* OCEAN slack tide

requerimiento: ~ **de corriente** *m* ELEC current requirement; ~ **de renuncia** *m* TEC PETR *licencias* relinquishment requirement; ~ **técnico** *m* METR technical requirement

requerimientos: ~ **de aeronavegabilidad apropiados** *m pl* TRANSP AÉR appropriate airworthiness requirements

requisito *m* INFORM&PD, SEG *condiciones* requirement; ~ **de almacenamiento** *m* INFORM&PD storage requirement; ~ **de ensayo** *m* PROD test requirement; ~ **de interfase** *m* INFORM&PD interface requirement; ~ **de interfaz** *m* INFORM&PD interface requirement; ~ **de memoria** *m* INFORM&PD storage requirement; ~ **de prueba** *m* SEG test requirement; ~ **técnico de seguridad** *m* SEG technical safety requirement

requisitos: ~ **de protección** *m pl* SEG security requirements; ~ **y supervisión de seguridad** *m pl* SEG safety requirements and supervision

re-radiación *f* ELECTRÓN scattering

re-rectificado *m* CARBÓN, PROD regrinding

re-registro *m* ACÚST rerecording

re-revelado *m* CINEMAT redevelopment

res: ~ **muerta** *f* AGRIC carcass

resaca *f* HIDROL backwash, OCEAN backrush, backwash, underset, undertow, TRANSP MAR *mar* surge, undertow

resalte *m* MECÁ flange; ~ **central** *m* METAL midrib

resalto *m* ING MECÁ ledge, stud; **~ hidráulico** *m* HIDROL hydraulic jump; **~ horizontal** *m* MINAS *fallas* floor heave; **~ de leva** *m* AUTO, TRANSP AÉR, VEH *motor* cam lobe

resbalabilidad *m* TRANSP *lubricantes* slipperiness

resbaladera *f* ING MECÁ guide bar; **~ de cojinete de bolas** *f* ING MECÁ ball-bearing guideway; **~ de la cruceta** *f* ING MECÁ slide bar

resbalamiento *m* CARBÓN *correa de transmisión*, PROD *correas* creep, TRANSP AÉR slip; **~ del embrague** *m* AUTO, VEH clutch slip

resbalarse *v refl* MECÁ slide

rescatar *vt* CARBÓN recover, SEG, TRANSP AÉR, TRANSP MAR rescue

rescate *m* D&A, SEG rescue, TRANSP salvaging, TRANSP AÉR recovery, rescue, TRANSP MAR rescue; **~ aeromarítimo** *m* D&A air-sea rescue

rescoldo *m* CARBÓN breeze

reseña *f* ING MECÁ outline

reserva[1]: **de ~** *adj* CALIDAD standby, INFORM&PD standby, backup; **en la ~** *adj* ING MECÁ, PROD, TRANSP AÉR, TRANSP MAR inactive; **en ~** *adj* D&A on standby

reserva[2] *f* IMPR resist, PROD *producción* stand-by; **~ de baterías** *f* PROD battery backup; **~ en caliente** *f* TEC ESP hot stand-by; **~ de calor** *f* PROD hot backup; **~ de carbón** *f* CARBÓN known coal deposit; **~ de combustible** *f* TRANSP AÉR fuel reserve; **~ de dispositivo** *f* INFORM&PD device reserve; **~ de flotabilidad** *f* TRANSP MAR *diseño de barcos* reserve buoyancy; **~ fotosensible** *f* COLOR photosensitive resist; **~ de fuente de energía** *f* TELECOM backup power supply; **~ genética** *f* AGRIC genetic reservoir; **~ de tinta** *f* IMPR ink tank

reservado: **~ a instrumentos** *adj* TRANSP AÉR instrument restricted; **~ para uso especial** *adj* PROD reserved for special use

reservas *f pl* (*cf existencias AmL*) MINAS *de carbón, minerales* offtake, TRANSP AÉR reserves; **~ comprobadas** *f pl* PETROL, TEC PETR *yacimientos* proven reserves; **~ posibles** *f pl* PETROL, TEC PETR *yacimientos* possible reserves; **~ probables** *f pl* PETROL, TEC PETR *yacimientos* probable reserves; **~ recuperables** *f pl* PETROL, TEC PETR *yacimientos* recoverable reserves

reservorio *m AmL* (*cf yacimiento Esp*) GEOL source rock, PETROL, TEC PETR *geología, yacimiento* reservoir; **~ oleífero** *m* MINAS *botánica* oil reservoir

resguardado: **~ del polvo** *adj* ELEC *equipo* dustproof

resguardar *vt* TEC ESP shield; **~ de** *vt* SEG guard against

resguardo *m* D&A shelter, ING MECÁ *esclusa*, MECÁ clearance, TEC ESP shelter, shield; **~ de la rampa de lanzamiento** *m* TEC ESP launch ramp shelter

residente: **~ en disco** *adj* INFORM&PD disc-resident (*BrE*), disk-resident (*AmE*); **~ en memoria** *adj* INFORM&PD memory-resident

residir *vi* INFORM&PD reside

residual *adj* ELEC residual, GEOL *de mineral o estructura* relict, ING ELÉC, METAL, PROD, QUÍMICA residual, TEC ESP relict, TEC PETR residual

residuo *m* AGRIC residue, scrap, AGUA residue, ALIMENT residue, waste, CONTAM *procesos industriales*, CONTAM MAR residue, INFORM&PD remainder, residue, MATEMÁT remainder, MINAS *procesos de afino* foot, QUÍMICA residue, TEC PETR *geología* sediment, *producción, refino* residue; **~ acuoso** *m* AGRIC *de la*

fermentación alcohólica de los granos grapple; **~ de alto nivel** *m* NUCL, RECICL high-level waste (*HLW*); **~ armónico** *m* ELECTRÓN harmonic content; **~ combustible** *m* CONTAM combustible waste; **~ de cosecha** *m* AGRIC crop residue; **~ de denudación** *m* GEOL lag deposit; **~ del despepitado** *m* AGRIC gin trash; **~ de destilación** *m* CARBÓN, MINAS bottom; **~ de destilería** *m* ALIMENT *industria cervecera* distillery residue; **~ de frigoríficos** *m* AGRIC tankage; **~ de mataderos** *m* AGRIC tankage; **~ de petróleo** *m* PETROL *Rusia* astatki; **~ radiactivo alcalino de media actividad** *m* NUCL alkaline medium-level radioactive waste; **~ seco** *m* ALIMENT pomace; **~ sulfurado** *m* GAS sulfurous residue (*AmE*), sulphurous residue (*BrE*); **~ del tamizado** *m* PROC QUÍ sieving residue

residuos *m pl* AGUA refuse, ALIMENT *fermentación* bottoms, *molienda, panadería, repostería* break tailings, CARBÓN scrap, refuse, *refino del petróleo* tailings, CONST tailings, CONTAM waste, EMB rubbish (*BrE*), refuse, garbage (*AmE*), METAL debris, PAPEL waste, PROD, RECICL waste, garbage (*AmE*), refuse, rubbish (*BrE*), TEC PETR *refino, perforación* bottoms; **~ de aceite** *m pl* CONTAM *contaminación por petróleo* oil waste; **~ de aceiterías** *m pl* AGRIC residues of oil manufacture; **~ de almidonería** *m pl* AGRIC residues of the starch industry; **~ de baja y mediana actividad** *m pl* (*RBMA*) NUCL low and intermediate level waste (*L&ILW*); **~ brutos** *m pl* AGUA, CONTAM, HIDROL, RECICL raw sewage; **~ combustibles semisólidos** *m pl* CONTAM semisolid combustible waste; **~ sin depurar** *m pl* AGUA, CONTAM, HIDROL, RECICL raw sewage; **~ deshidratados** *m pl* NUCL dewatered waste; **~ de dislocación** *m pl* METAL dislocation debris; **~ domésticos** *m pl* RECICL domestic sewage, household waste; **~ de escasa actividad** *m pl* RECICL low-level waste; **~ de escasa radiactividad** *m pl* RECICL low-level waste; **~ gaseosos** *m pl* NUCL waste gas; **~ glaciales no estratificados** *m pl* CARBÓN till; **~ de gran radioactividad** *m pl* NUCL, RECICL high-level waste (*HLW*); **~ industriales** *m pl* RECICL industrial waste; **~ del lavado de carbones** *m pl* CARBÓN silt; **~ líquidos** *m pl* RECICL liquid waste; **~ de matadero** *m pl* AGRIC offals; **~ no tratados** *m pl* AGUA, CONTAM, HIDROL, RECICL raw sewage; **~ nucleares** *m pl* CONTAM, NUCL, RECICL, SEG nuclear waste; **~ patógenos** *m pl* RECICL pathogenic waste; **~ patológicos** *m pl* RECICL pathological waste; **~ petrolíferos** *m pl* CONTAM *contaminación por petróleo* oil waste; **~ de piedras** *m pl* CONST *labrado* stone chippings; **~ radiactivos** *m pl* CONTAM, FÍS, FÍS RAD, NUCL radioactive waste; **~ radiactivos de actividad media** *m pl* NUCL intermediate level radioactive waste; **~ radiactivos de alta actividad** *m pl* NUCL high-level radioactive waste; **~ radiactivos de baja actividad** *m pl* NUCL low-level radioactive waste; **~ sépticos** *m pl* RECICL septic sludge; **~ sólidos** *m pl* CONTAM solid waste; **~ transuránicos** *m pl* NUCL transuranic waste; **~ de trilla** *m pl* AGRIC tailings

resiembra *f* AGRIC reseeding, resowing

resiliencia *f* INFORM&PD, ING MECÁ, P&C, PAPEL, PROD, TEXTIL *elasticidad* resilience

resina *f* EMB, MECÁ, P&C, QUÍMICA, TEXTIL, TRANSP MAR resin; **~ acetónica** *f* EMB, P&C acetone resin;

~ **acrílica** *f* EMB, MECÁ, P&C *plásticos, revestimientos, adhesivos* acrylic resin; ~ **alquílica** *f* P&C *aglutinante para pinturas* alkyd resin; ~ **alquílica de alto contenido en aceite** *f* P&C *para pintura* long-oil alkyd; ~ **alquílica de bajo contenido en aceite** *f* P&C *para pintura* short oil alkyd; ~ **alquílica corta en aceite** *f* P&C *para pintura* short-oil alkyd; ~ **alquílica larga en aceite** *f* P&C *para pintura* long-oil alkyd; ~ **amínica** *f* P&C *pinturas, adhesivos* amino resin; ~ **de anilina-formaldehído** *f* EMB aniline formaldehyde resin; ~ **de cresol** *f* P&C *polímero* cresol resin; ~ **de cumarona** *f* P&C *polímero* coumarone resin; ~ **epoxi** *f* CONST, ELEC, EMB, P&C, QUÍMICA epoxy resin; ~ **epoxi acrílica** *f* P&C *pinturas, adhesivos, plásticos* acrylated epoxy resin; ~ **epoxi curada con amina** *f* P&C amine-cured epoxy; ~ **de epoxia** *f* ELECTRÓN glass epoxy laminate; ~ **epóxica** *f* CONST, ELEC, EMB, P&C, QUÍMICA epoxy resin; ~ **epóxica acrílica** *f* P&C *pinturas, adhesivos, plásticos* acrylated epoxy resin; ~ **en estado A** *f* P&C A-stage resin; ~ **estérica** *f* DETERG ester resin; ~ **fenólica** *f* ELEC *aislación*, P&C *plásticos* phenolic resin; ~ **fluorocarbonada** *f* P&C *tipo de polímero* fluorocarbon resin; ~ **de formaldehido de urea** *f* ELEC *aislación*, P&C *polímero* urea-formaldehyde resin (*UF*); ~ **fundida en molde de cáscara** *f* P&C shell-molding resin (*AmE*), shell-moulding resin (*BrE*); ~ **gliptálica** *f* P&C *aglutinante de pintura* glyptal resin; ~ **de melamina** *f* ELEC *aislación, aislamiento* melamine resin; ~ **de melamina-formaldehído** *f* P&C melamine formaldehyde resin (*MF*); ~ **moldeada en cascarón** *f* P&C shell-molding resin (*AmE*), shell-moulding resin (*BrE*); ~ **poliéster** *f* QUÍMICA polyester resin; ~ **de poliuretano** *f* CONST polyurethane resin; ~ **de resorcinol** *f* P&C *tipo de polímero* resorcinol resin; ~ **sintética** *f* P&C *polímero*, TEC PETR *petroquímica* synthetic resin; ~ **termoendurecible en estado inicial** *f* P&C *fenólica* A-stage resin, resol; ~ **termoendurecible en su fase final** *f* P&C *tipo de polímero* C-stage resin, resite; ~ **termoplástica en estado blando** *f* P&C *tipo de polímero* B-stage resin; ~ **de trementina** *f* P&C *materia prima de la pintura* rosin; ~ **de urea-formaldehido** *f* (*UF*) ELEC *aislación*, P&C *polímero* urea-formaldehyde resin

resinas: ~ **que imparten resistencia en húmedo** *f pl* PAPEL wet-strength resins

resincronizar *vt* TELECOM resynchronize

resistencia[1]: **sin** ~ *adj* ING MECÁ frictionless

resistencia[2] *f* CONST, ELEC resistance, FÍS *por unidad de longitud* resistance, strength, FÍS FLUID, GAS resistance, INFORM&PD resilience, ING ELÉC resistance, ING MECÁ stiffness, resistance, METAL strength, METR durability, MINAS *de una mina* resistance, drag, P&C resistance coating, resist coating, PAPEL resistance, PROD *fuerza* strength, REVEST resistance coating, resist coating, TEC PETR, TELECOM, TERMO resistance, TEXTIL resistance, strength, TRANSP AÉR endurance;

~ a ~ **a la abrasión** *f* P&C *propiedad física, prueba*, PAPEL abrasion resistance; ~ **al ácido** *f* P&C *propiedad, prueba* acid resistance; ~ **de acoplamiento** *f* ELEC coupling resistance; ~ **acústica** *f* ACÚST acoustic resistance; ~ **del adhesivo al corte** *f* P&C *propiedad física, prueba* adhesive shear strength; ~ **aerodinámica** *f* FÍS, TEC ESP, TRANSP AÉR aerodynamic drag, drag; ~ **aerodinámica de la compresibilidad** *f* TRANSP AÉR compressibility drag; ~ **aerodinámica de forma** *f* TRANSP AÉR form

drag; ~ **aerodinámica parasítica** *f* TRANSP AÉR parasitic drag; ~ **aerodinámica del perfil** *f* TRANSP AÉR profile drag; ~ **aerodinámica por presión** *f* TRANSP AÉR pressure drag; ~ **aerodinámica por rozamiento** *f* TRANSP AÉR friction drag; ~ **aguas arriba** *f* PROD upstream resistance; ~ **de aislamiento** *f* ING ELÉC fault resistance, insulation resistance; ~ **del aislamiento** *f* FÍS insulation resistance; ~ **de alambre** *f* ING ELÉC inductive wirewound resistor; ~ **de alambre devanado** *f* ELEC wirewound resistor; ~ **a los álcalis** *f* P&C *propiedad, prueba* alkali resistance; ~ **alineal** *f* ELEC, TELECOM nonlinear resistance, nonlinear resistor; ~ **de amortiguación** *f* FÍS damping resistance; ~ **de la antena** *f* FÍS aerial resistance; ~ **antiparasitaria del distribuidor** *f* VEH *sistema de encendido* distributor suppressor; ~ **apareada** *f* ING ELÉC matched resistor; ~ **al aplastamiento** *f* ING MECÁ retroactive tenacity, PAPEL crushing resistance; ~ **para aplicaciones diversas** *f* ING ELÉC general-purpose resistor; ~ **al arrancado** *f* PAPEL pick resistance, picking resistance, surface bonding strength; ~ **a las arrugas** *f* TEXTIL crease resistance; ~ **autorreguladora** *f* ELEC ballast resistor, *estabilizador de tensión* barretter, FÍS barretter; ~ **autorreguladora de corriente** *f* ELEC ballast resistor; ~ **al avance** *f* ENERG RENOV *energía de viento*, FÍS FLUID *de una esfera* drag, TRANSP resistance to forward motion, TRANSP AÉR, TRANSP MAR *diseño de barcos* drag;

~ b ~ **a la baja temperatura** *f* P&C *propiedad física, prueba* low-temperature resistance; ~ **de base** *f* ING ELÉC base resistance; ~ **a las bases** *f* P&C alkali resistance; ~ **bobinada** *f* ELEC wirewound resistor;

~ c ~ **de CA** *f* ING ELÉC AC resistance; ~ **de caída de voltaje** *f* ING ELÉC voltage-dropping resistor; ~ **de caldeo** *f* ING ELÉC heater; ~ **de calefacción** *f* TRANSP AÉR heating resistor; ~ **de calentamiento** *f* ELEC, TRANSP AÉR heating resistor; ~ **al calor** *f* P&C *propiedad, prueba* heatsealing, TERMO hot strength, resistance to heat, TERMOTEC resistance to heat; ~ **a cambiar de color por acción de la luz** *f* P&C *propiedad física* light fastness; ~ **al cambio brusco de temperatura** *f* TERMO resistance to thermal shock; ~ **en campo** *f* AGRIC field resistance; ~ **de capa delgada** *f* ELECTRÓN thin-film resistor; ~**-capacidad** *f* (*RC*) ING ELÉC resistance-capacitance (*RC*); ~**-capacitancia** *f* (*RC*) ELEC, ELECTRÓN, ING ELÉC resistance-capacitance (*RC*); ~ **capacitiva** *f* ELEC, ING ELÉC capacitive resistance; ~ **de característica alineal** *f* ELEC *resistor* varistor, ING ELÉC varistance, varistor; ~ **de carbón** *f* ELEC, FÍS, ING ELÉC carbon composition resistor, carbon resistor; ~ **de carburo de silicio comprimido** *f* ING ELÉC silicon carbide varistor; ~ **de carga** *f* FÍS load resistance, ING ELÉC ballast resistor; ~ **de la cáscara** *f* ALIMENT *cáscara de huevo* shell strength; ~ **del casco** *f* TRANSP MAR *construcción naval* hull resistance; ~ **del casquillo** *f* ING ELÉC base resistance; ~ **de cátodo** *f* ING ELÉC bias resistor; ~ **de CC** *f* ING ELÉC DC resistance; ~ **al choque** *f* FÍS impact strength, ING MECÁ resistance to impact, resistance to shock, MECÁ impact strength; ~ **al choque térmico** *f* TERMO resistance to thermal shock; ~ **a los choques** *f* P&C *propiedad física, prueba* shock resistance; ~ **al cizallamiento** *f* CONST shearing stress, ING MECÁ, METAL shearing strength; ~ **de compensación** *f* ELEC balancing resistor, FÍS barretter, ING ELÉC bleeder resistor; ~ **compensada** f ING ELÉC barretter; ~ **compensadora** f ELEC balancing resistor, ballast resistor, *estabilizador de tensión* barretter; ~ **a la compresión** f EMB compression strength, ING MECÁ resistance to crushing, MECÁ compression strength, compression stress, P&C *propiedad física, prueba* compression strength; ~ **a compresión a la rotura** f P&C *propiedad física, prueba* bursting strength; ~**condensador** f (*RC*) ELECTRÓN resistance-capacitance (*RC*); ~ **conectada en paralelo** f ELEC parallel-connected resistance; ~ **conectada en serie** f ELEC series-connected resistance; ~ **de conexión perpendicular** f ING ELÉC radial-lead resistor; ~ **a la congelación-descongelación** f ALIMENT, PROC QUÍ, REFRIG freeze-thaw resistance; ~ **de contacto** f ELEC *entre dos metales,* ING ELÉC contact resistance; ~ **de contacto de escobillas** f ELEC *en máquina* brush-contact resistance; ~ **a la corona** f P&C *propiedad* corona resistance; ~ **a la corrosión** f CONST, P&C *propiedad, prueba* corrosion resistance; ~ **a la cortadura** f METAL shearing strength; ~ **al corte** f CARBÓN shear force, ING MECÁ shearing strength, resistance to shearing, P&C *propiedad física, prueba* shearing strength; ~ **crítica** f ELEC critical resistance;

~ **d** ~ **debida al viento** f FÍS FLUID wind resistance; ~ **a la deformación** f ING MECÁ stress; ~ **dependiente de voltaje** f ING ELÉC voltage-dependent resistor; ~ **de derivación** f ING ELÉC bias resistor; ~ **en derivación** f ING ELÉC shunt resistance; ~ **de derivación para instrumento** f ING ELÉC instrument shunt; ~ **derivada** f ING ELÉC tapped resistor; ~ **derivadora** f ING ELÉC bleeder resistor; ~ **de descarga** f ELEC discharge resistor; ~ **al desgarramiento** f CONTAM MAR resistance to tearing, P&C *propiedad física, prueba* tear strength; ~ **al desgarro** f IMPR *papel* tear strength, P&C *propiedad física, prueba* tear resistance, PAPEL, TEXTIL tear strength; ~ **al deslizamiento** f ING MECÁ resistance to sliding; ~ **devanada** f ELEC wirewound resistor; ~ **devanada de potencia** f ING ELÉC power wirewound resistor; ~ **devanada de precisión** f ING ELÉC precision wirewound resistor; ~ **diagonal** f CONST diagonal strength; ~ **dieléctrica** f ELEC *de capacitor* dielectric resistance, FÍS, ING ELÉC dielectric strength, disruptive strength; ~ **dinámica** f ELEC *de circuito resonante* dynamic resistance, TRANSP ram drag; ~ **directa** f ELEC *de semiconductor* forward resistance; ~ **discreta** f ING ELÉC discrete resistor; ~ **al doblado** f EMB folding strength; ~ **dúctil** f ING ELÉC flexible resistor;

~ **e** ~ **efectiva** f FÍS, ING ELÉC effective resistance; ~ **eléctrica** f AUTO resistor, ELEC, FÍS, ING ELÉC electric resistance, resistor; ~ **al encamado** f AGRIC resistance to lodging; ~ **a enfermedades** f AGRIC disease resistance; ~ **enrollada** f ING ELÉC reeled resistor; ~ **al ensuciado** f TEXTIL resistance to soiling; ~ **de entrada** f ING ELÉC input resistance; ~ **equivalente** f ELEC equivalent resistance; ~ **escalar** f ING ELÉC scalar resistor; ~ **de escape de electrones de la rejilla** f FÍS grid leak resistor; ~ **al esfuerzo cortante** f ING MECÁ resistance to shearing, shearing strength, P&C *propiedad física, prueba* shearing strength; ~ **específica** f ELEC specific resistance, FÍS resistivity, ING ELÉC specific resistance, PETROL, TEC PETR resistivity; ~ **específica de masas**

f ELEC mass resistivity; ~ **específica de volumen** f ING ELÉC volume resistivity; ~ **en estado húmedo** f P&C *propiedad física, prueba* wet strength; ~ **al estallido** f EMB, PAPEL, TEXTIL *tejido* bursting strength; ~ **al estrujamiento** f TEXTIL crush resistance; ~ **exterior** f ELEC, ING ELÉC external resistor; ~ **externa** f ELEC, ING ELÉC external resistance;

~ **f** ~ **a la fatiga** f CRISTAL fatigue strength, ING MECÁ stress, MECÁ, METAL, P&C fatigue strength, PROD, TEC ESP endurance, TRANSP AÉR fatigue strength; ~ **fija** f ELEC, FÍS, ING ELÉC fixed resistor; ~ **de filamento** f ELEC *de válvula, tubo* filament resistance; ~ **de filamento grueso** f ELECTRÓN thick-film resistor; ~ **de la flexibilidad** f TELECOM flexibility strength; ~ **a la flexión** f IMPR flexural strength, ING MECÁ resistance to bending, resistance to buckling, MECÁ flexural strength, NUCL bending strength, flexural strength, P&C *propiedad física, prueba* tending resistance, flexing resistance, flexing endurance, flexural strength, PAPEL bending strength, flexural strength; ~ **flexural** f IMPR, MECÁ, NUCL, P&C flexural strength; ~ **a la floculación** f CARBÓN *fueloil* stability; ~ **a la fluencia** f CARBÓN creep strength, P&C *propiedad, prueba* creep resistance; ~ **al flujo** f ACÚST *mecánica* flow resistance, FÍS FLUID resistance to flow; ~ **a la formación de olas** f TRANSP *buques* wave drag; ~ **por formación de olas** f TRANSP MAR *diseño de barcos* wave resistance; ~ **de fractura frágil** f METAL brittle-fracture resistance; ~ **de frenado** f TRANSP braking resistance; ~ **de fricción** f TRANSP MAR *diseño de barcos* frictional resistance; ~ **friccional** f CARBÓN skin resistance; ~ **al frío** f TERMO cold strength; ~ **en frío** f ING ELÉC cold resistance; ~ **al frotamiento en húmedo** f PAPEL wet rub resistance; ~ **al fuego** f P&C *pintura, plásticos, caucho* TERMO fire resistance; ~ **a las fuerzas de tensión** f ING MECÁ endurance tensile strength; ~ **de fuga** f FÍS leakage resistance;

~ **g** ~ **a la gasolina** f *Esp* (*cf resistencia a la nafta AmL*) P&C gas resistance (*AmE*), gasoline resistance (*AmE*), petrol resistance (*BrE*); ~ **a los golpes** f P&C *propiedad física, prueba* shock resistance; ~ **de gran disipación** f ING ELÉC power resistor; ~ **de gran valor** f ING ELÉC large-value resistor; ~ **a la grasa** f P&C *propiedad, prueba* powdering;

~ **h** ~ **de Hall** f FÍS Hall resistance; ~ **híbrida integrada** f ING ELÉC integrated hybrid resistor; ~ **hidrodinámica** f FÍS FLUID drag, TRANSP hydrodynamic drag; ~ **de hilo bobinado** f ELEC, ING ELÉC wirewound resistor; ~ **de hilo bobinado inductiva** f ING ELÉC inductive wirewound resistor;

~ **i** ~ **al impacto** f EMB impact resistance, ING MECÁ resistance to impact, NUCL, P&C impact strength, TRANSP impact resistance; ~ **al impacto con entalla** f P&C indentation hardness; ~ **al impacto con péndulo** f P&C pendulum hardness; ~ **inalámbrica** f CONST, ELEC non-wirewound resistor; ~ **inalámbrica no inductiva** f CONST, ELEC noninductive wirewound resistor; ~ **a inclemencias** f AGRIC hardiness; ~ **del inducido** f ELEC *máquina,* ING ELÉC armature resistance; ~ **inductiva** f ING ELÉC inductive resistor; ~ **a la intemperie** f CONST weathering resistance; ~ **interna** f ELEC *pila, fuente de alimentación* internal resistance; ~ **interna del inducido** f ELEC *máquina* armature resistance;

~ **j** ~ **de la junta** f EMB joint strength;

~ l ~ **de lámina** f ING ELÉC sheet resistance; ~ **al laminado** f TEXTIL laminating resistance; ~ **de laminado** f EMB laminating strength; ~ **a largo plazo** f P&C fatigue; ~ **limitadora** f ELEC limiting resistor; ~ **límite** f ING ELÉC terminating resistor; ~ **límite a la flexión** f ING MECÁ safe stress under bending; ~ **lineal** f FÍS linear resistance; ~ **líquida de arranque** f ELEC *resistor* liquid starter resistance, liquid starting resistance; ~ **a la luz solar** f P&C *propiedad, prueba* sunlight resistance;

~ m ~ **magnética** f ELEC, FÍS, ING ELÉC magnetic resistance, reluctance; ~ **magnética específica** f ELEC *magnetismo*, FÍS, ING ELÉC reluctivity; ~ **máxima** f ING MECÁ ultimate strength; ~ **máxima a la flexión** f ING MECÁ ultimate bending strength; ~ **máxima a la rotura por flexión** f ING MECÁ ultimate bending strength; ~ **máxima a la tracción** f ING MECÁ ultimate tensile strength; ~ **mecánica** f ACÚST mechanical resistance, PAPEL strength, PROD mechanical endurance; ~ **mecánica del papel** f IMPR, PAPEL paper strength; ~ **para medida de la corriente** f ING ELÉC current-sensing resistor; ~ **de la mina a la corriente de aire** f MINAS *ventilación* mine resistance; ~ **al movimiento** f ING MECÁ resistance to motion;

~ n ~ **a la nafta** f *AmL (cf resistencia a la gasolina Esp)* P&C *propiedad prueba* gas resistance *(AmE)*, gasoline resistance *(AmE)*, petrol resistance *(BrE)*; ~ **negativa** f ELEC, ELECTRÓN, ING ELÉC negative resistance; ~ **de nitruro de telurio** f ING ELÉC tellurium nitride resistor; ~ **no inductiva** f CONST, ELEC noninductive resistor; ~ **no lineal** f ELEC *resistor* varistor; ~ **del nudo** f TEXTIL knot strength;

~ o ~ **óhmica** f ELEC ohmic resistance, noninductive resistor, ING ELÉC ohmic resistance; ~ **orgánica** f ING ELÉC organic resistor; ~ **oscura** f NUCL *de una fotocélula* dark resistance; ~ **al ozono** f P&C *propiedad, prueba* ozone resistance;

~ p ~ **al pandeo** f ING MECÁ resistance to buckling; ~ **en paralelo** f ELEC bridge resistance, ING ELÉC parallel resistance, shunt resistor; ~ **en paralelo para galvanómetro** f ELEC, ING ELÉC galvanometer shunt; ~ **de película** f ING ELÉC film resistor; ~**película de carbón** f ING ELÉC carbon film resistor; ~ **de película metalizada** f ING ELÉC metal film resistor; ~ **a la perforación** f PAPEL puncture strength; ~ **a la pisada** f AGRIC resistance to trampling; ~ **al plegado** f EMB folding strength, PAPEL folding endurance; ~ **al pliegue** f TEXTIL crease resistance; ~ **de polarización** f FÍS bias resistor; ~ **de polarización de rejilla** f ING ELÉC bias resistor; ~ **a la propagación de grietas** f NUCL notch toughness; ~ **protectora** f ING ELÉC protection resistor; ~ **puerta-cátodo** f ING ELÉC gate-to-cathode resistor; ~ **en un punto** f CARBÓN point resistance; ~ **pura** f ING ELÉC pure resistance;

~ q ~ **química** f C&V, P&C *propiedad, prueba* chemical resistance;

~ r ~ **a la radiación** f FÍS radiation resistance, FÍS RAD radiation hardness, radiation resistance, ING ELÉC radiation hardness, NUCL radiation resistance, radiation hardness, TELECOM radiation hardness; ~ **al rasgado** f ING MECÁ resistance to tearing, P&C *propiedad física, prueba* tear strength; ~ **al rayado** f P&C *propiedad de las superficies, prueba* scratch resistance; ~ **real** f ELEC ohmic resistance; ~ **de**

realimentación f ING ELÉC feedback resistor; ~ **reflejada** f ING ELÉC, ÓPT reflected resistance; ~ **de reglaje** f ING ELÉC regulating resistance; ~ **reguladora** f ELEC ballast resistor, *estabilizador de tensión* barretter; ~ **reostrictiva** f ING ELÉC pinched resistor; ~ **de reposo** f ING ELÉC, NUCL *de una fotocélula* dark resistance; ~ **residual** f ELEC, ING ELÉC residual resistance; ~ **al resquebrajamiento** f P&C *propiedad, prueba* crack resistance; ~ **de retropulsión** f ING ELÉC feedback resistor; ~ **de rociado** f TRANSP AÉR spray drag; ~ **a la rodadura** f VEH rolling resistance, *neumático* road resistance; ~ **a la rotura** f CONST, ING MECÁ ultimate strength, MECÁ breaking strength, METAL rupture strength, P&C *propiedad física, prueba* breaking strength, crack resistance, TEXTIL breaking strength; ~ **a la rotura por flexión** f ING MECÁ ultimate bending strength; ~ **a la rotura por tensión** f MECÁ, P&C, TEXTIL breaking strength; ~ **a la rotura traccional** f ING MECÁ, MECÁ ultimate tensile strength; ~ **de rozamiento** f MECÁ frictional drag;

~ s ~ **de la secuencia de fases positivas** f ELEC positive-phase sequence resistance; ~ **del secundario** f ING ELÉC secondary resistance; ~ **de seguridad** f ING ELÉC snubber resistor; ~ **de semiconductor** f ING ELÉC semiconductor resistor; ~ **sensora** f ING ELÉC sensing resistor; ~ **en sentido directo** f ELEC *de semiconductor* forward resistance; ~ **a sequía** f AGRIC drought resistance; ~ **a ser desmenuzado** f EMB resistance to shattering; ~ **en serie** f ELEC, ING ELÉC series resistance; ~ **serie de colector** f ING ELÉC series collector resistance; ~ **serie del diodo** f ELECTRÓN diode series resistance; ~ **superficial** f PAPEL surface bonding strength; ~ **de la superficie** f IMPR *papel* surface strength; ~ **de superficie** f CARBÓN, TERMOTEC surface resistance;

~ t ~ **de la tapa de contacto** f ING ELÉC ferrule resistor; ~ **a la tensión** f ING MECÁ endurance tensile strength, P&C *propiedad física* tensile strength; ~ **a tensión** f ING MECÁ tensile strength; ~ **tensional** f FÍS tensile strength; ~ **térmica** f FÍS thermal resistance, TERMO, TERMOTEC thermal resistance, resistance to heat; ~ **termodependiente** f ING ELÉC temperature-dependent resistor; ~ **a la termofluencia** f CARBÓN creep strength; ~ **de tierra** f ELEC, ING ELÉC earth resistance *(BrE)*, ground resistance *(AmE)*; ~ **a la torsión** f ING MECÁ resistance to twisting, torsional strength; ~ **total al avance** f TRANSP total drag; ~ **a la tracción** f ING MECÁ resistance to tension, strain, tensile stress, MECÁ strain, tensile strength, METAL strain, TELECOM, TEXTIL tensile strength; ~ **a la tracción en la dirección z** f PROD z-direction tensile strength; ~ **a la tracción en húmedo** f PAPEL wet strength; ~ **a la tracción del secado al horno** f TEXTIL oven-dry tensile strength; ~ **a la tracción de vano cero** f PROD zero span tensile strength;

~ u ~ **última al corte** f ING MECÁ ultimate shear strength; ~ **última a la tracción** f ING MECÁ ultimate tensile strength; ~ **al uso** f TEXTIL wear resistance;

~ v ~ **del vapor** f TERMOTEC vapor resistance *(AmE)*, vapour resistance *(BrE)*; ~ **variable** f ELEC slide rheostat, variable resistor, FÍS variable resistor, ING ELÉC variable resistance; ~ **variable del arco** f ELEC arc rheostat; ~ **variable con la temperatura** f ING ELÉC thermistor; ~ **de variación lineal** f ELEC linear resistance; ~ **a ventilación en minas** *f* PROD drag; ~ **volumétrica** *f* ING ELÉC bulk resistivity

resistente *adj* CONST resistant, ELEC resistant, resistive, ING ELÉC resistive, ING MECÁ, P&C resistant, TELECOM resistive; ~ **a la abrasión** *adj* IMPR abrasion-resistant; ~ **a la acción de los mohos** *adj* EMB mold-resistant (*AmE*), mould-resistant (*BrE*); ~ **a la acción de productos químicos** *adj* EMB chemically resistant; ~ **a los ácidos** *adj* EMB acid-resistant, IMPR acid-proof; ~ **al aceite** *adj* P&C *propiedad, prueba* oil-resistant; ~ **al ácido** *adj* IMPR acid-resistant; ~ **al agua** *adj* EMB, TEXTIL water-resistant; ~ **al ataque de los ácidos** *adj* CONTAM acid-resistant; ~ **al ataque de los insectos** *adj* EMB insect-proof; ~ **al blanqueo** *adj* COLOR bleaching-resistant; ~ **al calor** *adj* EMB, INSTAL TERM, TERMO heat-proof, heat-resistant; ~ **al choque** *adj* ING MECÁ, METR resistant to impact; ~ **a descarga eléctrica** *adj* ING ELÉC lightning-resistant; ~ **a la descoloración** *adj* IMPR fade-proof, fade-resistant; ~ **al desgaste** *adj* EMB wear-resistant, ING MECÁ hard-faced; ~ **al envejecimiento** *adj* EMB ageing-resistant; ~ **al estrujamiento** *adj* TEXTIL crush-resistant; ~ **al fuego** *adj* COLOR, ELEC, EMB, P&C, QUÍMICA, REVEST, SEG, TERMO, TRANSP AÉR flame-resistant; ~ **a los gases** *adj* GAS, ING MECÁ, MECÁ, PROD, PROC QUÍ, QUÍMICA, TERMO *pinturas* gas-proof, gastight; ~ **a las grasas** *adj* EMB, PAPEL greaseproof; ~ **a heladas** *adj* AGRIC, CARBÓN *terreno* frost-hardy, frost-resistant; ~ **a la humedad** *adj* EMB moisture-resistant; ~ **a los líquidos** *adj* EMB liquid-proof; ~ **a la luz** *adj* COLOR light-fast; ~ **al moho** *adj* REVEST mildew-proof; ~ **a la pudrición** *adj* PAPEL rot-proof; ~ **al sol** *adj* COLOR sunfast;

resistir *vt* PROD *manipulación violenta* stand

resistividad *f* CARBÓN resistivity, ELEC resistivity, specific resistance, FÍS resistivity, ING ELÉC resistivity, specific resistance, P&C *propiedad eléctrica, prueba*, PETROL, TEC PETR, TELECOM, TERMOTEC resistivity; ~ **aparente** *f* PETROL *geofísica* apparent resistivity; ~ **eléctrica** *f* ELEC, P&C *propiedad, prueba* electrical resistivity; ~ **al flujo** *f* ELEC *metales* flow resistivity; ~ **de formación** *f* PETROL formation resistivity; ~ **de masas** *f* ELEC mass resistivity; ~ **superficial** *f* ING ELÉC surface resistivity; ~ **térmica** *f* FÍS, TERMOTEC thermal resistivity; ~ **del vapor** *f* TERMOTEC vapor resistivity (*AmE*), vapour resistivity (*BrE*)

resistivo *adj* ELEC, ING ELÉC, TELECOM resistive

resistómetro *m* FÍS ohmmeter

resistor *m* AUTO, ELEC, FÍS, ING ELÉC resistor; ~ **ajustable** *m* ELEC adjustable resistor; ~ **alineal** *m* ELEC, TELECOM nonlinear resistor; ~ **de amortiguación** *m* ELEC damping resistor; ~ **de aplanamiento** *m* ELEC smoothing resistor; ~ **de bobina** *m* FÍS wirewound resistor; ~ **de caldeo** *m* ELEC heating resistor; ~ **de capa delgada** *m* ELECTRÓN thin-film resistor; ~ **de capa gruesa** *m* ELECTRÓN thick-film resistor; ~ **de carga** *m* ELEC load resistor; ~ **de compensación** *m* ELEC balancing resistor; ~ **compensador** *m* ELEC balancing resistor; ~ **de filtro** *m* ELEC smoothing resistor; ~ **graduable** *m* ING ELÉC adjustable resistor; ~ **de hilo arrollado** *m* ELEC wirewound resistor; ~ **limitador** *m* ELEC limiting resistor; ~ **de metal vidriada** *m* ING ELÉC metal glaze resistor; ~ **metálico** *m* ELEC metallic resistor; ~ **no inductivo** *m* ELEC noninductive resistor; ~ **no regulable** *m* ELEC, FÍS, ING ELÉC fixed resistor;

~ **óhmico** *m* ELEC noninductive resistor; ~ **reductor de tensión** *m* ING ELÉC dropping resistor; ~ **regulador de la carga** *m* ELEC load resistor; ~ **en serie** *m* ELEC, ING ELÉC series resistor; ~ **variable** *m* ELEC variable resistor, adjustable resistor, ING ELÉC variable resistor; ~ **de variación lineal** *m* ELEC linear resistor

resita *f* P&C *tipo de polímero* C-stage resin, resite

resitol *m* P&C *tipo de polímero* B-stage resin

resma *f* EMB, IMPR *500 hojas* ream

resol *m* P&C *tipo de polímero* resol

resoldadura *f* PROD resoldering

resoldeo *m* PROD resoldering

resolución *f* GEN resolution; ~ **en alcance** *f* TEC ESP range resolution; ~ **de una ambigüedad** *f* TEC ESP ambiguity resolution; ~ **en distancia** *f* TEC ESP range resolution; ~ **espacial** *f* CONTAM spatial resolution; ~ **de frecuencia** *f* ELECTRÓN frequency resolution; ~ **geométrica del haz** *f* NUCL geometric beam resolution; ~ **horizontal** *f* TV horizontal resolution; ~ **de orden superior** *f* INFORM&PD high resolution (*HR*); ~ **pobre** *f* CINEMAT poor resolution; ~ **de regulación** *f* ING ELÉC trimming resolution; ~ **temporal** *f* CONTAM, FÍS RAD temporal resolution; ~ **de la trama** *f* IMPR screen resolution

resonador *m* ACÚST, ELECTRÓN, TELECOM resonator; ~ **acústico** *m* ACÚST, ELECTRÓN acoustic resonator, OCEAN *aparatos de comunicación* multibeam sounder; ~ **anular** *m* ELECTRÓN annular resonator; ~ **de cavidad** *m* ELECTRÓN, FÍS, TELECOM, TV cavity resonator; ~ **con cavidad de sintonización constante** *m* ELECTRÓN fixed-tuned cavity resonator; ~ **coaxial** *m* ELECTRÓN coaxial resonator; ~ **de cristal** *m* CRISTAL, ELECTRÓN crystal resonator; ~ **de cuarzo** *m* ELECTRÓN quartz resonator; ~ **dieléctrico** *m* TEC ESP, TELECOM dielectric resonator; ~ **eléctrico** *m* ELEC, ELECTRÓN electrical resonator; ~ **electromagnético** *m* ELEC, ELECTRÓN electromagnetic resonator; ~ **de entrada** *m* ELECTRÓN buncher resonator, input resonator; ~ **exterior** *m* ELECTRÓN open resonator; ~ **de haz de cesio** *m* ELECTRÓN caesium beam resonator (*BrE*), cesium beam resonator (*AmE*); ~ **de Helmholtz** *m* ACÚST, FÍS Helmholtz resonator; ~ **de microondas** *m* ELECTRÓN, TELECOM microwave resonator; ~ **con modo flexible** *m* ELECTRÓN flexure-mode resonator; ~ **de onda de medición de volumen** *m* ELECTRÓN bulk-wave resonator; ~ **óptico** *m* FÍS RAD, TELECOM optical resonator; ~ **piezoeléctrico** *m* ING ELÉC piezoelectric resonator; ~ **de rendija** *m* ACÚST slit resonator

resonancia *f* GEN resonance; ~ **aceptora** *f* ELEC *circuito de CA* acceptor resonance; ~ **de amplitud** *f* FÍS amplitude resonance; ~ **asistida** *f* ACÚST assisted resonance; ~ **de la barra de control** *f* TRANSP AÉR control-rod resonance; ~ **Briet-Wigner** *f* NUCL Briet-Wigner resonance; ~ **de cables eléctricos armados** *f* ING ELÉC ferroresonance; ~ **de cavidad** *f* ELECTRÓN, FÍS, TELECOM, TV cavity resonance; ~ **ciclotrónica** *f* TELECOM cyclotronic resonance; ~ **del cuadripolo** *f* NUCL quadrupole resonance; ~ **del espín del electrón** *f* FÍS, FÍS PART, NUCL electron-spin resonance; ~ **magnética** *f* FÍS, ING ELÉC magnetic resonance; ~ **magnética nuclear** *f* FÍS, FÍS RAD, NUCL, QUÍMICA, TEC PETR *evaluación de la formación* nuclear magnetic resonance (*NMR*); ~ **mecánica** *f*

FÍS mechanical resonance; ~ **óptica** *f* ELECTRÓN optical resonance; ~ **paralela** *f* ELECTRÓN, FÍS parallel resonance; ~ **paramagnética** *f* ELEC, ELECTRÓN, FÍS paramagnetic resonance; ~ **paramagnética electrónica** *f* ELECTRÓN, FÍS, NUCL electron paramagnetic resonance; ~ **parásita** *f* NUCL hangover; ~ **en serie** *f* FÍS, ING ELÉC series resonance; ~ **a un solo nivel** *f* NUCL single-level resonance; ~ **de tierra** *f* TRANSP AÉR ground resonance; ~ **de velocidad** *f* FÍS velocity resonance

resonante *adj* GEN resonant

resoplado *m* C&V *soplado del vidrio* blow back

resoplido *m* TEC ESP chuffing

resoplo *m* C&V puff

resorcílico *adj* QUÍMICA *aldehído* resorcylic

resorcinol *m* QUÍMICA resorcinol

resorte *m* FÍS, ING MECÁ, MECÁ, VEH spring; ~ **bajo flexión** *m* ING MECÁ spring subjected to bending; ~ **bajo torsión** *m* ING MECÁ spring subjected to torsion; ~ **cilíndrico** *m* PROD cylindrical spring; ~ **circular** *m* ING MECÁ, MECÁ circlip; ~ **circular de obturación** *m* ING MECÁ, MECÁ circlip; ~ **de coche** *m* ING MECÁ carriage spring; ~ **de compresión** *m* ING MECÁ open-coil spring, open-spiral spring, INSTAL HIDRÁUL compression spring, open-coil spring, open-spiral spring; ~ **cónico en espiral** *m* ING MECÁ volute spring; ~ **elíptico** *m* ING MECÁ elliptic spring; ~ **del embrague** *m* ING MECÁ, VEH clutch spring; ~ **de la escobilla** *m* ING ELÉC brush spring; ~ **espiral** *m* AUTO, ING MECÁ, MECÁ, VEH coil spring, coiled spring; ~ **espiral cónico** *m* ING MECÁ volute spring; ~ **de extensión** *m* ING MECÁ extension spring; ~ **de flexión** *m* ING MECÁ blade spring, flexion spring; ~ **flexionado** *m* ING MECÁ spring subjected to bending; ~ **de fricción** *m* ING MECÁ friction spring; ~ **en hélice** *m* MECÁ helical spring; ~ **helicoidal** *m* AUTO, ING MECÁ, MECÁ, VEH coil spring, coiled spring, helical spring; ~ **helicoidal cerrado** *m* ING MECÁ close-coil spring, close-spiral spring; ~ **de hojas** *m* ING MECÁ leaf spring, blade spring, carriage spring; ~ **de hojas múltiples** *m* ING MECÁ multiple blade spring; ~ **de horquilla** *m* ING MECÁ hairpin spring; ~ **de inserción** *m* AUTO insert spring; ~ **de lámina flexible** *m* ING MECÁ leaf spring; ~ **motor** *m* D&A, ING MECÁ main spring; ~ **múltiple** *m* ING MECÁ cluster spring; ~ **de palanca de desenganche** *m* AGRIC, AUTO release-lever spring; ~ **del percutor** *m* D&A striker spring; ~ **de planchuela** *m* ING MECÁ leaf spring; ~ **principal** *m* D&A, ING MECÁ main spring; ~ **de recuperación** *m* AUTO retracting spring; ~ **de recuperación del freno** *m* AUTO brake-release spring; ~ **recuperador** *m* D&A recoil spring; ~ **de retención** *m* ING MECÁ, MINAS retaining spring; ~ **de retorno** *m* AUTO, ING MECÁ, PROD, VEH return spring; ~ **de retracción** *m* ING MECÁ drawback spring; ~ **retráctil** *m* AUTO retracting spring; ~ **del ruptor** *m* AUTO breaker spring; ~ **de seguridad** *m* D&A safety spring; ~ **semielíptico** *m* ING MECÁ half-elliptic spring; ~ **de somier** *m* PROD hourglass spring; ~ **térmico** *m* TERMO thermal spring; ~ **de torsión** *m* ING MECÁ torsion spring; ~ **en V** *m* ING MECÁ hairpin spring

respaldado *adj* TERMO *fuego de horno* banked up

respaldar *vt* ING MECÁ back

respaldo *m* (*cf pared Esp*) AUTO backrest, FOTO back, backing paper, MINAS wall, VEH *de un asiento* backrest; ~ **para accesorios** *m* FOTO accessory shoe; ~ **para ampliaciones** *m* FOTO, IMPR enlarger support; ~ **antiestático** *m* FOTO antistatic backing; ~ **antihalo** *m* FOTO antihalation backing; ~ **de asiento** *m* VEH seat back; ~ **duro** *m* CONST hard shoulder; ~ **de fragua** *m* PROD forge back; ~ **de memoria** *m* PROD memory backup; ~ **de la película** *m* FOTO film backing

respiración: ~ **agitada** *f* OCEAN fast breathing; ~ **fluida** *f* OCEAN fluid breathing

respiradero *m* AUTO breather, breather pipe, CONST *entrada y salida de aire* vent, IMPR air vent, ING MECÁ breather, MECÁ vent, MINAS gas vent, air shaft, NUCL air vent, vent, PROD *ventilar* air hole, breather, *molde* sand vent, TEC ESP vent, TRANSP MAR air vent, TV aircheck, VEH *cárter* breather; ~ **del cárter motor** *m* AUTO, VEH crankcase breather

respirador: ~ **de oxígeno** *m* TEC ESP oxygen respirator

respiratorio *adj* CARBÓN, SEG respiratory

resplandeciente *adj* QUÍMICA, TERMO glowing

resplandor *m* CINEMAT flare, glare, TEC ESP radiance, TRANSP MAR *de luz* loom; ~ **blanco** *m* METEO whiteout; ~ **de la celda** *m* CINEMAT cell flare; ~ **marginal** *m* CINEMAT edge flare; ~ **del norte** *m* METEO aurora borealis; ~ **del objetivo** *m* CINEMAT lens flare; ~ **del sur** *m* METEO aurora australis

respondedor *m* TRANSP AÉR responder; ~ **para espacio profundo** *m* TEC ESP deep-space transponder

responsabilidad *f* CONTAM MAR liability, ING MECÁ responsibility, TELECOM accountability; ~ **del mensaje por intercambio de datos electrónicos** *f* TELECOM EDIM responsibility; ~ **por producto** *f* CALIDAD product liability

responsividad *f* TELECOM responsivity; ~ **espectral** *f* TELECOM spectral responsivity

respuesta *f* ACÚST response, ELECTRÓN reply, response, INFORM&PD feedback, reply, response, TELECOM, TV reply, response; ~ **acústica** *f* ELECTRÓN audio response, voice response; ~ **de amplitud** *f* ELECTRÓN amplitude response, amplitude-amplitude response; ~ **de amplitud de filtro** *f* ELECTRÓN filter-amplitude response; ~ **de amplitud de frecuencia** *f* ELECTRÓN amplitude-frequency response; ~ **asociada A** *f* TELECOM A-associate response (*AARE*); ~ **automática** *f* TELECOM answerback; ~ **de baja frecuencia** *f* ELECTRÓN low-frequency response; ~ **de color** *f* CINEMAT color response (*AmE*), colour response (*BrE*); ~ **de corriente** *f* ACÚST response to current; ~ **dada en casos de derrames de hidrocarburos** *f* CONTAM MAR oil spill response; ~ **dinámica** *f* TELECOM dynamic response; ~ **direccional** *f* ELECTRÓN directional response; ~ **de emergencia** *f* NUCL emergency response; ~ **de entrada** *f* ELECTRÓN input response; ~ **espacial** *f* ELECTRÓN spatial response; ~ **de fase** *f* ELECTRÓN phase response; ~ **de fase del filtro** *f* ELECTRÓN filter phase response; ~ **de filtro** *f* ELECTRÓN filter response; ~ **de frecuencia** *f* CINEMAT, ELEC, ELECTRÓN, FÍS ONDAS, ÓPT, TELECOM frequency response; ~ **de frecuencia de filtrado** *f* ELECTRÓN, FÍS RAD filter-frequency response; ~ **de función escalonada** *f* ELECTRÓN step function response; ~ **de impulso finito** *f* (*RIF*) ELECTRÓN finite impulse response (*FIR*); ~ **de impulso infinito** *f* ELECTRÓN, TELECOM infinite impulse response (*IIR*); ~ **de impulsos** *f* ÓPT, PETROL, TELECOM impulse response; ~ **indicativa** *f* ELECTRÓN indicial response; ~ **inicial** *f*

TV head response; ~ **de paso bajo** *f* ELECTRÓN low-pass response; ~ **paso banda** *f* ELECTRÓN *filtros* pass-band response; ~ **plana** *f* ELECTRÓN flat response; ~ **de potencia** *f* ACÚST response to power; ~ **de radar** *f* D&A, TRANSP, TRANSP AÉR, TRANSP MAR radar response; ~ **en régimen transitorio** *f* TV transient response; ~ **telefónica** *f* TELECOM voice response; ~ **térmica** *f* ING ELÉC temperature response; ~ **transitoria** *f* ELECTRÓN, ING ELÉC transient response; ~ **a transitorios** *f* TV surge characteristic; ~ **de voltaje** *f* ACÚST response to voltage

resquebrajadura: ~ **por fatiga** *f* METAL fatigue crack

resquebrajamiento *m* CALIDAD cracking, CONST cleaving, P&C *defecto de pinturas, barnices y revestimientos* alligatoring

resquebrajarse *v refl* MINAS split

resquicio *m* EMB eyelet

restablecedor: ~ **de cable** *m* ELEC, ING ELÉC cable repeater

restablecer *vt* CONTAM recover, ELECTRÓN repair

restablecimiento *m* CARBÓN recovery, restoration, CONTAM *contaminación del suelo* reclamation, recovery, rehabilitation, PROD restoration, RECICL reclamation; ~ **de una señal** *m* TELECOM signal restoration

restador: ~ **de un bit** *m* ELECTRÓN one-digit subtractor; ~ **digital** *m* ELECTRÓN digital subtractor; ~ **serie** *m* ELECTRÓN serial subtractor

restauración *f* CINEMAT reset, CONTAM *contaminación del suelo* recovery, reclamation, rehabilitation, INFORM&PD restore, reset, ING MECÁ, PROD reset, resetting, RECICL reclamation, TEC PETR *fondo marino* restoration, TV reset; ~ **de archivos** *f* INFORM&PD file restore; ~ **de ficheros** *f* INFORM&PD file restore; ~ **de imagen** *f* INFORM&PD image restoration; ~ **en memoria principal** *f* INFORM&PD roll-in; ~ **de pulsos** *f* TV pulse restoration; ~ **de suelo** *f* AGRIC soil building; ~ **de la superficie de corte** *f* PROD dressing

restaurador: ~ **de roscas** *m* ING MECÁ thread restorer

restaurar *vt* CONTAM recover, INFORM&PD restore; ~ **en memoria principal** *vt* INFORM&PD roll in

restinga *f* OCEAN shingle spit, *geomorfología de playas* beach ridge

resto *m* INFORM&PD, MATEMÁT remainder

restos: ~ **de crisoles viejos** *m pl* PROD pot scrap; ~ **flotantes** *m pl* TRANSP MAR flotsam; ~ **de naufragio** *m pl Esp* TRANSP MAR wreck; ~ **de náufrago** *m pl Esp* (*cf restos náufragos AmL*) TRANSP MAR wreck; ~ **náufragos** *m pl AmL* (*cf restos de náufragos Esp*) TRANSP MAR wreck; ~ **de uranio** *m pl* NUCL uranium scrap

restricción *f* TEC PETR confinement; ~ **calibrada** *f* TEC PETR *tuberías* gaged restriction (*AmE*), gauged restriction (*BrE*); ~ **dosificada** *f* TEC PETR *tuberías* gaged restriction (*AmE*), gauged restriction (*BrE*); ~ **del espacio aéreo** *f* TRANSP AÉR airspace restriction; ~ **de identificación de línea conectada** *f* TELECOM connected-line identification restriction (*COLR*); ~ **de identificación de línea de llamada** *f* TELECOM calling-line identification restriction (*CLIR*); ~ **tecnológica** *f* TEC ESP technological restriction

restrictor *m* TELECOM restrictor

restringir: ~ **banda** *vt* ELECTRÓN band limit

resultado *m* GEOM solution, METR *de una inspección* result; ~ **corregido** *m* METR *de errores asumidos sistemáticamente* corrected result; ~ **sin corregir** *m* METR uncorrected result; ~ **cuántico** *m* FÍS quantum yield; ~ **final** *m* TEC ESP payoff; ~ **de opinión media** *m Esp* TELECOM mean opinion score; ~ **de precipitación** *m* CONTAM *tratamiento de aguas y gases* precipitation event; ~ **de la subida** *m* TRANSP AÉR climb performance

resultante *f* FÍS, ING MECÁ, MATEMÁT resultant

resumen *m* ING MECÁ outline

resumidero *m AmL* (*cf sumidero Esp*) CONTAM *instalaciones industriales* sink

resumir *vt* INFORM&PD abstract, RECICL digest

resurgencia *f* AGUA, HIDROL *corriente de agua subterránea* resurgence

resurgente *adj* AGUA, HIDROL resurgent

retacado *m* (*cf atacado Esp*) MINAS *barrenos* stemming, TRANSP MAR *de hierro* caulking

retacador *m* ING MECÁ, MECÁ, PROD caulker, TRANSP MAR caulking iron, *de planchas* caulker; ~ **neumático** *m* TRANSP MAR caulking iron

retacar *vt* TRANSP MAR *mantenimiento, construcción de barcos* caulk

retales *m pl* CARBÓN *siderurgia, caucho* cuttings; ~ **del corte** *m pl* TEXTIL waste

retallo *m* CONST ledge, offset

retardación *f* ELEC, ING MECÁ, QUÍMICA, TEC ESP, TRANSP retardation; ~ **de la marea** *f* OCEAN *mareas máximas* tidal epoch

retardado *adj* INFORM&PD lagged, PROD on delay

retardador *m* AGUA inhibitor, CINEMAT, FOTO restrainer, ING ELÉC slug, ING MECÁ, MECÁ dashpot, P&C *materia prima, vulcanización, aditivo,* TRANSP *camiones* retarder; ~ **de aire** *m* TRANSP AÉR air dashpot; ~ **de fraguado** *m* CONST *hormigón* retarding agent; ~ **de llamas** *m* TRANSP AÉR flame holder

retardar[1] *vt* FÍS decelerate, *un aumento de temperatura* lag, MECÁ decelerate, QUÍMICA, TRANSP retard

retardar[2] *vi* ELEC *de relé, conmutador* delay, retard, ING MECÁ, TEC ESP retard

retardo[1]: **con** ~ *adj* PROD on delay

retardo[2] *m* AGUA lag, ELEC *corriente alterna* lag, *en relé, conmutador* delay, retardation, ELECTRÓN lag, delay, INFORM&PD delay, ING ELÉC lag, delay, ING MECÁ retardation, MECÁ lagging, drag, deceleration, MINAS *pega* drag, PROD lag, QUÍMICA retardation, TEC ESP *en circuitos* delay, retardation, TEC PETR *refino* lagging, TELECOM delay, time delay, TRANSP retardation, TRANSP AÉR drag, lagging, TV lag; ~ **en el cierre del temporizador** *m* PROD timer-off delay; ~ **de conmutación** *m* TELECOM switching delay; ~ **cromático** *m Esp* TV chroma delay; ~ **debido a la conmutación** *m* TELECOM switching delay; ~ **por derivación** *m* ELECTRÓN tapped delay; ~ **en desconexión** *m* PROD *temporizadores* off delay; ~ **después del discado** *m* TELECOM postdialing delay (*AmE*), postdialling delay (*BrE*); ~ **envolvente** *m* TEC ESP envelope delay; ~ **de escape** *m* INSTAL HIDRÁUL *distribuidor* exhaust lag, retarded release; ~ **de fase** *m* ELEC *corriente alterna,* ELECTRÓN, FÍS, OCEAN phase lag, TV phase delay; ~ **fijo** *m* ELEC *de conmutador de relé* fixed delay; ~ **de forma** *m* TRANSP AÉR form drag; ~ **grupal multimodo** *m* ÓPT, TELECOM multimode group delay; ~ **de grupo** *m* ING ELÉC, TEC ESP group delay; ~ **de**

grupo multimodal *m* ÓPT, TELECOM multimode group delay; **~ de la imagen** *m* TV image lag; **~ de imanación** *m* ING MECÁ magnetic lag; **~ de imantación** *m* ING MECÁ magnetic lag; **~ instantáneo** *m* ELECTRÓN high-order delay; **~ de la luminosidad** *m* TV luminance delay; **~ de magnetización** *m* ING MECÁ magnetic lag; **~ de la marea** *m* OCEAN *mareas máximas* tidal epoch; **~ del modo diferencial** *m* ÓPT, TELECOM differential mode delay; **~ de paquete** *m* INFORM&PD packet delay; **~ parasítico** *m* TRANSP AÉR parasitic drag; **~ postdiscado** *m* TELECOM postdialing delay (*AmE*), postdialling delay (*BrE*); **~ posterior al discado** *m* TELECOM postdialing delay (*AmE*), postdialling delay (*BrE*); **~ de propagación** *m* INFORM&PD, TEC ESP propagation delay; **~ de propagación de grupo** *m* TEC ESP *comunicaciones* envelope delay; **~ de la red** *m* INFORM&PD network delay; **~ por la resistencia del aire** *m* TRANSP AÉR pressure drag; **~ de respuesta de la puerta** *m* ELECTRÓN gate delay; **~ rotacional** *m* INFORM&PD, ÓPT rotational delay; **~ de señal** *m* ELECTRÓN, TELECOM *a través de una red* signal delay; **~ de señal entre circuitos** *m* ELECTRÓN intercircuit signal delay; **~ de señal entre pastillas electrónicas** *m* ELECTRÓN interchip signal delay; **~ en la señal de línea** *m* TELECOM dial-tone delay; **~ de la señal a través de la puerta** *m* ELECTRÓN gate delay; **~ térmico** *m* NUCL temperature lag, TERMO thermal lagging; **~ variable** *m* TELECOM variable delay

retén *m* (*cf arrancatubos AmL*) ING MECÁ detent, dog, grip, keeper, lock stop, ratchet, safety catch, trip catch, MECÁ catch, dog, ratchet, MINAS *entubación* dog, kep, PROD grip; **~ de cierre esférico** *m* ING MECÁ ball lock retainer; **~ de la compuerta del tren de aterrizaje** *m* TRANSP AÉR landing gear door latch; **~ del lubricante** *m* MECÁ seal; **~ de rebote** *m* AUTO rebound clip; **~ de resorte** *m* AUTO spring retainer; **~ de seguridad de muñón cuadrado** *m* ING MECÁ square-ball lock retainer; **~ de uña** *m* ING MECÁ pawl

retención *f* AGUA, HIDROL retention, ING ELÉC retention, sticking, ING MECÁ holding up, retention, INSTAL HIDRÁUL *agua residual* backing-up, PAPEL retention, PROD hold, QUÍMICA retention, TEC ESP hold, TEC PETR retention, TRANSP AÉR holding; **~ de agua** *f* AGUA water retention; **~ de derechos** *f* TEC PETR *licencias* retention of rights; **~ de la imagen** *f* TV image retention; **~ de llamadas** *f* TELECOM call holding; **~ de la respiración** *f* OCEAN breath-holding; **~ de visualización** *f* ELECTRÓN display retention

retenedor *m* ING MECÁ gib; **~ de la empaquetadura** *m* ING MECÁ packing retainer; **~ de grasa lubricante** *m* ING MECÁ grease retainer; **~ de la rueda del tren de aterrizaje** *m* TRANSP AÉR landing gear boot retainer

retener *vt* FERRO *tren* hold back, INFORM&PD, PROD hold, TELECOM *una llamada* book

retenida *f* TRANSP MAR *cabos* guy; **~ de botavara** *f* TRANSP MAR *yates* boom guy

retentividad *f* FÍS, TV retentivity

reticita *f* MINERAL rhaetizite

retícula *f* CARBÓN, TEC ESP mesh; **~ de arranque** *f* CARBÓN release mesh; **~ cartográfica mercatoriana** *f* OCEAN Mercator plotting chart; **~ óptica** *f* METR optical graticule; **~ de pantalla roja** *f* ELECTRÓN, TV red-screen grid

reticulación *f* CINEMAT reticulation, wrinkling, ELECTRÓN, FOTO, IMPR reticulation; **~ por irradiación** *f* FÍS RAD, P&C *polimerización* radiation cross-linking; **~ por radiación** *f* FÍS RAD, P&C radiation cross-linking

reticulado *m* AmL (*cf retículo Esp*) AGRIC reticulum, CINEMAT, ELECTRÓN graticule, raticle, IMPR cross web, INSTR graticule, raticle, MATEMÁT lattice, METR graticule, NUCL *pila atómica* lattice, ÓPT graticule, P&C network, TEC ESP *reactor nuclear* lattice, TELECOM array, TV graticule, grating; **~ de difracción** *m* FÍS ONDAS diffraction grating; **~ hexagonal compacto** *m* NUCL hexagonal close-packed lattice

reticular[1] *adj* TELECOM networked

reticular[2] *vt* P&C *polimerización, vulcanización* cross link

retículo *m* Esp (*cf reticulado AmL*) AGRIC reticulum, CINEMAT, ELECTRÓN graticule, reticle, IMPR cross web, INSTR graticule, reticle, MATEMÁT lattice, METR graticule, NUCL *pila atómica* lattice, ÓPT graticule, P&C network, TEC ESP *reactor nuclear* lattice, TELECOM array, TV graticule, grating; **~ de difracción** *m* FÍS ONDAS diffraction grating; **~ hexagonal compacto** *m* NUCL hexagonal close-packed lattice

retinalita *f* MINERAL retinalite, retinellite

retinasfalto *m* MINERAL retinasphalt

retinita *f* MINERAL retinite

retiración *f* IMPR back, perfecting, backing, backing-up

retirada *f* CONST removal, removing, CONTAM *tratamiento de aguas y gases* removal; **~ de las basuras** *f* AGUA scavenging; **~ de escombros** *f* TEC PETR debris removal; **~ de ganado de los pastos** *f* AGRIC removal from pasture; **~ de partículas contaminantes del aire** *f* CONTAM *tratamiento de gases* particulate removal of air pollutants; **~ del respaldo** *f* CINEMAT backing removal; **~ de ripios** *f* TEC PETR *perforación* cuttings removal

retirar *vt* CONST *andamiaje* take down, IMPR perfect, back, back up, ING MECÁ back, MINAS *terrenos de recubrimiento* remove, TRANSP push back

retirarse *v refl* HIDROL *las aguas* subside

retiro: **~ de pie con cabeza** *m* IMPR twelve ways back up

retocado *m* C&V, COLOR, FOTO, IMPR retouching, touching up

retocar *vt* C&V, COLOR, FOTO, IMPR retouch, touch up, ING MECÁ true up

retoque *m* C&V, COLOR, FOTO, IMPR retouching, touching up; **~ de negativos** *m* CINEMAT *usando barniz cubriente* opaquing

retorcedura *f* ING MECÁ twisting

retorcerse *v refl* PROD bend, bow, crook, curve

retorcimiento *m* FÍS FLUID *vorticidad de fondo*, ING MECÁ twisting

retornable: **no ~** *adj* ALIMENT nonreturnable

retornar *vti* INFORM&PD return

retorno[1]: **sin ~ a cero** *adj* INFORM&PD, TELECOM nonreturn-to-zero (*NRZ*); **sin ~ al cero invertido** *adj* TELECOM nonreturn-to-zero-inverted (*NRZI*)

retorno[2] *m* ELECTRÓN flyback, INFORM&PD return, TV flyback; **~ automático de la última palabra** *m* INFORM&PD word wrap, wrap around; **~ del carro** *m* IMPR carriage return; **~ combinado** *m* TELECOM common return; **~ del haz** *m* TV beam return, field flyback; **~ inactivo** *m* ING MECÁ idle return; **~ de línea** *m* TV line flyback; **~ de la llama** *m* CONST,

TERMO flashback; **~ por tierra** *m* GEOFÍS, ING ELÉC, PROD, TELECOM earth return (*BrE*), ground return (*AmE*); **~ de tierra del circuito de señal** *m* TELECOM signal ground; **~ al servicio** *m* TELECOM return to service; **~ en vacío** *m* ING MECÁ idle return

retorta *f* INSTAL TERM *química*, LAB *material de vidrio*, PROC QUÍ, QUÍMICA, TEC PETR *ingeniería de lodos* retort; **~ de destilación** *f* LAB, QUÍMICA, TERMO distillation retort; **~ de gas** *f* GAS, PROD gas retort; **~ refrigerada por agua** *f* INSTAL TERM water-cooled retort

retracción *f* PROD shrinking, *de fraguado* shrinkage

retractible *adj* MECÁ retractable

retráctil *adj* MECÁ retractable

retractilación: tras la ~ *adj* EMB after-shrinkage

retractiladora: ~ de bandejas *f Esp* (*cf retractiladora de pálets AmL*) EMB pallet wrapper; **~ de paletas** *f Esp* (*cf retractiladora de pálets AmL*) EMB pallet wrapper; **~ de pálets** *f AmL* (*cf retractiladora de paletas Esp*) EMB pallet wrapper

retraíble *adj* MECÁ retractable

retranquear *vt* CONST throw back into alignment

retranqueo *m* CONST *muro* setback, GEOFÍS *métodos de prospección* offset

retransmisión: ~ N *f* TELECOM N-relay; **~ televisiva** *f* ELECTRÓN, FÍS, ING ELÉC, TELECOM, TV television relay

retransmitir *vt* TELECOM rebroadcast, retransmit, TV rebroadcast

retrasar *vt* ELEC retard, ELECTRÓN *en transmisión de impulso* delay, ING MECÁ, QUÍMICA, TEC ESP, TRANSP retard

retraso *m* AGUA lag, ELEC *corriente alterna* lag, retardation, ELECTRÓN lag, ENERG RENOV *de las mareas* lagging, ING ELÉC lag, ING MECÁ retardation, MECÁ lagging, NUCL delay, PROD lag, QUÍMICA retardation, TEC ESP *procesos comunicaciones* delay, retardation, TEC PETR *refino* lagging, TELECOM *dispositivos temporizadores* time-out, TRANSP delay, retardation, TV lag; **~ de cosecha** *m* AGRIC cropping delay; **~ en la entrega** *m* EMB delivery delay; **~ de fase** *m* ELEC *corriente alterna* phase lag; **~ de la imagen** *m* TV image lag; **~ de la luminancia** *m* TV luminance delay; **~ de la marea** *m* OCEAN lag of the tide; **~ operacional** *m* TRANSP operational delay; **~ pirotécnico** *m* TEC ESP pyrotechnic delay; **~ de respuesta de la cuenca colectora** *m* HIDROL catchment area response lag; **~ rotacional** *m* INFORM&PD, ÓPT rotational delay

retratamiento *m* PROD retreat

retratar *vt* CARBÓN retreat, PROD reprocess, retreat, RECICL reprocess

retrete: ~ de expulsión controlada *m* FERRO *vehículos* controlled emission toilet

retroacción *f* GEN feedback

retroalimentación *f* GEN feedback; **~ anterior** *f* PROD last feedback; **~ negativa** *f* FÍS ONDAS negative feedback; **~ positiva** *f* FÍS RAD positive feedback; **~ táctil** *f* PROD tactile feedback

retrocabalgamiento *m* GEOL *geología estructural* back thrust

retroceder *vi* AGUA, HIDROL recede, INFORM&PD backspace, TRANSP AÉR backtrack

retroceso *m* AGUA recession, D&A *de un arma de fuego* recoil, ELECTRÓN retrace, flyback, HIDROL *del caudal* recession, IMPR backspace (*BS*), backward motion,

INFORM&PD backspace (*BS*), ING MECÁ reversing motion, recoil, reversing, MECÁ recoil, TERMO *cerámica* comeback, TRANSP AÉR push back, backing, TV flyback; **~ del fileteado** *m* ING MECÁ screw-cutting reverse; **~ de frecuencia** *m* ELECTRÓN frequency retrace; **~ de la hélice** *m* TRANSP MAR slip; **~ de la llama** *m* CONST, TERMO *soldadura* flashback; **~ transitorio de la llama en el soplete** *m* CONST backfire

retrocohete *m* TEC ESP, TRANSP retrorocket

retrocruza *f* AGRIC *mejoramiento* back-cross

retrocruzamiento *m* AGRIC *mejoramiento* back-cross

retrodifracción *f* CRISTAL back-reflection method

retrodifundir *vt* ÓPT backscatter

retrodifundirse *v refl* ÓPT backscatter

retrodifusión *f* NUCL back diffusion, ÓPT backscattering, TEC ESP backscatter

retrodispersar *vt* ÓPT backscatter

retrodispersión *f* FÍS RAD, TEC ESP, TELECOM backscattering

retroespacio *m* ACÚST backgap

retroesparcimiento *m* ÓPT backscattering

retroexcavadora *f* CONTAM MAR backhoe

retrogalvanostegia *f* PROD barrel plating

retrogradación *f* ALIMENT retrogradation

retrógrado *adj* ENERG RENOV, GEOL, TEC ESP retrograde

retroinstalación *f* INFORM&PD retrofit

retrometamorfismo *m* ENERG RENOV retrograde metamorphism, GEOL retrograde metamorphism, retrogressive metamorphism

retropaís *m* GEOL hinterland

retroplegamiento *m* GEOL backfolding

retroprogramación *f* PROD backward scheduling

retrorreflejar *vt* ÓPT backscatter

retrorreflexión *f* FÍS RAD retroreflection

retrosecuencia *f* TEC ESP retrosequence

retroseguimiento *m* IMPR, INFORM&PD backspace (*BS*), backtracking

retrotitulación *f* QUÍMICA back titration

retrotitular *vt* QUÍMICA back titrate

retrovaloración *f* QUÍMICA back titration

retrovalorar *vt* QUÍMICA back titrate

retrovisor: ~ lateral *m* VEH side mirror

reubicación *f* INFORM&PD relocation

reubicar *vt* INFORM&PD relocate

reunidora: ~ de cintas *f* ING MECÁ, TEXTIL sheeter box

reunión *f* ACÚST *combinación de varios registros sonoros* dubbing; **~ espacial** *f* TEC ESP rendezvous; **~ en la obra** *f* CONST site meeting; **~ en órbita** *f* TEC ESP rendezvous; **~ a pie de obra** *f* CONST site meeting; **~ de trabajo** *f* PROD workshop

reunir *vt* AGUA bulk

reutilizable *adj* CONTAM recyclable, EMB recyclable, INFORM&PD reusable, RECICL recyclable, reusable; **no ~** *adj* INFORM&PD nonreusable

reutilización *f* C&V, CARBÓN, CONST, CONTAM, PAPEL recycling, PROD reprocessing, recycling, QUÍMICA recycling, RECICL recycling, reprocessing, reuse; **~ de frecuencia** *f* ELECTRÓN, TEC ESP, TELECOM frequency reuse

reutilizar *vt* CONST, CONTAM, EMB recycle, RECICL recycle, reuse

revegetación *f* AGRIC regrazing

revelado[1]: **~ insuficiente** *adj* FOTO, IMPR underdeveloped

revelado[2] *m* FOTO processing; **~ de acoplador** *m* CINEMAT coupler development; **~ de un baño** *m*

IMPR one-bath development; **~ en color** *m* CINEMAT colour processing (*BrE*), FOTO color development (*AmE*), colour development (*BrE*), color processing (*AmE*), colour processing (*BrE*); **~ en cremallera y clavija** *m* CINEMAT rack-and-pin processing; **~ forzado** *m* CINEMAT forced development; **~ por inversión** *m* CINEMAT reversal development, reversal processing, FOTO reversal processing, reversal development, IMPR reversal processing; **~ químico** *m* FOTO, QUÍMICA chemical development; **~ superficial** *m* FOTO *de un grano* surface development; **~ en tambor** *m* CINEMAT drum development; **~ en tanque** *m* CINEMAT tank developing, FOTO tank development

revelador *m* CINEMAT, FOTO developer; **~ agotado** *m* FOTO exhausted developer; **~ de baño único** *m* FOTO single-bath developer; **~ para color** *m* FOTO color developer (*AmE*), colour developer (*BrE*); **~ compensador** *m* FOTO compensating developer; **~ para contraste tenue** *m* CINEMAT, FOTO developer for soft contrast; **~ de grano fino** *m* CINEMAT, FOTO fine-grain developer; **~ rápido** *m* FOTO fast developer; **~ suave** *m* FOTO soft developer

revelar *vt* CINEMAT develop, process, FOTO develop; **~ con exceso** *vt* CINEMAT, FOTO overdevelop

revenido[1] *adj* CRISTAL tempered, ING MECÁ drawn

revenido[2] *m* C&V *relajar las tensiones internas* stress relaxation, METAL artificial ageing, tempering, MINAS *lingotes* drawing, PROD *funderías* drawback, TERMO *metales* artificial ageing

revenir *vt* MECÁ anneal, PROD *metales en tubos* draw

reventar[1] *vt* CONST, ING MECÁ burst

reventar[2] *vi* INFORM&PD burst, PROD *de moldes* break out

reventazón *f* OCEAN breaking, *morfología de playas* surf

reventón *m* PETROL, TEC PETR *perforación* blowout

reverberación *f* ACÚST, FÍS, FÍS ONDAS reverberation

reversibilidad *f* FÍS, ING ELÉC, ING MECÁ, QUÍMICA, TRANSP reversibility

reversible *adj* FÍS, ING ELÉC, ING MECÁ, QUÍMICA, TRANSP reversible

reverso *m* CONST back, IMPR backside, reverse, verso, PAPEL *papel* backside

revesa *f* TRANSP MAR *agua, viento* eddy

revesas: ~ de la estela *f pl* OCEAN dead water

revestido[1] *adj* ELEC *conductor*, ING ELÉC covered, REVEST coated; **~ de algodón** *adj* ELEC, ING ELÉC, REVEST cotton-covered; **~ de caucho** *adj* P&C, REVEST rubber-coated; **~ electrolíticamente con estaño** *adj* REVEST electroplated with tin; **~ por emulsión** *adj* EMB, REVEST emulsion-coated; **~ de seda** *adj* TEXTIL faced with silk

revestido[2]: **~ por extrusión** *m* PAPEL extrusion coating; **~ por inmersión** *m* PAPEL dip coating; **~ mediante fusiones en caliente** *m* PAPEL hot-melt coating; **~ termosellable** *m* EMB heat-seal coating

revestidor *m* ING MECÁ *para cilindros y perforaciones* liner; **~ sin perforaciones** *m* TEC PETR *perforación* blank liner

revestimiento[1]: **con ~ doble** *adj* ING MECÁ double-skinned; **con ~ electrolítico** *adj* REVEST electroplatable; **con ~ metálico** *adj* ING MECÁ, MECÁ metalclad

revestimiento[2] *m* (*cf tabla de forro AmL*) CARBÓN liner, *de hornos* shell, CINEMAT coating, CONST coat-

ing, facing, housing, revetment, sheeting, veneering, EMB coating, FÍS cladding, coating, ING ELÉC armor (*AmE*), armour (*BrE*), sheath, ING MECÁ surfacing, INSTAL HIDRÁUL clothing, enclosed casing, cleading, casing, MECÁ *ejes* sheath, lagging, lining, casing, liner, MINAS *de un pozo de mina* lining, *pared* lagging, lagging piece, ÓPT *fibras ópticas* cladding, P&C *pintura, plásticos* coating, PETROL coating, liner, PROD *hornos* shirt, relining, *encofrados* sheathing, *horno alto* lining, REVEST sheathing, relining, lining, coating, TEC ESP shell, casing, TEC PETR coating, TELECOM cladding, serving, TRANSP AÉR liner; **~ abrasivo** *m* REVEST abrasive coating; **~ adaptado** *m* TELECOM matched cladding; **~ adhesivo** *m* EMB, REVEST cling film; **~ aislante** *m* ING ELÉC, PROD insulating covering, TERMO lagging; **~ ajustado** *m* ÓPT matched cladding; **~ de alta resistencia** *m* REVEST stress coating; **~ de aluminio** *m* REVEST aluminium coating (*BrE*), aluminum coating (*AmE*); **~ de amianto** *m* AmL (*cf revestimiento de asbesto Esp*) CONST asbestos sheeting; **~ anódico** *m* REVEST anodic coating; **~ antiabrasivo** *m* CINEMAT anti-abrasion coating; **~ antiácido** *m* COLOR acid-proof coat; **~ anticorrosión** *m* NUCL anticorrosion coating; **~ antideslizante** *m* REVEST nonskid coating; **~ antideslumbrante** *m* TELECOM antiglare coating; **~ antihalo** *m* *Esp* FOTO antihalo layer; **~ antinflamable** *m* REVEST, TERMO fire-resistant coating; **~ antirreflectante** *m* REVEST antireflective coating; **~ antirreflector** *m* C&V, CINEMAT, FOTO, ÓPT antireflection coating, TELECOM antireflection coating, antireflective coating; **~ antirreflejante** *m* ÓPT antireflection coating; **~ aplicado con brocha** *m* REVEST brush coating; **~ aplicado por lecho fluidizado** *m* PROC QUÍ, REVEST fluidized-bed coating; **~ de asbesto** *m* *Esp* (*cf revestimiento de amianto AmL*) CONST asbestos sheeting; **~ asfáltico** *m* CONST asphalt surfacing; **~ asfáltico para tuberías** *m* ING MECÁ bitumen pipe-coating; **~ por aspersión** *m* REVEST spray coating; **~ del avión** *m* TRANSP AÉR skin; **~ básico** *m* METAL basic lining; **~ basto** *m* CONST rough casting; **~ de betún** *m* REVEST bitumen coating; **~ bituminoso** *m* REVEST bitumen coating; **~ de bobinas** *m* REVEST roll coating; **~ del cable** *m* ELEC, ING ELÉC, P&C *caucho, plásticos* cable covering (*AmE*), cable sheathing (*BrE*); **~ de caldera** *m* INSTAL HIDRÁUL boiler jacketing; **~ calorifugado** *m* INSTAL HIDRÁUL *de un motor de vapor, TERMO de tuberías, calderas* lagging; **~ calorífugo** *m* PROD *de calderas, cilindros* lag; **~ de caucho** *m* P&C rubber coating, rubber lining, REVEST rubber coating, SEG rubber lining; **~ cerámico** *m* C&V ceramic coating, PROD dome cover, REVEST ceramic coating, dome casing; **~ de chocolate** *m* ALIMENT chocolate coating; **~ del cojinete** *m* MECÁ bearing lining; **~ de concreto** *m* AmL (*cf revestimiento de hormigón Esp*) COLOR concrete coating, CONST concrete lining, PETROL concrete coating, REVEST concrete coating, concrete lining; **~ de cromo duro** *m* REVEST hard chrome plating; **~ de cubierta** *m* TRANSP MAR *construcción naval* decking; **~ delgado** *m* MECÁ film; **~ doble** *m* ING MECÁ double casing; **~ de doble capa** *m* TELECOM double-layer coating; **~ electrolítico** *m* ELEC, ING ELÉC electroplated coating, PROD plating, REVEST electroplated coating; **~ de electrón** *m* TV M wrap; **~ del embrague** *m* AUTO,

MECÁ, VEH clutch lining; ~ **del embrague** *m* ING MECÁ clutch lining; ~ **de envés** *m* TEXTIL back-coating; ~ **exterior** *m* INSTAL TERM *calderas* jacket, TEC ESP cladding; ~ **de fábrica** *m* CARBÓN *pozos mina* stone tubbing; ~ **de fibras** *m* ÓPT, TELECOM fiber cladding (*AmE*), fiber coating (*AmE*), fibre cladding (*BrE*), fibre coating (*BrE*); ~ **del fondo** *m* C&V *hornos* bottom paving; ~ **de fosfato** *m* REVEST phosphate coating; ~ **fotoprotector** *m* ELECTRÓN photoresist coating; ~ **galvánico delgado** *m* TEC ESP *recubrimientos superficiales* film; ~ **de grafito** *m* NUCL *en superficie interna de la vaina*, REVEST graphite coating; ~ **hidrófugo** *m* REVEST water-repellent coat; ~ **homogéneo** *m* ÓPT homogeneous cladding; ~ **de hormigón** *m Esp* (*cf revestimiento de concreto AmL*) COLOR concrete coating, CONST concrete lining, PETROL concrete coating, REVEST concrete coating, concrete lining; ~ **húmedo** *m* PETROL wetsuit; ~ **ignífugo** *m* COLOR fireproof coat, NUCL fire-retardant coat, REVEST fire-resistant coating, fire-retardant coat, fireproof coat, flameproof coat, TERMO fire-resistant coating, fire-retardant coat, fireproof coat; ~ **impermeable** *m* REVEST water-repellent coat, waterproof coating; ~ **por inmersión** *m* REVEST dip coat; ~ **interior** *m* CONST lining, EMB interior coating, PAPEL wadding, PROD lining, REVEST subcoating, undercoating, TRANSP MAR *del casco* inner lining, inner skin; ~ **interior antiácido** *m* REVEST acid-proof lining; ~ **interior de superficies cilíndricas** *m* REVEST reverse-roll coating; ~ **intermedio** *m* REVEST intermediate coat; ~ **de la junta** *m* PETROL joint coating; ~ **de madera** *m* INSTAL HIDRÁUL wood lagging; ~ **de madera para pared** *m* REVEST woodchip wall covering; ~ **magnético** *m* TV magnetic coating; ~ **de mampostería** *m* CARBÓN *pozos mina* stone tubbing, MINAS *pared* stone tubbing, wall bearing; ~ **con material duro** *m* MECÁ hard facing; ~ **de material trenzado** *m* ING ELÉC *cables* braiding; ~ **con mercurio** *m* PROD *para espejos*, REVEST quicksilvering; ~ **de metal antifricción** *m* FERRO *vehículo* white metal packing; ~ **metálico** *m* EMB metal coating, metallic coating, ING ELÉC metal sheath, ING MECÁ, PROD, REVEST metal coating, metallic coating, TEC ESP *de alas, fuselaje* skin, TELECOM metal coating, metallic coating; ~ **metálico por inmersión en baño caliente** *m* REVEST hot-dip metal coating; ~ **metálico por inmersión en baño de metal húmedo** *m* REVEST hot-dip metal coating; ~ **de nitruro de titanio** *m* ING MECÁ titanium nitride coating; ~ **del objetivo** *m* FOTO lens coating; ~ **obtenido por oxidación catódica** *m* REVEST cathodic-oxide coating; ~ **Omega** *m* TV Omega wrap; ~ **orgánico** *m* REVEST organic coating; ~ **de papel** *m* REVEST paper lining; ~ **parcial** *m* REVEST partial coating; ~ **del piso** *m* CONST *solado*, SEG flooring; ~ **plástico** *m* REVEST plastic coating; ~ **de plata** *m* ACÚST, PROD silvering; ~ **en polvo** *m* COLOR powder coating; ~ **de polvo electrostático** *m* COLOR, P&C, REVEST electrostatic powder coating; ~ **posterior protector** *m* CINEMAT backside coating; ~ **primario** *m* CARBÓN precoating, ÓPT, TELECOM primary coating; ~ **protector** *m* FÍS protective coating, MECÁ overlay cladding, ÓPT protective coating, REVEST protective coating, protective sheathing; ~ **protector antióxido** *m* REVEST protective oxide coat; ~ **protector galvanizado** *m*

REVEST galvanized protective coating; ~ **protector interior** *m* MECÁ liner, lining; ~ **a prueba de ácidos** *m* COLOR acid-proof coat; ~ **químico** *m* NUCL, REVEST chemical coating; ~ **realizado con soplete** *m* MECÁ, REVEST, TERMO flame-spray coating; ~ **realizado utilizando llama** *m* MECÁ, REVEST, TERMO flame-spray coating; ~ **refractario** *m* INSTAL TERM refractory coating, refractory lining, REVEST, TERMO fire-resistant coating, refractory coating, refractory lining; ~ **de resina acrílica** *m* REVEST acrylic resin coating; ~ **resistente** *m* P&C resist coating, resistance coating, REVEST resist coating, *fibras ópticas* resistance coating; ~ **resistente a la abrasión** *m* REVEST abrasion resistant coating; ~ **resistente al calor** *m* REVEST heat-resistant coating; ~ **resistente al desgaste** *m* REVEST wear-resistant coating; ~ **resistente a los álcalis** *m* SEG alkaline-resistant lining; ~ **resistente a los productos químicos** *m* QUÍMICA, REVEST chemical-resistant coating; ~ **con reverso transflexivo** *m* ELECTRÓN transflective-back coating; ~ **por rociado** *m* REVEST spray coating; ~ **secundario** *m* ÓPT secondary coating; ~ **selectivo** *m* ENERG RENOV selective coating; ~ **del silenciador** *m* AUTO muffler shell; ~ **de sílice** *m* TELECOM silica coating; ~ **de silicona** *m* ING ELÉC silicone cladding; ~ **suelto, abierto** *m* TEC ESP unbonded skin; ~ **de superficie** *m* CONST hardfacing; ~ **del tambor** *m* PROD dome casing, dome cover; ~ **tosco** *m* CONST rough cast; ~ **transparente** *m* REVEST transparent coating; ~ **a vacío** *m* REVEST vacuum-coated film; ~ **de vapor** *m* INSTAL HIDRÁUL steam casing; ~ **de yeso** *m* REVEST plaster coating; ~ **de yute** *m* ING ELÉC jute covering; ~ **de zinc** *m* REVEST zinc coating

revestir[1]: **sin** ~ *adj* ING ELÉC bare

revestir[2] *vt* CINEMAT coat, CONST line, sheet, FÍS lag, MINAS *sondeos* tub, P&C *pintura*, PAPEL coat, PROD *hornos* line, REVEST coat, line, *lentes* coat; ~ **de estaño** *vt* PROD *caja* line with tin; ~ **de ladrillo** *vt* CONST brick; ~ **de madera** *vt* CONST timber, board; ~ **de mampostería** *vt* MINAS *pozos de minas* wall

revisar *vt* IMPR revise, SEG check, TEC ESP inspect, TRANSP AÉR, TRANSP MAR *mantenimiento* overhaul

revisión *f* IMPR review, ING MECÁ check, TEC ESP review; ~ **del anteproyecto** *f* (*PDR*) TEC ESP preliminary-design review (*PDR*); ~ **de bits** *f* PROD *ordenadores* bit examining; ~ **de la configuración según la cláusula primera** *f* TEC ESP *gestión de proyectos* first-article configuration review (*FACR*); ~ **crítica de diseño** *f* (*RCD*) TEC ESP critical design review (*CDR*); ~ **de datos** *f* INFORM&PD data vet; ~ **del diseño intermedio** *f* TEC ESP intermediate design review; ~ **del diseño del sistema** *f* TEC ESP system design review (*SDR*); ~ **de la disponibilidad de lanzamiento** *f* TEC ESP launch readiness review; ~ **de disponibilidad de vuelo** *f* TEC ESP flight readiness review (*FRR*); ~ **final del diseño** *f* TEC ESP final design review (*FDR*); ~ **general** *f* NUCL major overhaul; ~ **de las herramientas** *f* PROD *ingeniería* bit examining; ~ **de la máquina** *f* TRANSP MAR engine overhaul; ~ **menor** *f* TRANSP AÉR minor check; ~ **del motor** *f* TRANSP MAR engine overhaul; ~ **postvuelo** *f* TRANSP AÉR post-flying check; ~ **preliminar de diseño** *f* TEC ESP preliminary-design review (*PDR*); ~ **de la prueba de calificación** *f* TEC ESP qualification test review

revisores: ~ **secundarios** *m pl* C&V secondary checkers

revista: ~ **de diseño intermedio** *f* TEC ESP intermediate design review

revolución *f* GEOM revolution, ING MECÁ revolution, turn, TEC ESP, VEH *velocidad de giro del motor* rev *n*, revolution; ~ **del eje** *f* ING MECÁ shaft lap; ~ **verde** *f* AGRIC green revolution

revoluciones: ~ **por minuto** *f pl* AUTO, CINEMAT, FÍS, VEH revolutions per minute

revólver *m* D&A gun

revolver *vt* FOTO agitate

revoque *m* CONST *albañilería* rendering, REVEST coater; ~ **de inyección** *m* PETROL *perforación* mud cake; ~ **de lodo** *m* PETROL *perforación* mud cake

revuelta *f* HIDROL *de un río* sweep

rezón *m* ING MECÁ, TRANSP MAR grapnel

rezumado *m* CARBÓN seepage

rezumamiento *m* HIDROL *fango* ooze

rezumar *vi* HIDROL percolate, TEXTIL seep through, TRANSP MAR leak

RF *abr* ACÚST, ELEC, ELECTRÓN TELECOM, TRANSP AÉR, TV *(frecuencia de radio, radiofrecuencia)* RF *(radio frequency)*

Rf *abr (ruterfordio)* FÍS RAD Rf *(rutherfordium)*

RFMA *abr (radiofaro omnidireccional de muy alta frecuencia)* TRANSP VHFO *(very high-frequency omnirange)*

RGB *abr (rojo verde azul)* TV RGB *(red green blue)*

Rh *abr (rodio)* QUÍMICA *elemento* Rh *(rhodium)*

rhäticita *f* MINERAL rhaetizite

ría *f* OCEAN firth, TRANSP MAR sea loch, *geografía* firth

riachuelo *m* HIDROL rill, streamlet, *pequeño curso de agua* run

riada *f* HIDROL freshet; ~ **repentina** *f* HIDROL flash flood

ribazo *m* GEOL, GEOFÍS, OCEAN coastal ridge

ribera *f* AGUA riverbank, CONST *lago* shore, *río* riverbank, shoreline, strand, bank, HIDROL *río* bank, shoreline, shore, strand, riverbank, OCEAN shore, shoreline, strand, *río* riverbank, bank, TRANSP MAR *lago* shore, shoreline, shore, *río* bank, riverbank; ~ **posterior** *f* OCEAN backshore

ribereño[1] *adj* AGUA, HIDROL, OCEAN, TRANSP MAR riparian, riverside

ribereño[2] *m* AGUA, HIDROL, OCEAN, TRANSP MAR riverside

ribeteado: ~ **del armazón** *m* ING MECÁ frame edging

ribeteadora *f* EMB flanging machine, ING MECÁ trimming machine

ribetear *vt* CONST trim

riboflavina *f* ALIMENT, QUÍMICA riboflavin

ribonucleico *adj* QUÍMICA *ácido* ribonucleic

ricia *f* AGRIC volunteer

ricinoleico *adj* QUÍMICA *ácido* ricinoleic

riebeckita *f* MINERAL riebeckite

riego *m* AGRIC, AGUA irrigation; ~ **con aguas cloacales** *m* AGUA broad irrigation; ~ **con aguas negras** *m* AGUA broad irrigation; ~ **por amelgas** *m* AGUA border irrigation; ~ **por aspersión** *m* HIDROL spray irrigation; ~ **por bordes** *m* AGUA border irrigation; ~ **por compartimiento** *m* AGRIC basin irrigation; ~ **directo** *m* HIDROL direct irrigation; ~ **por goteo** *m* AGRIC dribble irrigation, drip irrigation; ~ **por gravedad** *m* AGRIC gravity irrigation; ~ **por inundación** *m* AGRIC, AGUA, HIDROL flood irrigation; ~ **por inundación controlada** *m* AGRIC, AGUA, HIDROL flood-control irrigation; ~ **permanente** *m* AGRIC permanent irrigation; ~ **por surcos** *m* AGRIC, AGUA furrow irrigation; ~ **por tablares** *m* AGRIC border irrigation, gravity-check irrigation

riel *m* CONST, FERRO, ING ELÉC rail, ING MECÁ plate; ~ **acanalado** *m* FERRO grooved rail; ~ **aislado** *m* FERRO insulated rail; ~ **continuo soldado** *m* FERRO *equipo inamovible* continuous welded rail *(CWR)*; ~ **descamado** *m* FERRO flaked rail; ~ **DIN** *m* PROD DIN rail; ~ **de doble seta** *m* FERRO double-headed rail; ~ **de grúa** *m* CONST crane rail; ~ **paramagnético** *m* TRANSP paramagnetic rail; ~ **de sustentación** *m* TRANSP bearing rail

riesgo[1]: **sin** ~ **de fallo** *adj* INFORM&PD fail-safe

riesgo[2] *m* CALIDAD risk, SEG danger, hazard, risk, safety hazard, safety risk; ~ **del ambiente** *m* CALIDAD environmental risk; ~ **biológico** *m* SEG biological hazard; ~ **causado por ondas radioeléctricas** *m* SEG radio-wave hazard; ~ **causado por ultrasonidos** *m* SEG ultrasonic hazard; ~ **climatológico** *m* METEO, SEG climatic hazard; ~ **económico** *m* CALIDAD economic risk; ~ **eléctrico** *m* ELEC, SEG electrical hazard; ~ **de exposición** *m* FÍS RAD exposure risk, *de radiación* risk of exposure, NUCL exposure risk, SEG *de radiación* exposure risk, risk of exposure; ~ **de incendio** *m* CONST, SEG, TERMO, TRANSP AÉR, TRANSP MAR fire hazard; ~ **individual** *m* CALIDAD individual risk; ~ **mayor** *m* CALIDAD, SEG major hazard; ~ **mecánico** *m* MECÁ, SEG mechanical hazard; ~ **microbiológico** *m* SEG microbiological hazard; ~ **no mecánico** *m* SEG *incluyendo polvos y accidentes eléctricos* nonmechanical hazard; ~ **originado por la maquinaria** *m* SEG machinery hazard; ~ **originado por maquinaria de carpintería** *m* SEG woodworking machinery hazard; ~ **originado por sierra de cinta** *m* SEG narrow-bandsaw hazard; ~ **originado por soldadura** *m* SEG welding hazard; ~ **originado por vibración** *m* SEG vibration hazard; ~ **originado por vibraciones transmitidas corporalmente** *m* SEG body-transmitted vibration hazard; ~ **patógeno** *m* SEG pathogenic hazard; ~ **potencial** *m* SEG potential hazard; ~ **profesional** *m* SEG occupational hazard; ~ **químico** *m* QUÍMICA, SEG chemical hazard; ~ **de radiación** *m* SEG, TELECOM radiation hazard; ~ **de radiación láser** *m* ELECTRÓN, FÍS RAD, SEG laser radiation hazard; ~ **relativo** *m* MATEMÁT relative risk; ~ **sanitario** *m* SEG health hazard; ~ **social** *m* CALIDAD societal risk

RIF *abr (respuesta de impulso finito)* ELECTRÓN FIR *(finite impulse response)*

rifle *m* D&A rifle

rigidez *f* ACÚST rigidity, stiffness, FÍS, ING MECÁ, PAPEL resistance to bending, rigidity, stiffness, TEXTIL *tejido* stiffness; ~ **acústica** *f* ACÚST, FÍS acoustic stiffness; ~ **de batimiento del cubo** *f* TRANSP AÉR hub-flapping stiffness; ~ **dieléctrica** *f* ELEC *de capacitor* dielectric strength; ~ **específica** *f* TEC ESP specific stiffness; ~ **a la flexión** *f* FÍS, IMPR, MECÁ, NUCL flexural rigidity

rigidización *f* CONST, TEC ESP stiffening

rigidizador *m* TEC ESP, TRANSP MAR *geografía* stiffener

rígido *adj* CINEMAT hard, ING MECÁ set

rigor *m* PROD exactness

rinde *m AmL* AGRIC yield

riñón *m* CONST *arquitectura* haunch

río[1]: ~ **abajo** *adv* CONST, FÍS, FÍS FLUID, GAS, MECÁ, TRANSP MAR downstream; ~ **arriba** *adv* CONST, FÍS, FÍS FLUID, GAS, MECÁ, TRANSP MAR upstream

río[2]: ~ **anastomosado** *m* GEOL braided river; ~ **canalizado** *m* AGRIC, AGUA, HIDROL, TRANSP MAR river channel; ~ **de cauces interconectados** *m* PETROL braided stream; ~ **costero** *m* AGUA coastal river; ~ **decapitado** *m* AGUA beheaded river; ~ **joven** *m* HIDROL young river; ~ **natural** *m* HIDROL mature river; ~ **navegable** *m* HIDROL, TRANSP MAR navigable river, waterway; ~ **subterráneo** *m* HIDROL subterranean river, underground river

riolita *f* PETROL rhyolite

riostra *f* AGUA diagonal brace, CARBÓN strut, CONST brace, bridging piece, *estructura* tie beam, strut, *vigas de celosía* stringer, *viga* stay, ING MECÁ bracing truss, TEC ESP brace; ~ **diagonal** *f* CONST angle brace; ~ **transversal** *f* TRANSP MAR cross brace, *construcción naval* cross tie; ~ **travesal guía** *f* ING MECÁ guide cross tie

ripidolita *f* MINERAL ripidolite

ripio *m* CONST *grava* gravel, *albañilería* chip, GEOL *vidrio volcánico* shard, ING MECÁ *de piedra* chip, PETROL rubble

ripios *m pl* (*cf desechos AmL*) GAS drill cuttings, attle, cuttings

riple *m* GEOL ripple mark; ~ **de corriente** *m* GEOL *estructura sedimentaria* current ripple, ripple mark, ~ **de interferencia** *m* GEOL *estructura sedimentaria* interference ripple

riqueza *f* CARBÓN *de un mineral* assay value, yield, *de un filón* strength, MINAS strength, *pozo* yield; ~ **en mineral** *f* MINAS ore contents; ~ **en peso** *f* MINAS weight strength (*WS*)

riser *m* TEC PETR *perforación costa-fuera* marine riser

ritmo *m* ACÚST rhythm, tempo, CINEMAT *de una secuencia* pace, TEC ESP pulse; ~ **de afluencia de oleadas de tráfico** *m* TRANSP tidal flow; ~ **de afluencia de tráfico** *m* TRANSP tidal flow; ~ **de agotamiento** *m* PETROL, TEC PETR depletion rate, pumping rate; ~ **de defecto** *m* CALIDAD failure rate; ~ **de digitalización** *m* ELECTRÓN, INFORM&PD, TELECOM digitizing rate; ~ **de falla** *m* CALIDAD failure rate; ~ **de fallas instantáneas** *m* CALIDAD instantaneous failure rate; ~ **de ganancia de peso** *m* AGRIC rate of gain; ~ **del lapso adiabático** *m* TRANSP AÉR adiabatic lapse rate; ~ **del patrón** *m* PROD clock rate; ~ **de transformación** *m* CONTAM *procesos químicos* transformation rate

rizado[1] *adj* TEXTIL *papel* crisp

rizado[2] *m* ACÚST, TEC ESP ripple, TEXTIL crimp, curling

rizamiento: ~ **foliar** *m* AGRIC leaf curl

rizar *vt* (*cf plegar*) C&V curl, crimp, *vela* reef

rizo *m* ELECTRÓN *de vector* curl, MINAS *geología* crimping, TEXTIL curl, TRANSP MAR *vela* reef; ~ **de agua** *m* CONST ripple

rizoma *m* AGRIC rhizome

rizósfera *f* AGRIC *suelos* rhizosphere

RLS *abr* (*radiobaliza de localización de siniestros*) TRANSP MAR EPIRB (*emergency position-indicating radio beacon*)

RMP *abr* (*pasta mecánica de refino*) PAPEL RMP (*refiner mechanical pulp*)

Rn *abr* (*radón*) FÍS, FÍS RAD, QUÍMICA Rn (*radon*)

roblón *m* CONST, ING MECÁ, MECÁ, PROD rivet

roblonado *m* CONST, ING MECÁ, MECÁ, PROD riveting

roblonadura *f* CONST, ING MECÁ, MECÁ, PROD riveting

robo *m* SEG burglary; ~ **de ciclo informático** *m* INFORM&PD cycle stealing

robos: contra ~ *adj* SEG antitheft

robot *m* INFORM&PD robot; ~ **industrial** *m* SEG *medidas de seguridad* industrial robot; ~ **industrial para manipulación** *m* ING MECÁ manipulating industrial robot; ~ **industrial manipulador** *m* ING MECÁ manipulating industrial robot; ~ **de montaje** *m* MECÁ assembly robot

robótica *f* INFORM&PD robotics

roca *f* CONST stone, MINAS stone, *mina de carbón* metal; ~ **ácida** *f* GEOL acid rock, acidic rock, QUÍMICA acidic rock; ~ **almacén** *f* GAS reservoir rock, GEOL host rock, reservoir rock; ~ **de almacenamiento** *f* GAS, GEOL reservoir rock; ~ **biogénica** *f* PETROL biogenic rock; ~ **de Boundstone** *f* GEOL *petrología química* boundstone; ~ **carbonatada** *f* GEOL *con más del 10% de granos de lodo en una micrita* wackestone; ~ **carbonatada de granos y matriz micrítica** *f* GEOL packstone; ~ **carbonatada micrítica** *f* GEOL mudstone; ~ **carbonífera** *f* CARBÓN, GEOL, MINAS coal-bearing rock; ~ **de cobertura impermeable** *f* GAS watertight caprock; ~ **cristalina laminada** *f* GEOL foliated crystalline rock; ~ **de cubierta** *f* ENERG RENOV, GEOL, MINAS, TEC PETR *geología* cap rock; ~ **desintegrada** *f* MINERAL rock meal; ~ **encajante** *f* GEOL country rock; ~ **escarpada** *f* GEOL scar; ~ **escarpada en pendiente** *f* GEOL slip scar; ~ **estéril** *f* CARBÓN dirt, MINAS *geología* dead ground, *preparación de minerales* waste rock, *Esp* (*cf roca de los respaldos AmL*) *filones* country rock; ~ **extrusiva** *f* ENERG RENOV, GEOL extrusive rock; ~ **extrusora** *f* ENERG RENOV, GEOL extrusive rock; ~ **ferruginosa** *f* GEOL ironstone; ~ **fosfática** *f* PETROL phosphate rock; ~ **fragmentaria** *f* GEOL fragmental rock; ~ **granular** *f* GEOL grainstone; ~ **hialoclástica** *f* GEOL hyaloclastic rock; ~ **hipoabisal** *f* GEOL hypabyssal rock; ~ **ígnea** *f* ENERG RENOV, GEOL, PETROL igneous rock; ~ **ígnea estratificada** *f* GEOL layered igneous rock; ~ **improductiva** *f* ENERG RENOV, MINAS, TEC PETR *recubre un filón* cap rock; ~ **intermedia** *f* GEOL intermediate rock; ~ **laminar** *f* GEOL flagstone; ~ **machacada** *f* CONST crushed stone; ~ **madre** *f* MINAS country rock, TEC PETR *geología* source rock; ~ **madre del diamante** *f* MINAS diamond matrix; ~ **metabásica** *f* GEOL metabasic rock; ~ **metamórfica** *f* ENERG RENOV metamorphic rock; ~ **no aurífera** *f* MINAS mullock; ~ **piroclástica** *f* GEOFÍS, GEOL, PETROL pyroclastic rock; ~ **de playa** *f* GEOL beach rock, OCEAN arenite, beach rock, sandstone; ~ **plutónica** *f* GEOFÍS *petrología* plutonic rock; ~ **productiva** *f* GAS, GEOL reservoir rock; ~ **recién extraída de la cantera** *f* CONST, GEOL greenstone; ~ **de los respaldos** *f AmL* (*cf roca estéril Esp, roca madre Esp*) MINAS *filones* stone, country rock; ~ **de sal** *f* GEOL salt rock; ~ **salina** *f* GEOL salt rock; ~ **sedimentaria** *f* CARBÓN, GEOL, TEC PETR sedimentary rock; ~ **silícea no detrítica** *f* GEOL siliceous nondetrital rock; ~ **siliciclástica** *f* GEOL siliciclastic rock; ~ **subyacente** *f* GEOL underlying rock; ~ **del techo** *f* CARBÓN *minas*, MINAS *de mina, de filón* roof; ~ **terrígena clástica** *f* GEOL clastic terrigenous rock;

~ **de trípoli** *f* GEOL tripoli stone; ~ **ultrabásica** *f* GEOL ultrabasic rock; ~ **ultramáfica** *f* GEOL ultramafic rock; ~ **verde** *f* CONST, GEOL greenstone; ~ **verdosa** *f* CONST, GEOL greenstone; ~ **volada** *f* CARBÓN blasted stone; ~ **volcánica** *f* TEC PETR volcanic rock; ~ **de yeso** *f* CONST plaster rock; ~ **zoogénica** *f* TEC PETR zoogenic rock

rocalla *f* AGUA rock fill, MINAS rock work

rocas: ~ **clásticas** *f pl* GEOL clastic rocks; ~ **detríticas** *f pl* GEOL clastic rocks; ~ **rudíticas** *f pl* GEOL rudaceous rocks

roce *m* CONST, ING MECÁ attrition, MECÁ fretting

rociado *m* AGUA, QUÍMICA sprinkling; ~ **con hormigón proyectado** *m* CONST spray with shotcrete; ~ **térmico** *m* REVEST thermal spraying; ~ **de tijeras** *m* C&V shear spray; ~ **de yeso para estucado** *m* REVEST plast spraying

rociador *m* AGUA sprinkler, ALIMENT *máquina para proceso de alimentos* sparge pipe, CONST *contraincendios* sprinkler, PAPEL sprayer, shower, QUÍMICA sprayer, SEG *incendios* sprinkler; ~ **de alquitrán** *m* CONST tar sprinkler; ~ **de alta presión** *m* PAPEL knock-off shower; ~ **antiestático** *m* INFORM&PD antistatic spray; ~ **antimaculante** *m* IMPR anti-offset spray; ~ **automático contraincendios** *m* TRANSP MAR automatic fire sprinkler; ~ **contra incendios** *m* SEG, TERMO fire sprinkler; ~ **giratorio** *m* PAPEL rotating shower; ~ **oscilante** *m* PAPEL oscillating shower; ~ **de refilar** *m* PAPEL trim shower

rociadura *f* AGUA sprinkling

rociar *vt* AGUA sprinkle, CINEMAT spray down, CONTAM MAR spray, QUÍMICA sprinkle; ~ **con bórax** *vt* PROD flux

rocío *m* METEO, REFRIG dew, TRANSP MAR spray; ~ **del mar** *m* METEO, OCEAN spindrift; ~ **salino** *m* P&C *prueba de pintura* salt spray

roda[1]: **de** ~ **limpia** *adj* TRANSP MAR *buque* bluff-bowed

roda[2] *f* TRANSP MAR *construcción naval* stem; ~ **con lanzamiento** *f* TRANSP MAR *construcción naval* raking stem

rodada *f* CONST *carreteras* rut

rodadura *f* TRANSP AÉR taxiing

rodaje *m* AUTO running-in, CINEMAT shooting, ING MECÁ breaking-in, TRANSP AÉR *sobre el suelo* taxiing; ~ **diario** *m* CINEMAT dailies; ~ **fuera del estudio** *m* CINEMAT location shooting; ~ **marcha atrás** *m* CINEMAT back run; ~ **de la película** *m* CINEMAT film shoot; ~ **con playback** *m* CINEMAT shooting to playback; ~ **con sincronización sonora a una banda previamente grabada** *m* CINEMAT shooting to playback; ~ **sincronizado** *m* CINEMAT synchronized shooting

rodamen: ~ **de apoyo** *m* TRANSP landing gear

rodamiento *m* CONST rolling, FÍS roll, TRANSP MAR engine bearing, VEH bearing; ~ **de agujas** *m* ING MECÁ needle bearing; ~ **de bolas** *m* ING MECÁ, VEH ball bearing; ~ **de bolas de construcción fuerte** *m* ING MECÁ heavy-duty ball bearing; ~ **de eje primario** *m* VEH *embrague* input shaft bearing; ~ **de empuje axial** *m* ING MECÁ axial thrust bearing; ~ **de rodillos** *m* ING MECÁ, VEH roller bearing; ~ **de rodillos de acción radial** *m* ING MECÁ radial cylindrical roller bearing; ~ **de rodillos cónicos** *m* VEH tapered roller bearing; ~ **de rueda** *m* VEH wheel bearing

rodapié *m* CONST base board (*AmE*), skirting board

(*BrE*), washboard; ~ **de madera** *m* CONST mop board

rodar[1] *vt* VEH *hacer funcionar un motor* run in

rodar[2] *vi* TRANSP AÉR *sobre el suelo* taxi

rodear *vt* CONST brace, ING MECÁ encompass

rodeo *m* AmL (*cf rebaño Esp*) AGRIC herd

rodete *m* ENERG RENOV *turbina*, ING MECÁ runner, PROD *turbina, bomba* runner, *ventiladores* wheel

rodicita *f* MINERAL rhodizite

rodillera *f* SEG knee pad

rodillo *m* AGRIC, C&V, CINEMAT, CONST, IMPR roller, INFORM&PD platen, MECÁ drum, roller, P&C *parte de equipo* roller, PAPEL roll, cylinder, PROD *tren laminador*, TEXTIL roll; ~ **abrillantador** *m* PAPEL glazing roller; ~ **acanalado** *m* ING MECÁ grooved roller; ~ **de acarreo** *m* C&V carrying roller; ~ **accionado** *m* PAPEL follower roll; ~ **de acondicionamiento de la tela** *m* PAPEL wire drive roll; ~ **afiligranador** *m* PAPEL dandy roll, watermark roll; ~ **con agujeros ciegos** *m* PAPEL blind drill roll; ~ **de alimentación** *m* ALIMENT *maquinaria para proceso de alimentos* feed roller, PAPEL feed roll, feeding roll; ~ **alimentador** *m* ALIMENT feed roll, IMPR infeed roller, ING MECÁ feed roll, feed roller; ~ **alisador** *m* PAPEL smoothing roll; ~ **de alisar** *m* CONST smooth roller; ~ **amortiguado** *m* CINEMAT pad roller; ~ **anilox** *m* IMPR *flexografía* anilox roll; ~ **aplicador** *m* EMB application roller, IMPR lay-on roller, PAPEL applicator roll; ~ **de apoyo** *m* IMPR backing roll, ING MECÁ carrying idler; ~ **de arrastre** *m* CINEMAT drag roller, IMPR pull roller; ~ **de avance** *m* ALIMENT feed roll, IMPR forwarding roller, ING MECÁ feed roll, feed roller; ~ **de bagazo** *m* ALIMENT bagasse roller; ~ **de barro** *m* C&V clay roller; ~ **de borde** *m* C&V edge roll; ~ **cabecero** *m* PAPEL breast roll; ~ **de la caja de entrada** *m* PAPEL roll headbox; ~ **de la calandra** *m* PAPEL calender bowl, calender roll; ~ **cepillador** *m* PAPEL brush-roller; ~ **de cojinetes de bolas** *m* PROD ball-bearing roller; ~ **de cola de la matriz** *m* IMPR pan roller; ~ **compactador** *m* AGRIC cultipacker, land packer; ~ **compensador del combado** *m* PAPEL swimming roll; ~ **conductor** *m* PAPEL guide roll, leading roll; ~ **conductor del fieltro** *m* PAPEL felt-carrying roll; ~ **de costuras** *m* ING MECÁ seam roller; ~ **de cuchilla** *m* PAPEL knife cylinder; ~ **curvado** *m* PAPEL bowed roll; ~ **de curvar** *m* C&V *método Libbey-Owens* bending roller, ING MECÁ bending roll; ~ **para curvar chapas** *m* ING MECÁ plate-bending roller; ~ **delantero** *m* PROD front-facing roller; ~ **desbrozador** *m* ING MECÁ breaking down; ~ **desgotador** *m* PAPEL dandy roll, watermark roll; ~ **desgotadora** *m* PAPEL dewatering roll; ~ **desmenuzador** *m* AGRIC land roller; ~ **desplegador** *m* PAPEL spreader roll, *fieltro* spiral felt roll; ~ **distribuidor** *m* PAPEL distributor roll, holey roll; ~ **de dos pestañas** *m* PROD *de puente-grúa* double-flanged traveling wheel (*AmE*), double-flanged travelling wheel (*BrE*); ~ **de empuje lateral** *m* PROD side-push roller; ~ **encolador** *m* PAPEL size roll; ~ **enfriador** *m* IMPR chill roll; ~ **de enfriamiento** *m* P&C *parte de equipo* chill roll; ~ **entintador** *m* IMPR ink-form roller; ~ **para entintar a mano** *m* IMPR brayer; ~ **de entrada** *m* IMPR, PAPEL leading-in roll; ~ **escurridor** *m* IMPR squeegee roller, PAPEL squeeze roll; ~ **para evitar el combado** *m* PAPEL antideflection roll; ~ **exprimidor** *m* PAPEL wringer roll, TEXTIL squeezing roller; ~ **para forjar redondo** *m* PROD forge roll; ~ **formador** *m* PAPEL forming roll; ~ **de fricción** *m* PAPEL, PROD friction roller; ~ **giratorio por gravedad** *m* EMB *sistemas transportadores* gravity roller; ~ **gofrador** *m* EMB, IMPR, PAPEL embossing roll; ~ **de gran potencia** *m* ING MECÁ heavy-section roll; ~ **de granulación** *m* PROC QUÍ granulating roller; ~ **guía** *m* CINEMAT guide roller, PAPEL guide roll, leading roll, PROD guide roller; ~ **guía del fieltro** *m* PAPEL fly roll; ~**-guía del fieltro** *m* PAPEL felt roll; ~ **guía de la tela** *m* PAPEL wire guide roll; ~**-guía de tela** *m* PAPEL wire roll; ~ **hacia atrás** *m* PROD rear-facing roller; ~ **para el hilo** *m* TEXTIL yarn roll; ~ **de huecograbado** *m* IMPR gravure roller; ~ **humectador** *m* IMPR dip roller, PAPEL dampener; ~ **igualador** *m* PAPEL *estucadoras* evener roll; ~ **impregnador** *m* TEXTIL quetch roller; ~ **impresor** *m* IMPR printing roller, P&C *parte de equipo* impression roller; ~ **inactivo** *m* TV idle roller; ~ **inferior** *m* ING MECÁ lower roll, PAPEL king roll; ~ **de leva** *m* MECÁ follower; ~ **de línea de contacto blanda** *m* PAPEL soft nip roll; ~ **liso** *m* CONST smooth roller; ~ **manchón aspirante** *m* PAPEL suction couch roll; ~ **para marcas de agua** *m* PAPEL dandy roll; ~ **mojador** *m* IMPR dipping roller; ~ **molturador** *m* ALIMENT *molienda, panadería, repostería* break roller; ~ **no radial** *m* PROD offset roller; ~ **oscilador** *m* IMPR vibrator; ~ **de palanca de horquilla** *m* PROD fork-lever roller; ~ **de pata de cabra** *m* CONST *maquinaria* sheepsfoot roller; ~ **perfilador** *m* CARBÓN *muelas abrasivas* crusher; ~ **perfilador giratorio** *m* ING MECÁ swinging crusher; ~ **de perfilar** *m* ING MECÁ profiling roller; ~ **perforado** *m* PAPEL holey roll; ~ **de pestaña de la rueda** *m* ING MECÁ wheel flange roller; ~ **plegador** *m* IMPR folding roller; ~ **de plomo** *m* IMPR lead roll; ~ **de la prensa** *m* IMPR, PAPEL press roll; ~ **de prensa húmeda** *m* PAPEL wet-press roll; ~ **de prensa simple** *m* PAPEL plain-press roll; ~ **prensador** *m* ACÚST, C&V, ING MECÁ pressure roller; ~ **prensor** *m* PAPEL rider roll; ~ **de presión** *m* ACÚST pressure roller, CINEMAT pinch roller, pressure roller, IMPR pinch roller, ING MECÁ pressure roller; ~ **pulverizador** *m* AGRIC soil pulverizer; ~ **ranurado** *m* PAPEL grooved roll; ~ **de rasqueta** *m* PAPEL doctor roll; ~ **de rebabas** *m* ING MECÁ seam roller; ~ **rectificador** *m* PAPEL *caja de cabeza de máquina* rectifier roll; ~ **regulador** *m* IMPR metering roller; ~ **con resaltes en espiral** *m* PAPEL worm roll; ~ **de residuo** *m* ALIMENT bagasse roller; ~ **de retorno** *m* ING MECÁ return roller; ~ **de retorno de tela** *m* PAPEL wire-return roll; ~ **de rodadura** *m* MECÁ, PROD runner; ~ **de rotorno de la tela** *m* PAPEL forward-drive roll; ~ **secador** *m* IMPR roller dryer; ~ **soplador** *m* PAPEL air roll, blow roll; ~ **soplante** *m* PAPEL air roll; ~ **soportante** *m* PAPEL *bobinadora* bed roll; ~ **de soporte** *m* PAPEL backing roll; ~ **superior** *m* PAPEL top roll; ~ **superior de la prensa del manchón** *m* PAPEL lump breaker roll; ~ **de tensión** *m* ING MECÁ binding pulley, jockey-roller, jockey-wheel, PAPEL tension roller; ~ **tensor** *m* CINEMAT tension roller, PAPEL hitch roll, stretch roll; ~ **del tintero** *m* IMPR ink fountain roller; ~ **de tiro** *m* PAPEL draw roll; ~ **tomador** *m* PAPEL pick-up roll; ~ **tomador en la forma redonda** *m* PAPEL *máquina de cartón* baby press; ~ **tramado** *m* IMPR screen roller; ~ **de trituración** *m* PROC QUÍ crushing roll; ~ **triturador** *m*

ALIMENT break roller, ING MECÁ, PROD crushing roll; ~ **de troncos** *m* AGRIC log roller; ~ **unidireccional** *m* PROD one-way roller; ~ **de vaciado** *m* C&V casting roller; ~ **de varias partes** *m* C&V split roller; ~ **de ventilación de las bolsas secadoras** *m* PAPEL pocket-dryer ventilation roll; ~ **vibratorio** *m* CONST vibrating roller

rodio *m* (*Rh*) QUÍMICA rhodium (*Rh*)

rodita *f* MINERAL rhodite

rodo *m* PROD *herramienta picafuegos* rake

rodocrosita *f* MINERAL rhodochrosite

rodonita *f* MINERAL rhodonite

ROE *abr* (*relación de onda estacionaria*) FÍS, TEC ESP, TELECOM SWR (*standing-wave ratio*)

roentgen *m* (*R*) FÍS, FÍS RAD roentgen (*R*)

rojo[1]: **en** ~ *adj* TRANSP red; **al** ~ **blanco** *adj* TERMO white-hot; **al** ~ **vivo** *adj* TERMO, TERMOTEC red-hot

rojo[2]: ~ **de Inglaterra** *m* PROD rouge; ~ **primario** *m* TV red primary; ~ **de pulir** *m* PROD *química* rouge; ~ **verde azul** *m* (*RGB*) INFORM&PD, TV red green blue (*RGB*)

rolada *f* TRANSP MAR *del viento* backing

rolado *m* C&V rolling process; ~ **de vela en el mármol** *m* C&V marvering

rolar *vi* TRANSP MAR *cargamento* shift, *viento* veer

roldana *f* ING MECÁ cone sheave, pulley, pulley sheave, pulley wheel, PROD sheaf, TRANSP MAR pulley, *accesorios* sheave; ~ **circular para soldadura continua** *f* ING MECÁ seam welding wheel blank; ~ **de eje** *f* CONST axle pulley; ~ **pivotante** *f* PROD *para muebles* caster; ~ **del tambor** *f* CONST drum washer

rollo *m* CINEMAT roll, CONTAM MAR reel, EMB coil, reel, FOTO cartridge, ING MECÁ, MECÁ coil, PAPEL *de papel* reel; ~ **de cable** *m* EMB cable reel; ~ **matriz** *m* CINEMAT parent roll; ~ **de papel gráfico** *m* ELEC *registro* chart strip; ~ **para rayos X** *m* INSTR *película* X-ray cartridge; ~ **de tramas** *m* CINEMAT matte roll

ROM[1] *abr* (*memoria de sólo lectura*) ELEC, INFORM&PD, ING ELÉC ROM (*read-only memory*)

ROM[2]: ~ **borrable eléctricamente** *f* (*ROMBE*) INFORM&PD electrically-erasable ROM (*EEROM*); ~ **óptica** *f* INFORM&PD optical ROM

romana *f* FÍS steelyard, ING MECÁ lever balance, lever scales, scale; ~ **de máquina herramienta** *f* ING MECÁ machine tool scales; ~ **de muelle** *f* ING MECÁ spring balance

romaneo *m* TRANSP MAR *de barco, cargamento* trim

ROMBE *abr* (*ROM borrable eléctricamente*) INFORM&PD EEROM (*electrically-erasable ROM*)

rómbico *adj* CRISTAL orthorhombic, GEOM rhomboidal

rombo *m* GEOM rhomb, rhombus

romboédrico *adj* CRISTAL rhombohedral, trigonal

romboidal *adj* GEOM rhomboidal

romboide[1] *adj* GEOM rhomboid

romboide[2] *m* GEOM rhomboid

romeíta *f* MINERAL romeite

romo *adj* ING MECÁ blunt

rompedor[1] *adj* MINAS *explosivos* disruptive

rompedor[2] *m* CARBÓN splitter; ~ **de camellón** *m* AGRIC ridge buster; ~ **de caminos** *m* CONST ripper; ~ **de concreto** *m* AmL (*cf rompedor de hormigón Esp*) CONST concrete breaker; ~ **de hormigón** *m* Esp (*cf rompedor de concreto AmL*) CONST concrete breaker

rompegrumos *m* AmL PAPEL lump breaker roll

rompehielos *m* OCEAN ice breakup, TRANSP MAR *buque* icebreaker

rompelingotes *m* METAL, PROD pig-iron breaker

rompeolas *m* AGUA breakwater, CONST, OCEAN, TRANSP MAR breakwater, jetty

romper¹ *vt* CONST break, rip, burst, METAL, QUÍMICA rupture, SEG break

romper² *vi* CONST break, fail, P&C *prueba* tear, TRANSP MAR *onda* break; ~ **la canasta** *vi* TRANSP MAR *bandera* break

romperse *v refl* CONST break open, ING MECÁ come away, come off, part, MECÁ, PROD fail, TRANSP MAR *onda* break; ~ **bruscamente** *v refl* PROD snap

rompetuercas *m* ING MECÁ nut splitter

rompevientos: ~ **de playa** *m* (*RDP*) OCEAN blow-out preventer (*BOP*)

rompevirutas *m* ING MECÁ chip breaker

rompiente *m* OCEAN, TRANSP MAR *oleaje* breaker, roller

rompimiento *m* (*cf apertura AmL*) CALIDAD fracture, MINAS forcing, *galerías* holing, OCEAN breaking

roña *f* ALIMENT *fitopatología* scab

roñada *f* TRANSP MAR *cuerdas* eyelet hole, grommet

ronquido *m* AUTO *del motor* rumble

ronzar *vt* TRANSP MAR *bolinas* haul in

rood *m* METR *medida de superficie equivalente a 0.25 acres* rood

ROP *abr* (*tasa de penetración*) TEC PETR ROP (*rate of penetration*)

ropa: ~ **aceitada** *f* TRANSP MAR diving suit, oilskin, wetsuit; ~ **blanca crujiente** *f* TEXTIL *del hogar* crisp linen; ~ **de hogar** *f* TEXTIL linen clothing; ~ **de laboratorio** *f* SEG laboratory clothing; ~ **de trabajo** *f* CONST working clothes, SEG industrial clothing, TEXTIL working clothes; ~ **de vestir** *f* TEXTIL formal wear

rosa: ~ **de la aguja** *f* TRANSP MAR compass card; ~ **de la brújula** *f* INSTR compass card; ~ **de las corrientes** *f* OCEAN current rose; ~ **de los vientos** *f* ENERG RENOV, METEO wind rose, TRANSP MAR compass card

rosario: ~ **de draga** *m* MINAS, TRANSP MAR dredge chain

rosca *f* ING MECÁ, MECÁ screw, thread, worm, PROD *de tornillo* fillet, thread; ~ **Acme** *f* ING MECÁ Acme thread; ~ **americana** *f* ING MECÁ Sellers thread; ~ **cónica NPT** *f* (*rosca cónica de tubos American National*) ING MECÁ American NPT (*American National Taper Pipe thread*); ~ **cónica de tubos American National** *f* (*rosca NPT*) ING MECÁ American National Taper Pipe thread (*American NPT*); ~ **derecha** *f* ING MECÁ right-hand thread; ~ **estándar de los Estados Unidos** *f* ING MECÁ US standard thread; ~ **de estándar internacional** *f* ING MECÁ international standard thread; ~ **exterior** *f* ING MECÁ outside thread; ~ **sin fin** *f* EMB continuous thread; ~ **fina de gas** *f* GAS, ING MECÁ, PROD gas thread; ~ **en forma de trapecio** *f* CONST buttress thread; ~ **de gas** *f* GAS, ING MECÁ, PROD gas thread; ~ **guía** *f* PROD guide stock; ~ **hacia la derecha** *f* ING MECÁ right-hand thread; ~ **hembra** *f* ING MECÁ female thread; ~ **hundida** *f* C&V depressed thread; ~ **interior** *f* ING MECÁ female thread, inside screw, inside thread; ~ **macho** *f* ING MECÁ male thread; ~ **metalizada** *f* REVEST metal-coated thread; ~ **métrica de paso fino** *f* ING MECÁ metric fine-pitch thread; ~ **NPT** *f* (*rosca cónica de tubos American National*) ING MECÁ American NPT (*American*

National Taper Pipe thread); ~ **de paso fino** *f* ING MECÁ close-fit thread; ~ **de paso fino unificado** *f* (*UNF*) ING MECÁ unified fine thread (*UNF*); ~ **de paso de precisión** *f* ING MECÁ close-fit thread; ~ **de perfil trapezoidal** *f* ING MECÁ buttress screw thread; ~ **de poca altura** *f* ING MECÁ thread undercut; ~ **de poco diámetro** *f* ING MECÁ thread undercut; ~ **Sellers** *f* ING MECÁ Sellers thread; ~ **de tornillo** *f* ING MECÁ screw thread; ~ **de tornillo en miniatura** *f* ING MECÁ miniature screw thread; ~ **de tornillo NPT** *f* ING MECÁ NPT screw thread; ~ **de tornillo de una pulgada** *f* ING MECÁ inch screw thread; ~ **trapezoidal** *f* ING MECÁ trapezoidal thread; ~ **de tubería** *f* CONST pipe threading; ~ **de tubería de gas** *f* GAS, ING MECÁ, PROD gas thread; ~ **para tuberías paralela** *f* ING MECÁ parallel pipe-thread; ~ **de tubo** *f* ING MECÁ pipe thread, screw thread for pipework; ~ **de tubo Brigg** *f* ING MECÁ American standard pipe thread, Brigg's pipe thread; ~ **de tubo cónico** *f* ING MECÁ taper pipe thread; ~ **para tubos cónicos** *f* ING MECÁ national pipe thread (*NPT*); ~ **de uso general** *f* ING MECÁ general-purpose screw thread; ~ **de 29 grados** *f* ING MECÁ Acme thread; ~ **Whitworth** *f* ING MECÁ Whitworth screw thread, Whitworth thread

roscado¹ *adj* MECÁ threaded

roscado² *m* ING MECÁ chasing, screw cutting, screw-cutting lathe, *con machos* tapping, threading, TEC PETR *tuberías* screwing; ~ **externo** *m* ING MECÁ external thread

roscadora *f* ING MECÁ screw machine, threader, MECÁ screw machine; ~ **-aterrajadora** *f* ING MECÁ screwing-and-tapping machine; ~ **automática** *f* MECÁ automatic screw machine; ~ **de pernos y tuercas** *f* PROD bolt-screwing and nut-tapping machine; ~ **de terrajas** *f* ING MECÁ die-threading machine; ~ **de tubos** *f* CONST pipe threader, ING MECÁ tube-screwing machine

roscadura *f* ING MECÁ screwing

roscar *vt* MECÁ screw, PROD tap, *de tornillos* thread, *remaches* chase; ~ **con macho** *vt* PROD tap

roseta *f* AGRIC sandbur, C&V rosette; ~ **para cielo raso** *f* ELEC ceiling rose

rosetón: ~ **de techo** *m* ING ELÉC ceiling rose

rotación *f* AUTO *de la rueda* spinning, CONST slewing, CRISTAL rotation, ELECTRÓN circular shift, FÍS, FÍS RAD rotation, GEOM rotation, revolution, INFORM&PD circular shift, ING MECÁ revolution, turn, ÓPT rotation, PROD *de mercancías* turnover, QUÍMICA *óptica, estereoquímica* rotation, TEC ESP revolution, rotation, spin; ~ **según las agujas del reloj** *f* ALIMENT, MECÁ clockwise rotation; ~ **anticiclónica** *f* METEO anticyclonic circulation, anticyclonic rotation; ~ **ciclónica** *f* METEO cyclonic rotation; ~ **de cultivos** *f* AGRIC crop rotation, rotation; ~ **en dirección contraria a las agujas del reloj** *f* MECÁ anticlockwise rotation, counterclockwise rotation, PROD ccw rotation, counterclockwise rotation; ~ **de ejes coordenados** *f* GEOM rotation of coordinate axes; ~ **específica** *f* FÍS *potencia de rotación* specific rotation; ~ **de las existencias** *f* PROD *comercio* inventory turnover, stock turnover; ~ **fuera del plano normal al eje de rotación** *f* TEC ESP wobble; ~ **de izquierda a derecha** *f* ALIMENT, MECÁ clockwise rotation; ~ **a izquierdas** *f* MECÁ anticlockwise rotation, counterclockwise rotation, PROD ccw rotation, counterclockwise rotation; ~ **del motor por**

medio de manivela *f* TRANSP AÉR cranking;
~ siniestrosa *f* MECÁ anticlockwise rotation, counterclockwise rotation, PROD ccw rotation,
counterclockwise rotation

rotacional[1] *adj* GEN rotational

rotacional[2] *m* FÍS, FÍS FLUID *vector* curl

rotador: **~ forzado** *m* AUTO positive-type valve rotator;
~ de válvula *m* AUTO valve rotator; **~ de válvula tipo
positivo** *m* AUTO positive-type valve rotator

rotámetro *m* AGUA, CARBÓN, CONST, ELEC flowmeter,
FÍS flowmeter, rotameter, HIDROL, ING ELÉC, INSTAL
HIDRÁUL flowmeter, INSTR rotameter, LAB flow-
meter, rotameter, PAPEL, PROC QUÍ flowmeter, TEC
PETR flowmeter, *tuberías* rotameter, TERMO flow-
meter

rotativa *f* IMPR reel-feed press, rotary printing
machine, rotary unit; **~ de bobina** *f* IMPR web press;
~ offset de bobina *f* IMPR web-fed offset rotary
press; **~ de periódicos** *f* IMPR newspaper rotary
press; **~ de varios colores** *f* IMPR multicolor rotary
printing machine (*AmE*), multicolour rotary printing
machine (*BrE*)

rotativo *adj* GEN rotational

rotatorio: **~ de ferrita** *m* ING ELÉC ferrite rotator

rotavapor *m* INSTR, LAB *material de vidrio* rotary
evaporator

rotenona *f* QUÍMICA rotenone

rotoazada *f* AGRIC rotary hoe, shovel cultivator

rotocriba *f* CARBÓN, CONST, MINAS, PROD rotary screen

rotofita *f* MINERAL rothoffite

rotonda *f* AUTO, CONST, TRANSP rotary (*AmE*), roundabout (*BrE*), traffic circle (*AmE*), VEH rotary (*AmE*),
roundabout (*BrE*)

rotopala *f* CONST bucket excavator, MINAS bucket-
chain excavator, bucket-wheel excavator

rotor *m* AUTO, CARBÓN, ELEC, FÍS rotor, ING ELÉC rotor,
armature, ING MECÁ runner, MECÁ impeller, PROD
tornos drum, TRANSP AÉR rotor, impeller, VEH *de
bomba* impeller, *encendido, generador* rotor; **~ alado
de motor a reacción** *m* TRANSP AÉR jet-flapped
rotor; **~ de anillo colector** *m* ELEC *generador* slip-
ring rotor; **~ de anillo rozante** *m* ELEC *generador*
slip-ring rotor; **~ de anillos colectores
cortocircuitados** *m* ELEC *generador* short-circuited
slip-ring rotor; **~ antipar** *m* TRANSP *helicópteros*
antitorque rotor; **~ antitorsión de cola** *m* TRANSP
AÉR antitorque propeller; **~ articulado** *m* TRANSP AÉR
articulated rotor; **~ auxiliar** *m* TRANSP AÉR auxiliary
rotor; **~ bifásico** *m* ING ELÉC two-phase rotor;
~ bobinado *m* ING ELÉC wound rotor; **~ centrífugo**
m LAB *separación* centrifuge rotor; **~ de cola** *m*
TRANSP AÉR tail rotor; **~ del compresor** *m* TRANSP
AÉR compressor rotor; **~ de compresor centrífugo**
m TRANSP AÉR impeller; **~ de control** *m* TRANSP AÉR
control rotor; **~ en cortocircuito** *m* ELEC *generador*
short-circuit rotor, ING ELÉC squirrel-cage rotor;
~ con devanado en fase *m* ELEC phase-wound rotor
motor; **~ del distribuidor** *m* AUTO, VEH *sistema de
encendido* distributor rotor; **~ de elevación** *m*
TRANSP AÉR lifting rotor; **~ enjaulado** *m* ING ELÉC
cage rotor; **~ equilibrado** *m* ENERG RENOV teetered
rotor; **~ libre** *m* TRANSP AÉR free rotor; **~ macizo** *m*
ELEC solid rotor; **~ no devanado** *m* CONST non-
wound rotor; **~ a palas de motor a reacción** *m*
TRANSP AÉR jet-flapped rotor; **~ principal** *m* TRANSP
AÉR main rotor; **~ de resistencia aumentada** *m*

ELEC *generador* increased-resistance rotor; **~ rígido** *m*
ING MECÁ, TRANSP AÉR rigid rotor; **~ trifásico** *m* ING
ELÉC three-phase rotor; **~ de turbina** *m* INSTAL
HIDRÁUL, TRANSP MAR turbine wheel; **~ único** *m*
TRANSP AÉR single rotor; **~ de ventilador** *m* SEG fan
wheel

rotos: **~ húmedos** *m pl* PAPEL wet broke

rotoscopio *m* CINEMAT rotoscope

rótula *f* *Esp* AUTO knuckle, CINEMAT, FOTO ball-and-
socket head, ING MECÁ gimbal joint, knee, knuckle,
MECÁ knuckle joint, TEC PETR *perforación* swivel,
TRANSP MAR knee; **~ cardan** *f* AUTO universal joint;
~ de dirección *f* AUTO steering knuckle; **~ esférica** *f*
FOTO ball-and-socket; **~ giratoria** *f* ING MECÁ swivel;
~ superior *f* ING MECÁ upper ball joint

rotulación *f* EMB labeling (*AmE*), labelling (*BrE*), IMPR
lettering, PROD *cables* labeling (*AmE*), labelling
(*BrE*); **~ con letras autoadhesivas** *f* IMPR dry
transfer lettering

rotulador *m* EMB labeler (*AmE*), labeller (*BrE*)

rotuladora: **~ mecánica** *f* EMB labeling machine
(*AmE*), labelling machine (*BrE*), PROD label stamper

rótulo *m* ELEC *de artefacto* nameplate, EMB label, IMPR
callout, ING MECÁ placard, PROD docket; **~ de aviso**
m EMB, SEG caution label

rotura *f* CONST breaking, breaking down, bursting,
failure, ENERG RENOV fracture, GAS failure, ING MECÁ
fracture, gash, MECÁ fracture, METAL failure, rupture,
PAPEL break, QUÍMICA rupture, TEXTIL break, *del
flock* crushing; **~ de la aislación** *f* ELEC insulation
breakdown; **~ del aislamiento** *f* ELEC insulation
breakdown; **~ de la banda** *f* IMPR web break;
~ catastrófica *f* ING ELÉC catastrophic failure;
~ por compresión *f* CONST crushing;
~ espontánea *f* C&V spontaneous breaking; **~ frágil**
f NUCL brittle failure; **~ en húmedo** *f* PAPEL wet
break; **~ de línea** *f* TV line tear; **~ de la línea de
vapor principal** *f* (*MSLB*) NUCL main-steamline
break (*MSLB*); **~ normal** *f* METAL normal rupture;
~ por presión *f* C&V bursting-off; **~ progresiva** *f*
CARBÓN progressive failure; **~ de rail** *f* FERRO rail
break; **~ de la trama** *f* TEXTIL weft break; **~ de tubos
en un generador de vapor** *f* (*SGTR*) NUCL steam
generator tube rupture (*SGTR*); **~ de la urdimbre** *f*
TEXTIL warp break; **~ de la vaina a finales de vida** *f*
NUCL end-of-life cladding rupture

roturadora *f* AUTO, CONST, VEH road plough (*BrE*),
road plow (*AmE*)

rovín: **~ peludo** *m* C&V hairy roving

roya *f* AGRIC rust; **~ foliar** *f* AGRIC leaf rust; **~ de la
hoja** *f* AGRIC leaf rust; **~ de hoja de cebada** *f* AGRIC
barley-leaf rust; **~ parda** *f* AGRIC rust of wheat;
~ vesicular *f* AGRIC blister rust

roza *f* CARBÓN *minas* cut, MINAS jad, kirving, kerving,
holing; **~ a cielo abierto** *f* MINAS surface cut; **~ para
estemple** *f* MINAS cut

rozado *adj* MINAS holed

rozadora *f* CARBÓN rotary heading machine, MINAS
channeler (*AmE*), channeling machine (*AmE*), channeller (*BrE*), channelling machine (*BrE*), bar coal-
cutting machine; **~ de cadena para entalladuras
verticales** *f* MINAS chain shearing machine; **~ para
carbón** *f* CARBÓN, MINAS coal cutter, coal-mining
machine; **~ para entalladuras verticales** *f* CARBÓN
minas shearing machine; **~ de galería de avance** *f*
MINAS heading machine; **~ de percusión** *f* CARBÓN

pick breaker; ~ **rotativa de galería de avance** *f*
CARBÓN rotary heading machine
rozadura *f* ING MECÁ attrition
rozamiento[1]: **de** ~ *adj* CONST frictional; **sin** ~ *adj* ING
MECÁ frictionless
rozamiento[2] *m* GEN friction, CARBÓN, CONST attrition,
TRANSP MAR *velas* chafing; ~ **debido al rodamiento**
m FÍS rolling friction; ~ **por deslizamiento** *m* FÍS
sliding friction; ~ **estático** *m* FÍS static friction;
~ **interno** *m* METAL internal friction; ~ **laminar** *m*
FÍS *mecánica de fluidos* skin friction; ~ **superficial**
negativo *m* CARBÓN *pilotes* negative skin friction
RPD *abr* (*red pública de datos*) INFORM&PD PDN
(*public data network*)
RPS *abr* (*sistema de protección del reactor*) NUCL RPS
(*reactor protection system*)
RT *abr* (*radiotelefonía*) TELECOM, TRANSP, TRANSP MAR
RT (*radiotelephony*)
Ru *abr* (*rutenio*) QUÍMICA Ru (*ruthenium*)
rubelita *f* MINERAL rubellite
ruberítrico *adj* QUÍMICA ruberythric
rubí *m* CONST, GEOL, MINERAL ruby; ~ **balas** *m*
MINERAL balas ruby; ~ **de Brasil** *m* MINERAL Brazilian ruby; ~ **de crisol** *m* C&V pot ruby
rubidio *m* (*Rb*) QUÍMICA rubidium (*Rb*)
rubor *m* IMPR blushing
ruditas *f pl* GEOL rudites
rueda *f* ING MECÁ wheel, OCEAN net roller, TRANSP, VEH
wheel; ~ **abrasiva** *f* C&V grinding wheel, *para corte de*
artículos de vidrio o cristal cutting wheel, ING MECÁ
grinding wheel, abrasive wheel, MECÁ abrasive wheel,
PROD abrading wheel, grinding wheel, SEG abrasive
wheel; ~ **abrasiva de aletas** *f* ING MECÁ abrasive flap
wheel; ~ **de acción** *f* PROD impulse wheel; ~ **de**
acero en sistema de rail de acero *f* TRANSP steel
wheel on steel rail system; ~ **de afilar** *f* MECÁ grinding
wheel; ~ **de aletas con eje** *f* ING MECÁ flap-wheel
with shaft; ~ **con aletas integral** *f* ENERG RENOV
turbinas integral runner; ~ **de alfarero** *f* C&V throw-
ing wheel; ~ **de alta velocidad específica** *f* ENERG
RENOV high specific speed wheel; ~ **de amolar** *f* PROC
QUÍ grinding cylinder; ~ **de arcaduces** *f* AGUA
overshot wheel; ~ **de arrastre** *f* ING MECÁ driving
wheel; ~ **auxiliar** *f* VEH *semirremolque* fifth wheel;
~ **de Barlow** *f* FÍS Barlow's wheel; ~ **bruñidora** *f*
PROD buffing wheel; ~ **de cadena** *f* INFORM&PD
sprocket, ING MECÁ chain pulley, chain sheave,
sprocket wheel; ~ **de cangilones** *f* CONST bucket
wheel; ~ **de Carborundo** *f* PROD Carborundum
wheel®; ~ **catalina** *f* ING MECÁ, MECÁ, PROD sprocket
wheel; ~ **cilíndrica** *f* ING MECÁ tub wheel; ~ **de cola** *f*
AGRIC *de arado* rear-furrow wheel, TRANSP AÉR tail
wheel; ~ **de colores** *f* COLOR, TEXTIL color wheel
(*AmE*), colour wheel (*BrE*); ~ **conducida** *f* ING MECÁ
follower, driven wheel; ~ **conductora** *f* ING MECÁ
driving wheel; ~ **de conexión con tornillo sin fin** *f*
ING MECÁ wheel which engages a worm; ~ **cónica** *f*
ING MECÁ bevel wheel, cone wheel; ~ **de control** *f*
C&V control wheel; ~ **corona** *f* AUTO crown wheel;
~ **de corte** *f* ING MECÁ cutting-off wheel; ~ **de corte**
de diamante *f* C&V diamond-slitting wheel; ~ **de**
corte para vidrio *f* C&V glass-cutting wheel; ~ **de**
costado *f* INSTAL HIDRÁUL breast wheel; ~ **dentada** *f*
AUTO, INFORM&PD sprocket, ING MECÁ *piñón* cogwheel, gear, gear wheel, rack wheel, toothed wheel,
MECÁ gear wheel, PROD toothed wheel; ~ **dentada de**

alimentación *f* CINEMAT feed sprocket; ~ **dentada**
de cadena *f* ING MECÁ chain sprocket; ~ **dentada**
para cadena *f* ING MECÁ chain gear, chain wheel;
~ **dentada para cadena articulada** *f* ING MECÁ,
PROD sprocket wheel, VEH sprocket; ~ **dentada**
cónica *f* ING MECÁ crown wheel, INSTR bevel;
~ **dentada de dientes rectos** *f* ING MECÁ spur wheel;
~ **dentada de doble hélice** *f* ING MECÁ herringbone
gear; ~ **dentada con fiador** *f* ING MECÁ ratch;
~ **dentada de los linguetes** *f* ING MECÁ pawl rim;
~ **dentada mayor** *f* ING MECÁ crown wheel; ~ **den-**
tada motriz *f* CINEMAT drive sprocket; ~ **dentada**
motriz para cadena *f* VEH *mando por cadena* drive
sprocket; ~ **dentada de tornillo sin fin** *f* ING MECÁ
worm gear; ~ **dentada con trinquete** *f* ING MECÁ
ratch; ~ **dentada de trinquete** *f* ING MECÁ click
wheel; ~ **con dentadura postiza de madera** *f* PROD
mortise wheel; ~ **de desbaste** *f* INSTR roughing
wheel; ~ **de dientes rectos** *f* ING MECÁ straight-
tooth wheel; ~ **de disco** *f* VEH disc centre wheel
(*BrE*), disk center wheel (*AmE*); ~ **disparadora** *f*
AUTO trigger wheel; ~ **de división** *f* ING MECÁ division
wheel; ~ **divisora** *f* ING MECÁ dividing wheel;
~ **elástica** *f* TRANSP elastic wheel; ~ **elevadora** *f*
ING MECÁ lifting wheel, elevating wheel; ~ **de enfo-**
que central *f* INSTR central focusing wheel; ~ **de**
engranaje *f* ING MECÁ pitch wheel, PROD toothed
wheel; ~ **de escape** *f* ING MECÁ rack wheel; ~ **para**
escombros *f* ING MECÁ rubbish wheel; ~ **de esmeril**
f MECÁ, PROD emery wheel; ~ **espiral** *f* ING MECÁ
spiral wheel; ~ **estabilizadora** *f* TRANSP stabilizing
wheel; ~ **de estrella** *f* ING MECÁ ratchet wheel, star
wheel; ~ **estrellada** *f* AUTO starwheel; ~ **fija** *f* ING
MECÁ fixed wheel, PROD fast wheel; ~ **de filetear** *f*
ING MECÁ change wheel; ~ **sin fin** *f* ING MECÁ worm
wheel; ~ **de fricción** *f* ING MECÁ friction wheel, gear
wheel, MECÁ gear wheel; ~ **frontal** *f* TRANSP front
wheel; ~ **de gobierno** *f* TRANSP MAR *construcción*
naval steering wheel, wheel; ~ **de guía** *f* TRANSP guide
wheel; ~ **helicoidal** *f* ING MECÁ screw wheel, spiral
wheel, worm wheel; ~ **hidráulica de admisión**
superior *f* AGUA overshot water-wheel; ~ **hidráulica**
de alimentación superior *f* AGUA overshot wheel;
~ **hidráulica de pera** *f* INSTAL HIDRÁUL danaide;
~ **hidráulica de reacción** *f* AGUA reaction water-
wheel; ~ **de impresión** *f* INFORM&PD print wheel;
~ **de impulsión** *f* ING MECÁ driving wheel, PROD
impulse wheel; ~ **de inercia** *f* TEC ESP momentum
wheel, reaction wheel; ~ **de inercia con rodamiento**
magnético *f* TEC ESP magnetic bearing momentum
wheel; ~ **intermedia** *f* ING MECÁ intermediate wheel,
runner, stud wheel, MECÁ idler; ~ **de levas** *f* MECÁ
crown; ~ **libre** *f* ING MECÁ freewheel; ~ **loca** *f* ING
MECÁ idler; ~ **maestra** *f* ING MECÁ leader, master
wheel; ~ **a mano** *f* MECÁ hand wheel; ~ **de mano** *f*
ING MECÁ hand wheel; ~ **de mariposa** *f* INFORM&PD
daisywheel; ~ **matriz** *f* ING MECÁ leader; ~ **motora** *f*
ING MECÁ driving wheel; ~ **motriz** *f* AUTO drive wheel,
CINEMAT, ING MECÁ driving wheel, MECÁ impeller,
VEH drive wheel; ~ **móvil** *f* CARBÓN *de turbina* rotor,
MECÁ, PROD *bomba centrífuga* impeller; ~ **orientada** *f*
VEH *dirección* steered wheel; ~ **de palas** *f* AGUA
paddle wheel; ~ **de paletas** *f* INSTAL HIDRÁUL flash
wheel, float water wheel, impeller, paddle, runner,
PROD *turbina*, *bomba* runner, TRANSP MAR *construc-*
ción naval impeller, paddle wheel; ~ **parásita** *f* ING

MECÁ runner, stud wheel; **~ parcialmente dentada** *f* ING MECÁ mutilated wheel; **~ Pelton** *f* PROD impulse turbine; **~ pequeña pivotante** *f* ING MECÁ castor; **~ de perforación** *f* IMPR perforation wheel; **~ de pestañas** *f* ING MECÁ flange wheel; **~ de piñón** *f* ING MECÁ pinion wheel; **~ pivotante** *f* ING MECÁ castor; **~ planetaria** *f* ING MECÁ planet wheel; **~ de plato compacto** *f* VEH disc centre wheel (*BrE*), disk center wheel (*AmE*); **~ portadora** *f* PROD *puente-grúa* traveling wheel (*AmE*), travelling wheel (*BrE*), TRANSP carrying wheel; **~ portadora embridada** *f* ING MECÁ flanged traveling wheel (*AmE*), flanged travelling wheel (*BrE*); **~ propulsora** *f* AUTO, VEH *rueda, en transmisión* drive wheel; **~ pulidora** *f* C&V polishing wheel, PROD buffing wheel; **~ de puntería azimutal** *f* D&A training wheel; **~ de radios** *f* VEH spoke wheel, spoked wheel; **~ con radios de alambre** *f* VEH wire wheel; **~ de reacción** *f* AGUA reaction wheel, *turbinas* pressure wheel, TEC ESP reaction wheel; **~ de recambio** *f* AUTO spare tire (*AmE*), spare tyre (*BrE*), spare wheel (*BrE*), ING MECÁ change wheel, VEH spare tire (*AmE*), spare tyre (*BrE*), spare wheel (*BrE*); **~ de recortar** *f* ING MECÁ cutoff wheel; **~ rectificadora** *f* ING MECÁ truing wheel; **~ de repuesto** *f* AUTO, VEH spare tire (*AmE*), spare tyre (*BrE*), spare wheel (*BrE*); **~ de retorno** *f* ING MECÁ return wheel; **~ retráctil** *f* AUTO, VEH retractable wheel; **~ de riel** *f* ING MECÁ rail wheel; **~ satélite** *f* ING MECÁ planet gear, planet wheel; **~ de soplador** *f* ING MECÁ blower-wheel; **~ de talón** *f* AGRIC rear-furrow wheel; **~ del timón** *f* TRANSP MAR *buque* helm, *construcción naval* steering wheel, wheel; **~ tipo Pelton** *f* ENERG RENOV Pelton wheel; **~ tipo Pelton con árbol vertical** *f* ENERG RENOV vertical-shaft Pelton wheel; **~ de tiras abrasivas con eje** *f* ING MECÁ flap-wheel with shaft; **~ de tornillo sinfín** *f* MECÁ worm gear; **~ de torno** *f* CONST gin wheel; **~ de tracción** *f* AUTO, VEH *rueda, en transmisión* drive wheel; **~ de transmisión** *f* TV drive sprocket; **~ transmisora** *f* ING MECÁ driving wheel; **~ de trapo de calicó** *f* PROD calico mop; **~ trasera** *f* AUTO, TRANSP, VEH rear wheel; **~ de traslación** *f* ING MECÁ rail wheel, running wheel, translating wheel, PROD *puente-grúa* traveling wheel (*AmE*), travelling wheel (*BrE*); **~ de traslación embridada** *f* ING MECÁ flanged traveling wheel (*AmE*), flanged travelling wheel (*BrE*); **~ de traslación de pestaña única** *f* ING MECÁ single-flanged traveling wheel (*AmE*), single-flanged travelling wheel (*BrE*); **~ de traslación plana** *f* ING MECÁ flat-running wheel; **~ de traslación simple** *f* ING MECÁ plain traveling wheel (*AmE*), plain travelling wheel (*BrE*); **~ del tren de aterrizaje del morro** *f* TRANSP AÉR nose gear wheel; **~ de trinquete** *f* ING MECÁ click wheel, dog wheel, ratchet wheel; **~ trituradora de nitruro de boro cúbico** *f* ING MECÁ cubic boron nitride grinding wheel; **~ de turbina** *f* INSTAL HIDRÁUL turbine wheel; **~ de turbina con paletas** *f* INSTAL HIDRÁUL turbine wheel with vanes; **~ de velocidad periférica** *f* ING MECÁ peripheral wheel speed

ruedas: **~ dobles** *f pl* TRANSP AÉR dual wheels; **~ gemelas** *f pl* TRANSP AÉR twin wheels

ruedecilla *f* CINEMAT thumbwheel, ING MECÁ wheel

ruedo *m* GEOM circle

rugosidad *f* ING MECÁ, MECÁ, PAPEL roughness, PETROL

rugosity; **~ de superficie** *f* CONST, IMPR, ING MECÁ surface roughness

ruido[1]: **sin ~** *adj* SEG noiseless

ruido[2] *m* GEN noise; **~ acústico** *m* TELECOM acoustic noise; **~ aditivo** *m* TELECOM additive noise; **~ aéreo** *m* ACÚST airborne noise; **~ aerodinámico** *m* ACÚST, TRANSP AÉR aerodynamic noise; **~ de aeronaves** *m* ACÚST aircraft noise; **~ de aeropuertos** *m* ACÚST airport noise; **~ de agitación térmica** *m* ÓPT, TEC ESP, TELECOM shot noise; **~ aleatorio** *m* ACÚST, ELECTRÓN, FÍS, GEOFÍS, TEC ESP, TELECOM random noise; **~ ambiental** *m* ACÚST ambient noise, environmental noise, SEG environmental noise; **~ ambiente** *m* ACÚST, TELECOM ambient noise; **~ artificial** *m* TEC ESP, TELECOM artificial noise; **~ atmosférico** *m* ELECTRÓN, TEC ESP atmospheric noise, TV static; **~ de aviones** *m* ACÚST airplane noise; **~ de banda ancha** *m* ELECTRÓN, TELECOM wideband noise; **~ de banda estrecha** *m* ELECTRÓN narrow-band noise; **~ básico** *m* ELECTRÓN basic noise; **~ de batido** *m* TEC ESP *válvulas de vacío* shot noise; **~ blanco** *m* ACÚST, ELECTRÓN, FÍS, INFORM&PD, TEC ESP, TELECOM white noise; **~ de canal** *m* ELECTRÓN channel noise; **~ casual** *m* ACÚST, ELECTRÓN, FÍS, GEOFÍS, TEC ESP, TELECOM random noise; **~ de certificación** *m* ACÚST certification noise; **~ de chorro** *m* ACÚST jet noise; **~ cíclico** *m* ACÚST cyclic noise; **~ de circuito** *m* ELECTRÓN, ING ELÉC line noise, circuit noise; **~ coherente** *m* GEOFÍS *prospección sísmica* coherent noise; **~ complejo** *m* ACÚST, ELECTRÓN, FÍS, GEOFÍS, TEC ESP, TELECOM random noise; **~ comunitario** *m* ACÚST community noise; **~ constante** *m* ACÚST steady noise; **~ continuo** *m* ACÚST steady noise; **~ cósmico** *m* TEC ESP cosmic noise; **~ cromático** *m* TV color noise (*AmE*), colour noise (*BrE*); **~ cuántico** *m* ELECTRÓN, FÍS, ÓPT, TELECOM quantum noise; **~ de cuantificación** *m* INFORM&PD, TEC ESP, TELECOM quantization noise; **~ de descarga** *m* ELECTRÓN, ÓPT, TEC ESP shot noise; **~ por diacromía** *m* TV cross-color noise (*AmE*), cross-colour noise (*BrE*); **~ de disparo** *m* FÍS, TEC ESP shot noise; **~ eléctrico** *m* ELECTRÓN electric noise; **~ de emisión secundaria** *m* ELECTRÓN secondary emission noise; **~ de empalme** *m* CINEMAT bloop; **~ de engranaje** *m* CINEMAT sprocket noise; **~ enmascarador** *m* ACÚST masking noise; **~ errático** *m* ACÚST, ELECTRÓN, FÍS, GEOFÍS, TEC ESP, TELECOM random noise; **~ de escintilación** *m* TEC ESP scintillation noise; **~ con espectro continuo y uniforme** *m* TEC ESP white noise; **~ estable** *m* ACÚST stable noise; **~ estadístico** *m* ACÚST, ELECTRÓN, FÍS, GEOFÍS, TEC ESP, TELECOM random noise; **~ de estática** *m* TV static; **~ estructural** *m* ACÚST structure noise; **~ exterior** *m* ACÚST outdoor noise; **~ fluctuante** *m* ACÚST fluctuating noise, ELECTRÓN flicker noise; **~ de fondo** *m* ACÚST background noise, ELECTRÓN background noise, ground noise, FÍS, GEOFÍS, TEC ESP background noise, TELECOM *telefonía* background noise, sidetone; **~ de fotones** *m* TELECOM photon noise; **~ fotónico** *m* ÓPT photon noise; **~ de frecuencia** *m* ELECTRÓN frequency noise; **~ de frenos** *m* ACÚST brake noise; **~ del fuselaje** *m* ACÚST airframe noise; **~ galáctico** *m* ELECTRÓN galactic noise; **~ de gas** *m* ELECTRÓN gas noise; **~ gausiano** *m* ACÚST, ELECTRÓN, INFORM&PD, TELECOM Gaussian noise; **~ granular** *m* ÓPT shot noise, TELECOM shot noise, granular noise; ~ **de impactos normalizado** *m* ACÚST standardized impact-sound; ~ **impulsivo** *m* ACÚST impulsive noise, *radio* impulse noise, TELECOM impulse noise; ~ **del impulso** *m* TEC ESP pulse noise; ~ **de impulsos** *m* INFORM&PD impulse noise; ~ **inducido por la corriente** *m* PROD induced noise current; ~ **industrial** *m* TEC ESP man-made noise; ~ **intermitente** *m* ACÚST intermittent noise; ~ **de intermodulación** *m* TEC ESP intermodulation noise; ~ **interno** *m* ELECTRÓN internal noise; ~ **intrínseco** *m* ELECTRÓN intrinsic noise; ~ **Johnson** *m* ELECTRÓN, FÍS Johnson noise; ~ **de línea** *m* ELECTRÓN, ING ELÉC line noise; ~ **llevado por el viento** *m* ING MECÁ *emitido por maquinaria de excavar, sierras de cadena* airborne noise; ~ **de mácula** *m* TELECOM speckle noise; ~ **del mar** *m* OCEAN *cartografía marina por sonar etología marina* sea noise; ~ **mezcla de múltiples frecuencias** *m* TEC ESP white noise; ~ **modal** *m* ÓPT, TELECOM modal noise; ~ **de modulación** *m* ACÚST, ELECTRÓN, TELECOM modulation noise; ~ **en modulación de frecuencia** *m* TELECOM frequency-modulation noise; ~ **modulado** *m* AmL (*cf ruido producido por la señal Esp*) TV modulation noise; ~ **por moteado** *m* ÓPT speckle noise; ~ **de los motores a reacción** *m* TEC ESP white noise; ~ **no gausiano** *m* TELECOM non-Gaussian noise; ~ **ocupacional** *m* ACÚST occupational noise; ~ **de operaciones** *m* ACÚST *aviación* operational noise; ~ **parásito** *m* ELECTRÓN interference, interference noise, TELECOM extraneous noise; ~ **de partición** *m* ELECTRÓN partition noise; ~ **de la portadora** *m* ELECTRÓN carrier noise; ~ **producido por la señal** *m Esp* (*cf ruido modulado AmL*) TV modulation noise; ~ **propio** *m* OCEAN *equipos hidrofónicos* self-noise; ~ **pseudoaleatorio** *m* TELECOM pseudo-random noise; ~ **radioeléctrico** *m* ELECTRÓN radio noise; ~ **de referencia** *m* ACÚST, ELECTRÓN reference noise; ~ **residual** *m* INFORM&PD residual noise; ~ **rosa** *m* ACÚST, FÍS pink noise; ~ **de Schottky** *m* FÍS Schottky noise; ~ **de secuencias de igual amplitud** *m* ELECTRÓN white noise; ~ **de señal de amplificación** *m* ELECTRÓN amplifier noise; ~ **sondeo** *m Esp* (*cf subterráneo AmL*) MINAS rumbling; ~ **de superficie** *m* ACÚST, ELECTRÓN surface noise; ~ **térmico** *m* ELECTRÓN, FÍS, TEC ESP thermal noise; ~ **transmitido a través del aire** *m* SEG airborne acoustical noise, airborne noise

rulemán *m* ING MECÁ roller bearing; ~ **de agujas** *m* ING MECÁ needle bearing; ~ **para trabajos pesados** *m* ING MECÁ heavy-duty ball bearing

ruleta *f* GEOM roulette

rumbo[1] *m* TEC ESP heading, vector, TRANSP AÉR, TRANSP MAR course, heading; ~ **de aguja** *m* TRANSP MAR compass heading; ~ **de la aguja** *m* TRANSP MAR magnetic course; ~ **del bote** *m* TRANSP MAR boat heading; ~ **de la brújula** *m* TRANSP AÉR compass heading; ~ **por círculo máximo** *m* TRANSP MAR great-circle course; ~ **corregido** *m* TRANSP MAR course made good; ~ **de dirección** *m* GEOL, MINAS *filones* line of strike; ~ **efectivo** *m* TRANSP MAR course made good; ~ **estimado** *m* OCEAN estimated reckoning; ~ **del filón** *m Esp* (*cf hallazgo AmL*) MINAS *de filones* run, *de un filón subterráneo* strike; ~ **franco** *m* TRANSP MAR steady bearing; ~ **del haz localizador** *m* TRANSP AÉR localizer beam heading; ~ **magnético** *m* TEC ESP magnetic bearing, TRANSP MAR magnetic

course; ~ **marcado por la brújula** *m* GEOL compass bearing; ~ **mixto** *m* TEC ESP mixed path; ~ **opuesto** *m* TRANSP MAR reciprocal course; ~ **radárico** *m* TRANSP, TRANSP AÉR, TRANSP MAR radar heading; ~ **real por el agua** *m* TRANSP MAR course through the water; ~ **con riesgo de abordaje** *m* TRANSP MAR collision course; ~ **de salida** *m* TRANSP AÉR outbound heading; ~ **sinuoso** *m* TRANSP weaving; ~ **verdadero** *m* TRANSP MAR true course; ~ **en zigzag** *m* TRANSP *buques* weaving

rumbo[2]: **estar a** ~ *vi* TRANSP MAR be on course

rumbos: ~ **de la aguja** *m pl* TRANSP MAR points of the compass

rumbotrón *m* ELECTRÓN rhumbatron

rumen *m* AGRIC rumen

ruptor *m* AUTO breaker, contact breaker, contact breaker-point, ELEC *conmutador* interrupter, ING ELÉC trembler, VEH contact breaker-point (*BrE*), contact breaker, points (*AmE*); ~ **del arco** *m* ING ELÉC arc breaker; ~ **a chorro de aire** *m* ELEC *cortocircuito* air blast breaker; ~ **doble** *m* AUTO two-system contact breaker, VEH two-system contact-breaker; ~ **de dos sistemas** *m* AUTO two-system contact breaker, VEH two-system contact-breaker; ~ **de posición** *m* ELEC position switch

ruptura *f* C&V *de la base del envase* breaking-off, CONST, ELEC breakdown, IMPR breakthrough, INFORM&PD truncation, break, ING MECÁ parting, fracture, MECÁ fracture, METAL rupture, PROC QUÍ breaking, PROD break, QUÍMICA *presión, esfuerzo* breakdown, *enlace, ligadura* rupture, TELECOM breakthrough; ~ **por avalancha** *f* ELEC *diodo Zener* avalanche breakdown; ~ **dieléctrica** *f* ELEC dielectric breakdown; ~ **del dieléctrico** *f* ELEC dielectric breakdown; ~ **entre patillas** *f* ING ELÉC pin-to-pin breakdown; ~ **de interrupción** *f* GEOL knickpoint (*BrE*), nickpoint (*AmE*); ~ **de pendiente** *f* GEOL knickpoint (*BrE*), nickpoint (*AmE*); ~ **de red** *f* TELECOM network breakdown; ~ **en los sedimentos** *f* OCEAN sediment break

rurbanización *f* AGRIC rurbanization

ruta *f* INFORM&PD route, TEC ESP course, TELECOM, TRANSP AÉR, TRANSP MAR *comercial, mercantil* route; ~ **de acceso** *f* INFORM&PD access path, path; ~ **aérea** *f* TRANSP AÉR airway; ~ **aérea internacional** *f* TRANSP AÉR international air route; ~ **alternativa** *f* TELECOM alternative route; ~ **de la banda** *f* IMPR web path; ~ **del cable** *f* ELEC, ING ELÉC *fuente de alimentación* cable run; ~ **de enlace** *f* INFORM&PD linkage path; ~ **de líneas sobre postes** *f* TELECOM pole route; ~ **marítima** *f* OCEAN *navegación comercial* sea lane, TRANSP MAR sea lane, sea route, shipping route; ~ **de navegación** *f* TRANSP lane; ~ **ortodrómica** *f* TRANSP AÉR great-circle route; ~ **perimétrica** *f* TRANSP AÉR perimeter track; ~ **de postes** *f* TELECOM pole route; ~ **de tratamiento** *f* PROD flow path

ruténico *adj* METAL, QUÍMICA ruthenic

rutenio *m* (*Ru*) METAL, QUÍMICA ruthenium (*Ru*)

ruterfordio *m* (*Rf*) FÍS RAD rutherfordium (*Rf*)

rutilante *adj* QUÍMICA, TERMO glowing

rutilo *m* MINERAL, QUÍMICA *mineral* rutile

rutina *f* INFORM&PD routine; ~ **de bifurcación** *f* PROD jump routine; ~ **de control** *f* INFORM&PD control routine

S

s *abr* ACÚST, CONST, FÍS (*segundo*) s (*second*), FÍS PART
(*espín*) s (*spin*), METR (*segundo*) s (*second, spin*)

S *abr* FÍS (*siemens, siemensio*) S (*siemens*), METAL
(*azufre*) S (*sulfur AmE, sulphur BrE*), METR
(*siemens, siemensio*) S (*siemens*), P&C, QUÍMICA, TEC
PETR (*azufre*) S (*sulfur AmE, sulphur BrE*)

sábana *f* TEXTIL *de la cama* sheet; **~ impermeable** *f*
TEXTIL waterproof sheet

SABD *abr* (*sistema de administración de base de datos*)
INFORM&PD, TELECOM DBMS (*database management
system*)

sabine *m* ACÚST sabin

sable *m* TRANSP MAR *vela* batten

sabor *m* ALIMENT flavor (*AmE*), flavour (*BrE*)

saborizante *m* ALIMENT flavoring (*AmE*), flavouring
(*BrE*)

saborizar *vt* ALIMENT flavor (*AmE*), flavour (*BrE*)

sacabocados *m* ING MECÁ, MECÁ punch; **~ cilíndrico
con cabeza cónica** *m* ING MECÁ round punch with
conical head; **~ dúplex** *m* ING MECÁ duplex punching
bear; **~ a golpe** *m* PROD socket punch; **~ mecánico**
m ING MECÁ machine punch; **~ revólver** *m* ING MECÁ
revolving punch pliers

sacabrocas *m* ING MECÁ drift

sacachavetas *m* ING MECÁ drift

sacaclavos: **~ de horquilla** *m* CONST claw bar

sacacorchos: **~ en espiral** *m* TEXTIL spiral corkscrew

sacacubos *m* ING MECÁ hub extractor, MECÁ hub
puller

sacado[1]: **~ del barril** *adj* ALIMENT drawn from the
wood

sacado[2] *m* C&V takeout

sacador *m* C&V takeout, *saca y da forma al vidrio
fundido* former; **~-alimentador** *m* C&V ladler; **~ de
molde** *m* C&V mold-emptier (*AmE*), mould-emptier
(*BrE*)

sacafusibles *m pl* PROD fuse puller

sacamecha *f* ING MECÁ center key (*AmE*), centre key
(*BrE*)

sacamuestras *m AmL* (*cf sacatestigos Esp*) MINAS
core catcher, core bit, core breaker, *sondeos* core
barrel; **~ de pared lateral** *m* TEC PETR sidewall
sampler

sacapajas *m* AGRIC straw carrier

sacapasador *m* ING MECÁ pin drift, pin extractor

sacapernos *m* ING MECÁ drift

sacapliegos *m* IMPR sheet deliverer

sacapuntas *m* ING MECÁ sharpener

sacar[1] *vt* CARBÓN extract, *agua* tap, CONST pull out,
extract, IMPR pull, ING MECÁ throw off, MINAS
develop, QUÍMICA extract; **~ con cucharón** *vt* C&V
ladle; **~ con espita** *vt* ALIMENT *procesamiento de
alimentos, fermentación* tap

sacar[2]: **~ las interlíneas** *vi* IMPR unlead; **~ la jaula** *vi*
MINAS cut a landing

sacarato *m* QUÍMICA saccharate

sacárico *adj* QUÍMICA saccharic

sacárido *m* QUÍMICA saccharide

sacarina *f* QUÍMICA saccharin

sacarosa *f* ALIMENT saccharose

sacatestigos *m Esp* MINAS (*cf sacamuestras AmL, cf
tubo sacatestigos Esp*) core extractor, core lifter, core
catcher, core bit, core breaker, *sondeos* core barrel

sacatornillos *m* CONST screw elevator

saco *m* AGRIC bag, EMB sack; **~ de basura** *m* EMB
refuse sack (*BrE*); **~ de boca abierta** *m* EMB open
mouth sack; **~ de carga** *m* CINEMAT changing bag;
~ cosido *m* EMB sewn sack; **~ para filtrar** *m* CARBÓN,
PROC QUÍ filter bag; **~ de fondo plano** *m* EMB flat-end
sack; **~ de fuelle** *m* EMB gusseted sack; **~ para
insertar** *m* EMB *en cajas, bidones* insertable sack;
~ negro *m* CINEMAT changing bag; **~ de papel** *m* EMB
paper sack; **~ pegado** *m* EMB pasted sack; **~ plano** *m*
EMB flat sack; **~ único** *m* PROD single bag; **~ de
varias hojas** *m* EMB multiply sack, multiwall sack

sacoblasto *m* QUÍMICA saccoblast

sacudida *f* CONTAM *descolmatación de electrodos de un
electrofiltro* rapping, PROD jar, TEC ESP shock, buffeting, nudging; **~ de la columna de control** *f* TRANSP
AÉR control-column whip; **~ eléctrica** *f* ELEC *seguridad* electric shock

sacudidas: **~ ligeras para desprender un modelo** *f*
PROD *moldeo* rapping

sacudimiento *m* ING MECÁ shake

SADO *abr* (*sistema de adquisición de datos oceánicos*)
OCEAN ODAS (*ocean-data acquisition system*)

saetín *m* AGUA headrace, leat, race, *molinos* mill tail,
mill course, raceway, mill race, CARBÓN *canal de
esclusa* flume, ENERG RENOV flume, *presas* chute
spillway, ING MECÁ raceway, INSTAL HIDRÁUL guide,
PROD *de molino* leat, *hidráulica* raceway; **~ de aguas
arriba** *m* AGUA head crown, head bay

safflorita *f* MINERAL safflorite

safranina *f* QUÍMICA safranin

safrol *m* QUÍMICA safrol

sagenita *f* MINERAL sagenite

sahlita *f* MINERAL sahlite

sal *f* QUÍMICA salt; **~ ácida** *f* ALIMENT acid salt; **~ de
ácido tartárico** *f* ALIMENT, ING ELÉC, QUÍMICA
Rochelle salt; **~ amoníaco** *f* QUÍMICA sal ammoniac;
~ básica *f* QUÍMICA basic salt; **~ en bloque** *f* ALI-
MENT block salt cake, salt cake; **~ en bruto** *f* C&V salt
cake; **~ de cebolla** *f* QUÍMICA onion salt; **~ de cocina**
f ALIMENT kitchen salt; **~ común** *f* MINERAL rock
salt; **~ gema** *f* GAS rock salt, MINERAL halite, rock
salt; **~ gema sedimentaria** *f* GAS sedimentary rock
salt; **~ gorda** *f* ALIMENT rock salt; **~ de hierro** *f*
DETERG iron salt; **~ de mar** *f* ALIMENT sea salt;
~ marina *f* ALIMENT sea salt; **~ normal** *f* QUÍMICA
normal salt; **~ reguladora** *f* ALIMENT buffer salt;
~ de roca *f* QUÍMICA rock salt; **~ de roca
sedimentaria** *f* GAS sedimentary rock salt; **~ de
Rochelle** *f* ALIMENT, ING ELÉC, QUÍMICA Rochelle
salt; **~ rosada** *f* QUÍMICA pink salt; **~ de sodio** *f*
QUÍMICA sodium salt

sala *f* CARBÓN *minas* room, stall, MINAS *hueco pequeño*
stall, *cámara* chamber, *habitación* room, P&C cabinet;
~ anecoica *f* FÍS anechoic room; **~ aséptica** *f* TEC

ESP clean room; ~ **blanca** *f* TEC ESP clean room, *acústica* white room; ~ **de bombas** *f* AGUA pump room; ~ **de calderas** *f* INSTAL HIDRÁUL boiler-room, stokehold, INSTAL TERM boiler-room; ~ **de cardado** *f* TEXTIL cardroom; ~ **central de control** *f* TV central control room; ~ **de congelación** *f* REFRIG, TERMO chill room; ~ **de control** *f* ELEC, NUCL, TV control room; ~ **de control de la cadena** *f* TV network control room (*NCR*); ~ **de control del estudio** *f* TV studio control room; ~ **de control de imagen** *f* TV vision control room; ~ **de control de luz e imagen** *f* TV lighting and vision control room; ~ **de control de producción** *f* TV production control room; ~ **de control de rayos X** *f* INSTR X-ray control room; ~ **de control de video** *AmL*, ~ **de control de vídeo** *f Esp* TV video control room; ~ **de curado** *f* REFRIG ageing room; ~ **de enfriamiento** *f* REFRIG, TERMO chill room; ~ **de equipaje** *f* TRANSP baggage room (*AmE*), luggage compartment (*BrE*); ~ **de espera** *f* TRANSP *de estación* concourse; ~ **esterilizada** *f* TEC ESP white room; ~ **fría** *f* ING MECÁ, REFRIG, TERMO coldroom; ~ **de gálibos** *f* TRANSP MAR mold loft (*AmE*), mould loft (*BrE*); ~ **insonorizada** *f* CINEMAT sound stage; ~ **limpia** *f* LAB, TEC ESP clean room; ~ **de mando** *f* ELEC control room; ~ **de máquinas** *f* PROD engine room; ~ **de montaje** *f* CINEMAT, TV editing room; ~ **de ordeño en espina de pescado** *f* AGRIC herringbone milking parlor (*AmE*), herringbone milking parlour (*BrE*); ~ **de pasajeros** *f* TRANSP *ferrocarril* passenger hall; ~ **para preparar las formulaciones de estucado** *f* REVEST coating kitchen; ~ **de proyección** *f* CINEMAT screening room, viewing theater (*AmE*), viewing theatre; ~ **de pruebas** *f* TELECOM test room; ~ **de pruebas acústicas** *f* ING MECÁ acoustic testing room; ~ **del reactor** *f* NUCL reactor hall; ~ **de redacción** *f* TV editorial newsroom, newsroom; ~ **de refrigeración** *f* REFRIG, TERMO chill room; ~ **de servicio** *f* TELECOM operations room; ~ **sorda** *f* ACÚST dead room; ~ **de teñido** *f* COLOR dye room; ~ **de trazado** *f* TRANSP MAR *arquitectura naval* drawing office; ~ **de turbinas** *f* NUCL turbine building, turbine house

salabardo *m* OCEAN scoop net

salacetol *m* QUÍMICA salacetol

salado *adj* AGUA salty, CONST brackish, HIDROL briny, salty, OCEAN briny, QUÍMICA, TRANSP MAR salty

salazón *f* ALIMENT salting, OCEAN salt curing

salbanda *f* GEOL selvage, MINAS selvage, wall, *filón* gouge; ~ **arcillosa** *f* MINAS *filón* pug

sales: ~ **impermeabilizadoras** *f pl* REVEST waterproofing salts; ~ **nutritivas** *f pl* HIDROL nutrient salts

salgar *m* AGRIC salt lick

salicilado *adj* QUÍMICA salicylated

salicilaldehído *m* QUÍMICA salicylaldehyde

salicilato *m* QUÍMICA salicylate

salicílico *adj* QUÍMICA salicylic

salicilo *m* QUÍMICA salicyl

salicina *f* QUÍMICA salicin

salida[1]: **de ~ inmediata** *adj* IMPR quick-release

salida[2] *f* AGUA, AUTO outlet, CONST way out, GAS output, IMPR delivery, output, INFORM&PD output, exit, ING MECÁ opening, output, discharge, starting, INSTAL HIDRÁUL release, MECÁ vent, NUCL *de agua* issue, PROD *máquinas*, TELECOM outlet, TRANSP AÉR *de aterrizaje frustrado* flare-out, TRANSP MAR *de barco* way out; ~ **de agua** *f* AGUA, REFRIG water outlet;

~ **del agujero** *f* TEC PETR *perforación* coming out of hole; ~ **de aire** *f* ING MECÁ air outlet, MECÁ air discharge, TEC PETR air exhaust; ~ **de aire deshelador** *f* TRANSP AÉR de-icing air outlet; ~ **de amplificador de potencia** *f* PROD line-driver output; ~ **de amplificador vertical** *f* ELECTRÓN vertical-amplifier output; ~ **analógica** *f* INSTR analog output; ~ **asimétrica** *f* ING ELÉC single-ended output, TEC ESP unbalanced output; ~ **de CC** *f* ING ELÉC DC output; ~ **de CC de bajo nivel** *f* PROD low-level DC output; ~ **de computadora en microfilm** *f AmL* (*cf salida de ordenador en microfilm Esp*) INFORM&PD computer output on microfilm (*COM*); ~ **controlada** *f* ING MECÁ controlled outlet; ~ **de corriente** *f* ELECTRÓN electrical output, ING ELÉC current output, electrical output; ~ **de corriente alterna** *f* ING ELÉC alternating-current output; ~ **de corriente digital** *f* ELECTRÓN digital output; ~ **de datos** *f* INFORM&PD, INSTR data output; ~ **desbalanceada** *f* TEC ESP unbalanced output; ~ **digital en serie** *f* INFORM&PD serial digital output; ~ **directa** *f* ELECTRÓN direct output; ~ **de emergencia** *f* CONST, SEG emergency exit, fire exit, TEC ESP emergency escape, TERMO fire exit, TRANSP AÉR emergency exit; ~ **de energía eléctrica** *f* ELECTRÓN, ING ELÉC electrical output; ~ **de energía óptica** *f* ING ELÉC optical output; ~ **espacial** *f* TEC ESP space step-out; ~ **por el fondo** *f* AGUA bottom outlet; ~ **frontal** *f* IMPR front delivery; ~ **de gases** *f* TRANSP gas outlet; ~ **del generador** *f* ELEC, ING ELÉC generator output; ~ **impresa** *f* INFORM&PD hard copy; ~ **de impresión** *f* TELECOM printout; ~ **de incendio** *f* CONST, SEG, TERMO fire-escape; ~ **inmediata** *f* PROD immediate output; ~ **del inyector** *f* FÍS nozzle exit; ~ **de línea** *f* TV line-out; ~ **máxima** *f* INFORM&PD fan-out; ~ **no retentiva** *f* PROD non iretentive output, nonretentive output; ~ **nominal** *f* INFORM&PD rated output; ~ **de octeto** *f* ELECTRÓN, INFORM&PD eight-bit output; ~ **de ordenador en microfilm** *f Esp* (*cf salida de computadora en microfilm AmL*) INFORM&PD computer output on microfilm (*COM*); ~ **de oxígeno** *f* PROD oxygen outlet; ~ **del pivote** *f* AUTO kingpin inclination; ~ **plana** *f* IMPR flat-sheet delivery; ~ **de la portadora de crominancia** *f Esp* (*cf salida de la subportadora de color AmL*) TV chrominance-carrier output; ~ **de potencia desequilibrada** *f* ING ELÉC unbalanced output; ~ **programada** *f* TRANSP MAR *embarcaciones* sailing, yachting; ~ **sin referencia a tierra** *f* ING ELÉC floating output; ~ **de residuos** *f* NUCL waste outlet; ~ **de rodaje** *f* TRANSP AÉR exit taxiway; ~ **de rosca para fijador** *f* ING MECÁ thread run-out for fastener; ~ **del secuenciador** *f* PROD sequencer output; ~ **del sistema** *f* INFORM&PD log-off, log-out; ~ **de la subportadora de color** *f AmL* (*cf salida de la portadora de crominancia Esp*) TV chrominance-carrier output; ~ **de tiempo de ejecución** *f* INFORM&PD run-time output; ~ **de tiempo de pasada** *f* INFORM&PD run-time output; ~ **en tiempo real** *f* CINEMAT, ELECTRÓN, INFORM&PD, TELECOM, TV real-time output; ~ **de tres estados** *f* ELECTRÓN three-state output; ~ **de vahos** *f* PAPEL *sección de secado* exhaust; ~ **de válvula** *f* ING MECÁ valve outlet; ~ **del viento** *f* ING MECÁ *soplante* air outlet; ~ **vocal** *f* INFORM&PD voice output

salidas: ~ **en abanico** *f pl* PROD fanning out; ~ **complementarias** *f pl* ELECTRÓN complementary outputs; **~ escalonadas** *f pl* FERRO flighted departure

salido: **~ de telar** *adj* TEXTIL loom state

saliente[1] *adj* TELECOM outgoing

saliente[2] *m* CONST corbel, projection, set-off, IMPR kern, ING MECÁ boss, flange, height above pinch line, ledge, offset, lug, point, tit, MECÁ lug, rib, PROD *sobre superficie* lug, ridge, TEC ESP *de pieza* lobe, VEH *carrocería* overhang, *de buje de rueda* boss; **~ del cerrojo** *m* ING MECÁ keeper; **~ de inclinación** *m* TV tip protrusion; **~ de leva** *m* AUTO, TRANSP AÉR, VEH cam lobe; **~ del macho** *m* PROD *fundición* seating, *moldería* core print; **~ posicionador** *m* MECÁ locating key; **~ del radiador** *m* AUTO, VEH radiator flange; **~ roscada** *m* ING MECÁ teat screw; **~ superficial** *m* PROD *en moldes* boss

saliente[3] *f* ING MECÁ ledge, height above pinch line, offset, point; **~ roscada** *f* ING MECÁ teat screw

salificable *adj* QUÍMICA salifiable

saligenol *m* QUÍMICA saligenin

salina *f* ALIMENT, OCEAN saltern

salinas *f pl* PROD salt refinery, salt works

salinidad *f* AGUA, ENERG RENOV, HIDROL, QUÍMICA, TEC PETR, TRANSP MAR salinity

salinización *f* HIDROL salinization

salino *adj* AGUA saline, GEOL salt-bearing, HIDROL saline

salinómetro *m* CARBÓN, INSTR, OCEAN salinometer

salir: **~ al mar** *vi* HIDROL *ríos* discharge into the sea, outfall to sea; **~ de madre** *vi* HIDROL *un río* burst its banks; **~ del sistema** *vi* INFORM&PD log off, log out

salirse *v refl* AGUA *líquidos* run out; **~ de** *v refl* TRANSP AÉR *la pista* overshoot, run off; **~ del cauce** *v refl* HIDROL *un río* burst its banks; **~ de la vertical** *v refl* ING MECÁ run out of the vertical

salita *f* MINERAL salite

salitre *m* ALIMENT, C&V, QUÍMICA saltpeter (*AmE*), saltpetre (*BrE*); **~ de Perú** *m* QUÍMICA nitratine, Peru saltpeter (*AmE*), Peru saltpetre (*BrE*)

salmer *m* CONST *arquitectura* springer, springer stone

salmonicultor *m* OCEAN salmon breeder, salmon farmer

salmonicultura *f* OCEAN salmon culture

salmuera *f* ALIMENT, GAS, HIDROL, NUCL, QUÍMICA, REFRIG brine

salmuerado *m* OCEAN *del pescado* brine pickling

salobre *adj* AGUA, ALIMENT, GEOL brackish, HIDROL, OCEAN briny, brackish

salol *m* QUÍMICA salol

salón *m* TRANSP MAR *buques* saloon

salpicadero *m* PROD splash guard, splash wing, splasher

salpicador: **~ de aceite** *m* ING MECÁ oil slinger

salpicadura *f* CONST spatter, splashback; **~ de metal fundido** *f* SEG *ropa protectora* molten-metal splash; **~ de tungsteno** *f* CONST tungsten spatter

salpicaduras: **~ de acero líquido** *f pl* METAL scattering

saltachispas: **~ protector** *m* ELEC protective spark discharger

saltación *f* HIDROL saltation

saltamontes *m* AGRIC grasshopper

saltar[1]: **~ del freno** *m* FERRO *vehículos* brakes off

saltar[2] *vt* CINEMAT, IMPR skip

saltar[3] *vi* ELEC *arco* jump, PROD *remaches de una*

costura fly off, fly, TEC ESP *de un vehículo en emergencia* bail out, TRANSP MAR eddy, *el viento* shift

saltarse *v refl* PROD *hilos* snap

saltillo *m* TRANSP MAR *buque* raised deck

salto *m* AGUA fall, CRISTAL step, INFORM&PD jump, skip, MINAS leap, TEC ESP judder; **~ de agua** *m* ENERG RENOV head of water, HIDROL waterfall, *nivel del río* fall; **~ de agua diferencial** *m* ENERG RENOV differential head; **~ de agua efectivo** *m* ENERG RENOV effective head; **~ de agua elevado** *m* ENERG RENOV high head; **~ de agua medio** *m* ENERG RENOV medium head; **~ de aguas arriba** *m* ENERG RENOV upstream head; **~ de arco** *m* ELEC arc-over; **~ bajo** *m* ENERG RENOV low head; **~ de la banda de energía** *m* NUCL energy-band gap; **~ de caída libre** *m* D&A free-fall jump; **~ de carnero** *m* CONST *ferrocarriles* sheep's leap; **~ de cinta** *m* INFORM&PD tape skip; **~ condicional** *m* INFORM&PD conditional jump; **~ cuántico** *m* FÍS, FÍS PART quantum leap; **~ de falla** *m* CRISTAL leap; **~ de fase** *m* TRANSP *control de tráfico* phase skipping; **~ incondicional** *m* INFORM&PD unconditional jump; **~ de intercambio** *m* INFORM&PD exchange jump; **~ de línea** *m* INFORM&PD line feed (*LF*); **~ medio** *m* ENERG RENOV medium head; **~ de modos** *m* ÓPT mode hopping, *de un láser* mode jumping, TELECOM mode hopping, mode jumping; **~ de papel** *m* INFORM&PD paper skip, paper slew (*AmE*), paper throw (*BrE*); **~ en paracaídas** *m* D&A parachute drop; **~ de paracaídas a gran altitud** *m* D&A high-altitude low-opening (*HALO*); **~ pequeño** *m* INFORM&PD jitter; **~ de tensión** *m* ELEC voltage jump; **~ vertical** *m* GEOL *geología estructural* vertical throw; **~ de voltaje** *m* ELEC voltage jump

salud: **~ pública** *f* AGUA public health

saludo: **~ inicial** *m* INFORM&PD handshake; **~ inicial entre equipos** *m* PROD hardware handshaking

salva *f* D&A round, volley, *de artillería* salvo

salvado *m* AGRIC *veterinaria* bran; **~ de acemite** *m* ALIMENT *molienda, repostería, panadería* middlings bran

salvaguardia *f* SEG *de edificios* safeguarding

salvamento *m* SEG rescue, TRANSP AÉR recovery, rescue, TRANSP MAR rescue, *de buque, de bienes perdidos* salvage, *de personas o cosas en el mar* recovery; **~ aeromarítimo** *m* D&A air-sea rescue; **~ con medios marítimos y aéreos** *m* TRANSP MAR air-sea rescue

salvar *vt* INFORM&PD save, SEG, TRANSP AÉR, TRANSP MAR *buque, bienes perdidos, personas* rescue; **~ el desnivel de** *vt* AGUA lock; **~ mediante un puente** *vt* CONST bridge

salvarsán *m* QUÍMICA salvarsan

salvavidas: **~ de herradura** *m* TRANSP MAR horseshoe lifebuoy

salvo: **~ una constante arbitraria** *fra* FÍS to within an arbitrary constant; **~ error u omisión** *fra* IMPR errors and omissions excepted

samario *m* (*Sm*) QUÍMICA samarium (*Sm*)

samarsquita *f* MINERAL samarskite

saneamiento *m* AGUA cleansing, sanitation, CONTAM, RECICL reclamation, TEC PETR purging

sanear *vt* AGUA cleanse, CONST drain

saneo: **~ de frentes** *m* MINAS *túneles* lifter hole

sangrador *m* INSTAL HIDRÁUL clapper valve

sangradura *f* MECÁ bleeding

sangrar[1] *vt* CARBÓN *horno alto* tap, IMPR indent, MECÁ, P&C *plastificación* bleed

sangrar[2] *vi* CINEMAT bleed; **~ la escoria** *vi* CARBÓN deslag

sangría *f* IMPR indent, MECÁ bleeding, PROD *horno* tap; **~ recta** *f* IMPR squared indentation; **~ de vapor** *f* MECÁ bleeding

sanidad: **~ animal** *f* AGRIC animal health; **~ pecuaria** *f* AGRIC cattle sanitation; **~ pública** *f* AGUA public health, sanitation, SEG public health

sanidina *f* MINERAL sanidine

sanitarizado *m* INFORM&PD sanitization

santónico *adj* QUÍMICA santonic

santonina *f* QUÍMICA santonin

sapeli *m* TRANSP MAR *construcción naval* sapele

sapogenina *f* QUÍMICA sapogenin

saponificación *f* QUÍMICA saponification

saponificador *m* QUÍMICA saponifier

saponificante *adj* QUÍMICA saponifying

saponificar *vt* QUÍMICA saponify

saponina *f* QUÍMICA saponin

saponita *f* MINERAL bowlingite, saponite, soapstone

sapropel *m* OCEAN sapropel

saquete: **~ para ennegrecer** *m* PROD *fundición* blacking bag; **~ de espolvorear** *m* PROD dusting bag

SAR *abr* (*búsqueda y rescate*) D&A, TELECOM, TRANSP AÉR SAR (*search and rescue*)

sarcina *f* QUÍMICA sarcine

sarcinita *f* MINERAL sarkinite

sarcoláctico *adj* QUÍMICA sarcolactic

sarcolita *f* MINERAL sarcolite

sarcosina *f* QUÍMICA sarcosine, *aminoácido* sarcosin

sarda *f* MINERAL sard

sardinal *m* OCEAN *red de pesca* sardine net

sardónica *f* MINERAL sardonyx

sarga *f* TEXTIL *ligamentos* twill

sarna *f* AGRIC blind sector without traffic rights

sarta *f* TEC PETR *perforación* string; **~ de cementación** *f* TEC PETR *perforación* cementing string; **~ de fondo** *f* TEC PETR *perforación* bottom-hole assembly (*BHA*); **~ de lavar** *f* TEC PETR *perforación* washover string; **~ de perforación** *f* TEC PETR drill string; **~ de producción** *f* TEC PETR *producción* production string

sartorita *f* MINERAL sartorite

sassolina *f* MINERAL sassolite

sassolita *f* MINERAL sassoline

SATCOM *abr* TELECOM, TRANSP MAR (*comunicaciones por satélite, telecomunicaciones vía satélite*) SATCOM (*satellite communications*)

satélite *m* FÍS, IMPR, TEC ESP satellite; **~ activo** *m* TEC ESP active satellite; **~ artificial** *m* TEC ESP orbiter, TELECOM artificial satellite; **~ artificial recuperable** *m* TEC ESP recoverable orbiter; **~ asíncrono** *m* TEC ESP non-synchronous satellite; **~ de carga** *m* TEC ESP cargo satellite; **~ en cautividad** *m* TEC ESP tethered satellite; **~ de comprobación** *m* TEC ESP monitoring satellite; **~ de comunicaciones** *m* TEC ESP communication satellite; **~ de contrastación** *m* TEC ESP monitoring satellite; **~ de detección a distancia** *m* TEC ESP remote sensing satellite; **~ de detección nuclear** *m* TEC ESP nuclear-detection satellite; **~ de detección remota** *m* TEC ESP remote sensing satellite; **~ de difusión en directo** *m* TV direct broadcast satellite (*DBS*); **~ de doble giro** *m* TEC ESP dual-spin satellite; **~ de doble rotación** *m* TEC ESP dual-spin satellite; **~ para exploración del ambiente** *m* TEC ESP

environment-survey satellite; **~ geoestacionario** *m* FÍS geostationary satellite, TEC ESP synchronous satellite, geostationary satellite, TELECOM, TRANSP MAR, TV geostationary satellite; **~ investigación de recursos terrestres** *m* TEC ESP earth-resources research satellite; **~ de mantenimiento en órbita** *m* TEC ESP station-keeping satellite; **~ marítimo** *m* TEC ESP maritime satellite; **~ meteorológico** *m* METEO, TEC ESP meteorological satellite; **~ no fijo** *m* TEC ESP non-synchronous satellite; **~ de observación** *m* TEC ESP observation satellite; **~ de observación militar** *m* D&A, TEC ESP military observation satellite; **~ de observación de recursos terrestres** *m* TEC ESP earth-resources research satellite; **~ de observación de la tierra** *m* TEC ESP earth remote-sensing satellite; **~ en órbita ecuatorial** *m* TEC ESP equatorial-orbiting satellite; **~ en órbita sincrónica con la Tierra** *m* TEC ESP earth-synchronous satellite, orbiting satellite; **~ pasivo** *m* TEC ESP, TV passive satellite; **~ de percepción remota** *m* TEC ESP earth remote-sensing satellite; **~ de radiodifusión** *m* TEC ESP broadcasting satellite, radio-broadcast satellite, TELECOM radio-broadcast satellite, TV broadcasting satellite; **~ de radiodifusión televisiva** *m* TEC ESP, TELECOM, TV television broadcast satellite; **~ de recogida de datos** *m* TEC ESP data-relay satellite (*DRS*); **~ para recolección de datos** *m* TEC ESP data-collection satellite; **~ reflector** *m* ÓPT, TEC ESP reflecting satellite; **~ repetidor** *m* TEC ESP, TV relay satellite, repeater satellite; **~ repetidor en tiempo real** *m* CINEMAT, ELECTRÓN, INFORM&PD, TELECOM, TV real-time repeater satellite; **~ de retransmisión** *m* TEC ESP, TV relay satellite; **~ retransmisor de datos** *m* TEC ESP data-relay satellite (*DRS*); **~ secundario en el lanzador** *m* TEC ESP piggyback satellite; **~ síncrono** *m* TEC ESP synchronous satellite; **~ síncronosolar** *m* TEC ESP sun-synchronous satellite; **~ solar síncrona** *m* TEC ESP sun-synchronous satellite; **~ subsíncrono** *m* TEC ESP subsynchronous satellite; **~ de telecomunicaciones** *m* TRANSP MAR communications satellite (*COMSAT*); **~ de teledetección** *m* TEC ESP remote sensing satellite; **~ de televisión directa** *m* TEC ESP, TELECOM, TV television broadcast satellite; **~ terrestre** *m* TEC ESP earth satellite; **~ terrestre con recepción a distancia** *m* TEC ESP earth remote-sensing satellite; **~ para transmisión de datos** *m* TEC ESP data-relay satellite (*DRS*); **~ transmisor de señales** *m* TEC ESP active satellite; **~ transportado sobre otro satélite** *m* TEC ESP piggyback satellite; **~ de utilidad** *m* TEC ESP utility satellite; **~ de vigilancia** *m* TEC ESP surveillance satellite

Satélite: **~ para Comunicaciones Internacionales** *m* (*INTELSAT*) TEC ESP International Communication Satellite (*INTELSAT*)

satén *m* TEXTIL satin

satinado[1]: **~ entre placas** *adj* PAPEL plate-glazed; **~ por fricción** *adj* PAPEL friction-glazed

satinado[2] *m* C&V, IMPR, MECÁ burnishing, PAPEL glaze, glazing, PROD glazing, gloss; **~ por cepillado** *m* PAPEL brush glazing; **~ entre placas** *m* PAPEL plate glazing; **~ por fricción** *m* PAPEL friction glazing; **~ en húmedo** *m* PAPEL water finishing

satinadora *f* EMB supercalender

satinar *vt* PROD *papel*, TEXTIL glaze

satinómetro *m* P&C *pinturas, instrumento* gloss meter

satnav *m* (*sistema de radionavegación por satélite*) TEC ESP, TRANSP MAR satnav (*satellite navigator*)

saturación *f* GEN saturation; **~ de aceite residual** *f* PETROL residual-oil saturation (*ROS*); **~ de agua** *f* PETROL water saturation; **~ de agua residual** *f* HIDROL, PETROL residual-water saturation; **~ del ánodo** *f* ING ELÉC anode saturation; **~ de bandas** *f* TV saturation banding; **~ de blanco** *f* TV white saturation; **~ de color** *f* ELECTRÓN color saturation (*AmE*), colour saturation (*BrE*); **~ de equilibrio** *f* PETROL equilibrium saturation; **~ de gas** *f* PETROL gas saturation; **~ de hidrocarburos** *f* PETROL hydrocarbon saturation; **~ irreducible de agua** *f* TEC PETR irreducible water saturation; **~ magnética** *f* ELEC, ING ELÉC magnetic saturation; **~ de petróleo** *f* PETROL oil saturation; **~ térmica** *f* ELECTRÓN temperature saturation; **~ del transistor** *f* ELECTRÓN transistor saturation

saturado *adj* CINEMAT, PAPEL, QUÍMICA *solución, compuesto*, REFRIG saturated; **no ~** *adj* ALIMENT, P&C, QUÍMICA, TEC PETR unsaturated

saturante *m* QUÍMICA saturant

saturar *vt* QUÍMICA impregnate

sazonar *vt* ALIMENT flavor (*AmE*), flavour (*BrE*)

Sb *abr* (*antimonio*) C&V, METAL, QUÍMICA Sb (*antimony*)

SBE *abr* (*semiconductor de bombardeo electrónico*) ELECTRÓN, FÍS PART, NUCL EBS (*electron-bombarded semiconductor*)

SBLOCA *abr* (*LOCA pequeño*) NUCL SBLOCA (*small break LOCA*)

SBM *abr* (*anclaje a boya simple*) TEC PETR SBM (*single-buoy mooring*)

SBR *abr* (*caucho butadieno-estireno*) P&C SBR (*styrene butadiene rubber*)

Sc *abr* METEO (*estratocúmulo*) Sc (*stratocumulus*), QUÍMICA (*escandio*) Sc (*scandium*)

scheelita *f* MINERAL scheelite

scheelitina *f* MINERAL scheelitine

scheererita *f* MINERAL scheererite

schefferita *f* MINERAL schefferite

schreibersita *f* MINERAL schreibersite

schwartzembergita *f* MINERAL schwartzembergite

schwatzita *f* MINERAL schwatzite

SCOM *abr* (*semiconductor complementario de óxido de metal*) ELECTRÓN, PROD CMOS (*complementary metal oxide semiconductor*)

SCPC *abr* TEC ESP, TELECOM (*canal único por portadora, portadora monocanal, un solo canal por portadora*) SCPC (*single channel per carrier*)

SCPO *abr* (*sistema para control de posición y órbita*) TEC ESP AOCS (*altitude and orbit control system*)

SDLC *abr* (*procedimiento de control de transmisión síncrona*) INFORM&PD SDLC (*synchronous data-link control*)

Se *abr* (*selenio*) ING ELÉC, QUÍMICA Se (*selenium*)

Seabee *m* TRANSP MAR Seabee carrier

sebácico *adj* QUÍMICA *ácido* sebacic

sebo *m* ALIMENT, QUÍMICA, TEXTIL tallow; **~ mineral** *m* MINERAL hatchetine, mineral tallow

seboso *adj* ALIMENT, QUÍMICA, TEXTIL tallowy

sec *abr* (*secante*) CONST, GEOM, INFORM&PD, MATEMÁT *trigonometría* sec (*secant*)

seca: **~~pastas** *f* PAPEL pulp machine

secadero *m* ALIMENT kiln, PROD drying floor, drying room, TERMO dryer; **~ de estanterías a vacío** *m* (*SCA*) ALIMENT vacuum shelf dryer; **~ de maíz** *m* AGRIC maize drying shed; **~ para quesos** *m* ALIMENT, REFRIG cheese-drying room; **~ de vapor** *m* INSTAL HIDRÁUL steam dryer

secado[1]: **~ al aire** *adj* ALIMENT, EMB air-dried; **~ por calor** *adj* IMPR heat-set; **~ en estufa** *adj* PAPEL oven-dry, TERMO oven-dried; **~ por goteo** *adj* TEXTIL *proceso de acabado* drip-dry; **~ en horno** *adj* PAPEL oven-dry, TERMO kiln-dried, oven-dried; **de ~ rápido** *adj* EMB, TEXTIL quick-drying

secado[2] *m* CARBÓN drying, CONST de-watering, seasoning, NUCL dry-out, P&C *de la pintura, operación*, PAPEL, QUÍMICA, TEC PETR, TERMO drying, TEXTIL drying, baking; **~ en bidón** *m* NUCL in-drum drying; **~ por centrífuga** *m* C&V centrifugal drawing; **~ por choque** *m* CINEMAT impingement drying; **~ por chorro de aire** *m* IMPR air drying; **~ por congelación** *m* ALIMENT, EMB freeze-drying, PROC QUÍ freeze-drying, lyophilization, REFRIG, TERMO freeze-drying; **~ por convección** *m* TERMO convection drying; **~ en espiga** *m* AGRIC ear-drying; **~ de espuma en lecho** *m* ALIMENT foam mat drying; **~ de espuma al vacío** *m* ALIMENT foam vacuum drying; **~ en estufa** *m* PROD kiln-drying, REFRIG oven dehydration; **~ mediante estufa** *m* PROD, REVEST stoving; **~ al horno** *m* CONST kiln drying, PROD kiln-drying; **~ en mazorca** *m* AGRIC ear-drying; **~ superficial** *m* PROD *fundería* skin-drying; **~ de la tinta** *m* IMPR ink setting

secador *m* IMPR, P&C, QUÍMICA, TERMO dryer; **~ de banda colgante** *m* PAPEL festoon drier, festoon dryer; **~ con circulación de aire caliente a través del papel** *m* PAPEL through-drying machine; **~ por congelación** *m* ALIMENT, PROC QUÍ, REFRIG freeze-dryer; **~ de convección** *m* TERMO convection dryer; **~ de correa** *m* PROC QUÍ belt dryer; **~ de doble cilindro** *m* ALIMENT double drum dryer; **~ eléctrico** *m* ELEC, INSTAL TERM electric dryer; **~ del fieltro** *m* PAPEL felt dryer; **~ por flotación de aire** *m* IMPR air flotation drier; **~ de infrarrojos** *m* IMPR infrared dryer, PAPEL radiant dryer; **~ de lecho fluidificado** *m* ALIMENT, INSTAL TERM, PROC QUÍ fluidized-bed dryer; **~ de lecho fluidizado** *m* ALIMENT, INSTAL TERM, PROC QUÍ fluidized-bed dryer; **~ de lecho fluido** *m* ALIMENT, INSTAL TERM, PROC QUÍ fluidized-bed dryer; **~ de película** *m* CINEMAT, FOTO film dryer; **~ radiante** *m* PAPEL radiant dryer; **~ de rodillos** *m* ALIMENT drum dryer; **~ rotativo** *m* PROD rotary dryer; **~ de tambor** *m* CARBÓN, TERMO drum dryer; **~ de tamices** *m* PROC QUÍ sieve dryer; **~ a vacío** *m* ALIMENT vacuum dryer, vacuum drying; **~ de vacío** *m* CARBÓN vacuum dryer; **~ de vapor** *m* INSTAL HIDRÁUL steam dryer; **~ Yankee** *m* PAPEL yankee dryer

secadora *f* AGRIC, CARBÓN dryer, EMB drying machine, ING MECÁ, TEXTIL dryer; **~ centrífuga** *f* CARBÓN, PROC QUÍ centrifugal dryer; **~ por centrifugado** *f* TEXTIL tumble dryer; **~ por choque de aire caliente** *f* TEXTIL hot-air impingement dryer; **~ cilíndrica** *f* TEXTIL cylinder drying machine; **~ esmaltadora** *f* FOTO dryer glazer; **~ de espigas** *f* AGRIC ear-dryer; **~ de flujo continuo** *f* AGRIC continuous-flow dryer; **~ de listones** *f* PAPEL slat dryer; **~ de mazorcas** *f* AGRIC ear-dryer; **~ para muela rectificadora** *f* ING MECÁ dryer for grinding wheel; **~ de pruebas** *f* FOTO print-dryer; **~ rotativa alimentada por gasóleo** *f* INSTAL TERM *planta de asfaltado* oil-fired rotary dryer; ~ **de tambor** *f* TEXTIL drum dryer; ~ **tubular** *f* TEXTIL tubular dryer; ~ **por volteo** *f* TEXTIL tumble dryer

SECAM *abr* (*sistema de color secuencial con memoria*) TV SECAM (*sequential color with memory system AmE, sequential colour with memory system BrE*)

secante¹ *adj* PAPEL, TERMO siccative

secante² *f* ALIMENT desiccant, siccative, CONSTR secant (*sec*), EMB drying agent, GEOM secant (*sec*), IMPR dryer, INFORM&PD, MATHEMÁT secant (*sec*), TERMO siccative

secar *vt* ALIMENT desiccate, CARBÓN, TERMO dry, TEXTIL bake; ~ **al aire** *vt* CONST weather; ~ **con aire frío** *vt* TERMO dry by cold air; ~ **por calor** *vt* TERMO dry by heat; ~ **al horno** *vt* PROD kiln; ~ **en horno** *vt* TERMO kiln-dry

secarse *v refl* AGUA riachuelos, canales run dry, CARBÓN dry, HIDROL, OCEAN run dry; ~ **al aire** *v refl* CONST weather

sección *f* GEOL, GEOM section, ING MECÁ compartment, run, section, segment, PETROL, PROD, TELECOM, TEXTIL section, TRANSP *tráfico* platoon; ~ **abocinada** *f* TEC ESP antenas, guías de onda flared section; ~ **de bobina** *f* ELEC coil section; ~ **de cable** *f* ELEC, ING ELÉC cable section; ~ **caliente** *f* TRANSP AÉR hot section; ~ **de carril** *f* FERRO vehículos rail section; ~ **central del ala** *f* TRANSP AÉR center wing section (*AmE*), centre wing section (*BrE*); ~ **circular** *f* ING ELÉC pie section; ~ **de congelación** *f* REFRIG en planta de tratamiento de alimentos freezing section; ~ **contraída** *f* INSTAL HIDRÁUL contracted section; ~ **crítica** *f* INFORM&PD critical section; ~ **decreciente** *f* ING ELÉC tapered section; ~ **delgada** *f* CRISTAL thin section; ~ **desmontable** *f* TRANSP collapsible section; ~ **de diámetro progresivamente mayor** *f* TEC ESP antenas, guías de ondas flared section; ~ **digital** *f* TELECOM digital section (*DS*); ~ **eficaz** *f* FÍS cross section, túnel de viento working section; ~ **eficaz de absorción** *f* FÍS RAD absorption cross-section; ~ **eficaz de absorción neutrónica cero** *f* NUCL zero neutron-absorption cross-section; ~ **eficaz atómica** *f* NUCL atomic cross-section; ~ **eficaz de difusión** *f* FÍS scattering cross-section; ~ **eficaz de fisión** *f* FÍS fission cross-section; ~ **eficaz macroscópica** *f* NUCL macroscopic cross-section; ~ **eficaz de Thomson** *f* FÍS Thomson cross-section; ~ **de empaquetado** *f* EMB packing station; ~ **de enfriamiento** *f* ALIMENT, REFRIG cooling section; ~ **de entrada** *f* PROD input section; ~ **específica del dominio** *f* TELECOM domain-specific part (*DSP*); ~ **de filtro** *f* ELECTRÓN filter section; ~ **final** *f* NUCL end section; ~ **fría** *f* TRANSP AÉR cold section; ~ **frontal abierta** *f* INSTAL HIDRÁUL open front; ~ **geológica** *f* GEOL geological section; ~ **de guía de ondas** *f* ING ELÉC waveguide section; ~ **de guíaondas** *f* ING ELÉC waveguide section; ~ **de herradura** *f* ING MECÁ horseshoe section; ~ **de intensidad máxima de esfuerzo** *f* ING MECÁ section of maximum intensity of stress; ~ **de intensidad máxima de fatiga** *f* ING MECÁ section of maximum intensity of stress; ~ **en L** *f* ELECTRÓN L section; ~ **laminada** *f* P&C plásticos, transformación laminated section; ~ **libre de paso de aire** *f* REFRIG entre las aletas de la parrilla del acondicionador de aire free area; ~ **de línea aérea** *f* TELECOM section overhead (*SOH*); ~ **longitudinal** *f* FERRO longitudinal section,

TRANSP MAR diseño naval buttock lines; ~ **media** *f* TRANSP MAR geometría del buque midship section; ~ **de múltiplex** *f* TELECOM multiplex section (*MS*); ~ **neutra** *f* FERRO línea de contacto neutral section; ~ **no presurizada** *f* TRANSP AÉR nonpressurized section; ~ **de oro** *f* GEOM golden section; ~ **de paso bajo** *f* ELECTRÓN low-pass section; ~ **plegable** *f* TRANSP collapsible section; ~ **de plegado** *f* IMPR folding station; ~ **de popa** *f* TRANSP MAR de barco aft section; ~ **de un programa** *f* INFORM&PD bead; ~ **de prueba** *f* FÍS túnel de viento test section; ~ **ranurada de guía de ondas** *f* ING ELÉC waveguide slotted section; ~ **ranurada de guíaondas** *f* ING ELÉC waveguide slotted section; ~ **del regenerador** *f* TELECOM de impulsos regenerator section; ~ **de resistencia uniforme** *f* ING MECÁ section of uniform strength; ~ **de RF** *f* ELEC, ELECTRÓN, ING ELÉC RF section; ~ **de secado** *f* ALIMENT drying section, PAPEL dryer section; ~ **de secado de varios cilindros** *f* PAPEL multicylinder drier section, multicylinder dryer section; ~ **de secadores superpuestos** *f* PAPEL stacked-drier section, stacked-dryer section; ~ **solapada** *f* FERRO infraestructura overlap section; ~ **en T** *f* ELECTRÓN T-section; ~ **total de rejilla** *f* REFRIG core area; ~ **de tramo** *f* FERRO infraestructura block section; ~ **transversal** *f* AGUA, CONST, GEOL cross section, ING MECÁ cross section, section, PAPEL, PROD cross section, TELECOM transverse section, TEXTIL cross section, TRANSP MAR diseño de barcos body section, cross section, station, ordinate, transverse section; ~ **transversal de la pala** *f* TRANSP AÉR blade cross-section; ~ **transversal de un puente** *f* CONST bridge shown in cross-section; ~ **transversal rectangular** *f* CONST rectangular cross-section; ~ **variable por condensador** *f* ING ELÉC variable-capacitor section; ~ **ventral** *f* FÍS ONDAS antinode; ~ **de vía** *f* FERRO infraestructura track section, ING ELÉC telegraph line

seccionado *adj* GEOM sectional

seccionador *m* ELEC red de distribución sectionalizing switch, *conmutador* line breaker, *interruptor de seccionamiento* disconnecting switch, ING ELÉC isolating switch; ~ **de barra colectora** *m* ELEC busbar-sectionalizing switch; ~ **de ferrita** *m* ING ELÉC ferrite isolator; ~ **de potencia** *m* ELEC load switch; ~ **tripolar** *m* ING ELÉC three-pole switch

seccional *adj* GEOM, ING MECÁ sectional

seccionamiento *m* ELEC red de distribución sectionalization, PROD slicing, TV tearing; ~ **total** *m* NUCL double-ended break

secciones¹: **en** ~ *adj* ING MECÁ sectional

secciones²: **por** ~ *adv* ING MECÁ sectional

secesión: ~ **convergente** *f* MATEMÁT convergent sequence

seco¹ *adj* ALIMENT dried, withered, CONST ladrillo dried, QUÍMICA drying, TERMO dry, dried, TEXTIL dry; **al** ~ **absoluto** *adj* PAPEL a $100°C$ bone-dry; ~ **al aire** *adj* PAPEL air-dried; **en** ~ **estampado** *adj* IMPR blind embossing

seco²: **en** ~ *adv* TRANSP MAR aground

secreto: ~ **de la comunicación** *m* TELECOM privacy; ~ **de datos** *m* INFORM&PD data privacy

sector¹: **de** ~ **duro** *adj* INFORM&PD disco hard-sectored

sector² *m* CONST, GEOM sector, INFORM&PD slice, TEC PETR, TRANSP MAR de una luz sector; ~ **agrícola** *m* AGRIC farming sector; ~ **angular** *m* MECÁ angle section; ~ **de bienes de consumo** *m* EMB consumer-goods sector; ~ **del cambio de marcha** *m* ING MECÁ quadrant, reversing link; ~ **ciego** *m* TRANSP AÉR blind sector; ~ **comercial** *m* GAS commercial sector; ~ **de la construcción** *m* CONST building trade; ~ **de control** *m* TRANSP AÉR control sector; ~ **dentado** *m* AUTO sector gear, ING MECÁ quadrant, rack, sector gear, sector wheel, segment gear, toothed sector, MECÁ rack, VEH *dirección* sector gear; ~ **de disco** *m* INFORM&PD disk sector; ~ **esférico** *m* GEOM spherical sector; ~ **marítimo** *m* TRANSP MAR maritime industry; ~ **muerto** *m* FERRO dead sector; ~ **redondo del muelle en espiral** *m* ING MECÁ round-section coil spring; ~ **redondo del muelle helicoidal** *m* ING MECÁ round-section coil spring; ~ **de Stephenson** *m* ING MECÁ reversing link; ~ **del timón** *m* TRANSP AÉR, TRANSP MAR rudder quadrant; ~ **de unidad binaria** *m* ELECTRÓN bit slice

sectorización: ~ **blanda** *f* INFORM&PD soft sectoring; ~ **dura** *f* INFORM&PD hard sectoring

sectorizado: ~ **blando** *adj* INFORM&PD soft-sectored

secuencia *f* ACÚST sequence, ELECTRÓN frame, GEOL *de los estratos* sequence, INFORM&PD sequencing, MATEMÁT sequence; ~ **de apilamiento** *f* CRISTAL stacking sequence; ~ **de aterrizaje** *f* TRANSP AÉR landing sequence; ~ **binaria** *f* INFORM&PD, TELECOM binary sequence; ~ **binaria pseudoaleatoria** *f* TELECOM pseudo-random binary sequence (*PRBS*); ~ **de capas** *f* CARBÓN layer sequence; ~ **de las capas** *f* CARBÓN seam sequence; ~ **de colación** *f* INFORM&PD collating sequence; ~ **de comprobación** *f* INFORM&PD sequence check, TELECOM check sequence; ~ **condensada** *f* GEOL condensed sequence; ~ **de conmutación** *f* ELEC switching sequence; ~ **de conmutación de colores** *f* TV color-sampling sequence (*AmE*), colour-sampling sequence (*BrE*); ~ **de control** *f* INFORM&PD control sequence; ~ **convergente** *f* MATEMÁT convergent sequence; ~ **cronometrada** *f* ELECTRÓN clocking sequence; ~ **de depósito** *f* GEOL depositional sequence; ~ **entrelazada** *f* TV interlace sequence; ~ **de escape** *f* INFORM&PD escape sequence; ~ **de estratos** *f* CARBÓN layer sequence; ~ **de exploración** *f* PROD scanning sequence; ~ **de fases** *f* ELEC, ELECTRÓN phase sequence; ~ **de fases positivas** *f* ELEC *generador de AC* positive-phase sequence; ~ **granocreciente** *f* GEOL coarsening-up sequence; ~ **granodecreciente** *f* GEOL fining-up sequence; ~ **de imagen** *f* TELECOM image sequence; ~ **de impulsos** *f* ELECTRÓN pulse sequence; ~ **de intercalación** *f* INFORM&PD collating sequence; ~ **de llamada** *f* INFORM&PD calling sequence, TELECOM call sequence; ~ **de marcación** *f* TELECOM marking sequence; ~ **negativa** *f* GEOL coarsening-up sequence; ~ **de operaciones** *f* PROD operations sequence; ~ **de paquetes** *f* INFORM&PD packet sequencing; ~ **positiva** *f* ELEC *CA* positive sequence, GEOL fining-up sequence; ~ **prioritaria** *f* INFORM&PD priority sequencing; ~ **de pseudorruido** *f* TRANSP MAR *comunicación por satélite* pseudonoise sequence; ~ **de pseudorruidos digitales** *f* TELECOM digital pseudonoise sequence; ~ **de retroceso** *f* PROD *en soldaduras* backstep sequence; ~ **de retrocohete** *f* TEC ESP, TRANSP retrorocket sequence; ~ **sedimentaria** *f* GEOL sedimentary sequence; ~ **de señales electrónicas entre equipos** *f* PROD hard-

ware handshaking; ~ **de soldadura** *f* CONST welding sequence; ~ **de somerización** *f* GEOL shallowing-up sequence; ~ **de vuelo** *f* TEC ESP, TRANSP AÉR flight sequence; ~ **en vuelo** *f* TEC ESP in-flight sequence

secuenciado *m* INFORM&PD sequencing

secuenciador *m* INFORM&PD, ING ELÉC, PROD, TELECOM sequencer; ~ **fijo** *m* TELECOM fixed sequencer; ~ **programable** *m* TELECOM programmable sequencer

secuencial *adj* INFORM&PD sequential, TELECOM serial, sequential

secuenciamiento *m* TELECOM *de serie de operaciones* sequencing

secuestrar *vt* QUÍMICA sequester

secuestreno *m* QUÍMICA sequestrene

secular *adj* FÍS, GEOFÍS secular

secundarios: ~ **de fase invertida** *m pl* ING ELÉC phase-reversed secondaries

seda *f* C&V, TEXTIL silk; ~ **para estarcir** *f* C&V stencil silk

sedán *m* TRANSP *automóviles* saloon (*BrE*), sedan (*AmE*)

sedentario *adj* CARBÓN sedentary

sedimentación *f* AGUA sedimentation, CARBÓN settlement, CONTAM *depuración* sedimentation, deposition, FÍS sedimentation, GEOL, HIDROL silting up, P&C *pigmentos, materiales de carga* sedimentation, QUÍMICA *depósito lodoso, azolve* silting, sedimentation, settling, RECICL, TEC PETR *geología* sedimentation; ~ **aluvial** *f* OCEAN alluviation; ~ **biogénica** *f* GEOL biogenic sedimentation; ~ **cíclica** *f* GEOL cyclic sedimentation; ~ **de contaminante** *f* CONTAM pollutant deposition; ~ **húmeda** *f* CONTAM *depuración* wet deposition; ~ **preliminar** *f* HIDROL *alcantarillado* preliminary sedimentation; ~ **primaria** *f* CARBÓN primary settlement; ~ **química** *f* GEOL, QUÍMICA chemical sedimentation; ~ **regulada** *f* CONTAM regulated deposition; ~ **seca** *f* CONTAM *depuración* dry deposition; ~ **secundaria** *f* CARBÓN secondary settlement; ~ **de la tinta** *f* IMPR ink lay-down; ~ **total** *f* CONTAM *depuración* total deposition

sedimentado *adj* AGUA, CARBÓN, PROC QUÍ, QUÍMICA, RECICL, TEC PETR sedimented

sedimentador: ~ **desbastador** *m* AGUA roughing tank

sedimentar *vt* PROC QUÍ settle

sedimentarse *v refl* PROC QUÍ thicken

sedimento *m* AGUA sediment, silt, CARBÓN deposit, sediment, settling, silt, GEOL deposit, HIDROL settling, sludge, OCEAN *redes planctónicas* bottom, PROC QUÍ, QUÍMICA sediment, RECICL sludge, sediment, TEC PETR *geología* sediment; ~ **ácido** *m* CONTAM *química* acid deposit; ~ **activado** *m* RECICL activated sludge; ~ **acuoígneo** *m* GEOL hydrothermal deposit; ~ **arcilloso** *m* AGUA clay silt; ~ **biológico** *m* HIDROL *aguas residuales* biological sludge; ~ **carbonoso del aceite** *m* AUTO oil-carbon deposit; ~ **concentrado** *m* CARBÓN concentrated sludge; ~ **costero** *m* AGUA coastal deposit; ~ **detrítico** *m* OCEAN detrital sediment; ~ **inicial** *m* CARBÓN initial settlement; ~ **marino** *m* OCEAN marine sediment; ~ **no tratado** *m* AGUA, CONTAM, RECICL raw sludge; ~ **sapropélico** *m* GEOL sapropelic deposit; ~ **sin tratar** *m* AGUA, CONTAM, RECICL raw sludge

sedimentos *m pl* CARBÓN cuttings, dirt; ~ **cíclicos estacionales** *m pl* GEOL varve; ~ **del fondo** *m pl*

GEOL *sedimentación deltáica* bottomset beds; ~ **del lecho** *m pl* HIDROL bed deposit

segador *m* AGRIC mower

segadora: ~-**atadora** *f* AGRIC reaper-binder; ~ **de césped** *f* CONST lawn mower; ~ **de cultivos en líneas** *f* AGRIC row-binder; ~ **de rastrillo** *f* AGRIC self-rake reaper; ~ **de rastrillos** *f* AGRIC reaper, sail reaper

segmentación *f* GEOM, INFORM&PD, QUÍMICA segmentation; ~ **y rearmado** *f* TELECOM segmentation and reassembly

segmental *adj* GEOM segmental

segmentar *vt* GEOM, INFORM&PD segment

segmentario *adj* GEOM segmental

segmento *m* AUTO ring, GEOM section, segment, INFORM&PD, ING MECÁ segment, MECÁ ring, NUCL, TEC ESP, TELECOM segment, VEH ring; ~ **UNA** *m* TELECOM UNA segment; ~ **de compresión** *m* AUTO, VEH *motores, pistones* compression ring; ~ **de control de engrase** *m* AUTO oil control ring, oil ring; ~ **dentado** *m* ING MECÁ toothed segment; ~ **de engrase** *m* AUTO oil control ring, oil ring, slotted oil control ring; ~ **de engrase con expansores** *m* AUTO oil-expander ring; ~ **espacial** *m* TEC ESP space segment; ~ **del inducido** *m* ING ELÉC armature bar; ~ **ISA** *m* TELECOM ISA segment; ~ **lunar** *m* TEC ESP moon segment; ~ **de Marte** *m* TEC ESP Mars segment; ~ **del pilar de en medio** *m* CARBÓN middle-pile segment; ~ **del pilote** *m* CARBÓN pile segment; ~ **de pistón** *m* MECÁ ring; ~ **del pistón** *m* AUTO, VEH *motor* piston ring; ~ **principal** *m* ING ELÉC main bar; ~ **rascador de aceite** *m* VEH *pistón* oil control ring; ~ **de recta** *m* GEOM line segment; ~ **de revestimiento** *m* CONST lining segment; ~ **terreno** *m* TEC ESP ground segment; ~ **terrestre** *m* TEC ESP ground segment; ~ **de tiempo** *m* TELECOM time slot; ~ **de transición** *m* TRANSP AÉR transition segment; ~ **UNB** *m* TELECOM UNB segment; ~ **UNH** *m* TELECOM UNH segment

segmentos: ~ **lisos** *m pl* TEC ESP *alojamientos del cierre de cañones* slot

segregación *f* C&V segregation

segregado *m* NUCL segregation

seguidor *m* ING MECÁ follower, TEC ESP tracker, TRANSP MAR *radar* tracker register; ~ **de aguja** *m* ING MECÁ needle follower; ~ **de cátodo** *m* FÍS cathode follower; ~ **de emisor** *m* FÍS, ING ELÉC emitter follower; ~ **de estrellas** *m* TEC ESP star tracker; ~ **láser** *m* ELECTRÓN laser tracker; ~ **de leva** *m* ING MECÁ, MECÁ, TRANSP AÉR cam follower

seguimiento *m* D&A, OCEAN, TEC ESP, TELECOM, TV tracking; ~ **de auditoría** *m* INFORM&PD audit trail; ~ **automático** *m* TEC ESP autotracking, TRANSP MAR *con el radar* automatic tracking, TV autotracking; ~ **de la desintegración del muón** *m* FÍS RAD muon decay track; ~ **espacial** *m* D&A, TEC ESP space tracking; ~ **de fabricación** *m* PROD expediting, manufacturing follow-up; ~ **por láser** *m* ELECTRÓN laser tracking; ~ **de órbita** *m* TEC ESP orbit tracking; ~ **con radar** *m* D&A, TRANSP, TRANSP AÉR, TRANSP MAR radar tracking (*RT*); ~ **por radar** *m* D&A, TRANSP, TRANSP AÉR, TRANSP MAR radar tracking (*RT*); ~ **de radiación** *m* FÍS RAD, TEC ESP radiation monitoring; ~ **radioeléctrico** *m* D&A radio tracking

seguir *vt* CINEMAT track

segunda: ~ **cubierta** *f* TRANSP MAR *transbordadores*

second deck; ~ **generación** *f* INFORM&PD second generation; ~ **generación filial** *f* AGRIC F2 generation; ~ **ley** *f* FÍS *de la termodinámica* second law; ~ **marcha** *f* AUTO second gear; ~ **mayor** *f* ACÚST major second; ~ **menor** *f* ACÚST minor second; ~ **parte** *f* ACÚST second (*s*); ~ **parte del ciclo de combustible nuclear** *f* NUCL back end of nuclear fuel cycle; ~ **pasada** *f* ING MECÁ *de macho, terraja, herramienta de corte* second cut; ~ **prueba** *f* IMPR second proof; ~ **unidad** *f* CINEMAT *filmación* second unit; ~ **zona telefónica de la central de línea privada** *f* TELECOM second-tier trunk exchange area

segundero *m* ING MECÁ second hand

segundo¹ *adj* ACÚST second

segundo² *m* (*s*) ACÚST *de tiempo*, CONST, FÍS, METR second (*s*); ~ **ánodo** *m* TV second anode; ~ **del arco** *m* FÍS arc second; ~ **armónico** *m* ELECTRÓN second harmonic; ~ **centro de conexiones telefónicas de línea privada** *m* TELECOM second-tier trunk switching center (*AmE*), second-tier trunk switching centre (*BrE*); ~ **cuartillo** *m* D&A second dog watch; ~ **erróneo** *m* TELECOM errored second (*ES*); ~ **fuera** *m* TELECOM OFS (*out-of-frame second*); ~ **fuera de cuadro** *m* TELECOM out-of-frame second (*OFS*); ~ **indisponible** *m* TELECOM unavailable second (*UAS*); ~ **inutilizable** *m* TELECOM unavailable second (*UAS*); ~ **macho de roscar** *m* ING MECÁ second tap; ~ **maquinista** *m* TRANSP MAR *a bordo* second engineer; ~ **momento de inercia del área** *m* ING MECÁ second moment of area; ~ **no disponible** *m* TELECOM unavailable second (*UAS*); ~ **de pérdida de trama** *m* TELECOM frame loss second (*FLS*); ~ **plano** *m* FOTO background; ~ **severamente erróneo** *m* TELECOM severely errored second (*SES*); ~ **vecino próximo** *m* CRISTAL second nearest neighbor (*AmE*), second nearest neighbour (*BrE*)

segundogrado *m* ACÚST *música* second (*s*)

seguridad¹: **de** ~ *adj* INFORM&PD backup

seguridad² *f* CALIDAD dependability, reliability, security, safety, CONTAM MAR, CRISTAL, ELEC, INFORM&PD, ING ELÉC, ING MECÁ reliability, MECÁ safety, METR reliability, PROD security, SEG *protección de edificios y personas* security, *para máquinas de manejo manual* safety, TEC ESP, TELECOM, TV reliability; ~ **activa de vehículos de motor** *f* TRANSP active motor vehicle safety; ~ **de archivos** *f* INFORM&PD file security; ~ **en la carretera** *f* CONST, SEG road safety; ~ **en caso de aglomeración** *f* SEG crowd safety; ~ **del ciclotrón** *f* FÍS, FÍS RAD, SEG cyclotron safety; ~ **de computadora** *f* AmL (*cf seguridad de ordenadores Esp*) INFORM&PD computer security; ~ **contra incendios** *f* D&A, SEG fire safety; ~ **de datos** *f* INFORM&PD data security; ~ **de ficheros** *f* INFORM&PD file security; ~ **física** *f* TELECOM physical security; ~ **fluvial** *f* SEG, TRANSP MAR river safety; ~ **de funcionamiento** *f* ING MECÁ dependability; ~ **en el hogar** *f* SEG safety in the home; ~ **industrial** *f* SEG industrial safety; ~ **informática** *f* INFORM&PD computer security; ~ **integrada** *f* TEC ESP integrated safety; ~ **en el mar** *f* SEG marine safety; ~ **marítima** *f* SEG, TRANSP MAR maritime safety; ~ **en los métodos de trabajo** *f* SEG safe methods of working; ~ **nuclear** *f* SEG nuclear safety; ~ **de ordenadores** *f* Esp (*cf seguridad de computadora AmL*) INFORM&PD computer security; ~ **pasiva de vehículos de motor** *f* TRANSP passive motor-vehicle safety; ~ **de servicio** *f*

ING MECÁ dependability; ~ **del sistema** *f* INFORM&PD system security; ~ **en el trabajo** *f* SEG occupational safety, safety at work, work safety; ~ **de transmisión** *f* INFORM&PD transmission security; ~ **en el transporte** *f* SEG transportation safety; ~ **vial** *f* CONST, SEG road safety; ~ **viaria** *f* CONST, SEG road safety; ~ **en vías férreas** *f* FERRO, SEG rail safety

seguro *m* D&A *de un arma de fuego* safety catch, ING MECÁ safety bolt, detent, pawl, MECÁ dog, stop; ~ **del casco** *m* TRANSP MAR hull insurance; ~ **laboral** *m* SEG industrial insurance; ~ **marítimo** *m* TRANSP MAR marine insurance

seibertita *f* MINERAL seybertite

seiche *f* OCEAN seiche

seísmo: ~ **de base de diseño** *m* NUCL design base earthquake; ~ **de parada segura** *m* NUCL safe shutdown earthquake; ~ **submarino** *m* GEOFÍS, OCEAN seaquake

selección *f* C&V grading, selection, PAPEL sorting, TEC ESP screening, TELECOM selecting; ~ **automática interurbana** *f* TELECOM subscriber trunk dialing (*AmE*) (*STD*), subscriber trunk dialling (*BrE*) (*STD*); ~ **de colores** *f* IMPR color separation (*AmE*), colour separation (*BrE*); ~ **de colores por proceso indirecto** *f* IMPR indirect color separation (*AmE*), indirect colour separation (*BrE*); ~ **de componentes** *f* TEC ESP component selection; ~ **a distancia por frecuencia vocal** *f* TELECOM voice dialing (*AmE*), voice dialling (*BrE*); ~ **de la dosificación** *f* CONST *hormigón* mix design; ~ **por frecuencia vocal** *f* TELECOM voice dialing (*AmE*), voice dialling (*BrE*); ~ **por fuerza** *f* PROD force selection; ~ **de llegadas** *f* GEOL selection of arrivals; ~ **a mano** *f* CARBÓN handscreening; ~ **de modo a distancia** *f* PROD remote-mode selection; ~ **por palanca basculante** *f* PROD toggle selection; ~ **reducida de artistas** *f* TV narrowcasting; ~ **de semitonos** *f* IMPR halftone selection

seleccionable *adj* ING ELÉC selectable; ~ **por conmutador** *adj* PROD switch-selectable

seleccionador: ~ **de normas** *m* TV standards selector

seleccionar *vt* CARBÓN faredice, MINAS select; ~ **a mano** *vt* CARBÓN sort by hand

selectividad *f* ELECTRÓN, FÍS selectivity; ~ **contra canales adyacentes** *f* TELECOM adjacent channel selectivity; ~ **direccional** *f* TV directional selectivity

selector *m* ELEC multiple contact switch, selector switch, tap selector, tap switch, INFORM&PD, ING ELÉC, TELECOM, VEH *transmisión automática* selector; ~ **de barras cruzadas** *m* ING ELÉC crossbar selector; ~ **de canales** *m* TELECOM, TV channel selector; ~ **de canales múltiples** *m* TV multichannel selector; ~ **de combustible** *m* INSTAL TERM fuel selector; ~ **a distancia por frecuencia vocal** *m* TELECOM voice dialer (*AmE*), voice dialler (*BrE*); ~ **doble** *m* ELECTRÓN *de magnetrón* double moding; ~ **de escala** *m* TRANSP MAR *radar* variable-range marker; ~ **de escobilla** *m* ELEC *en máquina* brush-selector; ~ **por frecuencia vocal** *m* TELECOM voice dialer (*AmE*), voice dialler (*BrE*); ~ **de frecuencias** *m* TELECOM frequency selector; ~ **de funciones** *m* ELEC *conmutador* function selector; ~ **de indicador de rumbo** *m* TRANSP AÉR, TRANSP MAR course indicator selector; ~ **del nivel de combustible** *m* TRANSP AÉR fuel-level selector; ~ **normalizado** *m* PROD standard knob selector; ~ **omnidireccional** *m*

TRANSP AÉR omnibearing selector; ~ **por palanca universal** *m* ELEC *conmutador* joystick selector; ~ **de pistas** *m* TV track selector; ~ **de programas** *m* TELECOM program selector (*AmE*), programme selector (*BrE*); ~ **de rango** *m* INSTR, METR measurement range selector; ~ **de rumbo** *m* TRANSP AÉR course selector, heading selector; ~ **de tiempo de vuelo** *m* FÍS RAD, NUCL time-of-flight velocity selector; ~ **de velocidades** *m* CINEMAT speed selector; ~ **de voltaje** *m* ING ELÉC voltage selector

selenato *m* QUÍMICA selenate

selénico *adj* QUÍMICA *ácido* selenic

selenio *m* (*Se*) ING ELÉC, QUÍMICA selenium (*Se*)

selenioso *adj* QUÍMICA selenous, *ácido* selenious

selenita *f* MINERAL, QUÍMICA *mineral* selenite

selenítico *adj* MINERAL, QUÍMICA selenitic

seleniuro *m* QUÍMICA selenide; ~ **de plata** *m* QUÍMICA silver selenide

selenocianato *m* QUÍMICA selenocyanate

selenociánico *adj* QUÍMICA selenocyanic

selenuro *m* QUÍMICA selenide

sellado[1] *adj* METEO, MINAS, PAPEL, PROD, QUÍMICA occluded

sellado[2] *m* C&V sealing, CONST *de grietas* sealing, stopping, IMPR stamping, METEO, PAPEL, QUÍMICA occlusion; ~ **en caliente** *m* PAPEL heat-sealing; ~ **en frío** *m* IMPR cold seal; ~ **hermético** *m* EMB hermetic seal, NUCL, TELECOM hermetic sealing; ~ **hermético al vacío** *m* EMB vacuum heat sealer; ~ **por inducción interna** *m* EMB induction inner seal; ~ **laleríntico** *m* ING MECÁ, NUCL labyrinth seal; ~ **de lámina metálica** *m* EMB foil sealing; ~ **con lechada** *m* CONST slurry seal; ~ **mediante una cuchilla caliente** *m* EMB *termosoldado* hot blade sealing; ~ **mediante lacado** *m* EMB lacquer sealing; ~ **mediante precinto** *m* EMB band sealing; ~ **térmico** *m* REVEST heat-sealing; ~ **térmico por laminación** *m* REVEST heat seal laminating; ~ **por termoinducción** *m* EMB heat induction sealing; ~ **ultrasónico** *m* EMB ultrasonic sealing, TERMO ultrasonic welding; ~ **en vertical al vacío** *m* EMB vertical vacuum sealer

sellador *m* C&V sealant, COLOR sealer, IMPR sealant, P&C *pintura* sealer, *preparación polimérica* sealant, QUÍMICA sealant; ~ **de bandejas** *m* EMB tray sealer; ~ **hermético al vacío** *m* EMB vacuum heat sealer

selladora: ~ **por inducción** *f* EMB induction sealer

sellaíta *f* MINERAL sellaite

sellamiento *m* TEC PETR sealing

sellar *vt* MINAS insulate, PAPEL seal, PROD insulate, stamp; ~ **herméticamente** *vt* PROD make a tight joint

sello *m* C&V, ING MECÁ, MECÁ, TEC PETR *tuberías, geología* seal; ~ **de aceite** *m* ING MECÁ, MECÁ oil seal; ~ **anular** *m* GAS ring seal; ~ **dédalo** *m* EMB labyrinth seal; ~ **del eje de válvulas** *m* AUTO valve shaft seal; ~ **de empaquetadura hidráulica** *m* ING MECÁ hydraulic packing seal; ~ **estructural** *m* TEC PETR *geología* structural tap; ~ **extruido** *m* ING MECÁ extruded seal; ~ **graduado** *m* C&V graded seal; ~ **de la inspección** *m* PROD inspection stamp; ~ **de labios** *m* ING MECÁ lip seal; ~ **de lubricación** *m* ING MECÁ, MECÁ oil seal; ~ **moldeado de neopreno** *m* P&C *caucho sintético* neoprene molded seal (*AmE*), neoprene moulded seal (*BrE*); ~ **pegado** *m* EMB bonded seal; ~ **vacuoobturador** *m* ING MECÁ vacuum seal

selva: ~ **tropical húmeda** *f* AGRIC rainforest

semáforo *m* TRANSP MAR signal mast, signal station

sembrado *m* NUCL seeding

sembrador: ~ **de minas** *m* D&A, TRANSP MAR mine-layer; ~ **de precisión** *m* AGRIC precision seeder

sembradora *f* AGRIC planter, seed drill, seeder; ~ **bajo cubierta** *f* AGRIC sod seeder; ~ **en camellones** *f* AGRIC ridge drill; ~ **centrífuga manual** *f* AGRIC sack type seeder; ~ **para cultivos en hilera** *f* AGRIC row-crop seeder; ~ **de granos en línea** *f* AGRIC semi-deep furrow drill; ~ **de hortalizas** *f* AGRIC vegetable seeder; ~ **en línea** *f* AGRIC grain drill; ~ **en líneas** *f* AGRIC deep-furrow drill; ~ **de maíz** *f* AGRIC corn drill (*AmE*), maize drill (*BrE*); ~ **de mano** *f* AGRIC seed darrow; ~ **manual de mochila** *f* AGRIC knapsack seeder; ~ **de papas** *f AmL* (*cf sembradora de patatas Esp*) AGRIC potato planter; ~ **de patatas** *f Esp* (*cf sembradora de papas AmL*) AGRIC potato planter; ~ **a voleo distribuidora de fertilizantes** *f* AGRIC field distributor

semejanza: ~ **dinámica** *f* FÍS FLUID dynamic similarity

semelín *f* MINERAL semeline

semelina *f* MINERAL semelin

semental *m* AGRIC sire

sementera *f* AGRIC seedbed

semiacabado *adj* PROD semifinished

semiacetal *m* QUÍMICA hemi-acetal

semialeatorio *adj* TELECOM pseudo-random

semialeta *f* ING MECÁ *tubos* half flange

semiantracita *f* CARBÓN semianthracite

semiautomático *adj* ING MECÁ semiautomatic

semibituminoso *adj* CARBÓN semibituminous

semibrida *f* ING MECÁ half clamp

semicaja: ~ **superior** *f* PROD *caja de moldeo, molde de arcilla* cope

semicarbazida *f* QUÍMICA semicarbazide

semicarbazona *f* QUÍMICA semicarbazone

semicercha *f* CONST half truss

semiciclo *m* ING ELÉC, NUCL half cycle

semicírculo *m* GEOM semicircle

semiconductor *m* ELECTRÓN, FÍS semiconductor; ~ **amorfo** *m* ELECTRÓN amorphous semiconductor; ~ **de bombardeo electrónico** *m* (*SBE*) ELECTRÓN, FÍS PART, NUCL electron-bombarded semiconductor (*EBS*); ~ **con caracteres completos** *m* ELECTRÓN i-type semiconductor; ~ **degenerado** *m* ELECTRÓN degenerate semiconductor; ~ **dopado** *m* ELECTRÓN doped semiconductor; ~ **electrónico** *m* NUCL electronic semiconductor; ~ **extrínseco** *m* ELECTRÓN, FÍS, INFORM&PD extrinsic semiconductor; ~ **intrínseco** *m* ELECTRÓN, FÍS, INFORM&PD intrinsic semiconductor; ~ **intrínseco por compensación** *m* ING ELÉC compensated semiconductor; ~ **ligeramente alterado** *m* ELECTRÓN lightly-doped semiconductor; ~ **de memoria** *m* ELECTRÓN bulk semiconductor; ~ **metal-óxido** *m* (*SOM*) ELECTRÓN metallic oxide semiconductor (*MOS*), INFORM&PD metal-oxide semiconductor (*MOS*); ~ **monocristalino** *m* ELECTRÓN semiconductor single crystal; ~ **de óxido de metal difuso** *m* (*SOMD*) ELECTRÓN diffuse metal oxide semiconductor (*DMOS*); ~ p^+ *m* ELECTRÓN p^+ semiconductor; ~ p^- *m* ELECTRÓN p^- semiconductor; ~ **policristalino** *m* ELECTRÓN poly-crystalline semiconductor; ~ **de separación indirecta** *m* ELECTRÓN indirect gap semiconductor; ~ **de un solo cristal** *m* ELECTRÓN single-crystal semiconductor; ~ **tipo n** *m* ELECTRÓN, FÍS n-type semiconductor; ~ **de tipo p** *m* ELECTRÓN p-type

semiconductor; ~ **tipo p** *m* FÍS p-type semiconductor; ~ **tipo n+** *m* ELECTRÓN n+-type semiconductor

semicopa *f* ING MECÁ half cup

semicorte *m* PROD *dibujos planos* half section

semicuadratín *m* IMPR en space

semidescremada *adj* ALIMENT *lechería* semiskimmed

semidina *f* QUÍMICA semidine

semi-dúplex *m* (*HDX*) INFORM&PD half-duplex (*HDX*)

semieje *m* VEH *transmisión* axle shaft, halfshaft; ~ **mayor** *m* TEC ESP semimajor axis; ~ **trasero** *m* AUTO, MECÁ, VEH rear-axle drive shaft

semielaborado *adj* PROD semifinished, REVEST *papel* unfinished

semifijo *adj* ING MECÁ semiportable

semigelatina *f* MINAS semigelatin

semiimpulso *m* ELECTRÓN half pulse

semilla *f* AGRIC, C&V seed, NUCL fission spike, spike; ~ **de algodón** *f* AGRIC cottonseed; ~ **base** *f* AGRIC foundation seed; ~ **certificada** *f* AGRIC certified seed; ~ **de césped** *f* AGRIC lawn seed; ~ **curada** *f* AGRIC cured seed; ~ **descascarada** *f* AGRIC hulled seed; ~ **de lino** *f* AGRIC linseed; ~ **del mejorador** *f* AGRIC breeder seed; ~ **oleaginosa** *f* AGRIC oilseed; ~ **original** *f* AGRIC original seed; ~ **peleteada** *f* AGRIC pelleted seed; ~ **de primera multiplicación** *f* AGRIC registered seed; ~ **registrada** *f* AGRIC registered seed

semillanta *f* ING MECÁ half flange

semimetal *m* QUÍMICA semimetal

semimetálico *adj* QUÍMICA semimetalic (*AmE*), semi-metallic (*BrE*)

semi-montaje: ~ **en derivación** *m* ING ELÉC half bridge arrangement

semionda *f* ELEC *corriente alterna* half wave

semipalabra *f* INFORM&PD half word

semipersonalizado *adj* INFORM&PD semicustom

semiplato *m* ING MECÁ half flange

semipolar *adj* QUÍMICA semipolar

semiportátil *adj* ING MECÁ semiportable

semi-precoz *adj* AGRIC medium-early

semipuente *m* ING ELÉC half bridge

semipuño *m* PROD *cepillo, garlopa* handle

semirremolque *m* VEH semitrailer; ~ **basculante** *m* TRANSP tilt-type semitrailer

semirrestador *m* ELECTRÓN one-digit subtractor

semisubstractor *m* ELECTRÓN, INFORM&PD half sub-tractor

semisumador *m* ELECTRÓN half adder, one-digit adder, INFORM&PD half-adder

semisumergible *adj* TRANSP hull-borne

semi-sustractor *m* INFORM&PD half-subtractor

semi-tardío *adj* AGRIC medium-late

semitono *m* ACÚST, FÍS semitone, IMPR halftone; ~ **cromático** *m* ACÚST chromatic semitone; ~ **diatónico** *m* ACÚST diatonic semitone; ~ **en dos colores** *m* IMPR duplex halftone; ~ **menor** *m* ACÚST minor semitone; ~ **sobreimpreso tapando** *m* IMPR blocked-out halftone; ~ **de trama fina** *m* IMPR fine-screen halftone

semitrailer: ~ **basculante** *m* TRANSP tilt-type semi-trailer

semitransportable *adj* ING MECÁ semiportable

semiunión *f* ING MECÁ half union

sen *abr* (*seno*) CONST, GEOM, INFORM&PD, MATEMÁT *trigonometría* sin (*sine*)

señal[1]: **de ~ en el aire** *adj* TELECOM off-air
señal[2] *f* ACÚST cueing, signal, CINEMAT cueing, D&A landmark, ELECTRÓN, FERRO, FÍS signal, GEOFÍS trace, INFORM&PD signal, token, PROD marking, TEC ESP, TRANSP MAR signal, TV cueing, VEH *carretera* sign;
~ a **~ de aceptación de llamada** *f* TELECOM call acceptance signal; **~ de activación** *f* INFORM&PD enabling signal; **~ de activación del transmisor** *f* TELECOM transmitter turn-on signal; **~ activada por peatones** *f* TRANSP pedestrian-actuated signal; **~ activada por el tráfico** *f* TRANSP traffic-actuated signal; **~ de actuación** *f* NUCL actuating signal; **~ acústica** *f* ELECTRÓN acoustic signal, audio signal, sound signal, tone signal, FERRO *vehículos* sound signal, ING MECÁ acoustic signal, TELECOM, TRANSP MAR sound signal; **~ acústica de evacuación de urgencia** *f* SEG audible emergency evacuation signal; **~ de advertencia** *f* TRANSP MAR warning signal; **~ de advertencia de hielo** *f* SEG, TRANSP ice warning sign; **~ de advertencia de incidente** *f* TRANSP accident advisory sign, incident warning sign; **~ de AF** *f* (*señal de alta frecuencia*) ELECTRÓN HF signal (*high-frequency signal*); **~ ajena** *f* ELECTRÓN aliased signal; **~ de ajuste de banda** *f* ELECTRÓN band-limited signal; **~ de ajuste de fase** *f* ELECTRÓN phasing signal; **~ de alarma** *f* PROD warning signal, TRANSP MAR alarm signal, emergency radio-call; **~ de alarma inferior final del pantógrafo** *f* FERRO *infraestructura* lower pantograph final warning sign; **~ de alarma navegacional** *f* TELECOM, TRANSP MAR navigation warning signal; **~ aleatoria** *f* ELECTRÓN, TELECOM, TRANSP MAR random signal; **~ para la alimentación del bucle "petición bucle acústico"** *f* TELECOM LCA (*loopback command "audio loop request"*); **~ para la alimentación del bucle "petición bucle por audio"** *f* TELECOM loopback command "audio loop request" (*LCA*); **~ para la alimentación del bucle "petición bucle digital"** *f* TELECOM loopback command "digital loop request"; **~ para la alimentación del bucle "petición bucle por video"** *f* TELECOM loopback command "video loop request" (*LCV*); **~ de alineación de trama** *f* TELECOM frame alignment signal (*FAS*); **~ de alta frecuencia** *f* (*señal de AF*) ELECTRÓN high-frequency signal (*HF signal*); **~ de alto nivel** *f* ELECTRÓN high-level signal; **~ analógica** *f* ELECTRÓN, FÍS, INFORM&PD, TELECOM analog signal; **~ artificial** *f* ELECTRÓN exotic signal; **~ audible** *f* TELECOM audible signal; **~ de audio** *f* TELECOM sound signal; **~ autorizada** *f* ELECTRÓN enable signal; **~ avanzada exterior** *f* FERRO *infraestructura* outer distant signal; **~ avanzada de precaución** *f* FERRO *infraestructura* distant caution signal; **~ de aviso** *f* TRANSP MAR warning signal; **~ de aviso navegacional** *f* TELECOM, TRANSP MAR navigation warning signal; **~ de aviso de retención** *f* TRANSP queue warning sign; **~ azul** *f* ELECTRÓN blue signal;
~ b **~ de baja amplitud** *f* ELECTRÓN low-amplitude signal; **~ de baja frecuencia** *f* ELECTRÓN low-frequency signal; **~ de bajo nivel** *f* ELECTRÓN, PROD low-level signal; **~ de banda ancha** *f* ELECTRÓN, TELECOM wideband signal; **~ de banda base** *f* ELECTRÓN baseband signal; **~ de banda estrecha** *f* ELECTRÓN, TELECOM narrow-band signal; **~ de banda lateral residual** *f* ELECTRÓN vestigial sideband signal; **~ binaria** *f* ELECTRÓN, INFORM&PD, TELECOM binary signal; **~ bipolar** *f* TELECOM bipolar signal; **~ de borrado** *f* TV blanking signal; **~ de borrado y sincronismo** *f* TV blanking and sync signal; **~ en bucle** *f* TELECOM looped signal; **~ B-Y** *f* TV B-Y signal;
~ c **~ de la cabina** *f* FERRO *vehículos* cab signal; **~ de caída** *f* TV drop-out switch signal; **~ de cambio** *f* CINEMAT, TV changeover cue; **~ de cambios en la ruta** *f* TRANSP variable route sign; **~ captadora** *f* ELECTRÓN sensor signal; **~ de carretera** *f* AUTO, CONST, TRANSP, VEH road sign; **~ en caso de peligro** *f* FERRO *señalización* signal at danger; **~ casual** *f* ELECTRÓN, TELECOM, TRANSP MAR random signal; **~ de cierre de rampa** *f* TRANSP ramp-closure sign; **~ de circulación** *f* ELECTRÓN traffic signal; **~ codificada** *f* ELECTRÓN encoded signal; **~ de código binario** *f* ELECTRÓN binary-coded signal; **~ con cohetes** *f* D&A, TEC ESP rocket signal; **~ de color** *f* TV color signal (*AmE*), colour signal (*BrE*); **~ de color compuesta** *f* TV composite color signal (*AmE*), composite colour signal (*BrE*); **~ de comando digital** *f* TELECOM digital command signal (*DCS*); **~ compensadora** *f* TV shading signal; **~ compleja** *f* ELECTRÓN complex signal; **~ comprimida** *f* ELECTRÓN compressed signal; **~ de comprobación** *f* ELECTRÓN calibration signal; **~ compuesta** *f* ELECTRÓN, TELECOM, TV composite signal; **~ de comunicaciones** *f* ELECTRÓN communications signal; **~ de confirmación** *f* TELECOM acknowledgement signal; **~ confusa** *f* ELECTRÓN deception signal; **~ continua** *f* ELECTRÓN continuous signal; **~ de control** *f* D&A command, ELECTRÓN control signal, INFORM&PD, TEC ESP command, TV control signal; **~ de control de errores** *f* TELECOM error-check signal; **~ de control remoto** *f* TRANSP remote control sign; **~ de cromeado** *f* ELECTRÓN chrominance signal; **~ de crominancia** *f* TEC ESP *colorimetría*, TV chrominance signal; **~ de cuadratura** *f* ELECTRÓN quadrature signal; **~ cuantificada** *f* ELECTRÓN quantized signal;
~ d **~ degradada** *f* TV degraded signal; **~ de densidad máxima** *f* TELECOM picture black; **~ de densidad mínima** *f* TELECOM picture white; **~ de deslizamiento** *f* METAL slip marking; **~ de destellos** *f* TRANSP MAR *luz* flashing signal; **~ de desviación** *f* TRANSP AÉR deviation signal; **~ detectable mínima** *f* ELECTRÓN minimum detectable signal; **~ detectible** *f* TELECOM blip; **~ detectora** *f* ELECTRÓN detector signal; **~ detonante** *f* FERRO detonator (*BrE*), torpedo (*AmE*); **~ en diente de sierra** *f* FÍS RAD sawtooth signal wave; **~ de diferencia de color** *f* TV color-difference signal (*AmE*), colour-difference signal (*BrE*); **~ de diferencia de luminancia** *f* TV luminance difference signal; **~ diferencial** *f* ELECTRÓN difference signal, differential signal; **~ digital** *f* ELECTRÓN, FÍS, INFORM&PD, TELECOM digital signal; **~ digital paralela** *f* ELECTRÓN parallel digital signal; **~ digital en serie** *f* ELECTRÓN serial digital signal; **~ digitalizada** *f* ELECTRÓN digitized signal; **~ de dirección** *f* CONST direction sign; **~ de dirección de la onda portadora** *f* TELECOM carrier sense signal; **~ de disco** *f* FERRO *infraestructura* disc signal (*BrE*), disk signal (*AmE*); **~ discreta** *f* ELECTRÓN discrete signal; **~ de duración muy corta** *f* TEC ESP *radio* pulse;
~ e **~ de eco** *f* ELECTRÓN echo signal; **~ eléctrica** *f*

ELEC electric signal, electrical signal, ELECTRÓN electrical signal, ING ELÉC electric signal, electrical signal; **~ emitida por banderas** *f* TRANSP MAR *comunicaciones* flag signal; **~ de encendido del transmisor** *f* TELECOM transmitter turn-on signal; **~ de entrada** *f* ELECTRÓN input signal, signal input, FERRO *infraestructura* home signal (*BrE*), home switch (*AmE*), TV input signal; **~ de entrada de circuito lógico** *f* ELECTRÓN logic input signal; **~ de entrada digital** *f* ELECTRÓN digital input signal; **~ de entrada hacia adelante** *f* TELECOM forward-input signal; **~ de entrada interna** *f* TELECOM internal input signal; **~ de entrada regresiva** *f* TELECOM backward-input signal; **~ entrante** *f* TELECOM incoming signal; **~ errónea** *f* ELECTRÓN, TELECOM false signal; **~ de error** *f* ELECTRÓN, TRANSP AÉR error signal; **~ de espectro ensanchado** *f* TRANSP MAR *comunicaciones por satélite* spread spectrum signal; **~ espúrea** *f* TEC ESP spurious signal; **~ estadística** *f* ELECTRÓN, TELECOM, TRANSP MAR random signal; **~ de estallido de acceso** *f* TELECOM access burst signal; **~ de exceso** *f* ELECTRÓN hard-over signal; **~ de excitación de base** *f* ELECTRÓN base drive signal; **~ de existencia de petróleo** *f* TEC PETR *perforación, exploración* oil show; **~ de exploración** *f* TV line drive signal; **~ externa** *f* ELECTRÓN, TELECOM external signal; **~ para extraer la hoja defectuosa** *f* EMB faulty-sheet ejection signal;

- f **~ falsa** *f* TEC ESP spurious signal; **~ de fase** *f* *AmL* (*cf* señal de puesta en fase *Esp*) TV phasing signal; **~ en fase** *f* ELECTRÓN in-phase signal; **~ de FI** *f* ELECTRÓN, TELECOM IF signal; **~ de fin de cinta** *f* CINEMAT run-out signal; **~ de fin de comunicación** *f* TELECOM end-of-communication signal; **~ de FMA** *f* ELECTRÓN VHF signal; **~ en forma de escalera** *f* TV staircase signal; **~ fortuita** *f* ELECTRÓN, TELECOM, TRANSP MAR random signal; **~ de frecuencia intermedia** *f* ELECTRÓN, TELECOM, TV intermediate frequency signal; **~ de frecuencia vocal** *f* TELECOM speech signal; **~ de frecuencia vocal no lineal** *f* TELECOM nonlinear digital speech;

- g **~ de gran amplitud** *f* ELECTRÓN large signal; **~ G-Y** *f* TV G-Y signal;

- h **~ de habilitación** *f* ELECTRÓN enable signal; **~ hablante de carretera** *f* TRANSP talking road sign; **~ del haz** *f* ELECTRÓN beam signal; **~ hidrográfica** *f* OCEAN hydrographic signal; **~ horaria** *f* TRANSP MAR time signal;

- i **~ de identificación** *f* CINEMAT pilotone, FERRO *infraestructura* signal identification plate, TELECOM who-are-you (*WRU*), TV identification signal; **~ de identificación de carretera** *f* AUTO road identification sign; **~ de identificación del programa** *f* TV program-identification signal (*AmE*), programme-identification signal (*BrE*); **~ de identificación de transmisor** *f* TRANSP transmitter identification signal; **~ identificativa de información de tráfico** *f* TRANSP traffic information identification signal; **~ de imagen** *f* ELECTRÓN image signal, picture signal, TEC ESP video signal; **~ de imagen efectiva** *f* TV effective picture signal; **~ de impulso** *f* ELECTRÓN beat signal, pulse signal; **~ incidente** *f* ELECTRÓN incident signal; **~ de indicación de alarma** *f* TELECOM alarm indication signal; **~ de indicación de alarma de la sección de múltiplex** *f* TELECOM multiplex-section alarm indication signal; **~ indica-**

dora variable *f* TRANSP variable route sign; **~ de información** *f* FERRO *infraestructura ferroviaria* information sign; **~ para informes programados** *f* TELECOM scheduled reporting signal; **~ inhibidora** *f* ELECTRÓN inhibiting signal; **~ de interferencia** *f* ELECTRÓN interference signal; **~ interferente** *f* ELECTRÓN jamming signal; **~ interna lógica** *f* PROD internal logic signal; **~ interrumpida** *f* ELECTRÓN chopped signal; **~ de interrupción** *f* ELECTRÓN gated signal, gating signal, INFORM&PD interrupt signal; **~ invariable con el tiempo** *f* ELECTRÓN time invariant signal;

- l **~ de largo alcance** *f* ELECTRÓN long-way signal; **~ de liberación de mando para varios puntos** *f* TELECOM multipoint command release token; **~ limitada** *f* ELECTRÓN companded signal, limited signal; **~ de línea** *f* ELECTRÓN line signal; **~ de línea visual** *f* ELECTRÓN line-of-sight signal; **~ de llamada** *f* CONST, ELECTRÓN, ING ELÉC ringing, TELECOM calling signal; **~ de llamada en espera** *f* TELECOM call-waiting signal; **~ de llegada** *f* ELECTRÓN incoming signal, FERRO home signal (*BrE*), home switch (*AmE*); **~ local del oscilador** *f* ELECTRÓN local oscillator signal; **~ lógica** *f* ELECTRÓN logic signal; **~ de luminancia** *f* TEC ESP, TV luminance signal (*Y-signal*); **~ luminosa** *f* D&A flare, ELECTRÓN, FERRO light signal, TRANSP AÉR flare; **~ luminosa de color** *f* FERRO *infraestructura* colored light signal (*AmE*), coloured light signal (*BrE*); **~ luminosa en la pantalla** *f* TELECOM *radar* blip; **~ luminosa de punto de aguja** *f* FERRO *infraestructura* switchpoint light;

- m **~ de mando asignada para varios puntos** *f* TELECOM multipoint command assign token (*MCA*); **~ de mando digital** *f* TELECOM DCS (*digital command signal*); **~ de mando de puerta** *f* ELECTRÓN gate-drive signal; **~ manual** *f* SEG *para carga mecánica*, TRANSP hand signal; **~ de marcha atrás** *f* FERRO *equipo inamovible* backup signal, *infraestructura* backing signal; **~ de marcha libre** *f* ELECTRÓN free-running signal; **~ marina** *f* TEC ESP sea mark; **~ de máxima** *f* ELECTRÓN maximum signal; **~ de mediciones estándar** *f* TV standard measuring signal; **~ de MF** *f* ELECTRÓN FM signal; **~ de microondas** *f* ELECTRÓN microwave signal; **~ mínima** *f* ELECTRÓN minimum signal; **~ en modo común** *f* ELECTRÓN, TELECOM common-mode signal; **~ de modo diferencial** *f* ELECTRÓN differential mode signal; **~ de modulación** *f* ELECTRÓN modulating signal; **~ modulada** *f* TELECOM modulated signal; **~ monocroma** *f* TEC ESP monochrome signal; **~ monoestable** *f* ELECTRÓN one-shot signal; **~ muestreada** *f* TELECOM sampled signal; **~ múltiple en la distribución del tiempo** *f* ELECTRÓN time-division multiplexed signal; **~ muy limitada** *f* ELECTRÓN hard limited signal;

- n **~ negativa de video** *AmL*, **~ negativa de vídeo** *f Esp* TV negative video signal; **~ de niebla** *f* TRANSP MAR fog signal; **~ del nivel de aceite** *f* VEH *lubricación* oil-level mark; **~ de nivel lógico** *f* PROD logic level signal;

- o **~ obligatoria** *f* SEG mandatory sign; **~ OC** *f* ING ELÉC CW signal; **~ oculta** *f* FERRO *infraestructura* concealed signal; **~ de ocupado** *f* INFORM&PD busy signal; **~ de OK** *f* TELECOM OK-signal; **~ óptica** *f* ELECTRÓN, TELECOM optical signal; **~ ortogonal** *f*

TELECOM orthogonal signal; ~ **oscilante** *f* FERRO *infraestructura* wig-wag signal;

~ p ~ **de parada** *f* FERRO home signal (*BrE*), home switch (*AmE*); ~ **de parada absoluta** *f* FERRO *infraestructura* absolute stop signal; ~ **de parada discrecional** *f* FERRO *infraestructura* semipermissive stop signal; ~ **parásita** *f* ELECTRÓN interfering signal, TEC ESP spurious signal, *radio* interfering signal; ~ **parlante de carretera** *f* TRANSP talking road sign; ~ **de peligro** *f* CONST warning sign, SEG danger signal; ~ **de peligro aislado** *f* TRANSP MAR *marcas de navegación* isolated danger mark; ~ **pequeña** *f* ELECTRÓN small signal; ~ **periódica** *f* ELECTRÓN, TELECOM periodic signal; ~ **perturbadora** *f* TEC ESP interfering signal; ~ **de picos** *f* TV peak signal; ~ **piloto** *f* TV pilot signal; ~ **de la pista de control** *f* ACÚST, TV control-track signal; ~ **de la pleamar** *f* HIDROL *mar* high-water mark; ~ **de portadora** *f* INFORM&PD carrier signal; ~ **posterior** *f* ELECTRÓN back signal; ~ **de precaución** *f* CONST warning sign, FERRO, SEG caution signal; ~ **pre-programada** *f* TRANSP *control de tráfico* pretimed signal; ~ **para prevenir accidentes** *f* SEG accident prevention advertising sign; ~ **de propagación del espectro** *f* ELECTRÓN spread-spectrum signal; ~ **protectora de trenes** *f* FERRO *infraestructura* train-protecting signal; ~ **de prueba** *f* ELECTRÓN sampled signal, test signal; ~ **de prueba de intervalo vertical** *f* TV vertical interval test signal; ~ **pseudoaleatoria** *f* TELECOM pseudorandom signal; ~ **de puesta en fase** *f* *Esp* (*cf señal de fase AmL*) TV phasing signal;

~ q ~ **Q** *f* ELECTRÓN, TV Q-signal;

~ r ~ **de radio** *f* ELECTRÓN radio signal, receiver signal, FÍS RAD receiver signal, ING ELÉC receiver signal, radio signal, TELECOM receiver signal; ~ **de realimentación** *f* ELECTRÓN feedback signal; ~ **de realimentación en circuito cerrado** *f* ELECTRÓN loop feedback signal; ~ **recibida** *f* ELECTRÓN, INFORM&PD, TELECOM received signal; ~ **de referencia** *f* ELECTRÓN reference signal; ~ **de referencia del color** *f* TV color reference signal (*AmE*), colour reference signal (*BrE*); ~ **reflejada** *f* ELECTRÓN, ING ELÉC, ÓPT reflected signal; ~ **de regeneración** *f* ELECTRÓN, INFORM&PD refresh signal; ~ **registradora** *f* ELECTRÓN sensor signal; ~ **de relleno DX** *f* TELECOM DX-stuffing signal; ~ **de reloj** *f* ELECTRÓN clock signal; ~ **repetida** *f* ELECTRÓN, TRANSP MAR repeated signal; ~ **repetidora** *f* ELECTRÓN repeater signal, FERRO *equipo inamovible* repeating signal, TRANSP MAR *equipo electrónico* repeater signal; ~ **repetidora reflectante** *f* FERRO *infraestructura* scotchlight signal; ~ **repetitiva** *f* ELECTRÓN repetitive signal; ~ **requerida** *f* ELECTRÓN wanted signal; ~ **de respuesta** *f* TELECOM answer signal; ~ **de retorno** *f* PROD feedback signal; ~ **de retroceso** *f* FERRO backing signal; ~ **roja** *f* ELECTRÓN, TV red signal; ~ **de ruido blanco** *f* ELECTRÓN white noise signal; ~ **de ruidos** *f* TV noise signal; ~ **de ruidos complejos** *f* ELECTRÓN, FÍS, TEC ESP, TELECOM random-noise signal; ~ **R-Y** *f* TV red colour difference signal (*R-Y signal*);

~ s ~ **de salida** *f* ING ELÉC, TELECOM, TV output signal; ~ **de salida de circuito lógico** *f* ELECTRÓN logic output signal; ~ **de salida digital** *f* ELECTRÓN digital output signal; ~ **de saturación** *f* ELECTRÓN saturating signal; ~ **de seguridad** *f* LAB, SEG safety sign; ~ **de seguridad fosforescente** *f* SEG phosphorescent safety-sign; ~ **de sincronismo de color** *f* TV burst; ~ **de sincronismo de color múltiple** *f* TV multiburst; ~ **de sincronismo de colores** *f* TV color sync signal (*AmE*), colour sync signal (*BrE*); ~ **de sincronismo compuesta** *f* TV composite synchronization signal; ~ **de sincronización** *f* CINEMAT sync beep; ~ **de sincronización de control** *f* TEC ESP control burst; ~ **sincronizada** *f* ELECTRÓN clock signal; ~ **sinusoidal** *f* ELECTRÓN, TELECOM sinusoidal signal; ~ **de socorro** *f* TELECOM distress alerting, TRANSP AÉR, TRANSP MAR distress signal; ~ **de una sola banda lateral** *f* ELECTRÓN single-sideband signal; ~ **de un solo impulso** *f* ELECTRÓN single-pulse signal; ~ **sonárica** *f* ELECTRÓN sonar signal; ~ **de sonido** *f* TELECOM sound signal; ~ **sonora de alarma por silbido** *f* INSTAL HIDRÁUL alarm whistle signal; ~ **sonora de identificación** *f* CINEMAT pilot-tone sound; ~ **de stop** *f* CONST halt sign; ~ **de subida rápida** *f* ELECTRÓN fast-rise signal; ~ **de subportadora de crominancia** *f* TV chrominance subcarrier signal; ~ **subportadora de fase en cuadratura** *f* TV quadrature phase subcarrier signal; ~ **supersíncrona** *f* TV supersync signal; ~ **de supresión** *f* ELECTRÓN blanking signal; ~ **de supresión retardada** *f* TV delayed blanking signal;

~ t ~ **de tecleado** *f* TV keying signal; ~ **telegráfica** *f* ELECTRÓN telegraph signal; ~ **televisiva** *f* ELECTRÓN, FÍS, ING ELÉC, TELECOM, TV television signal; ~ **de tiempo** *f* ELECTRÓN time signal, timing signal; ~ **total** *f* TELECOM aggregate signal; ~ **de tráfico** *f* AUTO, CONST road sign, TRANSP road sign, traffic sign, VEH road sign; ~ **de tráfico semi-activada** *f* TRANSP traffic semiactuated signal; ~ **de tramo cerrada** *f* FERRO *infraestructura* block signal locked; ~ **de transmisión FM** *f* ELECTRÓN, FÍS RAD, TELECOM frequency-modulation transmitting signal; ~ **de transmisión de frecuencia modulada** *f* ELECTRÓN, FÍS RAD, TELECOM frequency-modulation transmitting signal; ~ **transmitida por radio** *f* ELECTRÓN radio signal, receiver signal, FÍS RAD receiver signal, ING ELÉC radio signal, receiver signal, TELECOM receiver signal; ~ **de tres niveles** *f* ELECTRÓN three-level signal; ~ **triangular de peligro** *f* SEG warning triangle; ~ **tricromática** *f* TV tristimulus signal;

~ u ~ **de UHF** *f* ELECTRÓN UHF signal; ~ **umbral** *f* ELECTRÓN threshold signal;

~ v ~ **variable con el tiempo** *f* ELECTRÓN time-varying signal; ~ **de vía libre** *f* FERRO line-clear signal; ~ **de vía única** *f* FERRO single-line token; ~ **de video** *AmL*, ~ **de vídeo** *f* *Esp* CINEMAT camera signal, FÍS, TEC ESP video signal, TELECOM video signal, picture signal, TV camera signal, video signal; ~ **de video con borrado** *AmL*, ~ **de vídeo con borrado** *f* *Esp* TV video signal with blanking; ~ **de video compuesta** *AmL*, ~ **de vídeo compuesta** *f* *Esp* TV composite video signal; ~ **visual** *f* TEC ESP video signal; ~ **visual-audible** *f* INFORM&PD visual-audible signal; ~ **vocal** *f* TELECOM speech signal; ~ **vocal no lineal** *f* TELECOM nonlinear digital speech; ~ **vocal simulada** *f* TELECOM simulated speech;

~ y ~ **Y** *f* TV Y-signal (*luminance signal*)
señalado *adj* ING MECÁ set
señalador *m* IMPR, INFORM&PD flag; ~ **de recorrido** *m* TEC ESP tracker

señalamiento *m* INFORM&PD signaling (*AmE*), signalling (*BrE*)

señalar *vt* TRANSP MAR signal

señalero: ~ **primero** *m* TRANSP MAR yeoman of signals

señales: ~ **para el alzado** *f pl* IMPR back marks, collating marks; ~ **de arrastre** *f pl Esp* (*cf señales de excitación AmL*) TV driving signals; ~ **audibles** *f pl* SEG auditory signals; ~ **automáticas de tráfico** *f pl* FERRO *en pasos de nivel* automatic traffic light signals; ~ **de doble imagen** *f pl* TV multipath signals; ~ **de excitación** *f pl AmL* (*cf señales de arrastre Esp*) TV driving signals; ~ **de frecuencia de cierre** *f pl* ELECTRÓN close frequency signals; ~ **luminosas automáticas** *f pl* FERRO *en pasos a nivel* automatic light signals; ~ **de mando de operador** *f pl* TELECOM operator commands; ~ **de marea fósil** *f pl* HIDROL ripple mark; ~ **mediante código por campana** *f pl* FERRO *infraestructura* bell-code signaling (*AmE*), bell-code signalling (*BrE*); ~ **parásitas** *f pl* ELECTRÓN grass; ~ **parásitas en pantalla** *f pl* ELECTRÓN *en radar*, TEC ESP, TELECOM, TRANSP MAR clutter; ~ **de peligro contra incendio** *f pl* SEG fire safety sign; ~ **de presión negativa** *f pl* SEG negative pressure signs

señalización *f* INFORM&PD, TEC ESP, TELECOM, TRANSP MAR signaling (*AmE*), signalling (*BrE*), TV flagging; ~ **de accidente** *f* TRANSP accident advisory sign; ~ **accionada por vehículos** *f* AUTO vehicle-actuated signalization (*AmE*), vehicle-actuated signallization (*BrE*); ~ **asociada al canal** *f* TELECOM channel associated signaling (*AmE*) (*CAS*), channel associated signalling (*BrE*) (*CAS*); ~ **asociada a las vías de transmisión** *f* TELECOM channel associated signaling (*AmE*) (*CAS*), channel associated signalling (*BrE*) (*CAS*); ~ **de aviso de incidente** *f* TRANSP incident warning sign; ~ **de cabina** *f* TRANSP cab signaling (*AmE*), cab signalling (*BrE*); ~ **por canal común** *f* TELECOM common channel signaling (*AmE*) (*CCS*), common channel signalling (*BrE*) (*CCS*); ~ **de carreteras** *f* CONST road signposting; ~ **de conexión y desconexión** *f* TELECOM connect and disconnect signaling (*AmE*), connect and disconnect signalling (*BrE*); ~ **de corriente continua** *f* ELECTRÓN DC signaling (*AmE*), DC signalling (*BrE*); ~ **dentro de banda** *f* TEC ESP in-band signaling (*AmE*), in-band signalling (*BrE*); ~ **E y M** *f* TELECOM E and M signaling (*AmE*), E and M signalling (*BrE*); ~ **por frecuencias vocales** *f* TEC ESP in-band signaling (*AmE*), in-band signalling (*BrE*); ~ **fuera de banda** *f* TEC ESP out-of-band signaling (*AmE*), out-of-band signalling (*BrE*); ~ **de garita guarda-vías** *f* TRANSP cab signaling (*AmE*), cab signalling (*BrE*); ~ **del generador** *f* ING ELÉC generator signaling (*AmE*), generator signalling (*BrE*); ~ **de guardabarrera** *f* TRANSP grade crossing signposting (*AmE*), level crossing signposting (*BrE*); ~ **lateral de la vía** *f* FERRO trackside signaling (*AmE*), trackside signalling (*BrE*); ~ **manual** *f* TRANSP hand signal; ~ **matriz** *f* TRANSP matrix signalization (*AmE*), matrix signallization (*BrE*); ~ **con matriz luminosa** *f* TRANSP matrix signalization; ~ **primaria** *f* TRANSP matrix signalization; ~ **para taxis** *f* TRANSP cab signaling (*AmE*), cab signalling (*BrE*); ~ **de tonalidad** *f* ELECTRÓN tone signaling (*AmE*), tone signalling (*BrE*); ~ **de usuario a usuario** *f* TELECOM user-to-user signaling (*AmE*) (*UUS*), user-to-user signalling (*BrE*) (*UUS*)

señalizaciones: ~ **de canal** *f pl* TRANSP channel marks

señalizador *m* IMPR sentinel, TELECOM acknowledgement flag; ~ **de nuevos datos** *m* TELECOM new data flag (*NDF*)

senarmontita *f* MINERAL senarmontite

señas: ~ **auxiliares del destinatario** *f pl* TELECOM subaddress

sencillo *adj* PROD bare

senda *f* CONST bypath; ~ **crítica** *f* INFORM&PD critical path; ~ **de datos** *f* INFORM&PD data path

sendero *m* CONST byway, path; ~ **cortafuegos** *m* TERMO firebreak

seno *m* (*sen*) CONST, GEOM, INFORM&PD, MATEMÁT *trigonometría* sine (*sin*), PROD *de correas, de cuerdas* slack portion, slack length, TELECOM *tendido de cables* slack, TRANSP MAR *de cabo* loop, *de cuerda, de cinta* slack, *de la ola* bight, *de cabos* bight; ~ **central** *m* METR sine center (*AmE*), sine centre (*BrE*); ~ **natural** *m* GEOM natural sine

senoidal *adj* GEN sinusoidal

sensación *f* ACÚST sensation

sensibilidad *f* ACÚST, CARBÓN sensitivity, ELEC *reactancia* susceptibility, *instrumento* sensitivity, ING ELÉC *instrumentos*, INSTR, MINAS, ÓPT, TEC ESP, TELECOM sensitivity; ~ **al agrietamiento por tratamiento térmico** *f* METAL, TERMO heat treatment crack sensitivity; ~ **al calor** *f* P&C heat sensitivity; ~ **de desviación** *f* ELECTRÓN deflection sensitivity; ~ **DIN** *f* FOTO DIN speed; ~ **espectral** *f* CINEMAT spectral sensitivity, ÓPT spectral sensibility; ~ **del explosivo** *f* MINAS explosive sensitivity; ~ **a la humedad** *f* PAPEL hygrosensibility; ~ **a la iniciación** *f* MINAS sensitivity to initiation; ~ **a la luz** *f* FÍS RAD sensitivity to light; ~ **óptica** *f* ELECTRÓN, ÓPT optical sensing; ~ **de la película** *f* CINEMAT, FOTO film speed; ~ **de propagación** *f* MINAS propagation sensitivity; ~ **radioeléctrica** *f* FÍS RAD radio sensitivity; ~ **de referencia** *f* TELECOM reference sensibility; ~ **de señal tangencial** *f* FÍS tangential signal sensitivity; ~ **de sintonización electrónica** *f* ELECTRÓN, TELECOM electronic-tuning sensitivity

sensibilización *f* CINEMAT, FOTO, IMPR sensitization

sensibilizado: poco ~ *adj* IMPR undersensitized

sensibilizador *m* CINEMAT, FOTO, IMPR sensitizer

sensibilizar *vt* CINEMAT, FOTO, IMPR sensitize

sensible: ~ **al calor** *adj* P&C heat-sensitive; ~ **a contexto** *adj* INFORM&PD context-sensitive; ~ **a infrarrojos** *adj* FÍS RAD infrared sensitive

sensitividad *f* FÍS, ING ELÉC, TEC ESP sensitivity; ~ **de desplazamiento angular** *f* TRANSP AÉR angular displacement sensitivity

sensitometría *f* ACÚST, CINEMAT sensitometry

sensor *m* GEN sensor, IMPR feeler; ~ **acústico** *m* ELECTRÓN acoustic sensor, TEC ESP voice sensor; ~ **analógico** *m* METEO dial gage (*AmE*), dial gauge (*BrE*); ~ **angular** *m* MECÁ angle sensor; ~ **de apagado** *m* TEC ESP shutdown sensor; ~ **de bloque** *m* PROD block sensor; ~ **detector** *m* INSTR, TEC ESP sensor; ~ **a distancia** *m* CONST remote sensor; ~ **de la entrada** *m* PROD input sensor; ~ **estelar** *m* TEC ESP star sensor; ~ **del flujo de aire** *m* AUTO airflow sensor; ~ **de la frecuencia triaxial angular** *m* TRANSP AÉR angular three-axis rate sensor; ~ **del horizonte** *m* TEC ESP horizon sensor; ~ **de imagen** *m* TELECOM, TV image sensor; ~ **inercial** *m* TEC ESP inertial sensor; ~ **láser** *m* ELECTRÓN laser sensor; ~ **lógico** *m*

INFORM&PD logical sensor; ~ **luminoso** *m* ELECTRÓN light sensor; ~ **de nivel** *m* TEC ESP level sensor; ~ **óptico** *m* ELECTRÓN, ÓPT, TEC ESP optical sensor; ~ **pasivo** *m* ELECTRÓN, TEC ESP passive sensor; ~ **piezoeléctrico** *m* ELEC, ÓPT piezoelectric sensor; ~ **de posición** *m* TEC ESP attitude sensor; ~ **de la proporción de la velocidad angular** *m* TRANSP AÉR angular velocity rate sensor; ~ **de rayos infrarrojos** *m* D&A, TEC ESP *misiles* infrared sensor; ~ **de referencia** *m* AUTO reference sensor; ~ **remoto** *m* CONST remote sensor; ~ **de RF** *m* TEC ESP, TELECOM RF sensor; ~ **sidéreo** *m* TEC ESP star sensor; ~ **solar** *m* GEOFÍS sun sensor, TEC ESP solar sensor; ~ **de temperatura** *m* AUTO, INSTR, PROD, REFRIG, TERMO temperature sensor; ~ **terrestre** *m* GEOFÍS earth sensor; ~ **tubular** *m* PROD tubular sensor; ~ **de viento** *m* METEO wind sensor; ~ **de volumen** *m* TEC ESP volume sensor

sentencia *f* INFORM&PD *programación* sentence, statement; ~ **de asignación** *f* INFORM&PD assignment statement; ~ **compuesta** *f* INFORM&PD *programa* compound statement; ~ **declarativa** *f* INFORM&PD pseudo-instruction, *programación* declarative statement; ~ **ejecutable** *f* INFORM&PD *programa* executable statement; ~ **indefinida** *f* INFORM&PD undefined statement; ~ **LOOP** *f* INFORM&PD LOOP statement; ~ **única** *f* TEC ESP unique word (*UW*)

sentido[1]: **de ~ único** *adj* ELEC *corriente*, TELECOM unidirectional

sentido[2]: **en el ~ de la inclinación** *adv* GEOL downdip; **en ~ longitudinal** *adv* ING MECÁ lengthwise, TRANSP MAR fore and aft; **en ~ transversal** *adv* IMPR *papel* against the grain

sentido[3] *m* HIDROL *dirección de la corriente* drift; ~ **longitudinal** *m* IMPR grain direction, PAPEL machine direction; ~ **de máquina** *m* IMPR *longitudinal* machine direction; ~ **del paso del cableado** *m* ELEC *de componente de cable* direction of lay; ~ **de la torsión** *m* TEXTIL *hilado* direction of twist; ~ **del tráfico** *m* FERRO direction of traffic; ~ **transversal** *m* IMPR cross direction

sentina *f* TRANSP MAR bilge

señuelo *m* OCEAN lure

separabilidad *f* NUCL separability

separable *adj* CARBÓN separable, ELEC, FOTO, INFORM&PD removable, ING MECÁ detachable, removable, ÓPT removable, PROC QUÍ separable, PROD removable, QUÍMICA separable, TRANSP AÉR removable

separación *f* AGUA shutting-off, C&V puntying, CARBÓN separation, sorting, CONST *hormigón* segregation, CONTAM *tratamiento de aguas y gases* removal, ELEC gap, FÍS *entre los componentes de un multiplete* splitting, spacing, separation, *de un frente de ondas* division, gap, GEOL sorting, INFORM&PD gap, *de información* isolation, ING ELÉC clearance, gap, ING MECÁ pitch, parting, detachment, MECÁ espacement, gap, MINAS *de minerales en un yacimiento* removal, PROC QUÍ separation, QUÍMICA separation, segregation, TEC ESP *mecánica* gap, separation, TEC PETR *perforación, producción* separation, TRANSP, TRANSP AÉR spacing; ~ **del adhesivo** *f* P&C *caucho* bond separation; ~ **de aire** *f* CONTAM *procesos industriales* air shed, PROC QUÍ air separation; ~ **por aire** *f* ALIMENT air separation; ~ **angular** *f* TELECOM angular separation; ~ **de anillo** *f* AUTO, VEH ring gap; ~ **de banda** *f* ELECTRÓN band separation; ~ **de la basura** *f* RECICL waste sorting; ~ **de la capa límite** *f* FÍS FLUID boundary-layer separation; ~ **en capas** *f* P&C *cubierta de caucho* ply separation; ~ **por capas** *f* GEOL layering; ~ **de la carga** *f* ELEC *fuente de alimentación* load shedding; ~ **cinética** *f* NUCL kinetic separation; ~ **de colores** *f* CINEMAT, FOTO color separation (*AmE*), colour separation (*BrE*), color-separation overlay (*AmE*), colour-separation overlay (*BrE*), TV color break-up (*AmE*), colour break-up (*BrE*), color separation (*AmE*), colour separation (*BrE*), color-separation overlay (*AmE*), colour-separation overlay (*BrE*); ~ **de los componentes** *f* PROD component spacing; ~ **de contacto** *f* ELEC *relé* contact gap; ~ **de los contactos** *f* AUTO, VEH contact-points gap; ~ **por contracorriente de aire** *f* CONTAM *tratamiento de líquidos* air flotation; ~ **de la corriente de aire** *f* TRANSP AÉR airstream separation; ~ **cromática** *f* TV chromatic splitting; ~ **y cruzamiento de los ríos** *f* HIDROL braiding of river; ~ **del electrodo** *f* ELEC, ING ELÉC electrode gap; ~ **de los electrodos de la bujía** *f* AUTO, ELEC, VEH spark plug gap; ~ **energética** *f* FÍS energy gap; ~ **entre chispas** *f* FÍS spark gap; ~ **entre partículas** *f* METAL interparticle spacing; ~ **entre pistas** *f* INFORM&PD track pitch; ~ **espacial** *f* CONTAM spatial resolution; ~ **por espuma** *f* PROC QUÍ foam separation; ~ **de fases** *f* ELEC, ELECTRÓN phase splitting; ~ **por filtración** *f* PROC QUÍ filtering separation; ~ **por floculación** *f* PROC QUÍ separation by flocculation; ~ **en forma de escamas** *f* IMPR *capa de estucado* flaking; ~ **de franja** *f* FÍS, FÍS ONDAS, ÓPT fringe separation; ~ **de frecuencia** *f* ELECTRÓN frequency separation; ~ **de función** *f* ELECTRÓN mode separation; ~ **geométrica** *f* NUCL separation by geometry; ~ **de guiaderas** *f* MINAS rodding; ~ **hidráulica** *f* PROC QUÍ hydraulic separation; ~ **horizontal** *f* GEOL horizontal separation; ~ **horizontal normal** *f* GEOL normal horizontal separation; ~ **de impulsos** *f* ELECTRÓN pulse separation; ~ **incorrecta** *f* IMPR bad break; ~ **de isótopos** *f* FÍS isotope separation; ~ **laminar** *f* FÍS FLUID laminar separation; ~ **lateral** *f* GEOL lateral separation; ~ **magnética de isótopos** *f* FÍS, NUCL magnetic isotope separation; ~ **por medios densos** *f* CARBÓN, MINAS *minerales* jigging; ~ **de metales nobles** *f* QUÍMICA parting; ~ **de módem** *f* ELECTRÓN modem interfacing; ~ **de partículas** *f* CONTAM particulate removal, FÍS PART particle separation; ~ **de partículas contaminantes del aire** *f* CONTAM *tratamiento de gases* particulate removal of air pollutants; ~ **de los platinos** *f* AUTO, VEH contact-points gap; ~ **en pliegues** *f* P&C *cubierta de caucho* ply separation; ~ **de polarización** *f* TEC ESP polarization isolation; ~ **positiva** *f* CINEMAT positive separation; ~ **principal** *f* ING ELÉC main gap; ~ **del punto de rocío** *f* REFRIG dew point depression; ~ **secundaria** *f* NUCL secondary separation; ~ **temporal** *f* CONTAM temporal resolution; ~ **turbulenta** *f* FÍS FLUID turbulent separation; ~ **vertical** *f* *Esp* (*cf franqueo vertical AmL*) MINAS clearance, PROD vertical separation

separaciones: ~ **múltiples** *f pl* IMPR ganged separations

separado *adj* C&V split, TELECOM spaced-out

separador *m* ALIMENT separator, C&V spacer, CARBÓN

pilas liner, gage (*AmE*), gauge (*BrE*), CINEMAT splitter, CONTAM MAR, CRISTAL separator, EMB insert, IMPR decollator, INFORM&PD *papel continuo* decollator, delimiter, ING MECÁ divider, separator, MECÁ insert, MINAS separator, forcing, PROC QUÍ separator, PROD divider, *papel* spacer, TEC ESP spacer, limiter, TEC PETR *perforación, producción* trap, separator, TELECOM separator, TV buffer; ~ **de aceite** *m* AGUA oil trap, HIDROL oil separator, oil trap, REFRIG oil separator; ~ **de agua** *m* AGUA separator; ~ **de aire** *m* ING MECÁ air separator; ~ **de amplitud** *m* TELECOM clipper; ~ **de archivos** *m* INFORM&PD file separator; ~ **de audio** *m* TELECOM audio splitter; ~ **centrífugo** *m* PROC QUÍ centrifugal separator; ~ **de choque** *m* PROC QUÍ baffle-type separator; ~ **cilíndrico** *m* CARBÓN drum separator; ~ **de colores** *m* CINEMAT color splitter (*AmE*), colour splitter (*BrE*), color-separation overlay (*AmE*), colour-separation overlay (*BrE*), FOTO color-separation overlay (*AmE*), colour-separation overlay (*BrE*), TV color splitter (*AmE*), color-separation overlay (*AmE*), colour splitter (*BrE*), colour-separation overlay (*BrE*); ~ **cónico** *m* ING MECÁ conical spacer; ~ **de crominancia** *m* TV chromakey; ~ **del cubo** *m* TRANSP AÉR hub spacer; ~ **de elementos de datos** *m* TELECOM data element separator; ~ **por elementos de los datos de componentes** *m* TELECOM component data element separator; ~ **de fibras** *m* PAPEL fiberizer; ~ **de ficheros** *m* INFORM&PD file separator; ~ **de gas** *m* CARBÓN *pozo de petróleo* flume; ~ **de granos finos** *m Esp* (*cf desiltor AmL*) TEC PETR *perforación* desilter; ~ **de grupo** *m* (*SG*) INFORM&PD group separator (*GS*); ~ **de haces** *m* IMPR beam splitter; ~ **de impurezas** *m* REFRIG scale trap; ~ **inercial** *m* NUCL inertial separator; ~ **de información** *m* (*SI*) INFORM&PD information separator, (*IS*); ~ **por lavado** *m AmL* MINAS scrubber; ~ **de limolita** *m Esp* (*cf desiltor AmL*) TEC PETR *perforación* desilter; ~ **magnético** *m* CARBÓN, MINAS, NUCL, PROD magnetic separator; ~ **de mineral** *m* CARBÓN riffle sampler; ~ **de minerales** *m* MINAS ore separator; ~ **de módem** *m* ELECTRÓN modem interface; ~ **de modos** *m* ELECTRÓN, ÓPT, TELECOM mode stripper; ~ **de nata** *m* ALIMENT cream separator; ~ **de negativos** *m* CINEMAT breakdown operator; ~ **del negro** *m* TV black clipper; ~ **de nudos** *m* PAPEL strainer; ~ **de páginas** *m* INFORM&PD burst mode; ~ **de plato** *m* ING MECÁ dishpan spacer; ~ **de polvos** *m* CARBÓN dust catcher; ~ **de pulsos** *m* TV pulse separator; ~ **redondo** *m* ING MECÁ round distance piece; ~ **de registros** *m* INFORM&PD record separator (*RS*; ~ **de semillas** *m* AGRIC *zaranda* fanning mill; ~ **de sincronización** *m* TELECOM clipper; ~ **de sincronización cromática** *m* TV burst separator; ~ **de tabiques** *m* PROC QUÍ baffle-type separator; ~ **de tambor** *m* CARBÓN *de minerales* drum cobber; ~ **de tuercas** *m* ING MECÁ nut cage; ~ **de uretano** *m* ING MECÁ urethane buffer; ~ **de vapor** *m* INSTAL HIDRÁUL *tubería de vapor* steam separator, TEC PETR *refino* steam trap; ~ **de ventana** *m* CONST sash bar, window bar

separadora: ~ **de recortes** *f* IMPR stripper
separadores *m pl* PROD dividers; ~ **de humedad-recalentadores** *m pl* NUCL moisture separator-reheaters; ~ **de resortes** *m pl* ING MECÁ spring dividers

separanudos *m* PAPEL knotter screen
separar[1] *vt* CINEMAT lock off, CONST split, CONTAM, GAS separate, IMPR, INFORM&PD decollate, ING ELÉC isolate, PROC QUÍ separate, PROD *cortar* cut out, *brocas, taladros* clear, QUÍMICA separate, TEXTIL split; ~ **por congelación** *vt* REFRIG freeze out; ~ **por destilación** *vt* TERMO distil off (*BrE*), distill off (*AmE*)
separar[2] *vi* ING MECÁ come apart; ~ **por cristalización** *vi* CRISTAL, PROC QUÍ crystallize out; ~ **para impresión en colores** *vi* IMPR break for colors (*AmE*), break for colours (*BrE*)
separarse *v refl* CARBÓN separate, CONST branch off, break, PROC QUÍ, QUÍMICA separate; ~ **de la derrota** *v refl* TRANSP MAR *para evitar algo* sheer off
sepiolita *f* MINERAL meerschaum, sepiolite
séptima *f* ACÚST seventh; ~ **mayor** *f* ACÚST major seventh; ~ **menor** *f* ACÚST minor seventh
septivalente *adj* QUÍMICA septivalent
septoriasis: ~ **de la hoja** *f* AGRIC leaf blotch
sequedad *f* CARBÓN PAPEL, TEXTIL dryness; ~ **comercial teórica** *f* PAPEL *de la pasta* theoretical commercial dryness
sequía *f* CARBÓN, PAPEL, TEXTIL dryness
serial *adj* TELECOM serial
serialización *f* INFORM&PD, ING ELÉC serialization
serializador *m* ING ELÉC serializer
serializar *vt* INFORM&PD, ING ELÉC serialize
sericita *f* MINERAL sericite
serie[1]: **en** ~ *adj* ELEC, FÍS in series, IMPR, INFORM&PD serial, ING ELÉC serial form, MATEMÁT in series, MECÁ in-line, TELECOM serial; **en ~-paralelo** *adj* INFORM&PD serial-parallel
serie[2] *f* ELEC *circuito*, GEOL *grupo de capas* series, INFORM&PD series, suite, ING MECÁ series, MATEMÁT series, *de figuras, cifras* array, MECÁ batch, QUÍMICA *de reacciones* series, TEC ESP array, TELECOM *equipo* suite, TEXTIL range; ~ **de los actínidos** *f* FÍS RAD actinide series; ~ **del actinio** *f* FÍS RAD actinium series; ~ **alveal** *f* CARBÓN bed sequence; ~ **armónica** *f* ACÚST harmonic series; ~ **aromática** *f* QUÍMICA *relativos al benceno* aromatic series; ~ **de Balmer** *f* FÍS Balmer series; ~ **de barrenos** *f* MINAS *voladuras* round; ~ **de bits** *f* INFORM&PD bit string; ~ **de Brackett** *f* FÍS Brackett series; ~ **del curio** *f* FÍS RAD curium series; ~ **electroquímica** *f* ELEC, ING ELÉC, QUÍMICA electrochemical series; ~ **de elementos radiantes** *f* TEC ESP array aerial (*BrE*), array antenna (*AmE*); ~ **de enlaces** *f* TELECOM *conexión* series of links; ~ **de Fibonacci** *f* MATEMÁT Fibonacci sequence; ~ **de Fourier** *f* ACÚST, CRISTAL, ELECTRÓN, FÍS, MATEMÁT Fourier series; ~ **de fusibles** *f* ING ELÉC fuse array; ~ **geométrica** *f* MATEMÁT geometric series; ~ **homóloga** *f* TEC PETR *refino* homologous series; ~ **de impulsos** *f* INFORM&PD pulse train; ~ **isomórfica** *f* CRISTAL isomorphous series; ~ **de laminadores** *f* ING MECÁ battery of rolls; ~ **de láminas** *f* ING MECÁ leaf chain; ~ **literal** *f* INFORM&PD string literal; ~ **de Lyman** *f* FÍS, FÍS RAD Lyman series; ~ **del metano** *f* GAS methane series; ~ **de objetivos** *f* FOTO set of lenses; ~ **de objetivos auxiliares** *f* FOTO set of supplementary lenses; ~ **de ondas** *f* TEC ESP ripple; ~ **de parafina** *f* TEC PETR *petroquímica* paraffin series; ~ **de pilotes** *f* CARBÓN row of piles; ~ **radiactiva** *f* FÍS, FÍS RAD radioactive series; ~ **de símbolos** *f* INFORM&PD symbol string;

~ **del torio** *f* FÍS RAD thorium series; ~ **unitaria** *f* INFORM&PD unit string; ~ **vacía** *f AmL* (*cf cadena vacía Esp*) INFORM&PD empty string, null string; ~ **de válvulas** *f* VEH *motor* valve train

series: ~ **de Paschen** *f pl* FÍS Paschen series; ~ **de Pfund** *f pl* FÍS Pfund series

serif *f* IMPR serif; ~ **oblicua** *f* IMPR oblique serif

serigrafía *f* IMPR screen printing, serigraphy

serotonina *f* QUÍMICA serotonin

serpenteo *m* CONST meandering, ENERG RENOV serpentinization, TRANSP AÉR snaking

serpentín *m* CONST coil, ING ELÉC helixing, ING MECÁ coil, LAB *destilación* serpent coil, worm, TERMOTEC *intercambiadores de calor* coil; ~ **acumulador** *m* REFRIG hold-over coil; ~ **de calefacción** *m* REFRIG, TERMO, TERMOTEC heating coil; ~ **de enfriamiento** *m* GAS, NUCL, REFRIG cooling coil; ~ **de pared** *m* REFRIG wall coil; ~ **de parrilla** *m* PROD grid coil; ~ **plano** *m* PROD grid coil; ~ **refrigerador** *m* GAS, NUCL, REFRIG *del condensador* cooling coil; ~ **refrigerante** *m* ING MECÁ, REFRIG cooling spiral; ~ **de techo** *m* REFRIG ceiling coil; ~ **de vapor** *m* TERMOTEC steam coil

serpentina *f* MINERAL serpentine

serpentines: ~ **Fenske** *m pl* LAB *destilación* Fenske helices

serpeta: ~ **del manzano** *f* AGRIC oystershell scale

serpiente *f* C&V snake

serrado *m* CONST breaking down; ~ **con sierra vertical de vaivén** *m* CONST jigsawing

serrar: ~ **con sierra de vaivén** *vt* CONST jigsaw

serrería: ~ **de banda** *f* ING MECÁ band sawing machine; ~ **mecánica** *f AmL* (*cf aserradero Esp*) AGRIC, CONST, MECÁ, MINAS sawmill

serreta *f* TRANSP MAR *vela* batten

serrín *m* CARBÓN dust, CONST sawdust

serrodina *f* ELECTRÓN serrodyne

serrodinar *vt* ELECTRÓN serrodyne

serrucho *m* CONST handsaw, ING MECÁ, MECÁ saw; ~ **de calar** *m* CONST keyhole saw; ~ **de costilla** *m* CONST back saw; ~ **eléctrico** *m* CONST, ELEC electric saw; ~ **de marquetería** *m* CONST compass saw; ~ **para podar** *m* ING MECÁ pruning saw; ~ **de punta** *m* CONST compass saw, lock saw

servicio[1]: **en** ~ *adj* PROD in active service, in commission, in operation, in service, TRANSP MAR *buque* in commission; **para** ~ **pesado** *adj* DETERG, ING MECÁ, PROD heavy-duty

servicio[2] *m* CALIDAD, FERRO service, PROD duty, TEC ESP service, utility, TELECOM *de abonados* servicing, TRANSP AÉR overhaul; ~ **de abastecimiento** *m* ELEC *al consumidor* supply service; ~ **de abonado** *m* TELECOM subscriber service; ~ **de abonados ausentes** *m* TELECOM *telefonía* absent subscriber service; ~ **aeronáutico móvil por satélite** *m* TEC ESP *comunicaciones* aeronautical mobile satellite service; ~ **aeronáutico satelital móvil** *m* TEC ESP *comunicaciones* aeronautical mobile satellite service; ~ **de agua** *m* TRANSP MAR *circuitos* water system; ~ **de alarma de advertencia** *m* TELECOM reminder-alarm service; ~ **de almacenamiento y reexpedición** *m* TELECOM store-and-forward facility; ~ **alternativo de palabras** *m* TELECOM alternate speech service; ~ **de autobuses con horario según demanda** *m* TRANSP demand-scheduled bus service; ~ **automático de tarjetas de crédito** *m* TELECOM automatic credit card service; ~ **de bajo nivel** *m* TELECOM lower-level service; ~ **de calibrado** *m* METR calibration service; ~ **de la capa física** *m* TELECOM physical-layer service (*PLS*); ~ **de carga** *m* TRANSP AÉR all-freight service; ~ **de cómputo** *m* INFORM&PD computing facility; ~ **confirmado** *m* TELECOM confirmed service; ~ **no confirmado** *m* TELECOM nonconfirmed service; ~ **de conmutación de datos de multimegabits** *m* TELECOM switched multimegabit data service (*SMDS*); ~ **conmutado** *m* TELECOM switched service; ~ **de contestador** *m* TELECOM answering service; ~ **continuo** *m* ELEC *equipo* continuous duty; ~ **de control de aproximación** *m* TRANSP AÉR approach control service; ~ **de control de tráfico aéreo** *m* TRANSP AÉR air traffic control service; ~ **de conversión por almacenamiento y reexpedición** *m* TELECOM store-and-forward conversion facility; ~ **de correo electrónico** *m* ELECTRÓN, INFORM&PD, TELECOM electronic mail service (*e-mail service*); ~ **de corriente portadora** *m* TELECOM bearer service; ~ **de datos de banda ancha sin conexiones** *m* TELECOM connectionless broadband data service (*CBDS*); ~ **de dirección de pistas** *m* TRANSP AÉR apron management service; ~ **directo de paquetes** *m* FERRO *vehículos* through parcel service; ~ **de distribución principal** *m* TELECOM cab-and-pillar distribution service; ~ **doméstico** *m* TRANSP AÉR domestic service; ~ **de emergencia** *m* TELECOM emergency attention; ~ **de exploración de la tierra por satélite** *m* TEC ESP Earth exploration satellite service; ~ **de extensión** *m* AGRIC extension service; ~ **de ferry** *m* TRANSP, TRANSP MAR ferry service; ~ **fijo aeronáutico** *m* TRANSP AÉR aeronautical fixed service (*BrE*), aeronautical fixed system (*AmE*); ~ **fijo por satélite** *m* TEC ESP fixed-satellite service (*FSS*); ~ **físico** *m* TELECOM physical service (*PhS*); ~ **general** *m* TEC PETR *agua, gas, electricidad* utility; ~ **con horario fijo** *m* TRANSP scheduled service; ~ **de información** *m* ELECTRÓN directory enquiries (*BrE*), enquiries service (*AmE*), TELECOM directory assistance (*AmE*), directory enquiries (*BrE*), enquiries service (*AmE*); ~ **de información aeronáutico** *m* TRANSP AÉR aeronautical information service; ~ **de información de gestión común** *m* TELECOM common management information service (*CMIS*); ~ **de información de vuelo** *m* TRANSP AÉR flight information service; ~ **ininterrumpido** *m* ELEC *equipo* uninterrupted duty; ~ **de investigación espacial** *m* TEC ESP space research service; ~ **de larga distancia** *m* TRANSP, TRANSP AÉR long-haul service; ~ **de líneas conmutadas de datos de multimegabits** *m* TELECOM switched multimegabit data service (*SMDS*); ~ **de llamada virtual** *m* INFORM&PD virtual call service; ~ **de llamadas por especificación** *m* TELECOM custom calling service; ~ **médico de urgencia** *m* SEG, TEC ESP emergency service; ~ **menor de base** *m* TRANSP AÉR minor base check; ~ **de mercancías** *m* TRANSP AÉR all-freight service; ~ **meteorológico** *m* METEO weather bureau; ~ **de monitorización** *m* TELECOM monitoring service; ~ **móvil de comunicaciones por satélite** *m* TRANSP MAR mobile satellite communications; ~ **móvil marítimo por satélite** *m* TRANSP MAR *satcom, navcom* maritime mobile satellite service; ~ **móvil telefónico** *m* TELECOM mobile telephone service; ~ **de múltiples portadores** *m* TELECOM multibearer service; ~ **N** *m* TELECOM N-service; ~ **a nivel de red sin conexiones** *m* TELECOM connectionless network layer service (*CLNS*); ~ **de nivel superior** *m* TELECOM higher-level service; ~ **nocturno** *m* TELECOM night service; ~ **de observación** *m* TELECOM observation service; ~ **de operaciones internacionales** *m* TELECOM international operations service; ~ **de paginación** *m* TELECOM paging service; ~ **de paquetería urgente** *m* FERRO *vehículos* express parcel service; ~ **de pasajeros** *m* TRANSP passenger service; ~ **de portador** *m* TELECOM audio bearer service; ~ **portador del circuito virtual** *m* TELECOM virtual circuit bearer service; ~ **portador sin conexiones** *m* TELECOM connectionless bearer service; ~ **portador con conmutación de paquetes** *m* TELECOM packet-switched bearer service; ~ **portador del modo trama** *m* TELECOM frame-mode bearer service (*FMBS*); ~ **portador en modos por paquete** *m* TELECOM packet-mode bearer service; ~ **del portador sin restricciones** *m* TELECOM unrestricted bearer service; ~ **portador de señalización al usuario** *m* TELECOM user-signaling bearer service (*AmE*), user-signalling bearer service (*BrE*); ~ **postventa** *m* CALIDAD, CONST after sales service; ~ **programado** *m* CONST scheduled service; ~ **público de información registrada** *m* TELECOM recorded public information service; ~ **de puente aéreo** *m* TRANSP AÉR shuttle service; ~ **de puente aéreo de helicóptero** *m* TRANSP AÉR helicopter shuttle service; ~ **de radioaficionados** *m* TELECOM amateur radio service; ~ **de radionavegación aeronáutica** *m* TRANSP AÉR aeronautical radio navigation service; ~ **de red sin conexiones** *m* TELECOM connectionless mode network service; ~ **reforzado** *m* TELECOM enhanced service; ~ **de rescate e incendios** *m* SEG, TRANSP AÉR rescue and fire fighting service; ~ **de reserva** *m* TRANSP backup service; ~ **sin restricciones** *m* TELECOM unrestricted service; ~ **restringido** *m* TELECOM restricted service; ~ **restringido de transferencia de información** *m* TELECOM restricted information-transfer service; ~ **de satélite meteorológico** *m* METEO, TEC ESP *comunicaciones espaciales* meteorological satellite service; ~ **de satélite móvil** *m* TEC ESP mobile satellite service; ~ **entre satélites** *m* TEC ESP intersatellite service; ~ **de satélites de radiodifusión** *m* TEC ESP broadcasting satellite service; ~ **de seguridad** *m* TELECOM *de transmisiones* security service; ~ **sesenta y cuatro de desconexión** *m* TELECOM clear sixty-four service; ~ **de socorro** *m* TRANSP emergency aid; ~ **sólo para correos** *m* TRANSP AÉR all-mail service; ~ **de suministro** *m* ELEC supply service; ~ **suplementario** *m* TELECOM supplementary service; ~ **de tarifa plana** *m* TELECOM flat-rate service; ~ **técnico** *m* ING MECÁ technical service; ~ **por teléfono** *m* TELECOM teleservice; ~ **de transbordador** *m* TRANSP, TRANSP MAR ferry service; ~ **de transferencia de información no reconocido** *m* TELECOM unacknowledged information transfer service (*UITS*); ~ **de transferencia de información sin restricciones** *m* TELECOM unrestricted information transfer service; ~ **de transmisión de datos** *m* TELECOM data link service (*DLS*); ~ **de transmisión de la palabra** *m* TELECOM speech service; ~ **transparente del portador** *m* TELECOM transparent bearer service; ~ **no transparente del portador** *m* TELECOM non-

transparent bearer service; ~ **de transporte de carga** *m* TRANSP AÉR all-cargo service; ~ **de traslado de información reconocida** *m* TELECOM acknowledged information transfer service; ~ **de urgencia** *m* TRANSP emergency aid; ~ **de valor agregado** *m AmL* (*cf servicio de valor añadido Esp*) TELECOM value-added service; ~ **de valor añadido** *m Esp* (*cf servicio de valor agregado AmL*) TELECOM value-added service; ~ **de varios portadores** *m* TELECOM *RDSI* multibearer service; ~ **de vigilancia costera** *m* TRANSP MAR coastguard

servicios: ~ **de cabina** *m pl* TRANSP, TRANSP AÉR cabin services; ~ **de emergencia** *m pl* TELECOM emergency services; ~ **de equipo** *m pl* TRANSP AÉR equipment services; ~ **de radio** *m pl* TELECOM, TRANSP, TRANSP AÉR, TRANSP MAR radio facilities; ~ **de regulación del tráfico marítimo** *m pl* TRANSP MAR vessel traffic services; ~ **de la ruta aérea** *m pl* TRANSP AÉR air-route facilities; ~ **de salvamento** *m pl* SEG rescue services

servidor *m* INFORM&PD *redes* server, TRANSP AÉR servicer; ~ **de archivos** *m* INFORM&PD file server; ~ **de comunicaciones** *m* INFORM&PD communication server; ~ **sin conexiones** *m* TELECOM connectionless server (*CLS*); ~ **de correo** *m* TELECOM mail server; ~ **designado** *m* TELECOM name server; ~ **del fichero** *m* INFORM&PD file server; ~ **de ficheros** *m* INFORM&PD file server; ~ **de impresión** *m* INFORM&PD print-server; ~ **terminal** *m* INFORM&PD *redes* terminal server; ~ **de videotex** *m Esp* (*cf servidor de videtex AmL*) TELECOM videotex server; ~ **de videtex** *m AmL* (*cf servidor de videotex Esp*) TELECOM videotex server

servidumbre: ~ **de paso** *f* CONST right of way

serviola *f* TRANSP MAR *navegación* lookout

servo *m* TRANSP, VEH *frenos, dirección, embrague* servo; ~ **cabrestante** *m* TV capstan drive, servo capstan; ~ **de mando** *m* VEH *componente del automóvil* booster; ~ **óptico** *m* ING ELÉC optical servo; ~ **regulador de paso cíclico** *m* TRANSP AÉR cyclic-pitch servo trim; ~ **de segundo orden** *m* ING ELÉC second-order servo; ~ **de tensión** *m* TV tension servo

servoaltímetro *m* TRANSP AÉR servoaltimeter

servoamplificador *m* ELECTRÓN servo amplifier

servocontrol *m* CINEMAT, ING ELÉC, TRANSP AÉR servocontrol; ~ **auxiliar** *m* ING MECÁ auxiliary servo control

servodirección *f* AUTO, VEH power steering, power-assisted steering

servofreno *m* AUTO power brake, servobrake, ING MECÁ air brake, VEH brake servo, power-assisted brake, servobrake; ~ **hidráulico** *m* AUTO hydraulic brake servo

servomando *m* CINEMAT servocontrol

servomecanismo *m* FÍS, INFORM&PD, ING ELÉC, ING MECÁ servomechanism, TEC ESP servosystem, TV servomechanism; ~ **del cabezal móvil** *m* CINEMAT capstan servo; ~ **para control de la velocidad** *m* TV velocity control servo; ~ **de fijación del cabezal** *m* TV head servo lock; ~ **programado** *m* TEC ESP programmed servo-system

servomotor *m* ENERG RENOV, ING ELÉC servomotor, ING MECÁ relay, MECÁ actuator, NUCL servomotor; ~ **de CC** *m* ING ELÉC DC servomotor; ~ **de corriente alterna** *m* ING ELÉC alternating-current servomotor; ~ **de efecto simple** *m* ENERG RENOV single acting servomotor; ~ **de paleta guiadora** *m* ENERG RENOV guide vane servomotor

servorueda *f* TV servo wheel

servosistema *m* TEC ESP servo system; ~ **programado** *m* TEC ESP programmed servo-system

servoválvula *f* ING MECÁ servovalve, PROD servo valve; ~ **de múltiples espiras** *f* ING MECÁ multiturn valve actuator

servozoom *m* CINEMAT servozoom

sesgado *adj* CONST, INFORM&PD skew

sesgadura *f* CONST skewing, INSTR beveling (*AmE*), bevelling (*BrE*)

sesgar *vt* ING MECÁ chamfer

sesgo *m* CARBÓN bias, CONST skewing, GEOL pitch, INFORM&PD, MATEMÁT bias; ~ **de ordenación** *m* INFORM&PD ordering bias; ~ **positivo** *m* ING ELÉC positive bias

sesquicarbonato: ~ **de sodio** *m* DETERG sodium sesquicarbonate

sesquiterpenoide *adj* QUÍMICA sesquiterpenoid

set: ~ **de filtros** *m* FOTO filter set

seudorabia *f* AGRIC mad itch

s.e.u.o. *abr* (*salvo error u omisión*) IMPR E&OE (*errors and omissions excepted*)

severita *f* MINERAL severite

sexta *f* ACÚST *música* sixth; ~ **mayor** *f* ACÚST major sixth; ~ **menor** *f* ACÚST minor sixth

sextante *m* FÍS sextant, INSTR index mirror, ÓPT, TRANSP MAR sextant; ~ **periscópico** *m* TEC ESP periscopic sextant

sexteto *m* FÍS *espectroscopia* sextet

SG *abr* (*separador de grupo*) INFORM&PD GS (*group separator*)

SGTR *abr* (*rotura de tubos en un generador de vapor*) NUCL SGTR (*steam generator tube rupture*)

shunt: ~ **Ayrton** *m* ELEC *resistencia* universal shunt; ~ **de galvanómetro** *m* ELEC, ING ELÉC *instrumento, resistencia* galvanometer shunt; ~ **magnético** *m* ELEC *imán* keeper; ~ **universal** *m* ELEC *resistencia* universal shunt

Si *abr* GEOL (*sílice*) Si (*silica*), QUÍMICA (*silicio*) Si (*silicon*)

SI *abr* (*separador de información*) INFORM&PD IS (*information separator*), ELEC, FÍS, METR (*sistema internacional de unidades*) IS (*international system of units*), TELECOM (*sistema intermedio*) IS (*intermediate system*)

SIA *abr* (*sistema integrado de administración*) TELECOM MIS (*management information system*)

siberita *f* MINERAL siberite

sicotrina *f* QUÍMICA psychotrine

sicrómetro *m* LAB *humedad*, REFRIG psychrometer; ~ **de aspiración** *m* REFRIG aspiration psychrometer; ~ **aspirado** *m* REFRIG aspirated psychrometer; ~ **de honda** *m* REFRIG sling hygrometer

sidecar *m* *Esp* (*cf zapato AmL*) TRANSP sidecar

siderita *f* MINERAL chalybite, siderite, QUÍMICA ironstone

siembra *f* AGRIC planting, QUÍMICA *de cristales* seeding; ~ **en banda** *f* AGRIC band seeding; ~ **en cuadros** *f* AGRIC check row planting; ~ **en hilera** *f* AGRIC drill planting; ~ **a nivel** *f* AGRIC contour planting; ~ **de respaldo** *f* AGRIC background planting; ~ **al voleo** *f* AGRIC broadcast seeding

siemens *m* (*S*) FÍS, METR *unidad de conductancia eléctrica* siemens (*S*)

siemensio *m* (*S*) FÍS, METR siemens (*S*)

sienita *f* QUÍMICA, TEC PETR syenite

sierra *f* CONST handsaw, ING MECÁ, MECÁ saw; ~ **abrazadera** *f* PROD pitsaw; ~ **alternativa** *f* ING MECÁ alternating saw, reciprocating saw, PROD reciprocating saw; ~ **alternativa de mano** *f* PROD hand-power hacksaw; ~ **alternativa mecánica para metales** *f* ING MECÁ power hacksaw; ~ **alternativa para metales** *f* ING MECÁ hacksaw; ~ **alternativa para trocear madera en rollo** *f* PROD log cross-cutting machine; ~ **alternativa vertical** *f* CONST jigsaw; ~ **de arco** *f* ING MECÁ scroll saw; ~ **de arco circular** *f* ING MECÁ segmental circular saw; ~ **de banda** *f* ING MECÁ band saw; ~ **de bastidor** *f* PROD frame saw; ~ **de cable flexible** *f* ING MECÁ flexible wire saw; ~ **de cadena** *f* ING MECÁ chain saw; ~ **de calar** *f* ING MECÁ scroll saw, PROD fret saw; ~ **para chapa** *f* ING MECÁ plate saw; ~ **de cinta** *f* ING MECÁ continuous saw; ~ **de cinta para madera en rollo** *f* PROD log band mill; ~ **circular** *f* CONST circular saw, ING MECÁ annular saw, circular saw, MECÁ circular saw; ~ **circular de eje fijo** *f* PROD fixed-spindle circular-saw bench; ~ **circular de mesa** *f* CONST saw bench; ~ **para concreto** *f* AmL (*cf sierra para hormigón Esp*) CONST concrete saw; ~ **continua** *f* ING MECÁ continuous saw; ~ **de contornear** *f* CONST compass saw, ING MECÁ scroll saw; ~ **para cortar metales** *f* ING MECÁ cold saw; ~ **de corte poco profundo** *f* ING MECÁ undercutting saw; ~ **de diamante** *f* ING MECÁ diamond saw; ~ **con dientes** *f* ING MECÁ saw with teeth; ~ **de dientes** *f* CONST tenon saw; ~ **de dientes adiamantados** *f* ING MECÁ diamond saw; ~ **de dientes articulados** *f* ING MECÁ *madera* chain saw; ~ **de dientes biselados** *f* ING MECÁ fleam-tooth saw; ~ **eléctrica** *f* CONST, ELEC *herramienta* electric saw; ~ **sin fin** *f* ING MECÁ band saw; ~ **giratoria** *f* ING MECÁ turning saw; ~ **helizoidal flexible** *f* ING MECÁ flexible wire saw; ~ **de hojas múltiples** *f* PROD gang saw; ~ **para hormigón** *f* Esp (*cf sierra para concreto AmL*) CONST concrete saw; ~ **larga** *f* PROD long saw; ~ **para leña** *f* CONST cleaving saw; ~ **de madera** *f* CONST wood saw; ~ **de mano** *f* PROD handsaw; ~ **de mano vertical** *f* ING MECÁ vertical handsaw; ~ **de marquetería** *f* ING MECÁ scroll saw, PROD fret saw, fret saw blade; ~ **mecánica** *f* AGRIC power saw, ING MECÁ power saw, sawing machine; ~ **de mecánico** *f* ING MECÁ, MECÁ hacksaw; ~ **para metales** *f* MECÁ hacksaw, PROD metal saw; ~ **móvil** *f* SEG moving saw; ~ **de nitruro de boro cúbico** *f* ING MECÁ cubic boron nitride saw; ~ **ordinaria** *f* PROD frame saw; ~ **de pelo flexible** *f* ING MECÁ flexible wire saw; ~ **de perforación** *f* ING MECÁ hole saw; ~ **de perforación de filo de polvo de carburo al tungsteno** *f* ING MECÁ tungsten-carbide grit hole saw; ~ **de perforación con puntas de carburo** *f* ING MECÁ, MECÁ carbide-tipped hole saw; ~ **de punta** *f* PROD fret saw blade; ~ **de rebaje** *f* ING MECÁ undercutting saw; ~ **para resquebrajar** *f* CONST cleaving saw; ~ **segmentada** *f* ING MECÁ segmented saw; ~ **tronzadera** *f* PROD crosscut saw; ~ **de tumba** *f* PROD crosscut saw; ~ **de vaivén** *f* CONST jigsaw; ~ **vertical alternativa de varias hojas** *f* PROD log frame

sievert *m* (*Sv*) FÍS, FÍS RAD, NUCL sievert (*Sv*)

sifón *m* CONST *fontanería* trap, FÍS, HIDROL, LAB siphon, MECÁ gooseneck; ~ **en la acometida a la alcantarilla**

m HIDROL interceptor sewer; ~ **de drenaje** *m* CONST drain trap; ~ **espumador** *m* PROD *colada de horno alto* skimmer; ~ **de filtración** *m* PROC QUÍ filter siphon; ~ **guar-daolores** *m* CONST stink trap, *fontanería* stench trap; ~ **en S** *m* CONST S trap; ~ **térmico** *m* ENERG RENOV thermosiphon; ~ **en U** *m* CONST *fontanería* running trap; ~ **vertedero** *m* ENERG RENOV siphon spillway; ~ **vertedor** *m* ENERG RENOV siphon spillway

SIG *abr* (*sistema integrado de administración*) INFORM&PD MIS (*management information system*)

signaturas: ~ **plegadas sin perforar** *f pl* IMPR closed bolts

signo *m* INFORM&PD sign; ~ **&** *m* IMPR ampersand; ~ **igual** *m* IMPR parallel mark; ~ **indicador** *m* ING MECÁ index; ~ **de intercalación** *m* IMPR caret; ~ **de llamada** *m* IMPR fist; ~ **plus** *m* CONST plus sign; ~ **de polaridad** *m* ELEC *pila* polarity sign

signos: ~ **convencionales** *m pl* IMPR arbitrary signs; ~ **de corrección de pruebas** *m pl* IMPR proof-correction marks, proof marks; ~ **de puntuación** *m pl* IMPR punctuation marks

SIL *abr* (*nivel de interferencia con la palabra*) ACÚST SIL (*speech interference level*)

silano *m* P&C, QUÍMICA silane

silbato *m* ELECTRÓN whistle; ~ **de alarma** *m* ING MECÁ alarm whistle, INSTAL HIDRÁUL alarm whistle signal; ~ **de alarma regulado por flotador** *m* INSTAL HIDRÁUL float-controlled alarm whistle

silbido: ~ **del micrófono** *m* CINEMAT mike stew

silenciador *m* ACÚST, AUTO muffler (*AmE*), silencer (*BrE*), ING MECÁ noise absorption device, SEG *en tuberías* silencer, TRANSP, VEH muffler (*AmE*), silencer (*BrE*); ~ **de absorción** *m* SEG *aislamiento acústico* absorption silencer; ~ **de la admisión** *m* MECÁ inlet silencer; ~ **del aireador** *m* TRANSP AÉR aerator silencer; ~ **en ausencia de señal** *m* TELECOM squelch; ~ **catalítico** *m* AmL CONTAM catalytic muffler (*AmE*), catalytic silencer (*BrE*); ~ **del escape del motor** *m* AUTO, ING MECÁ, MECÁ engine muffler (*AmE*), engine silencer (*BrE*); ~ **de pantallas** *m* AUTO baffle silencer; ~ **de placas** *m* AUTO baffle silencer; ~ **de resonancia** *m* SEG *aislamiento contra ruidos* resonance silencer

silencio *m* ACÚST muteness

silentbloc *m* VEH *suspensión* damper

sílex *m* C&V flint, GEOL chert, flint; ~ **negro** *m* GEOL *microcuarcita grafítica* black chert; ~ **de radiolarios** *m* GEOL radiolarian chert

sílica: ~ **gel** *f* QUÍMICA *agente desecador, agente de empaque* silica gel

silicato *m* QUÍMICA silicate; ~ **de aluminio** *m* DETERG, QUÍMICA aluminium silicate (*BrE*), aluminum silicate (*AmE*); ~ **de calcio** *m* DETERG calcium silicate; ~ **de plomo** *m* C&V lead silicate; ~ **de sodio** *m* DETERG sodium silicate

sílice *f* (*Si*) GEOL silica (*Si*); ~ **coloidal** *f* DETERG colloidal silica; ~ **fundida** *f* C&V fused silica, METAL silica glass, ÓPT, TELECOM fused silica; ~ **precipitada** *f* P&C *carga, pigmento* precipitated silica; ~ **vítrea** *f* C&V, ÓPT, TELECOM vitreous silica

silíceo *adj* GEOFÍS siliceous, GEOL cherty, siliceous, QUÍMICA, TEC PETR siliceous

silicio *m* (*Si*) QUÍMICA silicon (*Si*); ~ **amorfo** *m* ELECTRÓN amorphous silicon; ~ **policristalino** *m* ELECTRÓN polycrystalline silicon; ~ **de tipo n** *m* ELECTRÓN n-type silicon; ~ **de tipo p** *m* ELECTRÓN p-type silicon; ~ **con zafiro** *m* (*SCZ*) ELECTRÓN silicon on sapphire (*SOS*)

siliciuret *m* QUÍMICA siliciuret

siliciuro *m* QUÍMICA silicide

silicoborato *m* QUÍMICA silicoborate

silicofenil *m* QUÍMICA silicophenyl

silicofluoruro *m* QUÍMICA silicofluoride

silicona *f* ELEC, ING ELÉC, P&C, QUÍMICA silicone; ~ **monocristalina** *f* TEC ESP monocrystalline silicon

silicotitanato *m* QUÍMICA silicotitanate

silicotungstato *m* QUÍMICA silicotungstate

silicotúngstico *adj* QUÍMICA silicotungstic

sill: ~ **mixto** *m* GEOL *intrusión ígnea* composite sill

silla *f* C&V chair; ~ **de montar de forma simétrica** *f* GEOM symmetric saddle shape

sillar: ~ **de esquina** *f* CONST *albañilería* quoin

silleta *f* Esp (*cf soporte AmL*) MINAS *de ejes suspendidos* bracket hanger, PROD *de eje* hanger; ~ **colgante** *f* AmL (*cf soporte colgante Esp*) MINAS pendant bracket; ~ **de piso** *f* MINAS *ejes de transmisiones* floor hanger; ~ **de transmisión** *f* PROD sling hanger

sillimanita *f* MINERAL sillimanite

silo *m* AGRIC elevator, silo, D&A silo, EMB bin, TEC ESP silo; ~ **cuba** *f* AGRIC pit silo; ~ **hecho con duelas de madera** *m* AGRIC stave silo; ~ **para minerales** *m* Esp MINAS ore hopper; ~ **parva** *m* AGRIC stack silo; ~ **torre** *m* AGRIC tower silo; ~ **trinchera** *m* AGRIC trench silo; ~ **de troceados** *m* PAPEL chips; ~ **vertical** *m* AGRIC upright silo

siloxano *m* QUÍMICA siloxane

silueta *f* ING MECÁ outline

silvanita *f* MINERAL sylvanite

silvestreno *m* QUÍMICA sylvestrene

silvinita *f* MINERAL sylvinite

silvita *f* MINERAL sylvine, sylvite

sima *f* GEOFÍS sima

símbolo *m* INFORM&PD, QUÍMICA symbol; ~ **abstracto** *m* INFORM&PD abstract symbol; ~ **gráfico** *m* SEG *planes de protección contra incendios* graphical symbol; ~ **de lógica** *m* AmL (*cf símbolo lógico Esp*) INFORM&PD logic symbol; ~ **lógico** *m* Esp (*cf símbolo de lógica AmL*) INFORM&PD logic symbol; ~ **de organigrama** *m* INFORM&PD flowchart symbol; ~ **de puesta a tierra de seguridad** *m* PROD safety earth symbol (*BrE*), safety ground symbol (*AmE*); ~ **separador** *m* INFORM&PD separator symbol; ~ **terminal** *m* INFORM&PD terminal symbol

simetría *f* GEOM, QUÍMICA *de molécula* symmetry; ~ **aguas arriba-aguas abajo** *f* FÍS FLUID upstream-downstream symmetry; ~ **de amperios** *f* ELEC *instrumento* ampere balance; ~ **axial** *f* GEOM, MATEMÁT axial symmetry; ~ **cristalina** *f* CRISTAL crystal symmetry; ~ **de grupos** *f* CRISTAL point group symmetry; ~ **paritaria de carga** *f* FÍS PART, FÍS RAD charge parity symmetry; ~ **de rotación** *f* (*cf simetría rotativa*) GEOM, MATEMÁT rotational symmetry; ~ **rotativa** *f* (*cf simetría de rotación*) GEOM, MATEMÁT rotational symmetry

simétrico *adj* GEOM, INFORM&PD, MATEMÁT, QUÍMICA symmetric, symmetrical, TELECOM *electricidad*, TRANSP push-pull

símil[1] *adj* PAPEL imitation

símil[2]: ~ **fibra** *m* CONST fiber-like material (*AmE*), fibre-like material (*BrE*)

simili *adj* PAPEL imitation

SIMM *abr* (*módulo simple de memoria en línea*) INFORM&PD SIMM (*single in-line memory module*)

simple *adj* C&V *vidrio no decorado* plain, TELECOM simplex

símplex[1] *adj* INFORM&PD, TELECOM simplex

símplex[2]: ~ **con dos frecuencias** *m* TELECOM two-frequency simplex

simulación *f* ELECTRÓN, GAS, INFORM&PD simulation; ~ **lógica** *f* ELECTRÓN logic simulation; ~ **de planificaciones** *f* PROD scenario; ~ **de la señal** *f* ELECTRÓN signal simulation; ~ **en tiempo real** *f* ELECTRÓN, INFORM&PD, TELECOM real-time simulation; ~ **de tráfico** *f* TRANSP simulation of traffic, traffic simulation

simulacro: ~ **de incendio** *m* PROD, SEG, TERMO fire drill

simulador *m* ELECTRÓN, INFORM&PD simulator, TEC ESP mock-up, TELECOM, TRANSP AÉR simulator; ~ **de combate aéreo** *m* D&A, TRANSP AÉR air combat simulator; ~ **de entrenamiento** *m* NUCL, TEC ESP, TRANSP AÉR training simulator; ~ **lógico** *m* ELECTRÓN logic simulator; ~ **de red** *m* INFORM&PD network simulator; ~ **de sondeo** *m* TEC ESP feel simulator; ~ **suave** *m* TEC ESP soft mock-up; ~ **en tiempo real** *m* ELECTRÓN, INFORM&PD, TELECOM real-time simulator; ~ **de tráfico** *m* TRANSP traffic simulator; ~ **de vuelo** *m* INFORM&PD, TRANSP AÉR flight simulator

simular *vt* ELECTRÓN, FÍS RAD, GAS, INFORM&PD, TELECOM, TRANSP simulate

simultaneidad *f* FÍS simultaneity

simultáneo *adj* INFORM&PD concurrent

sinantrosa *f* QUÍMICA synanthrose

sinápico *adj* QUÍMICA sinapic

sinapina *f* QUÍMICA sinapine

sinclinal *m* TEC PETR *geología* syncline; ~ **elongado** *m* GEOL canoe fold; ~ **periférico** *m* GEOL rim syncline

sinclinorio *m* GEOL *tectónica* synclinorium

síncope: ~ **del buzo** *m* OCEAN *problemas fisiológicos de inmersión* shallow water blackout

sincrociclotrón *m* FÍS synchrocyclotron

sincronía *f* ING MECÁ synchronism

sincrónica: ~ **ecuatorial** *f* TEC ESP *orbitografía* earth-synchronous orbit

sincronismo[1]: **en** ~ *adv* ING ELÉC in step

sincronismo[2] *m* ELECTRÓN, ING MECÁ synchronism, TELECOM timing

sincronismos: ~ **mixtos** *m pl* Esp (*cf mezcla de sincronismos AmL*) TV mixed syncs

sincronización *f* AUTO synchronization, timing, ELEC *corriente alterna* phasing, ELECTRÓN *oscilador* lock, phasing, timing, INFORM&PD timing, clocking, ING MECÁ, TEC ESP synchronization, TELECOM synchronization, timing, alignment, TRANSP, TV phasing; ~ **de apertura del circuito** *f* ING ELÉC break time; ~ **de la chispa** *f* AUTO spark timing; ~ **por correa dentada** *f* AUTO, VEH cogged belt timing; ~ **degradada** *f* TV degraded sync, degraded synchronization; ~ **del encendido** *f* AUTO timing of the ignition; ~ **externa** *f* ELECTRÓN external sync, external synchronization; ~ **de fase** *f* ELECTRÓN phase locking, TV phase lock; ~ **horizontal** *f* AmL (*cf sincronización de línea Esp*) TV horizontal sync, horizontal synchronization; ~ **de imagen** *f* TV field sync, field synchronization; ~ **labial** *f* CINEMAT *para el doblaje* lip sync, lip synchronization; ~ **de línea** *f* Esp (*cf sincronización*

horizontal AmL) TV horizontal sync, horizontal synchronization; ~ **de montaje** *f* CINEMAT, TV edit sync, edit synchronization; ~ **mutua** *f* TELECOM mutual sync, mutual synchronization; ~ **de la pista de control** *f* ACÚST, TV control-track time code; ~ **de la positivadora** *f* CINEMAT printer sync, printer synchronization; ~ **precisa** *f* TELECOM fine tuning; ~ **de pulsos** *f* TV pulse phasing, pulse sync, pulse synchronization; ~ **con la red** *f* TV locking; ~ **de trama** *f* TELECOM frame sync, frame synchronization

sincronizado[1]: **en** ~ *adj* CINEMAT in sync, in synchronization, IMPR in-step, ING ELÉC in step, PROD gated on, TV in sync; **no** ~ *adj* CINEMAT, TV, nonsync

sincronizado[2]: ~ **horizontal** *m* TV horizontal lock; ~ **de ignición** *m* AUTO ignition timing; ~ **por polea centrada** *m* VEH cogged belt timing; ~ **por polea dentada** *m* AUTO cogged belt timing

sincronizador *m* AUTO synchromesh, synchronizer, CINEMAT, ELEC, INFORM&PD, ING ELÉC synchronizer, TELECOM timer, TV phaser; ~ **de cabeceo** *m* TRANSP AÉR pitch synchro; ~ **sin cable** *m* CINEMAT cordless sync, cordless synchronization; ~ **de cadena y rueda dentada** *m* AUTO sprocket and chain timing; ~ **de correa dentada** *m* AUTO notched-belt timing; ~ **detector de cabeceo** *m* TRANSP AÉR pitch-detector synchro, pitch-detector synchronizer; ~ **detector del paso** *m* TRANSP AÉR pitch-detector synchro, pitch-detector synchronizer; ~ **de la imagen** *m* CINEMAT picture synchronizer; ~ **de paso** *m* TRANSP AÉR pitch synchro, pitch synchroniation; ~ **del paso del colectivo** *m* TRANSP AÉR collective-pitch synchronizer; ~ **del regulador de paso** *m* TRANSP AÉR pitch-throttle synchronizer; ~ **de rumbo** *m* TRANSP AÉR heading synchronizer; ~ **separador** *m* TV sync separator, synchronization separator; ~ **de trama** *m* TELECOM frame-synchronization control

sincronizar[1] *vt* CINEMAT marry up, synchronize (*sync*), PROD time, TV synchronize (*sync*)

sincronizar[2]: ~ **el rodaje diario** *vi* CINEMAT sync dailies

síncrono *adj* ELECTRÓN, INFORM&PD synchronous

sincrotransformador: ~ **sólido** *m* ING ELÉC solid synchrotransformer

sincrotrón *m* ELECTRÓN, FÍS, FÍS PART, NUCL electron accelerator, ion accelerator, synchrotron; ~ **de electrones** *m* ELECTRÓN, FÍS PART, NUCL electron synchrotron; ~ **de protones** *m* FÍS PART proton synchrotron (*PS*)

síndrome: ~ **nervioso de alta presión** *m* OCEAN high-pressure nerve syndrome, high-pressure nervous syndrome

sinéresis *f* P&C *defecto*, QUÍMICA *exudación de líquido contenido en gel* syneresis

sinergético *adj* QUÍMICA *efecto* synergetic

sinergía *f* DETERG synergy

sinergismo *m* DETERG synergism

sinergista *m* ALIMENT synergist

sinesquistoso *adj* GEOL synschistous

sinfín *m* AGRIC grain auger

singenita *f* MINERAL syngenite

singlete *m* FÍS, NUCL singlet

singularidad *f* FÍS singularity

siniestro *m* SEG, TRANSP MAR accident

siniforme *m* GEOL synform

sinorogénico *adj* GEOL synorogenic

sintáctico *adj* MINAS, QUÍMICA syntactic

sintaxis: ~ **de instrucciones** *f* PROD instruction syntax; ~ **de transferencia** *f* TELECOM transfer syntax

sintectónico *adj* GEOL syntectonic

sinterización *f* CARBÓN, INSTAL TERM, METAL, PROC QUÍ sintering; ~ **en lecho fluidizado** *f* PROC QUÍ fluidized-bed sintering; ~ **a presión** *f* PROC QUÍ sintering under pressure; ~ **en el vacío** *f* METAL vacuum sintering

sinterizado[1] *adj* C&V, CARBÓN, MECÁ, PROC QUÍ, QUÍMICA sintered

sinterizado[2] *m* C&V, CARBÓN, MECÁ, PROC QUÍ sintering

sinterizante *m* FERRO *vehículos* sintering

sinterizar *vt* CARBÓN, PROC QUÍ, QUÍMICA *polvo* sinter

síntesis *f* GAS, PROC QUÍ, QUÍMICA synthesis; ~ **acústica** *f* ELECTRÓN speech synthesis; ~ **aditiva** *f* FOTO, IMPR additive synthesis; ~ **de audición digital** *f* ELECTRÓN digital speech synthesis; ~ **de audiofrecuencia electrónica** *f* ELECTRÓN electronic-speech synthesis; ~ **del cuadripolo** *f* ELEC network synthesis; ~ **del filtro** *f* ELECTRÓN filter synthesis; ~ **de la forma de onda** *f* ELECTRÓN *generación de señal* waveform synthesis; ~ **de frecuencia** *f* ELECTRÓN frequency synthesis; ~ **de frecuencia directa** *f* ELECTRÓN direct frequency synthesis; ~ **de frecuencia indirecta** *f* ELECTRÓN indirect frequency synthesis; ~ **de Friedel-Crafts** *f* DETERG, PROC QUÍ Friedel-Crafts synthesis; ~ **de redes** *f* ELEC, ING ELÉC network synthesis; ~ **de la señal** *f* ELECTRÓN signal synthesis; ~ **de sonidos vocales** *f* ELECTRÓN, INFORM&PD, TELECOM speech synthesis; ~ **sustractiva** *f* FOTO subtractive synthesis; ~ **de tiempo** *f* ELECTRÓN time synthesis; ~ **de voz** *f* ELECTRÓN, INFORM&PD speech synthesis

sintético *adj* QUÍMICA *compuesto* synthetic

sintetizador *m* ELECTRÓN, TELECOM synthesizer; ~ **de colores** *m* TV color synthesizer (*AmE*), colour synthesizer (*BrE*); ~ **de frecuencia** *m* ELECTRÓN frequency synthesizer; ~ **de frecuencia constante** *m* ELECTRÓN fixed-frequency synthesizer; ~ **de frecuencia directa** *m* ELECTRÓN direct frequency synthesizer; ~ **de frecuencia indirecta** *m* ELECTRÓN indirect frequency synthesizer; ~ **de impulsos** *m* ELECTRÓN pulse synthesizer; ~ **de microondas** *m* ELECTRÓN microwave synthesizer; ~ **de tiempo** *m* ELECTRÓN time synthesizer; ~ **de video** *AmL*, ~ **de vídeo** *m* *Esp* TV video synthesizer; ~ **de voz** *m* INFORM&PD speech synthesizer

sintol *m* QUÍMICA *combustibles* synthol

sintonía *f* GEN tuning; ~ **por barra** *f* ING ELÉC slug tuning; ~ **por condensadores** *f* ING ELÉC capacitive tuning; ~ **nuclear** *f* ING ELÉC slug tuning; ~ **óptica** *f* TELECOM optical tuning; ~ **en tándem** *f* ING ELÉC ganged tuning

sintonina *f* QUÍMICA syntonin

sintonización *f* GEN tuning; ~ **adaptable** *f* ELECTRÓN adaptive tuning; ~ **en avance progresivo** *f* ELECTRÓN incremental tuning; ~ **constante** *f* ELECTRÓN fixed-tuning; ~ **continua** *f* ELECTRÓN continuous tuning; ~ **digital** *f* ELECTRÓN digital tuning; ~ **eléctrica** *f* ELECTRÓN electric tuning; ~ **electromagnética** *f* ELECTRÓN electromagnetic tuning; ~ **electrónica** *f* ELECTRÓN, TELECOM electronic tuning; ~ **de fase** *f* TELECOM phase tuning; ~ **de frecuencia** *f* ELECTRÓN, TELECOM frequency tuning; ~ **horizontal** *f* TV horizontal hold; ~ **con mando único** *f* ING ELÉC ganged tuning; ~ **de onda**

sinusoidal *f* ELECTRÓN sine wave tuning; ~ **óptica** *f* TELECOM optical tuning; ~ **de paso de banda** *f* TV banding; ~ **de varias octavas** *f* ELECTRÓN multi-octave tuning

sintonizado *adj* ACÚST, ELECTRÓN, FÍS ONDAS, TELECOM, TV tuned

sintonizador *m* ACÚST, ELECTRÓN, FÍS RAD, TELECOM, TV tuner; ~ **de guía de ondas** *m* ING ELÉC slug; ~ **de UHF** *m* TV UHF tuner; ~ **de UHF y VHF** *m* TV VHF and UHF tuner

sintonizar *vt* ACÚST, AUTO, ELECTRÓN, FÍS ONDAS, TELECOM, TV tune

sinuosidad *f* PETROL *de terreno* convolution

sinusoidal *adj* GEN sinusoidal

sinusoide *f* GEOM, INFORM&PD sinusoid, MATEMÁT sine curve, sinusoid

SIO *abr* (*entrada-salida secuencial*) INFORM&PD SIO (*sequential input-output, series input-output*)

sirena *f* PROD siren, *de fábrica* hooter, SEG *de advertencia* siren, TRANSP *buques* horn, TRANSP MAR *navegación* siren; ~ **de incendio** *f* SEG fire siren

siríngico *adj* QUÍMICA syringic

sisal *m* TRANSP MAR *cordaje* sisal

sísmica: ~ **de martillo** *f* GEOFÍS hammer seismics

sismicidad *f* CONST, GEOFÍS, TEC ESP seismicity

sísmico *adj* CONST, GEOFÍS, TEC PETR *geofísica* seismic

sismo: ~ **submarino** *m* GEOFÍS, OCEAN seaquake

sismogénico *adj* GEOFÍS seismogenic

sismógrafo *m* CONST seismograph, FÍS seismograph, seismometer, FÍS ONDAS seismic-profile recorder, GAS seismograph, GEOFÍS seismograph, seismometer, INSTR seismic-profile recorder; ~ **horizontal** *m* GEOFÍS horizontal seismograph; ~ **vertical** *m* GEOFÍS vertical seismograph

sismograma *m* TEC PETR seismogram

sismología *f* FÍS, GEOFÍS seismology

sismologista *m* FÍS, GEOFÍS seismologist

sismómetro *m* FÍS, GEOFÍS seismometer

sismotectónica *f* GEOFÍS seismotectonics

sistema *m* CONST, GEOL system, PROD setup, TELECOM system; ~ **abierto** *m* INFORM&PD, TELECOM, TERMO open system; ~ **de acceso continuo de transporte público** *m* TRANSP continuous-access public transport system; ~ **de acceso múltiple** *m* INFORM&PD multiaccess system; ~ **de accionamiento por motor** *m* TELECOM motor-driven system; ~ **de accionamiento motorizado** *m* TELECOM motor-driven system; ~ **activo** *m* ACÚST active system; ~ **acuático** *m* AGUA aquatic system; ~ **acústico** *m* ACÚST acoustic system; ~ **adaptivo** *m* INFORM&PD adaptive system; ~ **de adherencia** *m* TRANSP adhesion system; ~ **adiabático** *m* TERMO adiabatic system; ~ **de adición de boro** *m* NUCL rack railroad (*AmE*), rack railway (*BrE*); ~ **aditivo de color** *m* CINEMAT additive color system (*AmE*), additive colour system (*BrE*); ~ **de administración de base de datos** *m* (*SABD*) INFORM&PD, TELECOM database management system (*DBMS*); ~ **de admisión** *m* ING MECÁ intake system; ~ **de adquisición de datos oceánicos** *m* (*SADO*) OCEAN ocean-data acquisition system (*ODAS*); ~ **aéreo** *m* ING ELÉC overhead system; ~ **de aflojamiento rápido** *m* ING MECÁ, PROD quick-release clamping system; ~ **de agua** *m* TRANSP MAR *circuitos* water system; ~ **de agua de alimentación auxiliar** *m* (*AFWS*) NUCL auxiliary feedwater system (*AFWS*); ~ **de agua de alimentación principal** *m* (*MFWS*) NUCL main feedwater system (*MFWS*); ~ **de agua caliente de alta presión** *m* INSTAL TERM high-pressure hot-water system; ~ **de agua caliente a baja presión** *m* INSTAL TERM low-pressure hot-water system; ~ **de agua de circulación** *m* NUCL circulating water system; ~ **de agua de refrigeración de componentes** *m* (*CCWS*) NUCL component-cooling water system (*CCWS*); ~ **de aire comprimido** *m* ING MECÁ compressed-air system, TRANSP AÉR pressure-air system; ~ **de aire a presión** *m* ING MECÁ compressed-air system; ~ **aislado** *m* FÍS isolated system; ~ **de alarma** *m* SEG alarm system; ~ **de alarma acústica** *m* EMB audible warning system; ~ **de alarma contra gas** *m* *Esp* (*cf sistema avisador de desprendimiento de gases AmL*) MINAS gas alarm system; ~ **de alarma y registro** *m* SEG *aplicaciones industriales* alarm and logging system; ~ **de alcantarillado** *m* MINAS *saneamiento* drainage, RECICL sewer system; ~ **de alcantarillado separado** *m* AGUA separate sewerage system; ~ **de alerta anticolisiones** *m* TEC ESP, TRANSP AÉR collision warning system; ~ **de algodón modificado** *m* TEXTIL modified cotton system; ~ **de alimentación** *m* PAPEL feeding system, TEC ESP feed system; ~ **de alimentación por CA** *m* TEC ESP AC power system, alternating-current power system; ~ **de alimentación de emergencia** *m* ELEC emergency power supply; ~ **de alimentación del garfio** *m* CINEMAT claw-feed system; ~ **de alimentación a presión de la goma** *m* EMB pressurized glue feed; ~ **de alivio de vapor** *m* NUCL steam dumping system; ~ **de alivio de vapor al condensador** *m* NUCL turbine bypass system; ~ **de almacenado, mezcla y distribución** *m* EMB, PROD mix-and-dispense storage system; ~ **de almacenaje ambientalmente controlado** *m* EMB controlled-environment storage system, PROD controlled-environment storage system; ~ **de almacenamiento de energía térmica** *m* TRANSP thermal energy storage system; ~ **de almacenamiento subterráneo** *m* GAS underground storage system; ~ **ALOHA de línea ranurada** *m* TELECOM slotted ALOHA system; ~ **de alta tensión** *m* ELEC *fuente de alimentación* high-voltage system; ~ **de altavoces** *m* ACÚST loudspeaker system; ~ **de alumbrado** *m* GEN lighting system; ~ **de alumbrado de aproximación** *m* TRANSP AÉR approach lighting system; ~ **de alumbrado de vía estrecha** *m* TRANSP AÉR narrow-gage lighting system (*AmE*), narrow-gauge lighting system (*BrE*); ~ **de amortiguamiento** *m* TRANSP AÉR lagging system, shock absorbing system; ~ **de amplificación del circuito digital** *m* TELECOM digital circuit multiplication system (*DCMS*); ~ **de amplio alcance** *m* TELECOM wide area system; ~ **analógico** *m* TELECOM analog system; ~ **anamórfico** *m* CINEMAT anamorphic system; ~ **de ancho de banda ampliado** *m* TELECOM extended bandwidth system; ~ **de anclaje** *m* GEN anchorage system; ~ **de antenas** *m* TEC ESP *telecomunicaciones* aerial system (*BrE*), antenna system (*AmE*); ~ **antibloqueo** *m* (*ABS*) AUTO anti blocking system (*ABS*), ING ELÉC non-blocking network, VEH *frenos* antilock system; ~ **antibloqueo de frenos** *m* VEH antilock brake system; ~ **anticongelamiento** *m* TEC ESP, TRANSP AÉR anti-icing system; ~ **antihielo** *m* TEC ESP, TRANSP AÉR anti-icing system; ~ **de apoyo a la vida** *m* SEG,

TEC ESP, TRANSP AÉR life support system; ~ **de aprobación** *m* CALIDAD approval system; ~ **de aproximación controlado desde tierra** *m* TRANSP AÉR ground-controlled approach system; ~ **de apuntalamiento** *m* CONST propping, system of shoring; ~ **de archivo por discos ópticos** *m* INFORM&PD, ÓPT optical disk filing system; ~ **de archivo de documentos del disco óptico de datos** *m* INFORM&PD, ÓPT optical data disk document filing system; ~ **de armas** *m* D&A weapons system; ~ **de arranque con llave** *m* ING MECÁ key-locked starting system; ~ **de arrastre** *m* MINAS tram system; ~ **de arrollado** *m* TEXTIL take-up system; ~ **arterial** *m* AGUA arterial system; ~ **articulado** *m* ING MECÁ linkage, frame, MECÁ frame; ~ **de aspiración central** *m* SEG central vacuum cleaning system; ~ **de aspiración de polvos** *m* CARBÓN, PROD dust exhaust fan; ~ **de aterrizaje por instrumentos** *m* (*ILS*) TRANSP AÉR instrument landing system (*ILS*); ~ **de aterrizaje de microondas** *m* (*MLS*) TRANSP AÉR microwave landing system (*MLS*); ~ **de autobuses mediante aviso** *m* TRANSP on-call bus system; ~ **autoestructurado** *m* INFORM&PD self-organizing system; ~ **automático** *m* TEC ESP automatic system; ~ **automático de alimentación** *m* EMB, PROD automatic feeding system; ~ **automático de carga** *m* FOTO automatic loading system; ~ **automático de conmutadores rotativos** *m* TELECOM rotary system; ~ **automático contra incendios** *m* SEG *usando agua atomizada* automatic fire-fighting system; ~ **automático de detección de incendios** *m* SEG automatic fire detection system; ~ **automático para eliminar el abarquillado** *m* EMB automatic decurling; ~ **automático de interceptación** *m* TELECOM automatic intercept system; ~ **automático panel** *m* TELECOM panel system; ~ **autónomo** *m* TEC ESP independent system, stand-alone system; ~ **auto-organizado** *m* INFORM&PD self-organizing system; ~ **autorregulado** *m* TV self-controlling system; ~ **autorregulado de mantenimiento** *m* TRANSP self-regulating maintenance system; ~ **de autorreposición** *m* IMPR autoreclosing system; ~ **de autoventilación** *m* EMB self-venting system; ~ **auxiliar para urgencias** *m* TELECOM hot stand-by system; ~ **de avance del encendido** *m* VEH *encendido* advance mechanism; ~ **avanzado de apoyo de fuego aéreo** *m* TRANSP AÉR advanced airborne fire support system; ~ **avisador de desprendimiento de gases** *m* *AmL* (*cf sistema de alarma contra gas Esp*) MINAS gas alarm system; ~ **de aviso de colisión en nubes** *m* SEG, TRANSP AÉR cloud collision warning system; ~ **de aviso de peligro** *m* AUTO hazard warning system;

~ **b** ~ **de baipás de la turbina** *m* NUCL turbine bypass system; ~ **de balizado luminoso de aproximación** *m* TRANSP AÉR approach lighting system; ~ **de barra colectora** *m* ELEC *conexión* busbar system; ~ **de barras colectoras no regulado** *m* TEC ESP unregulated bus system; ~ **de barras cruzadas** *m* TELECOM crossbar system; ~ **de barrido** *m* PROD scavenge system; ~ **bifásico** *m* ELEC *red*, ING ELÉC two-phase system; ~ **bifilar** *m* ING ELÉC two-wire system; ~ **binario** *m* ELECTRÓN, INFORM&PD binary system; ~ **bipolar** *m* ELEC *de máquina eléctrica* two-pole system; ~ **de blocaje del paso** *m* TRANSP AÉR pitch-locking system; ~ **bloqueador controlado a mano** *m* FERRO lock-and-block; ~ **bloqueador del**

paso *m* TRANSP AÉR pitch-locking system; **~ de bloqueo** *m* SEG *mecánico, eléctrico, neumático, hidráulico* interlocking system; **~ de bloqueo con accionamiento de paso** *m* FERRO *infraestructura* lock-and-block system; **~ de bloqueo automático** *m* FERRO *equipo inamovible* lock-and-block system; **~ de bootstrap** *m* REFRIG bootstrap system; **~ de a bordo** *m* TEC ESP on-board system; **~ de buques remolcadores para transporte de gabarras cargadas con contenedores** *m* TRANSP containerized lighter aboard ship system (*CLASS*); **~ de búsqueda y rescate marítimos mediante satélite** *m* TELECOM satellite-aided maritime search and rescue system (*SAMSARS*);

~ c **~ de cabeza simple** *m* EMB, PROD single head system; **~ de cabina sobre rieles** *m* TRANSP cabin system on rail; **~ de cable coaxial** *m* TELECOM pipeline system; **~ de calefacción** *m* AUTO heater system, TERMO, VEH heating system; **~ de calefacción por agua caliente** *m* INSTAL TERM hot water heating system; **~ de calefacción por aire caliente** *m* INSTAL TERM hot air heating system, warm air heating system, SEG hot air radiation heating system; **~ de calefacción por alta presión** *m* INSTAL TERM high-pressure heating system; **~ de calefacción aumentador** *m* ENERG RENOV *calefacción solar y geotérmica* booster heating system; **~ de calefacción central por gasóleo** *m* INSTAL TERM, TERMOTEC oil-fired central heating system; **~ de calefacción a gas** *m* GAS, TERMO, TERMOTEC gas-heating system; **~ de calefacción solar** *m* ENERG RENOV solar heating system; **~ de calefacción por vapor** *m* INSTAL TERM steam heating system; **~ de calentamiento por camisa calefactora** *m* NUCL jacket heating system; **~ de calibrado modular** *m* METR modular gaging system (*AmE*), modular gauging system (*BrE*); **~ de calidad** *m* CALIDAD quality system; **~ de cambio de herramientas** *m* ING MECÁ tool-changing system; **~ de cambio de herramientas flexible** *m* ING MECÁ flexible tool-changing system; **~ de camión a camión** *m* TRANSP truck-to-truck system; **~ de camuflaje** *m* D&A camouflage system; **~ de canalización** *m* ING MECÁ pipework system; **~ de canalización circular** *m* ELEC, ING ELÉC ring supply system; **~ cardinal** *m* TRANSP MAR *marcas de navegación* cardinal system; **~ de carga rápida** *m* FOTO rapid-loading system; **~ de carreteras elevadas para tránsito rápido** *m* TRANSP elevated rapid-transit system; **~ cegesimal** *m* (*sistema cgs*) METR centimetre-gram-second system (*CGS system*); **~ celular** *m* TELECOM cellular system; **~ centrado** *m* FÍS centered system (*AmE*), centred system (*BrE*); **~ centralizado** *m* TELECOM centralized system; **~ centralizado para la separación de tráfico** *m* TELECOM centralized traffic division system; **~ Centrex®** *m* TELECOM Centrex system®; **~ de centro de masas** *m* MECÁ, NUCL center of mass system (*AmE*), centre of mass system (*BrE*); **~ cerrado** *m* TERMO closed system; **~ de certificación** *m* CALIDAD certification system; **~ cgs** *m* (*sistema cegesimal*) METR CGS system (*centimetre-gram-second system*); **~ de cierre** *m* CONST system of seals, EMB interlocking device, TEC ESP shutter system; **~ de cierre de la plancha** *m* IMPR plate lockup system; **~ de cierres de vapor** *m* NUCL gland steam system; **~ de circuito de anillo** *m* ELEC,

ING ELÉC ring supply system; **~ en circuito de anillo** *m* ELEC *red de distribución*, ING ELÉC ring-main system; **~ en circuito de canalización circular** *m* ELEC, ING ELÉC ring-main system; **~ de circulación** *m* ING MECÁ circulating system; **~ de circulación rápida** *m* FERRO *metro subterráneo, tren elevado* rapid-transit system (*RTS*); **~ circulante hidráulico** *m* TEC PETR *perforación* hydraulic circulation system; **~ de clasificación** *m* CALIDAD classification system; **~ de codificación de cuatro firmas** *m* TELECOM four-signature coding system; **~ de codificación y descodificación** *m* TELECOM, TRANSP code-decode system; **~ de codificación de mensajes** *m* TELECOM message handling system (*MHS*); **~ colchón de aire por cámara impelente** *m* TRANSP plenum-chamber air-cushion system; **~ de colector regulado** *m* TEC ESP regulated bus system; **~ de colectores** *m* CONTAM MAR manifold system; **~ de colocación de las bolsas** *m* EMB bag-placing system; **~ de color secuencial con memoria** *m* (*SECAM*) TV sequential color with memory system (*AmE*) (*SECAM*), sequential colour with memory system (*BrE*) (*SECAM*); **~ combinado local-interurbano** *m* TELECOM combined local-toll system; **~ de combustible** *m* TRANSP AÉR fuel system; **~ de combustible líquido** *m* TEC ESP liquid propellant systems; **~ de combustible del motor** *m* ING MECÁ engine fuel system; **~ de combustión controlada** *m* AUTO, TERMO, VEH controlled combustion system; **~ comercial** *m* TELECOM business system; **~ de comportamiento de trayectoria** *m* TEC ESP step track system; **~ de compresión** *m* REFRIG compression system; **~ de comunicación** *m* INFORM&PD communication system; **~ de comunicación auditiva en vehículo** *m* TRANSP in-vehicle aural communication system; **~ de comunicación por cable radiante** *m* FÍS RAD, TELECOM radiating-cable communication system; **~ de comunicación de defensa** *m* TELECOM defence communication system (*BrE*) (*DCS*), defense communication system (*AmE*) (*DCS*); **~ de comunicaciones comerciales** *m* TELECOM business communication system; **~ de comunicaciones entre redes** *m* INFORM&PD gateway; **~ de concentración de potencia** *m* TRANSP power-collection system; **~ condensado** *m* NUCL, PROC QUÍ condensed system, condensate system; **~ condensador** *m* FOTO condenser system; **~ conductor** *m* ING MECÁ drive system; **~ de conductor neutro** *m* ELEC third wire system; **~ de conexión bifilar** *m* ELEC double wire system; **~ de conexión de conductos eléctricos** *m* ELEC bus duct plug-in unit; **~ de conexiones** *m* TELECOM trunk system; **~ de conexiones en delta dobles** *m* ELEC double delta connection; **~ de conexiones en triángulo dobles** *m* ELEC double delta connection; **~ de conmutación** *m* ING ELÉC relay-switching system, TELECOM message-switching system, switching system, TV switching system; **~ de conmutación analógica** *m* TELECOM analog switching system; **~ de conmutación bifilar** *m* TELECOM two-wire switching system; **~ de conmutación bifilar doble** *m* TELECOM four-wire switching system; **~ de conmutación de circuitos** *m* TELECOM circuit switching system; **~ de conmutación de control común** *m* TELECOM common control switching system; **~ de conmutación digital** *m* TELECOM digital switching system; **~ de conmutación por**

división de frecuencias *m* TELECOM frequency-division switching system (*FDS system*); ~ **de conmutación por divisiones espaciadas** *m* TELECOM spaced division switching system; ~ **de conmutación de dos conductores** *m* TELECOM two-wire switching system; ~ **de conmutación de dos hilos** *m* TELECOM two-wire switching system; ~ **de conmutación electromecánica** *m* ELEC, TELECOM electromechanical-switching system; ~ **de conmutación electrónico** *m* TELECOM electronic switching system (*ESS*); ~ **de conmutación por modulación de impulsos codificados** *m* TELECOM PCM switching system; ~ **de conmutación de múltiple tarifa** *m* TELECOM multirate switching system; ~ **de conmutación óptica** *m* TELECOM optical-switching system; ~ **de conmutación de programa almacenado** *m* TELECOM stored program switching system (*AmE*), stored programme switching system (*BrE*); ~ **de conmutación programable cableado** *m* TELECOM hardwired programmable switching system; ~ **de conmutación de servicios múltiples** *m* TELECOM multiservice switching system; ~ **de conmutación de tarifas múltiples** *m* TELECOM multirate switching system; ~ **de conmutación temporal** *m* TELECOM time-division switching system; ~ **de conmutación en el tiempo** *m* TELECOM time-division switching system; ~ **de conmutación del videófono** *m* TELECOM videophone switching system; ~ **conmutador rotatorio** *m* TELECOM rotary system; ~ **de contabilizar llamadas** *m* TELECOM call accounting system; ~ **de contado de rejilla múltiple de alta velocidad** *m* EMB high-speed multirack counting system; ~ **de contención con amortiguación por aire** *m* TRANSP air cushion restraint system; ~ **de contención por bolsa de aire** *m* TRANSP air bag restraint system; ~ **de contenedores industriales para productos a granel** *m* EMB industrial bulk container system; ~ **de control** *m* GEN control system; ~ **de control adaptable** *m* ING ELÉC adaptive control system; ~ **de control autoadaptable** *m* INFORM&PD, MECÁ adaptive control system; ~ **de control automático** *m* PROD automatic control system; ~ **de control en bucle cerrado** *m* ELEC, ELECTRÓN, ING ELÉC closed-loop control system; ~ **de control centralizado** *m* TELECOM centralized control system; ~ **de control común** *m* TELECOM common control system; ~ **para control de las condiciones ambientales** *m* TEC ESP *vehículos espaciales* environmental-control system; ~ **de control del directorio** *m* TELECOM directory control system; ~ **de control distribuido** *m* PROD, TELECOM distributed control system; ~ **de control electrónico** *m* ELECTRÓN electronic control system; ~ **de control hidráulico** *m* AUTO hydraulic control system; ~ **de control por impulsos inversos** *m* TELECOM revertive control system; ~ **de control indirecto** *m* TELECOM indirect control system; ~ **de control del marcador** *m* TELECOM marker control system; ~ **de control de la pala** *m* TRANSP AÉR blade control system; ~ **para control de posición y órbita** *m* (*SCPO*) TEC ESP altitude and orbit control system (*AOCS*); ~ **de control de procesos** *m* INFORM&PD process control system; ~ **de control de programas por conexionado** *m* TELECOM wired-programme control system (*WPC system*); ~ **de control químico y de volumen** *m* NUCL chemical and volume control system (*CVCS*); ~ **de control por radio** *m* TELECOM radio control system; ~ **de control por registradores** *m* TELECOM register-controlled system; ~ **de control remoto** *m* SEG remote control system; ~ **de control totalmente distribuido** *m* TELECOM fully-distributed control system; ~ **de control de tráfico en circuito** *m* ING ELÉC closed-loop traffic control system; ~ **de control de tráfico en circuito cerrado** *m* ELEC, ELECTRÓN, TRANSP closed-loop traffic control system; ~ **de control de vuelo automático** *m* TRANSP AÉR automatic flight control system; ~ **de control de vuelo doble** *m* TRANSP AÉR dual flight-control system; ~ **de controlador programable** *m* PROD programmable-controller system; ~ **de coordenadas** *m* FÍS, MATEMÁT coordinate system; ~ **coordinado Cartesiano** *m* CONST, ELECTRÓN, FÍS, GEOM, MATEMÁT Cartesian coordinate system; ~ **de copiado de imágenes** *m* IMPR imaging system; ~ **de corriente CA** *m* TEC ESP AC power system; ~ **cristalino** *m* CRISTAL crystal system; ~ **cúbico** *m* METR cubic system; ~ **sin cubo** *m* PROD bucketless system; ~ **de cultivos** *m* AGRIC cropping system;

~ **d** ~ **de demanda flexible** *m* TRANSP demand-responsive system; ~ **de desagüe** *m* RECICL drainage system, sewerage system; ~ **de descarga atmosférica de vapor** *m* NUCL steam dumping system; ~ **de descarga de combustible** *m* TEC ESP fuel dumping system; ~ **de descarga eléctrica** *m* ING ELÉC lighting system; ~ **descentralizado** *m* TELECOM decentralized system; ~ **de descontaminación** *m* NUCL decontamination system; ~ **de desempañamiento** *m* VEH demister system; ~ **de desmontaje** *m* TEC ESP strapdown system; ~ **de desviación** *m* TV deflection system; ~ **de detección y alarma de incendios** *m* SEG fire detection and alarm system; ~ **de detección de elementos defectuosos** *m* NUCL failed element detection system; ~ **de detección en el exterior del vehículo** *m* TRANSP detection system outside the vehicle; ~ **de detección de incendios** *m* SEG, TRANSP AÉR, TRANSP MAR fire-detection system; ~ **de detección en el interior del vehículo** *m* TRANSP detection system inside the vehicle; ~ **de detección y seguimiento espacial** *m* D&A space detection and tracking system (*SPADATS*); ~ **difásico** *m* ELEC *red* two-phase system; ~ **de diferencias de la onda portadora** *m* TV carrier difference system; ~ **digestivo** *m* RECICL digestive system; ~ **de dirección** *m* AUTO steering system; ~ **de dirección obligatoria por medios físicos** *m* TRANSP system with compulsory guidance by physical means; ~ **de dirección de la onda portadora** *m* TELECOM carrier sense system; ~ **de direccionamiento** *m* INFORM&PD addressing system; ~ **de disparo** *m* FÍS RAD triggering system; ~ **de dispersión coloidal** *m* CONTAM *tratamiento de líquidos*, TEC PETR *ingeniería de lodos, perforación* colloid disperse system; ~ **de distribución** *m* AGUA, ELEC, GAS, ING ELÉC distribution system; ~ **de distribución en el tiempo** *m* TELECOM time-division system; ~ **distribuido multiantena** *m* TELECOM distributed multi-antenna system; ~ **distribuidor mediante bomba** *m* EMB, PROD pump dispenser system; ~ **dividido funcionalmente** *m* TELECOM functionally-divided system; ~ **de división de funciones** *m* TELECOM function-division system; ~ **de división de tiempo** *m* TELECOM time-division system; ~ **de división de tráfico** *m* TELECOM traffic division system; ~ **por divisiones espaciadas** *m* TELECOM spaced division system; ~ **divisor del haz** *m* CINEMAT beam-splitting system; ~ **de doble combustible** *m* TRANSP dual-fuel system; ~ **de doble procesador** *m* TELECOM dual-processor system; ~ **doméstico integrado** *m* INFORM&PD integrated home system (*IHS*); ~ **doméstico de video** *m* *AmL*, ~ **doméstico de vídeo** *m* *Esp* TV video home system (*VHS*); ~ **doméstico de video-compacto** *AmL*, ~ **doméstico de vídeo-compacto** *m* *Esp* TV video home system-compact (*VHS-C system*); ~ **de dos fases** *m* ELEC *red* two-phase system; ~ **de dos hilos** *m* ELEC *red de distribución*, TELECOM two-wire system; ~ **de dos nucleones** *m* NUCL two-nucleon system; ~ **de dos pozos** *m* ENERG RENOV *energía geotérmica*, TERMO doublet; ~ **de dos rodillos** *m* IMPR two-roll system; ~ **de drenaje** *m* CONST drainage structure; ~ **de drenaje pluvial** *m* CONST storm sewer system; ~ **dual de combustible** *m* TRANSP dual-fuel system;

■ **e** ~ **EBCS** *n* (*Sistema Europeo de Transporte de Barcazas*) TRANSP EBCS (*European barge-carrier system*); ~ **de ecuaciones** *m* MATEMÁT simultaneous equations; ~ **de efluentes activos** *m* NUCL active effluent system; ~ **de electrodos de conexión a tierra** *m* ELEC, ING ELÉC, PROD earthing electrode system (*BrE*), grounding electrode system (*AmE*); ~ **de electrodos de puesta a tierra** *m* ELEC, ING ELÉC, PROD earthing electrode system (*BrE*), grounding electrode system (*AmE*); ~ **electrónico antideslizante** *m* TRANSP electronic anti-skid system; ~ **electrónico de mensajes** *m* TELECOM electronic message system; ~ **electrosensible de seguridad** *m* SEG, TEC ESP electrosensitive safety system; ~ **de eliminación de polvo** *m* CONST dust suppression system; ~ **para eliminar los orillos** *m* PAPEL trim removal system; ~ **de embalaje desmontable y reutilizable** *m* EMB collapsible and reusable packaging system; ~ **de embalaje por termoformación** *m* EMB thermoforming packaging system; ~ **de embarque y desembarque autopropulsado** *m* TRANSP, TRANSP MAR roll-on/roll-off system (*ro-ro system*); ~ **de embolsado doble** *m* EMB twin-bagging system; ~ **de emergencia** *m* TRANSP AÉR emergency system; ~ **de empedernado del techo** *m* MINAS *galerías* resin roof bolting; ~ **de encendido de dos circuitos** *m* AUTO two-circuit ignition system; ~ **de encendido transistorizado** *m* AUTO transistorized ignition system; ~ **de enclavamiento controlado por computadora** *m* *AmL* (*cf sistema de enclavamiento controlado por ordenador Esp*) FERRO *equipo inamovible* computerized interlocking system; ~ **de enclavamiento controlado por ordenador** *m* *Esp* (*cf sistema de enclavamiento controlado por computadora AmL*) FERRO *equipo inamovible* computerized interlocking system; ~ **de enclavamiento eléctrico** *m* ELEC *seguridad*, PROD, SEG electric interlocking system; ~ **de end** *m* (*cf sistema de fin*) TELECOM end system; ~ **de energía terrestre** *m* TEC ESP ground power system; ~ **de enfriado por agua** *m* ING MECÁ water-cooled system; ~ **de enfriado por aire** *m* ING MECÁ air-cooled system; ~ **de enfriamiento** *m* AUTO, CONST, CONTAM, ELEC, TERMO, TEXTIL cooling system; ~ **de enfriamiento por expansión directa** *m* REFRIG direct expansion refrigeration system; ~ **de enfriamiento con refrigerante parcialmente**

recuperado *m* REFRIG partial-recovery refrigeration system; ~ **de enfriamiento con refrigerante perdido** *m* REFRIG total loss refrigeration system; ~ **de enfriamiento en ruta** *m* REFRIG over-the-road system; ~ **de enfriamiento por transmisión indirecta** *m* REFRIG indirect expansion refrigeration system; ~ **de engomado en frío** *m* EMB cold-glueing system; ~ **de engranaje helicoidal** *m* ING MECÁ spiral; ~ **de engranaje planetario** *m* AUTO planetary-gear system; ~ **de engranajes** *m* ING MECÁ train of gearing; ~ **de enlaces** *m* TELECOM trunk system, link system; ~ **de ensayo de materiales** *m* TEXTIL materials testing system; ~ **de ensayos no destructivos** *m* ING MECÁ nondestructive testing system; ~ **de entrega física** *m* TELECOM physical delivery system (*PDS*); ~ **para envolver con un material retráctil** *m* EMB shrink flow line wrappers; ~ **de equilibrio** *m* QUÍMICA equilibrium system; ~ **de escape** *m* AUTO, VEH exhaust system; ~ **de escape de lanzamiento** *m* TEC ESP launch escape system; ~ **de escape de la máquina** *m* ING MECÁ, TRANSP MAR engine exhaust system; ~ **de escape del motor** *m* ING MECÁ, TRANSP MAR engine exhaust system; ~ **de esclusas** *m* NUCL *de entrada en contención* airlock system; ~ **de eslabones** *m* ING MECÁ linkage; ~ **de espera de velódromo** *m* TRANSP AÉR racetrack holding pattern; ~ **en estrella** *m* ELEC, ING ELÉC radial system; ~ **de etiquetado mediante código de barras** *m* EMB bar-code labeling system (*AmE*), bar-code labelling system (*BrE*); ~ **de etiquetado modular** *m* EMB modular labeling system (*AmE*), modular labelling system (*BrE*); ~ **de etiquetado múltiple** *m* EMB multilane labeling system (*AmE*), multilane labelling system (*BrE*); ~ **de expedición de enlaces** *m* TELECOM trunked dispatch system; ~ **experto** *m* INFORM&PD expert system; ~ **de exploración por haz electrónico** *m* ELECTRÓN scanning electron-beam system; ~ **de explotación agrícola** *m* AGRIC farming system; ~ **de explotación de datos** *m* INFORM&PD data processing (*DP*); ~ **de extracción** *m* MINAS hoisting system, pithead works; ~ **de extracción de humos y calor** *m* SEG smoke and heat extraction system; ~ **de extracción de humos de soldadura** *m* SEG welding smoke extraction system; ~ **de extracción por jaulas** *m* MINAS cage winding system; ~ **de extracción no continua** *m* MINAS skip-winding system; ~ **de extracción del recorte tipo eyector** *m* EMB ejector-type trim exhaust system; ~ **de extracción de residuos** *m* CONTAM, EMB waste extraction system; ~ **extractor** *m* SEG extractor-fan system;

■ **f** ~ **de fichas** *m* INFORM&PD card system; ~ **de ficheros de raíz** *m* PROD root file system; ~ **fijo aeronáutico** *m* TRANSP AÉR aeronautical fixed service (*BrE*), aeronautical fixed system (*AmE*); ~ **de fin** *m* (*cf sistema de end*) TELECOM end system; ~ **de flap soplado** *m* TRANSP AÉR blown-flap system; ~ **de formación de imágenes** *m* ÓPT imaging system; ~ **de formación de imágenes térmicas** *m* D&A thermal imager; ~ **de frenado** *m* ING MECÁ braking system; ~ **de frenado por aire comprimido** *m* ING MECÁ pneumatic brake system; ~ **de frenado antideslizante** *m* AUTO antiskid braking system; ~ **de frenado de disco** *m* AUTO, CONST disc braking system (*BrE*), disk braking system (*AmE*); ~ **de frenado de emergencia** *m* TRANSP AÉR, VEH emer-

gency-brake system; ~ **de freno auxiliar** *m* VEH secondary brake system; ~ **de freno a las cuatro ruedas** *m* AUTO four-wheel brake system; ~ **de freno en las cuatro ruedas** *m* VEH four-wheel brake system; ~ **de freno de disco** *m* AUTO, CONST disc braking system (*BrE*), disk braking system (*AmE*); ~ **de freno hidráulico** *m* ING MECÁ hydraulic brake system; ~ **de frenos** *m* ING MECÁ, VEH brake system; ~ **frigorífico** *m* ING MECÁ refrigerating system, REFRIG refrigerating system, refrigeration system, TERMO refrigeration system; ~ **frigorífico de absorción** *m* REFRIG absorption refrigeration system; ~ **frigorífico autónomo** *m* REFRIG factory-assembled system; ~ **con frigorífico carburante** *m* REFRIG fuel-freeze system; ~ **frigorífico de eyección de vapor** *m* REFRIG ejector-cycle refrigeration system; ~ **de fuerza de fluido** *m* ING MECÁ fluid power system;
~ g ~ **de gas frío** *m* TEC ESP *aeronave* cold gas system; ~ **de gestión de base de datos** *m* INFORM&PD, TELECOM database management system (*DBMS*); ~ **Giorgi** *m* METR Giorgi system; ~ **de globo aerostático** *m* D&A balloon system; ~ **de globo aerostático para transporte de carga** *m* D&A load-carrying balloon system; ~ **de graduación láser** *m* CONST laser grading system; ~ **de grúa de hasta 20 toneladas** *m* ING MECÁ cranage up to 20 tonnes capacity; ~ **de grúa de 20 toneladas de capacidad** *m* ING MECÁ cranage of 20 tonnes capacity; ~ **de grúas** *m* ING MECÁ cranage; ~ **de guía de aparcamiento** *m* TRANSP AÉR docking guidance system; ~ **de guía automático** *m* ING ELÉC automatic guidance system; ~ **de guiado por adherencia** *m* TRANSP system with guidance by adhesion; ~ **de guiado de navegación** *m* TEC ESP guidance navigation system;
~ h ~ **de haz único** *m* TRANSP monobeam system; ~ **híbrido** *m* TELECOM, TRANSP hybrid system; ~ **hidrante contra incendio** *m* SEG, TEC PETR *refino, seguridad* hydrant system; ~ **hidráulico** *m* INSTAL HIDRÁUL hydraulic system, TEC PETR *mecánica de fluidos* hydraulics; ~ **de hidrorrefrigeración** *m* ING MECÁ water-cooled system; ~ **de hilatura** *m* TEXTIL spinning system; ~ **de hilo neutro** *m* ELEC third wire system; ~ **hiperbólico de determinación de la situación** *m* TRANSP MAR *navegación* hyperbolic position-fixing system; ~ **hiperbólico de navegación** *m* TRANSP MAR hyperbolic navigation;
~ i ~ **de ignición** *m* TEC ESP ignition system; ~ **de iluminación** *m* ING ELÉC lighting system; ~ **de iluminación de vía estrecha** *m* TRANSP AÉR narrow-gage lighting system (*AmE*), narrow-gauge lighting system (*BrE*); ~ **de imagen directa en una plancha** *m* IMPR direct-to-plate imaging system; ~ **de impresión sin impacto** *m* IMPR nonimpact printing system; ~ **de impulsión** *m* TEXTIL drive system; ~ **incremental-analógico** *m* INFORM&PD analog-incremental system; ~ **independiente** *m* TEC ESP independent system, stand-alone system; ~ **de indicación de cámara** *m* CINEMAT, TV camera-prompting system; ~ **indicador de fugas** *m* SEG leakage indicator system; ~ **indicador del nivel de la vasija del reactor** *m* NUCL reactor vessel level-indicating system (*RVLIS*); ~ **inductor** *m* FÍS inducing system; ~ **inercial** *m* NUCL inerting system; ~ **inercial de navegación** *m* TRANSP MAR inertial

navigation system (*INS*); ~ **de inertización** *m* NUCL inerting system; ~ **de información** *m* INFORM&PD information system (*IS*); ~ **de información de estación de barco** *m* TELECOM ship reporting system; ~ **de información oceanográfica** *m* OCEAN Oceanographic Data Acquisition System; ~ **de información de vuelo** *m* TEC ESP flight-data system; ~ **informático** *m* ELEC computer system, INFORM&PD computer system, data processing system, TELECOM computer system; ~ **insensible a fallos** *m Esp* INFORM&PD fault-tolerant system; ~ **de inspección visual del tren de aterrizaje** *m* TRANSP AÉR landing gear optical inspection system; ~ **instrumental de aterrizaje** *m* TEC ESP instrument landing system (*ILS*); ~ **integrado** *m* TELECOM integrated system; ~ **integrado de administración** *m* (*SIG*) INFORM&PD, TELECOM management information system (*MIS*); ~ **integrado de armamento** *m* TRANSP AÉR integrated weapon system; ~ **integrado de casa** *m AmL* INFORM&PD integrated home system (*IHS*); ~ **integrado de gestión** *m* INFORM&PD management information system; ~ **interactivo** *m* TELECOM interactive system; ~ **de intercalación** *m* EMB collating system; ~ **de intercambio de texto orientado a mensajes** *m* TELECOM Message-Oriented Text Interchange Standard (*MOTIS*); ~ **de intercomunicación** *m* TELECOM intercom; ~ **intermedio** *m* (*SI*) TELECOM intermediate system (*IS*); ~ **internacional de unidades** *m* (*SI*) ELEC, FÍS PART, METR international system of units (*SI*); ~ **de interrogación** *m* TELECOM polling system; ~ **de inyección de tinta** *m* IMPR, INFORM&PD inkjet printer, inkjet system
~ j ~ **jerárquico** *m* TELECOM hierarchical system;
~ l ~ **de laboratorio** *m* FÍS laboratory frame; ~ **de laboreo por grandes tajos** *m* CARBÓN *minas de carbón* longwall system; ~ **lasérico de medición** *m* FÍS RAD laser monitoring system; ~ **lasérico de monitorización** *m* FÍS RAD laser monitoring system; ~ **lasérico de vigilancia** *m* FÍS RAD laser monitoring system; ~ **lateral** *m* TRANSP MAR *navegación* lateral system; ~ **lateral de guiado electromagnético** *m* TRANSP electromagnetic lateral-guidance system; ~ **de lectura** *m* ING MECÁ read-out system; ~ **de lectura óptica de caracteres** *m* (*sistema OCR*) EMB optical character reading system (*OCR system*); ~ **Leitz** *m* INSTR Leitz system; ~ **de límites cruzados** *m* TELECOM cross-border system; ~ **de línea con alternación de fase** *m* TV phase alternation line (*PAL*); ~ **de línea coaxial** *m* FÍS, ING ELÉC coaxial-line system, TELECOM pipeline system, coaxial line system; ~ **de línea concéntrica** *m* TELECOM pipeline system; ~ **de línea ranurada** *m* TELECOM slotted system; ~ **lineal** *m* ELECTRÓN *de fotodetectores, diodos electroluminiscentes* linear array; ~ **de líneas** *m* TELECOM line system; ~ **de llamada de emergencia** *m* TELECOM, TRANSP emergency-call system; ~ **de llamada sin hilos** *m* TELECOM radio-paging system; ~ **de llamada selectiva** *m* TELECOM selective calling system (*SELCAL*); ~ **de llamadas por radio** *m* TELECOM radio-paging system; ~ **de llamadas de socorro radiotelefónicas** *m* TELECOM distress radio call system; ~ **de llave en mano** *m* INFORM&PD turnkey system; ~ **de llenado de cajas de cartón por los laterales y por la parte superior** *m* EMB horizontal and top loader cartoner; ~ **de**

llenado y cierre *m* EMB blow-fill-seal system; ~ **de llenado de la memoria de discos** *m* ÓPT jukebox filing system; ~ **de lodo** *m* TEC PETR *perforación, ingeniería de lodos* mud system; ~ **de lógica cableada** *m* TELECOM wired-logic system; ~ **de lubricación** *m* AUTO, ING MECÁ lubricating system, PROD lubrication system; ~ **de lubricación centralizado** *m* ING MECÁ centralized lubricating system; ~ **de lubricación de cojinetes** *m* PROD bearing lubrication system;

■ m ~ **de maclaje** *m* METAL twinning system; ~ **maestro-esclavo** *m* INFORM&PD master-slave system; ~ **de mando del acelerador** *m* VEH *carburador* accelerator linkage; ~ **de mando automático** *m* ELEC control system; ~ **de mando de la dirección** *m* VEH steering gear; ~ **de mando por radio** *m* TELECOM radio control system; ~ **de manejo de cables** *m* ELEC, ING ELÉC cable-handling system; ~ **de maniobra orbital** *m* (*OMS*) TEC ESP orbital maneuvering system (*AmE*) (*OMS*), orbital manoeuvring system (*BrE*) (*OMS*); ~ **de manipulación electrónica** *m* TELECOM electronic-key system; ~ **de manipulación de mensajes** *m* TELECOM Message-Oriented Text Interchange Standard (*MOTIS*), message handling system (*MHS*); ~ **de manivela** *m* MECÁ bellcrank system; ~ **manual** *m* TELECOM manual system; ~ **de manutención mediante transportador** *m* EMB conveyor handling system; ~ **del marcador** *m* TELECOM marker system; ~ **mecánico** *m* ACÚST, ING MECÁ mechanical system; ~ **de medición absoluta** *m* ELECTRÓN absolute measuring system; ~ **de medición analógica** *m* INFORM&PD analog measuring system; ~ **de medida** *m* ELEC *instrumento*, INSTR measuring system; ~ **de medida por puntos tipográficos** *m* IMPR point system; ~ **de megafonía** *m* SEG *por medio de radios* staff-calling installation, TRANSP AÉR public address system (*PA system*); ~ **de memoria** *m* TV memory system; ~ **de memoria compartida** *m* INFORM&PD shared memory system; ~ **de memoria virtual** *m* INFORM&PD virtual memory system (*VMS*); ~ **de mensajería por intercambio de datos electrónicos** *m* TELECOM EDI-messaging system (*EDIMS*); ~ **de mensajería interpersonal** *m* TELECOM interpersonal messaging system; ~ **meteorológico** *m* METEO weather system; ~ **métrico** *m* METR metric system; ~ **metro/kilogramo/segundo/amperio** *m* (*sistema MKSA*) METR meter/kilogram/second/ampere system (*AmE*) (*MKSA system*), metre/kilogram/second/ampere system (*BrE*) (*MKSA system*); ~ **de microondas** *m* TELECOM microwave system; ~ **MKSA** *m* (*sistema metro/kilogramo/segundo/amperio*) METR MKSA system (*meter/kilogram/second/ampere system, metre/kilogram/second/ampere system*); ~ **modificado** *m* TEXTIL modified system; ~ **de modulación digital** *m* TELECOM digital modulation system; ~ **con modulación por impulsos codificados** *m* TELECOM PCM system; ~ **modular de aspiración** *m* SEG point vacuum-cleaning system; ~ **de mojado por cortina de aire** *m* IMPR, PAPEL air doctor dampening system; ~ **de moldeo de la pasta** *m* EMB pulp-molding system (*AmE*), pulp-moulding system (*BrE*); ~ **monitor de despegue** *m* TRANSP AÉR takeoff-monitoring system; ~ **de monitorización** *m* FÍS RAD monitoring system; ~ **monoclínico** *m* METAL monoclinic system; ~ **monofilar** *m* ELEC single-wire

system; ~ **motor** *m* ING MECÁ drive system; ~ **multifilar** *m* ELEC multiple wire system; ~ **multimicroprocesador** *m* TELECOM multimicroprocessor system; ~ **de multinivel** *m* TELECOM multilevel system; ~ **con múltiples derivaciones** *m* ELEC *red de distribución* treed system; ~ **de multiplicación del circuito digital** *m* TELECOM digital circuit multiplication gain (*DCMG*); ~ **multiplinto** *m* TRANSP multiple skirt system, multiskirt system; ~ **multiprocesador** *m* TELECOM multiprocessor system; ~ **de multiprocesador segmentado** *m* TELECOM segmented multiprocessor system; ~ **de multiprocesamiento** *m* INFORM&PD multiprocessing system; ~ **de multiprogramación** *m* INFORM&PD multiprogramming system; ~ **multiusuario** *m* INFORM&PD multiuser system;

■ n ~ **de navegación por inercia** *m* TRANSP AÉR inertial navigation system (*INS*); ~ **neumático** *m* ING MECÁ compressed-air system; ~ **con neutro aislado** *m* ELEC isolated neutral system; ~ **con neutro puesto directamente a tierra** *m* ELEC solidly-earthed neutral system (*BrE*), solidly-grounded neutral system (*AmE*); ~ **con neutro a tierra** *m* ELEC earthed neutral system (*BrE*), grounded neutral system (*AmE*), impedance earthed neutral system (*BrE*), impedance grounded neutral system (*AmE*), ING ELÉC earthed neutral system (*BrE*), grounded neutral system (*AmE*); ~ **con neutro a tierra resonante** *m* ELEC resonant-earthed neutral system (*BrE*), resonant-grounded neutral system (*AmE*); ~ **no balanceado** *m* TELECOM unbalanced system; ~ **no jerárquico** *m* TELECOM nonhierarchical system; ~ **normalizado** *m* TELECOM Message-Oriented Text Interchange Standard (*MOTIS*); ~ **norteamericano de calibres de alambres y de chapas** *m* PROD American wire gage (*AmE*) (*AWG*), American wire gauge (*BrE*) (*AWG*); ~ **nuboso** *m* METEO cloud system; ~ **numérico** *m* INFORM&PD number system;

■ o ~ **para obtener pruebas** *m* IMPR proofing system; ~ **OCR** *m* (*sistema de lectura óptica de caracteres*) EMB OCR system (*optical character reading system*); ~ **de oficina integrado** *m* (*SOI*) INFORM&PD, TELECOM integrated office system (*IOS*); ~ **de onda portadora de uno más uno** *m* TELECOM one-plus-one carrier system; ~ **del operador** *m* TELECOM operator system; ~ **operativo** *m* (*SO*) INFORM&PD, TELECOM operating system (*OS*); ~ **operativo de discos** *m* (*DOS*) INFORM&PD disk operating system (*DOS*); ~ **operativo distribuido** *m* INFORM&PD distributed operating system; ~ **operativo de red** *m* (*NOS*) INFORM&PD network operating system (*NOS*); ~ **óptico** *m* ELECTRÓN optical system; ~ **óptico de haces iónicos** *m* FÍS RAD ion beam optical system; ~ **óptico Schmidt** *m* CINEMAT Schmidt optical system; ~ **de orden y control** *m* INFORM&PD command and control system (*CSM*); ~ **de órdenes** *m* INFORM&PD command system; ~ **de la organización del tráfico marítimo** *m* TRANSP MAR *navegación* routing system; ~ **oscilatorio** *m* ELECTRÓN, TELECOM, TV oscillatory system;

■ p ~ **de palanca acodada** *m* MECÁ bellcrank system; ~ **de palancas** *m* ING MECÁ leverage, rod linkage; ~ **Panavisión** *m* CINEMAT Panavision system; ~ **panel** *m* TELECOM panel system; ~ **de parada**

secundaria *m* NUCL secondary shutdown system; ~ **con paradas intermedias** *m* TRANSP system with intermediate stops; ~ **pasivo** *m* ACÚST, ENERG RENOV passive system; ~ **pasivo de cinturón de seguridad** *m* TRANSP passive seat-belt system; ~ **pasivo de sujeción del pasajero** *m* TRANSP passive occupant-restraint system; ~ **paso a paso** *m* TELECOM step-by-step system; ~ **patrón de trabajo** *m* FÍS working standard; ~ **de película de aire** *m* TRANSP air film system; ~ **de plegado de pala automático** *m* TRANSP AÉR automatic blade-folding system; ~ **de plinto múltiple** *m* TRANSP multiple skirt system, multiskirt system; ~ **de portador desplazado** *m* TV offset carrier system; ~ **de portadora analógico** *m* INFORM&PD analog carrier system; ~ **de portadora suprimida** *m* TV suppressed-carrier system; ~ **posicionador del haz** *m* TV beam-positioning system; ~ **de posicionamiento global** *m* (*GPS*) TRANSP AÉR, TRANSP MAR global positioning system (*GPS*); ~ **de potencia en tierra** *m* TEC ESP ground power system; ~ **de presión** *m* TRANSP AÉR pressure system; ~ **de prevención de colisión de a bordo** *m* TRANSP AÉR airborne collision avoidance system; ~ **de prevención de colisión de radiofaro activo** *m* TRANSP AÉR active beacon collision avoidance system; ~ **preventivo anticolisión de a bordo** *m* TRANSP AÉR airborne collision avoidance system; ~ **preventivo anticolisión de radiofaro activo** *m* TRANSP AÉR active beacon collision avoidance system; ~ **principal-subordinados** *m* INFORM&PD master-slave system; ~ **con procesador único y control común** *m* TELECOM single-processor common-control system; ~ **de procesamiento de datos** *m* INFORM&PD data processing system; ~ **de procesamiento múltiple** *m* INFORM&PD multiprocessing system; ~ **de procesamiento de tareas** *m* INFORM&PD job stream, job-processing system; ~ **de proceso de datos** *m* NUCL data processing system; ~ **de producción nuclear de vapor** *m* (*NSSS*) NUCL nuclear steam supply system (*NSSS*); ~ **de programación orientado al objeto** *m* INFORM&PD object-oriented programming system; ~ **progresivo** *m* TRANSP progressive system; ~ **progresivo limitado** *m* TRANSP limited progressive system; ~ **de propelentes sólidos** *m* TEC ESP solid-propellant system; ~ **de propulsión** *m* ING ELÉC power plant, TEC ESP propulsion system; ~ **de propulsión de hidrazina** *m* TEC ESP hydrazine propulsion system; ~ **propulsor** *m* TEC ESP, TRANSP AÉR, TRANSP MAR *motores* power plant; ~ **de protección colectiva** *m* D&A collective protective system; ~ **de protección discriminante** *m* ELEC *cortocircuito* discriminating protective system; ~ **de protección del reactor** *m* (*RPS*) NUCL reactor protection system (*RPS*); ~ **protector de sujeción** *m* SEG protective restraint system; ~ **público de altavoces** *m* TRANSP AÉR public address system (*PA system*); ~ **de puesta en fila de espera de llamadas** *m* TELECOM call-queueing facility; ~ **de puesta a tierra** *m* ELEC, ING ELÉC, PROD earthed system (*BrE*), grounded system (*AmE*); ~ **de puntería** *m* D&A sighting line; ~ **de pura pérdida** *m* PROD dead loss system; ~ **de purificación del condensado** *m* NUCL condensate polishing system;

`~ q` ~ **para quitar el polvo** *m* CONTAM, EMB, PROD dust removal system;

`~ r` ~ **de radiación dirigida** *m* TELECOM guided radiation system; ~ **radial** *m* ELEC, ING ELÉC radial system; ~ **de radio monocanal terrestre y aéreo** *m* D&A single channel ground and airborne radio system (*SINCGARS*); ~ **de radiocontrol** *m* TELECOM radio control system; ~ **de radiodeterminación por satélite** *m* TEC ESP, TRANSP AÉR, TRANSP MAR *navegación* radio-determination satellite system; ~ **de radionavegación por satélite** *m* (*satnav*) TEC ESP satellite navigator (*satnav*), TRANSP MAR satellite navigator (*satnav*), system of satellite navigation; ~ **ramificado** *m* ELEC *red de distribución* treed system; ~ **de ranurado en L** *m* AUTO L split system; ~ **de ranuras** *m* TEC ESP slit system; ~ **de rastreo** *m* TELECOM tracking system; ~ **de recogida de datos** *m* TELECOM data collection system (*DCS*); ~ **recuperación** *m* CONTAM recovery device, recovery system, NUCL recovery system; ~ **de recuperación de datos** *m* FÍS RAD data recovery system; ~ **de recuperación de información** *m* INFORM&PD information retrieval system; ~ **de referencia** *m* FÍS frame; ~ **de referencia de Galileo** *m* FÍS Galilean frame; ~ **de referencia inercial** *m* FÍS inertial frame, TEC ESP inertial reference system; ~ **de referencia de posición** *m* PETROL position reference system; ~ **de reflote** *m* TRANSP lift on-off system; ~ **de refrigeración** *m* AUTO, CONST, CONTAM, ELEC cooling system, REFRIG refrigeration system, TERMO refrigeration system, cooling system, TEXTIL cooling system; ~ **de refrigeración por agua** *m* ING MECÁ water-cooled system; ~ **de refrigeración autónomo** *m* AUTO closed-and-sealed cooling system; ~ **de refrigeración del blindaje** *m* NUCL shield cooling system; ~ **de refrigeración del blindaje biológico** *m* NUCL biological-protection cooling system; ~ **de refrigeración de chimenea** *m* CONST chimney cooler; ~ **de refrigeración de ciclo cerrado** *m* ELEC closed-cyle cooling system, NUCL, TERMO closed-cycle cooling system; ~ **de refrigeración hermético** *m* AUTO closed-and-sealed cooling system, sealed cooling system; ~ **de refrigeración de las máquinas** *m* TRANSP MAR engine-cooling system; ~ **refrigerante** *m* AUTO, CONST, CONTAM, ELEC cooling system, ING MECÁ, REFRIG refrigerating system, TERMO, TEXTIL cooling system; ~ **del refrigerante del reactor** *m* (*RCS*) NUCL reactor-coolant system (*RCS*); ~ **de refugio** *m* TEC ESP slit system; ~ **regional de advertencia por radio** *m* TELECOM, TRANSP regional radio-warning system; ~ **registrador** *m* TEC ESP sensor system; ~ **registrador activo** *m* TEC ESP active sensor; ~ **de regulación** *m* ELEC control system; ~ **de regulación por contrarreacción** *m* ELECTRÓN, ING ELÉC feedback control system; ~ **de regulación del tráfico** *m* TRANSP traffic regulation system; ~ **regulador de voltaje** *m* TRANSP AÉR dimmer system; ~ **de relé hertziano** *m* TELECOM relay system; ~ **de relé hertziano en dos frecuencias** *m* TELECOM two-frequency relay system; ~ **de relé de láminas** *m* ELEC, ING ELÉC, TELECOM reed-relay system; ~ **de remetido** *m* TEXTIL drafting system; ~ **de reparto de la carga** *m* TELECOM load-sharing system; ~ **de reparto de la carga de doble procesador** *m* TELECOM dual-processor load-sharing system; ~ **por repelado** *m* EMB peelable system; ~ **de repetidores sin amplificador de antena** *m* TELECOM nonboosted antenna repeater system; ~ **de repostaje en vuelo** *m* TRANSP, TRANSP AÉR refueling in-flight system

(*AmE*), refuelling in-flight system (*BrE*); ~ **de reserva** *m* TELECOM stand-by system; ~ **de reservaciones de pasajeros** *m AmL* (*cf sistema de reservas de pasajeros Esp*) TRANSP passenger reservation system; ~ **de reservas de pasajeros** *m Esp* (*cf sistema de reservaciones de pasajeros AmL*) TRANSP passenger reservation system; ~ **con una resolución temporal de 5 ns** *m* FÍS RAD system providing a temporal resolution of 5 ns; ~ **de retransmisión** *m* TELECOM relay system; ~ **de retransmisión en dos frecuencias** *m* TELECOM two-frequency relay system; ~ **retrovisor** *m* AUTO, TRANSP, VEH rear-view system; ~ **de revestimiento** *m* TRANSP AÉR lagging system; ~ **robotizado de paletización y retractilado** *m* EMB robotic palletizing and stretch system; ~ **de rociado de contención** *m* NUCL containment spray system; ~ **de rociado del núcleo de alta presión** *m* NUCL railroad semitrailer (*AmE*), railway semitrailer (*BrE*); ~ **rociador de agua** *m* NUCL water spray system; ~ **de rodillos para revestimiento** *m* REVEST coating roller system; ~ **rotativo de desequilibrio de masas** *m* ING MECÁ rotating system of out-of-balance masses; ~ **rotativo de llenado** *m* EMB rotary filling; ~ **de rotulado modular** *m* EMB modular labeling system (*AmE*), modular labelling system (*BrE*); ~ **s** ~ **de saneamiento** *m* CONST drainage structure; ~ **satelitario de radiodeterminación** *m* TEC ESP, TRANSP AÉR, TRANSP MAR *navegación* radio-determination satellite system; ~ **Schmidt** *m* INSTR Schmidt system; ~ **de secado** *m* GAS drying system; ~ **SECAM** *m* TV SECAM system; ~ **de seguimiento** *m* TELECOM tracking system; ~ **semiautomático** *m* TELECOM semiautomatic system; ~ **de señales** *m* TV signal complex; ~ **de señales por tramos de vía** *m* FERRO block system; ~ **de señalización** *m* TELECOM signaling system (*AmE*), signalling system (*BrE*); ~ **sensible a la demanda** *m* TRANSP demand-responsive system; ~ **de sensitividad de cabeceo del piloto automático** *m* TRANSP AÉR autopilot pitch sensitivity system; ~ **sensor** *m* TEC ESP sensor system; ~ **de separación de las palas** *m* TRANSP AÉR blade-spacing system; ~ **de separación de petróleo** *m* GAS oil-removing system; ~ **de servodirección integral** *m* AUTO self-contained power steering system; ~ **de servodirección de unión** *m* AUTO linkage power steering system; ~ **de servomecanismo** *m* TV servo system; ~ **silenciador de control continuo** *m* TELECOM continuous controlled squelch system (*CTCSS*); ~ **simple** *m* CINEMAT single system; ~ **simultáneo** *m* TRANSP *tráfico* simultaneous system; ~ **sincronizado del cambio de velocidades** *m* AUTO synchromesh; ~ **síncrono en paralelo** *m* TELECOM parallel synchronous system; ~ **de sintonía silenciosa** *m* TELECOM tone squelch system; ~ **solar activo** *m* ENERG RENOV active solar system; ~ **solar pasivo** *m* CONST passive solar system; ~ **soldado a gas** *m* ING MECÁ, MECÁ gas-welded system; ~ **de soldadura oxiacetilénica** *m* ING MECÁ, MECÁ gas-welded system; ~ **Strowger** *m* TELECOM Strowger system; ~ **subsidiario de una red de sistemas interconectados** *m* PROD peer-to-peer slave; ~ **subterráneo** *m* PROD undergrounded system; ~ **de suministro de electricidad** *m* CONST, ELEC, ING ELÉC, ING MECÁ electricity supply system; ~ **de supervivencia** *m* OCEAN life support system; ~ **de supresión** *m* TELECOM tone squelch system;

~ **suspendido** *m* TRANSP suspended system; ~ **de suspensión** *m* AUTO suspension system; ~ **t** ~ **de tarjetas** *m* INFORM&PD card system; ~ **de teleconmutación** *m* TELECOM, TV remote-switching system; ~ **de telefonía por interconexión de procesos** *m* TELECOM switching process interworking telephony event (*SPITE*); ~ **telefónico a prueba de incendios** *m* CONST fireproof telephone system, flameproof telephone system; ~ **telefónico con teclado** *m* TELECOM key telephone system; ~ **de telemando por radio** *m* TELECOM radio control system; ~ **de televisión de barrido lento** *m* TV slow-scan television system; ~ **de televisión por cable** *m* TELECOM, TV cable television system; ~ **de televisión color NTSC** *m* TV NTSC color television system (*AmE*), NTSC colour television system (*BrE*); ~ **de televisión en color PAL** *m* TV PAL color system (*AmE*), PAL colour system (*BrE*); ~ **de televisión de pequeña velocidad de escansión** *m* TV slow-scan television system; ~ **termodinámico** *m* TERMO thermodynamic system; ~ **tetrafilar** *m* ING ELÉC four-wire system; ~ **tetragonal** *m* METAL tetragonal system; ~ **en tiempo real** *m* CINEMAT, ELECTRÓN, TELECOM, TV real-time system; ~ **de tirahilos** *m* TEXTIL take-up system; ~ **tolerante a fallos** *m AmL* INFORM&PD fault-tolerant system; ~ **de toma** *m* ING MECÁ intake system; ~ **de trabajo y soporte** *m* MINAS face and support system; ~ **de tramas ranuradas** *m* TELECOM frame-slotted system; ~ **de transferencia de datos** *m* TELECOM data transfer system; ~ **de transferencia electrónica de fondos** *m* (*EFTS*) INFORM&PD electronic funds transfer system (*EFTS*); ~ **de transferencia de mensajes** *m* TELECOM message transfer system; ~ **de tránsito rápido elevado** *m* TRANSP elevated rapid-transit system; ~ **de tránsito rápido estabilizado al máximo** *m* TRANSP top-stabilized rapid transit system; ~ **de tránsito rápido sin paradas** *m* TRANSP nonstop rapid transit system; ~ **de transmisión** *m* ING ELÉC transmission system, ING MECÁ shafting; ~ **de transmisión bidireccional** *m* INFORM&PD operation duplex; ~ **de transmisión birideccional** *m* (*FDX*) INFORM&PD full duplex (*FDX*); ~ **de transmisión de energía** *m* ING MECÁ power-transmission system; ~ **de transmisión de fibra óptica** *m* ING ELÉC, ÓPT, TELECOM fiber-optic transmission system (*AmE*), fibre-optic transmission system (*BrE*); ~ **de transmisión por fibras ópticas** *m* ÓPT optical fiber transmission system (*AmE*), optical fibre transmission sýstem (*BrE*); ~ **de transmisión hidráulica** *m* ING MECÁ hydraulic transmission system; ~ **de transmisión mecánica** *m* ENERG RENOV, ING MECÁ mechanical transmission system; ~ **de transmisión neumática** *m* ING MECÁ pneumatic transmission system; ~ **de transmisión óptica** *m* ÓPT optical transmission system; ~ **de transporte** *m* CONST conveyor system, transportation system; ~ **de transporte automático** *m* TRANSP automatic transportation system; ~ **de transporte continuo** *m* TRANSP continuous transportation system; ~ **para transporte de mercancías** *m* TRANSP lift on-off system; ~ **de transporte múltiple** *m* TRANSP multiple mode transportation system; ~ **de transporte de película al vacío** *m* EMB vacuum film transport system; ~ **de transporte público controlado** *m* TRANSP guided public mass transportation system; ~ **de transporte rápido** *m* FERRO rapid-transit system (*RTS*); ~ **de transporte sobre raíles** *m* TRANSP track-guided transport system; ~ **de tratamiento de la información** *m* INFORM&PD, TELE-COM information-processing system; ~ **triclínico** *m* METAL triclinic system; ~ **trifásico** *m* ING ELÉC three-phase system; ~ **trifásico de cuatro hilos** *m* ING MECÁ three-phase four-wire system; ~ **trifásico desequilibrado** *m* ING MECÁ unbalanced three-phase system; ~ **trifásico tetrafilar** *m* ING MECÁ three-phase four-wire system; ~ **trifilar** *m* ING ELÉC three-wire system; ~ **de tuberías** *m* CONST pipage, pipeworks, *refinería, planta industrial* piping, tubing, ING MECÁ pipework system; ~ **de tuberías de hierro** *m* CONST iron piping;

■ **u** ~ **de unidades** *m* ELEC, FÍS, METR system of units; ~ **con unidades de transporte sin fin** *m* TRANSP system with endless transportation units; ~ **con unidades de transporte de longitud intermedia** *m* TRANSP system with transportation units of intermediate length; ~ **unifilar** *m* ELEC single-wire system;

■ **v** ~ **de vaciadero caliente** *m* ING MECÁ hot runner system; ~ **de vacuometría** *m* TV vacuum guide system; ~ **de vagonetas suspendidas** *m* FERRO *vehículos* suspended-load trolley set (*BrE*); ~ **de vapor principal** *m* NUCL main-steam system; ~ **de varias cubiertas** *m* TRANSP multidecking system; ~ **de varillas articuladas** *m* ING MECÁ rod linkage; ~ **de vehículo suspendido** *m* TRANSP suspended vehicle system (*SVS*); ~ **de vehículos subterráneos** *m* TRANSP tube vehicle system (*TVS*); ~ **de velocidad visualizada** *m* TRANSP displayed speed system; ~ **de ventilación para aspersores de pintura** *m* SEG paint-spraying apparatus ventilation system; ~ **de ventilación de la máquina** *m* TRANSP MAR engine ventilation system; ~ **de ventilación del motor** *m* TRANSP MAR engine ventilation system; ~ **de verificación** *m* TEC ESP *tecnología* checkout system; ~ **vibratorio** *m* FÍS ONDAS *que provoca ondas* vibrating system; ~ **de videodisco** *m* INFORM&PD, ÓPT video-disk system; ~ **de vigilancia** *m* FÍS RAD monitoring system; ~ **de vigilancia por láser** *m* FÍS RAD laser monitoring system; ~ **de vigilancia con radar** *m* D&A, TRANSP, TRANSP AÉR, TRANSP MAR radar-surveillance system; ~ **visual de lectura de caracteres** *m* EMB, INFORM&PD character-reading vision system; ~ **de voltaje medio** *m* ING ELÉC medium-voltage system

Sistema: ~ **Europeo de Transporte de Barcazas** *m* (*sistema EBCS*) TRANSP European barge-carrier system (*EBCS*); ~ **Mundial de Socorro y Seguridad Marítimos** *m* (*SMSSM*) TRANSP MAR *salvamento en la mar* Global Marine Distress and Safety System (*GMDSS*); ~ **de Transporte Espacial** *m* TEC ESP Space Transportation System (*STS*)

sistemas: ~ **acoplados** *m pl* FÍS coupled systems; ~ **interconectados** *m pl* ELEC interconnected systems

sistematización: ~ **a bordo** *f* TEC ESP on-board processing

sitio: ~ **de carga de baterías** *m* TRANSP battery loading point; ~ **de distribución de cables** *m* ELEC, ING ELÉC cable distribution point; ~ **exclusivo** *m* TRANSP exclusive site; ~ **de reacción** *m* QUÍMICA reactive site

situación *f* CONST location, ING MECÁ location,

positioning; ~ **astronómica** *f* TRANSP MAR *astronavegación* astronomical position; ~ **de balanceo** *f* TEC ESP roll attitude; ~ **del buque** *f* TRANSP MAR *navegación* ship's position; ~ **por dos marcaciones sucesivas** *f* TRANSP MAR *navegación* running fix; ~ **estimada** *f* OCEAN dead reckoning position, TRANSP MAR *navegación* dead reckoning position, estimated position; ~ **sin existencias** *f* PROD out-of-stock situation; ~ **inicial** *f* PROD start statement; ~ **libre** *f* PROD open condition; ~ **por marcaciones sucesivas** *f* TRANSP MAR *navegación* running fix; ~ **de mínima energía** *f* FÍS PART minimum energy level; ~ **observada** *f* TRANSP MAR *navegación* observed position; ~ **en paralelo** *f* ING ELÉC bank; ~ **de prueba** *f* PROD test condition; ~ **radio-goniométrica** *f* TELECOM, TRANSP AÉR, TRANSP MAR *navegación* radio fix; ~ **relativa** *f* TEC ESP fix; ~ **relativa de un astro** *f* TEC ESP astro fix; ~ **relativa de la caja** *f* IMPR lay of the case; ~ **del tráfico** *f* TRANSP traffic situation

situado: ~ **a tierra** *adj* ING ELÉC connected to ground (*AmE*), connected to earth (*BrE*)

situador: ~ **en posición** *m* ING MECÁ positioner

situar *vt* CONST lay, PROD, TRANSP MAR *barco, marca, hombre al agua* locate

sizigia *f* GEOM syzygy

skiatrón *m* ELECTRÓN skiatron

skid *m* TEC PETR *perforación* skid

skuttedurita *f* MINERAL skutterudite

SLTT: ~ **Schottky** *m* ELECTRÓN Schottky TTL

Sm *abr* (*samario*) QUÍMICA Sm (*samarium*)

smithsonita *f* CARBÓN, MINAS, MINERAL smithsonite

SMPTE *abr* (*Sociedad de Técnicos Cinematográficos*) CINEMAT SMPTE (*Society of Motion Picture Technicians*)

SMSSM *abr* (*Sistema Mundial de Socorro y Seguridad Marítimos*) TRANSP MAR GMDSS (*Global Marine Distress and Safety System*)

Sn *abr* (*estaño*) METAL, MINAS, PROD, QUÍMICA Sn (*tin*)

SO *abr* (*sistema operativo*) INFORM&PD, TELECOM OS (*operating system*)

sobrante *m* TEC PETR *producción, refino* residue

sobrealimentación *f* AGRIC overfeed, overfeeding, AUTO supercharging, ING MECÁ, REFRIG, TEXTIL overfeed, overfeeding, TRANSP MAR *motor diesel* turbocharging; ~ **antes del acoplamiento sexual** *f* AGRIC *zootecnia* flushing

sobrealimentado *m* TRANSP AÉR supercharge

sobrealimentador *m* AUTO supercharger, ELEC *transformador*, ELECTRÓN voltage booster, MECÁ supercharger, VEH *componente del automóvil* booster; ~ **centrífugo** *m* AUTO, VEH centrifugal supercharger; ~ **de potencia** *m* AUTO power booster

sobrealimentar *vt* AGRIC, ING MECÁ, PROD, REFRIG, TEXTIL *en el acabado* overfeed

sobreamperaje *m* ELEC overcurrent, ING ELÉC over-current, overload

sobreatenuación *f* FÍS overdamping

sobrecabeza *f* IMPR head cap

sobrecalefacción *f* QUÍMICA superheating

sobrecalentado *adj* TERMO *lingotes* scorched

sobrecalentador *m* INSTAL TERM, PROD, TERMOTEC superheater; ~ **por convección** *m* INSTAL TERM convection superheater, convective superheater, TERMO convection superheater; ~ **radiante** *m* INSTAL

TERM radiant superheater; **~ de vapor** *m* TERMOTEC steam superheater

sobrecalentamiento *m* C&V reboil, ING MECÁ overheating, PROD overheating, superheating, QUÍMICA superheating, TERMO overheating

sobrecalentar *vt* INSTAL TERM superheat, TERMO overheat, *lingotes* scorch, TERMOTEC superheat

sobrecapacidad *f* TRANSP AÉR overcapacity

sobrecarga *f* CARBÓN surcharge load, CONST overload, overloading, ELEC *de acumulador* overcharge, overloading, *corriente* overload, EMB overload, FERRO *vehículos* supercharging, FÍS overload, overloading, GAS overload, HIDROL *aguas negras* surcharge, INFORM&PD overload, ING ELÉC overload, overloading, METAL overstressing, PROD overloading, overload, overstress, overstressing, TEC ESP overload, TEC PETR *geología* overburden, TELECOM overloading, overload, *transmisión de datos* overhead, TRANSP *tráfico* overloading, TRANSP AÉR surge; **~ del compresor** *f* TRANSP AÉR compressor surge; **~ de corriente** *f* ELEC excess current, overcurrent; **~ DCME** *f* TELECOM DCME overload; **~ destructiva** *f* TEC ESP *instrumentos de medición* burnout; **~ dinámica** *f* CONST moving load; **~ en funcionamiento** *f* PROD operating overload; **~ de ganado por unidad de superficie** *f* AGRIC overstocking; **~ móvil** *f* CONST moving load; **~ de la sección múltiplex** *f* TELECOM multiplex section overhead; **~ de transmisión** *f* TELECOM transmission overload; **~ de voltaje** *f* ELEC, FÍS overvoltage

sobrecargado *adj* CONST, ELEC, FÍS overloaded, HIDROL surcharged, ING ELÉC, PROD, TELECOM, TRANSP overloaded

sobrecargar *vt* CONST overload, weigh down, ELEC, FÍS, ING ELÉC overload, ING MECÁ stress, PROD, TELECOM, TRANSP, TV overload

sobrecargo *m* TRANSP MAR *marina mercante* purser

sobrecarrera *f* PROD overtravel

sobrecorriente *f* ELEC excess current, overcurrent, ING ELÉC current surge, PROD surge current

sobrecorte *m* ACÚST overcutting

sobrecrecimiento *m* AGRIC overgrowth

sobrecubierta *f* HIDROL downstream face

sobredesviación *f* TEC ESP overdeviation

sobredimensión *f* CARBÓN oversize, ING MECÁ oversizing, MECÁ, MINAS oversize

sobredimensionado *adj* ING MECÁ oversize

sobredosificación *f* QUÍMICA overdosage

sobreelevación *f* CONST superelevation, ING MECÁ *vía férrea* cant

sobreembalado: ~ de la bobina *m* EMB reel overwrapper

sobreembarque *m* PROD overshipment

sobreenvoltura *f* ELEC *cable* oversheath, EMB overwrap

sobreescribir *vt* IMPR overwrite

sobreescritura *f* INFORM&PD overwriting

sobreespesor[1]**: de ~** *adj* ING MECÁ, MECÁ oversized

sobreespesor[2] *m* ING MECÁ, MECÁ oversizing; **~ para el mecanizado** *m* PROD *fundición* machining allowance, *maquinado* tooling allowance

sobreexploración *f* TV overscan

sobreexponer *vt* CINEMAT, FÍS RAD, FOTO overexpose

sobreexposición *f* CINEMAT, FÍS RAD, FOTO overexposure

sobreexpuesto *adj* FOTO *película, imagen* overexposed

sobreextracción *f* HIDROL *aguas subterráneas* overextraction

sobrefatiga *f* PROD *estructuras* overstress

sobreflexión *f* PETROL overbend

sobreflujo *m* INSTAL HIDRÁUL back current

sobrefunda *f* ELEC *cable* oversheath

sobrefusión *f* METAL supercooling

sobregrabación *f* INFORM&PD overwriting

sobrehilado *m* TEXTIL oversewing

sobreimpresión *f* CINEMAT, FOTO overprint, IMPR imprint, imprinting, overprinting, surprint, ING MECÁ, PROD overprinting; **~ tapando** *f* IMPR patch

sobreimprimir *vt* CINEMAT, FOTO overprint, IMPR double-print, imprint, overprint, patch; **~ tapando** *vt* IMPR block out, double-print, patch

sobreimpulso *m* ELEC *tensión* pulse spike, ING ELÉC overshoot

sobreintensidad *f* ELEC overcurrent, PROD surge current

sobremaduro *adj* AGRIC over ripe

sobremarcha *f* AUTO, MECÁ overdrive

sobremedida: de ~ *adj* ING MECÁ oversize

sobremodulación *f* TRANSP AÉR overshoot

sobremultiplicación *f* MECÁ overdrive

sobrenadante *adj* QUÍMICA supernatant

sobreorillar *vt* TEXTIL overlock

sobreoxidar *vt* QUÍMICA overoxidize

sobrepasado *adj* MECÁ overriding

sobrepasar *vt* FERRO *estación, señal de stop* overrun, MECÁ override

sobrepastoreo *m* AGRIC overgrazing

sobrepesca *f* OCEAN overfishing

sobrepeso *m* METR overweight

sobreponer *vt* CINEMAT, FOTO, TV superimpose

sobreposición *f* GEOL overprinting

sobrepresión *f* PROD overpressure, surge pressure, TEC PETR *geología*, TRANSP AÉR overpressure; **~ intersticial** *f* CARBÓN pore overpressure; **~ máxima de aire ambiente** *f* TRANSP AÉR free-air peak overpressure

sobreproducción *f* PROD overproduction

sobrequilla *f* TRANSP MAR *construcción naval* keelson, rider plate

sobrerrevelado *m* CINEMAT, FOTO overdevelopment

sobrerrefinado *m* PROD *metales* overrefining

sobresaturación *f* QUÍMICA supersaturation, TV oversaturation

sobresaturado *adj* CONTAM, FÍS super-saturated, GEOL oversaturated, QUÍMICA super-saturated

sobresaturar *vt* FÍS, QUÍMICA supersaturate

sobresecar *vt* PAPEL overdry

sobresoplado *m* PROD *producción de acero* afterblow

sobretamaño *m* ING MECÁ oversizing, MECÁ oversize

sobretemperatura *f* TERMO over temperature

sobreteñir *vt* TEXTIL overdye

sobretensión *f* ELEC excess voltage, overvoltage, ING ELÉC overvoltage, voltage surge, MECÁ impulse, METAL overshoot, PROD *electricidad* overpressure, surge, QUÍMICA overvoltage, TELECOM voltage surge; **~ de concentración** *f* TEC ESP concentration overvoltage; **~ por descarga atmosférica** *f* ELEC, GEOFÍS lightning surge; **~ dinámica** *f* ING ELÉC dynamic overvoltage; **~ de energía** *f* ING ELÉC power surge, TEC ESP activity overvoltage; **~ inducida por el rayo** *f* ELEC *conductor, equipo*, GEOFÍS *magnetismo terrestre* lightning surge; **~ inicial de encendido** *f* TELECOM voltage surge; ~ **pasajera** *f* ELEC transient voltage; ~ **transitoria** *f* ELEC *tensión*, ING ELÉC surge

sobretensionado *adj* ING ELÉC transient

sobrevaina *f* ELEC *cable* oversheath

sobrevelocidad *f* ING MECÁ, PROD overspeed; ~ **máxima del motor** *f* TRANSP AÉR maximum engine overspeed; ~ **del rotor** *f* TRANSP AÉR rotor overspeed

sobrevidriado *m* C&V overglazing

sobreviraje *m* AUTO, VEH oversteer

sobrevirar *vi* AUTO, VEH oversteer

sobrevoltaje *m* ELEC excess voltage, overvoltage, ING ELÉC impulse voltage, overvoltage, voltage surge, MECÁ impulse voltage, PROD overpressure, surge, QUÍMICA overvoltage, TELECOM voltage surge; ~ **de energía** *m* TEC ESP activity overvoltage

sobrevuelta: ~ **perforada** *f* EMB perforated overlap

sobrevulcanización *f* P&C *plásticos, caucho, defecto* overcure

socabado *m* IMPR *de líneas y puntos de la trama* undercut

socaire *m* TRANSP MAR lee, *navegación a vela* shelter

socava *f* MINAS kirving

socavación *f* AGRIC undercutting, AGUA scour, CARBÓN *minas* cut, ENERG RENOV *presas*, HIDROL *fuerza erosiva del agua* scour, MINAS fan cut, crater cut, OCEAN *erosión*, TEC PETR scouring; ~ **interna** *f* CARBÓN internal scour; ~ **paralela** *f* MINAS parallel cut

socavadora *f* CARBÓN, MINAS coal cutter

socavar *vt* AGUA *orillas de ríos* scour, CONST undermine, ENERG RENOV, HIDROL scour, MINAS cut, undermine

socave *m* MINAS kerving, kirving

socavón *m* CARBÓN adit, gangway, CONST undermining, MINAS cross drift, drift, undermining, day drift, day hole, day level, drainage level, level, cavern, hole, sapping, tunnel, OCEAN *del litoral por acción del oleaje* nip, slumping; ~ **por la acción del agua** *m* FERRO *del terraplén* washout; ~ **de cateo** *m* MINAS prospect tunnel

socaz *m* AGUA *turbina hidráulica*, HIDROL tailrace

sociedad: ~ **certificada** *f* CALIDAD certified company; ~ **de clasificación** *f* TRANSP MAR *seguros* classification society; ~ **flotadora** *f* TRANSP MAR charter company

Sociedad: ~ **Norteamericana de Ingenieros Mecánicos** *f* (*ASME*) American Society of Mechanical Engineers (*ASME*); ~ **de Técnicos Cinematográficos** *f* (*SMPTE*) CINEMAT Society of Motion Picture Technicians (*SMPTE*)

socio *m* TRANSP MAR co-owner, part owner

socorrista *m* SEG first aider, first-aid worker

sodalita *f* MINERAL sodalite

sodamida *f* QUÍMICA sodamide

sódico *adj* GEOL, QUÍMICA sodic

sodio *m* (*Na*) METAL sodium (*Na*), QUÍMICA natrium, sodium (*Na*); ~ **reducido** *m* ALIMENT reduced sodium

sofito *m* CONST *arquitectura* soffit

sofocar *vt* TERMO smother

software *m* ELEC, FÍS, INFORM&PD, TEC PETR software; ~ **de administración de transacciones** *m* INFORM&PD transaction management software; ~ **de aplicación** *m* INFORM&PD applications software; ~ **de comunicaciones** *m* INFORM&PD communications software; ~ **didáctico** *m* *Esp*

INFORM&PD courseware; ~ **de gestión de transacciones** *m* INFORM&PD transaction management software; ~ **orientado al problema** *m* INFORM&PD problem-oriented software; ~ **propio** *m* INFORM&PD in-house software; ~ **de ratón** *m* INFORM&PD mouse software; ~ **de red** *m* INFORM&PD network software; ~ **del sistema** *m* INFORM&PD systems software

soga *f* CONST *aparejo* stretcher, MECÁ, SEG rope; ~ **de retorno** *f* ING MECÁ return rope; ~ **de sisal** *f* ING MECÁ sisal rope; ~ **de tracción** *f* ING MECÁ traction rope

SOI *abr* (*sistema de oficina integrado*) INFORM&PD IOS (*integrated office system*)

sola: **de una ~ capa** *adj* REVEST single-layer; **de una ~ fase** *adj* ELEC *fuente de alimentación* monophase; **todo en una ~ pieza** *adj* ING MECÁ all in one piece

solado *m* P&C *caucho* soling

solador *m* CONST pavior (*AmE*), paviour (*BrE*), *operario* tiler

solanidina *f* QUÍMICA *alcaloide* solanidine

solapa *f* IMPR *encuadernación* overlap; ~ **externa** *f* EMB outer flap

solapado *m* REVEST *también solapamiento* overlapping; ~ **de líneas** *m* TV pairing

solapamiento *m* GEOL, INFORM&PD overlap, MECÁ overriding, METAL overlapping, TV overlap; ~ **expansivo** *m* GEOL onlap; ~ **de frecuencias** *m* TV frequency overlap; ~ **regresivo** *m* GEOL offlap

solapas: ~ **solapadas** *f pl* EMB overlapping flaps

solape *m* CONST lap, overlap, overlapping, PROD lapping, TRANSP MAR *construcción* overlap, *de planchas* harriscut

solar¹ *adj* CONST solar

solar² *m* CONST floor

solarímetro *m* ENERG RENOV solarimeter

solarización *f* CINEMAT Sabattier effect, solarization, FOTO solarization

soldabilidad *f* ING MECÁ, MECÁ, METAL weldability, PROD *con estaño* solderability

soldable *adj* PROD *con estaño* solderable

soldado *adj* ING MECÁ, MECÁ, METAL welded

soldador *m* ELEC *herramienta* soldering iron, ING MECÁ, MECÁ, METAL *persona* welder, solderer, PROD *herramienta* copper bit, soldering iron, SEG *persona* welder; ~ **del arco eléctrico** *m* MECÁ arc welder; ~ **de bronce** *m* CONST brass solder; ~ **de cobre** *m* PROD *herramienta* soldering bit, soldering copper; ~ **de fontanero** *m* *Esp* (*cf soldador de plomero AmL*) CONST plumber's solder; ~ **de plomero** *m* *AmL* (*cf soldador de fontanero Esp*) CONST plumber's solder

soldadora *f* MECÁ welder, PROD welding machine; ~ **por arco** *f* MECÁ arc welder; ~ **de plásticos** *f* ING MECÁ plastic-welding machine

soldadura¹: **sin ~** *adj* PROD weldless

soldadura² *f* AGRIC graft union, C&V welding, CONST soldering, ING ELÉC solder, ING MECÁ welding, wipe, MECÁ uphand welding, METAL, NUCL weld, P&C welding, PROD weld, welding, SEG welding, TERMO weld, welding; ~ **con aleación de cinc y cobre** *f* TERMO brazing; ~ **de almas** *f* ELEC *conexiones* cored solder; ~ **por alta frecuencia** *f* EMB high-frequency welding; ~ **aluminotérmica** *f* CONST, FERRO thermit welding, MECÁ aluminothermic welding; ~ **en ángulo** *f* MECÁ fillet weld; ~ **anular** *f* NUCL ring weld; ~ **por aproximación** *f* CONST, ING MECÁ, MECÁ butt welding, PROD butt welding, jump weld, jump welding; ~ **por arco** *f* CONST arc welding, fusion welding, ELEC, PROD arc welding, TERMO fusion welding; ~ **al arco automática** *f* ING MECÁ automatic arc welding; ~ **por arco en atmósfera de argón** *f* PROD argon arc welding; ~ **con arco eléctrico** *f* PROD arc welding; ~ **por arco eléctrico con fundente** *f* CONST flux-cored arc welding; ~ **con arco con electrodo de carbón** *f* PROD carbon arc welding; ~ **por arco con electrodo consumible** *f* ING MECÁ manual metal arc welding; ~ **por arco manual** *f* CONST, MECÁ manual arc welding; ~ **con arco metálico sumergido en gas inerte** *f* CONST gas-shielded metal arc welding; ~ **con arco a presión** *f* CONST flash welding; ~ **por arco sumergido** *f* CONST submerged arc welding; ~ **con arco sumergido y fundente en gas activo** *f* CONST flux-cored submerged arc welding with active-gas shielding; ~ **en atmósfera de gas inerte** *f* PROD inert gas welding; ~ **bajo agua** *f* TEC PETR underwater welding; ~ **con baño de sal** *f* CONST salt bath brazing; ~ **en bisel** *f* PROD scarf weld, scarf welding; ~ **blanda** *f* ELEC *conexión* soft solder, soft soldering; ~ **blanda con estaño y plomo** *f* ELEC *conexión* soft soldering; ~ **de boca abierta** *f* PROD split weld, split welding; ~ **de bronce** *f* CONST brass solder, bronze welding; ~ **en caliente** *f* EMB heat welding; ~ **por centelleo** *f* PROD flash welding; ~ **circunferencial** *f* MECÁ, NUCL girth weld; ~ **de cobre** *f* CONST braze welding; ~ **cóncava** *f* NUCL concave weld; ~ **por corriente de inducción** *f* PROD induction welding; ~ **de costura por resistencia** *f* CONST resistance seam welding; ~ **de costuras transversales** *f* PROD butt-seam welding; ~ **de descarga** *f* PROD discharge welding; ~ **por deslizamiento eléctrico** *f* NUCL electroslag welding; ~ **eléctrica** *f* ELEC, ING ELÉC electric welding; ~ **eléctrica con escorias** *f* CONST, MECÁ electroslag welding; ~ **eléctrica por resistencia por puntos** *f* ELEC, ING MECÁ resistance spot welding; ~ **con escoria eléctricamente conductora** *f* CONST, MECÁ electroslag welding; ~ **espaldar** *f* PROD 'back weld; ~ **de filete** *f* MECÁ fillet weld; ~ **de forja** *f* CONST, PROD forge welding; ~ **por forja** *f* ING MECÁ plastic welding; ~ **fría** *f* ELEC *circuitos*, ING ELÉC cold junction; ~ **de fricción** *f* CONST friction welding; ~ **fuerte** *f* INSTAL TERM brazing, PROD brazing, hard soldering, SEG *con bronce o latón*, TEC ESP, TERMO brazing; ~ **fuerte por arco** *f* PROD arc brazing; ~ **por fuerza magnética** *f* NUCL magnetic-force welding; ~ **por fundición líquida** *f* PROD *fundería* casting-on; ~ **por fusión** *f* CONST, PROD, TERMO fusion welding; ~ **por fusión líquida** *f* PROD burning; ~ **con gas activo de metal** *f* (*soldadura MAG*) CONST, TERMO metal active-gas welding (*MAG welding*); ~ **por gas inerte** *f* ING MECÁ inert gas welding; ~ **por gas noble** *f* ING MECÁ inert gas welding; ~ **de haz de electrones** *f* CONST, ELECTRÓN, NUCL electron-beam welding; ~ **por hidrógeno atómico** *f* NUCL atomic hydrogen welding; ~ **por inclusión eléctrica** *f* CONST electroslag welding; ~ **por inducción** *f* ELEC *proceso* induction welding; ~ **por inmersión** *f* CONST dip brazing; ~ **interna** *f* MECÁ inside welding; ~ **por láser** *f* CONST, ELECTRÓN laser welding; ~ **con latón** *f* TERMO brazing; ~ **de latón** *f* CONST braze welding, PROD brazing, hard soldering, TEC ESP hard soldering; ~ **de latón dulce** *f* PROD soft-brazing solder; ~ **de lengüeta** *f* PROD tongue weld, tongue welding;

~ **ligera** *f* NUCL light weld; ~ **con llama** *f* P&C *plásticos, operación* flame-welding; ~ **por llama de gas** *f* GAS, ING MECÁ, MECÁ, TERMO gas welding; ~ **MAG** *f* (*soldadura con gas activo de metal*) CONST, TERMO MAG welding (*metal active-gas welding*); ~ **manual con electrodo consumible** *f* ING MECÁ manual metal arc welding; ~ **con metal no ferroso** *f* INSTAL TERM brazing; ~ **de montaje** *f* NUCL field weld; ~ **de muesca** *f* PROD cleft weld, cleft welding; ~ **en obra** *f* NUCL site weld; ~ **ortogonal** *f* MECÁ fillet weld; ~ **oxiacetilénica** *f* CONST oxyacetylene welding, GAS, ING MECÁ gas welding, MECÁ gas welding, oxyacetylene welding, PROD oxyacetylene welding, TERMO gas welding, oxyacetylene welding; ~ **con oxiacetileno** *f* CONST, MECÁ, PROD, TERMO oxyacetylene welding; ~ **con pasador** *f* CONST stud welding; ~ **de presión** *f* CONST pressure welding; ~ **a presión en frío** *f* CONST, MECÁ cold pressure welding; ~ **pulida** *f* METAL bright field; ~ **a pulso** *f* MECÁ uphand welding; ~ **de punto por resistencia** *f* ELEC, ING MECÁ resistance spot welding; ~ **por puntos** *f* CONST, ELEC spot welding, MECÁ tack welding; ~ **de recubrimiento** *f* EMB, NUCL, PROD lap weld; ~ **por reflujo** *f* ING ELÉC reflow soldering; ~ **por resistencia** *f* CONST, ELEC, ING ELÉC, ING MECÁ, TERMO resistance welding; ~ **por resistencia eléctrica** *f* CONST, ELEC, ING ELÉC, ING MECÁ, TERMO resistance welding; ~ **de revés** *f* MECÁ, PROD backhand welding; ~ **con saliente** *f* CONST, PROD projection welding; ~ **de sellado** *f* NUCL sealing weld; ~ **solapada** *f* EMB, NUCL, PROD lap weld; ~ **a solape** *f* EMB, NUCL, PROD lap weld; ~ **con solvente** *f* P&C *operación* solvent welding; ~ **a soplete** *f* CONST torch brazing; ~ **con soplete** *f* PROD burning; ~ **por superposición** *f* PROD lap welding; ~ **de termita** *f* CONST thermit welding; ~ **a tope** *f* CONST, ING MECÁ, MECÁ, PROD butt weld, butt welding; ~ **a tope en ángulo recto** *f* PROD jump weld, jump welding; ~ **a tope por chispa** *f* CONST flash welding; ~ **a tope por resistencia** *f* CONST resistance butt welding; ~ **de tungsteno a gas inerte** *f* CONST tungsten inert gas welding (*TIG welding*); ~ **por ultraacústica** *f* TERMO ultrasonic welding; ~ **ultrasónica** *f* P&C ultrasonic welding; ~ **por ultrasónico** *f* EMB ultrasonic sealing; ~ **por ultrasonido** *f* P&C ultrasonic welding; ~ **al vacío** *f* CONST vacuum brazing; ~ **de viga por láser** *f* CONST laser beam welding

soldante: ~ **de almas** *m* ELEC *conexiones* cored solder; ~ **blando** *m* ELEC *conexión* soft solder

soldar *vt* CONST solder, wipe, ELEC run, solder, ELECTRÓN, ING ELÉC solder, MECÁ weld, PROD solder, TERMO weld; ~ **con latón** *vt* CONST, ING MECÁ, PROD, TERMO braze; ~ **con suelda fuerte** *vt* CONST, ING MECÁ, PROD, TERMO braze

soldeo *m* ING MECÁ, TERMO welding; ~ **con aleaciones de estaño y plomo** *m* PROD soldering; ~ **al arco en atmósfera de gas inerte con electrodo consumible** *m* CONST metal inert-gas welding (*MIG welding*); ~ **con arco eléctrico** *m* TERMO arc welding; ~ **por arco eléctrico** *m* ELEC, PROD, TERMO electric-arc welding; ~ **por arco con electrodo metálico** *m* TERMO metal arc welding; ~ **con electrodo consumible** *m* TERMO metal arc welding; ~ **con gas activo de metal** *m* CONST, TERMO metal active-gas welding (*MAG welding*); ~ **por llama de gas** *m* GAS, ING MECÁ, MECÁ, TERMO gas welding; ~ **MAG** *m*

CONST, TERMO MAG welding (*metal active-gas welding*); ~ **oxiacetilénico** *m* CONST oxyacetylene welding, GAS, ING MECÁ gas welding, MECÁ gas welding, oxyacetylene welding, PROD oxyacetylene welding, TERMO gas welding, oxyacetylene welding; ~ **de plásticos** *m* ING MECÁ, P&C plastic welding; ~ **por presión** *m* PROD, TERMO pressure welding; ~ **por recalcadura** *m* TERMO upset welding; ~ **submarino** *m* TERMO underwater welding; ~ **ultrasónico** *m* EMB ultrasonic sealing, TERMO ultrasonic welding

soleamiento *m AmL* (*cf insolación Esp*) ENERG RENOV insolation

solenoide *m* ELEC *bobina*, FÍS solenoid, ING ELÉC plunger, solenoid, PROD, REFRIG, TV, VEH *motor de arranque* solenoid; ~ **de cierre** *m* PROD lockout solenoid; ~ **de drenaje** *m* ING MECÁ drain solenoid; ~ **eléctrico** *m* PROD electrical solenoid; ~ **toroidal** *m* ELEC *bobina* toroid

solera *f* CONST curb, floor, mudsill, sole plate, sole, sole piece, *estructuras* sill, ING MECÁ bedplate, MINAS *entibación minas* sill, sole, groundsill, PROD *bocarte, trituración de mineral* sill, *de horno* sole; ~ **de frente** *f* PROD nose sill; ~ **de la puerta** *f* CONST door sill

solicitar *vt* INFORM&PD request, MECÁ apply for, PROD apply for, request, TELECOM request, TEXTIL *patentes, licencias* apply for

solicitud *f* INFORM&PD, TELECOM request; ~ **de administración de la red** *f* TELECOM network management application service element (*NM-ASE*); ~ **de capacidades de transacción** *f* TELECOM transaction capabilities application request; ~ **de compra** *f* PROD procurement request, procurement requisition; ~ **de conexión** *f* TELECOM connection request (*CR*); ~ **de cotización** *f* PROD request for quotation; ~ **para enviar** *f* INFORM&PD, TELECOM request to send; ~ **de mantenimiento y explotación** *f* TELECOM operating and maintenance application request; ~ **de servicio** *f* TELECOM request for service; ~ **de tarea** *f* INFORM&PD job request

solidario *adj* ING MECÁ integral

solidez *f* ENERG RENOV solidity, GAS *física* density, strength, MECÁ hardness, METAL strength, hardness, TEXTIL *de los colores de los tintes* fastness, TRANSPAÉR solidity; ~ **al lavado** *f* TEXTIL *tintura* fastness to washing; ~ **a la luz** *f* TEXTIL *de los colorantes* fastness to light; ~ **al roce** *f* TEXTIL *de la tela* fastness to rubbing; ~ **a la transpiración** *f* TEXTIL *del tejido y colores* fastness to perspiration; ~ **de unión** *f* P&C *adhesivos, caucho* bond strength

solidificación *f* CRISTAL setting, FÍS solidification, METAL freezing, solidifying, P&C set, PROC QUÍ setting; ~ **por compresión** *f* P&C *propiedad física* compression set; ~ **en frío** *f* P&C *adhesivos* cold setting; ~ **residual** *f* P&C residual set

solidificado: ~ **por congelación** *adj* ALIMENT, REFRIG, TERMO frozen solid

solidificador *m* CONTAM MAR solidifier

sólido[1] *adj* CONST solid, GEOL *textura* massive, QUÍMICA solid, TEXTIL *color, tinte* fast

sólido[2] *m* FÍS, GEOM, PROD solid; ~ **cilíndrico de una revolución** *m* GEOM cylindrical solid of revolution; ~ **policristalino** *m* CRISTAL polycrystalline solid; ~ **de revolución** *m* GEOM solid of revolution; ~ **rígido** *m* FÍS *mecánica analítica* rigid body

sólidos: ~ **de cacao** *m pl* ALIMENT cocoa solids;

~ **sedimentables** *m pl* HIDROL *aguas residuales* settleable solids, QUÍMICA settlings

sollado *m* TRANSP MAR *armada* mess, *entrepuente* orlop deck

soltador *m* ING MECÁ tripper

soltar *vt* AUTO, GAS release, ING MECÁ release, relieve, unclamp, INSTAL HIDRÁUL relieve, MECÁ release, relieve, PROD relieve, QUÍMICA release, TRANSP MAR *aro salvavidas* cast, *bandera* break, VEH *el freno* release

soltarse *v refl* ING MECÁ work loose

soltura *f* ING MECÁ slack

solubilidad *f* CARBÓN, P&C solubility, PROC QUÍ dissolubility, QUÍMICA *de sustancia* solubility, dissolubility; ~ **restringida** *f* METAL restricted solubility

solubilización *f* METAL precipitation

soluble *adj* QUÍMICA, TEC PETR soluble; ~ **en aceite** *adj* P&C *pintura*, QUÍMICA oil-soluble; ~ **en agua** *adj* ALIMENT, QUÍMICA, TEXTIL water-soluble

solución *f* MATEMÁT, PROC QUÍ, QUÍMICA solution; ~ **ácida** *f* ALIMENT acid solution, TRANSP MAR *mantenimiento de buques* pickling; ~ **aleatoria** *f* METAL random solution; ~ **amortiguadora** *f* PROD buffer solution; ~ **antirrayas** *f* CINEMAT antiscratch solution; ~ **de cloruro cálcico** *f* PROD bleach liquor; ~ **concentrada** *f* QUÍMICA concentrated solution; ~ **desengrasante** *f* DETERG scouring solution; ~ **estándar** *f* QUÍMICA standard solution; ~ **de Fehling** *f* QUÍMICA Fehling's solution; ~ **de glicol** *f* DETERG, QUÍMICA, REFRIG glycol solution; ~ **de grabado químico** *f* METAL etching solution; ~ **humectadora** *f* IMPR *impresión offset* dampening solution; ~ **normal** *f* QUÍMICA normal solution; ~ **patrón** *f* QUÍMICA standard solution; ~ **reforzadora** *f* CINEMAT, FOTO, TV replenishing solution; ~ **regeneradora** *f* CINEMAT, FOTO, TV replenishing solution; ~ **reguladora** *f* ALIMENT buffer solution, P&C *control de pH* buffer; ~ **retardadora** *f* PROD buffer solution; ~ **salina** *f* NUCL saline solution; ~ **saturada** *f* QUÍMICA saturated solution; ~ **sobresaturada** *f* QUÍMICA supersaturated solution; ~ **sólida** *f* CRISTAL, GEOL, QUÍMICA solid solution; ~ **sólida intersticial** *f* CRISTAL interstitial solid solution; ~ **sólida nucleada** *f* ING MECÁ, METAL coring; ~ **sólida ordenada** *f* METAL ordered solid solution; ~ **sólida sustitución** *f* CRISTAL substitutional solid solution; ~ **tope** *f* PROD buffer solution; ~ **de viraje** *f* FOTO toning solution

soluto[1] *adj* QUÍMICA solute

soluto[2] *m* ALIMENT solute

solvatación *f* QUÍMICA solvation

solvatado *adj* QUÍMICA solvated

solvato *m* QUÍMICA solvate

solvente[1] *adj* CONTAM, QUÍMICA *características químicas* solvent

solvente[2] *m* CARBÓN, CONTAM, DETERG, P&C, QUÍMICA solvent; ~ **no polar** *m* P&C nonpolar solvent; ~ **selectivo** *m* QUÍMICA selective solvent

SOM *abr* (*semiconductor metal-óxido*) ELECTRÓN, INFORM&PD MOS (*metallic oxide semiconductor, metal-oxide semiconductor*)

sombear *vt* PROD cross-hatch

sombra *f* C&V shade, CINEMAT shadow, FOTO, IMPR shade, shadow, P&C shading, TV shadowing; ~ **de**

presión f GEOL *petrología* pressure shadow; ~ **propia** f TEC ESP *vehículos espaciales* eigenshadow
sombreado[1] *adj* C&V, FOTO, IMPR shaded
sombreado[2] *m* TV shading; ~ **bajo letras** *m* IMPR drop shadow; ~ **mecánico** *m* IMPR benday; ~ **negro** *m* TV black shading; ~ **parabólico** *m* TV parabolic shading; ~ **de polo** *m* ELEC *generador* pole-shading; ~ **a rayas** *m* ING MECÁ hatching, cross hatching
sombrerete *m* CONST *chimenea* cowl, hood, ING MECÁ, INSTR *cojinetes* cap, cover, MECÁ, PETROL cap, PROD *de la cúpula* dome cap, TERMOTEC *chimenea* cowl; ~ **de biela** *m* AUTO, ING MECÁ, VEH connecting-rod cap; ~ **del eje** *m* AUTO axle cap; ~ **del pilote** *m* CARBÓN pile cap; ~ **del portacojinetes** *m* ING MECÁ pedestal cover; ~ **de protección** *m* REVEST protective bonnet; ~ **del soporte del cigüeñal** *m* AUTO, VEH crankshaft-bearing cap
sombrero: ~ **de copa** *m* CINEMAT *trípode enano de cámara* high hat; ~ **de protección** *m* REVEST protective bonnet
sombrilla f CINEMAT umbrella
SOMD *abr* (*semiconductor de óxido de metal difuso*) ELECTRÓN DMOS (*diffuse metal oxide semiconductor*)
somero *adj* TRANSP MAR *aguas* fleet
someter: ~ **a esfuerzo** *vt* ING MECÁ stress
sometido: ~ **a una fuerza** *fra* FÍS acted upon by a force
sonar *m* ACÚST, D&A, FÍS ONDAS, OCEAN, TELECOM *Esp* (*cf dar timbre AmL*), TRANSP MAR sonar; ~ **explorador** *m* OCEAN scanning sonar
sonda f CARBÓN auger, probe, CONST *perforación* boreholing plant, ELEC *medición* probe, HIDROL lead, MINAS *sondeo* auger, *medir* probe, swivel, OCEAN dragrope, probe, TEC ESP sensor, probe, TRANSP AÉR probe, TRANSP MAR *navegación* sounder, sounding, sound; ~ **para acoplamiento** f TEC ESP *de una nave* docking probe; ~ **activa** f TEC ESP active sensor; ~ **acústica** f D&A sonic depth finder, FÍS ONDAS echo sounding, PROD *marina* sounding, TRANSP MAR *navegación* echo sounding, sounder; '~ **de avión nodriza para aprovisionar en vuelo a otros aviones** f TRANSP AÉR refueling boom (*AmE*), refuelling boom (*BrE*); ~ **de barrera de diamantes** f PROD diamond drill; ~ **de cuchara** f CONST spoon auger; ~ **por eco** f ACÚST, FÍS ONDAS echo sounding, OCEAN echo sounding, TRANSP MAR echo sounder; ~ **ecoica** f ACÚST, FÍS ONDAS, OCEAN depth sounder; ~ **eléctrica** f CONST, ELEC, MECÁ electric drill; ~ **electrónica** f NUCL electron probe; ~ **espacial** f TEC ESP space probe; ~ **para el espacio profundo** f TEC ESP deep space mission, deep space probe; ~ **de formación de hielo** f TRANSP AÉR icing probe; ~ **del fotómetro** f FOTO light meter probe; ~ **de Hall** f FÍS, FÍS RAD Hall probe; ~ **de hielo** f TRANSP AÉR ice probe; ~ **de incidencia** f TRANSP AÉR incidence probe; ~ **interplanetaria** f TEC ESP interplanetary probe; ~ **interruptora de contacto** f ING MECÁ touch trigger probe; ~ **interruptora de toque** f ING MECÁ touch trigger probe; ~ **lambda** f AUTO lambda probe; ~ **de línea ranurada** f ING ELÉC slotted line probe; ~ **lunar** f TEC ESP lunar probe; ~ **magnética** f ELEC search coil, NUCL magnetic probe; ~ **medidora electrónica** f ING MECÁ electronic gaging probe (*AmE*), electronic gauging probe (*BrE*); ~ **para muestras** f *AmL* (*cf sonda para testigos Esp*) MINAS core borer; ~ **de percusión** f ING MECÁ percussive

drill, MINAS percussion drill, drill jar, jumper, *pozos petrolíferos* churn drill; ~ **pirométrica** f ELEC pyrometer probe; ~ **planetaria** f TEC ESP planetary probe; ~ **de profundidad** f OCEAN depth meter; ~ **de prueba** f ING ELÉC test prod; ~ **de reaprovisionamiento en vuelo** f TRANSP AÉR flight refueling probe (*AmE*), flight refuelling probe (*BrE*); ~ **de red** f OCEAN netsonde; ~ **de rejilla** f NUCL grid probe; ~ **de sedimentos** f OCEAN *geología submarina* sediment sounder; ~ **de temperatura** f TERMO temperature probe; ~ **de temperatura del aceite** f TRANSP AÉR oil temperature probe; ~ **de temperatura del aire exterior** f TRANSP AÉR outside air temperature probe; ~ **de temperatura del combustible** f TRANSP AÉR fuel temperature probe; ~ **para testigos** f *Esp* (*cf sonda para muestras AmL*) MINAS core borer, core drilling; ~ **de toque disparadora** f ING MECÁ touch trigger probe; ~ **con tubo contador** f NUCL counter-tube probe; ~ **ultrasónica** f CONST ultrasonic probe
sondador: ~ **acústico** *m* FÍS ONDAS echo sounder, OCEAN echo sounding, TRANSP MAR echo sounder; ~ **radar** *m* D&A, FÍS RAD, TRANSP, TRANSP AÉR, TRANSP MAR radar scanner; ~ **del subsuelo marino** *m* OCEAN *sedimentología marina* sub-bottom profiler
sondadora[1]: ~ **a base de diamantes** *m* TEC PETR *perforación* diamond core drill
sondadora[2]: ~ **de granalla de acero** f MINAS shot drill; ~ **de rotación** f MINAS shot drill
sondaje *m* PROD *marina* sounding; ~ **ultrasónico** *m* FÍS ONDAS ultrasonic sounding
sondaleza f TRANSP MAR lead line, sounding line
sondar[1] *vt* CARBÓN, MINAS, TECH PETR bore
sondar[2] *vi* CONST bore, sound
sondeado: ~ **de profundidad** *m* GEOFÍS, MINAS, OCEAN, TECH PETR, ING MECÁ deep-hole boring
sondeador: ~ **de haces múltiples** *m* OCEAN multi-beam sounder; ~ **lidar** *m* TEC ESP sounding lidar; ~ **de sedimentos** *m* OCEAN *geología submarina* sediment sounder
sondeadora f TEC PETR rig; ~ **de corona** f MINAS annular borer
sondear[1] *vt* INFORM&PD poll, ING MECÁ, MINAS bore, OCEAN *con la sondaleza* sound
sondear[2] *vi* TRANSP MAR sound
sondeo *m* CARBÓN boring, drilling, CONST *perforación* boreholing, boring tool, *carpintería* boring, GEOFÍS well log, HIDROL bore, INFORM&PD polling, ING MECÁ boring, MINAS *investigación minera* cutting, grab, boring, prospecting, OCEAN *acción o acto de sondar* sounding, PROD *marina* sounding, boring, TEC ESP tracking, TEC PETR *exploración* prospecting; ~ **acústico** *m* GEOFÍS acoustic log, OCEAN echo sounding; ~ **por aire** *m* TEC PETR *perforación* air drilling; ~ **en ángulo fijo** *m* OCEAN fixed angle sounding; ~ **de apreciación** *m* TEC PETR *perforación* appraisal drilling; ~ **de caída libre** *m* MINAS free fall; ~ **del concentrador** *m* INFORM&PD *redes* hub polling; ~ **por contracta** *m* *Esp* (*cf sondeo a tanto alzado AmL*) MINAS contract boring, contract drilling; ~ **con corona de diamantes** *m* MINAS, TEC PETR *perforación* diamond drilling; ~ **por corriente de agua** *m* TEC PETR wash boring; ~ **a la cuerda por percusión** *m* CARBÓN percussive-rope boring, percussive-rope drilling; ~ **dinámico** *m* CARBÓN dynamic sounding; ~ **estratigráfico** *m* MINAS *geología* core drilling, TEC PETR *exploración, perforación*

core drill; ~ **de evaluación** m TEC PETR *perforación* appraisal well; ~ **explorador** m TEC PETR wildcat; ~ **a gran profundidad** m MINAS deep borehole, deep boring; ~ **por granalla de acero** m MINAS shot drilling; ~ **hidráulico por rotación** m AmL (*cf sondeo a rotorpercusión hidráulica Esp*) MINAS hydraulic rotary percussion drilling; ~ **por inyección** m CARBÓN wash boring; ~ **en mar abierto** m TEC PETR offshore drilling; ~ **nominal** m INFORM&PD roll call polling; ~ **de percusión** m TEC PETR *perforación* percussion drilling; ~ **por percusión** m CARBÓN percussion drilling, percussive drilling; ~ **de producción** m TEC PETR *perforación* production drilling; ~ **profundo** m MINAS deep drilling; ~ **recíproco** m PETROL reciprocal probe; ~ **de roca** m CARBÓN rock sounding; ~ **por rotación** m MINAS shot drilling; ~ **rotativo** m CARBÓN, MINAS, PETROL, TEC PETR rotary drilling; ~ **rotativo con extracción continua de muestras** m MINAS rotary continuous-core drilling; ~ **a rotorpercusión hidráulica** m *Esp* MINAS hydraulic rotary percussion drilling; ~ **a tanto alzado** m AmL (*cf sondeo por contracta Esp*) MINAS contract boring, contract drilling; ~ **testiguero** m *Esp* (*cf sondeo estratigráfico AmL*) MINAS core drilling; ~ **con trépano de diamantes** m MINAS diamond drilling; ~ **con trépano de granalla de acero** m MINAS shot boring; ~ **con ultrasonidos** m ACÚST, FÍS ONDAS ultrasonic sounding

sondista m CARBÓN, MINAS, TEC PETR *perforación* borer, driller

sone m ACÚST, FÍS sone

sónico adj ACÚST sonic

sonido m ACÚST, FÍS sound, TEC ESP noise; ~ **aéreo** m ACÚST airborne sound; ~ **aerodinámico** m ACÚST aerodynamic sound; ~ **de alta fidelidad** m TV hi-fi sound; ~ **asincrónico** m CINEMAT asynchronous sound; ~ **de banda ancha** m ACÚST broadband sound; ~ **combinado** m ACÚST combination sound; ~ **estructural** m ACÚST structural sound, structure-borne sound; ~ **sobre imagen** m TV sound on vision; ~ **no sincronizado** m CINEMAT unmarried sound; ~ **óptico** m CINEMAT optical sound; ~ **perimétrico** m CINEMAT surround sound; ~ **pregrabado** m ACÚST playback; ~ **sincronizado** m CINEMAT married sound

sonio m ACÚST, FÍS sone

sonodetector m D&A sound detector

sonolocalizador m ACÚST sound locator

sonómetro m ACÚST sound level meter, FÍS, FÍS ONDAS sonometer

sonoridad f ACÚST, FÍS loudness; ~ **subjetiva** f ACÚST subjective loudness

sonorización f ACÚST *de película muda, de film mudo* postscoring

sopanda f CONST straining piece

soperas f pl C&V *vasija honda para servir sopa, caldo* hollow ware

soplado[1]: **no** ~ adj C&V *vidrio* not blown up; ~ **al aire** adj C&V *vidrio* free-blown

soplado[2] m C&V *vidrio* blowing; ~ **de aire** m ING MECÁ air blowing; ~ **de arco** m ING ELÉC arc quenching; ~ **con la boca** m C&V mouth blowing; ~ **de chispas** m ELEC *relé* spark blowout; ~ **de fondo con oxígeno** m METAL oxygen bottom blowing; ~ **magnético** m ING ELÉC magnetic blowout; ~ **con molde** m C&V mold blowing (*AmE*), mould blowing (*BrE*); ~ **en molde empastado** m C&V paste-mold blowing

(*AmE*), paste-mould blowing (*BrE*); ~ **de oxígeno** m PROD *aceros* oxygen lancing; ~ **con vacío** m C&V vacuum blowing; ~ **con vapor** m C&V steam blowing; ~ **de vidrio** m C&V glassblowing

soplador m ALIMENT, AUTO blower, C&V *vidrio soplado* blower, glass former, MECÁ blower, impeller; ~ **de aire caliente** m INSTAL TERM, LAB *calentamiento* hot air blower; ~ **del calentador** m INSTAL HIDRÁUL, TRANSP AÉR heater blower; ~ **de chispas** m ELEC *relé* spark extinguisher; ~ **para machos** m ING MECÁ core-blowing machine; ~ **manual** m C&V hand bellows; ~ **de vidrio** m C&V glass blower

sopladura f CARBÓN *pieza fundida* bubble, IMPR *planchas de estereotipia* honeycomb, PROD *en calderas* blowhole

soplante m PAPEL *campana y sección de secado*, PROD *ventilador impelente* blower; ~ **para despegue vertical** m TRANSP *aviación* lift fan

soplar vt ALIMENT puff, PROD blow

soplete m CONST blowpipe, torch, GAS torch, ING MECÁ blowpipe, MECÁ flame-cutter, torch, PROD blowpipe, torch, TERMO flame-cutter; ~ **para acetileno** m CONST acetylene blowpipe; ~ **de alta presión** m CONST high-pressure blowpipe; ~ **de baja presión** m CONST low-pressure blowpipe; ~ **de calentamiento** m CONST heating blowpipe; ~ **de cobre-soldadura** m CONST brazing blowpipe; ~ **cortador** m ING MECÁ cutting torch; ~ **de corte** m CONST gouging blowpipe, *soldadura* cutting blowpipe, ING MECÁ cutting torch; ~ **de corte bajo el agua** m CONST underwater cutting blowpipe; ~ **a gas** m PROD gas blowpipe; ~ **manual** m ING MECÁ manual blowpipe; ~ **oxiacetilénico** m CONST oxyacetylene blowpipe, MECÁ acetylene-oxygen torch, PROD oxyacetylene blowpipe; ~ **de oxiacetileno** m CONST *soldadura*, PROD oxyacetylene blowpipe; ~ **para oxicorte** m TERMO flame-cutting torch; ~ **oxídrico** m PROD compound blowpipe; ~ **de pantalla de aire** m PROD *producción, mantenimiento* air screen blow gun; ~ **perforador** m CONST *soldadura* oxygen lance; ~ **de soldadura** m CONST soldering blowpipe, welding blowpipe; ~ **para soldadura de acetileno** m CONST acetylene blowpipe; ~ **para soldadura MIG-MAG** m CONST torch for MIG-MAG welding; ~ **de soldadura con pasador** m CONST stud welding gun; ~ **para soldadura de plasma** m CONST torch for plasma welding; ~ **para soldadura TIG** m CONST torch for TIG welding; ~ **de soldar** m CONST, TERMOTEC blowlamp; ~ **de soldeo** m TERMO welding torch

sopleteado m CONST blowpiping

soplo: ~ **final** m C&V final blow; ~ **en molde empastado** m C&V paste-mold blowing (*AmE*), paste-mould blowing (*BrE*); ~ **primario** m C&V preblowing

soportabrazos m INSTR support arm

soportado adj INFORM&PD supported

soportar vt CONST *estructuras* carry, bear, support, INFORM&PD support, ING MECÁ bear, carry, support

soportatubos m ING MECÁ pipe support

soporte[1]: **con el** ~ **hacia abajo** adj CINEMAT base down

soporte[2] m CINEMAT base, stand, CONST bearing, bed, *escaleras* standard, support, *carpintería* upright, FERRO *infraestructura* upright, FOTO mounting, GEO-FÍS pillar, IMPR substrate, INFORM&PD medium, support, ING ELÉC holder, ING MECÁ standard, leg, mount, fitment, prop, follow rest, droop restrainer, upright, bearing, rack, rest, seating, pedestal, pillow block, mounting, INSTR base, cradle, LAB *equipo general* stand, MECÁ cradle, pedestal, rack, bearing, MINAS *de ejes suspendidos* bracket hanger, PAPEL *de revestimiento mural* base web, base stock, PROD hanger, *cilindro del calentador* pedestal, QUÍMICA stand, TELECOM rack, TRANSP AÉR pylon, backing, VEH *motocicleta* stand; **~ de acetato** *m* IMPR acetate base; **~ de acetato de celulosa** *m* CINEMAT cellulose-acetate base, P&C cellulose-nitrate base; **~ de amortiguador** *m* TRANSP cushion borne base; **~ de ángulo** *m* ING MECÁ angle bracket; **~ antivibratorio** *m* ING MECÁ spring clip; **~ de árbol** *m* ING MECÁ *fresadora* arbor support; **~ del arco de ladrillo** *m* CONST brick-arch bearer; **~ articulado** *m* ING MECÁ swivel bearing; **~ azimut-elevación** *m* TEC ESP Az-El mount; **~ de bajo encogimiento contracción** *m* CINEMAT low-shrink base; **~ basculante** *m* ING MECÁ swivel bearing, swivel hanger; **~ de base** *m* CONST foothold; **~ de batería** *m* AUTO battery cradle, PROD battery holder; **~ de la bolsa** *m* EMB bag holder; **~ del brazo del balancín** *m* AUTO, VEH rocker-arm support; **~ de bureta** *m* LAB burette stand; **~ del cable** *m* ING MECÁ, PROD, TRANSP cable support; **~ de la caja de cambios principal** *m* TRANSP AÉR main-gearbox support; **~ de la cámara** *m* CINEMAT, TV *para colocar accesorios* camera bracket; **~ de cámara de tres patas** *m* CINEMAT crowfoot; **~ del canalón** *m* CONST gutter bracket; **~ del carro** *m* ING MECÁ turning rest; **~ del carro basculante** *m* ING MECÁ swivel slide rest; **~ del carro portaherramientas** *m* ING MECÁ handrest socket; **~ de carro transversal y longitudinal** *m* ING MECÁ compound slide rest; **~ por catenaria** *m* FERRO *equipo inamovible* catenary support; **~ de cátodo** *m* ING ELÉC core; **~ central** *m* ING MECÁ *torno* center rest (*AmE*), centre rest (*BrE*); **~ para chip sin patillas** *m* ELECTRÓN, ING ELÉC leadless chip carrier; **~ de choque principal del tren de aterrizaje** *m* TRANSP AÉR landing gear main shock strut; **~ colgante** *m* ING MECÁ hanger, MINAS pendant bracket; **~ colgante con cojinetes** *m* ING MECÁ hanger with bearings; **~ colgante de manguera** *m* PROD hose hanger; **~ coloreado** *m* CINEMAT tinted base; **~ de la columna del tren de aterrizaje** *m* TRANSP AÉR landing gear leg support; **~ de comunicaciones** *m* INFORM&PD communication medium; **~ conductor** *m* IMPR carrying medium; **~ de consola** *m* CONST knee bracket; **~ de control cíclico lateral** *m* TRANSP AÉR lateral cyclic control support; **~ de corredera oscilante** *m* ING MECÁ swivel slide rest; **~ de cristal** *m* CRISTAL, ELECTRÓN crystal holder; **~ de cuchilla** *m* METR *del fiel de la balanza*, PROD knife edge; **~ de cuchilla central** *m* METR *balanza* center knife edge (*AmE*), centre knife edge (*BrE*); **~ de cuerda** *m* CONST line pin; **~ de cuerpo** *m* CINEMAT *para sujetar la cámara* body brace; **~ de datos** *m* INFORM&PD data medium; **~ de desplazamiento transversal y longitudinal** *m* ING MECÁ compound slide rest; **~ en diagonal** *m* INSTAL HIDRÁUL *de caldera* diagonal stay; **~ doble** *m* PROD *fundería* stud, stud chaplet; **~ del eje de afiladora** *m* ING MECÁ grinding-spindle carrier; **~ del eje de rectificadora** *m* ING MECÁ grinding-spindle carrier;

~ elástico *m* ING MECÁ mount, TEC ESP shock mount; **~ de embudo** *m* LAB *filtración* funnel stand; **~ encastrado** *m* FOTO *lentes de cámara* sunk setting; **~ de enmascaramiento** *m* FOTO masking frame; **~ de escuadra** *m* ING MECÁ angle plate; **~ de estar** *m* CINEMAT *del negativo* estar base; **~ de estribo** *m* CINEMAT stirrup hanger; **~ fijo** *m* ING MECÁ steady rest; **~ fijo para sistema de frío** *m* REFRIG cold static base; **~ de filtro** *m* LAB filter support; **~ del filtro** *m* CINEMAT, FOTO, TV filter holder; **~ giratorio** *m* CARBÓN martingale, ING MECÁ trunnion, MECÁ trunnion, turntable, VEH *junta universal* trunnion; **~ de goma del motor** *m* AUTO, P&C, VEH rubber engine mounting; **~ de grabación** *m* INFORM&PD recording medium; **~ para la grabación de información** *m* FOTO data recording back; **~ en herradura** *m* INSTR horseshoe mount, horseshoe mounting; **~ de herramientas** *m* ING MECÁ tool rest; **~ de hombro** *m* CINEMAT shoulder brace; **~ del horno de ladrillo** *m* CONST brick-arch bearer; **~ para iluminación** *m* FOTO lighting stand; **~ del inducido** *m* ELEC *máquina*, ING ELÉC armature spider; **~ para lámparas de flash** *m* FOTO flash bar; **~ lineal** *m* TERMOTEC linear bearing; **~ logable** *m* FOTO collapsible stand; **~ lógico** *m* INFORM&PD software; **~ lógico de análisis estructural** *m* TEC ESP structural analysis software; **~ lógico de comunicación** *m* INFORM&PD communications software; **~ lógico de un curso de enseñanza** *m* INFORM&PD courseware; **~ lógico inalterable** *m* INFORM&PD firmware; **~ lógico de producción propia** *m* INFORM&PD in-house software; **~ lógico de sistemas** *m* INFORM&PD systems software; **~ lógico software remotamente disponible** *m* INFORM&PD telesoftware; **~ para machos** *m* PROD *fundería* chaplet; **~ de madera** *m* CONST timber jack; **~ magnético** *m* INFORM&PD magnetic media; **~ de mandíbula** *m* ING MECÁ jaw holder; **~ de mano** *m* ING MECÁ handrest; **~ de la máquina** *m* MECÁ engine pedestal; **~ de memoria** *m* INFORM&PD storage medium; **~ de microtono direccional** *m* CINEMAT gunpod; **~ del molde** *m* C&V, P&C mold holder (*AmE*), mould holder (*BrE*); **~ de montaje** *m* FOTO mounting bracket; **~ de mordaza** *m* ING MECÁ jaw holder; **~ del motor** *m* AUTO engine mounting, engine support, TRANSP AÉR, VEH engine support; **~ móvil** *m* ING MECÁ follow rest, MECÁ carriage; **~ del muelle del mecanismo de tracción** *m* FERRO *vehículos* draw gear spring plate; **~ por muelles** *m* VEH *carrocería* sprung weight; **~ de la mufla** *m* C&V muffle support; **~ neumático** *m* MINAS air leg; **~ de nitrato** *m* CINEMAT, FOTO nitrate base; **~ de nitrato de celulosa** *m* CINEMAT cellulose-nitrate base; **~ óptico** *m* INFORM&PD optical medium; **~ de órbita** *m* TEC ESP orbit support; **~ de la pantalla de enfoque** *m* CINEMAT, FOTO, TV focusing-screen frame; **~ de pared** *m* *Esp* (*cf brazo mural AmL*) CINEMAT wall brace, MINAS wall bracket; **~ de película** *m* FOTO base, support; **~-pescante** *m* ING MECÁ cantilever; **~ de pipeta** *m* LAB burette stand; **~ para placas** *m* FOTO magazine back; **~ de plataforma de altura variable** *m* PROD stand with rising table; **~ plegable** *m* CINEMAT collapsible stand; **~ de poliéster** *m* CINEMAT polyester base; **~ portafresas en saliente** *m* PROD *fresadora* overhanging arm; **~ portátil de tornillo** *m* CONST portable-vice stand (*BrE*), portable-vise stand (*AmE*); **~ portátil de tornillo con tornillo**

de pie *m* CONST portable-vice stand with leg vice (*BrE*), portable-vise stand with leg vise (*AmE*); **~ del radiador** *m* AUTO, VEH radiator support; **~ del reflector** *m* CINEMAT, FOTO, TV reflector stand; **~ para refrentar** *m* ING MECÁ facing head; **~ de registro magnético** *m* TV magnetic recording medium; **~ regulable de altura** *m* LAB, VEH jack; **~ de remolque** *m* VEH towing bracket; **~ de resbaladeras** *m* ING MECÁ motion plate, guide bearer, guide yoke; **~ de reserva** *m* TEC ESP backup bearing; **~ de retorta** *m* LAB retort stand; **~ de rótula** *m* ING MECÁ swivel bearing; **~ de seguridad** *m* CINEMAT safety base; **~ sencillo** *m* PROD staple, *fundería* single stud; **~ suspendido** *m* ING MECÁ hanger; **~ suspendido con cojinetes** *m* ING MECÁ hanger with bearings; **~ telescópico** *m* INSTR telescope mount, telescopic support; **~ del torno** *m* MECÁ lathe saddle; **~ transparente** *m* CINEMAT clear base; **~ del transporte de carga** *m* TRANSP AÉR cargo-carrier support; **~ del tren de aterrizaje del morro** *m* TRANSP AÉR nose gear saddle; **~ de tres pies** *m* PROD tripod stand; **~ de triacetato** *m* CINEMAT triacetate base; **~ de triacetato de celulosa** *m* CINEMAT cellulosetriacetate base; **~ de tubo electrónico** *m* ING ELÉC tube socket; **~ vacío** *m* INFORM&PD empty medium; **~ de la vara** *m* CONST *topografía* staff holder; **~ virgen** *m* INFORM&PD virgin medium

soportes *m pl* CONST braces

sorbente *m* CONTAM MAR sorbent

sorbita *f* QUÍMICA sorbite

sorbitol *m* QUÍMICA sorbitol

sorbosa *f* QUÍMICA sorbose

sorción *f* AGUA, QUÍMICA sorption

sordera: ~ profesional *f* ACÚST professional deafness

sorgo: ~ forrajero *m* AGRIC broom millet, forage sorghum, grass sorghum; **~ granífero** *m* AGRIC grain sorghum; **~ lanudo** *m* AGRIC velvet grass; **~ negro** *m* AGRIC columbus grass

sosa *f* C&V, QUÍMICA soda; **~ cáustica** *f* DETERG caustic soda, QUÍMICA sodium hydroxide, caustic soda; **~ comercial** *f* DETERG soda ash; **~ para lavar** *f* QUÍMICA *carbonato sódico hidratado* washing soda

sostén *m* C&V jamb, CONST support, ING MECÁ leg, holder, prop; **~ del lente** *m* C&V lens holder

sostener *vt* ING MECÁ bear, hold up, support

sostenido *m* ACÚST *música* sharp

sostenimiento *m* CONST buttressing, PROD hold, upkeep

sótano *m* CONST vault, GEOL basement, PETROL, TEC PETR basement, cellar

sotavento[1]: **de ~** *adj* METEO, TRANSP MAR leeward

sotavento[2]: **a ~** *adv* METEO, TRANSP MAR *navegación* leeward

sotavento[3] *m* TRANSP MAR lee

soterramiento *m* CARBÓN landfill, CONTAM, RECICL landfill, landfilling

spaniolita *f* MINERAL spaniolite

sparker *m* GEOFÍS sparker

spin *m* FÍS spin

spooler *m* INFORM&PD spooler

spooling *m* INFORM&PD spooling

sprite *m* INFORM&PD sprite

Sr *abr* (*estroncio*) QUÍMICA Sr (*strontium*)

SSM *abr* TEC ESP (*espejo de superficie secundaria*) SSM (*second surface mirror*), (*misil superficie-superficie*) SSM (*surface-to-surface missile*), (*monitor del sistema de satélites*) SSM (*satellite system monitor*), TELECOM (*mensaje del segmento único*) SSM (*single-segment message*)

SSOG *abr* (*Guía de Explotación del Sistema de Satélites*) TEC ESP SSOG (*Satellite System Operation Guide*)

SSUS *abr* (*última fase rotatoria sólida*) TEC ESP SSUS (*Solid Spinning Upper Stage*)

St *abr* (*stratus*) METEO St (*stratus*)

stándard[1] *adj* FÍS *tipo de estadísitica* standard

stándard[2] *m* FÍS standard; **~ secundario** *m* FÍS secondary standard

standarización *f* FÍS, ING MECÁ, ING ELÉC, METAL, NUCL standardization

standpipe *m* TEC PETR *perforación* standpipe

stassfurtita *f* MINERAL stassfurtite

Steadicam® *m* CINEMAT *soporte* Steadicam; **~ de Panavision**® *m* CINEMAT panaglide

steinmannita *f* MINERAL steinmannite

sternbergita *f* MINERAL sternbergite

stock: ~ de peces *m* OCEAN stock of fish

stone *m* METR *unidad de peso* stone

stratocumulus *m* (*Sc*) METEO stratocumulus (*Sc*)

stratus *m* (*St*) METEO stratus (*St*); **~ fractus** *m* METEO stratus fractus

streamer *m* INFORM&PD streamer

stromeyerita *f* MINERAL stromeyerite

struvita *f* MINERAL struvite

suavizador *m* ING MECÁ strop; **~ de la luz** *m* CINEMAT light softener

suavizamiento: ~ de contrastes *m* TEC ESP *comunicaciones* de-emphasis

suavizante *m* TEXTIL softener; **~ del agua** *m* TEXTIL water softener; **~ de telas** *m* DETERG, TEXTIL fabric softener

subacetato *m* QUÍMICA subacetate

subalimentación *f* REFRIG *insuficiencia de refrigerante en el evaporador* starving

subantomórfico *adj* GEOL subantomorphic

subarmónica *f* ELECTRÓN subharmonic

subarmónico *adj* ACÚST subharmonic

subbase *f* CONST *carreteras* subbase

subcadena *f* INFORM&PD substring

subcampo: ~ inferior *m* TELECOM lower subfield

subcanal *m* ELECTRÓN subchannel

subcanalización *f* TEC PETR subduction

subcapa *f* FÍS subshell; **~ de convergencia** *f* TELECOM convergence sublayer (*CS*); **~ de convergencia de transmisión** *f* TELECOM transmission convergence sublayer; **~ electrónica** *f* FÍS RAD *de átomo* electronic subshell; **~ física media** *f* TELECOM physical medium sublayer; **~ de segmentación y rearmado** *f* TELECOM segmentation and reassembly sublayer

subcarbonato *m* QUÍMICA subcarbonate

subcentral *f* ELEC *red de distribución*, ING ELÉC substation

subcloruro *m* QUÍMICA subchloride

subcompactación *f* TEC PETR *geología* undercompaction

subcompactado *adj* TEC PETR *geología* undercompacted

subconjunto *m* INFORM&PD, MATEMÁT subset; **~ de caracteres** *m* INFORM&PD character subset; **~ correcto** *m* INFORM&PD proper subset

subdirección *f* TELECOM subaddress, subaddressing; **~ terminal** *f* TELECOM terminal subaddressing

subdivisión *f* PROD splitting
subdominante *adj* ACÚST subdominant
subducción *f* TEC PETR *geología* subduction
subenfriador *m* REFRIG subcooler
subenfriamiento: ~ **por evaporación súbita** *m* NUCL flash subcooling
suberato *m* QUÍMICA suberate
subérico *adj* QUÍMICA suberic
suberificación *f* QUÍMICA suberification
suberílico *adj* QUÍMICA suberylic
suberilo *m* QUÍMICA suberyl
suberina *f* QUÍMICA suberin
suberona *f* QUÍMICA suberone
subespacio: ~ **neumático** *m* CARBÓN pneumatic cell
subestación *f* CONST *electricidad*, ELEC *red de distribución*, ING ELÉC, PROD substation; ~ **de bifurcación** *f* ELEC, ING ELÉC tee-off substation; ~ **para la conversión de frecuencia** *f* ELEC, ING ELÉC substation for frequency conversion; ~ **de derivación** *f* ELEC, ING ELÉC tee-off substation; ~ **de distribución** *f* ELEC, ING ELÉC *red de distribución* distribution substation; ~ **de distribución de energía** *f* ELEC, ING ELÉC switching substation; ~ **eléctrica** *f* ELEC, ING ELÉC electric-power substation; ~ **rectificadora** *f* ELEC, ING ELÉC, MECÁ rectifier substation; ~ **con tomas** *f* ELEC, ING MECÁ tapped substation; ~ **de transformación** *f* ELEC, ING ELÉC transformer substation; ~ **transformadora** *f* ELEC, ING ELÉC transformer substation
subestándar *adj* CALIDAD substandard
subestructura *f* TEC PETR *perforación* substructure
subexcitado *adj* ING MECÁ underdriven
subexplotación *f* OCEAN *de una pesquería* underfishing
subexponer *vt* CINEMAT, FOTO, IMPR underexpose
subexposición *f* CINEMAT, FOTO, IMPR underexposure
subfusil *m* D&A sub-machine gun
subgalería *f* MINAS subdrift
subhédrico *adj* GEOL *textura ígnea* subhedral
subida *f* CONST *del magma* upright, MINAS hoisting, OCEAN ascent, PROD hoisting, TEC ESP climb, TRANSP AÉR climb, climb-out; ~ **capilar** *f* CARBÓN capillary rise; ~ **de crucero** *f* TRANSP AÉR cruise climb; ~ **en globo** *f* OCEAN *accidente de buceo* ballooning; ~ **del petróleo** *f* MINAS *pozos* rising; ~ **del pozo** *f* TEC PETR *perforación* coming out of hole; ~ **de rampas con tracción en cola** *f* FERRO *vehículos* banking, pusher operation
subido *adj* TEXTIL *tono* deep
subíndice *m* IMPR inferior character, inferior figure, inferior letter, subscript, INFORM&PD subscript
subir[1] *vt* MECÁ lift, PROD hoist; ~ **al exterior** *vt* MINAS bring to the surface; ~ **a la grada** *vt* TRANSP MAR slip
subir[2] *vi* TRANSP MAR *marea* flood, *presión* rise; ~ **a la superficie** *vi* TRANSP MAR *submarino* return to surface
sublimable *adj* QUÍMICA sublimable
sublimación *f* FÍS, QUÍMICA sublimation; ~ **catódica con descarga luminosa** *f* NUCL glow-discharge sputtering
sublimado[1] *adj* QUÍMICA sublimed
sublimado[2] *m* QUÍMICA, TEC PETR *química* sublimate
sublimador *m* QUÍMICA sublimatory
sublimar *vt* QUÍMICA sublime
sublote *m* PROD sublot
submarino[1] *adj* OCEAN, TRANSP MAR submarine, undersea

submarino[2] *m* OCEAN, TRANSP MAR *vehículo subacuático* submarine, submersible
submodulación *f* ELECTRÓN undermodulation
submúltiples: ~ **mínimos** *m pl* GEOM minimal submanifolds
submultiplexar *vt* TELECOM submultiplex
submúltiplo *m* METR *de la unidad de medida* submultiple
submuniciones *f pl* D&A submunitions
subnitrato *m* QUÍMICA subnitrate
subnivel *m* FÍS sublevel, MINAS counter, *galerías* sublevel
subnormal *f* GEOM subnormal
suboficial *m* TRANSP MAR petty officer
subpastoreo *m* AGRIC undergrazing
subpiso *m* MINAS substage, *galerías* subdrift, sublevel
subportadora *f* TEC ESP, TV subcarrier; ~ **cromática** *f* TV color subcarrier (*AmE*), colour subcarrier (*BrE*); ~ **de cromeado** *f* ELECTRÓN chrominance subcarrier; ~ **de crominancia** *f* TV chrominance subcarrier
subpresión *f* CARBÓN *presas* pore pressure, ING MECÁ underpressure, TERMO uplift
subprocesamiento: ~ **múltiple** *m* INFORM&PD multithreading
subproducto *m* CARBÓN, CONTAM, ING MECÁ by-product, PROC QUÍ side product, by-product, PROD by-product, QUÍMICA *de reacción* side product, by-product, TEC PETR by-product; ~ **gaseoso** *m* GAS by-product gas; ~ **de molienda de trigo** *m* AGRIC wheat shorts
subproductos: ~ **de molienda** *m pl* AGRIC fine-offals, milling by-products;
subprograma *m* INFORM&PD subprogram
subrayar *vt* IMPR underline; ~ **por arriba** *vt* IMPR overscore
subred *f* TELECOM subnetwork; ~ **de gestión de jerarquía digital síncrona** *f* TELECOM SDH management subnetwork (*SMS*)
subreflector *m* TEC ESP subreflector
subrevelar *vt* CINEMAT, FOTO underdevelop
subrutina *f* INFORM&PD subroutine
subsaturación *f* GEOL, METAL undersaturation
subsaturado *adj* GEOL, METAL undersaturated
subscriptor *m* TELECOM subscriber
subserie *f* INFORM&PD substring
subsidencia *f* CONTAM *nivel de aguas*, GEOL, METEO subsidence, MINAS *del techo* subsidence, *geología* sinking, TEC PETR subsidence; ~ **regional** *f* GEOL downwarp
subsistema *m* INFORM&PD, TEC ESP, TELECOM subsystem; ~ **de acceso** *m* TELECOM access subsystem; ~ **de empuje** *m* TEC ESP thrust subsystem; ~ **de gestión de la transacción** *m* TELECOM transaction management subsystem; ~ **de localización del vehículo** *m* TELECOM vehicle location subsystem; ~ **de potencia** *m* TEC ESP power sub-system; ~ **radioeléctrico** *m* TELECOM radio subsystem
subsolador *m* AGRIC breaker plough (*BrE*), breaker plow (*AmE*), chisel plough (*BrE*), chisel plow (*AmE*), subsoiler; ~ **rotativo** *m* AGRIC rotary subsoiler
substancia *f* C&V substance, ING MECÁ material, QUÍMICA substance; ~ **amalgamadora** *f* CARBÓN amalgamator; ~ **cancerígena** *f* SEG carcinogenic substance; ~ **contaminante** *f* CONTAM pollutant; ~ **contaminante del aire** *f* CONTAM air pollutant,

air-polluting substance; **~ contaminante atmosférica** *f* CONTAM air pollutant, air-polluting substance; **~ corrosiva** *f* SEG corrosive substance; **~ disipativa** *f* ELEC *dieléctrico, línea* lossy material; **~ fecal** *f* INSTAL HIDRÁUL, RECICL faecal matter (*BrE*), fecal matter (*AmE*); **~ irritante** *f* SEG irritant substance; **~ miscible** *f* CONTAM miscible substance; **~ nociva** *f* SEG harmful substance; **~ oxidante** *f* SEG oxidizing substance; **~ peligrosa** *f* QUÍMICA dangerous substance, SEG dangerous substance, hazardous substance, TRANSP dangerous substance; **~ radiactiva** *f* CONTAM, FÍS RAD, SEG radioactive substance; **~ tóxica** *f* SEG toxic substance

substituir: ~ a *vt* PROD override

substituto: ~ de sal *m* ALIMENT salt substitute

substituyente *m* QUÍMICA substituent

substractor *m* ELECTRÓN, INFORM&PD subtractor; **~ de un solo dígito** *m* INFORM&PD one-digit subtractor; **~ de un solo dígito** *m* ELECTRÓN, INFORM & PD one-digit subtractor

substraendo *m* INFORM&PD subtrahend

substrato *m* FÍS substrate, GEOL substratum, INFORM&PD, P&C substrate, TEC PETR *geología* substratum, TELECOM substrate; **~ aislante** *m* ING ELÉC, TEC ESP insulating substrate; **~ amorfo** *m* ELECTRÓN amorphous substrate; **~ de arseniuro de galio** *m* ELECTRÓN gallium arsenide substrate; **~ blindado** *m* ELECTRÓN metal-clad substrate; **~ de circuito híbrido de capa delgada** *m* ELECTRÓN thin-film hybrid circuit substrate; **~ de circuito híbrido de capa gruesa** *m* ELECTRÓN thick-film hybrid circuit substrate; **~ de circuito impreso** *m* ELECTRÓN printed-circuit substrate; **~ de circuito integrado** *m* ELECTRÓN integrated circuit substrate; **~ de doble capa** *m* ELECTRÓN double-sided substrate; **~ de microondas** *m* ELECTRÓN microwave substrate; **~ pasivo** *m* ELECTRÓN passive substrate; **~ piezoeléctrico** *m* ING ELÉC piezoelectric substrate; **~ semiaislante** *m* ELECTRÓN semi-insulating substrate; **~ de semiconductor** *m* ELECTRÓN semiconductor substrate; **~ de silicio** *m* ELECTRÓN silicon substrate; **~ de silicio de tipo p** *m* ELECTRÓN p-type silicon substrate; **~ tipo n** *m* ELECTRÓN n-type substrate; **~ de vidrio** *m* ELECTRÓN glass substrate; **~ de zafiro** *m* ELECTRÓN sapphire substrate

subsuelo *m* GEOL substratum, MINAS undersoil

subsulfato *m* QUÍMICA subsulfate (*AmE*), subsulphate (*BrE*)

subtangente *f* GEOM subtangent

subtarea *f* INFORM&PD subtask

subtender *vt* GEOM subtend

subtensión *f* NUCL underload

subterráneo[1] *adj* AGUA, CONST, CONTAM, ELEC underground, ELECTRÓN *canal* buried, GAS, GEOL, HIDROL, ING ELÉC, MINAS, TEC PETR, TELECOM, TERMO, TRANSP underground

subterráneo[2] *m* FERRO subway (*AmE*), tube, underground (*BrE*)

subtipo *m* INFORM&PD subtype

subtitulación *f* QUÍMICA undertitration

subtítulo *m* CINEMAT caption, IMPR bank

subtormenta *f* GEOFÍS substorm; **~ ionosférica** *f* GEOFÍS ionospheric substorm

subtracción: ~ binaria *f* INFORM&PD binary subtraction

subtractor: ~ digital *m* ELECTRÓN digital subtractor

subvariedades: ~ mínimas *f pl* GEOM minimal submanifolds

subvirador *m* VEH *dirección* understeer

subvoltaje *m* ING MECÁ underpressure, PROD undervoltage

subyacente *adj* GEOL underlying

succinato *m* QUÍMICA succinate

succínico *adj* QUÍMICA succinic

succinilo *m* QUÍMICA succinyl

succinimida *f* QUÍMICA succinimide

succinita *f* MINERAL succinite

succión *f* CONST draft (*AmE*), draught (*BrE*), ING MECÁ indraft (*AmE*), indraught (*BrE*), suction, MECÁ suction; **~ de arena** *f* CONST sand dredging; **~ ascendente** *f* AUTO updraft (*AmE*), updraught (*BrE*); **~ termohalina** *f* OCEAN thermohaline pumping

succionador *m* CONTAM, PROD *petróleo* skimmer; **~ de choque** *m* ING ELÉC *guías de ondas* choke plunger; **~ para la extracción de petróleo** *m* CONTAM *combustibles*, CONTAM MAR oil-recovery skimmer

sucedáneo: ~ de aguarrás *m* COLOR turpentine substitute

sucesión *f* MATEMÁT sequence, series; **~ de Fibonacci** *f* MATEMÁT Fibonacci series; **~ geométrica** *f* GEOM, MATEMÁT geometric sequence, geometric series

suceso *m* GEN event; **~ aleatorio** *m* ELECTRÓN random event; **~ iniciador** *m* NUCL initiating event; **~ simulado** *m* FÍS PART, FÍS RAD simulated event

suciedad *f* PAPEL bark specks, PROD *fundería* sullage; **~ impresa** *f* CINEMAT *manchas en la película* printed dirt

sucio *adj* ING MECÁ, PROD fouled

sucrasa *f* QUÍMICA invertase

sucrato *m* QUÍMICA sucrate

sudación *f* MINAS sweat

sudar *vt* ALIMENT sweat

suela *f* PROD *horno eléctrico* hearth; **~ antideslizante** *f* SEG nonslip sole, *calzado protector* slip resistant sole

suelda: ~ de estaño y plomo *f* PROD fine solder; **~ fuerte** *f* TERMO brazing; **~ de plata** *f* PROD silver solder; **~ de plomo** *f* PROD coarse solder

suelo *m* AGRIC ground, soil, CARBÓN earth (*BrE*), ground (*AmE*), floor, soil, bottom, CONST *piso* ground, floor, GEOFÍS soil, GEOL *de filón, yacimiento* footwall, MINAS sole, grass, bottom, *pozo* deck, *filones* footwall; **~ ácido** *m* CONTAM *contaminación terrestre caracterización de suelos* acid earth; **~ agua** *m* AGUA soil water; **~ antrópico** *m* AGRIC anthropic soil; **~ arcilloso** *m* AGRIC, CARBÓN clay soil; **~ de arrastre** *m* AGRIC drift soil; **~ de baldosas** *m* CONST tile flooring; **~ blando** *m* CONST soft ground; **~ de la cabina** *m* TRANSP AÉR cabin floor; **~ de cieno** *m* MINAS *piso de galerías* mud bottom; **~ cohesivo** *m* CARBÓN cohesive soil; **~ contraíble** *m* CARBÓN contractant soil, contractible soil; **~ endurecido** *m* GEOL hard ground; **~ ensamblado** *m* CONST framed floor; **~ flotante** *m* ACÚST floating floor; **~ forestal** *m* AGRIC forest floor; **~ grueso** *m* CARBÓN coarse soil; **~ helado** *m* CARBÓN, GEOL frozen ground; **~ incohesivo** *m* CARBÓN noncohesive soil; **~ con juntas en ángulo recto** *m* CONST square-jointed floor; **~ de lajas** *m* CONST flagstone, flagstone pavement; **~ liso** *m* CARBÓN even-grained soil; **~ de losas** *m* CONST tile floor; **~ margoso** *m* AGRIC marly soil; **~ móvil** *m* TRANSP moving floor; **~ nevado** *m*

METEO snow cover; ~ **orgánico** *m* CARBÓN organic soil; ~ **de parqué a espiga** *m* CONST herringbone pattern parquet flooring; ~ **presurizado** *m* TRANSP AÉR pressurized floor; ~ **refractario** *m* SEG fireproof floor, flameproof floor; ~ **semi turboso** *m* AGRIC half bog soil; ~ **sulfuroso** *m* CARBÓN sulfide soil (*AmE*), sulphide soil (*BrE*); ~ **de la torre** *m* TEC PETR *perforación* derrick floor

suelta: ~ **de carga** *f* TRANSP AÉR load release; ~ **manual del tren de aterrizaje** *f* TRANSP AÉR landing gear manual release

suelto *adj* ING MECÁ, PROD *tuerca* loose

suero *m* ALIMENT *industria láctea* whey; ~ **concentrado** *m* ALIMENT whey concentrate; ~ **en polvo** *m* ALIMENT whey powder

suerte: ~ **corrida por el derrame** *f* CONTAM MAR fate of oil

suertes *f pl* IMPR sorts

sufijo *m* INFORM&PD suffix

sufridor *m* PROD *remachado* dolly

sufrir: ~ **un accidente grave** *vi* SEG meet with a serious accident; ~ **un naufragio** *vi* TRANSP MAR be shipwrecked

sujeción *f* CONST *armadura de hormigón* fastening, ING ELÉC holding, ING MECÁ cramping, screwing, fastening, MECÁ, PROD fastening, TRANSP AÉR holding, TV cramping, VEH *dispositivo para remolcar* shackle; ~ **con correas de color** *f* EMB colored strapping (*AmE*), coloured strapping (*BrE*); ~ **con flejes de acero** *f* EMB steel band strapping; ~ **hidráulica** *f* ING MECÁ hydraulic clamping; ~ **magnética** *f* ING MECÁ magnetic holding; ~ **a rosca** *f* ING MECÁ screw fixing; ~ **de la subportadora** *f* TV subcarrier lock; ~ **de tubería auxiliar de revestimiento** *f* PETROL liner hanger; ~ **para tubería de bombeo** *f* PETROL tubing hanger; ~ **de tubería de revestimiento** *f* PETROL casing hanger; ~ **de viga** *f* TRANSP AÉR beam capture

sujetacable *m* ELEC, ING ELÉC cable clip

sujetacarril *m* FERRO *infraestructura* rail clip

sujetado *m* ING MECÁ tieback

sujetador *m* ELEC *cable* clamp, ING MECÁ clip, latch, fastener, MECÁ fastener, clip, MINAS *tacos* catch, PROD grip, fastener; ~ **de acero inoxidable anti-corrosivo** *m* ING MECÁ corrosion-resistant stainless steel fastener; ~ **avellanado** *m* ING MECÁ countersunk fastener; ~ **de borde** *m* C&V edge holder; ~ **de cabeza cruciforme** *m* ING MECÁ cruciform head fastener; ~ **de cabeza moleteada** *m* ING MECÁ knurled-head fastener; ~ **de cabeza con ranuras en cruz** *m* ING MECÁ recessed-head fastener; ~ **de cabeza rectangular** *m* ING MECÁ rectangle head-fastener; ~ **de correa** *m* PROD belt fastener; ~ **de desenganche rápido** *m* ING MECÁ, PROD quick-release fastener; ~ **embutido** *m* ING MECÁ countersunk fastener; ~ **de lámparas de carbón** *m* CINEMAT carbon holder; ~ **de la página** *m* IMPR page bearer; ~ **sobre platina** *m* LAB *microscopio* stage clip; ~ **para pulido** *m* ING MECÁ lapping fixture; ~ **de resortes** *m* ING MECÁ spring clip; ~ **de rosca externa** *m* ING MECÁ external-threaded fastener; ~ **roscado** *m* ING MECÁ threaded fastener

sujetafusible *m* PROD fuse clip

sujetapapeles *m* EMB paperclip

sujetar *vt* CONST anchor, ING MECÁ cramp, take in, MECÁ clamp, PROD grip, hold, hold fast; ~ **con mordazas** *vt* CONST clamp

sujetarse *v refl* CONST bind, lock

sujeto *adj* CONST, ELEC, EMB, ING MECÁ, MECÁ, PROD clamped

sulfafurazol *m* QUÍMICA sulfafurazole (*AmE*), sulphafurazole (*BrE*)

sulfamato *m* QUÍMICA sulfamate (*AmE*), sulphamate (*BrE*)

sulfámico *adj* QUÍMICA sulfamic (*AmE*), sulphamic (*BrE*)

sulfamida[1]: **sin** ~ *adj* QUÍMICA sulfa-free (*AmE*), sulpha-free (*BrE*)

sulfamida[2] *f* QUÍMICA sulfa drug (*AmE*), sulpha drug (*BrE*), sulfamide (*AmE*), sulphamide (*BrE*)

sulfanatos: ~ **de petróleo y calcio** *m pl* TEC PETR *petroquímica* calcium petroleum sulfanates (*AmE*), calcium petroleum sulphanates (*BrE*)

sulfanilamida *f* QUÍMICA sulfanilamide (*AmE*), sulphanilamide (*BrE*)

sulfanilato *m* QUÍMICA sulfanilate (*AmE*), sulphanilate (*BrE*)

sulfanílico *adj* QUÍMICA sulfanilic (*AmE*), sulphanilic (*BrE*)

sulfapiridina *f* QUÍMICA sulfapyridine (*AmE*), sulphapyridine (*BrE*)

sulfarsénico *adj* QUÍMICA sulfarsenic (*AmE*), sulpharsenic (*BrE*)

sulfarsenido *m* QUÍMICA sulfarsenide (*AmE*), sulpharsenide (*BrE*)

sulfarseniuro *m* QUÍMICA sulfarsenide (*AmE*), sulpharsenide (*BrE*)

sulfatación *f* DETERG, QUÍMICA sulfation (*AmE*), sulphation (*BrE*)

sulfatiazol *m* QUÍMICA sulfathiazole (*AmE*), sulphathiazole (*BrE*)

sulfatida *f* QUÍMICA sulfatide (*AmE*), sulphatide (*BrE*)

sulfato *m* CONTAM, QUÍMICA sulfate (*AmE*), sulphate (*BrE*); ~ **de aluminio** *m* HIDROL, QUÍMICA aluminium sulphate (*BrE*), aluminum sulfate (*AmE*); ~ **de aluminio y potasio** *m* QUÍMICA potash alum; ~ **cálcico** *m* ALIMENT, QUÍMICA calcium sulfate (*AmE*), calcium sulphate (*BrE*); ~ **de cobre** *m* FOTO, QUÍMICA copper sulfate (*AmE*), copper sulphate (*BrE*); ~ **de dimetilo** *m* QUÍMICA dimethyl sulfate (*AmE*), dimethyl sulphate (*BrE*); ~ **ferroso** *m* QUÍMICA ferrous sulfate (*AmE*), ferrous sulphate (*BrE*); ~ **de hierro y cobre** *m* MINERAL copperasine; ~ **de magnesio** *m* DETERG magnesium sulfate (*AmE*), magnesium sulphate (*BrE*); ~ **de plomo** *m* QUÍMICA *venenosa* lead sulfate (*AmE*), lead sulphate (*BrE*); ~ **plumboso** *m* QUÍMICA lead sulfate (*AmE*), lead sulphate (*BrE*); ~ **de potasio** *m* DETERG potassium sulfate (*AmE*), potassium sulphate (*BrE*); ~ **de potasio y hierro** *m* QUÍMICA iron alum; ~ **sódico** *m* QUÍMICA sodium sulfate (*AmE*), sodium sulphate (*BrE*); ~ **de sodio** *m* DETERG sodium sulfate (*AmE*), sodium sulphate (*BrE*)

sulfhidrador *m* AmL (*cf aparato de Kipp Esp*) LAB *generador de H$_2$S*, QUÍMICA Kipp's apparatus

sulfhidrato *m* QUÍMICA sulfhydrate (*AmE*), sulphhydrate (*BrE*)

sulfhídrico *adj* QUÍMICA sulfhydric (*AmE*), sulphhydric (*BrE*)

sulfínico *adj* QUÍMICA sulfinic (*AmE*), sulphinic (*BrE*)

sulfinilo *m* QUÍMICA sulfinyl (*AmE*), sulphinyl (*BrE*)

sulfito *m* QUÍMICA sulfite (*AmE*), sulphite (*BrE*)

sulfocloración *f* DETERG, QUÍMICA sulfochlorination (*AmE*), sulphochlorination (*BrE*)

sulfolano *m* QUÍMICA, TEC PETR sulfolane (*AmE*), sulpholane (*BrE*)

sulfoleno *m* QUÍMICA, TEC PETR sulfolene (*AmE*), sulpholene (*BrE*)

sulfona *f* DETERG, QUÍMICA sulfone (*AmE*), sulphone (*BrE*)

sulfonación *f* DETERG, QUÍMICA sulfonation (*AmE*), sulphonation (*BrE*)

sulfonado *adj* QUÍMICA sulfonated (*AmE*), sulphonated (*BrE*)

sulfonamida *f* QUÍMICA sulfonamide (*AmE*), sulphonamide (*BrE*)

sulfonato *m* QUÍMICA sulfonate (*AmE*), sulphonate (*BrE*); ~ **de alcano** *m* DETERG, QUÍMICA alkane sulfonate (*AmE*), alkane sulphonate (*BrE*)

sulfónico *adj* QUÍMICA sulfonic (*AmE*), sulphonic (*BrE*)

sulfonilo *m* QUÍMICA sulfonyl (*AmE*), sulphonyl (*BrE*)

sulfonio *m* QUÍMICA *ion* sulfonium (*AmE*), sulphonium (*BrE*)

sulfosalicílico *adj* QUÍMICA sulfosalicylic (*AmE*), sulphosalicylic (*BrE*)

sulfoxilato: ~ **de formaldehído** *m* ALIMENT, QUÍMICA formaldehyde sulfoxylate (*AmE*), formaldehyde sulphoxylate (*BrE*)

sulfuración *f* QUÍMICA sulfuration (*AmE*), sulphuration (*BrE*)

sulfúrico *adj* QUÍMICA sulfuric (*AmE*), sulphuric (*BrE*)

sulfurilo *m* QUÍMICA sulfuryl (*AmE*), sulphuryl (*BrE*)

sulfurización *f* QUÍMICA sulfurization (*AmE*), sulphurization (*BrE*)

sulfurizar *vt* QUÍMICA sulfurize (*AmE*), sulphurize (*BrE*)

sulfuro *m* QUÍMICA sulfide (*AmE*), sulphide (*BrE*); ~ **de antimonio** *m* QUÍMICA stibnite; ~ **de carbonilo** *m* CONTAM carbonyl sulfide (*AmE*), carbonyl sulphide (*BrE*); ~ **de hidrógeno** *m* ALIMENT hydrogen sulfide (*AmE*), hydrogen sulphide (*BrE*); ~ **de níquel capilar** *m* MINERAL hair pyrites; ~ **de petróleo** *m* TEC PETR petrosulfur (*AmE*), petrosulphur (*BrE*); ~ **de plomo** *m* QUÍMICA lead sulfide (*AmE*), lead sulphide (*BrE*)

sulfuroso *adj* QUÍMICA sulfurous (*AmE*), sulphurous (*BrE*)

sultam *m* QUÍMICA sultam

sultona *f* QUÍMICA sultone

suma *f* ING MECÁ, MATEMÁT addition; ~ **binaria** *f* ELECTRÓN binary addition; ~ **de control** *f* ELECTRÓN *informática* checksum, INFORM&PD checksum, control total; ~ **hidrográfica** *f* ENERG RENOV summation hydrograph; ~ **lógica** *f* ELECTRÓN logic addition

sumador *m* INFORM&PD adder, full adder; ~ **de acarreo variable** *m* *Esp* (*cf sumador de llevar por onda AmL*) INFORM&PD ripple-carry adder; ~ **binario** *m* ELECTRÓN binary adder; ~ **de un bit** *m* ELECTRÓN one-digit adder; ~ **de circuito electrónico en escalera** *m* ELECTRÓN ladder adder; ~ **completo** *m* INFORM&PD full adder; ~ **digital** *m* ELECTRÓN digital adder; ~ **de llevar por onda** *m* *AmL* (*cf sumador de acarreo variable Esp*) INFORM&PD ripple-carry adder; ~ **paralelo** *m* ELECTRÓN, INFORM&PD parallel adder; ~ **en serie** *m* ELECTRÓN, INFORM&PD serial adder; ~ **de un solo dígito** *m* ELECTRÓN one-digit adder

sumando *m* INFORM&PD addend

sumar *vt* INFORM&PD, MATEMÁT add

sumatrol *m* QUÍMICA sumatrol

sumergencia *f* QUÍMICA submergence

sumergible[1] *adj* OCEAN, TEC PETR submersible

sumergible[2]: ~ **autónomo** *m* OCEAN autonomous submersible; ~ **remolcado** *m* OCEAN towed submersible; ~ **telecomandado no tripulado** *m* OCEAN unmanned submersible

sumergido *adj* FÍS, QUÍMICA immersed

sumergir *vt* AGUA submerge, FÍS immerse, MECÁ quench, PROD dip, QUÍMICA immerse, TEXTIL dip

sumersión *f* OCEAN submergence, QUÍMICA submergence, submersion, TEC ESP *vehículos* dive

sumidero *m* *Esp* (*cf resumidero AmL*) AGUA catch basin, sump, CARBÓN drain, CONST catch pit, sink hole, sump man (*BrE*), CONTAM *instalaciones industriales* sink, GEOL *cárstico* swallow hole, sink hole, INSTAL HIDRÁUL sink, LAB *servicios* sump, MINAS sinkhole, PROD sump, TEC PETR *producción* sump, *refino* catchpit; ~ **de bomba** *m* AGUA pump sump; ~ **de calor** *m* ELEC heat sink, INFORM&PD sink, NUCL, PROD, TEC ESP, TERMO heat sink; ~ **de calor final** *m* NUCL ultimate heat sink; ~ **de contención** *m* NUCL containment sump; ~ **térmico** *m* ELEC, FÍS heat sink

suministrador *m* EMB deliverer; ~ **de combustible** *m* TEC PETR fueler (*AmE*), fueller (*BrE*); ~ **de servicios CMISE** *m* TELECOM CMISE service provider

suministrar[1] *vt* FÍS supply, ING MECÁ fit out

suministrar[2]: ~ **energía** *vi* PROD provide power; ~ **una reserva de oxígeno** *vi* FÍS, PROC QUÍ provide an oxygen supply

suministro *m* AGUA, GAS, GEOL supply, ING MECÁ delivery, PROD supplying; ~ **de agua** *m* AGUA water delivery, HIDROL, TEXTIL water supply; ~ **de agua asegurado** *m* AGUA assured water supply; ~ **de agua subterránea** *m* AGRIC, AGUA, CARBÓN, HIDROL ground-water supply; ~ **de aire** *m* TRANSP AÉR air supply; ~ **aislado** *m* ELEC *red* floating supply; ~ **del arco** *m* CINEMAT arc feed; ~ **automático** *m* PROC QUÍ, TEC PETR *material* automatic feed; ~ **de calor** *m* TERMO heat rise; ~ **calorífico** *m* TERMO heat supply; ~ **de CC** *m* ING ELÉC DC supply; ~ **sin conexiones** *m* ELEC *red* floating supply; ~ **de corriente** *m* FÍS current supply (*AmE*), mains supply (*BrE*); ~ **de electricidad** *m* CONST, ELEC, ING ELÉC, ING MECÁ electricity supply; ~ **eléctrico** *m* CINEMAT power supply, CONST, ELEC, ING ELÉC electricity supply, ING MECÁ, MECÁ electricity supply, power; ~ **eléctrico de tierra** *m* TRANSP AÉR ground power supply; ~ **de energía** *m* ELEC *red* energy supply, power supply, ENERG RENOV, ELECTRÓN, energy supply, FERRO power supply, ING ELÉC, ING MECÁ, MECÁ energy supply, power supply, NUCL, TEC ESP energy supply, TELECOM power feed, power supply, TERMO energy supply; ~ **de energía de alta tensión** *m* ING ELÉC high-tension power supply; ~ **de energía de alto voltaje** *m* ING ELÉC high-voltage power supply; ~ **de energía con armazón abierto** *m* ING ELÉC open-frame power supply; ~ **de energía bipolar** *m* ING ELÉC bipolar power supply; ~ **de energía para calentador** *m* ING ELÉC heater power supply; ~ **de energía por conmutación** *m* ING ELÉC switching power supply; ~ **de energía conmutando a armazón abierto** *m* ING ELÉC open-frame switching power supply; ~ **de energía de continuidad absoluta** *m* TEC ESP no-break power supply; ~ **de energía a**

distancia *m* ELEC, ING ELÉC remote power supply;
~ de energía eléctrica *m* ELEC, ELECTRÓN, ING ELÉC,
NUCL, TELECOM electrical-power supply; **~ de energía electrónica** *m* ELECTRÓN, ING ELÉC electronic
power supply; **~ de energía con estabilizador de
tensión** *m* ELEC voltage-stabilized power supply;
~ de energía con estabilizador de voltaje *m* ELEC
voltage-stabilized power supply; **~ de energía
ininterrumpida** *m* TEC ESP no-break power supply;
~ de energía lineal a bastidor abierto *m* ING ELÉC
open-frame linear power supply; **~ de energía
nuclear** *m* TEC ESP nuclear power supply; **~ de
energía positiva** *m* ING ELÉC positive-power supply;
~ de energía regulado *m* ING ELÉC regulated power
supply; **~ de energía regulado por RCS** *m* ING ELÉC
SCR-regulated power supply; **~ de energía regulado en serie** *m* ING ELÉC series-regulated power
supply; **~ de energía de reserva** *m* ING ELÉC backup
power supply; **~ de energía de retroalimentación** *m*
ING ELÉC kickback power supply; **~ de energía sin
transformador** *m* ELEC *red de distribución* transfor-
merless power supply; **~ flotante** *m* ELEC *red* floating
supply; **~ de fuerza de tierra** *m* TRANSP AÉR ground
power supply; **~ monofásico** *m* ELEC, ING ELÉC
single-phase supply; **~ de oxígeno** *m* TEC ESP oxygen
supply; **~ de potencia** *m* FÍS power supply; **~ de
potencia íntegra** *m* PROD integral power supply;
~ de potencia en modo de conmutación *m* PROD
switch-mode power supply; **~ de presión** *m* INSTAL
HIDRÁUL pressure delivery; **~ de presión hidráulica**
m INSTAL HIDRÁUL hydraulic pressure supply;
~ principal *m* TV current supply (*AmE*), mains
supply (*BrE*); **~ a la red** *m* ELEC, NUCL *de energía
eléctrica* delivery into the mains (*BrE*), delivery into
the utility network (*AmE*); **~ de la red** *m* ING ELÉC
current supply (*AmE*), mains supply (*BrE*); **~ de
reserva** *m* ELEC *red de distribución* stand-by supply;
~ de tensión muy elevada *m* TV EHT supply; **~ de
tensión negativa** *m* ING ELÉC negative-voltage
supply; **~ trifásico** *m* ING ELÉC three-phase supply;
~ único *m* ING ELÉC single supply; **~ de verdeo** *m*
AGRIC soiling; **~ de voltaje negativo** *m* ING ELÉC
negative-voltage supply
super: **~ sincrotrón de protones** *m* FÍS PART proton
supersynchrotron, superproton synchrotron (*SPS*);
~~transportador de crudo *m* TEC PETR ultralarge
crude carrier (*ULCC*), very large crude carrier
(*VLCC*)
superabundancia *f* ING MECÁ overflow
supercadenas *f pl* FÍS superstrings
supercalandra *f* PAPEL supercalender; **~ de bastidor
abierto** *f* PAPEL open-frame super calender
supercalandrado *m* PAPEL supercalenderizing
supercalentamiento *m* FÍS superheating
supercarbonato *m* QUÍMICA supercarbonate
supercomputación *f* AmL (*cf superinformática Esp*)
INFORM&PD supercomputing
supercomputadora *m* AmL (*cf superordenador Esp*)
INFORM&PD supercomputer
superconductividad *f* ELEC *conductores*, ELECTRÓN,
FÍS, INFORM&PD superconductivity; **~ a altas
temperaturas** *f* FÍS high-temperature superconductivity
superconductor *m* ELEC, ELECTRÓN, FÍS, TELECOM
superconductor; **~ débil** *m* ELEC, ELECTRÓN, FÍS soft
superconductor

supercrítico *adj* NUCL supercritical
superefecto: **~ doble** *m* TV double supereffect
superenfriado *adj* FÍS supercooled
superestructura *f* CONST, CRISTAL, TRANSP MAR *construcción naval* superstructure
superficial *adj* MINAS surface
superficie *f* CARBÓN surface, CONST area, *edificio*
floorspace, CONTAM *de cuerpos de agua, de tierra*
surface area, FÍS, GEOM surface, MATEMÁT area,
MINAS *de una parte especificada* content, TRANSP
AÉR skin; **~ de activación** *f* METAL activation area;
~ de adhesión *f* P&C adherend; **~ aerodinámica** *f*
TRANSP AÉR aerofoil (*BrE*), airfoil (*AmE*); **~ aerodinámica de control** *f* TEC ESP aerodynamic control
surface; **~ aleteada** *f* TERMOTEC *intercambiadores de
calor* finned surface; **~ de aplicación** *f* REVEST use-
surface; **~ de apoyo** *f* ING MECÁ bearing surface,
MECÁ, PROD seat; **~ de asiento** *f* CONST bedding
surface; **~ áspera** *f* ING MECÁ rough surface; **~ de
aterrizaje de helicópteros** *f* TRANSP AÉR helicopter
landing surface; **~ baja** *f* TRANSP AÉR lower surface;
~ de caldeo *f* TERMO, TERMOTEC heating surface;
~ de caldeo real *f* TERMO, TERMOTEC effective
heating-surface; **~ de caldeo del recalentador** *f*
PROD superheater heating surface; **~ de calefacción**
f TERMO heating surface, TERMOTEC heating surface;
~ calefactora *f* INSTAL TERM, TERMO, TERMOTEC
heating surface; **~ cáustica** *f* FÍS, QUÍMICA caustic
surface; **~ cóncava** *f* GEOM concave surface;
~ cónica *f* GEOM conical surface; **~ de contacto** *f*
ING MECÁ interface, surface of contact, MECÁ seat,
METAL, PETROL interface; **~ de contacto con el agua
de mar** *f* HIDROL seawater interface; **~ de contacto
con el agua salada** *f* HIDROL seawater interface; **~ de
contacto de chapado del núcleo** *f* ÓPT, TELECOM
core-cladding interface; **~ de contacto de la válvula**
f AUTO valve-mating surface; **~ continua** *f* FERRO
continuous surface; **~ de control** *f* FÍS, TRANSP AÉR
control surface; **~ de control equilibrada** *f* TRANSP
AÉR balanced control surface; **~ convexa** *f* C&V,
GEOM, ÓPT convex surface; **~ desgastada** *f* TEXTIL
wearing surface; **~ de desgaste** *f* ING MECÁ wearing
surface; **~ de deslizamiento** *f* CARBÓN slip surface;
~ equifásica *f* ING ELÉC equiphase surface;
~ equipotencial *f* ELEC, FÍS, GEOFÍS, ING ELÉC, TEC
ESP equipotential surface; **~ equivalente** *f* TEC ESP
integración matemática equivalent area; **~ específica**
f INSTR special surface, P&C *propiedad del pigmento*
specific surface area; **~ externa** *f* EMB exterior surface; **~ externa de contacto** *f* TEC ESP *mecánica*
outer contact area, *vehículos espaciales* external
interface; **~ de Fermi** *f* FÍS Fermi surface; **~ de
filtrado** *f* CARBÓN screening surface; **~ fría** *f* C&V
cold surface; **~ de frotamiento** *f* PROD rubbing
surface; **~ geométrica** *f* GEOM geometric surface;
~ de grabación *f* INFORM&PD recording surface;
~ hidrodinámica *f* TRANSP AÉR hydrofoil; **~ de
imposición** *f* IMPR imposing surface, slab;
~ inferior *f* FÍS lower surface; **~ inferior de la pala** *f*
TRANSP AÉR blade lower surface; **~ isócrona** *f* GEOL
isochronal surface; **~ de lente negativa** *f* INSTR, ÓPT
negative-lens surface; **~ de lente positiva** *f* INSTR
positive-lens surface; **~ libre** *f* FÍS free surface, MINAS
voladuras free face, TRANSP MAR *diseño de barcos* free
surface; **~ limitante** *f* ACÚST, ENERG RENOV, FÍS, FÍS
FLUID, MECÁ, METAL, OCEAN, REFRIG, REVEST,

TRANSP AÉR boundary layer; ~ **lisa** *f* ING MECÁ plain surface; ~ **magnética** *f* ING MECÁ magnetic face; ~ **maquinada** *f* MECÁ machined surface; ~ **mínima** *f* GEOM minimal surface; ~ **mínima helocoidal** *f* GEOM helicoid minimal surface; ~ **mojada** *f* TRANSP MAR *diseño de barcos* wetted surface; ~ **neta** *f* ENERG RENOV *de colector solar* net area; ~ **de onda** *f* ACÚST, ING ELÉC wave surface; ~ **ondulada** *f* FERRO *vía* running surface; ~ **ópticamente lisa** *f* FÍS optically-smooth surface; ~ **del piso** *f* EMB floor space; ~ **del pistón** *f* ING MECÁ piston surface; ~ **plan de granito** *f* ING MECÁ granite surface plate; ~ **plana** *f* ING MECÁ plane, INSTR flat surface, PROD *especificaciones de la máquina* plain surface; ~ **plana entre acanaladuras** *f* P&C *parte de equipo* land; ~ **poco curtida** *f* IMPR hungry surface; ~ **polar** *f* TV pole face; ~ **portante** *f* ING MECÁ bearing surface; ~ **posterior** *f* C&V back surface; ~ **de precipitación** *f* PROC QUÍ precipitation area; ~ **primitiva de rodadura** *f* ING MECÁ pitch surface; ~ **de referencia** *f* ÓPT, TELECOM reference surface; ~ **de registro** *f* INFORM&PD recording surface; ~ **de revolución** *f* GEOM surface of revolution; ~ **de rodadura** *f* CONST tread, FERRO *rueda de vehículo* running surface; ~ **de rodadura de la cubierta** *f* AmL P&C *caucho* tire tread (*AmE*), tyre tread (*BrE*); ~ **de rodadura del neumático** *f* P&C *caucho* tire tread (*AmE*), tyre tread (*BrE*); ~ **de rodamiento** *f* ING MECÁ wearing surface; ~ **de rodamiento del neumático** *f* TRANSP AÉR tread; ~ **de rozamiento** *f* ING MECÁ bearing surface, PROD rubbing surface; ~ **rugosa** *f* ING MECÁ rough surface, P&C *defecto de la pintura* orange peel; ~ **selectiva** *f* ENERG RENOV selective surface; ~ **de sellado** *f* C&V sealing surface; ~ **de separación** *f* ING MECÁ interface; ~ **de separación de agua dulce** *f* HIDROL freshwater interface; ~ **de separación entre el núcleo y el revestimiento** *f* ÓPT, TELECOM corecladding interface; ~ **superior** *f* FÍS *de viento*, TRANSP AÉR upper surface; ~ **superior de la pala** *f* TRANSP AÉR blade upper surface; ~ **sustentadora** *f* ENERG RENOV, TRANSP AÉR, TRANSP MAR *diseño de barcos* aerofoil (*BrE*), airfoil (*AmE*); ~ **sustentadora aerodinámica** *f* TRANSP MAR *diseño de barcos* aerofoil (*BrE*), airfoil (*AmE*); ~ **de termotransferencia** *f* TERMO heat transfer surface; ~ **terrestre** *f* TEC ESP terrestrial surface; ~ **total** *f* CONST total area; ~ **de trabajo** *f* PROD work surface; ~ **de trabajo de la mesa** *f* ING MECÁ table work surface; ~ **translúcida** *f* ÓPT translucent substance; ~ **de transmisión de calor** *f* REFRIG, TERMO heat transfer surface; ~ **transparente** *f* ÓPT transparent substance; ~ **útil** *f* CONST floorspace occupied, ING MECÁ useful surface
superficies: ~ **de acoplamiento** *f pl* ING MECÁ mating surfaces; ~ **coincidentes** *f pl* ING MECÁ mating surfaces; ~ **de medir interiores** *f pl* ING MECÁ inside-measuring faces
superfluidez *f* FÍS superfluidity
superfluido *m* FÍS superfluid
superfosfato *m* QUÍMICA superphosphate; ~ **de cal** *m* QUÍMICA *abono* superphosphate of lime
superfrío *m* FÍS supercooling
supergás *m* GEN liquid petroleum gas (*LPG*), liquefied petroleum gas (*LPG*)
supergrupo *m* TELECOM supergroup
superimpulso: ~ **de borrado** *m* TV super blanking pulse

superíndice *m* INFORM&PD superscript
superinformática *f* Esp (*cf supercomputación AmL*) INFORM&PD supercomputing
superintendente: ~ **de perforación** *m* PETROL drilling superintendent
superior *adj* GEOL *edad* late
superluminiscencia *f* ÓPT, TELECOM superluminescence
supermarcha *f* AUTO, VEH *caja de cambio* overdrive
supermini *m* INFORM&PD supermini
supernova *f* TEC ESP supernova
superordenador *m* Esp (*cf supercomputadora AmL*) INFORM&PD supercomputer
superoxigenar *vt* QUÍMICA superoxygenate
superpetrolero *m* TEC PETR *transporte* very large crude carrier (*VLCC*)), TRANSP supertanker
superplastificante *m* CONST *aditivo de hormigón* superplasticizer
superponer *vt* CINEMAT super, PROD fold over, TV overlay
superposición *f* CRISTAL stacking, FÍS overlapping, GEOL *geocronología estructural* overprinting, overplacement, IMPR overlap, overstrike, INFORM&PD overlap, overlay, ING ELÉC, METAL overlapping, MINAS *geología* overplacement; ~ **envuelta** *f* IMPR wrapped overlay; ~ **mecánica** *f* IMPR mechanical overlay; ~ **de renglones** *f* ING MECÁ overprinting; ~ **única** *f* ING MECÁ single overlap
superradiancia *f* ÓPT, TELECOM superradiance
superred *f* CRISTAL superlattice
superrefracción *f* TEC ESP super-refraction
supersal *f* QUÍMICA supersalt
supersónica *f* FÍS ONDAS ultrasonics
supersónico *adj* ACÚST supersonic, FÍS transonic
supertipo *m* IMPR supertype
supertónica *f* ACÚST *música* supertonic
supervisar *vt* INFORM&PD, PROD monitor
supervisión[1] *f* CALIDAD surveillance; ~ **de la calidad del agua** *f* AGUA water quality monitoring; ~ **de mensaje de llamada** *f* TELECOM call message supervisory (*CMS*); ~ **rigurosa** *f* SEG close supervision; ~ **del transcurso del intervalo de retardo** *f* TELECOM time-out supervision
supervisión[2]: ~ **y mantenimiento** *fra* TELECOM monitoring and maintenance
supervisor *m* CONST surveyor, CONTAM MAR, INFORM&PD supervisor, MINAS deputy, SEG supervisor; ~ **activo** *m* TELECOM *transmisión por paquetes* active supervisor; ~ **de la red** *m* TELECOM network supervisor; ~ **de reserva** *m* TELECOM backup supervisor; ~ **de turno** *m* NUCL shift supervisor; ~ **de unidad** *m* PROD unit supervisor
suplementario *adj* ING MECÁ additional
suplemento *m* IMPR addendum, back matter; ~ **nutritivo** *m* ALIMENT nutritional supplement; ~ **de perfil** *m* ING MECÁ form shim
supletorio *m* TELECOM extension
supraconductividad *f* ELEC superconductivity
supraconductor *m* ELEC superconductor
supresión[1]: **con** ~ **de corrientes momentáneas** *adv* PROD transient-suppressed
supresión[2] *f* CONST removal, IMPR deletion, INFORM&PD deletion, suppression, ING ELÉC suppression, TELECOM *radio* blackout; ~ **del arco** *f* ING ELÉC arc suppression; ~ **armónica** *f* ELECTRÓN harmonic rejection; ~ **de ceros** *f* INFORM&PD zero suppression;

~ **de la chispa** *f* ING ELÉC spark suppression; ~ **del color** *f* TV color kill (*AmE*), colour kill (*BrE*); ~ **del eco** *f* ELECTRÓN, INFORM&PD, TEC ESP, TELECOM echo suppression; ~ **del empuje** *f* TEC ESP thrust cutoff; ~ **de FI** *f* ELECTRÓN IF rejection; ~ **de frecuencia** *f* ELECTRÓN frequency rejection; ~ **de frecuencia intermedia** *f* ELECTRÓN intermediate frequency rejection, TELECOM intermediate frequency rejection, intermediate frequency stage; ~ **del haz** *f* ELECTRÓN beam blanking, TV beam blanking, gating; ~ **de la imagen** *f* TV horizontal blanking; ~ **de interfase** *f* PROD interface suppression; ~ **de las interferencias** *f* TRANSP MAR *radio, radar* interference rejection; ~ **de línea** *f* ELECTRÓN trace blanking, vertical blanking, TV line blanking; ~ **del modo común** *f* ELECTRÓN common-mode rejection; ~ **de la onda portadora** *f* ELECTRÓN, TELECOM carrier suppression; ~ **de la portadora** *f* ELECTRÓN carrier suppression; ~ **de ruido** *f* SEG noise abatement, TELECOM squelch; ~ **de transitorios** *f* ING ELÉC transient suppression

supresor *m* ELECTRÓN rejector, suppressor, ING ELÉC arrester, suppressor, PROD, TELECOM suppressor; ~ **de armónicos** *m* ELECTRÓN, FÍS RAD harmonic suppressor; ~ **de campo** *m* ELEC, ING ELÉC field suppressor; ~ **de la chispa** *m* ING ELÉC spark suppressor; ~ **del circuito** *m* ING ELÉC channel stopper; ~ **de color** *m* TV color kill (*AmE*), colour kill (*BrE*); ~ **del distribuidor** *m* VEH *sistema de encendido* distributor suppressor; ~ **de ecos** *m* ELECTRÓN, INFORM&PD, TEC ESP, TELECOM echo-suppressor; ~ **de modos en el revestimiento** *m* ÓPT *fibras ópticas* cladding mode stripper; ~ **de ruido del reactor** *m* TRANSP AÉR jet noise suppressor; ~ **de transitorios** *m* ING ELÉC transient suppressor

suprimir *vt* IMPR kill, delete, INFORM&PD delete, suppress

Sur: ~ **magnético** *m* GEOFÍS magnetic south

surco *m* ACÚST groove, AGRIC drill, furrow, AGUA furrow, CONST groove, GEOL furrow, MECÁ groove, ÓPT *de disco fonográfico* groove, track; ~ **del arrecife** *m* GEOL, OCEAN reef furrow; ~ **en blanco** *m* ACÚST blank groove; ~ **de cables** *m* ELEC, ING ELÉC *fuente de alimentación* cable trench; ~ **final** *m* ACÚST *disco gramófono* lead-out groove; ~ **finalizador** *m* ACÚST *disco gramófono* finishing groove; ~ **grueso** *m* ACÚST coarse groove; ~ **inicial** *m* ACÚST *disco gramófono* lead-in groove; ~ **intermedio** *m* ACÚST *disco gramófono* lead-over groove; ~ **modulado** *m* ACÚST modulated groove; ~ **muerto** *m* AGRIC dead furrow; ~ **sublitoral** *m* OCEAN offshore trough

surfactante *m* CARBÓN surfactant, CONTAM, DETERG, QUÍMICA surface-active agent, surfactant; ~ **aniónico** *m* DETERG, QUÍMICA anionic surface-active agent

surfactivo *m* QUÍMICA surfactant

surfear *vt* INFORM&PD surf

surfing *m* FÍS ONDAS surfing

surgencia *f* OCEAN upwelling, TEC PETR *perforación, producción* surge

surgente *adj* AGUA *pozo* artesian

surgidero *m* TRANSP MAR anchorage

surgir *vi* TRANSP MAR anchor

surtidor *m* AUTO jet, C&V spout, HIDROL spouting, LAB *análisis* dispenser; ~ **de aceleración** *m* AUTO acceleration jet; ~ **de alimentación** *m* VEH *carburador* accelerator jet; ~ **de arranque** *m* AUTO starter jet, starting jet; ~ **auxiliar** *m* VEH *carburador* auxiliary jet;

~ **compensador** *m* VEH *carburadores* compensating jet; ~ **de dosificación** *m* ING MECÁ metering jet; ~ **economizador** *m* VEH *carburador* economizer jet; ~ **de marcha lenta** *m* AUTO, VEH *carburador* idle jet; ~ **de mínimo** *m* VEH *carburador* idle jet; ~ **principal** *m* AUTO, VEH *del carburador* main jet; ~ **de ralentí** *m* AUTO idle jet; ~ **de regulación** *m* ING MECÁ metering jet; ~ **de rociado** *m* PROD *para ruedas abrasivas y similares* water can

surtir *vt* ING MECÁ fit out

surto: **estar** ~ **al ancla** *vi* TRANSP MAR ride at anchor

susceptancia *f* ELEC, FÍS, ING ELÉC susceptance

susceptibilidad *f* ELEC, FÍS susceptibility; ~ **diamagnética** *f* FÍS, FÍS RAD diamagnetic susceptibility; ~ **dieléctrica** *f* ELEC dielectric susceptib-il-ity; ~ **eléctrica** *f* ELEC, FÍS electric susceptibility; ~ **de errores** *f* TELECOM error susceptibility; ~ **a la helada** *f* CARBÓN frost susceptibility; ~ **magnética** *f* ELEC, FÍS, PETROL magnetic susceptibility

susceptible: ~ **de detonación** *adj* MINAS capable of detonation

suscriptor *m* INFORM&PD subscriber

suspendedor *m* MINAS suspension

suspender *vt* ING MECÁ hold up, MINAS *trabajo* cease, PROD hang

suspendido *adj* PROD pending

suspensión *f* FÍS, ING ELÉC suspension, ING MECÁ holding up, mount, knock-off, MINAS, P&C suspension, PAPEL slurry, PROD hanging, hanging-up, QUÍMICA suspension, TELECOM cease, VEH *motocicleta* antidive suspension; ~ **activa** *f* TRANSP active suspension; ~ **acuosa** *f* NUCL water slurry; ~ **acuosa espesa** *f* RECICL slurry; ~ **en arcilla** *f* AGUA clay suspension; ~ **articulada** *f* ING MECÁ hinged suspension; ~ **asimétrica** *f* TRANSP *monorrail* asymmetric suspension; ~ **bifilar** *f* ELEC *de instrumento*, FÍS bifilar suspension; ~ **a la cardán** *f* MECÁ gimbal; ~ **cardán** *f* ING MECÁ Cardan's suspension, TRANSP MAR gimbal mounting; ~ **cardánica** *f* ING MECÁ gimbal suspensión, TRANSP MAR gimbal mounting; ~ **de cargas** *f* PAPEL *cargas* slurry; ~ **sin contacto** *f* TRANSP non-contact suspension; ~ **delantera** *f* AUTO, VEH front suspension; ~ **delantera independiente** *f* AUTO, VEH independent front suspension; ~ **delantera suficiente** *f* VEH independent front suspension; ~ **de flujo nulo** *f* TRANSP *levitación de vehículos* null flux suspension; ~ **fluorescente** *f* ÓPT fluorescent substance; ~ **frontal de puntal MacPherson** *f* AUTO, VEH MacPherson strut front suspension; ~ **de la galería** *f* MINAS suspension of the rods; ~ **hidroelástica** *f* AUTO, VEH hydroelastic suspension; ~ **hidroneumática** *f* AUTO, VEH hydropneumatic suspension; ~ **de horquilla doble** *f* AUTO double wishbone suspension; ~ **independiente de las ruedas delanteras** *f* AUTO, VEH independent front suspension; ~ **independiente de las ruedas traseras** *f* AUTO, VEH independent rear suspension; ~ **con juntas** *f* ING MECÁ suspension with linkages; ~ **magnética** *f* TRANSP magnetic suspension; ~ **neumática** *f* TRANSP *por aire comprimido* pneumatic suspension; ~ **opaca** *f* ÓPT opaque substance; ~ **pendular** *f* TRANSP *de automóvil* pendulum suspension; ~ **de pigmentos** *f* PAPEL slurry; ~ **de procesos** *f* INFORM&PD process suspension; ~ **de la respiración** *f* OCEAN apnea (*AmE*), apnoea (*BrE*); ~ **secundaria** *f* TRANSP secondary suspension;

~ **súbita** *f* TEC ESP *comunicaciones* blowout; ~ **con superficie activa** *f* QUÍMICA surfactant; ~ **de tipo brazo paralelo** *f* AUTO parallel arm-type suspension, VEH parallel armtype suspension; ~ **de tipo brazo trapezoidal** *f* AUTO, VEH trapezoid arm-type suspension; ~ **trasera** *f* AUTO rear suspension, VEH rear suspensión, rear suspension; ~ **trasera independiente** *f* AUTO, VEH independent rear suspension; ~ **unifilar** *f* ELEC *instrumento* unifilar suspension; ~ **al vacío** *f* TRANSP vacuum suspension; ~ **de vuelos** *f* TRANSP AÉR grounding

suspenso: en ~ *adj* PROD held-up

sustancia: ~ **adhesiva** *f* IMPR bond, P&C bonding agent; ~ **antioxidante** *f* AUTO, MECÁ, TRANSP MAR, VEH rust inhibitor; ~ **biodegradable** *f* CONTAM biodegradable substance; ~ **colorante** *f* COLOR coloring substance (*AmE*), colouring substance (*BrE*); ~ **conservante** *f* AGRIC food preservative; ~ **fluorescente** *f* ELEC, FÍS, FÍS RAD fluorescent substance; ~ **fotoendurecible** *f* COLOR photoresistant pigment; ~ **opaca** *f* FÍS RAD opaque substance; ~ **productora de gas** *f* P&C *polímeros celulares* blowing-agent; ~ **retardante a las llamas** *f* P&C *pintura, plásticos, caucho*, REVEST, TERMO fireproof coating; ~ **con superficie activa** *f* FÍS surfactant; ~ **taponadora** *f* QUÍMICA sealant; ~ **translúcida** *f* FÍS RAD translucent substance; ~ **transparente** *f* FÍS RAD transparent substance; ~ **trazadora** *f* QUÍMICA tracer substance

sustentación *f* ENERG RENOV lift, FÍS FLUID *líquidos* uplift, ING MECÁ holding up, MINAS upright, TRANSP AÉR lift; ~ **del avión** *f* TRANSP AÉR aircraft lift; ~ **máxima** *f* TRANSP AÉR maximum lift; ~ **de la pala** *f* TRANSP AÉR blade lift; ~ **total** *f* TRANSP total lift

sustentáculo *m* MINAS rest

sustentadora: ~ **de la cinta** *f* TV tape lifter

sustitución: ~ **del componente gris** *f* IMPR gray component replacement (*AmE*), grey component replacement (*BrE*); ~ **de diálogo** *f* CINEMAT dialog replacement (*AmE*), dialogue replacement (*BrE*); ~ **de parámetros** *f* INFORM&PD parameter substitution

sustracción: ~ **binaria** *f* ELECTRÓN, INFORM & PD binary subtraction

sustractor: ~ **todo completo** *m* ELECTRÓN, INFORM&PD full subtractor

sustrato *m* FÍS, INFORM&PD, METAL, P&C, TELECOM substrate; ~ **de Ekman** *m* OCEAN Ekman layer

Sv *abr* (*sievert*) FÍS, FÍS RAD, NUCL Sv (*sievert*)

swivel *m Esp* TEC PETR *perforación* swivel

T

T¹ *abr* FÍS (*tesla*) T (*tesla*), METR (*tera, tesla*) T (*tesla, tera*)

T²: **en ~** *adj* GEOM, ING MECÁ T-shaped, tee-shaped

Ta *abr* ING ELÉC, METAL, QUÍMICA (*tantalio*) Ta (*tantalum*)

tabergita *f* MINERAL tabergite

tabique *m* ING MECÁ parting, MINAS midwall; **~ cortafuego** *m* TEC ESP fireproof bulkhead; **~ cortafuegos** *m* TERMO fire-resistant bulkhead; **~ divisorio** *m* CONST partition, MECÁ, TEC ESP bulkhead; **~ de ladrillo** *m* CONST noggin (*BrE*), nogging; **~ de lona** *m* Esp (*cf cañamazo AmL*) CONST canvas brattice; **~ de madera** *m* CONST plank partition; **~ para materias flotantes** *m* HIDROL aguas negras scum baffle; **~ con postes** *m* CONST stud partition; **~ de ventilación** *m AmL* (*cf compartimiento de ventilación Esp*) MINAS air brattice, stoping, brattice

tabiquería *f* CONST albañilería walling, arquitectura partitioning; **~ autoportante** *f* CONST edificaciones self-supporting partition

tabla *f* CARBÓN *diamante* table, CONST *madera* board, shelf, slab, strip, INFORM&PD table, ING MECÁ barrel, TRANSP MAR *construcción naval* plank; **~ de actividad del agua** *f* ALIMENT water table; **~ automática** *f* INFORM&PD autochart; **~ de avances** *f* ING MECÁ feed chart; **~ de barras** *f AmL* (*cf diagrama de barras Esp*) INFORM&PD bar chart; **~ de la bovedilla** *f* CONST counter lathe; **~ del canalón del alero** *f* CONST eaves board; **~ de cargas de capacidad bruta** *f* PROD rough-capacity load table; **~ de categoría de canales** *f* INFORM&PD channel status table; **~ de código de fallas** *f* PROD fault code chart; **~ de código de identificación de fallas** *f* PROD fault-identification code chart; **~ con códigos de comprobación** *f* INFORM&PD hash table; **~ de consulta** *f* INFORM&PD look-up table; **~ de contingencia** *f* MATEMÁT contingency table; **~ de datos** *f* INFORM&PD data tablet, PROD data table; **~ de datos inflados** *f* PROD expanded-data table; **~ de decisión** *f* INFORM&PD decision table; **~ de decisión lógica** *f* INFORM&PD truth table; **~ delantal** *f* CONST skirt board; **~ de entibación del techo** *f* MINAS headboard; **~ estadística** *f* CALIDAD statistical table; **~ de estado de canal** *f* Esp INFORM&PD channel status table; **~ de exposiciones** *f* FOTO exposure-calculating chart; **~ de forro** *f AmL* (*cf revestimiento Esp*) MINAS lagging piece; **~ de frontis** *f* CONST fascia board; **~ de grátil** *f* TRANSP MAR *navegación* headboard; **~ de imagen de salida** *f* PROD output image table; **~ de imágenes** *f* PROD image table; **~ de imágenes de entrada** *f* PROD input image table; **~ indicadora** *f* MECÁ index table; **~ de localización de averías** *f* ING MECÁ fault-finding table; **~ de localización de fallos** *f* ING MECÁ fault-finding table; **~ machihembrada** *f* CONST match-board; **~ de machos** *f* PROD *funderías* core board; **~ de mareas** *f* OCEAN, TRANSP MAR *navegación* tide table; **~ de materias** *f* MINAS contents; **~ de mortalidad** *f* MATEMÁT life table; **~ oficial de horarios** *f* FERRO official timetable; **~ de operaciones** *f* INFORM&PD operation table; **~ de páginas** *f* INFORM&PD page table; **~ de partículas** *f* MECÁ particle board; **~ periódica** *f* QUÍMICA periodic table; **~ de piso** *f* TEC PETR *perforación* footboard (*AmE*), monkey board (*BrE*); **~ para piso de madera** *f* CONST batten; **~ de quitapón** *f* AGUA flashboard; **~ de ripia** *f* CONST shake; **~ senoidal** *f* MECÁ sine table; **~ de símbolos** *f* INFORM&PD symbol table; **~ de tiro** *f* D&A *de artillería* range table; **~ de velas** *f* TRANSP MAR sailboard; **~ de verdad** *f* INFORM&PD truth table

tablaestaca *f* TRANSP MAR *duques de alba* pile

tablas: **~ de cosenos** *f pl* GEOM cosine tables; **~ de descompresión** *f pl* PETROL decompression tables

tablazón *m* CARBÓN planking

tablero *m* CONST floor, board, decking, *puente* cover, IMPR board, PROD *cintas transportadoras* apron, *mesa* basin, TEXTIL panel; **~ de acceso** *m* TEC ESP access panel; **~ acústico** *m* SEG acoustic board; **~ de aglomerado** *m* CONST, EMB hardboard; **~ aislado** *m* ELEC insulating board; **~ de alimentación** *m* IMPR feedboard; **~ para ampliaciones** *m* FOTO baseboard, enlarger baseboard; **~ de anuncios** *m* INFORM&PD bulletin board; **~ AZERTY** *m* INFORM&PD *teclado francés* AZERTY keyboard; **~ de bornas de potencia** *m* PROD power terminal bloc; **~ de bornes** *m* ELEC *conexión*, ING ELÉC terminal block; **~ de calibración del balancín** *m* METR beam caliper gage (*AmE*), beam calliper gauge (*BrE*); **~ de cierre** *m* ING MECÁ closing panel; **~ de circuito impreso** *m* (*PCI, tablero de CI*) ELEC, ELECTRÓN, INFORM&PD, TELECOM, TV printed-circuit board (*PC board, PCB*); **~ de conexiones** *m* INFORM&PD plugboard; **~ de conmutación** *m* TELECOM switchboard; **~ de conmutación con clavija y cordón** *m* TELECOM plug-and-cord switchboard; **~ conmutador de la centralita privada** *m* TELECOM PBX switchboard; **~ de contacto auxiliar** *m* PROD auxiliary contact deck; **~ de control** *m* CONST control panel, ELEC control board, INFORM&PD control panel, ING ELÉC pin-board, MECÁ, TEC ESP, TRANSP AÉR control panel; **~ de control del mecánico de vuelo** *m* TRANSP AÉR flight-engineer's panel; **~ de control principal** *m* Esp (*cf panel del control maestro AmL*) TV master control panel; **~ de control sonoro** *m* CINEMAT sound control desk; **~ para copias y reproducciones** *m* FOTO copy stand, copying stand; **~ de crujía** *m* TRANSP MAR *construcción naval* king plank; **~ de descarga de la barra colectora** *m* PROD bus-discharge board; **~ de dibujo** *m* CONST, MECÁ, TRANSP MAR *arquitectura naval* drawing board; **~ de distribución** *m* ELEC, VEH *sistema de encendido* distributing board; **~ de distribución sinóptica** *m* ELEC synoptical switchboard; **~ de distribución telefónica** *m* TELECOM telephone switchboard; **~ experimental para componentes electrónicos** *m* Esp (*cf tablero inicial para componentes electrónicos AmL*) INFORM&PD breadboard; **~ de gobierno de puertas** *m* PROD gate-drive board, gate-driving board; ~ **gráfico** *m* TV graphic tablet; ~ **inclinable** *m* FOTO tilting baseboard; ~ **a inglete** *m* CONST miter board (*AmE*), mitre board (*BrE*); ~ **inicial para componentes electrónicos** *m AmL* (*cf tablero experimental para componentes electrónicos Esp*) INFORM&PD breadboard; ~ **de instrumentos** *m* AUTO instrument panel, dashboard, PROD, TRANSP AÉR, TRANSP MAR instrument panel, VEH dashboard; ~ **de instrumentos del mecánico de vuelo** *m* TRANSP AÉR flight-engineer's panel; ~ **lógico de modulador** *m* PROD modulator logic board; ~ **de mandos** *m* TRANSP AÉR console; ~ **marcador** *m* IMPR feedboard; ~ **de mira** *m* CONST sighting board; ~ **multiuso** *m* ELECTRÓN general-purpose board; ~ **de partículas** *m* PAPEL particle board; ~ **de pasta mecánica** *m* EMB mechanical-pulp board; ~ **de progresión de vuelo** *m* TRANSP AÉR flight-progress board; ~ **de pruebas** *m* ELEC test board; ~ **terminal** *m* ELEC *conexión* terminal block; ~ **de trazado** *m* INFORM&PD plotting board

tablestaca *f* CARBÓN *construcción* sheet pile, CONST pile plank, sheeting pile, sheet pile, MINAS lagging piece; ~ **de acero** *f* CARBÓN steel pile; ~ **de cohesión** *f* CARBÓN, MINAS cohesion pile

tablestacado *m* CONST sheet piling

tableta *f* INFORM&PD tablet; ~ **de datos** *f* INFORM&PD data tablet; ~ **de digitalización** *f* INFORM&PD digitizing tablet; ~ **gráfica** *f* INFORM&PD graphics tablet; ~ **Rand** *f* INFORM&PD Rand tablet

tableteo *m* *AmL* (*cf trinqueteo Esp*) NUCL ratchetting

tablilla *f* CONST vane, *topografía* target, PROD board

tablita *f* CONST *madera* cleat

tablón *m* CONST plank, strickle board, *madera* deal, TRANSP MAR *construcción naval* plank; ~ **americano y canadiense** *m* CONST *madera* deal; ~ **del andamio** *m* CONST scaffold board; ~ **de anuncios** *m* CONST board, SEG general warning panel; ~ **de cierre** *m* AGUA stop plank; ~ **de etiquetas** *m* TEXTIL tagboard; ~ **de pie** *m* CONST *andamios* toe board

tabulación *f* INFORM&PD tabulation (*tab*); ~ **horizontal** *f* INFORM&PD horizontal parity, horizontal tabulation (*HT*); ~ **vertical** *f* INFORM&PD vertical tabulation (*VT*)

tabulador *m* INFORM&PD tabulator (*tab*); ~ **electrónico** *m* ELEC *mediciones*, INSTR logger; ~ **electrónico de datos** *m* INSTR data logger

tabular[1] *adj* INFORM&PD tabular

tabular[2] *vt* CARBÓN table, INFORM&PD tabulate (*tab*)

taburete *m* ING MECÁ seat; ~ **de laboratorio** *m* LAB *mobiliario* laboratory stool

tacha *f* ING MECÁ clout nail, MECÁ flaw

tacho *m* CARBÓN *evaporador de vacío en la fabricación de azúcar de caña* pan

tachuela *f* ING MECÁ tack, stud; ~ **de estaño** *f* ING MECÁ tin tack

taco *m* CONST plug, ING MECÁ heel, MINAS *jaula* prop, PROD *galga, bloque de frenado* chock; ~ **abrasivo** *m* ING MECÁ honing stone; ~ **abrasivo de cobre** *m* PROD copper lap; ~ **abrasivo de plomo** *m* PROD lead lap; ~ **autoavanzante** *m* *Esp* (*cf entibación auto-avanzante AmL*) MINAS self-advancing chock; ~ **elástico** *m* VEH *suspensión* damper; ~ **de goma** *m* AUTO, P&C rubber pad; ~ **limpiador** *m* PETROL pig; ~ **de limpiar** *m* PROD *pozos petrolíferos* pig; ~ **de madera que se inserta en la hendidura abierta por la rozadura** *m* MINAS holing prop; ~ **de plomo** *m*

PROD *para lapidar* lead lap; ~ **de presión** *m* AUTO, CINEMAT, FOTO pressure pad; ~ **para roca** CONST rock dowel; ~ **del sector de la excéntrica** *m* ING MECÁ link block; ~ **de suspensión del motor** *m* AUTO, VEH engine support lug

tacogenerador *m* ING MECÁ tachogenerator

tacógrafo *m* TRANSP tachograph

tacómetro *m* AUTO, CINEMAT tachometer, FÍS *medidor de rapidez* tachometer, velocimeter, ING MECÁ revolution counter, motion indicator, INSTR rev counter, revolution counter, tachometer, TRANSP motion detector, TV tachometer, VEH rev counter, tachometer; ~ **de alta precisión** *m* ING MECÁ high-sensitivity tachometer; ~ **de alta sensibilidad** *m* ING MECÁ high-sensitivity tachometer; ~ **de alta sensitividad** *m* ING MECÁ high-sensitivity tachometer; ~ **manual** *m* EMB hand tachometer; ~ **de precisión** *m* ING MECÁ precision tachometer

tacos: ~ **de jaula de extracción** *m pl* MINAS cage shuts

táctico *adj* D&A *vuelo* tactical

tacto: ~ **áspero** *m* TEXTIL *proceso de acabado* harsh handle; ~ **crujiente** *m* TEXTIL crisp handle; ~ **duro** *m* TEXTIL *tejido* hard handle; ~ **firme** *m* TEXTIL *tejido* firm handle; ~ **lleno** *m* TEXTIL *tejidos* full handle; ~ **sedoso** *m* TEXTIL silk-like handle; ~ **suave** *m* TEXTIL soft handle; ~ **vivo** *m* TEXTIL *tejido* lively handle

TADG *abr* (*generador de datos triaxial*) TEC ESP TADG (*three-axis data generator*)

tafetán *m* TEXTIL taffeta

tafrogénesis *f* GEOL taphrogenesis

tajada *f* INFORM&PD bit slice

tajadera *f* PROD *forja* hot set

tajamar *m* AGUA breakwater, CONST nosing, breakwater, OCEAN *ingeniería portuaria*, TRANSP MAR breakwater; ~ **de salida** *m* AGUA downstream cutwater

taje: ~ **de avance** *m* *AmL* (*cf tajo de avance Esp*) MINAS drift stope

tajo *m* CONST cut; ~ **de arranque** *m* MINAS stope; ~ **de avance** *m* *Esp* (*cf taje de avance AmL*) MINAS drift stope

tajuelo *m* ING MECÁ footstep, footstep bearing

tal: ~ **como se requiere** *fra* PROD as required

tala *f* AGRIC felling

taladrable *adj* TEC PETR *perforación* drillable

taladracorchos *m* LAB *herramienta* cork borer

taladrado *m* CARBÓN drilling, boring, CONST *carpintería* boring, ING MECÁ boring, drilling, MINAS *de rocas o piedras*, PROD boring; ~ **libre** *m* CONST free-fall drilling; ~ **de pozos** *m* MINAS *pega* breaking-in hole; ~ **transversal** *m* IMPR cross perforation

taladrador *m* ING MECÁ piercer; ~ **con núcleo de diamante** *m* PROD diamond core drill

taladradora *f* CARBÓN rock drill, CONST drilling tool, rock drill, ING MECÁ, MECÁ boring-and-turning mill, boring machine, boring tool, drill, drilling machine, MINAS boring tool, boring machine, rock drill, TEC PETR *perforación* driller; ~ **para ángulos** *f* ING MECÁ drilling machine; ~ **de banco** *f* ING MECÁ bench drill, bench-drilling machine, MECÁ bench drill; ~ **de billetes** *f* TRANSP ticket punch; ~ **de columna** *f* ING MECÁ column-type drilling machine, drill press, pillar-drilling machine, MECÁ drill press; ~ **de columna de banco** *f* ING MECÁ bench-pillar drilling machine; ~ **dúplex** *f* ING MECÁ duplex boring machine; ~ **eléctrica a pilas** *f* ING MECÁ cordless power drill; ~ **de fricción de precisión** *f* ING MECÁ sensitive friction drill; ~ **de láser** *f* ELECTRÓN laser drill; ~ **de machos** *f* PROD core box-boring machine; ~ **mecánica** *f* MINAS machine drilling; ~ **múltiple** *f* MECÁ multiple drilling machine, PROD gang drill, multiple drilling machine; ~ **múltiple de precisión** *f* ING MECÁ sensitive gang drill; ~ **múltiple sensible** *f* ING MECÁ sensitive gang drill; ~ **a pilas** *f* ING MECÁ cordless power drill; ~ **de plantillas** *f* MECÁ jig borer; ~ **de plantillas de CN** *f* ING MECÁ *control numérico* NC jig borer; ~ **de pliegos para su costura** *f* IMPR stabbing machine; ~ **de polihusillos** *f* PROD multiple spindle drill; ~ **portátil** *f* ING MECÁ drilling jig, portable drilling-machine; ~ **en punta de flecha** *f* ING MECÁ arrowhead drill; ~ **radial** *f* MECÁ radial drilling-machine; ~ **radial de brazo giratorio** *f* ING MECÁ swing-jib radial drill; ~ **de tickets** *f* TRANSP ticket punch; ~ **de traviesas** *f* FERRO *infraestructura* sleeper drilling machine

taladradura *f* CARBÓN drilling

taladrar *vt* CARBÓN bore, tap, CONST bite, bore, punch, hole, drill, ING MECÁ, MECÁ bore, punch, drill, MINAS bore, drill, *salida de líquidos* tap, PROD punch

taladro *m* CARBÓN auger, rock drill, drill, boring, CONST *perforación, sondeo* auger, hole, borehole, brace, boring tool, rock drill, *carpintería* boring, ING MECÁ drill socket, boring, screw auger, punching out, MECÁ drill, brace, MINAS rock drill, *de rocas o piedras* boring, PETROL drilling platform, bit, PROD boring, *con punzón sacabocados* perforator, punching out, TEC PETR *perforación* drilling rig, drilling platform, rig, bit; ~ **de albañilería rotativo por percusión** *m* ING MECÁ rotary percussive masonry drill; ~ **de albañilería rotopercutiente** *m* ING MECÁ rotary percussive masonry drill; ~ **de albañilería** *m* ING MECÁ masonry drill; ~ **de billetes** *m* TRANSP ticket punch; ~ **de cabezales múltiples** *m* PROD gang drill; ~ **de columna** *m* ING MECÁ pillar drill; ~ **de desvío** *m* AUTO bypass bore; ~ **eléctrico** *m* CARBÓN electrodrilling, CONST, ELEC, MECÁ electric drill; ~ **giratorio** *m* MINAS churn drill; ~ **de hormigón** *m* ING MECÁ masonry drill; ~ **de mano** *m* ING MECÁ fiddle drill, piercer, MECÁ hand drill; ~ **mecánico** *m* CONST power drill, MINAS jumper drill, *sondeo* drill; ~ **mecánico eléctrico** *m* CONST, ELEC, MECÁ electric drill; ~ **múltiple** *m* ING MECÁ multidrill head; ~ **neumático** *m* CONST pneumatic drill; ~ **de pecho** *m* ING MECÁ breast drill; ~ **de percusión** *m* CONST hammer drill; ~ **por percusión** *m* CARBÓN percussion drilling, percussive drilling; ~ **de plantilla** *m* ING MECÁ jig boring, jig-boring machine; ~ **rápido** *m* MECÁ high-speed drill; ~ **de sondeo** *m* MINAS jumper

talar *vt* AGRIC fell

talco *m* MINERAL steatite

tálico *adj* QUÍMICA thallic

talio *m* (*Tl*) QUÍMICA thallium (*Tl*)

talioso *adj* QUÍMICA thallous

talla *f* AmL (*cf corte Esp*) CARBÓN *de gemas* cut, *de diamantes* cutting, MINAS *de diamantes* cutting, PROD hobbing, TEXTIL *dimensiones* size; ~ **basta** *f* ING MECÁ, PROD rough cut; ~ **del diamante** *f* MINAS diamond cut, PROD diamond cutting; ~ **de engranajes** *f* ING MECÁ, MECÁ gear cutting

tallado[1]: ~ **en la barra** *adj* PROD *pernos, tuercas y*

tornillos cut from bar; ~ **en el primordio** *adj* PROD *engranajes* cut from the solid banks

tallado[2]: ~ **en brillante** *m* C&V brilliant cutting; ~ **de engranajes** *m* ING MECÁ hobbing; ~ **de engranajes con fresa matriz** *m* ING MECÁ hob cutting; ~ **de fragmentos** *m* MINAS shatter cut; ~ **en el primordio** *m* PROD *engranajes* cut from the solid

tallador: ~ **de diamante** *m* PROD *persona* diamond cutter

tallar *vt* MECÁ cut

talle *m* TEXTIL waist

taller *m* C&V shop, FERRO *infraestructura* car repair shop (*AmE*), workshop (*BrE*), MECÁ shop, workshop, PROD shop floor, works, workshop, shop, SEG workshop; ~ **de ajuste** *m* ING MECÁ fitting shop; ~ **asistido** *m* TEC ESP manned workshop; ~ **de bocartes** *m* MINAS battery house; ~ **de calderería** *m* INSTAL HIDRÁUL *fábrica de calderas* boiler shop, boiler works, flue shop; ~ **de cambio de aceite** *m* AUTO oil change shop; ~ **de chapas finas** *m* PROD sheet-iron works; ~ **de concentración de fangos** *m* MINAS *minerales* jig mill; ~ **de concentraciones** *m* MINAS *minerales* dressing works; ~ **de conformado** *m* TRANSP MAR plater's shop; ~ **de construcción** *m* MECÁ erecting shop; ~ **de corte** *m* C&V cutting shop; ~ **de desarenado** *m* PROD *en fundarías* dressing shop; ~ **de desarenar** *m* PROD *fundición* cleaning shop; ~ **de desmontaje** *m* CONST dismantling chamber; ~ **de encuadernación** *m* IMPR bindery; ~ **espacial** *m* TEC ESP space workshop; ~ **de fabricación** *m* MECÁ fabricating shop; ~ **de forja** *m* ING MECÁ blacksmith's shop, PROD forge; ~ **de herramientas** *m* MECÁ toolroom; ~ **de laminación** *m* ING MECÁ rolling mill, PROD flatting works, *establecimiento* rolling mill; ~ **de laminación de chapas** *m* PROD plate-works; ~ **de láminas** *m* ING MECÁ, PROD rolling mill; ~ **de lavado** *m* PROD flow shop, wash house; ~ **de mantenimiento** *m* CONST maintenance shop; ~ **de modelos** *m* PROD *fundición* pattern shop; ~ **de moldeo** *m* PROD molding floor (*AmE*), moulding floor (*BrE*), molding shop (*AmE*), moulding shop (*BrE*); ~ **de moldeo en arena** *m* PROD *fundería* sand floor; ~ **de montaje** *m* MECÁ assembly shop, PROD erecting shop, TRANSP MAR *construcción naval* assembly hall, erecting shop; ~ **de motores** *m* AUTO engine shop; ~ **orbital** *m* TEC ESP orbital workshop; ~ **organizado por secciones homogéneas** *m* PROD job shop; ~ **de prefabricación** *m* TRANSP MAR *construcción naval* erecting shop; ~ **de preparación de arenas** *m* PROD *fundería* sand shop; ~ **de preparación mecánica** *m* MINAS mill, *minerales* dressing works; ~ **de producción de alambres** *m* ING MECÁ wire mill; ~ **con protección respiratoria** *m* SEG respiratory-protection workshop; ~ **de pruebas** *m* ING MECÁ test shop; ~ **de pulido** *m* C&V polishing shop; ~ **para la reparación de vagones** *m* FERRO *infraestructura* car repair shop (*AmE*), workshop (*BrE*); ~ **de reparaciones** *m* CONST maintenance shop, PROD repair shop; ~ **de revestimientos en caliente** *m* REVEST hot coating shop; ~ **de teñido** *m* COLOR, TEXTIL dyehouse; ~ **de verificación** *m* ING MECÁ testing shop

tallo: ~ **de los cereales** *m* AGRIC culm; ~ **del maíz** *m* AGRIC corn stalk (*AmE*), maize stalk (*BrE*)

tallón *m* C&V scrub mark (*BrE*)

talocha *f* CONST *herramientas* float

talón *m* C&V heel, CONST ogee, TRANSP MAR *de palo* heel, VEH *neumático* bead; **~ de cable** *m* ELEC, ING ELÉC cable lug; **~ de formularios con intercalación de papel carbón** *m* IMPR snap-out; **~ de la llanta** *m* VEH *rueda* rim flange

talonar: ~ la aguja *vi* FERRO *infraestructura* force open the points

talonario: ~ doble *m* IMPR double stub

talónico *adj* QUÍMICA talonic

talosa *f* QUÍMICA talose

talud *m* CARBÓN slope, CONST bank, battering, earthwork embankment, slope, GEOFÍS *fisiografía terrestre* scree, talus, GEOL scree, MINAS *minería a cielo abierto* slope; **~ continental** *m* GEOFÍS, GEOL, OCEAN, TEC PETR *geología* continental slope

taluzar *vt Esp* (*cf amontonar AmL*) CARBÓN pile, MINAS bank up

tamaño[1]**: de ~ gigante** *adj* PROD jumbo-size; **de ~ reducido** *adj* EMB space-saving

tamaño[2] *m* C&V, CARBÓN size, CONST bulk, CRISTAL size, EMB, FOTO bulk, IMPR format, size, INFORM&PD bulk, ING MECÁ format, size, MECÁ bulk, format, size, METAL, METR, OCEAN, P&C, PROC QUÍ, QUÍMICA format, size, REVEST size, TV format; **~ A4** *m* IMPR A4 size; **~ apaisado** *m* IMPR landscape size; **~ de archivo** *m* INFORM&PD file size; **~ del bloque** *m* INFORM&PD block size; **~ bruto** *m* IMPR *papel* untrimmed size; **~ de control** *m* CARBÓN control size; **~ sin cortar** *m* IMPR *papel* untrimmed size; **~ DIN** *m* IMPR DIN size; **~ final** *m* EMB trimmed size; **~ final cortado** *m* IMPR *papel* trim size; **~ de fotografía** *m* FOTO picture size; **~ del grano** *m* CARBÓN, CRISTAL, METAL grain size; **~ de grano útil** *m* CARBÓN effective grain size; **~ de letra** *m* IMPR body size; **~ límite** *m* MECÁ limit size; **~ mal clasificado** *m* CARBÓN misplaced size; **~ de la malla** *m* MECÁ, OCEAN mesh size; **~ de malla** *m* CARBÓN mesh size; **~ máximo y mínimo especificado y ajustes** *m* ING MECÁ limits and fits; **~ de muestra** *m* METR sample size; **~ natural** *m* ING MECÁ full scale; **~ nominal** *m* EMB, ING MECÁ nominal size; **~ óptimo de triturado** *m* CARBÓN optimal crushing size; **~ de palabra** *m* INFORM&PD word size; **~ del papel** *m* IMPR paper size; **~ de partición** *m* CARBÓN partition size; **~ de la partícula** *m* CARBÓN *metalurgia*, CRISTAL grain size, METAL particle size, grain size, NUCL, P&C, PROC QUÍ, QUÍMICA particle size; **~ de la retícula** *m* CARBÓN mesh size; **~ de separación** *m* PROC QUÍ separation size; **~ total** *m* METR total size

tamaños: ~ B *m pl* IMPR B sizes; **~ de cama** *m pl* TEXTIL bed sizes; **~ de composición** *m pl* IMPR composition sizes; **~ de corte estándar** *m pl* C&V cut sizes; **~ grandes** *m pl* CARBÓN, MINAS *mineral quebrantado* oversizes; **~ seleccionados** *m pl* ING MECÁ selected sizes; **~ de vidrio plano de línea** *m pl* C&V stock sheets (*AmE*), stock sizes (*BrE*)

tambalearse *v refl* TEC ESP wobble

tambaleo *m* ELECTRÓN wobbulation

tambo *m AmL* (*cf granja lechera*) AGRIC dairy farm

tambor *m* AGUA, AUTO, C&V tumbler, CARBÓN, CINEMAT drum, CONST, DETERG, ELEC roller, EMB, FOTO, GEOFÍS, IMPR, INFORM&PD, ING ELÉC drum, ING MECÁ barrel, shell, INSTAL HIDRÁUL can, drum, tin, MECÁ, MINAS barrel, cylinder, drum, OCEAN *buques tendedores de cables telefónicos submarinos* cable drum, net

drum, P&C *contenedor, recipiente*, PAPEL, PROC QUÍ drum, PROD drum, *draga de rosario* tumbler, TEC PETR drum, TEXTIL *cilindro de carda* swift, TRANSP barrel, TRANSP MAR capstan drum, *accesorios de cubierta* winch drum, TV, VEH drum; **~ de la cabeza** *m Esp* (*cf tambor porta cabeza AmL*) TV head drum; **~ de cable** *m* ELEC cable drum; **~ del cable** *m* CONST wire reel; **~ para cables** *m* ELEC cable drum; **~ de chigre** *m* TRANSP MAR *accesorios de cubierta* warping drum; **~ de clasificación** *m* PROC QUÍ classifying drum; **~ cónico** *m* ING MECÁ cone drum, MINAS, PROD *extracción de minas* fuse wheel; **~ cribador** *m* PROC QUÍ sieve drum; **~ desarenador** *m* PROD *fundería* rattler, shaking barrel, shaking machine, shaking mill; **~ descortezador** *m* PAPEL debarking drum; **~ del embrague** *m* VEH clutch drum; **~ del embrague** *m* AUTO, ING MECÁ clutch drum; **~ enrollador** *m* PAPEL winding drum; **~ de la enrolladora** *m* PAPEL reel drum; **~ del espesador** *m* PROC QUÍ thickener drum; **~ de exploración** *m* TV drum scanner, scanning drum; **~ de extracción** *m* CARBÓN *minas* drum, MINAS *para cable plano* hoisting reel; **~ de fibras** *m* EMB fiber drum (*AmE*), fibre drum (*BrE*); **~ de filtración** *m* PROC QUÍ filter drum; **~ filtrante** *m* CARBÓN, HIDROL drum filter; **~ foto-conductor** *m* ÓPT photoconducting drum; **~ de frenado** *m* MECÁ, TEC ESP snubber; **~ del freno** *m* AUTO, ING MECÁ, MECÁ, VEH brake drum; **~ giratorio** *m* CINEMAT, GEOFÍS rotating drum, PROD tumbling barrel, tumbling box, tumbling drum, tumbling mill; **~ del guinche** *m* OCEAN winch drum; **~ de impresión** *m* INFORM&PD print drum; **~ de impulsos** *m* TRANSP driving drum; **~ izador** *m* MINAS hoisting drum; **~ lavador** *m* PAPEL washing drum; **~ de lentes** *m* INSTR drum lens; **~ de limpieza** *m* PROD tumbling barrel, *fundería* shaking barrel; **~ de madera contrachapada** *m* EMB plywood drum; **~ magnético** *m* INFORM&PD, ING ELÉC magnetic drum; **~ de la maquinilla** *m* OCEAN, TRANSP MAR *accesorios de cubierta* winch drum; **~ de metal** *m* EMB metal drum; **~ de molinete** *m* TRANSP MAR *accesorios de cubierta* warping drum; **~ de molturación** *m* CARBÓN balling drum; **~ motor** *m* TRANSP driving drum; **~ para película** *m* CINEMAT film drum; **~ de plegado** *m* IMPR folding drum; **~ porta cabeza** *m AmL* (*cf tambor de la cabeza Esp*) TV head drum; **~ receptor** *m* FOTO take-up drum, TEC PETR receiving drum; **~ de revelado** *m* CINEMAT developing drum, FOTO processing drum; **~ para revestimientos** *m* REVEST coating drum; **~ de secado** *m* CONST, DETERG drying drum; **~ de secado de película** *m* CINEMAT, FOTO film-drying drum; **~ del separador sedimentario** *m* PROC QUÍ thickener drum; **~ tamizador** *m* PROC QUÍ sieve drum; **~ de tiempos** *m* C&V *mecanismo de control de máquinas IS* timing drum; **~ de traza** *m* INFORM&PD barrel plotter (*BrE*), drum plotter (*AmE*); **~ de trituración** *m* PROC QUÍ grinding drum

tamborete: ~ racamento *m* TRANSP MAR *mástil* hound

tambucho *m* TRANSP MAR yates coachroof

tamiz *m* CARBÓN cribble, riddle, sieve, CONST sieve, screen, GEOL screen, HIDROL, LAB sieve, MECÁ, MINAS, PAPEL screen, PROC QUÍ sieve, PROD riddle, *bocarte para minerales* grate, screen, sieve, gauze, RECICL screen, sieve; **~ de alambre** *m* CONST wire strainer; **~ de bucarán** *m* TEXTIL *estarcir* scrim screen; **~ para**

criba *m* MINAS jigger screen; **~ de desagüe** *m* CARBÓN draining screen; **~ de malla ancha** *m* CARBÓN coarse screen; **~ metálico** *m* PROD gauze strainer; **~ de moldear** *m* PROD foundry riddle; **~ de prueba** *m* ING MECÁ test sieve; **~ de tela metálica** *m* PROD wire sieve

tamizado *m* CONST sifting, CONTAM MAR, PROC QUÍ sieving, PROD sifting, TEC ESP *rayos X* screening; **~ húmedo** *m* CARBÓN wet screening; **~ de prueba** *m* ING MECÁ test sieving

tamizador: ~ de gránulos *m* PROC QUÍ particle siever

tamizar *vt* ALIMENT sieve, sift, CARBÓN faredice, sieve, sift, CONST screen, sieve, sift, CONTAM MAR sift, PROC QUÍ sieve, PROD sieve, sift, QUÍMICA screen, RECICL sieve

tampón *m* IMPR dabber, P&C, QUÍMICA buffer; **~ variable** *m* ÓPT loose buffering

tanato *m* QUÍMICA tannate

tanda *f* CINEMAT, TEC ESP, TEXTIL, TRANSP batch; **~ comercial** *f* TV commercial; **~ publicitaria** *f* TV advertising slot

tangencia *f* FÍS, GEOM, MECÁ tangency

tangencial *adj* FÍS, GEOM, MECÁ tangential

tangente *f* (*tg*) CONST, GEOM, INFORM&PD, MATEMÁT *trigonometría* tangent (*tan*); **~ al círculo** *f* GEOM tangent to the circle

tangón: ~ de aspersión *m* CONTAM MAR spray boom; **~ del espí** *m* TRANSP MAR *palos* spinnaker boom

tánico *adj* QUÍMICA tannic

tanino[1]**: ~ catecútico** *adj* QUÍMICA catechutannic

tanino[2] *m* QUÍMICA tannin

tanque *m* C&V, CONST, CONTAM, GAS, MECÁ tank, PROD vat, tank, QUÍMICA vat, REFRIG, TEC ESP tank, TEC PETR *perforación* pit, TRANSP MAR *construcción naval* tank; **~ de aceite** *m* CONST, ING ELÉC oil tank; **~ de achique** *m* AGUA bailing tank; **~ de acumulación de hielo** *m* REFRIG ice bank tank; **~ acumulador** *m* TEC PETR *almacenaje* accumulator tank; **~ adicional** *m* TEC ESP additional tank; **~ de agua** *m* AGUA, TRANSP MAR water tank; **~ de agua caliente** *m* INSTAL TERM hot water tank; **~ de aire** *m* TEC PETR air tank; **~ de aireación** *m* RECICL aeration tank; **~ de alimentación** *m* PROD feed tank; **~ de almacenado** *m Esp* (*cf tanque de almacenaje AmL*) CONTAM, TEC PETR storage tank; **~ de almacenaje** *m AmL* (*cf tanque de almacenado Esp*) CONTAM, TEC PETR storage tank; **~ de almacenamiento** *m* CONTAM, PROD, TEC PETR storage tank; **~ de alta presión** *m* TEC ESP high-pressure tank; **~ anfibio** *m* D&A amphibious tank; **~ de barro** *m* C&V earthenware tank; **~ de buceo** *m* OCEAN diving tank; **~ de carga** *m* CONTAM *instalación industrial* cargo tank; **~ de carga para la luz diurna** *m* FOTO daylight loading tank; **~ para cieno húmico** *m* AGUA humus tank; **~ cintura de avispa** *m* C&V wasp-waisted tank; **~ de coagulación** *m* PROC QUÍ coagulating tank; **~ colector** *m* PETROL flow tank; **~ de combustible** *m* AUTO, MECÁ fuel tank, PROD fuel bunker, TEC ESP, TRANSP AÉR fuel tank, TRANSP MAR bunker; **~ compensador** *m* AGUA equalizing tank, GAS buffer tank; **~ de compresión** *m* ING MECÁ receiver; **~ criogénico** *m* TEC ESP cryogenic tank; **~ de cromatografía** *m* LAB *análisis* chromatography tank, QUÍMICA chromatography tank, paper-chromatography tank; **~ de cromatografía sobre papel** *m* LAB *análisis* paper-chromatography tank; **~ de**

decantación *m* AGUA sedimentation tank, PROC QUÍ decantation tank; **~ de decantación final** *m* AGUA final-settling tank; **~ decantador** *m* MINAS *metalurgia* dewaterer; **~ de desechos** *m* TEC ESP *vuelos tripulados* disposal tank; **~ diario** *m* NUCL *para aceite* day tank; **~ diario de combustible** *m* NUCL day fuel tank; **~ digestor** *m* HIDROL *aguas residuales*, TERMO digestion tank; **~ enfriador de granja** *m* REFRIG refrigerated farm tank; **~ de enfriamiento** *m* CINEMAT cooling tank; **~ de equilibrio** *m* ENERG RENOV surge tank; **~ del espesador** *m* PROC QUÍ thickener tank; **~ de espuma** *m* PAPEL foam tank; **~ de evaporación** *m* AGUA evaporation pan; **~ de expansión** *m* AUTO, INSTAL HIDRÁUL, MECÁ, TERMOTEC expansion tank; **~ de fango activado** *m* HIDROL *aguas residuales* activated sludge tank; **~ de filtración** *m* AGUA filter tank; **~ flexible para almacenamiento de combustible** *m* D&A flexible oil storage tank; **~ de flotabilidad** *m* TRANSP MAR *buques, botavaras flotantes* buoyancy tank; **~ de flotación** *m* TRANSP MAR flotation tank; **~ flotante flexible** *m* CONTAM MAR floating flexible tank; **~ flotante de material flexible** *m* CONTAM MAR floating flexible tank; **~ de fluido hidráulico** *m* ING MECÁ hydraulic fluid reservoir; **~ de gas de presurización** *m* TEC ESP pressurizing gas tank; **~ de hidrógeno** *m* TEC ESP hydrogen tank; **~ de humidificación** *m* CARBÓN conditioning tank; **~ incorporado** *m* ING MECÁ built-in tank, TRANSP *metanero* integrated tank; **~ inyector de ácido** *m* PROC QUÍ acid injecting tank; **~ de lastre** *m* TRANSP deep tank, TRANSP MAR *buques y submarinos* ballast tank; **~ de lavado** *m* FOTO wash tank; **~ de líquido hidráulico** *m* ING MECÁ hydraulic fluid reservoir; **~ de lodo** *m* TEC PETR *perforación* mud pit, mud tank; **~ de medición** *m* TEC PETR *producción* gaging tank (*AmE*), gauging tank (*BrE*); **~ con membrana elastómerica** *m* TEC ESP *vehículos* elastomeric membrane tank; **~ de mezclado** *m* CINEMAT, PROC QUÍ mixing tank; **~ de mezclar** *m* CINEMAT, PROC QUÍ mixing tank; **~ de petróleo** *m* CONST oil tank; **~ petrolero** *m* CONTAM *buques* bunker tank; **~ plástico de revelado** *m* FOTO plastic developing-tank; **~ de precipitación** *m* PROC QUÍ precipitating tank, precipitation tank; **~ de presión** *m* OCEAN pressure tank; **~ presurizado de agua caliente** *m* INSTAL TERM pressurized hot-water tank; **~ profundo** *m* TRANSP deep tank; **~ receptor** *m* PETROL flow tank; **~ de reextracción** *m* NUCL backwash tank; **~ de revelado** *m* CINEMAT developing tank, FOTO developing tank, development tank; **~ de revelado de unidades múltiples** *m* FOTO multiunit developing tank; **~ de revelado universal** *m* FOTO universal developing tank; **~ secundario** *m* TRANSP AÉR feeder tank; **~ de sedimentación** *m* CARBÓN thickener, PROC QUÍ, RECICL sedimentation tank; **~ de sedimentación a filtro** *m* CARBÓN filter thickener; **~ de señuelo** *m* D&A decoy tank; **~ del separador sedimentario** *m* PROC QUÍ thickener tank; **~ séptico** *m* AGUA, HIDROL, RECICL septic tank; **~ de succión** *m* TEC PETR *perforación* suction pit; **~ de suministro** *m* TEC PETR *refino, producción* feed tank; **~ de tensión superficial** *m* TEC ESP surface tension tank; **~ vertical** *m* TRANSP deep tank; **~ de vidrio** *m* C&V, LAB *material de vidrio* glass tank

tantalato *m* QUÍMICA tantalate

tantálico *adj* QUÍMICA tantalic

tantalio *m* (*Ta*) ING ELÉC, METAL, QUÍMICA tantalum (*Ta*)

tantalita *f* MINERAL tantalite

tapa *f* C&V cap, *de botellas* closure, CINEMAT cap, CONST cap, dome, CONTAM, EMB lid, FOTO hood, ING MECÁ head, cover, LAB *de contenedor* lid, MECÁ cover, lid, PETROL cap, TEC ESP, TEXTIL cover, TRANSP MAR *embocadura, chimenea* bonnet, VEH cover; **~ de abertura automática** *f* EMB key-opening lid; **~ de aguja** *f* PROD *cambio de vía* switch cover plate; **~ de agujero de hombre** *f* MECÁ manhole cover; **~ articulada** *f* EMB hinged lid; **~ de balancín** *f* EMB lever lid; **~ de balancines** *f* AUTO, VEH rocker cover; **~ de buje** *f* VEH *rueda* hubcap; **~ de la cámara para pilas** *f* FOTO battery chamber cover; **~ ciega** *f* TRANSP MAR *construcción naval* deadlight; **~ de cierre** *f* PROD locking cover; **~ del cilindro** *f* ING MECÁ head, cylinder head, INSTAL HIDRÁUL *motores, máquina de vapor* cylinder cover, cylinder head, PROD cylinder cover, TRANSP MAR *máquina de vapor* cylinder head; **~ del cilindro posterior** *f* INSTAL HIDRÁUL *motor alternativo* back-cylinder cover; **~ del cojinete** *f* AUTO, ING MECÁ bearing cap; **~ de color** *f* PROD *para pulsadores* color cap (*AmE*), colour cap (*BrE*); **~ del combustible** *f* AmL MECÁ filler cap; **~ del compartimiento para pilas** *f* FOTO battery-compartment cover; **~ del cono atrapador de vapor** *f* PROC QUÍ vapor-catching cone cap (*AmE*), vapour-catching cone cap (*BrE*); **~ corrediza** *f* MECÁ slide; **~ del depósito** *f* VEH *de gasolina, gas-oil* tank cap; **~ del depósito de llenado** *f* VEH *combustible* filler compartment flap; **~ deslizante** *f* EMB sliding lid; **~ del distribuidor** *f* AUTO distributor cap, timing cover; **~ del doble fondo** *f* TRANSP MAR *arquitectura naval* inner bottom; **~ embisagrada** *f* PROD hinge cover; **~ de escotilla** *f* TRANSP MAR *de barco* hatch cover; **~ frontal de cilindro** *f* INSTAL HIDRÁUL *máquina de vapor* front cylinder cover, *máquina horizontal* front cylinder head; **~ de fusible** *f* ELEC fuse cover; **~ de gancho** *f* EMB hooked lid; **~ guardapolvos** *f* ING MECÁ dust cap; **~ lengüeta para rasgar** *f* EMB *sistema de apertura* tear tab lid; **~ del maletero** *f* VEH boot lid (*BrE*), trunk lid (*AmE*); **~ del objetivo** *f* CINEMAT lens cap, FOTO cap, lens cap; **~ del objetivo de rosca** *f* FOTO screw-on lens cap; **~ de obturación** *f* TRANSP AÉR blanking cover; **~ para el ojo de la cerradura** *f* CONST keyhole guard; **~ con pestaña** *f* EMB flanged cap, flanged edge; **~ de pistón** *f* ING MECÁ follower; **~ posterior** *f* FOTO back cover; **~ posterior desmontable** *f* FOTO removable back; **~ posterior giratoria** *f* FOTO revolving back, swinging back; **~ de pozo** *f* CONST manhole door; **~ de prensaestopa** *f* NUCL stuffing box lid; **~ de protección** *f* MECÁ *electricidad* fender; **~ protectora** *f* ING MECÁ protective cap; **~ de registro** *f* MECÁ manhole cover; **~ de relleno** *f* AmL MECÁ filler cap; **~ de rosca sin fin** *f* EMB continuous-thread cap; **~ roscada** *f* C&V screw cap, EMB screw cap, screw lid, screw top; **~ del ruptor** *f* VEH *sistema de encendido* distributor cap; **~ de seguridad** *f* PROD safety cover; **~ superior** *f* ELECTRÓN top cap; **~ de terminales** *f* PROD terminal cover; **~ transparente contra toda falsificación** *f* PROD transparent antitamper cover; **~ de válvula** *f*

VEH *de neumático* valve cap; **~ del visor** *f* FOTO finder hood

tapacubos *m* AUTO, VEH *rueda* hubcap

tapadera: ~ del cabezal de pozo *f* TEC PETR *perforación* bradenhead cap for the casing

tapado: ~ por contacto *m* ELECTRÓN contact masking

tapador *m* MECÁ plug

tapafusibles *m* PROD fuse cover

tapagrietas *m* Esp (*cf capa de aparejo AmL*) COLOR filler coat

tapajuntas *m* VEH sealant

tapaporos *m* COLOR filler, CONST *pintura* primer, NUCL filler, VEH body filler

tapar[1] *vt* AGUA stop, CINEMAT cap up, CONST cap, MINAS *una junta* stem, PROD *agujero de colada* stop up, *remaches* close, TERMO smother; **~ con masilla** *vt* CONST stop with putty

tapar[2]: **~ el agujero de colada** *vi* PROD *fundición* boat up the furnace

tapas: ~ y envolturas de polietileno para paletas *f pl* EMB polyethylene pallet-covers and liners; **~ con lomo** *f pl* IMPR book case

tapiar *vt* CONST, MINAS wall up; **~ con ladrillos** *vt* CONST brick up

tapicería *f* TEXTIL upholstery, VEH seat upholstery, *asientos* upholstery

tapiolita *f* MINERAL tapiolite

tapón *m* ALIMENT bung, C&V cover tile (*BrE*), spout cover (*AmE*), feeder plunger, plug, stopper, CONST cover tile (*BrE*), spout cover (*AmE*), plug, LAB *cierre* bung, cork, MECÁ plug, PROD *del agujero de colada* boat, TEC PETR *tuberías, conducto*, VEH *de vaciado* plug; **~ aerosol** *m* EMB aerosol cap; **~ para aislamiento acústico** *m* SEG noise-protective plug; **~ de arcilla** *m* CARBÓN *de un barreno* ram pump; **~ de arcilla para parar la colada** *m* PROD *horno de cúpula* plug; **~ de la botella** *m* EMB bottle stopper; **~ de carga** *m* PROD filler; **~ del cárter** *m* AUTO, VEH engine support plug; **~ de caucho** *m* Esp (*cf tapón de hule AmL*) LAB, P&C rubber stopper; **~ de cementación superior** *m* TEC PETR *perforación* top-cementing plug; **~ de cemento** *m* TEC PETR *perforación* cement plug, cementing plug; **~ ciego** *m* CONST *fontanería* cap; **~ de cierre** *m* EMB corking plug; **~ de cierre de la tubería de ventilación** *m* CONST vent plug; **~ de configuración** *m* PROD configuration plug; **~ cónico para la piquera de escoria** *m* PROD *horno alto* bott; **~ corona** *m* EMB *botellas* crown closure; **~ corona con opérculo de corcho** *m* EMB crown cork; **~ del depósito** *m* AUTO tank filler cap; **~ del depósito de gasolina** *m* Esp (*cf tapón del depósito de nafta AmL*) AUTO, VEH gas tank cap (*AmE*), gasoline tank cap (*AmE*), petrol tank cap (*BrE*); **~ del depósito de nafta** *m* AmL (*cf tapón del depósito de gasolina Esp*) AUTO, VEH gas tank cap (*AmE*), gasoline tank cap (*AmE*), petrol tank cap (*BrE*); **~ de desagüe magnético** *m* ING MECÁ magnetic drain plug; **~ de drenaje** *m* AUTO, VEH *motor, transmisión* drain plug; **~ de drenaje del aceite** *m* AUTO oil drain plug; **~ esmerilado** *m* EMB, LAB *material de vidrio* ground stopper; **~ del filtro de aceite** *m* AUTO oil filter cap; **~ fusible** *m* INSTAL HIDRÁUL lead plug, plug, safety plug; **~ fusible para caldera de vapor** *m* INSTAL HIDRÁUL fusible plug for steam boiler; **~ térmico** *m* TERMO *calderas* heat plug; **~ hembra roscado** *m* ING MECÁ pipe cap; **~ hermético** *m* EMB sealing cap; ~ **de hongo** *m* C&V mushroom stopper; ~ **de hule** *m* *AmL* (*cf tapón de caucho Esp*) LAB, P&C rubber stopper; ~ **de ida** *m* MECÁ go plug; ~ **incandescente** *m* TERMO glow plug; ~ **inferior de cementación** *m* TEC PETR *perforación* bottom-cementing plug; ~ **insonorizador** *m* SEG soundproof plug; ~ **de légamo** *m* OCEAN silt plug; ~ **de llenado** *m* AUTO, VEH *depósito de combustible, de aceite* filler cap; ~ **macho roscado** *m* ING MECÁ pipe plug; ~ **obturador** *m* CONST plug, ING MECÁ obturating plug, PROD blanking plug; ~ **obturador de empalme** *m* PROD coupling blanking plug; ~ **para el oído** *m* ACÚST, CONST, LAB, SEG earplug; ~ **de plástico** *m* EMB plastic plug; ~ **de precintado** *m* ING MECÁ sealing plug; ~ **de precintar** *m* ING MECÁ sealing plug; ~ **de presión del radiador** *m* AUTO, VEH radiator-pressure cap; ~ **de purga de aceite** *m* MECÁ oil drain plug; ~ **de purga magnético** *m* ING MECÁ magnetic drain plug; ~ **del radiador** *m* AUTO, VEH *sistema de refrigeración* radiator cap, radiator filler cap; ~ **de relleno** *m* PROD filling plug; ~ **del repostador por gravedad** *m* TRANSP AÉR gravity filler plug; ~ **que se rompe al abrirse por primera vez** *m* EMB snap-on lid; ~ **de rosca** *m* EMB screw cap, screw lid, screw top; ~ **de rosca de garantía** *m* EMB guarantee cap; ~ **roscado** *m* CONST screw plug, EMB bottle cap, ING MECÁ, MECÁ cap nut, PROD screw plug; ~ **roscado del cárter de aceite** *m* VEH *lubricación* oil drain plug; ~ **roscado de limpieza** *m* PROD *caldera* mud plug; ~ **roscado de purga del aceite** *m* VEH *lubricación* oil drain plug; ~ **de seguridad** *m* INSTAL HIDRÁUL safety plug; ~ **superior** *m* NUCL upper end plug; ~ **superior del núcleo** *m* NUCL core head plug unit; ~ **térmico** *m* TERMO heat plug; ~ **de tolerancias** *m* ING MECÁ limit internal gage (*AmE*), limit internal gauge (*BrE*); ~ **de tubo** *m* ING MECÁ pipe plug; ~ **de vaciado** *m* AUTO drain plug, PROD emptying plug, VEH *motor, transmisión* drain plug; ~ **de vaciado magnético** *m* ING MECÁ magnetic drain plug; ~ **de vaciado del radiador** *m* AUTO, VEH *sistema de refrigeración* radiator drain tap; ~ **de vapor** *m* AUTO vapor lock (*AmE*), vapour lock (*BrE*); ~ **de vidrio** *m* C&V, LAB *material de vidrio* glass stopper

taponado *m* TEC PETR *perforación* plugging; ~ **de tubos en un generador de vapor** *m* NUCL steam generator tube plugging

taponamiento *m* ING MECÁ, TEC PETR *perforación* plugging; ~ **de la piquera** *m* PROD *fundición* bott

taponar *vt* ING MECÁ, MECÁ caulk, PROD stop, caulk, *agujero de colada* stop up, *tubo* close, TRANSP MAR *construcción naval* plug

taqué *m* AUTO tappet, VEH valve lifter

taquetes: ~ **de la jaula de minas** *m pl* MINAS cage shuts

taquihidrita *f* MINERAL tachhydrite

taquilla *f* TRANSP MAR *construcción naval* locker; ~ **de banderas** *f* TRANSP MAR signal locker

taquimetría *f* CONST tacheometry, *topografía* stadia surveying, FÍS tacheometry

taquímetro *m* CONST, FÍS, INSTR tacheometer, tachymeter

taquión *m* FÍS tachyon

tara *f* EMB, TEXTIL tare, TRANSP load-no charge ratio

tarado *m* GEN calibration, METR rating, scaling, setting

tarar *vt* METR measure, PROD gage (*AmE*), gauge (*BrE*)

tardar *vt* ELECTRÓN *en la orden de ejecución* delay

tardío *adj* GEOL *edad* late

tarea *f* INFORM&PD task, MINAS *trabajo* labor (*AmE*), labour (*BrE*); ~ **en primer plano** *f* INFORM&PD foreground job; ~ **prioritaria** *f* INFORM&PD foreground job

tareas: ~ **agrícolas** *f pl* AGRIC farm operations; ~ **de mantenimiento** *f pl* CALIDAD, INFORM&PD housekeeping

tarifa *f* ELEC, EMB, TELECOM tariff; ~ **de la carga** *f* EMB, TEC PETR freight rate (*AmE*), goods rate (*BrE*); ~ **del flete** *f* EMB, TEC PETR freight rate (*AmE*), goods rate (*BrE*); ~ **inicial** *f* TELECOM access charge rate; ~ **por línea** *f* IMPR line rate; ~ **de llamada** *f* TELECOM call charge rate; ~ **media** *f* TELECOM *tasa de conversación* medium rate; ~ **nocturna** *f* ELEC *fuente de alimentación* night tariff; ~ **reducida** *f* TELECOM reduced rate; ~ **reducida de llamadas** *f* TELECOM cheap call rate; ~ **según contador** *f* TELECOM metering rate; ~ **a tanto alzado** *f* ELEC *consumo* flat-rate tariff; ~ **con todo incluído** *f* ELEC *fuente de alimentación* all-in tariff; ~ **única** *f* ELEC *consumo* flat-rate tariff; ~ **uniforme** *f* ELEC *consumo* flat-rate tariff

tarima *f* C&V *para recipientes de vidrio* pallet, CONST floor

tarjeta *f* ELECTRÓN card, INFORM&PD *electrónica* board; ~ **de alarma** *f* TELECOM alarm card; ~ **de ampliación** *f* INFORM&PD expansion board, expansion card; ~ **para anotar la hora de entrada y salida** *f* PROD *talleres* time sheet; ~ **de apertura** *f* INFORM&PD aperture card; ~ **con chip** *f* INFORM&PD chip card; ~ **de circuito impreso** *f* (*PCI, tarjeta de CI*) ELEC, ELECTRÓN, INFORM&PD, TELECOM, TV printed-circuit board (*PC board, PCB*); ~ **colgante del precio** *f* EMB price tag; ~ **de contacto** *f* ELECTRÓN interface card; ~ **de control** *f* EMB control tag; ~ **de control del tiempo** *f* PROD *tableros* time card; ~ **de cronometraje** *f* CINEMAT timing card; ~ **de datos de fallos** *f* TRANSP AÉR failure data card; ~ **de expansión** *f* ELECTRÓN expansion card; ~ **explosiva** *f* D&A, TEC ESP *proyectiles balísticos* detonating card; ~ **de gestión** *f* PROD driver card; ~ **Hollerith** *f* INFORM&PD Hollerith card; ~ **de identificación** *f* INFORM&PD badge, SEG identification tag; ~ **de identificación de paquete** *f* TRANSP parcel registration card; ~ **inteligente** *f* INFORM&PD smart card; ~ **de interconexión** *f* ELECTRÓN interface card; ~ **de interconexión universal S** *f* TELECOM *red de transmisión digital de servicios integrados* S-universal interface card; ~ **de interfaz de red** *f* INFORM&PD network interface card; ~ **de inversión** *f* INFORM&PD turn-around card; ~ **láser** *f* ÓPT lasercard®; ~ **laseróptica** *f* ÓPT laser-optic card; ~ **de lógica** *f* *AmL* ELECTRÓN, INFORM&PD logic card; ~ **lógica** *f* ELECTRÓN, INFORM&PD logic card; ~ **de lote** *f* PROD batch card; ~ **maestra** *f* INFORM&PD master card; ~ **magnética** *f* IMPR mag card, INFORM&PD magnetic card; ~ **de memoria** *f* INFORM&PD memory card, PROD memory map; ~ **de memoria óptica** *f* ÓPT optical memory card; ~ **con microprocesador** *f* INFORM&PD smart card; ~ **de ocho columnas** *f* INFORM&PD eight-column card; ~ **de pago** *f* TELECOM chargecard; ~ **de la palanca de arranque** *f* TELECOM swipe card; ~ **perforada** *f* INFORM&PD punch card (*AmE*), punched card (*BrE*); ~ **perforada de doce columnas** *f* INFORM&PD twelve-row punched card;

~ **de pestaña** *f* IMPR tab card; ~ **postal ilustrada** *f* PAPEL illustrated postcard; ~ **de procesamiento** *f* TELECOM processing card; ~ **de recorrido** *f* PROD turnaround card; ~ **de remesa** *f* PROD batch card; ~ **telefónica** *f* TELECOM phonecard, telephone card; ~ **del título** *f* CINEMAT title card; ~ **de viaje redondo** *f* PROD turnaround card

tarmacadam *f* CONST tarmac, tarmacadam

tarro *m* ALIMENT jar; ~ **de vidrio** *m* C&V, EMB *conservas* glass jar

tartárico *adj* QUÍMICA tartaric

tártaro *m* QUÍMICA tartar

tartaro: ~ **bruto** *m* QUÍMICA argol

tartrado *adj* QUÍMICA tartrated

tartrato *m* QUÍMICA tartrate; ~ **ácido de potasio** *m* ALIMENT cream of tartar; ~ **sódico-potásico** *m* ALIMENT, ING ELÉC, QUÍMICA Rochelle salt

tartrónico *adj* QUÍMICA tartronic

tartronilurea *f* QUÍMICA tartronylurea

tarugo *m* CARBÓN *pozo de petróleo* slug, MECÁ bolt, plug

tas: ~ **de aplanar** *m* ING MECÁ planishing stake; ~ **de estampar** *m* PROD *fraguas* swage block

tasa *f* FÍS rate, ratio, INFORM&PD rate, MATEMÁT, METR, REFRIG ratio; ~ **de actividad de archivos** *f* INFORM&PD file-activity ratio; ~ **de actividad de ficheros** *f* INFORM&PD file-activity ratio; ~ **de agotamiento** *f* Esp (*cf* rata de agotamiento AmL) PETROL pumping rate, TEC PETR *producción, yacimiento* depletion rate, pumping rate; ~ **anual de radiación de fondo** *f* FÍS RAD annual natural background radiation; ~ **de aplicación** *f* CONTAM MAR application rate; ~ **básica de bits** *f* TELECOM basic bit rate; ~ **en baudios** *f* AmL (*cf* velocidad de línea en baudios Esp) INFORM&PD baud rate; ~ **de bits** *f* INFORM&PD, TEC ESP, TELECOM bit rate; ~ **de bits constante** *f* TELECOM constant bit rate; ~ **de bits erróneos** *f* TELECOM bit error rate (*BER*); ~ **de bits variable** *f* INFORM&PD, TELECOM variable-bit rate (*VBR*); ~ **de carga** *f* TELECOM charge rate; ~ **de caudal del refrigerante del núcleo** *f* NUCL core coolant flow rate; ~ **de colisiones** *f* TELECOM collision rate; ~ **de crecimiento** *f* AGRIC growth rate; ~ **de decantación** *f* PROC QUÍ decantation rate; ~ **de desintegración** *f* FÍS RAD decay rate; ~ **de dosis** *f* FÍS, FÍS RAD dose rate; ~ **de dosis absorbida** *f* FÍS, FÍS RAD absorbed dose rate; ~ **de dosis equivalente efectiva colectiva** *f* NUCL collective effective dose equivalent rate, SE; ~ **efectiva de transferencia de datos** *f* INFORM&PD effective data-transfer rate; ~ **de eficiencia bovina** *f* AGRIC cattle performance; ~ **de eliminación biológica** *f* NUCL biological clearance rate; ~ **de eliminación de calor** *f* NUCL heat rejection rate; ~ **de emergencia** *f* AGRIC rate of emergence; ~ **de error** *f* ELECTRÓN, INFORM&PD, TELECOM error rate; ~ **de errores binarios** *f* TELECOM binary error-rate; ~ **de errores en bloques** *f* INFORM&PD block error rate; ~ **de exposición** *f* FÍS exposure rate; ~ **de fallos** *f* INFORM&PD, ING ELÉC, TEC ESP, TELECOM failure rate; ~ **de fecundidad** *f* AGRIC fertility rate; ~ **de fluencia energética** *f* FÍS energy-fluence rate; ~ **de fluencia de partículas** *f* FÍS, FÍS PART particle-fluence rate; ~ **de flujo de energía radiante** *f* FÍS, FÍS RAD, TELECOM radiant-energy fluence rate; ~ **de fugas** *f* NUCL leak rate; ~ **de fugas metaestable** *f* FÍS RAD metastable loss rate; ~ **de giro** *f* FÍS rate gyro; ~ **de hallazgo** *f* CONTAM MAR encounter rate; ~ **de kerma** *f* FÍS kerma rate; ~ **legal de humedad** *f* TEXTIL *fibras* moisture regain; ~ **de llamadas del abonado** *f* TELECOM subscriber calling rate; ~ **máxima** *f* TELECOM peak rate; ~ **de mediación** *f* TELECOM metering rate; ~ **mínima** *f* TELECOM minimum rate; ~ **de muestreo** *f* INFORM&PD sampling rate; ~ **neta de asimilación** *f* AGRIC net assimilation rate; ~ **neta de reproducción** *f* NUCL net breeding rate; ~ **de ocupación** *f* CONST occupancy rate; ~ **de parición** *f* AGRIC calving rate; ~ **de penetración** *f* (*ROP*) TEC PETR *perforación* drilling rate, rate of penetration (*ROP*); ~ **de penetración normalizada** *f* TEC PETR *perforación* normalized drilling rate (*NDR*); ~ **de pérdidas metaestable** *f* FÍS RAD metastable loss rate; ~ **de recombinación** *f* ELECTRÓN, FÍS recombination rate; ~ **de relleno** *f* TELECOM stuffing rate; ~ **de repetición de pulsos** *f* FÍS pulse-repetition rate; ~ **de sedimentación** *f* TEC PETR *geología* sedimentation rate; ~ **de transferencia** *f* ÓPT transfer rate; ~ **de transferencia de calor** *f* INSTAL TERM, TERMO rate of heat transfer; ~ **volumétrica** *f* FÍS *de flujo* volume rate

tasación *f* FÍS rating, TELECOM metering; ~ **de los daños** *f* TRANSP MAR *seguros* damage assessment

tasador: ~ **de averías** *m* TRANSP MAR *seguros* average adjuster

tasas: ~ **por contaminación** *f pl* CONTAM polluter pays principle

tasmanita *f* CARBÓN gyttja, MINERAL tasmanite

tauón *m* FÍS tauon

taurina *f* QUÍMICA taurine

tauriscita *f* MINERAL tauriscite

taurocolato *m* QUÍMICA taurocholate

taurocólico *adj* QUÍMICA taurocholic

tautomerismo *m* QUÍMICA tautomerism; ~ **ceto-enólico** *m* QUÍMICA keto-enol tautomerism

tautomerización *f* QUÍMICA tautomerization

tautómero[1] *adj* QUÍMICA tautomeric

tautómero[2] *m* QUÍMICA tautomer

taxi *m* TRANSP cab; ~ **aéreo** *m* TRANSP AÉR air taxi; ~ **sin conductor** *m* TRANSP self-drive taxi; ~ **con radio** *m* TRANSP radio taxicab

taxímetro *m* TRANSP MAR *navegación* pelorus, VEH taximeter

taylorita *f* MINERAL taylorite

taza: ~ **del horno** *f* C&V tank furnace; ~ **de porcelana** *f* C&V porcelain cup

Tb *abr* (*terbio*) QUÍMICA Tb (*terbium*)

Tc *abr* (*tecnecio*) QUÍMICA Tc (*technetium*)

TCR *abr* TEC ESP (*control telemedida telemando y distancia*) TCR (*telemetry command and ranging subsystem*)

TDE *abr* (*tratamiento de datos electrónico*) ELECTRÓN EDP (*electronic data processing*)

TDI *abr* (*toluendiisocianato*) P&C TDI (*toluene diisocyanate*)

TDM *abr* ELECTRÓN (*teclado de desplazamiento mínimo*) MSK (*minimum-shift keying*), FÍS, INFORM&PD, TEC ESP, TELECOM (*transmisión simultánea por división del tiempo, modulación por división de tiempo, multiplexación por división de tiempo*) TDM (*time-division modulation, time-division multiplexing*)

te: ~ **de derivación** *f* ELEC *cable* tee joint

TE *abr* (*transversal eléctrico*) ING ELÉC TE (*transverse electric*)

Te *abr* (*telurio*) ING ELÉC, QUÍMICA Te (*tellurium*)

tebaína *f* QUÍMICA thebaine

TEC[1] *abr* (*transistor de efecto de campo*) ELECTRÓN, FÍS, INFORM&PD, ÓPT, TEC ESP FET (*field-effect transistor*)

TEC[2]: ~ **con aislante Schottky** *m* ELECTRÓN Schottky-barrier FET; ~ **de canal p** *m* ELECTRÓN p-channel FET; ~ **discreto de canal n** *m* ELECTRÓN n-channel discrete FET; ~ **heterounión** *m* ELECTRÓN hetero-junction FET; ~ **integrado de canal p** *m* ELECTRÓN p-channel integrated FET; ~ **mínimo** *m* ELECTRÓN depletion-mode FET; ~ **de silicio** *m* ELECTRÓN silicon FET; ~ **de unión integrada de canal p** *m* ELECTRÓN p-channel integrated-junction FET

teca *f* TRANSP MAR teak

TECCA *abr* (*transistor de efecto de campo con circuito aislado*) ELECTRÓN IGFET (*insulated grid field-effect transistor*)

techado *m* CONST roofing

techador *m* CONST roofer

techo *m* CONST cap, roof, FERRO *vehículos* roof, MINAS back, TEC ESP, TRANSP AÉR ceiling, TRANSP MAR deckhead, *construcción naval* tank top, VEH roof; ~ **de concreto** *m* AmL (*cf techo de hormigón Esp*) CONST concrete roofing tile; ~ **convertible** *m* VEH *carrocería* convertible top; ~ **de la cúpula** *m* CONST dome roof; ~ **descapotable** *m* VEH *coches* convertible top; ~ **giratorio** *m* TRANSP swiveling roof (*AmE*), swivelling roof (*BrE*); ~ **de hormigón** *m* Esp (*cf techo de concreto AmL*) CONST concrete roofing tile; ~ **móvil** *m* PROD *hornos* bung; ~ **de nubes** *m* METEO, TRANSP AÉR cloud ceiling; ~ **operacional** *m* TRANSP AÉR operational ceiling; ~ **operativo** *m* TEC ESP, TRANSP AÉR ceiling; ~ **del porche** *m* Esp (*cf porche AmL*) CONST porch roof; ~ **práctico** *m* TRANSP AÉR service ceiling; ~ **reforzado** *m* CONST trussed roof; ~ **de servicio** *m* TRANSP AÉR service ceiling; ~ **de vidrio** *m* C&V, CONST glass roof

techumbre *f* CONST roof

tecla *f* ACÚST, INFORM&PD, PROD key, TELECOM press-button, key, TV key; ~ **blanda** *f* INFORM&PD soft key; ~ **de búsqueda** *f* INFORM&PD search key; ~ **de ciclo único** *f* PROD *automatismo industrial* single scan key; ~ **para conectar interrogación** *f* PROD examine-on key; ~ **de control** *f* INFORM&PD control key; ~ **del cursor** *f* INFORM&PD cursor key; ~ **para desconectar interrogación** *f* PROD examine-off key; ~ **de escape** *f* (*ESC*) INFORM&PD escape key (*ESC*); ~ **de función** *f* INFORM&PD function key; ~ **indefinida** *f* INFORM&PD undefined key; ~ **de justificación** *f* IMPR justification key; ~ **de letras Baudot** *f* INFORM&PD letters shift; ~ **de mayúsculas derecha** *f* INFORM&PD right shift; ~ **programable** *f* INFORM&PD soft key; ~–**pulsadora momentánea** *f* PROD momentary push-key; ~ **de repetición** *f* INFORM&PD, TV repeat key; ~ **de retroceso** *f* IMPR, INFORM&PD backspace (*BS*); ~ **de salida** *f* (*ESC*) INFORM&PD escape key (*ESC*); ~ **de tabulación** *f* INFORM&PD tabulator key (*TAB*); ~ **tabuladora** *f* INFORM&PD tabulator key (*TAB*)

teclado *m* GEN keyboard; ~ **adicional** *m* IMPR additional keyboard; ~ **blando** *m* INFORM&PD soft keyboard; ~ **ciego** *m* IMPR blind keyboard; ~ **desconectable** *m* INFORM&PD detachable keyboard; ~ **desmontable** *m* PROD detachable keyboard; ~ **desplazable** *m* INFORM&PD, PROD detachable keyboard; ~ **de desplazamiento mínimo** *m* (*TDM*) ELECTRÓN minimum-shift keying (*MSK*); ~ **de membrana** *m* ING ELÉC, PROD membrane keyboard; ~ **numérico** *m* FÍS keypad, INFORM&PD *teléfono* keypad, numeric keypad, TELECOM, TV keypad; ~ **programable** *m* INFORM&PD soft keyboard; ~ **QWERTY** *m* INFORM&PD QWERTY keyboard

tecleado *m* PROD keying

teclear *vt* GEN keyboard, INFORM&PD, CINEMAT key in, TV *información, texto* key

tecnecio *m* (*Tc*) QUÍMICA technetium (*Tc*)

técnica *f* ING MECÁ mechanism; ~ **de acabado** *f* TEXTIL finishing technique; ~ **de activación con isótopos** *f* NUCL labeling technique (*AmE*), labelling technique (*BrE*); ~ **de automatización de bajo costo** *f* ING MECÁ low-cost automation technique; ~ **a baja temperatura** *f* TERMO low-temperature technique; ~ **de bajo nivel de ruido** *f* ING MECÁ low-noise engineering; ~ **de barrido de campo próximo** *f* ÓPT near-field scanning technique; ~ **celular** *f* TELECOM cellular technique; ~ **de codificación de frecuencia síncrona** *f* TELECOM synchronous frequency encoding technique (*SFET*); ~ **de compresión de datos** *f* ELECTRÓN, INFORM&PD, TELECOM data compression technique (*DCT*); ~ **cónica equilibrada** *f* TEC ESP matched conics technique; ~ **de construcción de túneles** *f* CONST, MINAS tunneling technique (*AmE*), tunnelling technique (*BrE*); ~ **de crisol doble** *f* ÓPT double crucible technique; ~ **de cultivos** *f* AGRIC farming technique; ~ **de deposición en fase de vapor** *f* PROC QUÍ vapor deposition technique (*AmE*), vapour deposition technique (*BrE*); ~ **de depósito axial por fase de vapor** *f* TELECOM vapor phase axial deposition technique (*AmE*), vapour phase axial deposition technique (*BrE*); ~ **de diferenciación de metales pesados** *f* NUCL heavy metal difference technique; ~ **de dirección de la calidad** *f* CALIDAD quality engineering; ~ **de distribución** *f* EMB distribution technique; ~ **de ensayo ambiental** *f* SEG environmental-testing procedure; ~ **de excavación** *f* CONST digging technique; ~ **de exploración de campo próximo** *f* TELECOM near-field scanning technique; ~ **de fabricación** *f* ELECTRÓN fabrication technique; ~ **de impresión** *f* FOTO printing technique; ~ **intercambiadora de iones** *f* TELECOM ion exchange technique; ~ **de intercambio iónico** *f* ÓPT ion exchange technique; ~ **invasiva** *f* FÍS RAD invasive technique; ~ **para levantar pesos a mano** *f* SEG *métodos seguros de trabajo* manual lifting technique; ~ **de marcación** *f* NUCL labeling technique (*AmE*), labelling technique (*BrE*); ~ **de marcadores radioisotópicos** *f* AGRIC radioisotope tracer technique; ~ **de medición de tensión** *f* ING MECÁ strain gage technique (*AmE*), strain gauge technique (*BrE*); ~ **de medida cortando la fibra** *f* ÓPT cutback fiber technique (*AmE*), cutback fibre technique (*BrE*); ~ **de mezclado** *f* PROC QUÍ mixing technique; ~ **de muestreo de aire** *f* SEG air-sampling technique; ~ **de la posluminiscencia pulsada** *f* FÍS RAD pulsed-afterglow technique; ~ **de precisión** *f* ING MECÁ precision engineering; ~ **de predistorsión** *f* TELECOM predistortion technique; ~ **de primeros auxilios** *f* SEG first-aid procedure; ~ **de primeros auxilios en casos de emergencia** *f* SEG emergency

first-aid procedure; **~ de programación** *f*
INFORM&PD software engineering; **~ de radioactiva-ción con isótopos** *f* NUCL labeling technique (*AmE*),
labelling technique (*BrE*); **~ de recuento de
impulsos** *f* TELECOM pulse-counting technique;
~ de retrodispersión *f* TELECOM backscattering
technique; **~ por retroesparcimiento** *f* ÓPT back-scattering technique; **~ de revisión y evaluación de
programas** *f* (*PERT*) TEC ESP program evaluation
and review technique (*AmE*), programme evaluation
and review technique (*BrE*); **~ sanitaria** *f* AGUA
sanitary engineering; **~ de seccionamiento** *f* NUCL
sectioning technique; **~ síncrona de codificación de
frecuencia** *f* TELECOM synchronous frequency
encoding technique (*SFET*); **~ de sinterización** *f*
PROC QUÍ sintering technique; **~ de sinterizado e
infiltración** *f* PROC QUÍ sintering and infiltration
technique; **~ de transmisión** *f* NUCL transmission
technique; **~ del trenzado** *f* TEXTIL braiding techni-que; **~ de varilla en tubo** *f* ÓPT, TELECOM rod-in-tube
technique

técnico *m* GEN technician; **~ de calefacción** *m* ING
MECÁ heating technician; **~ encargado** *m* ING MECÁ
operator; **~ frigorista de mantenimiento** *m* REFRIG
refrigeration service engineer; **~ de montaje** *m* TV
editor; **~ radigráfico** *m* FÍS, FÍS RAD, INSTR radio-grapher; **~ en reparaciones** *m* ING MECÁ
troubleshooter; **~ del servicio radioeléctrico** *m*
TRANSP AÉR, TRANSP MAR radio engineer; **~ en
supervivencia** *m* OCEAN life support technician;
~ de ventas *m* PROD sales engineer

tecnología *f* GEN technology; **~ biergológica** *f* TRANSP
bi-ergol technology; **~ bipolar** *f* ELECTRÓN bipolar
technology; **~ bipolar combinada** *f* ELECTRÓN
merged bipolar technology; **~ de canal n** *f* ELECTRÓN
n-channel technology; **~ de capa gruesa** *f* TEC ESP
thick-film technology; **~ de computadoras** *f* *AmL*
(*cf tecnología de ordenadores Esp*) INFORM&PD com-puter technology; **~ de decapado con ácidos** *f*
NUCL acid pickling technology; **~ al día** *f* CONST
state-of-the-art technique; **~ diergólica** *f* TEC ESP
diergol technology; **~ energética** *f* CONTAM energy
technology; **~ espacial** *f* TRANSP space technology;
~ de fibra óptica *f* ING ELÉC, ÓPT, TELECOM fiber-optic technology (*AmE*), fibre-optic technology
(*BrE*); **~ fibroóptica** *f* ING ELÉC, ÓPT, TELECOM
fiber-optic technology (*AmE*), fibre-optic technology
(*BrE*); **~ de los fluidos** *f* FÍS FLUID fluidics; **~ de haz
de partículas** *f* ELECTRÓN, FÍS PART particle-beam
technology; **~ de la información** *f* INFORM&PD, TELE-COM information technology (*IT*); **~ de
manipulación de sólidos voluminosos** *f* ING MECÁ
bulk-solids handling technology; **~ de microondas** *f*
TEC ESP microwave technology; **~ minera** *f* MINAS
mining engineering; **~ mixta** *f* ELECTRÓN mixed
technology; **~ de ordenadores** *f* *Esp* (*cf tecnología
de computadoras AmL*) INFORM&PD computer tech-nology; **~ de película gruesa** *f* TEC ESP thick-film
technology; **~ de procesos** *f* TEC PETR process
engineering; **~ de puerta de silicio** *f* ELECTRÓN
silicon gate technology; **~ que mejora la
resolución** *f* IMPR resolution enhancement technol-ogy (*RET*); **~ de reactores reproductores rápidos** *f*
FÍS, NUCL fast-breeder reactor technology; **~ de
revestimiento delgado** *f* ELECTRÓN thin-film tech-nology; **~ de revestimiento grueso** *f* ELECTRÓN

thick-film technology; **~ para salas de operaciones
de bolsa** *f* TELECOM dealer room technology; **~ de
semiconductor de óxido de metal** *f* ELECTRÓN
metal-oxide semiconductor technology; **~ de
semiconductores** *f* ELECTRÓN semiconductor tech-nology; **~ solar** *f* ENERG RENOV solar technology;
~ de SOM *f* ELECTRÓN MOS technology; **~ de
SOMD** *f* ELECTRÓN DMOS technology; **~ del vacío**
f ING MECÁ vacuum technology; **~ de visualización
con ELCD** *f* ELECTRÓN TFEL display technology

TECRA *abr* (*transistor de efecto de campo con rejilla
aislada*) ELECTRÓN IGFET (*insulated grid field-effect
transistor*)

tectogénesis *m* TEC PETR *geología* tectogenesis
tectógeno *m* GEOL downbuckle, tectogene
tectónica *f* GEOL, TEC PETR *geología* tectonics;
~ penetrante *f* GEOL indenter tectonics; **~ de
placas** *f* GEOL plate tectonics; **~ de rift** *f* GEOL rift
tectonics; **~ salina** *f* GEOL, TEC PETR *geología* salt
tectonics; **~ tangencial** *f* GEOL *geología estructural*
tangential tectonics

tectónico *adj* GEOL, TEC PETR *geología* tectonic
Teflón® *m* QUÍMICA Teflon®; **~ aluminizado**® *m* TEC
ESP aluminized Teflon®

tefroíta *f* MINERAL tephroite
teína *f* QUÍMICA theine
teja *f* CONST crest tile, tile, TEC ESP tile; **~ de albardilla** *f*
CONST *arquitectura* saddle tile; **~ de caballete** *f*
CONST hip tile, ridge tile; **~ canalón** *f* CONST pantile;
~ de chapa *f* PROD *cilindro de calderas* shell plate;
~ cimera *f* CONST crown tile; **~ convexa** *f* CONST
arched tile; **~ de esquina** *f* CONST corner tile; **~ de
expansión** *f* INSTAL HIDRÁUL *distribuidor máquina de
vapor* expansion plate, expansion slide, *válvula de
cierre* cutoff plate; **~ de lima** *f* CONST hip tile; **~ de
madera** *f* CONST shingle; **~ plana** *f* CONST plain tile;
~ de protección térmica *f* TEC ESP thermal protec-tion tile; **~ sillín** *f* CONST *arquitectura* saddle tile; **~ de
vidrio para techos** *f* C&V, CONST glass roof-tile

tejadillo *m* CONST lean-to roof, pent roof, shed roof
tejado *m* C&V tiling, CONST roof; **~ a un agua** *m* CONST
single-pitch roof; **~ sin armadura** *m* CONST untrussed
roof; **~ a cuatro aguas** *m* CONST hip roof, hipped
roof; **~ a cuatro aguas con cumbrera** *m* CONST hip
roof with ridge, hipped ridge roof; **~ a cuatro aguas
con lima hoya** *m* CONST hip-and-valley roof; **~ con
cumbrera a cuatro aguas** *m* CONST hip-and-ridge
roof; **~ dentado** *m* CONST saw tooth roof; **~ a dos
aguas** *m* CONST double pitch roof, gable roof, pitch
roof, ridged roof; **~ de una sola agua** *m* CONST lean-to; **~ tipo masandra** *m* CONST curb roof; **~ de
vertiente simple** *m* CONST single-pitch roof

tejador *m* CONST *operario* tiler
tejamanil *m* CONST *tejados* shingle
tejar *vt* CONST tile
tejedura *f* TEXTIL weaving; **~ de punto** *f* TEXTIL
knitting

tejer *vt* TEXTIL weave
tejido *m* ELEC *cable* braid, TEXTIL fabric, VEH *neumá-tico* ply; **~ para abrigos** *m* TEXTIL coating;
~ acabado *m* TEXTIL finished fabric; **~ de calada**
m TEXTIL woven fabric; **~ de calada de fantasía** *m*
TEXTIL fancy woven fabric; **~ de calada más grueso**
m TEXTIL coarser woven fabric; **~ para chaquetas** *m*
TEXTIL jacketing; **~ de chenilla** *m* TEXTIL chenille
fabric; **~ para cortinas** *m* TEXTIL casement cloth; **~ a**

cuadros *m* TEXTIL checks; ~ **en cuerda** *m* TEXTIL fabric in rope form; ~ **estampado** *m* TEXTIL printed fabric; ~ **de felpilla** *m* TEXTIL chenille fabric; ~ **de fibra de vidrio** *m* C&V, EMB glass fabric; ~ **de fondo** *m* TEXTIL ground fabric; ~ **impermeable** *m* REVEST waterproof tissue; ~ **impregnado** *m* EMB impregnated fabric; ~ **inflamable** *m* TEXTIL flamm yarn; ~ **jacquard** *m* TEXTIL jacquard fabric; ~ **de lana peinada** *m* TEXTIL combed-wool fabric; ~ **de ligamento por maquinita** *m* TEXTIL dobby weave fabric; ~ **ligero para prendas** *m* TEXTIL *de poco peso* lightweight apparel fabric; ~ **ligero para tapicería** *m* TEXTIL lightweight furnishing fabric; ~ **de lino** *m* TEXTIL linen; ~ **liso** *m* TEXTIL plain fabric; ~ **en paneles** *m* TEXTIL panel fabric; ~ **de punto** *m* TEXTIL knitted fabric; ~ **de punto charmés** *m* TEXTIL locknit; ~ **de punto circular** *m* TEXTIL circular-knitted fabric; ~ **de punto de una fontura** *m* TEXTIL single jersey; ~ **de punto por trama** *m* TEXTIL weft-knitted fabric; ~ **para sábanas** *m* TEXTIL sheeting; ~ **tafetán** *m* TEXTIL plain fabric; ~ **para tapicería** *m* TEXTIL furnishing fabric; ~ **en toda su longitud** *m* TEXTIL full-length cloth; ~ **tupido** *m* TEXTIL firmly-asset fabric; ~ **de vidrio** *m* TRANSP MAR *material* glass cloth
tejidos: no ~ **de filamentos** *m pl* TEXTIL spunbond
tejo *m* CARBÓN ingot
tejuelo *m* ING MECÁ pillow, step box
tela *f* PAPEL fabric, *de formación metálica o plástica* wire, PETROL fabric, TEC ESP *traje* fabric, *vestimenta de astronautas* cloth, TEXTIL fabric, cloth; ~ **aceitada** *f* P&C *material textil* oilskin; ~ **acompañante** *f* TEXTIL leader cloth; ~ **de alambre** *f* C&V, P&C *equipo* wire mesh; ~ **de alambre de hierro** *f* PROD iron wire gauze; ~ **alquitranada** *f* CONST tarpaulin; ~ **para camisas** *f* TEXTIL shirting; ~ **de colchón** *f* TEXTIL *tejido* mattress ticking; ~ **encogible** *f* PAPEL shrink sleeve; ~ **de esmeril** *f* ING MECÁ abrasive cloth, emery cloth, MECÁ, PROD emery cloth; ~ **de esmeril fina** *f* ING MECÁ fine emery cloth; ~ **de esmeril de óxido de hierro** *f* ING MECÁ crocus cloth; ~ **de fibra de vidrio** *f* PROD glass cloth; ~ **de filtración** *f* PROC QUÍ filtering cloth; ~ **filtrante** *f* AGUA, PROC QUÍ filter cloth; ~ **de filtrar** *f* AGUA, PROC QUÍ filter cloth; ~ **de filtro** *f* C&V filter cloth; ~ **de fondo** *f* TEXTIL ground cloth; ~ **de formación** *f* PAPEL Fourdrinier wire, *de material plástico* forming fabric; ~ **forrada** *f* P&C, REVEST, TEXTIL coated fabric; ~ **de forro** *f* TEXTIL lining fabric; ~ **con hilos mezclados** *f* TEXTIL union cloth; ~ **impermeable** *f* P&C *material textil* oilskin; ~ **de lana regenerada** *f* TEXTIL shoddy fabric; ~ **de lija** *f* ING MECÁ, REVEST coated abrasive; ~ **de mesa plana** *f* PAPEL Fourdrinier wire; ~ **metálica** *f* P&C *parte de equipo* wire mesh, PROD gauze, wire cloth; ~ **metálica de alambre de bronce** *f* CONST brass-wire gauze; ~ **metálica fina** *f* PROD wire gauze; ~ **a rayas** *f* TEXTIL stripe fabric; ~ **de refuerzo de envés** *f* TEXTIL *forros* backing fabric; ~ **revestida** *f* P&C, REVEST, TEXTIL coated fabric; ~ **revestida poromérica** *f* P&C poromeric coated fabric; ~ **para sacos o bolsas** *f* AGRIC bagging; ~ **de soporte** *f* PAPEL backing wire; ~ **de sostén** *f* TEXTIL backing; ~ **de tapeta** *f* TEXTIL facing fabric; ~ **de tejido liso** *f* PAPEL plain weave; ~ **para trajes** *f* TEXTIL suiting; ~ **de vidrio tejida** *f* C&V knitted glass fabric
telar *m* CONST *arquitectura* reveal, TEXTIL loom; ~ **de**

lanzadera de aletas *m* TEXTIL fly-shuttle loom; ~ **no automático** *m* TEXTIL nonautomatic loom
telaraña: la ~ *f* INFORM&PD the Web; ~ **mundial** *f* (*WWW*) INFORM&PD World Wide Web (*WWW*)
teleaccionamiento *m* GEN remote control
teleautógrafo *m* TELECOM telewriter, telewriting
telecine *m* TV telecine
telecomercialización *f* TELECOM telemarketing
telecomprobación *f* INFORM&PD, TELECOM, TV remote monitoring
telecomputación *f* TELECOM teleinformatics
telecomunicaciones *f pl* FÍS, INFORM&PD, TRANSP telecommunications; ~ **vía satélite** *f pl* (*SATCOM*) TRANSP MAR satellite communications (*SATCOM*)
teleconferencia *f* INFORM&PD, TELECOM teleconference
teleconmutación *f* INFORM&PD telecommuting, TELECOM, TV remote switching
telecontrol *m* GEN remote control; ~ **por radio** *m* TELECOM, TRANSP radio telecontrol
telecopia *f* IMPR, INFORM&PD facsimile, TELECOM facsimile, facsimile message
telecopiadora *f* IMPR, INFORM&PD, TELECOM facsimile machine
teledetección *f* CONTAM MAR, GEOL, INFORM&PD, TEC ESP remote sensing, TELECOM remote detection, remote sensing, TV remote sensing; ~ **aérea** *f* CONTAM MAR airborne remote sensing; ~ **desde el aire** *f* CONTAM MAR airborne remote sensing
teledinámico *adj* ING MECÁ teledynamic
teledirección *f* GEN remote control, TELECOM remote management
teledistribución *f* TV teledistribution
teleescritor *m* Esp (*cf teleprinter AmL*) INFORM&PD, TELECOM teletypewriter (*AmE*), teleprinter (*BrE*)
teleespectador *m* TELECOM, TV television viewer
teleférico *m* FERRO, ING MECÁ cable railway, TRANSP cable railway, cableway, tramway; ~ **con cabinas** *m* TRANSP gondola cableway; ~ **góndola** *m* TRANSP gondola cableway
telefonear *vti* TELECOM ring
telefonía *f* FÍS, INFORM&PD, TELECOM telephony; ~ **sin hilos** *f* TELECOM wireless telephony; ~ **inalámbrica** *f* TELECOM wireless telephony; ~ **simultánea de una única portadora** *f* TELECOM voice-operated device anti-singing (*VODAS*)
telefonista *m* TELECOM telephone operator, telephonist
teléfono *m* TELECOM telephone; ~ **de automóvil** *m* AmL (*cf teléfono de coche Esp*) TELECOM, TRANSP car telephone; ~ **de botonera** *m* AmL (*cf teléfono de teclado Esp*) TELECOM push-button telephone; ~ **de cabina** *m* TRANSP AÉR cabin telephone; ~ **de carro** *m* AmL (*cf teléfono de coche Esp*) TELECOM, TRANSP car telephone; ~ **de coche** *m* Esp (*cf teléfono de automóvil AmL, teléfono de carro AmL*) TELECOM, TRANSP car telephone; ~ **contestador** *m* TELECOM telephone-answering machine; ~ **de disco** *m* TELECOM dial telephone; ~ **de doble señalización** *m* TELECOM dual-signaling telephone (*AmE*), dual-signalling telephone (*BrE*); ~ **de emergencia** *m* TELECOM, TRANSP emergency telephone; ~ **de escucha** *m* TELECOM observation telephone; ~ **inalámbrico** *m* TELECOM cordless telephone; ~ **llamado** *m* TELECOM called telephone; ~ **de mezclado** *m* D&A scrambler telephone; ~ **monedero**

m TELECOM coin-operated payphone; ~ **multifrecuencia** *m* TELECOM multifrequency telephone (*MF telephone*), multiple frequency telephone; ~ **operado con monedas** *m* TELECOM coin-operated payphone; ~ **de operador** *m* TELECOM operator's telephone; ~ **de pago previo** *m* TELECOM payphone; ~ **de pago con tarjeta** *m* TELECOM card-operated payphone, cardphone; ~ **público** *m* TELECOM payphone, public telephone, *accionado por monedas* coin-operated payphone; ~ **que llama** *m* TELECOM calling telephone; ~ **de teclado** *m Esp* (*cf teléfono de botonera AmL*) TELECOM push-button telephone
telegestión *f* ELEC remote control
telegobierno *m* GEN remote control
telegrafía: ~ **en facsímil** *f* TELECOM facsimile telegraphy
telégrafo: ~ **de máquinas** *m* TRANSP MAR engine-room telegraph
telegrama *m* TELECOM wire
teleimpresor *m* INFORM&PD, TELECOM teleprinter (*BrE*), teletypewriter (*AmE*)
teleindicación *f* ELEC *medición* telemetry; ~ **de alarma** *f* TELECOM remote-alarm indication (*RAI*)
teleindicador *m* ELEC *medición* telemeter
teleinformática *f* TELECOM teleinformatics
teleinscriptor *m* TELECOM telewriter
telemando *m* AUTO, CINEMAT, CONST remote control, D&A command, remote control, ELEC, FOTO remote control, INFORM&PD command, ING MECÁ, MECÁ remote control, TEC ESP remote control, *cohetes y vehículos espaciales no tripulados* command, TELECOM, TRANSP, TV, VEH remote control
telemaniobra *f* GEN remote control
telemanipulador *m* OCEAN *sumergibles tripulados de recolección de muestras* remote manipulator
telemantenimiento *m* CALIDAD housekeeping telemetry, TELECOM remote maintenance
telemarketing *m Esp* (*cf telemercadeo AmL*) TELECOM telemarketing
telemática *f* INFORM&PD data communication, TELECOM telematics
telemático *adj* TELECOM telematic
telemecánica *f* ING MECÁ telemechanics
telemecanismo *m* ING MECÁ telemechanism
telemedición *f* ELEC *medición*, ING MECÁ, TELECOM telemetry
telemedida *f* ELEC *medición*, ING MECÁ, TELECOM telemetry
telemedidor *m* ELEC *medición*, ING MECÁ telemeter
telemercadeo *m AmL* (*cf telemarketing Esp*) TELECOM telemarketing
telemetrar *vt* D&A range to
telemetría *f* GEN telemetry; ~ **y control de rastreo** *f* (*TTC*) TEC ESP tracking telemetry and command (*TTC*); ~ **de datos de abordo** *f* TEC ESP housekeeping telemetry; ~ **por láser** *f* TELECOM laser telemetry; ~ **preparatoria** *f* TEC ESP housekeeping telemetry
telémetro *m* CINEMAT rangefinder, D&A rangefinder, telemeter, ELEC *medición* telemeter, ELECTRÓN, FOTO rangefinder, ING MECÁ telemeter, TV rangefinder; ~ **acoplado** *m* FOTO coupled rangefinder; ~ **de imagen dividida** *m* FOTO split image rangefinder; ~ **por láser** *m* ELECTRÓN laser rangefinder; ~ **de nubes** *m* TRANSP AÉR ceilometer; ~ **óptico** *m* FOTO optical rangefinder; ~ **de radar** *m* D&A, TRANSP, TRANSP AÉR, TRANSP MAR radar telemeter
telemetrógrafo *m* ING MECÁ telemetrograph
teleobjetivo *m* CINEMAT telephoto lens, FOTO telelens, telephoto lens
teleoperador *m* INFORM&PD, OCEAN, TELECOM remote operator
telepoint *m* TELECOM telepoint
teleprinter *m AmL* (*cf teleescritor Esp*) INFORM&PD teleprinter (*BrE*), teletypewriter (*AmE*)
teleprocesamiento: ~ **de datos** *m* INFORM&PD, TELECOM remote data processing
teleproceso *m* INFORM&PD teleprocessing; ~ **por lotes** *m* INFORM&PD remote batch processing
telera: ~ **transportadora** *f* PROD traveling apron (*AmE*), travelling apron (*BrE*)
telerregistrador *m* CINEMAT, TV telerecorder
telerregistro *m* CINEMAT, TV telerecording
telerregulación *f* GEN remote control
telescópico *adj* INSTR, MECÁ, ÓPT, TEC ESP telescopic
telescopio *m* CONST, FÍS, INSTR, ÓPT, TEC ESP telescope; ~ **de ajuste** *m* INSTR adjusting telescope; ~ **de alineación** *m* INSTR alignment telescope; ~ **de anteojo central** *m* INSTR transit telescope; ~ **articulado de cizalla** *m* INSTR shear-jointed telescope; ~ **astronómico** *m* FÍS, INSTR astronomical telescope; ~ **de Cassegrain** *m* FÍS, TEC ESP Cassegrain telescope; ~ **cenital** *m* INSTR, TEC ESP zenith telescope; ~ **sin cúpula** *m* INSTR domeless telescope; ~ **de fusil** *m* INSTR rifle telescope; ~ **de Galileo** *m* FÍS Galilean telescope; ~ **gregoriano** *m* FÍS Gregorian telescope; ~ **meridiano** *m* INSTR, ÓPT meridian telescope; ~ **de mira** *m* CONST sighting telescope; ~ **monocular** *m* INSTR monocular telescope; ~ **newtoniano** *m* FÍS Newtonian telescope; ~ **de noche** *m* INSTR night telescope; ~ **panorámico** *m* INSTR panoramic telescope; ~ **de prisma erector** *m* INSTR erecting-prism telescope; ~ **protónico** *m* INSTR proton telescope; ~ **reflector** *m* FÍS reflecting telescope, INSTR, ÓPT reflecting telescope, reflector telescope; ~ **de reflexión** *m* FÍS, INSTR, ÓPT reflecting telescope; ~ **de refracción** *m* FÍS, INSTR, ÓPT refracting telescope; ~ **refractante** *m* FÍS refracting telescope, INSTR, ÓPT refracting telescope, refractor telescope; ~ **refringente** *m* FÍS refracting telescope, INSTR, ÓPT refracting telescope, refractor telescope; ~ **solar** *m* INSTR solar telescope; ~ **terrestre** *m* INSTR terrestrial telescope; ~ **de torre** *m* INSTR tower telescope; ~ **visual** *m* INSTR sighting telescope
telescospio: ~ **óptico** *m* TEC ESP optical telescope
teleselección: ~ **por frecuencia vocal** *f* TELECOM voice dialing (*AmE*), voice dialling (*BrE*)
teleselector: ~ **por frecuencia vocal** *m* TELECOM voice dialer (*AmE*), voice dialler (*BrE*)
telesilla *f* TRANSP chair lift
telesoftware *m* INFORM&PD telesoftware
teletexto *m* INFORM&PD, TELECOM, TV teletext
teletipo *m* (*TTY*) INFORM&PD, TELECOM teletype (*TTY*)
teletipocomposición *f* IMPR teletypesetting
teletrabajo *m* TELECOM teleworking
televigilancia *f* INFORM&PD, TELECOM, TV remote monitoring
televisar *vt* TV telecast, televise
televisión *f* (*TV*) TELECOM, TV television (*TV*); ~ **para abonados** *f Esp* (*cf televisión por subscripción AmL*)

TELECOM pay television; **~ de la aeronave** *f* TV airborne television; **~ de alta definición** *f Esp* (*cf televisión por definición extendida AmL*) TELECOM high-definition television (*HDTV*), TV extended definition television (*EDTV*), high-definition television (*HDTV*); **~ en blanco y negro** *f* TV black and white television; **~ por cable** *f* TV piped television; **~ por cable privada** *f* TV pay cable; **~ en circuito cerrado** *f* TELECOM, TV closed-circuit television (*CCTV*); **~ en color** *f* TV color television (*AmE*), colour television (*BrE*); **~ por definición extendida** *f AmL* (*cf televisión de alta definición Esp*) TELECOM high-definition television (*HDTV*), TV extended definition television (*EDTV*), high-definition television (*HDTV*); **~ digital** *f* TV digital television; **~ experimental** *f* TV experimental television; **~ de exploración lenta** *f* TELECOM slow-scan television; **~ interactiva** *f* TV interactive television; **~ de pago** *f Esp* (*cf televisión por subscripción AmL*) TELECOM, TV pay television; **~ proyectada una pantalla** *f* TV projection television; **~ por subscripción** *f AmL* (*cf televisión para abonados Esp*) TELECOM pay television

télex *m* INFORM&PD, TELECOM telex; **~-positivo** *m* TELECOM telex-plus

telurato *m* QUÍMICA tellurate

telurhídrico *adj* QUÍMICA tellurhydric

telúrico *adj* ENERG RENOV, QUÍMICA telluric

telúrido *m* QUÍMICA telluride

telurio *m* (*Te*) ING ELÉC, QUÍMICA tellurium (*Te*); **~ foliado** *m* MINERAL foliated tellurium, nagyagite; **~ hojoso** *m* MINERAL foliated tellurium; **~ hojoso nagyagita** *m* MINERAL nagyagite

telurita *f* MINERAL telluric ocher (*AmE*), telluric ochre (*BrE*), tellurite, QUÍMICA tellurite

telurito *m* MINERAL, QUÍMICA tellurite

teluro: **~ negro** *m* MINERAL black tellurium

telurómetro *m* CONST *topografía* tellurometer

teluroso *adj* QUÍMICA tellurous

TEM *abr* (*electromagnético transversal*) ING ELÉC TEM (*transverse electromagnetic*)

temblor *m* INFORM&PD, TELECOM jitter; **~ de la imagen** *m* TELECOM picture bounce; **~ de tierra** *m* CONST earthquake, GEOFÍS earth tremor, earthquake; **~ de tierra ocasionado por el hombre** *m* CONTAM *daño geológico* man-made earth tremor, man-made earthquake

temperamento *m* ACÚST temperament; **~ uniforme** *m* ACÚST equal temperament

temperatura[1]: **con ~ controlada** *adj* TERMO temperature-controlled; **con ~ regulada** *adj* TERMO temperature-controlled

temperatura[2] *f* GEN temperature, GAS, OCEAN, TEC ESP heat; **~ absoluta** *f* FÍS, REFRIG, TERMO absolute temperature; **~ del aire de entrada** *f* REFRIG entering air temperature; **~ de almacenamiento** *f* PROD storage temperature; **~ ambiente** *f* EMB room temperature, FÍS ambient temperature, room temperature, GAS, INSTAL TERM, METAL, METR ambient temperature, TERMO ambient temperature, room temperature; **~ ambiente normal** *f* REFRIG standard ambient temperature; **~ de aspiración** *f* REFRIG suction temperature; **~ de autoestiramiento** *f* C&V self-sagging temperature; **~ en boca de pozo** *f* ENERG RENOV wellhead temperature; **~ de Boyle** *f* FÍS Boyle temperature; **~ de bulbo húmedo** *f* TERMOTEC wet-bulb temperature; **~ de bulbo seco** *f*

TERMOTEC dry-bulb temperature; **~ de la cabina** *f* TRANSP AÉR cabin temperature; **~ Celsius** *f* FÍS, MECÁ, METEO, METR, PROD, QUÍMICA Celsius temperature; **~ de cocido** *f* C&V firing temperature; **~ de color** *f* CINEMAT, FOTO color temperature (*AmE*), colour temperature (*BrE*); **~ de compensación** *f* TERMO equalizing temperature; **~ de congelación** *f* ALIMENT, C&V, EMB, FÍS, ING MECÁ, METAL, P&C, PROC QUÍ, REFRIG, TERMO, TRANSP MAR freezing point; **~ de contención de grietas** *f* METAL crack arrest temperature; **~ controlada** *f* TERMO, TERMOTEC controlled temperature; **~ crítica** *f* FÍS, REFRIG, TEC PETR, TERMO, TERMOTEC critical temperature; **~ de cuerpo negro** *f* FÍS black-body temperature; **~ de Curie** *f* FÍS RAD Curie temperature; **~ de Debye** *f* FÍS Debye temperature; **~ de descarga** *f* CONTAM *procesos industriales* outlet temperature, REFRIG, TERMO discharge temperature; **~ de descomposición** *f* EMB decomposition temperature; **~ de descongelación** *f* VEH *aceite* pour point; **~ de detonación** *f* MINAS *explosivos* detonating point, detonation temperature; **~ diferencial** *f* TEC PETR differential temperature; **~ efectiva** *f* TERMO effective temperature; **~ de Einstein** *f* FÍS Einstein temperature; **~ empírica** *f* TERMO empirical temperature; **~ de encendido** *f* AUTO, MECÁ ignition point; **~ de enturbiamiento** *f* DETERG cloud point; **~ de evaporación** *f* REFRIG evaporating temperature; **~ excesiva** *f* TERMO excess temperature; **~ exterior de diseño** *f* INSTAL TERM design outside temperature; **~ exterior proyectada** *f* INSTAL TERM *de proyecto* design outside temperature; **~ de filamento** *f* ELEC *de válvula, tubo* filament temperature; **~ de fluidez** *f* VEH *lubricante* pour point; **~ de formación** *f* EMB *envases* forming temperature; **~ funcional** *f* TEC ESP operating temperature; **~ de funcionamiento** *f* PROD operation temperature, REFRIG operating temperature; **~ de fusión** *f* C&V, P&C fusing point, TERMO fusing point, melting point; **~ de fusión del hielo** *f* FÍS ice point; **~ de fusión incipiente** *f* REFRIG initial fusion temperature; **~ del gas de escape** *f* AUTO, TRANSP, TRANSP AÉR exhaust-gas temperature; **~ de la gota o carga** *f* C&V gob temperature; **~ en grados Kelvin** *f* TERMO Kelvin temperature; **~ de ignición** *f* FÍS, TEC PETR ignition point, TERMO kindling point; **~ de inflamabilidad** *f* ALIMENT, AUTO, FÍS, P&C, QUÍMICA, REFRIG, TEC PETR, TERMO, TERMOTEC flash point; **~ de inflamabilidad espontánea** *f* SEG *aceites y petróleos* fire point; **~ de inflamación** *f* ALIMENT, AUTO, FÍS, P&C, QUÍMICA, REFRIG, TEC PETR flash point, TERMO kindling point, flash point, TERMOTEC flash point; **~ interna** *f* EMB internal temperature; **~ de interrupción de grietas** *f* METAL crack arrest temperature; **~ intrínseca** *f* ELECTRÓN intrinsic temperature; **~ de inversión** *f* ELEC *de termopar*, FÍS inversion temperature; **~ Kelvin** *f* CINEMAT, TEC ESP Kelvin temperature; **~ de licuación de soldadura** *f* TEC PETR flow point; **~ de maduración** *f* C&V maturing temperature; **~ del mar** *f* TRANSP MAR sea temperature; **~ máxima en el centro del combustible** *f* NUCL maximum fuel-central temperature; **~ máxima de la vaina** *f* NUCL peak cladding temperature (*PCT*); **~ media normal** *f* REFRIG average mean temperature; **~ mínima de revenido** *f* C&V lower annealing temperature; **~ mínima de templado** *f* C&V lower annealing temperature;

~ Nel *f* ELEC *antiferromagnetismo* Nel temperature; **~ normal** *f* TEC PETR standard temperature, STP; **~ de operación** *f* TRANSP AÉR operating temperature; **~ de punto de rocío** *f* NUCL, TERMO dew point temperature; **~ de recocido** *f* C&V annealing range; **~ de referencia** *f* METEO, METR reference temperature; **~ de reforzamiento** *f* GEOL blocking temperature; **~ de ruido** *f* ING ELÉC, TEC ESP noise temperature; **~ de ruido equivalente** *f* TEC ESP *telecomunicaciones* equivalent noise temperature; **~ de ruido espacial** *f* TEC ESP sky noise temperature; **~ de ruido de fondo** *f* TEC ESP sky noise temperature; **~ de saturación** *f* REFRIG saturation temperature, TERMO dew point temperature; **~ de seguridad para el planchado** *f* TEXTIL safe ironing temperature; **~ de sinterización** *f* PROC QUÍ sintering temperature; **~ de spin** *f* FÍS spin temperature; **~ stándard** *f* FÍS standard temperature (*STP*); **~ de superficies que se pueden tocar** *f* SEG temperature of touchable surfaces; **~ superior de revenido** *f* C&V upper annealing temperature; **~ termodinámica** *f* FÍS thermodynamic temperature; **~ de termosellado** *f* EMB heat seal temperature; **~ de trabajo** *f* C&V working temperature, TEC ESP operating temperature; **~ de transformación** *f* C&V transformation temperature; **~ de transición** *f* FÍS, METAL, NUCL, REFRIG transition temperature; **~ de transición de ductilidad nula** *f* NUCL nil-ductility transition temperature; **~ de transición del estado vítreo** *f* C&V, P&C *prueba* glass transition temperature; **~ de transición a la rotura frágil** *f* MECÁ brittle-fracture transition-temperature; **~ de transición vítrea** *f* C&V, P&C glass transition temperature; **~ de la tubería de salida** *f* TEC PETR flow-line temperature; **~ de los tubos del reactor** *f* TRANSP AÉR jet pipe temperature; **~ del vidrio** *f* C&V *en el horno* gathering temperature

temperismo *m* CARBÓN, GEOL weathering

tempestad *f* METEO storm; **~ tropical** *f* METEO tropical revolving storm

templa *f* P&C *pintura* distemper

templabilidad *f* CRISTAL, METAL hardenability

templado[1] *adj* MECÁ tempered; **~ en aceite** *adj* METAL, TERMO oil-hardened, oil-quenched; **~ con agua** *adj* TERMO water-hardened; **~ artificialmente** *adj* TERMO artificially aged; **~ en fueloil** *adj* METAL oil-hardened

templado[2] *m* C&V temper, CONST tempering; **~ en agua** *m* C&V, CRISTAL, QUÍMICA quenching; **~ brusco en agua** *m* METAL quench; **~ desigual** *m* C&V uneven temper; **~ rápido** *m* C&V rapid annealing

templador: **~ con banda transportadora** *m* C&V conveyor-belt lehr; **~ continuo** *m* C&V continuous recirculation lehr; **~ de convección forzada** *m* C&V forced-convection lehr; **~ de túnel** *m* C&V tunnel lehr

templar *vt* C&V, CRISTAL, MECÁ, METAL, PROD, QUÍMICA, TERMO quench, TRANSP MAR *cuerda* haul taut; **~ en aceite** *vt* METAL, TERMO *metales* oil quench; **~ al aire** *vt* PROD chill harden; **~ artificialmente** *vt* TERMO *funderías* artificially age; **~ instantáneamente** *vt* METAL, PROD, TERMO *metales* flash harden; **~ a la llama** *vt* TERMO flame-harden; **~ superficialmente con la llama** *vt* TERMO flame-harden

templarse *v refl* PROD harden

temple *m* CARBÓN *aceros* hydrogenation, CRISTAL, METAL hardening, NUCL tempering, PROD *pinturas* distemper, TEC ESP *acero* hardening; **~ al aire** *m* PROD chill hardening; **~ artificial** *m* METAL, TERMO *metales* artificial ageing; **~ en coquilla** *m* PROD *fundición* chilling; **~ por corrientes de inducción** *m* ING ELÉC induction hardening; **~ gamma** *m* NUCL gamma quench; **~ general instantáneo en aceite** *m* METAL, TERMO *metales* oil quenching; **~ instantáneo** *m* METAL quenching; **~ instantáneo al vapor** *m* METAL vapor quenching (*AmE*), vapour quenching (*BrE*); **~ isotérmico** *m* METAL isothermal quenching; **~ químico** *m* METAL, QUÍMICA chemical hardening; **~ del revestimiento** *m* TRANSP film cooling

temporada *f* AGRIC season

temporal *m* METEO gale; **~ con mucha mar** *m* TRANSP MAR heavy weather

temporización *f* ELECTRÓN, INFORM&PD, TELECOM timing; **~ del elemento de señal** *f* TELECOM signal element timing; **~ del elemento de señal del receptor** *f* TELECOM receiver-signal element timing; **~ del intervalo elemental** *f* TELECOM signal element timing; **~ del intervalo unitario** *f* TELECOM signal element timing; **~ lógica** *f* ELECTRÓN logic timing

temporizador *m* INFORM&PD, ING ELÉC, LAB timer, PROD time delay position switch, timer; **~ en cascada** *m* PROD cascading timer; **~ de circuito** *m* ELECTRÓN circuit delay; **~ de clavijas** *m* ING ELÉC plug-timer; **~ continuo** *m* PROD free-running timer; **~ controlador de secuencia** *m* PROD watchdog timer; **~ en demora** *m* PROD timer-on delay; **~ diferencial** *m* ELECTRÓN differential delay; **~ electrónico** *m* ELECTRÓN electronic timer; **~ de enchufe** *m* ING ELÉC plug-timer; **~ de intervalos** *m* INFORM&PD interval timer; **~ de intervalos programable** *m* (*TIP*) ELECTRÓN programmable interval timer (*PIT*); **~ no retentivo** *m* PROD non-retentive timer; **~ programable** *m* LAB *calentamiento* program timer (*AmE*), programme timer (*BrE*); **~ supervisor** *m* TELECOM supervisory timer; **~ de vigilancia** *m* INFORM&PD watchdog timer

temprano *adj* GEOL early

tenacidad *f* ING MECÁ tenacity, MECÁ, TELECOM tensile strength, TEXTIL tenacity; **~ de cizallamiento** *f* ING MECÁ shearing tenacity; **~ de corte** *f* ING MECÁ shearing tenacity; **~ a la entalla** *f* NUCL notch toughness; **~ a la fractura** *f* P&C *propiedad física, prueba* fracture toughness; **~ a la rotura** *f* P&C *propiedad física, prueba* fracture toughness; **~ torsional** *f* ING MECÁ torsional tenacity

tenacillas *f pl* ELEC *herramienta*, ING MECÁ pliers, LAB *equipo general* tweezers

tenaza: **~ dobladora** *f* ELEC, MECÁ crimping tool; **~ engarzadora** *f* ELEC, MECÁ crimping tool; **~ de junta de labios** *f* ING MECÁ lip-joint pliers

tenazas *f pl* C&V tweezers, CONST pincers, tongs, PROD *fundición* lifter; **~ para alambre de fijación** *f pl* ING MECÁ lockwire pliers; **~ de aros de pistón** *f pl* ING MECÁ piston-ring pliers; **~ articuladas para manejar sillares** *f pl* PROD lever grip tongs; **~ de boca curva** *f pl* PROD elbow tongs; **~ de boca plana** *f pl* ING MECÁ duckbill pliers; **~ de bujes** *f pl* PROD elbow tongs; **~ cónicas** *f pl* ING MECÁ cone pliers; **~ de corte** *f pl* CONST cutting nippers, PROD nippers; **~ de corte de extremos** *f pl* PROD end-cutting nippers; **~ estriadas** *f pl* ING MECÁ multigrip pliers; **~ para forja** *f pl* CONST smith's pliers; **~ de herrero** *f pl* ING MECÁ blacksmith's tongs; **~ de mufla** *f pl* LAB crucible tongs; **~ para pernos** *f pl* PROD bolt tongs; **~ planas** *f pl* ING

MECÁ end nippers; ~ **portacarriles** *f pl* FERRO *infraestructura* rail tongs; ~ **de sujeción** *f pl* ING MECÁ lock-grip pliers (*BrE*), locking pliers (*AmE*); ~ **de triscar** *f pl* ING MECÁ saw set; ~ **para tubería** *f pl* CONST pipe tongs

tendencia *f* TEXTIL trend; ~ **espacial** *f* CONTAM spatial trend; ~ **normal** *f* TEC PETR normal trend; ~ **a orzar** *f* TRANSP MAR *navegación* weather helm; ~ **de presión** *f* METEO pressure tendency; ~ **a producir burbujas** *f* C&V tendency to reboil; ~ **en el tiempo** *f* CONTAM temporal fluctuation, temporal variation, time trend; ~ **a vulcanizarse prematuramente** *f* P&C *vulcanización, prueba* scorching tendency

ténder *m* FERRO *vehículos* tender

tender *vt* ING MECÁ, PROD *líneas telefónicas* run

tendido *m* CONST laying, ELEC, ING ELÉC, TELECOM, TV *fuente de alimentación* cabling; ~ **del cable** *m* ELEC, ING ELÉC *fuente de alimentación* cable run; ~ **de cables** *m* ELEC *fuente de alimentación* cable laying, *conexión* wiring, ING ELÉC cable laying, wire; ~ **de carriles** *m* FERRO *infraestructura* rail laying; ~ **eléctrico** *m* ELEC *conexión* wiring; ~ **de tuberías** *m* CONST pipelaying, pipeworks; ~ **de vía** *m* CONST *ferrocarriles* tracklaying, FERRO *infraestructura* tracklaying, *vehículos* plate laying

tenedero *m* TRANSP MAR anchoring ground

tenedor: ~ **de carga** *m* C&V carrying-in fork

tener: ~ **firme** *vi* PROD hold fast; ~ **la medida** *vi* CONST be to size

teñido[1] *adj* COLOR dyed, tinted; ~ **en cable** *adj* TEXTIL tow-dyed; ~ **en canillas** *adj* COLOR cop-dyed; ~ **en corona** *adj* TEXTIL *hilatura* cake-dyed; ~ **de encoladora** *adj* TEXTIL slasher-dyed; ~ **en hilado** *adj* TEXTIL *después de hilar* spun-dyed; ~ **en la masa** *adj* COLOR mass-colored (*AmE*), mass-coloured (*BrE*); ~ **en mecha** *adj* TEXTIL slubbing-dyed; ~ **en peinado** *adj* TEXTIL top-dyed; ~ **en pieza** *adj* TEXTIL piece-dyed; ~ **en rama** *adj* TEXTIL stock-dyed

teñido[2] *m* ALIMENT coloring (*AmE*), colouring (*BrE*), COLOR dyeing, IMPR tinting, PAPEL color (*AmE*), colour (*BrE*), QUÍMICA staining, TEXTIL dyeing; ~ **azoico** *m* COLOR naphthol dyeing; ~ **al azúcar** *m* COLOR sugar dye; ~ **en bobinas** *m* COLOR package dyeing; ~ **en cinta peinada** *m* COLOR top dyeing; ~ **con colorante de tina** *m* COLOR vat dyeing; ~ **comparado** *m* COLOR comparative dyeing; ~ **continuo** *m* COLOR continuous dyeing; ~ **en cuerda** *m* COLOR rope dyeing; ~ **al disolvente** *m* COLOR solvent dyeing; ~ **de doble tono** *m* COLOR two-tone dyeing; ~ **en estrella** *m* COLOR star dyeing; ~ **de fibras en rama** *m* COLOR raw-stock dyeing; ~ **de fondo** *m* COLOR bottom dyeing; ~ **en frío** *m* COLOR cold dyeing; ~ **fuera de matiz** *m* COLOR off-shade dyeing; ~ **mediante impregnación** *m* COLOR pad dyeing; ~ **de inmersión** *m* COLOR dip dyeing; ~ **localizado** *m* COLOR localized dyeing; ~ **en madeja a presión** *m* COLOR pressure hank dyeing; ~ **con máquina de teñir al ancho** *m* COLOR jig dyeing; ~ **metacromo** *m* COLOR metachrome dyeing; ~ **monocromo** *m* COLOR monochrome dyeing; ~ **con mordente** *m* FOTO mordant dyeing; ~ **con nudos** *m* COLOR tie-dyeing; ~ **parejo** *m* COLOR level dyeing; ~ **pesado** *m* COLOR dead dyeing; ~ **con pigmentos** *m* COLOR pigment dyeing; ~ **a pincel** *m* COLOR brush dyeing; ~ **a presión** *m* COLOR pressure-dyeing; ~ **puro** *m* COLOR plain dyeing; ~ **suelto** *m* COLOR loose dyeing; ~ **en tambor** *m* COLOR drum dyeing; ~ **en tina** *m* COLOR kettle dyeing; ~ **en urdimbre** *m* COLOR warp dyeing

teñir[1]: **sin** ~ *adj* TEXTIL undyed

teñir[2] *vt* COLOR dye, stain, tincture, tinge, tint, FOTO dye, PAPEL color (*AmE*), colour (*BrE*), dye, stain, QUÍMICA stain, TEXTIL dye; ~ **de carmesí** *vt* COLOR crimson; ~ **en hilado** *vt* TEXTIL yarn-dye; ~ **en pieza** *vt* TEXTIL dye in the piece; ~ **a pistola** *vt* COLOR shot-dye; ~ **a presión** *vt* COLOR dye under pressure; ~ **con pulverizador** *vt* COLOR shot-dye; ~ **con rojo turco** *vt* COLOR dye turkey red

tennantita *f* MINERAL tennantite

tenor *m* ING MECÁ tenor

tenorita *f* MINERAL tenorite

tensador: ~ **de película** *m* CINEMAT, FOTO film holder

tensar[1] *vt* ING MECÁ strain, tighten

tensar[2]: ~ **con vientos** *vi* CONST, ING ELÉC guying

tensioactivo *adj* QUÍMICA surface-active

tensiómetro *m* LAB *instrumento* surface tension meter, OCEAN tensiometer

tensión[1]: **bajo** ~ *adj* ELEC live, *circuito* alive; **en** ~ *adj* PROD *electricidad* live

tensión[2] *f* C&V strain, stress, CARBÓN, CONST stress, ELEC voltage, FÍS strain, stress, tension, GEOL stress, ING ELÉC tension, voltage, ING MECÁ strain, stress, stretch, tension, tightening, MECÁ pressure, strain, stress, METAL strain, tension, P&C, TEC ESP, TEC PETR stress, TELECOM voltage, TEXTIL tension, VEH voltage; ~ **de aceleración** *f* ING ELÉC accelerating voltage; ~ **aceleradora** *f* ELEC *partículas cargadas electrizadas* accelerating voltage; ~ **acelerante** *f* ELEC *partículas cargadas electrizadas* accelerating voltage; ~ **aceleratriz** *f* ELEC *partículas cargadas electrizadas* accelerating voltage; ~ **activa** *f* ELEC *corriente alterna*, ING ELÉC active voltage; ~ **de activación** *f* ING ELÉC firing voltage; ~ **admisible** *f* ING ELÉC permissible voltage; ~ **de ajuste** *f* ELEC adjusting voltage; ~ **aleatoria** *f* ELEC, ELECTRÓN, ING ELÉC random voltage; ~ **de alimentación** *f* PROD supply voltage; ~ **alterna** *f* ELEC alternating voltage; ~ **alternativa** *f* ING ELÉC alternating voltage; ~ **de anulación** *f* ELEC null voltage; ~ **aplicada** *f* ELEC impressed voltage; ~ **en avalancha** *f* ELEC avalanche voltage; ~ **de bobinado** *f* CINEMAT winding tension; ~ **de carga** *f* ELEC, ING ELÉC *acumulador, batería* charging voltage; ~ **de CC** *f* ING ELÉC DC voltage; ~ **de CC de bajo nivel** *f* ELEC, PROD low-level DC voltage; ~ **de choque** *f* ELEC impulse voltage; ~ **de la cinta** *f* TV tape tension; ~ **de circuito abierto** *f* ELEC, ENERG RENOV open-circuit voltage; ~ **del circuito cortado** *f* ELEC *interruptor o disyuntor* recovery voltage; ~ **circunferencial** *f* METAL circumferential stress; ~ **compensadora** *f* ELEC compensating voltage; ~ **de compresión** *f* METAL compression stress; ~ **constante** *f* ELEC, ING ELÉC constant voltage; ~ **de contorneamiento en seco** *f* ELEC *de arco* dry flashover voltage; ~ **de correa** *f* ING MECÁ belt tension; ~ **de correa trapezoidal** *f* ING MECÁ V-belt tension; ~ **de correa en V** *f* ING MECÁ V-belt tension; ~ **de la corriente** *f* ELEC *de alimentación* current voltage (*AmE*), mains voltage (*BrE*), ING ELÉC, TV mains voltage (*BrE*), current voltage (*AmE*); ~ **de corte** *f* FÍS, P&C *propiedad física, prueba* shear stress; ~ **de cortocircuito a corriente nominal** *f* ELEC *de transformador* impedance voltage at rated current;

~ **de cortocircuito a corriente de régimen** *f* ELEC *de transformador* impedance voltage at rated current; ~ **en cresta** *f* ING ELÉC peak voltage; ~ **crítica** *f* ELEC, ING ELÉC critical voltage; ~ **en cuadratura** *f* ELEC *circuito de corriente alterna* quadrature voltage; ~ **de cuasicresta** *f* TELECOM *móvil aeronáutico* quasi-peak voltage; ~ **de desplazamiento de punto neutro** *f* ELEC *red de distribución* neutral point displacement voltage; ~ **de desprendimiento** *f* ING ELÉC drop-out voltage; ~ **diferencial** *f* ING ELÉC differential voltage; ~ **directa sin carga** *f* ELEC no-load direct voltage; ~ **directa en circuito abierto** *f* ELEC no-load direct voltage; ~ **directa en vacío** *f* ELEC no-load direct voltage; ~ **de disparo** *f* ELEC, ING ELÉC, ING MECÁ breakdown voltage; ~ **disruptiva** *f* ELEC disruptive voltage; ~ **disruptiva en seco** *f* ELEC *de arco* dry flashover voltage; ~ **efectiva** *f* CARBÓN effective stress, ING ELÉC effective voltage, TEC PETR *geología* effective stress; ~ **de ensayo** *f* ELEC test voltage; ~ **de entrada** *f* ELEC, ING ELÉC input voltage; ~ **entre fase y neutro** *f* ELEC star voltage, voltage to neutral; ~ **entre fases** *f* ELEC star voltage, *fuente de alimentación trifásica* line-to-line voltage, *sistema trifásico* phase-to-phase voltage; ~ **especificada** *f* ELEC, ING ELÉC, PROD rated voltage; ~ **en estrella** *f* ELEC voltage to neutral; ~ **excesiva** *f* ELEC, ING ELÉC, QUÍMICA overvoltage; ~ **de excitación** *f* ELEC alternator field voltage; ~ **de fase** *f* ELEC *sistema* phase voltage; ~ **de una fase** *f* ELEC star voltage; ~ **por fase** *f* ELEC *sistema* phase voltage; ~ **de una fase de la estrella** *f* ELEC star voltage; ~ **de funcionamiento** *f* ELEC working voltage, ING ELÉC operate voltage, PROD operating voltage; ~ **hexagonal** *f* ELEC *sistema de seis fases* hexagon voltage; ~ **hidrostática** *f* METAL hydrostatic stress; ~ **de impacto** *f* EMB impact stress; ~ **de inducción** *f* ING ELÉC induction voltage; ~ **inducida** *f* ELEC *electromagnetismo* induced voltage; ~ **del inductor** *f* ELEC *de máquina* field voltage; ~ **del inductor del alternador** *f* ELEC alternator field voltage; ~ **instantánea** *f* ELEC instantaneous voltage; ~ **interfacial** *f* CONTAM MAR interfacial tension; ~ **inversa** *f* ELEC, ING ELÉC reverse voltage; ~ **inversa de cresta** *f* ELEC peak inverse voltage; ~ **inversa inicial** *f* ELEC initial inverse voltage; ~ **limitada** *f* ING MECÁ limited tightness; ~ **de línea** *f* ELEC *red de distribución* line voltage, METAL line tension; ~ **de línea-neutro** *f* ELEC *red de distribución* line-to-neutral voltage; ~ **máxima** *f* ELEC maximum voltage; ~ **máxima de arco** *f* ELEC peak arc voltage; ~ **máxima inversa** *f* ELEC peak inverse voltage; ~ **máxima nominal en escalón** *f* ELEC *conmutador de tomas* maximum-rated step voltage; ~ **medida** *f* ELEC *corriente* measured voltage; ~ **mínima** *f* ELEC minimum voltage; ~ **momentánea** *f* ELEC transient voltage; ~ **muy elevada** *f* (*MAT*) TV extra-high tension (*EHT*); ~ **negativa** *f* ING ELÉC negative voltage; ~ **neutra a fase** *f* ELEC *sistema* phase-to-neutral voltage; ~ **no disruptiva** *f* ELEC *instalación* withstand voltage; ~ **no disruptiva de aislamiento** *f* ELEC insulation withstand voltage; ~ **nodal** *f* ELEC *de circuito* nodal voltage; ~ **nominal** *f* ELEC rated voltage, *de sistema* nominal voltage, ING ELÉC rated voltage, METAL nominal voltage, PROD nominal voltage; ~ **nominal de aislación** *f* ELEC, ING ELÉC, PROD rated insulation voltage; ~ **nominal por pasos** *f* ELEC *transformador* rated step voltage; ~ **normal** *f* CAR-

BÓN normal stress; ~ **de ondulación** *f* CONST, ELECTRÓN ripple voltage; ~ **ondulatoria** *f* CONST, ELECTRÓN ripple voltage; ~ **de perforación** *f* ELEC *de dieléctrico, diodo Zener,* ING ELÉC, ING MECÁ *de dieléctrico* breakdown voltage; ~ **periférica** *f* METAL circumferential stress; ~ **plena** *f* ELEC, ING ELÉC, PROD full voltage; ~ **primaria** *f* ELEC input voltage, *transformador* primary voltage, ING ELÉC primary voltage; ~ **de prueba** *f* ELEC test voltage; ~ **de puerta** *f* ING ELÉC gate voltage; ~ **puerta a generador** *f* ING ELÉC gate-to-source voltage; ~ **reactiva** *f* ELEC *circuito de CA,* ELECTRÓN, FÍS, ING ELÉC reactive voltage; ~ **de realimentación** *f* ING ELÉC feedback voltage; ~ **de rebobinado** *f* CINEMAT rewind tension; ~ **rectificada** *f* ELEC, ING ELÉC rectified voltage; ~ **rectilínea** *f* METAL true strain; ~ **reflejada** *f* ING ELÉC, ÓPT reflected voltage; ~ **de régimen** *f* ELEC, ING ELÉC, PROD rated voltage; ~ **residual** *f* CONST residual stress; ~ **de restablecimiento** *f* ELEC *interruptor o disyuntor* recovery voltage; ~ **de retroceso** *f* ING ELÉC kickback; ~ **de rotura** *f* ING MECÁ ultimate stress; ~ **de rotura del árido** *f* CONST aggregate crushing value; ~ **de ruptura** *f* ELEC *de dieléctrico, diodo Zener* breakdown voltage, ING ELÉC breakdown voltage, interrupting voltage, ING MECÁ breakdown voltage; ~ **de salida** *f* ELEC, TELECOM output voltage; ~ **de salida máxima** *f* ELEC output voltage; ~ **de salida regulada** *f* ING ELÉC regulated-output voltage; ~ **de salto arco con aislador seco** *f* ELEC *de arco* dry flashover voltage; ~ **secundaria** *f* ELEC *transformador* secondary voltage; ~ **de servicio** *f* ELEC on-load voltage, working voltage, *de sistema* operating voltage; ~ **sinusoidal** *f* ING ELÉC sinusoidal voltage; ~ **de sobrecarga** *f* ELEC *circuito, equipo* overload voltage; ~ **superficial** *f* CARBÓN, CONST, FÍS, P&C *propiedad física* surface tension; ~ **tangencial** *f* ING MECÁ tangential strain; ~ **a tierra** *f* ING ELÉC voltage to ground; ~ **de trabajo** *f* ELEC on-load voltage, working voltage, *de sistema* operating voltage, ING ELÉC driving potential; ~ **de tracción a la rotura** *f* P&C *propiedad física, prueba* ultimate tensile strength, load at break; ~ **transitoria** *f* ELEC transient voltage; ~ **umbral** *f* ING ELÉC threshold voltage; ~ **unimodal común** *f* ING ELÉC common-mode voltage

tensionar *vt* FÍS stress

tensoactivo *adj* QUÍMICA surface-active

tensor *m* CONST guy, stretcher, turnbuckle, FERRO *vagones* drawbar, ING MECÁ tightener, stretcher, strainer, tie rod, MATEMÁT tensor, MECÁ guy, OCEAN *aparejos* turnbuckle, PETROL tensioner, TEXTIL stretcher, tensioner, TRANSP MAR *aparejos* turnbuckle, VEH *cadena* tensioner; ~ **de acoplamiento** *m* ING MECÁ coupling nut; ~ **del cable** *m* ELEC, ING ELÉC cable tensioner; ~ **de cadena** *m* VEH *transmisión, motocicletas* chain tensioner; ~ **de correa** *m* AUTO belt idler; ~ **de cuerpo cerrado** *m* TRANSP MAR *aparejos* bottle screw; ~ **de empalme** *m* ING MECÁ coupling nut; ~ **de esfuerzos** *m* FÍS strain tensor; ~ **del fieltro** *m* PAPEL felt stretcher; ~ **de gancho y ojal** *m* ING MECÁ hook with eye; ~ **de las líneas guía** *m* PETROL guide-line tensioner; ~ **de la tela** *m* PAPEL wire stretcher; ~ **de tensiones** *m* FÍS stress tensor; ~ **de tornillo** *m* ING MECÁ screw tightener, turnbuckle, straining screw, MECÁ turnbuckle; ~ **del tubo de subida** *m* PETROL, TEC PETR riser tensioner

tensores: ~ **del pararrayos** *m pl* ING MECÁ lightning brace

tentáculo *m Esp* (*cf brazo AmL*) MINAS horn

tentemozo *m AmL* (*cf puntal de seguridad Esp*) MINAS safety drag bar

tenue *adj* PROD thin

tenuidad *f* PROD thinness

teobromina *f* QUÍMICA theobromine

teodolito *m* CONST *topografía* surveyor's transit, theodolite, transit theodolite, INSTR theodolite; ~ **digital electrónico** *m* CONST electronic digital theodolite; ~ **de un segundo** *m* INSTR one second theodolite

teofilina *f* QUÍMICA theophylline

teorema *m* ELEC, FÍS, FÍS PART, GEOM theorem; ~ **de Ampère** *m* ELEC *electromagnetismo* Ampère's theorem; ~ **de Ampère-Laplace** *m* ELEC *electromagnetismo* Ampère-Laplace theorem; ~ **de Bayes** *m* MATEMÁT Bayes' theorem; ~ **de Bernoulli** *m* FÍS Bernoulli's theorem; ~ **binomial** *m* MATEMÁT binomial theorem; ~ **del brillo** *m* ÓPT brightness theorem; ~ **de Carnot** *m* FÍS Carnot's theorem; ~ **central del límite** *m* MATEMÁT central limit theorem; ~ **de compensación** *m* FÍS compensation theorem; ~ **de Coulomb** *m* ELEC *campo eléctrico*, FÍS Coulomb's theorem; ~ **CPT** *m* FÍS, FÍS PART CPT theorem; ~ **cuadricolor** *m* GEOM four-color theorem (*AmE*), four-colour theorem (*BrE*); ~ **de los cuatro colores** *m* GEOM four-color theorem (*AmE*), four-colour theorem (*BrE*); ~ **de Earnshaw** *m* FÍS Earnshaw's theorem; ~ **de Gauss** *m* ELEC, FÍS Gauss's theorem; ~ **del límite central** *m* MATEMÁT central limit theorem; ~ **de muestreo** *m* TEC PETR sampling theorem; ~ **de Norton** *m* FÍS Norton's theorem; ~ **de Pitágoras** *m* GEOM Pythagoras' theorem; ~ **de Poynting** *m* FÍS Poynting's theorem; ~ **de reciprocidad** *m* FÍS, ING ELÉC reciprocity theorem; ~ **tetracromo** *m* GEOM four-color theorem (*AmE*), four-colour theorem (*BrE*); ~ **de Thevenin** *m* FÍS Thévenin's theorem; ~ **virial** *m* FÍS virial theorem

teoría: ~ **de Abbé** *f* FÍS Abbé theory; ~ **de adición de colores** *f* IMPR additive color theory (*AmE*), additive colour theory (*BrE*); ~ **de bandas** *f* FÍS band theory; ~ **de Bardeen-Cooper-Schrieffer** *f* (*teoría de BCS*) ELECTRÓN, FÍS *de superconductividad* Bardeen-Cooper-Schrieffer theory (*BCS theory*); ~ **de BCS** *f* (*teoría de Bardeen-Cooper-Schrieffer*) ELECTRÓN *de superconductividad*, FÍS BCS theory (*Bardeen-Cooper-Schrieffer theory*); ~ **del campo de ligandos** *f* QUÍMICA ligand field theory; ~ **del caos** *f* MATEMÁT chaos theory; ~ **cinética** *f* FÍS kinetic theory; ~ **de circuitos** *f* ING ELÉC circuit theory; ~ **de codificación** *f* INFORM&PD coding theory; ~ **de colas** *f* INFORM&PD queueing theory; ~ **de comunicación** *f* INFORM&PD communication theory; ~ **de conjuntos** *f* MATEMÁT set theory; ~ **de conmutación** *f* ELECTRÓN, INFORM&PD switching theory; ~ **cuántica** *f* ELECTRÓN, FÍS, FÍS PART, FÍS RAD, NUCL, QUÍMICA, TEC ESP, TELECOM, TV quantum theory; ~ **cuántica de campos** *f* FÍS, FÍS PART quantum-field theory; ~ **de decisiones** *f* MATEMÁT decision theory; ~ **de Deybe-Hückel** *f* QUÍMICA Deybe-Hückel theory; ~ **de duda** *f* INFORM&PD fuzzy theory; ~ **electrodébil** *f* FÍS, FÍS PART electroweak theory; ~ **electrónica de los metales** *f* FÍS RAD electron theory of metals; ~ **del ensanchamiento**

de rayas por colisión *f* FÍS RAD impact theory of line broadening; ~ **especial** *f* FÍS *de la relatividad* special theory; ~ **de esperar** *f* INFORM&PD queueing theory; ~ **de explotación agrícola** *f* AGRIC principles of farm management; ~ **del flujo laminar** *f* FÍS FLUID laminar flow theory; ~ **general de la relatividad** *f* FÍS general theory of relativity; ~ **de gráficas** *f* MATEMÁT graph theory; ~ **de grafos** *f* MATEMÁT graph theory; ~ **de la gran explosión** *f* FÍS, TEC ESP big bang theory; ~ **de gran unificación** *f* FÍS, FÍS PART grand unified theory (*GUT*); ~ **de grupos** *f* MATEMÁT group theory; ~ **de Huygen** *f* FÍS RAD Huygens' theory; ~ **de la información** *f* ELECTRÓN, INFORM&PD information theory; ~ **de la interacción electrodébil** *f* FÍS, FÍS PART electroweak theory; ~ **de juegos** *f* GEOM *estadísticas* game theory; ~ **de nebulosa** *f* INFORM&PD fuzzy theory; ~ **de nudos** *f* GEOM knot theory; ~ **de nudos que comprende varias curvas cerradas** *f* GEOM knot theory comprising several closed curves; ~ **de números** *f* MATEMÁT number theory, theory of numbers; ~ **de números trascendentales** *f* MATEMÁT theory of transcendental numbers; ~ **de números trascendentes** *f* MATEMÁT theory of transcendental numbers; ~ **ondulatoria de la luz** *f* FÍS ONDAS wave theory of light; ~ **de probabilidad** *f* MATEMÁT probability theory; ~ **del radio efectivo** *f* NUCL theory of effective radius; ~ **de redes** *f* ING ELÉC network theory; ~ **de la separación de Kynch** *f* NUCL Kynch's separation theory

TEP *abr* (*tetraetilplomo*) QUÍMICA TEL (*tetraethyl lead*)

tepetate *m AmL* (*cf ripios Esp*) MINAS *creación de paredes* attle

tera *pref* (*T*) METR tera (*T*)

terabyte *m* ÓPT terabyte

terapia: ~ **por infrarrojos** *f* FÍS RAD infrared therapy

térbico *adj* QUÍMICA terebic

terbio *m* (*Tb*) QUÍMICA terbium (*Tb*)

terc: ~-**butilbenceno** *m* QUÍMICA tert-butylbenzene

tercer: ~ **armónico** *m* ELECTRÓN third harmonic; ~ **carril** *m* FERRO third rail; ~ **macho** *m* ING MECÁ third tap; ~ **riel** *m* FERRO *equipo inamovible* third rail

tercera: ~ **cubierta** *f* TRANSP MAR *diseño de barcos* third deck; ~ **generación** *f* INFORM&PD third generation; ~ **ley** *f* FÍS *de la termodinámica* third law; ~ **marcha** *f* AUTO third gear; ~ **mayor** *f* ACÚST major third; ~ **menor** *f* ACÚST minor third

terciopelo *m* TEXTIL velvet

tercloruro *m* QUÍMICA terchloride

terebenteno *m* QUÍMICA terebenthene

tereftalato *m* QUÍMICA terephthalate; ~ **de polietileno** *m* ELEC *dieléctrico*, P&C polyethylene terephthalate

tereftálico *adj* QUÍMICA terephthalic

tergal *m* QUÍMICA, TEXTIL *fibras sintéticas de poliéster* tergal

terileno *m* QUÍMICA Terylene®, TRANSP MAR *lona* Terylene®

termalización: ~ **de neutrones** *f* FÍS RAD neutron thermalization

termalizar *vt* QUÍMICA *neutrones* thermalize

termas *f pl* OCEAN hot springs

térmicamente: ~ **aislado** *adj* TERMOTEC heat-insulated

térmico *adj* FÍS, REFRIG thermal, TERMO thermal, caloric, thermic

terminación *f* ACÚST, ELEC *conexión de cable*, IMPR,

INFORM&PD, PROD, TELECOM termination; ~ **anecoica** *f* ACÚST anechoic termination; ~ **anormal** *f* INFORM&PD abnormal termination; ~ **avanzada** *f* PROD advance termination; ~ **de bifurcación** *f* PROD branch close; ~ **de camino de menor orden** *f* TELECOM lower-order path termination (*LPT*); ~ **en clavija** *f* PROD pin termination; ~ **de conductores comunes** *f AmL* INFORM&PD bus network; ~ **2 de RDSI-BA** *f* (*terminación 2 de red digital de servicios integrados*) TELECOM NT2-LB (*B-ISDN network termination 2*); ~ **2 de red digital de servicios integrados** *f* (*terminación 2 de RDSI-BA*) TELECOM B-ISDN network termination 2 (*NT2-LB*); ~ **de línea** *f* (*TL*) IMPR, TELECOM line termination (*LT*); ~ **en massa** *f* ING ELÉC mass termination; ~ **en porcelana** *f* REVEST porcelain finish (*BrE*); ~ **a pozo abierto** *f* PETROL barefoot completion, open-hole completion; ~ **de la red** *f* TELECOM network termination (*NT*); ~ **de la red digital de servicios integrados** *f* (*T-RDSI*) TELECOM B-ISDN network termination (*NT-LB*); ~ **de la sección múltiplex** *f* TELECOM multiplex section termination (*MST*); ~ **de la sección del regenerador** *f* TELECOM *de impulsos* regenerator-section termination; ~ **de soldadura** *f* PROD solder termination; ~ **submarina** *f* TEC PETR *operaciones costa-fuera* subsea completion; ~ **de trayecto de orden superior** *f* TELECOM higher-order path termination (*HPT*); ~ **1 de RDSI-BA** *f* (*terminación 1 de red digital de servicios integrados*) TELECOM NT1-LB (*B-ISDN network termination 1*); ~ **1 de red digital de servicios integrados** *f* (*terminación 1 de RDSI-BA*) TELECOM B-ISDN network termination 1 (*NT1-LB*)

terminado *m* C&V dressing, finish

terminador *m* INFORM&PD, PROD terminator; ~ **de bus** *m* INFORM&PD bus network; ~ **del segmento** *m* TELECOM segment terminator

terminal[1] *m* GEN terminal, PROD *electricidad* stud; ~ **para accesorios** *m* CINEMAT accessory shoe; ~ **de alta tensión** *m* AUTO high-tension terminal; ~ **de batería** *m* AUTO, ING ELÉC battery terminal; ~ **de bloque** *m* TELECOM block terminal; ~ **de bolsillo** *m* TELECOM pocket terminal; ~ **de la bujía** *m* AUTO, ELEC, VEH spark plug terminal; ~ **del cabezal imperdible** *m* ELEC *conexión* captive head terminal; ~ **de cable** *m* TV shoe; ~ **de cableado** *m* PROD wiring terminal; ~ **de carga** *m* EMB freight terminal (*AmE*), goods terminal (*BrE*); ~ **del cárter** *m* PROD *engranajes* box lug; ~ **de central telefónica** *m* TELECOM exchange termination (*ET*); ~ **del circuito de entrada** *m* PROD input circuit terminal; ~ **de comunicación de datos** *m* INFORM&PD data communication terminal; ~ **de conexión** *m* ELEC *de cable* ING ELÉC connecting terminal; ~ **de datos** *m* INFORM&PD, TELECOM data terminal; ~ **de datos preparado** *m* TELECOM data terminal ready; ~ **de derivación** *m* ELEC *conexión* branch terminal; ~ **a distancia** *m* INFORM&PD remote terminal; ~ **de drenaje** *m* ING ELÉC drain terminal; ~ **emisor-receptor automático** *m* (*ASR*) INFORM&PD automatic send-receive (*ASR*); ~ **de entrada** *m* ELEC *conexión*, ING ELÉC input terminal; ~ **de entrada de datos en tiempo real** *m* PROD RTD output terminal; ~ **para estañosoldar** *m* PROD *cables* solder tag; ~ **extremo de excitación** *m* ING ELÉC drive end; ~ **de fase** *m* ELEC *conexión* phase terminal; ~ **inactivo** *m*

INFORM&PD dormant terminal; ~ **inteligente** *m* INFORM&PD intelligent terminal, smart terminal; ~ **interactivo** *m* TELECOM interactive terminal; ~ **lateral** *m* AUTO side-mounted terminal; ~ **de línea** *m* ELEC *conexión*, PROD *telegrafía*, TELECOM line terminal; ~ **de línea plesiosincrónico** *m* TELECOM plesiosynchronous line terminal; ~ **por lotes remoto** *m AmL* INFORM&PD remote batch terminal (*RBT*); ~ **de masa** *m* ELEC, ING ELÉC, PROD earth lug *BrE*, ground lug *AmE*; ~ **en modo de paquetes** *m* INFORM&PD packet-mode terminal; ~ **negativo** *m* AUTO minus terminal, negative terminal, ELEC, ING ELÉC, VEH *instalación eléctrica* negative terminal; ~ **neutro** *m* ELEC *conexión*, ING ELÉC neutral terminal; ~ **orientado hacia las tareas** *m* INFORM&PD job-oriented terminal; ~ **de pantalla** *m* TELECOM screen terminal; ~ **de pantalla de visualización** *m* (*VDU*) IMPR video display unit (*AmE*) (*VDU*), visual display unit (*BrE*) (*VDU*); ~ **para las pilas** *m* FOTO battery terminal; ~ **portátil** *m* TELECOM hand-held terminal, TRANSP MAR *comunicación por satélite, ordenadores* portable terminal; ~ **positivo** *m* AUTO plus terminal, ELEC *conexión*, VEH *batería* positive terminal; ~ **de potencia de entrada** *m* PROD incoming power terminal; ~ **principal** *m* AUTO main terminal; ~ **de procesado de datos fijos** *m* TELECOM fixed-data processing terminal equipment; ~ **sin procesador** *m* INFORM&PD dumb terminal; ~ **de pruebas** *m* ING ELÉC test terminal; ~ **en el punto de venta** *m* EMB point-of-sale terminal; ~ **punto de venta** *m* (*TPV*) INFORM&PD electronic point of sale (*EPOS*); ~ **remoto** *m* INFORM&PD remote terminal; ~ **de representación visual** *f* TEC ESP *informática, radar* display terminal; ~ **de salida** *m* ELEC, ING ELÉC output terminal; ~ **de salida de datos en tiempo real** *m* PROD RTD output terminal; ~ **del secundario** *m* ING ELÉC secondary terminal; ~ **simple** *m* INFORM&PD dumb terminal; ~ **de la sujeción del carro portaherramientas** *m* PROD saddle clamp terminal; ~ **TDMA** *m* TEC ESP TDMA terminal; ~ **teleaccionada** *m* INFORM&PD, TELECOM remote-operating terminal; ~ **telefónico** *m* TELECOM telephone terminal; ~ **telemandada** *m* INFORM&PD, TELECOM remote-operating terminal; ~ **telemaniobrada** *m* INFORM&PD, TELECOM remote-operating terminal; ~ **telemático** *m* INFORM&PD data communication terminal; ~ **de teleproceso por lotes** *m* INFORM&PD remote batch terminal (*RBT*); ~ **de tierra** *m* ELEC, ELECTRÓN, ING ELÉC earth lug (*BrE*), earth pin (*BrE*), earth terminal (*BrE*), ground lug (*AmE*), ground pin (*AmE*), ground terminal (*AmE*), PROD earth pin (*BrE*), ground pin (*AmE*), earth lug (*BrE*), ground lug (*AmE*); ~ **a tierra negativo** *m* AUTO negative earthed terminal (*BrE*), negative grounded terminal (*AmE*); ~ **de tornillo** *m* CINEMAT screw terminal, PROD terminal screw; ~ **de transmisión de datos** *m* INFORM&PD data communication terminal; ~ **de usuario local** *m* TELECOM local user terminal (*LUT*); ~ **de video** *AmL*, ~ **de vídeo** *m* *Esp* INFORM&PD video terminal; ~ **de videotexto** *m* TV viewdata terminal; ~ **virtual** *m* INFORM&PD virtual terminal (*VT*); ~ **virtual en red** *m* (*NVT*) INFORM&PD network virtual terminal (*NVT*); ~ **de visualización de datos** *m* TV data display terminal

terminal[2] *f* TRANSP AÉR *aeropuerto* terminal; ~ **aérea** *f*

TRANSP AÉR air terminal; ~ **de aeropuerto** *f* TRANSP AÉR airport terminal; ~ **de aire** *f* REFRIG air terminal device; ~ **de a bordo** *f* TRANSP MAR *telecomunicaciones* shipboard terminal; ~ **de carga** *f* PROD load terminal, TRANSP AÉR cargo terminal; ~ **de carga y correo** *f* TRANSP mail and cargo terminal; ~ **de carga y descarga platanera** *f* TRANSP banana-handling terminal; ~ **de cisternas** *f* TRANSP tanker terminal; ~ **de la ciudad** *f* TRANSP AÉR city terminal; ~ **de contenedores** *f* TRANSP container station, container terminal; ~ **de equipaje** *f* TRANSP baggage terminal; ~ **local de usuario** *f* TRANSP MAR *localización por satélite* local user terminal; ~ **de pasajeros** *f* TRANSP passenger terminal; ~ **de salidas** *f* TRANSP departure terminal

terminar[1] *vt* GEOL die out

terminar[2]: ~ **la sesión** *vi* INFORM&PD log off, log out

término *m* INFORM&PD, MATEMÁT term; ~ **de la vía** *m* FERRO rail head

términos: ~ **de embarque** *m pl* TRANSP MAR shipping terms; ~ **espectrales** *m pl* FÍS spectral terms; ~ **del mismo grado** *m pl* MATEMÁT terms of the same degree; ~ **de puesta en aplicación** *m pl* PROD effectivity terms

termión *m* TERMO thermion

termiónico *adj* TERMO thermionic

termistor *m* FÍS, ING ELÉC thermistor, PROD capillary-type temperature switch, temperature sensor, REFRIG, TELECOM thermistor, TERMO temperature sensor; ~ **de tipo bulbo** *m* PROD bulb-type temperature switch

termita *f* QUÍMICA thermite

termo *m* C&V vacuum flask, FÍS, LAB, PROC QUÍ Dewar flask; ~ **de agua por gas** *m* GAS, HIDROL, ING MECÁ, INSTAL TERM, TERMO gas-fired water heater

termoabsorción *f* REFRIG heat pick-up

termoacumulador: ~ **para desescarche** *m* REFRIG thermobank defrosting

termoadherir *vt* TERMO hot-bond

termoadhesión *f* TERMO hot bonding

termoagitación *f* ING ELÉC thermal agitation

termoaislación *f* MECÁ, TEC PETR, TERMO heat insulation

termoaislado *adj* MECÁ, TEC PETR, TERMO heat-insulated

termoaislador *m* MECÁ, TEC PETR, TERMO thermal insulation

termoaislamiento *m* MECÁ heat insulation, insulation, TEC PETR, TERMO heat insulation

termoaislante *adj* MECÁ, TEC PETR, TERMO heat-insulating

termoamperímetro *m* ELEC *instrumento* thermoammeter

termoanálisis *m* QUÍMICA, TERMO thermoanalysis

termobalanza *f* QUÍMICA thermobalance

termocambiador *m* PROD, TRANSP MAR *motor* heat exchanger; ~ **intermedio** *m* PROD *compresor de aire* intercooler

termocanjeador *m* ING MECÁ, MECÁ, TRANSP AÉR heat exchanger; ~ **primario** *m* TRANSP AÉR primary heat exchanger; ~ **de tubos** *m* ING MECÁ tube-type heat exchanger

termoclina *f* OCEAN thermocline

termocolorímetro *m* CINEMAT color temperature meter (*AmE*), colour temperature meter (*BrE*),

kelvinometer, FOTO color-temperature meter (*AmE*), colour temperature meter (*BrE*), kelvinometer

termocompensación *f* TERMO heat compensation

termoconductor[1] *adj* TERMO heat-conducting

termoconductor[2] *m* TERMO thermal conductor, thermoconductor

termoconformado *adj* TERMO heat-formed

termoconformar *vt* TERMO heat-form

termoconservante *adj* TERMO heat-retaining

termocontracción *f* MECÁ thermal contraction, TERMO heat shrinking, thermal contraction

termocontraer *vt* TERMO heat-shrink

termocontraíble *adj* MECÁ heat-shrunk, TERMO heat-shrinkable

termocontraído *adj* MECÁ, TERMO heat-shrunk

termocupla *f* FÍS, ING ELÉC thermocouple

termocurado *m* TERMO hot curing

termocurar *vt* TERMO heat-cure

termodeformación: ~ **en caliente** *f* TERMO *circuitos* hot creep; ~ **en frío** *f* TERMO *circuitos* cold creep; ~ **lenta** *f* TERMOTEC creep; ~ **plástica** *f* CARBÓN, MECÁ creep, METAL creep strain, P&C *propiedad física, prueba* creep

termodependiente *adj* TERMO temperature-dependent

termodesintegración: ~ **catalítica** *f* QUÍMICA, TEC PETR *refino* catalytic cracking

termodifusión *f* TERMO thermodiffusion

termodifusividad *f* TERMO, TERMOTEC rate of heat release

termodinámica *f* CONST, FÍS, ING MECÁ, QUÍMICA, TERMO thermodynamics; ~ **aplicada** *f* ING MECÁ applied thermodynamics; ~ **clásica** *f* TERMO classic thermodynamics; ~ **lineal** *f* TERMO linear thermodynamics

termodinámico *adj* CONST, FÍS, ING MECÁ, QUÍMICA, TERMO thermodynamic

termodisipación *f* TERMO heat dissipation

termodistorsión *f* TERMO heat distortion

termoelectricidad *f* FÍS, ING ELÉC, TERMO thermoelectricity

termoeléctrico *adj* FÍS, ING ELÉC, TERMO thermionic, thermoelectric, thermoelectrical

termoelectrón *m* TERMO thermion

termoencoger *vt* MECÁ, TERMO heat-shrink

termoencogible *adj* MECÁ heat-shrunk, TERMO heat-shrinkable

termoencogido *adj* MECÁ, TERMO heat-shrunk

termoendurecer *vt* IMPR bake, TERMO heat-harden

termoendurecible *adj* P&C, QUÍMICA, TERMO thermosetting

termoendurecido[1] *adj* MECÁ thermosetting, TERMO heat-hardened, thermosetting

termoendurecido[2] *m* IMPR burning-in; ~ **de las planchas** *m* IMPR plate-backing

termoendurecimiento *m* TERMO heat-hardening

termoendurente *adj* P&C thermosetting

termoesfuerzo *m* TERMO thermal stress

termoestabilizado *adj* TERMO heat-stabilized

termoestable *adj* P&C thermoset, QUÍMICA, TEC PETR thermostable, TERMO heat-stable, temperature-stable, thermosetting, thermostable

termoestables *m pl* ING MECÁ injection mold for thermosetting resins (*AmE*), injection mould for thermosetting resins (*BrE*)

termoestirado[1] *adj* TERMO hot-drawn

termoestirado[2] *m* TERMO hot drawing

termoestirar *vt* TERMO hot-draw

termofijado[1] *adj* TERMO heat-setting

termofijado[2] *m* TEXTIL *tintura* thermofixing

termofluencia: **~ secundaria** *f* METAL secondary creep

termófono *m* ACÚST thermophone

termoformación *f* GAS thermoforming; **~ automática partiendo del material en bobina** *f* EMB thermoforming automatically from the reel

termoformado[1] *adj* PAPEL thermoforming

termoformado[2] *m* TERMO heat-forming

termoformadora *f* EMB thermoform machinery; **~ a presión** *f* EMB pressure-forming machine

termofosforescencia *f* TERMO thermophosphorescence

termofraguable *adj* QUÍMICA thermosetting

termofraguables *m pl* ING MECÁ injection mold for thermosetting resins (*AmE*), injection mould for thermosetting resins (*BrE*)

termofraguado *m* MECÁ, TERMO thermosetting

termofraguante *adj* QUÍMICA thermosetting, TERMO heat-setting

termofugacia *f* FÍS, TERMO heat resistance

termófugo *adj* FÍS, TERMO heat-insulating

termofundible *adj* FÍS, TERMO heat-fusible

termogeneración *f* FÍS, TERMO heat generation

termogenerador[1] *adj* FÍS, TERMO heat-generating

termogenerador[2] *m* FÍS, TERMO heat generator, thermogenerator

termógrafo *m* FÍS, LAB, NUCL thermograph, TERMO recording thermometer, *instrumento* thermograph

termograma *m* TERMO *diagrama* thermograph

termogravimetría *f* P&C *prueba* thermogravimetry

termohigrógrafo *m* LAB *registro de temperatura y humedad atmosférica* thermohygrograph

termoimagen *f* TERMO heat image

termoimpresora *f* EMB thermal transfer printer

termoindurantes *m pl* ING MECÁ injection mold for thermosetting resins (*AmE*), injection mould for thermosetting resins (*BrE*)

termoinestabilidad *f* P&C, TELECOM, TERMO thermal instability

termoinmersor *m* ELEC *calefacción*, MECÁ immersion heater

termointercambiador *m* ING MECÁ, MECÁ, TEC PETR *refino* heat exchanger; **~ primario** *m* TRANSP AÉR primary heat exchanger

termoiónico *adj* TERMO thermionic

termolecular *adj* QUÍMICA thermolecular

termólisis *f* TERMO thermolysis

termoluminiscencia *f* FÍS, FÍS RAD, TERMO thermoluminescence

termoluminiscente *adj* FÍS, FÍS RAD, TERMO thermoluminescent

termomadurar *vt* TERMO heat-cure

termomagnético *adj* TERMO thermomagnetic

termomagnetismo *m* TERMO thermomagnetism

termometría *f* FÍS, TERMO thermometry

termométrico *adj* FÍS, TERMO thermometric

termómetro *m* GEN thermometer; **~ por acción remota** *m* INSTR, REFRIG, TERMOTEC remote temperature gage (*AmE*), remote temperature gauge (*BrE*); **~ de acumulación de calor** *m* REFRIG thermal storage thermometer; **~ de alcohol** *m* REFRIG alcohol thermometer; **~ avisador** *m* ING MECÁ alarm thermometer; **~ para baja temp-**

eratura *m* TERMO low-temperature thermometer; **~ de Beckmann** *m* FÍS Beckmann thermometer; **~ de bola seca** *m* REFRIG, TERMO, TERMOTEC dry-bulb thermometer; **~ de bulbo húmedo** *m* REFRIG wet-bulb thermometer, TERMO *psicometría* wet bulb thermometer, TERMOTEC wet-bulb thermometer; **~ de bulbo seco** *m* REFRIG, TERMO, TERMOTEC dry-bulb thermometer; **~ de cabina** *m* TRANSP AÉR cabin temperature indicator; **~ calorimétrico** *m* TERMO calorimetric thermometer; **~ de cuadrante** *m* REFRIG dial thermometer; **~ para cubetas** *f* FOTO dish thermometer; **~ diferencial** *m* REFRIG differential thermometer; **~ de dilatación de gases** *m* GAS, INSTR, REFRIG gas thermometer; **~ de dilatación de sólidos** *m* REFRIG solid expansion thermometer; **~ de gas** *m* FÍS, GAS, INSTR gas thermometer; **~ de lectura directa** *m* REFRIG indicating thermometer; **~ de lectura a distancia** *m* INSTR, REFRIG, TERMO remote-reading thermometer; **~ líquido** *m* TERMO liquid thermometer; **~ magnético** *m* REFRIG magnetic thermometer; **~ de máxima y mínima** *m* FÍS, LAB, TERMO, TERMOTEC maximum-minimum thermometer; **~ de mercurio** *m* FÍS, REFRIG mercury thermometer; **~ para pozos profundos** *m* PETROL deep well thermometer (*DWT*); **~ de presión a vapor** *m* REFRIG vapor pressure thermometer (*AmE*), vapour pressure thermometer (*BrE*); **~ de resistencia** *m* FÍS, INSTR, REFRIG, TERMO resistance thermometer; **~ de resistencia de platino** *m* FÍS platinum-resistance thermometer; **~ seco** *m* REFRIG, TERMO, TERMOTEC dry-bulb thermometer; **~ para el tanque de revelado** *m* CINEMAT, FOTO developing tank thermometer; **~ de termopar** *m* ING ELÉC thermocouple thermometer; **~-higrómetro** *m* TRANSP moisture-and-temperature detector

termonatrita *f* MINERAL thermonatrite

termoneutralidad *f* QUÍMICA thermoneutrality

termopar *m* ELEC *medición*, LAB *medida de temperatura*, NUCL, TEC PETR, TERMO thermocouple; **~ blindado** *m* NUCL sheathed thermocouple; **~ diferencial** *m* TERMO differential thermocouple; **~ de salida del núcleo** *m* NUCL core outlet thermocouple

termopermutador *m* ING MECÁ, MECÁ, PROD heat exchanger; **~ primario** *m* TRANSP AÉR primary heat exchanger; **~ refrigerante del combustible** *m* TRANSP AÉR fuel-coolant heat exchanger; **~ de tubos** *m* ING MECÁ tube-type heat exchanger

termopila *f* FÍS, ING ELÉC, TEC ESP thermopile

termoplasticidad *f* P&C *prueba* thermoplasticity

termoplástico[1] *adj* ING MECÁ, MECÁ, P&C, TEXTIL, TRANSP MAR *construcción de poliéster* thermoplastic

termoplástico[2] *m* ING MECÁ thermoplastics, MECÁ, P&C, TEXTIL, TRANSP MAR *construcción de poliéster* thermoplastic

termoquímica *f* QUÍMICA thermochemistry

termoresistente *adj* SEG heat-resistant

termorrecuperación *f* TERMO heat recovery

termorrecuperador *m* ING MECÁ, MECÁ heat exchanger, METAL, PROD checker (*AmE*), chequer (*BrE*), TEC ESP regenerator

termorresistencia *f* FÍS, TEC ESP, TERMO heat resistance

termorresistente *adj* FÍS heat-resistant, TEC ESP heat-resistant, heat-resisting, TERMO heat-proof, heat-resistant

termorresistor: ~ **de recalentamiento** *m* TRANSP AÉR overheat thermoresistor

termosellable *adj* TERMO heat-sealable

termosellado[1] *adj* EMB head-sealed, heat-sealing, TERMO heat-sealed

termosellado[2] *m* EMB heat-sealing, IMPR heat seal, thermosealing, TERMO heatsealing; ~ **por impulsos** *m* EMB impulse heat sealer

termoselladora: ~ **de cuchilla en caliente** *f* EMB hot blade sealing

termosellar *vt* EMB, TERMO heat-seal

termosensibilidad *f* P&C *prueba* heat sensitivity

termosensible *adj* P&C, TERMO heat-sensitive

termosfera *f* GEOFÍS thermosphere

termostática *f* TERMO thermostatics

termostático *adj* REFRIG, TERMO thermostatic

termostato *m* AUTO, CONST, ELEC, FÍS, INSTAL TERM, LAB thermostat, PROD temperature switch, REFRIG, TERMO, VEH *sistema de refrigeración* thermostat; ~ **de agua** *m* TERMO water thermostat; ~ **con alarma** *m* ING MECÁ alarm thermometer; ~ **ambiente** *m* REFRIG, TERMO room thermostat; ~ **de bulbo remoto** *m* REFRIG, TERMO remote-bulb thermostat; ~ **de habitación** *m* REFRIG, TERMO room thermostat; ~ **de seguridad de aceite** *m* REFRIG oil temperature cutout; ~ **de seguridad de impulsión** *m* REFRIG high-discharge temperature cut-out

termotécnica *f* ING MECÁ heat engineering

termoterapia *f* AGRIC heat treatment

termotransformación *f* TERMO heat transformation

termotransición *f* TERMO heat transition

termotransmisión *f* TERMO heat transmission

termotratable *adj* TERMO heat-treatable

termotratado *adj* TERMO *metales* heat-treated

termotratamiento *m* METAL heat treatment; ~ **para endurecimiento estructural** *m* PROD precipitation heat-treatment

termotratar *vt* TERMO heat-treat

ternario *adj* MATEMÁT, QUÍMICA ternary

ternera *f* AGRIC heifer calf

ternero: ~ **para engordar** *m* AGRIC feeder calf; ~ **de mala calidad con síndrome de fiebre de embarque** *m* AGRIC knot head

ternitrato *m* QUÍMICA ternitrate

terpadieno *m* QUÍMICA terpadiene

terpénico *adj* QUÍMICA terpenic

terpeno *m* QUÍMICA terpene

terpina *f* QUÍMICA terpin

terpineno *m* QUÍMICA terpinene

terpineol *m* QUÍMICA terpineol

terpinol *m* QUÍMICA terpinol

terpinoleno *m* QUÍMICA terpinolene

terpolímero *m* P&C, QUÍMICA terpolymer

terracota *f* C&V terracotta

terraja *f* ING MECÁ die, screw plate, screw stock, screwing die, stock, tube-screwing machine, PROD sweep, *fundería* spindle and sweep, sweep board, *moldeo* strickle, *moldeo en arcilla* templet; ~ **de apertura automática** *f* ING MECÁ self-opening die-head; ~ **enganchadora** *f* MINAS *sondeos* grabbing tap; ~ **giratoria** *f* PROD *moldeo* loam board; ~ **hembra** *f* ING MECÁ outside-threading tool; ~ **macho** *f* ING MECÁ inside-threading tool; ~ **de mano** *f* ING MECÁ tap plate; ~ **de peines** *f* ING MECÁ

chaser die screwing stock; ~ **de una pieza** *f* ING MECÁ one-part screwplate; ~ **de roscar** *f* ING MECÁ die; ~ **transportable** *f* PROD *moldeo* gig; ~ **para tubos de gas** *f* GAS, PROD gas tap

terrajas: ~ **concéntricas** *f pl* ING MECÁ self-centering dies (*AmE*), self-centring dies (*BrE*)

terral *m* METEO, TRANSP MAR land breeze

terramicina *f* QUÍMICA terramycin

terraplén *m* Esp (*cf malecón AmL*) AGUA breakwater, dike, mole (*mol*), CARBÓN, fill, landfill, embankment, CONST earthwork embankment, jetty, mole (*mol*), embankment, bank, breakwater, CONTAM landfill, FERRO, MINAS embankment, OCEAN, RECICL landfill, TRANSP terraplein, TRANSP MAR sea wall, breakwater, mole (*mol*)

terraplenado *m* CARBÓN, CONST, CONTAM, RECICL, VEH landfilling

terraplenar *vt* GEN bank up

terraza: ~ **de abrasión marina** *f* OCEAN *erosión de costas* wave-cut bench; ~ **de base angosta** *f* AGRIC narrow-based terrace; ~ **de borde** *f* AGRIC ridge terrace; ~ **continental** *f* OCEAN continental terrace; ~ **de desagüe** *f* AGRIC graded channel terrace, TRANSP AÉR drainage terrace; ~ **de fiorita** *f* GEOFÍS fiorite terrace; ~ **Magnum** *f* AGRIC Magnum terrace; ~ **a nivel** *f* AGRIC level terrace; ~ **de siliceo sinterizado** *f* GEOFÍS siliceous sinter terrace; ~ **sinterizada** *f* GEOFÍS sinter terrace; ~ **sinterizada de siliceo** *f* GEOFÍS siliceous sinter terrace

terremoto *m* CONST, GEOFÍS earthquake

terreno *m* AGRIC ground, soil, CARBÓN soil, CONST bank, ground, GEOL terrain; ~ **de abono** *m* MINAS dumping ground; ~ **de aluvión** *m* AGRIC flood plain, AGUA alluvial deposit; ~ **de aluviones** *m* MINAS placer ground; ~ **arcilloso duro** *m* CONST, GEOL gault; ~ **arrancado por una pala excavadora** *m* MINAS spoil heap; ~ **aurífero** *m* AmL (*cf terreno productivo Esp*) MINAS pay dirt; ~ **compactado** *m* AGRIC heavy soil; ~ **de dragado** *m* MINAS dredging ground; ~ **entregado al constructor para ejecutar la obra** *m* CONST delivered site; ~ **para escombros** *m* RECICL dumping ground; ~ **esponjoso** *m* CARBÓN swelling soil; ~ **estéril** *m* MINAS dead ground, waste ground, pillar; ~ **firme** *m* CONST firm ground; ~ **helado** *m* CARBÓN, GEOL frozen ground; ~ **inadecuado** *m* CONST bad ground; ~ **inclinado** *m* CARBÓN graded soil; ~ **liso** *m* AGRIC fine soil, CARBÓN even-grained soil, fine soil; ~ **llano** *m* CONST even ground; ~ **mineral** *m* CARBÓN, MINERAL mineral soil; ~ **mixto** *m* TELECOM mixed terrain; ~ **movedizo** *m* CARBÓN running ground, running soil; ~ **no susceptible de helarse** *m* CARBÓN non-frost-susceptible soil; ~ **pantanoso** *m* GEOL marshy environment; ~ **de pláceres** *m* MINAS placer ground; ~ **productivo** *m* Esp (*cf terreno aurífero AmL*) MINAS pay dirt; ~ **de recubrimiento** *m* MINAS rock capping, *filones* capping; ~ **saturado** *m* CARBÓN saturated soil; ~ **sedimentario** *m* CARBÓN sedimentary soil; ~ **suave** *m* CONST soft ground; ~ **zonal** *m* CARBÓN zonal soil

terrenos: ~ **ganados al mar** *m pl* AGUA land reclamation; ~ **de recubrimiento** *m pl* CARBÓN *minería, geología*, GEOL, MINAS overburden, overplacement

terrero: ~ **aluvial** *m* OCEAN alluvial deposit

terrestre *adj* GEOL terrigenous, TEC ESP terrestrial

terrígeno *adj* GEOL terrigenous

territorio: **~ de alimentación** *m* OCEAN feeding ground; **~ de engorde** *m* OCEAN fattening ground

terrón *m* CONST lump, *tierra* clod; **~ de arcilla** *m* PROD *fundición* loam cake; **~ de azúcar** *m* ALIMENT lump sugar

Terylene® *m* QUÍMICA Terylene®

tesar *vt* TRANSP MAR *cuerda* haul taut

tesla *f* (*T*) FÍS, METR tesla (*T*)

tesoro *m* INFORM&PD *léxico* thesaurus

tesselita *f* MINERAL tesselite

test: **~ eléctrico** *m* ELEC, ING ELÉC electrical test; **~ no destructivo** *m* FÍS, ING MECÁ nondestructive test; **~ tensional** *m* FÍS tensile test

testeador *m* METEO feeler gage (*AmE*), feeler gauge (*BrE*)

testera *f* ING MECÁ headstock

testero *m* CARBÓN *minas* breast, CONST back stope, PAPEL header

testificar *vt* GEOL *un sondeo*, TEC PETR *perforación* core

testigo *m* AGUA sample, CARBÓN, GAS core sample, GEOFÍS, GEOL core, INFORM&PD *red en anillo* token, MINAS *de perforación* core, *sondeos* bore core, core sample, TEC PETR *evaluación de la formación* core, *perforación* core sample; **~ lateral** *m* TEC PETR *evaluación de la formación* sidewall core; **~ orientado** *m* PETROL oriented core; **~ de pared** *m* TEC PETR sidewall core; **~ de la perforación** *m* MINAS *sondeos* drill core; **~ de profundidad** *m* METR, PROD *moldeo en tierra* depth gage (*AmE*), depth gauge (*BrE*); **~ de sondeo** *m* CONST core sampling, MINAS drill core

testiguero *m* AmL (*cf barril de testigos Esp*) TEC PETR *perforación* core barrel

tetón: **~ tubular** *m* ING MECÁ barrel nipple

tetrabásico *adj* QUÍMICA tetrabasic, quadribasic

tetraborato: **~ de sodio** *m* DETERG, QUÍMICA sodium tetraborate

tetrabromoetano *m* QUÍMICA tetrabromoethane

tetrabromoetileno *m* QUÍMICA tetrabromoethylene

tetrabromuro *m* QUÍMICA tetrabromide

tetracarbonilo *m* QUÍMICA tetracarbonyl

tetraceno *m* QUÍMICA tetrazene

tetraciclo *m* QUÍMICA quadricycle

tetracloretileno *m* QUÍMICA tetrachloroethylene

tetracloroetano *m* QUÍMICA tetrachlorethane

tetraclorometano *m* QUÍMICA tetrachloromethane

tetracloruro: **~ de carbono** *m* CINEMAT, QUÍMICA, carbon tetrachloride (*carbon tet*), tetrachloromethane

tetracordo *m* ACÚST diatonic tetrachord, tetrachord

tetradecanoico *adj* QUÍMICA tetradecanoic

tetradimita *f* MINERAL tetradymite

tetraédrico *adj* CRISTAL, GEOM tetrahedral

tetraedrita *f* MINERAL fahl ore, panabase

tetraedro *m* CRISTAL, GEOM tetrahedron; **~ regular** *m* GEOM regular tetrahedron

tetraetilo *m* QUÍMICA tetraethyl

teatrilplomo *m* (*TEP*) CHEM tetraethyl lead (*TEL*)

tetrafásico *adj* ELECTRÓN four-phase

tetragonal *adj* CRISTAL tetragonal

tetrágono *m* GEOM tetragon

tetrahédrico *adj* QUÍMICA tetrahedral

tetrahedrita *f* MINERAL tetrahedrite

tetrahidrobenceno *m* QUÍMICA tetrahydrobenzene

tetrahidroglioxalina *f* QUÍMICA tetrahydroglyoxaline

tetrahidronaftaleno *m* QUÍMICA tetrahydronaphthalene

tetrahidroquinona *f* QUÍMICA tetrahydroquinone

tetrahidroxiquinona *f* QUÍMICA tetrahydroxyquinone

tetrahidruro *m* QUÍMICA tetrahydride

tetralina *f* QUÍMICA tetralin

tetramérico *adj* QUÍMICA tetrameric

tetrametileno *m* QUÍMICA tetramethylene

tetrametílico *adj* QUÍMICA tetramethylic

tetramina *f* QUÍMICA tetramine

tetranitrol *m* QUÍMICA tetranitrol

tetraoxosulfato: **~ VI** *m* CONTAM sulfate (*AmE*), sulphate (*BrE*); **~ VI de hidrógeno** *m* CONTAM sulfuric acid (*AmE*), sulphuric acid (*BrE*)

tetrapak *m* ALIMENT tetrapak

tetrapolar *adj* ELEC *generador* four-polar

tetrasulfuro *m* QUÍMICA tetrasulfide (*AmE*), tetrasulphide (*BrE*)

tetratiónico *adj* QUÍMICA tetrathionic

tetratomicidad *f* QUÍMICA tetratomicity

tetravalencia *f* QUÍMICA tetravalence, tetravalency, quadrivalence

tetravalente *adj* QUÍMICA quadrivalent, tetravalent

tetrayodofluoresceína *f* QUÍMICA tetraiodofluorescein

tetrazina *f* QUÍMICA tetrazine

tetrazol *m* QUÍMICA tetrazole

tetrilo *m* QUÍMICA tetryl

tetrodo *m* ELECTRÓN, FÍS tetrode; **~ de gas** *m* ELECTRÓN gas tetrode

tetrólico *adj* QUÍMICA tetrolic

tetrosa *f* QUÍMICA tetrose

tetróxido *m* QUÍMICA tetroxide, quadroxide; **~ de plomo** *m* MINERAL, QUÍMICA minium

tevetina *f* QUÍMICA thevetin

texatita *f* MINERAL texasite

textil: **~ para el hogar** *m* TEXTIL home textile; **~ de producción casera** *m* TEXTIL home-produced textile

texto *m* IMPR matter; **~ de cabecera** *m* IMPR leading matter; **~ cifrado** *m* TELECOM ciphertext; **~ despejado** *m* TELECOM clear text; **~ a distancia** *m* INFORM&PD teletext; **~ del mensaje** *m* INFORM&PD message text; **~ opuesto** *m* IMPR opposite text; **~ de organigrama** *m* INFORM&PD flowchart text; **~ que abarca una página doble** *m* IMPR double spread; **~ superpuesto** *m* IMPR against text; **~ de transmisión** *m* TV transmission copy

textual *adj* INFORM&PD word-oriented

textura[1]: **de ~ fina** *adj* GEOL fine-textured

textura[2] *f* CARBÓN *metales* texture, grain, GEOL *petrología*, PETROL fabric, VEH *neumático* ply; **~ alveolar** *f* GEOL *agentes atmosféricos, inclinación* honeycomb texture; **~ deposicional** *f* TEC PETR depositional fabric; **~ direccional** *f* GEOL directional fabric; **~ dolerítica** *f* PETROL doleritic texture; **~ esferolítica** *f* GEOL spherulitic texture; **~ fibrosa** *f* CRISTAL fibrous texture, METAL fiber texture (*AmE*), fibre texture (*BrE*); **~ fluidal** *f* GEOL *de rocas ígneas* fluidal texture; **~ gráfica** *f* PETROL graphic texture; **~ gruesa** *f* GEOL coarse texture; **~ lenticular en fláser** *f* GEOL augen structure; **~ de la miga de pan** *f* ALIMENT crumb texture; **~ del molde** *f* ING MECÁ mold texturing (*AmE*), mould texturing (*BrE*); **~ de pan** *f* ALIMENT bread texture; **~ superficial** *f* ING MECÁ surface texture; **~ vidriosa** *f* C&V, PETROL glassy texture

tg *abr* (*tangente*) CONST, GEOM, INFORM&PD, MATEMÁT *trigonometría* tan (*tangent*)

Th *abr* (*torio*) FÍS RAD, QUÍMICA Th (*thorium*)

thenardita *f* MINERAL thenardite
thomselonita *f* MINERAL thomsenolite
thomsonita *f* MINERAL thomsonite
thulita *f* MINERAL thulite
thuringita *f* MINERAL thuringite
Ti *abr* (*titanio*) METAL, QUÍMICA, TEC ESP Ti (*titanium*)
tiacina *f* QUÍMICA thiazine
tiación *f* QUÍMICA thiation
tial *m* QUÍMICA thial
tialdina *f* QUÍMICA thialdine
tiazina *f* QUÍMICA thiazine, thioindamine
tiazol *m* QUÍMICA thiazole
tiazolina *f* QUÍMICA thiazoline
tibio *adj* TERMO tepid
ticket: **~ de trabajo** *m* PROD job ticket
tiemannita *f* MINERAL tiemannite
tiempo[1]: **de ~ real** *adj* CINEMAT, ELECTRÓN, INFORM&PD, ING ELÉC, TELECOM, TV real-time; **en ~ real** *adj* CINEMAT, ELECTRÓN, INFORM&PD, ING ELÉC, TELECOM, TV real-time
tiempo[2]:
~ a **~ de acceso** *m* IMPR, INFORM&PD, ING ELÉC, ÓPT access time; **~ de aceleración** *m* INFORM&PD acceleration time; **~ de activación** *m* ING ELÉC firing time, turn-on time; **~ de activación del emisor** *m* TELECOM transmitter turn-on time; **~ de adelanto** *m* ING MECÁ lead time; **~ de admisión** *m* VEH *motor* induction stroke; **~ en el aire** *m* TV airtime, on-air time; **~ de aireación** *m* HIDROL *aguas residuales* aeration time; **~ de amortiguamiento del impulso** *m Esp* TELECOM pulse decay-time; **~ de apertura** *m* ELEC *relé* break time, opening time, ING ELÉC opening time, break time; **~ de aprovisionamiento** *m* PROD lead time; **~ para aprovisionamiento** *m* PROD manufacturing lead time; **~ de aproximación** *m* TRANSP AÉR approach time; **~ de aproximación previsto** *m* TRANSP AÉR expected approach time; **~ aproximado de ejecución** *m* PROD approximate execution time; **~ de arranque** *m* INFORM&PD start time; **~ de arreglo** *m* IMPR make-ready time; **~ de ascenso** *m* FÍS, INFORM&PD rise time; **~ de asignación de canal** *m* TELECOM channel allocation time; **~ de avance** *m* ING MECÁ, TEXTIL *mecánica* lead time;
~ b **~ de bajada del impulso** *m* TELECOM pulse decay-time; **~ de bloque** *m* TRANSP AÉR block time; **~ de bloque a bloque** *m* TRANSP AÉR block-to-block time; **~ bonancible** *m* METEO, TRANSP AÉR clear weather, TRANSP MAR clear weather, moderate weather; **~ borrascoso** *m* METEO, TRANSP MAR rough weather; **~ de búsqueda** *m* INFORM&PD, ÓPT seek time;
~ c **~ de caída** *m* FÍS fall time, INFORM&PD decay time; **~ de caída del impulso** *m* TELECOM pulse decay-time; **~ de caldeo** *m* TERMO heating time; **~ de calentamiento** *m* TERMO heating-up time, warm-up time; **~ de cambio de sentido** *m* INFORM&PD turnaround time; **~ del ciclo** *m* INFORM&PD cycle time; **~ de cierre** *m* ELEC *relé* closing time; **~ de coherencia** *m* FÍS, ÓPT, TELECOM coherence time; **~ de combustión** *m* D&A burn time; **~ compartido** *m* INFORM&PD time-sharing, TELECOM time sharing; **~ de compilación** *m* INFORM&PD compilation time; **~ de compresión** *m* TERMO *motor* compression stroke; **~ de computadora** *m AmL* (*cf tiempo de ordenador Esp*) INFORM&PD machine time; **~ de comunicación** *m* PROD operating time; **~ de**

~ conexión *m* INFORM&PD connect time; **~ de configuración** *m* INFORM&PD setup time; **~ de confirmación** *m* TELECOM submission time; **~ de congelación nominal** *m* REFRIG nominal freezing time; **~ de conmutación** *m* ING ELÉC switching time;
~ d **~ de decaimiento** *m* INFORM&PD decay time; **~ de deceleración** *m* INFORM&PD deceleration time; **~ de declinación** *m* PETROL decay time; **~ de demora previsible** *m* PROD slack time; **~ de desaccionamiento** *m* ELEC *de relé* drop-out time; **~ de desactivación** *m* ELEC *de relé* drop-out time; **~ de descompresión** *m* PETROL decompression time; **~ de desconexión** *m* ING ELÉC turn-off time; **~ desde la orden de pedido hasta la entrega** *m* ING MECÁ lead time; **~ de desexcitación** *m* PROD drop-out time; **~ de desintegración** *m* ELECTRÓN *de impulso*, FÍS decay time; **~ de despacho** *m* TRANSP clearance period; **~ de desplazamiento** *m* PROD move time; **~ diferencial** *m* ELECTRÓN differential time; **~ de digestión** *m* HIDROL digestion time; **~ de disparo** *m* PROD trip time; **~ disponible** *m* INFORM&PD available time;
~ e **~ efectivo de congelación** *m* REFRIG effective freezing-time; **~ de ejecución** *m* INFORM&PD execution time, run time, PROD run time; **~ de elevación** *m* CONST rise time; **~ de emisión** *m* TRANSP AÉR release time; **~ en emisión** *m* TV on-air time; **~ de emisora** *m* TV station time; **~ de encendido del transmisor** *m* TELECOM transmitter turn-on time; **~ de endurecimiento** *m* P&C *polimerización* setting time; **~ entre percepción y reacción** *m* TRANSP *tráfico* perception-reaction time; **~ de escritura** *m* ELECTRÓN *protectores* writing time, INFORM&PD write time; **~ espacial continuo** *m* FÍS, TEC ESP space-time continuum; **~ de espera** *m* ALIMENT bench time, ELECTRÓN lead time, INFORM&PD standby time, timeout, PROD delivery lead time, lead time, turnaround time, TELECOM holding time, TV queueing time; **~ de espera en cola** *m Esp* (*cf tiempo de espera en línea AmL*) TELECOM queueing time; **~ de espera en línea** *m AmL* (*cf tiempo de espera en cola Esp*) TELECOM queueing time; **~ de estabilización** *m* ÓPT stabilization time, PROC QUÍ settling time; **~ estable** *m* METEO, TRANSP MAR settled weather; **~ de establecimiento** *m* PROD setup time; **~ de establecimiento del sintetizador** *m* TELECOM synthesizer settling time; **~ de examen** *m* PROD scan time; **~ de expansión** *m* TERMO *motor* expansion stroke; **~ de exploración de entrada/salida** *m* (*tiempo de exploración de E/S*) PROD input/output scan time (*I/O scan time*); **~ de exploración de E/S** *m* (*tiempo de exploración de entrada/salida*) PROD I/O scan time (*input/output scan time*); **~ de exploración del programa** *m* PROD programme scan time (*BrE*), program scan time (*AmE*); **~ de exposición** *m* CINEMAT time exposure, exposure time, FOTO exposure time, time exposure; **~ de extinción** *m* ING ELÉC, PETROL decay time;
~ f **~ de fabricación** *m* PROD production time; **~ del fallo** *m* INFORM&PD fault time; **~ focal** *m* CINEMAT, FOTO, TV focal time; **~ de formación** *m* PETROL *impulsos* rise time; **~ de formación de ráfaga** *m* TEC ESP, TRANSP AÉR, TRANSP MAR gust formation time; **~ de frenado** *m* TRANSP braking time; **~ fuera de funcionamiento** *m* TRANSP AÉR downtime; **~ fuera de servicio** *m* TELECOM out-of-service time; **~ fuera de servicio previsto** *m* TRANSP AÉR esti-

mated off-block time; **~ de funcionamiento** *m* ING ELÉC operate lag, operate time, PROD operating time, TRANSP running time; **~ de funcionamiento acumulativo** *m* PROD cumulative working time; **~ de funcionamiento por inercia del motor** *m* TRANSP AÉR engine coasting-down time; **~ de fusión** *m* TERMO melting time;

~ g ~ **de gelatinización** *m* P&C *propiedad física, polimerización* gel time; **~ de gelificación** *m* EMB gel time; **~ de giro total** *m* INFORM&PD turnaround time; **~ de grabación** *m* ELECTRÓN *protectores* writing time, ING ELÉC write time;

~ h ~ **hecho** *m* METEO, TRANSP MAR settled weather;

~ i ~ **de impresión** *m* FOTO printing time; **~ improductivo** *m* MECÁ, PROD, TELECOM downtime; **~ impropio** *m* FÍS improper time; **~ de impulso** *m* ELECTRÓN period pulse; **~ de inactividad** *m* INFORM&PD downtime, idle time, PROD downtime; **~ de inactividad por avería** *m* INFORM&PD fault time; **~ inactivo** *m* TV downtime; **~ de indisponibilidad** *m* NUCL unavailability time; **~ indisponible** *m* TELECOM unavailable time (*UAT*); **~ inefectivo en el aire** *m* TELECOM ineffective airtime; **~ de instalación** *m* PROD setup time; **~ de instrucción doble** *m* TRANSP AÉR dual instruction time; **~ de integración** *m* ELECTRÓN, ING ELÉC integration time; **~ de intercambio** *m* TELECOM interchange time; **~ interoperational** *m* PROD interoperation time; **~ interoperativo** *m* PROD interoperative time; **~ inutilizable** *m* TELECOM unavailable time (*UAT*); **~ de inutilización** *m* TEXTIL setup time; **~ de inversión** *m* FERRO, TELECOM *del sentido de transmisión* turnaround time; **~ de inversión en el terminal** *m* FERRO *vehículos* turnaround time at terminus;

~ l ~ **de lectura** *m* IMPR, INFORM&PD read time; **~ libre total** *m* PROD *proyectos* total float; **~ de lixiviación** *m* CARBÓN leaching time; **~ de luna nueva o llena en el ciclo de fases** *m* ENERG RENOV *astronomía* syzygy;

~ m ~ **de maniobra** *m* ING ELÉC operate time, PROD operating time; **~ para maniobrar** *m* TRANSP MAR *radar* time to manoeuvre; **~ de manipulación** *m* TEXTIL handling time; **~ de mantenimiento** *m* PROD move time; **~ de máquina** *m* INFORM&PD machine time; **~ medio entre desmontaje** *m* (*MTBR*) TEC ESP mean time between removals (*MTBR*); **~ medio entre fallos** *m* (*MTBF*) CALIDAD, INFORM&PD, ING ELÉC, MECÁ, TEC ESP mean time between failures (*MTBF*); **~ medio de parada** *m* TRANSP average stopped time; **~ medio de reparación** *m* (*MTTR*) INFORM&PD, ING ELÉC, MECÁ, TEC ESP mean time to repair (*MTTR*); **~ de memorización y búsqueda** *m* ING ELÉC storage time; **~ de mezcla** *m* CONST, PETROL mixing time; **~ de mezclado** *m* CONST, PETROL mixing time; **~ de montaje abierto** *m* P&C *adhesivos, transformación* open assembly time; **~ muerto** *m* ELECTRÓN dead time, EMB downtime, INFORM&PD dead halt, dead time, MECÁ, NUCL downtime; **~ de muestreo** *m* PROD sample time;

~ n ~ **no disponible** *m* TELECOM unavailable time (*UAT*);

~ o ~ **de observación** *m* PROD scan time; **~ de operación** *m* C&V running time; **~ de ordenador** *m* Esp (*cf tiempo de computadora AmL*) INFORM&PD machine time;

~ p ~ **de palabra** *m* INFORM&PD word time; **~ de parada** *m* INFORM&PD, MECÁ downtime, NUCL outage time, PROD downtime; **~ de pasada** *m* INFORM&PD, PROD run time; **~ pasivo** *m* INFORM&PD idle time; **~ de paso a la posición de reposo** *m* ELEC *de relé* drop-out time; **~ de penetración del calor** *m* TERMO heat penetration time; **~ de penetración térmica** *m* TERMO heat penetration time; **~ perdido** *m* TEC PETR *perforación* lost time; **~ de permanencia** *m* TEXTIL dwelling time; **~ de permanencia en el aire** *m* TELECOM air time; **~ de persistencia** *m* PETROL decay time, TELECOM hangover time; **~ por pieza** *m* PROD time per piece; **~ de posicionado radial** *m* ÓPT radial positioning time; **~ de preparación** *m* IMPR lead time, INFORM&PD, PROD setup time; **~ de procesamiento** *m* INFORM&PD processing time; **~ de producción realizado** *m* PROD realized production time; **~ productivo** *m* INFORM&PD productive time, uptime; **~ de programación programado** *m* TELECOM scheduled operating time; **~ de propagación** *m* TEC ESP travel time, TELECOM time delay, TRANSP AÉR rise time; **~ de propagación de la señal** *m* TELECOM *en un medio* signal delay; **~ propio** *m* FÍS *relatividad* proper time; **~ de protección** *m* PROD protection time;

~ q ~ **para que el vidrio esté libre de gases** *m* C&V seed-free time;

~ r ~ **de rampa a rampa total del viaje** *m* TRANSP AÉR ramp-to-ramp time; **~ de ranurado** *m* PROD slitting time; **~ de reacción** *m* TRANSP *seguridad en carretera, seguridad social* reaction time; **~ real** *m* CINEMAT, ELECTRÓN, INFORM&PD, TELECOM, TV real time; **~ real de intervalo** *m* TV real gap length; **~ real en línea** *m* (*OLRT*) INFORM&PD on line real time (*OLRT*); **~ de recarga** *m* FOTO recharge time; **~ de recirculación** *m* FOTO recycle time, recycling time; **~ de recorrido** *m* GEOFÍS *análisis de ondas* travel time, ING ELÉC, PROD transit time; **~ de recuperación** *m* IMPR clearing time; **~ de recuperación inversa** *m* ELECTRÓN reverse-recovery time; **~ reducido a alta temperatura** *m* (*TRAT*) ALIMENT high temperature short time; **~ de registro** *m* CINEMAT playing time, PETROL record time; **~ de relajación** *m* ELEC *dieléctrico*, METAL relaxation time; **~ de reparación** *m* INFORM&PD repair time; **~ de reposición** *m* ING ELÉC release lag; **~ de reposo** *m* AGRIC rest period; **~ requerido para iniciar algún proceso** *m* ING MECÁ lead time; **~ de respuesta** *m* ELECTRÓN response time, FERRO turnaround time, INFORM&PD, METR response time, PROD turnaround directive, *línea semidúplex* turnaround time, TELECOM response time, *del sentido de transmisión* turnaround time; **~ de retardo** *m* ELEC *de conmutador* delay time, ELECTRÓN time lag, time delay, TELECOM, TV delay time; **~ de retardo estimable** *m* PROD *informática* slack time; **~ de retención** *m* INFORM&PD hold time, ING ELÉC retention time, TEC ESP hold time; **~ de retorno** *m* PROD turnaround time, TV return interval; **~ de retroceso del borrado** *m* TV flyback blanking; **~ de revelado** *m* CINEMAT development time, developing time; **~ de reverberación** *m* ACÚST reverberation time; **~ de ruptura** *m* ELEC *relé* break time;

~ s ~ **de secado de la tinta** *m* IMPR ink-setting time; **~ de seguridad** *m* PROD safety time, TEC ESP guard time; **~ de servicio** *m* PROD, TELECOM service time;

~ de sinterización *m* PROC QUÍ sintering time; **~ solar medio** *m* TEC ESP mean solar time; **~ de solidificación** *m* P&C *polimerización* setting time; **~ de subida** *m* INFORM&PD rise time; **~ de suma-resta** *m* INFORM&PD add-subtract time;

~ t **~-temperatura-tolerancia** *m* (*t-t-t*) REFRIG *de alimentos congelados* time-temperature-tolerance (*t-t-t*); **~ total de viaje** *m* TRANSP overall travel time; **~ total del vuelo** *m* TRANSP AÉR block time; **~ transcurrido** *m* INFORM&PD, MECÁ elapsed time; **~ transcurrido aproximado** *m* TRANSP AÉR estimated elapsed time; **~ de transferencia** *m* PROD move time; **~ para transformación** *m* P&C *adhesivos, transformación, polimerización* pot life; **~ de transición de autorotación** *m* TRANSP AÉR autorotation transition time; **~ de transiente** *m* FÍS transient time; **~ de tránsito** *m* NUCL, PETROL, PROD transit time; **~ de tránsito integrado** *m* TEC PETR *geofísica, petrofísica* integrated transit time (*ITT*); **~ de transporte** *m* PROD move time, transportation time;

~ u **~ por unidad** *m* PROD time per unit; **~ unitario** *m* PROD unit time; **~ universal coordinado** *m* METEO, TELECOM universal time coordinates (*UTC*); **~ útil** *m* INFORM&PD uptime;

~ v **~ de variación** *m* ELECTRÓN time drift; **~ de ventolinas y brisas ligeras** *m* METEO, TRANSP MAR *meteorología* light weather; **~ de viaje** *m* TRANSP trip time; **~ de vuelo** *m* FÍS PART, FÍS RAD, NUCL time of flight (*TOF*), TRANSP AÉR flying time; **~ de vuelo previsto** *m* TRANSP AÉR estimated flight time; **~ de vuelo en solitario** *m* TRANSP AÉR *sin copiloto ni instructor* solo flight time; **~ de vulcanización** *m* P&C *polímeros, elastómeros* curing time

tiempo[3]: **~ que tarda el lodo desde el fondo del pozo hasta la superficie** *fra* TEC PETR *perforación* lag time

tienda: **~ protectora para trabajadores** *f* SEG workers' protective tents; **~ de teñido** *f* COLOR dye shop

tierra[1]: **de ~** *adj* TEC ESP land; **sobre ~** *adj* MINAS above-ground; **de ~ firme** *adj* TEC PETR onshore

tierra[2]: **a ~** *adv* TRANSP MAR ashore; **en ~** *adv* TRANSP MAR ashore; **hacia ~** *adv* GEOL landward, TRANSP MAR *navegación* landward, shoreward; **sobre ~** *adv* MINAS above-ground, TRANSP MAR onshore

tierra[3] *f* AUTO earth connection (*BrE*), ground connection (*AmE*), CARBÓN earth (*BrE*), ground (*AmE*), ELEC earth (*BrE*), ground (*AmE*), earth connection (*BrE*), ground connection (*AmE*), FÍS earth (*BrE*), ground (*AmE*), GEOFÍS soil, ING ELÉC, PROD earth (*BrE*), earth connection (*BrE*), ground connection (*AmE*), TELECOM earth (*BrE*), ground (*AmE*), TRANSP MAR land, VEH earth (*BrE*), earth connection (*BrE*), ground connection (*AmE*); **~ accidental** *f* ELEC, ING ELÉC, PROD earth fault (*BrE*), ground fault (*AmE*); **~ accidental doble** *f* ELEC double earth fault (*BrE*), double ground fault (*AmE*); **~ accidental a fase** *f* ELEC *corriente alterna* phase-to-earth fault (*BrE*), phase-to-ground fault (*AmE*); **~ ácida** *f* CONTAM *caracterización de suelos* acid earth; **~ para alfarería** *f* C&V potter's earth; **~ de aluvión** *f* CARBÓN dirt; **~ arcillosa** *f* AGRIC, CARBÓN clay soil; **~ arenácea** *f* CARBÓN sandy ground; **~ de batanero** *f* GEOL *arcilla enriquecida con esmectita* fuller's earth; **~ de blanqueo** *f* DETERG bleaching earth; **~ en bloques** *f* CARBÓN boulder soil; **~ en descanso** *f* AGRIC resting land; **~ de descombro** *f* CARBÓN *construcción*, MINAS overburden; **~ de**

diatomeas *f* GEOL, REFRIG diatomaceous earth; **~ doble** *f* ELEC double earth fault (*BrE*), double ground fault (*AmE*); **~ excavada** *f* CARBÓN dirt; **~ de infusorios** *f* MINERAL infusorial earth; **~ de ladrillos** *f* CONST brick earth; **~ de moldeo** *f* GEOL marly loam, PROD loam; **~ de la red** *f* ELEC *conexión* system earth (*BrE*), system ground (*AmE*); **~ rica** *f* CARBÓN saturated soil; **~ saneada** *f* AGRIC reclaimed land; **~ de seguridad** *f* ELEC safety earth (*BrE*), safety ground (*AmE*); **~ de servicio** *f* ELEC *conexión* system earth (*BrE*), system ground (*AmE*); **~ de sílice** *f* MINERAL tripolite; **~ vegetal** *f* CONST loam, topsoil, CONTAM unspoilt land; **~ verde** *f* MINERAL green earth; **~ virgen** *f* CONTAM unspoilt land

tierras: **~ raras** *f pl* QUÍMICA rare earths

tiffanyita *f* MINERAL tiffanyite

tifón *m* METEO typhoon, TRANSP twister

tíglico *adj* QUÍMICA tiglic

tijera *f* MECÁ shear; **~ para chapa** *f* ING MECÁ plate shears; **~ cortapernos** *f* ING MECÁ bolt cutter; **~ mecánica** *f* ING MECÁ shearing machine; **~ de podar** *f* AGRIC hedge shears; **~ punzonadora y azalladora de palanca** *f* PROD lever punching-and-shearing machine; **~ para setos** *f* ING MECÁ hedge trimmer

tijeras *f pl* C&V shears, ING MECÁ scissors, shears, TEXTIL scissors; **~ anulares** *f pl* ING MECÁ ring shears; **~ curvadas para planchas** *f pl* ING MECÁ corner-rounding cutters; **~ para desbastes** *f pl* PROD slab shears; **~ de espiga** *f pl* CONST pin shears; **~ de festonear** *f pl* ING MECÁ pinking shears; **~ de guillotina** *f pl* MECÁ guillotine shears, PROD guillotine shearing machine, guillotine shears; **~ de hojalatero** *f pl* PROD tinman's shears, tinman's snips; **~ para lupias** *f pl* ING MECÁ bloom shears; **~ para metales** *f pl* PROD metal shears; **~ picafestones** *f pl* ING MECÁ pinking shears

timbar *vt* PROD stamp

timbrado *m* C&V badge

timbrar *vt* PROD stamp, *calderas* badge

timbre *m* ACÚST timbre, CARBÓN *electrotecnia* bell, electric bell, FÍS ONDAS *calidad de la nota* timbre, tone, pitch, INSTAL HIDRÁUL *de caldera de vapor* test plate, working pressure, PROD *caldera* badge plate, TELECOM ringer; **~ de la caldera** *m* INSTAL HIDRÁUL boiler test-plate, working pressure; **~ intermitente** *m* ELEC trembler bell; **~ de llamada** *m* ING ELÉC call bell; **~ de sonido continuo** *m* ING ELÉC continuous ringing bell; **~ supletorio** *m* TELECOM extension bell; **~ de teléfono** *m* TELECOM telephone bell; **~ trepidante** *m* ELEC trembler bell

timol *m* QUÍMICA thymol

timolftaleína *f* QUÍMICA thymolphthalein

timón *m* FÍS rudder, TEC ESP *vehículos* flange mounting, TRANSP AÉR, TRANSP MAR *construcción naval* rudder; **~ de cola** *m* TRANSP AÉR aileron follow-up; **~ colgado** *m* TRANSP MAR spade rudder, underhung rudder; **~ compensado** *m* TRANSP MAR balanced rudder; **~ de dirección** *m* FÍS, TRANSP AÉR, TRANSP MAR rudder; **~ de dirección de arrastre** *m* TRANSP AÉR drag rudder; **~ de fortuna** *m* TRANSP MAR jury rudder; **~ hidráulico** *m* OCEAN hydrofoil rudder; **~ horizontal** *m* TRANSP MAR *construcción naval* fin, *submarinos* diving rudder, hydroplane; **~ del izador** *m* TRANSP AÉR elevator follow-up; **~ ordinario** *m* TRANSP MAR unbalanced rudder; **~ del paso del**

colectivo *m* TRANSP AÉR collective-pitch follow-up; **~ de plancha sencilla** *m* TRANSP MAR single-plate rudder; **~ de profundidad** *m* OCEAN depth rudder

timonel *m* TRANSP MAR helmsman; **~ de combate** *m* *AmL* (*cf* **cabo de mar** *Esp*) TRANSP MAR quartermaster

timones: ~ de dirección en serie *m pl* TRANSP AÉR serial rudders

timpa *f* PROD *horno alto* tymp

timpanismo *m* AGRIC bloat

tímpano *m* ACÚST eardrum, CONST *arquitectura* spandrel, IMPR tympan, INSTAL HIDRÁUL tympanum

tina *f* ALIMENT vat, LAB *material de vidrio* trough, MINAS trunk, PAPEL vat, chest, PROD tub, TEXTIL vat; **~ de blanqueo** *f* PAPEL bleaching chest; **~ de desgotado** *f* *AmL* PAPEL straining chest; **~ de desgote** *f* PAPEL drainage chest; **~ de máquina** *f* PAPEL machine chest; **~ de mezcla** *f* PAPEL blending-chest, mixing chest, service chest, stock chest; **~ de pasta** *f* PAPEL stuff chest

tinaja *f* ALIMENT vat

tinción *f* P&C *pintura* tinting, QUÍMICA staining

tingladillo: de ~ *adj* IMPR bicycling, TRANSP MAR *botes* clinker-built

tinglado *m* CONST *edificación* siding, TRANSP MAR dock warehouse

tinkal *m* MINERAL tinkal

tinkalconita *f* MINERAL tincalconite

tinta *f* COLOR, IMPR, INFORM&PD, INSTR, PAPEL ink; **~ al aceite** *f* IMPR oil-based ink; **~ de aceite** *f* COLOR oleic ink; **~ al alcohol** *f* IMPR alcohol ink; **~ de anilina** *f* COLOR aniline ink; **~ autográfica** *f* COLOR autographic ink; **~ de baja densidad** *f* IMPR light-bodied ink; **~ de base acuosa** *f* IMPR aqueous ink; **~ a base de disolventes** *f* IMPR solvent-based ink; **~ brillante** *f* COLOR gloss ink; **~ de bronce** *f* COLOR bronze ink; **~ al carbón** *f* COLOR carbon ink; **~ china** *f* COLOR Chinese ink, Indian ink, IMPR Indian ink; **~ de China** *f* REVEST china water; **~ de copiar** *f* COLOR copying ink; **~ para cuatricromía** *f* IMPR four-color process ink (*AmE*), four-colour process ink (*BrE*); **~ densa** *f* IMPR heavy ink; **~ de diamante** *f* COLOR diamond ink; **~ de dibujo** *f* COLOR, IMPR drawing ink; **~ en dos colores** *f* COLOR duplex ink; **~ de dos tonos** *f* COLOR bitone ink, double tone ink, duotone ink; **~ dúplex** *f* COLOR duplex ink; **~ electrográfica** *f* COLOR, ELEC electrographic ink; **~ de elevada densidad** *f* IMPR heavy-bodied ink; **~ ferrogálica** *f* COLOR ferrogallic ink; **~ flexográfica** *f* IMPR flexographic ink; **~ de fraguado al vapor** *f* COLOR moisture-set ink; **~ para grabado en plancha de cobre** *f* COLOR, IMPR copperplate engraving ink; **~ de grabar** *f* COLOR etching ink; **~ hipsométrica** *f* COLOR hypsometric tint; **~ para huecograbado** *f* COLOR gravure ink, photogravure ink; **~ para ilustraciones** *f* IMPR halftone ink; **~ de imprenta** *f* IMPR printer's ink, printing ink; **~ para impresión de huecograbado** *f* COLOR gravure printing ink; **~ de imprimir** *f* P&C printing ink; **~ para imprimir sobre hojas de aluminio** *f* IMPR aluminium foil printing ink (*BrE*), aluminum foil printing ink (*AmE*); **~ indeleble** *f* COLOR indelible ink; **~ invisible** *f* COLOR invisible ink, sympathetic ink; **~ de libro** *f* COLOR book ink; **~ magnética** *f* IMPR, INFORM&PD magnetic ink; **~ de marcar** *f* COLOR marking ink; **~ mate** *f* COLOR mat ink; **~ metálica** *f* COLOR metallic ink; **~ metálica de bronce** *f* IMPR bronze-powder ink; **~ normal** *f* IMPR standard ink; **~ oléica** *f* COLOR oleic ink; **~ opaca** *f* CINEMAT blooping ink; **~ en pasta** *f* COLOR paste ink; **~ permanente** *f* COLOR permanent ink; **~ en polvo** *f* COLOR dry ink; **~ de protección** *f* COLOR resistor ink; **~ pulverizada** *f* COLOR powdering ink; **~ que seca por absorción** *f* IMPR absorptive-type ink; **~ que seca por calor** *f* IMPR thermosetting ink; **~ que seca por infrarrojos** *f* IMPR infrared process ink; **~ de registro** *f* COLOR record ink; **~ para remiendo** *f* IMPR jobbing ink; **~ de retocar** *f* COLOR retouching ink; **~ de secado en frío** *f* COLOR cold-set ink; **~ de secado por presión** *f* COLOR pressure-set ink; **~ de secado rápido por calor** *f* COLOR heat-set ink; **~ simpática** *f* COLOR sympathetic ink; **~ para tricromía** *f* IMPR process ink

tintada *f* COLOR tint

tinte *m* C&V tint, COLOR dyer, tint, FOTO tint, QUÍMICA dye; **~ ácido** *m* TEXTIL acid dye; **~ directo** *m* COLOR substantive dye; **~ encapsulado** *m* COLOR encapsulated dye; **~ indofenólico** *m* COLOR indophenol dye; **~ para lana** *m* COLOR woollen dyer; **~ monoazoico** *m* COLOR monoazo dye; **~ de muestra** *m* COLOR swatch dyer; **~ reactivo** *m* COLOR, TEXTIL reactive dye; **~ de trazado** *m* PROD layout dye

tintero *m* COLOR ink duct, ink fountain, IMPR fountain, ink ductor, ink fountain; **~ tipo conducto** *m* COLOR ductor-type ink fountain

tintinear *vi* ING MECÁ clatter

tintóreo *adj* COLOR tinctorial

tintura *f* COLOR dyeing, stainer, tincture, QUÍMICA dye, TEXTIL dyeing; **~ de base** *f* IMPR base tint; **~ bronceadora** *f* COLOR bronzing tincture; **~ de colorante vegetal** *f* COLOR water stain; **~ de efecto retardado** *f* COLOR vital stain; **~ en foulard** *f* TEXTIL pad dyeing; **~ en hilado** *f* TEXTIL yarn dyeing; **~ de igualación** *f* TEXTIL level dyeing; **~ para madera** *f* COLOR wood stain; **~ de mechas** *f* TEXTIL roving dyeing; **~ en plegador** *f* TEXTIL beam dyeing; **~ en polvo** *f* COLOR powdered dye; **~ de yodo** *f* COLOR tincture of iodine

tinturar *vt* COLOR tinge

tio- *pref* QUÍMICA thio-

tioacético *adj* QUÍMICA thioacetic

tioácido *m* QUÍMICA thioacid

tioalcohol *m* QUÍMICA thioalcohol

tioaldehído *m* QUÍMICA thioaldehyde

tioamida *f* QUÍMICA thioamide

tioarsénico *adj* QUÍMICA thioarsenic

tiocarbamida *f* QUÍMICA thiocarbamide

tiocarbanilida *f* QUÍMICA thiocarbanilide

tiocarbonato *m* QUÍMICA thiocarbonate

tiocarbónico *adj* QUÍMICA thiocarbonic

tiocetona *f* QUÍMICA thioketone

tiocianato *m* QUÍMICA thiocyanate

tiociánico *adj* QUÍMICA thiocyanic

tiodifenilamina *f* QUÍMICA thiodiphenylamine

tioéter *m* QUÍMICA thioether

tiofeno *m* QUÍMICA thiophene

tiofenol *m* QUÍMICA thiophenol

tioflavina *f* QUÍMICA thioflavin

tiofosgeno *m* QUÍMICA thiophosgene

tioglicólico *adj* QUÍMICA thioglycolic

tiogoma *f* QUÍMICA thioplast

tioindamina *f* QUÍMICA thioindamine

tioíndigo *m* QUÍMICA thioindigo
tiol *m* CONTAM *producto químico*, QUÍMICA thiol
tiolato *m* QUÍMICA thiolate
tiona *f* QUÍMICA thione
tionación *f* QUÍMICA thionation
tionaftaleno *m* QUÍMICA thionaphthene
tionafteno *m* QUÍMICA thionaphthene
tionato *m* QUÍMICA thionate
tioneína *f* QUÍMICA thioneine
tiónico *adj* QUÍMICA thionic
tionilo *m* QUÍMICA thionyl
tionina *f* QUÍMICA thionine
tiopental *m* QUÍMICA thiopental
tioplasto *m* QUÍMICA thioplast
tiosulfato *m* QUÍMICA thiosulfate (*AmE*), thiosulphate (*BrE*); **~ sódico** *m* QUÍMICA sodium thiosulfate (*AmE*), sodium thiosulphate (*BrE*)
tiosulfúrico *adj* QUÍMICA thiosulfuric (*AmE*), thiosulphuric (*BrE*)
tiotetanol: **~ etílico** *m* TEC PETR *petrofísica* ethylthioethanol
tiourea *f* QUÍMICA thiourea
tioxantona *f* QUÍMICA thioxanthone
tioxeno *m* QUÍMICA thioxene
típico *adj* ING MECÁ standard
tipificación *f* TELECOM standardization
tipo[1]: **de ~ abierto** *adj* ING MECÁ open-type
tipo[2] *m* EMB grade, TELECOM standard; **~ de acción rápida** *m* PROD pop-up type; **~ adornado** *m* IMPR fancy type; **~ de amplificador** *m* ELECTRÓN amplifier class; **~ antiguo** *m* IMPR antique face, old style (*AmE*); **~ booleano** *m* INFORM&PD Boolean type; **~ burbuja** *m* TEXTIL bubble type; **~ de carácter** *m* IMPR, INFORM&PD character type; **~ de codificación** *m* TELECOM coding type; **~ comercial** *m* IMPR commercial type; **~ complejo** *m* INFORM&PD complex type; **~ común** *m* IMPR body type; **~ de datos** *m* INFORM&PD data type; **~ de datos resumidos** *m* INFORM&PD abstract data type; **~ desgatado** *m* IMPR blunt; **~ de empaquetado** *m* PROD packing type; **~ encapsulado** *m* INFORM&PD encapsulated type; **~ entero** *m* INFORM&PD integer type; **~ de enumeración** *m* INFORM&PD enumeration type; **~ escalar** *m* INFORM&PD scalar type; **~ estrecho** *m* IMPR condensed face; **~ estructurado** *m* INFORM&PD structured type; **~ de exactitud** *m* METR class of accuracy; **~ extra ancha** *m* IMPR extended type; **~ del flete** *m* EMB, TEC PETR *transporte, comercio* freight rate (*AmE*), goods rate (*BrE*); **~ gótico** *m* IMPR gothic face; **~ de imprenta** *m* IMPR printing type; **~ de información** *m* TELECOM information type (*IT*); **~ de información codificada** *m* TELECOM encoded information type (*EIT*); **~ de latón** *m* IMPR brass type; **~ de letra** *m* IMPR typeface; **~ de letra dimensionable** *m* IMPR scalable font, scalable fount; **~ de letra fino** *m* IMPR light face; **~ de letra para libros** *m* IMPR bookfont; **~ de letra mecanográfica** *m* IMPR typewriter face; **~ de letra negrita** *m* IMPR bold type; **~ de letras** *m* IMPR, INFORM&PD font; **~ lógico** *m* INFORM&PD logical type; **~ de madera** *m* IMPR woodtype; **~ de madera de gran tamaño** *m* IMPR block letter; **~ mejorado** *m* PROD improved type; **~ de muestras** *m* IMPR swatch type; **~ n** *m* ELECTRÓN n-type; **~ negrilla** *m* PROD bold type; **~ de número** *m* TELECOM type of number (*TON*); **~ de ojo pleno** *m* IMPR full-face type; **~ de palo seco** *m* IMPR

sanserif, sanserif face; **~ palo seco plano** *m* IMPR slab serifs; **~ en pantalla** *m* IMPR display type; **~ privado** *m* INFORM&PD private type; **~ de producto** *m* CALIDAD product type; **~ para publicidad** *m* IMPR ad face; **~ de puerta** *m* CONST door brand; **~ de quark** *m* FÍS PART flavor of quark (*AmE*), flavour of quark (*BrE*); **~ de reactor** *m* NUCL reactor design; **~ real** *m* INFORM&PD real type; **~ de remiendo** *m* IMPR jobbing face, jobbing type; **~ de roca** *m* CARBÓN, PETROL rock type; **~ de segmento** *m* TELECOM segment type (*ST*); **~ de terreno intermedio** *m* CARBÓN intermediate type of soil; **~ para titulares** *m* IMPR display face
tipografía *f* IMPR letterpress, letterpress printing, relief printing, typography
tipográfico *adj* IMPR letterpress
tipógrafo *m* IMPR comp
tipos: **~ raros** *m pl* IMPR pi types
tira *f* CINEMAT, EMB, FÍS, FOTO strip, IMPR string, PROD strap, strip, TRANSP AÉR strip, TRANSP MAR davit fall; **~ bimetálica** *f* FÍS bimetallic strip (*BrE*); **~ de cubierta de la nervadura de la raíz** *f* TRANSP AÉR cover strip of root rib; **~ de ensayo** *f* CINEMAT, FOTO test strip; **~ del estimulador de la entrada** *f* PROD input simulator strip; **~ de identificación** *f* PROD identification strip; **~ de imágenes** *f* IMPR picture strip; **~ sin imágenes que precede al filme** *f* TV leader; **~ laminada** *f* P&C *producto semielaborado* laminated strip; **~ de muestra** *f* TEXTIL sample strip; **~ de prueba** *f* FOTO test print; **~ de rasgado** *f* EMB tear tape; **~ para rasgar** *f* EMB *sistema de apertura* tear strip; **~ de referencia** *f* CINEMAT reference strip; **~ del simulador de la entrada** *f* PROD input simulator strip; **~ de toma a tierra** *f* ELEC earthing strip (*BrE*), grounding strip (*AmE*); **~ de toma de tierra** *f* ING ELÉC earthing strip (*BrE*), grounding strip (*AmE*); **~ de unión** *f* PROD bonding tab
tirabuzón *m* TRANSP AÉR spin
tirachinas *m* TEC ESP slingshot
tirada *f* IMPR presswork, printing; **~ de combinación** *f* IMPR combination run; **~ corta** *f* IMPR short run; **~ de doble producción** *f* IMPR straight run; **~ excesiva** *f* IMPR overrun
tirador *m* CONST *puertas* button, handle, locking handle, TEC ESP slingshot; **~ con cerradura** *m* CONST locking handle; **~ emboscado** *m* D&A sniper; **~ de la puerta** *m* CONST door bolt, door handle, doorknob
tirafondo *m* CONST lag screw, FERRO *equipo inamovible* screw spike, *infraestructura* sleeper screw (*BrE*), tie screw (*AmE*), ING MECÁ coach screw; **~ de traviesa** *m* FERRO *infraestructura* sleeper screw (*BrE*), tie screw (*AmE*); **~ de traviesas** *m* FERRO railroad tie screw (*AmE*), railway sleeper screw (*BrE*)
tirafondos: **~ de mano** *m* ING MECÁ hand spike; **~ manual** *m* ING MECÁ hand spike
tirahilo: **~ de aguja** *m* TEXTIL take-up
tiraje: **~ de alta velocidad** *m* Esp (*cf duplicación de alta velocidad AmL*) TV high-speed duplication; **~ de pruebas** *m* FOTO printing, printing stage
tiramina *f* QUÍMICA tyrosamine
tiramollar *vt* TRANSP MAR *mantenimiento* overhaul
tirante *m* CARBÓN *ferrocarril* brace, CONST brace, stay rod, *edificación* stay, guy rope, *estructura* tie rod, CONTAM MAR *mecánica* tension member, ING MECÁ tie, tie bolt, brace strut, tie bar, link, MECÁ guy, anchor, MINAS stretcher piece, TEC ESP brace, TRANSP

MAR *construcción de motor* tie rod; ~ **de armadura** *m* CONST *estructuras* truss rod; ~ **de arrastre** *m* TRANSP AÉR drag brace; ~ **articulado** *m* ING MECÁ flexible stay bolt; ~ **de caldera** *m* ING MECÁ boiler stay; ~ **de freno** *m* ING MECÁ brake strap; ~ **MacPherson** *m* AUTO, VEH *suspensión* MacPherson strut; ~ **de placa de guarda** *m* FERRO *vehículos* axle guide stay; ~ **de reglaje** *m* ING MECÁ pull rod; ~ **de separación** *m* TEXTIL separating rod; ~ **transversal** *m* VEH *dirección* track rod; ~ **de tubería** *m* ING MECÁ pipe anchor

tirantería *f* VEH *de dirección, embrague, frenos* linkage

tirantez *f* ING MECÁ stretch, tightness

tirar¹ *vt* CARBÓN dump, IMPR print, RECICL dump

tirar² *vi* CONST draw

tirar³: **no** ~ *fra* EMB *instrucciones de uso* do not throw, not to be thrown

tiras *f pl* TRANSP MAR boat falls; ~ **de metal** *f pl* ING MECÁ strip shim

tiratrón *m* ELECTRÓN, QUÍMICA thyratron; ~ **de hidrógeno** *m* ELECTRÓN hydrogen thyraton

tireta *f* PROD lace; ~ **para coser correas** *f* PROD belt lace

tiristor *m* ELEC, FÍS, ING ELÉC, TELECOM thyristor; ~ **bidireccional** *m* ELEC, ING ELÉC triac; ~ **en estado de desconexión** *m* ELECTRÓN off thyristor; ~ **de potencia** *m* TELECOM power thyristor

tiro *m* CINEMAT *de cámara o proyección* throw, CONST *chimenea* draft (*AmE*), shaft, draught (*BrE*), D&A shot, EMB *resistencia a despegarse*, IMPR *de la tinta* tack, ING MECÁ pull, pulling, MINAS shot, PAPEL *en la máquina* draw, PROD *chimenea* draught (*BrE*), draft (*AmE*), TEC ESP *propulsión* casting, TEC PETR *geofísica* shot; ~ **de artillería** *m* D&A artillery fire; ~ **por aspiración** *m* ING MECÁ induced draft (*AmE*), induced draught (*BrE*); ~ **aspirado** *m* ING MECÁ *chimeneas* induced draft (*AmE*), induced draught (*BrE*); ~ **de barrera** *m* D&A *artillería* barrage; ~ **combinado** *m* IMPR collect run; ~ **de enfilada** *m* D&A raking fire; ~ **espacial** *m* TEC ESP space launch, space shot; ~ **forzado** *m* CONST *chimenea*, ING MECÁ forced draft (*AmE*), forced draught (*BrE*); ~ **de horquilla** *m* D&A ranging fire; ~ **indirecto** *m* D&A indirect fire; ~ **inducido** *m* ING MECÁ induced draft (*AmE*), induced draught (*BrE*); ~ **irregular** *m* D&A ragged fire; ~ **de protección** *m* D&A covering fire; ~ **rápido** *m* D&A rapid fire; ~ **de reglaje** *m* D&A ranging fire, registration fire; ~ **de repetición** *m* D&A repetition fire

tirón *m* ING MECÁ lug, strain, MECÁ, PROD lug, TRANSP MAR *pesca* haul

tironina *f* QUÍMICA thyronine

tirosamina *f* QUÍMICA tyrosamine

tirosina *f* QUÍMICA tyrosine

tiroxina *f* QUÍMICA thyroxine

titanato *m* QUÍMICA titanate

titánico *adj* QUÍMICA titanic

titanilo *m* QUÍMICA titanyl

titanio *m* (*Ti*) METAL, QUÍMICA, TEC ESP titanium (*Ti*)

titanita *f* MINERAL eucolite, titanite

titanoso *adj* QUÍMICA titanous

titilación *f* ACÚST flutter, TELECOM scintillation

titrar *vt* CARBÓN *química, microbiología*, DETERG, LAB, QUÍMICA titrate

titulación *f* LAB, QUÍMICA titration; ~ **automática** *f* *AmL* (*cf valoración automática Esp*) LAB *análisis* automatic titration; ~ **complejométrica** *f* QUÍMICA *análisis* complexometric titration; ~ **conductimétrica** *f* QUÍMICA *análisis* conductimetric titration; ~ **electrométrica** *f* QUÍMICA electrometric titration; ~ **potenciométrica** *f* QUÍMICA electrometric titration; ~ **por retorno** *f* QUÍMICA back titration

titulador *m* CARBÓN, DETERG, LAB, QUÍMICA titrator

tituladora *f* CINEMAT title printer; ~ **de tambor** *f* CINEMAT drum titler

titular¹: ~ **de un abono** *m* TELECOM subscriber; ~ **principal de la cuenta** *m* TELECOM major account holder

titular² *vt* CARBÓN, DETERG, LAB, QUÍMICA titrate

título *m* CINEMAT credit, title, IMPR title, INFORM&PD heading, PROD grade, QUÍMICA *análisis* titer (*AmE*), titre (*BrE*), TEXTIL *hilado* count; ~ **de calibre** *m* PROD *galga de alambres* gage number (*AmE*), gauge number (*BrE*); ~ **de capitán de la marina mercante** *m* TRANSP MAR master's certificate; ~ **de contraportada** *m* IMPR bastard title; ~ **a dos o más columnas** *m* IMPR spreadhead; ~ **fino** *m* TEXTIL *hilatura* fine count; ~ **grueso** *m* TEXTIL *hilado* coarse count; ~ **de la lana** *m* TEXTIL *números* wool count; ~ **a lo ancho de la página** *m* IMPR shoulder head; ~ **del lomo** *m* IMPR back title; ~ **molar** *m* REFRIG *de una disolución* mole titer (*AmE*), mole titre (*BrE*); ~ **provisional** *m* CINEMAT working title; ~ **recuadrado** *m* IMPR boxed head; ~ **repetido** *m* IMPR running title; ~ **a toda plana** *m* IMPR streamer; ~ **de vapor** *m* TERMOTEC steam quality

títulos *m pl* CINEMAT, TV credits; ~ **de crédito** *m pl* CINEMAT, TV credits; ~ **finales** *m pl* CINEMAT, TV closing credits, end credits

tixotropía *f* CARBÓN, FÍS thixotropy, P&C *propiedad física* thixotropic, PETROL, QUÍMICA thixotropy

tixotrópico *adj* CARBÓN, FÍS, P&C, QUÍMICA, TEC PETR thixotropic

tiza *f* C&V chalk

tizón *m* AGRIC smut, *patología vegetal* blight, CONST *albañilería* perpend, bonder

Tl *abr* (*talio*) QUÍMICA Tl (*thallium*)

TL *abr* (*terminación de línea*) TELECOM LT (*line termination*)

TLV¹ *abr* (*valor límite umbral*) CONTAM *legislación, toxicología* TLV (*threshold limit value*)

TLV²: ~ **en áreas de trabajo** *m* CONTAM TLV at place of work; ~ **en el lugar de trabajo** *m* CONTAM *legislación, toxicología* TLV at place of work; ~ **en el medio ambiente** *m* CONTAM *legislación, toxicología* TLV in the free environment; ~ **ocupacional** *m* CONTAM *legislación* occupational TLV

TM *abr* (*magnético transversal*) ING ELÉC TM (*transverse magnetic*)

Tm *abr* (*tulio*) QUÍMICA Tm (*thulium*)

TMAD *abr* (*tráfico medio anual por día*) TRANSP AADT (*average annual daily traffic*)

TNI *abr* (*índice de ruido de tráfico*) ACÚST TNI (*traffic noise index*)

TNT *abr* (*trinitrotolueno*) QUÍMICA TNT (*trinitrotoluene*)

toallero *m* INSTAL TERM *eléctrico* towel rail

toba *f* CONST, GEOL *roca volcánica* tufa, PETROL tufa, tuff, TEC PETR tuff, tufa; ~ **calcárea** *f* GEOL calcareous tufa; ~ **caliza** *f* GEOL calcareous tufa, PETROL call-tufa; ~ **cristalina** *f* CRISTAL, GEOL *roca piroclástica* crystal tuff; ~ **lítica** *f* GEOL *tipo de roca* *piroclástica* lithic tuff; ~ **silícea** *f* TEC PETR siliceous sinter; ~ **vítrea** *f* GEOL *cenizas petrificadas de grano fino* vitric tuff; ~ **volcánica** *f* TEC PETR volcanic tuff; ~ **volcánica soldada** *f* GEOL welded tuff

tobera *f* CARBÓN nozzle, FÍS *alto horno* tue iron, tuyere, nozzle, ING MECÁ, INSTAL HIDRÁUL, MECÁ nozzle, NUCL end section, PROC QUÍ nozzle, PROD nozzle, *alto horno* tuyere, tue iron, TEC ESP nozzle, TEC PETR *perforación* jet, TRANSP AÉR nozzle, *alto horno* tue iron, tuyere; ~ **de aire** *f* CARBÓN, ING MECÁ air nozzle; ~ **ajustable** *f* ING MECÁ adjustable nozzle; ~ **de atomización** *f* PROC QUÍ atomizer nozzle, atomizing cone; ~ **de chorro** *f* TEC ESP jet nozzle; ~ **cilíndrica** *f* INSTAL HIDRÁUL cylindrical mouthpiece; ~ **de combustible** *f* AUTO, ING MECÁ, VEH fuel nozzle; ~ **convergente** *f* ING MECÁ combining cone, combining tube; ~ **coronada** *f* TRANSP AÉR notched nozzle; ~ **de decantación** *f* CARBÓN thickening cone; ~ **dentada** *f* TRANSP AÉR notched nozzle; ~ **de descarga** *f* NUCL *de una bomba* discharge nozzle; ~ **divergente** *f* FÍS divergent nozzle, PROD *inyectores* delivery cone, delivery tube; ~ **de empuje vectorial** *f* TEC ESP, TRANSP AÉR thrust-vectoring nozzle; ~ **de entrada de aire** *f* ING MECÁ air inlet nozzle; ~ **de entrada de alimentación** *f* NUCL *de un generador de vapor* feedwater-inlet nozzle; ~ **de entrada del primario** *f* NUCL reactor-coolant inlet nozzle, REFRIG reactor-coolant pump; ~ **de entrada del refrigerante del reactor** *f* NUCL reactor-coolant inlet nozzle, REFRIG reactor-coolant pump; ~ **de escape** *f* NUCL vent nozzle, TRANSP AÉR exhaust nozzle; ~ **de expansión** *f* TRANSP AÉR expansion nozzle; ~ **del extremo de la pala** *f* TRANSP AÉR blade-tip nozzle; ~ **giratoria** *f* TEC ESP rotatable nozzle; ~ **de inyección** *f* TEC ESP thrust cone, VEH *de gasolina* injection nozzle; ~ **de inyección del atomizador** *f* PROC QUÍ atomizer injection nozzle; ~ **de inyector** *f* ING MECÁ cone; ~ **inyectora** *f* ING MECÁ combining cone; ~ **de mezcla** *f* ING MECÁ combining cone; ~ **orientable de empuje** *f* TEC ESP, TRANSP AÉR thrust-vectoring nozzle; ~ **propulsora** *f* TRANSP AÉR propelling nozzle; ~ **del reactor** *f* TRANSP AÉR jet nozzle; ~ **de salida del aire** *f* MECÁ air discharge nozzle; ~ **secundaria** *f* TRANSP AÉR secondary nozzle; ~ **soplante** *f* PROD blast nozzle; ~ **superior** *f* NUCL top fitting, *de un elemento de combustible* upper end fitting; ~ **de suspensión** *f* TEC ESP construction nozzle; ~ **de tipo Venturi** *f* MECÁ venturi nozzle; ~ **de tubo múltiple** *f* TRANSP AÉR multitube nozzle; ~ **de vapor** *f* INSTAL HIDRÁUL *inyector* steam cone, steam nozzle; ~ **de ventilación** *f* CARBÓN air nozzle

tobería: ~ **de caudal** *f* ING MECÁ flow nozzle

tobogán *m* SEG *aviones*, TRANSP AÉR toboggan; ~ **de carga** *m* TRANSP freight chute (*AmE*), goods chute (*BrE*); ~ **de emergencia** *m* SEG, TRANSP AÉR emergency slide; ~ **de escape** *m* TEC ESP escape chute; ~ **para hielo** *m* REFRIG ice dump table; ~ **inflable** *m* TRANSP AÉR inflatable slide; ~ **de mercancías** *m* TRANSP freight chute (*AmE*), goods chute (*BrE*); ~ **de paquetes** *m* TRANSP parcels chute

tocadiscos *m* ACÚST *gramófono* record player, turntable, ING ELÉC record player

tocar[1]: ~ **contra** *vt* ING MECÁ, MECÁ impinge on; ~ **fondo** *vt* OCEAN *vehículos y objetos subacuáticos* bottom, TRANSP MAR *del barco* touch bottom

tocar[2]: ~ **el timbre** *vi Esp* TELECOM ring

TOCC *abr* (*Centro de Control Técnico y de Operaciones*) TEC ESP TOCC (*Technical and Operational Control Center AmE*)

tocho *m* CARBÓN, MECÁ, PROD ingot; ~ **para tornillo** *m* ING MECÁ, PROD screw blank

tocoferol *m* ALIMENT tocopherol

tocón *m* CONST *madera* stump

tojino *m* TRANSP MAR *construcción naval* lug

Tokamac: ~ **JET** *m* FÍS RAD, NUCL JET Tokamac

tokamak *m* FÍS tokamak

tolerancia *f* CARBÓN analysis error, ING ELÉC clearance, ING MECÁ margin, allowance, ply, tolerance, MECÁ allowance, tolerance, METR *mecánica* tolerance, PROD allowance, TEXTIL tolerance; ~ **al ácido** *f* CONTAM *características o propiedades químicas* acid tolerance; ~ **de ajuste positiva** *f* INSTAL HIDRÁUL *mecanismo de pistón y cilindro* clearance, clearance space; ~ **del árbol de levas** *f* AUTO, VEH camshaft clearance; ~ **del carril al desgaste** *f* FERRO *infraestructura* rail-wear tolerance; ~ **de concentricidad** *f* MECÁ concentricity tolerance; ~ **por contracción** *f* ING MECÁ allowance for shrinkage; ~ **del diámetro del alma** *f* ÓPT, TELECOM core-diameter tolerance; ~ **del diámetro del núcleo** *f* ÓPT, TELECOM core-diameter tolerance; ~ **del diámetro del revestimiento** *f* ÓPT *fibras ópticas* cladding-diameter tolerance; ~ **de espesor** *f* ING MECÁ thickness margin; ~ **de fabricación** *f* ING MECÁ limit; ~ **de fallo** *f* ING ELÉC fault tolerance; ~ **general** *f* ING MECÁ general tolerance; ~ **en más** *f* ING MECÁ margin over; ~ **máxima y mínima** *f* ING MECÁ top and bottom clearance; ~ **en menos** *f* ING MECÁ margin under; ~ **del pistón** *f* AUTO, VEH *motor* piston clearance; ~ **de planeidad** *f* ING MECÁ flatness tolerance; ~ **de válvula** *f* ING MECÁ valve clearance; ~ **en la verticalidad** *f* C&V verticality tolerance

tolerancias: ~ **geométricas** *f pl* ING MECÁ geometric tolerancing; ~ **y perfiles de los filetes de roscas** *f pl* ING MECÁ screw thread profiles and tolerances

tolerante: ~ **de defecto** *adj* ING ELÉC fault-tolerant; ~ **a fallos** *adj AmL* INFORM&PD fault-tolerant

tolilénico *adj* QUÍMICA tolylene

tolilo *m* QUÍMICA tolyl

tolua *f Esp* (*cf alcancía AmL*) CARBÓN chute, ore pass, MINAS chute, *mineral* coarse concentration mill; ~ **principal** *f Esp* (*cf alcancía principal AmL*) CARBÓN, MINAS main chute

toluato *m* QUÍMICA toluate

toluendiisocianato *m* (*TDI*) P&C toluene diisocyanate (*TDI*)

tolueno *m* DETERG, P&C, QUÍMICA, TEC PETR *petrofísica* toluene

toluico *adj* QUÍMICA toluic

toluidina *f* QUÍMICA toluidine

toluileno *m* QUÍMICA toluylene

toluilo *m* QUÍMICA toluyl

toluldehído *m* QUÍMICA toluldehyde

tolunitrilo *m* QUÍMICA tolunitrile

toluol *m* QUÍMICA toluol

toluquinolina *f* QUÍMICA toluquinoline

tolva *f* AGRIC grain feeder, hopper, seed box, C&V *para alimentar una máquina* boot, CONTAM tank, ING MECÁ hopper, MECÁ bin, hopper, PROD *horno alto* cone, hopper, TEXTIL hopper; ~ **de aire** *f* ING MECÁ air scoop; ~ **de alimentación** *f* CONST, EMB, MECÁ, PROD feed hopper; ~ **de carga** *f* CONTAM cargo tank, MINAS loading chute; ~ **decantadora** *f* CARBÓN settling cone; ~ **de destino** *f* PROD destination bin; ~ **para minerales** *f* MINAS ore hopper; ~ **vibratoria** *f* EMB vibratory hopper

toma[1] *f* AGUA collection, inlet, intake, uptake, CINEMAT shot, take, ELEC *devanado* tap, *conexión* tapping, FOTO shot, HIDROL intake, ING ELÉC *de corriente* collection, ING MECÁ pick-up, intake, meshing, INSTAL HIDRÁUL *de vapor* drawing off, intake, MECÁ meshing, intake, TELECOM seizure, plug, TRANSP AÉR *de aire* intake; ~ **aérea** *f* CINEMAT aerial shot; ~ **de agua** *f* AGUA hydrant, offtake, CONST, ENERG RENOV, HIDROL water intake; ~ **de agua contraincendios** *f* AGUA, CONST, SEG, TERMO, TRANSP MAR fire hydrant (*BrE*), fireplug (*AmE*); ~ **de aire** *f* ING MECÁ air inlet, air intake, PROD *horno alto* belly pipe, *colada* air hole, TRANSP AÉR intake, TRANSP MAR air intake; ~ **de aire cuya sección puede cambiarse en vuelo** *f* TRANSP variable geometry intake; ~ **de aire dinámica** *f* ING MECÁ *motores* air scoop; ~ **de aire de geometría variable en vuelo** *f* TRANSP AÉR variable geometry inlet, variable geometry intake; ~ **de aire del motor** *f* TRANSP AÉR engine air-intake; ~ **de aire a presión** *f* ING MECÁ compressed-air socket; ~ **de alimentación** *f* ING ELÉC lead; ~ **de altura** *f* TEC ESP climb, TRANSP AÉR climb, climb-out; ~ **de altura inicial** *f* TRANSP AÉR initial climb-out; ~ **angular** *f* CINEMAT angle shot; ~ **en ángulo** *f* CINEMAT angle shot; ~ **de ángulo bajo** *f* FOTO low-angle shot; ~ **de archivo** *f* CINEMAT stock shot; ~ **con boom** *f* CINEMAT boom shot; ~ **central** *m* ELEC *transformador* center tap (*AmE*), centre tap (*BrE*); ~ **central secundaria** *f* ING ELÉC secondary center tap (*AmE*), secondary centre tap (*BrE*); ~ **combinada** *f* CINEMAT composite shot; ~ **de combustible** *f* CONTAM MAR *embarcaciones* refueling (*AmE*), refuelling (*BrE*), TRANSP MAR bunkering; ~ **de conexión** *f* ING ELÉC lead; ~ **de contacto con tierra** *f* TRANSP AÉR touchdown; ~ **de corriente** *f* ELEC *conexión* socket, tapping, FERRO *vehículos* current collector, ING ELÉC outlet, current collector, collector, PROD *electricidad* outlet connection; ~ **de corriente múltiple** *f* TV multiple outlet plug; ~ **en corrientes de agua** *f* AGUA tapping stream; ~ **crepuscular** *f* FOTO twilight shot; ~ **de datos** *f* FÍS ONDAS scan, FÍS RAD scanning; ~ **desde un ángulo alto** *f* CINEMAT high-angle shot; ~ **desde un ángulo bajo** *f* CINEMAT low shot, low-angle shot; ~ **directa** *f* ING MECÁ direct drive; ~ **a distancia** *f* CINEMAT long shot; ~ **con dolly** *f* CINEMAT dolly shot; ~ **de enchufe** *m* ELEC *conexión* plug receptacle; ~ **de ensayo** *f* FOTO test shot; ~ **especular** *f* CINEMAT mirror shot; ~ **exterior de día** *f* CINEMAT day exterior; ~ **de flujo** *f* TRANSP AÉR inflow; ~ **de gas lateral** *f* PROD *horno alto* downcomer, downtake; ~ **de la hoja por aspiración** *f* PAPEL suction pickup transfer; ~ **inclinada** *f* CINEMAT tilted shot; ~ **de inducción** *f* ING MECÁ induction pickup; ~ **inicial** *f* CINEMAT opening shot; ~ **de inserción** *f* CINEMAT insert shot; ~ **instantánea** *f* IMPR shot; ~ **intermedia** *f* ELEC *conexión* tapping; ~ **intermedia principal** *f* ELEC *conexión* principal tapping; ~ **larga** *f* CINEMAT long take; ~ **macho de dispositivo de conexión múltiple** *m* TV multiple outlet plug; ~ **con maquetas** *f* CINEMAT model shot; ~ **de mar** *f* TRANSP MAR *construcción naval* seacock; ~ **a masa** *f* AUTO, ELEC, ING ELÉC, PROD, VEH earth connection (*BrE*), ground connection (*AmE*); ~ **con máscara** *f* CINEMAT mask

shot; ~ **de muestras** *f* AGUA, CARBÓN, TEC PETR *geología* sampling; ~ **múltiple** *f* TELECOM multiple seizure; ~ **negativa** *f* ELEC minus tapping; ~ **panorámica** *f* CINEMAT pan; ~ **con panorámica vertical** *f* CINEMAT tilt shot; ~ **picada** *f* CINEMAT tilt shot; ~ **a ras de suelo** *f* CINEMAT, TV ground angle shot; ~ **reducida de energía** *f* ELEC reduced power tapping; ~ **reducida de potencia** *f* ELEC reduced power tapping; ~ **de seguimiento** *f* CINEMAT tracking shot; ~ **de sonido** *f* ACÚST, CINEMAT sound take; ~ **sin sonido** *f* CINEMAT mute shot; ~ **sin sonido directo** *f* CINEMAT wild shot; ~ **submarina** *f* CINEMAT underwater shot; ~ **a tierra** *f* AUTO earth connection (*BrE*), ground connection (*AmE*), ING ELÉC connection to ground (*AmE*); ~ **de tierra** *f* ELEC earth (*BrE*), earth connection (*BrE*), earth connector (*BrE*), ground (*AmE*), ground connection (*AmE*), ground connector (*AmE*), earth plate (*BrE*), ground plate (*AmE*), FÍS earth (*BrE*), ground (*AmE*), ING ELÉC earth (*BrE*), earth connection (*BrE*), earth plate (*BrE*), ground connection (*AmE*), ground plate (*AmE*), PROD earth (*BrE*), earth connection (*BrE*), earth connector (*BrE*), ground (*AmE*), ground connection (*AmE*), ground connector (*AmE*), TEC ESP touchdown, TELECOM earth (*BrE*), ground (*AmE*), VEH *instalación eléctrica* earth (*BrE*), earth connection (*BrE*), ground (*AmE*), ground connection (*AmE*); ~ **del transformador** *f* ING ELÉC transformer tap; ~ **de transición** *f* CINEMAT bridging shot; ~ **a través de un cristal** *f* CINEMAT glass shot; ~ **de vapor** *f* INSTAL HIDRÁUL drawing off, input, input of steam, throttle; ~ **de vapor del inyector** *f* INSTAL HIDRÁUL injection cock, injector throttle; ~ **con viñeta** *f* CINEMAT *para efectos especiales* matte shot; ~ **de vistas imagen por imagen** *f* CINEMAT stop motion

toma[2]: **en la ~** *fra* CINEMAT in shot

tomacorriente *m* CINEMAT plug, ELEC *conexión* power outlet, socket, ING ELÉC tap, ING MECÁ outlet, *taladro mecánico* socket, MECÁ, PROD, TV socket; ~ **de clavija** *m* ELEC *conexión* plug receptacle; ~ **de clavija tipo enchufe** *m* ELEC *conexión* plug-type outlet; ~ **enclavado de múltiples conductores** *m* ELEC *conexión* multiconductor locking plug; ~ **enclavado policonductor** *m* ELEC *conexión* multiconductor locking plug; ~ **múltiple** *m* ELEC *conexión* multiple socket; ~ **mural** *m* ELEC *conexión* wall outlet; ~ **de pared** *m* ELEC *conexión* wall outlet; ~ **sencillo** *m* ING MECÁ plain socket; ~ **trabado policonductor** *m* ELEC *conexión* multiconductor locking plug

tomador *m* PAPEL *sieltro* pick-up, TRANSP MAR *para aferrar velas* gasket

tomamuestras *m* CARBÓN sampler, LAB water sampler, PROD *máquina*, TELECOM sampler; ~ **automático** *m* CARBÓN *sondeos* automatic sampler; ~ **basculante** *m* CARBÓN tip sampler

tomar[1] *vt* FERRO *curva* negotiate, TRANSP MAR bunker; ~ **muestras de** *vt* TELECOM sample; ~ **un primer plano de** *vt* CINEMAT, FOTO close up

tomar[2]: ~ **agua** *vi* AGUA water; ~ **una boya de amarre** *vi* TRANSP MAR *maniobra, yates* pick up a mooring; ~ **cubierta** *vi* TRANSP MAR land; ~ **una marcación** *vi* TRANSP MAR *navegación* take a bearing; ~ **un muerto** *vi* TRANSP MAR *maniobra, yates* pick up a mooring; ~ **rizos** *vi* TRANSP MAR *vela* reef; ~ **salida** *vi* TRANSP

MAR gather way; ~ **una vuelta al cabo con el chigre**
vi TRANSP MAR make a turn round winch with line
tomar[3]: ~ **y trazar marcaciones** *fra* TRANSP MAR plot
a bearing
tomas[1]: **sin** ~ *adj* ELEC *bobina* untapped
tomas[2]: ~ **en el secundario** *f pl* ING ELÉC secondary
tap
tome: ~ **a tierra** *f* ING ELÉC connection to earth (*BrE*),
connection to ground (*AmE*)
tomografía *f* ING MECÁ tomography
tomógrafo: ~ **axial computarizado por rayos x** *m* FÍS
RAD scanner
tonalidad *f* CINEMAT key, COLOR tonality, IMPR tone;
~ **máxima** *f* TEXTIL *en intensidad de color* overall
shade; ~ **oscura** *f* TEXTIL dark shade
tonalita *f* PETROL tonalite
tonel *m* PROD tub, TEC ESP *aviones* roll, TRANSP barrel,
TRANSP AÉR roll; ~ **desarenador** *m* PROD *fundición*
rattling barrel, rumbler; ~ **para desarenar** *m* PROD
fundición cleaning barrel
tonelada *f* METR ton, *métrica* tonne; ~ **de arqueo** *f*
TRANSP MAR measured ton, register ton; ~ **de arqueo**
bruto *f* TRANSP MAR gross ton; ~ **corta** *f* METR short
ton; ~ **de desplazamiento** *f* TRANSP MAR ton of
displacement; ~ **de flete** *f* TRANSP MAR shipping ton;
~ **larga** *f* METR gross ton, long ton; ~ **métrica** *f*
METR, TEC PETR *unidades de medida* metric ton,
metric tonne; ~ **neta** *f* METR net ton
toneladas: ~ **de peso muerto** *f pl* TEC PETR *transporte*
marítimo tonnes dead weight, dead weight tons
tonelaje *m* METR *peso en toneladas* tonnage; ~ **de**
carga *m* TEC PETR *transporte marítimo* dead weight
tonnage; ~ **inactivo** *m* TRANSP MAR idle shipping;
~ **inmovilizado** *m* TRANSP MAR idle shipping; ~ **neto**
m TEC PETR *transporte* net tonnage; ~ **de peso**
muerto *m* TEC PETR *transporte marítimo* dead weight
tonnage; ~ **de registro bruto** *m* TEC PETR gross
tonnage
tóner *m* COLOR dry ink, IMPR *tinta en polvo* toner;
~ **para offset** *m* IMPR dry offset ink
tongada *f* CONST, TRANSP MAR layer
tónica *f* ACÚST leading note, *música* keynote
tonificar *vt* IMPR tone up
tono[1]: **de** ~ **verdoso** *adj* GEOL *silex alterado* green-
stained
tono[2] *m* ACÚST *música* tone, C&V shade, COLOR tone, FÍS
pitch, FÍS ONDAS timbre, pitch, FOTO tone, hue, IMPR
shade, hue, ÓPT pitch, TELECOM tone; ~ **de alerta**
para terceros *m* TELECOM third-party warning tone;
~ **audible** *m* CINEMAT beep tone, bleep, blip tone;
~ **claro** *m* COLOR light tone, TEXTIL light shade; ~ **de**
color *m* COLOR, P&C *pintura* color tone (*AmE*), colour
tone (*BrE*); ~ **de combinación** *m* ACÚST combination tone; ~ **combinado** *m* ACÚST *música*
combination tone; ~ **complejo** *m* ACÚST complex
tone; ~ **completo** *m* ACÚST whole tone; ~ **completo**
mayor *m* ACÚST major whole tone; ~ **completo**
menor *m* ACÚST minor whole tone; ~ **continuo** *m*
IMPR continuous tone; ~ **cromático** *m* COLOR color
hue (*AmE*), colour hue (*BrE*); ~ **de frecuencia**
audible *m* FÍS ONDAS beat-note pitch; ~ **de frecuen-**
cia variable *m* ACÚST warble; ~ **fundamental** *m*
ACÚST, FÍS ONDAS fundamental tone; ~ **intenso de**
color *m* COLOR, P&C *pintura* deep color tone (*AmE*),
deep colour tone (*BrE*); ~ **lateral** *m* TELECOM *tele-*
fonía sidetone; ~ **modulado** *m* ACÚST warble tone;

~ **natural** *m* TEXTIL natural shade; ~ **neutro** *m* COLOR
neutral tint, TEXTIL neutral shade; ~ **de nota de**
batido *m* FÍS ONDAS beat-note pitch; ~ **oscuro** *m*
COLOR dark tone; ~ **periódico** *m* ACÚST periodic
tone; ~ **piloto** *m* TV pilot tone; ~ **puro** *m* ACÚST pure
tone; ~ **relativo** *m* ACÚST relative tone; ~ **subido** *m*
COLOR, TEXTIL deep shade; ~ **subjetivo** *m* ACÚST
subjective tone; ~ **supervisor** *m* TELECOM supervisory tone; ~ **a tres cuartos** *m* IMPR three-quarter
tone
tonos: ~ **conjuntos** *m pl* ACÚST conjoined pitches;
~ **medios** *m pl* IMPR middletones
tonsura *f* ING MECÁ shearing
tonsurar *vt* ING MECÁ shear
TOP[1] *abr* (*tubo de ondas progresivas*) ELECTRÓN, FÍS,
TEC ESP, TELECOM TWT (*traveling-wave tube AmE,*
travelling-wave tube BrE)
TOP[2]: ~ **de banda X** *m* ELECTRÓN X-band TWT
topacio *m* MINERAL topaz; ~ **rosa** *m* MINERAL pink
topaz, Brazilian ruby
topazolita *f* MINERAL andradite, melanite, topazolite
tope *m* AmL (*cf sujetador Esp*) CONST abutment, butt,
butt end, stop, stub, FERRO buffer, IMPR lay, ING MECÁ
scotch, *ferrocarril* limiter, buffer, MECÁ stop, MINAS
prop, *jaula de minas* kep, *tacos* catch, TRANSP MAR *del*
palo masthead; ~ **de aguja** *m* FERRO *infraestructura*
bearing; ~ **del banco** *m* CONST bench stop; ~ **de**
choque *m* FERRO *vehículos* buffer-stop block; ~ **del**
cierre del motor *m* TRANSP AÉR engine shut-off stop;
~ **compensador** *m* FERRO *infraestructura* compensating buffer; ~ **del control del regulador** *m* TRANSP
AÉR governor-control stop; ~ **esférico** *m* ING MECÁ
ball stop; ~ **del expulsor** *m* ING MECÁ ejector stop-
piece; ~ **de extremo mecánico** *m* ELEC *cambiador de*
toma mechanical end stop; ~ **del eyector** *m* ING
MECÁ ejector stop-piece; ~ **fijo** *m* ING MECÁ fixed
stop; ~ **final** *m* ING MECÁ end stop; ~ **frontal** *m* IMPR
front stop; ~ **graduable** *m* ING MECÁ, MECÁ adjustable stop; ~ **de inclinación del cubo** *m* TRANSP AÉR
hub tilt stop; ~ **indicador** *m* ING MECÁ indicating
stop; ~ **de líquido** *m* CARBÓN liquid limit; ~ **magné-**
tico para la puerta *m* CONST magnetic doorstop;
~ **movible** *m* ING MECÁ moveable stop; ~ **móvil** *m*
ING MECÁ moveable stop; ~ **de parada** *m* FERRO
equipo inamovible buffer stop; ~ **del paso** *m* TRANSP
AÉR pitch stop; ~ **del regulador de relentí** *m* TRANSP
AÉR idle throttle stop; ~ **de resorte** *m* FERRO
infraestructura spring buffer (*BrE*), spring bumper
(*AmE*), ING MECÁ spring stop; ~ **de retención** *m*
IMPR backstop; ~ **seco** *m* FERRO *infraestructura*
spring buffer (*BrE*), spring bumper (*AmE*);
~ **trasero** *m* TRANSP tailback
topes *m pl* MINAS overwind gear, overwinding gear;
~ **de fin de carrera** *m pl* MINAS prop, *jaula* dog
topetazo *m* AUTO, ING MECÁ, VEH knocking
topo *m* MINAS roadheader
topografía *f* AGRIC land surveying, CONST land
surveying, surveying, topography; ~ **de memoria** *f*
AmL INFORM&PD memory map; ~ **de rayos X** *f*
CRISTAL, METAL X-ray topography; ~ **submarina** *f*
OCEAN bottom topography, submarine topography;
~ **subterránea** *f* GEOL buried topography
topógrafo *m* CONST surveyor, topographer
topología *f* GEOM, INFORM&PD topology; ~ **de anillo** *f*
INFORM&PD *red* ring topology; ~ **de árbol** *f*
INFORM&PD tree topology; ~ **de bus** *f* INFORM&PD

bus topology; ~ **de conductores comunes** f AmL
INFORM&PD bus topology; ~ **diferencial** f GEOM
differential topology; ~ **en estrella** f INFORM&PD
snowflake topology, star topology; ~ **en forma de
copo de nieve** f INFORM&PD snowflake topology;
~ **de interconexión** f INFORM&PD interconnection
topology; ~ **de red** f INFORM&PD network topology

topólogo m GEOM, INFORM&PD topologist

toque: ~ **de queda** m ACÚST curfew; ~ **del tren de
aterrizaje con la pista** m TRANSP AÉR touchdown

torbellino m CARBÓN vortex, FÍS FLUID eddy, vortex,
TEC ESP vortex, TRANSP twister, TRANSP MAR *construcción naval* slip; ~ **descendiente** m TRANSP AÉR
downwash; ~ **hacia abajo** m TRANSP AÉR downwash;
~ **de la hélice** m TRANSP AÉR propeller wash, slipstream; ~ **de la punta del ala** m TRANSP AÉR wing-tip
vortex; ~ **de la punta de la pala** m TRANSP AÉR bladetip vortex

torbellinos: ~ **readheridos** m pl FÍS FLUID attached
eddies

torbernita f MINERAL torbernite, NUCL copper uranite,
torbernite

torcedura f C&V warpage, *del vidrio óptico* warp, ING
MECÁ torsion, set, MECÁ torsion, PROD kink; ~ **de
hojas** f IMPR cockle

torcer: ~ **hasta soltar** vt PROD twist to release, twistrelease

torcerse v *refl* MECÁ buckle

torcido[1] *adj* ING MECÁ out-of-true, MECÁ warped

torcido[2] m C&V *de tubos* twisting

torianita f MINERAL thorianite

tórico *adj* QUÍMICA thoric

torio m (Th) FÍS RAD, QUÍMICA thorium (Th)

torita f MINERAL thorite

tormenta[1] f METEO storm, thunderstorm; ~ **eléctrica** f
METEO electrical storm; ~ **magnética** f GEOFÍS, TEC
ESP magnetic storm; ~ **solar** f GEOFÍS solar storm

tormenta[2]: ~ **más fuerte de los últimos 100 años** *fra*
TEC PETR *operaciones costa-fuera* hundred year storm

tornado m METEO tornado; ~ **efímero** m METEO shortlived tornado

tornasol m QUÍMICA litmus

tornasolado[1] *adj* C&V iridescent

tornasolado[2] m C&V iridizing

torneado m ING MECÁ turning; ~ **al aire** m ING MECÁ
facing; ~ **de cilindros** m ING MECÁ straight turning;
~ **máximo** m ING MECÁ swing; ~ **en el plato liso** m
ING MECÁ turning on the face plate; ~ **entre puntos**
m ING MECÁ turning between centers (*AmE*), turning
between centres (*BrE*)

torneadora f ING MECÁ boring-and-turning mill

torneaduras f pl PROD turnings

tornear vt ING MECÁ turn; ~ **al aire** vt PROD *sin puntos*
face

torneo m CARBÓN contest

tornero m C&V *alfarero* thrower, MECÁ lathe operator

tornillería f PROD screw works

tornillo m CONST box screw, screw, ING MECÁ bolt, male
screw, screw, screw bolt, MECÁ screw, P&C *parte de
equipo* mandrel, mandril, screw, TRANSP MAR
construcción, VEH screw;

-a ~ **de ajustar** m TRANSP MAR *construcción naval*
rigging screw; ~ **de ajuste** m FÍS *guía de ondas* tuning
screw, IMPR adjusting screw, ING MECÁ temper screw,
setscrew, adjusting screw, MECÁ adjusting screw; ~ **de
ajuste elevador** m INSTR elevation adjusting screw;

~ **de ajuste manual** m PROD thumbscrew; ~ **de
ajuste de marcha lenta** m AUTO idle adjustment
screw; ~ **de ajuste del ralentí** m AUTO idle adjustment screw; ~ **de ajuste de la resistencia
aerodinámica** m INSTR windage adjustment screw;
~ **alado** m ING MECÁ wing screw, winged screw;
~ **alimentador** m ING MECÁ temper screw; ~ **de
apriete** m ING MECÁ setscrew, PROD clamping screw;
~ **de apriete manual** m ING MECÁ thumbscrew; ~ **de
Arquímedes** m INSTAL HIDRÁUL Archimedean screw,
spiral conveyor; ~ **de arrastre** m ING MECÁ drag
screw; ~ **autoroscante plano** m ING MECÁ flat and
self-tapping screw; ~ **autosuficiente plano** m ING
MECÁ flat and self-tapping screw; ~ **de avance** m ING
MECÁ lead screw, feed screw, MECÁ, PROD lead screw;
~ **de avance esférico** m ING MECÁ ball circulating
lead screw; ~ **de avance esférico circulante** m ING
MECÁ circulating ball lead screw;

-b ~ **de banco** m CONST bench screw, ING MECÁ vice
(*BrE*), vise (*AmE*), MECÁ clamp, vice (*BrE*), vise
(*AmE*), PROD vice bench (*BrE*), vise bench (*AmE*);
~ **de banco con cola** m ING MECÁ tail vice (*BrE*), tail
vise (*AmE*); ~ **de banco giratorio** m ING MECÁ swivel
vice (*BrE*), swivel vise (*AmE*); ~ **de banco de
movimiento paralelo** m ING MECÁ parallel vice
(*BrE*), parallel vise (*AmE*); ~ **de banco paralelo** m
ING MECÁ parallel vice (*BrE*), parallel vise (*AmE*);
~ **con bridas** m ING MECÁ vice with clamp (*BrE*), vise
with clamp (*AmE*); ~ **de bronce** m ING MECÁ brass
screw;

-c ~ **sin cabeza** m ING MECÁ grub screw, MECÁ
headless screw; ~ **de cabeza avellanada** m CONST,
ING MECÁ, MECÁ countersunk-head screw, PROD
crosshead-terminal screw; ~ **de cabeza biselada** m
CONST bevel-headed bolt; ~ **de cabeza cilíndrica
ranurada** m PROD fillister-head screw; ~ **de cabeza
cuadrada** m ING MECÁ square-head screw; ~ **de
cabeza elevada** m ING MECÁ raised-head screw;
~ **de cabeza embutida** m CONST, ING MECÁ, MECÁ
countersunk-head screw; ~ **con cabeza gota de
cebo** m ING MECÁ round-headed screw; ~ **de cabeza
hendida** m ING MECÁ grub screw; ~ **de cabeza
hexagonal** m MECÁ hexagon-head screw, PROD
hexagon head screw; ~ **de cabeza hexagonal hueca
con punto plano** m ING MECÁ hexagon socket screw
with flat point; ~ **de cabeza hueca hexagonal** m
ING MECÁ hexagon-socket-head cap screw, hexagon
socket screw; ~ **de cabeza de martillo** m ING MECÁ
hammer head screw; ~ **de cabeza moleteada** m ING
MECÁ knurled-head fastener, knurled screw, thumbscrew; ~ **de cabeza oculta** m PROD shrouded-cover
screw; ~ **de cabeza ovalada** m ING MECÁ oval-head
screw; ~ **de cabeza perdida** m CONST, ING MECÁ,
MECÁ countersunk-head screw; ~ **de cabeza Phillips**
m ING MECÁ Phillips screw; ~ **con cabeza plana** m
CONST cheese-head screw; ~ **con cabeza redonda** m
CONST capstan-headed screw; ~ **de cabeza redonda**
m ING MECÁ cup-head bolt, round-headed screw;
~ **de cabeza semiesférica** m ING MECÁ cup-head
bolt, round-headed screw; ~ **de cabeza semiredonda** m ING MECÁ saucer-head screw; ~ **de
cabeza troncocónica** m CONST, ING MECÁ, PROD
pan-head screw; ~ **de cabeza troncocónica de
cruz hendida** m ING MECÁ cross-recessed pan head
screw; ~ **de cadena** m ING MECÁ chain vice (*BrE*),
chain vise (*AmE*); ~ **de carpintero** m CONST bench

screw; ~ **de carpintero de mano** *m* CONST pin vice (*BrE*), pin vise (*AmE*); ~ **de carpintero con mordaza** *m* CONST bench-vice with clamp (*BrE*), bench-vise with clamp (*AmE*); ~ **cautivo de cabeza de cruceta** *m* PROD captive cross-head terminal screw; ~ **de centrar** *m* ING MECÁ centering screw (*AmE*), centring screw (*BrE*); ~ **del cierre** *m* D&A breech screw; ~ **concéntrico** *m* ING MECÁ self-centering vise (*AmE*), self-centring vice (*BrE*); ~ **de conexión** *m* ING MECÁ connecting screw; ~ **de corrección** *m* ING MECÁ adjusting screw;

~ d ~ **dentado** *m* CONST barbed bolt; ~ **deslizante de barras paralelas** *m* ING MECÁ vice sliding between parallel bars (*BrE*), vise sliding between parallel bars (*AmE*); ~ **deslizante con guías paralelas** *m* ING MECÁ vice sliding between parallel bars (*BrE*), vise sliding between parallel bars (*AmE*); ~ **dextrógiro** *m* ING MECÁ right-hand screw; ~ **diferencial** *m* ING MECÁ compound screw;

~ e ~ **elevador** *m* ING MECÁ elevating screw; ~ **embutido** *m* CONST, ING MECÁ countersunk screw; ~ **embutido para madera** *m* CONST countersunk woodscrew; ~ **esférico** *m* ING MECÁ ball screw; ~ **de extracción** *m* ING MECÁ draw screw; ~ **extractor** *m* ING MECÁ jackscrew;

~ f ~ **fiador** *m* ING MECÁ grub screw; ~ **de fijación** *m* CONST fixing screw, lag screw, ING MECÁ lock screw, fixing screw, setscrew, PROD binding-head screw, clamp screw, set screw; ~ **de fijación del distribuidor** *m* AUTO distributor-clamp bolt; ~ **fijador** *m* PROD set screw; ~ **de filetes convergentes** *m* PROD hourglass screw; ~ **sin fin** *m* ING MECÁ endless screw, perpetual screw, tangent screw, worm, worm screw, MECÁ worm; ~ **de fricción** *m* ING MECÁ friction screw;

~ g ~ **giratorio** *m* ING MECÁ swivel vice (*BrE*), swivel vise (*AmE*); ~ **glóbico** *m* PROD hourglass screw; ~ **globoide** *m* PROD hourglass screw; ~ **de graduar** *m* ING MECÁ temper screw; ~ **de guía** *m* ING MECÁ guide clamp; ~ **guía** *m* ING MECÁ guide screw; ~ **de guía esférico** *m* ING MECÁ ball circulating lead screw;

~ h ~ **hecho a máquina** *m* ING MECÁ machine bolt; ~ **hembra** *m* PROD internal screw; ~ **hidráulico** *m* INSTAL HIDRÁUL Archimedean screw, Archimedes screw; ~ **de hilo cuadrado** *m* ING MECÁ square-thread screw, square-thread tap; ~ **de hilo único** *m* ING MECÁ single-threaded screw; ~ **con hilo en V** *m* ING MECÁ V-threaded screw; ~ **de hueco hexagonal** *m* MECÁ hexagon-socket screw;

~ i ~ **imperdible** *m* ING MECÁ captive screw; ~ **interior** *m* ING MECÁ inside screw; ~ **interno** *m* C&V internal friction; ~ **de inversión** *m* ING MECÁ reversing screw;

~ l ~ **de latón** *m* ING MECÁ brass screw; ~ **limitador** *m* FOTO stop screw;

~ m ~ **macho** *m* ING MECÁ outside screw, PROD external screw; ~ **para madera** *m* CONST wood screw; ~ **de madera con cabeza redonda de bronce** *m* CONST brass round-head woodscrew; ~ **con mandíbulas insertadas** *m* ING MECÁ vice with inserted jaws (*BrE*), vise with inserted jaws (*AmE*); ~ **con mandíbulas intercambiables** *m* ING MECÁ vice with detachable jaws (*BrE*), vise with detachable jaws (*AmE*); ~ **de mano** *m* CONST pin vice (*BrE*), pin vise (*AmE*), ING MECÁ hand vice (*BrE*), hand vise (*AmE*), tail vice (*BrE*), tail vise (*AmE*), vice clamp (*BrE*), vise

clamp (*AmE*); ~ **de máquina de cabeza cilíndrica ranurada** *m* ING MECÁ, PROD fillister-head machine screw; ~ **de máquina de cabeza redonda** *m* ING MECÁ, PROD fillister-head machine screw; ~ **de mariposa** *m* CONST butterfly screw, ING MECÁ butterfly screw, thumbscrew, wing screw, PROD *sin destornillador* thumbscrew; ~ **de martillo** *m* ING MECÁ hammer drive screw; ~ **mecánico** *m* ING MECÁ machine screw; ~ **de mesa con brida** *m* ING MECÁ table vice with clamp (*BrE*), table vise with clamp (*AmE*); ~ **de mesa con pie** *m* PROD leg vice (*BrE*), leg vise (*AmE*); ~ **para metales** *m* ING MECÁ machine screw; ~ **micrométrico** *m* ING MECÁ, INSTR micrometer screw; ~ **de micrómetro de paso fino** *m* ING MECÁ finely-threaded micrometer screw; ~ **de micrómetro de paso de precisión** *m* ING MECÁ finely-threaded micrometer screw; ~ **de montaje** *m* CONST fixing screw, ING MECÁ mounting bolt, fixing screw;

~ n ~ **neumático de frenado** *m* TRANSP braking airscrew; ~ **de nivelación** *m* CONST *topografía, carpintería, albañilería* bubble nut; ~ **de nivelado** *m* ING MECÁ, INSTR leveling screw (*AmE*), levelling screw (*BrE*); ~ **nivelador** *m* ING MECÁ leveling screw (*AmE*), levelling screw (*BrE*), footscrew, jackscrew; ~ **nivelador de cabeza autoajustable** *m* ING MECÁ jackscrew with self-adjusting head; ~ **nivelante** *m* ING MECÁ, INSTR leveling screw (*AmE*), levelling screw (*BrE*);

~ o ~ **de ojo** *m* ING MECÁ, MECÁ eyebolt; ~ **opresor** *m* ING MECÁ stud bolt; ~ **de orejas** *m* ING MECÁ wing bolt; ~ **de orejetas** *m* CONST butterfly screw, ING MECÁ butterfly screw, thumbscrew, wing screw, PROD thumbscrew;

~ p ~ **de palomilla** *m* ING MECÁ thumbscrew, wing screw; ~ **paralelo de combinación** *m* ING MECÁ combined parallel vice (*BrE*), combined parallel vise (*AmE*); ~ **paralelo combinado** *m* ING MECÁ combined parallel vice (*BrE*), combined parallel vise (*AmE*); ~ **paralelo desplazable** *m* ING MECÁ sliding parallel vice (*BrE*), sliding parallel vise (*AmE*); ~ **pasante** *m* CONST, ING MECÁ through bolt; ~ **de paso ancho** *m* CONST coarse-pitch screw; ~ **de paso angular** *m* MECÁ angular-thread screw; ~ **de paso cuadrado** *m* ING MECÁ square-thread screw; ~ **de paso grande** *m* ING MECÁ long screw, long-pitch screw; ~ **de paso pequeño** *m* CONST fine-pitch screw; ~ **de paso rápido** *m* ING MECÁ long screw, long-pitch screw; ~ **de paso unificado** *m* ING MECÁ unified screw thread; ~ **de pasos contrarios** *m* ING MECÁ compound screw, right-and-left screw; ~ **patrón** *m* ING MECÁ, MECÁ, PROD lead screw; ~ **con pernos protegidos** *m* ING MECÁ vice with protected screw (*BrE*), vise with protected screw (*AmE*); ~ **Phillips** *m* ING MECÁ Phillips screw; ~ **con pie** *m* PROD standing vice (*BrE*), standing vise (*AmE*), staple vice (*BrE*), staple vise (*AmE*); ~ **de plancha** *m* ING MECÁ sheet-metal screw; ~ **plurirroscas** *m* PROD multiple-threaded screw; ~ **Posidriv** *m* ING MECÁ Posidriv screw; ~ **de precisión horizontal** *m* INSTR *topografía* horizontal-tangent screw; ~ **de precisión vertical** *m* INSTR vertical-tangent screw; ~ **de presión** *m* ING MECÁ clamp, grub screw, pressure screw, setscrew; ~ **prisionero** *m* ING MECÁ setscrew; ~ **puntiagudo de cabeza cuadrada** *m* CONST pointed square-head coach-screw; ~ **de purga** *m* ING MECÁ bleed screw;

~ r ~ **de regulación** *m* ING MECÁ regulating screw, MECÁ adjusting screw; ~ **de regulación de la mariposa** *m* VEH *del carburador* throttle stop screw; ~ **de regulación del mínimo** *m* VEH *carburador* idle adjustment screw; ~ **de regulación del ralentí** *m* VEH *carburador* idle adjustment screw; ~ **regulador** *m* ING MECÁ lead screw, regulating screw, temper screw, MECÁ lead screw, MINAS temper screw, PROD lead screw; ~ **regulador esférico** *m* ING MECÁ ball circulating lead screw; ~ **regulador esférico circulante** *m* ING MECÁ circulating ball lead screw; ~ **de rosca cuadrada** *m* ING MECÁ square-thread screw, square-threaded screw; ~ **de rosca a derechas** *m* ING MECÁ right-handed screw; ~ **de rosca dextrógira** *m* ING MECÁ right-handed screw; ~ **con rosca en V** *m* ING MECÁ V-threaded screw; ~ **de roscas a derechas y a izquierdas** *m* ING MECÁ right-and-left screw; ~ **rotatorio de banco de sujeción instantánea** *m* ING MECÁ sudden-grip rotary bench vice (*BrE*), sudden-grip rotary bench vise (*AmE*);

~ s ~ **de sangrado** *m* AUTO bleeder screw; ~ **de seguridad** *m* ING MECÁ lock screw, protected screw; ~ **sinfín curvado** *m* ING MECÁ curved worm gear; ~ **sinfín tractor** *m* ING MECÁ draw screw; ~ **de sintonización** *m* ING ELÉC tuning screw; ~ **de sujeción** *m* PROD set screw; ~ **de sujeción del cojinete de roscar** *m* ING MECÁ tap with metric thread;

~ t ~ **de taladradora de plantilla** *m* ING MECÁ jig-boring vice (*BrE*), jig-boring vise (*AmE*); ~ **tensor** *m* ING MECÁ stretching screw, tension screw, tightening screw; ~ **tensor para terminales** *m* PROD terminal-tensioning screw; ~ **de tetón** *m* ING MECÁ tit screw; ~ **del tintero** *m* IMPR fountain screw; ~ **de traba** *m* ING MECÁ lock screw; ~ **de tres entradas** *m* ING MECÁ three-threaded screw; ~ **de trípode** *m* CINEMAT tripod screw; ~ **de tubos** *m* ING MECÁ tube vice (*BrE*), tube vise (*AmE*); ~ **para tubos** *m* CONST pipe vice (*BrE*), pipe vise (*AmE*);

~ u ~ **de unión** *m* ING MECÁ connecting screw

tornillos: ~ **de mano** *m pl* ING MECÁ vice grips (*BrE*), vise grips (*AmE*); ~ **reguladores del tintero** *m pl* IMPR fountain keys

torno *m* C&V *para mangos* machine, CONST winch, ING MECÁ turning lathe, MECÁ, PROD *de tornear* lathe; ~ **al aire** *m* ING MECÁ face lathe, pole lathe; ~ **de aire comprimido** *m* ING MECÁ air hoist; ~ **para el acabado de bronce** *m* CONST brass-finisher's lathe; ~ **acoplado directamente** *m* ING MECÁ plain ungeared lathe; ~ **de alfarería** *m* C&V potter's wheel; ~ **de alfarero** *m* C&V potter's wheel; ~ **de arco** *m* ING MECÁ bow lathe; ~ **automático** *m* MECÁ automatic lathe; ~ **auxiliar** *m* PETROL cathead; ~ **de bancada** *m* ING MECÁ bed lathe; ~ **de bancada prismática** *m* ING MECÁ gantry lathe; ~ **de bancada recta** *m* ING MECÁ plain-bed lathe; ~ **de banco** *m* ING MECÁ bench lathe; ~ **para banco** *m* MECÁ bench lathe; ~ **de barrenar y cilindrar de tipo carrusel** *m* ING MECÁ floor-type surfacing-and-boring lathe; ~ **de barrenar y cilindrar de tipo vertical** *m* ING MECÁ floor-type surfacing-and-boring lathe; ~ **de barrenas horizontal** *m* ING MECÁ turning mill; ~ **con cabezal revólver** *m* PROD monitor lathe; ~ **de cable** *m*

TRANSP cable winch; ~ **de centrar** *m* ING MECÁ centering lathe (*AmE*), centring lathe (*BrE*); ~ **cerrado** *m* PROD close lathing; ~ **de cilindrar** *m* ING MECÁ plain lathe; ~ **de cilindrar y taladrar** *m* ING MECÁ surfacing-and-boring lathe; ~ **cilíndrico** *m* ING MECÁ shafting lathe; ~ **de cilindros** *m* ING MECÁ plain turning-lathe; ~ **copiador** *m* ING MECÁ tracing lathe, copying lathe; ~ **de copiar automático** *m* ING MECÁ automatic copying lathe; ~ **de cortador** *m* C&V cutter's lathe; ~ **de despojar** *m* ING MECÁ backing-off lathe, relieving lathe; ~ **de destalonar** *m* ING MECÁ backing-off lathe, relieving lathe; ~ **de dos ejes** *m* ING MECÁ twin-spindle lathe; ~ **de dos husillos de rosca** *m* ING MECÁ twin-screw lathe; ~ **para ejes** *m* ING MECÁ axle lathe; ~ **eléctrico** *m* ELEC, MECÁ electric hoist; ~ **elevador de pared** *m* PROD gypsy winch; ~ **para elevar pesos** *m* MECÁ crab; ~ **de embutir** *m* ING MECÁ chasing lathe; ~ **de extracción** *m* MINAS mine hoist, PROD hoisting crab, tackle; ~ **extractor** *m* ING MECÁ tackle; ~ **de filetear** *m* ING MECÁ screw-cutting lathe; ~ **fresador copiador** *m* ING MECÁ copy-milling lathe; ~ **para grabar** *m* C&V engraving lathe; ~ **de herramentista** *m* ING MECÁ toolmaker's lathe; ~ **de herramientas múltiples** *m* ING MECÁ multiple tool lathe, multitool lathe; ~ **hidráulico para izar pesos** *m* MINAS hydraulic winch; ~ **de hilar** *m* TEXTIL spinning wheel; ~ **de husillos gemelos de roscado** *m* ING MECÁ twin-screw lathe; ~ **de izar** *m* PROD crab winch; ~ **de izar pesos** *m* MECÁ hoist; ~ **para machos** *m* PROD *fundición* core lathe; ~ **para madera** *m* CONST wood-turner's lathe, wood-turning lathe; ~ **de maniobras** *m* CONST shunting winch; ~ **de mano** *m* PROD gypsy winch; ~ **neumático** *m* Esp (*cf winchy AmL*) ING MECÁ, TEC PETR air hoist; ~ **paralelo automático** *m* ING MECÁ self-acting sliding lathe; ~ **de plato** *m* ING MECÁ face lathe; ~ **de plato automático** *m* ING MECÁ automatic chucking lathe; ~ **de plato combinado** *m* ING MECÁ combination lathe; ~ **de plato horizontal** *m* ING MECÁ boring-and-turning mill, MECÁ boring mill; ~ **de polea** *m* ING MECÁ pulley-turning lathe; ~ **para poleas** *m* ING MECÁ pulley lathe; ~ **para producción pesada** *m* ING MECÁ heavy-duty lathe; ~ **de puente** *m* ING MECÁ gap lathe; ~ **pulidor** *m* ING MECÁ polishing lathe; ~ **para pulidoras** *m* ING MECÁ polishing head; ~ **de puntos** *m* ING MECÁ center lathe (*AmE*), centering lathe (*AmE*), centring lathe (*BrE*), centre lathe (*BrE*), pole lathe; ~ **de puntos fijos** *m* ING MECÁ dead center (*AmE*), dead centre (*BrE*); ~ **rápido** *m* ING MECÁ speed lathe; ~ **para redondos** *m* ING MECÁ bar turning-tool; ~ **de repetición** *m* ING MECÁ forming lathe; ~ **revólver** *m* ING MECÁ, MECÁ capstan lathe, turret lathe; ~ **de roscar** *m* ING MECÁ screw machine, screw-cutting lathe, MECÁ screw machine; ~ **para roscar** *m* ING MECÁ chasing lathe; ~ **de roscar automático** *m* MECÁ automatic screw machine; ~ **de rueda doble** *m* ING MECÁ double-wheel lathe; ~ **simple** *m* ING MECÁ plain lathe; ~ **de sondeo** *m* MINAS *sondeos* drilling winch, hoist; ~ **de taladrar y cilindrar de tipo carrusel** *m* ING MECÁ floor-type surfacing-and-boring lathe; ~ **de taladrar y cilindrar de tipo vertical** *m* ING MECÁ floor-type surfacing-and-boring lathe; ~ **de taladrar y desbastar tipo bancada** *m* ING MECÁ bed-type surfacing-and-boring lathe; ~ **de torre hexagonal** *m* ING MECÁ hexagon-turret lathe; ~ **troceador** *m*

ING MECÁ cutting-off lathe; ~ **troceador y de perfiles** m ING MECÁ cutting-off and forming lathe; ~ **troceador y de repetición** m ING MECÁ cutting-off and forming lathe; ~ **de velocidad única** m ING MECÁ single-geared lathe

toro m FÍS RAD *física del plasma* torus, GEOM doughnut, torus, TEC ESP annulus; ~ **de dos agujeros** m GEOM two-hole torus; ~ **reproductor** m AGRIC bull for service; ~ **de tres agujeros** m GEOM three-hole torus

toroide m ELEC *bobina* toroid, torus, NUCL toroid

torón m ÓPT *de cable*, PROD *cuerdas, cables*, TEC ESP strand

torpedero m TRANSP MAR torpedo boat; ~ **cañonero** m D&A, TRANSP MAR torpedo gunboat

torpedo m D&A, MINAS, P&C, TRANSP MAR torpedo

torque m FÍS torque, ING MECÁ moment, torque

torr m FÍS torr

torre f CONST tower, ELEC *red de distribución* pylon, tower, ING ELÉC pylon, PETROL *perforación de sondeos* derrick; ~ **de absorción** f ALIMENT, CARBÓN, PROC QUÍ absorption tower, TEC PETR absorption tower, *refino* absorber, absorption column; ~ **de amarre** f ELEC *red de distribución* anchoring tower; ~ **basculante** f ENERG RENOV *energía del viento* tiltable tower; ~ **de blanqueo** f PAPEL bleaching tower; ~ **de carga** f C&V batch tower; ~ **de celosía** f CONST, ELEC *línea de alimentación aérea* lattice tower; ~ **de control** f TRANSP AÉR control tower; ~ **depuradora** f MINAS scrubber; ~ **de destilación** f PROC QUÍ distillation tower, distilling tower, TEC PETR *refino*, TERMO distillation tower; ~ **de difusión** f PROC QUÍ diffusion tower; ~ **de elevación de agua** f CONST *contraincendios* water tower; ~ **de enfriado** f *Esp* (*cf enfriador de chimenea AmL*) MINAS cooling tower; ~ **de enfriamiento** f CONST cooling tower, IMPR chilling tower, ING MECÁ cooling tower, MINAS water tower, NUCL, PROC QUÍ, PROD, REFRIG, TERMO, TERMOTEC cooling tower; ~ **de enfriamiento seco** f REFRIG dry-cooling tower; ~ **para escape de emergencia** f TEC ESP *instalaciones en tierra* emergency-escape tower; ~ **estabilizadora** f TEC PETR *refino* stabilizer tower; ~ **de la estación de clasificación** f TRANSP classification yard tower; ~ **de estirado** f C&V drawing tower; ~ **de extracción** f MINAS winding tower, TEC PETR absorption tower; ~ **extrema** f ELEC *red de distribución* dead end tower; ~ **de fraccionamiento** f PROC QUÍ fractionating tower; ~ **de lanzamiento** f TEC ESP launching tower; ~ **de lavado** f PROC QUÍ washing tower; ~ **metálica de electricidad** f CONST, ELEC electricity pylon; ~ **de microondas** f TELECOM microwave tower; ~ **de observación** f D&A observation tower; ~ **de perforación** f TEC PETR *perforación* derrick; ~ **de perforación sobre barcaza** f TEC PETR *perforaciones costa-fuera* drill barge; ~ **de perforación flotante** f TEC PETR floating rig; ~ **de perforación fuera de la costa** f TEC PETR accommodation rig; ~ **del RDP** f OCEAN BOP stack; ~ **de refrigeración** f CONST, ING MECÁ, MINAS, NUCL, PROC QUÍ, PROD, REFRIG, TERMO, TERMOTEC cooling tower; ~ **revólver** f PROD *tornos* monitor; ~ **de secado** f ALIMENT drying tower; ~ **de separación** f LAB, PROC QUÍ separating tower; ~ **solar** f ENERG RENOV solar tower, *energía solar* power tower; ~ **de sondeo** f TEC PETR *perforación* derrick; ~ **de sondeo de exploración** f TEC PETR *perforación* exploration rig; ~ **de soporte** f

ELEC *red de distribución* supporting pylon; ~ **de telescopio solar** f INSTR solar telescope tower; ~ **de terminales** f ELEC *red de distribución* terminal tower; ~ **de toma de agua** f HIDROL *de un embalse* intake tower; ~ **de transmisión** f ELEC *red de distribución* transmission tower; ~ **de transposición** f ELEC *red de distribución* transposition tower

torrefacción f PROD torrefaction

torrefactor m PROD torrefier

torrencial adj METEO torrential

torrente m CONST *carpintería* rush, HIDROL flood, torrent, *crecida* spate

torrero m TEC PETR derrick man

torreta f CINEMAT *de cámara*, D&A, turret, ING MECÁ turret head, turret, MECÁ turret, TRANSP cable support; ~ **autopropulsada** f D&A self-propelled turret; ~ **de control** f TRANSP MAR *submarino* conning tower; ~ **de cuatro herramientas** f ING MECÁ four-tool turret; ~ **de observación** f OCEAN *investigación submarina* survey diving bell; ~ **portaherramientas revolver de cuatro caras** f ING MECÁ four-tool turret; ~ **portaobjetivos** f ÓPT lens turret; ~ **rotativa** f CINEMAT lens turret; ~ **triple** f CINEMAT three-lens turret

torrotito m TRANSP MAR jack flag

torsiómetro m INSTR, LAB *resistencia de materiales* torsion meter

torsión[1]: **de ~** adj MECÁ torsional

torsión[2] f FÍS torsion, ING MECÁ torsion, twisting, MECÁ, METAL torsion, TEXTIL twist; ~ **aerodinámica** f TRANSP AÉR aerodynamic twist; ~ **del alambre fijador** f ING MECÁ lock wire twist; ~ **cero** f TEXTIL zero twist; ~ **de rosca lubricada** f ING MECÁ lubricated thread torque; ~ **suplementaria** f TEC ESP *cuerdas* hardening

torsional adj MECÁ torsional

torsionamiento m ING MECÁ twisting

torta: ~ **amarilla** f NUCL uranium concentrate, yellow cake; ~ **de cachaza** f CARBÓN filter cake; ~ **de extracción** f ALIMENT expeller cake; ~ **de filtración** f PROC QUÍ filter cake; ~ **de filtro** f CARBÓN filter cake; ~ **de linaza** f AGRIC linseed cake; ~ **oleaginosa** f AGRIC oilcake

tortas: ~ **de hielo** f pl OCEAN pancake ice

tortuga f CINEMAT *pantalla* turtle

tortuosidad: ~ **de núcleo de aire** f TEC ESP air core winding

torulina f QUÍMICA torulin

torzal m TEXTIL organzine, *de seda* rove

tosca f GEOL calcrete

tosco adj CONST *piedra* rubbly, ING MECÁ coarse; ~ **de soldadura** adj PROD as welded

tosilo m QUÍMICA tosyl

tostadero m ALIMENT kiln

tostado[1] adj TERMO scorched

tostado[2] m ALIMENT roasting

tostadura f PROD torrefaction

tostar vt TERMO scorch

total[1] adj ING MECÁ overall

total[2] m INFORM&PD *suma* tally; ~ **de batch** m INFORM&PD, PROC QUÍ batch processing; ~ **de control** m ELECTRÓN checksum, INFORM&PD control total, checksum; ~ **de lotes** m INFORM&PD batch processing, batch total, PROC QUÍ batch processing; ~ **de nutrientes digestibles** m AGRIC total digestible nutrients (*TDN*)

totalizador *m* INFORM&PD full adder, TRANSP integrator; **~ de etapa** *m* TEC ESP stage integrator

totalmente: **~ automático** *adj* ING MECÁ, INSTR, PROD fully automatic; **~ pegado** *adj* EMB full glueing

toxicidad *f* CONTAM MAR, P&C, QUÍMICA, SEG, TEC PETR toxicity

tóxico[1]: **~** *adj* CARBÓN, SEG, TEC PETR toxic; **no ~** *adj* SEG nontoxic

tóxico[2]: **~ de contacto** *m* AGRIC contact poison

toxicología *f* SEG toxicology

toxina *f* CALIDAD toxin; **~ fúngiga** *f* ALIMENT mycotoxin

toxisterol *m* QUÍMICA toxisterol

TPV *abr* (*terminal punto de venta*) INFORM&PD EPOS (*electronic point of sale*)

t-quark *m* FÍS PART top quark

TR *abr* (*transmisor-receptor*) ELECTRÓN TR (*transmitter-receiver*)

traba *f* CONST window bar, *cantería, piedras* joggle joint, ING MECÁ cramp

trabado *adj* ING MECÁ, MECÁ clogged

trabajabilidad *f* ING MECÁ machinability

trabajable *adj* CONST workable

trabajado: **~ en la máquina** *adj* MECÁ machined

trabajador: **~ en cajón de hinca** *m* AmL (*cf trabajador en cámara hiperbárica Esp*) OCEAN caisson worker; **~ en cámara hiperbárica** *m* Esp (*cf trabajador en cajón de hinca AmL*) OCEAN caisson worker; **~ de gran destreza** *m* PROD highly-skilled worker; **~ de la industria gráfica** *m* IMPR printing-trade worker; **~ migratorio** *m* AGRIC migrant worker

trabajar[1] *vt* ING MECÁ work, MECÁ machine

trabajar[2] *vi* ING MECÁ operate, MINAS *en realce* rise; **~ bajo tierra** *vi* CONST work underground; **~ contra veteado** *vi* CONST work against the grain; **~ con trípode** *vi* FOTO work with tripod; **~ a plena capacidad** *vi* ING MECÁ work to full capacity; **~ a toda su potencia** *vi* ING MECÁ work to its full capacity

trabajo[1]: **para ~ pesado** *adj* MECÁ heavy-duty; **de ~ de reserva** *adj* TELECOM stand-by work

trabajo[2] *m* FÍS work, MINAS labor (*AmE*), labour (*BrE*), PROD *servicio* duty; **~ en el banco** *m* CINEMAT bench work; **~ del bronce** *m* CONST *metalurgia* brassing; **~ en caliente** *m* C&V, METAL hot working; **~ en campaña** *m* D&A field work; **~ con la caña de soplo** *m* C&V making on blowpipe; **~ en circuito abierto** *m* ELEC *equipo* open-circuit operation; **~ constante** *m* CONST permanent work; **~ de construcción** *m* CONST construction work; **~ en curso de ejecución** *m* PROD work in hand, work-in-progress by employee; **~ encargado** *m* PROD commissioned work; **~ de enrejado** *m* CONST lattice work; **~ de estudio fotográfico** *m* FOTO studio work; **~ excesivo** *m* PROD overworking; **~ explotado en retirada** *m* MINAS retreat mining; **~ de forja artística** *m* CONST art metal work; **~ con la gubia** *m* PROD gouging; **~ hecho a máquina** *m* ING MECÁ machine work; **~ de impresión** *m* IMPR printjob; **~ de instalación eléctrica** *m* ELEC, ELECTRÓN, TRANSP MAR electrical-installation work; **~ en el interior** *m* MINAS bottom; **~ intermitente** *m* ELEC *de equipo* intermittent duty; **~ mal hecho** *m* PROD bad work; **~ a mano** *m* PROD handwork, manual labor (*AmE*), manual labour (*BrE*); **~ manual** *m* PROD handwork, handworking, manual labor (*AmE*), manual labour (*BrE*);

~ mecánico *m* ING MECÁ, MECÁ machining; **~ de muelles** *m* PROD *puertos* dock work; **~ a pala** *m* CONST shovel work; **~ perjudicial** *m* ING MECÁ loss; **~ preferencial** *m* INFORM&PD foreground job; **~ preparatorio** *m* MINAS mine opening, dead work, *de una cantera* opening; **~ en proceso** *m* PROD work in process; **~ realizado por encargo** *m* PROD *rendimiento de servicios* commissioned work; **~ en roca** *m* MINAS rock work; **~ en serie** *m* PROD repetitive work; **~ subterráneo** *m* MINAS underground operation, underground working, underground workings; **~ teórico** *m* FÍS theoretical work; **~ por testeros** *m* MINAS overhand stoping; **~ de torno** *m* PROD lathe work; **~ tridimensional** *m* ING MECÁ three-dimensional form working; **~ tridimensional a máquina** *m* ING MECÁ three-dimensional machining; **~ por turnos** *m* PROD shift work; **~ útil** *m* ING MECÁ effective power

trabajos[1]: **para ~ fuertes** *adj* ING MECÁ heavy-duty; **para ~ pesados** *adj* ING MECÁ heavy-duty

trabajos[2]: **~ en bronce** *m pl* CONST *metalurgia* brassworks; **~ en capa** *m pl* CARBÓN, MINAS seam work; **~ con cemento** *m pl* CONST cement works; **~ de exploración** *m pl* MINAS exploration work; **~ en manto** *m pl* CARBÓN, MINAS seam work; **~ en roca** *m pl* MINAS mullocking; **~ temporales** *m pl* CONST temporary works; **~ de trazado y de acceso** *m pl* MINAS development work; **~ con tuercas y pernos** *m pl* CONST nut-and-bolt works

trabar *vt* ING MECÁ clog, cramp, MECÁ fasten, jam, lock, TEXTIL interlock

trabarse *v refl* MECÁ jam

traca *f* CARBÓN *buques*, TRANSP MAR *construcción naval* strake; **~ de aparadura** *f* TRANSP MAR *construcción naval* garboard strake; **~ de cinta** *f* TRANSP MAR *construcción naval* sheerstrake; **~ del pantoque** *f* TRANSP MAR *construcción naval* bilge strake; **~ de quilla** *f* TRANSP MAR *construcción naval* keel strake

tracas: **~ de cubierta** *f pl* TRANSP MAR *construcción naval* deck plating

tracción[1]: **~ por animal** *adj* AGRIC animal-drawn

tracción[2] *f* CINEMAT pull-down, CONST, FERRO traction, ING MECÁ pull, pulling, thrust, tension, traction, PAPEL draw, TEC ESP pull, TELECOM *cable* pulling-in; **~ axial** *f* METAL tensile axis; **~ a las cuatro ruedas** *f* AUTO, VEH *transmisión* four-wheel drive; **~ delantera** *f* MECÁ, TRANSP front-wheel drive; **~ diesel-eléctrica** *f* TRANSP diesel-electric drive; **~ efectiva de ruedas** *f* AUTO, TRANSP, VEH rear-wheel drive; **~ de la hélice** *f* TRANSP propeller thrust; **~ inversa** *f* TRANSP AÉR reverse thrust; **~ máximal** *f* IMPR maximal stress; **~ del papel** *f* IMPR paper pull; **~ del rotor** *f* TRANSP AÉR rotor thrust; **~ a las ruedas delanteras** *f* AUTO front-wheel drive; **~ a ruedas y ejes** *f* TRANSP *ferrocarril* wheel-and-axle drive; **~ a las ruedas traseras** *f* AUTO, TRANSP, VEH rear-wheel drive; **~ por ruedas traseras con motor trasero** *f* AUTO rear-engine rear-wheel drive, VEH rear-engine rear-weel drive; **~ trasera** *f* AUTO, TRANSP, VEH rear-wheel drive

traceo *m* PROD cutting-off

trackball *m* INFORM&PD trackball, trackerball

tractor *m* AGRIC tractor, AUTO road tractor, CONST, D&A, INFORM&PD, MECÁ, TRANSP tractor, VEH road tractor, tractor; **~ de carriles** *m* AGRIC tracklayer; **~ por cultivos en línea** *m* AGRIC root-crop tractor;

~ monosurco *m* AGRIC one-furrow tractor; **~ oruga** *m* AGRIC, D&A caterpillar tractor, MECÁ crawler; **~ de orugas** *m* CONST *maquinaria obras públicas* tracked tractor; **~ porta-aperos** *m* AGRIC implement carrier; **~ de zancas** *m* AGRIC high-clearance tractor

traducción *f* INFORM&PD translation; **~ asistida por ordenador** *f Esp* (*cf traducción auxiliada por computadora LAm*) INFORM&PD computer-assisted translation; **~ auxiliada por computadora** *f AmL* (*cf traducción auxiliada por ordenador Esp*) INFORM&PD computer-assisted translation; **~ a dibujo** *f* PROD translation on a drawing; **~ dinámica de direcciones** *f* INFORM&PD dynamic address translation

traducir *vt* INFORM&PD, TELECOM translate; **~ a la forma digital** *vt* ELEC digitize; **~ en valor digital** *vt* ELEC digitize; **~ en valor numérico** *vt* ELEC digitize

traductor *m* INFORM&PD, TELECOM translator; **~ de lenguaje** *m* INFORM&PD language translator

traer *vt AmL* (*cf extraer Esp*) INFORM&PD fetch

tráfico *m* CONST transit, FERRO, TELECOM, TRANSP traffic, TRANSP AÉR transit, VEH traffic; **~ aéreo** *m* TRANSP AÉR air traffic; **~ aéreo del aeropuerto** *m* TRANSP AÉR airport traffic; **~ de agrupamiento** *m* FERRO *vehículos* groupage traffic; **~ de alta velocidad** *m* TRANSP superhigh-speed traffic; **~ de cabotaje** *m* TRANSP MAR coastal trade, *marina mercante* coastwise trade; **~ de cargas parciales** *m* FERRO part-load traffic; **~ de cercanías** *m* TRANSP commuter traffic; **~ de circunvalación** *m* TRANSP turning traffic; **~ comercial** *m* TRANSP business traffic, commercial traffic; **~ conducido** *m* TELECOM traffic carried; **~ constante** *m* TRANSP stationary traffic; **~ desviable** *m* TRANSP bypassable traffic; **~ diario** *m* TELECOM day traffic; **~ efectivo** *m* TELECOM effective traffic; **~ de entrada** *m* TRANSP entering traffic, inward traffic; **~ entrante** *m* TELECOM incoming traffic; **~ exterior** *m* TRANSP external traffic, outbound traffic; **~ por ferrocarril de vía estrecha** *m* FERRO *vehículos* light rail transit (*AmE*); **~ fluvial** *m* TRANSP MAR river traffic; **~ de giro a la derecha** *m* TRANSP right-turning traffic; **~ de giro a la izquierda** *m* TRANSP left-turning traffic; **~ de grupaje** *m* FERRO *vehículos* groupage traffic; **~ del hogar al trabajo** *m* TRANSP home-to-work traffic; **~ hora punta** *m* FERRO rush-hour traffic, TRANSP peak-period traffic, rush-hour traffic; **~ de hora punta** *m* FERRO, TRANSP peak-hour traffic; **~ ideal** *m* TELECOM pure-chance traffic; **~ interior** *m* TRANSP inbound traffic; **~ intermodal** *m* FERRO *vehículos* intermodal traffic; **~ interno** *m* TELECOM internal traffic; **~ interurbano** *m* TRANSP *larga distancia* through traffic; **~ para ir al trabajo** *m* TRANSP work traffic; **~ lanzadera** *m* TRANSP shuttle traffic; **~ local** *m* CONST, TRANSP local traffic; **~ marítimo** *m* TRANSP MAR sea trade, seaborne trade, shipping; **~ medio anual por día** *m* (*TMAD*) TRANSP average annual daily traffic (*AADT*); **~ de mercancías de larga distancia** *m* TRANSP long-distance freight traffic (*AmE*), long-distance goods traffic (*BrE*); **~ mínimo por hora** *m* TRANSP lowest hourly traffic; **~ en movimiento** *m* TRANSP moving traffic; **~ multimodal** *m* FERRO multimodal traffic; **~ normal** *m* TRANSP normal traffic; **~ ofrecido** *m* TELECOM traffic offered; **~ de origen** *m* TELECOM originating traffic; **~ de parada y continuación alternativas** *m* TRANSP stop-and-go traffic; **~ perdido** *m* TELECOM lost traffic; **~ pesado** *m* TRANSP peak-load traffic; **~ Poisson** *m* TELECOM Poisson traffic; **~ de procedencia** *m* TRANSP originating traffic; **~ puramente aleatorio** *m* TELECOM pure-chance traffic; **~ regularizado** *m* TELECOM smooth traffic; **~ de remolques sobre vagón de ferrocarril** *m* FERRO piggyback traffic; **~ rodado** *m* AUTO, TRANSP, VEH road traffic; **~ de salida** *m* TRANSP outward traffic; **~ saliente** *m* TELECOM outgoing traffic; **~ suburbano** *m* TRANSP suburban traffic; **~ terminal** *m* TELECOM, TRANSP terminating traffic; **~ de tránsito** *m* TELECOM transit traffic; **~ en tránsito** *m* TRANSP through traffic; **~ ultra rápido** *m* TRANSP ultrahigh-speed traffic; **~ urbano** *m* TRANSP urban traffic; **~ de vehículos pesados** *m* TRANSP heavy freight vehicle traffic (*AmE*), heavy goods vehicle traffic (*BrE*), HGV traffic; **~ en vías preferentes** *m* TRANSP straight-through traffic

tragacanto: ~ de la India *m* QUÍMICA karaya gum

tragaluz *m* CONST, TRANSP MAR *equipamiento de cubierta* skylight

tragante *m* CARBÓN *alto horno* slope top, ING MECÁ inlet, MINAS *hornos* mouth, PROD *horno alto* throat, *horno horno* tunnel, *hornos, convertidores* mouth; **~ de chimenea** *m* CONST chimney flue

traída *f AmL* (*cf extracción Esp*) INFORM&PD fetch; **~ de la instrucción** *f* INFORM&PD instruction-fetching

trailer *m* CONST trailer, bogie (*BrE*), truck (*AmE*), FERRO *vehículos* bogie (*BrE*), truck (*AmE*), ING MECÁ truck (*AmE*), bogie (*BrE*), MECÁ bogie (*BrE*), TRANSP caravan (*BrE*), trailer (*AmE*), VEH caravan (*BrE*), trailer, bogie (*BrE*); **~ multiuso** *m* TRANSP all-purpose trailer; **~ de uso múltiple** *m* TRANSP all-purpose trailer

traílla *f* CARBÓN *excavaciones* pan, CONST *maquinaria* scraper, MINAS *excavaciones* pan; **~ portadora** *f* MINAS slusher

traje: ~ abierto *m* OCEAN *buceo* wetsuit; **~ anti-G** *m* TRANSP AÉR G-suit; **~ de buceo** *m* OCEAN diving suit; **~ de buzo** *m* TRANSP MAR diving suit, wetsuit; **~ para contrarrestar de la aceleración** *m* TEC ESP anti-g suit; **~ espacial** *m* TEC ESP spacesuit; **~ estanco** *m* TRANSP MAR diving suit, wetsuit; **~ de inmersión articulado** *m AmL* (*cf escafandra articulada Esp*) OCEAN articulated diving suit; **~ de protección total** *m* SEG overall protective suit; **~ protector** *m* SEG protective suit

trama *f* CINEMAT matte, ELECTRÓN raster, IMPR raster, screen, INFORM&PD pattern, raster, ING ELÉC *guiaondas* grating, frame, PAPEL *fieltro, tela formación* weft, TELECOM board, TEXTIL *del tejido* weft, TV line, raster; **~ de borde rígido** *f* CINEMAT hard-edged matte; **~ de color** *f* IMPR color screen (*AmE*), colour screen (*BrE*); **~ complementaria** *f* CINEMAT complementary matte; **~ sin contraste** *f* IMPR flat screen; **~ y contratrama** *f* CINEMAT matte and counter-matte; **~ directa** *f* IMPR direct screen; **~ de distribución de alta frecuencia** *f* TELECOM high-frequency distribution frame; **~ de distribución combinada** *f* TELECOM combined distribution frame (*CDF*); **~ de distribución de grupo** *f* TELECOM group distribution frame; **~ de distribución intermedia** *f* TELECOM intermediate distribution frame (*IDF*); **~ de distribución principal de repetición** *f* TELECOM main repeater distribution frame; ~ **exploradora** *f* TV raster pitch; ~ **fina** *f* IMPR fine screen; ~ **gruesa** *f* IMPR coarse screen; ~ **de identificación digital** *f* TELECOM digital identification frame; ~ **impresa transparente adhesiva** *f* IMPR Zip-a-tone; ~ **de lana regenerada** *f* TEXTIL shoddy-type filling (*AmE*), shoddy-type weft (*BrE*); ~ **pintada** *f* CINEMAT painted matte; ~ **principal de distribución** *f* TELECOM main distribution frame (*MDF*); ~ **salida de telar** *f* TEXTIL loom state weft; ~ **de traveling** *f* CINEMAT traveling matte (*AmE*), travelling matte (*BrE*)

tramado: ~ **sin contraste** *m* IMPR flat-screening

tramo *m* AGUA reach, section, CONST *carretera*, FERRO *equipo inamovible* section, ING MECÁ run, section, NUCL *de una tubería* leg, TRANSP MAR reach; ~ **de cabecera** *m* AGUA headwater reach; ~ **de canal entre dos esclusas** *m* AGUA, CARBÓN pond; ~ **de cañería doble** *m* PETROL loop; ~ **de ensayos** *m* CONST *carreteras* trial strip; ~ **de escalera** *m* CONST flight of stairs; ~ **a favor del viento** *m* TRANSP AÉR downwind leg; ~ **inferior** *m* HIDROL *de un río* lower track; ~ **de pruebas** *m* CONST *carreteras* test section; ~ **superior** *m* HIDROL *de un río* upper part, VEH *del generador de vapor, por encima de la planta de operaciones* upper shell assembly; ~ **de tiempo** *m* ELECTRÓN time slice; ~ **de vía** *m* FERRO *infraestructura* track section; ~ **con viento de cola** *m* TRANSP AÉR downwind leg

tramoyista *m* CINEMAT grip

trampa *f* GEN *producción* trap; ~ **por acuñamiento** *f* TEC PETR *geología* pinch-out trap; ~ **de adsorción** *f* NUCL adsorption trap; ~ **anticlinal** *f* TEC PETR *geología* anticlinal trap; ~ **de arena** *f* AGUA, TEC PETR *perforación* sand trap; ~ **en la aspiración** *f* REFRIG dead end trap; ~ **para bomba de vacío** *f* LAB trap for vacuum pump; ~ **de calor** *f* NUCL heat trap; ~ **de condensación** *f* LAB *material de vidrio* condensing trap; ~ **de diablos** *f* PETROL launching trap; ~ **de discordancia** *f* TEC PETR *geología* unconformity trap; ~ **estratigráfica** *f* TEC PETR *geología* stratigraphic trap; ~ **por falla** *f* TEC PETR *geología* fault trap; ~ **de fangos** *f* AGUA silt trap; ~ **de hidrocarburo** *f* TEC PETR *geología* hydrocarbon trap; ~ **de iones** *f* TV ion trap; ~ **de limo** *f* AGUA silt trap; ~ **de luz** *f* CINEMAT light trap; ~ **de luz aterciopelada** *f* FOTO velvet light trap; ~ **de petróleo** *f* TEC PETR *geología* oil trap; ~ **de raspatubos** *f* PETROL launching trap; ~ **receptora** *f* PETROL receiving trap; ~ **de retroceso** *f* C&V check; ~ **de terciopelo** *f* CINEMAT velvet trap; ~ **térmica enfriada por agua** *f* CINEMAT water-cooled heat trap; ~ **de vapor** *f* PROC QUÍ vapor trap (*AmE*), vapour trap (*BrE*), TEC PETR *refino*, TERMOTEC steam trap; ~ **de velocidad** *f* TRANSP *tráfico* speed trap

trampilla *f* CONST flap door, trap door; ~ **de descarga** *f* INSTAL HIDRÁUL expansion trap; ~ **de inspección** *f* MECÁ inspection door

trancanil *m* TRANSP MAR deck stringer, *cubierta* stringer

tranche: ~ **nuclear** *m* ELEC, NUCL nuclear tranche

tranco *m* PROD *hogar, hornos* fire bridge

transacción *f* INFORM&PD transaction; ~ **de entrada aceptada para ejecución** *f* TELECOM input transaction accepted for delivery (*ITD*); ~ **de entrada rechazada** *f* TELECOM input transaction rejected (*ITR*); ~ **de existencias** *f* PROD stock transaction

transatlántico *m* TRANSP MAR ocean liner

transbordador *m* ING MECÁ transporter, TRANSP, TRANSP MAR *barco* ferry; ~ **aéreo** *m* TRANSP AÉR air ferry; ~ **de automóviles** *m* TRANSP, TRANSP MAR *comercio marítimo* car ferry; ~ **sobre carriles** *m* TRANSP rail ferry; ~ **de contenedores** *m* TRANSP MAR feeder ship; ~ **espacial** *m* TEC ESP shuttle, space shuttle, TELECOM space shuttle; ~ **portabarcazas HDW** *m* TRANSP HDW barge carrier; ~ **Tierra-órbita** *m* TEC ESP, TELECOM earth-to-orbit shuttle; ~ **para trenes** *m* TRANSP MAR train ferry; ~ **de trenes y pasaje** *m* TRANSP MAR ferryboat

transbordar *vt* TEC PETR, TRANSP MAR *pasajeros, cargamento* transship

transbordo *m* FERRO interchange, TEC ESP transfer, TEC PETR *transporte* transshipment, TRANSP MAR ferrying, transshipment; ~ **desde el tanque** *m* TEC ESP transfer from tank

transcendental *adj* MATEMÁT transcendental

transcendente *adj* MATEMÁT transcendental

transceptor *m* INFORM&PD, TELECOM, TRANSP MAR transceiver, TV transmitter-receiver

transcodificación *f* TELECOM transcoding

transcodificador *m* TELECOM, TV transcoder

transcomputadora *m* AmL (*cf transordenador Esp*) INFORM&PD transputer

transconductancia *f* FÍS, ING ELÉC transconductance

transcontenedor *m* TRANSP transcontainer

transcribir *vt* INFORM&PD transcribe, transliterate

transcripción *f* ACÚST rerecording

transcurso *m* TEC ESP range

transductor *m* ACÚST, ELEC transducer, FÍS transducer, INFORM&PD, ING ELÉC, ING MECÁ, METR transducer, PROD transducer, transductor, TEC ESP transductor, TELECOM transducer; ~ **activo** *m* ELEC, ING ELÉC active transducer; ~ **autogenerador** *m* ING ELÉC self-generating transducer; ~ **bidireccional** *m* ING ELÉC bidirectional transducer, bilateral transducer; ~ **bilateral** *m* ING ELÉC reversible transducer; ~ **de CC** *m* ING ELÉC DC transducer; ~ **sin contactos** *m* ELEC contactless pickup (*BrE*), pointless pickup (*AmE*); ~ **de conversión heterodina** *m* ING ELÉC heterodyne conversion transducer; ~ **diferencial** *m* ING ELÉC differential transducer; ~ **a diodo** *m* ELECTRÓN diode transducer; ~ **eléctrico** *m* ELEC, ING ELÉC electric transducer; ~ **electroacústico** *m* ACÚST, ING ELÉC electroacoustic transducer; ~ **electromecánico** *m* ACÚST, ELEC, ING ELÉC electromechanical transducer; ~ **del embrague** *m* TRANSP AÉR clutch pick-off; ~ **de entrada** *m* ING ELÉC input transducer; ~ **de equilibrio de fuerzas** *m* ING ELÉC force-balance transducer; ~ **de fibra óptica** *m* ING ELÉC fiber-optic transducer (*AmE*); ~ **fibroóptico** *m* ING ELÉC, ÓPT, TELECOM fiber-optic transducer (*AmE*), fibre-optic transducer (*BrE*); ~ **fotoactivo** *m* ING ELÉC photoactive transducer; ~ **fotoeléctrico** *m* ING ELÉC photoelectric transducer; ~ **de frecuencia** *m* ELEC, ING ELÉC frequency transducer; ~ **interdigital** *m* TELECOM interdigital transducer (*IDT*); ~ **lineal** *m* ACÚST linear transducer; ~ **magnetoestrictivo** *m* ING ELÉC magnetostrictive transductor; ~ **de medida** *m* METR measuring transducer; ~ **para medir presiones** *m* PROD pressure transducer; ~ **optoelectrónico** *m* ING ELÉC optoelectronic transducer; ~ **pasivo** *m* ING ELÉC passive transducer; ~ **de película** *m* TEC ESP film transducer; ~ **piezoeléctrico** *m* ING ELÉC piezoelectric transducer; ~ **recíproco** *m* ACÚST reciprocal transducer; ~ **reversible** *m* ACÚST reversible transducer; ~ **de salida** *m* ING ELÉC output transducer; ~ **simétrico** *m* ING ELÉC symmetrical transducer; ~ **sonar** *m* ACÚST sonar transducer; ~ **de transparencias a televisión** *m* TV slide pickup; ~ **unidireccional** *m* ING ELÉC unidirectional transducer

transesterificación *f* QUÍMICA cross-esterification

transferencia *f* GEN transfer; ~ **de archivos** *f* INFORM&PD file transfer; ~ **automática** *f* ELEC *conmutador* automatic changeover; ~ **automática de la carga** *f* ELEC *fuente de alimentación* automatic load transfer; ~ **automática de llamadas** *f* TELECOM automatic call transfer; ~ **de la banda sonora de una grabación a otra** *f* ACÚST rerecording; ~ **bidireccional de bloques** *f* PROD bidirectional block transfer; ~ **de bits en paralelo** *f* INFORM&PD bit parallel transfer; ~ **en blanco** *f* C&V blank transfer; ~ **de bloques** *f* INFORM&PD block transfer; ~ **de calor** *f* FÍS, INSTAL TERM heat transfer; ~ **de calor por radiación** *f* FÍS RAD, TEC ESP radiative heat transfer; ~ **de calor radioactivo** *f* FÍS RAD, TEC ESP radiative heat transfer; ~ **de carga** *f* FÍS RAD, NUCL charge transfer; ~ **de cargas** *f* ELEC *fuente de alimentación*, TELECOM load transfer; ~ **cerámica** *f* C&V ceramic transfer; ~ **de cinta a película** *f* TV tape-to-film transfer; ~ **de cinta de vídeo a película** *f* CINEMAT tape-to-film transfer; ~ **controlada** *f* TELECOM transfer controlled (*TFC*); ~ **de datos** *f* INFORM&PD data transfer; ~ **de datos solicitada** *f* TELECOM data transfer requested; ~ **por difusión** *f* IMPR photomechanical transfer; ~ **electrónica de fondos** *f* TELECOM electronic funds transfer (*EFT*); ~ **electrónica de fondos en punto de venta** *f* TELECOM electronic funds transfer at point of sale (*EFTPOS*); ~ **electrostática de la tinta** *f* ELEC, IMPR electrostatic ink transfer; ~ **de energía** *f* FÍS, TERMO energy transfer; ~ **de energía lineal** *f* FÍS, FÍS RAD linear energy transfer; ~ **de energía resonante** *f* FÍS RAD resonant-energy transfer; ~ **de energía por vibraciones** *f* FÍS ONDAS *onda progresiva* energy transference by vibration; ~ **de ficheros** *f* INFORM&PD file transfer; ~ **de humedad** *f* REFRIG, TEXTIL moisture transfer; ~ **de imagen** *f* TELECOM image transfer; ~ **de licencia** *f* CALIDAD transfer of licence (*BrE*), transfer of license (*AmE*); ~ **lineal de energía** *f* FÍS, FÍS RAD linear energy transfer; ~ **de llamada** *f* TELECOM call transfer; ~ **magnética** *f* CINEMAT magnetic transfer; ~ **de mando** *f* TRANSP AÉR override control; ~ **de masa** *f* GAS mass transfer; ~ **de mensajes** *f* INFORM&PD, TELECOM message transfer (*MT*); ~ **óptica** *f* CINEMAT optical transfer; ~ **de un ordenador a otro** *f* INFORM&PD, TELECOM downloading; ~ **paralela de bit** *f* INFORM&PD bit parallel transfer; ~ **en paralelo** *f* INFORM&PD parallel transfer; ~ **de película** *f* CINEMAT, TV film transfer; ~ **de película a cinta** *f* CINEMAT *vídeo*, TV film-to-tape transfer; ~ **de periféricos** *f* INFORM&PD peripheral transfer; ~ **permitida** *f* TELECOM transfer allowed (*TFA*); ~ **prohibida** *f* TELECOM transfer prohibited (*TFP*); ~ **radiativa** *f* FÍS RAD radiative transfer; ~ **restringida** *f* TELECOM transfer restricted (*TFR*); ~ **sin sacudidas** *f* PROD bumpless transfer; ~ **en serie** *f* INFORM&PD serial transfer; ~ **en serie de bit** *f* INFORM&PD bit serial transfer; ~ **sonora** *f*

CINEMAT sound transfer; ~ **térmica** *f* GAS thermal transfer, P&C heat transfer; ~ **de la tinta** *f* IMPR ink transfer

transferible *adj* INFORM&PD transportable

transferido: ~ **por octeto** *m* PROD *automatismo industrial* put byte

transferir *vt* IMPR, INFORM&PD transfer, ING MECÁ *fuerza motriz* convey

transfinito *adj* MATEMÁT transfinite

transformación *f* CARBÓN process, CONTAM reclamation, ELEC conversion, ELECTRÓN conversion, transform, INFORM&PD transform, ING ELÉC conversion, ING MECÁ refitting, MATEMÁT transform, PROD reprocessing, RECICL reclamation, reprocessing; ~ **adiabática** *f* METAL adiabatic transformation, TERMO adiabatic change; ~ **afín** *f* METAL affine transformation; ~ **coordinada** *f* ELECTRÓN coordinate transformation; ~ **discreta Fourier** *f* ELECTRÓN discrete Fourier transform; ~ **de electricidad** *f* ELEC *red de distribución* transformation of electricity; ~ **de energía** *f* ELEC, ING ELÉC, TERMO energy transformation; ~ **de estrella a delta** *f* ELEC *conexión* star-to-delta transformation; ~ **de estrella a triángulo** *f* ELEC *conexión* star-to-delta transformation; ~ **estrella-delta** *f* FÍS star-delta transformation; ~ **eutéctica** *f* METAL, QUÍMICA eutectic transformation; ~ **de fase** *f* TERMO phase transformation; ~ **de Fourier** *f* ACÚST, CRISTAL, ELECTRÓN, FÍS Fourier transformation; ~ **de Galileo** *f* FÍS Galilean transformation; ~ **de Laplace** *f* ELECTRÓN Laplace transformation, FÍS Laplace transformation, Laplace transform, MATEMÁT Laplace transform; ~ **de Lorentz** *f* FÍS Lorentz transformation; ~ **martensítica** *f* CRISTAL martensitic transformation; ~ **orden-desorden** *f* CRISTAL order-disorder transformation; ~ **polimórfica** *f* CRISTAL polymorphic transformation; ~ **radioactiva** *f* FÍS RAD radioactive transformation; ~ **sedimentaria** *f* ENERG RENOV diagenesis; ~ **térmica** *f* TERMO heat transformation; ~ **termodinámica** *f* TERMO thermodynamic transformation; ~ **en ventanas** *f* INFORM&PD window transformation; ~ **de visualización** *f* INFORM&PD viewing transformation

transformada *f* ACÚST, CRISTAL, ELECTRÓN, FÍS, INFORM&PD, MATEMÁT transform; ~ **discreta de Fourier** *f* ELECTRÓN discrete Fourier transform; ~ **de Fourier** *f* ACÚST, CRISTAL, ELECTRÓN, FÍS Fourier transform; ~ **de Laplace** *f* FÍS, MATEMÁT Laplace transform; ~ **rápida de Fourier** *f* (*TRF*) ELECTRÓN, INFORM&PD fast Fourier transform (*FFT*)

transformador *m* ELEC converter, *corriente AC/CC* transformer, ELECTRÓN converter, FÍS transformer, ING ELÉC transformer, converter, INSTR, NUCL, TELECOM transformer; ~ **de acoplamiento** *m* ELEC, ING ELÉC coupling transformer; ~ **acorazado** *m* ELEC, ING ELÉC shell-type transformer; ~ **de adaptación** *m* ELEC, ING ELÉC matching transformer; ~ **de adaptación de impedancias** *m* ING ELÉC impedance-matching transformer; ~ **de aire** *m* ELEC air transformer; ~ **de aislamiento** *m* ING ELÉC, PROD isolation transformer; ~ **ajustable** *m* ELEC adjustable transformer; ~ **de alimentación** *m* ELEC current transformer (*AmE*), mains transformer (*BrE*), feeding transformer, ELECTRÓN, ING ELÉC current transformer (*AmE*); ~ **de alta frecuencia** *m* ING ELÉC high-frequency transformer; ~ **de alta tensión** *m* ELEC high-voltage transformer; ~ **de amperaje** *m*

ING ELÉC mains transformer (*BrE*); ~ **de amperaje constante** *m* ING ELÉC constant-current transformer; ~ **de amplificación** *m* FÍS step-up transformer; ~ **para aparatos de medida** *m* ELEC instrument transformer; ~ **de arranque** *m* ING ELÉC starting transformer; ~ **de arrollamiento múltiple** *m* ELEC multiwinding transformer; ~ **auxiliar** *m* ELEC auxiliary transformer; ~ **auxiliar de la unidad** *m* NUCL unit auxilliary transformer; ~ **en baño de aceite** *m* ELEC oil transformer, ING ELÉC oil-immersed transformer; ~ **de barras** *m* ELEC bar-type transformer; ~ **bidireccional** *m* ING ELÉC bidirectional transducer; ~ **bifilar** *m* ING ELÉC double-wound transformer; ~ **blindado** *m* ING ELÉC shielded transformer; ~ **de una bobina** *m* ELEC one-coil transformer; ~ **de CA directa** *m* ELEC direct AC converter; ~ **de CC** *m* ELEC DC converter, direct DC converter, ING ELÉC DC transducer, DC transformer; ~ **a chorro de aire** *m* ELEC air blast transformer; ~ **de cinco circuitos derivados** *m* ELEC five-legged transformer; ~ **de columnas** *m* ELEC, ING ELÉC core-type transformer; ~ **concentrador** *m* MINAS *electrotermia* concentrator; ~ **de corriente** *m* ELEC, ELECTRÓN current transformer (*AmE*), mains transformer (*BrE*), ING ELÉC current transformer (*AmE*); ~ **de corriente continua** *m* ELEC direct current transformer; ~ **desfasador** *m* ELEC phase-shifting transformer; ~ **de devanado independiente** *m* ELEC separate winding transformer; ~ **de devanado múltiple** *m* ELEC multiwinding transformer; ~ **diferencial** *m* ELEC, ING ELÉC differential transformer; ~ **de distribución para montaje en poste** *m* ELEC *línea aérea* pole-mounted transformer; ~ **de dos devanados** *m* ELEC double-wound transformer; ~ **elevador** *m* ELEC booster transformer, voltage booster, step-up transformer, ELECTRÓN voltage booster, booster transformer, ING ELÉC step-up transformer; ~ **elevador de tensión** *m* ELECTRÓN, ING ELÉC booster transformer; ~ **elevador de voltaje** *m* ELEC, ELECTRÓN booster transformer, TV step-up transformer; ~ **EMF** *m* FÍS transformer EMF; ~ **de encendido** *m* ELEC, ING ELÉC ignition transformer; ~ **de energía** *m* ING ELÉC, PROD power transformer, TEC ESP power converter; ~ **enfriado por aceite** *m* ELEC, ING ELÉC oil-cooled transformer; ~ **enfriado por aire** *m* ELEC air-cooled transformer, ING ELÉC oil-cooled transformer; ~ **de enfriamiento por agua** *m* ELEC water-cooled transformer, ING ELÉC air-cooled transformer; ~ **de enlace** *m* ELEC, ING ELÉC interstage transformer; ~ **de entrada** *m* FÍS, ING ELÉC input transformer; ~ **estático** *m* ELEC, ING ELÉC static transformer; ~ **para filamentos** *m* ELEC, ING ELÉC filament transformer; ~ **de frecuencia** *m* ELEC, ING ELÉC frequency transformer; ~ **de fuerza** *m* PROD power transformer; ~ **giratorio** *m* ELEC, ING ELÉC dynamotor; ~ **de gran potencia** *m* ELEC, ING ELÉC high-power transformer; ~ **de guía de ondas** *m* ELEC, ING ELÉC grating converter, waveguide transformer; ~ **de guíaondas** *m* ELEC, ING ELÉC waveguide transformer; ~ **ideal** *m* ELEC, ING ELÉC ideal transformer; ~ **de imagen** *m* INSTR image converter; ~ **de impedancias** *m* FÍS impedance transformer; ~ **de impulsos** *m* ELEC, ING ELÉC pulse transformer; ~ **de instrumentos** *m* ELEC, ING ELÉC instrument transformer; ~ **de intensidad** *m* ELEC, ELECTRÓN, ING ELÉC current transformer (*AmE*), mains transformer

(*BrE*); ~ **de intensidad secuencial** *m* PROD sequence current transformer; ~ **de mando** *m* ELEC, ING ELÉC control transformer; ~ **de medida** *m* ELEC, ING ELÉC instrument transformer; ~ **monofásico** *m* ELEC, ING ELÉC single-phase transformer; ~ **para montaje en poste** *m* ELEC, ING ELÉC *línea aérea* pole-mounted transformer; ~ **de núcleo** *m* ELEC, ING ELÉC core transformer; ~ **sin núcleo** *m* ELEC, ING ELÉC air core transformer; ~ **de núcleo abierto** *m* ELEC, ING ELÉC open-core transformer; ~ **de núcleo de aire** *m* ELEC, ING ELÉC air core transformer; ~ **de núcleo ferromagnético** *m* ELEC, ING ELÉC iron core transformer; ~ **de núcleo de hierro** *m* ELEC iron core transformer, core-type transformer, ING ELÉC core-type transformer, iron core transformer; ~ **sin núcleo magnético** *m* ELEC, ING ELÉC air core transformer; ~ **de núcleo saturable** *m* ELEC, ING ELÉC peaking transformer; ~ **de par** *m* ING MECÁ torque converter; ~ **polifásico** *m* ELEC, ING ELÉC polyphase transformer; ~ **para poste** *m* ELEC, ING ELÉC pole-type transformer; ~ **de potencial** *m* ELEC, ING ELÉC potential transformer, voltage transformer; ~ **principal** *m* NUCL main transformer; ~ **de prueba** *m* TELECOM test transformer; ~ **a prueba de descargas atmosféricas** *m* ELEC, ING ELÉC lightning-proof transformer; ~ **a prueba de rayos** *m* ELEC, ING ELÉC lightning-proof transformer; ~ **de puesta a tierra trifásico** *m* ELEC, ING ELÉC three-phase earthing transformer (*BrE*), three-phase grounding transformer (*AmE*); ~ **de pulsos** *m* FÍS pulse transformer; ~ **de rayos X** *m* INSTR X-ray transformer; ~ **del RCS** *m* ELEC, ING ELÉC SCR trimmer transformer; ~ **rebajador** *m* ELEC, ING ELÉC step-down transformer; ~ **rebajador de tensión** *m* ELEC, ING ELÉC step-down transformer; ~ **rebajador de voltaje** *m* ELEC, ING ELÉC step-down transformer; ~ **rebajador del voltaje** *m* ELEC, ING ELÉC negative booster; ~ **rectificador** *m* ELEC, ING ELÉC rectifier transformer; ~ **de la red de corriente** *m* ELECTRÓN current transformer (*AmE*); ~ **de la red de corriente eléctrica** *m* ELEC, ING ELÉC current transformer (*AmE*), mains transformer (*BrE*); ~ **de la red principal** *m* ELEC current transformer (*AmE*), mains transformer (*BrE*), ELECTRÓN current transformer (*AmE*), ING ELÉC mains transformer (*BrE*), current transformer (*AmE*); ~ **de reducción** *m* FÍS step-down transformer; ~ **reductor** *m* ELEC *inductor* choke, step-down transformer, ING ELÉC choke, PROD step-down transformer; ~ **reductor-elevador** *m* PROD step transformer; ~ **refrigerado por aceite** *m* ELEC, ING ELÉC oil-cooled transformer; ~ **refrigerado por agua** *m* ELEC, ING ELÉC water-cooled transformer; ~ **refrigerado por aire** *m* ELEC, ING ELÉC air-cooled transformer; ~ **con refrigerante de aire** *m* ELEC, ING ELÉC air-cooled transformer; ~ **regulable** *m* ELEC, ING ELÉC variable transformer; ~ **de regulación** *m* ELEC, ING ELÉC regulation transformer; ~ **regulador** *m* ING ELÉC booster transformer, LAB regulating transformer; ~ **regulador de la tensión** *m* ELEC, ING ELÉC voltage-regulating transformer; ~ **de relación de voltaje regulable** *m* ELEC, ING ELÉC variable-ratio transformer; ~ **de retracción** *m* *Esp* (*cf transformador vertical de alta tensión AmL*) TV flyback transformer; ~ **de RF** *m* ELECTRÓN, ING ELÉC RF transformer; ~ **rotatorio** *m* ELEC, ING ELÉC rotary transformer; ~ **de salida** *m* ELEC, FÍS, ING ELÉC output transformer; ~ **saturable** *m* ELEC, ING ELÉC saturable transformer; ~ **saturado** *m* ELEC, ING ELÉC saturated transformer; ~ **en seco** *m* ING ELÉC dry-type power transformer; ~ **sellado** *m* ELEC, ING ELÉC sealed transformer; ~ **en serie** *m* ELEC series transformer, ING ELÉC current transformer (*AmE*), mains transformer (*BrE*); ~ **simétrico** *m* ELEC, ELECTRÓN, TELECOM push-pull transformer; ~ **de sincronización** *m* PROD synchronization transformer; ~ **sintonizado** *m* ELEC, ING ELÉC tuned transformer; ~ **de soldar** *m* ELEC, ING ELÉC welding transformer; ~ **de tensión** *m* ELEC, ING ELÉC potential transformer, voltage transformer; ~ **de tensión de ánodos** *m* ELEC, ING ELÉC high-voltage transformer; ~ **de timbre** *m* ELEC, ING ELÉC bell transformer; ~ **de tipo seco** *m* ELEC, ING ELÉC dry-type transformer; ~ **de tipo seco no encapsulado** *m* ELEC, ING ELÉC nonencapsulated-winding dry-type transformer; ~ **con tomas** *m* ELEC, ING ELÉC tapped transformer; ~ **con tomas de regulación** *m* ELEC, ING ELÉC tapped transformer; ~ **toroidal** *m* ELEC, ING ELÉC toroidal transformer; ~ **toroidal saturado** *m* ELEC, ING ELÉC saturated toroidal transformer; ~ **de tres devanados** *m* ELEC, ING ELÉC three-winding transformer; ~ **tridevanado** *m* ELEC, ING ELÉC three-winding transformer; ~ **trifásico** *m* ELEC, ING ELÉC three-phase transformer; ~ **de triple arrollamiento** *m* ELEC, ING ELÉC triple-wound transformer; ~ **vertical de alta tensión** *m* *AmL* (*cf transformador de retracción Esp*) TV flyback transformer; ~ **de voltaje** *m* ING ELÉC potential transformer, voltage transformer, PROD potential transformer; ~ **de voltaje constante** *m* ELEC, PROD constant-voltage transformer

transformar *vt* ELEC, ELECTRÓN convert, INFORM&PD transform, ING ELÉC convert, PROD, RECICL reprocess; ~ **en diesel** *vt* TRANSP convert to diesel

transgresión *f* GEOL *incursión del agua en el continente* transgression

transición *f* TELECOM crossover; ~ **Auger** *f* FÍS RAD Auger effect; ~ **de bajo a alto** *f* ELECTRÓN low-to-high transition; ~ **cónica** *f* ÓPT tapered transition; ~ **del cuadrípolo eléctrico** *f* ELEC, FÍS RAD, ING ELÉC electric quadrupole transition; ~ **débil del positrón** *f* NUCL weak positron transition; ~ **desde el estado fundamental** *f* FÍS, FÍS PART, FÍS RAD, QUÍMICA ground-state transition; ~ **por desenfoque** *f* CINEMAT defocus transition; ~ **desfavorecida** *f* NUCL unfavored transition (*AmE*), unfavoured transition (*BrE*); ~ **de dipolo magnético** *f* FÍS RAD magnetic dipole transition; ~ **donde una imagen quita a otra de la pantalla** *f* CINEMAT, TV push-off wipe; ~ **donde una imagen quita a otra de la pantalla verticalmente** *f* CINEMAT, TV vertical wipe; ~ **donde una imagen reemplaza otra girando** *f* CINEMAT spin wipe; ~ **electrónica dipolar permitida** *f* FÍS RAD *reglas de selección* allowed electron dipole transition; ~ **sin emisión de radiación** *f* FÍS RAD radiationless transition; ~ **espontánea** *f* FÍS RAD spontaneous transition; ~ **de estado frágil a estado correoso** *f* METAL tough-brittle transition; ~ **de falso a verdadero** *f* PROD false-to-true transition; ~ **frágil-dúctil** *f* METAL, NUCL brittle ductile transition; ~ **de guía de ondas** *f* ING ELÉC waveguide transition; ~ **de guíaondas** *f* ING ELÉC waveguide transition; ~ **de la inversión del espín** *f* FÍS PART spin reversal transition; ~ **isomérica** *f* FÍS RAD isomeric transition;

~ **lasérica** *f* FÍS RAD laser transition; ~ **magnética** *f* METAL magnetic transition; ~ **óptica** *f* FÍS RAD optical transition; ~ **permitida** *f* FÍS, FÍS RAD, NUCL allowed transition; ~ **de primer orden** *f* FÍS first-order transition; ~ **progresiva** *f* TELECOM progressive transition; ~ **prohibida** *f* FÍS, FÍS RAD, NUCL forbidden transition; ~ **sin radiación** *f* FÍS RAD radiationless transition; ~ **radiativa** *f* FÍS RAD radiative transition; ~ **rectangular a circular** *f* ING ELÉC rectangular-to-circular transition; ~ **de la rotura dúctil a la rotura frágil** *f* METAL ductile-brittle transition; ~ **de segundo orden** *f* FÍS second-order transition; ~ **suave** *f* CINEMAT soft cut; ~ **térmica** *f* TERMO heat transition; ~ **de verdadero a falso** *f* PROD true-to-false transition; ~ **de vuelo** *f* TRANSP AÉR flight transition

transiente[1] *adj* FÍS, ING ELÉC, NUCL transient
transiente[2] *m* FÍS, ING ELÉC, NUCL transient
transistor *m* ELECTRÓN transistor, transistor chip, FÍS, INFORM&PD, ÓPT, TEC ESP transistor; ~ **de aleación por difusión** *m* ELECTRÓN diffused alloy transistor; ~ **de alta frecuencia** *m* ELECTRÓN high-frequency transistor; ~ **de amplificación** *m* ELECTRÓN amplifying transistor; ~ **de amplificación de potencia** *m* ELECTRÓN power-amplifier transistor; ~ **autoalineado** *m* ELECTRÓN self-aligned transistor; ~ **de bajo nivel** *m* ELECTRÓN low-level transistor; ~ **en base común** *m* ELECTRÓN common-base transistor; ~ **BICMOS** *m* INFORM&PD BICMOS transistor; ~ **bipolar** *m* ELECTRÓN, FÍS, INFORM&PD, TELECOM bipolar transistor; ~ **bipolar discreto** *m* ELECTRÓN discrete bipolar transistor; ~ **bipolar de gran potencia** *m* ELECTRÓN high-power bipolar transistor; ~ **bipolar integrado** *m* ELECTRÓN integrated bipolar transistor; ~ **bipolar de silicio** *m* ELECTRÓN silicon bipolar transistor; ~ **bipolar de tecnología planar** *m* ELECTRÓN planar bipolar transistor; ~ **bipolar vertical** *m* ELECTRÓN vertical bipolar transistor; ~ **de capa delgada** *m* ELECTRÓN thin-film transistor; ~ **en chip** *m* ELECTRÓN on-chip transistor; ~ **de colector múltiple** *m* ELECTRÓN multicollector transistor; ~ **de conmutación** *m* ELECTRÓN switching circuit; ~ **de conmutación de potencia** *m* ELECTRÓN power-switching transistor; ~ **de consumo común** *m* ELECTRÓN common-drain transistor; ~ **de contacto de punta** *m* ELECTRÓN point-contact transistor; ~ **de corriente elevada** *m* ELECTRÓN high-current transistor; ~ **cuántico** *m* ELECTRÓN quantum transistor; ~ **difuso** *m* ELECTRÓN double-diffused transistor; ~ **de diodo conectado** *m* ELECTRÓN diode-connected transistor; ~ **en drenaje común** *m* ELECTRÓN common-drain transistor; ~ **de efecto de campo** *m* (*TEC*) ELECTRÓN, FÍS, INFORM&PD, ÓPT, TEC ESP field effect transistor (*FET*); ~ **de efecto de campo con circuito aislado** *m* (*TECCA*) ELECTRÓN insulated grid field effect transistor (*IGFET*); ~ **de efecto campo de empobrecimiento** *m* ELECTRÓN enhancement mode FET, enhancement mode field-effect transistor; ~ **de efecto de campo con rejilla aislada** *m* (*TECRA*) ELECTRÓN insulated grid field effect transistor (*IGFET*); ~ **de efecto de campo vertical** *m* ELECTRÓN vertical field-effect transistor; ~ **de emisión común** *m* ELECTRÓN common-emitter transistor; ~ **de emisor múltiple** *m* ELECTRÓN multi-emitter transistor; ~ **emisor-captador por difusión** *m* ELEC-

TRÓN diffused emitter-collector transistor; **~ de enganche** *m* ELECTRÓN latching transistor; **~ estabilizado Schottky** *m* ELECTRÓN Schottky clamped transistor; **~ de filamento fibroso** *m* ELECTRÓN cat's whisker transistor; **~ en fuente común** *m* ELECTRÓN common-source transistor; **~ funcionando como inversor** *m* ELECTRÓN inverting transistor; **~ de germanio** *m* ELECTRÓN germanium transistor; **~ de gran capacidad de conmutación** *m* ELECTRÓN high-speed switching transistor; **~ inactivo** *m* ELECTRÓN off transistor; **~ laminar** *m* ELECTRÓN laminar transistor; **~ lateral** *m* ELECTRÓN lateral transistor; **~ de lectura** *m* ELECTRÓN read transistor; **~ libre** *m* ELECTRÓN uncommitted transistor; **~ de memoria** *m* ELECTRÓN memory transistor; **~ mesa** *m* ELECTRÓN mesa transistor; **~ de mezcla** *m* ELECTRÓN mixing transistor; **~ de microondas** *m* ELECTRÓN microwave transistor; **~ no afectado** *m* ELECTRÓN uncommitted transistor; **~ npn** *m* ELECTRÓN npn transistor; **~ parásito** *m* ELECTRÓN parasitic transistor; **~ pasivado** *m* ELECTRÓN passivated transistor; **~ de paso** *m* ELECTRÓN pass transistor; **~ de pasos en serie** *m* ELECTRÓN series-pass transistor; **~ de pequeña señal** *m* ELECTRÓN small-signal transistor; **~ planar epitáxico de silicio** *m* ELECTRÓN silicon epitaxial planar transistor; **~ p-n-p** *m* ELECTRÓN, FÍS p-n-p transistor; **~ de posición secundaria con indicación diversa** *m* (*transistor de PSID*) ELECTRÓN multipoint indication secondary-status transistor (*MIS transistor*); **~ de potencia** *m* ELECTRÓN power transistor; **~ de potencia bipolar** *m* ELECTRÓN bipolar power transistor; **~ de potencia de conmutación** *m* ELECTRÓN fast-switching power transistor; **~ de potencia de microondas** *m* ELECTRÓN microwave power transistor; **~ de potencia de pasos en serie** *m* ELECTRÓN series-pass power transistor; **~ de potencia de semiconductor de óxido de metal** *m* ELECTRÓN metal-oxide semiconductor power transistor; **~ de potencia SOM** *m* ELECTRÓN MOS power transistor; **~ de PSID** *m* (*transistor de posición secundaria con indicación diversa*) ELECTRÓN MIS transistor (*multipoint indication secondary-status transistor*); **~ en puerta común** *m* ELECTRÓN common-gate transistor; **~ con puerta de silicio** *m* ELECTRÓN silicon gate transistor; **~ de RF** *m* ELEC, ELECTRÓN, ING ELÉC RF transistor; **~ saturado** *m* ELECTRÓN saturated transistor; **~ de SCOM** *m* ELECTRÓN CMOS transistor; **~ selector** *m* ELECTRÓN gating transistor; **~ de semiconductor de óxido de metal** *m* ELECTRÓN metal-oxide semiconductor transistor, MOS transistor; **~ de una sola unión** *m* ELECTRÓN unijunction transistor; **~ de una sola unión programable** *m* ELECTRÓN programmable uni-junction transistor; **~ SOM** *m* ELECTRÓN MOS transistor; **~ SOM de arseniuro de galio** *m* ELECTRÓN gallium arsenide MOS transistor; **~ de SOM con canal corto** *m* ELECTRÓN short channel MOS transistor; **~ SOM de canal N** *m* ELECTRÓN NMOS transistor; **~ de SOM integrado** *m* ELECTRÓN integrated MOS transistor; **~ SOM integrado de canal n** *m* ELECTRÓN n-channel integrated MOS transistor; **~ SOM en modo de intensificación del canal p** *m* ELECTRÓN p-channel enhancement-mode MOS transistor; **~ SOM en modo de transición a canal p** *m* ELECTRÓN p-channel depletion-mode MOS transistor; **~ SOM de potencia vertical** *m*

ELECTRÓN vertical power MOS transistor; **~ SOM vertical** *m* ELECTRÓN vertical MOS transistor; **~ de SOMD** *m* ELECTRÓN DMOS transistor; **~ de tetrodo** *m* ELECTRÓN tetrode transistor; **~ de tracción con canal n** *m* ELECTRÓN n-channel pulldown transistor; **~ de unión de aleación** *m* ELECTRÓN alloy junction transistor; **~ de uniones** *m* ELECTRÓN, FÍS junction transistor; **~ de vector mayoritario** *m* ELECTRÓN majority-carrier transistor

transistores: ~ complementarios *m pl* ELECTRÓN complementary transistors; **~ emparejados** *m pl* ELECTRÓN matched transistors

transitario *m* FERRO, TRANSP, TRANSP MAR forwarding agent (*BrE*), freight agent (*AmE*), goods agent (*BrE*)

tránsito *m* CONST, NUCL transit, TRANSP traffic, TRANSP MAR *navegación* passage; **~ de material** *m* EMB material flow; **~ privado rápido automatizado** *m* TRANSP automated personal rapid transit

transitorio[1] *adj* NUCL, TELECOM transient

transitorio[2] *m* FÍS, ING ELÉC, NUCL transient

transitorios: ~ previstos sin parada de emergencia *m pl* (*ATWS*) NUCL anticipated transient without scream (*ATWS*)

translación: ~ de un punto a otro en un intervalo de tiempo *f* TEC ESP *aviones, misiles* vector

translador *m* TELECOM *telefonía* translator; **~ de anillo** *m* TELECOM ring translator

transliterar *vt* INFORM&PD transliterate

translucidez *f* FÍS RAD, ÓPT, QUÍMICA translucence

translúcido *adj* FÍS RAD, ÓPT, QUÍMICA translucent

transluciente *adj* FÍS RAD, ÓPT, QUÍMICA translucent

transmisibilidad *f* ACÚST, TELECOM transmissibility

transmisión[1]: **de ~ directa** *adj* MECÁ gearless

transmisión[2] *f* AUTO, ELEC, ENERG RENOV, FÍS, GAS transmission (*TX*), INFORM&PD transmission (*TX*), sending, ING ELÉC, ING MECÁ shafting, transmission (*TX*), MECÁ, ÓPT, TEC PETR transmission (*TX*), TELECOM transmittance, keying, TERMOTEC, TV, VEH transmission (*TX*); **~ accesoria** *f* TRANSP AÉR accessory drive; **~ acústica** *f* TELECOM sound transmission; **~ por almacenamiento y reexpedición** *f* TELECOM store-and-forward transmission; **~ analógica** *f* TELECOM analog transmission; **~ de ángulo** *f* MECÁ angle drive; **~ anisócrona** *f* INFORM&PD anisochronous transmission; **~ asíncrona** *f* INFORM&PD, TELECOM asynchronous transmission; **~ asincrónica** *f* INFORM&PD asynchronous transmission; **~ automática** *f* AUTO, MECÁ, TELECOM, VEH automatic transmission; **~ por balancín** *f* ING MECÁ beam drive; **~ de banda ancha** *f* TELECOM, TV wideband transmission; **~ por una banda lateral** *f* TELECOM, TV single-sideband transmission; **~ en banda lateral doble** *f* TELECOM, TV double-sideband transmission; **~ por banda lateral única** *f* FÍS, TELECOM, TV single-sideband transmission; **~ de bloques** *f* TELECOM block transmission; **~ por cable** *f* TV cable transmission, cablecast, wired broadcasting; **~ por cadena** *f* ING MECÁ, MECÁ, VEH chain drive; **~ de cadena y rueda dentada** *f* AUTO chain and sprocket drive; **~ de calor** *f* P&C, REFRIG heat transfer, TERMOTEC heat transmission; **~ de la cámara** *f* CINEMAT, TV camera drive; **~ codificada** *f* TELECOM coded transmission; **~ coherente** *f* ÓPT, TELECOM coherent transmission; **~ sin conexiones** *f* TELECOM connectionless mode transmission; **~ por conos de fricción** *f* ING MECÁ cone gear; ~ **por correa** *f* ING MECÁ, MECÁ, PROD belt drive; ~ **de correa trapezoidal** *f* ING MECÁ V-belt drive; ~ **por correa trapezoidal múltiple** *f* ING MECÁ multiple V-belt drive; ~ **de correa en V** *f* ING MECÁ V-belt drive; ~ **por correa en V múltiple** *f* ING MECÁ multiple V-belt drive; ~ **por corredera** *f* ING MECÁ link gear; ~ **por cremallera** *f* TRANSP rack gearing; ~ **de datos** *f* INFORM&PD data transmission, data communication, TELECOM data link; ~ **de datos simétricos de anulación de mando para varios puntos** *f* TELECOM multipoint command negating MCS (*MCN*); ~ **de datos simétricos de mando para varios puntos** *f* TELECOM multipoint command symmetrical data transmission; ~ **despejada** *f* TELECOM clear transmission; ~ **por desplazamiento coherente de fase** *f* TELECOM coherent phase-shift keying (*CPSK*); ~ **por desplazamiento de fase cuaternaria** *f* TELECOM phase-shaped QPSK; ~ **digital** *f* TELECOM digital transmission; ~ **digital comprimida** *f* TELECOM compressed digital transmission; ~ **directa** *f* AUTO positive drive, ING MECÁ direct drive; ~ **de doble portadora** *f* TELECOM dual-carrier transmission; ~ **con dos bandas laterales** *f* TV double-sideband transmission; ~ **de dos velocidades** *f* AUTO two-speed final drive; ~ **con ejes de acero en compresión** *f* ING MECÁ compressed steel shafting; ~ **por ejes y engranajes** *f* AUTO positive drive; ~ **eléctrica** *f* ELEC *red de distribución* electric linkage (*AmE*), electric transmission (*BrE*); ~ **de electricidad** *f* ELEC, ELECTRÓN, TERMO electricity transmission; ~ **de elevada potencia** *f* TELECOM high-power transmission; ~ **de energía** *f* ING ELÉC power transmission, transmission of energy, TERMO energy transmission; ~ **de energía de alta tensión en CC** *f* ING ELÉC DC high-tension power transmission; ~ **de energía por correa** *f* ING MECÁ power transmission by belt drive; ~ **de energía eléctrica** *f* ELEC, ING ELÉC, TERMO electric-power transmission; ~ **por engranajes** *f* CINEMAT sprocket drive, ING MECÁ, MECÁ gear drive; ~ **de engranajes deslizantes** *f* AUTO sliding-gear transmission; ~ **de fibra óptica** *f* ING ELÉC, ÓPT, TELECOM fiber-optic transmission (*AmE*), fibre-optic transmission (*BrE*), optical fiber transmission (*AmE*), optical fibre transmission (*BrE*); ~ **por fibras ópticas** *f* ÓPT optical fiber transmission (*AmE*), optical fibre transmission (*BrE*); ~ **final** *f* AUTO final drive; ~ **flexible** *f* ING MECÁ flexible joint; ~ **por un fluido** *f* MECÁ fluid drive; ~ **por fricción** *f* CINEMAT friction drive, ING MECÁ friction drive, friction gear; ~ **hidráulica** *f* ING MECÁ hydraulic drive, MECÁ fluid drive; ~ **hidrostática** *f* MECÁ hydrostatic transmission; ~ **Hotchkiss** *f* VEH Hotchkiss drive; ~ **de imagen** *f* TELECOM image transmission; ~ **independiente** *f* ING MECÁ individual drive; ~ **intermedia** *f* ING MECÁ counterdriving motion, countergear, countershaft, countershafting; ~ **isócrona** *f* INFORM&PD isochronous transmission; ~ **de una llamada terminal** *f* TELECOM terminal call forwarding; ~ **manual** *f* AUTO manual transmission; ~ **con matrices** *f* TV matrixing; ~ **mecánica** *f* ENERG RENOV mechanical transmission, ING MECÁ mechanical drive, mechanical transmission; ~ **de mensaje** *f* TEC ESP, TELECOM pass-along message (*PAM*); ~ **de mensajes hablados** *f* TELECOM voice messaging; ~ **del movimiento lateral** *f* VEH *diferencial* side gear;

~ **múltiple digital** *f* ELECTRÓN digital multiplex, digital multiplexing; ~ **múltiple con división de tiempo** *f* (*TMDT*) ELECTRÓN time-division modulation (*TDM*), time-division multiplexing (*TDM*); ~ **en múltiplex** *f* TV multiplex transmission; ~ **neutral** *f* INFORM&PD neutral transmission; ~ **de la onda** *f* TELECOM wave transmission; ~ **de paquetes** *f* INFORM&PD packet transmission; ~ **por paquetes** *f* INFORM&PD, ING ELÉC packet switching, TELECOM packet transmission, packet switching; ~ **en paralelo** *f* INFORM&PD, TELECOM parallel transmission; ~ **de parámetros** *f* INFORM&PD parameter passing; ~ **con portadora interrumpida** *f* TELECOM voice-operated transmission; ~ **por portadora suprimida** *f* ELECTRÓN suppressed-carrier transmission; ~ **de potencia máxima** *f* ING ELÉC maximum power transmission; ~ **principal** *f* TEC PETR *mecanismos* transmission main; ~ **de prueba** *f* TV test transmission; ~ **rápida por desplazamiento de frecuencia** *f* TELECOM fast frequency-shift keying; ~ **por rueda de fricción** *f* ING MECÁ friction gearing; ~ **por satélite** *f* TELECOM satellite transmission; ~ **secreta** *f* TELECOM scrambled message, scrambled transmission, scrambling; ~ **semiautomática** *f* AUTO VEH semiautomatic transmission; ~ **de la señal** *f* TELECOM signal transmission; ~ **de señales** *f* TEC ESP, TELECOM signaling (*AmE*), signalling (*BrE*); ~ **en serie** *f* INFORM&PD serial transmission; ~ **simultánea** *f* TEC ESP, TELECOM multiplexing; ~ **simultánea de datos** *f* ELECTRÓN, TELECOM data multiplexing; ~ **simultánea por división del tiempo** *f* (*TDM*) FÍS, INFORM&PD, TEC ESP, TELECOM time-division modulation (*TDM*), time-division multiplexing (*TDM*); ~ **simultánea de MCI** *f* ELECTRÓN, TELECOM PCM multiplexing; ~ **simultánea de muchas informaciones** *f* TEC ESP, TELECOM multiplexing; ~ **simultánea sobre la misma onda** *f* TEC ESP, TELECOM *radio, TV* multiplex; ~ **simultánea sobre el mismo hilo** *f* TEC ESP, TELECOM *telefonía, telegrafía alámbrica* multiplex; ~ **simultánea óptica** *f* ELECTRÓN, TV optical multiplexing; ~ **simultánea de la señal** *f* ELECTRÓN, TELECOM signal multiplexing; ~ **simultánea de varias longitudes de onda** *f* ÓPT wavelength multiplexing; ~ **síncrona** *f* INFORM&PD, TELECOM synchronous transmission; ~ **sincrónica** *f* CINEMAT synchronous drive, INFORM&PD clocking; ~ **sincrónica binaria** *f* INFORM&PD binary synchronous transmission (*BISYNC, BSC*); ~ **sincronizada** *f* AUTO synchromesh transmission, synchronized transmission; ~ **de sonido** *f* ACÚST, TELECOM sound transmission; ~ **del sonido** *f* ACÚST sound transmission; ~ **sonora** *f* TELECOM sound transmission; ~ **sucesiva de señales** *f* TV time multiplex; ~ **superior** *f* ING MECÁ overhead transmission; ~ **térmica** *f* TERMOTEC thermal transmission; ~ **por tornillo sinfín** *f* AUTO worm gear final drive; ~ **transparente** *f* INFORM&PD data link escape (*DLE*), transparent transmission; ~ **de velocidad por correa trapezoidal** *f* ING MECÁ V-belt speed transmission; ~ **de video** *AmL*, ~ **de vídeo** *f* *Esp* TELECOM video transmission

transmisor *m* ACÚST, ELECTRÓN, FÍS, FÍS ONDAS, ING ELÉC transmitter, ING MECÁ drive, driving, INSTAL HIDRÁUL, TEC ESP transmitter, TELECOM transmission sender, sender, transmitter, TRANSPAÉR, TRANSP MAR, TV transmitter; ~ **Argos** *m* TRANSP MAR *investigación* *oceanográfica y del medio ambiente* Argos transmitter; ~ **de correa plana** *m* ING MECÁ flat belt drive; ~ **direccional de un haz** *m* ELECTRÓN, TELECOM, TV directional beam transmitter; ~ **fibroóptico** *m* ING ELÉC, ÓPT, TELECOM fiber-optic transmitter (*AmE*), fibre-optic transmitter (*BrE*); ~ **de fuerza motriz** *m* ING MECÁ transmitter of motive power; ~ **de imágenes** *m* TV pick-up transmitter; ~ **del indicador del combustible** *m* AUTO, TRANSP, TRANSP AÉR, VEH fuel gage transmitter (*AmE*), fuel gauge transmitter (*BrE*); ~ **de localización de siniestros** *m* TRANSP MAR *para aeronaves* emergency-locator transmitter (*ELT*); ~ **localizador de emergencia** *m* TELECOM emergency locator transmitter (*ELT*); ~ **de modem** *m* ELECTRÓN modem transmitter; ~ **del nivel de combustible** *m* TRANSP AÉR fuel gage transmitter (*AmE*), fuel gauge transmitter (*BrE*), fuel-level transmitter; ~ **de paso de la pala** *m* TRANSP AÉR blade-pitch transmitter; ~ **de película** *m* CINEMAT, TV film transmitter; ~ **de polea escalonada** *m* ING MECÁ cone-pulley drive; ~ **con portadora suprimida** *m* ELECTRÓN suppressed-carrier transmitter; ~ **portátil** *m* TELECOM, TV portable transmitter; ~ **de presión** *m* INSTAL HIDRÁUL pressure transmitter; ~ **de radiodifusión** *m* TELECOM broadcast transmitter; ~ **de radiodifusión de sonido** *m* TELECOM sound broadcast transmitter; ~**-receptor** *m* (*TR*) ELECTRÓN transmitter-receiver, FÍS transponder, INFORM&PD sender-receiver, TELECOM, TRANSP MAR, TV transmitter-receiver; ~**-receptor mediante teclado** *m* INFORM&PD keyboard send-receive (*KSR*); ~**-receptor multifrecuencia** *m* TELECOM MF sender-receiver; ~**-receptor a teclado** *m* (*KSR*) INFORM&PD keyboard send-receive (*KSR*); ~ **de rejilla** *m* ELECTRÓN mask carrier; ~ **repetidor** *m* TELECOM, TV relay transmitter; ~ **de socorro de baja potencia** *m* TELECOM low-power distress transmitter (*LPDT*); ~ **telefónico** *m* ACÚST, TELECOM telephone transmitter; ~ **de televisión** *m* ELECTRÓN, FÍS, ING ELÉC, TELECOM, TV television transmitter

transmitancia *f* FÍS, FÍS ONDAS, ÓPT, PROC QUÍ, TELECOM transmittance; ~ **espectral** *f* FÍS spectral transmittance; ~ **del filtro** *f* PROC QUÍ filter transmittance

transmite: que no ~ *adj* TV off-air

transmitir *vt* FÍS impart, FÍS ONDAS broadcast, INFORM&PD send, transmit, ING MECÁ impart, drive, TRANSP MAR *radio* transmit; ~ **por cable coaxial** *vt* TV pipe; ~ **energía a** *vt* FÍS, TERMO impart energy to; ~ **simultáneamente** *vt* ELECTRÓN, TELECOM multiplex; ~ **por submúltiplex** *vt* TELECOM submultiplex

transmultiplexor *m* TELECOM transmultiplexer (*TMUX*)

transmutación *f* FÍS, FÍS RAD, NUCL transmutation; ~ **radiactiva** *f* FÍS RAD radioactive transmutation

transordenador *m* *Esp* (*cf transcomputadora AmL*) INFORM&PD transputer

transparencia *f* CINEMAT back projection, FÍS RAD translucence, FOTO, IMPR, P&C transparency, TELECOM *red* transparency; ~ **nubosa** *f* PAPEL nepheloid transparency

transparente[1] *adj* METEO clear, PAPEL look-through, TELECOM transparent

transparente[2]: ~ **irregular** *m* PAPEL wild formation; ~ **nuboso** *m* PAPEL wild lookthrough; ~ **a los rayos Ultra Violeta** *m* C&V UV-transmitting

transpiración: ~ **de gases** *f* GAS transpiration of gases

transplantador *m* AGRIC garden trowel

transplante: ~ **de macollos** *m* AGRIC tufting

transpondedor *m* TEC ESP, TELECOM transponder; ~ **para espacio profundo** *m* TEC ESP *comunicaciones* deep-space transponder; ~ **interrogador** *m* TELECOM interrogator transponder

transportado: ~ **por agua** *adj* QUÍMICA waterborne; ~ **por mar** *adj* TRANSP MAR seaborne

transportador *m* AGRIC, C&V conveyor, CARBÓN follower, CONST conveyor, ELECTRÓN carrier, EMB conveyor, GEOM protractor, ING MECÁ, INSTR conveyor, METR *mecánica* protractor, PROD conveyor, TEC ESP, TEXTIL *maquinaria* carrier; ~ **accionado por motor** *m* PROD motor-driven conveyor; ~ **acústico** *m* ELECTRÓN acoustic carrier; ~ **aéreo** *m* EMB aerial conveyor, PROD overhead runway; ~ **de ángulos** *m* TRANSP MAR *navegación* protractor; ~ **de banda articulada** *m* EMB, ING MECÁ apron conveyor, PROD traveling apron (*AmE*), travelling apron (*BrE*), TRANSP apron conveyor; ~ **de banda sinfín** *m* AGRIC band conveyor; ~ **de bandejas** *m* *Esp* (*cf transportador de pálets AmL*) AGRIC pallet truck; ~ **de botellas** *m* EMB bottle carrier; ~ **de cable** *m* PAPEL cable conveyor; ~ **por cable** *m* FERRO, ING MECÁ, TRANSP cable railway; ~ **de cadena** *m* EMB apron conveyor, ING MECÁ chain conveyor; ~ **de cangilones** *m* CONST bucket conveyor; ~ **de carga** *m* MINAS load conveyor, TRANSP loading conveyor; ~ **de carga externa** *m* TRANSP AÉR external-load carrier; ~ **de cinta** *m* AGRIC, EMB, ING MECÁ, MINAS, PROD belt conveyor; ~ **de cinta cóncava** *m* PROD trough conveyor; ~ **de cinta cóncava parabólico** *m* (*PTC*) INSTR parabolic trough conveyor (*PTC*); ~ **circular** *m* GEOM circular protractor; ~ **de combustible** *m* MINAS bunker conveyor; ~ **de correa** *m* AGRIC, EMB, ING MECÁ, MINAS, PROD belt conveyor; ~ **de correa con polea guía** *m* ING MECÁ belt conveyor with carrying idlers; ~ **de correa sinfín** *m* EMB, PROD band conveyor; ~ **de cuba basculante** *m* TRANSP tipping bucket conveyor; ~ **de cubetas** *m* PROD trough conveyor; ~ **elevado** *m* EMB overhead conveyor, aerial conveyor; ~ **de estera** *m* ING MECÁ *equipo continuo de transporte* apron conveyor; ~ **para el frente de arranque blindado** *m* *AmL* (*cf panzer Esp*) MINAS armored face conveyor (*AmE*), armoured face conveyor (*BrE*); ~ **de graneles** *m* CONST bulk carrier; ~ **de hilos** *m* TEXTIL yarn carrier; ~ **con listones** *m* PROD flight conveyor; ~ **de mandil** *m* ING MECÁ *equipo continuo de transporte*, TRANSP apron conveyor; ~ **mediante tubería neumática** *m* TRANSP pneumatic-pipe conveyor; ~ **monocarril** *m* MINAS, TRANSP monorail conveyor; ~ **monocarril eléctrico aéreo** *m* ING MECÁ overhead electrical monorail conveyor; ~ **monorrail** *m* MINAS, TRANSP monorail conveyor; ~ **en muelle** *m* TRANSP quayside conveyor; ~ **neumático** *m* TRANSP airlift; ~ **de nivel universal** *m* METR universal level protractor; ~ **de objetos apilados** *m* EMB stacking conveyor; ~ **oscilante** *m* ING MECÁ oscillating conveyor; ~ **de paletas** *m* *Esp* (*cf transportador de pálets AmL*) AGRIC pallet truck, PROD flight conveyor, push-plate conveyor; ~ **con paletas rascadoras** *m* MINAS scraper chain conveyor, PROD scraper conveyor; ~ **de pálets** *m* *AmL* (*cf transportador de bandejas Esp*) AGRIC pallet truck; ~ **peatonal** *m* TRANSP *convoy turístico* pedestrian conveyor; ~ **de pinzas e hilado** *m* TEXTIL grippers and yarn carrier; ~ **de plataforma** *m* CONST apron conveyor; ~ **para plateado** *m* C&V conveyor for silvering; ~ **de raederas** *m* MINAS scraper chain conveyor, PROD scraper conveyor; ~ **de rasquetas** *m* MINAS scraper chain conveyor, PROD scraper conveyor; ~ **de rastras** *m* PROD flight conveyor; ~ **de rodillo por gravedad** *m* EMB gravity-roller conveyor; ~ **de rollizos** *m* PROD log conveyer; ~ **de sacos** *m* TRANSP bag conveyor; ~ **por sacudidas** *m* PROD shaker conveyor; ~ **con soportes colgantes** *m* PROD sling hanger; ~ **de tornillo** *m* ING MECÁ, PAPEL screw conveyor; ~ **de tornillo sin fin** *m* EMB, ING MECÁ, PROD screw conveyor; ~ **de tornillo helicoidal** *m* PROD screw conveyor; ~ **a tornillo sinfin** *m* AGRIC auger; ~ **vertical** *m* CONST, MECÁ elevator (*AmE*), lift (*BrE*); ~ **vertical de cangilones** *m* MINAS dredge elevator; ~ **vertical de cargas unitarias** *m* PROD unit load vertical conveyor; ~ **vibratorio** *m* ALIMENT vibration conveyor

transportar[1] *vt* CONST carry, HIDROL *cuerpos flotantes* transport, ING MECÁ carry, MECÁ lift, TELECOM, TEXTIL carry, TRANSP transport, remove; ~ **en recipientes cerrados** *vt* TRANSP containerize; ~ **por tren y carretera** *vt* TRANSP transport by rail and road

transportar[2]: ~ **cargas con seguridad** *vi* SEG carry the load safely; ~ **fuerza motriz** *vi* ING MECÁ convey power

transporte *m* ELECTRÓN *aritmética* carry, EMB carriage, INFORM&PD transport, MINAS *de obreros* transportation, TEXTIL carrying; ~ **A** *m* TRANSP *que contiene motor* carriage A; ~ **acuático** *m* TRANSP MAR water transport; ~ **acuático interior** *m* TRANSP inland water transport; ~ **de agua** *m* TRANSP water transport, *aljibes* water carriage; ~ **bajo temperatura controlada** *m* REFRIG transport under controlled temperature; ~ **en barca** *m* TRANSP boat carriage; ~ **en barco** *m* TRANSP boat carriage; ~ **en buque** *m* TRANSP boat carriage; ~ **de cabestrillo** *m* TRANSP AÉR sling transport; ~ **por canal** *m* TRANSP canal transport; ~ **de carga en bandeja** *m* Esp (*cf transporte de carga en palét AmL*) EMB, TRANSP palletized-cargo carrier; ~ **de carga en palét** *m* AmL (*cf transporte de carga en bandeja Esp*) EMB, TRANSP palletized-cargo carrier; ~ **por carretera** *m* AUTO road haulage, road transport, CONST road transport, TRANSP, VEH road haulage, road transport; ~ **en carretilla** *m* CONST barrowing; ~ **de cercanías** *m* TRANSP short-distance transport; ~ **a cielo abierto** *m* MINAS surface conveyance; ~ **de cinta** *m* INFORM&PD tape transport; ~ **combinado** *m* REFRIG intermodal transport, TRANSP combination bulk carrier; ~ **combinado crudo-mineral** *m* TRANSP oil-ore carrier (*OO carrier*); ~ **combinado petróleo-mineral** *m* TRANSP oil-ore carrier (*OO carrier*); ~ **continuo** *m* TRANSP continuous transport; ~ **entre ciudades** *m* TRANSP intercity transport; ~ **de eslinga** *m* TRANSP AÉR sling transport; ~ **por ferrocarril de vía estrecha** *m* FERRO *vehículos* light rail transport (*BrE*) (*LRT*); ~ **frigorífico** *m* ALIMENT, REFRIG, TRANSP refrigerated transport; ~ **en gabarras** *m* TRANSP MAR lighterage; ~ **de gas de larga distancia** *m* TRANSP long-distance gas transport; ~ **de gas natural licuado** *m* TRANSP liquid-natural-gas carrier (*LNG carrier*); ~ **por helicóptero** *m* TRANSP AÉR heli-lifting, TRANSP MAR transport by

helicopter; ~ **hidráulico** *m* AGUA sluicing; ~ **industrial** *m* TRANSP industrial carrier; ~ **interurbano** *m* TRANSP intercity transport; ~ **de larga distancia** *m* TRANSP long-haul carriage; ~ **a largas distancias** *m* TRANSP long-range transport; ~ **litográfico** *m* IMPR retransfer; ~ **por mar** *m* EMB carriage by sea, TRANSP MAR sea carriage, carriage by sea; ~ **marítimo** *m* TRANSP MAR sea transport; ~ **marítimo combinado** *m* TRANSP *petróleo, granel, mineral* oil-bulk-ore carrier (*OBO carrier*); ~ **de material petrolífero** *m* TRANSP oil-ore carrier (*OO carrier*); ~ **de mercancías** *m* TRANSP merchant haulage; ~ **de mercancías a granel** *m* EMB, TRANSP bulk transport; ~ **mercante** *m* TRANSP merchant haulage; ~ **de metano con tanques autoportantes** *m* TRANSP methane carrier with self-supporting tank; ~ **mixto** *m* TRANSP *petróleo, carbón, mineral* oil-coal-ore carrier (*OCO carrier*); ~ **mixto a granel** *m* TRANSP combination bulk carrier; ~ **mixto por mar** *m* TRANSP *petróleo, granel, mineral* oil-bulk-ore carrier (*OBO carrier*); ~ **neumático de materiales a granel** *m* ING MECÁ pneumatic conveying of bulk materials; ~ **por oleoducto** *m* TRANSP pipeline transportation; ~ **de pasajeros** *m* TRANSP passenger transport; ~ **de la película** *m* CINEMAT, FOTO, TV film transport; ~ **privado rápido** *m* TRANSP personal rapid transport (*PRT*); ~ **de puerta a puerta** *m* TRANSP point-to-point transport; ~ **puerta a puerta** *m* FERRO door-to-door delivery; ~ **rápido automático** *m* TRANSP rapid automatic transport; ~ **de remolques en vagón plataforma** *m* TRANSP piggyback transport; ~ **por rodillo** *m* CINEMAT, TRANSP roller transport; ~ **de sedimentos** *m* HIDROL bottom transport; ~ **de sólidos del fondo** *m* HIDROL bed-load transport; ~ **terrestre de alta velocidad** *m* TRANSP high-speed ground transportation; ~ **de troncos cortados** *m* PROD logging; ~ **de tropas acorazado** *m* D&A armored personnel carrier (*AmE*), armoured personnel carrier (*BrE*) (*APC*); ~ **por tubería** *m* TRANSP *de conducción* pipeline transportation; ~ **por tubería a baja presión** *m* TRANSP transport in low-pressure tube; ~ **por tuberías** *m* CONST pipage; ~ **tubular** *m* TRANSP tubular transportation; ~ **urbano sin paradas** *m* TRANSP nonstop urban transportation; ~ **con vagón de carga parcial** *m* FERRO less-than-carload freight shipment (*AmE*), less-than-carload goods shipment (*BrE*); ~ **por vía férrea** *m* FERRO, TRANSP rail transport; ~ **por vía marítima** *m* EMB, TRANSP MAR carriage by sea; ~ **por vías navegables interiores** *m* TRANSP inland water transport

transportes: ~ **y comunicaciones** *m pl* TRANSP transport and communications

transportista *m* AUTO road hauler (*AmE*), road haulier (*BrE*), TEC ESP carrier, TRANSP *oleoductos* pipeliner, *relacionado con una red de distribución* loader, road hauler (*AmE*), road haulier (*BrE*), TRANSP MAR sea carrier, VEH road hauler (*AmE*), road haulier (*BrE*); ~ **de mercancías en grupaje** *m* FERRO *vehículos* groupage traffic forwarder

transposición *f* ACÚST, ELEC *de cables aislados* transposition; ~ **de Beckmann** *f* QUÍMICA Beckmann rearrangement

transuránico *adj* FÍS, QUÍMICA transuranic

transvase: ~ **de combustible** *m* TRANSP AÉR fuel transfer

transversal¹ *adj* TRANSP MAR *diseño de barcos* transverse

transversal²: ~ **eléctrico** *m* (*TE*) ING ELÉC transverse electric (*TE*)

transversal³ *f* GEOM transversal, MINAS *Esp* (*cf galería de labor atravesada AmL*) offset drive, (*cf pasadizo AmL*) *túnel* passageway

transversalmente *adv* IMPR crosswise

tranvía *m* TRANSP streetcar (*AmE*), tram (*BrE*), tramway; ~ **metropolitano** *m* TRANSP tramway metro; ~ **subterráneo** *m* TRANSP underground tramway; ~ **urbano** *m* TRANSP tramway metro

trapa: ~ **de la botavara** *f* TRANSP MAR *brazo de izado* kicking strap, *del yate* boom vang

trapecio *m* GEOM trapezium

trapezoidal *adj* GEOM trapezoidal

trapezoide *m* GEOM trapezium, trapezoid

trapo *m* PAPEL *materia prima* rag, TEC ESP *navegación acuática* cloth, TRANSP MAR *vela, cuerda* sheet, *velamen* canvas

traquetear *vi* ING MECÁ clatter

traquiandesita *f* PETROL trachyandesite

traquibasalto *m* PETROL trachybasalt

traquita *f* PETROL basalt glass, trachyte

trasdós *m* CONST extrados, *arco, muro* back

trasiego *m* ALIMENT racking, REFRIG bleed-off

traslación *f* FÍS *desplazamiento*, GEOM, ING MECÁ translation, PROD *de puente-grúa* traveling (*AmE*), travelling (*BrE*), traversing, TEC PETR *movimientos de equipos flotantes* heave

trasladar¹ *vt* HIDROL remove, INFORM&PD transfer

trasladar²: ~ **el carro de un puente grúa** *vi* ING MECÁ rack the carriage of a traveling crane (*AmE*), rack the carriage of a travelling crane (*BrE*); ~ **fuerza motriz** *vi* ING MECÁ convey power

traslado *m* INFORM&PD transfer, ING MECÁ removal, movement, transfer, translation

traslape *m* C&V lap

trasmallo *m* OCEAN *tecnología de pesca* flue net, trammel net

traspaso: ~ **de acciones** *m* PROD stock transfer; ~ **de energía** *m* GEOFÍS energy transfer

trasplaya *f* OCEAN backshore

tratado: ~ **criogénicamente** *adj* ING MECÁ cryogenically treated; ~ **con película antirreflectora** *adj* C&V *lentes* bloomed; ~ **térmicamente** *adj* TERMO *metales* heat-treated; ~ **al vapor** *adj* TRANSP MAR *madera* steamed

tratamiento *m* AGUA, C&V treatment, CARBÓN *aceros* air conditioning, CINEMAT, CONST treatment, CONTAM *de desperdicios*, DETERG, HIDROL treatment, conditioning, INFORM&PD, ING MECÁ handling, NUCL, TEC PETR, TEXTIL treatment; ~ **de acontecimientos** *m* INFORM&PD event handling; ~ **adicional** *m* PROD retreat; ~ **del agua** *m* AGUA water conditioning, NUCL, PROC QUÍ water treatment; ~ **del agua de alimentación** *m* HIDROL feedwater treatment; ~ **del agua de calderas** *m* INSTAL TERM boiler-water treatment; ~ **de aguas** *m* CARBÓN water treatment; ~ **de aguas negras** *m* AmL (*cf tratamiento de aguas residuales Esp*) CONTAM wastewater treatment; ~ **de aguas residuales** *m* Esp (*cf tratamiento de aguas negras AmL*) CONTAM, HIDROL, RECICL wastewater treatment; ~ **del aire** *m* REFRIG air treatment; ~ **por aire comprimido** *m* CARBÓN pneumatic treatment; ~ **alcalino** *m* CONST *suelos*

liming, DETERG alkali treatment; ~ **anaeróbico** *m* RECICL anaerobic treatment; ~ **antirreflector** *m* C&V antireflecting treatment; ~ **de la basura** *m* RECICL waste processing, waste treatment; ~ **biológico** *m* CONTAM biological treatment; ~ **biológico de agua** *m* AGUA biological water treatment; ~ **por calentamiento en el vacío** *m* METAL vacuum heat treatment; ~ **por cargas** *m* PROD batch process; ~ **con chatarra** *m* PROD *fabricación de acero* scrap process; ~ **de datos electrónico** *m* (*TDE*) ELECTRÓN electronic data processing (*EDP*); ~ **de depuración de los efluentes** *m* RECICL effluent-purification process; ~ **de depuración de las emanaciones** *m* RECICL effluent-purification process; ~ **de desechos** *m* RECICL waste treatment; ~ **digital** *m* ELECTRÓN digital processing; ~ **de los efluentes** *m* RECICL waste disposal; ~ **final** *m* EMB final treatment, NUCL tailend process, PROD last-stage treatment; ~ **físico del agua** *m* AGUA physical water treatment; ~ **con gas** *m* QUÍMICA *por metal* gassing; ~ **del haz de electrones** *m* ELECTRÓN electron-beam processing; ~ **de hidrofugación** *m* CONST damp-proof course; ~ **húmedo** *m* CARBÓN, ING MECÁ wet treatment; ~ **de la imagen** *m* ELECTRÓN image processing; ~ **de imágenes** *m* ELECTRÓN image processing; ~ **de impermeabilización** *m* CONST damp-proof course; ~ **inencogible** *m* TEXTIL antishrink treatment, nonshrink treatment; ~ **de la información** *m* (*PI*) ELECTRÓN, INFORM&PD information processing (*IP*); ~ **de la información en microplaca** *m* ELECTRÓN on-chip processing; ~ **por irradiación** *m* FÍS RAD, NUCL radiation processing; ~ **por láser** *m* ELECTRÓN laser treatment, lasing; ~ **de llamadas** *m* TELECOM call handling; ~ **del lodo** *m* RECICL sludge processing; ~ **por lotes** *m* PROD batch process; ~ **con material antirreflector** *m* CINEMAT blooming; ~ **con mineral** *m* PROD ore process; ~ **de minerales** *m* MINAS ore treatment; ~ **de minerales de uranio** *m* NUCL uranium milling; ~ **mixto de aguas negras y residuales** *m* AGUA mixed sewage and waste water treatment; ~ **de la pastilla** *m* ELECTRÓN *semiconductores* wafer processing; ~ **por precipitación** *m* METAL, TERMO *metales* artificial ageing; ~ **preliminar** *m* AGUA preliminary treatment, CARBÓN pretreatment; ~ **previo** *m* CARBÓN pretreatment; ~ **primario** *m* HIDROL *de aguas residuales* primary treatment; ~ **por el procedimiento húmedo-seco** *m* TEXTIL wet-dry processing treatment; ~ **químico** *m* CARBÓN, QUÍMICA chemical treatment; ~ **químico de agua** *m* AGUA, HIDROL, QUÍMICA chemical water treatment; ~ **de residuos** *m* RECICL waste treatment; ~ **de sedimentos** *m* HIDROL *aguas residuales* sludge conditioning; ~ **de la señal** *m* ELECTRÓN signal processing; ~ **de señal de ámbito de frecuencia** *m* ELECTRÓN frequency-domain signal processing; ~ **de señal digital** *m* (*TSD*) ELECTRÓN digital signal processing (*DSP*); ~ **de la señal en dominio temporal** *m* ELECTRÓN time-domain signal processing; ~ **de señal electrónica** *m* ELECTRÓN electronic-signal processing; ~ **de señal electroóptico** *m* ELECTRÓN, ÓPT electro-optical signal processing; ~ **de señal óptica** *m* ELECTRÓN optical signal processing; ~ **de la señal en tiempo real** *m* ELECTRÓN, INFORM&PD real-time signal processing, TELECOM real-time processing, real-time signal processing, TV real-time signal processing; ~ **superficial** *m* PAPEL surface application; **~ superficial bituminoso doble** *m* CONST double bituminous surface treatment (*DBST*); **~ superficial bituminoso simple** *m* CONST *carreteras* single bituminous surface treatment (*SBST*); **~ de la superficie** *m* C&V surface treatment, CONST *obra civil* surfacing (*AmE*); **~ térmico** *m* CARBÓN, METAL, TERMO heat treatment; **~ de texto** *m* (*WP*) INFORM&PD text processing, word processing (*WP*); **~ de textos con opciones gráficas** *m* INFORM&PD desktop publishing (*DTP*); **~ de urgencia** *m* SEG emergency treatment; **~ de vidrio** *m* C&V, ELECTRÓN glass passivation

tratante: **~ de ganado** *m* AGRIC animal dealer
tratar¹: **sin ~** *adj* HIDROL *agua* untreated
tratar² *vt* CARBÓN *aceros, aire* condition, treat, PROD handle, QUÍMICA process; **~ con cloro** *vt* DETERG, HIDROL, QUÍMICA chlorinate; **~ de nuevo** *vt* CARBÓN, PROD retreat; **~ con silicio** *vt* ELECTRÓN commit to silicon
trauma: **~ acústico** *m* ACÚST acoustic trauma
traverso: **~ de viga maestra** *m* CONST *edificación* summer tree
travertino *m* ENERG RENOV travertine, GEOL calcrete
través¹: **a ~ del objetivo** *adj* CINEMAT through-the-lens
través²: **por el ~** *adv* TEC ESP abeam, TRANSP MAR abeam, abreast; **a ~ de proa** *adv* TRANSP MAR across the bow
través³ *m* TRANSP MAR *amarres* breast line; **~ de popa** *m* TRANSP MAR *para rescate* breeches buoy; **~ de proa** *m* TRANSP MAR *para rescate* breeches buoy
través⁴: **a ~ de las caras** *fra* ING MECÁ across flats
travesaño *m* CONST bolster, crossbeam, transom, bridging piece, crossbar, rail, crosspiece, header, *arquitectura* cap, traverse, ING MECÁ headstock, MECÁ spreader, MINAS *vagonetas* headstock, cap piece, PROD *marco de fundición* stay, cross bearer, *de lecho de torno* crossbar, TRANSP AÉR *edificación* crossbar, TRANSP MAR crossbeam, VEH *carrocería* cross member; **~ del armazón** *m* ING MECÁ frame cross-member, frame crossbeam; **~ del bastidor** *m* ING MECÁ frame cross-member; **~ de encuentro** *m* CONST checkrail; **~ de escala de relés** *m* PROD relay ladder rung; **~ guía** *m* ING MECÁ guide cross tie; **~ intermedio** *m* CONST *puertas* middle rail; **~ portapolea** *m* TEC PETR crown block; **~ posterior** *m* FERRO *vehículos, locomotoras* drawbar; **~ de tope** *m* FERRO *vehículos* buffer beam
travesaños: **~ diagonales** *m pl* MINAS dividers
travesero *m* ING MECÁ cleat
travesia *f* FERRO splice bar
travesía *f* TRANSP MAR *mar* crossing, *navegación* passage, *viaje marítimo* sailing, yachting; **~ de ida** *f* TRANSP MAR outward passage; **~ de regreso** *f* TRANSP MAR homeward passage
traviesa *f* FERRO cross member, *equipo inamovible* tie (*AmE*), *vehículos* tie bar, sleeper (*BrE*), TRANSP track girder, VEH *carrocería* cross member; **~ del armazón** *f* ING MECÁ frame cross-member, frame crossbeam; **~ del bastidor** *f* ING MECÁ frame cross-member; **~ colocada en sentido longitudinal** *f* FERRO *equipo inamovible* stringer; **~ de concreto** *f* *AmL* (*cf traviesa de hormigón Esp*) CONST, FERRO concrete sleeper (*BrE*), concrete tie (*AmE*), *equipo inamovible* monobloc concrete sleeper (*BrE*), monobloc concrete tie (*AmE*); **~ en forma de T invertida** *f* TRANSP inverted

T-shaped track girder; **~ de hormigón** *f* *Esp* (*cf traviesa de concreto AmL*) CONST, FERRO concrete sleeper (*BrE*), concrete tie (*AmE*), *equipo inamovible* monobloc concrete sleeper (*BrE*), monobloc concrete tie (*AmE*); **~ hueca de sección rectangular** *f* TRANSP box-section track-girder; **~ hueca de vía** *f* TRANSP hollow-type track girder; **~ de madera** *f* *Esp* (*cf durmiente de madera AmL*) CONST *ferrocarril* wooden sleeper (*BrE*), wooden tie (*AmE*); **~ oscilante** *f* FERRO *equipo inamovible* dancing sleeper (*BrE*), dancing tie (*AmE*), pumping sleeper (*BrE*); **~ partida** *f* FERRO *equipo inamovible* split railroad tie (*AmE*), split sleeper (*BrE*); **~ de piso** *f* CONST girt; **~ del pivote del bogie** *f* FERRO *vehículos* bogie bolster (*BrE*), truck bolster (*AmE*); **~ del pivote del carretón** *f* FERRO bogie bolster (*BrE*), truck bolster (*AmE*); **~ de sección rectangular hueca** *f* TRANSP box-section track-girder; **~ de separación** *f* CONST *estructura* tie bar; **~ de vía en U** *f* TRANSP U-shaped track girder
trayecto *m* INFORM&PD, ING ELÉC path, ING MECÁ section, PROD path, TELECOM route, TRANSP AÉR path; **~ de aproximación segmentado** *m* TRANSP AÉR segmented approach path; **~ de la corriente** *m* ING ELÉC current path; **~ de la puesta a tierra** *m* ING ELÉC, PROD grounding path; **~ rápido de frecuencia** *m* ELECTRÓN fast frequency hopping; **~ recíproco** *m* TRANSP AÉR reciprocal leg
trayectografía *f* FÍS RAD *detectores de partículas*, TEC ESP trajectography
trayectoria *f* CINEMAT path, D&A trajectory, ELEC, FÍS path, trajectory, GEOL line of bearing, GEOM path, ING ELÉC trace, PROD path, QUÍMICA way, TEC ESP course, path, trajectory, TELECOM route, TRANSP, TRANSP AÉR, TRANSP MAR *del barco* path; **~ del agua** *f* CONTAM watercourse; **~ aleatoria** *f* *Esp* (*cf caminata aleatoria AmL*) INFORM&PD random walk; **~ de aproximación** *f* TRANSP AÉR approach path; **~ de aproximación final** *f* TRANSP AÉR final-approach path; **~ de aproximación inicial** *f* TRANSP AÉR initial approach path; **~ asignada de vuelo** *f* TRANSP AÉR assigned flight path; **~ de aspersión** *f* CONTAM MAR spraypath; **~ de aterrizaje** *f* TRANSP landing path; **~ balística** *f* D&A, TEC ESP ballistic trajectory; **~ de cita** *f* TEC ESP rendezvous trajectory; **~ de colisión** *f* TEC ESP collision course; **~ de la comprobación de la seguridad** *f* TELECOM *de transmisiones* security audit trail; **~ curvada de aproximación por acimuz** *f* TRANSP AÉR curved azimuth approach path; **~ descendente** *f* TEC ESP descent path; **~ de descenso** *f* TEC ESP descent orbit; **~ directa** *f* ING ELÉC forward path; **~ electrónica** *f* ELECTRÓN, NUCL electron trajectory; **~ de enhebrado** *f* CINEMAT threading path; **~ de enlace** *f* INFORM&PD linkage path; **~ de espera** *f* TRANSP AÉR holding path; **~ del haz de iluminación** *f* INSTR illumination beam path; **~ de la iluminación** *f* INSTR illumination path; **~ de impacto** *f* TEC ESP collision course; **~ media** *f* D&A mean trajectory; **~ mixta** *f* TEC ESP mixed path; **~ múltiple** *f* TEC ESP multiple path; **~ óptica** *f* TELECOM optical path; **~ óptima** *f* TELECOM optimal path; **~ ortodrómica** *f* TEC ESP great-circle path; **~ de planeo** *f* TRANSP glide path, TRANSP AÉR glide path, glide slope; **~ de planeo instrumental** *f* TEC ESP *aeropuertos* ILS glide path; **~ de planeo con instrumentos** *f* TEC ESP *aeropuertos*

ILS glide path; ~ **de planeo mínima** *f* TRANSP AÉR
minimum glide path; ~ **rasante** *f* D&A flat trajectory;
~ **de un rayo** *f* ELEC *conductor* lightning path; ~ **de
rayos** *f* PETROL ray path; ~ **del satélite** *f* TEC ESP
satellite track; ~ **según un círculo máximo** *f* TEC ESP
great-circle path; ~ **sísmica** *f* TEC PETR *geofísica*
seismic path; ~ **de vuelo** *f* TRANSP AÉR flight path;
~ **de vuelo actual** *f* TRANSP AÉR actual flight path;
~ **de vuelo de despegue** *f* TRANSP AÉR takeoff flight
path; ~ **de vuelo garantizada** *f* TRANSP AÉR guaranteed flight path; ~ **de vuelo indicada** *f* TRANSP AÉR
indicated flight path; ~ **de vuelo prevista** *f* TRANSP
AÉR intended flight path; ~ **de vuelo requerida** *f*
TRANSP AÉR required flightpath
trayectorias: ~ **de partículas sacudidas por moléculas turbulentas** *f pl* FÍS RAD particle trajectories
buffeted by turbulent molecules
traza *f* AmL (*cf huella Esp*) GAS trace, ING MECÁ
outline, MATEMÁT *de matriz* trace, NUCL *detectores*,
ÓPT track, PETROL trace; ~ **axial** *f* GEOL axial trace;
~ **de neutrones** *f* FÍS prompt neutron; ~ **radárica** *f*
TRANSP AÉR radar blip
trazado *m* *Esp* (*cf plano AmL*) CONST alignment,
drawing, *de planos* running, sketch, ELECTRÓN *con
lasers* scribing, GEOFÍS *sismograma* trace, GEOM
drawing, IMPR layout, INFORM&PD plotting, plot,
ING MECÁ drawing, layout, outline, MECÁ layout,
MINAS development, PROC QUÍ drawing, PROD layout,
TELECOM *vía utilizada* routing, route, TRANSP AÉR
plot, TRANSP MAR *arquitectura naval* drawing; ~ **de la
derrota** *m* TRANSP MAR *navegación* plot, track chart;
~ **por láser** *m* ELECTRÓN laser scribing; ~ **de la mina**
m MINAS mine development; ~ **de organigrama** *m*
INFORM&PD flowcharting; ~ **de planos** *m* D&A plotting; ~ **de poligonal** *m* CONST *topografía* traverse;
~ **de secciones** *m* TRANSP MAR *proyecto del buque*
sectional drawing; ~ **de secciones longitudinales y
transversales** *m* TRANSP MAR *proyecto del buque*
sectional drawing; ~ **del terreno** *m* CONST lie of the
land; ~ **transversal** *m* PETROL crossplot; ~ **de la
trayectoria** *m* TELECOM *satélite* tracking
trazador *m* C&V, CONTAM MAR, D&A tracer, ELEC
computadoras, INFORM&PD plotter, INSTR plotter,
tracer, NUCL, PETROL, TRANSP AÉR tracer; ~ **de base
plana** *m* INFORM&PD flat-bed plotter; ~ **bioquímico**
m CONTAM biochemical tracer; ~ **de curvas** *m* TEC
ESP X-Y plotter; ~ **digital** *m* INFORM&PD digital
plotter; ~ **electrostático** *m* ELEC, INFORM&PD electrostatic plotter; ~ **gráfico** *m* INFORM&PD graph
plotter, graphics plotter; ~ **gráfico X-Y** *m* TEC ESP
X-Y plotter; ~ **de gráficos digital** *m* INFORM&PD
digital plotter; ~ **incremental** *m* INFORM&PD incremental plotter; ~ **isotópico** *m* NUCL isotopic tracer;
~ **por láser** *m* ELECTRÓN laser scriber; ~ **plano** *m*
INFORM&PD flat-bed plotter; ~ **de plumas** *m*
INFORM&PD pen plotter (*AmE*), pen recorder (*BrE*);
~ **radiactivo** *m* FÍS, FÍS RAD radioactive tracer; ~ **de
rodillo** *m* INFORM&PD barrel plotter (*BrE*), drum
plotter (*AmE*); ~ **de rumbo** *m* TRANSP AÉR course
tracer; ~ **de tambor** *m* INFORM&PD barrel plotter
(*BrE*), drum plotter (*AmE*)
trazar *vt* CONST line, mark out, plot, *carpintería* scribe,
GEOL lay down, GEOM trace, plot, INFORM&PD trace,
ING MECÁ run, MATEMÁT trace, MECÁ design, MINAS
develop, *pozo* lay out, *un filón* chase, TRANSP MAR *el
rumbo* plot; ~ **a escala** *vt* PROD *dibujos, planos* scale

trazo *m* IMPR stroke, ING MECÁ line, PROD mark, *dibujos*
line; ~ **descendente** *m* IMPR *de una letra* descender;
~ **extrafino** *m* IMPR hairline; ~ **fino** *m* IMPR hair-
stroke; ~ **focal** *m* ELECTRÓN focal spot; ~ **luminoso
catódico** *m* ELECTRÓN, ING ELÉC cathode spot;
~ **oblicuo** *m* GEOM oblique stroke; ~ **del
osciloscopio** *m* ELECTRÓN oscilloscope trace; ~ **de
referencia** *m* PROD guiding line; ~ **con remates** *m*
IMPR serif; ~ **del tiempo** *m* GEOFÍS *prospección
sísmica* time trace
TRC[1] *abr* (*tubo de rayos catódicos*) ELEC, ELECTRÓN,
IMPR, INFORM&PD, SEG, TV CRT (*cathode-ray tube*)
TRC[2]: ~ **electrostático** *m* ELECTRÓN electrostatic
CRT; ~ **de haces múltiples** *m* ELECTRÓN multibeam
CRT; ~ **intensificador** *m* ELECTRÓN postaccelerator
CRT
T-RDSI *abr* (*terminación de la red digital de servicios
integrados*) TELECOM NT-LB (*B-ISDN network termination*)
trefilado *adj* ING MECÁ *alambre* drawn
trefilador *m* PROD *persona* wiredrawer; ~ **de carburo al
tungsteno** *m* ING MECÁ tungsten-carbide wire-drawing die
trefilar *vt* ING MECÁ draw, draw out, PROD draw out of a
kiln
trefilería *f* ING MECÁ wire mill, PROD wire works
trehalosa *f* QUÍMICA trehalose
trementina *f* COLOR, QUÍMICA turpentine
tremolita *f* MINERAL calamite, tremolite
trémolo *m* ACÚST *música* tremolo
tremorina *f* QUÍMICA tremorine
tren *m* ING MECÁ battery, cluster, MINAS *de vagonetas*
rake; ~ **de accionamiento por cremallera y piñón** *m*
NUCL rack-and-pinion drive gear; ~ **de adherencia
total** *m* TRANSP total adherence train; ~ **aéreo** *m*
TRANSP aerotrain; ~ **aerodeslizador** *m* TRANSP
hovercraft train; ~ **aeromagnético** *m* TRANSP aeromagnetic train; ~ **de alambre** *m* PROD *laminador*
looping mill; ~ **alternativo** *m* ING MECÁ *máquinas*
alternating motion; ~ **de amortiguación magnética**
m TRANSP magnetic-cushion train; ~ **de aterrizaje** *m*
TEC ESP landing gear, undercarriage, TRANSP *aviación*
undercarriage, TRANSP AÉR landing gear, undercarriage; ~ **de aterrizaje del morro** *m* TRANSP AÉR nose
gear; ~ **de aterrizaje principal** *m* TRANSP AÉR main
landing gear; ~ **de aterrizaje de ruedas dobles en
tándem** *m* TRANSP AÉR dual tandem wheel undercarriage; ~ **de aterrizaje semiactivo** *m* TRANSP AÉR
semiactive landing-gear; ~ **de aterrizaje en triciclo**
m TRANSP AÉR tricycle landing-gear; ~ **de auxilio** *m*
TRANSP wrecker; ~ **blindado** *m* D&A armored train
(*AmE*), armoured train (*BrE*); ~ **de bombas de
mina** *m* MINAS lift; ~ **de burbujas** *m* C&V train of
bubbles; ~ **camión** *m* TRANSP road-rail; ~ **de carga**
m FERRO *vehículos* freight train (*AmE*), goods train
(*BrE*); ~ **de cargas** *m* CONST rolling load; ~ **de
cercanías** *m* FERRO stopping train, local train, VEH
local train; ~ **de chapas** *m* ING MECÁ plate mill; ~ **de
coches-cama** *m* FERRO car sleeper train (*AmE*),
wagon sleeper train (*BrE*); ~ **de conexión** *m* FERRO
connecting train; ~ **correo** *m* FERRO mail train;
~ **desbastador** *m* ING MECÁ billetting roll, PROD
cogging mill, roughing mill; ~ **de desbaste** *m*
CARBÓN, PROD rougher; ~ **de dispersión** *m* NUCL
dispersion train; ~ **dúo** *m* PROD two-high mill, two-
high roll, two-high train; ~ **de embotellado** *m* EMB

bottling line; ~ **de engranaje epicíclico** *m* ING MECÁ, MECÁ epicyclic-gear train; ~ **de engranaje planetario** *m* ING MECÁ planetary-gear train; ~ **de engranajes** *m* AUTO set of gears, ING MECÁ gear assembly, gear train, gearing, gear work, train of gearing, VEH gear train, set of gears; ~ **epicíclico** *m* ING MECÁ, MECÁ epicyclic train; ~ **expreso** *m* FERRO *vehículos* express train; ~ **de fuerza** *m* AUTO, TRANSP AÉR power train; ~ **de hormigonado** *m* FERRO *vehículos* concreting train; ~ **de impulsos** *m* ELECTRÓN, INFORM&PD pulse train; ~ **de impulsos digitales** *m* TELECOM digital-pulse stream; ~ **inteligente** *m* TRANSP brain train; ~ **de inundación del núcleo** *m* NUCL core-flooding train; ~ **de inyección de seguridad** *m* NUCL core-flooding train; ~ **de laminación de desbastes planos** *m* PROD slabbing mill; ~ **laminador** *m* FERRO rolling-mill train, ING MECÁ rolling-mill train, rolls, PROD rolling-mill train; ~ **de laminador doble dúo** *m* PROD four-high rod mill; ~ **laminador pequeño para perfiles** *m* PROD *laminación* small bar mill; ~ **de laminadores** *m* ING MECÁ battery of rolls; ~ **de laminar redondos** *m* CARBÓN rod mill; ~ **de laminar redondos oscilantes** *m* CARBÓN vibrating rod mill; ~ **local** *m* FERRO local train, slow train, stopping train, VEH local train; ~ **local de carga** *m* FERRO *vehículos* way freight train (*AmE*), way goods train (*BrE*); ~ **local de mercancías** *m* FERRO *vehículos* way freight train (*AmE*), way goods train (*BrE*); ~ **de mercancías** *m* FERRO *vehículos* freight train (*AmE*), goods train (*BrE*); ~ **moderno de viajeros** *m* FERRO advanced passenger train (*APT*); ~ **de montaje** *m* ING MECÁ, TRANSP MAR *construcción naval* production line; ~ **motor** *m* AUTO power train; ~ **de movimiento** *m* ING MECÁ train of action; ~ **de ondas** *m* ACÚST, FÍS wave train, FÍS ONDAS wave train, wavelet; ~ **ondulador** *m* PAPEL corrugator, fluter; ~ **de palancas** *m* ING MECÁ leverage; ~ **portacontenedores** *m* FERRO *vehículos* freightliner train; ~ **prototipo** *m* FERRO preproduction train; ~ **de prueba** *m* FERRO *vehículos* test train; ~ **de pudelaje** *m* PROD puddle rolls; ~ **de pudelar** *m* PROD puddle train; ~ **de pudelar barras** *m* PROD puddle-bar train; ~ **a punto de salir** *m* FERRO train about to depart; ~ **de puntos de parada limitados** *m* FERRO limited train (*AmE*), semifast train (*BrE*); ~ **que produce ingresos** *m* FERRO *vehículos* revenue-earning train; ~ **que viaja comtinuamente entre dos puntos** *m* FERRO shuttle train; ~ **de ramificación** *m* FERRO *vehículos* feeder train; ~ **rápido** *m* FERRO *vehículos* fast train; ~ **de recogida de carga** *m* FERRO pick-up freight train (*AmE*), pick-up goods train (*BrE*); ~ **de rectificar carriles** *m* FERRO *vehículos* rail-grinding train; ~ **rentable** *m* FERRO *vehículos* revenue-earning train; ~ **para rociar herbicidas sobre la vía** *m* FERRO weed-killing train; ~ **de rodadura** *m* MECÁ carriage; ~ **de rodaje** *m* ING MECÁ running gear; ~ **rodante** *m* ING MECÁ running gear; ~ **de rodillos** *m* ING MECÁ train of rolls, PROD roll train; ~ **de ruedas** *m* ING MECÁ wheel train; ~ **de secado** *m* PAPEL dryer section; ~ **semidirecto** *m* FERRO limited train (*AmE*), semifast train (*BrE*); ~ **semiexpreso** *m* FERRO limited train (*AmE*), semifast train (*BrE*); ~ **semirrápido** *m* FERRO limited train (*AmE*), semifast train (*BrE*); ~ **de socorro** *m* FERRO *vehículos* relief train; ~ **de sondeo** *m* MINAS

drilling rig, PROD *equipamiento* rig; ~ **de sondeos** *m* MINAS core drilling; ~ **de suministro de energía** *m* TRANSP AÉR power train; ~ **termostático** *m* REFRIG power system; ~**-tranvía** *m* TRANSP rail motor unit; ~ **trío** *m* ING MECÁ three-high mill; ~ **con vacío de gravedad** *m* TRANSP, VEH gravity-vacuum transit train (*GVT train*); ~ **de válvulas en cabeza** *m* AUTO I-head valve train; ~ **de válvulas en cabeza en T** *m* AUTO T-head valve train

trencilla *f* ING ELÉC braid

trenza *f* ELECTRÓN, ING ELÉC, PROD, TEXTIL braid; ~ **de algodón** *f* ING ELÉC cotton braid; ~ **de cáñamo** *f* PROD gasket, hemp gasket; ~ **de cobre** *f* ELEC, ELECTRÓN, PROD copper braid; ~ **de filamentos** *f* FÍS strand

trenzado[1] *adj* TELECOM twisted together

trenzado[2] *m* ELEC *cable* braid, TEXTIL braiding

trenzar *vt* ELEC, ING ELÉC *cables*, TEXTIL braid

treosa *f* QUÍMICA threose

trepanador *m* ING MECÁ trepanner

trépano *m* CARBÓN *AmL* (*cf barreno Esp*) churn drill bit, drill jar, drilling bit, *berbiquí, pistolete de mina* chisel, *sondeos* jackhammer, bit, *perforación* rotary bit, three-cone bit, rock bit, PETROL bit; ~ **adiamantado** *m* PROD diamond drill; ~ **de aletas** *m* TEC PETR *perforación* blade bit; ~ **de cabeza** *m* *Esp* (*cf trépano compuesto AmL*) ING MECÁ, MINAS *sondeos* boring head; ~ **de chorro** *m* TEC PETR *perforación* jet bit; ~ **en cola de carpa** *m* TEC PETR *perforación* fishtail bit, roller bit; ~ **compuesto** *m* AmL (*cf trépano de cabeza Esp*) ING MECÁ, MINAS *sondeos* boring head; ~ **cónico** *m* AmL (*cf barrena cónica Esp*) TEC PETR *perforación* cone bit; ~ **de corona** *m* MINAS star bit, TEC PETR *perforación* crown bit; ~ **de corona adiamantada** *m* MINAS diamond crown set; ~ **cortante** *m* CARBÓN *bisel, sondeos* cutting bit, MINAS *barrenación* chisel bit, *sondeos* chopping bit, cutting bit, TEC PETR *perforación* chisel bit; ~ **de corte en cruz** *m* MINAS star bit; ~ **de cuatro aletas** *m* TEC PETR four-wing bit, *perforación* four-way bit; ~ **de cuatro puntas** *m* TEC PETR *perforación* four-way bit; ~ **para escariado** *m* TEC PETR *perforación* reaming bit; ~ **en espiral** *m* TEC PETR *perforación* spiral bit; ~ **en estrella** *m* MINAS star bit; ~ **excéntrico** *m* MINAS eccentric bit, eccentric chisel, TEC PETR *perforación* eccentric bit; ~ **para formación dura** *m* TEC PETR *perforación* hard of formation bit; ~ **iniciador** *m* TEC PETR *perforación* spudding bit; ~ **de insertos** *m* AmL (*cf barrena de insertos Esp*) TEC PETR insert bit; ~ **de perforación** *m* TEC PETR *perforación* drill bit; ~ **piloto** *m* TEC PETR *perforación* pilot bit; ~ **plano** *m* ING MECÁ, MINAS *sondeos* flat chisel; ~ **para roca** *m* MINAS *sondeos* drill bit; ~ **sacatestigos** *m* TEC PETR *perforación* core bit; ~ **de sondar** *m* MINAS earth-borer; ~ **de sondeo** *m* AmL (*cf tricono Esp*) MINAS boring chisel, drilling bit; ~ **para sondeo al cable** *m* *Esp* MINAS *sondeo* churn drill; ~ **para sondeo a la cuerda** *m* AmL (*cf trépano para sondeo al cable Esp*) MINAS *sondeo* churn drill; ~ **para testigos** *m* AmL (*cf corona testiguera Esp*) MINAS *sondeos* core cutter; ~ **tetracónico** *m* AmL (*cf barrena tetracónica Esp*) TEC PETR *perforación* quadricone bit; ~ **tricónico** *m* TEC PETR *perforación* tricone bit

trepar *vt* CARBÓN, CONST bore

trepidación *f* ACÚST flutter, PROD jarring, shaking, TEC

ESP judder, TRANSP AÉR flutter; ~ **por huelgo** *f* ING MECÁ shake

trepidante *adj* PROD jarring

tres[1]: **de ~ fases** *adj* ELEC *red de distribución* three-phase; **de ~ vías** *adj* MECÁ three-way

tres[2]: ~ **pliegues** *m pl* IMPR quarter fold

tres[3]: ~ **herramientas** *f pl* ING MECÁ tool trias

TRF *abr* (*transformada rápida de Fourier*) ELECTRÓN, INFORM&PD FFT (*fast Fourier transform*)

triac *m* ELEC *tiristor* triac

triacetato *m* TEXTIL triacetate; ~ **de celulosa** *m* (*CTA*) CINEMAT, P&C cellulose triacetate

triacetina *f* QUÍMICA triacetin

triacetonamina *f* QUÍMICA triacetonamine

triácido *m* QUÍMICA triacid

tríada *f* QUÍMICA triad

triamilo *m* QUÍMICA triamyl

triangulación *f* CONST *topografía*, GEOM, METR, TEC ESP triangulation

triangular[1] *adj* C&V, GEOM triangular

triangular[2] *vt* GEOM triangulate

triangularidad *f* C&V, GEOM triangularity

triángulo *m* CONST, FÍS, GEOM triangle, ING ELÉC delta, LAB triangle; ~ **agudo** *m* GEOM acute triangle; ~ **de ángulo recto** *m* GEOM right-angle triangle; ~ **básico** *m* METAL basic triangle; ~ **de colores** *m* FÍS color triangle (*AmE*), colour triangle (*BrE*); ~ **congruente** *m* GEOM congruent triangle; ~ **equiangular** *m* GEOM equiangular triangle; ~ **equilátero** *m* GEOM equiangular triangle; ~ **escaleno** *m* GEOM scalene triangle; ~ **esférico** *m* GEOM spherical triangle; ~ **de fuerzas** *m* CONST, FÍS triangle of forces; ~ **de gran capacidad** *m* ELECTRÓN high-speed mesh; ~ **isósceles** *m* GEOM isosceles triangle; ~ **oblicuo** *m* GEOM oblique triangle; ~ **obtuso** *m* GEOM obtuse triangle; ~ **plano** *m* GEOM plane triangle; ~ **polar** *m* GEOM polar triangle; ~ **rectángulo** *m* GEOM right-angle triangle; ~ **de tubo de arcilla** *m* LAB *soporte para crisol* pipeclay triangle

triar *vt* MINAS sort

triazoico *adj* QUÍMICA triazoic

triazol *m* QUÍMICA triazole

tribásico *adj* QUÍMICA tribasic

triboelectricidad *f* ELEC frictional electricity, FÍS, TRANSP triboelectricity

triboeléctrico *adj* ELEC *detector*, FÍS, TRANSP *detector* triboelectric

triboluminiscencia *f* FÍS triboluminescence

tributario[1] *adj* HIDROL, TELECOM tributary

tributario[2] *m* HIDROL, TELECOM tributary

tributirina *f* QUÍMICA tributyrin

tricarbalílico *adj* QUÍMICA tricarballylic

tricíclico *adj* QUÍMICA tricyclic

triclínico *adj* CRISTAL, QUÍMICA triclinic

tricloroacético *adj* QUÍMICA trichloroacetic

tricloroetileno *m* DETERG, QUÍMICA trichlorethylene

tricloruro *m* QUÍMICA trichloride

tricono *m* *Esp* (*cf barrena tricónica Esp*) MINAS *sondeo petrolífero* drilling bit, TEC PETR three-cone bit, tricone bit

tricopirita *f* MINERAL hair pyrites

tricosano *m* QUÍMICA tricosane

tricotado *m* TEXTIL knitting; ~ **de medias** *m* TEXTIL hose knitting

tricotar *vt* TEXTIL knit

tricotosa: ~ **rectilínea** *f* TEXTIL flat-knitting machine

tricresilo *m* QUÍMICA tricresyl

tricresol *m* QUÍMICA tricresol

tricroico *adj* FÍS trichroic

tricroismo *m* FÍS trichroism

tricromático *adj* INSTR trichromatic

tricromía *f* CINEMAT three-color process (*AmE*), three-colour process (*BrE*)

tridimensional *adj* FÍS three-dimensional

tridimita *f* MINERAL tridymite

tridireccional *adj* MECÁ three-way

triédrico *adj* GEOM trihedral

triedro[1] *adj* GEOM trihedral

triedro[2] *m* GEOM trihedron

triestearato: ~ **de glicerilo** *m* ALIMENT, QUÍMICA glyceryl tristearate; ~ **de glicerina** *m* QUÍMICA tristearin

triestearina *f* QUÍMICA tristearin

triéster *m* QUÍMICA triester

trietanolamina *f* DETERG, QUÍMICA triethanolamine

trifana *f* MINERAL triphane

trifásico *adj* ELEC *red de distribución*, FÍS three-phase, ING ELÉC triphase, three-phase, QUÍMICA triphasic

trifenilmetilo *m* QUÍMICA triphenylmethyl

trifenol *m* QUÍMICA triphenol

trifílico *adj* QUÍMICA triphilic

trifilina *f* MINERAL triphyline, triphylite

trifosfato: ~ **de adenosina** *m* (*ATP*) ALIMENT, PETROL, QUÍMICA adenosine triphosphate (*ATP*)

trifurcador *m* ELEC *accesorio de cable* trifurcator

trifurcante *m* ELEC *accesorio de cable* trifurcator

trigatrón *m* ELECTRÓN trigatron

trigger: ~ **de Schmitt** *m* ELECTRÓN, FÍS Schmitt trigger

triglicérido *m* DETERG, QUÍMICA triglyceride

trigo: ~ **duro** *m* AGRIC, ALIMENT durum wheat; ~ **sarraceno** *m* AGRIC, ALIMENT buckwheat

trigonal *adj* CRISTAL trigonal

trigonometría *f* GEOM trigonometry; ~ **plana** *f* GEOM plane trigonometry

trigonométrico *adj* GEOM trigonometric, trigonometrical

trihidrato *m* QUÍMICA trihydrate

trihídrico *adj* QUÍMICA trihydric

trihidrol *m* QUÍMICA trihydrol

trihidruro: ~ **de nitrógeno** *m* CONTAM *producto químico*, QUÍMICA ammonia (NH_3)

trilátero *m* GEOM trilateral

trilla: ~ **de retorno** *f* AGRIC rethreshing

trilladora: ~ **de maíz** *f* AGRIC corn thresher (*AmE*), maize thresher (*BrE*)

trimerizar *vt* QUÍMICA trimerize

trímero *m* P&C trimer

trimésico *adj* QUÍMICA trimesic

trimetilbenceno *m* QUÍMICA trimethylbenzene

trimetilcarbinol *m* QUÍMICA trimethylcarbinol

trimetileno *m* QUÍMICA trimethylene

trimetilpiridina *f* QUÍMICA trimethylpyridine

trimiristina *f* QUÍMICA trimyristin

trimolecular *adj* QUÍMICA trimolecular

trimórfico *adj* QUÍMICA trimorphic

trimorfismo *m* QUÍMICA trimorphism

trimorfo *adj* QUÍMICA trimorphous

trinca *f* TRANSP MAR *estibaje* lashing

trincar *vt* TRANSP MAR *carga* make fast, *cordaje* lash

trincha *f* MECÁ chisel

trinchera *f* AGUA ditch, CARBÓN ditch, trench, D&A, MINAS trench; ~ **de cateo** *f* AmL (*cf trinchera de* *exploración Esp*) MINAS exploration trench; **~ de exploración** *f Esp* (*cf trinchera de cateo AmL*) MINAS exploration trench; **~ noruega** *f* TEC PETR *geología del Mar Norte* Norwegian trench

trineo: ~ de aparejo *m* PROD *explotación forestal* pig

trinitrado *adj* QUÍMICA trinitrated

trinitrato *m* QUÍMICA trinitrate

trinitrina *f* QUÍMICA trinitrin

trinitrobenceno *m* QUÍMICA trinitrobenzene

trinitrocompuesto *m* QUÍMICA trinitro-compound

trinitrocresol *m* QUÍMICA trinitrocresol

trinitrofenol *m* QUÍMICA trinitrophenol

trinitroresorcinol *m* QUÍMICA styphnic acid

trinitrotolueno *m* (*TNT*) QUÍMICA trinitrotoluene (*TNT*)

trino *m* ACÚST *música* trill

trinomio *m* MATEMÁT trinomial

trinquetar *vt* ING MECÁ ratch

trinquete *m* AmL (*cf garfio Esp*) CONST *carpintería* ratchet, ING MECÁ tripper, click, dog, keeper, ratchet, pawl, MECÁ stop, ratchet, dog, MINAS prop, *jaula* kep, *jaula, accesorio de extracción* catch, NUCL *del mecanismo de accionamiento de las barras de control* latch; **~ de aparcamiento** *m* AUTO parking pawl; **~ de elevación** *m* NUCL lift gripper; **~ estacionario** *m* NUCL stationary gripper; **~ del freno de estacionamiento** *m* AUTO parking pawl; **~ de mandril** *m* ING MECÁ lathe dog; **~ móvil** *m* NUCL moving gripper; **~ reversible** *m* ING MECÁ reversible ratchet; **~ de tubo abierto** *m* ING MECÁ open-socket ratchet

trinqueteo *m* Esp (*cf tableteo AmL*) NUCL ratchetting

trinquetilla *f* TRANSP MAR forestay sail

triodo *m* ELECTRÓN, FÍS triode; **~ doble** *m* ELECTRÓN double triode; **~-exodo** *m* ELECTRÓN triode-hexode; **~ de gas** *m* ELECTRÓN gas triode; **~ planar** *m* ELECTRÓN planar triode; **~ termiónico** *m* ELECTRÓN thermionic triode; **~ de vacío** *m* ELECTRÓN vacuum triode

triol *m* QUÍMICA triol

trioleína *f* QUÍMICA triolein

triosa *f* QUÍMICA triose

trioxano *m* QUÍMICA trioxane

trióxido *m* QUÍMICA teroxide, trioxide; **~ de arsénico** *m* QUÍMICA arsenic trioxide; **~ de azufre** *m* CONTAM sulfuric anhydride (*AmE*), sulphuric anhydride (*BrE*), QUÍMICA sulfur trioxide (*AmE*), sulphur trioxide (*BrE*); **~ de dinitrógeno** *m* CONTAM *producto químico*, QUÍMICA nitrous oxide (N_2O)

trioximetileno *m* QUÍMICA trioxymethylene

trioxipurina *f* QUÍMICA trioxypurine

trioxosulfato: ~ IV de hidrógeno *m* CONTAM, QUÍMICA sulfurous acid (*AmE*), sulphurous acid (*BrE*)

tripa: ~ del cartón *f* PAPEL board liner; **~ para ondular** *f* PAPEL *cartón ondulado* fluting corrugating medium, fluting corrugating paper, fluting medium, fluting paper

tripalmitato: ~ de glicerilo *m* QUÍMICA tripalmitin

tripalmitina *f* QUÍMICA tripalmitin

triparanol *m* QUÍMICA triparanol

triple[1] *m* Esp (*cf pareja AmL*) TEC PETR *perforación* stand of pipe; **~ enlace** *m* QUÍMICA triple bond

triple[2]: **~ ligadura** *f* QUÍMICA triple bond

triplete *m* FÍS, QUÍMICA *espectroscopia*, NUCL triplet

triplita *f* MINERAL triplite

trípode *m* CINEMAT tripod, CONST shear legs, FOTO

camera stand, tripod, INSTR, LAB, METR, MINAS, PROD tripod; **~ ajustable** *m* METR adjustable tripod, MINAS adjustable stilt; **~ alto** *m* CINEMAT standard legs; **~ bajo** *m* CINEMAT baby tripod; **~ con cabezal panorámico horizontal** *m* FOTO pan-head tripod; **~ cardánico** *m* CINEMAT gimbal tripod; **~ copa** *m* CINEMAT high hat; **~ de mesa** *m* CINEMAT table-top tripod, FOTO table tripod; **~ de patas cortas** *m* CINEMAT baby legs; **~ de patas extensibles** *m* CINEMAT, FOTO extension tripod; **~ de un pie** *m* CINEMAT, FOTO unipod; **~ telescópico** *m* CINEMAT telescopic tripod, extension tripod, FOTO extension tripod

trípodo: ~ bajo *m* CINEMAT baby legs

trípoli *m* MINERAL diatomite, tripolite, PETROL tripolite

tripropulsante *m* TEC ESP, TERMO tripropellant; **~ líquido** *m* TERMO liquid tripropellant

tripropulsor *m* TEC ESP tripropellant

triptano *m* QUÍMICA triptane

tríptico *adj* QUÍMICA tryptic

triptomin *m* QUÍMICA tryptomin

triptomina *f* QUÍMICA tryptomine

tripulación[1]: **sin ~** *adj* TEC ESP unmanned; **con ~ insuficiente** *adj* TRANSP undermanned

tripulación[2] *f* TEC ESP *de una nave*, TRANSP AÉR crew, TRANSP MAR *del buque* crew, ship's hands; **~ de cabina** *f* TRANSP, TRANSP AÉR cabin crew; **~ instrumento** *f* TEC ESP payload; **~ de relevo** *f* TRANSP AÉR relief crew; **~ de vuelo** *f* TRANSP AÉR flight crew

tripular *vt* TRANSP MAR *embarcación* crew, man

triquita *f* METAL whisker, PETROL trichite

triscado *m* ING MECÁ set

triscador *m* ING MECÁ, PROD *sierras* jumper, swage

triscadora *f* ING MECÁ setting machine

triscadura *f* ING MECÁ setting

triscamiento *m* ING MECÁ set

triscar *vt* ING MECÁ, PROD *sierras* jump

trisección *f* GEOM trisection

trisilicato *m* QUÍMICA trisilicate

trisnitrato *m* QUÍMICA trisnitrate

trisódico *adj* QUÍMICA trisodium

trisustituido *adj* QUÍMICA trisubstituted

tritano *m* QUÍMICA tritan

tritiado *adj* NUCL tritiated

triticina *f* QUÍMICA triticin

tritilo *m* QUÍMICA trityl

tritio *m* FÍS, QUÍMICA tritium

tritiónico *adj* QUÍMICA trithionic

tritóxido *m* QUÍMICA tritoxide

triturabilidad *f* PROC QUÍ grindability

trituración *f* ALIMENT grinding, CARBÓN, CONST crushing, grinding, HIDROL comminution, MINAS *minerales* crushing, grinding, NUCL comminution, P&C *operación* crushing, powdering, mastication, PROC QUÍ comminution, PROD grinding, crushing, stamping, QUÍMICA trituration, grinding; **brecha de ~** *f* GEOL crush breccia **~ basta** *f* CARBÓN, MINAS coarse grinding; **~ fina** *f* CARBÓN, PROD fine crushing, fine grinding; **~ de finos** *f* PROD fine crushing; **~ gruesa** *f* CARBÓN, MINAS coarse grinding; **~ de grueso** *f* CARBÓN, MINAS, PROD coarse crushing; **~ a mano** *f* MINAS ragging; **~ primaria** *f* CARBÓN precrushing

triturado: ~ de mineral *m* MINAS *preparación* ore crusher; **~ terciario** *m* CARBÓN tertiary crushing

triturador *m* ALIMENT *molienda, panadería* kibbler,

CARBÓN stamp mill, HIDROL *sedimentos* comminutor, ING MECÁ crusher, triturator, NUCL, PROC QUÍ comminutor, PROD crusher; ~ **de bolas** *m AmL* (*cf molino de bolas Esp*) ALIMENT, C&V, CARBÓN, MINAS, P&C ball mill, PROC QUÍ ball mill, milling ball; ~ **giratorio** *m* ALIMENT *maquinaria para proceso de alimentos* gyratory crusher, ING MECÁ swinging crusher; ~ **de grueso** *m* MINAS, PROD coarse crusher; ~ **de hielo** *m* REFRIG ice crusher; ~ **de mineral** *m AmL* (*cf troquel Esp*) MINAS die; ~ **de minerales** *m* MINAS ore stamp; ~ **de muelas horizontales** *m* PROD roller mill; ~ **oscilante** *m* ING MECÁ swinging crusher; ~ **primario** *m* MINAS primary crusher; ~ **de pulpa** *m* PAPEL kneader pulper

trituradora *f* AGRIC *trilladora* feedgrinder, CARBÓN crusher, CONST *maquinaria obras públicas* stone crusher, EMB shredding machine, ING MECÁ breaker, grinding machine, MECÁ grinding machine, PROC QUÍ grinding device, breaker, crushing machine, PROD mill; ~ **de Bradford** *f* ING MECÁ Bradford breaker; ~ **de carbón** *f* CARBÓN, MINAS coal breaker, coal crusher; ~ **de carbón y roca** *f* MINAS coal and stone breaker; ~ **de cilindros** *f* CARBÓN roll crusher, PROC QUÍ crushing cylinder; ~ **de cilindros anulares** *f* CARBÓN ring-roll crusher; ~ **de conos** *f* CARBÓN cone crusher; ~ **de coque** *f* CARBÓN, MINAS coke breaker; ~ **y criba combinadas** *f* ING MECÁ, PROD combined grinder and sieve; ~ **en fino** *f* PROC QUÍ fine-crushing mill; ~ **de finos** *f* CARBÓN, PROD fine crusher; ~ **gruesa de rodillos** *f* PROC QUÍ coarse crushing mill; ~ **de grueso** *f* CARBÓN, PROD coarse crusher; ~ **de gruesos** *f* MINAS coarse crusher; ~ **de hielo** *f* ING MECÁ crushing machine for ice; ~ **de mandíbulas** *f* CARBÓN, PROD jaw crusher; ~ **de martillos** *f* CARBÓN hammer crusher; ~ **mezcladora** *f* PROD mixing mill; ~ **de mineral** *f* MINAS ore crusher; ~ **móvil** *f* CARBÓN mobile crusher; ~ **de rodillos** *f* CARBÓN bowl mill crusher, ring-roll crusher; ~ **de tortas oleaginosas** *f* AGRIC oilcake breaker

triturar *vt* AGRIC crush, ALIMENT grind, CARBÓN crush, grind, CONST crush, ING MECÁ, MECÁ grind, MINAS *preparación* squeeze, PROC QUÍ grind, PROD crush, grind, pulverize, stamp, QUÍMICA triturate, grind

tritutador: ~ **de grueso** *m* CARBÓN coarse crusher

trivalencia *f* QUÍMICA tervalence, trivalence, trivalency

trivalente *adj* QUÍMICA tervalent, trivalent

triyoduro *m* QUÍMICA triiodide

troceado *m* PAPEL chip

troceador: ~ **binario** *m* INFORM&PD binary chop

troceadora *f* PAPEL chipper

trocear *vt* PROD chip, *herramientas* cut off

troceo *m* PROD cutting

trocha: ~ **de vía** *f AmL* CONST gage (*AmE*), gauge (*BrE*)

troctolita *f* PETROL troctolite, troctolyte

trogerita *f* MINERAL troegerite

troilita *f* MINERAL troilite

trole *m* CARBÓN *explotación forestal*, MINAS buggy, TRANSP *transporte público* trolley; ~ **eléctrico** *m* ELEC, TRANSP electric trolley

trolebús *m* TRANSP trolleybus; ~ **articulado** *m* AUTO articulated trolleybus; ~ **subterráneo** *m* TRANSP underground trolleybus (*BrE*), underground trolley-car (*AmE*)

tromba *f* METEO spout, waterspout; ~ **de agua** *f* AGUA

flush; ~ **marina** *f* METEO, OCEAN, TRANSP MAR *estado de la mar* waterspout

tromel *m* CARBÓN trommel screen, *metalurgia* revolving screen, MINAS revolving screen; ~ **para finos** *m* PROD fine trommel

trompa: ~ **acústica** *f* ACÚST horn; ~ **de Eustaquio** *f* ACÚST *audición* Eustachian tube; ~ **de vacío** *f Esp* (*cf bomba filtrante AmL*) LAB filter pump

trompeta: ~ **del eje trasero** *f* AUTO, MECÁ, VEH rear-axle flared tube

trona *f* MINERAL trona

tronar *vi* METEO thunder

troncal *m* TELECOM trunk

tronchador *m* MECÁ cutter

tronchadura *f* ING MECÁ shearing

tronco *m* GEOM *de un sólido* frustum, INFORM&PD trunk (*AmE*), link (*BrE*), PAPEL log, TELECOM trunk; ~ **para aserrar** *m* CONST saw log

tronera *f* MECÁ vent

tronido *m* METEO thunderclap

tronzador *m* ING MECÁ parting tool

troostita *f* METAL, MINERAL troostite

tropas: ~ **transportadas por camión** *f pl* D&A lorried troops (*BrE*), trucked troops (*AmE*)

trópico *adj* QUÍMICA tropic

tropina *f* QUÍMICA tropine

tropopausa *f* METEO tropopause

troposfera *f* METEO, TEC ESP troposphere

troquel *m* CINEMAT die, die-casting die, press tool, driver; ~ **de acuñar** *m* ING MECÁ coining die; ~ **de cinta perforada** *m* INFORM&PD tape punch; ~ **combinado** *m* ING MECÁ compound die; ~ **compuesto** *m* ING MECÁ compound die; ~ **de corte** *m* ING MECÁ shearing die; ~ **encabezador** *m* ING MECÁ heading die; ~ **estampador** *m* MECÁ die-stamper; ~ **de explosión** *m* ING MECÁ explosion die; ~ **de extrusión** *m* ING MECÁ extrusion die; ~ **extrusor** *m* ING MECÁ extrusion die; ~ **de forjado en caliente** *m* ING MECÁ hot forging die; ~ **de forjado en frío** *m* ING MECÁ, PROD, TERMO cold-forging die; ~ **de forjar** *m* ING MECÁ forging die; ~ **de martillo** *m Esp* (*cf triturador de mineral AmL*) ING MECÁ hammer die, MINAS die; ~ **de una parte** *m* ING MECÁ one-part die; ~ **partido** *m* ING MECÁ split die; ~ **para perforar** *m* ING MECÁ die for punching; ~ **progresivo** *m* ING MECÁ progression die; ~ **de punzonar** *m* ING MECÁ piercing-die; ~ **de ranurar** *m* ING MECÁ notching die; ~ **de suelta rápida** *m* ING MECÁ, PROD quick-release die

troquelado[1] *adj* MECÁ die-cast

troquelado[2] *m* IMPR die cut, die cutting, ING MECÁ dying, punching, PROD *con el punzón sacabocados* punching; ~ **en frío** *m* ING MECÁ cold hobbing; ~ **de pestañas** *m* IMPR tabbing

troqueladora *f* PAPEL cutting die, PROD *láminas de metal* stamping machine

troquelar *vt* IMPR die-cut, ING MECÁ die-cast, MINAS *preparación* squeeze

trotilo *m* QUÍMICA trotyl

trozo: ~ **de ladrillo** *m* CONST bat

trueno *m* METEO thunder

trueque *m* PROD trade-in

trulla *f* CONST brick trowel; ~ **de albañil** *f* CONST bricklayer's trowel

truncación *f* INFORM&PD truncation

truncamiento *m* GEOM truncating, truncation, PETROL beveling (*AmE*), bevelling (*BrE*)

truncar *vt* GEOM, INFORM&PD truncate

truth: ~ **quark** *m* FÍS PART top quark

truxílico *adj* QUÍMICA truxillic

truxilina *f* QUÍMICA truxilline

tscheffkinita *f* MINERAL tscheffkinite, tschewkinite

tschermigita *f* MINERAL tschermigite

tschernakita *f* MINERAL tschermakite

TSD *abr* (*tratamiento de señal digital*) ELECTRÓN DSP (*digital signal processing*)

TSP *abr* (*fosfato trisódico*) DETERG TSP (*trisodium phosphate*)

tsunami *m* GEOFÍS, OCEAN tsunami

TTC *abr* (*telemetría y control de rastreo*) TEC ESP TTC (*tracking telemetry and command*)

t-t-t *abr* (*tiempo-temperatura-tolerancia*) REFRIG t-t-t (*time-temperature-tolerance*)

TTY *abr* (*teletipo*) INFORM&PD, TELECOM TTY (*teletype*)

tubería *f* AGUA conduit, water supply line, CARBÓN *presas* piping, CONST pipe, line, conduit pipe, pipeline, CONTAM pipeline, ENERG RENOV pipeline, *perforación* tubing, FÍS FLUID pipe, GAS piping, ING MECÁ pipeline, MECÁ pipe, TEC PETR tubing, TEC PETR pipe; ~ **de abastecimiento** *f* CONST supply pipe; ~ **de abastecimiento de agua** *f* AGUA water supply pipe; ~ **absorbente** *f* PROC QUÍ absorber tubing; ~ **de acero** *f* CONST steel pipe; ~ **de agua** *f* AGUA water pipe; ~ **para agua pluvial** *f* CONST rainwater pipe; ~ **de agua de refrigeración** *f* AGUA, ING MECÁ, REFRIG cooling water pipe; ~ **de aire** *f* ING MECÁ air pipeline; ~ **de alimentación** *f* NUCL feeder pipe, PROD feed pipe; ~ **de arcilla con juntas abiertas** *f* CONST open-jointed clayware pipe; ~ **ascendente** *f* ELEC *fuente de alimentación* rising main; ~ **de aspiración** *f* AGUA, HIDROL suction pipe, PROD suction line, TRANSP MAR *drenajes* suction pipe; ~ **aspirante** *f* INSTR vacuum manifold; ~ **de barro** *f* CONST earthenware pipe; ~ **con brida** *f* CONST flange pipe; ~ **de cementación** *f* TEC PETR *perforación* cementing string; ~ **de combustible** *f* AUTO, VEH fuel line; ~ **concéntrica** *f* GAS concentric pipe; ~ **de concreto** *f* AmL (cf *tubería de hormigón Esp*) CONST concrete pipe; ~ **para conductores** *f* ELEC *fuente de alimentación* conduit; ~ **de desagüe** *f* CONST waste pipe, RECICL outfall pipe; ~ **de descarga** *f* INSTAL HIDRÁUL *turbina hidráulica* suction tube, aspiring tube, draught tube (*BrE*), draft tube (*AmE*), RECICL discharge pipe; ~ **de descarga de gases de combustión** *f* NUCL flue tube; ~ **de desviación** *f* PROC QUÍ deflecting tubing; ~ **de distribución** *f* AGUA distributing pipe; ~ **elevada** *f* CONST lift pipe; ~ **de enchufe** *f* CONST socket pipe; ~ **de engrase** *f* PROD *lubricación* oil pipe; ~ **enterrada** *f* GAS buried pipeline; ~ **de entrada de vapor** *f* INSTAL HIDRÁUL steam supply pipe; ~ **de entubación** *f* GAS casing, MINAS drill casing, *perforación* casing pipe; ~ **de escape** *f* TERMO, TRANSP AÉR, TRANSP MAR exhaust pipe; ~ **de evacuación** *f* AGUA soil pipe; ~ **exterior** *f* GAS outside pipe; ~ **flexible** *f* ING MECÁ flexible pipe; ~ **de fluidos** *f* ING MECÁ fluid pipeline; ~ **de flujo** *f* PETROL, TEC PETR flow line; ~ **forzada** *f* AGUA penstock, INSTAL HIDRÁUL full pipe; ~ **de freno** *f* VEH brake line; ~ **de fuego** *f* TERMO fire riser; ~ **de fundición** *f* CONST cast-iron pipe; ~ **de fundición con**

brida *f* CONST flanged cast-iron pipe; ~ **de hierro forjado** *f* CONST wrought-iron pipe; ~ **de hierro fundido** *f* ING MECÁ cast-iron pipeline; ~ **de hierro gris** *f* ING MECÁ gray-iron pipe (*AmE*), grey-iron pipe (*BrE*); ~ **de hormigón** *f Esp* (cf *tubería de concreto AmL*), CONST concrete pipe; ~ **horquillada** *f* CONST forked pipe; ~ **de humos** *f* TERMO fire riser; ~ **de impulsión** *f* MINAS plunger lift; ~ **interconecta** *f* PROD interconnecting pipework; ~ **de llegada del vapor** *f* INSTAL HIDRÁUL steam supply pipe; ~ **de llenado** *f* TRANSP MAR filling pipe; ~ **marina** *f* TEC PETR *perforación costa-fuera* marine riser; ~ **de material polivalente** *f* TRANSP multipurpose material pipeline; ~ **metálica flexible** *f* ING MECÁ flexible metal conduit; ~ **de muestreo** *f* CALIDAD *agua* sampling line; ~ **neumática** *f* ING MECÁ airline; ~ **de palastro** *f* CONST sheet-iron pipe; ~ **de perforación** *f* MINAS, PETROL, TEC PETR *producción* drill pipe; ~ **perforada** *f* PETROL perforated pipe; ~ **de plomo** *f Esp* CONST lead piping; ~ **porosa con juntas abiertas** *f* CONST open-jointed porous pipe; ~ **sin presión** *f* ING MECÁ nonpressure pipeline; ~ **principal** *f* AGUA current, mains (*BrE*), CONST main; ~ **principal del agua** *f* AGUA water main; ~ **principal del viento** *f* ING MECÁ air main, PROD blast main; ~ **de producción** *f* PETROL production string, TEC PETR *producción* production string, production tubing; ~ **de revestimiento** *f* AGUA casing pipe, ENERG RENOV *geotérmica* well casing, GAS casing, MINAS drill casing, *perforación* casing pipe, PETROL casing, TEC PETR *perforación* casing, liner, conductor; ~ **rígida** *f* ING MECÁ rigid pipe; ~ **de salida** *f* AGUA outlet pipe; ~ **secundaria** *f* CONST branch pipe; ~ **de succión** *f* PROD suction line; ~ **de suministros** *f* AUTO delivery pipe; ~ **de suspensión** *f* MECÁ hanger pipe; ~ **en T** *f* CONST junction pipe; ~ **de toma del vapor** *f* INSTAL HIDRÁUL *máquinas* steamway; ~ **de vapor** *f* INSTAL HIDRÁUL steam line, steam loop, steam pipe, steam piping, vapor line (*AmE*), vapour line (*BrE*); ~ **vástago** *f* MINAS, PETROL, TEC PETR *producción* drill pipe; ~ **de ventilación** *f* CONST vent pipe, MINAS ventilation pipe; ~ **vertical** *f* CONST riser pipe; ~ **vertical de subida** *f* PETROL, TEC PETR *producción costa-fuera* riser pipeline; ~ **del viento** *f* PROD *alto horno* blast pipe; ~ **en Y** *f* CONST forked pipe

tuberías: ~ **y accesorios** *f pl* CONST pipes and fittings; ~ **flexibles de acero** *f pl* ING MECÁ flexible steel piping; ~ **metálicas flexibles** *f pl* ING MECÁ flexible metal piping; ~ **de plástico** *f pl* ING MECÁ plastic pipeline; ~ **del sistema de frenado** *f pl* AUTO brake line; ~ **con uniones de cordón y grifo** *f pl Esp* CONST spigot-and-faucet joint pipes; ~ **con uniones de enchufe y cordón** *f pl AmL* CONST spigot-and-faucet joint pipes

tuberina *f* QUÍMICA tuberin

tubing *m* TEC PETR *producción* production tubing

tubista *m* ING MECÁ pipelayer

tubo *m* CONST barrel, spigot, *cerradura* tube, ELECTRÓN, ENERG RENOV, FÍS, GAS tube, ING MECÁ duct, LAB tube, MECÁ duct, pipe, MINAS *agotamiento* stock, ÓPT, TEC ESP tube, TEC PETR pipe, TEXTIL tube; ~ **en U** *m* CONST channel iron, ING MECÁ hairpin tube, LAB *material de vidrio* U tube; ~ **en Y** *m* ING MECÁ, MECÁ, PROD breeches pipe;

[a] ~ **abductor** *m* INSTAL HIDRÁUL feed pipe;

~ **abocardado** *m* PROD flared tube; ~ **de absorción** *m* LAB *material de vidrio* absorption tube; ~ **de aceleración** *m* FÍS, FÍS PART, PROC QUÍ *acelerador de partículas* accelerating tube; ~ **de acero al carbono de baja tensión de rotura** *m* INSTAL TERM, METAL low-tensile carbon steel tube; ~ **de acero inoxidable** *m* ING MECÁ stainless steel tube; ~ **acodado** *m* LAB *material de vidrio*, MECÁ, NUCL elbow; ~ **de acometida a la alcantarilla** *m* AGUA soil pipe; ~ **de acoplamiento** *m* ING MECÁ coupling tube; ~ **activado** *m* ELECTRÓN fired tube; ~ **de admisión** *m* INSTAL HIDRÁUL admission pipe, induction pipe; ~ **de agua** *m* AUTO, TERMOTEC water tube; ~ **ahorquillado** *m* ING MECÁ, MECÁ, PROD breeches pipe; ~ **de aire** *m* ING MECÁ air pipe; ~ **de aire comprimido** *m* CONST, ING MECÁ compressed-air line; ~ **de aireación** *m* ING MECÁ breather, MECÁ funnel, PROD breather, funnel, VEH breather; ~ **de aislamiento** *m* ING MECÁ insulation pipe; ~ **aislante** *m* ING ELÉC conduit; ~ **ajustable para montaje del parasol** *m* CINEMAT adjustable lens hood barrel; ~ **de alcance principal** *m* INSTR main scope tube; ~ **de aletas** *m* MECÁ finned tube; ~ **de alimentación** *m* PROD delivery tube, feed pipe, filler; ~ **alimentador** *m* INSTAL HIDRÁUL feed pipe, flow pipe; ~ **de almacenamiento** *m* ELECTRÓN storage tube; ~ **de almacenamiento de carga** *m* ELECTRÓN charge-storage tube; ~ **de almacenamiento de la grabación** *m* ELECTRÓN recording storage tube; ~ **de almacenamiento de imagen** *m* ELECTRÓN image storage tube; ~ **de almacenamiento mallado** *m* ELECTRÓN mesh storage tube; ~ **de almacenamiento de material aislante** *m* ELECTRÓN barrier-grid storage tube; ~ **de almacenamiento de memoria** *m* INFORM&PD storage tube; ~ **de almacenamiento de persistencia variable** *m* ELECTRÓN variable-persistence storage tube; ~ **de almacenamiento de un solo cañón** *m* ELECTRÓN single-gun storage tube; ~ **de alta definición** *m* INSTR fine focus sleeve; ~ **de alto vacío** *m* ELECTRÓN hard tube, high vacuum tube; ~ **de amplificación regulable** *m* ELECTRÓN variable-mu tube; ~ **amplificador** *m* ELECTRÓN amplifier tube; ~ **de amplificador de microonda** *m* ELECTRÓN microwave-amplifier tube; ~ **de amplificador de potencia** *m* ELECTRÓN power-amplifier tube; ~ **amplificador de rayos X** *m* INSTR X-ray amplifier tube; ~ **de anillos** *m* TEXTIL ring tube; ~ **de ánodo giratorio** *m* CONST rotating-anode tube; ~ **ascendente** *m* PETROL riser, TEC PETR *perforación costa-fuera* marine riser, riser; ~ **de aspiración** *m* AGUA tailpipe, ENERG RENOV draft tube (*AmE*), draught tube (*BrE*), INSTAL HIDRÁUL aspiring tube, draft tube (*AmE*), draught tube (*BrE*), suction tube; ~ **atirantador** *m* PROD *calderas* stay tube; ~ **autopolarizado** *m* ELECTRÓN self-biased tube;

~b ~ **de bajada de aguas** *m* CONST downpipe (*BrE*), downspout (*AmE*); ~ **de banda ancha** *m* ELECTRÓN wideband tube; ~ **de banda estrecha** *m* ELECTRÓN narrow-band tube; ~ **de barro** *m* C&V clay pipe, earthenware pipe; ~ **bifurcado** *m* AGUA, CONST branch pipe, ING MECÁ breeches pipe, MECÁ breeches pipe, branch pipe, PROD breeches pipe; ~ **de la bobina plana** *m* TEXTIL cheese tube; ~ **Bourdon** *m* FÍS, TEC PETR *aparatos de medida* Bourdon gage (*AmE*), Bourdon gauge (*BrE*); ~ **con bridas** *m* ING

MECÁ flanged pipeline; ~ **de burbuja** *m* CONST bubble tube;

~c ~ **de caldera** *m* ING MECÁ boiler tube; ~ **de calibre ancho** *m* LAB *material de vidrio* wide-bore tube; ~ **de calibre angosto** *m* LAB *material de vidrio* narrow-bore tube; ~ **de calor** *m* TEC ESP *satélites* heat pipe; ~ **de cámara** *m* CINEMAT, ELECTRÓN, TV camera tube; ~ **de cámara con gran proporción de rayos gamma** *m* ELECTRÓN high-gamma camera tube; ~ **de cámara de televisión** *m* ELECTRÓN, FÍS, ING ELÉC, TELECOM, TV television camera tube; ~ **de campo cruzado** *m* ELECTRÓN, FÍS, TELECOM crossed-field tube; ~ **capilar** *m* FÍS, LAB, TEC PETR capillary tube; ~ **capilar blindado** *m* PROD armored capillary (*AmE*), armoured capillary (*BrE*); ~ **captador** *m* TV pick-up tube; ~ **de capuchón** *m* ELECTRÓN acorn tube; ~ **de cartulina** *m* EMB, PAPEL cardboard tube; ~ **de cátodo caliente** *m* ELEC *de radio* thermionic tube (*AmE*), thermionic valve (*BrE*); ~ **de cátodo frío** *m* ELECTRÓN cold-cathode tube; ~ **de cátodo incandescente** *m* ELECTRÓN hot cathode tube; ~ **de caucho** *m* LAB, P&C rubber tubing; ~ **de cebado** *m* AGUA priming pipe; ~ **de centrífuga** *m* LAB *separación* centrifuge tube; ~ **de charco de mercurio** *m* ELECTRÓN, ING ELÉC mercury pool tube; ~ **de chimenea** *m* CONST chimney stack; ~ **de cobre forjado sin costura** *m* ING MECÁ seamless wrought-copper tube; ~ **de cobre fundido sin costura** *m* ING MECÁ seamless wrought-copper tube; ~ **de cola** *m* CONST tailpipe; ~ **colector** *m* MECÁ header pipe; ~ **con collarín** *m* ING MECÁ flanged pipeline; ~ **de combustible** *m* AUTO, VEH fuel pipe; ~ **conductor** *m* MINAS *perforación* conductor pipe; ~ **congelador** *m* PROD freezing tube; ~ **cónico** *m* ING MECÁ taper pipe, TEXTIL cone tube; ~ **de conmutación** *m* ELECTRÓN switching tube; ~ **de conmutación de gas** *m* ELEC, ELECTRÓN, GAS gas-filled switching tube; ~ **de conmutación lleno de gas** *m* ELEC, ELECTRÓN, GAS gas-filled switching tube; ~ **con cono metálico** *m* ELECTRÓN metal cone tube; ~ **contador** *m* ELECTRÓN, NUCL counter tube; ~ **contador de cátodo frío** *m* ELECTRÓN, ING ELÉC cold-cathode counter tube; ~ **contador de irradiación** *m* ELECTRÓN, FÍS RAD radiation counter tube; ~ **contador de neutrones** *m* NUCL neutron counter tube; ~ **contador de radiación** *m* ELECTRÓN, FÍS RAD radiation counter tube; ~ **convertidor** *m* INSTR converter tube; ~ **de convertidor de imagen** *m* CINEMAT, ELECTRÓN image converter tube; ~ **de corte rápido** *m* ELECTRÓN sharp-cut-off tube; ~ **con costura** *m* ING MECÁ seamed pipe; ~ **sin costura** *m* ING MECÁ seamless pipe; ~ **de Crookes** *m* ELECTRÓN Crookes tube; ~ **de cuello de cisne** *m* ING MECÁ gooseneck pipe; ~ **cuentagotas** *m* LAB *material de vidrio* dropping tube; ~ **curvado** *m* MECÁ bend;

~d ~ **de deflexión** *m* FÍS RAD deflection tube; ~ **con depósito de mercurio** *m* ELECTRÓN, ING ELÉC mercury pool tube; ~ **desactivado** *m* ELECTRÓN unfired tube; ~ **de desagüe** *m* AUTO overflow pipe, CONST drainpipe, VEH *sistema de refrigeración* overflow pipe; ~ **de desbordamiento** *m* AGUA overflow, overflow pipe; ~ **de descarga** *m* AGUA delivery pipe, discharge pipe, offtake, ELECTRÓN, FÍS discharge tube, LAB *material de vidrio* spout, MINAS offtake, PROD spout, RECICL discharge pipe; ~ **de descarga del arco** *m* ING ELÉC arc discharge tube; ~ **de**

descarga de cenizas *m* PROD cinder chute; ~ **de descarga luminiscente** *m* ELEC, ELECTRÓN, ING ELÉC glow-discharge tube; ~ **de descarga de semilla** *m* AGRIC seed tube; ~ **de desconexión remoto** *m* ELECTRÓN remote-cut-off tube; ~ **desecador** *m* FOTO, LAB *material de vidrio* potash bulb; ~ **de deslizamiento** *m* ELECTRÓN, ING MECÁ drift tube; ~ **desmontable** *m* EMB collapsible tube; ~ **de destellos** *m* ELECTRÓN flash tube; ~ **de destilación** *m* PROC QUÍ, QUÍMICA distilling tube; ~ **de dilatación** *m* LAB *material de vidrio* expansion tube; ~ **de diodo** *m* ELECTRÓN diode tube (*AmE*), diode valve (*BrE*); ~ **de diodo conectado** *m* ELECTRÓN diode-connected tube; ~ **disector** *m* TV dissector tube; ~ **de doble cañón** *m* TV double gun tube; ~ **donde se montan las lentes** *m* CINEMAT, FOTO lens barrel; ~ **de drenaje** *m* C&V gutter tile, CONST drainpipe;

~ e ~ **del eje motor** *m* AUTO, VEH torque drive tube; ~ **del eje propulsor del motor** *m* AUTO, VEH torque tube; ~ **electrométrico** *m* ELEC, ELECTRÓN, ING ELÉC electrometer tube; ~ **electrónico** *m* ELECTRÓN electron tube, electronic tube, PROD, TEC ESP valve; ~ **electrónico industrial** *m* ELECTRÓN industrial electronic tube; ~ **electrónico memorizador** *m* INFORM&PD storage tube; ~ **electrónico del microscopio** *m* INSTR electron microscope tube; ~ **elevador de calor** *m* AUTO heat riser tube; ~ **embridado** *m* ING MECÁ flanged pipeline; ~ **de emisión secundaria** *m* ELECTRÓN secondary emission tube; ~ **de enfoque secundario** *m* INSTR secondary viewing tube; ~ **enfriado con agua** *m* ELECTRÓN water-cooled tube; ~ **enrollado en espiral** *m* EMB spirally-wound tube; ~ **de ensayo** *m* C&V, LAB, PROD, QUÍMICA *equipo* test tube; ~ **de entrada** *m* MECÁ inlet, PROD inlet pipe; ~ **de entrada de aire** *m* ING MECÁ air inlet pipe; ~ **para envíos postales** *m* EMB mailing tube (*AmE*), postal tube (*BrE*); ~ **de ER** *m* ELECTRÓN TR tube; ~ **de escape** *m* AUTO tailpipe, ING MECÁ, PROD blowpipe, TERMO exhaust pipe, TERMOTEC exhaust, TRANSP AÉR, TRANSP MAR exhaust pipe, VEH *sistema de escape de gases del motor* tailpipe; ~ **de escape de bomba de aire** *m* ING MECÁ air pump exhaust pipe; ~ **de escaso vacío** *m* ELECTRÓN soft tube; ~ **estabilizador de voltaje** *m* ING ELÉC voltage stabilizer tube; ~ **para estibado** *m* TEC PETR *perforación* racking pipe; ~ **de exhaustación** *m* INSTAL HIDRÁUL aspiring tube, draft tube (*AmE*), draught tube (*BrE*), suction tube; ~ **de expansión** *m* LAB *material de vidrio* expansion tube; ~ **de exploración por punto deslizante** *m* TV flying-spot tube scanner; ~ **exponencial** *m* ELECTRÓN exponential tube; ~ **de expulsión** *m* TERMOTEC exhaust; ~ **de extensión** *m* LAB *material de vidrio* extension tube;

~ f ~ **flexible** *m* ING MECÁ flexible pipe, MINAS flexible hose, PROD, TEC ESP hose; ~ **flexible de freno de aire comprimido** *m* Esp (*cf manga flexible de freno de aire comprimido AmL*) ING MECÁ air-brake hose; ~ **flexible de gasolina** *m* Esp (*cf tubo flexible de nafta AmL*) AUTO gas hose (*AmE*), gasoline hose (*AmE*), petrol hose (*BrE*), P&C, VEH *combustible* petrol hose (*BrE*), gasoline hose (*AmE*), gas hose (*AmE*); ~ **flexible de nafta** *AmL* (*cf tubo flexible de gasolina Esp*) *m* AUTO, P&C petrol hose (*BrE*), gasoline hose (*AmE*), gas hose (*AmE*), VEH gas hose (*AmE*), gasoline hose (*AmE*), petrol hose (*BrE*); ~ **flexible**

ondulado *m* CINEMAT, FOTO, ING MECÁ, MECÁ, PROD, TEC ESP *esclusas, trajes espaciales* bellows; ~ **fluorescente** *m* ELEC *alumbrado*, FÍS, FÍS RAD, GAS, ING ELÉC fluorescent tube; ~ **fluorescente de neón** *m* FÍS RAD neon fluorescent tube; ~ **focalizador** *m* INSTR focusing sleeve; ~ **de formación de imagen térmica** *m* ELECTRÓN thermal-imaging tube; ~ **fotoeléctrico** *m* ELECTRÓN photoelectric tube, phototube; ~ **fotoeléctrico de cesio** *m* ELECTRÓN, INSTR caesium phototube (*BrE*), cesium phototube (*AmE*); ~ **fotosensible** *m* ELECTRÓN photosensitive tube; ~ **para frecuencias ultraelevadas** *m* ING MECÁ microwave tube; ~ **para freno neumático** *m* Esp (*cf manga para freno neumático AmL*) P&C air-brake hose;

~ g ~ **de gas** *m* CONST gas pipe, ELECTRÓN gas tube, GAS gas pipe, gas tube, TERMO, TRANSP gas tube; ~ **de gas de cátodo incandescente** *m* ELECTRÓN hot-cathode gas tube; ~ **de gas raro** *m* ELECTRÓN, GAS, QUÍMICA rare-gas tube; ~ **Geiger-Müller** *m* FÍS Geiger-Müller tube; ~ **Geissler** *m* ELECTRÓN, FÍS Geissler tube; ~ **de goteo** *m* LAB *material de vidrio* dropping tube; ~ **de gran potencia** *m* ELECTRÓN high-power tube; ~ **guía** *m* ING MECÁ guide tube, MINAS guide tube, *perforación* conductor pipe; **~~guía** *m* NUCL thimble; **~~guía de grafito** *m* NUCL graphite guide tube; ~ **guiador** *m* ING MECÁ, MINAS guide tube;

~ h ~ **de haz de electrones** *m* ELECTRÓN electron-beam tube; ~ **de haz lineal** *m* ELECTRÓN linear-beam tube; ~ **de haz ondulado** *m* ELECTRÓN re-entrant beam tube; ~ **de haz perfilado** *m* ELECTRÓN shaped beam tube; ~ **de haz periódico** *m* TV gated beam tube; ~ **de hermeticidad de disco** *m* ELECTRÓN disc seal tube (*BrE*), disk seal tube (*AmE*); ~ **de hierro dúctil** *m* ING MECÁ ductile iron pipe; ~ **hincado** *m* MINAS drive pipe; ~ **de hiperfrecuencias** *m* ING MECÁ microwave tube; ~ **hueco** *m* CARBÓN *entubado de pozos* shell liner; ~ **de humos** *m* PROD *de calderas*, TERMO flue, TERMOTEC smoke tube;

~ i ~ **de imagen** *m* ELECTRÓN, TELECOM *TV* picture tube; ~ **de imagen en color de tres cañones** *m* ELECTRÓN three-gun color picture tube (*AmE*), three-gun colour picture tube (*BrE*); ~ **de imagen en color de tres haces** *m* ELECTRÓN three-beam color picture tube (*AmE*), three-beam colour picture tube (*BrE*); ~ **de imagen de máscara de sombra** *m* TV shadow mask tube; ~ **de imagen televisiva** *m* ELECTRÓN, FÍS, ING ELÉC, TELECOM, TV television picture tube; ~ **de impulsión** *m* AGUA discharge pipe; ~ **inclinado de agar** *m* ALIMENT agar slant; ~ **indicador** *m* ELECTRÓN indicator tube, TV index tube; ~ **intensificador** *m* ELECTRÓN intensifier tube; ~ **intensificador de imagen** *m* ELECTRÓN, INSTR, TV image intensifier tube; ~ **de interacción ampliado** *m* ELECTRÓN extended-interaction tube; ~ **de interacción corta** *m* ELECTRÓN short-interaction tube; ~ **de inyección** *m* INSTAL HIDRÁUL injection pipe; ~ **inyector** *m* TEC ESP jet pipe; ~ **isotérmico** *m* TEC ESP, TRANSP heat pipe;

~ k ~ **de Kundt** *m* ACÚST, FÍS Kundt's tube;

~ l ~ **laminado** *m* EMB laminated tube; ~ **lanzatorpedos** *m* D&A torpedo-launching tube; ~ **de lavado** *m* LAB *material de vidrio* washing tube; ~ **lineal** *m* ELECTRÓN linear tube; ~ **de llenado del aceite** *m* AUTO oil filler pipe; ~ **lleno de gas** *m* ELEC,

ELECTRÓN, GAS gas-filled tube; ~ **loctal** *m* ELECTRÓN loctal tube; ~ **de lubricación** *m* PROD oil pipe; ~ **luminoso** *m* ING ELÉC light pipe;

~ m ~ **de machos** *m* PROD *fundición* core tube; **~-máscara con compensación de temperatura** *m* TV temperature-compensated shadow mask mount; ~ **para máscara de sombra** *m* ELECTRÓN shadow mask tube; ~ **medidor** *m* C&V gage glass (*AmE*), gauge glass (*BrE*); ~ **de memoria** *m* ELECTRÓN memory tube; ~ **memorizador de expansión** *m* ELECTRÓN expansion storage tube; ~ **metálico** *m* ELECTRÓN metal tube; ~ **metálico flexible** *m* ING MECÁ flexible metal conduit; ~ **mezclador** *m* ELECTRÓN, ING MECÁ mixer tube, INSTR converter tube; ~ **de microondas** *m* ELECTRÓN, ING MECÁ microwave tube; ~ **de microondas tipo O** *m* ELECTRÓN O-type microwave tube; ~ **de microondas tipo M** *m* ELECTRÓN M-type microwave tube; ~ **sin montar** *m* CONST unmounted tubing; ~ **de muestreo** *m* LAB *material de vidrio* sampling tube; ~ **multielectródico** *m* ELECTRÓN multielectrode tube; ~ **multiplicador electrónico** *m* ELECTRÓN, FÍS RAD, INSTR electron-multiplier tube;

~ n ~ **de neón** *m* ELEC *alumbrado*, FÍS, ING ELÉC neon tube; ~ **con nervios** *m* PROD gilled tube; ~ **de nivel** *m* C&V water glass, PROD *calderas* gage glass (*AmE*), gauge glass (*BrE*); ~ **del nivel de agua** *m* PROD gage glass (*AmE*), gauge glass (*BrE*); ~ **Nixie** *m* ELECTRÓN Nixie tube;

~ o ~ **octal** *m* ELECTRÓN octal tube; ~ **de onda lenta** *m* ELECTRÓN slow-wave tube; ~ **de onda milimétrica** *m* ELECTRÓN millimeter-wave tube (*AmE*), millimetre-wave tube (*BrE*); ~ **de onda de propagación milimétrica** *m* ELECTRÓN millimeter-wave traveling-wave tube (*AmE*), millimetre-wave travelling-wave tube (*BrE*); ~ **de onda rápida** *m* ELECTRÓN, ING ELÉC fast-wave tube; ~ **de onda regresiva** *m* ELECTRÓN, FÍS backward-wave tube; ~ **de ondas electrónicas** *m* ELECTRÓN, FÍS ONDAS electron-wave tube; ~ **de ondas estacionarias** *m* ACÚST stationary-wave tube; ~ **de ondas progresivas** *m* (*TOP*) ELECTRÓN, FÍS, TEC ESP, TELECOM traveling-wave tube (*AmE*) (*TWT*), travelling-wave tube (*BrE*) (*TWT*); ~ **de ondas progresivas de banda X** *m* ELECTRÓN X-band traveling wave tube (*AmE*), X-band travelling wave tube (*BrE*); ~ **de ondas progresivas de helicoidal** *m* ELECTRÓN helix traveling-wave tube (*AmE*), helix travelling-wave tube (*BrE*); ~ **de órgano** *m* ACÚST organ pipe; ~ **de oscilador local** *m* ELECTRÓN local oscillator tube; ~ **de oscilador de microondas** *m* ELECTRÓN microwave oscillator tube; ~ **de osciloscopio** *m* ELECTRÓN oscilloscope tube;

~ p ~ **de perforación** *m* MINAS *sondeos* drill pipe, drive pipe, PETROL, TEC PETR drill pipe; ~ **de Pitot** *m* FÍS, TRANSP AÉR Pitot tube; ~ **de Pitot presostático** *m* CARBÓN *aviones* pressure head; ~ **de Pitot con toma estática** *m* CARBÓN *aviones* pressure head; ~ **plano** *m* CONST plain tube, EMB lay-flat tubing; ~ **plano con fuelle** *m* EMB gusseted layflat tubing; ~ **plegable según el eje vertical** *m* EMB center folding tubing (*AmE*), centre folding tubing (*BrE*); ~ **de porcelana** *m* C&V porcelain tube; ~ **portaocular** *m* INSTR *microscopio compuesto* drawtube; ~ **portaviento** *m* ING MECÁ *alto horno* blowpipe, TRANSP MAR air intake; ~ **de potencia** *m* ELECTRÓN power tube; ~ **de presión** *m* AUTO pressure tube, ING MECÁ pressure pipe; ~ **de presión con bridas** *m* ING MECÁ flanged pressure pipe; ~ **de presión con collarín** *m* ING MECÁ flanged pressure pipe; ~ **de propagación de ondas milimétricas** *m* ELECTRÓN millimeter-wave traveling-wave tube (*AmE*), millimetre-wave travelling-wave tube (*BrE*); ~ **de protección de cables** *m* ING ELÉC conduit; ~ **protector del pirómetro** *m* SEG pyrometer protection-tube; ~ **de proyección de boca pequeña** *m* ELECTRÓN short-neck projection tube; ~ **de purificación por contacto** *m* HIDROL *aguas residuales* contact column;

~ q ~ **quemador** *m* PROD *lámpara de solar* flame tube; ~ **del quinqué** *m* C&V lamp chimney;

~ r ~ **de radar** *m* ELECTRÓN, FÍS RAD radar tube; ~ **del radiador** *m* AUTO, TRANSP AÉR, VEH *sistema de refrigeración* radiator hose; ~ **de rayos catódicos** *m* (*TRC*) ELEC *presentación visual, osciloscopio*, ELECTRÓN, IMPR, INFORM&PD, SEG, TV cathode-ray tube (*CRT*); ~ **de rayos catódicos de alto vacío** *m* TV high-vacuum cathode ray tube; ~ **de rayos catódicos de doble haz** *m* ELECTRÓN dual-beam cathode ray tube; ~ **de rayos catódicos electrostático** *m* (*TRC electrostático*) ELECTRÓN electrostatic cathode ray tube (*electrostatic CRT*); ~ **de rayos catódicos de exploración de la imagen** *m* ELECTRÓN raster-scan cathode ray tube; ~ **de rayos catódicos de exploración vectorial** *m* ELECTRÓN vector-scan cathode ray tube; ~ **de rayos catódicos de haces múltiples** *m* (*TRC de haces múltiples*) ELECTRÓN multibeam cathode ray tube (*multibeam CRT*); ~ **de rayos catódicos de haz dividido** *m* TV split-beam cathode ray tube; ~ **de rayos catódicos con monoacelerador** *m* ELECTRÓN monoaccelerator CRT; ~ **de rayos catódicos de un solo haz** *m* ELECTRÓN single-beam cathode ray tube; ~ **de rayos catódicos para televisión** *m* ELECTRÓN, FÍS, ING ELÉC, TELECOM, TV television tube; ~ **de rayos X** *m* CRISTAL, ELECTRÓN, FÍS RAD, INSTR X-ray tube; ~ **de rebose** *m* AUTO overflow pipe, LAB *material de vidrio* spout, VEH overflow pipe; ~ **recalentador** *m* PROD superheater pipe; ~ **receptor** *m* ELECTRÓN pick-up tube, receiving tube; ~ **rectificador** *m* ING ELÉC rectifier tube; ~ **reductor** *m* GEOFÍS, LAB *material de vidrio* reduction tube; ~ **de referencia de voltaje** *m* ING ELÉC voltage reference tube; ~ **refractario** *m* INSTAL TERM, TERMO refractory tube; ~ **de refrigeración** *m* ING MECÁ, REFRIG cooling tube; ~ **de refrigeración por aire** *m* ELECTRÓN air-cooled tube; ~ **regulado por rejilla** *m* ELECTRÓN grid-controlled tube; ~ **regulador de voltaje** *m* ING ELÉC voltage regulator tube; ~ **de rejilla de pantalla** *m* ELECTRÓN screen grid tube; ~ **de rejilla única** *m* ELECTRÓN single-grid tube; ~ **de rejillas múltiples** *m* ELECTRÓN multigrid tube; ~ **para la reproducción directa de la imagen** *m* ELECTRÓN direct-view storage tube; ~ **de respiración** *m* AUTO breather pipe; ~ **de respiración atmosférica** *m* OCEAN snorkel; ~ **de retención de imagen** *m* ELECTRÓN display storage tube; ~ **de retorno** *m* PROD return tube; ~ **revestidor de fondo** *m* PETROL liner; ~ **de revestimiento** *m* MINAS drive pipe, *sondeos petrolíferos* lining tube; ~ **de revestimiento perforado** *m* PETROL perforated casing; ~ **rojo** *m* ELECTRÓN red tube;

~ s ~ **en S** *m* ING MECÁ, MECÁ gooseneck pipe;

~ **sacatestigos** m MINAS *sondeos* core barrel; ~ **de salida** m AGUA outlet pipe; ~ **de sección iluminada** m INSTR light section tube; ~ **de seguridad** m LAB *material de vidrio* safety tube; ~ **con una sola salida** m ELECTRÓN single-ended tube; ~ **de un solo ánodo** m ELECTRÓN single-anode tube; ~ **de un solo haz** m ELECTRÓN single-beam tube; ~ **de sondeo** m MINAS drive pipe; ~ **de subida** m PETROL, TEC PETR riser; ~ **de succión** m INSTAL HIDRÁUL *de turbina* draft box (*AmE*), draft tube (*AmE*), draught box (*BrE*), draught tube (*BrE*); ~ **de supervivencia** m OCEAN *equipos de buceo* umbilical;

~ **t** ~ **telescópico** m CINEMAT extension tube, FOTO extension tube, telescopic tube; ~ **de termocambiador** m ING MECÁ heat exchanger tube; ~ **termoiónico** m ELEC *de radio*, ELECTRÓN thermionic tube (*AmE*), thermionic valve (*BrE*); ~ **de termopermutador** m ING MECÁ heat exchanger tube; ~ **testigos** m MINAS core tube; ~ **de tetrodo** m ELECTRÓN tetrode tube (*AmE*), tetrode valve (*BrE*); ~ **de Thiele** m LAB *punto de fusión* Thiele tube; ~ **tipo O** m ELECTRÓN O-type tube; ~ **tipo M** m ELECTRÓN M-type tube; ~ **para toma de aire libre** m OCEAN *equipos para buceo* snorkel; ~ **tomavistas** m TV pick-up tube; ~ **de torsión** m AUTO, VEH *motor* torque tube; ~ **de transferencia** m PROD transfer tube; ~ **de tres electrodos** m ELECTRÓN three-electrode tube; ~ **de tres rejillas** m ELECTRÓN three-grid tube; ~ **de triodo** m ELECTRÓN triode tube (*AmE*), triode valve (*BrE*); ~ **de la tubería de impulsión** m MINAS *desagüe* tree;

~ **u** ~ **de unidad múltiple** m ELECTRÓN multiple unit tube;

~ **v** ~ **de vacío** m ELECTRÓN, FÍS, INFORM&PD, TV vacuum tube; ~ **de vacío elevado** m ELECTRÓN, FÍS high-vacuum tube; ~ **de variación** m ELECTRÓN drift tube; ~ **de velocidad modulada** m ELECTRÓN velocity-modulated tube; ~ **de venteo** m NUCL vent pipe; ~ **de venteo de aire** m NUCL air vent; ~ **de ventilación** m ING MECÁ air duct, LAB vent, TRANSP MAR *botavara* gooseneck; ~ **de Venturi** m FÍS, ING MECÁ venturi tube; ~ **vertical** m AGUA, TEC PETR standpipe; ~ **vertical de bajada** m PROD *alto horno*, TEC PETR downcomer; ~ **vertical de bajada en una torre de destilación** m TEC PETR *refino* downcomer; ~ **vidicón** m ELECTRÓN vidicon tube; ~ **de vidrio** m C&V, ELECTRÓN, LAB *material de vidrio* glass tube, glass tubing; ~ **de vidrio para soplado** m LAB *material de vidrio* blown-glass tube; ~ **en virolas** m CONST conical tubing; ~ **de visualización** m ELECTRÓN display tube

tubos: ~ **adaptados** m pl ELECTRÓN matched tubes; ~ **conductores** m pl CARBÓN piping; ~ **de descarga fluorescente** m pl FÍS RAD fluorescent-discharge tubes; ~ **detectores para muestreo a corto plazo** m pl SEG detector tubes for short-term sampling; ~ **del sistema de frenado** m pl AUTO brake line

tubulador: ~ **de admisión** m ING MECÁ inlet manifold (*BrE*), intake manifold (*AmE*)

tubuladora: ~ **de impulsión** f AGUA head pipe

tubular *adj* PROD tubular

tuerca f CONST nut, ING MECÁ knurling, nut, MECÁ, PROD, VEH nut; ~ **abovedada** f ING MECÁ dome nut; ~ **de acoplamiento** f ING MECÁ coupling nut; ~ **de agujeros** f PROD holed nut; ~ **de ajuste** f ING MECÁ set nut; ~ **alada** f ING MECÁ wing nut; ~ **de alas** f ING

MECÁ thumbnut, wing nut; ~ **almenada** f Esp (*cf tuerca castilla AmL*) CONST castle nut, ING MECÁ castellated nut, castle nut, notched nut; ~ **de anclaje** f ING MECÁ anchor nut; ~ **de apriete** f ING MECÁ locknut, maiden nut, press nut, PROD locknut; ~ **aterrajada** f ING MECÁ tapped nut; ~ **de aterrajar hexagonal** f ING MECÁ hexagonal die nut; ~ **auto-trabadora con tope de nylon** f ING MECÁ nylstop self-locking nut; ~ **en bóveda** f ING MECÁ dome nut; ~ **castilla** f AmL (*cf tuerca almenada Esp*) CONST castle nut, ING MECÁ castellated nut, castle nut, notched nut; ~ **de centrado** f ING MECÁ centering nut (*AmE*), centring nut (*BrE*); ~ **de centrar** f ING MECÁ centering nut (*AmE*), centring nut (*BrE*); ~ **cerillada** f ING MECÁ milled nut; ~ **ciega** f ING MECÁ box nut, cap nut, MECÁ cap nut, TRANSP AÉR acorn nut; ~ **cilíndrica** f ING MECÁ round nut; ~ **en cúpula** f ING MECÁ dome nut; ~ **en domo** f ING MECÁ dome nut; ~ **elástica** f ING MECÁ, PROD elastic nut; ~ **entallada** f CONST castellated nut, castle nut, ING MECÁ castle nut; ~ **esférica circulante** f ING MECÁ ball circulating nut; ~ **estriada** f ING MECÁ milled nut, knurled nut; ~ **de fijación** f ING MECÁ locknut, set nut, PROD locknut; ~ **de fijación de la rueda** f VEH wheel nut; ~ **guía** f ING MECÁ guide nut; ~ **hecha a máquina** f ING MECÁ machine-made nut; ~ **hexagonal** f ING MECÁ hexagon nut, hexagonal nut, MECÁ hex nut; ~ **hexagonal común** f ING MECÁ ordinary hexagonal nut; ~ **hexagonal ordinaria** f ING MECÁ ordinary hexagonal nut; ~ **hexagonal de torsión dominante** f ING MECÁ prevailing torque-type hexagon-nut; ~ **del husillo** f ING MECÁ clasp nut; ~ **inaflojable** f ING MECÁ, PROD locknut; ~ **de inmovilización** f ING MECÁ jam nut, set nut; ~ **de inserción** f MECÁ insert nut; ~ **manual** f ING MECÁ thumbnut; ~ **de mariposa** f CONST butterfly nut, ING MECÁ thumbnut, wing nut, PROD hand nut, VEH wing nut; ~ **moleteada** f ING MECÁ knurled nut; ~ **normalizada** f ING MECÁ standard nut; ~ **ochavada** f ING MECÁ octagonal nut; ~ **octogonal** f ING MECÁ octagonal nut; ~ **de orejeta** f ING MECÁ wing nut; ~ **de orejetas** f AUTO lug nut (*AmE*), wheelnut (*BrE*), CONST finger nut, ING MECÁ thumbnut, PROD fly nut, hand nut; ~ **de palomilla** f AUTO lug nut (*AmE*), wheelnut (*BrE*), ING MECÁ thumbnut, wing nut, VEH wing nut; ~ **de paro elástica** f ING MECÁ, PROD elastic stop-nut; ~ **pesada** f ING MECÁ heavy nut; ~ **profunda** f CONST deep nut; ~ **rayada** f ING MECÁ milled nut; ~ **reforzada** f ING MECÁ heavy nut; ~ **reglamentaria** f ING MECÁ standard nut; ~ **de regulación** f ING MECÁ regulating nut; ~ **para repasar roscas** f PROD die nut; ~ **de repasar roscas hexagonal** f ING MECÁ hexagonal die nut; ~ **roscada** f ING MECÁ tapped nut, threaded nut; ~ **sin roscar** f PROD blank nut; ~ **de la rueda** f AUTO lug nut (*AmE*), wheelnut (*BrE*); ~ **de seguridad** f ING MECÁ locknut, checknut, safety nut, PROD locknut; ~ **de sombrerete** f ING MECÁ box nut, cap nut, MECÁ cap nut; ~ **de sujeción** f ING MECÁ checknut; ~ **de sujeción flotante** f ING MECÁ floating anchor nut; ~ **tapón** f ING MECÁ flanged nut; ~ **de tope elástica** f ING MECÁ, PROD elastic stop-nut; ~ **de un tornillo** f ING MECÁ screw box; ~ **torsiométrica** f PROD torque nut; ~ **de unión** f ING MECÁ coupling nut

tuercas: ~ **y pernos** fra CONST nuts and bolts

tueste m INSTAL HIDRÁUL calcin

tulio *m* (*Tm*) QUÍMICA thulium (*Tm*)

tulipa *f* ING ELÉC lampshade

tumbador *m* PROD *cerraduras* tumbler

tumbar *vi* ING MECÁ overturn

tundido *adj* TEXTIL cropped

tundir *vt* TEXTIL *acabado de la tela* crop

túnel *m* GEN tunnel; ~ **para acoplamiento** *m* TEC ESP docking tunnel; ~ **aerodinámico** *m* CONST, TEC ESP, TRANSP AÉR wind tunnel; ~ **aerodinámico de garganta cerrada** *m* FÍS closed-throat wind tunnel; ~ **del árbol de transmisión** *m* AUTO, VEH drive-shaft tunnel (*AmE*), propeller-shaft tunnel (*BrE*); ~ **de calefacción** *m* EMB heating tunnel; ~ **de comunicación** *m* CONST *excavación* drift; ~ **de conexión** *m* TEC ESP connecting tunnel; ~ **de congelación** *m* REFRIG freezing tunnel; ~ **de control** *m* ELECTRÓN drift tunnel; ~ **de curado** *m* CONST curing tunnel; ~ **de derivación** *m* HIDROL *presas* diversion tunnel; ~ **de descarga** *m* AGUA, HIDROL *presas* tailrace tunnel; ~ **de endurecimiento** *m* REFRIG *de helados* hardening tunnel; ~ **de enfriamiento** *m* ALIMENT, REFRIG cooling tunnel; ~ **de exploración** *m* MINAS drift; ~ **ferroviario** *m* CONST, FERRO railroad tunnel (*AmE*), railway tunnel (*BrE*); ~ **hipersónico de arco** *m* TRANSP AÉR hot shot wind tunnel; ~ **a presión** *m* ENERG RENOV *energía hidroeléctrica* pressure tunnel; ~ **de prospección** *m* MINAS prospect tunnel; ~ **de refrigeración** *m* ALIMENT, REFRIG cooling tunnel; ~ **de retractilación para el enfundado** *m* EMB shrink tunnel for sleeving; ~ **de retractilación para el sellado de fundas** *m* EMB shrink tunnel for sleeve sealing; ~ **de secado** *m* ALIMENT, EMB drying tunnel; ~ **de variación** *m* ELECTRÓN drift tunnel; ~ **de ventilación** *m* MINAS ventilation drive; ~ **de viento** *m* FÍS, TRANSP AÉR wind tunnel; ~ **para voladura** *m* MINAS gopher hole

tunelización *f* MINAS *construcciones* tunneling (*AmE*), tunnelling (*BrE*)

tungstato *m* QUÍMICA tungstate

tungsteno *m* (*W*) ING ELÉC, METAL, QUÍMICA tungsten (*W*), wolfram (*W*)

túngstico *adj* QUÍMICA tungstic

tungstosilicato *m* QUÍMICA tungstosilicate

turba *f* CARBÓN, QUÍMICA peat; ~ **fósil** *f* CARBÓN bog butter, fibrous peat; ~ **de pantanos** *f* CARBÓN bog peat

turbidez *f* GAS cloudiness, HIDROL, QUÍMICA turbidity

turbidimetría *f* QUÍMICA turbidimetry

turbidímetro *m* LAB *instrumento, análisis, medida de turbidez* turbidity meter, OCEAN turbidity meter, turbidimeter, QUÍMICA turbidimeter

turbidita *f* GEOL turbidite

turbiedad *f* ENERG RENOV turbidity

turbina *f* GEN turbine, MINAS, REFRIG, TERMO, TERMOTEC cylinder; ~ **de acción** *f* INSTAL HIDRÁUL action turbine, MECÁ, PROD impulse turbine; ~ **de acción y de impulsión** *f* AGUA, INSTAL HIDRÁUL reaction and impulse turbine; ~ **de aire** *f* ING MECÁ air turbine; ~ **de aire caliente de ciclo cerrado** *f* TRANSP closed-cycle hot-air turbine; ~ **de aire comprimido** *f* ING MECÁ air turbine; ~ **anegada** *f* INSTAL HIDRÁUL drowned turbine, submerged turbine; ~ **atmosférica** *f* ING MECÁ air turbine; ~ **axial** *f* INSTAL HIDRÁUL journal turbine, parallel flow turbine; ~ **de bomba** *f* ENERG RENOV pump turbine; ~ **de**

caudal paralelo *f* INSTAL HIDRÁUL axial-flow turbine, parallel flow turbine; ~ **centrífuga** *f* INSTAL HIDRÁUL outward-flow turbine; ~ **centrípeta** *f* PROD inward-flow turbine; ~ **de chorro** *f* TRANSP jet turbine; ~ **de combustión** *f* GAS, PROD, TEC ESP, TERMO combustion-gas turbine, gas generator; ~ **de combustión interna** *f* GAS, MECÁ, PROD, TERMO, TRANSP AÉR, TRANSP MAR gas turbine, internal combustion turbine; ~ **de condensación** *f* INSTAL TERM condensing turbine; ~ **de contraflujo** *f* ENERG RENOV reverse-flow turbine; ~ **de doble chorro** *f* TERMO, TRANSP, TRANSP AÉR fan jet turbine; ~ **de enfriamiento** *f* ING MECÁ cooling turbine; ~ **engranada** *f* MECÁ geared turbine; ~ **eólica** *f* ENERG RENOV, ING MECÁ wind turbine; ~ **de escape** *f* TERMOTEC exhaust turbine; ~ **de expansión** *f* REFRIG expansion turbine, turboexpander, TRANSP AÉR expansion turbine; ~ **de expansión de impulsión** *f* INSTAL HIDRÁUL *rueda hidráulica entre dos paredes verticales* velocity stage turbine; ~ **de flujo axial** *f* INSTAL HIDRÁUL axial-flow turbine, journal turbine; ~ **de gas** *f* GAS, MECÁ, PROD, TERMO gas turbine, TRANSP jet turbine engine, TRANSP AÉR, TRANSP MAR *motor* gas turbine; ~ **de gas de ciclo abierto** *f* TRANSP open-cycle gas turbine; ~ **de gas de ciclo cerrado** *f* TRANSP closed-cycle gas turbine; ~ **a gas de pistón libre** *f* ING MECÁ free-piston gas turbine; ~ **de gases** *f* GAS, PROD, TEC ESP, TERMO gas generator; ~ **de gases de escape** *f* AUTO exhaust-gas turbine; ~ **de hélice** *f* ENERG RENOV propeller turbine; ~ **hidráulica** *f* ENERG RENOV water turbine, INSTAL HIDRÁUL drowned turbine, TERMO water turbine; ~ **de impulsión** *f* ENERG RENOV, MECÁ, PROD impulse turbine; ~ **de inyección total** *f* INSTAL HIDRÁUL full-injection turbine; ~ **libre** *f* TRANSP AÉR free turbine; ~ **de limitación** *f* INSTAL HIDRÁUL limit turbine; ~ **lineal** *f* TRANSP linear turbine; ~ **mixta** *f* AGUA reaction and impulse turbine, INSTAL HIDRÁUL combined-flow turbine, mixed-flow turbine, reaction and impulse turbine; ~ **monoepática** *f* INSTAL HIDRÁUL single-stage turbine; ~ **monoexpansiva** *f* INSTAL HIDRÁUL single-stage turbine; ~ **motriz** *f* INSTAL HIDRÁUL impulse turbine; ~ **de reacción** *f* AGUA reaction-turbine, ENERG RENOV reaction turbine, INSTAL HIDRÁUL pressure turbine, reaction turbine, MECÁ reaction turbine, TRANSP AÉR turbojet; ~ **de rueda Pelton con árbol horizontal** *f* ENERG RENOV horizontal-shaft Pelton Wheel; ~ **de succión** *f* MINAS draft engine (*AmE*), draught engine (*BrE*); ~ **tipo Francis** *f* AGUA reaction-turbine, ENERG RENOV Francis turbine, reaction turbine, INSTAL HIDRÁUL, MECÁ reaction turbine; ~ **tipo Pelton** *f* ENERG RENOV impulse turbine, Pelton turbine; ~ **de vapor** *f* INSTAL HIDRÁUL, INSTAL TERM, TRANSP MAR *motor* steam turbine; ~ **a vapor de escape** *f* TERMOTEC *de la caldera* exhaust steam turbine; ~ **de vapor de mercurio** *f* ING MECÁ mercury vapor turbine (*AmE*), mercury vapour turbine (*BrE*); ~ **de viento** *f* ENERG RENOV, ING MECÁ wind turbine; ~ **de viento con eje vertical** *f* ENERG RENOV, ING MECÁ vertical-axis wind turbine

turbio *adj* AGUA turbid

turboalimentador: ~ **de escape** *m* AUTO exhaust turbocharger

turboalternador *m* ELEC *generador* turbo-alternator, ING ELÉC turbo-alternator, turbogenerator

turbobomba *f* AGUA turbopump, INSTAL HIDRÁUL turbine pump, TEC ESP, TRANSP MAR turbopump

turbocompresor *m* AUTO turbosupercharger, TRANSP MAR turbocompressor, VEH *motor* turbocharger; **~ radial y axial** *m* ING MECÁ radial and axial turbocompressor

turbocrucero *m* TRANSP MAR turbocruiser

turbodínamo *m* ING ELÉC turbogenerator

turboeléctrico *adj* ING ELÉC, TRANSP turboelectric

turbogenerador *m* ING ELÉC turbogenerator

turbohélice *f* TRANSP turbofan, TRANSP AÉR turbo-prop, turbopropeller

turbomáquina *f* TRANSP MAR turbine engine

turbomezclador *m* PROD *producción* turbomixer

turbomolecular *adj* ING MECÁ turbomolecular

turbomotor *m* ING ELÉC, TRANSP MAR turbine engine; **~ con ventilador entubado** *m* TRANSP ducted-fan turbo engine

turbonada *f* METEO, OCEAN squall; **~ negra** *f* OCEAN black squall

turbonave *f* TRANSP MAR turbine vessel

turboperforación *f* PETROL, TEC PETR turbodrilling

turbopropulsión *f* TRANSP MAR turbine propulsion

turborreactor *m* TERMO turbojet, turboreactor, TRANSP AÉR turbojet; **~ a postcombustión** *m* TRANSP turbo-ramjet; **~ con soplante turbofan** *m* TRANSP AÉR turbofan

turboseparación *f* ALIMENT turboseparation

turbosobrealimentador *m* AUTO turbosupercharger, ING MECÁ turbocharger

turbosoplante *m* ING MECÁ turbocharger, PROD blower, TRANSP MAR *motor* turbocharger

turbotren *m* FERRO, TRANSP turbotrain

turbulencia *f* AUTO, ENERG RENOV, METEO, TEC ESP, TRANSP AÉR turbulence; **~ de aire claro** *f* TRANSP AÉR clear-air turbulence; **~ de aire limpio** *f* TRANSP AÉR clear-air turbulence; **~ convectiva** *f* TRANSP AÉR convective turbulence; **~ espiral** *f* FÍS FLUID spiral turbulence; **~ de la estela** *f* TRANSP AÉR wake turbulence; **~ del extremo de la pala** *f* TRANSP AÉR blade-tip vortex; **~ homogénea isótropa** *f* FÍS FLUID homogeneous isotropic turbulence; **~ isótropa** *f* FÍS FLUID isotropic turbulence; **~ en rejilla** *f* FÍS FLUID grid turbulence

turbulento *adj* FÍS FLUID, METEO turbulent

turgita *f* MINERAL turgite

turgor *m* ALIMENT turgor

turmalina *f* FÍS, MINERAL tourmaline; **~ azul** *f* MINERAL blue tourmaline; **~ negra** *f* MINERAL schorl; **~ parda** *f* MINERAL brown tourmaline; **~ roja** *f* MINERAL rubellite

turno *m* PROD *de trabajo* shift, TEC PETR *perforación* crew, TEXTIL *de trabajo* shift; **~ de trabajo** *m* TEC PETR *perforación* shift (*BrE*), tour (*AmE*)

turquesa *f* MINERAL turquoise; **~ oriental** *f* MINERAL oriental turquoise

tutor *m* CARBÓN *de planta* prop

tuyano *m* QUÍMICA thujane

tuyeno *m* QUÍMICA thujene

tuyona *f* QUÍMICA thujone

TV[1] *abr* (*televisión*) TELECOM, TV TV (*television*)

TV[2]: **~ paga** *f* TV pay TV

TVC *abr* (*control del vector de empuje*) TEC ESP TVC (*thrust vector control*)

U

U *abr* (*uranio*) FÍS RAD, NUCL, QUÍMICA U (*uranium*)

UA *abr* AGRIC (*unidad animal*) AU (*animal unit*), INFORM&PD (*unidad aritmética*) AU (*arithmetic unit*)

UAL *abr* (*unidad aritmética y lógica*) INFORM&PD ALU (*arithmetic and logic unit*)

ubicación *f* CONST locating, location, INFORM&PD *memoria* location, ING ELÉC locating, ING MECÁ position, PETROL location, PROD, TELECOM, TRANSP MAR locating; **~ en el almacenamiento** *f* INFORM&PD storage location; **~ direccionable** *f* INFORM&PD addressable location; **~ por eco** *f* OCEAN echo location; **~ de existencias** *f* PROD stock location; **~ de memoria** *f* INFORM&PD memory location; **~ de la obra** *f* CONST construction site; **~ protegida** *f* AmL (*cf posición protegida Esp*) INFORM&PD protected location; **~ de la tabla de imagen de salida** *f* PROD output image table location

ubitrón *m* ELECTRÓN ubitron

UCP *abr* (*procesador central, unidad central de proceso*) INFORM&PD, TELECOM CPU (*central processing unit*)

udómetro *m* ENERG RENOV rain gage (*AmE*), rain gauge (*BrE*)

ued *m* HIDROL wadi

UHF *abr* (*frecuencia ultra-alta, frecuencia ultraelevada, hiperfrecuencia, ultra-alta frecuencia*) ELECTRÓN, TELECOM, TV UHF (*ultrahigh frequency*)

UHT *abr* (*uperizada*) ALIMENT UHT (*ultra heat treated*)

ULA *abr* (*unidad de llamada automática*) INFORM&PD ACU (*automatic calling unit*)

ulexina *f* QUÍMICA ulexine

ulexita *f* MINERAL ulexite

úlmico *adj* QUÍMICA ulmic

ulmina *f* QUÍMICA ulmin

ulmoso *adj* QUÍMICA ulmous

última: **~ fase** *f* TEC ESP *cohetes* upper stage; **~ fase inercial** *f* TEC ESP *cohetes* inertial upper stage (*IUS*); **~ fase rotatoria sólida** *f* (*SSUS*) TEC ESP Solid Spinning Upper Stage (*SSUS*); **~ mano** *f* CONST *pintura* finishing coat, MECÁ finish, P&C *pintura* finishing coat

último[1]: **~ grito** *m* TEXTIL *de la moda* craze; **~ peldaño incondicional** *m* PROD unconditional end rung

último[2]: **~ en entrar, primero en salir** *fra* (*LIFO*) INFORM&PD, PROD last-in-first-out (*LIFO*)

ultra[1]: **~ pequeño** *adj* PROD ultrasmall

ultra[2]: **~-alta frecuencia** *f* (*UHF*) ACÚST, CRISTAL, ELECTRÓN, FÍS, FÍS ONDAS, TELECOM ultrahigh frequency (*UHF*)

ultraacústica *f* ACÚST, CRISTAL, ELECTRÓN, FÍS, FÍS ONDAS ultrasonics

ultracentrífuga *f* FÍS, LAB, NUCL, QUÍMICA ultracentrifuge

ultracentrifugación *f* QUÍMICA ultracentrifugation

ultracentrifugado *m* NUCL ultracentrifugation

ultracongelación *f* EMB deep-freezing

ultracromatografía *f* LAB, QUÍMICA ultrachromatography

ultrafiltración *f* PROC QUÍ ultrafiltration

ultrafiltrado *m* QUÍMICA ultrafiltrate, ultrafiltration

ultrafiltro *m* QUÍMICA ultrafilter

ultramafita *f* GEOL ultramafite

ultramar: **de ~** *adj* TRANSP MAR *comercio* overseas

ultramarino[1] *adj* TRANSP MAR *comercio* overseas

ultramarino[2] *m* QUÍMICA ultramarine

ultramicroanálisis *m* INSTR, LAB, NUCL ultramicroanalysis

ultramicroscopía *f* FÍS, LAB, METAL, ÓPT, QUÍMICA ultramicroscopy

ultramicroscópico *adj* FÍS, LAB, METAL, ÓPT, QUÍMICA ultramicroscopic

ultramicroscopio *m* FÍS, LAB, METAL, ÓPT, QUÍMICA ultramicroscope

ultramilonita *f* GEOL ultramylonite

ultrapasteurización *f* ALIMENT ultrapasteurization

ultraquímico *adj* QUÍMICA ultrachemical

ultrarrápido *adj* MECÁ high-speed

ultrarrefracción *f* TEC ESP super-refraction

ultrasónica *f* FÍS ONDAS ultrasonics

ultrasónico *adj* FÍS ONDAS ultrasonic

ultrasonido *m* ACÚST, ELECTRÓN, FÍS, FÍS ONDAS, LAB ultrasound

ultratraza *f* QUÍMICA ultratrace

ultravacío *m* TERMO ultravacuum

ultraveloz *adj* MECÁ high-speed

ultravioleta *m* (*UV*) FÍS, FÍS RAD, ÓPT, TEC ESP ultraviolet (*UV*); **~ cercano** *m* FÍS, FÍS RAD near ultraviolet; **~ lejano** *m* FÍS, FÍS RAD far ultraviolet; **~ en vacío** *m* TEC ESP vacuum ultraviolet

ululación *f* TELECOM warble

UMA *abr* (*unidad de masa atómica*) FÍS AMU (*atomic mass unit*), AWU (*atomic weight unit*)

umbélico *adj* QUÍMICA umbellic

umbilical *adj* GEOL *paleontología* umbilical

umbra *f* FÍS, ÓPT umbra

umbral *m* ACÚST threshold, AGUA sill, CONST ground-sill, *puerta* threshold, *puertas* sill, ELECTRÓN, FÍS, GEOL, INFORM&PD threshold, MINAS *entibación*, OCEAN *hidrología* sill, PROD *puerta contraincendios* sill plate, TEC ESP threshold; **~ absoluto** *m* TV absolute threshold; **~ de acción láser** *m* FÍS, ÓPT, TELECOM lasing threshold; **~ de amplitud** *m* ELECTRÓN amplitude threshold; **~ de asiento** *m* AGUA clap sill; **~ de audibilidad** *m* CONTAM *legislación* threshold of audibility, threshold of hearing; **~ de audición normalizado** *m* ACÚST standardized threshold hearing; **~ auditivo normal** *m* ACÚST normal hearing threshold; **~ del color** *m* TV color threshold (*AmE*), colour threshold (*BrE*); **~ de compuerta** *m* AGUA gate sill; **~ de detección** *m* ÓPT, TELECOM detection threshold; **~ diferencial** *m* ACÚST *audición* difference limen, differential threshold; **~ diferencial de frecuencia** *m* ACÚST differential threshold of frequency; **~ diferencial de nivel de presión del sonora** *m* ACÚST differential threshold of sound pressure level; **~ diferencial relativo** *m* ACÚST relative differential threshold; **~ efectivo del boro** *m* NUCL effective boron cut-off; **~ de efecto láser** *m* FÍS, ÓPT laser-effect threshold; **~ de energía geomagnética**

m TEC ESP geomagnetic cutoff energy; ~ **de esclusa** *m* AGUA lock sill; ~ **fotoeléctrico** *m* FÍS photoelectric threshold; ~ **de iluminación visible** *m* TRANSP AÉR visual threshold of illumination; ~ **de ionización** *m* FÍS, GAS ionization threshold; ~ **de luminiscencia** *m* FÍS, ÓPT, TV luminescence threshold; ~ **negro** *m* TV black porch; ~ **normal de audición dolorosa** *m* ACÚST normal threshold of painful hearing; ~ **observado** *m* ELECTRÓN, FÍS, NUCL, ÓPT observed threshold; ~ **de pista** *m* TRANSP AÉR runway threshold; ~ **posterior** *m* TV back porch; ~ **de la puerta** *m* CONST door sill; ~ **de la señal** *m* ELECTRÓN, FÍS, TEC ESP, TELECOM signal threshold; ~ **del voltaje** *m* TV threshold voltage

uña *f* ING MECÁ finger, pawl, ratchet, MECÁ pawl, ratchet, TRANSP MAR *del ancla* anchor fluke, fluke; ~ **de apriete** *f* ING MECÁ dog hook; ~ **de arrastre** *f* ING MECÁ catch pin, drive pin

unario *adj* INFORM&PD unary

UNC *abr* (*hilo de paso ancho unificado*) ING MECÁ UNC (*unified coarse thread*)

undecano *m* QUÍMICA undecane

undecanoico *adj* QUÍMICA undecanoic

undecilénico *adj* QUÍMICA undecylenic

UNF *abr* (*rosca de paso fino unificado*) ING MECÁ UNF (*unified fine thread*)

uniáxico *adj* CRISTAL, FÍS, ÓPT, QUÍMICA uniaxial

único: ~ **de borde posterior** *m* PROD trailing-edge one-shot

unidad[1]: **de** ~ **completa** *adj* ING MECÁ, PROD complete-assembly

unidad[2] *f* CINEMAT, ELEC, FÍS, GEOL unit, INFORM&PD drive, unit, device, deck, ING MECÁ item, NUCL, QUÍMICA, TELECOM unit;
▪ **a** ~ **de acabado de gas** *f* GAS gas completion unit; ~ **de acceso** *f* TELECOM access unit (*AU*); ~ **de acceso al intercambio de datos electrónicos** *f* TELECOM EDI-UA; ~ **de acceso a la entrega física** *f* TELECOM physical-delivery access unit (*PDAU*); ~ **de acceso télex** *f* TELECOM telex access unit; ~ **para acoplamiento** *f* TEC ESP docking unit; ~ **de activación de armas y seguridad a distancia** *f* TEC ESP remote arming and safety unit; ~ **administrativa** *f* TELECOM administrative unit (*AU*); ~ **agrícola** *f* AGRIC farming unit; ~ **de alimentación** *f* CINEMAT, ING ELÉC power pack, PROD *electricidad* power-supply unit (*PSU*); ~ **de alimentación portátil** *f* TV portable pack; ~ **de almacenamiento** *f* PROD storage unit; ~ **de almacenamiento central** *f* AmL INFORM&PD main storage (*AmE*), main store (*BrE*); ~ **de almacenamiento principal** *f* INFORM&PD main storage (*AmE*), main store (*BrE*); ~ **de almacenamiento y reexpedición** *f* TELECOM store-and-forward unit (*SFU*); ~ **de alquilación** *f* DETERG, PROC QUÍ alkylation unit; ~ **amplificadora de distribución de impulsos** *f* TELECOM pulse-distribution amplifier unit (*PDAU*); ~ **animal** *f* (*UA*) AGRIC animal unit (*AU*); ~ **antideslizante** *f* TRANSP AÉR antiskid unit; ~ **de área** *f* METR unit of area; ~ **aritmética** *f* (*UA*) INFORM&PD arithmetic unit (*AU*); ~ **aritmética y lógica** *f* (*UAL*) INFORM&PD arithmetic and logic unit (*AUL*); ~ **de armadura balanceada** *f* ELEC *motor* balanced-armature unit; ~ **de armadura equilibrada** *f* ELEC *motor* balanced-armature unit; ~ **de arranque automático** *f* TRANSP AÉR automatic starting unit; ~ **de autoposición** *f*

TRANSP AÉR autopositioning unit; ~ **autosituadora** *f* TRANSP AÉR autopositioning unit; ~ **auxiliar de conmutación** *f* TELECOM auxiliary switching unit;
▪ **b** ~ **base** *f* TELECOM base unit (*BU*); ~ **básica** *f* ING MECÁ fundamental unit; ~ **binaria** *f* ELECTRÓN bit; ~ **bioestratigráfica** *f* GEOL biozone; ~ **de bombeo** *f* CONTAM MAR pumping unit, MINAS *pozo petrolífero* jack, PROD pumping unit;
▪ **c** ~ **del cabezal de video** *AmL*, ~ **del cabezal de vídeo** *Esp f* TV video head assembly; ~ **de captación** *f* PETROL capture unit; ~ **de captura** *f* PETROL capture unit; ~ **de cartucho** *f* INFORM&PD cartridge drive; ~ **de CD-ROM** *f* INFORM&PD, ÓPT CD-ROM drive; ~ **central** *f* INFORM&PD mainframe; ~ **central de conmutación** *f* TELECOM central switching unit; ~ **central de proceso** *f* (*UCP*) INFORM&PD, TELECOM central processing unit (*CPU*); ~ **de cinta** *f* INFORM&PD tape deck, tape drive, tape unit; ~ **de cinta para copia de seguridad** *f* INFORM&PD streaming tape drive; ~ **de cinta magnética** *f* INFORM&PD magnetic tape unit; ~ **de cola doble** *f* TRANSP AÉR twin-tail unit; ~ **compresora de aire** *f* ING MECÁ air compressor set; ~ **conectable** *f* ING ELÉC plug-in unit; ~ **de conexión** *f* TELECOM junctor; ~ **de conexión de líneas** *f* TELECOM line connection unit; ~ **de conmutación** *f* TELECOM switching unit; ~ **de conmutación de circuitos** *f* TELECOM circuit switching unit; ~ **de conmutación electromecánica** *f* ELEC, TELECOM electromechanical-switching unit; ~ **conmutadora** *f* TELECOM switching unit; ~ **de construcción modular** *f* ELEC *equipo*, ING MECÁ, PROD module; ~ **de control** *f* CINEMAT, ELEC, INFORM&PD, ING ELÉC, ING MECÁ, TELECOM control unit; ~ **de control de aceleración** *f* TRANSP AÉR acceleration control unit; ~ **de control de la cámara** *f* CINEMAT, TV camera control unit (*CCU*); ~ **de control de combustible** *f* TRANSP AÉR fuel-control unit; ~ **de control a distancia** *f* TELECOM remote control unit (*RCU*); ~ **de control electrónico** *f* AUTO electronic control unit; ~ **de control móvil** *f* CINEMAT, TV mobile control unit; ~ **de control del piloto automático** *f* TRANSP AÉR autopilot control unit; ~ **para control de posición de vuelo** *f* TEC ESP attitude control unit; ~ **de control y representación visual** *f* TEC ESP control and display unit; ~ **de control de temperatura** *f* MINAS temperature-monitoring unit; ~ **de control transistorizado** *f* AUTO transistor control unit; ~ **de control del tren de aterrizaje** *f* TRANSP AÉR landing gear control unit; ~ **conveniente** *f* ING MECÁ practical unit; ~ **de conversión operacional** *f* TRANSP AÉR operational conversion unit (*OCU*); ~ **copiadora** *f* ING MECÁ copying unit; ~ **correctora de cabeceo** *f* TRANSP AÉR pitch-correcting unit; ~ **cronostratigráfica** *f* GEOL *estratigrafía* chronostratigraphic unit;
▪ **d** ~ **de datos** *f* INFORM&PD item; ~ **de datos protocolarios de la subcapa de convergencia** *f* TELECOM convergence-sublayer protocol data unit (*CSPDU*); ~ **de datos de protocolo** *f* TELECOM protocol data unit (*PDU*); ~ **de datos de protocolo N** *f* TELECOM *para técnicos* N-service data unit; ~ **de datos del protocolo de la red** *f* TELECOM network protocol data unit (*NPDU*); ~ **de datos del protocolo de sesión** *f* TELECOM session protocol data unit (*SPDU*); ~ **de datos de servicio** *f* TELECOM service data unit (*SDU*); ~ **datos de servicio AAL** *f* TELE-

COM AAL service data unit (*AAL-SDU*); ~ **derivada** *f* FÍS, ING MECÁ derived unit; ~ **de deshidratación de helio** *f* NUCL helium dehydrator unit; ~ **de destilación de crudo** *f* TEC PETR crude distillation unit (*CDU*); ~ **digital de acceso al abonado** *f* TELECOM digital subscriber access unit; ~ **de disco** *f* IMPR, INFORM&PD, TELECOM disk drive, disk unit, floppy disk drive; ~ **de disco borrable** *f* INFORM&PD erasable disk drive; ~ **de disco de CD-ROM** *f* ÓPT CD-ROM disc drive; ~ **de disco compacto** *f* INFORM&PD CD drive; ~ **de disco compacto de música** *f* INFORM&PD compact music disk drive; ~ **de disco flexible** *f* IMPR, INFORM&PD, TELECOM floppy disk drive; ~ **de disco óptico** *f* INFORM&PD optical disk drive, ÓPT optical disc drive (*BrE*), optical disk drive (*AmE*); ~ **de disco óptico registrable** *f* ÓPT writable optical disc drive (*BrE*), writable optical disk drive (*AmE*); ~ **de display de video** *AmL*, ~ **de display de vídeo** *Esp f* TV video display unit (*VDU*); ~ **de disquete** *f* IMPR, INFORM&PD, TELECOM diskette drive, floppy disk drive;

▪ **e**   ~ **electrolítica** *f* ELEC, FÍS, ING ELÉC electrolytic unit; ~ **electromagnética** *f* ELEC electromagnetic unit; ~ **de elementos amovibles** *f* ELEC draw-out unit; ~ **de embrague y desembrague** *f* TRANSP AÉR freewheel and clutch unit; ~ **de empuje para torcer y soltar** *f* PROD twist-to-release push unit; ~ **de encendido automático** *f* TRANSP AÉR automatic starting unit; ~ **de encendido transistorizado** *f* AUTO transistor ignition unit; ~ **enchufable** *f* ELEC plug-in unit, *conexión* plug pin; ~ **enchufable del conducto portacable de la barra colectora** *f* ELEC *conexión* bus duct plug-in unit; ~ **de energía** *f* D&A power-supply unit (*PSU*), ING MECÁ power unit, TERMO unit of energy; ~ **de energía auxiliar** *f* D&A auxiliary power unit (*APU*); ~ **de entropía** *f* TERMO unit of entropy; ~ **esclava** *f* TV slave unit, slaving; ~ **de esfuerzo pesquero** *f* OCEAN unit of fishing effort; ~ **de existencias** *f* PROD stock unit; ~ **de exposición** *f* FÍS RAD unit of exposure; ~ **de extensión** *f* AGRIC field unit; ~ **extractora de polvo** *f* CONTAM, SEG *para gas de chimenea* dust removal plant; ~ **extraíble** *f* ELEC draw-out unit;

▪ **f**   ~ **de faro hermético** *f* AUTO sealed beam unit; ~ **de flash alimentada por pilas** *f* FOTO battery-powered flash unit; ~ **para formar, llenar y cerrar bolsitas** *f* EMB sachet form fill seal unit; ~ **forrajera** *f* AGRIC feed unit; ~ **de fuelle de extensión** *f* CINEMAT, FOTO extension bellows unit; ~ **de fuerza** *f* FÍS, ING MECÁ force unit, unit of force; ~ **de fuerza del timón de dirección** *f* TRANSP AÉR, TRANSP MAR rudder-power unit; ~ **funcional** *f* INFORM&PD, TELECOM functional unit (*FU*); ~ **funcional entrelazada** *f* TELECOM interworking functional unit (*IFU*); ~ **fundamental** *f* ELEC, FÍS, ING MECÁ fundamental unit;

▪ **g**   ~ **ganadera** *f* AGRIC cattle unit; ~ **de ganado mayor** *f* AGRIC large animal unit; ~ **giroscópica triaxial** *f* TEC ESP three-axis gyro unit; ~ **del giróscopo del piloto automático** *f* TRANSP AÉR autopilot gyro unit; ~ **de glaseado doble** *f* C&V double glazing unit; ~ **de grabación** *f* INSTR cartridge recorder, recording unit; ~ **gravimétrica** *f* PETROL gravity unit;

▪ **h**   ~ **herméticamente cerrada** *f* ING ELÉC, ING MECÁ, PROC QUÍ, PROD, SEG hermetically-sealed unit; ~ **del husillo** *f* ING MECÁ spindle nose;

▪ **i**   ~ **independiente** *f* PROD stand-alone unit; ~ **inercial** *f* TEC ESP inertial unit; ~ **de información** *f* INFORM&PD data item, TEC ESP *comunicaciones* frame; ~ **interconectada** *f* TELECOM cross-connect unit; ~ **de interfase** *f* INFORM&PD interface unit; ~ **de interfaz** *f* INFORM&PD interface unit; ~ **internacional** *f* ELEC, FÍS international unit;

▪ **l**   ~ **de lectura-escritura** *f* INFORM&PD, ÓPT read-write drive; ~ **limpia** *f* ELECTRÓN *fabricación de C.I.* clean room; ~ **de llamada automática** *f* (*ULA*) INFORM&PD automatic calling unit (*ACU*); ~ **de llenado volumétrica** *f* EMB volumetric-filling unit; ~ **lógica** *f* ELEC *control* logic unit; ~ **de longitud** *f* METR unit of length;

▪ **m**   ~ **de maniobra asistida** *f* TEC ESP manned maneuvering unit (*AmE*), manned manoeuvring unit (*BrE*); ~ **de mantenimiento** *f* PROD, TRANSP AÉR maintenance unit; ~ **de masa atómica** *f* (*UMA*) FÍS atomic mass unit (*AMU*), atomic weight unit (*AWU*); ~ **de mediación de combustible** *f* TEC ESP fuel measuring unit; ~ **de medida** *f* METR, TEC PETR measurement unit, unit of measurement; ~ **de medida de la fabricación** *f* PROD production unit of measure; ~ **del mezclador** *f* PROC QUÍ, PROD, TRANSP AÉR mixing unit; ~ **modular** *f* ELECTRÓN module set, ING MECÁ modular unit; ~ **monitora** *f* CINEMAT, INFORM&PD, TELECOM, TV monitor unit; ~ **de motor** *f* ING MECÁ, PROD, TRANSP power unit; ~ **motora** *f* PROD, TRANSP driving unit; ~ **motorizada** *f* D&A motorized unit; ~ **motriz** *f* ING MECÁ, PROD, TRANSP power unit; ~ **móvil** *f* CINEMAT, D&A mobile unit; ~ **móvil de registro** *f* GAS mobile logging unit;

▪ **n**   ~ **N de datos del protocolo** *f* TELECOM N-protocol data unit; ~ **de nivelación** *f* TRANSP AÉR leveling unit (*AmE*), levelling unit (*BrE*);

▪ **o**   ~ **óptica** *f* INFORM&PD, ÓPT optical drive; ~ **óptica borrable** *f* INFORM&PD, ÓPT erasable optical drive; ~ **óptica de lectura y escritura** *f* INFORM&PD, ÓPT read-write optical drive; ~ **óptica de lectura-escritura** *f* INFORM&PD, ÓPT read-write optical drive;

▪ **p**   ~ **para-autóctona** *f* GEOL *materias transportadas a corta distancia* para-autochthonous unit; ~ **de pasajeros** *f* TRANSP passenger car unit (*PCU*); ~ **de pedido** *f* PROD ordering unit; ~ **periférica** *f* INFORM&PD peripheral, peripheral unit; ~ **de permeabilidad** *f* HIDROL darcy; ~ **de peso** *f* METR weight unit; ~ **de potencia** *f* FOTO power-pack unit, ING MECÁ, PROD power unit; ~ **de potencia de tierra** *f* (*GPU*) TRANSP AÉR ground power unit (*GPU*); ~ **práctica** *f* ING MECÁ practical unit; ~ **prefabricada** *f* CONST precast unit, prefabricated unit; ~ **de preformado de émbolo** *f* ING MECÁ piston-type preforming unit; ~ **de presentación de datos del protocolo** *f* TELECOM presentation protocol data unit (*PPDU*); ~ **de presentación visual** *f* (*VDU*) ELECTRÓN, INFORM&PD, INSTR, TELECOM visual display unit (*VDU*); ~ **principal** *f* INFORM&PD mainframe, ING MECÁ main unit; ~ **de programa** *f* INFORM&PD program unit, programming unit; ~ **de programador** *f* INFORM&PD programmer unit; ~ **de propulsión** *f* TEC ESP, TRANSP AÉR propulsion unit; ~ **de protocolo** *f* TELECOM protocol unit (*PU*); ~ **de puerta de tiro duro** *f* PROD hard-fired gate drive; ~ **de puesta en fase** *f* TRANSP AÉR phasing unit; ~ **pulidora** *f* C&V polishing unit;

~ r ~ **de radar** *f* TRANSP, TRANSP AÉR, TRANSP MAR radar unit; ~ **de la radiación** *f* FÍS RAD *medida* radiation unit; ~ **de radioterapia** *f* FÍS RAD *medida* radiation unit; ~ **de rayos X** *f* INSTR X-ray unit; ~ **recambiable** *f* ELEC plug-in unit; ~ **rectificadora** *f* ELEC, ELECTRÓN, TELECOM rectifier unit; ~ **de recursos** *f* PROD resource unit; ~ **reemplazable en línea** *f* ING ELÉC line replaceable unit (*LRU*); ~ **de referencia para posición de vuelo** *f* TEC ESP attitude reference unit; ~ **de referencia vertical** *f* TEC ESP vertical reference unit; ~ **de refinado de crudo** *f* TEC PETR crude distillation unit (*CDU*); ~ **de relé de la hélice** *f* TRANSP AÉR propeller-relay unit; ~ **remota** *f* INFORM&PD, TELECOM, TV remote unit (*RU*); ~ **de repetición** *f* QUÍMICA recurring unit; ~ **de representación visual** *f* (*VDU*) ELECTRÓN, INFORM&PD, INSTR visual display unit (*VDU*); ~ **de reserva** *f* D&A, ING ELÉC, PROD standby unit; ~ **de respuesta de audio** *f* ELECTRÓN, INFORM&PD audio response unit; ~ **de respuesta oral** *f* ELECTRÓN, INFORM&PD audio response unit; ~ **de respuestas telefónicas** *f* TELECOM voice response unit; ~ **de rodaje fuera del estudio** *f* CINEMAT location unit;

~ s ~ **de seguridad** *f* PROD, TEC ESP safety unit; ~ **de sellado** *f* NUCL seal unit; ~ **de señal de mensajes** *f* TELECOM message signal unit (*MSU*); ~ **separadora** *f* AGRIC *paja*, CARBÓN, ING MECÁ, PROC QUÍ, PROD separating unit; ~ **SI** *f* (*unidad del sistema internacional*) ELEC, FÍS, FÍS PART, METR SI unit (*international system unit*); ~ **de sintonización** *f* ELECTRÓN, TELECOM, TV tuner; ~ **del sistema internacional** *f* (*unidad SI*) ELEC, FÍS, FÍS PART, METR international system unit (*SI unit*); ~ **de sistema de referencia de avance y vertical** *f* TEC ESP heading and vertical reference unit system; ~ **del sistema de vacío** *f* ING MECÁ vacuum unit; ~ **sustituible en línea** *f* ING ELÉC line replaceable unit (*LRU*);

~ t ~ **técnica** *f* PROD engineering unit; ~ **de teleconmutación** *f* TELECOM, TV remote-switching unit (*RSU*); ~ **tensora** *f* PROD take-up unit; ~ **térmica británica** *f* (*BTU*) GAS, ING MECÁ, PETROL, TERMOTEC British thermal unit (*BTU*); ~ **de tiempo** *f* INFORM&PD time slice; ~ **de tiro para torcer y soltar** *f* PROD twist-to-release pull unit; ~ **para torcer hasta soltar** *f* PROD twist-to-release unit; ~ **de tracción** *f* TRANSP tractive unit, tractor unit; ~ **tractora** *f* TRANSP tractive unit, tractor unit; ~ **de tráfico** *f* TELECOM traffic unit (*TU*); ~ **de transferencia de datos del protocolo** *f* TELECOM transport protocol data unit (*TPDU*); ~ **de transmisión** *f* TELECOM transmission unit; ~ **de transmisión principal** *f* ING MECÁ, INSTR, PROD main drive unit; ~ **de transporte no motora** *f* TRANSP passive transport unit; ~ **de transporte y refrigeración** *f* ING MECÁ, REFRIG, TRANSP cooling and conveying unit; ~ **de tratamiento de energía** *f* TEC ESP power-conditioning unit; ~ **de tratamiento de potencia** *f* TEC ESP power-conditioning unit; ~ **tributaria** *f* TELECOM tributary unit (*TU*);

~ v ~ **de vaciado** *f* C&V casting unit; ~ **de vapor** *f* TRANSP steam car; ~ **de viajeros** *f* TRANSP passenger car unit (*PCU*); ~ **de vidriado múltiple** *f* C&V multiple glazing unit; ~ **de visualización** *f* (*VDU*) INFORM&PD, TELECOM visual display unit (*VDU*); ~ **de visualización virtual** *f* (*VDU*) INFORM&PD

visual display unit (*VDU*); ~ **de volumen** *f* ACÚST volume unit

unidad[3]: **por ~ de área** *fra* FÍS per unit area; **por ~ de longitud** *fra* FÍS per unit length; **por ~ de masa** *fra* FÍS per unit mass; **por ~ de volumen** *fra* FÍS per unit volume

unidades: ~ **intercambiables** *f pl* FERRO *vehículos* swap bodies

unidimensional *adj* FÍS one-dimensional

unidireccional *adj* ACÚST *micrófono*, ELEC *corriente*, ING ELÉC unidirectional, TELECOM *circuito de transmisión* simplex, unidirectional

unido *adj* ING MECÁ connected; ~ **a tierra** *adj* ELEC, ING ELÉC, TELECOM earthed (*BrE*), grounded (*AmE*)

unificación *f* TEC PETR *concesiones, licencias* unitization

unifinación *f* QUÍMICA unifining

uniformación: ~ **de métodos de prueba** *f* ING MECÁ, INSTR, LAB standardization of test methods

uniforme *adj* ELEC, FÍS uniform, GEOL even, HIDROL, ING ELÉC uniform, ING MECÁ standard, METAL, ÓPT, QUÍMICA uniform

uniformidad *f* INFORM&PD consistency; ~ **del pavimento** *f* CONST pavement surface evenness; ~ **del tono** *f* IMPR hue consistency

uniformización *f* GEN standardization

uniformizar *vt* GEN standardize

unilateral *adj* ELEC *corriente* unidirectional

unión *f* C&V splice, CINEMAT join, CONST binder, connection, bond, fastening, meeting, union, ELEC *conexión* joint, junction, ELECTRÓN conjunction, EMB bond, INFORM&PD union, junction, jump instruction, ING MECÁ coupling link, tie, linkage, link, fastening, coupling, couple, connection, MECÁ link, linkage, coupling, fastening, bonding, METAL binding, NUCL bond, joint, jointing, ÓPT joint, P&C *enlace químico* bond, linkage, *adhesivos* bonding, PETROL joint, collar, PROD fastening, jointing, junction, QUÍMICA bond, TEC ESP attachment, TEC PETR collar, TELECOM splice, junction, TRANSP AÉR coupling, TRANSP MAR *construcción* bonding; ~ **abrupta** *f* ELECTRÓN abrupt junction; ~ **acodada** *f* ING MECÁ cranked link; ~ **de asiento cónico** *f* ING MECÁ conical clamping connection; ~ **de barra colectora** *f* ELEC *conmutador* busbar coupler; ~ **de bayoneta** *f* ELEC bayonet joint; ~ **biselada** *f* CONST bevel joint; ~ **de bordes** *f* ELECTRÓN edge latching; ~ **de bridas** *f* ING MECÁ flanged union; ~ **de cables** *f* ELEC, ING ELÉC *conexión*, TELECOM cable coupling, cable joint; ~ **caliente** *f* ELEC *termopar* PROD hot junction; ~ **por calor** *f* P&C *transformación* heat bonding; ~ **por chavetas** *f* ING MECÁ, PROD keying; ~ **cónica** *f* ING MECÁ cone union body, MECÁ taper; ~ **covalente** *f* METAL, QUÍMICA covalent bond; ~ **cruzada** *f* ING MECÁ cross union; ~ **de la cubierta al casco** *f* TRANSP MAR deck-hull bonding; ~ **del cubrejunta** *f* PROD butt-strap joint; ~ **por difusión** *f* ELECTRÓN diffused junction; ~ **de la dislocación** *f* METAL dislocation junction; ~ **doble** *f* PETROL double joint; ~ **eléctrica** *f* TEC ESP *circuitos eléctricos* bonding; ~ **emisor base** *f* ELECTRÓN emitter-base junction; ~ **de entramado** *f* CONST stud union; ~ **de escotadura** *f* CONST notch joint; ~ **de espiga doble** *f* CONST double tenon joint; ~ **de espiga y mortaja doble** *f* CONST double mortise-and-tenon joint; ~ **estacionaria** *f* ING MECÁ stationary link; ~ **de fibra óptica** *f* TELECOM optical fiber splice

(*AmE*), optical fibre splice (*BrE*); ~ **de fieltro y espuma** *f* CONST felt and foam joint; ~ **en forma de T** *f* ELEC cable T-joint, tee joint; ~ **giratoria** *f* ING MECÁ swivel joint, TEC PETR swivel; ~ **de hierro fundido** *f* CONST cast-iron joint; ~ **por hilo** *f* ING ELÉC wire bonding; ~ **en L** *f* CONST union elbow; ~ **de lengüeta postiza** *f* CONST loose-tongue joint; ~ **local** *f* TELECOM local junction; ~ **localmente oxidada** *f* ELECTRÓN locally-oxided junction; ~ **de manguera** *f* MINAS flexible-hose union; ~ **de la manguera de alimentación** *f* TEC PETR *perforación* feed-hose union; ~ **mecánica** *f* ÓPT, TELECOM, TV mechanical splice; ~ **metal-semiconductor** *f* ELECTRÓN metal-semiconductor junction; ~ **multifibra** *f* ING ELÉC, ÓPT, TELECOM multifiber joint (*AmE*), multifibre joint (*BrE*); ~ **de múltiples fibras** *f* ING ELÉC, ÓPT, TELECOM multifiber joint (*AmE*), multifibre joint (*BrE*); ~ **de nuez** *f* ING MECÁ swivel joint; ~ **de paro** *f* ING MECÁ knock-off link; ~ **con pasador** *f* CONST cottered joint; ~ **de pasador** *f* CONST *tuberías* stud coupling; ~ **con pernos** *f* CONST stud link, bolting, stud union, TEC ESP bolted connection; ~ **de una pieza en T** *f* CONST T-piece union, tee-piece union; ~ **plegable** *f* CONST folding joint; ~ **posterior** *f* ELEC *conexión de cable* breeches joint; ~ **con ranura de encaje** *f* CONST fillistered joint; ~ **ranurada y embadurnada** *f* CONST ploughed-and-feathered joint (*BrE*), plowed-and-feathered joint (*AmE*); ~ **ranurada y con lengüeta** *f* CONST ploughed-and-tongued joint (*BrE*), plowed-and-tongued joint (*AmE*); ~ **de reactancias** *f* FERRO reactance bond; ~ **repulsiva** *f* METAL repulsive junction; ~ **a rosca** *f* ING MECÁ screw joint; ~ **roscada** *f* CONST, ING MECÁ nipple, NUCL threaded joint; ~ **de rótula** *f* ING MECÁ swivel joint; ~ **con solape** *f* CONST overlap joint; ~ **en T** *f* CONST *fontanería* T-joint, union T, ING MECÁ elbow joint; ~ **en T para tubos** *f* ING MECÁ pipe tee; ~ **de taza** *f* PROD *tubería de plomo* cup joint; ~ **de TEC** *f* ELECTRÓN junction FET; ~ **térmica** *f* P&C, PROD, TERMO thermal bonding; ~ **a tierra** *f* AUTO earth connection (*BrE*), ground connection (*AmE*), CONST earthing (*BrE*), grounding (*AmE*), ELEC earth connection (*BrE*), earthing (*BrE*), ground connection (*AmE*), grounding (*AmE*), ELECTRÓN, FERRO earthing (*BrE*), grounding (*AmE*), ING ELÉC connection to earth (*BrE*), connection to ground (*AmE*), earth connection (*BrE*), earth connector (*BrE*), earthing (*BrE*), ground (*AmE*), ground connection (*AmE*), ground connector (*AmE*), grounding (*AmE*), PROD, VEH earth connection (*BrE*), ground connection (*AmE*); ~ **transversal** *f* ING MECÁ cross union; ~ **triple** *f* AGUA, CONST, GAS, ING MECÁ, METAL, PROC QUÍ, TEC PETR triple junction; ~ **de tubería** *f* CONST pipe union; ~ **de tuberías** *f* AGUA, CONST, GAS, ING MECÁ, METAL, PROC QUÍ, TEC PETR pipe connection; ~ **de tubo** *f* ING MECÁ pipe junction; ~ **de tubo flexible** *f* MINAS flexible-hose union; ~ **de tubos** *f* ING MECÁ pipe union; ~ **en U** *f* ING MECÁ, MECÁ clevis link; ~ **en Y** *f* ELEC y-joint

Unión: ~ **Europea de Brandis y Licores** *f* ALIMENT European Alcohol Brandy and Spirit Union

unipolar *adj* ELEC *dinamo* unipolar, *fuente de alimentación* monopolar, *terminal* single-pole, ELECTRÓN, INFORM&PD, ING ELÉC, NUCL unipolar, TV single-pole

unir *vt* CINEMAT join, CONST join, make a joint, match, ING MECÁ attach, hook on, MECÁ lock, PROD inter-face, QUÍMICA bond, unite, TEC ESP *vehículos* attach, TELECOM splice; ~ **con la cremallera** *vt* TEXTIL *forro* zip in; ~ **en frío** *vt* PROD, TERMO cold-bond; ~ **a masa** *vt* AUTO, ELEC, FÍS, ING ELÉC, PROD, TELECOM, VEH *instalación eléctrica* earth (*BrE*), ground (*AmE*); ~ **mediante puentes** *vt* CONST bridge; ~ **con pernos** *vt* CONST bolt; ~ **a tope** *vt* CONST butt-joint

unirse *vi* ING MECÁ lock

uniselector *m* ING ELÉC selector switch, uniselector

unísono *m* ACÚST unison

universo *m* MATEMÁT universal set

untar: ~ **la tinta** *vi* COLOR, IMPR roll ink on

untuosidad *f* TRANSP slipperiness

UO₂ *abr* (*óxido de uranio*) QUÍMICA UO_2 (*uranium oxide*)

up: ~ **quark** *m* FÍS PART up quark

uperización *f* ALIMENT uperization

uperizada *adj* (*UHT*) ALIMENT ultra heat treated (*UHT*)

uralita *f* MINERAL uralite

uramido *adj* QUÍMICA uramid

uranato *m* QUÍMICA uranate

uránico *adj* QUÍMICA uranic

uranífero *adj* FÍS RAD, MINERAL, QUÍMICA uranium-bearing

uranilo *m* QUÍMICA uranyl

uraninita *f* MINERAL pitchblende, uraninite, NUCL uranium black

uranio *m* (*U*) FÍS RAD, NUCL, QUÍMICA uranium (*U*); ~ **enriquecido** *m* FÍS, NUCL enriched uranium; ~ **sin sus hijos** *m* NUCL uranium free from its daughters

uranoso *adj* QUÍMICA uranous

uranuro *m* QUÍMICA uranide

urazol *m* QUÍMICA urazole

urdido *m* TEXTIL warping; ~ **seccional** *m* TEXTIL section warping, sectional warping

urdidor *m* TEXTIL warper; ~ **seccional** *m* TEXTIL sectional warping machine

urdimbre *f* TEXTIL warp; ~ **encolada** *f* TEXTIL sized warp

urea¹: **de la** ~ *adj* QUÍMICA ureal

urea² *f* QUÍMICA urea

ureico *adj* QUÍMICA ureic

ureido *m* QUÍMICA ureide

ureotélico *adj* QUÍMICA ureotelic

uretano *m* QUÍMICA urethane

úrico *adj* QUÍMICA uric

uridina *f* QUÍMICA uridine

urna *f* METR *de balanza de precisión* case

urobilinógeno *m* QUÍMICA urobilinogen

urónico *adj* QUÍMICA uronic

uropterina *f* QUÍMICA uropterin

urotropina *f* QUÍMICA urotropine

uroxánico *adj* QUÍMICA uroxanic

US: ~ **NRC** *abr* (*Comisión Reguladora Nuclear de los Estados Unidos*) NUCL US NRC (*United States Nuclear Regulatory Commission*)

usado *adj* MECÁ worn

usar *vt* ING MECÁ wear, wear off, wear out, TEXTIL *ropa y calzado* wear; ~ **fuera de temporada** *vt* PROD deseasonalize; ~ **de nuevo** *vt* RECICL reuse

usina *f* ENERG RENOV powerhouse

uso¹: **de** ~ **corriente** *adj* ING MECÁ standard; **de** ~ **eficiente de combustible** *adj* TERMO fuel-efficient; **de** ~ **general** *adj* DETERG, ING ELÉC, ING MECÁ, MECÁ, PROD general-purpose; **de** ~ **universal** *adj*

DETERG, ING ELÉC, ING MECÁ, MECÁ, PROD general-purpose; **para todo** ~ *adj* DETERG, ING ELÉC, ING MECÁ, MECÁ, PROD general-purpose

uso[2] *m* ING MECÁ, MECÁ, TEXTIL wear; ~ **y desgaste** *m* ING MECÁ, PROD wear and tear; ~ **diario** *m* PROD daily use; ~ **de dos o más tipos de tintas en impresión multicolor** *m* IMPR sandwich; ~ **final** *m* CALIDAD, ING MECÁ, PROD, SEG, TEXTIL end use; ~ **de originales en varias obras** *m* IMPR bicycling; ~ **de palabras** *m* PROD word usage; ~ **repetido** *m* RECICL reuse; ~ **sin riesgos** *m* SEG *explosivos* safe use

usual *adj* ING MECÁ standard

usuario *m* GEN user; ~ **en conjunto** *m* INFORM&PD, TELECOM joint user; ~ **del directorio** *m* TELECOM directory user; ~ **final** *m* AGUA, CONST, INFORM&PD, ING ELÉC, TELECOM, TRANSP end user; ~ **de intercambio de datos electrónicos** *m* TELECOM EDI user; ~ **de mensajería por intercambio de datos electrónicos** *m* TELECOM EDI-messaging user, EDIMG user; ~ **de un río** *m* AGRIC, AGUA, HIDROL, TRANSP MAR river user; ~ **del servicio de calidad de transmisión de CMISE** *m* TELECOM performing CMISE service user; ~ **del servicio telefónico** *m* TELECOM telephone user; ~ **del teléfono** *m* TELECOM telephone user

utensilio *m* CONST utensil, ING MECÁ implement, tool, METR *verificación y calibrado* jig; ~ **para el hogar** *m* ELEC domestic appliance; ~ **de porcelana** *m* C&V porcelain utensil

utensilios: ~ **de lavado** *m pl AmL* (*cf herramientas de lavado Esp*) MINAS flushing tools

útil *m* ING MECÁ, MECÁ tool; ~ **acodado** *m* ING MECÁ bent tool; ~ **de ensayo** *m* CALIDAD, LAB, TEC ESP test tool; ~ **para fresar** *m* ING MECÁ milling jig; ~ **giratorio para soldar** *m* ING MECÁ positioner; ~ **de montaje** *m* ING MECÁ assembly jig

utilidad *f* TEC ESP service, utility

utilización *f* AGUA harnessing, MECÁ application; ~ **aparente de oxígeno** *f* OCEAN apparent utilization of oxygen; ~ **del centro de trabajo** *f* PROD work center utilization (*AmE*), work centre utilization (*BrE*); ~ **en circuito abierto** *f* ELEC *equipo* open-circuit operation; ~ **compartida de ficheros** *f* INFORM&PD file sharing; ~ **en común de automóviles** *f AmL* (*cf aparcamiento de coches Esp*) TRANSP car pooling; ~ **manual** *f* TEC ESP hand-operated; ~ **de nuevo** *f* RECICL reuse; ~ **de los residuos para abono de tierras** *f* RECICL sewage farming; ~ **por semana** *f* PROD week utilization

utilizar[1] *vt* FOTO operate, PROD operate, run; ~ **de nuevo** *vt* RECICL reuse

utilizar[2]: ~ **un preflash** *vi* CINEMAT preflash; ~ **el zoom** *vi* CINEMAT, FOTO, INFORM&PD, INSTR, TV zoom

utillaje: ~ **de chapistería** *m* ING MECÁ car body tooling; ~ **para emblistado** *m* ING MECÁ blister-pack tooling; ~ **para envasado** *m* ING MECÁ canning tooling; ~ **del sacatestigos** *m Esp* (*cf cortanúcleos AmL*) TEC PETR *perforación* coring tool

utillería: ~ **que puede destruirse rápidamente** *f* CINEMAT breakaway prop

UV *abr* (*ultravioleta*) FÍS, FÍS RAD, ÓPT, TEC ESP UV (*ultraviolet*)

uvarovita *f* MINERAL uvarovite

uvítico *adj* QUÍMICA uvitic

V

V[1] *abr* ELEC, FÍS, ING ELÉC, METR (*voltio*) V (*volt*), QUÍMICA (*vanadio*) V (*vanadium*)

V[2]: **en ~** *adj* GEOM, ING MECÁ V-shaped

V[3]: **~ de mecánico** *f* ING MECÁ V-block

vaca: **~ grávida** *f* AGRIC bred cow; **~ lechera** *f* AGRIC dairy cow, milk cow; **~ preñada** *f* AGRIC bred cow; **~ que acaba de parir** *f* AGRIC fresh cow

vacante *m* CRISTAL vacancy

vaccinina *f* QUÍMICA vacciniin

vaciable *adj* C&V castable

vaciada *f* C&V casting

vaciadero: **~ municipal** *m* CONTAM, RECICL municipal dump; **~ de productos de dragado** *m* CONTAM dumping ground

vaciado *m* *AmL* (*cf escape Esp*) C&V cast, pour, tapping, IMPR pouring, INFORM&PD dump, MINAS *del aire* exhausting, PROD *metales* casting, drawing the charge, SEG *fundición* pouring, TRANSP *contenedor* unstuffing, destuffing, TRANSP AÉR dumping; **~ centrifugado** *m* C&V centrifugal casting; **~ de contenedores** *m* TRANSP container destuffing, container unstuffing; **~ continuo** *m* C&V continuous casting; **~ a fondo** *m* ING ELÉC deep depletion; **~ en mesa** *m* C&V table casting; **~ de metal fundido** *m* PROD *fundería* teeming; **~ del nitrógeno** *m* TEC ESP *del tanque de la aeronave* nitrogen purging; **~ de rescate** *m* INFORM&PD rescue dump; **~ semi-continuo** *m* C&V semicontinuous casting

vaciadura: **~ ilegal** *f* AGUA, CONTAM illegal dumping

vaciar *vt* AGUA pump out, run off, C&V *vidrio plano* cast, *el vidrio del horno de fundición* tap, teem, CONST pump out, run, shoot, ING MECÁ, MECÁ, PROD drain, evacuate, hollow, hollow out, pump out, RECICL drain, TEC ESP *combustible* dump

vacilación *f* CINEMAT, INFORM&PD, TV flicker

vacío[1]: **en ~** *adj* ING MECÁ light, PROD empty; **~** *adj* INFORM&PD empty, TRANSP AÉR no-load; **de ~ casi perfecto** *adj* MECÁ high-vacuum

vacío[2] *m* GEN, P&C *defecto*, QUÍMICA *destilación y otros procesos* hollowness, vacuum, void; **bajo ~** *m* FÍS low vacuum; **~ absoluto** *m* REFRIG absolute vacuum, TERMO ultimate vacuum; **~ aproximado** *m* FÍS rough vacuum; **~ natural** *m* GAS natural void; **~ preliminar** *m* FÍS fore vacuum; **~ ultra-alto** *m* FÍS, REFRIG ultrahigh vacuum

vacío[3]: **bajo ~** *fra* TERMO under vacuum

vacka *f* GEOL wacke

vacunos: **~ jóvenes alimentados normalmente** *m pl* AGRIC stocker cattle; **~ en periodo final de engorde** *m pl* AGRIC finishing cattle; **~ en terminación de engorde** *m pl* AGRIC finishing cattle; **~ terminados con concentrados** *m pl* AGRIC long-fed animals

vacuoaislamiento *m* ING MECÁ vacuum insulation

vacuofreno *m* FERRO *vehículos*, ING MECÁ vacuum brake; **~ automático dúplex** *m* ING MECÁ duplex automatic vacuum brake

vacuofusión *f* METAL vacuum melting

vacuohorno *m* ING MECÁ vacuum furnace

vacuoingeniería *f* ING MECÁ vacuum engineering

vacuolado *adj* TEC PETR *geología* vuggy

vacuolar *adj* TEC PETR *geología* vugular

vacuómetro *m* ING MECÁ, INSTR, PROD, TEC PETR vacuum gage (*AmE*), vacuum gauge (*BrE*), TV vacuum guide

vacuotécnica *f* ING MECÁ vacuum engineering

vacuotecnología *f* ING MECÁ vacuum technology

vadear *vt* CONST, HIDROL *río* ford

vado *m* CONST fording, *río* ford, HIDROL *ríos* ford, fording

vagón *m* ING MECÁ carriage, MINAS car, PROD carriage, TRANSP car, carriage, carload (*AmE*), wagonload (*BrE*), VEH lorry (*BrE*), truck (*AmE*); **~ abierto** *m* FERRO gondola car (*AmE*), open wagon (*BrE*); **~ de agrupamiento** *m* FERRO groupage car (*AmE*), groupage wagon (*BrE*); **~ autobasculante** *m* FERRO self-tipping car (*AmE*), self-tipping wagon (*BrE*); **~ de auto-descarga con caja abierta** *m* TRANSP bogie open self-discharge wagon (*BrE*), truck open self-discharge car (*AmE*); **~ averiado** *m* FERRO damaged car (*AmE*), damaged wagon (*BrE*); **~ del bar** *m* FERRO *vehículos* bar coach; **~ basculante** *m* TRANSP dump car (*AmE*), dump wagon (*BrE*); **~ basculante con descarga hacia atrás** *m* FERRO *vehículos* back-discharge car (*AmE*), back-discharge wagon (*BrE*), rear-dump car (*AmE*); **~ basculante hacia atrás** *m* FERRO *vehículos* back-discharge car (*AmE*), back-discharge wagon (*BrE*), rear-dump car (*AmE*); **~ basculante lateralmente** *m* FERRO *vehículos* side dump car (*AmE*), side dump wagon (*BrE*); **~ basculante pequeño** *m* FERRO *vehículos* spoil car (*AmE*), spoil wagon (*BrE*); **~ bastidor** *m* FERRO *vehículos* trestle car (*AmE*), trestle wagon (*BrE*); **~ batea** *m* FERRO *laterales bajos* gondola car (*AmE*), open wagon (*BrE*); **~ batea de plataforma rebajada** *m* FERRO *vehículos* depressed deck car (*AmE*), depressed deck wagon (*BrE*); **~ de bordes altos** *m* FERRO *vehículos* box wagon (*BrE*), boxcar (*AmE*); **~ de bordes bajos** *m* FERRO gondola car (*AmE*), open wagon (*BrE*); **~ butanero** *m* TEC PETR *transporte*, TRANSP butane carrier; **~ de carga** *m* CONST carrier wagon; **~ sin carga colocado entre dos vagones cargados** *m* FERRO *vehículos* idler; **~ cerrado** *m* FERRO *vehículos* box wagon (*BrE*), boxcar (*AmE*), covered car (*AmE*), covered wagon (*BrE*); **~ cisterna** *m* FERRO rail tank car (*AmE*), rail tank wagon (*BrE*), *vehículos* tank car (*AmE*), tank wagon (*BrE*), MINAS tank car (*AmE*), tank wagon (*BrE*), TRANSP rail tank car (*AmE*), rail tank wagon (*BrE*); **~ de clase turista** *m* TRANSP coach class; **~ para la colocación de catenaria** *m* FERRO overhead drum car (*AmE*), overhead drum wagon (*BrE*); **~ contenedor** *m* FERRO, TRANSP container car (*AmE*), container wagon (*BrE*); **~ con costados revestidos** *m* FERRO sheeted car (*AmE*), sheeted wagon (*BrE*); **~ cuba** *m* FERRO *vehículos*, MINAS tank car (*AmE*), tank wagon (*BrE*); **~ de descarga automática** *m* FERRO self-discharging car (*AmE*), self-discharging wagon (*BrE*); **~ escudo** *m* FERRO *vehículos* cushion car

(*AmE*), cushion wagon (*BrE*); ~ **de ferrocarril** *m* FERRO rail coach; ~ **con fondo en tolva** *m* TRANSP saddle-bottomed car; ~ **frenado** *m* FERRO *vehículos* braked car (*AmE*), braked wagon (*BrE*); ~ **freno** *m* FERRO *vehículos* caboose (*AmE*), guard's van (*BrE*); ~ **con freno posterior** *m* FERRO *vehículos* rear-brake van; ~ **para ganado** *m* FERRO, TRANSP cattle car (*AmE*), cattle wagon (*BrE*); ~ **góndola** *m* FERRO *vehículos* well car (*AmE*), well wagon (*BrE*); ~ **de grupaje** *m* FERRO *vehículos* groupage car (*AmE*), groupage wagon (*BrE*); ~ **incontrolado** *m* FERRO *vehículos* runaway car (*AmE*), runaway wagon (*BrE*); ~ **de inspección de vía** *m* CONST railroad inspection trolley (*AmE*), railway inspection trolley (*BrE*); ~ **isotermo** *m* REFRIG insulated lorry (*BrE*), insulated truck (*AmE*); ~ **de mercancías** *m* CONST bogie (*BrE*), truck (*AmE*), FERRO *vehículos* bogie (*BrE*), car (*AmE*), carriage, truck (*AmE*), wagon (*BrE*), ING MECÁ, MECÁ bogie (*BrE*), truck (*AmE*), TRANSP freight car (*AmE*), car (*AmE*), freight wagon (*BrE*), goods wagon (*BrE*), VEH bogie (*BrE*); ~ **de mercancías auto-descargable** *m* TRANSP self-discharge freight car (*AmE*), self-discharge freight wagon (*BrE*); ~ **de mina** *m* MINAS mine truck; ~ **minero** *m* TRANSP mine car; ~ **normal de carga parcial** *m* FERRO *vehículos* regular part-load car (*AmE*), regular part-load wagon (*BrE*); ~ **particular** *m* FERRO *vehículos* private car (*AmE*), private wagon (*BrE*); ~ **de pasajeros** *m* TRANSP carriage, passenger car; ~ **pendular** *m* TRANSP tilting body coach, tilting car (*AmE*), tilting wagon (*BrE*); ~ **perforador** *m* MINAS *para hacer barrenos* jumbo including boom; ~ **plataforma** *m* FERRO *vehículos* flat car (*AmE*), flat wagon (*BrE*), platform car, TRANSP passive flat car; ~ **presurizado** *m* FERRO pressure-sealed car (*AmE*), pressure-sealed wagon (*BrE*); ~ **privado** *m* FERRO *vehículos* private car (*AmE*), private wagon (*BrE*); ~ **que se ha cortado de un tren en marcha** *m* FERRO runaway car (*AmE*), runaway wagon (*BrE*); ~ **refrigerado** *m* FERRO *vehículos*, REFRIG refrigerated car (*AmE*), refrigerated wagon (*BrE*); ~ **refrigerante con depósito de hielo en el techo** *m* REFRIG overhead-bunker refrigerated truck; ~ **refrigerante con depósito de hielo en los testeros** *m* REFRIG end-bunker refrigerated truck; ~ **de remolque** *m* TRANSP trailer wagon (*AmE*), wagon car (*BrE*); ~ **restaurante** *m* FERRO *vehículos* buffet car; ~ **Schnabel** *m* FERRO *vehículos* Schnabel car (*BrE*); ~ **de servicio con generador diesel** *m* FERRO *vehículos* diesel generator unit service wagon; ~ **de socorro** *m* FERRO *equipo inamovible* breakdown car (*AmE*), breakdown wagon (*BrE*); ~ **con suspensión pendular** *m* TRANSP *monorrail* pendulum vehicle suspension; ~ **tolva** *m* FERRO *vehículos* hopper car (*AmE*), hopper wagon (*BrE*), skip wagon; ~ **tolva autodescargable** *m* TRANSP saddle-bottomed self-discharging car; ~ **de transporte** *m* MINAS transit car; ~ **para transporte de carriles** *m* FERRO rail-carrying car (*AmE*), rail-carrying wagon (*BrE*); ~ **para transporte de ganado** *m* AGRIC, TRANSP cattle lorry (*BrE*), cattle truck (*AmE*); ~ **para el transporte de troncos** *m* FERRO *vehículos* timber car (*AmE*), timber wagon (*BrE*); ~ **para uso interno en la estación** *m* FERRO, VEH car for internal yard use (*AmE*), wagon for internal yard use (*BrE*); ~ **a vapor**

m TRANSP steam car; ~ **de viajeros** *m* TRANSP *ferrocarril* coach

vagoneta *f* CARBÓN mine car, tip, tip car, CONST bucket, manrider, MINAS barrow, hurley, hutch, larry car, mine car, mine truck, mining bucket, tub; ~ **basculadora** *f* CARBÓN tip box car; ~ **basculante** *f* CARBÓN dump truck (*AmE*), tipper (*BrE*), FERRO *vehículos, volquete* tip car (*AmE*), tip wagon (*BrE*); ~ **basculante pequeña** *f* FERRO small tip car (*AmE*), small tip wagon (*BrE*); ~ **basculante de pico** *f* TRANSP scoop dump car (*AmE*), scoop dump wagon (*BrE*); ~ **Baum** *f* CARBÓN Baum box; ~ **para combustible** *f* MINAS bunker car; ~ **de mina** *f* MINAS tram; ~ **de minas basculante** *f* MINAS tipping mine car; ~ **sobre monocarril** *f* TRANSP monorail grab trolley; ~ **de remolque** *f* TRANSP trailer wagon (*AmE*), wagon car (*BrE*)

vagra *f* TRANSP MAR *construcción naval* girder; ~ **lateral** *f* TRANSP MAR *construcción naval* side girder; ~ **del pantoque** *f* TRANSP MAR *construcción naval* bilge stringer

vaguada *f* CONST talweg, thalweg, watercourse, METEO trough; ~ **barométrica** *f* METEO barometric trough

vaina *f* AGRIC sheath, *fruto* pod, ING ELÉC *cables eléctricos* sheathing, *en la vela* batten pocket, *bandera* hoist, *perforación AmL* (cf *hueco rata Esp, hueco ratón Esp*) mouse hole, rat hole; ~ **aislante** *f* ING ELÉC insulating sheath; ~ **con aletas disipadoras de calor** *f* NUCL finned can; ~ **del cable** *f* ELEC, ING ELÉC, ING MECÁ cable sheath; ~ **del combustible** *f* NUCL fuel cladding; ~ **de la fibra** *f* ÓPT fiber jacket (*AmE*), fibre jacket (*BrE*); ~ **foliar** *f* AGRIC leaf sheath; ~ **metálica** *f* ELEC metallic sheath; ~ **metálica del cartucho** *f* D&A cartridge case; ~ **de plomo** *f* ING ELÉC lead sheath; ~ **reventada** *f* NUCL burst can; ~ **de zircaloy** *f* NUCL zircaloy cladding

vainillina: ~ **de etilo** *f* ALIMENT ethyl vanillin

vaivén[1]: **de ~** *adj* ING MECÁ, MECÁ reciprocating

vaivén[2] *m* ELEC *máquinas* hunting, ING MECÁ seesawing, swaying, TEXTIL roving frame

valdosa: ~ **vitrificada** *f* CONST glazed tile

valencia[1]: **sin ~** *adj* QUÍMICA avalent

valencia[2] *f* FÍS, METAL valence, QUÍMICA valency, valence

valentinita *f* MINERAL valentinite

valeramida *f* QUÍMICA valeramide

valerato *m* QUÍMICA valerate

valérico *adj* QUÍMICA valeric

valerileno *m* QUÍMICA valerylene

valerilo *m* QUÍMICA valeryl

validación *f* CALIDAD, INFORM&PD validation; ~ **de datos** *f* INFORM&PD data validation

validar *vt* INFORM&PD, TEC ESP validate

validez *f* TELECOM validity

valla *f* CONST barricade, enclosure, pale; ~ **de contención de nieve** *f* CONST snow barrier; ~ **limítrofe** *f* CONST boundary fence; ~ **protectora para máquina** *f* SEG machine fence

vallar: **sin ~** *adj* CONST unfenced

valle: ~ **anegado** *m* HIDROL drowned valley; ~ **árido** *m* HIDROL dry valley; ~ **energético** *m* NUCL *diagramas* energy valley; ~ **intermedio** *m* GEOL median valley; ~ **de un río inundado** *m* OCEAN drowned river valley; ~ **del silicio** *m* ELECTRÓN *California* silicon valley; ~ **submarino** *m* OCEAN submarine valley; ~ **tectónico** *m* GEOL rift valley

vallicultura *f* OCEAN sea ranch
valor[1]: ~ **agregado** *adj AmL* (*cf valor añadido Esp*)
INFORM&PD value-added; ~ **añadido** *adj Esp* (*cf valor
agregado AmL*) INFORM&PD value-added
valor[2] *m* TELECOM value; ~ **de abrasión de los áridos**
m CONST *hormigón* aggregate abrasion value (*AA*);
~ **absoluto** *m* MATEMÁT, QUÍMICA absolute value;
~ **ácido** *m* ALIMENT acid value; ~ **ácido total** *m*
(*VAT*) QUÍMICA total acid number (*TAN*); ~ **en
aduana** *m* TRANSP MAR bonded value; ~ **agregado**
m AmL (*cf valor añadido Esp*) ING MECÁ value-added;
~ **añadido** *m Esp* (*cf valor agregado AmL*) ING MECÁ
value-added; ~ **de una barra de control** *m* NUCL
control rod worth; ~ **booleano** *m* INFORM&PD Boolean value; ~ **del brillo** *m AmL* (*cf valor de la
luminosidad Esp*) TV brightness value; ~ **calórico
neto** *m* TERMOTEC net calorific value; ~ **calorífico**
m FÍS, GEOL, TEC PETR, TERMO, TERMOTEC calorific
value; ~ **calorífico bruto** *m* TERMOTEC gross calorific
value; ~ **de la carga** *m* TRANSP *tráfico* load value;
~ **de causa** *m* TELECOM cause value; ~ **cereal** *m*
AGRIC grain equivalent; ~ **de cresta** *m* ELEC *de
corriente, tensión, etc*, TELECOM peak value; ~ **cuadrático medio** *m* CONST root-mean-square value,
ELEC mean square value, ÓPT root-mean-square
value; ~ **por defecto** *m* INFORM&PD default; ~ **de
depósito** *m* CONTAM deposition value; ~ **deseado** *m*
NUCL desired value; ~ **desmultiplicador** *m* PROD
scaling value; ~ **diferencial de una barra de
control** *m* NUCL differential control rod worth;
~ **efectivo** *m* CONST, ÓPT *de una magnitud sinusoidal*
root-mean-square value; ~ **eficaz** *m* CONST root-
mean-square value, ELEC mean square value, ING
ELÉC effective value, ÓPT *de una magnitud sinusoidal*
root-mean-square value; ~ **eigen** *m* INFORM&PD
eigenvalue; ~ **de emisión fijado por la ley** *m*
CONTAM *legislación* emission standard; ~ **de
entrada** *m* FÍS RAD input; ~ **en la escala de grises**
m CINEMAT, FOTO, INFORM&PD, TV gray-scale value
(*AmE*), grey-scale value (*BrE*); ~ **estérico** *m* DETERG
ester value; ~ **de exposición** *m* CINEMAT, FOTO
exposure value; ~ **indicado** *m* NUCL indicated value;
~ **de índice** *m* NUCL index value; ~ **instantáneo** *m*
ELEC *de tensión*, FÍS instantaneous value; ~ **de
interrupción** *m* PROD interrupt value; ~ **límite** *m*
NUCL limit setting, TELECOM limiting value; ~ **límite
umbral** *m* CONTAM *legislación, toxicología* threshold
limit value (*TLV*); ~ **límite umbral en el medio
ambiente** *m* CONTAM *legislación, toxicología* threshold limit value in the free environment; ~ **límite
umbral ocupacional** *m* CONTAM *legislación* occupational threshold limit value; ~ **lineal total** *m* PROD
total line value; ~ **lógico** *m* INFORM&PD logical value;
~ **de la luminosidad** *m Esp* (*cf valor del brillo AmL*)
TV brightness value; ~ **luminoso** *m* FOTO light value;
~ **en lux** *m* FOTO lux value; ~ **máximo** *m* ELEC *de
corriente, tensión*, ELECTRÓN, FÍS peak value, NUCL
maximum, TELECOM peak value; ~ **medio** *m* ELEC,
FÍS average value; ~ **medio de dosis letal** *m* CONTAM,
FÍS RAD median lethal dose; ~ **medio de un flujo
turbulento** *m* FÍS FLUID average in turbulent flow;
~ **de muestra** *m* ELECTRÓN sampled value;
~ **nominal** *m* ELEC nominal value, *aparato* rated
value; ~ **nominal del amperaje** *m* PROD amp rating;
~ **numérico** *m* METR numerical value; ~ **nutritivo** *m*
AGRIC, ALIMENT food value; ~ **objetivo de quemado**

m NUCL target burn-up; ~ **óhmico** *m* ING ELÉC ohmic
value; ~ **por omisión** *m* INFORM&PD default, TELE-
COM default value; ~ **de parámetro** *m* TELECOM
parameter value (*PV*); ~ **de pH** *m* AGRIC, HIDROL pH
value; ~ **de pico** *m* ELECTRÓN peak value; ~ **pico a
pico** *m* ACÚST peak-to-peak value; ~ **prefijado** *m*
PROD preset value; ~ **promedio** *m* ELEC average
value, FÍS mean value; ~ **propio** *m* ELECTRÓN, FÍS
eigenvalue; ~ **real** *m* IMPR actual value; ~ **de
sonoemisión** *m* ING MECÁ noise emission value;
~ **de taponamiento del filtro** *m* PROC QUÍ filter
plugging value; ~ **térmico** *m* TERMO thermal value;
~ **de la tierra** *m* AGRIC land value; ~ **tonal** *m* FOTO
tonal value; ~ **de la tonalidad** *m* IMPR tone value;
~ **umbral** *m* ELECTRÓN threshold value; ~ **venal** *m*
MINAS *de una roca* commercial value

valoración *f* CARBÓN, DETERG, LAB titration, MATEMÁT
estimation, QUÍMICA titration; ~ **automática** *f Esp* (*cf
titulación automática AmL*) LAB *análisis* automatic
titration; ~ **de la calidad** *f* METR quality assessment;
~ **complejométrica** *f* QUÍMICA *análisis* complexometric titration; ~ **conductimétrica** *f* QUÍMICA
análisis conductimetric titration; ~ **del error** *f* MATE-
MÁT error estimation; ~ **potenciométrica** *f* QUÍMICA
electrometric titration; ~ **de los riesgos** *f* CALIDAD,
CONTAM, CONTAM MAR, SEG risk assessment, risk
evaluation

valorar *vt* QUÍMICA titrate
valores: ~ **eléctricos nominales** *m pl* PROD electrical
ratings
valuación: ~ **de existencias** *f* PROD inventory valuation, stock valuation
válvula *f* GEN valve, CONST trap, ING MECÁ cap, INSTAL
HIDRÁUL *bombas* cap, tube, INSTR cap, MECÁ cap TEC
ESP tube;

~ **de acción instantánea** *f* REFRIG snap-action
valve; ~ **de acción termostática** *f* ING MECÁ thermostatic valve; ~ **accionada por aire comprimido** *f* ING
MECÁ pneumatically-operated valve; ~ **accionada
hidráulicamente** *f* ING MECÁ hydraulically-operated
valve; ~ **accionada por leva** *f* AUTO poppet valve;
~ **accionada por muelle** *f* INSTAL HIDRÁUL spring-
loaded valve; ~ **accionada por resorte** *f* INSTAL
HIDRÁUL spring-loaded valve; ~ **de admisión** *f*
AUTO inlet valve, ING MECÁ admission valve, throttle
valve, INSTAL HIDRÁUL admission valve, induction
valve, inlet valve, intake valve, throttle valve, VEH
inlet valve, *motor* intake valve; ~ **de admisión de
aire** *f* ING MECÁ air-inlet valve; ~ **de admisión de la
turbina** *f* NUCL turbine stop valve; ~ **de admisión del
vapor** *f* INSTAL HIDRÁUL steam valve; ~ **aerosol** *f*
EMB aerosol valve; ~ **de agua** *f* CONST water valve;
~ **de aguja** *f* AUTO needle valve, CONST pin valve,
ENERG RENOV, ING MECÁ, INSTAL HIDRÁUL, LAB,
NUCL, QUÍMICA, REFRIG needle valve, VEH *carburador*
needle; ~ **de aguja de alimentación visible** *f* ING
MECÁ sight-feed needle valve; ~ **de aguja del
carburador** *f* AUTO, VEH carburetor needle (*AmE*),
carburettor needle (*BrE*); ~ **de aire** *f* ING MECÁ air
valve, breather, PROD, VEH breather; ~ **de aire
caliente** *f* TRANSP AÉR hot air valve; ~ **de
aislamiento** *f* LAB *equipo general*, PROD isolating
valve, TEC ESP isolation valve; ~ **de aislamiento
automático** *f* TEC PETR *tuberías, bombas* automatic
isolating valve; ~ **de aislamiento del indicador** *f*
PROD gage isolating valve (*AmE*), gauge isolating

valve (*BrE*); ~ **de aislamiento de vapor principal** *f* NUCL main-steam isolation valve; ~ **de ajuste** *f* GAS adjusting valve; ~ **de alimentación** *f* PROD feed cock, TEC ESP *vehículos* filling valve, TEC PETR *refino* feed valve; ~ **de alimentación cruzada de combustible** *f* TRANSP AÉR fuel cross-feed valve; ~ **de alivio** *f* AUTO, ENERG RENOV relief valve, ING MECÁ air-snifting valve, pressure-relief valve, INSTAL HIDRÁUL, MECÁ, PROD, REFRIG, TEC PETR, VEH relief valve; ~ **de alivio del presionador** *f* NUCL pressurizer relief-valve; ~ **Allen** *f* ING MECÁ Allen valve; ~ **de amortiguación** *f* ING MECÁ dashpot valve; ~ **de amortiguación bidireccional** *f* AUTO, ING MECÁ two-way damper valve; ~ **de amortiguación de doble paso** *f* AUTO, ING MECÁ two-way damper valve; ~ **de amortiguación de dos vías** *f* AUTO, ING MECÁ two-way damper valve; ~ **del amortiguador** *f* ING MECÁ dashpot valve; ~ **antiagitación** *f* TRANSP AÉR antisurge valve; ~ **antisobretensión** *f* TRANSP AÉR antisurge valve; ~ **antisobrevoltaje** *f* TRANSP AÉR antisurge valve; ~ **de asiento cónico** *f* AUTO poppet valve, INSTAL HIDRÁUL *distribución de vapor* conical-seat valve, poppet valve, PROD poppet valve; ~ **de asiento sencillo** *f* INSTAL HIDRÁUL single-seated valve; ~ **de aspiración** *f* AUTO suction valve, ING MECÁ foot valve, INSTAL HIDRÁUL inlet valve, intake valve, suction valve, REFRIG suction valve; ~ **automática** *f* REFRIG, TEC PETR *tuberías, bombas* automatic valve; ~ **automática de agua** *f* REFRIG water-regulating valve; ~ **auxiliar** *f* ING MECÁ auxiliary valve, pilot valve;

▪ b ~ **de bloqueo** *f* TRANSP AÉR lockout valve; ~ **de boca de pozo** *f* ENERG RENOV wellhead valve; ~ **de bola** *f* ING MECÁ ball valve, INSTAL HIDRÁUL ball valve, hinged valve, open flap valve, swing valve, MECÁ ball valve; ~ **de bola de bridas** *f* ING MECÁ flanged ball valve; ~ **de bomba** *f* TEC PETR pump valve; ~ **de la bomba de lodo** *f* TEC PETR mud pump valve; ~ **de bóveda** *f* PROD *horno alto* chimney valve;

▪ c ~ **calibrada** *f* ING MECÁ calibrated valve; ~ **de cambio de marcha** *f* INSTAL HIDRÁUL reversing valve; ~ **de campana** *f* INSTAL HIDRÁUL bell valve; ~ **de la cápsula madura** *f* AGRIC *algodón* lock; ~ **de carga** *f* TEC ESP filling valve; ~ **de carga de aire** *f* TRANSP AÉR air-charging valve; ~ **de caudal sobrante** *f* INSTAL HIDRÁUL overflow valve; ~ **cerrada** *f* NUCL valve off; ~ **de charnela** *f* CONST flap valve, ING MECÁ clack valve, clapper, flap valve, INSTAL HIDRÁUL clack valve, clapper, flap valve, hinged valve, open flap valve, swing valve, leaf valve, PROD flap valve; ~ **de cierre** *f* CONST cutoff valve, stop valve, LAB *control de fluido* faucet (*AmE*), stopcock (*BrE*), PROD shutoff valve, TRANSP AÉR lockout valve; ~ **de cierre atornillada** *f* CONST screw-down stop valve; ~ **de cierre rápido** *f* AGUA quick-closing valve; ~ **de cierre de vapor** *f* INSTAL HIDRÁUL steam stop valve; ~ **de combustible sin paso de aire** *f* TRANSP AÉR air-no-fuel vent valve; ~ **de compuerta** *f* AGUA sluice valve, water gate, ING MECÁ, INSTAL HIDRÁUL gate valve, sluice valve, PROD gate; ~ **de comunicación con el mar** *f* TRANSP MAR *construcción naval* seacock; ~ **cónica** *f* AUTO, ENERG RENOV, ING MECÁ needle valve, INSTAL HIDRÁUL cone valve, needle valve, LAB, NUCL, QUÍMICA, REFRIG needle valve; ~ **de contención** *f* AGUA retaining valve, AUTO, ENERG RENOV check valve, INSTAL

HIDRÁUL *tuberías* check valve, cutoff valve, font valve, foot valve, reflux valve, retaining valve, stop valve; ~ **contra el retroceso de la llama** *f* SEG, TERMO flashback preventer; ~ **contra vacío** *f* ENERG RENOV *hidroelectricidad* antivacuum valve; ~ **de contrapresión** *f* INSTAL HIDRÁUL back valve; ~ **de control** *f* AUTO proportioning valve, INSTAL HIDRÁUL control valve, controlling valve, maneuvering valve (*AmE*), manoeuvring valve (*BrE*), MECÁ control valve; ~ **de control del accionador** *f* TRANSP AÉR actuator-control valve; ~ **de control de dirección** *f* ING MECÁ direction-control valve; ~ **de control direccional de cuatro polos** *f* ING MECÁ four-port directional control valve; ~ **de control electrónico** *f* CONST, ELECTRÓN, ING MECÁ electronically-controlled valve; ~ **de control de explosión** *f* TEC PETR blowout preventer (*BOP*); ~ **de control de flujo** *f* ING MECÁ flow-control valve; ~ **de control proporcional** *f* PROD proportional-control valve; ~ **de control de vacío** *f* AUTO vacuum check valve; ~ **controlada de caudal** *f* ING MECÁ flow-control valve; ~ **corredera** *f* CONST *fontanería* running trap; ~ **de corredera** *f* ING MECÁ gate valve, INSTAL HIDRÁUL gate valve, sluice valve, *máquina alternativa de vapor* slide valve, VEH *de motor* slide valve; ~ **de corredera antihielo del motor** *f* TRANSP AÉR engine anti-icing gate valve; ~ **de corredera de avance a la evacuación** *f* INSTAL HIDRÁUL inside-lead slide valve; ~ **de corredera compensada** *f* ING MECÁ balanced slide valve; ~ **de corredera con huelgo interior** *f* INSTAL HIDRÁUL *máquina alternativa de vapor* inside-clearance slide valve; ~ **de corredera con recubrimiento al escape** *f* INSTAL HIDRÁUL *distribuidores* inside-lap slide valve; ~ **de corredera con retardo a la admisión** *f* INSTAL HIDRÁUL late-admission slide valve; ~ **de corredera con retardo al escape** *f* INSTAL HIDRÁUL *motores* late-release slide valve; ~ **corrediza del tren de aterrizaje** *f* TRANSP AÉR landing gear sliding valve; ~ **de cuatro pasos** *f* AGUA four-way cock; ~ **de cuello** *f* INSTAL HIDRÁUL main steam pipe; ~ **en culata** *f* AUTO, METAL, PETROL, TEC PETR, TERMO *mecánica* overhead valve (*OHV*);

▪ d ~ **D** *f* INSTAL HIDRÁUL *máquina de vapor* D-valve; ~ **del depósito** *f* INSTAL HIDRÁUL tank valve; ~ **del depósito de combustible** *f* TEC ESP fuel dump valve; ~ **del depósito de vapor** *f* INSTAL HIDRÁUL cap of steam chest; ~ **de derivación** *f* LAB *equipo general* bypass valve; ~ **de derrame** *f* FERRO, INSTAL HIDRÁUL overflow valve; ~ **de desagüe** *f* AGUA sluice valve, ING MECÁ drain valve; ~ **de desahogo** *f* AGUA bleeding-valve, ING MECÁ air-snifting valve, pressure-relief valve, INSTAL HIDRÁUL bleeding-valve, pet valve, pressure-relief valve, sifter valve; ~ **de descarga** *f* AUTO, ENERG RENOV relief valve, ING MECÁ discharge valve, unloading valve, INSTAL HIDRÁUL delivery valve, discharge valve, escape valve, release valve, relief valve, unloading valve, air valve, breather, MECÁ, PROD relief valve, REFRIG discharge valve, relief valve, TEC PETR, VEH relief valve; ~ **de deshielo de la superficie aerodinámica** *f* TRANSP AÉR aerofoil de-icing valve (*BrE*), airfoil de-icing valve (*AmE*); ~ **deslizante balanceada** *f* ING MECÁ balanced slide valve; ~ **de despresurización** *f* ING MECÁ depressurization valve; ~ **desviadora** *f* INSTAL HIDRÁUL deflecting valve; ~ **de desvío** *f* INSTAL HIDRÁUL cross valve, switch valve, three-way valve, triple valve; ~ **de diafragma** *f* ING MECÁ diaphragm valve; ~ **directriz** *f* PROD director valve; ~ **de disco** *f* INSTAL HIDRÁUL *bombas, compresores* circular slide valve, deck valve, disc valve (*BrE*), disk valve (*AmE*); ~ **de disco con movimiento vertical** *f* AUTO poppet valve; ~ **de disco rotatorio** *f* AUTO, VEH *motor de dos tiempos* rotary-disc valve (*BrE*), rotary-disk valve (*AmE*); ~ **de disparo de accionamiento rápido** *f* NUCL fast-acting trip valve; ~ **de distribución** *f* ING MECÁ distribution valve; ~ **distribuidora cilíndrica** *f* INSTAL HIDRÁUL *máquina de vapor* circular slide valve, deck valve, disc valve (*BrE*), disk valve (*AmE*); ~ **de doble asiento** *f* INSTAL HIDRÁUL cup valve, double beat valve; ~ **de doble efecto** *f* PROD shuttle valve; ~ **de dos presiones** *f* PROD dual-pressure valve; ~ **de dosificación** *f* AUTO proportioning valve; ~ **dosificadora** *f* ING MECÁ metering valve; ~ **de drenaje** *f* ING MECÁ drain valve;

▪ **e** ~ **electrónica** *f* ELECTRÓN tube, *tubo* electronic valve, PROD valve; ~ **de elevación** *f* AUTO poppet valve; ~ **de entrada** *f* AUTO inlet valve, INSTAL HIDRÁUL induction valve; ~ **de entrada de aire** *f* ING MECÁ air-inlet valve, air-snifting valve; ~ **de entrada de combustible** *f* AUTO fuel-inlet valve; ~ **equilibrada** *f* AGUA bleeding-valve, ENERG RENOV cylindrical balanced valve, INSTAL HIDRÁUL *bombas* balanced valve, bleeding-valve, pet valve, pressure-relief valve, sifter valve; ~ **de equilibrio** *f* INSTAL HIDRÁUL equalizing valve, equilibrated valve, pet valve; ~ **de escape** *f* AUTO exhaust valve, ING MECÁ reducing valve, return valve, INSTAL HIDRÁUL delivery valve, discharge valve, escape valve, eduction valve, exhaust valve, outlet valve, LAB exhaust valve, *control de gas* reducing valve, PROD reducing valve, REFRIG blow-off valve, VEH *motor* exhaust valve; ~ **de escape de aire** *f* TRANSP AÉR air-vent valve; ~ **de esclusa** *f* ING MECÁ, INSTAL HIDRÁUL gate valve; ~ **de esfera** *f* REFRIG globe valve; ~ **esférica** *f* ING MECÁ float valve, INSTAL HIDRÁUL ball cock, ball valve, float valve, globe valve, MECÁ ball valve, REFRIG float valve; ~ **esférica de retención** *f* NUCL ball check valve; ~ **de espiga** *f* AUTO, ENERG RENOV, ING MECÁ, INSTAL HIDRÁUL, LAB, NUCL, QUÍMICA, REFRIG needle valve; ~ **de estrangulación** *f* AmL (*cf válvula reguladora Esp*) AUTO throttle valve, metering valve, CARBÓN choker valve, throttle valve, ING MECÁ throttle valve, INSTAL HIDRÁUL maneuvering valve (*AmE*), manoeuvring valve (*BrE*), regulating valve, governor valve, controlling valve, ballast valve, control valve, LAB regulating valve, MECÁ control valve, MINAS, PROD throttle valve, TRANSP AÉR throttle, VEH throttle valve; ~ **de estrangulamiento de la aspiración** *f* AUTO suction-throttling valve; ~ **por estrechamiento** *f* QUÍMICA pinch cock; ~ **de evacuación** *f* INSTAL HIDRÁUL eduction valve, escape valve, outlet valve; ~ **de exhalación** *f* TRANSP AÉR exhalation valve; ~ **de expansión** *f* ING MECÁ, INSTAL HIDRÁUL, REFRIG expansion valve; ~ **de expansión con conexión soldada** *f* REFRIG sweat-type expansion valve; ~ **de expansión manual** *f* REFRIG hand-expansion valve; ~ **de expansión termostática** *f* REFRIG thermostatic expansion valve; ~ **expulsora** *f* ING MECÁ ejector valve; ~ **de extracción de aire** *f* ALIMENT air-bleed valve; ~ **del eyector** *f* ING MECÁ ejector valve;

▪ **f** ~ **de flotador** *f* ING MECÁ, INSTAL HIDRÁUL ball valve, float valve, MECÁ ball valve, REFRIG float valve; ~ **de flotador de alta presión** *f* REFRIG high-pressure float valve; ~ **de flotador de baja presión** *f* REFRIG low-pressure float valve; ~ **de flujo magnético** *f* TRANSP AÉR flux gate (*AmE*), flux valve (*BrE*); ~ **de frenado** *f* FERRO *vehículos* brake valve, ING MECÁ, MECÁ check valve; ~ **de freno** *f* FERRO *vehículos* brake valve; ~ **de fuelle** *f* REFRIG bellows valve;

▪ **g** ~ **de gas** *f* MINAS gas vent, PROD gas cock, gas valve; ~ **giratoria** *f* ENERG RENOV, INSTAL HIDRÁUL rotary valve; ~ **de gobierno termostático** *f* ING MECÁ thermostatic valve;

▪ **h** ~ **de haz de luz** *f* CINEMAT light valve; ~ **hidráulica** *f* REFRIG hydraulic-operated valve; ~ **de hongo** *f* AUTO mushroom valve;

▪ **i** ~ **impide reventones** *f* AmL PETR TECH blowout preventer (*BOP*); ~ **de impulsión** *f* INSTAL HIDRÁUL *bombas* delivery valve, discharge valve; ~ **de impulsión de retroceso** *f* INSTAL HIDRÁUL *bombas* back-pressure valve; ~ **de incremento de presión** *f* AUTO pressure-boost valve; ~ **indicador** *f* TV index tube; ~ **de inducción** *f* AUTO induction valve; ~ **industrial** *f* ING MECÁ industrial valve; ~ **de interrupción de vapor** *f* INSTAL HIDRÁUL steam stop valve; ~ **de inversión** *f* INSTAL HIDRÁUL reversing valve; ~ **inversora** *f* PROD changeover valve; ~ **invertida** *f* INSTAL HIDRÁUL *motores* drop valve; ~ **de inyección** *f* LAB *cromatografía de gases* injection valve; ~ **de inyección de líquido** *f* REFRIG liquid injection valve;

▪ **l** ~ **de lanzamiento** *f* TRANSP AÉR jettison valve; ~ **limitadora** *f* REFRIG restrictor valve; ~ **de llenado** *f* TEC ESP *vehículos* filling valve;

▪ **m** ~ **maestra de climatización** *f* ING MECÁ air-conditioning master valve; ~ **de mando** *f* ING MECÁ pilot valve; ~ **de manguito** *f* INSTAL HIDRÁUL equalizing valve; ~ **de maniobra** *f* INSTAL HIDRÁUL control valve, controlling valve, maneuvering valve (*AmE*), manoeuvring valve (*BrE*), MECÁ control valve; ~ **de maniobra direccional** *f* PROD directional control valve; ~ **manorreductora** *f* ING MECÁ pressure-reducing valve, INSTAL HIDRÁUL pressure reducer, pressure-reducing valve, PROD pressure-reducing valve, TEC ESP pressure reducer; ~ **de mar** *f* TRANSP MAR *construcción naval* seacock; ~ **de mariposa** *f* AUTO throttle valve, CONST flap valve, ENERG RENOV *presas* balanced disc valve (*BrE*), balanced disk valve (*AmE*), butterfly valve, ING MECÁ throttle valve, butterfly valve, flap valve, INSTAL HIDRÁUL butterfly valve, flap valve, pivot valve, butterfly throttle valve, MINAS *para el tiro de chimenea* throttle valve, PROD flap valve, VEH throttle valve; ~ **de membrana** *f* ING MECÁ diaphragm valve; ~ **mezcladora** *f* ING MECÁ mixer tube; ~ **del motor** *f* ING MECÁ engine valve; ~ **de movimiento vertical** *f* INSTAL HIDRÁUL lift valve, lifting valve;

▪ **n** ~ **neumática** *f* ING MECÁ pneumatic valve; ~ **neumática por control remoto** *f* ING MECÁ remote-controlled pneumatic valve; ~ **neumática por telemando** *f* ING MECÁ remote-controlled pneumatic valve; ~ **neumoaccionada** *f* ING MECÁ pneumatically-operated valve; ~ **no restringida** *f* CONST nonrestricted valve;

▪ **o** ~ **obturada por fuelle** *f* ING MECÁ bellows valve; ~ **oscilante** *f* PROD swinging valve;

▪ **p** ~ **de palanca** *f* INSTAL HIDRÁUL lever valve; ~ **de**

palomilla *f* AUTO, ING MECÁ, VEH throttle valve; ~ **de parada** *f* PROD shutoff valve; ~ **de parada accionada por solenoide** *f* PROD solenoid-operated shutoff valve; ~ **de parrilla** *f* PROD gridiron valve; ~ **de paso** *f* ING MECÁ throttle valve; ~ **de paso directo** *f* ING MECÁ, INSTAL HIDRÁUL gate valve; ~ **de paso recto** *f* ENERG RENOV straight-flow valve; ~ **de paso de vapor** *f* INSTAL HIDRÁUL steam valve; ~ **del pedal del freno** *f* AUTO, ING MECÁ treadle brake valve; ~ **piloto** *f* ING MECÁ pilot valve; ~ **pirotécnica** *f* TEC ESP pyrotechnic valve; ~ **del pistón** *f* ING MECÁ piston valve; ~ **de presión diferencial** *f* AUTO pressure-differential warning valve; ~ **presostática** *f* REFRIG pressure-controlled valve; ~ **presurizada** *f* TRANSP AÉR pressurizing valve; ~ **principal** *f* INSTAL TERM main valve; ~ **principal de aislamiento** *f* NUCL main isolating valve; ~ **principal del vapor** *f* INSTAL HIDRÁUL main steam pipe; ~ **prioritaria** *f* INSTAL HIDRÁUL priority valve; ~ **proporcional** *f* PROD proportional valve; ~ **de purga** *f* AGUA bleeding-valve, ING MECÁ bleed valve, drain valve, INSTAL HIDRÁUL bleeding-valve, TEC PETR *tuberías, bombas* bleed valve, TERMOTEC *de calderas* blow-down valve, TRANSP MAR drain valve; ~ **de purga de aire** *f* ING MECÁ air-vent valve, NUCL air-drain valve, TRANSP AÉR air-bleed valve; ~ **de purga cruzada de aire** *f* TRANSP AÉR air-cross bleed valve; ~ **de purgado de aire** *f* TRANSP AÉR air-vent valve; ~ **purgadora** *f* ENERG RENOV *equipo de perforación geotérmica* purge valve;

■ **q** ~ **quemada** *f* VEH *del motor* burnt valve;
■ **r** ~ **de rebose** *f* FERRO, INSTAL HIDRÁUL overflow valve; ~ **rectificadora de alimentación** *f* ELEC power rectifier; ~ **rectificadora de MAT** *f* TV EHT rectifier; ~ **de reducción de presión** *f* PROD pressure-relief valve; ~ **de reducción de presión proporcional** *f* PROD proportional pressure-reducing valve; ~ **reductora** *f* ING MECÁ, LAB reducing valve, PROD reducing valve, restrictor; ~ **reductora de presión** *f* CONST, ING MECÁ, REFRIG pressure-reducing valve; ~ **reductora de presión de gas** *f* GAS, INSTAL TERM, TERMO gas-pressure reducing valve; ~ **de refrigeración por sodio** *f* AUTO sodium-cooled valve; ~ **de regulación** *f* AUTO proportioning valve; ~ **de regulación de la aspiración** *f* AUTO suction-throttling valve; ~ **de regulación del caudal** *f* PROD flow-control valve; ~ **regulada termostáticamente** *f* PROD thermostatically-controlled valve; ~ **reguladora** *f* *Esp* (*cf válvula de estrangulación AmL*) AUTO metering valve, throttle valve, CARBÓN choker valve, throttle valve, ING MECÁ throttle valve, INSTAL HIDRÁUL ballast valve, control valve, controlling valve, governor valve, maneuvering valve (*AmE*), manoeuvring valve (*BrE*), regulating valve, LAB *control de fluidos* regulating valve, MECÁ control valve, MINAS, PROD *tiro de chimeneas* throttle valve; ~ **reguladora del agua** *f* REFRIG water-regulating valve; ~ **reguladora de la corriente** *f* AGUA flow-regulating valve; ~ **reguladora de escape libre** *f* AUTO, ELEC, ING ELÉC regulator cutout; ~ **reguladora de mariposa** *f* INSTAL HIDRÁUL *red de tuberías* throttle valve; ~ **reguladora principal** *f* AUTO main regulator valve; ~ **reguladora del temporizador controlador de secuencia** *f* PROD watchdog-timer set valve; ~ **de resorte** *f* INSTAL HIDRÁUL spring valve; ~ **de resortes** *f* AUTO, PROD poppet valve;

~ **del retardador** *f* ING MECÁ dashpot valve; ~ **de retención** *f* AGUA check valve, retaining valve, AUTO *tuberías* check valve, CONST nonreturn valve, ENERG RENOV check valve, nonreturn valve, ING MECÁ check valve, nonreturn valve, clack valve, retention valve, INSTAL HIDRÁUL *tuberías* reflux valve, nonreturn valve, stop valve, foot valve, cutoff valve, check valve, shutoff valve, retaining valve, fort valve, MECÁ check valve, PROD nonreturn valve, TEC ESP *fluidos*, VEH *lubricación* check valve; ~ **de retención accionada por piloto** *f* PROD pilot-operated check valve; ~ **de retención criptográfica** *f* TELECOM cryptographic check valve; ~ **de retención de vapor** *f* INSTAL HIDRÁUL steam stop valve; ~ **de retenida** *f* CONST, ENERG RENOV, ING MECÁ, INSTAL HIDRÁUL, PROD nonreturn valve; ~ **roscada** *f* CONST screwdown valve; ~ **rotativa** *f* ENERG RENOV, INSTAL HIDRÁUL rotary valve;

■ **s** ~ **de salida** *f* AUTO, INSTAL HIDRÁUL outlet valve; ~ **de sangrado** *f* AUTO, ING MECÁ, TEC PETR *tuberías, bombas* bleed valve; ~ **de sangrado de aire** *f* ALIMENT air-bleed valve; ~ **satélite** *f* AUTO satellite valve; ~ **de seccionamiento** *f* PROD isolating valve, TRANSP MAR *motor* shutoff valve; ~ **secuencial** *f* ING MECÁ sequence valve; ~ **de seguridad** *f* AUTO, ENERG RENOV relief valve, GAS safety valve, ING MECÁ pressure-relief valve, INSTAL HIDRÁUL blow-valve, safety-relief valve, pressure-relief valve, relief valve, safety valve, release valve, MECÁ relief valve, safety valve, PROD relief valve, pressure-relief valve, safety valve, REFRIG relief valve, SEG safety valve, TEC PETR *tuberías, bombas*, VEH *lubricación* relief valve; ~ **de seguridad de agujero de perforación** *f* TEC PETR downhole safety valve; ~ **de seguridad de cartucho** *f* PROD *de dinamita* cartridge-relief valve; ~ **de seguridad por diferencia de presión** *f* AUTO pressure-differential warning valve; ~ **de seguridad para el fondo del pozo** *f* TEC PETR *producción* downhole safety valve; ~ **de seguridad de resorte** *f* REFRIG spring-loaded pressure-relief valve; ~ **de seguridad de vapor** *f* INSTAL HIDRÁUL steam-relief valve; ~ **selectora de todo o nada** *f* ING MECÁ hit-or-miss selector valve; ~ **servoaccionada** *f* REFRIG servo-operated valve; ~ **servomoduladora** *f* AUTO servo-modulator valve; ~ **de seta** *f* AUTO mushroom valve, poppet valve; ~ **simuladora de percepción** *f* TRANSP AÉR feel-simulator valve; ~ **del sistema contraincendios** *f* ING MECÁ fire valve, water valve for firefighting, SEG, TERMO fire valve; ~ **de sobrante** *f* INSTAL HIDRÁUL overflow valve; ~ **del sobrante** *f* ING MECÁ return valve; ~ **de sobrepresión** *f* AUTO pressure-boost valve; ~ **de sobrepresión del estrangulador** *f* AUTO throttle pressure-boost valve; ~ **de sobrepresión de la mariposa** *f* AUTO throttle pressure-boost valve; ~ **de solenoide** *f* INSTAL HIDRÁUL *tuberías*, PROD, REFRIG solenoid valve; ~ **de suministro** *f* INSTAL HIDRÁUL delivery valve, discharge valve, head valve, pressure valve;

■ **t** ~ **tarada** *f* ING MECÁ calibrated valve; ~ **termiónica** *f* ELEC *de radio* thermionic tube (*AmE*), thermionic valve (*BrE*), ING ELÉC thermionic rectifier; ~ **termoelectrónica** *f* ELEC *de radio* thermionic tube (*AmE*), thermionic valve (*BrE*); ~ **termostática** *f* REFRIG thermostatically-controlled valve; ~ **de toma** *f* ING MECÁ admission valve; ~ **de toma de aire** *f* ING MECÁ air-inlet valve, TRANSP AÉR air-intake valve; ~ **de tornillo** *f* LAB *fluidos* screw valve; ~ **de tres vías** *f* INSTAL HIDRÁUL cross valve, switch valve, three-way valve, triple valve;

~ v ~ **de vaciado rápido de depósitos** *f* TEC ESP *aviones* jettison valve; ~ **de vacío** *f* AUTO vacuum valve, ELECTRÓN vacuum phototube, INFORM&PD vacuum tube, VEH *sistema de refrigeración* vacuum valve; ~ **de vástago** *f* AUTO poppet valve; ~ **de viento** *f* PROD *horno alto* cold-blast valve

válvulas: ~ **y accesorios de seguridad** *f pl* SEG safety valves and fittings; ~ **en culata** *f pl* AUTO, VEH *motor* overhead valves (*OHV*)

vanadato *m* QUÍMICA vanadate

vanádico *adj* QUÍMICA vanadic

vanadífero *adj* QUÍMICA vanadiferous

vanadilo *m* QUÍMICA vanadyl

vanadinita *f* MINERAL vanadinite

vanadio *m* (*V*) METAL, QUÍMICA vanadium (*V*)

vanadiolita *f* MINERAL vanadiolite

vanadito *m* QUÍMICA vanadite

vanadoso *adj* QUÍMICA vanadous

vano *m* CONST distance piece, span, ING MECÁ opening; ~ **de la puerta** *m* CONST door opening, doorway

vapoacumulador *m* INSTAL HIDRÁUL steam accumulator

vapofreno *m* INSTAL HIDRÁUL steam brake

vapohermético *adj* INSTAL HIDRÁUL steamtight

vapor[1] *m* AUTO vapor (*AmE*), vapour (*BrE*), CARBÓN collier, ELECTRÓN, ENERG RENOV, FÍS steam, vapor (*AmE*), vapour (*BrE*), GAS vapor (*AmE*), vapour (*BrE*), INSTAL TERM steam, METAL vapor (*AmE*), vapour (*BrE*), MINAS collier, NUCL, PROC QUÍ, QUÍMICA, REFRIG vapor (*AmE*), vapour (*BrE*), TEC PETR *perforación, refino* mist, vapor (*AmE*), vapour (*BrE*), TELECOM, TERMO, TERMOTEC vapor (*AmE*), vapour (*BrE*), TEXTIL steam; ~ **de agua** *m* INSTAL HIDRÁUL steam; ~ **de agua pesada** *m* NUCL heavy water vapor (*AmE*), heavy water vapour (*BrE*); ~ **de azufre** *m* GAS sulfur vapor (*AmE*), sulfur vapour (*BrE*); ~ **de escape** *m* INSTAL HIDRÁUL dead steam, exhaust steam, waste steam, TERMO dead steam, TERMOTEC exhaust steam; ~ **de exhaustación** *m* INSTAL HIDRÁUL waste steam; ~ **expulsado** *m* ALIMENT exhaust steam; ~ **de hélices gemelas** *m* TRANSP MAR twin-screw steamer; ~ **húmedo** *m* ENERG RENOV wet steam; ~ **inflamable** *m* SEG flammable vapor (*AmE*), flammable vapour (*BrE*); ~ **instantáneo** *m* REFRIG flash gas; ~ **de inversión** *m* INSTAL HIDRÁUL reversing steam; ~ **de mercurio** *m* CONST, ING MECÁ, QUÍMICA mercury vapor (*AmE*), mercury vapour (*BrE*); ~ **perdido** *m* ALIMENT exhaust steam; ~ **a presión** *m* INSTAL HIDRÁUL live steam, pressurized steam, NUCL, TERMO live steam; ~ **principal** *m* NUCL main steam; ~ **saturado** *m* FÍS saturated vapor (*AmE*), saturated vapour (*BrE*), INSTAL TERM, TERMO, TERMOTEC saturated steam; ~ **seco** *m* ENERG RENOV, INSTAL TERM dry steam; ~ **de servicios** *m* NUCL service steam; ~ **sobrecalentado** *m* INSTAL TERM, TERMOTEC superheated steam; ~ **supercalentado** *m* FÍS superheated steam; ~ **turístico** *m* TRANSP excursion steamer; ~ **vivo** *m* INSTAL HIDRÁUL pressurized steam, live steam, NUCL, TERMO live steam; ~ **de zinc** *m* CARBÓN zinc vapor (*AmE*), zinc vapour (*BrE*)

vapor[2]: **a ~** *fra* ING MECÁ under steam

vaporización *f* GEN, PAPEL *madera, troceados* evapora-tion, vaporization; ~ **instantánea** *f* IMPR flash-off; ~ **por láser** *f* FÍS RAD laser vaporization; ~ **rápida** *f* COLOR flash ageing

vaporizado[1] *adj* AUTO, FÍS, GAS, PROC QUÍ, QUÍMICA, TERMO vaporized

vaporizado[2]: ~ **húmedo** *m* PROC QUÍ ageing

vaporizador *m* ALIMENT, C&V, INSTAL TERM, LAB, NUCL, PROC QUÍ evaporator, vaporizer, PROD evaporator, vaporator, QUÍMICA, REFRIG, TERMO, VEH evaporator, vaporizer

vaporizar *vt* AUTO, FÍS, GAS, PROC QUÍ, QUÍMICA, TERMO vaporize, TEXTIL *someter al vapor* steam

vaporoso *adj* FÍS, GAS, QUÍMICA vaporous

Vapotrón *m* ELECTRÓN Vapotron

vapoturbina *m* INSTAL TERM steam turbine

vaquillona *f* AmL AGRIC heifer; ~ **castrada** *f* AmL AGRIC spayed heifer; ~ **sin servicio** *f* AmL AGRIC maiden heifer

vaquita: ~ **de San Antonio** *f* AGRIC ladybird

vara *f* C&V stick; ~ **de agrimensor** *f* CONST ranging pole, ranging rod, *topografía* range pole; ~ **de calibrado** *f* CONST *perforación, sondeo* bore rod; ~ **de empaque** *f* C&V packing stick; ~ **de gas** *f* ING MECÁ gas poker; ~ **de hierro** *f* PROD iron rod; ~ **de medir** *f* CONST *topografía* rod

varactor *m* ELECTRÓN varactor

varadero *m* TRANSP MAR *construcción naval* slip, *reparaciones* slipway; ~ **del ancla** *m* TRANSP MAR anchor well; ~ **carenero** *m* OCEAN camber

varado *adj* TRANSP MAR *del barco* aground, *navegación* stranded

varadura *f* CONTAM MAR grounding

varar[1] *vt* TRANSP MAR slip, *embarcación* beach

varar[2] *vi* OCEAN, TRANSP MAR go aground, ground, run aground, run ashore

varenga *f* TRANSP MAR *construcción naval* floor, floor plate

variabilidad: ~ **espacial** *f* CONTAM spatial variability

variable[1] *adj* ING MECÁ adjustable, MATEMÁT variable, MECÁ floating

variable[2] *f* INFORM&PD, MATEMÁT variable; ~ **aleatoria** *f* ELECTRÓN, INFORM&PD, MATEMÁT random variable; ~ **booleana** *f* INFORM&PD Boolean variable; ~ **controlada** *f* ING ELÉC controlled variable; ~ **dependiente** *f* MATEMÁT dependent variable; ~ **de desplazamiento** *f* TEC ESP *vehículos* displacement variable; ~ **de desviación** *f* TEC ESP displacement variable; ~ **eléctrica** *f* ELEC electric variable; ~ **estadística** *f* ELECTRÓN, INFORM&PD, MATEMÁT random variable; ~ **global** *f* INFORM&PD global variable; ~ **local** *f* INFORM&PD local variable; ~ **lógica** *f* INFORM&PD logical variable; ~ **reológica** *f* METAL rheological variable

variables: ~ **macroscópicas** *f pl* FÍS macroscopic variables

variación[1]: **sin ~** *adj* ELECTRÓN jitter-free

variación[2] *f* ELEC fluctuation, FÍS FLUID *del patrón de flujo* change, GEOFÍS *magnetometría*, ING MECÁ variation, METR rate, TRANSP AÉR hunting; ~ **absoluta de velocidad** *f* ELEC *máquina* absolute speed variation; ~ **adiabática** *f* TERMO adiabatic change; ~ **de la aguja** *f* TRANSP MAR compass variation, magnetic variation; ~ **anisoelástica** *f* TEC ESP *giróscopos* anisoelastic drift; ~ **anual media** *f* TRANSP MAR *de la marea* mean annual variation; ~ **de autoinductancia** *f* ELEC self-inductance variation;

~ barométrica *f* METEO barometrical variation; **~ de la capa freática** *f* HIDROL water table fluctuation; **~ de la carga** *f* ELEC load fluctuation; **~ del cero** *f* INSTR drift; **~ cíclica primaria** *f* TRANSP AÉR primary cyclic variation; **~ de corriente** *f* ELEC current fluctuation; **~ del coste** *f* PROD cost variance; **~ diaria** *f* TRANSP MAR diurnal variation; **~ de fase** *f* ELEC *corriente alterna* phase shift, phase variation, TEC ESP phase shift; **~ de fase digital** *f* ELECTRÓN digital phase shifting; **~ de fase por efecto de retardo** *f* (*VFE de retardo*) ELECTRÓN delayed modified phase-shift keying (*delayed MPSK*); **~ de flujo macroscópica** *f* NUCL macroscopic flux variation; **~ de frecuencia** *f* ELECTRÓN frequency pulling; **~ de la frecuencia por cristal** *f* CRISTAL, ELECTRÓN crystal frequency drift; **~ de la frecuencia por efecto Doppler** *f* PETROL Doppler shift; **~ de ganancia** *f* ELECTRÓN gain drift; **~ inversa** *f* MATEMÁT inverse variation; **~ lateral** *f* GEOL *de facies* lateral variation; **~ en la línea** *f* C&V in-line variation; **~ magnética** *f* GEOFÍS, TRANSP MAR magnetic variation; **~ magnética diaria** *f* GEOFÍS magnetic daily variation; **~ magnética lunar diaria** *f* GEOFÍS magnetic lunar daily variation; **~ magnética solar de un día en calma** *f* GEOFÍS magnetic solar-quiet-day variation; **~ magnética solar de un día tranquilo** *f* GEOFÍS magnetic solar-quiet-day variation; **~ en la octava fase** *f* ELECTRÓN eight-phase shift keying; **~ de paso de la pala** *f* TRANSP AÉR blade-pitch variation; **~ pequeña del voltaje** *f* TEC ESP ripple; **~ periódica** *f* GEOFÍS periodic variation; **~ permisible** *f* ING MECÁ tolerance; **~ de potencia** *f* REFRIG *de un compresor* capacity control; **~ de presión** *f* FÍS ONDAS *en onda estacionaria*, GAS pressure variation; **~ por saltos en la frecuencia** *f* TELECOM frequency hopping; **~ secular** *f* GEOFÍS secular variation, secular change; **~ secular geomagnética** *f* GEOFÍS geomagnetic secular variation; **~ de temperatura** *f* REFRIG temperature variation; **~ temporal** *f* CONTAM temporal fluctuation, temporal variation, time trend; **~ transitoria** *f* TELECOM transient fluctuation; **~ de la velocidad corriente y tensión** *f* ING ELÉC *motores sincrónicos* hunting

variador: **~ de fase** *m* ELEC *corriente alterna*, TELECOM phase shifter; **~ de fase de diodo** *m* ELECTRÓN diode phase shifter; **~ de fase de diodo NIP** *m* ELECTRÓN PIN-diode phase shifter

variamina *f* QUÍMICA variamine

variancia *f* MATEMÁT variance

variante: **~ de producto** *f* PROD product variant; **~ temporal** *f* PETROL time variant

varianza *f* FÍS, INFORM&PD, MATEMÁT variance

variar[1] *vt* ELECTRÓN *velocidad de electrones* modulate; **~ desde** *vt* TELECOM range from

variar[2] *vi* ELEC fluctuate; **~ la fase** *vi* ELECTRÓN, FÍS ONDAS, ING ELÉC, TELECOM phase-shift

varias: **~ de ~ capas** *adj* QUÍMICA multilayer

variedad *f* GEOM manifold; **~ resistente** *f* AGRIC resistant variety; **~ temprana** *f* AGRIC early variety; **~ de tres dimensiones** *f* GEOM three-dimensional manifold; **~ tri-dimensional** *f* GEOM three-dimensional manifold

varilla *f* C&V *de vidrio óptico* rod, CARBÓN *paraguas, abanicos* rib, CONST *escayola* lath, *soldadura* rod, IMPR, INFORM&PD wand, ING MECÁ link, tie, LAB *agitador* rod, TEC PETR *unidad de presión*, TEXTIL *banda* bar; **~ absorbente** *f* NUCL absorber rod; **~ absorbente negra** *f* NUCL black absorber-rod; **~ de acabado a mano** *f* ING MECÁ hand-finishing stick; **~ accionadora de señales** *f* FERRO *infraestructura* signal-operating rod; **~ de accionamiento de la mariposa** *f* INSTAL HIDRÁUL *válvulas* throttle-reach rod; **~ de acoplamiento** *f* ING MECÁ link rod; **~ para agitar los baños** *f* CINEMAT stirring rod; **~ de amarre** *f* C&V tie rod; **~ de apriete del freno** *f* VEH brake connecting-rod; **~ de arrastre** *f* TEC PETR *perforación* kelly; **~ de bomba** *f* AGUA spear; **~ de bombeo** *f* AGUA pump rod; **~ de centrar** *f* ING MECÁ centering rod (*AmE*), centring rod (*BrE*); **~ de combustible** *f* FÍS *reactor nuclear* fuel rod; **~ de combustible defectuosa** *f* NUCL defective fuel rod; **~ de combustible segmentada** *f* NUCL segmented fuel rod; **~ de combustible venteado** *f* NUCL vented fuel rod; **~ compactadora** *f* CONST *ensayos* tamping rod; **~ de conexión** *f* AUTO tie rod, ING MECÁ connecting rod, tie rod, MECÁ connecting rod; **~ de conexión en U** *f* ING MECÁ, MECÁ clevis link; **~ de conexión de longitud infinita** *f* ING MECÁ connecting rod of infinite length; **~ de contacto** *f* ING MECÁ prod; **~ de control de la palomilla** *f* AUTO throttle control rod; **~ de control de precisión** *f* ING MECÁ fine-control rod; **~ desenganchada** *f* MINAS *sondeos* uncoupling rod; **~ de desenganche** *f* AUTO, VEH release rod; **~ de embrague** *f* AUTO clutch release fork, clutch rod, VEH clutch rod; **~ empujadora** *f* AUTO, ING MECÁ, MECÁ push rod; **~ empujadora de válvulas** *f* AUTO valve push rod; **~ empujaválvulas** *f* AUTO valve push rod; **~ de empuje** *f* AUTO, ING MECÁ, MECÁ push rod; **~ de enclavamiento** *f* FERRO locking bar; **~ de enfoque** *f* CINEMAT, TV focusing lever; **~ excéntrica** *f* ING MECÁ, MECÁ eccentric rod; **~ de expulsión** *f* ING MECÁ knock-out rod; **~ de extracción** *f* ING MECÁ knock-out rod; **~ de eyección** *f* ING MECÁ knock-out rod; **~ de fijación aislante** *f* REFRIG breaker strip; **~ de freno** *f* VEH brake rod; **~ de gatear** *f* PROD jack rod; **~ de hierro** *f* PROD iron rod; **~ del horno** *f* C&V buckstay; **~ de ignición** *f* ING MECÁ ignition poker; **~ impulsora** *f* AUTO, ING MECÁ, MECÁ push rod; **~ de levantamiento de la válvula** *f* INSTAL HIDRÁUL valve tappet; **~ de llamada** *f* ING MECÁ drawbar; **~ maestra** *f* ING MECÁ main rod; **~ de mando** *f* AUTO control rod; **~ de maniobra** *f* FERRO *agujas* connecting rod, ING MECÁ connecting rod, pull rod, MECÁ connecting rod; **~ de la mariposa** *f* INSTAL HIDRÁUL throttle rod; **~ para medición del nivel** *f* AUTO, VEH *lubricación* dipstick; **~ para medir la profundidad del líquido** *f* METR *depósitos* measuring rod; **~ de metal de aportación** *f* CONST, MECÁ, PROD *soldadura* filler rod; **~ del mezclador** *f* TRANSP AÉR mixer rod; **~ del nivel del aceite** *f* AUTO, VEH *lubricación* dipstick, oil dipstick, oil-level stick; **~ de obturación** *f* ING MECÁ rod seal; **~ de ojal** *f* PROD eyed rod; **~ de pararrayos** *f* ING ELÉC lightning rod; **~ pararrayos** *f* GEOFÍS lightning rod; **~ de paro** *f* ING MECÁ knock-out rod; **~ de perforación** *f* MINAS, PETROL drill rod; **~ para perforación** *f* CONST *sondeos* boring rod; **~ del pistón** *f* ING MECÁ sucker rod; **~ plana** *f* ING MECÁ flat rod; **~ de pretensado** *f* NUCL tendon; **~ de refuerzo** *f* PROD *ganchos de moldería* dabber, *molde de arena* prod; **~ de reglaje** *f* AUTO control rod;

~ reguladora *f* ING MECÁ governor rod; **~ del relé** *f* AUTO relay rod; **~ separadora** *f* C&V punty; **~ de sonda** *f* MINAS drill rod; **~ de sondeo** *f* ING MECÁ boring bar, MINAS, PETROL, TEC PETR *perforación* drill pipe; **~ de suspensión** *f* MINAS hanging rod; **~ de tensión** *f* ING MECÁ tie rod; **~ de tierra** *f* ELEC *conexión* earth rod (*BrE*), earthing rod (*BrE*), ground rod (*AmE*), grounding rod (*AmE*), FERRO earthing paddle (*BrE*), grounding paddle (*AmE*), ING ELÉC earth rod (*BrE*), ground rod (*AmE*); **~ de toma de tierra** *f* ELEC *conexión*, ING ELÉC earth rod (*BrE*), ground rod (*AmE*); **~ de tracción** *f* ING MECÁ pull rod; **~ de transmisión** *f* ING MECÁ transmission rod; **~ transversal** *f* FERRO crossbar; **~ de unión** *f* AUTO tie rod, ING MECÁ connecting rod, tie bar, MECÁ connecting rod, tie rod; **~ de unión del pantógrafo** *f* FERRO pantograph tie-bar; **~ de vidrio** *f* C&V, LAB glass rod; **~ de zoom** *f* CINEMAT zoom lever

varillaje *m* ING MECÁ linkage, rod linkage, MINAS bracket crab, drill pipe, PETROL, TEC PETR drill pipe, VEH *de frenos* rods; **~ del acelerador** *m* AUTO, VEH *carburador* accelerator linkage; **~ de carburador** *m* AUTO, VEH carburetor linkage (*AmE*), carburettor linkage (*BrE*); **~ de la dirección** *m* AUTO, VEH steering linkage; **~ del freno** *m* AUTO, ING MECÁ, VEH brake linkage; **~ de mando** *m* AUTO, VEH *de dirección, embrague, frenos* linkage; **~ de mando del embrague** *m* AUTO, VEH clutch linkage; **~ de mando de la mariposa** *m* AUTO, VEH *del carburador* throttle linkage

varillas: **~ de sonda** *f pl* MINAS rods

variómetro: **~ magnético** *m* GEOFÍS magnetic variometer

variscita *f* MINERAL variscite

varistancia *f* ELEC *resistor* varistor

varistor *m* ELEC *resistor* varistor, ING ELÉC varistance, varistor, PROD varistor; **~ de carburo de silicio** *m* ING ELÉC silicon carbide varistor; **~ de óxido de zinc** *m* ING ELÉC zinc oxide varistor

varita: **~ de zahorí** *f* HIDROL dowser

varva *f* GEOL varve

vaselina ® *f* PETROL, QUÍMICA, TEC PETR Vaseline ®, mineral jelly, petrolatum, petroleum jelly; **~ líquida** *f* TERMO liquid paraffin

vasija *f* FÍS vessel, ING MECÁ pan, LAB *material de vidrio* basin; **~ de acero inoxidable** *f* LAB stainless steel beaker; **~ cruda** *f* C&V unfired pot; **~ de evaporación** *f* LAB evaporating basin; **~ interior del calentador** *f* INSTAL TERM *calentadores de agua* heating body; **~ de presión** *f* ING MECÁ, MECÁ pressure vessel; **~ de presión criogénica** *f* ING MECÁ cryogenic pressure vessel; **~ del reactor** *f* NUCL reactor vessel; **~ de recogida de fugas** *f* NUCL leakage interception vessel; **~ rotatoria** *f* C&V rotating bowl

vasijas: **~ de barro** *f pl* C&V pottery, stoneware

vaso *m* C&V glass, ELEC *pilas y baterías eléctricas* cell, EMB cup; **~ acumulador** *m* AUTO accumulator cell; **~ de batería** *m* AUTO battery cell; **~ de boca ancha** *m* ALIMENT, C&V beaker; **~ cónico** *m* LAB *material de vidrio* conical beaker; **~ de decantación** *m* ALIMENT, CONTAM decanting glass, PROC QUÍ decantation glass, decanting glass, QUÍMICA, TEC PETR decanting glass; **~ Dewar** *m* FÍS, LAB, PROC QUÍ Dewar flask; **~ del embalse** *m* AGUA reservoir basin; **~ del filtro** *m* VEH *de carburador* filter bowl; **~ de laboratorio** *m* ALIMENT beaker; **~ de pila** *m* C&V battery jar; **~ de**

precipitados *m* LAB *material de vidrio* beaker, PROC QUÍ precipitation vessel; **~ de precipitados con pico** *m* LAB *material de vidrio* beaker with spout

vasopresina *f* QUÍMICA vasopressin

vasos: **~ comunicantes** *m pl* LAB communicating vessels

vástago *m* ING MECÁ, MECÁ, PROD pin; **contra ~** *m* AGUA tailrod, CONST rod, stem, ING MECÁ ram, shank, INSTAL HIDRÁUL *de válvula* pin, rod, stem, MECÁ pin, MINAS rod; **~ de aislador** *m* PROD insulator pin; **~ de bomba** *m* AGUA spear; **~ para broca postiza** *m* MINAS drill rod; **~ del distribuidor** *m* FERRO *vehículos* valve rod, INSTAL HIDRÁUL slide rod, valve rod; **~ elevador** *m* INSTAL HIDRÁUL *de mecanismo de válvula de Corliss* lifting rod; **~ del émbolo de extremo doble** *m* ING MECÁ double-ended piston rod cylinder; **~ de estampa** *m* PROD stamp stem; **~ impulsor** *m* ING MECÁ bumper rod; **~ del levantaválvulas** *m* AUTO tappet stem; **~ de perforación** *m* PETROL drill stem, TEC PETR kelly; **~ del pistón** *m* AUTO, FERRO, ING MECÁ, PROD, VEH piston rod; **~ del pistón del cilindro** *m* ING MECÁ cylinder piston rod; **~ del pistón de extremo doble** *m* ING MECÁ double-ended piston rod cylinder; **~ pulido** *m* ENERG RENOV *bomba de molino de viento* polished rod; **~ rojo** *m* ENERG RENOV *bomba de molino de viento* red rod; **~ de tierra** *m* GEOFÍS earth rod (*BrE*), ground rod (*AmE*); **~ de válvula** *m* INSTAL HIDRÁUL valve rod, valve spindle, valve stem, VEH *motor* valve stem; **~ de válvula de compuerta** *m* INSTAL HIDRÁUL gate stem; **~ de válvula de mariposa** *m* INSTAL HIDRÁUL throttle stem

VAT *abr* (*valor ácido total*) QUÍMICA TAN (*total acid number*)

vatiaje *m* ING ELÉC wattage

vatihora *m* ELEC *unidad de trabajo* watt-hour

vatímetro *m* ELEC wattmeter; **~ dinamométrico** *m* INSTR dynamometer wattmeter; **~ ferrodinámico** *m* ELEC, INSTR ferrodynamic wattmeter; **~ de hilo caliente** *m* ELEC hot-wire wattmeter; **~ de paleta** *m* ELEC, FÍS, ING ELÉC, INSTR vane wattmeter; **~ térmico** *m* ELEC hot-wire wattmeter

vatio *m* (*W*) ELEC, FÍS, ING ELÉC, METR watt (*W*); **~~-hora** *m* ELEC *unidad de trabajo*, FÍS, ING ELÉC watt-hour

vauquelinita *f* MINERAL vauquelinite

VCL[1] *abr* (*visualización en cristal líquido*) ELEC, ELECTRÓN, FÍS, INFORM&PD, TELECOM, TV LCD (*liquid crystal display*)

VCL[2]: **~ dicroico** *m* ELECTRÓN dichroic LCD; **~ reflectante** *m* ELECTRÓN reflective LCD

VDU *abr* (*unidad de presentación visual, unidad de representación visual, unidad de visualización*) ELECTRÓN, IMPR, INFORM&PD, INSTR TELECOM, TV VDU (*visual display unit*)

vecinal *adj* QUÍMICA vicinal

vecindad: **~ más próxima** *f* METAL *coordinación* next nearest neighbors (*AmE*), next nearest neighbours (*BrE*); **~ próxima al origen** *f* METAL first-nearest neighbors (*AmE*), first-nearest neighbours (*BrE*)

vecino *adj* QUÍMICA vicinal

vecinos: **~ más cercanos** *m pl* CRISTAL nearest neighbors (*AmE*), nearest neighbours (*BrE*)

vector *m* GEN vector; **~ base** *m* FÍS basis vector; **~ de Burgers** *m* CRISTAL Burgers vector; **~ de campo** *m* ING ELÉC field vector; **~ de corriente** *m* ELEC *forma* *de onda* phasor; **~ desplazamiento** *m* ELEC *campo eléctrico* displacement vector; **~ eigen** *m* INFORM&PD eigenvector; **~ de empuje** *m* TEC ESP thrust vector; **~ de interrupción** *m* INFORM&PD interrupt vector; **~ mayoritario** *m* ELECTRÓN majority carrier; **~ de onda** *m* FÍS, ING ELÉC wave vector; **~ de ondas de Fermi** *m* FÍS Fermi-wave vector; **~ de posición** *m* FÍS, GEOM position vector; **~ de Poynting** *m* FÍS Poynting vector; **~ propio** *m* ELECTRÓN, FÍS eigenvector; **~ de salida** *m* TRANSP AÉR trailing vector; **~ unitario** *m* FÍS unit vector

vectorial *adj* GEOM vectorial

vectorización: **~ de empuje en vuelo** *f* TEC ESP *reactor* in-flight thrust vectoring

vectorizado *adj* TEC ESP vectored

vectorscopio *m* TV vectorscope

vegetación: **~ de playa** *f* OCEAN *botánica* beach growth

vehículo *m* AGUA medium, IMPR *tinta* vehicle, ING MECÁ carriage, MECÁ, QUÍMICA, TEC ESP vehicle, TRANSP vehicle, carriage; **~ accionado por acumuladores** *m* ING ELÉC battery vehicle; **~ acorazado de recuperación** *m* D&A armored recovery vehicle (*AmE*) (*ARV*), armoured recovery vehicle (*BrE*) (*ARV*); **~ aislado de apoyo** *m* TRANSP contactless support vehicle; **~ con amortiguación por aire** *m* TRANSP, TRANSP AÉR TRANSP MAR air cushion vehicle (*ACV*); **~ anfibio** *m* D&A amphibious vehicle, TRANSP MAR amphibian; **~ blindado de combate sobre orugas** *m* D&A tracked armored fighting vehicle (*AmE*), tracked armoured fighting vehicle (*BrE*) (*AFV*); **~ de cadenas** *m* TRANSP tracked vehicle; **~ de carga** *m* TEC ESP cargo vehicle; **~ de carga pesada** *m* TEC ESP heavy-lift vehicle; **~ cisterna** *m* AUTO vehicle tanker; **~ con colchón de aire tipo aerostático** *m* TRANSP aerostatic-type air cushion vehicle; **~ comercial** *m* VEH commercial vehicle; **~ contra incendios** *m* SEG, TERMO firefighting vehicle; **~ de doble uso** *m* VEH dualpurpose vehicle; **~ de efecto en superficie** *m* TRANSP surface-effect vehicle (*SEV*); **~ eléctrico** *m* ELEC, TRANSP, VEH electric vehicle; **~ eléctrico accionado por acumuladores** *m* TRANSP batterypowered electric vehicle; **~ eléctrico comercial** *m* TRANSP commercial electric vehicle; **~ eléctrico polivalente** *m* TRANSP general-purpose electric vehicle; **~ eléctrico de transporte** *m* ELEC, TRANSP electric transport vehicle; **~ eléctrico urbano** *m* TRANSP urban electric vehicle; **~ elevador** *m* EMB lifting vehicle; **~ espacial** *m* TEC ESP space vehicle, spaceship; **~ experimental de seguridad** *m* SEG, TRANSP experimental safety vehicle (*ESV*); **~ frigorífico** *m* REFRIG, TRANSP, VEH mechanically refrigerated vehicle; **~ guiado por raíles** *m* TRANSP track-guided vehicle; **~ híbrido** *m* TRANSP hybrid vehicle; **~ impulsado por baterías** *m* ING ELÉC battery vehicle; **~ isoeléctrico** *m* TRANSP isoelectric vehicle; **~lanzadera espacial** *m* TEC ESP spacelab; **~ de lanzamiento** *m* TEC ESP launch vehicle; **~ de lanzamiento de carga pesada** *m* TEC ESP heavy lift launch vehicle (*HLLV*); **~ logístico lunar** *m* (*LLV*) TEC ESP lunar logistics vehicle (*LLV*); **~ de mercancías pesadas multieje** *m* TRANSP multiaxle heavy freight vehicle (*AmE*), multiaxle heavy goods vehicle (*BrE*); **~ montacargas de horquilla** *m* AGRIC, AUTO, CONST, EMB, ING MECÁ, PROD, VEH forklift truck;

~ con motor de gasolina *m* *Esp* (*cf vehículo con motor de nafta AmL*) AUTO, CONTAM, VEH gas-engine vehicle (*AmE*), gasoline-engine vehicle (*AmE*), petrol-engine vehicle (*BrE*); **~ con motor de nafta** *m* *AmL* (*cf vehículo con motor de gasolina Esp*) AUTO, CONTAM, VEH gas-engine vehicle (*AmE*), gasolineengine vehicle (*AmE*), petrol-engine vehicle (*BrE*); **~ de motor semi-trailer** *m* TRANSP semitrailer motor vehicle; **~ orbital** *m* TEC ESP orbital vehicle; **~ oruga** *m* VEH tracked vehicle; **~ pilotado a distancia** *m* D&A, TRANSP AÉR remote-piloted vehicle; **~ de recogida de basura** *m* RECICL, VEH refuse-collection vehicle; **~ de reconocimiento** *m* D&A reconnaissance vehicle; **~ de reentrada** *m* TEC ESP, VEH re-entry vehicle; **~ refrigerado** *m* REFRIG, TRANSP, VEH refrigerated vehicle; **~ remolcado** *m* AUTO tow vehicle; **~ remolcador** *m* VEH towing vehicle; **~ remolcador semi-trailer** *m* TRANSP semitrailer towing vehicle; **~ para repostar** *m* TEC ESP fueling vehicle (*AmE*), fuelling vehicle (*BrE*); **~ de rescate** *m* SEG, TRANSP AÉR, TRANSP MAR rescue vehicle; **~ de rescate de la tripulación** *m* TEC ESP crew rescue vehicle; **~ rígido auto-portante** *m* TRANSP self-supporting rigid vehicle; **~ de salvamento** *m* TRANSP salvage lorry (*BrE*), salvage truck (*AmE*); **~ de servicio** *m* TRANSP AÉR service vehicle; **~ subacuático** *m* TRANSP MAR underwater vehicle; **~ submarino** *m* TEC PETR underwater vehicle; **~ subterráneo** *m* TRANSP tube vehicle; **~ suspendido por aspiración** *m* TRANSP suctionsuspended vehicle; **~ tanque de carreta** *m* CONTAM MAR road tanker; **~ terrestre** *m* TEC ESP land vehicle; **~ terrestre de captura** *m* TEC ESP *equipos en tierra* earth-capture vehicle; **~ terrestre levitante por la reacción de chorros de aire sobre el terreno** *m* TRANSP, VEH ground effect machine (*GEM*); **~ todo terreno** *m* AUTO all-terrain vehicle (*AT vehicle*); **~ con tracción a las cuatro ruedas** *m* AUTO, VEH four-wheel drive vehicle; **~ de transferencia entre órbitas** *m* (*OTV*) TEC ESP orbital transfer vehicle (*OTV*); **~ para transporte de personal** *m* MINAS manriding car; **~ de transporte de tropas** *m* D&A troop-carrying vehicle; **~ de uso general** *m* D&A utility vehicle; **~ utilitario** *m* VEH utility vehicle; **~ ventilado** *m* REFRIG ventilated vehicle

vejiga *f* TRANSP AÉR blister

vela[1]: **a ~ llena** *adv* TRANSP MAR *navegación* full and by

vela[2] *f* C&V gathering bubble, gob, parison, CARBÓN *puntal grueso de túnel* prop, minas, *túneles* strut, MINAS punch prop, TRANSP MAR sail; **~ de capa** *f* TRANSP MAR storm sail, *navegación a vela* trysail; **~ doble** *f* C&V double gobbing; **~ mayor** *f* TRANSP MAR mainsail; **~ solar** *f* TEC ESP solar sail; **~ trinquete** *f* TRANSP MAR foresail

velado: **~ por la luz** *adj* CINEMAT light-struck

velamen *m* TEC ESP *buques de vela* cloth

velar *vt* CINEMAT, FOTO fog

velas: **con las ~ cargadas** *fra* TRANSP MAR *navegación a vela* ahull; **con las ~ desplegadas** *fra* TRANSP MAR *navegación a vela* under canvas

velero: **~ cón motor auxiliar** *m* TRANSP MAR auxiliary engine sailing ship

veleta *f* METEO wind inclination meter, TRANSP MAR wind telltale; **~ bidireccional** *f* METEO bidirectional wind vane

vello *m* TEXTIL *del tejido* nap

vellosidad *f* TEXTIL hairiness

velo *m* ACÚST fog, CINEMAT veil, fog, FOTO fog, IMPR fog, *en planchas offset* scum, TEC ESP film; ~ **de la base** *m* FOTO base fog; ~ **dicroico** *m* CINEMAT, FOTO dichroic fog; ~ **marginal** *m* CINEMAT edge fogging; ~ **por la plata** *m* CINEMAT silver fog; ~ **químico** *m* CINEMAT chemical fogging; ~ **por revelado** *m* CINEMAT development fog

velocidad *f* GEN velocity, CINEMAT, FÍS ONDAS speed, INFORM&PD rate, ING MECÁ gear, TEXTIL, TRANSP speed;

~ a ~ **absoluta** *f* MECÁ absolute velocity, TRANSP AÉR ground speed; ~ **absoluta de agua** *f* ENERG RENOV absolute water velocity; ~ **de absorción** *f* AGUA rate of absorption; ~ **de acceso** *f* ING ELÉC access speed; ~ **de acercamiento** *f* TEC ESP approach speed; ~ **de acumulación de dosis** *f* FÍS RAD dose rate; ~ **acústica de referencia** *f* ACÚST reference sound velocity; ~ **de admisión** *f* INSTAL HIDRÁUL *de turbinas* inlet velocity; ~ **adquirida** *f* FÍS, ING MECÁ, MECÁ momentum; ~ **de aferrar** *f* ENERG RENOV *energía de viento* Furling speed; ~ **de agua relativa** *f* AGUA, ENERG RENOV, HIDROL relative water velocity; ~ **del aire** *f* INSTAL TERM air velocity; ~ **al extinguirse el motor** *f* TEC ESP *cohetes* burnout velocity; ~ **de alcance** *f* TERMO rate of spread; ~ **de alimentación** *f* CARBÓN feed rate; ~ **de amortiguamiento** *f* TV decay rate; ~ **anemométrica indicada** *f* TRANSP AÉR indicated airspeed; ~ **anemométrica verdadera** *f* TRANSP AÉR true airspeed (*TAS*); ~ **angular** *f* ELEC *fuente de alimentación de CA* angular frequency, ENERG RENOV, FÍS, ING ELÉC, ING MECÁ, TEC ESP angular velocity; ~ **angular constante** *f* INFORM&PD, ÓPT constant angular velocity (*CAV*); ~ **angular de precesión** *f* ENERG RENOV angular velocity of precession; ~ **aparente** *f* GEOFÍS apparent velocity; ~ **de aproximación** *f* TRANSP AÉR approach speed, TRANSP MAR *navegación* closing speed; ~ **de aproximación de aterrizaje** *f* TEC ESP, TRANSP AÉR landing approach speed; ~ **de aproximación de aterrizaje de referencia** *f* TRANSP AÉR reference landing approach speed; ~ **ascensional** *f* TEC ESP, TRANSP AÉR rate of climb (*RC*); ~ **de ascenso** *f* TEC ESP climb rate, climb speed, rate of climb (*RC*), TRANSP AÉR climb speed, rate of climb (*RC*); ~ **de aterrizaje** *f* TRANSP AÉR landing speed; ~ **de aterrizaje de diseño** *f* TRANSP AÉR design landing speed; ~ **de avance** *f* CARBÓN feed rate, TEC PETR *perforación* drilling rate, TRANSP progression speed; ~ **de avance por el agua** *f* TRANSP MAR *del barco* speed through the water; ~ **de avance del carro portaherramienta** *f* ING MECÁ saddle-feed rate; ~ **de avance normalizada** *f* TEC PETR *perforación* normalized drilling rate (*NDR*); ~ **axial** *f* ENERG RENOV axial velocity;

~ b ~ **de bandas** *f* TV velocity banding; ~ **de barrido** *f* TRANSP AÉR scanning speed; ~ **de bloque** *f* TRANSP AÉR block speed;

~ c ~ **de cabeza a cinta** *f* TV head-to-tape speed; ~ **de caída** *f* ACÚST decay rate, rate of decay; ~ **de caída adiabática** *f* TERMO adiabatic lapse rate; ~ **de calentamiento** *f* INSTAL TERM, TERMO rate of heating; ~ **calibrada mínima** *f* TRANSP AÉR minimum calibrated speed; ~ **de la cámara** *f* CINEMAT, TV camera speed; ~ **de cambio de fase** *f* TELECOM phase-change velocity; ~ **de cangilón** *f* ENERG RENOV bucket velocity; ~ **del chip** *f* TELECOM chip rate; ~ **de cierre** *f* ELEC *relé* closing speed; ~ **de cinta** *f* ACÚST, TV tape speed; ~ **circunferencial** *f* TRANSP AÉR circumferential speed, rotor tip velocity; ~ **de combustión** *f* TERMO, TERMOTEC rate of combustion; ~ **de congelación** *f* REFRIG, TERMO freezing rate; ~ **de conmutación** *f* ELECTRÓN switching speed; ~ **constante** *f* FÍS, ING MECÁ, MECÁ constant speed, TELECOM constant rate, TRANSP constant speed; ~ **de control mínima en el aire** *f* TRANSP AÉR minimum control speed in the air; ~ **de control mínima en tierra** *f* TRANSP AÉR minimum control speed on the ground; ~ **de corriente libre** *f* FÍS, FÍS FLUID free-stream velocity; ~ **de corriente libre fuera de la capa límite** *f* FÍS, FÍS FLUID free-stream velocity outside the boundary layer; ~ **de corrientes** *f* TRANSP MAR current rate; ~ **de corte** *f* ING MECÁ cutting speed, P&C shear rate, *propiedad física* rate of shear; ~ **cósmica** *f* TEC ESP cosmic velocity; ~ **de crecimiento** *f* ING ELÉC rate of rise; ~ **crítica** *f* FÍS critical velocity, MECÁ, TRANSP, TRANSP AÉR critical speed; ~ **de crucero** *f* TRANSP AÉR, TRANSP MAR cruising speed; ~ **de crucero de diseño** *f* TRANSP AÉR design cruising speed; ~ **de cuadro** *f* CINEMAT, TV frame rate; ~ **cubicada del viento** *f* ENERG RENOV wind velocity cubed;

~ d ~ **de decantación** *f* PROC QUÍ decantation rate; ~ **de decisión** *f* TRANSP AÉR decision speed (*V1*); ~ **de deposición** *f* CONTAM *depuración* deposition rate, deposition speed, deposition velocity; ~ **de derivación** *f* METAL drift velocity; ~ **de derrape** *f* TEC ESP yaw rate; ~ **de desagüe** *f* CONTAM washout rate; ~ **de descarga** *f* CARBÓN rate of flow, HIDROL discharge velocity, RECICL discharge rate; ~ **de descenso** *f* NUCL, TEC ESP, TRANSP AÉR rate of descent; ~ **de descenso límite en la toma de tierra** *f* TRANSP AÉR limit rate of descent at touchdown; ~ **de desintegración radiactiva** *f* FÍS PART, FÍS RAD radioactive-decay rate; ~ **de despegue** *f* TRANSP AÉR liftoff speed, takeoff speed, unstick speed; ~ **de despegue mínima** *f* TRANSP AÉR minimum unstick speed; ~ **de desplazamiento** *f* NUCL *de una barra de control* rate of travel; ~ **de desplazamiento de la cruceta** *f* FÍS crosshead displacement rate; ~ **de desprendimiento de calor** *f* TERMO, TERMOTEC rate of heat release; ~ **de detonación** *f* MINAS *explosivos* velocity of detonation (*VOD*); ~ **de difusión** *f* CONST rate of spread; ~ **de diseño con flaps extendidos** *f* TRANSP AÉR design flaps-extended speed; ~ **de diseño para ráfagas de máxima intensidad** *f* TRANSP AÉR design speed for maximum gust intensity; ~ **de diseño de turbulencia de aire** *f* TRANSP AÉR design rough air speed; ~ **de dislocación** *f* METAL dislocation velocity;

~ e ~ **económica** *f* TRANSP MAR economical speed; ~ **efectiva de transmisión de datos** *f* INFORM&PD effective data-transfer rate; ~ **de elevación** *f* TRANSP AÉR liftoff speed; ~ **de embalamiento** *f* ENERG RENOV *turbinas* runaway speed; ~ **en emersión** *f* TRANSP MAR *submarinos* surface speed; ~ **de la emulsión** *f* CINEMAT emulsion speed; ~ **de endurecimiento** *f* P&C, TERMO rate of cure; ~ **de enfriamiento** *f* REFRIG rate of cooling, temperature gradient, TERMO rate of cooling; ~ **de enhebrado** *f* PROD *laminadores* thread speed; ~ **de ensanchamiento** *f* METAL drift velocity; ~ **de entrada** *f* INSTAL HIDRÁUL inlet velocity; ~ **escalar de ascenso** *f* TEC ESP, TRANSP AÉR climbing speed;

~ de escaneo _f_ INFORM&PD scan rate; **~ para escapar de la Tierra** _f_ TEC ESP earth-escape velocity; **~ de escape** _f_ FÍS escape velocity; **~ específica** _f_ ENERG RENOV specific speed; **~ específica de ionización** _f_ FÍS PART ionization rate; **~ específica máxima** _f_ VEH maximum design speed; **~ de espera** _f_ TRANSP AÉR holding speed; **~ de evaporación** _f_ PROC QUÍ, REVEST evaporation rate; **~ excesiva** _f_ ING MECÁ, MECÁ, PROD overspeed; **~ de exploración** _f_ INFORM&PD scan rate, TRANSP spot speed, TRANSP AÉR scanning rate, TV scanning speed; **~ de extracción** _f_ ALIMENT extraction rate;

■ **f** **~ de fase** _f_ FÍS, PETROL phase velocity; **~ de filmación** _f_ CINEMAT filming speed; **~ de flujo** _f_ GAS FÍS FLUID flow speed, P&C, REFRIG rate of flow; **~ del flujo de aire** _f_ INSTAL TERM airflow rate; **~ del flujo de fusión** _f_ FÍS FLUID melt flow rate; **~ de fuga** _f_ FÍS escape velocity; **~ de funcionamiento** _f_ ING MECÁ working speed, TRANSP operating speed; **~ de funcionamiento real** _f_ MECÁ actual running speed;

■ **g** **~ del gasto** _f_ AGRIC rate of flow; **~ del generador** _f_ ENERG RENOV generator speed; **~ global del viaje** _f_ TRANSP overall travel speed; **~ de grabación** _f_ ELECTRÓN _tubos de almacenamiento_ writing speed; **~ de grabado** _f_ ACÚST recorded velocity; **~ de gradiente** _f_ CINEMAT gradient speed; **~ de grupo** _f_ ACÚST, FÍS, GEOFÍS, ÓPT group velocity, TELECOM envelope velocity, group velocity; **~ de guiñada** _f_ TEC ESP yaw rate;

■ **h** **~ hipersónica** _f_ FÍS hypersonic speed; **~ del husillo** _f_ ING MECÁ wheel spindle speed;

■ **i** **~ ideal** _f_ ENERG RENOV ideal velocity; **~ de impacto** _f_ METAL impact velocity; **~ de impresión** _f_ INFORM&PD print speed; **~ de infiltración** _f_ INSTAL TERM infiltration rate; **~ de información obligada** _f_ TELECOM committed information rate (_CIR_); **~ infrasónica** _f_ FÍS infrasonic speed; **~ inicial** _f_ TEC ESP initial velocity; **~ de intervalo** _f_ TEC PETR _geofísica_ interval velocity;

■ **l** **~ de lectura** _f_ INFORM&PD reading rate; **~ lenta para recibir el tocho** _f_ PROD _laminadores_ thread speed; **~ de liberación de calor** _f_ TERMO, TERMOTEC rate of heat release; **~ de la línea** _f_ EMB line speed; **~ de línea en baudios** _f_ _Esp_ (_cf tasa en baudios AmL_) INFORM&PD, PROD _teleproceso_ baud rate; **~ lineal constante** _f_ ÓPT constant linear velocity (_CLV_); **~ de la luz** _f_ FÍS, FÍS ONDAS speed of light;

■ **m** **~ de maniobra de diseño** _f_ TRANSP AÉR design maneuvering speed (_AmE_), design manoeuvring speed (_BrE_); **~ de la máquina** _f_ C&V machine speed; **~ de marcha** _f_ TRANSP running speed; **~ máxima** _f_ AGUA peak velocity, TRANSP, TRANSP AÉR, TRANSP MAR, VEH maximum speed; **~ máxima del árbol** _f_ ENERG RENOV _energía de viento_ maximum shaft speed; **~ máxima con flaps extendidos** _f_ TRANSP AÉR maximum flap-extended speed; **~ máxima con el tren de aterrizaje extendido** _f_ TRANSP AÉR maximum landing gear-extended speed; **~ máxima con el tren de aterrizaje operando** _f_ TRANSP AÉR maximum landing gear-operating speed; **~ máxima en vuelo nivelado con potencia nominal** _f_ TRANSP AÉR maximum speed in level flight with rated power; **~ media** _f_ TRANSP mean speed; **~ media de cruce** _f_ TELECOM average crossing rate; **~ media de funcionamiento** _f_ IMPR average running speed; **~ media incluyendo paradas** _f_ PROD average speed

including stoppages; **~ media de marcha** _f_ TRANSP average running speed; **~ media en un punto** _f_ TRANSP average spot speed; **~ media del viento** _f_ ENERG RENOV, METEO average wind speed, mean wind speed; **~ mínima de control durante aproximación al aterrizaje** _f_ TRANSP AÉR minimum control speed during landing approach; **~ mínima de un motor** _f_ MECÁ idling speed; **~ mínima de seguridad de despegue** _f_ TRANSP AÉR minimum takeoff safety speed; **~ del motor** _f_ TRANSP AÉR, VEH engine speed; **~ de movimiento** _f_ CARBÓN rate of flow;

■ **n** **~ negativa** _f_ ING MECÁ retarded velocity; **~ nominal del motor** _f_ AUTO, VEH rated engine speed; **~ nominal del viento** _f_ ENERG RENOV, METEO rated wind speed;

■ **o** **~ del obturador** _f_ FOTO shutter speed; **~ de las olas** _f_ OCEAN wave velocity; **~ de onda** _f_ FÍS ONDAS wave velocity; **~ de operación permisible máxima** _f_ TRANSP AÉR maximum permissible operating speed; **~ orbital** _f_ FÍS orbital velocity;

■ **p** **~ de las paletas** _f_ ENERG RENOV blade speed, vane velocity; **~ parabólica** _f_ TEC ESP parabolic velocity; **~ de partículas** _f_ ACÚST, FÍS PART particle velocity; **~ de paso** _f_ CARBÓN rate of flow; **~ de la película** _f_ CINEMAT, FOTO film speed; **~ periférica** _f_ ENERG RENOV peripheral velocity, TRANSP AÉR rotor tip velocity; **~ de picado de diseño** _f_ TRANSP AÉR design diving speed; **~ de precesión** _f_ TEC ESP precession rate; **~ del programa** _f_ TRANSP schedule speed; **~ de propagación** _f_ ACÚST propagation speed, FÍS ONDAS propagation velocity, propagation speed, speed of propagation, TELECOM propagation velocity, TERMO _de la llama_ rate of spread; **~ de propagación de una grieta** _f_ METAL crack velocity, NUCL crack propagation rate; **~ de proyección** _f_ CINEMAT projection speed; **~ en las pruebas** _f_ TRANSP MAR trial speed; **~ punta del motor** _f_ VEH peak engine speed; **~ de las puntas de las palas del rotor** _f_ TRANSP AÉR rotor tip velocity; **~ del punto luminoso** _f_ TRANSP spot speed;

■ **r** **~ radial** _f_ ENERG RENOV, FÍS RAD radial velocity; **~ de ráfaga derivada** _f_ TRANSP AÉR derived gust velocity; **~ de ráfaga nominal** _f_ TRANSP AÉR nominal gust velocity; **~ de ráfaga vertical equivalente** _f_ TRANSP AÉR equivalent vertical gust speed; **~ de reacción** _f_ METAL, QUÍMICA reaction rate; **~ del reactor** _f_ TEC ESP jet velocity; **~ de rebobinado** _f_ CINEMAT, FOTO, INFORM&PD, TV rewind speed; **~ recomendada** _f_ AUTO, TRANSP, VEH recommended speed; **~ de regeneración** _f_ ELECTRÓN, INFORM&PD refresh rate; **~ de régimen** _f_ TRANSP design speed; **~ regulable** _f_ ING MECÁ variable velocity; **~ relativa** _f_ FÍS, ING MECÁ relative velocity, TRANSP AÉR airspeed; **~ relativa calibrada** _f_ TRANSP AÉR calibrated air speed; **~ relativa de diseño** _f_ TRANSP AÉR design airspeed; **~ con respecto al aire** _f_ TRANSP AÉR airspeed; **~ con respecto al aire equivalente** _f_ TRANSP AÉR equivalent airspeed; **~ con respecto al fondo** _f_ TRANSP MAR _del barco_ speed over the ground; **~ de retardo** _f_ ING MECÁ retarded velocity; **~ de retroalimentación anterior** _f_ PROD last-feedback rate; **~ de rotación** _f_ ENERG RENOV rotational speed, TEC PETR rotation speed, _perforación_ rotating speed, TRANSP rotation speed; **~ rotatoria** _f_ TEC PETR _perforación_, TRANSP AÉR rotation speed; **~ del rotor** _f_ TRANSP AÉR rotor speed; **~ del rotor angular** _f_

TRANSP AÉR angular rotor speed; ~ **de rotor máxima** *f* TRANSP AÉR maximum rotor speed;

■ s ~ **de salida** *f* INSTAL HIDRÁUL angle of departure, emergent velocity, exit velocity; ~ **de salto** *f* PROD slew speed; ~ **de secado** *f* TERMO rate of drying; ~ **de sedimentación** *f* CARBÓN settling speed, CONTAM *depuración* deposition rate, deposition speed, deposition velocity, PROC QUÍ settling rate, RECICL sedimentation rate; ~ **sincrónica** *f* ELEC, ENERG RENOV synchronous speed; ~ **de sincronismo** *f* ELEC, ENERG RENOV synchronous speed; ~ **sísmica** *f* GEOL, TEC PETR seismic velocity; ~ **sobremultiplicada** *f* VEH *caja de cambio* overdrive; ~ **del sonido** *f* ACÚST sound velocity, FÍS speed of sound, FÍS ONDAS *en un medio* seep of sound, velocity of sound, TRANSP AÉR speed of sound; ~ **de Stokes** *f* C&V Stokes velocity; ~ **de subida** *f* TEC ESP, TRANSP AÉR climb speed; ~ **en suelo** *f* TRANSP AÉR ground speed; ~ **supersónica** *f* FÍS supersonic speed; ~ **del surco** *f* ACÚST groove speed;

■ t ~ **de tamizado** *f* PROC QUÍ sieving rate; ~ **tangencial** *f* ENERG RENOV tangential velocity; ~ **del telar** *f* TEXTIL loom speed; ~ **terminal** *f* FÍS terminal velocity; ~ **de toma de contacto** *f* TRANSP AÉR touchdown speed; ~ **de trabajo** *f* ING MECÁ working speed; ~ **de transferencia** *f* INFORM&PD, TELECOM transfer rate; ~ **de transferencia de datos** *f* INFORM&PD, TELECOM data transfer rate; ~ **de transferencia de la información** *f* TELECOM information transfer rate; ~ **de transformación** *f* CONTAM *procesos químicos* transformation rate; ~ **de transmisión** *f* INFORM&PD transfer rate, baud rate, transmission speed, ING ELÉC transmission rate, TELECOM transmission speed; ~ **de transmisión de bits** *f* INFORM&PD, TEC ESP, TELECOM bit rate; ~ **de transmisión de datos** *f* INFORM&PD data rate, data transfer rate, TELECOM data transfer rate; ~ **transónica** *f* FÍS transonic speed; ~ **de traslación** *f* TRANSP translation speed; ~ **de traslado** *f* INFORM&PD transfer rate; ~ **de trituración** *f* PAPEL breakage rate;

■ u ~ **de umbral demostrada mínima** *f* TRANSP AÉR minimum demonstrated threshold speed; ~ **de umbral máxima** *f* TRANSP AÉR maximum threshold speed;

■ v ~ **de vaciado** *f* CONTAM *procesos industriales* washout rate; ~ **en vacío** *f* MECÁ idling speed; ~ **variable** *f* ING MECÁ variable velocity; ~ **variable continua de correa trapezoidal** *f* ING MECÁ endless variable speed of V-belt; ~ **de variación de fase** *f* TELECOM phase-change velocity; ~ **verdadera** *f* TRANSP AÉR true airspeed (*TAS*); ~ **vertical** *f* TEC ESP *de la nave* vertical speed; ~ **de la vibración** *f* TV flutter rate; ~ **del viento** *f* ENERG RENOV, METEO wind speed, wind velocity; ~ **de viento de cortar** *f* ENERG RENOV cutout wind speed; ~ **de viento de intercalar** *f* ENERG RENOV cut-in wind speed; ~ **de viento de supervivencia** *f* ENERG RENOV survival wind speed; ~ **de volumen** *f* ACÚST volume velocity; ~ **de vulcanización** *f* P&C, TERMO rate of cure

velocimetría *f* FÍS velocimetry

velocímetro *m* AUTO, VEH *accesorio* speedometer

vena *f* AmL (*cf rumbo del filón Esp*) MINAS *de filones* run; ~ **cruzada** *f* GEOL intersecting vein; ~ **gaseosa** *f* GAS gaseous vein

venasquita *f* MINERAL venasquite

vencerse *v refl* CONST give way

vencimiento *m* TV deadline

vendedor: ~ **de herramientas de construcción** *m* CONST builders' hardware merchant

vendimia *f* AGRIC grape harvest

veneno *m* CONTAM, NUCL, PROC QUÍ, SEG poison; ~ **de catalizador** *m* CONTAM *procesos químicos* catalyst poison; ~ **de un catalizador** *m* PROC QUÍ catalytic poison; ~ **consumible** *m* NUCL burnable poison

venir: ~ **de cierto puerto** *vi* TRANSP MAR *buque* hail from a port

venta: ~ **de cuero al por menor** *f* PROD leather cutting

ventaja: ~ **competitiva** *f* PROD competitive edge

ventana *f* CONST, INFORM&PD window, SEG *de una careta* visor; ~ **de abertura** *f* ELECTRÓN aperture grill (*AmE*), aperture grille (*BrE*); ~ **de balancín** *f* CONST centre-hung window (*AmE*), centre-hung window (*BrE*); ~**-basculante** *f* CONST center-hung window (*AmE*), centre-hung window (*BrE*); ~ **de buharda** *f* CONST dormer; ~ **ciega** *f* CONST blind window; ~ **corredera** *f* (*cf ventana corrediza*) CONST pivot-hung window, sliding window; ~ **corrediza** *f* (*cf ventana corredera*) CONST pivot-hung window, sliding window; ~ **desplegable** *f* INFORM&PD pop-up window; ~ **desprendible** *f* TRANSP AÉR jettisonable window; ~ **con emplomado** *f* C&V stained glass window; ~ **de enfoque automático** *f* FOTO autofocus window; ~ **especial de lanzamiento** *f* TEC ESP firing window; ~ **espectral** *f* ÓPT spectral window; ~ **francesa** *f* C&V French window; ~ **de guillotina** *f* CONST sash window; ~ **de hoja basculante** *f* CONST pivot-hung window, sliding window; ~ **de inspección** *f* EMB inspection window; ~ **lanzable** *f* TRANSP AÉR jettisonable window; ~ **de lanzamiento** *f* TEC ESP launch window, *de cosmonaves* firing window; ~ **metálica bloqueada** *f* TRANSP AÉR flag window; ~ **en mosaico** *f* INFORM&PD tiled window; ~ **de observación** *f* INSTR viewing window; ~ **óptica** *f* ÓPT optical window; ~ **oval** *f* ACÚST *audición* oval window; ~ **practicable** *f* CONST casement window, French doors (*AmE*), French windows (*BrE*); ~ **de protección** *f* SEG security window; ~ **de prueba espectral** *f* TELECOM spectral window; ~ **de radio** *f* FÍS, FÍS RAD radio window; ~ **radioeléctrica** *f* FÍS, FÍS RAD radio window; ~ **redonda** *f* ACÚST *audición* round window; ~ **de la sala de control** *f* TV control-room window; ~ **de sílice fundida** *f* TEC ESP fused-silica window; ~ **de sincronización** *f* TEC ESP synchronization window; ~ **del telémetro** *f* AmL (*cf visor del telémetro Esp*) CINEMAT, ELECTRÓN, FOTO, TV rangefinder window; ~ **del tiempo** *f* TEC PETR *operaciones costa-fuera* weather window; ~ **de tiro** *f* TEC ESP firing window; ~ **de transmisión** *f* ÓPT transmission window; ~ **de ventilación** *f* TRANSP AÉR vent window; ~ **de visualización** *f* Esp (*cf ventanilla de vista AmL*) INFORM&PD viewing window

ventanilla *f* AUTO quarter light (*BrE*), quarter window (*AmE*), CINEMAT, TRANSP AÉR quarter light (*BrE*), quarter window (*AmE*), VEH *carrocería* window; ~ **de cámara** *f* CINEMAT, TV camera gate; ~ **húmeda** *f* CINEMAT wet gate; ~ **de impresión** *f* CINEMAT printing gate; ~ **líquida** *f* CINEMAT liquid gate; ~ **para la película** *f* CINEMAT film gate; ~ **posterior** *f* AUTO, VEH rear window; ~ **de presión** *f* CINEMAT pressure gate; ~ **de proyección** *f* CINEMAT picture gate, projection port; ~ **de vista** *f* AmL (*cf* *ventana de visualización Esp*) INFORM&PD viewing window

ventanillo *m* TEC ESP peephole

ventarrón *m* METEO gale; **~ moderado** *m* METEO high wind, moderate gale, moderate wind

ventas: ~ por teléfono *f pl* TELECOM telesales

venteadura *f* CONST *madera* honeycombing

ventear *vt* PROD *agujeros* pierce

venteo *m* ING MECÁ air outlet, air vent, NUCL venting, PROD piercing; **~ de aire** *m* ING MECÁ air vent

ventilación *f* GEN aeration, ventilation; **~ artificial** *f* SEG artificial ventilation; **~ por aspiración** *f* ING MECÁ exhaust draft (*AmE*), exhaust draught (*BrE*); **~ autónoma** *f* MINAS separate ventilation; **~ del cárter motor** *f* AUTO, MECÁ, VEH *motor* crankcase ventilation; **~ cerrada de cárter** *f* AUTO positive crankcase ventilation (*PCV*); **~ por extracción** *f* ING MECÁ exhaust draft (*AmE*), exhaust draught (*BrE*); **~ por falso techo** *f* REFRIG underfloor ventilation; **~ forzada** *f* ELEC, ING ELÉC forced ventilation, ING MECÁ forced draft (*AmE*), forced draught (*BrE*), forced ventilation, PROD forced ventilation; **~ mecánica por insuflación** *f* ING MECÁ forced draft (*AmE*), forced draught (*BrE*), pressure draft (*AmE*), pressure draught (*BrE*); **~ natural** *f* NUCL natural ventilation; **~ positiva del cárter** *f* AUTO positive crankcase ventilation (*PCV*); **~ a presión** *f* ELEC, ING ELÉC, ING MECÁ, PROD forced ventilation

ventiladero *m* TEC PETR air vent

ventilado *adj* GEN ventilated; **no ~** MINAS dead

ventilador *m* ALIMENT blower, AUTO blower, fan, C&V *vidrio soplado* blower, CINEMAT wind machine, CONST ventilating fan, fan, ventilator, ELEC *de artefacto*, ELECTRÓN, FÍS, ING ELÉC, ING MECÁ, INSTAL TERM, fan, MECÁ blower, fan, MINAS ventilating fan, fan, PROD fan, blower, REFRIG fan, ventilator, SEG ventilator, TERMOTEC, TRANSP fan, TRANSP MAR *equipamiento de cubierta* ventilator, VEH *refrigeración* fan, *sistema de refrigeración del motor* blower; **~ de alto rendimiento** *m* REFRIG high-performance fan; **~ aspirador** *m* ING MECÁ induced-draft fan (*AmE*), induced-draught fan (*BrE*); **~ aspirante** *m* ING MECÁ exhauster, suction fan, LAB *seguridad* extraction fan; **~ aspirante colocado sobre una ventana** *m* ING MECÁ porthole fan; **~ axial** *m* SEG axial ventilator; **~ axial contrarrotativo** *m* TRANSP counter-revolving axial fan; **~ axial de elevación** *m* TRANSP axial flow lift fan; **~ axial de hélices contrarrotativas** *m* TRANSP counter-revolving axial fan; **~ axial de paletas** *m* REFRIG vane axial fan; **~ de cable entretejido** *m* ING ELÉC laced cable fan; **~ centrífugo** *m* ING MECÁ fan blower, MINAS screw fan, REFRIG, TERMO, TERMOTEC centrifugal fan, TRANSP centrifugal flow fan; **~ centrífugo helicoidal** *m* REFRIG mixed-flow fan; **~ sin conductos** *m* TRANSP AÉR unducted fan; **~ de corriente inducida** *m* ING MECÁ induced-draft fan (*AmE*), induced-draught fan (*BrE*); **~ de doble entrada** *m* ING MECÁ, REFRIG double inlet fan; **~ de dos oídos** *m* ING MECÁ, REFRIG double inlet fan; **~ eléctrico** *m* AUTO power fan; **~ de enfriamiento** *m* AUTO, CINEMAT, REFRIG cooling fan; **~ de una entrada** *m* ING MECÁ single inlet fan; **~ entubado** *m* ING MECÁ, TRANSP ducted fan; **~ estático** *m* PROD displacement fan; **~ de extracción** *m* ING MECÁ extractor fan, exhaust fan, MECÁ, PAPEL, TERMOTEC

exhaust fan; **~ extractor** *m* ING MECÁ induced-draft fan (*AmE*), induced-draught fan (*BrE*); **~ de flujo axial** *m* TERMOTEC axial flow fan; **~ de flujo cruzado** *m* TERMOTEC crossflow fan; **~ de flujo radial** *m* TERMOTEC radial-flow fan; **~ para fundarías** *m* PROD foundry blower; **~ de gas axial de combustión** *m* ING MECÁ, TERMOTEC combustion axial-gas fan; **~ de hélice** *m* ING MECÁ, TERMOTEC propeller fan; **~ helicoidal** *m* ING MECÁ propeller fan, MINAS screw fan, REFRIG propeller fan; **~ impelente** *m* ALIMENT, AUTO, C&V blower, ING MECÁ plenum ventilator, plenum fan, compressing fan, MECÁ blower; **~ de impulsión** *m* REFRIG blower; **~ impulsor de flujo axial** *m* TRANSP axial flow lift fan; **~ de mina** *m* MINAS mine fan; **~ del motor** *m* AUTO engine fan, TRANSP fan motor; **~ de un oído** *m* ING MECÁ, REFRIG single inlet fan; **~ de paletas aerodinámicas** *m* REFRIG aerofoil fan (*BrE*), airfoil fan (*AmE*); **~ portátil** *m* ING MECÁ portable fan; **~ de recirculación** *m* REFRIG *de aire*, TERMO circulating fan; **~ de refrigeración** *m* AUTO, CINEMAT, REFRIG, VEH cooling fan; **~ rotatorio** *m* TRANSP air propeller; **~ secundario** *m* MINAS booster ventilation fan; **~ de sentinas** *m* TRANSP MAR bilge blower; **~ del soplante** *m* PROD blower-fan; **~ soplante** *m* PROD force fan; **~ de superficie** *m* MINAS surface ventilating fan; **~ de tiro inducido** *m* ING MECÁ induced-draft fan (*AmE*), induced-draught fan (*BrE*); **~ con turbina eólica** *m* REFRIG wind turbine fan; **~ ventilado** *m* REFRIG ventilated fan

ventilar *vt* CARBÓN ventilate, RECICL aerate, SEG, TERMO ventilate, TRANSP MAR *sollado, pantoques* air

ventilocalefactor *m* INSTAL TERM fan heater

ventiloconvector *m* REFRIG fan coil unit

ventímetro *m* TRANSP MAR ventimeter

ventisca *f* METEO blizzard; **~ de nieve** *f* METEO blowing snow, drifting snow

ventolina *f* METEO, TRANSP MAR *meteorología* catspaw

ventolinas *f pl* METEO, TRANSP MAR *meteorología* light airs

ventosa *f* INSTR sucker, PROD *hornos* draft hole (*AmE*), draught hole (*BrE*); **~ de aire** *f* ING MECÁ air vent

ventoso *adj* METEO windy

venturi *m* AUTO venturi, CONTAM venturi scrubber, VEH *carburador* venturi

venturímetro *m* ING MECÁ venturi tube

venturina *f* C&V aventurine

ver *vt* CINEMAT view; **~ antes de transmitir** *vt* TV preview

veratramina *f* QUÍMICA veratramine

verátrico *adj* QUÍMICA veratric

veratrina *f* QUÍMICA veratrine

veratrol *m* QUÍMICA veratrol

verbo *m* INFORM&PD *COBOL* verb

verdadero: ~ espesor del valor mitad *m* FÍS RAD true half-width

verde: ~ cromo *m* C&V, COLOR chrome green; **~ de cromo** *m* C&V, COLOR chrome green; **~ malaquita** *m* QUÍMICA green mineral; **~ primario** *m* TV green primary

verdeo *m* AGRIC soilage

verduguillo *m* TRANSP MAR rubbing strake, *construcción naval* molding (*AmE*), moulding (*BrE*)

vereda *f* AmL (*cf acera Esp*) CONST bypath, causeway, pavement (*BrE*), sidewalk (*AmE*); **~ móvil** *f* AmL (*cf acera móvil Esp*) TRANSP moving pavement (*BrE*), traveling sidewalk (*AmE*); ~ **móvil capotada** *f AmL* (*cf acera móvil capotada Esp*) TRANSP cabin-type moving pavement (*BrE*), cabin-type moving sidewalk (*AmE*); ~ **para transporte** *f AmL* (*cf acera para transporte Esp*) CONST moving platform

verga: ~ **de señales** *f* TRANSP MAR *a bordo* signal mast

verificación *f* CALIDAD checking, CARBÓN testing, EMB checking, INFORM&PD check, verification, ING MECÁ check, gaging (*AmE*), gauging (*BrE*), testing, verification, MECÁ inspection, PROD monitoring, inspection, TEC ESP monitoring, checkout (*AmE*), test, TEC PETR *yacimientos* gaging (*AmE*), gauging (*BrE*); ~ **automática** *f* CALIDAD *de datos cuantitativos* automatic verification, INFORM&PD machine check; ~ **de base** *f* TRANSP AÉR base check; ~ **de datos** *f* INFORM&PD data vet; ~ **por eco** *f* INFORM&PD echo check; ~ **de errores** *f* TELECOM error check; ~ **estática** *f* TEC ESP static test; ~ **de fábrica** *f* MECÁ factory acceptance; ~ **inicial** *f* METR initial verification; ~ **lógica** *f* ELECTRÓN logic test; ~ **de longitud periférica** *f* ING MECÁ peripheral-length checking; ~ **marginal** *f* INFORM&PD marginal check; ~ **meticulosa** *f* CONST, SEG meticulous inspection; ~ **de paridad** *f* INFORM&PD parity check; ~ **de paridad par-impar** *f AmL* INFORM&PD odd-even check; ~ **de programa** *f* INFORM&PD program verification; ~ **de redundancia** *f* INFORM&PD, TEC ESP redundancy check; ~ **de redundancia cíclica** *f* INFORM&PD, MATEMÁT, TELECOM cyclic redundancy check (*CRC*); ~ **de redundancia longitudinal** *f* INFORM&PD longitudinal redundancy check (*LRC*); ~ **de redundancia vertical** *f* INFORM&PD vertical redundancy check (*VRC*); ~ **de la resistencia de la soldadura** *f* PROD weld strength check; ~ **por saltos** *f* INFORM&PD leapfrog test; ~ **en tierra** *f* TEC ESP, TRANSP AÉR ground test; ~ **de validez** *f* INFORM&PD, TELECOM validity check; ~ **verdadero-falso** *f* PROD true-false check; ~ **visual de la máquina** *f* EMB machine vision verification; ~ **de vuelco** *f* INFORM&PD dump check

verificaciones: ~ **previas a la puesta en marcha** *f pl* NUCL precommissioning checks

verificador *m* INFORM&PD verifier, ING MECÁ gage (*AmE*), gauge (*BrE*), PROD checker, QUÍMICA assayer; ~ **de aislamiento** *m* ELEC *instrumento* insulation tester; ~ **automático de velocidad** *m* ING MECÁ automatic speed checker; ~ **combinado de planos** *m* ING MECÁ combination surface gage (*AmE*), combination surface gauge (*BrE*); ~ **de cortocircuitos** *m* ELEC *instrumento*, INSTR growler; ~ **de electroaislación** *m* ELEC *instrumento* insulation tester; ~ **lógico** *m* ELECTRÓN logic tester; ~ **de pieza** *m* ING MECÁ detail gage (*AmE*), detail gauge (*BrE*); ~ **de secuencia** *m* INFORM&PD sequence check; ~ **de superficies planas** *m* ING MECÁ scribing block; ~ **de la velocidad de la cámara** *m* CINEMAT, TV camera-speed checker

verificar *vt* CINEMAT, CONST check, INFORM&PD verify, METR *medidas* check, PROD monitor, *tablas de datos* adjust, TEC ESP *ensayos, gestión* check out; ~ **de nuevo** *vt* METR recheck

verjuras *f pl* PAPEL laid lines

vermiculita *f* MINERAL, REFRIG, TERMOTEC vermiculite

vernier *m* ING MECÁ vernier

veronal *m* QUÍMICA veronal

verosimilitud *f* INFORM&PD, MATEMÁT likelihood

versalitas *f pl* IMPR level small caps, small cap

versatilidad *f* TEC ESP versatility

versene *m* QUÍMICA versene

versión *f* INFORM&PD version, *software* release; ~ **doblada** *f* CINEMAT dubbed version; ~ **sintáctica** *f* TELECOM syntax version

vertedera *f* AGRIC moldboard (*AmE*), mouldboard (*BrE*), turn plough (*BrE*), turn plow (*AmE*), CARBÓN *dragas* chute, PROD *cuchara de fundición* nose

vertedero *m Esp* (*cf vertedor AmL*) AGUA outfall, spillway, weir, ALIMENT dispenser, C&V weir, CONST *tierras, escombros* dump (*AmE*), dump site, disposal site, *presas* spillway, CONTAM *instalaciones industriales* sink, ENERG RENOV chute spillway, MINAS dump, landing, RECICL dump, dump site, dumping ground, tip (*BrE*), waste dump, waste heap; ~ **de aforo** *m Esp* (*cf vertedor de aforo AmL*) AGUA, HIDROL measuring weir, INSTAL HIDRÁUL weir; ~ **de aguas sobrantes** *m Esp* (*cf vertedor de aguas sobrantes AmL*) NUCL leaping weir; ~ **de aguja** *m Esp* (*cf vertedor de aguja AmL*) AGUA pin weir; ~ **al aire libre** *m Esp* (*cf vertedor al aire libre AmL*) RECICL open dump; ~ **anegado** *m Esp* (*cf vertedor anegado AmL*) INSTAL HIDRÁUL submerged weir; ~ **de basura** *m Esp* (*cf vertedor de basura AmL*) AGUA refuse dump, RECICL garbage chute (*AmE*), refuse dump, rubbish chute (*BrE*); ~ **con contracción** *m Esp* (*cf vertedor con contracción AmL*) AGUA contracted weir; ~ **de control** *m Esp* (*cf vertedor de control AmL*) AGUA control weir; ~ **de crecidas** *m Esp* (*cf vertedor de crecidas AmL*) AGUA *hidráulica* spillway, *presas* weir; ~ **de cresta aguda** *m Esp* (*cf vertedor de cresta aguda AmL*) AGUA sharp-crested weir, INSTAL HIDRÁUL narrow weir, sharp-crested weir, thin-edged weir; ~ **de cresta ancha** *m Esp* (*cf vertedor de cresta ancha AmL*) INSTAL HIDRÁUL flat-crested weir; ~ **de cresta delgada** *m Esp* (*cf vertedor de cresta delgada AmL*) AGUA sharp-crested weir, INSTAL HIDRÁUL narrow weir, sharp-crested weir, thin-edged weir; ~ **del efluente** *m Esp* (*cf vertedor del efluente AmL*) NUCL effluent weir; ~ **inflable** *m Esp* (*cf vertedor inflable AmL*) AGUA inflatable weir; ~ **de mineral** *m Esp* (*cf cancha de mineral AmL, vertedor de mineral AmL*) MINAS ore dump; ~ **municipal** *m Esp* (*cf vertedor municipal AmL*) CONTAM, RECICL municipal dump; ~ **de palanca** *m Esp* (*cf vertedor de palanca AmL*) AGUA lever weir; ~ **en pared delgada** *m Esp* (*cf vertedor en pared delgada AmL*) AGUA sharp-crested weir, INSTAL HIDRÁUL *presas* narrow weir, sharp-crested weir, thin-edged weir; ~ **de pared gruesa** *m Esp* (*cf vertedor de pared gruesa AmL*) AGUA broad-crested weir, INSTAL HIDRÁUL broad-crested weir, flat-crested weir; ~ **de pozo** *m Esp* (*cf vertedor de pozo AmL*) AGUA *presas* shaft spillway; ~ **de productos de dragado** *m Esp* (*cf vertedor de productos de dragado AmL*) MINAS dumping ground; ~ **proporcional** *m Esp* (*cf vertedor proporcional AmL*) HIDROL proportional weir; ~ **público** *m Esp* (*cf vertedor público AmL*) CARBÓN, CONTAM, RECICL landfill; ~ **de rodillos** *m Esp* (*cf vertedor de rodillos AmL*) AGUA roller weir; ~ **sumergido** *m Esp* (*cf vertedor sumergido AmL*) AGUA, HIDROL drowned weir, INSTAL HIDRÁUL drowned weir, flush weir, submerged orifice, submerged overfall, submerged weir; ~ **de superficie** *m Esp* (*cf vertedor de superficie AmL*) ING MECÁ overflow; ~ **totalmente libre** *m Esp* (*cf vertedor totalmente libre AmL*) HIDROL free overall weir

vertedor *m AmL* (*cf vertedero Esp*) ALIMENT dispenser, C&V weir, ENERG RENOV *presas* spillway, chute spillway

verter[1] *vt AmL* (*cf rellenar Esp*) CARBÓN, CONTAM MAR dump, MINAS land, fill, RECICL discharge, dump; **~ por la cresta** *vt* HIDROL *dique* overtop

verter[2]: **~ hormigón** *vi* CONST concrete

verterse *v refl* ING MECÁ run

vertical *f* CONST plumb

vértice *m* (*nodo*) CONST crown, CRISTAL vertex, ELECTRÓN node, FÍS vertex, GEOM apex, vertex, MINAS *topografía* hub, TEC ESP *geometría* apex, TELECOM vertex; **~ de la lente** *m* FOTO lens vertex

vertido *m* RECICL discharge, dumping, tipping (*BrE*); **~ de basuras** *m* AGUA, RECICL refuse dumping (*AmE*), refuse tipping (*BrE*); **~ controlado** *m* CARBÓN landfill, CONTAM, RECICL controlled dumping, controlled tipping, landfill; **~ de efluentes** *m* RECICL effluent discharge; **~ de emanaciones** *m* RECICL effluent discharge; **~ de envergadura** *m* RECICL fly tipping; **~ de escombros** *m* CONST spoil disposal; **~ indiscriminado** *m* RECICL indiscriminate dumping

vertiente *f* AGUA watershed, CONST slope, GAS seepage, HIDROL watershed; **~ calibrada** *f* CONTAM calibrated watershed

vesicular *adj* GEOL vesicular

vestíbulo *m* CONST *arquitectura* portal, FERRO hall, vestibule (*BrE*), INSTAL TERM *de un horno* vestibule

vestido: **~~camisa** *m* TEXTIL *prenda de vestir* shift, shirt-dress

vestiduras *f pl* PAPEL *telas, fieltros* clothing, paper-machine clothings, TEXTIL clothing, garments

vestigio *m* PETROL trace

vestimenta *f* SEG clothing; **~ protectora** *f* D&A protective clothing; **~ de trabajo** *f* CONST working clothes

vesubiana *f* MINERAL vesuvianite, idocrase; **~ azul** *f* MINERAL cyprine

vesubina *f* QUÍMICA vesuvin

veta *f* CONST *madera, piedra* vein, MINAS vein, lode, seam, lead; **~ de arcilla entre la roca madre y las paredes** *f* MINAS *filones* pug; **~ atravesada** *f* MINAS cross lode; **~ explotada** *f* MINAS worked-out vein; **~ gaseosa** *f* GAS gaseous vein; **~ madre** *f AmL* (*cf veta principal Esp*) MINAS *filones* main lode; **~ pequeña** *f* MINAS *de mineral* streak; **~ principal** *f Esp* (*cf veta madre AmL*) MINAS *filones* main lode; **~ transversal** *f* MINAS counter, counterlode

vetas: **~ de color** *f pl* C&V color streaks (*AmE*), color striking (*AmE*), colour streaks (*BrE*), colour striking (*BrE*)

VFE: **~ de retardo** *f* (*variación de fase por efecto de retardo*) ELECTRÓN delayed MPSK (*delayed modified phase-shift keying*)

VGA *abr* (*adaptador de gráficos de video AmL, adaptador de gráficos de vídeo Esp*) INFORM&PD VGA (*video graphics adaptor*)

VHF *abr* (*frecuencia muy alta*) GEN VHF (*very high frequency*)

VHFO *abr* (*frecuencia muy alta omnidireccional*) TRANSP AÉR VHFO (*very high-frequency omnirange*)

vía[1]: **por ~ marítima** *adv* TRANSP MAR by sea

vía[2] *f* CINEMAT, FERRO track, MINAS roadway, TELECOM path, route, routing, TRANSP AÉR *del tren de aterrizaje* tread; **~ de acceso** *f* CONST access road (*AmE*), approach, INFORM&PD path; **~ de acceso a los datos** *f* INFORM&PD data path; **~ aérea** *f* PROD overhead runway, overhead track; **~ aerífera** *f* MINAS air passage; **~ de agua** *f* AGUA waterway; **~ de alcance** *f* CONST, TRANSP overtaking lane; **~ de anchura mixta** *f* FERRO, TRANSP mixed gage track (*AmE*), mixed gauge track (*BrE*); **~ apartadero** *f* FERRO turnout, passing track, *equipo inamovible* loop line, *vehículos* shunt; **~ apartadero en la estación** *f* FERRO passing track in station; **~ apartadero sin personal** *f* FERRO unmanned turnout; **~ de aporte** *f* GEOL feeder; **~ auxiliar** *f* FERRO auxiliary track; **~ auxiliar de paso** *f* FERRO relief track; **~ sin balasto** *f* TRANSP ballastless track; **~ para bicicletas** *f* CONST, TRANSP cycle track; **~ de circulación** *f* MINAS gangway, OCEAN, TRANSP MAR *navegación* traffic lane; **~ de clasificación** *f* FERRO, FERRO *vehículos* marshalling track (*AmE*), marshalling track (*BrE*), switching track, TRANSP classification track; **~ de clasificación cubierta por una señal** *f* FERRO advance classification track; **~ de comunicaciones** *f* ING ELÉC, TELECOM communications circuit; **~ de conversación** *f* TELECOM speech path; **~ crítica** *f Esp* (*cf camino crítico AmL*) NUCL critical pathway; **~ de cruzamiento** *f* FERRO *equipo inamovible* crossover; **~ de derivación** *f* FERRO loop line; **~ de desintegración del muón** *f* FÍS RAD muon decay track; **~ de desmonte** *f* FERRO cutout; **~ de desviación** *f* TRANSP bypass line; **~ de desvío** *f* TRANSP bypass line; **~ diagonal de enlace** *f* FERRO *equipo inamovible* crossover; **~ diagonal de unión** *f* FERRO *haz de vías paralelas* scissors crossing, *infraestructura* scissor crossing; **~ de la dolly** *f* CINEMAT dolly track; **~ elevada** *f* CONST elevated runway, FERRO elevated track; **~ de enlace** *f* CONST *ferrocarril* crossover; **~ de enlace doble** *f* CONST double crossover; **~ férrea** *f* FERRO, MINAS, TRANSP rail track; **~ férrea principal** *f* TRANSP articulated railroad (*AmE*), articulated railway (*BrE*), arterial railway (*BrE*); **~ férrea de tránsito rápido** *f* FERRO, NUCL, TRANSP rapid-transit railroad (*AmE*), rapid-transit railway (*BrE*); **~ férrea troncal** *f* CONST main line; **~ de ferrocarril de ancho normal** *f* FERRO standard rail gage (*AmE*), standard rail gauge (*BrE*); **~ fluvial** *f* AGUA waterway; **~ fuera de servicio** *f* FERRO *equipo inamovible* out-of-service track; **~ de gran tráfico marítimo** *f* OCEAN *navegación comercial* shipping corridor; **~ de intercambio** *f* FERRO *infraestructura férrea* interchange track; **~ lateral** *f* FERRO *equipo inamovible* branch line, feeder line; **~ libre** *f* FERRO *infraestructura* line clear; **~ marítima** *f* OCEAN *navegación comercial* sea lane, seaway route, TRANSP MAR sea lane, shipping lane; **~ métrica** *f* FERRO meter gage (*AmE*), metre gauge (*BrE*); **~ de un metro** *f* FERRO meter gage (*AmE*), metre gauge (*BrE*); **~ minera** *f* MINAS mine opening; **~ muerta** *f* CONST, FERRO siding, MINAS dead end switch; **~ de navegación** *f* TRANSP MAR waterway; **~ de navegación interior** *f* TRANSP MAR inland waterway; **~ normal** *f* FERRO *equipo inamovible* standard track; **~ ocupada** *f* FERRO line occupied, *infraestructura* occupied track; **~ en pendiente** *f* CONST climbing lane; **~ principal** *f* FERRO main line, main track, TELECOM trunk channel; **~ principal de una red** *f* TELECOM trunk channel; ~ **de propagación** *f* ING ELÉC propagation path; ~ **de pruebas** *f* TRANSP test track; ~ **pública** *f* CONST public road, *carreteras* thoroughfare; ~ **que da servicio a un apartadero** *f* FERRO line serving a siding; ~ **rápida de autobús** *f* TRANSP busway for rapid transit; ~ **rápida de bus** *f* TRANSP busway for rapid transit; ~ **al ras del pavimento** *f* FERRO *infraestructura* sunken track; ~ **recta** *f* FERRO *infraestructura* straight track; ~ **para reparaciones de vagones** *f* FERRO *infraestructura* repair track (*BrE*), rip track (*AmE*); ~ **de retorno** *f* TELECOM return path; ~ **secundaria** *f* FERRO *equipo inamovible* branch line, feeder line; ~ **en servicio** *f* FERRO track in service, *equipo inamovible* line in service; ~ **de tráfico** *f* TRANSP lane; ~ **de transbordo** *f* FERRO transfer track (*AmE*), transshipment track (*BrE*); ~ **de transmisión** *f* TELECOM path, transmission highway, transmission path, channel; ~ **de transmisión aérea** *f* TELECOM path overhead (*POH*); ~ **de transmisión óptica** *f* TELECOM optical path; ~ **de transmisión óptima** *f* TELECOM optimal path; ~ **de transmisión virtual** *f* TELECOM virtual path (*VP*); ~ **del tranvía** *f* TRANSP streetcar track (*AmE*), tram track (*BrE*); ~ **del tren de aterrizaje** *f* TRANSP AÉR landing gear track; ~ **en trinchera** *f* FERRO cutout, *infraestructura* cutting

viabilidad *f* MINAS feasibility

viada *f* TRANSP MAR headway, *de barco* way out

viaducto *m* CONST viaduct

viajar *vt* FÍS trip, TRANSP travel

viaje[1]: **en ~ de ida** *adj* TRANSP MAR *buque, cargamento* outward bound; **en ~ de regreso** *adj* TRANSP MAR *buque, cargamento* homeward-bound

viaje[2]: **en ~ de ida** *adv* TRANSP *buque, cargamento*, TRANSP MAR outward-bound; **en ~ de regreso** *adv* TRANSP, TRANSP MAR homeward-bound

viaje[3] *m* TEC PETR *perforación* trip; ~ **espacial** *m* TEC ESP space travel; ~ **de ida** *m* TRANSP MAR outward passage; ~ **de ida y vuelta** *m* TEC ESP, TRANSP, TRANSP MAR round trip; ~ **inaugural** *m* TRANSP MAR *de un buque* maiden voyage; ~ **interplanetario** *m* TEC ESP interplanetary travel

vías[1]: **en ~ de calibrado** *adv* METR in process gaging (*AmE*), in process gauging (*BrE*)

vías[2]: ~ **de maniobra** *f pl* FERRO, TRANSP marshaling track (*AmE*), marshalling track (*BrE*), switching track; ~ **en terraplén** *f pl* FERRO railroad embankment (*AmE*), railway embankment (*BrE*)

vibración *f* ACÚST vibration, ELEC oscillation, *corriente alterna* oscillation, ELECTRÓN oscillation, FÍS beat, vibration, FÍS ONDAS pulse, GEOFÍS vibration, ING MECÁ shaking, swinging, oscillation, vibration, shake, INSTR, MECÁ, METAL vibration, PROD shaking, jar, chattering, jarring, REFRIG, SEG vibration, TEC ESP *ondas* pulse, vibration, TRANSP AÉR flutter, buffeting, vibration, TV flutter; ~ **acústica** *f* ACÚST acoustic vibration; ~ **aeroelástica** *f* TRANSP AÉR flutter; ~ **de alta velocidad** *f* TRANSP AÉR high-speed buffeting; ~ **armónica** *f* FÍS harmonic; ~ **aural** *f* ACÚST aural flutter; ~ **de baja frecuencia** *f* ACÚST *gramófonos* rumble; ~ **del contacto** *f* ELEC *relé* contact chatter; ~ **de los contactos** *f* AUTO contact chatter; ~ **estructural** *f* TEC ESP *aviones* buffeting; ~ **forzada** *f* ACÚST, FÍS forced vibration; ~ **de frecuencia** *f* ACÚST frequency flutter; ~ **lateral** *f* ING MECÁ whirling of shafts; ~ **libre** *f* ACÚST, FÍS free

vibration; ~ **mecánica** *f* ACÚST mechanical vibration; ~ **del motor** *f* TRANSP AÉR engine vibration; ~ **de paleta guiadora** *f* ENERG RENOV guide vane vibration; ~ **de los platinos** *f* AUTO contact chatter; ~ **torsional** *f* ING MECÁ whirling of shafts; ~ **transmitida a través de la mano** *f* SEG hand-transmitted vibration; ~ **de una válvula** *f* REFRIG valve flutter

vibraciones: ~ **amortiguadas** *f pl* FÍS ONDAS damped vibrations; ~ **de curvatura** *f pl* FÍS RAD bending vibrations; ~ **de flexión** *f pl* FÍS RAD bending vibrations

vibrado: ~ **de imagen** *m* TV flopover

vibrador *m* ACÚST *audición* vibrator, CONST *hormigón* poker vibrator, vibrator, ING ELÉC, TEC ESP vibrator; ~ **de combustible** *m* MINAS bunker vibrator; ~ **de llenado** *m* EMB filling vibrator; ~ **óseo** *m* ACÚST bone vibrator

vibradora: ~ **de tamices** *f* LAB sieve shaker, PROC QUÍ sieve jigger

vibradores *m pl* TEC PETR *perforación* shale shaker

vibrante *adj* ELEC, ELECTRÓN, FÍS ONDAS oscillating, PROD jarring

vibrar[1] *vt* CONST *hormigón* vibrate, ELECTRÓN oscillate, MINAS jump

vibrar[2] *vi* PROD chatter

vibrato *m* ACÚST vibrato

vibrocompactación *f* NUCL vibrocompaction

vibrocorrosión *f* NUCL chafing

vibrocultivador *m* AGRIC spring-tooth harrow

vibroenfriador: ~ **de lecho fluidizado** *m* PROC QUÍ, REFRIG fluidized-bed vibro-cooler

vida *f* CALIDAD, ING MECÁ life, TEC ESP lifetime; ~ **atmosférica** *f* CONTAM *radiactividad* atmospheric lifetime; ~ **de la explotación** *f* Esp (*cf longevidad AmL*) MINAS life, lifetime; ~ **en funcionamiento** *f* ING ELÉC operating life; ~ **mecánica** *f* ING ELÉC, PROD mechanical life; ~ **media** *f* FÍS *cuerpos radiactivos* average life, half-life, mean life, FÍS PART *de partícula* half-life, lifetime, GEOL, QUÍMICA *cuerpos radiactivos* half-life; ~ **media de la luminosidad** *f* FÍS RAD luminosity lifetime; ~ **media radiactiva** *f* FÍS RAD, NUCL half-life, radioactive half-life; ~ **media radiativa** *f* FÍS RAD radiotive lifetime; ~ **de la pala** *f* TRANSP AÉR blade life; ~ **de percentil Q** *f* NUCL Q-percentile life; ~ **de servicio** *f* NUCL operating lifetime; ~ **útil** *f* C&V working life, CARBÓN, CONST service life, EMB service life, useful life, working life, FOTO working life, ING ELÉC lifetime, service life, ING MECÁ life, NUCL service life, P&C *adhesivos, termoestables* pot life, TEC ESP service life; ~ **útil de la cabeza** *f* TV head life; ~ **útil comercial** *f* TEC ESP commercial life; ~ **útil en depósito** *f* CINEMAT shelf life; ~ **útil de la herramienta** *f* ING MECÁ tool life

video *AmL ver vídeo Esp*

vídeo *Esp m* INFORM&PD, TELECOM, TV video; ~ **de bajo nivel** *m* TV low-level video; ~ **grabador** *m* (*cf videograbador AmL, videograbadora AmL*) TV video recorder, videocassette recorder (*VCR*), videotape recorder (*VTR*); ~ **inverso** *m* INFORM&PD inverse video, reverse video; ~ **invertido** *m* INFORM&PD reverse video; ~ **de larga duración** *m* ÓPT long-playing video (*VLP*)

videoamplificación *f* ELECTRÓN video amplification

videoamplificador *m* ELECTRÓN video amplifier

videocasete *m* TV videocassette

videocinta *f* TELECOM videotape

videoclip *m* TV videoclip

videoconferencia *f* TEC ESP, TELECOM videoconference; **~ de exploración lenta** *f* TELECOM slow-scan videoconferencing; **~ de movimiento total** *f* TELECOM full-motion videoconferencing

videodisco *m* ÓPT, TV videodisk; **~ de alta densidad** *m* AmL ÓPT video high-density disc (*BrE*) (*VHD*), video high-density disk (*AmE*) (*VHD*); **~ analógico** *m* ÓPT analog videodisk; **~ codificado digitalmente** *m* ÓPT digitally-encoded videodisc (*BrE*), digitally-encoded videodisk (*AmE*); **~ digital** *m* ÓPT digital videodisc (*BrE*), digital videodisk (*AmE*); **~ de escritura superficial** *m* ÓPT surface-written videodisc (*BrE*), surface-written videodisk (*AmE*); **~ interactivo** *m* ÓPT interactive videodisc (*BrE*), interactive videodisk (*AmE*); **~ láser** *m* ÓPT laser videodisc (*BrE*), laser videodisk (*AmE*); **~ óptico** *m* INFORM&PD, ÓPT optical videodisc (*BrE*), optical videodisk (*AmE*); **~ síncrono** *m* ÓPT synchronous videodisc (*BrE*), synchronous videodisk (*AmE*); **~ de visión láser** *m* ÓPT laservision videodisc (*BrE*), laservision videodisk (*AmE*)

videofrecuencia *f* ELECTRÓN, TELECOM video frequency

videograbador *m* AmL (*cf* vídeo grabador Esp) TV video recorder, videocassette recorder (*VCR*), videotape recorder (*VTR*)

videograbadora *f* AmL (*cf* vídeo grabador Esp) TV video recorder, videocassette recorder (*VCR*), videotape recorder (*VTR*)

videografía *f* TELECOM videography; **~ interactiva** *f* TELECOM interactive videography

videoindicador: **~ activo** *m* TELECOM video indicator active (*VIA*); **~ listo para ser activado** *m* AmL TELECOM video indicator ready-to-activate (*VIR*); **~ suprimido** *m* AmL TELECOM video indicator suppressed (*VIS*)

videoseñal *f* ELECTRÓN, TELECOM video signal

videoteca *f* TV videotape library

videoteléfono *m* TELECOM videophone

videotex *m* Esp (*cf* videtex AmL) INFORM&PD, TELECOM videotex; **~ interactivo** *m* TELECOM interactive videotex

videtex *m* AmL (*cf* videotex Esp) INFORM&PD, TELECOM videotex

vidicón: **~ intensificador** *m* ELECTRÓN intensifier vidicon

vidriado[1]: **~** *adj* C&V *loza* glazed; **no ~** *adj* C&V, QUÍMICA unglazed

vidriado[2] *m* C&V glaze, ENERG RENOV *de colector con placas planas* glazing, QUÍMICA vitrification; **~ absorbente de calor** *m* C&V heat-absorbing glazing; **~ burdo rolado** *m* C&V rough-rolled glazing; **~ cerámico** *m* C&V ceramic glaze; **~ de colchoneta** *m* C&V matt glaze; **~ común** *m* REVEST salt glaze; **~ con fritas** *m* C&V fritted glaze; **~ transparente** *m* C&V transparent glaze

vidriar *vt* C&V, PROD glaze

vidriera *f* AmL (*cf* escaparate Esp) C&V picture glass, CONST glazing

vidriero *m* C&V, CONST glass cutter, glazier

vidrio *m* GEN glass; **~ absorbente de calor** *m* C&V heat-absorbing glass; **~ acorazado y pulido** *m* C&V polished wired glass; **~ agrietado** *m* C&V crackled glass; **~ ahumado** *m* C&V smoked glass; **~ para aislamiento acústico** *m* ACÚST, SEG noise-protective insulating glass; **~ aislante contra incendios** *m* SEG insulating glass for fire protection; **~ aislante insonorizador** *m* ACÚST, SEG soundproof insulating glass; **~ alambrado con malla hexagonal** *m* C&V hexagonal mesh-wired glass; **~ ambar neutro** *m* C&V neutral amber glass; **~ antisolar** *m* C&V solar control glass; **~ de aumento** *m* FÍS, LAB, MECÁ magnifying glass; **~ base** *m* C&V base glass; **~ biselado** *m* C&V satin finish glass; **~ blindado** *m* C&V armored glass (*AmE*), armoured glass (*BrE*); **~ de borosilicato** *m* C&V, LAB, TERMOTEC borosilicate glass; **~ cabal** *m* C&V cabal glass; **~ de calibrar** *m* TEC PETR *refino* gage glass (*AmE*), gauge glass (*BrE*); **~ celular** *m* TERMOTEC cellular glass; **~ cementado** *m* C&V cemented glass; **~ cerámico** *m* C&V, NUCL glass ceramic; **~ en cilindros** *m* C&V blown sheet (*AmE*), cylinder glass (*BrE*); **~ de color** *m* C&V tinted glass; **~ de color laminado** *m* C&V tinted laminated glass; **~ coloreado** *m* C&V colored glass (*AmE*), coloured glass (*BrE*), REVEST waterglass color (*AmE*), waterglass colour (*BrE*); **~ común** *m* CONST window glass; **~ corrugado** *m* C&V corrugated glass; **~ de Crookes** *m* C&V Crookes glass; **~ crown** *m* C&V *óptico de bajo índice de refracción* crown glass; **~ de deshecho** *m* C&V cullet; **~ de dosímetro** *m* C&V dosimeter glass; **~ duro** *m* C&V hard glass, short glass; **~ con elementos en forma metálica** *m* C&V chalcogenide glass; **~ emplomado** *m* C&V cathedral glass, X-ray protective glass; **~ para encapsular** *m* C&V encapsulating glass; **~ encorvado** *m* C&V bent glass; **~ endurecido** *m* C&V, CONST, TRANSP toughened glass; **~ endurecido por zonas** *m* C&V zone-toughened glass; **~ para envases** *m* C&V container glass; **~ escarchado** *m* C&V frosted glass, ice-patterned glass; **~ espumado** *m* C&V foamed glass; **~ estirado** *m* C&V antique drawn glass; **~ estructural** *m* C&V structural glass; **~ flotado** *m* C&V float glass; **~ de fondo** *m* C&V *botellas* bottom glass; **~ fotosensible** *m* C&V photosensitive glass; **~ fundido** *m* C&V molten glass; **~ granulado** *m* C&V granulated glass; **~ hilado** *m* C&V spun glass; **~ para horticultura** *m* C&V horticultural glass; **~ incoloro** *m* C&V colorless glass (*AmE*), colourless glass (*BrE*); **~ laminado** *m* TRANSP laminated glass; **~ láser** *m* C&V laser glass; **~ luminiscente** *m* C&V luminescent glass; **~ martillado** *m* C&V hammered glass; **~ mitad plateado y mitad claro** *m* INSTR *sextante* horizon glass; **~ molido** *m* C&V cullet crush, powdered glass; **~ con muchas semillas** *m* C&V highly seeded glass, non-reflective glass; **~ neutro** *m* C&V neutral glass; **~ neutro blanco** *m* C&V neutral white glass; **~ neutro de color** *m* C&V neutral-tinted glass; **~ no reflejante** *m* C&V non-reflective glass, nonreflecting glass; **~ obscuro** *m* C&V dark glass; **~ opaco** *m* C&V opaque glass; **~ opaco a la radiación** *m* C&V, FÍS RAD, SEG radiation-shielding glass; **~ opalescente** *m* C&V opalescent glass; **~ opalino** *m* C&V fluoride opal glass, opal glass; **~ ópticamente plano** *m* CINEMAT optical flat; **~ óptico** *m* C&V spectacle glass; **~ orgánico** *m* C&V organic glass; **~ pirorresistente** *m* TERMO heat-resisting glass; **~ plano** *m* C&V flat glass, sheet glass; **~ plano extra delgado** *m* C&V extra-thin sheet glass; **~ plano grueso** *m* C&V thick sheet glass (*BrE*), thick window glass (*AmE*); **~ plano, grueso y vaciado en bruto** *m* C&V thick roughcast plate glass; **~ plano de línea** *m* C&V stock sheets (*AmE*), stock sizes (*BrE*); ~ **plano pulido** *m* CONST plate glass; ~ **plano pulido por ambos lados** *m* C&V twin-ground plate; ~ **plano, pulido y grueso** *m* C&V thick polished plate glass; ~ **plano sencillo** *m* C&V single-thickness sheet glass (*BrE*), single-thickness window glass (*AmE*), thin sheet glass; ~ **plano soplado** *m* C&V blown sheet (*AmE*), cylinder glass (*BrE*); ~ **prensado** *m* C&V pressed glass (*BrE*), pressware (*AmE*); ~ **prismático** *m* C&V prismatic glass; ~ **pristino** *m* C&V pristine glass; ~ **producido con 100% de fundante** *m* C&V glass melted from cullet; ~ **producido con 100% materia prima** *m* C&V glass melted from batch only; ~ **protector** *m* SEG protective glass; ~ **protector contra rayos X** *m* SEG X-ray protective glass; ~ **a prueba de explosiones** *m* C&V explosion-proof glazing; ~ **a prueba de fuego** *m* C&V fireproof glass, flameproof glass; ~ **a prueba de hornos** *m* TERMO oven-proof glass; ~ **recubierto** *m* C&V coated glass; ~ **reforzado** *m* TRANSP toughened glass; ~ **refractario** *m* TERMO heat-resisting glass, oven-proof glass; ~ **de reloj** *m* *AmL* (*cf cristalizador Esp*) CRISTAL crystallizing dish, LAB watch glass, dish, crystallizing dish, QUÍMICA dish; ~ **resistente al calor** *m* C&V, LAB borosilicate glass, heat-resisting glass, TERMOTEC borosilicate glass, heat-resistant glass; ~ **resistente a substancias químicas** *m* C&V chemically-resistant glass; ~ **rolado** *m* C&V rolled glass, figured rolled glass (*BrE*), patterned glass (*AmE*), corrugated glass; ~ **rolado acorazado** *m* C&V wired cast glass; ~ **rolado alambrado** *m* C&V corrugated wired glass; ~ **rolado sin grabado** *m* C&V plain-rolled glass; ~ **rosa** *m* C&V pink glass; ~ **ruby selénico** *m* C&V selenium ruby glass; ~ **sandwich** *m* C&V ply glass, laminated glass; ~ **de seguridad** *m* SEG safety glass; ~ **de seguridad acorazado** *m* C&V, SEG wired safety glass; ~ **de seguridad laminado** *m* C&V, SEG laminated safety glass; ~ **de señales** *m* C&V signal glass; ~ **de sílice** *m* METAL silica glass; ~ **sinterizado** *m* C&V sintered glass; ~ **soda** *m* C&V crown glass; ~ **soluble** *m* DETERG soluble glass, QUÍMICA waterglass; ~ **soplado a mano** *m* C&V hand-blown glass, mouth-blown glass; ~ **suave** *m* C&V soft glass; ~ **suelto** *m* C&V loose glass; ~ **templado** *m* C&V safety glass, tempered glass, TRANSP tempered glass; ~ **para termómetro** *m* C&V thermometer glass; ~ **tosco** *m* C&V blank glass; ~ **translúcido** *m* C&V translucent glass; ~ **transmisor de calor** *m* C&V heat-transmitting glass; ~ **transparente** *m* C&V colorless glass (*AmE*), colourless glass (*BrE*), TEC PETR refino sight glass; ~ **trifocal** *m* C&V trifocal glass; ~ **para tubos** *m* C&V tubing glass; ~ **vaciado** *m* C&V cast glass; ~ **vaciado para horticultura** *m* C&V horticultural cast glass; ~ **de una o varias capas** *m* SEG single and multilayer glass; ~ **verde** *m* C&V bottle glass; ~ **volcánico** *m* MINERAL pitchstone; ~ **de Woods** *m* C&V Wood's glass

vidrioso *adj* C&V, ÓPT, QUÍMICA, REVEST, TELECOM vitreous

viento[1]: **al ~** *adv* TRANSP MAR *vela* upwind; **con el ~** *adv* TRANSP MAR downwind; **contra el ~** *adv* TRANSP MAR upwind; **de ~ de cola** *adv* TRANSP AÉR downwind; **con ~ libre a popa del través** *adv* TRANSP MAR on a broad reach; **de ~ en popa** *adv* ENERG RENOV downwind; **con el ~ a popa del través** *adv* TRANSP MAR *navegación a vela* off the wind; **con el ~ por la proa** *adv* TRANSP MAR *navegación a vela* in stays

viento[2] *m* CONST guy rope, METEO, TEXTIL wind, TRANSP MAR *cabos* guy; ~ **alisio** *m* METEO trade wind; ~ **en altitud** *m* METEO upper wind; ~ **aparente** *m* TRANSP MAR *navegación* apparent wind; ~ **arremolinado** *m* OCEAN eddy wind; ~ **de la atmósfera alta** *m* METEO upper wind; ~ **de cara** *m* METEO, TRANSP MAR headwind; ~ **de cola** *m* METEO tailwind, TRANSP AÉR downwind; ~ **en contra** *m* TRANSP AÉR headwind; ~ **contrario** *m* METEO, TRANSP MAR foul wind; ~ **de la costa** *m* METEO land breeze, offshore breeze, TRANSP MAR land breeze; ~ **cruzado** *m* TRANSP AÉR crosswind; ~ **descendente** *m* METEO downwind; ~ **dominante** *m* METEO prevailing wind; ~ **duro** *m* METEO gale; ~ **del este** *m* METEO, TRANSP MAR easterly wind; ~ **de frente** *m* METEO, TRANSP AÉR headwind; ~ **fresco** *m* METEO moderate gale, moderate wind; ~ **fuerte** *m* METEO high wind; ~ **geostrófico** *m* METEO geostrophic wind; ~ **geotrópico** *m* GEOFÍS geotropic wind; ~ **hecho** *m* TRANSP MAR steady breeze; ~ **a un largo** *m* TRANSP MAR quarter wind; ~ **muy duro** *m* METEO strong gale; ~ **del noroeste** *m* TRANSP MAR northwest wind; ~ **del norte** *m* GEOL, METEO, TRANSP MAR north wind; ~ **en popa** *m* ENERG RENOV, TRANSP MAR downwind; ~ **en popa cerrado** *m* TRANSP MAR following wind; ~ **predominante** *m* METEO prevailing wind; ~ **prevaleciente** *m* METEO prevailing wind; ~ **a ráfagas** *m* METEO baffling wind, gusting wind, squalling wind; ~ **rafagoso** *m* METEO baffling wind, gusting wind, squalling wind; ~ **solar** *m* FÍS, GEOFÍS, TEC ESP solar wind; ~ **de superficie** *m* METEO surface wind; ~ **terral** *m* TRANSP MAR offshore wind; ~ **a la tierra** *m* TRANSP MAR onshore wind; ~ **de tierra** *m* METEO land breeze, onshore wind, TRANSP MAR offshore wind, land breeze; ~ **verdadero** *m* TRANSP MAR true wind

viento[3]: **con ~ a proa del través sin ceñir todo** *fra* TRANSP MAR on a close reach

vientos: ~ **de montaña y de valle** *m pl* METEO mountain and valley winds; ~ **del oeste** *m pl* METEO westerlies; ~ **periódicos** *m pl* METEO periodical winds

vientre *m* ACÚST antinode, AGRIC third stomach, C&V *del horno de cuba* belly, CARBÓN *alto horno* breast, ELEC *onda* antinode, PROD *caja de moldear* belly; ~ **del alto horno** *m* CARBÓN shaft kiln; ~ **de tensión** *m* ELEC *tensión* potential loop; ~ **de voltaje** *m* ELEC *voltaje* potential loop

vierteaguas *m* CARBÓN weathering, CONST *ventana* wash, weathering; ~ **calibrado** *m* CONTAM *procesos industriales* calibrated watershed

viga *f* CONST beam, stretcher, joist, rafter, FÍS, ING MECÁ beam, TRANSP AÉR jib, TRANSP MAR *construcción naval* girder, ship girder; ~ **en U** *f* CONST channel iron; ~ **de acero** *f* CONST steel beam; ~ **de ala ancha** *f* CONST broad-flange girder; ~ **de alma abierta** *f* CONST open-web girder; ~ **de alma llena** *f* CONST plate-web girder; ~ **de anclaje** *f* C&V *hornos* buckstay; ~ **apoyada** *f* CONST supported beam; ~ **sin apoyo** *f* CONST unsupported beam; ~ **armada** *f* CONST built-up girder, plate girder, trussed beam; ~ **de armazón** *f* CONST skeleton girder; ~ **atirantada** *f* CONST trussed beam; ~ **balancín** *f* ING MECÁ walking beam; ~ **cajón** *f* CONST box girder, cased beam; ~ **de cambio de paso** *f* TRANSP AÉR pitch-change beam; ~ **carrera** *f* CONST curb plate; ~**-casco** *f* TRANSP MAR *construcción naval* hull girder; ~ **de**

celosía f CONST lattice beam, lattice girder, trussed girder, MECÁ lattice girder; **~ de celosía de madera reforzada** f CONST trussed wooden beam; **~ compuesta** f CONST composite girder, compound girder; **~ continua** f CONST continuous beam; **~ de cruce** f FERRO *equipo inamovible* crossing timber; **~ de cumbrera** f CONST *arquitectura* ridge beam; **~ de deformación** f CONST straining beam; **~ en doble T** f CONST I-beam; **~ del eje del engranaje principal** f TRANSP AÉR main-gear axle beam; **~ elástica** f CONST bowstring girder; **~ empotrada** f CONST fixed beam, ING MECÁ built-in beam; **~ empotrada en los dos extremos** f ING MECÁ built-in beam; **~ encastrada por una extremidad** f ING MECÁ beam fixed at one end; **~ de falso tirante** f CONST collar truss; **~ en H** f CONST H-beam; **~ de hierro** f CONST plate iron girder; **~ libre** f CONST *vigas* free beam; **~ de madera** f CONST balk (*AmE*), baulk (*BrE*); **~ de madera con arriostramiento de hierro** f CONST *estructuras* timber truss with iron bracing; **~ maestra** f CONST girder, main beam, summer, summer beam; **~ muestra** f CONST breastsummer; **~ oculta** f CONST encased beam; **~ del piso** f TRANSP AÉR floor beam; **~ principal** f CONST girder, main girder; **~ principal en doble T** f CONST I-girder; **~ principal en H** f CONST H-girder; **~ principal de hierro** f CONST iron girder; **~ rebajada** f CONST trimmer beam; **~ de recalzo** f CONST needle; **~ de rellano** f CONST *escalera* landing trimmer; **~ simplemente apoyada** f CONST *estructuras* simple beam; **~ de sujeción** f CONST binding beam; **~ en T** f CONST T-beam; **~ de trabazón** f ING MECÁ collar beam; **~ transversal** f CONST, TRANSP MAR *construcción naval* crossbeam; **~ transversal del armazón** f ING MECÁ frame crossbeam; **~ tubular** f CONST box girder; **~ tubular cerrada** f TRANSP closed box girder; **~ voladiza** f ING MECÁ cantilever; **~ en voladizo** f CONST cantilever, cantilever beam, FÍS cantilever beam; **~ en voladizo con carga en la punta** f CONST cantilever loaded at free end

vigía m TRANSP MAR *navegación* lookout

vigilancia f FERRO *infraestructura* vigilance, METR *de la cantidad* surveying, survey, PROD monitoring, SEG *de trabajadores* surveillance, TEC ESP monitoring, TELECOM alerting; **~ aérea** f CONTAM MAR aerial surveillance; **~ a distancia** f INFORM&PD, TELECOM, TV remote monitoring; **~ electrónica** f TELECOM electronic surveillance; **~ con radar** f D&A, TRANSP, TRANSP AÉR, TRANSP MAR radar surveillance; **~ por radar** f D&A, TRANSP, TRANSP AÉR, TRANSP MAR radar monitoring; **~ de radiación** f FÍS RAD *en una zona*, TEC ESP radiation monitoring; **~ remota de temperatura** f REFRIG, TERMO remote temperature monitoring

vigilante: ~ de vía m FERRO lineman

vigor: ~ germinativo m AGRIC rate of germination

viguería f CONST joisting

vigueta f CONST girt, joist, ING MECÁ beam; **~ de izada de pesos** f PROD *grúas* lifting beam; **~ de piso** f CONST floor joist; **~ de techo** f CONST ceiling joist; **~ de unión** f CONST *arquitectura* trimmed joist

vinagre: ~ blanco m ALIMENT white vinegar

vinatera f TRANSP MAR *cabos* becket

vincular vt INFORM&PD bind

vínculo m ING MECÁ link, MECÁ link, linkage, TEC ESP

cement; **~ con el prestador de servicios** m TELECOM service provider link (*SPL*)

viñeta f FOTO, IMPR vignette

viñeteado m FOTO, IMPR vignetting

vínico adj QUÍMICA vinic

vinilacetileno m QUÍMICA vinylacetylene

vinilación f QUÍMICA vinylation

vinilbenceno m QUÍMICA vinylbenzene

vinilideno m QUÍMICA vinylidene

vinilita f QUÍMICA vinylite

vinilo m QUÍMICA vinyl

vinílogo[1] adj QUÍMICA vinylogous

vinílogo[2] m QUÍMICA vinylog

vinilpiridina f QUÍMICA vinylpyridine

viocid m QUÍMICA viocid

violación: ~ de código f TELECOM code violation (*CV*); **~ de paridad** f FÍS nonconservation of parity

violencia f HIDROL *de la caída del agua* force

violento adj METEO *viento* blustery

violeta: ~ de genciana f FÍS FLUID gentian violet

violín m TRANSP MAR *buque* fiddle, *palos* jumper strut

violúrico adj QUÍMICA violuric

virada f TRANSP MAR *navegación a vela* tacking

virado m IMPR tone, toning

virador m NUCL turning gear, TRANSP MAR *cabos* messenger, *motor* turning gear

viraje m CINEMAT dye toning, toning, CONST turning, FOTO toning; **~ de doble tono** m FOTO two-bath toning; **~ de emergencia** m TRANSP MAR emergency turn; **~ estándar de aterrizaje** m TRANSP AÉR landing pattern turn; **~ en oro** m FOTO gold toning; **~ plano** m TRANSP AÉR flat turn; **~ pronunciado** m TRANSP AÉR steep turn; **~ en sepia** m FOTO sepia toning; **~ en subida** m TRANSP AÉR climb turn; **~ con sulfuros** m FOTO sulfide toning (*AmE*), sulphide toning (*BrE*)

virar[1] vt IMPR tone, ING MECÁ swing, TRANSP MAR *anclas* heave, *bolinas* haul in, *cuerdas de amarre* heave in the mooring ropes

virar[2] vi ING MECÁ swing, TEC ESP turn; **~ por avante** vi TRANSP MAR tack, *navegación a vela* go about; **~ por redondo** vi TRANSP MAR *buques* veer, wear

virazón f METEO onshore breeze

virgen adj INFORM&PD, PAPEL virgin

viridina f QUÍMICA viridine

virola f CONST hoop, ING MECÁ ferrule, hoop, burr, MECÁ burr, ferrule, NUCL shell course, ÓPT ferrule, PROD *calderas* strake, *calderas cilíndricas* shell plate, TELECOM ferrule; **~ de tornillo** f ING MECÁ screw ferrule

virotillo: ~ articulado m ING MECÁ flexible stay bolt

viruelas f pl C&V pocking

viruta f ING MECÁ, MECÁ chip; **~ de silicio** f ELEC silicon chip; **~ con sobrecorte** f ING MECÁ built-up edge

virutas f pl CARBÓN chippings, CONST wood shavings, PROD filings; **~ para embalar** f pl EMB excelsion; **~ metálicas** f pl D&A chaff, PROD *de torno, cepilladora* swarf; **~ de perforación** f pl GAS drill cuttings; **~ de taladrar** f pl CARBÓN drillings; **~ de torno** f pl PROD turnings

visagra f ING MECÁ joint

viscoelasticidad f P&C viscoelasticity

viscoestático adj QUÍMICA viscostatic

viscofluencia: ~ a altas temperaturas f METAL high-temperature creep; **~ de estado permanente** f METAL steady-state creep

viscómetro *m* FÍS, FÍS FLUID, LAB, P&C viscometer; ~ **de Ostwald** *m* LAB, PROC QUÍ Ostwald viscometer

viscoplástico *adj* GAS viscoplastic

viscosa *f* QUÍMICA viscose

viscosidad *f* ALIMENT *panadería* ropiness, CARBÓN, FÍS viscosity, FÍS FLUID viscosity, viscidity, GAS viscosity, IMPR bodying, P&C, QUÍMICA, TEC PETR, TERMO, VEH viscosity; ~ **de Brookfield** *f* P&C *prueba* Brookfield viscosity; ~ **cinemática** *f* ENERG RENOV, FÍS, FÍS FLUID, MECÁ, REFRIG kinematic viscosity; ~ **dinámica** *f* ENERG RENOV, FÍS, FÍS FLUID, P&C dynamic viscosity; ~ **intrínseca** *f* FÍS FLUID intrinsic viscosity; ~ **de Mooney** *f* FÍS, P&C *prueba* Mooney viscosity; ~ **relativa** *f* FÍS FLUID relative viscosity

viscosímetro *m* FÍS, FÍS FLUID, LAB viscometer, viscosimeter, TERMO viscosity meter; ~ **de bolas** *m* *Esp* (*cf viscosímetro de esfera de caída AmL*) INSTR, LAB *viscosidad de líquidos* falling sphere viscometer; ~ **capilar** *m* LAB *flujo de líquido*, PROC QUÍ capillary viscometer, capillary viscosimeter; ~ **de esfera de caída** *m* *AmL* (*cf viscosímetro de bolas Esp*) INSTR, LAB *viscosidad de líquidos* falling sphere viscometer; ~ **de Ostwald** *m* LAB, PROC QUÍ Ostwald viscometer

viscoso *adj* ALIMENT, FÍS viscous, FÍS FLUID viscid, viscous, MECÁ heavy, viscous, METAL viscous, QUÍMICA *solución* thick, viscid, viscous, VEH viscous

visera *f* CINEMAT *de un foco de luz* barndoor, INSTR, SEG eye-shade, TRANSP AÉR coaming; ~ **antideslumbrante** *f* VEH *accesorio* antidazzle visor (*BrE*), antiglare vizor (*AmE*); ~ **contra el sol** *f* *AmL* (*cf parasol Esp*) AUTO, VEH *accesorio* sun visor (*BrE*), sun vizor (*AmE*); ~ **del ocular** *f* INSTR, SEG eye-shields

visibilidad *f* FÍS, METEO visibility, TEC ESP *meteorología aeronáutica* ceiling, TRANSP visibility, TRANSP AÉR ceiling, TRANSP MAR visibility; ~ **en tierra** *f* TRANSP AÉR ground visibility; ~ **en vuelo** *f* TRANSP AÉR flight visibility

visión: ~ **en estereo** *f* TV stereovision; ~ **óptica** *f* INSTR optical sight; ~ **telescópica** *f* INSTR telescopic sight

visita *f* TRANSP MAR *barco* boarding; ~ **de cortesía** *f* TRANSP MAR *entre buques de guerra, de inspectores* boarding party; ~ **de inspección** *f* TRANSP MAR boarding party

visitar *vt* TRANSP MAR *barco* board

viso *m* TEXTIL *brillo de la seda* sheen

visor *m* CINEMAT finder, viewfinder, FOTO viewfinder, INSTR finder, sight, REFRIG sight glass, SEG *visera ocular* visor, TEC ESP viewfinder; ~ **alimentado por pilas** *m* FOTO battery-powered viewer; ~ **angular** *m* CINEMAT angular viewfinder; ~ **de ángulo recto** *m* FOTO right-angle finder; ~ **de cámara oscura** *m* CINEMAT, FOTO, ÓPT reflecting viewfinder; ~ **con caperuza** *m* *AmL* (*cf visor con tapa Esp*) FOTO viewfinder with hood; ~ **de copias** *m* CINEMAT print-viewer; ~ **de cristal líquido transmisivo** *m* ING ELÉC transmissive LCD; ~ **descentrado** *m* CINEMAT offset viewfinder; ~ **de diapositivas** *m* FOTO slide viewer; ~ **directo** *m* CINEMAT direct viewfinder; ~ **para el director** *m* CINEMAT director's finder; ~ **enderezador** *m* CINEMAT, FOTO reversal finder; ~ **de esfuerzos** *m* C&V strain viewer; ~ **estereoscópico** *m* FOTO stereo viewer; ~ **facial** *m* SEG face visor (*BrE*), face vizor (*AmE*); ~ **para fotografías deportivas** *m* FOTO sports finder; ~ **del**

horizonte *m* TEC ESP horizon sensor; ~ **de imágenes directas** *m* FOTO erect-image viewfinder; ~ **de inversión** *m* CINEMAT, FOTO reversal finder; ~ **lateral** *m* CINEMAT side finder; ~ **de líquido** *m* REFRIG liquid-flow indicator; ~ **luminoso** *m* FOTO bright-line viewfinder; ~ **multifocal** *m* CINEMAT multifocal finder; ~ **de negativos** *m* FOTO negative viewer; ~ **al nivel de cintura intercambiable** *m* FOTO interchangeable waist-level finder; ~ **óptico** *m* IMPR optical viewfinder; ~ **orientable** *m* CINEMAT orientable viewfinder; ~ **sin paralaje** *m* CINEMAT parallax-free viewfinder, parallel-free viewfinder; ~ **plegable** *m* FOTO collapsible finder; ~ **reflex** *m* CINEMAT, FOTO, TV reflex viewfinder; ~ **con tapa** *m* *Esp* (*cf visor con caperuza AmL*) FOTO viewfinder with hood; ~ **del telémetro** *m* *Esp* (*cf ventana del telémetro AmL*) CINEMAT, ELECTRÓN, FOTO, TV rangefinder window; ~ **transmisivo de cristal líquido** *m* ING ELÉC transmissive LCD; ~ **universal** *m* CINEMAT, FOTO universal viewfinder; ~ **de zoom** *m* CINEMAT zoom viewfinder

vista *f* INSTR sight, TEXTIL *de las prendas* facing; ~ **cercana** *f* CINEMAT, TV close-up, TEC ESP *instrumentos ópticos, televisión* close-up; ~ **de corte** *f* ING MECÁ cutaway view, sectional view; ~ **de la costa** *f* OCEAN *navegación* seascape; ~ **desarrollada** *f* MECÁ exploded view; ~ **desde arriba** *f* ING MECÁ plan view; ~ **detallada** *f* ING MECÁ exploded view; ~ **de frente** *f* CONST *topografía* foresight, ING MECÁ front elevation; ~ **frontal** *f* ING MECÁ front elevation; ~ **lateral** *f* INSTR lateral view; ~ **panorámica** *f* INSTR panoramic sight; ~ **parcial** *f* ING MECÁ section drawing; ~ **en planta** *f* ING MECÁ plan view; ~ **recortada** *f* ING MECÁ cutaway view; ~ **de sección** *f* ING MECÁ sectional view; ~ **seccional** *f* ING MECÁ sectional view

vistaclara *m* TRANSP MAR clearview screen

visto: ~ **bueno** *m* METR *después de una inspección* approval sign, VEH *normas* approval

visual *f* CONST *topografía* sight; ~ **de frente** *f* CONST *topografía* minus sight; ~ **inversa** *f* CONST *topografía* backsight

visualización *f* ELEC display, IMPR viewing, INFORM&PD display, visualization; ~ **de barrido** *f* INFORM&PD raster display; ~ **en colores** *f* INFORM&PD color display (*AmE*), colour display (*BrE*); ~ **en cristal líquido** *f* (*VCL*) ELEC, ELECTRÓN, FÍS, INFORM&PD, TELECOM, TV liquid crystal display (*LCD*); ~ **electrolítica** *f* ELECTRÓN electrolytic display; ~ **electroluminiscente** *f* ELEC, ELECTRÓN, INFORM&PD electroluminescent display; ~ **de indicación para varios puntos** *f* TELECOM multipoint indication visualization (*MIV*); ~ **de pantalla** *f* *AmL* INFORM&PD screen display; ~ **de plasma** *f* INFORM&PD plasma display; ~ **por puntos** *f* INFORM&PD raster display; ~ **de tramas** *f* INFORM&PD raster display; ~ **en vehículo** *f* TRANSP in-vehicle visual display

visualizador *m* TELECOM display; ~ **alfageométrico** *m* TELECOM alphageometric display; ~ **alfanumérico** *m* INFORM&PD, TELECOM alphanumeric display; ~ **ampliador de diapositivas** *m* INSTR magnifying picture viewer; ~ **electroluminiscente de capa gruesa** *m* ELECTRÓN thick-film electroluminescent display; ~ **de haz dirigido** *m* ELECTRÓN directed beam display; ~ **de identificación de línea de**

llamada *m* TELECOM calling line identification display (*CLID*); ~ **matricial** *m* TELECOM matrix display
visualizar *vt* ELECTRÓN, INFORM&PD, TELECOM display
visualrecíproca *f* CONST *topografía* folding sight
vitamina: ~ **E** *f* ALIMENT tocopherol
vitela *f* PAPEL wove
vitelina *f* QUÍMICA vitellin
viticultor *m* AGRIC grape grower
viticultura *f* AGRIC grape growing
vitola *f* ING MECÁ standard gage (*AmE*), standard gauge (*BrE*)
vitral *m* C&V stained glass window
vítreo *adj* C&V vitreous, GEOL hyaline, ÓPT, QUÍMICA, REVEST, TELECOM vitreous
vitrificación *f* GEOL, QUÍMICA vitrification
vitrificado *adj* COLOR, REVEST enameled (*AmE*), enamelled (*BrE*)
vitrificar *vt* C&V glaze, REVEST vitrify
vitrina *f* C&V display window, P&C cabinet, REFRIG display case; ~ **frigorífica** *f* REFRIG refrigerated counter, refrigerated display case, refrigerated showcase
vitriolización *f* QUÍMICA vitriolization
vitriolo *m* QUÍMICA vitriol; ~ **azul** *m* QUÍMICA blue vitriol; ~ **verde** *m* QUÍMICA green vitriol
vitrofibra: ~ **de núcleo grande** *f* ING ELÉC large-core glass fiber (*AmE*), large-core glass fibre (*BrE*)
vituallar *vt* TRANSP MAR *barco* supply with provisions
viuda *f* IMPR widow
vivero: ~ **de mariscos** *m* OCEAN shellfish farm; ~ **de semillas** *m* AGRIC seed nursery
vivianita *f* MINERAL vivianite
vivo *adj* PROD *vapor* live, TEXTIL *colores de telas* bright, TV live
vocoder *m* INFORM&PD, TELECOM vocoder; ~ **de canales** *m* TELECOM channel vocoder; ~ **formante** *m* TELECOM formant vocoder; ~ **de tono excitado** *m* TELECOM pitch-excited vocoder
vocodificador *m* INFORM&PD, TELECOM vocoder
vogesita *f* TEC PETR vogesite (*AmE*), vogesyte (*BrE*)
voile *m* TEXTIL voile
voladizo[1]: **en** ~ *adj* FÍS, MECÁ cantilevered
voladizo[2] *m* CONST cantilevering, corbel, overhanging, FÍS, ING MECÁ cantilevering, TRANSP AÉR pylon
volado *adj* IMPR *letra más alta que las demás* high in line
voladura *f* CARBÓN blast, blasting, CONST *explosivos* blasting, GEOL round, MINAS blasting; ~ **de muros suave** *f* MINAS smooth wall blasting; ~ **de pozos profundos** *f* MINAS long-hole blasting; ~ **por rotación** *f* MINAS rotation firing; ~ **secundaria** *f* MINAS secondary blasting; ~ **suave** *f* MINAS smooth blasting
voladuras: ~ **sucesivas** *f pl* MINAS multiple row blasting
volante *m* AUTO flywheel, *dirección* steering wheel, ING MECÁ flywheel, driving wheel, wheel, hand wheel, MECÁ flywheel, PROD *regulador de paletas* fly governor, REFRIG flywheel, TEXTIL *vestimenta* ruffle, TRANSP driving wheel, VEH *del motor* flywheel; ~ **del árbol de levas** *m* AUTO, VEH camshaft drive; ~ **del balancín** *m* PROD *de bomba* bob; ~ **en el contorno de busto** *m* TEXTIL bustline ruffle; ~ **de control del alerón** *m* TRANSP AÉR aileron control wheel; ~ **de control de dirección de la rueda del morro** *m* TRANSP AÉR nose-wheel steering control

wheel; ~ **cruciforme** *m* ING MECÁ split-set collar; ~ **de dirección** *m* VEH steering wheel; ~ **elevador** *m* ING MECÁ elevating wheel; ~ **de fricción** *m* ING MECÁ friction wheel; ~ **de inercia** *m* AUTO inertia drive, TEC ESP momentum wheel, reaction wheel; ~ **macizo experimental** *m* ING MECÁ experimental solid-disc flywheel (*BrE*), experimental solid-disk flywheel (*AmE*); ~ **de mando** *m* ING MECÁ operating handwheel; ~ **de maniguetas radiales** *m* ING MECÁ pilot wheel; ~ **de maniobra** *m* ING MECÁ hand wheel, operating handwheel; ~ **de mano** *m* ING MECÁ, MECÁ hand wheel; ~ **del motor** *m* ING MECÁ engine flywheel; ~ **de puntería en elevación** *m* D&A *cañones* elevating wheel; ~ **regulador** *m* CINEMAT balance wheel; ~ **del ventilador** *m* ING MECÁ fan wheel
volar[1] *vt* CARBÓN blast, CONST *vertedero* shoot, TERMO *mina* explode
volar[2] *vi* MINAS mine; ~ **en círculo** *vi* TEC ESP orbit
volátil: ~ *adj* AGRIC, CARBÓN, INFORM&PD, P&C, QUÍMICA, TEXTIL volatile; **no** ~ *adj* CONST, INFORM&PD, P&C nonvolatile
volátiles *m pl* CARBÓN volatile bodies
volatilidad *f* METAL, P&C, QUÍMICA, TEXTIL volatility
volatilización *f* QUÍMICA volatilization; ~ **catódica en el vacío** *f* METAL cathode sputtering
volatilizar *vt* QUÍMICA volatilize
volatilizarse *v refl* QUÍMICA, TEXTIL volatilize
volborthita *f* MINERAL volborthite
volcado: ~ **de memoria** *m* INFORM&PD memory dump
volcador *m* CONST tipping device, PROD tilter; ~ **de moldes de hielo** *m* REFRIG ice tip; ~ **de vagones** *m* CARBÓN dump truck (*AmE*), *de vagonetas* tipper (*BrE*); ~ **de vagonetas** *m* CARBÓN dump truck (*AmE*), tipper (*BrE*)
volcán: ~ **en actividad** *m* GEOL active volcano; ~ **apagado** *m* GEOL extinct volcano; ~ **compuesto** *m* GEOL composite volcano; ~ **de escudo** *m* GEOL shield volcano; ~ **estratificado** *m* GEOFÍS stratovolcano; ~ **extinto** *m* GEOL extinct volcano; ~ **de lodo** *m* ENERG RENOV, TEC PETR *geología* mud volcano; ~ **subterráneo** *m* GEOFÍS subterranean volcano; ~ **de tipo Hawaiano** *m* GEOL Hawaiian-type volcano
volcánico *adj* GEOL, TEC PETR *geología* volcanic
volcanismo *m* GEOL volcanism
volcar[1] *vt* TRANSP MAR *embarcación* capsize
volcar[2] *vi* CONST tip up, ING MECÁ overturn, SEG overbalance
volcar[3]: **no** ~ *fra* EMB not to be dropped, *instrucciones de uso* do not drop
volquete *m* CARBÓN dump cart, tip box car, CONST dumper, *verter tierras* dump (*AmE*), tip (*BrE*), CONTAM MAR skip, TRANSP dumper (*AmE*), tipper (*BrE*); ~ **al extremo** *m* CONST end dump
volt *m* ELEC volt (*V*)
voltaje *m* ELEC, FÍS voltage, ING ELÉC potential, tension, voltage, TEC ESP potential, TELECOM, VEH *instalación eléctrica, encendido* voltage; **bajo** ~ *m* ELEC low voltage, *tensión* low tension, ING ELÉC low tension, low voltage, TELECOM low voltage; ~ **acelerador** *m* ING ELÉC accelerating voltage; ~ **activo** *m* ELEC *corriente alterna*, FÍS, ING ELÉC active voltage; ~ **de ajuste** *m* ELEC adjusting voltage; ~ **de alimentación** *m* ELEC, ING ELÉC *fuente de alimentación* current voltage (*AmE*), mains voltage (*BrE*), PROD input voltage, supply voltage, TV current voltage (*AmE*), mains voltage (*BrE*); ~ **alternativo** *m*

ING ELÉC alternating voltage; ~ **alterno** *m* ELEC alternating voltage; ~ **del ánodo** *m* ING ELÉC anode voltage; ~ **antinodal** *m* ING ELÉC voltage antinode; ~ **aplicado** *m* ELEC impressed voltage; ~ **en avalancha** *m* ELEC avalanche voltage; ~ **en los bornes** *m* TV output voltage; ~ **de borrado** *m* TV blanking voltage; ~ **de CA rectificado de onda completa** *m* ELEC, ING ELÉC, PROD full-wave rectified AC voltage; ~ **de captación** *m* ING ELÉC pick-up voltage; ~ **de carga** *m* ELEC, ING ELÉC *acumulador, batería* charging voltage; ~ **en carga** *m* ING ELÉC on-load voltage; ~ **de CC de pequeña intensidad** *m* ELEC, PROD low-level DC voltage; ~ **de cebado** *m* ING ELÉC firing voltage; ~ **cero** *m* ING ELÉC zero voltage; ~ **del circuito inducido** *m* ING ELÉC secondary voltage; ~ **compensador** *m* ELEC compensating voltage; ~ **de contorneamiento en seco** *m* ELEC *de arco* dry flashover voltage; ~ **de control** *m* ELEC, PROD control voltage; ~ **de la corriente** *m* ELEC, ING ELÉC, TV current voltage (*AmE*), mains voltage (*BrE*); ~ **en cuadratura** *m* ELEC *circuito de corriente alterna* quadrature voltage; ~ **de cuasicresta** *m* TELECOM *móvil aeronáutico* quasi-peak voltage; ~ **de la descarga disruptiva** *m* ING ELÉC disruptive voltage; ~ **diente de sierra** *m* ELEC, FÍS, ING ELÉC, TRANSP MAR ramp voltage; ~ **en diente de sierra** *m* ING ELÉC sawtooth voltage; ~ **diferencial** *m* ING ELÉC differential voltage; ~ **de diodo** *m* ELECTRÓN diode voltage; ~ **de disparo** *m* AmL TV triggering voltage; ~ **disruptivo** *m* ING ELÉC flashover voltage; ~ **de entrada** *m* ING ELÉC, PROD input voltage, TV threshold voltage; ~ **de entrada de línea** *m* ELEC, PROD line voltage in; ~ **entre fase y neutro** *m* ELEC voltage to neutral; ~ **entre fases** *m* ELEC *fuente de alimentación trifásica* line-to-line voltage; ~ **de entretenimiento** *m* ING ELÉC keep-alive voltage; ~ **excesivo** *m* ELEC, ING ELÉC, QUÍMICA overvoltage; ~ **de excitación** *m* TV drive voltage; ~ **exterior** *m* ELEC, ING ELÉC external voltage; ~ **en fase de corriente** *m* ING ELÉC active voltage; ~ **del filamento** *m* ING ELÉC *tubos* heater voltage; ~ **de funcionamiento** *m* ELEC *de sistema,* ING ELÉC operating voltage; ~ **de Hall** *m* FÍS Hall voltage; ~ **del haz electrónico** *m* ELECTRÓN, TV electron-beam voltage; ~ **de impedancia** *m* ING ELÉC impedance voltage; ~ **impulsivo no disruptivo** *m* ELEC, PROD withstand impulse voltage; ~ **inducido** *m* TELECOM induced voltage; ~ **del inductor del alternador** *m* ELEC alternator field voltage; ~ **de interrupción** *m* ING ELÉC interrupting voltage; ~ **inverso** *m* ELEC, ING ELÉC reverse voltage, PROD inverse voltage; ~ **inverso máximo** *m* ELEC peak inverse voltage; ~ **de línea de entrada** *m* PROD incoming line voltage; ~ **de línea a tierra** *m* ELEC *red de distribución* line-to-earth voltage (*BrE*), line-to-ground voltage (*AmE*); ~ **máximo** *m* ELEC full voltage, FÍS peak voltage, ING ELÉC peak voltage, full voltage, PROD full voltage, TV peak voltage; ~ **medido** *m* ELEC *corriente* measured voltage; ~ **medio** *m* ING ELÉC medium voltage, *electrodos* bias; ~ **medio del electrodo** *m* ING ELÉC electrode-bias voltage; ~ **momentáneo** *m* ELEC transient voltage; ~ **negativo** *m* ING ELÉC negative voltage; ~ **de nivel lógico** *m* ELEC, PROD logic level voltage; ~ **no disruptivo** *m* ELEC *instalación* withstand voltage; ~ **no disruptivo de aislación** *m* ELEC, PROD insulation withstand voltage; ~ **no regulado** *m* ING ELÉC

unregulated voltage; ~ **nominal de aislamiento** *m* ELEC, ING ELÉC, PROD rated insulation voltage; ~ **normal** *m* CONST normal voltage; ~ **nulo** *m* ELEC, ING ELÉC no-voltage; ~ **de onda cuadrada** *m* TV square-wave voltage; ~ **ondulado** *m* CONST, ELECTRÓN ripple voltage; ~ **de perturbación del suministro** *m* ELEC, FÍS, ING ELÉC breakdown voltage; ~ **de polarización** *m* FÍS bias voltage; ~ **de polarización de rejilla** *m* ING ELÉC grid bias; ~ **de prueba** *m* ING MECÁ proof pressure; ~ **reactivo** *m* ELEC, ELECTRÓN, FÍS, ING ELÉC reactive voltage; ~ **de la red** *m* ELEC, ING ELÉC, TV current voltage (*AmE*), mains voltage (*BrE*); ~ **de referencia** *m* ELEC, ING ELÉC reference voltage; ~ **de régimen** *m* ELEC on-load voltage, rated voltage, working voltage, *de sistema* operating voltage, ING ELÉC rated voltage, PROD operating voltage, rated voltage; ~ **regulado** *m* ING ELÉC regulated voltage; ~ **relativamente negativo** *m* ING ELÉC relatively-negative voltage; ~ **relativamente positivo** *m* ING ELÉC relatively-positive voltage; ~ **de ruido** *m* ING ELÉC noise voltage; ~ **de ruptura** *m* ELEC *de dieléctrico, diodo Zener,* FÍS, ING ELÉC, ING MECÁ, TELECOM breakdown voltage; ~ **de salida** *m* ING ELÉC, TV output voltage; ~ **de salto arco con aislador seco** *m* ELEC dry flashover voltage; ~ **de saturación** *m* ING ELÉC saturation voltage; ~ **del secundario** *m* ING ELÉC secondary voltage; ~ **de servicio** *m* ELEC closed-circuit voltage, on-load voltage, operating voltage, working voltage, ING ELÉC closed-circuit voltage, PROD operating voltage; ~ **sinusoidal** *m* FÍS sinusoidal voltage; ~ **de suministro doble** *m* ING ELÉC dual-supply voltage; ~ **de suministro único** *m* ING ELÉC single-supply voltage; ~ **a tierra** *m* ING ELÉC voltage to earth (*BrE*), voltage to ground (*AmE*); ~ **de trabajo** *m* ELEC *de sistema* operating voltage; ~ **en vacío** *m* ING ELÉC open-circuit voltage; ~ **de variación lenta** *m* ING ELÉC slowly varying voltage; ~ **Zener** *m* FÍS Zener voltage; ~ **zumbador** *m* ING ELÉC hum voltage; ~ **con zumbido** *m* ING ELÉC hum voltage

voltámetro *m* ELEC, FÍS, ING ELÉC, INSTR, METR coulombmeter, coulometer

volteador: ~ **de hojas** *m* IMPR sheet-turning device

voltear *vt* ING MECÁ overturn

voltejear *vi* TRANSP MAR *navegación a vela* beat

voltejeo *m* TRANSP MAR *navegación a vela* tacking

voltiamperímetro *m* ING ELÉC multimeter, voltameter (*BrE*), voltammeter (*AmE*), voltmeter, TV multimeter

voltímetro *m* ELEC, FÍS voltmeter; ~~**amperímetro** *m* FÍS, QUÍMICA voltameter (*BrE*), voltammeter (*AmE*); ~ **analógico** *m* ING ELÉC analog voltmeter; ~ **astático** *m* ELEC astatic voltmeter; ~ **de CA** *m* ING ELÉC AC voltmeter; ~ **Cardew** *m* ELEC Cardew voltmeter; ~ **de CC** *m* ING ELÉC DC voltmeter; ~ **de corriente alterna** *m* ELEC alternating-current voltmeter; ~ **de cresta** *m* ELEC peak voltmeter; ~ **de cuadro móvil** *m* ELEC moving-coil voltmeter; ~ **diferencial** *m* ELEC, INSTR differential voltmeter; ~ **digital** *m* ING ELÉC digital voltmeter; ~ **estático** *m* ELEC, INSTR static voltmeter; ~ **de hilo caliente** *m* ELEC hot wire voltmeter; ~ **indicador del valor de cresta** *m* ELEC peak voltmeter; ~ **con núcleo de hierro** *m* ELEC iron core voltmeter; ~ **de núcleo de hierro** *m* ELEC iron core voltmeter; ~ **selectivo** *m* ELEC selective voltmeter; ~ **térmico** *m* ELEC hot wire voltmeter

voltio *m* (*V*) ELEC, FÍS, ING ELÉC, METR volt (*V*)

volumen *m* ACÚST *radio* volume, AGUA turnover, volume, CARBÓN content, CONST bulk, FÍS bulk, volume, GEOM volume, INFORM&PD volume, bulk, ING MECÁ holding capacity, MATEMÁT volume, MECÁ bulk, cubic capacity, METR cubic capacity, QUÍMICA, TEXTIL volume, TRANSP MAR *de los espacios de estiba* cubic capacity, *del tanque* volume; ~ **de agua superficial** *m* HIDROL surface water load; ~ **atómico** *m* NUCL atomic volume; ~ **base** *m* TRANSP base volume; ~ **de capacidad** *m* PETROL *buque* bulk volume; ~ **de carga** *m* MINAS loading density; ~ **conductor** *m* ELECTRÓN conducting state; ~ **crítico** *m* TERMOTEC critical volume; ~ **de diseño** *m* TRANSP design volume; ~ **divergente** *m* TRANSP diverging volume; ~ **específico** *m* FÍS specific volume, PAPEL bulk; ~ **específico del aire húmedo** *m* REFRIG humid volume; ~ **de flujo de saturación** *m* ELECTRÓN saturation output state; ~ **de la fusión** *m* TRANSP merge volume; ~ **global de consumo** *m* PROD throughput; ~ **de hoja** *m* PAPEL bulk; ~ **en la incorporación** *m* TRANSP merge volume; ~ **de llamadas** *m* TELECOM call flow; ~ **de llenado** *m* EMB volume filling; ~ **máximo por hora** *m* TRANSP maximum hourly volume; ~ **de modal efectivo** *m* ÓPT effective mode volume; ~ **de modo efectivo** *m* TELECOM effective mode volume; ~ **de modos** *m* ÓPT, TELECOM mode volume; ~ **molar** *m* FÍS molar volume; ~ **del movimiento de tierras** *m* CONST earthwork cubature, earthwork cubing; ~ **de la muestra** *m* ELECTRÓN sample size; ~ **de negocios** *m* PROD turnover; ~ **de polvos** *m* CONTAM dust content; ~ **de poro** *m* CARBÓN *hormigón, terrenos*, PETROL pore volume; ~ **de referencia** *m* ACÚST reference volume; ~ **de rotación** *m* GEOM volume of rotation; ~ **de sedimentación** *m* PROC QUÍ sedimentation volume; ~ **de servicio** *m* TRANSP service volume; ~ **sonoro** *m* ACÚST *radio* volume; ~ **total** *m* EMB gross volume; ~ **total de almacenamiento** *m* GAS total storage volume; ~ **de tráfico** *m* CONST, TELECOM traffic flow, TRANSP traffic volume; ~ **de tráfico en horas punta** *m* TRANSP peak traffic flow, peak traffic volume; ~ **de tráfico remoto** *m* TELECOM remote loading; ~ **vacío** *m* TEXTIL void volume; ~ **de vapor** *m* INSTAL HIDRÁUL *de caldera* bulk of steam, head of steam, steam space; ~ **de ventas** *m* PROD turnover

volumétrico *adj* ING MECÁ, METAL *ecuación* volumetric

volúmetro *m* ACÚST volumeter

voluta *f* ING MECÁ helix

volutina *f* QUÍMICA volutin

volvedor *m* ING MECÁ tap wrench; ~ **de chapas** *m* PROD *laminación* turnover

volver *vt* INFORM&PD return; ~ **a cargar** *vt* CINEMAT, FOTO, INFORM&PD, PROD, TV reload; ~ **a cero** *vt* ING MECÁ set to zero; ~ **a comenzar** *vt* INFORM&PD restart; ~ **a comprobar** *vt* METR recheck; ~ **a editar** *vt* CINEMAT, TV re-edit; ~ **a ejecutar** *vt* INFORM&PD rerun; ~ **a encuadrar** *vt* CINEMAT reframe; ~ **a montar** *vt* CONST reassemble; ~ **a poner a cero** *vt* ING MECÁ *instrumento* reset to zero; ~ **a rodar** *vt* CINEMAT *escena*, TV reshoot; ~ **a usar** *vt* RECICL reuse; ~ **a utilizar** *vt* RECICL reuse

vomicina *f* QUÍMICA vomicine

vórtex *m* CARBÓN vortex

vórtice *m* CARBÓN, FÍS, MECÁ, TEC ESP vortex; ~ **de la**
punta del ala *m* TRANSP AÉR wing-tip vortex; ~ **de la punta de la pala** *m* TRANSP AÉR blade-tip vortex

vorticidad *f* ACÚST, FÍS, FÍS FLUID vorticity; ~ **localmente grande** *f* FÍS FLUID locally-high vorticity

voz *f* INFORM&PD, TELECOM voice; ~ **artificial** *f* ACÚST artificial voice; ~ **digitalizada** *f* ELECTRÓN, TELECOM digitalized speech, digitized speech

VSI *abr* (*indicador de la velocidad vertical*) TEC ESP, TRANSP AÉR VSI (*vertical speed indicator*)

vuelco *m* AGRIC lodging, CONST dumping, INFORM&PD dump, PROD dumping, tilting, SEG overturning, TRANSP *vehículos* overturning, roll-over; ~ **binario** *m* INFORM&PD binary dump; ~ **de cambios** *m* INFORM&PD change dump; ~ **en cinta** *m* INFORM&PD tape dump; ~ **en cinta magnética** *m* INFORM&PD tape dump; ~ **dinámico** *m* INFORM&PD dynamic allocation, dynamic dump; ~ **estático** *m* INFORM&PD static dump; ~ **instantáneo** *m* INFORM&PD snapshot dump; ~ **de memoria** *m* INFORM&PD memory dump; ~ **de la memoria central** *m* INFORM&PD core dump; ~ **de pantalla** *m* INFORM&PD screen dump; ~ **selectivo** *m* INFORM&PD selective dump

vuelo *m* CONST *arquitectura* nosing, projection, jetty, *grúa* span, TRANSP AÉR flight; ~ **de adiestramiento** *m* TRANSP AÉR training flight; ~ **de aeroremolque** *m* TRANSP AÉR aerotow flight (*BrE*), airtow flight (*AmE*); ~ **en ascendencia térmica** *m* TERMO thermal soaring; ~ **asistido** *m* TEC ESP manned flight; ~ **de autorotación** *m* TRANSP AÉR autorotation flight; ~ **autorotativo** *m* TRANSP AÉR autorotative flight; ~ **de autotraslado** *m* TRANSP AÉR ferry flight; ~ **de calibración** *m* TRANSP AÉR calibration flight; ~ **cautivo** *m* TEC ESP captive flight; ~ **charter de carga** *m* TRANSP AÉR all-cargo charter flight; ~ **a ciegas** *m* TRANSP AÉR blind flight; ~ **de contacto** *m* TRANSP AÉR contact flight; ~ **controlado** *m* TRANSP AÉR controlled flight; ~ **de la cosmonave utilizando el campo de gravitación** *m* TEC ESP swing-by; ~ **de demostración** *m* TRANSP AÉR proving flight; ~ **a la deriva** *m* TRANSP AÉR drifting flight; ~ **directo** *m* TRANSP AÉR direct flight; ~ **doméstico** *m* TRANSP AÉR domestic flight; ~ **sin escalas** *m* TRANSP AÉR nonstop flight; ~ **espacial** *m* TEC ESP space flight; ~ **estacionario** *m* TRANSP AÉR hovering; ~ **experimental** *m* TEC ESP test flight; ~ **de exploración** *m* TEC ESP sweep; ~ **de familiarización de ruta** *m* TRANSP AÉR route-familiarization flight; ~ **en formación** *m* TRANSP AÉR formation flight; ~ **hacia atrás** *m* TRANSP AÉR backward flight; ~ **horizontal** *m* TRANSP AÉR level flight; ~ **inaugural** *m* TRANSP AÉR maiden flight; ~ **por instrumentos** *m* TRANSP AÉR instrument flying; ~ **interno** *m* TRANSP AÉR domestic flight; ~ **interplanetario** *m* TEC ESP interplanetary flight; ~ **de larga distancia** *m* TRANSP AÉR long-distance flight; ~ **libre** *m* TEC ESP free flight; ~ **mediante instrumentos** *m* TRANSP AÉR instrument flight; ~ **nivelado** *m* TRANSP AÉR level flight; ~ **nocturno** *m* TRANSP AÉR night flight; ~ **en órbita espacial tripulado** *m* TEC ESP manned orbital space flight; ~ **orbital** *m* TEC ESP orbital flight; ~ **de planeo** *m* TRANSP AÉR gliding flight; ~ **de pruebas** *m* TRANSP AÉR proving flight, test flight; ~ **radiodirigido** *m* D&A, TRANSP AÉR radio-guided flight; ~ **radiogoniométrico hacia la estación de destino** *m* TEC ESP homing; ~ **radioguiado** *m* D&A, TRANSP AÉR radio-guided flight; ~ **de recepción** *m* TRANSP AÉR acceptance flight; ~ **regular** *m* TRANSP AÉR scheduled flight; ~ **teledirigido** *m* D&A, TRANSP AÉR remote-controlled flight; ~ **telemaniobrado** *m* D&A remote-controlled flight; ~ **tripulado** *m* TEC ESP manned flight; ~ **uniforme** *m* TRANSP AÉR steady flight; ~ **sin visibilidad** *m* TEC ESP *astronáutica* blind navigation

vuelta[1] *f* FÍS *del bobinado* turn, GEOM revolution, INFORM&PD *secuencia* run, ING MECÁ revolution, turn, TEC ESP spin, revolution, TRANSP MAR *cabos* hitch, VEH *régimen de giro del motor* rev; ~ **a la base** *f* TRANSP AÉR homing; ~ **de braza** *f* TRANSP MAR *nudo* timber hitch; ~ **del eje** *f* ING MECÁ shaft lap; ~ **encontrada** *f* TRANSP MAR *navegación* reciprocal course; ~ **espiral** *f* ING MECÁ spire; ~ **de gancho** *f* TRANSP MAR *nudo* blackwall hitch; ~ **de gancho doble** *f* TRANSP MAR *nudo* double blackwall hitch; ~ **de gancho simple** *f* TRANSP MAR *nudo* single blackwall hitch; ~ **inactiva de vacío** *f* ING MECÁ noncutting return; ~ **redonda y medio cote doble** *f* TRANSP MAR *nudo* round turn and two half hitches; ~ **de rezón** *f* TRANSP MAR *nudo* fisherman's bend; ~ **por tierra** *f* ING ELÉC, PROD ground return; ~ **de vinatera** *f* TRANSP MAR *nudo* becket bend, sheet bend

vuelta[2]: ~**redonda y medio cote doble** *fra* TRANSP MAR round turn and two half hitches

vueltas: ~ **por metro** *f pl* TEXTIL turns per meter; ~ **por pulgadas** *f pl* TEXTIL turns per inch

vulcanicidad *f* GEOFÍS vulcanicity

vulcanismo: ~ **entre placas** *m* GEOL intraplate volcanism

vulcanita *f* ELEC, MECÁ, P&C, QUÍMICA vulcanite

vulcanización *f* ELEC, MECÁ, P&C, QUÍMICA cure, curing, vulcanization, TEC ESP *del caucho* metalization (*AmE*), metallization (*BrE*), TERMO curing, vulcanization; ~ **continua** *f* P&C continuous vulcanization; ~ **en frío** *f* PROD, TERMO cold curing; ~ **prematura** *f* P&C scorch; ~ **rápida** *f* P&C fast curing

vulcanizado: *adj* TERMO vulcanized; **no** ~ *adj* ELEC, MECÁ, P&C, QUÍMICA, TERMO uncured, unvulcanized; ~ **en frío** *adj* PROD, TERMO cold-cured

vulcanizar *vt* ELEC, MECÁ, P&C, QUÍMICA cure, vulcanize, TERMO heat-cure, vulcanize; ~ **en frío** *vt* PROD, TERMO cold-cure

vulnerabilidad: ~ **electromagnética** *f* ING ELÉC electromagnetic vulnerability

vulneraria *f* AGRIC kidney welch

V1 *abr* (*velocidad de decisión*) TRANSP AÉR V1 (*decision speed*)

W

W *abr* ELEC, FÍS (*vatio*) W (*watt*), FÍS ONDAS, FÍS RAD (*frecuencia angular*) W (*angular frequency*), ING ELÉC, METR (*vatio*) W (*watt*), QUÍMICA (*tungsteno, wolframio*) W (*tungsten, wolfram*)

wadi *m* HIDROL wadi

wagnerita *f* MINERAL wagnerite

walchowita *f* MINERAL walchowite

WANO *abr* (*Asociación Mundial de Operadores Nucleares*) NUCL WANO (*World Association of Nuclear Operators*)

warfarina *f* QUÍMICA warfarin

wataje *m* QUÍMICA wattage

wavellita *f* MINERAL wavellite

Wb *abr* (*weber, weberio*) ELEC, FÍS, ING ELÉC, METR Wb (*weber*)

Web: **el ~** *m* INFORM&PD the Web

weber *m* (*Wb*) ELEC, FÍS, ING ELÉC, METR weber (*Wb*)

weberio *m* (*Wb*) ELEC, FÍS, ING ELÉC, METR weber (*Wb*)

wernerita *f* MINERAL scapolite, wernerite

whewelita *f* MINERAL whewellite

whipstock *m* TEC PETR *perforación* whipstock

whitneyita *f* MINERAL whitneyite

willemita *f* MINERAL willemite

williamsita *f* MINERAL williamsite

wiluíta *f* MINERAL wiluite

winchy *m* AmL (*cf torno neumático Esp*) ING MECÁ, TEC PETR air hoist

withamita *f* MINERAL withamite

witherita *f* MINERAL witherite

wittichenita *f* MINERAL wittichenite

wohlerita *f* MINERAL wohlerite

wolastonita *f* MINERAL wollastonite

wolfrámico *adj* QUÍMICA wolframic

wolframio *m* ING ELÉC (*cf tungsteno, W*), METAL (*cf tungsteno, W*), QUÍMICA (*cf tungsteno, W*) tungsten (*W*), wolfram (*W*)

wolframita *f* MINERAL, QUÍMICA wolframite

wolfsbergita *f* MINERAL chalcostibite, wolfsbergite

World: **~ Wide Web** *m* (*WWW*) INFORM&PD World Wide Web (*WWW*)

WOW *abr* (*espera por mal tiempo*) TEC PETR WOW (*waiting on weather*)

WP *abr* INFORM&PD (*procesado de texto, procesamiento de texto, tratamiento de texto*) WP (*word processing*), (*procesador de textos*) WP (*word processor*)

wulfenita *f* INFORM&PD, MINERAL wulfenite

wurtzita *f* MINERAL wurtzite

WWW *abr* (*World Wide Web*) INFORM&PD WWW (*World Wide Web*)

WYSIWYG *abr* (*lo que se ve es lo que se obtiene*) INFORM&PD WYSIWYG (*what you see is what you get*)

X

xantato *m* QUÍMICA xanthate
xanteína *f* QUÍMICA xanthein
xanteno *m* QUÍMICA xanthene
xántico *adj* QUÍMICA xanthic
xantidrol *m* QUÍMICA xanthydrol
xantilo *m* QUÍMICA xanthyl
xantina *f* QUÍMICA xanthine
xantocreatinina *f* QUÍMICA xanthocreatinine
xantófila *f* AGRIC xanthophyl
xantofilita *f* MINERAL xanthophyllite
xantogénico *adj* QUÍMICA xanthogenic
xantomoniasis *f* AGRIC bacterial streak
xantona *f* QUÍMICA xanthone
xantoproteico *adj* QUÍMICA xanthoproteic
xantosina *f* QUÍMICA xanthosine
xantoso *adj* QUÍMICA xanthous
xantoxilina *f* QUÍMICA xanthoxylin
Xe *abr* (*xenón*) QUÍMICA Xe (*xenon*)
xenocristal *m* GEOL xenocryst
xenolito *m* GEOL xenolith

xenomórfico *adj* GEOL xenomorphic
xenón *m* (*Xe*) QUÍMICA xenon (*Xe*)
xenotima *f* MINERAL xenotime
xerofita *f* AGRIC xerophyte
xerografía *f* ELEC, ELECTRÓN, GAS, INFORM&PD xerography
xilantita *f* QUÍMICA xylanthite
xilenetiol *m* QUÍMICA xylenethiol
xileno *m* DETERG, P&C, QUÍMICA, TEC PETR xylene
xilenol *m* QUÍMICA xylenol
xilidina *f* QUÍMICA xylidine
xilileno *m* QUÍMICA xylylene
xililo *m* QUÍMICA xylyl
xilita *f* ALIMENT *repostería y confituras* xylitol
xilitol *m* QUÍMICA xylitol
xilografía *f* IMPR xylograph
xilol *m* QUÍMICA xylol
xilonita *f* P&C xylonite
xilosa *f* QUÍMICA xylose
xilotila *f* MINERAL xylotile

Y

Y *abr* (*itrio*) METAL, QUÍMICA Y (*yttrium*)
yacimiento *m* CARBÓN bed, layer, CONST deposit, GAS pool, GEOL accumulation, ore, reservoir, HIDROL lens, MINAS mineral deposit, PETROL pool, field, TEC PETR field; **~ ácido** *m* CONTAM, MINAS acid deposit; **~ aluvial** *m* AGUA, MINAS alluvial deposit; **~ de aluvión** *m* GEOL, MINAS *placeres* placer deposit; **~ aurífero** *m* MINAS gold field; **~ de caliza** *m* AGUA chalk formation; **~ de carbón** *m* CARBÓN coal bed, coal measure, GEOL coal measure, cola bed, MINAS coal bed, coal measure; **~ desprendido del filón** *m* CARBÓN *aluvión, acarreos,* MINAS float; **~ de gas** *m* GAS gas field; **~ de gas condensado** *m* GAS condensate gas reservoir; **~ de gas natural** *m* GAS natural gas deposit; **~ gasífero** *m* GAS gas field; **~ de hierro** *m* MINAS iron deposit; **~ hullero** *m* CARBÓN, GEOL, MINAS coal deposit; **~ irregular** *m* MINAS bunchy; **~ metalífero** *m* METAL, MINAS, NUCL ore deposit; **~ mineral** *m* Esp (*cf criadero mineral AmL*) MINAS mineral deposit; **~ de petróleo** *m* PETROL reservoir, TEC PETR reservoir, storage lake; **~ petrolífero** *m* PETROL oil reservoir, oilfield, petroleum field, TEC PETR oil reservoir, *producción* oilfield; **~ de plomo** *m* MINAS lead deposit; **~ vertical de oro dentro del cuarzo** *m* MINAS goldreef, reef
yacimientos *m pl* MINAS *excavaciones* diggings
yarda *f* METR yard; **~ cuadrada** *f* METR square yard; **~ cúbica** *f* METR cubic yard
yardaje *m* METR yardage
yate: **~ para cruceros** *m* TRANSP MAR cruiser; **~ de recreo** *m* TRANSP cruiser; **~ velero motorizado** *m* TRANSP MAR motor sailer
Yb *abr* (*iterbio*) METAL, QUÍMICA Yb (*ytterbium*)
yema: **~ axilar** *f* AGRIC axillary bud; **~ foliar** *f* AGRIC leaf bud
yesar *m* MINAS gypsum quarry
yesífero *adj* GEOL gypsiferous, gypsum-bearing
yeso *m* AGRIC gypsum, CONST plaster, GEOL, MINERAL, PETROL, QUÍMICA, TEC PETR *geología* gypsum; **~ cristalizado** *m* QUÍMICA selenite; **~ defectuoso y quebrado** *m* CONST fractured-and-faulted chalk; **~ fino** *m* CONST plaster of Paris; **~ floculento** *m* TERMOTEC flocculent gypsum; **~ de París** *m* CONST plaster of Paris

yesoso *adj* QUÍMICA gypseous
yodar *vt* PROC QUÍ, QUÍMICA iodize
yodargirita *f* MINERAL iodargyrite
yodato *m* QUÍMICA iodate
yodembolita *f* MINERAL iodembolite
yodhidrato *m* QUÍMICA hydriodide
yodhídrico *adj* QUÍMICA hydriodic
yódico *adj* QUÍMICA iodic
yodo *m* (*I*) ELECTRÓN, FOTO, QUÍMICA, TEXTIL iodine (*I*); **~ radioactivo** *m* FÍS RAD, QUÍMICA radioiodine
yodoaurato *m* QUÍMICA iodoaurate
yodobenceno *m* QUÍMICA iodobenzene
yodobromito *m* MINERAL iodobromite
yodoformo *m* QUÍMICA iodoform
yodohidrina *f* QUÍMICA iodohydrin
yodomercurato *m* QUÍMICA iodomercurate
yodometría *f* QUÍMICA iodometry
yodométrico *adj* QUÍMICA iodometric
yodonio *m* QUÍMICA iodonium
yodopsina *f* QUÍMICA iodopsin
yodoso *adj* QUÍMICA iodous
yodosobenceno *m* QUÍMICA iodosobenzene
yodrita *f* MINERAL iodyrite
yoduro *m* QUÍMICA iodide; **~ de acetilo** *m* QUÍMICA acetyl iodide; **~ de metileno** *m* QUÍMICA methylene iodide; **~ de metilo** *m* QUÍMICA methyl iodide; **~ de plata** *m* FOTO, QUÍMICA silver iodide
yola *f* TRANSP MAR yawl
yugo *m* ELEC, ING MECÁ, NUCL, QUÍMICA yoke, TV scanning yoke; **~ de apriete** *m* ING MECÁ *maquinaria* gripping yoke; **~ doble de conexión** *m* TRANSP AÉR connecting twin-yoke; **~ de escobilla** *m* ELEC brush yoke; **~ de sujeción** *m* ING MECÁ *maquinaria* gripping yoke
yunque *m* ACÚST *audición*, IMPR anvil, ING MECÁ *martillo pilón* anvil die, anvil, MECÁ, MINAS anvil; **~ de tornillo de banco** *m* ING MECÁ anvil vice (*BrE*), anvil vise (*AmE*)
yunta *f* ING MECÁ yoke
yute *m* TEXTIL jute
yuxtaposición: **no ~ de las líneas** *f* TV underlap
yuyo: **~ bola** *m* AGRIC tumbleweed; **~ colorado** *m* AGRIC rough pigweed

Z

Z *abr* (*número atómico*) FÍS, FÍS PART, QUÍMICA Z (*atomic number*)

zabordar *vi* OCEAN, TRANSP MAR *de barco* run ashore

zafar: ~ **el ancla** *vi* TRANSP MAR trip anchor

zafarse *v refl* ING MECÁ come away, come off, part, TRANSP MAR claw off

zafirino *m* MINERAL sapphirine

zafirita *f* MINERAL sapphire

zafiro *m* ELECTRÓN sapphire; ~ **azul** *m* MINERAL blue sapphire

zafra *f* AGRIC cane crop

zahine *f* AGRIC durra

zahorra *f* TRANSP MAR *playa, orilla* shingle bank

zallarse *v refl* TRANSP MAR *bote* swing away

zampeado *m* AGUA floor, CARBÓN *esclusas* invert

zanca *f* CONST string piece, *escalera* stringer, *escaleras* string, ELEC *sistema* spur; ~ **cerrada** *f* CONST close string, cut string, cut wall string; ~ **exterior** *f* CONST face string, outside string; ~ **a la francesa** *f* CONST *escaleras* housed string; ~ *f* CONST open string; ~ **inglesa** *f* CONST open string; ~ **interior** *f* CONST wall string

zanja *AmL f* (*cf trinchera Esp*) AGUA ditch, CARBÓN berm, ditch, drain, trench, CONST ditch, drain, trench, CONTAM MAR trench, HIDROL ditch, MINAS, OCEAN, PETROL trench; ~ **de agotamiento** *f* AGUA drainage ditch; ~ **antisocavación** *f* AGUA, OCEAN *hidráulica* cutoff trench; ~ **para avenamiento** *f* AGUA, HIDROL drainage ditch; ~ **de desagüe** *f* AGUA drainage channel; ~ **de drenaje** *f* AGRIC drainage channel, AGUA drainage ditch; ~ **de exploración** *f* MINAS costean trench; ~ **interceptora** *f* AGUA, OCEAN cutoff trench; ~ **de lixiviación** *f* AGUA leaching trench; ~ **de oxidación** *f* AGUA, HIDROL *aguas negras* oxidation ditch; ~ **pequeña de desagüe** *f* CARBÓN grip; ~ **perimetral** *f* CONST cutoff ditch; ~ **de préstamo** *f* CARBÓN *excavaciones*, CONST *movimiento de tierra* borrow pit, PROD borrow; ~ **de saneamiento** *f* CONST berm

zanjadora *f* CONST ditcher; ~ **de cangilones** *f* CONST trench cutter

zapa *f* MINAS *excavación a cielo abierto* mining

zapapico *m* CONST, PROD mattock, pick mattock, pickax (*AmE*), pickaxe (*BrE*)

zapata *f* CONST footing, shoe, ELEC *conmutador* collector shoe, FERRO *vehículos* shoe, GEOL footing, MECÁ shoe, drag, MINAS *colector* headboard, *retén o gancho agarrador de la vagoneta* drag, NUCL shoe, TEC ESP sole plate, TRANSP MAR *construcción naval* skeg, VEH *freno* shoe; ~ **de AP** *f* PROD HP shoe; ~ **de aterrizaje** *f* MECÁ skid; ~ **del carril** *f* FERRO *infraestructura* rail foot; ~ **de cimentación** *f* CARBÓN *muros* pad foundation; ~ **de contacto para flash** *f* FOTO hot-shoe flash contact; ~ **cortante** *f* MINAS drive shoe; ~ **de la cruceta** *f* ING MECÁ crosshead slipper; ~ **encajadora** *f* MINAS drive shoe; ~ **de formación** *f* PAPEL forming shoe; ~ **de freno** *f* *Esp* (*cf guarnición de freno AmL*) AUTO brake lining, brake shoe, C&V brake lining, FERRO *vehículos* brake block, ING MECÁ brake lining, brake shoe, brake block, MECÁ brake lining, brake calliper (*BrE*), brake caliper (*AmE*), brake shoe, VEH brake shoe, brake lining, *de bicicleta* brake block; ~ **de freno sin pestaña** *f* FERRO *vehículos* flangeless brake block; ~ **para golpes** *f* AGRIC *arado* shock shoe; ~ **de guía** *f* MINAS guide shoe; ~ **de pilón** *f* PROD *bocarte* stamp shoe; ~ **sobre pilotes** *f* CARBÓN *pilares* pile footing; ~ **posterior** *f* VEH *de freno* trailing shoe; ~ **primaria** *f* AUTO primary shoe; ~ **propulsora** *f* MINAS drive shoe; ~ **protectora** *f* ING MECÁ wear pad; ~ **de rodillo** *f* PROD roller shoe; ~ **de la toma colectora** *f* FERRO, ING ELÉC collector shoe

zapato *m* AmL (*cf sidecar Esp*) TRANSP sidecar; ~ **de entubación** *m* GAS casing shoe

zaranda *f* AGRIC angle sieve, MECÁ screen, TEC PETR *perforación* shale shaker; ~ **de granos limpios** *f* AGRIC shoe sieve; ~ **inferior de granos** *f* AGRIC lower sieve

zaratita *f* MINERAL zaratite

zarcillo: ~ **foliar** *m* AGRIC leaf tendril

zarpa *f* CONST *arquitectura* set-off

zarpar *vi* TRANSP MAR lift anchor

zeína *f* P&C, QUÍMICA zein

zenit *m* ENERG RENOV, TEC ESP, TRANSP MAR *navegación celeste* zenith

zeolita *f* DETERG, MINERAL, QUÍMICA zeolite

zeunerita *f* MINERAL zeunerite

zietrisikita *f* QUÍMICA zietrisikite

zimasa *f* QUÍMICA zymase

zímico *adj* QUÍMICA zymic

zimina *f* QUÍMICA zymin

zinc (*Zn*) *m* GEN zinc (*Zn*)

zincado *m* P&C *pintura* zinc coating

zincita *f* MINERAL zincite

zincoso *adj* QUÍMICA zincous

zingibereno *m* QUÍMICA zingiberene

zinkenita *f* MINERAL zinckenite

zinnwaldita *f* MINERAL zinnwaldite

zircaloy *m* NUCL zircaloy

zircón *m* C&V zircon

zirconato *m* QUÍMICA zirconate

zircónico *adj* QUÍMICA zirconic

zirconifluoruro *m* QUÍMICA zirconifluoride

zirconilo *m* QUÍMICA zirconyl

zirconio (*Zr*) *m* METAL, QUÍMICA zirconium (*Zr*)

Zn (*zinc*) *abr* GEN Zn (*zinc*)

zócalo *m* CONST base board (*AmE*), footing block, skirting, skirting board (*BrE*), ELEC electric socket, GEOL basement, oldland, ING MECÁ bedplate, LAB *red de distribución* electric socket, PETROL *geología, estratigrafía* basement, PROD *cilindro del calentador* pedestal, TEC PETR basement; ~ **de bayoneta** *m* CINEMAT *luz* bayonet socket; ~ **para casquillo de bayoneta** *m* ELEC *luz* bayonet socket; ~ **conector acoplador** *m* ELEC coupler socket-connector; ~ **de tomacorriente portátil** *m* ELEC *conexión* portable-socket outlet; ~ **de tubo** *m* ING ELÉC tube socket

zoisita *f* MINERAL zoisite

zona f GEN area, zone;

~a ~ **de acción de seguridad** f TV safe action area; ~ **de acidez** f CONTAM *característica química* acidic area; ~ **de acondicionamiento** f C&V conditioning zone; ~ **activa** f ELECTRÓN *de substrato semiconductor* active region, NUCL *del elemento de combustible* active length; ~ **de acumulación de petróleo** f TEC PETR *yacimientos* zone of petroleum accumulation; ~ **administrativa** f CONST administrative area; ~ **afectada por el calor** f TERMO heat-affected zone; ~ **algodonera** f AGRIC cotton belt; ~ **de almacenaje** f CONST yard; ~ **de almacenamiento** f CONST storage area; ~ **de alta presión** f METEO anticyclone, high-pressure area, high-pressure zone; ~ **aluvial** f CONST alluvial area; ~ **de amortiguación** f NUCL *de una barra de control* dashpot; ~ **de aparcamiento** f *Esp* (*cf plaza de estacionamiento AmL*) CONST parking area; ~ **de aplanamiento del flujo** f NUCL flux-flattened region; ~ **de aterrizaje** f D&A landing zone (*LZ*); ~ **atonal** f ACÚST atonal space; ~ **auroral** f GEOFÍS auroral belt, TEC ESP auroral zone;

~b ~ **de baja presión** f METEO depression, low-pressure area; ~ **de bajo rendimiento** f ELECTRÓN low-yield region; ~ **de base** f ELECTRÓN base region; ~ **de Benioff** f ENERG RENOV *recursos geotérmicos* Benioff zone; ~ **de Brillouin** f FÍS, METAL Brillouin zone; ~ **de búsqueda** f INFORM&PD seek area;

~c ~ **de caída** f CINEMAT, TV footprint; ~ **de caldeo** f TERMO heating zone; ~ **de calefacción** f TERMO heating zone; ~ **de calentamiento** f TERMO heating zone; ~ **de calmas ecuatoriales** f METEO, OCEAN, TRANSP MAR doldrums; ~ **de captación** f ENERG RENOV catchment area; ~ **de carga** f NUCL charge area; ~ **de carga del reactor** f NUCL reactor-charging face; ~ **de cargamento** f MINAS tip area (*BrE*), dump area (*AmE*); ~ **de central urbana** f TELECOM local exchange area; ~ **de cizalla** f GEOL shear zone; ~ **clara de la imagen** f IMPR highlight; ~ **de clasificación** f TRANSP classification yard (*AmE*), marshalling yard (*BrE*), shunting yard (*BrE*); ~ **climática ecuatorial** f METEO equatorial climate; ~ **de cobertura** f TEC ESP coverage area; ~ **de cobertura por satélite** f TV satellite coverage area; ~ **del colector de corriente** f ING ELÉC collector region; ~ **común** f INFORM&PD common area; ~ **de conducción** f ING ELÉC conducting zone; ~ **de contacto** f ELECTRÓN interface; ~ **contigua** f OCEAN contiguous zone; ~ **de contracción** f MINAS crumple zone; ~ **controlada** f NUCL controlled zone; ~ **de coordinación** f TEC ESP coordination area; ~ **costera** f GEOL, HIDROL, OCEAN, TRANSP MAR coastal zone; ~ **de crestas** f OCEAN crest province;

~d ~ **de deformaciones elásticas** f MECÁ elastic range; ~ **de derrumbamiento** f MINAS crumple zone; ~ **de desagüe** f CONST drainage area; ~ **de desbordamiento** f INFORM&PD overflow area; ~ **de descarga** f MINAS dump site; ~ **de descenso** f TEC ESP *vehículos* drop zone; ~ **de desechos** f *AmL* (*cf área de estériles Esp*) MINAS tailings area; ~ **desnuda** f METAL denuded zone; ~ **de despegue** f TRANSP AÉR takeoff area; ~ **despejada** f INFORM&PD clear area; ~ **desplazada** f GEOFÍS offset; ~ **de desplazamientos** f NUCL displacement spike; ~ **desplomada** f GEOL, MINAS caved area; ~ **detrás de una barrera de arrecifes** f GEOL *en dirección al*

continente back reef; ~ **de dirección final** f PROD end address field; ~ **de dislocación** f CRISTAL dislocation line; ~ **de dragado** f MINAS dredging field;

~e ~ **ecuatorial** f METEO equatorial zone, torrid zone; ~ **de ejercicios navales** f D&A, OCEAN naval maneuvers zone (*AmE*), naval manoeuvres zone (*BrE*); ~ **de embalamiento** f TRANSP AÉR runup area; ~ **de empaquetado** f PROD packaging area; ~ **de enfriamiento** f C&V, REFRIG cooling zone; ~ **de engorde** f AGRIC feeding area; ~ **enrejada** f CONST grate area; ~ **de enriquecimiento** f GEOL ore shoot; ~ **sin ensayar** f CONST unproved area; ~ **epicéntrica** f FÍS, GEOFÍS, GEOL epicentral area; ~ **erosionada** f AGRIC, GEOL galled spot; ~ **de escobenes** f TRANSP MAR *buque* hawse; ~ **de la estación** f FERRO infraestructura station area; ~ **de estacionamiento** f AmL (*cf plaza de aparcamiento Esp*) CONST parking area; ~ **de estériles** f MINAS tailings area; ~ **de estiraje** f TEXTIL draught zone (*BrE*); ~ **de estiramiento** f TEXTIL stretching zone; ~ **de exclusión** f NUCL exclusion zone; ~ **exclusiva de pesca** f OCEAN exclusive fishing zone; ~ **explorada** f TEC ESP *comunicaciones* frame; ~ **de extinción** f GAS extinction zone;

~f ~ **fértil** f NUCL blanket; ~ **forestal** f AGRIC, CONST forest area; ~ **forrajera principal** f AGRIC main forage area; ~ **de fractura** f GEOL crush belt; ~ **franca** f TRANSP MAR *puerto* bonded area; ~ **de Fresnel** f TELECOM Fresnel zone; ~ **de freza** f OCEAN *peces* spawning ground;

~g ~ **de generación de olas** f OCEAN wave-generating area; ~ **generadora de aguas frías** f OCEAN cold water down-welling zone; ~ **de Guinier-Preston** f CRISTAL Guinier-Preston zone (*GP*);

~h ~ **habilitada para aterrizaje de helicópteros** f CONTAM MAR helicopter pad;

~i ~ **de la imagen** f IMPR image area; ~ **de impresión** f IMPR printing area; ~ **de incendio** f NUCL fire area; ~ **de inclemencias meteorológicas** f CONST area of deep weathering; ~ **de influencia de la fuente** f CONTAM *acústica* source area; ~ **interfacial coherente** f METAL coherent interface; ~ **interfacial externa** f TEC ESP *equipos* external interface; ~ **de interferencia** f TV interference area; ~ **intermareal** f CONTAM MAR intertidal zone; ~ **de inundación** f HIDROL food plain; ~ **invadida** f TEC PETR invaded zone;

~l ~ **de lanzamiento** f D&A *con paracaídas* drop zone, dropping zone; ~ **lavada** f PETROL flushed zone; ~ **libre de obstáculos** f ING MECÁ clearway; ~ **litoral** f GEOL *de sedimentación* littoral zone; ~ **de lluvias** f AGUA, METEO rainfall area; ~ **lluviosa** f AGUA, HIDROL, METEO rain area;

~m ~ **de madera en la planchada del taladro** f TEC PETR *perforación* monkey board (*BrE*); ~ **maicera** f AGRIC corn belt (*AmE*), maize belt (*BrE*); ~ **de mal tiempo** f OCEAN bad-weather zone; ~ **de mar** f TRANSP MAR sea area; ~ **media del río** f HIDROL midstream; ~ **metamórfica** f GEOL metamorphic zone; ~ **minera** f MINAS mining area; ~ **muerta** f CINEMAT dead spot, INFORM&PD dead zone, MINAS *servomecanismos* backlash;

~n ~ **de navegación** f OCEAN, TRANSP MAR navigation zone; ~ **nerítica** f GEOL neritic zone; ~ **neutra** f ING ELÉC neutral zone; ~ **no compacta** f TEC PETR *geología* undercompacted zone; ~ **no minada** f

MINAS mined-out area; ~ **no saturada** *f* HIDROL unsaturated zone; ~ **no saturada de agua** *f* HIDROL unsaturated zone water; ~ **núcleo** *f AmL* AGRIC corn belt (*AmE*), maize belt (*BrE*);

~ o ~ **oculta** *f* PETROL blind zone; ~ **de onda** *f* FÍS RAD wave zone; ~ **de operaciones** *f* NUCL operation area; ~ **orogénica** *f* GEOL orogenic zone; ~ **oscura** *f* IMPR shadow;

~ p ~ **pantanosa** *f* HIDROL marsh; ~ **de parada** *f* TRANSP AÉR stopway; ~ **de la parcela que florece antes** *f* AGRIC hot spot; ~ **peatonal** *f* TRANSP pedestrian area, pedestrian zone; ~ **de peligro** *f* SEG danger zone; ~ **peligrosa** *f* SEG *instalación eléctrica* hazardous zone; ~ **periférica del núcleo** *f* NUCL outer fueled zone (*AmE*), outer fuelled zone (*BrE*); ~ **de pesca** *f* OCEAN fishing zone; ~ **de practicaje** *f* OCEAN pilotage waters, pilotage zone, TRANSP MAR pilot waters; ~ **de producción neta** *f* PETROL net-pay zone; ~ **productiva** *f* PETROL pay zone; ~ **con prohibición de fumar** *f* SEG smokeless zone; ~ **prohibida** *f* D&A, TRANSP MAR prohibited area; ~ **protegida** *f* CONST protected area;

~ q ~ **que no imprime en los procesos offset** *f* IMPR low spot;

~ r ~ **de raíces** *f* GEOL root zone; ~ **de reacción** *f* CARBÓN reaction zone; ~ **de rebote** *f* C&V spring zone (*AmE*); ~ **de recepción insuficiente** *f* TV poor-reception area; ~ **de recepción de mensajes** *f* PROD message store area; ~ **receptora** *f* CONTAM receptor region; ~ **de recorrido** *f* TRANSP AÉR runup area; ~ **de refino** *f* C&V refining zone; ~ **de remolque** *f* CONST *canales y ríos* towpath; ~ **de reunión** *f* TRANSP MAR *emergencia a bordo* assembly area; ~ **de rift axial** *f* GEOL axial rift zone;

~ s ~ **de salida** *f* INFORM&PD output area; ~ **de saturación** *f* ELECTRÓN saturation region, HIDROL zone of saturation; ~ **de secado** *f* ALIMENT drying area; ~ **de seguridad** *f* INFORM&PD clear zone, TV safe action area; ~ **de seguridad autorizada** *f* SEG approved safety area; ~ **de seguridad de final de pista** *f* TRANSP AÉR runway-end safety area; ~ **de sensación auditiva normal** *f* ACÚST normal auditory sensation area; ~ **de silencio** *f* FÍS ONDAS *radio* skip distance; ~ **sísmica** *f* GEOFÍS seismic zone; ~ **de la soldadura** *f* NUCL weld region; ~ **de sombra** *f* TV shadow area; ~ **de sondeos** *f* MINAS drilling site; ~ **de subducción** *f* TEC PETR *geología* subduction zone; ~ **subsuperficial** *f* OCEAN subsurface zone; ~ **superior** *f* HIDROL *de un río* headwater;

~ t ~ **telefónica de conexiones** *f* TELECOM trunk-switching exchange area; ~ **templada** *f* METEO temperate zone; ~ **en tensión** *f* C&V *alrededor de una piedra* stressed zone; ~ **termoafectada** *f* MECÁ, TERMO heat-affected zone; ~ **de tolerancias** *f* CALIDAD tolerance interval; ~ **de toma de contacto con tierra** *f* TRANSP AÉR touchdown zone; ~ **tórrida** *f* METEO equatorial zone, torrid zone; ~ **de trabajo** *f* INFORM&PD *memoria* scratch pad; ~ **de trabajo de gran riesgo** *f* SEG high-risk area of work; ~ **sin tráfico rodado** *f* TRANSP blind sector without traffic rights; ~ **de transición** *f* PETROL, TEC PETR *geología* transition zone; ~ **de trazado** *f* CONST plotting ground;

~ v ~ **de vientos del oeste** *f* METEO westerlies

zonación *f* GEOL *de terreno metamórfico* zonation

zonificación *f* GEOL *de un mineral* zoning

zoom *m* INFORM&PD zoom, zooming, zoom-in, INSTR zoom lens; ~ **óptico** *m* CINEMAT optical zoom

zoosterol *m* QUÍMICA zoosterol

zootaxía *f* QUÍMICA zootaxy

zootáxico *adj* QUÍMICA zootaxic

zootecnia *f* AGRIC animal production

zooxantina *f* QUÍMICA zooxanthine

zorgita *f* MINERAL zorgite

zozobra *f* SEG overturning, TRANSP *barcos* overturning, roll-over, TRANSP MAR capsizing

zozobrar *vi* SEG overturn, TRANSP roll over, TRANSP MAR capsize

Zr *abr* (*zirconio*) METAL, QUÍMICA Zr (*zirconium*)

zulaque *m* PETROL, TRANSP MAR mastic

zumaca: ~ **pesquera** *f* TRANSP MAR fishing smack

zumbador *m* ING ELÉC buzzer, TELECOM buzzer, howler

zumbido *m* ACÚST, ING ELÉC hum, TEC ESP *máquinas* drone, TRANSP AÉR drone, hum; ~ **de arranque** *m* TV starting hum; ~ **de corriente** *m* ELEC *fuente de alimentación*, ING ELÉC current hum (*AmE*), mains hum (*BrE*); ~ **de corriente alterna** *m* ELEC *amplificador* alternating-current hum; ~ **de la red** *m* ELEC *fuente de alimentación*, ING ELÉC current hum (*AmE*), mains hum (*BrE*); ~ **del relé** *m* ING ELÉC relay hum

zunchado *m* ELEC *cable* reinforcement, PROD hooping; ~ **con bandas de color** *m* EMB colored strapping (*AmE*), coloured strapping (*BrE*)

zunchar *vt* MECÁ hoop

zuncho[1]: **con ~ de hierro** *adj* PROD ironbound

zuncho[2] *m* CONST hoop, band, ING MECÁ band, hoop, clamp, MECÁ clamp, PROD hoop; ~ **de arraigada de los obenques** *m* TRANSP MAR *mástil* hound; ~ **de pilote** *m* CONST pile ferrule

zuñita *f* MINERAL zunyite

zurcir *vt* TEXTIL *calcetines* darn

zuro *m* AGRIC cob; ~ **del maíz** *m* AGRIC corn cob (*AmE*), maize cob (*BrE*)

Conversion tables /
Tablas de conversión

1 Length / Longitud

		metre *metro*	inch *pulgada*	*foot *pie*	*yard *yarda*	*rod *rod*	*mile *milla*
1 metre *metro*	=	1	39.37	3.281	1.093	0.1988	6.214×10^{-4}
1 inch *pulgada*	=	2.54×10^{-2}	1	0.083	0.02778	5.050×10^{-3}	1.578×10^{-5}
1 foot *pie*	=	0.3048	12	1	0.3333	0.0606	1.894×10^{-4}
1 yard *yarda*	=	0.9144	36	3	1	0.1818	5.682×10^{-4}
1 rod *rod*	=	5.029	198	16.5	5.5	1	3.125×10^{-3}
1 mile *milla*	=	1609	63360	5280	1760	320	1

1 imperial standard yard = 0.914 398 41 metre / 1 yarda inglesa = 0,914 398 41 metros
1 yard (scientific) = 0.9144 metre (exact) / 1 yarda científica internacional = 0,9144 metros (exactamente)
1 US yard = 0.914 401 83 metre / 1 yarda americana = 0,914 401 83 metros
1 English nautical mile = 6080 ft = 1853.18 metres / 1 milla marina inglesa = 6080 pies = 1853,18 metros
1 international nautical mile = 1852 metres = 6076.12 ft / 1 milla marina internacional = 1852 metros = 6076,12 pies

* not recognized officially in Spain / unidad sin valor oficial en España

2   Area / Superficie

		sq. metre $metro^2$	sq. inch $pulgada^2$	*sq. foot pie^2	*sq. yard $yarda^2$	*acre $acre$	*sq. mile $milla^2$
1 sq. metre $metro^2$	=	1	1550	10.76	1.196	2.471×10^{-4}	3.861×10^{-7}
1 sq. inch $pulgada^2$	=	6.452×10^{-4}	1	6.944×10^{-3}	7.716×10^{-4}	1.594×10^{-7}	2.491×10^{-10}
1 sq. foot pie^2	=	0.0929	144	1	0.1111	2.296×10^{-5}	3.587×10^{-8}
1 sq. yard $yarda^2$	=	0.8361	1296	9	1	2.066×10^{-4}	3.228×10^{-7}
1 acre $acre$	=	4.047×10^3	6.273×10^6	4.355×10^4	4840	1	1.563×10^{-3}
1 sq. mile $milla^2$	=	259.0×10^4	4.015×10^9	2.788×10^7	3.098×10^6	640	1

1 are = 100 sq. metres = 0.01 hectare / 1 rea = 100 metros cuadrados = 0,01 hectáreas
1 circular mil = 5.067×10^{-10} sq. metre / 1 milipulgada circular = $5,067 \times 10^{-10}$ metros cuadrados
$\qquad\qquad$ = 7.854×10^{-7} sq. inch / $\qquad\qquad\qquad$ = $7,854 \times 10^{-7}$ pulgadas cuadradas
1 acre (statute) = 0.4047 hectare / 1 acre = 0,4047 hectáreas

* not recognized officially in Spain / unidad sin valor oficial en España

3   Volume / Volumen

		cubic metre $metro\ cúbico$	*cubic inch $pulgada\ cúbica$	*cubic foot $pie\ cúbico$	*UK gallon $galón\ imperial$	*US gallon $galón\ americano$
1 cubic metre $metro\ cúbico$	=	1	6.102×10^4	35.31	220.0	264.2
1 cubic inch $pulgada\ cúbica$	=	1.639×10^{-5}	1	5.787×10^{-4}	3.605×10^{-3}	4.329×10^{-3}
1 cubic foot $pie\ cúbico$	=	2.832×10^{-2}	1728	1	6.229	7.480
1 UK gallon† $galón\ imperial$	=	4.546×10^{-3}	277.4	0.1605	1	1.201
1 US gallon‡ $galón\ americano$	=	3.785×10^{-3}	231.0	0.1337	0.8327	1

† volume of 10lb of water at 62°F / volumen de 10 libras de agua a 62°F
‡ volume of 8.328 28 lb of water at 60°F / volumen de 8,328 28 libras de agua a 60°F
1 cubic metre = 999.972 litres / 1 metro cúbico = 999,972 litros
1 acre foot = 271 328 UK gallons = 1233 cubic metres / 1 acre-pie = 271 328 galones imperiales = 1233 metros cúbicos

* not recognized officially in Spain / unidad sin valor oficial en España

until 1976 the litre was equal to 1000.028cm^3 (the volume of 1kg of water at maximum density) but then it was revalued to be 1000cm^3 exactly

hasta 1976 el litro era igual a 1000,028cm^3 (volumen de 1kg de agua a máxima densidad) pero ha sido objeto de nueva definición y ahora vale 1.000cm^3 exactamente

4   Angle / Ángulo

		degree *grado*	minute *minuto*	second *segundo*	radian *radián*	revolution *revolución*
1 degree *grado*	=	1	60	3600	1.745×10^{-2}	2.778×10^{-3}
1 minute *minuto*	=	1.677×10^{-2}	1	60	2.909×10^{-4}	4.630×10^{-5}
1 second *segundo*	=	2.778×10^{-4}	1.667×10^{-2}	1	4.848×10^{-6}	7.716×10^{-7}
1 radian *radián*	=	57.30	3438	2.063×10^{5}	1	0.1592
1 revolution *revolución*	=	360	2.16×10^{4}	1.296×10^{6}	6.283	1

1 mil = 10^{-3} radian / 1 milésima angular (artillería) = 10^{-3} radianes

5   Time / Tiempo

		year *año*	solar day *día solar medio*	hour *hora*	minute *minuto*	second *segundo*
1 year *año*	=	1	365.24*	8.766×10^{3}	5.259×10^{5}	3.156×10^{7}
1 solar day *día solar medio*	=	2.738×10^{-3}	1	24	1440	8.640×10^{4}
1 hour *hora*	=	1.141×10^{-4}	4.167×10^{-2}	1	60	3600
1 minute *minuto*	=	1.901×10^{-6}	6.944×10^{-4}	1.667×10^{-2}	1	60
1 second *segundo*	=	3.169×10^{-8}	1.157×10^{-5}	2.778×10^{-4}	1.667×10^{-2}	1

1 year = 366.24 sidereal days / 1 año = 366,24 días sidéreos
1 sidereal day = 86 164,090 6 seconds / 1 día sidéreo = 86 164.090 6 segundos

* exact figure = 365.242 192 64 in A.D. 2000 / cifra exacta = 365,242 192 64 en el año 2000 de nuestra era

6   Mass / Masa

		kilogram *kilogramo*	pound *libra*	*slug *slug*	*UK ton *tonelada larga*	*US ton *tonelada corta*	u
1 kilogram *kilogramo*	=	1	2.205	6.852×10^{-2}	9.842×10^{-4}	11.02×10^{-4}	6.024×10^{26}
1 pound *libra*	=	0.4536	1	3.108×10^{-2}	4.464×10^{-4}	5.000×10^{-4}	2.732×10^{26}
1 slug *slug*	=	14.59	32.17	1	1.436×10^{-2}	1.609×10^{-2}	8.789×10^{27}
1 UK ton *tonelada larga*	=	1016	2240	69.62	1	1.12	6.121×10^{29}
1 US ton *tonelada corta*	=	907.2	2000	62.16	0.8929	1	5.465×10^{29}
1 u	=	1.660×10^{-27}	3.660×10^{-27}	1.137×10^{-28}	1.634×10^{-30}	1.829×10^{-30}	1

1 imperial standard pound = 0.453 592 338 kilogram/1 libra imperial patrón = 0,453 592 338 kilogramos
1 US pound = 0.453 592 427 7 kilogram / 1 libra americana = 0,453 592 427 7 kilogramos
1 international pound = 0.453 592 37 kilogram / 1 libra internacional = 0,453 592 37 kilogramos
1 tonne = 10^3 kilograms / 1 tonelada = 10^3 kilogramos
1 troy pound = 0.373 242 kilogram / 1 libra troy = 0,373 242 kilogramos

* not recognized officially in Spain / unidad sin valor oficial en España

7   Force / Fuerza

		dyne *dina*	newton *neutonio*	*pound force *libra fuerza*	*poundal *poundal*	gram force *gramo fuerza*
1 dyne *dina*	=	1	10^{-5}	2.248×10^{-6}	7.233×10^{-5}	1.020×10^{-3}
1 newton *neutonio*	=	10^5	1	0.2248	7.233	102.0
1 pound force *libra fuerza*	=	4.448×10^5	4.448	1	32.17	453.6
1 poundal *poundal*	=	1.383×10^4	0.1383	3.108×10^{-2}	1	14.10
1 gram force *gramo fuerza*	=	980.7	980.7×10^{-5}	2.205×10^{-3}	7.093×10^{-2}	1

* not recognized officially in Spain / unidad sin valor oficial en España

8   Power / Potencia

		*Btu per hr *Ucb por hora*	*ft lb s^{-1} *libra-pie por segundo*	kg metre s^{-1} *kilogramo-metro por segundo*	cal s^{-1} *caloría por segundo*	HP *caballo de vapor*	watt *vatio*
1 Btu per hour *Ucb por hora*	=	1	0.2161	2.987×10^{-2}	6.999×10^{-2}	3.929×10^{-4}	0.2931
1 ft lb per second *libra-pie por segundo*	=	4.628	1	0.1383	0.3239	1.818×10^{-3}	1.356
1 kg metre per second *kilogramo-metro por segundo*	=	33.47	7.233	1	2.343	1.315×10^{-2}	9.807
1 cal per second *caloría por segundo*	=	14.29	3.087	4.268×10^{-1}	1	5.613×10^{-3}	4.187
1 HP *caballo de vapor*	=	2545	550	76.04	178.2	1	745.7
1 watt *vatio*	=	3.413	0.7376	0.1020	0.2388	1.341×10^{-3}	1

1 international watt = 1.000 19 absolute watt / 1 vatio internacional = 1,000 19 vatios absolutos

* not recognized officially in Spain / unidad sin valor oficial en España

9 Energy, work, heat / Energía, trabajo, calor

		*Btu Ucb	joule julio	*ft lb libra-pie	cm⁻¹ cm⁻¹	cal caloria	kW h kilovatio-hora	electron volt electrón-voltio	kg kilogramo	u†
1 Btu *Ucb*	=	1	1.055×10^{3}	778.2	5.312×10^{25}	252	2.930×10^{-4}	6.585×10^{21}	1.174×10^{-14}	7.074×10^{12}
1 joule *julio*	=	9.481×10^{-4}	1	7.376×10^{-1}	5.035×10^{22}	2.389×10^{-1}	2.778×10^{-7}	6.242×10^{18}	1.113×10^{-17}	6.705×10^{9}
1 ft lb *libra-pie*	=	1.285×10^{-3}	1.356	1	6.828×10^{22}	3.239×10^{-1}	3.766×10^{-7}	8.464×10^{18}	1.507×10^{-17}	9.092×10^{9}
1 cm⁻¹ *cm⁻¹*	=	1.883×10^{-26}	1.986×10^{-23}	1.465×10^{-23}	1	4.745×10^{-24}	5.517×10^{-30}	1.240×10^{-4}	2.210×10^{-40}	1.332×10^{-13}
1 cal 15°C *caloria 15°C*	=	3.968×10^{-3}	4.187	3.088	2.108×10^{23}	1	1.163×10^{-6}	2.613×10^{19}	4.659×10^{-17}	2.807×10^{10}
1 kW h *kilovatio-hora*	=	3412	3.600×10^{6}	2.655×10^{6}	1.813×10^{29}	8.598×10^{5}	1	2.247×10^{25}	4.007×10^{-11}	2.414×10^{16}
1 electron volt *electrón-voltio*	=	1.519×10^{-22}	1.602×10^{-19}	1.182×10^{-19}	8.066×10^{3}	3.827×10^{-20}	4.450×10^{-26}	1	1.783×10^{-36}	1.074×10^{-9}
1 kg *kilogramo*	=	8.521×10^{13}	8.987×10^{16}	6.629×10^{16}	4.525×10^{39}	2.147×10^{16}	2.497×10^{10}	5.610×10^{35}	1	6.025×10^{2}
1 u	=	1.415×10^{-13}	1.492×10^{-10}	1.100×10^{-10}	7.513×10^{12}	3.564×10^{-11}	4.145×10^{-17}	9.31×10^{8}	1.660×10^{-27}	1

† from the mass-energy relationship $E = mc^2$ / partiendo de la relación energía-masa $E = mc^2$

* not recognized officially in Spain / unidad sin valor oficial en España

	standard atmosphere *atmósfera patrón*	kg force cm⁻² *kg fuerza cm⁻²*	dyne cm⁻² *dina cm⁻²*	pascal *pascal*	*pound force in.⁻² *libra fuerza pulgada⁻²*	*pound force ft⁻² *libra fuerza pie⁻²*	millibar *milibar*	torr *torr*	barometric in. Hg *pulgada barométrica de mercurio*
1 standard atmosphere *atmósfera patrón*	= 1	1.033	1.013×10^6	1.013×10^5	14.70	2116	1013	760	29.92
1 kg force cm⁻² *kg fuerza cm⁻²*	= 0.9678	1	9.804×10^5	9.804×10^4	14.22	2048	980.7	735.6	28.96
1 dyne cm⁻² *dina cm⁻²*	= 9.869×10^{-7}	10.20×10^{-7}	1	0.1	14.50×10^{-6}	2.089×10^{-3}	10^{-3}	750.1×10^{-6}	29.53×10^{-6}
1 pascal *pascal*	= 9.869×10^{-6}	10.20×10^{-6}	10	1	14.50×10^{-5}	2.089×10^{-2}	10^{-2}	750.1×10^{-5}	29.53×10^{-5}
1 pound force in.⁻² *libra fuerza pulgada⁻²*	= 6.805×10^{-2}	7.031×10^{-2}	6.895×10^4	6.895×10^3	1	144	68.95	51.71	2.036
1 pound force ft⁻² *libra fuerza pie⁻²*	= 4.725×10^{-4}	4.882×10^{-4}	478.8	47.88	6.944×10^{-3}	1	47.88×10^{-2}	0.3591	14.14×10^{-3}
1 millibar *milibar*	= 0.9869×10^{-3}	1.020×10^{-3}	10^3	10^2	14.50×10^{-3}	2.089	1	0.7500	29.53×10^{-3}
1 torr *torr*	= 1.316×10^{-3}	1.360×10^{-3}	1.333×10^2	1.333×10^3	1.934×10^{-2}	2.784	1.333	1	3.937×10^{-2}
1 barometric in. Hg *pulgada barométrica de mercurio*	= 3.342×10^{-2}	3.453×10^{-2}	3.386×10^4	3.386×10^3	4.912×10^{-1}	70.73	33.87	25.40	1

1 torr = 1 barometric mmHg density 13.5951 g cm⁻³ at 0°C and acceleration due to gravity 980.665 cm/s⁻²
1 torr = 1 milímetro de mercurio de una densidad de 13,5951 g cm⁻³ a 0°C y para una aceleración de la gravedad de 980,665 cm/s⁻²
1 dyne cm⁻² = 1 barad
* not recognized officially in Spain / unidad sin valor oficial en España

11 Magnetic flux / Flujo de inducción magnética

		maxwell *maxvelio*	kiloline *kilolino*	weber *weberio*
1 maxwell (1 line) *maxvelio (1 lino)*	=	1	10^{-3}	10^{-8}
1 kiloline *kilolino*	=	10^3	1	10^{-5}
1 weber *weberio*	=	10^8	10^5	1

12 Magnetic flux density / Inducción magnética

		gauss *gausio*	weber m^{-2} (tesla) *weberio m^{-2} (tesla)*	gamma *gamma*	maxwell cm^{-2} *maxvelio cm^{-2}*
1 gauss (line cm^{-2}) *gausio (lino cm^{-2})*	=	1	10^{-4}	10^5	1
1 weber m^{-2} (tesla) *weberio m^{-2} (tesla)*	=	10^4	1	10^9	10^4
1 gamma *gamma*	=	10^{-5}	10^{-9}	1	10^{-5}
1 maxwell cm^{-2} *maxvelio cm^{-2}*	=	1	10^{-4}	10^5	1

13 Magnetomotive force / Fuerza magnetomotriz

		*abamp turn *abamperio-vuelta*	amp turn *amperio-vuelta*	gilbert *gilbertio*
1 abampere turn *abamperio-vuelta*	=	1	10	12.57
1 ampere turn *amperio-vuelta*	=	10^{-1}	1	1.257
1 gilbert *gilbertio*	=	7.958×10^{-2}	0.7958	1

* not recognized officially in Spain / unidad sin valor oficial en España

14 Magnetic field strength / Intensidad de campo magnético

		amp turn cm^{-1} *amperio-vuelta cm^{-1}*	amp turn m^{-1} *amperio-vuelta m^{-1}*	oersted *oerstedio*
1 amp turn cm^{-1} *amperio-vuelta cm^{-1}*	=	1	10^2	1.257
1 amp turn m^{-1} *amperio-vuelta m^{-1}*	=	10^{-2}	1	1.257×10^{-2}
1 oersted *oerstedio*	=	0.7958	79.58	1

15 Illumination / Iluminación

		lux *lux*	phot *fotio*	*foot-candle *bujía-pie*
1 lux (1m m^{-2}) *lux (1m m^{-2})*	=	1	10^{-4}	9.29×10^{-2}
1 phot (1m cm^{-2}) *fotio (1m cm^{-2})*	=	10^4	1	929
1 foot-candle (1m ft^{-2}) *bujía-pie (1m pie^{-2})*	=	10.76	10.76×10^{-4}	1

* not recognized officially in Spain / unidad sin valor oficial en España

16 Luminance / Luminancia

		nit *nit*	stilb *estilbio*	*cd ft^{-2} *candela-pie^{-2}*	apostilb *apostilbio*	lambert *lambertio*	*foot-lambert *lambertio-pie*
1 nit (cd m^{-2}) *nit (candela m^{-2})*	=	1	10^{-4}	9.29×10^{-2}	π	$\pi \times 10^{-4}$	0.292
1 stilb (cd cm^{-2}) *estilbio (candela cm^{-2})*	=	10^{4}	1	929	$\pi \times 10^{4}$	π	2920
1 cd ft^{-2} *candela-pie^{-2}*	=	10.76	1.076×10^{-3}	1	33.8	3.38×10^{-3}	π
1 apostilb (1m m^{-2}) *apostilbio (1m m^{-2})*	=	$1/\pi$	$1/(\pi \times 10^{4})$	2.96×10^{-2}	1	10^{-4}	9.29×10^{-2}
1 lambert (1m cm^{-2}) *lambertio (1m cm^{-2})*	=	$1/(\pi \times 10^{-4})$	$1/\pi$	296	10^{4}	1	929
1 foot lambert *or* equivalent foot candle *lambertio-pie* o *bujía-pie equivalente*	=	3.43	3.43×10^{-4}	$1/\pi$	10.76	1.076×10^{-3}	1

luminous intensity of candela = 98.1% that of international candle
intensidad luminosa de la candela = 98,1% de la bujía internacional
1 lumen = flux emitted by 1 candela into unit solid angle
1 lumen = flujo luminoso emitido por una fuente puntual uniforme, situada en el vértice de un ángulo
 sólido de 1 estereorradián y cuya intensidad es de 1 candela

* not recognized officially in Spain / unidad sin valor oficial en España

Chemical elements / Elementos químicos

Symbol/Símbolo	Element	Elemento	Atomic number/ Número atómico
Ac	Actinium	Actinio	89
Ag	Silver	Plata	47
Al	Aluminium *(BrE)* Aluminum *(AmE)*	Aluminio	13
Am	Americium	Americio	95
Ar	Argon	Argón	18
As	Arsenic	Arsénico	33
At	Astatine	Astato	85
Au	Gold	Oro	79
B	Boron	Boro	5
Ba	Barium	Bario	56
Be	Beryllium	Berilio	4
Bi	Bismuth	Bismuto	83
Bk	Berkelium	Berquelio	97
Br	Bromine	Bromo	35
C	Carbon	Carbono	6
Ca	Calcium	Calcio	20
Cd	Cadmium	Cadmio	48
Ce	Cerium	Cerio	58
Cf	Californium	Californio	98
Cl	Chlorine	Cloro	17
Cm	Curium	Curio	96
Co	Cobalt	Cobalto	27
Cr	Chromium	Cromo	24
Cs	Caesium *(BrE)* Cesium *(AmE)*	Cesio	55
Cu	Copper	Cobre	29
Dy	Dysprosium	Disprosio	66
Er	Erbium	Erbio	68
Es	Einsteinium	Einsteinio	99
Eu	Europium	Europio	63
F	Fluorine	Flúor	9
Fe	Iron	Hierro	26
Fm	Fermium	Fermio	100
Fr	Francium	Francio	87
Ga	Gallium	Galio	31
Gd	Gadolinium	Gadolinio	64
Ge	Germanium	Germanio	32
H	Hydrogen	Hidrógeno	1
He	Helium	Helio	2
Hf	Hafnium	Hafnio	72
Hg	Mercury	Mercurio	80
Ho	Holmium	Holmio	67
I	Iodine	Yodo Iodo	53
In	Indium	Indio	49
Ir	Iridium	Iridio	77
K	Potassium	Potasio	19
Kr	Krypton	Cripton	36
La	Lanthanum	Lantano	57
Li	Lithium	Litio	3

Lr	Lawrencium	Laurencio	103
Lu	Lutetium	Lutecio	71
Md	Mendelevium	Mendelevio	101
Mg	Magnesium	Magnesio	12
Mn	Manganese	Manganeso	25
Mo	Molybdenum	Molibdeno	42
N	Nitrogen	Nitrógeno	7
Na	Sodium	Sodio	11
Nb	Niobium	Niobio	41
Nd	Neodymium	Neodimio	60
Ne	Neon	Neón	10
Ni	Nickel	Níquel	28
No	Nobelium	Nobelio	102
Np	Neptunium	Neptunio	93
O	Oxygen	Oxígeno	8
Os	Osmium	Osmio	76
P	Phosphorus	Fósforo	15
Pa	Protactinium	Protactinio	91
Pb	Lead	Plomo	82
Pd	Palladium	Paladio	46
Pm	Promethium	Promecio	61
Po	Polonium	Polonio	84
Pr	Praseodymium	Praseodimio	59
Pt	Platinum	Platino	78
Ra	Radium	Radio	88
Rb	Rubidium	Rubidio	37
Re	Rhenium	Renio	75
Rh	Rhodium	Rodio	45
Rn	Radon	Radón	86
Ru	Ruthenium	Rutenio	44
S	Sulphur *(BrE)* Sulfur *(AmE)*	Azufre	16
Sb	Antimony	Antimonio	51
Sc	Scandium	Escandio	21
Se	Selenium	Selenio	34
Si	Silicon	Silicio	14
Sm	Samarium	Samario	62
Sn	Tin	Estaño	50
Sr	Strontium	Estroncio	38
Ta	Tantalum	Tantalio	73
Tb	Terbium	Terbio	65
Tc	Technetium	Tecnecio	43
Te	Tellurium	Telurio	52
Th	Thorium	Torio	90
Ti	Titanium	Titanio	22
Tl	Thallium	Talio	81
Tm	Thulium	Tulio	69
U	Uranium	Uranio	92
V	Vanadium	Vanadio	23
W	Tungsten Wolfram	Tungsteno Wolframio	74
Xe	Xenon	Xenón	54
Y	Yttrium	Itrio	39
Yb	Ytterbium	Iterbio	70
Zn	Zinc	Cinc Zinc	30
Zr	Zirconium	Circonio Zirconio	40